6th edition CPCR®

1990

6th edition CPCR®

1990

Crop Protection Chemicals Reference

CHEMICAL AND PHARMACEUTICAL PRESS

a joint venture of

JOHN WILEY & SONS

New York • Chichester • Brisbane • Toronto • Singapore

and

CHEMICAL AND PHARMACEUTICAL PUBLISHING CORP.,

a subsidiary of OVP, Paris

CPCR® has been made possible through the courtesy of agricultural chemical manufacturers and the information concerning each product has been prepared by participating manufacturers. Neither the inclusion nor exclusion of any product in CPCR should be construed as a recommendation for or against the use of such product by the Publisher.

Although great care has been taken by the Publisher in compiling and checking the information and data contained in CPCR to insure that it is current and accurate at the time of publication, the Publisher disclaims any and all liability for any errors, omissions, or inaccuracies in such information and data, whether attributable to inadvertence or otherwise, and for any consequences arising therefrom. The Publisher further disclaims any and all liability for inaccuracies in the information contained in CPCR due to revisions in the product labels, from which the information contained in the CPCR is compiled. The body of the book contains information current as of September 15, 1989. The update of last minute information covers changes and additional information received before November 10, 1989. READERS OF CPCR ARE CAUTIONED TO CONTACT MANUFACTURERS FOR THE MOST CURRENT PRODUCT INFORMATION, INCLUDING SUPPLEMENTAL LABELING AND SLN REGISTRATIONS. REFERENCE SHOULD ALWAYS BE MADE TO THE PRODUCT LABEL PRIOR TO THE USE OF ANY CROP PROTECTION CHEMICAL. IT IS A VIOLATION OF FEDERAL LAW TO USE ANY CROP PROTECTION CHEMICAL DESCRIBED IN CPCR IN A MANNER INCONSISTENT WITH ITS LABELING. It is further understood that THE PUBLISHER MAKES NO REPRESENTATIONS OR WARRANTIES OF ANY KIND, INCLUDING, BUT NOT LIMITED TO, THE WARRANTIES OF FITNESS FOR A PARTICULAR PURPOSE OR MERCHANTABILITY NOR ARE ANY SUCH REPRESENTATIONS OR WARRANTIES TO BE IMPLIED WITH RESPECT TO THE DATA FURNISHED, AND THE PUBLISHER TAKES NO RESPONSIBILITY WITH RESPECT TO THE USE OF THE INFORMATION AND DATA FURNISHED. THE PUBLISHER SHALL NOT BE LIABLE FOR ANY SPECIAL, CONSEQUENTIAL, OR EXEMPLARY DAMAGES RESULTING, IN WHOLE OR IN PART, FROM CUSTOMER'S USE OF THE DATA. Further, no one should administer or recommend medical treatment in reliance upon the data contained in CPCR without first verifying such treatment with a knowledgeable medical professional.

Library of Congress Cataloging in Publication Data:

Library of Congress Catalog Number: 85–642883
ISBN 0–471–51641–4

Printed in the United States of America

10 9 8 7 6 5 4 3 2 1

Foreword

This 6th edition of the CPCR could be the most important ever. Crop protection products are in the news and the public debate more than ever before. In that light, the CPCR is a more valuable reference than it always has been in the past.

Those of us who make, sell, or use crop protection chemicals will find ourselves using the CPCR more and more as we are called upon to discuss the relationship between crop protection chemicals and the bounty of food and fiber we enjoy.

The makeup of our industry and our products continue to change, but the CPCR's ability to reflect that change remains the same. The information in this edition is presented in the usual well ordered and usable fashion.

The 6th edition will occupy a valuable spot on my reference shelf and I'm sure it will on yours, too.

JACK D. EARLY, Ph.D.

President
National Agricultural Chemicals Association
1155 15th Street, N.W.—Suite 900
Washington, D.C. 20005

All of us at the Chemical and Pharmaceutical Press would like to thank our Advisory Panel for their invaluable advice and support.

Kathleen Brody, Environmental Protection Agency
Jack D. Early, National Agricultural Chemicals Association
William Engel, Brandt Chemical Company
David Harris, School of Medicine, SUNY Stony Brook, New York
Mort McDonald, American Cyanamid Company
Hilton Withers, Southern States Cooperative
Virgil Young, Tenkoz, Inc.

Contents

PRODUCT LABELS

SUPPLEMENT

As of press time, we are aware of the following products included in the supplement. Please check for others that may have arrived after the CPCR press date.

American Cyanamid Company
- Tri-Scept

Ciba-Geigy Corporation
- Tilt
- Supracide 2E
- Ridomil 2E

E.I. Du Pont de Nemours & Co. (Inc.)
- Manzate 200 DF

Hoechst-Roussel Agri-Vet Company
- Hoelon 3EC

Brand Name Quick Index

In this section, each manufacturers' brand name products are listed in alphabetical sequence (including many of the products in the supplement). The manufacturer of each product follows the brand name in parentheses.

*** Restricted Use Only** **† Only Registered for Turf, Ornamentals and Forestry**

Always read and follow label directions carefully.

*** Restricted Use Only** **† Only Registered for Turf, Ornamentals and Forestry**

* Restricted Use Only   † Only Registered for Turf, Ornamentals and Forestry

Always read and follow label directions carefully.

T

V

W

Z

*** Restricted Use Only** **† Only Registered for Turf, Ornamentals and Forestry**

Manufacturer Index

Listed here are manufacturers of crop protection chemicals in the United States, that are participating in this edition of CPCR. Included are the manufacturer's name and address, branch or regional offices, the names of individuals or departments to whom product inquiries may be directed, as well as general and emergency telephone numbers.

Also listed in this index are the brand name products of each manufacturer that are included in this book. Some of the products in the supplement may not be listed.

ABBOTT LABORATORIES 251
14th Street & Sheridan Road
North Chicago, IL 60064
Address inquiries to: Chemical & Agricultural Products Division
(800) 323-9597
For emergencies contact:
J. Berdahl, Dept. 836
(312) 937-6100

Branch Offices

Abbott Laboratories
1540 East Shaw
Suite 123
Fresno, CA 93710
(209) 227-2926

Abbott Laboratories
310 S. Dillard St.
Suite 410
Winter Garden, FL 32787
(407) 656-8779

Products Listed in CPCR

DeVine
DiPel ES East
DiPel ES West
DiPel 2X
DiPel 4L
†DiPel 8AF
†DiPel 8L
DiPel 10G
ProGibb Plus 2X
ProGibb 4% Liquid
ProGibb Spraying Guide
Promalin
†ProShear
ProVide
VectoBac-12AS
VectoBac-G

AMERICAN CYANAMID COMPANY 271
Agricultural Division
One Cyanamid Plaza
Wayne, NJ 07470
(201) 831-2000
Emergency Number: (201) 835-3100

Branch Offices

Eastern Region
P.O. Box 590
650 East Carmel Drive
Suite 360
Carmel, IN 46032
(317) 844-4952

Central Region
655 First Bank Drive
Suite A
Palatine, IL 60067
(312) 991-9340

Western Region
Forest Ridge Park
Suite 102
Forest Ridge Road
Bedford, TX 76022
(817) 571-3535

Midwest Region
2910 Westown Parkway
Suite 310
West Des Moines, IA 50265
(515) 224-0202

Mid-South Region
P.O. Box 381347
Suite 200
7640 Highway 72
Poplar Avenue
Germantown, TN 38183-1347
(901) 755-4000

Products Listed in CPCR

*AAStar
Abate 1-SG
Abate 2-CG
Abate 5-CG
Abate 4-E
Amdro
Arsenal
Arsenal Forestry Granule
Arsenal Applicators Concentrate
Arsenal 0.5 G
Assert
Avenge
†Chopper
*Counter
Cycocel
Cygon 400
Cyprex 65-W
Cythion ULV
Cythion 57% EC
Cythion RTU
†Event
†Image
Prowl
Prozine 70DF Non-CRop
Prozine 70DF
Pursuit
Pursuit Plus
Scepter
Scepter 70DG
Scepter OT
Squadron
†Stomp
*Thimet 15-G
*Thimet 20-G
Tri-Scept

BASF CORPORATION 379

Chemicals Division
100 Cherry Hill Road
Parsippany, NJ 07054
(201) 316-3000
For Emergencies Contact:
CHEMTREC 800-424-9300
BASF Corporation 201-316-3000
Address inquiries to:
BASF Corporation Chemicals Division
Agricultural Chemicals
100 Cherry Hill Road
Parsippany, NJ 07054

Branch Offices:

BASF Corporation
Chemicals Division
Agricultural Chemicals
5420 W. Southern Avenue
Suite 105
Indianapolis, Indiana 46241
(317) 248-8511

Products Listed in CPCR

Basagran
Blazer
Dash
Galaxy
Laddok
Pix
Poast
Poast Plus
Pyramin DF
Pyramin FL
Ronilan WP
Ronilan FL
Storm

CIBA-GEIGY CORPORATION AGRICULTURAL DIVISION 491

P.O. Box 18300
Greensboro, NC 27419
For emergencies contact:
(919) 292-7100

Products Listed in CPCR

AAtrex 4L
AAtrex 80W
AAtrex Nine-0
Apron 25W
Aquazine 90 WDG
Atratol 8P
Atratol 90
Bicep 6L
Caparol 4L
Concep II 70W
Cotoran 80W
Cotoran 4L
Cotoran DF
*Curacron 6E
*Curacron 8E
D•z•n diazinon 4E
D•z•n diazinon 14G
D•z•n diazinon 50W
D•z•n diazinon AG500
Dual
Dual 25G
Evik 80W
Orbit
Pramitol 25E
Pramitol 5PS
Prime+
Princep 4G
Princep 4L
Princep 80W
Princep Caliber 90
Ridomil 2E
Ridomil MZ58
Ridomil 5G
Ridomil PC 11G
Ridomil Bravo 81W
Sequestrene 138 Fe
Sequestrene 330 Fe
*Supracide 2E
*Swat 8E
Tilt
Trigard 75W

THE DOW CHEMICAL COMPANY 623

Midland, MI 48640
Address inquiries to: Agricultural Products Department 1-800-258-2436 (CHEM)
For emergencies contact:
(517) 636-4400

Products Listed In CPCR

†Confront
Crossbow
Curtail
Curtail M
Lorsban 4E
Lorsban 15G
Lorsban 50W
N-Serve 24 for Stabilized Nitrogen
N-Serve 24E for Stabilized Nitrogen
*Pathway
Stinger
Tandem
*Telone II
*Telone C-17
*Tordon 22K
†Tordon RTU
†Turflon Amine
†Turflon II Amine
†Turflon D
†Turflon Ester

E. I. DU PONT DE NEMOURS & CO., (INC.) 695

Agricultural Products Department
Walker's Mill, Barley Mill Plaza
Wilmington, DE 19898
Address inquiries to: Product Information:
(800) 441-7515
For emergencies contact: Transportation Emergency -
Call Chemtrec (800) 424-9300
Medical Emergency - Du Pont
(800) 441-3637

Products Listed in CPCR

Ally
*Asana
*Asana XL
Assure
Atrazine 4L
Benlate
Benlate 50 DF
*Bidrin 8
*Bladex 4L
*Bladex 90DF
Canopy
Classic
Escort
Express
*Extrazine II 4L
*Extrazine II DF
Finesse
Gemini
Glean
Harmony
Harmony Extra
Hyvar X
Hyvar X-L
Hyvar L
Karmex DF
Krenite S
Krovar I DF
Krovar II DF
Lannate
*Lannate L
*Lannate LV
Lexone DF
Lexone 4L
Londax
Londax CA only
Lorox DF
Lorox Plus
Lorox L
Manzate 200 DF
Oust
Pinnacle
Preview
*Pydrin 2.4
Savey
Sinbar
†Telar
†Tersan 1991 DF
†Tupersan
Velpar
Velpar L
Velpar ULW
Vendex 4L
Vendex 50WP
*Vydate C
*Vydate L

ELANCO PRODUCTS COMPANY 921

Ag Chem Marketing
4040 Vincennes Circle, Suite 400
Indianapolis, IN 46268 USA Address inquiries to:
J. E. Meeker, Director of Marketing (317) 276-5903 or J. E. Moody, Marketing Services (317) 276-3071

*** Restricted Use Only** **† Only Registered for Turf, Ornamentals and Forestry**

Always read and follow label directions carefully.

For emergencies contact:
Rocky Mountain Poison Control Center
1-800-722-8977
Elanco Products Co. (317) 276-2000
For Human Ingestion: Dr. F. W. Price (317) 276-3665
Dr. R. A. Browne (317) 276-6626
Dr. J. T. Callaghan (317) 276-4793
Animal ingestion: Robert Stanley (317) 276-3164, Tucker Thomson (317) 467-4246
Chemical Spill: Stephen Nash, (317) 276-3537, Bert Gorman (317) 276-2181, William Fields (317) 276-2570

Products Listed in CPCR

Balan E.C.
Balan Dry Flowable
Paarlan E.C.
Rubigan E.C.
Sonalan E.C.
Surflan A.S.
Treflan E.C.
Treflan M.T.F.
Treflan 5
Treflan TR-10

FERMENTA ASC CORPORATION 991

5966 Heisley Rd.
P.O. Box 8000
Mentor, OH 44061-8000
Address inquiries to: E. R. Sabala, Mgr., Corporate Communications
(216) 357-4100
For emergencies contact: G. L. Eilrich
Vice President Technology (216) 357-4145

Branch Offices

Northern Area
5966 Heisley Rd.
P.O. Box 8000
Mentor, OH 44061-8000
(216) 357-4610
Southern Area
1517 Johnson Ferry Rd., Suite 275
Marietta, GA 30062
(404) 578-9990
Western Area
1318 East Shaw Avenue, Suite 305
Fresno, CA 93710
(209) 224-9661

Products Listed in CPCR

Ansar 6.6
Bravo 500
Bravo W-75
Bravo 720
Bravo 90DG
Bueno 6
Dacamine 4D
Daconate 6
†Daconil 2787 Flowable
†Daconil 2787 Fungicide
†Daconil 2787 WDG
Dacthal W-75
†Dacthal W-75 Turf
DSMA Liquid
Frigate
†2 Plus 2 (MCPP + 2,4-D Amine)

FMC CORPORATION 1035

2000 Market Street
Philadelphia, PA 19103
Address inquiries to:
D. Hopper/K. Gaskins
2000 Market St.
Philadelphia, PA 19103
(215) 299-6000
For emergency contact:
(716) 735-3765

Branch Offices

FMC Corporation
6065 Roswell Road NE
Atlanta, GA 30328
(404) 256-9333
FMC Corporation
4922 E. Yale Box 1669
Fresno, CA 93717
(209) 252-1641
FMC Corporation
11128 John Galt Blvd.
Omaha, NE 68137
(402) 592-5090
FMC Corporation
1563 E. County Line Road
Suite 402 MS 39211
(601) 956-8081

Products Listed in CPCR

*Ammo 2.5 EC
*Ammo WSB
*Capture 2EC
Carbamate WDG
Command 4EC
Commence EC
Dusting Sulfur
Elgetol
Ethion 4 Miscible
Funginex 1.6 EC
*Furadan 3G
*Furadan 4F
*Furadan 5G
*Furadan 15G
Golden Leaf Tobacco Spray
*Methyl Parathion 2 Thiodan 3 EC
Option 1 EC
*Pounce 1.5 G
*Pounce 3.2 EC
*Pounce 25 WP
†Talstar 10 WP
Thiodan 2 CO EC
Thiodan 3 EC
Thiodan 50 WP
Wettable Sulfur
Ziram 76 WDG

HOECHST-ROUSSEL AGRI-VET COMPANY 1103

Rt. 202-206 North
Somerville, NJ 08876
Address inquiries to: Technical Training Manager, Agricultural Chemicals
(201) 231-3520
For emergencies contact:
(201) 231-2000

Products Listed in CPCR

†Acclaim 1EC
*Hoelon 3EC
Horizon 1EC
Rely
*Scout X-TRA
Tiller
Whip 1EC

ICI AMERICAS INC. 1123

Agricultural Products
Wilmington, DE 19897
(302) 886-1000
Address inquiries to:
Agricultural Products
(800) 759-2500
Emergency Number: 1-800-327-8633

Regional Offices

Agricultural Products-Southern Region
6625 The Corners Parkway
Suite 250
Norcross, GA 30092
(404) 441-2977
Agricultural Products-Northeast Region
Suite 101, Fortune Park 10
3905 Vincennes Rd.
Indianapolis, IN 46268
(317) 872-5052
Agricultural Products-Western Region
2200-B Douglas Blvd.
Suite 200
Roseville, CA 95661
(916) 783-7467
Agricultural Products-Midwestern Region
2630 South 156th Circle
Omaha, NE 68130
(402) 333-0450

Products Listed in CPCR

Actellic 5E
*Ambush
*Ambush 25W
*Ambush 25W-WSP
Arrosolo 3-3E
†Betasan 2.9-E
†Betasan 4-E LF
†Betasan 3.6-G
†Betasan 7-G
†Betasan 12.5-G
*Cyclone
*Cymbush 3E
Demon EC
Demon TC
†Demon WP
Demon WP Packet (4 oz.)
†Demon WP Water Soluble Packets

Devrinol 2-E
Devrinol 2-G
Devrinol 5-G
Devrinol 10-G
Devrinol 50-WP
†Devrinol 50-WP—Ornamentals
Devrinol 50 DF
*Dyfonate 4-EC
Dyfonate II 10-G
*Dyfonate II 20-G
†Eptam 2.3-G
†Eptam 5-G
Eptam 7-E
Eptam 10-G
Eradicane Extra
Eradicane 6.7-E
*Force 1.5G
Fusilade 2000
*Gramoxone Extra
*Gramoxone Super
Imidan 50-WP
*Karate
Ordram 8-E Southern Region
Ordram 8-E Southwest
Ordram 8-E California
Ordram 10-G California
Ordram 15-G
Prefar 4-E
Reflex 2LC
Ro-Neet 6-E
Sulfur Alfa
Sulfur Signal
Sulfur Special Electric
Surpass 6.7-E
Sutan + 6.7-E
Sutan + 10-G
Sutazine +18-6 G
Sutazine +
Talon-G
Talon G Bait Packs
Talon G Bait Pack Mini-Pellets
Talon G Mini Pellet
Tillam 6-E
Torpedo
Vapam
†Vapam—Garden Soils
Vernam 7-E
WeatherBlok

MOBAY CORPORATION 1337

8400 Hawthorn Road
Kansas City, MO 64120
Address inquiries to: Product Management
(816) 242-2000

Branch Offices

Mobay Corporation
1900 N. Gateway Blvd.
Suite 152
Fresno, CA 93727-1694
(209) 252-5121

Mobay Corporation
407 N. 117th Street
Suite 2
Omaha, NE 68154
(402) 333-0325

Mobay Corporation
6077 Primacy Parkway
Suite 310
Memphis, TN 38119-5799
(901) 761-9711

Mobay Corporation
8125 Knue Road
Suite 127
Indianapolis, IN 46250-1936
(317) 845-9890

Mobay Corporation
1587 Phoenix Blvd. #6
Atlanta, GA 30349-5503
(404) 997-7466

Products Listed in CPCR

Bayleton 50% WP PVA
Bayleton 50% DF
*Baythroid 2
*Bolstar 6
Def 6
*Di-Syston 15% G
*Di-Syston 8
Dylox 80% SP
Dyrene
*Furadan 3G
*Furadan 4F
*Guthion 2S
*Guthion 2L
Guthion 35% WP
Guthion 50% WP Water Soluble Packets
*Mesurol 75% WP
*Metasystox-R
*Monitor 4
Morestan 25% WP
Morestan 25% WP Water Soluble Packets
*Nemacur 15% G
*Nemacur 15%-citrus
*Nemacur 3
*Nemacur 3-citrus
Salute 4EC
Sencor DF
Sencor 4
Turbo 8EC

MONSANTO AGRICULTURAL COMPANY 1493

800 North Lindbergh Blvd.
St. Louis, MO 63167
For emergencies contact:
(314) 694-4000

Branch Offices

Monsanto Company
11550 N. Meridian
Suite 200
Carmel, IN 46032
(317) 843-4186

Monsanto Company
1025 Ashworth Rd. #110
West Des Moines, IA 50265
(515) 225-0412

Monsanto Company
9785 Maroon Circle
Suite 110
Englewood, CO 80112
(303) 799-4155

Monsanto Company
1515 Johnson Ferry Road, #200
Marietta, GA 30062
(404) 565-6759

Products Listed in CPCR

*Bronco
Buckle
*Bullet
*Cannon
Fallow Master
Far-Go
Far-Go Granular
*Freedom
Honcho
Landmaster BW
Landmaster II
*Lariat
*Lasso
*Lasso II
*Lasso-Micro-Tech
Pondmaster
Ramrod Flowable
Ramrod and Atrazine
Ramrod 20 Granular
Ranger
Roundup
Roundup RT
Showdown

NOR-AM CHEMICAL COMPANY 1623

P.O. Box 7495
3509 Silverside Road
Wilmington, DE 19803
Address inquiries to: (302) 575-2000
For emergencies contact:
Emergency (800) 424-9300
Customer Service (800) 992-9909

Branch Offices

NOR-AM Chemical Company
730 Sunrise Boulevard
Suite 251
Roseville CA 95661
(916) 782-9606

NOR-AM Chemical Company
14041 Burnhaven Drive
Suite 102
Burnsville MN 55337
(612) 435-2575

NOR-AM Chemical Company
6262 Poplar Ave., Suite 201
Memphis, TN 38119
(901) 683-1593

NOR-AM Chemical Company
134 W. University Dr.
Suite 101A
Rochester, MI 48063
(313) 656-1000

Products Listed in CPCR

Antor 4ES
Apollo
Betamix
Betanex
Botec

Botran 75W
Carzol SP
Dropp 50WP
Mitac WP
Nortron EC
Rotate 2 1/2 G
*Vorlex

PENNWALT CORPORATION/ AGCHEM DIVISION 1651

Three Parkway, Room 619
Philadelphia, PA 19102
Address inquiries to: Benjamin VanDeest
Director of Marketing and Sales (215) 587-7770
For emergencies contact:
(409) 779-0060

Branch Offices

AgChem Division
7506 E. Independence Blvd.
Suite 127
Quorum Business Park
Charlotte, NC 28227
(704) 568-6892

AgChem Division
8777 Purdue Road
Suite 330
Indianapolis, IN 46268
(317) 875-5803

AgChem Division
516 W. Shaw Avenue
Suite 107
Fresno, CA 93704
(209) 226-8400

Products Listed in CPCR

Accelerate
Aquathol K
Aquathol Granular
Des-I-Cate
*Desiccant L-10
†Endothal Turf Herbicide
Herbicide 273
Hydrothol 191
Hydrothol 191 Granular
Knox Out 2FM
Kryocide
Maneb 80
Maneb Plus Zinc F4
Microthiol Special
*Penncap-M
Penncozeb
Penncozeb DF
†Penncozeb Turf & Ornamental
Topsin M 70W
†Topsin M Turf & Ornamentals
Topsin M 4.5F
†Topsin M 4.5 F Turf & Ornamentals
Ziram F-4
Ziram W-76

RHÔNE-POULENC AG COMPANY 1701

2 T. W. Alexander Drive
Research Triangle Park, NC 27709
Address inquiries to: Communications Dept., P.O. Box 12014
Research Triangle Park, NC 27709
For PRODUCT USE information call 800-334-9745
For EMERGENCIES ONLY call 24 hours a day—1-800-334-7577

Branch Offices

Rhone Poulenc Ag Company
115 E. Main Street, Suite 19
P.O. Box 591
Williamston, NC 27892
(919) 792-8122

Rhone Poulenc Ag Company
11900 W. 87th St. Pkwy., Suite 250
Lenexa, KS 66215
(913) 894-1004

Rhone Poulenc Ag Company
2491 West Shaw Ave., Suite 123
Fresno, CA 93711
(209) 224-8484

Products Listed in CPCR

Aliette
Amiben Granular
Amiben DS
Amid Thin W
*Amitrol T
*Amizol Industrial
Aqua Kleen
Asulox
*Bronate
*Buctril
*Buctril + Atrazine
Butyrac 175
Butyrac 200
Cerone
Chiptox MCPA Sodium
Envert 171
Esteron 99
Ethrel Plant Regulator For Flue-Cured Tobacco
Ethrel Plant Regulator For Pineapple and Sugarcane
Ethrel Plant Regulator
Folex 6EC
Formula 40
Fruitone N
Fruitone CPA
Larvin 3.2
*Liquid Amizine
Mocap 10%
*Mocap 20%
*Mocap EC
*Mocap Plus
*Mocap Plus 4-2 EC
NAA-800
Prep
Rhomene
Rhonox
Rovral
Rovral 4 Flowable
Sevin XLR Plus
Sevin 4-Oil
Sevin 80S
Sevin 50W
Sevin 4F
Tackle Plus
*Temik 15G
*Temik 15G for Citrus
*Temik 15G for CA & AZ Only
*Temik TSX
Thistrol
†Tre-Hold Tree Wound Dressing
Tre-Hold Sprout Inhibitor A112
Tre-Hold Sprout Inhibitor for Citrus
Weedar Emulsamine E-3
Weedar 64
Weedar 64-A
Weedone LV4
Weedone LV6
Weedone 2,4-DP
Weedone 170
Weedone 638
Weedone CB

ROHM AND HAAS COMPANY 1911

Independence Mall West
Philadelphia, PA 19105
Address inquiries to:
Fred W. Belledin, Sales and Marketing Manager, (215) 592-3627
(215) 592-2797 (Fax)
For emergencies contact:
Medical Emergencies (215) 592-3000
CHEMTREC (Transporting Emergencies) (800) 424-9300

Branch Offices

North Central Region
5750 W. Jarvis Avenue
Niles, IL 60648
(312) 774-9100

Eastern Region
Independence Mall West
Philadelphia, PA 19105
(215) 592-3655

South Central Region
1355 Lynnfield Road
Suite 227
Memphis, TN 38119
(901) 682-1533

Western Region
1520 E. Shaw Avenue
Suite 119
Fresno, CA 93710
(209) 224-0158

Canada
2 Manse Road
West Hill, Ontario
Canada M1E 3T9
(416) 284-4711

Corporate Headquarters
Independence Mall West
Philadelphia, PA 19105
(215) 592-3000

Products Listed in CPCR

Dikar
Dithane DF

Dithane F-45
Dithane M-45
†Fore
Goal 1.6E
Karathane LC
Karathane WD
Kelthane MF
Kelthane 35
*Kerb 50-W
Nova
Prostar 4E
Rally 40W AK, ID, MT, NV, OR, UT, WA, WY
Rally 40W AZ, CA, HI
Stam M-4
Stampede CM
Strel 4E
Triton AG-44M Compatibility and Buffering (Acidifying) Agent
Triton AG-98 Spreader Activator
Triton B-1956 Spreader Sticker
Triton CS-7 Spreader Binder

SANDOZ CROP PROTECTION CORPORATION **1981**
1300 E. Touhy Ave.
Des Plaines, IL 60018
(312) 699-1616
Address inquiries to: (312) 699-1616
Emergency Number: (312) 699-1616
(24 hr. answering service)

Branch Offices
Western Region (Ag)
Suite 3700, 4949 Syracuse Pkwy.
Denver, CO 80237
(303) 779-9722
Eastern Region (Ag)
1906 Exeter Rd.
Germantown, TN 38138
(901) 756-8717
Central Region (Ag)
1300 E. Touhy Ave.
Des Plaines, IL 60018
(312) 390-3701

Products Listed in CPCR
Banvel
Banvel SGF
Evital
Javelin WG
Marksman
Probe
Solicam DF
Thiolux DF
Trident
Weedmaster
Zorial Rapid 80

UNIROYAL CHEMICAL COMPANY, INC. **2015**
Middlebury, CT 06749
Address inquiries to:
Judith A. Seweryn (203) 573-2400
For emergencies contact: Uniroyal Chemical Group (203) 723-3670

Products Listed in CPCR
Alanap-L
Apron–Terraclor
†B-Nine SP
†Casoron 2G
Casoron 4G
Casoron 10G
Casoron 50W
Comite
Comite for Citrus
*Dimilin 25W
*Dimilin 25W for Ornamentals
Harvade 5F
Omite 30W
Omite 30W WSP
Omite-CR
Omite-CR for California
Omite-6E
†Ornamite
†Plantvax 75W
Reldan 3%
Rescue
Royal MH-30–Tobacco
Royal MH-30–Potatoes & Onions
Royal MH-30SG—Potatoes & Onions
Royal Slo Gro
Sulv
Terraclor 2 lb. Emulsifiable
*Terraclor 6.5% Plus Di-Syston 6.5% Granular
Terraclor 10% Granular
Terraclor 75% WP
Terraclor with Mocap 10 Plus 3
Terraclor Super-X Emulsifiable
Terraclor Super-X Granular
*Terraclor Super-X/Di-Syston
Terra-Coat LT-2N–Gustafson
†Terrazole 25% Emulsifiable
†Terrazole 35% WP
†Turfcide 10% Granular
†Turfcide 2lb Emulsifiable
Vitavax-200
Vitavax Pour On

VALENT USA CORPORATION **2071**
1333 North California Blvd.
P.O. Box 8025
Walnut Creek, CA 94596-8025
Address inquiries to: See below for nearest regional office
For emergencies contact:
Emergency
Telephone (415) 233-3737

Regional Sales Offices
Valent U.S.A. Corporation
Western Region Office
1052 East Shaw Avenue
Fresno, CA 93710
(209) 224-2513
Valent U.S.A. Corporation
Eastern Region Office
6075 Poplar Avenue, Suite 322
Memphis, TN 38119-4743
(901) 684-4800

Products Listed in CPCR
Bolero 8EC
Bolero 10G
Cobra Herbicide
Dibrom 8 Emulsive
Dibrom 14 Concentrate
Diquat Herbicide H/A
*Monitor 4 Spray
Orthene 75S Soluble Powder
Orthene 80 Seed Protectant
Orthene 90S
Orthene Specialty Concentrate
Orthene Tobacco Insect Spray
†Orthene Turf, Tree & Ornamental Spray
Volck Supreme Spray

6th edition CPCR®

1990

Common and Chemical Name Index

Products included in CPCR are listed here under common and chemical name headings according to each product's active ingredient. Combination products are listed under all appropriate active ingredients.

The headings under which products are listed were determined by CPCR in cooperation with participating manufacturers.

Most products included only in the supplement are not indexed here. Please check the supplement also.

* Restricted Use Only † Only Registered for Turf, Ornamentals and Forestry

* Restricted Use Only † Only Registered for Turf, Ornamentals and Forestry

Always read and follow label directions carefully.

*** Restricted Use Only** **† Only Registered for Turf, Ornamentals and Forestry**

*** Restricted Use Only** **† Only Registered for Turf, Ornamentals and Forestry**

Product Category Index

In this index, products included in CPCR are listed alphabetically under the following categories:

- Desiccants
- Fungicides
- Herbicides
- Insecticides
- Insect Growth Regulators
- Miticides
- Nematicides
- Plant Growth Regulators
- Rodenticides
- Seed Protectants
- Soil Fumigants
- Adjuvants
- Other

The category under which each product is listed was determined by CPCR in cooperation with participating manufacturers.

Most products included only in the supplement are not indexed here. Please check the supplement also.

DESICCANTS

FUNGICIDES

HERBICIDES

*** Restricted Use Only** **† Only Registered for Turf, Ornamentals and Forestry**

* Restricted Use Only † Only Registered for Turf, Ornamentals and Forestry

INSECTICIDES

*** Restricted Use Only** **† Only Registered for Turf, Ornamentals and Forestry**

INSECT GROWTH REGULATORS

MITICIDES

NEMATICIDES

PLANT GROWTH REGULATORS

RODENTICIDES

SEED PROTECTANTS

SOIL FUMIGANTS

ADJUVANTS

OTHER

*** Restricted Use Only** **† Only Registered for Turf, Ornamentals and Forestry**

Crop and Non-Crop Use Index

Products included in CPCR are listed alphabetically under appropriate crop and non-crop headings. The headings under which products are listed were determined by CPCR in cooperation with participating manufacturers. SLN registrations are not identified in this index. Refer to product container label for uses in your area.

The letter code appearing with each product indicates to which pesticide category it belongs:

(D)	Desiccant	(IGR)	Insect Growth Regulator	(R)	Rodenticide
(F)	Fungicide	(M)	Miticide	(SP)	Seed Protectant
(H)	Herbicide	(N)	Nematicide	(SF)	Soil Fumigant
(I)	Insecticide	(PGR)	Plant Growth Regulator	(A)	Adjuvant
				(O)	Other

Most product information included in the supplement is not indexed here. Please check the supplement also.

* **Restricted Use Only** † **Only Registered for Turf, Ornamentals and Forestry**

Always read and follow label directions carefully.

Artichokes

Asparagus

Avocados

B

Bananas

Black-eyed peas

Blueberries

Boysenberries

Broccoli

* **Restricted Use Only** † **Only Registered for Turf, Ornamentals and Forestry**

C

Cabbage

* **Restricted Use Only** † **Only Registered for Turf, Ornamentals and Forestry**

Celery

Chemfallow

Cherries (sweet)

* Restricted Use Only   † Only Registered for Turf, Ornamentals and Forestry

Cocoa

Coffee

Collards

Conifers

Container-grown Nursery Crops (all varieties)

Corms

Corn (field)

Corn (pop)

Corn (sweet)

Cotton

Cowpeas

Crabapples

Cranberries

Crenshaw

Cress (upland)

Cress (water)

Cucumbers

* Restricted Use Only † Only Registered for Turf, Ornamentals and Forestry

* Restricted Use Only † Only Registered for Turf, Ornamentals and Forestry

Forest Nursery and Plantings

Forestry (non-selective)

G

Garlic

* **Restricted Use Only** † **Only Registered for Turf, Ornamentals and Forestry**

Grasses (grown for seed)

Grasses (lawn and turf; not grazed)

Grasses (pasture, forage and rangeland)

*** Restricted Use Only** **† Only Registered for Turf, Ornamentals and Forestry**

* Restricted Use Only † Only Registered for Turf, Ornamentals and Forestry

Lentils

Lespedeza

Lettuce

Limes

*** Restricted Use Only** **† Only Registered for Turf, Ornamentals and Forestry**

P

Papayas

Parsley

Parsnips

Passion Fruit

Peaches

* Restricted Use Only † Only Registered for Turf, Ornamentals and Forestry

Pecans

Peppermint

Peppers

*** Restricted Use Only** **† Only Registered for Turf, Ornamentals and Forestry**

Ponds

Pools

Potatoes

Prunes

* Restricted Use Only † Only Registered for Turf, Ornamentals and Forestry

Soybeans

Standing Water

Stored Products

Strawberries

Sugar Beets

Sugar Cane

* Restricted Use Only   † Only Registered for Turf, Ornamentals and Forestry

Trefoil (big)

Trefoil (birdsfoot)

Turf

(see under grasses: lawn and turf)

Turnips

Turnip Greens

V

Velvet Beans

Vetch

W

Walnuts (black)

*** Restricted Use Only** **† Only Registered for Turf, Ornamentals and Forestry**

Wheat (Spring planted)

Wild Rice

Wineries

Y

Youngberries

Pest Use Index

Products included in CPCR are listed alphabetically under appropriate pest headings. The major pests and pest groups were selected as sub-headings. The headings and sub-headings under which products are listed were determined by CPCR in cooperation with participating manufacturers.

Latin names have been added to most of the insect, mite, nematode and weed headings in an effort to minimize misunderstanding. The Latin names of the plant diseases were significantly more difficult to identify properly, and are not included in this edition.

It should be noted that individual products may be approved for use against pests other than those listed in this index. Additional or more specific pest listings may be found on the product label. SLN registrations are not identified in this index. Refer to product container label for uses in your area.

Most product information included in the supplement is not indexed here. Please check the supplement also.

The following major pest headings are included in this index:

DISEASES

* Restricted Use Only   † Only Registered for Turf, Ornamentals and Forestry

Always read and follow label directions carefully.

Dithane F-45 (Rohm and Haas) p 1920
Dithane M-45 (Rohm and Haas) p 1927
Manzate 200 DF (DuPont) p Suppl.
Nova (Rohm and Haas) p 1965
Penncozeb (Pennwalt) p 1676
Penncozeb DF (Pennwalt) p 1678
Rally 40W AK, ID, MT, NV, OR, UT, WA, WY (Rohm and Haas) p 1968
Rally 40W AZ, CA, HI (Rohm and Haas) p 1970
Terraclor 75% WP (Uniroyal Chemical) p 2056
Topsin M 70W (Pennwalt) p 1683
Topsin M 4.5F (Pennwalt) p 1690
†Turfcide 10% Granular (Uniroyal Chemical) p 2066

Black spot

Carbamate WDG (FMC) p 1041
†Daconil 2787 Flowable (Fermenta) p 1014
†Daconil 2787 Fungicide (Fermenta) p 1016
†Daconil 2787 WDG (Fermenta) p 1019
Dithane F-45 (Rohm and Haas) p 1920
Dithane M-45 (Rohm and Haas) p 1927
Maneb 80 (Pennwalt) p 1665
Maneb Plus Zinc F4 (Pennwalt) p 1667
Manzate 200 DF (DuPont) p Suppl.
†Penncozeb Turf & Ornamental (Pennwalt) p 1681
Sulfur Alfa (ICI) p 1286
Sulfur Signal (ICI) p 1287
Sulfur Special Electric (ICI) p 1288
†Tersan 1991 DF (DuPont) p 894
Thiolux DF (Sandoz) p 2006
†Topsin M Turf & Ornamentals (Pennwalt) p 1689
†Topsin M 4.5 F Turf & Ornamentals (Pennwalt) p 1695
Ziram F-4 (Pennwalt) p 1697
Ziram 76 WDG (FMC) p 1101
Ziram W-76 (Pennwalt) p 1698

Blast

Bravo 500 (Fermenta) p 992
Bravo W-75 (Fermenta) p 996
Bravo 720 (Fermenta) p 999
Bravo 90DG (Fermenta) p 1004

Blight

(see also under individual species)
Aliette (Rhone–Poulenc) p 1701
Benlate (DuPont) p 724
Benlate 50 DF (DuPont) p 727
Botran 75W (NOR-AM) p 1633
†Daconil 2787 Flowable (Fermenta) p 1014
†Daconil 2787 Fungicide (Fermenta) p 1016
†Daconil 2787 WDG (Fermenta) p 1019
Dithane DF (Rohm and Haas) p 1913
Dithane F-45 (Rohm and Haas) p 1920
Dyfonate II 10-G (ICI) p 1187
†Fore (Rohm and Haas) p 1935
Maneb 80 (Pennwalt) p 1665
Maneb Plus Zinc F4 (Pennwalt) p 1667
Manzate 200 DF (DuPont) p Suppl.
Penncozeb (Pennwalt) p 1676
Penncozeb DF (Pennwalt) p 1678
†Penncozeb Turf & Ornamental (Pennwalt) p 1681
Rovral (Rhone–Poulenc) p 1808
Rovral 4 Flowable (Rhone–Poulenc) p 1812
Sulfur Alfa (ICI) p 1286
Terraclor 75% WP (Uniroyal Chemical) p 2056
†Tersan 1991 DF (DuPont) p 894
†Topsin M Turf & Ornamentals (Pennwalt) p 1689
†Topsin M 4.5 F Turf & Ornamentals (Pennwalt) p 1695

Blossom blight

Bayleton 50% WP PVA (Mobay) p 1337
Bayleton 50% DF (Mobay) p 1340
Benlate (DuPont) p 724
Benlate 50 DF (DuPont) p 727
Botran 75W (NOR-AM) p 1633
Bravo 500 (Fermenta) p 992
Bravo 720 (Fermenta) p 999
Bravo 90DG (Fermenta) p 1004
Carbamate WDG (FMC) p 1041
†Daconil 2787 Flowable (Fermenta) p 1014
†Daconil 2787 Fungicide (Fermenta) p 1016
†Daconil 2787 WDG (Fermenta) p 1019
Dithane DF (Rohm and Haas) p 1913
Dithane F-45 (Rohm and Haas) p 1920
Dithane M-45 (Rohm and Haas) p 1927
†Fore (Rohm and Haas) p 1935
Funginex 1.6 EC (FMC) p 1063
Maneb 80 (Pennwalt) p 1665
Maneb Plus Zinc F4 (Pennwalt) p 1667
†Penncozeb Turf & Ornamental (Pennwalt) p 1681
Ronilan WP (BASF) p 480
Ronilan FL (BASF) p 484
Rovral (Rhone–Poulenc) p 1808
Rovral 4 Flowable (Rhone–Poulenc) p 1812
Sulfur Alfa (ICI) p 1286
Sulfur Signal (ICI) p 1287
Sulfur Special Electric (ICI) p 1288
Topsin M 70W (Pennwalt) p 1683
Topsin M 4.5F (Pennwalt) p 1690
Ziram F-4 (Pennwalt) p 1697
Ziram 76 WDG (FMC) p 1101
Ziram W-76 (Pennwalt) p 1698

Blotch

(see under individual species)

Blue mold

Benlate (DuPont) p 724
Benlate 50 DF (DuPont) p 727
Carbamate WDG (FMC) p 1041
Ridomil 2E (Ciba-Geigy) p Suppl.

Branch blight

†Daconil 2787 Flowable (Fermenta) p 1014
†Daconil 2787 Fungicide (Fermenta) p 1016
†Daconil 2787 WDG (Fermenta) p 1019

Branch canker

Aliette (Rhone–Poulenc) p 1701

Brown blotch

Tilt (Ciba-Geigy) p Suppl.

Brown fruit spot

Bravo 500 (Fermenta) p 992
Bravo 720 (Fermenta) p 999
Bravo 90DG (Fermenta) p 1004

Brown leaf spot

Benlate (DuPont) p 724
Benlate 50 DF (DuPont) p 727
Bravo 500 (Fermenta) p 992
Bravo 720 (Fermenta) p 999
Bravo 90DG (Fermenta) p 1004
Cyprex 65-W (Cyanamid) p 297
Tilt (Ciba-Geigy) p Suppl.

Brown mold

Elgetol (FMC) p 1061

Brown patch

†Daconil 2787 Flowable (Fermenta) p 1014
†Daconil 2787 Fungicide (Fermenta) p 1016
†Daconil 2787 WDG (Fermenta) p 1019
Dithane DF (Rohm and Haas) p 1913
Dithane F-45 (Rohm and Haas) p 1920
†Fore (Rohm and Haas) p 1935
Maneb Plus Zinc F4 (Pennwalt) p 1667
Manzate 200 DF (DuPont) p Suppl.
†Penncozeb Turf & Ornamental (Pennwalt) p 1681
Terraclor 2 lb. Emulsifiable (Uniroyal Chemical) p 2052
Terraclor 75% WP (Uniroyal Chemical) p 2056
†Tersan 1991 DF (DuPont) p 894
†Topsin M Turf & Ornamentals (Pennwalt) p 1689
†Topsin M 4.5 F Turf & Ornamentals (Pennwalt) p 1695
†Turfcide 10% Granular (Uniroyal Chemical) p 2066
†Turfcide 2lb Emulsifiable (Uniroyal Chemical) p 2067

Brown rot

Benlate (DuPont) p 724
Benlate 50 DF (DuPont) p 727
Botran 75W (NOR-AM) p 1633
Carbamate WDG (FMC) p 1041
Cyprex 65-W (Cyanamid) p 297
Dusting Sulfur (FMC) p 1060
Funginex 1.6 EC (FMC) p 1063
Maneb 80 (Pennwalt) p 1665
Maneb Plus Zinc F4 (Pennwalt) p 1667
Manzate 200 DF (DuPont) p Suppl.
Ronilan WP (BASF) p 480
Ronilan FL (BASF) p 484
Rovral (Rhone–Poulenc) p 1808
Rovral 4 Flowable (Rhone–Poulenc) p 1812
Sulfur Alfa (ICI) p 1286
Sulfur Signal (ICI) p 1287
Sulfur Special Electric (ICI) p 1288
Thiolux DF (Sandoz) p 2006
Topsin M 70W (Pennwalt) p 1683
Topsin M 4.5F (Pennwalt) p 1690
Wettable Sulfur (FMC) p 1099
Ziram F-4 (Pennwalt) p 1697
Ziram 76 WDG (FMC) p 1101
Ziram W-76 (Pennwalt) p 1698

Bubbles

Elgetol (FMC) p 1061

Bud blight

Dithane DF (Rohm and Haas) p 1913
Dithane F-45 (Rohm and Haas) p 1920
Dithane M-45 (Rohm and Haas) p 1927
†Fore (Rohm and Haas) p 1935

* Restricted Use Only † Only Registered for Turf, Ornamentals and Forestry

*** Restricted Use Only** **† Only Registered for Turf, Ornamentals and Forestry**

* **Restricted Use Only** † **Only Registered for Turf, Ornamentals and Forestry**

*** Restricted Use Only** **† Only Registered for Turf, Ornamentals and Forestry**

INSECTS

A

* Restricted Use Only † Only Registered for Turf, Ornamentals and Forestry

* Restricted Use Only † Only Registered for Turf, Ornamentals and Forestry

*** Restricted Use Only** **† Only Registered for Turf, Ornamentals and Forestry**

*** Restricted Use Only** **† Only Registered for Turf, Ornamentals and Forestry**

*** Restricted Use Only** † **Only Registered for Turf, Ornamentals and Forestry**

Green fruitworm (*Lithophane antennata*)
*Ambush (ICI) p 1134
*Ambush 25W (ICI) p 1124
*Ambush 25W-WSP (ICI) p 1129
*Asana (DuPont) p 700
*Asana XL (DuPont) p 706
*Guthion 2S (Mobay) p 1372
Guthion 35% WP (Mobay) p 1389
Guthion 50% WP Water Soluble Packets (Mobay) p 1397
Imidan 50-WP (ICI) p 1266
Lannate (DuPont) p 822
*Lannate L (DuPont) p 827
*Lannate LV (DuPont) p 833
Lorsban 50W (Dow) p 645
*Penncap-M (Pennwalt) p 1670
*Pounce 3.2 EC (FMC) p 1080
*Pounce 25 WP (FMC) p 1084
*Pydrin 2.4 (DuPont) p 880
Sevin XLR Plus (Rhone–Poulenc) p 1816
Sevin 80S (Rhone–Poulenc) p 1830
Sevin 50W (Rhone–Poulenc) p 1841
Sevin 4F (Rhone–Poulenc) p 1848
Thiodan 2 CO EC (FMC) p 1091
Thiodan 3 EC (FMC) p 1093
Thiodan 50 WP (FMC) p 1096

Green stink bug (*Acrosternum hilare*)
*Asana (DuPont) p 700
*Guthion 2S (Mobay) p 1372
Guthion 35% WP (Mobay) p 1389
Guthion 50% WP Water Soluble Packets (Mobay) p 1397
*Karate (ICI) p 1268
*Methyl Parathion 2 Thiodan 3 EC (FMC) p 1075
Sevin XLR Plus (Rhone–Poulenc) p 1816
Sevin 80S (Rhone–Poulenc) p 1830
Sevin 50W (Rhone–Poulenc) p 1841
Sevin 4F (Rhone–Poulenc) p 1848

Grubs
(see under White grub)

Gypsy moth (*Lymantria dispar*)
*Dimilin 25W (Uniroyal Chemical) p 2028
DiPel 2X (Abbott) p 254
DiPel 4L (Abbott) p 257
†DiPel 8AF (Abbott) p 259
†DiPel 8L (Abbott) p 259
Imidan 50-WP (ICI) p 1266
Kryocide (Pennwalt) p 1663
Orthene Specialty Concentrate (Valent) p 2093
Sevin XLR Plus (Rhone–Poulenc) p 1816
Sevin 4-Oil (Rhone–Poulenc) p 1827
Sevin 80S (Rhone–Poulenc) p 1830
Sevin 50W (Rhone–Poulenc) p 1841

H

Harlequin bug (*Murgantia histrionica*)
D•z•n diazinon 50W (Ciba-Geigy) p 541
D•z•n diazinon AG500 (Ciba-Geigy) p 546
*Methyl Parathion 2 Thiodan 3 EC (FMC) p 1075
Sevin XLR Plus (Rhone–Poulenc) p 1816
Sevin 80S (Rhone–Poulenc) p 1830
Sevin 50W (Rhone–Poulenc) p 1841
Sevin 4F (Rhone–Poulenc) p 1848
Thiodan 2 CO EC (FMC) p 1091
Thiodan 3 EC (FMC) p 1093
Thiodan 50 WP (FMC) p 1096

Hessian fly (*Mayetiola destructor*)
*Di-Syston 15% G (Mobay) p 1351
*Di-Syston 8 (Mobay) p 1355
*Thimet 15-G (Cyanamid) p 368
*Thimet 20-G (Cyanamid) p 372

Hickory shuckworm (*Laspeyresia caryana*)
*Ammo 2.5 EC (FMC) p 1035
*Ammo WSB (FMC) p 1037
*Asana (DuPont) p 700
*Asana XL (DuPont) p 706
*Cymbush 3E (ICI) p 1150
*Guthion 2S (Mobay) p 1372
*Guthion 2L (Mobay) p 1380
Guthion 35% WP (Mobay) p 1389
Guthion 50% WP Water Soluble Packets (Mobay) p 1397
Imidan 50-WP (ICI) p 1266
Lorsban 4E (Dow) p 632
Lorsban 50W (Dow) p 645
*Pydrin 2.4 (DuPont) p 880
Sevin XLR Plus (Rhone–Poulenc) p 1816
Sevin 80S (Rhone–Poulenc) p 1830
Sevin 50W (Rhone–Poulenc) p 1841
Sevin 4F (Rhone–Poulenc) p 1848
*Supracide 2E (Ciba-Geigy) p Suppl.

Honey bee (*Apis mellifera*)
Demon EC (ICI) p 1152
Demon TC (ICI) p 1154

Horn fly (*Haematobia irritans*)
Dibrom 8 Emulsive (Valent) p 2078
Dibrom 14 Concentrate (Valent) p 2080

I

Imported cabbageworm (*Artogeia rapae*)
*Ambush (ICI) p 1134
*Ambush 25W (ICI) p 1124
*Ambush 25W-WSP (ICI) p 1129
*Asana (DuPont) p 700
*Asana XL (DuPont) p 706
D•z•n diazinon 50W (Ciba-Geigy) p 541
D•z•n diazinon AG500 (Ciba-Geigy) p 546
Dibrom 8 Emulsive (Valent) p 2078
DiPel 2X (Abbott) p 254
DiPel 4L (Abbott) p 257
*Guthion 2S (Mobay) p 1372
*Guthion 2L (Mobay) p 1380
Guthion 35% WP (Mobay) p 1389
Guthion 50% WP Water Soluble Packets (Mobay) p 1397
Javelin WG (Sandoz) p 1995
Kryocide (Pennwalt) p 1663
Lannate (DuPont) p 822
*Lannate L (DuPont) p 827
*Lannate LV (DuPont) p 833
Lorsban 50W (Dow) p 645
*Methyl Parathion 2 Thiodan 3 EC (FMC) p 1075
*Monitor 4 (Mobay) p 1412
*Monitor 4 Spray (Valent) p 2083
Orthene 75S Soluble Powder (Valent) p 2085
*Penncap-M (Pennwalt) p 1670
*Pounce 3.2 EC (FMC) p 1080
*Pounce 25 WP (FMC) p 1084
*Pydrin 2.4 (DuPont) p 880
Sevin XLR Plus (Rhone–Poulenc) p 1816
Sevin 80S (Rhone–Poulenc) p 1830
Sevin 50W (Rhone–Poulenc) p 1841
Sevin 4F (Rhone–Poulenc) p 1848
Thiodan 2 CO EC (FMC) p 1091
Thiodan 3 EC (FMC) p 1093
Thiodan 50 WP (FMC) p 1096

Imported fire ant (*Solenopsis*)
Lorsban 4E (Dow) p 632
Orthene 75S Soluble Powder (Valent) p 2085
†Orthene Turf, Tree & Ornamental Spray (Valent) p 2095
Sevin XLR Plus (Rhone–Poulenc) p 1816
Sevin 80S (Rhone–Poulenc) p 1830
Sevin 50W (Rhone–Poulenc) p 1841

Indian meal moth (*Plodia interpunctella*)
Actellic 5E (ICI) p 1123
D•z•n diazinon 4E (Ciba-Geigy) p 536
DiPel 2X (Abbott) p 254
DiPel 4L (Abbott) p 257
Reldan 3% (Uniroyal Chemical) p 2042

J

Japanese beetle (*Popillia japonica*)
*Ambush (ICI) p 1134
*Ambush 25W (ICI) p 1124
*Ambush 25W-WSP (ICI) p 1129
*Asana (DuPont) p 700
*Asana XL (DuPont) p 706
Cythion ULV (Cyanamid) p 298
D•z•n diazinon 14G (Ciba-Geigy) p 538
Imidan 50-WP (ICI) p 1266
†Orthene Turf, Tree & Ornamental Spray (Valent) p 2095
*Penncap-M (Pennwalt) p 1670
*Pounce 3.2 EC (FMC) p 1080
*Pydrin 2.4 (DuPont) p 880
Sevin XLR Plus (Rhone–Poulenc) p 1816
Sevin 4-Oil (Rhone–Poulenc) p 1827
Sevin 80S (Rhone–Poulenc) p 1830
Sevin 50W (Rhone–Poulenc) p 1841
Sevin 4F (Rhone–Poulenc) p 1848
†Talstar 10 WP (FMC) p 1088
*Vydate L (DuPont) p 914

June beetle (*Phyllophaga spp.*)
D•z•n diazinon 50W (Ciba-Geigy) p 541
D•z•n diazinon AG500 (Ciba-Geigy) p 546
Dylox 80% SP (Mobay) p 1361
Sevin XLR Plus (Rhone–Poulenc) p 1816
Sevin 80S (Rhone–Poulenc) p 1830
Sevin 50W (Rhone–Poulenc) p 1841
Sevin 4F (Rhone–Poulenc) p 1848

L

Lace bug (*Tingidae*)
Cythion 57% EC (Cyanamid) p 303
*Di-Syston 15% G (Mobay) p 1351
*Guthion 2S (Mobay) p 1372
†Orthene Turf, Tree & Ornamental Spray (Valent) p 2095
Sevin XLR Plus (Rhone–Poulenc) p 1816

*** Restricted Use Only** **† Only Registered for Turf, Ornamentals and Forestry**

Imidan 50-WP (ICI) p 1266
*Penncap-M (Pennwalt) p 1670
Sevin XLR Plus (Rhone–Poulenc) p 1816
Sevin 80S (Rhone–Poulenc) p 1830
Sevin 50W (Rhone–Poulenc) p 1841
Sevin 4F (Rhone–Poulenc) p 1848
Thiodan 2 CO EC (FMC) p 1091
Thiodan 3 EC (FMC) p 1093
Thiodan 50 WP (FMC) p 1096

Peach tree borer (*Synanthedon exitiosa*)

*Guthion 2S (Mobay) p 1372
*Guthion 2L (Mobay) p 1380
Guthion 35% WP (Mobay) p 1389
Guthion 50% WP Water Soluble Packets (Mobay) p 1397
Lorsban 4E (Dow) p 632
Lorsban 50W (Dow) p 645
*Penncap-M (Pennwalt) p 1670
*Pounce 25 WP (FMC) p 1084
*Pydrin 2.4 (DuPont) p 880
Thiodan 2 CO EC (FMC) p 1091
Thiodan 3 EC (FMC) p 1093
Thiodan 50 WP (FMC) p 1096

Peach twig borer (*Anarsia lineatella*)

*Ambush (ICI) p 1134
*Ambush 25W (ICI) p 1124
*Ambush 25W-WSP (ICI) p 1129
D•z•n diazinon 50W (Ciba-Geigy) p 541
D•z•n diazinon AG500 (Ciba-Geigy) p 546
Dibrom 8 Emulsive (Valent) p 2078
*Guthion 2L (Mobay) p 1380
Guthion 35% WP (Mobay) p 1389
Guthion 50% WP Water Soluble Packets (Mobay) p 1397
Imidan 50-WP (ICI) p 1266
Lorsban 4E (Dow) p 632
Lorsban 50W (Dow) p 645
*Penncap-M (Pennwalt) p 1670
*Pounce 3.2 EC (FMC) p 1080
*Pounce 25 WP (FMC) p 1084
*Pydrin 2.4 (DuPont) p 880
Sevin XLR Plus (Rhone–Poulenc) p 1816
Sevin 80S (Rhone–Poulenc) p 1830
Sevin 50W (Rhone–Poulenc) p 1841
Sevin 4F (Rhone–Poulenc) p 1848
*Supracide 2E (Ciba-Geigy) p Suppl.
Thiodan 2 CO EC (FMC) p 1091
Thiodan 3 EC (FMC) p 1093
Thiodan 50 WP (FMC) p 1096

Pealeaf weevil (*Sitona lineatus*)

*Guthion 2S (Mobay) p 1372
*Guthion 2L (Mobay) p 1380
Imidan 50-WP (ICI) p 1266
*Penncap-M (Pennwalt) p 1670
Sevin 50W (Rhone–Poulenc) p 1841
Sevin 4F (Rhone–Poulenc) p 1848

Pear plant bug (*Lygocoris communis*)

*Penncap-M (Pennwalt) p 1670

Pear psylla (*Psylla pyricola*)

*Ambush (ICI) p 1134
*Ambush 25W (ICI) p 1124
*Ambush 25W-WSP (ICI) p 1129
Cythion 57% EC (Cyanamid) p 303
D•z•n diazinon 50W (Ciba-Geigy) p 541
Dikar (Rohm and Haas) p 1911
Imidan 50-WP (ICI) p 1266
Lorsban 4E (Dow) p 632
Mitac WP (NOR-AM) p 1641
Morestan 25% WP (Mobay) p 1417
*Pounce 3.2 EC (FMC) p 1080
*Pounce 25 WP (FMC) p 1084
*Pydrin 2.4 (DuPont) p 880
Sevin XLR Plus (Rhone–Poulenc) p 1816
Sevin 80S (Rhone–Poulenc) p 1830
Sevin 50W (Rhone–Poulenc) p 1841
Sevin 4F (Rhone–Poulenc) p 1848
*Supracide 2E (Ciba-Geigy) p Suppl.
Thiodan 2 CO EC (FMC) p 1091
Thiodan 3 EC (FMC) p 1093
Thiodan 50 WP (FMC) p 1096
Volck Supreme Spray (Valent) p 2099

Pear slug

D•z•n diazinon 4E (Ciba-Geigy) p 536
D•z•n diazinon 50W (Ciba-Geigy) p 541
D•z•n diazinon AG500 (Ciba-Geigy) p 546
*Pydrin 2.4 (DuPont) p 880
Sevin XLR Plus (Rhone–Poulenc) p 1816
Sevin 50W (Rhone–Poulenc) p 1841
Sevin 4F (Rhone–Poulenc) p 1848

Pecan weevil (*Curculio caryae*)

*Ammo 2.5 EC (FMC) p 1035
*Ammo WSB (FMC) p 1037
*Asana (DuPont) p 700
*Asana XL (DuPont) p 706
*Cymbush 3E (ICI) p 1150
Imidan 50-WP (ICI) p 1266
Lorsban 50W (Dow) p 645
*Pydrin 2.4 (DuPont) p 880
Sevin XLR Plus (Rhone–Poulenc) p 1816
Sevin 80S (Rhone–Poulenc) p 1830
Sevin 50W (Rhone–Poulenc) p 1841
Sevin 4F (Rhone–Poulenc) p 1848

Pepper maggot (*Zonosemata electa*)

Thiodan 2 CO EC (FMC) p 1091
Thiodan 3 EC (FMC) p 1093
Thiodan 50 WP (FMC) p 1096

Pepper weevil (*Anthonomus eugenii*)

*Ambush (ICI) p 1134
*Ambush 25W (ICI) p 1124
*Ambush 25W-WSP (ICI) p 1129
*Asana XL (DuPont) p 706
Kryocide (Pennwalt) p 1663
*Pounce 3.2 EC (FMC) p 1080
*Pounce 25 WP (FMC) p 1084
*Pydrin 2.4 (DuPont) p 880
*Vydate L (DuPont) p 914

Phylloxera (*Phylloxeridae*)

*Asana XL (DuPont) p 706
Cythion 57% EC (Cyanamid) p 303
*Furadan 4F (FMC) p 1066
*Furadan 15G (FMC) p 1070
Lorsban 4E (Dow) p 632
Lorsban 50W (Dow) p 645
Sevin 50W (Rhone–Poulenc) p 1841
Sevin 4F (Rhone–Poulenc) p 1848
Thiodan 2 CO EC (FMC) p 1091
Thiodan 3 EC (FMC) p 1093
Thiodan 50 WP (FMC) p 1096

Pickleworm (*Diaphania nitidalis*)

*Ambush (ICI) p 1134
*Ambush 25W (ICI) p 1124
*Ambush 25W-WSP (ICI) p 1129
*Asana (DuPont) p 700
*Asana XL (DuPont) p 706
Dibrom 8 Emulsive (Valent) p 2078
Kryocide (Pennwalt) p 1663
Lannate (DuPont) p 822
*Lannate L (DuPont) p 827
*Lannate LV (DuPont) p 833
*Pounce 25 WP (FMC) p 1084
*Pydrin 2.4 (DuPont) p 880
Sevin XLR Plus (Rhone–Poulenc) p 1816
Sevin 80S (Rhone–Poulenc) p 1830
Sevin 50W (Rhone–Poulenc) p 1841
Sevin 4F (Rhone–Poulenc) p 1848
Thiodan 2 CO EC (FMC) p 1091
Thiodan 3 EC (FMC) p 1093
Thiodan 50 WP (FMC) p 1096

Pillbug (*Crustacea*)

Demon EC (ICI) p 1152
Demon TC (ICI) p 1154
†Demon WP (ICI) p 1156
Demon WP Packet (4 oz.) (ICI) p 1159
†Demon WP Water Soluble Packets (ICI) p 1158

Pineapple scale (*Diaspis bromeliae*)

D•z•n diazinon 50W (Ciba-Geigy) p 541
D•z•n diazinon AG500 (Ciba-Geigy) p 546

Pink bollworm (*Pectinophora gossypiella*)

*Ambush (ICI) p 1134
*Ambush 25W (ICI) p 1124
*Ambush 25W-WSP (ICI) p 1129
*Ammo 2.5 EC (FMC) p 1035
*Ammo WSB (FMC) p 1037
*Asana XL (DuPont) p 706
*Baythroid 2 (Mobay) p 1344
*Bolstar 6 (Mobay) p 1346
*Capture 2EC (FMC) p 1039
*Cymbush 3E (ICI) p 1150
DiPel 2X (Abbott) p 254
DiPel 4L (Abbott) p 257
*Guthion 2S (Mobay) p 1372
*Guthion 2L (Mobay) p 1380
*Karate (ICI) p 1268
Larvin 3.2 (Rhone–Poulenc) p 1780
Lorsban 4E (Dow) p 632
Orthene 75S Soluble Powder (Valent) p 2085
Orthene 90S (Valent) p 2091
*Pounce 3.2 EC (FMC) p 1080
*Pounce 25 WP (FMC) p 1084
*Scout X-TRA (Hoechst) p 1115
Sevin XLR Plus (Rhone–Poulenc) p 1816
Sevin 80S (Rhone–Poulenc) p 1830
Sevin 50W (Rhone–Poulenc) p 1841
Sevin 4F (Rhone–Poulenc) p 1848
*Supracide 2E (Ciba-Geigy) p Suppl.

Plum curculio (*Conotrachelus nenuphar*)

*Ambush (ICI) p 1134
*Ambush 25W (ICI) p 1124
*Ambush 25W-WSP (ICI) p 1129
*Asana (DuPont) p 700
*Asana XL (DuPont) p 706
Cythion 57% EC (Cyanamid) p 303

* Restricted Use Only   † Only Registered for Turf, Ornamentals and Forestry

*** Restricted Use Only** **† Only Registered for Turf, Ornamentals and Forestry**

Always read and follow label directions carefully.

S

*** Restricted Use Only** **† Only Registered for Turf, Ornamentals and Forestry**

* Restricted Use Only   † Only Registered for Turf, Ornamentals and Forestry

Tomato hornworm (*Manduca quinquemaculata*)
*Asana (DuPont) p 700
*Asana XL (DuPont) p 706
Dibrom 8 Emulsive (Valent) p 2078
DiPel 2X (Abbott) p 254
DiPel 4L (Abbott) p 257
Dylox 80% SP (Mobay) p 1361
Lannate (DuPont) p 822
*Pounce 25 WP (FMC) p 1084
*Pydrin 2.4 (DuPont) p 880
Sevin XLR Plus (Rhone–Poulenc) p 1816
Sevin 80S (Rhone–Poulenc) p 1830
Sevin 4F (Rhone–Poulenc) p 1848
Thiodan 50 WP (FMC) p 1096

Tomato pinworm (*Keiferia lycopersicella*)
*Ambush (ICI) p 1134
*Ambush 25W (ICI) p 1124
*Ambush 25W-WSP (ICI) p 1129
*Asana XL (DuPont) p 706
DiPel 2X (Abbott) p 254
DiPel 4L (Abbott) p 257
*Guthion 2S (Mobay) p 1372
Kryocide (Pennwalt) p 1663
Lannate (DuPont) p 822
*Lannate L (DuPont) p 827
*Lannate LV (DuPont) p 833
*Penncap-M (Pennwalt) p 1670
*Pounce 3.2 EC (FMC) p 1080
*Pounce 25 WP (FMC) p 1084
*Pydrin 2.4 (DuPont) p 880
Sevin XLR Plus (Rhone–Poulenc) p 1816
Sevin 80S (Rhone–Poulenc) p 1830
Sevin 50W (Rhone–Poulenc) p 1841
Sevin 4F (Rhone–Poulenc) p 1848

Tufted apple budmoth (*Platynota idaeusalis*)
*Asana (DuPont) p 700
*Asana XL (DuPont) p 706
DiPel 2X (Abbott) p 254
DiPel 4L (Abbott) p 257
Imidan 50-WP (ICI) p 1266
Lannate (DuPont) p 822
*Lannate L (DuPont) p 827
*Lannate LV (DuPont) p 833
*Penncap-M (Pennwalt) p 1670
*Pydrin 2.4 (DuPont) p 880

Tussock moth (*Lymantriidae*)
*Dimilin 25W (Uniroyal Chemical) p 2028
DiPel 2X (Abbott) p 254
DiPel 4L (Abbott) p 257
†DiPel 8AF (Abbott) p 259
†DiPel 8L (Abbott) p 259
*Guthion 2L (Mobay) p 1380
Guthion 35% WP (Mobay) p 1389
Guthion 50% WP Water Soluble Packets (Mobay) p 1397
Lannate (DuPont) p 822
*Lannate L (DuPont) p 827
*Lannate LV (DuPont) p 833
Orthene Specialty Concentrate (Valent) p 2093
†Orthene Turf, Tree & Ornamental Spray (Valent) p 2095
Sevin XLR Plus (Rhone–Poulenc) p 1816
Sevin 4-Oil (Rhone–Poulenc) p 1827
Sevin 80S (Rhone–Poulenc) p 1830
Sevin 50W (Rhone–Poulenc) p 1841

Twostriped grasshopper (*Melanoplus bivittatus*)
Sevin 4F (Rhone–Poulenc) p 1848

V

Variegated cutworm (*Peridroma saucia*)
DiPel 2X (Abbott) p 254
DiPel 4L (Abbott) p 257
Dylox 80% SP (Mobay) p 1361
Kryocide (Pennwalt) p 1663
Lannate (DuPont) p 822
*Lannate L (DuPont) p 827
*Lannate LV (DuPont) p 833
Lorsban 4E (Dow) p 632
Sevin XLR Plus (Rhone–Poulenc) p 1816
Sevin 80S (Rhone–Poulenc) p 1830
Sevin 50W (Rhone–Poulenc) p 1841
Sevin 4F (Rhone–Poulenc) p 1848

Vegetable weevil (*Listroderes costirostris obliquus*)
*Penncap-M (Pennwalt) p 1670

Velvetbean caterpillar (*Anticarsia gemmatalis*)
*Ambush (ICI) p 1134
*Ambush 25W (ICI) p 1124
*Ambush 25W-WSP (ICI) p 1129
*Asana (DuPont) p 700
*Asana XL (DuPont) p 706
*Bolstar 6 (Mobay) p 1346
*Dimilin 25W (Uniroyal Chemical) p 2028
DiPel ES East (Abbott) p 251
DiPel 2X (Abbott) p 254
DiPel 4L (Abbott) p 257
*Guthion 2S (Mobay) p 1372
*Guthion 2L (Mobay) p 1380
Guthion 35% WP (Mobay) p 1389
Guthion 50% WP Water Soluble Packets (Mobay) p 1397
Lannate (DuPont) p 822
*Lannate L (DuPont) p 827
*Lannate LV (DuPont) p 833
Larvin 3.2 (Rhone–Poulenc) p 1780
Lorsban 4E (Dow) p 632
Orthene 75S Soluble Powder (Valent) p 2085
*Penncap-M (Pennwalt) p 1670
*Pounce 3.2 EC (FMC) p 1080
*Pounce 25 WP (FMC) p 1084
*Pydrin 2.4 (DuPont) p 880
*Scout X-TRA (Hoechst) p 1115
Sevin XLR Plus (Rhone–Poulenc) p 1816
Sevin 80S (Rhone–Poulenc) p 1830
Sevin 50W (Rhone–Poulenc) p 1841
Sevin 4F (Rhone–Poulenc) p 1848

W

Walnut caterpillar (*Datana integerrima*)
D•z•n diazinon 50W (Ciba-Geigy) p 541
D•z•n diazinon AG500 (Ciba-Geigy) p 546
DiPel 2X (Abbott) p 254
DiPel 4L (Abbott) p 257
*Guthion 2S (Mobay) p 1372
*Guthion 2L (Mobay) p 1380
Guthion 35% WP (Mobay) p 1389
Guthion 50% WP Water Soluble Packets (Mobay) p 1397
Sevin XLR Plus (Rhone–Poulenc) p 1816
Sevin 80S (Rhone–Poulenc) p 1830
Sevin 50W (Rhone–Poulenc) p 1841
Sevin 4F (Rhone–Poulenc) p 1848

Walnut husk fly (*Rhagoletis completa*)
*Ambush (ICI) p 1134
*Ambush 25W (ICI) p 1124
*Ambush 25W-WSP (ICI) p 1129
*Asana (DuPont) p 700
*Asana XL (DuPont) p 706
*Pounce 3.2 EC (FMC) p 1080
*Pounce 25 WP (FMC) p 1084
*Pydrin 2.4 (DuPont) p 880

Wasps
Demon EC (ICI) p 1152
Demon TC (ICI) p 1154
†Demon WP (ICI) p 1156
Demon WP Packet (4 oz.) (ICI) p 1159
†Demon WP Water Soluble Packets (ICI) p 1158

Weevil
(see also under individual species)
*Asana XL (DuPont) p 706
Carzol SP (NOR-AM) p 1637
*Furadan 4F (FMC) p 1066
*Furadan 4F (Mobay) p 1367
*Furadan 15G (FMC) p 1070
*Guthion 2L (Mobay) p 1380
Imidan 50-WP (ICI) p 1266
Knox Out 2FM (Pennwalt) p 1662
Kryocide (Pennwalt) p 1663
Lannate (DuPont) p 822
*Lannate L (DuPont) p 827
*Lannate LV (DuPont) p 833
†Orthene Turf, Tree & Ornamental Spray (Valent) p 2095
*Pounce 3.2 EC (FMC) p 1080
*Pydrin 2.4 (DuPont) p 880
Sevin XLR Plus (Rhone–Poulenc) p 1816
Sevin 80S (Rhone–Poulenc) p 1830
Sevin 50W (Rhone–Poulenc) p 1841
Sevin 4F (Rhone–Poulenc) p 1848
*Supracide 2E (Ciba-Geigy) p Suppl.

Western bean cutworm (*Loxagrotis albicosta*)
*Ambush 25W (ICI) p 1124
*Ambush 25W-WSP (ICI) p 1129
*Asana (DuPont) p 700
*Asana XL (DuPont) p 706
Larvin 3.2 (Rhone–Poulenc) p 1780
Lorsban 4E (Dow) p 632
*Penncap-M (Pennwalt) p 1670
*Pydrin 2.4 (DuPont) p 880
Sevin XLR Plus (Rhone–Poulenc) p 1816
Sevin 4-Oil (Rhone–Poulenc) p 1827
Sevin 80S (Rhone–Poulenc) p 1830
Sevin 50W (Rhone–Poulenc) p 1841
Sevin 4F (Rhone–Poulenc) p 1848

Western field wireworm (*Limonius infuscatus*)
*Dyfonate 4-EC (ICI) p 1184
Dyfonate II 10-G (ICI) p 1187
*Dyfonate II 20-G (ICI) p 1191

Western tent caterpillar (*Malacosoma californicum*)
DiPel 2X (Abbott) p 254
DiPel 4L (Abbott) p 257
†DiPel 8AF (Abbott) p 259

MITES

A

B

*** Restricted Use Only** **† Only Registered for Turf, Ornamentals and Forestry**

NEMATODES

* Restricted Use Only   † Only Registered for Turf, Ornamentals and Forestry

RODENTS

M

R

WEEDS

A

* Restricted Use Only   † Only Registered for Turf, Ornamentals and Forestry

* Restricted Use Only † Only Registered for Turf, Ornamentals and Forestry

* Restricted Use Only   † Only Registered for Turf, Ornamentals and Forestry

* Restricted Use Only † Only Registered for Turf, Ornamentals and Forestry

* Restricted Use Only † Only Registered for Turf, Ornamentals and Forestry

* Restricted Use Only † Only Registered for Turf, Ornamentals and Forestry

Creeping speedwell (*Veronica filiformis*) (*Veronica filiformis*)

Creeping woodsorrel (*Oxalis corniculata*)

Creosotebush (*Larrea tridentata*)

Crotalaria (*Crotalaria*)

Croton spp. (*Croton spp.*)

Crowfootgrass (*Dactyloctenium aegyptium*)

Cupgrass (*Eriochloa*)

*** Restricted Use Only** † **Only Registered for Turf, Ornamentals and Forestry**

G

H

* Restricted Use Only † Only Registered for Turf, Ornamentals and Forestry

* Restricted Use Only   † Only Registered for Turf, Ornamentals and Forestry

Johnsongrass (*Sorghum halepense*)

Johnsongrass (Seedling only) (*Sorghum halepense*)

Junglerice (*Echinochloa colonum*)

M

* Restricted Use Only   † Only Registered for Turf, Ornamentals and Forestry

* Restricted Use Only   † Only Registered for Turf, Ornamentals and Forestry

O

P

*** Restricted Use Only** **† Only Registered for Turf, Ornamentals and Forestry**

*** Restricted Use Only** **† Only Registered for Turf, Ornamentals and Forestry**

* Restricted Use Only   † Only Registered for Turf, Ornamentals and Forestry

* Restricted Use Only † Only Registered for Turf, Ornamentals and Forestry

* Restricted Use Only † Only Registered for Turf, Ornamentals and Forestry

* Restricted Use Only † Only Registered for Turf, Ornamentals and Forestry

Always read and follow label directions carefully.

* Restricted Use Only † Only Registered for Turf, Ornamentals and Forestry

* Restricted Use Only † Only Registered for Turf, Ornamentals and Forestry

Always read and follow label directions carefully.

* Restricted Use Only † Only Registered for Turf, Ornamentals and Forestry

Always read and follow label directions carefully.

*** Restricted Use Only** **† Only Registered for Turf, Ornamentals and Forestry**

*** Restricted Use Only** **† Only Registered for Turf, Ornamentals and Forestry**

Always read and follow label directions carefully.

* Restricted Use Only   † Only Registered for Turf, Ornamentals and Forestry

6th edition CPCR®
1990

APPENDIX I

Calibrations and Conversions
General Information on Handling and Storage of Pesticides
Storage and Temperature Requirements by Product
DOT Information Table

In this section, calibrations and conversion tables are provided to insure the application of proper dosages of crop protection chemicals. General agricultural chemical storage and handling guidelines are also provided for the safe handling of products in warehouses and shipping facilities. These guidelines have been prepared by, and are printed with permission from the National Agricultural Chemicals Association, Washington, D.C.

Also included is a list of each product that appears in CPCR with pertinent storage and temperature information. The information was extracted from the product labels and reviewed by each manufacturer. Each product label is specific in giving instructions and precautions for the material in its container and all readers are cautioned to read the label before using any crop protection chemical.

The DOT Information Table contains information about each product contained in CPCR. This index contains as complete information as was available to us from the manufacturers when this edition of CPCR went to press. Even though not all products require each type of information, we advise that you consult other sources of DOT information when shipping or invoicing a product for which incomplete information has been supplied herein.

CALIBRATIONS AND CONVERSIONS

Sprayer Calibration

Calibration of your spraying equipment is very important. It should be done at least every other day of operation to insure application of the proper dosages. This is probably the most important step in your whole spraying operation since applying incorrect amounts may do much more damage than good. Ecah nozzle must be checked so that all nozzles are delivering the same amount of material for a uniform spray pattern.

If a lower rate is desired it may be obtained by increasing the speed, reducing the pressure or changing to a smaller nozzle or a combination of the three. The opposite is true if a higher rate is desired.

General Calibration

I. Method I.
- A. Measure out 660 feet.
- B. Determine the amount of spray put out in traveling this distance at the desired speed.
- C. Use this formula:
  $$\text{gallons/acre} = \frac{\text{gallons used in 660 feet} \times 66}{\text{swath width in feet}}$$
- *D. Fill the tank with the desired concentration.

II. Method II.
- A. Fill spray tank and spray a specified number of feet.
- B. After spraying refill tank measuring the quantity of material needed for refilling.
- C. Use this formula:
  $$\text{gallons/acre} = \frac{43560 \times \text{gallons delivered}}{\text{swath length (ft.)} \times \text{swath width (ft.)}}$$
- *D. Fill the tank with the desired concentrate.

III. Method III.
- A. Measure 163 feet in the field.
- B. Time tractor in 163 feet. Make two passes to check accuracy.
- C. At edge of field adjust pressure valve until you catch 2 pints (32 ounces) of spray in the same amount of time it took to run the 163 feet. Be sure tractor is at the same throttle setting. You are now applying 20 gallons per acre on a 20 inch nozzle boom spacing.
- D. For each inch of nozzle spacing on boom, increase time by 5% or reduce the volume by 5%.

* If you have calibrated your rig and it is putting out 37 gpa, the required dosage is 4 lbs actual/acre. Therefore, for every 37 gallons of carrier (water, oil, etc.) in the spray tank you add 4 lbs of active material.

Useful Formulae

1. To determine the amount of active ingredient needed to mix in the spray tank.
   No. of gallons or pounds =
   $$\frac{\text{No. of acres to be sprayed} \times \text{pounds active ingredient required per Acre}}{\text{pounds active ingredient per gallon or per pound}}$$
2. To determine the amount of pesticide needed to mix a spray containing a certain percentage of the active ingredient.
   No. of gallons or pounds =
   $$\frac{\text{gallons of spray desired} \times \text{\% active ingredient wanted} \times 8.345}{\text{lbs. active ingredient per gallon or pound} \times 100}$$
3. To determine the percent active ingredient in a spray mixture.
   Percent =
   $$\frac{\text{lbs. or gallons of concentrate used (not just active ingredient)} \times \text{\% active ingredient in the concentrate}}{\text{gallons of spray} \times 8.345 \text{ (weight of water/gallon)}}$$

4. To determine the amount of pesticide needed to mix a dust with a given percent active ingredient.
pounds material =
$$\frac{\text{\% active ingredient wanted} \times \text{lbs of mixed dust wanted}}{\text{\% active ingredient in pesticide used}}$$

5. To determine the size of pump needed to apply a given number of gallons/acre.
Pump capacity =
$$\frac{\text{Gallons/acre desired} \times \text{boom width (feet)} \times \text{mph}}{495}$$

6. To determine the nozzle capacity in gallons per minute at a given rate/ acre and miles/hour.
Nozzle capacity =
$$\frac{\text{gallons/acre} \times \text{nozzle spacing (inches)} \times \text{mph}}{5940}$$

7. To determine the acres per hour sprayed.
Acres per hour =
$$\frac{\text{swath width (inches)} \times \text{mph}}{100}$$

8. To determine the rate of speed in miles per hour.
   1. Measure off a distance of 300 to 500 feet
   2. Measure in seconds the time it takes the tractor to go the marked off distance.
   3. Multiply .682 times the distance traveled in feet and divide product by the number of seconds.
   MPH =
$$\frac{.682 \times \text{distance}}{\text{seconds}}$$

9. To determine the nozzle flow rate.
Time the seconds necessary to fill a pint jar from a nozzle.
Divide the number of seconds into 7.5
$$\text{gallons/minute/nozzle} = \frac{7.5}{\text{seconds}}$$

10. To determine the gallons per minute per boom.
Figure out the gallons/minute/nozzle and multiply by the number of nozzles.

11. To determine the gallons per acre delivered.
$$\frac{5940 \times \text{gallons/minute/nozzle} = \text{gpa}}{\text{nozzle spacing (inches)} \times \text{mph}}$$

12. To determine the acreage sprayed per hour.
$$\text{acres sprayed/hour} = \frac{\text{boom width (feet)} \times \text{mph}}{12}$$
This allows 30% of time for filling, turning, etc.

13. Sprayer Tank Capacity
Calculate as follows:
   1. Cylindrical Tanks:
   Multiply the length in inches times the square of the diameter in inches and multiply the product by .0034.
   length × diameter squared × .0034 = number per gallons.
   2. Elliptical Tanks:
   Multiply the length in inches times the short diameter in inches times the long diameter in inches times .0034.
   length × short diameter × long diameter × .0034 = number of gallons.
   3. Rectangular Tanks:
   Multiply the length times the width times the depth in inches and multiply the product by .004329.
   length × width × depth × .004329 = number of gallons.

14. To determine the acres in a given area.
Multiply the length in feet times the width in feet times 23. Move the decimal point 6 places to the left to give the actual acres.

Conversion Tables (U.S.)

Linear Measure—
1 foot = 12 inches
1 yard = 3 feet
1 rod = 5.5 yards = 16.5 feet
1 mile = 320 rods = 1760 yards = 5280 feet

Square Measure—
1 square foot (sq. ft.) = 144 square inches (sq. inch)
1 square yard (sq. yd.) = 9 sq. feet
1 square rod (sq. rd.) = 272.25 sq. ft. = 30.25 sq. yd.
1 acre = 43,560 sq. ft. = 4840 sq. yds. = 160 sq. rds.
1 square mile = 640 acres

Cubic Measure—
1 cubic foot (cu. ft.) = 1728 cubic inches (cu. in.) 29.922 liquid quarts = 7.48 gallons
1 cubic yard = 27 cubic feet

Liquid Capacity Measure—
1 tablespoon = 3 teaspoons
1 fluid ounce = 2 tablespoons
1 cup = 8 fluid ounces
1 pint = 2 cups = 16 fluid ounces
1 quart = 2 pints = 32 fluid ounces
1 gallon = 4 quarts = 8 pints = 128 fluid ounces

Weight Measure—
1 pound (lb.) = 16 ounces
1 hundredweight (cwt.) = 100 pounds
1 ton = 20 cwt. = 2000 pounds

Rates of Application—
1 ounce/sq. ft. = 2722.5 lbs/acre
1 ounce/sq. yd. = 302.5 lbs/acre
1 ounce/100 sq. ft. = 27.2 lbs/acre
1 pound/100 sq. ft. = 435.6 lbs/acre
1 pound/1000 sq. ft. = 43.6 lbs/acre
1 gallon/acre = ⅓ ounces/1000 sq. ft.
5 gallons/acre = 1 pint/1000 sq. ft.
100 gallons/acre = 2.5 gallons/1000 sq. ft. = 1 quart/100 sq. ft.
100 lbs/acre = 2.5 lbs/1000 sq. ft.

Important Facts—
Volume of sphere = diameter^3 × .5236
Diameter = circum. × .31831
Area of circle = dia^2 × .7854
Area of elipse = prod. of both dia × .7854
Vol. of cone = area of base × ⅓ ht.
1 cu. ft. water = 7.5 gallons = 62.5 lbs.
Pressure in psi = ht. (ft.) × .434
1 acre = 209 feet square
ppm = % × 10,000
$$\% = \frac{\text{ppm}}{10{,}000}$$
1% by volume = 10,000 ppm

Table of Conversion Factors

To Convert From	To	Multiply By
Cubic feet	gallons	7.48
Cubic feet	liters	28.3
Gallons	milliliters	3785
Grams	pounds	.0022
Grams/liter	parts/million	1000
Grams/liter	pounds/gallon	.00834
Liters	cubic feet	.0353
Milligrams/liter	parts/million	1
Milliliters/gallons	gallons	.0026
Ounces	grams	28.35
Parts/million	grams/liter	.001
Parts/million	pounds/million gallons	8.34
Pounds	grams	453.59
Pounds/gallon	grams/liter	111.83

1 gram = .035 ounce
1 kilogram = 2.2 lbs
1 quintal = 100 kg = 221 lbs
1 metric ton = 1000 kg = 2,205 lbs
1 hectare = 2.5 acres
1 meter = 39.4 inches
1 kilometer = .6 mile

Conversion Table

1 kilogram (kg) = 1000 grams (g) = 2.2 lbs
1 gram (g) = 1000 milligrams (mg) = .035 ounce
1 liter = 1000 milliliters (ml) or cubic centimeters (cc) = 1.058 quarts
1 milliliter or cubic centimeter = .034 fluid ounce
1 milliliter or cubic centimeter of water weighs 1 gram
1 liter of water weighs 1 kilogram

1 lb = 453.6 grams
1 ounce = 28.35 grams
1 pt. of water weighs approximately 1 lb
1 gallon of water weighs approximately 8.34 lbs

1 gallon = 4 qts = 3.785 liters
1 qt = 2 pts = .946 liters
1 pt = .473 liters
1 fluid ounce = 29.6 milliliters or cubic centimeters

1 part per million (ppm) = 1 milligram/liter
= 1 milligram/kilogram
= .0001 per cent
= .013 ounces in 100 gallons of water

1 percent = 10.000 ppm
= 10 grams per liter
= 10 grams per kilogram
= 1.33 ounces by weight per gallon of water
= 8.34 pounds/100 gallons of water

.1 percent = 1000 ppm = 1000 milligrams/liter
.01 percent = 100 ppm = 100 milligrams/liter
.001 percent = 10 ppm = 10 milligrams/liter
.0001 percent = 1 ppm = 1 milligram/liter

Chemical Elements

Name	Symbol	Atomic Weight	Valance
Aluminum	Al	26.97	3
Antimony	Sb	121.76	3, 5
Arsenic	As	74.91	3, 5
Barium	Ba	137.36	2
Bismuth	Bi	209.00	3, 5
Boron	B	10.82	3, 0
Bromine	Br	79.916	1, 3, 5, 7
Cadmium	Cd	112.41	2
Calcium	Ca	40.08	2
Carbon	C	12.01	2, 4
Chlorine	Cl	35.457	1, 3, 5, 7
Cobalt	Co	58.94	2, 3
Copper	Cu	63.57	1, 2
Fluorine	F	19.00	1
Hydrogen	H	1.008	1
Iodine	I	126.92	1, 3, 5, 7
Iron	Fe	55.85	2, 3
Lead	Pb	207.21	2, 4
Magnesium	Mg	24.32	2
Mercury	Hg	200.61	1, 2
Molybdenum	Mo	95.95	3, 4, 6
Nickel	Ni	58.69	2, 3
Nitrogen	N	14.008	3, 5
Oxygen	O	16.00	2
Phosphorus	P	30.98	3, 5
Potassium	K	39.096	1
Selenium	Se	78.96	3
Silicon	Si	28.06	4
Silver	Ag	107.88	1
Sodium	Na	22.997	1
Sulfur	S	32.06	2, 4, 6
Thallium	Tl	204.29	1, 3
Tin	Sn	118.70	2, 4
Titanium	Ti	47.90	3, 4
Uranium	U	238.17	4, 6
Zinc	Zn	65.38	2

Width Of Area Covered To Acres Per Mile Traveled

Width of Strip (feet)	Acres/mile
6	.72
10	1.21
12	1.45
16	1.93
18	2.18
20	2.42
25	3.02
30	3.63
50	6.04
75	9.06
100	12.1
150	18.14
200	24.2
300	36.3

Temperature Conversion Table Relationship Of Centigrade And Fahrenheit Scales

°C	°F	°C	°F
−40	−40	25	77
−35	−31	30	86
−30	−22	35	95
−25	−13	40	104
−20	−4	45	113
−15	5	50	122
−10	14	55	131
−5	23	60	140
0	32	80	176
5	41	100	212
10	50		
15	59		
20	68		

Proportionate Amounts Of Dry Materials

Water	Quantity of Material				
100 gallons	1 lb.	2 lbs.	3 lbs.	4 lbs.	5 lbs.
50 gallons	8 oz.	1 lb.	24 oz.	2 lbs.	2½ lbs.
5 gallons	3 tbs.	1½ oz.	2½ oz.	3¼ oz.	4 oz.
1 gallon	2 tsp.	3 tsp.	1½ tbs.	2 tbs.	3 tbs.

Proportionate Amounts Of Liquid Materials

Water	Quantity of Material		
100 gallons	1 qt.	1 pt.	1 cup
50 gallons	1 pt.	1 cup	½ cup
5 gallons	3 tbs.	5 tbs.	2½ tsp.
1 gallon	2 tsp.	1 tsp.	½ tsp.

Miles Per Hour Converted To Feet Per Minute

MPH	fpm
1	88
2	176
3	264
4	352

Emulsifiable Concentrate Percent Ratings In Pounds Actual Per Gallon

%EC	lbs/Gallon
10–12	1
15–20	1.5
25	2
40–50	4
60–65	6
70–75	8
80–100	10

Conversion Table For Liquid Formulations*

Concentration of Active Ingredient in Formulations, lbs/gal.

Rate Desired Lbs/A	1	2	2.5	3	4	5	6
	(ml of formulation per 100 square feet)						
1	8.67	4.33	3.47	2.89	2.17	1.73	1.44
2	17.3	8.67	6.93	5.78	4.33	3.47	2.89
3	26.0	13.0	10.4	8.67	6.50	5.20	4.33
4	34.8	17.4	13.9	11.6	8.69	6.95	5.79
5	43.4	21.7	17.4	14.5	10.0*	8.68	7.24
6	52.1	26.0	20.8	17.4	13.0	10.4	8.68
7	60.8	30.4	24.3	20.3	15.2	12.2	10.1
8	69.4	34.7	27.8	23.1	17.4	13.9	11.6
9	78.1	39.0	31.2	26.0	19.5	15.6	13.0
10	86.7	43.3	34.7	28.9	21.7	17.3	14.4

*Example: To put out a 100 sq. ft. plot at the rate of 5 lbs/A active ingredient using a formulation containing 4 lb/gal active ingred., use 10.7 ml of the 4 lbs/gal form. & distribute evenly.

Conversion Table For Dry Formulations

Concentration of Active Ingredient in Formulation

Rate Desired Lbs/A	100%	90%	80%	75%	70%	60%	50%	40%	30%	25%	20%	10%	5%
	(Grams of formulation per 100 square feet)												
1	1.04	1.16	1.30	1.39	1.49	1.74	2.08	2.60	3.47	4.17	5.21	10.4	20.8
2	2.08	2.31	2.60	2.78	2.98	3.47	4.17	5.21	6.94	8.33	10.4	20.8	41.7
3	3.12	3.47	3.90	4.17	4.46	5.20	6.25	7.81	10.4	12.5	15.6	31.2	62.5
4	4.17	4.63	5.21	5.55	5.95	6.94	8.33*	10.4	13.9	16.7	20.8	41.7	83.3
5	5.21	5.79	6.51	6.94	7.44	8.68	10.4	13.0	17.4	20.8	26.0	52.1	104
6	6.25	6.94	7.81	8.33	8.93	10.4	12.5	15.6	20.8	25.0	31.2	62.5	125
7	7.29	8.10	9.11	9.72	10.4	12.1	14.6	18.2	24.3	29.2	36.4	72.9	146
8	8.33	9.26	10.4	11.1	11.9	13.9	16.7	20.8	27.8	33.3	41.7	83.3	167
9	9.37	10.4	11.7	12.5	13.4	15.6	18.7	23.4	31.2	37.5	46.9	93.7	187
10	10.4	11.6	13.0	13.9	14.9	17.4	20.8	26.0	34.7	41.7	52.1	104	208

*Example: To put out a 100 sq. ft. plot at the rate of 4 lbs/A active ingredient using a formulation containing 50% active ingredient, use 8.33 grams of the 50% formulation and distribute evenly over the 100 sq. ft.

Conversion Table For Granular Formulations

Concentration of Active Ingredient in Formulation

Rate Desired Lbs/A	20%	15%	10%	7.5%	5%	4%	3%	2%	1%
	(Grams of formulation per 100 square feet)								
1	5.2	6.94	10.4	13.86	20.8	26.0	34.66	52.0	104.0
2	10.4	13.9	20.8	27.7	41.7	52.0	69.3	104.0	208.0
3	15.6	20.8	31.2	41.6	62.5	78.0	103.9	156.0	312.0
4	20.8	27.8	41.7	55.4	83.3	104.0	138.6	208.0	416.0
5	26.0	34.7	52.1	69.3	104.0*	130.0	173.3	260.0	520.0
6	31.2	41.6	62.5	83.2	125.0	156.0	207.9	312.0	624.0
7	36.4	45.6	72.9	97.0	146.0	182.0	242.6	364.0	728.0
8	41.7	55.5	83.3	110.9	167.0	208.0	277.3	416.0	832.0
9	46.9	62.5	93.7	124.7	187.0	234.0	311.9	468.0	936.0
10	52.1	69.4	104.0	168.6	208.0	260.0	346.6	520.0	1,040.0
15	78.0	104.1	156.0	207.9	312.0	390.0	519.2	780.0	1,560.0
20	104.0	138.8	208.0	277.2	416.0	520.0	693.2	1,040.0	2,080.0
25	130.0	173.5	260.0	346.5	520.0	650.0	866.5	1,300.0	2,600.0
30	156.0	208.2	312.0	415.8	624.0	780.0	1,039.8	1,560.0	3,120.0

*Example: To put out a 100 sq. ft. plot at the rate of 5 lbs/A active ingredient using a formulation containing 5% active material, use 104.0 grams of the 5% formulation and distribute it evenly over the 100 sq. ft.

Quick Conversions

Temperature °C	Temperature °F	Length cm	Length inch	Volume liters	Volume quarts
100	212	2.5	1	1	1.1
90	194	5	2	2	2.1
80	176	10	4	3	3.2
70	158	20	8	4	4.2
60	140	30	12	5	5.3
50	122	40	16	6	6.3
40	104	50	20	7	7.4
35	95	60	24	8	8.5
30	86	70	28	9	9.5
25	77	80	32		
20	68	90	36		
15	59	100	39		
10	50	200	79		
5	41		feet		
0	32	300	10		
−5	23	400	13		
−10	14	500	16		
−15	5	1,000	33		
−20	−4				
−25	−13				
−30	−22				
−40	−40				

Quick Conversions

kg/ha	lb/A
1.	0.9
2.	1.8
3.	2.7
4.	3.6
5.	4.5
10.	9
20.	18
30.	27
40.	36
50.	45
60.	54
70.	62
80.	71
90.	80
100.	89
200.	180
300.	270
400.	360
500.	450
600.	540
700.	620
800.	710
900.	800
1,000.	890
2,000.	1,800
	ton/A
3,000	1¼
4,000	1¾
5,000	2¼
6,000	2¾
7,000	3
8,000	3½
9,000	4
10,000	4½
11,000	5
12,000	5½
13,000	5¾
14,000	6¼
15,000	6¾
16,000	7
17,000	7½
18,000	8
19,000	8½
20,000	9

Conversion Factors Used In Calculation

Convert	To	By
gram (gm.)	kilogram (kg.)	move decimal 3 places to left
Example 2000 gm. = 2.0 kg.		
gram (gm.)	milligram (mg.)	move decimal 3 places to right
Example 2.0 gm. = 2000 mg.		
gram (gm.)	pound (lb.)	divide by 454
Example 658 gm./ ÷ 454 = 1.45 lb.		
gram/pound	percent (%)	divide by 4.54
Example 90 gm./lb. ÷ 4.54 = 19.8%		
gram/ton	percent	multiply by 11, move decimal 5 places to left
Example 45 gm./ton × 11 = 495 = .00495%		
kilogram (kg.)	gram	move decimal 3 places to right
Example 5.5 kg. = 5500 gm.		
milligram (mg.)	gram	move decimal 3 places to left
Example 95 mg. = 0.095 gm.		
percent	gram/pound	multiply by 4.54
Example 25 × 4.54 = 113.5 gm./lb		
percent	parts/million (ppm)	move decimal 4 places to right
Example .025% = 250 ppm.		
percent	gram/ton	divide by 11, move decimal 5 places to right
Example .011 ÷ 11 = .001 = 100 gm./ton		
pound	gram	multiply by 454
Example 0.5 lb. × 454 = 227 gm.		
ppm	percent	move decimal 4 places to left
Example 100 ppm = 0.01%		

Determine The Number of Rows To The Acre

Rows/Acre	Length of Rows 32″	36″	38″	40″	60″
1	16,355	14,520	13,756	13,068	8,712
2	8,168	7,260	6,878	6,534	4,356
3	5,445	4,840	4,585	4,356	2,904
4	4,084	3,630	3,439	3,267	2,178
5	3,267	2,904	2,751	2,614	1,742
6	2,723	2,420	2,293	2,178	1,452
7	2,334	2,074	1,965	1,867	1,245
8	2,042	1,815	1,719	1,634	1,089
9	1,815	1,613	1,528	1,452	968
10	1,634	1,452	1,376	1,307	871
11	1,485	1,320	1,251	1,188	792
12	1,361	1,210	1,156	1,089	726
13	1,257	1,117	1,058*	1,005	670
14	1,167	1,037	982	933	622
15	1,089	968	917	871	581
16	1,021	908	860	817	545
17	961	854	809	769	512
18	908	807	764	726	484
19	860	764	724	688	459
20	817	726	688	653	436
21	778	691	655	622	415
22	743	660	625	594	396
23	710	631	598	568	379
24	681	605	573	544	363
25	653	581	550	523	348
26	628	558	529	503	335
27	605	538	509	484	323
28	583	519	491	467	311
29	563	501	474	450	300
30	545	484	459	436	290

*Example: A growers field is 1,058 feet long furrowed out on 38-inch centers. Therefore, every 13 rows across the field represents an acre.

Grams/Gallon Table

Gallons PPM	5	10	15	20	25	50	75	100	150	200	300	400
5	0.1	0.2	0.3	0.4	0.5	1.0	1.4	1.9	2.8	3.8	5.7	7.6
10	0.2	0.4	0.6	0.8	1.0	1.9	2.8	3.8	5.7	7.6	11.0	15.0
15	0.3	0.6	0.9	1.1	1.4	2.8	4.3	5.7	8.5	11.0	17.0	23.0
20	0.4	0.8	1.1	1.5	1.9	3.8	5.7	7.6	11.0	15.0	23.0	30.0
25	0.5	0.9	1.4	1.9	2.4	4.7	7.1	9.5	14.0	19.0	28.0	38.0
50	0.9	1.9	2.8	3.8	4.7	9.5	14.0	19.0	28.0	38.0	57.0	76.0
75	1.4	2.8	4.3	5.7	7.1	14.0	21.0	28.0	43.0	57.0	85.0	114.0
100	1.9	3.8	5.7	7.6	9.5	19.0	28.0	38.0	57.0	76.0	114.0	151.0
125	2.4	4.7	7.1	9.5	12.0	24.0	36.0	47.0	71.0	95.0	142.0	189.0
150	2.8	5.7	8.5	11.0	14.0	28.0	43.0	57.0	85.0	114.0	170.0	227.0
175	3.3	6.6	9.9	13.0	17.0	33.0	50.0	66.0	99.0	133.0	199.0	265.0
200	3.8	7.6	11.0	15.0	19.0	38.0	57.0	76.0	114.0	151.0	227.0	303.0
250	4.7	9.5	14.0	19.0	24.0	47.0	71.0	95.0	142.0	189.0	284.0	379.0
300	5.7	11.0	17.0	23.0	28.0	57.0	85.0	114.0	170.0	227.0	341.0	454.0
400	7.6	15.0	23.0	30.0	38.0	76.0	114.0	151.0	227.0	303.0	454.0	606.0

Standard Measurements

Measure of Length (Linear Measure)

4 inches = 1 hand
9 inches = 1 span
12 inches = 1 foot
3 feet = 1 yard
6 feet = 1 fathom
5½ yards - 16½ feet = 1 rod
40 poles = 1 furlong
8 furlongs = 1 mile
5,280 feet = 1,760 yards = 320 rods = 1 mile
3 miles = 1 league

Measure of Surface (area)

144 square inches = 1 square foot
9 square feet = 1 square yard
30¼ square yards = 1 square rod
160 square rods = 1 acre
43,560 square feet = 1 acre
640 square acres = 1 square mile
36 square miles = 1 township

Surveyor's Measure

7.92 inches = 1 link
25 links = 1 rod
4 rods = 1 chain
10 square chains = 160 square rods = 1 acre
640 acres = 1 square mile
80 chains = 1 mile
1 Gunters chain = 66 feet

Metric Length

1 inch = 2.54 centimeters
1 foot = .305 meter
1 yard = .914 meter
1 mile = 1.609 kilometers
1 fathom = 6 feet
1 knot = 6,086 feet
3 knots = 1 league
1 centimeter = .394 inch
1 meter = 3.281 feet
1 meter = 1.094 yards
1 kilometer = .621 mile

Troy Weight

24 grains = 1 pennyweight
20 pennyweight = 1 ounce
12 ounces = 1 pound

Liquid Measure

2 cups = 1 pint
4 gills = 1 pint
16 fluid ounces = 1 pint
2 pints = 1 quart
4 quarts = 1 gallon
31½ gallons = 1 barrel
2 barrels = 1 hogshead
1 gallon = 231 cubic inches
1 cubic foot = 7.48 gallons
1 teaspoon = .17 fluid ounces (⅙ oz.)
3 teaspoons (level) = 1 tablespoon (½ oz.)
2 tablespoons = 1 fluid ounce
1 cup (liquid) = 16 tablespoons (8 oz.)
1 teaspoon = 5 to 6 cubic centimeters
1 tablespoon = 15 to 16 cubic centimeters
1 fluid ounce = 29.57 cubic centimeters

Cubic Measure (Volume)

1,728 cubic inches = 1 cubic foot
27 cubic feet = 1 cubic yard
2,150.42 cubic inches = 1 standard bushel
231 cubic inches = 1 standard gallon (liquid)
1 cubic foot = ⅘ of a bushel
128 cubic feet = 1 cord (wood)
7.48 gallons = 1 cubic foot
1 bushel = 1.25 cubic feet

Dry Measure

2 pints = 1 quart
8 quarts = 1 peck
4 pecks = 1 bushel
36 bushels = 1 chaldron

Apothecaries' Weight

20 grains = 1 scruple
3 scruples = 1 dram
8 drams = 1 ounce
12 ounces = 1 pound
$27\frac{11}{32}$ grains = 1 dram
16 drams = 1 ounce
16 ounces = 1 pound
2,000 pounds = 1 ton (short)
2,240 pounds = 1 ton (long)

Conversion Factors

Degree C. = 5/9 (Degree F. −32).
Degree F. = 9/5 (Degree C. +32).
Degrees absolute (Kelvin) = Degrees centigrade +273.16.
Degrees absolute (Rankine) = Degrees fahrenheit +459.69.

Multiply	*By*	*To Obtain*
Diameter circle	3.1416	Circumference circle
Diameter circle	0.8862	Side of equal square
Diameter circle squared	0.7854	Area of circle
Diameter sphere squared	3.1416	Area of sphere
Diameter sphere cubed	0.5236	Volume of sphere
U.S. Gallons	0.8327	Imperial gallons (British)
U.S. Gallons	0.1337	Cubic feet
U.S. Gallons	8.330	Pounds of water (20° C.)
Cubic feet	62.427	Pounds of water (4°C.)
Feet of water (4° C.)	0.4335	Pounds per sq. inch
Inch of mercury (0° C.)	0.4912	Pounds per sq. inch
Knots	1.1516	Miles per hour

Figuring Grain Storage Capacity

1 bu. ear corn = 70 lbs. 2.5 cu. ft. (15.5% moisture)
1 bu. shelled corn = 56 lbs. 1.25 cu. ft. (15.5% moisture)
1 cu. ft. = 1/2.50 = .4 bu. of ear corn
1 cu. ft. = 1/1.25 = .8 bu. of shelled corn;

Bu. × 1.25 ft.^{3}, Ft.^{3} × .8 bu.

Ft.^{3} = Bu. × 1.25

Bu. = Ft.^{3} × .8

Rectangular or square cribs or bins
cu. ft. = width × height × length (W × H × L)
Round cribs, bins or silos (= 3.1416)

Volume = $R^2H = D^2H/r$
cu. ft. = × radius × radius × height = (R × R × H)
or × $\frac{\text{diameter} \times \text{diameter} \times \text{height} = (D \times D \times H)}{4}$

or $\frac{3.1416 \times D \times D \times H}{4}$ = .785 × D × D ×H

Examples:

1. Crib — ear corn — 6' wide by 12' high by 40' long
   a. 6 × 12 × 40 = 2880 cu. ft. × 4 bu./cu. ft. = 1152 bu.
   b. 6 × 12 × 1 = 72 cu. ft. × .4 28.8 bu/ft. of length × 40' = 1152 bu.
2. Round crib — ear corn — 14' diameter by 14' high
   a. .785 × 14' × 14' × 14' × .8 = 1722 bushel
   b. .785 × 14' × 1 × .4 6.15 bu/ft. × 14 = 861 bushel
3. Round Bin or Silo — shell corn — 14' diameter by 14' high
   a. .785 × 14' × 14' × 14' × .8 = 1722 bushel
   b. .785 × 14' × 14' × 14' × 1 × .8 – 123 bu/ft. × 14' = 1722 bushel

Metric Equivalents

Metric Weight

1 grain	=	.065 gram
1 apothecaries' scruple	=	1.296 grams
1 avoirdupois ounce	=	28.350 grams
1 troy ounce	=	31.103 grams
1 avoirdupois pound	=	.454 kilogram
1 troy pound	=	.373 kilogram
1 gram	=	15.432 grains
1 gram	=	.772 apothecaries' scruple
1 gram	=	.035 avoirdupois ounce
1 gram	=	.032 troy ounce
1 kilogram	=	2.205 avoirdupois pounds
1 kilogram	=	2.679 troy pounds

Capacity

1 U.S. fluid ounce	=	29.573 milliliters
1 U.S. liquid quart	=	.946 liter
1 U.S. dry quart	=	1.101 liters
1 U.S. gallon	=	3,785 liters
1 U.S. bushel	=	.3524 hectoliters
1 cubic inch	=	16.4 cubic centimeters
1 liter	=	1,000 milliliters or 1,000 cubic centimeters
1 cubic foot water	=	7.48 gallons or 62½ pounds
231 cubic inches	=	1 gallon
1 milliliter	=	.034 U.S. fluid ounce
1 liter	=	1.057 U.S. liquid quarts
1 liter	=	.908 U.S. dry quart
1 liter	=	.264 U.S. gallon
1 hectoliter	=	2.838 U.S. bushels
1 cubic centimeter	=	.061 cubic inch

Miscellaneous Equivalents

9 in. equal 1 span.
6 ft. equal 1 fathom.
6,080 ft. equal 1 nautical mile.
1 board ft. equals 144 cu. in.
1 cylindrical ft. contains 4⅞ gals.
1 cu. ft. equals .8 bushel.
12 dozen (doz.) equals 1 gross (gr.).
1 gal. water weighs about 8⅓ lbs.
1 gal. milk weighs about 8.6 lbs.
1 gal. cream weighs about 8.4 lbs.
46½ qts. of milk weighs 100 lbs.
1 cu. ft. water weighs 62½ lbs., contains 7½ gals.
1 gal. kerosene weighs about 6½ lbs.
1 bbl. cement contains 3.8 cu. ft.
1 bbl. oil contains 42 gals.
1 standard bale cotton weighs 480 lbs.
1 keg of nails weighs 100 lbs.
4 in. equal 1 hand in measuring horses.

Useful Measurements

LENGTH
1 mile = 80 chains = 8 furlongs = 1,760 yards = 5,280 ft. = 1.6 kilometres
1 chain = 22 yards = 4 rods, poles or perches = 100 links

WEIGHT
1 long ton = 20 cwt. = 2,240 lbs.
1 lb. = 16 oz. = 454 grams = 0.454 kilograms
1 short ton = 2,000 lbs.
1 metric ton = 2,204 lbs. = 1,000 kilograms

AREA
1 acre = 10 sq. chains = 4,840 sq. yards = 43,560 sq. ft. = 0.405 hectares
1 sq. mile = 640 acres = 2.59 kilometres
1 hectare = 2.471 acres

VOLUME
1 gal. = 4 quarts = 8 pints = 128 fluid ozs. = 3.785 litres
1 fluid oz. = 2 tablespoons = 4 dessertspoons = 8 teaspoons = 28.4 c.c.'s
1 litre = 1,000 c.c.'s = 0.22 Imperial gallons = 1.76 pints

CAPACITIES
Cylinder-diameter^2/x depth × 0.785 = cubic feet
Rectangle-breadth × depth × length = cubic feet
Cubic Feet × 6.25 = gallons

QUICK CONVERSIONS

1 pint/acre	= 1 fluid oz./242 sq. yards
1 gal./acre	= 1 pint/605 sq. yards
1 lb./acre	= 1 oz./300 sq. yards
1 cwt./acre	= 0.37 oz./sq. yard
1 m.p.h.	= 88 ft./minute
3 m.p.h.	= 1 chain/15 sec.
1 litre/hectare	= 0.089 gal./acre
1 kilogram/hectare	= 0.892 lb./acre
1 c.c/100 litres	= 0.16 fl. oz./100 gallons
125 c.c./100 litres	= 1 pint/100 gallons
1 gm./100 litres	= 0.16 oz./100 gallons

A strip 3 ft. wide × 220 chains
A strip 4 ft. wide × 165 chains   1 acre
A strip 5 ft. wide × 132 chains

CAPACITIES OF ROUND BINS AND WIRE CRIBS

Diameter	Height	Cu. ft.	Capacities: Small grain	Capacities: Ear corn
feet	feet		bushels	bushels
8	6	301	241	120
8	8	400	320	160
8	10	502	400	200
10	8	628	502	251
10	10	785	628	314
10	12	942	753	376
12	10	1130	904	452
12	12	1356	1084	542
12	14	1582	1265	632
14	12	1846	1476	738
14	14	2154	1723	861
14	16	2462	1969	984
16	14	2814	2251	1125
16	16	3216	2572	1286
16	18	3618	2894	1447
16	20	4029	3216	1608

To find the capacity of a circular grain bin in bushels, multiple the radius (½ diameter) times the radius, times 3.1416, times height, times 0.8 for small grain and shelled corn.

For a circular crib, use 0.4 in place of 0.8 to find bushels of ear corn. (All measurements must be in feet.)

CAPACITIES OF SQUARE GRAIN BINS

Size	Capacity per foot of depth	Capacity of bin 10′ deep
feet	bushels	bushels
5×5	20	200
6×6	28.8	288
8×8	51.2	512
10×10	80	800

To find the capacity of a square or rectangular grain bin in bushels, multiply length times width, times height, times 0.8 (All measurements must be in feet.)

Mich. State Univ. Coop. Ext. Service

TO DETERMINE TRACTOR FIELD SPEED IN MILES PER HOUR

1. Stake off a distance of 88 feet (1/60 of a mile) in the field to be sprayed.
2. Select tractor gear and throttle setting to be used.
3. Clock the seconds required to drive 88 feet from a running start.
4. Divide the time in seconds into 60; the answer is miles per hour.

CAPACITIES OF ROUND SILOS IN TONS

Depth of settled silage	Weight silage per cu. ft. at this depth	Average weight of silage per cu. ft. to this depth	Total quantity of settled silage, from the top to the depth indicated in silos having a diameter of— 10 feet	12 feet	14 feet	16 feet	18 feet	20 feet
Ft.	Lbs.	Lbs.	Tons	Tons	Tons	Tons	Tons	Tons
2	29.2	23.5	2	3	4	5	6	7
4	37.2	29.5	5	7	9	12	15	18
6	41.8	33.3	8	11	15	20	25	31
8	45.0	36.0	11	16	22	29	37	45
10	47.1	38.1	15	22	29	38	49	60
12	48.7	39.8	19	27	37	48	61	75
14	49.9	41.2	23	33	44	58	73	90
16	51.0	42.4	27	38	52	68	86	107
18	51.8	43.5	31	44	60	79	100	123
20	52.4	44.3	35	50	68	89	113	139
22	53.0	45.1	39	56	76	100	126	156
24	53.5	45.8	43	62	85	110	140	173
26	53.9	46.4	47	68	92	121	154	189
28	54.2	46.9	52	74	101	132	167	206
30	54.4	47.4	56	80	110	143	181	224
32	54.6	47.9	60	87	118	154	195	241
34	54.7	48.3	65	93	126	165	209	258
36	—	—	—	—	135	176	223	275
38	—	—	—	—	143	187	237	292
40	—	—	—	—	152	198	251	310
42	—	—	—	—	—	209	265	327
44	—	—	—	—	—	220	279	344
46	—	—	—	—	—	—	293	361
48	—	—	—	—	—	—	307	378
50	—	—	—	—	—	—	320	396
52	—	—	—	—	—	—	335	413
54	—	—	—	—	—	—	349	430
56	—	—	—	—	—	—	363	448
58	—	—	—	—	—	—	377	466
60	—	—	—	—	—	—	392	484

Silage table and material on settled silage content of silos taken from USDA circular 603. The weight of a cubic foot of silage is so variable that tables or rules for estimating may be very inaccurate. The following rule will give approximate results: Diameter squared × .7854 × depth of settled silage will give number of cubic feet of silage. The number of cubic feet times the weight per cubic foot, divided by 2,000 will give tons of silage.

CONVERTING BROADCAST RATE TO BAND RATE

Unless specified otherwise, directions for the application of preemergence herbicides are given **for broadcast coverage.** If the herbicide is to be **band-applied,** the actual amount of solution to use per acre is calculated by the following formula:

$$\frac{\text{band width (inches)}}{\text{row spacing (inches)}} \times \text{broadcast rate in gallons per acre}$$

Example: $\frac{12 \text{ in.}}{40 \text{ in.}} \times$ 20 gal./A (broadcast) = 6 gal./A (banded)

AIRCRAFT CALIBRATION

Acres covered per minute

Speed mph	Swath width in feet									
	30	**35**	**40**	**45**	**50**	**75**	**100**	**200**	**300**	**500**
75	4.5	5.2	6.0	6.7	7.5	11.2	15.0	30.0	45.0	75.0
80	4.8	5.6	6.4	7.2	8.0	12.0	16.0	32.0	48.0	80.0
85	5.1	5.9	6.8	7.6	8.5	12.7	17.0	34.0	51.0	85.0
90	5.4	6.3	7.2	8.1	9.0	13.5	18.0	36.0	54.0	90.0
95	5.7	6.6	7.6	8.5	9.5	14.2	19.0	38.0	57.0	95.0
100	6.0	7.0	8.0	9.0	10.0	15.0	20.0	40.0	60.0	100.0
110	6.6	7.7	8.8	9.9	11.0	16.5	22.0	44.0	66.0	110.0
120	7.2	8.4	9.6	10.8	12.0	18.0	24.0	48.0	72.0	120.0
130	7.8	9.1	10.4	11.7	13.0	19.5	26.0	52.0	78.0	130.0
140	8.4	9.8	11.2	12.6	14.0	21.0	28.0	56.0	84.0	140.0
150	9.0	10.5	12.0	13.5	15.0	22.5	30.0	60.0	90.0	150.0

The rate of application in gallons or pounds per minute is calculated by multiplying the acres per minute by the number of gallons or pounds per acre to be applied. Take for example a 100-mile-per-hour aircraft that has a 50 foot effective swath. The chart indicates that the plane has a coverage of 10.0 acres per minute. If spray is to be applied at a rate of 2 gallons per acre, the unit should be calibrated to dispense 20 gallons per minute ($2 \times 10 = 20$). If 7.5 pounds of dry material is to be applied per acre, the unit should be calibrated to dispense 75 pounds per minute ($10 \times 7.5 = 75$).

The basic formula for calculating acres per minute is:

$$\frac{\text{swath width} \times 2 \times \text{miles per hour}}{1{,}000}$$

NOZZLE OUTPUTS (20-INCH NOZZLE SPACING*)

Output	Ground speed, miles per hour					
	3	4	5	6	7	8
			5 gallons per acre			
Gallons per minute	.05	.067	.084	.101	.118	.135
Ounces per minute	6.4	8.6	10.8	12.9	15.1	17.2
			6 gallons per acre			
Gallons per minute	.06	.081	.101	.121	.141	.162
Ounces per minute	7.8	10.3	12.9	15.5	18.1	20.6
			7 gallons per acre			
Gallons per minute	.071	.094	.118	.141	.165	.189
Ounces per minute	9.0	12.1	15.1	18.1	21.1	24.2
			8 gallons per acre			
Gallons per minute	.081	.108	.135	.162	.189	.216
Ounces per minute	10.3	13.8	17.2	20.7	24.1	27.6
			9 gallons per acre			
Gallons per minute	.091	.121	.152	.182	.212	.242
Ounces per minute	11.6	15.5	19.4	23.3	27.1	31.0
			10 gallons per acre			
Gallons per minute	.101	.135	.168	.202	.236	.269
Ounces per minute	12.9	17.2	21.5	25.9	30.1	34.4
			15 gallons per acre			
Gallons per minute	.152	.202	.253	.303	.354	.404
Ounces per minute	19.4	25.9	32.3	38.8	45.2	51.8
			20 gallons per acre			
Gallons per minute	.202	.269	.337	.404	.472	.538
Ounces per minute	25.9	34.5	43.1	51.8	60.2	69.0

* Data also applies if every other outlet is plugged and two nozzles are used per drop, as is often the case in post-emergence herbicide application.
The table may be used to:

1. Determine gallonage per acre at given ground speed and nozzle output.
2. Select proper ground speed when operating at a given nozzle pressure and output.
3. Select new nozzles when range of gallons per acre and ground speed is known.

COMPUTING PLANT POPULATIONS AND YIELDS

Row Width	Length of Row in 1/1000 Acre
40 inches	**13-ft., 1-in.**
38 inches	**13-ft., 9-in.**
36 inches	**14-ft., 6 in.**
30 inches	**17-ft., 5-in.**
20 inches	**26-ft., 2-in.**
15 inches	**34-ft., 10-in.**

Determining Plant Population*

1. Measure off 1/1000 acre row.
2. Count number of plants; then multiply by 1000 to determine plants per acre.

Determining Yields*

1. Harvest grain from 1/1000 acre row.
2. Weigh harvested grain, then multiply by 1000 to get pounds of grain per acre.
3. Determine moisture content, then divide pounds per acre by pounds per bushel for adjusted moisture.

* Average three or more 1/1000 acre rows for more accurate estimate of field populations and yields.

PLANT POPULATION AT ROW WIDTHS USED FOR SOYBEANS

Row Width (inches)	Plants per foot of row					
	12	**9**	**8**	**6**	**4**	**3**
7	896,088	672,066	597,392	448,044	298,696	224,022
10	627,264	470,448	418,176	313,632	209,088	156,816
14	448,044	336,033	298,696	224,022	149,348	112,011
20	313,632	235,224	209,088	156,816	104,544	78,408
21	298,692	224,019	199,128	149,346	99,564	74,673
24	261,360	196,020	174,240	130,680	87,120	65,340
28	224,028	168,021	149,352	112,014	74,676	56,007
30	209,088	156,816	139,392	104,544	69,696	52,272
36	174,240	130,680	116,160	87,120	58,080	43,560
40	156,816	117,612	104,544	78,408	52,272	39,204

PLANTING CAPACITY IN ACRES PER 10-HOUR DAY FOR VARIOUS PLANTERS

Planter	Speed and field efficiency											
	3 mph				4 mph				5 mph			
	50%	60%	70%	80%	50%	60%	70%	80%	50%	60%	70%	80%
	acres per 10 hours				acres per 10 hours				acres per 10 hours			
2-row 40 inches	12	15	17	19	16	19	23	26	20	24	28	32
4-row 20 inches	12	15	17	19	16	19	23	26	20	24	28	32
4-row 28 inches	17	20	24	27	23	27	32	36	28	34	40	45
4-row 30 inches	18	22	25	29	24	29	34	39	30	36	42	49
4-row 36 inches	22	26	31	35	29	35	41	47	36	44	51	58
4-row 38 inches	23	28	32	37	31	37	43	49	38	46	54	62
4-row 40 inches	24	29	34	39	32	39	45	52	40	48	57	65
8-row 20 inches	24	29	34	39	32	39	45	52	40	48	57	65
6-row 28 inches	25	31	36	41	34	41	48	54	42	51	59	68
6-row 30 inches	27	33	38	44	36	44	51	58	45	55	64	73
6-row 36 inches	32	39	46	52	43	52	61	70	54	66	77	88
8-row 28 inches	34	41	48	54	45	54	63	73	57	68	79	91
6-row 40 inches	36	44	51	58	48	58	68	78	61	73	85	97
8-row 30 inches	36	44	51	58	48	58	68	78	61	73	85	97
12-row 20 inches	36	44	51	58	48	58	68	78	61	73	85	97

Field efficiency indicates how much field time is spent planting. The greater the time spent in turning, filling seed hoppers, making minor adjustments and performing other jobs, the lower the field efficiency. Source: Purdue University

GENERAL INFORMATION ON HANDLING AND STORAGE OF PESTICIDES

Introduction

All agricultural chemicals should be handled and stored with care.

This publication suggests the proper and safe handling of agricultural chemicals in warehousing and in transportation.

Extreme care is required when handling many types of agricultural chemicals. Some are highly toxic. Not all agricultural chemicals have the same degree of hazard. Some products, if inhaled or allowed to remain on the skin for a short period of time, can cause serious illness or possible death. Contaminated food, clothing or any other material likely to come into intimate contact with humans, or contaminated animal feed could cause damage to either humans or animals who are far removed from the place of contamination. *ALL AGRICULTURAL CHEMICALS SHOULD BE TREATED WITH CAUTION.*

The need for careful handling and inspection of all empty and full agricultural chemical containers cannot be overemphasized. Damaged or leaking containers can create a serious hazard if not properly handled. The main cause of ruptured containers is poor handling or abuse. When handling containers always be alert to detect possible leakers so you can protect yourself and others, and avoid the contamination of other goods.

Manufacturers of agricultural chemicals have technical staff skilled in dealing with emergencies. Expert advice and on-site assistance is available by calling CHEMTREC, 1-800-424-9300. Many manufacturers include an emergency number on their bill of lading and on shipping container labels. *CALL THE SHIPPER'S NUMBER FIRST.*

Agricultural chemical workers have three main responsibilities:

1. Prevent emergencies by careful handling of agricultural chemicals.
2. Check for spillage—*immediately* bring it to the supervisor's attention.
3. Know the proper first aid and emergency steps to take until expert assistance is obtained.

Personnel Precautions

1. Do not store food, feedstuffs, beverages, tobacco, eating utensils, or smoking equipment in the work area, or where agricultural chemicals are stored.
2. Do not drink, eat, smoke, or use tobacco in the work area or where agricultural chemicals are stored.
3. Wear rubber or neoprene gloves or other gloves directed by the product label while handling any agricultural chemical, even if the container seems tightly closed.
4. Do not put your fingers to your ears, mouth, or other parts of body, or rub your eyes while working with or around agricultural chemicals.
5. After handling agricultural chemicals, always wash your hands and face before eating, drinking, smoking, or using the toilet.
6. When cleaning up spills, wear appropriate eye and respiratory protection and impervious footwear (rubber or other footwear as directed by label). Depending upon the magnitude of the spill, follow label directions for other protective equipment, such as coveralls, rainsuits, etc. Leather footwear or leather outer clothing should *not* be worn as they do not provide adequate protection and cannot be decontaminated. Ensure that capabilities and limitations of respiratory protection are known prior to cleanup.
7. Avoid breathing vapors from leaking containers, and exercise caution to avoid fire in the area.
8. Never place agricultural chemicals in an unmarked container, particularly a new container that originally contained a beverage or food.

Containers

Many types and varieties of containers are used for the packaging and shipment of agricultural chemicals. Containers are designed to safely and effectively contain the products. Containers cannot always withstand abusive treatment, such as protruding nails, top loading or sharp objects, and puncture by forklift or other mechanical equipment. *HANDLE WITH CARE.*

Health Hazards

Routes of Exposure

Even a small amount of certain agricultural chemicals taken into the body can cause serious illness—or death. Harmful amounts can enter the body by (1) absorption through the skin, (2) inhalation, or (3) ingestion. So, after handling an agricultural chemical, make it a standard practice to thoroughly wash hands and face with soap and water. Always shower at end of shift. If accidentally contaminated, remove any contaminated clothing and shower immediately.

This product can kill you if touched by hands or spilled or splashed on skin, in eyes or on clothing (liquid goes through clothes).

This product can kill you if vapors, spray mist or dust are breathed.

This product can kill you if swallowed even in small amounts; spray mist or dust may be fatal if swallowed.

First Aid in Case of Contact or Exposure

The product labels on all agricultural chemical products classified as Poison B contain first aid statements which clearly outline the procedure to be followed in case of contact or exposure. Statements on the labels are specific for the products in that container. READ THE LABEL. Medical doctors should call emergency numbers for additional information.

Even slightly contaminated clothing should be laundered separately, twice with great care *before being reused.* Use a very strong detergent and liquid chlorine bleach; rinse twice and dry completely. Clothing and shoes which have been badly contaminated should be destroyed by burning or buried by qualified personnel. Rubber footwear is preferable, since leather absorbs organic materials readily and canot be decontaminated.

Labels

The entire product label should always be read and totally understood before handling the product. Follow all precautions! By law, agricultural chemical containers are required to bear the label registered with the U.S. Environmental Protection Agency. Outer shipping containers for products meeting the criteria of the U.S. Department of Transportation (DOT) for a Poison B, in either a liquid or solid form, must display the DOT poison label (Appendix I). Outer shipping containers for other hazardous materials must also be labeled according to their respective classification; for example, flammable liquids should carry a DOT flammable liquid label (Appendix I). Certain agricultural chemicals may require more than one DOT label.

A look at the product label will show that extreme care has been taken to correctly describe the product, the hazards involved, methods to be used in handling the product, and what to do in case of poisoning. The product labels of highly toxic products carry the word "POISON" and the "skull and crossbones" symbol. Generally, there will be a list of warning statements headed by the word "DANGER." These statements describe the primary hazards, i.e. highly toxic or irritating or corrosive to eyes, skin, nose, and throat, and what should be done to avoid them. In addition, there is often information for the physician to assist him in treating a case of poisoning, should one occur. The labels on all POISON and other toxic category products are informative, and should contain all the basic information needed to handle the product safely. The chemical name and an ingredient statement are always shown on the label.

Signal Words for Labeling

The EPA toxicity rating system for agricultural chemical labeling requires the following signal words to appear on labels:

Signal Words	Toxicity of Material	Signal Words/ Symbol
Danger	Highly Toxic	Danger-Poison Skull & Crossbones
Warning	Moderately Toxic	Warning
Caution	Slightly Toxic	Caution

ALWAYS READ THE LABEL

HANDLING SHIPMENTS

Agricultural Chemicals Classified by DOT as Hazardous Materials

Department of Transportation's Hazardous Materials Regulations require that those agricultural chemicals which are defined by DOT as hazardous materials, when offered for transportation, be identified on shipping papers by their proper hazardous materials shipping name, hazard class, appropriate four-digit identification number preceded by the designation 'UN' or 'NA', as applicable, and the total quantity. Additional information may also be required to be included on shipping papers. DOT regulations describe the exact manner in which the shipping papers are to be completed. The Federal regulations also prescribe container marking requirements and specific specifications for the containers in which hazardous products must be packaged and shipped. Materials labeled as Poison B must never be carried in the same vehicle with foodstuffs, food ingredients, feeds or any other edible material intended for consumption by humans or animals. In general, when tendering packaged hazardous materials in amounts of or greater than 1,000 pounds gross weight per vehicle to a motor carrier, the shipper must provide appropriate placards. Placarding of vehicles carrying less than 1,000 pounds gross weight is permissible, but is not required. For specific information, consult the DOT Hazardous Materials Regulations.

During transportation of agricultural chemicals, stick to the following rules:

1. STOP ALL WORK IF A LEAK OR SPILL IS FOUND IN A SHIPMENT.
2. CLOSE AND LOCK THE VEHICLE TO ISOLATE THE SHIPMENT.
3. NOTIFY THE SHIPPER, OWNER, OR CONSIGNOR.
4. POST THE IMMEDIATE AREA AND GUARD SO THAT NO UNAUTHORIZED PERSON CAN COME IN CONTACT WITH THE SHIPMENT.

A suspected contaminated vehicle should not be used for any purpose until it has been released by the shipper.

If leaking containers are discovered after receipt, read the label (avoid contact with the material), and always consult shipper for necessary procedures before cleaning up a spill as follows:

1. Care should be taken to avoid direct contact with the material.
2. Damaged containers should be removed from the vehicle and segregated from the undamaged containers.
3. Damaged containers should be placed in suitable overpack containers, and placed in a secure location.
4. Residue material must be thoroughly cleaned from the vehicle, using the approved clean-up method.
5. Shipper or consignee should be notified of leakage.

Occasionally, a portion of the containers in a shipment of agricultural chemicals may be damaged during transit. **IN SUCH CASES, PLACE THE MATERIAL IN AN ISOLATED AREA, AND INFORM THE SHIPPER OR OWNER OF THE DAMAGE.**

Storage

Agricultural chemicals should be stored in a dry, well-ventilated, separate room, building, or covered area dedicated exclusively to agricultural chemical storage, and where fire protection is provided. The following precautions should be taken:

1. When possible, the entire storage facility should be secured by climb-proof fences. Doors and gates should be kept locked to prevent unauthorized entry.
2. Identification signs may be placed on rooms, buildings, and fences to advise of the contents and warn of their hazardous nature.
3. All items of movable equipment used for handling agricultural chemicals at the storage site should be dedicated for this purpose. If it becomes necessary that they be used for other purposes, they should be thoroughly cleaned and decontaminated before they leave the site.
4. Pallets used for agricultural chemicals should not be used for other products.
5. Where possible, agricultural chemicals should not be stored on the upper floors of buildings. Neither should they be stored in basements, unless proper fire protection is provided. Ground floor storage is preferable.
6. A separate, *secure* storage area should be designated for leaking or damaged containers.

Disposal

Before disposing of any agricultural chemical wastes, consult the requirements of the Resource Conservation and Recovery Act (RCRA), as published by the U.S. Environmental Protection Agency, to determine if the waste is defined as a "hazardous" waste. If so, its disposal is subject to stringent requirements under this law (RCRA).

Helpful Operating Hints

Agricultural chemical containers should be stored with the label plainly visible from operating aisles. If containers are not in good condition when received, they should be placed in a suitable overpack container and properly relabeled after consulting the shipper. If granular agricultural chemicals are received in bags that are damaged, the bag and the contents should be placed in suitable containers and properly labeled. (Fibre-packs or transparent heavy duty plastic bags are ideal.) Metal or rigid plastic containers should be inspected carefully to make sure that the lids and bungs are tight.

Classification and Separation

Each agricultural chemical formulation should be stored separately. Rigid containers should be stored in an upright position. All containers should be stored off the ground in an orderly way, so as to permit ready access and inspection. They should be placed in rows or stacking position with labels visible, and with lanes to provide easy access. A complete inventory should be maintained, indicating the number and identity of containers in each storage position row.

Container Inspection and Maintenance

Containers should be checked regularly for corrosion and leaks. Leaking containers should be isolated, the leak controlled if possible (by inverting container or placing in suitable overpack), and identified. The manufacturer should be consulted for further instructions. Materials such as adsorptive clay, hydrated lime, and sodium hypochlorite (household bleach) should be kept on hand for treatment of spills or leaks. Soda ash and liquid detergent are useful in neutralizing organic phosphate liquids.

Safety Precautions

Rules for personal safety and accident prevention similar to those listed below should be posted in areas where workers congregate. Examples are:

1. Inspect all containers of agricultural chemicals for leaks before handling them. Always wear impervious gloves. Wear personal protective equipment when handling leakers.
2. Handle containers carefully to prevent emergencies.

3. Permit only authorized persons in the storage area.
4. Store agricultural chemicals separately from materials used or consumed by humans or animals.
5. Clean up residue before and after loading or unloading.
6. Inspect all vehicles for placarding and road worthiness.
7. Never ship damaged containers.

Fire control

Where large amounts of agricultural chemicals are stored or where conditions may otherwise warrant, the owner of stored agricultural chemicals should inform the local fire department, in writing, of the hazards that such agricultural chemicals may present in the event of a fire. A floor plan of the storage area, indicating where different agricultural chemical classifications are regularly stored, should be provided. The fire chief should be furnished with the home telephone numbers of those who are responsible for the agricultural chemical storage facility. Annual or semi-annual inspections by the fire department should be encouraged. Pre-fire plans should be formulated and reviewed annually, at least. A listing of agricultural chemicals inventory should be maintained in an area separate from the area of storage, which could be involved in a fire, and easily accessible for emergency purposes.

ALWAYS READ THE LABEL

NACA—Pesticide Safety Team Network

The Pesticide Safety Team Network (PSTN) of the National Agricultural Chemicals Association (NACA) represents a joint effort of technically qualified manufacturers to respond to emergency situations where the involuntary release of agricultural chemicals has occurred. The PSTN can be reached by telephone 24 hours a day through CHEMTREC at 1-800-424-9300.

Suggestions for Fire Hazard Abatement

Where applicable, plainly label the outside of each storage area with "DANGER-POISON" "PESTICIDE STORAGE." Consult with the local fire department regarding the use of the current hazard signal system of the National Fire Protection Association.

In case of fire, make certain that all persons are evacuated immediately to an upwind location. Avoid exposure to smoke and possible toxic fumes. Provide the fire department with as accurate a list as possible of the types and amounts of stored agricultural chemicals. This will provide guidance to determine the best method of attack, and alert firemen to the specific hazards which the stored material(s) may present.

Suggestions for fire fighting procedures and precautions:

1. Wear air-supplied breathing apparatus and rubber clothing.
2. Avoid breathing or otherwise contacting toxic smoke and fumes.
3. Remove clothing and bathe completely as soon as possible after encountering smoke and fumes. Clothing should be properly laundered or destroyed.
4. Contain the water used in fire fighting by a diking system, if possible, to catch all drainage from the area.
5. Evacuate persons who may come in contact with smoke or fumes or contaminated surfaces.

BASIC ELEMENTS OF SAFE PESTICIDE USE

1. Always read the label with care. This is the first step in selecting the right material for the job. Never rely on your memory. Before opening the container, pay strict attention to warnings and cautions printed on the label.
2. Keep spray and dust materials out of the reach of children, pets and irresponsible persons. Storage outside of the home, away from food and feed, and under lock and key is the safest method.
3. Store only in the original container and keep tightly closed.
4. NEVER smoke, eat or drink while applying pesticides.
5. Avoid inhalation or direct contact. Always use personal protective equipment prescribed by the label.
6. Avoid spills. If spills occur, take *immediate action* to remove contaminated clothing and wash thoroughly.
7. After each application, bathe and change to clean clothing. Wash clothing after each use. Always use fresh clothing when starting new application.
8. Avoid contamination of fish ponds and water supplies. Cover feed and water containers when treating around livestock or pet areas.
9. Keep separate equipment for use with hormone-type herbicides to avoid accidental injury to susceptible plants. Also avoid applications under wind conditions which could create drift to non-target areas.
10. Rinse empty containers 3 times before disposing of them. Add the rinse to the spray tank and dispose of containers according to local regulations to avoid hazards to humans, animals and the environment.
11. Follow label directions for mixing and application to keep residues within the limits prescribed by law.
12. Plan ahead. Discuss with your physician the materials you will be using during the season so that he can be prepared to provide the appropriate treatment in case of accidental exposure. If symptoms of illness occur, call the physician or get the patient to a hospital immediately. Always provide the medical personnel with as much information as possible.

Modern pesticide chemicals are developed and tested under strict protocols designed to provide maximum safety both to growers and consumers when used as directed. As with other tools and equipment, however, care must be exercised in handling and using them. Thus, the attitude of the user is a key factor in the exercise of safety procedures.

PESTICIDE SPRAYS AND DUSTS ARE EFFECTIVE FARMING TOOLS IF YOU FOLLOW THESE SIMPLE RULES OF SAFETY

STORAGE AND TEMPERATURE REQUIREMENTS BY PRODUCT

ABBOTT LABORATORIES

DeVine: Store material under refrigerated conditions but DO NOT FREEZE.

DiPel ES: Avoid excessive storage temperatures and direct sunlight (maintain below 110°F).

DiPel 2X: Avoid excessive storage temperatures and direct sunlight. Store in dry area, keep material dry and container tightly closed.

DiPel 4L: Avoid excessive storage temperatures (stay below 110°F) and direct sunlight.

DiPel 8AF: Keep containers tightly closed when not in use. At temperatures below 32°F or above 100°F, store under cover.

DiPel 8L: Avoid excessive storage temperatures and direct sunlight (maintain below 110°F).

DiPel 10G: Avoid excessive storage temperatures and direct sunlight. Store in dry area, keep material dry and container tightly closed.

ProGibb Plus 2X: Avoid excessive storage temperatures and direct sunlight. Store in dry area, keep material dry and container tightly closed.

ProGibb Plus 4% Liquid: Avoid excessive storage temperatures and direct sunlight.

ProGibb Spraying Guide: Flammable. Keep away from heat and open flame.

Promalin: Avoid excessive storage temperatures and direct sunlight.

ProShear: Avoid excessive storage temperatures and direct sunlight.

ProVide: Avoid excessive storage temperatures and direct sunlight.

VectoBac–12AS: Avoid excessive storage temperatures and direct sunlight.

VectoBac–G: Avoid excessive storage temperatures and direct sunlight. Store in dry area, keep material dry and container tightly closed.

AMERICAN CYANAMID COMPANY

Aastar: Store below 120°F to maintain physical properties. Not for use or storage in or around the home.

Abate 1–SG: Segregate from incompatible materials listed in Reactivity Data Section.

Abate 2–CG: Segregate from incompatible materials listed in Reactivity Data Section.

Abate 5–CG: Segregate from incompatible materials listed in Reactivity Data Section.

Abate 4–E: Because of this material's flash point, areas containing this material should conform to the codes (such as NFPA 30 and NFPA 70) governing Class III combustible liquids. Segregate from incompatible materials listed in Reactivity Data Section.

Amdro: Store in a cool, dry, secure place and keep container tightly closed. Use within 3 months after opening container.

Arsenal: Segregate from incompatible materials listed in Reactivity Data Section. Corrosive to mild steel and brass. Store only in areas where flooding is unlikely.

Arsenal Forestry Granule: Store in a secure, dry, well-ventilated separate room, building or covered area. Keep away from sources of ignition and protect from exposure to fire and heat.

Arsenal Concentrate: No special storage requirements needed.

Arsenal 0.5G: Store only in areas where flooding is unlikely. Corrosive to mild steel and brass.

Assert: Do not mix or store in unlined containers. If ice occurs from prolonged storage at $<5°F$ ($<-15°C$), place container at room temperature until ice melts.

Avenge: Do not store below 32°F. Vessels and piping should not be constructed of aluminum.

Chopper: Segregate from incompatible materials listed in Reactivity Data Section. Corrosive to mild steel and brass.

Counter: Do not store in opened bags.

Cycocel: Do not store below freezing temperatures. Because of this material's flash point, areas containing this material should conform to the codes (such as NFPA 30 and NFPA 70) governing Class III combustible liquids.

Cygon 400: Keep away from sources of ignition and protect from exposure to fire and heat. Because of this material's flash point, areas containing this material should conform to the codes (such as NFPA 30 and NFPA 70) governing Class III combustible liquids. To prevent freezing, do not store below 10°C.

Cyprex 65 W: This material has a hazard class St–1 for dust deflagration. Maintain good housekeeping to control accumulation of dust.

Cythion: Because of this material's flash point, areas containing this material should conform to the codes (such as NFPA 30 and NFPA 70) governing Class III combustible liquids. Segregate from incompatible materials listed in Reactivity Data Section.

Cythion 57% EC: Keep away from sources of ignition and protect from exposure to fire and heat.

Cythion RTU: Material rapidly decomposes above 150°C.

Event: Keep from freezing. Do not store below 32°F.

Image: Keep from freezing. Do not store below 32°F.

Pursuit: Do not store below 32°F. If product becomes crystalline, store at room temperature to thaw.

Pursuit Plus: Keep from freezing. Do not store below 32°F.

Prowl: Because of this material's flash point, areas containing this material should conform to the codes (such as NFPA 30 and NFPA 70) governing Class I flammable liquids. Segregate from incompatible materials listed in Reactivity Data Section.

Prozine 70 DF: Store in a dry location.

Prozine 70 DF Non-Crop: Store in a dry location.

Scepter: Do not store below 32°F.

Scepter 70DG: Store in a secure, dry, well-ventilated separate room, building or covered area. Not for use or storage in or around the home. Keep away from exposure to fire and heat.

Scepter OT: Keep from freezing. Do not store below 32°F.

Squadron: Do not store below 32°F. Do not use, pour, spill, or store near heat or open flame. Do not use in the vicinity of pilot lights.

Stomp: Because of this material's flash point, areas containing this material should conform to the codes (such as NFPA 30 and NFPA 70) governing Class I flammable liquids. Segregate from incompatible materials listed in Reactivity Data Section.

Thimet 15–G: Store below 120°F to maintain physical properties.

Thimet 20–G: Store below 120°F to maintain physical properties.

Tri–Scept: Store in a secure, dry, well-ventilated separate room, building, or covered area. Keep away from sources of ignition and protect from exposure to fire or heat. Segregate from oxidizers of fluorides, nitrogen and carbon and incompatible materials listed in Section VI.

BASF CORPORATION

Basagran: Do not allow to freeze.

Blazer: Keep from freezing. Store above 32°F.

Dash: Store above 40°F.

Galaxy: Keep from freezing. Store above 40°F.

Laddok: Do not allow to freeze. Store above 15°F.

Pix: Do not allow to freeze.

Poast: Do not allow to freeze.

Poast Plus: Do not contaminate water, food or feed by storage or disposal.

Pyramin DF: Avoid moisture.

Pyramin FL: Do not allow to freeze.

Ronilan WP: Avoid moisture.

Ronilan FL: Do not allow to freeze.

Storm: Protect from freezing.

CIBA–GEIGY CORPORATION

Aatrex 4L: Store the material in a well-ventilated, secure area.

Aatrex 80W: Store the material in a well-ventilated, secure area.

Aatrex Nine–O: Store the material in a well-ventilated, secure area.

Apron 25W: Store the material in a well-ventilated, secure area. Do not use or store near heat or open flame.

Aquazine 90WDG: Store the material in a well-ventilated, secure area.

Atratol 8P: Store the material in a well-ventilated, secure area. If exposed to intense heat, Sodium chlorate may release toxic fumes of Sodium oxide, Chlorine and Chlorine dioxide.

Atratol 90: Store the material in a well-ventilated, secure area.

Bicep 6L: Store the material in a well-ventilated, secure area.

Caparol 4L: Store the material in a well-ventilated, secure area.

Concep II 70W: Store the material in a well-ventilated, secure area. Do not store near ignition sources such as electrical sparks, flames, or heated surfaces. Do not smoke while using this product. Follow good housekeeping practices to avoid dust accumulations. Ground all metallic surfaces.

Cotoran 4L: Store the material in a well-ventilated, secure area.

Cotoran 80W: Store the material in a well-ventilated, secure area.

Cotoran DF: Store the material in a well-ventilated, secure area.

Curacron 6E: Store the material in a well-ventilated, secure area. Curacron 6E is a Combustible material (DOT regulations).

Curacron 8E: Store the material in a well-ventilated, secure area. Curacron 8E is a Combustible material (DOT regulations). Do not store near heat or open flame.

Dual Herbicide: Store the material in a well-ventilated, secure area.

DZN 4E: Store the material in a well-ventilated, secure area. Do not use, pour, spill, or store near heat or open flame.

DZN 14G: Store the material in a well-ventilated, secure area.

DZN 50W: Store the material in a well-ventilated, secure area. Do not use, pour, spill, or store near heat or open flame.

DZN AG500: Store the material in a well-ventilated, secure area. Forms flammable mixtures with air at temperatures above 41°C. DZN AG500 is a Combustible material (DOT and an NFPA Class I) flammable liquid. Do not use, pour, spill or store near heat or open flame.

Dual 25G: Store the material in a well-ventilated, secure area.

Evik 80W: Store the material in a well-ventilated, secure area.

Orbit: Store the material in a well-ventilated, secure area. Forms flammable mixtures with air at temperatures above 43°C (109°F). Do not use, pour, spill, or store near heat or open flame.

Pramitol 25E: Store the material in a well-ventilated, secure area. Forms flammable mixtures above 77°F. Do not use, pour, spill, or store near heat or open flame. Store at temperatures above 32°F.

Pramitol 5PS: Store the material in a well-ventilated, secure area. Intense heat may release toxic fumes of sodium oxide, chlorine, and chlorine dioxide.

Prime +: Store the material in a well-ventilated, secure area. Prime +is a DOT Combustible liquid and an NFPA Class II Combustible liquid.

Princep 4G: Store the material in a well-ventilated, secure area.

Princep 4L: Store the material in a well-ventilated, secure area. Princep 4L is an NFPA Class IIIB Liquid.

Princep 80W: Store the material in a well-ventilated, secure area. Presents a possible dust explosion hazard at temperatures above 110°C.

Princep Caliber 90: Store the material in a well-ventilated, secure area.

Ridomil 2E: Store the material in a well-ventilated, secure area. Ridomil 2E is an NFPA Class II Combustible Liquid. Do not use or store near heat or open flame.

Ridomil MZ58: Store the material in a well-ventilated, secure area. Exposure to excessive heat or ignition source presents a definite ignition hazard. Fire can occur in closely packed, unventilated piles of bags. Exposure to moisture promotes decomposition which can cause spontaneous combustion.

Ridomil 5G: Store the material in a well-ventilated, secure area.

Ridomil PC 11G: Store the material in a well-ventilated, secure area.

Ridomil Bravo 81W: Store the material in a well-ventilated, secure area.

Sequestrene 138 Fe: Store the material in a well-ventilated, secure area.

Sequestrene 330 Fe: Store the material in a well-ventilated, secure area.

Supracide 2E: Store the material in a well-ventilated, secure area. Supracide 2E is an NFPA Class II Combustible liquid. Do not use, pour, spill, or store near heat or open flame.

Swat: Store the material in a well-ventilated, secure area. Flammable. Fire hazard is greater as liquid temperatures rise above 76°F. Keep away from heat, open flame, sparks, or hot surfaces. Use only in well-ventilated area. Keep container closed. Clean up spills immediately. This product presents fire hazard. Liquid quickly evaporates and forms vapor (fumes) which can catch fire and burn with explosive violence. Invisible vapor spreads easily and can set on fire by many sources such as pilot lights, welding equipment and electrical motors and switches. Do not heat the container. Replace the cap or bung. Emptied container still contains hazardous or explosive vapor or liquid.

Tilt: Store the material in a well-ventilated, secure area. Forms flammable mixtures with air at temperatures above 43°C (109°F). Tilt is an NFPA Class II Combustible liquid. Do not use, pour, spill, or store near heat or open flame.

Trigard 75W: Store the material in a well-ventilated, secure area. This product presents a slight dust explosion hazard.

THE DOW CHEMICAL COMPANY

Confront: Store above 28°F or agitate before use.

Crossbow: Do not use or store near heat or open flame. Store above 10°F or agitate before use.

Curtail: Store above 10°F or warm and agitate before use.

Curtail M: Keep container closed. Do not ship or store with food, feed or drugs or clothing.

Lorsban 4E: Do not use or store near heat or open flame. Do not store above 100°F. Store in a dry area.

Lorsban 15G: No special storage requirement stated.

Lorsban 50W: Store in a cool/dry place. Do not expose to temperatures above 115°F for any extended period of time.

N–Serve 24 for Stabililized Nitrogen: Do not use, pour, spill or store near heat or open flame. Store above 18°F to keep product in solution. Do not use stainless steel for long-term storage unless product and vapor space is kept free of moisture.

N–Serve 24 E for Stabilized Nitrogen: Do not use, pour, spill or store near heat or open flame. Store above 18°F to keep product in solution. Do not use stainless steel for long-term storage unless product and vapor space is kept free of moisture.

Stinger: Keep container closed. Do not ship or store with food, feed, drugs or clothing.

Tandem: Do not use or store near heat or open flame.

Telone II: Do not use, pour, spill, or store near heat or open flame. Store in tightly closed container in a cool place away from dwellings.

Telone C–17: Do not use, pour, spill or store near heat or open flame. Store in tightly closed container in a cool place away from dwellings.

Tordon 22K: No special storage requirements stated.

Tordon RTU: Keep away from heat and open flame.

Turflon Amine: Do not use or store near heat or open flame. Store above 28°F.

Turflon II Amine: Store above 28°F.

Turflon D: Do not use or store near heat or open flame. Store above 10° F.

Turflon Ester: Do not use or store near heat or open flame. Store above 28°F.

E.I. DUPONT DE NEMOURS & CO. (INC.)

Ally: Store product in original container only, away from other pesticides, fertilizer, food or feed.

Asana: Store in a secure, dry temperate area. Avoid contact with water.

Asana XL: Store in a secure, dry and temperate area. Store in original container. Keep container closed when not in use. Do not store near food or feed. Do not use or store around the home. Avoid contact with water.

Assure: Store in well ventilated area. Keep container tightly closed. Do not subject to temperatures below 32°F.

Atrazine 4L: Store in well ventilated area. Store in cool place. Keep container tightly closed. Avoid contact with water.

Benlate: Store in well ventilated area. Never allow to become wet during storage. Keep container tightly closed.

Benlate 50 DF: Store in a well ventilated area. Never allow product to get wet during storage. Keep container tightly closed.

Bidrin 8: Store in well ventilated area. Keep container tightly closed.

Bladex 4L: Store in well ventilated area. Store in cool place. Keep container tightly closed. Avoid contact with water.

Bladex 90DF: Store in well ventilated area. Store in cool place. Avoid contact with water.

Canopy: Store product in original container. Keep container tightly closed.

Classic: Store product in original container. Keep container tightly closed.

Escort: Store product in original container. Keep container tightly closed.

Extrazine II 4L: Store in well ventilated area. Store in cool place. Keep container tightly closed.

Extrazine II DF: Store in well ventilated area. Store in cool place. Keep container tightly closed.

Express: Store product in original container.

Finesse: Store product in original container. Keep container tightly closed.

Gemini: Store product in original container. Keep container tightly closed.

Glean: Store product in original container only. Keep container tightly closed.

Harmony: Store product in original container only. Keep container tightly closed.

Harmony Extra: Store product in original container only, away from other pesticides, fertilizer, food or feed.

Hyvar X: Store product in original container. Keep container tightly closed.

Hyvar X–L: Store product in original container. Keep container tightly closed.

Hyvar L: Store product in original container. Keep container tightly closed. Do not store below 32°F.

Karmex DF: Store product in original container. Keep container tightly closed.

Krenite S: Store product in original container. Keep container tightly closed.

Krovar I DF: Store product in original container. Keep container tightly closed.

Krovar II DF: Store product in original container. Keep container tightly closed.

Lannate: Store in well ventilated area. Keep container tightly closed. Handle carefully to prevent breakage of inner bag when stored at low temperature.

Lannate L: Store in well ventilated area. Do not subject to temperatures below 32°F.

Lannate LV: Store in well ventilated area. Keep container tightly closed. Do not subject to temperatures below 32°F.

Lexone DF: Store product in original container. Keep container tightly closed.

Lexone 4L: Store product in original container. Keep container tightly closed.

Londax: Store in well ventilated area.

Londax California Only: Store product in original container only, away from other pesticides, fertilizer, food or feed. Not for use or storage in or around the home. Keep container closed.

Lorox DF: Store product in original container. Keep container tightly closed. Keep from freezing.

Lorox L: Store product in original container. Keep container tightly closed. Keep from freezing.

Lorox Plus: Store product in original container. Keep container tightly closed.

Manzate 200 DF: Never allow the product to become wet during storage. Store in well ventilated area. Keep container tightly closed.

Oust: Store product in original container. Keep container tightly closed.

Pinnacle: Store product in original container.

Preview: Store product in original container. Keep container tightly closed.

Pydrin 2.4E: Store in well ventilated area. Keep container tightly closed. Avoid contact with water.

Savey: Store product in original container.

Sinbar: Store product in original container.

Telar: Store product in original container. Keep container tightly closed.

Tersan 1991: Store in well ventilated area. Keep container tightly closed. Never allow product to become wet during storage.

Tupersan: Store product in original container.

Velpar: Store product in original container. Keep container tightly closed.

Velpar L: Store in well ventilated area. Keep container tightly closed. Do not store below 32°F.

Velpar ULW: Store in well ventilated area. Keep container tightly closed.

Vendex 4L: Store in secure, dry and temperate area. Keep container tightly closed. Avoid contact with water.

Vendex 50 WP: Store in secure, dry and temperate area. Keep container tightly closed. Avoid contact with water.

Vydate L: Store in well ventilated area. Keep container tightly closed. Do not subject product to temperatures below 32°F.

Vydate C: Flammable. Keep away from heat, sparks, and open flame. Keep container closed. Use with adequate ventilation. Do not subject to temperatures below 32°F. Store the product in original container only, away from pesticides, fertilizers, food or feed. Not for use or storage in or around the home.

ELANCO PRODUCTS COMPANY

Balan EC: Do not store near heat or flame. Store in original container only.

Balan Dry Flowable: Do not store above 120°F. Store in original container only.

Paarlan EC: Do not store near heat or open flame. Store in original container only.

Rubigan EC: Do not store near heat or open flame. Store only in original container.

Sonalan EC: Flammable liquid. Do not use or store near heat or open flame. Tanks that have been used for phenoxy-type herbicides should not be used for Sonalan. Store in original container.

Surflan AS: Store in original container only.

Treflan EC: Avoid freezing. Store above 40°F.

Treflan MTF: Do not store near heat or open flame. Store in original container only.

Treflan 5: Do not store near heat or open flame. Store in original container only.

Treflan TR–10: Avoid extreme cold or heat.

FMC CORPORATION

Ammo WSB: Keep out of reach of children and animals. Store in original containers only. Store in a cool dry place and avoid excess heat. Do not store at temperatures below 32°F (0°C). Rough handling may cause breakage, especially at low temperatures. Allow to warm above 50 (10°C) for use. Do not allow inner bags to become wet during storage. Do not handle inner bag with wet hands or wet gloves. Do not put concentrate or dilute material into food or drink containers. Do not contaminate other pesticides, fertilizers, water, food or feed by storage or disposal.

Ammo 2.5 EC: Store in a cool, dry, well-ventilated location. Avoid excess heat. Do not store below 10°F, (–12°C).

Capture 2EC: Do not freeze. Do not store below 40°F. If crystals are observed, warm material to above 60°F by placing container in warm location. Shake or roll container periodically to redissolve solids.

Carbamate WDG: Keep away from fire and sparks. Store in a cool, dry place and avoid excess heat.

Command 4EC: Do not store below 40°F. Do not freeze. If solid crystals are observed, warm material to above 60°F by placing container in warm location. Store in original container only.

Commence EC: Store in a cool, dry, well-ventilated location. Avoid excess heat.

Dusting Sulfur: Keep away from fire and sparks. Store in a cool, dry place and avoid excess heat.

Elgetol: Store in a cool, dry, well-ventilated location. Avoid excess heat.

Ethion 4 Miscible: Store in a cool, dry, well-ventilated place. Do not use or store near heat, open flame or hot surfaces. Store in original containers only.

Funginex 1.6EC: Store in a cool, dry, well-ventilated place. Don't store below 32°F. Do not use or store near heat, open flame or hot surfaces. Store in original containers only.

Furadan 3G: Store in a cool, dry place and avoid excess heat.

Furadan 4F: Store in a cool, dry, well-ventilated location. Avoid excess heat. Store in original container only.

Furadan 5G: Store in a cool, dry place and avoid excess heat.

Furadan 15G: Store in a cool, dry, well-ventilated location. Avoid excess heat.

Golden Leaf Tobacco Spray: Do not store in or around the home. Do not store below 20°F, (–7°C). Do not use or store near heat, open flame or hot surfaces. Store in a cool, dry place and avoid excess heat.

Methyl Parathion 2 Thiodan 3EC: Do not store below 20°F, (−7°C). Do not use or store near heat, open flame or hot surfaces. Store in a cool, dry place and avoid excess heat.

Option 1EC: Do not store over 100°F or below 10°F. Do not use or store near heat or open flame.

Pounce 1.5G: Store in original containers only. Store in a cool, dry place and avoid excess heat.

Pounce 3.2 EC: Store in a cool, dry, well-ventilated location. Avoid excess heat.

Pounce 25 WP: Store in original containers only. Store in a cool, dry place and avoid excess heat.

Talstar 10 WP: Store in a cool, dry, well-ventilated place. Do not freeze or store at temperatures below 40°F. Avoid excess heat. Store in original containers only.

Thiodan 2 CO EC: Do not store below 0°F, (−18°C). Do not use or store near heat, open flame or hot surfaces.

Thiodan 3EC: Do not store in or around the home. Do not store below 20°F, (−7°C). Do not use or store near heat, open flame or hot surfaces. Store in a cool, dry place and avoid excess heat.

Thiodan 50 WP: Store in a cool, dry, well-ventilated place. Avoid excess heat. Store in original containers only.

Wettable Sulfur: Store in a cool/dry place and avoid excess heat. Keep away from fire and sparks. Store in original containers only.

Ziram 76 WDG: Keep away from fire and sparks. Store in a cool, dry place and avoid excess heat.

HOECHST-ROUSSEL AGRIVET COMPANY

Acclaim 1 EC: Do not store over 100°F or below 10°F. Do not use or store near heat or open flame.

Hoelon 3EC: Do not use or store near heat or open flame. Do not store below 20°F.

Horizon 1EC: Do not store over 100°F or below 10°F. Do not use or store near heat or open flame.

Rely: Store above 28°F or agitate before use. Do not use or store near heat or open flame.

Scout X-TRA: Do not store over 100°F or below 10°F. Do not use or store near heat or open flame.

Tiller: Do not store over 100°F or below 10°F. Do not use or store near heat or open flame.

Whip 1EC: Do not store over 100°F or below 10°F. Do not use or store near heat or open flame.

ICI AMERICAS, INC.

Actellic 5E: Store in a cool place and protect from freezing. Do not store near heat or open flame. Keep container closed when not in use.

Ambush 2E: Protect from freezing. Do not store near heat or flame. Keep container closed when not in use.

Ambush 25W: Do not store diluted material. Keep container closed when not in use.

Ambush 25W WSP: Keep water soluble pack in original foil bag. Store in a cool dry place. Do not store below 32°F. Handle carefully when stored at temperatures less than 50°F to avoid breakage of pack. Keep away from heat and flame.

Arrosolo 3-3E: Containers should be stored in a cool, dry, well-ventilated area away from flammable materials and sources of heat or flame. Use lacquer-lined steel containers.

Betasan 2.9E: Do not store near heat or open flame. Protect from temperatures below 32°F. Keep container closed when not in use.

Betasan 3.6G: Store only in original container in a cool dry place.

Betasan 4-EL F: Do not store near heat or open flame. Protect from temperatures below 42°F.

Betasan 7-G: Store in a cool dry place.

Betasan 12.5-G: Store in a cool dry place.

Captan 50-WP: Keep pesticide in original container. Keep container tightly closed when not in use. Protect from excessive heat. Store in a cool dry place.

Cyclone: Store above 32°F in original container.

Cymbush 3E: Protect from freezing. Do not store near heat or open flame. Keep container closed when not in use.

Demon EC: Do not store near heat or open flame. Protect from freezing. Keep container closed when not in use.

Demon TC: Do not store near heat or open flame. Protect from freezing. Keep container closed when not in use.

Demon WP: Do not store diluted material. Keep container closed when not in use.

Demon WP Packet: Do not store diluted material. Keep water-soluble packets in original foil package and store in a cool dry place. Do not store below 32°F. Handle carefully when stored at temperatures less than 50°F to avoid breakage of packets.

Demon WP WSP: Keep water-soluble packets in original foil package and store in cool/dry place. Do not store below 32°F. Handle carefully when stored at temperatures less than 50°F to avoid breakage of packets.

Devrinol 2-E: Store in a cool dry place. Do not store below 20°F. Do not store near heat or open flame. Keep container closed.

Devrinol 2-G: Store in a cool dry place. Keep container closed when not in use.

Devrinol 5-G: Store in a cool dry place. Keep container closed when not in use.

Devrinol 10-G: Store in a cool, dry, well-ventilated area away from flammable materials and sources of heat or flame. Keep container closed when not in use.

Devrinol 50-DF: Store in a cool dry place. Keep container closed when not in use.

Devrinol 50-WP: Store in a cool, dry, well-ventilated area away from flammable materials and sources of heat and flame. Keep container closed when not in use.

Devrinol 50-WP Ornamental: Store in a cool dry place. Keep container closed when not in use.

Dyfonate 4EC: Store in a cool, dry well-ventilated area away from flammable materials and sources of heat or flame. Minimum storage temperature: MINUS 40°F.

Dyfonate II 10-G: Store in a cool, dry well-ventilated area away from flammable materials and sources of heat or flame.

Dyfonate II 20-G: Not for storage in or around the home.

Eptam 2.3-G: Keep container closed when not in use.

Eptam 5G: Store in a cool dry place. Keep container closed when not in use.

Eptam 7-E: Store in a cool, dry, well-ventilated area away from flammable materials and sources of heat or flame. Can be stored at temperatures as low as minus 50°F. Keep container tightly closed when not in use.

Eptam 10-G: Store in a cool, dry place. Keep container closed when not in use.

Eradicane Extra: Store in a cool, dry, well-ventilated area away from heat or flame. Can be stored at temperatures as low as minus 50°F. Keep container tightly closed when not in use.

Eradicane 6.7E: Store in a cool, dry, well-ventilated area. Store away from flammable materials and sources of heat and flame. Can be stored as low as minus 50°F.

Force 1.5G: Do not store in or around the home. Store in original container only.

Fusilade 2000: Store in original container only. Do not store near heat or open flame. Keep container closed when not in use.

Gramoxone Extra: Do not store in containers made of aluminum or having aluminum fittings. Store above 32°F.

Gramoxone Super: Do not store in containers made of aluminum or having aluminum fittings. Store above 32°F.

Imidan 50-WP: Store in a cool, dry, well-ventilated area away from flammable materials and sources of heat or flame.

Karate: Store in original containers only. Do not use or store near heat or open flame. Keep container closed when not in use.

Ordram 8-E Southern Region: Keep container closed when not in use.

Ordram 8-E Southwest: No special storage requirement stated.

Ordram 8-E California: Keep container closed when not in use.

Ordram 10-G California: Keep container closed when not in use.

Ordram 15-G: Keep container closed when not in use.

Prefar 4-E: Protect from temperatures below 42°F. Do not store near heat or open flame. Keep container closed when not in use.

Reflex 2LC: Store above 32°F in original containers only. Keep container closed when not in use.

Ro-Neet 6-E: Protect from temperatures below 20°F but do not store near sources of heat or open flame.

Sulfur Alfa: Store in a cool dry place. Keep away from heat, sparks and flame. Keep container closed when not in use.

Sulfur Signal: Store in a cool dry place. Keep away from heat, sparks and flame. Keep container closed when not in use.

Sulfur Special Electric: Store in a cool dry place. Keep away from heat, sparks and flame. Keep container closed when not in use.

Surpass 6.7-E: Can be stored at temperatures as low as minus 50°F.

Sutan +6.7-E: Can be stored at temperatures as low as minus 50°F.

Sutan +10-G: Keep container closed when not in use.

Sutazine +18.6-G: Keep container closed when not in use.

Sutazine Can be stored at temperatures as low as minus 30°F. Keep container tightly closed when not in use.

Talon-G Pellets: Store in original container only in a dry place inaccessible to children and pets.

Talon-G Bait Pack: Store in original container only in a dry place inaccessible to children and pets.

Talon G Bait Mini: Store in original container only in a dry place inaccessible to children and pets.

Talon G Mini: Store in original container only in a dry place inaccessible to children and pets.

Tillam 6-E: Keep away from heat and open flame.

Torpedo: Protect from freezing. Do not store near heat or open flame. Keep container closed when not in use.

Vapam: Do not store below 0°F. Keep containers tightly closed when not in use.

Vapam Garden Soil: Do not store below 0°F. Keep container tightly closed when not in use.

Vernam 7-E: Can be stored at temperatures as low as minus 50°F. Keep away from heat and flame. Keep container tightly closed when not in use.

Weather Blok Bait: Store in original container only in a dry place.

MOBAY CORPORATION

Bayleton 50 WP PVA: Store in a cool, dry place. Don't store below freezing (32°F).

Bayleton 50% DF: Store in a cool, dry place.

Bolstar 6: Store in a cool, dry area.

Baythroid 2: Store in a cool, dry area.

DEF 6: Store in a cool dry area.

Di-Syston 15G: Store in a cool dry area.

Di-Syston 8: Store in a cool dry area.

Dylox: Store in a cool dry area.

Dyrene: Store in a cool dry area.

Furadan 3G: Store in a cool dry area.

Furadan 4F: Store in a cool dry area. Store between temperatures of 32°F and 120°F. Store the liquid formulations away from excessive heat and open flame.

Guthion 2S: Store in a cool dry area away from excessive heat and open flame.

Guthion 2L: Store in a cool dry area away from excessive heat and open flame.

Guthion 35% WP: Store in a cool dry area away from heat sources.

Guthion 35% WP PVA: Store in a cool dry place. Do not store below freezing (32°F). Exposure to moisture or excessive handling of water soluble packets may cause breakage. Store packets in original container.

Guthion 50% WP PVA: Store in a cool dry place. Do not store below freezing (32°F). Protect from heat. Keep away from open flame. Do not heat.

Mesurol 75 WP: Store in a cool dry area away from excessive heat and open flame.

Metasystox R: Store in a cool dry place. Keep from heat, open flame.

Monitor 4: Store in a cool dry area.

Morestan 25 WP: Store in a cool dry area.

Morestan 25 WP PVA: Store in a cool dry place but not below 32°F. Protect from rain/moisture.

Nemacur 15G: Store in a cool dry area away from excessive heat and open flame.

Nemacur 15G citrus: Store in a cool dry place. Do not store in front of space heaters. Protect from rain/moisture.

Nemacur 3: Store in a cool dry area.

Nemacur 3 Citrus: Store below 120°F, but not below 32°F. Do not store in front of space heaters. Protect from rain/moisture.

Salute: Store in a cool, dry area away from excessive heat and open flame.

Sencor DF: Store in a cool dry area. Store the liquid formulations away from excessive heat and open flame. Store away from hormone-type herbicides and insecticides.

Sencor 4: Protect from rain and moisture. Store in a cool dry place. Do not store in front of a space heater.

Turbo 8 EC: Store in a cool dry area. Store the liquid formulations away from excessive heat and open flame. Store away from hormone-type herbicides and insecticides.

MONSANTO AGRICULTURAL PRODUCTS COMPANY

Bronco: Store above 40°F to keep product in solution. Do not store near heat or flame.

Buckle: Keep bag closed to prevent spills and contamination.

Bullet: Store above −4°F to keep product from freezing.

Cannon: Store above 32°F to keep product in solution.

Fallow Master: Store above 40°F to keep product in solution.

Far-Go: Store above 32°F to keep from freezing. Do not use, pour, spill or store near heat or flame.

Far-Go Granular: Keep bag closed to prevent spills and contamination.

Freedom: Store above 32°F to keep product in solution.

Honcho: Store above 10°F (−12°C) to keep from crystallizing.

Landmaster BW: Store above 40°F to keep product in solution.

Landmaster II: Store above 40°F to keep product in solution.

Lariat: Store above 40°F. Do not use or store near heat or flame.

Lasso: Store above 32°F to keep product in solution. Do not use or store near heat or flame.

Lasso II: No special storage requirements stated.

Lasso Micro-Tech: Store above −4°F to keep product from freezing.

Pondmaster: Store in secured place in original container. Do not store this product or spray solutions of this product in galvanized steel or unlined steel (except stainless steel) containers or spray tanks.

Ramrod Flowable: Store above 20°F and below 100°F to keep product in suspension.

Ramrod and Atrazine: Store above 20°F and below 100°F to keep product in suspension.

Ramrod 20 Granular: No special storage requirements stated.

Ranger: At temperatures below 50°F, the viscosity of this product may increase to a point that pouring and mixing could be difficult.

Roundup: Store above 10°F. Store only in stainless steel, aluminum, fiberglass, plastic and plastic-lined steel containers.

Roundup RT: Store only in stainless steel, aluminum, fiberglass, plastic and plastic-lined steel containers.

Showdown: Keep bag closed to prevent spills and contamination.

NOR-AM CHEMICAL COMPANY

Antor 4ES: Do not use, pour or store near heat or open flame. Use with adequate ventilation. Do not store below 36°F. Keep container closed.

Apollo: Store in original container in a cool and dry place away from foodstuffs.

Betamix: Do not use or store near heat or open flame. If exposed to subzero temperatures, Betamix gradually thickens, the liquid returns to its original consistency when placed in a room (over 50°F) for several days.

Betanex: Store in original container and keep closed. Store in a cool, dry place. Do not use or store near heat or open flame. If exposed to subzero temperatures Betanex gradually thickens, the liquid returns to its original consistency when placed in a room (over 50°F) for several days.

Botec: Store in original container and keep closed. Store in a cool, dry place.

Botran 75W: Store in original container and keep closed. Store in a cool, dry place.

Carzol SP: Store in a cool/dry place in original container. As with all pesticides, this product should be stored in a separate location under lock. Humid conditions should be avoided during long-term storage.

Dropp 50WP: Do not store under wet conditions. Handle carefully when stored at temperature of less than 50°F.

Mitac WP: Store in a cool/dry place.

Nortron EC: Do not use, pour, spill or store near heat or open flame. Keep containers cool and dry away from foodstuffs. Store only in designated areas for highly flammable liquids.

Rotate 2 1/2G: Store in a cool, dry place away from foodstuffs. Do not store near materials with strong odors.

Vorlex: Store in a tightly closed original container in a cool place away from dwellings. In outside storage, store drums on sides to avoid accumulation of rainwater in top or bottom of recessed areas. Make sure bungs are tightly closed. Store only in areas designated for highly flammable liquids.

PENNWALT CORPORATION

Accelerate: Store in original container.

Aquathol K: Store in original container. Keep container closed.

Aquathol Granular: Store in original container. Keep container closed.

Des-i-cate: Store in original container. Keep container closed.

Desiccant L-10: Don't store in galvanized metal container. Store in original container. Keep container container closed.

Endothal: Store in original container. Keep container closed.

Herbicide 273: Store in original container. Keep container closed.

Hydrothol 191: Store in original container. Keep container closed.

Hydrothol 191 Granular: Store in original container. Keep container closed.

Knox-Out 2 FM: Store in original container. Keep container closed.

Kryocide: Store in original container. Keep container closed.

Maneb 80: Store in original container in a dry area. Keep away from sources of ignition (e.g. sparks and open flame). Close bag when not in use.

Maneb Plus Zinc F4: Keep away from heat, sparks, and open flames.

Penncap-M: Store in original container. Keep container closed.

Penncozeb: Store in original container. Keep container closed.

Penncozeb DF: Store in original container, preferably in a locked storage area. Keep container closed.

Penncozeb Turf and Ornamental: Store in original container. Keep container closed.

Topsin M: Store in original container. Keep container closed.

Topsin M 70W: Store in original container. Keep container closed.

Topsin M 4.5F: Store in original container. Keep container closed.

Topsin M 4.5 Turf and Ornamentals: Store in original container. Keep container closed.

Ziram F4: Store in original container in a frost-free room. Store in the shade. Keep container closed.

Ziram W-76: Store in original container in a cool dry place. Keep away from sources of ignition. Keep container closed.

RHONE-POULENC AG COMPANY

Aliette: Do not store near food, feedstuffs, fertilizers, or seed. Do not contaminate water, food or feed by storage and disposal.

Amiben Granular: Do not store near food, feedstuffs, seeds or fertilizers.

Amiben DS: Do not store near food, feedstuffs, seeds or fertilizers. Do not contaminate water, food or feed by storage or disposal.

Amid-Thin W: Do not store near food, feedstuffs, seeds or fertilizers. Do not contaminate water, food or feed by storage or disposal.

Amitrol T: Do not store near food, feedstuffs, seeds or fertilizers. Do not contaminate water, food or feed by storage or disposal.

Amizol Industrial: Do not store near food, feedstuffs, seeds or fertilizers. Avoid wet or excessively humid conditions. Do not contaminate water, food or feed by storage or disposal.

Aqua-Kleen: Do not store near food, feedstuffs, seeds or fertilizers. Do not contaminate water, food or feed by storage or disposal.

Asulox: Store at temperatures greater than 20°F (−6.6°C). Do not store near food, feedstuffs, fertilizers, or seed. Do not contaminate water, food or feed by storage or disposal.

Bronate: Keep away from heat, sparks, open flame and other ignition sources. Store at temperatures above 3°F (−16°C). If allowed to freeze, remix before using. Do not store near food, feedstuffs, fertilizers, or seed. Do not contaminate water, food or feed by storage or disposal.

Buctril: Store in closed containers in a cool well ventilated area. Stable in storage at 55°C (131°F) for eight weeks in original container. Discolors but not pit steel. Keep away from heat, sparks, open flame and other ignition sources. Store at temperatures above 3°F (−16°C). If allowed to freeze, remix before using. Do not store near food, feedstuffs, fertilizers, or seed. Do not contaminate water, food or feed by storage or disposal.

Buctril + Atrazine: Store at temperatures above 32°F. If exposed to freezing temperatures, store at temperatures above 55°F for 24 hours or until completely thawed. Shake well before using. Do not store near food, feedstuffs, fertilizers, or seed. Do not contaminate water, food or feed by storage or disposal.

Butyrac 175: Do not store near food, feedstuffs, seeds or fertilizers. Do not contaminate water, food or feed by storage or disposal.

Butyrac 200: Do not store near food, feedstuffs, seeds or fertilizers. Do not contaminate water, food or feed by storage or disposal.

Cerone: Do not store at temperatures in excess of 75°C (167°F). Do not store near food, feedstuffs, fertilizers, or seed. Do not contaminate water, food or feed by storage and disposal.

Chiptox MCPA Sodium: Protect from freezing (33°F, 0°C). Do not store near food, feedstuffs, fertilizers, or seed. Do not contaminate water, food or feed by storage or disposal.

Envert 171: Keep away from heat, sparks, open flame. Do not store near food, feedstuffs, fertilizers, or seed. Do not contaminate water, food or feed by storage and disposal.

Esteron 99: Do not cut or weld container. Keep container tightly closed when not in use. This product can be stored in an unheated building.

Ethrel for Flue Cured Tobacco: Do not store near food, feedstuffs, seeds or fertilizers. Do not contaminate water, food or feed by storage and disposal.

Ethrel Plant Regulator (for pineapple): Do not store at temperatures in excess of 75°C (167°F). Do not store near food, feedstuffs, fertilizers, or seed. Do not contaminate water, food or feed by storage and disposal.

Ethrel Plant Regulator: Do not store at temperatures in excess of 75°C (167°F). Do not store near food, feedstuffs, fertilizers, or seed. Do not contaminate water, food or feed by storage and disposal.

Folex 6EC: Keep away from heat, sparks, open flame and other ignition sources. Do not store at temperatures less than 0°F or 30 day average not to exceed 100°F. Do not store near food, feedstuffs, fertilizers, or seed. Do not contaminate water, food or feed by storage and disposal.

Formula 40: Do not cut or weld container. Keep container tightly closed when not in use.

Fruitone N: Do not store near food, feedstuffs, seeds or fertilizers.

Fruitone CPA: Store above 40°F (4.4°C). Do not store near food, feedstuffs, fertilizers, or seed. Do not contaminate water, food or feed by storage and disposal.

Larvin 3.2: Do not store in or around the home. Store unused product in a cool, ventilated, dry, locked area. Do not allow prolonged storage in areas where temperatures frequently exceed 115°F (46°C). NEVER TRANSFER THIS PRODUCT TO ANOTHER CONTAINER FOR STORAGE. Do not contaminate water, food or feed by storage and disposal.

Liquid Amizine: Keep from freezing. Do not store near fertilizers, seeds, insecticides or fungicides.

Mocap 10% G: Do not store near food, feedstuffs, seeds or fertilizers. Do not contaminate water, food or feed by storage and disposal.

Mocap 20% G: Do not store near food, feedstuffs, seeds or fertilizers. Do not contaminate water, food or feed by storage and disposal.

Mocap EC: Keep away from heat, sparks, open flame and other ignition sources. Do not store near food, feedstuffs, fertilizers, or seed. Do not contaminate water, food or feed by storage and disposal.

Mocap Plus: Do not store near food, feedstuffs, seeds or fertilizers. Do not contaminate water, food or feed by storage and disposal.

Mocap Plus 4−2 EC: Keep away from heat, sparks, open flame and other ignition sources. Do not store near food, feedstuffs, fertilizers, or seed. Do not contaminate water, food or feed by storage and disposal.

NAA−800: FLAMMABLE! Keep away from heat, sparks, open flame and other ignition sources. Do not contaminate water, food or feed by storage and disposal.

Rhonox: Always store pesticides in a secured warehouse or storage building. Do not store near seeds, fertilizers, insecticides or fungicides. Containers should be opened in well ventilated areas. Keep container tightly sealed when not in use. Do not stack cardboard cases more than 2 pallets high. Do not contaminate water, food or feed by storage or disposal. COMBUSTIBLE! Keep away from heat, sparks, open flame, and other ignition sources.

Prep: Do not store near food, feedstuffs, seeds or fertilizers. Do not contaminate water, food or feed by storage and disposal. Do not store at temperatures in excess of 75°C (167°F).

Rhomene: Do not store near seeds, fertilizers, insecticides or fungicides. Do not contaminate water, food or feed by storage or disposal. Open dumping is prohibited. Avoid freezing. If stored below 20°F (−6.6°C), rewarm container and agitate.

Rovral: Do not store near food, feedstuffs, seeds or fertilizers. Do not contaminate water, food or feed by storage or disposal.

Rovral 4 Flowable: Do not store near food, feedstuffs, seeds or fertilizers. Do not contaminate water, food or feed by storage or disposal.

Sevin XLR Plus: Store unused product in original container only, in cool, dry area out of reach of children and animals. Do not store in areas where temperatures frequently exceed 100°F (37.8 C). Do not contaminate water, food or feed by storage or disposal.

Sevin 4-Oil: Store unused product in original container only, in cool, dry area out of reach of children and animals. Do not store in areas where temperatures frequently exceed 100°F. Keep away from heat, sparks, open flame and other ignition sources. Do not contaminate water, food or feed by storage or disposal.

Sevin 80S: Store unused product in original container only, in cool, dry area out of reach of children and animals. Do not store in areas where temperatures frequently exceed 100°F (37.8°C). Do not contaminate water, food or feed by storage or disposal.

Sevin 50W: Store unused product in original container only, in cool, dry area out of reach of children and animals. Do not store in areas where temperatures frequently exceed 100°F. Do not contaminate water, food or feed by storage or disposal.

Sevin 4F: Store unused product in original container only, in cool, dry area out of reach of children and animals. Do not store in areas where temperatures frequently exceed 100°F (37.8°C). This product may be used following several freeze-thaw cycles. Do not contaminate water, food or feed by storage or disposal.

Tackle Plus: Protect from freezing. Store above 32°F (0°C). Do not store near food, feedstuffs, fertilizers, or seed. Do not contaminate water, food or feed by storage or disposal.

Temik 15G: Store unused product in original container only, in well-ventilated, clean, dry area out of reach of children and animals. Do not store in areas where temperatures frequently exceed 115°F (46°C) or greater. Do not store in or around home or home garden. Do not store near food, feedstuffs, fertilizers, or seed. Do not contaminate water, food or feed by storage or disposal. If container is broken, handle with rubber gloves.

Temik 15G Citrus: Store unused product in original container only, in well-ventilated, clean, dry area out of reach of children and animals. Do not store in areas where temperatures frequently exceed 115°F (46°C) or greater. Do not store in or around home or home garden. Do not store near food, feedstuffs, fertilizers, or seed. Do not contaminate water, food or feed by storage or disposal. If container is broken, handle with rubber gloves.

Temik 15G CA-AZ: Store unused product in original container only, in well-ventilated, clean, dry area out of reach of children and animals. Do not store in areas where temperatures frequently exceed 115°F (46°C) or greater. Do not store in or around home or home garden. Do not store near food, feedstuffs, fertilizers, or seed. Do not contaminate water, food or feed by storage or disposal. If container is broken, handle with rubber gloves.

Temik TSX: Store unused product in original container only, in well-ventilated, clean, dry area out of reach of children and animals. Do not store in areas where temperatures frequently exceed 115°F (46°C) or greater. Do not store in or around home or home garden. Do not store near food, feedstuffs, fertilizers, or seed. Do not contaminate water, food or feed by storage or disposal. If container is broken, handle with rubber gloves.

Thistrol: Protect from freezing. If allowed to freeze, hold in warm temperatures for several days and remix before using. Do not store near food, feedstuffs, fertilizers, or seed. Do not contaminate water, food or feed by storage or disposal.

Tre-Hold Tree Wound Dressing: Contents under pressure. Do not puncture, incinerate or store above 120°F (49°C). Keep at room temperature. Avoid exposure to direct sunlight and other sources of heat. Do not contaminate water, food or feed by storage or disposal.

Tre-Hold Sprout Inhibitor A112: FLAMMABLE. Keep away from heat, sparks, open flame and other ignition sources. Do not store near food, feedstuffs, fertilizers, or seed. Do not contaminate water, food or feed by storage or disposal.

Tre-Hold Citrus: Do not store near food, feedstuffs, seeds or fertilizers. Do not contaminate water, food or feed by storage or disposal.

Weedar Emulsamine E-3: Keep away from heat, sparks, open flame and other ignition sources. Do not store near food, feedstuffs, fertilizers, or seed. Do not contaminate water, food or feed by storage or disposal.

Weedar 64: Do not store near food, feedstuffs, seeds or fertilizers. Do not contaminate water, food or feed by storage or disposal.

Weedar 64A: Do not store near food, feedstuffs, seeds or fertilizers. Do not contaminate water, food or feed by storage or disposal.

Weedone LV4: Do not store near heat, sparks, open flame, or other ignition sources. Do not store near food, feedstuffs, fertilizers, or seed. Do not contaminate water, food or feed by storage or disposal.

Weedone LV6: Keep away from heat, sparks, open flame, or other ignition sources. Do not store near food, feedstuffs, fertilizers, or seed. Do not contaminate water, food or feed by storage or disposal.

Weedone 2,4-DP: Keep away from heat, sparks, open flame, or other ignition sources. Do not store near food, feedstuffs, fertilizers, or seed. Do not contaminate water, food or feed by storage or disposal.

Weedone 170: Keep away from heat, sparks, open flame, or other ignition sources. Do not store near food, feedstuffs, fertilizers, or seed. Do not contaminate water, food or feed by storage or disposal.

Weedone 638: Keep away from heat, sparks, open flame, or other ignition sources. Do not store near food, feedstuffs, fertilizers, or seed. Do not contaminate water, food or feed by storage or disposal.

Weedone CB: Keep away from heat, sparks, open flame and other ignition sources. Do not store near food, feedstuffs, fertilizers, or seed. Do not contaminate water, food or feed by storage or disposal.

ROHM AND HAAS COMPANY

Dikar: Store in a cool/dry, well-ventilated area. Keep away from fire and sparks. Do not allow it to become wet or overheated.

Dithane DF: Store in a cool, dry, well-ventilated area. Do not allow to become wet or overheated. Keep away from fire and sparks. Store above 32°F.

Dithane F-45: Do not store below 32°F. Store in a cool, well-ventilated area. Keep away from fire and sparks. Do not allow to become overheated in storage.

Dithane M-45: Store in a cool, dry, well-ventilated area. Do not allow to become wet or overheated. Keep away from fire and sparks. Store above 32°F.

Fore: Store in a cool, dry, well-ventilated area. Keep away from fire and sparks. Do not allow to become wet or overheated.

Goal 1.6E: Store above 32°F. Do not store near heat or open flame, combustible.

Karathane LC: Do not store near heat or open flame. Do not store above 90°F for periods longer than 48 hours. Flammable.

Karathane WD: Do not store above 90°F for more than 48 hours.

Kelthane MF: Do not store near heat or open flame. Keep from freezing. Store at 32°F-110°F.

Kelthane 35: Avoid temperatures above 100°C/212°F.

Kerb 50-W: Do not store below 32°F (0°C). Store in a cool, dry place. Do not remove package from container except for immediate use.

Nova: Store in a cool dry place. Do not store below 32°F. Do not remove package from container except for immediate use.

Prostar 4E: Protect from freezing. Do not store near heat or open flame. Combustible.

Rally 40W: Store in a cool dry place. Do not store below 32°F. Do not remove package from container except for immediate use.

Rally 40W AZ: Store in a cool dry place but not below 32°F. Do not remove package from container except for immediate use.

Stam M-4: Protect from freezing. Do not store near heat or open flame. Combustible.

Stampede CM: Do not store below 15°F. Do not store near heat or open flame. Combustible.

Strel 4E: Protect from freezing. Do not store near heat or open flame. Combustible.

Triton AG-44M: Keep away from heat and flame. Combustible.

Triton AG-98: Keep away from heat, sparks and flame. Combustible.

Triton B-1956: Keep away from heat and flame.

Triton CS-7: Keep away from heat, sparks and flame. Combustible.

SANDOZ CROP PROTECTION CORPORATION

Banvel: Store in cool, ventilated, area.

Banvel SGF: Store in a well-ventilated area.

Evital: Store in a cool, dry place.

Javelin WG: Store in a cool area, protect from freezing. Activity may be impaired by storage at temperature above 90°F.

Marksman: Store in a cool, ventilated area.

Probe: Store in original container in a well-ventilated area.

Solicam DF: Store in a cool, dry place. Avoid extreme heat conditions.

Thiolux DF: Store in a cool, dry place.

Trident: Store in a cool area. Activity may be impaired by storage at temperatures above 90°F.

Weedmaster: Store in a cool, ventilated area.

Zorial Rapid 80: Store in a cool, dry place. Avoid extreme heat conditions.

UNIROYAL CHEMICAL COMPANY, INC.

Alanap-L: Store away from sources of direct heat in a dry area. Store below 32°F (0°C).

Apron-Terraclor: Store in a cool, dry place. Do not store in direct sunlight. Protect from extreme temperatures.

B-Nine SP: Store away from sources of direct heat in a dry area.

Casoron 2G: Store away from sources of direct heat in a dry area.

Casoron 4G: Store away from sources of direct heat in a dry area.

Casoron 10G: Store away from sources of direct heat in a dry area.

Casoron 50W: Store away from sources of direct heat in a dry area.

Comite: Store away from sources of direct heat in a dry area.

Comite for Citrus: Flammable, keep away from heat and open flame.

Dimilin 25W: Store away from sources of direct heat in a dry area.

Dimilin Ornamentals: Store in a dry location.

Harvade 5F: Store away from sources of direct heat in a dry area.

Omite 30W East: Store away from sources of direct heat in a dry area.

Omite 30W West: Store away from sources of direct heat in a dry area.

Omite 30W WSP: Flammable, keep away from heat and open flame. Store in a dry location.

Omite CR: Store away from sources of direct heat in a dry area.

Omite CR California: Store in a dry location.

Omite 6E: Store away from sources of direct heat in a dry area.

Ornamite: Store away from sources of direct heat in a dry area.

Plantvax 75W: Store away from sources of direct heat in a dry area.

Reldan 3%: Store in a cool, dry location. Do not store in direct, hot sunlight.

Rescue: Store away from sources of direct heat in a dry area. Store above 32°F (0°C).

Royal MH-30 Tobacco: Store away from sources of direct heat in a dry area; store below 134°F (56°C).

Royal MH-30 Potatoes and Onions: Store away from sources of direct heat in a dry area. Store below 134°F (56°C).

Royal MH-30 SG: Store away from sources of direct heat in a dry area.

Royal Slo-Gro: Store away from sources of direct heat in a dry area. Store below 134°F (56°C).

SULV: Do not contaminate water, food or feed by storage or disposal.

Terraclor 2 lb. Emulsifiable: Store away from sources of direct heat in a dry area.

Terraclor 6.5 + Di-Syston 6.5: Store in a dry location.

Terraclor 10% Granular: Store away from sources of direct heat in a dry area.

Terraclor 75% WP: Store away from sources of direct heat in a dry area.

Terraclor with Mocap: Do not contaminate water, food or feed by storage or disposal. Do not store in or around the home.

Terraclor Super-X Emulsifiable: Store away from sources of direct heat in a dry area.

Terraclor Super-X Granular: Store away from sources of direct heat in a dry area.

Terraclor Super-X/Di-Syston: Store away from sources of direct heat in a dry area.

Terraclor Super-X/Thimet: Store away from sources of direct heat in a dry area.

Terra-Coat LT-2N: Do not use, pour, spill or store near heat or open flame. Store in a cool place. Do not store in direct sunlight. Protect from temperatures below 20°F.

Terrazole 25% Emulsifiable: Store away from sources of direct heat in a dry area.

Terrazole 35% WP: Store away from sources of direct heat in a dry area.

Turfcide 10% Granular: Store away from sources of direct heat in a dry area.

Turfcide 2 lb. Emulsifiable: Store away from sources of direct heat in a dry area.

Vitavax 200: Do not store near heat or open flame. Store in a cool/dry place. Do not store in direct sunlight. Protect from extreme temperatures.

Vitavax Pour On: Do not use or store near heat or open flame.

VALENT USA CORPORATION

Bolero 8EC: Do not use or store near flame, sparks or hot surfaces. Use only in well ventilated areas. Keep container closed.

Bolero 10G: Keep pesticide in original container.

Cobra Herbicide: Do not use or store near flame, sparks or hot surfaces. Use only in well ventilated areas. Keep container closed.

Dibrom 8 Emulsive: Do nôt use or store near flame, sparks or hot surfaces. Use only in well ventilated areas. Keep container closed. Keep pesticide in original container.

Dibrom 14 Concentrate: Do not use or store near flame, sparks or hot surfaces. Use only in well ventilated areas. Keep container closed. Keep pesticide in original container.

VALENT Diquat Herbicide H/A: Keep pesticide in original container.

Monitor 4 Spray: Do not use or store near flame, sparks or hot surfaces. Use only in well ventilated areas. Keep container closed. Keep pesticide in original container.

Orthene 75S Soluble Powder: Store in a cool/dry place. Protect from excessive heat.

Orthene 80 Seed Protectant: Store in a cool, dry place. Protect from excessive heat. Keep pesticide in original container.

Orthene 90S: Store in a cool/dry place. Protect from excessive heat.

Orthene Specialty Concentrate: Store in a cool/dry place. Protect from excessive heat.

Orthene Tobacco Insect Spray: Store in a cool/dry place. Protect from excessive heat.

Orthene Turf, Tree & Ornamental Spray: Store in a cool/dry place. Protect from excessive heat.

Volck Supreme Spray: Keep pesticide in original container.

Product Brand Name	EPA Registration	Hazardous Materials Descriptions and Proper Shipping Names	Hazard Class	Id no.	Labels
		ABBOTT LABORATORIES			
DeVine	275-39	Not regulated			
DiPel 10G	275-55	Not regulated			
DiPel 2X	275-37	Not regulated			
DiPel 4L	275-36	Not regulated			
DiPel 8 AF	275-67				
DiPel 8L	275-51	Not regulated			
DiPel ES	275-65	Not regulated			
DiPel ES East	275-65	Not regulated			
DiPel ES West	275-65	Not regulated			
ProGibb 4%	275-61	Isopropanol	Flammable Liquid	UN 1219	Flammable
ProGibb Plus 2X	275-62	Not regulated			
ProGibb Spraying Guide					
ProVide	275-2	Combustible Liquid, N.O.S.	Combustible Liquid	NA 1993	
Promalin	275-32	Combustible Liquid, N.O.S.	Combustible Liquid	NA 1993	
VectoBac-12 AS	275-66	Not regulated			
VectoBac-G	275-50	Not regulated			
		AMERICAN CYANAMID COMPANY			
Aastar	241-279	Organic Phosphate Compound Mixture, Dry	Poison B	NA-2783, DOT-E-9277	Poison B
Abate 1-SG	241-174AA	Insecticides or Fungicides, Agricultural, N.O.I. Other than Liquid			
Abate 2-CG	241-151AA	Insecticides or Fungicides, Agricultural, N.O.I. Other than Liquid			
Abate 4-E	241-132AA	Insecticides or Fungicides, Agricultural, N.O.I.			
Abate 5-CG	241-150AA	Insecticides or Fungicides, Agricultural, N.O.I. Other than Liquid			
Amdro	241-260	Insecticides or Fungicides, Agricultural, N.O.I. Other than Liquid			
Arsenal	241-273	Not regulated			
Arsenal 0.5G	241-295	Weed Killing Compound, Liquid			
Arsenal Applicator Concentrate	241-299				
Arsenal Forestry Granule	241-308	Not regulated			
Assert	241-285	Not regulated			
Avenge	241-266AA	Weed Killing Compound, Liquid			
Chopper	241-296	Weed Killing Compound, Liquid			
Counter 15G	241-238	Organic Phosphate Compound, Mixture, Dry	Poison B	NA-2783, DOT-E-9277	Poison
Cycocel	241-74AA	Not regulated			
Cygon 400	241-333ZA	Insecticide, Liquid, N.O.S., Combustible Liquid, RQ (Dimethoate)	Combustible Liquid	NA 1993	
Cyprex 65-W	241-51AA	Insecticides or Fungicides, Agricultural, N.O.I. Other than Liquid			
Cythion	241-208AA	Malathion, RQ	ORM-A	NA 2783	
Cythion 57% EC	241-47	Insecticide, Liquid, N.O.S. (Malathion), RQ	Flammable Liquid	NA 1993	Flammable
Cythion RTU	241-282	Malathion, RQ	ORM-A	NA 2783	
Event	241-317				
Image	241-289				
Prowl	241-243	30 or 55 Gal. Drum (bulk): Compound, Weed Killing Liquid, RQ; 5 Gal. Pail: Compound, Weed Killing Liquid	Flammable Liquid	NA 1993	Flammable
	241-243ZA	4 x 1 Gal. Drum Case, Compound, Weed Killing Liquid, Limited Quantities	Flammable Liquid	NA 1993	Flammable
Prozine 70DF	241-302	Compound, Weed Killing, Dry			
Prozine 70DF Non-Crop	241-302	Compound, Weed Killing, Dry			
Pursuit	241-310	Weed Killing Compound, N.O.I.			
Pursuit Plus	241-315				
Scepter	241-289	Compound, Weed Killing, Liquid			
Scepter 70 DG	241-306	Weed Killing Compound, N.O.I.			

Product Brand Name	EPA Registration	Hazardous Materials Descriptions and Proper Shipping Names	Hazard Class	Id no.	Labels
Scepter OT	241-321				
Squadron	241-297	Compound, Weed Killing, Liquid	Flammable Liquid	NA 1993	Flammable
Stomp	241-305				
Thimet 15-G	241-145AA	Organic Phosphate Compound Mixture, Dry	Poison B	NA 2783, DOT-E-9277	Poison
Thimet 20-G	241-257AA	Organic Phosphate Compound Mixture, Dry, RQ (Phorate)	Poison B	NA 2783, DOT-E-9277	Poison
Tri-Scept	241-307	Weed Killing Compound, N.O.I.			
BASF CORPORATION					
Basagran	7969-45	Compounds, Tree or Weed Killing, N.O.I. (Herbicides)	None	None	None
Blazer	7969-79	Compounds, Tree or Weed Killing, N.O.I. (Herbicides) (Freezable product-product from freezing-maintain above 40 F);	None	Not applicable	None
Dash		Naphthalene, Adjuvants, spreader or sticker	None	Not applicable	None
Galaxy	7969-77	Compounds, Tree or Weed Killing, N.O.I. (Herbicides), NOIBN	None	None	None
Laddok		Compounds, Tree or Weed Killing, N.O.I. (Herbicides)	None	None	None
Pix	7969-52	Plant Growth Inhibitor Modifier or Regulator (Keep from freezing)	None	None	None
Poast	7969-58	Bulk-Compounds, Weed Killing, Liquid (contains Xylene); Pkg-Compounds, Tree or Weed Killing, (Herbicides), NOIBN	Bulk-Combustible; Pkg-None	NA 1993	Bulk-Combustible; Pkg-None
Poast Plus	7969-88				
Pyramin DF	7969-81	Compounds, Tree or Weed Killing (Herbicides), N.O.I.B.N.	None	None	None
Pyramin FL	7969-50	Compounds, Tree or Weed Killing, N.O.I. (Herbicides)	None	None	None
Ronilan FL	7969-62	Fungicide, Agricultural, N.O.I.B.N.	None	None	None
Ronilan WP	7969-53	Fungicide, Agricultural, N.O.I.B.N.	None	None	None
Storm	7969-76	Compounds, Tree or Weed Killing (Herbicides), N.O.I.	None	N/A	None
CIBA-GEIGY CORPORATION					
Aatrex 4L	100-497	Compound, Weed Killing (Herbicide) N.O.S.	Not applicable		No label or placard required
Aatrex 80W	100-439	Compound, Weed Killing (Herbicide) N.O.S.	Not applicable		No label or placard required
Aatrex Nine-O	100-585	Compound, Weed Killing (Herbicide) N.O.S.	Not applicable		No label or placard required
Apron 25W	100-639	Insecticide or Fungicide, Agricultural N.O.S.	Not applicable		No label or placard required
Aquazine 90WDG	100-650	Compound, Weed Killing (Herbicide) N.O.S.	Not applicable		No label or placard required
Atratol 8P	100-475	Compound, Weed Killing (Herbicide) N.O.S.	Not applicable		No label or placard required
Atratol 90	100-622	Compound, Weed Killing (Herbicide) N.O.S.			No label or placard required
Bicep 6L	100-645	Compound, Tree and Weed Killing, N.O.S.	Not applicable		No label or placard required
Caparol 4L	100-620	Compound, Weed Killing (Herbicide), N.O.S.	Not applicable		No label or placard required
Concep II 70W		Compound, Seed Treating	Not applicable		No label or placard required
Cotoran 4L	100-642	Compound, Weed Killing (Herbicides), N.O.S.	Not applicable		No label or placard required
Cotoran 80W	100-609	Compound, Weed Killing (Herbicides), N.O.S.	Not applicable		No label or placard required
Cotoran DF	100-704				
Curacron 6E	100-599	Insecticides or Fungicides, Agricultural, N.O.S.	Not applicable		No label or placard required
Curacron 8E	100-669	Insecticides or Fungicides, Agricultural, N.O.S.	Not applicable		No label or placard required
DZN 14G	100-469	Diazinon Mixture, RQ Insecticides or Fungicides, Agricultural, N.O.S.	ORM-A	NA 2783	
DZN 4E	100-463	Diazinon Solution, RQ Insecticides or Fungicides, Agricultural, N.O.S.	ORM-A	NA 2783	
DZN 50W	100-460	Diazinon Mixture, RQ Insecticides or Fungicides, Agricultural, N.O.S.	ORM-A	NA 2783	
DZN AG500	100-461	Diazinon Mixture, RQ Insecticides or Fungicides, Agricultural, N.O.S.	ORM-A	NA 2783	
Dual	100-673				
Dual 25G	100-638	Compound, Weed Killing (Herbicide), N.O.S.	Not applicable		No label or placard required
Evik 80W	100-473	Compound, Weed Killing (Herbicide), N.O.S.	Not applicable		No label or placard required
Orbit	100-617	Compound, Insecticides or Fungicides, Agricultural, N.O.S.	Not applicable		No label or placard required
Pramitol 25E	100-443	Triazine Pesticide, Liquid, Compound, Weed Killing (Herbicide), N.O.S.	Flammable	UN 2764	Flammable
Pramitol 5PS	100-479	Insecticides or Fungicides, N.O.S.	Not applicable		No label or placard required
Prime +	100-640	Plant Growth Inhibitor, Modifer, or Regulator	Not applicable		No label or placard required
Princep 4G	100-435	Compound, Weed Killing (Herbicide), N.O.S.	Not applicable		No label or placard required

Product Brand Name	EPA Registration	Hazardous Materials Descriptions and Proper Shipping Names	Hazard Class	Id no.	Labels
Princep 4L	100-526	Compound, Weed Killing (Herbicide), N.O.S.	Not applicable		No label or placard required
Princep 80W	100-437	Compound, Weed Killing (Herbicide), N.O.S.	Not applicable		No label or placard required
Princep Caliber 90	100-603	Compound, Weed Killing (Herbicide), N.O.S.	Not applicable		No label or placard required
Ridomil 2E	100-607	Insecticides or Fungicides, Agricultural, N.O.S.	Not applicable		No label or placard required
Ridomil 5G	100-628	Insecticides or Fungicides, Agricultural, N.O.S.	Not applicable		No label or placard required
Ridomil MZ58	100-629	Insecticides or Fungicides, Agricultural, N.O.S.	Not applicable		No label or placard required
Ridomil PC 11G	100-664	Insecticides or Fungicides, Agricultural, N.O.S.	Not applicable		No label or placard required
Ridomil/Bravo 81W	100-658	Poison B, Solid, N.O.S., Insecticides or Fungicides, Agricultural, N.O.S.	Poison B	UN 2811	Poison B
Sequestrene 138 Fe		Fertilizing Compound, (Manufactured Fertilizer), N.O.S.	Not applicable		No label or placard required
Sequestrene 330 Fe		Fertilizing Compound, (Manufactured Fertilizer), N.O.S.	Not applicable		No label or placard required
Supracide 2E	100-501	Organophosphorus Pesticide, Liquid Insecticides or Fungicides, Agricultural, N.O.S.	Poison B	UN 2783 DOT E-9416	Poison B
Swat	100-668	Organophosphorus Pesticide Liquid, N.O.S. (Phosphamidon), Insecticides or Fungicides, Agricultural, N.O.S.	Poison B, Flammable Liquid	UN 2784	Poison B, Flammable Liquid
Tilt	100-617	Compound, Insecticides or Fungicides, Agricultural, N.O.S.	Not applicable		No label or placard required
Trigard 75W	100-654	Insecticides or Fungicides, Agricultural, N.O.S.	Not applicable		No label or placard required

THE DOW CHEMICAL COMPANY

Product Brand Name	EPA Registration	Hazardous Materials Descriptions and Proper Shipping Names	Hazard Class	Id no.	Labels
Confront					
Crossbow	464-59	RQ, Compound, Weed Killing, Liquid, (2,4-D Ester)	Combustible Liquid	NA 1993	None
Curtail	464-563	Not regulated by DOT			
Curtail M	464-625	Not regulated by DOT			
Lorsban 15G	464-523	RQ, Chlorpyrifos Solution	ORM-A	NA 2783	None
Lorsban 4E	464-448	RQ, Chlorpyrifos Solution	ORM-A	NA 2783	None
Lorsban 50W	464-552	RQ, Chlorpyrifos Mixture	ORM-A	NA 2783	None
N-Serve 24 Nitrogen Stabilizer	464-433	RQ, Combustible Liquid, N.O.S. (xylene). Under 110 gal. - Not DOT regulated	Combustible Liquid	NA 1993	None
N-Serve 24E Nitrogen Stabilizer	464-432	RQ, Combustible Liquid, N.O.S. (xylene). Under 110 gal. - Not DOT regulated	Combustible Liquid	NA 1993	None
Pathway	464-510				
Stinger					
Tandem	464-615	Not regulated by D.O.T. as hazardous material			
Telone C-17	464-379	RQ, Chloropicrin Mixture, Flammable, (1,3-Dichloropropene)	Poison B	UN 2929	Poison and Flammable
Telone II	464-511	RQ, Dichloropropene, (1,3-Dichloropropene)	Flammable Liquid	UN 2047	Flammable
Tordon 22K	464-323	Not regulated by D.O.T. as hazardous material			
Tordon RTU	464-510	Not regulated by D.O.T. as hazardous material in containers 55 gal. or less			
Turflon Amine	464-546	Not regulated by D.O.T. in less than 110 gallon containers			
Turflon D	464-589	55 gal. drum quantities: RQ, Compound, Weed Killing Liquid (2,4-D Ester). Not regulated by D.O.T. in less than 55 gal. drum quantities	Combustible Liquid	NA 1993	None
Turflon Ester	464-554	Not regulated by D.O.T. in less than 110 gal. containers			
Turflon II Amine	464-606	55 gal. drum quantities: RQ, Hazardous Substance Liquid, N.O.S. (2,4-D Salt). Not regulated by D.O.T. in less than 55 gal. drum quantities	ORM-E	NA 9188	None

E.I. DUPONT DE NEMOURS & CO. (INC.)

Product Brand Name	EPA Registration	Hazardous Materials Descriptions and Proper Shipping Names	Hazard Class	Id no.	Labels
Ally	352-435	Not regulated by DOT, Herbicide Preparation, Other			
Asana	352-502	Not regulated by DOT unless in containers > 110 gal. cap., Agricultural Insecticide, Liquid			
Asana XL	352-515	Not regulated by DOT when in containers < 110 gal., Agricultural Insecticide, Liquid			
Assure	352-441	Not regulated by DOT, Herbicide Preparations, Other			
Atrazine 4L	352-490	Not regulated by DOT, Herbicide Preparations, Other			
Benlate	352-354	Not regulated by DOT, Fungicide Preparations, Other			

DOT Information for CROP PROTECTION CHEMICALS REFERENCE 6th Edition (cont'd.)

Product Brand Name	EPA Registration	Hazardous Materials Descriptions and Proper Shipping Names	Hazard Class	Id no.	Labels
Benlate 50DF	352-447	Not regulated by DOT, Fungicide Preparations, Other			
Bidrin 8	352-466	Organic Phosphate Compound, Liquid (82% Dicrotophos)	Poison, Class B	NA 2783	Poison B
Bladex 4L	352-470	Not regulated by DOT, Herbicide Preparation, Other			
Bladex 90DF	352-495	Not regulated by DOT, Herbicide Preparations, Other			
Canopy	352-444	Not regulated by DOT, Herbicide Preparations, Other			
Classic	352-436	Not regulated by DOT, Herbicide Preparations, Other			
Escort	352-439	Not regulated by DOT, Herbicide Preparations, Other			
Express	352-509	Not regulated by DOT, Herbicide Preparations, Other			
Extrazine II 4L	352-500	Not regulated by DOT, Herbicide Preparations, Other			
Extrazine II DF	352-497	Not regulated by DOT, Herbicide Preparations, Other			
Finesse	352-445	Not regulated by DOT, Herbicide Preparations, Other			
Gemini	352-443	Not regulated by DOT, Herbicide Preparations, Other			
Glean	352-404	Not regulated by DOT, Herbicide Preparations, Other			
Harmony	352-446	Not regulated by DOT, Herbicide Preparations, Other			
Harmony Extra	352-538	Not regulated by DOT, Herbicide Preparations, Other			
Hyvar L Citrus	352-425	Not regulated by DOT, Herbicide Preparations, Other			
Hyvar X	352-287	Not regulated by DOT, Herbicide Preparations, Other			
Hyvar X-L	352-346	Not regulated by DOT, Herbicide Preparations, Other			
Karmex DF	352-508	Not regulated if <125 lbs/pkg. If 125 lbs. or more per package: RQ Hazardous Substance, Solid, N.O.S. (Diuron)	ORM-E	NA 9188	
Krenite S	352-395	Not regulated by DOT, Herbicide Preparations, Other			
Krovar I DF	352-505	Not regulated if <100 lbs. Diuron per pkg. If 100 lbs. Diuron or more: RQ Hazardous Substance, Solid, N.O.S. (Diuron)	ORM-E	NA 9188	
Krovar II DF	352-440	Not regulated if <100 lbs. Diuron per pkg. If 100 lbs. Diuron or more: RQ Hazardous Substance, Solid, N.O.S. (Diuron)	ORM-E	NA 9188	
Lannate	352-342	Carbamate Pesticide, Solid, N.O.S. (90% Methomyl); Add RQ for 100 lbs or more methomyl per pkg	Poison B	UN 2757	Poison B
Lannate L	352-370	Carbamate Pesticide Liquid, N.O.S. (24% Methomyl in Methanol); Add RQ for 100 lbs or more methomyl per pkg	Flammable Liquid, Poison B	UN 2758	Flammable Liquid, Poison B
Lannate LV	352-384	Carbamate Pesticide, Liquid, N.O.S. (29% Methomyl in Cyclohexanone); Add RQ for 100 lbs. or more per pkg.	Poison B	UN 2757	Poison
Lexone 4L	352-382	Not regulated by DOT, Herbicide Preparations, Other			
Lexone DF	352-390	Not regulated by DOT, Herbicide Preparations, Other			
Londax DF	352-506	Not regulated by DOT, Herbicide Preparations, Other			
Londax DF California Only	352-506	Not regulated by DOT, Herbicide Preparations, Other			
Lorox DF	352-394	Not regulated by DOT, Herbicide Preparations, Other			
Lorox L	352-391	Not regulated by DOT, Herbicide Preparations, Other			
Lorox Plus		Not regulated by DOT, Herbicide Preparations, Other			
Manzate 200 DF	352-449	Not regulated except for water. For water: Pesticides, water reactive	ORM-C (water shipments only)	NA 2210 (water shipments only)	
Oust	352-401	Not regulated by DOT, Herbicide Preparations, Other			
Pinnacle	352-525	Not regulated by DOT, Herbicide Preparations, Other			
Preview	352-448	Not regulated by DOT, Herbicide Preparations, Other			
Pydrin 2.4	352-485	Not regulated by DOT unless shipped in portable tanks >110 gal. capacity			
Savey	352-531	Not regulated by DOT, Pesticide Preparations, Other			
Sinbar	352-317	Not regulated by DOT, Herbicide Preparations, Other			
Telar	352-404	Not regulated by DOT, Herbicide Preparations, Other			
Tersan 1991 DF	352-507	Not regulated by DOT, Fungicide Preparations, Other			
Tupersan	352-307	Not regulated by DOT, Herbicide Preparations, Other			
Velpar	352-378	Not regulated by DOT, Herbicide Preparations, Other			
Velpar L	352-392	Ethanol Solution	Flammable Liquid	UN 1170	Flammable Liquid

Product Brand Name	EPA Registration	Hazardous Materials Descriptions and Proper Shipping Names	Hazard Class	Id no.	Labels
Velpar ULW	352-450	Not regulated by DOT, Herbicide Preparations, Other			
Vendex 4L	352-493	Not regulated by DOT, Agricultural Insecticide, Liquid, N.O.I.B.N.			
Vendex 50WP	201-369	Not regulated by DOT, Agricultural Insecticide, N.O.I.B.N.			
Vydate C	352-372	Flammable Liquid, Poisonous, N.O.S. (24% Oxamyl in Methanol)	Flammable Liquid, Poison B	UN 1992	Flammable Liquid and Poison B
Vydate L	352-372	Flammable Liquid, Poisonous, N.O.S. (24% Oxamyl in Methanol)	Flammable Liquid, Poison B	UN 1992	Poison B
		ELANCO PRODUCTS COMPANY			
Balan Dry Flowable	1471-140 (ID5529)	Not regulated			
Balan E.C.	1471-55 (ID5023) 1471-77 (ID5805)	Compound, Weed Killing, Liquid (xylene)	Flammable Liquid	NA 1993	Flammable Liquid
Paarlan E.C.	1471-79	Compound, Weed Killing, Liquid (xylene)	Combustible Liquid	NA 1993	Combustible Liquid
Rubigan E.C.	1471-146	Combustible Liquid, N.O.S. (propylene glycol monomethyl ether, naphthalene)	Combustible Liquid	NA 1993	Combustible Liquid
Sonalan E.C.	1471-122	Compound, Weed Killing, Liquid (xylene)	Flammable Liquid	NA 1993	Flammable Liquid
Surflan A.S.	1471-112 (ID5500) 1471-113 (ID5949) 1471-154 (ID5834)	Not regulated			
Treflan 5	1471-120	Compound, Weed Killing, Liquid (monochlorotoluene)	Combustible Liquid	NA 1993	Combustible Liquid
Treflan E.C.	1471-35 (ID5011) 1471-65 (ID5902) 1471-72 (ID5811)	Compound, Weed Killing, Liquid (xylene)	Combustible Liquid	NA 1993	Combustible Liquid
Treflan M.T.F.	1471-116	Compound, Weed Killing, Liquid (xylene, monochlorotoluene)	Combustible Liquid	NA 1993	Combustible Liquid
Treflan TR-10	1471-143	Not regulated			
		FERMENTA ASC CORPORATION			
2 Plus 2 (MCPP + 2,4-D Amine)	50534-11	Hazardous Substances, Liquid, N.O.S. RQ (2,4-D salts and esters)	ORM-E	NA 9188	
Ansar 6.6	50534-16	Not regulated			
Bravo 500	50534-8	Not regulated			
Bravo 720	50534-188	Not regulated			
Bravo 90 DG	50534-157	Not regulated			
Bravo W-75	50534-23	Not regulated			
Bueno 6	50534-6	Not regulated			
DSMA Liquid	50534-27	Not regulated			
Dacamine 4D	50534-13	2,4-Dichlorophenoxyacetic acid, > =30 gals. solution, RQ	ORM-A	NA 2765	
Daconate 6	50534-5	Not regulated			
Daconil 2787	50534-4	Not regulated			
Daconil 2787 Flowable	50534-9	Not regulated			
Daconil 2787 WDG	50534-195	Not regulated			
Dacthal W-75	50534-1	Not regulated			
Dacthal W-75 Turf	50534-20	Not regulated			
Frigate/Lo Dose		Flammable Liquid, N.O.S.	Flammable	UN 1993	Flammable Liquid
		FMC CORPORATION			
Ammo 2.5 EC	279-3027	Insecticides, Agricultural Liquid (Not regulated)			
Ammo WSB	279-3084				
Capture 2 EC	279-3069	Pesticide, Liquid, Toxic, Flammable, N.O.S. (25.1% Bifenthrin containing Xylene)		UN 2903	
Carbamate WDG	279-388	Fungicides, Agricultural Dry			
Command 4EC	279-3053	(Not regulated) Compound, Weed Killing (Herbicide) N.O.I.			
Commence EC	279-3074	(Not regulated) Compound, Weed Killing (Herbicide) N.O.I.			
Dusting Sulfur	279-1721ZB	Fungicide, Agricultural, Dry (Not regulated)			
Elgetol	279-1853	Substituted Nitrophenol Pesticide, Liquid, N.O.S.	Poison B	UN 2779	

DOT Information for CROP PROTECTION CHEMICALS REFERENCE 6th Edition (cont'd.)

Product Brand Name	EPA Registration	Hazardous Materials Descriptions and Proper Shipping Names	Hazard Class	Id no.	Labels
Ethion 4 Miscible	279-1254	RQ, Insecticide, Liquid, N.O.S. (Ethion)	Combustible Liquid	NA 1993	None
Funginex 1.6 EC	21137-4-279	Fungicide, Agricultural, Liquid (Not regulated)			
Furadan 15G	279-3023	Insecticides, Agricultural, Dry (Not regulated)			
Furadan 3G	279-2792	Insecticides, Agricultural, Dry (Not regulated)			
Furadan 4F	279-2876	RQ, Carbofuran, Poison B	NA 2757		Poison
Furadan 5G	279-2874	Insecticides, Agricultural, Dry (Not regulated)			
Golden Leaf Tobacco Spray	279-2924	RQ, Insecticide Liquid, N.O.S. (Endosulfan)	Combustible Liquid	NA 1993	None
Methyl Parathion 2 Thiodan 3 EC		RQ Methyl Parathion Mixture (Containing 20.4% Methyl Parathion) (Endosulfan)	Poison B	NA 2783	
Option 1 EC	8340-23-279	Compound, Weed Killing Liquid	Flammable Liquid	NA 1993	Flammable
Pounce 1.5G	279-3059	Insecticides, Agricultural, Dry (Not regulated)			
Pounce 25 WP	279-3051	Insecticides, Agricultural, Dry (Not regulated)			
Pounce 3.2 EC	279-3014	(Not regulated) Insecticide, Agricultural, Liquid			
Talstar 10 WP	279-3057	(Not regulated) Insecticide, Agricultural, Dry			
Thiodan 2 CO EC	279-2659				
Thiodan 3 EC	279-2924	RQ, Insecticide Liquid, N.O.S. (Endosulfan)	Combustible Liquid	NA 1993	None
Thiodan 50 WP	279-1380	RQ, Hazardous Substance, Solid, N.O.S. (Endosulfan)	ORM-E	NA 9188	None
Wettable Sulfur	279-387	Fungicides, Agricultural, Dry (Not regulated)			
Ziram 76 WDG	279-268	Fungicides, Agricultural, Dry (Not regulated)			
		HOECHST-ROUSSEL AGRI-VET COMPANY			
Acclaim 1EC		Compound, Weed Killing, Liquid	Flammable Liquid	UN 1993	Flammable
Hoelon 3EC	8340-20-54382	Compound, Weed Killing, Liquid	Combustible Liquid	NA 1993	None
Horizon 1EC		Compound, Weed Killing, Liquid	Flammable Liquid	UN 1993	Flammable
Rely					
Scout X-TRA	34147-3-54382	Not regulated by DOT			
Tiller		Not regulated unless shipped in containers of 110 gals. or more. Compound, Weed Killing, Liquid		NA 1993	
Whip 1EC		Compound, Weed Killing, Liquid	Flammable Liquid	UN 1993	Flammable
		ICI AMERICAS INC.			
Actellic 5E	10182-79	(Pirimiphos-methyl: Toxic to Fish), Insecticides or Fungicides, Agricultural, N.O.I., Liquid			
Ambush	10182-18	(Peremethrin-methyl: Toxic to Fish), Insecticides or Fungicides, Agricultural, N.O.I., Liquid			
Ambush 25W	10182-35	(Peremethrin-methyl: Very Toxic to Fish), Insecticides or Fungicides, Agricultural, N.O.I., Other than Liquid			
Ambush 25W WSP	10182-110	Insecticides or Fungicides, Agricultural, N.O.I., Other than Liquid			
Arrosolo 3-3E	10182-260	Compound, Tree or Weed Killing, N.O.I.			
Betasan 12.5-G	10182-175	Not regulated by DOT/IMO/ICAO. Do not load with food or animal feed			
Betasan 2.9E	10182-170	Compound, Tree or Weed Killing, N.O.I. (For overland, packaged truck only)			
Betasan 3.6-G	10182-183	Compounds, Weed Killing (Herbicides), N.O.I.			
Betasan 4-E LF	10182-163	Compound, Tree or Weed Killing, N.O.I. (For overland, packaged truck only)			
Betasan 7-G	10182-240	Compound, Tree or Weed Killing, N.O.I.			
Captan —					
Cyclone	10182-103	Bipyridilium Pesticide, Liquid, N.O.S. (Paraquat)	Poison B	UN 2781	Poison
Cymbush 3E	10182-65	(Cypermethrin: Toxic to Fish), Insecticides or Fungicides, Agricultural, N.O.I., Liquid			
Demon EC	10182-104	Insecticides or Fungicides, Agricultural, N.O.I., Liquid			
Demon TC	10182-107	Insecticides or Fungicides, Agricultural, N.O.I., Liquid			

Product Brand Name	EPA Registration	Hazardous Materials Descriptions and Proper Shipping Names	Hazard Class	Id no.	Labels
Demon WP	10182-71	6 x 1 lb. jars - (Cypermethrin: Very Toxic to Fish), Insecticides or Fungicides, Agricultural, N.O.I., Liquid, Other than Liquid			
Demon WP Packet	10182-100	(Cypermethrin: Very Toxic to Fish), Insecticides, Agricultural, N.O.I., Other than Liquid			
Demon WP WSP	10182-100	(Cypermethrin: Very Toxic to Fish), Insecticides, Agricultural, N.O.I., Other than Liquid			
Devrinol 10-G	10182-253	Compound, Tree or Weed Killing, N.O.I.			
Devrinol 2-E	10182-219	Compounds, Weed Killing, Liquid	Flammable Liquid	NA 1993	Flammable Liquid
Devrinol 2-G	10182-243	Compound, Tree or Weed Killing, N.O.I.			
Devrinol 5-G	10182-242	Compound, Tree or Weed Killing, N.O.I.			
Devrinol 50 DF	10182-258	Compound, Tree or Weed Killing, N.O.I.			
Devrinol 50-WP	10182-205	Compound, Tree or Weed Killing, N.O.I.			
Devrinol 50-WP Ornamentals	10182-231	Compound, Tree or Weed Killing, N.O.I.			
Dyfonate 10-G	10182-180	Insecticides or Fungicides, Agricultural, N.O.I., Other than Liquid			
Dyfonate 20-G	476-2028	(50 lbs.) Organophosphorus Pesticide, Solid, N.O.S. (Fonofos), DOT-E7987	Poison B	UN 2783	Poison
Dyfonate 4EC	10182-212	(5 gal.) Organophosphorus Pesticide, Liquid, N.O.S. (Fonofos)	Poison B	UN 2783	Poison
Eptam 10-G	10182-160	Compound, Tree or Weed Killing, N.O.I.			
Eptam 2.3-G	10182-172	Compound, Tree or Weed Killing, N.O.I.			
Eptam 5-G	10182-155	Compound, Tree or Weed Killing, N.O.I.			
Eptam 7-E	10182-220	(2 x 2 1/2 gal., 5 gal., 30 gal., 110 gal.,) Compounds, Weed Killing (Herbicides) N.O.I.			
	10182-220	Bulk (tank truck) Compound, Weed Killing, Liquid, (Contains Kerosene)	Combustible Liquid	NA 1993	
Eradicane 6.7-E	10182-223	Bulk (tank truck) Compound, Weed Killing, Liquid, (Contains Kerosene)	Combustible Liquid	NA 1993	
	10182-223	Compound, Tree or Weed Killing, N.O.I.			
Eradicane Extra	10182-244	Bulk (tank truck) Compound, Weed Killing, Liquid, (Contains Kerosene)	Combustible Liquid	NA 1993	
	10182-244	Compound, Tree or Weed Killing, N.O.I.			
Force		Insecticides or Fungicides, Agricultural, N.O.I., Other than Liquid			
Fusilade 2000 1E	10182-88	Compounds, Weed Killing (Herbicides), N.O.I.			
Gramoxone Extra	10182-280	Bipyridilium Pesticide, Liquid, N.O.S. (Paraquat)	Poison B	UN 2781	Poison
Gramoxone Super	10182-103	Bipyridilium Pesticide, Liquid, N.O.S. (Paraquat)	Poison B	UN 2781	Poison
Imidan 50-WP	10182-173	Insecticides or Fungicides, Agricultural, N.O.I., Other than Liquid			
Karate	10182-96	Insecticides or Fungicides, Agricultural, N.O.I., Liquid			
Ordram 10-G California	476-1932	Compound, Tree or Weed Killing, N.O.I.			
Ordram 15-G	10182-274	Compound, Tree or Weed Killing, N.O.I.			
Ordram 8-E	10182-204	Compound, Tree or Weed Killing, N.O.I.			
Ordram 8-E California	10182-204	Compound, Tree or Weed Killing, N.O.I.			
Ordram 8-E Southwest	10182-204	Compound, Tree or Weed Killing, N.O.I.			
Prefar 4E	10182-184	Compound, Tree or Weed Killing, N.O.I.			
Reflex 2LC		Compound, Tree or Weed Killing, N.O.I.			
Ro-Neet 6-E	10182-178	Compound, Tree or Weed Killing, N.O.I.			
Sulfur Alfa	10182-144	Sulfur, refined, OT crude	ORM-C	UN 1350	
Sulfur Signal	10182-143	Sulfur, Solid	ORM-C	UN 1350	
Sulfur Special Electric	10182-142	Sulfur, Solid	ORM-C	UN 1350	
Surpass 6.7E	10182-235	Compound, Tree or Weed Killing, N.O.I.			
Sutan + 10-G	10182-182	Compound, Tree or Weed Killing, N.O.I.			
Sutan + 6.7E	10182-222	Bulk, (tank truck) Compound, Weed Killing Liquid, RQ (Calcium Dodecylbenzene Sulfonate)	Combustible Liquid	NA 1993	
	10182-222	Compound, Tree or Weed Killing, N.O.I.			
Sutazine +	10182-248	Bulk, (tank truck) Hazardous Substance, Liquid, N.O.S. RQ (Calcium Dodecylbenzene Sulfonate)	ORM-E	NA 9188	
	10182-248	Compound, Tree or Weed Killing, N.O.I.			

Product Brand Name	EPA Registration	Hazardous Materials Descriptions and Proper Shipping Names	Hazard Class	Id no.	Labels
Sutazine + 18-6.G	10182-201	Compound, Tree or Weed Killing, N.O.I.			
Talon G	10182-38	Exterminators, Vermin, N.O.I.			
Talon G Bait Mini-Pellet	10182-41				
Talon G Bait Pack Mini-Pellet	10182-40				
Talon G Bait Pack Pellet	10182-39	Not regulated by DOT			
Tillam 6-E	10182-158	Compound, Tree or Weed Killing, N.O.I.			
Torpedo	10182-18	(Permethrin: Toxic to Fish), Insecticides or Fungicides, Agricultural, N.O.I, Liquid			
Vapam	10182-150	Compound, Tree or Weed Killing, N.O.I. Not DOT regulated			
Vapam Garden Soils	10182-229	Compound, Tree or Weed Killing, Liquid			
Vernam 7-E	10182-221	Compound, Tree or Weed Killing, N.O.I.			
Weather Blok Bait	10182-48	Exterminators, Vermin, N.O.I.			
MOBAY CORPORATION					
Bayleton 50% DF	3125-320	Triadimefon Fungicide, Agricultural, Other than Liquid	Not applicable	Not applicable	Not applicable
Bayleton 50% WP WSP	3125-340	Triadimefon Fungicide, Agricultural, N.O.I., Other than Liquid	Not applicable	Not applicable	None
Baythroid 2	3125-351	Cyfluthrin Insecticides, Agricultural, Liquid	Not applicable	Not applicable	Not applicable
	3125-351	Cyfluthrin, Over 110 gals.: Combustible Liq. NOS, NA 1993. Bulk quantities of 7,300 lbs. (Xylenes) or 37,000 lbs. (Ethyl Benzene) or more: add RQ after Haz. Id. No., show (Xylenes) or (Xylenes and Ethyl Benzene)			
Bolstar 6	3125-321	Sulprofos Insecticides, Agricultural, Liquid	Not applicable	Not applicable	Not applicable
	3125-321	Sulprofos, Over 110 gals.: Combustible Liq. NOS, NA 1993. Bulk quantities of 20,000 lbs. or more: add RQ after Hazard Id. No., then show (Xylenes)			
DEF 6	3125-282	S,S,S, Tributyl phosphorotrithioate, Over 110 gals.: Combustible Liquid, N.O.S.; Bulk quantities of 20,800 lbs. or more: add RQ after Hazard Id. No., then show: (Xylenes). Defoliant, Plant, N.O.I.	Combustible Liquid - if > 110 gals.	NA 1993	None
Di-Syston 15% G	3125-172	Disulfoton, Mixture, Dry, R.Q., Insecticides, Agricultural, Other than Liquid	Poison B	NA 2783	Poison
Di-Syston 8	3125-307	Disulfoton, Liquid, RQ, Insecticides, Agricultural, Liquid	Poison B	NA 2783	Poison
	3125-307	Over 110 gals.: Combustible Liquid	Combustible Liquid - if > 100 gals.		
Dylox 80% SP	3125-184	Trichlorfon, If shipped in quantities of 125 lbs. or more: Hazardous Substance, Solid, N.O.S.	ORM-E	NA 9188	None
Dyrene 50% WP	3125-50	Anilazine, Fungicide, Agricultural, Other than Liquid	Not applicable	Not applicable	None
Furadan 3G	279-2792-3125	Carbofuran, Insecticide, Agricultural, Other than Liquid	Not applicable	Not applicable	None
Furadan 4F	279-2876-ZC-3125	Carbofuran, Mixture, Liquid, RQ, Insecticide, Agricultural, Liquid	Poison B	NA 2757	Poison
Guthion 2L	3125-102	Azinphos Methyl, Mixture, Liquid, Insecticide, Agricultural, Liquid	Poison B (Combustible Liquid-bulk only)	NA 2783	Poison
Guthion 2S	3125-123	Azinphos Methyl, Mixture, Liquid, Insecticide, Agricultural, Liquid	Poison B (Combustible Liquid-bulk only)	NA 2783	Poison
Guthion 35% WP	3125-378	Azinphos Methyl, Solid, NA 2783, RQ, Insecticide, Agricultural, Other than Liquid	Poison B	NA 2783	Poison
Guthion 35% WP WSP	3125-379	Azinphos Methyl, Solid, NA 2783, RQ, Insecticide, Agricultural, Other than Liquid	Poison B	NA 2783	Poison
Guthion 50% WP WSP	3125-301				
Mesurol 75% WP	3125-288	Mercaptodimethur, Insecticide, Agricultural, Other than Liquid	Not applicable	Not applicable	None
Metasystox-R SC	3125-111	Oxydemetonmethyl, Organophosphorus Pesticide, Mixture, Liquid, N.O.S., Insecticide, Agricultural, Liquid; if 10,000 lbs. or more RQ must be added to the end of proper shipping name	Flammable Liquid	UN 2784	Flammable
Monitor 4	3125-280-AA	Methamidophos, Organophosphorus Pesticide, Mixture, Liquid N.O.S.; Insecticide, Agricultural, Liquid	Poison B	Un 2783	Poison
Morestan 25% WP	3125-117	Chinomethionat, Insecticide, Agricultural, Other than Liquid	Not applicable	Not applicable	Not applicable

DOT Information for CROP PROTECTION CHEMICALS REFERENCE 6th Edition (cont'd.)

Product Brand Name	EPA Registration	Hazardous Materials Descriptions and Proper Shipping Names	Hazard Class	Id no.	Labels
Morestan 25% WP WSP	3125-302-AA	Chinomethionat, Insecticide, Agricultural, Other than Liquid	Not applicable	Not applicable	Not applicable
Nemacur 15% G	3125-236	Fenamiphos, Organophosphorus Pesticide, Mixture, Solid, N.O.S., RQ Insecticide, Agricultural, Other than Liquid	Poison B	UN 2783	Poison
	3125-236	Organophosphorus Pesticide, Mixture, Solid, N.O.S., RQ; Insecticide, Agricultural, Liquid	Poison B	UN 2783	Poison
Nemacur 3	3125-283	Fenamiphos, Bulk quantities of 11,300 lbs. or more: add RQ after Hazard Identification Number, then show: (Xylenes); If > 110 gals. in one container is Combustible Liquid			
	3125-283	Organophosphorus Pesticide, Mixture, Liquid, N.O.S., Insecticide, Agricultural, Liquid	Poison B	UN 2783	Poison, if > 110 gals. add Combustible placard after Poison
Nemacur 3 Citrus	3125-283	Fenamiphos, Bulk quantities of 11,300 lbs. or more: add RQ after Hazard Identification Number, then show: (Xylenes); If > 110 gals. in one container is Combustible Liquid			
	3125-283	Organophosphorus Pesticide, Mixture, Liquid, N.O.S., Insecticide, Agricultural, Liquid	Poison B	UN 2783	Poison, if > 110 gals. add Combustible placard after Poison
Salute 4EC	3125-375	Compounds, Tree or Weed Killing, N.O.I.	Not applicable	Not applicable	None
Sencor 4	3125-314	Metribuzen, Compounds, Tree or Weed Killing, Liquid	Not applicable	Not applicable	None
Sencor DF	3125-325	Metribuzen, Compounds, Tree or Weed Killing, Other than Liquid	Not applicable	Not applicable	None
Turbo 8EC	3125-366	Metribuzen-Metolachlor, Compounds, Tree or Weed Killing, Other than Liquid; If shipped in quantities of > 110 gals., Combustible Liquid, N.O.S.	Not applicable, if > 110 gal. Combustible Liquid	Not applicable, if > 110 gal, UN 1993	None, if > 110 gals., Combustible
		MONSANTO AGRICULTURAL COMPANY			
Bronco	524-341-AA	650 lbs. or more, Weed Killing Compound, Liquid (Chlorobenzene), Combustible Liquid, NA 1993, RQ	Combustible Liquid	NA 1993	Not applicable
	524-341-AA	< 650 lbs. Weed Killing Compound, Liquid	Not applicable	Not applicable	Not applicable
Bullet		Weed Killing Compound, NOIBN	Not applicable	Not applicable	Not applicable
Cannon	524-412	110 gals. or more: Weed Killing Compound, Liquid	Combustible Liquid	NA 1993	Not applicable
	524-412	< 110 gals. Weed Killing Compound, Liquid	Not applicable	Not applicable	Not applicable
Fallow Master	524-390	Weed Killing Compound, NOIBN	Not applicable	Not applicable	Not applicable
Far-Go	524-145	110 gals. or more, Weed Killing Compound, Liquid	Combustible Liquid	NA 1993	Not applicable
	524-145	< 110 gals. Weed Killing Compound, Liquid, NOIBN	Not applicable	Not applicable	Not applicable
Far-Go Granular	524-292	Weed Killing Compound, NOIBN	Not applicable	Not applicable	Not applicable
Freedom	524-422	110 gals. or more: Weed Killing Compound, Liquid	Combustible Liquid	NA 1993	Not applicable
	524-422	< 110 gals. Weed Killing Compound, Liquid	Not applicable	Not applicable	Not applicable
Granular Buckle		Weed Killing Compound, NOIBN	Not applicable	Not applicable	Not applicable
Honcho	524-326	Weed Killing Compound, NOIBN	Not applicable	Not applicable	Not applicable
Landmaster BW	524-351	485 lbs. or more, RQ, ORM-A, N.O.S., NA 1693, (2,4-Dichlorophenoxyacetic acid salt) Weed Killing Compound, NOIBN	ORM-A	NA 1693	Not applicable
	524-351	< 485 lbs. Weed Killing Compound, NOIBN	None	None	Not applicable
Landmaster II	524-376	900 lbs. or more, RQ, ORM-A, N.O.S. NA 1693, (2,4-Dichlorophenoxyacetic acid salt) Weed Killing Compound, NOIBN	ORM-A	NA 1693	Not applicable
	524-376	< 900 lbs. Weed Killing Compound, NOIBN	None	None	Not applicable
Lariat Flowable	524-329-AA	600 lbs. or more: Weed Killing Compound, Liquid, (Chlorobenzene)	Combustible Liquid	NA 1993	Not applicable
	524-329-AA	< 600 lbs.: Weed Killing Compound, Liquid	None	None	Not applicable
Lasso	524-314-AA	370 lbs. or more: Weed Killing Compound, Liquid, Combustible Liquid, NA 1993, (Chlorobenzene) RQ	Combustible Liquid	NA 1993	Not applicable
	524-314-AA	< 370 lbs.: Weed Killing Compound, Liquid	None	None	Not applicable
Lasso II Granular	524-296-AA	Weed Killing Compound, NOIBN	Not applicable	Not applicable	Not applicable
Lasso Micro-Tech	524-344	Weed Killing Compound, NOIBN	Not applicable	Not applicable	Not applicable
Pondmaster	524-308	Weed Killing Compound, NOIBN	Not applicable	Not applicable	Not applicable
Ramrod 20 Granular	524-152	Weed Killing Compound, NOIBN	Not applicable	Not applicable	Not applicable

Product Brand Name	EPA Registration	Hazardous Materials Descriptions and Proper Shipping Names	Hazard Class	Id no.	Labels
Ramrod Flowable	524-331	Weed Killing Compound, NOIBN, ORM-A, N.O.S., Via Air,	ORM-A	NA 1693 Via Air	Not applicable
Ramrod and Atrazine	524-328	Weed Killing Compound, NOIBN, ORM-A, N.O.S., Via Air	ORM-A	NA 1693 Via Air	Not applicable
Ranger		Weed Killing Compound, NOIBN	Not applicable	Not applicable	Not applicable
Roundup	524-308	Weed Killing Compound, NOIBN	Not applicable	Not applicable	Not applicable
Roundup RT	524-308	Weed Killing Compound, NOIBN	Not applicable	Not applicable	Not applicable
Showdown	524-416	Weed Killing Compound, NOIBN	Not applicable	Not applicable	Not applicable
		NOR-AM CHEMICAL COMPANY			
Antor 4ES	45639-54	Compounds, Tree or Weed Killing, N.O.I.	None	None	None
Apollo SC	45639-135	Insecticides, Agricultural, Liquid, N.O.I.			
Betamix	45639-87	Compounds, Turf or Weed Killing, N.O.I. (Herbicides)	None	None	None
Betanex	45639-86	Compound, Tree or Weed Killing, N.O.I.	None	None	None
Botec	45639-121	Insecticides or Fungicides, Agricultural, Dry, N.O.I.	None	None	None
Botran 75W	45639-110	Fungicides, Agricultural, N.O.I., Other than Liquid	None	None	None
Carzol SP	45639-74	Agricultural Insecticide, Other than Liquid, N.O.I., Carbamate Pesticide, Solid, N.O.S., (Formetanate Hydrochloride)	Poison B	UN 2757	Poison B
Dropp 50WP	45639-89	Defoliant, Plant, N.O.I.	None	None	None
Mitac WP	45639-61	Insecticides or Fungicides, Agricultural, Other than Liquid, N.O.I.	None	None	None
Nortron EC	45639-5	Compound, Tree or Weed Killing, Liquid, N.O.I.	Flammable Liquid	NA 1993	Flammable Liquid
Rotate 2 1/2 G	45639-100	Insecticides, Other than Agricultural, Dry, N.O.I.	None	None	None
Vorlex	45639-90	Agricultural Insecticides or Fungicides, Liquid, NOI; 5 Gal., 20 Liter and 115 Liter: Insecticide, Liquid, NOS; 30 Gal., 55 Gal. and 200 Liter: RQ, Insecticide, Liquid, NOS (1,3-dichloropropene)	Flammable Liquid	NA 1993	Flammable Liquid, Corrosive
		PENNWALT CORPORATION			
Accelerate	4581-284	Not regulated by DOT	None	None	
Aquathol Granular	4581-201	Not regulated by DOT	Not regulated by DOT	Not regulated by DOT	
Aquathol K	4581-204	Compound, Tree or Weed Killing, Liquid (Endothall)	Poison B	NA 2810	Poison 2810
Des-i-cate	4581-206	Not regulated by DOT	None	None	
Desiccant L-10	4581-231	Arsenic Acid Solution	Poison B	UN 1553	
Endothal Turf	4581-79	Not regulated by DOT	Not regulated by DOT	Not regulated by DOT	
Herbicide 273	4581-223	Compound, Tree or Weed Killing, Liquid (Endothall)	Poison B	NA 2810	
Hydrothol 191	4581-174	Compound, Tree or Weed Killing, Liquid (Endothall)	Poison B	NA 2810	
Hydrothol 191 G	4581-172	Not regulated by DOT	Not regulated by DOT	Not regulated by DOT	
Knox Out 2 FM	4581-335	RQ Hazardous Substance, Liquid, N.O.S. (Diazinon)	ORM-E	NA 9188	
Kryocide	4581-116	Not regulated by DOT	Not regulated by DOT	Not regulated by DOT	
Maneb + Zinc F4		Not regulated by DOT			
Maneb 80	4581-255	For water shipment: Pesticide, Water reactive	ORM-C (for water shipment)	NA 2210 (for water shipment)	
Microthiol Special					
Penncap-M	4581-292	RQ Hazardous Substance, Liquid, N.O.S. (Methyl Parathion) (tank truck only)	ORM-E (tank truck only)	NA 9188 (tank truck only)	
Penncozeb	4581-358	For water shipment: Pesticide, Water Reactive	ORM-C (for water shipment)	NA 2210 (for water shipment)	
Penncozeb DF	4581-370	For water shipment: Pesticide, Water Reactive	ORM-C (for water shipment)	NA 2210 (for water shipment)	
Penncozeb Turf & Ornamental	4581-357	For water shipment: Pesticide, Water Reactive	ORM-C (for water shipment)	NA 2210 (for water shipment)	
Topsin M 4.5F	4581-352	Not regulated by DOT	Not regulated by DOT	Not regulated by DOT	
Topsin M 4.5F Turf & Ornamental	4581-352				
Topsin M 70% WP	4581-322	Not regulated by DOT	Not regulated by DOT	Not regulated by DOT	

Product Brand Name	EPA Registration	Hazardous Materials Descriptions and Proper Shipping Names	Hazard Class	Id no.	Labels
Topsin M Turf & Ornamental	4581-322	Not regulated by DOT	Not regulated by DOT	Not regulated by DOT	
Ziram F4		Not regulated by DOT	Not regulated by DOT	Not regulated by DOT	
Ziram W76		Dithiocarbamate Pesticide, Solid, N.O.S.	Poison B	UN 2771	
RHÔNE-POULENC AG COMPANY					
Aliette	264-467	Not regulated			
Amiben DS	264-305	Not regulated			
Amiben Granular	264-167 AA	Not regulated			
Amid Thin W	264-137 AA	Not regulated			
Amitrol T	264-135	Hazardous Substance, Liquid, N.O.S. RQ (Amitrole)	ORM-E	NA 9188	
Amizol Industrial	264-119	Hazardous Substance, Solid, N.O.S. (RQ Amitrole)	ORM-E	NA 9188	
Aqua Kleen	264-109	Not regulated			
Asulox	259-662	Not regulated			
Bronate	264-438	Compounds, Tree or Weed Killing, N.O.S. (only regulated in containers larger than 110 gal.)	Combustible Liquid	NA 1993	
Buctril	264-437	Compounds, Tree or Weed Killing, N.O.S. (only regulated in containers larger than 110 gal.)	Combustible Liquid	NA 1993	
Buctril + Atrazine	264-477	Not regulated			
Butyrac 175	264-164	Not regulated			
Butyrac 200	264-105	Not regulated			
Cerone	264-377	Corrosive Liquid, N.O.S.	Corrosive Material	UN 1760	Corrosive
Chiptox	264-55AA	Not regulated			
Envert 171	264-223	Not regulated			
Esteron 99	464-349-264	RQ, Tree or Weed Killing, Liquid (2,4-Dichlorophenoxyacetic acid ester). Regulated only when 2,4-D Ester > 100 lbs./container	Combustible Liquid	NA 1993	
Ethrel Pineapple Growth Regulator	264-257	Corrosive Liquid, N.O.S.	Corrosive Material	UN 1760	Corrosive
Ethrel Plant Regulator	264-267	Corrosive Liquid, N.O.S.	Corrosive Material	UN 1760	Corrosive
Ethrel Plant Regulator For Flue-Cured Tobacco	264-292	Corrosive Liquid, N.O.S.	Corrosive Material	UN 1760	Corrosive
Folex 6EC	264-498				
Formula 40	464-1-264				
Fruitone CPA	264-338	Not regulated			
Fruitone N	264-141 AA	Not regulated			
Larvin 3.2	264-379	Not regulated			
Liquid Amizine	264-196	Hazardous Substance, Liquid, N.O.S. RQ (Amitrole) (only regulated if package contains 1 lb. or more Amitrole)	ORM-E	NA 9188	
Mocap 10% G	264-465	Not regulated			
Mocap 20%G	264-469	Organophosphorus pesticide, Solid, N.O.S.	Poison B	UN 2783	Poison
Mocap EC	264-458	Organophosphorus pesticide, Liquid, N.O.S.	Poison B	UN 2783	Poison
Mocap Plus	264-459	RQ, Disulfoton Mixture, Dry	Poison B	NA 2783	Poison
Mocap Plus 4-2 EC	264-464	RQ, Disulfoton Mixture, Liquid	Poison B	NA 2783	Poison
NAA-800	264-142 AA	Ethyl Alcohol Solution	Flammable Liquid	UN 1170	Flammable Liquid
Prep	264-418	Corrosive Liquid, N.O.S.	Corrosive Liquid	UN 1760	Corrosive
Rhomene	264-47 AA	Not regulated			
Rhonox	228-156-264	Not regulated			
Rovral	359-685	Not regulated			
Rovral 4	359-724				
Sevin 4-Oil	264-323	For packaging of 110 gals. or more, RQ, Combustible Liquid, N.O.S. (Carbaryl)	Combustible Liquid	NA 1993	
	264-323	RQ, Carbaryl Mixture, (Not regulated when < 100 lbs. of Carbaryl/package)	ORM-A	NA 2757	

Product Brand Name	EPA Registration	Hazardous Materials Descriptions and Proper Shipping Names	Hazard Class	Id no.	Labels
Sevin 4F	264-349	RQ, Carbaryl Mixture, (Not regulated when < 100 lbs. of Carbaryl/package)	ORM-A	NA 2757	
Sevin 50W	264-314	RQ, Carbaryl Mixture, (Not regulated when < 100 lbs. of Carbaryl/package)	ORM-A	NA 2757	
Sevin 80S	264-316	RQ, Carbaryl Mixture, (Not regulated when < 100 lbs. of Carbaryl/package)	ORM-A	NA 2757	
Sevin XLR Plus	264-333	RQ, Carbaryl Mixture, (Not regulated when < 100 lbs. of Carbaryl/package)	ORM-A	NA 2757	
Tackle Plus	7969-76-264	Not regulated			
Temik 15G (Gypsum)	264-330	Carbamate Pesticide, Solid, N.O.S. RQ (Aldicarb)	Poison B	UN 2757	Poison
Temik 15G CA-AZ	264-426	Carbamate Pesticide, Solid, N.O.S. RQ (Aldicarb)	Poison B	UN 2757	Poison
Temik 15G Citrus	264-417	Carbamate Pesticide, Solid, N.O.S. RQ (Aldicarb)	Poison B	UN 2757	Poison
Temik TSX	264-319	Carbamate Pesticide, Solid, N.O.S. RQ (Aldicarb, Pentachloronitrobenzene)	Poison B	UN 2757	Poison
Thistrol	264-181	Not regulated			
Tre-Hold Tree Sprout Inhibitor A112	264-336	Flammable liquid, N.O.S.	Flammable Liquid	UN 1993	Flammable Liquid
Tre-Hold Tree Sprout Inhibitor for Citrus	264-397	Not regulated			
Tre-Hold Tree Wound Dressing	264-431	Compressed Gas, N.O.S., Limited Quantity	Flammable Gas	UN 1954	Flammable Gas
Weedar 64	264-179 AA	Hazardous substance, Liquid, N.O.S., RQ (2,4-D Salt) (only regulated when exceeding 2,4-D Salt RQ of 100 lbs.)	ORM-E	NA 9188	
Weedar 64A	264-143	Hazardous substance, N.O.S., RQ (2,4-D Salt). Not regulated when < 100 lbs. of 2,4-D Salt/package	ORM-E	NA 9188	ORM-E
Weedar Emulsamine E-3	264-179	Hazardous substance, Liquid, N.O.S., RQ (2,4-D Salt) (only regulated in packagings containing 100 lbs. or more of 2,4-D Salt)	ORM-E	NA 9188	
Weedone 170	264-222 ZB	RQ, Compounds, Tree or Weed Killing, Liquid (2,4 Dichlorophenoxy Acetic Acid Ester); Not regulated when < 100 lbs. of 2,4-D/package	Combustible liquid	NA 1993	
Weedone 2,4-DP	264-231	Not regulated			
Weedone 638	264-37	Not regulated			
Weedone CB	264-393	RQ, Compounds, Tree or Weed Killing Liquid (2,4-Dichlorophenoxyacetic acid ester (Not regulated when < 100 lbs. of 2,4-D Ester/package)	Combustible Liquid	NA 1993	
Weedone LV4	264-20	RQ, Compounds, Tree or Weed Killing, Liquid (2,4 Dichlorophenoxy Acetic Acid Ester); Not regulated when < 100 lbs. of 2,4-D/package	Combustible Liquid	NA 1993	
Weedone LV6	264-271	RQ, Compounds, Tree or Weed Killing, Liquid (2,4 Dichlorophenoxy Acetic Acid Ester); Not regulated when < 100 lbs. of 2,4-D/package	Combustible Liquid	NA 1993	

ROHM AND HAAS COMPANY

Dikar	707-93	Fungicide, Agricultural, NOIBN, Dry			
Dithane DF	707-180				
Dithane F-45	707-156	Fungicide, Agricultural, NOIBN, Liquid			
Dithane M-45	707-78	Fungicide, Agricultural, NOIBN, Dry			
Fore	707-87	Fungicide, Agricultural, NOIBN, Dry			
Goal 1.6E	707-174	(2 1/2 gal. Jug, 2/case) Weed Killing Compound, Liquid		NA 1993, UN 1993	Combustible
Karathane LC	707-71	RQ, Insecticide Liquid, N.O.S., Contains Xylene, Flammable, NA 1993		NA 1993, UN 1993	Flammable
Karathane WD	707-53	Fungicide, Agricultural, NOIBN, Dry			
Kelthane 35 WP	707-205	RQ, ORM-E Solid, N.O.S. Kelthane, ORM-E, NA 9188			
Kelthane MF	707-202	(30 gal. drum, 5 gal. pail, 1 gal. jug, 4/case) RQ, ORM-E, Insecticide Liquid, N.O.S., Kelthane NA 9188	ORM-E	NA 9188	
Kerb 50-W	707-159	Weed Killing Compound, Dry			
Nova	707-221	Fungicide, Agricultural, NOIBN, Dry			
Prostar 4E	707-109	RQ, Compound Weed Killing Liquid (Isophorone, Methyl Isobutyl Keton)	Combustible Liquid	NA 1993	
Rally 40W	707-221	Fungicide, Agricultural, NOIBN, Dry			
Rally 40W AZ, CA, HI	707-215	Fungicide, Agricultural, NOIBN, Dry			
Stam M-4	707-109	(35 gal. drum) Weed Killing Compound, Liquid		NA 1993, UN 1993	Combustible
Stampede CM	707-182	(2 1/2 gal jug, 2/case) Weed Killing Compound, Liquid		NA 1993, UN 1993	Combustible

DOT Information for CROP PROTECTION CHEMICALS REFERENCE 6th Edition (cont'd.)

Product Brand Name	EPA Registration	Hazardous Materials Descriptions and Proper Shipping Names	Hazard Class	Id no.	Labels
Strel 4E	707-109	(35 gal. drum) Weed Killing Compound, Liquid			
Triton AG-44M Compatibility Agent	CA Reg 707-50040	(1 gal. jug, 4/case) Adhesive, Adjuvant Spreader, Liquid		NA 1993, UN 1993	Combustible
Triton AG-98 Spreader Activator	CA Reg 707-50036	(1 gal. jug, 4/case) Adhesive, Adjuvant Spreader, Liquid		NA 1993, UN 1993	Combustible
Triton B-1956 Spreader Sticker	CA Reg 707-50032	(1 gal. jug, 4/case) Adhesive, Adjuvant Spreader, Liquid			
Triton CS-7 Spreader Binder	CA Reg 707-50033	(1 gal. jug, 4/case) Adhesive, Adjuvant Spreader, Liquid		NA 1993, UN 1993	Combustible
		SANDOZ CROP PROTECTION CORPORATION			
Banvel	55947-1	RQ Hazardous Substance, Liquid, N.O.S.	ORM-E	NA 9188 (Dicamba)	
Banvel SGF	55947-28	RQ Hazardous Substance, Liquid, N.O.S.	ORM-E	NA 9188 (Dicamba)	
Evital 5% Granular	55947-57		Non-hazardous		None Required
Javelin WG	55947-136				
Marksman	55947-39	RQ, Hazardous Substance, Liquid, N.O.S.	ORM-E	NA 9188 (Dicamba)	
Probe	55947-23				
Solicam DF	55947-78				None Required
Thiolux DF	55947-48		Non-hazardous		None Required
Trident	55947-131				
Weedmaster	55947-24				
Zorial Rapid 80	55947-77	Agricultural Insecticide, NOIBN			None Required
		UNIROYAL CHEMICAL COMPANY, INC.			
Alanap L	400-49	Not regulated by DOT			
Apron-Terraclor Dust Seed	7501-126				
B-Nine SP	400-110	Not regulated by DOT			
Casoron 10G	400-178	Not regulated by DOT unless over 1000 pounds per package: RQ, Hazardous Substance, Solid, N.O.S. (Dichlobenil)	ORM-E	NA 9188	ORM-E
Casoron 2G	400-170	Not regulated by DOT unless over 5000 pounds per package: RQ, Hazardous Substance, Solid, N.O.S. (Dichlobenil)	ORM-E	NA 9188	ORM-E
Casoron 4G	400-168	Not regulated by DOT unless over 2500 pounds per package: RQ, Hazardous Substance, Solid, N.O.S. (Dichlobenil)	ORM-E	NA 9188	ORM-E
Casoron 50W	400-169	Not regulated by DOT unless over 200 pounds per package: RQ, Hazardous Substance, Solid, N.O.S. (Dichlobenil)	ORM-E	NA 9188	ORM-E
Comite	400-104	RQ, Insecticide, Liquid, N.O.S. (Propargite) Flammable Liquid, N.O.S.	Flammable Liquid	NA 1993	Flammable Liquid
Comite for Citrus	400-104				
Ded-Weed SULV			Combustible Liquid	NA 1993	
Dimilin 25W	37100-1-400	Not regulated by DOT			
Dimilin for Ornamentals	37100-8-400				
Harvade 5F	400-155	Not regulated by DOT			
Omite 30W	400-82	Not regulated by DOT unless over 33 pounds per package, RQ, Hazardous Substance, Solid, N.O.S. Propargite	ORM-E	NA 9188	ORM-E
	400-82	RQ, Propargite Mixture	ORM-E	NA 2765 when quantity exceeds 33 lbs.	
Omite 30W WSP	400-82AA				
Omite CR	400-82	Only when over 33 pounds per package: RQ, Hazardous Substance, Solid, N.O.S.	ORM-E	NA 9188 (Propargite)	ORM-E
	400-82	RQ, Propargite Mixture	ORM-E	NA 9188 when net quantitiy exceeds 33 lbs.	

Product Brand Name	EPA Registration	Hazardous Materials Descriptions and Proper Shipping Names	Hazard Class	Id no.	Labels
Omite CR California	400-82				
Omite-6E	400-144-AA	RQ, Insecticide Liquid, N.O.S. Propargite	Combustible Liquid	NA 1993	None
Ornamite	400-82-AA		ORM-E	NA 9188 when net quantity exceeds 33 lbs.	
Plantvax 75W	400-144-AA	Not regulated by DOT			
Reldan 3% Gustafson	7501-99				
Rescue	400-166	Not regulated by DOT			
Royal MH-30	400-84	Not regulated by DOT unless over 7500 gallons per package: RQ, Hazardous Substance, Liquid, N.O.S. (Maleic Hydrazide)	ORM-E	NA 9188	ORM-E
Royal MH-30 SG	400-165	Not regulated by DOT unless over 8344 pounds per package: RQ, Hazardous Substance, Solid, N.O.S. (Maleic Hydrazide)	ORM-E	NA 9188	ORM-E
Royal Slo-Gro	400-94	Not regulated by DOT unless over 7500 gallons per package: RQ, Hazardous Substance, Liquid, N.O.S. (Maleic Hydrazide)			
Terra Coat LT-2N	7501-55				
Terraclor 10% G	400-402	Not regulated by DOT			
Terraclor 2EC	400-400	RQ, Flammable Liquid, N.O.S. (Xylene)	Flammable Liquid	UN 1993	Flammable Liquid
Terraclor 6.5 + Di-Syston 6.5	400-411				
Terraclor 75% WP	400-399	Not regulated by DOT			
Terraclor Super X Emulsifiable	400-405	RQ, Flammable Liquid, N.O.S. (Xylene)	Flammable Liquid	UN 1993	Flammable Liquid
Terraclor Super X Granular	400-406				
Terraclor Super X with Disyston	400-408	RQ, Organic Phosphate Compound Mixture, Dry (Disulfoton)	Poison B	NA 2783	Poison
Terraclor Super X with Thimet	400-409				
Terraclor with Mocap	264-475-400	Not regulated by DOT			
Terrazole 25% EC	400-417	Flammable Liquid, N.O.S. (Xylene)	Flammable Liquid	UN 1993	Flammable Liquid
Terrazole 35% WP	400-416	Not regulated by DOT			
Turfcide 10% G	400-407	Not regulated by DOT			
Turfcide Emulsifiable (2 lbs.)	400-404	Flammable Liquid, N.O.S. (Naphtha Solvent)	Flammable Liquid	UN 1993	Flammable Liquid
Vitavax 200	400-112	RQ, Thiram Mixture	ORM-A	NA 2771	ORM-A
Vitavax Pour On	400-156				

VALENT USA CORPORATION

Product Brand Name	EPA Registration	Hazardous Materials Descriptions and Proper Shipping Names	Hazard Class	Id no.	Labels
Bolero 10 G	239-2449-AA	Not regulated by DOT			
Bolero 8 EC	239-2450	Not regulated by DOT			
Cobra	239-2585	In packages, not regulated by DOT; In bulk: Compound, Tree or Weed Killing, Liquid	Combustible Liquid	NA 1993	Not applicable
Dibrom 14 Conc.	239-1721-AA	RQ, Corrosive Liquid, N.O.S. (Naled)	Corrosive Liquid	UN 1760	Corrosive
Dibrom 8 Emulsive	239-1281-AA	RQ, Corrosive Liquid, N.O.S. (Naled)	Corrosive Liquid	UN 1760	Corrosive
Diquat Herbicide (H/A)	239-1663-ZA	In packages by truck, rail, water; none. In packages by air: ORM-B, N.O.S.	ORM-B	By air: NA 1760	
Monitor 4 Spray	239-2404-AA	Organophosphorus Pesticide, Liquid, N.O.S.	Poison B	UN 2783	Poison
Orthene 75 S Soluble Powder	239-2418-AA	Not regulated by DOT			
Orthene 80 Seed Protectant	239-2434	Not regulated by DOT			
Orthene 90 S	239-2584	Not regulated by DOT			

DOT Information for CROP PROTECTION CHEMICALS REFERENCE 6th Edition (cont'd.)

Product Brand Name	EPA Registration	Hazardous Materials Descriptions and Proper Shipping Names	Hazard Class	Id no.	Labels
Orthene Specialty Concentrate	239-2486-AA	Not regulated by DOT			
Orthene Tobacco Insect Spray	239-2419	Not regulated by DOT			
Orthene Turf, Tree & Ornamental Spray	239-2418	Not regulated by DOT			
Volck Supreme Spray	239-1736-AA	Not regulated by DOT			

APPENDIX II

Poison Control Centers
Solid and Hazardous Waste Agencies by State
Pesticide Control Officials by State
State Extension Pesticide Leaders/Coordinators

The list of Poison Control Centers was generously donated by EMERGENCY MEDICINE Magazine, from their "Poisoning Hotlines" section of the August 15, 1989 issue. Following are some guidelines that will make using this list easier.

Listed in **boldface:**	Each state or regional poison control center, and other centers that provide a 24-hour poison control staff.
Boldface with an asterisk*:	Those 24-hour centers that are also certified by the American Association of Poison Control Centers.
Listed in lightface:	Centers that can give out limited poisoning information. You should specify that you are calling with a poison emergency.

Please note: each manufacturer has an emergency telephone number, which is listed in the manufacturer index.

The Solid and Hazardous Waste agencies provide information about the disposal of crop protection chemicals and containers. This list is provided by the Environmental Protection Agency's Office of Solid Waste.

The list of Pesticide Control Officials was kindly supplied by the Association of American Pesticide Control Officials and the Pesticide Leader/Coordinator listing supplied by the U.S.D.A.

POISON CONTROL CENTERS

ALABAMA

Birmingham
The Children's Hospital
of Alabama Regional Poison Control Center*
1600 Seventh Ave., S. 35233
(800) 292-6678 (Alabama only)
(205) 933-4050
939-9201
939-9202

Tuscaloosa
Alabama Poison Center*
809 University Blvd., E. 35401
(800) 462-0800 (Alabama only)
(205) 345-0600

Chattanooga Poison Control Center
Chattanooga, Tenn.
(615) 778-6100 (northern Alabama)

ALASKA

Anchorage
Anchorage Poison Center
Providence Hospital
3200 Providence Dr.
P.O. Box 196604 99519-6604
(800) 478-3193
(907) 261-3193

ARIZONA

Phoenix
Samaritan Regional Poison Center*
Good Samaritan Medical Center
1130 E. McDowell Rd., Ste. A-5 85006
(602) 253-3334

Tucson
Arizona Poison and Drug
Information Center*
University of Arizona
Arizona Health Sciences Center
1501 N. Campbell Ave., Rm. 3204-K
85724
(800) 362-0101 (Arizona only)
(602) 626-6016

ARKANSAS

Little Rock
Statewide Poison Control Drug
Information Center
University of Arkansas for Medical
Sciences

College of Pharmacy
Slot 522 (internal mailing)
4301 W. Markham St. 72205
(800) 482-8948 (Arkansas only)
(501) 666-5532 (Pulaski County)

CALIFORNIA

Fresno
Fresno Regional Poison Control Center*
Fresno Community Hospital
and Medical Center
Fresno & R Sts., P.O. Box 1232 93715
(800) 346-5922 (California only)
(209) 445-1222

Los Angeles
Los Angeles County Medical Association
Regional Poison Control Center*
1925 Wilshire Blvd. 90057
(800) 777-6476 (public)
825-2722 (MDs only)
(213) 484-5151
644-2121 (MDs and hospitals)

Orange
UC Irvine Regional Poison Center
University of California, Irvine Medical Center
101 City Dr. S, Rte. 78 92668
(800) 544-4404 (southern California only)
(714) 634-5988

Sacramento
UC Davis Medical Center
Regional Poison Control Center*
2315 Stockton Blvd., Rm. 1511 95817
(800) 342-9293 (northern California only)
(916) 453-3692
453-3414 (business information)

San Diego
San Diego Regional Poison Center*
University of California
San Diego Medical Center
225 Dickinson St., H-925 92103-1990
(800) 876-4766
(619) 543-6000

San Francisco
San Francisco Bay Area
Regional Poison Center*
San Francisco General Hospital
1001 Potrero Ave., Rm 1 E 86 94110
(800) 523-2222 (415 and 707 area codes only)
(415) 476-6600

San Jose
Regional Poison Center
Santa Clara Valley Medical Center
751 S. Bascom Ave. 95128
(800) 662-9886
(408) 299-5112

COLORADO

Denver
Rocky Mountain Poison Center*
645 Bannock St. 80204-4507
(800) 332-3073 (Colorado only)
525-5042 (Montana only)
442-2702 (Wyoming only)
(303) 629-1123

Mid-Plains Poison Control Center*
Omaha, Neb.
(800) 228-9515

CONNECTICUT

Farmington
Connecticut Poison Control Center
University of Connecticut
Health Center 06302
(800) 343-2722 (Connecticut only)
(203) 679-3473 (administration)
679-4346 (TDD)

DISTRICT OF COLUMBIA

Washington
National Capital Poison Center*
Georgetown University Hospital
3800 Reservoir Rd. , NW 20007
(202) 625-3333
784-4660 (TTY)

FLORIDA

Jacksonville
St. Vincent's Medical Center
1800 Barrs St.
P.O. Box 2982 32203
(904) 378-7500
378-7499 (TTY)

Tallahassee
Tallahassee Memorial Regional
Medical Center
1300 Miccosukee Rd. 32308
(904) 681-5411

Tampa
The Florida Poison Information Center
at the Tampa General Hospital*
Davis Islands
P.O. Box 1289 33601
(800) 282-3171 (Florida only)
(813) 253-4444

GEORGIA

Atlanta
Georgia Regional Poison Control Center*
Grady Memorial Hospital
80 Butler St., S.E. 30335
(800) 282-5846 (Georgia only)
(404) 589-4400
525-3323 (TTY)

Macon
Regional Poison Control Center
Medical Center of Central Georgia
777 Hemlock St. 31201
(912) 744-1427
744-1146
744-1000

Savannah
Savannah Regional Poison Control
Center
Department of Emergency Medicine
Memorial Medical Center
4700 Waters Ave. 31405
(912) 355-5228 (southeastern Georgia and southwestern South Carolina)

Chattanooga Poison Control Center
Chattanooga, Tenn.
(615) 778-6100 (northern Georgia)

HAWAII

Honolulu
Hawaii Poison Center
Kapiolani-Children's Medical Center
1319 Punahou St. 96826
(800) 362-3585
(808) 941-4411

IDAHO

Boise
Idaho Poison Center
St. Alphonsus Regional Medical Center
1055 N. Curtis Rd. 83706
(800) 632-8000
(208) 378-2707

Spokane Poison Center
Spokane, Wash.
(800) 541-5624 (northern Idaho)

ILLINOIS

Chicago
Chicago & Northeastern Illinois
Regional Poison Control Center
Rush-Presbyterian-St. Luke's Medical
Center
1753 W. Congress Pkwy. 60612
(800) 942-5969 (Illinois only)
(312) 942-5969

Springfield
Central and Southern Illinois
Regional Poison Resource Center
St. John's Hospital
800 E. Carpenter St. 62769
(800) 252-2022
(217) 753-3330

Cardinal Glennon Children's Hospital
Regional Poison Center*
St. Louis Mo.
(800) 366-8888 (western Illinois)

INDIANA

Indianapolis
Indiana Poison Center
Methodist Hospital of Indiana
1701 N. Senate Blvd. 46206
(800) 382-9097 (Indiana only)
(317) 929-2323
929-2336 (TTY)

Kentucky Regional Poison Center of
Kosair Children's Hospital*
Louisville, Ky.
(502) 589-8222 (southern Indiana)

IOWA

Des Moines
Variety Club Poison and Drug Information Center
Iowa Methodist Medical Center
1200 Pleasant St. 50309
(800) 362-2327 (Iowa only)
(515) 283-6254

Iowa City
Poison Control Center
University of Iowa Hospitals and Clinics
52242
(800) 272-6477 (Iowa only)
(319) 356-2922

McKennan Hospital Poison Center
Sioux Falls, S. Dak.
(800) 843-0505

Mid-Plains Poison Control Center*
Omaha, Neb.
(800) 228-9515

KANSAS

Kansas City
Mid-America Poison Control Center
University of Kansas Medical Center
3900 Rainbow, Rm. B-400 66103
(800) 332-6633 (Kansas only)
(913) 588-6633

Wichita
Wesley Medical Center
550 N. Hillside Ave. 67214
(316) 688-2277

Cardinal Glennon Children's Hospital Regional Poison Center*
St. Louis, Mo.
(800) 366-8888 (Topeka)

Mid-Plains Regional Poison Center*
Omaha, Neb.
(800) 228-9515

KENTUCKY

Louisville
Kentucky Regional Poison Center of Kosair Children's Hospital*
P.O. Box 35070 40232-5070
(800) 722-5725 (Kentucky only)
(502) 589-8222 (metropolitan Louisville and southern Indiana)

Knoxville Poison Control Center
Knoxville, Tenn.
(615) 544-9400 (southern Kentucky)

MAINE

Portland
Maine Poison Control Center at Maine Medical Center
22 Bramhall St. 04102
(800) 442-6305 (Maine only)
(207) 871-2950 (ER)

MARYLAND

Baltimore
Maryland Poison Center*
University of Maryland School of Pharmacy
20 N. Pine St. 21201
(800) 492-2414 (Maryland only)
(301) 528-7701

MASSACHUSETTS

Boston
Massachusetts Poison Control System*
300 Longwood Ave.
02115
(800) 682-9211 (Massachusetts only)
(617) 232-2120
277-3323 (TTY)

MICHIGAN

Detroit
Poison Control Center*
Children's Hospital of Michigan
3901 Beaubien Blvd. 48201
(800) 462-6642 (Michigan only)
(313) 745-5711

Grand Rapids
Blodgett Regional Poison Center*
Blodgett Memorial Medical Center
1840 Wealthy St., S.E. 49506
(800) 632-2727 (Michigan only)
(616) 774-7851 (administration)

Kalamazoo
Bronson Poison Center
Bronson Methodist Hospital
252 E. Lovell St. 49007
(800) 442-4112 (Michigan only)
(616) 341-6409

MINNESOTA

Minneapolis
Hennepin Regional Poison Center*
Hennepin County Medical Center
701 Park Ave. 55415
(612) 347-3141
347-6219 (TTY)

St. Paul
Minnesota Regional Poison Center*
St. Paul-Ramsey Medical Center
640 Jackson St. 55101
(800) 222-1222 (Minnesota only)
(612) 221-2113
221-3096 (administration)

McKennan Hospital Poison Center
Sioux Falls, S. Dak.
(800) 843-0505

North Dakota Poison Information Center
Fargo, N. Dak.
(701) 234-5575 (northwestern Minnesota—call collect)

St. Luke's Midland Regional Medical Center
Poison Control Center
Aberdeen, S. Dak.
(800) 592-1889

MISSISSIPPI

Jackson
Regional Poison Control Center
University Medical Center
2500 N. State St. 39216
(601) 354-7660

MISSOURI

Kansas City
The Children's Mercy Hospital
24th and Gillham Rd. 64108
(800) 366-8888 (Missouri Poison System)
(816) 234-3000

St. Louis
Cardinal Glennon Children's Hospital Regional Poison Center*
1465 S. Grand Blvd. 63104
(800) 392-9111 (Missouri only)
366-8888 (western Illinois and Topeka, Kan.)
(314) 772-5200

Mid-Plains Poison Control Center*
Omaha, Neb.
(800) 228-9515

MONTANA

Rocky Mountain Poison Center*
Denver, Colo.
(800) 525-5042

Spokane Poison Center
Spokane, Wash.
(800) 541-5624 (western Montana)

NEBRASKA

Omaha
Mid-Plains Poison Control Center*
Children's Memorial Hospital
8301 Dodge St. 68114
(800) 642-9999 (Nebraska only)
228-9515 (Colorado, Iowa, Kansas, Missouri, South Dakota, Wyoming)
(402) 390-5400

McKennan Hospital Poison Center
Sioux Falls, S. Dak.
(800) 843-0505

NEW HAMPSHIRE

Hanover
New Hampshire Poison Information Center
Dartmouth/Hitchcock Medical Center
2 Maynard St. 03756
(800) 562-8236 (New Hampshire only)
(603) 646-5000

NEW JERSEY

Newark
New Jersey Poison Information and Education System*
Newark Beth Israel Medical Center
201 Lyons Ave. 07112
(800) 962-1253 (New Jersey only)
(201) 923-0764
926-8008 (TTY)
926-7443 (administration)

NEW MEXICO

Albuquerque
New Mexico Poison and Drug Information Center*
University of New Mexico
87131
(800) 432-6866 (New Mexico only)
(505) 843-2551
277-4261 (administration)

El Paso Poison Control Center
El Paso, Tex.
(915) 533-1244 (southern New Mexico)

NEW YORK

Buffalo
Western New York Regional Poison Control Center at Children's Hospital of Buffalo
219 Bryant St. 14222
(800) 888-7655
(716) 878-7654
878-7655

East Meadow
Long Island Regional Poison Control Center*
Nassau County Medical Center
2201 Hempstead Tpk. 11554
(516) 542-2324
542-2325
542-2323 (TTY)

New York
New York City Poison Control Center*
455 First Ave., Rm. 123 10016
(212) 340-4494
764-7667

Nyack
Hudson Valley Regional Poison Center
Nyack Hospital
N. Midland Ave. 10960
(914) 353-1000

Rochester
Life Line/The Finger Lakes Regional Poison Control Center
University of Rochester Medical Center
Box 777 14642
(800) 333-0542
(716) 275-5151
275-4354 (office of director)
275-2700 (TTY)

Schenectady
Ellis Hospital Poison Center
1101 Nott St. 12308
(518) 382-4039
382-4309

Syracuse
Central New York Poison Control Center
University Hospital at Syracuse
750 E. Adams St. 13210
(800) 252-5655 (outside Onondaga County)
(315) 476-4766

Northwest Regional Poison Center
Erie, Pa.
(800) 822-3232 (southwestern New York)

Vermont Poison Center
Burlington, Vt.
(802) 658-3456

NORTH CAROLINA

Asheville
Western NC Poison Control Center
Memorial Mission Hospital
509 Biltmore Ave. 28801
(800) 542-4225
(704) 255-4490

Charlotte
Mercy Hospital Poison Control Center
2001 Vail Ave. 28207
(704) 379-5827

Durham
Duke University Regional Poison Control Center*
Duke University Medical Center
P.O. Box 3007 27710
(800) 672-1697 (North Carolina only)
(919) 684-8111

Greensboro
Moses H. Cone Memorial Hospital
Triad Poison Center
1200 N. Elm St. 27401-1020
(800) 722-2222 (North Carolina only)
(919) 379-4105

Hickory
Catawba Memorial Hospital
Poison Control Center
810 Fairgrove Church Rd., S.E. 28602
(704) 322-6649
322-0363 (administration)

Tidewater Poison Center
Norfolk, Va.
(800) 552-6337 (northeastern North Carolina)

NORTH DAKOTA

Fargo
North Dakota Poison Information Center
St. Luke's Hospitals
720 Fourth St., N. 58122
(800) 732-2200 (North Dakota only)
(701) 234-5575 (local; northwestern Minnesota—call collect)

St. Luke's Midland Regional Medical Center
Poison Control Center
Aberdeen, S. Dak.
(800) 592-1889

OHIO

Akron
Akron Regional Poison Control Center
Children's Hospital Medical Center of Akron
281 Locust St. 44308
(800) 362-9922
(216) 379-8562
379-8446 (TTY)
379-8447 (FAX)

Cincinnati
Regional Poison Control System and Drug & Poison Information Center*
University of Cincinnati College of Medicine
231 Bethesda Ave., M.L. 144 45267-0144
(800) 872-5111 (Ohio only)
(513) 558-5111

Cleveland
Greater Cleveland Poison Control Center
2101 Adelbert Rd. 44106
(216) 231-4455

Columbus
Central Ohio Poison Center*
Children's Hospital
700 Children's Dr. 43205
(800) 682-7625 (Ohio only)
(614) 228-1323

Dayton
Western Ohio Poison and Drug Information Center
Children's Medical Center
1 Children's Plaza 45404-1815
(800) 762-0727 (Ohio only)
(513) 222-2227

Lorain
Lorain Community Hospital
3700 Kolbe Rd. 44053
(800) 821-8972
(216) 282-2220

Toledo
Poison Information Center
Medical College of Ohio Hospital
3000 Arlington Ave. 43614
(419) 381-3897

Youngstown
Mahoning Valley Poison Center
St. Elizabeth Hospital Medical Center
1044 Belmont Ave. 44501
(800) 426-2348 (Mahoning, Trumbull, Columbiana counties in Ohio; Mercer and Lawrence in Pennsylvania)
(216) 746-2222
746-5510 (TTY)

Northwest Regional Poison Center
Erie, Pa.
(800) 822-3232 (northeastern Ohio)

OKLAHOMA

Oklahoma City
Oklahoma Poison Control Center
Oklahoma Children's Memorial Hospital
940 N.E. 13th St.
P.O. Box 26307 73126
(800) 522-4611 (Oklahoma only)
(405) 271-5454

OREGON

Portland
Oregon Poison Center*
Oregon Health Sciences University
3181 S. W. Sam Jackson Park Rd.
97201
(800) 452-7165 (Oregon only)
(503) 279-8968

Spokane Poison Center
Spokane, Wash.
(800) 541-5624 (northwestern Oregon)

PENNSYLVANIA

Allentown
Lehigh Valley Poison Center
Allentown Hospital
17th & Chew Sts. 18102
(215) 433-2311

Altoona
Keystone Region Poison Center
Mercy Hospital
2500 Seventh Ave. 16603
(814) 946-3711

Danville
Susquehanna Poison Center
Geisinger Medical Center
N. Academy Ave. 17821
(717) 271-6116
275-6119

Erie
Northwest Regional Poison Center
Saint Vincent Health Center
232 W. 25th St. 16544
(814) 452-3232
(800) 822-3232 (northwestern Pennsylvania, northeastern Ohio, southwestern New York)

Hershey
Capital Area Poison Center
University Hospital
Milton S. Hershey Medical Center
University Dr. 17033
(717) 531-6111
531-6039

Philadelphia
Delaware Valley Regional Poison Control Center*
1 Children's Ctr.
34th & Civic Center Blvd. 19104
(215) 386-2100
386-2066 (administration)

Pittsburgh
Pittsburgh Poison Center*
1 Children's Pl.
3705 Fifth Ave. at DeSoto St. 15213
(412) 681-6669
692-5600 (administration/consultation)

Mahoning Valley Poison Center
Youngstown, Ohio
(800) 426-2348 (Mercer and Lawrence counties)

RHODE ISLAND

Providence
Rhode Island Poison Center*
Rhode Island Hospital
593 Eddy St. 02903
(401) 277-5727
277-8062 (TDD)

SOUTH CAROLINA

Columbia
Palmetto Poison Center
University of South Carolina
College of Pharmacy
29208
(800) 922-1117 (South Carolina only)
(803) 765-7359

Savannah Regional Poison Control Center
Savannah, Ga.
(912) 355-5228 (southwestern South Carolina)

SOUTH DAKOTA

Aberdeen
St. Luke's Midland Regional Medical Center
Poison Control Center
305 S. State St. 57401
(800) 592-1889 (South Dakota, North Dakota, Minnesota, Wyoming)
(605) 225-1880

Sioux Falls
McKennan Hospital Poison Center
800 E. 21 St.
P.O. Box 5045 57117-5045
(800) 952-0123 (South Dakota only)
843-0505 (Iowa, Minnesota, Nebraska)
(605) 336-3894

Mid-Plains Poison Control Center*
Omaha, Neb.
(800) 228-9515

TENNESSEE

Chattanooga
Chattanooga Poison Control Center
T.C. Thompson Children's Hospital
910 Blackford St. 37403
(615) 778-6100 (southeastern Tennessee, northern Georgia, northern Alabama)

Knoxville
Knoxville Poison Control Center
University of Tennessee Memorial Research Center and Hospital
1924 Alcoa Hwy. 37920
(615) 544-9400 (eastern Tennessee and southern Kentucky)

Memphis
Southern Poison Center, Inc.
848 Adams Ave. 38103
(901) 528-6048
522-5985 (administration)

TEXAS

Dallas
North Texas Poison Center*
P.O. Box 35926 75235
(800) 441-0040 (Texas only)
(214) 590-5000

El Paso
El Paso Poison Control Center
R.E. Thomason General Hospital
4815 Alameda Ave. 79905
(915) 533-1244 (southwestern Texas and southern New Mexico)

Galveston
Texas State Poison Center*
University of Texas Medical Branch
Eighth & Mechanic Sts. 77550-2780
(800) 392-8548 (Texas only)
(409) 765-1420 (Galveston)
(713) 654-1701 (Houston)
(512) 478-4490 (Austin)

UTAH

Salt Lake City
Intermountain Regional Poison Control Center*
50 N. Medical Dr., Building 428 84132
(800) 456-7707 (Utah only)
(801) 581-2151

VERMONT

Burlington
Vermont Poison Center
Medical Center Hospital of Vermont
Colchester Ave. 05401
(802) 658-3456 (Vermont and bordering parts of New York)
656-2721 (education programs)

VIRGINIA

Charlottesville
Blue Ridge Poison Center
University of Virginia Hospital
22908
(800) 451-1428 (Virginia only)
446-9876 (outside Virginia; TTY; MDs only)
(804) 924-5543

Norfolk
Tidewater Poison Center
150 Kingsley Lane 23505
(800) 552-6337 (southeastern Virginia and northeastern North Carolina)
(804) 489-5288

Richmond
Central Virginia Poison Center
Medical College of Virginia
Box 522, MCV Station 23298
(804) 786-9123 (call collect)
786-4780 (administration)

Roanoke
Southwest Virginia Poison Center
Roanoke Memorial Hospitals
Belleview at Jefferson St.
P.O. Box 13367 24033
(703) 981-7336

WASHINGTON

Seattle
Seattle Poison Center
Children's Hospital and
Medical Center
4800 Sand Point Way, N.E.
P.O. Box C-5371 98105-0371
(800) 732-6985 (Washington only)
(206) 526-2121

Spokane
Spokane Poison Center
S. 715 Cowley, Ste. 132 99202
(800) 572-5842 (Washington only)
541-5624 (northern Idaho, western Montana, northwestern Oregon)
(509) 747-1077 (TTY)

Tacoma
Mary Bridge Poison Center
Mary Bridge Children's Hospital
311 S. L St.
P.O. Box 5299 98405-0987
(800) 542-6319 (Washington only)
(206) 594-1414

Yakima
Central Washington Poison Center
Yakima Valley Memorial Hospital
2811 Tieton Dr. 98902
(800) 572-9176 (Washington only)
(509) 248-4400

WEST VIRGINIA

Charleston
West Virginia Poison Center*
West Virginia University
School of Pharmacy
3110 MacCorkle Ave., S.E. 25304
(800) 642-3625 (West Virginia only)
(304) 348-4211

WISCONSIN

Green Bay
Green Bay Poison Control Center
St. Vincent Hospital
P.O. Box 13508 54307-3508
(414) 433-8100

Madison
University of Wisconsin Hospital
Regional Poison Control Center
600 Highland Ave. 53792
(608) 262-3702

Milwaukee
Milwaukee Poison Center
Children's Hospital of Wisconsin
9000 W. Wisconsin Ave.
P.O. Box 1997 53201
(414) 266-2222

WYOMING

Cheyenne
Wyoming Poison Center
Hathaway Bldg., Rm. 527 82001
(800) 442-2702 (Wyoming only)

Mid-Plains Poison Control Center*
Omaha, Neb.
(800) 228-9515

Rocky Mountain Poison Center*
Denver, Colo.
(800) 442-2702

St. Luke's Midland Regional Medical Center
Poison Control Center
Aberdeen, S. Dak.
(800) 592-1889

SOLID AND HAZARDOUS WASTE AGENCIES BY STATE

ALABAMA

Land Division
Alabama Dept. of Environmental Management
1751 Federal Drive
Montgomery, Alabama 36130
CML (205) 271-7730

ALASKA

Alaska Dept. of Environ. Conserv.
Division of Environ. Quality
P.O. Box O
Juneau, Alaska 99801
CML (907) 465-2666

AMERICAN SAMOA

Environmental Quality Commission
Government of American Samoa
Pago Pago, American Samoa 96799

Overseas Operator
(Commercial Call 663-2304)

ARIZONA

Office of Waste and Water Quality Management
Arizona Department of Environmental Quality
2005 N. Central Avenue, Room 304
Phoenix, Arizona 85004
CML (602) 257-2305

ARKANSAS

Hazardous Waste Division
Arkansas Department of Pollution Control and Ecology
P.O. Box 9583
Little Rock, Arkansas 72219
CML (501) 562-7444 Ext. 504

CALIFORNIA

Department of Health Services
P.O. Box 942732
400 P Street
Sacramento, California 95814
CML (916) 323-2913

Toxic Substances Control Division
Department of Health Services
CML (916) 324-1826

COLORADO

Waste Management Division
Colorado Department of Health
4210 E. 11th Ave.
Denver, Colorado 80220
CML (303) 331-4830

COMMONWEALTH OF NORTHERN MARIANA ISLANDS

Division of Environmental Quality
Department of Public Health and Environmental Services
Commonwealth of the Northern Mariana Islands
Office of the Governor
Saipan, Mariana Islands 96950

Overseas Operator: 6984
Cable address: GOV. NMI Saipan

CONNECTICUT

Hazardous Material Management Unit
Department of Environmental Protection
State Office Building
165 Capitol Ave.
Hartford, Connecticut 06106
CML (203) 566-4924

DELAWARE

Director of Air & Waste Management
Division of Air & Waste Management
Department of Natural Resources and Environmental Control,
P.O. Box 1401, 89 Kings Highway
Dover, Delaware 19903
CML (302) 736-3672

DISTRICT OF COLUMBIA

Pesticides & Hazardous Materials Division/
Superfund Department of Consumer and Regulatory Affairs
5010 Overlook Ave., S.W. Room 114
Washington, DC 20032
CML (202) 783-3194

FLORIDA

Division of Waste Management
Underground Storage Tanks (UST)
Department of Environmental Regulations
Twin Towers Office Building
2600 Blair Stone Rd.
Tallahassee, Florida 32301
CML (904) 488-0190

GEORGIA

Land Protection Branch
Industrial and Hazardous Waste Management Program
Floyd Towers East/Rm 1154
205 Butler St., S.E.
Atlanta, Georgia 30334
CML (404) 656-2833

GUAM

Haz. Waste Management Prog.
Guam Environmental Protection Agency
P.O. Box 2999
Agana, Guam 96910

Overseas Operator
CML (671) 646-8863

HAWAII

Department of Health
Hazardous Waste Program
P.O. Box 3378
Honolulu, Hawaii 96801

CML (808) 548-2270

IDAHO

Hazardous Materials Bureau
Dept. of Health & Welfare
Idaho State House/450 W. State St.
Boise, Idaho 83720
FTS 8-554-5879
CML (208) 334-5879

ILLINOIS

Illinois Environmental Protection Agency
2200 Churchill Road
Springfield, Illinois 62706
CML (217) 782-6760

Division of Land Pollution Control
Environmental Protection Agency
2200 Churchill Rd., Room A-104
Springfield, Illinois 62706
CML (217) 782-6760

INDIANA

Indiana Department of Environmental Management
105 S. Meridian Street,
P.O. Box 6015
Indianapolis, Indiana 46225
CML (317) 232-3210

IOWA

Air Quality & Solid Waste Protection
Depart. of Water, Air & Waste Manage.
900 East Grand Ave.
Henry A. Wallace Building
Des Moines, Iowa 50319-0034

KANSAS

Bureau of Waste Management
Dept. of Health & Environment
Forbes Field, Bldg. 321
Topeka, Kansas 66620
CML (913) 862-9360 Ext. 290

KENTUCKY

Division of Waste Management
Dept. of Environmental Protection
Cabinet for Natural Resources and Environmental Protection
Ft. Boone Plaza, Bldg. #2
Frankfort, Kentucky 40601
CML (502) 564-6716 Ext. 214

LOUISIANA

Hazardous Waste Division
Office of Solid Waste and Hazardous Waste
Louisiana Dept. of Environmental Quality
P.O. Box 44307/625 N. 4th St.
Baton Rouge, Louisiana 70804
CML (504) 342-9079

MAINE

Bureau of Oil and Hazardous Materials Control
Dept. of Environmental Protection
State House Station #17
Augusta, Maine 04333
CML (207) 289-2651

MARYLAND

Maryland Waste Management Administrator
National Resources Planner
Dept. of Health and Mental Hygiene
201 W. Preston Street, Room 212
Baltimore, Maryland 21201
CML (301) 225-5647

Hazardous and Solid Waste Management Administration
Maryland Dept. of the Environment
201 West Preston St., Room 212
Baltimore, Maryland 21201
CML (301) 225-5647

MASSACHUSETTS

Division of Solid and Hazardous Waste
Massachusetts Department of Environmental Quality Engineering
One Winter Street, 5th Floor
Boston, Massachusetts 02108
CML (617) 292-5589

MICHIGAN

Waste Management Division
Environmental Protection Bureau
Dept. of Natural Resources
Box 30038
Lansing, Michigan 48909
CML (517) 373-2730

Permits Unit
Technical Services Sec.
Waste Management Div.
Dept. of Natural Resources
Box 30038
Lansing, Michigan 48909
CML (517) 373-2730

MINNESOTA

Solid & Hazardous Waste Division
Minnesota Pollution Control Agency
520 Lafayette Road, North
St. Paul, Minnesota 55155
CML (612) 296-7282

Program Development Sec.
Solid and Hazardous Waste
Minnesota Pollution Control Agency
520 Lafayette Road, North
St. Paul, Minnesota 55155
CML (612) 296-7444

MISSISSIPPI

Division of Solid & Waste Management
Bureau of Pollution Control
Dept. of Natural Resources
P.O. Box 10385
Jackson, Mississippi 39209
CML (601) 961-5062

MISSOURI

Waste Management Program
Dept. of Natural Resources
Jefferson Building
205 Jefferson St. (13/14 floor)
P.O. Box 176
Jefferson City, Missouri 65102
CML (314) 751-3176

MONTANA

Solid & Hazardous Waste Bureau
Dept. of Health & Environmental Sciences
Cogswell Bldg., Room B-201
Helena, Montana 59620
CML (406) 444-2821

NEBRASKA

Hazardous Waste Management Section
Dept. of Environmental Control
State House Station
P.O. Box 94877
Lincoln, Nebraska 68509
CML (402) 471-2186

NEVADA

Waste Management Program
Division of Environmental Protection
Dept. of Conservation & Natural Resources
Capitol Complex
201 South Fall Street
Carson City, Nevada 89710
CML (702) 885-4670

NEW HAMPSHIRE

Division of Public Health Services
Office of Waste Management
Dept. of Health & Welfare
Health & Welfare Bldg.
6 Hazen Drive
Concord, New Hampshire 03301
CML (603) 271-4662

NEW JERSEY

Division of HQ Waste Management
Dept. of Environmental Protection
401 East State St./CN 028
Trenton, New Jersey 08625
CML (609) 292-1250

NEW MEXICO

Groundwater & Hazardous Waste Bureau
Environmental Improvement Division
New Mexico Health & Environment Department
P.O. Box 968
Santa Fe, New Mexico 87504-0968
CML (505) 827-2924

Groundwater and Hazardous Waste Bureau
New Mexico Health and Environment Dept.
P.O. Box 968
Santa Fe, New Mexico 87504-0968
CML (505) 827-2924

NEW YORK

Division of Hazardous Substance Regulation
Dept. of Environmental Conservation
50 Wolfe Rd., Room 209
Albany, New York 12233
CML (518) 457-6603

NORTH CAROLINA

Solid & Hazardous Waste Mgmt. Branch
Division of Health Services
Dept. of Human Resources
P.O. Box 2091
Raleigh, North Carolina 27602
CML (919) 733-2178

NORTH DAKOTA

Division of Hazardous Waste Management & Special Studies
Dept. of Health
1200 Missouri Ave., Room 302
Bismarck, North Dakota 58502-5520
CML (701) 224-2366

OHIO

Division of Solid and Hazardous Waste Management
Ohio EPA
1800 Watermark Drive/P.O. Box 1049
Columbus, Ohio 43266-0149
CML (614) 466-7220

OKLAHOMA

Waste Management Service
Oklahoma State Dept. of Health
P.O. Box 53551
1000 N.E. 10th St.
Oklahoma City, Oklahoma 73152
CML (405) 271-5338

OREGON

Hazardous and Solid Waste Division
Dept. of Environmental Quality
811 Southwest 6th Avenue
Portland, Oregon 97204
CML (503) 229-5356

PENNSYLVANIA

Bureau of Solid Waste Management
Pennsylvania Dept. of Environmental Resources
P.O. Box 2063/Fulton Building
Harrisburg, Pennsylvania 17120
CML (717) 787-9870

PUERTO RICO

Environmental Quality Board
Santurce, Puerto Rico 00910-1488
CML (809) 725-0439

RHODE ISLAND

Solid Waste Management Program
Dept. of Environmental Management
204 Canon Bldg.
75 Davis St.
Providence, Rhode Island 02908
CML (401) 277-2797

SOUTH CAROLINA

Bureau of Solid & Hazardous Waste Management
Dept. of Health & Environmental Control
2600 Bull Street
Columbia, South Carolina 29201
CML (803) 758-5681

SOUTH DAKOTA

Office of Air Quality & Solid Waste
Dept. of Water & Natural Resources
523 E. Capitol
Foss Bldg., Room 416
Pierre, South Dakota 57501
CML (605) 773-3153

TENNESSEE

Division of Solid Waste Mgmt.
Tennessee Dept. of Public Health
701 Broadway
Customs House, 4th Floor
Nashville, Tennessee 37219-5403
CML (615) 741-3424

TEXAS

Hazardous and Solid Waste Division
Texas Water Commission
P.O. Box 13087, Capitol Station
Austin, Texas 78711-3087
CML (512) 463-7760

UTAH

Bureau of Solid and Hazardous Waste Management
Department of Health
P.O. Box 16700
288 North 1460 West Street
Salt Lake City, Utah 84116-0700
CML (801) 533-4145

VERMONT

Waste Management Division
Agency of Environmental Conservation
103 South Main Street
Waterbury, Vermont 05676
CML (802) 244-8702

VIRGIN ISLANDS

Dept. of Conservation & Cultural Affairs
P.O. Box 4399, Charlotte
St. Thomas, Virgin Islands 00801
CML (809) 774-6420

VIRGINIA

Division of Technical Ser.
Va. Dept. of Waste Mgmt.
Monroe Building, 11th Floor
101 North 14th Street
Richmond, Virginia 23219
CML (804) 225-2667

WASHINGTON

Solid & Hazardous Waste Mgmt. Division
Dept. of Ecology
Mail Stop PV-11
Olympia, Washington 98504
CML (206) 459-6316

WEST VIRGINIA

Waste Management Division
Dept. of W. Va. Natural Resources
1260 Greenbrier Street
Charleston, West Virginia 25311
CML (304) 348-5935

WISCONSIN

Bureau of Solid Waste
Dept. of Natural Resources
P.O. Box 7921
Madison, Wisconsin 53707
CML (608) 266-1327

WYOMING

Solid Waste Mgmt. Program
State of Wyoming
Dept. of Environmental Quality
122 West 25th Street
Herschler Bldg.
Cheyenne, Wyoming 82002
CML (307) 777-7752

CML - Commercial Line

PESTICIDE CONTROL OFFICIALS BY STATE

ALABAMA

J.A. Bloch
Ag Chemistry/Plant Protection Division
Alabama Dept. of Agriculture
& Industries
P.O. Box 3336
1445 Federal Dr.
Montgomery, AL 36193
(205) 261-2631

ALASKA

Douglas Donegan
Environmental Health
Dept. of Environmental Conservation
Pouch O
Juneau, AK 99811
(907) 465-2696

ARIZONA

Les Davis
Agricultural Chemicals Environmentals
Services Division,
Commission of Agriculture
and Horticulture,
1688 W. Adams, Suite 415
Phoenix, AZ 85007
(602) 542-3578

ARKANSAS

Tim Jessup
Div. of Feeds, Fertilizer & Pesticides
Arkansas State Plant Board
P.O. Box 1069
Little Rock, AR 72203
(501) 225-1598

CALIFORNIA

Doug Okumura
Pesticide Enforcement Branch
Division of Pest Management
California Dept. of Food & Agriculture
1220 N Street, Room A-170
Sacramento, CA 95814
(916) 445-5343

CANADA

S.W. Ormrod
Agriculture Canada
Sir John Carling Building,
930 Carling Avenue,
Ottowa, Ontario, Canada K1A OC5,
(613) 992-2114

COLORADO

Linda Coulter
Pesticide Applicator Section
Division of Plant Industry
Colorado Dept. of Agriculture
1525 Sherman St., 4th Floor
Denver, CO 80203
(303) 866-2838

CONNECTICUT

Gregory J. Piontek
Pesticide Control Section
Dept. of Environmental Protection
State Office Building
Hartford, CT 06106
(203) 566-5148

DELAWARE

H. Grier Stayton
Delaware Dept. of Agriculture
2320 S. Dupont Highway
Dover, DE 19901
(302) 736-4811

DISTRICT OF COLUMBIA

Angelo C. Tompros
Pesticide & Hazardous Waste Mgmt. Branch
Environmental Control Division
Dept. of Consumer & Regulatory Affairs
5010 Overlook Ave., SW, Rm. 114
Washington, DC 20032
(202) 727-7419

FLORIDA

Roger Inman
Bureau of Product Data Evaluation
Division of Inspection
Florida Dept. of Agriculture &
Consumer Services
Mayo Building, Rm. 211
Tallahassee, FL 32399-0800
(904) 487-0532

GEORGIA

J. Ron Conley
Entomology & Pesticide Division
Georgia Dept. of Agriculture
Capitol Square
Atlanta, GA 30334
(404) 656-4958

HAWAII

Robert Boesch
Pesticides Branch,
Plant Industry Division
Hawaii Dept. of Agriculture
1428 S. King Street
Honolulu, HI 96814
(808) 548-7124

IDAHO

Rodney A. Awe
Bureau of Pesticides
Division of Plant Industries
Idaho Dept. of Agriculture
P.O. Box 790
Boise, ID 83701-0790
(208) 334-3243

ILLINOIS

Tom A. Walker
Bureau of Plant & Apiary Protection
Illinois Dept. of Agriculture
State Fairgrounds
Springfield, IL 62794-9281
(217) 785-2427

INDIANA

David E. Scott
Indiana State Chemist Office
Dept. of Biochemistry
Purdue University
West Lafayette, IN 47907
(317) 494-1587

IOWA

Charles A. Eckermann
Pesticide Section
Iowa Dept. of Agriculture
Wallace Building
Des Moines, IA 50319
(515) 281-8591

KANSAS

Dale Lambley
Plant Health Division
Kansas Dept. of Agriculture
109 SW Ninth Street
Topeka, KS 66612-1281
(913) 296-2263

KENTUCKY

Ron Egnew
Kentucky Dept. of Agriculture
Capitol Plaza Tower
Frankfort, KY 40601
(502) 564-7274

LOUISIANA

H.F. "Butch" Calhoun
Pesticide & Environmental Programs
Louisiana Dept. of Agriculture
P.O. Box 44153, Capitol Station
Baton Rouge, LA 70804-4153
(504) 925-3763

MAINE

Robert I. Batteese, Jr.
Board of Pesticide Control
Maine Dept. of Agriculture
State House Station, #28
Augusta, ME 04333
(207) 289-2731

MARYLAND

Charles Puffinberger
Office of Plant Industries and Resource
Conservation
50 Harry S. Truman Parkway
Annapolis, MD 21401
(301) 841-5870

MASSACHUSETTS

Jeffrey Carlson
Pesticides Bureau
Massachusetts Dept. of Food & Agriculture
100 Cambridge St., 21st Floor
Boston, MA 02202
(617) 727-7712

MICHIGAN

Robert L. Mesecher
Pesticide & Plant Management Division
Michigan Dept. of Agriculture
P.O. Box 30017
Lansing, MI 48909
(517) 373-1087

MINNESOTA

Michael Fresvik
Pesticide Regulatory Program
Agronomy Services Division
Minnesota Dept. of Agriculture
90 W. Plato Boulevard
St. Paul, MN 55107
(612) 296-8547

MISSISSIPPI

Robert H. McCarty
Division of Plant Industry,
Mississippi Dept. of Agriculture & Commerce
P.O. Box 5207
Mississippi State, MS 39762
(601) 325-3390

MISSOURI

John R. Hagan
Bureau of Pesticide Control
Plant Industries Division
Missouri Dept. of Agriculture
P.O. Box 630
Jefferson City, MO 65102-0630
(314) 751-2462

MONTANA

Steven F. Baril
Field Services Bureau
Environmental Management Division
Montana Dept. of Agriculture
Agriculture/Livestock Bldng.
Capitol Station
Helena, MT 59620-0205
(406) 444-2944

NEBRASKA

Donald Kemper
Bureau of Plant Industry
Nebraska Dept. of Agriculture
301 Centennial Mall
Lincoln, NE 68509
(402) 471-2341

NEVADA

Robbin Rose
Nevada Dept. of Agriculture
P.O. Box 11100
350 Capitol Hill Ave.
Reno, NV 89510-1100
(702) 789-0180

NEW HAMPSHIRE

Murray L. McKay
Division of Pesticide Control
New Hampshire Dept. of Agriculture
Caller Box 2042,
Concord, NH 03301
(603) 271-3550

NEW JERSEY

Raymond A. Ferrarin
Bureau of Pesticide Control
New Jersey Dept. of Environmental Protection
401 E. State Street, CN 411
Trenton, NJ 08625
(609) 530-4123

NEW MEXICO

Barry Patterson
Division of Agricultural and Environmental Services,
New Mexico Dept. of Agriculture
P.O. Box 30005, Dept. 3150
Las Cruces, NM 88003
(505) 646-3208

NEW YORK

Marilyn Dubois
Bureau of Pesticide Management
New York Dept. of Environmental Conservation
50 Wolf Rd., Room 404
Albany, NY 12233-7254
(518) 457-7482

NORTH CAROLINA

John L. Smith
Food & Drug Protection Division
North Carolina Dept. of Agriculture
P.O. Box 27647
Raleigh, NC 27611-0647
(919) 733-3556

NORTH DAKOTA

Peri Dura
Registration Division
North Dakota Department of Health and Consolidated Labs
P.O. Box 937
Bismarck, ND 58502
(701) 221-6149

OHIO

Oren W. Spilker
Pesticide Regulation
Division of Plant Industry
Ohio Dept. of Agriculture
8995 E. Main Street
Reynoldsburg, OH 43068-3399
(614) 866-6361

OKLAHOMA

Robert L. Chada
Pesticide Management Group
Oklahoma Dept. of Agriculture
2800 N. Lincoln Boulevard
Oklahoma City, OK 73105-4298
(405) 521-3864

OREGON

Christopher K. Kirby
Program Coordinator
Plant Division
Oregon Dept. of Agriculture
635 Capitol Street, NE
Salem, OR 97310
(503) 378-3776

PENNSYLVANIA

John Longenecker
Bureau of Plant Industry
Division of Agronomic Services
Pennsylvania Dept. of Agriculture
2301 N. Cameron Street
Harrisburg, PA 17110
(717) 787-4843

PUERTO RICO

Arline Gonzales
Analysis & Registration of Agricultural Materials
Puerto Rico Dept. of Agriculture
P.O. Box 10163
Santurce, PR 00908
(809) 796-1710

RHODE ISLAND

Kenneth D. Ayars
Division of Agriculture & Marketing
Rhode Island Dept. of Environmental Management
22 Hayes Street
Providence, RI 02908
(401) 277-2782

SOUTH CAROLINA

V.H. McCaskill
Department of Fertilizer and Pesticide Control,
256 Poole Agricultural Center
Clemson, SC 29634-0394
(803) 656-3171

SOUTH DAKOTA

Brenda Puepke
Pesticide Activity
Division of Regulatory Services
South Dakota Dept. of Agriculture
Anderson Building
445 E. Capitol
Pierre, SD 57501
(605) 773-3724

TENNESSEE

Douglas A. Hite
Division of Plant Industries
Tennessee Dept. of Agriculture
P.O. Box 40627, Melrose Station
Nashville, TN 37204
(615) 360-0130

TEXAS

Susan Rieff
Agr. Resources Protection Division
Texas Dept. of Agriculture
P.O. Box 12847
Austin, TX 78711
(512) 463-7443

UTAH

Joseph M. Beckstrand
Pesticide & Fertilizer Inspection
Utah State Agricultural Dept.
350 N. Redwood Rd.
Salt Lake City, UT 84116
(801) 533-5916

VERMONT

Jeffrey Comstock
Plant Industry Section
Vermont Dept. of Agriculture
116 State St.
State Office Building
Montpelier, VT 05602
(802) 828-2443

VIRGIN ISLANDS

Donald C. Francois
Div. of Natural Resources Mgmt.
Dept. of Conserv. & Cultural Affairs
P.O. Box 4340
St. Thomas, VI 00801
(809) 774-6420

VIRGINIA

William E. "Billy" Walls
Office of Pesticide Regulation
Virginia Dept. of Agriculture & Consumer Services
P.O. Box 1163
Richmond, VA 23209
(804) 786-3798

WASHINGTON

Glenn E. Smerdon
Washington Dept. of Agriculture
406 General Administration Building, AX-41
Olympia, WA 98504
(206) 753-5064

WEST VIRGINIA

Robert Frame
Regulatory & Inspection Division
West Virginia Dept. of Agriculture
Capitol Building
Charleston, WV 25305
(304) 348-2212

WISCONSIN

Nick Neher
Groundwater & Regulatory Services Section
Wisconsin Dept. of Agriculture, Trade and Consumer Protection
P.O. Box 8911
801 W. Badger Rd.
Madison, WI 53708
(608) 266-7137

WYOMING

Jim Bigelow
Technical Services
Wyoming Dept. of Agriculture
2219 Carey Ave.
Cheyenne, WY 82002-0100
(307) 777-6590

STATE EXTENSION PROGRAM LEADERS/COORDINATORS PESTICIDE CHEMICALS EDUCATION, SAFETY AND TRAINING

ALABAMA

G. Talmadge Balch
Pesticide Education Specialist
203 Extension Hall
Auburn University
Auburn, AL 36849
(205) 826-4940

*Steve L. Brown
Pesticide Education Specialist
206 Extension Hall
Auburn University
Auburn, AL 36849
(205) 826-4940

ALASKA

Wayne Vandre
Horticulture Specialist,
Pesticide Coordinator
Cooperative Extension Service
2221 E. Northern Lights Blvd. #240
Anchorage, AK 99508
(907) 279-6575

AMERICAN SAMOA

Agnes Vargo
American Samoa Community College
Land Grant Program
Mapusaga Campus
P.O. Box 2609
Pago Pago, AS 96779
011 (684) 699-1394

ARIZONA

Paul Baker
State Pesticide Coordinator
University of Arizona
1109 E. Helen Street
Tucson, AZ 85719
(602) 621-4012

ARKANSAS

J. Ples Spradley
Cooperative Extension Service
P.O. Box 391
Little Rock, AR 72203
(501) 373-2640

CALIFORNIA

Michael W. Stimmann
Statewide Pesticide Coord.
Cooperative Extension
OPIC
University of California
Davis, CA 95616
(916) 752-7011

*Patrick Marer
Pesticide Training Coordinator
IPM Education and Publications
University of California
Davis, CA 95616
(916) 752-7694

COLORADO

Bert L. Bohmont
Coordinator
Pesticide Programs
College of Agriculture
116 Weed Science Laboratory
Colorado State University
Fort Collins, CO 80523
(303) 491-5237

CONNECTICUT

Candace L. Bartholomew
Pesticide Coordinator
Cooperative Extension Service
1800 Asylum Avenue
West Harford, CT 06117
(203) 241-4940

DELAWARE

Susan Whitney
Pesticide Coordinator
Townsend Hall
Rm: 254
University of Delaware
Newark, DE 19711-1301
(302) 451-2526
FAX: (302) 292-3651

DISTRICT OF COLUMBIA

*M. Shaeed Khan
Coordinator
Pesticide Chemicals/Extension Specialist
Cooperative Extension Service
University of D.C.
901 Newton Street, NE
Washington, D.C. 20017
(202) 576-7419

W. B. Easley
Program Leader
Agriculture & Natural Resources
Cooperative Extension Service
University of D.C.
901 Newton St. NE
Washington, D.C. 20017
(202) 576-6951

FLORIDA

O. Norman Nesheim
Pesticide Coordinator
University of Florida
Building 847
Gainesville, FL 32611
(904) 392-4721
FAX: (904) 392-6532

GEORGIA

Burton R. Evans
Coordinator
Pesticide Applicator Training
Cooperative Extension Service
University of Georgia
Athens, GA 30602
(404) 542-3685

GUAM

R. Muniappan
Pesticide Coordinator
UOG Station
Mangilao, GU 96923
FTS Phone: (671) 734-2575

HAWAII

Barry M. Brennan
Extension Specialist—
Pesticide Chemicals
Department of Agricultural
Biochemistry
University of Hawaii
1800 East-West Road
Honolulu, HI 96822
(808) 948-8352
FAX: (808) 948-5037

IDAHO

*Hugh W. Homan
Extension Entomologist
Department of Plant, Soil
& Entomology Sciences
University of Idaho
Moscow, ID 83843
(208) 734-3600

Gene P. Carpenter
Pesticide Coordinator
Department of Plant, Soil
& Entomology Sciences
University of Idaho
Moscow, ID 83843
(208) 885-7541

ILLINOIS

*Donald E. Kuhlman
Pesticide Coordinator
Office of Agricultural Entomology
University of Illinois
172 Natural Resources Building
607 E. Peabody Drive
Champaign, IL 61820
(217) 333-6494
FAX: (217) 333-1952

Loren Bode
Chairman
Pesticide Applicator Training
Committee
Department of Agricultural Engineering
360-L Agriculture Engineering Science
Building
1304 W. Pennsylvania
Urbana, IL 61801
(217) 333-3000

INDIANA

Thomas N. Jordan
Pesticide Coordinator
Purdue Pesticide Programs
Rm: G316
Lilly Hall of Life Sciences
West Lafayette, IN 47907
(317) 494-4629

IOWA

Wendy Wintersteen
Extension Entomologist
Pest Applicator Trng. Coord.
109 Insectary Building
Iowa State University
Ames, IA 50011
(515) 294-1101
FAX: (515) 294-0907

KANSAS

Donald C. Cress
Extension Pesticide Coordinator
Department of Entomology
Waters Hall
Kansas State University
Manhattan, KS 66506
(913) 532-5891

KENTUCKY

James R. Martin, Coordinator
Pesticide Chemicals and Weeds Specialist
P.O. Box 469
Princeton, KY 42445
(502) 365-7541

*C.M. Christensen
Coordinator
Pesticide Applicator Training
S-225, Agriculture Science Center-N
University of Kentucky
Lexington, KY 40546
(606) 257-5956

LOUISIANA

Mary L. Grodner
Pesticide Chemicals Coordinator
Cooperative Extension Service
Knapp Hall
Louisiana State University
Baton Rouge, LA 70803
(504) 388-2180

MAINE

James F. Dill
Pest Management Specialist
UMCES—Pest Management Office
491 College Avenue
Orono, ME 04473
(207) 581-3879

MARYLAND

Amy E. Brown
Pesticide Coordinator
Department of Entomology
University of Maryland
College Park, MD 20742
(301) 454-7127

MASSACHUSETTS

*Jeffrey J. Jenkins
Coordinator—Pesticides Program
Pesticide Chemicals
Department of Entomology
Fernald Hall
University of Massachusetts
Amherst, MA 01003
(413) 545-2284

Natalia P. Clifton—Ext. Specialist
Pesticide Chemicals
Department of Entomology
Fernald Hall
University of Massachusetts
Amherst, MA 01003
(413) 545-2284

MICHIGAN

*Larry G. Olson
Pesticide Education Coordinator
Cooperative Extension Service
11 Agricultural Hall
Michigan State University
East Lansing, MI 48824-1039
(517) 355-0117

Robin R. Rosenbaum
Extension Associate/IPM
Cooperative Extension Service
11 Agricultural Hall
Michigan State University
East Lansing, MI 48824-1039
(517) 355-0117

MINNESOTA

Richard A. Meronuck
Extension Plant Pathologist
Plant Pathology Department
495 Borlaug Hall
1991 Buford Circle
University of Minnesota
St. Paul, MN 55108
(612) 625-6290

*Dean Herzfeld
Assistant Extension Specialist
Plant Pathology Department
495 Borlaug Hall
1991 Buford Circle
University of Minnesota
St, Paul, MN 55108
(612) 625-6290

MISSISSIPPI

Edna R. Morgan
Pesticide Coordinator
P.O. Box 5446
Mississippi State, MS 39762
(601) 325-3371

MISSOURI

*Daryl Sanders
IPM Coordinator/Pesticides
University of Missouri
45 Agriculture Building
Columbia, MO 65211
(314) 882-7181
Dialcom #: AGS3500

Mahlon Fairchild
IPM Pesticides
University of Missouri
45 Agriculture Building
Columbia, MO 65211
(314) 882-5099
Dialcom #: AGS3500

MONTANA

Greg Johnson
Pesticide Education Specialist
Entomology Research Laboratory
Montana State University
Bozeman, MT 59717
(406) 994-3518

NEBRASKA

*Larry D. Schulze
Extension Pesticide Coordinator
Natural Resources Hall, Room: 101
University of Nebraska
Lincoln, NE 68583-0818
(402) 472-1632
FAX: (402) 472-3574

Edward F. Vitzthum
Associate Coordinator
Environmental Programs
Natural Resources Hall, Room: 101
University of Nebraska
Lincoln, NE 68583-0818
(402) 472-1446

Roger E. Gold
Coordinator, Env. Prog.
University of Nebraska
103 Natural Resources Hall
Lincoln, NE 68583-0844
(402) 472-3305

NEVADA

Duam Flom, Acting Coord. PAT
Lyon County Extension
P.O. Box 811
Yerington, NV 89447
(702) 463-3341, Ext. 236

NEW HAMPSHIRE

Stanley R. Swier
Pesticide Coordinator & Extension Entomology
University of New Hampshire
Nesmith Hall
Durham, NH 03824
(603) 862-1159

NEW JERSEY

George C. Hamilton
Extension Pesticide Coordinator
Rutgers University
J.B. Smith Hall, Room: 108
P.O. Box 231
New Brunswick, NJ 08903
(201) 932-9801

NEW MEXICO

L. Michael English
Coordinator
Pesticide Chemicals & Entomology
Box 3AE
New Mexico State University
Las Cruces, NM 88003
(505) 646-2546

NEW YORK

*Donald Rutz
Coordinator
Chemicals—Pest Program
5123 Comstock Hall
Cornell University
Ithaca, NY 14853
(607) 255-1866

W.G. Smith
Associate Coordinator
Chemicals—Pesticide Program
5123 Comstock Hall
Cornell University
Ithaca, NY 14853
(607) 255-1866

NORTH CAROLINA

John H. Wilson Jr.
Pesticide Coordinator
Dept. of Horticultural Science
North Carolina State University
Box 7609
Raleigh, NC 27695-7609
(919) 737-3113

NORTH DAKOTA

Gregory K. Dahl
Pesticide Program Coordinator
NDSU Pesticide Programs
Department of Horticulture
North Dakota State University
P.O. Box 5658
Fargo, ND 58105
(701) 237-7180

OHIO

*Acie C. Waldron
Pesticide Coordinator
Extension Entomology
Ohio State University
1991 Kenny Road
Columbus, OH 43210-1090
(614) 292-4077

Joanne Kick-Raack
Training Specialist
Ohio State University
1991 Kenny Road
Columbus, OH 43210-1090
(614) 292-5274

OKLAHOMA

Jim T. Criswell
Pesticide Coordinator
Pesticide Program Specialist
Department of Entomology
Oklahoma State University
501 Life Science West
Stillwater, OK 74078
(405) 744-5531

OREGON

Terry L. Miller
Coordinator
Pesticide Applicator Training
Department of Agricultural Chemistry
Oregon State University
Corvallis, OR 97331
(503) 754-2906

PENNSYLVANIA

*Winand K. Hock
Pesticide Coordinator
417 Agricultural Administration Building
Pennsylvania State University
University Park, PA 16802
(814) 863-0263

Scott A. Harrison
Pesticide Education Specialist
419 Agricultural Administration Building
Pennsylvania State University
University Park, PA 16802
(814) 863-0263

PUERTO RICO

Hipolito O'Farrill-Nieves
State Pesticide Training Coordinator
College of Agricultural Sciences
Mayaguez Campus
University of Puerto Rico
Mayaguez, PR 00708
(809) 832-4040 Ext. 3481
Toll Free: (809) 137-800-462-2100

RHODE ISLAND

Steven Alm
Pesticide Coordinator
316 Woodward Hall
Plant Science Department
University of Rhode Island
Kingston, RI 02881
(401) 792-5998

SOUTH CAROLINA

*Paul MacHorton
Pesticide & Chemical Coordinator
Clemson University
Department of Entomology
103 Long Hall
Clemson, SC 29634-0365
(803) 656-5042

Bob Bellinger
Clemson University
Department of Entomology
103 Long Hall
Clemson, SC 29634-0365
(803) 656-5056

SOUTH DAKOTA

Jim Wilson
Pesticide Applicator Training
South Dakota State University
Agriculture Hall, 152 B
P.O. Box 2207D
Brookings, SD 57007
(605) 688-4752

TENNESSEE

*Gene Burgess
Pesticide Coordinator
Entomology & Plant Pathology Section
University of Tennessee
P.O. Box 1071
Knoxville, TN 37901
(615) 974-7138

Charles H. Hadden
Professor & Leader
Entomology & Plant Pathology Section
University of Tennessee
P.O. Box 1071
Knoxville, TN 37901
(615) 974-7138

TEXAS

Jack D. Price
Agricultural Chemist
115 Agronomy Field Laboratory
Texas A&M University
College Station, TX 77843-2474
(409) 845-3849

UTAH

Howard M. Deer
Extension Pesticide Coordinator
Utah State University
Logan, UT 84322-4649
(801) 750-1600

VERMONT

Gordon R. Nielson
Coordinator
Pesticide Chemicals
217 Hills Building
University of Vermont
Burlington, VT 05405-0082
(802) 656-0487

VIRGINIA

Michael J. Weaver
Extension Coordinator
Chemical, Drug & Pesticide Unit
Virginia Polytechnic Institute & State University
Blacksburg, VA 24061-0409
(703) 231-6543

VIRGIN ISLANDS

David Farrar
Coordinator
Pesticide Chemicals
Virgin Islands Extension Service
P.O. Box 598
St. Croix, VI 00850
(809) 778-0246

WASHINGTON

*Gary L. Thomasson
Extension Pesticide Education Specialist
315 Johnson Hall
Washington State University
Pullman, WA 99164-6432
(509) 335-2876

Richard Maxwell
Cooperative Extension
317 Johnson Hall
Washington State University
Pullman, WA 99164-6432
(509) 335-2995

WEST VIRGINIA

Jack Baniecki
Pesticide Coordinator
West Virginia University
414 Brooks Hall
P.O. Box 6057
Morgantown, WV 26506-6057
(304) 293-3911

WISCONSIN

*Ronald E. Doersch
Coordinator
University of Wisconsin
1575 Linden Drive
Madison, WI 53706
(608) 262-1392

Roger Flashinski
Specialist
University of Wisconsin
1575 Linden Drive
Madison, WI 53706
(608) 263-6358

WYOMING

Mark A. Ferrell
Ext. Pesticide Coordinator
Department of Plant, Soil & Insect Science
University of Wyoming
P.O. Box 3354, Univ. Station
Union Station
Laramie, WY 82070
(307) 766-5381

EXTENSION SERVICE—USDA

Bonnie L. Poli
Program Leader—
Pesticide Education
USDA Extension Service
Room: 3347—South Building
Washington, DC 20250
(202) 447-6506
FAX: (202) 475-5289

An asterisk (*) indicates the contact person in the State where more than one coordinator is listed.

APPENDIX III

Safety Information

This section was kindly provided by the National Agricultural Chemicals Association, Washington, D.C.

SAFE HANDLING PRECAUTIONS

Proper handling of crop protection chemicals by those involved in mixing, loading, and application is essential in order to minimize the risk of injury or intoxication. Therefore, safe handling procedures and body protection to reduce potential exposure to crop protection chemicals are absolutely essential.

RISK = TOXICITY × EXPOSURE

When dealing with chemical products, a simple formula to remember is:

Risk = Toxicity (Hazard) × Exposure.

The toxicity (hazard) is inherent in the product. The operator can't change that part of the formula. But one can **reduce exposure** and thereby **reduce risk.** Reduction of exposure to reduce risk is a principle that fits with every farm operation, from handling and applying chemicals to driving a farm tractor, an automobile or handling farm animals.

Methods and procedures to reduce exposure when handling crop protection chemicals take little time and in the long run pay great dividends. Knowledge of the toxicity of a product and the potential for exposure is essential for developing risk management strategies. For example, gasoline is extremely toxic and flammable. This is well understood, and we have learned how to handle it safely. No matter how toxic a substance may be, if the exposure is low enough, the risk can be minimized to an acceptable level.

EXPOSURE ROUTES

Exposure occurs via four routes: (1) **oral** (mouth), (2) **dermal** (skin), (3) **eyes** and (4) **inhalation** (lungs). Each route of exposure can occur as the result of accidents, spills, broken hoses, poor personal hygiene (transporting chemicals to the mouth, eyes and pubic areas via the hands when eating, handling tobacco products and toileting). Each route of exposure can be significantly reduced by simply following basic principles for protection.

For example, field applicator exposure studies show the primary source of exposure is dermal (the skin) with the forearms and hands receiving the overwhelming majority. Published dermal exposure data in Table I show deposition of pesticide residues on various parts of the body when measured under field conditions. Further, the data demonstrate the wearing of rubber gloves can reduce potential dermal exposure to the mixer, loader and applicator. Thus, by simply wearing appropriate chemical resistant gloves, and washing the gloves and hands following chemical handling, potential exposure by the dermal route can be significantly reduced. At the same time, transport of chemical residues to the mouth and eyes by unwashed hands can be minimized or essentially eliminated.

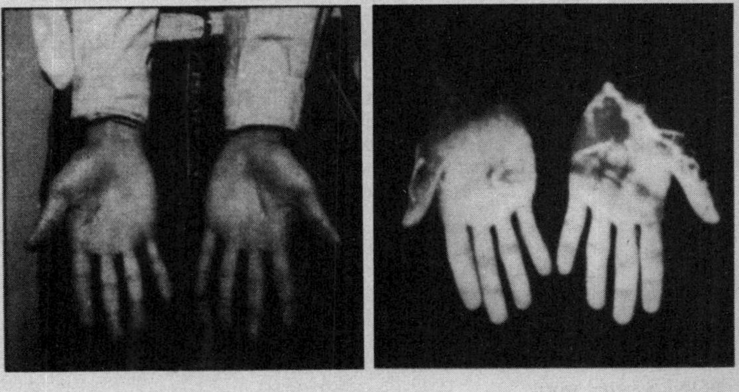

Simulated Spray Exposure—Fluorescent dye was sprayed onto sleeved arm and bare hands (left), then photographed under ultraviolet light. Observe spray exposure on unprotected hands (right).

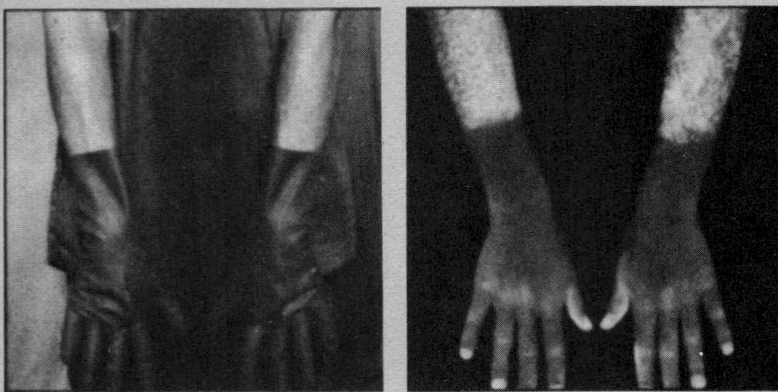

Simulated Spray Exposure—Fluorescent dye was sprayed onto bare arms and gloved hands (left), then photographed under ultraviolet light. Observe spray exposure on unprotected arms (right). Long sleeves and rubber gloves reduce potential dermal exposure to hands and arms.

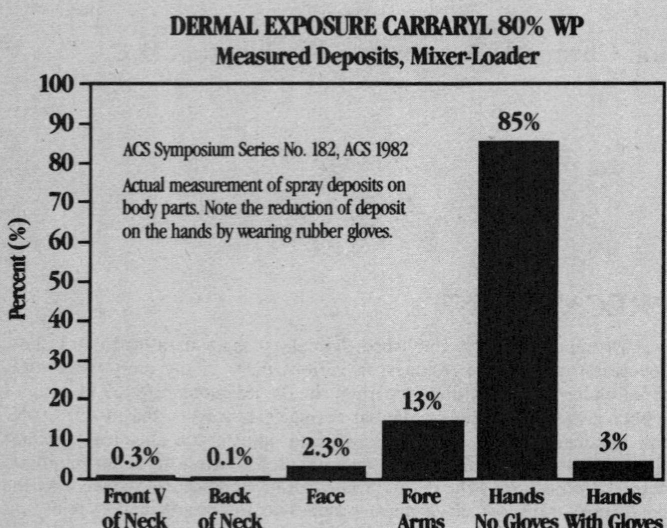

Chart 1—Percentage of measured spray deposits on various parts of the body. Note amount on hands and without rubber gloves. Rubber gloves on hands reduced potential dermal exposure by over 95 percent.

READ THE PESTICIDE LABEL

The signal words DANGER, WARNING, and CAUTION are boldly printed on the pesticide label for one purpose: that is to warn the user of the toxicity level of the compound that is being handled. From the signal word, kinds and types of protective measures or garments are determined for exposure reduction and protection.

The word **CAUTION** appears on pesticide products that have a relatively low level of toxicity. It means that while the chemical, based on laboratory animal studies, is not very toxic some basic handling precautions are important. Common sense methods of holding down exposure and safe use, such as wearing long sleeves, long pants, shoes, socks, chemical resistant gloves and a hat should be considered.

Labels for more toxic materials will carry the words **WARNING** or **DANGER** and sometimes **POISON.** These signal words suggest that very specific protective measures *must be followed.* Label or labeling information must be read and directions specifically followed. Take time to be careful; carelessness may invite injury, serious illness and sometimes death, particularly if the product carries **DANGER-POISON, SKULL** and **CROSSBONES** signal words.

PROTECTIVE CLOTHING/EQUIPMENT

Toxicity categories, routes of potential exposure and application procedures dictate the kinds and types of protection required for various pesticidal products.

Obviously fumigants require a self-contained breathing apparatus if the fumigation takes place in an enclosed environment. When the label requires a respirator, the equipment filtering cartridge must be approved for pesticide use by the National Institute for Occupational Safety and Health (NIOSH) and the Mine Safety Health Administration (MSHA).

Eye protection becomes a first priority as there is always the potential for splashing or misting of chemicals into the eyes. Wear goggles as standard practice during the transfer or handling of all chemicals. If the user already wears eyeglasses, then a plastic-type face shield would be the equipment of choice.

Rubber or chemical resistant gloves, aprons, boots, shoe covers and pants may be very uncomfortable at times when worn, but in the event of a spill, broken hose, splash or need for equipment repair, the use of such protection is critical and worthwhile. Long sleeved garments further reduce the exposure potential and should be standard wearing apparel when handling chemicals.

Leather or cloth gloves, leather boots or shoes absorb chemicals. They are very difficult to decontaminate, and **should not be worn.** Rubber covering over leather footwear will provide protection.

WATER IN THE FIELD AND AT THE LOADING SITE

The availability of clean water is absolutely essential. Washing of gloves and hands after handling farm chemicals and the flushing of body parts or equipment after a spill or an accident is critical if exposure is to be reduced. Water is inexpensive, available and effective as a safety tool. Water should always be on equipment at the mixing and loading site, and available for emergencies.

IN SUMMARY

Following product label instructions and common sense will go far in allowing crop protection products to be safely used for purposes intended. The signal words CAUTION, WARNING or DANGER tell about the product toxicity. The formula "RISK = TOXICITY × EXPOSURE" provides the principle for risk reduction. When **exposure is reduced** by wearing proper protective equipment, **risk is reduced.** It is that simple.

While protective equipment is not always beautiful, comfortable or convenient, it is necessary. Always follow label instructions and utilize basic safety procedures. Provide and wear rubber gloves, wash gloves and hands after handling farm chemicals, wear eye protection and provide plenty of clean water for washing and rinsing of gloves and hands after use. Use other protective procedures and equipment as required. Don't take chances. Truly, crop protection chemical safety should be an attitude as well as a meaningful practice.

Dr. John F. McCarthy
Vice President for Scientific and Regulatory Affairs
National Agricultural Chemicals Association
1155 Fifteenth Street, N.W., Washington, D.C. 20005

GOOD WAREHOUSE PRACTICES CHECKLIST FOR AGRICULTURAL CHEMICALS

NOTICE CONCERNING USE OF THIS CHECKLIST

This checklist is presented as self-inspection guidelines for employers to use in evaluating their warehousing environmental, health and safety programs. The use of this document is strictly a voluntary matter. Any actions taken in response to this checklist does not necessarily assure complete compliance with federal, state, or local law. Should assistance concerning the management of effective risk control programs be needed, professional technical or legal expertise is recommended.

PREFACE

This document was prepared by the Environmental Management Committee of the National Agricultural Chemicals Association, a non-profit trade organization, whose members manufacture or formulate virtually all the agricultural chemicals used in the United States and a large percentage used abroad. The checklist contained herein represents our industry's suggestions with respect to self-inspection procedures for enhancing good warehouse practices for agricultural chemicals. NACA is hopeful that the dissemination of this document will be helpful in further strengthening existing warehousing environmental, health and safety programs.

HAZARD AWARENESS

There are many facets to consider when evaluating the environmental, health and safety hazards associated with the warehousing of agricultural chemicals. Like any other chemicals, recognition of the key hazards, e.g., flammability, toxicity, corrosivity, etc., associated with the products you store and handle is the first and essential step to an effective risk control program.

Yes No

1. Do you store hazardous materials *labeled* as flammable, corrosives, reactives, poisons (toxics)? — —
2. Do you have Materials Safety Data Sheets on each product stored? — —
   Are they available to warehouse personnel?
3. Do you have a listing of supplier contacts for obtaining additional information? — —

EMPLOYEE INDOCTRINATION AND TRAINING

Effective employee indoctrination and training is critical if your risk control program is to be successful. Employees must be thoroughly trained in hazards recognition and the appropriate methods for protecting themselves and your property from such hazards.

Yes No

1. Are you subject to federal, state or community Right-to-Know regulations? — —
2. Are your employees trained in the proper handling of agricultural chemicals? — —
   Is special training given to employees in the proper handling of hazardous materials? — —
3. Who (by name) conducts the training? — —
   Is the trainer experienced and knowledgeable in the proper handling of hazardous materials? — —
4. Are training programs reviewed and updated on an annual basis? — —
5. Are newly hired/assigned employees given initial training and annual reviews? — —
6. Do you have a forklift safety training program? — —
7. Is all training documented in writing by employees' signatures? — —

PRODUCT STORAGE AND HANDLING

How products are stored and handled should be determined according to their levels of hazard and potentials for accidental release. Certainly precaution and care should focus on those products presenting the greatest hazards and potentials for release.

Yes No

1. Are agricultural chemicals stored in a separate warehouse, room or portion of a warehouse to prevent possible contamination of animal feed, grain, etc.? — —
2. Are flammable/combustible materials segregated in a well-ventilated area of the warehouse away from any ignition source? — —
3. Are agricultural chemicals stored in bulk? — —
   If so, are storage facilities properly diked, contained, and secured? — —
4. Does a trained employee operate bulk handling operations? — —
5. Are materials stacked in a stable manner? — —
   Do you abide by stacking height limitations as recommended by suppliers? — —
6. Do you maintain sufficient aisle space for product inspections and forklift maneuverability? — —
7. Is smoking confined to specifically designated areas? — —
   Are "No Smoking" signs posted? — —
8. Do you make routine inspections of all inventory to check for leaks and spills? — —
9. Do you have recovery drums in which to place damaged or leaking materials? — —
   When used, do you properly label the drums as to content? — —

GOOD HOUSEKEEPING

Good housekeeping practices are a common sense component of an effective risk control program. The physical appearance of your warehouse is a good indicator of your commitment to your business and your risk control program.

Yes No

1. Are storage areas in the warehouse neat and orderly? — —
2. Are floors swept and trash containers emptied regularly? — —
3. Are leaks/spills immediately cleaned up and residues properly contained? — —
4. Are damaged goods properly labeled and stored in designated areas? — —
5. Are exits marked and unobstructed? — —
6. Is a clean lunchroom available separate from material storage areas? — —
7. Are clean, properly supplied restrooms available? — —

EMERGENCY PREPAREDNESS

In the event that emergencies or accidents do take place, your chances of minimizing their consequence are much better if you are prepared for them in advance. Preparation means to identify responses to potential emergency situations in advance and *practice* the roles and actions of designated personnel, both internal and external to your facility.

Yes No

1. Have you evaluated your entire operation to see what/where spill, leak, or fire potentials exist? — —
2. Do you have a warehouse pre-emergency plan? — —
   Does the fire department have a copy of the plan? — —
3. Did the fire department assist in the preparation of the pre-emergency plan? — —
4. Is the fire department invited to tour your warehouse on an annual basis? — —
5. Do your fire extinguishers comply with OSHA specifications:
   — Sufficient number? — —
   — Properly sized? — —
   — Visually inspected monthly and marked by location personnel? — —
   — Professionally inspected annually? — —
6. Are the fire extinguishers properly identified and accessible? — —
7. If you have a sprinkler system, is it:
   — Inspected monthly? — —
   — Serviced annually? — —
8. If you have fire alarm systems, are they:
   — Heat sensors? — —
   — Smoke sensors? — —
   — Audible alarm? — —
   — Direct connection to fire department? — —
   Are your employees familiar with their location and use? — —
9. Do you know the location of fire hydrants within

the vicinity of your warehouse? — —

10. Are your material Safety Data Sheets:
    - — Available to employees? — —
    - — Available to fire department? — —
    - — Available to medical facilities? — —
11. Are your employees trained in emergency response? — —
12. Do you have contacts and telephone numbers of surrounding residences and/or businesses for emergency notification? — —
13. Do you have an accessible list of company personnel to call in the event of an emergency? — —
14. Do you have an evacuation plan? — —
    Are your employees familiar with it? — —
15. Do you have location contacts (manager, assistant, etc.) with telephone numbers *posted* at your warehouse visible to the public? — —
16. Do you have a list of government agencies to call in the event of a reportable spill, leak or fire? — —
17. Do you have someone assigned to coordinate fire, police, medical and other local emergency response groups? — —
18. In the event of a fire, can the contaminated firewater be contained? — —
    If not, are you aware of where runoff will go? — —
19. Do you have plastic sheeting and sandbags accessible to block runoff into drains and/or sewers? — —
20. Do you have copies of relevant emergency information away from your place of business that would be readily accessible to emergency response personnel if needed? — —
21. Do you conduct pre-emergency preparedness practices? — —
22. Do you have a source arranged in advance to supply large equipment and services for emergencies (i.e., large spill clean-up, earth moving equipment, etc.)? — —

SPILLS

Product spills can be prevented or substantially reduced, both in numbers and consequences, through effective spill prevention and control programs. Loss of product not only has serious economical considerations, but if not properly contained and cleaned up, can also result in serious environmental consequences, adding further to your economic losses.

Yes No

1. Do you have a spill prevention and control program? — —
2. Are spills immediately cleaned up and residues properly disposed? — —
3. Do you have the means to contain large spills? — —
   If not, where will runoff go? — —
4. Do you have floor drains? — —
   If so, where do they discharge? — —
5. Can drains be manually blocked/closed off in the event of a spill? — —
6. Do you maintain a "spill response kit" (e.g., recovery drums, absorbent material, tools, personal protection equipment etc.)? — —
7. Is contaminated personal protective equipment (boots, gloves, etc.) properly cleaned or disposed? — —

DISPOSAL

Federal and state laws, as well as financial liabilities arising from past hazardous waste disposal practices, are presenting potentially serious economic risks to businesses who process and store chemicals. To minimize these risks to your warehousing operation strict hazardous waste accountability is essential.

Yes No

1. Do you generate any waste considered hazardous according to federal or state regulations? — —
2. Did the previous owners generate waste? — —
3. Has any waste been disposed of on site? — —
4. Who is designated (by name) to be responsible for waste disposal decisions? — —
5. Does your off-site hazardous waste treatment/storage/disposal facility:
   - — Have the required operating permits? — —
   - — Have any current or potential problems? — —
   - — Incurred any fines for non-compliance with Federal or State regulations? — —
   - — Had any adverse publicity? — —
6. Does your hazardous waste transporter:
   - — Have the required permits? — —
   - — Have the required insurance coverage? — —
   - — Have an accident record? — —
7. If you ship hazardous wastes off-site, are required manifests used? — —
8. Do your have a procedure for handling damaged product? — —
   - — Is it repackaged on site? — —
   - — Is it returned to the supplier? — —
   - — Is is disposed of properly? — —
9. Do you dispose of waste in the municipal landfill? — —
   Could hazardous waste be sent unknowingly? — —
10. Do you vacuum, sweep, or wash the floor of your warehouse? — —
    - — Is clean-up material properly disposed? — —
    - — Is clean-up material utilized in any way? — —

SECURITY

Good security practices have become essential to the protection of your warehouse property and against accidental or intentional emergency situations resulting from entry of unauthorized personnel. A careful evaluation of practices or conditions in or surrounding your warehouse which may compromise security protection can significantly help to minimize these types of losses to your property.

Yes No

1. Do you have an electronic security alarm? — —
   - — Audible alarm? — —
   - — Silent alarm wired to a security service, local police or sheriff's office? — —
2. Do you have a guard service contract? — —
3. Do you encourage observation of your warehouse by local police and neighbors during off hours? — —
4. Are deadbolt locks used on all entry doors? — —
5. Do you have adequate lighting in all areas in and around the warehouse? — —
6. Do you conduct regular inventories and periodic shipment checks? — —
7. Do you prohibit unauthorized personnel from entering the warehouse? — —
8. Are windows protected with wire mesh, bars, locks, etc.? — —
9. Are the walls of your warehouse properly secured or "double-walled" to discourage entry? — —
10. Is your employee parking area away from the warehouse? — —
11. Does the lack of adequate housekeeping practices promote places for hiding around the warehouse? — —
12. Do you have procedures for securing areas while unattended during working hours and at closing time? — —

LAWS AND REGULATIONS

The various operations involved in warehousing of agricultural chemicals can make you legally liable for complying with federal, state and local regulations. Do you know what regulations affect your warehouse?

Yes No

1. Do you repackage pesticides? — —
   If so, do you have an EPA Establishment number? — —
   # ______________________
2. Do you have an EPA/State hazardous waste generator ID number? — —
   # ______________________
   If not, should you? — —
3. Are you required to have operating permits for air, water or solid wastes? — —
4. Have you ever incurred any citations from a Federal, State or local agency? — —
5. Do you have above or below ground storage tanks that are required to be registered by federal or state laws? — —

WAREHOUSE SITE CONSIDERATIONS

The physical characteristics of buildings and warehouse locations can have significant impacts on the success of risk control programs, especially as they relate to emergency response/preparedness and community relations. Before purchasing future facilities/sites, a thorough exposure assessment can help prevent future problems.

	Yes	No
1. Are you familiar with the history of the site?	—	—
2. Have you checked on local zoning laws and regulations?	—	—
3. Are safe traffic entrances and exits to your warehouse available?	—	—
4. Are existing structures appropriate for warehousing operations?	—	—
5. Have any waste materials been buried on site?	—	—
6. Have you considered the following potential hazards?:		
— Surrounding occupancies?	—	—
— Proximity to schools, hospitals, or high population areas?	—	—
— Location and capabilities of local fire and emergency response units?	—	—
— Proximity to water wells?	—	—
— Quantity and location of stormwater discharge points from the property?	—	—
— Relationship of the property regarding earthquakes, floods, etc.?	—	—

ATTENTION!

The next page in CPCR is page 251, which begins with Abbott Laboratories. There are no pages 209-250.

Abbott Laboratories

Chemical & Agricultural Products Division
14th & Sheridan Road
North Chicago, IL 60064
(800) 323-9597
(800) 642-1959 (in Illinois)

DEVINE®

BIOLOGICAL HERBICIDE

ACTIVE INGREDIENT:
Live Chlamydospores of *Phytophthora palmivora* MWV.

3.2×10^8 per pint. (6.7×10^5	W/W
live chlamydospores per ml)	0.8%
INERT INGREDIENTS:	99.2%
TOTAL	100.0%

EPA Reg. No. 275-39 EPA Est. No. 275-IL-1

LABEL EFFECTIVE DATE: 2/88

CAUTION

KEEP OUT OF REACH OF CHILDREN

KEEP REFRIGERATED—DO NOT FREEZE

FOR USE IN FLORIDA ONLY

DO NOT USE IN CLAY, GULF, LIBERTY OR GADSDEN COUNTIES

DEVINE® is a mycoherbicide for control of *Morrenia odorata*, strangler or milkweed vine, in citrus groves. This fungus will initiate a root infection in milkweed vine plants that kills the vine in two to ten weeks following application, depending on the size and maturity of the vine. Populations of the vine generally continue to decrease over the period of a year after treatment. Complete control may not be achieved in the first year of treatment. DEVINE will not harm citrus trees.

PRECAUTIONARY STATEMENTS

Environmental Hazard-Keep out of lakes, ponds and streams. Do not contaminate water by cleaning of equipment or disposal of wastes.

DIRECTIONS FOR USE

It is a violation of federal law to use this product in a manner inconsistent with its labeling.

Do not apply this product through any type of irrigation system.

STORAGE AND DISPOSAL

Do not contaminate water, food, or feed by storage or disposal.

STORAGE: KEEP REFRIGERATED. DO NOT FREEZE.

Failure to maintain refrigeration causes rapid product deterioration and bottle leakage. Return unopened, still refrigerated, out-of-date product to your supplier immediately upon expiration for disposal.

PESTICIDE DISPOSAL:

Wastes resulting from the use of this product may be disposed of on site or at an approved waste disposal facility.

CONTAINER DISPOSAL:

Do not reuse empty containers. Triple rinse (or equivalent). Puncture and dispose of in a sanitary landfill, or incinerate, or, if allowed by state and local authorities, by burning. If burned, stay out of smoke.

MIXING DIRECTIONS

Triple flush the tank and lines of the herbicide boom sprayer system. Adjust nozzles for coverage of all the soil where the vine can grow. Use 50 mesh or coarser screens. Half-fill the tank with water, add the required amount of DEVINE® and agitate as the tank is filled.

APPLICATION DIRECTIONS

Use rate: 1 pint DEVINE per treated acre.

DEVINE may be applied in any type of citrus grove from May through September after the weed has germinated or is actively growing. Only one treatment per season is necessary.

Apply with a herbicide boom sprayer to achieve uniform coverage of the soil under the tree canopy. Use at least 50 gallons of spray water per treated acre. The surface of the soil *must* be wet at the time of application. Adequate moisture can be achieved with either 2 inches of water, rainfall or irrigation, prior to treatment or 1 inch water pretreatment followed by ½ inch on the third day after treatment.

APPLICATION RESTRICTIONS

DO NOT:

—mix with chlorinated water

—tank mix with a wetting agent, fertilizer, or any other pesticide

—apply in wind conditions favoring drift

—apply where the susceptible plants watermelon and periwinkle are grown

—apply within 100 ft. of the following plants: cucumber, squash, Begonia, Bougainvilla, boxwood, Hibiscus, oak areca palm, Pittosporum, snapdragon, Washingtonia and coconut palm, and hybrid Rhododendron

—apply more than 1 pint per acre per season

NOTICE TO USER

Seller makes no warranty, express or implied, of merchantability, fitness or otherwise concerning the use of this product other than as indicated on the label. User assumes all risks of use, storage or handling not in strict accordance with accompanying directions.

This is a specimen label. Abbott Laboratories is not responsible for the accuracy of the information contained herein. As labels are subject to revision, always carefully read and follow the label on the product container.

DIPEL® ES

(EAST OF THE ROCKY MOUNTAINS)
WORM KILLER
EMUSIFIABLE SUSPENSION
BIOLOGICAL INSECTICIDE

ACTIVE INGREDIENT:
Bacillus thuringiensis, var. kurstaki,
17,600 International Units of Potency per

mg (64 Billion International Units per gallon)	3.5%
INERT INGREDIENTS:	96.5%
TOTAL	100.0%

E.P.A. Registration No. 275-65 E.P.A. Est. No. 33762-IA-1

CAUTION

KEEP OUT OF REACH OF CHILDREN

PRECAUTIONARY STATEMENTS

HAZARDS TO HUMANS (AND DOMESTIC ANIMALS)

CAUTION

Avoid contacts with skin, eyes or clothing. In case of contact immediately flush eyes or skin with plenty of water. Get medical attention if irritation persists.

BENEFICIAL INSECTS

Honeybees foraging treated areas are not harmed by DIPEL ES use.

DIPEL ES does not interrupt the activities of beneficial and predacious arthropods in pest management programs.

STORAGE AND DISPOSAL

Do not contaminate water, food or feed by storage or disposal.

STORAGE

Keep containers tightly closed when not in use. At temperatures less than 0°F and greater than 100°F, DIPEL ES should be stored under cover. Roll or shake the container before dispensing.

PESTICIDE DISPOSAL

Wastes resulting from the use of this product may be disposed of on site or at an approved waste disposal facility.

CONTAINER DISPOSAL

Triple rinse (or equivalent). Then puncture and dispose of in a sanitary landfill, or by other procedures approved by state and local authorities.

Days to harvest: There are no restrictions on applying DIPEL ES up to the time of harvest.

DIRECTIONS FOR USE

It is a violation of Federal law to use this product in a manner inconsistent with its labeling.

DIPEL ES is a highly selective insecticide for use against listed caterpillars (larvae) of lepidopterous insects. Close scouting and early attention to infestations is highly recommended. Larvae must eat deposits of DIPEL ES to be affected. Always follow these directions:

- Treat when larvae are young (early instars) before extensive damage has occurred.
- Larvae must be actively feeding on treated, exposed plant parts.
- Thorough spray coverage is needed to provide a uniform deposit of DIPEL ES at the site of larval feeding.
- Under heavy pest population pressure, use the higher label rates and/ or raise gallonage to improve spray coverage.
- If attempting to control a pest with a single spray, make the treatment when egg hatch is essentially complete, but before extensive crop damage occurs.

- A spreader-sticker or surfactant which has been approved for use on growing and harvested crops should be added to improve the effectiveness of DIPEL ES (ground or aerial application).

After eating a lethal dose of DIPEL ES, larvae stop feeding within the hour and will die within several days. Dying larvae move slowly, discolor, then shrivel, blacken and die.

DIPEL ES may be applied in conventional ground, aerial equipment or center pivot irrigation systems with quantities of water sufficient to provide thorough coverage of infested plant parts. The amount of water needed per acre will depend on weather, spray equipment, and local experience. Fill the mix tank or plane hopper with the desired quantity of water. Start the mechanical or hydraulic agitation to provide moderate circulation before adding DIPEL ES. Always add the surfactant or spreader-sticker to the water prior to the addition of DIPEL ES. Add the desired volume of DIPEL ES to the tank or plane hopper and continue circulation. Include rinse water from containers. Never add DIPEL ES to the mix tank before introducing the desired quantity of water. Maintain the suspension while loading and spraying. Do not mix more DIPEL ES than can be used in a 6-day period. **CAUTION:** Rinse and flush spray equipment thoroughly following each use.

DIPEL ES For Field Crops (East of the Rockies)

Crop	Pest	Pints/Acre (Ground Equipment)	Pints/Acre (Aerial Application)	Pints/Acre (Center Pivot Irrigation)
Alfalfa	Alfalfa	¼	¼	
(Hay &	Caterpillar	2 to 2½	2 to 2½	
Seed)	Armyworm	½	½	
Hay and	European	1	1	
Other	Skipper			
Forage	Loopers			
Crops				
Corn: Field Corn, Popcorn, Seed Corn, Sweet Corn	European Cornborer & Southwestern Cornborer (First generation population)	1½ to 2½		1½ to 2½
	European Cornborer & Southwestern Cornborer (Second generation population)	1½ to 2½	1½ to 2½	1½ to 2½
Cotton	Armyworm	2 to 2½	2 to 2½	2 to 2½
	Cotton Bollworm	1 to 2½	1 to 2½	1 to 2½
	Loopers	1	1	1
	Tobacco Budworm	1 to 2½	1 to 2½	1 to 2½
Soybeans	Armyworm	2	2	
	Green Cloverworm	½	½	
	Loopers	1	1	
	Podworm (Heliothis zea)	1	1	
	Velvetbean Caterpillar	⅔	⅔	

CORN APPLICATION TIMING

Applications should be made when young larvae are present for first or second generation corn borers. One application against the first generation of larvae should provide economic control. Two or more applications may be required against second generation borers if there is an extended period of egg deposition.

Treatments should be made before extensive damage has occurred while larvae are still actively feeding on exposed plant parts. Thorough spray coverage is needed to provide a uniform deposit at the site of larval feeding.

Under heavy pest poulation pressure, use a higher label rates increased spray volume, and/or multiple applications.

Cool weather may cause cornborer larvae to seek protected areas of the corn plant and to reduce the amount of feeding normally done on exposed plant parts. This alteration in feeding behavior will hamper the effectiveness of DIPEL ES.

CORN APPLICATION INSTRUCTIONS

Mixing Recommendations: Important — do not add DIPEL ES to the mix tank before introducing the desired quantity of water. Start the mechanical or hydraulic agitation to provide moderate circulation before adding DIPEL ES. When using a surfactant or speader-sticker, add it to the water prior to addition of DIPEL ES. Add the desired volume of DIPEL ES to the tank mix and continue circulation. Include rinse water from the containers. For chemigation, keep the ratio at 3 parts water to 1 part DIPEL ES. Maintain the suspension while loading and spraying. When using a non-emulsifiable oil or another pesticide, add it after the DIPEL ES. Do not mix more DIPEL ES than can be used in a 6-day period. Rinse and flush spray equipment thoroughly following each use. Selection of fluid to flush the application system will depend on what type of mixture was used during the application period. Use 20-mesh screens.

Undiluted injection for chemigation: Flush and clean nurse tank, lines, screen canister, and pump with diesel fuel or a non-emulsifiable oil until they are water free before and after application. Use a 20-mesh screen. Continue agitation during injection.

Spray Volume: For aerial application use at least 3 gallons of total volume per acre in water based sprays. For ground application, use at least 7 gallons of volume per acre. For chemigation, use irrigation levels of 0.15 to 0.5 inches of water per water. Up to 1 inch of irrigation water may be used, but efficacy may be reduced. Apply DIPEL ES only through sprinkler irrigation systems such as center-pivot, lateral move, end-tow, side (wheel) roll, traveler, solid set or hand move. Do not apply this product through any other type of irrigation system.

COTTON, SOYBEANS AND ALFALFA APPLICATION CONSIDERATIONS

DIPEL ES may be used to control small armyworms (1st and 2nd instar) when populations are light and full coverage sprays are applied. Repeat treatment as necessary. If mature worms or heavy populations are present a contact insecticide should be added to enhance control.

COTTON APPLICATION TIMING

Use DIPEL ES to suppress light to moderate populations of newly hatched budworms and bollworms in pest management programs. Use under close scouting when beneficial insects are active or building. Repeat treatments at 4 to 5 day intervals or as long as necessary and results are acceptable. Use only in combination with registered ovicides at labeled rates. For armyworms and loopers follow general directions for use.

COTTON APPLICATION INSTRUCTIONS

Mixing Recommendations: Use general guidelines given for corn.

Spray Volume: For aerial application use at least 3 gallons of total volume per acre in water based sprays. For ground equipment, use at least 7 gallons volume per acre. Use 20-mesh screens.

SOYBEANS APPLICATION TIMING

DIPEL ES suppresses early instar larvae of Podworms and Armyworms when populations are light to moderate.

SOYBEANS APPLICATION INSTRUCTIONS

Mixing Recommendations: Use general guidelines for corn.

Spray Volume: For aerial application use at least 3 gallons of total volume per acre in water based sprays. For ground equipment, use at least 7 gallons volume per acre. Use 20-mesh screens.

ALFALFA APPLICATION TIMING

DIPEL ES may be used to control small armyworms (1st and 2nd instar) when populations are light and full coverage sprays are applied. Repeat treatment as necessary. If mature worms or heavy populations are present a contact insecticide should be added to enhance control.

ALFALFA APPLICATION INSTRUCTIONS

Mixing Recommendations: Use general guidelines given for corn.

Spray Volume: For aerial application use at least 3 gallons of total volume per acre in water based sprays. For ground equipment, use at least 7 gallons volume per acre. Use 20-mesh screens.

CHEMIGATION

Apply this product only through sprinkler systems such as center pivot, lateral move, end tow, side (wheel) roll, traveler, solid set, or hand move. Do not apply this product through any other type of irrigation system.

Crop injury, lack of effectiveness, or illegal pesticide residues in the crop can result from nonuniform distribution of treated water.

If you have questions about calibration, you should contact State Extension Service specialists, equipment manufacturers or other experts.

Do not connect an irrigation system (including greenhouse systems) used for pesticide application to a public water system unless the pesticide label-prescribed safety devices for public water systems are in place.

A person knowledgeable of the chemigation system and responsible for its operation, or under the supervision of the responsible person, shall shut the system down and make necessary adjustments should the need arise.

The active ingredient in DIPEL ES will settle in the tank and injection lines; adequate agitation must be provided before and during the injection period. Use only in systems that apply uniformly and have appropriate check valves. Do not apply when wind speed favors drift beyond the area intended for treatment. When application is complete, thoroughly flush the injection system and sprinkler lines.

The system must contain a functional check valve, vacuum relief valve, and low pressure drain appropriately located on the irrigation pipeline to prevent water source contamination from backflow.

The pesticide injection pipeline must contain a functional, automatic, quick-closing check valve to prevent the flow of fluid back toward the injection pump.

The pesticide injection pipeline must also contain either a functional, normally closed, solenoid operated valve connected to the system interlock located on the intake side of the injection pump and to prevent fluid from being withdrawn from the supply tank when the irrigation system is either automatically or manually shut down.

The system must contain functional interlocking controls to automatically shut off the pesticide injection pump when the water pump motor stops.

The irrigation line or water pump must include a functional pressure switch which will stop the water pump motor when the water pressure decreases to the point where pesticide distribution is adversely affected.

Systems must use a metering pump, such as a positive displacement injection pump (e.g., diaphragm pump) effectively designed and constructed of materials that are compatible with pesticides and capable of being fitted with a system interlock.

NOTICE TO USER

Seller makes no warranty, express or implied, of merchantability, fitness or otherwise concerning use of this product other than as indicated on the label. User assumes all risks of use, storage or handling not in strict accordance with accompanying directions.

This is a specimen label. Abbott Laboratories is not responsible for the accuracy of the information contained herein. As labels are subject to revision, always carefully read and follow the label on the product container.

DIPEL® ES

(WEST OF THE ROCKY MOUNTAINS) EMULSIFIABLE SUSPENSION

Biological Insecticide

ACTIVE INGREDIENT:
Bacillus thuringiensis, var. *kursatki*,
17,600 International Units of Potency per mg
(64 billion International Units per gallon) . . . 3.5%
INERT INGREDIENTS . . . 96.5%
TOTAL . . . 100.00%

EPA Registration No. 275-65 EPA Est. No. 33762-IA-1

CAUTION

KEEP OUT OF REACH OF CHILDREN

PRECAUTIONARY STATEMENTS

HAZARDS TO HUMANS AND DOMESTIC ANIMALS

CAUTION

Avoid contacts with skin, eyes or clothing. In case of contact, immediately flush eyes or skin with plenty of water. Get medical attention if irritation persists.

BENEFICIAL INSECTS

Honeybees foraging treated areas are not harmed by DIPEL ES use.

DIPEL ES does not interrupt the activities of beneficial and predacious arthropods in pest management programs.

STORAGE AND DISPOSAL

Do not contaminate water, food or feed by storage or disposal.

STORAGE

Keep containers tightly closed when not in use. At temperatures less than 0°F and greater than 100°F, DIPEL ES should be stored under cover. Roll or shake the container before dispensing.

PESTICIDE DISPOSAL

Wastes resulting from the use of this product may be disposed of on site or at an approved waste disposal facility.

CONTAINER DISPOSAL

Triple rinse (or equivalent). Then puncture and dispose of in a sanitary landfill, or by other procedures approved by state and local authorities.

DAYS TO HARVEST

There are no restrictions on applying DIPEL ES up to the time of harvest.

DIRECTIONS FOR USE

It is a violation of Federal law to use this product in a manner inconsistent with its labeling.

DIPEL ES is a highly selective insecticide for use against listed caterpillars (larvae) of lepidopterous insects. Close scouting and early attention to infestations is highly recommended. Larvae must eat deposits of DIPEL ES to be affected. Always follow these directions:

- Treat when larvae are young (early instars) before extensive damage has ocurred.
- Larvae must be actively feeding on treated, exposed plant parts.
- Thorough spray coverage is needed to provide a uniform deposit of DIPEL ES at the site of larval feeding.
- Under heavy pest population pressure, use the higher label rates and/or raise gallonage to improve spray coverage.
- If attempting to control a pest with a single spray, make the treatment when egg hatch is essentially complete, but before extensive crop damage occurs.
- A spreader-sticker or surfactant which has been approved for use on growing and harvested crops should be added to improve the effectiveness of DIPEL ES (ground or aerial application).

After eating a lethal dose of DIPEL ES larvae stop feeding within the hour and will die within several days. Dying larvae move slowly, discolor, then shrivel, blacken and die.

DIPEL ES may be applied in conventional ground or aerial equipment with quantities of water sufficient to provide thorough coverage of infested plant parts. The amount of water needed per acre will depend on weather, spray equipment, and local experience. Unless otherwise indicated, use at least 4 gallons of water by air: except in the far west, 5 to 10 gallons is the usual minimum. Fill the mix tank or plane hopper with the desired quantity of water. Start the mechanical or hydraulic agitation to provide moderate circulation before adding DIPEL ES. Always add the surfactant or spreader-sticker to the water prior to the addition of DIPEL ES. Add the desired volume of DIPE ES to the tank or plane hopper and continue circulation. Include rinse water from containers. Never add DIPEL ES to the mix tank before introducing the desired quantity of water. Maintain the suspension while loading and spraying. Do not mix more DIPEL ES than can be used in a 6-day period. **CAUTION:** Rinse and flush spray equipment thoroughly following each use.

COTTON AND ALFALFA APPLICATION CONSIDERATIONS

DIPEL ES may be used to control small armyworms (1st and 2nd instar) when populations are light and full coverage sprays are applied. Repeat treatment as necessary. If mature worms or heavy populations are present a contact insecticide should be added to enhance control.

DIPEL ES FOR COTTON

Crop	Pest	Pints/Acre (Ground Equipment)	Pints/Acre (Aerial Application)
Cotton	Armyworms	2 to 2½	2 to 2½
	Cotton Bollworms*	1½ to 2½	1½ to 2½
	Loopers	1	1
	Tobacco Budworm*	1½ to 2½	1½ to 2½

*Not registered in California

COTTON APPLICATION TIMING

Use DIPEL ES to suppress light to moderate populations of newly hatched budworms and bollworms in pest management programs. Use under close scouting when beneficial insects are active or building. Repeat treatments at 4 to 5 day intervals or as long as necessary and results are acceptable. Use only in combination with registered ovicides at labeled rates. For armyworms and loopers follow general directions for use.

COTTON APPLICATION INSTRUCTIONS

Mixing Recommendations: Use general guidelines given for alfalfa.

Spray Volume: For aerial application use at least 3 gallons of total volume per acre in water based sprays. For ground equipment, use at least 7 gallons volume per acre. Use 20-mesh screens.

DIPEL ES FOR ALFALFA

Crop	Pest	Pints/Acre (Ground Equipment)	Pints/Acre (Aerial Application)
Alfalfa (Hay and Seed), Hay and Other Forage Crops	Alfalfa Caterpillar*	¼	¼
	Armyworms	2 to 2½	2 to 2½
	European Skipper*	½	½
	Loopers*	1	1

*Not registered in California

ALFALFA APPLICATION TIMING

DIPEL ES may be used to control small armyworms (1st and 2nd instar) when populations are light and full coverage sprays are applied. Repeat treatment as necessary. If mature worms or heavy populations are present a contact insecticide should be added to enhance control.

ALFALFA APPLICATION INSTRUCTIONS

Mixing Recommendations: Important — do not add DIPEL ES to the mix tank before introducing the desired quantity of water. Start the mechanical or hydraulic agitation to provide moderate circulation before adding DIPEL ES. When using a surfactant or spreader-sticker, add it to the water prior to addition of DIPEL ES. Add the desired volume of DIPEL

ES to the mix tank and continue circulation. Include rinse water from the containers. Maintain the suspension while loading and spraying. When using a non-emulsifiable oil or another pesticide, add it after the DIPEL ES. Do not mix more DIPEL ES than can be used in a 6-day period. Rinse and flush spray equipment thoroughly following each use. Selection of fluid to flush the application system will depend on what type of mixture was used during the application period. Use 20-mesh screens.

Spray Volume: For aerial application use at least 3 gallons of total volume per acre in water based sprays. For ground equipment, use at least 7 gallons volume per acre. Use 20-mesh screens.

NOTICE TO USER

Seller makes no warranty, express or implied, of merchantability, fitness or otherwise concerning use of this product other than as indicated on the label. User assumes all risks of use, storage or handling not in strict accordance with accompanying directions.

This is a specimen label. Abbott Laboratories is not responsible for the accuracy of the information contained herein. As labels are subject to revision, always carefully read and follow the label on the product container.

DIPEL® 2X

WETTABLE POWDER
BIOLOGICAL INSECTICIDE

ACTIVE INGREDIENT:
Bacillus thuringiensis, var. *kurstaki*, 32,000 International Units of Potency per mg.
(14.52 Billion International Units per pound) 6.4%
INERT INGREDIENTS: 93.6%
TOTAL 100.0%
E.P.A. Reg. No. 275-37 E.P.A. Est. No. 33762-IA-1

KEEP OUT OF REACH OF CHILDREN
CAUTION

STORAGE AND DISPOSAL

Do not contaminate water, food or feed by storage or disposal.

BENEFICIAL INSECTS

Honeybees foraging treated areas are not harmed by DIPEL 2X use.

DIPEL 2X does not interrupt the activities of beneficial and predacious arthropods in integrated pest management programs.

DIRECTIONS FOR USE

It is a violation of Federal law to use this product in a manner inconsistent with its labeling.

Do not apply this product through any type of irrigation system.

Days to harvest: There are no restrictions on applying DIPEL 2X up to the time of harvest.

Sites: DIPEL 2X may be used for any labeled pest in both field and greenhouse uses.

DIPEL 2X is a highly selective insecticide for use against listed caterpillars (larvae) of lepidopterous insects. Close scouting and early attention to infestations is highly recommended. Larvae must eat deposits of DIPEL 2X to be affected. Always follow these directions:

- Treat when larvae are young (early instars) before the crop is extensively damaged.
- Larvae must be actively feeding on treated, exposed plant parts.
- Thorough spray coverage is needed to provide a uniform deposit of DIPEL 2X at the site of larval feeding. For some crops directed drop nozzles by ground machine are required.
- Under heavy pest population pressure, use the higher label rates, shorten the spray interval, and/or raise the gallonage to improve spray coverage.
- Repeat applications at an interval sufficient to maintain control, usually 3 to 14 days depending on plant growth rate, moth activity, rainfall after treating, and other factors. If attempting to control a pest with a single spray, make the treatment when egg hatch is essentially complete, but before extensive crop damage occurs.
- A spreader-sticker which has been approved for use on growing and harvested crops should be added for hard-to-wet crops such as cole crops, or to improve weather-fastness of the spray deposits.
- DIPEL 2X is a non-restricted use pesticide and does not require a restricted use permit for purchase and use.
- DIPEL 2X may be tank mixed with other labeled insecticides to enhance control.
- Before tank mixing DIPEL 2X with other labeled products, check for tank-mix compatability.

After eating a lethal dose of DIPEL 2X, larvae stop feeding within the hour, and will die within several days. Dying larvae move slowly, discolor, then shrivel, blacken and die.

DIPEL 2X may be applied in conventional ground or aerial equipment with quantities of water sufficient to provide thorough coverage of infested plant parts. The amount of water needed per acre will depend on crop size, weather, spray equipment, and local experience. Unless otherwise indicated, use at least 2 gallons of water by air; except in the Western U.S., where 5 to 10 gallons is the usual minimum. Add water to the spray or mixing tank at the level that provides maximum agitation. With the agitator running, slowly sprinkle in the DIPEL 2X. Continue agitation. Then add other spray materials (if any). Add the balance of the water and agitate until mixed. Maintain the suspension while loading and spraying. Do not mix more DIPEL 2X than can be used in a 12-hour period.

For Smaller Spray Volumes: If Rate is	Use this Amount Per Gallon
1/4 lb./acre or 100 gals.	1/2 tsp.
1/2 lb./acre or 100 gals.	1 tsp.
1 lb./acre or 100 gals.	2 tsps.
2 lb./acre or 100 gals.	4 tsps.

APPLICATION CONSIDERATIONS: DIPEL IS EXEMPT FROM THE REQUIREMENTS OF A TOLERANCE ON ALL RAW AGRICULTURAL COMMODITIES AND HAS NO PLANTBACK OR REENTRY RESTRICTIONS.

APPLICATION RATE
(For all states except California)

Crop Group (Typical Crops)	Pest	Pounds/Acre
Root and Tuber such as Carrots, Potatoes, Beets and Sugarbeets **Bulb** such as Onions (green and bulb) and Garlic **Leafy and Cole Crops** such as Lettuce (head and leaf) Kale, Celery, Spinach, Broccoli, Cabbage, Mustard Greens, Brussels Sprouts, Cauliflower, Collards, Chinese Cabbage, Endive, Kohirabi and Parsley	Loopers	1/2-1
	Omnivorous Leafroller	1/2-1
	Hornworms	1/2-1
	Imported Cabbageworm	1/2-1
	Diamondback Moth	1/2-1
	Green Cloverworm	1/2-1
	Webworm	1/2-1
	Saltmarsh Caterpillar	1/2-1
	Armyworms*	1/2-2
	Cutworms	1/2-1
Fruiting Vegetables such as Tomatoes Peppers and Eggplant	Loopers	1/2-1
	Hornworms	1/2-1
	Tomato Fruitworm	1/2-1
	Variegated Cutworm	1/2-1
	Saltmarsh Caterpillar	1/2-1
	Armyworms*	1/2-2
Cucurbit Vegetables such as Melons, Cucumbers and Squash	Loopers	1/2-1
	Melonworms	1/2-1
	Rindworm complex	1/2-1
	Armyworms*	1/2-2
Legume Vegetables Such as Beans, Peas, Lentils and Soybeans	Loopers	1/2-1
	Soybean Looper	1/2-1
	Green Cloverworm	1/2-1
	Velvetbean Caterpillar	1/2-1
	Armyworms*	1/2-2
	Podworms*	1/2-1

Other Crops:

Tobacco	Tobacco Budworm	1/2-1
	Hornworms	1/4-1/2
	Loopers	1/2-1

DIPEL 2X may be tank mixed with sucker control products at recommended rates. Follow the application directions on the sucker control labels. Observe all label precautions for all products used.

Crop Group (Typical Crops)	Pest	Pounds/Acre
Safflower	Loopers	1/2-1
	Saltmarsh Caterpillar	1/2-1
	Armyworms*	1/2-2
Sunflowers	Loopers	1/2-1
	Head Moth	1/2-1
Peanuts	Loopers	1/2-1
	Velvetbean Caterpillar	1/2-1
	Green Cloverworm	1/2-1
	Podworms*	1/2-1
Alfalfa (Hay and Seed)	Loopers	1/2-1

APPLICATION RATE (For all states except California)		
Crop Group (Typical Crops)	**Pest**	**Pounds/Acre**
Hay and Other Forage Crops	Alfalfa Caterpillar European Skipper (Essex Skipper) Armyworms*	½-1 ½-1 ½-2
Cotton	Tobacco Budworm** Cotton Bollworm** Loopers Saltmarsh Caterpillar Armyworms*	½-2 ½-2 ½-1 ½-1 ½-2

** Use to control light to moderate populations of newly hatched worms in integrated pest management conditions. Repeat treatments at 4 to 5-day intervals as long as necessary and results are acceptable. Use in combination with ovicidal rates of labeled *Heliothis* ovicides.

Avacodo	Amorbia Moth Omnivorous Leafroller Omnivorous Looper Orange Tortix Spanworm	½-2 ½-2 ½-2 ½-2 ½-2
Malanga	Saltmarsh Caterpillar Armyworms*	½-1 ½-2
Watercress	Loopers Armyworms*	½-1 ½-2
Kiwi Fruit	Omnivorous Leafroller	½-2
Hops	Loopers Armyworms*	½-1 ½-2
Bananas	Banana Skipper	½-1
Asparagus	Armyworms*	½-2
Corn (Sweet and Field) and Sorghum	Armyworms* Headworms	½-2 ½-1
Rape	Looper Armyworms* Heliothis	1½-2 ½-2 1½-2
Herbs, Spices, and Mints such as Basil, Chives, Dill, Leeks and Peppermint	Looper Saltmarsh Caterpillar Armyworms*	½-1 ½-1 ½-2
Pineapple	Gummosos-Batrachedra comosae (Hodges) Thecla-Thecla basilides (Geyr)	¼-½
Turf	Sod Webworm	1-2
Flowers, Bedding Plants and Ornamentals (Ground Only)	Loopers Tobacco Budworm Omnivorous Looper Omnivorous Leafroller Diamondback Moth Armyworms* Ello Moth (Hornworm) Io Moth Oleander Moth Azalea Caterpillar	¼-½ ¼-½ ¼-½ ¼-½ ¼-½ ½-2 ¼-½ ¼-½ ½-½ ¼-½
Greenhouse/Shadehouse and Outdoor Nursery Crops such as Leafy, Herbs, Brassica and Fruiting groups	Loopers Heliothis	½-1 ½-1

DIPEL 2X for Small Fruits

Crop	Pest	Pounds/Acre
Small Fruit and Berries such as Grapes, Strawberries, Blackberries and Cranberries	Grapeleaf Skeletonizer (ground only) Grape Leafroller Achema Sphinx Moth (Hornworm) Saltmarsh Caterpillar (ground only) Omnivorous Leafroller (ground only) Loopers Orange Tortrix Oblique Banded Leafroller Armyworms* Tobacco Budworm	½-1 ½-1 ½-1 ½-1 ½-1 ½-1 ½-1 ½-1 ½-2 ½-1

DIPEL 2X for Tree Fruits, Nuts, Citrus and Tropical Fruit

Crop	Pest	Pounds/Acre
Stone Fruit such as Cherries, Plum, Peach, Prune and Nectarine **Pome Fruit** such as Apples and Pears **Tree Nuts** such as Almonds, Pecan, Walnut and Filbert **Pomegrantes**	Redhumped Caterpillar Tent Caterpillars Omnivorous Leafroller Fall Webworm Walnut Caterpillar Cankerworms Gypsy Moth Variegated Leafroller Redbanded Leafroller Tufted Apple Budmoth Fruittree Leafroller Filbert Leafroller Obliquebanded Leafroller	½-2 ½-2 ½-2 ½-2 ½-2 ½-2 ½-2 ½-2 ½-2 ½-2 ½-2 ½-2 ½-2
Citrus	Fruittree Leafroller Orangedog Citrus Cutworm*	½-2 ¼-2 ½-2
	* Apply to light to moderate populations of newly-heatched worms	
Tropical Fruits	Hornworms Leafrollers Omnivorous Looper Loopers	½-2 ½-2 ½-2 ½-2

DIPEL 2X FOR STORED AGRICULTURAL COMMODITIES

GRAINS, SOYBEANS, SUNFLOWER SEED, CROP SEED, CONDIMENTAL SEEDS, SPICES, HERBS, BIRDSEED and POPCORN.

Pest	Rate
Indian Meal Moth, Almond Moth	⅜ lb/100 bu (undiluted and diluted)*

* As a surface treatment, apply ½ lb. DIPEL 2X in 5-10 gal. of water per 500 sq. ft. of grain surface area; mix into top 4 inches.**

For the control and prevention of these pests, apply DIPEL 2X in a constantly agitated water suspension to the top four inch surface layer of grain in the bin. Use a sprinkler can or sprayer to apply the dosage into the grain stream as the last (top) four inch layer is augered into the bin. Mix 1/20 lb. DIPEL 2X per gallon of water. Apply 0.6 pint in this mixture per bushel as grain is augered into storage. Or, sprinkle the dosage into the surface of the grain in the bin and mix thoroughly with a scoop or rake to the depth of four inches. More thorough coverage may be achieved by dividing the recommended dosage into three applications and mixing the grain between applications.

For the protection of bagged grain including popcorn, apply the dosage to the entire grain mass and mix thoroughly prior to bagging.

Treatments can be applied to stored grain at any time, but for best results, make application immediately after harvest before moth activity occurs. In areas where late fall harvested grain is not subject to infestation because of low temperatures, application can be delayed until late winter or early spring before moth activity begins. Control for a full storage season should normally be expected: however, repeat application if infestation recurs.

This treatment controls the moth larvae. If an infestation is present when the grain is treated, moth emergence may continue for several days. If immediate control of severe infestations is desired, grain should be fumigated prior to application of this treatment. **DIPEL 2X will not control weevils or other beetles.**

DIPEL 2X is exempt from tolerance restrictions and the treated grain can be used at any time after treatment for any use.

** For commodities coarser than shelled corn, increase depth of treatment according to the habit of the pest.

PEANUTS

Pest	Rate
Indian Meal Moth, Almond Moth	¼ lb./ton*

* Apply this rate to the top four to eight feet of nuts when filling the warehouse.

To prevent and control these pests, spray an even coating of DIPEL 2X on the farmer stock peanuts while filling the warehouse. To make the spray solution, mix 3¾ lbs. DIPEL 2X per 5 gallons of water. Apply to 15 tons of commodity. Do not pre-mix more spray solution than will be used within 12 hours. Keep the spray suspension agitated during application, and use pressures and nozzles sufficient to handle this suspension.

Before filling the warehouse, clean thoroughly, then spray interior of the facility with a DIPEL 2X suspension at the rate of ½ lb. DIPEL 2X per 100 gallons water. Spray enough suspension to wet all cracks and crevices.

For bagged peanuts, treat the whole mass of commodity at the rate indicated above.

FLUE-CURED TOBACCO

Pest	Rate
Tobacco Moth	0.2 oz./100 lbs.*

* Apply 0.2 ounce (approximately 2½ teaspoonsful) of DIPEL 2X in one quart of

water per 100 pounds of tobacco as a fine mist spray. Avoid overwetting. Tobacco should have just enough moisture to be handled without shattering at the time of application.

Tobacco to be stored up to Twelve Months

Spray loose leaves as the tobacco is being bundled from the curing barn. For tobacco on sticks, treat both side of leaves.

Stored Tobacco

For tobacco which is to be carried over, rebundle or restack sticks, fluff up tobacco and spray loose leaves.

For tobacco that has been stored over three weeks, apply at first signs of infestation, promptly open bundles, spray loose leaves, then bundle.

Treatment of Storage Barns

If tobacco has been treated, or is going to be treated, treatment of the floors and walls may be made to aid in control. Sweep out the area, especially cracks and corners, and all of the loose tobacco pieces in which the moth might breed. Make a spray mixture containing 1/2 oz. DIPEL 2X per 2 1/2 gallons of water. Apply this at a rate of 1/2 gallon per 1000 sq. ft. of surface area. Be sure to spray into cracks and between floorboards.

* DIPEL 2X may be used to control small armyworms (1st and 2nd instar) and/or podworms when populations are light and full coverage sprays are applied. Repeat treatment as necessary. If mature worms or heavy populations are present, a contact insecticide should be added to enhance control.

DIPEL 2X for Trees and Forests

Crop	Pest	Pounds/ 100 gallons* (Ground Equipment)	Pounds/ Acre (Aerial** Application)
Forest Shade, Sugar Maple Trees and Ornamentals	Gypsy Moth	1/4-3/4	1/2-1 1/2
	Bagworm	1/4-3/4	1/2-1 1/2
	Redhumped Caterpillar	1/4-3/4	1/2-1 1/2
	Spring & Fall Cankerworm	1/4-3/4	1/2-1 1/2
	Fall Webworm	1/4-3/4	1/2-1 1/2
	Elm Spanworm	1/4-3/4	1/2-1 1/2
	Tent Caterpillars	1/4-3/4	1/2-1 1/2
	California Oakworm	1/4-3/4	1/2-1 1/2
	Pine Butterlfy	1/4-3/4	1/2-1 1/2
	Spruce Budworms	1/4-3/4	1/2-1 1/2
	Saddle Prominent Caterpillar	1/4-3/4	1/2-1 1/2
	Douglas Fir Tussock Moth	1/4-3/4	1/2-1 1/2
	Western Tussock Moth	1/4-3/4	1/2-1 1/2
	Fruittree Leafroller	1/4-3/4	1/2-1 1/2
	Blackheaded Budworm	1/4-3/4	1/2-1 1/2
	Mimosa Webworm	1/4-3/4	1/2-1 1/2
	Jack Pine Budworm	1/4-3/4	1/2-1 1/2
	Saddleback Caterpillar	1/4-3/4	1/2-1 1/2
	Greenstriped Mapleworm	1/4-3/4	1/2-1 1/2

* Rate for hydraulic sprayer. For mist blowers, mix the applicable amount (lbs.) in 10 gallons of water.

** For aerial application, use in one to five gallons of water depending on type and density of trees. For best results, spray systems which deliver droplet size of LESS THAN 150 microns should be used.

NOTICE TO USER

Seller makes no warranty, express or implied, of merchantability, fitness or otherwise concerning the use of this product other than as indicated on the label. User assumes all risks of use, storage or handling not in strict accordance with accompanying directions.

Before using any tank mix, read the product labels for all environmental and use cautions.

List 5209

APPLICATION RATE (For California only)

Crop	Pest	Pounds/Acre
Cole Crops and Vegetables including Leafy, Root, Stalk, and Seed and Pod Types (Peas, Beans, Lentils, etc., dry or succulent) Potatoes, Cucurbits	Loopers	1/4 to 1/2
	Hornworms	1/8 to 1/4
	Imported Cabbageworm	1/8 to 1/4
	Diamondback Moth	1/6 to 1/3
	Green Cloverworm	1/8 to 1/4
	Webworm	1/4 to 1/2
	Armyworms*	1/2 to 1

* DIPEL 2X may be used to suppress small armyworms (1st and 2nd instar) when populations are light and full coverage ground sprays are applied.

Tomatoes, Peppers, Eggplant	Loopers	1/4 to 1/2
	Hornworms	1/8 to 1/4
	Tomato Fruitworm*	1/2
	Variegated Cutworm	1/2
	Saltmarsh Caterpillars	1/4 to 1/2
	Armyworms**	1/2 to 1

* Apply weekly in a preventive program.

** DIPEL 2X may be used to suppress small worms (1st and 2nd instar) when populations are light and full coverage ground sprays are applied.

Crop	Pest	Pounds/Acre
Cucurbits, Melons, Squash, Cucumbers, etc.	Rindworm Complex*	1/4 to 1/2
	Melonworms	1/2 to 1

* Apply weekly in a preventive program.

Crop	Pest	Pounds/Acre
Safflower, Sugar beets, Mints	Loopers	1/4 to 1/2
	Saltmarsh Caterpillar	1/4 to 1/2
Alfalfa, Hay, and Small Grains	Loopers	1/4 to 1/2
	Alfalfa Caterpillar	1/16 to 1/8
	European Skipper (Essex Skipper)	1/8 to 1/4
Cotton	Tobacco Budworm*	1/8 to 1/2
	Cotton Bollworm*	1/8 to 1/2
	Loopers**	1/4 to 1/2

* Use to suppress light to moderate populations of newly hatched worms in pest management conditions. Use under close scouting when beneficial insects are active or building. Repeat treatments at 4 to 5-day intervals as long as necessary and results are acceptable. If 15% or more terminals have eggs, add 1/8 lb. AI Chlordimeform[1] (Galecron® or Fundal®) per acre to 1/8 lb. DIPEX 2X. Increase chlordimeform to 1/4 lb. AI if needed. Once beneficial insects are no longer a factor, DIPEL 2X (at 1/8 to 3/8 lb./acre) may be tank mixed with methomyl (Lannate® or Nudrin®) at .225 to .675 lbs. AI per acre to control light to moderate infestations of early instar larvae. Observe any rate, frequency, or boll load labels restrictions for methomyl in your area. Methomyl will likely cause reddening of cotton foliage. Discontinue this tank mix if reddening becomes excessive. Before using any tank mix, read the product labels for all environmental and usage cautions. Chlordimeform is restricted to aerial application and to states where registered.

** Full plant coverage needed.

[1] The mixture of **B.t.** with chlordimeform for inhibiting the growth of lepidopterous larvae on plants is covered by Abbott U.S. patent 3,937,813.

Crop	Pest	Pounds/Acre
Turf	Sod Webworm	1 to 2

Crop	Pest	Lbs./100 Gals.
Flowers and Ornamentals (ground only)	Loopers	1/4 to 1/2
	Tobacco Budworm	1/4 to 1/2
	Omnivorous Looper	1/4 to 1/2
	Omnivorous Leafroller	1/4 to 1/2
	Diamondback Moth	1/6 to 1/3
	Armyworms*	1/2 to 1
	Ello Moth (Hornworm)	1/8 to 1/4
	Io Moth	1/4 to 1/2
	Oleander Moth	1/4 to 1/2

* DIPEL 2X may be used to suppress small armyworms (1st and 2nd instar) when populations are light and full coverage ground sprays are applied.

DIPEL 2X for Fruits, Nuts, Trees, Ornamentals

Crop	Pest	Pounds/Acre
Grapes, Small Fruits, Cane and Bush Berries	Grapeleaf Skeletonizer (ground only)	1/3 to 1/2
	Grape Leaffolder	1/4 to 3/8
	Achema Sphinx Moth (Hornworm)	1/6 to 3/8
	Saltmarsh Caterpillar (ground only)	1/2
	Omnivorous Leafroller (ground only)	1/4 to 1/2
	Loopers	1/4 to 1/2

Crop	Pest	Pounds/ 100 Gallons* (Ground Equipment)	Pounds/ Acre (Aerial** Application)
Stone Fruits, Nut Trees, Pomegranates	Redhumped Caterpillar	1/8 to 1/4	1/6 to 1/2
	Tent Caterpillars	1/8 to 1/4	1/4 to 1/2
	Omnivorous Leafroller	1/4 to 1/2	—
	Fall Webworm	1/8 to 1/4	1/4
	Walnut Caterpillars	1/4 to 1/2	1/4 to 1/2

Crop	Pest	Pounds/ 100 Gallons* (Ground Equipment)	Pounds/ Acre (Aerial** Application)
	Cankerworms	1/8 to 1/4	1/4 to 1/2
Pome Fruits[1]	Cankerworms	1/8 to 1/4	1/4 to 1/2
	Gypsy Moth	1/4 to 1/2	1/2
	Variegated Leafroller	1/4 to 1/2	—
	Redbanded Leafroller	1/4 to 1/2	—
	Tufted Apple Budmoth	1/4 to 1/2	—
	Tent Caterpillars	1/8 to 1/4	1/4 to 1/2
Citrus[2]	Fruittree Leafroller	1/4 to 1/2	—
	Orangedog	1/8 to 1/4	—
Tropical Fruits[3]	Hornworm	1/4 to 1/2	—
	Leafrollers	1/4 to 1/2	—
	Omnivorous Looper	1/4 to 1/2	—
	Loopers	1/4 to 1/2	—
Forest, Shade, Sugar Maple Trees and Ornamentals	Gypsy Moth	1/4 to 1/2	1/2
	Bagworm	1/4 to 1/2	1/4 to 1/2
	Redhumped Caterpillar	1/8 to 1/4	1/8 to 1/2
	Spring and Fall Cankerworm	1/8 to 1/4	1/8 to 1/4
	Fall Webworm	1/4	1/4
	Elm Spanworm	1/4 to 1/2	1/4 to 1/2
	Tent Caterpillars	1/8 to 1/4	1/4
	California Oakworm	1/8 to 1/4	1/8 to 1/4
	Pine Butterfly	1/2	1/2
	Spruce Budworms[4]	1/2	1/2
	Saddle Prominent Caterpiller	1/4 to 1/2	—
	Douglas Fir Tussock Moth	1/2	1/2
	Western Tussock Moth	1/4 to 1/2	—
	Fruittree Leafroller	1/4 to 1/2	—
	Blackheaded Budworm	1/2	—
	Mimosa Webworm	1/4 to 1/2	—
	Jack Pine Budworm	1/4 to 1/2	3/8 to 1/2
	Saddleback Caterpillar	1/4 to 1/2	—

* Rate for hydraulic sprayer. For mist blowers, mix the applicable amount (lbs) in 10 gallons of water.

** For aerial application, use in one to ten gallons of water depending on type and density of trees. For best results, spray systems which deliver droplet size of 200 microns should be used.

[1] Use 400 to 450 gallons/acre to suppress listed pests in pest management programs.

[2] Use 100 to 400 gallons/acre.

[3] Use at least 200 gallons/acre.

[4] Use against populations below 35 larvae per 18-inch branch tip. Use a minimum of 1/2 gallon/acre by air.

DIPEL 2X FOR STORED AGRICULTURAL COMMODITIES GRAINS, SOYBEANS, SUNFLOWER SEED, CROP SEED, CONDIMENTAL SEEDS, SPICES and HERBS.

Pest	Rate
Indian Meal Moth Almond Moth	3/8 lb. 100 bu. (undiluted and diluted)*

* As a surface treatment, apply 1/2 lb. DIPEL 2X in 5-10 gal. of water per 500 sq. ft. of grain surface area, mix into top 4 inches.**

For the control and prevention of these pests, apply DIPEL 2X in a constantly agitated water suspension to the top four inch surface layer of grain in the bin. Use a sprinkler can or sprayer to apply the dosage into the grain stream as the last (top) four inch layer is augered into the bin. Mix 1/20 lb. DIPEL 2X per gallon of water. Apply 0.6 pint in this mixture per bushel as grain is augered into storage. Or, sprinkle the dosage into the surface of the grain in the bin and mix thoroughly with a scoop or rake to the depth of four inches. More thorough coverage may be achieved by dividing the recommended dosage into three applications and mixing the grain between applications.

For the protection of bagged grain including popcorn, apply the dosage to the entire grain mass and mix thoroughly prior to bagging.

Treatments can be applied to stored grain at any time, but for best results, make application immediately after harvest before moth activity occurs. In areas where late fall harvested grain is not subject to infestation because of low temperatures, application can be delayed until late winter or early spring before moth activity begins. Control for a full storage season should normally be expected: however, repeat application if infestation recurs.

This treatment controls the moth larvae. If an infestation is present when the grain is treated, moth emergence may continue for several days. If immediate control of severe infestations is desired, grain should be fumigated prior to application of this treatment. DIPEL 2X will not control weevils or other beetles.

DIPEL 2X is exempt from tolerance restrictions and the treated grain can be used at any time after treatment for any use.

** For commodities coarser than shelled corn, increased depth of treatment according to the habit of the pest.

This is a specimen label. Abbott Laboratories is not responsible for the accuracy of the information contained herein. As labels are subject to revision, always carefully read and follow the label on the product container.

DIPEL® 4L

EMULSIFIABLE SUSPENSION BIOLOGICAL INSECTICIDE

ACTIVE INGREDIENT:
Bacillus thuringiensis, var. *kurstaki,*
8,800 International Units of Potency per mg.
(32 billion International Units per gallon) 1.76%
INERT INGREDIENTS: 98.24%
TOTAL 100.00%

EPA Reg. No. 275-36 EPA Est. No. 33762-IA-1

CAUTION

KEEP OUT OF REACH OF CHILDREN

STORAGE AND DISPOSAL

See Container Label.

CAUTION

Avoid contact with skin, eyes or clothing. In case of contact immediately flush eyes or skin with plenty of water. Get medical attention if irritation persists.

Beneficial Insects: Honeybees foraging treated areas are not harmed by DIPEL 4L use.

DIPEL 4L does not interrupt the activities of beneficial and predacious arthropods in pest management programs.

Days to Harvest: There are no restrictions on applying DIPEL 4L up to the time of harvest.

Sites: DIPEL 4L may be used for any labeled pest in both field and greenhouse uses.

For Smaller Spray Volumes: If Rate is	Use this Amount Per Gallon
1/2 pt./acre or 100 gals.	1/2 tsp.
1 pt./acre or 100 gals.	1 tsp.
1 qt./acre or 100 gals.	2 tsps.
2 qts./acre or 100 gals.	4 tsps.

DIRECTIONS FOR USE

It is a violation of Federal law to use this product in a manner inconsistent with its labeling.

Do not apply this product through any type of irrigation system.

DIPEL 4L is a highly selective insecticide for use against listed catepillars (larvae) of lepidopterous insects. Close scouting and early attention to infestations is highly recommended. Larvae must eat deposits of DIPEL 4L to be affected. Always follow these directions:

- Treat when larvae are young (early instars) before the crop is extensively damaged.
- Larvae must be actively feeding on treated, exposed plant parts.
- Thorough spray coverage is needed to provide a uniform deposit of DIPEL 4L at the site of larval feeding. For some crops directed drop nozzles by ground machine are required.
- Under heavy pest population pressure, use the higher label rates, shorten the spray interval, and/or raise gallonage to improve spray coverage.
- Repeat applications at an interval sufficient to maintain control, usually 3 to 14 days depending on plant growth rate, moth activity, rainfall after treating, and other factors. If attempting to control a pest with a single spray, make the treatment when egg hatch is essentially complete, but before extensive crop damage occurs.
- A spreader-sticker which has been approved for use on growing and harvested crops should be added for hard-to-wet crops such as cole crops, or to improve weather-fastness of the spray deposits.

After eating a lethal dose of DIPEL 4L, larvae stop feeding within the hour and will die within several days. Dying larvae move slowly, discolor, then shrivel, blacken and die.

DIPEL 4L may be applied in conventional ground or aerial equipment with quantities of water sufficient to provide thorough coverage of infested plant parts. The amount of water needed per acre will depend on crop size, weather, spray equipment, and local experience. Unless otherwise indicated, use at least 2 gallons of water by air; except in the far west, 5 to 10 gallons is the usual minimum. Add water to the spray or mixing tank at the level that provides maximum agitation. With the agitator running, add the DIPEL 4L. Continue agitation. Then add other spray materials (if any). Add the balance of the water. Maintain the suspension while loading and spraying. Do not mix more DIPEL 4L than can be used in a 72-hour period. **CAUTION:** Rinse and flush spray equipment thoroughly following each use.

CAUTION

DIPEL 4L should not be used in combination with Comite® Bravo® (except celery), Difolatan, Captan (except seed) or Dyrene®.

(ALL STATES EXCEPT CALIFORNIA)

The effects of combining DIPEL 4L with fungicides (other than benomyl, maneb, mancozeb or captan) and miticides (other than chlordimeform) are unknown.

FOR CALIFORNIA ONLY

The effects of combining DIPEL 4L with insecticides, fungicides and other spray materials are unknown. Tank mixes should be avoided except where specific recommendations are made in the label for use in California.

APPLICATION RATE

Crop	Pest	Pints/Acre
Cole Crops*	Loopers	1 to 2
Celery[1]	Imported Cabbageworm	½ to 1
Lettuce	Diamondback Moth	⅔ to 1⅓
Spinach[1]		

[1] For all states except California.

* Do not apply by air to plants after transplant or other stress before 6 weeks in the field. Use more than 25 gallons of water per acre by ground and 10 gallons of water per acre by air.

Soybeans	Loopers	1 to 2
Peanuts[1]	Green Cloverworm	½ to 1
Peas[1]	Velvetbean Caterpillar	⅔ to 1⅓
	Podworm (Heliothis zea)*	1 to 2
	Armyworms*,[1]	2 to 4

* DIPEL 4L suppresses early instar larvae of Podworms and Armyworms when populations are light to moderate.

[1] For all states except California.

Tobacco[1]	Tobacco Budworm	1
	Hornworms	½ to 1
	Loopers	1

[1] For all states except California.

Tomatoes	Loopers	1 to 2
	Tomato Fruitworm*	2
	Variegated Cutworm	2
	Saltmarsh Caterpillar	1 to 2
	Hornworms	½ to 1
	Armyworms**	2 to 4

* Apply weekly in a preventive program.

** DIPEL 4L may be used to suppress small Armyworms (1st and 2nd instar) when populations are light and full coverage ground sprays are applied.

DIPEL 4L[1] may be tank-mixed with benomyl, maneb or mancozeb for application on tomatoes. Observe all label precautions for all products used.

[1] For all states except California.

Crop	Pest	Pints/Acre
Alfalfa, Hay, and Small Grains	Loopers	1 to 2
	Alfalfa Caterpillar	¼ to ½
	European Skipper (Essex Skipper)	½ to 1
Cotton	Tobacco Budworm*	½ to 2
	Cotton Bollworm*	½ to 2
	Loopers**	1 to 2

* Use to suppress light to moderate populations of newly hatched worms in pest management conditions. Use under close scouting when beneficial insects are active or building. Repeat treatments at 4 to 5-day intervals as long as necessary and results are acceptable. If 15% or more terminals have eggs, add ⅛ lb. AI chlordimeform[1] (Galecron® or Fundal®, or ⅛ lbs. AI methomyl[2] (Lannate® or Nudrin®) per acre to ½ pt. DIPEL 4L. Increase chlordimeform to ¼ lb. AI if needed.

Once beneficial insects are no longer a factor, DIPEL 4L (at ½ to 1½ pts./acre) may be tank mixed with methomyl (Lannate® or Nudrin®) at .225 to .675 lbs. AI per acre to control light to moderate infestations of early instar larvae. Observe any rate, frequency, or boll load label restrictions for methomyl in your area. Methomyl will likely cause reddening of cotton foliage. Discontinue this tank mix if reddening becomes excessive.

Before using any tank mix, read the product labels for all environmental and usage cautions. Chlordimeform is restricted to aerial application and to states where registered.

DIPEL 4L[2] may be tank mixed with vegetable or cottonseed oil at the ratio of 1 pint 4L to 1 pint oil. DIPEL 4L may also be mixed at a ratio of 40 parts 4L to 60 parts water, and applied in total spray volume not to exceed 80 fl. oz. per acre.

** Full plant coverage needed.

[1] The mixture of B.t. with chlordimeform for inhibiting the growth of lepidopterous larvae on plants is covered by Abbott U.S. patent 3,937,813.

[2] All states except California.

Crop	Pest	Pints/Acre
Avocados[1]	Amorbia Moth	Not more than 8
	Omnivorous Leafroller	
	Omnivorous Looper	
	Orange Tortix	

Use a minimum of 200 gallons water per acre by ground rig or 10 gallons by aircraft. Maximum of two applications per season.

[1] For all states except California.

CROP SEED[1] Pest	Rate
Indian Meal Moth	1½ pts/100 bu.
Almond Moth	

Mix thoroughly prior to bagging.

DIPEL 4L is compatible with common seed treatments including captan, methoxychlor, carboxin (Vitavax) and malathion. Fumigation has not been found to decrease the effectiveness of DIPEL 4L.

[1] For all states except California.

DIPEL 4L FOR TREES AND SHRUBS[2]

Crop	Pest	Pints/100 Gallons* (Ground Equipment)	Pints/Acre (Aerial** Application)
Forest, Shade Sugar Maple Trees and Shrubs[2]	Gypsy Moth	1[1] to 4	2[1] to 4
	Bagworm	1 to 2	1 to 2
	Redhumped Caterpillar	½ to 1	½ to 1
	Spring & Fall Cankerworm	½ to 1	½ to 1
	Fall Webworm	1	1
	Elm Spanworm	1 to 2	1 to 2
	Tent Caterpillars	½ to 1	1[1] to 2
	California Oakworm	½ to 1	½ to 1
	Pine Butterfly	2	2
	Spruce Budworms[3]	2[1] to 4	2[1] to 4
	Saddled Prominent Caterpillar	1 to 2	1 to 2[2]
	Douglas Fir Tussock Moth	2	2
	Western Tussock Moth	1 to 2	—
	Fruittree Leafroller	1 to 2	—
	Blackheaded Budworm	2	—
	Mimosa Webworm	1 to 2	—
	Jack Pine Budworm	1 to 2	1½ to 2
	Saddleback Caterpillar	1 to 2	—

* Rate for hydraulic sprayer in California. In other states, less water may be used if adequate coverage is provided. For mist blowers, mix the applicable amount (pints) in zero to 10 gallons of water: except in California, mix the applicable amount (pints) in 10 gallons of water.

** For aerial application, use in zero to 10 gallons (in California, against all pests except Spruce Budworms, use in 1 to 10 gallons) of water depending on type of density of trees. For best results, spray systems which deliver droplet size of 200 microns or less should be used.

[1] Use only the low rate in California.

[2] All states except in California.

[3] Use rates greater than 2 pints in Northern states for heavy populations. In California, tank mix ratio of water to DIPEL 4L must be no less than 50:50.

NOTICE TO USER

Seller makes no warranty, express or implied, of merchantability, fitness or otherwise concerning use of this product other than as indicated on the label. User assumes all risks of use, storage or handling not in strict accordance with accompanying directions.

®Fundal, Comite, Galecron, Lannate, Nudrin, Bravo, Vitavax, and Dyrene are trademarks of companies other than Abbott Laboratories.

This is a specimen label. Abbott Laboratories is not responsible for the accuracy of the information contained herein. As labels are subject to revision, always carefully read and follow the label on the product container.

DIPEL® 8AF

AQUEOUS FLOWABLE BIOLOGICAL INSECTICIDE

ACTIVE INGREDIENT:
Bacillus thuringiensis, var. *kurstaki*,
14,500 International Units of Potency per mg
(64 billion International Units per gallon) 2.9%
INERT INGREDIENTS 97.1%

TOTAL 100.0%

E.P.A. Registration No. 275-67 E.P.A. Est. No. 33762-IA-1

KEEP OUT OF REACH OF CHILDREN
CAUTION

PRECAUTIONARY STATEMENTS

HAZARDS TO HUMANS
(and Domestic Animals)
CAUTION

Avoid contact with skin, eyes or clothing. In case of contact immediately flush eyes or skin with plenty of water. Get medical attention if irritation persists.

BENEFICIAL INSECTS

Honeybees foraging treated areas are not harmed by DIPEL 8AF use.

DIPEL 8AF does not interrupt the activities of beneficial and predacious arthropods in pest management programs.

STORAGE AND DISPOSAL

Do not contaminate water, food or feed by storage or disposal.

STORAGE: Keep containers tightly closed when not in use. At temperatures below 32°F or above 100°F, DIPEL 8AF should be stored under cover.

PESTICIDE DISPOSAL: Wastes resulting from the use of this product may be disposed of on site or at an approved waste disposal facility.

CONTAINER DISPOSAL: Triple rinse (or equivalent). Then puncture and dispose of in a sanitary landfill, or by other procedures approved by state and local authorities.

DIRECTIONS FOR USE

It is a violation of Federal law to use this product in a manner inconsistent with its labeling.

DIPEL 8AF is a highly selective insecticide for use against listed caterpillars (larvae) of lepidopterous insects. Close scouting and early attention to infestations is highly recommended. Larvae must eat deposits of DIPEL 8AF to be affected. Always follow these directions:

- Treat when larvae are young (early instars) before extensive damage has occurred.
- Larvae must be actively feeding on treated, exposed plant parts.
- Thorough spray coverage is needed to provide a uniform deposit of DIPEL 8AF at the site of larval feeding.
- Under heavy pest population pressure, use the higher label rates and/or consider a second application.
- If attempting to control a pest with a single spray, make the treatment when egg hatch is essentially complete, but before extensive tree damage occurs.
- An approved spreader-sticker may be added to improve weather-fastness of spray deposits. Do not add sticker to the undiluted product.

After eating a lethal dose of DIPEL 8AF, larvae stop feeding within the hour and will die within several days. Dying larvae move slowly, discolor, then shrivel, blacken and die.

DIPEL 8AF is completely water miscible and may be applied through conventional ground or aerial equipment with quantities of water sufficient to provide thorough coverage of infested plant parts. The volume of water needed per acre will depend on weather, spray equipment and plant canopy type. It is recommended that DIPEL 8AF be added to water and not in reverse order. Fill the mix tank with the appropriate quantity of water. Start the mechanical or hydraulic agitation to provide moderate circulation before adding DIPEL 8AF. Include rinse water from containers. Always add the sticker to the water before adding DIPEL 8AF. Add the required volume of DIPEL 8AF to the mix tank or plane hopper and agitate until uniformly suspended. Mild agitation is sufficient to maintain mixture suspension during loading and spraying. Do not mix more DIPEL 8AF that can be used in a 144 hour period.

CAUTION: Rinse and flush spray equipment thoroughly with water following each use.

DIPEL 8AF can also be applied undiluted from aircraft for control of Spruce Budworms, Jack Pine Budworm and Gypsy Moth. It is recommended that rotary or other atomizers be used to provide droplet Volume Mean Diameters (VMD) of 20 to 80 micorns for needle conifers and 50-150 microns for deciduous hardwoods. Do not spray undiluted DIPEL 8AF under less than 50% relative humidity. After prolonged storage, undiluted DIPEL 8AF should be recirculated once to redistribute prior to use. During loading and spraying, agitation of the product is unnecessary and should be avoided.

DIPEL 8AF for Trees and Shrubs

Crop	Pest	Ounces/ Acre* (Ground Equipment)	Ounces/ Acre (Aerial** Application)	BIU's/ Acre
Forest, Shade, Sugar Maple, Trees & Shrubs	Bagworm	8 to 16	8 to 16	4 to 8
	Blackheaded Budworm	16	—	8
	Browntail Moth	16 to 40	16 to 40	8 to 20
	California Oakworm	8	8	4
	Douglas Fir Tussock Moth	16 to 32	16 to 32	8 to 16
	Elm Spanworm	8 to 16	8 to 16	4 to 8
	Fall Webworm	8	8	4
	Fruittree Leafroller	8 to 16	—	4 to 8
	Green Striped Maple Worm	16 to 24	16 to 24	8 to 12
	Gypsy Moth	16 to 40	16 to 40	8 to 20
	Jack Pine Budworm	16 to 16	16 to 32	8 to 16
	Mimosa Webworm	8 to 16	—	4 to 8
	Oak Leaf Tier	16 to 24	16 to 24	8 to 12
	Pine Butterfly	16	16	8
	Redhumped Caterpillar	8 to 16	8 to 16	4 to 8
	Saddleback Caterpillar	8 to 16	—	4 to 8
	Saddled Prominent Caterpillar	8 to 16	8 to 16	4 to 8
	Spring & Fall Cankerworm	8 to 16	8 to 16	4 to 8
	Spruce Budworms[1]	16 to 40	16 to 40	8 to 20
	Tent Caterpillars	8 to 16	8 to 16	4 to 8
	Western Tussock Moth	8 to 16	—	4 to 8

* Water dilution rate for hydraulic sprayer may be varied depending on coverage. For mist blowers, mix the applicable amount (pts.) in up to 10 gallons of water

** For diluted aerial application, use in up to 10 gallons of water depending on type and density of trees. For best results, spray systems which deliver droplet size of 150 microns or less should be used.

[1] Use rates greater than 1 pint in Northern states for heavy populations.

NOTICE TO USER

Seller makes no warranty, express or implied, of merchantability, fitness or otherwise concerning use of this product other than as indicated on the label. User assumes all risks of use, storage or handling not in strict accordance with accompanying directions.

This is a specimen label. Abbott Laboratories is not responsible for the accuracy of the information contained herein. As labels are subject to revision, always carefully read and follow the label on the product container.

DIPEL® 8L

EMULSIFIABLE SUSPENSION BIOLOGICAL INSECTICIDE

ACTIVE INGREDIENT:
Bacillus thuringiensis, var. *kurstaki*,
17,600 International Units of Potency per mg
(64 Billion International Units per gallon) 3.5%
INERT INGREDIENTS: 96.5%

TOTAL 100.00%

EPA Reg. No. 275-51 EPA Est. No. 33762-IA-1

CAUTION
KEEP OUT OF REACH OF CHILDREN

PRECAUTIONARY STATEMENTS

HAZARDS TO HUMANS (AND DOMESTIC ANIMALS)

CAUTION

Avoid contact with skin, eyes or clothing. In case of contact immediately flush eyes or skin with plenty of water. Get medical attention if irritation persists.

BENEFICIAL INSECTS

Honeybees foraging treated areas are not harmed by DIPEL 8L use. DIPEL 8L does not interrupt the activities of beneficial and predacious arthropods in pest management programs.

STORAGE AND DISPOSAL

Do not contaminate water, food or feed by storage or disposal.

STORAGE

Keep containers tightly closed when not in use. Do not store at temperatures below 0°F or above 100°F. Roll or shake the drum before dispensing.

PESTICIDE DISPOSAL

Wastes resulting from the use of this product may be disposed of on site or at an approved waste disposal facility.

CONTAINER DISPOSAL

Triple rinse (or equivalent). Then puncture and dispose of in a sanitary landfill, or by other procedures approved by state and local authorities.

DIRECTIONS FOR USE

It is a violation of Federal Law to use this product in a manner inconsistent with its labeling.

DIPEL 8L is a highly selective insecticide for use against listed caterpillars (larvae) of lepidopterous insects. Close scouting and early attention to infestations is highly recommended. Larvae must eat deposits of DIPEL 8L to be affected. Always follow these directions:

- Treat when larvae are young (early instars) before extensive damage has occurred.
- Larvae must be actively feeding on treated, exposed plant parts.
- Thorough spray coverage is needed to provide a uniform deposit of DIPEL 8L at the site of larval feeding.
- Under heavy pest population pressure, use the higher label rates and/or consider a second application.
- If attempting to control a pest with a single spray, make the treatment when egg hatch is essentially complete, but before extensive crop damage occurs.
- An approved spreader-sticker may be added to diluted tank mixes to improve weather-fastness of the spray deposits. Do not add sticker to the undiluted product.

After eating a lethal dose of DIPEL 8L, larvae stop feeding within the hour and will die within several days. Dying larvae move slowly, discolor, then shrivel, blacken and die.

DIPEL 8L may be applied in conventional ground or aerial equipment with quantities of water sufficient to provide thorough coverage of infested plant parts. The amount of water needed per acre will depend on weather, spray equipment and plant canopy type. DIPEL 8L should always be mixed with at least an equal amount of water for diluted applications. Fill the mix tank or plane hopper with the desired quantity of water. Include rinse water from containers. Start the mechanical or hydraulic agitation to provide moderate circulation before adding DIPEL 8L. Always add the sticker to the water prior to the addition of DIPEL 8L. Add the desired volume of DIPEL 8L to the tank or plane hopper and continue circulation. **Never add DIPEL 8L to the mix tank before introducing the desired quantity of water.** Maintain the suspension while loading and spraying. Do not mix more DIPEL 8L than can be used in a 144-hour period.

DIPEL 8L can also be applied undiluted from aircraft for control of Spruce Budworms, Jack Pine Budworm and Gypsy Moth. It is recommended that rotary or other atomizers be used to provide droplet Volume Mean Diameters (VMD) of 20-80 microns for needle conifers and 50-150 microns for deciduous hardwoods. **CAUTION:** Rinse and flush spray equipment thoroughly following each use. Use oil solvent for flushing and rinsing undiluted DIPEL 8L.

DIPEL 8L for Trees and Shrubs

Crop	Pest	Ounces/Acre* (Ground Equipment)	Ounces/Acre (Aerial** Application)	BIU's/Acre
Forest, Shade, Sugar Maple, Trees & Shrubs	Bagworm	8 to 16	8 to 16	4 to 8
	Blackheaded Budworm	16	—	8
	Browntail Moth	16 to 40	16 to 40	8 to 20
	California Oakworm	8	8	4
	Douglas Fir Tussock Moth	16 to 32	16 to 32	8 to 16
	Elm Spanworm	8 to 16	8 to 16	4 to 8
	Fall Webworm	8	8	4
	Fruittree Leafroller	8 to 16	—	4 to 8
	Green Striped Maple Worm	16 to 24	16 to 24	8 to 12
	Gypsy Moth	16 to 40	16 to 40	8 to 20
	Jack Pine Budworm	16 to 16	16 to 32	8 to 16
	Mimosa Webworm	8 to 16	—	4 to 8
	Oak Leaf Tier	16 to 24	16 to 24	8 to 12
	Pine Butterfly	16	16	8
	Redhumped Caterpillar	8 to 16	8 to 16	4 to 8
	Saddleback Caterpillar	8 to 16	—	4 to 8
	Saddled Prominent Caterpillar	8 to 16	8 to 16	4 to 8
	Spring & Fall Cankerworm	8 to 16	8 to 16	4 to 8
	Spruce Budworms[1]	16 to 40	16 to 40	8 to 20
	Tent Caterpillars	8 to 16	8 to 16	4 to 8
	Western Tussock Moth	8 to 16	—	4 to 8

* Water dilution rate for hydraulic sprayer may be varied depending on coverage. For mist blowers, mix the applicable amount (pts.) in up to 10 gallons of water.

** For aerial application, use in up to 10 gallons of water depending on type and density of trees. For best results, spray systems which deliver droplet VMD of 150 microns or less should be used. DIPEL 8L should always be mixed with at least an equal amount of water for diluted applications.

[1] Use rates greater than 1 pint in Northern states for heavy populations. DIPEL 8L may be sprayed undiluted for the control of Spruce Budworms, Jack Pine Budworm and Gypsy Moth.

NOTICE TO USER

Seller makes no warranty, express or implied, of merchantability, fitness or otherwise concerning use of this product other than as indicated on the label. User assumes all risks of use, storage or handling not in strict accordance with accompanying directions.

This is a specimen label. Abbott Laboratories is not responsible for the accuracy of the information contained herein. As labels are subject to revision, always carefully read and follow the label on the product container.

DIPEL® 10G

BIOLOGICAL INSECTICIDE

ACTIVE INGREDIENT:
Bacillus thuringiensis, var. *kurstaki*,
1600 International Units of Potency per mg* 0.32%
INERT INGREDIENTS: 99.68%
TOTAL 100.00%

*This product contains 0.726 Billion IU/lb.

EPA Reg. No. 275-55

CAUTION

KEEP OUT OF REACH OF CHILDREN

PRECAUTIONARY STATEMENTS

HAZARDS TO HUMANS

CAUTION

Avoid inhalation or contact with eyes.

ENVIRONMENTAL HAZARDS

Keep out of lakes, ponds or streams. Do not contaminate water by cleaning of equipment or disposal of wastes.

DIRECTIONS FOR USE

It is a violation of Federal law to use this product in a manner inconsistent with its labeling.

GENERAL INFORMATION

DIPEL® 10G is a granular formulation of **Bacillus thuringiensis,** variety **kurstaki,** designed specifically as a selective insecticide for control of certain leaf-eating caterpillars (Lepidoptera). For practical determinations of active ingredient, DIPEL 10G can be expressed as a 10% formula of DIPEL WP.

The active ingredient in DIPEL 10G is a stomach poison and therefore must be eaten by larvae to be effective. Climatic extremes such as heavy rain or strong wind shortly after treatment may act to remove DIPEL 10G from the feeding zone making re-treatment necessary.

STORAGE AND DISPOSAL

Do not contaminate water, food, or feed by storage or disposal.

		Application:	
Crop	Pest	Rate	Comments
Field Corn Popcorn, Seed Corn, and Sweet Corn	European Corn Borer, Southwestern Corn Borer, First and Second Generation	10 lbs/acre or 5-10 lbs/acre banded over-row with ground application	**Ground or aerial equipment** First Generation: Apply DIPEL 10G directly to corn whorls. Treat seed corn when 25 percent of plants show shot hole feeding in the whorls and field corn when 35 percent of whorls show feeding. Larvae should be small and present at this time. Satisfactory control cannot be expected if treatment is delayed to higher injury levels or after larvae have begun stalk boring. Second Generation: Apply DIPEL 10G when a field count shows 50 egg masses per 100 plants and the first hatch is occurring or follow your State Extension Service recommendation. Under conditions of inten-

Application:

Crop	Pest	Rate	Comments
			sive population pressure, a second application may be necessary.
	Fall Armyworm	10 lbs/acre	**Ground or aerial equipment** Apply DIPEL 10G directly to corn whorls. Treat sweet corn when 10-12 percent of the plants show feeding in the whorls. Field corn should be treated when 75 percent of the plant whorls show feeding damage. Treatment thresholds vary for different types of corn so consult State Agricultural Extension specialists for specific recommendations. Under high pest pressure, a second application may be necessary.
Tobacco	Tobacco Budworm	5-10 lbs./acre	**Ground equipment** Apply DIPEL 10G by hand or properly calibrated mechanical ground equipment.

STORAGE

Store in a cool, dry place.

PESTICIDE DISPOSAL

Wastes resulting from the use of this product may be disposed of on site or at an approved waste disposal facility.

CONTAINER DISPOSAL

Completely empty bag in application equipment. Then dispose of empty bag in a sanitary landfill or by incineration, or, if allowed by State and local authorities, by burning. If burned, stay out of smoke.

NOTICE TO USER

Local conditions may affect the use of DIPEL 10G. Consult State Agricultural Extension or Experiment Station Specialist for specific recommendations related to local crop protection problems. Abbott makes no warranty, express or implied, including the warranties of commercially acceptable quality and/or fitness for any particular purpose concerning this material, other than those which are contained on this label.

This is a specimen label. Abbott Laboratories is not responsible for the accuracy of the information contained herein. As labels are subject to revision, always carefully read and follow the label on the product container.

PROGIBB® PLUS 2X

SOLUBLE POWDER
(Gibberellic Acid)

ACTIVE INGREDIENT:	W/W
Gibberellin A	20%
INERT INGREDIENTS:	80%
TOTAL	100%

EPA Reg. No. 275-62 EPA Est. No. 48477-CA-02

LABEL EFFECTIVE DATE: 10/87

CAUTION
KEEP OUT OF REACH OF CHILDREN

PRECAUTIONARY STATEMENTS

HAZARDS TO HUMANS (AND DOMESTIC ANIMALS)

CAUTION

Powder causes eye irritation. In case of contact with eyes flush thoroughly with water.

Do not apply this product in such a manner as to directly or through drift expose workers or other persons. The area being treated must be vacated by unprotected persons.

REENTRY STATEMENT

Do not enter treated areas without protective clothing until sprays have dried.

Because certain states may require more restrictive reentry intervals for various crops treated with this product, consult your State Department of Agriculture for further information.

ENVIRONMENTAL HAZARDS

Keep out of lakes, ponds or streams. Do not contaminate water by cleaning of equipment or disposal of wastes.

DIRECTIONS FOR USE

GENERAL CLASSIFICATION

It is a violation of Federal law to use this product in a manner inconsistent with its labeling.

Do not apply this product through any type of irrigation system.

IMPORTANT

Before application, read accompanying PROGIBB Spray Guide carefully and use only as directed.

STORAGE AND DISPOSAL

Do not contaminate water, food or feed by storage or disposal.

STORAGE

Keep containers tightly closed when not in use.

PESTICIDE DISPOSAL

Wastes resulting from the use of this product may be disposed of on site or at an approved waste disposal facility.

CONTAINER DISPOSAL

Do not reuse empty containers. Triple rinse (or equivalent). Then puncture and dispose of in a sanitary landfill, or by incineration, or, if allowed by state and local authorities, by burning. If burned, stay out of smoke.

NOTICE TO USER

Seller makes no warranty, express or implied, of merchantability, fitness or otherwise concerning the use of this product other than as indicated on the label. User assumes all risks of use, storage or handling not in strict accordance with accompanying directions.

This is a specimen label. Abbott Laboratories is not responsible for the accuracy of the information contained herein. As labels are subject to revision always carefully read and follow the label on the product container.

PROGIBB® 4% LIQUID

CONCENTRATE
(Gibberellic Acid)

ACTIVE INGREDIENT:	W/W
Gibberellic Acid	4%
INERT INGEDIENTS:	96%
TOTAL	100%

EPA Reg. No. 275-61 EPA Est. No. 35347-CA-1

LABEL EFFECTIVE DATE: 10/87

CAUTION
KEEP OUT OF REACH OF CHILDREN

PRECAUTIONARY STATEMENTS

HAZARDS TO HUMANS (AND DOMESTIC ANIMALS)

CAUTION

PROGIBB 4% Liquid Concentrate may cause eye irritation and is harmful if swallowed or if vapors are breathed for prolonged periods. In case of contact with eyes, flush thoroughly with water.

Do not apply this product in such a manner as to directly or through drift expose workers or other persons. The area being treated must be vacated by unprotected persons.

REENTRY STATEMENT

Do not enter treated areas without protective clothing until sprays have dried. Because certain states may require more restrictive reentry intervals for various crops treated with this product, consult your State Department of Agriculture for further information.

ENVIRONMENTAL HAZARDS

Keep out of lakes, ponds or streams. Do not contaminate water by cleaning of equipment or disposal of wastes.

PHYSICAL OR CHEMICAL HAZARDS

FLAMMABLE: Keep away from heat and open flame.

DIRECTIONS FOR USE

GENERAL CLASSIFICATION

It is a violation of Federal law to use this product in a manner inconsistent with its labeling.

Do not apply this product through any type of irrigation system.

IMPORTANT
Before application, read accompanying PROGIBB Spray Guide carefully and use only as directed.

STORAGE AND DISPOSAL
Do not contaminate water, food or feed by storage or disposal.

STORAGE
Keep containers tightly closed when not in use. Keep away from heat and open flame.

PESTICIDE DISPOSAL
Wastes resulting from the use of this product may be disposed of on site or at an approved waste disposal facility.

CONTAINER DISPOSAL
Do not reuse empty containers. Triple rinse (or equivalent). Puncture and dispose of in a sanitary landfill, or by incineration or, if allowed by state and local authorities, by burning. If burned, stay out of smoke.

NOTICE TO USER
Seller makes no warranty, express or implied, of merchantability, fitness or otherwise concerning the use of this product other than as indicated on the label. User assumes all risks of use, storage or handling not in strict accordance with accompanying directions.

This is a specimen label. Abbott Laboratories is not responsible for the accuracy of the information contained herein. As labels are subject to revision, always carefully read and follow the label on the product container.

PROGIBB®

SPRAYING GUIDE

PROGIBB Plus 2X Soluble Powder (160 grams/bottle)

ACTIVE INGREDIENT:
Gibberellic Acid* . . . . 20% W/W

Equivalent to 32 grams* of Gibberellic Acid per bottle.

EPA Reg. No.: 275-62

PROGIBB 4% Liquid Concentrate (20 fl. oz./bottle and 128 fl. oz./bottle)

ACTIVE INGREDIENT:
Gibberellic Acid* . . . . 4% W/W

Equivalent to approximately 1.0 gram* of Gibberellic Acid per fluid ounce of product.

EPA Reg. No.: 275-61

(Gibberellic Acid)

CAUTION

KEEP OUT OF REACH OF CHILDREN

PRECAUTIONARY STATEMENTS

HAZARDS TO HUMANS (AND DOMESTIC ANIMALS)
May cause eye irritation and is harmful if swallowed or if vapors are breathed for prolonged periods.

STATEMENT OF PRACTICAL TREATMENT

IF IN EYES: Immediately flush eyes with plenty of water. Get medical attention if irritation persists.
IF SWALLOWED: Seek medical aid.
IF ON SKIN: Immediately flush skin with plenty of water. Get medical attention if irritation persists.
IF INHALED: Move to fresh air.

PHYSICAL OR CHEMICAL HAZARDS
FLAMMABLE! Keep away from heat and open flame.

ENVIRONMENTAL HAZARDS
Keep out of lakes, ponds or streams. Do not contaminate water by disposal of waste or cleaning of equipment.

DIRECTIONS FOR USE
It is a violation of Federal law to use this product in a manner inconsistent with its labeling.

Do not apply this product in such a manner as to directly or through drift expose workers or other persons. The area being treated must be vacated by unprotected persons.

Do not apply this product through any type of irrigation system.

STORAGE AND DISPOSAL
See container label.

REENTRY STATEMENT
Do not enter treated areas without protective clothing until sprays have dried.

Written or oral warnings must be given to workers who are expected to be in a treated area or in an area about to be treated with this product. Oral warnings must include the following information:

> Inform workers of areas or fields that must not be entered without appropriate protective clothing until sprays have dried. In case of accidental exposure, wash with plenty of water. If there is any irritation in eyes after washing, get medical attention.

When oral warnings are given, warnings shall be given in language customarily understood by workers. Oral warnings must be given if there is reason to believe that written warnings cannot be understood by workers. Written warnings must include the following information:

> Area treated with PROGIBB on (date of application). Do not enter without appropriate protective clothing until sprays have dried. In case of accidental exposure, wash with plenty of water. If there is any irritation in eyes after washing, get medical attention.

NOTE:
Gibberellic Acid is an extremely potent plant growth regulator. For best results, read all directions for use thoroughly. Consult your local experiment station specialist, distributor, or the Abbott agricultural specialist in your area for the spray schedule best suited to your conditions.

GENERAL DIRECTIONS FOR USE
Discard any unused spray material at the end of each day. Prepare solution concentrations by mixing the required amount of product with water only in a clean, empty spray tank. For best results, applications should be made during cooler parts of the day.

Use only as directed. Good spray practices should be followed. The label should be read thoroughly and understood before making applications. Effectiveness requires that all parts of plant or crop must receive spray or desired result will not occur, so spray thoroughly. When a range of rates is indicated, use the concentration and spray volume recommended locally.

Gibberellic Acid is a naturally occurring compound, produced by Abbott Laboratories in a biological process.

Data concerning the compatibility of PROGIBB with other agricultural compounds except Dipel® 2X is not available.

SPRAY GUIDELINES FOR GRAPES
For all grapes, application is recommended by ground sprayer. Use 100 to 500 gallons as a dilute spray according to foliage density, or 30 to 80 gallons as a concentrate spray, unless specified otherwise. Do not exceed maximum rates. It is important to wet all berries thoroughly.

THOMPSON SEEDLESS
For cluster elongation ("Stretch"), looser cluster forms, and reducing cost of thinning, when used in conjunction with established girdling and thinning practices.

- **Guide:** Apply 8 to 16 grams*/A before bloom when flower clusters are 3 to 5 inches long.

For decreased berry set ("Thinning"), reducing hand-thinning costs, and hastened maturity.

- **Guide:** Apply 8 to 16 grams*/A during bloom as one application or as two applications of equal amounts when the bloom period is extended with the second made 3 to 7 days after the first application.

For larger berries ("Sizing") and larger clusters when used in conjunction with established girdling and thinning practices.

- **Guide:** Apply 32 to 80 grams*/A when average berry size is 4 to 5 millimeters in diameter or as two applications of equal amounts with the first made at or 2 to 3 days after shatter, followed during the next two weeks by the second application. Timing of the second spray will be dictated by experience in the vineyard to be sprayed and temperatures occurring during the interim between sprays. Potential effect will be reduced if the second spray occurs more than two weeks after the first application.

THOMPSON SEEDLESS FOR RAISINS
For decreasing berry set, with increased raisin quality, and hastened maturity.

- **Guide:** Apply 0.75 to 6 grams*/A when most bunches are in 60% to 80% bloom.

FLAME SEEDLESS
For decreased berry set ("Thinning") and reducing hand-thinning costs.

- **Guide:** Apply 3 to 7.5 grams*/A during bloom.

Higher amounts may cause an excess of shot berries or overthinning.

For larger berries ("Sizing") and larger clusters when used in conjunction with established girdling and thinning practices.

- **Guide:** Apply 20 to 48 grams*/A as one application when berry diameter reaches 6 to 8 millimeters, or as two applications of equal amounts with the first made when berry diameter reaches 6 to 8 millimeters, followed during the next 5 to 10 days by the second application. Timing of the second spray will be dictated by experience in the vineyard to be sprayed and rate of berry growth during the interim between sprays.

PERLETTES
For larger berries ("Sizing") and larger clusters when used in conjunction with established girdling and thinning practices.

• **Guide:** Apply 32 to 80 grams*/A when average berry size is 4 to 5 millimeters in diameter or as two applications of equal amounts with the first made at or 2 to 3 days after shatter, followed during the next two weeks by the second application. Timing of the second spray will be dictated by experience in the vineyard to be sprayed and temperatures occurring during the interim between sprays. Potential effect will be reduced if the second spray occurs more than two weeks after the first application.

OTHER SEEDLESS VARIETIES SUCH AS SEEDLESS TOKAY, INTERLOCKEN SERIES AND RELATED HYBRIDS

For larger berries and larger clusters when used in conjunction with established girdling and thinning practices.

• **Guide:** Apply 8 to 48 grams*/A as one application at or just after shatter (usually 2 to 3 days later) or as two applications of equal amounts not to exceed a total of 48 grams*/A, with the first made at or just after shatter, followed during the next two weeks by the second application. Timing of the second spray with split application will be dictated by experience in the vineyard to be sprayed and temperatures occurring during the interim between sprays. Potential effect will be reduced if the second spray occurs more than two weeks after the first application.

EMPEROR

For reducing berry shrivel. This use can also increase berry size.

• **Guide:** Apply 20 grams*/A as one application in 200 to 250 gallons/A approximately two weeks after completion of shatter following bloom. This timing should correspond to a period when the predominant berry diameter ranges from 10 to 15 millimeters.

BLACK CORINTH (ZANTE CURRANT)

For improving berry size.

• **Guide:** Apply spray containing 1 to 8 grams*/A 3 to 5 days after full bloom, but before shatter begins.

CONCORD

(Arkansas, Michigan, New York, Ohio and Pennsylvania)

For cluster elongation (Stretch), looser cluster forms, increased berry size, reduced number of green berries, increased soluble solids content, and increased yields, when used in conjunction with established girdling and thinning practices and a first-bloom application of daminozide (Alar®-85) to increase berry set.

• **Guide:** Apply 40-80 grams*/A in a postbloom spray at the berry shatter stage. Grape vines should have received a first-bloom application of daminozide (Alar®-85) at the recommended rate of 1 lb./A Alar®-85. See current Alar®-85 label for precautionary statements and other specific recommendations.

Applications should not be made to vines considered to be in low vigor.

Apply in sufficient water to give uniform and complete coverage.

®Alar is the trademark of a company other than Abbott Laboratories.

SPRAY GUIDELINES FOR CITRUS

NAVEL ORANGES (California)

To delay aging of the rind and reduce rind disorders (e.g., rind staining, water spotting, sticky or tacky surface, puffy rind and rupture under pressure) and to produce a more orderly harvesting pattern.

EARLY SPRAY (before any color change).

Apply to groves where harvest is not anticipated before March 1. The delay in rind aging is greatest when the early spray is applied before a color change. This spray timing produces the firmest rind possible.

• **Guide:** Apply one spray two weeks prior to color break which normally occurs August to November. Apply until February 15 in San Joaquin Valley, March 1 in Southern California. On large mature trees, apply 10 to 40 grams*/A in 400 to 500 gallons/A dilute, or 50 to 100 gallons/A concentrate.

NOTE: Do not apply to groves that may be harvested before March 1 as a reduction in grade may result due to the delayed coloring. Do not apply in white wash sprays in which lime or other caustic material has produced a high pH in the spray tank.

LATE SPRAY (after marketable color is reached).

Apply to groves where harvest may be before March 1 (or not known).

• **Guide:** Apply one spray just after marketable color has developed which is normally from October through December. On large mature trees, apply 16 to 48 grams*/A in 400 to 500 gallons/A dilute, or 50 to 100 gallons/A concentrate.

NOTE: Sprays applied in late January/February may cause reduced production the following year. Do not apply within 10 days of harvest. Do not spray navel orange trees between February 15 and August 1.

VALENCIA ORANGES (California)

To reduce rind creasing and to delay aging and softening of the rind.

• **Guide:** Apply a single spray in August or September to trees with a target crop of young fruit. On large mature trees, apply 40 to 80 grams*/A in approximately 500 gallons/A dilute or 100 gallons/A concentrate.

NOTE: Some increased regreening, or slower color development, should be expected in the target crop. Some increased regreening of mature fruit, if present, may occur.

* Refers to actual Gibberellic Acid. See Conversion Table to convert to amount of formulated Progibb needed.

LEMONS (California)

To decrease the amount of small tree ripe fruit and to produce a more desirable production pattern in relation to market demand.

• **Guide:** Apply in a single spray when target crop is 1/2 to 3/4 full size but still green. Typically October through December. Use 20 Grams* in 500-700 gallons/A on large mature trees.

When applied two years in a row, an even larger difference in harvest pattern and maturity occurs.

NOTE: Do not apply within one month of harvest. Do not apply in spring or summer.

TANGERINE HYBRIDS (Florida)

To increase fruit set and yields on tangerine hybrids with pollination problems such as the Orlando, Robinson, and Minneola.

• **Guide:** Apply spray during full bloom. Be sure to wet the leaves sufficiently.

Fruits are generally seedless. Use 8 to 30 grams* in 400 to 500 gallons/A on large mature trees.

NOTE: A slight increase in mature leaf drop occurs at concentrations above 25 ppm. Fruit sizes may be reduced and color development slightly retarded.

(California)

To delay disorders associated with rind aging of the Minneola tangelo; e.g., puffiness and softening, and to increase peel strength.

• **Guide:** Apply 20 to 40 grams*/A in 400-500 gallons dilute spray two weeks prior to color break.

NOTE: Do not apply if early harvest is planned. Do not apply after coloring as pre-harvest rind staining may occur. Application during coloring may cause variation in rind color development.

GRAPEFRUIT (Florida and Texas)

To delay disorders associated with rind aging: e.g., puffiness, softening, and orange coloration, to prevent preharvest drop of mature fruit, and to increase peel strength and reduce water loss during storage.

• **Guide:** Apply a single spray to fully colored fruit during the November through January period. Use 20 to 56 grams* in 500 to 700 gallons/A containing a suitable non-ionic surfactant at the manufacturer's recommended rate. It is advisable to spot pick heavy crops to aid early marketing and to avoid reduction of yields which generally follow late held crops.

NOTE: Applications made after January or when trees begin to break dormancy may adversely affect new crop. Do not use concentrate sprays. Results may vary season to season depending on environmental conditions.

GRAPEFRUIT, STAR RUBY VARIETY (Texas)

To reduce early-season drop of small fruit of Star Ruby Variety thereby increasing yields.

• **Guide:** Apply a single spray during the bloom period.

Use PROGIBB 4% Liquid Concentrate. Use 25 fluid ounces (1 1/4 20-ounce bottles) (25 grams*) in 250 gallons water final spray mixture per acre. A suitable surfactant may be used to enhance efficacy.

NOTE: Do not tank-mix with other chemicals.

Do not apply concentrated solution.

Results may vary season to season depending on environmental conditions.

Maintain a well-balanced fertilization and watering program.

SPRAY GUIDELINES FOR FRUIT CROPS

BLUEBERRIES

For improving fruit set. For set problems due to insufficient natural honeybee pollination on varieties such as Coville, Jersey, Stanley, Earlieblue, Weymouth and others.

• **Guide:** Make a single foliage spray application at full bloom (when over 75 percent of all flowers are fully open). For Weymouth, application can be delayed up to two weeks after full bloom to affect sizing of shot berries.

Use PROGIBB 4% Liquid Concentrate. Mix 80 fluid ounces in 100 gallons of water. Use of a spreader-sticker is recommended. Apply to the point of run-off, thoroughly wetting all parts of the plant. Total gallonage will depend on size and density of the plants.

NOTE: Do not exceed 300 gallons/A. Although some varieties bloom closer to harvest than others—in no case should application be made closer than 40 days before harvest. Do not apply to plants in a low state of vigor.

SWEET CHERRIES

To delay harvesting, to produce a brighter colored, firmer fruit, and to increase size.

• **Guide:** Apply spray when the fruit is light green to straw colored. Apply spray to thoroughly wet the entire tree. Use 16 to 48 grams* in 400 to 600 gallons/A on large mature trees.

NOTE: Do not apply within one week of harvest.

RED TART CHERRIES (All states except California)

To maintain and extend high fruiting capacity of bearing tart cherry trees and reduce occurrence of "blind" nodes by stimulating lateral vegetative buds to develop a more productive balance of lateral shoots and spurs.

- **Guide:** Apply a foliar spray containing 4 to 8 fluid ounces of PROGIBB 4% in 100 gallons finished spray from 14 to 28 days after bloom (or up to 14 days after shuck split). Use full coverage sprays of 50 to 150 gal/A on medium to large bearing trees. Be sure entire trees receive good coverage. Use of a good horticultural wetting agent at the manufacturer's recommended rate will aid foliar wetting. PROGIBB must be applied annually to insure vegetative development and subsequent yield improvement year after year.

Note: PROGIBB works by affecting lateral bud differentiation which is apparent the year *after* application. Therefore, changes in shoot, spur, and flower production will not be evident until 2 or 3 years after program initiation. Once this period is satisfied, response will be yearly provided annual applications have been made.

NOTE: Do not spray within one month of harvest. Adjust PROGIBB rate to complement vigor of trees. If trees are vigorous, use lowest recommended rates. Use higher rate for trees low in vigor and showing weak shoot and spur production. Excessive application rates on any tree will increase vegetative growth at the expense of fruit production the following year.

PROGIBB will *not* improve growth of trees under stress (nutritional, moisture, winter injury) or other factors inhibiting normal growth and development resulting from physical damage or unsound orchard practices. Best results from PROGIBB will be obtained when combined with good cultural practices.

YOUNG TART AND SWEET CHERRY TREES

(All states except California)

To reduce flowering and fruiting in young tart and sweet cherry trees to minimize the competitive effect of early fruiting on tree development.

- **Guide:** Apply PROGIBB Liquid Concentrate two to four weeks after bloom. Mix 20 to 40 ounces of PROGIBB 4% Liquid Concentrate in 100 gallons of water. Apply a foliar spray of 25 to 50 gallons per acre, assuming a tree density of 100 trees per acre equivalent, or apply about one quart of spray volume per tree.

* Refers to actual Gibberellic Acid. See Conversion Table to convert to amount of formulated PROGIBB needed.

Under conditions of low vigor two applications are recommended. If two spray applications are made, allow at least a seven-day interval between sprays.

NOTE: DO NOT SPRAY TREES IN THE FIRST YEAR. Treat in the second season for reduction of flowering in the third season, and again in the third season if reduction of flowering and fruiting is desired in the fourth season.

OLYMPUS STRAWBERRIES

(N.W. US Only; propagation stock)

To increase runner production of mother plants of the Olympus cultivar.

- **Guide:** Apply a single spray to mother plants 10 to 30 days after planting. At the time of spraying, plants should have 1 to 6 leaves. Apply 100 gallons/A to thoroughly wet new foliage to the point of run-off. Use 20 grams*/A.

NOTE: Not for use on fruiting plants. Treatments may not be effective on plantings set out after mid-May.

FORCING RHUBARB

To increase yield of marketable forced rhubarb and to break dormancy on plants receiving insufficient chilling.

- **Guide:** Apply 2 fluid ounces (60 ml) of a solution containing 20 grams* in 10 gallons to each cleaned crown, when the rest period is not completely broken. When the rest period is broken by cold weather, apply 2 fluid ounces (60 ml) of a solution containing 10 grams* in 10 gallons.

NOTE: Keep forcing house temperatures at 40° to 50°F for 24 hours after application. If house is warmer than 50°F, the crowns should be covered with plastic. Temperatures in the forcing house above 50°F will result in lower yields and poor stalk color.

SPRAY GUIDELINES FOR VEGETABLE CROPS

ARTICHOKES (California)

To accelerate maturity of artichokes and to shift the harvest to an earlier date.

- **Guide:** Apply spray at bud initiation time, normally six weeks prior to anticipated harvest. Be sure the entire plant (leaves, stems and buds) are covered to point of run-off. Use 10 grams* in 100 to 125 gallons/A.

NOTE: Do not apply within seven days of harvest.

CELERY

To increase plant height and yield and overcome stress due to cold weather conditions, or saline soils and to obtain earlier maturity.

- **Guide:** Apply spray one to four weeks prior to harvest. Lower concentrations are applied at the three to four-week interval. Higher concentrations at the one to two-week interval. Use 2.5 to 10 grams* in 25 to 50 gallons/A.

NOTE: Do not apply earlier than four weeks before harvest as Gibberellic Acid may induce bolting (seed stalk formation).

Applications made less than one week preharvest may result in residues.

Celery plants must be harvested when mature to ensure quality.

LETTUCE FOR SEED

To obtain uniform bolting and increase seed production.

- **Guide:** Apply the following spray schedule:

Growth Stage	ppm*	g.*/A	Gallon/Acre
4 leaf stage	10	0.4	10
8 leaf stage	10	1.6	40
12 leaf stage	10	4	100

NOTE: Do not feed crop wastes to livestock.

SEED POTATOES

To stimulate uniform sprouting—for maximum production, more uniform development, fewer late maturing plants, and to break dormancy of newly harvested potatoes that have not had a full rest period.

- **Guide:** Dip freshly dug seed pieces in a solution containing 0.2 to 0.4 gram* in 100 gallons prior to planting.

NOTE: If soil temperature is very high, avoid treating rested seed and use the minimum concentration for dormant seed.

SPINACH (All states except California)

To facilitate harvest, increase yield and improve quality of fall and over-winter spinach.

- **Guide:** Apply a single spray 10 to 14 days before each anticipated harvest on fall or over-winter spinach, ideally when daytime temperatures are 40° to 70°F and during early morning hours when dew is present on crop.

Use PROGIBB 4% Liquid Concentrate.

Mix 6 to 8 fluid ounces/A (6 to 8 grams*/A) in 10 to 50 gallons/A by ground sprayer or in a minimum of 5 to 10 gallons/A by air.

Maximum benefit from PROGIBB is obtained when below normal temperatures predominate following application and growth would be otherwise slowed in untreated spinach.

NOTE: Since Gibberellic Acid can promote bolting, do not apply to spinach after the mid-winter period or if temperatures may be expected to exceed 75°F within several days of application. Do not apply on spring-planted spinach.

SPRAY GUIDELINES FOR FLORICULTURE CROPS

POMPOM CHRYSANTHEMUMS (Florida)

For elongating peduncles on pompom chrysanthemums.

- **Guide:** Apply a single spray 4 to 5 weeks after initiation of short day conditions.

Use PROGIBB 4% Liquid Concentrate. Use 1/2 to 1 fluid ounce (1/2 to 1 gram*) in 12 gallons for application to 1,000 sq. ft. of bed (20 to 40 fluid ounces equivalent to 20 to 40 grams* in 500 gallons/A).

Apply with overhead nozzles directing the spray to the flower buds.

NOTE: Overuse or incorrect timing may cause long, spindly, and weak stems.

STATICE (Florida)

To promote earlier flowering and to increase flower yield.

- **Guide:** Apply a single drench spray when plants are more than 10 inches in diameter (approximately 90 to 110 days after normal seeding time). Use 40 to 50 grams* in 25 gallons to provide 10 ml (5 mg*) solution per plant.

NOTE: Do not exceed specified rates. Do not apply repeated sprays. Accelerated flowering is influenced by extended photoperiod, adequate nutrition, and reduced night temperature. Treatment with gibberellins lessens the requirement for the cold requirement and/or the long photoperiod.

SPRAY GUIDELINES FOR ADDITIONAL CROPS

BERMUDAGRASS GOLD TURF (Florida)

To initiate or maintain growth and prevent color change during periods of cold stress and light frosts on golf course Bermudagrass (e.g., Tifdwarf, Tifgreen, etc.)

- **Guide:** Apply 10 grams* weekly or 25 grams* biweekly in 25 to 100 gallons/A.

* Refers to actual Gibberellic Acid. See Conversion Table to convert to amount of formulated PROGIBB needed.

Use PROGIBB 4% Liquid Concentrate. Mix 1/4 to 2/3 fluid ounce (1/4 to 2/3 gram*) in approximately 6 gallons appropriate for the spray equipment for application to 1,000 sq. ft. (10 1/2 to 26 1/2 fluid ounces/A equivalent to 10 to 25 grams*/A in 25 to 100 gallons/A).

NOTE: Do not exceed specified rates.

Do not apply during extended warm periods where night temperatures exceed 65°F.

Maintain adequate moisture and proper fertilization programs recommended in local area.

Discontinue treatments if thinning is observed.

Do not apply the high rate more frequently than every two weeks. More frequent mowing may be necessary.

Do not use on dormant turf.

HOPS

For seeded and seedless Fuggle hops and similar varieties adapted to Oregon and the Northwest.

To increase yield and pickability.

- **Guide:** Apply spray when vine growth is five to eight feet in length. Use 4 to 6 grams* in 100 to 150 gallons/A.

NOTE: Do not apply within three weeks of harvest.

SUGAR CANE (Hawaii)

Use PROGIBB Plus 2X Soluble Powder

For Increase in sucrose yield.

- **Guide:** Apply 28 to 84 grams* in 7 to 10 gallons/A of spray by airplane. Uniform coverage is essential for maximum response. Use 56 grams as a single treatment, or 28 grams two or three times in separate applications with 30 to 45-day intervals. Application may be made to cane during the first and/or second year of culture. Young cane should be at least three months old to avoid possible tiller reduction. Application should not be made less than 4 months prior to harvest.

Application should be made when growth rate is depressed by temperature. Cane growth below 1,500 feet elevation will benefit from applications made during November through March.

Cane should be treated when there is sufficient soil moisture from rain or irrigation to sustain a high growth rate for at least 30 days following each treatment. Lack of water will negate treatment effect.

Two forms of PROGIBB brand Gibberellic Acid are now available to better serve the needs of individual growers.

CONVERSION TABLE
GRAMS OF ACTUAL GIBBERELLIC ACID PER ACRE TO AMOUNT OF PROGIBB® FORMULATION PER ACRE

Desired Actual Gibberellic Acid Concentration (Grams[1]) in Finished Spray (Per Acre)	PROGIBB Plus 2X Soluble Powder Contains (2.0 Grams[1]/10 Grams Formulated Product)	PROGIBB 4% Liquid Contains (1.0 Gram[1]/ Fluid Ounce of Formulated Product
0.5	2.5 grams	0.5oz.
1.0	5 grams	1oz.
2.0	10 grams	2oz.
4.0	20 grams (1/8 bottle)	4oz.
5.0	25 grams	5oz.
8.0	40 grams (1/4 bottle)	8oz.
10.0	50 grams	10oz.
12.0	60 grams	12oz.
16.0	80 grams (1/2 bottle)	16oz.
20.0	100 grams	20oz.
25.0	125 grams	25oz.
32.0	160 grams (1 bottle)	32oz.
40.0	200 grams	40oz.
48.0	240 grams (1-1 1/2 bottles)	48oz.
50.0	250 grams	50oz.

[1] Refers to actual Gibberellic Acid.

NOTICE TO USER

Seller makes no warranty, express or implied, of merchantability, fitness or otherwise concerning the use of this product other than as indicated on the label. User assumes all risks of use, storage or handling not in strict accordance with accompanying directions.

* Refers to actual Gibberellic Acid. See Conversion Table to convert to amount of formulated PROGIBB needed.

This is a specimen label. Abbott Laboratories is not responsible for the accuracy of the information contained herein. As labels are subject to revision, always carefully read and follow the label on the product container.

PROMALIN®

PLANT GROWTH REGULATOR

ACTIVE INGREDIENTS:	W/W
N-(phenylmethyl)-1Hpurine-6-amine	1.8%
Gibberellins A_4 A_7	1.8%
INERT INGREDIENTS:	96.4%
TOTAL	100.0%

EPA Reg. No. 275-32 EPA Est. No. 35347-CA-1

LABEL EFFECTIVE DATE: 1/88

WARNING

KEEP OUT OF REACH OF CHILDREN

Causes eye irritation. Do not get in eyes. Harmful if swallowed. In case of contact, immediately flush eyes with plenty of water. Call a physician.

PROMALIN® is a plant growth regulator for use on apples*. When applied at specific spray timing and concentration, PROMALIN will provide the following effects:

1. PROMALIN improves the shape of Delicious* applies through elongation of fruit and development of more prominent calyx lobes. These desirable effects will be most evident in years when natural typiness is limited.

2. On Delicious and other varieties PROMALIN may increase weight of individual fruit and yield per tree. Some thinning may occur from the use of PROMALIN.

3. On red apple varieties, e.g. Delicious, Jonathan, Jonamac, MacIntosh, Paulared, Spartan, Empire. PROMALIN may be combined (in all States except California) with Alar®-85 to enhance expected desirable effects from Alar-85, including improved color, earlier coloring, increased fruit pressure and prevention of preharvest fruit drop.

**4. On most apple varieties, PROMALIN will increase lateral bud break, shoot growth, and improve branch angles on *non-bearing* apple trees, including nursery stock. This effect will provide a better tree framework for early cropping.

*Only Red Delicious apples in California.

**All States except California.

DIRECTIONS FOR USE

It is a violation of Federal law to use this product in a manner inconsistent with its labeling.

Do not apply this product through any type of irrigation system.

Keep out of lakes, ponds or streams. Do not contaminate water by cleaning of equipment or disposal of wastes.

STORAGE AND DISPOSAL

Do not contaminate water, food or feed by storage or disposal.

STORAGE

Store Below 75°F (23°C). Keep containers tightly closed when not in use.

PESTICIDE DISPOSAL

Wastes resulting from the use of this product may be disposed of on site or at an approved waste disposal facility.

CONTAINER DISPOSAL

Do not reuse empty containers. Triple rinse (or equivalent). Puncture and dispose of in a sanitary landfill, or by incineration or, if allowed by state and local auathorities, by burning. If burned, stay out of smoke.

A. RATE AND TIMING OF APPLICATION-FRUIT DEVELOPMENT

1. Single Application for Improved Type.

Western U.S.:A single application of PROMALIN should be made from early king bloom (king blossoms opening and side blossoms in the balloon stage) to the early stages of petal fall of the king blossoms.

***PROMALIN-ELGETOL®** (DNOC, 19% AI) (for type and thinning on spur Delicious strains): Apply a tank mix of PROMALIN and Elgetol at the recommended Elgetol timing, i.e. 80% full king bloom on the north side of the tree. Apply 2 pints of PROMALIN plus 4 pints of Elgetol in 200 gallons of water per acre. Observe all precautionary statements on Elgetol label.

*All States except California.

Eastern U.S.: A single application of PROMALIN should be made from early king bloom to the early stages of petal fall of side blossoms.

For both eastern and western U.S. apply 1 to 2 pints of PROMALIN per acre in 50 to 200 gallons of spray mixture as a fine mist spray.

2. Split Applications for Improved Type and Size.

A split application may be utilized to improve typing and sizing responses. The first application should be made from early king bloom to the early stages of petal fall of the king bloom. The second application should be made 7 to 21 days later. Apply 0.5 to 1 pint of PROMALIN per acre in 50 to 200 gallons of spray mixture in each application. Do not, however, exceed the maximum recommended rate of a total of 2 pints per acre for the combined sprays.

*3. Split Applications to Improve "Typiness" in Locations where the Bloom Period is Prolonged or Occurs over a Period of Several Days.

Make two PROMALIN applications three or more days apart. The first spray should be timed during the first flush of bloom. A second application should be made when the remainder of the tree canopy comes into bloom.

The time between PROMALIN applications may vary from orchard to orchard within a given growing area because of the influence of the local microclimate. In most instances a period from 3 to 7 days between sprays is adequate. The time interval between sprays should be lengthened should cool weather further prolong bloom. For the most efficient use of PROMALIN, bloom progression within individual orchards should be assesssed at least every 48 hours.

Apply PROMALIN twice as two concentrate sprays each containing 0.5-1 pint of PROMALIN in 100 gpa. The spray machine should be calibrated to deliver 100 gpa and nozzled to direct the spray mist to those portions of the tree canopy currently in bloom. An appropriate wetting agent should be included so as to insure even thorough coverage of open blossoms.

*All States except California.

*4. PROMALIN and ALAR®-85 (Plant Growth Regulants) for Commercial Applications to Apples.

*PROMALIN and ALAR-85 may be combined as a tank mix application during mid-summer (60-80 days pre-harvest) on apples. This tank mix may enhance the expected desirable effect from Alar-85 application, including improved color, earlier coloring, increased fruit pressure, and prevention of preharvest fruit drop.

Use 1 to 2 pints of PROMALIN and 2 to 4 pounds of Alar-85 per acre in 100-400 gallons of spray solution. Alar should be added to the tank mix prior to the addition of PROMALIN.

The PROMALIN and Alar-85 tank mix should be applied as a foliar spray in a manner that insures uniform and complete coverage. Do not exceed the per acre recommended rate. See current Alar-85 label for precautionary statements and other specific recommendations.

*All States except California.

Mixing Instructions

Add the required amount of PROMALIN to a spray tank about half-filled with water. Agitate while bringing the total volume of water to the required level. New solutions should be mixed only in a clean, empty spray tank, and used within 24 hours. A nonionic wetting agent e.g. Tween®-20, Glyodin® or Triton X-100, may improve spray coverage and enhance absorption.

Application Considerations

1. Since PROMALIN is a plant growth regulator that must be absorbed by the plant to be effective, best performance can be expected with good absorption conditions and healthy, vigorous trees. Do not spray the recommended tank mixes on any low vigor trees or trees under stress as from drought, low fertility, winter injury, etc.

2. Apply PROMALIN spray mixture as a uniform fine mist spray.

3. Use a properly calibrated and adjusted sprayer that insures uniform and complete coverage of all foliage and blossoms. Spray volume should be adjusted to tree size and density.

4. Approximately 85% of the spray volume should be directed into the upper two-thirds of the trees.

5. To minimize excessive spray drift, apply when air is calm.

6. High relative humidity and slow drying conditions favor maximum absorption of PROMALIN. Application in morning or evening may be preferable. Conditions favoring rapid drying of the spray deposit should be avoided.

7. Rainfall within six hours after spraying may reduce activity of PROMALIN.

8. Do not apply PROMALIN when air temperatures are lower than 40°F or greater than 90°F.

9. Do not apply PROMALIN in more than 200 gallons of spray per acre.

B. Rate and Timing of Application—

***Lateral branching and tree development**

A single foliar application of PROMALIN or a PROMALIN-latex paint spot application may be applied to apples, non-bearing pears, and non-bearing sweet cherry trees, including nursery stock, to increase lateral bud break, shoot growth, improve branch angles, and provide a better tree framework for early cropping.

Foliar Applications

Foliar applications have proven to be efficacious in increasing bud break and branch angles on apples, non-bearing pears, and sweet cherries being grown in nurseries.

Directions for Use

Crop	Rate	Timing
Apples (Nursery and Orchard)	125–500 PPM (.25–1 pt. PROMALIN/5 gal. of spray solution)	For orchard trees apply at 1–3 inches of new terminal growth
Non-bearing Pears (Nursery and Orchard)	250–1000 PPM (.5–2 pts. PROMALIN/5 gal. of spray solution)	Nursery stock should be treated after trees have reached a terminal height at which lateral branching is desired
Non-bearing Sweet Cherries (Nursery only)	See Pears	

A buffered wetting agent, e.g. buffer X or non-ionic wetting agent, e.g. Tween 20, Triton X-100, Glyodin, etc. should be added to the tank at a rate of .2–3% (1–2 oz./5 gal.) prior to the addition of PROMALIN. The final spray solution should not exceed a pH of 8.

Uniformly apply PROMALIN with a pressurized hand sprayer, hand gun attachment to an airblast sprayer, or an airblast sprayer to thoroughly wet the bark and foliage surfaces. Total spray volume needed to achieve thorough wetting will vary with the equipment used and the size of the trees to be treated. Generally, 50–100 gallons of spray mixture will treat an acre of nursery trees planted on a 4 foot by 4 inch spacing (2200 trees). Approximately 5–10 gallons of spray mixture applied with a pressurized hand sprayer will treat 200–300 non-bearing orchard trees 1–4 years old.

Latex Applications

PROMALIN applied in latex paint has proven effective in increasing bud break and branch angles when applied to one year old wood on apples and orchard grown non-bearing sweet cherries.

Crop	Rate	Timing
Apples (Nursery and Orchard) Sweet Cherries (Orchard only)	5000–7500 PPM (.2–.33 pt. PROMALIN per pt. of latex paint)	Apply in the spring when terminal buds begin to swell but before shoots emerge. NOTE: Do not apply after buds break. Applications after buds have broken may cause some injury to tender shoot tips and fail to promote shoot growth from that point

A buffered or non-ionic wetting agent should be added to the PROMALIN and latex paint mixture at the rate of .5–1% (.1–.15 oz. per pint of paint) prior to the addition of PROMALIN. The addition of a wetting agent will improve the dispersion of PROMALIN in the latex paint and will also improve wetting and absorption through the waxy cuticular layer of the bark surface.

Uniformly apply the PROMALIN-latex paint mixture with a brush or sponge to thoroughly cover the bark surface. NOTE: Apply only to one year old wood.

Tank Mix Compatibility

1. PROMALIN may be applied in combination with plant nutrients.

2. A buffer should be added to alkaline spray water to reduce the pH below 8.

*3. Fungicides.
Benlate®, Ferbam®, Polyram®, Thiram®, Glyodin®, Dithane® M-45, Orthocide®, wettable sulfur may be tank mixed with PROMALIN at the recommended timing associated with bloom applications as described under "Rate and Timing of Application" for PROMALIN. When using fungicides, refer to respective labels for directions and precautionary statements.

Elgetol, Alar, Tween, Benomyl, Ferbam, Polyram, Thiram, *Dithane M-45, *Triton, Glyodin and Orthocide are trademarks of companies other than Abbott Laboratories.

*All States except California.

NOTICE TO USER

Seller makes no warranty, express or implied, of merchantability, fitness or otherwise concerning the use of this product other than as indicated on the label. User assumes all risks of use, storage or handling not in strict accordance with accompanying directions.

This is a specimen label. Abbott Laboratories is not responsible for the accuracy of the information contained herein. As labels are subject to revision, always carefully read and follow the label on the product container.

PROSHEAR®

LIQUID CONCENTRATE BIORATIONAL PLANT GROWTH REGULATOR

ACTIVE INGREDIENT:	W/W
N-[Phenylmethyl]-1H-purine-6-amine	2.0%
INERT INGREDIENTS:	98.0%
TOTAL	100.0%

EPA Reg. No. 275-57 EPA Est. No. 52889-IL-01

WARNING

KEEP OUT OF REACH OF CHILDREN

PRECAUTIONARY STATEMENTS

HAZARDS TO HUMANS (AND DOMESTIC ANIMALS)

Causes eye irritation. Do not get in eyes. Harmful if swallowed. In case of contact, immediately flush eyes with plenty of water and call a physician.

ENVIRONMENTAL HAZARDS

Keep out of lakes, ponds or streams. Do not contaminate water by cleaning of equipment or disposal of wastes.

DIRECTIONS FOR USE

It is a violation of Federal law to use this product in a manner inconsistent with its labeling.

Do not apply this product through any type of irrigation system.

STORAGE AND DISPOSAL

Do not contaminate water, food or feed by storage or disposal.

STORAGE: Keep containers tightly closed.

PESTICIDE DISPOSAL. Wastes resulting from the use of this product may be disposed of on site or at an approved waste disposal facility.

CONTAINER DISPOSAL. Do not reuse empty containers. Triple rinse (or equivalent). Puncture and dispose of in a sanitary landfill, or by incineration, or, if allowed by state and local authorities by burning. If burned, stay out of smoke.

PROSHEAR® is a plant growth regulator for use on white pine to increase lateral bud set the year of application and branch development the year following application. Application of PROSHEAR will result in a "fuller" or more dense tree and produce a very desirable full top due to the enhanced lateral shoot development at and below the current year's whorl. Optimal results will be obtained when PROSHEAR is used in conjunction with normal shearing practices and other standard cultural practices.

Application of PROSHEAR to older trees (5–9 years) should be made at least one year prior to intended harvest to allow for full phenotypic expression in top growth. Applications of PROSHEAR to younger trees can be made annually to stimulate bud development and lateral shoot growth during the rotation period.

APPLICATION RATE

—White Pine

The selection of rate is dependent on the total spray volume applied per acre and the degree of vigor of the trees. The optimal rate is in the range of 500–1000 ppm Active Ingredient in the final spray solution. Highly vigorous trees will generally respond better to treatment.

Based on the degree of response desired for White Pine, use one (1) to two (2) pints of PROSHEAR per five gallons of aqueous spray solution (3.2 to 6.4 fluid ounces of PROSHEAR per gallon of spray solution) (25–50 ml of PROSHEAR per liter) to attain a concentration of 500 to 1000 ppm Active Ingredient in the final spray solution. Add the required volume of PROSHEAR to about one-half the final volume of water, begin stirring or agitating and add the remainder of the water to the final volume. It is not necessary to add a wetting agent to the spray solution. The spray solution should be stirred or agitated prior to application. Do not store spray solutions more than 24 hours.

METHOD OF APPLICATION

Uniformly apply the spray mixture as a fine spray. Best results have been achieved using a pressurized hand sprayer equipped with a fine nozzle orifice, such an as 8001-8002 flat fan or D2-12 to D2-25 hollow cone nozzle operating at approximately 30 psi. The nozzle should be directed to thoroughly wet the new terminal leader growth. Approximately 3–8 gallons of spray mixture will treat an acre of trees (10–30 ml/leader) planted on a spacing 6 ft x 6 ft (1210 trees/acre). Larger trees and/or closer tree spacing may require greater spray volumes. The 10–30 ml/tree is an average spray volume for white pine trees set 3–6 years in the field.

TIME OF APPLICATION

A single annual application of PROSHEAR can be made from 14 days before to 14 days following shearing of white pine. Best results have been obtained with sprays applied from 1–14 days after shearing. This period generally occurs from mid-June to mid-July for most North American growing areas. Do not apply to white pine more than 21 days before or more than 21 days after shearing.

OTHER APPLICATION CONSIDERATIONS

1. High relative humidity and slow drying conditions favor the maximum absorption or PROSHEAR. Conditions favoring the rapid drying of the spray desposit should be avoided.
2. Rainfall within six hours after spraying may reduce the activity of PROSHEAR and if this occurs trees should be retreated.
3. Do not apply PROSHEAR when air temperatures are lower than 40°F (5°C) or greater than 100°F (38°C)
4. Do not apply PROSHEAR on any low vigor trees or trees under stress from drought, low fertility, winter injury, etc.

NOTICE TO USER

Seller makes no warranty, express or implied, or merchantability, fitness or otherwise concerning the use of this product other than as indicated on the label. User assumes all risks of use, storage or handling not in strict accordance with accompanying directions.

This is a specimen label. Abbott Laboratories is not responsible for the accuracy of the information contained herein. As labels are subject to revision, always carefully read and follow the label on the product container.

PROVIDE®

(GIBBERELLINS A_4A_7)

PLANT GROWTH REGULATOR

ACTIVE INGREDIENT:	
Gibberellins A_4A_7	2.0% w/w
INERT INGREDIENTS:	98.0% w/w
TOTAL	100.0% w/w

(Contains 21 grams active ingredient per liter or 9.93 grams per pint.)

EPA Reg. No. 275-2 EPA Est. No. 35347-CA-1

KEEP OUT OF REACH OF CHILDREN

CAUTION

Net Contents: One (1) Pint (16 fluid ounces, 473 ml)

PRECAUTIONARY STATEMENTS

HAZARDS TO HUMANS (AND DOMESTIC ANIMALS)

Causes eye irritation. In case of contact flush skin and eyes with plenty of water. In case of irritation contact a physician.

Causes eye irritation. Do not get in eyes. In case of contact, immediately flush eyes with plenty of water and call a physician.

ENVIRONMENTAL HAZARDS

Keep out of lakes, ponds or streams. Do not contaminate water by cleaning of equipment or disposal of wastes.

DIRECTIONS FOR USE

It is a violation of Federal Law to use this product in a manner inconsistent with its labeling.

Do not apply this product through any type of irrigation system.

DIRECTIONS FOR USE

PROVIDE is for suppression of fruit russeting of Delicious apples (Goldens and Reds). PROVIDE is a plant growth regulator containing gibberellins A_4A_7 which may reduce the development of physiological fruit russeting in Delicious apples when applied in a post-bloom spray program. Used in years when russet conditions are present, PROVIDE may boost overall fresh market grades of the crop. Physiological russeting is generally associated with the presence of water on the fruit surface during the first 30–40 days of fruit development. Russet severity at a given location is therefore specifically influenced by the weather conditions during and following the bloom period. The key conditions influencing russet include precipitation, humidity, and temperature. PROVIDE may act only on that russet caused by climatic factors active early in fruit development and will not reduce damage associated with mechanical forces such as limb or pedicel rub, or insect damage.

APPLICATION RATE

Apply 26 to 40 ounces per acre for a total seasonal rate.

TIMING OF APPLICATIONS

Apply PROVIDE as two to four consecutive sprays beginning at petal fall, and continuing at 7 to 10-day intervals. This spray timing corresponds to the most sensitive russet development period shown to occur in Delicious apples.

METHOD OF APPLICATION

Apply 10 ounces* (15 ppm) to 13 ounces (20 ppm) of PROVIDE in 100 gallons of water per acre, per application for 2 to 4 total applications. Number and rate of applications should be based on both tree size and local weather conditions. In general, as tree size and overall rainfall increase, the number and rate of applications should be increased. This increase is necessary to maintain adequate coverage of the developing fruit surface during the most sensitive russet period. NOTE: No more than 40 ounces of PROVIDE should be applied per season.

* Best russet control may be expected with four 10-ounce (15 ppm) per acre applications of PROVIDE.

APPLICATION CONSIDERATIONS

1. Do not use with spray surfactants, spreader stickers, etc. since these materials have been shown to induce russet.
2. Do not use in excessive spray volumne, (i.e. greater than 100 gal/acre) since excess moisture has been shown to induce russet.
3. Use a properly calibrated and adjusted sprayer that insures uniform and complete coverage of all foliage and fruitlets.
4. Approximately 85% of the spray volume should be directed into the upper two-thirds of the trees.
5. To minimize excessive spray drift, apply when air is calm.
6. Rainfall within six hours after spraying may reduce activity of PROVIDE.
7. PROVIDE may be applied in combination with other materials: refer to respective levels for directions and precautionary statements.

STORAGE AND DISPOSAL

Do not contaminate water, food or feed by storage or disposal.

STORAGE: Store below 78°F (25°C). Keep containers tightly closed when not in use.

PESTICIDE DISPOSAL:

Wastes resulting from the use of this product may be disposed of on site or at an approved waste disposal facility.

CONTAINER DISPOSAL:

Do not reuse empty containers. Triple rinse, puncture and dispose of in a sanitary landfill, or by incineration, or, if allowed by state and local authorities by burning. If burned, stay out of smoke.

NOTICE TO USER

Seller makes no warranty, express or implied, or merchantability, fitness or otherwise concerning the use of this product other than as indicated on the label. User assumes all risks of use, storage or handling not in strict accordance with accompanying directions.

This is a specimen label. Abbott Laboratories is not responsible for the accuracy of the information contained herein. As labels are subject to revision, always carefully read and follow the label on the product container.

VECTOBAC®-12 AS
Aqueous Suspension

BIOLOGICAL MOSQUITO AND BLACKFLY LARVICIDE

ACTIVE INGREDIENT:
Bacillus thuringiensus. Serotype H-14, 1200 International Toxic Units (ITU) Per Milligram (Equivalent to 4.84 billion ITU per gallon: 1.279 billion ITU per liter) 1.2%
INERT INGREDIENTS: 98.8%
TOTAL 100.0%

EPA Reg. No. 275-66 EPA Est. No. 33762-IA-1

CAUTION
KEEP OUT OF REACH OF CHILDREN

PRECAUTIONARY STATEMENTS

HAZARD TO HUMANS

As a precautionary measure in case of contact, flush eyes with plenty of water. In case of irritation, contact a physician.

DIRECTIONS FOR USE—GENERAL CLASSIFICATION

It is a violation of Federal Law to use this product in a manner inconsistent with its labeling.

CHEMIGATION

Refer to supplemental labeling for use directions for chemigation. Do not apply this product through any irrigation system unless the supplemental labeling on chemigation is followed.

STORAGE AND DISPOSAL

Do not contaminate water, food, or feed by storage or disposal.

STORAGE

Store in a cool place.

PESTICIDE DISPOSAL

Wastes resulting from the use of this product may be disposed of on site or at an approved waste disposal facility.

CONTAINER DISPOSAL

Triple rinse (or equivalent). Then puncture and dispose of in a sanitary landfill, or by incineration, or, if allowed by state and local authorities by burning. If burned, stay out of smoke.

VECTOBAC®-12 AS is a highly selective insecticide for use against mosquito, blackfly and larvae.

APPLICATION DIRECTIONS

Mosquitoes Habitat	Suggested Range Rate*
(Such as the following examples): Irrigation ditches, roadside ditches, flood water, standing ponds, woodland pools, snow melt pools, pastures, catch basins, storm water retention areas, tidal water, salt marshes and rice fields.	0.25 pt–1 pt/acre
Polluted water (such as sewage lagoons, animal waste lagoons)	1 pt–2 pts/acre

* Use higher rate in polluted water and when late 3rd and early 4th instar larvae predominate, mosquito populations are high, water is heavily polluted, and/or algae are abundant.

Blackflies Habitat	Suggested Range Rate
Streams	0.5–2.5 mg/liter stream water** (= ppm)

** Use higher rate when stream contains high concentration of organic material, or dense aquatic vegetation.

GROUND AND AERIAL APPLICATION

VECTOBAC-12 AS may be applied in conventional ground or aerial application equipment with quantities of water sufficient to provide uniform coverage of the target area. The amount of water needed per acre will depend on weather, spray equipment, and mosquito habitat characteristics. Do not mix more VECTOBAC-12 AS than can be used in a 2-week period.

For most ground spraying, apply in 5-100 gallons per acre using hand-pump, airblast, mist blower, etc., spray equipment.

For aerial application, VECTOBAC-12 AS may be applied either undiluted or diluted with water. For undiluted applications, apply 0.25 to 2.0 pts./acre of VECTOBAC-12 AS through fixed wing or helicopter aircraft equipped with either conventional boom and nozzle systems or rotary atomizers.

For diluted application, fill the mix tank or plane hopper with the desired quantity of water. Start the mechanical or hydraulic agitation to provide moderate circulation before adding the VECTOBAC-12 AS. Maintain the agitation while loading or spraying. For all aerial applications, a droplet spectrum in the range of 50–100 microns (VMD) will produce best results.

Rinse and flush spray equipment throughly following each use.

For blackfly aerial applications, VECTOBAC-12 AS can be applied undiluted via fixed wing or helicopter aircraft equipped with either conventional boom and nozzle systems or open pipes. Rate of application will be determined by the stream discharge and the required amount of VECTOBAC-12 AS necessary to maintain a 0.5–25 ppm concentration in the stream water. VECTOBAC-12 AS can also be applied diluted with similar spray equipment. Do not mix more VECTOBAC-12 AS than can be used in a 2-week period.

RICE—FLOOD (BASIN) CHEMIGATION

Systems using a gravity flow pesticide dispensing systems must meter the pesticide into the water at the head of the field and downstream of a hydraulic discontinuity such as a drop structure or weir box to decrease potential for water source contamination from backflow if water flow stops.

VECTOBAC is metered or dripped into rice floodwater at application stations positioned at the point of introduction (levee cut) of water into each rice field or pan. Two pints of VECTOBAC AS (or 1 pt of VECTOBAC 12AS) are diluted in water to a final volume of 5 gallons. The diluted solution is contained in a 5 gallon container and metered or dispersed into the irrigation water using a constant flow device at the rate of 80 ml per minute. Introduction of the solution should begin when 1/3 to 1/2 of the pan or field is covered with floodwater. Delivery of the solution should continue for a period of approximately 4½ hours. Floodwater depth should not exceed 10–12 inches to prevent excessive dilution of VECTOBAC which could result in reduced larval kill. Agitation is not required during the period in which the VECTOBAC solution is being dispersed.

Application of VECTOBAC into rice floodwater is not permitted using a pressurized water and pesticide injection system.

NOTICE TO USER

Seller makes no warranty, express or implied, of merchantability, fitness or otherwise concerning the use of this product other than as indicated on the label. User assumes all risks of use, storage or handling not in strict accordance with accompanying directions.

This is a specimen label. Abbott Laboratories is not responsible for the accuracy of the information contained herein. As labels are subject to revision, always carefully read and follow the label on the product container.

VECTOBAC®-G Granules

BIOLOGICAL LARVICIDE

ACTIVE INGREDIENT:
Bacillus thuringiensis, Serotype H-14
200 International Toxic Units (ITU) per milligram
(Equivalent to 0.091 billion ITU's per pound) 0.2%
INERT INGREDIENTS: 99.8%

TOTAL 100.0%

EPA Registration No. 275-50 EPA Est. No. 2393-IL-3

Filled by weight, not volume
List No. 5108

CAUTION
KEEP OUT OF REACH OF CHILDREN

APPLICATION DIRECTIONS

MOSQUITOES

Habitat	Suggested Range Rate*
(Such as the following examples): Irrigation ditches, roadside ditches, flood water, standing ponds, woodland pools, snow melt pools, pastures catch basins, storm water retention areas, tidal water, salt marshes and rice fields.	2.5–10 lb acre

* Use 10–20 lbs/acre when late 3rd and early 4th instar larvae predominate, mosquito populations are high, water is heavily polluted, (sewage lagoons, animal waste lagoons), and/or algae are abundant.

Apply uniformly by aerial or ground conventional equipment.

A 7 to 14 day interval between applications should be employed. Where non-target aquatic invertebrate parasites and predators are present longer periods of suppression may result since these beneficials would be conserved to aid in mosquito population management.

NOTICE TO USER

Seller makes no warranty, express or implied, of merchantability, fitness or otherwise concerning the use of this product other than as indicated on the label. User assumes all risks of use, storage or handling not in strict accordance with accompanying directions.

VECTOBAC®-G is a highly selective insecticide for use against mosquito larvae.

DIRECTIONS FOR USE

It is a violation of Federal Law to use this product in a manner inconsistent with its labeling.

STORAGE AND DISPOSAL

Do not contaminate potable water, food, or feed by storage or disposal.

STORAGE: Store in a cool, dry place.

PESTICIDE DISPOSAL:

Wastes resulting from the use of this product may be disposed of on site or at an approved waste disposal facility.

CONTAINER DISPOSAL:

Completely empty bag into application equipment. Then dispose of empty bag in a sanitary landfill or by incineration, or, if allowed by State and local authorities, by burning. If burned, stay out of smoke.

INTRODUCTION

VECTOBAC®-G biological larvicide is a highly selective insecticide for control of most species of mosquito larvae. It contains spores and delta-endotoxin crystals of *Bacillus thuringiensis,* Serotype H-14 (*B.t.* H-14). Activity against mosquito larvae is dependent upon the delta-endotoxin crystal. *B.t.* H-14 is well-suited for use in integrated pest management (IPM) control programs and environmentally sensitive areas, since toxicity testing has not indicated any harmful effect on mammals (and, by analogy, humans), birds, fish, or aquatic invertebrate predators and parasites.

The rapid rate of kill, typically within 24 hours, is in harmony with the inspection and evaluation programs currently used by most mosquito control operators.

The fermentation and formulation expertise acquired by Abbott Laboratories (North Chicago, Illinois, USA) in developing *B.t.* products for agriculture and forestry is being applied to the production of VECTOBAC® brand *B.t.* H-14, thereby insuring high quality, stability, and performance.

FORMULATION

VECTOBAC® -G is a granular formulation of *Bacillus thuringiensis,* Serotype H-14, containing 200 International Toxic Units (ITU) per milligram (equivalent to 0.091 billion ITU's per pound).

This formulation has been designed with unique penetration and deposition properties for use in breeding areas with moderate to heavy vegetation, such as flooded pastures, salt marshes, ponds, ditches, and rice fields.

SELECTIVITY

The insecticidal activity of VECTOBAC®-G is primarily limited to larvae of mosquitoes (Family Culicidae). Tests have been conducted to determine the safety of *B.t.* H-14 to non-target species of vertebrates and invertebrates. Results of these tests have shown that B.t. H-14 provides excellent control of mosquito larvae, while demonstrating no harmful effects on mammals (and, by analogy, humans), fish and most non-target aquatic organisms and beneficial insects.

RECOMMENDATIONS FOR USE

Application rates of VECTOBAC®-G mosquito larvicide are related to surface area of water to be treated, not depth. Recommended application rates range from 2.5 to 20 pounds of VECTOBAC® granules per acre, depending upon habitat and larval stage.

GROUND APPLICATION

VECTOBAC® -G can be applied through manual and mechanically driven ground-rig granule applicators, e.g. cyclone seeder, Ortho™ Whirlybird Seeder, and air-blast ground equipment, e.g. Buffalo turbines. Tests to date have shown VECTOBAC®-G application through a horn seeder to be unacceptable.

AERIAL APPLICATION

VECTOBAC®-G can be applied through conventional fixed wing and helicopter granule application equipment.

Higher application rates may be needed to control late 3rd and 4th instar larvae and in highly organic waters.

NOTE: Alternate liquid formulations, VECTOBAC®-AS and VECTOBAC®-12AS, are available for use against mosquito and black fly larvae.

This is a specimen label. Abbott Laboratories is not responsible for the accuracy of the information contained herein. As labels are subject to revision, always carefully read and follow the label on the product container.

American Cyanamid Company

Agricultural Division
One Cyanamid Plaza
Wayne, NJ 07470
(201) 831-2000

AASTAR®

SOIL AND SYSTEMIC INSECTICIDE

RESTRICTED USE PESTICIDE

Due to acute oral and dermal toxicity; and bird toxicity.

For retail sale to and use only by Certified Applicators or persons under their direct supervision and only for those uses covered by the Certified Applicator's certification.

American Cyanamid Company endorses Certification to promote the responsible use of pesticides to insure the protection of man and the environment.

FOR USE IN FIELD CORN

ACTIVE INGREDIENTS:	
phorate (O,O-diethyl S-[(ethylthio)methyl] phosphorodithioate) .	15.00%
flucythrinate [(±)-cyano(3-phenoxyphenyl) methyl(+)-4-(difluoromethoxy)-alpha-(1-methylethyl)benzeneacetate	0.75%
INERT INGREDIENTS:	84.25%
TOTAL	100.00%

EPA Reg. No. 241-279

® Registered Trademark of American Cyanamid Company

7/88

KEEP OUT OF REACH OF CHILDREN

DANGER–POISON

¡PELIGRO!

PRECAUCION AL USUARIO: Si usted no lee ingles, no use este producto hasta que la etiqueta le haya sido explicada ampliamente.

FIRST AID

IF SWALLOWED, Drink one or two glasses of water and induce vomiting by touching back of throat with finger. Do not induce vomiting or give anything by mouth to an unconscious person. Avoid alcohol. Get medical attention.

IF INHALED, remove to fresh air. If not breathing, give artificial respiration, preferably mouth-to-mouth. If breathing is difficult, give oxygen. Get medical attention.

IF ON SKIN, wash thoroughly with soap and water. Remove contaminated clothing and shoes. Wash clothing and decontaminate shoes before reuse.

IF IN EYES, immediately flush eyes with plenty of water. Get medical attention.

Repeated inhalation or skin contact may, without symptons, progressively increase susceptibility to poisoning.

Antidote: Atropine is an antidote.

CALL A PHYSICIAN AT ONCE IN ALL CASES OF SUSPECTED POISONING

NOTE TO PHYSICIANS: Warning symptoms include weakness, headache, tightness in chest, blurred vision, nonreactive pinpoint pupils, salivation, sweating, nausea, vomiting, diarrhea and abdominal cramps. Give atropine intramuscularly or intravenously, depending on severity of poisoning, 2 to 4 milligrams every ten minutes until fully atropinized as shown by dilated pupils, dry flushed skin and tachycardia. Twenty to thirty milligrams, or more, may be required during the first 24 hours. Never give opiates or phenothiazine tranquilizers. Clear chest by postural drainage. Artificial respiration of oxygen administration may be necessary. Observe patient continuously for at least 48 hours. Allow no further exposure to any cholinesterase inhibitor until cholinesterase regeneration has taken place as determined by blood tests.

Pralidoxime chloride (2-PAM; PROTOPAM chloride) may be effective as an adjunct to atropine. Use according to label directions.

PRECAUTIONARY STATEMENTS

HAZARDS TO HUMANS AND DOMESTIC ANIMALS

DANGER!

Fatal if swallowed, inhaled, or absorbed through the skin. Do not breathe dust. Do not get in eyes, on skin, or on clothing.

Wear freshly laundered, long-sleeved work clothing daily. While transferring from package to equipment, wear a clean cap, gloves (rubber or cotton) and goggles. If cotton gloves are used, they must be laundered or discarded after each day's use. Rubber gloves should be washed with soap and water after each use. Do not wear the same gloves for other work. Destroy and replace gloves frequently.

In case of contact, immediately remove contaminated clothing and wash skin thoroughly with soap and water. Launder clothing and decontaminate shoes before reuse. Wash thoroughly with soap and water before eating or smoking. Bathe at the end of the work day and change clothing.

Do Not Breathe Dust

Wear a face mask or other respiratory equipment while emptying bags of product into a hopper. While emptying bags into equipment, pour downwind and allow as little free fall as possible. Do not pour at face level and do not allow dust to reach the breathing zone.

Do Not Contaminate Food or Feed Products

Once a bag has been opened, use it completely. Make sure the hoppers are emptied while still in the field. Refer to STORAGE AND DISPOSAL statement for further instructions.

Keep All Unprotected Persons Out of Operating Areas

Do not apply this product in such a manner as to directly or through drift expose workers or other persons.

Keep Out of Reach of Domestic Animals

Not for Use or Storage In or Around the Home

REENTRY

Do not enter treated areas without protective clothing until treatments have been completed.

Because certain states may require more restrictive reentry intervals for various crops treated with this product, consult your State Department of Agriculture for further information.

Written or oral warnings must be given to workers who are expected to be in a treated area or in an area about to be treated with this product. When oral warnings are given, warnings shall be given in a language customarily understood by workers. Oral warnings must be given if there is reason to believe that written warnings cannot be understood by workers. Written warnings must include the following information: DANGER. Area treated with AASTAR on (date of application). Do not enter without protective clothing.

Fields may be reentered on day of treatment with soil applications of AASTAR.

ENVIRONMENTAL HAZARDS

This product is toxic to fish, shrimp, crab, birds, and other wildlife. Birds and other wildlife in treated areas may be killed. Keep out of lakes, streams, ponds, tidal marshes and estuaries. Do not apply where runoff is likely to occur. Do not apply when weather conditions favor drift from treated areas. Do not contaminate water by cleaning of equipment or disposal of wastes. Shrimp and crab may be killed at application rates recommended on this label. Do not apply where these are important resources. Apply this product only as specified on this label.

This pesticide is highly toxic to bees exposed to direct application. Application should be timed to coincide with periods of minimum bee activity, usually between late evening and early morning.

DISCLAIMER

The label instructions for the use of this product reflect the opinion of experts based on field use and tests. The directions are believed to be reliable and should be followed carefully. However, it is impossible to eliminate all risks inherently associated with use of this product. Crop injury, ineffectiveness or other unintended consequences may result because of such factors as weather conditions, presence of other materials, or the use or application of the product contrary to label instructions, all of which are beyond the control of American Cyanamid Company. All such risks shall be assumed by the USER.

American Cyanamid Company warrants only that the material contained herein conforms to the chemical description on the label and is reasonably fit for the use therein described when used in accordance with the directions for use, subject to the risks referred to above.

Any damages arising from a breach of this warranty shall be limited to direct damages and shall not include consequential commercial damages such as loss of profits or values or any other special or indirect damages.

American Cyanamid Company makes no other express or implied warranty, including any other express or implied warranty of FITNESS or of MERCHANTABILITY.

DIRECTIONS FOR USE

It is a violation of Federal law to use this product in a manner inconsistent with its labeling.

This label must be in the possession of the user at the time of pesticide application.

STORAGE AND DISPOSAL

STORAGE

Store pesticide products in a secure locked area where children, unauthorized persons and animals cannot enter. Do not store in the same area with food or feed. Do not store opened bags.

PROHIBITIONS

Do not contaminate water, food or feed by storage or disposal. Open dumping is prohibited.

PESTICIDE DISPOSAL

Pesticide wastes are acutely hazardous. Improper disposal of excess pesticide, spray mixture, or rinsate is a violation of Federal law. If these wastes cannot be disposed of by use according to label instructions, contact your state pesticide or Environmental Control Agency, or the Hazardous Waste Representative at the nearest EPA Regional Office for guidance.

CONTAINER DISPOSAL

Completely empty bag into application equipment. Then dispose of empty bag in a sanitary landfill or by incineration, or, if allowed by state and local authorities, by burning. If burned, stay out of smoke.

GENERAL

Consult federal, state or local disposal authorities for approved alternative procedures such as limited open burning.

USE RATE AS POUNDS PER ACRE BASED ON VARIOUS ROW SPACINGS

Application Rate 8 oz./1,000 ft. of Row

Row spacing (inches)	30	32	34	36	38	40
Lbs./acre	8.7	8.2	7.7	7.3	6.9	6.5

BEFORE USING, READ PRECAUTIONARY STATEMENTS

AASTAR should be applied with a granular pesticide applicator properly calibrated to assure accurate placement and proper dosage. Cover granules that may be exposed on the ends of the treated rows and turns and loading areas by deep discing immediately after treating fields.

Crop	Pests Controlled	Rate of AASTAR	Application	Remarks
FIELD CORN At Planting	Cutworms Corn rootworms Wireworms White grubs Seedcorn maggots Seedcorn beetles Flea beetles Sod Webworm	8 oz. per 1,000 ft. of row (minimum 30-inch row spacing)	Place granules in a 6-8 inch bend over the row directly behind or in front of the press wheel	Do not place granules in direct contact with seed. Do not graze or cut for forage within 60 days of treatment.

IN CASE OF AN EMERGENCY ENDANGERING LIFE OR PROPERTY INVOLVING THIS PRODUCT
CALL COLLECT, DAY OR NIGHT, AREA CODE 201-836-3100.

ABATE® 1-SG

INSECTICIDE

FOR CONTROL OF MOSQUITO AND MIDGE LARVAE

ACTIVE INGREDIENT:
Temephos ((O,O′-(thiodi-4,1-phenylene) O,O,O′,O′-tetramethyl phosphorothioate) 1%
INERT INGREDIENTS: 99%
TOTAL 100%

EPA Reg. No. 241-174AA
®Registered trademark of American Cyanamid Company
9/87

KEEP OUT OF REACH OF CHILDREN
CAUTION

PRECAUCION

PRECAUCION AL USUARIO: Si usted no lee ingles, no use este producto hasta que la etiqueta le haya sido explicada ampliamente.

STATEMENT OF PRACTICAL TREATMENT

If in Eyes: Flush with plenty of water. Get medical attention.

If on Skin: Wash with plenty of soap and water after handling. Get medical attention if irritation persists.

NOTE TO PHYSICIANS: This product may cause cholinesterase inhibition. Atropine is antidotal. Pralidoxime chloride (2-PAM; PROTOPAM chloride) may be effective as an adjunct to atropine.

PRECAUTIONARY STATEMENTS

HAZARDS TO HUMANS AND DOMESTIC ANIMALS

CAUTION

Avoid breathing dust. Avoid contact with skin. Wash thoroughly with soap and water after handling. Do not contaminate food or feed products.

ENVIRONMENTAL HAZARDS

This product is toxic to birds and fish. Fish and other aquatic organisms in water treated with this product may be killed. You must consult your State Fish and Game Agency before applying this product to waters or wetlands. Do not contaminate water when disposing of equipment washwaters.

DIRECTIONS FOR USE

It is a violation of Federal law to use this product in a manner inconsistent with its labeling.

STORAGE AND DISPOSAL

PROHIBITIONS: Do not contaminate water, food, or feed by storage or disposal.

PESTICIDE DISPOSAL: Wastes resulting from the use of this product may be disposed of on site or at an approved waste disposal facility.

CONTAINER DISPOSAL: Completely empty bag into application equipment. Then dispose of empty bag in a sanitary landfill or by incineration, or, if allowed by State and local authorities, by burning. If burned, stay out of smoke.

GENERAL: Consult Federal, State or local disposal authorities for approved alternative procedures.

For control of mosquito and midge larvae, apply ABATE 1-SG as specified below:

Areas of Treatment	Lbs. Per Acre
Standing water, shallow ponds, lakes, woodland pools, catch basins	5-10
Tidal water, marshes, swamps and waters high in organic content	10-20

May be applied as a spot treatment to catch basins and similar areas where mosquitoes may breed.

Apply uniformly and repeat applications as necessary. Do not use on crops used for food, forage or pasture.

DISCLAIMER

The label instructions for the use of this product reflect the opinion of experts based on field use and tests. The directions are believed to be reliable and should be followed carefully. However, it is impossible to eliminate all risks inherently associated with use of this product. Crop injury, ineffectiveness or other unintended consequences may result because of such factors as weather conditions, presence of other materials, or the use or application of the product contrary to label instructions, all of which are beyond the control of American Cyanamid Company. All such risks shall be assumed by the user.

American Cyanamid Company warrants only that the material contained herein conforms to the chemical description on the label and is reasonably fit for the use therein described when used in accordance with the directions for use, subject to the risks referred to above.

Any damages arising from a breach of this warranty shall be limited to direct damages and shall not include consequential commercial damages such as loss of profits or values or any other special or indirect damages.

American Cyanamid Company makes no other express or implied warranty, including any other express or implied warranty of FITNESS or of MERCHANTABILITY.

In case of an emergency endangering life or property involving this product, call collect, day or night, Area Code 201-835-3100.

ABATE® 2-CG

INSECTICIDE

FOR CONTROL OF MOSQUITO AND MIDGE LARVAE

ACTIVE INGREDIENT:
Temephos (O,O′-(thiodi-4,1-phenylene) O,O,O′,O′-tetramethyl phosphorothioate) 2%
INERT INGREDIENTS: 98%

TOTAL 100%

EPA Reg. No.241-151AA EPA Est. No. 8329-1L-01

®Registered trademark of American Cyanamid Company

24680-20 D-42 SEPTEMBER 1987

KEEP OUT OF REACH OF CHILDREN

CAUTION

PRECAUCION

PRECAUCION AL USUARIO: Si usted no lee ingles, no use este producto hasta que la etiqueta le haya sido explicada ampliamente.

See Below For Additional Precautionary Statements

STATEMENT OF PRACTICAL TREATMENT

If in Eyes: Flush with plenty of water. Get medical attention.

If on Skin: Wash with plenty of soap and water after handling. Get medical attention if irritation persists.

NOTE TO PHYSICIANS: This product may cause cholinesterase inhibition. Atropine is antidotal. Pralidoxime chloride (2-PAM; PROTOPAM chloride) may be effective as an adjunct to atropine.

PRECAUTIONARY STATEMENTS

HAZARDS TO HUMANS AND DOMESTIC ANIMALS

CAUTION

Avoid breathing dust. May cause eye irritation. Avoid contact with eyes, skin or clothing. Wash thoroughly with soap and water after handling. Do not contaminate food or feed products.

ENVIRONMENTAL HAZARDS

This product is toxic to birds and fish. Fish and other aquatic organisms in water treated with this product may be killed. You must consult your State Fish and Game Agency before applying this product to waters or wetlands. Do not contaminate water when disposing of equipment washwaters.

DIRECTIONS FOR USE

It is a violation of Federal law to use this product in a manner inconsistent with its labeling.

STORAGE AND DISPOSAL

PROHIBITIONS: Do not contaminate water, food, or feed by storage or disposal.

PESTICIDE DISPOSAL: Wastes resulting from the use of this product may be disposed of on site or at an approved waste disposal facility.

CONTAINER DISPOSAL: Completely empty bag into application equipment. Then dispose of empty bag in a sanitary landfill or by incineration, or, if allowed by State and local authorities, by burning. If burned, stay out of smoke.

GENERAL: Consult Federal, State or local disposal authorities for approved alternative procedures.

In case of emergency endangering life or property involving this product, call collect, day or night, Area Code 201-835-3100.

Areas of Treatment	Insects (Larvae)	Lbs./ Acre	Application
Standing water, shallow ponds, lakes, woodland pools, catch basins	Mosquito and Midge	2.5-5	Apply uniformly by air or ground equipment. Repeat applications as necessary.
Tidal waters, marshes, swamps and water high in organic content		10	
Highly polluted waters		25	
Marsh lands, margins of streams, intertidal zones of sandy beaches	Gnat Punkie Sandfly	5-10	Apply uniformly by ground or air equipment. Repeat applications as necessary.

Do not use on crops used for food, forage or pasture.

DISCLAIMER

The label instructions for the use of this product reflect the opinion of experts based on field use and tests. The directions are believed to be reliable and should be followed carefully. However, it is impossible to eliminate all risks inherently associated with use of this product. Crop injury, ineffectiveness or other unintended consequences may result because of such factors as weather conditions, presence of other materials, or the use or application of the product contrary to label instructions, all of which are beyond the control of American Cyanamid Company. All such risks shall be assumed by the user.

American Cyanamid Company warrants only that the material contained herein conforms to the chemical description on the label and is reasonably fit for the use therein described when used in accordance with the directions for use, subject to the risks referred to above.

Any damages arising from a breach of this warranty shall be limited to direct damages and shall not include consequential commercial damages such as loss of profits or values or any other special or indirect damages.

American Cyanamid Company makes no other express or implied warranty, including any other express or implied warranty of FITNESS or of MERCHANTABILITY.

ABATE® 5-CG

INSECTICIDE

FOR CONTROL OF MOSQUITO AND MIDGE LARVAE

ACTIVE INGREDIENT:
Temephos (O, O′-(thiodi-4,1-phenylene) O, O, O′, O′-tetramethyl phosphorothioate) 5%
INERT INGREDIENTS: 95%

TOTAL 100%

EPA Reg. No. 241-150AA

®Registered trademark of American Cyanamid Company

4/88

KEEP OUT OF REACH OF CHILDREN

DANGER/PELIGRO

PRECAUCION AL USUARIO: Si usted no lee ingles, no use este producto hasta que la etiqueta le haya sido explicada ampliamente.

STATEMENT OF PRACTICAL TREATMENT

If in Eyes: Hold eyelids open and flush with a steady, gentle stream of water for 15 minutes.

If Swallowed: Drink promptly a large quantity of milk, egg white, gelatin solution, or, if these are not available, large quantities of water. Avoid alcohol. Do not induce vomiting or give anything by mouth to an unconscious person.

NOTE TO PHYSICIANS: This product may cause cholinesterase inhibition. Atropine is antidotal. Pralidoxime chloride (2-PAM; PROTOPAM chloride) may be effective as an adjunct to atropine.

PRECAUTIONARY STATEMENTS

HAZARDS TO HUMANS AND DOMESTIC ANIMALS

In case of an emergency endangering life or property involving this product, call collect, day or night,
Area Code 201-835-3100.

DANGER

Corrosive: Causes irreversible eye damage. Do not get in eyes or on clothing. Wear goggles. Wash thoroughly with soap and water after handling. Remove contaminated clothing and wash before reuse.

ENVIRONMENTAL HAZARDS

This product is toxic to birds and fish. Fish and other aquatic organisms in water treated with this product may be killed. You must consult your State Fish and Game Agency before applying this product to waters or wetlands. Do not contaminate water when disposing of equipment washwaters.

DIRECTIONS FOR USE

It is a violation of Federal law to use this product in a manner inconsistent with its labeling.

STORAGE AND DISPOSAL

PROHIBITIONS: Do not contaminate water, food, or feed by storage or disposal.

PESTICIDE DISPOSAL: Pesticide wastes are acutely hazardous. Improper disposal of excess pesticide, spray mixture, or rinsate is a violation of Federal law. If these wastes cannot be disposed of by use according to label instructions, contact your State Pesticide or Enviromental Control Agency, or the Hazardous Waste representative at the nearest EPA Regional Office for guidance.

CONTAINER DISPOSAL: Completely empty bag into application equipment. Then dispose of empty bag in a sanitary landfill or by incineration, or, if allowed by State and local authorities, by burning. If burned, stay out of smoke.

GENERAL: Consult Federal, State or local disposal authorities for approved alternative procedures.

For control of mosquito and midge larvae, apply ABATE 5-CG as specified below:

Areas of Treatment	Lbs. Per Acre
Standing water, shallow ponds, lakes, woodland pools	2
Tidal waters, marshes, swamps and waters high in organic content	4
Highly-polluted waters	10

May be applied as a spot treatment to catch basins and similar areas where mosquitoes may breed.

Apply uniformly and repeat applications as necessary. Do not use on crops used for food, forage or pasture.

DISCLAIMER

The label instructions for the use of this product reflect the opinion of experts based on field use and tests. The directions are believed to be reliable and should be followed carefully. However, it is impossible to eliminate all risks inherently associated with use of this product. Crop injury, ineffectiveness or other unintended consequences may result because of such factors as weather conditions, presence of other materials, or the use or application of the product contrary to label instructions, all of which are beyond the control of American Cyanamid Company. All such risks shall be assumed by the user.

American Cyanamid Company warrants only that the material contained herein conforms to the chemical description on the label and is reasonably fit for the use therein described when used in accordance with the directions for use, subject to the risks referred to above.

Any damages arising from a breach of this warranty shall be limited to direct damages and shall not include consequential commercial damages such as loss of profits or values or any other special indirect damages.

American Cyanamid Company makes no other express or implied warranty, including any other express or implied warranty of FITNESS or of MERCHANTABILITY.

ABATE® 4-E

INSECTICIDE

FOR CONTROL OF MOSQUITO LARVAE

ACTIVE INGREDIENT:

Temephos (O,O′-(thiodi-4,1-phenylene) O, O, O′, O′-tetramethyl phosphorothioate)	44.6%
INERT INGREDIENTS:*	55.4%
TOTAL	100.0%

(1 gallon contains 4 lbs. of Temephos)

*contains petroleum distillate

EPA Reg. No. 241-132AA

®Registered trademark of American Cyanamid Company

9/87

KEEP OUT OF REACH OF CHILDREN

WARNING/AVISO

PRECAUCION AL USUARIO: Si usted no lee inglés, no use este producto hasta que la etiqueta le haya sido explicada ampliamente.

STATEMENT OF PRACTICAL TREATMENT

If Swallowed: Do not induce vomiting. Get medical attention.

If in Eyes: Flush with plenty of water. Get medical attention.

If on Skin: Wash with plenty of soap and water after handling. Get medical attention if irritation persists.

NOTE TO PHYSICIANS: This product may cause cholinesterase inhibition. Atropine is antidotal. Pralidoxime chloride (2-PAM; PROTOPAM chloride) may be effective as an adjunct to atropine.

PRECAUTIONARY STATEMENTS

HAZARDS TO HUMANS AND DOMESTIC ANIMALS

WARNING

Harmful if swallowed or absorbed through skin. Causes skin and eye irritation. Wash thoroughly with soap and water after handling. Avoid breathing spray mist. Do not contaminate food or feed products.

ENVIRONMENTAL HAZARDS

This product is toxic to birds and fish. Fish and other aquatic organisms in water treated with this product may be killed. You must consult your State Fish and Game Agency before applying this product to waters or wetlands. Do not contaminate water when disposing of equipment washwaters. This product is toxic to bees and should not be applied when bees are actively visiting the treatment area.

DIRECTIONS FOR USE

It is a violation of Federal law to use this product in a manner inconsistent with its labeling.

STORAGE AND DISPOSAL

PROHIBITIONS: Do not contaminate water, food, or feed by storage or disposal. Do not use or store near heat or open flame.

PESTICIDE DISPOSAL: Wastes resulting from the use of this product may be disposed of on site or at an approved waste disposal facility.

CONTAINER DISPOSAL: Triple rinse (or equivalent). Then offer for recycling or reconditioning, or puncture and dispose of in a sanitary landfill, or by incineration or, if allowed by State and local authorities by burning. If burned, stay out of smoke.

GENERAL: Consult Federal, State or local disposal authorities for approved alternative procedures.

MOSQUITO LARVAE CONTROL

Apply ABATE 4-E as a uniform spray in sufficient water for good coverage at the rate of 0.5 to 1.5 fluid ounces per acre. Apply to standing water, moist areas, woodland pools, shallow ponds, around the edges of lakes, swamps, marshes, tidal waters, and catch basins. Use 0.5 fluid ounces per acre in clean open water and 1.0 to 1.5 fluid ounces per acre in waters high in organic matter content. The higher rate should be used in areas known to have organophosphate resistant mosquitoes. Repeat applications as necessary.

Do not use on crops used for food, forage or pasture.

DISCLAIMER

The label instructions for the use of this product reflect the opinion of experts based on field use and tests. The directions are believed to be reliable and should be followed carefully. However, it is impossible to eliminate all risks inherently associated with use of this product. Crop injury, ineffectiveness or other unintended consequences may result because of such factors as weather conditions, presence of other materials, or the use or application of the product contrary to label instructions, all of which are beyond the control of American Cyanamid Company. All such risks shall be assumed by the user.

American Cyanamid Company warrants only that the material contained herein conforms to the chemical description on the label and is reasonably fit for the use therein described when used in accordance with the directions for use, subject to the risks referred to above.

Any damages arising from a breach of this warranty shall be limited to direct damages and shall not include consequential commercial damages such as loss of profits or values or any other special or indirect damages.

American Cyanamid Company makes no other express or implied warranty, including any other express or implied warranty of FITNESS or of MERCHANTABILITY.

In case of an emergency endangering life or property involving this product, call collect, day or night, Area Code 201-835-3100.

AMDRO®

INSECTICIDE

ACTIVE INGREDIENT:	BY WT.
Hydramethylnon: Tetrahydro-5,5-dimethyl-2(1*H*)-pyrimidinone(3-[4-(trifluoromethyl)phenyl]-1-(2-[4-(trifluoromethyl)phenyl]-ethenyl)-2-propenylidene)hydrazone	0.88%
INERT INGREDIENTS:	99.12%
TOTAL	100.00%

EPA Reg. No. 241-260

®Registered Trademark of American Cyanamid Company

9/88

KEEP OUT OF REACH OF CHILDREN

CAUTION

PRECAUCION

PRECAUCION AL USUARIO: Si usted no lee inglés, no use este producto hasta que la etiqueta le haya sido explicada ampliamente.

In case of an emergency endangering life or property involving this product, call collect, day or night, Area Code 201-835-3100.

PRECAUTIONARY STATEMENTS

HAZARDS TO HUMANS AND DOMESTIC ANIMALS

CAUTION!

MAY BE HARMFUL IF SWALLOWED

Avoid any prolonged contact with skin or eyes. Wash thoroughly after handling.

STATEMENT OF PRACTICAL TREATMENT

IF SWALLOWED, drink one or two glasses of water and induce vomiting by touching back of throat with finger. Do not induce vomiting or give anything by mouth to an unconscious person. Get medical attention.

ENVIRONMENTAL HAZARDS

This product is toxic to fish. Do not apply directly to water. Do not contaminate water when disposing of equipment washwaters. Do not apply when weather conditions favor drift from target areas.

DIRECTIONS FOR USE

It is a violation of Federal law to use this product in a manner inconsistent with its labeling.

STORAGE AND DISPOSAL

PROHIBITIONS: Do not contaminate water, food or feed by storage or disposal. Open dumping is prohibited.

STORAGE: STORE IN COOL, DRY, SECURE PLACE AND KEEP CONTAINER TIGHTLY CLOSED. AMDRO is formulated in an oil bait that functions as an attractant to ants. Prolonged exposure to air may turn oil rancid and reduce the attractiveness of the bait. This product may be an attractant to pets and rodents. Use within 3 months after opening.

PESTICIDE DISPOSAL: Wastes resulting from the use of this product may be disposed of on site or at an approved waste disposal facility.

CONTAINER DISPOSAL: Completely empty bag into application equipment. Then dispose of empty bag in a sanitary landfill or by incineration, or, if allowed by State and local authorities, by burning. If burned, stay out of smoke.

GENERAL: Consult Federal, State, or local disposal authorities for approved alternative procedures.

AMDRO is for use on lawns, turf, sod, pasture, range grass, noncropland, and non-food bearing nursery stock for the control of fire ants, harvester ants and big-headed ants.

Treatment	Rate	Comments
Individual mound treatment: fire ants harvester ants big-headed ants	5 level tablespoons	Distribute bait uniformly around the base of the mound. Do not disturb ant mounds. Do not apply directly to the top of the mound. Do not apply more than 1½ lbs of bait per acre. Do not contaminate kitchen utensils by use or storage.
Broadcast treatment: fire ants harvester ants	1-1½ lb/acre ¼-⅓ lb/ 10,000 square feet	Broadcast bait uniformly with ground equipment (spinning disk types) or fixed or rotary wing aerial equipment.
Broadcast treatment: big-headed ants	1-2 lb/acre ¼-½ lb/ 10,000 square feet	Broadcast bait uniformly with ground equipment (spinning disk types) or fixed or rotary wing aerial equipment.

Apply when ants are active (typically when soil temperature is greater than 60°F).

AMDRO is a bait which is highly attractive to fire ants, harvester ants and big-headed ants. It works by becoming part of the ant food chain. Worker ants will find the AMDRO bait, carry it back to the mound and pass it along to the queen and the other ants in the mound.

An effective ant insecticide must be slow-acting so that it can be passed by the workers throughout the ant mound and eventually to the queen. AMDRO is a slow-acting insecticide and is especially effective against the queen. Typically, in 1-4 weeks the queen and a number of ants are killed. Within 2-8 weeks, a visible reduction in mound activity is observed. Very large mounds may continue to be active for a period of time even though the queen is dead and no young are being produced. Retreatment may be desirable under these circumstances.

Further information regarding ants may be obtained from your Cooperative Agricultural Extension Service.

DISCLAIMER

The label instructions for the use of this product reflect the opinion of experts based on field use and tests. The directions are believed to be reliable and should be followed carefully. However, it is impossible to eliminate all risks inherently associated with use of this product. Crop injury, ineffectiveness or other unintended consequences may result because of such factors as weather conditions, presence of other materials, or the use or application of the product contrary to label instructions all of which are beyond the control of American Cyanamid Company. All such risks shall be assumed by the user. American Cyanamid Company warrants only that the material contained herein conforms to the chemical description on the label and is reasonably fit for the use therein described when used in accordance with the directions for use, subject to the risks referred to above.

Any damages arising from a breach of this warranty shall be limited to direct damages and shall not include consequential commercial damages such as loss of profits or values or any other special or indirect damages.

American Cyanamid Company makes no other express or implied warranty, including any other express or implied warranty of FITNESS or of MERCHANTABILITY.

ARSENAL®

HERBICIDE

ACTIVE INGREDIENT:
Isopropylamine salt of Imazapyr (2-[4,5-dihydro-4-methyl-4-(1-methylethyl)-5-oxo-1*H*-imidazol-2-yl]-3-pyridinecarboxylic acid)* 27.6%
INERT INGREDIENT: 72.4%
TOTAL 100.0%

*Equivalent to 22.6% 2-[4,5-dihydro-4-methyl-4-(1-methylethyl)-5-oxo-1*H*-imidazol-2-yl]-3-pyridinecarboxylic acid or 2 pounds acid per gallon.

EPA Reg. No. 241-273

®Registered Trademark of American Cyanamid Company

9/88

KEEP OUT OF REACH OF CHILDREN

CAUTION!

PRECAUCION

PRECAUCION AL USUARIO: Si usted no lee ingles, no use este producto hasta que la etiqueta le haya sido explicado ampliamente.

In case of emergency endangering life or property involving this product, call collect, day or night, Area Code 201-835-3100.

PRECAUTIONARY STATEMENTS

HAZARDS TO HUMANS

CAUTION!

Avoid contact with skin, eyes or clothing. Avoid breathing spray mist. Wash thoroughly with soap and water after handling. Remove contaminated clothing and wash before reuse.

FIRST AID

IF ON SKIN: Wash with plenty of soap and water.

IF IN EYES: Flush with plenty of water. Get medical attention if irritation persists.

PHYSICAL AND CHEMICAL HAZARDS

Spray solutions of ARSENAL should be mixed, stored and applied only in stainless steel, fiberglass, plastic and plastic-lined steel containers.

DO NOT mix, store or apply ARSENAL or spray solutions of ARSENAL in unlined steel (except stainless steel) containers or spray tanks.

ENVIRONMENTAL HAZARDS

DO NOT apply directly to water or wetlands. **DO NOT** contaminate water when disposing of equipment washwaters.

IMPORTANT

DO NOT use on food or feed crops. **DO NOT** treat irrigation ditches. Keep from contact with fertilizers, insecticides, fungicides and seeds. **DO NOT** apply or drain or flush equipment on or near desirable trees or other plants, or on areas where their roots may extend, or in locations where the chemical may be washed or moved into contact with their roots. **DO NOT** use on lawns, walks, driveways, tennis courts, or similar areas. **DO NOT** side trim desirable vegetation with this product. Prevent drift of spray to desirable plants. **DO NOT USE** in California.

Clean application equipment after using this product by thoroughly flushing with water.

GENERAL INFORMATION

ARSENAL herbicide is an aqueous solution to be mixed in water and applied as a spray for control of most annual and perennial grasses and broadleaf weeds on noncropland areas.

ARSENAL may be applied either preemergence or postemergence to the weeds; however, postemergence application is the method of choice in most situations, particularly for control of perennials. For maximum activity, weeds should be growing vigorously at the time of postemergence applications. The preemergence activity of ARSENAL will provide residual control of most weed species following a postemergence application.

ARSENAL is readily absorbed through foliage and roots and is translocated rapidly throughout the plant, with accumulation in the meristematic regions. Treated plants stop growing soon after spray application. Chlorosis appears first in the newest leaves, and necrosis spreads from this point.

In perennials, the herbicide is translocated into and kills underground storage organs, thus preventing regrowth. Chlorosis and tissue necrosis may not be apparent in some plant species until two weeks after application. Complete kill of plants may not occur for several weeks.

DISCLAIMER

The label instructions for the use of this product reflect the opinion of experts based on field use and tests. The directions are believed to be reliable and should be followed carefully. However, it is impossible to eliminate all risks inherently associated with use of this product. Ineffectiveness or other unintended consequences may result because of such factors as weather conditions, presence of other materials, or the use or application of the product contrary to label instructions, all of which are beyond the control of American Cyanamid Company. All such risks shall be assumed by the user.

American Cyanamid Company warrants only that the material contained herein conforms to the chemical description on the label and is reasonably fit for the use therein described when used in accordance with the directions for use, subject to the risks referred to above.

Any damages arising from a breach of this warranty shall be limited to direct damages and shall not include consequential commercial damages such as loss of profits or values or any other special or indirect damages.

American Cyanamid Company makes no other express or implied warranty, including other express or implied warranty of FITNESS or of MERCHANTABILITY.

DIRECTIONS FOR USE

It is a violation of Federal law to use this product in a manner inconsistent with its labeling.

ARSENAL herbicide should be used only in accordance with recommendations on the leaflet label attached to the container. Keep containers closed to avoid spills and contamination.

A postemergence use of ARSENAL is recommended for control of most annual and perennial grasses and broadleaf weeds on noncropland areas such as railroad, utility, highway and pipeline rights-of-way, utility plant sites, petroleum tank farms, pumping installations, fence rows, storage areas, non-irrigation ditch banks, and other similar areas. ARSENAL is recommended for the establishment and maintenance of wildlife openings. ARSENAL may also be used for the release of improved bermudagrass (see specific labeling).

STORAGE AND DISPOSAL

PROHIBITIONS: DO NOT store below 10 degrees F. DO NOT contaminate water, food or feed by storage or disposal.

PESTICIDE DISPOSAL: Wastes resulting from the use of this product may be disposed of on site or at an approved waste disposal facility.

CONTAINER DISPOSAL: Triple rinse (or equivalent). Then offer for recycling or reconditioning, or puncture and dispose of in an approved sanitary landfill, or by incineration, or, if allowed by state and local authorities, by burning. If burned, stay out of smoke.

FOR CONTROL OF UNDESIRABLE WEEDS IN UNIMPROVED BERMUDAGRASS

For use on unimproved bermudagrass turf such as roadsides, utility rights-of-way and other non-cropland industrial sites. The application of ARSENAL on established Common and Coastal Bermudagrass provides control of labeled broadleaf and grass weeds. Competition from these weeds is eliminated, releasing the bermudagrass. Treatment of bermudagrass with ARSENAL results in a compacted growth habit and seedhead inhibition.

Uniformly apply with properly calibrated ground equipment using at least 20 gallons of water per acre with a spray pressure 20 to 50 psi.

DOSAGE RATES AND TIMING

EARLY SPRING—DORMANT: Apply ARSENAL at 6 to 12 fluid oz. per acre when the bermudagrass is still dormant and has not initiated new growth.

SPRING—UP TO 25% GREEN-UP: Apply ARSENAL at 6 to 8 fluid oz. per acre after the bermudagrass has initiated green-up but has not exceeded 25% green-up.

WEEDS CONTROLLED

Bedstraw (*Galium* spp.)
Bishopweed *(Ptilimnium capillaceum)*
Buttercup *(Ranunculos parviflorus)*
Carolina geranium *(Geranium carolinianum)*
Fescue (*Festuca* spp.)
Foxtail (*Setaria* spp.)
Little barley *(Hordeum pusillum)*
Seedling Johnsongrass *(Sorghum halepense)*
Wild carrot *(Daucus carota)*
White clover *(Trifolium repens)*
Yellow woodsorrel *(Oxalis stricta)*

1. DO NOT APPLY to bermudagrass during its first growing season.
2. DO NOT APPLY to bermudagrass that is under stress from drought, disease, insects, or other causes.
3. Temporary yellowing of bermudagrass may occur when treatment is made after growth commences.
4. DO NOT add a surfactant.

APPLICATION EQUIPMENT AND TECHNIQUES

ARSENAL herbicide may be applied with the following application equipment:

Aerial—fixed wing and helicopter.

Boom—conventional boom mounted, manifold mounted, and off-center nozzles.

Low-volume hand-held spray equipment-backpack, knapsack and other pump-up type pressure sprayers and backpack mist blowers used to direct application to weed foliage.

High-volume spray equipment — pressure handguns and vehicle mounted high-volume directed spray equipment.

AERIAL EQUIPMENT

Uniformly apply the recommended amount of ARSENAL with properly calibrated aerial equipment in 5 to 30 gallons of water per acre. All precautions should be taken to minimize or eliminate spray drift. Aerial equipment designed to minimize spray drift, such as a helicopter equipped with a MICROFOIL boom, THRU-VALVE boom or raindrop nozzles, must be used. Applications should not be made under gusty conditions or when wind velocity exceeds 5 mph. Except when applying with a MICROFOIL boom, a drift control agency may be added at the recommended label rate. A foam reducing agent may be added at the recommended label rate, if needed.

IMPORTANT: DO NOT make applications by fixed wing aircraft unless appropriate buffer zones can be maintained to insure that drift does not occur off the target area. Thoroughly clean application equipment, including landing gear, immediately after use. Prolonged exposure of this product to uncoated steel (except stainless steel) surfaces may result in corrosion and failure of the exposed part.

BOOM EQUIPMENT

Mix the recommended amount of ARSENAL in 15 to 60 gallons of water per acre in the spray tank with the agitator running. A foam reducing agent may be added at the recommended label rate, if needed, If desired, a spray pattern indicator may be added at the recommended label rate. Check for even distribution in spray pattern.

IMPORTANT: To minimize drift, select proper nozzles to avoid spraying a fine mist, DO NOT exceed spray pressure of 50 psi, and DO NOT spray under windy or gusty conditions.

Clean application equipment after using this product by thoroughly flushing with water.

LOW-VOLUME HAND-HELD SPRAY EQUIPMENT

Thoroughly mix a 1/2 to 1 percent solution of ARSENAL in water. To determine the proper percent solution of ARSENAL to use, see the "WEEDS CONTROLLED" section of this label and the "PERCENT SOLUTION RATE GUIDE" below.

Percent Solution Rate Guide

Arsenal Rate Per Acre	Percent Solution to Mix
2–3 pints	½%
3–4 pints	¾%
4–6 pints	1%

For best results, uniformly cover the foliage of the vegetation to be controlled with the spray solution.

DO NOT over apply and cause runoff from the treated foliage.

To mix the spray solution, add the volume of ARSENAL indicated in the table below to the desired amount of water.

Spray Solution Mixing Guide

Solution Volume	Amount of Arsenal to Use (fluid ounces)		
	½%	¾%	1%
1 gallon	⅔ fl. oz.	1 fl. oz.	1⅓ fl. oz.
5 gallons	3⅓ fl. oz.	5 fl. oz.	6½ fl. oz.
10 gallons	6⅔ fl. oz.	10 fl. oz.	13 fl. oz.
25 gallons	1 pint	1½ pints	2 pints

2 tablespoons = 1 fluid ounce

It is not necessary to add additional surfactants or wetting agents.

IMPORTANT: DO NOT exceed recommended dosage rate per acre. DO NOT side-trim desirable vegetation with this product. Clean application equipment after using this product by thoroughly flushing with water.

HIGH-VOLUME SPRAY EQUIPMENT

ARSENAL may be applied using high-volume spray equipment. For best results, apply ARSENAL using the least amount of water practical to obtain uniform coverage of the vegetation foliage. Using excessive spray volumes which cause runoff from the plant foliage may result in reduced performance.

When using spray volumes greater than 60 gallons per acre, additional nonionic surfactant such as Ortho* 1 X-77 must then be added at the rate of 1 quart per 100 gallons of spray solution to provide optimum wetting and/or contact activity. A foam reducing agent may be added at the recommended label rate, if needed. If desired, a spray pattern indicator may be added at the recommended label rate.

To mix the spray solution, determine the proper ARSENAL pints per acre rate from the "WEEDS CONTROLLED" section of this label and mix according to the table below.

Spray Solution Mixing Guide

Spray Volume (Gal/Acre)	Pints Arsenal to Mix Per 100 Gallons Water			
	3 Pts/Acre	4 Pts/Acre	5 Pts/Acre	6 Pts/Acre
50	6	8	10	12
100	3	4	5	6
150	2	2¾	3¼	4

IMPORTANT: DO NOT side-trim desirable vegetation with this product. Clean application equipment after using this product by thoroughly flushing with water.

*Registered trademark of Chevron Chemical Company.

FOR CONTROL OF UNDESIRABLE WEEDS UNDER PAVED SURFACES

ARSENAL herbicide can be used under asphalt, pond liners and other paved areas, ONLY in industrial sites or where the pavement has a suitable barrier along the permimeter that prevents encroachment of roots of desirable plants.

ARSENAL should be used only where the area to be treated has been prepared according to good construction practices. If rhizomes, stolons, tubers or other vegetative plant parts are present in the site, they should be removed by scalping with a grader blade to a depth sufficient to insure their complete removal.

Application Directions:

Applications should be made to the soil surface only when final grade is established. Do not move soil following ARSENAL application.

Uniformly apply ARSENAL to the area to be surfaced, including the shoulder areas at a rate of 6 pts. per acre.

Apply ARSENAL in sufficient water (at least 100 gal. per acre) to insure thorough wetting of the soil surface. Add the recommended amount of ARSENAL to clean water in the spray tank during the filling operation. Agitate before spraying.

For Herbicide Activation:

On Moist Subsoils: Apply ARSENAL after final grading and immediately before laying asphalt or liner surface. Apply uniformly, using at least 100 gallons of water per acre.

If Moisture Is Not Present: Incorporation of ARSENAL is needed for herbicide activation. ARSENAL can be incorporated into the soil to a depth of 4 to 6 inches using a rototiller or disc. Rainfall or irrigation of 1 inch will also provide uniform incorporation. Do not allow treated soil to wash or move into untreated areas.

IMPORTANT

Paving should follow ARSENAL applications as soon as possible. DO NOT apply where the chemical may contact the roots of desirable trees or other plants.

The product is not recommended for use under pavement on residential properties such as driveways or parking lots, nor is it recommended for use in recreational areas such as under bike or jogging paths, golf cart paths, or tennis courts, or where landscape plantings could be anticipated.

Injury or death of desirable plants may result if this product is applied where roots are present or where they may extend into the treated area. Roots of trees and shrubs may extend a considerable distance beyond the branch extremities or so called drip line.

WEEDS CONTROLLED

ARSENAL herbicide will provide postemergence control with residual control of the following target vegetation species at the rates listed. ARSENAL herbicide should be used only in accordance with the recommendation on this label.

GRASSES

Apply 2-3 pints per acre*

Annual blue grass *(Poa annua)*
Broadleaf signalgrass *(Brachiaria platphylla)*
Canada bluegrass *(Poa compressa)*
Downy brome *(Bromus tectorum)*
Fescue *(Festuca* spp.)
Foxtail *(Setaria* spp.)
Italian ryegrass *(Lolium multiflorum)*
Johnsongrass *(Sorghum halepense)*
Kentucky bluegrass *(Poa pratensis)*
Lovegrass *(Eragrostis* spp.)
Orchardgrass *(Dactylis glomerata)*
Paragrass *(Brachiaria mutica)*
Quackgrass *(Agropyron repens)*
Saltgrass *(Distichlis stricta)*
Sandbur *(Cenchrus* spp.)
Smooth brome *(Bromus inermis)*
Vaseygrass *(Paspalum urvillei)*
Wild oats *(Avena fatua)*
Witchgrass *(Panicum capillare)*

Apply 3-4 pints per acre*

Beardgrass *(Andropogon* spp.)
Cheat *(Bromus secalinus)*
Crabgrass *(Digitaria* spp.)
Fall panicum *(Panicum dichotomiflorum)*
Goosegrass *(Eleusine indica)*
Prairie threeawn *(Aristida oligantha)*
Reed canary grass *(Phalaris arundinacea)*
Torpedograss *(Panicum repens)*
Wild barley *(Hordeum* spp.)

Apply 4-6 pints per acre*

Bahiagrass *(Paspalum notatum)*
Bermudagrass *(Cynodon dactylon)*
Big bluestem *(Andropogon geradi)*
Cattail *(Typha* spp.)
Dallisgrass *(Paspalum dilatatum)*
Feathertop *(Pennisetum villosum)*
Guineagrass *(Panicum maximum)*
Phragmites *(Phragmites australis)*
Prairie cordgrass *(Spartina pectinata)*
Sand dropseed *(Sporobolus cryptandrus)*
Timothy *(Phleum pratense)*
Wirestem muhly *(Muhlenbergia frondosa)*

BROADLEAF WEEDS

Apply 2-3 pints per acre*

Burdock *(Arctium* spp.)
Camphorweed *(Heterotheca subaxillaris)*
Carpetweed *(Mullugo verticillata)*
Carolina geranium *(Geranium carolinianum)*
Clover *(Trifolium* spp.)
Common chickweed *(Stellaria media)*
Common ragweed *(Ambrosia artemislifolia)*
Dandelion *(Taraxacum officinale)*
Dogfennel *(Eupatorium capillifolium)*
Filaree *(Erodium* spp.)
Fleabane *(Erigeron* spp.)
Hoary vervain *(Verbena stricta)*
Horseweed *(Conyza canadensis)*
Indian mustard *(Brassica juncea)*
Kochia *(Kochia scoparia)*
Lambsquarters *(Chenopodium album)*
Lespedeza *(Lespedeza* spp.)
Miners lettuce *(Montia perfoliata)*
Mulleim *(Verbascum* spp.)
Nettleleaf goosefoot *(Chenopodium murale)*
Oxeye daisy *(Chrysanthemum leucanthemum)*
Pepperweed *(Lepidium* spp.)
Pigweed *(Amaranthus* spp.)
Plantain *(Plantago* spp.)
Russian thistle *(Salsola kali)*
Smartweed *(Polygonum* spp.)
Sorrell *(Rumex* spp.)
Sunflower *(Helianthus* spp.)
Sweet clover *(Melilotus* spp.)
Tansymustard *(Descurainia pinnata)*
Western ragweed *(Ambrosia psilostachya)*
Wild carrot *(Daucus carota)*
Wild lettuce *(Lactuca* spp.)
Wild parsnip *(Pastinaca sativa)*
Wild turnip *(Brassica campestris)*
Wollyleaf bursage *(Ambrosia grayi)*
Yellow woodsorrel *(Oxalis stricta)*

Apply 3-4 pints per acre*

Bull thistle *(Circium volgare)*
Cocklebur *(Xanthium strumarium)*
Dock *(Rumex* spp.)
Goldenrod *(Solidago* spp.)
Pokeweed *(Phytolacca americana)*
Purslane *(Portulaca* spp.)
Saltbush *(Atriplex* spp.)
Yellow starthistle *(Centaurea solstitialis)*

Apply 4-6 pints per acre*

Arrowwood *(Pluche sericea)*
Canada thistle *(Circium arvense)*
Giant ragweed *(Ambrosia trifida)*
Little mallow *(Malva parviflora)*
Milkweed *(Asclepias* spp.)
Primrose *(Oenothera kunthiana)*
Silverleaf nightshade *(Solanum elaeagnifolium)*
Sowthistle *(Sonchus* spp.)
Texas thistle *(Circium texanum)*

VINES AND BRAMBLES

Apply 1 pint per acre		
Field bindweed *(Convolvulus arvensis)*		Hedge bindweed *(Calystegia sequium)*
Apply 2-3 pints per acre*		
Wild buckwheat *(Polygonum convolvulus)*		
Apply 3-4 pints per acre*		
Greenbriar *(Smilax spp.)*	Morningglory *(Ipomoea spp.)*	Wild rose *(Rosa spp.)*
Honeysuckle *(Lonicera spp.)*	Poison Ivy *(rhus radicans)*	Including: Multiflora rose *(Rosa multiflora)*
	Redvine *(Brunnichia cirrhosa)*	Macartney rose *(Rosa bractreata)*
Apply 4-6 pints per acre*		
Blackberry *(Rubus spp.)***	Kudzu *(Pueraria lobata)*	Virginia creeper *(Parthenocissus quinquefolia)*
Dewberry *(Rubus spp.)***	Trumpetcreeper *(Campsis radicans)*	Wild grape *(Vitis spp.)*

BRUSH SPECIES

Apply 4-6 pints per acre*		
American beech *(Fagus grandifolia)*	Dogwood *(Cornus spp.)*	Red alder *(Alnus rubra)*
Ash *(Fraxinus spp.)*	Hawthorn *(Crataegus spp.)*	Red Maple *(Acer rubrum)*
Bald cypress *(Taxodium distichum)*	Hickory *(Carya spp.)*	Saltcedar *(Tamarix ramosissima)*
Bigleaf maple *(Acer macrophylum)*	Maple *(Acer spp.)*	Sassafras *(Sassafras albidum)*
Blackgum *(Nyssa sylvatica)*	Mulberry *(Morus spp.)*	Sourwood *(Oxydendrum arboreum)*
Boxelder *(Acer negundo)*	Oak *(Quercus spp.)*	Sumac *(Rhus spp.)*
Cherry *(Prunus spp.)*	Persimmon *(Diospyros virginiana)*	Sweetgum *(Liquidambar styraciflua)*
Chinaberry *(Melia azadarach)*	Poplar *(Populos spp.)*	Willow *(Salix spp.)*
Chinese tallow-tree *(Sapium sebiferum)*	Privet *(Ligustrum vulgare)*	Yellow poplar *(Liriodendron tulipifera)*

*The higher rates should be used where heavy or well established infestations occur.

**The degree of control is species dependent. Some Rubus species may not be controlled.

ARSENAL®
HERBICIDE
FORESTRY GRANULE

ACTIVE INGREDIENT:

Imazapyr (2-[4,5-dihydro-4-methyl-4-(1-methylethyl)-5-oxo-1*H*-imidazol-2-yl]-3-pyridinecarboxylic acid)	5.0%
INERT INCREDIENTS:	95.0%
TOTAL	100.0%

EPA Reg. No. 241-308
8/89

KEEP OUT OF REACH OF CHILDREN
CAUTION!

PRECAUCION

AL USUARIO: Si usted no lee ingles, no use este producto hasta que la etiqueta la haya sido explicada ampliamente.

In case of an emergency endangering life or property involving this product, call collect, day or night, Area Code 201-835-3100

PRECAUTIONARY STATEMENTS

HAZARDS TO HUMANS
CAUTION!

Avoid contact with skin, eyes or clothing. Avoid breathing dust. Wash thoroughly with soap and water after handling. Remove contaminated clothing and wash before reuse.

FIRST AID

IF ON SKIN: Wash with plenty of soap and water.

IF IN EYES: Flush with plenty of water. Get medical attention if irritation persists.

PHYSICAL AND CHEMICAL HAZARDS

ARSENAL granules should be mixed and stored only in lined paper bags, stainless steel, fiberglass, plastic and plastic-lined steel containers.

DO NOT mix or store ARSENAL granules in unlined steel (except stainless steel) containers.

ENVIRONMENTAL HAZARDS

DO NOT apply directly to water or wetlands. DO NOT contaminate water when disposing of equipment washwaters.

IMPORTANT

DO NOT use on food or feed crops. DO NOT apply on ditches used to transport irrigation water. DO NOT apply where runoff or irrigation water may flow onto agricultural land as injury to crops may result. Keep from contact with fertilizers, insecticides, fungicides and seeds. DO NOT apply or drain or flush equipment on or near desirable trees or other plants, or on areas where their roots may extend, or in locations where the chemical may be washed or moved into contact with their roots. DO NOT use on lawns, walks, driveways, tennis courts, or similar areas. DO NOT use in California.

GENERAL INFORMATION

ARSENAL herbicide forestry granule is applied for control of most annual and perennial grasses, broadleaf weeds, brush and hardwood trees under forestry conditions.

ARSENAL herbicide forestry granule is readily absorbed through the roots and is translocated rapidly throughout the plant, with accumulation in the meristematic regions. Treated plants stop growing soon after application. Chlorosis appears first in the newest leaves, and necrosis spreads from this point. In perennials, the herbicide is translocated into and kills underground storage organs, thus preventing regrowth. Complete kill of plants may not occur for several weeks.

DISCLAIMER

The label instructions for the use of this product reflect the opinion of experts based on field use and tests. The directions are believed to be reliable and should be followed carefully. However, it is impossible to eliminate all risks inherently associated with use of this product. Ineffectiveness or other unintended consequences may result because of such factors as weather conditions, presence of other materials, or the use or application of the product contrary to label instructions, all of which are beyond the control of American Cyanamid Company. All such risks shall be assumed by the user.

American Cyanamid Company warrants only that the material contained herein conforms to the chemical description on the label and is reasonably fit for the use therein described when used in accordance with the directions of use, subject to the risks referred to above.

Any damages arising from a breach of this warranty shall be limited to direct damages and shall not include consequential commercial damages such as loss of profits or values of any other special or indirect damages.

American Cyanamid Company makes no other express or implied warranty, including other express or implied warranty of FITNESS or of MERCHANTABILITY.

DIRECTIONS FOR USE

It is a violation of Federal law to use this product in a manner inconsistent with its labeling.

An application of ARSENAL herbicide forestry granule is recommended for control of most annual and perennial grasses and broadleaf weeds. ARSENAL herbicide forestry granule may be applied anytime during the growing season.

STORAGE AND DISPOSAL

PROHIBITIONS: DO NOT store below 10°F. DO NOT contaminate water, food or feed by storage or disposal.

PESTICIDE DISPOSAL: Wastes resulting from the use of this product may be disposed of on site or at an approved waste disposal facility.

CONTAINER DISPOSAL: Completely empty bag into application equipment. Then dispose of empty bag in a sanitary landfill or by incineration, or, if allowed by state and local authorities, by burning. If burned, stay out of smoke.

APPLICATION INSTRUCTIONS

Uniformly apply with properly calibrated aerial or ground equipment at a rate of 0.5 to 1.5 lbs. ae/A (10 to 30 lbs. of product per acre).

SITE PREPARATION

ARSENAL herbicide forestry granule may be used to control labeled grasses, broadleaf weeds, vines and brambles, and woody brush and trees on forest sites in advance of regenerating. Herbaceous grass and broadleaf weeds killed will provide fuel to facilitate a controlled fire, if desired, to control residual pines.

Apply 10 to 30 pounds of ARSENAL herbicide forestry granule per acre (0.5 to 1.5 lb ae/Acre) when making helicopter or mechanical ground applications. Use the higher label rates of ARSENAL herbicide forestry granule when controlling particularly dense, multi-layered, canopies of hardwood stands, or difficult to control species.

DO NOT plant loblolly pine seedling on sites treated with ARSENAL herbicide forestry granule for three months following application or injury may occur.

CONIFER RELEASE

ARSENAL herbicide forestry granule may be applied as a broadcast application for control of labeled weed species in loblolly pine. Some minor conifer growth inhibition may be observed when release treatments are made during periods of most active conifer growth. To minimize potential conifer injury, release treatments may be made after formation of final conifer resting buds in the fall.

FOR RELEASE OF LOBLOLLY PINE: Apply 10 to 30 pounds of ARSENAL herbicide forestry granule per acre (0.5 to 1.5 lb. ai/Acre) when making helicopter or mechanical ground spray applications. Use the higher label rates of ARSENAL herbicide forestry granule when controlling particularly dense stands or difficult to control species. Do not make broadcast applications of ARSENAL herbicide forestry granule to loblolly pine stands younger than three years old. Directed applications may be made in loblolly plantations of all ages by targeting the unwanted vegetation and avoiding direct application near the loblolly pine.

WEEDS CONTROLLED

ARSENAL herbicide forestry granule may be used for the control and residual control of the following target vegetation species when applied at the recommended rates. ARSENAL herbicide forestry granule should be used only in accordance with the recommendation on this label.

GRASSES

The species of annual and perennial grasses controlled by ARSENAL herbicide forestry granule include:

ANNUAL BLUEGRASS (*Poa annua*)
BAHIAGRASS (*Paspalum notatum*)
BEARDGRASS (*Andropogon* spp.)
BERMUDAGRASS (*Cynodon dactylon*)
BIG BLUESTEM (*Andropogon geradi*)
BROADLEAF SIGNALGRASS (*Brachiaria platphylla*)
CANADA BLUEGRASS (*Poa compressa*)
CATTAIL (*Typha* spp.)
CHEAT (*Bromus secalinus*)
CRABGRASS (*Digitaria* spp.)
DALLISGRASS (*Paspalum dilatatum*)
DOWNY BROME (*Bromus tectorum*)
FALL PANICUM (*Panicum dichotomiflorum*)
FEATHERTOP (*Pennisetum vitiosum*)
FESCUE (*Festuca* spp.)
FOXTAIL (*Setaria* spp.)
GOOSEGRASS (*Eleusine indica*)
GUINEAGRASS (*Panicum maximum*)
ITALIAN RYEGRASS (*Lolium multiflorum*)
JOHNSONGRASS (*Sorghum halepense*)
KENTUCKY BLUEGRASS (*Poa pratensis*)
LOVEGRASS (*eragrostis* spp.)
ORCHARDGRASS (*Dactylis glomerata*)
PARAGRASS (*Brachiaria mutica*)
PHRAGMITES (*Phragmites australis*)
PRAIRIE CORDGRASS (*Spartina pectiata*)
PRAIRIE THREEAWN (*Aristida oligantha*)
QUACKGRASS (*Agropyron repens*)
REED CANARY GRASS (*Phalaris arundinacea*)
SALTGRASS (*Distichlis stricta*)
SAND DROPSEED (*Sporobolus crptandrus*)
SANDBUR (*Cenchrus* spp.)
SMOOTH BROME (*Bromus inermis*)
TIMOTHY (*Phleum pratense*)
TORPEDOGRASS (*Ponicum repens*)
VASEYGRASS (*Paspalum urvillei*)
WILD BARLEY (*Hordeum* spp.)
WILD OATS (*Avena fatua*)
WIRESTEM MUHLY (*Muhlenbergia frondosa*)
WITCHGRASS (*Panicum capillare*)

BROADLEAF WEEDS

The species of annual and perennial broadleaf weeds controlled by ARSENAL herbicide forestry granule include:

BULL THISTLE (*Cirsium vulgare*)
BURDOCK (*Arctium* spp.)
CAMPHORWEED (*Heterotheca subaxillaris*)
CANADA THISTLE (*Cirsium arvense*)
CAROLINA GERANIUM (*Geranium carolinianum*)
CARPETWEED (*Mullugo verticillata*)
CLOVER (*Trifolium* spp.)
COCKLEBUR (*Xanthium strumarium*)
COMMON CHICKWEED (*Stellaria media*)
COMMON RAGWEED (*Ambrosia artemisiifolia*)
DANDELION (*Taraxacum officinale*)
DOCK (*Rumex* spp.)
DOGFENNEL (*Eupatorium capillifolium*)
FILAREE (*Erodium* spp.)
FLEABANE (*Erigeron* spp.)
GIANT RAGWEED (*Ambrosia trifida*)
HOARY VERVAIN (*Verbena stricta*)
HORSEWEED (*Conyza canadensis*)
INDIAN MUSTARD (*Brassica juncea*)
KOCHIA (*Kochia scoparia*)
LAMBSQUARTERS (*Chenopodium album*)
LESPEDEZA (*Lespedeza* spp.)
LITTLE MALLOW (*Malvapar viflora*)
MILKWEED (*Asclepias* spp.)
MINERS LETTUCE (*Montia perfoliata*)
MULLEIN (*Verbascum* spp.)
NETTLELEAF GOOSEFOOT (*Chenopodium murale*)
OXEYE DAISY (*Chrysanthemus leucanthemum*)
PEPPERWEED (*Lepidium* spp.)
PIGWEED (*Amaranthus* spp.)
PLANTAIN (*Plantago* spp.)
POKEWEED (*Phytolacca americana*)
PRIMROSE (*Oenothera kunthiana*)
PURSLANE (*Portulaca* spp.)
SILVERLEAF NIGHTSHADE (*Solanum elaeagnifolium*)
SMARTWEED (*Polygonum* spp.)
SORREL (*Rumex* spp.)
SOWTHISTLE (*Sonchus* spp.)
SUNFLOWER (*Helianthus* spp.)
SWEET CLOVER (*Melilotus* spp.)
TANSYMUSTARD (*Descurainia pinnata*)
TEXAS THISTLE (*Cirsium texanum*)
WESTERN RAGWEED (*Ambrosia psilostachya*)
WILD CARROT (*Caucus carota*)
WILD LETTUCE (*Lactuca* spp.)
WILD PARSNIP (*Pastinaca sativa*)
WILD TURNIP (*Brassica campestris*)
WOOLLYLEAF BURSAGE (*Ambrosia grayi*)
YELLOW STARTHISTLE (*Centaurea solstitialis*)
YELLOW WOODSORREL (*Oxalis stricta*)

VINES AND BRAMBLES

The species of vines and brambles controlled by ARSENAL herbicide forestry granule include:

BLACKBERRY (*Rubus* spp.)*
DEWBERRY (*Rubus* spp.)*
FIELD BINDWEED (*Convolvulus arvensis*)
GREENBRIAR (*Smilax* spp.)
HEDGE BINDWEED (*Calystegia sequium*)
HONEYSUCKLE (*Lonicera* spp.)
KUDZU (*Pueraria lobata*)
MORNINGGLORY (*Ipomoea* spp.)
POISON IVY (*Rhus radicans*)
REDVINE (*Brunnichia cirrhosa*)
TRUMPETCREEPER (*Campsis radicans*)
VIRGINIA CREEPER (*Parthenocissus quinquefolia*)
WILD BUCKWHEAT (*Polyqonum convolvulus*)
WILD GRAPE (*Vitis* spp.)
WILD ROSE (*Rosa* spp.)
Including:
MULTIFLORA ROSE (*Rosa multifloria*)
MACARTNEY ROSE (*Rosa bractreata*)

*The degree of control may be species dependent. Some *Rubus* species may not be controlled.

WOODY BRUSH AND TREES

The species of woody brush and trees controlled by ARSENAL herbicide forestry granule include:

ASH (*Fraxinus* spp.)
BLACKGUM (*Nyssa sylvatica*)
BOXELDER (*Acer negundo*)
CHERRY (*Runus* spp.)
CHINESE TALLOW-TREE (*Sapium sebiferum*)
DOGWOOD (*Cornus* spp.)
HAWTHORN (*Crataegus* spp.)
HICKORY (*Carya* spp.)
MAPLE (*Acer* spp.)
MULBERRY (*Morus* spp.)*
OAK (*Quercus* spp.)
PERSIMMON (*Diospyros virginiana*)
POPLAR (*Populus* spp.)
PRIVET (*Liqustrum vulgare*)
RED MAPLE (*Acer rubrum*)
SASSAFRAS (*Sassafras albidum*)
SUMAC (*Rhus* spp.)
SWEETGUM (*Liquidambar styraciflua*)
WILLOW (*Salix* spp.)
YELLOW POPLAR (*Liriodendron tulipifera*)

The degree of control may be species dependent.

® Registered Trademark of American Cyanamid Company

ARSENAL®

HERBICIDE
APPLICATORS CONCENTRATE

ACTIVE INGREDIENT:
Isopropylamine salt of Imazapyr (2-[4,5-dihydro-4-methyl-4-(1-methylethyl)-5-oxo-1*H*-imidazol-2-yl]-3-pyridinecarboxylic acid)* 53.1%
INERT INGREDIENTS: 46.9%
TOTAL 100.0%

*Equivalent to 43.3% (2-[4,5-dihydro-4-methyl-4-(1-methylethyl)-5-oxo-1*H*-imidazol-2-yl]-3-pyridinecarboxylic acid) or 4 pounds acid per gallon.

EPA Reg. No. 241-299

® Registered Trademark of American Cyanamid Company

9/88

KEEP OUT OF REACH OF CHILDREN
CAUTION!

PRECAUCION

PRECAUCION AL USUARIO: Si usted no lee inglés, no use este producto hasta que la etiqueta le haya sido explicada ampliamente.

In case of emergency endangering life or property involving this product, call collect, day or night, Area Code 201-835-3100.

PRECAUTIONARY STATEMENTS

HAZARDS TO HUMANS

CAUTION!

Avoid contact with skin, eyes, or clothing. Avoid breathing spray mist. Wash thoroughly with soap and water after handling. Remove contaminated clothing and wash before reuse.

FIRST AID

IF ON SKIN: Wash with plenty of soap and water.

IF IN EYES: Flush with plenty of water. Get medical attention if irritation persists.

ENVIRONMENTAL HAZARDS

DO NOT apply directly to water or wetlands. DO NOT contaminate water when disposing of equipment washwaters.

PHYSICAL AND CHEMICAL HAZARDS

Spray solutions of ARSENAL should be mixed, stored, and applied only in stainless steel, fiberglass, plastic, and plastic-lined steel containers.

DO NOT mix, store, or apply ARSENAL or spray solutions of ARSENAL in unlined steel (except stainless steel) containers or spray tanks.

IMPORTANT

DO NOT use on food or feed crops. Keep from contact with fertilizers, insecticides, fungicides, and seeds. **DO NOT** apply or drain or flush equipment on or near desirable trees or other plants, or on areas where their roots may extend, or in locations where the chemical may be washed or moved into contact with their roots. **DO NOT** side trim desirable vegetation with this product. Prevent drift of spray to desirable plants. **DO NOT USE** in California.

Clean application equipment after using this product by thoroughly flushing with water.

GENERAL INFORMATION

ARSENAL herbicide APPLICATORS CONCENTRATE is an aqueous solution to be mixed in water and applied as a spray for control of most annual and perennial grasses, broadleaf weeds, vines and brambles, and hardwood trees for site preparation and release for forest management. ARSENAL herbicide APPLICATORS CONCENTRATE is also recommended for control of undesirable vegetation along forest roads, nonirrigation ditchbanks, and the establishment and maintenance of wildlife openings. ARSENAL may also be mixed in water and used for injection and cut surface treatment for control of unwanted woody vegetation.

ARSENAL is readily absorbed through foliage and roots and is translocated rapidly throughout the plant, with accumulation in the meristematic regions. Treated plants stop growing soon after spray application. Chlorosis appears first in the newest leaves, and necrosis spreads from this point. In perennials, the herbicide is translocated into and kills the underground storage organs, thus preventing resprouting. Chlorosis and tissue necrosis may not be apparent in some plant species for several weeks after application. Complete kill of some brush species may not occur for several months.

DIRECTIONS FOR USE

It is a violation of Federal law to use this product in a manner inconsistent with its labeling.

ARSENAL herbicide APPLICATORS CONCENTRATE should be used only in accordance with recommendations on this label. Keep containers closed to avoid spills and contamination.

A postemergence application of ARSENAL is recommended for control of most annual and perennial grasses and broadleaf weeds, vines and brambles, and woody brush and tree species for the site preparation and release for forest management. ARSENAL may be applied using helicopters, ground operated sprayers, and low-volume hand-operated spray equipment such as back-pack and pump-up sprayers and injection equipment.

ARSENAL may be applied for the control of annual and perennial weeds and brush along drainage ditches and forest roads in forest management areas. ARSENAL may also be used in the establishment and maintenance of wildlife openings. Thoroughly mix 1 to 2½ pints of ARSENAL in 15 to 60 gallons of water per acre. A suitable nonionic surfactant such as Ortho X-77* must be added at a minimum rate of ¼ percent by volume to the spray solution. DO NOT apply ARSENAL directly to water. Only the area above the water surface in ditches may be treated. DO NOT treat any ditch into which roots of desirable trees or other plants may extend, as injury may result.

STORAGE AND DISPOSAL

PROHIBITIONS:

DO NOT store below 10°F. DO NOT contaminate water, food, or feed by storage or disposal.

PESTICIDE DISPOSAL:

Wastes resulting from the use of this product may be disposed of on site or at an approved waste disposal facility.

CONTAINER DISPOSAL:

Triple rinse (or equivalent). Then offer for recycling or reconditioning, or puncture and dispose of in an approved sanitary landfill, or incineration, or, if allowed by state and local authorities, by burning. If burned, stay out of smoke.

MIXING AND APPLICATION INSTRUCTIONS

HELICOPTER SPRAY EQUIPMENT

Thoroughly mix the recommended amount of ARSENAL in 5 to 30 gallons of water per acre and uniformly apply with properly calibrated aerial equipment. A suitable nonionic surfactant such as Ortho X-77* must be added to the spray solution to insure proper control of undesirable vegetation. All precautions should be taken to minimize or eliminate spray drift. Applications should not be made under gusty conditions or when wind velocity exceeds 5 mph. Except when applying with a MICROFOIL boom, a drift control agent may be added at the recommended label rate. A foam reducing agent may be added at the recommended label rate, if needed.

*Registered Trademark of Chevron Chemical Company

IMPORTANT: DO NOT make applications by fixed wing aircraft. Maintain adequate buffer zones to insure that drift does not occur off the target area. Thoroughly clean application and mixing equipment, including landing gear, immediately after use. Prolonged exposure of this product to uncoated steel (except stainless steel) surfaces may result in corrosion and failure of the exposed part.

GROUND OPERATED SPRAY EQUIPMENT

Thoroughly mix and apply the recommended amount of ARSENAL in 15 to 60 gallons of water per acre. A suitable surfactant such as Ortho X-77* must be added to the spray solution to insure proper control of undesirable vegetation. A drift control agent and a foam reducing agent may be added at the recommended label rates, if needed. If desired, a spray pattern indicator may be added at the recommended label rate.

For best results, uniformly cover the foliage of the vegetation to be controlled with the spray solution.

IMPORTANT: DO NOT spray under windy or gusty conditions. Maintain adequate buffer zones to insure that drift does not occur off the target area. Clean application and mixing equipment after using this product by thoroughly flushing with water.

LOW-VOLUME HAND-OPERATED SPRAY EQUIPMENT

When using back-pack sprayers or similar equipment, thoroughly mix a solution of ½ percent ARSENAL herbicide APPLICATORS CONCENTRATE and ¼ percent of a nonionic surfactant such as Ortho X-77* in water.

To mix the spray solution, add the volume of ARSENAL herbicide and nonionic surfactant indicated in the table below to the desired amount of water.

Spray Solution Mixing Guide

Solution Volume	Arsenal	Surfactant
1 gallon	⅔ oz.	⅓ oz.
5 gallons	3⅓ oz.	1⅔ oz.
10 gallons	6⅔ oz.	3⅓ oz.
25 gallons	1 pint	8 oz.
100 gallons	4 pints	2 pints

2 tablespoons = 1 ounce

For best results, uniformly cover the foliage of the vegetation to be controlled with the spray solution.

* Registered Trademark of Chevron Chemical Company

IMPORTANT: DO NOT over apply causing runoff from the treated foliage. Avoid direct application to desired plant species as injury may occur. Clean application and mixing equipment after using this product by thoroughly flushing with water.

CUT SURFACE AND INJECTION TREATMENTS

ARSENAL APPLICATORS CONCENTRATE may be used to control undesirable hardwoods in forest management by treating cut stumps or use of injection techniques.

Thoroughly mix 4.0 to 6.0 fluid ounces of ARSENAL in one gallon of water.

For cut surface treatments: Spray or brush the solution onto the cambium area of the freshly cut stump surface. Insure that the solution thoroughly wets the entire cambium area (the wood next to the bark of the stump).

For tree injection treatments: Using standard injection equipment, apply 1 milliliter of solution at each injection site around the tree with no more than 1 inch intervals between cut edges. Insure that the injector completely penetrates the bark at each injection site.

For frill or girdle treatments: Using a hatchet, machete, or similar device, make cuts through the bark at intervals around the tree with no more than 2 inch intervals between cut edges. Spray or brush the solution into each cut until thoroughly wet.

NOTE: Injury may occur to non-target or desirable hardwoods if they extend from the same root system or their root systems are grafted to those of the treated tree.

SITE PREPARATION TREATMENTS

ARSENAL may be used to control labeled grasses, broadleaf weeds, vines and brambles, and woody brush and trees on forest sites in advance of regeneration.

Herbaceous grass and broadleaf weeds are killed and will provide fuel to facilitate a controlled fire, if desired, to control residual pines.

For site preparation in advance of regeneration with the following species:

Species	Rate (pints/A)
Loblolly Pine *(Pinus taeda)*	1½–2½
Virginia Pine *(Pinus virginiana)*	1½–2½
Slash Pine *(Pinus elliotti)*	1½–2½
Ponderosa Pine *(Pinus ponderosa)*	¾–2
Douglas Fir *(Pseudotsuga menziesii)*	¾–2
White Spruce *(Picea glauca)*	¾–2

Apply recommended rate of ARSENAL per acre when making helicopter or mechanical ground spray applications. Refer to mixing and application instructions for proper spray volumes. Use a minimum of ¼ percent by volume of a nonionic surfactant. Use the higher label rates of ARSENAL and higher spray volumes when controlling particularly dense or multi-layered canopies of hardwood stands, or difficult to control species.

DO NOT plant seedlings on sites treated with ARSENAL for three months following application or injury may occur.

Tank mixes may be necessary for chemical control of residual pines in certain cases.

RELEASE TREATMENTS

For herbaceous competition release of the following specie:

Loblolly Pine *(Pinus taeda)*

ARSENAL herbicide may be applied as a broadcast or directed spray application for release of loblolly pine from herbaceous weeds. Some minor conifer growth inhibition may be observed when release treatments are made during periods of active conifer growth. For release of loblolly pine from herbaceous weeds, apply 6 to 10 fluid ounces of ARSENAL per acre when making helicopter or ground spray applications. Refer to mixing and application instructions for proper spray volumes. A suitable nonionic surfactant should be added at a rate of ¼ percent by volume in water. For best results, applications should be made to newly emerged weeds. For difficult to control weeds, use the higher rates of ARSENAL.

For hardwood competition release of the following species:

Species	Rate (pints/A)
Loblolly Pine *(Pinus taeda)*	1–2
Virginia Pine *(Pinus virginiana)*	1–2

ARSENAL may be applied as a broadcast or directed spray application for control of labeled weed species. Some minor conifer growth inhibition may be observed when release treatments are made during periods of active conifer growth. To minimize potential conifer injury, release treatments may be made after formation of final conifer resting buds in the fall.

Apply recommended rate of ARSENAL per acre when making helicopter or mechanical ground spray applications. Refer to mixing and application instructions for proper spray volumes. Addition of a nonionic surfactant at ¼ percent by volume is required. Use the higher label rates of ARSENAL when controlling particularly dense stands or difficult to control species. To minimize injury, do not make applications to pine stands before formation of resting buds at the end of the second growing season. Directed spray applications may be made with low-volume hand-operated sprayers in plantations of all ages by targeting the unwanted vegetation and avoiding direct application to the pine.

For spot treatment of undesirable hardwood vegetation.

ARSENAL may be used to control undesirable brush and hardwoods in the management of commercial conifer species as a directed spray application using low-volume hand-operated spray equipment. Refer to mixing and application instructions for proper use rates and application techniques.

Avoid direct application to desired plant species as injury may occur. Injury may occur to non-target or desirable hardwoods if they extend from the same root system or their root systems are grafted to those of the treated tree.

WEEDS CONTROLLED

ARSENAL herbicide APPLICATORS CONCENTRATE will provide postemergence control and some residual control of the following target vegetation species when applied at the recommended rates. ARSENAL herbicide should be used only in accordance with the recommendation on this label.

GRASSES

The species of annual and perennial grasses controlled by ARSENAL herbicide include the following:

Annual bluegrass *(Poa annua)*
Bahiagrass *(Paspalum notatum)*
Beardgrass *(Andropogon* spp.*)*
Bermudagrass *(Cynodon dactylon)*
Big bluestem *(Andropogon geradi)*
Broadleaf signalgrass *(Brachiaria platphylla)*
Canada bluegrass *(Poa compressa)*
Cattail *(Typha* spp.*)*
Cheat *(Bromus secalinus)*
Crabgrass *(Digitaria* spp.*)*
Dallisgrass *(Paspalum dilatatum)*
Downy brome *(Bromus tectorum)*
Fall panicum *(Panicum dichotomiflorum)*
Feathertop *(Pennisetum villosum)*
Fescue *(Festuca* spp.*)*
Foxtail *(Setaria* spp.*)*
Goosegrass *(Eleusine indica)*
Guineagrass *(Panicum maximum)*
Italian ryegrass *(Lolium multiflorum)*
Johnsongrass *(Sorghum halepense)*
Kentucky bluegrass *(Poa pratensis)*
Lovegrass *(Eragrostis* spp.*)*
Orchardgrass *(Dactylis glomerata)*
Paragrass *(Brachiaria mutica)*
Phragmites *(Phragmites australis)*
Prairie cordgrass *(Spartina pectinata)*
Prairie threeawn *(Aristida oligantha)*
Quackgrass *(Agropyron repens)*
Reed canary grass *(Phalaris arundinacea)*
Saltgrass *(Distichlis stricta)*
Sand dropseed *(Sporobolus cryptandrus)*
Sandbur *(Cenchrus* spp.*)*
Smooth brome *(Bromus inermis)*
Timothy *(Phleum pratense)*
Torpedograss *(Panicum repens)*
Vaseygrass *(Paspalum urvillei)*
Wild barley *(Hordeum* spp.*)*
Wild oats *(Avena fatua)*
Wirestem muhly *(Muhlenbergia frondosa)*
Witchgrass *(Panicum capillare)*

BROADLEAF WEEDS

The species of annual and perennial broadleaf weeds controlled by ARSENAL herbicide include the following:

Bull thistle *(Circsium volgare)*
Burdock *(Arctium* spp.*)*
Camphorweed *(Heterotheca subaxillaris)*
Canada thistle *(Cirsium arvense)*
Carolina geranium *(Geranium carolinianum)*
Carpetweed *(Mullugo verticillata)*
Clover *(Trifolium* spp.*)*
Cocklebur *(Xanthium strumarium)*
Common chickweed *(Stellaria media)*
Common ragweed *(Ambrosia artemislifolia)*

Dandelion *(Taraxacum officinale)*
Dock *(Rumex* spp. *)*
Dogfennel *(Eupatorium capillifolium)*
Filaree *(Erodium* spp. *)*
Fleabane *(Erigeron* spp. *)*
Giant ragweed *(Ambrosia trifida)*
Goldenrod *(Solidago* spp. *)*
Hoary vervain *(Verbena stricta)*
Horseweed *(Conyza canadensis)*
Indian mustard *(Brassica juncea)*
Kochia *(Kochia scoparia)*
Lambsquarters *(Chenopodium album)*
Lespedeza *(Lespedeza* spp. *)*
Little mallow *(Malva parviflora)*
Milkweed *(Asclepias* spp. *)*
Miners lettuce *(Montia perfoliata)*
Mullein *(Verbascum* spp. *)*
Nettleleaf goosefoot *(Chenopodium murale)*
Oxeye daisy *(Chrysanthemum leucanthemum)*
Pepperweed *(Lepidium* spp. *)*
Pigweed *(Amaranthus* spp. *)*
Plantain *(Plantago* spp. *)*
Pokeweed *(Phytolacca americana)*
Primrose *(Oenothera kunthiana)*
Purslane *(Portulaca* spp. *)*
Silverleaf nightshade *(Solanum elaeagnifolium)*
Smartweed *(Polygonum* spp. *)*
Sorrell *(Rumex* spp. *)*
Sowthistle *(Sonchus* spp. *)*
Sunflower *Helianthus* spp. *)*
Sweet clover *(Melilotus* spp. *)*
Tansymustard *(Descurainia pinnata)*
Texas thistle *(Cirsium texanum)*
Western ragweed *(Ambrosia psilostachya)*
Wild carrot *(Daucus carota)*
Wild lettuce *(Lactuca* spp. *)*
Wild parsnip *(Pastinaca sativa)*
Wild turnip *(Brassica campestris)*
Woollyleaf bursage *(Ambrosia grayi)*
Yellow starthistle *(Centaurea solstitialis)*
Yellow woodsorrel *(Oxalis stricta)*

VINES AND BRAMBLES

The species of vines and brambles controlled by ARSENAL herbicide include the following:

Blackberry *(Rubus* spp. *)**
Dewberry *(Rubus* spp. *)**
Field bindweed *(Convolvulus arvensis)*
Greenbriar *(Smilax* spp. *)*
Hedge bindweed *(Calystegia sequium)*
Honeysuckle *(Lonicera* spp. *)*
Kudzu *(Pueraria lobata)*
Morningglory *(Ipomoea* spp. *)*
Poison ivy *(Rhus radicans)*
Redvine *(Brunnichia cirrhosa)*
Trumpetcreeper *(Campsis radicans)*
Virginia creeper *(Parthenocissus quinquefolia)*
Wild buckwheat *(Polygonum convolvulus)*
Wild grape *(Vitis* spp. *)*
Wild rose *(Rosa* spp. *)*
  Including: Multiflora rose *(Rosa multiflora)*
    Macartney rose *(Rosa bractreata)*

* The degree of control may be species dependent. Some *Rubus* species may not be controlled.

** Control of established stands may require multiple applications.

WOODY BRUSH AND TREES

The species of woody brush and trees controlled by ARSENAL herbicide include the following:

American Beech *(Fagus grandifolia)*
Ash *(Fraxinus* spp. *)*
Bald Cypress *(Taxodium distichum)*
Bigleaf Maple *(Lacer macrophyllum)*
Blackgum *(Nyssa sylvatica)*
Boxelder *(Acer negundo)*
Cherry *(Prunus* spp. *)*
Chinaberry *(Melia azedarach)*
Chinese tallow-tree *(Sapium sebiferum)*
Dogwood *(Cornus* spp. *)*
Hawthorn *(Crataegus* spp. *)*
Hickory *(Carya* spp. *)*
Maple *(Acer* spp. *)*
Mulberry *(Morus* spp. *)**
Oak *(Quercus* spp. *)*
Persimmon *(Diospyros virginiana)*
Poplar *(Populus* spp. *)*
Privet *(Ligustrum vulgare)*
Red Alder *(Alnus Ruba)*
Red Maple *(Acer rubrum)*
Sassafras *(Sassafras albidum)*
Sourwood *(Oxydendrum arboreum)*
Sumac *(Rhus* spp. *)*
Sweetgum *(Liquidambar styraciflua)*
Willow *(Salix* spp. *)*
Yellow poplar *(Liriodendron tulipifera)*

* The degree of control may be species dependent.

DISCLAIMER

The label instructions for the use of this product reflect the opinion of experts based on field use and tests. The directions are believed to be reliable and should be followed carefully. However, it is impossible to eliminate all risks inherently associated with use of this product. Ineffectiveness or other unintended consequences may result because of such factors as weather conditions, presence of other materials, or the use or application of the product contrary to label instructions, all of which are beyond the control of American Cyanamid Company. All such risks shall be assumed by the user.

American Cyanamid Company warrants only that the material contained herein conforms to the chemical description on the label and is reasonably fit for the use therein described when used in accordance with the directions for use, subject to the risks referred to above.

Any damages arising from a breach of this warranty shall be limited to direct damages and shall not include consequential commercial damages such as loss of profits or values or any other special or indirect damages.

American Cyanamid Company makes no other express or implied warranty, including other express or implied warranty of FITNESS or of MERCHANTABILITY.

ARSENAL®

HERBICIDE

0.5 Granule

FOR USE IN NONCROP AREAS

ACTIVE INGREDIENT:

Imazapyr (2-[4,5-dihydro-4-methyl-4-(1-methylethyl)-5-oxo-1H-imidazol-2-yl]-3-pyridinecarboxylic acid)	0.5%
INERT INGREDIENT:	99.5%
TOTAL	100.0%

EPA Reg. No. 241-295

® Registered Trademark of American Cyanamid Company

7/87

KEEP OUT OF REACH OF CHILDREN

CAUTION!

PRECAUCION

AL USUARIO: Si usted no lee inglés, no use este producto hasta que la etiqueta le haya sido explicada ampliamente.

In case of an emergency endangering life or property involving this product, call collect, day or night, Area Code 201-835-3100

PRECAUTIONARY STATEMENTS

HAZARDS TO HUMANS

CAUTION!

Avoid contact with skin, eyes or clothing. Avoid breathing dust. Wash thoroughly with soap and water after handling. Remove contaminated clothing and wash before reuse.

FIRST AID

IF ON SKIN: Wash with plenty of soap and water.

IF IN EYES: Flush with plenty of water. Get medical attention if irritation persists.

PHYSICAL AND CHEMICAL HAZARDS

ARSENAL granules should be mixed and stored only in lined paper bags, stainless steel, fiberglass, plastic and plastic-lined steel containers.

DO NOT mix or store ARSENAL granules in unlined steel (except stainless steel) containers.

ENVIRONMENTAL HAZARDS

DO NOT apply directly to any body of water. DO NOT contaminate water when disposing of equipment washwaters.

IMPORTANT

DO NOT use on food or feed crops. **DO NOT** apply on ditches used to transport irrigation water. **DO NOT** apply where runoff or irrigation water may flow onto agricultural land as injury to crops may result. Keep from contact with fertilizers, insecticides, fungicides and seeds. **DO NOT** apply or drain or flush equipment on or near desirable trees or other plants, or on areas where their roots may extend, or in locations where the chemical may be washed or moved into contact with their roots. **DO NOT** use on lawns, walks, driveways, tennis courts or similar areas. **DO NOT** use in California.

GENERAL INFORMATION

ARSENAL® herbicide 0.5 Granule is applied for control of most annual and perennial grasses and broadleaf weeds.

ARSENAL herbicide 0.5 Granule is readily absorbed through the roots and is translocated rapidly throughout the plant, with accumulation in the meristematic regions. Treated plants stop growing soon after spray application. Chlorosis first in the newest leaves, and necrosis spreads from this point. In perennials, the herbicide is translocated into and kills underground storage organs, thus preventing regrowth. Chlorosis and tissue necrosis may not be apparent in some plant species until two weeks after application. Complete kill of plants may not occur for several weeks.

DIRECTIONS FOR USE

It is a violation of Federal law to use this product in a manner inconsistent with its labeling.

A preemergence application of ARSENAL granular is recommended for control of most annual and perennial grasses and broadleaf weeds. ARSENAL granular may be applied anytime during the growing season before weeds have emerged.

STORAGE AND DISPOSAL

PROHIBITIONS:

DO NOT store below 10°F. DO NOT contaminate water, food or feed by storage or disposal.

PESTICIDE DISPOSAL:

Wastes resulting from the use of this product may be disposed of on site or at an approved waste disposal facility.

CONTAINER DISPOSAL:

Triple rinse (or equivalent). Then offer for recycling or dispose of in a sanitary landfill, or by incineration, or if allowed by state and local authorities, by burning. If burned, stay out of smoke.

DISCLAIMER

The label instructions for the use of this product reflect the opinion of experts based on field use and tests. The directions are believed to be reliable and should be followed carefully. However, it is impossible to eliminate all risks inherently associated with use of this product. Ineffectiveness or other unintended consequences may result because of such factors as weather conditions, presence of other materials, or the use or application of the product contrary to label instructions, all of which are beyond the control of American Cyanamid Company. All such risks shall be assumed by the user.

American Cyanamid Company warrants only that the material contained herein confirms to the chemical description on the label and is reasonably fit for the use therein described when used in accordance with the directions of use, subject to the risks referred to above.

Any damages arising from a breach of this warranty shall be limited to direct damages and shall not include consequential commercial damages such as loss of profits or values of any other special or indirect damages.

American Cyanamid Company makes no other express or implied warranty, including other express or implied warranty of FITNESS or of MERCHANTABILITY.

APPLICATION INSTRUCTIONS

Uniformly apply with properly calibrated aerial or ground equipment at a rate of 1.0 lbs. ae/A (200 lbs. of product per acre).
ARSENAL herbicide will provide preemergence control with residual control of the following target vegetation species at 1.0 lb. ae/A.

BROADLEAF WEEDS	Bull thistle *(Circium vulgare)*	Mullein *(Verbascum* spp.)
	Burdock *(Arctium* spp.)	Nettleleaf goosefoot *(Chenopodium murale)*
	Camphorweed *(Heterotheca subzxillaris)*	Oxeye daisy *(Chrysanthemus leucanthemum)*
	Canada thistle *(Circium arvense)*	Pepperweed *(Lepidium* spp.)
	Carpetweed *(Mullugo verticillata)*	Pigweed *(Amaranthus* spp.)
	Carolina geranium *(Geranium carolinianum)*	Plantain *(Plantago* spp.)
	Clover *(Trifolium* spp.)	Pokeweed *(Phytolacca americana)*
	Cocklebur *(Xanthium strumarium)*	Primrose *(Oenothera kunthiana)*
	Common chickweed *(Stellaria media)*	Purslane *(Portulaca* spp.)
	Common ragweed *(Ambrosia artemisiifolia)*	Silverleaf nightshade *(Solanum elaeagnifolium)*
	Dandelion *(Taraxacum officinale)*	Smartweed *(Ploygonum* spp.)
	Dock *(Rumex* spp.)	Sorrel *(Rumex* spp.)
	Dogfennel *(Eupatorium capillifolium)*	Sowthistle *(Sonchus* spp.)
	Filaree *(Erodium* spp.)	Sunflower *(Helianthus* spp.)
	Fleabane *(Erigeron* spp.)	Sweet clover *(Melilotus* spp.)
	Giant ragweed *(Ambrosia trifida)*	Tansymustard *(Descurainia pinnata)*
	Hoary vervain *(Verbena stricta)*	Texas thistle *(Circium texanum)*
	Horseweed *(Conyza canadensis)*	Western ragweed *(Ambrosia psilostachya)*
	Indian mustard *(Brassica juncea)*	Wild carrot *(Caucus carota)*
	Kochia *(Kochia scoparia)*	Wild lettuce *(Lactuca* spp.)
	Lambsquarters *(Chenopodium album)*	Wild parsnip *(Pastinaca sativa)*
	Lespedeza *(Lespedeza* spp.)	Wild turnip *(Brassica campestris)*
	Little mallow *(Malvapar viflora)*	Woollyleaf bursage *(Ambrosia grayl)*
	Milkweed *(Asclepias* spp.)	Yellow starthistle *(Centaurea solstitialis)*
	Miners lettuce *(Montia perfoliata)*	Yellow woodsorrel *(Oxalis stricta)*
GRASSES	Annual bluegrass *(Poa annua)*	Johnsongrass *(Sorghum halepense)*
	Bahiagrass *(Paspalum notatum)*	Kentucky bluegrass *(Poa pratensis)*
	Beardgrass *(Andropogon* spp.)	Lovegrass *(Eragrostis* spp.)
	Bermudagrass *(Cynodon dactylon)*	Orchardgrass *(Cactylis glomerata)*
	Big bluestem *(Andropogon geradi)*	Paragrass *(Brachiaria mutica)*
	Broadleaf signalgrass *(Brachiaria platphylla)*	Prairie cordgrass *(Spartina pectnata)*
	Canada bluegrass *(Poa compressa)*	Prairie threeawn *(Aristida oligantha)*
	Cattail *(Typha* spp.)	Quackgrass *(Agropyron repens)*
	Cheat *(Bromus secalinus)*	Sandbur *(Cenbchrus* spp.)
	Crabgrass *(Digitaria* spp.)	Sand dropseed *(Sporobolus cyptandrus)*

Dallisgrass *(Paspalum dilatatum)*
Downey brome *(Bromus tectorum)*
Fall panicum *(Panicum dichotomiflorum)*
Fescue *(Festuca* spp.*)*
Foxtail *(Setaria* spp.*)*
Goosegrass *(Eleusine indica)*
Guineagrass *(Panicum maximum)*
Italian ryegrass *(Lolium multiflorum)*
Smooth brome *(Bromus inermis)*
Timothy *(Phleum pratense)*
Torpedograss *(Ponicum repens)*
Vaseygrass *(Paspalum urvillei)*
Wild barley *(Hordeum* spp.*)*
Wild oats *(Avena fatua)*
Wiresterm muhly *(Muhlenbergia frondosa)*
Witchgrass *(Panicum capillare)*

ASSERT®

HERBICIDE

FOR USE IN WHEAT, BARLEY AND SUNFLOWERS

ACTIVE INGREDIENT:

m-Toluic acid, 6-(4-isopropyl-4-methyl-5-oxo-2-imidazolin-2-yl)-, methyl ester and *p*-Toluic acid, 2-(4-isopropyl-4-methyl-5-oxo-2-imidazolin-2-yl)methyl ester	27.0%
INERT INGREDIENTS:	73.0%
TOTAL	100.0%

(One gallon contains 2.5 lbs of active ingredient)

EPA Reg. No. 241-285

® Registered Trademark of American Cyanamid Company

9/88

KEEP OUT OF REACH OF CHILDREN

DANGER!/PELIGRO!

PRECAUCION AL USUARIO: Si usted no lee inglés, no use este producto hasta que la etiqueta le haya sido explicada ampliamente.

STATEMENT OF PRACTICAL TREATMENT

If in eyes: Hold eyelids open and flush with a steady gentle stream of water for 15 minutes. Get medical attention.

If swallowed: Drink promptly a large quantity of milk, egg white, gelatin solution, or if these are not available, large quantities of water. Avoid alcohol. Get medical attention.

If on skin: Wash thoroughly with soap and water. Get medical attention if irritation persists.

Note to Physician: Probable mucosal damage may contraindicate the use of gastric lavage.

In case of an emergency endangering life or property involving this product, call collect, day or night (201) 835-3100.

PRECAUTIONARY STATEMENTS

HAZARDS TO HUMANS AND DOMESTIC ANIMALS

DANGER

CORROSIVE, CAUSES IRREVERSIBLE EYE DAMAGE

DANGER! Do not get in eyes. Wear goggles and gloves when handling. Avoid contact with skin and clothing. Avoid breathing vapor or spray mist. Use with adequate ventilation. Wash thoroughly after handling with soap and water. Remove contaminated clothing and wash before reuse.

ENVIRONMENTAL HAZARDS

DO NOT apply directly to any body of water or wetlands.

DO NOT contaminate water when disposing of equipment washwaters.

DIRECTIONS FOR USE

It is a violation of Federal law to use this product in a manner inconsistent with its labeling.

ASSERT should be used only in accordance with recommendations on this label.

Do Not apply this product through any type of irrigation system.

Observe all cautions and limitations on this label and on the labels of products in combination with ASSERT. American Cyanamid Company will not be responsible for losses or damages resulting from the use of this product in any manner not specifically recommended by American Cyanamid. User assumes all risks associated with such non-recommended use. Keep container closed to avoid spills and contamination.

STORAGE AND DISPOSAL

Keep from freezing. Do not store below 32°F. Ice will form in the container at 5°F (−15°C). If ice occurs from prolonged storage at temperatures of 5°F (−15°C) or below, place container at room temperature until ice melts. Stability of ASSERT is not effected by freezing or thawing.

DO NOT contaminate water, food or feed by storage or disposal. **DO NOT** mix or store in unlined containers.

PESTICIDE DISPOSAL: Pesticide wastes are acutely hazardous. Improper disposal of excess pesticide, spray mixture, or rinsate is a violation of Federal law. If these wastes cannot be disposed of by use according to label instructions, contact your State Pesticide or Environmental Control Agency, or the Hazardous Waste representative at the nearest EPA Regional Office for guidance.

CONTAINER DISPOSAL: Triple rinse (or equivalent). Then offer for recycling or reconditioning, or puncture and dispose of in a sanitary landfill, or by incineration, or if allowed by state and local authorities, by burning. If burned, stay out of smoke.

EQUIPMENT APPLICATION INSTRUCTIONS

The spray equipment must be clean and properly calibrated before treatments are applied. For best results, use flat fan nozzles, and a minimum pressure of 20 to 40 psi to achieve uniform spray distribution and minimize drift. DO NOT use flood jet nozzles. Keep the by-pass line on or near the bottom of the tank to minimize foaming. An antifoaming agent may be added to the tank if needed.

In ground applications, check nozzles for even distribution of spray droplets.

In aerial applications, it is recommended that a flagman be located at each end of the field, or an automatic mechanical flagging unit be attached to the aircraft to ensure uniform aerial application.

MIXING INSTRUCTIONS

Fill tank one-half to three-quarters full with clean water.

When using ASSERT alone, add ASSERT to the partially filled tank while agitating. When ASSERT is thoroughly mixed, fill remainder of tank with water while agitating.

When using ASSERT with a broadleaf herbicide, add the broadleaf herbicide first to the partially filled tank while agitating. When the broadleaf herbicide is thoroughly mixed, add ASSERT to the tank, then complete filling the tank with water while agitating. **Maintain continuous agitation until spraying is completed.**

The addition of a non-ionic surfactant of at least 80% active ingredient is required. For **each** gallon/acre of spray solution, add 0.3 fluid ounces of non-ionic surfactant. Use a recommended surfactant registered for use on crops. An antifoaming agent may be added to the tank if needed.

APPLICATIONS WITH LIQUID FERTILIZERS

GENERAL

Liquid fertilizers can be applied with ASSERT alone or in combination with MCPA ester, 2,4-D ester, or Bromoxynil + MCPA ester[3] (Bronate[4]) to wheat and barley. Follow all ASSERT label recommendations regarding timing of application, special instructions and precautions.

Apply the ASSERT/liquid fertilizer combination with a minimum of 5 gallons of spray solution per acre with aircraft or a minimum of 10 gallons per acre with ground equipment.

NOTE: Herbicides can increase contact burn of fertilizers on plant foliage. Reduced gallons of fertilizer per acre may decrease leaf burn. DO NOT allow ASSERT to remain overnight in a liquid fertilizer solution.

All individual state regulations relating to fluid fertilizer mixing, registration, labeling and application are the responsibility of the individual and/or company selling the ASSERT/liquid fertilizer mixture.

LIQUID FERTILIZER COMPATIBILITY DETERMINATIONS

If a liquid fertilizer and herbicide(s) mixture separates in the spray tank, clogged equipment and uneven application can result, which can cause poor weed control and crop injury. Always predetermine the compatibility of ASSERT alone or with other herbicides in the specific liquid fertilizer to be used according to the following directions:

1. Add 1 pint of liquid fertilizer to each of 2 one-quart jars.
2. Add 1/2 teaspoon of adjuvant* to one jar.
3. (a) When using ASSERT alone, add to each jar the correct amount of ASSERT as specified in the table below. (b) When using ASSERT tank mixtures, first add the specified quantity of MCPA ester, or 2,4-D ester, or Bronate and then add the correct amount of ASSERT.
4. Close both jars and shake thoroughly for 10 seconds. Let them stand for 30 minutes and then observe the results. Look for signs of separation, an oily layer or globules, sludge, flakes or other precipitates.
5. Determine compatibility.
   (a) If the mixture without adjuvant does not separate, use this mixture in your spray tank.
   (b) If the mixture with adjuvant does not separate, but the one with-

out adjuvant separates, use the adjuvant mixture in your spray tank. Add the adjuvant to the liquid fertilizer as directed on the manufacturer's label.
(c) If either mixture separates, but mixes readily with shaking, the mixture can be used providing good agitation is maintained in the spray tank.
(d) If separation of the mixture occurs, and agitation and/or adjuvant does not correct this problem, **DO NOT** use the herbicide(s) in that specific liquid fertilizer.

Teaspoons of Specific Herbicide to be Added to 1 pint of Liquid Fertilizer Solution

Gallons of Liquid Fertilizer plus Water to be Applied per acre	ASSERT	MCPA ester	2,4-D ester	Bronate
5	3	2	2	2
10	1½	1	1	1
15	¾	½	½	½

SPRAYING INSTRUCTIONS

Uniformly apply the recommended ASSERT or ASSERT tank-mixture in 8 or more gallons of water per acre with ground equipment or in 5 or more gallons of water per acre by aircraft. Avoid overlapping, and shut off spray booms while starting, turning, slowing or stopping, or injury to the crop or following crops may result. **DO NOT** allow spray to drift onto adjacent crops as injury may occur.

IMPORTANT

DO NOT make more than one application of ASSERT per growing season.

DO NOT graze treated fields or cut treated forage for silage or hay.

Wheat or barley straw may be fed or used for bedding.

FOLLOW CROP RESTRICTIONS

The following rotational crops may be planted after applying ASSERT at recommended rates in wheat, barley and sunflowers:

Barley	Edible bean	Safflower	Sunflowers
Corn	Potatoes	Soybean	Wheat

If ASSERT and GLEAN,[5] or ASSERT and ALLY[5] or ASSERT and HARMONY[5] are applied during the same year, follow the most restrictive precautions, directions and recropping/rotation limitations that appear on the respective product labels.

DO NOT plant sugar beets for at least 20 months following an ASSERT application.

DO NOT plant other rotational crops for at least 15 months following an ASSERT application. Land should be deep plowed prior to planting other crops.

WHEAT AND BARLEY

GENERAL INFORMATION

ASSERT is a selective herbicide for the postemergence control of wild oats and certain broadleaved weeds in wheat and barley. Apply ASSERT postemergence to wheat (including durum) and barley from the 2-leaf stage, but before development of the first internode (jointing).

ASSERT alone or in combination can be applied by ground equipment or aircraft.

Maximum weed control with ASSERT alone or in combination is obtained when temperature, moisture, fertility and cultural practices provide favorable conditions for active plant growth. Thorough uniform spray coverage is required to maximize weed control.

ASSERT rapidly inhibits growth of susceptible weeds; however, typical symptoms (discoloration) of dying weeds may not be noticeable for 2-4 weeks after application. Weeds hardened by cold weather or drought stress may not be fully controlled and regrowth may occur.

WEEDS CONTROLLED:

Wild Oats *(Avena fatua)*
Wild Mustard *(Brassica kaber)*
Tansy Mustard *(Descurainia pinnata)*
London Rocket *(Sisymbrium irio)*
Roughstalk bluegrass *(poa trivialis)*
Flixweed *(Descurainia sophia)*
Field Pennycress, Fanweed *(Thlaspi arvense)*
*Catchweed bedstraw *(Galium aparine)*
*Tartary buckwheat *(Fagopyrum tataricum)*, up to 3 leaf stage
*Wild buckwheat *(Polygonum convolvulus)*, up to 3 leaf stage

* When ASSERT is applied up to the 3 leaf stage plants stop growing, resulting in non-competitive plant.

HERBICIDE COMBINATIONS:

Additional weeds can be controlled in wheat and barley when ASSERT is tank-mixed with MCPA ester, 2,4-D ester, bromoxynil + MCPA ester (BRONATE), GLEAN, ALLY, HARMONY, or CURTAILM[6] at manufacturer's recommended label rates and application timing.

NOTE: FOLLOW THE MOST RESTRICTIVE PRECAUTIONS, DIRECTIONS AND RECROPPING/ROTATION LIMITATIONS THAT APPEAR ON THE RESPECTIVE PRODUCT LABELS. TANK-MIXTURES MUST BE APPLIED PRIOR TO THE DEVELOPMENT OF THE FIRST INTERNODE OF THE CROP.

DO NOT tank-mix ASSERT with 2,4-D ester unless the crop is fully tillered.

DO NOT tank-mix ASSERT with 2,4-D ester unless crop is fully tillered.

DO NOT tank-mix ASSERT with BANVEL[7] or amine formulations of MCPA or 2,4-D.

RATES AND TIMING OF APPLICATION

For wild oat populations in excess of 25 plants per square foot, use a minimum spray volume of 15 gallons per acre.

EAST OF THE ROCKY MOUNTAINS

Use ASSERT at 1.2-1.5 pints per acre (19-24 fl.oz/A) alone or in tank-mixtures with MCPA ester, 2,4-D ester, bromoxynil + MCPA ester (BRONATE), GLEAN, ALLY, HARMONY, or CURTAILM at manufacturer's recommended label rates when wild oat plants are in the 1-4 leaf stage of growth. Use the high rate and increase spray volume if the wild oats have initiated tillering.

WEST OF THE ROCKY MOUNTAINS

Use ASSERT at 1.5 pints per acre (24 fl.oz./A) alone or in tank-mixtures with MCPA ester, 2,4-D ester, bromoxynil + MCPA ester (BRONATE), GLEAN, ALLY, HARMONY, or CURTAILM at manufacturer's recommended label rates when wild oat plants are in the 1-4 leaf stage of growth.

SUNFLOWERS

GENERAL INFORMATION

ASSERT herbicide is a selective herbicide for the postemergence control of wild mustard and wild oats in sunflower. Apply ASSERT postemergence to sunflowers from the 2 to 8-leaf stage of growth (sunflowers must be less than 15 inches high). Maximum weed control with ASSERT is obtained when temperature, moisture, fertility and cultural practices provide favorable conditions for active plant growth. Thorough uniform spray coverage is required to maximize weed control.

WEEDS CONTROLLED:

Wild mustard *(Brassica kaber)*
Wild oats *(Avena fatua)*

TIMING AND RATES OF APPLICATION

For optimum **wild mustard** control apply **ASSERT at 0.6 to 0.8 pints per acre (9.6-12.8 fl. oz./A)** when the majority of wild mustard plants are in the rosette stage and prior to bloom.

For wild oat control apply **ASSERT at 1.2 pints per acre (19 fl. oz./A)** when the majority of wild oat plants are in the 1-4 leaf stage of growth.

ASSERT may be applied following application of a soil applied herbicide registered for use in sunflowers such as AMIBEN[4], EPTAM[8], PROWL[9], Treflan[10].

[1]MCPA ester is sold under various brand names.
[2]2,4-D ester is sold under various brand names.
[3]Bromoxynil + MCPA ester is sold under various brand names.
[4]BRONATE and AMIBEN are trademarks of Rhône Poulenc.
[5]GLEAN, ALLY, and HARMONY are trademarks of E.I. Du Pont de Nemours and Company, Inc.
[6]CURTAILM is a trademark of Ciba-Geigy Corporation.
[7]BANVEL is a trademark of Sandoz Crop Protection Corp.
[8]EPTAM is a trademark of ICI Americas, Inc.
[9]PROWL is a trademark of American Cyanamid Company.
[10]TREFLAN is a trademark of ELANCO.

DISCLAIMER

The label instructions for the use of this product reflect the opinion of experts based on field use and tests. The directions are believed to be reliable and should be followed carefully. However, it is impossible to eliminate all risks inherently associated with use of this product. Crop injury, ineffectiveness or other unintended consequences may result because of such factors as weather conditions, presence of other materials, or the use or application of the product contrary to label instructions, all of which are beyond the control of American Cyanamid Company. All such risks shall be assumed by the user.

American Cyanamid Company warrants only that the material contained herein conforms to the chemical description on the label and is reasonably fit for the use therein described when used in accordance with the directions for use, subject to the risks referred to above.

Any damages arising from breach of this warranty shall be limited to direct damages and shall not include consequential commercial damages such as loss of profits or values or any other special or indirect damages.

American Cyanamid Company makes no other expresss or implied warranty, including other express or implied warranty of FITNESS or of MERCHANTABILITY.

AVENGE®

WILD OAT HERBICIDE

FOR USE IN BARLEY AND WHEAT

ACTIVE INGREDIENT:
difenzoquat (1,2-dimethyl-3,5-diphenyl-1*H*-pyrazolium methyl sulfate) 31.2%
INERT INGREDIENTS: 68.8%
TOTAL 100.00%
(1 gallon contains 2 lbs. of difenzoquat cation)

EPA Reg. No. 241-266

® Registered Trademark of American Cyanamid Company

3/88

KEEP OUT OF REACH OF CHILDREN

DANGER! PELIGRO!

PRECAUCION AL USUARIO: Si usted no lee ingles, no use este producto hasta que la etiqueta le haya sido explicada ampliamente.

STATEMENT OF PRACTICAL TREATMENT

If in eyes: Hold eyelids open and flush with a steady, gentle stream of water for 15 minutes. Get medical attention.

If swallowed: Drink promptly a large quantity of milk, egg white, gelatin solution, if these are not available, large quantities of water. Avoid alcohol.

If on skin: Wash with plenty of soap and water. Get medical attention.

NOTE TO PHYSICIAN: Probable mucosal damage may contraindicate the use of gastric lavage.

In case of an emergency endangering life or property involving this product, call collect, day or night, Area Code 201-835-3100.

PRECAUTIONARY STATEMENTS

HAZARDS TO HUMANS

DANGER!

CORROSIVE: CAUSES IRREVERSIBLE EYE DAMAGE

HARMFUL IF SWALLOWED OR ABSORBED THROUGH THE SKIN

Do not get in eyes, on skin, or on clothing. Wear goggles or face shield and rubber gloves when handling. Wash thoroughly with soap and water after handling. Remove contaminated clothing and wash before reuse.

Do not apply this product in such a manner as to directly or through drift expose workers or other persons. The area being treated must be vacated by unprotected persons. Do not enter treated areas without protective clothing until sprays have dried.

ENVIRONMENTAL HAZARDS

Do not apply directly to water or wetlands. Do not apply when weather conditions favor run-off and/or drift from target area. Do not contaminate water when disposing of equipment washwaters.

GENERAL INFORMATION

AVENGE is a selective herbicide for the postemergence control of wild oats in barley and wheat. See the section DIRECTIONS FOR USE for varietal restrictions.

If broadleaf weeds must also be controlled in these crops, AVENGE may be tank-mixed with MCPA amine or ester, bromoxynil, MCPA plus bromoxynil, 2,4-D amine or ester, Glean[5] or Ally[6]. AVENGE alone or in combination can be applied postemergence by ground equipment or aircraft.

Maximum control of wild oats in barley and wheat with AVENGE alone or in combination is obtained when temperature, moisture, fertility and cultural practices provide favorable conditions for plant growth. When applied under poor conditions for plant growth such as cold/wet or hot/dry weather or low soil fertility, AVENGE can produce yellowing or slight tip burn of barley or wheat. However, recovery is rapid when good growth conditions return. A slight reduction of yield may result under conditions described above.

Observe all cautions and limitations on this label and the labels of products used in tank mixtures with AVENGE. The use of AVENGE not consistent with this label can result in injury to crops, animals or persons. Keep containers closed to avoid spills and contamination.

EQUIPMENT APPLICATION INSTRUCTIONS

The spray equipment must be clean and properly calibrated before treatments are applied. For best results, use flat fan nozzles, at pressures of 20 to 40 psi to achieve uniform spray distribution and minimize drift. DO NOT use flood jet nozzles. Keep the by-pass line on or near the bottom of the tank to minimize foaming. In ground applications, check nozzles for even distribution of spray droplets. In aerial applications, it is recommended that a flagman be located at each end of the field, or an automatic mechanical flagging unit be attached to the aircraft to ensure uniform aerial application and to avoid overlapping and possible crop injury.

DIRECTIONS FOR USE

It is a violation of Federal law to use this product in a manner inconsistent with its labeling.

DO NOT apply this product through any type of irrigation system.

This label must be in the possession of the user at the time of pesticide application.

WHERE TO USE AVENGE: AVENGE alone or in combination can be applied to:

BARLEY: All Varieties

WHEAT: SEE TABLE BELOW

Wheat Type	Yes	No	
Fall and Winter Seeded	All	EXCEPT	Borah, WS 1809, WS 1877, WS 1859, Klasic, Probrand 771, Tadinia, and Mexican durums
Soft-White Spring Seeded	All		
Spring Seeded durums	All	EXCEPT	Edmore, Fjord Kyle, Laker, Lakota, Vic, Wakooma, Wascona, Westbred 803, and Westbred 881
Hard Red Spring Seeded	Apex, Benito, Bronze Chief, Buckshot, Butte, Butte 86, Canuck, Centa, Chester, Columbus, Coteau, Courtney, Era, Erik, Fortuna, Glenlea, Glennman, HY320, Katepwa, Kitt, Kodiak, Leader, Leif, McKay, Marberg, Marshall, Newana, Neepawa, NK(PB) 751, Norak, Norana, Olaf, Oslo, Pioneer 2369, Pondera, Probrand 711, Probrand 715, Prodax, Selkirk, Solar, Stoa, Success, Victory, Walera, Wheaton and Yecoro rojo		DO NOT apply to varieties of hard red spring wheat not listed.

WHEN TO USE AVENGE: AVENGE or AVENGE tank-mix combinations are to be applied when the majority of wild oat plants are in the 3- to 5-true leaf stage of growth (tillering). The timing of AVENGE application to wild oats frequently coincides with barley in the 2- to 7-leaf stage, spring-seeded wheat in the 5- to 6-leaf stage, and fall-seeded wheat in the 4-leaf to tillered stage of growth. To achieve maximum control, the growth stage of wild oats in each field to be treated should be examined frequently to determine the 3- to 5-leaf stage (tillering) for proper timing of AVENGE application. Density of wild oats per square foot should also be determined to select the correct AVENGE application rate.

AVENGE APPLICATION

3-true leaf stage

4-true leaf stage

5-true leaf stage

AVENGE APPLICATION

DO NOT apply AVENGE when the flagleaf is exposed or crop injury may occur. DO NOT apply AVENGE if rain is predicted within 6 hours. DO NOT spray AVENGE when freezing temperatures are forecast or when plants are wet with dew or rain. For spray volumes in excess of 10 gallons of spray per acre (gpa) the addition of a surfactant is required. Use any non-ionic surfactant of at least 80% active ingredient cleared for application to growing crops at a rate of 0.6 fluid ounces for each gallon of spray in excess of 10 gpa.

Example: For a 15 gpa spray add 3 fluid ounces of non-ionic surfactant (0.6 ounce for each gallon of water over 10 gpa). Use only recommended surfactants cleared for application to growing crops. DO NOT apply AVENGE or AVENGE tank-mix combinations under severe conditions for plant growth such as drought, extreme temperature fluctuation, low soil fertility and plant disease.

AVENGE
Rate Recommendation Table

Wild Oat Population	Wild Oat Density Plants Per Square Foot	AVENGE Rate Per Acre (pints)
Low	1-10	$2\frac{1}{2}$
Moderate	11-25	3
High	more than 25	4

MIXING INSTRUCTIONS

HOW TO MIX AVENGE ALONE

1. Fill tank one-half to three-quarters full with clean water and start agitation.
2. When using AVENGE alone, add AVENGE to the partially filled tank while agitating. When AVENGE is thoroughly mixed, fill remainder of tank with water while agitating.
3. Allow mixture to agitate and recycle for at least five minutes before spraying.
4. Agitation must be continued until spraying is completed.

HOW TO MIX AVENGE PLUS BROADLEAF HERBICIDES

1. Fill tank one-half to three-quarters full with clean water and start agitation.
2. When using AVENGE with a broadleaf herbicide, add the broadleaf herbicide first to the partially filled tank while agitating.
3. When the broadleaf herbicide is thoroughly mixed, add the AVENGE to the tank, then complete filling the tank with water while agitating.
4. Allow mixture to agitate and recycle for at least five minutes before spraying.
5. Agitation must be continued until spraying is completed.

HOW TO USE AVENGE ALONE

If wild oats are growing vigorously, select the proper application rate from the AVENGE Recommendation Table that corresponds to the population of wild oat plants per square foot in the field to be treated.

Apply AVENGE in 5 to 20 gallons of water per acre by ground equipment or 3 to 10 gallons of water per acre by aircraft. DO NOT let spray drift onto adjacent crops.

HOW TO USE AVENGE PLUS BROADLEAF HERBICIDES

Select the proper AVENGE application rate as described in the AVENGE Recommendation Table. DO NOT apply AVENGE tank mix combinations under stress conditions. In order to assure optimum broadleaf weed control follow label recommendations of the broadleaf herbicides used with AVENGE. Follow the Mixing Instructions and use the herbicide as follows:

AVENGE PLUS MCPA[1]: Use only amine or ester formulations of MCPA at the rate of 0.25 to 1.0 lb. acid equivalent per acre. Apply in 5 to 20* gallons per acre by ground equipment or 5 to 10 gallons per acre by aircraft.

AVENGE PLUS BROMOXYNIL[2]: Use bromoxynil at the rate of 0.375 to 0.5 lb. active ingredient per acre. Apply in 10 to 20* gallons per acre by ground equipment or 5 to 10 gallons per acre by aircraft.

AVENGE PLUS MCPA PLUS BROMOXYNIL[3]: Use MCPA at the rate of 0.5 lb. acid equivalent per acre plus bromoxynil at 0.25 to 0.5 lb. active ingredient per acre. Apply in 10 to 20* gallons per acre by ground equipment or 5 to 10 gallons per acre by aircraft.

AVENGE PLUS 2,4-D[4]: Use only amine or ester formulations of 2,4-D at the rate of 0.25 to 0.75 lb. acid equivalent per acre. Apply in 5 to 20* gallons of spray per acre by ground equipment and 5 to 10 gallons per acre by aircraft. Apply tank-mix after crops are 6 inches tall, after the crops are well tillered, or after the bloom stage. DO NOT apply in seedling stage or early boot to dough stage.

AVENGE PLUS GLEAN[5]: Use Glean at the rate of $\frac{1}{6}$ to $\frac{1}{2}$ ounce of the 75% dry flowable formulation per acre. Apply in 5 to 20* gallons of spray per acre by ground equipment and 3 to 10 gallons per acre by aircraft. Prior to use of Glean, determine soil pH by laboratory analysis using a 1:1 soil:water suspension. DO NOT use Glean on soils with a pH greater than 7.5. DO NOT exceed $\frac{1}{2}$ oz/acre/year on soils with a pH of 6.5 or lower, or $\frac{1}{3}$ oz/acre/year on soils with a pH of 6.6 to 7.5. Follow the most restrictive precautions, directions, and recropping/rotation limitations that appear on both products' labels.

AVENGE PLUS ALLY[6]: Use Ally at the rate of $\frac{1}{10}$ ounce of the 60% dry flowable formulation per acre. Apply in 5 to 20* gallons of spray per acre by ground equipment and 3 to 10 gallons of spray per acre by aircraft. DO NOT use Ally on highly calcareous soils (soil pH of greater than 8.0). DO NOT apply more than $\frac{1}{10}$ ounce per acre per year. DO NOT apply to irrigated land where tail water will be used to irrigate other crop land. Follow the most restrictive precautions, directions, and recropping/rotation limitations that appear on both product labels.

NOTE: DO NOT MAKE MORE THAN ONE APPLICATION OF AVENGE PER GROWING SEASON. DO NOT GRAZE TREATED FIELDS OR CUT TREATED FORAGE FOR SILAGE OR HAY. BARLEY OR WHEAT CAN BE REPLANTED IF LOSS OF CROP OCCURS. BARLEY AND WHEAT GRAIN AND STRAW FROM TREATED FIELDS CAN BE FED TO LIVESTOCK. ROTATION TO OTHER CROPS CAN BE MADE THE FOLLOWING YEAR.

* When applying more than 10 gallons of spray per acre, follow instructions relating to the addition of surfactant described above.

[1] The amine and ester formulations of MCPA are sold under various names. Consult manufacturer's label recommendation before use.

[2] Bromoxynil is sold under various brand names. Consult manufacturer's label recommendation before use.

[3] There are several brand names for MCPA + bromoxynil. Consult manufacturer's label recommendation before use.

[4] The amine and ester formulations of 2,4-D are sold under various brand names. Consult manufacturer's label recommendation before use.

[5] Glean is a trademark of E. I. duPont de Nemours and Company, Inc.

[6] Ally is a trademark of E. I. duPont de Nemours and Company, Inc.

STORAGE AND DISPOSAL

Ice will form in the container at 15°F (−10°C). If ice occurs from prolonged storage at temperatures 15°F (−10°C) or below, place container at room temperature until ice melts.

PROHIBITIONS: Do not contaminate water, food, or feed by storage or disposal. Open dumping is prohibited. Do not reuse empty container.

PESTICIDE DISPOSAL: Pesticide wastes are acutely hazardous. Improper disposal of excess pesticide, spray mixture, or rinsate is a violation of Federal law. If these wastes cannot be disposed of by use according to label instructions, contact your State Pesticide or Environmental Control Agency, or the hazardous waste representative at the nearest EPA Regional Office for guidance.

CONTAINER DISPOSAL: Triple rinse (or equivalent). Then offer for recycling or reconditioning, or puncture and dispose of in a sanitary landfill, or by incineration, or, if allowed by state and local authorities, by burning. If burned, stay out of smoke.

GENERAL: Consult federal, state, or local authorities for approved alternative procedures.

DISCLAIMER

The label instructions for the use of this product reflect the opinion of experts based on field use and tests. The directions are believed to be reliable and should be followed carefully. However, it is impossible to eliminate all risks inherently associated with use of this product. Crop injury, ineffectiveness or other unintended consequences may result because of such factors as weather conditions, presence of other materials, or the use or application of the product contrary to label instructions all of which are beyond the control of American Cyanamid Company. All such risks shall be assumed by the user.

American Cyanamid Company warrants only that the material contained herein conforms to the chemical description on the label and is reasonably fit for the use therein described when used in accordance with the directions for use, subject to the risks referred to above.

Any damages arising from a breach of this warranty shall be limited to direct damages and shall not include consequential commercial damages such as loss of profits or values or any other special or indirect damages.

American Cyanamid Company makes no other express or implied warranty, including any other express or implied warranty of FITNESS or of MERCHANTABILITY.

CHOPPER®

HERBICIDE

ACTIVE INGREDIENT:

Isopropylamine salt of Imazapyr (2-[4,5-dihydro-4-methyl-4-(1-methylethyl)-5-oxo-1*H*-imidazol-2-yl]-3-pyridinecarboxylic acid)*	27.6%
INERT INGREDIENTS:	72.4%
TOTAL:	100.0%

* Equivalent to 22.6% 2-[4,5-dihydro-4-methyl-4-(1-methylethyl)-5-oxo-1*H*-imidazol-3-yl]-3-pyridinecarboxylic acid or 2 pounds acid per gallon.

EPA Reg. No. 241-296

KEEP OUT OF REACH OF CHILDREN
CAUTION!

PRECAUCION AL USUARIO: Si usted no lee ingles, no use este producto hasta que la etiqueta le haya sido explicada ampliamente.

In case of an emergency endangering life or property involving this product, call collect, day or night, Area Code 201-835-3100.

10/88

® Registered Trademark of American Cyanamid Company

PRECAUTIONARY STATEMENTS

HAZARDS TO HUMANS

CAUTION!

May be a skin sensitizer. Avoid contact with skin, eyes or clothing. Avoid breathing spray mist. Wash thoroughly with soap and water after handling. Remove contaminated clothing and wash before reuse.

FIRST AID

IF ON SKIN: Wash with plenty of soap and water.

IF IN EYES: Flush with plenty of water. Get medical attention if irritation persists.

PHYSICAL AND CHEMICAL HAZARDS

Spray solutions of CHOPPER should be mixed, stored and applied only in stainless steel, fiberglass, plastic and plastic-lined steel containers.

DO NOT mix, store or apply CHOPPER or spray solutions of CHOPPER in unlined steel (except stainless steel) containers or spray tanks.

ENVIRONMENTAL HAZARDS

DO NOT apply directly to water or wetlands. **DO NOT** contaminate water when disposing of equipment washwaters.

IMPORTANT

DO NOT use on food or feed crops. **DO NOT** apply on ditches used to transport irrigation water. **DO NOT** apply where runoff water may flow onto agricultural land as injury to crops may result. Keep from contact with fertilizers, insecticides, fungicides and seeds. **DO NOT** apply or drain or flush equipment on or near desirable trees or other plants, or on areas where their roots may extend, or in locations where the chemical may be washed or moved into contact with their roots. **DO NOT** use on lawns, walks, driveways, tennis courts, or similar areas. **DO NOT** use in California.

Thoroughly clean application equipment after use. Flush tank, pump, hoses and boom with several changes of water after removing nozzle tips and screens (clean these parts separately).

GENERAL INFORMATION

CHOPPER herbicide is an emulsifiable concentrate mixable with water, fuel or penetrating oils. CHOPPER is to be mixed with a penetrating oil and applied as a spray to cut stumps, the basal areas of brush and trees and frilling cuts for the control of brush in noncrop areas. **Prior to actual tank mixing with other products, compatibility testing in small containers is recommended.**

CHOPPER herbicide is readily absorbed through bark and roots and is translocated rapidly throughout the plants. In dormant plants, leaf-out is prevented and dieback occurs as the plant depletes its energy stores.

DIRECTIONS FOR USE

It is a violation of Federal law to use this product in a manner inconsistent with its labeling.

STORAGE AND DISPOSAL

PROHIBITIONS: DO NOT store below 10°F. **DO NOT** contaminate water, food or feed by storage or disposal.

PESTICIDE DISPOSAL: Wastes resulting from the use of this product may be disposed of on site or at an approved waste disposal facility.

CONTAINER DISPOSAL: Triple rinse (or equivalent). Then offer for recycling or reconditioning, or puncture and dispose of in an approved sanitary landfill, or by incineration, or, if allowed by state and local authorities, by burning. If burned, stay out of smoke.

CHOPPER herbicide should be used only in accordance with recommendations in this label. Keep containers closed to avoid spills and contamination.

An application of CHOPPER is recommended for control of brush in noncropland areas such as railroad, utility, highway and pipeline rights-of-way, utility plant sites, petroleum tank farms, pumping installations, fence rows, storage areas, nonirrigation ditchbanks and other similar areas. CHOPPER is also recommended for woody vegetation control in forestry sites and wildlife habitats.

CUT STUMP TREATMENTS

Mix 8.0-12.0 fluid ounces of CHOPPER in one gallon of water*, diesel oil, or a penetrating oil such as CIDE-KICK[1], CIDE-KICK II[2] or ARBORCHEM BASAL OIL[3].

Spray or brush the CHOPPER solution onto the cambium area of the freshly cut stump surface and the bark of the cut stump. Insure that the CHOPPER solution thoroughly wets the cambium area, the wood next to the bark of the stump.

DO NOT OVER APPLY CAUSING PUDDLING.

* Note: Use water as a diluent only when temperatures are sufficient to prevent freezing.

TREE INJECTION TREATMENTS

Mix 8.0-12.0 fluid ounces of CHOPPER in one gallon of water*, diesel oil, or a penetrating oil such as CIDE-KICK, CIDE-KICK II or ARBORCHEM BASAL OIL.

Using standard injection equipment, apply 1 mil. of CHOPPER solution at each injection site around the tree with no more than 1 inch intervals between cut edges. Insure that the injector completely penetrates the bark at each site.

* Note: Use water as a diluent only when temperatures are sufficient to prevent freezing.

FRILL OR GIRDLE TREATMENTS

Mix 8.0-12.0 fluid ounces of CHOPPER in one gallon of water*, diesel oil, or a penetrating oil such as CIDE-KICK, CIDE-KICK II or ARBORCHEM BASAL OIL.

Using an axe or machete, make cuts through the bark and completely around the tree with no more than 2 inch intervals between cut edges. Spray or brush the CHOPPER solution into each cut until thoroughly wet.

* Note: Use water as a diluent only when temperatures are sufficient to prevent freezing.

LOW VOLUME BASAL BARK TREATMENTS

Mix 8.0-12.0 fluid ounces of CHOPPER in one gallon of diesel oil, or a penetrating oil such as CIDE-KICK, CIDE-KICK II or ARBORCHEM BASAL OIL.

To control mixed brush species up to 4 inches DBH (tree diameter at breast height) spray to wet the lower 12-18 inches of the plant stem with the CHOPPER oil mixture.

DO NOT over apply causing dripping or puddling.

CHOPPER herbicide will control the following species:

Ash (*Fraxinus* spp.)	Poplar (*Populus* spp.)
Cherry (*Prunus* spp.)	Prickly wild rose (*Rosa arkansana*)
Dogwood (*Cornus* spp.)	Privet (*Ligustrum vulgare*)
Hawthorn (*Crataegus* spp.)	Red Maple (*Acer rubrum*)
Hickory (*Carya* spp.)	Sassafras (*Sassafras albidum*)
Maple (*Acer* spp.)	Sumac (*Rhus* spp.)
Mulberry (*Morus* spp.)	Sweetgum (*Liquidambar styraciflua*)
Multiflora rose (*Rosa multiflora*)	Willow (*Salix* spp.)
Oak (*Quercus* spp.)	

[1,2] Registered Trademark of JLB International Chemical, Inc.
[3] Registered Trademark of Arborchem Products Co.

DISCLAIMER

The label instructions for the use of this product reflect the opinion of experts based on field use and tests. The directions are believed to be reliable and should be followed carefully. However, it is impossible to eliminate all risks inherently associated with use of this product. Ineffectiveness or other unintended consequences may result because of such factors as weather conditions, presence of other materials, or the use or application of the product contrary to label instructions, all of which are beyond the control of the American Cyanamid Company. All such risks shall be assumed by the user.

American Cyanamid Company warrants only that the material contained herein conforms to the chemical description on the label and is reasonably fit for the use therein described when used in accordance with the directions of use, subject to the risks referred to above.

Any damages arising from a breach of this warranty shall be limited to direct damages and shall not include consequential commercial damages such as loss of profits or values of any other special or indirect damages.

American Cyanamid Company makes no other express or implied warranty, including other express or implied warranty of FITNESS or of MERCHANTABILITY.

COUNTER®

SYSTEMIC INSECTICIDE NEMATICIDE

RESTRICTED USE PESTICIDE

Due to Acute Oral and Dermal Toxicity.

For retail sale to, and use only by Certified Applicators or persons under the direct supervision of a Certified Applicator, and only for those uses covered by the Certified Applicator's certification.

American Cyanamid Company endorses Certification to promote the responsible use of pesticides to insure the protection of man and the environment.

FOR USE IN FIELD CORN, POPCORN, SWEET CORN, SUGAR BEETS AND GRAIN SORGHUM

ACTIVE INGREDIENT:	
terbufos (S-[[(1,1-dimethylethyl)thio]methyl]O,O-diethyl phosphorodithioate)	15.0%
INERT INGREDIENTS:	85.0%
TOTAL	100.0%

EPA Reg. No. 241-238
7/89

KEEP OUT OF REACH OF CHILDREN

DANGER–POISON

DANGER/¡PELIGRO!

PRECAUCION AL USUARIO: Si usted no lee inglés, no use este producto hasta que la etiqueta le haya sido explicada ampliamente.

FIRST AID

If swallowed, drink one or two glasses of water and induce vomiting by touching back of throat with finger. Do not induce vomiting or give anything by mouth to an unconscious person. Avoid alcohol. Get medical attention.

If inhaled, remove to fresh air. If not breathing, give artificial respiration, preferably mouth-to-mouth. If breathing is difficult, give oxygen. Get medical attention.

If on skin, wash thoroughly with soap and water. Remove contaminated clothing and shoes. Wash clothing and decontaminate shoes before reuse.

If in eyes, immediately flush eyes with plenty of water. Get medical attention.

Repeated inhalation or skin contact may, without symptoms, progressively increase susceptibility to poisoning. **Antidote:** Atropine is an antidote.

CALL A PHYSICIAN AT ONCE IN ALL CASES OF SUSPECTED POISONING

NOTE TO PHYSICIANS: Warning symptoms include weakness, headache, tightness in chest, blurred vision, nonreactive pinpoint pupils, salivation, nausea, vomiting, diarrhea and abdominal cramps. Give atropine intramuscularly or intravenously, depending on severity of poisoning, 2 to 4 milligrams every 10 minutes until fully atropinized as shown by dilated pupils, dry flushed skin and tachycardia. Twenty to thirty milligrams, or more, may be required during the first 24 hours. Never give opiates or phenothiazine tranquilizers. Clear chest by postural drainage. Artificial respiration or oxygen administration may be necessary. Observe patient continuously for at least 48 hours. Allow no further exposure to any cholinesterase inhibitor until cholinesterase regeneration has taken place as determined by blood tests.

Pralidoxime chloride (2-PAM; PROTOPAM chloride) may be effective as an adjunct to atropine. Use according to label directions.

PRECAUTIONARY STATEMENTS

DANGER!

HAZARDS TO HUMANS AND DOMESTIC ANIMALS

Fatal if swallowed, inhaled, or absorbed through the skin. Do not breathe dust. Do not get in eyes, on skin, or on clothing.

Wear freshly laundered, long-sleeved work clothing daily. While transferring from container to equipment, wear a clean cap, gloves (rubber or cotton) and goggles. If cotton gloves are used, they must be laundered or discarded after each day's use. Rubber gloves should be washed with soap and water after each use. Do not wear the same gloves for other work. Destroy and replace gloves frequently.

In case of contact, immediately remove contaminated clothing and wash skin thoroughly with soap and water. Launder clothing and decontaminate shoes before reuse. Wash thoroughly with soap and water before eating or smoking. Bathe at the end of the work day and change clothing.

Do Not Breathe Dust

While emptying bags into equipment, pour downwind and allow as little free fall as possible. Do not pour at face level and do not allow dust to reach the breathing zone.

Do Not Contaminate Food or Feed Products

Once a bag has been opened, use it completely. Make sure the hoppers are emptied while still in the field. Refer to STORAGE AND DISPOSAL statement for further instructions.

Keep All Unprotected Persons Out of Operating Areas

Do not apply this product in such a manner as to directly or through drift expose workers or other persons.

Keep Out of Reach of Domestic Animals
Not For Use or Storage In or Around the Home

RE-ENTRY STATEMENT

Do not enter treated areas without protective clothing until treatments have been completed.

Because certain states may require more restrictive re-entry intervals for various crops treated with this product, consult your State Department of Agriculture for further information.

Written or oral warnings must be given to workers who are expected to be in a treated area or in an area about to be treated with this product. When oral warnings are given, warnings shall be given in a language customarily understood by workers. Oral warnings must be given if there is reason to believe that written warnings cannot be understood by workers. Written warnings must include the following information: DANGER. Area treated with COUNTER on (date of application). Do not enter without protective clothing.

WORK SAFETY RULES

REPEATED EXPOSURES TO CHOLINESTERASE INHIBITORS SUCH AS ARE CONTAINED IN THIS PRODUCT MAY, WITHOUT WARNING, CAUSE PROLONGED SENSITIVITY TO VERY SMALL DOSES OF ANY CHOLINESTERASE INHIBITOR.

Persons working with this product should have frequent blood tests of their cholinesterase levels. If the cholinesterase level falls below a critical point, no further exposure should be allowed until it has been determined by means of blood tests that the cholinesterase level has returned to normal. Before using this product, consult the National Pesticide Telecommunication Network for recommendations regarding such blood tests, poisoning management, emergency treatment, and other information regarding the toxicity of terbufos. The toll-free number for the National Pesticide Telecommunication Network is 1-800-858-7378.

If handled indoors, provide mechanical exhaust ventilation.

Keep all unprotected persons, children, livestock, and pests away from treated areas or where there is danger of drift.

Do not rub eyes or mouth with hands. If you feel sick in any way, STOP work and get help right away. See First Aid section.

WEAR THE FOLLOWING PROTECTIVE CLOTHING DURING LOADING, APPLICATION, EQUIPMENT REPAIR, EQUIPMENT CLEANING, AND DISPOSAL OF THE PESTICIDE.

Wear a protective suit of one or two pieces that covers all parts of the body except the head, hands and feet. Wear chemical-resistant gloves and a pesticide respirator approved by the National Institute for Occupational Safety and Health (NIOSH) and the Mine Safety and Health Administration (MSHA) at all times except during application.

If application is performed using an enclosed cab or cockpit, the following protective clothing may be worn as an alternate: long-sleeved shirt and long-legged pants; shoes and socks. Chemical-resistant gloves must be available in the cab or cockpit and must be worn while exiting. All other protective clothing required for use during application must be available in the cab and must be worn when exiting the cab into treated areas.

IMPORTANT! Before removing gloves, wash them with soap and water. Always wash hands, face, and arms with soap and water before smoking, drinking, or toileting.

IMMEDIATELY AFTER COMPLETING WORK WITH THE PESTICIDE, take off all clothing and shoes. Shower using soap and water. Do not reclothe with contaminated clothing and shoes. Wash protective clothing and protective equipment with soap and water after each use. Respirators must be cleaned and filters replaced according to instructions included with the respirators. Personal clothing worn during use must be laundered separately from household articles. Clothing and protective equipment heavily contaminated with terbufos must be destroyed according to State and local regulations. HEAVILY CONTAMINATED CLOTHING (CLOTHING ON WHICH THE PRODUCT HAS SPILLED) CANNOT BE ADEQUATELY DECONTAMINATED.

DIRECTIONS FOR USE

It is a violation of Federal law to use this product in a manner inconsistent with its labeling.

This label must be in the possession of the user at the time of pesticide application.

BEFORE USING, READ PRECAUTIONARY STATEMENTS

COUNTER should be applied with a granular pesticide applicator properly calibrated to assure accurate placement and proper dosage. Cover granules that may be exposed on the ends of the treated rows and turns and loading areas by deep discing immediately after treating fields.

CROPS	PESTS CONTROLLED	RATES OF COUNTER	APPLICATION	REMARKS
FIELD CORN, POPCORN, AND SWEET CORN **At Planting**	Corn rootworms Wireworms White grubs Seedcorn maggots Seedcorn beetles Corn flea beetles Maize billbugs Southern corn billbugs Thrips Symphylans Nematodes: Lance Lesion Root Knot Spiral Stunt Sting Stubby Root Dagger Suppression of: Cutworms Lesser corn-stalk borers	**Banded or In-Furrow** 8 oz. per 1,000 ft. of row for any row spacing. Do not exceed 13.1 pounds per acre.	**Banded** Place granules in a 7-inch band over the row, in front of or behind the presswheel and lightly incorporate. **In-Furrow** Place granules directly in the seed furrow behind the planter shoe.	For use on conventional and conservation tillage corn. In situations where crop debris or winds can prevent proper placement of granules, in-furrow applications are recommended. In-furrow applications reduce the potential for insecticide runoff resulting from excessive rain. Under dry soil conditions or heavy infestations, it may be necessary to apply an insecticide rescue treatment with another registered insecticide after corn emergence to control surviving cutworm larvae. If application is made at planting, do not make post-emergence or cultivation time treatments of COUNTER.
Post-emergence Incorporated **(Same as crops above)**	Maize billbugs Southern corn billbugs	**Banded** 12 oz. per 1,000 ft. of row for any row spacing (minimum 30-inch row spacing).	Apply in a 7-inch band over the row of seeding corn plants and lightly incorporate into the soil when billbugs or damage are observed, usually in the 1-6 leaf stage of growth. Use cultivators or other suitable implements to lightly incorporate granules into soil.	Only one at planting, postemergence incorporated or cultivation time treatment of COUNTER may be used. Do not graze or cut for forage within 30 days of treatment. Consult your state experiment station, state extension service or certified pest management specialist for proper rate and timing of application.
At Cultivation **(Same as crops above)**	Corn rootworms	8 oz. per 1,000 ft. of row spacing (minimum 30-inch row spacing).	Apply granules to base of plants or over the top of plants just ahead of cultivation shovels so as to cover granules with soil.	Only one at planting postemergence incorporated or cultivation time treatment of COUNTER may be used. Do not graze or cut for forage within 30 days of treatment.
SUGAR BEETS **At Planting**	Sugar beet root maggots (1) Wireworms (1) White grubs (1)	**Modified in-furrow or Banded** 4-8 oz. per 1,000 ft. of row for any row spacing (minimum 20-inch row spacing).	**Modified in-furrow** Apply in-furrow at planting time 2-3 inches behind the seed drop zone after some soil has covered the seed.	Only one application of COUNTER per year may be made.
SUGAR BEETS **Post-emergence**	Sugar beet root maggots	**Banded** 4-8 oz. per 1,000 ft. of row for any row spacing (minimum 20-inch row spacing).	**Banded** Apply in a 5 to 7-inch band over the row and lightly incorporate into the soil.	**At Planting** Do not place granules in direct contact with the seed as crop injury may occur. Power incorporation may be used. Do not incorporate deeper than 2 inches.
SUGAR BEETS **At Planting**	Suppression of Cutworms	**Banded** 8 oz. per 1,000 ft of row for any row spacing (minimum 20-inch row spacing).		**Postemergence** Apply at the time of first fly emergence. Lightly incorporate into the soil.

CROPS	PESTS CONTROLLED	RATES OF COUNTER	APPLICATION	REMARKS
	Suppression of: Sugar Beet cyst nematode Control of: Beet leafhopper	**Knifed-in** 18 oz. per 1,000 ft. of row for any row spacing (minimum 20-inch row spacing) or no more than 29 lbs. pr acre.	**Knifed-in** Drill granules 2 inches to the side of the seed and 2-4 inches below the seed.	[1] If especially heavy infestations are expected use the 8 oz. rate.
GRAIN SORGHUM **At Bedding**	Greenbugs Corn leaf aphids	**Knifed-in** 8-16 oz. per 1,000 ft. of row for any row spacing (minimum 20-inch row spacing) or no more than 26 lbs. per acre.	**Knifed-in** Drill granules 1-4 inches directly below the seed OR 1-4 inches below the seed and up to 5 inches to the side of the seed.	Do not place granules in direct contact with seed as crop injury may occur.
GRAIN SORGHUM **At Planting**	Greenbugs Corn leaf aphids	**Knifed-in or Banded** **Knifed-in** 8-16 oz. per 1,000 ft. of row for any row spacing (minimum 20-inch row spacing) or no more than 26 lbs. per acre. **Banded** 8 oz. per 1,000 ft. of row for any row spacing (minimum 20-inch row spacing) or no more than 13.1 lbs. per acre.	**Knifed-in** Drill granules 1-4 inches directly below the seed OR 1-4 inches below the seed and up to 5 inches to the side of the seed. **Banded**[2] Place granules in a 5-7 inch band directly behind the planter shoe in front of the press wheel.	[2] Do not use banded applications for aphid control in New Mexico, West Texas and the Panhandle of Oklahoma. **Only one application of COUNTER per year may be used.**
GRAIN SORGHUM **At Planting**	Corn rootworms Wireworms White grubs Nematodes	**Banded** 8 oz. per 1,000 ft. of row for any row spacing (minimum 20-inch row spacing) or no more than 13.1 lbs. per acre.	**Banded** Place granules in a 5-7 inch band directly behind the planter shoe in front of the press wheel.	

ENVIRONMENTAL HAZARDS

This product is extremely toxic to fish, and wildlife. Birds and wild mammals utilizing treated fields may be killed.

Do not apply directly to water or wetlands (including swamps, marshes, bogs, and potholes). Runoff and drift from treated areas may be hazardous to aquatic organisms in adjacent aquatic sites. Do not contaminate water when disposing of equipment washwaters. Cover or incorporate granules that are spilled during mixing and loading.

IN CASE OF AN EMERGENCY ENDANGERING LIFE OR PROPERTY INVOLVING THIS PRODUCT, CALL COLLECT, DAY OR NIGHT, AREA CODE 201-835-3100.

STORAGE AND DISPOSAL

Storage

Store pesticide products in a secure locked area where children, unauthorized persons, and animals cannot enter. Do not store in the same area with food or feed. Do not store opened bags.

Prohibitions

Do not contaminate water, food, or feed by storage or disposal. Open dumping is prohibited.

Pesticide Disposal

Pesticide wastes are acutely hazardous. Improper disposal of excess pesticide, spray mixture, or rinsate is a violation of Federal Law. If these wastes cannot be disposed of by use according to label instructions, contact your state pesticide or Environmental Control Agency, or the hazardous waste representative at the nearest EPA Regional Office for guidance.

Container Disposal

Completely empty bag into application equipment. Then dispose of empty bag in a sanitary landfill or by incineration, or, if allowed by state and local authorities, by burning. If burned, stay out of smoke.

General

Consult federal, state or local disposal authorities for approved alternative procedures such as limited open burning.

USE RATES AS POUNDS PER ACRE* (BASED ON VARIOUS ROW SPACINGS AND APPLICATION RATES) APPLICATION RATES (OZ./1,000 FT. OF ROW)

ROW WIDTH (inches)	4 oz.	6 oz.	8 oz.	10 oz.	12 oz.	14 oz.	16 oz.
40″	3.3	4.9	6.5	8.2	9.8	11.4	13.1
38″	3.4	5.2	6.9	8.6	10.3	12.0	13.7
36″	3.6	5.4	7.3	9.1	10.9	12.7	14.5
34″	3.9	5.6	7.7	9.7	11.6	13.5	15.4
32″	4.1	6.1	8.2	10.2	12.3	14.3	16.3
30″	4.4	6.5	8.7	10.9	13.1	15.2	17.4
28″	4.7	7.1	9.5	11.8	14.2	16.6	18.9
26″	5.0	7.6	10.2	12.7	15.3	17.9	20.4
24″	5.4	8.2	10.9	13.6	16.3	19.1	21.8
22″	5.9	8.9	11.9	14.9	17.9	21.0	24.0
20″	6.5	9.8	13.1	16.3	19.6	22.8	26.1

* See directions for specific use rates on each crop.

DISCLAIMER

The label instructions for the use of this product reflect the opinion of experts based on field use and tests. The directions are believed to be reliable and should be followed carefully. However, it is impossible to eliminate all risks inherently associated with use of this product. Crop injury, ineffectiveness or other unintended consequences may result because of such factors as weather conditions, presence of other materials, or the use or application of the product contrary to label instructions, all of which are beyond the control of American Cyanamid Company. All such risks shall be assumed by the user.

American Cyanamid Company warrants only that the material contained herein conforms to the chemical description on the label and is reasonably fit for the use therein described when used in accordance with the directions for use, subject to the risks referred to above.

Any damages arising from a breach of this warranty shall be limited to direct damages and shall not include consequential commercial damages such as loss of profits or values or any other special or indirect damages.

American Cyanamid Company makes no other express or implied warranty, including any other express or implied warranty of FITNESS or of MERCHANTABILITY.

CALIBRATION INFORMATION

FIRST READ THE LABEL

It is important that applicator equipment be properly set to deliver the labeled rate. This chart will help determine the desired rate of application.

When label rate is 8 ounces per 1,000 feet of row, use these suggested starting gauge settings, regardless of row spacing:

PLANTING SPEED

APPLICATOR	3 mph	5 mph	7 mph
John Deere Max Emerge 2 (Models 7200 and 7300)	16	23	35
John Deere Max-Emerge (Odd Notches 5-15-25-35) (Models 7000 and 7100)	11	15	19
John Deere Max-Emerge (Even Notches 0-10-20-30) (Models 7000 and 7100)	13	19	25
John Deere 71 Flexi-Planter (Metal Hoppers)	1/28	2/6	2/15
International Harvester	1/7.0	2/1.5	2/7.5
New International Harvester	1/9.0	2/6.0	3/2.0
Gandy	18	24	28
New Noble	8	12	16
Old Noble	11	17	23
Allis Chalmers (78 and 79 Series)		All Speeds—Gauge 7	
Allis Chalmers #385	10	14	18
Buffalo (Gauged)	9	15	20
Buffalo (Fluted-Roll Feed)		All speeds—5 1/2 Revolutions Hexagonal nut	
White		Refer to Nobel Settings	

Note: These settings should be used as starting points only. Continually check the amount of COUNTER used against a known length of row or acreage and make further adjustments accordingly. Also, check calibration occasionally to make sure equipment wear, changing moisture conditions, etc., have not caused a change in flow rate.

® Trademark of American Cyanamid Company

SUPPLEMENTAL LABELING

COUNTER® SYSTEMIC INSECTICIDE-NEMATICIDE

RESTRICTED USE PESTICIDE

Due to acute oral and dermal toxicity

For retail sale to, and use only by Certified Applicators or persons under the direct supervision of a Certified Applicator, and only for those uses covered by the Certified Applicator's certification. American Cyanamid Company endorses Certification to promote the responsible use of pesticides to insure the protection of man and the environment.

EPA Reg. No. 241-238

® Registered Trademark of American Cyanamid Company

4/88

FOR DISTRIBUTION AND USE ONLY WITHIN THE STATES OF LOUISIANA, GEORGIA, ALABAMA, MISSISSIPPI, FLORIDA, TEXAS FOR USE ON FIELD CORN

DIRECTIONS FOR USE

BEFORE USING, READ PRECAUTIONARY STATEMENTS ON BAG

It is a violation of Federal law to use this product in a manner inconsistent with its labeling. COUNTER should be applied with a granular pesticide applicator properly calibrated to assure placement and proper dosage. See label for specific instructions.

Cover granules that may be exposed on the ends of the treated rows and turns and loading areas by deep disking immediately after treating fields.

Crop	Pests Controlled	Rate of Counter	Application	Remarks
Field Corn At Planting	Chinch Bug	In-furrow 8 oz/1,000 ft. of row for any row spacing (minimum 30 inch row spacing)	In-furrow Apply in-furrow at planting time. Place granules directly in the seed furrow behind the planter shoe.	For early season control of light to moderate infestations.

THIS LABEL SHOULD BE IN THE POSSESSION OF THE USER AT THE TIME OF PESTICIDE APPLICATION

CYCOCEL®

PLANT GROWTH REGULANT

FOR USE ON AZALEAS, RED POINSETTIAS, GERANIUMS AND HIBISCUS

ACTIVE INGREDIENT:

chlormequat (2-chloroethyl)trimethylammonium chloride	11.8%
INERT INGREDIENTS:	88.2%
TOTAL	100.0%

(1 gallon contains 1 pound (2-chloroethyl)trimethylammonium chloride)

EPA Reg. No. 241-74-AA

® Registered Trademark of American Cyanamid Company

4/88

KEEP OUT OF REACH OF CHILDREN

CAUTION!

PRECAUCION!

PRECAUCION AL USUARIO. Si usted no lee inglés, no use este producto hasta que la etiqueta le haya sido explicada ampliamente.

STATEMENT OF PRACTICAL TREATMENT

If Swallowed: Call a physician or Poison Control Center. Drink 1 or 2 glasses of water and induce vomiting by touching back of the throat with a finger. DO NOT induce vomiting or give anything by mouth to an unconscious person. Avoid alcohol.

If in Eyes: Flush eyes with plenty of water. Call a physician if irritation persists.

If on Skin: Flush with plenty of water. Get medical attention if irritation persists.

Note to Physician: The use of Atropine is contraindicated.

PRECAUTIONARY STATEMENTS

HAZARDS TO HUMANS AND DOMESTIC ANIMALS

CAUTION!

Harmful if swallowed or absorbed through the skin.
Avoid contact with skin, eyes, or clothing.
Wash thoroughly with soap and water after handling.
Remove contaminated clothing and wash before reuse.

ENVIRONMENTAL HAZARDS

This product is toxic to wildlife. Keep out of lakes, streams and ponds. Do not contaminate water when disposing of equipment washwaters.

In case of emergency endangering life or property involving this product, call collect, day or night, Area Code 201-835-3100.

DIRECTIONS FOR USE

It is a violation of Federal law to use this product in a manner inconsistent with its labeling.

DO NOT apply this product through any type of irrigation system.

STORAGE AND DISPOSAL

DO NOT STORE BELOW FREEZING TEMPERATURES

Storage: DO NOT contaminate water, food, or feed by storage.

Pesticide Disposal: Wastes resulting from the use of this product may be disposed of on site or at an approved waste disposal facility.

Container Disposal: Triple rinse (or equivalent). Then offer for recycling or reconditioning, or puncture and dispose of in a sanitary landfill, or incineration, or, if allowed by state and local authorities, by burning. If burned, stay out of smoke.

AZALEAS

OBTAIN EXPERIENCE IN SMALL-SCALE TRIALS BEFORE EXTENSIVE USE

I. TO PRODUCE EARLY BUDDED, COMPACT, SYMMETRICAL PLANTS—Apply as a uniform foliar spray at the rate of 2 to 3 ounces CYCOCEL per gallon of water in early July when regrowth following pinching or shearing is 3/4 to 1 inch long. Repeat application 1 week later. Use the higher rate in the South. In conjunction with optimum cultural practices, this treatment has induced development of multiple buds on many varieties forced for early blooming in the greenhouse. CYCOCEL has been used effectively on the following early blooming varieties: Alaska, Ambrosia, Ambrosius, Chimes, Coral Bells, Dorothy Gish, Easter Parade, Firelight, Gloria, Hexe, Hinodigiri, Red Wing, Mrs. L. C. Fischer, Salmon Queen, Snow, Snowbank, Sunset, Triumphe.

Do not apply CYCOCEL on Dawn or other Pericat varieties as leaf injury may result. Over treatment with CYCOCEL may cause temporary leaf yellowing of some varieties forced for early blooming in the greenhouse. Experience indicates that such yellowing usually disappears during storage.

CYCOCEL treatment is not a substitute for any cultural practice (light, watering, fertilizer, temperature). Optimum cultural practices for greenhouse forcing of high quality plants vary with variety and geographical location. Consult your experiment station horticulturist for proper timing and duration of light intensity, day length, cool storage and greenhouse conditions for forcing early blooming azaleas. CYCOCEL may slightly delay flowering of some varieties exposed to certain cool storage conditions.

II. TO CONTROL UNDESIRABLE VEGETATIVE GROWTH AND PREVENT LATE SEASON GROWTH IN NURSERY STOCK—When plants have reached desired head size following pinching or shearing, apply as a uniform foliar spray at the rate of 3 ounces CYCOCEL per gallon of water. Repeat application 1 week later.

Apply spray to well watered, vigorously growing plants.

Foliage should be dry at time of treatment.

Repeat application if rain occurs or foliage is washed within 24 hours after treatment.

The compatibility of CYCOCEL when mixed with insecticides, acaricides, fungicides and fertilizers is not known. Therefore, do not use CYCOCEL in combination with these materials.

RED POINSETTIAS

CYCOCEL is recommended for use on all red poinsettia varieties. Plants treated with CYCOCEL are more compact, with stronger stems, deeper green foliage and richer red bracts.

Dilute CYCOCEL by adding 1 quart to 10 gallons of water. Apply the dilute solution as a uniform soil drench using the amounts listed in the table.

Pot Diameter (Inches)	Fluid Ounces of Dilute Solution Per Pot	1 Quart of CYCOCEL Treats (Number of Pots)
2 1/4 to 3*	2	640
4	3	425
5	4	300
6	6	200
8	8	150

* Two applications will be necessary to apply the recommended volume in 2 1/4 inch pots.

Plants should be well rooted and established in the container before treatment. CYCOCEL may be applied to plants rooted directly in pots or when rooted cuttings are established after transplanting. Most effective response is obtained when application is made during periods of active root growth. Treatments during late July and early September are preferred to later applications to produce compact plants on early propagated stock. CYCOCEL may be applied at one of the following stages of growth:

1. In pots or pans, treat after cuttings or rooted cuttings have established a well-developed root system. Do not apply later than September 15 in the North and October 1 in the South.

2. After panning 2 1/4" to 3" stock, treat when active root development is observed. Do not apply later than October 15 in the North and November 1 in the South.

3. After pinching plants or within two weeks thereafter. If 2 1/4" to 3" stock is pinched and shifted to pans, wait until root system is well established before treating. Do not apply later than October 15 in the North and November 1 in the South.

Proper rates of application for use on pink and white poinsettias are still to be determined. Do not apply CYCOCEL after the dates recommended since over-dwarfing and crinkled bracts may result. Avoid application when plants are pot bound.

Clean application equipment by rinsing thoroughly with water.

GERANIUMS

CYCOCEL is recommended to induce early flowering of seed geraniums grown under greenhouse cultural conditions.

Apply CYCOCEL at 1500 ppm using a foliar application. Make two applications of CYCOCEL at 35 and 42 days after seeding. Plants treated with CYCOCEL show a decrease in days to flowering, vegetative height, and an increase in number of breaks.

To mix the spray solution, add the volume of CYCOCEL indicated in the table below to the desired amount of water.

Spray Solution Mixing Guide

Solution Volume	Amount of CYCOCEL to Use 1500 PPM
1 Gallon	1.6 Fl. Oz.
5 Gallons	8.0 Fl. Oz.
10 Gallons	16.0 Fl. Oz.

HIBISCUS

CYCOCEL is recommended to reduce shoot growth and to improve flowering of Hibiscus spp. grown under greenhouse cultural conditions.

Apply CYCOCEL at 460 ppm using a foliar application after pruning to new shoots when they are 4-5 cm long. Make two or three applications 14 days apart. Plants treated with CYCOCEL show a dwarfing shoot growth and an increase in flowering.

To mix the spray solution, add the volume of CYCOCEL indicated in the table below to the desired amount of water.

Spray Solution Mixing Guide

Solution Volume	Amount of CYCOCEL to use 460 PPM
1 Gallon	0.5 Fl. Oz.
5 Gallons	2.5 Fl. Oz.
10 Gallons	5.0 Fl. Oz.

DISCLAIMER

The label instructions for use of this product reflect the opinion of experts based on field use and tests. The directions are believed to be reliable and should be followed carefully. However, it is impossible to eliminate all risks inherently associated with use of this product. Crop injury, ineffectiveness or other unintended consequences may result because of such factors as weather conditions, presence of other materials, or the use or application of the product contrary to label instructions, all of which are beyond the control of American Cyanamid Company. All such risks shall be assumed by the user. American Cyanamid Company warrants only that the material contained herein conforms to the chemical description on the label and is reasonably fit for the use therein described when used in accordance with the directions for use, subject to the risks referred to above.

Any damages arising from a breach of this warranty shall be limited to direct damages and shall not include consequential commercial damages such as loss of profits or values or any other special or indirect damages.

American Cyanamid Company makes no other express or implied warranty, including any other express or implied warranty of FITNESS or MERCHANTABILITY.

CYGON® 400

SYSTEMIC INSECTICIDE-MITICIDE

ACTIVE INGREDIENT:
Dimethoate (O,O-dimethyl S-[N-(methylcarbamoyl) methyl] phosphorodithioate) 43.5%
INERT INGREDIENTS: 56.5%
TOTAL 100.0%
(1 Gallon contains 4.0 pounds of Dimethoate)

EPA Reg. No. 241-233-ZA

® Registered Trademark of American Cyanamid Company

2/89

KEEP OUT OF REACH OF CHILDREN

WARNING/AVISO

PRECAUCION AL USUARIO: Si usted no lee ingles, no use este producto hasta que la etiqueta le haya sido explicada ampliamente.

FIRST AID

In case of contact with eyes, flush eyes with plenty of water. Get medical attention.

NOTE TO PHYSICIANS: This product may cause cholinesterase inhibition. Atropine is antidotal. Pralidoxime chloride (2-PAM: PROTOPAM chloride) may be effective as an adjunct to atropine. Use according to label directions.

See Next Page For Additional Precautionary Statements See Directions For Use Inside.

In case of an emergency endangering life or property involving this product, call collect, day or night, Area Code 201-835-3100.

PRECAUTIONARY STATEMENTS

HAZARDS TO HUMANS AND DOMESTIC ANIMALS

WARNING

- Harmful or fatal if swallowed.
- Vapor harmful—concentrated material.
- Causes eye irritation.
- Avoid breathing vapor or spray mist.
- Use only with adequate ventilation.
- Keep container closed.
- Avoid contact with skin and eyes.
- Wash thoroughly after handling.
- Do not contaminate food or feed products.

Required Clothing and Equipment for Application

All applicators, including homeowners, flaggers and all personnel involved with the mixing, loading, and transferring operations must wear the protective clothing and equipment enumerated below. Pilots are exempt from this requirement. The protective clothing and equipment to be worn is as follows:

a. Impermeable gloves (for example, rubber or plastic covered reinforced gloves.)

b. Boots or boot covers.

c. Long-sleeved shirt and long pants.

d. Wide-brimmed hat.

e. Respirators must be worn by flaggers and mixed/loaders.

ENVIRONMENTAL HAZARDS

This pesticide is toxic to wildlife and aquatic invertebrates. Do not apply directly to water or wetlands. Runoff from treated areas may be hazardous to aquatic organisms in neighboring areas. Do not contaminate water when disposing of equipment washwaters.

This pesticide is highly toxic to bees exposed to direct treatment or residues on blooming crops or weeds. Do not apply this product or allow it to drift to blooming crops or weeds if bees are visiting the treatment area. Protective information may be obtained from your Agricultural Extension Service.

PHYSICIAL/CHEMICAL HAZARDS

Combustible liquid and vapor.
Do not use, pour, spill or store near heat or open flame.

DIRECTIONS FOR USE

It is a violation of Federal law to use this product in a manner inconsistent with its labeling.

Do not apply this product through any type of irrigation system.

Do not apply this product in such a manner as to directly or through drift expose workers or other persons. The area being treated must be vacated by unprotected persons.

STORAGE AND DISPOSAL

DO NOT STORE BELOW 45°F

Prohibitions: Do not contaminate water, food or feed by storage or disposal. Open dumping is prohibited.

Pesticide Disposal: Pesticide wastes are acutely hazardous. Improper disposal of excess pesticide, spray mixture, or rinsate is a violation of Federal law. If these wastes cannnot be disposed of by use according to label instructions, contact your State Pesticide or Environmental Control Agency, or the Hazardous Waste representative at the nearest EPA Regional Office for guidance.

Container Disposal:

Triple rinse (or equivalent). Then offer for recycling or reconditioning, or puncture and dispose of in a sanitary landfill, or by other approved state and local procedures.

GENERAL WORKER PROTECTION

Do not enter treated areas for 4 days unless appropriate protective clothing is worn.

Protective clothing means, at least, a hat or other suitable head covering, a long sleeved shirt and long legged trousers or a coverall type garment (all of closely woven fabric covering the body, including the arms and legs), shoes and socks.

Because certain states may require more restrictive reentry intervals for various crops treated with this product, consult your State Department of Agriculture for further information.

Written or oral warnings must be given to workers who are expected to be in a treated area or in an area about to be treated with this product. Inform workers of areas or fields that may not be entered without specific protective clothing, period of time field must be vacated and appropriate actions to take in case of accidental exposure. An example of such information is given under written warnings. When oral warnings are given, warnings shall be given in a language customarily understood by workers. Oral warnings must be given if there is reason to believe that written warnings cannot be understood by workers. Written warnings must include the following information:

WARNING

Area treated with CYGON® spray (insert date). Do not enter without appropriate protective clothing for 4 days (insert State Department of Agriculture's re-entry interval, if more restrictive).

In case of accidental exposure to pesticide spray, wash the skin thoroughly with soap and water. Remove contaminated clothing and wash before reuse. If in eyes, flush with plenty of water. If inhaled, go to an area where the pesticide has not been applied. Get medical attention if needed.

DIRECTIONS FOR APPLICATION

This product is intended for use in conventional hydraulic sprayers, ground applicators or aerial sprayers. Do not apply when weather conditions favor drift of spray from treated areas. Repeat applications as necessary unless otherwise specified. Consult your state experiment station or state extension service for proper timing of applications.

CYGON 400 has systemic and contact activity against a broad spectrum of piercing, sucking and chewing insects; however, it may not control certain organophosphate resistant species.

COMPATIBILITY*: CYGON® 400 is compatible in spray tank mixes with most insecticides, miticides, and fungicides, provided they are not alkaline in reaction. Field experience indicates that CYGON has been satisfactorily mixed with SEVIN[1/], DIAZINON[2/], PYDRIN[3/], GUTHION[4/], CYTHION®, 'The Premium Grade Malathion'®, parathion, KELTHANE[5/], CYPREX® Fruit Fungicide, captan, zineb, and thiram. Because uniform dispersibility and sprayability may be influenced by pesticide combinations used, it is recommeded that compatability be determined before adding pesticides to the spray tank. The addition of X-77 Spreader[6/] at recommended rates will usually eliminate any incompatibility noted.

FOR PROPER MIXING SPRAY TANK SHOULD BE AT LEAST THREE-QUARTERS FILLED WITH WATER BEFORE ADDING CYGON 400. MECHANICAL AGITATION OR RECIRCULATION THROUGH PUMP BYPASS TO TANK IS USUALLY SUFFICIENT FOR MAINTAINING A GOOD DISPERSION.

Spray tank mixtures of CYGON with alkaline insecticides and fungicides should be applied promptly.

ODOR: CYGON formulations may produce a distinctive odor during the spray operation, but under normal conditions this odor does not persist.

Aerial Applications: Apply at least one gallon of finished spray per acre. Apply at least 5 gallons finished spray per acre in California.

Automatic flagging devices should be used whenever feasible.

If human flaggers are employed, they must wear the protective clothing and respirator specified on this label.

Ground Applications: Use water for dilution and apply at least 5 gallons of finished spray per acre.

[1/] SEVIN is a registered trademark of Rhone Poulenc.

[2/] DIAZINON is a registered trademark of Ciba-Geigy Corporation.

3/ PYDRIN is a registered trademark of E.I. du Pont de Nemours and Company.
4/ GUTHION is a registered trademark of the parent company of Farbenfabriken Bayer GmbH Leverkusen.
5/ KELTHANE is a registered trademark of Rohm and Haas Company.
6/ X-77 Spreader is a registered trademark of Chevron Chemical Company.
® Registered Trademark of American Cyanamid Company.
* Tank mixes not registered in California

FIELD CROPS

CROPS	PESTS CONTROLLED	RATE	INTERVAL (DAYS) BETWEEN LAST APPLICATION AND HARVEST
Alfalfa	Aphids, grasshoppers, leafhoppers, plant bugs including Lygus, reduction of alfalfa weevil larvae	½ to 1 pt./acre	10 Do not apply if the crop or weeds in the treatment areas are in bloom. Do not apply within 10 days of harvest or pasturing. Make only one application per cutting. Effective only on cutting to which applied.
Field Corn	Banks grass mites (excluding Trans-Pecos area of Texas), aphids, bean beetle, corn rootworm adult	⅔ to 1 pt./acre	14 Apply as necessary. Make no more than three applications per year. Do not feed or graze within 14 days of last application. Do not apply to corn during the pollen-shed period if bees are actively foraging in the treated area.
	Grasshoppers	1 pt./acre	
Cotton (grown in California and Arizona)	Leafhoppers, fleahoppers, plant bugs including Lygus	½ to 1 pt./acre	14 Repeat applications should not be made at intervals closer than 14 days. Make only 2 applications per season at the higher rate. Do not feed treated forage or graze livestock on treated fields.
Cotton (except Arizona and California)	Aphids, mites, thrips, fleahoppers	¼ to ½ pt./acre	14 *When water is used for dilution:* Repeat applications should not be made at intervals closer than 14 days. Do not feed treated forage or graze livestock on treated fields. 40 *When once refined vegetable oil is used for dilution:* Repeat applications should not be made at intervals closer than 40 days. Make only 2 applications per season at the higher rate. Apply at least one quart of finished spray/acre. Do not feed treated forage or graze livestock on treated fields.
	Plant bugs including Lygus	½ pt./acre	
Safflower (grown in California and Arizona)	Aphids, leafhoppers, plant bugs including Lygus, thrips	½ to 1 pt./acre	14 Repeat applications should not be made at intervals closer than 14 days. Make only 2 applications per season at the higher rate.
Sorghum (milo)	Aphids	½ to 1 pt./acre	Do not feed or graze within 28 days of last application. Make no more than 3 applications as needed per season. Do not apply after heading.
	Banks grass mites (excluding Trans-Pecos area of Texas), spider mites	1 pt./acre	
	Grasshoppers	1 pt./acre	
	Sorghum midge	¼ to ½ pt./acre	
Soybeans	Mexican bean beetle, spider mites, bean leaf beetle, leafhoppers, three-cornered alfafa hopper*	1 pt./acre	21 Do not feed or graze within 5 days of last application.
	Grasshoppers	1 pt./acre	
Wheat	Aphids	½ to ¾ pt./acre	Do not apply within 14 days of grazing immature plant.
	Brown wheat mite	⅓ to ½ pt./acre	Do not harvest grain within 60 days of last application.
	Grasshoppers	¾ pt./acre	Do not make more than 2 applications per season.

* Not registered in California.

SEED CROPS

CROP	PESTS CONTROLLED	RATE	INTERVAL (DAYS) BETWEEN LAST APPLICATION AND HARVEST
Alfalfa	Aphids, leafhoppers, Lygus bugs, grasshoppers, reduction of alfalfa weevil larvae	½ to 1 pt./acre	Do not apply if the crop or weeds in the treatment areas are in bloom. Do not feed or graze livestock in treated crops, hay, threshings or stubble within 10 days of application.

ATTENTION

DO NOT USE ON SEED ONIONS, SEED CARROTS, OR SEED BERMUDA GRASS.

VEGETABLE CROPS

CROP	PESTS CONTROLLED	RATE	INTERVAL (DAYS) BETWEEN LAST APPLICATION AND HARVEST
Beans (green, lima, snap, dry)	Aphids, grasshoppers, leafhoppers, leaf miners, Lygus bugs, mites, bean leaf beetle, Mexican bean beetle	½ to 1 pt./acre	Beans may be harvested on day of application. Do not feed treated vines. Do not apply if the crop or weeds in the treatment area are in bloom.
Broccoli Cauliflower	Aphids	½ to 1 pt./acre	7
Cabbage	Aphids	½ to 1 pt./acre	3
Celery (Florida)	Leaf miners	1 pt./acre	7
Head Lettuce	Aphids, leafhoppers, leaf miners	½ pt./acre	7
Leaf Lettuce Spinach Collards Kale Turnip (greens and roots)	Aphids, leafhoppers, leaf miners	½ pt./acre	14

CROP	PESTS CONTROLLED	RATE	INTERVAL (DAYS) BETWEEN LAST APPLICATION AND HARVEST
Mustard Greens Swiss Chard Endive (Escarole)			
Melons (except watermelons)	Aphids, leafhoppers, leaf miners, thrips	1 pt./acre	3
Watermelons	Aphids, leaf miners, leafhoppers	½ to 1 pt./acre	3
Lupine	Aphids, lygus bugs	½ to 1 pt./acre	Apply when aphids first appear. Make only 2 applications per season. Lupine may be harvested on day of application. Do not feed or graze forage or hay. Do not apply if the crop or weeds in the treatment area are in bloom.
Peas Lentils	Aphids	⅓ pt./acre	Peas or lentils may be harvested on day of application. Do not feed or graze hay within 21 days after last application. Do not make more than one application per season. Do not apply if the crop or weeds in treatment area are in bloom.
Peppers	Aphids, leaf miners, maggots	½ to ⅔ pt./acre	Peppers may be harvested on day of application.
Potatoes	Aphids, grasshoppers, leaf miners, leafhoppers	½ to 1 pt./acre	Potatoes may be harvested on day of application.
Tomatoes	Aphids, leaf miners, leafhoppers	½ to 1 pt./acre	7

FRUIT AND NUT CROPS

CROP	PESTS CONTROLLED	RATE	INTERVAL (DAYS) BETWEEN LAST APPLICATION AND HARVEST
Apples	Apple maggot Codling moth*	1 pt./100 gals. water	28 Do not apply when trees or substantial numbers of weeds in the orchard are in bloom. Apply at petal-fall and every 10 to 14 days thereafter until control is achieved. Under heavy infestations, some sting injury may occur. *Midwest and eastern states only. Do not graze livestock in treated orchards.
Apples Pears	Aphids, leafhoppers, mites (except rust mites)	½ to 1 pt./100 gals. water	28 Do not apply when trees or substantial numbers of weeds in the orchard are in bloom. Do not graze livestock in treated orchards.
Grapefruit Lemons Oranges Tangerines	Aphids	½ to 1 pt./100 gals. water Aerial: 1 to 2 qts./acre in 5 to 10 gals. water	15
	Mites (except rust mites)	½ to 1 pt./100 gals. water	15
	Scales (except black or snow)	1 to 1-½ pts./100 gals. water. Apply as a thorough coverage spray.	45
	Thrips	½ to 1 pt./100 gals. water Aerial: 1 to 2 qts./acre in 5 to 10 gals. water	15
	Whiteflies	1 pt./100 gals. water	15
	Do not apply when trees or substantial number of weeds in the grove are in bloom. Do not use on citrus seedlings. Make no more than 2 applications to mature fruit. Do not enter treated groves within 4 days of last application. Do not graze livestock in treated orchards.		
CITRUS: (California, Arizona) Non-bearing and nursery stock	Aphids, thrips	Foliar Spray: 1 pt./100 gals. water	Repeat applications as necessary. May be applied in the year grapefruit, lemon, orange and tangerine trees begin to bear fruit. Do not enter treated groves within 4 days of last application. Do not graze livestock in treated groves.
		Soil Drench (trees 1 to 3 years old): 2 qts./acre	Apply in the furrow or basin around the base of tree. Apply when insect injury to new growth appears. Do not apply to trees that will bear fruit within one year. Do not graze livestock in treated groves.
NUTS Pecans	Aphids, mites, leafhoppers	⅔ pt./acre	21 Do not graze livestock in treated groves.

DISCLAIMER

The label instructions for the use of this product reflect the opinion of experts based on field use and tests. The directions are believed to be reliable and should be followed carefully. However, it is impossible to eliminate all risks inherently associated with use of this product. Crop injury, ineffectiveness or other unintended consequences may result because of such factors as weather conditions, presence of other materials, or the use or application of the product contrary to label instructions, all of which are beyond the control of American Cyanamid Company. All such risks shall be assumed by the user.

American Cyanamid Company warrants only that the material contained herein conforms to the chemical description on the label and is reasonably fit for the use therein described when used in accordance with the directions for use, subject to the risks referred to above.

Any damages arising from a breach of this warranty shall be limited to direct damages and shall not include consequential commercial damages such as loss of profits or values or any other special or indirect damages.

American Cyanamid Company makes no other express or implied warranty, including any other express or implied warranty of FITNESS or of MERCHANTABILITY.

CYPREX® 65-W

FRUIT FUNGICIDE

FOR CONTROL OF CERTAIN DISEASES OF APPLES, CHERRIES, PEACHES, PEARS AND PECANS.

ACTIVE INGREDIENT:	
Dodine (n-dodecylguanidine acetate)	65.0%
INERT INGREDIENTS:	35.0%
TOTAL	100.0%

EPA Reg. No. 241-51

®Registered trademark of American Cyanamid Company

4/88

KEEP OUT OF REACH OF CHILDREN

DANGER!/¡PELIGRO!

¡PELIGRO!

PRECAUCION AL USUARIO: Si usted no lee ingles, no use este producto hasta que la etiqueta le haya sido explicada ampliamente.

FIRST AID

IF SWALLOWED: Drink 1 or 2 glasses of water and induce vomiting by touching back of throat with finger. Do not induce vomiting or give anything by mouth to an unconscious person. Get medical attention.

IF ON SKIN: Wash with plenty of soap and water.

IF IN EYES: Flush with plenty of water. Get medical attention.

In case of an emergency endangering life or property involving this product, call collect, day or night, Area Code 201-835-3100.

PRECAUTIONARY STATEMENTS

HAZARDS TO HUMANS (AND DOMESTIC ANIMALS)

DANGER

HARMFUL IF SWALLOWED

CAUSES EYE BURNS

Do not get in eyes or on skin. Wear gloves and goggles when handling. Avoid breathing of dust. Keep container closed. Use with adequate ventilation. Avoid contact with eyes. Wash thoroughly after handling. Do not contaminate food or feed products.

ENVIRONMENTAL HAZARDS

This product is toxic to fish. Do not apply directly to water. DO NOT apply when weather conditions favor drift from target area. DO NOT contaminate water when disposing of equipment washwaters.

STORAGE AND DISPOSAL

DIRECTIONS

Prohibitions: Do not contaminate water, food or feed by storage and disposal. Open dumping is prohibited. Store in a secure cool dry place and keep container tightly closed.

Pesticide Disposal: Pesticide wastes are acutely hazardous. Improper disposal of excess pesticide, spray mixture, or rinsate is a violation of Federal law. If these wastes cannot be disposed of by use according to label instructions, contact your State Pesticide or Environmental Control Agency, or the Hazardous Waste representative at the nearest EPA Regional Office for guidance.

Container Disposal: Completely empty bag into application equipment. Then dispose of empty bag in a sanitary landfill or by incineration, or, if allowed by State and local authorities, by burning. If burned, stay out of smoke.

DILUTE AND CONCENTRATE SPRAYS: CYPREX 65-W may be used effectively in dilute and concentrate sprays when applied properly to deliver the recommended rate (lb. per acre). Trees must be covered uniformly. Fruit and foliage should not be excessively drenched with dilute sprays. Avoid heavy deposits when using concentrate sprays.

INSTRUCTIONS FOR MIXING CYPREX 65-W

CYPREX 65-W, a fine wettable powder, should be added directly to the spray tank when half filled with water. Keep agitator running during filling and spraying. Do not wash wettable powder through filter screen into spray tank.

COMPATIBILITY

Experience indicates that CYPREX 65-W can be used effectively at recommended dosages with most of the commonly used fungicides, insecticides and acaricides such as:

Cygon® dimethoate	Imidan®
diazinon	Omite®
Karathane® dinocap	sulfurs
ferbam	zineb

—and wettable powder formulations of—

Guthion® azinphosmethyl	malathion
Sevin® carbaryl	Dimecron® phosphamidon
Mitox® chlorbenside	
Kelthane® dicofol	parathion
Thiodan® endusulfan	ryania

While CYPREX 65-W has been used satisfactorily with emulsifiable concentrate formulations and flowable (water-based) emulsions, it is desirable to check the compatibility of this type combination before materials are mixed in the spray tank. CYPREX 65-W is not compatible with Acaraben® chlorobenzilate, Bordeaux mixture, fixed copper, Fundal®, Galecron®, lime or lime sulfur. Observe all cautions and limitations on labeling of all products used in mixtures.

CYPREX 65-W AND OIL

Tank-mix combinations of CYPREX 65-W and oils of 60-75 viscosity have been used effectively as a delayed dormant application on apples. Oils of the slow-breaking type are suggested for tank mixing with CYPREX 65-W. Liquid concentrate or flowable (water-based) emulsion formulations of the phosphate insecticides have been used satisfactorily in CYPREX 65-W oil systems. Wettable powder formulations of parathion, Guthion® azinphosmethyl, Trithion® carbophenothion and certain other insecticides have appeared suitable in limited trials. CYPREX 65-W oil mixtures are not compatible with wettable powder malathion, Sevin® carbaryl, zineb, ferbam and sulfurs.

Because uniform dispersibility and sprayability may be influenced by the particular oil or phosphate insecticide formulation used or by the local water supply, it is recommended that compatibility be determined before adding pesticides to the spray tank.

PROCEDURE FOR TANK-MIXING CYPREX 65-W, INSECTICIDE AND OIL

Maintain continuous agitation from the time CYPREX 65-W is added to the spray tank until mixture is sprayed out:

1. CYPREX 65-W should be added to the tank when one-quarter to one-half full with water and allowed to mix thoroughly.
2. If insecticide is used, add as tank is filling.
3. Add self-emulsifying oil last when tank is nearly three-quarters full.

Consult your state agricultural experiment station or state extension service for information on self-emulsifying oils and pesticide formulations for tank mixing with CYPREX 65-W.

DIRECTIONS FOR USE

It is a violation of Federal law to use this product in a manner inconsistent with its labeling.

Do not apply this product through any type of irrigation system.

Consult your state experiment station or state extension service for proper timing of sprays. Do not graze cover crops in treated orchards.

Do not apply this product in such a manner as to directly or through drift expose workers or other persons. The area being treated must be vacated by unprotected persons.

Do not enter treated areas without protective clothing until sprays have dried.

Because certain states may require more restrictive reentry intervals for various crops treated with this product, consult your State Department of Agriculture for further information.

Written or oral warnings must be given to workers who are expected to be in a treated area or in an area about to be treated with this product. Inform workers of areas or fields that may not be entered without specific protective clothing, period of time field must be vacated and appropriate actions to take in case of accidental exposure. An example of such information is given under written warnings. When oral warnings are given, warnings shall be given in a language customarily understood by workers. Oral warnings must be given if there is reason to believe that written warnings cannot be understood by workers. Written warnings must include the following information:

DANGER

Area treated with CYPREX® 65-W (insert date). Do not enter without appropriate protective clothing until sprays have dried.

In case of accidental exposure to pesticide spray or dust, wash the skin thoroughly with soap and water. Remove contaminated clothing and wash before reuse. If in eyes, flush with plenty of water. If inhaled, go to an area where the pesticide has not been applied. Get medical attention if needed.

APPLE: Scab—Protection Schedule (Ground Application)—Use 1 to 2 lbs. per acre CYPREX 65-W at 5 to 7-day intervals or as needed to maintain scab control from prebloom through the first cover applications. Use 1 lb. per acre CYPREX 65-W in subsequent cover sprays as needed.

CYPREX 65-W may be used in a delayed dormant oil schedule. Use 1 to 2 lbs. per acre CYPREX 65-W with 2 gallons of oil of 60-75 viscosity. For mixing instructions, refer to "CYPREX 65-W and Oil" section on label.

(Air Application)—Use 1½ lbs. CYPREX 65-W per acre in 5 to 7 gallons of spray at 5 to 7-day intervals or as needed to maintain scab control from prebloom through the first cover sprays. Repeat application in subsequent cover sprays as needed.

"After-infection" Applications—After an infection period has occurred, use 3 lbs. per acre CYPREX 65-W during or following a rain from prebloom through first cover applications and, in later cover sprays, as needed. While timing of the spray may vary considerably depending upon temperatures, it is recommended that application be made within 36 hours from the beginning of the rain or wet period which caused the infection.

Do not apply within 7 days of harvest.

Wettable sulfur can be used in combination with CYPREX 65-W for control of powdery mildew on apples. Consult your state extension service for proper rates and timing of sprays.

CYPREX 65-W TREATED APPLES SHOULD NOT BE USED IN THE MANUFACTURE OF APPLE POMACE FOR USE IN LIVESTOCK FEEDS. SUCH USE MAY RESULT IN ILLEGAL RESIDUES IN MILK OR MEAT.

Fruit color and finish on most apple varieties sprayed with CYPREX 65-W at recommended dosages have frequently been outstanding.

CYPREX 65-W may cauase russetting of Golden Delicious and Grimes Golden apple varieties if applied during cool, slow drying weather conditions. Do not use on these varieties under these conditions.

Freezing or near-freezing temperatures, particularly when accompanied by slow-drying conditions may injure apple fruit or predispose the fruit to chemical injury. Do not exceed 2 lbs. CYPREX 65-W per acre immediately before, during or immediately after occurence of such conditions.

CHERRY (Sour and Sweet): Leaf Spot—Use 1 to 2 lbs. CYPREX 65-W per acre at petal fall or when first leaves unfold. Repeat application thereafter at 7 to 10-day intervals. If conditions favor **leaf spot,** continue applications through harvest.

Post-Harvest Sprays: Apply 1 to 2 lbs. CYPREX 65-W per acre immediately after harvest and as needed. **Blossom Brown Rot** (Western States Only): Use 2 lbs. CYPREX 65-W per acre at the time petals begin to open. Repeat application at full bloom and again at petal fall.

Wettable sulfur can be used in combination with CYPREX 65-W for control of powdery mildew on cherries. Consult your state extension service for proper rates and timing of sprays.

Applications of CYPREX 65-W to cherries when temperatures and humidity are excessively high may cause spotting to foliage or fruit.

PEACH: Bacterial Spot—Under light to moderate disease conditions, use 2 lbs. CYPREX 65-W and 4 lbs. captan (50%) per acre (use ½ lb. CYPREX 65-W and 1 lb. captan (50%) per 100 gallons of spray) at shuck-split and repeat application at 7 to 10-day intervals through the cover sprays. Shorten intervals during wet periods as needed to maintain maximum protection against infection.

Do not use concentrate sprays higher than 3X.

Do not apply within 15 days of harvest.

Captan or the CYPREX-captan mixture is toxic to fish. Do not contaminate any body of water.

When it is necessary to use an insecticide with the combination, the wettable powder formulation of parathion is suggested. Limited experience indicates that wettable powder formulations of Guthion© azinphosmethyl, Kelthane© dicofol, Tedion© tetradifon, and Thiodan© endosulfan are compatible with the mixture. Do not combine CYPREX 65-W and captan with solvent formulations of insecticides or sulfur. Spray injury may result if the combination is applied with, immediately before, or closely after an oil or zinc sulfate-lime spray.

PEACH: Leaf Curl (Western States Only)—Use 4 lbs. CYPREX 65-W per acre just before buds swell in the spring. If leaf curl was moderate to severe the previous season, use 4 lbs. CYPREX 65-W per acre 1 month before buds swell in the spring and again just prior to bud swelling. **Blossom Brown Rot** (Western States Only)—Use 2 lbs. CYPREX 65-W at the time petals begin to open. Repeat application at full bloom and again at petal fall.

Do not apply after petal fall.

PEAR: Scab (Western States Only)—Use 3 lbs. CYPREX 65-W per acre of spray at 5 to 7-day intervals or as needed to maintain scab control from prebloom through petal fall and again in first cover. Use 2 to 3 lbs. CYPREX 65-W in subsequent sprays as needed.

Do not apply within 7 days of harvest.

(Air Application-Oregon Only)—Use 2 to 3 lbs. CYPREX 65-W per acre in 5 to 7 gallons of spray at 5 to 7-day intervals or as needed to maintain scab control from prebloom through the first cover sprays. Repeat application in subsequent cover sprays as needed.

Do not apply within 7 days of harvest.

PECANS: Scab, Liver Spot, Brown Leaf Spot, Downy Leaf Spot, leaf blotch (Ground Application)—Use 2 to 4 lbs. CYPREX 65-W per acre of spray. Begin applications when first leaves are showing and repeat at 10 to 14-day intervals through first cover. Repeat applications at 2 to 3-week intervals or as needed to maintain disease control.

(Air Application)—Use 2 lbs. CYPREX 65-W per acre in a minimum of 20 gallons of spray. Begin applications when first leaves are showing and repeat at 10 to 14-day intervals through first cover. Repeat application at 2 to 3-week intervals or as needed to maintain disease control. Under severe scab conditions, use 4 lbs. CYPREX 65-W per acre in a minimum of 20 gallons of spray.

Do not apply after shucks have started to open.

Do not graze meat or dairy animals in treated groves.

CYPREX 65-W has been used effectively on most pecan varieties. Severe foliage injury has been experienced on Moore and Van Deman in the Southeast.

STRAWBERRY: Leaf Scorch, Leaf Spot and Leaf Blight—Use 1½ to 2 lbs. CYPREX 65-W per acre in sufficient water (250-300 gallons) for thorough plant coverage. Make first application as soon as new growth starts and repeat sprays at weekly intervals.

Do not apply within 14 days of harvest.

BLACK WALNUT: Anthracnose (Leaf Blight)—Use 4 lbs. CYPREX 65-W per acre when first leaves unfold. Repeat application at 10 to 14-day intervals. Do not graze meat or dairy animals in treated groves.

DISCLAIMER

American Cyanamid Company warrants only that the material contained herein conforms to the chemical description on the label and is reasonably fit for the use therein described when used in accordance with the directions for use.

Any damages arising from a breach of this warranty shall be limited to direct damages, and shall not include consequential commercial damage such as loss of profits or values, etc.

American Cyanamid Company makes no other express or implied warranty, including any other express or implied warranty of FITNESS or of MERCHANTABILITY.

User assumes the risk of any use contrary to label instructions or under abnormal conditions, or under conditions not reasonably foreseeable by American Cyanamid Company.

CYTHION® ULV

INSECTICIDE

ACTIVE INGREDIENT:

Malathion: O,O-dimethyl phosphorodithioate of diethyl mercaptosuccinate	91.0%
INERT INGREDIENTS:	9.0%
TOTAL	100.0%

(One gallon contains 9.33 pounds of malathion)

EPA Reg. No. 241-208

® Registered Trademark of American Cyanamid Company

6/89

KEEP OUT OF REACH OF CHILDREN

CAUTION/PRECAUCION

PRECAUCION AL USUARIO: Si usted no lee inglés, no use este producto hasta que la etiqueta le haya sido explicada ampliamente.

IN CASE OF AN EMERGENCY ENDANGERING LIFE OR PROPERTY INVOLVING THIS PRODUCT CALL COLLECT, DAY OR NIGHT, AREA CODE 201-835-3100

STATEMENT OF PRACTICAL TREATMENT

IF SWALLOWED: Call a physician or Poison Control Center. Drink 1 or 2 glasses of water and induce vomiting by touching back of throat with finger. Do not give anything by mouth to an unconscious person.

IF ON SKIN: Wash with plenty of soap and water. Get medical attention.

IF IN EYES: Flush with plenty of water. Call a physician if irritation persists.

NOTE TO PHYSICIAN: This product may cause cholinesterase inhibition. Atropine is antidotal. 2-PAM may be effective as an adjunct to atropine.

PRECAUTIONARY STATEMENTS

HAZARDS TO HUMANS (AND DOMESTIC ANIMALS)

CAUTION

Harmful if swallowed, inhaled or absorbed through skin. Avoid breathing spray mist. Avoid contact with skin, eyes, or clothing. Wash thoroughly with soap and water after handling. Remove contaminated clothing and wash before reuse.

ENVIRONMENTAL HAZARDS

This product is toxic to fish and aquatic invertebrates. Do not apply directly to water except as specified on this label. Drift and runoff may be hazardous to aquatic organisms in areas near the application site. Do not contaminate water when disposing of equipment washwaters. This product is highly toxic to bees exposed to direct treatment on blooming crops or weeds. Do not apply this product or allow it to drift to blooming crops or weeds if bees are visiting the treatment area.

DIRECTIONS FOR USE

It is a violation of Federal law to use this product in a manner inconsistent with its labeling.

Do not apply this product through any type of irrigation system.

CYTHION ULV insecticide may only be used in accordance with the directions in this leaflet or approved supplementary labeling. Read all directions carefully before using.

Undiluted spray droplets of CYTHION ULV insecticide, will permanently damage automobile paint. If accidental exposure does occur, the vehicle should be washed immediately.

STORAGE AND DISPOSAL

Do not store in any premise occupied by humans or animals.
Do not contaminate water, food or feed by storage or disposal.

Pesticide Disposal: Wastes resulting from the use of this product may be disposed of on site or at an approved waste disposal facility.

Container Disposal: Triple rinse (or equivalent). Then offer for recycling or reconditioning, or puncture and dispose of in a sanitary landfill, or by other procedures approved by state and local authorities.

Before using, read the directions contained in this label for the proper methods and procedures which must be followed to achieve effective insect control and avoid permanent damage to automobile and other paint finishes.

DISCLAIMER

The label instructions for the use of this product reflect the opinion of experts based on field use and tests. The directions are believed to be reliable and should be followed carefully. However, it is impossible to eliminate all risks inherently associated with use of this product. Crop injury, automobile paint damage, ineffectiveness or other unintended consequences may result because of such factors as weather conditions, presence of other materials, or the use or application of the product contrary to label instructions, all of which are beyond the control of American Cyanamid Company. All such risks shall be assumed by the user.

American Cyanamid Company warrants only that the material contained herein conforms to the chemical description on the label and is reasonably fit for the use therein described when used in accordance with the directions for use, subject to the risks referred to above.

Any damages arising from a breach of this warranty shall be limited to direct damages and shall not include consequential commercial damages such as loss of profits or values or any other special or indirect damages.

American Cyanamid Company makes no other express or implied warranty, including any other express or implied warranty of FITNESS or of MERCHANTABILITY.

The sale of this product does not include a license under any patent owned by the American Cyanamid Company.

AGRICULTURAL USES—OPERATING INSTRUCTIONS

General Worker Protection Statements

Do not apply this product in such a manner as to directly or through drift, expose workers or other persons. The area being treated must be vacated by unprotected persons.

Do not enter treated areas without protective clothing until sprays have dried. Because certain states may require more restrictive reentry intervals for various crops treated with this product, consult your State Department of Agriculture for further information.

Written or oral warnings must be given to workers who are expected to be in a treated area or in an area about to be treated with this product. When oral warnings are given, warnings shall be given in a language customarily understood by workers. Oral warnings must be given if there is reason to believe that written warnings cannot be understood by workers. Written warnings must include the following information: CAUTION. Area treated with CYTHION ULV insecticide (date of application). Do not enter without appropriate protective clothing.

CYTHION ULV is used undiluted in specially designated aircraft or ground equipment capable of applying ultra low volumes for control of the insects indicated. Aerial applications are most effective when made at a boom height of 5 feet and a swath width of 50 feet. Apply only when weather conditions are favorable. Wind and rising air currents may cause undesirable spray drift and reduce insect control.

Mist blowers and boom sprayers utilizing a controlled air flow to facilitate particle size and spray deposition may be used at a vehicle speed of 4 to 10 mph.

Mist blowers with a pump capable of producing up to 40 psi and blower speeds of 2600 rpm are satisfactory. Use flat fan nozzles, 8001 to 8002, place 30° into air blast or rotary atomizers into the air blast that produce an efficient spray particle with a mass median diameter of 40 to 100 microns. Swath widths should not exceed 30 feet. Apply only when weather conditions are favorable. Wind and rising air currents may cause undesirable spray drift and reduce insect control.

Boom sprayers with a filtered rotary air compressor, either PTO or gas engine driven or an air pump capable of producing at least 12 psi are satisfactory. Use air pressure on chemical tanks and an accurate metering valve to assure a calibrated flow of the pesticide. Air should be regulated with relief valve and gauge for proper air and liquid mixture. Pneumatic-type spray nozzles, as suggested by equipment manufacturer, should be used for spray particles with mass median diameter of 30 to 100 microns. Apply only when weather conditions are favorable. Wind and rising air currents may cause undesirable spray drift and reduce insect control.

Repeat applications should be made as necessary unless otherwise specified.

OTHER AGRICULTURAL USES:

Alfalfa, Clover, Pasture and Range Grass, Grass and Grass Hay, Grain Crops, Beans, Rice, Tomatoes and Nonagricultural Lands (wasteland): Adult mosquitoes and flies—Apply CYTHION ULV at the rate of 2 to 4 fl. oz. for control of adult mosquitoes and at 6 to 8 fl. oz. per acre for control of adult flies. Repeat applications as necessary. On alfalfa, clover, pasture and range grass, grass and grass hay, may be applied on day of harvest or grazing. Do not apply to alfalfa and clover in bloom. Do not use on seed alfalfa, On Grain crops, make no application within 7 days of harvest or forage use; on corn, within 5 days of harvest or forage; on rice, within 7 days of harvest; on beans and tomatoes, within 1 day of harvest.

IMPORTANT—Undiluted spray droplets of CYTHION ULV WILL PERMANENTLY damage automobile paint. If accidental exposure does occur, the vehicle should be washed immediately.

Consult your state experiment station or state extension service for proper timing of sprays.

This product is highly toxic to bees exposed to direct treatment or residues on crops. Protective information may be obtained from your Cooperative Agricultural Extension Service.

CROP	PESTS CONTROLLED	FLUID OUNCES PER ACRE	DAYS TO HARVEST or GRAZING and COMMENTS
Alfalfa	Alfalfa Caterpillar Western Yellow Striped Armyworm	8	0 day. Apply when larvae are small.
		12	5 days. Apply when larvae are large or when foliage is dense.
	Alfalfa Weevil Larvae	16	5 days. Apply when day temperatures are expected to exceed 65°F. and when 50-75% of leaves show feeding damage.
	Beet Armyworm	8	0 day. Apply when larvae are small.
		16	5 days. Apply when larvae are large or when foliage is dense.
	Grasshoppers	8	0 day.
	Do not apply to alfalfa in bloom. Do not apply to seed alfalfa.		
Barley	Grasshoppers Cereal Leaf Beetle	8 4-8	7 days 7 days.
Beans (lima, green, snap, Navy, red kidney, wax, dry, blackeye)	Mexican Bean Beetle, Leafhoppers, Green Cloverworm, Japanese Beetle, Lygus Bug	8	1 day.
Blueberries	Blueberry Maggot	10	0 day.

CROP	PESTS CONTROLLED	FLUID OUNCES PER ACRE	DAYS TO HARVEST or GRAZING and COMMENTS
Cherries	Cherry Fruit Fly	12-16	1 day. Apply by aircraft only. Use higher rate when foliage is heavy or infestation is severe. Make first application as soon as flies appear.
Clover, Pasture and Range Grass, Grass Hay	Blackgrass Bugs, Grasshoppers	8-12	0 day. Do not apply to clover in bloom.
Corn	Adult Corn Rootworm Grasshoppers Cereal Leaf Beetle	4 8 4-8	5 days. 5 days. 5 days.
Cotton†	Aphids (Green Peach and Cotton)	14-16	
	Boll Weevil	8-16	Early to late season.
	Grasshoppers	8	
	Fleahoppers, Leafhoppers	4-8	
	Lygus Bugs including Tarnished Plant Bugs	8-12	Moderate Populations.
		16	Very heavy or migrating populations.
	Thrips	4-8	

† For use on cotton:
CYTHION ULV can be used alone as a ULV concentrate spray or diluted in once-refined cottonseed or vegetable oil sufficient to make at least one quart of finished spray per acre.

CROP	PESTS CONTROLLED	FLUID OUNCES PER ACRE	DAYS TO HARVEST or GRAZING and COMMENTS
Flax	Grasshoppers	8	Apply once per season. Do not apply within 45 days of harvest. Do not graze or feed forage.
Grain Sorghum	Grasshoppers Sorghum Midge	8 8-12	7 days. 7 days. Apply during the bloom stage.
Grasses	Blackgrass Bugs Grasshoppers	8-12	0 day.
	Cereal Leaf Beetle	4-8	
Oats	Cereal Leaf Beetle Grasshoppers	4-8 8	7 days. 7 days.
Peas (Northwest)	Pea Weevil	8	14 days.
Rice	Grasshoppers	8	7 days.
Rice-Grain Form (Louisiana, Texas)	Rice Stink Bug	8	7 days. Apply by aircraft only. Apply during early milk and dough stage of growing rice.
Rye	Grasshoppers	8	7 days.
Safflower	Grasshoppers Lygus Bugs	8	3 days of harvesting seeds.
Soybeans	Mexican Bean Beetle, Grasshoppers, Japanese Beetle, Green Cloverworm	8	7 days.
Sugar Beets	Grasshoppers Sugar Beet Root Maggot Adults	8	0 day. 7 days. If tops are to be used for food or feed.
Wheat	Cereal Leaf Beetle Grasshoppers	4-8 8	7 days. 7 days.
Nonagricultural Lands (wastelands, roadsides)	Beet Leafhopper (on wild host plants)	8	0 day.
	Blackgrass Bugs Grasshoppers	8-12	0 day.
Beef Cattle-Feed Lots and Holding Pens	Adult Flies and Mosquitoes	6-8	0 days.

IMPORTANT

Consult your state experiment station or state extension service for proper timing of sprays.

These products are highly toxic to bees exposed to direct treatment or residues on crops. Protective information may be obtained from your Cooperative Agricultural Extension Service.

All applicable directions, restrictions and precautions on the EPA registered labels are to be followed.

Operator should inspect and calibrate equipment to assure that the proper amounts of pesticide are being applied, and that uniform coverage is obtained.

This label must be in the possession of the user at time of pesticide application.

FOREST INSECTS

Apply with aircraft equipped for ultra low volume application. Make application when air is calm and temperature is below 68°F. Do not allow spray to contact ferns, hickory and maples as injury may result. Do not spray on elms under extreme heat, drought or disease conditions.

TREE	PESTS CONTROLLED	FLUID OUNCES PER ACRE	DIRECTIONS
Douglas Fir, True Fir, Spruce	Spruce Budworm	13	Apply when highest percentage of larvae are in the fifth instar.
Hemlock	Hemlock Looper	8	Apply when most larvae are in the third and fourth instar.
Pines	European Pine Sawfly	10	Apply when larvae are in the first or second instar or before they reach ½ inch in length.
	Saratoga Spittlebug		Apply when 95% of the population has become adult.
Larch	Larch Casebearer	8	Apply in spring as soon as larvae break hibernation and begin feeding on new foliage.

TREE	PESTS CONTROLLED	FLUID OUNCES PER ACRE	DIRECTIONS
Nonagricultural Lands (Wastelands, roadsides)	Beet Leafhopper (on wild host plants)	8	0 day.

MOSQUITO CONTROL IN POPULATED AND RURAL AREAS

IMPORTANT NOTICE

TO BE APPLIED ONLY BY TRAINED PERSONNEL OR PUBLIC HEALTH ORGANIZATIONS, MOSQUITO ABATEMENT DISTRICTS OR PEST CONTROL OPERATORS.

AERIAL APPLICATION

ADULT MOSQUITO CONTROL OVER CITIES, TOWNS, AND OTHER AREAS WHERE AUTOMOBILES, TRAILERS, TRUCKS AND PLEASURE BOATS ARE PRESENT: Apply 2.6 to 3.0 fluid ounces of CYTHION ULV per acre. Apply only when weather conditions are favorable. Wind and rising air currents may cause undersirable spray drift and reduce insect control.

IMPORTANT—Undiluted spray droplets of CYTHION ULV will permanently damage vehicle paint finishes unless the aircraft used for the ultra low volume application meets all of the specifications listed below.

FIXED WING AIRCRAFT

1. Aircraft is operated at 150 mph or more.
2. There are no leaks in the ultra low volume spray system.
3. Nozzles are placed on the boom at a 45° angle down and into the wind.
4. Diaphragm check valves are used on all nozzles to insure positive cut-off of the spray.
5. Dosage of CYTHION ULV does not exceed 3 fluid ounces per acre.
6. The spray system produces droplets of this product in the 50 to 60 mass median diameter (MMD) micron range, with no more than 10% of the droplets exceeding 100 microns, as determined by readings made from microscope slides coated with DRI-FILM[1] or TEFLON[2].

[1] Trademark of General Electric Company

[2] Registered Trademark of E.I. duPont de Nemours & Co., Inc.

HELICOPTER

Equipment specifications

1. Rotary nozzle equivalent to Beecomist Spray Head Assembly Model: No. 350 equipped with:

a. a direct reading RPM tachometer or low RPM signal light readily visible to operator;

b. a stainless steel porous metal sleeve, 20 micron pore size, dynamically balanced to the nozzle;

c. a diaphragm check valve as near to the rotary nozzle as possible to insure positive cut off of the spray;

d. nozzle on-off switch separate from main switch and pump switch.

2. Minimum no-load nozzle speed of 10,500 RPM.

3. A continuous nonpulsating metered flow must be maintained by a variable speed metering pump equipped with:

a. a positive cut off valve between tank and pump;

b. a flow gauge or tachometer visible to operator;

c. a pump on-off switch separate from main switch and nozzle switch.

4. Maximum flow rate of 0.5 gallon per minute per nozzle.

5. Rotary nozzle must be mounted behind and below the boom with the sleeve directed toward the rear of the aircraft and parallel to the ground during flight. Nozzle must be positioned to minimize air turbulence and the collection of CYTHION ULV droplets on mounting brackets, feed lines, fittings, etc., or any part of the aircraft.

Operating Procedures

1. CYTHION ULV must be prefiltered through a 10 micron filter prior to transfer into helicopter tank. A 50 mesh stainless steel line strainer must be installed in the pump feed line.

2. Entire system, including tank, pump, nozzle and feed lines, to be used only for application of CYTHION ULV.

3. Entire system must be inspected daily to insure that there are no leaks.

4. Sleeve must be removed and cleaned immediately after each use by washing with hot water and blowing dry from outside in with clean air.

5. Rotating nozzle must be turned on and operating before turning on pump. For shut off, pump must be shut off and lines cleared prior to stopping nozzle rotation.

Dosage of CYTHION ULV does not exceed 3 fluid ounces per acre.

7. The spray system must produce droplets of CYTHION ULV with a mass median diameter (MMD) of less than 50 microns, with no more than 2.5% of the droplets exceeding 100 microns, as determined by readings made from microscope slides coated with DRI-FILM or TEFLON.

GROUND APPLICATION

Thermal Aerosols or Fogs

For control of adult mosquitoes with thermal aerosols or fogs, apply CYTHION ULV at the rate of 6-8 oz. actual/gallon (3.9-5.2 gallons CYTHION ULV in 100 gallons finished solution*) by ground equipment delivering 40 gallons per hour at a vehicle speed of 5 miles per hour to treat a swath width of 300-400 feet.

* There is a great variation in the chemical composition of fuel oils which may be used as thermal fog solvents. These differences may cause sludge and/or affect the solubility of the CYTHION ULV. For more complete details on tests for sludge formation and solubility in thermal fog solutions, write American Cyanamid Company, One Cyanamid Plaza, Wayne, New Jersey 07470.

Nonthermal Aerosols

Over a 300-foot swath can be produced using the nonthermal ultra low volume aerosol method with CYTHION ULV.

Use the following rates at the indicated vehicle speeds:

Vehicle Speed Miles per Hour	Flow Rate of CYTHION ULV Fluid Ounces per Minute	Maximum Flow Rate per Hour
5	1.0 to 2.1 fluid ounces	1 gallon
10	2.0 to 4.3 fluid ounces	2 gallons
15	3.0 to 6.45 fluid ounces	3 gallons
20	4.0 to 8.6 fluid ounces	4 gallons

For control of adult stable fly in populated and rural areas with nonthermal aerosols of CYTHION ULV using the ultra low volume method, use the following flow rates at the indicated speeds:

Vehicle Speed Miles per Hour	Flow Rate of CYTHION ULV Fluid Ounces per Minute	Maximum flow Rate per Hour
5	2.1 fluid ounces	1 gallon
10	4.3 fluid ounces	2 gallons

DROPLET SIZE

1. The Mass Median Diameter (MMD) of the droplets should not exceed 17 microns. The MMD is the drop diameter which divides the spray volume into two equal parts; i.e., 50% of the volume is in the drop sizes below the MMD and 50% is above the MMD.

2. Spray droplets should not exceed 32 microns in size. Three percent of the spray droplets (6 droplets out of 200) can exceed 32 microns providing the MMD does not exceed 17 microns and no droplets exceed a maximum of 48 microns. Larger droplets, when transported by natural air currents, impinge more readily on objects in their pathway and will permanently damage automobile-type paints.

3. More than one-half of the total spray mass must consist of droplets in the 6 to 18 micron range to achieve adequate dispersal of insecticide over a 300-foot swath.

4. A minimum of two-thirds, preferably four-fifths of the total spray mass must consist of droplets not exceeding 24 microns in range.

OPERATING EQUIPMENT

Each Nonthermal Aerosol Generator used for dispersal of CYTHION ULV to control adult mosquitoes must have minimum capability of producing the droplet spectrum described under DROPLET SIZE. The initial determination of droplet size is made after the unit is installed in a vehicle and prior to its use in mosquito control operations. The unit should be rechecked as frequently as necessary to insure that proper droplet size is maintained for each operation. Determination of droplet size every two months is usually sufficient if the unit has been maintained in good operating condition. Equipment manufacturer's instructions setting forth cleaning and maintenance of the unit must be followed. The unit must be inspected before each operation to correct any leaks or obstructions in the spray system; to detect whether the nozzle, hoses, or other parts are worn and need replacement; to insure that the flow meter is properly calibrated; and to determine that the pressure recommended by the manufacturer is being maintained.

Flow Rate—must be regulated by accurate flow meter.
—not greater than 1 gallon per hour at 5 mph, 2 gallons per hour at 10 mph, 3 gallons per hour at 15 mph or 4 gallons per hour at 20 mph.

Nozzle Direction—rear of the vehicle.
—upward at an angle of 45° or more.

Vehicle Speed—not greater than 20 miles per hour.
—shut off spray equipment when vehicle is stopped.

IMPORTANT—Spray droplets of undiluted CYTHION ULV will permanently damage automobile paint unless all the conditions described and recommended in this label are met.

Directions for Determining the Droplet Size of CYTHION ULV Nonthermal Aerosols

Permanent records of each droplet size determination must be kept and made available to American Cyanamid upon request.

1. Preparation of Slides with DRI-FILM

CYTHION ULV droplet sizes are determined by depositing a sample of the aerosol on a coated glass slide and measuring the droplets under a high-power microscope. Ordinary 3" x 1" glass slides must be coated with silicone (General Electric SC-87 DRI-FILM) prior to sampling to prevent excessive spreading or coalescence of the droplets. The slides are dipped into a 10 percent solution of DRI-FILM in toluene, drained and dried at about 200°F. for 30 minutes, after which they are dipped in acetone, allowed to dry and stored in a tight slide box. Coating solution must be freshly prepared. Do not store coating solution because it will deteriorate. Slides are lightly polished with a soft tissue before using to remove any foreign particles.

2. Deposition of CYTHION ULV Droplets on Slides

Droplets should be collected under ideal operating conditions to insure representative sampling of droplets in the aerosol. A sample of the CYTHION ULV aerosol is deposited on a slide by waving the slide as rapidly as possible perpendicular through the aerosol cloud at a distance of 25 feet from the point of discharge. The slide velocity may be increased by attaching it to a 3 or 4 foot stick by means of a spring paper clip. At least two slides should be exposed to insure an adequate sample. Store slides in a tight slide box for transfer to a location where measurements can be made. Avoid excessive heat during transit and store in a cool place until measurements can be made.

Although label specifications require the aerosol nozzle to be angled upward at 45° or more during operation it is more convenient to position the nozzle parallel to the ground for droplet sampling. If this is not possible it will be necessary to be positioned at a sufficient height to obtain a representative sample of the aerosol.

Figure 1.
Percentage of the total volume of aerosol samples below each stated droplet size (from Table 1). The Mass Median Diameter is determined from the 50 percent point on the line. The Mass Median Diameter (MMD) = 9.2 divisions times 1.75 = 16.1 microns.

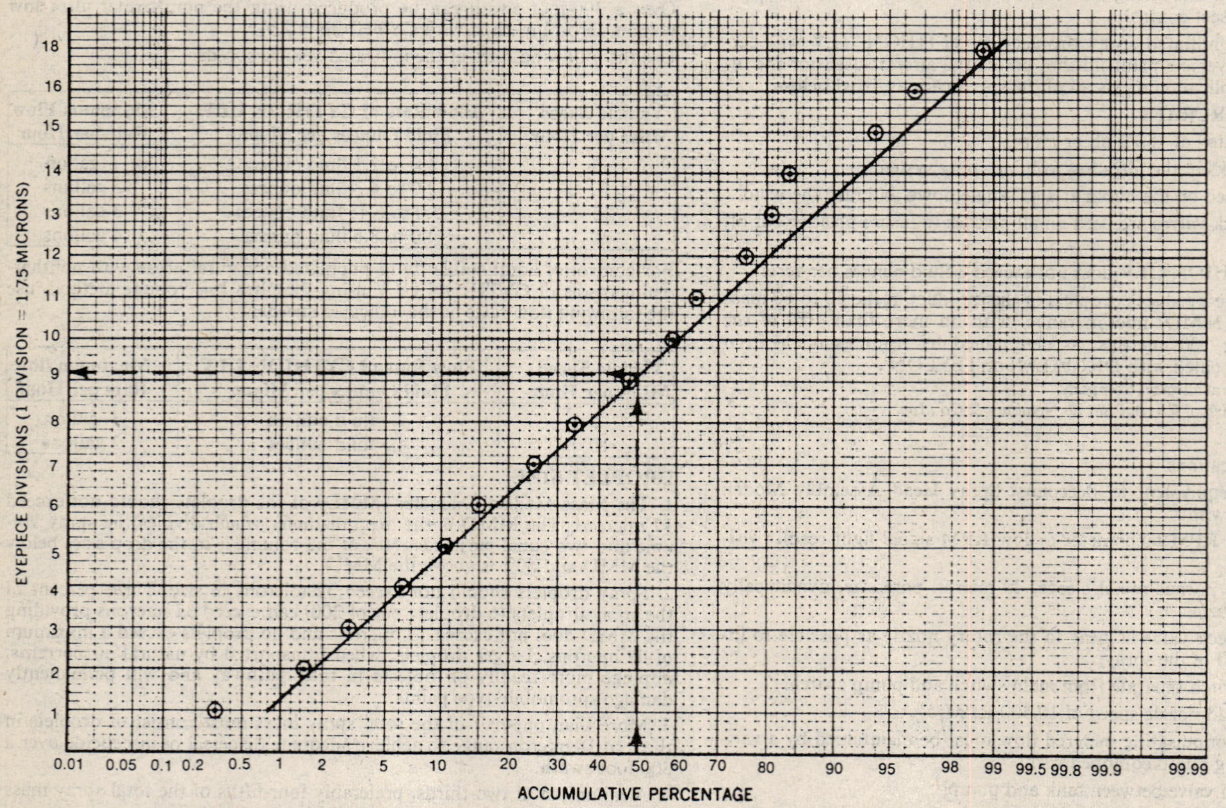

3. Determination of CYTHION ULV Droplet Sizes

A microscope with mechanical stage and an eyepiece micrometer are used to determine the size of the individual aerosol droplets. Prior to taking measurements, the divisions of the eyepiece micrometer must be calibrated into microns by means of a stage micrometer. In the example represented in Table 1, droplets were measured at 400x magnification. At that magnification each division of the eyepiece was calibrated to equal 3.5 microns.

At least 200 droplets should be measured. Usually this is easily accomplished on one slide. An accurate method is to measure all droplets that pass through the micrometer scale as the slide is moved from one edge to the other by using the mechanical stage. Measurements should not be taken along the margins of the slide. It is more convenient to measure in terms of the divisions of the eyepiece micrometer and then convert these divisions into microns.

The measurements converted into microns must then be corrected for the amount of spread that occurred on the slide. The CYTHION ULV spread factor for silicone-coated slides is 0.5. Therefore, in Table 1 each division of the eyepiece actually equals 1.75 microns (3.5 microns times the 0.5 spread factor).

The spread factor for TEFLON-coated slides is 0.69. The following procedure as given for silicone-coated slides, would be the same for TEFLON-coated slides once the value for each eyepiece division has been determined.

The measurements are tabulated and processed as in Table 1. The Maximum Diameter is calculated by converting the diameter of the largest droplet measured into microns. In Table 1, the largest droplet measured had a diameter of 19 eyepiece divisions. Therefore, the Maximum Diameter is 33.3 microns (19x1.75 = 33.3).

To determine the Mass Median Diameter (MMD), the accumulative percentages from the last column in Table 1 are plotted against the eyepiece divisions (D) on semi-logarithmic paper as in Figure 1. Directly across from the 50 percent point on the line is the median droplet size in eyepiece divisions which must be converted to microns. In Figure 1, 9.2 eyepiece divisions times the conversion factor of 1.75 equals a Mass Median Diameter of 16.1 microns.

Table 1—Representative Count of CYTHION ULV Aerosol Droplets Impinged on Microscope Slides Coated with DRI-FILM.

Eyepiece Divisions (D*)	Number of Droplets (N)	DxN	% of Total DxN Σ (DxN)	Accumulative Percentages
1	5	5	0.31	0.31

Eyepiece Divisions (D*)	Number of Droplets (N)	DxN	% of Total DxN Σ (DxN)	Accumulative Percentages
2	10	20	1.22	1.53
3	9	27	1.65	3.18
4	12	48	2.93	6.11
5	15	75	4.58	10.69
6	12	72	4.40	15.09
7	25	175	10.70	25.79
8	14	112	6.85	32.64
9	28	252	15.40	48.04
10	19	190	11.61	59.65
11	14	154	9.41	65.06
12	10	120	7.33	76.39
13	6	78	4.77	81.16
14	4	56	3.42	84.58
15	11	165	10.09	94.67
16	2	32	1.96	96.63
18	2	36	2.20	98.83
19	1	19	1.16	99.99
Total	199	1636		

* Measurements were taken at 400x magnification. Each eyepiece division equals 1.75 microns (3.5 microns times the 0.5 spread factor).

Also for use in accordance with the recommendations and instructions issued by the United States Department of Agriculture for quarantine programs. To be used only by or under the direction of Federal/State personnel for quarantine treatments.

Also for use in accordance with the recommendations and instructions issued by the United States Department of Agriculture for quarantine programs. To be used only by or under the direction of Federal/State personnel for quarantine treatments.

CYTHION® 57% EC

INSECTICIDE

'The Premium Grade Malathion' ®

57% Emulsifiable Concentrate

Before Using Read Directions in this Leaflet.

ACTIVE INGREDIENT:

Malathion: O,O-dimethyl phosphorodithioate of diethyl mercaptosuccinate 57.0%

INERT INGREDIENTS:** 43.0%

TOTAL 100.0%

Contains Xylene

(1 gallon contains 5 pounds of Malathion)

12/88

EPA Reg. No. 241-47

KEEP OUT OF REACH OF CHILDREN

WARNING/AVISO

FLAMMABLE LIQUID AND VAPOR

HARMFUL IF SWALLOWED

Keep away from heat, sparks, and flame. Avoid breathing spray mist.

PRECAUCION AL USUARIO: Si usted no lee inglés, no use este producto hasta que la etiqueta le haya sido explicada ampliamente.

FIRST AID

IF SWALLOWED: DO NOT induce vomiting. Get medical attention!

IF IN EYES: Immediately flush eyes with plenty of water for at least 15 minutes. Call a physician.

IF ON SKIN: Immediately wash with plenty of soap and water while removing contaminated clothing and shoes. Get medical attention.

IF INHALED: Remove victim to fresh air. If not breathing give artificial respiration, preferably mouth-to-mouth. Get medical attention.

NOTE TO PHYSICIAN: This product is a cholinesterase inhibitor. Atropine is antidotal. 2-PAM may be effective as an adjunct to atropine.

PRECAUTIONARY STATEMENTS

HAZARDS TO HUMANS

WARNING

Harmful if swallowed, inhaled or absorbed through the skin. Keep container closed. Use with adequate ventilation. Avoid breathing spray mist. Avoid contact with skin or clothing. Wash skin thoroughly after handling. Remove and wash contaminated clothing before reuse. May be a skin sensitizer. Causes eye irritation.

ENVIRONMENTAL HAZARDS

This pesticide is toxic to fish and aquatic invertebrates. Do not apply directly to water except as specified on this label. Drift and runoff may be hazardous to aquatic organisms in areas near the application site. Do not contaminate water when disposing of equipment washwaters.

This product is highly toxic to bees exposed to direct treatment on blooming crops or weeds. Do not apply this product or allow it to drift to blooming crops or weeds if bees are visiting the treatment area.

PHYSICAL/CHEMICAL HAZARDS

Flammable: Do not use or store near heat sources or open flame. Do not use in the vicinity of pilot lights.

DISCLAIMER

American Cyanamid Company warrants only that the material contained herein conforms to the chemical description on the label and is reasonably fit for the use therein described when used in accordance with the directions for use.

Any damages arising from a breach of this warranty shall be limited to direct damages, and shall not include consequential commercial damage such as loss of profits or values, etc.

American Cyanamid Company makes no other express or implied warranty, including any other express or implied warranty of FITNESS or of MERCHANTABILITY.

User assumes the risk of any use contrary to label instructions or under abnormal conditions, or under conditions not reasonably foreseeable by American Cyanamid Company.

In case of an emergency endangering life or property involving this product, call collect, day or night, Area Code 201-835-3100.

GENERAL WORKER PROTECTION STATEMENTS

Do not apply this product in such a manner as to directly or through drift expose workers or other persons. The area being treated must be vacated by unprotected persons.

Do not enter treated areas without protective clothing until sprays have dried. Because certain states may require more restrictive reentry intervals for various crops treated with this product, consult your State Department of Agriculture for further information.

Written or oral warnings must be given to workers who are expected to be in a treated area or in an area about to be treated with this product. When oral warnings are given, warnings shall be given in a language customarily understood by workers. Oral warnings must be given if there is reason to believe that written warnings cannot be understood by workers. Written warnings must include the following information: WARNING. Area treated with CYTHION insecticide on (date of application). Do not enter without appropriate protective clothing.

DIRECTIONS FOR USE

It is a violation of Federal law to use this product in a manner inconsistent with its labeling. Refer to label directions for use. Read all directions carefully before using. Do not apply this product through any type of irrigation system.

STORAGE AND DISPOSAL

PROHIBITIONS: Do not contaminate water, food, or feed by storage or disposal. Open dumping is prohibited.

PESTICIDE DISPOSAL: Pesticide wastes are toxic. Improper disposal of excess pesticide, spray mixture, or rinsate is a violation of Federal Law. If these wastes cannot be disposed of by use according to label instructions, contact your State Pesticide or Environmental Control Agency, or the Hazardous Waste representative at the nearest EPA Regional Office for guidance.

CONTAINER DISPOSAL: Triple rinse (or equivalent). Then offer for recycling or reconditioning, or puncture and dispose of in a sanitary landfill, or by other approved state and local procedures.

GENERAL: Consult Federal, State or local disposal authorities for approved alternative procedures such as limited open burning.

VEGETABLES: Apply in a minimum of 30 gals. by ground equipment or 5 gals. by air equipment unless otherwise specified.

CROP	PESTS CONTROLLED	RATE/A	INTERVAL (DAYS) BETWEEN LAST APPLICATION AND HARVEST
Asparagus	Asparagus aphid Asparagus beetle	2 pts.	Do not apply within 1 day of harvest.
	Thrips	1½-2 pts.	
Beans (Green Beans, Lima Beans, Navy Beans, Red Kidney Beans, Snap Beans, Wax Beans, Cowpeas and Blackeyed Peas)	Aphids *Bean leaf beetles *Blister beetles Cucumber beetles Potato leafhoppers Lygus bugs	2 pts.	Do not apply within 1 day of harvest.
	Japanese beetle Mexican bean beetle	1½-2 pts.	
	Red Spider mites	1½ pts.	
Dry Beans (California and Northwest only)	Lygus bugs	1½-2 pts. (ground) 1½ pts. (air) in 10 gallons of water	Do not apply within 1 day of harvest.
Beets (Garden and Table)	Aphids	1½-2 pts.	Do not apply within 7 days of harvest if tops are to be used as food or feed.
Broccoli	Aphids Cabbage looper Imported cabbageworm	1-2 pts.	Do not apply within 3 days of harvest.
Brussels Sprouts	Aphids	1-2 pts.	Do not apply within 7 days of harvest.
Cabbage	Aphids Cabbage looper Imported cabbageworm	1-2 pts.	Do not apply within 7 days of harvest. For control of caterpillars on summer and fall plantings in the South, begin when true leaves appear. On other plantings, and for control of other insects, begin when insects appear.
	Diamondback moth Webworm	2 pts.	
Carrots	Aphids	1½-2 pts.	Do not apply within 7 days of harvest.
	Leafhoppers	2½ pts.	
Cauliflower	Diamondback moth	2-4 pts.	Do not apply within 7 days of harvest.
	Aphids	1-2 pts.	
Celery, Anise	Aphids Spider mites	1½ pts.	To be applied to fresh leaves and stalks only. Do not use on crops grown for seed and oil. Do not apply within 7 days of harvest
Collards	Aphids	1½-2 pts.	Do not apply within 7 days of harvest. For control of caterpillars on summer and fall plantings in the South, begin when true leaves appear. On other plantings, and for control of other insects, begin when insects appear.
	Harlequin cabbage bug	1 pt.	
Cucumbers	Aphids Pickleworms Spider mites	1½-2 pts.	Do not apply to cucumbers unless plants are dry. Do not apply within 1 day of harvest.
	Squash vine borer	3 pts.	
	Cucumber beetle Leaf miner	2 pts.	
Dandelion, Parsley, Parsnip, Swiss Chard, Watercress	Aphids	1½-2 pts.	Do not apply within 21 days of harvesting parsley or within 7 days of harvesting dandelion, parsnip, Swiss chard or watercress
Eggplant	Aphids Spider mites	1 pt.	Do not apply within 3 days of harvest.
	Lacebugs	3 pts.	
Endive	Aphids Spider mites	1½-2 pts.	Do not apply within 7 days of harvest.
Garlic, Shallots	Aphids Thrips	1½-2 pts.	Do not apply within 3 days of harvest.
Hops	Aphids Spider mites	1 pt.	Do not apply within 10 days of harvest.

CROP	PESTS CONTROLLED	RATE/A	INTERVAL (DAYS) BETWEEN LAST APPLICATION AND HARVEST
Kale	Aphids Cabbage looper Imported cabbageworm	1-2 pts.	Do not apply within 7 days of harvest. For control of caterpillars on summer and fall plantings in the South, begin when true leaves appear. On other plantings, and for control of other insects, begin when insects appear.
	Webworm Diamondback moth	2 pts.	
Kohlrabi	Aphids	1-2 pts.	Do not apply within 7 days of harvest.
	Cabbage looper Diamondback moth Imported cabbageworm Webworm	2 pts.	
Leeks	Aphids	1½-2 pts.	Do not apply within 3 days of harvest.
	Onion maggot flies	2½ pts.	
	Thrips	3 pts.	
Lentils	Cowpea aphid Pea aphid	1½ pts.	Do not apply within 3 days of harvest.
Lettuce	Aphids Leafhoppers Spider mites	2 pts.	Do not apply within 7 days of harvesting head lettuce or within 14 days of harvesting leaf lettuce.
	Cabbage looper	3 pts.	
Melons (Cantaloupe, Casaba, Crenshaw, Honeydew Melons, Honey Balls, Muskmelons, Persian Melons and hybrids of these, Watermelons and their hybrids)	Aphids Spider mites	1½-2 pts.	Do not apply to melons unless plants are dry. Do not apply within 1 day of harvest.
	Cucumber beetles	2-3 pts.	
	Leaf miners	2 pts.	
	Leafhoppers	1½-2 pts.	
	Pickleworms	2 pts.	
	Squash vine borer	3 pts.	
Mushrooms	Mites Phorid and Sciarid flies	2½ pts. in 130 gallons of water; or, 2 tablespoons in 3 gallons of water per 1000 sq. ft. of bed	Make thorough applications as soon after picking as possible. Repeat applications as necessary, usually twice a week. Do not apply within 1 day of harvest.
Mustards	Aphids Cabbage looper Imported cabbageworm	2 pts.	For control of caterpillars on summer and fall plantings in the South, begin when true leaves appear. On other plantings, and for control of other insects, begin when insects appear. Do not apply within 7 days of harvest.
	Flea beetles	1½-2 pts.	
Okra	Aphids	1½ pts.	Do not apply within 1 day of harvest.
	Japanese beetle	2 pts.	
Onions	Onion maggots	1½-2½ pts.	Do not apply within 3 days of harvest.
	Onion thrips	1½-2 pts.	
Peas	Aphids	1½-2 pts.	Do not apply within 7 days of harvest if vines are to be fed, or within 3 days of harvest if vines are not to be fed.
	Grasshoppers	2 pts.	
	Leafhoppers	1¼-4 pts.	
Peppermint, Spearmint	Aphids Flea beetles Leafhoppers Spider mites	1½ pts.	Do not apply within 7 days of harvest.
Peppers	Aphids	1-2½ pts.	Do not apply within 3 days of harvest.
	Pepper maggots	2½ pts.	
Potatoes	Aphids Grasshoppers Leafhoppers	2 pts.	May be applied on the day of harvest.
	False chinch bugs	1½ pts.	
	Mealybugs	2-2½ pts.	
Pumpkins	Aphids Pickleworms Spider mites	2 pts.	Do not apply to pumpkins unless plants are dry. Do not apply within 3 days of harvest.
	Leafhoppers	1½-2 pts.	
	Cucumber beetles Squash vine borer	3 pts.	

CROP	PESTS CONTROLLED	RATE/A	INTERVAL (DAYS) BETWEEN LAST APPLICATION AND HARVEST
Radish, Horseradish	Aphids	1½-2 pts.	Do not apply within 7 days of harvest.
Rice	Rice leaf miner	2½ pts.	Make first application shortly after the first rice blades appear on the surface of the water and repeat if necessary. Do not apply within 7 days of harvest.
	Rice stink bugs	1-1½ pts.	Apply by airplane in 2 gallons of water per acre during the early milk and dough stage of growing rice. Repeat applications as necessary. Do not apply within 7 days of harvest.
Rutabaga	Aphids	1½ pts.	Do not apply within 3 days of harvest.
Spinach	Aphids	2 pts.	Do not apply within 7 days of harvest.
Squash	Aphids Leaf miners Spider mites Pickleworm	2 pts.	Do not apply to squash unless plants are dry. Do not apply within 1 day of harvest.
	Cucumber beetles Squash vine borer	3 pts.	
Strawberries	Aphids Spider mites	1½ pts.	Do not apply within 3 days of harvest.
	Field crickets Lygus bugs Spittlebugs Thrips	1½-3 pts.	
	Potato leafhopper Strawberry leafroller Strawberry root weevil Whiteflies	1½-2½ pts.	
Sweet Corn	Japanese beetle	2 pts.	Do not apply within 5 days of harvest. Injury may occur in the whorl or to the silks.
Sweet Potatoes	Leafhoppers	1½-2 pts.	Do not apply within 3 days of harvest.
	Morninglory leaf miner	2½-3 pts.	
Tomatoes	Aphids	1½ pts.	Do not apply within 1 day of harvest.
	Drosophila	2½ pts.	
	Spider mites	1½ pts.	
	Armyworms Fruitworms (California only)	2¾ qts.	Do not apply within 3 days of harvest.
Turnip	Aphids Cabbage looper Imported cabbageworm	1-2 pts.	For control of caterpillars on summer and fall plantings in the South, begin when true leaves appear. On other plantings, and for control of other insects, begin when insects appear. Do not apply within 7 days of harvest if tops are to be used for food or feed.

VEGETABLES GROWN IN COMMERCIAL GREENHOUSES: Apply to give complete cpverage.

CROPS	PESTS CONTROLLED	RATE/100 GALS.	INTERVAL (DAYS) BETWEEN LAST APPLICATION AND HARVEST
Cucumbers	Aphids Armyworms Cabbage loopers Earwigs Garden fleahoppers Mealybugs Spider mites Thrips Whiteflies	1½-2 pts.	Spray plants thoroughly. Do not apply to cucumber unless plants are dry. Do not apply within 1 day of harvest.
Endive	Aphids Armyworms Cabbage loopers Serpentine leaf miners Thrips Whiteflies	1½-2 pts.	Spray plants thoroughly. Do not apply within 7 days of harvest.
Lettuce	Aphids Armyworms Cabbage loopers Serpentine leaf miners Thrips Whiteflies	1½-2 pts.	Spray plants thoroughly. Do not apply within 14 days of harvesting leaf lettuce or within 7 days of harvesting head lettuce.
Radish	Aphids Flea beetles	1½-2 pts.	Spray plants thoroughly. Do not apply within 7 days of harvest.
Tomato	Aphids Armyworms Cabbage loopers Cutworms Drosophilia Garden fleahoppers Mealybugs Serpentine leaf miners Spider mites	1½-2 pts.	Spray plants thoroughly. Do not apply within 1 day of harvest.

CROPS	PESTS CONTROLLED	RATE/100 GALS.	INTERVAL (DAYS) BETWEEN LAST APPLICATION AND HARVEST
	Thrips Tomato russet mites Whiteflies		
Watercress	Aphids Serpentine leaf miners Thrips	1½-2 pts.	Spray plants thoroughly. Do not apply within 7 days of harvest.

FRUIT AND NUT

CROP	PESTS CONTROLLED	RATE/ 100 GALS.	RATE/A	INTERVAL (DAYS) BETWEEN LAST APPLICATION AND HARVEST
Almonds	Aphids Spider mites	1-2 pts.	2-4 pts.	May be applied on the day of harvest.
	Peach twig borer	1-1½ pts.	2-3 pts.	Apply during the petal fall period or during the period appropriate for spraying larvae of the May brood. May be applied on the day of harvest.
	The rates for use on almonds are based on a standard of 200 gallons per acre dilute spray for mature trees. Do not apply more than 8 lbs actual CYTHION per acre to almond trees.			
Apples	Budmoth Forbes scale Wooly apple aphid	1 pt.	4 pts.	Do not apply within 3 days of harvest.
	Eastern tent caterpillar	1-1½ pts.	4-6 pts.	
	Green apple aphid Rosy apple aphid	1½ pts.	6 pts.	
	Codling moth Plum curculio Red-banded leafroller	2 pts.	8 pts.	
	Mealybug Mites, such as Clover mite, European red mite, Twospotted spider mite, Willamette mite	1-2 pts.	4-8 pts.	
	The rates for use on apples are based on a standard of 400 gallons per acre dilute spray for mature trees pruned 20 to 22 feet high in rows 40 feet apart.			
	CYTHION Emulsifiable Concentrate may cause fruit injury to McIntosh and Cortland varieties if spray is applied within 4 weeks of harvest.			
Dormant or Delayed Dormant Sprays	Aphids Mites Red-banded leafrollers	1 pts. + 1 gal. superior oil	—	Make full coverage dormant or delayed dormant applications only.
Apricots	Aphids Codling moth European fruit lecanium Orange tortrix Soft brown scale Terrapin scale	1½-2 pts.	4½-6 pts.	Do not apply within 7 days of harvest.
	The rate for use on apricots is based on a standard of 300 gallons per acre dilute spray for mature trees.			
Avocado	Greenhouse thrips Latania scale Omnivorous looper Orange tortrix Soft brown scale	1½ pts.	7½ pts.	Do not apply within 7 days of harvest.
	The rate for use on avocado is based on a standard of 500 gallons per acre dilute spray for mature trees.			
Blackberry Boysenberry Dewberry Loganberry Raspberry	Aphids Rose scale chafers Japanese beetle Leafhoppers Mites Thrips	1½ pts.	3 pts.	Do not apply within 1 day of harvest.
	The rates for use on brambles are based on a standard of 200 gallons per acre dilute spray.			
Blueberries	Blueberry maggots	1 pt. + 1½ qts. Staley's Sauce Base No. 7	—	For use in the Northeast only. Apply in 100 gallons of water per acre. Do not apply within 8 hours of harvest.
	Cherry fruitworm	1 pt.	2 pts.	Do not apply within 1 day of harvest.
	Cranberry fruitworm	1 pt.	2 pts.	Make first application at egg hatch and repeat applications every 4 or 5 days until a total of four applications has been made. Do not apply within 1 day of harvest.
	Japanese beetle	¾ pt.	1½ pts	Make application when first berries turn blue and at 10 day intervals as long as infestation persists. Do not apply within 1 day of harvest.
	The rates for use on blueberries are based on a standard of 200 gallons per acre dilute spray.			
Cherry	Black cherry aphid Fruit-tree leafroller	1½ pts.	6 pts.	Do not apply within 3 days of harvest.
	Cherry fruit fly Japanese beetle	1 pt.	4 pts.	
	The rates for use on cherries are based on a standard of 400 gallons per acre dilute spray for mature trees. Do not apply more than 8 lbs of actual CYTHION per acre to cherry trees.			
	Injury may occur on certain varieties of sweet cherries, particularly in the Northwest.			

CROP	PESTS CONTROLLED	RATE/ 100 GALS.	RATE/A	INTERVAL (DAYS) BETWEEN LAST APPLICATION AND HARVEST
Citrus Grapefruit Kumquat Lemon Lime Orange Tangerine Tangelo	Black scale (single and off-brooded California red scale Citricola scale Purple scale Soft brown scale Yellow scale	1-1½ pts.	—	Do not apply when trees are in bloom. Do not apply within 7 days of harvest.
	Florida purple scale Florida red scale	2 pts.	—	
	Citrus red mites Thrips	—	2½ pts.	Apply in 200 gallons of water per acre. Do not apply when trees are in bloom. Do not apply within 7 days of harvest.
	The rate for use on citrus is based on a standard of 500 gallons per acre dilute spray for mature trees.			
Currants Gooseberries	Japanese beetles	1½ pts.	3 pts.	Do not apply within 3 days of harvest.
	Rose chafers Mites	1 pt.	2 pts.	
	The rates for use on currants and gooseberries are based on a standard of 200 gallons per acre dilute spray.			
Figs	Dried fruit beetles Vinegar flies	—	2 qts. + 1-2 gals. unsulfured molasses	Do not apply within 3 days of harvest.
Filberts	Apple mealybug Eye-spotted bud moth Filbert aphids Tingids	1 pt.	4 pts.	Do not apply within 3 days of harvest.
	Scales (crawler stages)	1 pt.	—	
	The rates for use on filberts are based on a standard of 400 gallons per acre dilute spray for mature trees. Do not apply more than 3 lbs of actual CYTHION per acre to filbert trees.			
Grapes	Leafhoppers Spider mites	1½ pts.	3 pts.	Do not apply within 3 days of harvest.
	European fruit lecanium	1½ pts.	—	Make full coverage applications when newly hatched nymphs are migrating over vines, usually shortly after bloom. Do not apply within 3 days of harvest.
	Mealybugs	—	1½ pts.	Apply in 50-100 gallons of water per acre. Do not apply within 3 days of harvest.
	The rates for use on grapes are based on a standard of 200 gallons per acre dilute spray.			
	Injury may occur on grapes of Almeria, Cardinal, Italia and Ribier varieties when sprays containing CYTHION are applied after clusters appear.			
Grape Vines (Overwintering on nursery stock only)	Grape phylloxera	3 pts.	—	Remove excess soil from the roots and dip in the CYTHION solution. Submerge the entire root system for 5 minutes. Keep the solution agitated at all times.
Macadamia Nuts	Green stink bugs	1½ pts.	—	Apply when insects start to feed on the nuts and repeat as necessary. May be applied during harvest.
	No more than 15 lbs of actual CYTHION per acre should be applied to macadamia nut trees.			
Nectarines	Spider mites	1-2 pts.	3-6 pts.	Do not apply within 7 days of harvest.
	Plum curculio	2 pts.	6 pts.	
	The rates for use on nectarines are based on a standard of 300 gallons per acre dilute spray for mature trees.			
	CYTHION may cause fruit spotting on nectarines.			
Papaya	Aphids Mealybugs	1½-2 pts.	—	May be applied on the day of harvest.
Peaches	Black cherry aphid Black peach aphid European red mite Green peach aphid Rusty plum aphid Japanese beetle Spider mites	1 pt.	3 pts	Do not apply within 7 days of harvest.
	Oriental fruit moth Plum curculio	2 pts.	3 qts.	
	Cottony peach scale European fruit lecanium Terrapin scale	2 pts.	—	Make full coverage applications at completion of hatching, about the last of June. Do not apply within 7 days of harvest.
	The rates for use on peaches are based on a standard of 300 gallons per acre dilute spray for mature trees. Do not apply more than 9 pounds of actual CYTHION per acre to peach trees.			
Pears	Aphids	1 pt.	3 pts.	Do not apply within 1 day of harvest.
	Mealybug Mites Pear psylla	1-2 pts.	3-6 pts.	
	Codling moth Fruit-tree leafroller Plum curculio Red-banded leafroller	2 pts.	6 pts.	
	The rates for use on pears are based on a standard of 300 gallons per acre dilute spray for mature trees.			
	Injury may occur under certain conditions in the Northeast on Bosc pears.			
Pecans	Aphids Spider mites Walnut aphid	1½-2 pts.	7½-10 pts.	May be applied on the day of harvest.
	European red mite	1¼ pts.	6¼ pts.	May be applied on the day of harvest.
	Pecan phylloxera	1¼ pts.	6¼ pts.	Apply when buds begin to develop and repeat as necessary. May be applied on the day of harvest.
	Pecan nut casebearer Walnut husk fly	1¼ pts.	6¼ pts.	Apply after first generation eggs begin to hatch. May be applied on the day of harvest.
	The rates for use on pecans are based on a standard of 500 gallons per acre dilute spray for mature trees 25-35 feet high.			
Pineapple	Mealybug	1 pt.	8 pts.	Do not apply within 7 days of harvest.

CROP	PESTS CONTROLLED	RATE/ 100 GALS.	RATE/A	INTERVAL (DAYS) BETWEEN LAST APPLICATION AND HARVEST
Plums and Prunes	Mealy plum aphid Plum curculio	1-1½ pts.	3 pts.	Do not apply within 3 days of harvest. Make full coverage applications to trunks and scaffold limbs 4 times at 21-day intervals beginning with emergence in June. In Georgia, two applications each 4 weeks apart for first (April and May) and second (August and September) broods. Do not apply within 3 days of harvest.
	The rates for use on plums and prunes are based on a standard of 300 gallons per acre dilute spray for mature trees.			
Quince	Codling moth Oriental fruit moth Plum curculio	2 pts.	6 pts.	Do not apply within 3 days of harvest.
	Spider mites	1-2 pts.	3-6 pts.	
	Forbes scale	1 pt.	—	Make full coverage applications when crawlers are present, about first or second cover and last half of July. Do not apply within 3 days of harvest.
	The rates for use on quince are based on a standard of 300 gallons per acre dilute spray for mature trees.			
Walnuts	Aphids European red mite Walnut aphid	½ pt.	1½-3 pts.	May be applied on the day of harvest.
	Walnut husk fly	½ pt.	1½-3 pts.	Apply with conventional sprayer. For bait sprays, combine Staley's Sauce Base No. 2 or No. 7 at the rate of 2 qts. per acre with CYTHION. May be applied on the day of harvest.
		—	3-4 pts.	Apply in 500 gallons of water per acre with air-carrier type sprayer. For bait sprays, combine Staley's Sauce Base No. 2 or No. 7 at the rate of 2 qts. per acre with CYTHION. May be applied on the day of harvest.

FIELD CROPS AND PASTURE AND RANGE GRASSES

CROP	PESTS CONTROLLED	RATE/A	INTERVAL (DAYS) BETWEEN LAST APPLICATION AND HARVEST
Alfalfa	Alfalfa weevil larvae Aphids Grasshoppers Lygus bugs Potato leafhoppers Spider mites Spittlebugs Stinkbugs	1½-2 pts.	Apply to alfalfa in bloom only in the evening or early morning when bees are not working in the field or are not hanging on outside of hives. May be applied on the day of harvest.
	Armyworms	2 pts.	
	Clover leaf weevil	1½ pts.	
	Vetch bruchid	2-2½ pts.	
Clover	Alfalfa weevil larvae Aphids Grasshoppers Lygus bugs Potato leafhoppers Spider mites Spittlebugs Stink bugs	1½-2 pts.	Do not apply to clover in bloom. May be applied on the day of harvest.
	Armyworms	2 pts.	
	Clover leaf weevil	1½ pts.	
Corn (Grain or Forage)	Aphids Corn earworm Corn rootworm adults Grasshoppers Sap beetle Thrips	1½ pts.	For control of corn earworm and sap beetles, begin treatments when 10% of the ears show silk. Repeat applications at 3-5 day intervals until 4-5 applications have been made. Do not apply within 5 days of harvest or forage use.
	Armyworms	1½-2 pts.	
	Injury may occur in the whorl and silk stages with CYTHION.		
Cotton	Brown cotton leafworm Cotton aphid Cotton leafworm Cotton leafperforator Desert spider mite Leafhoppers Lygus bugs Thrips Whiteflies	½-2 pts.	May be applied on the day of harvest. Consult local agricultural authorities for exact time of application.
	Boll weevil	2-4 pts.	
	Cotton fleahoppers	1-1½ pts.	
	Fall armyworms Garden webworms Grasshoppers	1½-3 pts.	May be applied on the day of harvest.
	Lygus bugs Thrips	1-4 pts.	
Grasses (Barnyardgrass, Canarygrass, Fescue,	Cereal leaf beetle	1-1½ pts.	May be applied on the day of harvest or grazing.

CROP	PESTS CONTROLLED	RATE/A	INTERVAL (DAYS) BETWEEN LAST APPLICATION AND HARVEST
Orchardgrass, Red top, Timothy, Yellow foxtail)			
Pasture and Range Grass, Grass and Grass Hay	Aphids Grasshoppers Leafhoppers	1½-2 pts. OR 1½ pts. in 1 gal. of diesel fuel oil per acre	May be applied on the day of harvest or grazing.
	Armyworms	2 pts. OR 1½ pts. in 1 gal. of diesel fuel oil per acre	
Peanuts	Potato leafhoppers Thrips	1½ pts.	May be harvested or grazed on the day of application.
Safflower	Aphid Grasshoppers Lygus bugs	1½-2 pts.	Do not apply within 3 days of harvesting seeds.
Small Grains (Barley, Oats, Rye, Wheat)	Cereal leaf beetle	1-1½ pts.	Do not apply within 7 days of harvest or forage use.
	English grain aphids Grasshoppers Greenbugs	1½ pts.	
	Winter grain mite	2 pts.	
Sorghum	Greenbugs	1½ pts.	Do not apply within 7 days of harvest or forage use.
Soybeans	Green cloverworms Mexican bean beetle	3 pts.	May be applied on the day of harvest or forage use.
Sugar Beets	Aphids Spider mites	1½-2 pts.	Do not apply within 7 days of harvest if tops are to be used as feed or forage.
	Grasshoppers	3 pts.	
Tobacco	Aphids	1½-2½ pts.	May be applied on the day of harvest. Use maximum dosage in plant beds.
Vetch	Omniverous leaf tier Pea aphid Vetch bruchid	1½-2 pts.	May be applied on the day of harvest or grazing.

STORED PRODUCTS

SITE	PESTS CONTROLLED	RATE	DIRECTIONS FOR USE
Stored in-shell Almonds	Indian meal moth Merchant grain beetle	4 fl. oz. in sufficient water to treat 10,000 lbs of almonds.	Use a suitable mechanical spray applicator that regulates the rate application to the flow of the almonds. Avoid spraying with a fine mist that drifts away. Shield the nozzle against wind and air currents.
Stored Peanuts	Confused flour beetle Flat grain beetle Granary weevil Indian meal moth Lesser grain borer Red flour beetle Rice weevil Rusty grain beetle Saw-toothed grain beetle	1 pt. in 2½ gal. of water OR 8 pts. in 19 gal. of water.	Residual Warehouse Spray - Before Storing Peanuts: Apply as a coarse spray at the rate of 2 gal. per 1,000 sq. ft. of surface or to run-off. Clean warehouse thoroughly of trash and remains of old peanuts 1 to 2 weeks before new peanut crop is stored. Then, thoroughly spray the interior of the empty warehouse, especially cracks and protected places. treat outside walls to a height of 6 to 8 feet and the ground to a distance of 6 feet from the warehouse.
		2½ pts. in 5 gal. of water for each 15 tons of farmers stock peanuts.	Bulk Spray Treatment—Peanuts Going Into Storage: Use good spray equipment. Apply coarse spray uniformly. Preferably, use a suitable mechanical spray applicator that regulates the rate of application to the flow water for each 15 tons of peanuts. Adjust the operating pressure of spray pump and size of nozzle opening to correlate the amount of spray delivery with the rate of flow of peanuts being treated. Avoid spraying with a fine mist that drifts away, by using low nozzle pressure. Shield the nozzle against wind and air currents.
Nonmedicated cattle feed concentrate blocks	Cigarette beetles	8 pts. in 25 gal. water	Before storing feed blocks thoroughly clean storage areas and remove and burn all debris and sweepings. Apply as a thorough application for a residual spray.
		4 fl. oz. in 1 qt. water	Treat paper on the side next to the feed concentrate at a rate of 100 mg/sq. ft., or 1 qt. of diluted CYTHION per 710 sq. ft.
Bagged citrus pulp Residual warehouse spray	Almond moth Angoumois grain beetle Cigarette beetle Confused flour beetle Flat grain beetle Indian meal moth Mediterranean flour moth Red flour beetle Saw-toothed grain beetle	1 pt. in sufficient water to make 2½ gals of spray OR 8 pts. per 19 gal of water.	Apply spray at the rate of 2 gallons per 1,000 sq. ft. of surface or to run-off. Before bagged citrus pulp is stored, thoroughly clean warehouses by removing and burning all debris and sweepings. Thoroughly spray with sufficient pressure, interior of empty warehouse (including cracks and protected places), outside walls to a height of 6 to 8 feet and ground to a distance of about 6 ft. from warehouse.
	Do not use treated burlap bags other than for dried citrus pulp.		

SITE	PESTS CONTROLLED	RATE	DIRECTIONS FOR USE
Stored Grains (Barley, corn, grain sorghum, oats, rice, rye, wheat) and field or garden seeds	Cereal leaf beetle Confused flour beetle Flat grain beetle Granary weevil Indian meal moth Lesser grain borer Maize weevil Red flour beetle Rice weevil Rusty grain beetle Saw-toothed grain beetle	8 pts. per 25 gal of water	Residual Spray Before Storing Grains: Before applying spray, clean elevators, box cars, etc. thoroughly. Remove and burn all sweepings and debris. For a residual wall, floor and machinery spray in grain elevators in treating truck beds, box cars and ships' holds, before loading grain make a thorough application.
			Grains Going Into Storage: Apply as the grain is being loaded or turned into final storage.
		1 pt. in 2 to 5 gal of water per 1,000 bushels	
		½ pt. in 1 to 2 gal of water per 1,000 sq ft of grain surface area.	As a Surface Treatment After Grains are Stored: Apply the spray evenly over the surface of the grain. Apply immediately after grain is loaded into storage and repeat if necessary.
Warehouse	Khapra Beetle	8 pts. per 20 gallons of water	Apply spray at the rate of 2 gallons per 1,000 sq ft of surface or to run-off. Thoroughly spray with sufficient pressure, interior of empty warehouse including cracks and protected places.

Use only with adequate ventilation. After using this product indoors, ventilate thoroughly before occupying enclosed spaces. Do not allow contact with treated surface until sprays have dried.

FLY AND MOSQUITO CONTROL

Fly Control: For use in and around buildings which house domestic animals, around yards, homes and meat and food-processing plants. Do not use in edible products areas of food processing plants, restaurants or other areas where food is commercially prepared or processed. Do not use in serving areas while food is exposed.

PESTS CONTROLLED	RATE	DIRECTIONS FOR USE
Adult flies	Straight Sprays: 5 Tablespoons + 1 gal water OR 1 cup + 2½ gal water OR 1 quart + 12 gal water.	Apply as a spray at the rate of 1 gallon per 1,000 sq. ft. on painted surfaces and 2 gals per 1,000 sq. ft. on unpainted surfaces where flies alight or congregate, such as walls, ceilings, stanchions, windows in dairy barns, fences, around garbage cans, etc.
Adult flies Fly maggots	Bait Sprays: 5 Tablespoons + 7 Tablespoons sugar or molasses (unsulfurized) or corn syrup + 1 gal water OR 1 cup + 1 cup sugar or molasses (unsulfurized) or corn syrup + 2½ gal water OR 1 quart + 2½ lbs sugar or 1 quart molasses (unsulfurized) or 1 quart corn syrup + 12 gal water.	Apply as a bait spray over the surface of manure or poultry droppings. In loafing sheds, spray the dry bedding within 18 inches of the walls and around upright braces. For effective control in and around dairy barns, fly-breeding sites such as manure and other waste material, should be eliminated. Do not apply to freshly whitewashed surfaces. Wait 14 days after whitewashing before applying.

Repeat applications as necessary.
Avoid contamination of milk, milk equipment and water.
Avoid contamination of feed and food products, also drinking fountains and feed troughs.
Remove lactating dairy animals from buildings before treating. Also remove animals under one month of age before treating.
Avoid applying oil-based formulations to valuable ornamental plants as injury may occur.

Mosquito larvae	13 fl oz per acre	For use in standing water (intermittently flooded areas, stagnant water, temporary rain pools). Mix in sufficient water or oil when applied by air or ground equipment.
Mosquito adults	1 part to 28 parts water, fuel oil or diesel oil.	Spray building foundations, shrubs, low trees and lawn areas.

LIVESTOCK PEST CONTROL

ANIMAL	PEST	RATE	REMARKS
Hogs	Lice	1 gal per 100 gals of water OR 6½ oz per 5 gals of water.	Apply complete coverage spray to animals, pens and litter. One treatment may be sufficient. Repeat only if needed.
	Sarcoptic mange	1 gal per 100 gals of water OR 6½ ozs per 5 gals of water.	Apply complete coverage spray to all animals in herd, bedding and walls. Use extreme care to thoroughly cover all body surfaces of the animal, including inside of ears. One thorough spraying will usually control sarcoptic mange of swine, however, extensive cases will required second treatment about 10 days after first application.
			After spraying, swine should be kept out of sun and wind for a few hours.
			Residue tolerance: 4 ppm in meat, fat, and meat by-products.
			Avoid contamination of feed, food containers and watering troughs.
Sheep and Goats	Lice Ticks Keds	1 gal per 100 gals of water OR 6½ ozs per 5 gals of water.	Spray animals thoroughly. Repeat application after 2 or 3 weeks if needed. Do not apply to milk goats. Do not treat animals under one month of age.
			Residue tolerance: 4 ppm in meat, fat and meat by-products.
Horses, beef and non-milking cattle	Lice	1 gal per 100 gals of water OR 6½ ozs per 5 gals of water.	Apply complete coverage spray. One treatment may be sufficient. Repeat only if needed.
		BACK-RUBBING DEVICES: To reduce lice apply a mixture of 2% CYTHION (using 57% Emulsifiable Concentrate) in fuel oil. There may also be a reduction in hornflies. These devices should be made continuously accessible, one to each 35-45 head of cattle. Back-rubbing devices must be retreated every 2 to 3 weeks.	
	Ticks	1-2 gals per 100 gals of water OR 6½-13 ozs per 5 gals of water.	Apply complete coverage spray. Repeat at 2 week intervals if needed.

ANIMAL	PEST	RATE	REMARKS
	Hornflies	1-1½ gals per 100 gals of water OR 6½-10 ozs per 5 gals of water.	Apply complete coverage spray. Repeat at 2 week intervals if needed. Do not apply to lactating dairy animals, or nonlactating dairy animals within 2 weeks of freshening. Do not treat animals under one month of age. Residue tolerance: 4 ppm in meat, fat and meat by-products.
Poultry: Chickens, Ducks Geese, Turkeys	Northern fowl mite Poultry lice Chicken red mite	2 tbs per 1 gal water per 100-150 birds	Repeat application in 4-8 weeks or when necessary. As a supplement to premise treatment for chicken red mite.

ANIMAL	PEST	SITE	RATE	REMARKS
	Northern fowl mite Chicken body lice Shaft lice	Tail dipping	8½ ozs per 15 gals water per 400 birds	Hold bird by wings and dip 3 to 4 inches of tail into solution. Treat vent and surrounding areas. Repeat in 7-10 days if necessary.
	Chicken red mite Poultry lice	Roost paint	2-7 ozs per 1 gal water	Brush on at rate of 1 pt per 150 ft of roost.
	Northern fowl mite Chicken red mite Poultry lice Flies	Premise treatments	4 tbs per 1 gal water	Apply liberally to litter, walls, ceilings, roost nests and adjacent areas. Force spray into cracks and crevices.
	Poultry ticks	Premise treatments	6-7 ozs per 1 gal water	Apply liberally to walls, ceilings and adjacent areas. Force spray into cracks and crevices.
	Chiggers	Range treatments	1-1½ pt per acre	Treat range thoroughly the day before placing poultry on range. Repeat every 2-3 weeks.
	Residue tolerance: 4 ppm in or on meat and meat by-products Eggs 0.1 ppm (from application to poultry)			
Domestic pets (dogs and cats)	Fleas Lice Ticks	—	1 oz/gal	Wet animal thoroughly. Repeat in 2 to 3 weeks if necessary.
	Fleas Ticks	Premise Treatment:	5 oz/gal	Apply per 1,000 sq ft of surface to pet quarters, yards and lawns. Remove manure or debris before treating. Repeat treatment in 3 to 4 weeks if necessary.

OUTDOOR ORNAMENTALS
FLOWERS, SHADE TREES, SHRUBS

Injury may occur on Ferns, Hickory, Viburnum, Lantana, Crassula and Canareti Juniper following the use of CYTHION Emulsifiable Concentrate, slight injury has also been reported on Boston, Pteris, and Maidenhair Ferns, Petunias, Small-Leaf Spirea, White Pine and Maples. Under extreme heat, drought and disease conditions the emulsifiable concentrates may cause slight damage to elms.

PEST	RATE	REMARKS
Aphids Spider mites	1½ pts per 100 gal OR 2 teaspons per gal	—
Bagworms Birch leaf miner Boxwood leaf miner	2 pts per 100 gal OR 2 teaspoons per gal	—
European pine shoot moth Four-lined leaf bug Japanese beetle adult Potato leafhopper Rose leafhopper Tarnished plant bug Thrips	1½ pts per 100 gal OR 2 teaspoons per gal	—
Lace bug	1 pt per 100 gal OR 2 teaspoons per gal	—
Mealybugs Whiteflies	1½ pts per 100 gal OR 2 teaspoons per gal	—
Oak kermes	2 pts per 100 gals OR 2 teaspoons per gal	Apply when scale crawlers have settled on foliage.
Tent caterpillar	2 pts per 100 gals OR 2 teaspoons per gal	—
Oyster shell scale	1 pt per 100 gal OR 2 teaspoons per gal	Apply when scale crawlers have settled on foliage.
Euonymus scale Scurfy scale	1½ pt per 100 gal OR 2 teaspoons per gal	—
Azalea scale Magnolia scale Pine leaf scale	2 pts per 100 gal OR 2 teaspoons per gal	—
Fletcher scale	2 pts per 100 gal OR 2 teaspoons per gal	Apply when scale crawlers have settled on foliage.
Florida red scale Juniper scale	2 pts per 100 gal OR 2 teaspoons per gal	Apply when scale crawlers have settled on foliage.
Black scale crawlers Soft brown scale	2¼ pts per 100 gal OR 2 teaspoons per gal	—
Monterey pine scale	2½ pts per 100 gal OR 2 teaspoons per gal	—

PEST	RATE	REMARKS
Pine needle scale	4 pts per 100 gal OR 4 teaspoons per gal	—
Wax scale	4 pts per 100 gal OR 4 teaspoons per gal	Apply in spring when crawlers are active. Repeat 1 or 2 full-coverage applications at 10-day intervals.

FOREST TREES

TREE	PEST	RATE	REMARKS
Deciduous Forest and Shade trees	Tent caterpillar	2 pts/100 gal water	Make thorough coverage application before larvae reach third instar.
Pines	Pine needlesheath miner	1½ pts/acre	Apply by ground or aircraft equipment in 25 gals of water. Make application when air is calm and temperature is below 68°F. Apply when 75% of larvae have left overwintering mines.
Eastern pines	Pine tortoise scale	1½ pts/acre	Apply by aircraft in 2 gals of water per acre when crawlers emerge. Repeat application in 1 week under conditions of heavy infestations. Make application when air is calm and temperature is below 68°F.
Red pine	Red-headed pine sawfly	0.8 pt/acre	Apply by mist blower in 2 gallons of water per acre when egg hatch is complete.

AROUND THE HOME

SITE	PEST	RATE	REMARKS
In and around green-house and gardens	Millipedes Sawbugs Springtails	1 tsp/gal	Apply to 150 square feet of soil surface or where insects congregate. Repeat at 7 to 10 day intervals as needed.
In and around homes	Bedbugs	2 to 4 tablespoons per gallon of deodorized kerosene	Apply lightly to all mattress surfaces in sufficient quantity to "mist" the fabric and generously to beds and woodwork, with special care taken to wet all possible hiding places.
Lawns	Ant mounds	1½ pts/100 gal	Spray ant hills thoroughly so that they are well soaked. For other small ants in flower beds, lawns, around trees, spray lightly in the infested areas. Repeat in 10 to 15 days if ants return.
	Ground pearls	3 to 4 qts/100 gal	Make full coverage to soil surface when ground pearl nymphs are in the pink, "crawler" or active stage and immediately wash into soil with additional water.

ON AND AROUND CULL FRUIT AND VEGETABLE DUMPS

PEST	RATE	REMARKS
Drosophila flies Dried fruit beetles	1½ gal per 100 gal water	Apply as a drench using 8 to 10 gal of spray per 100 sq ft. For best results, dumps should not be over 18 inches deep. DO NOT FEED TREATED FRUIT AND VEGETABLES.

IN AND AROUND WINERIES AND PROCESSING PLANTS

Drosophila flies	3½ oz in 1 qt water	Paint all doors and window screens. AVOID CONTAMINATION OF WINE, FOOD, UTENSILS, EQUIPMENT AND WATER.

PLANTS PROCESSING DRY MILK

Black carpet beetles Trogoderma species	1 pt per 2½ gal	Clean premises thoroughly before applying and maintain good sanitation at all times. Use spray equipment and nozzles that will produce a coarse spray. Application must be made only by an experienced or trained person. Apply as a residual spray to all sections of the plant and warehouses where insects hide or crawl such as cracks, corners, edges of floors, lower parts of walls, floors under storage platforms and underneath and behind protected places. Avoid contamination of milk, dry milk, equipment, utensils, work surfaces, containers and liners. Repeat application as necessary.

CAUTION: CYTHION sprays may damage finished surfaces and fabrics. Avoid contamination of food, utensils, milk, milk equipment, and water. Do not use in milk processing rooms.

CRACK AND CREVICE TREATMENT IN FOOD HANDLING ESTABLISHMENTS

IN FOOD AREAS: Includes areas for receiving, serving, storage, packing (canning, bottling, wrapping, boxing), preparing, edible waste storage and enclosed processing systems (mills, dairies, edible oils, syrups).

PEST		RATE	DIRECTIONS FOR USE
Ants Carpet beetles Cigarette beetles Clover mites Confused flour beetles Crickets Drug store beetles Flat grain beetles Granary weevils	Indian meal moths Red flour beetles Rice weevils Roaches Rusty grain beetles Saw-toothed grain beetles Silverfish Spiders	1 pt/2½ gal OR 1 part in 19 parts mixture of 4 parts kerosene-type solvent and 1 part aromatic hydrocarbon-type solvent.	Apply in small amounts directly into cracks and crevices using a pin stream of insecticide into points such as expansion joints between different elements of construction or between equipment bases and the floor, wall voids, motor housing, junction boxes or switch boxes, conduits or hollow equipment legs where the named insects hide. Care should be taken to avoid depositing the product onto exposed surfaces or introducing the material into the air. Avoid contamination of food or food processing surfaces. Repeat application as necessary. Application must be made only by a commercial pest control operator.

IN NONFOOD AREAS: Includes garbage rooms, lavatories, flood drains (to sewers), entries and vestibules, offices, locker rooms, machine rooms, boiler rooms, garages, mop closets and storage.

PEST		RATE	DIRECTIONS FOR USE
Ants Carpet beetles Cigarette beetles Clover mites Confused flour beetles Crickets Drug store beetles Flat grain beetles Granary weevils	Indian meal moths Red flour beetles Rice weevils Roaches Rusty grain beetles Saw-toothed grain beetles Silverfish Spiders	1 pt/2 1/2 gal OR 1 part in 19 parts mixture of 4 parts kerosene-type solvent and 1 part aromatic hydrocarbon-type solvent.	Apply as a coarse spray to baseboard areas, around water pipes, surfaces behind and beneath sinks, lockers, tables, pallets and similar areas where insects hide or through which they may enter. Repeat application as necessary. Application must be made only by a commercial pest control operator.

CYTHION® RTU

INSECTICIDE

ACTIVE INGREDIENT:
Malathion O,O-dimethyl phosphorodithioate of diethyl mercaptosuccinate. 46.2%
INERT INGREDIENTS: 53.8%
TOTAL 100.0%
(One gallon contains 4.1 pounds of malathion)

APRIL 1988

® Registered Trademark of American Cyanamid Company

EPA Reg. No. 241-282 2/88

KEEP OUT OF REACH OF CHILDREN

CAUTION!

PRECAUCION

PRECAUCION AL USUARIO: Si usted no lee ingles, no use este producto hasta que la etiqueta le haya sido explicada ampliamente.

STATEMENT OF PRACTICAL TREATMENT

IF SWALLOWED: Call a physician or poison controller. Drink 1 to 2 glasses of water and induce vomiting by touching back of throat with finger. Do not induce vomiting or give anything by mouth to an unconscious person.

IF INHALED: Remove victim to fresh air. Give artificial respiration if indicated.

IF ON SKIN: Wash with plenty of soap and water.

IF IN EYES: Flush with plenty of water. Get medical attention if irritation persists.

NOTE TO PHYSICIAN: This product may cause cholinesterase inhibition. Atropine is antidotal. 2-PAM may be effective as an adjunct to atropine.
In case of an emergency endangering life or property involving this product call collect, day or night, Area Code 201-835-3100

PRECAUTIONARY STATEMENTS

HAZARDS TO HUMANS (AND DOMESTIC ANIMALS)
CAUTION!

HARMFUL IF SWALLOWED, INHALED OR ABSORBED THROUGH SKIN

Avoid breathing spray mist.
Avoid contact with skin, eyes, or clothing.
May be a skin sensitizer.
Wash thoroughly with soap and water after handling.
Remove contaminated clothing and wash before reuse.

ENVIRONMENTAL HAZARDS

This product is toxic to fish and aquatic invertebrates. Do not apply directly to water and wetlands (i.e., swamps, bogs, marshes, and potholes). Drift and runoff may be hazardous to aquatic organisms in areas near the application site. Do not contaminate water when disposing of equipment washwaters.

This product is highly toxic to bees exposed to direct treatment on blooming crops or weeds. Do not apply this product or allow it to drift to blooming crops or weeds if bees are visiting the treatment area.

DIRECTIONS FOR USE

It is a violation of Federal law to use this product in a manner inconsistent with its labeling.

Before using, read the use directions for the proper methods and procedures which must be followed to achieve effective insect control and avoid permanent damage to automobile and other paint finishes.

Do not apply this product in such a manner as to directly or through drift expose workers or other persons. The area being treated must be vacated by unprotected persons.

Do not apply this product through any type of irrigation system.

Do not enter treated areas without protective clothing until sprays have dried.

STORAGE AND DISPOSAL

Do not contaminate water, food, or feed by storage or disposal.

Pesticide Disposal: Wastes resulting from the use of this product may be disposed of on site or at an approved waste disposal facility.

Container Disposal: Triple rinse (or equivalent). Then offer for recycling or reconditioning, or puncture and dispose of in a sanitary landfill, or by other procedures approved by state and local authorities.

IMPORTANT—Undiluted spray droplets of CYTHION RTU will permanently damage automobile paint. If accidental exposure does occur, the vehicle should be washed immediately.

Apply CYTHION RTU insecticide uniformly at the volumes and dosages rates below.

Aerial Applications: If water is used for dilution, apply at least one gallon of finished spray per acre. CYTHION RTU can be used in specially designed aircraft or ground equipment capable of applying ultra low volumes for control of cotton insects indicated. When CYTHION RTU is applied ultra low volume, use a sufficient amount of once refined vegetable oil to make at least one quart of finished spray solution per acre.

Conventional Ground Application: Use water for dilution and apply at least 5 gallons of finished spray per acre.

Apply only when weather conditions are favorable. Wind and rising air currents may cause undesirable spray drift and reduce insect control.

Cover foliage uniformly! Time treatments and set dosages according to insect populations or damage. Repeat treatments as necessary to maintain control. Consult your local extension service or professional consultant for treatment thresholds.

Crop	Pests Controlled	Fluid Ounces Per Acre	Days to Harvest or Grazing and Comments
Cotton	Aphids (Green Peach and Cotton)	32-36	
	Boll Weevil	18-36	Early to midseason
		27-36	Late Season
	Grasshoppers	18	0
	Fleahoppers Leafhoppers	9-18	Day
	Lygus Bugs including	18-27	Moderate populations
	Tarnished Plant Bugs	36	Very heavy or migrating populations
	Thrips	9-18	

TANK-MIX INSTRUCTIONS

For a complete cotton spray program, this product can be combined with other registered cotton insecticides. Carefully follow all precautions and directions in the labeling of any product used in combination with CYTHION RTU. Do not use registered products at higher dosages than is recommended on the label. Do not use any insecticide in combination with CYTHION RTU with label prohibitions against such use. The use of CYTHION RTU in combination with products containing carbaryl may cause injury to cotton.

DISCLAIMER

The label instructions for the use of this product reflect the opinion of experts based on field use and tests. The directions are believed to be reliable and should be followed carefully. However, it is impossible to eliminate all risks inherently associated with use of this product. Crop injury, automobile paint damage, ineffectiveness, or other unintended consequences may result because of such factors as weather conditions, presence of other materials, or the use or application of the product contrary to label instructions, all of which are beyond the control of American Cyanamid Company. All such risks shall be assumed by the user. American Cyanamid Company warrants only that the material contained herein conforms to the chemical description on the label and is reasonably fit for the use therein described when used in accordance with the directions for use, subject to the risks referred to above.

Any damages arising from a breach of this warranty shall be limited to direct damages and shall not include consequential commercial damages such as loss of profits or values or any other special or indirect damages.

Amercan Cyanamid Company makes no other express or implied warranty, including any other express or implied warranty of FITNESS or of MERCHANTABILITY.

The sale of this product does not include a license under any patent owned by the Amercan Cyanamid Company.

EVENT™

GRASS GROWTH REGULATOR

For Use on Tall Fescues, Perennial Ryegrasses, Bluegrasses, and Bahiagrasses

ACTIVE INGREDIENTS:

Ammonium salt of imazethapyr [(±)-2-[4,5-dihydro-4-methyl-4-(1-methylethyl-5-oxo-1-*H*-imidazol-2-yl]-5-ethyl-3-pyridinecarboxylic acid]*	17.26%
Ammonium salt of imazapyr [2-[4,5-dihydro-4-methyl-4-(1-methylethyl)-5-oxo- 1-*H*-imidazol-2-yl]-3-pyridinecarboxylic acid]]	0.64%
INERT INGREDIENTS	82.10%
TOTAL	100.00%

*Equivalent to 16.3% imazethapyr and 0.6% imazapyr or 1.46 pounds acid per gallon.

EPA Reg. No. 241-317

®Trademark of American Cyanamid Company

4/89

KEEP OUT OF REACH OF CHILDREN

CAUTION!/¡PRECAUCION!

PRECAUCION AL USUARIO: Si usted no lee inglés, no use este producto hasta que la etiqueta le haya sido explicada ampliamente.

STATEMENT OF PRACTICAL TREATMENT

IF ON SKIN: Wash with plenty of soap and water. Get medical attention if irritation persists.

IF IN EYES: Flush with plenty of water.

In case of emergency endangering life or property involving this product, call collect, day or night, area code 201-835-3100.

PRECAUTIONARY STATEMENTS

HAZARDS TO HUMANS

CAUTION!

Harmful if inhaled or absorbed through skin. Avoid breathing spray mist. Avoid contact with skin, eyes, or clothing. Wash thoroughly with soap and water after handling. Remove contaminated clothing and wash before reuse.

ENVIRONMENTAL HAZARDS

DO NOT apply directly to water or wetlands.

DO NOT contaminate water when disposing of equipment washwaters.

GENERAL INFORMATION

EVENT reduces the foliar growth of established tall fescue, perennial ryegrass, bluegrass, and bahiagrass. EVENT controls seedhead production by tall fescue and controls or suppresses seedhead production by perennial ryegrass, bluegrass, and bahiagrass.

EVENT may be used on limited care-low maintenance areas, such as roadsides, airports, fairgrounds, and golf course roughs, and limited wear areas such as industrial, institutional, and cemetery grounds.

DIRECTIONS FOR USE

It is a violation of Federal law to use this product in a manner inconsistent with its labeling.

Observe all cautions and limitations on this label and on the labels of products used in combination with EVENT. When EVENT is used in tank mixture with another product, refer to each label for rates, methods of application, proper timing, limitations and precautions. **ALWAYS USE** in accordance with the most restrictive label limitations. EVENT cannot be mixed with any product having a label prohibiting against such mixing. Keep containers closed to avoid spills and contamination.

Do not apply this product through any type of irrigation system.

STORAGE AND DISPOSAL

PROHIBITIONS: KEEP FROM FREEZING

DO NOT store below 32°F.

DO NOT contaminate water, food, or feed by storage or disposal.

PESTICIDE DISPOSAL: Wastes resulting from the use of this product may be disposed of on site or at an approved waste disposal facility.

CONTAINER DISPOSAL: Triple rinse (or equivalent). Then offer for recycling or reconditioning, or puncture and dispose of in a sanitary landfill, by incineration or, if allowed by State and local authorities by burning. If burned, stay out of smoke.

DOSAGE RATES AND TIMING

Apply EVENT at a rate of 8 to 10 fluid ounces per acre. Use the higher rate for more esxtended periods of foliar growth suppression. Apply after grasses are at 100 percent greenup and have at least two inches of vertical growth. On finer turfgrass areas where more uniform quality is desired, apply after two mowings. For seedhead suppression, apply prior to the time of seed stalk elongation. EVENT may be applied to mixed grass-legume (e.g., crown vetch) stands for legume release.

ADDITION OF SURFACTANTS

THE ADDITION OF A SURFACTANT TO EVENT SPRAYS IS REQUIRED. Add a nonionic surfactant containing at least 80 percent active ingredient at 0.25 percent v/v of the spray (2 pints per 100 gallons of spray mixture). An antifoam agent may be added to the tank, if needed.

SPRAYING INSTRUCTIONS

Uniformly apply EVENT sprays with a properly calibrated sprayer in sufficient volume to insure adequate coverage. **AVOID DRIFT** onto the foliage of desirable nontarget plants or injury may result. **AVOID** overlaps when spraying. A spray indicator dye is useful for preventing overlaps. **DO NOT** apply when wind velocity is greater than 10 mph.

HERBICIDE COMBINATIONS

EVENT may be used in combination with EMBARK* or TELAR** for broad-spectrum grass seedhead suppression. To use, mix 2 to 4 ounces of EVENT with either 8 ounces of EMBARK or 1/8 to 1/4 ounce of TELAR per acre.

Where control of broadleaf weeds is desirable, EVENT may be tank-mixed with most broadleaf herbicides currently registered for use in turf. DO NOT tank-mix with sulfonylurea (EXCEPT TELAR) or imidazolinone herbicides. EVENT may be tank-mixed with methylarsonate herbicides such as MSMA or DSMA. DO NOT tank-mix with other postemergence grass herbicides labeled for use in turfgrasses. EVENT may be used prior to or sequential to all currently registered herbicides EXCEPT there must be a three month interval of grass growth between applications of EVENT and sulfonylurea and imidazolinone herbicide applications.

*Registered Trademark of P.B.I. Gordon

**Registered Trademark of DuPont

MIXING INSTRUCTIONS

Fill the spray tank one-half to three-quarters full with clean water. While agitating, add the required measured amount of EVENT. When thoroughly mixed, add the required measured amount of nonionic surfactant. Fill the tank to required volume and add an antifoam agent, if needed.

When tank mixing EVENT with recommended herbicides, add the herbicide first and make sure it is thoroughly mixed. Then add EVENT and proceed as above.

SPECIAL PRECAUTIONS

USE ONLY on well-established non-stressed grass stands.

DO NOT use on newly established stands less than 1-year old.

DO NOT use on highly managed turfgrasses.

DO NOT reseed before three months following an application of EVENT.

DO NOT use on intense wear areas such as athletic fields, playgrounds, or footpaths.

DO NOT use on ornamental plants as injury may result.

DO NOT graze or feed clippings of treated grasses to livestock.

DO NOT apply aerially.

IMPORTANT: Thoroughly clean application equipment after use. Flush tank, pump, hoses, and boom with several changes of water after removing nozzle tips and screens (clean these parts separately).

DISCLAIMER

The label instructions for the use of this product reflect the opinion of experts based on research and field use. The directions are believed to be reliable and should be followed carefully. However, it is impossible to eliminate all risks inherently associated with use of this product. Turf injury, ineffectiveness or other unintended consequences may result because of such factors as weather conditions, presence of other materials, or the use of, or application of the product contrary to label instructions, all of which are beyond the control of American Cyanamid Company. All such risks shall be assumed by the user.

American Cyanamid Company warrants only that the material contained herein conforms to the chemical description on the label and is reasonably fit for the use therein described when used in accordance with the directions for use, subject to the risks referred to above.

Any damages arising from breach of this warranty shall be limited to direct damages and shall not include consequential commercial damages such as loss of profits or values or any other special or indirect damages.

American Cyanamid Company makes no other express or implied warranty, including other express or implied warranty of FITNESS or of MERCHANTABILITY.

IMAGE®

HERBICIDE 1.5 LC

A Selective Herbicide for Use in Established:
Bermudagrass St. Augustinegrass
Centipedegrass Zoysiagrass

ACTIVE INGREDIENT:

Ammonium salt of imazaquin (2-[4,5-dihydro-4-methyl-4-(1-methylethyl-5-oxo-1 *H*-imidazol-2-yl] -3-quinolinecarboxylic acid)*	17.3%
INERT INGREDIENTS	82.7%
TOTAL	100.0%

*Equivalent to 16.4% 2-[4,5-dihydro-4-methyl-4-(1-methylethyl-5-oxo-1*H*-imidazol-2-yl]-3-quinolinecarboxylic acid.

(1 gallon contains 1.5 pounds of active ingredient as the free acid)

EPA Reg. No. 241-289

®Registered Trademark of American Cyanamid Company 4/89

In case of emergency endangering life or property involving this product, call collect, day or night, Area Code 201-835-3100.

KEEP OUT OF REACH OF CHILDREN

CAUTION!

PRECAUCION!

PRECAUCION AL USUARIO: Si usted no lee inglés, no use este producto hasta que la etiqueta le haya sido explicada ampliamente.

See Next Page For Additional Precautionary Statements.

STATEMENT OF PRACTICAL TREATMENT

If on skin: Wash with plenty of soap and water. Get medical attention.

If in eyes: Flush eyes with plenty of water.

PRECAUTIONARY STATEMENTS

HAZARDS TO HUMANS AND DOMESTIC ANIMALS

CAUTION!

Harmful if inhaled or absorbed through skin. Avoid breathing spray mist. Avoid contact with skin, eyes or clothing. Wash thoroughly with soap and water after handling. Remove contaminated clothing and wash before reuse.

ENVIRONMENTAL HAZARDS

DO NOT apply directly to water or wetlands (swamps, bogs, marshes and potholes).

DO NOT contaminate water when disposing of equipment washwaters.

DIRECTIONS FOR USE

It is a violation of Federal law to use this product in a manner inconsistent with its labeling.

Do not apply this product through any irrigation system.

Observe all cautions and limitaions on this label and on the labels of products used in combination with IMAGE. The use of IMAGE not consistent with this label may result in injury to turf. Keep container closed to avoid spills and contamination.

STORAGE AND DISPOSAL

PROHIBITIONS:
KEEP FROM FREEZING
DO NOT store below 32°F.
DO NOT contaminate water, food, or feed by storage or disposal.

PESTICIDE DISPOSAL: Wastes resulting from the use of this product may be disposed of on site or at an approved waste disposal facility.

CONTAINER DISPOSAL: Triple rinse (or equivalent). Then offer for recycyling or reconditioning, or puncture and dispose of in a sanitary landfill, or by incineration, or if allowed by state and local authorities, by burning. If burned, stay out of smoke.

GENERAL INFORMATION

IMAGE is a unique versatile herbicide that will control or aid in the control of several important weeds in warm-season turfgrasses. IMAGE may be used at weed control rates of application on established Bermudagrasses, Centipedegrasses, St. Augustinegrasses and Zoysiagrasses. IMAGE will effectively control susceptible weeds when applied to the foliage of actively growing weeds (postemergence treatment).

IMAGE-treated turfgrasses may have a compacted growth habit and formation of seed heads may be inhibited; vigor is not impaired.

HERBICIDE COMBINATIONS

IMAGE may be tank mixed with registered preemergence and broadleaf postemergence herbicides and with methylarsonate herbicides, such as MSMA, to control weeds not controlled by IMAGE alone. DO NOT tank mix IMAGE with other postemergence herbicides labeled for grass weed control in turf. IMAGE may be applied prior to or sequential to the use of all currently labeled preemergence and postemergence herbicides.

When IMAGE is used in tank mixture with another herbicide refer to each label for rates, methods of application, proper timing, weeds controlled, limitations and precautions. ALWAYS USE in accordance with the more restrictive label limitations and precautions. IMAGE cannot be mixed with any product containing a label prohibition against such mixing.

MIXING INSTRUCTIONS

Fill the spray tank one-half to three-quarters full with clean water. While agitating add the required amount of IMAGE and then fill the remainder of the tank with water. Maintain agitation while spraying to ensure a uniform spray mixture.

When tank mixing IMAGE with recommended herbicides, add the other herbicide to the spray tank first and make sure it is thoroughly mixed before adding IMAGE.

ADDITION OF SURFACTANTS

For postemergence application of IMAGE, the addition of a nonionic surfactant, at 0.25% v/v (2 pints per 100 gallons of spray mixture) of the spray volume is required. Add the nonionic surfactant to the spray mixture after herbicide(s) have mixed. An antifoaming agent may be added to the tank, if needed.

SPRAYING INSTRUCTIONS

Uniformly apply IMAGE sprays with a properly calibrated sprayer in sufficient volume to insure adequate coverage (40-200 gallons/acre). AVOID drift onto vegetables, flowers, ornamental shrubs and other desirable plants or injury may result. AVOID overlaps when spraying. Add a spray indicator dye to prevent overlaps. DO NOT apply when wind velocity is greater than 10 mph.

SPECIAL PRECAUTIONS

USE ONLY on well established non-stressed turfgrass with a dense and uniform stand. TEMPORARY YELLOWING may occur.

DO NOT use on golf or bowling greens.

DO NOT use on unlabeled turfgrass species, including dichondra.

DO NOT mow St. Augustine grasses until 48 hours after application.

DO NOT use on tall fescue or mixed stands of tall fescue and Bermudagrass.

DO NOT reseed or winter overseed or sprig turfgrass for 1½ months after treatment.

DO NOT apply to newly planted or sprigged lawns.

DO NOT use on winter overseeded turfgrass. IMAGE may injure Ryegrass, Bentgrass or *Poa trivialis.*

DO NOT use on ornamentals as injury may result.

DO NOT graze or feed clippings of treated turfgrasses.

DO NOT apply aerially.

IMAGE USE RATES AND WEEDS CONTROLLED

Rates: Use 0.5 to 1.0 fluid ounce per 1000 square feet of turf. One gallon treats 3-6 acres. The high rate should be used where heavy weed infestations occur. IMAGE can be applied at any time during the year except at turfgrass transition from winter dormancy. (See Note 4 below.)

		Preemergence*	Postemergence
Bluegrass, annual	*Poa annua*	X	XE
Burweed, lawn	*Soliva pterosperma*	X	XE
Chickweed, common	*Stellaria media*	X	XE
Chickweed, mouseear	*Cerastium vulgatum*	X	XE
Dollarweed (Pennywort, water)	*Hydrocotyle umbellata*		XM
Flatsedge, rice	*Cyperus iria*	X	X
Garlic, wild	*Allium vineale*		X
Henbit	*Lamium amplexicaule*	X	XE
Kyllinga, green	*Kyllinga brevifolia*		X
Nutsedge, purple	*Cyperus rotundus*		X
Nutsedge, yellow	*Cyperus esculentus*		X**
Onion, wild	*Allium canadense*		X
Parsley-piert	*Alchemilla arvensis*	X	XE
Ryegrass, perennial (overseeded)	*Lolium perenne*	X	X
Sandbur, field	*Cenchrus incertus*	X	X
Sedge, globe	*Cyperus globosus*	X	X
Torpedograss	*Panicum repens*		X***

X* Application timed just before germination.

X* Application timed just before germination.
X** Use the high rate of application and/or tank mixes with a methyl arsonate herbicide.
X*** Use the 0.75 or 1.0 fl. oz. rate in combination with MSMA at 2.0 lb. ai/A.
XE Application timed early postemergence.
XM IMAGE aids in the control of this hard-to-control weed. Use only the 0.75 or 1.0 fluid ounce rates.

Notes:
1. A repeat application may be required for preferred postemergence control of weeds such as dollarweed, kyllinga, the nutsedges and torpedograss.
2. Addition of a methylarsonate herbicide, such as MSMA, at labeled rates will result in more rapid and improved sedge control in methylarsonate tolerant turfgrasses.
3. Bahiagrass, *Paspalum notatum*, is injured by IMAGE applications and the height and numbers of seedheads are suppressed.
4. Tolerant turfgrasses detoxify IMAGE. Best tolerance occurs under optimal growth conditions. When the turfgrasses are growing under marginal conditions, such as low temperatures, drought stress, compaction, etc., temporary yellowing or browning can occur and recovery could take 2-4 weeks.

DISCLAIMER

The label instructions for the use of this product reflect the opinion of experts based on field use and tests. The directions are believed to be reliable and should be followed carefully. However, it is impossible to eliminate all risks inherently assiciated with use of this product. Turf injury, ineffectiveness or other unintended consequences may result because of such factors as weather conditions, presence of other materials, or the use or application of the product contrary to label instructions, all of which are beyond the control of American Cyanamid Company. All such risks shall be assumed by the user.

American Cyanamid Company warrants only that the material contained herein conforms to the chemical description on the label and is reasonably fit for the use therein described when used in accordance with the directions for use, subject to the risks referred to above.

Any damages arising from breach of this warranty shall be limited to direct damages and shall not include consequential commercial damages such as loss of profits or values or any other special or indirect damages.

American Cyanamid Company makes no other express or implied warranty, including other express or implied warranty of FITNESS or of MERCHANTABILITY.

PROWL®

HERBICIDE

FOR CONTROL OF MOST ANNUAL GRASSES AND CERTAIN BROADLEAF WEEDS

FOR USE IN COTTON, EDIBLE BEANS, FIELD CORN, FORAGE LEGUMES, GARLIC, GRAIN SORGHUM, NONBEARING FRUIT, NUT CROPS, AND VINEYARDS; PEANUTS, POTATOES, RICE, SOYBEANS, SUNFLOWERS, SWEET CORN, SWEET LUPINES, AND TOBACCO

ACTIVE INGREDIENT:	
pendimethalin (N-(1-ethylpropyl)-3,4-dimethyl-2,6-dinitrobenzenamine)	42.3%
INERT INGREDIENTS:	57.7%
TOTAL	100.0%

(1 gallon contains 4 lbs. of pendimethalin)

EPA Reg. No. 241-243-ZA

® Registered Trademark of American Cyanamid Company.

8/89

KEEP OUT OF REACH OF CHILDREN

WARNING!/¡AVISO!

PRECAUCION AL USUARIO: Si usted no lee inglés, no use este producto hasta que la etiqueta le haya sido explicada ampliamente.

STATEMENT OF PRACTICAL TREATMENT

If in eyes, hold eyelids open and flush with a steady gentle stream of water for 15 minutes.

If swallowed, DO NOT induce vomiting. Call a physician or Poison Control Center immediately.

For skin contact, wash thoroughly with soap and water. Call a physician if eye or skin irritation persists.

This product contains monochlorobenzene.

NOTE TO PHYSICIAN: Because of increased risk of chemical pneumonia or pulmonary edema caused by aspiration of the hydrocarbon solvent (monochlorobenzene), vomiting should be induced only under professional supervision.

PRECAUTIONARY STATEMENTS

HAZARDS TO HUMANS

WARNING!

CAUSES SUBSTANTIAL BUT TEMPORARY EYE INJURY

HARMFUL IF SWALLOWED, ABSORBED THROUGH THE SKIN, OR INHALED.

DO NOT get in eyes or on clothing. Wear goggles or face shield. Avoid breathing spray mist. Avoid contact with skin. Wash skin thoroughly with soap and water after handling. Use with adequate ventilation. Keep container closed. Remove contaminated clothing and wash before reuse.

ENVIRONMENTAL HAZARDS

This product is toxic to fish. **DO NOT** apply directly to water or wetlands (swamps, bogs, marshes, potholes). Drift and runoff from treated areas may be hazardous to aquatic organisms in adjacent sites. **DO NOT** contaminate water when disposing of equipment washwaters.

PHYSICAL HAZARDS

FLAMMABLE LIQUID AND VAPOR

DO NOT use, pour, spill, or store near heat or open flame. DO NOT use in the vicinity of pilot lights.

DISCLAIMER

The label instructions for the use of this product reflect the opinion of experts based on field use and tests. The directions are believed to be reliable and should be followed carefully. However, it is impossible to eliminate all risks inherently associated with use of this product. Crop injury, ineffectiveness or other unintended consequences may result because of such factors as weather conditions, presence of other materials, or the use or application of the product contrary to label instructions, all of which are beyond the control of American Cyanamid Company. All such risks shall be assumed by the user.

American Cyanamid Company warrants only that the material contained herein conforms to the chemical description on the label and is reasonably fit for the use therein described when used in accordance with the directions for use, subject to the risks referred to above.

Any damages arising from a breach of this warranty shall be limited to direct damages, and shall not include consequential commercial damages such as loss of profits or values or any other special or indirect damages.

American Cyanamid Company makes no other express or implied warranty, including any other express or implied warranty of FITNESS or MERCHANTABILITY.

DIRECTIONS FOR USE

It is a violation of Federal law to use this product in a manner inconsistent with its labeling.

This labeling must be in the possession of the user at the time of pesticide application.

Observe all cautions and limitations in this booklet and the labels of products used in combination with PROWL. The use of PROWL not consistent with this label can result in injury to crops, animals, or persons. Keep containers closed to avoid spills and contamination.

GENERAL WORKER PROTECTION STATEMENTS

Do not apply this product in such a manner as to directly or through drift expose workers or other persons. The area being treated must be vacated by unprotected persons.

Do not enter treated areas without protective clothing until sprays have dried. Because certain states may require more restrictive re-entry intervals for various crops treated with this product, consult your State Department of Agriculture for further information.

Written or oral warnings must be given to workers who are expected to be in a treated area or in an area about to be treated with this product. When oral warnings are given, warnings shall be given in a language customarily understood by workers. Oral warnings must be given if there is reason to believe that written warnings cannot be understood by workers. Written warnings must include the following information: WARNING. Area Treated with PROWL on (date of application). Do not enter without appropriate protective clothing.

STORAGE AND DISPOSAL

STORAGE:

DO NOT STORE BELOW 40°F. Extended storage at temperatures below 40°F can result in the formation of crystals on the bottom of the container. If crystallization does occur, store the container on its side at room temperature (70°F) and rock occasionally until crystals re-dissolve.

DO NOT contaminate water, food, or feed by storage or disposal.

DISPOSAL:

Pesticide wastes are toxic. Improper disposal of excess pesticide, spray mixture, or rinsate is a violation of Federal law. If these wastes cannot be disposed of by use according to label instructions, contact your State Pesticide or Environmental Control Agency, or the Hazardous Waste representative at the nearest EPA Regional Office for guidance.

CONTAINER DISPOSAL:
Triple rinse (or equivalent). Then offer for recycling or reconditioning, or puncture and dispose of in a sanitary landfill, or incineration, or, if allowed by state and local authorities, by burning. If burned, stay out of smoke.

GENERAL INFORMATION

Special Precautions
PROWL herbicide controls most annual grasses and certain broadleaf weeds as they germinate, but it will not control established weeds. Destroy existing weeds before applying PROWL (except as recommended in specific postemergence combination treatments). Unusually cold, excessively wet, or hot and dry conditions that delay germination or extend germination over a long period of time can reduce weed control.

Applied according to label directions and under normal growing conditions, PROWL or PROWL tank-mix combinations will not cause crop injury. Over-application can result in crop stand loss, crop injury, or soil residues. Uneven application or improper soil incorporation can decrease weed control or cause crop injury. Soil incorporation deeper than recommended can reduce weed control.

Seedling diseases, cold weather, excessive moisture, deep planting, low or high soil pH, high soil salt concentration, or drought can weaken seedlings and plants and increase the possibility of crop damage from PROWL. Under these conditions, crop yields can be reduced.

MIXING INSTRUCTIONS

1. Fill tank one-half to three-quarters full with clean water or liquid fertilizer and agitate. Prior to mixing PROWL or PROWL tank mixtures in liquid fertilizer, refer to appropriate label sections (see table of contents for page number) for recommended uses in liquid fertilizer, application instructions, and compatibility determinations.
2. PROWL Alone
   When using PROWL alone, add PROWL to the partially-filled tank while agitating and then fill the remainder of the tank with water or liquid fertilizer.
3. PROWL Tank Mixtures
   Add the tank mixture ingredients in the order listed below prior to adding PROWL. (For tank mixtures with Gramoxone Super or Roundup see mixing instructions in PROWL tank mixtures in minimum and no-till systems subsection of **Soybeans section** [see table of contents for page number]):

(a) Wettable powder formulations
Make a slurry of the wettable powder in water (1 part WP + 2 parts water). Add the slurry slowly into the partially-filled tank while agitating. If tank mixing more than one wettable powder formulation, add the first one while agitating. Continue agitation and add second wettable powder only after complete mixing of the first wettable powder has occurred.

(b) Flowable formulations
Premix one part of the flowable and one part water and add this diluted mixture into the partially-filled tank while agitating.

(c) Water soluble concentrate formulations
Add the water soluble concentrate to the partially-filled tank while agitating.

(d) Dry flowable/water dispersible granule formulations
Add the granules to the partially-filled tank while agitating.

(e) Emulsifiable concentrate formulations
Add the emulsifiable concentrate to the partially-filled tank while agitating.

After complete mixing, add PROWL to the tank.

Fill the remainder of the tank with water or liquid fertilizer while agitating.

4. MAINTAIN CONTINUOUS AGITATION WHILE ADDING HERBICIDES AND UNTIL SPRAYING IS COMPLETED. If the spray mixture is allowed to settle for any period of time, thorough agitation is essential to resuspend the mixture before spraying is resumed. Continue agitation while spraying.

FORMULATION CONVERSION TABLES

The rate tables in this label for PROWL tank mixtures with atrazine, Bladex, prometryn, SCEPTER, Sencor/Lexone and Lorox list rates for one formulation of the respective herbicides. Use the following conversion tables to determine rates when using other formulations.

Atrazine, Bladex, prometryn Conversion Tables

Rate of Bladex Atrazine or prometryn 80W	Rate of Bladex Atrazine or prometryn 4L	Rate of Bladex or Atrazine 90DF
1.25 lbs.	2.0 pts.	1.11 lbs.
1.50 lbs.	2.4 pts.	1.33 lbs.
2.00 lbs.	3.2 pts.	1.78 lbs.
2.50 lbs.	4.0 pts.	2.22 lbs.
3.00 lbs.	4.8 pts.	2.67 lbs.

Sencor/Lexone Conversion Table

Rate of Sencor/Lexone DF	Rate of Sencor/Lexone 50WP	Rate of Sencor 4F/Lexone 4L
0.33 lb.	0.50 lb.	0.50 pt.
0.50 lb.	0.75 lb.	0.75 pt.
0.67 lb.	1.00 lb.	1.00 pt.

Lorox Conversion Table

Rate of Lorox L	Rate of Lorox 50W
1.0 pt.	1.0 lb.
1.5 pts.	1.5 lbs.
2.0 pts.	2.0 lbs.
2.5 pts.	2.5 lbs.

GENERAL APPLICATION INSTRUCTIONS

Uniformly apply recommended PROWL or PROWL tank mixture treatments in 10 or more gallons of water or 20 or more gallons of liquid fertilizer per acre with ground equipment. Refer to **Applications with Liquid Fertilizers** section (see table of contents for page number) for liquid fertilizer recommended uses, application instructions and compatibility determinations. Apply in 5 or more gallons of water per acre with aircraft. Refer to **Spraying Instructions** section (see table of contents for page number) for spraying directions. Refer to **Preplant Incorporated Applications With Dry Bulk Fertilizers** section (see table of contents for page number) for PROWL/dry bulk fertilizer applications.

CULTURAL PRACTICES FOLLOWING APPLICATION

Should weeds develop, a shallow cultivation or rotary hoeing will generally result in better weed control. A PROWL treatment may be followed by any registered herbicide to control weeds not listed on the PROWL label.

RATES

Follow all recommendations in specific crop sections of this label regarding rates per acre of PROWL and herbicides used in combination with PROWL. When a rate range is recommended, unless other directions are given in a crop section, the high rate of the range should be used if heavy weed populations are anticipated or if extensive crop residues were present prior to seedbed preparation.

SOIL TEXTURES

The rate tables for recommended PROWL treatments for each crop refer to coarse, medium, and fine soils. Soil type should be determined since rates for PROWL treatments will change with soil texture. The following table lists soil types for coarse, medium and fine soils:

COARSE	MEDIUM	FINE
sands	sandy clay loams*	silty clay loams*
loamy sands	sandy clays	silty clays
sandy loams	loams	clay loams
	silt loams	clays
	silts	

* Sometimes considered transitional soils and may be classified as either medium or fine textured soils.

DO NOT use PROWL on peat or muck soils.

CHEMIGATION

PROWL may be applied through sprinkler irrigation systems in cotton, field corn, grain sorghum, peanuts, potatoes, soybeans, and sunflowers. Follow all label recommendations for these crops regarding rates per acre, timing of application, special instructions, and precautions.

Apply this product only through a sprinkler including center pivot, lateral move, end tow, side (wheel) roll, traveler, big gun, solid set, or hand move irrigation system. Do not apply this product through any other type of irrigation system.

Crop injury, lack of effectiveness, or illegal pesticide residues in the crop can result from nonuniform distribution of treated water.

The system must be properly calibrated (with water only) to ensure that the amount of PROWL applied corresponds to the recommended rate.

If you have questions about calibration, you should contact State Extension Service specialists, equipment manufacturers, or other experts.

Do not connect an irrigation system (including greenhouse systems) used for pesticide application to a public water system unless the pesticide label prescribed safety devices for public water systems are in place.

A person knowledgeable of the chemigation system and responsible for its operation, or under the supervision of the responsible person, shall shut the system down and make necessary adjustments should the need arise.

The system must be free of leaks and clogged nozzles.

The pesticide must be supplied continuously for the duration of the aqueous application. An uneven application may cause injury to the crop or poor weed control.

Agitation must be maintained in the nurse tank.

The sprinkler-chemigation system must contain a functional check valve, vacuum relief valve, and low pressure drain appropriately located on the irrigation pipeline to prevent water source contamination from backflow.

The pesticide injection pipeline must also contain a functional, normally closed, solenoid-operated valve located on the intake side of the injection pump and connected to the system interlock to prevent fluid from being withdrawn from the supply tank when the irrigation system is either automatically or manually shut down.

The system must contain functional interlocking controls to automatically shut off the pesticide injection pump when the water pump motor stops, or in cases where there is no water pump, when the water pressure decreases to the point where pesticide distribution is adversely affected.

The irrigation line or water pump must include a functional pressure switch which will stop the water pump motor when the water pressure decreases to the point where pesticide distribution is adversely affected.

Systems must use a metering pump such as a positive displacement injection pump (e.g., diaphragm pump) effectively designed and constructed of materials that are compatible with pesticides and capable of being fitted with a system interlock.

Do not apply when wind speed favors drift beyond the area intended for treatment.

Apply PROWL in 1/2 to 3/4 inches of water during the first sprinkler set. When application is complete, flush the system with water.

Chemigation Systems Connected to Public Water Systems

Public water system means a system for the provision to the public of piped water for human consumption if such system has at least 15 service connections or regularly serves an average of at least 25 individuals daily at least 60 days out of the year.

Chemigation systems connected to public water systems must contain a functional, reduced-pressure zone, backflow preventer (RPZ) or the functional equivalent in the water supply line upstream from the point of pesticide introduction. As an option to the RPZ, the water from the public water system should be discharged into a reservoir tank prior to pesticide introduction. There shall be a complete physical break (air gap) between the outlet end of the fill pipe and the top or overflow rim of the reservoir tank of at least twice the inside diameter of the fill pipe.

The pesticide injection must contain a functional, automatic, quick-closing check valve to prevent the flow of fluid back toward the injection pump.

Do not connect an irrigation system (including greenhouse systems) used for pesticide application to a public water system unless the pesticide label-prescribed safety devices for public water systems are in place.

A person knowledgeable of the chemigation system and responsible for its operation, or under the supervision of the responsible person, shall shut the system down and make necessary adjustments should the need arise.

WEED SPECIES CONTROLLED

The following grass and broadleaf weeds are controlled by treatments of PROWL at the rates recommended for soil texture in the respective crop section:

GRASSES CONTROLLED

Barnyardgrass *(Echinochloa crus-galli)*
Crabgrass *(Digitaria* spp.)
Crowfootgrass *(Dactyloctenium aegyptium)*
Fall panicum *(Panicum dichotomiflorum)*
Field sandbur *(Cenchrus incertus)*
Foxtail (giant) *(Setaria faberi)*
Foxtail (green) *(Setaria viridis)*
Foxtail (yellow) *(Setaria glauca)*
Goosegrass *(Eleusine indica)*
Johnsongrass (from seed) *(Sorghum halepense)*
Signalgrass *(Brachiara platyphylla)*
Texas Panicum *(Panicum texanum)*
Witchgrass *(Panicum capillare)*

BROADLEAF WEEDS CONTROLLED

Carpetweed *(Mollugo verticillata)*
Florida pusley *(Richardia scabra)*
Kochia *(Kochia scoparia)*
Lambsquarters *(Chenopodium album)*
Pigweed *(Amaranthus* spp.)
Purslane *(Portulaca oleracea)*
Spurge, annual *(Euphorbia* spp.)

PROWL will aid in the control of and reduce competition from:

Pennsylvania smartweed *(Polygonum pensylvanicum)*
Velvetleaf (Buttonweed) *(Abutilon theophrasti)*

PROWL will provide control or aid in the control of and reduce competition from the following hard-to-control weeds (refer to appropriate section for use instructions; see table of contents for page number):

Itchgrass *(Rottboellia exaltata)*, (soybeans)
Red rice *(Oryza sativa)*, (soybeans)
Rhizome johnsongrass *(Sorghum halepense)*, (cotton), (soybeans)
Shattercane *(Sorghum bicolor)*, (soybeans), (field corn, grain sorghum),
Wild proso millet *(Panicum miliaceum)*, (field and sweet corn, grain sorghum)

When applied preemergence in corn, PROWL will control the following weeds in addition to those listed above:

Pennsylvania smartweed *(Polygonum pensylvanicum)*
Velvetleaf (Buttonweed) *(Abutilon theophrasti)*
Woolly cupgrass *(Eriochloa villosa)*

When applied as directed in potatoes, PROWL will control the following weed in addition to those listed above:

Stinging nettle *(Urtica urens)*

SPRAYING INSTRUCTIONS

GROUND APPLICATIONS

Use a properly calibrated low-pressure (20 to 40 psi) sprayer equipped with 8002 or larger size Tee-jet or comparable flat fan nozzles to achieve uniform spray distribution and minimize drift. Keep the bypass line on or near the bottom of the tank to minimize foaming. Nozzle screens must be no finer than 50 mesh. DO NOT apply PROWL during periods of gusty winds or when wind velocity is greater than 10 mph.

BROADCAST TREATMENT—Uniformly apply in 10 or more gallons of water for 20 or more gallons of liquid fertilizer as specified in the appropriate sections of this leaflet.

BAND TREATMENT—Uniformly apply the broadcast equivalent rate and volume per acre. To determine these:

$$\frac{\text{Band width in inches}}{\text{Row width in inches}} \times \text{Broadcast RATE per acre} = \text{Band RATE per acre}$$

$$\frac{\text{Band width in inches}}{\text{Row width in inches}} \times \text{Broadcast VOLUME per acre} = \text{Band VOLUME per acre}$$

AERIAL APPLICATIONS

Uniformly apply in 5 or more gallons of water per acre. To minimize drift, DO NOT apply during periods of gusty winds or winds in excess of 5 mph. It is recommended that a flagman or an automatic mechanical flagging unit on the aircraft be used to avoid overlapping and possible crop injury.

INCORPORATION INSTRUCTIONS

Break up clods before a preplant incorporated application.

1. Flat-planted crop

PROWL is to be incorporated prior to planting and within 7 days of application. When PROWL is applied to flat land that will not be bedded, mechanical incorporation should be into the top 1 or 2 inches of soil. Mechanical incorporation of PROWL into the top 1 or 2 inches of soil can be achieved by the following:

(a) Disk harrow set to cut 3 to 4 inches deep and operated in two different directions at 4 to 6 mph.
(b) PTO-driven equipment (tillers, cultivators, hoes) set to cut 2 inches deep and operated one time at 4 mph or less.
(c) Rolling cultivator set to cut 2 to 3 inches deep and operated two times at 6 to 8 mph. Use only on coarse and medium textured soils.
(d) Field cultivator must have 3 or 4 rows of sweeps, spaced at intervals of 7 inches or less and staggered so that no soil is left unturned. Do not use chisel points. Set implement to cut 3 inches deep and make two passes at more than 5 mph. The second pass should be made at an angle to the first.
(e) LELY-ROTERRA[1] set and operated according to the manufacturer's directions to thoroughly incorporate PROWL to a depth of 1 to 2 inches.
(f) Do-All set and operated according to the manufacturer's directions to thoroughly incorporate PROWL to a depth of 1 to 2 inches.

NOTE: PROWL can be incorporated in a single pass when the soil is of good tilth with moderate moisture and relatively free of clods and trash. Implements must be properly set up to thoroughly incorporate PROWL into the top 1 or 2 inches of soil. Recommended implements include: (1) C-shank or S-Shank field cultivators equipped with flextine drag or rolling basket attachment and (2) the combination disk/field cultivator implement equipped with flextine drag or rolling basket attachment. Field cultivators must have 3 or 4 rows of sweeps, spaced at intervals of 7 inches or less and staggered so that no soil is left unturned. Implements must be operated at 6 to 8 miles per hour and set to cut 3 to 4 inches deep.

[1] LELY-ROTERRA is a trademark of Lely.

2. Bedded crop—Application Over Beds

PROWL is to be incorporated prior to planting and within 7 days of application. When PROWL is applied over beds, mechanical incorporation should be into the top 1 or 2 inches of soil. Mechanical incorporation of PROWL into the top 1 or 2 inches of soil can be achieved by the following:

(a) Rolling cultivator set to cut 2 to 3 inches deep and operated two times at 6 to 8 mph. Use only on coarse and medium textured soils.
(b) PTO-driven equipment (tillers, cultivators, hoes) set to cut 2 inches deep and operated one time at 4 mph or less.
(c) Bed conditioner set to cut 2 to 3 inches deep and operated one time at 4 to 6 mph.

During planting, or if beds must be reshaped prior to planting, avoid tillage that will bring untreated soil to the surface or expose untreated soil in the seedbed or in the furrow.

3. Bedded Crop—Application Prior to Bedding
PROWL is to be incorporated prior to planting and within 7 days of application. When PROWL is applied to flat land prior to bedding, mechanical incorporation should be of sufficient depth so that bedding does not bring up untreated soil. Mechanical incorporation of PROWL can be achieved by the following:

(a) Disk harrow set to cut 4 to 6 inches deep and operated in two different directions at 4 to 6 mph.

(b) PTO-driven equipment (tillers, cultivators, hoes) set to cut 3 to 4 inches deep and operated one time at 4 mph or less. During planting, or if beds must be reshaped prior to planting, avoid tillage that will bring untreated soil to the surface or expose untreated soil on the seedbed or in the furrow.

APPLICATIONS WITH LIQUID FERTILIZERS

GENERAL

PROWL alone and preemergence applications of PROWL plus atrazine, Bladex, Marksman, Bicep, Lariat, Bullet, Dual 8E or Lasso can be applied in liquid fertilizers, as can preplant incorporated or preemergence applications of PROWL plus SCEPTER, PURSUIT, Canopy or Preview. PROWL plus Cotoran can be applied in nitrogen solutions.

Follow all PROWL label recommendations regarding registered crops, rates per acre, timing of application, special instructions and precautions. Apply treatments in 20 or more gallons of liquid fertilizer per acre with ground equipment. DO NOT apply PROWL postemergence in liquid fertilizers.

All individual state regulations relating to fluid fertilizer mixing, registration, labeling and application are the responsibility of the individual and/or company selling the PROWL/liquid fertilizer mixture.

LIQUID FERTILIZER COMPATIBILITY DETERMINATIONS

If liquid fertilizer/herbicide(s) mixture separates in the spray tank, clogged equipment and uneven application can result, which can cause poor weed control and crop injury. Always predetermine the compatibility of PROWL alone or with other herbicides in the specific liquid fertilizer to be used according to the following directions:

1. Add 1 pint of fertilizer to each of 2 one-quart jars.
2. Add ½ teaspoon of adjuvant* to one jar.
3. (a) When using PROWL alone, add to each jar the correct amount of PROWL as specified in the table below.
   (b) When using PROWL tank mixtures, first add the specified quantity of product(s) to be tank mixed with PROWL (see table) and then add the correct amount of PROWL.
4. Close both jars and shake thoroughly for 10 seconds. Let them stand for 30 minutes and then observe the results. Look for signs of separation, an oily layer or globules, sludge, flakes or other precipitates.
5. Determine compatibility.

(a) If the herbicide(s) and liquid fertilizer mixture without adjuvant does not separate, use this mixture in your spray tank.

(b) If the mixture with adjuvant does not separate, but the one without adjuvant separates, use the adjuvant mixture in your spray tank. Add the adjuvant to the liquid fertilizer as directed on the manufacturer's label.

(c) If either mixture separates, but mixes readily with shaking, the mixture can be used providing good agitation is maintained in the spray tank.

(d) If separation of the mixture occurs and agitation and/or adjuvant does not correct this problem, DO NOT use PROWL with that specific liquid fertilizer.

*Compex-Kalo Laboratories Incorporated, Kansas City, MO; Sponto 168-D-Witco Chemical Company, Houston, TX; Unite-Hopkins Agricultural Chemical Company, Madison, WI; or other comparable adjuvants.

Teaspoons of Specified Herbicide to be Added to 1 Pint of Liquid Fertilizer Solution**

Gallons of Liquid Fertilizer to be Applied per Acre	PROWL	Bladex 4L or Atrazine 80W	AAtrex Nine-O or Atrazine 4L or Marksman or Dual 8E	Bullet or Lariat or Bladex 80W	Bladex 90DF or Cotoran or Bicep 6L or Lasso	Canopy or Preview	Scepter	Pursuit
20	2½	3	2	4½	4	1	½	¼
30	1¾	2	1½	3	2¾	¾	⅓	⅙
40	1¼	1½	1	2¼	2	½	¼	⅛

** Based on highest per acre rate used in the leaflet for each product.

PREPLANT INCORPORATED APPLICATIONS WITH DRY BULK FERTILIZERS

PROWL may be impregnated on dry bulk fertilizers. When applied as directed, PROWL/dry bulk fertilizer mixtures provide weed control equal to that provided by the same rates of PROWL applied in water.

Follow all PROWL label recommendations regarding rates per acre, registered crops, incorporation, special instructions and precautions. Apply PROWL/dry bulk fertilizer mixtures only for preplant incorporated uses with registered crops. Apply PROWL/dry bulk fertilizer mixtures only with ground equipment.

All individual state regulations relating to dry bulk fertilizer blending, registration, labeling, and application are the responsibility of the individual and/or company selling the PROWL/dry bulk fertilizer mixture.

A minimum of 200 pounds and a maximum of 450 pounds of dry bulk fertilizer impregnated with the recommended amount of PROWL must be applied per acre.

DO NOT impregnate PROWL onto coated ammonium nitrate or limestone because these materials will not absorb the herbicide. Dry fertilizer blends containing mixtures of ammonium nitrate or limestone may be impregnated with PROWL. A minimum of 200 pounds of impregnated dry bulk fertilizer, excluding the weight of ammonium nitrate or limestone, must be applied per acre.

Refer to appropriate crop section in PROWL booklet to determine the rate of PROWL per acre. Use the following table to determine the amount of PROWL to be impregnated on a ton of dry bulk fertilizer based on the rate of fertilizer which will be applied per acre.

Rate Chart for Impregnation of Dry Bulk Fertilizers with PROWL (Pints of PROWL per Ton of Fertilizer)

Fertilizer Rate lbs/acres	PROWL Rate per Acre (recommended rate for crop and soil texture)				
	1 pint	1½ pints	2 pints	2½ pints	3 pints
200	10 pts/ton	15 pts/ton	20 pts/ton	25 pts/ton	30 pts/ton
250	8 pts/ton	12 pts/ton	16 pts/ton	20 pts/ton	24 pts/ton
300	6⅔ pts/ton	10 pts/ton	13⅓ pts/ton	16⅔ pts/ton	20 pts/ton
350	5¾ pts/ton	8½ pts/ton	11½ pts/ton	14¼ pts/ton	17 pts/ton
400	5 pts/ton	7½ pts/ton	10 pts/ton	12½ pts/ton	15 pts/ton
450	4½ pts/ton	6⅔ pts/ton	9 pts/ton	11 pts/ton	13⅓ pts/ton

For those rates not listed in this table, calculate the pints of PROWL to be impregnated on a ton of dry bulk fertilizer using the following formula:

$$\frac{2000}{\text{Pounds of dry fertilizer per acre}} \times \text{Pints of PROWL per acre (recommended rate for crop and soil texture)} = \text{pints of PROWL per ton of fertilizer}$$

To impregnate PROWL on bulk fertilizer, use a closed rotary-drum mixer or other commonly used dry bulk fertilizer blender equipped with suitable spray equipment. Spray nozzles must be placed to provide uniform coverage of PROWL onto the fertilizer during mixing.

Apply the PROWL/dry bulk fertilizer mixture with an accurately calibrated dry fertilizer spreader. The PROWL/dry bulk fertilizer mixture must be spread uniformly on the soil surface. Uneven spreading can cause poor weed control and crop injury.

Refer to **Incorporation Instructions** section (see table of contents for page number) for incorporation directions.

FOLLOW CROP RESTRICTIONS

1. Land treated with PROWL may be planted to other crops the following year. See restrictions below for sugar beets, red beets, and spinach.

2. Winter wheat, winter barley
Winter wheat and winter barley may be planted in the fall 120 days after a PROWL application in any registered crop. Winter wheat and winter barley may be planted in the fall 90 days after a PROWL CULTI-SPRAY™ application in irrigated field corn or grain sorghum. The treated crop must be grown to maturity and harvested before planting wheat or barley.

In areas where irrigation is necessary to produce the crop treated with PROWL, DO NOT plant winter wheat or winter barley as follow crops if crop failure/destruction occurs and land is fallowed during the summer as crop injury may result.

DO NOT plant winter wheat or winter barley as follow crops in treated land if the follow crop is planted using no-tillage procedures as crop injury may result.

DO NOT plant winter wheat or winter barley as follow crops in treated land if higher PROWL rates were used for rhizome johnsongrass or red rice control in soybeans or itchgrass suppression, or for weed control in no-till cotton or rhizome johnsongrass control in cotton.

DO NOT feed forage or graze livestock for 75 days after planting wheat or barley in treated land.

3. Sugar beets, red beets, spinach
To avoid crop injury, do not plant sugar beets, red beets or spinach for 12 months following a PROWL application. Land should be plowed to a depth of 12 inches prior to planting these crops.

4. When PROWL is used in tank-mix or sequential combinations, refer to labels of other herbicides for additional follow crop restrictions.

™Trademark of American Cyanamid Company

CORN
(Field, Sweet)

PROWL may be applied in field corn as follows:

Preemergence (surface treatment after planting)	Early Postemergence
PROWL + atrazine	PROWL + atrazine
PROWL + Bladex	PROWL + Bladex 80W or 90DF
PROWL + atrazine + Bladex	PROWL + atrazine + Bladex 80W or 90DF
PROWL alone	PROWL + Marksman
PROWL + Banvel	PROWL + Marksman + Bladex 80W
PROWL + Marksman	PROWL + Bicep
PROWL + Marksman + Bladex	PROWL + Lariat or Bullet
PROWL + Bicep	PROWL + Lasso
PROWL + Lariat or Bullet	
PROWL + Dual 8E	
PROWL + Lasso	

DO NOT INCORPORATE

PROWL may be applied in sweet corn (processing varieties ONLY) in Illinois, Minnesota, New York, and Wisconsin ONLY as follows:

Preemergence (surface treatment after planting)	Early Postemergence
PROWL + atrazine	PROWL + atrazine
PROWL + Bladex	PROWL + Bladex 80W or 90DF
PROWL + atrazine + Bladex	PROWL + atrazine + Bladex 80W or 90DF
PROWL alone	PROWL + Bicep
PROWL + Bicep	PROWL + Lariat or Bullet
PROWL + Lariat or Bullet	PROWL + Lasso
PROWL + Dual 8E	
PROWL + Lasso	

DO NOT INCORPORATE

Refer to "Culti-Spray" section (see table of contents for page number) for instructions on CULTI-SPRAY applications (4 inch stage to layby treatment) in field corn.

DO NOT use PROWL on peat or muck soils.

GENERAL INSTRUCTIONS

With the exception of minimum tillage systems (see below) plant into a seed bed that is firm and free of clods and trash. Use only where adequate tillage is practiced to provide good soil coverage of the corn seed. Plant corn AT LEAST 1½ inches deep to ensure good seed coverage.

PROWL or PROWL tank-mix combination treatments are most effective in controlling weeds when adequate rainfall or overhead irrigation is received within 7 days after application. If cultivation is necessary because of soil crusting, soil compaction or weed germination before rain or irrigation, use shallow tillage (such as rotary hoe), and make certain corn seeds are below the tilled area. For maximum effectiveness, wait at least 7-10 days before cultivating early postemergence treatments.

Field Corn Grown In Minimum Tillage Systems

PROWL alone and PROWL tank mixtures with atrazine, atrazine plus Bladex, Bicep, Dual, Lariat or Bullet, Marksman or Marksman plus Bladex may be used in field corn in minimum tillage systems where adequate tillage is practiced to provide a seedbed without living vegetation which allows for good seed coverage. Plant corn AT LEAST 1½ inches deep to ensure good seed coverage. The use of no-till planters in minimum tillage corn under conditions which do not allow good soil coverage of the corn seed can result in reduced crop stand or injury if PROWL contacts the germinating corn seed. Check equipment to ensure good seed coverage. Apply in 20 or more gallons of spray solution per acre.

Applications of PROWL Alone

Apply PROWL alone preemergence (surface treatment), after planting, but before weeds and crop emerge. Apply with ground equipment or aircraft.

DO NOT INCORPORATE or serious corn injury can result. DO NOT apply PROWL alone postemergence.

Weeds Controlled

When applied as directed, PROWL preemergence treatments in corn will control the following weeds in addition to the weeds listed previously as being controlled by PROWL:

Pennsylvania smartweed *(Polygonum pensylvanicum)*
Velvetleaf (Buttonweed) *(Abutilon theophrasti)*
Woolly cupgrass *(Eriochloa villosa)*

PROWL will aid in the control of and reduce competition from wild proso millet *(Panicum miliaceum)*

Broadcast Rate per Acre of PROWL Alone Preemergence In Corn (PROWL may be applied in sweet corn in Illinois, Minnesota, New York, and Wisconsin only.)

Soil Texture	PROWL (pints) Less than 1.5% Organic Matter	PROWL (pints) 1.5 to 3.0% Organic Matter	PROWL (pints) More than 3.0% Organic Matter
COARSE	1.5 to 2.0	3.0	3.0
MEDIUM	2.0 to 3.0	3.0	3.0 to 4.0
FINE	2.0 to 3.0	3.0 to 4.0	4.0

NOTE: When PROWL alone is used, if corn loss occurs due to weather conditions, corn, cotton or soybeans can be replanted the same year without adverse effects. If corn is replanted, seeding depth must be below re-tilled area.

RESTRICTIONS FOR EARLY POSTEMERGENCE APPLICATIONS

To avoid serious crop injury with early postemergence applications observe the following restrictions:

DO NOT apply following PROWL preemergence treatments.
DO NOT apply PROWL alone postemergence.
DO NOT INCORPORATE.
DO NOT apply in liquid fertilizer.

DO NOT apply PROWL plus Bladex 80W or 90DF if 5th corn leaf is visible.

DO NOT use Bladex 4L postemergence on corn.
DO NOT apply PROWL plus Bladex 80W or 90DF under extended cold (below 50°F), wet weather conditions, or when cold, wet conditions are predicted.

PROWL Plus Atrazine or PROWL Plus Bladex or PROWL Plus Atrazine Plus Bladex

PREEMERGENCE (surface treatment after planting)

Apply PROWL plus atrazine or PROWL plus Bladex or PROWL plus atrazine plus Bladex preemergence, after planting, but before weeds or crop emerge. Apply in water or liquid fertilizer with ground equipment or in water with aircraft. Refer to Application with Liquid Fertilizers section (see table of contents for page number) for liquid fertilizer compatibility determinations.

DO NOT INCORPORATE or serious corn injury can result.

EARLY POSTEMERGENCE

Apply PROWL plus atrazine or PROWL plus Bladex 80W or 90DF or PROWL plus atrazine plus Bladex 80W or 90DF, after corn emergence but no later than when corn is in the 4-leaf stage of growth and when weeds are no more than 1 inch tall except for Texas panicum which must be no larger than the 2 leaf stage. Apply with ground equipment or aircraft in water ONLY. For ground applications, flat fan nozzles are preferred.

Weeds Controlled

When applied as directed, PROWL plus atrazine and PROWL plus Bladex will control the following weeds in addition to the weeds controlled by PROWL.

Black nightshade *(Solanum nigrum)*
Common ragweed *(Ambrosia artemisiifolia)*
Mustard *(Brassica* spp.)
Venice mallow *(Hibiscus trionum)*
Smartweed *(Polygonum pensylvanicum)*
Velvetleaf (Buttonweed) *(Abutilon theophrasti)*
Woolly cupgrass *(Eriochloa villosa)*

PROWL plus atrazine or PROWL plus Bladex will aid in the control of and reduce competition from:

Cocklebur *(Xanthium strumarium)*
Annual morningglory *(Ipomoea* spp.)
Jimsonweed *(Datura stramonium)*
Shattercane *(Sorghum bicolor)*
Wild proso millet* (see below) *(Panicum miliaceum)*

* Effective management of wild proso millet can be achieved with a program involving herbicide applications and mechanical cultivation to eliminate weed escapes. PROWL early postemergence (no later than when corn is in the 4-leaf stage) treatments have provided the most consistent suppression of wild proso millet because: (1) early postemergence treatments provide a burn-down of the initial germination flush of wild proso millet and (2) early postemergence treatments provide residual activity later into the growing season. Additional wild proso millet reduction can be obtained when PROWL treatments follow effective registered preplant incorporated herbicide treatments.

Broadcast Rate per Acre of PROWL Plus Atrazine Preemergence and Early Postemergence Tank Mixtures in Corn (PROWL may be applied in sweet corn in Illinois, Minnesota, New York, and Wisconsin only.)

Soil Texture	Less than 1.5% Organic Matter PROWL + atrazine 80W[1]	1.5 to 3.0% Organic Matter PROWL + atrazine 80W[1]	More than 3.0% Organic Matter PROWL + atrazine 80W[1]
COARSE	1.5 to 2.0 pts + 1.25 lbs.	2.0 pts. + 1.25 lbs.	3.0 pts. + 1.25 lbs.
MEDIUM	2.0 pts. + 1.25 to 1.5 lbs.	3.0 pts. + 1.25 to 1.5 lbs.	3.0 pts. + 1.5 to 2.0 lbs.
FINE	2.0 pts. + 1.25 to 1.5 lbs.	3.0 pts. + 1.5 to 2.0 lbs.	3.0 pts. + 1.5 to 2.0 lbs.

[1] When using atrazine 4L or AAtrex Nine-O use equivalent rates. One quart of atrazine 4L equals 1.25 lbs. of atrazine 80W. One pound of AAtrex Nine-O equals 1.12 lbs. of atrazine 80W. See conversion chart in Formulation Conversion Tables section (see Table of Contents for page number).

NOTE: If corn crop loss occurs due to weather conditions and corn is replanted, seeding depth must be below retilled area. Follow cropping restrictions on atrazine labels. Observe all precautions and limitations on atrazine labels.

Broadcast Rate per Acre of PROWL Plus Bladex 90 DF Preemergence and Early Postemergence Tank Mixtures in Corn (PROWL may be applied in sweet corn in Illinois, Minnesota, New York, and Wisconsin only.)

Soil Texture	Less than 1.5% Organic Matter PROWL + Bladex 90DF[1]	1.5 to 3.0% Organic Matter PROWL + Bladex 90DF[1]	More than 3.0% Organic Matter PROWL + Bladex 90DF[1]
COARSE[2]	1.5 to 2.0 pts. + 1.11 lbs. to 1.33 lbs.	2.0 pts. + 1.78 lbs.	3.0 pts. + 1.78 to 2.22 lbs.
MEDIUM	2.0 pts. + 1.11 to 1.33 lbs.	3.0 pts. + 1.78 to 2.22 lbs.	3.0 pts. + 2.22 to 2.67 lbs.[3]
FINE	2.0 pts. + 1.78 to 2.22 lbs.	3.0 pts. + 2.22 to 2.67 lbs.[3]	3.0 pts. + 2.22 to 2.67 lbs.[3]

[1] When using Bladex 4L or 80W use equivalent rates. One pound of Bladex 90DF equals 1.12 lbs. of Bladex 80W. One quart of Bladex 4L equals 1.11 pounds of Bladex 90DF. DO NOT use Bladex 4L early postemergence. See conversion chart in Formulation Conversion Tables section (see Table of Contents for page number).

[2] DO NOT use this combination on sands or loamy sands with less than 1% organic matter.

[3] The MAXIMUM RATE of Bladex 90DF for early postemergence application is 2.22 lbs. per acre.

NOTE: If corn crop loss occurs due to weather conditions and corn is replanted, seeding depth must be below retilled area. Observe all precautions and limitations on Bladex and atrazine labels.

Broadcast Rate per Acre of PROWL plus Atrazine plus Bladex Preemergence and Early Postemergence Tank Mixtures in Field and Sweet Corn (PROWL may be applied in sweet corn in Illinois, Minnesota, New York, and Wisconsin only.)

Soil Texture	Less than 1.5% Organic Matter PROWL + Atrazine 80W** + Bladex 80WP**	1.5 to 3.0%* Organic Matter PROWL + Atrazine 80W** + Bladex 80WP**	More than 3.0%* Organic Matter PROWL + Atrazine 80W** + Bladex 80WP**
COARSE[1]	1.5 to 2.0 pts. + 0.5 lb. + 1.0 lb.	2 pts. + 0.75 lb. + 0.75 to 1.25 lbs.	3 pts. + 0.75 lb. + 0.75 to 1.5 lbs.
MEDIUM	2 pts. + 0.5 lb. + 1.0 lb.	3 pts. + 0.75 lb. + 1.0 to 1.75 lbs.	3 pts. + 1.0 lb. + 1.5 to 2.0 lbs.
FINE	2 pts. + 0.75 lb. + 1.0 to 1.75 lbs.	3 pts. + 1.0 lb. + 1.5 to 2.0 lbs.	3 pts. + 1.0 lb. + 1.5 to 2.0 lbs.

[1] **DO NOT** use this combination for preemergence applications in the Atlantic Coastal Plain. **DO NOT** use this combination for postemergence applications on sands or loamy sands with less than 1% organic matter. **DO NOT** use on peat or muck soils.

* The high rate for each soil texture, where listed, should be used if heavy infestations of grass or broadleaf weeds are anticipated, if hard-to-control weed species are anticipated, or if extensive crop residues were present following seedbed preparation, or if soil organic matter content falls in high end of range.

** When using Bladex 4L, Bladex 80W, atrazine 4L, or Aatrex Nine-O use equivalent rates. One quart of Bladex 4L or atrazine 4L equals 1.25 lb. of Bladex 80WP or atrazine 80W. One pound of Aatrex Nine-O or Bladex 90DF equals 1.12 lb. of atrazine 80W or Bladex 80W. One quart of Bladex 4L or atrazine 4L equals 1.1 pounds of Bladex 90DF or AAtrex Nine-O. **DO NOT use Bladex 4L postemergence.** See conversion chart in Formulation Conversion Tables section (see Table of Contents for page number).

NOTE: If corn crop loss occurs due to weather conditions and corn is replanted, seeding depth must be below retilled area. Observe all precautions and limitations on Bladex and atrazine labels.

FOLLOW CROP RESTRICTIONS

1. Plant only corn, peanuts, sorghum, or soybeans the year following the use of this mixture.
2. Injury may occur to soybeans planted the year following application on soils having a calcareous surface layer.
3. If applied after June 10, do not rotate with crops other than corn or sorghum the next year or injury may occur.
4. In the high plains and intermountain areas of the West where rainfall is sparse and erratic or where irrigation is required, use only when corn or sorghum is to be planted the following year, or a crop of corn or sorghum not treated with atrazine is to precede other rotational crops.
5. All crops may be planted 18 months after application.
6. Refer to atrazine and Bladex labels for other follow crop restrictions.

PROWL Plus Marksman Or PROWL Plus Marksman Plus Bladex

Observe all precautionary statements in the Marksman and Bladex labels before using.

Apply only on level planted corn. Plant corn at least 1½ inches deep to insure good seed coverage.

AVOID drift of these mixtures to soybeans or other sensitive crops.

DO NOT use in California.

PREEMERGENCE (surface treatment after planting)

Apply preemergence, after planting, but before weeds or crop emerge. Apply in water or liquid fertilizer with ground equipment or in water with aircraft. Refer to Application with Liquid Fertilizers section (see table of contents for page number) for liquid fertilizer compatibility determinations.

DO NOT use on furrow irrigated corn.

DO NOT incorporate or serious corn injury can result.

EARLY POSTEMERGENCE

Apply PROWL plus Marksman or PROWL plus Marksman plus Bladex after corn emergence but no later than when corn is in the 2-leaf stage of growth and when weeds are no larger than 1 inch tall.

Apply with ground equipment or aircraft in water only. For ground applications flat fan nozzles are preferred.

WEEDS CONTROLLED

When applied as directed, PROWL plus Marksman or PROWL plus Marksman plus Bladex will control the following weeds in addition to those controlled by PROWL alone:

- Annual clover (*Trifolium* spp.)
- Black nightshade (*Solanum nigrum*)
- Burcucumber (*Sicyos angulatus*)
- Chickweed, common (*Stellaria media*)
- Jimsonweed (*Datura stramonium*)
- Mallaow
  - common (*Malva* spp.)
  - Venice (*Hibiscus trionum*)
- Mustard (*Brassica* spp.)
- Puncturevine (*Tribulus terrestris*)
- Ragweed
  - common* (*Ambrosia artemisiifolia*)
  - giant (*Ambrosia trifida*)
- Sicklepod (*Cassia obtusifolia*)
- Sida, prickly (Teaweed), (*Sida spinosa*)
- Smartweed
  - ladysthumb (*Polygonum persicaria*)
- Spurge
  - prostrate (*Euphorbia humistrata*)
- Sunflower, common (*Helianthus annuus*)
- Tansymustard (*Descurainia pinnata*)
- Thistle, Russian (*Salsola kali*)
- Velvetleaf (*Abutilon theophrasti*)
- Wild buckwheat (*Polygonum convolvulus*)
- Wild cucumber (*Echinocystis lobata*)

PROWL plus Marksman or PROWL plus Marksman plus Bladex will aid in the control of and reduce competition from the following annual weed species. Consult the Marksman label for perennial weed species suppressed by these tank mixtures.

- Annual morningglory (*Ipomoea* spp.)
- Shattercane (*Sorghum bicolor*)
- Wild proso millet (*Panicum miliaceum*)

*PROWL plus Marksman plus Bladex tank mixtures only.

Broadcast Rates per Acre of PROWL Plus Marksman Preemergence and Early Postemergence Tank Mixtures in Field Corn

PROWL (pints) + Marksman (pints)

Soil Texture	Less than 1.5% Organic Matter PROWL + Marksman	1.5% to 3.0% Organic Matter PROWL + Marksman	More than 3.0% Organic Matter PROWL + Marksman
COARSE**	1.5 to 2 + 2	2 + 2	3 + 2
MEDIUM	2 + 3.5	3 + 3.5	3 + 3.5
FINE	2 + 3.5	3 + 3.5	3 + 3.5

DO NOT apply preemergence on coarse textured soils or any soil with less than 2.5% organic matter.

NOTE: If crop loss occurs due to weather conditions and corn is replanted, seeding depth must be below retilled area. Observe all cropping restrictions on the Marksman and Bladex labels.

Broadcast Rates per Acre of PROWL Plus Marksman plus Bladex Preemergence and Early Postemergence Tank Mixtures In Field Corn

PROWL (pints) + Marksman (pints) + Bladex 90DF[1,2] (pounds)

Soil Texture	Less than 1.5% Organic Matter PROWL + Marksman + Bladex 90DF	1.5 to 3.0% Organic Matter PROWL + Marksman + Bladex 90DF	More than 3.0 Organic Matter PROWL + Marksman + Bladex 90DF
COARSE**	1.5 to 2 + 2 + 1	2 + 2 + 0.67 to 1.11	3 + 2 + 6.7 to 1.3
MEDIUM	2 + 3.5 + 1	3 + 3.5 + 1 to 1.5	3 + 3.5 + 1.3 to 1.8

Soil Texture	Less than 1.5% Organic Matter PROWL + Marksman + Bladex 90DF	1.5 to 3.0% Organic Matter PROWL + Marksman + Bladex 90DF	More than 3.0 Organic Matter PROWL + Marksman + Bladex 90DF
FINE	2 + 3.5 + 1 to 1.5	3 + 3.5 + 1.3 to 2.2	3 + 3.5 + 1.3 to 2.2

[1] When using Bladex 4L or Bladex 80W use equivalent rates. One quart of Bladex 4L equals 1.11 pounds of Bladex 90DF. One pound of Bladex 90DF equals 1.12 pounds of Bladex 80W. See conversion chart in Formulation Conversion Tables section (see Table of Contents for page number).

[2] Apply only Bladex 80W in postemergence tank mixtures with Marksman.

** DO NOT apply preemergence on coarse textured soils or any soil with less than 2.5% organic matter.

NOTE: If crop loss occurs due to weather conditions and corn is replanted, seeding depth must be below retilled area. Observe all cropping restrictions on the Marksman and Bladex labels.

PROWL Plus Bicep

PROWL may be applied in sweet corn (processing varieties) in Illinois, Minnesota, New York and Wisconsin ONLY.

Observe all precautionary statements in the Bicep label before using.

Add PROWL to the spray tank prior to the addition of Bicep.

DO NOT use in California.

PREEMERGENCE (surface treatment after planting)

Apply PROWL plus Bicep preemergence, after planting, but before weeds or crop emerge. Apply in water or liquid fertilizer with ground equipment. Preemergence applications in field corn may be made through center pivot irrigation systems. Refer to Application with Liquid Fertilizers section (see table of contents for page number) for liquid fertilizer compatibility determinations.

DO NOT incorporate or serious crop injury can result.

EARLY POSTEMERGENCE

Apply after corn emergence but no later than when corn is in the 4-leaf stage of growth and when weeds are no larger than the 1-2 leaf growth stage. Apply with ground equipment in water ONLY. Flat fan nozzles are preferred. Use PROWL rates listed in the following table with the Bicep rates listed under the "More than 3% Organic Matter" section of the table. For best results, destroy emerged weeds by cultivation prior to application.

WEEDS CONTROLLED

When applied preemergence as directed, PROWL plus Bicep will control the following weeds in addition to those controlled by PROWL alone:

- Black nightshade (*Solanum nigrum*)
- Browntop panicum (*Panicum fasciculatum*)
- Cocklebur (*Xanthium strumarium*)
- Galinsoga (*Galinsoga* spp.)
- Hairy nightshade (*Solanum sarrachoides*)
- Jimsonweed (*Datura Stramonium*)
- Morningglory (*Ipomoea* spp.)
- Ragweed (*Ambrosia* spp.)
- Red rice (*Oryza sativa*)
- Smartweed (*Polygonum pensylvanicum*)
- Southwestern cupgrass (*Eriochloa gracilis*)
- Velvetleaf (*Abutilon theophrasti*)
- Woolly cupgrass (*Eriochloa villosa*)
- Yellow nutsedge (*Cyperus esculentus*)

Preemergence tank mixtures of PROWL plus Bicep will aid in the control of and reduce competition from:

- Shattercane (*Sorghum bicolor*)
- Sicklepod (*Cassia obstusifolia*)
- Volunteer sorghum (*Sorghum vulgare*)
- Wild proso millet (*Panicum miliaceum*)

When applied early postemergence as directed, PROWL plus Bicep will control the following weeds in addition to those controlled by PROWL alone:

- Jimsonweed (*Datura stramonium*)
- Mustard (*Brassica* spp.)
- Prickly sida (*Sida spinosa*)
- Ragweed (*Ambrosia* spp.)
- Smartweed (*Polygonum pensylvanicum*)
- Velvetleaf (*Abutilon theophrasti*)

Early postemergence tank mixtures of PROWL plus Bicep will aid in the control of and reduce competition from:

- Cocklebur (*Xanthium strumarium*)
- Morningglory (*Ipomoea* spp.)
- Wild proso millet (*Panicum miliaceum*)
- Woolly cupgrass (*Eriochloa villosa*)
- Yellow nutsedge (*Cyperus esculentus*)

Broadcast Rates per Acre of PROWL Plus Bicep Preemergence and Early Postemergence Tank Mixtures In Corn
(PROWL may be applied in sweet corn in Illinois, Minnesota, New York and Wisconsin ONLY.)

PROWL (pints) + Bicep 6L** (quarts)

Soil Texture	Less than 1.5% Organic Matter PROWL + Bicep 6L	1.5% to 3.0% Organic Matter PROWL+ Bicep 6L	More than 3.0% Organic Matter PROWL + Bicep 6L
COARSE	1.5 to 2+1.5	2+1.5	3+1.8
MEDIUM	2+1.8	3+1.8	3+2.4
FINE	2+2.4	3+2.4	3+2.4 to 3

**When using Bicep 4.5L use equivalent rates. One quart of Bicep 6L equals 1.33 quarts of Bicep 4.5L.

NOTE: If crop loss occurs due to weather conditions and corn is replanted, seeding depth must be below retilled area.

Observe all cropping restrictions on the Bicep label.

PROWL Plus Lariat Or PROWL Plus Bullet

PROWL may be applied in sweet corn (processing varieties) in Illinois, Minnesota, New York and Wisconsin ONLY.

Observe all precautionary statements in the Lariat and Bullet labels before using.

DO NOT graze treated area or feed treated forage to livestock for 21 days following application.

DO NOT use in California.

PREEMERGENCE (surface treatment after planting)

Apply PROWL plus Lariat or Bullet after planting, but before weeds or crop emerge. Apply in water or liquid fertilizer with ground equipment. Preemergence applications in field corn may be made through center pivot irrigation systems. Refer to Application with Liquid Fetilizers section (see table of contents for page number) for liquid fertilizer compatibility determinations.

DO NOT incorporate or serious corn injury can result.

EARLY POSTEMERGENCE

Apply after corn emergence but no later than when corn is in the 4-leaf stage of growth and when weeds are no larger than the 2-leaf growth stage. Apply with ground equipment in water ONLY. Flat fan nozzles are preferred.

DO NOT apply postemergence to sweet corn.

WEEDS CONTROLLED

When applied as directed, PROWL plus Lariat or PROWL plus Bullet will control the following weeds in addition to those controlled by PROWL alone:

- Black nightshade (*Solanum nigrum*)
- Browntop panicum (*Panicum fasciculatum*)
- Common ragweed (*Ambrosia artemisiifolia*)
- Hairy nightshade (*Solanum sarrachoides*)
- Jimsonweed (*Datura stramonium*)
- Mustard (*Brassica* spp.)
- Red sprangletop (*Leptochloa filiformis*)
- Smartweed
  - Ladysthumb (*Polygonum persicaria*)
- Velvetleaf (*Abutilon theophrasti*)
- Wild proso millet (*Panicum millaceum*)
- Woolly cupgrass (*Eriochloa villosa*)

Broadcast Rates per Acre of PROWL Plus Lariat or PROWL Plus Bullet
Preemergence and Early Postemergence Tank Mixtures In Corn
(PROWL may be applied in sweet corn in Illinois, Minnesota, New York and Wisconsin ONLY)

PROWL (pints) + Lariat or Bullet (quarts)

Soil Texture	Less than 1.5% Organic Matter PROWL + Lariat or Bullet	1.5 to 3.0% Organic Matter PROWL+ Lariat or Bullet	More than 3.0% Organic Matter PROWL + Lariat or Bullet
COARSE	1.5 to 2 + 2.5	2 + 2.5	3 + 3
MEDIUM	2 + 3	3 + 3	3 + 3.75
FINE	2 + 3.75	3 + 3.75	3 + 3.75 to 4.5

NOTE: If crop loss occurs due to weather conditions, corn or grain sorghum may be planted the same year without adverse effects. Seeding depth must be below the retilled area.

Observe all cropping restrictions on the Lariat or Bullet labels.

PROWL Plus Dual 8E

PROWL may be applied in sweet corn (processing varieties) in Illinois, Minnesota, New York and Wisconsin ONLY.

Observe all precautionary statements in the Dual 8E label before using.

DO NOT use in California.

PREEMERGENCE (surface treatment after planting)

Apply PROWL plus Dual 8E preemergence, after planting, but before weeds or crop emerge. Apply in water or liquid fertilizer with ground equipment or in water with aircraft. Preemergence applications in field corn may be made through center pivot irrigation systems. Refer to Application with Liquid Fertilizers section (see table of contents for page number) for liquid fertilizer compatibility determinations.

DO NOT incorporate or serious corn injury can result.

WEEDS CONTROLLED

When applied as directed, PROWL plus Dual 8E will control the following weeds in addition to those controlled by PROWL alone:

- Black nightshade (*Solanum nigrum*)
- Florida beggarweed* (*Desmodium tortuosum*)
- Galinsoga (*Galinsoga* spp.)
- Hairy nightshade (*Solanum sarrachoides*)
- Red rice (*Oryza sativa*)
- Southwestern cupgrass (*Eriochloa gracilis*)
- Volunteer sorghum (*Sorghum vulgare*)
- Yellow nutsedge (*Cyperus esculentus*)

*Use a minimum of 2 pints of Dual 8E per acre.

Broadcast Rates per Acre of PROWL Plus Dual 8E Preemergence Tank Mixture In Corn

(PROWL may be applied in sweet corn in Illinois, Minnesota, New York and Wisconsin ONLY)

PROWL (pints) + Dual 8E (quarts)

Soil Texture	Less than 1.5% Organic Matter PROWL + Dual 8E	1.5 to 3.0% Organic Matter PROWL + Dual 8E	More than 3.0% Organic Matter PROWL + Dual 8E
COARSE	1.5 to 2 + 1.25 to 1.5	2 + 1.25 to 1.5	3 + 1.5
MEDIUM	2 + 1.5 to 2	3 + 1.5 to 2	3 + 2
FINE	2 + 2	3 + 2	3 + 2 to 2.5

NOTE: If crop loss occurs due to weather conditions corn, potatoes or soybeans may be planted the same year without adverse effects.

Observe all cropping restrictions on the Dual 8E label.

PROWL Plus Lasso

PROWL may be applied in sweet corn (processing varieties) in Illinois, Minnesota, New York and Wisconsin ONLY.

Observe all precautionary statements in the Lasso label before using.

DO NOT use in California.

PREEMERGENCE (surface treatement after planting)

Apply preemergence, after planting, but before weeds or crop emerge. Apply in water or liquid fertilizer with ground equipment. Preemergence applications in field corn may be made through center pivot irrigation systems. Refer to Application with Liquid Fertilizers section (see table of contents for page number) for liquid fertilizer compatibility determinations. This treatment must be applied within 5 days of the last preplant tillage.

DO NOT INCORPORATE or serious corn injury can result.

EARLY POSTEMERGENCE

Apply after corn emergence but before corn exceeds 5 inches in height. Apply with ground equipment in water ONLY. EMERGED WEEDS WILL NOT BE CONTROLLED BY THIS TANK MIX. Destroy emerged weeds by cultivation prior to application.

WEEDS CONTROLLED

When applied as directed, PROWL plus Lasso will control the following weeds in addition to those controlled by PROWL alone:

- Galinsoga (*Galinsoga* spp.)
- Pennsylvania smartweed (*Polygonum pensylvanicum*)
- Velvetleaf (*Abutilon theophrasti*)

Broadcast Rates per Acre of PROWL Plus Lasso Preemergence and Early Postemergence Tank Mixtures In Corn

(PROWL may be applied in sweet corn in Illinois, Minnesota, New York and Wisconsin ONLY)

PROWL (pints) + Lasso (quarts)

Soil Texture	Less than 1.5% Organic Matter PROWL + Lasso	1.5 to 3.0% Organic Matter PROWL + Lasso	More than 3.0% Organic Matter PROWL + Lasso
COARSE	1.5 to 2 + 1.5 to 2.5	2 + 1.5 to 2.5	3 + 2 to 3
MEDIUM	2 + 1.75 to 2.5	3 + 1.75 to 2.5	3 + 2 to 3
FINE	2 + 2.25 to 3	3 + 2.25 to 3	3 + 3 to 4

NOTE: If crop loss occurs due to weather conditions corn, potatoes or soybeans may be planted the same year without adverse effects. If corn is replanted, seeding depth must be below retilled area.

Preemergence Applications of PROWL Plus Banvel Tank Mixture in Field Corn

Apply PROWL plus Banvel preemergence (surface treatment) only, after planting, but before weeds or crop emerge. Apply in water with GROUND EQUIPMENT ONLY.

DO NOT INCORPORATE or serious corn injury can result.

APPLY ONLY ON LEVEL PLANTED CORN. Plant corn AT LEAST 1½ inches deep to ensure good seed coverage.

Weeds Controlled

When applied as directed, PROWL plus Banvel will control the following weeds in addition to the weeds controlled by PROWL.
Cocklebur *(Xanthium strumarium)*
Venice mallow *(Hibiscus trionum)*

Broadcast Rate per Acre of PROWL Plus Banvel Preemergence Tank Mixture in Field Corn

Soil Texture[1]	PROWL + Banvel (pints) 1.5 to 3.0% Organic Matter	PROWL + Banvel (pints) More than 3.0% Organic Matter
MEDIUM	DO NOT USE	3.0 + 0.75
FINE	3.0 + 0.75	3.0 + 0.75

[1] DO NOT use on sands, loamy sands, or sandy loams.

DO NOT use on medium soils containing less than 3% organic matter.

AVOID drift of this mixture to soybeans or other sensitive crops.

DO NOT use on furrow irrigated corn.

DO NOT apply this tank mixture by aircraft.

When applying this tank mixture at planting, the spray nozzles must be far enough behind the planter to avoid contact of the spray mixture with the seeds to avoid crop injury.

NOTE: Follow cropping restrictions on Banvel label. Observe all precautions and limitations on Banvel label.

CULTI-SPRAY™ (POSTEMERGENCE INCORPORATED) FIELD CORN—GRAIN SORGHUM

PROWL CULTI-SPRAY (postemergence incorporated) can be applied in field corn as follows:

PROWL alone — PROWL + atrazine

PROWL CULTI-SPRAY (postemergence incorporated) can be applied in grain sorghum as follows:

PROWL alone — PROWL + atrazine

Winter wheat and winter barley may be planted in the fall 90 days after an application with PROWL CULTI-SPRAY in irrigated field corn or grain sorghum or 120 days after an application with PROWL CULTI-SPRAY in dry land field corn or grain sorghum. The treated crop must be grown to maturity and harvested before planting wheat or barley. See complete follow crop instructions given previously in this label (see table of contents for page number). DO NOT use PROWL on peat or muck soils.

GENERAL INFORMATION

PROWL or PROWL plus atrazine tank mixtures applied as a CULTI-SPRAY (postemergence incorporated) treatment will control most late season annual grasses and certain broadleaf weeds in field corn or grain sorghum. PROWL treatments can be applied from the 4-inch growth stage to as late as the last cultivation (layby) of field corn or grain sorghum. PROWL treatments will not control established weeds. DESTROY EMERGED WEEDS BY CULTIVATION PRIOR TO PROWL APPLICATION.

Additional Herbicide Applications

CULTI-SPRAY (postemergence incorporated) applications of PROWL or PROWL plus atrazine tank mixture can be applied in field corn or grain sorghum previously treated with herbicides registered for use in field corn or grain sorghum. Consult the labels of those herbicides for suggested treatments, rates to be used, and precautions or restrictions for use in field corn or grain sorghum and for follow crop restrictions.

CULTI-SPRAY (Postemergence Incorporated) APPLICATIONS

Cultivate and destroy all emerged weeds prior to application. Cultivate with a sweep-type or rolling cultivator operated at sufficient speed to throw at least one inch of soil over the bases of the field corn or grain sorghum plants. This will kill small weed seedlings growing in the field corn or grain sorghum row and will prevent direct contact of the zone of brace root formation by PROWL during application.

Apply PROWL or PROWL plus atrazine tank mixture with ground sprayer. Cover soil uniformly. Use drop nozzles if field corn or grain sorghum foliage will prevent uniform coverage of the soil surface within the rows.

PROWL treatments can be applied from the 4-inch growth stage to as late as the last cultivation (layby) of field corn or grain sorghum.

Thoroughly and uniformly incorporate PROWL treatments into the soil (1) with a sweep-type or rolling cultivator set to provide thorough incorporation in the top 1 inch of soil OR (2) with adequate irrigation water or rainfall. For best results, incorporate PROWL treatments as soon as possible after application. Incorporation must be completed within 7 days after application. If adequate moisture is not received within 7 days after application, incorporate PROWL with a sweep-type or rolling cultivator.

Under situations of low rainfall or soil moisture, when deep germinating weeds such as shattercane or field sandbur are anticipated, mechanical incorporation will provide best results.

If cultivation is needed after application and incorporation of PROWL, the depth of cut should be no deeper than the depth of cut used to incorporate.

DO NOT APPLY PROWL in field corn or grain sorghum more than once per crop season.

Weeds Controlled

When applied as directed, PROWL CULTI-SPRAY (postemergence incorporated) will control the following weeds in addition to the weeds previously listed as being controlled by PROWL.

Shattercane *(Sorghum bicolor)*
Wild proso millet *(Panicum miliaceum)*

When applied as directed, PROWL plus atrazine CULTI-SPRAY (postemergence incorporated) tank mixture will control the following weeds in addition to those controlled by PROWL alone:

Pennsylvania smartweed *(Polygonum pensylvanicum)*
Velvetleaf (Buttonweed) *(Abutilon theophrasti)*
Venice mallow *(Hibiscus trionum)*

FIELD CORN

DO NOT APPLY PROWL preplant incorporated in field corn as serious crop injury can result.

Broadcast Rate Per Acre of PROWL CULTI-SPRAY (Postemergence Incorporated) in Field Corn

	PROWL (pints)	PROWL (pints)
Soil Texture	Southern States[1]	Northern States[1]
COARSE	1.0 to 1.5	1.5 to 2.0
MEDIUM	1.5 to 2.0	2.0 to 3.0
FINE	1.5 to 3.0	2.0 to 3.0

[1] See map at end of this label for specific states.

NOTE: Livestock can graze or be fed forage from treated field corn after 21 days following application.

Broadcast Rate Per Acre of PROWL Plus Atrazine CULTI-SPRAY (Postemergence Incorporated) Tank Mixture in Field Corn

	PROWL + atrazine 80W[1]	
Soil Texture	Southern States[2]	Northern States[2]
COARSE	1.0 to 1.5 pts. + 1.25 lbs.	1.5 to 2.0 pts. + 1.25 lbs.
MEDIUM	1.5 to 2.0 pts. + 1.25 to 1.5 lbs.	2.0 to 3.0 pts. + 1.25 to 1.5 lbs.
FINE	1.5 to 3.0 pts. + 1.25 to 1.5 lbs.	2.0 to 3.0 pts. + 1.25 to 1.5 lbs.

[1] When using atrazine 4L or AAtrex Nine-O use equivalent rates. One quart of atrazine 4L equals 1.25 lbs. of atrazine 80W. One pound of AAtrex Nine-O equals 1.12 lbs. of atrazine 80W. See conversion chart in **Formulation Conversion Tables** section (see table of contents for page number).

[2] See map at end of this label for specific states.

NOTE: Observe all precautions and limitations on atrazine labels. Follow cropping restrictions on atrazine labels.
Livestock can graze or be fed forage from treated field corn after 21 days following application.

GRAIN SORGHUM

DO NOT APPLY PROWL in grain sorghum preplant incorporated or preemergence as serious crop injury can result.

DO NOT APPLY PROWL in grain sorghum which is planted in double row beds since this cultural practice does not permit adequate soil coverage of the bases of the grain sorghum plants with cultivation.

Broadcast Rate Per Acre of PROWL CULTI-SPRAY (Postemergence Incorporated) in Grain Sorghum

	PROWL (pints)	PROWL (pints)
Soil Texture	Southern States[1]	Northern States[1]
COARSE	1.0 to 1.5	1.5 to 2.0
MEDIUM	1.5 to 2.0	2.0 to 3.0
FINE	1.5 to 3.0	2.0 to 3.0

[1] See map at end of this label for specific states.

NOTE: Livestock can graze or be fed forage from treated grain sorghum after 21 days following application.

Broadcast Rate Per Acre of PROWL Plus Atrazine CULTI-SPRAY (Postemergence Incorporated) Tank Mixture in Grain Sorghum

	PROWL + atrazine 80W[1]	
Soil Texture	Southern States[2]	Northern States[2]
COARSE	DO NOT USE	DO NOT USE
MEDIUM	1.5 to 2.0 pts. + 1.25 lbs.	2.0 to 3.0 pts. + 1.25 lbs.
FINE	1.5 to 3.0 pts. + 1.25 lbs.	2.0 to 3.0 pts. + 1.25 lbs.

[1] When using atrazine 4L or AAtrex Nine-O use equivalent rates. One quart of atrazine 4L equals 1.25 lbs. of atrazine 80W. One pound of AAtrex Nine-O equals 1.12 lbs. of atrazine 80W. See conversion chart in **Formulation Conversion Tables** section (see table of contents for page number).

[2] See map at end of this label for specific states.

NOTE: Observe all precautions and limitations on atrazine labels. Follow cropping restrictions on atrazine labels. Livestock can graze or be fed forage from treated grain sorghum fields after 21 days following application.

GRAIN SORGHUM

PROWL may be applied early postemergence in grain sorghum grown in states east of the Mississippi River and in Arkansas, East Texas, Louisiana, and the Missouri 'bootheel' as follows:

PROWL + atrazine
DO NOT INCORPORATE
DO NOT USE ON PEAT OR MUCK SOILS

Application Instructions

The seedbed should be firm and free of clods and trash. Use only where adequate tillage is practiced to provide good seed coverage. Plant grain sorghum AT LEAST 1½ inches deep to ensure good seed coverage.

Uniformly apply PROWL plus atrazine tank mix treatment in water by ground equipment or by aircraft. Refer to **Mixing Instructions** section (see table of contents for page number) for complete mixing directions. Apply PROWL plus atrazine tank mixture only after grain sorghum has reached the 2-leaf stage and when weeds are no more than 1 inch tall.

DO NOT INCORPORATE or serious grain sorghum injury can result.

DO NOT apply in liquid fertilizer.

PROWL plus atrazine treatments are most effective in controlling weeds when adequate rainfall or overhead irrigation is received within 7 days after application. If cultivation is necessary because of soil crusting, soil compaction, or weed germination before rain or irrigation, use shallow tillage (such as a rotary hoe), and make certain grain sorghum seeds are below the tilled area. Wait 7 to 10 days after application before cultivating.

Weed Species Controlled

When applied as directed, PROWL plus atrazine will control the following weeds in addition to those controlled by PROWL alone:

Black nightshade (*Solanum nigrum*)
Common ragweed (*Ambrosia artemisiifolia*)
Mustard (*Brassica* spp.)
Smartweed (*Polygonum pensylvanicum*)
Velvetleaf (Buttonweed) (*Abutilon theophrasti*)

Recommended tank mixtures of PROWL plus atrazine will reduce competition from the following hard-to-control annual broadleaf weeds:

Cocklebur (*Xanthium strumarium*)
Annual morningglory (*Ipomoea* spp.)
Jimsonweed (*Datura stramonium*)

Broadcast Rate per Acre for Early Postemergence Application for PROWL Plus Atrazine in Grain Sorghum[1]

Soil Texture	PROWL + atrazine 80W[2]
COARSE	DO NOT USE
MEDIUM	1.5 to 2.0 pts. + 1.25 to 1.5 lbs.
FINE	2.0 pts. + 1.25 to 1.5 lbs.

[1] For use in states east of the Mississippi River, plus Arkansas, East Texas, Louisiana, and the Missouri 'bootheel' only.

[2] When using atrazine 4L or AAtrex Nine-O use equivalent rates. One quart of atrazine 4L equals 1.25 lbs. of atrazine 80W. One pound of AAtrex Nine-O equals 1.12 lbs of atrazine 80W. See conversion chart under **Formulation Conversion Tables** section (see Table of Contents for page number).

The high rate for each soil texture above should be used if heavy weed populations are anticipated.

DO NOT use on peat or muck soils.

NOTE: DO NOT replant grain sorghum if crop loss occurs due to weather conditions. Observe all precautions and limitations on atrazine labels. Follow cropping restrictions on atrazine labels. Livestock can graze or be fed forage from PROWL plus atrazine treated grain sorghum fields after 21 days following application.

COTTON

PROWL can be applied in cotton as follows:

Preplant Incorporated	Sequential	Preemergence
PROWL alone	PROWL preplant incorporated followed by Cotoran preemergence	PROWL alone
PROWL + prometryn	PROWL preplant incorporated followed by Karmex preemergence	PROWL + Cotoran
PROWL + Zorial		PROWL + Zorial

Refer to Rhizome Johnsongrass subsection (see table of contents for page number) for rhizome johnsongrass control program.

DO NOT use PROWL on peat or muck soils.

Preplant Incorporated Application of PROWL Alone

Apply PROWL with ground equipment or aircraft immediately before planting or up to 60 days prior to planting. Incorporate prior to planting and within 7 days of application. Refer to **Incorporation Instructions** section (see table of contents for page number) for incorporation directions.

DO NOT APPLY PROWL AS A POSTEMERGENCE SPRAY IN COTTON.

Broadcast Rate Per Acre of PROWL Preplant Incorporated in Cotton

Soil Texture	PROWL (pints)
COARSE	1.0
MEDIUM	1.5
FINE	2.0

For heavy clay soils, apply PROWL at the broadcast rate of 3 pints per acre.

When heavy weed populations are anticipated, apply PROWL at the broadcast rate of 1.5 pints on coarse soils, 2.0 pints on medium soils, and 3.0 pints on fine soils.

NOTE: DO NOT feed forage or graze livestock in treated cotton fields. If loss of cotton crop occurs due to adverse weather conditions, cotton or any crop registered for PROWL preplant incorporated use can be replanted the same year into treated soil without adverse effects. If replanting is necessary, DO NOT rework the soil deeper than the treated zone.

Fall Application

PROWL may be applied for weed control in cotton in the fall, after October 15 (up to 140 days prior to planting cotton) in Arizona, California, Louisiana, New Mexico, Mississippi, Oklahoma and Texas. Apply PROWL at the broadcast rate per acre of 1.5 pints on coarse soils, 2.0 pints on medium soils and 3.0 pints on fine soils.

Preplant Incorporated Application of PROWL Followed by Preemergence Application of Cotoran or Karmex

PROWL preplant incorporated followed by an overlay preemergence application of Cotoran 80W or Karmex may be applied for weed control in cotton.

Refer to the Cotoran or Karmex label for specific directions regarding rates, mixing, spraying and application. Use the Cotoran or Karmex alone preemergence rates. Follow all precautions and limitations on the Cotoran or Karmex label.

When applied as directed, in addition to those weeds controlled by PROWL alone, PROWL preplant incorporated followed by Cotoran or Karmex preemergence will control those additional broadleaf weeds listed on the Cotoran or Karmex label.

Preplant Incorporated Application of PROWL Plus Prometryn Mixtures

PROWL plus prometryn 4L or 80W tank mixture may be applied preplant incorporated for weed control in cotton grown in Arizona, California, New Mexico and the Upper and Lower El Paso Valley of Texas.

Apply PROWL plus prometryn with ground equipment prior to listing or over partially finished or finished beds. Incorporation should immediately follow application. Refer to **Incorporation Instructions** section (see table of contents for page number) for incorporation directions. Maintain good agitation at all times until spraying is completed to keep the material in suspension. Mechanical agitation should be used in Arizona and California.

Do not apply PROWL plus prometryn 4L in California.

Weeds Controlled

When applied as directed, PROWL plus prometryn will control the following weeds in addition to those controlled by PROWL alone:

GRASSES CONTROLLED

Junglerice *(Echinochloa colonum)*
Wild oat *(Avena fatua)*

BROADLEAF WEEDS CONTROLLED

Annual morningglory *(Ipomoea* spp.)
Black nightshade *(Solanum nigrum)*
Common ragweed *(Ambrosia artemisiifolia)*
Groundcherry *(Physalis* spp.)
Malva *(Malva* spp.)
Mustard *(Brassica* spp.)
Pennsylvania smartweed *(Polygonum pensylvanicum)*
Prickly sida *(Sida spinosa)*
PROWL plus prometryn will control shallow-germinating seedlings of:
Cocklebur *(Xanthium strumarium)*
Coffeeweed *(Daubentonia texana)*
Sandbur *(Cenchrus* spp.)

Broadcast[1] Rate Per Acre of PROWL Plus Prometryn Preplant Incorporated Tank Mixture in Cotton

Soil Texture		PROWL + Prometryn 4L[3]
COARSE	Sands, loamy sands	DO NOT USE
	Sandy loams	1.0 to 1.5 pts. + 2.4 to 3.2 pints
MEDIUM	Loams	1.5 to 2.0 pts. + 3.2 pints
	Silt loams, silts, sandy clay loams[2]	1.5 to 2.0 pts. + 3.2 to 4.8 pints
FINE	Silty clay loams[2], clay loams, clays	1.5 to 3.0 pts. + 3.2 to 4.8 pints

[1] For calculation of band treatment rates, see **Spraying Instructions** section (see table of contents for page number).
[2] Sometimes considered transitional soils and may be classified as either medium or fine textured soils.
[3] When using prometryn 80W, use equivalent rates. One pound of prometryn 80W equals 1.6 pints of prometryn 4L. See conversion chart in Formulation Conversion Tables section (see Table of Contents for page number).

Use the 3 pint rate of PROWL for heavy clay soils.

NOTE: DO NOT feed forage or graze livestock in treated cotton fields. If replanting is necessary, cotton may be planted in soil previously treated with PROWL plus prometryn. DO NOT rework the soil. Follow cropping restrictions on prometryn label.

PRECAUTIONS:

DO NOT use this tank mixture in cut areas of newly leveled fields, in areas of excess salt or in areas where flooding over the bed is likely to occur as crop injury may result. Cotton should not be planted in tractor wheel depressions or crop injury may result. DO NOT use this tank mixture when cotton is irrigated up as crop injury may result. On mulch-planted cotton, water back only after cotton seedlings are well established. Follow all precautions and limitations on prometryn label.

Preplant Incorporated Application of PROWL Plus Zorial

PROWL plus Zorial 80WP tank mix may be applied preplant incorporated for weed control in cotton grown in all states except Texas, Oklahoma, New Mexico, Arizona and California.

Uniformly apply recommended tank mixtures of PROWL plus Zorial by ground equipment immediately before or up to 30 days before planting. DO NOT incorporate deeper than 2-3 inches with commonly used equipment. Incorporate within seven days after application. Refer to **Incorporation Instructions** section for complete directions (see table of contents for page number).

Weed Species Controlled

When applied as directed, PROWL plus Zorial preplant incorporated tank mix will control the following weeds in addition to those controlled by PROWL alone:
Prickly sida *(Sida spinosa)*
Common ragweed *(Ambrosia artemisiifolia)*
Smartweed *(Polygonum pensylvanicum)*
Sicklepod *(Cassia obtusifolia)*
Spurred anoda *(Anoda cristata)*
Tank mixtures of PROWL plus Zorial will reduce competition from the following hard-to-control weeds:
Cocklebur *(Xanthium strumarium)*
Annual morningglory *(Ipomoea* spp.)
Nutsedge *(Cyperus* spp.)

Broadcast Rates Per Acre of PROWL Plus Zorial Preplant Incorporated Tank Mixture[1]

Soil Texture	PROWL + ZORIAL 80WP
COARSE	1.0 to 1.5 pts. + 1.25 lbs.
MEDIUM	1.5 to 2.0 pts. + 1.9 lbs.
FINE	1.5 to 3.0 pts. + 2.5 lbs.

[1] Not for use in Texas, Oklahoma, New Mexico, Arizona or California.

Use the high rate of PROWL for each soil texture above if heavy weed populations are anticipated or if extensive crop residues were present prior to seedbed preparation. Use the 3 pint rate of PROWL for heavy clay soils.

NOTE: If the cotton plant is stressed during early development, application of PROWL plus Zorial at the label rate may result in temporary bleaching or chlorosis of the leaves from which the plant will recover. If cotton crop loss occurs due to weather conditions, cotton can be replanted. If replanting is necessary, a light reworking of the soil will retain the herbicidal activity of PROWL plus Zorial. DO NOT feed forage or graze livestock in treated cotton fields. Observe all precautions and limitations on the Zorial label.

Preemergence Application of PROWL Alone

Uniformly apply PROWL in water or in liquid fertilizer. Refer to **Applications with Liquid Fertilizers** section (see table of contents for page number) for liquid fertilizer compatibility determinations.

Apply PROWL preemergence at planting or up to 2 days after planting. Apply to a seedbed which is firm and free of clods and trash. DO NOT APPLY POSTEMERGENCE. Preemergence treatments are most effective in controlling weeds when adequate rainfall or overhead irrigation is received within 7 days after application.

A shallow cultivation is recommended if soil crusting or soil compaction occurs. If weeds begin to germinate or adequate moisture is not received within 7 days after application, use shallow tillage (rotary hoe or light harrow) and make certain cotton seeds are below the tilled area.

A shallow, thorough mixing of the soil after herbicide application with a rotary hoe or similar implement will improve consistency of performance and reduce dependency on rainfall for activation.

Broadcast Rate per Acre for PROWL Preemergence In Cotton

Soil Texture	PROWL (pints)
COARSE	1 to 1.5
MEDIUM	1.5 to 2.0
FINE	1.5 to 3.0

DO NOT USE on peat or muck soils.

The high rates of PROWL for each soil texture, as stated above, should be used if heavy weed populations are anticipated or if extensive crop residues were present prior to seedbed preparation.

NOTE: DO NOT feed forage or graze livestock in treated cotton fields. If loss of cotton crop occurs due to adverse weather conditions, cotton or soybeans can be replanted the same year into treated soil without adverse effects. If replanting is necessary, a light reworking of the soil will retain the herbicidal activity of PROWL.

Preemergence Application of PROWL Plus Cotoran

PROWL plus Cotoran 80WP or 4L preemergence tank mixture may be applied immediately after planting for weed control in cotton grown in all states except West Texas, Oklahoma, New Mexico, Arizona and California.

Apply PROWL plus Cotoran in water or nitrogen solution with ground equipment immediately after planting. Make sure cotton seeds are placed 1 inch or deeper below soil surface. Apply PROWL plus Cotoran immediately after planting or crop injury may result. Refer to **Applications with Liquid Fertilizers** section (see table of contents for page number) for liquid fertilizer compatibility determinations. Maintain good agitation at all times until spraying is completed.

PROWL plus Cotoran 80WP or 4L is most effective in controlling weeds when adequate rainfall or overhead irrigation is received within 7 days after application. A shallow cultivation is recommended if soil crusting or soil compaction occurs. If weeds begin to germinate or adequate moisture is not received within 7 days after application, use shallow tillage (rotary hoe or light harrow) and make certain cotton seeds are below the tilled area.

Weeds Controlled

When applied as directed, PROWL plus Cotoran will control the following weeds in addition to those controlled by PROWL alone:
Jimsonweed *(Datura stramonium)*
Prickly sida *(Sida spinosa)*
Cocklebur* *(Xanthium strumarium)*
Hemp Sesbania* *(Sesbania exaltata)*
Morningglory* *(Ipomoea* spp.*)*
Common ragweed* *(Ambrosia artemisiifolia)*
Pennsylvania Smartweed* *(Polygonum pensylvanicum)*
Sicklepod* *(Cassia obtusifolia)*
Velvetleaf* *(Abutilon theophrasti)*

* Cotoran 4L only

PROWL plus Cotoran will aid in the control of and reduce competition from:
Cocklebur *(Xanthium strumarium)*
Annual morningglory *(Ipomoea* spp.*)*

Broadcast Rate[1] Per Acre of PROWL Plus Cotoran Preemergence Tank Mixture in Cotton

Soil Texture/States	PROWL + Cotoran 80WP	PROWL + Cotoran 4L
COARSE Arkansas, Louisiana and Mississippi	1.0 to 1.5 pts. + 1.0 lb.	1.0 to 1.5 pts. + 1.6 pts.
Other states, except West Texas, Oklahoma, New Mexico, Arizona, and California	1.0 to 1.5 pts. + 1.25 lbs.	1.0 to 1.5 pts. + 2.0 pts.
MEDIUM All states, except West Texas, Oklahoma, New Mexico Arizona, and California	1.5 to 2.0 pts. + 1.25 to 2.0 lbs.	1.5 to 2.0 pts. + 2.0 pts. to 3.2 pts.
FINE All states, except West Texas, Oklahoma, New Mexico, Arizona, and California	1.5 to 3.0 pts. + 2.0 to 2.5 lbs.	1.5 to 3.0 pts. + 3.2 pts. to 4.0 pts.

[1] For calculation of band treatment rate, see **Spraying Instructions** section (see table of contents for page number).

Use the 3 pint rate of PROWL for heavy clay soils.

NOTE: The use of Cotoran 80WP or 4L as a preemergence application, following the use of a systemic insecticide at planting, may result in injury to cotton. DO NOT plant crops other than cotton within 6 months of the last application of Cotoran or injury may result. DO NOT feed foliage from treated cotton plants or gin trash to livestock. Observe all precautions and limitations on the Cotoran label.

Preemergence Application of PROWL Plus Zorial

PROWL plus Zorial 80WP or 4L preemergence tank mixture may be applied immediately after planting for weed control in cotton grown in all states except Texas, Oklahoma, New Mexico, Arizona and California.

Apply PROWL plus Zorial with ground equipment immediately after planting. Make sure cotton seeds are placed 1 inch or deeper below soil surface. Apply PROWL plus Zorial immediately after planting or crop injury may result. Maintain good agitation at all times until spraying is completed.

PROWL plus Zorial is most effective in controlling weeds when adequate rainfall or overhead irrigation is received within 7 days after application. A shallow cultivation is recommended if soil crusting or soil compaction occurs. If weeds begin to germinate or adequate moisture is not received within 7 days after application, use shallow tillage (rotary hoe or light harrow) and make certain cotton seeds are below the tilled area.

Weeds Controlled

When applied as directed, PROWL plus Zorial will control the following weeds in addition to those controlled by PROWL alone:
Common ragweed *(Ambrosia artemisiifolia)*
Pennsylvania smartweed *(Polygonum pensylvanicum)*
Prickly sida *(Sida spinosa)*
Sicklepod *(Cassia obtusifolia)*
Spurred anoda *(Anoda cristata)*

PROWL plus Zorial will aid in the control of and reduce competition from:
Cocklebur *(Xanthium strumarium)*
Annual morningglory *(Ipomoea* spp.*)*
Nutsedge *(Cyperus* spp.*)*

Broadcast Rate[1] Per Acre of PROWL Plus Zorial Preemergence Tank Mixture in Cotton

Soil Texture	PROWL + Zorial 80WP
COARSE	1.0 to 1.5 pts. + 1.25 lbs.
MEDIUM	1.5 to 2.0 pts. + 1.9 lbs.
FINE	1.5 to 3.0 pts. + 2.5 lbs.

[1] For calculation of band treatment rate, see **Spraying Instructions** section (see table of contents for page number).

Use the 3 pint rate of PROWL for heavy clay soils.

NOTE: If the cotton plant is stressed during early development, application of PROWL plus Zorial at the label rate may result in temporary bleaching or chlorosis of the leaves from which the plant will recover. If cotton crop loss occurs due to weather conditions, cotton can be replanted. If replanting is necessary, a light reworking of the soil will retain the herbicidal activity of PROWL plus Zorial. DO NOT feed forage or graze livestock in treated cotton fields. Observe all precautions and limitations on the Zorial label.

PROWL and PROWL Tank Mixtures in Minimum and No-till Systems

PROWL alone or in a tank mixture may be applied preplant up to 14 days before planting or preemergence up to 2 days after planting in minimum or no-till cotton as follows:

- PROWL alone
- PROWL + Roundup
- PROWL + Gramozone Super

Roundup and Gramoxone Super are used to kill existing vegetative growth, if present at application. DO NOT apply PROWL or PROWL tank mixtures postemergence to cotton. No-till applications are most effective in controlling weeds when adequate rainfall or overhead irrigation is received within 7 days after application. DO NOT use in no-till cotton in California.

Broadcast Rate per Acre for No-till Cotton

Soil Texture (Up to 3% organic matter[1])	PROWL (pints)
COARSE	1.5 to 2.0
MEDIUM	2.0 to 3.0
FINE	3.0 to 4.0

* Use the high rates of PROWL for each soil texture, as stated above, if heavy weed populations are anticipated, or if extensive crop residues are present.

[1] This use is not recommended for soils with more than 3% organic matter.

Preplant Incorporated Application of PROWL for Control of Rhizome Johnsongrass

PROWL applied preplant incorporated for two consecutive years will provide control of rhizome johnsongrass *(Sorghum halepense)* in cotton at the rates recommended for soil textures listed below. This use is not recommended for Arizona, New Mexico and California. Rhizome johnsongrass will be suppressed after the first year and controlled after the second year.

Before application, use a chisel plow or similar implement to bring johnsongrass rhizomes to the surface. Chop rhizomes into small pieces with a disk harrow set to cut 4 to 6 inches deep and operated in two different directions at 4 to 6 mph.

PROWL is to be incorporated into the soil within 7 days after application prior to planting. For maximum control of rhizome johnsongrass, incorporate PROWL as soon as possible after application. Deep and thorough incorporation of PROWL is necessary for control of rhizome johnsongrass. Mechanical incorporation can be achieved by the following methods:

a) Disk harrow set to cut 4 to 6 inches deep and operated in two different directions at 4 to 6 mph.

b) PTO-driven equipment (tillers, cultivators, hoes) set to cut 3 to 4 inches deep and operated one time at 4 mph or less.

For johnsongrass escapes during the crop season, cultivation and/or application of registered postemergence herbicides are recommended. Follow the directions for use on the labels of the respective herbicides.

Broadcast Rate Per Acre of PROWL Preplant Incorporated for Control of Rhizome Johnsongrass in Cotton
APPLY FOR TWO CONSECUTIVE YEARS
PROWL (pints)

Soil Texture—Up to 3% Organic Matter[1]	
COARSE	2.0
MEDIUM	3.0
FINE	4.0

[1] This use is not recommended for soils with more than 3% organic matter.

NOTE: DO NOT feed forage or graze livestock in treated cotton fields.

If cotton crop loss occurs due to weather conditions, cotton or soybeans can be replanted the same year into treated soil without adverse effects. If replanting is necessary, DO NOT rework the soil deeper than the treated zone.

SOYBEANS

PROWL can be applied in soybeans as follows (DO NOT USE PROWL IN SOYBEANS IN CALIFORNIA):

Preplant Incorporated

PROWL alone
PROWL + SCEPTER
PROWL + Sencor/Lexone
PROWL + Amiben
PROWL + Command
PROWL + Canopy
PROWL + Amiben + Sencor/Lexone
PROWL + Lasso
PROWL + Dual
PROWL + PURSUIT
PROWL + Preview

Sequential

PROWL preplant incorporated followed by preemergence applications of:
Sencor/Lexone
Amiben
Lorox

PROWL preplant incorporated or preemergence followed by postemergence applications of:
SCEPTER
SCEPTER O.T.
PURSUIT

Preemergence

PROWL alone
PROWL + SCEPTER
PROWL + Sencor/Lexone
PROWL + Lasso
PROWL + Dual
PROWL + Canopy
PROWL + Lorox
PROWL + Amiben + Sencor/Lexone
PROWL + Amiben
PROWL + PURSUIT
PROWL + Preview

Refer to **PROWL Tank Mixtures in Minimum and No-Till Systems** subsection for use of PROWL in minimum and no-till soybeans.

PROWL will provide control or aid in the control of and reduce competition from the following special weeds as detailed in the appropriate subsection:
Itchgrass
Red rice
Rhizome johnsongrass
Shattercane

DO NOT use PROWL on peat or muck soils.

Preplant Incorporated Applications of PROWL Alone

Apply PROWL with ground equipment or aircraft immediately before planting or up to 60 days prior to planting. Incorporate prior to planting and within 7 days of application. Refer to **Incorporation Instructions** section (see table of contents for page number) for incorporation directions.

The following rates for PROWL alone preplant incorporated should be used when no tank mixtures or sequential preemergence treatments are planned. Use PROWL alone rates when sequential postemergence broadleaf herbicide treatments are planned.

Broadcast Rate Per Acre of PROWL Alone Preplant Incorporated in Soybeans

SOUTHERN STATES (Including Eastern Coastal Plans)[1]	
Soil Texture	**PROWL (pints)**
COARSE	1.0
MEDIUM	1.5
FINE	2.0

[1] See map at the end of this label for specific Southern states. DO NOT use in California.

For heavy clay soils, apply PROWL at the broadcast rate of 3 pints per acre.

When heavy weed populations are anticipated, apply PROWL at the broadcast rate per acre of 1.5 pints on coarse soils, 2.0 pints on medium soils, and 3.0 pints on fine soils.

Broadcast Rate Per Acre of PROWL Alone Preplant Incorporated in Soybeans

NORTHERN STATES[1]		
	PROWL (pints)	**PROWL (pints)**
Soil Texture	**Up to 3% Organic Matter**	**More than 3% Organic Matter**
COARSE	1.0 to 2.0	2.0
MEDIUM	1.5 to 2.5	2.5 to 3.0
FINE	2.0 to 3.0	3.0

[1] See map at end of this label for specific Northern states.

Use the 3 pint rate for heavy clay soils.

NOTE: Livestock can graze or be fed soybean forage from treated fields. If soybean crop loss occurs due to weather conditions, soybeans or any crop registered for PROWL preplant incorporated use can be replanted. If replanting is necessary, DO NOT rework the soil deeper than the treated zone.

PROWL Preemergence General Instructions

Use the following instructions when applying recommended PROWL preemergence tank mixtures in soybeans. Apply PROWL preemergence tank mixtures at planting or up to 2 days after planting. Apply to a seedbed which is firm and free of clods and trash. DO NOT APPLY POSTEMERGENCE. Preemergence treatments are most effective in controlling weeds when adequate rainfall or overhead irrigation is received within 7 days after application.

If moisture is insufficient to activate the herbicides, a shallow cultivation (preferably with a rotary hoe) should be made after emergence of soybeans, but while weeds are small enough to be controlled by mechanical cultivation.

A shallow, thorough mixing of soil after herbicide application with a rotary hoe or similar implement will improve consistency of performance and reduce dependency on rainfall to activate herbicides.

DO NOT make applications of PROWL preemergence north of Interstate 80.

Broadcast Rates per Acre for PROWL Preemergence in Soybeans

	PROWL (pints)	**PROWL (pints)**
Soil Texture	**Up to 3% Organic Matter**	**More than 3% Organic Matter**
COARSE	1 to 1.5	1.5
MEDIUM	1.5 to 2.0	1.5 to 2.0
FINE	1.5 to 2.0	2.0 to 2.5

The high rates of PROWL for each soil texture, as stated above, should be used if heavy weed populations are anticipated or if extensive crop residues were present prior to seedbed preparation.

NOTE: Livestock can graze or be fed forage from treated soybean fields. In the event of a crop loss due to weather conditions, soybeans can be replanted without adverse effects the same year. If replanting is necessary, a light reworking of the soil will retain the herbicidal activity of PROWL.

PROWL Plus SCEPTER

The following PROWL Plus SCEPTER herbicide treatments can be used in soybeans. Apply PROWL plus SCEPTER treatments with ground equipment or aircraft. Use PROWL and SCEPTER rates listed in table below. See SCEPTER label for use restrictions. See Rotational Crop Restrictions below.

Uniformly apply PROWL plus SCEPTER with properly calibrated equipment in 10 or more gallons of water or 20 or more gallons of liquid fertilizer per acre. Use sprayers equipped with nozzles that provide accurate and uniform application. To minimize drift, use a maximum spray pressure of 40 psi. Do not apply SCEPTER 70DG in liquid fertilizers.

For aerial applications, apply PROWL plus SCEPTER with properly calibrated aerial equipment in 5 or more gallons of water per acre. To avoid injury to sensitive crops from drift, aerial applicators must adhere to the SPECIAL AERIAL USE DIRECTIONS AND PRECAUTIONS found in the SCEPTER label.

1. Preplant Incorporated tank mixture of PROWL plus SCEPTER

Incorporate into the soil within 7 days after application, and plant soybeans within 45 days after application. Refer to **Incorporation Instructions** section (see table of contents for page number) for incorporation directions.

2. Preemergence tank mixture of PROWL plus SCEPTER

Apply at planting or up to 2 days after planting. DO NOT APPLY postemergence to soybeans. As with other herbicides applied preemergence, rainfall or irrigation is necessary to activate PROWL plus SCEPTER in the soil. If sufficient moisture to activate PROWL plus SCEPTER is not received within 7 days after application, a thorough, shallow tillage or cultivation is recommended (preferably with a rotary hoe) for optimum weed control.

DO NOT make applications of PROWL preemergence north of Interstate 80.

3. Sequential applications of PROWL preplant incorporated or preemergence followed by SCEPTER postemergence.

Refer to the appropriate section of this label for PROWL use rates and instructions. See the SCEPTER label for use rates, instructions, and weeds controlled.

Weeds Controlled

When applied as directed, tank mixtures of PROWL plus SCEPTER will control the following weeds in addition to those controlled by PROWL alone:

Shattercane* (*Sorghum bicolor*)
Bristly starbur (*Acanthospermum hispidum*)
Burcucumber (*Sicyos angulatus*)
Cupgrass, Woolly* (*Eriochloa villosa*)
Cocklebur, common (*Xanthium strumarium*)
Galinsoga (*Galinsoga* spp.)
Jimsonweed (*Datura stramonium*)
Mallow, Venice (*Hibiscus trionum*)
Morningglory
- Palmleaf (*Ipomoea wrightii*)
- Pitted (*Ipomoea lacunosa*)
- Smallflower (*Jacquemontia tamnifolia*)

Mustard Species (*Brassica* spp.)
Nightshade, Eastern black* (*Solanum ptycanthum*)
Poinsettia, Wild (*Euphorbia heterophylla*)
Ragweed,
- Common (*Ambrosia artemisiifolia*)
- Giant* (*Ambrosia trifida*)

Sida, Prickly (Teaweed) (*Sida spinosa*)
Smartweed
- Ladysthumb (*Polygonum persicaria*)
- Pennsylvania (*Polygonum pensylvanicum*)

Sunflower, common (*Helianthus annuus*)
Velvetleaf* (*Abutilon theophrasti*)

PROWL plus SCEPTER soil applied treatments will aid in the control and reduce competition from:

Beggarweed, Florida (*Desmodium tortuosum*)
Mexicanweed (*Caperonia castaniflia*)
Morningglory**
- Entireleaf (*Ipomoea hederacea var. integriuscula*)
- Ivyleaf (*Ipomoea hederacea*)
- Tall (*Ipomoea purpurea*)

Nutsedge, Yellow* (*Cyperus esculentus*)
Sicklepod (*Cassia obstusifolia*)

* Preplant incorporated applications only. Use the high rates of PROWL if heavy weed populations are anticipated.
** For best activity on morningglory species, use a preplant incorporated application.

Broadcast Rates of PROWL Plus Scepter, Preplant Incorporated or Preemergence Tank Mix Applications in Soybeans[1]

	Less than 3% organic matter		More than 3% organic matter	
Soil Texture	PROWL (pints)	SCEPTER[2] (pints)	PROWL (pints)	SCEPTER[2] (pints)
Coarse	1-1.5	⅔	1.5	⅔
Medium	1.5-2	⅔	1.5-2	⅔
Fine	.5-2	⅔	2-2.5	⅔

NOTE: Use the higher rate of PROWL if heavy weed populations are anticipated or if shattercane will be present.

[1] See USE AREA section of the SCEPTER label for states where PROWL plus SCEPTER can be applied.
[2] When using SCEPTER 70DG use equivalent rates. A SCEPTER 70DG rate of 5.6 ounces per acre is equivalent to ⅔ pint per acre of SCEPTER.

ROTATIONAL CROP RESTRICTIONS

Follow all rotational crop restrictions on the PROWL and SCEPTER labels. Always follow the most restrictive label.

In the event of a crop loss due to weather conditions, soybeans can be replanted. DO NOT work the soil deeper than 2 inches.

DO NOT GRAZE OR FEED TREATED SOYBEAN FORAGE, HAY OR STRAW TO LIVESTOCK

PROWL Followed by SCEPTER O.T.

Preplant incorporated or preemergence applications of PROWL in soybeans may be followed by postemergence applications of SCEPTER O.T. See SCEPTER O.T. label for use restrictions and precautions.

Refer to the appropriate section of this label for PROWL use rates and instructions. See the SCEPTER O.T. label for use rates, instructions, and weeds controlled.

Follow all rotational crop restrictions on the PROWL and SCEPTER O.T. labels. Always follow the most restrictive label.

PROWL Plus PURSUIT

The following PROWL plus PURSUIT herbicide treatments can be used in soybeans. Apply PROWL plus PURSUIT treatments with ground equipment. Use PROWL and PURSUIT rates listed in table below. See PURSUIT label for use restrictions. See Rotational Crop Restrictions below.

Uniformly apply PROWL plus PURSUIT with properly calibrated equipment in 10 or more gallons of water or 20 or more gallons of liquid fertilizer per acre. Use sprayers equipped with nozzles that provide accurate and uniform application. To minimize drift, use a maximum spray pressure of 40 psi.

1. **Preplant Incorporated tank mixture of PROWL plus PURSUIT**
   Incorporate into the soil within 7 days after application, and plant soybeans within 45 days after application. Refer to **Incorporation Instructions** section (see table of contents for page number) for incorporation directions.
2. **Preemergence tank mixture of PROWL plus PURSUIT**
   Apply at planting or up to 2 days after planting. DO NOT APPLY postemergence to soybeans. As with other herbicides applied preemergence, rainfall or irrigation is necessary to activate PROWL plus PURSUIT in the soil. If sufficient moisture to activate PROWL plus PURSUIT is not received within 7 days after application, a thorough, shallow tillage or cultivation is recommended (preferably with a rotary hoe) for optimum weed control.

DO NOT make applications of PROWL preemergence north of Interstate 80.

Sequential applications of PROWL preemergence or preplant Incorporated followed by PURSUIT postemergence

Refer to the appropriate section of this label for PROWL use rates and instructions. See the PURSUIT label for use rates and instructions.

Weeds Controlled

When applied as directed, PROWL plus PURSUIT will control the following weeds in addition to those controlled by PROWL alone. See the PURSUIT label for weed height limitations for PURSUIT postemergence applications.

Artichoke, Jerusalem[1] (*Helianthus tuberosus*)
Cocklebur, common[2,3] (*Xanthium strumarium*)
Foxtail
- giant green (*Setaria* sp.)
- robust purple (*Setaria viridis, var. robusta-purp*)
- robust white (*Setaria viridis, var. robusta-alba*)

Galinsoga (*Galinsoga* spp.)
Itchgrass[4,5] (*Rottboellia exaltata*)
Jimsonweed[3] (*Datura stramonium*)
Johnsongrass, rhizome[5,6] (*Sorghum halepense*)
Lambsquarters, common[3] (*Chenopodium album*)
Mallow, Venice[1] (*Hibiscus trionum*)
Morningglory
- entireleaf[3] (*Ipomoea hederacea var. integriuscula*)
- ivyleaf[3] (*Ipomoea hederacea*)
- pitted[3] (*Ipomoea lacunosa*)
- smallflower (*Jacquemontia tamnifolia*)
- tall[3] (*Ipomoea purpurea*)

Mustard species (*Brassica* spp.)
Nutsedge, yellow[5] (*Cyperus esculentus*)
Nightshade
- black (*Solanum nigrum*)
- Eastern black (*Solanum ptycanthum*)
- hairy (*Solanum sarrachoides*)

Puncturevine (*Tribulus terrestris*)
Ragweed

common (*Ambrosia artemisiifolia*)
giant[3] (*Ambrosia trifida*)
Shattercane[6] (*Sorghum bicolor*)
Sida, prickly (teaweed)[3] (*Sida spinosa*)
Smartweed
ladysthumb (*Polygonum persicaria*)
Pennsylvania (*Polygonum pensylvanicum*)
Spurge
prostrate (*Euphorbia humistrata*)
spotted (*Euphorbia maculata*)
Sunflower, common[3] (*Helianthus annuus*)
Velvetleaf[3] (*Abutilon theophrasti*)
Waterhemp, tall (*Amaranthus tuberculatos*)
Wild proso millet[4,5] (*Panicum miliaceum*)
Woolly cupgrass[6] (*Eriochloa villosa*)

[1] PROWL preplant incorporated or preemergence followed by PURSUIT postemergence only.
[2] Heavy infestations will only be controlled by PROWL preplant incorporated or preemergence followed by PURSUIT postemergence.
[3] For soil applied tank mixtures, preplant incorporated applications give better results than preemergence applications.
[4] PROWL plus PURSUIT preplant incorporated only.
[5] Suppression only.
[6] PROWL plus PURSUIT preplant incorporated, or PROWL soil applied followed by PURSUIT postemergence only.

Broadcast Rates of PROWL Plus Pursuit, Preplant Incorporated or Preemergence Tank Mix Applications in Soybeans[1]

	Less than 3% organic matter		More than 3% organic matter	
Soil Texture	PROWL (pints)	PURSUIT (pints)	PROWL (pints)	PURSUIT (pints)
Coarse	1-1.5	1/4	1.5	1/4
Medium	1.5-2	1/4	1.5-2	1/4
Fine	1.5-2	1/4	2-2.5	1/4

NOTE: Use the higher rate of PROWL if heavy weed populations are anticipated or if shattercane will be present.

[1] See USE AREA section of the PURSUIT label for states where PROWL plus PURSUIT can be applied.

DO NOT GRAZE OR FEED TREATED SOYBEAN FORAGE, HAY, OR STRAW TO LIVESTOCK.

ROTATIONAL CROP RESTRICTIONS

Follow all rotational crop restrictions on the PROWL and PURSUIT labels. Always follow the most restrictive label.

PROWL Plus Sencor/Lexone

The following PROWL plus Sencor/Lexone treatments may be used in soybeans. Apply PROWL plus Sencor/Lexone treatments with ground equipment or aircraft. Refer to **Incorporation Instructions** section (see table of contents for page number) for incorporation directions. Use PROWL and Sencor/Lexone rates listed in table below. See special precautions below.

1. PROWL Preplant Incorporated Followed by Sencor/Lexone Preemergence

Apply PROWL preplant incorporated. Follow with a preemergence treatment of Sencor/Lexone as directed on the Sencor or Lexone label.

2. Preplant Incorporated Tank Mixture of PROWL plus Sencor/Lexone

PROWL plus Sencor/Lexone must be incorporated into the soil within 7 days after application and soybeans must be planted no later than 7 days after application. For best results, incorporation should be in the top 1 to 2 inches of soil.

3. Preemergence Tank Mixture of PROWL plus Sencor/Lexone

Apply PROWL plus Sencor/Lexone preemergence tank mixture at planting or up to 2 days after planting. Refer to **PROWL Preemergence General Instructions** subsection above. DO NOT APPLY POSTEMERGENCE.

Special Precautions for PROWL Plus Sencor or Lexone Combinations

Soybean seed should be planted at least 1 1/2 inches below soil surface. Applications should not be made to soils having calcareous surface or a pH of 7.5 or higher. Soil incorporation deeper than recommended will reduce weed control and can result in crop injury.

The PROWL plus Sencor or Lexone treatment can be applied only once per cropping season. DO NOT USE on sensitive soybean varieties—Altona, Coker 102 and 156, Govan, Semmes, Tracy or Vansoy.

Follow cropping restrictions on Sencor and Lexone labels.

Weeds Controlled

PROWL plus Sencor/Lexone will control the following weeds in addition to those controlled by PROWL alone:

Common ragweed (*Ambrosia artemisiifolia*)
Jimsonweed (*Datura stramonium*)
Pennsylvania smartweed (*Polygonum pensylvanicum*)
Prickly sida (*Sida spinosa*)
Hemp Sesbania (*Sesbania exaltata*)
Velvetleaf (Buttonweed) (*Abutilon theophrasti*)
Venice mallow (*Hibiscus trionum*)
Wild mustard (*Brassica kaber*) (medium and fine soils only)

PROWL plus Sencor/Lexone will aid in the control of and reduce competition from cocklebur (*Xanthium strumarium*). Use either a preplant incorporated application of PROWL followed by a preemergence application of Sencor/Lexone or a preemergence tank mixture of PROWL plus Sencor/Lexone where severe cocklebur infestations are anticipated. See appropriate subsection below for PROWL plus Sencor/Lexone use for shattercane control.

Broadcast Rates Per Acre of PROWL Plus Sencor/Lexone Preplant Incorporated or Preemergence Tank Mixtures or PROWL Preplant Incorporated followed by Sencor/Lexone Preemergence

	SOUTHERN STATES (Including Eastern Coastal Plains)[1]	NORTHERN STATES[1]	
	PROWL + Sencor/Lexone DF[2]	PROWL + Sencor/Lexone DF[2]	PROWL + Sencor/Lexone DF[2]
Soil Texture[3]		1/2 to 3% Organic Matter	More Than 3% Organic Matter
COARSE	1.0 to 1.5 pts. + 0.33 lb.	1.0 to 1.5 pts. + 0.5 lb.	1.5 pts. + 0.5 lb.
MEDIUM	1.5 to 2.0 pts. + 0.5 lb.	1.5 to 2.0 pts. + 0.5 lb.	1.5 to 2.0 pts. + 0.5 to 0.67 lb.
FINE	1.5 to 2.0 pts. + 0.5 to 0.67 lb.	1.5 to 2.0 pts. + 0.5 to 0.67 lb.	2.0 to 2.5 pts. + 0.67 lb.

[1] See map at end of this label for specific states.
[2] When using Sencor 4F, Sencor 50WP, Lexone 4L, or Lexone 50WP, use equivalent rates. One pint of the flowable formulation or 1 lb. of the wettable powder formulation equals 0.67 lb. of Sencor DF or Lexone DF. See conversion chart in **Formulation Conversion Tables** section (see table of contents for page number).
[3] DO NOT USE on sands. DO NOT USE on loamy sands or on sandy loams that contain less than 1% organic matter, nor on other soils with less than 1/2% organic matter as crop injury may result. This tank mixture is not recommended for use on soils with less than 2% organic matter in the Coastal Plain of New Jersey or the Delmarva Peninsula.

NOTE: In the event of a crop loss due to weather conditions, soybeans can be replanted without adverse effects the same year. If replanting is necessary, DO NOT rework the soil. Livestock can graze or be fed soybean forage from treated fields 40 days after application. Observe all precautions and limitations on Sencor and Lexone labels.

PROWL plus Amiben

The following PROWL plus Amiben treatments may be used in soybeans in the Northern States. See the map at the end of this label for specific states. Apply PROWL plus Amiben treatments with ground equipment or aircraft. Refer to **Incorporation Instructions** section (see table of contents for page number) for incorporation directions. Use PROWL and Amiben rates listed in table below.

1. PROWL Preplant Incorporated Followed by Amiben Preemergence

Apply PROWL preplant incorporated. Follow with a preemergence treatment of Amiben as directed on the Amiben label.

2. Preplant Incorporated Tank Mixture of PROWL plus Amiben

PROWL plus Amiben must be incorporated into the soil within 7 days after application and soybeans must be planted no later than 7 days after application. For best results, incorporation should be in the top 1 to 2 inches of soil.

3. Preemergence Tank Mixture of PROWL plus Amiben

Apply PROWL plus Amiben preemergence tank mixture at planting or up to 2 days after planting. Refer to previous subsection on **PROWL Preemergence General Instructions.** DO NOT APPLY POSTEMERGENCE.

Weeds Controlled

When applied as directed, PROWL plus Amiben will control the following weeds in addition to those controlled by PROWL alone:

Common ragweed (*Ambrosia artemisiifolia*)
Mustard (*Brassica* spp.)
Pennsylvania smartweed (*Polygonum pensylvanicum*)
Velvetleaf (Buttonweed) (*Abutilon theophrasti*)

Broadcast Rates Per Acre of PROWL Plus Amiben Preplant Incorporated or Preemergence Tank Mixtures or PROWL Preplant Incorporated Followed by Amiben Preemergence

	PROWL + Amiben	PROWL + Amiben
Soil Texture	Up to 3% Organic Matter	More Than 3% Organic Matter
COARSE	DO NOT USE	DO NOT USE
MEDIUM	1.5 to 2.0 pts. + 1.0 gallon	1.5 to 2.0 pts. + 1.0 gallon
FINE	1.5 to 2.0 pts. + 1.0 gallon	2.0 to 2.5 pts. + 1.0 gallon

The high rate of PROWL for each soil texture above should be used if velvetleaf or Pennsylvania smartweed infestations are anticipated.

NOTE: In the event of crop loss due to weather conditions, soybeans can be replanted without adverse effects the same year. If replanting is necessary, DO NOT rework the soil. Livestock can graze or be fed soybean forage from treated fields. Observe all precautions and limitations on Amiben label.

PROWL Plus Amiben Plus Sencor/Lexone

The following PROWL plus Amiben plus Sencor/Lexone treatments may be used in soybeans in the Northern States. Refer to map at the end of this label for specific states. Apply PROWL plus Amiben plus Sencor/Lexone treatments with ground equipment or aircraft. When mixing, add Sencor or Lexone first, then Amiben followed by PROWL. Refer to **Mixing Instructions** section (see table of contents for page) for complete mixing directions. Refer to **Incorporation Instructions** section (see table of contents for page number) for directions for incorporation. Use PROWL plus Amiben plus Sencor/Lexone rates listed in table following. See special precautions stated previously for PROWL plus Sencor/Lexone combinations.

1. Preplant Incorporated Tank Mixture of PROWL plus Amiben plus Sencor/Lexone

PROWL plus Amiben plus Sencor/Lexone must be incorporated into the soil within 7 days after application and soybeans must be planted no later than 7 days after application. For best results, incorporation should be in the top 1 to 2 inches of soil.

2. Preemergence Tank Mixture of PROWL plus Amiben plus Sencor/Lexone

Apply PROWL plus Amiben plus Sencor/Lexone preemergence tank mixture at planting or up to 2 days after planting. Refer to **PROWL Preemergence General Instructions** subsection above. DO NOT APPLY POSTEMERGENCE.

Weeds Controlled

When applied as directed, PROWL plus Amiben plus Sencor/Lexone will control the following weeds in addition to those controlled by PROWL alone:

Common ragweed (*Ambrosia artemisiifolia*)
Mustard (*Brassica* spp.)
Pennsylvania smartweed (*Polygonum pensylvanicum*)
Prickly sida (*Sida spinosa*)
Velvetleaf (Buttonweed) (*Abutilon theophrasti*)
Venice mallow (*Hibiscus trionum*)

PROWL plus Amiben plus Sencor/Lexone preemergence will aid in the control of and reduce competition from cocklebur (*Xanthium strumarium*).

PROWL plus Amiben plus Sencor/Lexone preplant incorporated or preemergence will aid in the control of and reduce competition from jimsonweed (*Datura stramonium*).

Broadcast Rates Per Acre of PROWL Plus Amiben Plus Sencor/Lexone Preplant Incorporated or Preemergence Tank Mixtures

	PROWL + Amiben + Sencor/Lexone DF[1]	PROWL + Amiben + Sencor/Lexone DF[1]
Soil Texture[2]	½ to 3% Organic Matter	More Than 3% Organic Matter
COARSE	1.5 pts. + 6.0 pts. + 0.33 lb.	1.5 pts. + 6.0 pts. + 0.33 lb.
MEDIUM	1.5 to 2.0 pts. + 6.0 to 8.0 pts. + 0.33 to 0.5 lb.	1.5 to 2.0 pts. 6.0 to 8.0 pts. + 0.33 to 0.5 lb.
FINE	1.5 to 2.0 pts. + 6.0 to 8.0 pts. + 0.33 to 0.5 lb.	2.0 to 2.5 pts. + 8.0 pts. + 0.5 lb.

[1] When using Sencor 4F, Sencor 50WP, Lexone 4L, or Lexone 50WP, use equivalent rates. One pint of the flowable formulation or 1 lb. of the wettable powder formulation equals 0.67 lb. of Sencor DF or Lexone DF. See conversion chart in **Formulation Conversion Tables** section (see table of contents for page number).

[2] DO NOT USE on sands. DO NOT USE on loamy sands or on sandy loams that contain less than 1% organic matter, nor on other soils with less than ½% organic matter as crop injury may result.

The high rates of PROWL, Amiben and Sencor/Lexone for each soil texture, as stated above, should be used if cocklebur, jimsonweed, Pennsylvania smartweed or velvetleaf populations are expected.

NOTE: In the event of a crop loss due to weather conditions, soybeans can be replanted without adverse effects the same year. If replanting is necessary, DO NOT rework the soil. DO NOT use foliage for livestock feed or forage. Observe all precautions and limitations on Amiben, Sencor and Lexone labels.

PROWL Plus Lorox

The following PROWL plus Lorox treatments may be used in soybeans. Apply PROWL plus Lorox with ground equipment or aircraft. Refer to **Incorporation Instructions** section for incorporation directions. Use PROWL and Lorox rates listed in table below.

1. PROWL Preplant Incorporated Followed by Lorox Preemergence.

Apply PROWL preplant incorporated. Follow with a preemergence treatment of Lorox as directed on the Lorox label.

2. Preemergence Tank Mixtures of PROWL plus Lorox.

Apply PROWL plus Lorox preemergence tank mixture at planting or up to 2 days after planting. Refer to previous subsection on **PROWL Preemergence General Instructions** subsection above. DO NOT APPLY POSTEMERGENCE.

Weed Species Controlled

When applied as directed PROWL plus Lorox will control the following weeds in addition to those controlled by PROWL alone:
Mustard (*Brassica* spp.)
Pennsylvania smartweed (*Polygonum pensylvanicum*)
Velvetleaf (Buttonweed) (*Abutilon theophrasti*)

PROWL plus Lorox will aid in the control of and reduce competition from:
Common ragweed (*Ambrosia artemisiifolia*)

Broadcast Rate Per Acre of PROWL Plus Lorox Preemergence Tank Mixtures or PROWL Preplant Incorporated Plus Lorox Preemergence

	PROWL + Lorox L[1] (pints)	PROWL + Lorox L[1] (pints)
Soil Texture[2]	½ to 3% Organic Matter	More Than 3% Organic Matter
COARSE Sandy loams only	1.0 to 1.5 + 1.0	1.5 + 1.0 to 1.5
MEDIUM	1.5 to 2.0 + 1.0 to 1.5	1.5 to 2.0 + 1.5 to 2.0
FINE	1.5 to 2.0 + 1.5 to 2.0	2.0 to 2.5 + 2.0 to 2.5

[1] When using Lorox 50W use equivalent rates. One pint of the flowable formulation equals 1 lb. of the wettable powder formulation. See conversion chart in **Formulation Conversion Tables** section (see table of contents for page number).

[2] DO NOT USE on sands, loamy sands, or gravelly soils, or on soils containing less than ½% organic matter.

NOTE: Livestock may graze or be fed forage from treated soybean fields. If soybean crop loss occurs due to weather conditions, soybeans may be replanted. If replanting is necessary, thoroughly rework the soil before replanting. Observe all precautions and limitations on Lorox label.

PROWL Plus Lasso

The following PROWL plus Lasso treatments may be used in soybeans. Apply PROWL plus Lasso treatments with ground equipment. Refer to **Incorporation Instructions** section for incorporation directions. Use PROWL and Lasso rates listed in table below.

1. Preplant Incorporated Tank Mixture of PROWL Plus Lasso

PROWL plus Lasso must be incorporated into the soil within 7 days after application and soybeans must be planted no later than 7 days after application. Incorporate into the top 1 to 2 inches of soil.

Incorporation on coarse soils may reduce length of control due to leaching of Lasso with rainfall or irrigation. Incorporation is not recommended on coarse soils in the Southeastern States.

2. Preemergence Tank Mixtures of PROWL Plus Lasso

Apply PROWL plus Lasso preemergence tank mixture at planting or up to 2 days after planting. Refer to previous subsection on **PROWL Preemergence General Instructions** above. DO NOT APPLY POSTEMERGENCE. This treatment must be applied within 5 days of the last preplant tillage.
High intensity or excessive rainfall or excessive irrigation after application may reduce weed control.

Weeds Controlled

When applied as directed, PROWL plus Lasso will control the following weeds in addition to those controlled by PROWL alone:

Black nightshade (*Solanum nigrum*)
Hairy nightshade (*Solanum sarrachoides*)
Red rice (*Oryza sativa*)*
Yellow nutsedge (*Cyperus esculentus*)*

* Apply preplant incorporated only for control of these weeds.

Broadcast Rates of PROWL Plus Lasso Preplant Incorporated or Preemergence Tank Mixtures

	PROWL + Lasso	PROWL + Lasso
Soil Texture	Up to 3% Organic Matter	More Than 3% Organic Matter
COARSE	1.0 to 1.5 pts. + 2.5 qts.	1.5 pts. + 2.5 qts.

MEDIUM	1.5 to 2.0 pts. + 3.0 to 4.0 qts.	1.5 to 2.0 pts. + 3.0 to 4.0 qts.
FINE	1.5 to 2.0 pts. + 3.0 to 4.0 qts.	2.0 to 2.5 pts. + 3.0 to 4.0 qts.

Observe all precautions and limitations on the Lasso label.

PROWL Plus Dual

The following PROWL plus Dual treatments may be used in soybeans. Apply PROWL plus Dual treatments with ground equipment or aircraft. Refer to **Incorporation Instructions** section (see table of contents for page number) for incorporation directions. Use PROWL and Dual rates listed in following table.

1. Preplant Incorporated Tank Mixture of PROWL Plus Dual

PROWL plus Dual must be incorporated into the soil within 7 days after application and soybeans must be planted no later than 14 days after application. Incorporate into the top 1 to 2 inches of soil.

Use a preplant incorporated application if furrow irrigation is used or when a period of dry weather after application is expected. If soybeans are planted on beds, apply and incorporate after bed formation.

2. Preemergence Tank Mixtures of PROWL Plus Dual

Apply PROWL plus Dual preemergence tank mixture at planting or up to 2 days after planting. Refer to previous subsection for **PROWL Preemergence General Instructions** above. DO NOT APPLY POSTEMERGENCE.

Weeds Controlled

When applied as directed, PROWL plus Dual will control the following weeds in addition to those controlled by PROWL alone:
Red rice *(Oryza sativa)*
Yellow nutsedge *(Cyperus esculentus)*
Black nightshade *(Solanum nigrum)*

Broadcast Rates Per Acre of PROWL Plus Dual Preplant Incorporated or Preemergence Tank Mixtures

	PROWL + Dual 8E (pints)	
Soil Texture	Up to 3% Organic Matter	More Than 3% Organic Matter
COARSE	1.0 to 1.5 + 1.5 to 2.0	1.5 + 2.0
MEDIUM	1.5 to 2.0 + 2.0 to 2.5	1.5 to 2.0 + 2.0 to 2.5
FINE	1.5 to 2.0 + 2.0 to 2.5	2.0 to 2.5 + 2.5 to 3.0

Observe all precautions and limitations on the Dual label

PROWL Plus Command 4EC

PROWL plus Command 4EC tank mixture may be applied preplant incorporated for weed control in soybeans. See the Command label for states where PROWL plus Command may be applied.

Uniformly apply PROWL plus Command 4EC with properly calibrated ground equipment in 10 to 40 gallons of water per acre. Use sprayers equipped with nozzles that provide accurate and uniform application. NOTE: the use of an agriculturally approved drift reducing agent is required at finished spray volumes of 10 to 15 gallons per acre.

Apply PROWL plus Command 4EC to the soil surface and uniformly incorporate. See the Command label for incorporation requirements and instructions. DO NOT apply this tank mixture to overly moist or wet soils.

Special Precautions

Command 4EC is a volatile compound. Off-site movement of spray drift or vapors of Command can cause foliar whitening or yellowing of some plants. Prior to using Command, read and stictly follow all precautions and applications instructions on the Command label.

Weeds Controlled

When applied as directed, PROWL plus Command 4EC will control the following weeds in addition to those controlled by PROWL alone:
Velvetleaf (*Abutilon theophrasti*)

Tank mixtures of PROWL plus Command 4EC will reduce competition from:

Cocklebur (*Xanthium strumarium*)

Follow all rotational crop restrictions on the PROWL and Command labels. Always follow the most restrictive label.

Broadcast Rates Per Acre of PROWL Plus Command 4EC Preplant Incorporated

	PROWL (pints) + Command (pints)	PROWL (pints) + Command (pints)
Soil Texture	Up to 3% Organic Matter	More Than 3% Organic Matter
COURSE	1.0 to 1.5 + 0.75	1.5 + 0.75
MEDIUM	1.5 to 2.0 + 1.0	1.5 to 2.0 + 1.0
FINE	1.5 to 2.0 + 1.5	2.0 to 2.5 + 1.5

When heavy weed populations are anticipated, apply PROWL at the high rate for each soil texture.

NOTE: Do not allow livestock to graze on soybean vines or use the vines for feed. Cover crops may be planted anytime but stand reductions may occur. Do not graze or use the cover crop for feed.

PROWL Plus Canopy

Observe all precautionary statements on the Canopy label before using.

The following PROWL plus Canopy treatments may be used in soybeans. Apply PROWL plus Canopy with a properly calibrated low pressure (20 to 40 psi) fixed-boom sprayer equipped with fan type nozzles and screens no finer than 50 mesh. Continuous agitation in the spray tank is required to keep the material in suspension. Avoid overlap and shut off spray booms while starting, turning, slowing or stopping, or crop injury may result.

DO NOT apply on peat or muck soils or soils having a calcareous surface layer.

DO NOT apply to soils with a pH greater than 7.0, except as directed on the Canopy label.

Consult the Canopy label for sensitive soybean varieties. Before use on any soybean variety, tolerance to Canopy must first be determined.

1. **Preplant Incorporated Tank Mixture of PROWL plus Canopy**
   Apply immediately before planting or up to two weeks prior to planting. Incorporate prior to planting and within 7 days of application. Refer to **Incorporation Instructions** section (see table of contents for page number) for incorporation directions.
2. **Preemergence Tank Mixture of PROWL plus Canopy**
   Apply at planting or up to 2 days after planting, but before weeds and crop emerge. DO NOT APPLY POSTEMERGENCE. Preemergence treatments are most effective in controlling weeds when adequate rainfall of overhead irrigation is received within 7 days after application.

If moisture is insufficient to activate the herbicides, a shallow cultivation (preferably with a rotary hoe) should be made after emergence of soybeans, but while weeds are still small enough to be controlled by mechanical cultivation.

DO NOT make applications of PROWL preemergence north of Interstate 80.

WEEDS CONTROLLED

When applied as directed, PROWL plus Canopy will control the following weeds in addition to those controlled by PROWL alone:

Cocklebur* (*Xanthium strumarium*)
Florida beggarweed (*Desmodium tortuosum*)
Hemp sesbania (*Sesbania exaltata*)
Hophornbeam copperleaf (*Acalypha ostryaefolia*)
Jimsonweed (*Datura stramonium*)
Morningglory*
Entireleaf (*Ipomoea hederacea* var. intergriuscula)
Ivyleaf (*Ipomoea hederacea*)
Pitted (*Ipomoea lacunosa*)
Smallflower (*Jacquemontia tamnifolia*)
Tall (*Ipomoea purpurea*)
Prickly sida (teaweed) (*Sida spinosa*)
Ragweed
Common (*Ambrosia artemisiifolia*)
Giant* (*Ambrosia trifida*)
Sicklepod* (*Cassia obstusifolia*)
Smartweed (*Polygonum pensylvanicum*)
Sunflower (*Helianthus annuus*)
Velvetleaf (*Abutilon theophrasti*)

Tank mixtures of PROWL plus Canopy will aid in the control of and reduce competition from:

Itchgrass (*Rottboellia exaltata*)
Red rice (*Oryza sativa*)
Shattercane (*Sorghum bicolor*)

* Cultivation or application of a postemergence herbicide may be required for season long control.

Broadcast Rates Per Acre of PROWL Plus Canopy Preplant Incorporated and Preemergence Tank Mixtures

	PROWL (pints) + Canopy (ounces)	
	0.5%* to 3% Organic Matter	More Than 3% Organic Matter
Soil Texture	PROWL + Canopy	PROWL + Canopy
COARSE	1 + 6 to 10	1.5 + 6 to 10
MEDIUM	1.5 + 8 to 12	1.5 to 2 + 8 to 12
FINE	1.5 + 10 to 14	2 to 2.5 + 10 to 14

* DO NOT apply to soil with less than 0.5% organic matter.

Use the higher rates on soils with higher organic matter or heavy weed pressure.

NOTE: If crop loss occurs due to weather conditions and corn is replanted, seeding depth must be below retilled area.

Observe all cropping restrictions on the Canopy label.

PROWL Plus Preview

Observe all precautionary statements on the Preview label before using.

The following PROWL plus Preview treatments may be used in soybeans. Apply PROWL plus Preview with a properly calibrated low pressure (20 to 40 psi) fixed-boom sprayer equipped with fan type nozzles and screens no finer than 50 mesh.

Continuous agitation in the spray tank is required to keep the material in suspension. Avoid overlap and shut off spray booms while starting, turning, slowing or stopping, or crop injury may result.

DO NOT apply on peat or muck soils, soils having a calcareous surface layer, or soils with a pH greater than 6.8.

Consult the Preview label for sensitive soybean varieties. Before use on any soybean variety, tolerance to Preview must first be determined.

1. **Preplant Incorporated Tank Mixture of PROWL Plus Preview**
   Apply immediately before planting or up to two weeks prior to planting. Incorporate prior to planting and within 7 days of application. Refer to Incorporation Instructions section (see table of contents for page number) for incorporation directions.
2. **Preemergence Tank Mixture of PROWL Plus Preview**
   Apply at planting or up to 2 days after planting, but before weeds and crop emerge. DO NOT APPLY POSTEMERGENCE. Preemergence treatments are most effective in controlling weeds when adequate rainfall or overhead irrigation is received within 7 days after application.

If moisture is insufficient to activate the herbicides, a shallow cultivation (preferably with a rotary hoe) should be made after emergence of soybeans, but while weeds are still small enough to be controlled by mechanical cultivation.

DO NOT make applications of PROWL preemergence north of Interstate 80.

Weed Species Controlled

When applied as directed, PROWL plus Preview will control the following weeds in addition to those controlled by PROWL alone:

Cocklebur* (*Xanthium strumarium*)
Hophornbeam copperleaf (*Acalypha ostryaefolia*)
Jimsonweed (*Datura stramonium*)
Mustards (*Brassica* spp.)
Prickly sida (teaweed) (*Sida spinosa*)
Ragweed, common (*Ambrosia artemisiifolia*)
Smartweed (*Polygonum pensylvanicum*)
Sunflower* (*Helianthus annuus*)
Velvetleaf (*Abutilon theophrasti*)

Tank mixtures of PROWL plus Preview will aid in the control of and reduce competition from:

Black nightshade (*Solanum nigrum*)
Burcucumber (*Sicyos angulatus*)
Itchgrass (*Rottboellia exaltata*)
Morningglory*
  Entireleaf (*Ipomoea hederacea* var. intergriuscula)
  Ivyleaf (*Ipomoea hederacea*)
  Pitted (*Ipomoea lacunosa*)
  Tall (*Ipomoea purpurea*)
Nutsedge (*Cyperus* spp.)
Ragweed, giant* (*Ambrosia trifida*)
Red rice (*Oryza sativa*)
Shattercane (*Sorghum bicolor*)

* Cultivation may be required to achieve season long control.

Broadcast Rates Per Acre of PROWL Plus Preview Preplant Incorporated and Preemergence Tank Mixtures

	PROWL (pints) + Preview (ounces)	
	0.5%* to 3% Organic Matter	More Than 3% Organic Matter
Soil* Texture	PROWL + Preview	PROWL + Preview
COARSE	1 + 6	1.5 + 7
MEDIUM	1.5 + 7	1.5 to 2 + 8
FINE	1.5 + 8	2 to 2.5 + 9 to 10

* DO NOT apply to soil with less than 0.5% organic matter.

NOTE: If crop loss occurs due to weather conditions and corn is replanted, seeding depth must be below retilled area.

Observe all cropping restrictions on the Preview label.

Special Weeds

1. Shattercane

Preplant incorporated treatments of PROWL alone, PROWL plus Sencor/Lexone preplant incorporated tank mixture, or preplant incorporated PROWL applications followed by registered sequential preemergence herbicides will control shattercane (wild cane) (*Sorghum bicolor*).

Shattercane is difficult to control in many soybean growing areas because of its ability to germinate from a greater depth in the soil than most other weeds. Shattercane can emerge throughout the growing season depending on soil temperature, moisture, and seed dormancy.

Thoroughly mix previous crop residues into the soil and destroy any existing vegetation prior to herbicide application. Apply PROWL or PROWL combinations at the recommended rates listed in tables following. See **Incorporation Instructions** section (see table of contents for page number) for incorporation directions. Thoroughly and uniformly mechanically incorporate PROWL into TOP 2 INCHES of soil. Acceptable results CANNOT be expected if PROWL is applied preemergence. Cultivation may be necessary to control escapes and late germinating shattercane during the crop season.

Broadcast Rates Per Acre of PROWL Preplant Incorporated for Control of Shattercane in Soybeans

	PROWL (pints)	PROWL (pints)
Soil Texture	Up to 3% Organic Matter	More Than 3% Organic Matter
COARSE	2.0	2.0
MEDIUM	2.5	3.0
FINE	3.0	3.0

Broadcast Rate Per Acre of PROWL Plus Sencor/Lexone Preplant Incorporated Tank Mixture for Control of Shattercane in Soybeans[1]

	PROWL + Sencor/ Lexone DF[2]	PROWL + Sencor/ Lexone DF[2]
Soil Texture[3]	½ to 3% Organic Matter	More Than 3% Organic Matter
COARSE	1.5 pts. + 0.5 lb.	1.5 pts. + 0.5 lb.
MEDIUM	2.0 pts. + 0.5 lb.	2.0 pts. + 0.5 to 0.67 lb.
FINE	2.0 pts. + 0.5 to 0.67 lb.	2.5 pts. + 0.67 lb.

[1] Refer to **Special Precautions for PROWL Plus Sencor or Lexone Combinations** subsection above.

[2] When using Sencor 4F, Sencor 50WP, Lexone 4L, or Lexone 50WP use equivalent rates. One pint of the flowable formulation or 1 lb. of the wettable powder formulation equals 0.67 lb of Sencor DF or Lexone DF. See conversion chart in **Formulation Conversion Tables** section (see table of contents for page number).

[3] DO NOT USE on sands. DO NOT USE on loamy sands or on sandy loams that contain less than 1% organic matter, nor on other soils with less than ½% organic matter as crop injury may result.

2. Red Rice, Itchgrass

PROWL applied preplant incorporated will provide control of red rice (*Oryza sativa*) and aid in the control of and reduce competition from itchgrass (*Rottboellia exaltata*) at the rates listed in the table below.

Broadcast Rate Per Acre of PROWL Preplant Incorporated for Control of Red Rice and Suppression of Itchgrass in Soybeans

	PROWL (pints)
Soil Texture Up to 3% Organic Matter[1]	
COARSE	2.0 to 3.0 [2]
MEDIUM	3.0
FINE	4.0

[1] This use is not recommended for soils with more than 3% organic matter.

[2] The higher rate should be used if heavy red rice or itchgrass populations are anticipated.

NOTE: Livestock can graze or be fed from treated soybean fields. If soybean crop loss occurs due to weather conditions, cotton or soybeans can be replanted the same year into treated soil without adverse effects. If replanting is necessary, DO NOT rework the soil deeper than the treated zone.

3. Rhizome Johnsongrass

PROWL applied preplant incorporated for two consecutive years will provide control of rhizome johnsongrass (*Sorghum halepense*) in soybeans at the rates recommended for soil textures listed in the following table. This use is not recommended for Arizona, New Mexico and California. Rhizome johnsongrass will be suppressed after the first year and controlled after the second year.

Before application, use a chisel plow or similar implement to bring johnsongrass rhizomes to the surface. Chop rhizomes into small pieces with a disk harrow set to cut 4 to 6 inches deep and operated in two different directions at 4 to 6 mph.

PROWL is to be incorporated into the soil within 7 days after application prior to planting. For maximum control of rhizome johnsongrass, incorporate PROWL as soon as possible after application. Deep and thorough incorporation of PROWL is necessary for control of rhizome johnsongrass. Mechanical incorporation can be achieved by the following methods.

(a) Disk harrow set to cut 4 to 6 inches deep and operated in two different directions at 4 to 6 mph.
(b) PTO-driven equipment (tillers, cultivators, hoes) set to cut 3 to 4 inches deep and operated one time at 4 mph or less.

For johnsongrass escapes during the crop season, cultivation and/or application of registered postemergence herbicides are recommended. Follow the directions for use on the labels of the respective herbicides.

Broadcast Rate Per Acre of PROWL Preplant Incorporated for Control of Rhizome Johnsongrass
APPLY FOR TWO CONSECUTIVE YEARS

	PROWL (pints)
Soil Texture—Up to 3% Organic Matter[1]	
COARSE	2.0
MEDIUM	3.0
FINE	4.0

[1] This use is not recommended for soils with more than 3% organic matter.

NOTE: Livestock can graze or be fed from treated soybean fields. If soybean crop loss occurs due to weather conditions, cotton or soybeans can be replanted the same year into treated soil without adverse effects. If replanting is necessary, DO NOT rework the soil deeper than the treated zone.

PROWL TANK MIXTURES IN MINIMUM AND NO-TILL SYSTEMS

PROWL may be applied as a preemergence tank mixture treatment in minimum or no-till soybeans as follows:

PROWL + PURSUIT + Roundup
PROWL + PURSUIT + Gramoxone Super
PROWL + SCEPTER + Roundup
PROWL + SCEPTER + Gramoxone Super
PROWL + Sencor/Lexone + Roundup
PROWL + Sencor/Lexone + Gramoxone Super
PROWL + Lorox + Roundup
PROWL + Lorox + Gramoxone Super
PROWL + Canopy + Roundup
PROWL + Canopy + Gramoxone Super
PROWL + Preview + Roundup
PROWL + Preview + Graxomone Super

Gramoxone Super and Roundup are used to kill existing vegetative growth. Gramoxone Super or Roundup should be deleted from the tank mixture if vegetation is absent at time of application.

Application Instructions

Uniformly apply the recommended PROWL tank mixture in 20 to 60 gallons of water per acre with ground equipment. Use the higher gallonage for fields with heavy weed infestations or excessive crop residues.

Apply recommended PROWL tank mixtures at planting or up to 2 days after planting. DO NOT APPLY POSTEMERGENCE to soybeans since soybeans exposed at time of application will be killed.

These treatments are most effective in controlling weeds when adequate rainfall or overhead irrigation is received within 7 days after application.

Mixing Instructions

Fill tank one-half to three-quarters full with clean water. MAINTAIN CONTINUOUS AGITATION/WHILE ADDING HERBICIDES. Add PURSUIT, SCEPTER, Sencor Lexone, Lorox, Canopy or Preview formulation first, as follows:

When using Sencor/Lexone or Lorox Flowable Formulation: Premix one part of the flowable and one part water and add this diluted mixture into the partially filled tank while agitating.

When using Sencor/Lexone, Canopy or Preview Dry Flowable Formulation: Add the granules to the partially filled tank while agitating, and allow the granules to disperse.

When using Sencor/Lexone or Lorox Wettable Powder Formulation: Make a slurry of the wettable powder in water (1 part WP + 2 parts water). Add the slurry slowly into the partially filled tank while agitating.

After PURSUIT, SCEPTER, Sencor/Lexone, Lorox, Canopy or Preview herbicide has been added to the tank and mixing is complete, add PROWL to tank while agitating.

After complete mixing of PROWL, continue filling the sprayer with water and add Gramoxone Super or Roundup near the end of the filling process.

If Gramoxone Super is included in the tank mixture, add 8 oz. of Ortho X-77 spreader or similar non-ionic surfactant per 100 gallons of total spray mixture as the last ingredient in the tank.

Fill the remainder of the tank with water.

MAINTAIN CONTINUOUS AGITATION UNTIL SPRAYING IS COMPLETED. If the spray mixture is allowed to settle for any period of time, thorough agitation is essential to resuspend the mixture before spraying is resumed. Continue agitation while spraying.

Tank Mixtures with Gramoxone Super

Apply 1⅓ to 2⅔ pints of Gramoxone Super per acre in recommended tank mixtures. Use the 2⅔ pint rate if weeds are 4 to 6 inches tall. Weeds over 6 inches tall will not be controlled with this treatment.

The rate of Gramoxone Super is based on products containing 2 pounds active ingredient per gallon.

When Gramoxone Super is included in a recommended tank mixture, add 8 oz. of Ortho X-77 spreader or similar non-ionic surfactant per 100 gallons of spray mixture. See mixing instructions above.

Gramoxone Super will control most annual emerged weeds and suppress many emerged perennials. Refer to Gramoxone Super label for specific use recommendations.

Observe all precautions and limitations on Gramoxone Super label.

Tank Mixtures with Roundup

Annual Weeds

Apply 1.0 to 1.5 quarts of Roundup per acre in recommended tank mixtures. Use the 1.5 quart rate if annual weeds are more than 6 inches tall.

Roundup will control most annual emerged weeds. Refer to Roundup label for specific use recommendations.

Observe all precautions and limitations on Roundup label.

Perennial Weeds

At normal application times in minimum or no-till systems, perennial weeds may not be at the proper stage or growth for control with Roundup. Refer to Roundup label for the proper stage of growth for perennial weeds. Use of 2 to 4 quarts of Roundup per acre in recommended tank mixtures under these conditions provides top kill and reduces competition from many emerged perennial grass and broadleaf weeds. Refer to Roundup label for emerged perennial weeds controlled. To obtain control of perennial weeds, follow recommendations on Roundup label for stage of growth and rate of application.

NOTE: When using these tank mixtures, DO NOT exceed 4 quarts of Roundup per acre.

USE OF THESE TANK MIXTURES FOR BERMUDAGRASS OR JOHNSONGRASS CONTROL IS NOT RECOMMENDED.

Observe all precautions and limitations on Roundup label.

PROWL Plus PURSUIT

See the PROWL plus PURSUIT subsection of this label for use rates and weeds controlled. The high rates of PROWL for each soil texture should be used if heavy weed populations are anticipated or if field is particularly trashy or cloddy.

PROWL Plus SCEPTER

See the PROWL plus SCEPTER subsection of this label for use rates and weeds controlled. The high rates of PROWL for each soil texture should be used if heavy weed populations are anticipated or if field is particularly trashy or cloddy.

PROWL Plus Sencor/Lexone

See **Special Precautions for PROWL Plus Sencor or Lexone Combinations** subsection above.

Weeds Controlled

When applied as directed, PROWL plus Sencor/Lexone will control the following weeds in addition to those controlled by PROWL alone:

Common ragweed *(Ambrosia artemisiifolia)*
Jimsonweed *(Datura stramonium)*
Pennsylvania smartweed *(Polygonum pensylvanicum)*
Prickly sida *(Sida spinosa)*
Velvetleaf (Buttonweed) *(Abutilon theophrasti)*
Venice mallow *(Hibiscus trionum)*
Wild mustard *(Brassica kaber)* (Medium and fine soils only)

PROWL plus Sencor/Lexone will aid in the control of and reduce competition from cocklebur *(Xanthium strumarium)*

Broadcast Rates Per Acre of PROWL Plus Sencor/Lexone Preemergence Tank Mixture in Minimum and No-Till Soybeans

	SOUTHERN STATES (Including Eastern Coastal Plains)[1]	NORTHERN STATES[1]	
	PROWL + Sencor/Lexone DF[2]	PROWL + Sencor/Lexone DF[2]	PROWL + Sencor/Lexone DF[2]
Soil Texture[3]		½ to 3% Organic Matter	More Than 3% Organic Matter
COARSE	1.0 to 1.5 pts. + 0.33 lbs.	1.0 to 1.5 pts. + 0.5 lb.	1.5 pts. + 0.5 lbs.
MEDIUM	1.5 to 2.0 pts. + 0.5 lb.	1.5 to 2.0 pts. + 0.5 lb.	1.5 to 2.0 pts. + 0.5 to 0.67 lb.
FINE	1.5 to 2.0 pts. + 0.67 lb.	1.5 to 2.0 pts. + 0.5 to 0.67 lb.	2.0 to 2.5 pts. + 0.67 lb.

[1] See map at end of this label for specific states.

[2] When using Sencor 4F, Sencor 50WP, Lexone 4L, or Lexone 50WP, use equivalent rates. One pint of the flowable formulation or 1 lb. of the wettable powder formulation equals 0.67 lb. of Sencor DF or Lexone DF. See conversion chart in **Formulation Conversion Tables** section (see table of contents for page number).

[3] DO NOT USE on sands. DO NOT USE on loamy sands or on sandy loams that contain less than 1% organic matter, nor on other soils with less than ½% organic matter as crop injury may result. This tank mixture is not recommended for use on soils with less than 2% organic matter in the Coastal Plain of New Jersey or the Delmarva Peninsula.

The high rates of PROWL and Sencor/Lexone for each soil texture, as stated above, should be used if heavy weed populations are anticipated or if field is particularly trashy or cloddy.

NOTE: Livestock may graze or be fed forage from treated fields 40 days after application. Observe all precautions and limitations on Sencor and Lexone labels.

PROWL PLUS LOROX

See the PROWL plus Lorox subsection of this label for use rates and weeds controlled.

The high rates of PROWL and Lorox for each soil texture should be used if heavy weed populations are anticipated or if field is particularly trashy or cloddy.

NOTE: Livestock may graze or be fed forage from treated fields. Observe all precautions and limitations on Lorox label.

PROWL Plus Canopy

See the PROWL plus Canopy subsection of this label for use rates and weeds controlled.

The high rates of PROWL and Canopy for each soil texture should be used if heavy populations are anticipated or if field is particularly trashy or cloddy.

PROWL Plus Preview

See the PROWL plus Preview subsection of this label for use rates and weeds controlled.

The high rates of PROWL and Preview for each soil texture should be used if heavy weed populations are anticipated or if field is particularly trashy or cloddy.

RICE

PROWL plus Stam M-4 early postemergence tank mixture may be used in dry-seeded rice.
DO NOT use this treatment in water-seeded rice. DO NOT use in California.
DO NOT use PROWL on peat or muck soils.
This pesticide is toxic to fish and aquatic organisms. Fish may be killed at application rates recommended on the label. DO NOT contaminate water by cleaning of equipment or disposal of wastes.
DO NOT apply to rice fields if fields are used for Fish Production, especially catfish or crayfish farming.

GENERAL INFORMATION

PROWL plus Stam M-4 is a postemergence treatment which combines the direct contact action of Stam M-4 and the residual activity of PROWL. Since the residual activity of PROWL provides preemergence control of certain annual grasses which can germinate after this tank-mix treatment is applied, flooding after application can be delayed.

For maximum weed control with this treatment, it is important to carefully follow the directions below for (1) adequate spray coverage of weeds and soil and (2) proper timing of application, when barnyardgrass (watergrass) is in the 1 to 3 leaf stage of growth with an occasional 4 leaf plant (make application when sprangletop is less than 1/2 inch in height).

SPRAYING INSTRUCTIONS

AERIAL APPLICATIONS

For aerial application, apply the recommended rate in 10 to 12 gallons of water per acre to ensure adequate coverage. To minimize drift (see Stam M-4 label for sensitive crops), DO NOT apply during periods of gusty winds or winds in excess of 5 mph. Uniform spray distribution may not be achieved when wind velocity is greater than 5 mph. It is recommended that a flagman or an automatic mechanical flagging unit on the aircraft be used to avoid overlapping and possible crop injury.

GROUND APPLICATIONS

For ground equipment, apply the recommended rate in 15 to 25 gallons of water per acre to ensure adequate coverage. Use a properly calibrated low-pressure (20 to 40 psi) sprayer equipped with 8002 or larger size Tee-Jet or comparable nozzles to achieve uniform spray distribution and minimize drift. Keep the bypass line on or near the bottom of the tank to minimize foaming. Nozzle screens must be no finer than 50 mesh. DO NOT apply PROWL during periods of gusty winds or when wind velocity is greater than 20 mph.

Early Postemergence Applications

The seedbed should be firm and free of clods and trash. The seedbed must be prepared to allow for good seed coverage. Previous crop residues should be thoroughly mixed into the soil to a depth of 4 to 6 inches by plowing or disking before planting rice.

Uniformly apply recommended PROWL plus Stam M-4 treatment by aircraft or ground equipment after rice emergence, according to spraying instructions above, when barnyardgrass is in the 1 to 3 leaf growth stage with an occasional 4 leaf plant (make application when sprangletop is less than 1/2 inch). THE GROWER SHOULD INSPECT FIELDS FREQUENTLY TO CHECK GROWTH OF BARNYARDGRASS AND/OR SPRANGLETOP TO DETERMINE PROPER APPLICATION TIMING. Timing of applications should be based on the growth stage of barnyardgrass and/or sprangletop and not on the growth stage of rice. If rice is to small to maintain a flood on the field, the treatment can still be applied since flooding can be delayed because of the residual activity of PROWL.

While the residual activity of PROWL allows flooding to be delayed, proper water management practices must be followed for normal rice growth. Flooding should not be delayed if weeds begin to develop after application.

Since soil and weeds must be completely exposed to spray coverage, no flood water should be on field at time of application. If necessary, fields may be flushed prior to treatment to produce vigorous rice and weed growth.

Since the residual activity of PROWL is activated by moisture, PROWL is most effective in controlling emerging weeds when adequate rainfall or irrigation (flush) is received within 7 days after application.

For maximum direct contact activity of Stam M-4, delay application if there is a chance of rain within 6 to 8 hours.

Weeds Controlled

PROWL plus Stam M-4 tank mixture provides direct contact control of weeds listed on the Stam M-4 label. The following grass species are controlled by the residual activity of PROWL plus Stam M-4 tank mixture at the rates recommended for each soil texture listed below:

Barnyardgrass (watergrass) *(Echinochloa crus-galli or Echinochloa colonum)*
Crabgrass *(Digitaria* spp.)
Signalgrass *(Brachiaria platyphylla)*
Sprangletop *(Leptochloa* spp.)

Broadcast Rate Per Acre of PROWL Plus Stam M-4 Tank Mixture in Rice

Soil Texture	PROWL + Stam M-4 (pints)
COARSE	1.5 + 6.0 to 8.0
MEDIUM	1.5 to 2.0 + 6.0 to 8.0
FINE	1.5 to 2.0 + 6.0 to 8.0

Use the high rate of Stam M-4 if the following situations exist at the time of application: (1) unseasonably cool weather, (2) barnyardgrass has progressed to predominantly the 3 to 4 leaf stage, or (3) emerged sprangletop (less than 1/2 inch) is present.

DO NOT apply this tank mixture within 14 days before or after insecticide applications because serious damage to rice may occur.

DO NOT apply this tank mixture in liquid fertilizer.

DO NOT make more than one application of PROWL per season.

DO NOT bale or use rice straw from treated fields for feed or bedding.

Observe all precautions and limitations on the Stam M-4 label.

Pending the submission of rotational crop data, DO NOT apply pendimethalin on rice fields in which crayfish or catfish farming are included in the cultural practices, and DO NOT plant crops in pendimethalin-treated fields unless pendimethalin is registered for use on those crops.

DO NOT use water containing pendimethalin residues from rice cultivation to irrigate food or feed crops which are not registered for use with pendimethalin.

POTATOES

PROWL may be applied in potatoes as follows:

Preemergence	Early Postemergence	Preemergence Incorporated
PROWL alone	PROWL alone	PROWL alone
PROWL + Sencor/Lexone	PROWL + Sencor/Lexone	PROWL + Eptam
PROWL + Lorox	PROWL + Eptam	

DO NOT use PROWL on peat or muck soils.

Applications of PROWL Alone

Apply PROWL with ground equipment, aircraft or sprinkler irrigation. The seedbed should be firm and free of clods and trash. DO NOT apply prior to planting crop.

1. **Preemergence Applications**
   PROWL may be applied after-planting but before potatoes and weeds emerge or after drag-off where this operation is practiced.
   PROWL applied preemergence is most effective in controlling weeds when adequate rainfall or irrigation is received within 7 days after application.
2. **Preemergence Incorporated**
   Apply PROWL and incorporate after planting but before potatoes and weeds emerge. Where drag-off is practiced, apply PROWL and incorporate before, at or after drag-off but before potatoes and weeds emerge. PROWL is to be incorporated within 7 days after application.
   PROWL must be thoroughly and uniformly incorporated into the top 1 to 2 inches of soil. Care must be taken so that incorporation equipment does not damage seed pieces or elongating sprouts.

Mechanical incorporation is not required if adequate rainfall for good crop and weed emergence occurs or irrigation is received within 7 days after application.

3. Early Postemergence
PROWL may be applied early postemergence in potatoes. DO NOT apply to sweet potatoes or yams. DO NOT apply PROWL postemergence if potatoes are under stress from cold/wet or hot/dry conditions or crop injury may occur. PROWL may be applied from crop emergence to the 6-inch stage of growth. PROWL treatments will not control established weeds. Emerged weeds must be destroyed prior to application. DO NOT make more than one application per season. PROWL treatments are effective in controlling weeds when adequate rainfall or overhead irrigation is received within 7 days after application.

Applications of PROWL Alone-Sprinkler Irrigation Systems

PROWL may be applied preemergence after planting but before potatoes and weeds have emerged or after drag-off where this operation is practiced or early postemergence through sprinkler irrigation systems. See **Chemigation** section for complete instructions (see table of contents for page number).

Weeds Controlled

When applied as directed, PROWL will control the following weeds in addition to the weeds listed previously as being controlled by PROWL:

Stinging nettle (*Urtica urens*)

Broadcast Rate Per Acre of PROWL Alone in Potatoes

Soil Texture	PROWL (pints) Less Than 3% Organic Matter	PROWL (pints) More Than 3% Organic Matter
COARSE	1.5	1.5
MEDIUM	2.0	3.0
FINE	3.0	3.0

PROWL Plus Sencor/Lexone

PROWL plus Sencor/Lexone tank mixture may be applied preemergence after planting or after drag-off where this operation is practiced or early postemergence. Apply with ground equipment or aircraft. The seedbed should be firm and free of clods and trash.

This treatment is most effective in controlling weeds when adequate rainfall or irrigation is received within 7 days after application.

PROWL plus Sencor/Lexone postemergence tank mixture may be applied from crop emergence to the 6-inch stage of growth. For optimum control of weeds in potatoes apply before weeds are 1-inch tall. DO NOT make more than one application of this tank mixture per season. DO NOT apply within 24 hours of application of other pesticides. DO NOT apply within three days after periods of cool, wet or cloudy weather or crop injury may occur. DO NOT use on early maturing smooth-skinned white or red-skinned varieties of potatoes.

Weeds Controlled

When applied as directed, PROWL plus Sencor/Lexone will control the following weeds in addition to those controlled by PROWL alone:

Cocklebur (*Xanthium strumarium*)
Common ragweed (*Ambrosia artemisiifolia*)
Pennsylvania smartweed (*Polygonum pensylvanicum*)
Velvetleaf (Buttonweed) (*Abutilon theophrasti*)
Russian thistle* (*Salsola Kali*)

PROWL plus Sencor/Lexone will aid in the control of and reduce competition from:

Black nightshade (*Solanum nigrum*)
Hairy nightshade (*Solanum sarrachoides*)
* early postemergence only

Broadcast Rate Per Acre of PROWL Plus Sencor/Lexone Preemergence or Early Postemergence Tank Mixture in Potatoes

Soil Texture	PROWL + Sencor/Lexone DF[1] Less than 3% Organic Matter[2]	PROWL + Sencor/Lexone DF[1] More than 3% Organic Matter
COARSE	1.0 to 1.5 pts. + 0.33 to 0.5 lb.[3]	1.0 to 1.5 pts. + 0.33 to 0.5 lb.[3]
MEDIUM	1.5 to 2.0 pts. + 0.5 to 0.67 lb.	2.0 to 3.0 pts. + 0.5 to 0.67 lb.
FINE	2.0 to 3.0 pts. + 0.5 to 0.67 lb.	3.0 pts. + 0.5 to 0.67 lb.

[1] When using Sencor 4F, Sencor 50WP, Lexone 4L, or Lexone 50WP, use equivalent rates. One pint of the flowable formulation or 1 lb. of the wettable powder formulation equals 0.67 lb. of Sencor DF or Lexone DF. See conversion chart in **Formulation Conversion Tables** section (see table of contents for page number).
[2] Observe organic matter restrictions on Sencor and Lexone labels.
[3] Preemergence only, use 0.5 to 0.67 lb. of Sencor/Lexone on coarse soils in Wisconsin.

NOTE: Observe all precautions and limitations on Sencor and Lexone labels. Follow cropping restrictions on Sencor and Lexone labels.

PROWL Plus Lorox

PROWL plus Lorox preemergence tank mixture may be used in potatoes grown East of the Rocky Mountains. Apply after planting but before potatoes and weeds emerge or after drag-off or hilling where these operations are practiced. Apply with ground equipment or aircraft. The seedbed should be firm and free of clods and trash. Plant seed pieces at least two inches deep.

This treatment is most effective in controlling weeds when adequate rainfall or overhead irrigation is received within 7 days after application.

Weeds Controlled

When applied as directed, PROWL plus Lorox will control the following weeds in addition to those controlled by PROWL alone:

Chickweed (*Stellaria media*)
Mustard (*Brassica* spp.)
Common ragweed (*Ambrosia artemisiifolia*)
Pennsylvania smartweed (*Polygonum pensylvanicum*)

Broadcast Rate Per Acre of PROWL Plus Lorox Preemergence Tank Mixture in Potatoes

Soil Texture[2]	PROWL + Lorox L[1] (pints) 1 to 3% Organic Matter	PROWL + Lorox L[1] (pints) 3 to 5% Organic Matter
COARSE Sandy loams only	1.0 to 1.5 + 1.5 to 2.5	1.0 to 1.5 + 2.0 to 3.0
MEDIUM	1.5 to 2.0 + 1.5 to 3.0	2.0 to 3.0 + 2.5 to 4.0
FINE	2.0 to 3.0 + 2.5 to 4.0	3.0 + 3.0 to 4.0

Wisconsin-Central Sands Area Only: PROWL plus Lorox tank mixture may be used on sands and loamy sands. **Sands**—Apply 1.0 to 1.5 pints of PROWL plus 1 pint of Lorox L per acre. **Loamy sand**—Apply 1.0 to 1.5 pints of PROWL plus 2 pints of Lorox L per acre.

[1] When using Lorox 50W use equivalent rates. One pint of the flowable formulation equals 1 lb. of the wettable powder formulation. See conversion chart in **Formulation Conversion Tables** section (see table of contents for page number).
[2] DO NOT use on sands and loamy sands (except in Wisconsin-Central Sands Area), gravelly soils or exposed subsoils or on soils containing less than 1% organic matter as injury to the treated crop or subsequent crops may result.

NOTE: Observe all precautions and limitations on Lorox label.

PROWL Plus Eptam

PROWL plus Eptam tank mixture may be applied and incorporated after planting but before potatoes and weeds emerge. The seedbed should be firm and free of clods and trash. In areas where potatoes are normally dragged-off, apply and incorporate following drag-off but before potatoes and weeds emerge. Incorporate immediately after application.

PROWL plus Eptam must be thoroughly and uniformly incorporated into the top 1 or 2 inches of soil. Whenever possible, application and incorporation should be done in the same operation. Care must be taken so that incorporation equipment does not damage seed pieces or elongating sprouts.

PROWL plus Eptam postemergence tank-mixture may be applied through SPRINKLER IRRIGATION SYSTEMS ONLY from crop emergence to the 6-inch stage of growth. PROWL plus Eptam will not control established weeds. EMERGED WEEDS MUST BE DESTROYED PRIOR TO APPLICATION. DO NOT make more than one application per season of this tank-mixture. DO NOT apply if potatoes are under stress from cold/wet or hot/dry conditions or crop injury may occur.

Weeds Controlled

When applied as directed, PROWL plus Eptam will control the following weeds in addition to those controlled by PROWL alone:

Purple nutsedge (*Cyperus rotundus*)
Yellow nutsedge (*Cyperus esculentus*)
Black nightshade (*Solanum nigrum*)
Hairy nightshade (*Solanum sarrachoides*)

Broadcast Rate Per Acre of PROWL Plus Eptam Preemergence Incorporated in Potatoes[1]

Soil Texture	PROWL + Eptam 7E (pints) Less than 3% Organic Matter	PROWL + Eptam 7E (pints) More than 3% Organic Matter
COARSE	1.0 to 1.5 + 3.0 to 3.5	1.0 to 1.5 + 3.0 to 3.5
MEDIUM	1.5 to 2.0 + 3.0 to 3.5	3.0 + 3.0 to 3.5

	PROWL + Eptam 7E (pints)	PROWL + Eptam 7E (pints)
Soil Texture	Less than 3% Organic Matter	More than 3% Organic Matter
FINE	3.0 + 3.0 to 3.5	3.0 + 3.0 to 3.5

[1] For nutsedge control, use 4.5 to 7.0 pints of Eptam per acre.

NOTE: The Superior variety potato is sensitive to Eptam and under stress conditions, early season stunting may occur. Observe all precautions and limitations on Eptam label.

TOBACCO

PROWL may be applied in transplanted tobacco as follows:
- Preplant incorporated application
- Layby application

DO NOT use PROWL on peat or muck soils.

Preplant Incorporated Application

Apply PROWL with ground sprayer immediately before transplanting or up to 60 days prior to transplanting tobacco. Incorporate prior to planting and within 7 days of application. Refer to Incorporation Instructions section (see table of contents below) for incorporation directions.

Broadcast Rate Per Acre of PROWL Preplant Incorporated In Transplanted Tobacco

REGION	SOIL TEXTURE	PROWL (pints)
Maryland, Virginia, North Carolina, South Carolina, Georgia, Florida	COARSE	1.5 to 2.0
	MEDIUM sandy clay loams, loams	1.5 to 2.0
	silt loams, silts	2.0 to 2.5
	FINE	2.0 to 2.5
Other tobacco-growing states	COARSE	1.5 to 2.0
	MEDIUM	2.5 to 3.0
	FINE	2.5 to 3.0

NOTE: Applied according to directions and under normal growing conditions, PROWL will not harm transplanted tobacco. Under stress conditions for plant growth such as cold/wet or hot/dry weather, PROWL can produce a temporary retardation of tobacco development.

If loss of tobacco crop occurs due to adverse weather conditions, transplanted tobacco or any crop registered for PROWL preplant incorporated use can be replanted the same year into treated soil without adverse effects. If replanting is necessary, DO NOT rework the soil deeper than the treated zone.

Layby Applications

PROWL may be applied as a directed spray following the last normal cultivation (layby), usually 4 to 6 weeks after transplanting tobacco.

Layby applications can be applied in tobacco previously treated with herbicides registered for use in tobacco. Consult the labels of those herbicides for suggested treatments, rates to be used, and precautions or restrictions for use in tobacco and for follow crop restrictions.

Apply PROWL in a 16 to 24-inch band in the middle of the row between the crop rows. The spray should not contact tobacco plants. If the spray nozzles on the ends of the spray boom pass over the same row middle twice, use nozzles which apply one-half (1/2) the normal number of gallons per acre to prevent over-application.

PROWL is most effective in controlling weeds when adequate rainfall or irrigation is received within 7 days after application.

PROWL will not control established weeds. DESTROY EMERGED WEEDS PRIOR TO APPLICATION.

DO NOT APPLY AS A BROADCAST SPRAY OVER TOP OF TOBACCO LEAF. CONTACT MAY CAUSE MALFORMED LEAVES.

Broadcast[1] Rate Per Acre of PROWL Layby Application in Transplanted Tobacco

Soil Texture	PROWL[2] (pints)
COARSE	1.0 to 1.5
MEDIUM	1.5 to 2.0
FINE	1.5 to 2.0

[1] For calculation of band treatment rate, see **Spraying Instructions** section (see table of contents for page number).

[2] The high rate for each soil texture above should be used if a registered herbicide treatment was not applied prior to layby.

PEANUTS

PROWL may be applied preplant incorporated in peanuts as follows:
- PROWL alone
- PROWL + Vernam
- PROWL + Dual

DO NOT use in California.
DO NOT use PROWL on peat or muck soils.

Preplant Incorporated Applications of PROWL Alone

Apply PROWL with ground equipment or aircraft immediately before planting or up to 60 days prior to planting. Incorporate prior to planting and within 7 days of application. Refer to **Incorporation Instructions** section for incorporation directions.

Broadcast Rate Per Acre of PROWL Preplant Incorporated in Peanuts

REGION	PROWL (pints)
Texas, Oklahoma and New Mexico	1.0 to 1.5
Other peanut-growing states	1.5 to 2.0

For heavy weed infestations, especially of Texas panicum, up to 3 pints of PROWL can be used in Alabama, Georgia, or Florida.

If loss of peanut crop occurs due to adverse weather conditions, peanuts or any crop registered for PROWL preplant incorporated use can be replanted the same year into treated soil without adverse effects. If replanting is necessary, DO NOT rework the soil deeper than the treated zone.

PROWL Plus Vernam

PROWL plus Vernam tank mixture may be applied preplant incorporated in peanuts.

1. **Application Instructions**
   Apply PROWL plus Vernam with ground sprayer only.
2. **Incorporation Instructions**
   PROWL plus Vernam must be incorporated IMMEDIATELY (within minutes) after application to prevent loss of Vernam herbicide. Whenever possible, application and incorporation should be done in the same operation.

   Use power-driven cultivation equipment set to cut to a depth of 2 to 3 inches or tandem disks set to cut to a depth of 4 to 6 inches, operated at 4 to 6 mph, followed by a spike-toothed harrow or some other leveling device which extends beyond the ends of the disk. For thorough mixing, disk in two different directions (cross disk). Prior to second disking, raise the disk to prevent cutting deeper than 4 to 6 inches.
3. **Planting and Weed Control Instructions**
   DO NOT plant seed deeper than 2 inches. Seeding can be accomplished immediately after application or up to 10 days after application. If it is necessary to rework the soil after incorporation, but before planting, do not work the soil deeper than the depth of incorporation.

   For nutsedge control, existing stands must be turned under and chopped up thoroughly before application. Cultivation is suggested in addition to PROWL plus Vernam treatment in fields with moderate or heavy nutsedge infestations. Do not cultivate deeper than the depth of incorporation.

Weeds Controlled

When applied as directed, PROWL plus Vernam will control the following weeds in addition to those controlled by PROWL alone:

Purple nutsedge (*Cyperus rotundus*)
Yellow nutsedge (*Cyperus esculentus*)

Broadcast Rates Per Acre of PROWL Plus Vernam Preplant Incorporated Tank Mixture in Peanuts

REGION	SOIL TEXTURE	PROWL + Vernam 7E (pints)
Texas, Oklahoma and New Mexico	sands, loamy sands	1.0 to 1.5 + 2 1/3
	sandy loams and finer soils	1.0 to 1.5 + 3
Other peanut-growing states	sands, loamy sands	1.5 to 2.0 + 2 1/3
	sandy loams and finer soils	1.5 to 2.0 + 3

Follow all precautions and limitations on Vernam label.

PROWL Plus Dual

PROWL plus Dual tank mixture may be applied preplant incorporated in peanuts.

Apply PROWL plus Dual with ground equipment or aircraft within 14 days prior to planting. Incorporate into the top 1 to 2 inches of soil prior to planting and within 7 days of application. If crop will be planted on beds, apply and incorporate after bed formation. Refer to **Incorporation Instructions** section for incorporation directions.

Weeds Controlled

When applied as directed, PROWL plus Dual will control the following weeds in addition to those controlled by PROWL alone:

Black nightshade (*Solanum nigrum*)
Nutsedge (yellow) (*Cyperus esculentus*)
Prairie cupgrass (*Eriochloa contracta*)
Red rice (*Oryza sativa*)
Southwestern cupgrass (*Eriochloa gracilis*)

Broadcast Rate Per Acre of PROWL Plus Dual 8E Preplant Incorporated Tank Mixture in Peanuts

REGION	SOIL TEXTURE	PROWL + Dual 8E (pints)
Texas, Oklahoma and New Mexico	sands, loamy sands	1.0 to 1.5 + 1.25
	sandy loams	1.0 to 1.5 + 1.25 to 1.5
	finer soils	1.0 to 1.5 + 2.0 to 2.5
Other peanut-growing states	sands, sandy loams, loamy sands	1.5 to 2.0 + 1.5 to 2.0
	finer soils	1.5 to 2.0 + 2.0 to 2.5

DO NOT use in California.

Follow all precautions and limitations on Dual 8E label.

SUNFLOWERS

PROWL may be applied preplant incorporated in the spring in sunflowers as follows:
PROWL alone PROWL + Amiben
PROWL may be applied preplant incorporated in the fall in sunflowers as follows:
PROWL alone (North Dakota, South Dakota and Minnesota only)
PROWL + Eptam (North Dakota and Minnesota only)
DO NOT use PROWL on peat or muck soils.

Preplant Incorporated Applications of PROWL Alone

Apply PROWL with ground equipment or aircraft immediately before planting or up to 60 days prior to planting unless applying in the fall (see subsection below for fall application instructions). Incorporate prior to planting and within 7 days of application unless making a fall application (see subsection below for fall incorporation instructions). Refer to **Incorporation Instructions** section (see table of contents for page number) for incorporation directions.

Broadcast Rate Per Acre of PROWL Preplant Incorporated in Sunflowers

	PROWL (pints)	PROWL (pints)	
	SOUTHERN STATES[1]	NORTHERN STATES[1]	
Soil Texture	No Organic Matter Restrictions	Up to 3% Organic Matter	More than 3% Organic Matter
COARSE	1.0 to 1.5	1.0 to 2.0	2.0
MEDIUM	1.5 to 2.0	1.5 to 2.5	2.5 to 3.0
FINE	1.5 to 3.0	2.0 to 3.0	3.0

[1] See map at end of this label for specific states.

Use the 3 pint rate for heavy clay soils.

NOTE: DO NOT feed forage or graze livestock in treated sunflower fields. If sunflower crop loss occurs due to weather conditions, sunflowers or any crop registered for PROWL preplant incorporated use can be replanted. If replanting is necessary, DO NOT rework the soil deeper than the treated zone.

PROWL Plus Amiben

The following PROWL plus Amiben treatments may be applied in sunflowers grown in Northern States. See map at end of this label for specific states. Apply PROWL or PROWL plus Amiben tank mixture with ground equipment or aircraft. Refer to **Incorporation Instructions** section (see table of contents for page number) for incorporation directions. Use PROWL and Amiben rates listed below.

1. **PROWL Preplant Incorporated Followed by Amiben Preemergence** Apply PROWL preplant incorporated. Follow with a preemergence treatment of Amiben as directed on the Amiben label.
2. **Preplant Incorporated Tank Mixture of PROWL Plus Amiben** PROWL plus Amiben must be incorporated into the soil within 7 days after application and sunflowers must be planted no later than 7 days after application. For best results, incorporation should be in the top 1 to 2 inches of soil.

When applied as directed, PROWL plus Amiben will control the following weeds in addition to those controlled by PROWL alone:

Common ragweed (*Ambrosia artemisiifolia*)
Mustard (*Brassica* spp.)
Pennsylvania smartweed (*Polygonum pensylvanicum*)
Velvetleaf (Buttonweed) (*Abutilon theophrasti*)

Broadcast Rate per Acre of PROWL Plus Amiben Preplant Incorporated Tank Mixture or PROWL Preplant Incorporated Plus Amiben Preemergence in Sunflowers

	PROWL + Amiben	PROWL + Amiben
Soil Texture	Up to 3% Organic Matter	More than 3% Organic Matter
COARSE	DO NOT USE	DO NOT USE
MEDIUM	1.5 to 2.0 pts. + 1 gallon	1.5 to 2.0 pts. + 1 gallon
FINE	1.5 to 2.0 pts. + 1 gallon	2.0 to 2.5 pts. + 1 gallon

The high rate of PROWL for each soil texture above should be used if velvetleaf or smartweed infestations are anticipated.

NOTE: DO NOT feed forage or graze livestock in treated sunflower fields. If sunflower loss occurs due to weather conditions, sunflowers can be replanted. If replanting is necessary, DO NOT rework the soil. Observe all precautions and limitations on Amiben label.

Fall Application of PROWL® or PROWL Plus Eptam 7E

GENERAL INFORMATION

PROWL or PROWL plus Eptam 7E tank mix may be applied and incorporated in late fall prior to planting sunflowers the following spring.

Destroy existing weeds before applying PROWL or PROWL plus Eptam 7E. The soil should be dry enough to permit good incorporation.

DIRECTIONS FOR USE

PROWL alone may be applied by aircraft, but PROWL plus Eptam tank mixtures may only be applied by ground equipment.

Application should be made in the late fall when soil temperatures are 45°F or below but before the ground freezes. DO NOT apply PROWL or PROWL plus Eptam 7E tank mixes when the air temperature is below 45°F.

PROWL or PROWL plus Eptam 7E tank mixture should be incorporated immediately after application. Refer to **Incorporation Instructions** section (see table of contents for page number) and Eptam label for incorporation directions.

Prior to sunflower planting in the spring, fields treated with PROWL or PROWL plus Eptam 7E should receive at least one shallow additional incorporation. Spring incorporation should be at an angle to the last tillage operation.

Refer to the appropriate sections of this label for specific directions regarding mixing, spraying, and liquid fertilizer compatibility determinations.

Broadcast Rate Per Acre of PROWL for Fall Preplant Incorporated Application in Sunflowers[1]

	PROWL (pints)	PROWL (pints)
Soil Texture	Up to 3% Organic Matter	More than 3% Organic Matter
COARSE	1½ to 2½	2½
MEDIUM	2 to 3	3 to 3½
FINE	2½ to 3½	3½

[1] For use in North Dakota, South Dakota, and Minnesota only.

DO NOT USE on peat or muck soils.

Use the high rate for each soil texture if heavy weed pressure is anticipated.

NOTE: DO NOT feed forage or graze livestock in treated sunflower fields. If sunflower crop loss occurs due to weather conditions, sunflowers or any crop registered for PROWL preplant incorporated use can be replanted. If replanting is necessary, DO NOT rework the soil deeper than the treated zone.

PROWL Plus Eptam

Weeds Controlled

Tank mixtures of PROWL plus Eptam will control the following weeds in addition to those controlled by applications of PROWL alone:

Annual bluegrass (*Poa annua*)
Annual Morningglory (*Ipomoea purpurea*)
Black Nightshade (*Solanum nigrum*)
Corn spurry (*Spergula arvensis*)

Hairy nightshade (*Solanum villosum*)
Lovegrass (Stinkgrass) (*Eragrostis ciliianensis*)
Nettleleaf goosefoot (*Chenopodium murale*)
Wild oats (*Avena fatua*)

Broadcast Rate per Acre for PROWL Plus Eptam 7E Preplant Incorporated Tank Mixtures for Fall Application in Sunflowers[1]

Soil Texture	PROWL + EPTAM 7E (pints) Up to 3% Organic Matter	PROWL + EPTAM 7E (pints) More than 3% Organic Matter
COARSE	1 to 2 + 2¼	2 + 2¼
MEDIUM	1½ to 2½ + 2¼	2½ to 3 + 2¼
FINE	2 to 3 + 2¼	3 + 2¼

[1] For use in North Dakota or Minnesota only.

DO NOT USE on peat or muck soils.

Use the high rate for each soil texture if heavy weed pressure is anticipated.

NOTE: DO NOT feed forage or graze livestock in treated sunflower fields. If sunflower crop loss occurs due to weather conditions, sunflowers or any crop registered for PROWL + Eptam preplant incorporated use can be replanted. If replanting is necessary, DO NOT rework the soil deeper than the treated zone. Observe all precautions and limitations on the Eptam 7E label.

BEANS (DRY, LIMA, SNAP)

PROWL may be applied preplant incorporated in dry beans as follows:
PROWL alone PROWL + Eptam PROWL + Dual 8E
PROWL may be applied preplant incorporated in lima and snap beans as follows:
PROWL alone PROWL + Dual 8E
DO NOT use PROWL on peat or muck soils.

Preplant Incorporated Applications of PROWL Alone

Apply PROWL with ground equipment or aircraft immediately before planting or up to 60 days prior to planting. Incorporate prior to planting and within 7 days of application. Refer to **Incorporation Instructions** section (see table of contents for page number) for incorporation directions.

Broadcast Rate Per Acre of PROWL Preplant Incorporated in Dry, Lima or Snap Beans

Soil Texture	PROWL (pints) SOUTHERN STATES[1]	PROWL (pints) NORTHERN STATES[1] Up to 3% Organic Matter	PROWL (pints) NORTHERN STATES[1] More than 3% Organic Matter
COARSE	1.0 to 1.5	1.0 to 2.0	2.0
MEDIUM	1.5 to 2.0	1.5 to 2.5	2.5 to 3.0
FINE	1.5 to 3.0	2.0 to 3.0	3.0

[1] See map at end of this label for specific states.

Use the 3 pint rate for heavy clay soils.

NOTE: If loss of dry, lima or snap bean crop occurs due to adverse weather conditions, dry, lima or snap beans, or any crop registered for PROWL preplant incorporated use can be replanted the same year into treated soil without adverse effects. If replanting is necessary, DO NOT rework the soil deeper than the treated zone.

PRECAUTION: DO NOT APPLY PROWL PREEMERGENCE (surface treatment) to dry, lima and snap beans after planting or serious crop injury can result. PROWL must be applied preplant incorporated.

PROWL Plus Dual 8E

Observe all precautionary statements in the PROWL and Dual 8E leaflet labels before using.

PROWL plus Dual 8E tank mixture may be applied preplant incorporated in dry, lima, and snap beans. Uniformly apply PROWL plus Dual 8E with properly calibrated ground equipment in 10 to 40 gallons of water per acre, or by aircraft in a minimum of 5 gallons of water per acre. Use sprayers equipped with nozzles that provide accurate and uniform application.

Apply immediately before planting, or up to 14 days prior to planting. Incorporate uniformly into the top 2 inches of soil within 7 days after application. See **Incorporation Instructions** section of this label for acceptable incorporation techniques.

WEEDS CONTROLLED

When applied as directed, PROWL plus Dual 8E tank mix will control the following weeds in addition to those controlled by PROWL alone:

Galinsoga (*Galinsoga* spp.)
Red rice (*Oryza sativa*)
Southwestern cupgrass (*Eriochloa gracilus*)
Yellow nutsedge (*Cyperus esculentus*)

Tank mixtures of PROWL plus Dual 8E will aid in the control of and reduce competition from:

Hairy nightshade (*Solanum sarrachoides*)
Volunteer sorghum (*Sorghum vulgare*)

Broadcast Rates Per Acre of PROWL Plus Dual 8E Preplant Incorporated Tank Mixture in Dry, Lima or Snap Beans

Soil Texture	Northern States[1] PROWL (pints) + Duel 8E (pints) Up to 3% Organic Matter PROWL + Dual 8E	Northern States[1] PROWL (pints) + Duel 8E (pints) More than 3% Organic Matter PROWL + Dual 8E
COARSE	1 to 2 + 1.5 to 2	2 + 2
MEDIUM	1.5 to 2.5 + 2 to 2.5	2.5 to 3 + 2 to 2.5
FINE	2 to 3 + 2 to 2.5	3 + 2.5 to 3

Soil Texture	Southern States[1] PROWL (pints) + Duel 8E (pints) Up to 3% Organic Matter PROWL + Dual 8E	Southern States[1] PROWL (pints) + Duel 8E (pints) More than 3% Organic Matter PROWL + Dual 8E
COARSE	1 to 1.5 + 1.5 to 2	1 to 1.5 + 2
MEDIUM	1.5 to 2 + 2 to 2.5	1.5 to 2 + 2 to 2.5
FINE	1.5 to 3 + 2 to 2.5	1.5 to 3 + 2.5 to 3

[1] See map at end of this label for specific states.

NOTE: If loss of dry, lima or snap bean crop occurs due to weather conditions, any crop registered for preplant incorporated use on the PROWL and Dual 8E labels may be replanted the same year without adverse effects. If replanting is necessary, DO NOT rework the soil deeper than the treated zone.

PRECAUTION:

DO NOT APPLY PREEMERGENCE (surface treatment) to dry, lima and snap beans or serious crop injury can result.

DO NOT graze or feed forage or fodder.

Observe all rotational restrictions on the Dual 8E label.

PROWL Plus Eptam

PROWL plus Eptam tank mixture may be applied preplant incorporated in dry beans. DO NOT use this tank mixture on Adzuki beans, cowpeas (blackeye peas, blackeye beans), soybeans, lima beans or other flat podded beans.

Apply PROWL plus Eptam with ground sprayer up to 2 days before planting. Incorporate immediately (within minutes) after application to prevent loss of Eptam herbicide. Whenever possible, application and incorporation should be done in the same operation.

Incorporation Instructions

Use power-driven cultivation equipment set to cut to a depth of 2 to 3 inches or tandem disks set to cut to a depth of 4 to 6 inches, operated at 4 to 6 mph, followed by a spike-toothed harrow or some other leveling device which extends beyond ends of the disk. For thorough mixing, disk in two different directions (cross disk).

Weeds Controlled

When applied as directed, PROWL plus Eptam will control the following weeds in addition to those controlled by PROWL alone:

Black nightshade (*Solanum nigrum*)
Deadnettle (Henbit) (*Lamium amplexicaule*)
Hairy nightshade (*Solanum sarrachoides*)
Purple nutsedge (Nutgrass) (*Cyperus rotundus*)
Yellow nutsedge (Nutgrass) (*Cyperus esculentus*)

Broadcast Rate Per Acre of PROWL Plus Eptam Preplant Incorporated Tank Mixture in Dry Beans

Soil Texture	SOUTHERN STATES[1] PROWL + Eptam 7E (pints)	NORTHERN STATES[1] PROWL + Eptam 7E (pints) Up to 3% Organic Matter	NORTHERN STATES[1] PROWL + Eptam 7E (pints) More than 3% Organic Matter
COARSE[2]	1.0 to 1.5 + 2.5 to 4.5	1.0 to 2.0 + 2.5 to 4.5	2.0 + 2.5 to 4.5
MEDIUM	1.5 to 2.0 + 3.0 to 4.5	1.5 to 2.5 + 3.0 to 4.5	2.5 to 3.0 + 3.0 to 4.5
FINE	1.5 to 3.0 + 3.0 to 4.5	2.0 to 3.0 + 3.0 to 4.5	3.0 + 3.0 to 4.5

[1] See map at end of this label for specific states.

[2] DO NOT exceed 3.5 pints Eptam 7E per acre on small white beans on coarse-textured soils.

For nutsedge control, use 4.5 pints of Eptam 7E per acre.

DO NOT feed bean hay, vines and forage or graze livestock in treated bean fields.

SWEET LUPINES

PROWL maybe applied preplant incorporated or preemergence in sweet lupines.

Refer to **General Application Instructions** and **Weed Species Controlled** sections (see table of contents for page number).

BROADCAST RATE PER ACRE

	SOUTHERN STATES[1]	NORTHERN STATES[1]	
Soil Texture	No Organic Matter Restrictions (pints)	Up to 3% Organic Matter (pints)	More than 3% Organic Matter (pints)
COARSE	1.0 to 1.5	1.0 to 2.0	2.0
MEDIUM	1.5 to 2.0	1.5 to 2.5	2.5 to 3.0
FINE	1.5 to 3.0	2.0 to 3.0	3.0

[1] See map at end of label for specific states.

The high rate for each soil texture above, where listed, should be used if heavy weed populations are anticipated or if extensive crop residues are present prior to seedbed preparation. Use the 3 pint rate for heavy clay soils.

DO NOT USE on peat or muck soils.

NOTE: If loss of sweet lupine crop occurs due to adverse weather conditions, any crop registered for PROWL preplant incorporated or preemergence use can be replanted the same year into PROWL treated soil without adverse effects. If replanting is necessary, DO NOT rework the soil deeper than the PROWL treated zone or 2 inches deep if surface applied.

DO NOT feed lupine hay and forage or graze livestock in treated lupine fields.

GARLIC

PROWL APPLIED PREEMERGENCE, POSTEMERGENCE OR PREEMERGENCE FOLLOWED BY POSTEMERGENCE IN OR ON GARLIC IN CALIFORNIA, NEVADA OR OREGON ONLY

Refer to **Mixing, Application** and **Incorporation Instructions, Weeds Controlled** and **Follow Crop Restrictions** sections (see table of contents for page numbers).

Apply PROWL at a broadcast rate of 1.5 to 3 pints per acre. Uniformly apply PROWL in a minimum of 10 gallons of water per acre. Prowl may be applied in the following ways:
(a) Preemergence after planting but before crop and weeds emerge.
(b) Postemergence to the garlic at the 1 to 5 true leaf growth stage.
(c) As a split application, apply at both preemergence and postemergence timings.

Do not exceed 3 pints per acre per crop. Do not apply within 5 months of harvest. PROWL treatments are most effective when adequate rainfall or overhead irrigation is received within 7 days after application.

DO NOT USE on peat or muck soils.

NOTE: If loss of garlic crop occurs due to adverse weather conditions, any crop registered for PROWL preplant incorporated or preemergence use can be replanted the same year into PROWL treated soil without adverse effects. If replanting is necessary, DO NOT rework the soil deeper than the PROWL treated zone or 2 inches deep if surface applied.

DO NOT feed or graze this crop.

FORAGE LEGUMES

USED AS COVER CROP IN FEDERAL SET-ASIDE OR CONSERVATION RESERVE PROGRAM AREAS. PROWL may be Applied Preplant Incorporated or Preemergence For Weed Control in Legume Cover Crops.

Refer to **Mixing, Application** and **Incorporation Instructions, Weeds Controlled** and **Follow Crop Restrictions** sections (see table of contents for page numbers).

Soil Texture	Broadcast Rate per Acre (pints)
COARSE	1.0 to 1.5
MEDIUM	1.5 to 2.0
FINE	2.0 to 2.5

NOTE: If loss of cover crop occurs due to adverse weather conditions, any crop registered for PROWL use and application method can be replanted the same year into PROWL treated soil without adverse effects. If replanting is necessary, DO NOT rework the soil deeper than the PROWL treated zone. Some stand reduction of the legume cover crop may occur with this use. Consult local county extension service or the local ASC committee for recommended cover crops.

DO NOT feed or graze legume cover crops established following PROWL application. The cover crop residue should ultimately be destroyed by tillage or left on the surface to retard erosion or as direct by the local ASC committee.

NONBEARING FRUIT AND NUT CROPS AND VINEYARDS

PROWL may be applied for preemergence control of most annual grasses and certain broadleaf weeds in several nonbearing fruit and nut crops and vineyards. PROWL may be used on the following nonbearing crops:

Almond	Peach
Apple	Pear
Apricot	Pistachio
Cherry	Plum
Citrus	Prune
Grape	English walnut
Nectarine	

Uniformly apply in 20 or more gallons of water per acre (broadcast basis) with ground equipment. Applications may be band or broadcast.

PROWL controls weeds as they germinate but will not control established weeds. Destroy emerged weeds with tillage prior to PROWL application OR use PROWL in combination with a contact herbicide registered for use in specific nonbearing crop to remove existing vegetation. Consult contact herbicide label for all directions, precautions and restrictions.

PROWL treatments are most effective in controlling weeds when adequate rainfall or irrigation is received within 7 days after application.

PRECAUTIONS: Apply the spray directly to the ground beneath the trees or vines.

DO NOT apply to newly transplanted trees or vines until ground has settled and no cracks are present.

DO NOT feed forage or graze livestock in treated fields.

FOR NEWLY TRANSPLANTED AND ONE YEAR OLD GRAPEVINES: Apply only to dormant grapevines. DO NOT apply if buds have started to swell.

Weeds Controlled

In addition to the weeds previously listed as being controlled by PROWL, the following weeds are controlled by preemergence treatments of PROWL at the rates recommended below:

GRASSES CONTROLLED

Bluegrass (annual) (*Poa annua*)
Browntop panicum (*Panicum fasciculatum*)
Junglerice (*Echinochlora colonum*)
Lovegrass (*Eragrostis* spp.)
Mexican sprangletop (*Leptochloa uninervia*)
Red sprangletop (*Leptochloa filiformis*)
Signalgrass (*Brachiaria platyphylla*)
Woolly cupgrass (*Eriochloa villosa*)

BROADLEAF WEEDS CONTROLLED

Chickweed (common) (*Stellaria media*)
Fiddleneck (*Amsinckia intermedia*)
Henbit (*Lamium amplexicaule*)
Knotweed (prostrate) (*Polygonum aviculare*)
London rocket (*Sisymbrium irio*)
Pennsylvania smartweed (*Polygonum pensylvanicum*)
Puncturevine (*Tribulus terrestris*)
Sheperdspurse (*Capsella bursa-pastoris*)
Velvetleaf (Buttonweed) (*Abutilon theophrasti*)

RECOMMENDED BROADCAST RATES

Short-term control (4 months): 2 quarts PROWL per acre
Long-term control (6-8 months): 4 quarts PROWL per acre
DO NOT use PROWL on peat or muck soils.

REGIONAL MAP FOR RATE DETERMINATION

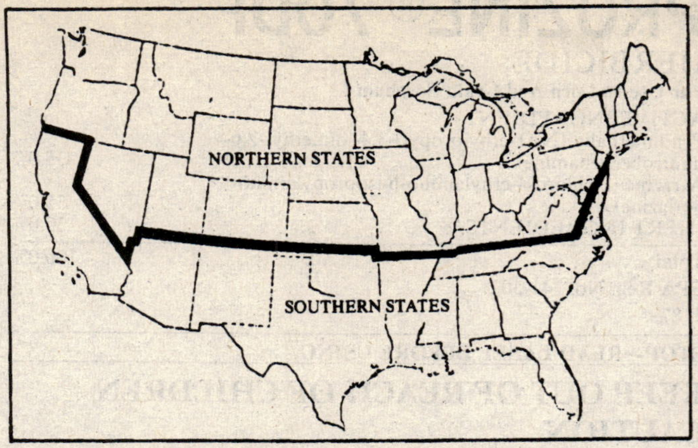

AAtrex is a leading brand of atrazine. AAtrex and AAtrex Nine—O are trademarks of Ciba-Geigy Corporation.
Bladex is a trademark of E.I. du Pont de Nemours and Company.
Marksman is a trademark of Sandoz Crop Protection Inc.
Bicep is a trademark of Ciba-Geigy Corporation.
Lariat is a trademark of Monsanto Agricultural Products Company.
Bullet is a trademark of Monsanto Agricultural Products Company.
Dual is a trademark of Ciba-Geigy Corporation.
Lasso is a trademark of Monsanto Agricultural Products Company.
Banvel is a trademark of Sandoz Crop Protection Inc.
Cotoran is a trademark of Ciba-Geigy Corporation.
Karmex is a trademark of E.I. du Pont de Nemours and Company.
Zorial is a trademark of Sandoz Crop Protection Inc.
SCEPTER is a registered trademark of American Cyanamid.
PURSUIT is a registered trademark of American Cyanamid.
Sencor is a trademark of Farbenfabriken Bayer GmbH, Leverkusen, West Germany.
Lexone is a trademark of E.I. du Pont de Nemours and Company.
Amiben is a trademark of Rhone Poulenc.
Lorox is a trademark of E.I. du Pont de Nemours and Company.
Stam is a trademark of Rohm and Hass Company.
Eptam is a trademark of ICI Americas, Inc.
Vernam is a trademark of ICI Americas, Inc.
Command is a trademark of FMC Corporation.
Canopy is a trademark of E.I. du Pont de Nemours and Company.
Preview is a trademark of E.I. du Pont de Nemours and Company.
TM Trademark of American Cyanamid Company.

PROZINE® 70DF

HERBICIDE

FOR USE IN NONCROP WEED CONTROL

ACTIVE INGREDIENTS:	
Pendimethalin (N-(1-ethylpropyl)-3,4-dimethyl-2,6-dinitrobenzenamine)	35.0%
Atrazine (2-chloro-4-ethylamino-6-isopropylamino-s-triazine)	35.0%
INERT INGREDIENTS:	30.0%
Total	100.0%

EPA Reg. No. 241-302

STOP—READ LABEL BEFORE USING

8/87

KEEP OUT OF REACH OF CHILDREN

CAUTION

STATEMENT OF PRACTICAL TREATMENT

IF ON SKIN: Remove contaminated clothing and wash with plenty of soap and water.

IF IN EYES: Flush eyes with plenty of water. Get medical attention if irritation persists.

IF INHALED: Remove victim to fresh air and apply respiration if indicated.

In case of fire or other emergency, call collect, day or night, AREA CODE 201-835-3100

PRECAUTIONARY STATEMENTS

HAZARDS TO HUMANS AND DOMESTIC ANIMALS

CAUTION!

Harmful if swallowed, inhaled, or absorbed through skin. Do not breathe vapors or spray mist. Avoid contact with eyes, skin or clothing. Remove contaminated clothing and wash before reuse. Wash thoroughly with soap and water after handling.

Do not apply this product in such a manner as to directly or through drift expose workers or other persons. The area being treated must be vacated by unprotected persons.

ENVIRONMENTAL HAZARDS

Atrazine can travel (seep or leach) through soil and can enter groundwater which may be used as drinking water. Atrazine has been found in ground water. Users are advised not to apply atrazine to sand and loamy sand soils where the water table (ground water) is close to the surface and where these soils are very permeable, i.e., well-drained. Your local agricultural agencies can provide further information on the type of soil in your area and the location of groundwater.

This product is toxic to fish and aquatic invertebrates. DO NOT apply directly to water or wetlands. DO NOT apply where runoff is likely to occur. DO NOT apply when weather conditions favor drift from treated areas. DO NOT contaminate water when disposing of equipment washwaters.

DIRECTIONS FOR USE

It is a violation of Federal law to use this product in a manner inconsistent with its labeling.

STORAGE AND DISPOSAL

DO NOT CONTAMINATE WATER, FOOD, OR FEED BY STORAGE OR DISPOSAL

STORAGE

Store in a dry location away from children, animals, foods, feeds, seeds, or other agricultural chemicals. Handle in accordance with information given under PRECAUTIONARY STATEMENTS. In the event of spillage, scrape up spillage. If spillage is not suitable for use, dispose of in accordance with information given under DISPOSAL. Repackage and relabel useable product in a sound container.

DISPOSAL

Pesticide Disposal: Wastes resulting from the use of this product may be disposed of on site or at an approved waste disposal facility.

Container Disposal: Completely empty bag into application equipment. Then dispose of bag in a sanitary landfill or by incineration or, if allowed by state and local authorities, by burning. If burned, stay out of smoke.

GENERAL INFORMATION

PROZINE controls many annual broadleaf and grass weeds. It may be applied before or after weeds emerge. The effectiveness of this product depends on moisture to move it into the root zone. PROZINE is recommended for preemergence control of most annual grasses and certain broadleaf weeds as they germinate on noncropland areas such as railroad, utility and pipeline rights-of-way, highway guardrails, delineators, and sign posts, utility substations, petroleum tank farms, pumping installations, fence rows, storage areas, and other similar areas.

PROZINE will not control established weeds. Therefore, areas to be treated should be free of established weeds at the time of treatment, or PROZINE may be used in conjunction with herbicides registered for post-emergence use in noncropland areas. Consult the labels of those herbicides for suggested treatments, rates to be used and precautions or restrictions for use in noncropland areas.

APPLICATION PROCEDURES

GROUND APPLICATION. Use conventional ground sprayers equipped with nozzles that provide accurate and uniform application. Unless otherwise specified, use a minimum of 20 gallons of spray mixture per acre for preemergence and postemergence applications with ground equipment.

AERIAL APPLICATION. Apply in a minimum of 10 gallons of water per acre. Avoid application to humans, animals, or nontarget areas. Flagmen and loaders should avoid inhalation of spray mist and prolonged contact with skin and should wash thoroughly before eating and at the end of each days operation.

MIXING PROCEDURES—ALL USES. Fill tank 1/4 full with clean water. Start agitation. Be certain that the agitation system is working properly and creates a rippling or rolling action on the liquid surface. Pour product directly from container into tank. Let it wet and settle into solution. Continue filling tank until 90% full. Increase agitation if necessary to maintain surface action. Finish filling tank. Maintain agitation to avoid separation of materials. Clean sprayer thoroughly after use by flushing system with water.

NONCROPLAND WEED CONTROL

For preemergence control of the weed species listed, apply PROZINE at the following rates:

Length of Control	PROZINE Rate
Short Term Control (2-4 months)	5 1/2 lb./Acre
Long Term Control (6-8 months)	11 1/2 lb./Acre

PROZINE will not control established weeds.

For post-emergence control of weeds, tank-mix combinations of PROZINE plus ARSENAL(R) herbicide are recommended. Recommended rates for the tank-mix compounds should be determined from the product labels of both PROZINE and ARSENAL prior to use.

Do not tank mix with ARSENAL in California.

WEED SPECIES CONTROLLED

GRASSES CONTROLLED

Barnyardgrass (Echinochloa crus-galli)
Bluegrass, Annual (Poa annua)
Crabgrass (Digitaria spp.)
Crowfootgrass (Dactyloctenium aegyptium)
Foxtail, Giant (Setaria faberi)
Foxtail, Green (Setaria viridis)
Foxtail, Yellow (Setaria lutescens)
Goosegrass (Eleusine indica)
Itchgrass (Rottboellia exaltata)
Johnsongrass (from seed) (Sorghum halepense)
Junglerice (Echinochloa colonum)
Lovegrass (Eragrostis spp.)
Panicum, Browntop (Panicum fasciculatum)
Panicum, Fall (Panicum dichotomiflorum)
Panicum, Texas (Panicum texanum)
Sandbur, Field (Cenchrus incertus)
Signalgrass (Brachiaria platyphylla)
Sprangletop, Mexican (Leptochloa uninervia)
Sprangletop, Red (Leptochloa filiformis)
Witchgrass (Panicum capillare)
Woolly Cupgrass (Eriochloa villosa)

BROADLEAF WEEDS CONTROLLED

Black nightshade (Solanum nigrum)
Carpetweed (Mollugo verticillata)
Chickweed, Common (Stellaria media)
Fiddleneck, Coast (Amsinckia intermedia)
Filaree (Erodium spp.)
Henbit (Lamium amplexicaule)
Knotweed, Prostrate (Polygonum aviculare)
Kochia (Kochia scoparia)
Lambsquarters, Common (Chenopodium album)
Mallow, Venice (Hibiscus trionum)
Mustard (Brassica spp.)
Pigweed (Amaranthus spp.)
Puncturevine (Tribulus terrestris)
Purslane, Common (Portulaca oleracea)
Pusley, Florida (Richardia scabra)
Ragweed, Common (Ambrosia artemisiifolia)
Rocket, London (Sisymbrium irio)
Sheperdspurse (Capsella bursa-pastoris)
Smartweed, Pennsylvania (Polygonum pensylvanicum)
Spurge (Euphorbia spp.)
Velvetleaf (Buttonweed) (Abutilon theophrasti)

PROZINE will aid in the control of the following weeds:

Annual Morningglory (Ipomoea spp.)
Cocklebur, Common (Xanthium strumarium)
Groundcherry (Physalis spp.)
Jimsonweed (Datura stramonium)
Millet, Wild Proso (Pancium miliaceum)
Shattercane (Sorghum bicolor)
Wild oats (Avena fatua)

DISCLAIMER

The label instructions for the use of this product reflect the opinion of experts based on field use and tests. The directions are believed to be reliable and should be followed carefully. However, it is impossible to eliminate all risks inherently associated with use of this product. Crop injury, ineffectiveness, or other unintended consequences may result because of such factors as weather conditions, presence of other materials, or the use or application of the product contrary to the label instructions, all of which are beyond the control of American Cyanamid Company. All such risks shall be assumed by the user.

American Cyanamid Company warrants only that the material contained herein conforms to the chemical description on the label and is reasonably fit for the use herein described when used in accordance with the directions for use, subject to the risks referred to above.

Any damages arising from a breach of this warranty shall be limited to direct damages, and shall not include any other express or implied warranty of FITNESS or MERCHANTABILITY.

PROZINE® 70DF

HERBICIDE

For Use In Corn And Grain Sorghum

ACTIVE INGREDIENT

Pendimethalin (N-(1-ethylpropyl)-3,4- dimethyl-2,6-dinitrobenzenamine)	35.0%
Atrazine (2-chloro-4-ethylamino-6-isopropylamino-s-triazine)	35.0%
INERT INGREDIENTS:	30.0%
Total	100.0%

EPA Reg. No. 241-302

1/87

STOP—READ LABEL BEFORE USING

KEEP OUT OF REACH OF CHILDREN
CAUTION

STATEMENT OF PRACTICAL TREATMENT

IF ON SKIN: Remove contaminated clothing and wash with plenty of soap and water.

IF IN EYES: Flush eyes with plenty of water. Get medical attention if irritation persists.

IF INHALED: Remove victim to fresh air and apply respiration if indicated.

In case of fire or other emergency, call collect, day or night, AREA CODE 201-835-3100

PRECAUTIONARY STATEMENTS

HAZARDS TO HUMANS AND DOMESTIC ANIMALS

CAUTION

Harmful if swallowed, inhaled, or absorbed through skin. Do not breathe vapors or spray mist. Avoid contact with eyes, skin or clothing. Remove contaminated clothing and wash before reuse. Wash thoroughly with soap and water after handling.

Do not apply this product in such a manner as to directly or through drift expose workers or other persons. The area being treated must be vacated by unprotected persons.

ENVIRONMENTAL HAZARDS

Atrazine can travel (seep or leach) through soil and can enter groundwater which may be used as drinking water. Atrazine has been found in groundwater. Users are advised not to apply atrazine to sand and loamy sand soils where the water table (groundwater) is close to the surface and where these soils are very permeable, i.e., well-drained. Your local agricultural agencies can provide further information on the type of soil in your area and the location of groundwater.

Plus see other PROZINE 70DF label.

This product is toxic to fish. Keep out of lakes, streams, or ponds. DO NOT apply when weather conditions favor drift from target area. DO NOT contaminate water by cleaning of equipment or disposal of wastes.

DIRECTIONS FOR USE

It is a violation of Federal law to use this product in a manner inconsistent with its labeling.

STORAGE AND DISPOSAL

DO NOT CONTAMINATE WATER, FOOD, OR FEED BY STORAGE OR DISPOSAL.

STORAGE

Store in a dry location away from children, animals, foods, feeds, seeds, or other agricultural chemicals. Handle in accordance with information given under PRECAUTIONARY STATEMENTS. In the event of spillage, scrape up spillage. If spillage is not suitable for use, dispose of in accordance with information given under DISPOSAL. Repackage and relabel useable product in a sound container.

DISPOSAL

Pesticide Disposal: Wastes resulting from the use of this product may be disposed of on site or at an approved waste disposal facility.

Container Disposal: Completely empty bag into application equipment. Then dispose of bag in a sanitary landfill, or by incineration or, if allowed by state and local authorities, by burning. If burned, stay out of smoke.

RE-ENTRY STATEMENT (FOR USE IN SWEET CORN IF HAND LABOR IS INVOLVED)

Do not enter treated areas without protective clothing until sprays have dried. Protective clothing means, at least, a hat or other suitable head covering, a long-sleeved shirt and long-legged trousers or a coverall-type garment (all of closely woven fabric covering the body, including the arms and legs), impermeable gloves and boots. Because certain states may require more restrictive re-entry intervals for various crops treated with this product, consult your State Department of Agriculture for further information.

Written or oral warnings must be given to workers who are expected to be in a treated area or in an area about to be treated with this product. When oral warnings are given, warnings shall be given in a language customarily understood by workers. Oral warnings must be given if there is reason to believe that written warnings cannot be understood by workers. Written warnings must include the following information: "CAUTION. Area treated with PROZINE on (date of application). Do not enter without appropriate protective clothing."

GENERAL INFORMATION

PROZINE controls many annual broadleaf and grass weeds in corn and sorghum. It may be applied before or after weeds emerge. The effectiveness of this product depends on moisture to move it into the root zone. If weeds develop, a shallow cultivation or rotary hoeing generally result in better weed control. Unusually cold, excessively wet, or hot and dry conditions that delay germination or extend germination over a long period of time can reduce weed control.

Seedling diseases, cold weather, excessive moisture, deep planting, low or high soil pH, high soil salt concentration, or drought can weaken seedlings and plants and increase the possibility of crop damage from this herbicide. Under these conditions, crop yields can be reduced.

Application in Water or Liquid Fertilizer: Nitrogen solution or complete liquid fertilizer may replace all or part of the water as a carrier for preemergence on corn. Check the compatibility of this product with liquid fertilizer and/or nitrogen solution as shown below before use. Do not apply in nitrogen solution or complete liquid fertilizer after corn emerges.

To determine the compatibility of PROZINE with liquid fertilizer, pour the product into a small container of liquid fertilizer in the proportions shown below. For different spray volume or herbicide rate, make appropriate changes in the ingredients.

For each pound of PROZINE to be applied per acre, add 1 heaping teaspoon of the product to one pint of liquid fertilizer (assuming a volume of 25 gallons per acre).

After thorough mixing, let stand 5 minutes. If the herbicide/fertilizer combination remains mixed, or can be remixed readily, make a slurry with the herbicide before adding fertilizer and use the mixture only if this test shows it is compatible.

WEED SPECIES CONTROLLED

GRASSES CONTROLLED

Barnyardgrass *(Echinochloa crus-galli)*
Crabgrass *(Digitaria* spp.*)*
Crowfootgrass *(Dactyloctenium aegyptium)*
Fall panicum *(Panicum dichotomiflorum)*
Field sandbur *(Cenchrus incertus)*
Foxtail (giant) *(Setaria faberi)*
Foxtail (green) *(Setaria viridis)*
Foxtail (yellow) *(Setaria glauca)*
Goosegrass *(Eleusine indica)*
Johnsongrass (from seed) *(Sorghum halepense)*
Signalgrass *(Brachiaria platyphylla)*
Texas panicum *(Panicum texanum)*
Witchgrass *(Panicum capillare)*

BROADLEAF WEEDS CONTROLLED

Black nightshade *(Solanum nigrum)*
Carpetweed *(Mollugo verticillata)*
Common ragweed *(Ambrosia artemisiifolia)*
Florida pusley *(Richardia scabra)*
Kochia *(Kochia scoparia)*
Lambsquarters *(Chenopodium album)*
Mustard *(Brassica* spp.*)*
Pennsylvania smartweed *(Polygonum pensylvanicum)*
Pigweed *(Amaranthus* spp.*)*
Purslane *(Portulaca oleracea)*
Spurge, annual *(Euphorbia* spp.*)*
Velvetleaf (Buttonweed) *(Abutilon theophrasti)*
Venice mallow *(Hibiscus trionum)*

PROZINE will aid in the control and reduce competition from:

Annual morningglory *(Ipomoea* spp.*)*
Cocklebur *(Xanthium strumarium)*
Groundcherry *(Physalis* spp.*)*
Jimsonweed *(Datura stramonium)*
Shattercane *(Sorghum bicolor)*
Wild oats *(Avena fatua)*
Wild proso millet *(Panicum miliaceum)*

ROTATIONAL CROPS: (1) Do not rotate to any crop except corn or sorghum until the following year, as injury may occur. (2) In the High Plains and intermountain areas of the West where rainfall is sparse and erratic or where irrigation is required, use only when corn or sorghum is to follow corn or sorghum or when a crop of untreated corn or sorghum is to precede other rotational crops. (3) Injury may occur to soybeans planted in north-central and northwest Iowa, south-central and southwest Minnesota, northeast Nebraska, southeast South Dakota, and other areas the year following application on soils having a calcareous surface layer. (4) Do not plant sugarbeets, tobacco, vegetables (including dry beans), spring-seeded small grains, or small-seeded legumes and grasses the year following application as injury may occur.

APPLICATION PROCEDURES

GROUND APPLICATION: Use conventional ground sprayers equipped with nozzles that provide accurate and uniform application. Be certain that nozzles are uniformly spaced and are the same size. Calibrate sprayer before use and recalibrate at the start of each season and when changing carriers. Unless otherwise specified, use a minimum of 10 gallons of spray mixture per acre for preemergence and postemergence applications with ground equipment.

Use a pump with capacity to (1) maintain 35-40 psi at nozzles, (2) provide sufficient agitation in tank to keep mixture in suspension, and (3) to provide a minimum of 20% bypass at all times. Use centrifugal pumps which provide propeller shear action for dispersing and mixing this product. The pump should provide a minimum of 10 gallons per minute per 100 gallons tank size circulated through a correctly positioned sparger tube or jets.

Use screens to protect the pump and to prevent nozzles from clogging. Screens placed on the suction side of the pump should be 16-mesh or coarser. Do not place a screen in the recirculation line. Use 50-mesh or coarser screens between the pump and boom and, where required, at the nozzles. Check nozzle manufacturer's recommendations.

AERIAL APPLICATION: Apply in a minimum of 5 gallons of water per acre. Avoid application to humans, animals, or non-target areas. Flagmen and loaders should avoid inhalation of spray mist and prolonged contact with skin and should wash thoroughly before eating and at the end of each day's operation.

MIXING PROCEDURES—ALL USES: Be sure sprayer is clean and not contaminated with any other materials, or crop injury or sprayer clogging may result. Fill tank 1/4 full with clean water, nitrogen solution, or complete liquid fertilizer. Start agitation. Be certain that the agitation system is working properly and creates a rippling or rolling action on the liquid surface. Pour product directly from container into tank. Let it wet and settle into solution. Continue filling tank until 90% full. Increase agitation if necessary to maintain surface action. Finish filling tank. Maintain agitation to avoid separation of materials. Clean sprayer thoroughly immediately after use by flushing system with water containing a detergent.

DISCLAIMER

The label instructions for the use of this product reflect the opinion of experts based on field use and tests. The directions are believed to be reliable and should be followed carefully. However, it is impossible to eliminate all risks inherently associated with use of this product. Crop injury, ineffectiveness, or other unintended consequences may result because of such factors as weather conditions, presence of other materials, or the use or application of the product contrary to the label instructions, all of which are beyond the control of American Cyanamid Company. All such risks shall be assumed by the user.

American Cyanamid Company warrants only that the material contained herein conforms to the chemical description on the label and is reasonably fit for the use therein described when used in accordance with the directions for use, subject to the risks referred to above.

Any damages arising from a breach of this warranty shall be limited to direct damages, and shall not include any other express or implied warranty of FITNESS or MERCHANTABILITY.

FIELD CORN AND GRAIN SORGHUM

POSTEMERGENCE INCORPORATED APPLICATIONS

GENERAL INFORMATION: Winter wheat and winter barley may be planted in the fall 90 days after an application with PROZINE postemergence incorporated in irrigated field corn or grain sorghum or 120 days after an application in dry land field corn or grain sorghum. The treated crop must be grown to maturity and harvested before planting wheat or barley. See GENERAL INFORMATION for complete rotational instructions.

PROZINE applied as a postemergence incorporated treatment will control most late season annual grasses and certain broadleaf weeds in field corn or grain sorghum. Treatments can be applied from the 4-inch growth stage to as late as the last cultivation (layby) of field corn or grain sorghum. Treatments will not control established weeds. **EMERGED WEEDS MUST BE DESTROYED BY CULTIVATION PRIOR TO APPLICATION.**

APPLICATION TECHNIQUES: The field must be cultivated and all emerged weeds must be destroyed prior to application. Cultivation should be accomplished with a sweep-type or rolling cultivator operated at sufficient speed to throw at least one inch of soil over the bases of the field corn or grain sorghum plants. This will kill small weed seedlings growing in the field corn or grain sorghum row and will prevent direct contact of the zone of brace root formation during application.

Apply with ground sprayer. Uniform soil coverage must be obtained. Drop nozzles should be used if field corn or grain sorghum foliage will prevent uniform coverage of the soil surface within the rows.

Treatments must be thoroughly and uniformly incorporated into the soil (1) with a sweep-type or rolling cultivator set to provide thorough incorporation in the top 1 inch of soil or (2) with adequate irrigation water or rainfall. For best results, treatments should be incorporated as soon as possible after application but incorporation must be completed within 7 days after application. If adequate moisture is not received within 7 days after application, PROZINE must be incorporated with a sweep-type or rolling cultivator.

Under situations of low rainfall or soil moisture, when deep germinating weeds such as shattercane or field sandbur are anticipated, mechanical incorporation will provide best results.

If cultivation is needed after application and incorporation, the depth of cut should be no deeper than the depth of cut used to incorporate.

DO NOT APPLY PROZINE in field corn or grain sorghum more than once per crop season.

WEEDS CONTROLLED: When applied as directed, PROZINE will control shattercane (*Sorghum bicolor*) and wild proso millet (*Panicum miliaceum*) in addition to the WEED SPECIES CONTROLLED previously listed.

ADDITIONAL HERBICIDE APPLICATION: Postemergence incorporated applications of PROZINE can be applied in field corn or grain sorghum previously treated with herbicides registered for use in field corn or grain sorghum. Consult the labels of those herbicides for suggested treatments, rates to be used, and precautions or restrictions for use in field corn or grain sorghum and for restrictions on following crops.

FIELD CORN

DO NOT APPLY PROZINE preplant incorporated in field corn as serious crop injury can result.

Broadcast Rate Per Acre

Soil Texture	Southern States[1]	Northern States[1]
COARSE	2.0 lbs.	3.0 lbs.
MEDIUM	3.0 lbs.	4.25 lbs.
FINE	4.25 lbs.	4.25 lbs.

[1] See map below.

GRAIN SORGHUM

DO NOT APPLY in grain sorghum preplant incorporated or preemergence as serious crop injury can result.

DO NOT APPLY in grain sorghum which is planted in double row beds since this cultural practice does not permit adequate soil coverage of the bases of the grain sorghum plants with cultivation.

Broadcast Rate per Acre

Soil Texture	Southern States[1]	Northern States[1]
COARSE	DO NOT USE	DO NOT USE
MEDIUM	3.0 lbs.	4.25 lbs.
FINE	4.25 lbs.	4.25 lbs.

[1] See map below.
NOTE: Observe all precautions, limitations, and cropping restrictions referred to in GENERAL INFORMATION section. Livestock can graze or be fed forage from treated sorghum fields 21 days following application.

CORN—FIELD AND SWEET

GENERAL INSTRUCTIONS

Use only where adequate tillage is practiced to provide good soil coverage of the corn seed. Plant corn AT LEAST 1½ inches deep to ensure good seed coverage.

PROZINE is most effective in controlling weeds when adequate rainfall or overhead irrigation is received within 7 days after application. If cultivation is necessary because of soil crusting, soil compaction or weed germination before rain or irrigation, use shallow tillage (such as rotary hoe), and make certain corn seeds are below the tilled area. For maximum effectiveness, wait at least 7 to 10 days before cultivating early postemergence treatments.

Field Corn Grown in Minimum Tillage Systems

PROZINE may be used in field corn in minimum tillage systems where adequate tillage is practiced to provide a seedbed without living vegetation and which allows for good seed coverage. The use of no-till planters in minimum tillage corn under conditions which do not allow good soil coverage of the corn seed can result in reduced crop stand or injury if PROZINE contacts the germinating corn seed. Check equipment to ensure good seed coverage. Apply in 20 or more gallons of spray solution per acre.

PREEMERGENCE (Surface Treatment After Planting)

Apply PROZINE preemergence after planting, but before weeds or crop emerge. Apply in water or liquid fertilizer with ground equipment or in water with aircraft. Refer to section on application in water or liquid fertilizer for liquid fertilizer compatibility determinations. DO NOT INCORPORATE or serious corn injury will result.

EARLY POSTEMERGENCE

Apply PROZINE after corn emergence but no later than when corn is in the 4-leaf stage of growth and when weeds are no more than 1 inch high, except for Texas panicum which must be no larger than the 1 to 2 leaf stage. Apply with ground equipment or aircraft in water ONLY. For ground applications, flat fan nozzles are preferred. DO NOT apply following PROWL preemergence treatment.

RESTRICTIONS FOR EARLY POSTEMERGENCE APPLICATIONS

To avoid serious crop injury with early postemergence applications, observe the following restrictions:

DO NOT INCORPORATE

DO NOT apply in liquid fertilizer.

Broadcast Rate Per Acre of PROZINE
Preemergence and Early Postemergence in Corn[1]

Soil Texture	Less than 1.5% Organic Matter	1.5 to 3.0% Organic Matter	More than 3% Organic Matter
COARSE	3.0 lbs.	3.0 lbs.	4.25 lbs.
MEDIUM	3.0 lbs.	4.25 lbs.	4.25 lbs.
FINE	3.0 lbs.	4.25 lbs.	4.25 lbs.

[1] PROZINE may be applied in sweet corn (processing varieties) in Illinois, Minnesota, Wisconsin, and New York only.
NOTE: If corn crop loss occurs due to weather conditions and corn is replanted, seeding depth must be below retilled area.
NOTE FOR ALL APPLICATIONS TO CORN: DO NOT graze or feed forage from treated areas for 21 days following application.

REGIONAL MAP FOR RATE DETERMINATION

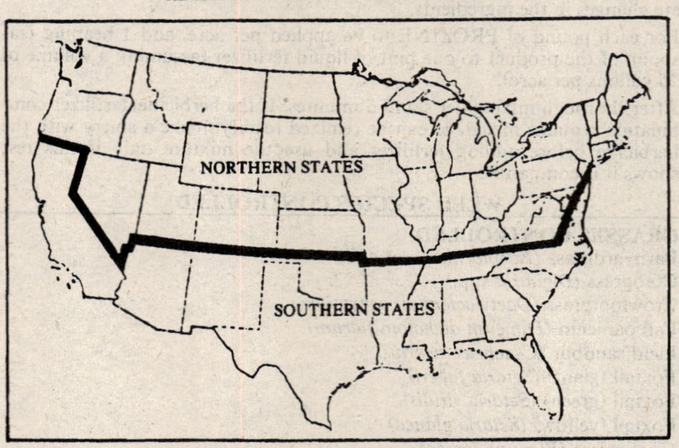

PURSUIT®

HERBICIDE

FOR USE IN SOYBEANS

ACTIVE INGREDIENT:

Ammonium salt of imazethapyr (±)-2-[4,5-dihydro-4-methyl-4-(1-methylethyl)-5-oxo-1*H*-imidazol-2-yl]-5-ethyl-3-pyridinecarboxylic acid* 22.87%
INERT INGREDIENTS 77.13%
TOTAL 100.00%

*Equivalent to 21.6% (±)-2-[4,5-dihydro-4-methyl-4-(1-methylethyl-5-oxo-1*H*-imidazol-2-yl]-5-ethyl-3-pyridinecarboxylic acid
(1 gallon contains 2.0 pounds of active ingredient as the free acid)

EPA Reg. No. 241-310 3/89
®Registered Trademark of American Cyanamid Company

KEEP OUT OF REACH OF CHILDREN

CAUTION!/¡PRECAUCION!

PRECAUCION AL USUARIO: Si usted no lee inglés, no use este producto hasta que la etiqueta le haya sido explicada ampliamente.

STATEMENT OF PRACTICAL TREATMENT

IF ON SKIN: Wash with plenty of soap and water. Get medical attention if irritation persists.

IF IN EYES: Flush with plenty of water.

In case of an emergency endangering life or property involving this product, call collect, day or night, area code 201-835-3100

PRECAUTIONARY STATEMENTS

HAZARDS TO HUMANS
CAUTION!

Harmful if inhaled or absorbed through skin. Avoid breathing spray mist. Avoid contact with skin, eyes or clothing. Wash thoroughly with soap and water after handling. Remove contaminated clothing and wash before reuse.

ENVIRONMENTAL HAZARDS

DO NOT apply directly to water or wetlands (swamps, bogs, marshes and potholes).

DO NOT clean equipment or dispose of equipment washwaters in a manner that will contaminate water resources.

DIRECTIONS FOR USE

It is a violation of Federal law to use this product in a manner inconsistent with its labeling.

Observe all cautions and limitations on this label and on the labels of products used in combination with PURSUIT. The use of PURSUIT not consistent with this label may result in injury to crops. Keep containers closed to avoid spills and contamination.

DO NOT apply this product through any type of irrigation system.

STORAGE AND DISPOSAL

PROHIBITIONS: KEEP FROM FREEZING

DO NOT store below 32°F.
DO NOT contaminate water, food or feed by storage or disposal.

PESTICIDE DISPOSAL: Wastes resulting from the use of this product may be disposed of on site or at an approved waste disposal facility.

CONTAINER DISPOSAL: Triple rinse (or equivalent). Then offer for recycling or reconditioning, or puncture and dispose of in a sanitary landfill, by incineration or, if allowed by State and local authorities by burning. If burned, stay out of smoke.

DISCLAIMER

The label instructions for the use of this product reflect the opinion of experts based on research and field use. The directions are believed to be reliable and should be followed carefully. However, it is impossible to eliminate all risks inherently associated with use of this product. Crop injury, ineffectiveness or other unintended consequences may result because of such factors as weather conditions, presence of other materials, or the use of, or application of the product contrary to label instructions, all of which are beyond the control of American Cyanamid Company. All such risks shall be assumed by the user.

American Cyanamid Company warrants only that the material contained herein conforms to the chemical description on the label and is reasonably fit for the use therein described when used in accordance with the directions for use, subject to the risks referred to above.

Any damages arising from breach of this warranty shall be limited to direct damages and shall not include consequential commercial damages such as loss of profits or values or any other special or indirect damages.

American Cyanamid Company makes no other express or implied warranty, including other express or implied warranty of FITNESS or of MERCHANTABILITY.

GENERAL INFORMATION

PURSUIT can be applied as an early preplant, preplant incorporated, preemergence or early postemergence treatment in soybeans. The application method of choice will depend on the anticipated weed spectrum and the preference of the applicator.

The mode of weed killing activity involves uptake of PURSUIT by weed roots and/or foliage and rapid translocation to the growing points. Adequate soil moisture is important for optimum PURSUIT activity. When adequate soil moisture is present, PURSUIT will provide residual control of susceptible germinating weeds; activity on established weeds will depend on the weed species and the location of its root system in the soil.

Occasionally, internode shortening and/or temporary yellowing of soybean plants may occur following PURSUIT applications. This will not affect soybean yields.

MIXING INSTRUCTIONS

Fill the spray tank one-half to three-quarters full with clean water. Use a calibrated measuring device to measure the required amount of PURSUIT. Add PURSUIT to the spray tank while agitating. Fill the remainder of the tank with water.

For postemergence applications, add a nonionic surfactant and a liquid fertilizer solution (10-34-0, 28 or 32% nitrogen) as the last ingredients in the tank. A liquid fertilizer solution is required for optimum weed control. Maintain agitation while spraying to ensure a uniform spray mixture. An antifoaming agent may be added to the tank if needed.

When tank mixing PURSUIT with recommended herbicides, add the other herbicide to the spray tank first and make sure it is thoroughly mixed before adding PURSUIT. When tank mixing Bronco[1], Roundup[1] or Gramoxone Super[2] with PURSUIT and PROWL®, trifluralin, Lasso[1] or Dual[3] add the herbicides in the following order: (1) PROWL, trifluralin, Lasso[1] or Dual[3], (2) Roundup or Gramoxone Super, (3) PURSUIT, and (4) nonionic surfactant.

To avoid injury to sensitive crops, spray equipment used for PURSUIT applications must be drained and thoroughly cleaned with water before being used to apply other products.

SPRAYING INSTRUCTIONS

GROUND APPLICATIONS: Uniformly apply with properly calibrated ground equipment in 10 or more gallons of water per acre. A spray pressure of 20 to 40 psi is recommended.

To ensure thorough coverage, use a minimum of 20 gallons of water per acre when applying PURSUIT to minimum or no-till soybeans. Use higher gallonage for fields with dense vegetation or heavy crop residues.

DO NOT apply with ground equipment when wind velocity is greater than 10 mph, or when spray may be carried to sensitive crops. Sensitive crops include leafy vegetables, sugar beets and cotton.

DO NOT apply by aerial application.

Avoid overlaps when spraying.

[1]Bronco, Lasso and Roundup are trademarks of Monsanto Agricultural Products Company.
[2]Gramoxone Super is a trademark of ICI Americas, Inc.
[3]Dual is a trademark of Ciba Geigy Corporation.
®Registered trademark of American Cyanamid Company.

PURSUIT HERBICIDE USE AREA

PURSUIT may be applied only in the states or parts of states shaded in the map below:

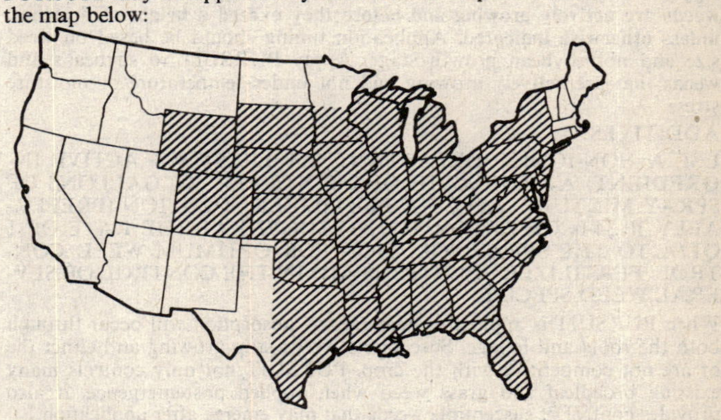

USE RATE
(4 OUNCES PER ACRE)

Apply PURSUIT at a broadcast rate of 4 ounces per acre (1/4 pint) for all methods of application: early preplant, preplant incorporated, preemergence (including minimum and no-till) and postemergence. At this broadcast rate, one gallon of PURSUIT will treat 32 acres of soybeans.

NOTE: Only one application of PURSUIT may be made during the season.

WEEDS CONTROLLED POSTEMERGENCE

BROADLEAF WEEDS

Weeds Controlled 1-2 leaves (1-2 inches)	Weeds Controlled 1-4 leaves (1-3 inches)	Weeds Controlled up to 8 leaves
Lambsquarters, common	Jimsonweed	Artichoke,
Morningglory species	Kochia	Jerusalem (6-10″)
entireleaf	Mallow, Venice	Cocklebur (1-8″)
ivyleaf	Morningglory	Pigweed (1-8″)
pitted	smallflower	palmer
tall	Mustard species	redroot
	Nightshade	smooth
	black	spiny
	Eastern black	Waterhemp, tall
	hairy	
	Ragweed	
	common	
	giant	
	Smartweed	
	ladysthumb	
	Pennsylvania	
	Spurge	
	prostrate	
	spotted	
	Sunflower, common	
	Velvetleaf	

GRASSES

Weeds controlled 1-3 leaves (1-3 inches)	Weeds Controlled 1-6 leaves (1-8 inches)
Barnyardgrass	Johnsongrass
Crabgrasss	seedling
large	Shattercane
smooth	
Cupgrass*, wooly	
Foxtail	
giant	
green	
giant green	
robust purple	
robust white	
yellow	

POSTEMERGENCE applications of PURSUIT will aid in the control and reduce competition from broadleaf signalgrass, Yellow nutsedge and rhizome Johnsongrass.

*PURSUIT controls emerged woolly cupgrass only.

APPLICATION INSTRUCTIONS

POSTEMERGENCE

Apply PURSUIT herbicide as an early postemergence treatment when weeds are actively growing and before they exceed a height of 3 inches, unless otherwise indicated. Application timing should be based on weed size and not soybean growth stage. Apply PURSUIT to soybeans and weeds that are actively growing and not under temperature or moisture stress.

ADDITIVES:

USE A NON-IONIC SURFACTANT (AT LEAST 80% ACTIVE INGREDIENT) AT THE RATE OF 2 PINTS PER 100 GALLONS OF SPRAY MIXTURE. APPLY A FERTILIZER SOLUTION (PREFERABLY 10-34-0, OR 28% OR 32% NITROGEN) AT THE RATE OF 1 QT/A TO THE SPRAY SOLUTION FOR OPTIMUM WEED CONTROL. FERTILIZER SOLUTIONS AID IN THE CONTROL OF SEVERAL WEED SPECIES.

When PURSUIT is applied postemergence, absorption will occur through both the roots and foliage. Susceptible weeds stop growing and either die or are not competitive with the crop. PURSUIT not only controls many existing broadleaf and grass weed when applied postemergence, it also provides control of susceptible weeds that may emerge after application.

A timely cultivation aids in the control of certain weeds when moisture is not received following application. Cultivation should be shallow. After postmergence applications, wait at least 10 days before cultivating.

PURSUIT may be used in combination with some other registered broadleaf herbicides. (See Page 6 for restrictions.)

PURSUIT may be followed by postemergence applications of a selective postemergence grass herbicides such as Fusilade 2000[4], Poast[5], or Assure[6]. DO NOT tank mix PURSUIT with postemergence grass herbicides.

[4]Fusilade 2000 is a trademark of ICI Americas, Inc.

[5]Poast is a trademark of BASF Corporation.

[6]Assure is a trademark of E. I. duPont de Nemours & Co., Inc.

WEEDS CONTROLLED BY SOIL APPLICATIONS OF PURSUIT ALONE

When applied as directed, PURSUIT soil-applied will control the following weeds:

BROADLEAF WEEDS

Carpetweed	Eastern black	Smartweed
Cocklebur, common**	hairy	ladysthumb
Galinsoga	Pigweed	Pennsylvania
Jimsonweed*	palmer	Spurge
Kochia	redroot	prostrate
Lambsquarters, common*	smooth	spotted
Morningglory	spiny	Sunflower,
smallflower	Puncturevine	common*
Mustard species	Purslane, common	Velvetleaf*
Nightshade	Pusley, Florida	Waterhemp, tall
black	Sida, prickly* (teaweed)	

GRASSES*

Foxtail	robust white
giant	yellow
giant green	Johnsongrass
green	seedling
robust purple	

*Common lambsquarters, jimsonweed, prickly sida, velvetleaf, sunflower and grass species are more consistently controlled by preplant incorporated treatments.

** Light to moderate infestations only. Must be preplant incorporated for best results.

PURSUIT soil applied will aid in the control of and reduce competition from:

BROADLEAF WEEDS	GRASSES*
Morningglory***	Barnyardgrass***
entireleaf	Crabgrass
ivyleaf	large
pitted	smooth
tall	Goosegrass
Ragweed	Panicum***
common	fall
giant***	Texas
	Shattercane
	Signalgrass, broadleaf

*** More consistent control can be obtained from preplant incorporated treatments.

WEEDS CONTROLLED BY PREPLANT AND PREEMERGENCE COMBINATION TREATMENTS

When applied as directed, PURSUIT preplant incorporated or preemergence combination treatments with PROWL, Treflan/trifluralin, Lasso or Dual will control the weeds listed in following table, in addition to those controlled by PURSUIT alone.

GRASSES	PROWL[a]	trifluralin[b]	Lasso	Dual
Barnyardgrass	X	X	X	X
Crabgrass, smooth	X	X	X	X
Crabgrass, large	X	X	X	X
Crowfootgrass	X	X		
Goosegrass	X	X	X	X
Panicum, fall	X	X	X	X
Panicum, Texas	X	X		
Sandbur, field	X	X		
Shattercane	X[b]	X[b]		
Signalgrass, broadleaf	X	X	X	X
Witchgrass	X	X	X	X

[a] Preplant incorporated tankmix applications of PURSUIT plus PROWL will suppress the growth of intchgrass, wild proso millet and rhizome Johnsongrass.

[b] Preplant incorporated treatments only.

SOIL APPLICATIONS

PURSUIT can be applied as an early preplant, preplant incorporated, or preemergence treatment in soybeans. It can also be applied in conventional, minimum tillage and no-till production systems. The application method of choice will depend on the anticipated weed spectrum and the preference of the applicator.

When PURSUIT is applied to the soil, susceptible weeds may emerge, growth will stop and the weeds will either die or are not competitive with the crop.

HERBICIDE COMBINATIONS

PURSUIT is active against many grass species. However, when heavy grass pressure is anticipated, PURSUIT should be used in combination with one of the following registered soil-applied grass herbicides; PROWL herbicide, trifluralin, Lasso or Dual.

When PURSUIT is used in combination with another herbicide, refer to the respective label for rates, methods of application, proper timing, weeds controlled, restrictions and precautions. Always use in accordance with the more restrictive label restrictions and precautions. No label dosages should be exceeded. PURSUIT cannot be mixed with any product containing a label prohibiting such mixtures.

SOIL APPLICATIONS WITH LIQUID FERTILIZERS

PURSUIT can be applied to the soil in liquid fertilizers, alone or in combination with PROWL, trifluralin, Lasso or Dual. Follow all PURSUIT label recommendations regarding incorporation, timing of application, special instructions and precautions. Apply treatments in 20 or more gallons of liquid fertilizer per acre with ground equipment. Always test the compatibility of PURSUIT with the liquid fertilizer before mixing in the spray tank.

PREEMERGENCE (SURFACE APPLICATIONS)

PURSUIT offers flexibility in that it can be utilized in all production tillage systems. It can be applied prior to planting or at planting in conventional, reduced tillage or no-till production systems.

PURSUIT controls weeds by uptake by weed roots and translocation to the growing points where it stops weed growth.

Adequate soil moisture is required for optimum activity. For surface applications, rainfall or overhead irrigation is necessary to move PURSUIT into the weed germination zone. The amount of rainfall or irrigation required following application depends on existing soil moisture, soil texture and organic matter content. Sufficient water to moisten the soil to a depth of 2 inches is normally adequate. If adequate moisture is not received within 7 days after a surface-applied treatment, then a cultivation is recommended to control escaped weeds. When adequate moisture is received after dry conditions, PURSUIT will provide residual control of susceptible germinating weeds; activity on established weeds will depend on the weed species and the location of its root system in the soil.

CONVENTIONAL:

PURSUIT may be applied after planting soybeans, but before crop emergence or up to 45 days prior to planting. If moisture is insufficient, a thorough shallow tillage or cultivation (preferably with a rotary hoe) is recommended to improve weed control.

NO-TILL OR REDUCED TILLAGE:

Apply PURSUIT treatments before, during or after planting but before crop emergence. To ensure thorough coverage, use a minimum of 20 gallons of water per acre. Use higher gallonage for fields with dense vegetation or heavy crop residues.

For maximum grass control, tank mix PURSUIT with PROWL, Lasso or Dual. To kill existing vegetation, Gramoxone Super or Roundup may be tank mixed with PURSUIT alone or in combination with PROWL, Lasso or Dual. Gramoxone Super or Roundup should be deleted from the tank mixture if vegetation is absent at the time of application. Refer to the respective labels for use rates and precautions.

NOTE: Adjust planters to ensure adequate seed coverage.

PREPLANT INCORPORATED APPLICATIONS

PURSUIT may be applied following land preparation and should be thoroughly incorporated to a depth of 1 to 2 inches. Application may be made up to 45 days prior to planting. Mechanical incorporation can be achieved by one of the following:

(a) Disk harrow set to cut 3 to 4 inches deep and operated at 4 to 6 mph with the second pass at an angle to the first.

(b) PTO-driven equipment (tillers, cultivators, hoes) set to cut 2-3 inches deep and operated one time at 4 mph or less.

(c) Rolling cultivator set to cut 2 to 4 inches deep and operated two times at 6 to 8 mph with the second pass at an angle to the first. Use only on coarse textured soils.

(d) Field cultivator with 3 or 4 rows of sweeps (Do not use chisel points), spaced at intervals of 7 inches or less and staggered so that no soil is left unturned, set to cut 3 inches deep, and operated two times at more than 5 mph with the second pass at an angle to the first.

(e) LELY-ROTERRA[7] set and operated according to the manufacturer's directions to thoroughly incorporate PURSUIT to a depth of 1 to 3 inches.

(f) Do-All set and operated according to the manufacturer's directions to incorporate PURSUIT 1 to 3 inches deep and operated at 4 to 6 mph.

If soybeans are planted on beds, apply and incorporate after bed formation using PTO-driven equipment or a rolling cultivator.

[7]LELY-ROTERRA is a trademark of Lely.

ROTATIONAL CROPS

The following rotational crops may be planted after applying PURSUIT at the recommended rate in soybeans:

1. Four months after last PURSUIT application:
   - Barley
   - Wheat
2. Nine and one-half months after last PURSUIT application:
   - Edible beans (snap, lima, dry)
   - Field corn
   - Peanuts
   - Peas
   - Tobacco

DO NOT plant other rotational crops prior to 18 months after a PURSUIT application.

If soybeans are furrow irrigated, till the soil prior to planting winter wheat. The beds should be broken up and the soil mixed to a depth of 4-6 inches deep.

Only rotational crops harvested at maturity may be used for feed or food.

In the event of a crop loss due to weather, soybeans can be replanted. DO NOT work the soil deeper than 2 inches.

There should be an interval of at least 85 days between an application of PURSUIT and soybean harvest.

Do not graze or feed treated soybean forage, hay or straw to livestock.

DO NOT apply products containing chlorimuron ethyl (e.g. Classic, Canopy, Gemini, Lorox Plus, Preview[8], etc.) or imazaquin (i.e. SCEPTER®, SQUADRON®, TRI-SCEPT®) the same year as PURSUIT or injury to follow crops may occur.

[8]Classic, Canopy, Gemini, Lorox Plus and Preview are trademarks of E. I. duPont de Nemours and Company.

®Registered Trademarks of American Cyanamid Company

PURSUIT® PLUS

HERBICIDE

FOR USE IN SOYBEANS

ACTIVE INGREDIENTS:

Ammonium salt of imazethapyr (±) -2-[4,5-dihydro-4-methyl-4-(1-methylethyl)-5-oxo-1*H*-imidazol-2-yl]-5-ethyl-3-pyridinecarboxylic acid*	2.39%
pendimethalin (N-1-ethylpropyl)-3,4-dimethyl-2,6-dinitrobenzenamine	31.66%
INERT INGREDIENTS	65.95%
TOTAL	100.00%

*Equivalent to 2.26%, (±)-2-[4,5-dihydro-4-methyl-4-(1-methylethyl)-5-oxo-1*H*-imidazol-2-yl]-5-ethyl-3-pyridinecarboxylic acid. PURSUIT PLUS contains 3.0 punds of active ingredients per gallon (2.8 pounds ai of pendimethalin and 0.2 pounds acid equivalent of imazethapyr).

EPA Reg. No. 241-315 3/89

®Registered Trademark of American Cyanamid Company

KEEP OUT OF REACH OF CHILDREN

CAUTION!/¡PRECAUCION!

PRECAUCION AL USUARIO: Si usted no lee inglés, no use este producto hasta que la etiqueta le haya sido explicada ampliamente.

STATEMENT OF PRACTICAL TREATMENT

IF ON SKIN: Wash with plenty of soap and water. Get medical attention if irritation persists.

In case of an emergency endangering life or property involving this product, call collect, day or night, Area Code 201-835-3100.

PRECAUTIONARY STATEMENTS

HAZARDS TO HUMANS (AND DOMESTIC ANIMALS)
CAUTION

Harmful if inhaled or absorbed through skin. Avoid breathing vapor or spray mist. Avoid contact with skin, eyes or clothing. Wash thoroughly with soap and water after handling. Remove contaminated clothing and wash before reuse.

ENVIRONMENTAL HAZARDS

Drift or runoff may adversely affect nontarget plants. This product is toxic to fish. DO NOT apply directly to water or wetlands. Drift and runoff from treated areas may be hazardous to fish in neighboring areas. DO NOT contaminate water when disposing of equipment washwaters.

DIRECTIONS FOR USE

It is a violation of Federal law to use this product in a manner inconsistent with its labeling.

This labeling must be in the possession of the user at the time of pesticide application.

Observe all cautions and limitations in this leaflet. The use of PURSUIT PLUS not consistent with this label can result in injury to crops, animals, or persons. Keep container closed to avoid spills and contamination. DO NOT apply this product through any type of irrigation system.

STORAGE AND DISPOSAL

STORAGE: KEEP FROM FREEZING. DO NOT STORE BELOW 32°F. SHAKE WELL BEFORE USING. DO NOT contaminate water, food, or feed by storage or disposal.

PESTICIDE DISPOSAL: Wastes resulting from the use of this product may be disposed of on site or at an approved waste disposal facility.

CONTAINER DISPOSAL: Triple rinse (or equivalent). Then offer for recycling or reconditioning, or puncture and dispose of in a sanitary landfill, or by incineration, or, if allowed by State and local authorities, by burning. If burned, stay out of smoke.

DISCLAIMER

The label instructions for the use of this product reflect the opinion of experts based on field use and tests. The directions are believed to be reliable and should be followed carefully. However, it is impossible to eliminate all risks inherently associated with use of this product. Crop injury, ineffectiveness or other unintended consequences may result because of such factors as weather conditions, presence of other materials, or the use or application of the product contrary to label instructions, all of which are beyond the control of American Cyanamid Company. All such risks shall be assumed by the user.

American Cyanamid Company warrants only that the material contained herein conforms to the chemical description on the label and is reasonably fit for the use therein described when used in accordance with the directions for use.

Any damages arising from a breach of this warranty shall be limited to direct damages, and shall not include consequential commercial damages such as loss of profits or values, etc.

American Cyanamid Company makes no other express or implied warranty, including any other express or implied warranty of FITNESS or MERCHANTIBILITY. USER assumes the risk of any use contrary to label instructions, or under abnormal conditions, or under conditions not reasonably foreseeable by American Cyanamid Company.

GENERAL INFORMATION

PURSUIT PLUS may be applied in conventional tillage or in minimum or no-till soybean practices. Apply PURSUIT PLUS preplant incorporated from 45 days prior to planting up to planting. PURSUIT PLUS may be surface applied (up to 45 days) prior to planting soybeans. After PURSUIT PLUS is applied, some susceptible weeds emerge, growth stops, and the weeds either die or are not competitive with the crop.

The weed killing activity of PURSUIT PLUS involves herbicide uptake by weed roots and rapid translocation to the growing points. Therefore, adequate soil moisture is important for optimum PURSUIT PLUS activity. When adequate soil moisture is present, PURSUIT PLUS will provide residual control of susceptible germinating weeds.

Some dwarfing of soybean plants may occur with PURSUIT PLUS applications. This is the result of shortened plant internodes and has no effect on soybean yields.

Replanting: If replanting is necessary in a field previously treated with PURSUIT PLUS, the field may be replanted to soybeans. Rework the soil no deeper than the treated zone. Do not apply a second treatment of PURSUIT PLUS.

MIXING INSTRUCTIONS

Fill the spray tank one-half to three-quarters full with clean water. While agitating add the required amount of product and then fill the remainder of the tank with water. Maintain agitation while spraying to ensure a uniform spray mixture.

When PURSUIT PLUS is tank mixed with paraquat or Roundup[1], add PURSUIT PLUS to the tank first, next add PROWL® (if used), then add paraquat or Roundup. When paraquat is included in a tank mixture, add 8 ounces of non-ionic surfactant per 100 gallons of spray mixture as the last ingredient in the tank.

SPRAYING INSTRUCTIONS

GROUND APPLICATIONS: Uniformly apply with properly calibrated ground equipment in 10 to 40 gallons of water per acre. A spray pressure of 20 to 40 psi is recommended.

To ensure thorough coverage, use a minimum of 20 gallons of water per acre when applying PURSUIT PLUS to minimum or no-till soybeans. Use higher gallonage for fields with dense vegetation or heavy crop residues.

DO NOT apply with ground equipment when wind velocity is greater than 10 mph.

APPLICATIONS WITH LIQUID FERTILIZER:

PURSUIT PLUS can be applied to the soil in liquid fertilizers, alone or in combination with PROWL®, trifluralin, Lasso[1] or Dual[2]. Follow all PURSUIT PLUS label recommendations regarding incorporation, timing of application, special instructions and precautions. Apply treatments in 20 or more gallons of liquid fertilizer per acre with ground equipment.

[1]Roundup and Lasso are trademarks of Monsanto Agricultural Products Company.
[2]Dual is a trademark of Ciba Geigy Corporation.
®Registered Trademark of American Cyanamid Company.

PURSUIT PLUS HERBICIDE USE AREA

PURSUIT PLUS may be applied to soybeans only in the states or parts of states shaded in the map below:

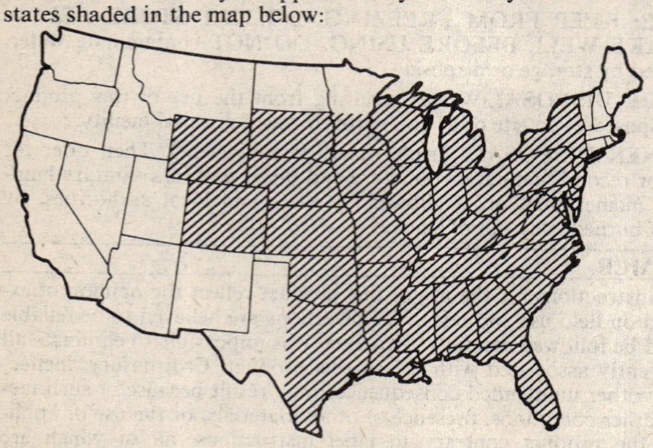

USE RATE
(2½ PINTS PER ACRE)

Apply PURSUIT PLUS at a broadcast rate of 2½ pints per acre for all methods of application: preplant incorporated and preemergence (including minimum and no-till). At this broadcast rate, one gallon of PURSUIT PLUS will treat 3.2 acres of soybeans.

NOTE: Only one application of PURSUIT PLUS may be made during the season.

WEEDS CONTROLLED

The following grass and broadleaf weeds are controlled by recommended treatments of PURSUIT PLUS:

GRASSES

Barnyardgrass	Goosegrass
Crabgrass	Johnsongrass, seedling
large	Panicum
smooth	browntop
Crowfootgrass	fall
Cupgrass, woolly**	Texas
Foxtail	Sandbur, field
giant	Shattercane**
green	Signalgrass, broadleaf**
giant green	Witchgrass
robust purple	
robust white	
yellow	

BROADLEAF WEEDS

Carpetweed	redroot
Cocklebur, common*	smooth
Galinsoga	spiny
Jimsonweed**	Puncturevine
Kochia	Purslane, common
Lambsquarters, common	Pusley, Florida
Mallow, Venice	Sida, prickly (Teaweed)
Morningglory	Smartweed
smallflower	ladysthumb
Mustard Species	Pennsylvania
Nightshade	Spurge
black	prostrate
Eastern black	spotted
hairy	Sunflower, common
Pigweed	Velvetleaf
palmer	Waterhemp, tall

PURSUIT PLUS will aid in the control and reduce competition from:

Itchgrass**	Morningglory	Ragweed
Johnsongrass,	entireleaf	common
rhizome**	ivyleaf	giant**
Millet, wild proso**	pitted	Nutsedge, yellow**
	tall	

*Preplant incorporated treatments only for best results and for light to moderate infestations.
**Preplant incorporated treatments only.

SOIL APPLICATION INSTRUCTIONS

PURSUIT PLUS may be soil applied prior to planting (up to 45 days) in conventional, reduced, or no-till soybeans. DO NOT apply after planting soybeans as crop injury may occur. PURSUIT PLUS may be surface applied or incorporated into the soil prior to planting.

PURSUIT PLUS controls weeds by uptake by weed roots and translocation to the growing points where it stops weed growth.

Adequate soil moisture is required for optimum activity. For surface applications, rainfall or overhead irrigation is necessary to move PURSUIT PLUS into the weed germination zone. The amount of rainfall or irrifation required following application depends on existing soil moisture, soil texture and organic matter content. Sufficient water to moisten the soil to a depth of 2 inches is normally adequate. If adequate moisture is not received within 7 days after a surface-applied treatment, then a cultivation is recommended to control escaped weeds. When adequate moisture is received after dry conditions, PURSUIT PLUS will provide residural control of susceptible germinating weeds; activity on established weeds will depend on the weed species and the location of its root system in the soil.

PREPLANT INCORPORATED APPLICATIONS

PURSUIT PLUS may be applied following land preparation and should be thoroughly incorporated to a depth of 1 to 2 inches. Application may be made up to 45 days prior to planting. Incorporate prior to soybean planting and within 7 days of application. Mechanical incorporation can be achieved by the following:

(a) Disk harrow set to cut 3 to 4 inches deep and operated at 4 to 6 mph with the second pass at an angle to the first.
(b) PTO-driven equipment (tillers, cultivators, hoes) set to cut 2-3 inches deep and operated one time at 4 mph or less.
(c) Rolling cultivator set to cut 2 to 4 inches deep and operated two times at 6 to 8 mph. Use only on coarse textured soils.

(d) Field cultivator with 3 to 4 rows of sweeps (Do not use chisel points), spaced at intervals of 7 inches or less and staggered so that no soil is left unturned, set to cut 3 inches deep, and operated two times at more than 5 mph with the second pass made at an angle to the first.

(e) LELY-Roterra[3] set and operated according to the manufacturer's directions to thoroughly incorporate PURSUIT PLUS to a depth of 1 to 3 inches.

(f) Do-All set and operated according to the manufacturer's directions to thoroughly incorporate PURSUIT PLUS to a depth of 1 to 3 inches.

(g) Do-All set and operated according to the manufacturer's directions to incorporate PURSUIT PLUS 1 to 3 inches deep and operated at 4 to 6 mph.

If soybeans are planted on beds, apply and incorporate after bed formation using PTO-driven equipment or a rolling cultivator.

[3]LELY-Roterra is a trademark of Lely.

PREEMERGENCE APPLICATIONS

SURFACE APPLICATIONS MADE BEFORE PLANTING

PURSUIT PLUS may be surface applied prior to soybean planting (up to 45 days). If sufficient rain does not occur before planting to activate PURSUIT PLUS, mechanically incorporate before planting.

SURFACE APPLICATIONS MADE AFTER PLANTING (preemergence)

PURSUIT PLUS may be surface applied (preemergence application) up to 2 days after soybean planting south of interstate highway I-80 only. DO NOT APPLY PURSUIT PLUS AFTER SOYBEAN PLANTING NORTH OF INTERSTATE HIGHWAY I-80.

NOTE: Surface applications made prior to soybean planting and not incorporated within 7 days of application are considered preemergence applications with respect to weeds controlled.

NO-TILL OR REDUCED TILLAGE:

Apply PURSUIT PLUS treatments from 45 days prior to planting, up to soybean planting but before crop emergence. To ensure thorough coverage, use a minimum of 20 gallons of water per acre. Use higher gallonage for fields with dense vegetation or heavy crop residues. Use ground equipment only.

If vegetation is present, apply paraquat or Roundup in combination with PURSUIT PLUS. Refer to the respective labels for rates, methods of application, proper timing, weeds controlled, restrictions, and precautions. Use ground equipment only.

When PURSUIT PLUS is tank mixed with paraquat or Roundup, add PURSUIT PLUS to the tank first then add paraquat or Roundup. When paraquat is included in a tank mixture, add 8 ounces of non-ionic surfactant per 100 gallons of spray mixture as the last ingredient in the tank.

NOTE: Adjust planters to ensure adequate seed coverage.

ROTATIONAL CROP RESTRICTIONS

The following rotational crops may be planted after applying PURSUIT PLUS at the recommended rate in soybeans:

1. Four months after application:
   - Barley
   - Wheat
2. Nine and one-half months after application:
   - Field Corn
   - Edible beans (snap, lima, dry)
   - Peanuts
   - Peas
   - Tobacco

DO NOT plant other rotational crops prior to 18 months after a PURSUIT PLUS application.

If soybeans are furrow irrigated, till the soil prior to planting winter wheat. The beds should be broken up and the soil mixed to a depth of 3-4 inches.

Only rotational crops harvested at maturity may be used for feed or food.

In the event of a crop loss due to weather, soybeans can be replanted. DO NOT work the soil deeper than 2 inches.

There should be an interval of at least 85 days between an application of PURSUIT PLUS and soybean harvest.

DO NOT graze or feed treated soybean forage, hay or straw to livestock.

DO NOT apply products containing chlorimuron ethyl (e.g. Classic, Canopy, Gemini, Lorox Plus, Preview[4], etc.) or imazaquin (i.e. SCEPTER®, SQUADRON®, or TRI-SCEPT®) the same year as PURSUIT PLUS or injury to follow crops may occur.

[4]Classic, Canopy, Gemini, Lorox Plus and Preview are trademarks of E.I. duPont de Nemours and Company.

®Registered Trademarks of American Cyanamid Company.

SCEPTER®

HERBICIDE

For Use in Soybeans

ACTIVE INGREDIENT:

Ammonium salt of imazaquin (2-[4,5-dihydro-4-methyl-4-(1-methylethyl)-5-oxo-1*H*-imidazol-2-yl]-3-quinolinecarboxylic acid)	17.3%
INERT INGREDIENTS:	82.7%
TOTAL	100.0%

Equivalent to 16.4% 2-[4,5-dihydro-4- methyl-4-(1- methylethyl)-5-oxo-1*H*-imidazol-2-yl]-3-quinolinecarboxylic acid.
(1 gallon contains 1.5 pounds of active ingredient as the free acid)

EPA Reg. No. 241-289 EPA Est. No.

® Registered Trademark of American Cyanamid Company

10/89

KEEP OUT OF REACH OF CHILDREN

CAUTION! ¡PRECAUCION!

PRECAUCION AL USUARIO: Si usted no lee inglés, no use este producto hasta que la etiqueta le haya sido explicada ampliamente.

STATEMENT OF PRACTICAL TREATMENT

If on skin: Wash with plenty of soap and water.

If in eyes: Flush eyes with plenty of water.

In case of an emergency endangering life or property involving this product, call collect, day or night, Area Code 201-835-3100.

PRECAUTIONARY STATEMENTS

HAZARDS TO HUMANS

CAUTION!

Harmful if inhaled or absorbed through skin. Avoid breathing spray mist. Avoid contact with skin, eyes or clothing. Wash thoroughly with soap and water after handling. Remove contaminated clothing and wash before reuse.

ENVIRONMENTAL HAZARDS

DO NOT apply directly to water or wetlands (swamps, bogs, marshes and potholes).

DO NOT contaminate water when disposing of equipment washwaters.

DIRECTIONS FOR USE

It is a violation of Federal law to use this product in a manner inconsistent with its labeling.

Observe all cautions and limitations on this label and on the labels of products used in combination with SCEPTER. Do not use SCEPTER other than in accordance with the instructions set forth on this label. The use of SCEPTER not consistent with this label may result in injury to crops. Keep container closed to avoid spills and contamination.

Do not apply this product through any type of irrigation system.

STORAGE AND DISPOSAL

PROHIBITIONS:

KEEP FROM FREEZING

DO NOT store below 32°F.

DO NOT contaminate water, food, or feed by storage or disposal of wastes.

PESTICIDE DISPOSAL: Wastes resulting from the use of this product may be disposed of on site or at an approved waste disposal facility.

CONTAINER DISPOSAL:

Triple rinse (or equivalent). Then offer for recyling or reconditioning, or puncture and dispose of in a sanitary landfill, or by incineration, or if allowed by state and local authorities, by burning. If burned, stay out of smoke.

DISCLAIMER

The label instructions for the use of this product reflect the opinion of experts based on field use and tests. The directions are believed to be reliable and should be followed carefully. However, it is impossible to eliminate all risks inherently associated with use of this product. Crop injury, ineffectiveness or other unintended consequences may result because of such factors as weather conditions, presence of other materials, or the use or application of the product contrary to label instructions, all of which are beyond the control of American Cyanamid Company. All such risks shall be assumed by the user.

American Cyanamid Company shall not be responsible for losses or damages resulting from use of this product in any manner not set forth on this label. User assumes all risks associated with the use of this product in any manner not specifically set forth on this label.

American Cyanamid Company warrants only that the material contained herein conforms to the chemical description on the label and is reasonably fit for the use therein described when used in accordance with the directions for use, subject to the risks referred to above.

American Cyanamid's liability will be limited to a refund of the purchase price of the SCEPTER. In no case shall American Cyanamid or the seller be liable for consequential, special or indirect damages resulting from the use of handling of this product.

AMERICAN CYANAMID COMPANY MAKES NO OTHER EPXRESS OR IMPLIED WARRANTY, INCLUDING OTHER EXPRESS OR IMPLIED WARRANTY OF FITNESS OR OF MERCHANTABILITY.

GENERAL INFORMATION

SCEPTER is a uniquely flexible broad-spectrum herbicide which may be applied preplant incorporated, preemergence or postemergence in soybeans.

When SCEPTER is applied to the soil, susceptible weeds emerge, growth stops, and then the weeds either die or are not competitive with the crop.

When SCEPTER is applied postemergence, absorption occurs through both the foliage and roots. Susceptible weeds stop growing and either die or are not competitive with the crop.

A timely cultivation may aid in the control of certain weeds or improve general weed control when adequate moisture is not received after application. Cultivation should be shallow. After postemergence treatments, wait at least 10 days before cultivating.

The mode of weed killing activity involves uptake of SCEPTER by weed roots and/or foliage and rapid translocation to growing points. Therefore, adequate soil moisture is important for optimum SCEPTER activity. The amount of rainfall or irrigation required following application depends on existing soil moisture, soil texture and organic matter content. Sufficient water to moisten the soil to a depth of 2 inches is normally adequate. If adequate moisture is not received within 7 days after treatment, then a cultivation is recommended to control escaped weeds. When adequate moisture is received after dry conditions, SCEPTER will provide residual control of susceptible germinating weeds; activity on established weeds will depend on the weed species and the location of its root system in the soil.

A SCEPTER soil treatment followed by a SCEPTER postemergence treatment will control certain problem weeds (See SEQUENTIAL PROGRAM section). Use sequential treatments in recommended states only.

Occasionally, internode shortening of soybean plants may be observed with SCEPTER applications. This has no effect on soybean yields.

There must be an interval of at least 90 days between the last SCEPTER application and soybean harvest. DO NOT graze or feed treated soybean forage, hay or straw to livestock.

PRECAUTIONS:

Do not use on crops other than soybeans. Crops other than soybeans, such as cotton, corn and vegetables, may be injured by spray drift or other indirect contact with SCEPTER.

To avoid injury to sensitive crops from spray drift, follow all use directions and precautions in SPRAYING INSTRUCTIONS on page 5.

To avoid injury to sensitive crops, spray equipment used for SCEPTER applications must be drained and thoroughly cleaned with water before being used to apply other products to these crops. See page 11 for ROTATIONAL CROP RESTRICTIONS.

WEEDS CONTROLLED
SCEPTER Alone

When applied as directed, SCEPTER soil-applied treatments will control the following weeds:

BROADLEAF WEEDS

BRISTLY STARBUR	(*Acanthospermum hispidum*)
BURCUCUMBER*	(*Sicyos angulatus*)
COCKLEBUR, COMMON	(*Xanthium strumarium*)
JIMSONWEED	(*Datura stramonium*)
LAMBSQUARTERS, COMMON	(*Chenopodium album*)
MALLOW, VENICE	(*Hisbiscus trionum*)
MORNINGGLORY	
PALMLEAF	(*Ipomoea wrightii*)
PITTED	(*Ipomoea lacunosa*)
SMALLFLOWER	(*Jacquemontia tamnifolia*)
MUSTARD SPECIES	(*Brassica* spp.)
NIGHTSHADE, EASTERN BLACK*	(*Solanum ptycanthum*)
PIGWEED	
PALMER	(*Amaranthus palmeri*)
REDROOT	(*Amaranthus retroflexus*)
SMOOTH	(*Amaranthus hybridus*)
SPINY	(*Amaranthus spinosus*)
WATERHEMP, TALL	(*Amaranthus tuberculatus*)
POINSETTIA, WILD	(*Euphorbia heterophylla*)
PUNCTUREVINE	(*Tribulus terrestris*)
PUSLEY, FLORIDA	(*Richardia scabra*)
RAGWEED,	
COMMON	(*Ambrosia artemisiifolia*)
GIANT*	(*Ambrosia trifida*)
REDWEED*	(*Melochia corchorifolia*)
SIDA, PRICKLY (TEAWEED)	(*Sida spinosa*)
SMARTWEED	
LADYSTHUMB	(*Polygonum persicaria*)
PENNSYLVANIA	(*Polygonum pensylvanicum*)
SUNFLOWER, COMMON	(*Helianthus annuus*)
VELVETLEAF*	(*Abutilon theophrasti*)

GRASSES

FOXTAIL	
GIANT	(*Setaria faberi*)
GREEN	(*Setaria viridis*)
YELLOW	(*Setaria glauca*)
JOHNSONGRASS, SEEDLING	(*Sorghum halepense*)

SCEPTER soil-applied treatments will aid in the control and reduce competition from:

BARNYARDGRASS	(*Echinochloa crus-galli*)
BEGGARWEED, FLORIDA****	(*Desmondium tortuosum*)
MEXICANWEED****	(*Caperonia castanifolia*)
GOOSEGRASS	(*Eleusine indica*)
MORNINGGLORY**	
ENTIRELEAF	(*Ipomoea hederacea* var. *intergriuscula*)
IVYLEAF	(*Ipomoea hederacea*)
TALL	(*Ipomoea purpurea*)
NUTSEDGE, YELLOW***	(*Cyperus esculentus*)
SHATTERCANE	(*Sorghum bicolor*)
SICKLEPOD****	(*Cassia obtusifolia*)
SIGNALGRASS, BROADLEAF	(*Brachiaria platyphylla*)
SPURGE, SPOTTED	(*Euphorbia maculata*)

* Eastern black nightshade, giant ragweed, burcucumber, redweed and velvetleaf are controlled by preplant incorporated treatments only.

** For best activity on morningglory species, use preplant incorporated treatments.

*** Preplant incorporated treatments only.

**** A sequential treatment of SCEPTER applied preplant incorporated or preemergence followed by SCEPTER postemergence will control these weeds. See pages 26 and 27 for more information on the sequential program.

MIXING INSTRUCTIONS

Fill the spray tank one-half to three-quarters full with clean water. While agitating add the required amount of SCEPTER and then fill the remainder of the tank with water. For postemergence applications, add the nonionic surfactant or the crop oil concentrate as the last ingredient in the tank. An antifoaming agent may be added to the tank if needed. Maintain agitation while spraying to ensure a uniform spray mixture.

When tank mixing SCEPTER with recommended herbicides, add the other herbicide to the spray tank first and make sure it is thoroughly mixed before adding SCEPTER.

When tank mixing Roundup or Gramoxone Super with SCEPTER and PROWL, Lasso or Dual, add herbicides to the tank in the following order: (1) PROWL, Lasso or Dual, (2) Roundup or Gramoxone Super, (3) SCEPTER, and (4) nonionic surfactant.

SPRAYING INSTRUCTIONS

NOTE: DO NOT make applications when spray may be carried by wind to sensitive crops.

Avoid overlaps when spraying.

GROUND APPLICATIONS:

Uniformly apply with properly calibrated ground equipment in 10 or more gallons of water per acre. Use sprayers equipped with nozzles that provide accurate and uniform application. To minimize drift, use a maximum spray pressure of 40 psi.

To ensure thorough coverage, use a minimum of 20 gallons of water per acre when applying SCEPTER to minimum or no-till soybeans or when applying as a postemergence treatment. Use higher gallonage for fields with dense vegetation or heavy crop residues.

DO NOT apply with ground equipment when wind velocity is greater than 10 mph.

For band treatment, apply the broadcast equivalent rate and volume per acre. To determine these:

$$\frac{\text{Band width in inches}}{\text{Row width in inches}} \times \text{Broadcast Rate per acre} = \text{Band Rate per acre}$$

$$\frac{\text{Band width in inches}}{\text{Row width in inches}} \times \text{Broadcast Volume per acre} = \text{Band Volume per acre}$$

AERIAL APPLICATIONS:

Uniformly apply with properly calibrated aerial equipment in 5 or more gallons of water per acre.

To avoid injury to sensitive crops from drift, aerial applicators must adhere to the following SPECIAL AERIAL USE DIRECTIONS AND PRECAUTIONS:

- Use nozzles which produce a coarse spray.
- Nozzle height above ground must be a maximum of 10 feet.
- Nozzles must be pointed toward the rear of the aircraft. The downward angle of the nozzle should not be greater than 20 degrees.
- To minimize wing-tip vortex roll, nozzles or spray boom must not be located any closer to end of wing or rotor than three-fourths the distance from the center of the aircraft.
- Use a maximum spray pressure of 40 psi.
- A buffer zone must be established between the area to be sprayed and sensitive crops.
- **DO NOT** spray when wind velocity is greater than 5 mph.

Applicator is responsible for any loss or damage which results from spraying SCEPTER in a manner other than recommended in this label. In addition, applicator must follow all applicable state and local regulations and ordinances in regard to spraying.

USE AREA

SCEPTER can be applied only in the states or parts of states shaded in the following map:

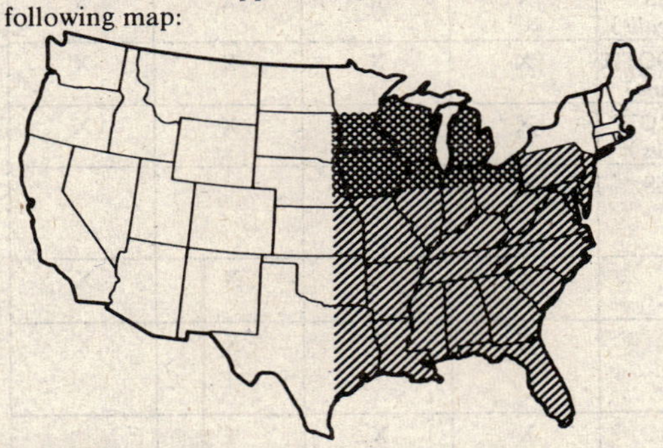

= Northern Use Area (includes: Nebraska, east of U.S. 81, north of U.S. 34; South Dakota, east of U.S. 81; Illinois, north of S. R. 116 for area west of Peoria, north of U.S. 24 for area east of Peoria; Indiana, north of U.S. 24 for area west of Huntington, north of U.S. 224 for area east of Huntington; Ohio, north of U.S. 224; and all other states or portions of states shaded in the Northern use area map).

= Southern Use Area (includes: Kansas, east of U.S. 81; southern portions of Illinois, Indiana, Ohio and eastern Nebraska not covered by Northern Use Area description; and all other states or portions of states shaded in the Southern Use Area map).

NOTE: See USE RATES section and ROTATIONAL CROP RESTRICTIONS section for recommendations applying to the Northern vs. Southern Use Areas.

NOTE: See SEQUENTIAL PROGRAM section in the label for geographical restrictions for this method of application.

USE RATES

Apply SCEPTER at a broadcast rate of 2/3 pint per acre for preplant incorporated and preemergence (including minimum and no-till systems) applications. At this broadcast rate, one gallon of SCEPTER will treat 12 acres of soybeans. Apply SCEPTER at a broadcast rate of 1/3 to 2/3 pint per acre for postemergence applications.

NOTE: For preplant incorporated and preemergence applications on sands and loamy sands, 1/2 pint per acre can be used without a substantial reduction in the control of only cocklebur and pigweed.

NOTE: In the Northern Use Area as defined by the Use Area section of the label, apply SCEPTER prior to July 1.

APPLICATION INSTRUCTIONS

PREPLANT INCORPORATED APPLICATIONS

Apply SCEPTER before planting and incorporate uniformly into the top 1 to 2 inches of soil. Mechanical incorporation can be achieved by the following:

(a) Disk harrow set to cut 3 to 4 inches deep and operated in two different directions at 4 to 6 mph.

(b) PTO-driven equipment (tillers, cultivators, hoes) set to cut 2 inches deep and operated one time at 4 mph or less.

(c) Rolling cultivator set to cut 2 to 3 inches deep and operated two times at 6 to 8 mph. Use only on coarse textured soils.

(d) Field cultivator with 3 or 4 rows of sweeps (Do not use chisel points), spaced at intervals of 7 inches or less and staggered so that no soil is left unturned, set to cut 3 inches deep, and operated two times at more than 5 mph with the second pass made at an angle to the first.

(e) LELY-Roterra[1] set and operated according to the manufacturer's directions to thoroughly incorporate SCEPTER to a depth of 1 to 2 inches.

(f) Do-All set and operated according to the manufacturer's directions to thoroughly incorporate SCEPTER to a depth of 1 to 2 inches.

[1] LELY-ROTERRA is a trademark of Lely.

SCEPTER may be applied immediately before planting or up to 30 days prior to planting in Alabama, Arkansas, Florida, Georgia, Louisiana, Mississippi, North Carolina, Oklahoma, South Carolina, Tennessee and Texas only. For the remaining Southern and Northern Use Areas, SCEPTER may be applied 45 days prior to planting. Incorporate within 7 days after application. Adequate moisture is required for activation of SCEPTER.

If soybeans are planted on beds, apply and incorporate after bed formation using PTO-driven equipment or rolling cultivator.

APPLICATIONS WITH LIQUID FERTILIZERS

GENERAL

SCEPTER can be applied to the soil in liquid fertilizers, alone or in combination with PROWL, trifluralin, Lasso or Dual. Follow all SCEPTER label recommendations regarding incorporation, timing of application, special instructions and precautions. Apply treatments in 20 or more gallons of liquid fertilizer per acre with ground equipment.

All individual state regulations relating to fluid fertilizer mixing, registration, labeling and application are the responsibility of the individual and/or company selling the SCEPTER/liquid fertilizer mixture.

LIQUID FERTILIZER COMPATIBILITY DETERMINATIONS

If a liquid fertilizer and herbicide(s) mixture separates in the spray tank, clogged equipment and uneven application can result, which can cause poor weed control and crop injury. Always predetermine the compatibility of SCEPTER alone or with other herbicides in the specific liquid fertilizer to be used according to the following directions:

1. Add 1 pint fertilizer to each of 2 one-quart jars.
2. Add 1/2 teaspoon of adjuvant to one jar. See recommended adjuvants below.

NOTE: Compex-Kalo Laboratories Incorporated, Kansas City, MO; Sponto-168 D-Witco-Retzloff Chemical Company, Houston, TX; Unite-Hopkins Agricultural Chemical Company, Madison, WI; or other compatible adjuvants.

3. (a) When using SCEPTER alone, add to each jar the correct amount of SCEPTER as specified in the table below.
   (b) When using SCEPTER tank mixtures, first add the specified quantity of PROWL, trifluralin, Lasso or Dual and then add the correct amount of SCEPTER.
4. Close both jars and shake thoroughly for 10 seconds. Let them stand for 30 minutes and then observe the results. Look for signs of separation, an oily layer or globules, sludge, flakes or other precipitates.
   (a) If the mixture without adjuvant does not separate, use this mixture in your spray tank.
   (b) If the mixture with adjuvant does not separate, but the one without adjuvant separates, use the adjuvant mixture in your spray tank. Add the adjuvant to the liquid fertilizer as directed on the manufacturer's label.
   (c) If either mixture separates, but mixes readily with shaking, the mixture can be used providing good agitation is maintained in the spray tank.
   (d) If separation of the mixture occurs, and agitation and/or adjuvant does not correct this problem, DO NOT use the herbicide(s) in that specific liquid fertilizer.

Teaspoons of Specific Herbicide to be added to 1 Pint of Liquid Fertilizer Solution

Gallons of Liquid Fertilizer to be Applied per Acre	SCEPTER	PROWL	Trifluralin	Lasso	Dual
20	½	2½	2½	5	2½
30	¼	1¾	1¾	3½	1¾
40	¼	1¼	1¼	2½	1¼

PREPLANT INCORPORATED APPLICATIONS WITH DRY BULK FERTILIZERS

SCEPTER may be impregnated on dry bulk fertilizers. When applied as directed, SCEPTER/dry bulk fertilizer mixtures provide weed control equal to that provided by the same rates of SCEPTER applied in water.

Follow all SCEPTER label recommendations regarding incorporation, special instructions and precautions. Apply SCEPTER/dry bulk fertilizer mixtures only with ground equipment.

All individual state regulations relating to dry bulk fertilizer blending, registration, labeling, and application are the responsibility of the individual and/or company selling the SCEPTER/dry bulk fertilizer mixture.

A minimum of 200 pounds and a maximum of 450 pounds of dry bulk fertilizer impregnated with the recommended amount of SCEPTER must be applied per acre.

DO NOT impregnate SCEPTER onto coated ammonium nitrate or limestone because these materials will not absorb the herbicide. Dry fertilizer blends containing mixtures of ammonium nitrate or limestone may be impregnated with SCEPTER. A minimum of 200 pounds of impregnated dry bulk fertilizer, excluding the weight of ammonium nitrate or limestone, must be applied per acre.

Use the following table to determine the amount of SCEPTER to be impregnated on a ton of dry bulk fertilizer based on the rate of fertilizer which will be applied per acre.

RATE CHART FOR IMPREGNATION OF DRY BULK FERTILIZER WITH SCEPTER

(Pints of SCEPTER per Ton of Fertilizer)

SCEPTER Rate Per Acre	Fertilizer Rate lbs/acre	Pints per Ton
⅔ Pints	200	6⅔
	250	5⅓
	300	4½
	350	3¾
	400	3⅓
	450	3

For those rates not listed in this table, calculate the pints of SCEPTER to be impregnated on a ton of dry bulk fertilizer using the following formula:

$$\frac{2000}{\text{Pounds of dry fertilizer per acre}} \times \begin{array}{c}\text{⅔ pint of}\\ \text{SCEPTER per acre}\\ \text{(recommended rate)}\end{array} = \begin{array}{c}\text{Pints of}\\ \text{SCEPTER per ton}\\ \text{of fertilizer}\end{array}$$

To impregnate SCEPTER on bulk fertilizer, use a closed rotary-drum mixer or other commonly used dry bulk fertilizer blender equipped with suitable spray equipment. Spray nozzles must be placed to provide uniform coverage of SCEPTER onto the fertilizer during mixing.

Apply the SCEPTER/dry bulk fertilizer mixture with an accurately calibrated dry fertilizer spreader. The SCEPTER/dry bulk fertilizer mixture must be spread uniformly on the soil surface. Uneven spreading can cause poor weed control and crop injury.

Refer to the Application Instructions section of this label under Preplant Incorporated applications on page 3 for incorporation directions.

PREEMERGENCE APPLICATIONS

Apply SCEPTER during or after planting but before the crop emerges. As with other herbicides applied preemergence, rainfall or irrigation is necessary to activate SCEPTER in the soil. If sufficient rainfall or irrigation to activate SCEPTER is not received within 7 days after application, a thorough shallow tillage or cultivation is recommended (preferably with a rotary hoe) for control of emerged weeds.

HERBICIDE COMBINATIONS

A soil treatment of SCEPTER is active against many grass species, but for maximum grass control SCEPTER should be tank mixed with PROWL® herbicide, trifluralin, Lasso or DUAL. SCEPTER may also be applied as a sequential preemergence or postemergence treatment following application of one of the above grass herbicides.

For no-till uses, SCEPTER may be tank mixed with Gramoxone Super, Roundup or Bronco.

SCEPTER may be followed by a postemergence application of Fusilade[6] or Poast[7]. DO NOT tank mix SCEPTER with postemergence grass herbicides.

When used in combination with another herbicide, SCEPTER should be used only in accordance with recommendations on this label. Always follow the more restrictive of label limitations and precautions. SCEPTER cannot be mixed with any product containing a label prohibition against such mixing.

Soil Texture

Although the rate of SCEPTER does not vary with soil texture, the recommended rates of other herbicides used with this product generally do vary with soil texture. Combination rate tables in this label refer to three soil texture groups: COARSE, MEDIUM and FINE. The following table lists soil textures included in each of these three groups:

COARSE	MEDIUM	FINE
sands	sandy clay loams*	silty clay loams*
loamy sands	sandy clays	silty clays
sandy loams	loams	clay loams
	silt loams	clays
	silts	

* Sometimes considered transitional soils and may be classified as either medium or fine textured soils.

[6] Fusilade is a trademark of ICI Americas, Inc.

[7] Poast is a trademark of BASF Wyandotte Corporation.

WEEDS CONTROLLED

PREPLANT AND PREEMERGENCE COMBINATION TREATMENTS

When applied as directed, SCEPTER preplant incorporated or preemergence combination treatments with PROWL, trifluralin, Lasso or Dual will control, as indicated, weeds listed in the following table in addition to those controlled by SCEPTER alone.

WEED	PROWL	Trifluralin[a]	Lasso	Dual
GRASSES				
BARNYARDGRASS (*Echinochloa crus-galli*)	X	X	X	X
CRABGRASS, SMOOTH (*Digiteria ischaemum*)	X	X	X	X
CRABGRASS, LARGE (*Digiteria sanguinalis*)	X	X	X	X
CROWFOOT GRASS (*Dactyloctenium aegyptium*)	X	X		
GOOSEGRASS (*Eleusine indica*)	X	X	X	X
PANICUM, FALL (*Panicum dichotomiflorum*)	X	X[c]	X	X
PANICUM, TEXAS (*Panicum texanum*)	X	X		
SANDBUR, FIELD (*Cenchrus incertus*)	X	X		
SHATTERCANE (*Sorghum bicolor*)	X[a,b]	X[c]		
SIGNALGRASS (*Brachiaria platyphylla*)	X	X	X	X
WITCHGRASS (*Panicum capillare*)	X	X	X	X
BROADLEAF WEEDS				
CARPETWEED (*Mollugo verticillata*)	X	X	X	X
GALINSOGA (*Galinsoga* spp.)			X	X
KOCHIA (*Kochia scoparia*)	X	X		
PURSLANE (*Portulaca oleracea*)	X	X	X	
SPURGE, ANNUAL (*Euphorbia* spp.)	X			

[a] Preplant incorporated treatments only.

[b] See NOTE on page 5.

[c] See NOTE on page 5.

SCEPTER Plus PROWL Tank Mixture

Apply SCEPTER plus PROWL preplant incorporated or preemergence using rates in the table below. Follow use directions under APPLICATION INSTRUCTIONS in this label. For preplant incorporated applications, incorporate into the soil within 7 days after application, and plant soybeans within 45 days after application. For preemergence applications, apply at planting or up to 2 days after planting. DO NOT apply if soybeans have emerged.

Broadcast Rate of SCEPTER Plus PROWL

Soil Texture	Less Than 3% Organic Matter: SCEPTER pints/acre +	PROWL pints/acre	3% or More Organic Matter: SCEPTER pints/acre +	PROWL pints/acre
Coarse	⅔	1 to 1½	⅔	1½
Medium	⅔	1 to 1½	⅔	1½ to 2
Fine	⅔	1 to 1½	⅔	2 to 2½

NOTE: Use the higher rate of PROWL if heavy weed populations are anticipated or if shattercane will be present.

Observe all precautions and limitations on PROWL label.

SCEPTER Plus trifluralin Tank Mixture

Apply SCEPTER plus trifluralin preplant incorporated using rates in the table below. Follow use directions under APPLICATION INSTRUCTIONS in this label. Incorporate into the soil within 24 hours after application, and plant soybeans within 45 days after application. Use equipment that provides uniform 2 inch incorporation.

Broadcast Rate of SCEPTER Plus trifuralin*

Soil Texture	SCEPTER pints/acre	Trifluralin* pints/acre
Coarse	⅔	1
Medium	⅔	1½
Fine	⅔	2

NOTE: If fall panicum and shattercane will be present, refer to trifluralin label for higher rates to control these weeds.
Observe all precautions and limitations on trifluralin label.
*Based on 4 pounds active ingredient per gallon.

SCEPTER Plus Lasso Tank Mixture

Apply SCEPTER plus Lasso preplant incorporated or preemergence using rates in the table below. Follow use directions under APPLICATION INSTRUCTIONS in this label. For preplant incorporated applications, apply and incorporate within 7 days prior to planting.

Broadcast Rate of SCEPTER Plus Lasso

Soil Texture	Less than 3% Organic Matter: SCEPTER (pints/acre)	Lasso (quarts/acre): Preplant Incorporated	Lasso (quarts/acre): Preemergence	3% or More Organic Matter: SCEPTER (pints/acre)	Lasso (quarts/acre): Preplant Incorporated	Lasso (quarts/acre): Preemergence
Coarse	⅔	2 to 2½	2	⅔	2 to 2½	2
Medium	⅔	2½ to 3	2	⅔	2½ to 3	2
Fine	⅔	2½ to 3	2*	⅔	2½ to 3	2½

* On silty clay and heavy clay soils of the Mississippi Delta, use 2 to 2½ quarts of Lasso.

NOTE: For preplant incorporated applications, use the higher rate of Lasso if heavy weed populations are anticipated.

Observe all precautions and limitations on the Lasso label.

SCEPTER Plus Dual Tank Mixture

Apply SCEPTER plus Dual preplant incorporated or preemergence using rates in table below. Follow use directions under APPLICATION INSTRUCTIONS in this label. For preplant incorporated applications, apply up to 29 days prior to planting. Incorporate within 7 days after application.

Broadcast Rate of SCEPTER Plus Dual

Soil Texture	Less Than 3% Organic Matter: SCEPTER pints/acre	Dual pints/acre
Coarse	⅔	1¼
Medium	⅔	1½
Fine	⅔	2

Soil Texture	3% or More Organic Matter: SCEPTER pints/acre	Dual pints/acre
Coarse	⅔	1½
Medium	⅔	2
Fine	⅔	2 to 2½*

Use the higher rate of Dual if heavy weed populations are anticipated.

NOTE: Observe all precautions and limitations on Dual label.

SCEPTER Plus PROWL Plus Command Tank Mixtures

Observe all precautionary statements and limitations in the PROWL and Command labels before using. The SCEPTER plus PROWL plus Command 4EC preplant incorporated tank mixture may be used when heavy infestations of velvetleaf are anticipated. Uniformly apply the tank mixture with properly calibrated ground equipment in 10 to 40 gallons of water per acre. Use sprayers equipped with nozzles that provide accurate and uniform application.

NOTE: The use of an agriculturally approved drift reducing agent is required at finished spray volumes of 10 to 15 gallons per acre.

Apply SCEPTER plus PROWL plus Command 4EC to the soil surface and uniformly incorporate. Refer to the Command 4EC label for incorporation instructions. DO NOT apply aerially or through irrigation equipment. DO NOT apply this tank mixture to overly moist or wet soils.

SPECIAL PRECAUTION

Command 4EC is a volatile compound. Off-site movement of spray drift or vapors of Command can cause foliar whitening or yellowing of some plants. Prior to using Command, read and strictly follow all precautions and application instructions on the Command label.

WEED SPECIES CONTROLLED

The SCEPTER plus PROWL plus Command tank mixture controls heavy infestations of velvetleaf as well as all the weeds listed in this label for combination preplant incorporated treatments of SCEPTER plus PROWL.

Broadcast Rate Per Acre of SCEPTER Plus PROWL Plus Command 4EC

Soil Texture	Less than 3% Organic Matter: SCEPTER (pints/acre)	PROWL (pints/acre)	Command (pints/acre)	3% or More Organic Matter: SCEPTER (pints/acre)	PROWL (pints/acre)	Command (pints/acre)
Coarse	⅔	1 to 1½	¼ to ½	⅔	1 to 1½	¼ to ½
Medium	⅔	1½ to 2	¼ to ½	⅔	1½ to 2	¼ to ½
Fine	⅔	1½ to 2	¼ to ½	⅔	2 to 2½	¼ to ½

When heavy grass populations are anticipated, apply PROWL at the high rate for each soil texture.

ROTATIONAL CROP RESTRICTIONS: Follow all rotational crop restrictions on the SCEPTER, PROWL and Command labels.

ALWAYS FOLLOW THE MOST RESTRICTIVE LABEL.

SCEPTER Plus Trifluralin Plus Command Tank Mixture

Observe all precautionary statements and limitations in the trifluralin and Command labels before using. The SCEPTER plus trifluralin plus Command 4EC preplant incorporated tank mixture may be used when heavy infestations of velvetleaf are anticipated. Uniformly apply the tank mixture with properly calibrated ground equipment in 10 to 40 gallons of water per acre. Use sprayers equipped with nozzles that provide accurate and uniform application.

NOTE: The use of an agriculturally approved drift reducing agent is required at finished spray volumes of 10 to 15 gallons per acre.

Apply SCEPTER plus trifluralin plus Command 4EC to the soil surface and uniformly incorporate. Refer to the Command 4EC label for incorporation instructions. DO NOT apply aerially or through irrigation equipment. DO NOT apply this tank mixture to overly moist or wet soils.

SPECIAL PRECAUTION

Command 4EC is a volatile compound. Off-site movement of spray drift or vapors of Command can cause foliar whitening or yellowing of some plants. Prior to using Command, read and strictly follow all precautions and application instructions on the Command label.

WEEDS SPECIES CONTROLLED

The SCEPTER plus trifluralin plus Command tank mixture controls heavy infestations of velvetleaf as well as all the weeds listed in this label for combination preplant incorporated treatments of SCEPTER plus trifluralin.

Broadcast Rates Per Acre of SCEPTER Plus Trifluralin* Plus Command 4EC

Soil Texture	SCEPTER (pints/acre)	Trifluralin (pints/acre)	Command (pints/acre)
Coarse	⅔	1	¼ to ½
Medium	⅔	1½	¼ to ½
Fine	⅔	2	¼ to ½

Use 1½ pints per acre of trifluralin on coarse soils with 2 to 5% organic matter. If fall panicum or shattercane infestations are anticipated, refer to the trifluralin label for higher rates to control these weeds.

*Based on 4 pounds active ingredient per gallon.

ROTATIONAL CROP RESTRICTIONS: Follow all rotational crop restrictions on the SCEPTER, trifluralin and Command labels.

ALWAYS FOLLOW THE MOST RESTRICTIVE LABEL.

MINIMUM AND NO-TILL SYSTEMS

SCEPTER may be applied in minimum or no-till soybeans.

For maximum grass control, tank mix SCEPTER with PROWL, Lasso or Dual. To kill existing vegetation, Gramoxone Super or Roundup may be tank mixed with SCEPTER alone or in combination with PROWL, Lasso or Dual. Gramoxone Super or Roundup should be deleted from the tank mixture if vegetation is absent at the time of application. SCEPTER may also be tank mixed with Bronco, a product which gives both preemergence control of annual grasses and postemergence control of existing vegetation.

Apply SCEPTER treatments immediately before, during or after planting but before the crop emerges. To ensure thorough coverage, use a minimum of 20 gallons of water per acre. Use higher gallonage for fields with dense vegetation or heavy crop residues. Use ground equipment only.

SCEPTER preemergence treatments are most effective in controlling weeds when adequate moisture is received within 7 days after application.

NOTE: Plant soybeans at least one inch deep and adjust planters to ensure adequate seed coverage.

Tank Mixtures with Prowl, Lasso or Dual

1. PROWL

Use the rates in table on page 5 for SCEPTER plus PROWL preemergence tank mixtures. Use the higher rate of PROWL when previous crop residues are extremely dense. Apply immediately before, during or up to 2 days after planting. DO NOT apply if soybeans have emerged.

Observe all precautions and limitations on PROWL label.

2. Lasso

Use the following rates of Lasso for SCEPTER plus Lasso preemergence tank mixtures in minimum and no-till systems:

Soil Texture	Lasso Rate (Quarts/acre)
Coarse	2
Medium	2 to 2½
Fine	2½ to 3

Use the higher rate of Lasso (1) when heavy grass infestations are anticipated, (2) when fall panicum or crabgrass will be present or (3) when previous crop residues are extremely dense.

Observe all precautions and limitations on Lasso label.

3. Dual

Use the following rates of Dual for SCEPTER plus Dual preemergence tank mixtures in minimum and no-till systems:

Soil Texture	Dual Rate (pints/acre)
Coarse	1½
Medium	2
Fine	2½

Observe all precautions and limitations on Dual label.

Tank Mixtures with Gramoxone Super

Apply 1½ to 2½ pints of Gramoxone Super per acre in SCEPTER tank mixtures. Use the 2½ pint rate if weeds are 4 to 6 inches tall. Weeds over 6 inches tall may not be controlled with this treatment.

When Gramoxone Super is included in a tank mixture, add a nonionic spreader surfactant at a rate of 8 fl. oz. per 100 gallons of spray mixture as the last ingredient in the tank.

Gramoxone Super will control most annual emerged weeds and suppress many emerged perennials. Refer to Gramoxone Super label for specific use recommendations and weeds controlled.

Observe all precautions and limitations on Gramoxone Super label.

Tank Mixtures With Roundup

Annual Weeds

Apply 1 to 1½ quarts of Roundup per acre in recommended tank mixtures. Use the 1½ quart rate if annual weeds are more than 6 inches tall.

Roundup will control most annual emerged weeds. Refer to Roundup label for specific use recommendations and weeds controlled.

Perennial Weeds

At normal application times in minimum or no-till systems, perennial weeds may not be at the proper stage of growth for control with Roundup. Refer to Roundup label for the proper stage of growth for perennial weeds. Use of 2 to 4 quarts of Roundup per acre in tank mixtures under these conditions provides top kill and reduces competition from many emerged perennial grass and broadleaf weeds. Refer to Roundup label for emerged perennial weeds controlled. To obtain control of perennial weeds, follow recommendations on Roundup label for stage of growth and rate of application.

NOTE: When using these tank mixtures, DO NOT exceed 4 quarts of Roundup per acre.

USE OF THESE ROUNDUP TANK MIXTURES FOR BERMUDAGRASS OR JOHNSONGRASS CONTROL IS NOT RECOMMENDED.

Observe all precautions and limitations on Roundup label.

Tank Mixtures With Bronco

Use the following rates of Bronco for SCEPTER plus Bronco tank mixtures:

Soil Texture	Bronco (quarts/acre)
Coarse	3 to 5
Medium	4 to 5
Fine	4 to 5

Use the higher rate of Bronco (1) when previous crop residues are extremely dense, (2) when emerged weeds are over 6 inches tall, (3) when heavy infestations of weeds are expected, and (4) when soils contain more than 3% organic matter.

Observe all precautions and limitations on Bronco label.

POSTEMERGENCE APPLICATIONS

Apply SCEPTER as a postemergence treatment after crop emergence and in accordance with the weeds controlled, weed heights, and rates of application information listed in the table below.

SCEPTER FOR POSTEMERGENCE WEED CONTROL

Weeds Controlled	Weed Height (inches)	SCEPTER* (pints/acre)
Cocklebur, Common *(Xanthium strumarium)*	up to 8	⅓**
	9 to 12	⅔
Pigweed		
Palmer *(Amaranthus palmeri)*	up to 12	⅔
Redroot *(Amaranthus retroflexus)*	up to 4	⅓**
	5 to 12	⅔
Smooth *(Amaranthus hybridus)*	up to 4	⅓**
	5 to 12	⅔
Spiny *(Amaranthus spinosus)*	up to 12	⅔
Tall waterhemp *(Amaranthus tuberculatus)*	up to 12	⅔
Poinsettia, Wild *(Euphorbia heterophylla)*	up to 6	⅔
Sunflower, Common *(Helianthus annuus)*	up to 4	⅓**
	5 to 8	⅔**
Volunteer Corn *(Zea mays)*	up to 8	⅓**
	9 to 12	⅔**

Apply when weeds are actively growing. DO NOT apply SCEPTER postemergence when soybeans and weeds have been subjected to stress conditions such as temperature and moisture extremes.

*Use a nonionic surfactant or crop oil concentrate. The nonionic surfactant should contain at least 80% active ingredient and should be applied at a rate of 2 pints per 100 gallons of spray mixture. Apply the crop oil concentrate (COC) at the rate stated on the COC label.

**For control of emerged weeds.

There must be an interval of at least 90 days between the last SCEPTER application and soybean harvest.

SEQUENTIAL PROGRAM

An additional treatment of SCEPTER may be applied postemergence after an initial preplant incorporated or preemergence treatment to control certain problem weeds.

Apply SCEPTER at ⅔ pint per acre as a sequential postemergence treatment following either a preplant incorporated or preemergence treatment of SCEPTER at ⅔ pint per acre. The total amount of SCEPTER applied should not exceed 1⅓ pints per acre in one season. **The sequential postemergence treatment should be applied before weed species exceed the 1 to 2 true leaf stage.**

Follow use directions for preplant incorporated or preemergence (see pages 6–7) and postemergence treatements (See above).

These sequential applications can be used in Alabama, Arkansas, Florida, Georgia, Louisiana, Mississippi, North Carolina, Oklahoma, South Carolina, Tennessee, and Texas only.

If grass species are anticipated, use one of the following registered soil-applied herbicides: PROWL® herbicide, trifluralin, Lasso or Dual. See pages 5-6 for specific use directions.

When applied as directed, SCEPTER preplant incorporated or preemergence followed by SCEPTER postemergence will control the following broadleaf weeds:

BEGGARWEED, FLORIDA	*(Desmodium tortuosum)*
MEXICANWEED	*(Caperonia castaniifolia)*
SICKLEPOD	*(Cassia obtusifolia)*
TEXASWEED	*(Caperonia palustria)*

ROTATIONAL CROP RESTRICTIONS

The following rotational crops may be planted after applying SCEPTER at recommended rates in soybeans:

1. Wheat may be planted in the fall, four months after last SCEPTER application in soybeans. Rice may be planted in the spring of the year following SCEPTER application. (See Northern Use Area Restrictions*).
2. Eleven months after last SCEPTER application:

Barley (See Northern Use Area Restrictions)*
Oats (See Northern Use Area Restrictions)*
Field Corn (See Northern Use Area* or Southern Use Area Restrictions**)
Edible beans

Grain sorghum
Peanuts
Tobacco

3. Sugarbeets may be planted 26 months after last SCEPTER application in soybeans.

* **NORTHERN USE AREA RESTRICTIONS:**
For the Northern Use Area as defined in the USE AREA section of the label; field corn, wheat, barley or oats may be planted as a rotational crop in the fall or in the spring of the year following SCEPTER application ONLY if SCEPTER was applied as a postemergence application at 1/3 pint per acre AND at least 10 inches of rainfall or irrigation has been received from the date of application through October of the same year.

** **SOUTHERN USE AREA RESTRICTIONS:**
For the Southern Use Area as defined in the USE AREA section of the label; field corn may be planted in the spring of the year following SCEPTER application unless extreme drought conditions develop or less than 15 inches of rainfall or irrigation is received from the date of application through October of the same year. A minimum of 10 inches of rainfall is needed for a postemergence application of SCEPTER only if the total amount of SCEPTER applied in one season does not exceed 1/3 pint per acre. For eastern Oklahoma, Arkansas, the bootheel of Missouri, Tennessee, North Carolina and states south: field corn may be planted in the spring of the year following SCEPTER application unless extreme drought conditions develop and less than 15 inches of rainfall or irrigation is received within 6 months following the date of application. A minimum of 10 inches of rainfall is needed for a postemergence application of SCEPTER only if the total amount of SCEPTER applied in one season does not exceed 1/3 pint per acre.

Only rotational crops harvested at maturity may be used for feed or food.

DO NOT plant other rotational crops prior to 18 months after last SCEPTER application.

Use of SCEPTER herbicide in accordance with label directions is expected to result in normal growth of rotational crops in most situations; however, various environmental and agronomic factors make it impossible to eliminate all risks associated with the use of this product and, therefore, rotational crop injury is always possible.

DO NOT apply products containing chlorimuron ethyl (e.g. Classic, Canopy, Gemini, Lorox Plus, Preview[9], etc.) or imazethapyr (e.g. PURSUIT® or PURSUIT® PLUS) the same year as SCEPTER or injury to follow crops may occur.

In the event of crop loss due to weather conditions, soybeans can be replanted. DO NOT work the soil deeper than 2 inches.

DO NOT graze or feed treated soybean forage, hay or straw to livestock.

[9] Classic, Canopy, Gemini, Lorox Plus and Preview are trademarks of E.I. du Pont de Nemours and Company.

SCEPTER® 70 DG

HERBICIDE

FOR USE IN SOYBEANS

ACTIVE INGREDIENT:

Imazaquin (2-[4,5-dihydro-4-methyl-4-(1-methylethyl)-5-oxo-1-imidazol-2-yl]-3-quinolinecarboxylic acid)	70.0%
INERT INGREDIENTS	30.0%
TOTAL	100.0%

EPA Reg. No. 241-306 3/89
® Registered Trademark of American Cyanamid Company

KEEP OUT OF REACH OF CHILDREN
CAUTION!/¡PRECAUCION!

PRECAUCION AL USUARIO: Si usted no lee inglés, no use este producto hasta que la etiqueta le haya sido explicada ampliamente.

STATEMENT OF PRACTICAL TREATMENT

If on skin: Wash with plenty of soap and water.

If in eyes: Flush eyes with plenty of water. Call a physician if irritation persists.

In case of an emergency endangering life or property involving this product, call collect, day or night, Area Code 201-835-3100.

PRECAUTIONARY STATEMENTS

HAZARDS TO HUMANS

CAUTION!

Causes eye irritation. Avoid contact with skin, eyes, or clothing. Harmful if absorbed through skin. Wash thoroughly with soap and water after handling.

ENVIRONMENTAL HAZARDS

DO NOT apply directly to water or wetlands (swamps, bogs, marshes and potholes). DO NOT contaminate water when disposing of equipment washwaters.

DIRECTIONS FOR USE

It is a violation of Federal law to use this product in a manner inconsistent with its labeling.

Observe all cautions and limitations on this label and on the labels of products used in combination with SCEPTER 70DG. SCEPTER 70DG should be used only in accordance with directions on this label. The use of SCEPTER 70DG not consistent with this label may result in injury to crops. Keep container closed to avoid spills and contamination.

DO NOT apply this product through any type of irrigation system.

STORAGE AND DISPOSAL

PROHIBITIONS: DO NOT contaminate water, food, or feed by storage or disposal.

PESTICIDE DISPOSAL: Wastes resulting from the use of this product may be disposed of on site or at an approved disposal facility.

CONTAINER DISPOSAL: Triple rinse (or equivalent). Then offer for recycling or reconditioning, or puncture and dispose of in a sanitary landfill, or by incineration, or if allowed by state and local authorities, by burning. If burned, stay out of smoke.

DISCLAIMER

The label instructions for the use of this product reflect the opinion of experts based on field use and tests. The directions are believed to be reliable and should be followed carefully. However, it is impossible to eliminate all risks inherently associated with use of this product. Crop injury, ineffectiveness or other unintended consequences may result because of such factors as weather conditions, presence of other materials, or the use or application of the product contrary to label instructions, all of which are beyond the control of American Cyanamid Company. All such risks shall be assumed by the user.

American Cyanamid Company will not be responsible for losses or damages resulting from use of this product in any manner not specifically recommended by American Cyanamid Company. User assumes all risk associated with such nonrecommended use.

American Cyanamid Company warrants only that the material contained herein conforms to the chemical description on the label and is reasonably fit for the use therein described when used in accordance with the directions for use, subject to the risks referred to above.

Any damages arising from breach of this warranty shall be limited to direct damages and shall not include consequential commercial damages such as loss of profits or values or any other special or indirect damages.

American Cyanamid Company makes no other express or implied warranty, including other express or implied warranty of FITNESS or of MERCHANTABILITY.

GENERAL INFORMATION

SCEPTER 70DG is a uniquely flexible broadspectrum herbicide which may be applied preplant incorporated, preemergence or postemergence in soybeans.

When SCEPTER 70DG is applied to the soil, susceptible weeds emerge, growth stops, and then the weeds either die or are not competitive with the crop.

When SCEPTER 70DG is applied postemergence, absorption occurs through both the foliage and roots. Susceptible weeds stop growing and either die or are not competitive with the crop.

A timely cultivation may aid in the control of certain weeds or improve general weed control when adequate moisture is not received after application. Cultivation should be shallow. After postemergence treatments, wait at least 10 days before cultivating.

The mode of weed killing activity involves uptake of SCEPTER 70DG by weed roots and/or foliage and rapid translation to growing points. Therefore, adequate soil moisture is important for optimum SCEPTER 70DG activity. The amount of rainfall or irrigation required following application depends on existing soil moisture, soil texture and organic matter content. Sufficient water to moisten the soil to a depth of 2 inches is normally adequate. If adequate moisture is not received within 7 days after treatment, then a cultivation is recommended to control escaped weeds. When adequate moisture is received after dry conditions, SCEPTER 70DG will provide residual control of susceptible germinating weeds; activity on established weeds will depend on the weed species and the location of its root system in the soil.

A SCEPTER 70DG soil treatment followed by a SCEPTER 70DG postemergence treatment will control certain problem weeds (See SEQUENTIAL PROGRAM section). Use sequential treatments in recommended states only.

Occasionally, internode shortening of soybean plants may be observed with SCEPTER 70DG applications. This has no effect on soybean yields.

There must be an interval of at least 90 days between the last SCEPTER 70DG application and soybean harvest. DO NOT graze or feed treated soybean forage, hay or straw to livestock.

PRECAUTIONS

Do not use on crops other than soybeans. Crops other than soybeans, such as cotton, corn and vegetables, may be injured by spray drift or other indirect contact with SCEPTER 70DG.

To avoid injury to sensitive crops from spray drift, follow all use directions and precautions in SPRAYING INSTRUCTIONS on pages 6 and 7.

To avoid injury to sensitive crops, spray equipment used for SCEPTER 70DG applications must be drained and thoroughly cleaned with water before being used to apply other products to these crops. See page 15 for ROTATIONAL CROP RESTRICTIONS.

WEEDS CONTROLLED

SCEPTER 70DG Alone

When applied as directed, SCEPTER 70DG soil-applied treatments will control the following weeds:

BROADLEAF WEEDS

Bristly Starbur (*Acanthospermum hispidum*)
Burcucumber* (*Sicyos angulatus*)
Cocklebur, Common (*Xanthium strumarium*)
Jimsonweed (*Datura stramonium*)
Lambsquarters, Common (*Chenopodium album*)
Mallow, Venice (*Hibiscus trionum*)
Morningglory,
  Palmleaf (*Ipomoea wrightii*)
  Pitted (*Ipomoea lacunosa*)
  Smallflower (*Jacquemontia tamnifolia*)
Mustard species (*Brassica* spp.)
Nightshade, Eastern Black* (*Solanum ptycanthum*)
Pigweed,
  Palmer (*Amaranthus palmeri*)
  Redroot (*Amaranthus retroflexus*)
  Smooth (*Amaranthus hybridus*)
  Spiny (*Amaranthus spinosus*)
  Waterhemp, Tall (*Amaranthus tuberculatus*)
Poinsettia, Wild (*Euphorbia heterophylla*)
Puncturevine (*Tribulus terrestris*)
Pusley, Florida (*Richardia scabra*)
Ragweed,
  Common (*Ambrosia artemisiifolia*)
  Giant* (*Ambrosia trifida*)
Redweed* (*Melochia corchorifolia*)
Sida, Prickly (teaweed) (*Sida spinosa*)
Smartweed,
  Ladysthumb (*Polygonum persicaria*)
  Pennsylvania (*Polygonum pensylvanicum*)
Sunflower, Common (*Helianthus annuus*)
Velvetleaf* (*Abutilon theophrasti*)

*Eastern black nightshade, giant ragweed, burcucumber, redweed and velvetleaf are controlled by preplant incorporated treatments only.

GRASSES

Foxtail,
  Giant (*Setaria faberi*)
  Green (*Setaria viridis*)
  Yellow (*Setaria glauca*)
Johnsongrass, seedling (*Sorghum halepense*)

SCEPTER 70DG soil-applied treatments will aid in the control and reduce competition from:

Beggarweed, Florida**** (*Desmodium tortuosum*)
Barnyardgrass (*Echinochloa crus-galli*)
Mexicanweed**** (*Caperonia castanifolia*)
Goosegrass (*Eleusine indica*)
Morningglory**
  Entireleaf (*Ipomoea hederacea var. intergriuscula*)
  Ivyleaf (*Ipomoea hederacea*)
  Tall (*Ipomoea purpurea*)
Nutsedge, Yellow*** (*Cyperus esculentus*)
Shattercane (*Sorghum bicolor*)
Sicklepod**** (*Cassia obtusifolia*)
Signalgrass, Broadleaf (*Brachiaria platyphylla*)
Spurge, Spotted (*Euphorbia maculata*)

**For best activity on morningglory species, use preplant incorporated treatments.
***Preplant Incorporated treatments only.
****A sequential treatment of SCEPTER 70DG applied preplant incorporated or preemergence followed by SCEPTER 70DG postemergence will control these weeds. See page 15 for more information on the sequential program.

MIXING INSTRUCTIONS

Fill the spray tank one-half to three-quarters full with clean water. While agitating add the required amount of SCEPTER 70DG and then fill the remainder of the tank with water. For postemergence applications, add the nonionic surfactant or the crop oil concentrate as last ingredient in the tank. An antifoaming agent may be added to the tank if needed. Maintain agitation while spraying to ensure a uniform spray mixture.

When tank mixing SCEPTER 70DG with recommended herbicides, add the SCEPTER 70DG to the spray tank first and make sure it is thoroughly mixed before adding the other herbicide.

When tank mixing Roundup or Gramoxone Super with SCEPTER 70DG and PROWL, Lasso or Dual, add herbicides to the tank in the following order: (1) SCEPTER 70DG, (2) PROWL, Lasso or Dual, (3) Roundup or Gramoxone Super, and (4) nonionic surfactant.

SPRAYING INSTRUCTIONS

NOTE: **DO NOT** make applications when spray may be carried by wind to sensitive crops. Avoid overlaps when spraying.

GROUND APPLICATIONS:

Uniformly apply with properly calibrated ground equipment in 10 or more gallons of water per acre. Use sprayers equipped with nozzles that provide accurate and application. Nozzle screens must be no finer than 50 mesh. To minimize drift, use a maximum spray pressure of 40 psi.

To ensure thorough coverage, use a minimum of 20 gallons of water per acre when applying SCEPTER 70DG to minimum or no-till soybeans or when applying as a postemergence treatment. Use higher gallonage for fields with dense vegetation or heavy crop residues.

DO NOT apply with ground equipment when wind velocity is greater than 10 mph.

For band treatment, apply the broadcast equivalent rate and volume per acre. To determine these:

$$\frac{\text{Band width in inches}}{\text{Row width in inches}} \times \text{Broadcast RATE per acre} = \text{Band RATE per acre}$$

$$\frac{\text{Band width in inches}}{\text{Row width in inches}} \times \text{Broadcast VOLUME per acre} = \text{Band VOLUME per acre}$$

AERIAL APPLICATIONS:

Uniformly apply with properly calibrated aerial equipment in 5 or more gallons of water per acre.

To avoid injury to sensitive crops from drift, aerial applicators must adhere to the following SPECIAL AERIAL USE DIRECTIONS AND PRECAUTIONS:

—Use nozzles which produce a coarse spray.

—Nozzle height above ground must be a maximum of 10 feet.

—Nozzles must be pointed toward the rear of the aircraft. The downward angle of the nozzle should not be greater than 20 degrees.

—To minimize wing-tip vortex roll, nozzles or spray boom must not be located any closer to end of wing or rotor than three-fourths the distance from the center of the aircraft.

—Use a maximum spray pressure of 40 psi.

—A buffer zone must be established between the area to be sprayed and sensitive crops.

DO NOT spray when wind velocity is greater than 5 mph.

Applicator is responsible for any loss or damage which results from spraying SCEPTER 70DG in manner other than recommended in this label. In addition, applicator must follow all applicable state and local regulations and ordinances in regard to spraying.

USE AREA

SCEPTER 70DG can be applied only in the states or parts of states shaded in the following map:

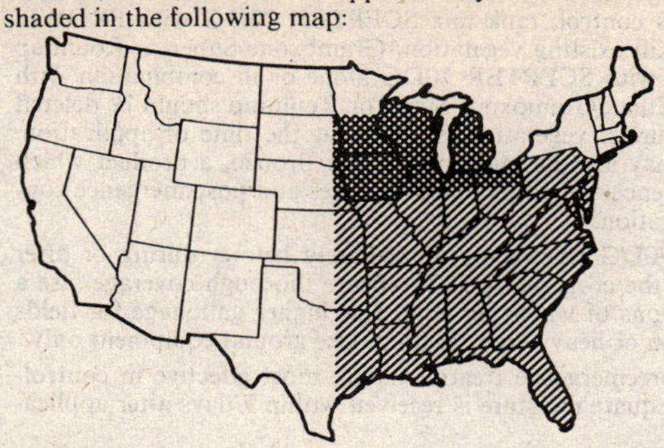

▓ = Northern Use Area (includes: Nebraska, east of U.S. 81, north of U.S. 34; South Dakota, east of U.S. 81; Illinois, north of S. R. 116 for area west of Peoria, north of U.S. 24 for area east of Peoria; Indiana, north of U.S. 24 for area west of Huntington, north of U.S. 224 for area east of Huntington; Ohio, north of U.S. 224; and all other states or portions of states shaded in the Northern use area map).

▨ = Southern Use Area (includes: Kansas, east of U.S. 81; southern portions of Illinois, Indiana, Ohio and eastern Nebraska not covered by Northern Use Area description; and all other states or portions of states shaded in the Southern Use Area map).

NOTE: See USE RATES section and ROTATIONAL CROP RESTRICTIONS section for recommendations applying to the Northern vs. Southern Use Areas.

NOTE: See SEQUENTIAL PROGRAM section in the label for geographical restrictions for the method of application.

USE RATES

Apply SCEPTER 70DG at a broadcast rate of 2.8 ounces per acre for preplant incorporated and preemergence (including minimum and no-till systems) applications. At this broadcast rate, one pound of SCEPTER 70DG will treat 5.7 acres of soybeans. Apply SCEPTER 70DG at a broadcast rate of 1.4 to 2.8 ounces per acre for postemergence applications.

NOTE: For preplant incorporated and preemergence applications on sands and loamy sands, 2.1 ounces per acre can be used without a substantial reduction in the control of only cocklebur and pigweed.

NOTE: In the Northern Use Area as defined by the Use Area section of the label, apply SCEPTER 70DG prior to July 1.

APPLICATION INSTRUCTIONS

PREPLANT INCORPORATED APPLICATIONS

Apply SCEPTER 70DG before planting and incorporate uniformly into the top 1 to 2 inches of soil. Mechanical incorporation can be achieved by the following:

(a) Disk harrow set to cut 3 to 4 inches deep and operated in two different directions at 4 to 6 mph.

(b) PTO-driven equipment (tillers, cultivators, hoes) set to cut 2 inches deep and operated one time at 4 mph or less.

(c) Rolling cultivator set to cut 2 to 3 inches deep and operated two times at 6 to 8 mph. Use only on coarse textured soils.

(d) Field cultivator with 3 to 4 rows of sweeps (Do not use chisel points), spaced at intervals of 7 inches or less and staggered so that no soil is left unturned, set to cut 3 inches deep, and operated two times at more than 5 mph with the second pass made at an angle to the first.

(e) LELY-Roterra[6] set and operated according to the manufacturer's directions to thoroughly incorporate SCEPTER 70DG to a depth of 1 to 2 inches.

(f) Do-All set and operated according to the manufacturer's directions to thoroughly incorporate SCEPTER 70DG to a depth of 1 to 2 inches.

[6]LELY-ROTERRA is a trademark of Lely.

SCEPTER 70DG may be applied immediately before planting or up to 30 days prior to planting in AL, AR, FL, GA, LA, MS, NC, OK, SC, TN, and TX only, For the rest of the Lower Use Area and all of Upper Use Area, SCEPTER 70DG may be applied up to 45 days prior to planting. Incorporate within 7 days after application. Adequate moisture is required for activation of SCEPTER 70DG.

If soybeans are planted on beds, apply and incorporate after bed formation using PTO-driven equipment or rolling cultivator.

PREEMERGENCE APPLICATIONS

Apply SCEPTER 70DG during or after planting but before the crop emerges. As with other herbicides applied preemergence, rainfall or irrigation is necessary to activate SCEPTER 70DG in the soil. If sufficient rainfall or irrigation to activate SCEPTER 70DG is not received within 7 days after application, a thorough shallow tillage or cultivation is recommended (preferably with a rotary hoe) for control of emerged weeds.

HERBICIDE COMBINATIONS

A soil treatment of SCEPTER 70DG is active against many grass species, but for maximum grass control SCEPTER 70DG should be tank mixed with PROWL® herbicide, trifluralin, Lasso or Dual. SCEPTER 70DG may also be applied as a sequential preemergence or postemergence treatment following application of one of the above grass herbicides.

For no-till uses, SCEPTER 70DG may be tank mixed with Gramoxone Super, Roundup or Bronco.

SCEPTER 70DG may be followed by a postemergence application of Fusilade[7] or Poast[8]. DO NOT tank mix SCEPTER 70DG with postemergence grass herbicides.

When used in combination with another herbicide, SCEPTER 70DG should be used only in accordance with directions on this label. SCEPTER 70DG cannot be mixed with any product containing a label prohibition against such mixing.

[7] Fusilade is a trademark of ICI Americas, Inc.
[8] Poast is a trademark of BASF Wyandotte Corporation.

SOIL TEXTURE

Although the rate of SCEPTER 70DG does not vary with soil texture, the recommended rates of other herbicides used with this product generally do vary with soil texture. Combination rate tables in this label refer to three soil texture groups: COARSE, MEDIUM, and FINE. The following table lists soil textures included in each of these three groups:

Coarse	Medium	Fine
sands loamy sands sandy loams	sandy clay loams* sandy clays loams silt loams silts	silty clay loams* silty clays clay loams clays

* Sometimes considered transitional soils and may be classified as either medium or fine textured soils.

WEEDS CONTROLLED PREPLANT AND PREEMERGENCE COMBINATION TREATMENTS

When applied as directed, SCEPTER 70DG preplant incorporated or preemergence combination treatments with PROWL, trifluralin, Lasso or Dual will control, as indicated, weeds listed in the following table in addition to those controlled by SCEPTER 70DG alone.

Weed	PROWL	Trifluralin[a]	Lasso	Dual
Grasses				
BARNYARDGRASS (*Echinochloa crus-galli*)	X	X	X	X
CRABGRASS, SMOOTH (*Digitaria ischaemum*)	X	X	X	X
CRABGRASS, LARGE (*Digiteria sanguinalis*)	X	X	X	X
CROWFOOT GRASS (*Dactylocterium aegyptium*)	X	X		
GOOSEGRASS (*Eleusine indica*)	X	X	X	X
PANICUM, FALL (*Panicum dichotomiflorum*)	X	X[c]	X	X
PANICUM, TEXAS (*Panicum texanum*)	X	X		
SANDBUR, FIELD (*Cenchrus incertus*)	X	X		
SHATTERCANE (*Sorghum bicolor*)	X[a, b]	X[c]		
SIGNALGRASS (*Brachiaria platyphylla*)	X	X	X	X
WITCHGRASS (*Panicum capillare*)	X	X	X	X
Broadleaf Weeds				
CARPETWEED (*Mollugo verticillata*)	X	X	X	X
GALINSOGA (*Galinsoga* spp.)		X	X	
KOCHIA (*Kochia scoparia*)	X	X		
PURSLANE (*Portulaca oleracea*)	X	X	X	
SPURGE, ANNUAL (*Euphorbia* spp.)	X			

[a]Preplant incorporated treatments only.
[b]See Note on page 10.
[c]See Note on page 11.

SCEPTER 70DG Plus PROWL Tank Mixture

Apply SCEPTER 70DG plus PROWL preplant incorporated or preemergence using rates in the table below. Follow use directions under APPLICATION INSTRUCTIONS in this label. For preplant incorporated applications, incorporate into the soil within 7 days after application, and plant soybeans within 45 days after application. For preemergence applications, apply at planting or up to 2 days after planting. DO NOT apply if soybeans have emerged.

Broadcast Rate of SCEPTER 70DG Plus PROWL

Soil Texture	SCEPTER 70DG + ounces/acre	PROWL pints/acre
	Less Than 3% Organic Matter	
Coarse	2.8	1 to 1½
Medium	2.8	1½ to 2
Fine	2.8	1½ to 2
	3% or More Organic Matter	
Coarse	2.8	1½
Medium	2.8	1½ to 2
Fine	2.8	2 to 2½

NOTE: Use the higher rate of PROWL if heavy weed populations are anticipated or if shattercane will be present. Observe all precautions and limitations on PROWL label.

SCEPTER 70DG Plus trifluralin Tank Mixture

Apply SCEPTER 70DG plus trifluralin preplant incorporated using rates in the table below. Follow use directions under APPLICATION INSTRUCTIONS in this label. Incorporate into the soil within 24 hours after application, and plant soybeans within 45 days after application. Use equipment that provides uniform 2 inch incorporation.

Broadcast Rate of SCEPTER 70DG Plus trifluralin*

Soil Texture	SCEPTER 70DG ounces/acre	Trifluralin* pints/acre
Coarse	2.8	1
Medium	2.8	1½
Fine	2.8	2

NOTE: If fall panicum or shattercane will be present, refer to trifluralin label for higher rates to control these weeds.
Observe all precautions and limitations on the trifluralin label.
*Based on 4 pounds active ingredient per gallon.

SCEPTER 70DG Plus Lasso Tank Mixture

Apply SCEPTER 70DG plus Lasso preplant incorporated or preemergence using rates in the table below. Follow use directions under APPLICATION INSTRUCTIONS in this label. For preplant incorporated applications, apply and incorporate within 7 days prior to planting.

Broadcast Rate of SCEPTER 70DG Plus Lasso

Soil Texture	SCEPTER 70DG (ounces/ acre)	Lasso (quarts/acre) Preplant Incorporated	Preemergence
	Less than 3% Organic Matter		
Coarse	2.8	2 to 2½	2
Medium	2.8	2½ to 3	2
Fine	2.8	2½ to 3	2*
	3% or More Organic Matter		
Coarse	2.8	2 to 2½	2
Medium	2.8	2½ to 3	2
Fine	2.8	2½ to 3	2½

*On silty clay and heavy clay soils of the Mississippi Delta, use 2 to 2½ quarts of Lasso.

NOTE: For preplant incorporated applications, use the higher rate of Lasso if heavy weed populations are anticipated.
Observe all precautions and limitations on the Lasso label.

SCEPTER 70DG Plus Dual Tank Mixture

Apply SCEPTER 70DG plus Dual preplant incorporated or preemergence using rates in table below. Follow use directions under APPLICATION INSTRUCTIONS in this label. For preplant incorporated applications, apply up to 29 days prior to planting. Incorporate within 7 days after application.

Broadcast Rate of SCEPTER 70DG Plus Dual

Soil Texture	SCEPTER 70DG ounces/acre	DUAL pints/acre
	Less Than 3% Organic Matter	
Coarse	2.8	1¼
Medium	2.8	1½
Fine	2.8	2
	3% or More Organic Matter	
Coarse	2.8	1½
Medium	2.8	2
Fine	2.8	2 to 2½*

*Use the higher rate of Dual if heavy weed populations are anticipated.

NOTE: Observe all precautions and limitations on Dual label. SCEPTER 70DG plus Command 4EC and SCEPTER 70DG plus Trifluralin plus Command 4EC Tank Mixtures Sections.

MINIMUM AND NO-TILL SYSTEMS

SCEPTER 70DG may be applied in minimum or no-till soybeans.

For maximum grass control, tank mix SCEPTER 70DG with PROWL, Lasso or Dual. To kill existing vegetation, Gramoxone Super or Roundup may be tank mixed with SCEPTER 70DG alone or in combination with PROWL, Lasso or Dual. Gramoxone Super or Roundup should be deleted from the tank mixture if vegetation is absent at the time of application. SCEPTER 70DG may also be tank mixed with Bronco, a product which gives both preemergence control of annual grasses and postemergence control of existing vegetation.

Apply SCEPTER 70DG treatments immediately before, during or after planting but before the crop emerges. To ensure thorough coverage, use a minimum of 20 gallons of water per acre. Use higher gallonage for fields with dense vegetation or heavy crop residues. Use ground equipment only.

SCEPTER 70DG preemergence treatments are most effective in controlling weeds when adequate moisture is received within 7 days after application.

NOTE: Plant soybeans at least one inch deep and adjust planters to ensure adequate seed coverage.

Tank Mixtures with PROWL, Lasso or Dual

1. PROWL

Use the rates in table on page 10 for SCEPTER 70DG plus PROWL preemergence tank mixtures. Use the higher rate of PROWL when previous crop residues are extremely dense. Apply immediately before, during or up to 2 days after planting. DO NOT apply if soybeans have emerged.

Observe all precautions and limitations on PROWL label.

2. Lasso

Use the following rates of Lasso for SCEPTER 70DG plus Lasso preemergence tank mixtures in minimum and no-till systems:

Soil Texture	Lasso Rate (Quarts/acre)
Coarse	2
Medium	2 to 2½
Fine	2½ to 3

Use the higher rate of Lasso (1) when heavy grass infestations are anticipated, (2) when fall panicum or crabgrass will be present or (3) when previous crop residues are extremely dense.
Observe all precautions and limitations on Lasso label.

3. Dual

Use the following rates of Dual for SCEPTER 70DG plus Dual preemergence tank mixtures in minimum and no-till systems:

Soil Texture	Dual Rate (pints/acre)
Coarse	1½
Medium	2
Fine	2½

Observe all precautions and limitations on Dual label.

Tank Mixtures With Gramoxone Super

Apply 1½ to 2½ pints of Gramoxone Super per acre in SCEPTER 70DG tank mixtures. Use the 2½ pint rate if weeds are 4 to 6 inches tall. Weeds over 6 inches tall may not be controlled with this treatment.

When Gramoxone Super is included in a tank mixture, add a nonionic spreader surfactant at a rate of 8 fl. oz. per 100 gallons of spray mixture as the last ingredient in the tank.

Gramoxone Super will control most annual emerged weeds and suppress many emerged perennials. Refer to the Gramoxone Super label for specific use recommendations and weeds controlled.

Observe all precautions and limitations on Gramoxone Super label.

Tank Mixtures With Roundup

Annual Weeds

Apply 1 to 1½ quarts of Roundup per acre in recommended tank mixtures. Use the 1½ quart rate if annual weeds are more than 6 inches tall.

Roundup will control most annual emerged weeds. Refer to Roundup label for specific use recommendations and weeds controlled.

Perennial Weeds

At normal application times in minimum or no-till systems, perennial weeds may not be at the proper stage of growth for control with Roundup. Refer to Roundup label for the proper stage of growth for perennial weeds. Use of 2 to 4 quarts of Roundup per acre in tank mixtures under these conditions provides top kill and reduces competition from many emerged perennial grass and broadleaf weeds. Refer to Roundup label for emerged perennial weeds controlled. To obtain control of perennial weeds, follow recommendations on Roundup label for stage of growth and rate of application.
NOTE: When using these tank mixtures, DO NOT exceed 4 quarts of Roundup per acre.

USE OF THESE ROUNDUP TANK MIXTURES FOR BERMUDAGRASS OR JOHNSONGRASS CONTROL IS NOT RECOMMENDED.

Observe all precautions and limitations on Roundup label.

Tank Mixtures With Bronco

Use the following rates of Bronco for SCEPTER 70DG plus Bronco tank mixtures:

Soil Texture	Bronco (quarts/acre)
Coarse	3 to 5
Medium	4 to 5
Fine	4 to 5

Use the higher rate of Bronco (1) when previous crop residues are extremely dense, (2) when emerged weeds are over 6 inches tall, (3) when heavy infestations of weeds are expected, and (4) when soils contain more than 3% organic matter.

Observe all precautions and limitations on Bronco label.

POSTEMERGENCE APPLICATIONS

Apply SCEPTER 70DG as a postemergence treatment after crop emergence and in accordance with the weeds controlled, weed heights, and rates of application information listed in the table below.

SCEPTER 70DG FOR POSTEMERGENCE WEED CONTROL

Weeds Controlled	Weed Height (inches)	SCEPTER 70DG* (ounces/acre)
Cocklebur, Common	up to 8	1.4**
(*Xanthium strumarium*)	9 to 12	2.8
Pigweed		
Palmer	up to 12	2.8
(*Amaranthus palmeri*)		
Redroot	up to 4	1.4**
(*Amaranthus retroflexus*)	5 to 12	2.8
Smooth	up to 4	1.4**
(*Amaranthus hybridus*)	5 to 12	2.8
Spiny	up to 12	2.8
(*Amaranthus spinosus*)		
Tall waterhemp	up to 12	2.8
(*Amaranthus tuberculatus*)		
Poinsettia, Wild	up to 6	2.8
(*Euphorbia heterophylla*)		
Sunflower, Common	up to 4	1.4**
(*Helianthus anuus*)	5 to 8	2.8.**
Volunteer Corn	up to 8	1.4**
(*Zea mays*)	9 to 12	2.8**

Apply when weeds are actively growing. DO NOT apply SCEPTER 70DG postemergence when soybeans and weeds have been subjected to stress conditions such as temperature and moisture extremes.

*Use a nonionic surfactant or crop oil concentrate.

The nonionic surfactant should contain at least 80% active ingredient and should be applied at a rate of 2 pints per 100 gallons of spray mixture. Apply the crop oil concentrate (COC) at the rate stated on the COC label.

**For control of emerged weeds, only.

There must be an interval of at least 90 days between the last SCEPTER 70DG application and soybean harvest.

SEQUENTIAL PROGRAM

An additional treatment of SCEPTER 70DG may be applied postemergence after an initial preplant incorporated or preemergence treatment to control certain problem weeds.

Apply SCEPTER 70DG at 1.4 to 2.8 ounces per acre as a sequential postemergence treatment following either a preplant incorporated or preemergence treatment of SCEPTER 70DG at 2.8 ounces per acre. The total amount of SCEPTER 70DG applied should not exceed 5.6 ounces per acre in one season. The sequential postemergence treatment should be applied before weed species exceed the 1 to 2 true leaf stage.

Follow use directions for preplant incorporated or preemergence (see page 8) and postemergence treatments (See above).

If grass species are anticipated, use one of the following registered soil-applied grass herbicides: PROWL herbicide, trifluralin, Lasso or Dual. See pages 9-12 for specific use directions.

This sequential application can be used in Alabama, Arkansas, Florida, Georgia, Louisiana, Mississippi, North Carolina, Oklahoma, South Carolina, Tennessee, and Texas only.

When applied as directed, SCEPTER 70DG preplant incorporated or preemergence followed by SCEPTER 70DG postemergence will control the following broadleaf weeds:

Beggarweed, Florida (*Desmodium tortuosum*)

Mexicanweed (*Caperonia castaniifolia*)

Sicklepod (*Cassia obtusifolia*)

ROTATIONAL CROP RESTRICTIONS

The following rotational crops may be planted after applying SCEPTER 70DG at recommended rates in soybeans:

1. Wheat may be planted in the fall, four months after last SCEPTER 70DG application in soybeans. Rice may be planted in the spring of the year following SCEPTER 70DG application (See Upper Use Area Restrictions*).
2. Eleven months after last SCEPTER 70DG application:
   Barley (See Upper Use Area Restrictions*)
   Oats (See Upper Use Area Restrictions*)
   Field Corn (See Upper Use Area or Lower Use Area Restrictions**)
   Edible beans
   Grain sorghum
   Peanuts
   Tobacco

*UPPER USE AREA RESTRICTIONS:

For the Upper Use Area as defined in the USE AREA section of the label; field corn, wheat, barley or oats may be planted as a rotational crop in the fall or in the spring of the year following SCEPTER 70DG application ONLY if SCEPTER 70DG was applied as a postemergence application at 1.4 ounces per acre AND at least 10 inches of rainfall or irrigation has been received from the date of application through October of the year of application.

**LOWER USE AREA RESTRICTIONS:

For the Lower Use Area as defined by in the USE AREA section of the label; field corn may be planted in the spring of the year following SCEPTER 70DG application unless extreme drought conditions develop and less than 15 inches (10 inches for 1.4 oz/acre postemergence applications) of rainfall or irrigation is received from the date of application through October of the year of application. For eastern Oklahoma, Arkansas, the bootheel of Missouri, Tennessee, North Carolina and states south; field corn may be planted in the spring of the year following SCEPTER 70DG application unless extreme drought conditions develop and less than 15 inches (10 inches for 1.4 oz/acre postemergence applications) of rainfall or irrigation is received within 6 months following the date of application.

Only rotational crops harvested at maturity may be used for feed or food.

DO NOT plant other rotational crops prior to 18 months after last SCEPTER 70DG application.

DO NOT apply products containing chlorimuron ethyl (e.g., Classic, Canopy, Gemini, Lorox Plus, Preview[9], etc.) or imazethapyr (e.g., PURSUIT® or PURSUIT® PLUS) the same year as SCEPTER 70DG or injury to follow crops may occur.

In the event of a crop loss due to weather conditions, soybeans can be replanted. DO NOT work the soil deeper than 2 inches.

DO NOT graze or feed treated soybean forage, hay or straw to livestock.

[9]Classic, Canopy, Gemini, Lorox Plus and Preview are trademarks of E.I. du Pont de Nemours and Company.

SUPPLEMENTAL LABELING

SCEPTER 70 DG herbicide

EPA Reg. No. 241-306

FOR USE IN SOYBEANS

SCEPTER 70DG Plus PROWL® herbicide Plus Command[1] Tank Mixture

OBSERVE ALL PRECAUTIONARY STATEMENTS, USE AREAS, APPLICATION INFORMATION, AND ROTATIONAL CROP RESTRICTIONS IN THE SCEPTER 70DG, PROWL, AND COMMAND LEAFLET LABELS BEFORE USING. ALWAYS FOLLOW THE MORE RESTRICTIVE LABEL.

DIRECTIONS FOR USE

It is a violation of Federal law to use this product in a manner inconsistent with its labeling. DO NOT apply this product through any type of irrigation system.

The SCEPTER 70DG plus PROWL plus Command preplant incorporated tank mixture may be used when heavy infestations of velvetleaf are anticipated.

MIXING INSTRUCTIONS

Fill the spray tank one-half to three quarters full with clean water. While agitating add the required amount of SCEPTER 70DG. Continue agitating until complete mixing of the SCEPTER 70DG has occurred. Next add PROWL, and then the Command. Finally, fill the remainder of the tank with water. Maintain agitation while spraying to ensure a uniform spray mixture.

SPRAYING INSTRUCTIONS

Unformly apply the tank mixture with properly calibrated ground equipment in 10 to 40 gallons of water per acre. Use sprayers equipped with nozzles that provide accurate and uniform application. NOTE: The use of an agriculturally approved drift reducing agent is required at finished spray volumes of 10 to 15 gallons per acre.

Apply SCEPTER 70DG plus PROWL plus Command to the soil surface and uniformly incorporate to a depth of 1 to 2 inches. Incorporate within 3 hours of application. DO NOT apply aerially or through irrigation equipment. DO NOT apply this tank mixture to overly moist or wet soils.

SPECIAL PRECAUTIONS

Command is a volatile compound. Off-site movement of spray drift or vapors of Command can cause foliar whitening or yellowing of some plants. Prior to using Command, read and strictly follow all precautions and application instructions on the Command label.

WEED SPECIES CONTROLLED

The SCEPTER 70DG plus PROWL plus Command tank mixture controls heavy infestations of velvetleaf as well as all the weeds listed on the SCEPTER 70DG label for combination preplant incorporated treatments of SCEPTER 70DG plus PROWL.

This label must be in the possession of the user at the time of herbicide application.

[1] Command is a Trademark of FMC Corporation
® Registered Trademarks of American Cyanamid Company
PE-12111 4/28/89

SUPPLEMENTAL LABELING

SCEPTER 70DG herbicide

EPA Reg. No. 241-306

FOR USE IN SOYBEANS

SCEPTER 70DG Plus Trifluralin Plus Command[1] Tank Mixture

OBSERVE ALL PRECAUTIONARY STATEMENTS, USE AREAS, APPLICATION INFORMATION, AND ROTATIONAL CROP RESTRICTIONS IN THE SCEPTER 70DG, TRIFLURALIN AND COMMAND LEAFLET LABELS BEFORE USING. ALWAYS FOLLOW THE MORE RESTRICTIVE LABEL.

DIRECTIONS FOR USE

It is a violation of Federal law to use this product in a manner inconsistent with its labeling. DO NOT apply this product through any type of irrigation system.

The SCEPTER 70DG plus trifluralin plus Command preplant incorporated tank mixture may be used when heavy infestations of velvetleaf are anticipated.

MIXING INSTRUCTIONS

Fill the spray tank one-half to three quarters full with clean water. While agitating add the required amount of SCEPTER 70DG. Continue agitating until complete mixing of the SCEPTER 70DG has occurred. Next add trifluralin, and then the Command. Finally, fill the remainder of the tank with water. Maintain agitation while spraying to ensure a uniform spray mixture.

SPRAYING INSTRUCTIONS

Uniformly apply the tank mixture with properly calibrated ground equipment in 10 to 40 gallons of water per acre. Use sprayers equipped with nozzles that provide accurate and uniform application. NOTE: The use of an agriculturally approved drift reducing agent is required at finished spray volumes of 10 to 15 gallons per acre.

Apply SCEPTER 70DG plus trifluralin plus Command to the soil surface and uniformly incorporate to a depth of 1 to 2 inches. Incorporate within 3 hours of application. DO NOT apply aerially or through irrigation equipment. DO NOT apply this tank mixture to overly moist or wet soils.

SPECIAL PRECAUTIONS

Command is a volatile compound. Off-site movement of spray drift or vapors of Command can cause foliar whitening or yellowing of some plants. Prior to using Command, read and strictly follow all precautions and application instructions on the Command label.

WEED SPECIES CONTROLLED

The SCEPTER 70DG plus trifluralin plus Command tank mixture controls heavy infestations of velvetleaf as well as all the weeds listed on the SCEPTER 70DG label for combination preplant incorporated treatments of SCEPTER 70DG plus trifluralin.

This label must be in the possession of the user at the time of herbicide application.

[1] Command is a Trademark of FMC Corporation
® Registered Trademark of American Cyanamid Company
PE-12112 5/1/89

SCEPTER® O.T.®

HERBICIDE

FOR USE IN SOYBEANS

ACTIVE INGREDIENTS:

Sodium salt of imazaquin (2-[4,5-dihydro-4-methyl-4-(1-methyl ethyl)-5-oxo-1-*H*-imidazol-2-yl]-3-quinolinecarboxylic acid)*	5.61%
Sodium salt of acifluorfen: Sodium 5-(2-chloro-4-(trifluoromethyl) phenoxy-2-nitrobenzoate)	20.93%
INERT INGREDIENTS	73.46%
TOTAL	100.00%

*Equivalent to 5.24%, 2-[4,5-dihydro-4-methyl-4-(1- methylethyl)5-oxo-1*H*-imidazol-2-yl]-3-quinolinecarboxylic acid. SCEPTER O.T. contains 2.5 pounds of active ingredients per gallon (0.5 pounds active ingredient of imazaquin and 2.0 pounds active ingredient of acifluorfen per gallon).

EPA Reg. No. 241-321

® Registered Trademarks of American Cyanamid Company

3/89

KEEP OUT OF REACH OF CHILDREN
DANGER!/PELIGRO!

PRECAUCION AL USUARIO: Si usted no lee inglés, no use este producto hasta que la etiqueta le haya sido explicada ampliamente.

In case of emergency endangering life or property involving this product, call collect, day or night, Area Code 201-835-3100.

STATEMENT OF PRACTICAL TREATMENT

IF IN EYES: Hold eyelids open and flush with a steady, gentle stream of water for 15 minutes. Get medical attention.

IF SWALLOWED: Drink promptly a large quantity of milk, egg white, gelatin solution, or if these are not available, large quantities of water. Avoid alcohol. Get medical attention.

IF ON SKIN: Wash with plenty of soap and water. Get medical attention if irritation persists.

IF INHALED: Remove victim to fresh air.

NOTE TO PHYSICIAN: Probable mucosal damage may contraindicate the use of gastric lavage.

PRECAUTIONARY STATEMENTS

HAZARDS TO HUMANS

DANGER!

CAUSES IRREVERSIBLE EYE DAMAGE
HARMFUL IF SWALLOWED,
ABSORBED THROUGH THE SKIN, OR INHALED

Avoid contact with eyes, skin, or clothing. Wear goggles or face shield. Avoid breathing vapor or spray mist. Wash thoroughly with soap and water after handling. Remove contaminated clothing and wash before reuse.

ENVIRONMENTAL HAZARDS

This product is toxic to fish. **DO NOT** apply directly to water or wetlands (swamps, bogs, or marshes). Drift or runoff from treated areas may be hazardous to aquatic organisms in neighboring areas. **DO NOT** contaminate water when disposing of equipment washwaters.

DIRECTIONS FOR USE

It is a violation of Federal law to use this product in a manner inconsistent with its labeling.

This labeling must be in the possession of the user at the time of pesticide application.

Observe all cautions and limitations in this leaflet. SCEPTER O.T. should be used only in accordance with directions on this label. The use of SCEPTER O.T. not consistent with this label can result in injury to crops, animals, or persons. Keep container closed to avoid spills and contamination.

DO NOT apply this product through any type of irrigation system.

STORAGE AND DISPOSAL

STORAGE: KEEP FROM FREEZING. DO NOT STORE BELOW 32°F. SHAKE WELL BEFORE USING.

DISPOSAL: Pesticide wastes are acutely hazardous. DO NOT contaminate water, food, or feed by storage or disposal. Improper disposal of excess pesticide, spray mixture, or rinsate is a violation fo Federal law. Dispose of excess or waste pesticide by use according to label directions, or contact your State Pesticides or Environmental Control Agency or the Hazardous Waste representative at the nearest EPA Regional Office for guidance.

CONTAINER DISPOSAL: Triple rinse (or equivalent). Then offer for recycling or reconditioning, or puncture and dispose of in a sanitary landfill, or by incineration, or, if allowed by state and local authorities, by burning. If burned, stay out of smoke.

DISCLAIMER

The label instructions for the use of this product reflect the opinion of experts based on field use and tests. The directions are believed to be reliable and should be followed carefully. However, it is impossible to eliminate all risks inherently associated with use of this product. Crop injury, ineffectiveness or other unintended consequences may result because of such factors as weather conditions, presence of other materials, or the use or application of the product contrary to label instructions, all of which are beyond the control of American Cyanamid Company. All such risks shall be assumed by the user.

American Cyanamid Company will not be responsible for losses or damages resulting from use of this product in any manner not specifically recommended by American Cyanamid Company. User assumes all risk associated with such nonrecommended use.

American Cyanamid Company warrants only that the material contained herein conforms to the chemical description on the label and is reasonably fit for the use therein described when used in accordance with the directions for use.

Any damages arising from a breach of this warranty shall be limited to direct damages, and shall not include consequential commercial damages such as loss of profits or values, etc.

American Cyanamid Company makes no other express or implied warranty, including any other express or implied warranty of FITNESS or MERCHANTABILITY. USER assumes the risk of any use contrary to label instructions, or under abnormal conditions, or under conditions not reasonably foreseeable by American Cyanamid Company.

GENERAL INFORMATION

SCEPTER O.T. is a selective postemergence herbicide for the control of important broadleaf weeds in soybeans. For best results, apply SCEPTER O.T. to small actively growing weeds.

CULTIVATION

A timely cultivation may aid in the control of certain weeds or improve general weed control. After postemergence treatments, wait at least 10 days before cultivating.

BIOLOGICAL ACTIVITY

When SCEPTER O.T. is applied postemergence, it is absorbed through the foliage and roots (root uptake occurs if adequate moisture is received from rainfall or overhead irrigations). Once absorbed through the foliage SCEPTER O.T. acts in two ways:

1. As a contact herbicide producing rapid necrosis of the plant tissue.
2. And as a translocatable herbicide with movement to the growing points (roots and leaves) of the weed.

Susceptible weeds treated with SCEPTER O.T. stop growing and rapidly become chlorotic and necrotic especially the growing points and either die or become non-competitive with the crop.

ADJUVANT USE

Use a nonionic surfactant at least 80% active at 1 quart per 100 gallons or 0.25% volume per volume (v:v) to enhance the wetting, contact, and uptake activity of SCEPTER O.T. on the plant foliage.

CROP RESPONSE

SCEPTER O.T. may cause some necrosis, cupping, speckling, and bronzing of the soybean leaves. Plants quickly recover without any effect on soybean yields.

There must be an interval of 90 days between the last SCEPTER O.T. application and soybean harvest. **DO NOT** graze or feed treated soybean forage hay or straw to livestock.

NOTE: Do not use on crops other than soybeans. Crops other than soybeans, such as cotton, corn, and vegetables, may be injured by spray or other indirect contact with SCEPTER O.T.

To avoid injury to sensitive crops from spray drift, follow all use directions and precautions in Spraying Instructions on page 3.

To avoid injury to sensitive crops, spray equipment used for SCEPTER O.T. applications must be drained and thoroughly cleaned with water before being used to apply other products to these crops.

See page 4 for Rotational Crop Restrictions.

MIXING INSTRUCTIONS

Fill the spray tank one-half to three-quarters full with clean water. While agitating add the required amount of SCEPTER O.T. and as the last ingredient in the tank add the nonionic surfactant. An antifoaming agent may be added to the tank if needed. Maintain agitation while spraying to ensure a uniform spray mixture.

SPRAYING INSTRUCTIONS

GROUND APPLICATIONS:

Uniformly apply with properly calibrated ground equipment in 10 to 30 gallons of water per acre and maintain a sprayer pressure of 30-40 PSI. Use higher gallonage for fields with dense vegetation. Use sprayers equipped for postemergence applications. Use nozzles that provide accurate and uniform application (tapered flat fan spray tips with 80-110 degree spray angles, and tip sizes should not exceed -06 to insure small droplet size of the spray solution; hollow cone nozzle tips may also be used).

A second application may be necessary if additional weeds emerge after treatment (total SCEPTER O.T. applied per season can not exceed 2 pints per acre).

DO NOT apply with ground equiment when wind velocity is greater than 10 mph or when spray may be carried by wind to sensitive crops. Avoid overlaps when spraying.

AERIAL APPLICATIONS:

Uniformly apply with properly calibrated aerial equipment in 5 or more gallons of water per acre.

To avoid injury to sensitive crops from drift, aerial applicators must adhere to the following SPECIAL USE DIRECTIONS AND PRECAUTIONS:

-Use diaphram-type nozzles which produce cone or flat spray patterns.

-Use a maximum spray pressure of 40 PSI.

-Nozzle height above ground must be a maximum of 10 feet.

-Nozzles must be pointed toward the rear of the aircraft. The downward angle of the nozzle should not be greater than 20 degrees.

-To minimize wing-tip vortex roll, nozzles or spray boom must not be located any closer to end of wing or rotor than three-fourths the distance from the center of the aircraft.

-A buffer zone must be established between the area to be sprayed and sensitive crops.

-DO NOT spray when wind velocity is greater than 5 mph.

Applicator is responsible for any loss or damage which results from spraying SCEPTER O.T. in manner other than recommended in this label. In addition, applicator must follow all applicable state and local regulations and ordinances in regard to spraying.

USE RATE RECOMMENDATION

Apply SCEPTER O.T. at a broadcast rate of 1 pint per acre for postemergence applications.

NOTE: In the Upper Use Area as defined by the Use Area section of the label, apply SCEPTER O.T. prior to July 1.

POSTEMERGENCE APPLICATIONS

Apply SCEPTER O.T. as a postemergence treatment after crop emergence and in accordance with the weeds controlled, weed height, and rates of application information listed in the table below.

PRODUCT	RATES	WEEDS CONTROLLED	WEED HEIGHT (inches)	MAXIMUM LEAF STAGE
SCEPTER O.T. + nonionic surfactant	1.0 pint/a + 0.25% v:v 1 quart per 100 gallons	cocklebur *(Xanthium strumarium)*	up to 4	4
		morningglories, ivyleaf *(Ipomoea hederacea)*	up to 2	3
		entireleaf *(Ipomoea hederacea var. intergriuscula)*	up to 2	3
		tall *(Ipomoea purpurea)*	up to 2	3
		pitted *(Ipomoea lacunosa)*	up to 2	3
		palmleaf *Ipomoea wrightii)*	up to 2	3
		smallflower *(Jacquemontia tamnifolia)*	up to 2	3
		redroot pigweed *(Amaranthus retroflexus)*	up to 6	6
		smooth pigweed *(Amaranthus hybridus)*	up to 4	4
		tall waterhemp *(Amaranthus tuberculatus)*	up to 4	4
		hemp sesbania *(Sesbania exaltata)*	up to 12	12

Apply when weeds are actively growing. DO NOT apply SCEPTER O.T. when weeds have been subjected to stress conditions such as temperature and moisture extremes.

There must be an interval of at least 90 days between the last SCEPTER O.T. application and soybean harvest.

SEQUENTIAL PROGRAM

An additional treatment of SCEPTER O.T. may be applied (in the Lower Use Area only) but the total amount of product can not exceed 2 pints per acre per year.

A single postemergence application of SCEPTER O.T. may be applied at 1 pint per acre following a preplant incorporated or preemergence treatment of SCEPTER®, SQUADRON®, or TRI-SCEPT® in Alabama, Arkansas, Florida, Georgia, Louisiana, Mississippi, North Carolina, Oklahoma, South Carolina, Tennessee, and Texas only.

Follow use directions for preplant incorporated or preemergence applications in the SCEPTER, SQUADRON, or TRI-SCEPT labels, and the use directions for SCEPTER O.T. contained in this label.

®Registered Trademarks of American Cyanamid Company.

SCEPTER O.T.
ROTATIONAL CROP RESTRICTIONS

The following rotational crops may be planted after applying SCEPTER O.T. in soybeans:

1. Wheat may be planted in the fall, four months after a SCEPTER O.T. application in soybeans. Rice may be planted in the spring of the year following SCEPTER O.T. application (See Upper Use Area Restrictions*)
2. Eleven months after SCEPTER O.T. application:
   Barley (See Upper Use Area Restrictions*)
   Oats (See Upper Use Area Restrictions*)
   Field corn (See Upper Use Area* or Lower Use Area Restrictions**)
   Edible beans
   Grain sorghum
   Peanuts
   Tobacco

* UPPER USE AREA RESTRICTIONS:
For the Northern Use Area as defined in the USE AREA section of this label; field corn, wheat, barley, or oats may be planted as a rotational crop in the fall or in the spring of the year following SCEPTER O.T. application ONLY if SCEPTER O.T. is applied as a single postemergence application at 1 pint per acre and 10 inches of rainfall or irrigation has been received from the date of application through October of the year of application.

** LOWER USE AREA RESTRICTIONS:
For the Lower Use Area, as defined in the USE AREA section of this label; field corn may be planted in the spring of the year following SCEPTER O.T. application unless extreme drought conditions develop and less than 15 inches of rainfall or irrigation is received from the date of SCEPTER O.T. application through October of the year of application. For eastern Oklahoma, Arkansas, the bootheel of Missouri, Tennessee, North Carolina and states south: field corn may be planted in the spring of the year following SCEPTER O.T. unless extreme drought conditions develop and less than 15 inches (10 inches for a single application of 1 pint/acre) of rainfall or irrigation is received within 6 months following the date of application.

ONLY rotational crops harvested at maturity may be used for feed or food.

DO NOT plant other rotational crops prior to 18 months after last SCEPTER O.T. application.

DO NOT apply products containing chlorimuron ethyl (e.g. Classic, Canopy, Gemini, Lorox Plus, Preview[1], etc.), imazaquin (e.g. SCEPTER, SQUADRON, TRI-SCEPT, etc. [refer to the SEQUENTIAL PROGRAM of this label for exceptions]), or imazethapyr (e.g. PURSUIT®, PURSUIT® PLUS, etc.) the same year as SCEPTER O.T. or injury to follow crops may occur.

DO NOT graze or feed treated soybean forage, hay, or straw to livestock.

Classic, Canopy, Gemini, Lorox Plus, and Preview are trademarks of E.I. duPont de Nemours and Company.

®Registered Trademarks Of American Cyanamid Company.

USE AREA

SCEPTER O.T. can be applied only in the states or parts shaded in the following map:

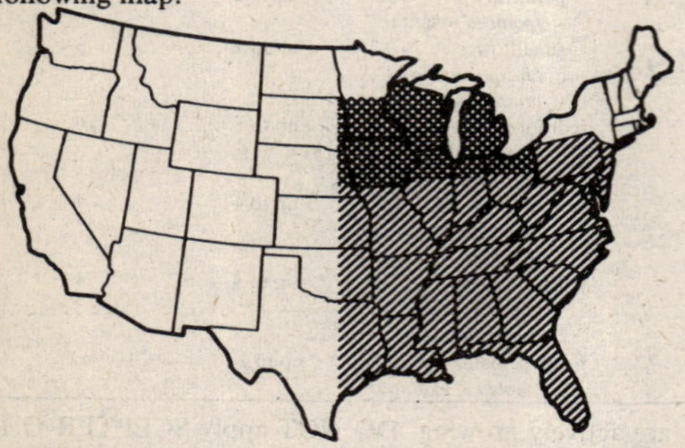

= Northern Use Area (includes: Nebraska, east of U.S. 81, north of U.S. 34; South Dakota, east of U.S. 81; Illinois, north of S. R. 116 for area west of Peoria, north of U.S. 24 for area east of Peoria; Indiana, north of U.S. 24 for area west of Huntington, north of U.S. 224 for area east of Huntington; Ohio, north of U.S. 224; and all other states or portions of states shaded in the Northern use area map).

= Southern Use Area (includes: Kansas, east of U.S. 81; southern portions of Illinois, Indiana, Ohio and eastern Nebraska not covered by Northern Use Area description; and all other states or portions of states shaded in the Southern Use Area map).

NOTE: See ROTATIONAL CROP RESTRICTIONS section for recommendations applying to the Upper vs. Lower Use Areas.

NOTE: See SEQUENTIAL PROGRAM section in the label for use area restrictions for this method of application.

SQUADRON®
HERBICIDE

FOR USE IN SOYBEANS

ACTIVE INGREDIENTS:

Ammonium salt of imazaquin (2-[4,5-dihydro-4-methyl-4-(1-methylethyl) -5-oxo-1*H*-imidazol-2-yl] -3-quinolinecarboxylic acid)*	3.84%
pendimethalin (N-1-ethylpropyl)-3,4-dimethyl-2,6-dinitrobenzenamine	21.85%
INERT INGREDIENTS:	74.31%
TOTAL	100.0%

*Equivalent to 3.64%, 2-[4,5-dihydro-4-methyl-4-(1-methylethyl)-5-oxo-1*H*-imidazol-2-yl] -3-quinolinecarboxylic acid.

SQUADRON contains 2.33 pounds of active ingredients per gallon (2.00 pounds ai of pendimethalin and 0.33 pounds ae of imazaquin).

EPA Reg. No. 241-297 EPA Est. No.

®Registered Trademark of American Cyanamid Company

10/89

OCTOBER 1989

KEEP OUT OF REACH OF CHILDREN
DANGER/¡PELIGRO!

PRECAUCION AL USUARIO; Si usted no lee inglés, no use este producto hasta que la etiqueta le naya sido explicada ampliamente.

STATEMENT OF PRACTICAL TREATMENT

If in eyes; hold eyelids open and flush with a steady, gentle stream of water for 15 minutes. Call a physician.

If swallowed; DO NOT induce vomiting. Call a physician or Poison Control Center immediately.

For skin contact; wash thoroughly with soap and water. Call a physician if skin irritation persists.

This product contains monochlorobenzene.

NOTE TO PHYSICIAN: Because of increased risk of chemical pneumonia or pulmonary edema caused by aspiration of the hydrocarbon solvent (monochlorobenzene), vomiting should be induced only under professional supervision.

In case of an emergency endangering life or property involving this product, call collect, day or night, Area Code 201-835-3100.

PRECAUTIONARY STATEMENTS

HAZARDS TO HUMANS

DANGER

CAUSES IRREVERSIBLE EYE DAMAGE

HARMFUL IF SWALLOWED, ABSORBED THROUGH THE SKIN, OR INHALED.

DO NOT get in eyes or on clothing. Wear goggles or face shield. Avoid breathing spray mist. Avoid contact with skin. Wash skin thoroughly with soap and water after handling. Use with adequate ventilation. Keep container closed. Remove contaminated clothing and wash before reuse.

ENVIRONMENTAL HAZARDS

This product is toxic to fish. DO NOT apply directly to water or wetlands (swamps, bogs, marshes, or potholes). Drift and runoff from treated areas may be hazardous to aquatic organisms in adjacent aquatic sites. DO NOT contaminate water when disposing of equipment washwaters.

PHYSICAL HAZARDS
FLAMMABLE LIQUID AND VAPOR

DO NOT use, pour, spill, or store near heat or open flame.
DO NOT use in the vicinity of pilot lights.

DISCLAIMER

The label instructions for the use of this product reflect the opinion of experts based on field use and tests. The directions are believed to be reliable and should be followed carefully. However, it is impossible to eliminate all risks inherently associated with use of this product. Crop injury, ineffectiveness or other unintended consequences may result because of such factors as weather conditions, presence of other materials, or the use or application of the product contrary to label instructions, all of which are beyond the control of American Cyanamid Company. All such risks shall be assumed by the user.

American Cyanamid Company shall not be responsible for losses or damages resulting from use of this product in any manner not set forth on this label. User assumes all risks associated with the use of this product in any manner not set forth on this label.

American Cyanamid Company warrants only that the material contained herein conforms to the chemical description on the label and in reasonably fit for the use therein described when used in accordance with the directions for use.

Any damages arising from a breach of this warranty shall be limited to direct damages, and shall not include consequential commercial damages such as loss of profits or values, etc. Cyanamid's liability will be limited to a refund of the purchase price of the SQUADRON. In no case shall Cyanamid or the seller be liable for consequential, special or indirect damages resulting from the use or handling of this product.

American Cyanamid Company makes no other express or implied warranty, including any other express or implied warranty of FITNESS or MERCHANTABILITY. USER assumes the risk of any use contrary to label instructions, or under abnormal conditions, or under conditions not reasonably foreseeable by Amercian Cyanamid Company.

DIRECTIONS FOR USE

It is a violation of Federal law to use this product in a manner inconsistent with its labeling.

This labeling must be in the possession of the user at the time of pesticide application.

Observe all cautions and limitations in this leaflet. Do not use SQUADRON other than in accordance with the instructions set forth on this label. The use of SQUADRON not consistent with this label can result in injury to crops, animals, or persons. Keep container closed to avoid spills and contamination.

DO NOT apply this product through any type of irrigation system.

STORAGE AND DISPOSAL

STORAGE:

KEEP FROM FREEZING. DO NOT STORE BELOW 32°F. SHAKE WELL BEFORE USING.

DO NOT contaminate water, food, or feed by storage or disposal.

DISPOSAL: Pesticide wastes are acutely hazardous. Improper disposal of excess pesticide, spray mixture, or rinsate is a violation of Federal law. If these wastes cannot be disposed of by use according to label instructions, contact your State Pesticide or Environmental Control Agency, or the Hazardous Waste Representative at the nearest EPA Regional Office for guidance.

CONTAINER DISPOSAL: Triple rinse (or equivalent). Then offer for recycling or reconditioning, or puncture and dispose of in a sanitary landfill, or by incineration, or, if allowed by state and local authorities, by burning. If burned, stay out of smoke.

GENERAL INFORMATION

For broad spectrum control of grass and broadleaf weeds in soybeans, SQUADRON may be applied as a preplant incorporated or preemergence treatment.

DO NOT APPLY SQUADRON OR ITS TANK MIXTURES PREEMERGENCE IN SOUTH DAKOTA.

After SQUADRON is applied, some susceptible weeds emerge, growth stops, and then the weeds either die or are not competitive with the crop.

A timely cultivation may aid in the control of certain weeds or improve general weed control when adequate moisture is not received after application. Cultivation should be shallow.

SQUADRON reaches the growing points of susceptible weeds either by direct contact in the soil, or by root uptake and rapid translocation to the growing points. Therefore, adequate soil moisture is important for optimum SQUADRON activity. The amount of rainfall or irrigation required following application depends on existing soil moisture, soil texture and organic matter content. Sufficient water to moisten the soil to a depth of 2 inches is normally adequate. If adequate moisture is not received within 7 days after treatment, then a cultivation is recommended to control escapted weeds. When adequate moisture is received after dry conditions, SQUADRON will provide residual control of susceptible germinating weeds; activity on established weeds will depend on the weed species and the location of its root system in the soil.

A SQUADRON soil treatment followed by a SCEPTER® or SCEPTER® O.T.® postemergence treatment will control certain problem weeds (see Sequential Program section). Use sequential treatments in recommended states only.

Occasionally, internode shortening of soybean plants may be observed with SQUADRON applications. This has no effect on soybeans yields.

There should be an interval of 90 days between the SQUADRON application and soybean harvest. DO NOT feed plants or straw to livestock.

Use of SQUADRON herbicide in accordance with label directions is expected to result in normal growth of rotational crops in most situations; however, various environmental and agronimic factors make it impossible to eliminate all risks associated with use of this product and therefore, rotational crop injury is always possible.

NOTE: DO NOT apply SQUADRON postemergence to soybeans as crop injury may occur.

Do not use SQUADRON other than in accordance with the instructions set forth on this label.

Do not use SQUADRON in any tank mixes other than those specifically provided in this label.

Crops other than soybeans, such as cotton, corn, grain sorghum, rice and vegetables, may be injured by spray drift or other indirect contact with SQUADRON.

To avoid injury to sensitive crops from spray drift, follow all use directions and precautions in Spraying Instructions on page 2.

To avoid injury to sensitive crops, spray equipment used for SQUADRON applications must be drained and thoroughly cleaned with water before being used to apply other products to these crops.

Apply SQUADRON prior to July 1 in the Northern Use Area, as defined in the USE AREA section of this label.

See page 5 for Rotational Crop restrictions.

Replanting: If replanting is necessary in a field previously treated with SQUADRON, the field may be replanted to soybeans. Rework the soil no deeper than the treated zone. DO NOT APPLY A SECOND TREATMENT OF SQUADRON.

MIXING INSTRUCTIONS

Fill the spray tank one-half to three-quarters full with clean water or liquid fertilizer. While agitating add the required amount of product and then fill the remainder of the tank with water or liquid fertilizer. Maintain agitation while spraying to ensure a uniform spray mixture.

When tank mixing SQUADRON with recommended herbicides, add the SQUADRON to the spray tank first and make sure it is thoroughly mixed before adding the other herbicide.

For mixing instructions for tank mixtures with Roundup or Gramoxone Super[1] see the NO-TILL SYSTEMS section of this leaflet.

SPRAYING INSTRUCTIONS

NOTE: DO NOT make applications when spray may be carried by wind to sensitive crops.

Avoid overlaps when spraying.

GROUND APPLICATIONS:

Uniformly apply with properly calibrated ground equipment in 10 to 40 gallons of water, or 20 or more gallons of liquid fertilizer per acre. A spray pressure of 20 to 40 psi is recommended.

To ensure thorough coverage, use a minimum of 20 gallons of water or liquid fertilizer per acre when applying SQUADRON to minimum or no-till soybeans. Use higher gallonage for fields with dense vegetation or heavy crop residues.

DO NOT apply with ground equipment when wind velocity is greater than 10 mph.

AERIAL APPLICATIONS:

Uniformly apply with properly calibrated aerial equipment in 5 or more gallons of water per acre.

To avoid injury to sensitive crops from drift, aerial applicators must adhere to the following SPECIAL AERIAL USE DIRECTIONS AND PRECAUTIONS:

- Use nozzles which produce a coarse spray.
- Nozzle height above ground must be a maximum of 10 feet.
- Nozzles must be pointed toward the rear of the aircraft. The downward angle of the nozzle should not be greater than 20 degrees.
- To minimize wing-tip vortex roll, nozzles or spray boom must not be located any closer to end of wing or rotor than three-fourths the distance from the center of the aircraft.
- Use a maximum spray pressure of 40 PSI
- A buffer zone must be established between the area to be sprayed and sensitive crops.
- DO NOT spray when wind velocity is greater than 5 mph.

Applicator is responsible for any loss or damage which results from spraying SQUADRON in a manner other than recommended in this label. In addition, applicator must follow all applicable state and local regulations and ordinances in regard to spraying.

[1] Gramoxone Super is a trademark of ICI Americas, Inc.

APPLICATIONS WITH LIQUID FERTILIZERS

SQUADRON can be applied in liquid fertilizers, alone or in combination with PROWL or Command. Follow all SQUADRON label recommendations regarding incorporation, timing of application, special instructions and precautions. Apply treatments in 20 or more gallons of liquid fertilizer per acre with ground equipment.

All individual state regulations relating to fluid fertilizer mixing, registration, labeling and application are the responsibility of the individual and/or company selling the SQUADRON/liquid fertilizer mixture.

LIQUID FERTILIZER COMPATIBILITY DETERMINATIONS

If a liquid fertilizer and herbicide(s) mixture separates in the spray tank, clogged equipment and uneven application can result, which can cause poor weed control and crop injury. Always predetermine the compatibility of SQUADRON alone or with PROWL or Command 4EC in the specific liquid fertilizer to be used according to the following directions:

1. Add 1 pint of fertilizer to each of 2 one-quart jars.
2. Add 1/2 teaspoon of adjuvant* to one jar.
3. (a) When using SQUADRON alone, add to each jar the correct amount of SQUADRON as specified in the table below.
   (b) When using SQUADRON plus PROWL or Command, first add the specified quantity of SQUADRON, then add the correct amount of PROWL or Command.
4. Close both jars and shake thoroughly for 10 seconds. Let them stand for 30 minutes and then observe the results. Look for signs of separation, an oily layer or globules, sludge, flakes or their precipitates.
5. Determine Compatibility
   (a) If the mixture without adjuvant does not separate, use this mixture in your spray tank.
   (b) If the mixture with adjuvant does not separate, but the one without adjuvant separates, use the adjuvant mixture in your spray tank. Add the adjuvant to the liquid fertilizer as directed on the manufacturer's label.
   (c) If either mixture separates, but mixes readily with shaking, the mixture can be used provding good agitation is maintained in the spray tank.
   (d) If separation of the mixture occurs, and agitation and/or adjuvant does not correct this problem, DO NOT use the herbicide(s) in that specific liquid fertilizer.

* Compex-Kalo Laboratories Inc., Kansas City, MO; Sponto 168 D-Witco Chemical Co., Houston, TX; Unite-Hopkins Agricultural Chemical Co., Madison, WI; or other comparable adjuvants.

Teaspoons of Specified Herbicide to be Added to 1 Pint of Liquid Fertilizer Solution

Gallons of Liquid Fertilizer to Be Applied per Acre	PROWL	SQUADRON	Command
20	2/3	2	1/3
30	1/2	1 1/2	1/4
40	1/3	1	1/6

PREPLANT INCORPORATED APPLICATIONS WITH DRY BULK FERTILIZERS

SQUADRON may be impregnated on dry bulk fertilizers. When applied as directed, SQUADRON/dry bulk fertilizer mixtures provide weed control equal to that provided by the same rates of SQUADRON applied in water or liquid fertilizer.

Follow all SQUADRON label recommendations regarding incorporation, special instructions and precautions. Apply SQUADRON/dry bulk fertilizer mixtures only with ground equipment.

All individual state regulations relating to dry bulk fertilizer blending, registration, labeling, and application are the responsibility of the individual and/or company selling the SQUADRON/dry bulk fertilizer mixture.

A minimum of 200 pounds and a maximum of 450 pounds of dry bulk fertilizer impregnated with the recommended amount of SQUADRON must be applied per acre.

DO NOT impregnate SQUADRON onto coated ammonium nitrate or limestone because these materials will not absorb the herbicide. Dry fertilizer blends containing mixtures of ammonium nitrate or limestone may be impregnated with SQUADRON. A minimum of 200 pounds of impregnated dry bulk fertilizer, excluding the weight of ammonium nitrate or limestone, must be applied per acre.

Apply SQUADRON at the rate of 3 pints per acre. Use the following table to determine the amount of SQUADRON to be impregnated on a ton of dry bulk fertilizer based on the rate of fertilizer which will be applied per acre.

Rate Chart for Impregnation of Dry Bulk Fertilizer with SQUADRON (Pints of SQUADRON per Ton of Fertilizer)

SQUADRON Rate Per Acre	Fertilizer Rate lbs/acre	Pints Per Ton
3 Pints	200	30
	250	24
	300	20
	350	17
	400	15
	450	13 1/3

For those rates not listed in this table, calculate the pints of SQUADRON to be impregnated on a ton of dry bulk fertilizer using the following formula:

$$\frac{200}{\text{Pounds of dry fertilizer per acre}} \times \text{3 Pints of SQUADRON per acre (recommended rate)} = \text{Pints of SQUADRON per ton of fertilizer}$$

To impregnate SQUADRON on bulk fertilizer, use a closed rotary-drum mixer or other commonly used dry bulk fertilizer blender equipped with suitable spray equipment. Spray nozzles must be placed to provide uniform coverage of SQUADRON onto the fertilizer during mixing.

Apply the SQUADRON/dry bulk fertilizer mixture with an accurately calibrated dry fertilizer spreader. The SQUADRON/dry bulk fertilizer mixture must be spread uniformly on the soil surface. Uneven spreading can cause poor weed control and crop injury.

Refer to Preplant Incorporated Applications section of this label for incorporation directions.

USE RATE

Apply SQUADRON at a rate of 3.0 pints per acre.

WEEDS CONTROLLED

The following grass and broadleaf weeds are controlled by recommended treatments of SQUADRON:

GRASSES

Barnyardgrass (*Echinochloa crus-galli*)
Crabgrass (*Digitaria* spp.)
Crowfootgrass (*Dactyloctenium aegyptium*)
Cupgrass, Woolly** (*Eriochloa villosa*)
Foxtail
  Giant (*Setaria faberi*)
  Green (*Setaria viridis*)
  Robust (*Setaria* spp.)
  Yellow (*Setaria glauca*)
Goosegrass (*Eleusine indica*)
Johnsongrass, seedling (*Sorghum halepense*)
Panicum
  Browntop (*Panicum fasciculatum*)
  Fall (*Panicum dichotomiflorum*)
  Texas (*Panicum texanum*)
Sandbur, Field (*Cenchrus incertus*)
Shattercane*** (*Sorghum bicolor*)
Signalgrass, Broadleaf (*Brachiaria platyphylla*)
Witchgrass (*Panicum capillare*)

BROADLEAF WEEDS

Bristly Starbur (*Acanthospermum hispidium*)
Burcucumber* (*Sicyos angulatus*)
Carpetweed (*Mollugo verticillata*)
Cocklebur, Common (*Xanthium strumarium*)
Jimsonweed (*Datura stramonium*)
Kochia (*Kochia scoparia*)
Lambsquarters, Common (*Chenopodium album*)
Mallow, Venice (*Hibiscus trionum*)
Morningglory
  Palm Leaf (*Ipomoea wrightii*)
  Pitted (*Ipomoea lacunosa*)
  Smallflower (*Jacquemontia tamnifolia*)
Mustard Species (*Brassica* spp.)
Nightshade, Eastern Black* (*Solanum ptycanthum*)
Pigweed
  Palmer (*Amaranthus palmeri*)
  Redroot (*Amaranthus retroflexus*)
  Smooth (*Amaranthus hybridus*)
  Spiny (*Amaranthus spinosus*)
  Waterhemp, Tall (*Amaranthus tuberculatus*)
Poinsettia, Wild (*Euphorbia heterophylla*)
Puncturevine (*Tribulus terrestris*)
Purslane (*Portulaca oleracea*)
Pusley, Florida (*Richardia scabra*)
Ragweed
  Common (*Ambrosia artemisiifolia*)
  Giant* (*Ambrosia trifida*)
Redweed* (*Melochia corchorifolia*)
Sida, Prickly (Teaweed) (*Sida spinosa*)
Smartweed
  Ladysthumb (*Polygonum persicaria*)
  Pennsylvania (*Polygonum pensylvanicum*)
Spurge
  Annual (*Euphorbia* spp.)
  Prostrate (*Euphorbia humistrata*)
  Spotted (*Euphorbia maculata*)

Sunflower, Common (*Helianthus annuus*)
Velvetleaf* (*Abutilon theophrasti*)

SQUADRON will aid in the control and reduce competition from:

Beggarweed, Florida (*Desmodium tortuosum*)
Mexicanweed (*Caperonia castanifolia*)
Morningglory
Entireleaf (*Ipomea hederacea var. intergriuscula*)
Ivyleaf (*Ipomoea hederacea*)
Tall (*Ipomoea purpurea*)
Nutsedge, Yellow* (*Cyperus esculentus*)
Sicklepod (*Cassia obtusifolia*)

* Preplant incorporated treatments only.
** Preplant incorporated treatments only; for best results, use a SQUADRON plus PROWL tank mixture.
*** Preplant incorporated treatments only; see SQUADRON plus PROWL section of this label for use directions.

ROTATIONAL CROP RESTRICTIONS

The following rotational crops may be planted after applying SQUADRON in soybeans:

1. Wheat may be planted in the fall, four months after SQUADRON application in soybeans. Rice may be planted in the spring of the year following SQUADRON application. (See Northern Use Area Restrictions*).
2. Eleven months after SQUADRON application:
   Barley (See Northern Use Area Restrictions*)
   Edible Beans
   Field Corn (See Northern Use Area* or Southern Use Area Restrictions**)
   Grain Sorghum
   Oats (See Northern Use Area Restrictions*)
   Peanuts
   Tobacco
3. Sugarbeets may be planted 26 months after SQUADRON application in soybeans.

***NORTHERN USE AREA RESTRICTIONS:**

For the Northern Use Area as defined in the USE AREA section of this label, DO NOT plant field corn, wheat, barley, or oats as rotational crops in the fall or in the spring of the year following SQUADRON application.

****SOUTHERN USE AREA RESTRICTIONS:**

For the Southern Use Area, as defined in the USE AREA section of this label; field corn may be planted as a rotational crop in the spring of the year following SQUADRON application unless extreme drought conditions develop or less than 15 inches of rainfall or irrigation is received from the date of SQUADRON application through October of the year of application. For eastern Oklahoma, Arkansas, the bootheel of Missouri, Tennesseee, North Carolina and states south: field corn may be planted in the spring of the year following SQUADRON application unless extreme drought conditions develop and less than 15 inches of rainfall or irrigation is received within 6 months following the date of application. If postemergence applications of SCEPTER or SCEPTER O.T. follow a SQUADRON treatment, the 6 month interval begins after the last postemergence application.

ONLY rotational crops harvested at maturity may be used for feed or food.

DO NOT plant other rotational crops prior to 18 months after last SQUADRON application.

Use of SQUADRON herbicide in accordance with label directions is expected to result in normal growth of rotational crops in most situtations; however, various environmental and agronomic factors make it impossible to eliminate all risks associated with use of this product and, therefore, rotational crop injury is always possible.

DO NOT apply products containing chlorimuron ethyl (e.g. Classic, Canopy, Gemini, Lorox Plus, Preview[2], etc.), imazaquin (e.g. SCEPTER®, ALA-SCEPT®, TRI-SCEPT®, etc [refer to the Sequential Program section of this label for exceptions]), or imazethapyr (e.g. PURSUIT® or PURSUIT® PLUS) the same year as SQUADRON or injury to follow crops may occur.

DO NOT graze or feed treated soybean forage, hay or straw to livestock.

[2] Classic, Canopy, Preview, Gemini and Lorox Plus are trademarks of E.I. du Pont de Nemours and Company.

APPLICATION INSTRUCTIONS

EARLY PREPLANT APPLICATIONS

SQUADRON may be applied up to 30 days prior to planting in Alabama, Arkansas, Florida, Georgia, Louisiana, Mississippi, North Carolina, Oklahoma, South Carolina, Tennessee, and Texas. For the rest of the Southern Use Area and all of the Northern Use Area, SQUADRON may be applied up to 45 days prior to planting. Incorporate uniformly into the top 1 to 2 inches of soil within 7 days after application. See PREPLANT INCORPORATED APPLICATIONS section for incorporation instructions. Adequate moisture is required for activation of SQUADRON.

If soybeans are planted on beds, apply and incorporate after bed formation.

PREPLANT INCORPORATED APPLICATIONS

Apply SQUADRON or its PROWL tank mixture before planting and incorporate uniformly into the top 1 to 2 inches of soil. Mechanical incorporation can be achieved by the following:

(a) Field cultivator with 3 to 4 rows of sweeps (Do not use chisel points), spaced at intervals of 7 inches or less and staggered so that no soil is left unturned, set to cut 3 inches deep, and operated two times at more than 5 mph with the second pass made at an angle to the first.
(b) PTO-driven equipment (tillers, cultivators, hoes) set to cut 2 inches deep and operated one time at 4 mph or less.
(c) Rolling cultivator set to cut 2 to 3 inches deep and operated two times at 6 to 8 mph. Use only on coarse textured soils.
(d) LELY-Roterra[3] set and operated according to the manufacturer's directions to thoroughly incorporate SQUADRON to a depth of 1 to 2 inches.
(e) Do-All set and operated according to the manufacturer's directions to thoroughly incorporate SQUADRON to a depth of 1 to 2 inches.
(f) Disk harrow set to cut 3 to 4 inches deep and operated in two different directions at 4 to 6 mph.

SQUADRON may be applied immediately before planting or up to 30 days prior to planting in Alabama, Arkansas, Florida, Georgia, Louisiana, Mississippi, North Carolina, Oklahoma, South Carolina, Tennessee, and Texas. For the rest of the Southern Use Area and all of the Northern Use Area, SQUADRON may be applied up to 45 days prior to planting. Incorporate within 7 days after application. Adequate moisture is required for activation of SQUADRON.

If soybeans are planted on beds, apply and incorporate after bed formation.

[3] LELY-ROTERRA is a trademark of Lely.

PREEMERGENCE APPLICATIONS

MINIMUM AND CONVENTIONAL TILLAGE:

Apply SQUADRON or its PROWL tank mixture during or up to 2 days after planting. DO NOT apply postemergence to soybeans. As with other herbicides applied preemergence, rainfall or irrigation is necessary to activate SQUADRON in the soil. If sufficient moisture to activate SQUADRON is not received within 7 days after application, a thorough shallow tillage or cultivation is recommended (preferably with rotary hoe) for optimum weed control.

NO-TILL SYSTEMS:

Apply SQUADRON or its PROWL tank mixture during or up to 2 days after planting. DO NOT apply postemergence to soybeans. If needed, apply Gramoxone Super or Roundup[4] in combination with SQUADRON. Refer to the respective labels for rates, methods of application, proper timing, weeds controlled restrictions, and precautions. Use ground equipment only.

When SQUADRON is tank mixed with Gramoxone Super or Roundup, add SQUADRON to the tank first, next add PROWL (if used), then add Gramoxone Super or Roundup. When Gramoxone Super is included in a tank mixture, add 8 oz. of non-ionic surfactant per 100 gallons of spray mixture as the last ingredient in the tank.

[4] Roundup is a trademark of Monsanto Agricultural Products Co.

HERBICIDE COMBINATIONS

SQUADRON should be used only in accordance with the directions on this label. Do not use SQUADRON in any tank mixes other than those specifically provided in this label.

SQUADRON plus PROWL® herbicide

Tank Mixture

PROWL can be tank mixed with SQUADRON and applied preplant incorporated or preemergence if heavy or difficult to control infestations are anticipated from grasses listed in this leaflet. SQUADRON plus PROWL must be applied preplant incorporated if shattercane infestations are anticipated.

On coarse textured soils add PROWL at 0.5 pint per acre and on medium and fine textured soil and PROWL at 0.5 to 1 pint per acre.

See the PROWL label for soil types classified as coarse, medium and fine textured.

Apply this tank mixture preplant incorporated north of Interstate 80.

Observe all precautions and limitations on the PROWL Label.

® Registered trademarks of Amercian Cyanamid Company

SQUADRON plus Command[5] 4EC

Tank Mixture

Observe all precautionary statements in the Command label before using.

The SQUADRON plus Command 4EC preplant incorporated tank mixture may be used when heavy infestations of velvetleaf are anticipated.

Uniformly apply the tank mixture with properly calibrated ground equipment in 10 to 40 gallons of water per acre. Use sprayers equipped with nozzles that provide accurate and uniform application.

NOTE: The use of an agriculturally approved drift reducing agent is required at spray volumes of 10 to 15 gallons per acre.

Apply 3 pints per acre of SQUADRON plus 0.25 to 0.5 pints per acre of Command 4EC to the soil surface and uniformly incorporate. Refer to the Command label for incorporation instructions. DO NOT apply aerially or through irrigation systems. DO NOT apply this tank mixture to overly moist or wet soils.

Command 4EC is a volatile compound. Off-site movement of spray drift or vapors of Command can cause foliar whitening or yellowing of some plants. Prior to using Command, read and strictly follow all precautions and application instructions on the Command 4 EC label.

When applied as directed, tank mixtures of SQUADRON plus Command control heavy infestations of velvetleaf as well as the weeds listed in the WEEDS CONTROLLED section of this label.

Apply SQUADRON at a rate of 3 pints per acre in combination with Command 4EC at a rate of 0.25 to 0.5 pints per acre.

Follow all rotational crop restrictions on the SQUADRON and Command labels. Always follow the most restrictive label.

In the event of a crop loss due to weather conditions, soybeans can be replanted. DO NOT work the soil deeper than 2 inches.

[5] Command is a trademark of FMC Corporation.

SEQUENTIAL PROGRAM

SQUADRON followed by SCEPTER® herbicide Postemergence

A preplant incorporated or preemergence application of SQUADRON may be followed by a single postemergence application of SCEPTER to control certain problem weeds.

Apply SCEPTER at 1/3 to 2/3 pints per acre as a sequential postemergence treatment following either a preplant incorporated or preemergence treatment of SQUADRON. The sequential postemergence treatment should be applied before weed species exceed the 1 to 2 true leaf stage.

Follow use directions for postemergence treatments on the SCEPTER label.

This sequential application can be used in Alabama, Arkansas, Florida, Georgia, Louisiana, Mississippi, North Carolina, Oklahoma, South Carolina, Tennessee, and Texas only.

When applied as directed, SQUADRON preplant incorporated or preemergence followed by SCEPTER postemergence at a rate of 2/3 pints per acre will control the following broadleaf weeds:

- Beggarweed, Florida *(Desmodium tortuosum)*
- Mexicanweed *(Caperonia castaniifolia)*
- Sicklepod *(Cassia obtusifolia)*
- Texasweed *(Caperonia palustris)*

SQUADRON followed by SCEPTER® O.T.® Postemergence

A preplant incorporated or preemergence application of SQUADRON may be followed by a single postemergence application of SCEPTER O.T. to control morningglories and hemp sesbania *(Sesbania exaltata).*

Apply SCEPTER O.T. at 1 pint per acre as a sequential postemergence treatment following either a preplant incorporated or preemergence treatment of SQUADRON.

Follow use directions on the SCEPTER O.T. label. See the SCEPTER O.T. label for weed height limitations.

This sequential application can be used in Alabama, Arkansas, Florida, Georgia, Louisiana, Mississippi, North Carolina, Oklahoma, South Carolina, Tennessee, and Texas only.

For postemergence control of other weeds following a soil application of SQUADRON, refer to the SCEPTER or SCEPTER O.T. labels in the Postemergence Applications section for use rates, weeds controlled, and growth stages.

USE AREA

SQUADRON can be applied only in the states or parts of states shaded in the following map:

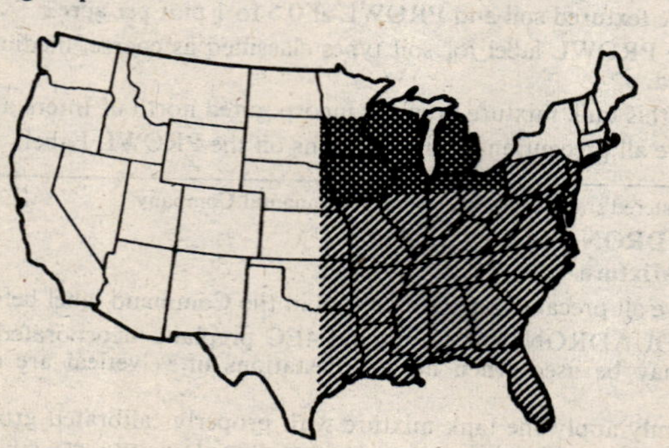

= Northern Use Area (includes: Nebraska, east of U.S. 81, north of U.S. 34; South Dakota, east of U.S. 81; Illinois, north of S. R. 116 for area west of Peoria, north of U.S. 24 for area east of Peoria; Indiana, north of U.S. 224 for area west of Huntington, north of U.S. 224 for area east of Huntington; Ohio, north of U.S. 224; and all other states or portions of states shaded in the Northern use area map).

= Southern Use Area (includes: Kansas, east of U.S. 81; and Cloud, Ellsworth, Harvey, Jewell, Lincoln, McPherson, Mitchell, Ottawa, Republic, Saline, Sedgewick and Sumner counties; southern portions of Illinois, Indiana, Ohio and eastern Nebraska not covered by Northern Use Area description; and all other states or portions of states shaded in the Southern Use Area map).

NOTE: See ROTATIONAL CROP RESTRICTIONS section for recommendations applying to the Northern vs. Southern Use Areas.

STOMP®

HERBICIDE

FOR CONTROL OF MOST ANNUAL GRASSES AND CERTAIN BROADLEAF WEEDS IN NONCROPLAND AREAS AND CHRISTMAS TREE PLANTINGS

ACTIVE INGREDIENT:	
pendimethalin, N-(1-ethylpropyl)-3,4-dimethyl-2,6-dinitrobenzenamine	42.3%
INERT INGREDIENTS:	57.7%
TOTAL	100.0%

(1 gal. contains 4 lbs. of pendimethalin)

EPA Reg. No. 241-305

® Registered Trademark of American Cyanamid Company

2/88

KEEP OUT OF REACH OF CHILDREN
WARNING!/AVISO!

PRECAUCION AL USUARIO: Si usted no lee inglés, no use este producto hasta que la etiqueta le haya sido explicada ampliamente.

FIRST AID: If in eyes, hold eyelids open and flush with steady, gentle stream of water for 15 minutes. If swallowed, DO NOT induce vomiting. Call a physician or Poison Control Center immediately. For skin contact, wash thoroughly with soap and water. Call a physician if eye or skin irritation persists.

This product contains monochlorobenzene.

NOTE TO PHYSICIAN: Because of increased risk of chemical pneumonia or pulmonary endema caused by aspiration of the hydrocarbon solvent (monochlorobenzene), vomiting should be induced only under professional supervision.

PRECAUTIONARY STATEMENTS

HAZARDS TO HUMANS

WARNING!

CAUSES SUBSTANTIAL BUT TEMPORARY EYE INJURY

HARMFUL IF SWALLOWED, ABSORBED THROUGH THE SKIN, OR INHALED

DO NOT get in eyes or on clothing. Wear goggles or face shield. Avoid breathing spray mist. Avoid contact with skin. Wash skin thoroughly with soap and water after handling. Use with adequate ventilation. Keep container closed. Remove contaminated clothing and wash before reuse.

ENVIRONMENTAL HAZARDS

This product is toxic to fish. DO NOT apply directly to water. Drift and runoff from treated areas may be hazardous to fish in neighboring areas. DO NOT contaminate water when disposing of equipment washwaters.

PHYSICAL HAZARDS

FLAMMABLE LIQUID AND VAPOR

DO NOT use, pour, spill, or store near heat or open flame.
DO NOT use in the vicinity of pilot lights.

GENERAL INFORMATION

STOMP is recommended for preemergence control of most annual grasses and certain broadleaf weeds as they germinate on noncropland areas such as railroad, utility and pipeline rights-or-way, highway guardrails, delineators, and sign posts, utility substations, petroleum tank farms, pumping installations, fence rows, storage areas, and other similar areas. STOMP can also be used in such areas without injury to the ornamental species listed on this label. STOMP may also be used for preemergence weed control in Christmas tree plantings. STOMP treatments are most effective in controlling weeds when adequate rainfall is received within 30 days after application.

STOMP will not control established weeds. Therefore, areas to be treated should be free of established weeds at the time of treatment, or STOMP may be used in conjuction with herbicides registered for postemergence use in noncropland areas. Consult the labels of those herbicides for suggested treatments, rates to be used and precautions or restrictions for use in noncropland areas.

DIRECTIONS FOR USE

It is a violation of Federal law to use this product in a manner inconsistent with its labeling. This labeling must be in the possession of the user at time of pesticide application. DO NOT apply this product through any type of irrigation system.

Observe all cautions and limitations in this label and the labels of products used in combination with STOMP. The use of STOMP not consistent with this label can result in injury to crops, animals, or persons. Keep containers closed to avoid spills and contamination.

STORAGE AND DISPOSAL

STORAGE: DO NOT STORE BELOW 40°F. Extended storage at temperature below 40°F can result in the formation of crystals on the bottom of container. If crystallization does occur, store the container on its side at room temperature (70°F) and rock occasionally until crystals redissolve.

DO NOT contaminate water, food, or feed by storage or disposal.

DISPOSAL: Pesticide wastes are toxic. Improper disposal of excess pesticide, spray mixture, or rinsate is a violation of Federal law. If these wastes cannot be disposed of by use according to label instructions, contact your State Pesticide or Environmental Control Agency, or the Hazardous Waste representative at the nearest EPA Regional Office for guidance.

CONTAINER DISPOSAL: Triple rinse (or equivalent). Then offer for recycling or reconditioning, or puncture and dispose of in a sanitary landfill, by incineration or, if allowed by state and local authorities, by burning. If burned, stay out of smoke.

DISCLAIMER

The label instructions for use of this product reflect the opinion of experts based on field use and tests. The directions are believed to be reliable and should be followed carefully. However, it is impossible to eliminate all risks inherently associated with use of this product. Crop injury, ineffectiveness or other unintended consequences may result because of such factors as weather conditions, presence of other materials, or the use or application of the product contrary to label instructions, all of which are beyond the control of Americam Cynanamid Company. All such risks shall be assumed by the user.

American Cyanamid Company warrants only that the material contained herein conforms to the chemical description on the label and is reasonably fit for the use therein described when used in accordance with the directions for use, subject to the risk referred to above.

Any damages arising from a breach of this warranty shall be limited to direct damages and shall not include consequential commercial damages such as loss of profits or values or any other special or indirect damages.

American Cyanamid Company makes no other express or implied warranty, including other express or implied warranty of FITNESS or of MERCHANTABILITY.

MIXING INSTRUCTIONS

1. Fill tank one-half to three-quarters full with clean water.
2. Add STOMP to the partially filled tank while agitating and then fill the remainder of the tank with water.
3. MAINTAIN CONTINUOUS AGITATION WHILE ADDING STOMP AND UNTIL SPRAYING IS COMPLETED. If the spray mixture is allowed to settle for any period of time, thorough agitation is essential to resuspend the mixture before spraying is resumed. Continue agitation while spraying.
4. If STOMP is to be used in tank mixtures with other registered herbicides, then follow directions on the labels of those products which recommend tank mixing.

SPRAYING INSTRUCTIONS

Uniformly apply with properly calibrated ground equipment in sufficient water per acre to uniformly treat the area with a spray pressure of 25 to 50 psi. Maintain continuous agitation during spraying with good mechanical or bypass agitation.

NONCROPLAND WEED CONTROL

For preemergence control of the weed species listed, apply STOMP at the following rates:

Length of Control	STOMP Rate
Short Term Control (*2-4 months*)	2 Quarts/Acre
Long Term Control (*6-8 months*)	4 Quarts/Acre

STOMP will not control established weeds.

For post-emergence control of weeds, tank-mix combinations of STOMP plus ARSENAL® herbicide are recommended. **DO NOT** tank mix with ARSENAL in California. Recommended rates for the tank-mix compounds should be determined from the product labels of both STOMP and ARSENAL prior to use.

WEED CONTROL IN AND AROUND ORNAMENTAL PLANTINGS ON INDUSTRIAL AND RIGHTS OF WAY SITES

STOMP may be used for the preemergence control of the weed species listed in and around the following established ornamental plants:

Scientific	Common Name
Arctotheca calendula	Capeweed
Carpobrotus edule	Large Leaf Iceplant
Gazania splendens	
Hedera helix	English Ivy
Hypericum calycinum	Rose-of-Sharon
Myoporum parviflorum	Myoporum
Osteospermum fruitcosum	African Trailing Daisy
Potentilla verna	

PRECAUTION!

Apply only to established plantings.

DO NOT apply STOMP in ornamentals nurseries.

WEED CONTROL IN CHRISTMAS TREE PLANTINGS

Do not use in California. STOMP is recommended for preemergence control of the weed species listed on this label as they germinate in areas planted with the following Christmas trees:

- Austrian Pine (*Pinus nigra*)
- Balsam Fir (*Abies balsamae*)
- Colorado Blue Spruce (*Picea pungens*)
- Douglas Fir (*Pseudotsuga taxifolia*)
- Scotch Pine (*Pine sylvestris*)
- Virginia Pine (*Pinus virginiana*)
- White Fir (*Abies concolor*)
- White Spruce (*Picea glauca*)

DO NOT apply STOMP to Christmas tree seed or nursery beds. Apply STOMP only to established Christmas tree plantings.

STOMP will not control established weeds. Therefore, areas to be treated should be free of established weeds at the time of treatment, or STOMP may be used in conjunction with herbicides registered for postermergence use in Christmas trees. Consult the labels of those herbicides for suggested treatments, rates to be used and precautions or restrictions for use in Christmas trees.

For preemergence control of the weed species listed, apply STOMP at the following rates:

Length of Control	STOMP Rate
Short Term Control (*2-4 months*)	2 Quarts/Acre
Long Term Control (*6-8 months*)	4 Quarts/Acre

STOMP will not control established weeds.

The efficacy of STOMP will be improved if the application is followed by one-half inch of rainfall or its equivalent in sprinkler irrigation.

WEED SPECIES CONTROLLED

The following grass and broadleaf weeds are controlled by preemergence treatments of STOMP herbicide at the rates recommended:

GRASSES CONTROLLED

- Barnyardgrass (*Echinochloa crus-galli*)
- Bluegrass, Annual (*Poa annua*)
- Crabgrass (*Digitaria spp.*)
- Crowfootgrass (*Dactyloctenium aegyptium*)
- Foxtail, Giant (*Setaria faberi*)
- Foxtail, Green (*Setaria viridis*)
- Foxtail, Yellow (*Setaria lutescens*)
- Goosegrass (*Eleusine indica*)
- Itchgrass (*Rottboellia exaltata*)
- Johnsongrass (from seed) (*Sorghum halepense*)
- Junglerice (*Echinochloa colonum*)
- Lovegrass (*Eragrostis* spp.)
- Panicum, Browntop (*Panicum fasciculatum*)
- Panicum, Fall (*Panicum dichotomiflorum*)
- Panicum, Texas (*Panicum texanum*)
- Sandbur, Field (*Cenchrus incertus*)
- Signalgrass (*Brachiaria platphyila*)
- Sprangletop, Mexican (*Leptochloa uninervia*)
- Sprangletop, Red (*Leptochloa filiformis*)
- Witchgrass (*Panicum capillare*)
- Woolly Cupgrass (*Eriochloa villosa*)

BROADLEAF WEEDS CONTROLLED

Carpetweed (*Mollugo verticillata*)
Chickweed, Common (*Stellaria media*)
Chickweed, Mouseear (*Cerastium vulgatum*)
Clover, Hop (*Trifolium procumbens*)
Cudweed (*Gnaphalium* spp.)
Fiddleneck (*Amsinckia intermedia*)
Filaree (*Erodium* spp.)
Henbit (*Lamium amplexicaule*)
Knotweed (prostrate) (*Polygonum aviculare*)
Kochia (*Kochia scoparia*)
Lambsquarters (*Chenopodium album*)
Pigweed (*Amaranthus* spp.)
Primrose, Evening (*Oenothera* spp.)
Puncturevine (*Tribulus terrestris*)
Purslane (*Portulaca oleracea*)
Pusley, Florida (*Richardia scabra*)
Rocket, London (*Sisymbrium irio*)
Sheperdspurse (*Capsella bursa-pastoris*)
Smartweed, Pennsylvania (*Polygonum pensylvanicum*)
Spurge, Annual (*Euphorbia* spp.)
Spurge, Prostrate (*Euphorbia supina*)
Woodsorrel, Yellow (*Oxalis stricta*)
Velvetleaf (Buttonweed) (*Abutilon therphrasti*)

THIMET® 15-G

SOIL AND SYSTEMIC INSECTICIDE

RESTRICTED USE CLASSIFICATION

Due to acute oral and dermal toxicity and avian hazards

For retail sale to, and use only by Certified Applicators or persons under their direct supervision and only for those uses covered by the Certified Applicator's certification.

American Cyanamid Company endorses Certification to promote the responsible use of pesticides to insure the protection of man and the environment.

ACTIVE INGREDIENT:	
Phorate (O,O-diethyl S-[(ethylthio) methyl] phosphorodithioate)	15.0%
INERT INGREDIENTS:	85.0%
TOTAL	100.0%

EPA Reg. No. 241-145
7/89

® Registered Trademark of American Cyanamid Company

KEEP OUT OF REACH OF CHILDREN

DANGER–POISON

¡PELIGRO!

PRECAUCION AL USUARIO: Si usted no lee inglés, no use este producto hasta que la etiqueta le haya sido explicada ampliamente.

FIRST AID

If swallowed, drink one or two glasses of water and induce vomiting by touching back of throat with finger. Do not induce vomiting or give anything by mouth to an unconscious person. Avoid alcohol. Get medical attention.

If inhaled, remove to fresh air. If not breathing, give artificial respiration, preferably mouth-to-mouth. If breathing is difficult, give oxygen. Get medical attention.

If on skin, wash thoroughly with soap and water. Remove contaminated clothing and shoes. Wash clothing and decontaminate shoes before reuse.

If in eyes, immediately flush eyes with plenty of water. Get medical attention. Antidote: Atropine is an antidote.

NOTE TO PHYSICIANS: Warning symptoms include weakness, headache, tightness in chest, blurred vision, nonreactive pinpoint pupils, salivation, nausea, vomiting, diarrhea, and abdominal cramps. Give atropine intramuscularly or intravenously, depending on severity of poisoning, 2 to 4 milligrams every 10 minutes until fully atropinized as shown by dilated pupils, dry flushed skin and tachycardia. Twenty to thirty milligrams, or more, may be required during the first 24 hours. Never give opiates or phenothiazine tranquilizers. Clear chest by postural drainage. Artificial respiration or oxygen administration may be necessary. Observe patient continuously for at least 48 hours. Allow no further exposure to any cholinesterase inhibitor until cholinesterase regeneration has taken place as determined by blood tests.

Pralidoxime chloride (2-PAM; PROTOPAM chloride) may be effective as an adjunct to atropine. Use according to label directions.

IN CASE OF AN EMERGENCY ENDANGERING LIFE OR PROPERTY INVOLVING THIS PRODUCT, CALL COLLECT, DAY OR NIGHT, AREA CODE 201-835-3100.

PRECAUTIONARY STATEMENTS

HAZARDS TO HUMANS AND DOMESTIC ANIMALS

DANGER!

Fatal if swallowed, inhaled or absorbed through the skin. Do not breathe dust. Do not get in eyes, on skin, or on clothing.

Wear freshly laundered, long-sleeved work clothing daily. While transferring from package to equipment, wear a clean cap, gloves (rubber or cotton) and goggles. If cotton gloves are used, they must be laundered or discarded after each day's use. Rubber gloves should be washed with soap and water after each use. Do not wear the same gloves for other work. Destroy and replace gloves frequently.

In case of contact, immediately remove contaminated clothing and wash skin thoroughly with soap and water. Launder clothing and decontaminate shoes before reuse. Wash thoroughly with soap and water before eating or smoking. Bathe at the end of the work day and change clothing.

Do Not Breathe Dust

Wear a face mask or other respiratory equipment while emptying bags of product into a hopper. While emptying bags into equipment, pour downwind and allow as little free fall as possible. Do not pour at face level and do not allow dust to reach the breathing zone.

Do Not Contaminate Food or Feed Products

Once a bag has been opened, use it completely. Make sure the hoppers are emptied while still in the field. Refer to STORAGE AND DISPOSAL statement for further instructions.

Keep all unprotected persons out of operating areas.

Do not apply this product in such a manner as to directly or through drift expose workers or other persons.

Keep Out of Reach of Domestic Animals.

Not for Use or Store In or Around the Home.

ENVIRONMENTAL HAZARDS

This pesticide is extremely toxic to fish and wildlife. Birds feeding in treated areas may be killed. Do not apply directly to water or wetlands (swamps, bogs, marshes, and pot-holes). Drift and runoff may be hazardous to aquatic organisms in neighboring areas. Collect or incorporate granules that are spilled during loading or are visible on soil surface in turn areas. Do not contaminate water when disposing of equipment washwaters.

REENTRY

After dusts have settled, do not enter or allow entry into the treated area until the 48 hour reentry interval has expired, unless the person entering the treated area is wearing the personal protective equipment listed below for early reentry.

For early reentry into treated areas wear a protective suit of one or two pieces covering all parts of the body except head, hands, and feet; chemical-resistant gloves and chemical resistant shoes, coverings, or boots.

Because certain states may require more restrictive re-entry intervals for various crops treated with this product, consult your State Department of Agriculture for further information.

Written or oral warnings must be given to workers who are expected to be in a treated area or in an area about to be treated with this product. When oral warnings are given, warnings shall be given in a language customarily understood by workers. Oral warnings must be given if there is reason to believe that written warnings cannot be understood by workers. Written warnings must include the following information: DANGER. Area treated with THIMET on (date of application). Do not enter without protective clothing.

PERSONAL PROTECTIVE EQUIPMENT

WEAR THE FOLLOWING PROTECTIVE CLOTHING DURING LOADING, APPLICATION, EQUIPMENT REPAIR/CLEANING, AND DISPOSAL OF THE PESTICIDE. A protective suit of one or two pieces covering all parts of the body except head, hands, and feet; chemical resistant gloves; chemical resistant shoes, shoe covering, or boots.

Unless a chemical resistant protective suit is otherwise required, mixers/loaders must wear a chemical resistant apron and goggles or a face shield.

IF APPLICATION IS PERFORMED USING AN ENCLOSED CAB OR COCKPIT, THE FOLLOWING PROTECTIVE CLOTHING AND EQUIPMENT MAY BE WORN AS AN ALTERNATE: Long-sleeved shirt and long-legged pants; shoes and socks. Chemical resistant gloves must be available in the cab or cockpit and must be worn when exiting. This clothing is inadequate to protect you during equipment repair or cleaning, reentry, or pesticide disposal work. Do not allow contaminated clothing to come in contact with the insides of the cab.

IMPORTANT! If pesticide comes in contact with skin, wash off with soap and water. Always wash hands, face, and arms with soap and water before smoking, eating, drinking, or using the toilet.

AFTER WORK: Before removing gloves, wash them with soap and water. Take off all work clothes and shoes. Shower using soap and water. Wear clean clothes. Do not reuse contaminated clothing. Personal clothing worn during work must be laundered separately from household articles. Store protective clothing separately from personal clothing. Clean or launder protective clothing after each use. Clothing and protection equipment heavily contaminated must be destroyed according to state and local regulations. HEAVILY CONTAMINATED CLOTHING CANNOT BE ADEQUATELY DECONTAMINATED.

WORK SAFETY RULE

REPEATED EXPOSURES TO CHOLINESTERASE INHIBITORS SUCH AS ARE CONTAINED IN THIS PRODUCT MAY, WITHOUT WARNING, CAUSE PROLONGED SUSCEPTIBILITY TO VERY SMALL DOSES OF ANY CHOLINESTERASE INHIBITOR.

Persons working with this product should have frequent blood tests of their cholinesterase levels. If the cholinesterase level falls below a critical point, no further exposures should be allowed until it has been determined by means of blood tests that the cholinesterase level has return to normal. Before using this product consult the National Pesticide Telecommunications Network for recommendations regarding such blood tests, poisoning management, emergency treatment, and other information regarding the toxicity of phorate. The toll free number for the National Pesticide Telecommunication Network is 1-800-858-7378.

If handled indoors provide mechanical exhaust ventilation.

Keep all unprotected persons, children, livestock, and pets away from treated areas or where there is danger of drift.

Do not rub eyes or mouth with hands. If feel sick in any way, STOP work and get help right away, see First Aid section.

CALL A PHYSICIAN AT ONCE IN ALL CASES OF SUSPECTED POISONING.

DIRECTIONS FOR USE

It is a violation of Federal law to use this product in a manner inconsistent with its labeling.

This label must be in the possession of the user at the time of pesticide application.

STORAGE AND DISPOSAL

Storage Store pesticide products in a secure locked area where children, unauthorized persons and animals cannot enter. Do not store in the same area with food or feed. Do not store opened bags.

Prohibitions Do not contaminate water, food or feed by storage or disposal. Open dumping is prohibited. Cover or incorporate spills.

Pesticide Disposal Pesticide wastes are acutely hazardous. Improper disposal of excess pesticide, spray mixture, or rinsate is a violation of Federal law. If these wastes cannot be disposed of by use according to label instructions, contact your state pesticide or Environmental Control Agency, or the hazardous waste representative at the nearest EPA Regional Office for guidance.

Container Disposal Completely empty bag into application equipment. Then dispose of empty bag in a sanitary landfill or by incineration, or, if allowed by State and local authorities, by burning. If burned, stay out of smoke.

General Consult Federal, State or local disposal authorities for approved alternative procedures such as limited open burning.

DISCLAIMER

The label instructions for the use of this product reflect the opinion of experts based on field use and tests. The directions are believed to be reliable and should be followed carefully. However, it is impossible to eliminate all risks inherently associated with use of this product. Crop injury, ineffectiveness or other unintended consequences may result because of such factors as weather conditions, presence of other materials, or the use or application of the product contrary to label instructions, all of which are beyond the control of American Cyanamid Company. All such risks shall be assumed by the user.

American Cyanamid Company warrants only that the material contained herein conforms to the chemical description on the label and is reasonably fit for the use therein when used in accordance with the directions for use, subject to the risk referred to above. Any damages arising from a breach of this warranty shall be limited to direct damages and shall not include consequential commercial damages such as loss of profits or values or any other special or indirect damages.

American Cyanamid Company makes no other express or implied warranty, including any other express or implied warranty of FITNESS or of MERCHANTABILITY.

BEFORE USING, READ PRECAUTIONARY STATEMENTS.

THIMET should be applied with a granular pesticide applicator properly calibrated to assure accurate placement and proper dosage. Cover granules that may be exposed on the ends of the treated rows and turns and loading areas by deep discing immediately after treating fields.

CROPS

CROP	PESTS CONTROLLED	RATE OF THIMET 15-G	APPLICATION	REMARKS
BEANS At Planting	Mexican bean beetles	6-12.5 ozs. per 1,000 ft. of row for any row (minimum 30-inch spacing)	Drill granules to the side of the seed.	Do not place THIMET granules in direct contact with seed. Do not feed the foliage of treated beans within 60 days of treatment.
	Leafhoppers Aphids Lygus bugs Thrips Mites Seedcorn maggots	6-9.3 ozs. per 1,000 ft. of row for any row spacing (minimum 30-inch spacing)		
	Seedcorn maggots		Apply granules in a band over the row.	
FIELD CORN AND SWEET CORN (excluding popcorn) At Planting	Corn rootworms Wireworms White grubs Seedcorn maggots Seedcorn beetles Flea beetles Mites	8 ozs. per 1,000 ft. of row for any row spacing (minimum 30-inch row spacing)	Place granules in a 7-inch band over the row directly behind the planter shoe in front of or behind the press wheel and lightly incorporate.	Do not use in-furrow application. One additional application can be applied at cultivation OR over corn later in the season.
FIELD CORN AND SWEET CORN (excluding popcorn) At Cultivation	Corn rootworms		Apply granules to base of plants or over the top of plants just ahead of cultivation shovels so as to cover granules with soil.	Do not make any applications of THIMET after cultivation treatment.

CROP	PESTS CONTROLLED	RATE OF THIMET 15-G	APPLICATION	REMARKS
FIELD CORN (excluding sweet and popcorn) At Cultivation	Chinch bug nymphs (Colorado, Kansas, and Nebraska)		Apply granules to base of plants or over the top of plants just ahead of cultivation shovels so as to cover granules with soil. Apply as soon as chinch bug nymphs are observed at base of plants. Granules must be placed at the infested region of the plants and adequately covered with soil.	Later infestations of chinch bugs may not be controlled. Do not make any applications of THIMET after cultivation treatment.
FIELD CORN (excluding sweet and popcorn) Over the Plant	European corn borer (1st brood)	6.7 lbs. per acre	Apply granule into the whorl of the plant prior to tassle emergence.	Do not graze or cut for forage within 30 days of treatment. Do not make more than one application over the plant. Do not apply under prolonged drought conditions. Consult your state experiment or state extension service for proper timing of application.
	Corn leaf aphids Mites		Broadcast the granules evenly over the top of plants by air or ground equipment.	
COTTON At Planting	Early season control of: Thrips Mites Leaf miners	4.0 ozs. per 1,000 ft. of row for any row spacing (minimum 36-inch spacing)	Distribute the granules evenly in the furrow at planting time. Soil temperature, moisture and tilth should be favorable for good germination and emergence.	Suggested for areas where a limited period of protection against early season pests is necessary for good cotton growth. Use 2.0 lbs. THIMET 15-G per acre for hill-dropped cotton.
	Mites Thrips Aphids Black cutworms Leafhoppers White flies Leaf miners	8-12 ozs. per 1,000 ft. of row for any row spacing (minimum 36-inch spacing)	Distribute the granules evenly in the furrow at planting time. Soil temperature, moisture and tilth should be favorable for good germination and emergence. Under adverse conditions plant stand may be affected.	Recommended for areas requiring an extended period of mite control as well as early season protection against other cotton pests. The higher rate is suggested for heavier soils.
COTTON Side-Dressing (Irrigated Cotton Only)	Mites from spring to early summer	12-16 ozs. per 1,000 ft. of row for any row spacing (minimum 36-inch spacing)	Incorporate granules into the soil as a side-dress application deep enough to avoid disturbance by future cultivations.	Irrigate as soon as possible after treatment. Do not apply later than 60 days before harvest.
PEANUTS At Planting	Thrips Leafhoppers	7.3 ozs. per 1,000 ft. of row for any row spacing (minimum 24-inch spacing)	Distribute the granules evenly in the furrow at planting time. For best results, plant peanuts when soil temperatures are favorable for establishing good peanut stands.	Do not graze or feed treated hay or forage to livestock.
PEANUTS At Pegging	Southern corn rootworms and Leafhoppers	14.6 ozs. per 1,000 ft. of row for any row spacing (minimum 24-inch spacing)	Distribute the granules as a band over the fruiting zone at pegging time. Work into the top few inches of soil immediately.	Do not graze or feed treated hay or forage to livestock.
POTATOES At Planting	Aphids Leafhoppers Leaf miners Psyllids Wireworms Flea beetle larvae and reduction of flea beetle adults	Light or Sandy Soils: 15 ozs. per 1,000 ft. of row for any spacing (minimum 32-inch spacing) Heavy or Clay Soils: 23 ozs. per 1,000 ft. of row for any row spacing (minimum 32-inch spacing)	Distribute the granules evenly in the furrow or band on each side of the row.	Wait 90 days after treatment before harvesting potatoes. Do not use for Colorado potato beetle control in Long Island, New York and the Del Marva Peninsula.
	Colorado potato beetle (early season control)	Heavy or Clay Soils: 23 ozs. per 1,000 ft. of row for any row spacing (minimum 32-inch spacing)		
SORGHUM At Planting	Greenbugs	8 ozs. per 1,000 ft. of row for any row spacing (minimum 30-inch spacing)	Place granules in a 7-inch band over the row OR drill 1 inch below and 2 inches to the side of the seed.	Do not place THIMET granules in direct contract with seed. A second application of granules may be applied topically later in the season.
SORGHUM At Cultivation	Chinch bug nymphs (Colorado, Kansas, Nebraska)		Apply granules to base of plants or over the top of plants just ahead of cultivation shovels so as to cover granules with soil. Apply as soon as chinch bug nymphs are observed at base of plants. Granules must be placed at the infested region of the plants and adequately covered with soil.	Later infestation of chinch bugs may not be controlled. Do not make any application of THIMET after cultivation treatment. Do not feed foliage before grain harvest.
SORGHUM Over the Plant	Greenbugs Bank grass mites	6.5 lbs. per acre	Broadcast granules into the whorls of the plants by air or ground equipment.	Make only one application after plant emergence. Do not apply within 28 days of harvest.
SOYBEANS At Planting	For the early season control of: Mexican bean beetles Leafhoppers Lygus bugs Seedcorn maggots Thrips Aphids Mites	12 ozs. per 1,000 ft. of row for any row spacing (minimum 30-inch row spacing)	Drill granules to the side of the seed.	Do not place THIMET granules in direct contract with seed. Do not feed the foliage of treated soybeans. Crop injury may result when preplant incorporated or preemergence applications of metribuzine herbicides are used in conjunction with this product or other soil-applied organo-phosphate pesticides.
	Seedcorn maggots		Place granules in a 7-inch band over the row directly behind the planter shoe and in front of the press wheel.	

CROP	PESTS CONTROLLED	RATE OF THIMET 15-G	APPLICATION	REMARKS
SUGAR BEETS	Aphids Beets root maggots Leafhoppers Mites Leaf miners	6 ozs. per 1,000 ft. of row for any row spacing (minimum 20-inch spacing)	Drill granules to the side of the seed OR place granules in a band over the row.	Do not place THIMET granules in direct contact with the seed. Do not apply within 30 days of harvest. Do not feed sugar beet tops or silage to dairy cattle. Consult your state experiment station or state extension service for proper timing of application in your area.
SUGAR BEETS Post-emergence	Mites Aphids	6.5-10 lbs. per acre	Apply to foliage when plants are dry. Use the higher rate on dense foliage for good coverage. Foliar application by air may be used for aphid control when wet soil prohibits its ground application.	
SUGAR CANE (Florida)	Wireworms	26 lbs. per acre	Apply in a 10-12 inch band directly on the seed piece and surrounding soil in the open furrow immediately before covering.	
WHEAT At Planting	Grasshoppers (1) Aphids (Western and North Central States only) Hessian fly	1.6 ozs. per 1,000 ft. of row for any row spacing (minimum 8-inch spacing)	Apply the granules in the seed furrow.	Do not feed or graze foliage within 45 days of treatment. Do not make any later applications after planting time treatment. (1) Grasshoppers controlled in winter wheat only.
WHEAT Over the Plant	Aphids (Western States only)	6.5 lbs. per acre	Broadcast the granules evenly over the field by air equipment when aphids first appear.	Do not apply over plants if application was made at planting time. Do not make more than one application over the plants. Do not apply within 70 days of harvest of grain. Do not feed or graze foliage within 28 days of treatment.

TYPICAL APPLICATION RATES FOR VARIOUS ROW SPACES

	RATE OZ./1,000 FT. OF ROW												
	1.6 oz.	2 oz.	4 oz.	6 oz.	8 oz.	10 oz.	12 oz.	14 oz.	16 oz.	18 oz.	20 oz.	22 oz.	24 oz.
40″				4.9	6.5	8.2	9.8	11.4	13.1	14.7	16.3	18.0	19.6
38″				5.2	6.9	8.6	10.3	12.0	13.7	15.5	17.2	18.9	20.6
36″				5.4	7.3	9.1	10.9	12.7	14.5	16.3	18.2	20.0	21.8
34″				5.6	7.7	9.7	11.6	13.5	15.4	17.4	19.3	21.3	23.2
32″				6.1	8.2	10.2	12.3	14.3	16.3	18.3	20.5	22.4	24.6
30″				6.5	8.7	10.9	13.1	15.2	17.4	19.6	21.8	24.0	26.1
26″				7.1	9.5	11.8	14.2	16.6	18.9	21.3	23.7	26.0	28.4
24″			5.4	8.2	10.9	13.6	16.3	19.1	21.8	24.5	27.2	30.0	32.7
22″			5.9	8.9	11.9	14.9	17.9	21.0	24.0	26.9	29.9	33.0	35.9
20″			6.5	9.8	13.1	16.3	19.6	22.8	26.1	29.3	32.6	35.9	39.1
16″			8.2	12.3	16.3	20.5	24.6	28.7	32.7	36.8	40.9		
12″		5.4	10.9	16.3	21.8	27.2	32.7	38.1	43.6				
8″	6.5	8.2	16.3	24.6	32.7	40.9							
6″	8.7	10.9	21.8	32.7	43.6								

EQUIVALENT POUNDS PER ACRE

CALIBRATION INFORMATION

FIRST READ THE LABEL

It is important that applicator equipment be properly set to deliver the labeled rate. The table below contains suggested starting gauge settings for calibration at planting speeds of 5 or 7 mph and application rates of 4.5, 6.0 or 9.0 oz./1,000 ft. of row.

NOTE: The settings in this table SHOULD ONLY BE USED AS STARTING POINTS. Continually check the amount of THIMET 15-G used against a known length of row and make further adjustments accordingly. Also check calibration occasionally to make sure equipment wear, changing moisture conditions, etc. have not caused a change in flow rate.

SUGGESTED GRANULAR APPLICATOR SETTING[1]

APPLICATOR	PLANTER SPEED 5 MPH			PLANTER SPEED 7 MPH		
	4.5 oz.	6.0 oz.	9.0 oz.	4.5 oz.	6.0 oz.	9.0 oz.
John Deere MaxEmerge 2	14	17	23	17	21	33
John Deere Max-Emerge (Odd Notches 5-15-25-35)	9	11	14	11	13	17
John Deere Max-Emerge (Even Notches 0-10-20-30)	9	13	20	13	19	26
John Deere 71 Flexi-Planter (John Deere Metal Hoppers)	1/23	1/26	2/3	1/26	1/32	2/14
New International Harvester (2 gauges/hopper)	1/9	2/2	2/6	2/2	2/5	3/3
Old International Harvester (1 gauge/hopper)	1/6	1/8	2/3	1/9	2/2	2/9
New Gandy	15	18	22	18	21	25
New Noble	6	8	11	8	11	14
Old Noble	8	11	16	12	15	22
Allis Chalmers (78 and 79 Series)	4	5	7	4	5	7

SUGGESTED GRANULAR APPLICATOR SETTING[1]						
APPLICATOR	PLANTER SPEED 5 MPH			PLANTER SPEED 7 MPH		
	4.5 oz.	6.0 oz.	9.0 oz.	4.5 oz.	6.0 oz.	9.0 oz.
Allis Chalmers #385	10	11	14	11	13	18

[1] Suggested starting point only.

THIMET® 20-G

SOIL AND SYSTEMIC INSECTICIDE

RESTRICTED USE CLASSIFICATION

Due to Acute Oral and Dermal Toxicity and Avian Hazards

For retail sale to and use only by Certified Applicators or persons under their direct supervision, and only for those uses covered by the Certified Applicator's certification.

American Cyanamid Company endorses Certification to promote the responsible use of pesticides to insure the protection of man and the environment.

ACTIVE INGREDIENT:
Phorate (0,0-diethyl S-[(ethylthio) methyl] phosphorodithioate . . 20.0%
INERT INGREDIENTS: 80.0%

TOTAL 100.0%

EPA Reg. No. 241-257

® Registered Trademark of American Cyanamid Company
7/89

KEEP OUT OF REACH OF CHILDREN

DANGER–POISON

DANGER/

¡PELIGRO! POISON.

PRECAUCION AL USUARIO: Si usted no lee inglés, no use este producto hasta que la etiqueta le haya sido explicada ampliamente.

FIRST AID

If swallowed, drink one or two glasses of water and induce vomiting by touching back of throat with finger. Do not induce vomiting or give anything by mouth to an unconscious person. Avoid alcohol. Get medical attention.

If inhaled, remove to fresh air. If not breathing, give artifical respiration, preferably mouth-to-mouth. If breathing is difficult, give oxygen. Get medical attention.

If on skin, wash thoroughly with soap and water. Remove contaminated clothing and shoes. Wash clothing and decontaminate shoes before reuse.

If in eyes, immediately flush eyes with plenty of water. Get medical attention.

Antidote: Atropine is an antidote

NOTE TO PHYSICIANS: Warning symptoms include weakness, headache, tightness in chest, blurred vision, nonreactive pinpoint pupils, salivation, sweating, nausea, vomiting, diarrhea and abdominal cramps. Give atropine intramuscularly or intravenously, depending on severity of poisoning, 2 to 4 milligrams every 10 minutes until fully atropinized as shown by dilated pupils, dry flushed skin and tachycardia. Twenty to thirty milligrams, or more, may be required during the first 24 hours. Never give opiates or phenothiazine tranquilizers. Clear chest by postural drainage. Artificial respiration or oxygen administration may be necessary. Observe patient continuously for at least 48 hours. Allow no further exposure to any cholinesterase inhibitor until cholinesterase regeneration has taken place as determined by blood tests.

Pralidoxime chloride (2-PAM; PROTOPAM chloride) may be effective as an adjunct to atropine. Use according to label directions.

In case of an emergency endangering life or property involving this product, call collect, day or night, Area Code 201-835-3100.

PRECAUTIONARY STATEMENTS

HAZARDS TO HUMANS AND DOMESTIC ANIMALS

DANGER!

Fatal if swallowed, inhaled, or absorbed through the skin. Do not breathe dust. Do not get in eyes, on skin, or on clothing.

Wear freshly laundered, long-sleeved work clothing daily. While transferring from package to equipment, wear a clean cap, gloves (rubber or cotton) and goggles. If cotton gloves are used, they must be laundered or discarded after each day's use. Rubber gloves should be washed with soap and water after each use. Do not wear the same gloves for other work. Destroy and replace gloves frequently.

In case of contact, immediately remove contaminated clothing and wash skin thoroughly with soap and water. Launder clothing and decontaminate shoes before reuse. Wash thoroughly with soap and water before eating or smoking. Bathe at the end of the work day and change clothing.

Do Not Breathe Dust

Wear a face mask or other respiratory equipment while emptying bags of product into a hopper. While emptying bags into equipment, pour downwind and allow as little free fall as possible. Do not pour at face level and do not allow dust to reach the breathing zone.

Do Not Contaminate Food or Feed Products

Once a bag has been opened, use it completely. Make sure the hoppers are emptied while still in the field. Refer to STORAGE AND DISPOSAL statement for further instructions.

Keep all Unprotected Persons Out of Operating Areas.

Do not apply this product in such a manner as to directly or through drift expose workers or other persons.

Keep Out of Reach of Domestic Animals

Not for use or Storage In or Around the Home

ENVIRONMENTAL HAZARDS

This pesticide is extremely toxic to fish and wildlife. Birds feeding in treated areas may be killed. Do not apply directly to water or wetlands (swamps, bogs, marshes, and pot-holes). Drift and runoff may be hazardous to aquatic organisms in neighboring areas. Collect or incorporate granules that are spilled during loading or are visible on soil surface in turn areas. Do not contaminate water when disposing of equipment washwaters.

REENTRY

After dusts have settled, do not enter or allow entry into the treated area until the 48 hour reentry interval has expired, unless the person entering the treated area is wearing the personal protective equipment listed below for early reentry.

For early reentry into treated areas wear a protective suit of one or two pieces covering all parts of the body except head, hands, and feet; chemical-resistant gloves and chemical resistant shoes, coverings, or boots.

Because certain states may require more restrictive re-entry intervals for various crops treated with this product, consult your State Department of Agriculture for further information.

Written or oral warnings must be given to workers who are expected to be in a treated area or in an area about to be treated with this product. When oral warnings are given, warnings shall be given in a language customarily understood by workers. Oral warnings must be given if there is reason to believe that written warnings cannot be understood by workers. Written warnings must include the following information: DANGER. Area treated with THIMET on (date of application). Do not enter without protective clothing.

PERSONAL PROTECTIVE EQUIPMENT

WEAR THE FOLLOWING PROTECTIVE CLOTHING DURING LOADING, APPLICATION, EQUIPMENT REPAIR/CLEANING, AND DISPOSAL OF THE PESTICIDE. A protective suit of one or two pieces covering all parts of the body except head, hands, and feet; chemical resistant gloves; chemical resistant shoes, shoe covering, or boots.

Unless a chemical resistant protective suit is otherwise required, mixers/loaders must wear a chemical resistant apron and goggles or a face shield.

IF APPLICATION IS PERFORMED USING AN ENCLOSED CAB OR COCKPIT, THE FOLLOWING PROTECTIVE CLOTHING AND EQUIPMENT MAY BE WORN AS AN ALTERNATE: Long-sleeved shirt and long-legged pants; shoes and socks. Chemical resistant goves must be available in the cab or cockpit and must be worn when exiting. This clothing is inadequate to protect you during equipment repair or cleaning, reentry, or pesticide disposal work. Do not allow contaminated clothing to come in contact with the insides of the cab.

IMPORTANT! If pesticide comes in contact with skin, wash off with soap and water. Always wash hands, face, and arms with soap and water before smoking, eating, drinking, or using the toilet.

AFTER WORK: Before removing gloves, wash them with soap and water. Take off all work clothes and shoes. Shower using soap and water. Wear clean clothes. Do not reuse contaminated clothing. Personal clothing worn during work must be laundered separately from household articles. Store protective clothing separately from personal clothing. Clean or launder protective clothing after each use. Clothing and protection equipment heavily contaminated must be destroyed according to state and local regulations. HEAVILY CONTAMINATED CLOTHING CANNOT BE ADEQUATELY DECONTAMINATED.

WORK SAFETY RULE

REPEATED EXPOSURES TO CHOLINESTERASE INHIBITORS SUCH AS ARE CONTAINED IN THIS PRODUCT MAY, WITHOUT WARNING, CAUSE PROLONGED SUSCEPTIBILITY TO VERY SMALL DOSES OF ANY CHOLINESTERASE INHIBITOR.

Persons working with this product should have frequent blood tests of their cholinesterase levels. If the cholinesterase level falls below a critical point, no further exposures should be allowed until it has been determined by means of blood tests that the cholinesterase level has return to normal. Before using this product consult the National Pesticide Telecommunications Network for recommendations regarding such blood tests, poisoning management, emergency treatment, and other information regarding the toxicity of phorate. The toll free number for the National Pesticide Telecommunication Network is 1-800-858-7378.

If handled indoors provide mechanical exhaust ventilation.

Keep all unprotected persons, children, livestock, and pets away from treated areas or where there is danger of drift.

Do not rub eyes or mouth with hands. If feel sick in any way, STOP work and get help right away, see First Aid section.

CALL A PHYSICIAN AT ONCE IN ALL CASES OF SUSPECTED POISONING.

DIRECTIONS FOR USE

It is a violation of Federal law to use this product in a manner inconsistent with its labeling.
This label must be in the possession of the user at the time of pesticide application.

STORAGE AND DISPOSAL

Storage

Store pesticide products in a secure locked area where children, unauthorized persons and animals cannot enter. Do not store in the same area with food or feed. Do not store opened bags.

Prohibitions

Do not contaminate water, food or feed by storage or disposal. Open dumping is prohibited. Cover or incorporate spills.

Pesticide Disposal

Pesticide wastes are acutely hazardous. Improper disposal of excess pesticide, spray mixture, or rinsate is a violation of Federal law. If these wastes cannot be disposed of by use according to label instructions, contact your state pesticide or Environmental Control Agency, or the hazardous waste representative at the nearest EPA Regional Office for guidance.

Container Disposal

Completely empty bag into application equipment. Then dispose of empty bag in a sanitary landfill or by incineration, or, if allowed by State and local authorities, by burning. If burned, stay out of smoke.

General

Consult Federal, State or local disposal authorities for approved alternative procedures such as limited open burning.

DISCLAIMER

The label instructions for the use of this product reflect the opinion of experts based on field use and tests. The directions are believed to be reliable and should be followed carefully. However, it is impossible to eliminate all risks inherently associated with use of this product. Crop injury, ineffectiveness or other unintended consequences may result because of such factors as weather conditions, presence of other materials, or the use or application of the product contrary to label instructions, all of which are beyond the control of American Cyanamid Company. All such risks shall be assumed by the user.

American Cyanamid Company warrants only that the material contained herein conforms to the chemical description on the label and is reasonably fit for the use therein when used in accordance with the directions for use, subject to the risk referred to above.

Any damages arising from a breach of this warranty shall be limited to direct damages and shall not include consequential commercial damages such as loss of profits or values or any other special or indirect damages.

American Cyanamid Company makes no other express or implied warranty, including any other express implied warranty of FITNESS or of MERCHANTABILITY.

BEFORE USING, READ PRECAUTIONARY STATEMENTS.

THIMET should be applied with a granular pesticide applicator properly calibrated to assure accurate placement and proper dosage. Cover granules that may be exposed on the ends of the treated rows and turns and loading areas by deep discing immediately after treating fields.

CROPS

CROP	PESTS CONTROLLED	RATE OF THIMET 20-G	APPLICATION	REMARKS
BEANS **At Planting**	Mexican bean beetles	4.9-9.4 ozs. per 1,000 ft. of row for any row (minimum 30-inch spacing).	Drill granules to the side of the seed.	Do not place THIMET granules in direct contact with seed. Do not feed the foliage of treated beans within 60 days of treatment.
	Leafhoppers Aphids Lygus bugs Thrips Mites Seedcorn maggots	4.5-7.0 ozs. per 1,000 ft. of row for any row spacing (minimum 30-inch spacing)		
	Seedcorn maggots		Apply granules in a band over the row.	
FIELD CORN AND SWEET CORN (excluding popcorn) **At Planting**	Corn rootworms Wireworms White grubs Seedcorn maggots Seedcorn beetles Flea beetles Mites	6 ozs. per 1,000 ft. row for any row spacing (minimum 30-inch row spacing)	Place granules in a 7-inch band over the row directly behind the planter shoe in front of or behind the press wheel and lightly incorporate.	Do not use in-furrow application. One additional application can be applied at cultivation OR over corn later in the season.
FIELD CORN AND SWEET CORN (excluding popcorn) **At Cultivation**	Corn rootworms		Apply granules to base of plants or over the top of plants just ahead of cultivation shovels so as to cover granules with soil	Do not make any applications of THIMET after cultivation treatment.
FIELD CORN (excluding sweet and popcorn) **At Cultivation**	Chinch bug nymphs (Colorado, Kansas, and Nebraska)		Apply granules to base of plants or over the top of plants just ahead of cultivation shovels so as to cover granules with soil. Apply as soon as chinch bug nymphs are observed at base of plants. Granules must be placed at the infested region of the plants and adequately covered with soil.	Later infestations of chinch bugs may not be controlled. Do not make any applications of THIMET after cultivation treatment.
FIELD CORN (excluding sweet and popcorn) **Over the Plant**	European corn borer (1st brood)	5.0 lbs. per acre	Apply granules into the whorl of the plant prior to tassle emergence	Do not graze or cut for forage within 30 days of treatment. Do not make more than one application over the plant. Do not apply under prolonged drought conditions. Consult your state experiment or state extension service for proper timing of application.
	Corn leaf aphids Mites		Broadcast the granules evenly over the top of plants by air or ground equipment.	

CROP	PESTS CONTROLLED	RATE OF THIMET 20-G	APPLICATION	REMARKS
COTTON **At Planting**	Early season control of: Thrips Mites Leaf miners	3.0 ozs. per 1,000 ft. row for any row spacing (minimum 36-inch spacing)	Distribute the granules evenly in the furrow at planting time. Soil temperature, moisture and tilth should be favorable for good germination and emergence.	Suggested for areas where a limited period of protection against early season pests is necessary for good cotton growth. Use 1.5 lbs. THIMET 20-G per acre for hill-dropped cotton.
	Mites Thrips Aphids Black cutworms Leafhoppers White flies Leaf miners	6.0-9.0 ozs. per 1,000 ft. of row for any row spacing (minimum 36-inch spacing)	Distribute the granules evenly in the furrow at planting time. Soil temperature, moisture and tilth should be favorable for good germination and emergence. Under adverse conditions plant stand may be affected.	Recommended for areas requiring an extended period of mite control as well as early season protection against other cotton pests. The higher rate is suggested for heavier soils.
COTTON **Side-Dressing (Irrigated Cotton Only)**	Mites from spring to early summer	9.0-12.0 ozs. per 1,000 ft. of row for any row spacing (minimum 36-inch spacing)	Incorporate granules into the soil as a side-dress application deep enough to avoid disturbance by future cultivations.	Irrigate as soon as possible after treatment. Do not apply later than 60 days before harvest.
PEANUTS **At Planting**	Thrips Leafhoppers	5.5 ozs. per 1,000 ft. of row for any row spacing (minimum 24-inch spacing)	Distribute the granules evenly in the furrow at planting time. For best results, plant peanuts when soil temperatures are favorable for establishing good peanut stands.	Do not graze or feed treated hay or forage to livestock.
PEANUTS **At Pegging**	Southern corn rootworms and Leafhoppers	11.0 ozs. per 1,000 ft. of row for any row spacing (minimum 24-inch spacing)	Distribute the granules as a band over the fruiting zone at pegging time. Work into the top few inches of soil immediately.	Do not graze or feed treated hay or forage to livestock.
POTATOES **At Planting**	Aphids Leafhoppers Leaf miners Psyllids Wireworms Flea beetle larvae and reduction of flea beetle adults	Light or Sandy Soils: 11.3 ozs. per 1,000 ft. of row for any spacing (minimum 32-inch spacing) Heavy or Clay Soils: 17.3 ozs. per 1,000 ft. of row for any row spacing (minimum 32-inch spacing)	Distribute the granules evenly in the furrow or band on each side of the row.	Wait 90 days after treatment before harvesting potatoes. Do not use for Colorado potato beetle control in Long Island, New York and the Del Marva Peninsula.
	Colorado potato beetle (early season control)	Heavy or Clay Soils: 17.3 ozs. per 1,000 ft. of row for any row spacing (minimum 32-inch spacing)		
SORGHUM **At Planting**	Greenbugs	6.0 ozs. per 1,000 ft. of row for any row spacing (minimum 30-inch spacing)	Place granules in a 7-inch band over the row OR drill 1 inch below and 2 inches to the side of the seed.	Do not place THIMET granules in direct contact with seed. A second application of granules may be applied topically later in the season.
SORGHUM **At Cultivation**	Chinch bug nymphs (Colorado, Kansas, Nebraska)		Apply granules to base of plants or over the top of plants just ahead of cultivation shovels so as to cover granules with soil. Apply as soon as chinch bug nymphs are observed at base of plants. Granules must be placed at the infested region of the plants and adequately covered with soil.	Later infestations of chinch bugs may not be controlled. Do not make any application of THIMET after cultivation treatment. Do not feed foliage before grain harvest.
SORGHUM **Over the Plant**	Greenbugs Banks grass mites	4.9 lbs. per acre	Broadcast granules into the whorls of the plants by air or ground equipment.	Make only one application after plant emergence. Do not apply within 28 days of harvest.
SOYBEANS **At Planting**	For the early season control of: Mexican bean beetles Leafhoppers Lygus bugs Seedcorn maggots Thrips Aphids Mites	9.0 ozs. per 1,000 ft. of row for any row spacing (minimum 30-inch row spacing)	Drill granules to the side of the seed	Do not place THIMET granules in direct contact with seed. Do not feed the foliage of treated soybeans. Crop injury may result when preplant incorporated or preemergence applications of metribuzine herbicides are used in conjunction with this product or other soil-applied organophosphate pesticides.
	Seedcorn maggots		Place granules in a 7-inch band over the row directly behind the planter shoe and in front of the press wheel.	
SUGAR BEETS	Aphids Beet root maggots Leafhoppers Mites Leaf miners	4.5 ozs. per 1,000 ft. row for any row spacing (minimum 20-inch spacing)	Drill granules to the side of the seed OR place granules in a band over the row.	Do not place THIMET granules in direct contact with the seed. Do not apply within 30 days of harvest. Do not feed sugar beet tops or silage to dairy cattle. Consult your state experiment station or state extension service for proper timing of application in your area.
SUGAR BEETS **Post-emergence**	Mites Aphids	4.9-7.5 lbs. per acre	Apply to foliage when plants are dry. Use the higher rate on dense foliage for good coverage. Foliar application by air may be used for aphid control when wet soil prohibits ground application.	

CROP	PESTS CONTROLLED	RATE OF THIMET 20-G	APPLICATION	REMARKS
SUGAR CANE (Florida)	Wireworms	19.5 lbs. per acre	Apply in a 10-12 inch band directly on the seed piece and surrounding soil in the open furrow immediately before covering.	
WHEAT **At Planting**	Grasshoppers (1) Aphids (Western and North Central States only) Hessian fly	1.2 ozs. per 1,000 ft. row for any row spacing (minimum 8-inch spacing)	Apply the granules in the seed furrow.	Do not feed or graze foliage within 45 days of treatment. Do not make any later applications after planting time treatment. (1) Grasshoppers controlled in winter wheat only.
WHEAT **Over the Plant**	Aphids (Western States only)	4.9 lbs. per acre	Broadcast the granules evenly over the field by air equipment when aphids first appear.	Do not apply over plants if application was made at planting time. Do not make more than one application over the plants. Do not apply within 70 days of harvest of grain. Do not feed or graze foliage within 28 days of treatment.

TYPICAL APPLICATION RATES FOR VARIOUS ROW SPACINGS
Rate Oz./1,000 Ft. of Row

	1.2 oz.	3 oz.	4.5 oz.	6 oz.	8 oz.	10 oz.	12 oz.	14 oz.	16 oz.	18 oz.
40″			3.7	4.9	6.5	8.2	9.8	11.4	13.1	14.7
38″			3.9	5.2	6.9	8.6	10.3	12.0	13.7	15.5
36″			4.1	5.4	7.3	9.1	10.9	12.7	14.5	16.3
34″			4.3	5.8	7.7	9.6	11.5	13.5	15.4	17.3
32″			4.6	6.1	8.2	10.2	12.2	14.3	16.3	18.4
30″			4.9	6.5	8.7	10.9	13.1	15.2	17.4	19.6
26″			5.7	7.5	10.0	12.6	15.1	17.6	20.1	22.6
24″			6.1	8.2	10.9	13.6	16.3	19.1	21.8	24.5
20″		4.9								
18″		5.4								
16″		6.1								
12″		8.2								
8″	4.9	12.2								
6″	6.5	16.3								

EQUIVALENT POUNDS PER ACRE

CALIBRATION INFORMATION

FIRST READ THE LABEL

It is important that applicator equipment be properly set to deliver the labeled rate. The table below contains suggested starting gauge settings for calibration at planting speeds of 5 or 7 mph and application rates of 4.5, 6.0 or 9.0 oz./1000 ft. of row.

NOTE: The settings in this table SHOULD ONLY BE USED AS STARTING POINTS. Continually check the amount of THIMET-20-G used against a known length of row and make further adjustments accordingly. Also check calibration occasionally to make sure equipment wear, changing moisture conditions, etc. have not caused a change in flow rate.

Suggested Granular Applicator Setting[1]

Applicator	Planter Speed At 5 MPH			Planter Speed At 7 MPH		
	4.5 oz.	6.0 oz.	9.0 oz.	4.5 oz.	6.0 oz.	9.0 oz.
John Deere MaxEmerge 2	14	17	23	17	21	33
John Deere Max-Emerge (Odd Notches 5-15-25-35)	9	11	14	11	13	17
John Deere Max-Emerge (Even Notches 0-10-20-30)	9	13	20	13	19	26
John Deere 71 Flexi-Planter (John Deere Metal Hoppers)	1/23	1/26	2/3	1/26	1/32	2/14
New International Harvester (2 gauges/hopper)	1/9	2/2	2/6	2/2	2/5	3/3
Old International Harvester (1 gauge/hopper)	1/6	1/8	2/3	1/9	2/2	2/9
New Gandy	15	18	22	18	21	25
New Noble	6	8	11	8	11	14
Old Noble	8	11	16	12	15	22
Allis Chalmers (78 and 79 series)	4	5	7	4	5	7
Allis Chalmers #385	10	11	14	11	13	18

[1]Suggested starting point only.

TRI-SCEPT®

HERBICIDE

FOR USE IN SOYBEANS

For use only in Alabama, Arkansas, Delaware, Florida, Georgia, Illinois, Indiana, Iowa, Eastern Kansas, Kentucky, Louisiana, Maryland, Southern Michigan, Mississippi, Missouri, Eastern Nebraska, New Jersey, North Carolina, Ohio, Eastern Oklahoma, Pennsylvania, South Carolina, Tennessee, Eastern Texas, Virginia, West Virginia and Southern Wisconsin.

For South Dakota use only in the following counties: Clay, Lincoln, Turner, Union, and Yankton.

Refer to "Supplement" for revised label, or if not in "Supplement" contact American Cyanamid (201) 831-2000. Ask for Product Mgr.

ACTIVE INGREDIENTS:

Ammonium salt of imazaquin (2-[4,5-dihydro-4-methyl-4-(1-methylethyl)-5-oxo-1*H*-imidazol-2-yl]-3-quinolinecarboxylic acid)* 5.09%

trifluralin: α,α,α-trifluoro-2,6-dinitro-N,N- dipropyl-p-toluidine . 28.99%

INERT INGREDIENTS: 65.92%

TOTAL 100.00%

Contains petroleum distillates

*Equivalent to 4.83%, 2-[4,5-dihydro-4-methyl-4-(1-methylethyl)5-oxo-1*H*-imidazol-2-yl]-3-quinolinecarboxylic acid.

TRI-SCEPT contains 3.0 pounds of active ingredients per gallon (0.43 pounds active ingredient of imazaquin and 2.57 pounds active ingredient of trifluralin per gallon).

EPA Reg. No. 241-307

DECEMBER 1989

® Registered Trademark of American Cyanamid Company

D41 12/88

KEEP OUT OF REACH OF CHILDREN

WARNING!/¡AVISO!

PRECAUCION AL USUARIO: Si usted no lee inglés, no use este producto hasta que la etiqueta le haya sido explicada ampliamente.

STATEMENT OF PRACTICAL TREATMENT

If in eyes: Hold eyelids open and flush with a steady, gentle stream of water for 15 minutes. Get medical attention if irritation persists.

If on skin: Wash with plenty of soap and water. Get medical attention if irritation persists.

If inhaled: Remove victim to fresh air.

In case of emergency endangering life or property involving this product, call collect, day or night, area code 201-835-3100.

PRECAUTIONARY STATEMENTS

HAZARDS TO HUMANS

WARNING!

Causes substantial but temporary eye injury.

HARMFUL IF ABSORBED THROUGH THE SKIN OR INHALED

DO NOT get in eyes or on clothing. Wear goggles or face shield. Avoid breathing spray mist. Avoid contact with skin. Wash skin thoroughly with soap and water after handling. Remove contaminated clothing and wash before reuse.

ENVIRONMENTAL HAZARDS

This product is toxic to fish. DO NOT apply directly to water or wetlands (swamps, bogs, or marshes). Drift or runoff from treated areas may be hazardous to aquatic organisms in neighboring areas. DO NOT contaminate water when disposing of equipment washwaters.

DIRECTIONS FOR USE

It is a violation of Federal law to use this product in a manner inconsistent with its labeling.

This labeling must be in the possession of the user at the time of pesticide application.

Observe all cautions and limitations in this leaflet. The use of TRI-SCEPT not consistent with this label can result in injury to crops, animals, or persons. Keep container closed to avoid spills and contamination.

DO NOT apply this product through any type of irrigation system.

STORAGE AND DISPOSAL

STORAGE: KEEP FROM FREEZING. DO NOT STORE BELOW 32°F. SHAKE WELL BEFORE USING. DO NOT contaminate water, food, or feed by storage or disposal.

DISPOSAL: Wastes resulting from the use of this product may be disposed of on site or at an approved waste disposal facility.

CONTAINER DISPOSAL: Triple rinse (or equivalent). Then offer for recycling or reconditioning, or puncture and dispose of in a sanitary landfill, or by incineration, or, if allowed by state and local authorities, by burning. If burned, stay out of smoke.

DISCLAIMER

The label instructions for the use of this product reflect the opinion of experts based on field use and tests. The directions are believed to be reliable and should be followed carefully. However, it is impossible to eliminate all risks inherently associated with use of this product. Crop injury, ineffectiveness or other unintended consequences may result because of such factors as weather conditions, presence of other materials, or the use or application of the product contrary to label instructions, all of which are beyond the control of American Cyanamid Company. All such risks shall be assumed by the user.

American Cyanamid Company warrants only that the material contained herein conforms to the chemical description on the label and is reasonably fit for the use therein described when used in accordance with the directions for use.

Any damages arising from a breach of this warranty shall be limited to direct damages, and shall not include consequential commercial damages such as loss of profits or values, etc.

American Cyanamid Company makes no other express or implied warranty, including any other express or implied warranty of FITNESS or MERCHANTABILITY. USER assumes the risk of any use contrary to label instructions, or under abnormal conditions, or under conditions not reasonably foreseeable by American Cyanamid Company.

GENERAL INFORMATION

For control of grass and broadleaf weeds, apply TRI-SCEPT as a preplant incorporated treatment.

After TRI-SCEPT is applied, some susceptible weeds emerge, growth stops, and then the weeds either die or are not competitive with the crop.

Some dwarfing of soybean plants may occur with TRI-SCEPT applications. This is the result of shortened plant internodes and has no effect on soybean yields.

There should be an interval of 90 days between the TRI-SCEPT application and soybean harvest.

NOTE: DO NOT apply TRI-SCEPT postemergence to soybeans as crop injury may occur. Crops other than soybeans, such as cotton, corn and vegetables, may be injured by spray drift or other indirect contact with TRI-SCEPT.

To avoid injury to sensitive crops from spray drift, follow all use directions and precautions in the Spraying Instructions on page 2.

To avoid injury to sensitive crops, spray equipment used for TRI-SCEPT applications must be drained and thoroughly cleaned with water before being used to apply other products to these crops.

See page 3 for Rotational Crop Restrictions.

Replanting: If replanting is necessary in field previously treated with TRI-SCEPT, the field may be replanted to soybeans. Rework the soil no deeper than the treated zone. DO NOT apply a second treatment of TRI-SCEPT.

MIXING INSTRUCTIONS

Fill the spray tank one-half to three-quarters full with clean water. While agitating add the required amount of product and then fill the remainder of the tank with water. Maintain agitation while spraying to ensure a uniform spray mixture.

SRAYING INSTRUCTIONS

NOTE: DO NOT make applications when spray may be carried by wind to sensitive crops.

Avoid overlaps when spraying.

GROUND APPLICATIONS

Uniformly apply with properly calibrated ground equipment in 10 to 40 gallons of water per acre. A spray pressure of 20 to 40 psi is recommended.

DO NOT apply with ground equipment when wind velocity is greater than 10 mph.

AERIAL APPLICATIONS:

Uniformly apply with properly calibrated aerial equipment in 5 or more gallons of water per acre.

DO NOT ALLOW TRI-SCEPT TO DRIFT OUTSIDE OF THE INTENDED TREATED FIELD.

DO NOT SPRAY WHEN ENVIRONMENTAL CONDITIONS COULD CAUSE DRIFT.

Applicator is responsible for any loss or damage which results from spraying TRI-SCEPT in a manner other than recommended in this label. In addition, applicator must follow all applicable state and local regulations and ordinances in regard to spraying.

USE RATE

Apply TRI-SCEPT at a rate of 2.33 pints per acre.

APPLICATION INSTRUCTIONS

PREPLANT INCORPORATED APPLICATIONS:

Apply TRI-SCEPT before planting and thoroughly incorporate. Incorporation implements should be set to thoroughly incorporate TRI-SCEPT into the top 2 inches of soil. A second pass must be made at an angle to the first pass to ensure thorough incorporation. PTO driven equipment should be set to cut 2 inches deep and operate one time at 4 mph or less. Incorporate into the soil within 24 hours after application. TRI-SCEPT may be applied during land preparation up to 45 days prior to planting or immediately prior to planting. Adequate moisture is required for activation of TRI-SCEPT.

If soybeans are planted on beds, apply and incorporate after bed formation.

WEEDS CONTROLLED

The following grass and broadleaf weeds are controlled by recommended treatments of TRI-SCEPT:

GRASSES

Barnyardgrass (*Echinochloa crus-galli*)
Crabgrass (*Digitaria* spp.)
Crowfootgrass (*Dactyloctenium aegyptium*)
Cupgrass, Woolly (*Eriochloa villosa*)
Foxtail
- Giant (*Setaria faberi*)
- Green (*Setaria viridis*)
- Robust (*Setaria* sp.)
- Yellow (*Setaria glauca*)

Goosegrass (*Eleusine indica*)
Johnsongrass, seedling (*Sorghum halepense*)
Panicum
- Fall* (*Panicum dichotomiflorum*)
- Texas (*Panicum texanum*)

Sandbur, Field (*Cenchrus incertus*)
Shattercane* (*Sorghum bicolor*)
Signalgrass, Broadleaf (*Brachiaria platyphylla*)

BROADLEAF WEEDS

Bristly Starbur (*Acanthospermum hispidum*)
Burcucumber (*Sicyos angulatus*)
Carpetweed (*Mollugo verticillata*)
Cocklebur, Common (*Xanthium strumarium*)
Jimsonweed (*Datura stramonium*)
Kochia (*Kochia scoparia*)
Lambsquarters, Common (*Chenopodium album*)
Mallow, Venice (*Hibiscus trionum*)

Morningglory
- Palm Leaf (*Ipomoea wrightii*)
- Pitted (*Ipomoea lacunosa*)
- Smallflower (*Jacquemontia tamnifolia*)

Mustard Species (*Brassica* spp.)
Nightshade, Eastern Black (*Solanum ptycanthum*)
Pigweed
- Palmer (*Amaranthus palmeri*)
- *Redroot (Amaranthus retroflexus)*
- Smooth (*Amaranthus hybridus*)
- Spiny (*Amaranthus spinosus*)
- Waterhemp, Tall (*Amaranthus tuberculatus*)

Poinsettia, Wild (*Euphorbia heterophylla*)
Puncturevine (*Tribulus terrestris*)
Purslane (*Portulaca oleracea*)
Pusley, Florida (*Richardia scabra*)
Ragweed
Common (*Ambrosia artemisiifolia*)
- Giant (*Ambrosia trifida*)

Redweed (*Melochia corchorifolia*)
Sida, Prickly (Teaweed)(*Sida spinosa*)
Smartweed
- Ladysthumb (*Polygonum persicaria*)
- Pennsylvania (*Polygonum pensylvanicum*)

Sunflower, Common (*Helianthus annuus*)
Velvetleaf (*Abutilon theophrasti*)
TRI-SCEPT will aid in the control and reduce competition from:
Morningglory
- Entireleaf (*Ipomoea hederacea var. intergriuscula*)
- Ivyleaf (*Ipomoea hederacea*)
- Tall (*Ipomoea purpurea*)

Nutsedge, Yellow (*Cyperus esculentus*)
Spurge, Spotted (*Euphorbia maculata*)

*See TRI-SCEPT plus trifluralin or PROWL® herbicide Tank Mixture section below for use directions.
® Registered Trademark of Amercian Cyanamid Company

TRI-SCEPT

ROTATIONAL CROP RESTRICTIONS

The following rotational crops may be planted after applying TRI-SCEPT in soybeans:

1. Wheat may be planted in the fall, four months after TRI-SCEPT application in soybeans. Rice may be planted in the spring of the year following TRI-SCEPT application. (See Northern Use Area Restrictions*).
2. Eleven months after TRI-SCEPT application:

   Barley (See Northern Use Area Restrictions*)
   Oats (See Northern Use Area Restrictions*) (***)
   Field corn (See Northern Use Area Restrictions* or Southern Use Area Restrictions**)
   Edible Beans
   Grain sorghum***
   Peanuts
   Tobacco

* NORTHERN USE AREA RESTRICTIONS:
For the Northern Use Area as defined in the USE AREA section of this label, DO NOT plant field corn, wheat, barley, oats as a rotational crop in the fall or in the spring of the year following TRI-SCEPT application.

** SOUTHERN USE AREA RESTRICTIONS:
For the Southern Use Area, as defined in the USE AREA section of this label; field corn may be planted in the spring of the year following TRI-SCEPT application unless extreme drought conditions develop and less than 15 inches of rainfall or irrigation is received from the date of application through October of the year of application. For eastern Oklahoma, Arkansas, the bootheel of Missouri, Tennessee, Northern Carolina and states south: field corn may be planted in the spring of the year following TRI-SCEPT application unless extreme drought conditions develop and less than 15 inches of rainfall or irrigation is received within 6 months following the date of application.

*** In those portions of Kansas, Nebraska, Oklahoma, and Texas where at least 20 inches of irrigation and/or rainfall (total) was used to produce the crop, grain sorghum or oats should not be planted for 12 months following TRI-SCEPT application. If less than 20 inches of water was used to produce the crop, DO NOT plant grain sorghum or oats for 18 months after application of TRI-SCEPT.

ONLY rotational crops harvested at maturity may be used for feed or food.

DO NOT plant other rotational crops prior to 18 months after last TRI-SCEPT application.

DO NOT apply products containing chlorimuron ethyl (e.g. Classic, Canopy, Gemini, Lorox Plus, Preview[1], etc.), imazaquin (e.g. SCEPTER® [refer to the SCEPTER label for exceptions], SQUADRON®, etc.), or imazethapyr (e.g. PURSUIT®, PURSUIT® PLUS, etc.) the same year as TRI-SCEPT or injury to follow crops may occur.

DO NOT graze or feed treated soybean forage, hay, or straw to livestock.

[1] Classic, Canopy, Gemini, Lorox Plus, and Preview are trademarks of E.I. duPont de Nemours and Company.
®Registered Trademark of American Cyanamid Company.

TRI-SCEPT Plus trifluralin or PROWL® Tank Mixture For Shattercane and Fall Panicum Infestations in Soybeans

Trifluralin or PROWL must be tank mixed with TRI-SCEPT and applied preplant incorporated if infestations of shattercane or fall panicum are anticipated.

For shattercane control, add trifluralin 4 EC or PROWL at a rate of 0.5 pint per acre on medium textured soils or 1.0 pint per acre on fine textured soils. For control of fall panicum, add 0.5 pint per acre of trifluralin or PROWL for all soil types.

Observe all precautions and limitations on the trifluralin label.

Observe all precautions and limitations on the PROWL label.

TRI-SCEPT PLUS Command[2] 4 EC Preplant Incorporated Tank Mixture For Use When Heavy Infestations of Velvetleaf are Anticipated

Observe all precautionary statements on the Command label before using. Use sprayers equipped with nozzles that provide accurate and uniform application.

NOTE: The use of an agriculturally approved drift reducing agent is required at spray volumes of 10 to 15 gallons per acre.

Apply TRI-SCEPT plus Command 4 EC to the soil surface and uniformly incorporate to a depth of 2 inches. Incorporate within 3 hours of application. DO NOT apply aerially or through irrigation systems. DO NOT apply this tank mixture to overly moist or wet soils.

SPECIAL PRECAUTION

Command 4 EC is a volatile compound. Off-site movement of spray drift or vapors of Command can cause foliar whitening or yellowing of some plants. Prior to using Command, read and strictly follow all precautions and application instructions on the Command 4 EC label.

WEED SPECIES CONTROLLED

The TRI-SCEPT plus Command tank mixture controls heavy infestations of velvetleaf as well as the weeds listed in the **WEEDS CONTROLLED** section of the label.

USE RATES

Apply TRI-SCEPT at a rate of $2\frac{1}{3}$ pints per acre in tank mixture with Command 4 EC at $\frac{1}{4}$ to $\frac{1}{2}$ pints per acre.

ROTATIONAL CROP RESTRICTIONS

Follow all rotational crop restrictions on the TRI-SCEPT and Command labels. Always follow the most restrictive label.

In the event of a crop loss due to weather conditions, soybeans can be replanted. DO NOT work the soil deeper than 2 inches.

[2] Command is a trademark of FMC Corporation

USE AREA

TRI-SCEPT can be applied only in the states or parts shaded in the following map:

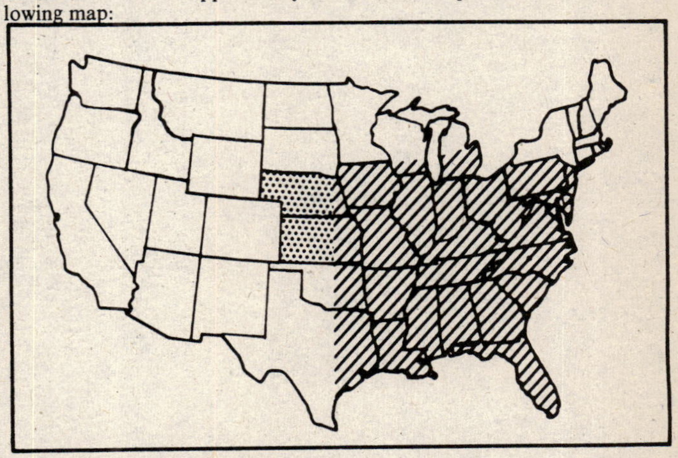

= for irrigated soybeans only

NOTE: See ROTATIONAL CROP RESTRICTIONS section for recommendations applying to the Northern vs. Southern Use Areas.

BASF Corporation

Agricultural Chemicals Division
100 Cherry Hill Road
Parsippany, NJ 07054
(201) 316-3000

BASAGRAN®

POSTEMERGENCE HERBICIDE

A soluble liquid formulation containing:

Active Ingredient:	
Sodium salt of bentazon*	42.0%
Inert Ingredients:	58.0%

* Equivalent to 4 pounds per gallon bentazon (3-(1-methylethyl)-1*H*-2,1,3-benzothiadiazin-4 (3*H*)-one 2,2-dioxide)

EPA Reg. No. 7969-45

KEEP OUT OF REACH OF CHILDREN.

CAUTION

STATEMENT OF PRACTICAL TREATMENT

Avoid contact with eyes or skin. In case of contact, immediately flush eyes or skin with plenty of water. Get medical attention if irritation persists. May cause allergic skin response.

First Aid: If contacted, flush eyes immediately with water for at least 15 minutes. Call a physician.

ENVIRONMENTAL HAZARDS

Do not apply directly to waters or wetlands **except** as noted under rice use directions.

Do not contaminate water when disposing of equipment wash waters.

Notice: It is a violation of Federal law to use any pesticide in a manner that results in the death of an endangered species or in adverse modification of their habitat.

The use of this product may pose a hazard to certain federally designated endangered species known to occur in specific areas within the **California** counties of Merced, Sacramento, and Solano. Before using this product in these counties, you must obtain the EPA Endangered Species Bulletin specific for these areas. The bulletin (EPA/ES-85-8) is available from either your County Agricultural Extension Agent, the Endangered Species Specialist in your State Wildlife Agency Headquarters, or the Regional Office of the U.S. Fish and Wildlife Service (Portland, Oregon). **The use of this product is prohibited in these counties unless specified otherwise in the bulletin.**

STORAGE AND DISPOSAL

Do not allow product to freeze.

Do not contaminate water, food, or feed by storage or disposal.

Wastes resulting from the use of this product may be disposed of on site or at an approved waste disposal facility.

Triple rinse container (or equivalent). Then offer for recycling or reconditioning, or puncture and dispose of in a sanitary landfill, or by incineration, or, if allowed by state and local authorities, by burning. If burned, stay out of smoke.

Do not re-use empty container.

In case of emergency

In case of large-scale spillage regarding this product, call:

CHEMTREC	(800) 424-9300
BASF Corporation	(201) 316-3000

In case of medical emergency regarding this product, call:

1. Your local doctor for immediate treatment.
2. Your local poison control center (hospital).
3. BASF Corporation (201) 316-3000.

DIRECTIONS FOR USE—ALL CROPS

It is a violation of Federal law to use this product in a manner inconsistent with its labeling.

Read the precautionary statement, environmental hazards, storage and disposal statements, and Conditions of sale and warranty statement appearing on the container label.

GENERAL INFORMATION

BASAGRAN® herbicide is intended for selective postemergence control of certain broadleaf weeds and sedges. (See **Directions for use** for specific crops and weeds.) BASAGRAN does not control grasses. BASAGRAN is effective mainly through contact action; therefore, weeds must be thoroughly covered with spray. Large crop-and-weed leaf canopies shelter smaller weeds and prevent adequate spray coverage. Labeled crops are tolerant to BASAGRAN; however, some leaf-speckling and leaf-bronzing may occur under certain conditions. (See **Restrictions and limitations** for each crop.)

TIMING OF APPLICATIONS

Make postemergence applications of BASAGRAN early, when weeds are small and actively growing and before weeds reach the maximum size listed in the application rate tables for the individual crops.

Early application to weeds produces the most beneficial effect on weed control (exceptions: yellow nutsedge and Canada thistle), allows use of the lower rate (depending on weed species), and makes it easier to obtain thorough spray coverage. Delay in application which permits weeds to exceed the maximum size stated will result in inadequate control.

Do not cultivate within five days before or after application of BASAGRAN in the following northern and western states: AZ, CA, CO, CT, IA, ID, IL, IN, KS, KY, MA, ME, MI, MN, MO, MT, NE, ND, NH, NJ, NV, NY, OH, OR, PA, RI, SD, UT, VT, WA, WI, WV, WY.

Water volume and spray pressure

Apply recommended rates of BASAGRAN as follows:

Ground equipment: Use a minimum of 20 gallons of water per broadcast acre and a minimum of 40 psi pressure (measured at the boom-not at the pump or in the line). When crop and weed foliage is dense, use up to 50 gallons of water and up to 80 psi pressure. Use standard high pressure pesticide hollow cone or flat fan nozzles spaced 20 inches apart. Do not use flood, whirl chamber, or controlled droplet application (CDA) nozzles.

Air equipment: Use a minimum of 5 gallons of water per acre (except 10 gallons for rice) and a maximum of 40 psi pressure. Use only diaphragm-type nozzles producing cone or fan spray patterns.

Aerial application—special directions

To obtain uniform coverage and to avoid drift hazards, the following application equipment and practices should be used:

Nozzle height: Maximum of 10 feet above crop.

Nozzle orientation: Nozzles must be oriented so as to discharge straight back with the air stream (opposite the direction of travel of the aircraft) or at some angle between straight back and straight down. For optimal coverage when applying BASAGRAN by air in rice, orient all nozzles straight down.

Nozzles must not be located farther out than three-fourths the distance from the center of the aircraft to the end of the wing or rotor.

Water volume and spray pressure: See Air equipment.

Do not apply BASAGRAN by aircraft when wind is blowing at a velocity above 10 mph (except above 5 mph in California). Coarse sprays (large droplets) are less likely to drift.

Do not apply BASAGRAN by air if ornamental or sensitive nontarget crops such as cotton, sugar beets, sunflowers or okra are within 200 feet downwind.

In California, do not apply BASAGRAN by air within 200 feet of ornamental or sensitive nontarget crops such as cotton, sugar beets, sunflowers or okra.

Applicator must follow the most restrictive use cautions to avoid drift hazards, including those found in this labeling as well as applicable state and local regulations and ordinances.

Special information for irrigated areas

In irrigated areas, it may be necessary to irrigate prior to treatment with BASAGRAN® herbicide to ensure that weeds are growing actively. Weeds growing under drought conditions usually are not satisfactorily controlled.

Addition of oil concentrate

A nonphytotoxic oil concentrate (commonly referred to as oil concentrate) should be added to the spray tank for certain weed problems as recommended in the directions for specific crops. The oil concentrate must contain either a petroleum or vegetable oil base and must meet the following criteria: 1) be nonphytotoxic, 2) contain only EPA-exempt ingredients, 3) provide good mixing quality in the jar test (see the following page), and 4) be successful in local experience.

The exact composition of suitable products will vary; however, vegetable and petroleum oil concentrates should contain emulsifiers which provide good mixing quality. For vegetable oil concentrates, it has been observed that highly refined vegetable oils are more satisfactory than unrefined vegetable oils. For additional information see **Jar test for estimating suitability of oil concentrates** at the end of this section.

With the addition of oil concentrate to BASAGRAN on soybeans, beans, and peanuts, a slight leaf burn may occur, but all new growth is normal and crop vigor is not reduced. The potential for leaf burn is increased when relative humidity and temperature are high. A few oil concentrates have exhibited excessive leaf burn. Refer to your supplier of BASAGRAN for information concerning successful local experience prior to purchasing any oil concentrate.

Do not add UAN (nitrogen) solution to BASAGRAN plus Blazer®/Tackle® herbicides when oil concentrate is included in the spray tank.

Do not add oil concentrate to BASAGRAN for use on peas.

Rate of oil concentrate:

Ground application—

2 pints/acre (maximum).
Air application—1 pint/acre (maximum).

California—refer to additional information under the specific crop (beans, and corn/grain sorghum).

MIXING/SPRAYING:

Fill tank of a thoroughly clean sprayer one-half to two-thirds full with clean water. Start agitation and add BASAGRAN; allow to mix thoroughly. Add oil concentrate and remaining volume of water. Maintain constant agitation during application.

Jar test for estimating suitability of oil concentrates

1. **Water supply:** Use only water from intended source and at the source temperature.
2. **Amount of water in jar:** Ground application—For 20 gal./A spray volume use 3⅓ cups (800 ml) of water.
   Air application—For 5 gal./A spray volume use ⅚ cup (200 ml) of water, or for 10 gal./A spray volume use 1⅔ cups (400 ml) of water.
   For other spray volumes, adjust proportionately to above.
3. **Amount of herbicide(s) and oil concentrate to add:** Add herbicides and oil concentrate at the rate of 1 teaspoon (5 ml) for each pint of recommended label rate.
4. **Add components in following sequence,** gently mixing between component additions:
   1) Dry products (dry flowables and wettable powders) when applicable.
   2) BASAGRAN, and when applicable, other water miscible products (such as **Blazer**), liquid fertilizers and/or liquid flowables.
   3) Oil concentrate.
   4) Poast® herbicide or other emulsifiable concentrates when applicable.
5. **Cap jar,** invert 10 cycles, let stand for 15 minutes, evaluate.
6. **Evaluation:** An ideal tank mix combination will be uniform; thus, the suitability of the oil concentrate is questionable if any of the following are observed:
   Free oil at the surface—film or globules.
   Flocculation—fine particles which may be suspended in the liquid or found as a precipitated layer at the bottom of the jar.
   Clabbering—thickening texture (coagulated) resembling yogurt or a curd-like texture as with cottage cheese.

RESTRICTIONS AND LIMITATIONS

Do not apply BASAGRAN to crops listed on this labeling that have been subject to stress conditions such as hail damage, flooding, drought, injury from other herbicides or widely fluctuating temperatures, as crop injury may result.

Do not apply BASAGRAN if crops listed on this label show injury (leaf phytotoxicity and/or plant stunting) produced by any other prior herbicide applications, because this injury may be enhanced and/or prolonged.

Do not apply BASAGRAN during prolonged periods of drought or during unseasonably cold weather, as unsatisfactory weed control may result.

Rainfall or overhead irrigation soon after application (within 8 hours) may nullify the effectiveness of BASAGRAN.

Do not mix or apply BASAGRAN with any other pesticide or with fertilizer except as specifically recommended on this labeling or approved supplemental labeling.

Clean sprayer thoroughly prior to application of BASAGRAN, particularly if a herbicide was used which has the potential to injure the crop to be sprayed with BASAGRAN.

Do not apply this product through any type of irrigation system.

Do not apply directly to water or wetlands **except** as noted under rice use directions.

Do not contaminate water when disposing of equipment wash waters.

SOYBEANS—DIRECTIONS FOR USE

Application of BASAGRAN herbicide should be made when weeds are small and actively growing and before weeds reach the maximum size listed in **Table 1, Application rate table for soybeans**. Such applications generally correspond to the soybean growth stages of unifoliate to two expanded trifoliate leaves. Soybeans are tolerant to BASAGRAN at all stages of growth. Slight yellowing, bronzing, speckling, or burning of leaves may occur under certain conditions. Soybean plants generally outgrow this condition within 10 days.

Restrictions and limitations (partial list)

Do not apply more than a total of 4 pints of BASAGRAN per acre in one season.

TABLE 1 APPLICATION RATE TABLE FOR SOYBEANS

WEEDS CONTROLLED	APPLICATION RATES FOR WEED GROWTH STAGES					
	1 PINT PER ACRE*		1½ PINTS PER ACRE		2 PINTS PER ACRE	
	LEAF STAGE	MAXIMUM HEIGHT	LEAF STAGE	MAXIMUM HEIGHT	LEAF STAGE	MAXIMUM HEIGHT
Balloonvine	—	—	2–4	2″	4–6	3″
Beggarticks	—	—	Up to 6	6″	6–8	8″
Bristly Starbur	—	—	Up to 4	2″	4–6	3″
Cocklebur[a]	2–4	4″	2–6	6″	6–10	10″
Coffee Senna	—	—	—	—	Up to 1** Pinnate	2″
Common Lambsquarters[b]	Up to 4**	1″	Up to 6**	1½″	4–8**	2″
Common Purslane	—	—	Up to 4	1″	4–6	2″
Common Ragweed	—	—	—	—	4–6**	3″
Dayflower	—	—	Up to 6	4″	6–10	8″
Devilsclaw	—	—	—	—	Up to 6**	3″
Galinsoga	—	—	—	—	Cotyledon to 6**	2″
Giant Ragweed[c]	—	—	—	—	Up to 4	6″
Jimsonweed	Up to 4	4″	Up to 6	6″	6–10	10″
Ladysthumb	Up to 4	4″	Up to 6	6″	6–10	10″
Marshelder	—	—	Up to 4	2″	Up to 8	4″
Pennsylvania Smartweed	Up to 4	4″	Up to 6	6″	6–10	10″
Prickly Sida or Teaweed	—	—	Up to 6	3″	6–8	4″
Redweed	—	—	4–6	6″	6–10	8″
Sesbania	—	—	—	—	3–5**	3″
Shepherdspurse[d]	—	—	Up to 6	4″	6–10	8″
Spurred Anoda	—	—	Up to 6	3″	6–8	4″
Tropic Croton	—	—	Up to 2	2″	2–4	4″
Velvetleaf[e]	Up to 3	2″	Up to 4	2″	4–6	5″
Venice Mallow	Up to 4	2″	Up to 6	2″	6–10	4″
Wild Buckwheat	—	—	Up to 4	3″	4–6	5″
Wild Mustard	Up to 4	2″	Up to 6	4″	6–10	8″
Wild Poinsettia	—	—	2–4**	4″	4–8**	6″
Wild Sunflower	Up to 2	3″	Up to 4	5″	4–6	8″

For additional weeds see **Special Directions** section following.

* Apply before weeds reach the maximum size or leaf stage indicated. If regrowth develops, make a second application of 1 pint 7 to 14 days after the first application. (This rate not applicable in California.)

** Add 2 pints per acre of oil concentrate. See **Addition of oil concentrate.**

a. Do not treat earlier than leaf stage shown and do not count cotyledon leaves.
b. Control may be partial or inconsistent.
c. If a second flush occurs, retreat field according to this rate table.
d. Do not treat rosette before seed stalk appears.
e. Add UAN solution according to the **Special Directions for Other Weed Problems** (See page 6) or add oil concentrate according to the section **Addition of oil concentrate.**

SPECIAL DIRECTIONS FOR OTHER WEED PROBLEMS IN SOYBEANS

Annual Morningglories

South: (AL, AR, FL, GA, LA, MS, NC, OK, SC, TN, TX, VA). To control smallflower and cypressvine morningglories, apply a single application of either 1½ pints of BASAGRAN® herbicide per acre to plants not larger than 4 true leaves and 4 inches in height, **or** 2 pints of BASAGRAN per acre to plants not larger than 6 true leaves and 6 inches in height. Add oil concentrate to the spray solution with BASAGRAN (see section **Addition of oil concentrate**).

To control palmleaf, pitted, tall (common), entireleaf, purple moonflower, and ivyleaf morningglories, apply 1½ pints of BASAGRAN per acre to plants not larger than 4 true leaves and 4 inches in height (14 to 18 days after morningglory emergence). Make a second application at the same rate 5 to 14 days later.

All states other than the South (see above): Apply 2 to 3 pints of BASAGRAN per acre to annual morningglories not larger than 4 true leaves. Control may be partial or inconsistent. Add oil concentrate to the spray solution of BASAGRAN (see section **Addition of oil concentrate**).

Because morningglories grow very rapidly, it is important to watch the growth stage carefully and to be certain that BASAGRAN is applied to morningglories before they exceed the maximum size recommended on this label.

Canada Thistle

Apply 2 pints of BASAGRAN per acre when plants are from 8 inches tall to bud stage. Make a second application at the same rate 7 to 10 days later.

Yellow Nutsedge

Two applications are preferred for best results. Apply 1½ to 2 pints of BASAGRAN per acre when plants are 6 to 8 inches tall. If needed, make a second application at the same rate 7 to 10 days later. Add oil concentrate to the spray solution of BASAGRAN (see section **Addition of oil concentrate**).

Field and Hedge Bindweed in KY, IL, IN, MI, OH only

For suppression of field and hedge bindweed, apply 2 to 3 pints of BASAGRAN per acre when vines are a maximum of 10 inches long. Add oil concentrate to the spray solution of BASAGRAN (see section **Addition of oil concentrate**).

Late Cocklebur Rescue Treatment

This treatment is intended to provide only partial control of cocklebur in the event early postemergence treatments were not made. Very thorough spray coverage is essential. Apply a single application of 2 to 3 pints of BASAGRAN per acre to plants up to 24 inches tall or, for best results, apply 1½ pints of BASAGRAN per acre to plants up to 24 inches tall and repeat 10 to 14 days later.

Late Velvetleaf Rescue Treatment

In the event early postemergence treatments were not made, partial control can be obtained. Very thorough spray coverage is essential. Apply a single application of 3 pints per acre of BASAGRAN plus 1 quart of oil concentrate and 1 gallon of UAN solution to velvetleaf plants up to 12". For best results, apply 1½ pints per acre of BASAGRAN plus 1 quart of oil concentrate plus 1 gallon of UAN solution followed in 4–7 days with the same treatment.

Addition of Urea Ammonium Nitrate (UAN) Solution for Velvetleaf and Other Weeds*

In soybeans,UAN solution (commonly referred to as 28%, 30% or 32% nitrogen solution) may be added to BASAGRAN in place of oil concentrate for improved control of velvetleaf. Improvement in the control of cocklebur, common sunflower, Pennsylvania smartweed, devilsclaw, venice mallow and wild mustard may also be attained.

The UAN solution is an agricultural grade fertilizer used by local dealers for agricultural applications. It may be added to the tank with BASAGRAN when velvetleaf is the primary target weed. However, BASAGRAN plus UAN solution will not provide adequate control of common ragweed and common lambsquarters. If these weeds or other weeds requiring oil concentrate (see **Table 1**) are present in addition to velvetleaf, then oil concentrate should also be used.

With the addition of UAN solution or UAN solution plus oil concentrate to BASAGRAN on soybeans, a slight leaf burn may occur, but the new growth is normal and crop vigor is not reduced. Refer to your supplier of BASAGRAN for information concerning successful local experience prior to using UAN solution. Do not use brass or aluminum nozzles when spraying BASAGRAN plus UAN solution.

For information on adding UAN solution to BASAGRAN® and Blazer®/Tackle® herbicides tank mix, see **Table 2, 3 and 4, pages 8-9.**

Do not add UAN solution to BASAGRAN for use on rice, peanuts, peas, mint, or turf.

Rate of UAN solution:

Ground application—½-1 gallon/acre
Air application—½ gallon/acre

* Not applicable in California.

SOYBEANS—TANK MIXES WITH BASAGRAN

Use the following chart as a guide to determine broadleaf weeds and grasses controlled by BASAGRAN® herbicide alone and various tank mixes with BASAGRAN.

BASAGRAN TANK MIXES*
GUIDE TO WEEDS CONTROLLED

BASAGRAN CONTROLS THE WEEDS LISTED BELOW	ADDITIONAL WEEDS CONTROLLED BY TANK MIXING VARIOUS HERBICIDES WITH BASAGRAN	REFER TO TABLE LISTED BELOW FOR RATE, WEED SIZE AND ADDITIVE INFORMATION
Balloonvine Beggarticks Cocklebur Coffee Senna Common Lambsquarters Common Purslane Common Ragweed Cypressvine Morningglory Canada Thistle** Dayflower Devilslaw Galinsoga Giant Ragweed Jimsonweed Ladysthumb Marshelder Pennsylvania Smartweed Prickly Side or Teaweed Redweed Shepherdspurse Smallflower Morningglory Spurred Anoda Tropic Croton Velvetleaf Venice Mallow Wild Buckwheat Wild Mustard Wild Pinsettia Wild Sunflower Yellow Nutsedge	**Blazer®/Tackle® herbicides**	
	Common Ragweed Tall Waterhemp Redroot Pigweed Smooth Pigweed Black Nightshade Sesbania Morningglories Crotalaria	BASAGRAN + Blazer/Tackle Tables 2, 3, and 4 Pages 8-9
	Reflex® 2LC herbicide	
	Common Ragweed Tall Waterhemp Redroot Pigweed Smooth Pigweed Black Nightshade Sesbania Morningglories Crotalaria	BASAGRAN + Reflex 2LC Page 10
	2,4-DB	
	Morningglories (Ivyleaf, tall, and entireleaf) Vines up to 6" long	BASAGRAN + 2,4-DB Table 5 Page 11
	Scepter® herbicide	
	Redroot Pigweed Smooth Pigweed Tall Waterhemp Wild Sunflower	BASAGRAN + Scepter Table 6 Page 12
	Poast® herbicide	

BASAGRAN TANK MIXES*
GUIDE TO WEEDS CONTROLLED

BASAGRAN CONTROLS THE WEEDS LISTED BELOW	ADDITIONAL WEEDS CONTROLLED BY TANK MIXING VARIOUS HERBICIDES WITH BASAGRAN	REFER TO TABLE LISTED BELOW FOR RATE, WEED SIZE AND ADDITIVE INFORMATION
	Wild Proso Millet Barnyardgrass Broadleaf Signalgrass Fall Panicum Giant Foxtail Green Foxtail Yellow Foxtail Seedling Johnsongrass Junglerice Red Sprangletop Texas Panicum Witchgrass Woolly Cupgrass Goosegrass Large Crabgrass Smooth Crabgrass	BASAGRAN + Poast Table 7 Page 13
	Poast® + Blazer®/Tackle® herbicides	
	See weeds listed above for Poast and Blazer/Tackle	BASAGRAN + Poast + Blazer/Tackle Table 8 Page 15
	Early Spot Spray	
	See weeds listed above for Poast and Blazer/Tackle	BASAGRAN + Poast + Blazer/Tackle Table 9 Page 15

* Tank mixes not applicable in California.

** Canada thistle requires two applications of BASAGRAN in accordance with this label.

BASAGRAN AND BLAZER/TACKLE TANK MIXES*-SOYBEANS

General and application information, Restrictions and Limitations for Tables 2, 3 and 4

General information

For postemergence broadleaf weed control, refer to **Tables 2, 3** and **4** as determined by weed problems and geographical area.

Table 2—Northern States

BASAGRAN® herbicide:
1-2 pints/A

Blazer® herbicide/Tackle:
½ pint/A

Additional weeds controlled:
pigweeds (redroot and smooth)
and tall waterhemp.

Table 3—All states (except California)

BASAGRAN: +1-2 pints/A

Blazer/Tackle: 1 pint/A

Additional weeds controlled:
Listed in Table 3.

Table 4—Southern States

BASAGRAN: 1 pint/A

Blazer/Tackle: 1 pint/A

Weeds controlled:
Listed in Table 4.

Time of application

The timing of all applications of BASAGRAN should be in accordance with the weed growth stages indicated in **Table 1** and when weeds are actively growing. With Blazer/Tackle in the tank mix, the timing should be in accordance with the weed growth stages indicated in **Tables 2, 3** and **4** and when weeds are actively growing. If weeds are not at the correct stage of growth for treatment at the same time, then separate applications should be made. Delay in application which permits weeds to exceed the maximum size stated will result in inadequate control.

Water volume and spray pressure

Ground equipment: Use a minimum of 20 gallons of water per broadcast acre and a minimum of 40 psi pressure (measured at the boom-not at the pump or in the line). When crop and weed foliage is dense, use up to 50 gallons of water and up to 80 psi pressure. Use standard high pressure pesticide hollow cone or flat fan nozzles spaced 20 inches apart. Do not use flood or whirl chamber or controlled droplet application (CDA) nozzles.

Air equipment: Use a minimum of 10 gallons of total spray solution per acre.

Mixing

Fill the spray tank half full with water and add the recommended amount of product in the following order—BASAGRAN, Blazer/Tackle, oil concentrate (or UAN solution)—while the agitator is running. Then add the remaining quantity of water.

Coverage

Thorough coverage of actively growing weeds is essential. Large crop-and-weed leaf canopies shelter smaller weeds and can prevent adequate spray coverage. Soybeans are tolerant to the above tank mixes; however, under certain conditions soybeans may burn, crinkle and bronze.

Restrictions and limitations (partial list)

Read and follow restrictions and limitations on the BASAGRAN and Blazer/Tackle labels. The most restrictive labeling applies in tank mixes.

Do not apply Blazer or Tackle within 50 days of harvest (see Blazer or Tackle label).

Do not graze treated soybean fields and do not feed treated soybean forage, ensilage or hay to livestock (see labels for Blazer and Tackle).

Do not include oil concentrate with UAN solution when tank mixing BASAGRAN with Blazer/Tackle.

* Tank mixes not applicable in California.

TABLE 2
NORTHERN STATES[a]
BASAGRAN + BLAZER/TACKLE TANK MIX–SOYBEANS
RATE AND TIME OF APPLICATION

PRODUCT	PRODUCT RATE	WEEDS CONTROLLED/WEED SIZE			ADDITIVE (RATE)
BASAGRAN	1-2 pints/A according to weed species and size (See Table 1, page 5)	Balloonvine Beggarticks Bristly Starbur Canada Thistle[b] Cocklebur Coffee Senna Common Lambsquarters Common Purslane Common Ragweed Cypressvine Morningglory Dayflower Devilsclaw Jimsonweed Ladysthumb	Marshelder Pennsylvania Smartweed Prickly Sida or Teaweed Redweed Shepherdspurse Smallflower Morningglory Spurred Anoda Tropic Croton Velvetleaf Venice Mallow Wild Buckwheat Wild Mustard Yellow Nutsedge		Oil Concentrate (2 pints/A) or UAN solution[c] (1/2-1.0 gallon/A) if velvetleaf is the primary weed target and lambsquarters or common ragweed are not a problem.
plus	plus		**Leaf Stage**	**Max. Height**	
Blazer/Tackle	1/2 pint/A	Pigweeds* (Redroot and smooth) Tall Waterhemp	Up to 4	2"	**Note:** Do not include Oil Concentrate with UAN solutions when tank mixing BASAGRAN with Blazer/Tackle.

* See Table 4 for control of additional weeds.

[a] Northern states, for the purpose of this table, are the following states and those to the north thereof: MD, WV, OH, IN, IL, KS and MO (except southeastern MO, Jefferson Co. and south).

[b] Requires two applications of BASAGRAN in accordance with this label.

[c] See Special Directions for Other Weed Problems in Soybeans page 6.

TABLE 3
ALL STATES (EXCEPT CALIFORNIA)
BASAGRAN + BLAZER/TACKLE TANK MIX—SOYBEANS
RATE AND TIME OF APPLICATION

PRODUCT	PRODUCT RATE	WEEDS CONTROLLED/WEED SIZE			ADDITIVE (RATE)
BASAGRAN	1 1/2-2 pints/A according to weed species and size (See Table 1, Page 5)	Balloonvine Beggarticks Bristly Starbur Canada Thistle[b] Cocklebur Coffee Senna Common Lambsquarters Common Purslane Common Ragweed Cypressvine Morningglory Dayflower Devilsclaw Galinsoga Giant Ragweed Jimsonweed Ladysthumb	Marshelder Pennsylvania Smartweed Prickly Sida or Teaweed Redweed Shepherdspurse Smallflower Morningglory Spurred Anoda Tropic Croton Velvetleaf Venice Mallow Wild Buckwheat Wild Mustard Wild Poinsettia Wild Sunflower Yellow Nutsedge		Oil Concentrate[a] (2 pints/A) (or) UAN solution[c] (1/2-1 gallon/A) if velvetleaf is the primary weed target and lambsquarters or common ragweed are not a problem. **Note:** Do not include Oil Concentrate with UAN solutions when tank mixing BASAGRAN with Blazer/Tackle.
plus	plus		**Leaf Stage**	**Max. Height**	
Blazer/Tackle	1 pint/A	Common Ragweed	Up to 6	6"	
		Black Nightshade	Up to 6	2"	
		Morningglories[d]	Up to 4	4"	
		Crotalaria	Up to 6	6"	
		Sesbania	Up to 4 Pinnate	6"	
		Redroot Pigweed	Up to 6	3"	
		Smooth Pigweed	Up to 6	3"	
		Tall Waterhemp	Up to 6	3"	

[a] Add oil concentrate to the tank mix according to recommendations in Table 1 Application Rate Table for Soybeans, page 5.

[b] Requires two applications of BASAGRAN® herbicide in accordance with this label.

[c] See Special Directions for Other Weed Problems in soybeans, page 6.

[d] For consistent control of common (tall) morningglory use the 1 1/2 pints rate of BASAGRAN.

TABLE 4
SOUTHERN STATES[a]
BASAGRAN + BLAZER/TACKLE TANK MIX—SOYBEANS RATE AND TIME OF APPLICATION

PRODUCT	PRODUCT RATE	WEEDS CONTROLLED	LEAF STAGE	WEED SIZE MAXIMUM HEIGHT	ADDITIVE (RATE)
BASAGRAN	1 pint/A	Black Nightshade	Up to 6	2"	Oil
+	+	Bristly Starbur	4–6	3"	Concentrate
Blazer/Tackle	1 pint/A	Cocklebur[b]	2–6	6"	(1 pint/A)
		Common Labmsquarters*	4–6	2"	
		Common Ragweed	4–6	3"	
		Crotalaria	Up to 6	6"	
		Giant Ragweed	Up to 4	6"	
		Jimsonweed	Up to 6	6"	
		Ladysthumb	Up to 6	6"	
		Morningglories[c]	Up to 4	4"	
		Pennsylvania Smartweed	Up to 6	6"	
		Prickly Sida (Teaweed)*	Up to 4	2"	
		Redroot Pigweed	Up to 6	3"	
		Redweed	2–4	3"	
		Sesbania	Up to 4	6"	
		Smooth Pigweed	Up to 6	3"	
		Spurred Anoda*	Up to 4	2"	
		Tall Waterhemp	Up to 6	3"	
		Velvetleaf*	Up to 4	2"	
		Venice Mallow	Up to 6	2"	
		Wild Mustard	Up to 6	4"	

* For more consistent control, increase rate of BASAGRAN to 1½ pints.
[a] Southern states, for the purpose of this table, are AL, AR, FL, GA, KY, LA, MS, NC, OK, SC, TN, TX, VA and southeastern MO (Jefferson Co. and south).
[b] Do not treat earlier than the two leaf stage and do not count cotyledon leaves.
[c] For common (tall) morningglory, increase rate of BASAGRAN to 1½ pints.

BASAGRAN + REFLEX 2LC TANK MIXES—SOYBEANS

General and application information, Restrictions and limitations

General information

A tank mix of BASAGRAN® herbicide and Reflex® 2LC herbicide may be applied for postemergence control of the major troublesome broadleaf weed species in soybeans.

BASAGRAN and Reflex 2LC are selective postemergence herbicides which control annual broadleaf weeds. The tank mix should be applied to actively growing weeds. Refer to the BASAGRAN and Reflex 2LC labels for defined environmental conditions that promote active growth; recommended rates, sizes and growth stages for susceptible weed species are described on these labels. The most restrictive labeling applies to the tank mixtures.

Water volume and spray pressure

Ground application: Use sufficient spray volume and pressure to ensure complete coverage of the target weeds. A minimum of 20 gallons per acre of spray mixture should be used with spray pressures of 40 to 60 psi (pressure) and 20-50 gallons per acre to ensure coverage of weed foliage. Use only hollow cone or flat fan nozzles. The sprayer must be calibrated to provide the proper volume and rate per acre. In addition, the boom and nozzle height must be adjusted to provide complete coverage of all weeds. **Do not use flood type, or other spray nozzles which deliver coarse sprays, or controlled droplet applicator (CDA) nozzles.**

Mixing

Fill the clean sprayer tank ½ full with clean water. Begin agitation and add the recommended amount of BASAGRAN and Reflex 2LC. Add the appropriate amount of oil concentrate to be used. Complete filling the tank with water to the needed volume. Allow the spray mixture to agitate and recycle 5-10 minutes.

Cultivation

Cultivation within 7 days prior to application is not recommended. Cultivation may put weeds under stress and reduce control obtained. Timely cultivation 2-3 weeks after applying the tank mix of BASAGRAN and Reflex 2LC will often assist weed control.

Rate

BASAGRAN may be used in the tank mixtures at rates of 1-2 pts. per acre in each of the regions listed for Reflex 2LC. Refer to tank mix table below for the recommended use rate of Reflex 2LC in Regions 1, 2, and 3. Geographic description of these regions is included in the Reflex 2LC label.

Application Rates for Basagran and Reflex 2LC in Tank Mix

Region*	Basagran**	Reflex 2LC**	Oil Concentrate
1	1-2 pts./A	1-1½ pts./A	1 qt./A
2	1-2 pts./A	¾-1¼ pts./A	1 qt./A
3	1-2 pts./A	¾-1 pt./A	

*See the Reflex label for states or parts of states included in regions.
**Consult labels for each product for specific weeds controlled.

Restrictions and limitations
(partial list)

- Always read and follow the restrictions and limitations for each product. The most restrictive labeling applies in tank mixes.
- Reflex 2LC can be applied only in the states or parts of states included in Regions 1, 2, & 3 as prescribed on the Reflex label. **Do not apply Reflex 2LC to any field in Regions 2 & 3 more than once every two years.**
- A maximum of 1.5 pints (0.375 lb. active) per acre of Reflex 2LC may be applied per growing season for soybeans in Region 1. A maximum of 1.25 pts. (0.313 lb. active) per acre may be applied in alternate years in Region 2. A maximum of 1.0 pt. (0.25 lb. active) per acre may be applied in alternative years in Region 3.
- Refer to Reflex 2LC label for recommendations concerning crop rotation.
- Do not apply a total of more than 4 pints of BASAGRAN per acre in one season on soybeans.
- Applications of the tank mix of BASAGRAN/Reflex 2LC should not be made beyond 3 weeks after the soybeans have emerged (see Reflex label).
- **Do not make more than one application of the BASAGRAN/Reflex 2LC tank mix in a single season.**
- BASAGRAN/Reflex 2LC requires an 8-hour rain-free period. Do not apply the tank mix if rain is threatening.
- Use of tank mix of BASAGRAN/Reflex 2LC during periods of dry weather when crop and weeds are under stress and not actively growing may result in reduced weed control. Do not apply to drought stressed weeds or weeds which have gone through an extended dry period.
- In the event of a crop loss due to weather conditions, only soybeans can be replanted (see Reflex label).
- Avoid drift to all other crops and nontarget areas. Crops other than soybeans may be severely injured by drift.
- Do not graze treated areas or harvest for forage or hay (see Reflex label).

BASAGRAN + 2,4-DB TANK MIX*—SOYBEANS

General and application information, Restrictions and limitations

General information

These directions are intended to provide the user of BASAGRAN® herbicide with instructions for tank mixing with 2,4-DB (such as Butyrac® 200 herbicide or Butoxone® 200 herbicide) to control entireleaf, tall (common), and ivyleaf morningglories. Weeds must be actively growing and at the recommended growth stages. Delay in application which permits weeds to exceed the maximum size stated will result in inadequate control.

Water volume and spray pressure

Refer to section entitled **Directions for use—all crops** for additional information.

Ground equipment: Use a minimum of 20 gallons of water per broadcast acre and a minimum of 40 psi pressure (measured at the boom-not at the pump or in the line). When crop and weed foliage is dense, use up to 50 gallons of water and up to 80 psi pressure. Use standard high pressure pesticide hollow cone or flat fan nozzles spaced 20 inches apart. Do not use flood, whirl chamber or controlled droplet application (CPA) nozzles.

Mixing: Fill the spray tank half full with water and add the recommended amount of BASAGRAN and 2,4-DB while the agitator is running. Then add the remaining quantity of water.

Coverage

The tank mix is effective partly through contact action. Therefore, weeds must be thoroughly covered with spray.

Large crop-and-weed leaf canopies shelter smaller weeds and prevent adequate spray coverage.

Restrictions and limitations
(partial list)

Read and follow the restrictions and limitations on the labels for BASAGRAN and 2,4-DB. The most restrictive labeling applies in tank mixes.

Use only amine formulations of 2,4-DB.

Do not add oil or any other additive (including UAN solution) to tank mix with 2,4-DB.

Rainfall soon after application (within 8 hours) may nullify the effectiveness of the tank mix.

Do not apply more than 1 application of the tank mix per season.

Do not apply within 60 days of harvest (see label for 2,4-DB).

The use of this tank mix will cause soybean foilage injury (such as burning, bronzing or crinkling) and may reduce yields.

Do not use this tank mix on soybeans that show symptoms of disease such as phytophthora root rot (see label for 2,4-DB).

* Tank mix not applicable in California.

TABLE 5
BASAGRAN + 2,4-DB TANK MIX—SOYBEANS
RATE AND TIME OF APPLICATION

PRODUCT	PRODUCT RATE	WEEDS CONTROLLED/WEED SIZE		ADDITIVE (RATE)
BASAGRAN	1½-2 pints/A according to weed species and size. (See Table 1 Page 5)	Balloonvine Beggarticks Bristly Starbur Canada Thistle* Cocklebur Coffee Senna Common Lambsquarters Common Purslane Common Ragweed Cypressvine Morningglory Dayflower Devilsclaw Galinsoga Giant Ragweed Jimsonweed	Ladysthumb Marshelder Pennsylvania Smartweed Prickly Sida or Teaweed Redweed Shepherdspurse Smallflower Morningglory Spurred Anoda Tropic Croton Velvetleaf Venice Mallow Wild Buckwheat Wild Mustard Wild Pointsettia Wild Sunflower Yellow Nutsedge	**Do not** add Oil Concentrate or any other additives (including UAN solution) to this tank mix.
plus 2,4-DB (amine formulation)	**plus** 2 fl. oz./A of Butoxone 200 or Butyrac 200. (0.03 pound ae**/A.)	Morningglories: Ivyleaf Tall (common) Entireleaf	Vines up to 6″long	

* Requires two applications of BASAGRAN in accordance with this label.
** Acid equivalent.

BASAGRAN + SCEPTER TANK MIX*—SOYBEANS NORTHERN STATES ONLY

General and application information, Restrictions and limitations

General information

The tank mix of BASAGRAN® herbicide plus Scepter® will control pigweeds, in addition to those weeds controlled by BASAGRAN. Weeds must be actively growing and at the recommended growth stages.

Water volume and spray pressure

Large crop-and-weed leaf canopies shelter smaller weeds and prevent adequate spray coverage. When this occurs, or the weed population is very high, use a higher than minimum spray volume and pressure.

Ground equipment: Use a minimum of 20 gallons of total spray mixture per acre (broadcast basis) and 40 psi pressure with standard high pressure hollow cone or flat fan nozzles spaced 20 inches apart.

Air equipment: Use a minimum of 5 gallons of water per acre.

Mixing

Fill spray tank half full with water, and add the recommended amount of product in the following order—BASAGRAN, Scepter, oil concentrate—while the agitator is running. Then add the remaining quantity of water.

Restrictions and limitations
(partial list)

Read and follow the restrictions and limitations on the labels for BASAGRAN and Scepter. The most restrictive labeling applies in tank mixes.

Observe all geographical and rotational crop restrictions on the label for Scepter.

* Tank mix not applicable in California.

TABLE 6
NORTHERN STATES (STATES WHERE SCEPTER IS APPROVED*)
BASAGRAN + SCEPTER TANK MIX–SOYBEANS
RATE AND TIME OF APPLICATION

PRODUCT	RATE	WEEDS CONTROLLED/WEED SIZE			ADDITIVE (RATE)
BASAGRAN	1-2 pints/A according to weed species and size (See Table 1 Page 5)	Balloonvine Beggarticks Bristly Starbur Canada Thistle** Cocklebur Coffee Senna Common Lambsquarters Common Purslane Common Ragweed Cypressvine Morningglory Dayflower Devilsclaw Galinsoga Giant Ragweed Jimsonweed	Ladysthumb Marshelder Pennsylvania Smartweed Prickly Sida or Teaweed Redweed Shepherdspurse Smallflower Morningglory Spurred Anoda Tropic Croton Velvetleaf*** Venice Mallow Wild Buckwheat Wild Mustard Wild Poinsettia Wild Sunflower Yellow Nutsedge		Oil Concentrate (2 pints/Acre)
			Leaf Stage	Max. Height	
plus Scepter	**plus** ⅓ pint/A	Redroot Pigweed Smooth Pigweed Tall Waterhemp Wild Sunflower	Up to 6	3″	

* Northern states, for the purpose of this table, are the following states: IA, S. MI, S. WI, PA, NJ, DE, NE, KS, MD, WV, OH, IN, IL, and MO (except southeastern: Jefferson Co. and south). See label for Scepter for list of approved states and parts of states.

** Requires two applications of BASAGRAN in accordance with this label.
*** For optimum control of velvetleaf, ½-1 gallon of UAN may be used in place of oil concentrate.

BASAGRAN + POAST TANK MIX*—SOYBEANS

General and application information, Restrictions and limitations

General information

BASAGRAN® and Poast® herbicides may be tank mixed for postemergence control of the broadleaf and grass weeds shown in **Table 7.** Weeds must be actively growing and at the recommended growth stages.

Separate applications should be made if: a) all weeds to be controlled are not at the correct growth stage for treatment at the same time, or b) grasses to be controlled include rhizome johnsongrass, quackgrass, bermudagrass, wirestem muhly, shattercane, volunteer cereals, wild oats, red rice or itchgrass. See **Table 10, Separate Applications of BASAGRAN** page 16.

Water volume and spray pressure

Ground equipment: Use 20 gallons of total spray solution per acre (broadcast basis) and a minimum of 40 psi pressure. Use standard high pressure hollow cone or flat fan nozzles spaced 20 inches apart. Do not use flood or whirl chamber nozzles.

Air equipment: Use a minimum of 5 gallons of total spray solution per acre.

Additives

At the low rate of Poast (1 pt./A) the additive Dash™ spray adjuvant plus UAN (or ammonium sulfate) must be used. For control of the additional grasses listed in **Table 7** use the higher rate of Poast (1½ pts./A) and either Dash or oil concentrate. To enhance weed control UAN (or ammonium sulfate) may also be added.

Mixing

Fill spray tank half full with water, and add the recommended amount of product in the following order—BASAGRAN, UAN or ammonium sulfate, oil concentrate, Poast—while the agitator is running. Then add the remaining quantity of water.

Coverage

Thorough coverage of actively growing weeds is essential. Large crop-and-weed leaf canopies shelter smaller weeds and can prevent adequate spray coverage. Soybeans are tolerant to the tank mix; however, under certain conditions soybeans may burn, crinkle and bronze. Soybeans at all stages of growth are tolerant to BASAGRAN and Poast.

Restrictions and limitations
(partial list)

Read and follow the restrictions and limitations on the labels for BASAGRAN and Poast. The most restrictive labeling applies in tank mixes.

Do not apply tank mix within 90 days of harvest (see label for Poast).

Do not graze treated soybean fields and do not feed treated soybean forage (green, succulent) or ensilage to livestock. Treated soybean hay may be fed (see label for Poast).

* Tank mix not applicable in California.

TABLE 7
BASAGRAN + POAST TANK MIX–SOYBEANS
RATE AND TIME OF APPLICATION

PRODUCT	RATE	WEEDS CONTROLLED/WEED SIZE				ADDITIVE (RATE/ACRE): DASH OR OIL CONCENTRATE	ADDITIVE (RATE/ACRE): UAN SOLUTION OR AMMONIUM SULFATE
		Broadleaves and Sedge					
BASAGRAN	1-2 pints/A according to weed species and size (See Table 1, Page 5)	Balloonvine		Ladysthumb		—	—
		Beggarticks		Marshelder			
		Bristly Starbur		Pennsylvania Smartweed			
		Canada Thistle*		Prickly Sida or Teaweed			
		Cocklebur		Redweed			
		Coffee Senna		Shepherdspurse			
		Common Lambsquarters		Smallflower Morningglory			
		Common Purslane		Spurred Anoda			
		Common Ragweed		Tropic Croton			
		Cypressvine Morningglory		Velvetleaf			
		Dayflower		Venice Mallow			
		Devilsclaw		Wild Buckwheat			
		Galinsoga		Wild Mustard			
		Giant Ragweed		Wild Poinsettia			
		Jimsonweed		Wild Sunflower			
				Yellow Nutsedge			
plus		**Annual Grasses****					
Poast	1 pint/A	Wild Proso Millet***	4-10"	Green Foxtail	3-8"	Dash (2 pts.) plus	½-1 gallon UAN or 2½ lbs. AMS
		Fall Panicum	3-8"	Witchgrass	3-8"		
		Giant Foxtail	3-8"	Woolly Cupgrass	3-8"		
				Volunteer Corn	1-12"		
or							
Poast	1½ pints/A****	Barnyardgrass	3-8"	Junglerice	3-8"	Dash (2 pts.) or Oil Concentrate (2 pts.)	½-1 gallon UAN or 2½ lbs. AMS may be added to this tank mixture.
		Broadleaf Signalgrass	3-8"	Red Sprangletop	3-8"		
		Yellow Foxtail	3-8"	Texas Panicum	3-8"		
		Seedling Johnsongrass	3-8"	Goosegrass	3-6"		
				Large Crabgrass	3-6"		
				Smooth Crabgrass	3-6"		

* Requires two applications of BASAGRAN for control.
** Tank mix does not control rhizome johnsongrass, quackgrass, bermudagrass, wirestem muhly, shattercane, volunteer cereals, wild oats, red rice, or itchgrass.
*** For control of wild proso millet only, include Poast in the tank mix at ¾ pint/A.
**** The 1½ pts./A rate of Poast will also control all grasses listed at the 1 pt./A rate.

BASAGRAN + BLAZER/TACKLE + POAST TANK MIX*—SOYBEANS

General and application information, Restrictions and limitations

General information

BASAGRAN®, Poast® and Blazer®/Tackle herbicides may be tank mixed for postemergence control of broadleaf and grass weeds. Weeds must be actively growing and at the recommended growth stages.

Separate applications should be made if: a) all weeds to be controlled are not at the correct growth stage for treatment at the same time, or b) grasses to be controlled include rhizome johnsongrass, quackgrass, bermudagrass, wirestem muhly, volunteer corn, shattercane, volunteer cereals, wild oats, red rice or itchgrass. See **Table 10, Separate Applications of BASAGRAN.**

Water volume and spray pressure

Ground equipment: For the tank mix of BASAGRAN + Poast + Blazer/Tackle, use 20 gallons of total spray solution per acre (broadcast basis) and a minimum of 40 psi pressure. Use standard high pressure hollow cone or flat fan nozzles spaced 20 inches apart. Do not use flood or whirl chamber nozzles.

Air equipment: Use a minimum of 10 gallons of total spray solution per acre.

Mixing

Fill the spray tank half full with water and add the recommended amount of product in the following order—BASAGRAN, Blazer/Tackle, oil concentrate, Poast—while the agitator is running. Then add the remaining quantity of water.

Early spot spray
When using knapsack sprayers or high volume equipment utilizing handguns (or other suitable nozzle arrangements), prepare spray solution according to **Table 9.** Apply to the foliage on a spray-to-wet basis. Complete coverage of all foliage is essential for control. Control of perennial weeds may be limited to burnoff of exposed foliage.

Observe all safety precautions when spot spraying BASAGRAN + Blazer/Tackle + Poast.

Coverage
Thorough coverage of actively growing weeds is essential. Large crop-and-weed leaf canopies shelter smaller weeds and can prevent adequate spray coverage. Soybeans are tolerant to tank mix; however, under certain conditions soybeans may burn, crinkle and bronze.

Restrictions and limitations
(partial list)
Read and follow the restrictions and limitations on the labels for BASAGRAN, Poast and Blazer/Tackle. The most restrictive labeling applies in tank mixes.

Do not apply tank mix within 90 days of harvest (see label for Poast).

Do not graze treated soybean fields and do not feed treated soybean forage, ensilage or hay to livestock (see labels for Blazer/Tackle and Poast).

Do not include UAN solution (or ammonium sulphate) when tank mixing BASAGRAN, Blazer/Tackle and Poast.

*Tank mix not applicable in California.

TABLE 8
BASAGRAN + BLAZER/TACKLE + POAST TANK MIX
SOYBEANS
RATE AND TIME OF APPLICATION

PRODUCT	RATE	WEEDS CONTROLLED/WEED SIZE***				ADDITIVE (RATE)
BASAGRAN	1-2 pints/A according to weed species and size. (See Table 1, Page 5)	Balloonvine		Ladysthumb		Oil Concentrate (2 pints/A)
		Beggarticks		Marshelder		
		Bristly Starbur		Pennsylvania Smartweed		
		Canada Thistle*		Prickly Sida or Teaweed		Note: Do not include UAN solution or ammonium sulfate when tank mixing oil concentrate with BASAGRAN and Blazer/Tackle.
		Cocklebur		Redweed		
		Coffee Senna		Shepherdspurse		
		Common Lambsquarters		Smallflower Morningglory		
		Common Purslane		Spurred Anoda		
		Common Ragweed		Tropic Croton		
		Cypressvine Morningglory		Velvetleaf		
		Dayflower		Venice Mallow		
		Devilsclaw		Wild Buckwheat		
		Galinsoga		Wild Mustard		
		Giant Ragweed		Wild Poinsettia		
		Jimsonweed		Wild Sunflower		
				Yellow Nutsedge		
plus Poast	plus 1½ pints/A	Wild Proso Millet**	4-10"	Junglerice	3-8"	
		Barnyardgrass	3-8"	Red Sprangletop	3-8"	
		Broadleaf Signalgrass	3-8"	Texas Panicum	3-8"	
		Fall Panicum	3-8"	Witchgrass	3-8"	
		Giant Foxtail	3-8"	Woolly Cupgrass	3-8"	
		Green Foxtail	3-8	Goosegrass	3-6"	
		Yellow Foxtail	3-8"	Large Crabgrass	3-6"	
		Seedling Johnsongrass	3-8"	Smooth Crabgrass	3-6"	
				Leaf Stage	**Max. Height**	
plus Blazer/Tackle	plus ½-1 pint/A	Common Ragweed		Up to 10	6"	
		Black Nightshade		Up to 6	2"	
	Use ½ pint for pigweed (up to 2") only; 1 pint if other weeds at right are present.	Morningglories****		Up to 4	4"	
		Crotalaria		Up to 6	6"	
		Sesbania		Up to 4	6"	
				Pinnate		
		Tall Waterhemp		Up to 6	3"	
		Redroot Pigweed		Up to 6	3"	
		Smooth Pigweed		Up to 6	3"	

* Requires two applications of BASAGRAN® herbicide in accordance with this label.
** For control of wild proso millet only, include Poast® herbicide in tank mix at ¾ pint/A.
*** Tank mix does not control rhizome johnsongrass, quackgrass, bermudagrass, wirestem muhly, volunteer corn, shattercane, volunteer cereals, wild oats, red rice or itchgrass.
**** For consistent control of common (tall) morningglory use the 1½ pints rate of BASAGRAN.

Table 9
BASAGRAN + Blazer/Tackle + Poast Tank Mix
Soybeans
Spot Treatment Application Table

	Concentration in Spray Solution			
	BASAGRAN	**Blazer/Tackle**	**Poast**	**Oil Concentrate**
See annual grasses and broadleaves listed in Table 8.	1%	1%	1%	1%

Desired Spray Solution Volume	Amount to be Added to Obtain a 1% Solution	
	Poast	**Oil Concentrate**
1 Gallon	1¼ Fl. Oz.*	1¼ Fl. Oz.
25 Gallons	1 Qt.	1 Qt.
50 Gallons	2 Qts.	2 Qts.
100 Gallons	4 Qts.	4 Qts.

* 2 tablespoons = 1 Fl. Oz.

SOYBEANS—SEPARATE APPLICATIONS OF BASAGRAN OR BASAGRAN + BLAZER/TACKLE TANK MIX* PRECEDED OR FOLLOWED BY POAST.

Applications of BASAGRAN® herbicide or BASAGRAN tank mixed* with Blazer® /Tackle® herbicides can be preceded or followed by Poast® herbicide to obtain broad spectrum control of weeds listed on the respective product labels (refer to this label and labels for Poast and BASAGRAN + Blazer/Tackle tank mix). Also refer to these product labels for timing, rate and other information for ground and aerial applications.

For best results when making separate applications, a minimum period of time is recommended between applications, depending upon their order, according to the table on this page.

*Tank mixes not applicable in California.

Table 10
Separate Applications of BASAGRAN or BASAGRAN + Blazer/Tackle Tank Mix* Preceded or Followed by Poast

Order of Application		Minimum Time Between Applications
First Product(s) Applied	**Second Products(s) Applied**	
BASAGRAN	Poast	24 hours
BASAGRAN +	Poast	7 days

Order of Application		Minimum Time Between Applications
First Product(s) Applied	Second Products(s) Applied	
Blazer/Tackle		
Poast	BASAGRAN or BASAGRAN + Blazer/Tackle	24 hours
* Tank mixes not applicable in California.		

CORN, GRAIN SORGHUM—DIRECTIONS FOR USE

Applications of BASAGRAN® herbicide should be made when weeds are small and actively growing and before weeds reach the maximum size listed in the **Application Rate Table for Corn, Grain Sorghum.** Such applications generally correspond to the crop growth stages of one to five leaves. Corn is tolerant to BASAGRAN at all stages of growth. Grain sorghum is tolerant to BASAGRAN at all stages of growth up to and including early boot stage. Very slight leaf-speckling of corn and grain sorghum may occur but plants generally outgrow this condition within 10 days. Corn types included are field, sweet, and popcorn; and corn grown for seed or silage.

Restrictions and limitations

Do not apply more than a total of 4 pints per acre in one season in corn or 2 pints per acre in one season in grain sorghum.

Seed producers should consult the seed company regarding tolerance of seed production inbred lines to BASAGRAN.

Do not apply to grain sorghum that is heading or blooming.

Do not graze treated corn fields for at least 12 days after the last treatment with BASAGRAN.

California only: Not recommended for control of yellow nutsedge in corn or grain sorghum.

TABLE 11
APPLICATION RATE TABLE FOR CORN, GRAIN SORGHUM

WEEDS CONTROLLED	APPLICATION RATES FOR WEED GROWTH STAGES*			
	1½ PINTS PER ACRE		2 PINTS PER ACRE	
	LEAF STAGE	MAX. HEIGHT	LEAF STAGE	MAX. HEIGHT
Beggarticks	Up to 6	6″	6–8	8″
Bristly Starbur	Up to 4	2″	4–6	3″
Cocklebur	2–6*	6″	6–10	10″
Common Lambsquarters†	—	—	4–8**	2″
Common Ragweed	—	—	4–6**	3″
Dayflower	Up to 6	4″	6–10	8″
Galinsoga	—	—	Cotyledon to 6**	2″
Giant Ragweed††	—	—	Up to 4	6″
Jimsonweed	Up to 6	6″	6–10	10″
Ladysthumb	Up to 6	6″	6–10	10″
Pennsylvania Smartweed	Up to 6	6″	6–10	10″
Prickly Sida or Teaweed	Up to 6	3″	6–8	4″
Spurred Anoda	Up to 6	3″	6–8	4″
Tropic Croton	Up to 2	2″	2–4	4″
Velvetleaf†††	Up to 4†††	2″	4–6†††	5″
Venice Mallow	Up to 6	2″	6–10	4″
Wild Buckwheat	Up to 4	3″	4–6	5″
Wild Mustard	Up to 6	4″	6–10	8″
Wild Sunflower	Up to 4	5″	4–6	8″

For additional weeds see Special Directions section following.

* Do not treat earlier than leaf stage shown and do not count cotyledon leaves.
** Add oil concentrate according to section Addition of oil concentrate, page 4. UAN solution may be substituted for oil concentrate for all weeds except common lambsquarters, common ragweed, and galinsoga.
† Control may be partial or inconsistent.
†† If after the first application a second weed flush develops, retreat according to this rate table (corn only).
††† Add UAN solution according to the Special Directions for Other Weed Problems (see page 18).

SPECIAL DIRECTIONS FOR OTHER WEED PROBLEMS IN CORN

Annual Morningglories

South: (AL, AR, FL, GA, LA, MS, NC, OK, SC, TN, TX, VA). To control smallflower and cypressive morningglories apply a single application of either 1½ pints of BASAGRAN® herbicide per acre to plants not larger than 4 true leaves and 4 inches in height, **or** 2 pints of BASAGRAN per acre to plants not larger than 6 true leaves and 6 inches in height.

To control palmleaf, pitted, tall (common), entireleaf, ivyleaf and purple moonflower morningglories, apply 1½ pints of BASAGRAN per acre to plants not larger than 4 true leaves and 4 inches in height (14 to 18 days after morningglory emergence). Make a second application at the same rate 5 to 14 days later.

All states other than the South (see above): Apply 2 to 3 pints of BASAGRAN per acre to annual morningglories not larger than 4 true leaves. Control may be partial or inconsistent. Add oil concentrate to the spray solution of BASAGRAN/water.

Because morningglories grow very rapidly, it is important to watch the growth stage carefully and to be certain that BASAGRAN is applied to morningglories before they exceed the maximum size recommended on this label.

Canada Thistle

Apply 2 pints of BASAGRAN per acre when plants are from 8 inches tall to the bud stage. Make a second application at the same rate 7 to 10 days later.

Yellow Nutsedge

Two applications are preferred for best results. Apply 1½ to 2 pints of BASAGRAN per acre when plants are 6 to 8 inches tall. If needed, make a second application at the same rate 7 to 10 days later. Add oil concentrate to the spray solution of BASAGRAN/water for each application (see section **Addition of oil concentrate**).

Field and Hedge Bindweed in KY, IL, IN, MI, OH only

For suppression of field and hedge bindweed, apply 2 to 3 pints of BASAGRAN per acre when vines are a maximum of 10 inches long.

Add oil concentrate to the spray solution of BASAGRAN/water according to the section **Addition of oil concentrate.**

Late Cocklebur Rescue Treatment

This treatment is intended to provide only partial control of cocklebur in the event early postemergence treatments were not made. Very thorough spray coverage is essential. Apply a single application of 2 to 3 pints of BASAGRAN per acre to plants up to 24 inches tall or for best results, apply 1½ pints of BASAGRAN per acre to plants up to 24 inches tall and repeat 10 to 14 days later.

SPECIAL DIRECTIONS FOR OTHER WEED PROBLEMS FOR CORN AND GRAIN SORGHUM

Addition of UAN (Urea Ammonium Nitrate) Solution for Velvetleaf and Other Weeds*

In corn and grain sorghum, UAN solution (commonly referred to as 28%, 30% or 32% nitrogen solution) may be added to BASAGRAN in place of oil concentrate for improved control of velvetleaf. Improvement in control of cocklebur, common sunflower, Pennsylvania smartweed, devilsclaw, venice mallow and wild mustard may also be attained.

The UAN solution is an agricultural grade fertilizer used by local dealers for agricultural applications. It should be added to the tank with BASAGRAN when velvetleaf is the primary target weed. However, BASAGRAN plus UAN solution will not provide adequate control of common ragweed and common lambsquarters. If these weeds are present in addition to velvetleaf, then oil concentrate should be used.

With the addition of UAN solution to BASAGRAN, a slight leaf tip burn or yellowing and speckling may occur, but the new growth is normal and crop vigor is not reduced. Refer to your supplier of BASAGRAN for information concerning successful local experience prior to using UAN solution. Do not use brass or aluminum nozzles when spraying BASAGRAN and UAN solution.

Do not add UAN solution to BASAGRAN for use on rice, peanuts, beans, peas, mint or turf.

Rate of UAN solution:
Ground application: 1/2-1 gallon/A.
Air application: 1/2 gallon/A.

*Not applicable in California.

SPECIAL DIRECTIONS FOR OTHER WEED PROBLEMS IN GRAIN SORGHUM

Annual Morningglories

Apply 2 pints of BASAGRAN® herbicide per acre to annual morningglories not larger than 4 true leaves. Control may be partial or inconsistent. Add oil concentrate to the spray solution of BASAGRAN/water, according to the section **Addition of Oil Concentrate,** page 4.

Because morningglories grow very rapidly, it is important to watch the growth stage carefully and to be certain that BASAGRAN is applied to morningglories before they exceed the maximum size recommended (see page 20).

Canada Thistle

Apply 2 pints of BASAGRAN per acre when plants are from 8 inches tall to the bud stage. Control may be partial or inconsistent.

Yellow Nutsedge

Apply 1 1/2 to 2 pints of BASAGRAN per acre when plants are 6 to 8 inches tall. Add oil concentrate to the spray solution of BASAGRAN/water, for each application according to the section **Addition of Oil Concentrate.** Control may be partial or inconsistent.

TANK MIX WITH ATRAZINE AND UAN SOLUTION OR OIL CONCENTRATE FOR POSTEMERGENCE APPLICATIONS IN CORN, GRAIN SORGHUM

The tank mix of BASAGRAN with atrazine and UAN solution or oil concentrate effectively controls a broad spectrum of broadleaf weeds included on the labeling of both products. For the control of annual morningglories, Canada thistle and yellow nutsedge, refer to the section entitled **Special Directions for Other Weed Problems in Corn and Special Directions for Other Weed Problems in Grain Sorghum.**

Atrazine products compatible with BASAGRAN include AAtrex® 80W, AAtrex® 4L, and AAtrex® Nine-O herbicides. Refer to the respective atrazine labels for additional directions and limitations. **Always add** UAN solution or oil concentrate according to sections regarding addition of oil concentrate and addition of UAN solution.

Addition of UAN Solution (Urea Ammonium Nitrate) (Not for use in California)

In corn and sorghum, UAN solution (commonly referred to as 28%, 30% or 32% nitrogen) may be added to BASAGRAN/Atrazine tank mixes. If common ragweed is the predominant weed to be controlled then one quart of oil concentrate should be used instead of UAN solution. With the addition of UAN solution to the BASAGRAN + Atrazine tank mix on corn or sorghum, a slight leaf tip burn or yellowing and speckling may occur. New growth is normal and crop vigor is not reduced.

Do not use brass or aluminum nozzles when spraying BASAGRAN + Atrazine + UAN tank mix solution.

Mixing and spray equipment: Use intake, in-line, or nozzle screens no finer than 50 mesh. Fill tank of a thoroughly clean sprayer one-half to two-thirds full of clean water. Start agitation. Add atrazine and allow to wet and mix thoroughly. Maintain agitation and add BASAGRAN and UAN solution, or oil concentrate if substituted for UAN solution; allow to mix. Last, add the remaining quantity of water and mix thoroughly. Maintain constant agitation during application. Avoid allowing the mixture to stand overnight. Always clean sprayer thoroughly immediately after use by flushing the system with water and a strong detergent. Do not allow cleaning water to contaminate any streams or ponds.

Time and rate of application: Apply tank mix when weeds are small and actively growing and before weeds reach the maximum size listed in **Table 12, Application Rate Table for Tank Mix of BASAGRAN + Atrazine for Corn and Grain Sorghum** (page 20). Such applications generally correspond to the crop growth stages of one to five leaves.

Corn is tolerant to the tank mix at all stages of growth. Grain sorghum is tolerant to the tank mix at all stages of growth up to and including early boot stage.

Very light leaf speckling may occur in corn and grain sorghum, but plants generally outgrow this condition within 10 days.

Refer to the **Acreage Conversion Table** on page 20 for application rates depending on formulation. A cultivation may be necessary if all weeds are not controlled or if regrowth of weeds occurs.

Restrictions and limitations for tank mix with atrazine (partial list)

Read and follow the restrictions and limitations on the labels for BASAGRAN and AAtrex. The most restrictive labeling applies in tank mixes.

Do not use tank mix when crop is under stress from prolonged cold, wet weather, poor fertility, or other factors or when crop is wet and succulent from recent rainfall as crop injury may occur.

Seed producers should consult the seed company regarding tolerance of seed production inbred lines to tank mix.

Do not apply to grain sorghum that is heading out or blooming.

Do not make more than one application of tank mix per season.

Do not apply more than 4 pints of BASAGRAN per acre in one season in corn or 2 pints of BASAGRAN per acre in one season in grain sorghum.

Do not graze treated area or feed treated forage to livestock for 21 days following application (see label for AAtrex).

Acreage Conversion Table

Tank Mix Rate Recommendation (lb ai/A)*	Amount of Formulated Product									
	Basagran	Atrazine (AAtrex)								
	1 Acre	1 Acre			10 Acres			50 Acres		
	Pts.	80W Lbs.	Nine-0 Lbs.	4L Pts.	80W Lbs.	Nine-0 Lbs.	4L Pts.	80W Lbs.	Nine 0 Lbs.	4L Pts.
1/2 + 1/2	1	5/8	6/10	1	6 1/4	6	10	31 1/2	30	50
3/4 + 3/4	1 1/2	1	9/10	1 1/2	10	9	15	50	45	75

* According to weed growth stage indicated in table below.

Table 12
Application Rate Table for Tank Mix of BASAGRAN + Atrazine for Corn and Grain Sorghum

Weeds Controlled	Application Rates for Weed Growth Stages*			
	1/2 + 1/2 lb. ai/A*		3/4 + 3/4 lb ai/A*	
	Leaf Stage	Max. Height	Leaf Stage	Max. Height
Cocklebur	2-6**	4"	6-10	8"
Common Lambsquarters	4-8	2"	8-12	4"
Common Ragweed	Up to 4***	2"	4-6***	3"
Giant Ragweed	Up to 4	4"	4-6	6"
Jimsonweed	Up to 6	4"	6-10	8"
Ladysthumb	Up to 6	4"	6-10	8"
Morningglory (tall and pitted)	4	4"	5-6	6"
Pennsylvania Smartweed	Up to 6	4"	6-10	8"
Redroot Pigweed	4-8	2"	8-12	4"
Smooth Pigweed	4-8	2"	8-12	4"
Tall Waterhemp	4-8	2"	4-8	2"
Velvetleaf	Up to 4	2"	4-6	5"
Venice Mallow	Up to 6	2"	6-10	4"
Wild Buckwheat	Up to 4	3"	4-6	5"
Wild Mustard	Up to 6	4"	6-10	8"
Wild Sunflower	Up to 4	4"	4-6	8"

Other weeds: Other weeds listed on the label for BASAGRAN® herbicide at the 3/4 pound rate will also be controlled with the 3/4 plus 3/4 pound BASAGRAN + atrazine tank mix. Refer to page 17.

* Refer to **Acreage Conversion Table** for recommended rate of formulated product per acre. Add one gallon of UAN solution or one quart of oil concentrate for all weeds except common ragweed. When common ragweed predominates, use oil concentrate. See section **Addition of UAN solution** or **Addition of oil concentrate.**

** Do not treat earlier than leaf stage shown and do not count cotyledon leaves.

*** Add one quart per acre of oil concentrate, not UAN solution, when this weed predominates.

RICE—DIRECTIONS FOR USE (NOT FOR USE IN CALIFORNIA)

Make postemergence applications of BASAGRAN® herbicide early, before weeds exceed the maximum size listed in **Table 13 Application Rate Table for Rice.**

Early application produces the most beneficial effect on crop yields, allows use of the lower rate (depending on weed species), and makes it easier to obtain thorough spray coverage. Delay in application which permits weeds to exceed the maximum size for a given rate will result in inadequate control.

BASAGRAN has no adverse effect on rice when used according to directions and may be used on the first and second (ratoon) crops. If grasses are a problem, use propanil in tank mixture with BASAGRAN.

For optimal coverage when applying BASAGRAN by air in rice, orient all nozzles straight down. For additional aerial application information, refer to Directions for use.

Alternate flooding culture: In Texas, Louisiana, Arkansas and Mississippi, weed growth stages generally correspond to rice that is tillering (stooling) and occur prior to the permanent flood. Application of BASAGRAN **must** be made when there is no water on the field and 24 hours or more prior to flooding. If BASAGRAN cannot be applied until after flooding see directions under **Continuous flooding culture.**

Continuous flooding culture: In states using continuous flooding culture or, when treating after permanent flood, treatment should be made only when weeds are above the surface of the water. Weeds submerged at the time of application are not adequately controlled.

For early treatment, water may be partly or completely drained to expose more weed growth to spray applications of BASAGRAN. Do not raise water level for at least 24 hours after application or unsatisfactory control may result. Do not use ground equipment for applications of flooded fields because splashing will wash BASAGRAN off weed leaf surfaces and ineffective control may result.

Restrictions and limitations Rice straw may be fed to livestock.

Do not apply BASAGRAN to rice with ground equipment when field is flooded because splashing will wash BASAGRAN off weed leaf surfaces and ineffective control may result.

Do not apply more than 6 pints of BASAGRAN per acre in one season. (Maximum of 4 pints per acre in first crop and 2 pints per acre in second [ratoon] crop.)

Do not use BASAGRAN on rice fields in which the commercial cultivation of catfish or crayfish is practiced.

Do not use water containing BASAGRAN residues from rice cultivation to irrigate crops used for food or feed unless BASAGRAN is registered for use on these crops.

Do not contaminate water when disposing of equipment washwaters.

Tank mix with propanil

Use a tank mix of BASAGRAN + propanil by ground or air for the control of mixed populations of grasses, sedges and broadleaf weeds listed as susceptible on the two product labels. Prepare tank mix by adding BASAGRAN to half the final volume of water with agitator running. Then add propanil and bring mix to final volume. Agitation must be continuous from time of mixing through spraying.

Apply BASAGRAN at the rate of 1½ pints per acre plus either 3 quarts (4 lbs./gal.) or 4 quarts (3 lbs./gal.) of propanil formulation* in the spray volume specified on this labeling.

Apply this tank mix only to drained fields.

Restrictions and limitations

Do not use propanil on second crop (ratoon) rice. When applying tank mix of BASAGRAN/propanil by air, orient all nozzles straight back in accordance with the propanil label.

Observe all restrictions and limitations on this and the propanil labels. In tank mixes the most restrictive labeling applies.

*Propanil products compatible with BASAGRAN are STAM F-34 (3 lbs./gal.); STAM M-4 (4 lbs./gal.); Helena 4 (4 lbs./gal.) and Crystal 4E (4 lbs./gal.).

TABLE 13
APPLICATION RATE TABLE FOR RICE—DRAINED FIELDS

WEEDS CONTROLLED (ALL STATES)	APPLICATION RATES FOR WEED GROWTH STAGES			
	1½ PTS. PER ACRE*		2 PTS. PER ACRE*	
	LEAF STAGE	MAX. HEIGHT	LEAF STAGE	MAX. HEIGHT
Cocklebur	2–10	10"	10–15	15"
Dayflower	2–10	6"	10–15	10"
Ducksalad	—	—	6–10**	6"
Gooseweed	4–6	4"	6–10	8"
Redstem	Up to 6	4"	6–10	8"
Redweed	4–6	6"	6–10	8"
Smartweed	2–10	6"	10–15	10"
Spikerush	2–6	6"	6–8	8"
Water Plantains				
Arrowhead	—	—	Up to 4	7"
Common Water Plantain	—	—	Up to 4	7"
Yellow Nutsedge	4–6	6"	6–8	10"

* If after the first application a second weed flush develops, retreat according to this rate table.
** Control may be partial or inconsistent.

TABLE 14
APPLICATION RATE TABLE FOR RICE—FLOODED FIELDS

WEEDS CONTROLLED ALL STATES	APPLICATION RATES FOR WEED GROWTH STAGES			
	1½ PTS. PER ACRE*		2 PTS. PER ACRE*	
	MAXIMUM HEIGHT ABOVE SOIL	MINIMUM HEIGHT RANGE ABOVE WATER LEVEL	MAXIMUM HEIGHT ABOVE SOIL	MINIMUM HEIGHT RANGE ABOVE WATER LEVEL
Cocklebur	10"	3"–6"	15"	6"–10"
Dayflower	6"	3"–5"	10"	5"–8"
Ducksalad	—	—	—	—
Gooseweed	—	—	—	—
Redstem	4"	2"–3"	8"	4"–6"
Redweed	—	—	—	—
Smartweed	6"	2"–5"	10"	5"–8"
Spikerush	—	—	—	—
Water Plantains				
Arrowhead	—	—	7"	5"–6"
Common Water Plantain	—	—	7"	5"–6"
Yellow Nutsedge	6"	4"–5"	10"	6"–8"

* If after the first application a second weed flush develops, retreat according to this rate table.

PEANUTS—DIRECTIONS FOR USE

Applications of BASAGRAN® herbicide should be made when weeds are small and actively growing and before weeds reach the maximum size listed in the **Application Rate Table for Peanuts**. Such applications may occur from peanut cracking through pegging.

Peanuts are tolerant to BASAGRAN at all stages of growth, but slight leaf-speckling may occur under certain conditions (see **Restrictions and limitations**). Peanut plants generally outgrow this condition within 10 days.

Restrictions and limitations Do not apply BASAGRAN if peanuts show injury (leaf phytotoxicity and/or plant stunting) produced by any prior herbicide applications (preplant incorporated, preemergence, cracking and/or postemergence), because this injury may be enhanced and/or prolonged. In the Southeast, in-furrow treatments of insecticides/nematicides may predispose the peanuts to injury from BASAGRAN.

Do not apply more than a total of 4 pints of BASAGRAN per acre in one season.

Peanut hay and forage may be fed to livestock.

Do not graze treated peanut fields for at least 50 days after the last treatment of BASAGRAN.

TABLE 15
APPLICATION RATE TABLE FOR PEANUTS

WEEDS CONTROLLED	APPLICATION RATES FOR WEED GROWTH STAGES†					
	1 PINT PER ACRE††		1½ PINTS PER ACRE		2 PINTS PER ACRE	
	LEAF STAGE	MAX. HEIGHT	LEAF STAGE	MAX. HEIGHT	LEAF STAGE	MAX. HEIGHT
Balloonvine	—	—	2–4	2″	4–6	3″
Beggarticks	—	—	Up to 6	6″	6–8	8″
Bristly Starbur	—	—	Up to 4	2″	4–6	3″
Cocklebur	2–4*	4″	2–6*	6″	6–10	10″
Coffee Senna	—	—	—	—	Up to 1** Pinnate	2″
Common Ragweed	—	—	—	—	4–6**	3″
Dayflower	—	—	Up to 6	4″	6–10	8″
Devilsclaw	—	—	—	—	Up to 6**	3″
Giant Ragweed†	—	—	—	—	Up to 4	6″
Jimsonweed	Up to 4	4″	Up to 6	6″	6–10	10″
Ladysthumb	Up to 4	4″	Up to 6	6″	6–10	10″
Pennsylvania Smartweed	Up to 4	4″	Up to 6	6″	6–10	10″
Prickly Sida or Teaweed	—	—	Up to 6	3″	6–8	4″
Spurred Anoda	—	—	Up to 6	3″	6–8	4″
Tropic Croton	—	—	Up to 2	2″	2–4	4″
Velvetleaf	—	—	Up to 4	2″	4–6	5″
Wild Sunflower	—	—	Up to 4	5″	4–6	8″

For additional weeds see Special Directions section following.

* Do not treat earlier than leaf stage shown and do not count cotyledon leaves.

** Add oil concentrate according to section Addition of oil concentrate page 4.

† If a second flush occurs, retreat acording to this rate table.

†† Apply before weeds reach the maximum size or leaf stage indicated. If regrowth develops, makes a second application of 1 pint 7 to 14 days after the first application.

SPECIAL DIRECTIONS FOR OTHER WEED PROBLEMS IN PEANUTS

Annual Morningglories

To control smallflower and cypressvine morningglories apply a single application of either 1½ pints of BASAGRAN per acre to plants not larger than 4 true leaves and 4 inches in height, **or** 2 pints of BASAGRAN per acre to plants not larger than 6 true leaves and 6 inches in height.

To control palmleaf, pitted, common, entireleaf, purple moonflower and ivyleaf morningglories, apply 1½ pints of BASAGRAN per acre to plants not larger than 4 true leaves and 4 inches in height (14 to 18 days after morningglory emergence). Make a second application at the same rate 5 to 14 days later.

Because morningglories grow very rapidly, it is important to watch the growth stage carefully and to be certain that BASAGRAN is applied to morningglories before they exceed the maximum size recommended on this label.

Yellow Nutsedge

Two applications are preferred for best results. Apply 1½ to 2 pints of BASAGRAN per acre when plants are 6 to 8 inches tall. In Texas and Oklahoma use 2 pints.

If needed, make a second application at the same rate 7 to 10 days later. Add oil concentrate to the spray solution of BASAGRAN/water, according to the section Addition of oil concentrate, page 4.

Late Cocklebur Rescue Treatment

This treatment is intended to provide only partial control of cocklebur in the event early postemergence treatments were not made. Very thorough spray coverage is essential. Apply a single application of 2 to 3 pints of BASAGRAN per acre to plants up to 24 inches tall or, for best results, apply 1½ pints of BASAGRAN per acre to plants up to 24 inches tall and repeat 10 to 14 days later.

BASAGRAN + 2,4-DB TANK MIX*—PEANUTS

General and application information, Restrictions and limitations

General information

These directions are intended to provide the user of BASAGRAN® herbicide with instructions for tank mixing with 2,4-DB (such as Butyrac® 200 herbicide or Butoxone® 200 herbicide) to control entireleaf, tall (common), and ivyleaf morningglories in addition to all the other weeds listed in Table 15. Weeds must be actively growing and at the recommended growth stages. Delay in application, which permits weeds to exceed the maximum size stated, will result in inadequate control.

Water volume and spray pressure
(Refer to section entitled "Directions for use" for additional information.)

Ground Equipment: For broadcast application, use a minimum of 20 gallons of total spray solution per acre and 40 psi pressure with flat fan or hollow cone nozzles spaced 20 inches apart.

Mixing: Fill the spray tank half full with water and add the recommended amount of BASAGRAN and 2,4-DB while the agitator is running, then finish filling. Add the remaining quantity of water.

Coverage

The tank mix is effective partly through contact action. Therefore, weeds must be thoroughly covered with spray. If a treatment is made to morningglories larger than 10″, control will be inadequate.

Large crop-and-weed leaf canopies shelter smaller weeds and prevent adequate spray coverage.

Peanuts are tolerant to the tank mix of BASAGRAN + 2,4-DB; however, under certain conditions peanuts may have a white, bleached appearance and the leaves may be slightly elongated.

Restrictions and limitations for tank mix with 2,4-DB
(partial list)

Read and follow the restrictions and limitations on the labels for BASAGRAN and 2,4-DB. The most restrictive labeling applies in tank mixes.

Use only amine formulations of 2,4-DB.

Do not apply to or allow drift to any other adjacent crop.

Do not add oil or any other additives to tank mix.

Do not apply tank mix if peanuts show injury (leaf phytotoxicity and/or plant stunting) produced by any other prior herbicide treatment or by disease because this injury may be enhanced and/or prolonged.

Do not apply tank mix during prolonged periods of drought or during unseasonably cold weather, as unsatisfactory weed control may result.

Rainfall soon after application (within 8 hours) may nullify the effectiveness of the tank mix.

Do not apply tank mix to peanuts that have been subject to stress conditions such as hail damage, flooding, drought, injury from other herbicides, or unseasonably cold or widely fluctuating temperatures because injury may result.

Do not apply more than 2 applications of the tank mix per season.

Do not apply within 30 days of harvest in Oklahoma, Texas and New Mexico or 45 days in the Virginia-Carolina area. (See label for 2,4-DB.)

Do not feed treated peanut vines and peanut hay to livestock. (See label for 2,4-DB.)

* Tank mix not applicable in California.

TABLE 16
BASAGRAN + 2,4-DB TANK MIX—PEANUTS
RATE AND TIME OF APPLICATION

PRODUCT	RATE	WEEDS CONTROLLED/WEED SIZE		ADDITIVES
BASAGRAN	1½-2 pints/A according to weed species and size. (See Table 15, Page 23)	Balloonvine Beggarticks Bristly Starbur Cocklebur Coffee Senna Common Ragweed Cypressvine Morningglory Dayflower Devilsclaw Giant Ragweed	Jimsonweed Ladysthumb Pennsylvania Smartweed Prickly Sida or Teaweed Smallflower Morningglory Spurred Anoda Tropic Croton Velvetleaf Wild Sunflower Yellow Nutsedge*	Do not add Oil Concentrate or any other additives (including UAN solution) to this tank mix.
plus 2,4-DB (amine formulation)	plus 8 fl. oz./A of Butoxone 200 or Butyrac 200. (0.125 pound ae**/A.)	Morningglories: Ivyleaf Tall (Common) Entireleaf	Vines up to 10" long	

* May requires two applications of BASAGRAN in accordance with this label.
** Acid equivalent.

BASAGRAN + BLAZER TANK MIXES*—PEANUTS
General and application information, Restrictions and limitations

General information

The tank mix of BASAGRAN® + Blazer® herbicides will control the weeds listed in Tables 17 and 18.

Table 17—All states
BASAGRAN 1 pint/A
Blazer 1 pint/A

Table 18—All states for additional weeds or larger sizes
BASAGRAN 1½–2 pint/A
Blazer 1 pint/A

Time of application

The timing of application should be in accordance with weed growth stages indicated in the respective tables and when weeds are actively growing. Such applications generally occur from peanut cracking through pegging.

Delay in application which permits weeds to exceed the maximum size stated will result in inadequate control.

Water volume and spray pressure

Ground equipment: Use a minimum of 20 gallons of water per broadcast acre and a minimum of 40 psi pressure (measure at the boom—not at the pump or in the line). When crop and weed foliage is dense, use up to 50 gallons of water and up to 80 psi pressure.

Use standard high pressure pesticide hollow cone or flat fan nozzles spaced 20 inches apart. Do not use flood, whirl chamber or controlled droplet application (CDA) nozzles.

Air equipment: Use a minimum of 10 gallons of total spray solution per acre.

Mixing:

Fill the spray tank half full with water and add the recommended amount of product in the following order: BASAGRAN, Blazer, oil concentrate—while the agitator is running. Then add the remaining quantity of water.

Coverage

Through coverage of actively growing weeds is essential. Large crop-and-weed leaf canopies shelter smaller weeds and can prevent adequate spray coverage. Peanuts are tolerant to the above tank mixes; however, under certain conditions peanuts may burn, crinkle and bronze.

Restrictions and limitations (partial list) for tank mix with Blazer

Observe all applicable directions, restrictions and precautions on this label and the label fror Blazer. The most restrictive labeling applies in tank mixes.

Do not apply tank mix if peanuts show injury (leaf phytotoxicity and/or plant stunting) produced by any other prior herbicide treatment because this injury may be enhanced and/or prolonged.

Do not apply tank mix during prolonged periods of drought or during unseasonably cold weather, as unsatisfactory weed control may result.

Do not apply tank mix to peanuts that have been subject to stress conditions such as hail damage, flooding, drought, or unseasonably cold or widely fluctuating temperatures because injury may result.

Do not add a surfactant or oil concentrate except where specifically recommended.

* Tank mix not applicable in California.

TABLE 17
ALL STATES
BASAGRAN + BLAZER TANK MIX—PEANUTS
RATE AND TIME OF APPLICATION

PRODUCT	RATE	WEEDS CONTROLLED/WEED SIZE			ADDITIVE (RATE)
			LEAF STAGE	MAX. HEIGHT	
BASAGRAN + Blazer	1 pint/A + 1 pint/A[c]	Black Nightshade	Up to 6	2"	Oil Concentrate (1 pt./A)
		Bristly Starbur	4–6	3"	
		Cocklebur	2–6	6"	
		Common Lambsquarters	4–6	2"	
		Common Ragweed[a]	4–6	3"	
		Crotalaria**	Up to 6	6"	
		Jimsonweed	Up to 6	6"	
		Morningglories[b]	Up to 4	4"	
		Pennsylvania Smartweed	Up to 6	6"	
		Prickly Sida (Teaweed)*	Up to 4	2"	
		Redroot Pigweed	Up to 6	3"	
		Sesbania**	Up to 4	6"	
		Smooth Pigweed	Up to 6	3"	
		Spurred Anoda*	Up to 4	2"	
		Velvetleaf*	Up to 4	2"	
		Wild Mustard	Up to 6	4"	

[a] For common ragweed up to 6 inches tall and 10 leaves use 1½ pints of BASAGRAN with 1 pint of Blazer.
[b] For common (tall) morningglory, increase rate of BASAGRAN to 1½ pts.
[c] Blazer may also be included in the tank mix at a rate of up to 2 pints per acre; however, this will increase the severity and/or frequency with which peanut injury is observed.
* For more consistent control, increase the rate of BASAGRAN to 1½ pints/A.
** If crotalaria or sesbania are present, add Triton AG-98 at the rate of ½ pint per 100 gallons of spray solution; but do not combine Triton AG-98 with oil concentrate.

TABLE 18
ALL STATES FOR ADDITIONAL WEEDS OR LARGER WEED SIZES
BASAGRAN + BLAZER TANK MIX—PEANUTS
RATE AND TIME OF APPLICATION

PRODUCT	RATE	WEEDS CONTROLLED/WEED SIZE			ADDITIVE RATE
BASAGRAN	1½-2 pints/A according to weed species and size (See Table 1, Page 5)	Balloonvine Beggarticks Bristly Starbur Cocklebur Coffee Senna[b] Common Ragweed[b] Cypressvine Morningglory Dayflower Devilsclaw[b] Giant Ragweed Jimsonweed	Ladysthumb Marshelder Pennsylvania Smartweed Prickly Sida or Teaweed Smallflower Morningglory Spurred Anoda Tropic Croton Velvetleaf[b] Wild Sunflower Yellow Nutsedge[b]		Oil Concentrate[b]
plus	plus		**Leaf Stage**	**Max. Height**	
Blazer	1 pint/A	Common Ragweed[b]	Up to 10	6"	
		Black Nightshade	Up to 6	2"	
		Morningglories	Up to 4	4"	
		Crotalaria[b]	Up to 6	6"	
		Sesbania[b]	Up to 4 Pinnate	6"	
		Tall Waterhemp	Up to 6	3"	
		Redroot Pigweed	Up to 6	3"	
		Smooth Pigweed	Up to 6	3"	

[a] Choose the rate of BASAGRAN® herbicide (1½ or 2 pints per acre) according to the size and species of the weeds to be controlled with BASAGRAN alone (see Application Rate Table for Peanuts). Then add Blazer® herbicide at the rate of 1 pint per acre, if needed, to control the additional weed species, up to the maximum size, as shown in the tank mix time of application table above. Blazer may also be included in the tank mix at a rate of up to 2 pints per acre; however, this will increase the severity and/or frequency with which peanut injury is observed.

[b] Add oil concentrate to the tank mix according to recommendations in Table 15, Application Rate Table for Peanuts, page, 23. The addition of oil concentrate may increase the severity and frequency of peanut injury. If crotalaria or sesbania are present add Triton AG-98 at the rate of ½ pint per 100 gallons of spray solution. But do not mix Triton AG-98 with oil concentrate.

BASAGRAN + GRAMOXONE SUPER TANK MIX*—PEANUTS

General and application information, Restrictions and limitations

The tank mix of BASAGRAN® herbicide plus Gramoxone Super® herbicide will also control certain weeds not controlled by BASAGRAN alone (see Table 19 below).

Since this tank mix is effective mainly through contact action, thorough coverage of weeds is essential for effective weed control. Large crop-and-weed leaf canopies shelter smaller weeds and prevent adequate spray coverage. Crop foliage present at application may bronze or crinkle, but the plants will soon outgrow these effects and develop normally.

Time and rate of application

The rates of application and weed sizes for the use of this tank mix are given in the Rate and Time Application Table. This tank mix should be applied at the ground crack stage of peanuts to control an early flush of weeds. A second application may be applied up to 28 days after ground crack stage. Do not make more than two applications of this tank mix to the same crop.

Apply the BASAGRAN + Gramoxone Super tank mix to weeds which are actively growing and before weeds reach the maximum size listed in the Application Recommendation Table.

Application to weeds which exceed the maximum size stated may result in inadequate control.

Spray additivies

Always add a nonionic surfactant containing at least 50% surface active agent at the rates listed in the Tank Mix Recommendation Table below. Do not use crop oil concentrate or any other oil-based additive with this tank mix.

Water volume and spray pressure

Use a minimum of 20 gallons of total spray mixture per acre (broadcast basis) and 30-50 psi pressure and standard high pressure hollow cone or flat fan nozzles spaced 20 inches apart. Use only ground equipment to apply this tank mix.

Mixing

Fill the spray tank half full with water and add the recommended amount of BASAGRAN, Gramoxone Super and nonionic surfactant while the agitator is running. Then add the remaining quantity of water.

Restrictions and limitations
(partial list)

Read and follow the restrictions and limitations on the labels for BASAGRAN and Gramoxone Super. The most restrictive labeling applies in tank mixes.

Do not apply tank mix if peanuts show injury (leaf phytotoxicity and/or plant stunting) produced by any other prior herbicide treatment because this injury may be enhanced and/or prolonged.

Do not apply tank mix during prolonged periods of drought or during unseasonably cold weather, as unsatisfactory weed control may result.

Do not apply tank mix to peanuts that have been subjected to stress conditions such as hail damage, flooding, drought, or unseasonably cold or widely fluctuating temperatures because injury may result.

Avoid drift to all other crops and non-target areas. Crops other than peanuts may be severely injured by drift.

*Tank mix not applicable in California.

TABLE 19
BASAGRAN + GRAMOXONE SUPER TANK MIX—PEANUTS
RATE AND TIME OF APPLICATION RECOMMENDATION TABLE

PRODUCT	RATE	WEEDS CONTROLLED	WEED GROWTH STAGES		ADDITIVE
			LEAF STAGE	MAX. HEIGHT	
BASAGRAN	1 pint/A	Balloonvine	2–4	2″	Use suitable nonionic surfactant at 8 oz./100 gallons water or as directed on respective label.
		Beggarticks	Up to 6	6″	
		Bristly Starbur	Up to 4	2″	
		Cocklebur	2–6*	6″	
		Coffee Senna	Up to 1 Pinnate	2″	
		Common Ragweed	Up to 6	3″	
		Dayflower	Up to 6	4″	
		Devilsclaw	Up to 6	3″	
		Giant Ragweed	Up to 4	6″	
		Jimsonweed	Up to 6	6″	
		Ladysthumb	Up to 6	6″	
		Pennsylvania Smartweed	Up to 6	6″	
		Prickly Sida or Teaweed	Up to 6	3″	
		Spurred Anoda	Up to 6	3″	
		Tropic Croton	Up to 2	2″	
		Velvetleaf	Up to 4	2″	
		Wild Sunflower	Up to 4	5″	
plus Gramoxone Super	plus ⅔ pint/A	Smooth Pigweed	Up to 6	4″	
		Redroot Pigweed	Up to 6	4″	
		Tall Waterhemp	Up to 6	4″	
		Sicklepod	Up to 4	4″	
		Florida Beggarweed	Up to 4	4″	
		Morningglories	Up to 6	4″	

* Do not treat earlier than leaf stage shown and do not count cotyledon leaves.

Beans (dry or succulent)—directions for use

Apply BASAGRAN® herbicide early postemergence when weeds are small and actively growing and before weeds reach the maximum size listed in the **Application Rate Table for Beans.** Such weed growth stages generally correspond to bean stages of greater than one expanded trifoliate leaf.

Beans are tolerant to BASAGRAN after the first trifoliate leaf has fully expanded. Snap bean injury can be very pronounced. Even at the tolerant stages yellowing, bronzing, speckling or burning of leaves may occur under certain conditions (see **Restrictions and limitations**). This temporary injury is generally outgrown without delaying podset or maturity or reducing yield. The use of oil with BASAGRAN may increase injury and may reduce yields.

Tolerant bean types are navy, pinto, pinks, great northern, kidney, red, white, cranberry, black turtle soup, small limas, large limas and snap beans.

TABLE 20
APPLICATION RATE TABLE FOR BEANS (DRY OR SUCCULENT)

WEEDS CONTROLLED	APPLICATION RATES FOR WEED GROWTH STAGES					
	1 PINT PER ACRE*		1½ PINTS PER ACRE		2 PINTS PER ACRE	
	LEAF STAGE	MAX. HEIGHT	LEAF STAGE	MAX. HEIGHT	LEAF STAGE	MAX. HEIGHT
Cocklebur (PNW)	2–4**	4″	2–6**	6″	6–10	10″
Common Lambsquarters†	Up to 4	1″	—	—	4–8***	2″
Common Purslane	—	—	Up to 4	1″	4–6	2″
Common Ragweed	—	—	—	—	4–6	3″
Devilsclaw	—	—	—	—	Up to 6***	3″
Galinsoga	—	—	—	—	Cotyledon Up to 6***	2″
Giant Ragweed††	—	—	—	—	2–4	6″
Hairy Nightshade****	—	—	—	—	2–6	4″
Jimsonweed	—	—	Up to 6	6″	6–10	10″
Ladysthumb	—	—	Up to 6	6″	6–10	10″
Marshelder	—	—	Up to 4	2″	4–8	4″
Pennsylvania Smartweed	Up to 4	4″	Up to 6	4″	6–10	10″
Prickly Sida or Teaweed	—	—	Up to 6	3″	6–8	4″
Shepherdspurse†††	—	—	Up to 6	4″	6–10	8″
Velvetleaf[a]	Up to 3	2″	Up to 4	2″	4–6***	5″
Venice Mallow	Up to 4	2″	Up to 6	2″	6–10	4″
Wild Mustard (PNW)	Up to 4	2″	Up to 6	4″	6–10	10″
Wild Sunflower	Up to 2	3″	Up to 4	5″	4–6	8″

* Apply before weeds reach the maximum size or leaf stage indicated. If regrowth develops, make a second application of 1 pint 7 to 14 days after the first application. (This rate is not applicable in California.)
** Do not treat earlier than leaf stage shown and do not count cotyledon leaves.
*** Add oil concentrate according to Directions for use—all crops.
**** BASAGRAN does not adequately control black nightshade.
† Control may be partial or inconsistent.
†† If after the first application a second weed flush develops, retreat according to this rate table.
††† Do not treat rosette before seed stalk appears.
PNW —See special directions for Pacific Northwest.
[a] See page 29 for UAN addition.

Western irrigated areas

In the Western irrigated areas, it may be necessary to irrigate prior to treatment with BASAGRAN® herbicide to ensure that weeds are growing actively. Weeds that are growing under moisture stress are not actively growing and are not satisfactorily controlled.

Avoid application of BASAGRAN during prolonged periods of cold weather (day temperature below 75°F and night temperature below 55°F for 2 to 5 days) because weed control may be nullified.

Addition of urea ammonium nitrate (UAN) solution for velvetleaf and other weeds*

In beans, UAN solution (commonly referred to as 28%, 30%, 32% nitrogen solution) may be added to BASAGRAN in place of oil concentrate for improved control of velvetleaf. Improvement in the control of cocklebur, wild sunflower, Pennsylvania smartweed, devilsclaw, venice mallow and wild mustard may also be attained.

The UAN solution is an agricultural grade fertilizer used by local dealers for agricultural applications. It should be added to the tank mix with BASAGRAN when velvetleaf is the primary target weed. However, BASAGRAN plus UAN solution will not provide adequate control of common ragweed and common lambsquarters. If these weeds or other weeds requiring oil concentrate (see **Table 1**) are present in addition to velvetleaf, then oil concentrate should be used.

With the addition of UAN solution or UAN solution plus oil concentrate to BASAGRAN on beans, a slight leaf burn may occur, but the new growth is normal and crop vigor is not reduced. Refer to your supplier of BASAGRAN for information concerning successful local experience prior to using UAN solution. Do not use brass or aluminum nozzles when spraying BASAGRAN + UAN solution.

Rate of UAN solution:

Ground application— 1/2-1 gallon/acre

Air application— 1/2 gallon/acre

SPECIAL DIRECTIONS FOR OTHER WEED PROBLEMS IN BEANS

Yellow Nutsedge

Two applications are preferred for best results. Apply 1 1/2 to 2 pints (except Pacific Northwest) of BASAGRAN per acre when plants are 6 to 8 inches tall. If needed, make a second application at the same rate 7 to 10 days later.

Add oil concentrate to the spray solution of BASAGRAN/water for each application according to the **Directions for use—all crops.**

In California: apply 2 pints of BASAGRAN per acre when plants are 6 to 8 inches tall. Make a second application at the same rate 10 to 14 days later. The use of oil concentrate with BASAGRAN may increase injury and may reduce yields.

Canada Thistle

Apply 2 pints of BASAGRAN per acre when plants are from 8 inches tall to the bud stage. Make a second application at the same rate 7 to 10 days later.

Field and Hedge Bindweed in KY, IL, IN, MI, OH only

For suppression of field and hedge bindweed, apply 2 or 3 pints of BASAGRAN per acre when vines are a maximum of 10 inches long. Add oil concentrate to the spray solution of BASAGRAN/water, according to the **Directions for use—all crops.**

Pacific Northwest (ID, OR, WA)

For control of cocklebur, yellow nutsedge, and wild mustard, use only the 2 pints rate.

For cocklebur, treat when plants are in the 2 to 10 leaf stage and a maximum height of 10 inches.

For yellow nutsedge, follow the directions indicated above using only the 2 pints rate.

For wild mustard, treat when plants are up to the 10 leaf stage and a maximum height of 10 inches.

Restrictions and limitations (partial list)

Do not apply BASAGRAN to bean fields until beans have at least the first trifoliate leaf fully expanded because severe crop damage may occur.

Do not apply BASAGRAN to blackeyes grown in California or to garbanzo beans at any stage of growth, as severe crop damage may occur.

Do not apply more than a total of 4 pints of BASAGRAN per acre in one season.

In succulent beans only, do not apply BASAGRAN within 30 days of harvest.

*Not for use in California.

PEAS (DRY OR SUCCULENT)—DIRECTIONS FOR USE

Apply BASAGRAN® herbicide early postemergence when weeds are small and actively growing and before weeds reach the maximum size listed in the **Application Rate Table for Peas.** Such weed growth stages generally correspond to pea stages of greater than 3 pairs of leaves (or 4 nodes).

Peas are tolerant to BASAGRAN after 3 pairs of leaves (or 4 nodes) are present. Pea injury can be very pronounced. Even at the tolerant stages yellowing, bronzing, speckling or burning of leaves may occur under certain conditions (see **Restrictions and limitations**). This temporary injury is generally outgrown without delaying podset or maturity or reducing yield.

Tolerant pea types are garden peas and southern peas.

Western irrigated areas

In the Western irrigated areas, it may be necessary to irrigate prior to treatment with BASAGRAN to ensure weeds are growing actively. Weeds that are growing under moisture stress are not actively growing and are not satisfactorily controlled.

Avoid application of BASAGRAN during prolonged periods of cold weather (day temperature below 75°F and night temperature below 55°F for 2 to 5 days) because weed control may be nullified.

Restrictions and limitations (partial list)

Do not apply BASAGRAN to pea fields until peas have at least 3 pairs of leaves (or 4 nodes) because severe crop damage may occur.

Do not apply BASAGRAN to peas under stress from root rot.

In the Southeast, in-furrow treatments of insecticides/nematicides may possibly predispose the peas to injury from BASAGRAN.

Do not apply BASAGRAN to blackeyes grown in California or to garbanzo beans at any stage of growth, as severe crop damage may occur.

Do not apply more than a total of 4 pints of BASAGRAN per acre in one season.

Do not add oil to BASAGRAN for use on peas.

Table 21
Application Rate Table for Peas (Dry or Succulent)

Weeds Controlled	Application Rates for Weed Growth Stages			
	1½ Pints Per Acre		2 Pints Per Acre	
	Leaf Stage	Max. Height	Leaf Stage	Max. Height
Cocklebur (PNW)	2–4*	6″	6–10	10″
Common Purslane	Up to 4	1″	4–6	2″
Giant Ragweed†	—	—	2–4	6″
Hairy Nightshade**	—	—	2–6	4″
Jimsonweed	Up to 6	6″	6–10	10″
Ladysthumb	Up to 6	6″	6–10	10″
Marshelder	Up to 4	2″	4–8	4″
Pennsylvania Smartweed	Up to 6	4″	6–10	10″
Prickly Sida or Teaweed	Up to 6	3″	6–8	4″
Shepherdspurse††	Up to 6	4″	6–10	8″
Venice Mallow	Up to 6	2″	6–10	4″
Wild Mustard (PNW)	Up to 6	4″	6–10	10″
Wild Sunflower	Up to 4	5″	4–6	8″

For additional weeds see **Special Directions** section following.

* Do not treat earlier than leaf stage shown and do not count cotyledon leaves.

** BASAGRAN does not adequately control black nightshade.

† If after the first application a second weed flush develops, re-treat according to this rate table.

†† Do not treat rosette before seed stalk appears.

PNW—See special directions for **Pacific Northwest.**

SPECIAL DIRECTIONS FOR OTHER WEED PROBLEMS IN PEAS

Canada Thistle

Apply 2 pints of BASAGRAN per acre when plants are from 8 inches tall to the bud stage. Make a second application at the same rate 7 to 10 days later.

Pacific Northwest (ID, OR, WA)

For control of cocklebur and wild mustard, use only the 2 pint rate.

For cocklebur, treat when plants are in the 2 to 10 leaf stage and a maximum height of 10 inches.

For wild mustard, treat when plants are up to the 10 leaf stage and a maximum height of 10 inches.

ESTABLISHED PEPPERMINT AND SPEARMINT—DIRECTIONS FOR USE

Apply BASAGRAN® herbicide early postemergence when weeds are small and actively growing and before weeds reach the maximum size listed in the **Application Rate Table for Peppermint and Spearmint.**

Peppermint and spearmint are tolerant to BASAGRAN; however, some leaf-burning may occur under certain conditions, such as when plants are growing very actively and have extensive new, succulent tissue. Mint plants generally outgrow this condition within 10 days.

Irrigated areas

In irrigated areas it may be necessary to irrigate prior to treatment with BASAGRAN to ensure that weeds are growing actively. Weeds growing under drought conditions or unseasonably cold weather usually are not satisfactorily controlled.

Restrictions and limitations

Do not apply more than a total of 8 pints of BASAGRAN per acre in one season.

Table 22
Application Rate Table for Peppermint and Spearmint

Weeds Controlled	2 Pints per Acre		4 Pints per Acre	
	Leaf Stage	Max. Height	Leaf Stage	Max. Height
Common Lambsquarters*	4–8**	2″	—	—
Common Ragweed	4–6**	3″	—	—
Hairy Nightshade***	2–6	4″	6–10	6″
Kochia	NA	4″**	NA	6″**
Ladysthumb	6–10	10″	—	—
Pennsylvania Smartweed	6–10	10″	—	—
Wild Mustard	6–10	8″	—	—

For additional weeds see **Special Directions** section below.
* Control may be partial or inconsistent.
** Add oil concentrate according to the **Directions for use—all crops.**
*** BASAGRAN does not adequately control black nightshade.
NA = not applicable.

SPECIAL DIRECTIONS FOR OTHER WEED PROBLEMS IN PEPPERMINT AND SPEARMINT

Yellow Nutsedge
Apply 2 pints of BASAGRAN per acre when plants are 6 to 8 inches tall. Make a second application at the same rate 7 to 10 days later. Add oil concentrate to the spray solution of BASAGRAN/water for each application according to the **Directions for use—all crops.**

Canada Thistle
Apply 4 pints of BASAGRAN per acre when plants are from 8 inches tall to the bud stage. Make a second application at the same rate 7 to 10 days later.

Common Groundsel
Apply 2 to 3 pints of BASAGRAN per acre when plants are less than 3 inches tall. Add oil concentrate to the spray solution of BASAGRAN/water, according to the **Directions for use—all crops.**

NON-FOOD USE

ESTABLISHED TURF
DIRECTIONS FOR USE

BASAGRAN® herbicide may be used on established bluegrass, fescue, bentgrass, bermudagrass, bahiagrass, centipedegrass, zoysiagrass, ryegrass, and St. Augustine grass.

Timing of application
Apply BASAGRAN postemergently to yellow nutsedge when actively growing and under good soil moisture conditions. If desired control is not obtained with the first application, make additional applications at intervals of 10 to 14 days. Do not apply more than 6 pints per acre in one season.

In the northern United States, yellow nutsedge can emerge from May through July; however, in the southern United States, yellow nutsedge can emerge throughout the year. Therefore, initial applications should be planned when most plants have emerged. If new yellow nutsedge plants emerge later in the season, make additional applications of BASAGRAN in accordance with the label directions. In unmowed turf, make first application after emergence but before yellow nutsedge is 8 inches tall. Thorough spray coverage of yellow nutsedge is essential for maximum control.

For optimum control do not mow 3 to 5 days before or after application.

Application conditions

Water volume: Minimum of 1 gal./1000 square feet or 40 gallons per acre.

Spray pressure: Minimum of 40 psi.

Equipment: Use any spray equipment customarily used for spraying pesticides on turf. Clean equipment carefully before using BASAGRAN. Rinse with water after use.

Sprayer calibration suggestions for homeowners
Hand sprayers:
1. Stake off a 400 sq. ft. area of turf for practice. This is an area 20′ (7 steps) × 20′.
2. Add a measured quantity (1½ gallons for example) of water to the sprayer and uniformly spray the 400 sq. ft. area. Measure water remaining and thereby determine the amount applied per 400 sq. ft. (**Note:** A minimum of 3 pints/400 sq. ft. is recommended).
3. Prepare spray solution according to **Application Rate Table for Ornamental Turf.**

Example: Assume that in Step 2 the 400 sq. ft. area was uniformly covered with ½ gallon of water. Referring to the table, add BASAGRAN at the rate of 2 to 4 teaspoons per ½ gallon of water for each 400 sq. ft. of turf to be sprayed. (**Note:** Use of this mixture for spot spraying of individual nutsedge plants may result in an excessive dosage and possible turf injury.)

Hose-end applicators: A procedure similar to the above may be followed for calibrating hose-end sprayers. Half-fill container with water to an even mark on the "Gallons" scale and note the gallonage level. Spray the 400 sq. ft. area, noting the new gallonage reading, and thereby determine the amount of water used to spray the area. Then proceed as Step 3 above.

Application Rate Table for Turf

Weed Controlled—Yellow Nutsedge Application Rate*			
Teaspoons Per 200 Sq. Ft.	Teaspoons Per 400 Sq. Ft.	Fluid Oz. Per 1000 Sq. Ft.	Pints per Acre
1 to 2	2 to 4	¾ to 1½	2 to 4

* If needed, make subsequent applications at 10-14 day intervals until yellow nutsedge is eliminated. Apply no more than 6 pints per acre in one season.
In perennial ryegrass, apply no more than 2 pints per acre at one time and make subsequent applications not less than 21 days later.

Restrictions and limitations
(partial list)
Do not apply BASAGRAN to any newly seeded or newly sprigged turf until seedlings or sprigs are well established, as injury may result.

Do not apply BASAGRAN to golf course greens or collars.

In perennial ryegrass, apply no more than 2 pints per acre at one time and make subsequent applications not less than 21 days later.

When treating turf with BASAGRAN, avoid over-the-top spraying of adjacent ornamental trees, shrubs, and flowers. Spraying near the base of established ornamental trees, shrubs, and flowers should not result in injury.

APPENDIX

The following are scientific names for the weeds listed in this section. For specific recommendations on control of these weeds, refer to the major crop and/or tank mix sections.

Grasses

Common Name	Scientific Name
Barnyardgrass	*Echinochloa crus-galli*
Bermudagrass	*Cynodon dactylon*
Broadleaf Signalgrass	*Brachiaria platyphylla*
Crabgrass, Large,	*Digitaria sanguinalis*
Smooth	*Digitaria ischaemum*
Cupgrass, Woolly	*Eriochloa villosa*
Foxtail, Giant,	*Setaria faberi*
Green,	*Setaria virdis*
Yellow	*Setaria lutescens*
Goosegrass	*Eleusine indica*
Itchgrass	*Rottboellia exaltata*
Johnsongrass	*Sorghum halepense*
Junglerice	*Echinochloa colonum*
Lovegrass	*Eragrostis cilianensis*
Panicum, Fall,	*Panicum dichotomiflorum*
Texas	*Panicum texanum*
Quackgrass	*Agropyron repens*
Red Rice	*Oryza sativa*
Red Sprangletop	*Leptochloa filiformis*
Volunteer, Barley,	*Hordeum vulgare*
Corn,	*Zea mays*
Oats,	*Avena sativa*
Rye	*Secale cereale*
Wheat	*Triticum aestivum*
Wildcane/Shattercane	*Sorghum bicolor*
Wild Oats	*Avena fatua*
Wild Proso Millet	*Panicum miliaceum*
Wirestem Muhly	*Muhlenbergia frondosa*
Witchgrass	*Panicum capillare*

Broadleaf Weeds

Common Name	Scientific Name
Arrowhead	*Sagittaria* spp.
Balloonvine	*Cardiospermum halicacabum*
Beggarticks	*Bidens frondosa*
Bindweed, Field,	*Convolvulus arvensis*
Hedge	*Convolvulus sepium*
Bristly Starbur	*Acanthospermum hispidum*
Butterprint (see Velvetleaf)	
Buttonweed (see Velvetleaf)	
Canada Thistle	*Cirsium arvense*
Cocklebur	*Xanthium strumarium*
Coffee Senna	*Cassia occidentalis*
Common Lambsquarters	*Chenopodium album*
Common Purslane	*Portulaca oleracea*
Crotalaria	*Crotalaria spectabilis*
Dayflower	*Commelina* spp.
Devilsclaw	*Probiscidea louisianica*
Ducksalad	*Heteranthera limosa*
Galinsoga	*Galinsoga* spp.
Goldenrod, Western	*Solidago occidentalis*
Gooseweed	*Sphenoclea zeylanica*

Common Name	Scientific Name
Groundsel, Common	*Senecio vulgaris*
Jimsonweed	*Datura stramonium*
Kochia	*Kochia scoparia*
Ladysthumb	*Polygonum persicaria*
Marshelder	*Iva xanthiofolia*
Morningglory, Tall (Common),	*Ipomoea purpurea*
Cypressvine,	*Ipomoea quamoclit*
Entireleaf,	*Ipomoea hederacea var. intergruscula*
Ivyleaf,	*Ipomoea hederacea*
Palmleaf,	*Ipomoea wrightii*
Pitted,	*Ipomoea lacunosa*
Purple Moonflower,	*Ipomoea muricata*
Smallflower	*Jacquemontia tamnifolia*
Nightshade, Black,	*Solanum nigrum*
Hairy	*Solanum sarachoides*
Pennsylvania Smartweed	*Polygonum pensylvanicum*
Pigweed, Redroot,	*Amaranthus retroflexus*
Smooth	*Amaranthus hybridis*
Prickly Sida or Teaweed	*Sida spinosa*
Ragweed, Common,	*Ambrosia artemisiifolia*
Giant	*Ambrosia trifida*
Redstem	*Ammannia* spp.
Redweed	*Melochia corchorifolia*
Sesbania	*Sesbania exaltata*
Shepherdspurse	*Capsella bursa-pastoris*
Spurred Anoda	*Anoda caristata*
Tropic Croton	*Croton glandulosus*
Velvetleaf	*Abutilon theophrasti*
Venice Mallow	*Hibiscus trionum*
Waterhemp, Tall	*Amaranthus tuberculatus*
Waterplantain, Common	*Alisma triviale*
Wild Buckwheat	*Polygonum convolvulus*
Wild Mustard	*Sinapsis arvensis*
Wild Poinsettia	*Euphorbia heterophylla*
Wild Sunflower	*Helianthus annuus*

Sedges

Common Name	Scientific Name
Bulrush, River,	*Scirpus fluviatilis*
Roughseed	*Scirpus mucronatus*
Spikerush	*Eleocharis macrostachya*
Umbrellaplant, Smallflower	*Cyperus difformis*
Yellow Nutsedge	*Cyperus esculentus*

AAtrex is a registered trademark of Ciba-Geigy.
BASAGRAN and Poast are registered trademarks of BASF AG.
Blazer is a registered trademark and Dash is a trademark of BASF Corporation.
Butoxone is a registered trademark of Vertac Chemical Corporation.
Butyrac is a registered trademark of Rhone-Poulenc Inc.
Reflex is a registered trademark of ICI Americas Inc.
STAM and Triton are registered trademarks of Rohm and Haas Company.
Scepter is a registered trademark of American Cyanamid Company.
Tackle is a registered trademark of Rhone-Poulenc Inc.
The purchase price of Basagran® herbicide includes a royalty for a license to practice the method of U.S. Patent 3,708,277.

CONDITIONS OF SALE AND WARRANTY

The Directions for Use of this product reflect the opinion of experts based on field use and tests. The directions are believed to be reliable and should be followed carefully. However, it is impossible to eliminate all risks inherently associated with use of this product. Crop injury, ineffectiveness or other unintended consequences may result because of such factors as weather conditions, presence of other materials, or use of the product in a manner inconsistent with its labeling, all of which are beyond the control of BASF CORPORATION ("BASF") or the Seller. All such risks shall be assumed by the Buyer.

BASF warrants that this product conforms to the chemical description on the label and is reasonably fit for the purposes referred to in the Directions for use, subject to the inherent risks referred to above. BASF MAKES NO OTHER EXPRESS OR IMPLIED WARRANTY OF FITNESS OR MERCHANTABILITY OR ANY OTHER EXPRESS OR IMPLIED WARRANTY. IN NO CASE SHALL BASF OR THE SELLER BE LIABLE FOR CONSEQUENTIAL, SPECIAL OR INDIRECT DAMAGES RESULTING FROM THE USE OR HANDLING OF THIS PRODUCT. BASF and the Seller offer this product, and the Buyer and User accept it, subject to the foregoing Conditions of sale and warranty which may be varied only by agreement in writing signed by a duly authorized representative of BASF.

BG 1 013 JJ6361 A0821088

BLAZER®

POSTEMERGENCE HERBICIDE

For use on soybeans, peanuts and rice.

ACTIVE INGREDIENT:*
Sodium salt of acifluorfen
Sodium 5-[2-chloro-4-(trifluoromethyl) phenoxy]-2-nitrobenzoate 20.1%*
INERT INGREDIENTS: 79.9%
TOTAL 100.0%

Equivalent to 2 pounds active ingredient per gallon.

EPA Reg. No. 7969-79
U.S. Pat. 31455

KEEP OUT OF REACH OF CHILDREN.

DANGER

PRECAUTIONARY STATEMENTS

HAZARDS TO HUMANS AND DOMESTIC ANIMALS

Causes eye damage. Harmful if swallowed, inhaled or absorbed through the skin. Do not get in eyes. Wear goggles or face shield when handling. For protection during application wear a hat, long sleeved shirt and trousers. Mixer-loaders must wear rubber gloves. Avoid breathing vapor or spray mist and contact with skin or clothing. In case of contact, immediately remove contaminated clothing and shoes. Wash contaminated clothing with soap and hot water before re-use.

STATEMENT OF PRACTICAL TREATMENT

If in eyes: Flush with large amounts of water for at least 15 minutes. Get medical attention.

If on skin: Wash with plenty of soap and water. Consult a physician if irritation persists.

If swallowed: Dilute by giving 2 glasses of water to drink and call a physician. Never give anything by mouth to an unconscious person.

Note to physician: Emesis is recommended.

ENVIRONMENTAL HAZARDS

Do not apply directly to water or wetlands except as specified on this label for application to rice.

Do not contaminate water by cleaning of equipment or disposal of wastes.

Do not apply when weather conditions favor drift from target area.

RE-ENTRY AND WORKERS' PROTECTION STATEMENTS

Do not apply this product in such a manner as to directly or through drift, expose workers or other persons. The area being treated must be vacated by unprotected persons. Do not enter treated areas without protective clothing until sprays have dried. Because certain states may require more restrictive re-entry intervals for various crops treated with this product, consult your State Department of Agriculture for further information.

DIRECTIONS FOR USE

It is a violation of Federal law to use this product in a manner inconsistent with its labeling.

Read the precautionary statement, environmental hazard statements, storage and disposal statements, and Conditions of sale and warranty statement appearing on the container label.

STORAGE AND DISPOSAL

Do not allow product to freeze. Store above 32 degrees F. Do not contaminate water, food or feed by storage or disposal. Pesticide wastes are acutely hazardous. Improper disposal of excess pesticide spray mixture, or rinsate is a violation of Federal law. If these wastes cannot be disposed of by use according to label instructions, contact your State Pesticide or Environmental Control Agency, or the Hazardous Waste representative at the nearest EPA Regional Office for guidance.

Container disposal

Triple rinse (or equivalent). Puncture and dispose of in a sanitary landfill, or by incineration, or if allowed by state and local authorities, by burning. If burned, stay out of smoke. Do not re-use empty container.

In case of emergency

In case of large scale spillage regarding this product, call:
CHEMTREC 800-424-9300
BASF Corporation 201-316-3000

In case of medical emergency regarding this product, call:
1. Your local doctor for immediate treatment.
2. Your local poison control center (hospital).
3. BASF Corporation 201-316-3000.

Steps to be taken in case material is released or spilled

Dike and contain spill with inert material (sand, earth, etc.) and transfer liquid and solid diking material to separate containers for disposal. Remove contaminated clothing and wash affected skin areas with water.

Wash clothing before re-use. Keep spill out of all sewers and open bodies of water.

GENERAL INFORMATION

BLAZER® herbicide is intended for selective postemergence control of certain broadleaf weeds and grasses (See Directions for use for specific crops and weeds). BLAZER is effective through contact action; therefore weeds must be thoroughly covered with spray. Large crop-and-weed leaf canopies shelter smaller weeds and prevent adequate spray coverage. Labeled crops are tolerant to BLAZER; however, leaf-speckling and leaf bronzing may occur under certain conditions, particularly on the youngest leaves present at time of application. Exposed stems may also exhibit external spotting and bronzing. (See **Restrictions and limitations** for each crop.)

Time of application

Make postemergence application of BLAZER early, when weeds are small and actively growing and before weeds reach the maximum size listed in the application rate tables for the individual crops.

After the weeds exceed the maximum leaf stage of development listed in the weed table, see use instructions for tank mix with Basagran (soybeans) and tank mix with 2,4-DB (soybeans and peanuts).

Early application to weeds results in improved weed control, allows use of the lower rate (depending on weed species), and makes it easier to obtain thorough spray coverage. Delay in application which permits weeds to exceed the maximum size stated will result in inadequate control.

Do not cultivate within 5 days before or 3 to 7 days after application of BLAZER.

Water volume and spray pressure

Apply recommended rates of BLAZER as follows:

Ground equipment: Use a minimum of 20 gals. of water per broadcast acre and a minimum of 40 psi pressure (measured at the boom—not at the pump or in the line). When crop and weed foliage is dense, use up to 50 gals. of water and up to 80 psi pressure. Use standard high pressure pesticide hollow cone or flat fan nozzles spaced 20 inches apart. Do not use flood, whirl chamber or controlled droplet application (CDA) nozzles. Adjust the height of the boom above the crop to give complete coverage of all weeds. The high gallonage and high pressure will promote necessary coverage of weeds. For further information on optimum spray pressures for specific nozzles, refer to manufacturers' charts for recommendations. Maintain sufficient agitation during mixing and spraying to insure a uniform spray mixture.

Note: Cultivation before or during the application is not recommended. Cultivation may put weeds under stress, thus making control more difficult to obtain. Timely cultivation 3 to 7 days after application will usually assist in weed control. When row banding equipment is employed, it should be adjusted to provide maximum coverage of weeds in the row.

Thorough coverage of the weeds can be obtained with two nozzles directed from either side of the crop row toward the weeds in the center rows. Recommended minimum band width is 15 inches with a minimum of 15 gallons of water per acre on this band. Application with a single nozzle over the top of weeds in the center row is not recommended.

Air equipment: In general, use a minimum of 10 gallons of water per acre and a maximum of 40 psi pressure. However, BLAZER applied in 5 gallons per acre has been effective for control of hemp sesbania.

Use only diaphragm-type nozzles producing cone or fan spray patterns.

Nozzle place and orientation: Nozzles should point to the rear of the aircraft and not be pointed downward more than 20 degrees. Nozzles on spray booms should not be placed any closer to wing tips than 3/4 of the wing span; this will minimize the formation of spray or wing-tip vortice roll. A height of 6 to 10 feet over the crop is recommended.

Drift hazard: When spraying labeled crops, care must be exercised to prevent spray drift which could result in damage to other crops. Aerial spraying when other crops are closer than 100 yards downwind or 50 yards upwind from the point of application is not recommended. The use of any cleared drift control agent may reduce this hazard; however, the drift control agent may also decrease the weed control activity. Do not apply BLAZER by aircraft when wind velocity exceeds 10 mph.

Important: Aerial applicators must be familiar with the EPA registered label and follow the use precautions. Spraying of BLAZER in a manner other than as recommended is done at the user's risk. Users are responsible for all loss or damage which results from such spraying. In addition, aerial applicators should follow all applicable state and local regulations and ordinances. In interpreting the label and local regulations, the most restrictive situations should apply to avoid drift hazards.

Spray additives

Adjuvants are needed with BLAZER to achieve consistent weed control. The standard label recommendation is 1 pint of an 80% active nonionic spray adjuvant per 100 gallons of water. For certain weeds, a higher spray adjuvant rate is recommended.

Urea Ammonium Nitrate (UAN) commonly referred to as 28%, 30% or 32% nitrogen solution, may be added in place of other spray adjuvants for improved velvetleaf control in soybean. The standard use rate of 1/2 to 1 gallon per acre is recommended.

Nonphytotoxic oil concentrate should be added to the spray tank for certain tank mix combinations as recommended in the directions for specific crops. The oil concentrate must contain either a petroleum or vegetable oil base and must meet the following criteria: 1) be nonphytotoxic, 2) contain only EPA exempt ingredients, 3) provide good mixing quality in the jar test (see the following section), and 4) be successful in local experience.

The exact composition of suitable products will vary; however, vegetable and petroleum oil concentrates should contain emulsifiers which provide good mixing quality. For vegetable oil concentrates, it has been observed that highly refined vegetable oils are more satisfactory than unrefined oils. For additional information refer to the **Jar test for estimating suitability of oil concentrates.**

With the addition of oil concentrate, the potential for leaf burn is increased especially when relative humidity and temperatures are high.

Jar test for estimating suitability of oil concentrates

If BLAZER is mixed with herbicides requiring the addition of a crop oil concentrate, the following jar test for estimating suitability of oil concentrate should be carried out.

1. **Water supply:** Use only water from intended source and at the source temperature.
2. **Amount of water in jar:** Ground application—for 20 gal./A spray volume use 3 1/3 cups (800 ml) of water. Air application—for 5 gal./A spray volume use 5/6 cup (200 ml) of water, or for 10 gal./A spray volume use 1 2/3 cups (400 ml) of water. For other spray volumes, adjust proportionately to above.
3. **Amount of herbicide(s) and oil concentrate to add:** Add herbicides and oil concentrate at the rate of 1 teaspoon (5 ml) for each pint of recommended label rate.
4. **Add components in following sequence,** gently mixing between component additions:

a. Dry products (dry flowables and wettable powders) when applicable.
b. BLAZER, and when applicable, other water miscible products (such as Basagran® herbicide), liquid fertilizers and/or liquid flowables.
c. Oil concentrate.
d. Emulsifiable concentrates, such as Poast® herbicide, when applicable.

5. **Cap jar, invert 10 cycles, let stand for 15 minutes, evaluate.**
6. **Evaluation:** An ideal tank mix combination will be uniform; thus, the suitability of the oil concentrate is questionable if any of the following are observed:

Free oil at the surface-film or globules.
Flocculation-fine particles which may be suspended in the liquid or found as a precipated layer at the bottom of the jar.
Clabbering-thickening texture (coagulated) resembling yogurt or a curd-like texture as with cottage cheese.

Mixing

Fill half of the spray tank with clean water and add the recommended amount of BLAZER followed by a spray adjuvant while the agitator is running. After thorough mixing, add the remaining quantity of water. For the mixing sequence of tank mix combination see labeling of respective compounds.

Restrictions and limitations

Do not apply BLAZER to crops listed on this label that have been subject to stress conditions, such as drought; flooding, frost or hail damage; high temperature stress or wilt; injury from herbicides or excessive fertilizer or soil salts; wind injury; widely fluctuating temperatures; stress symptoms from disease, nematodes or insects; cold temperatures when maximum day temperature is below 70°F or soil temperature is below 60°F; as weeds will not be actively growing and control may be reduced.

Crop rotation restriction: Root crops (such as carrots, turnips, sweet potatoes, etc.) must not be planted in fields treated with BLAZER herbicide for a period of 18 months following treatment.

In case of crop failure, only peanuts or soybeans may be immediately re-planted.

Do not use treated plants for feed or forage.

Avoid drift to all other crops and non-target areas.

Rainfall and overhead irrigation soon after application (within 6 hours) may nullify the effectiveness of BLAZER. Do not apply if rain is threatening.

Do not mix or apply BLAZER with any other pesticide or with liquid fertilizer, except as specifically recommended on this label or on other BASF-approved supplemental labeling.

Clean sprayer thoroughly prior to application of BLAZER. Thoroughly flush spray equipment (tank, hoses, boom, pump) with water after each use.

Do not apply this product through any type of irrigation system.

Attention

Clean sprayer thoroughly before and after application of herbicides.

Failure to clean sprayer thoroughly after a herbicide application may result in injury to other crops if sprayed with the same equipment. Fill the sprayer with clean water and add a commercial spray tank cleaner or a surfactant/adjuvant at the recommended rate on its label. Circulate through entire sprayer system. Spray approximately half the tank solution through the hoses, boom, and nozzles to clean these parts. Drain the tank and rinse the total system thoroughly several times with clean water.

For use directions on specific crops see following pages.

SOYBEANS

DIRECTIONS FOR USE

Applications of BLAZER should be made when weeds are small and actively growing and before they reach the maximum size listed in **Table 1, Application Rate Table for Soybeans.** In solid seeded narrow-row soybean plantings, BLAZER herbicide should be applied when soybeans are in the 1 to 2 trifoliate leaf stage in order to insure good spray coverage of weeds.

Restrictions and limitations for use in soybeans (partial list)

Do not apply BLAZER within 50 days of harvest for soybeans.

Do not apply more than 4 pints per acre of BLAZER herbicide per growing season for soybeans.

Do not use treated plants for feed or forage.

In the case of crop failure, only soybeans or peanuts may be immediately replanted.

Crop rotation restriction: Root crops (such as carrots, turnips, sweet potatoes, etc.) must not be planted in fields treated with BLAZER for a period of 18 months following treatment.

TABLE 1
APPLICATION RATE TABLE FOR SOYBEANS

Rate: The rate for broad spectrum postemergence weed control is 2 pints of BLAZER per acre plus 1 pint of a spray adjuvant per 100 gallons of spray mix.

	APPLICATION RATE TABLE FOR SOYBEANS							
	1.0 pt./A		1.5 pts./A		2.0 pts./A		3.0 pts./A	
	MAXIMUM[a]		MAXIMUM[a]		MAXIMUM[a]		MAXIMUM[a]	
WEEDS CONTROLLED	LEAF STAGE	HEIGHT INCHES	LEAF STAGE	HEIGHT INCHES	LEAF STAGE	HEIGHT INCHES	LEAF STAGE	HEIGHT INCHES
Amaranth, Palmer,	—	—	4	1-2	6	3	—	—
Spiny	—	—	—	—	6	2-3	—	—
Ballonvine	—	—	—	—	4	2	—	—
Buckwheat, Wild	—	—	—	—	4	3	—	—
Buffalobur	—	—	—	—	3[b]	2	—	—
Burgherkin	—	—	—	—	4[b]	3	—	—
Carpetweed	multi. 3″ dia.	2	multi. 6″ dia.	2	multi. 8″ dia.	2	—	—
Citron (Wild Watermelon)	—	—	—	—	4[b]	2	—	—
Cocklebur, Common,	—	—	—	—	4[b]	4	8[b]	8
Heartleaf	—	—	—	—	4[b]	4	8[b]	8
Copperleaf,	—	—	2	1-2	4	2	—	—
Hophornbeam, Virginia	—	—	—	—	4	<2	—	—
Cowpea, Volunteer	—	—	—	—	1[b]	<2	—	—
Crotalaria, Showy	—	—	—	—	—	—	—	—
Croton, Tropic, Woolly	1-2	<2	2	2	2	2	—	—
	1-2	<2	2	2	2	2	—	—
Cucumber, Wild Spiny	—	—	—	—	4[b]	3	—	—
Devilsclaw	—	—	—	—	2	1	—	—
Galinsoga, Hairy,	—	—	—	—	4	<2	6	2-3
Smallflower	—	—	—	—	4	<2	6	2-3
Gourd, Texas	—	—	—	—	3[b]	3	—	—
Groundcherry, Cutleaf,	—	—	2	1	4	<2	6	2-3
Lanceleaf	—	—	—	—	4	<2	6	2-3
Indigo, Hairy	—	—	—	—	3	2	—	—
Jimsonweed	4	3	6	6	8	8	12	12
Ladysthumb	—	—	—	—	4	4	—	—
Lambsquarters	—	—	—	—	3[b]	<1	—	—
Morningglory, Cypressvine,	—	—	—	—	4	2	6	2-4
Entireleaf,	—	—	—	—	3	<2	4	2
Ivyleaf,	—	—	—	—	3	<2	4	2
Purple Moonflower,	—	—	4	2	4	2	8	2-4
Scarlet,	—	—	—	—	4	2	6	2-3
Smallflower,	—	—	—	—	4	2	8	2-4
Small White (Pitted),	2	2	4	2-3	4	2-3	8	5
Tall (Common),	—	—	—	—	3	<2	4	2-3
Willowleaf (Palmleaf)	—	—	—	—	4	2	6	2-3
Mustard, Wild	2	1-2	4	3	6	4	10	8
Nightshade, Eastern Black,	—	—	—	—	4	2	8	2-4
Black	2	<2	2-3	2	4	2	8	2-4
Pigweed, Prostrate,	—	—	—	—	6	3	8	2-4
Redroot,	2	<2	4	2	6	3	8	4
Smooth	—	—	4	2	6	3	8	4
Poinsettia, Wild	—	—	—	—	2[c]	2		—
Poorjoe	—	—	—	—	4	2	—	3
Purslane, Common	—	—	multi. 6″ dia.	1	multi. 8″ dia.	1	—	—
Pusley, Florida	—	—	—	—	2	<2	4	2
Ragweed, Common,	2	<2	4	2	6	3	8	4
Giant	—	—	—	—	4	4		
Senna, Coffee	—	—	—	—	—	—	2[b]	<2
Sesbania, Hemp	—	—	—	—	—	—	c	—
Smartweed, Pennsylvania	—	—	—	—	4	4	8	8
Smellmelon	—	—	—	—	4[b]	3	—	—
Spurge, Prostrate,	—	—	—	—	multi. 0.5″ dia.	—	—	—
Spotted	—	—	—	—	multi. 0.5″ dia.	—	—	—
Starbur, Bristly	—	—	—	—	—	—	2[b]	<2
Velvetleaf[c]	—	—	—	—	c	c	c	c
Waterhemp, Tall	—	—	—	—	4	2	6	3
Annual Grasses								

APPLICATION RATE TABLE FOR SOYBEANS								
	1.0 pt./A		1.5 pts./A		2.0 pts./A		3.0 pts./A	
	MAXIMUM[a]		MAXIMUM[a]		MAXIMUM[a]		MAXIMUM[a]	
WEEDS CONTROLLED	LEAF STAGE	HEIGHT INCHES	LEAF STAGE	HEIGHT INCHES	LEAF STAGE	HEIGHT INCHES	LEAF STAGE	HEIGHT INCHES
Foxtail, Giant,	—	—	—	—	3[b]	2	4[b]	3
Green,	—	—	—	—	3[b]	2	4[b]	3
Yellow	—	—	—	—	3[b]	2	4[b]	3
Johnsongrass, Seedling	—	—	—	—	3[b]	2	4[b]	3
Panicum, Fall	—	—	—	—	3[b]	2	4[b]	2
Shattercane	—	—	—	—	3[b]	2	4[b]	3
Volunteer Small Grains	—	—	—	—	3[b]	2	4[b]	3
Perennial Weeds								
Bindweed, Field,	—	—	—	—	multi[b]	—	—	—
Hedge	—	—	—	—	multi[b]	—	—	—
Milkweed, Climbing,	—	—	—	—	multi[b]	—	—	—
Common	—	—	—	—	multi[b]	—	—	—
Redvine	—	—	—	—	multi[b]	—	—	—
Thistle, Canada	—	—	—	—	multi[b]	—	—	—
Trumpetcreeper	—	—	—	—	multi[b]	—	—	—

[a] Do not count leaves as pairs . . . count each leaf separately. Do not count cotyledon leaves. Spraying weeds in the cotyledon growth stage is not recommended.
[b] See **Special Use Directions** for these weed problems.
[c] See **Special Use Directions** for rate.
Note: Weed height will vary depending on environmental conditions and is only given as a guide. Emphasis should be placed on leaf stages.

SPECIAL USE DIRECTIONS FOR ADDITIONAL WEED PROBLEMS IN SOYBEANS

Buffalobur
Partial control of buffalobur can usually be obtained when the seedlings have less than 4 true leaves. Use BLAZER at 2 pints in 30 gallons of water per acre. The use of 2 pints of a spray adjuvant per 100 gallons of spray mix is recommended. If regrowth or new seedlings appear, repeat application. Regrowth usually occurs when day or night temperatures fall below 70°F.

Cocklebur, Common
Cocklebur, Heartleaf
The most consistent control of cocklebur is obtained when 2 pints of BLAZER is applied per acre to weeds in the 2 leaf stage which are actively growing under conditions of high soil moisture and high relative humidity. The use of 1 pint of spray adjuvant per 100 gallons of spray mix is recommended. As cocklebur approaches the 8 leaf stage, the use of 3 pints per acre of BLAZER is suggested. Control of cocklebur may be partial or inconsistent when sprayed during periods of dry weather or other stress conditions.

Note: Use the 3 pint rate for soybeans only.

Cowpea, Volunteer (Southern pea, Black eye pea)
Volunteer cowpea germinates over a long period of time, thus making control difficult. Partial control can be obtained by applying 2 pints of BLAZER per acre just past the cowpea cotyledon growth stage. Do not wait for the first leaves to fully expand. Use 2 pints of spray adjuvant per 100 gallons of spray mix for maximum activity. A repeat application is normally required for new seedling weeds.

Sesbania, Hemp
Crotalaria, Showy
Sesbania and crotalaria are very sensitive to BLAZER. The use of 2 pints of spray adjuvant per 100 gallons of spray mix is required for control. Effective control can be obtained at just about all plant heights. It is important, however, that BLAZER be applied prior to bloom. Applications after bloom are usually not effective and therefore not recommended. The use of 1 pint per acre has given excellent control of both sesbania and crotalaria. During or after periods of dry weather, control may be erratic. Best control is then obtained when 1½ pints are applied to each acre. Application for control of these weeds should be timed to occur after maximum weed emergence has taken place. Care must be exercised to make certain that crops do not shade this weed from spray deposits. Waiting for the sesbania to break through the crop canopy may be advisable for control of late season infestations.

Cucurbits: **Burgherkin**
Citron (Wild Watermelon)
Cucumber, Wild Spiny
Gourd, Texas
Smellmelon
Members of the cucumber family germinate over an extended period of time. Control is therefore difficult to obtain with a single spray. In order for BLAZER to be effective, initial application should be made to weeds no later than the 4 leaf growth stage. Use 2 pints of BLAZER per acre plus 2 pints of spray adjuvant per 100 gallons of spray mix. If subsequent weed flushes occur in severely infested soybean fields, a second application should be timed much like the first application. Waiting for several weed flushes to appear and then spraying will result in poor control. The most effective control is obtained when weeds have no more than 4 true leaves.

Lambsquarters, Common
BLAZER, at the 2 pint rate per acre, will usually cause spotting, stunting or death of many seedlings not exceeding 3 true leaves. The use of 2 pints of spray adjuvant per 100 gallons of spray mix is **required** for maximum kill of seedlings. Cultivation 3 to 7 days after the BLAZER/spray adjuvant application will usually assist in control.

Poinsettia, Wild
BLAZER, at the 2 pint rate per acre plus 2 pints of a spray adjuvant per 100 gallons of spray mix, will usually kill or severely stunt wild poinsettia. Application must be made prior to the formation of the third true leaf. In addition, the seedling must be actively growing. This treatment will usually result in a height differential between soybeans and surviving wild poinsettia, thus allowing post-directed applications for additional control.

Starbur, Bristly
Senna, Coffee
BLAZER, at the 3 pint rate per acre plus 2 pints of a spray adjuvant per 100 gallons of spray mix, will kill or suppress seedlings that are not past the 2 leaf stage. Applications after the 2 leaf stage are usually ineffective.

Note: Use the 3 pints rate for soybeans only.

Velvetleaf
BLAZER can be used to control velvetleaf in several ways:
1. Apply 3 pints of BLAZER per acre when velvetleaf does not exceed the 4 true leaf stage (maximum 3" to 4" height) in combination with 2 pints of nonionic spray adjuvant per 100 gallons of spray mix.
2. Apply 2 pints of BLAZER per acre in combination with UAN at the rate of ½ to 1 gallon per acre when velvetleaf does not exceed the 4 leaf stage (maximum 3" to 4" height). The UAN will function as a spray adjuvant.
3. Or, apply BLAZER and Basagran in tank mix combination with UAN. For further details refer to page 12 of this label and the Basagran label.

Annual Grasses: **Foxtail, Giant**
Foxtail, Green
Foxtail, Yellow
Johnsongrass, Seedling
Panicum, Fall
Shattercane
BLAZER must not be the basic component of a grass management program. For additional control of escaped grasses following a preplant incorporated or preemergence herbicide, use BLAZER at 2 pints per acre. The use of 2 to 4 pints of a spray adjuvant per 100 gallons of spray mix is required. Grasses should not exceed the 2 to 3 leaf stage for this treatment. When escaped grasses are in the 4 leaf growth stage use 3 pints of BLAZER herbicide per acre plus 2 pints of a spray adjuvant per 100 gallons of spray. Activity is dependent upon good soil moisture during and following spray application.

Note: Use the 3 pints rate for soybeans only.

Volunteer Small Grains (Barley, Oats, Rye, Wheat)
When BLAZER is applied to emerging volunteer small grains in the 2 to 3 leaf stage, many plants will die or remain stunted. BLAZER should be applied at 2 pints per acre plus 2 to 4 pints of a spray adjuvant per 100 gallons of spray mix. When volunteer small grains are in the 4 leaf growth stage use 3 pints of BLAZER per acre plus 2 to 4 pints of spray adjuvant per 100 gallons of spray. Activity is dependent upon good soil moisture during and following spray applications.

Note: Use the 3 pints rate for soybeans only.

Perennial Weeds: **Bindweed, Field**
Bindweed, Hedge
Milkweed, Climbing
Milkweed, Common
Redvine
Trumpetcreeper

Growth of perennial weeds from underground rootstocks is very difficult to control. BLAZER at 2 pints per acre, plus 2 to 4 pints of spray adjuvant per 100 gallons of spray solution applied under favorable environmental conditions, will burn-back the above-ground plant parts and retard regrowth. A second application may be applied for additional burndown. BLAZER will not kill the underground rootstocks of these weeds.

Thistle, Canada

BLAZER at 3 pints per acre plus a spray adjuvant at 1 pint per 100 gallons will provide added burn-back and growth retardation. Do not apply more than a single application when using the 3 pints per acre rate. When applied before bud stage, Canada Thistle will be severely stunted. The terminal bud of the plant will be killed.

Note: Use the 3 pints rate for soybeans only.

TABLE 2
SOYBEANS—TANK MIXES WITH BLAZER[a]

Use the following chart as a guide to determine broadleaf weeds and grasses controlled by BLAZER alone and various tank mixes with BLAZER.

BLAZER CONTROLS THE WEEDS LISTED BELOW	ADDITIONAL WEEDS CONTROLLED BY TANK MIXING VARIOUS HERBICIDES WITH BLAZER	REFER TO TABLES LISTED BELOW FOR RATE, WEED SIZE AND ADDITIONAL INFORMATION
Annual Broadleaf Weeds	**Basagran® herbicide**	
Amaranth, Palmer, Spiny Balloonvine Buckwheat, Wild Buffalobur Burgherkin Carpetweed Citron (Wild Watermelon) Cocklebur, Common, Heartleaf Copperleaf, Hophornbeam, Virginia Cowpea, Volunteer Crotalaria, Showy Croton, Tropic, Woolly Cucumber, Wild Spiny Devilsclaw Galinsoga, Hairy, Smallflower Gourd, Texas Groundcherry, Cutleaf, Lanceleaf Indigo, Hairy Jimsonweed Ladysthumb Lambsquarters Morningglory, Cypressvine, Entireleaf, Ivyleaf, Purple Moonflower, Scarlet, Smallflower, Small White (Pitted), Tall (Common), Willowleaf (Palmleaf) Mustard, Wild Nightshade, Eastern Black, Black Pigweed, Prostrate, Redroot, Smooth Poinsettia, Wild Poorjoe Purslane, Common Pusley, Florida Ragweed, Common, Giant Senna, Coffee Sesbania, Hemp Smartweed, Pennsylvania Smellmelon Spurge, Prostrate, Spotted Starbur, Bristly Velvetleaf[b] Waterhemp, Tall	Anoda, Spurred Balloonvine Beggarticks Cocklebur (large) Dayflower Galinsoga Lambsquarters Mallow, Venice Nutsedge, Yellow Poinsettia, Wild Ragweed, Giant Redweed Senna, Coffee Sida, Prickly (Teaweed) Shepherdspurse Starbur, Bristly Sunflower, Wild Thistle, Canada Velvetleaf	BLAZER + Basagran Tables 3, 4, 5 Pages 13 to 15
	2,4-DB	
	Cocklebur Morningglory (large) Pigweed, Redroot	BLAZER + 2,4-DB Table 6 Page 17
	Classic® herbicide	
	See Classic label for weed species controlled.	BLAZER + Classic Page 26
	Poast® herbicide	
	Barnyardgrass Crabgrass, Large, Smooth Cupgrass, Woolly Foxtail, Giant, Green, Yellow Johnsongrass, Seedling Junglerice Millet, Wild Proso Panicum, Fall, Giant, Texas Signalgrass, Broadleaf Sprangletop, Red Witchgrass	BLAZER + Poast Table 10 Page 23
	Scepter® herbicide	
	Cocklebur (large) Poinsettia, Wild	BLAZER + Scepter Table 7 Page 19
	Basagran® + Poast® herbicides	
	See weeds listed above for Basagran + Poast	BLAZER + Basagran + Poast Tables 8 & 9 Page 21
	Fusilade® herbicide	
	See Fusilade 2000 1E label for weed species controlled.	BLAZER + Fusilade Tables 11, 12 Pages 24 to 25
	Amiben® herbicide	
	See Amiben label for weed species controlled.	BLAZER + Amiben Page 26
Annual Grasses	**Rescue® herbicide (Salvage treatment)**	
Foxtail, Giant, Green, Yellow Johnsongrass, Seedling	Cocklebur (large) Jimsonweed Morningglory Pigweed Ragweed, Common, Giant	BLAZER + Rescue Table 13 Page 28

BLAZER CONTROLS THE WEEDS LISTED BELOW	ADDITIONAL WEEDS CONTROLLED BY TANK MIXING VARIOUS HERBICIDES WITH BLAZER	REFER TO TABLES LISTED BELOW FOR RATE, WEED SIZE AND ADDITIONAL INFORMATION
Panicum, Fall Shattercane Volunteer Small Grains		
Perennial Weeds		
Bindweed, Field, Hedge Milkweed, Climbing, Common Redvine Thistle, Canada Trumpetcreeper		

[a] Note: Tank mixes are not applicable in California.
[b] See **Special Use Directions** for rate.

BLAZER + BASAGRAN TANK MIX IN SOYBEANS GENERAL AND APPLICATION INFORMATION, RESTRICTIONS AND LIMITATIONS

General information
BLAZER can be tank mixed with Basagran for postemergence control of the major broadleaf weed species in soybeans.

Time of application
Application timing should be in accordance with weed sizes outlined in Tables 1, 3, 4 & 5. A delay in application will permit weeds to exceed the maximum size stated, resulting in inadequate control. (Refer to section entitled Directions for use, pages 2 & 3 for additional information.)

Rate
Use BLAZER at the rate of ½ to 2 pints per acre tank mixed with Basagran at the rate of 1 to 2 pints per acre. In order to determine the correct application rate of BLAZER to use in the tank mix, see the BLAZER use rate in **Table 1.**

For control of cocklebur up to the 6 leaf stage and velvetleaf up to the 4 leaf stage, add 1 pint per acre of Basagran to BLAZER. For prickly sida (teaweed) up to the 6 leaf stage, add 1½ pints per acre of Basagran to BLAZER.

Spray additive
The addition of a spray adjuvant is recommended and should be used at a minimum rate of 1 pint per 100 gallons of spray mixture. For control of certain weeds such as escaped grasses and hemp sesbania with the tank mix, the addition of 2 to 4 pints of a spray adjuvant per 100 gallons of spray mix is required. See the BLAZER Special Use Directions for additional weeds and specific amounts of spray adjuvant.

Special instructions for the use of Urea Ammonium Nitrate (UAN)
For improved velvetleaf control in soybeans, a UAN solution (commonly referred to as 28%, 30% or 32% nitrogen solution) may be added. The UAN is an agricultural grade fertilizer used by dealers for agricultural applications. It may be added to the tank when velvetleaf is one of the primary target weeds. Apply ½ to 1 gallon per acre depending on weed size and environmental conditions.

With the addition of UAN, a leaf burn on soybeans may occur, but the new growth is normal and crop vigor is not reduced. Refer to your supplier of Basagran/BLAZER for information concerning successful local experience prior to using UAN. Do not use brass or aluminum nozzles when spraying with the UAN solution.

Use rate
Ground application—½ to 1 gallon/acre
Air application—not recommended

Water volume and spray pressure
For additional information refer to the section entitled **Directions for use,** pages 2 & 3.

Ground equipment:
For the tank mix of BLAZER + Basagran, use a minimum of 20 gallons of total spray solution per acre (broadcast basis) and a minimum of 40 psi pressure. Use standard high pressure hollow cone or flat fan nozzles spaced 20 inches apart. Do not use flood or whirl chamber nozzles.

Air equipment:
Use a minimum of 10 gallons of total spray solution per acre.

Mixing
Fill half the spray tank with water and add the recommended amount of product in the following order—BLAZER, Basagran, spray adjuvant—while the agitator is running; then add the remaining quantity of water.

Coverage
Thorough coverage of actively growing weeds is essential. Large crop-and-weed leaf canopies shelter smaller weeds and can prevent adequate spray coverage. Soybeans are tolerant to the above tank mixes; however, under certain conditions soybeans may burn, crinkle and bronze.

Restrictions and limitations (partial list)
Read and follow restrictions and limitations on the BLAZER and Basagran labels. The most restrictive labeling applies to tank mixes.

Do not apply BLAZER within 50 days of harvest.

Do not use treated plants for feed or forage.

TABLE 3
BLAZER + BASAGRAN TANK MIX IN SOYBEANS RATE AND TIME OF APPLICATION TABLE

PRODUCT	RATE	WEEDS CONTROLLED			ADDITIVE INFORMATION
		Annual Broadleaf Weeds			
BLAZER	1-2 pts. according to weed species and size. (See **Table 1** Pages 5 & 6)	Amaranth, Palmer, Spiny Balloonvine Buckwheat, Wild Buffalobur Burgherkin Carpetweed Citron (Wild Watermelon) Cocklebur, Common, Heartleaf Copperleaf, Hophornbeam, Virginia Cowpea, Volunteer Crotalaria, Showy Croton, Tropic, Woolly Cucumber, Wild Spiny Devilsclaw	Galinsoga, Hairy, Smallflower Gourd, Texas Groundcherry, Cutleaf, Lanceleaf Indigo, Hairy Jimsonweed Ladysthumb Lambsquarters Morningglory, Cypressvine, Entireleaf, Ivyleaf, Purple Moonflower, Scarlet, Smallflower, Small White (Pitted), Tall (Common), Willowleaf (Palmleaf) Mustard, Wild	Nightshade, Eastern Black, Black Pigweed, Prostrate, Redroot, Smooth Poinsettia, Wild Poorjoe Purslane, Common Pusley, Florida Ragweed, Common, Giant Senna, Coffee Sesbania, Hemp Smartweed, Pennsylvania Smellmelon Spurge, Prostrate, Spotted Starbur, Bristly Waterhemp, Tall	Spray adjuvant 1 pt./100 gal. **or** UAN for velvetleaf (½ to 1 gal./A)[b]

PRODUCT	RATE	WEEDS CONTROLLED			ADDITIVE INFORMATION
		Annual Grasses			
		Foxtail, Giant, Green, Yellow; Johnsongrass, Seedling; Panicum, Fall; Shattercane; Volunteer Small Grains			
		Perennial Weeds			
		Bindweed, Field, Hedge; Milkweed, Climbing, Common; Redvine Trumpetcreeper; Thistle, Canada			
			Application Rates		
		Weeds Controlled	**1½ pts./A**	**2 pts./A**	
plus Basagran	**plus** 1½-2 pts.	Anoda, Spurred	3"	4"	
		Beggarticks	6"	8"	
		Cocklebur (large)	6"	10"	
		Dayflower	4"	8"	
		Mallow, Venice	2"	4"	
		Nutsedge, Yellow	[a]	[a]	
		Ragweed, Giant	—	6"	
		Shepherdspurse	4"	8"	
		Sunflower, Wild	5"	8"	
		Thistle, Canada	[a]	[a]	
		Velvetleaf	2"	5"	

[a] See **Special Directions for other Weed Problems in Soybeans** on the Basagran label.
[b] Do not include oil concentrate with UAN solutions when tank mixing BLAZER and Basagran.

Special use instructions
For postemergence weed control with BLAZER + Basagran, the following tank mix combinations can be recommended based on weed problems and geographic area:

Table 4—Northern States
BLAZER: ½ pint
Basagran: 1 to 2 pints
Weeds controlled: Listed in **Table 4**

TABLE 4
NORTHERN STATES[a]
BLAZER + BASAGRAN TANK MIX IN SOYBEANS

PRODUCT	PRODUCT RATE	WEEDS CONTROLLED/WEED SIZE	LEAF STAGE	MAXIMUM HEIGHT	ADDITIVE INFORMATION
BLAZER	½ pt./A	Pigweed[b] (Redroot and Smooth)	Up to 4	2"	
		Tall Waterhemp	Up to 4	2"	
plus Basagran	**plus** 1-2 pts./A according to weed species and size[c]	Anoda, Spurred; Balloonvine; Beggarticks; Buckwheat, Wild; Cocklebur; Croton, Tropic; Dayflower; Devilsclaw; Galinsoga; Jimsonweed; Ladysthumb; Lambsquarters, Common; Mallow, Venice; Morningglory, Cypressvine, Smallflower	Mustard, Wild; Nutsedge, Yellow; Poinsettia, Wild; Purslane, Common; Ragweed, Common, Giant; Redweed; Senna, Coffee; Shepherdspurse; Sida, Prickly (Teaweed); Smartweed, Pennsylvania; Starbur, Bristly; Sunflower, Wild; Thistle, Canada[c]; Velvetleaf		Oil concentrate (2 pts./A or UAN Solution (½-1 gal./A)[d]

[a] Northern States, for the purpose of this table, are the following states and those to the north thereof: IL, IN, KS, MD, MO (except southeastern MO, Jefferson County and south), NY and OH.
[b] See **Table 5** for control of additional weeds.
[c] Requires two applications of Basagran in accordance with the Basagran label.
[d] Do not include oil concentrate with UAN solutions when tank mixing BLAZER with Basagran.
[e] See and follow the Basagran label for additional information.

Table 5—Southern States
BLAZER: 1 pint
Basagran: 1 pint
Weeds controlled: Listed in **Table 5**

TABLE 5
SOUTHERN STATES[a]
BLAZER + BASAGRAN TANK MIX IN SOYBEANS

PRODUCT	PRODUCT RATE	WEEDS CONTROLLED/WEED SIZE	Leaf Stage	Maximum Height	ADDITIVE INFORMATION
		Anoda, Spurred	Up to 4	2"	Oil
BLAZER	1 pt./A	Carpetweed	—	2"	concentrate
		Cocklebur[c]	2-6	6"	(1 pt./A)
plus	plus	Crotalaria	Up to 6	6"	
		Croton, Tropic,	2	<2"	
Basagran	1 pt./A	Woolly	2	<2"	
		Ladysthumb	Up to 6	6"	
		Lambsquarters, Common[b]	4-6	2"	
		Jimsonweed	Up to 6	6"	
		Mallow, Venice	Up to 6	2"	
		Morningglories	Up to 4	4"	
		Mustard, Wild	2	2"	
		Nightshade, Black	Up to 6	2"	
		Pigweed, Redroot,	6	3"	
		Smooth	Up to 6	3"	
		Ragweed, Common,	4-6	3"	
		Giant	Up to 4	6"	
		Redweed	2-4	3"	
		Sesbania, Hemp	4	6"	
		Sida, Prickly (Teaweed)[b]	Up to 4	2"	
		Smartweed, Pennsylvania	Up to 6	6"	
		Starbur, Bristly	4-6	3"	
		Velvetleaf[b]	Up to 4	2"	
		Waterhemp, Tall	4-6	3"	

[a] Southern states, for the purpose of this table are, AL, AR, FL, GA, KY, LA, MS, NC, OK, SC, TN, TX, VA and southeastern MO (Jefferson Co. and south).

[b] For more consistent control increase rate of Basagran to 1½ pints.

[c] Do not treat earlier than the two leaf stage and do not count cotyledon leaves.

BLAZER + 2,4-DB TANK MIX IN SOYBEANS GENERAL AND APPLICATION INFORMATION, RESTRICTIONS AND LIMITATIONS

General information

A tank mix of BLAZER plus 2,4-DB is recommended for control of morningglory, cocklebur, common ragweed, redroot pigweed, jimsonweed, burgherkin and citron in soybeans when the weed size exceeds that specified on the BLAZER label.

Time of application

For optimum control apply the tank mix to actively growing weeds up to 12 inches in height or length. Applications at later stages will result in partial control or suppression. See **Table 6** for information on dosage rates and weed sizes.

The use of this tank mix will cause foliage injury and may reduce yields. Applications at the third or greater trifoliate leaf stage will assist in minimizing foliar injury.

Rate

Mix 2 fluid ounces of Butyrac® 200, or 2 fluid ounces of Butoxone®, with 1½ to 2 pints of BLAZER for each acre being treated. For additional control of cocklebur, add ½ pint Basagran per acre to the tank mixture.

Spray additive

Add 1 pint of a spray adjuvant per 100 gallons to increase control of weeds. Do not add crop oils to the tank mixture. The addition of a spray adjuvant will increase the hormonal 2,4-DB crop response.

Water volume and spray pressure

For additional information refer to the section entitled **Directions for use,** pages 2 & 3.

Ground equipment: For best results, the tank mix should be applied with ground equipment. For thorough coverage of weeds apply with flat fan or hollow cone nozzles spaced 20 inches apart in a minimum of 20 gallons of water per acre with a spray pressure of 40 psi.

Air equipment: Use a minimum of 10 gallons of total spray solution per acre.

Mixing

Fill half the spray tank with water and add the recommended amount of product in the following order—BLAZER, 2,4-DB and spray adjuvant—while the agitator is running; then add the remaining quantity of water.

Coverage

Control with this mixture may decrease with increasing weed size or density of weed or soybean canopy, due to poor spray coverage.

Large crop-and-weed leaf canopies shelter small weeds and prevent adequate spray coverage.

Drift hazard

Care must be taken when applying the tank mixture to prevent drift to all non-target crops. Tobacco, ornamentals, mustards, sugar beets, potatoes, vegetables, and cotton are a few of the crops known to be sensitive to this tank mixture. Hormone type injury in non-target crops can result from trace amounts of 2,4-DB drift. The use of any cleared drift control agent may reduce this hazard; however, the drift control agent may also decrease the weed control activity.

Restrictions and limitations (partial list)

Read and follow all directions and use restrictions on BLAZER and 2,4-DB labels.

Do not use rates of BLAZER or 2,4-DB in excess of that recommended on this label, or excessive injury and possible yield reduction could result.

Do not apply the tank mixture within 60 days of harvest for soybeans.

Do not apply more than one application of the tank mixture to soybeans per growing season.

Do not mix oils, liquid fertilizers or other pesticides with the tank mix except as specifically directed on this label.

Do not apply the tank mixture when soybeans are exhibiting injury from previously applied pesticides or are exhibiting stress symptoms from disease, nematodes, insects; excessive fertilizer of soil salts; wind injury; frost damage or high temperature stress or wilt; as increased crop response will result.

Aerial applicators must be familiar with the EPA registered labels and follow the use precautions. In addition, aerial applicators should follow all applicable state and local regulations. In interpreting the label and local regulations, the most restrictive situations should apply to avoid drift hazards.

TABLE 6
BLAZER + 2,4-DB TANK MIX IN SOYBEANS RATE AND TIME OF APPLICATION TABLE

PRODUCT	PRODUCT RATE	WEEDS CONTROLLED		ADDITIVE INFORMATION
		Annual Broadleaf Weeds		
BLAZER	1½-2 pts./A according to weed species and size. (See **Table 1** Pages 2 & 3)	Amaranth, Palmer, Spiny Balloonvine Buckwheat, Wild Buffalobur Burgherkin Carpetweed Citron (Wild Watermelon) Cocklebur, Common, Heartleaf Copperleaf, Hophornbeam, Virginia Cowpea, Volunteer Crotalaria, Showy Croton, Tropic, Woolly Cucumber, Wild Spiny Devilsclaw Galinsoga, Hairy, Smallflower Gourd, Texas Groundcherry, Cutleaf, Lanceleaf Indigo, Hairy Jimsonweed Ladysthumb Lambsquarters	Morningglory, Cypressvine, Entireleaf, Ivyleaf, Purple Moonflower, Scarlet, Smallflower, Small White (Pitted), Tall (Common), Willowleaf (Palmleaf) Mustard, Wild Nightshade, Eastern Black, Black Pigweed, Prostrate, Redroot, Smooth Poinsettia, Wild Poorjoe Purslane, Common Pusley, Florida Ragweed, Common, Giant Senna, Coffee Sesbania, Hemp Smartweed, Pennsylvania Smellmelon Spurge, Prostrate, Spotted Starbur, Bristly Velvetleaf[a] Waterhemp, Tall	Spray adjuvant[b] 1 pt./100 gal.
		Annual Grasses		
		Foxtail, Giant, Green, Yellow Johnsongrass, Seedling	Panicum, Fall Shattercane Volunteer Small Grain	
		Perennial Weeds		
		Bindweed, Field, Hedge Milkweed, Climbing, Common	Redvine Thistle, Canada Trumpetcreeper	
plus	plus	**Weeds Controlled**	**Weed Height**	
Butyrac 200[c] or Butoxone[c]	2 fl. oz./A	Burgherkin[c] Citron Cocklebur Jimsonweed Morningglory Pigweed, Redroot Ragweed, Common	Up to 12" height or length of vine	

[a] See **Special Use Directions** for rate.
[b] The addition of a nonionic spray adjuvant will increase the hormonal 2,4-D crop response.
[c] When size exceeds that specified on the BLAZER label.

BLAZER + SCEPTER TANK MIX IN SOYBEANS GENERAL AND APPLICATION INFORMATION, RESTRICTIONS AND LIMITATIONS

General information
BLAZER may be tank mixed with Scepter for improved control of cocklebur and wild poinsettia in soybeans.

Time of application
Application should be in accordance with weed sizes outlined in Tables 1 and 7. A delay in application will permit weeds to exceed maximum size stated resulting in inadequate control.

Rate
Use BLAZER at the rate of 1½ to 2 pints per acre. In order to determine the correct application rate of BLAZER to use in the tank mix, see the BLAZER use rate in **Table 1.**

For improved control of common cocklebur (up to 6 leaf), add Scepter at the rate of ⅓ to ⅔ pint per acre to BLAZER. For control of wild poinsettia (up to 6 leaf), add Scepter at a rate of ⅔ pint per acre. Timely cultivations will usually assist in weed control.

Spray additive
Add 2 pints of a nonionic spray adjuvant per 100 gallons of spray mixture. For the control of certain weeds such as escaped grasses, the addition of up to 4 pints of spray adjuvant per 100 gallon spray mix is required.

Water volume and spray pressure
For additional information refer to the section entitled **Directions for use,** pages 2 & 3.

Ground equipment: For broadcast application, use a minimum of 20 gallons of total spray solution per acre and 40 psi pressure with flat fan or hollow cone nozzles spaced 20 inches apart.

Mixing
Fill half the spray tank with water and add the recommended amount of product in the following order—BLAZER, Scepter and a spray adjuvant—while the agitator is running; then add the remaining quantity of water.

Coverage
Thorough coverage of weeds is essential. Control with this mixture may decrease with increasing weed size or density of weed or soybean canopy.

Large crop-and-weed leaf canopies shelter small weeds and prevent adequate spray coverage.

Scepter preplant followed by BLAZER + Scepter tank mix
When Scepter is applied postemergence following a Scepter preplant incorporated or preemergence application as described in the "Sequential Program" section of the Scepter herbicide label for the control of Florida Beggarweed, Mexican weed and sicklepod, the addition of BLAZER herbicide at the rate of 1½ to 2 pints per acre will provide control of annual morningglory and other major broadleaf weed species in soybeans.

Restriction and limitations (partial list)

Read and follow restrictions and limitations on the BLAZER and Scepter labels. The most restrictive labeling applies in tank mixtures.

Do not apply Scepter within 90 days of harvest.

Observe all geographic and rotational crop restrictions on the Scepter label.

TABLE 7
BLAZER + SCEPTER TANK MIX IN SOYBEANS
RATE AND TIME OF APPLICATION TABLE

PRODUCT	PRODUCT RATE	WEEDS CONTROLLED		ADDITIVE INFORMATION
		Annual Broadleaf Weeds		
BLAZER	1½-2 pts. according to weed species and size. (See **Table 1** Pages 2 & 3)	Amaranth, Palmer, Spiny Balloonvine Buckwheat, Wild Buffalobur Burgherkin Carpetweed Citron (Wild Watermelon) Cocklebur, Common, Heartleaf Copperleaf, Hophornbeam, Virginia Cowpea, Volunteer Crotalaria, Showy Croton, Tropic, Woolly Cucumber, Wild Spiny Devilsclaw Galinsoga, Hairy, Smallflower Gourd, Texas Groundcherry, Cutleaf, Lanceleaf Indigo, Hairy Jimsonweed Ladysthumb Lambsquarters	Morningglory, Cypressvine, Entireleaf, Ivyleaf, Purple Moonflower, Scarlet, Smallflower, Small White (Pitted), Tall (Common), Willowleaf (Palmleaf) Mustard, Wild Nightshade, Eastern Black, Black Pigweed, Prostrate, Redroot, Smooth Poinsettia, Wild Poorjoe Purslane, Common Pusley, Florida Ragweed, Common, Giant Senna, Coffee Sesbania, Hemp Smartweed, Pennsylvania Smellmelon Spurge, Prostrate, Spotted Starbur, Bristly Velvetleaf[a] Waterhemp, Tall	Spray adjuvant 2 pts./100 gal.
		Annual Grasses		
		Foxtail, Giant, Green, Yellow Johnsongrass, Seedling	Panicum, Fall Shattercane Volunteer Small Grains	
		Perennial Weeds		
		Bindweed, Field, Hedge, Milkweed, Climbing, Common	Redvine Thistle, Canada Trumpetcreeper	
plus	**plus**	**Weeds Controlled**	**Weed Height**	
Scepter	⅓ to ⅔ pt.	Cocklebur Wild Poinsettia	Up to 6 leaf (6")	

[a] See **Special Use Directions** for rate.

BLAZER + SCEPTER + 2,4-DB TANK MIX IN SOYBEANS

General Information

The addition of 2 fluid ounces of 2,4-DB (Butyrac 200, or Butoxone) to the BLAZER plus Scepter tank mix is recommended for improved control of morningglory, common ragweed, giant ragweed, redroot pigweed, jimsonweed, burgherkin, and citron in soybeans when the weed size (up to 8 leaf) exceeds that specified on the BLAZER label.

Control with this mixture may decrease with increasing weed size or density of weed or soybean canopy, due to poor spray coverage. Add 1 pint of spray adjuvant per 100 gallons of spray solution to increase control of weeds. The addition of surfactant will increase the hormonal 2,4-DB crop response.

For information on water volume, spray pressure, mixing and application, refer to pages 2 & 3.

Restrictions and limitations (partial list)

Always read and follow the restrictions and limitations for all products whether used alone or in a tank mix. The most restrictive labeling applies in tank mixes.

BLAZER + BASAGRAN + POAST APPLICATIONS IN SOYBEANS GENERAL AND APPLICATION INFORMATION, RESTRICTIONS AND LIMITATIONS

General information

BLAZER, Basagran and Poast may be tank mixed or applied sequentially for postemergence control of broadleaf and grass weeds. Weeds must be actively growing and at the recommended growth stages.

Sequential applications

Sequential applications should be made if: a) all weeds to be controlled are not at the correct growth stage for treatment at the same time, or b) grasses to be controlled include rhizome johnsongrass, quackgrass, bermudagrass, wirestem muhly, volunteer corn, shattercane, volunteer cereals, wild oats, red rice or itchgrass. For further information on sequential applications see **Table 9** (page 21).

Time of application

Applications should be made in accordance with weed species outlined in **Table 8.**

Rate

BLAZER at ½ to 1 pint per acre may be tank mixed with Basagran at 1 to 2 pints per acre and 1½ pints of Poast for postemergence control of selected annual broadleaf/grass weeds in soybeans. The rate of Poast recommended in the tank mix is 50% greater than the rate of Poast used alone; see the Poast label. In order to determine the correct application rate of BLAZER to use in the tank mixture, see the BLAZER use rate in **Table 1.**

Spray additive
Always add Poast herbicide in the tank mixture with 2 pints per acre of a recommended nonphytotoxic oil concentrate. Oil concentrate must be used with the tank mixture in place of a spray adjuvant. The addition of a crop oil concentrate may increase the crop response.

Water volume and spray pressure
For additional information refer to the section entitled **Directions for use,** pages 2 & 3.

Ground equipment: For the tank mix of BLAZER + Basagran + Poast use 20 gallons of total spray solution per acre (broadcast basis) and a minimum of 40 psi pressure. Use standard high pressure hollow cone or flat fan nozzles spaced 20 inches apart. Do not use flood or whirl chamber nozzles.

Air equipment: Use a minimum of 10 gallons of total spray solution per acre.

Mixing
Fill half the spray tank with water and add the recommended amount of product in the following order—BLAZER, Basagran, oil concentrate, Poast—while the agitator is running, then add the remaining quantity of water.

Coverage
Thorough coverage of actively growing weeds is essential. Large crop-and-weed leaf canopies shelter smaller weeds and can prevent adequate spray coverage. Soybeans are tolerant to the tank mix; however, under certain conditions soybeans may burn, crinkle and bronze.

Restrictions and limitations (partial list)
Read and follow restrictions and limitations on the BLAZER, Basagran and Poast labels. The most restrictive labeling applies in tank mixes.

TABLE 8
BLAZER + BASAGRAN + POAST TANK MIX IN SOYBEANS
RATE AND TIME OF APPLICATION

PRODUCT	RATE	WEEDS CONTROLLED				ADDITIVE INFORMATION
				Leaf Stage	Maximum Height	
BLAZER	½ pt./A	Pigweed		Up to 4	2″	
		Waterhemp		Up to 4	2″	
	1 pt.	Crotalaria		Up to 6	6″	
		Morningglory[a]		Up to 4	4″	
		Nightshade, Black		Up to 6	6″	
		Pigweed, Redroot,		Up to 6	3″	
		Smooth		Up to 6	3″	
plus	plus					
Basagran	1-2 pts./A according to weed species and size (see Basagran Label)	Anoda, Spurred		Mustard, Wild		Oil concentrate (2 pts./A.) Do not include UAN[b]
		Balloonvine		Nutsedge, Yellow		
		Beggarticks		Poinsettia, Wild		
		Buckwheat, Wild		Purslane, Common		
		Cocklebur		Ragweed, Common,		
		Croton, Tropic		Giant		
		Dayflower		Redweed		
		Devilsclaw		Senna, Coffee		
		Galinsoga		Shepherdspurse		
		Jimsonweed		Sida, Prickly (Teaweed)		
		Ladysthumb		Smartweed, Pennsylvania		
		Lambsquarters, Common		Starbur, Bristly		
		Mallow, Venice		Sunflower, Wild		
		Morningglory, Cypressvine,		Thistle, Canada[e]		
		Smallflower		Velvetleaf		
plus	plus					
Poast	1½ pts./A	Barnyardgrass	3-8″	Johnsongrass, Seedling	3-8″	
		Crabgrass, Large,	3-6″	Junglerice	3-8″	
		Smooth	3-6″	Millet, Wild Proso[c]	3-8″	
		Cupgrass, Woolly	3-8″		4-10″	
		Foxtail, Giant,	3-8″	Panicum, Fall	3-8″	
		Green,	3-8″	, Texas	3-8″	
		Yellow	3-8″	Signalgrass, Broadleaf	3-8″	
		Goosegrass	3-6″	Sprangletop, Red	3-8″	
				Witchgrass[d]	3-8″	

[a] For consistent control of morningglory species use sequential applications.
[b] Do not include UAN solution or Ammonium Sulfate when tank mixing oil concentrate with BLAZER, Basagran and Poast.
[c] For control of wild proso millet only, include Poast in tank mix at ¾ pt./A.
[d] Tank mix does not control rhizome johnsongrass, quackgrass, bermudagrass, wirestem muhly, volunteer corn, shattercane, volunteer cereals, wild oats, red rice or itchgrass.
[e] Requires two applications of Basagran in accordance with this label.

Sequential applications—(BLAZER, Basagran, Poast)
When making sequential and/or tank mix applications of BLAZER, Basagran and Poast consult the following table for order of application and minimum time between application.

Table 9

Order of Applications		
First Product(s) Applied	**Second Product(s) Applied**	**Minimum Time Between Applications**
BLAZER	Poast	7 days
BLAZER + Basagran	Poast	7 days
Poast	BLAZER or BLAZER + Basagran	24 hours

BLAZER + POAST APPLICATIONS IN SOYBEANS GENERAL AND APPLICATION INFORMATION, RESTRICTIONS AND LIMITATIONS.

General information
BLAZER and Poast may be tank mixed or applied sequentially for post-emergence control of broadleaf and grass weeds. Weeds must be actively growing and at the recommended growth stages.

It is important that grass previously sprayed with BLAZER has resumed active growth before spraying with Poast. This waiting period is important in achieving maximum activity with Poast.

Time of application
For optimum control apply the tank mix to actively growing weeds at the sizes indicated in Tables 1 & 10.

Sequential applications should be made if: a) all weeds to be controlled are not at the correct growth stage for treatment at the same time, or b) grasses to be controlled include rhizome johnsongrass, quackgrass, bermudagrass, wirestem muhly, volunteer corn, shattercane, volunteer cereals, wild oats, red rice or itchgrass. For further information on sequential applications see **Table 9** (page 20).

Rate

BLAZER at 1½ to 2 pints per acre may be tank mixed with Poast for postemergence control of selected annual broadleaf/grass weeds in soybeans. Use 1 pint of Poast with 2 pints of crop oil concentrate per acre with the appropriate rate of BLAZER to control the following annual grasses: broadleaf signalgrass, fall panicum, giant foxtail, junglerice and texas panicum. For all other annual grasses on the Poast label, increase the rate of Poast by 50%. In order to determine the correct application rate of BLAZER to use in the tank mix, see the BLAZER use rate in **Table 1.**

Spray additive

Oil concentrate must be used in this tank mix.

Water volume and spray pressure

For additional information refer to the section entitled **Directions for use,** pages 2 & 3.

Ground equipment: For the tank mix of BLAZER + Poast, use 20 gallons of total spray solution per acre (broadcast basis) and a minimum of 40 psi pressure. Use standard high pressure hollow cone or flat fan nozzles spaced 20 inches apart. Do not use flood or whirl chamber nozzles.

Air equipment: Use a minimum of 10 gallons of total spray solution per acre.

Mixing

Fill half the spray tank with water and add the recommended amount of product in the following order—BLAZER, oil concentrate, Poast—while the agitator is running, then add the remaining quantity of water.

Coverage

Thorough coverage of actively growing weeds is essential. Large crop-and-weed leaf canopies shelter smaller weeds and can prevent adequate spray coverage. Soybeans are tolerant to the tank mix; however, under certain conditions soybeans may burn, crinkle and bronze.

Restrictions and limitations (partial list)

Always read and follow the restrictions for all products when used alone, in a tank mix or a sequential application. The most restrictive labeling must apply in a tank mixture.

Do not apply BLAZER within 50 days of harvest and do not apply Poast within 90 days of harvest.

Do not use treated plants for feed or forage.

Do not apply BLAZER or Poast when soybeans and weeds exhibit injury or stress from previously applied pesticides, excessive fertilizer, soil salts; wind injury; frost damage; high temperature; disease, nematodes, insects or wilt; as increased crop response; and decreased weed control may result.

Do not add UAN solution or ammonium sulfate to a tank mix of BLAZER, Poast and oil concentrate.

TABLE 10
BLAZER + POAST TANK MIX—SOYBEANS
RATE AND TIME OF APPLICATION TABLE

PRODUCT	RATE	WEEDS CONTROLLED		ADDITIVE INFORMATION
		Annual Broadleaf Weeds		
BLAZER	1½-2 pts./A according to weed species and size. (See **Table 1** Pages 2 & 3)	Amaranth, Palmer, Spiny Balloonvine Buckwheat, Wild Carpetweed Cocklebur, Common, Heartleaf Copperleaf, Hophornbeam, Virginia Crotalaria, Showy Croton, Tropic, Woolly Cucumber, Wild Spiny Devilsclaw Galinsoga, Hairy, Smallflower Gourd, Texas Groundcherry, Cutleaf, Indigo, Hairy Jimsonweed Ladysthumb Lambsquarters Morningglory, Cypressvine, Entireleaf, Ivyleaf, Purple Moonflower, Scarlet, Smallflower, Small White (Pitted), Tall (Common), Willowleaf (Palmleaf)	Mustard, Wild Nightshade, Eastern Black, Black Pigweed, Prostrate, Redroot, Smooth Poinsettia, Wild Poorjoe Purslane, Common Pusley, Florida Ragweed, Common, Giant Senna, Coffee Sesbania, Hemp Smartweed, Pennsylvania Smellmelon Spurge, Prostrate, Spotted Starbur, Bristly Velvetleaf[a] Waterhemp, Tall	Oil concentrate (2 pts./A) Do not add UAN or ammonium sulfate
plus		**Weeds Controlled**	**Sizes**	
		Annual Grasses[b]		
Poast	¾ pt.	Wild Proso Millet	4-10″	
	1 pt.	Foxtail, Giant Junglerice Panicum, Fall, Texas Signalgrass, Broadleaf	3-8″ 3-8″ 3-8″ 3-8″ 3-8″	
	1½ pts.	Barnyardgrass Crabgrass, Large, Smooth Cupgrass, Woolly Foxtail, Green, Yellow Goosegrass Johnsongrass, Seedling Sprangletop, Red Witchgrass	3-8″ 3-6″ 3-6″ 3-6″ 3-8″ 3-8″ 3-6″ 3-8″ 3-8″ 3-8″	

[a] See **Special Use Directions** for rate.

[b] Tank mix does not control rhizome johnsongrass, quackgrass, bermudagrass, wirestem muhly, volunteer corn, shattercane, volunteer cereals, wild oats, red rice or itchgrass.

BLAZER + FUSILADE 2000 1E APPLICATIONS IN SOYBEANS GENERAL AND APPLICATION INFORMATION, RESTRICTIONS AND LIMITATIONS

General information

BLAZER and Fusilade 2000 1E may be applied sequentially or in a tank mix for postemergence broad spectrum weed control. The growth stage of weeds at the time of application will determine which method of application will provide the most satisfactory results. Both BLAZER and Fusilade should be applied to actively growing weeds.

Time of application

Applications should be made to actively growing weeds approximately 2 to 3 weeks after planting. A delay will permit weeds to exceed the maximum size, resulting in inadequate control. The growth stage of weeds at the time of application should govern the application system used for optimum weed control. For additional information see Tables 11 & 12.

Rate

Use BLAZER at the rate of 1½ to 2 pints per acre. Fusilade 2000 1E rates will vary depending on region and weed species from ¾ to 3 pints per acre. For details see the Fusilade 2000 1E label.

Spray additive

When applying the BLAZER + Fusilade tank mixture use a nonionic spray adjuvant at the rate of 2 pints per 100 gallon spray solution. If Fusilade is applied alone, either a crop oil concentrate at the rate of 1% of the finished spray volume or a nonionic spray adjuvant at the recommended rate should be added.

Water volume and spray pressure

For additional information refer to the section entitled Directions for use, pages 2 & 3.

Ground equipment: For the tank mix of BLAZER + Fusilade 2000 1E use 20 gallons of total spray solution per acre (broadcast basis) and a minimum of 40 psi pressure. Use standard high pressure flat fan or hollow cone nozzles spaced 20 inches apart. Do not use flood fan or whirl chamber nozzles.

Air equipment: Use a minimum of 10 gallons of total spray solution per acre.

Mixing

Fill half the spray tank with water and add the recommended amount of product in the following order—BLAZER, Fusilade, and a spray adjuvant—while the agitator is running, then add the remaining quantity of water.

Coverage

Thorough coverage of actively growing weeds is essential. Large crop-and-weed leaf canopies shelter smaller weeds and can prevent adequate spray coverage. Soybeans are tolerant to the tank mix; however, under certain conditions soybeans may burn, crinkle and bronze.

Restrictions and limitations (partial list)

Always read and follow the restrictions and limitations for all products whether used alone or in a tank mix. The most restrictive labeling applies in tank mixtures.

Do not apply more than a total of 4 pints of Fusilade 2000 1E per acre per season to soybeans.

Make the last Fusilade 2000 1E application before soybean bloom.

Do not make more than one application of BLAZER + Fusilade 2000 1E tank mix in a single season.

Do not plant rotational crops other than cotton and soybeans within 60 days after the last Fusilade 2000 1E application.

Do not apply BLAZER within 50 days of harvest for soybeans.

BLAZER requires a six hour rain-free period for best results. Do not apply BLAZER or the BLAZER + Fusilade 2000 1E tank mix if rain is threatening.

Table 11
Postemergence Application Systems of BLAZER and Fusilade 2000 1E

Sequential		Tank Mix
Fusilade Followed by BLAZER	**BLAZER Followed by Fusilade**	**BLAZER + Fusilade**
Apply Fusilade 3 to 5 days prior to BLAZER to enable adequate translocation of Fusilade in grasses.	Following the BLAZER application, grasses must have resumed active growth with development of new leaves.	Apply combination when weeds are at proper stage of growth as per individual label.

For dosage rates and growth stages consult individual labels prior to applications.

The growth stage of weeds at the time of application will determine which application will provide most satisfactory results.

Note: Tank mix applications sometimes have resulted in reduced grass weed control and possible increase in crop injury as compared to either product used alone. If grass regrowth occurs following an application of the tank mix or an additional flush of grasses emerge, make a second application of Fusilade to actively growing annual grass weeds, as per the label recommendations. A tank mix application is not recommended if perennial grass weeds are the predominant species to be controlled.

TABLE 12
BLAZER + FUSILADE 2000 1E TANK MIX IN SOYBEANS
RATE AND TIME OF APPLICATION TABLE

PRODUCT	PRODUCT RATE	WEEDS CONTROLLED			ADDITIVE INFORMATION
		Annual Broadleaf Weeds			
BLAZER	1½-2 pts./A according to weed species and size. (See **Table 1** Pages 2 & 3)	Amaranth, Palmer, Spiny Balloonvine Buckwheat, Wild Buffalobur Burgherkin Carpetweed Citron (Wild Watermelon) Cocklebur, Common, Heartleaf Copperleaf, Hophornbeam, Virginia Cowpea, Volunteer Crotalaria, Showy Croton, Tropic, Woolly Cucumber, Wild Spiny Devilsclaw Galinsoga, Hairy, Smallflower	Gourd, Texas Groundcherry, Cutleaf, Lanceleaf Indigo, Hairy Jimsonweed Ladysthumb Lambsquarters Morningglory, Cypressvine, Entireleaf, Ivyleaf, Purple Moonflower, Scarlet, Smallflower, Small White (Pitted), Tall (Common), Willowleaf (Palmleaf) Mustard, Wild Nightshade, Eastern Black, Black	Pigweed, Prostrate, Redroot, Smooth Poinsettia, Wild Poorjoe Purslane, Common Pusley, Florida Ragweed, Common, Giant Senna, Coffee Sesbania, Hemp Smartweed, Pennsylvania Smellmelon Spurge, Prostrate, Spotted Starbur, Bristly Velvetleaf[a] Waterhemp	Spray adjuvant 1 pt./100 gal.
		Annual Grasses			
		Foxtail, Giant, Green, Yellow	Johnsongrass, Seedling Panicum, Fall	Shattercane Volunteer Small Grains	
		Perennial Weeds			
		Bindweed, Field, Hedge	Milkweed, Climbing, Common Redvine	Thistle, Canada Trumpetcreeper	
plus	plus	**Weeds Controlled Size**			

PRODUCT	PRODUCT RATE	WEEDS CONTROLLED					ADDITIVE INFORMATION
			Region A[b]		Region B[b]		
		Annual Grass	Height	Leaves	Height	Leaves	
Fusilade 2000 1E	³/₄ to 3 pts./A	Barnyardgrass	2-3"	3	1-2"	3	
		Crabgrass	1-2"	4	1-2"	3	
		Johnsongrass, Seedling	2-8"	4	2-4"	3	
		Junglerice Volunteer	2-3"	3	2-3"	3	
		Cereals: Barley	2-6"	6	2-4"	3	
		Milo	6-12"	4	2-4"	4	
		Oats	2-6"	6	2-4"	3	
		Wheat	2-6"	6	2-4"	3	

[a] See Special Use Directions for rate.
[b] Region A—All U.S. soybean growing areas with the exception of Region B
B—Western TX and OK production area
Fusilade may be applied sequentially or in a tank mix with BLAZER. See and follow the Fusilade 2000 1E label for additional information on dosage, weed species and size.

BLAZER + Amiben tank mix in soybeans
General and application information, Restrictions and limitations

General information
When applied to young, actively growing broadleaf weed seedlings, the tank mix of BLAZER plus Amiben will control all weeds listed on the BLAZER label and will also provide preemergence control of those grasses and broadleaf weeds listed in the General information section of the Amiben label which have not yet germinated at time of treatment.

Time of application
Apply the tank mixture when broadleaf weeds have 2 to 4 leaves, usually 14 to 21 days after planting.

Rate
Apply a tank mixture of 1¹/₂ to 2 pints of BLAZER per acre plus 5 to 6 quarts of Amiben or 3 to 3³/₅ pounds Amiben DS per acre. In order to determine the correct application rate of BLAZER in the tank mix, see the BLAZER use rate in **Table 1.**

Spray additive
Spray adjuvant should be added at the rate of 2 to 4 pints per 100 gallons of spray for maximum control of certain susceptible weeds such as escaped grasses, sesbania and crotalaria.

Water volume and spray pressure
For additional information refer to the section entitled **Directions for use,** pages 2 & 3.

Ground equipment: For the tank mix of BLAZER + Amiben, use a minimum of 20 gallons of total spray solution per acre (broadcast basis) and a minimum of 40 psi pressure. Use standard high pressure hollow cone or flat fan nozzles spaced 20 inches apart. Do not use flood or whirl chamber nozzles.

Air equipment: Use a minimum of 10 gallons of total spray solution per acre.

Mixing
Fill the spray tank half with water and add the recommended amount of product in the following order—BLAZER, Amiben, spray adjuvant—while the agitator is running; then add the remaining quantity of water.

Coverage
Thorough coverage of actively growing weeds is essential. Large crop-and-weed leaf canopies shelter smaller weeds and can prevent adequate spray coverage. Soybeans are tolerant to the above tank mixes; however, under certain conditions soybeans may burn, crinkle and bronze.

Restrictions and limitations (partial list)
Follow all use restrictions detailed on the BLAZER and Amiben labels. The most restrictive labeling applies in tank mixes.

Do not add crop oil when using the tank mix.

Do not make applications later than 33 days after soybean planting.

BLAZER + CLASSIC TANK MIX IN SOYBEANS GENERAL AND APPLICATION INFORMATION, RESTRICTIONS AND LIMITATIONS

General information
BLAZER herbicide may be tank mixed with Classic for postemergence control of the major broadleaf weed species in soybeans.

Time of application
For optimum postemergence control apply the tank mix to actively growing weeds.

Rate
Use BLAZER at the rate of 1¹/₂ to 2 pints per acre. For improved control of the following weeds (up to 5 leaf stage): cocklebur, bristly starbur, florida beggarweed, sunflower and yellow nutsedge, add ¹/₂ to ³/₄ ounce per acre of Classic. In order to determine the correct application rate of BLAZER in the tank mixture, see the BLAZER use rate in **Table 1.**

Spray additive
The use of spray adjuvant at the rate of 2 pints per 100 gallons of spray mixture is recommended. For control of certain weeds, such as escaped grasses with the tank mix, the addition of up to 4 pints of surfactant per 100 gallons of spray mix is required.

Water volume and spray pressure
For additional information refer to the section entitled **Directions for use,** pages 2 & 3.

Ground equipment: For broadcast application, use a minimum of 20 gallons of total spray solution per acre and 40 psi pressure with flat fan or hollow cone nozzles spaced 20 inches apart.

Mixing
Follow recommendations for water volume and spray pressure, as per the BLAZER label. Fill the spray tank with half the amount of required water and add the recommended amount of Classic. Once thoroughly mixed in the spray tank, add the recommended amounts of BLAZER and a spray adjuvant while the agitator is running. Then add the remaining quantity of water.

Coverage
Thorough coverage is essential. For best results follow good herbicide application practices.

Restrictions and limitations (partial list)
Always read and follow the restrictions and limitations for all products whether used alone or in a tank mix. The most restrictive labeling applies in tank mix.
Do not use crop oil concentrate or vegetable oil, as severe crop injury may result.

BLAZER + RESCUE TANK MIX IN SOYBEANS (MID- TO LATE SEASON POSTEMERGENCE WEED CONTROL) GENERAL AND APPLICATION INFORMATION, RESTRICTIONS AND LIMITATIONS

General information
BLAZER may be tank mixed with Rescue for mid-to-late season postemergence weed control in soybeans. Rescue is a systemic herbicide and therefore requires 20-30 days to obtain maximum effect. The reduction of weed competition to the soybeans begins immediately after the tank mix of BLAZER plus Rescue is applied even though the weeds are not killed immediately.

Weed size will determine the most effective rate for the tank mix of BLAZER plus Rescue. Control with this tank mix will decrease with increasing weed size and density of the soybean canopy. Thorough coverage of the weed is essential for maximum control.

Most soybeans are tolerant to Rescue when used according to label instructions. Some upper soybean plant drooping, leaf wrinkling or twisting may occur under certain conditions; however, soybeans will usually outgrow this condition and continue to develop normally. Soybean height reduction may occur due to the shortening of the soybean stem internodes and has been shown to have no direct relationship to yield. Mitchell, Corsoy and Forrest varieties are more sensitive to Rescue than other varieties.

Time of application
When applied according to label instructions, the tank mix of BLAZER plus Rescue will control or suppress certain broadleaf weeds. See **Table 13** for detailed time of application information.

Rescue may be applied to determinate or indeterminate soybeans which are at least 14" tall or if blooming has begun. In the upper midwest and high plains (Minnesota, North Dakota, South Dakota), Rescue applications should be delayed until the group 0, I, or II soybeans have begun to bloom.

Rate
For application rates of BLAZER + Rescue refer to **Table 13** for details.

Spray additive
An 80% active nonionic surfactant, or crop oil concentrate should be used at the rate recommended on the Rescue label. Soybean crop response will increase with the addition of a crop oil concentrate with the tank mix.

Water volume and spray pressure
For additional information refer to the section entitled **Directions for use,** pages 2 & 3.

Ground equipment: For the tank mix of BLAZER + Rescue apply the suggested label rates (See **Table 13**) in 10-25 gallons of water per acre; maintain a high pressure (40-50 psi) during application to insure better overall coverage. The spray boom should be equipped with standard high pressure hollow cone or flat fan nozzles spaced 20 inches apart.

Air equipment: Apply the suggested label rates of the BLAZER plus Rescue tank mix as indicated, in a minimum of 5 gallons of water per acre.

Mixing
Fill half the spray tank with water and add the recommended amount of product in the following order—BLAZER, Rescue, spray adjuvant-while the agitator is running; then add the remaining quantity of water.

Coverage
Thorough coverage of actively growing weeds is essential. Large crop-and-weed leaf canopies shelter smaller weeds and can prevent adequate spray coverage.

Restrictions and limitations (partial list)
Before applying a tank mix of Rescue and BLAZER, read both labels and follow precautionary statements on each label.

Do not harvest soybeans earlier than 60 days after treatment.

Do not graze or feed forage from treated soybeans to livestock.

Avoid applications to drought, heat or disease stressed soybeans. Less activity should be expected if applied on stressed weeds.

Do not apply if rain is expected within 6 hours as effectiveness may be reduced.

Applications should not be made when weather conditions favor drift. Do not apply by air within 500 feet of susceptible crops such as cotton, tomatoes, tobacco or sunflower.

TABLE 13
BLAZER + RESCUE TANK MIX—SOYBEANS
RATE AND TIME OF APPLICATION TABLE

WEED	GROWTH STAGE FOR OPTIMUM RESULTS	RESCUE APPLICATION RATE/A	BLAZER APPLICATION RATE/A	COMMENTS	EXPECTED RESPONSES[a]
Cocklebur	Up to 12" (but before flowering begins)	2-3 qts.	1 pt.	Use 2 qts. of Rescue rate in midsouth and when open canopy exists	A, B, C, E, F
Cocklebur	Up to 24" (but before flowering begins)	3 qts.	1 pt.		A, B, C, G, F
Giant Ragweed	Up to 36" (but before flowering begins)	2-3 qts	1 pt.	Use 3 qts. of Rescue where heavy infestations occur or when soybeans cover middles	A, B, E, F
Morningglory	Up to 8 leaf	2 qts.	1½ pts.	Spray solutions should be applied before closure of crop to facilitate good coverage	A, F, H
(Annual Tall, Ivy-leaf, Entireleaf)	8 leaf but before vining	3 qts.	1½ pts.		A, F, H
Common Ragweed	Up to 24"	3 qts.	1 pt.		E, F, I
	24"or taller but before flowering	3 qts.	1½ pts.		B, E, G, I
Jimsonweed	Up to 18"	2 qts.	1 pt.		D, E, F, I
Pigweed	18" or taller but before flowering	3 qts.	1 pt.		A, E, F, I

[a] Letter codes correspond to the following expected responses:
A–Stem twisting
B–Growth termination
C–Reduction of viable seed
D–Eventual desiccation if soybeans canopy over weeds
E–Reduce harvest losses
F–Faster growth termination and better weed desiccation
G–Plants 24-36" will be suppressed and may not die unless soybeans canopy over them
H–Terminate growth and flowering
I –Reduce competition to crop

PEANUTS

Directions for use
BLAZER is a selective broad spectrum herbicide recommended for preemergence, cracking (initiation of soil cracking, but prior to peanut emergence from the soil) and postemergence applications to peanuts to control susceptible weeds. Optimum weed control is achieved when young actively growing weed seedlings are treated. It is important to cover all weed parts thoroughly, as BLAZER works primarily by contact action. Failure to follow the suggested dosages on maximum weed size may result in unsatisfactory control.

Restrictions and limitations in peanuts
Do not apply BLAZER within 75 days of harvest.

Do not apply more than a total of 8 pints of BLAZER per acre from combined preemergence and postemergence applications. Do not apply more than 2 pints of BLAZER postemergence during the peanut growing season.

Do not use treated plants for feed or forage.

In the case of crop failure, only peanuts or soybeans may be immediately replanted.

Crop rotation restriction: Root crops (such as carrots, turnips, sweet potatoes, etc.) must not be planted in fields treated with BLAZER for a period of 18 months following treatment.

Table 14
BLAZER Weed Control and Use Rate Table—Peanuts
Rate: The recommended rate for broad spectrum postemergence control is 2 pints of BLAZER per acre plus 1 pint of a nonionic spray adjuvant per 100 gallons of spray mix. When this rate is used, do not apply more than one postemergence application of BLAZER to peanuts.

WEEDS CONTROLLED	APPLICATION RATES FOR WEED GROWTH STAGES					
	1.0 PT./A MAXIMUM[a]		1.5 PTS./A MAXIMUM[a]		2.0 PTS./A MAXIMUM[a]	
Annual Broadleaf Weeds	LEAF STAGE	HEIGHT INCHES	LEAF STAGE	HEIGHT INCHES	LEAF STAGE	HEIGHT INCHES
Amaranth, Palmer,	—	—	4	1-2"	6	3"
Spiny	—	—	—	—	6	2-3"
Balloonvine	—	—	—	—	4	2"
Beggarweed, Florida,	—	—	—	—	3[b]	2"
Buffalobur	—	—	—	—	3[b]	2"
Burgherkin	—	—	—	—	4[b]	3"
Carpetweed	multi. 3" dia.	2"	multi. 6" dia.	< 2"	multi 8" dia.	< 2" —
Citron (Wild Watermelon)	—	—	—	—	4[b]	2"
Cocklebur, Common,	—	—	—	—	4[b]	4"
Heartleaf	—	—	—	—	4[b]	4"
Croton, Tropic,	2	< 2"	2-4	2"	—	—
Woolly	2	< 2"	2-4	2"	—	—
Cucumber, Wild Spiny	—	—	—	—	4[b]	3"
Galinsoga, Hairy,	—	—	—	—	4	< 2"
Smallflower	—	—	—	—	4	< 2"
Gourd, Texas,	—	—	—	—	3[b]	2"
Groundcherry, Cutleaf,	—	—	2	1"	4	2"
Lanceleaf	—	—	—	—	4	2"
Indigo, Hairy	—	—	—	—	3	—
Jimsonweed	3	3"	6	6"	8	8"
Lambsquarters, Common	—	—	—	—	3[b]	1"
Mallow, Venice	—	—	—	—	3	1-2"
Morningglory, Cypressvine,	—	—	—	—	4	2"
Entireleaf,	—	—	—	—	3	< 2"
Ivyleaf,	—	—	—	—	3	1-2"
Purple Moonflower,	—	—	4	2"	4	2"
Scarlet,	—	—	—	—	4	2"
Smallflower,	—	—	—	—	4	2"
Small White (Pitted),	2	1-2"	4	2"	4	2"
Tall (Common),	—	—	—	—	3	< 2"
Willowleaf, (Palmleaf)	—	—	—	—	4	2"
Mustard, Wild	—	—	4	3"	6	4"
Nightshade, Eastern Black,	—	—	—	—	4	2"
Black	—	—	—	—	4	< 2"
Pigweed, Prostrate,	—	—	—	—	6	3"
Redroot,	2	< 2"	4	2"	6	3"
Smooth	—	—	4	2"	6	3"
Poinsettia, Wild	—	—	—	—	2	2"
Poorjoe	—	—	—	—	4	—
Purslane, Common	—	—	multi. 6" dia.	1"	multi. 8" dia.	1"
Pusley, Florida	—	—	—	—	2	< 2"
Ragweed, Common,	2	< 2"	4	2"	6	3"
Giant	—	—	—	—	4	4"
Sesbania, Hemp	—	—	b	b	b	b
Smartweed, Pennsylvania	—	—	—	—	4	4"
Smellmelon	—	—	—	—	4[b]	3"
Spurge, Prostrate,	—	—	—	—	multi ½" dia.	—
Spotted	—	—	—	—	multi ½" dia.	— —
Annual Grasses						
Foxtail, Giant,	—	—	—	—	3[b]	2"
Green,	—	—	—	—	3[b]	2"
Yellow,	—	—	—	—	3[b]	2"
Johnsongrass, Seedling	—	—	—	—	3[b]	2"
Panicum, Fall	—	—	—	—	3[b]	2"
Shattercane	—	—	—	—	3[b]	2"
Volunteer Small Grains	—	—	—	—	3[b]	2"
Perennial Grasses						
Bindweed, Field,	—	—	—	—	multi[b]	—
Hedge	—	—	—	—	multi[b]	—
Milkweed, Climbing,	—	—	—	—	multi[b]	—
Common	—	—	—	—	multi[b]	—
Redvine	—	—	—	—	multi[b]	—
Thistle, Canada	—	—	—	—	multi[b]	—
Trumpetcreeper	—	—	—	—	multi[b]	—

[a] Do not count leaves as pairs . . . count each leaf separately. Do not count cotyledon leaves. Spraying weeds in the cotyledon growth stage is not recommended.
[b] See **Special Use Directions** for these weed problems.

Special Use Directions for Additional Weed Problems in Peanuts

Annual Weeds

Beggarweed, Florida

Control of Florida beggarweed is difficult due to the weed's long germination season. An early application (preemergence to cracking) of 2 pints of BLAZER per acre plus 2 pints of spray adjuvant per 100 gallons of spray mix should be applied when beggarweed seedlings have no more than 3 young expanding true leaves. Weeds at this time will not be more than 1½ inches high. It is important to obtain maximum control of the earliest weed flush. Cultivation should also be well timed to give maximum control of regrowth or secondary weed flushes. BLAZER will suppress or partially control weeds growing under conditions of high soil moisture and high relative humidity.

Cocklebur, Common

Cocklebur, Heartleaf

The most consistent control of cocklebur is obtained when 2 pints per acre of BLAZER are applied to seedlings in the 2 leaf stage which are actively growing under conditions of high soil moisture and high relative humidity. The use of 1 pint of spray adjuvant per 100 gallons of spray mix is recommended. Control of cocklebur may be partial or inconsistent when sprayed during periods of dry weather or other stress conditions.

Cowpea, Volunteer

Volunteer cowpea germinates over a long period of time, thus making control difficult. Partial control can be obtained by applying 2 pints per acre of BLAZER just past the cowpea cotyledon growth stage. Do not wait for the first leaves to fully expand. Use 2 pints of nonionic spray adjuvant per 100 gallons of spray mix for maximum activity.

Crotalaria, Showy

Sesbania, Hemp

Sesbania and crotalaria are very sensitive to BLAZER. The use of 2 pints of spray adjuvant per 100 gallons of spray mix is suggested. Effective control can be obtained at most plant heights. It is important, however, that BLAZER be applied prior to bloom. Applications after bloom are usually not effective and therefore not recommended. The use of 1 pint of BLAZER per acre has given excellent control of both sesbania and crotalaria. During or after a period of dry weather, control may be erratic. Best control is then obtained when $1^1/_2$ pints are applied to each acre. Application for control of these weeds should be timed to occur after maximum weed emergence has taken place. Care must be exercised to make certain that crops do not shade this weed from spray deposits. Waiting for the sesbania to break through the crop canopy may be advisable for control of late season infestations.

Cucurbits,
Burgherkin,
Citron (Wild Watermelon),
Cucumber,
Wild Spiny,
Smellmelon,
Texas Gourd

Members of the cucumber family germinate over an extended period of time. Control is therefore difficult to obtain with a single spray. In order for BLAZER to be effective, the application should be made to weeds no later than the 4 leaf growth stage. Use 2 pints of BLAZER per acre plus 2 pints of spray adjuvant per 100 gallons of spray mix.

Waiting for several weed flushes to appear and then spraying will result in poor control. The most effective control is obtained when weeds have no more than 4 true leaves.

Lambsquarters, Common

BLAZER, at the 2 pint per acre rate, will usually cause spotting, stunting or death of many seedlings not exceeding 3 true leaves. The use of 2 pints of nonionic surfactant per 100 gallons of spray mix is suggested for maximum kill of seedlings. Cultivation 3 to 7 days after the application will usually assist in control.

Poinsettia, Wild

BLAZER, at the 2 pint rate per acre plus 2 pints of spray adjuvant per 100 gallons of spray mix, will usually kill or severely stunt wild Poinsettia. Application must be made prior to the formation of the third true leaf. In addition, the seedlings must be actively growing.

Annual Grasses
Foxtail, Giant
Foxtail, Green
Foxtail, Yellow
Johnsongrass, Seedling
Panicum, Fall
Shattercane

BLAZER must not be the basic component of a grass management program. For additional control of escaped grasses following a preplant incorporated or preemergence herbicide, use BLAZER at 2 pints per acre. The use of 2 to 4 pints of spray adjuvant per 100 gallons of spray mix is required. Grasses should not exceed the 2 to 3 leaf stage for this treatment. Activity is dependent upon good soil moisture during and following the spray solution.

Volunteer Small Grains: Barley, Oats, Rye, Wheat

When BLAZER is applied to emerging volunteer small grains in the 2 to 3 leaf stage, many plants will die or remain stunted. BLAZER should be applied at 2 pints per acre plus 2 to 4 pints of spray adjuvant per 100 gallons of spray mix. Activity is dependent upon good soil moisture during and following the spray applications.

Perennial Weeds
Bindweed, Field
Bindweed. Hedge
Milkweed, Climbing
Milkweed, Common
Redvine
Thistle, Canada
Trumpetcreeper

Growth of perennial weeds resulting from underground rootstocks is very difficult to control. BLAZER at the 2 pints rate per acre, plus 1-2 pints of spray adjuvant per 100 gallons of spray mix applied under favorable environmental conditions, will burn-back the aboveground plant parts and retard regrowth. BLAZER will not kill the underground rootstocks of these weeds.

BLAZER can be applied alone or in combination with various herbicides. For tank mix combination refer to Tables 15 and 16.

Preemergence weed control

General information

For weed control from a preemergence treatment use 4 to 8 pints of BLAZER per acre. The highest rate is used where dense or severe weed populations exist.

Susceptible broadleaf species: cocklebur (common), copperleaf (Virginia), croton (tropic) groundcherry species, (seedling), jimsonweed, lambsquarters (common), morningglory (smallflower), mustard (black), pigweed (redroot), pusley (Florida), ragweed (common), smartweed (Pennsylvania)

Grass species (suppression only): crabgrass (large), crabgrass (smooth), foxtail (giant), johnsongrass (seedling), panicum (fall), sandbur (field)

Time of application

Apply BLAZER preemergence to fields which have been well worked with smooth seed beds in firm condition. Avoid applications to poorly prepared seedbeds.

Rate

Use 4 to 8 pints of BLAZER per acre on a broadcast basis after seeding.

Water volume and spray pressure

For additional information refer to section entitled **Directions for use,** pages 2 & 3.

Ground equipment: Application should be made using standard herbicide sprayers equipped with flat fan nozzles. Spray equipment should be calibrated to deliver a minimum of 20 gallons of spray mixture per acre. Flood nozzles are not recommended.

Mixing

Fill the spray tank about half full with water, than add the required amount of BLAZER, mix thoroughly and continue filling the spray tank.

Coverage

Uniform application is essential for satisfactory weed control.

Restrictions and limitations (partial list)

Do not apply more than a total of 8 pints of BLAZER per acre from combined preemergence and postemergence applications. See postemergence use restrictions.

Do not use treated plants for feed or forage.

Crop rotation restriction: Root crops (such as carrots, turnips, sweet potatoes, etc.) must not be planted in fields treated with BLAZER for a period of 18 months following treatment.

In the case of crop failure, only peanuts may be replanted.

Table 15

Tank Mixture Recommendations for Weed Control in Peanuts Using Different Application Timings

At Cracking; Postemergence	Postemergence	Preemergence; At Cracking
BLAZER + Basagran see pages 35 to 36	BLAZER + 2,4-DB see pages 37 to 38	BLAZER + Amiben BLAZER + Dual BLAZER + Lasso see pages 39 to 40

Note: Cracking stage is defined in this label as the initiation of soil cracking, but prior to peanut plant emergence. Applications of tank mix after the peanuts are past the cracking stage may result in crop response.

TABLE 16
PEANUTS—TANK MIXES WITH BLAZER[a]

Use the following chart as a guide to determine broadleaf weeds and grasses controlled by BLAZER alone and various tank mixes with BLAZER.

BLAZER CONTROLS THE WEEDS LISTED BELOW	ADDITIONAL WEEDS CONTROLLED BY TANK MIXING VARIOUS HERBICIDES WITH BLAZER		REFER TO TABLES LISTED BELOW FOR RATE, WEED SIZE AND ADDITIONAL INFORMATION
Annual Broadleaf Weeds	**Basagran® herbicide**		
Amaranth, Palmer, Spiny Balloonvine Buffalobur Burgherkin Carpetweed Citron (Wild Watermelon) Cocklebur, Common, Heartleaf Cowpea, Volunteer Crotalaria, Showy Croton, Tropic, Woolly Cucumber, Wild Spiny Galinsoga, Hairy, Smallflower Gourd, Texas Groundcherry, Cutleaf, Lanceleaf Indigo, Hairy Jimsonweed Lambsquarters, Common Mallow, Venice Morningglory, Cypressvine, Entireleaf Ivyleaf, Purple Moonflower, Scarlet Smallflower, Small White (Pitted), Tall (Common), Willowleaf (Palmleaf) Mustard, Wild Nightshade, Eastern Black, Black Pigweed, Prostrate, Redroot, Smooth Poinsettia, Wild Poorjoe Purslane, Common Pusley, Florida Ragweed, Common, Giant Sesbania, Hemp Smartweed, Pennsylvania Smellmelon Spurge, Prostrate, Spotted	Anoda, Spurred Balloonvine Beggarticks Cocklebur (Large) Dayflower Galinsoga Lambsquarters Mallow, Venice Nutsedge, Yellow Poinsettia, Wild Ragweed, Giant	Redweed Senna, Coffee Shepherdspurse Sida, Prickly (Teaweed) Starbur, Bristly Sunflower, Wild Thistle, Canada Velvetleaf	BLAZER + Basagran **TABLE 17** Page 36
	2,4-DB		
	Burgherkin Citron Cocklebur Morningglory (Large) Pigweed, Redroot		BLAZER + 2,4-DB **TABLE 18** Page 38
	Amiben® herbicide		
	See Amiben label for weed species controlled at cracking.		BLAZER + Amiben Page 39
	Dual® 8E herbicide		
	See Dual 8E label for annual grasses controlled at cracking.		BLAZER + Dual 8E Page 39
	Lasso® herbicide		
	See Lasso label for annual grasses controlled at cracking.		BLAZER + Lasso Page 40
Annual Grasses			
Foxtail, Giant, Green, Yellow Johnsongrass, Seedling Panicum, Fall Shattercane Volunteer Small Grains			
Perennial Weeds			
Bindweed, Field, Hedge Milkweed, Climbing, Common Redvine Thistle, Canada Trumpetcreeper			

[a]Note: Tank mixes are not applicable in California.

AT CRACKING AND/OR POSTEMERGENCE TANK MIXTURE RECOMMENDATIONS
BLAZER + BASAGRAN TANK MIX IN PEANUTS

General and application, Restrictions and limitations

General information

BLAZER may be tank mixed with Basagran for postemergence control of the major broadleaf weed species in peanuts.

Time of application

The timing of application should be in accordance with weed growth stages indicated in the respective tables, and when weeds are actively growing (See Tables 14 & 17).

Delay in application which permits weeds to exceed the maximum size stated will result in inadequate control.

Rate

Use BLAZER at the rate of 1 to 2 pints per acre tank mixed with Basagran at the rate of 1 to 2 pints per acre. The tank mix of 1 pint of BLAZER with 1 pint of Basagran will provide postemergence control (up to 4 leaf) of common cocklebur, hemp sesbania, carpetweed, wild mustard, jimsonweed, common ragweed, Pennsylvania smartweed, redroot pigweed, smooth pigweed, cypressvine morningglory, purple moonflower morningglory, scarlet morningglory, small white (pitted) morningglory, willowleaf (palmleaf) morningglory, smallflower morningglory, and showy crotalaria. In order to determine the correct application rate of BLAZER to use in all other applications of the tank mix, see the BLAZER use rate table.

For control of common cocklebur, up to the 6 leaf stage, add 1 pint per acre of Basagran to BLAZER. For the additional control of spurred anoda, beggarticks, dayflower, redweed and prickly sida up to the 6 leaf stage and bristly starbur up to the 4 leaf stage, add 1½ pints per acre of Basagran to BLAZER. Add 1 pint of a spray adjuvant for each 100 gallons of the tank mix spray solution.

Spray additive

One pint of a nonionic spray adjuvant should be added per 100 gallons of the spray mixture.

Water volume and spray pressure

For additional information refer to section entitled **Directions for use,** pages 2 & 3.

Ground equipment: For the tank mix of BLAZER + Basagran, use a minimum of 20 gallons of total spray solution per acre (broadcast basis) and a minimum of 40 psi pressure. Use standard high pressure hollow cone or flat fan nozzles spaced 20 inches apart. Do not use flood or whirl chamber nozzles.

Air equipment: Use a minimum of 10 gallons of total spray solution per acre.

Mixing

Fill half the spray tank with water and add the suggested amount of—BLAZER, Basagran, spray adjuvant—while the agitator is running, then add the remaining quantity of water.

Coverage

Thorough coverage of actively growing weeds is essential. Large crop-and-weed leaf canopies shelter smaller weeds and can prevent adequate spray coverage. Peanuts are tolerant to the above tank mixes; however, under certain conditions peanuts may burn, crinkle and bronze.

Restrictions and limitations (partial list)

Read and follow all applicable directions and use restrictions on the BLAZER herbicide and Basagran labels.

Do not apply more than 2 pints of BLAZER postemergence during the peanut growing season.

Do not apply the tank mix within 75 days of harvest for peanuts.

Do not apply more than a total of 4 pints of Basagran per acre per season.

Do not use treated plants for feed or forage.

Do not add crop oil concentrate when using the tank mix.

TABLE 17
BLAZER + BASAGRAN TANK MIX IN PEANUTS
RATE AND TIME OF APPLICATION TABLE
ADDITIVES

PRODUCT	RATE	WEEDS CONTROLLED		ADDITIVE INFORMATION
		Annual Broadleaf Weeds		
BLAZER	1-2 pts./A according to weed species and size. (See **Table 14** Pages 28 & 29)	Amaranth, Palmer, Spiny Balloonvine Buffalobur Burgherkin Carpetweed Citron (Wild Watermelon) Cocklebur, Common, Heartleaf Cowpea, Volunteer Crotalaria, Showy Croton, Tropic, Woolly Cucumber, Wild Spiny Galinsoga, Hairy, Smallflower Gourd, Texas Goundcherry, Cutleaf Lanceleaf Indigo, Hairy Jimsonweed Lambsquarters	Morningglory, Cypressvine, Entireleaf, Ivyleaf, Purple Moonflower, Scarlet, Smallflower, Small White (Pitted), Tall (Common) Willowleaf (Palmleaf) Mustard, Wild Nightshade, Eastern Black, Black Pigweed, Prostrate, Redroot, Smooth Poinsettia, Wild Poorjoe Purslane, Common Pusley, Florida Ragweed, Common, Giant Senna, Coffee Sesbania, Hemp Smartweed, Pennsylvania Smellmelon Spurge, Prostrate, Spotted Starbur, Bristly	Spray adjuvant 1 pt./100 gal.
		Annual Grasses		
		Foxtail, Giant, Green, Yellow Johnsongrass, Seedling	Panicum, Fall Shattercane Volunteer Small Grains	
		Perennial Weeds		
		Bindweed, Field, Hedge Milkweed, Climbing, Common	Redvine Thistle, Canada Trumpetcreeper	

PRODUCT	WEEDS CONTROLLED/WEED SIZE	DOSAGE 1 ½ PTS./A		DOSAGE 2 PTS./A	
		LEAF STAGE	MAX. HEIGHT	LEAF STAGE	MAX. HEIGHT
plus Basagran	Anoda, Spurred	Up to 6	3″	6–8	4″
	Beggarticks	Up to 6	6″	6–8	8″
	Cocklebur[a]	2–6″	6″	6–10	10″
	Dayflower	Up to 6	4″	6–10	8″
	Nutsedge, Yellow	c	c	c	c
	Ragweed, Giant	Not recommended		Up to 4	6″
	Senna, Coffee[b]	Not recommended		Up to 1 pinnate	2″
	Sida, Prickly (Teaweed)	Up to 6	3″	6–8	4″
	Sunflower, Wild	Up to 4	5″	4–6	8″
	Starbur, Bristly	Up to 4	2″	4–6	3″

PRODUCT	RATE	WEEDS CONTROLLED					ADDITIVE INFORMATION
		Velvetleaf	Up to 4	2"	4–6	5"	

[a]Do not treat earlier than leaf stage shown, do not count cotelydon leaves.
[b]Add oil concentrate according to the Directions for Use.
[c]See **Special Directions for Other Weed Problems in Peanuts** in the Basagran label.

BLAZER + 2,4-DB TANK MIX IN PEANUTS GENERAL AND APPLICATION INFORMATION, RESTRICTIONS AND LIMITATIONS

General information

A tank mix of BLAZER plus 2,4-DB is recommended for control of morningglory, cocklebur, common ragweed, redroot pigweed, jimsonweed, burgherkin and citron in peanuts when the weed size exceeds that specified on the BLAZER label. Control with this mix may decrease with increasing weed size or density of weed or crop canopy due to poor spray coverage. For control of other weeds refer to Tables 14 & 18, pages 28-29 & 38.

Do not apply the tank mix when peanuts are exhibiting injury from previously applied pesticides or are exhibiting stress symptoms from disease, nematodes, insects; excessive fertilizer or soil salts; wind injury; frost damage or high temperature stress or wilt; as increased crop response will result.

Time of application

For optimum control apply BLAZER plus 2,4-DB tank mix to actively growing weeds up to the 8 leaf stage, usually 2 to 12 weeks after planting. Applications at later weed stages will result in partial control or suppression.

Peanuts should be at least 2 weeks old when using a tank mix of BLAZER herbicide and 2,4-DB. Do not use after pod-filling stage begins.

Rate

Mix 1 pint of Butyrac 200, or 1 pint of Butoxone with $1^1/_2$ to 2 pints of BLAZER for each acre being treated.

Spray additives

Add 1 pint of spray adjuvant per 100 gallons to increase control of weeds. Do not add crop oils to the tank mixture. The addition of surfactant will increase the hormonal 2,4-DB crop response.

Water volume and spray pressure

For additional information refer to the section entitled **Directions for use,** pages 2 & 3.

Ground equipment: For best results, the tank mix should be applied with ground equipment. For broadcast application and thorough coverage of weeds apply with flat fan or hollow cone nozzles spaced 20 inches apart in a minimum of 20 gallons of water per acre with a spray pressure of 40 psi.

Air equipment: Use a minimum of 10 gallons of total spray solution per acre. Aerial applicators should review **Restrictions and limitations** and **Drift hazards.**

Mixing

Fill half the spray tank with water and add the recommended amount of—BLAZER, 2,4-DB, spray adjuvant—while the agitator is running, then add the remaining quantity of water.

Drift Hazards

Care must be taken when applying the tank mix to prevent drift to all non-target crops. Tobacco, ornamentals, mustards, sugar beets, potatoes, vegetables, and cotton are a few of the crops known to be sensitive to this tank mix. Hormone-type injury in non-target crops can result from trace amounts of 2,4-DB drift. The use of any cleared drift control agent may reduce this hazard; however, the drift control agent may also decrease the weed control activity.

Restrictions and limitations (partial list)

Read and follow all directions and use restrictions on BLAZER and 2,4-DB labels.

Do not apply the tank mixture within 75 days of harvest for peanuts.

Do not apply more than one application of the tank mixture to peanuts per growing season.

Do not use rates of BLAZER or 2,4-DB in excess of those recommended on this label, or excessive injury and possible yield reduction could result.

Do not mix oils, liquid fertilizers or other pesticides with this tank mix except as specifically directed on this label or on other approved supplemental labeling.

Aerial applicators must be familiar with the EPA registered labels and follow the use precautions. In addition, aerial applicators should follow all applicable state and local regulations. In interpreting the label and the local regulations, the most restrictive situations should apply to avoid drift hazards.

TABLE 18
BLAZER + 2,4-DB TANK MIX IN PEANUTS RATE AND TIME OF APPLICATION TABLE

PRODUCT	PRODUCT RATE	WEEDS CONTROLLED		ADDITIVE INFORMATION
		Annual Broadleaf Weeds		
BLAZER	$1^1/_2$-2 pts./A according to weed species and size. (See **Table 14** Pages 28 & 29)	Amaranth, Palmer, Spiny Balloonvine Buffalobur Burgherkin Carpetweed Citron (Wild Watermelon) Cocklebur, Common, Heartleaf Cowpea, Volunteer Crotalaria, Showy Croton, Tropic, Woolly Cucumber, Wild Spiny Galinsoga, Hairy, Smallflower Gourd, Texas Groundcherry, Cutleaf, Lanceleaf Indigo, Hairy Jimsonweed Lambsquarters	Morningglory, Cypressvine, Entireleaf, Ivyleaf, Purple Moonflower, Scarlet, Smallflower, Small White (Pitted), Tall (Common) Willowleaf (Palmleaf) Mustard, Wild Nightshade, Eastern Black, Black Pigweed, Prostrate, Redroot, Smooth Poinsettia, Wild Poorjoe Purslane, Common Pusley, Florida Ragweed, Common, Giant Senna, Coffee Sesbania, Hemp Smartweed, Pennsylvania Smellmelon Spurge, Prostrate, Spotted Starbur, Bristly	Surfactant 1 pt./100 gal.[b] Do not add crop oil to the tank mixture.
		Annual Grasses		

PRODUCT	PRODUCT RATE	WEEDS CONTROLLED		ADDITIVE INFORMATION
		Foxtail, Giant, Green, Yellow Johnsongrass, Seedling	Panicum, Fall Shattercane Volunteer Small Grains	
		Perennial Weeds		
		Bindweed, Field, Hedge Milkweed, Climbing, Common	Redvine Thistle, Canada Trumpetcreeper	
		Weeds Controlled/Weed Size		
plus Butyrac 200[a] or Butoxone[a]	**plus** 1 pt./A	Burgherkin[a] Citron Cocklebur Jimsonweed Morningglory Pigweed, Redroot Ragweed, Common	Up to 12" height or length of vine	

[a] When size exceeds that specified on the BLAZER label.
[b] The addition of a nonionic spray adjuvant will increase the hormonal, 2,4-DB crop response.

Preemergence/to at cracking treatments
BLAZER + Amiben tank mix in peanuts

General information
In addition to the major broadleaf weed species controlled by BLAZER postemergence, the tank mix of BLAZER with Amiben 2E or Amiben DF will provide preemergence control of many annual grasses (as listed on the Amiben label).

Time of application
Applications of this tank mix may be made immediately after planting up to the initiation of soil cracking. Crop stunting may occur with the application of the tank mix of BLAZER plus Amiben, although yields are not adversely affected.

Rate
Use BLAZER at the rate of 1½ to 2 pints per acre, tank mixed with either Amiben DF at the rate of 2⅖ to 3⅗ pounds per acre or Amiben 2E at the rate of 4 to 6 quarts per acre. In order to determine the correct application rate of BLAZER to use in the tank mix, see **Table 14,** pages 29 & 30.

Spray additive
The addition of a spray adjuvant is recommended and should be used at the rate of 1 pint per 100 gallons of spray mix.

Water volume and spray pressure
For additional information, refer to the section entitled **Directions for use,** pages 2 & 3.

Ground equipment: For best results, the tank mix should be applied with ground equipment. For thorough coverage of weeds apply with flat fan or hollow cone nozzles spaced 20 inches apart in a minimum of 20 gallons of water per acre with a spray pressure of 40 psi.

Mixing
Fill half the spray tank with water and add the suggested amount of—BLAZER, Amiben, spray adjuvant—while the agitator is running; then finish filling the tank.

Restrictions and limitations (partial list)
Read and follow all applicable directions and use restrictions on the BLAZER and Amiben labels.

Do not apply more than 3⅗ pounds per acre of Amiben DS or 6 quarts per acre of Amiben 2E herbicide to the same field of peanuts in the same season.

Do not apply the tank mix after the peanuts are past the cracking stage (initiation of soil cracking, but prior to peanut emergence from the soil) as severe injury will result.

BLAZER + DUAL 8E TANK MIX IN PEANUTS

General information
In addition to the major broadleaf weeds species controlled postemergence by BLAZER, the tank mix of BLAZER with Dual 8E will provide preemergence control of many annual grasses (as listed on the Dual 8E label).

This tank mix can be used as a sequential application after Vernam®, Balan®, or Treflan® herbicides.

Time of application
Applications may be made immediately after planting up to the initiation of soil cracking. Crop stunting may occur with the application of the tank mix of BLAZER plus Dual 8E, although yields are not adversely affected.

Rate
Use BLAZER at the rate of 1½ to 2 pints per acre tank mixed with Dual 8E at the rate of 1½ to 2 pints per acre. In order to determine the correct application rate of BLAZER in the tank mix, see **Table 14,** pages 29 & 30.

Spray additive
The addition of spray adjuvant is recommended and should be used at the rate of 1 pint per 100 gallons of spray mix.

Water volume and spray pressure
For additional information, refer to the section entitled **Directions for use,** pages 2 & 3.

Ground equipment: For best results, the tank mix should be applied with ground equipment. For thorough coverage of weeds apply with flat fan or hollow cone nozzles spaced 20 inches apart in a minimum of 20 gallons of water per acre with a spray pressure of 40 psi.

Mixing
Fill half the spray tank with water and add the suggested amount of—BLAZER, Dual 8E, spray adjuvant—while the agitator is running; then finish filling the tank.

Restrictions and limitations (partial list)
Read and follow all applicable directions and use restrictions on the BLAZER and Dual labels.

Do not apply the tank mix after the peanuts are past the cracking stage (initiation of soil cracking, but prior to peanut emergence from the soil) as severe injury will result.

BLAZER + LASSO 4E TANK MIX IN PEANUTS

General information
In addition to the major broadleaf weed species controlled postemergence by BLAZER, the tank mix of BLAZER with Lasso 4E will provide preemergence control of many annual grasses (as listed on the Lasso 4E label).

This tank mix can be used as a sequential application after Vernam, Balan, or Treflan.

Time of application
Applications of this tank mix may be made immediately after planting up to the initiation of soil cracking. Crop stunting may occur with the application of the tank mixture of BLAZER plus Lasso 4E, although yields are not adversely affected.

Rate
Use BLAZER at the rate of 1½ to 2 pints per acre tank mixed with Lasso 4E at the rate of 2 to 4 quarts per acre. In order to determine the correct application rate of BLAZER in the tank mixture, see **Table 14,** pages 29 & 30.

Spray additive
The addition of spray adjuvant is recommended and should be used at the rate of 1 pint per 100 gallons of spray mix.

Water volume and spray pressure
For additional information refer to the section entitled **Directions for use,** pages 2 & 3.

Ground equipment: For best results, the tank mix should be applied with ground equipment. For thorough coverage of weeds apply with flat fan or hollow cone nozzles spaced 20 inches apart in a minimum of 20 gallons of water per acre with a spray pressure of 40 psi.

Mixing

Fill half the spray tank with water and add the suggested amount of—BLAZER, Lasso 4E, spray adjuvant—while the agitator is running. Then finish filling the tank.

Restrictions and limitations (partial list)

Read and follow all applicable directions and use restrictions on the BLAZER and Lasso labels.

Do not apply the tank mix after the peanuts are past the cracking stage (initiation of soil cracking, but prior to peanut emergence from the soil) as severe injury will result.

RICE

Directions for use*

BLAZER is a selective broad spectrum herbicide recommended for post-emergence application in rice to control hemp sesbania prior to flowering. Optimum weed control is achieved when actively growing weeds are treated. Good coverage is important, as BLAZER works primarily by contact action. Failure to follow the suggested dosage will result in unsatisfactory control. When applied at the recommended growth stages and suggested dosage rates, rice is tolerant to postemergence applications of BLAZER. Do not apply BLAZER after the rice reaches the boot stage.

Time of application

BLAZER should be applied to actively growing hemp sesbania plants, but before sesbania is in the flowering stage. Best results are obtained when the sesbania growth extends above the rice. BLAZER may be applied when rice is at the late tillering stage up to the early boot stage, which normally occurs in June/July.

Rate

Apply 1/2 pint of BLAZER to actively growing hemp sesbania plants. A second application of 1/2 pint BLAZER can be made to control later germinating sesbania.

Spray additive

Two (2) pints of a nonionic spray adjuvant should be added per 100 gallons of the spray mixture. The use of a spray adjuvant is important for effective control of hemp sesbania.

Water volume and spray pressure

For additional information refer to the section entitled **Directions for use,** pages 2 & 3.

Ground equipment: For best results use a minimum of 20 gallons of water per acre and 40 psi. Use standard flat fan or hollow cone nozzles spaced 20 inches apart.

Air equipment: Apply BLAZER by aircraft using nozzling to deliver from 5 to 10 gallons of spray per acre.

Mixing

When mixing BLAZER and spray adjuvant, follow the standard mixing procedure, outlined on page 5.

Coverage

Thorough coverage of actively growing weeds is essential. Large crop-and-weed leaf canopies shelter smaller weeds and can prevent adequate spray coverage.

Drift hazard

When spraying labeled crops, care must be exercised to prevent spray drift which could result in damage to other crops. Spraying when other crops are closer than 100 yards downwind or 50 yards upwind from the point of application is not recommended. The use of any cleared drift control agent may reduce this hazard; however, the drift control agent may also decrease the weed control activity.

Restrictions and limitations (partial list)

Do not apply BLAZER to rice after it reaches the boot stage, or within 50 days of harvest.

Do not apply more than two applications to rice per season nor exceed 1 pint per acre per season.

Do not use treated plants for feed or forage.

Crop rotation restriction: Root crops (such as carrots, turnips, sweet potatoes, etc.) must **not** be planted in fields treated with BLAZER for a period of 18 months following treatment.

Do not harvest crayfish from treated rice areas for food.

Do not use water from treated rice fields for irrigation purposes for other than BLAZER labeled crops.

Avoid drift to all other crops and non-target areas.

Do not add crop oil concentrates.

*Not for use in California.

BLAZER + STAM M-4 TANK MIX IN RICE GENERAL AND APPLICATION INFORMATION, RESTRICTIONS AND LIMITATIONS

General information

When rice has at least 3 leaves, a tank mix of BLAZER® and Stam® M-4 herbicides can be used for the control of hemp sesbania and all weeds on the Stam M-4 label plus suppression of Northern Jointvetch and 4 to 6 leaf annual morningglories.

When using the tank mix, an increase in foliage burn may be noticed.

Time of application

The BLAZER + Stam M-4 tank mix combination should be applied after draining the rice fields when rice has at least 3 leaves.

Rate

Apply 1 pint of BLAZER plus 3 to 4 quarts of Stam M-4 per acre.

Spray additive

Two pints of nonionic spray adjuvant should be added per 100 gallons of the spray mixture.

Water volume and spray pressure

For additional information refer to the section entitled **Directions for use,** pages 2 & 3.

Ground equipment: For the tank mix of BLAZER + Stam M-4, use a minimum of 20 gallons of total spray solution per acre (broadcast basis) and a minimum of 40 psi pressure. Use standard high pressure hollow cone or flat fan nozzles spaced 20 inches apart. Do not use flood or whirl chamber nozzles.

Air equipment: Use 5 to 10 gallons of total spray solution per acre.

Mixing

Load half the spray tank with water and add the suggested amount of—BLAZER, Stam M-4, nonionic spray adjuvant—while the agitator is running; then add the remaining quantity of water.

Coverage

Thorough coverage of actively growing weeds is essential. A large weed canopy may shelter smaller weeds and can prevent adequate coverage.

Drift hazard

When spraying labeled crops, care must be exercised to prevent spray drift which could result in damage to other crops. Spraying when other crops are closer than 100 yards downwind or 50 yards upwind from the point of application is not recommended. The use of any cleared drift control agent may reduce this hazard; however, the drift control agent may also decrease the weed control activity.

Restrictions and limitations (partial list)

Always read and follow the restrictions and limitations for all products whether used alone or in a tank mix. The most restrictive labeling applies in tank mixtures.

To avoid excessive residue at harvest, do not apply Stam M-4 in a tank mix with BLAZER after the end of tillering.

BLAZER + COLLEGO TANK MIX IN RICE GENERAL AND APPLICATION INFORMATION, RESTRICTIONS AND LIMITATIONS

General information

A postemergence tank mix of BLAZER® and Collego® herbicides should be applied at the recommended rates and growth stages as described on the respective labels for the control of northern jointvetch and hemp sesbania in rice.

Time of application

Application should be made when hemp sesbania is 12 to 60 inches in height and northern jointvetch averages 8 to 24 inches tall. Applications should be made prior to the bloom stage but after plants have emerged through the rice canopy.

Rate

Apply Collego at labeled rates + BLAZER at 1/2 to 1 pint per acre. (For details see the Collego Label.)

Spray additive

The use of nonionic spray adjuvant at the rate of 2 pints per 100 gallons of spray mixture is recommended.

Water volume and spray pressure

For additional information refer to the section entitled **Directions for use,** pages 2 & 3.

Ground equipment: For the tank mix of BLAZER + Collego use a minimum of 20 gallons of total spray solution per acre (broadcast basis) and a minimum of 40 psi pressure. Use standard high pressure hollow cone or flat fan nozzles spaced 20 inches apart. Do not use flood or whirl chamber nozzles.

Air equipment: Use at least 10 gallons of total spray solution per acre.

Mixing

Fill the spray tank with half the amount of required water and add the recommended amount of Collego. Once thoroughly mixed in the spray tank, add the recommended amounts of BLAZER and spray adjuvant while the agitator is running. Add the remaining quantity of water.

Coverage

Thorough coverage of actively growing weeds is essential. A large weed canopy may shelter smaller weeds and can prevent adequate coverage.

Restrictions and limitations (partial list)

Always read and follow the restrictions and limitations for all products whether used alone or in a tank mix. The most restrictive labeling applies in tank mixes.

Table 19
Appendix

The following are scientific names for the weeds listed on this label. For specific recommendations on control of these weeds, refer to the major crop and/or tank mix sections.

Broadleaf Weeds

Common Name	Scientific Name
Amaranth, Palmer,	*Amaranthus palmeri*
Spiny	*Amaranthus spinosus*
Anoda, Spurred	*Anoda cristata*
Beggarweed, Florida	*Desmodium tortuosum*
Balloonvine	*Cardiospermum halicacabum*
Beggarticks	*Bidens frondosa*
Bindweed, Field,	*Convolvulus arvensis*
Hedge	*Convolvulus sepium*
Buckwheat, Wild	*Polygonum convolvulus*
Buffalobur	*Solanum rostratum*
Burgherkin	*Cucumis anguria*
Carpetweed	*Mollugo verticillata*
Citron (Wild Watermelon)	*Citrullus vulgaris*
Cocklebur, Common,	*Xanthium pensylvanicum*
Heartleaf	*Xanthium strumarium*
Copperleaf, Hophornbean,	*Acalypha ostryaefolia*
Virginia	*Acalypha virginica*
Cowpea, Volunteer	*Vigna sinensis*
Crotalaria, Showy	*Crotalaria spectabilis*
Croton, Tropic,	*Croton glandulosus*
Woolly	*Croton capitatus*
Cucumber, Wild Spiny	*Cucumis dipsaceus*
Dayflower	*Commelina* spp.
Devilsclaw	*Proboscidea lousianica*
Galinsoga, Hairy,	*Galinsoga ciliata*
Smallflower	*Galinsoga parviflora*
Gourd, Texas	*Cucurbita texana*
Groundcherry, Cutleaf,	*Physalis angulata*
Lanceleaf	*Physalis lanceifolia*
Indigo, Hairy	*Indigofera hirsuta*
Jimsonweed	*Datura stramonium*
Jointvetch, Northern	*Aeschynomene virginica*
Ladysthumb	*Polygonum persicaria*
Lambsquarters	*Chenopodium album*
Mallow, Venice	*Hibiscus trionum*
Mexicanweed	*Caperonia palustris*
Milkweed, Climbing,	*Sarcostemma cyanchoides*
Common	*Asclepias syriaca*
Morningglory, Cypressvine,	*Ipomoea quamoclit*
Entireleaf,	*Ipomoea hederacea var. integruscula*
Ivyleaf,	*Ipomoea hederacea var. hederacea*
Purple Moonflower,	*Ipomoea muricata*
Scarlet,	*Ipomoea coccinea*
Smallflower,	*Jacquemontia tamnifolia*
Small White (Pitted),	*Ipomoea lacunosa*
Tall (Common),	*Ipomoea purpurea*
Willowleaf (Palmleaf)	*Ipomoea wrightii*
Mustard, Wild	*Brassica kaber*
Nightshade, Eastern Black,	*Solanum ptycanthum*
Black	*Solanum nigrum*
Pigweed, Prostrate,	*Amaranthus blitoides*
Redroot,	*Amaranthus retroflexus*
Smooth	*Amaranthus hybridus*
Poinsettia, Wild	*Euphorbia heterophylla*
Poorjoe	*Diodia teres*
Purslane, Common	*Portulaca oleracea*
Pusley, Florida	*Richardia scabra*
Ragweed, Common,	*Ambrosia artemisiifolia*
Giant	*Ambrosia trifida*
Redvine	*Brunnichia cirrhosa*
Redweed	*Melochia corchorifolia*
Senna, Coffee	*Cassia occidentalis*
Sesbania, Hemp	*Sesbania exaltata*
Shepherdspurse	*Capsella bursa-pastoris*
Sicklepod	*Cassia obtusifolia*
Sida, Prickly (Teaweed)	*Sida spinosa*
Smartweed, Pennsylvania	*Polygonum pensylvanicum*
Smellmelon	*Cucumis melo*
Spurge, Prostrate	*Euphorbia supina*
Spotted	*Euphorbia maculata*
Starbur, Bristly	*Acanthospermum hispidum*
Sunflower, Wild	*Helianthus annuus*
Teaweed (See Sida, Prickly)	*Sida spinosa*
Thistle, Canada	*Cirsium arvense*
Trumpetcreeper	*Campsis radicans*
Velvetleaf	*Abutilon theophrasti*
Venice Mallow	*Hibiscus trionum*
Waterhemp, Tall	*Amaranthus ruberculatos*

Grasses

Common Name	Scientific Name
Barnyardgrass	*Echinochloa crus-galli*
Bermudagrass	*Cynodon dactylon*
Crabgrass, Large,	*Digitaria sanguinalis*
Smooth	*Digitaria ischaemum*
Cupgrass, Woolly	*Eriochloa villosa*
Foxtail, Giant,	*Setaria faberi*
Green,	*Setaria viridis*
Yellow	*Setaria lutescens*
Goosegrass	*Eleusine indica*
Itchgrass	*Rottboellia exaltata*
Johnsongrass, Seedling,	*Sorghum halepense*
Rhizome	*Sorghum halepense*
Junglerice	*Echinochloa colonum*
Millet, Wild Proso	*Panicum miliaceum*
Muhly, Wirestem	*Muhlenbergia frondosa*
Panicum, Fall,	*Panicum dichotomiflorum*
Texas	*Panicum texanum*
Quackgrass	*Agropyron repens*
Rice, Red	*Oryza rufipogon*
Sanbur, Field	*Cenchrus pauciflorus*
Shattercane	*Sorghum bicolor*
Signalgrass, Broadleaf	*Brachiaria platyphylla*
Sprangletop, Red	*Leptochloa filiformis*
Volunteer, Barley	*Hordeum vulgare*
Corn,	*Zea mays*
Oats,	*Avena sativa*
Rye,	*Secale cereale*
Wheat	*Triticum aestivum*
Wirestem Muhly	*Muhlenbergia frondosa*
Witchgrass	*Panicum capillare*

Sedges

Common Name	Scientific Name
Yellow Nutsedge	*Cyperus esculentus*

CONDITIONS OF SALE AND WARRANTY

The Directions for use of this product reflect the opinion of experts based on field use and tests. The directions are believed to be reliable and should be followed carefully. However, it is impossible to eliminate all risks inherently associated with use of this product. Crop injury, ineffectiveness or other unintended consequences may result because of such factors as weather conditions, presence of other materials, or use of the product in a manner inconsistent with its labeling, all of which are beyond the control of BASF CORPORATION ("BASF") or the Seller. All such risks shall be assumed by the Buyer.

BASF warrants that this product conforms to the chemical description on the label and is reasonably fit for the purposes referred to in the Directions for use, subject to the inherent risks referred to above. BASF MAKES NO OTHER EXPRESS OR IMPLIED WARRANTY OF FITNESS OR MERCHANTABILITY OR ANY OTHER EXPRESS OR IMPLIED WARRANTY. IN NO CASE SHALL BASF OR THE SELLER BE LIABLE FOR CONSEQUENTIAL, SPECIAL OR INDIRECT DAMAGES RESULTING FROM THE USE OR HANDLING OF THIS PRODUCT. BASF and the Seller offer this product, and the Buyer and User accept it, subject to the foregoing Conditions of Sale and Warranty which may be varied only by agreement in writing signed by a duly authorized representative of BASF.

Amiben and Butyrac are registered trademarks of Rhone-Poulenc Ag Products Co.
Balan and Treflan are registered trademarks of Elanco Products Company.
Basagran and Poast are registered trademarks of BASF AG.
BLAZER is a registered trademark of BASF Corporation.

Butoxone is a registered trademark of Vertac Chemical Corporation.
Collego is a registered trademark of NOR-AM Chemical Company.
Dual is a registered trademark of Ciba-Geigy.
Fusilade and Vernam are registered trademarks of ICI.
Lasso is a registered trademark of Monsanto.
Scepter is a registered trademark of American Cyanamid Company.
Stam is a registered trademark of Rohm and Haas Company.
Rescue is a registered trademark of Uniroyal Chemical Company, Inc.
Classic is a registered trademark of E.I. Dupont De Nemours & Co. (Inc.)

BL 2 005 A00908189 JJ6508

DASH®

Spray tank adjuvant for use with Poast® herbicide

Functioning agents 99%
Petroleum hydrocarbons
Alkyl esters & acids
Anionic surfactants
Ineffective constituents 1%

Note: All ingredients exempt from requirements of residue tolerances under Title 40 of the Code of Federal Regulations 180.1001

KEEP OUT OF REACH OF CHILDREN.

WARNING

USE WITH ADEQUATE VENTILATION.

Causes eye and skin irritation. Do not get into eyes or on skin. Harmful if swallowed or inhaled. Avoid breathing spray mists or vapors.

In case of eye contact immediately flush with plenty of water for at least 15 minutes. If irritation persists contact a physician. In case of skin contact, wash affected areas with soap and water. Remove contaminated clothing and launder before re-use.

Store above 40°F.

Shake well before using.

Net contents 2½ gallons

GENERAL INFORMATION

Soybeans, sugar beets, alfalfa, cotton, peanuts, flax and sunflowers

Dash® spray tank adjuvant was developed by BASF Corporation to be applied in water-based sprays to enhance the effectiveness of **Poast® herbicide.** This product is to be used as a replacement for oil concentrate for **Poast** whether alone or in a tank mix with **Basagran® herbicide** in soybeans. **Dash** enhances the performance of **Poast** by improving grass control, especially tough to control grasses such as volunteer corn, rhizome johnsongrass, wild oats and quackgrass.

Tank mix of Poast® and Basagran® herbicides in soybeans

Dash plus UAN or ammonium sulfate enhances the performance of **Poast** in a tank mix with **Basagran.** No longer do you need to increase the rate of **Poast** in the tank mix. Just add **Dash** and UAN solution or ammonium sulfate to the spray tank. UAN solution (Urea Ammonium Nitrate) is also commonly referred to as 28%, 30% or 32% nitrogen solution.

Rate Chart for Dash® Spray Tank Adjuvant

Herbicide Treatment*	Dash Rate/A	Oil Concentrate	UAN Solution (Rate/A)**	Benefits
Poast	2 pints	None	1 gallon on volunteer corn	Improved performance of **Poast**
Poast + Basagran Tank Mix	2 pints	None	1 gallon	Increase in rate of **Poast** not required

* See labels for weed sizes and rates of **Poast** and **Basagran.**
** Ammonium Sulfate at a rate of 2½ lbs. per acre may be substituted for UAN solution.

DIRECTIONS FOR USE

Use **Dash** at the rate of 2 pints per acre. For proper filling of the spray tank and mixing of the spray solution, refer to the mixing/spraying section of the label for **Poast.** Use **Dash** instead of crop oil concentrate. Refer to the label for **Poast** for additional recommendations restrictions.

For additional and more consistent control, UAN solution, at a rate of 1 gallon per acre, may be added to the spray mixture containing **Dash.** High quality ammonium sulfate may be used instead of UAN solution. Ammonium sulfate may be added, at a rate of 2½ pounds per acre. To determine the solubility of available ammonium sulfate, refer to the **Addition of ammonium sulfate** section of the label for **Poast.**

In case of emergency

In case of large-scale spillage regarding this product call:
Chemtrec 800-424-9300
BASF Corporation 201-316-3000

In case of medical emergency regarding this product, call:

1. Your local doctor for immediate treatment.
2. Your local poison control center (hospital).
3. BASF Corporation 201-316-3000.

Read the entire label. Use only according to label instructions. Read "Conditions of sale and warranty" before buying or using. If terms are not acceptable, return product at once, unopened.

CONDITIONS OF SALE AND WARRANTY

The Directions for use of this product reflect the opinion of experts based on field use and tests. The directions are believed to be reliable and should be followed carefully. However, it is impossible to eliminate all risks inherently associated with use of this product. Crop injury, ineffectiveness or other unintended consequences may result because of such factors as weather conditions, presence of other materials, or use of the product in a manner inconsistent with its labeling, all of which are beyond the control of BASF CORPORATION ("BASF") or the Seller. All such risks shall be assumed by the Buyer.

BASF warrants that this product conforms to the chemical description on the label and is reasonably fit for the purposes referred to in the Directions for use, subject to the inherent risks referred to above. BASF MAKES NO OTHER EXPRESS OR IMPLIED WARRANTY OF FITNESS OR MERCHANTABILITY OR ANY OTHER EXPRESS OR IMPLIED WARRANTY. IN NO CASE SHALL BASF OR THE SELLER BE LIABLE FOR CONSEQUENTIAL, SPECIAL OR INDIRECT DAMAGES RESULTING FROM THE USE OR HANDLING OF THIS PRODUCT. BASF and the Seller offer this product, and the Buyer and User accept it, subject to the foregoing Conditions of sale and warranty which may be varied only by agreement in writing signed by a duly authorized representative of BASF.

Dash is a trademark of BASF Corporation.
Poast and **Basagran** are registered trademarks of BASF AG.

DA 1 002 JJ6446 A0991288

GALAXY™

POSTEMERGENCE HERBICIDE

For broad spectrum control in soybeans
A soluble liquid formulation containing:

ACTIVE INGREDIENTS*
Sodium salt of bentazon 33.4%
Sodium salt of acifluorfen 6.8%
INERT INGREDIENTS: 59.8%
TOTAL 100%

* Equivalent to 3.00 pounds per gallon bentazon 3-(isopropyl)-1*H*-2,1,3-benzothiadiazin-4 (3*H*)-one 2, 2-dioxide and 0.67 pounds per gallon of sodium acifluorfen: sodium (5-[2-chloro-4-(trifluoromethyl) phenoxy]-2-nitrobenzoate).

EPA Reg. No. 7969-77

KEEP OUT OF REACH OF CHILDREN.

DANGER/PELIGRO

PRECAUCIÓN AL USUARIO: Si usted no lee ingles, no use este producto hasta que la etiqueta haya sido explicada ampliamente.

HAZARDS TO HUMANS

DANGER/PELIGRO

Causes eye damage. Harmful if swallowed, inhaled or absorbed through skin. Do not get in eyes. Wear goggles or face shield when handling. Avoid breathing vapor or spray mist and contact with skin or clothing. In case of contact, immediately remove contaminated clothing and shoes. Wash contaminated clothing with soap and hot water before re-use. May cause allergic skin response.

STATEMENT OF PRACTICAL TREATMENT

If in eyes: Flush with large amounts of water for at least 15 minutes. Get medical attention.

If on skin: Wash with plenty of soap and water. Consult a physician if irritation persists.

If swallowed: Dilute by giving 2 glasses of water to drink and call a physician. Never give anything by mouth to an unconscious person.

Note to physician: Emesis is recommended.

ENVIRONMENTAL HAZARDS

Do not apply directly to lakes, ponds or streams. Do not contaminate water when disposing of equipment washwaters. Do not apply when weather conditions favor drift from target area.

RE-ENTRY AND WORKERS' PROTECTION STATEMENTS

Do not apply this product in such a manner as to directly or through drift, expose workers or other persons. The area being treated must be vacated by unprotected persons. Do not enter treated areas without protective clothing until sprays have dried. Because certain states may require more restrictive re-entry intervals for various crops treated with this product, consult your State Department of Agriculture for further information.

DIRECTIONS FOR USE

It is a violation of Federal law to use this product in a manner inconsistent with its labeling.

STORAGE AND DISPOSAL

Keep from freezing. Store above 40°F.

Do not contaminate water, food or feed by storage or disposal.

Pesticide wastes are acutely hazardous. Improper disposal of excess pesticide spray mixture, or rinsate is a violation of Federal law. If these wastes cannot be disposed of by use according to label instructions, contact your State Pesticide or Environmental Control Agency, or the Hazardous Waste representative at the nearest EPA Regional Office for guidance.

Triple rinse container (or equivalent). Puncture and dispose of in a sanitary landfill, or by incineration, or if allowed by state and local authorities, by burning. If burned, stay out of smoke.

Do not re-use empty container.

In case of emergency

In case of large scale spillage regarding this product, call:

CHEMTREC	800-424-9300
BASF Corporation	201-316-3000

In case of medical emergency regarding this product, call:
1. Your local doctor for immediate treatment.
2. Your local poison control center (hospital).
3. BASF Corporation 201-316-3000.

Steps to be taken in case material is released or spilled.

Dike and contain spill with inert material (sand, earth, etc.) and transfer liquid and solid diking material to separate containers for disposal. Remove contaminated clothing and wash affected skin areas with water. Wash clothing before re-use. Keep spill out of all sewers and open bodies of water.

GENERAL INFORMATION

Galaxy™ herbicide is intended for selective postemergence control of certain broadleaf weeds. Galaxy is effective mainly through contact action; therefore, weeds must be thoroughly covered with spray. Galaxy may cause some soybean leaf-speckling and leaf-bronzing to occur under certain conditions. (See **Restrictions and limitations.)**

Timing of applications

Make postemergence applications of Galaxy early, when weeds are small and actively growing and before weeds reach the maximum size listed in the **Application Rate Table.** Delay in application which permits weeds to exceed the maximum size stated will result in inadequate control.

Cultivation before or during the application is not recommended. Cultivation may put weeds under stress, thus making control more difficult to obtain. Timely cultivation after application will usually assist in weed control.

Water volume and spray pressure

Apply recommended rates of Galaxy as follows:

Ground equipment: Use a minimum of 20 gallons of water per broadcast acre and a minimum of 40 psi pressure (measured at the boom, not the pump or in the line). When crop and weed foliage is dense, use up to 50 gallons of water and up to 80 psi pressure. Use standard high pressure pesticide hollow cone or flat fan nozzles spaced 20 inches apart. Do not use flood, whirl chamber, or controlled droplet applicator (CDA) nozzles.

Air equipment: Use a minimum of 10 gallons of water per acre and a maximum of 40 psi pressure. Use only diaphragm-type nozzles producing cone or fan spray patterns.

Aerial application–special directions

To obtain uniform coverage and to avoid drift hazards, the following application equipment and practices should be used:

Nozzle height: 6 to 10 feet above crop.

Nozzle orientation: Nozzles must be oriented so as to discharge straight back with the air stream (opposite the direction of travel of the aircraft) and not more than 20 degrees downward.

Nozzles must not be located farther out than three-fourths the distance from the center of the aircraft to the end of the wing or rotor.

Water volume and spray pressure: See Air equipment.

Do not apply Galaxy by aircraft when wind is blowing at a velocity above 10 mph. Coarse sprays (larger droplets) are less likely to drift.

Do not apply Galaxy by air if ornamentals or sensitive nontarget crops, such as cotton, sugar beets, sunflowers or okra are within 200 feet downwind.

Applicator must follow the most restrictive use cautions to avoid drift hazards, including those found in this labeling as well as applicable state and local regulations and ordinances.

Spray additives

A nonphytotoxic oil concentrate (commonly referred to as oil concentrate) should be added to the spray tank with Galaxy. The oil concentrate must contain either a petroleum or vegetable oil base and must meet the following criteria: 1) be nonphytotoxic, 2) contain only EPA-exempt ingredients, 3) provide good mixing quality in the jar test (see below), and 4) prove beneficial in local experience.

The exact composition of suitable products will vary; however, vegetable and petroleum oil concentrates should contain emulsifiers which provide good mixing quality. For vegetable oil concentrates, it has been observed that highly refined vegetable oils are more satisfactory than unrefined vegetable oils. For additional information see **Jar test for estimating suitability of oil concentrates** at the end of this section.

With the addition of oil concentrate to Galaxy on soybeans, some leaf burn may occur, but all new growth is normal and crop vigor is not reduced. The potential for leaf burn is increased when relative humidity and temperature are high. A few oil concentrates have exhibited excessive leaf burn. Refer to your supplier for information concerning successful local experience prior to purchasing any oil concentrate.

Rate of oil concentrate: Ground application–2 pints/acre (maximum). Air application–1 pint/acre (maximum).

Urea ammonium nitrate (UAN), commonly referred to as 28%, 30% or 32% nitrogen solution, may be added in place of other spray additives for improved pigweed and velvetleaf control. The standard use rates are: ½ to 1 gallon of UAN per acre is recommended for ground application; air application: ½ gallon/acre

Mixing/spraying

Fill tank of a thoroughly clean sprayer half to two-thirds full with clean water. Start agitation and add Galaxy; allow to mix thoroughly. Add spray additive and remaining volume of water. Maintain constant agitation during application.

Jar test for estimating suitability of oil concentrates

1. **Water supply:** Use on water from intended source and at the source temperature.
2. **Amount of water in jar:** Ground application—For 20 gal./A spray volume use 3⅓ cups (800 ml) of water. Air application—For 10 gal./A spray volume use 1⅔ cups (400 ml) of water. For other spray volumes, adjust proportionately to above.
3. **Amount of Galaxy and oil concentrate to add:** Add Galaxy and oil concentrate at the rate of 1 teaspoon (5 ml) for each pint of recommended label rate.
4. **Add components in following sequence,** gently mixing between component additions:
a. Galaxy
b. Oil concentrate
5. **Cap jar, invert 10 cycles, let stand for 15 minutes, evaluate.**
6. **Evaluation:** An ideal tank mix combination will be uniform; thus, the suitability of the oil concentrate is questionable if any of the following are observed:
Free oil (film or globules) at the surface.
Flocculation—fine particles which may be suspended in the liquid or found as a precipitated layer at the bottom of the jar.
Clabbering—thickening texture (coagulated) resembling yogurt or curd-like texture as with cottage cheese.

Application Rate Table

Applications of Galaxy™ herbicide should be made when weeds are small and actively growing and before weeds reach the maximum size listed below. Such applications generally correspond to the soybean growth stages of unifoliate to two expanded trifoliate leaves. Soybeans may experience slight yellowing, bronzing, speckling, or burning of leaves under certain conditions. Soybean plants generally outgrow this condition within 10 days

	Application Rates for Weed Growth Stages			
Weeds Controlled	Leaf Stage	Max. Height	Rate of Galaxy per Acre	Spray Additives Rate/A
Anoda, Spurred	Up to 6	3"		
Beggarticks	Up to 6	6"		
Buckwheat, Wild	Up to 4	3"		
Cocklebur[b]	2-6	6"		
Dayflower	Up to 6	4"		
Devilsclaw[a]	Up to 6	3"		
Galinsoga[a]	Cotyledon to 6	2"	2 pints	2
Jimsonweed	Up to 6	6"		pints
Ladysthumb	Up to 6	6"		oil
Lambsquarters Common[c]	4-6	2"		concentrate or
Mallow, Venice	Up to 6	4"		½ to 1 gal.
Morningglories[c]	Up to 4	2"		UAN
Mustard, Wild	Up to 6	4"		
Nightshade, Black	2	<2"		
Nutsedge, Yellow[a]	—	6-8"		
Pigweed, Redroot	Up to 4	2"		

Weeds Controlled	Application Rates for Weed Growth Stages: Leaf Stage	Max. Height	Rate of Galaxy per Acre	Spray Additives Rate/A
Pigweed, Smooth	Up to 4	2"		
Poinsettia, Wild	2-4	4"		
Purslane, Common	Up to 4	1"		
Ragweed, Common	4-6	3"		
Ragweed, Giant	Up to 4	6"		
Redweed	4-6	6"		
Shepherdspurse[d]	Up to 6	4"		
Sida, Prickly or Teaweed	Up to 6	3"		
Smartweed, Pennsylvania	Up to 6	6"		
Starbur, Bristly	Up to 4	2"		
Sunflower, Wild	Up to 4	5"		
Thistle, Canada[a]	—	8" to bud stage		
Velvetleaf	Up to 4	2"		
Waterhemp, Tall	4	2"		

[a] Control may be inconsistent with this rate of Galaxy. A later application of Basagran® herbicide may be necessary (see label for Basagran).

[b] Do not treat earlier than leaf stage shown and do not count cotyledon leaves.

[c] Control may be partial or inconsistent with this rate of Galaxy. A later application of Blazer® herbicide may be necessary (see label for Blazer).

[d] Do not treat rosette before seed stalk appears.

Restrictions and limitations

Do not apply more than a total of 3 pints of Galaxy per acre per season. Do not apply more than 1¾ pints of Basagran or 2 pints Blazer following an application of 3 pints of Galaxy per acre per season.

Do not apply Galaxy to soybeans that have been subject to stress conditions such as hail damage, flooding, drought, injury from other herbicides, or widely fluctuating temperatures, as crop injury may result.

Do not apply Galaxy to soybeans that show injury (leaf phytotoxicity and/or plant stunting) produced by any other prior herbicide applications, because this injury may be enhanced and/or prolonged.

Do not apply Galaxy during prolonged periods of drought or during unseasonably cold weather, as unsatisfactory weed control may result.

Rainfall or overhead irrigation soon after application (within 8 hours) may nullify the effectiveness of Galaxy.

Do not mix or apply Galaxy with any other pesticide or any fertilizer except as specifically recommended on this labeling.

Do not apply Galaxy within 50 days of soybean harvest.

Root crops (such as carrots, turnips, sweet potatoes, etc.) must not be planted in fields treated with Galaxy for a period of 18 months following treatment.

In case of crop failure, only peanuts or soybeans may be immediately replanted.

Do not use treated plants for feed or forage.

Do not apply Galaxy through any type of irrigation equipment.

Appendix

The following are scientific names for the weeds listed in this section. For specific recommendations on control of these weeds, refer to the **Application Rate Table.**

Broadleaf Weeds

Common Name	Scientific Name
Anoda, Spurred	*Anoda cristata*
Beggarticks	*Bidens frondosa*
Buckwheat, Wild	*Polygonum convolvulus*
Buttonweed (see Velvetleaf)	—
Butterprint (see Velvetleaf)	—
Cocklebur	*Xanthium strumarium*
Dayflower	*Commelina spp.*
Devilsclaw	*Probiscidea louisianica*
Galinsoga	*Galinsoga spp.*
Jimsonweed	*Datura stramonium*
Ladysthumb	*Polygonum persicaria*
Lambsquarters, Common	*Chenopodium album*
Mallow, Venice	*Hibiscus trionum*
Morningglory, Common (tall)	*Ipomoea purpurea*
Cypressvine	*Ipomoea quamoclit*
Entireleaf	*Ipomoea hederacea*
Ivyleaf	*Ipomoea hederacea*
Palmleaf	*Ipomoea wrightii*
Pitted	*Ipomoea lacunosa*
Purple Moonflower	*Ipomoea muricata*
Smallflower	*Jacquemontia tamnifolia*
Mustard, Wild	*Sinapsis arvensis*
Nightshade, Black	*Solanum nigrum*
Pigweed, Redroot	*Amaranthus retroflexus*
Smooth	*Amaranthus hybridis*
Poinsettia, Wild	*Euphorbia heterophylla*
Purslane, Common	*Portulaca oleracea*
Ragweed, Common	*Ambrosia artemisiifolia*
Giant	*Ambrosia trifida*
Redweed	*Melochia corchorifolia*
Shepherdspurse	*Capsella bursa-pastoris*
Sida Prickly or Teaweed	*Sida spinosa,*
Smartweed, Pennsylvania	*Polygonum pensylvanicum*
Starbur, Bristly	*Acanthospermum hispidum*
Sunflower, Wild	*Helianthus annuus*
Thistle, Canada	*Cirsium arvense*
Velvetleaf	*Hibiscus trionum*
Waterhemp, Tall	*Amaranthus tuberculatus*

Sedges

Common Name	Scientific Name
Nutsedge, Yellow	*Cyperus esculentus*

CONDITIONS OF SALE AND WARRANTY

The Directions for use of this product reflect the opinion of experts based on field use and tests. The directions are believed to be reliable and should be followed carefully. However, it is impossible to eliminate all risks inherently associated with use of this product. Crop injury, ineffectiveness or other unintended consequences may result because of such factors as weather conditions, presence of other materials, or use of the product in a manner inconsistent with its labeling, all of which are beyond the control of BASF CORPORATION ("BASF") or the Seller. All such risks shall be assumed by the Buyer.

BASF warrants that this product conforms to the chemical description on the label and is reasonably fit for the purposes referred to in the Directions for use, subject to the inherent risks referred to above. BASF MAKES NO OTHER EXPRESS OR IMPLIED WARRANTY OF FITNESS OR MERCHANTABILITY OR ANY OTHER EXPRESS OR IMPLIED WARRANTY. IN NO CASE SHALL BASF OR THE SELLER BE LIABLE FOR CONSEQUENTIAL, SPECIAL OR INDIRECT DAMAGES RESULTING FROM THE USE OR HANDLING OF THIS PRODUCT. BASF and the Seller offer this product, and the Buyer and User accept it, subject to the foregoing Conditions of Sale and Warranty which may be varied only by agreement in writing signed by a duly authorized representative of BASF.

Basagran is a registered trademark of BASF AG.
Galaxy is a trademark and Blazer is a registered trademark of BASF Corporation.

GL 1 001 JJ6612 A0350489

LADDOK®

POSTEMERGENCE FLOWABLE HERBICIDE

For Selective Postemergence Broadleaf Weed Control in Field Corn, Seed Corn, Silage Corn, Sweet Corn and Popcorn; and in Grain Sorghum. Not intended for use in California.

ACTIVE INGREDIENTS:	
Sodium salt of bentazon* [3-(isopropyl) 1*H*-2, 1, 3-benzothiadiazin-4(3*H*)-one, 2, 2-dioxide]	18.52%
Atrazine*(2-chloro-4-ethylamino-6-isopropylamino-s-triazine)	16.96%
INERT INGREDIENTS:	64.52%
TOTAL	100.0%

*Equivalent to 1.66 pounds per gallon each of bentazon and atrazine.

EPA REGISTRATION NO. 7969-54

KEEP OUT OF REACH OF CHILDREN.

DANGER/PELIGRO

PRECAUCIÓN AL USUARIO: Si usted no lee ingles, no use este producto hasta que la etiqueta haya sido explicada ampliamente.

SHAKE WELL BEFORE USING.

PRECAUTIONARY STATEMENTS

HAZARDS TO HUMANS

DANGER

Corrosive. Causes eye damage and skin irritation. Do not get in eyes, on skin or on clothing. Wear goggles or face shield and rubber gloves when handling. Harmful if swallowed. May cause allergic skin response.

If in eyes, immediately flush eyes with plenty of water. Get medical attention. If on skin, immediately flush skin with plenty of water. Get medical attention if irritation persists.

ENVIRONMENTAL HAZARDS

Atrazine, which is present in this product, can travel (seep or leach) through soil and can enter ground water which may be used as drinking water. Atrazine has been found in ground water. Users are advised not to apply LADDOK to sand and loamy sand soils where the water table (ground water) is close to the surface and where these soils are very permeable, i.e., well-drained. Your local agricultural agencies can provide further information on the type of soil in your area and the location of ground water.

This pesticide is toxic to aquatic invertebrates. Do not apply directly to water or wetlands. Do not apply where runoff is likely to occur. Do not apply when weather conditions favor drift from treated areas.
Runoff and drift from treated areas may be hazardous to aquatic organisms in neighboring areas. Do not contaminate water by cleaning of equipment or disposal of wastes.

ENDANGERED SPECIES RESTRICTIONS

Before using this pesticide on corn and sorghum in the counties listed below, you must obtain the Pesticide Use Bulletin for Protection of Endangered Species for the county in which the product is to be used. The bulletin is available from your County Extension Agent, State Fish and Game Office, or your pesticide dealer. **Use of this product in a manner inconsistent with the Pesticide Use Bulletin for Protection of Endangered Species is a violation of Federal laws.**

Alabama: Colbert, Greene, Jackson, Lamar, Lauderdale, Limestone, Madison, Marshall, Morgan, Pickens and Sumter

Arizona: Graham, Maricopa, Mohave, Pima, Pinal and Santa Cruz

Arkansas: Benton, Clay, Clark, Cross, Lawrence, Lee, Poinsette, Polk, Randolph, Sharp and St. Francis

Florida: Broward, Dade, Glades and Palm Beach

Kentucky: Ballard, Butler, Edmundson, Green, Hart, Jackson, Laurel, Livingston, Marshall, McCracken, McCreary, Pulaski, Rockcastle, Taylor, Warren and Wayne

Mississippi: Claiborne, Copiah, Hinds, Itawamba, Lowndes, Monroe and Noxubee

Missouri: Barry, Benton, Camden Christian, Dallas, Greene, Hickory, Jasper, Lawrence, Miller, Newton, Osage, Polk, St. Clair, Stone and Webster

Nevada: Clark

New Mexico: Chaves, Debaca and Eddy

North Carolina: Edgecombe, Nash and Pitt

Ohio: Pickaway

Oklahoma: Delaware, McCurtain and Pushmataha

Oregon: Lake

Tennessee: Bedford, Blount, Clairborne, Decatur, Franklin, Hancock, Hardin, Hickman, Knox, Lawrence, Lincoln, Loudon, Marshall, Maury, Meigs, Monroe, Rhea, Roane, Scott, Sequatchie, Smith, Sullivan and Wayne

Texas: Bastrop, Burleson, Comal, Harris, Hays, Jeff Davis, Pecos and Reeves

Utah: Utah and Washington

Virginia: Lee, Russell, Scott, Smyth, Tazewell, Washington and Wise

RE-ENTRY AND WORKERS' PROTECTION STATEMENTS

Do not apply this product in such a manner as to directly or through drift, expose workers or other persons. The area being treated must be vacated by unprotected persons. Do not enter treated areas without protective clothing until sprays have dried. Because certain states may require more restrictive re-entry intervals for various crops treated with this product, consult your State Department of Agriculture for further information.

DIRECTIONS FOR USE

It is a violation of Federal law to use this product in a manner inconsistent with its labeling.

GENERAL INFORMATION

LADDOK® herbicide is selective and intended for the postemergence control of a broad spectrum of broadleaf weeds. LADDOK does not control grasses. LADDOK is effective mainly through contact action; therefore, weeds must be thoroughly covered with spray. Large crop - and - weed leaf canopies shelter smaller weeds and prevent adequate spray coverage.

Corn is tolerant to LADDOK at all stages of growth. Very slight leaf speckling of corn may occur but plants generally outgrow this condition within 10 days. Corn types included are field, sweet and popcorn; and corn grown for seed or silage.

Always add UAN solution or oil concentrate according to the section entitled **Additive information.**

Timing of Applications

Apply LADDOK early postemergence when weeds are small and actively growing and before weeds reach the maximum size listed in the **Application Rate Table for Corn and Grain Sorghum.** Such applications generally correspond to the crop growth stages of one to seven leaves.

Early application to weeds produces the most beneficial effect on weed control, allows use of the lower rate (depending on weed species) and makes it easier to obtain thorough spray coverage. Delay in application which permits weeds to exceed the maximum size stated will result in inadequate control.

Do not cultivate within 5 days before or after application of LADDOK in the following states: AZ, CO, CT, ID, IL, IN, IA, KS, KY, ME, MA, MI, MN, MO, MT, NE, NV, NH, NJ, NY, ND, OH, OR, PA, RI, SD, UT, VT, WA, WV, WI, WY. A cultivation 5 or more days after application may be necessary if all weeds are not controlled or if a second flush of weeds occurs.

Water volume and spray pressure

Apply recommended rates of LADDOK as follows:

Ground equipment: Use a minimum of 20 gals. of water per broadcast acre and a minimum of 40 psi pressure (measured at the boom, not at the pump or in the line). When crop and weed foliage is dense use up to 50 gals. of water and up to 80 psi pressure. Use standard high pressure pesticide hollow cone or flat fan nozzles spaced 20 inches apart. Do not use flood or whirl chamber nozzles.

Air equipment Use a minimum of 5 gals. of water per acre and a maximum of 40 psi pressure. Use only diaphragm-type nozzles producing cone or fan spray patterns.

Aerial application-special directions

To obtain uniform coverage and to avoid drift hazards the following application equipment and practices should be used.

Nozzle height: Maximum of 10 feet above crop.

Nozzle orientation: Nozzles must be oriented so as to discharge straight back with the air stream (opposite the direction of travel of the aircraft) or at some angle between straight back and straight down.

Nozzles must be located no farther out than three-fourths the distance from the center of the aircraft to the end of the wing or rotor.

Water volume and spray pressure: See Air equipment

Do not apply LADDOK by aircraft when wind is blowing at a velocity above 10 mph. Coarse sprays (larger droplets) are less likely to drift.

Do not apply LADDOK by air within 200 feet upwind of ornamental or sensitive nontarget crops such as soybeans, peanuts, cotton, sugar beets, sunflowers or okra.

Applicator must follow the most restrictive use cautions to avoid drift hazards, including those found in this labeling as well as applicable state and local regulations and ordinances.

Special information for irrigated areas

In irrigated areas, it may be necessary to irrigate prior to treatment with LADDOK to ensure weeds are growing actively. Weeds growing under drought conditions usually are not satisfactorily controlled.

Additive information

UAN solution: In corn and grain sorghum UAN (Urea Ammonium Nitrate) solution may be added to LADDOK in the place of oil concentrate for most weeds. Consult the **Application Rate Table for Corn and Grain Sorghum.** UAN solution is commonly referred to as 28, 30 or 32% N. With the addition of UAN solution to LADDOK on corn or grain sorghum, a slight leaf tip burn or yellowing and speckling may occur. New growth is normal and vigor is not reduced. Do not use brass or aluminum nozzles when spraying LADDOK with UAN solution.

Rate of UAN solution

Ground application—1 gallon per acre
Air application—1/2 gallon per acre

Oil concentrate

If UAN solution is not used, a nonphytotoxic oil concentrate (commonly referred to as oil concentrate) should be added to the spray tank. Use oil concentrate if Canada thistle, yellow nutsedge, or field bindweed are to be controlled. The oil concentrate must contain either a petroleum or vegetable oil base and must meet the following criteria: 1) be nonphytotoxic, 2) contain only EPA-exempt ingredients, 3) provide good mixing quality in the jar test (see below), and 4) be successful in local experience.

The exact composition of suitable products will vary; however, vegetable and petroleum oil concentrates should contain emulsifiers which provide good mixing quality. For vegetable oil concentrates, it has been observed that highly refined vegetable oils are more satisfactory than unrefined vegetable oils. For additional information see **Jar test for estimating suitability of oil concentrates** at the end of this section.

Rate of oil concentrate
Ground application-2 pints/acre (maximum)
Air application-1 pint/acre (maximum)

Jar test for estimating suitability of oil concentrates

1. **Water supply:** Use only water from intended source and at the source temperature.
2. **Amount of water in jar:**
   Ground application–For 20 gal./A spray volume use 3 1/3 cups (800 ml) of water.
   Air application–For 5 gal./A spray volume use 5/6 cup (200 ml) of water, or
   For 10 gal./A spray volume use 1 2/3 cups (400 ml) of water.
   For other spray volumes, adjust propotionately to above.
3. **Amount of herbicide and oil concentrate or UAN solution to add:** Add herbicide and oil concentrate or UAN solution at the rate of 1 teaspoon (5 ml) for each pint of recommended label rate.
4. **Add components in the following sequence,** gently mixing between component additions:
   1.) UAN solution (if used)
   2.) LADDOK
   3.) Oil concentrate (if used)
5. **Cap jar,** invert 10 cycles, let stand for 15 minutes, evaluate.
6. **Evaluation:** An ideal tank mix combination will be uniform; thus, the suitability of the oil concentrate is questionable if any of the following are observed:
   Free oil at the surface-film or globules.
   Flocculation-fine particles which may be suspended in the liquid or found as a precipitated layer at the bottom of the jar.
   Clabbering-thickening texture (coagulated) resembling yogurt or a curd-like texture as with cottage cheese.

Mixing/spraying
Fill tank of a thoroughly clean sprayer half to two-thirds full with clean water. Start agitation. Add UAN solution then add LADDOK; allow to mix thoroughly. If UAN solution is not used add oil concentrate and remaining volume of water. Maintain constant agitation during application. Avoid allowing the mixture to stand overnight. Always clean sprayer thoroughly immediately after use by flushing the system with water and a strong detergent. Do not allow cleaning water to contaminate any streams or ponds.

APPLICATION RATE TABLE FOR CORN AND GRAIN SORGHUM

WEEDS CONTROLLED	APPLICATION RATES FOR LADDOK					
	2 PINTS*		2½ PINTS*		3½ PINTS	
	LEAF STAGE	MAX. HEIGHT	LEAF STAGE	MAX. HEIGHT	LEAF STAGE	MAX. HEIGHT
Cocklebur	2-4**	3"	2-10**	8"	—	—
Common Lambsquarters	—	—	Up to 8	5"	8-12	8"
Common Ragweed	—	—	Up to 5	4"	4-7	5"
Giant Ragweed	—	—	Up to 4	4"	4-6	6"
Jimsonweed	2-4	3"	Up to 6	4"	6-10	8"
Ladysthumb	2-6	4"	Up to 10	9"	10-14	12"
Pennsylvania Smartweed	2-6	4"	Up to 10	9"	10-14	12"
Redroot Pigweed	—	—	Up to 10	6"	—	—
Smooth Pigweed	—	—	Up to 10	6"	—	—
Tall Waterhemp	—	—	Up to 8	2"	6-9	4"
Velvetleaf	2-4	3"	Up to 6	5"	6-8	8"
Venice Mallow	—	—	Up to 8	4"	—	—
Wild Buckwheat	—	—	Up to 4	3"	4-6	5"
Wild Mustard	—	—	Up to 6	4"	6-10	8"
Wild Sunflower	—	—	Up to 5	6"	4-6	8"
Beggarticks	—	—	—	—	Up to 6	6"
Bristly Starbur	—	—	—	—	Up to 4	2"
Dayflower	—	—	—	—	Up to 6	4"
Prickly Sida or Teaweed	—	—	Up to 4	2"	Up to 6	3"
Spurred Anoda	—	—	—	—	Up to 6	3"
Annual Morningglory	—	—	Up to 4	4"	4-6	6"
Smallflower Morningglory	—	—	Up to 4	4"	4-6	6"
Yellow Nutsedge	See **Special directions** below.					
Canada Thistle	See **Special directions** below.					
Field Bindweed	See **Special directions** below.					

* Always add one gallon per acre of UAN solution or one quart per acre of oil concentrate to LADDOK. See **Additive information** section.
** Do not treat earlier than leaf stage shown and do not count cotyledon leaves.

Special directions for other weed problems

Corn

Canada Thistle
For suppression of Canada thistle, apply 3 1/2 pints of LADDOK per acre when plants are from 8-10 inches tall to the bud stage. Add oil concentrate for each application according to sections **Additive information** and **Mixing/spraying.**

Yellow Nutsedge
A single application of LADDOK at 3 1/2 pints per acre will provide suppression of yellow nutsedge. Add oil concentrate for each application according to sections **Additive information** and **Mixing/spraying.**

Field Bindweed
For suppression of field bindweed apply 3 1/2 pints of LADDOK per acre when vines are 8 to 10 inches long. Add oil concentrate for each application according to sections **Additive information** and **Mixing/spraying.**

Sorghum

Canada Thistle
Yellow Nutsedge
Field Bindweed
For suppression of these weeds, apply 3 1/2 pints of LADDOK per acre when Canada thistle plants are from 8-10 inches tall up to the bud stage, field bindweed vines are 8-10 inches long or yellow nutsedge is 1-4 inches tall. Add oil concentrate according to sections **Additive information** and **Mixing/spraying.** For best results, cultivate 7-14 days after application.

Appendix—Scientific names
The following are scientific names for the weeds listed in this label. For specific recommendations on control of these weeds, refer to the **Application Rate Table.**

Broadleaf Weeds

Common Name	Scientific Name
Beggarticks	*Bidens frondosa*
Bristly Starbur	*Acanthospermum hispidum*
Canada Thistle	*Circium arvense*
Cocklebur	*Xanthium strumarium*
Common Lambsquarters	*Chenopodium album*
Common Ragweed	*Ambrosia artemisiifolia*
Dayflower	*Commelina* spp.
Field Bindweed	*Convolvulus arvensis*
Giant Ragweed	*Ambrosia trifida*
Jimsonweed	*Datura stramonium*
Ladysthumb	*Polygonum persicaria*
Morningglory, Annual	*Ipomea spp.*
Pennsylvania Smartweed	*Polygonum pensylvanicum*
Prickly Sida or Teaweed	*Sida spinosa*
Redroot Pigweed	*Amaranthus retroflexus*
Morningglory, Smallflower	*Jacquemontia tamnifolia*

Common Name	Scientific Name
Smooth Pigweed	*Amaranthus hybridis*
Spurred Anoda	*Anoda cristata*
Tall Waterhemp	*Amaranthus tuberculatus*
Velvetleaf	*Abutilon theophrasti*
Venice Mallow	*Hibiscus trionum*
Wild Buckwheat	*Polygonum convolvulus*
Wild Mustard	*Sinapis arvensis*
Wild Sunflower	*Helianthus annuus*
Yellow Nutsedge	*Cyperus esculentus*

Restrictions and limitations

Do not apply LADDOK when crop is under stress from prolonged cold, wet weather, poor fertility, or other factors, or when crop is wet and succulent from recent rainfall as crop injury may occur.

Do not apply LADDOK if crop shows injury (leaf phytotoxicity and/or plant stunting) produced by any other prior herbicide applications, because this injury may be enhanced and/or prolonged.

Do not apply LADDOK during prolonged periods of drought or during unseasonably cold weather, as unsatisfactory weed control may result.

Seed corn producers should consult the seed corn company regarding tolerance of seed population inbred lines to LADDOK.

Do not apply to grain sorghum that is heading out or blooming.

Do not make more than one application of LADDOK per season.

Do not apply to grain sorghum that is heading out or blooming.

Do not graze treated area or feed treated forage to livestock for 21 days following application.

Rainfall or overhead irrigation soon after application (within 8 hours) may nullify the effectiveness of LADDOK.

Do not mix or apply LADDOK with any other pesticide or with fertilizer except as specifically recommended on this labeling.

Do not apply this product through any type of irrigation system.

STORAGE AND DISPOSAL

Store above 15 degrees F.

Do not contaminate water, food or feed by storage or disposal.

Pesticide wastes are acutely hazardous. Improper disposal of excess pesticide, spray mixture, or rinsate is a violation of Federal law. If these wastes cannot be disposed of by use according to label instructions, contact your State Pesticide or Environmental Control Agency, or the Hazardous Waste representative at the nearest EPA Regional Office for guidance.

Triple rinse container (or equivalent). Then offer for recycling or reconditioning, or puncture and dispose of in a sanitary landfill, or by incineration, or if allowed by state and local authorities, by burning. If burned, stay out of smoke.

In case of emergency

In case of large scale spillage regarding this product: Avoid contact, isolate area and keep out animals and unprotected persons.
Confine spill and call:
Chemtrec 800-424-9300
BASF 201-316-3000 (Collect)

In case of medical emergency regarding this product, call:
1. Your local doctor for immediate treatment.
2. Your local poison control center (hospital).
3. BASF Corporation 201-316-3000 (collect).

CONDITIONS OF SALE AND WARRANTY

The Directions for use of this product reflect the opinion of experts based on field use and tests. The directions are believed to be reliable and should be followed carefully. However, it is impossible to eliminate all risks inherently associated with use of this product. Crop injury, ineffectiveness or other unintended consequences may result because of such factors as weather conditions, presence of other materials, or use of the product in a manner inconsistent with its labeling, all of which are beyond the control of BASF CORPORATION ("BASF") or the Seller. All such risks shall be assumed by the Buyer.
BASF warrants that this product conforms to the chemical description on the label and is reasonably fit for the purposes referred to in the Directions for use, subject to the inherent risks referred to above. BASF MAKES NO OTHER EXPRESS OR IMPLIED WARRANTY OF FITNESS OR MERCHANTABILITY OR ANY OTHER EXPRESS OR IMPLIED WARRANTY. In no case shall BASF or the Seller be liable for consequential, special or indirect damages resulting from the use or handling of this product. BASF and the Seller offer this product, and the Buyer and User accept it, subject to foregoing Conditions of Sale and Warranty which may be varied only by agreement in writing signed by a duly authorized representative of BASF.
LADDOK is a registered trademark of BASF AG.

LD 2 001 JJ6360 A0811088

PIX®

PLANT REGULATOR

ACTIVE INGREDIENT:*

Mepiquat-chloride: N,N-dimethylpiperidinium chloride	4.2%
INERT INGREDIENTS:	95.8%
TOTAL	100.0%

*Equivalent to 0.35 pounds active ingredient per gallon.

EPA Reg. No. 7969-52

KEEP OUT OF REACH OF CHILDREN
CAUTION

Avoid contact with eyes. In case of contact immediately flush eyes or skin with plenty of water. Get medical attention if irritation persists.

Keep out of lakes, ponds or streams. Do not contaminate water when disposing of equipment washwaters.

DIRECTIONS FOR USE—COTTON

It is a violation of Federal law to use this product in a manner inconsistent with its labeling.

STORAGE AND DISPOSAL

Do not contaminate water, food, or feed by storage or disposal.

Pesticide wastes are toxic. Improper disposal of excess pesticide, spray mixture, or rinsate is a violation of Federal law. If these wastes cannot be disposed of by use according to label instructions, contact your State Pesticide or Environmental Control Agency, or the Hazardous Waste representative at the nearest EPA Regional Office for guidance.

Triple rinse container (or equivalent). Then offer for recycling or reconditioning, or puncture and dispose of in a sanitary landfill, or by incineration, or, if allowed by state and local authorities, by burning. If burned, stay out of smoke.

In case of emergency

In case of large scale spillage regarding this product, call:

CHEMTREC	800-424-9300
BASF Corporation	201-316-3000

In case of medical emergency regarding this product, call:
1. Your local doctor for immediate treatment.
2. Your local poison control center (hospital).
3. BASF Corporation 201-316-3000.

GENERAL INFORMATION

Pix® plant regulator is a foliar applied plant regulator which modifies the cotton plant in several beneficial ways. It is the only such compound which allows the grower to manage the cotton plant for earlier maturity, leading to reduced risk of yield and quality loss due to delayed and prolonged harvest. The use of **Pix** will also result in one or more of the following: height reduction and more open canopy, better boll retention, less boll rot, improved defoliation, reduced trash, and darker green leaf color. Some of these effects may favorably influence the yield potential of the cotton plant.

On both short-staple and Pima cotton the grower has the option of single, dual, or up to four low-rate multiple applications, which greatly facilitates his management flexibility. One of the major factors affecting cotton growth and development is soil moisture. If drought stress occurs when **Pix** is applied, or occurs after a full rate (½-1 pint) is applied, results will not be optimal. The low-rate multiple option gives the producer the ability to discontinue usage of **Pix** if any significant stresses occur after an earlier application; in such a case the total quantity of **Pix** used over a season may be reduced. If stress is relieved, the grower has the option of continuing treatments with **Pix.** In addition, the rate range indicated allows the grower to tailor his usage of **Pix** to his particular cultural or varietal conditions. In areas where insecticides/miticides are frequently applied, the timings are such that tank mixing is often possible. (See **Restrictions and limitations).**

Fields should be carefully scouted and **Pix** should not be applied if plants are under severe stress from weather factors, mite, insect or nematode damage, disease stress, herbicide injury, or fertility stress. In the absence of these stresses, up to four low-rate multiple applications should be made each season, generally using ⅛-¼ pt./A each time. After the first application (at matchhead square in the absence of stress) the rate and timing of subsequent application(s) will depend partially on growing conditions; under good growing conditions additional treatments (⅛-¼ pt./A) should be made at 7-14 day intervals. However, if new growth at any time is excessive, higher rates of **Pix** (¼-½ pt./A) can be used. Total use per season should not exceed 1½ pts./A.

Spray volume

Water as diluent: Ground and aerial application

When applying **Pix** with ground equipment in all states but California, use a minimum of 10 gallons of water per acre; in California, use a minimum of 20 gallons/A. For aerial application in all states but California, use a minimum of 3 gallons of water per acre; in California, use a minimum of 5 gallons/A. Regardless of method or gallonage of application, thorough coverage of the cotton foliage is required.

Oil as diluent: Ultra low volume (ULV) aerial application.

Only permitted in the following states: AL, AR, FL, GA, LA, MO, MS, NC, OK, SC, TN, and TX. A nonphytotoxic oil concentrate should be used which must contain either a petroleum or vegetable oil base and must meet the following criteria: 1) be nonphytotoxic, 2) contain only EPA-exempt ingredients, and 3) be successful in local experience. The exact composition of suitable products will vary; however, vegetable and petroleum oil concentrates should contain emulsifiers which provide good mixing quality. It is recommended that vegetable oil concentrates be highly refined.

Under constant agitation, pour approximately one-half of the required volume of oil (total final oil volume at least 2 pints/A) into the spray tank, and then simultaneously pour in the **Pix** and the remainder of the oil. Constant agitation is required during and after mixing and during application.

Rain safe period

If rain is expected within 8 hours, use of a high quality, EPA-exempt surfactant is recommended, which will result in a rain safe period of 4 hours.

Compatibility

Pix has an aqueous base, and as such is compatible with most insecticides and miticides. If in doubt, test compatibility by adding a teaspoon of the insecticide or miticide to a pint of ready-to-use spray solution of **Pix.** You may combine **Pix** with foliar fertilizers if your prior experience has shown the combination to be compatible and non-injurious under your conditions.

Restrictions and limitations

Insect or mite damage before, at, or after application of **Pix** can lead to yield decreases or other undesirable effects.

Do not apply a single application of 1/2 to 1 pint of **Pix** to cotton that is drought stressed; i.e., stressed due to lack of soil moisture. If using the low-rate multiple option, discontinue use until the moisture stress is alleviated.

Do not use **Pix** on the short-staple variety Delta & Pineland 70.

Do not apply more than 1 1/2 pints of **Pix** per acre per season.

Do not apply **Pix** within 30 days of harvest.

Do not graze or feed cotton forage to livestock within 30 days of application, or after applying **Pix** in oil as a ULV application by air.

The pink color of **Pix** may disappear instantly in spray tanks if chlorinated water is used; however, effectiveness is not related to color of spray solution.

Do not tank mix with other products other than those mentioned under **Compatibility.**

Time and rate application: short-staple and long-staple (Pima) cotton

The times and rates of application have been carefully researched and the Directions for use should be observed as specified below.

I. Single or dual applications

Geographic Area	Time of Application	Rate per Acre
AL, AR, AZ, CA, FL, GA, LA, MO, MS, NM, NC, SC, TN, VA	**First application:** Apply when cotton is actively growing and is between 20" and 30" tall, provided cotton is not more than 7 days beyond early bloom stage (5-6 blooms per 25 row feet). If cotton is 24" tall and has no blooms, apply **PIX® plant regulator.** Use 1/2 pint per acre on cotton where excessive vegetative growth is not likely to be a problem, and 1 pint per acre in areas tending to have excessive vegetative growth. See **Restrictions and limitations.**	1/2 to 1 pint
	Second application for control of excessive vegetative growth: If the cotton field has a history of excessive growth, and/or conditions after the first application of **Pix** favor excessive growth, apply a second application 3 to 4 weeks after the first application.	1/2 pint
OK, TX, (except Rio Grande Valley)	Areas where excessive vegetative growth is **not** a problem: **First application:** Apply when cotton is in early bloom stage (5-6 blooms per 25 row feet) and actively growing. If no blooms are present and the cotton is 24" tall and actively growing, apply **Pix.** See **Restrictions and limitations.**	1/2 pint
	Second application: If conditions after the first application of **Pix** favor excessive growth, apply a second application 3 to 4 weeks after the first application.	1/2 pint
OK, TX	Areas where excessive vegetative growth **is** a problem: **First application:** Optimum results will be achieved when plants are in early bloom stage (5-6 blooms per 25 row feet) and an average of 24" tall. Treatments can also be made when cotton height averages a minimum of 20" and a maximum of 28" provided cotton is not more than 7 days beyond early bloom as defined above. See **Restrictions and limitations.**	1 pint
	Second application for control of excessive vegetative growth: If cotton field has a history of excessive growth, and/or conditions after the first application of **Pix** favor excessive growth, apply a second application 3 to 4 weeks after the first application.	1/2 pint

Low-Rate Multiple Applications

Geographic Area (States)	Time of Application	Avg. Historical Use Rate = 1/2 Pint (for new users of PIX)	Avg. Historical Use Rate = 1 Pint (for fully Irrigated)
AL, AR AZ, CA, FL, GA, LA, MO, MS, NC, NM, OK, SC, TN, TX, VA	**First application:** Optimal results will be achieved when plants are in the match-head square[a] stage of growth.	1/8 pint	1/4 pint
	Second application: 7-14 days later, or when regrowth occurs.	1/8 to 1/4 pint*	1/4 to 1/2 pint*
	Third application: 7-14 days later, or when regrowth occurs.	1/8 to 1/4 pint*	1/4 to 1/2 pint*
	Fourth application: 7-14 days later, or when regrowth occurs.	1/8 to 1/4 pint*	1/4 to 1/2 pint*

* Use higher rate if previous application was not made or if growing conditions are conducive to excessive growth.

[a] Match-head square is when the first square of a typical cotton plant is 1/8-1/4" in diameter. The first application should go on when 50% of the plants have one or more match-head squares.

CONDITIONS OF SALE AND WARRANTY

The Directions for use of this product reflect the opinion of experts based on field use and tests. The directions are believed to be reliable and should be followed carefully. However, it is impossible to eliminate all risks inherently associated with use of this product. Crop injury, ineffectiveness or other unintended consequences may result because of such factors as weather conditions, presence of other materials, or use of the product in a manner inconsistent with its labeling, all of which are beyond the control of BASF CORPORATION ("BASF") or the Seller. All such risks shall be assumed by the Buyer.

BASF warrants that this product conforms to the chemical description on the label and is reasonably fit for the purposes referred to in the Directions for use, subject to the inherent risks referred to above. BASF MAKES NO OTHER EXPRESS OR IMPLIED WARRANTY OF FITNESS OR MERCHANTABILITY OR ANY OTHER EXPRESS OR IMPLIED WARRANTY. IN NO CASE SHALL BASF OR THE SELLER BE LIABLE FOR CONSEQUENTIAL, SPECIAL OR INDIRECT DAMAGES RESULTING FROM THE USE OR HANDLING OF THIS PRODUCT. BASF and the Seller offer this product, and the Buyer and User accept it, subject to the foregoing Conditions of sale and warranty which may be varied only by agreement in writing signed by a duly authorized representative of BASF.

Pix is a registered trademark of BASF AG.

PX 2 013 JJ6506 A0070189

POAST®

POSTEMERGENCE GRASS HERBICIDE

For use on alfalfa, cotton flax, peanuts, soybeans, sugar beets, sunflowers, and certain nonbearing and nonfood crops.

ACTIVE INGREDIENT: 2-[1-(ethoxyimino)butyl-5-[2-(ethylthio) propyl]-3-hydroxy-2-cyclohexen-1-one*	18.0%
INERT INGREDIENTS:	82.0%
TOTAL	100.0%

*Equivalent to 1.5 pounds per gallon

EPA Reg. No. 7969-58

KEEP OUT OF REACH OF CHILDREN.

WARNING

Causes substantial but temporary eye injury. Do not get into eyes or on clothing. Wear safety glasses. Wash thoroughly with soap and water after handling. Remove contaminated clothing and launder before re-use. Harmful if swallowed.

If in eyes: Hold eyelids open and flush with a steady stream of water for 15 minutes. If swallowed: Drink promptly a large quantity of milk, egg whites, gelatin solution or, if these are not available, large quantities of water. Avoid alcohol.

ENVIRONMENTAL HAZARDS

Do not apply directly to water.

Do not contaminate water by cleaning of equipment or disposal of wastes.

ENDANGERED SPECIES CONCERNS

The use of any pesticide in a manner that may kill or otherwise harm an endangered or threatened species or adversely modify their habitat is a violation of Federal law.

The use of this product is controlled to prevent death or harm to Solano grass which occurs in Solano County, **California.** Before using this product in this county you must obtain the EPA Endangered Species Bulletin (EPA/ES-85-13) available from either your County Agricultural Extension Agent, the Endangered Species Specialist in the California Department of Fish and Game, or the Regional Offices of the U.S. Fish and Wildlife Service (Portland, Oregon) or the U.S. Environmental Protection Agency (San Francisco, California). THIS BULLETIN MUST BE REVIEWED PRIOR TO PESTICIDE USE. THE USE OF THIS PRODUCT IS PROHIBITED IN THIS COUNTY UNLESS SPECIFIED OTHERWISE IN THE BULLETIN.

DIRECTIONS FOR USE—ALL CROPS

It is a violation of Federal law to use this product in a manner inconsistent with its labeling.

GENERAL INFORMATION

POAST® herbicide is a selective broad spectrum postemergence herbicide for control of annual and perennial grass weeds. POAST does not control sedges or broadleaf weeds.

Since all grass crops such as sorghum, corn, small grains and rice, as well as some ornamental grasses such as turf, are susceptible to POAST, avoid all direct or indirect contact with any desired grass plant.

Control symptoms: POAST rapidly enters the plant through the foliage and translocates throughout the plant. Control symptoms exhibited by the grass plant progress from a slowing and stopping of growth (generally within two days), to reddening of foliage, and to leaf tip burn. Subsequently, burn-back of the foliage occurs. These symptoms will generally be observed within three weeks, depending on environmental conditions.

Application information

Apply POAST to actively growing grasses when they are at the proper growth stage as specified in the **Recommendations for Use** tables.

Do not make applications to grasses under stress, such as stress due to lack of moisture, herbicide injury, mechanical injury or cold temperatures, since unsatisfactory control will probably result.

Nozzle selection: Thorough spray coverage of grass foliage is essential. For broadcast application use standard high pressure pesticide hollow cone or flat fan nozzles. Do not use flood or whirl chamber nozzles. Application of POAST with control drop applicator (CDA) nozzles is not recommended due to erratic coverage which causes inconsistent weed control.

Spray volume: Under most conditions 10 gallons per acre of spray volume is optimal. A minimum volume of 5 gallons and a maximum volume of 20 gallons of spray solution per acre for broadcast applications may be used. In the Western Region a minimum of 10 gallons per acre is recommended. In the high plains of Texas, Oklahoma, and eastern New Mexico a maximum of 10 gallons per acre is recommended.

Spray pressure: When using standard high pressure hollow cone or flat fan nozzles adjust pressure to a minimum of 40 psi and a maximum of 60 psi measured at the nozzle.

Boom height: Always adjust spray pressure, spray volume and height of spray boom to ensure penetration of plant canopy and thorough coverage of grasses to be controlled. When tall weeds, such as volunteer corn, are to be controlled, the boom height should be high enough to cover the entire plant. This may be as much as 20 inches above the weed. Refer to the nozzle manufacturer's directions for recommended height.

Tall crop applications: When a crop, such as cotton, is 24 or more inches in height and the grasses may be below the crop canopy, drop nozzles should be used to ensure good coverage of the grass species. Good coverage is essential for maximum control.

Band applications: Banding of POAST may be used to control annual grasses. Grasses which are not covered or only partly covered by the spray mixture will not be adequately controlled. All recommendations are on a broadcast basis unless otherwise stated. When banding, rates of POAST, additives and water should be reduced in proportion to the area sprayed.

Other spray equipment: Do not use selective application equipment such as recirculating sprayers, wiper applicators, or shielded applicators.

Air equipment: Thorough spray coverage of grass foliage is essential. Use a minimum of 5 gallons of water per acre. Increase water volume to 10 gallons per acre if grass foliage and/or crop canopy is dense.

Cultivation information

Do not cultivate within 5 days prior to application of POAST or within 7 days following application.

A timely cultivation after 7 days may aid in providing season-long control. For control of quackgrass a cultivation 14 to 21 days after application will aid in control.

Addition of Dash™ spray adjuvant* or oil concentrate

A nonphytotoxic oil concentrate (commonly referred to as oil concentrate) should be added to the spray tank. The oil concentrate must contain either a petroleum or vegetable oil base and must meet the following criteria: 1) be nonphytotoxic, 2) contain only EPA-exempt ingredients, 3) provide good mixing quality in the jar test (see at right), and 4) be successful in local experience.

The exact composition of suitable products will vary, however, vegetable and petroleum oil concentrates should contain emulsifiers which provide good mixing quality. For vegetable oil concentrates, it has been observed that highly refined vegetable oils are more satisfactory than unrefined vegetable oils. For additional information see **Jar test for estimating suitability of oil concentrates** at the end of this section.

Dash™ spray adjuvant may be used as a direct substitute for an oil concentrate with some exceptions. In some crops and tank mixes **Dash** is not recommended (see individual tables).

Rate of Dash* or oil concentrate

Ground and air application-2 pints/acre.

Addition of Urea Ammonium Nitrate Solution (UAN) or Ammonium Sulfate (AMS)**

Addition of UAN solution or AMS is recommended only for soybeans, alfalfa, sunflowers, peanuts, cotton and sugar beets. UAN solution is commonly referred to as 28%, 30% or 32% nitrogen, and is a water solution of urea and ammonium nitrate. When ammonium sulfate is used, three quarts of liquid ammonium sulfate (8.0.0 analysis) may be substituted for 2½ lb. solid ammonium sulfate.

For best control of volunteer corn, large crabgrass, wild oat, volunteer cereals, and quackgrass, add ammonium sulfate at 2½ pounds per acre plus oil concentrate, or ½—1 gallon of UAN solution plus oil concentrate.

* Dash is not to be used in California.

** Ammonium Sulfate is not to be used in California.

In some areas use of a nitrogen additive has improved control of rhizome johnsongrass; consult your local BASF representative for recommendations for your area.

Always include **Dash** or oil concentrate with the UAN solution or ammonium sulfate when applying POAST herbicide. UAN solution or ammonium sulfate does not replace **Dash** or oil concentrate.

Since most nitrogen solutions are mildly corrosive to galvanized, mild steel and brass spray equipment, rinse the entire spray system with water after use.

Note about ammonium sulfate (AMS): It is important to use high quality ammonium sulfate to avoid plugging of spray nozzles. The ammonium sulfate must be readily soluble in water and contain no insoluble materials. Local sources of high quality fine feed grade ammonium sulfate may be better than fertilizer grade. Low quality ammonium sulfate may contain material that will not readily dissolve which could result in nozzle tip plugging. To determine quality, perform a jar test adding ⅓ cup of ammonium sulfate to 1 gallon of water and agitate for 1 minute. If undissolved sediment is observed, predissolve the ammonium sulfate in water and filter prior to addition to the spray tank. If ammonium sulfate is added directly to the spray tank, add slowly with agitation. Adding too quickly may clog outlet lines. Ensure that ammonium sulfate is completely dissolved in the spray tank before adding other products.

	Ground Application	Air Application
Rate per Acre of UAN solution or Ammonium Sulfate	1/2-1 gallon 2 1/2 lbs.	1/2 gallon 2 1/2 lbs.

Mixing/spraying: Fill tank of a thoroughly clean sprayer one-half to two-thirds full with clean water. Start agitation and add oil concentrate; allow to mix thoroughly. Add POAST and remaining volume of water. When tank mixing, See **Jar test** (step 4) which follows for adding sequence. **Apply POAST soon after mixing.** Maintain constant agitation during application.

Jar test for estimating suitability of oil concentrates

1. Water supply: Use only water from intended source and at the source temperature.

2. Amount of water in jar: For 20 gal./A spray volume use 3 1/3 cups (800 ml) of water. For 10 gal./A spray volume use 1 2/3 cups (400 ml) of water. For 5 gal./A spray volume use 5/6 cup (200 ml) of water. For other spray volumes, adjust proportionately to above.

3. Amount of herbicides(s) and oil concentrate to add: Add herbicide(s) and oil concentrate at rate of 1 teaspoon (5 ml) for each pint of recommended label rate.

4. Add components in following sequence, gently mixing between component additions: 1) Water miscible or soluble products (such as **Basagran, Blazer,** ammonium sulfate, UAN solution) when applicable. 2) Oil concentrate. 3) POAST (and other emulsifiable concentrates when applicable).

5. Cap jar, invert 10 cycles, let stand for 15 minutes, evaluate.

6. Evaluation: An ideal tank mix combination will be uniform; thus, the suitability of the oil concentrate is questionable if any of the following are observed: Free oil at the surface—film or globules. Flocculation—fine particles which may be suspended in the liquid or found as a precipitated layer at the bottom of the jar. Clabbering—thickening texture (coagulated) resembling yogurt or a curd-like texture as with cottage cheese.

Spot or small area treatment

When using knapsack sprayers or high volume spray equipment utilizing hand guns or other suitable nozzle arrangements, prepare a 1% solution of POAST in water. **Dash** or a recommended oil concentrate must also be used at concentration of 1%.

Apply to foliage of grasses on a spray-to-wet basis. Spray coverage should be uniform and complete. Do not spray to point of runoff.

Prepare the desired volume of spray solution by mixing the amount of POAST and the amount of oil concentration in water according to the table below.

Desired Spray Solution Volume	Amount to be Added to Obtain a 1% Solution: POAST	Dash or Oil Concentrate
1 gallon	1 1/4 fl. oz.*	1 1/4 fl. oz.
25 gallons	1 qt.	1 qt.
50 gallons	2 qts.	2 qts.
100 gallons	4 qts.	4 qts.

* 2 tablespoons = 1 fl. oz.

In soybeans and cotton, spot or small area treatments should not exceed 1/10 of an acre in size, and no more than 10% of any given acre should be treated.

In soybeans, do not make more than one spot or small area treatment to the same area within the same growing season. Also in soybeans, do not apply both broadcast and spot or small area treatments to the same area within the same growing season.

In cotton, do not make more than two spot or small area treatments in the same area within the same growing season.

General restrictions and limitations

Do not make applications to grasses under stress, such as stress due to lack of moisture, herbicide injury, mechanical injury or cold temperatures, since unsatisfactory control will probably result.

Do not apply if rainfall is expected within one hour following application as grass control will probably be unsatisfactory.

Do not mix or apply POAST with any other pesticide, additive, or fertilizer except as specifically recommended on this labeling, or EPA approved BASF supplemental labeling.

Do not apply POAST as a preplant or preemergent treatment prior to corn, milo, millet or sorghum.

Do not apply POAST through any type of irrigation system.

PROCEDURE FOR CLEANING SPRAY EQUIPMENT

Attention! Clean sprayer thoroughly before and after application of POAST

Clean sprayer thoroughly prior to application of POAST, particulary if a herbicide was used which has the potential to injure crops.

The steps listed below are suggested for cleaning of spray equipment prior to or following applications of POAST.

Step 1: Hose down thoroughly the inside as well as the outside of equipment while filling the spray tank half full of water. Flush by operating sprayer until the system is purged of this rinse water.

Step 2: Refill tank with water while adding 1 gallon household ammonia or 1 pint household dishwashing detergent per 100 gallons of water. Or add a commercial sprayer cleaner according to the manufacturer's directions. Operate the pump to circulate the detergent solution through the sprayer system for 5 to 10 minutes and discharge a small amount of solution through the boom and nozzles. Let the solution stand for 24 hours.

Step 3: Flush the detergent solution out of the spray tank through the boom.

Step 4: Remove the nozzles and screens and flush the system with two tankfuls of water.

STORAGE AND DISPOSAL

Do not contaminate water, food, or feed by storage or disposal.

Pesticide wastes are toxic. Improper disposal of excess pesticide spray mixture, or rinsate is a violation of Federal law. If these wastes cannot be disposed of by use according to label instructions, contact your State Pesticide or Environmental Control Agency, or the Hazardous Waste representative at the nearest EPA Regional Office for guidance.

Triple rinse container (or equivalent). Then offer for recycling or reconditioning, or puncture and dispose of in a sanitary landfill, or by incineration, or, if allowed by state and local authorities, by burning. If burned, stay out of smoke.

DIRECTIONS FOR USE—SPECIFIC CROPS

Soybeans—Recommendations for grass control—all regions

- Apply to actively growing grasses at the sizes indicated below.
- Soybeans at all stages of growth are tolerant to POAST.
- Always follow recommendations given in **Application information** section.
- Always adjust spray pressure, spray volume and height of spray boom to ensure penetration of plant canopy and thorough coverage of grasses to be controlled.
- **Always add 2 pints of Dash™ or oil concentrate per acre.**

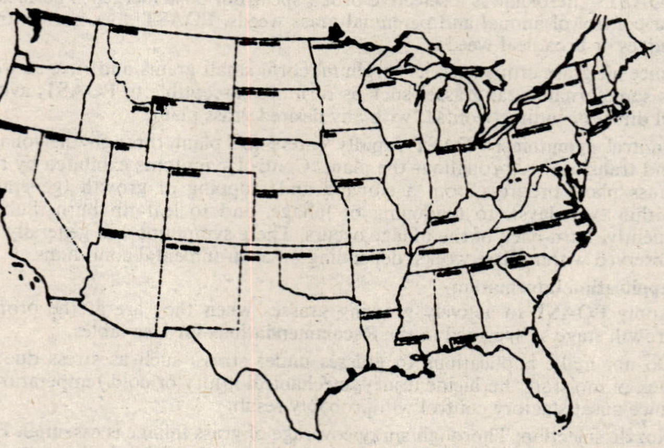

SOYBEANS—ANNUAL GRASSES*—SPECIAL RATE FOR EARLY TREATMENT

GROUP	GRASS	TIME OF APPLICATION	RATE OF POAST PER ACRE	ADDITIVES (RATE PER ACRE): DASH OR OIL CONCENTRATE (GROUND & AIR)	UAN SOLUTION OR AMMONIUM SULFATE (GROUND & AIR)
A	Wild Proso Millet	4-10"	1/2 pt. (16 acres/gal.)	2 pts.	—
B	Goosegrass	1-3"			

GROUP	GRASS	TIME OF APPLICATION	RATE OF POAST PER ACRE	ADDITIVES (RATE PER ACRE) DASH OR OIL CONCENTRATE — GROUND & AIR	ADDITIVES (RATE PER ACRE) UAN SOLUTION OR AMMONIUM SULFATE — GROUND & AIR
	Barnyardgrass** (Midwest only) Broadleaf Signalgrass Fall Panicum Texas Panicum Foxtails: Giant, Green	1-4"	¾ pt. (10.7 acres/gal.)	2 pts.	—
	Volunteer Corn	1-12"		2 pts. plus	Always add ½-1 gallon or 2½ lbs. AMS

* For broad spectrum control of annual grasses in Groups A & B, use ¾ pint of POAST per acre. If additional applications are needed, apply at the same rate and at the recommended stage of growth.

** In these states use 1 pint per acre as recommended in the next table: AL, AR, FL, GA, LA, MS, NC, SC, TN, TX, VA.

SOYBEANS—ANNUAL GRASSES—STANDARD RECOMMENDATIONS*

GROUP	GRASS	TIME OF APPLICATION	RATE OF POAST PER ACRE	ADDITIVES (RATE PER ACRE) DASH OR OIL CONCENTRATE — GROUND & AIR	ADDITIVES (RATE PER ACRE) UAN SOLUTION OR AMMONIUM SULFATE — GROUND & AIR
A	Wild Proso Millet	4-10"	½ pt. (16 acres/gal.)	2 pts.	—
B	Wild Oats	Up to 4"		2 pts. plus	½-1 gallon UAN or 2½ lbs. AMS is recommended
	Goosegrass Smooth Crabgrass	Up to 6"		2 pts.	—
	Large Crabgrass	Up to 6"		2 pts. plus	½-1 gallon UAN or 2½ lbs. AMS is recommended
	Barnyardgrass Broadleaf Signalgrass Browntop Panicum Fall Panicum Foxtails: Giant, Green, Yellow Johnsongrass, Seedling Junglerice Red Sprangletop Ryegrass, Annual Texas Panicum Witchgrass Woolly Cupgrass	Up to 8"	1 pt. (8 acres/gal.)	2 pts.	—
	Shattercane/Wildcane If needed, retreat at the same rate and stage of growth.	6-18"		2 pts.	—
	Volunteer Corn Maintain sufficient boom height above volunteer corn plants for best spray coverage.	Up to 20"		2 pts. plus	Always add ½-1 gallon UAN or 2½ lbs. AMS
C	Field Sandbur (Midwest only)	Before tillering up to 3"	1¼ pts. (6.4 acres/gal.)	2 pts.	—
D	Volunteer Cereals Barley, Oats Rye, Wheat Not recommended for spring control of volunteer cereals that emerged the previous fall.	Before tillering, Up to 4" and prior to over-wintering	1½ pts. (5.3 acres/gal.)	2 pts. plus	½-1 gallon UAN or 2½ lbs. AMS is recommended
E	Itchgrass Red Rice	2-4"	2 pts. (4 acres/gal.)	2 pts.	—

* For broad spectrum control of annual grasses in Group B (above), use 1 pint of POAST per acre. When weed populations include additional grasses in Group C, D or E, increase the rate of POAST herbicide as indicated. If later flushes of annual grasses emerge after the first application, make additional applications at the same rate and at the same recommended stage of growth.

Rescue treatment for controlling selected annual grasses

For best results, always apply POAST herbicide to annual grasses at the growth stage and rate specified in the above table (**Annual Grasses—Standard Recommendations**). However, if POAST herbicide cannot be applied at the recommended time, larger annual grasses can be controlled with a later application by increasing the rate of POAST. Apply to actively growing grasses at the rates and sizes indicated below.

Soybeans—Annual Grasses—Rescue Treatment

Grass	Time of Application	Rate of POAST per Acre	Dash or Oil Concentrate (Rate/A)
Wild Proso Millet	10-24"	1 pt. (8 acres/gal.)	

Foxtails: Giant, Green, Yellow Johnsongrass, Seedling	8-16"	1½ pts. (5.3 acres/gal.)	2 pts.
Barnyardgrass Broadleaf Signalgrass Fall Panicum Texas Panicum	8-12"		
Large Crabgrass Smooth Crabgrass Goosegrass	6-8"		

SOYBEANS—PERENNIAL GRASSES

GRASS	TIME OF APPLICATION	RATE OF POAST PER ACRE	ADDITIVES (RATE PER ACRE) DASH OR OIL CONCENTRATE	UAN SOLUTION OR AMMONIUM SULFATE
			GROUND & AIR	
Bermudagrass • First Application	Before stolon length exceeds 6".	1½ pts. (5.3 acres/gal.)	2 pts.	—
• Second Application If regrowth occurs or new plants emerge.	1-4" length of new plants or growth.	1 pt. (8 acres/gal.)	2 pts.	—
Johnsongrass, Rhizome • First Application Use 5-10 gallons of spray solution per acre. Maintain a ground speed of no more than 6 miles per hour. For best results rhizomes should be thoroughly fragmented (less than 6"). (When using 11-20 gallons of spray solution per acre use 1½ pints of POAST herbicide.)	15-25" (15-20" in no-till culture)	1 pt. (8 acres/gal.)	2 pts.	see page 5
• Second Application When regrowth occurs or new plants emerge.	6-12"	1 pt.	2 pts.	see page 5
Quackgrass • First Application For best results, rhizomes should be thoroughly fragmented (less than 6").	6-8"	1½ pts. (5.3 acres/gal.)	2 pts.	plus Always add 1 gal. UAN or 2½ lbs. AMS.
• Second Application If regrowth occurs or new plants emerge. Depending upon environmental conditions and crop cultural system, season-long control may not always be obtained. However, competition of quackgrass with the crop will be reduced. **Note:** In conventional wide-row soybeans, a cultivation no sooner than 14 days after application but within 21 days of application will aid in control.	6-8"	1 pt. (8 acres/gal.)	2 pts.	plus Always add ½-1 gal. UAN or 2½ lbs. AMS.
Wirestem Muhly If regrowth occurs, retreat at the same rate and stage of growth.	Up to 6"	1¼ pts. (6.4 acres/gal.)	2 pts.	—

Burndown application*
(soybeans only)

For the control of emerged annual and perennial grasses prior to soybean emergence. POAST herbicide may be applied before, during, or after planting in accordance with the **Directions for use.** Apply to actively growing grasses up to the maximum sizes indicated in the rate table for soybeans. Also see recommendation for broad spectrum weed control; a tank mixture of POAST and 2,4-D low volatile ester, pages 17-18. Subsequent applications of POAST may be made after soybean emergence in accordance with this label provided that no more than a total of 5 pints of POAST is applied per acre in a single season.

Restrictions and limitations for soybeans

Do not apply POAST to soybeans within 90 days of harvest.

Do not apply more than a total of 5 pints of POAST per acre to soybeans in one season (including application before or after planting).

Do not graze treated soybean fields and do not feed treated soybean forage (green succulent) or ensilage to livestock. Treated soybean hay may be fed.

Classic® herbicide may cause antagonism of POAST when sprayed from 7 days prior to application, to 1 day after POAST application. This antagonism is more likely to occur under stress conditions.

Do not apply POAST as a preplant or preemergence treatment prior to corn, milo, millet, or sorghum.

* Not applicable in California.

Soybeans—Tank Mixes with POAST

Use the following chart as a guide to determine grasses and broadleaf weeds controlled by POAST alone and various tank mixes with POAST.

TABLE 1
POAST TANK MIXES* AND GUIDE TO WEEDS CONTROLLED

POAST CONTROLS THE WEEDS LISTED BELOW	ADDITIONAL WEEDS CONTROLLED BY TANK MIXING VARIOUS HERBICIDES WITH POAST		REFER TO TABLE LISTED BELOW FOR RATE, WEED SIZE AND ADDITIVE INFORMATION
Barnyardgrass Broadleaf Signalgrass Fall Panicum Giant Foxtail Goosegrass Green Foxtail Junglerice Large Crabgrass Red Sprangletop Seedling Johnsongrass Smooth Crabgrass Texas Panicum Witchgrass Woolly Cupgrass Wild Proso Millet Yellow Foxtail	**BASAGRAN® HERBICIDE** Balloonvine Beggarticks Cocklebur Coffee Senna Common Lambsquarters Common Purslane Common Ragweed Cypressvine Morningglory Canada Thistle* * Dayflower Devilsclaw Galinsoga Giant Ragweed Jimsonweed Ladysthumb	Pennsylvania Smartweed Prickly Sida or Teaweed Redweed Shepherdspurse Smallflower Morningglory Spurred Anoda Tropic Croton Velvetleaf Venice Mallow Wild Buckwheat Wild Mustard Wild Poinsettia Wild Sunflower Yellow Nutsedge	POAST + Basagran* Table 2 Page 13
	BASAGRAN & BLAZER®/TACKLE® HERBICIDES In addition to weeds listed above for Basagran: Black Nightshade Common Ragweed Crotalaria Morningglories	Redroot Pigweed Sesbania Smooth Pigweed Tall Waterhemp	POAST + Basagran + Blazer* herbicides Tables 3, 4, and 6 Pages 14-16
Barnyardgrass Broadleaf Signalgrass Fall Panicum Giant Foxtail Goosegrass Green Foxtail Junglerice Large Crabgrass Red Sprangletop Seedling Johnsongrass Smooth Crabgrass Texas Panicum Witchgrass Woolly Cupgrass Wild Proso Millet Yellow Foxtail	**BLAZER® HERBICIDE** Balloonvine Black Nightshade Bristly Starbur Buffalobur Burgherkin Canada Thistle Carpetweed Citron Climbing Milkweed Cocklebur Coffee Senna Common Milkweed Common Ragweed Cutleaf Groundcherry Devilsclaw Eastern Black Nightshade Field Bindweed Florida Pusley Giant Ragweed Hairy Indigo Hairy Galinsoga Hedge Bindweed Hemp Sesbania Hophornbeam Copperleaf Jimsonweed Ladysthumb Lambsquarters Lanceleaf Groundcherry Morningglory, Common (Tall) Morningglory, Cypressvine Morningglory, Entireleaf	Morningglory, Ivyleaf Morningglory, Pitted (Small White) Morningglory, Purple Moonflower Morningglory, Scarlet Morningglory, Smallflower Morningglory, Willowleaf (Palmleaf) Palmer Amaranth Pennsylvania Smartweed Poorjoe Prostrate Pigweed Prostrate Spurge Purslane Redroot Pigweed Redvine Showy Crotalaria Smallflower Galinsoga Smellmelon Smooth Pigweed Spiney Amaranth Spotted Spurge Tall Waterhemp Texas Gourd Tropic Croton Trumpetcreeper Virginia Copperleaf Volunteer Cowpea Wild Buckwheat Wild Mustard Wild Poinsettia Wild Spiny Cucumber Woolly Croton	POAST + Blazer* Table 5 Pages 15-16
Barnyardgrass Broadleaf Signalgrass Fall Panicum Giant Foxtail Goosegrass Green Foxtail Junglerice Large Crabgrass Red Sprangletop Seedling Johnsongrass Smooth Crabgrass Texas Panicum Witchgrass Woolly Cupgrass Wild Proso Millet Yellow Foxtail	**2,4-D (LVE) BURNDOWN (PREPLANT)*** Canada Thistle Common Cocklebur Common Dandelion Common Lambsquarters Common Ragweed Field Bindweed Field Pennycress Giant Ragweed Horseweed	Pennsylvania Smartweed Prickly Lettuce Redroot Pigweed Shepherdspurse Velvetleaf White Cockle Wild Buckwheat Wild Mustard Yellow Rocket	POAST + 2,4-D (LVE) Table 7 Page 18

* Tank mixes not applicable in California.
* * Requires two applications of Basagran in accordance with the label for Basagran.

POAST + BASAGRAN TANK MIX—SOYBEANS* GENERAL AND APPLICATION INFORMATION, RESTRICTIONS AND LIMITATIONS

General information

POAST and Basagran may be tank mixed for postemergence control of the broadleaf and grass weeds shown in **Table 2.** Weeds must be actively growing and at the recommended growth stages.

Separate applications should be made if: a) all weeds to be controlled are not at the correct growth stage for treatment at the same time, or b) grasses to be controlled include rhizome johnsongrass, quackgrass, Bermudagrass, wirestem muhly, shattercane, volunteer cereals, wild oats, red rice or itchgrass. See rate tables on page 16 for POAST; and **Table 6, Separate Applications of POAST.**

Water volume and spray pressure

Ground equipment: Use 20 gallons of total spray solution per acre (broadcast basis) and a minimum of 40 psi pressure. Use standard high pressure hollow cone or flat fan nozzles spaced 20 inches apart. Do not use flood or whirl chamber nozzles.

Air equipment: Use a minimum of 5 gallons of total spray solution per acre.

Mixing

Fill spray tank half full with water, and add the recommended amount of product in the following order—Basagran, UAN or ammonium sulfate, oil concentrate, POAST—while the agitator is running. Then add the remaining quantity of water.

Additives

At the low rate of POAST (1 pt./A), the additives Dash® spray adjuvant **plus** UAN (or ammonium sulfate), must be used.

At the higher rate of POAST either Dash or oil concentrate must be used. UAN (or ammonium sulfate) may be added.

Coverage

Thorough coverage of actively growing weeds is essential. Large crop-and-weed leaf canopies shelter smaller weeds and can prevent adequate spray coverage. Soybeans are tolerant to the tank mix; however, under certain conditions soybeans may burn, crinkle and bronze. Soybeans at all stages of growth are tolerant to Basagran and POAST.

Restrictions and limitations (partial list)

Read and follow the restrictions and limitations on the labels for POAST and Basagran herbicides. The most restrictive labeling applies in tank mixes.

* Tank mix not applicable in California.

POAST + BASAGRAN + BLAZER/TACKLE TANK MIX*—SOYBEANS GENERAL AND APPLICATION INFORMATION, RESTRICTIONS AND LIMITATIONS

General information

POAST, Basagran and Blazer may be tank mixed for post-emergence control of broadleaf and grass weeds. Weeds must be actively growing and at the recommended growth stages.

Separate applications should be made if: a) all weeds to be controlled are not at the correct growth stage for treatment at the same time, or b) grasses to be controlled include rhizome johnsongrass, quackgrass, bermudagrass, wirestem muhly, volunteer corn, shattercane, volunteer cereals, wild oats, red rice or itchgrass. See rate tables on page 16 for POAST and **Table 6, Separate Applications of POAST.**

Water volume and spray pressure

Ground equipment: For the tank mix of POAST + Basagran + Blazer, use 20 gallons of total spray solution per acre (broadcast basis) and a minimum of 40 psi pressure. Use standard high pressure hollow cone or flat fan nozzles spaced 20 inches apart. Do not use flood or whirl chamber nozzles.

Air equipment: Use a minimum of 10 gallons of total spray solution per acre.

Mixing

Fill the spray tank half full with water and add the recommended amount of product in the following order—Basagran, Blazer, oil concentrate, POAST—while the agitator is running. Then add the remaining quantity of water.

Additive

Oil concentrate must be used in this tank mix; Dash is not recommended.

Early spot spray

When using knapsack sprayers or high volume equipment utilizing handguns (or other suitable nozzle arrangements), prepare spray solution according to **Table 4.** Apply to the foliage on a spray-to-wet basis. Complete coverage of all foliage is essential for control. Control of perennial weeds may be limited to burnoff of exposed foliage.

Coverage

Thorough coverage of actively growing weeds is essential. Large crop-and-weed leaf canopies shelter smaller weeds and can prevent adequate spray coverage. Soybeans are tolerant to the tank mix; however, under certain conditions soybeans may burn, crinkle and bronze.

Restrictions and limitations (partial list)

Read and follow the restrictions and limitations on the labels for POAST, Basagran, and Blazer or Tackle herbicides. The most restrictive labeling applies in tank mixes.

Do not add UAN solution or ammonium sulfate to a tank mix of POAST + Basagran + Blazer or Tackle + oil concentrate.

* Tank mix not applicable in California.

TABLE 2
POAST + BASAGRAN TANK MIX—SOYBEANS
RATE AND TIME OF APPLICATION TABLE

PRODUCT	PRODUCT RATE	WEEDS CONTROLLED/WEED SIZE				ADDITIVE (RATE/ACRE): DASH OR OIL CONCENTRATE		ADDITIVE (RATE/ACRE): UAN SOLUTION OR AMMONIUM SULFATE
		ANNUAL GRASSES*						
POAST	1 pt./A	Wild Proso Millet**	4-10"	Green Foxtail	3-8"	Dash (2 pts.)	plus	½-1 gallon UAN or 2½ lbs. AMS
		Fall Panicum	3-8"	Witchgrass	3-8"			
		Giant Foxtail	3-8"	Woolly Cupgrass	3-8"			
				Volunteer Corn	1-12"			
POAST	1½ pts./A	Barnyardgrass	3-8"	Junglerice	3-8"	Dash (2 pts.) or oil concentrate (2 pts.)	plus	½-1 gallon UAN or 2½ lbs. AMS may be added to this tank mixture.
		Broadleaf Signalgrass	3-8"	Red Sprangletop	3-8"			
		Yellow Foxtail	3-8"	Texas Panicum	3-8"			
		Seedling Johnsongrass	3-8"	Goosegrass	3-6"			
				Large Crabgrass	3-6"			
				Smooth Crabgrass	3-6"			
plus	plus							
		BROADLEAVES AND SEDGE						
Basagran	1-2 pts./A according to weed species and size (see label for Basagran).	Balloonvine		Ladysthumb				
		Beggarticks		Pennsylvania Smartweed				
		Bristly Starbur		Prickly Sida or Teaweed				
		Canada Thistle***		Redweed				
		Cocklebur		Shepherdspurse				
		Coffee Senna		Smallflower Morningglory				
		Common Lambsquarters		Spurred Anoda				
		Common Purslane		Tropic Croton				
		Common Ragweed		Velvetleaf				
		Cypressvine Morningglory		Venice Mallow				
		Dayflower		Wild Buckwheat				
		Devilsclaw		Wild Mustard				
		Galinsoga		Wild Poinsettia				
		Giant Ragweed		Wild Sunflower				
		Jimsonweed		Yellow Nutsedge				

* Tank mix does not control rhizome johnsongrass, quackgrass, bermudagrass, wirestem muhly, shattercane, volunteer cereals, wild oats, red rice, or itchgrass.

** For control of wild proso millet only, include POAST® herbicide in the tank mix at ¾ pint/A.

*** Requires two applications of Basagran® herbicide in accordance with the label for control.

TABLE 3
POAST + BASAGRAN + BLAZER TANK MIX—SOYBEANS
RATE AND TIME OF APPLICATION TABLE

PRODUCT	PRODUCT RATE	WEEDS CONTROLLED/WEED SIZE				ADDITIVE (RATE/ACRE)
		ANNUAL GRASSES*				
POAST	1½ pts./A	Wild Proso Millet**	4-10″	Junglerice	3-8″	Oil concentrate (2 pts./A)
		Barnyardgrass	3-8″	Red Sprangletop	3-8″	
		Broadleaf Signalgrass	3-8″	Texas Panicum	3-8″	
		Fall Panicum	3-8″	Witchgrass	3-8″	
		Giant Foxtail	3-8″	Woolly Cupgrass	3-8″	
		Green Foxtail	3-8″	Goosegrass	3-6″	
		Yellow Foxtail	3-8″	Large Crabgrass	3-6″	
		Seedling Johnsongrass	3-8″	Smooth Crabgrass	3-6″	
plus	**plus**	**BROADLEAVES AND SEDGE**				
Basagran	1-2 pts./A according to weed species and size. (See label for Basagran)	Balloonvine		Ladysthumb		(Dash is not recommended) Do not add UAN solution or ammonium sulfate.
		Beggarticks		Pennsylvania Smartweed		
		Bristly Starbur		Prickly Sida or Teaweed		
		Canada Thistle***		Redweed		
		Cocklebur		Shepherdspurse		
		Coffee Senna		Smallflower Morningglory		
		Common Lambsquarters		Spurred Anoda		
		Common Purslane		Tropic Croton		
		Common Ragweed		Velvetleaf		
		Cypressvine Morningglory		Venice Mallow		
		Dayflower		Wild Buckwheat		
		Devilsclaw		Wild Mustard		
		Galinsoga		Wild Poinsettia		
		Giant Ragweed		Wild Sunflower		
		Jimsonweed		Yellow Nutsedge		
plus	**plus**		**LEAF STAGE**	**MAX. HEIGHT**		
Blazer	½-1 pt./A Use ½ pint for pigweed (up to 2″) only; 1 pint if other weeds (at right) are present.	Common Ragweed	Up to 10	6″		
		Black Nightshade	Up to 6	2″		
		Morningglories	Up to 4	4″		
		Crotalaria	Up to 6	6″		
		Sesbania	Up to 4 Pinnate	6″		
		Tall Waterhemp	Up to 6	3″		
		Redroot Pigweed	Up to 6	3″		
		Smooth Pigweed	Up to 6	3″		

* Tank mix does not control rhizome johnsongrass, quackgrass, bermudagrass, wirestem muhly, volunteer corn, shattercane, volunteer cereals, wild oats, red rice or itchgrass.

** For control of wild proso millet only, include POAST® herbicide in tank mix at ¾ pint/A.

*** Requires two applications of Basagran® herbicide in accordance with the label for Basagran.

Note: Tackle may be substituted for the same rate of Blazer® herbicide as recommended above.

Table 4
Basagran + Blazer & POAST Mix—Soybeans
Spot Treatment Application Table

	Concentration in Spray Solution			
	Basagran	Blazer	POAST	Oil Concentrate**
See annual grasses and broadleaves listed in **Table 3.**	1%	1%	1%	1%

Desired Spray Solution Volume	Amount to Be Added To Obtain 1% Solution	
	POAST	**Oil Concentrate**
1 Gallon	1¼ fl. oz.*	1¼ fl. oz.
25 Gallons	1 qt.	1 qt.
50 Gallons	2 qts.	2 qts.
100 Gallons	4 qts.	4 qts.

* 2 Tablespoons = 1 fl. oz.

** Dash is not recommended.

Note: Tackle may be substituted for the same rate of Blazer as recommended above.

POAST + BLAZER TANK MIX* – SOYBEANS GENERAL AND APPLICATION INFORMATION, RESTRICTIONS AND LIMITATIONS

General information

POAST and Blazer® herbicides may be tank mixed for postemergence control of broadleaf and grass weeds. Weeds must be actively growing and at the recommended growth stages.

Separate applications should be made if: a) all weeds to be controlled are not at the correct growth stage for treatment at the same time, or b) grasses to be controlled include rhizome johnsongrass, quackgrass, bermudagrass, wirestem muhly, volunteer corn, shattercane, volunteer cereals, wild oats, red rice or itchgrass. See rate tables on page 16 for POAST and **Table 6, Separate Applications of POAST.**

Water volume and spray pressure

Ground equipment: For tank mix of POAST + Blazer use 20 gallons of total spray solution per acre (broadcast basis) and a minimum of 40 psi pressure. Use standard high pressure hollow cone or flat fan nozzles spaced 20 inches apart. Do not use flood or whirl chamber nozzles.

Air equipment: Use a minimum of 10 gallons of total spray solution per acre.

Mixing

Fill the spray tank half full with water and add the recommended amount of product in the following order—Blazer, oil concentrate, POAST—while the agitator is running. Then add the remaining quantity of water.

Additive

Oil concentrate must be used with this tank mix; Dash is not recommended.

Coverage

Thorough coverage of actively growing weeds is essential. Large crop-and-weed leaf canopies shelter smaller weeds and can prevent adequate spray coverage. Soybeans are tolerant to the tank mix; however, under certain conditions soybeans may burn, crinkle and bronze.

Restrictions and limitations (partial list)

Read and follow the restrictions and limitations on the labels for POAST and Blazer or Tackle herbicides. The most restrictive labeling applies in tank mixes.

TABLE 5
POAST AND BLAZER TANK MIX—SOYBEANS
RATE AND TIME OF APPLICATION

PRODUCT	PRODUCT RATE PER ACRE	WEEDS CONTROLLED AND WEED SIZE				ADDITIVE (RATE/ACRE)
		Annual Grasses*				
	3/4 pt.	Wild Proso Millet	4-10"			
POAST	1 pt.	Broadleaf Signalgrass	3-8"	Giant Foxtail	3-8"	Oil Concentrate (2 pts.) (Dash is not recommended.)
		Fall Panicum	3-8"	Junglerice	3-8"	
				Texas Panicum	3-8"	
	1½ pts.	Barnyardgrass	3-8"	Green Foxtail	3-8"	
		Yellow Foxtail	3-8"	Seedling Johnsongrass	3-8"	
		Red Sprangletop	3-8"	Witchgrass	3-8"	
		Woolly Cupgrass	3-6"	Large Crabgrass	3-6"	
		Goosegrass	3-6"	Smooth Crabgrass	3-6"	
plus	plus	BROADLEAF WEEDS AND VINES				
Blazer	1½-2 pts./A according to weed species and size. (See label for Blazer)	Balloonvine	Jimsonweed	Redroot Pigweed		Do not add UAN or Ammonium sulfate.
		Black Nightshade	Ladysthumb	Showy Crotalaria		
		Bristly Starbur	Lambsquarters	Smallflower Galinsoga		
		Buffalobur	Lanceleaf Groundcherry	Smellmelon		
		Burgherkin	Morningglory, Common (Tall)	Smooth Pigweed		
		Carpetweed	Cypressvine,	Spiny Amaranth		
		Citron	Entireleaf,	Spotted Spurge		
		Cocklebur	Ivyleaf,	Tall Waterhemp		
		Coffee Senna	Pitted (Small White),	Texas Gourd		
		Common Ragweed	Purple Moonflower,	Tropic Croton		
		Cutleaf Groundcherry	Scarlet,	Virgina Copperleaf		
		Devilsclaw	Smallflower,	Volunteer Cowpea		
		Eastern Black Nightshade	Willowleaf (Palmleaf)	Wild Buckwheat		
		Florida Pusley	Palmer Amaranth	Wild Mustard		
		Giant Ragweed	Pennsylvania Smartweed	Wild Poinsettia		
		Hairy Indigo	Poorjoe	Wild Spiny Cucumber		
		Hairy Galinsoga	Prostrate Pigweed	Woolly Croton		
		Hemp Sesbania	Prostrate Spurge			
		Hophornbeam Copperleaf	Purslane			
		PERENNIAL WEEDS				
		Canada Thistle	Field Bindweed	Redvine		
		Climbing Milkweed	Hedge Bindweed	Trumpetcreeper		
		Common Milkweed				

* Tank mixture does not control rhizome johnsongrass, quackgrass, bermudagrass, wirestem muhly, volunteer corn, shattercane, volunteer cereals, wild oats, red rice or itchgrass.

Note: Tackle may be substituted for the same rate of Blazer as recommended above.

SOYBEANS—SEPARATE APPLICATIONS OF POAST, PRECEDED OR FOLLOWED BY BASAGRAN OR BASAGRAN + BLAZER TANK MIX*

Applications of POAST® herbicide can be preceded or followed by Basagran® herbicide and/or Blazer® herbicide to obtain broad spectrum control of weeds listed on the respective product labels (refer to this label and the labels for POAST, Basagran and Basagran + Blazer tank mix). Also refer to these product labels for timing, rate and other information for ground and aerial applications. Make separate applications of Tackle in the same manner as Blazer.

For best results when making separate applications, a minimum period of time is recommended between applications, depending upon their order, according to **Table 6** below.

TABLE 6
SEPARATE APPLICATIONS

Order of Application		Minimum Time Between Applications
First Product(s) Applied	Second Product(s) Applied	
Basagran	POAST	24 hours
Basagran + Blazer*	POAST	7 days
POAST	Blazer or Basagran or Basagran + Blazer*	24 hours
Blazer	POAST	7 days

* Tank mixes not applicable in California.

POAST + 2,4-D LOW VOLATILE ESTER FOR USE AS A BURNDOWN PRIOR TO PLANTING SOYBEANS.

General information

For broad spectrum postemergence weed control a tank mix application of POAST with 2,4-D low volatile ester (LVE) may be made for control of emerged broadleaf and grass weeds before planting soybeans.

This tank mix does not control sedges or provide season-long control of hard-to-kill perennial weeds.

If grasses are larger than indicated **Table 7** then use rate of POAST as recommended in **Soybeans—Annual Grasses—Standard Recommendations,** page 8.

Seeding

Soybeans seed should be planted to a minimum depth of 3/4 inch and thoroughly covered by soil. Soybean injury may be observed if seed placement is too shallow or seeds are not thoroughly covered by soil.

Do not apply this tank mix during or following planting or after soybean emergence; severe soybean injury will result.

Water volume and spray pressure

Ground equipment only: Thorough spray coverage of weed foliage is essential. For broadcast application use standard high pressure pesticide flat fan nozzles. Do not use hollow cone, flood or whirl chamber nozzles, or control droplet applicator (CDA) nozzles. Use a minimum volume of 5 gallons and a maximum volume of 20 gallons of spray solution per acre. When using standard high pressure flat fan nozzles adjust pressure to a minimum of 40 psi and a maximum of 60 psi at the nozzle.

Additive

Dash or oil concentrate must be used with this tank mix.

Mixing

Fill tank of a thoroughly clean sprayer one-half to two-thirds full with clean water. Start agitation and add oil concentrate; allow to mix thoroughly. Add POAST, then 2,4-D (LVE), then the remaining volume of water. Maintain constant agitation during application.

Selection of 2,4-D (LVE) formulation

Use only low volatile ester formulations of 2,4-D such as 2,4-D isooctyl ester. Note that the recommended rate of 2,4-D is calculated on an acid equivalent basis. Make adjustments for the concentration of the 2,4-D formulation used. Since the exact composition of suitable products will vary, it is advised to conduct a **Jar Test for Estimating Suitability of Oil Concentrates and 2,4-D (LVE)** with each 2,4-D (LVE) formulation used. See this procedure on page 5.

Restrictions and limitations (partial list)

Do not plant soybeans until 3 months after treatment or until the 2,4-D (LVE) has disappeared from the soil.

Do not apply if rainfall is expected within 6 hours following application, as weed control will probably be unsatisfactory.

Since all crops such as sorghum, corn, small grains, cotton, soybeans, rice, sugar beets, trees, shrubs, as well as ornamental grasses such as turf, are extremely susceptible to POAST + 2,4-D (LVE) tank mix, avoid all direct or indirect postemergence contact with any desired plant.

Do not spray if the wind is blowing toward desired plants, or at any time when the wind exceeds 6 miles per hour (refer to 2,4-D (LVE) label).

Observe all restrictions and limitations specified on labels for 2,4-D (LVE) and POAST. The most restrictive labeling applies in tank mixes.

TABLE 7
POAST + 2,4-D (LVE)—SOYBEANS
PREPLANT BURNDOWN RATE AND TIME OF APPLICATION TABLE

WEED SPECIES		TIME OF APPLICATION	RATE OF POAST PER ACRE	DASH OR OIL CONCENTRATE PER ACRE	2,4-D (A.E.) RATE PER ACRE**	UAN SOLUTION OR AMMONIUM SULFATE
GRASSES						
Wild Proso Millet		Up to 4"				
Barnyardgrass Broadleaf Signalgrass Fall Panicum Foxtails: Giant, Green, Yellow Johnsongrass, Seedling	Witchgrass Woolly Cupgrass Large Crabgrass Smooth Crabgrass	Up to 3"				
BROADLEAVES						
Pennsylvania Smartweed		Up to 2"				
Field Bindweed* Wild Buckwheat*		Vine Length Up to 6"	½ pt.	2 pt.	½ lb	½-1 gal. UAN or 2½ lbs. AMS
Canada Thistle* Common Cocklebur Common Dandelion Common Lambsquarters Common Ragweed Field Pennycress Giant Ragweed Horseweed	Prickly Lettuce Redroot Pigweed Shepherdspurse Velvetleaf White Cockle* Wild Mustard Yellow Rocket	Up to 10"				

* Control may be partial or inconsistent.
** A.E. rate based on 2,4-D acid equivalent. See section entitled **Selection of 2,4-D (LVE) formulation.**

RECOMMENDATIONS FOR USE IN ALFALFA, COTTON, FLAX, PEANUTS, SUGAR BEETS, AND SUNFLOWERS.

General use recommendations—all regions

- Apply to actively growing grasses at the sizes indicated.
- Always follow recommendations given in **Application information section.**
- Always adjust spray pressure, spray volume and height of spray boom to ensure penetration of plant canopy and thorough coverage of grasses to be controlled.
- Do not apply to drought-stressed grass or grass which has gone through an extended dry period.
- In irrigated areas it may be necessary to irrigate prior to treatment with POAST to ensure weeds are growing actively.
- **Always add 2 pints Dash or oil concentrate per acre.**

Regional use map

All recommendations are based on growing region. Refer to the map below. Follow the recommendations for grass control for your region only.
Midwest, South and Northeast (see pages 20-27)
High and Rolling Plains of Texas, Oklahoma and Eastern New Mexico (see pages 28-33)
Western and Mountain States (see pages 34-39)

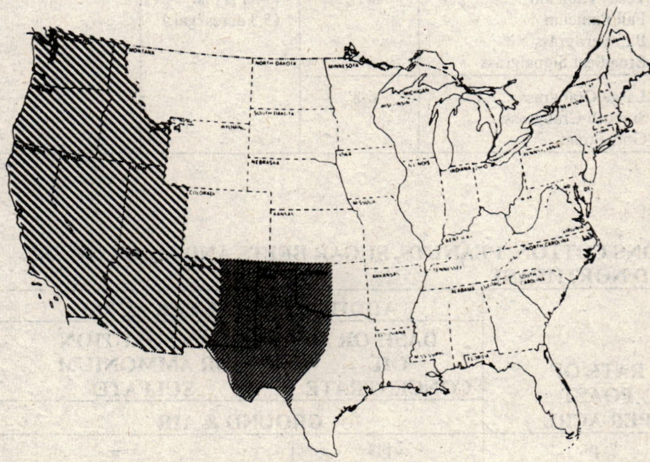

Restrictions and limitations—all regions

Alfalfa

- Do not apply POAST® herbicide within 7 days of grazing, feeding, or cutting for (undried) forage, or within 20 days of cutting alfalfa for (dry) hay.
- Do not apply more than a total of 5 pints of POAST per acre in one season.
- POAST may be applied to alfalfa by ground equipment only.
- Do not apply POAST and 2,4-DB as a tank mix unless the 60-day feeding, grazing and harvesting restrictions on the 2,4-DB label can be observed.

Cotton

- Do not apply POAST within 40 days of harvest.
- Do not apply more than a total of 7½ pints of POAST per acre in one season.
- Do not graze treated cotton fields and do not feed treated forage to livestock.

Flax

- Do not apply POAST within 75 days of harvest.
- Do not apply more than a total of 4 pints per acre in one season to flax.
- Under weather conditions of high humidity and/or temperature, some flowers may be damaged or dislodged by applications of POAST and oil concentrate.
- Do not graze or feed treated flax forage to livestock.
- Read and follow restrictions and limitations on bromoxynil or MCPA labels when tank mixing with POAST.
- Do not apply POAST by air equipment to flax.

Peanuts

- Do not apply POAST within 40 days of harvest.
- Do not apply more than a total of 2½ pints of POAST per acre in one season.
- Do not feed treated peanut forage or hay to livestock.

Sugar Beets

- Do not apply more than a total of 5 pints of POAST per acre in one season.
- If sugar beet tops are to be fed to livestock do not make more than two (2) applications of POAST, and do not make the last application sooner than 100 days before harvest.

Sunflower

- Do not apply POAST within 70 days of harvest.
- Do not apply more than a total of 2½ pints of POAST per acre in one season.
- Do not feed treated sunflower forage to livestock.
- Commercially released varieties of sunflower are tolerant to POAST at all stages of growth; however, leaf speckling has been occasionally observed on sunflower with no corresponding reduction in vigor or growth. POAST is not recommended for use on sunflower inbred lines grown for seed because crop safety of these lines has not been adequately established.

MIDWEST, SOUTH, NORTHEAST—RECOMMENDATIONS FOR USE

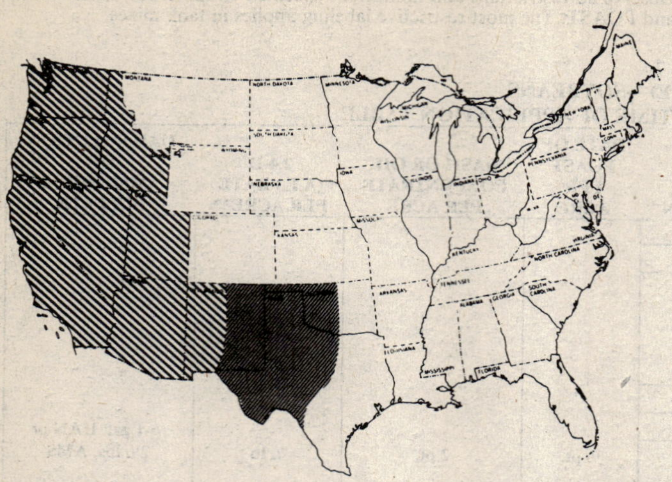

ROW CROPS—ANNUAL GRASSES—SPECIAL RATE FOR EARLY TREATMENT COTTON, PEANUTS, SUGAR BEETS, SUNFLOWERS MIDWEST, SOUTH AND NORTHEAST

GROUP	GRASS	TIME OF APPLICATION	RATE OF POAST PER ACRE	ADDITIVES (RATE/ACRE) DASH OR OIL CONCENTRATE (GROUND & AIR)		ADDITIVES (RATE/ACRE) UAN SOLUTION OR AMMONIUM SULFATE (GROUND & AIR)
A	Wild Proso Millet	4-10"	½ pt. (16 acres/gal.)	2 pts.		—
B	Goosegrass	1-3"	¾ pt. (10.7 acres/gal.)	2 pts.		—
	Barnyardgrass** (Midwest only) Broadleaf Signalgrass Fall Panicum Texas Panicum Foxtails: Giant, Green	1-4"				
	Volunteer Corn	1-12"		2 pts.	plus	Always add ½-gal. UAN or 2½ lbs. AMS

* For broad spectrum control of annual grasses in Groups A & B, use ¾ pint of POAST per acre. If additional applications are needed, apply at the same rate and at the recommended stage of growth.

** In the following states, for barnyardgrass use 1 pint per acre as recommended in *Standard recommendations* table: AL, AR, FL, GA, LA, MS, NC, SC, TN, TX, VA.

Rescue treatment for controlling selected annual grasses in row crops

For best results, always apply POAST to annual grasses at the growth stage and rate specified in the above table. However, if POAST cannot be applied at the recommended time, larger annual grasses can be controlled with a later application by increasing the rate of POAST. Apply to actively growing grasses at the rates and sizes indicated.

ROW CROPS—ANNUAL GRASSES—RESCUE TREATMENT

Grass	Time of Application (Weed Size)	Rate of POAST per Acre	Dash or Oil Concentrate Rate/Acre Ground & Air
Wild Proso Millet	10-24"	1 pt. (8 acres/gal.)	2 pts.
Foxtails: Giant, Green, Yellow Johnsongrass, Seedling	8-16"	1½ pts. (5.3 acres/gal.)	2 pts.
Texas Panicum Fall Panicum Barnyardgrass Broadleaf Signalgrass	8-12"		
Large Crabgrass Smooth Crabgrass Goosegrass	6-8"		

ROW CROPS—ANNUAL GRASSES—STANDARD RECOMMENDATIONS COTTON, PEANUTS, SUGAR BEETS AND SUNFLOWERS MIDWEST, SOUTH AND NORTHEAST

GROUP	GRASS	TIME OF APPLICATION	RATE OF POAST PER ACRE	ADDITIVES (RATE PER ACRE) DASH OR OIL CONCENTRATE (GROUND & AIR)		ADDITIVES (RATE PER ACRE) UAN SOLUTION OR AMMONIUM SULFATE (GROUND & AIR)
A	Wild Proso Millet	4-10"	½ pt. (16 acres/gal.)	2 pts.		—
B	Wild Oats	Up to 4"		2 pts.	plus	½-1 gal. UAN or 2½ lbs. AMS is recommended
	Goosegrass Smooth Crabgrass	Up to 6"		2 pts.		—

GROUP	GRASS	TIME OF APPLICATION	RATE OF POAST PER ACRE	ADDITIVES (RATE PER ACRE) DASH OR OIL CONCENTRATE		UAN SOLUTION OR AMMONIUM SULFATE
				GROUND & AIR		
	Large Crabgrass	Up to 6"	1 pt. (8 acres/gal.)	2 pts.	plus	½-1 gal. UAN or 2½ lbs. AMS is recommended
	Barnyardgrass, Broadleaf Signalgrass, Browntop Panicum, Fall Panicum Foxtails: Giant, Green, Yellow, Johnsongrass, Seedling, Junglerice, Red Sprangletop, Ryegrass, Annual, Texas Panicum, Witchgrass, Woolly Cupgrass	Up to 8"		2 pts.		—
	Shattercane/Wildcane If needed, retreat at the same rate and stage of growth.	6-18"		2 pts.		—
	Volunteer Corn Maintain sufficient boom height above volunteer corn plants for best spray coverage.	Up to 20"		2 pts.	plus	Always add ½-1 gal. UAN or 2½ lbs. AMS
C	Field Sandbur (Midwest only)	Up to 3"	1¼ pts. (6.4 acres/gal.)	2 pts.		—
D	Volunteer Cereals Barley, Oats, Rye, Wheat Not recommended for spring control of volunteer cereals that emerged the previous fall.	Before tillering, Up to 4" and prior to over-wintering	1½ pts. UAN (5.3 acres/gal.)	2 pts.	plus	½-1 gal. UAN or 2½ lbs. AMS is recommended
E	Itchgrass, Red Rice	2-4"	2 pts. (4 acres/gal.)	2 pts.		—

* For broad spectrum control of annual grasses in Group B (above), use 1 pint of POAST® herbicide per acre. When weed populations include additional grasses in Group C, D or E, increase the rate of POAST as indicated. If later flushes of annual grasses emerge after the first application, make additional applications at the same rate and at the same recommended stage of growth.

ROW CROPS—PERENNIAL GRASSES COTTON, PEANUTS, SUGAR BEETS, SUNFLOWERS MIDWEST, SOUTH, AND NORTHEAST

GRASS	TIME OF APPLICATION	RATE OF POAST PER ACRE	ADDITIVES (RATE/ACRE) DASH OR OIL CONCENTRATE		UAN SOLUTION OR AMMONIUM SULFATE
			GROUND & AIR		
Bermudagrass • First Application	Before stolon length exceeds 6".	1½ pts. (5.3 acres./gal.)	2 pts.		—
• Second Application If regrowth occurs or new plants emerge.	1-4" length of new plants or growth.	1 pt. (8 acres/gal.)	2 pts.		—
Johnsongrass, Rhizome • First Application Use 5-10 gallons of spray solution per acre. Maintain a ground speed of no more than 6 miles per hour. For best results rhizomes should be thoroughly fragmented (less than 6"). (When using 11-20 gallons of spray solution per acre use 1½ pints of POAST)	15-25" (15-20" in no-till culture)	1 pt. (8 acres/gal.)	2 pts.		See page 5
• Second Application When regrowth occurs or new plants emerge.	6-12"	1 pt.	2 pts.		See page 5
Quackgrass • First Application For best results, rhizomes should be thoroughly fragmented (less than 6").	6-8"	1½ pts. (5.3 acres/gal.)	2 pts.	plus	Always add ½ gal. UAN or 2½ lbs. AMS
• Second Application If regrowth occurs or new plants emerge. Depending upon environmental conditions and crop cultural system, season-long control may not always be obtained. However, competition of quackgrass with the crop will be reduced. **Note:** In conventional wide row crops, a cultivation no sooner than 14 days after application but within 21 days of application will aid in control.	6-8"	1 pt. (8 acres/gal.)	2 pts.	plus	Always add ½-1 gal. UAN or 2½ lbs. AMS
Wirestem Muhly If regrowth occurs, retreat at the same rate and stage of growth.	Up to 6"	1¼ pts. (6.4 acres/gal.)	2 pts.		—

USE RECOMMENDATIONS FOR POAST IN ALFALFA—MIDWEST, SOUTH AND NORTHEAST

POAST may be applied to seedling or established alfalfa grown for hay, silage, green chop, direct grazing or for seed. See the **Restrictions and limitations** on page 19 for the minimum length of time between application and harvest.

The effectiveness of POAST is dependent on the absorption and movement throughout the plant. For this to occur there must be enough leaf surface area to absorb the herbicide and the grass must be actively growing to move or translocate POAST to the roots and buds. Any stress conditions that slow the growth of the grass may decrease control or reduce the speed of control. These stress conditions include mowing, lack of moisture, herbicide injury, mechanical injury or cold temperatures.

Mowing

Best control of annual grasses can be achieved if applications are made before grass weeds are mowed. Once a grass is mowed it becomes tougher to control. This is because much of the leaf surface may be removed by the mowing and the grass is put under stress. In areas without a killing frost, some annuals can over-winter after having been mowed a number of times. These grasses can form large crowns which contain many viable buds. A large crown, even if it is an annual grass, may require repeated applications of POAST for control and depending on the species and size of the crown, control may not be complete.

Irrigated alfalfa

Irrigation practices can be very critical to the successful use of POAST. This is because an irrigation may be necessary to start the grass weeds growing again. Generally, applications 2-4 days after an irrigation are most effective. This is because: 1) grasses resume active growth, 2) grasses have less chance to grow too large, 3) by waiting later, the alfalfa begins to canopy and interferes with spray coverage. Irrigation shortly (2 days) after application has been effective, but more consistent grass control is obtained when the irrigation is made before the application. In large fields it may take several days for irrigation equipment to be moved across a field. Grasses must not be allowed to grow too large on the part of the field which is to be irrigated first. In these situations the field should be irrigated, then sprayed in segments, to obtain best results.

Annual grass control

Apply POAST at the grass size and rate indicated in the following tables. If a grass has been cut, apply POAST after the regrowth reaches the minimum height (so there will be enough leaf area to absorb POAST) and before it exceeds the maximum height indicated in the table. Applications should be made before alfalfa canopies over the grasses and interferes with spray coverage. Also, applications made after an alfalfa cutting may need to be timed to follow an irrigation or a rainfall which will allow the grasses to regrow to a treatable size.

Some annual grasses are spring and summer germinating, while others are fall germinating, and the time they are actively growing and most susceptible to POAST may vary from area to area. Also some annuals germinate over a long period of time, and since control of small grasses is desired, applications after each weed flush may be needed. As a general guideline, spray spring and summer germinating grasses as early in the season as possible. Optimum application timing may occur very early in the spring after initial green-up. Spray fall germinating weeds in the fall soon after they begin growing but before any killing frosts. This is because the weeds are more susceptible to POAST when they begin growth in the fall and control is more complete. Late fall applications may be less effective due to environmental changes, such as frosts, or due to the onset of flowering.

Inter-seeded oats

Oats inter-seeded with alfalfa may be killed back with an application of POAST. Their removal allows the seedling alfalfa to grow with less competition. This application should be made before the oats get too large. Applications made in the boot stage or later will not be as effective as when an application is made on young oats.

Perennial grass control

POAST effectively controls or suppresses perennial grasses such as Bermudagrass, Johnsongrass, quackgrass, wirestem muhly and perennial ryegrass. However, their growth characteristics are such that they are more difficult to control than annual grasses, especially in a perennial crop such as established alfalfa. A program consisting of repeated applications is usually necessary for best results.

The most economical way of controlling perennial grasses is to do so in the year of stand establishment before rhizomes or stolons become large and difficult to kill. The field should be disked before seeding to thoroughly fragment rhizomes or stolons.

In summer and fall seedings, cool season grasses (quackgrass, wirestem muhly, perennial ryegrass) can become very competitive under cool fall conditions. Fall applications of POAST will reduce late season grass growth and limit the ability of grasses to accumulate nutrient reserves in roots and rhizomes.

In established stands it is important to begin applications in the spring when conditions favor active growth and before storage tissues have increased their nutrient reserves. Additional applications should be made on any grass regrowth in later cuttings.

FORAGE CROPS—ANNUAL GRASSES—STANDARD RECOMMENDATIONS* ALFALFA MIDWEST, SOUTH AND NORTHEAST

				ADDITIVES (RATE PER ACRE)	
				DASH OR OIL CONCENTRATE	UAN SOLUTION OR AMMONIUM SULFATE
GROUP	GRASS	TIME OF APPLICATION	RATE OF POAST PER ACRE	GROUND	
A	Wild Proso Millet	4-10"	1/2 pt. (16 acres/gal.)	2 pts.	—
B	Inter-seeded Oats (Tame Oats)	2-8"	3/4 pt. (10.6 acres/gal.)	2 pts.	—
C	Goosegrass Smooth Crabgrass	2-4"	1 pt. (8 acres/gal.)	2 pts.	—
	Large Crabgrass Wild Oats	2-4"		2 pts.	plus 1/2-1 gal. UAN or 2 1/2 lbs. AMS is recommended.
	Barnyardgrass, Broadleaf Signalgrass, Browntop Panicum, Fall Panicum, Foxtails: Giant, Green, Yellow, Johnsongrass, Seedling, Junglerice, Red Sprangletop, Ryegrass, Annual, Texas Panicum, Witchgrass, Woolly Cupgrass	3-8"		2 pts.	—
	Inter-seeded Oats (Tame Oats)	4-10"		2 pts.	—
	Shattercane/Wildcane If needed, retreat at the same rate and stage of growth.	6-18"		2 pts.	—
	Volunteer Corn Maintain sufficient boom height above volunteer corn plants for best spray coverage.	6-20"		2 pts.	plus Always add 1/2-1 gal. UAN or 2 1/2 lbs. AMS.
D	Field Sandbur (Midwest only)	1-3"	1 1/2 pts. (5.3 acres/gal.)	2 pts.	—
	Volunteer Cereals Barley Oats Rye Wheat Not recommended for spring control of volunteer cereals that emerged the previous fall.	Before tillering, 2-4" and prior to over-wintering		2 pts.	plus 1/2-1 gal. UAN or 2 1/2 lbs. AMS. is recommended
E	Itchgrass Red Rice	2-4"	2 pts. (4 acres/gal.)	2 pts.	—

* For broad spectrum control of annual grasses in Group C (above), use 1 pint of POAST per acre. When weed populations include additional grasses in Group D or E, increase the rate of POAST as indicated. If later flushes of annual grasses emerge after the first application, make additional applications at the same rate and at the same recommended stage of growth.

FORAGE CROPS—PERENNIAL GRASSES—STANDARD RECOMMENDATIONS
ALFALFA
MIDWEST, SOUTH, AND NORTHEAST

GRASS	TIME OF APPLICATION	RATE OF POAST PER ACRE	ADDITIVES (RATE/ACRE) DASH OR OIL CONCENTRATE (GROUND)		ADDITIVES (RATE/ACRE) UAN SOLUTION OR AMMONIUM SULFATE (GROUND)
Bermudagrass • First Application	Before stolon length is 3-6".	2½ pts. (3.2 acres./gal.)	2 pts.		—
• Second Application If regrowth occurs or new plants emerge.	1-4" length of regrowth or new plants.	2½ pts.	2 pts.		—
Johnsongrass, Rhizome • First Application In established alfalfa, control may be partial or inconsistent; johnsongrass growth will be suppressed.	15-25"	2½ pts.	2 pts.		See page 5
• Second Application If regrowth occurs or new plants emerge.	6-12"	2½ pts.	2 pts.		See page 5
Quackgrass • First Application In established alfalfa, control may be partial or inconsistent; quackgrass growth will be suppressed.	6-8"	2½ pts.	2 pts.	plus	Always add ½ gal. UAN or 2½ lbs. AMS.
• Second Application If regrowth occurs or new plants emerge.	6-8"	2½ pts.	2 pts.	plus	Always add ½ gal. UAN or 2½ lbs. AMS.
Ryegrass, Perennial If regrowth occurs, retreat at the same rate and stage of growth.	3-8"	2 pts. (4 acres/gal.)	2 pts.		—
Wirestem Muhly If regrowth occurs, retreat at the same rate and stage of growth.	3-6"	1¼ pts. (6.4 acres/gal.)	2 pts.		—

Tank mix of POAST® herbicide with 2,4-DB for grass and broadleaf weed control in alfalfa

Use a tank mix of POAST +2,4-DB for the control of mixed populations of grasses and broadleaf weeds listed as susceptible on the two product labels. Prepare the tank mixture by adding 2,4-DB to half the final volume of water with the agitator running. Then add oil concentrate and then POAST; bring mixture to final volume. Agitation must be continuous from time of mixing through spraying. Include 2,4-DB as a tank mix with POAST according to rates recommended on the individual product labels. 2,4-DB formulations include Buxtone® herbicide (1.75 lb./gal.); Butoxone Ester (2.0 lb./gal.): Butyrac® 200 herbicide (2.0 lb./gal.); Butyrac Ester (2.0 lb./gal.).

Some leaf yellowing and burning of the alfalfa may occur with this tank mixture. Use of 2,4-DB ester formulations may increase the severity of leaf injury. Additionally, in established alfalfa, 2,4-DB alone may cause twisting of stems and malformation of leaves. (Refer to 2,4-DB label.) Alfalfa plants will generally outgrow these temporary leaf injuries.

Restrictions and limitations (partial list)

Observe all cautions and limitations on the label of both products. The most restrictive labeling applies to tank mixes.

Do not apply POAST and 2,4-DB as a tank mix unless all feeding, grazing and harvesting restrictions on the 2,4-DB label can be observed.

Do not add Dash. UAN solution or ammonium sulfate to a POAST plus 2,4-DB tank mixture.

USE RECOMMENDATIONS FOR FLAX

Flax competes poorly with weeds. It is important to control grass weeds before the flax stand is reduced and the crop vigor suffers. Where flax stands are poor or when flax is growing slowly, germination of new grass, following POAST application, may occur. If additional applications are needed apply at the same rate and recommended time of application. Apply POAST to actively growing grasses at the sizes indicated in the following tables.

FLAX—ANNUAL GRASSES—STANDARD RECOMMENDATIONS*

GROUP	GRASS	TIME OF APPLICATION	RATE OF POAST PER ACRE	ADDITIVES (RATE/ACRE) DASH OR OIL CONCENTRATE RATE/ACRE		ADDITIVES (RATE/ACRE) UAN SOLUTION OR AMMONIUM SULFATE
A	Wild Proso Millet	4-10"	½ pt.	2 pts.		—
B	Barnyardgrass** Fall Panicum Foxtails: Giant, Green, Yellow Witchgrass Woolly Cupgrass	Up to 4"	1 pt.	2 pts.		—
	Wild Oats	Up to 4"	1 pt.	2 pts.	plus	½-1 gal. UAN or 2½ lbs. AMS is recommended.
	Shattercane/Wildcane If needed, retreat at the same rate and stage of growth.	2-8"	1 pt.	2 pts.		—

GROUP	GRASS	TIME OF APPLICATION	RATE OF POAST PER ACRE	ADDITIVES (RATE/ACRE) DASH OR OIL CONCENTRATE RATE/ACRE	ADDITIVES (RATE/ACRE) UAN SOLUTION OR AMMONIUM SULFATE
	Volunteer Corn	2-8″	1 pt.	2 pts.	—
C	Volunteer Cereals* Barley Rye Oats Wheat POAST is not recommended for spring control of volunteer cereals that emerged in the previous fall.	Before tillering, up to over 6″ and prior to over-wintering	1½ pts.	2 pts.	plus ½-1 gal. UAN Solution or 2½ lbs. AMS is recommended

* For broad spectrum control of annual grasses in Groups A & B, use 1 pint of POAST per acre. When weed populations include additional grasses in Group C increase the rate of POAST to 1½ pints. If a second flush of annual grasses emerges after the first application, make an additional application at the same rate and at the recommended stage of growth.

Rescue treatment for controlling selected annual grasses

For best results, always apply POAST to annual grasses at the growth stage and rate specified in the above table. However, if POAST cannot be applied at the recommended time, larger annual grasses can be controlled with a later application by increasing the rate of POAST. Apply to actively growing grasses at the rates and sizes indicated below.

FLAX—ANNUAL GRASSES-RESCUE TREATMENT

Grass	Time of Application	Rate of POAST per Acre	Oil Concentrate Rate/Acre
Foxtails: Giant, Green, Yellow Barnyardgrass Wild Oats	Up to 8″	1½ pts.	2 pts.

TANK MIXTURES FOR FLAX

Tank mix of POAST® herbicide with bromoxynil and MCPA for grass and broadleaf weed control

Use a tank mix of POAST plus MCPA or POAST plus bromoxynil for the control of mixed populations of grasses and broadleaf weeds listed as susceptible on the respective product labels. Prepare the tank mixture by adding water soluble forms of herbicides (such as MCPA amine) to half the final water volume. Then add oil concentrate (not Dash) then POAST, then emulsifiable herbicides (such as bromoxynil esters) and bring the mixture to the final volume. Agitation must be continuous from the time of mixing through spraying. Include bromoxynil or MCPA with POAST according to the rates recommended on the respective product labels, up to a maximum of 0.25 lb. bromoxynil equivalent per acre or up to a maximum of 0.25 lb. MCPA acid equivalent per acre.

Do not delay spraying broadleaf weeds even though grassy weeds are not in correct stage for treatment. Bromoxynil or MCPA applied with POAST may cause leaf burn, retarded growth and delayed maturity of the crop. Some reduced grass control may be experienced with the above tank mixtures.

Do not add Dash, ammonium sulfate or UAN solution to a tank mixture of POAST plus bromoxynil or POAST plus MCPA.

Follow all restrictions detailed on the MCPA or bromoxynil labels that apply to use in flax. The most restrictive labeling must apply to a tank mix.

HIGH AND ROLLING PLAINS OF TEXAS, OKLAHOMA AND EASTERN NEW MEXICO RECOMMENDATIONS FOR USE

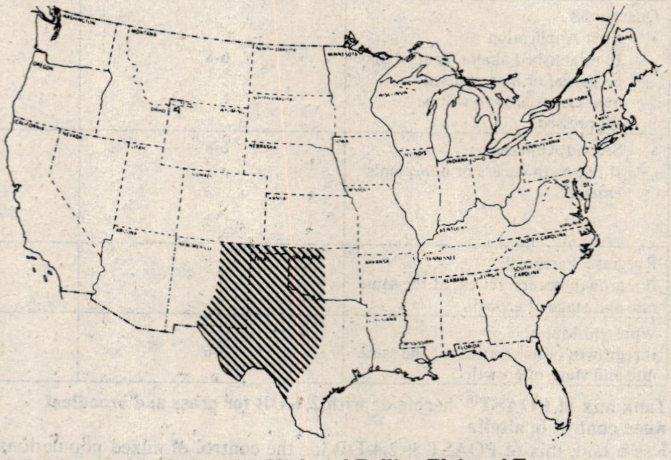

Special recommendations for High and Rolling Plains of Texas, Oklahoma, and Eastern New Mexico

POAST works best when applied to actively growing grasses. In nonirrigated areas of High and Rolling Plains, grasses often are stressed when limited rainfall is followed by periods of hot windy weather. In these rainfed areas the grasses may have gone through several periods of drought and growth, and even if they are actively growing, they will require the higher rate of POAST than is recommended for other parts of Texas and Oklahoma. For Western New Mexico see **Western and mountain states** starting on page 34.

ROW CROPS—ANNUAL GRASSES—STANDARD RECOMMENDATIONS
COTTON, PEANUTS, SUGAR BEETS AND SUNFLOWERS
HIGH AND ROLLING PLAINS OF TEXAS, OKLAHOMA AND EASTERN NEW MEXICO

GROUP	GRASS	TIME OF APPLICATION	RATE OF POAST PER ACRE	ADDITIVES (RATE PER ACRE) DASH OR OIL CONCENTRATE — GROUND & AIR	ADDITIVES (RATE PER ACRE) UAN SOLUTION OR AMMONIUM SULFATE — GROUND & AIR
A	Goosegrass Smooth Crabgrass	Up to 4″		2 pts.	—
	Large Crabgrass	Up to 4″		2 pts	plus ½-1 gal UAN or 2½ lbs. AMS is recommended
	Barnyardgrass Broadleaf Signalgrass Browntop Panicum Fall Panicum Foxtails: Giant, Green, Yellow Johnsongrass, Seedling Junglerice	Up to 8″	1½ pts. (5.3 acres/gal.)	2 pts.	—

GROUP	GRASS	TIME OF APPLICATION	RATE OF POAST PER ACRE	ADDITIVES (RATE PER ACRE) DASH OR OIL CONCENTRATE	UAN SOLUTION OR AMMONIUM SULFATE
				GROUND & AIR	
	Red Sprangletop Texas Panicum Witchgrass				
	Shattercane/Wildcane If needed, retreat at the same rate and stage of growth.	6-18″		2 pts.	—
	Volunteer Corn Maintain sufficient boom height above volunteer corn plants for best spray coverage.	Up to 20″		2 pts. plus	Always add ½-1 gal. UAN or 2½ lbs. AMS
B	Volunteer Cereals Barley Oats Rye Wheat Not recommended for spring control of volunteer cereals that emerged the previous fall.	Before tillering, up to 4″ and prior to over-wintering	2 pts. (4 acres/gal.)	2 pts. plus	½-1 gal. UAN or 2½ lbs. AMS is recommended

* For broad spectrum control of annual grasses in Group A (above), use 1½ pints of POAST® herbicide per acre. When weed populations include additional grasses in Group B, increase the rate of POAST as indicated. If later flushes of annual grasses emerge after the first application, make additional applications at the same rate and at the same recommended stage of growth.

ROW CROPS—PERENNIAL GRASSES
COTTON, PEANUTS, SUGAR BEETS, AND SUNFLOWERS
HIGH AND ROLLING PLAINS OF TEXAS, OKLAHOMA AND EASTERN NEW MEXICO

GRASS	TIME OF APPLICATION	RATE OF POAST PER ACRE	DASH OR OIL CONCENTRATE (RATE/ACRE) GROUND & AIR
Bermudagrass • First Application (In peanuts & sunflowers season-long control may not be obtained.)	6″ of stolon length	2 pts. (4 acres/gallon) (Use 1½ pts. in peanuts and sunflowers.)	2 pts.
• Second Application	21 days after first application	1½ pts. (5.3 acres/gal.) (Use 1 pt. in peanuts and sunflowers.)	2 pts.
• Third Application (Do not make a third application in peanuts & sunflowers.)	1-4″ length of regrowth or new plants	1½ pts.	2 pts.
Johnsongrass, Rhizome For best results, rhizomes should be thoroughly fragmented (less than 6″). Adjust volume of spray solution to a maximum of 10 gallons and a minimum of 5 gallons per acre while maintaining a ground speed of no more than 6 miles per hour. • First Application	6-10″ (5.3 acres/gal.)	1½ pts.	2 pts.
• Subsequent applications when regrowth occurs or new plants emerge.	4-8″	1 pt. (8 acres/gal.)	2 pts.

USE RECOMMENDATIONS FOR POAST IN ALFALFA—HIGH AND ROLLING PLAINS OF TEXAS, OKLAHOMA AND EASTERN NEW MEXICO

POAST® herbicide may be applied to seedling or established alfalfa grown for hay, silage, green chop, direct grazing or for seed. See **Restrictions and limitations** on page 19 for the minimum length of time between application and harvest.

The effectiveness of POAST is dependent on the absorption and movement throughout the plant. For this to occur there must be enough leaf surface area to absorb the herbicide and the grass must be actively growing in order to translocate POAST to the roots and buds. Any stress conditions that slow the growth of the grass may decrease control or reduce the speed of control. These stress conditions include mowing, lack of moisture, herbicide injury, mechanical injury or cold temperatures.

Mowing

Best control of annual grasses can be achieved if an application of POAST is made before grass weeds are mowed off. Once a grass is mowed it becomes more difficult to control. Because much of the leaf surface may be removed by the mowing and the grass is stressed. In areas without a killing frost some annuals can over-winter after having been mowed a number of times. These grasses can form large crowns which contain many viable buds. A large crown, even if it is an annual grass, may require repeated applications of POAST for control and depending on the species and size of the crown, control may not be complete.

Irrigation

Irrigation practices can be very critical to the successful use of POAST. This is because an irrigation may be necessary to start the grass weeds growing again. Generally, applications 2-4 days after an irrigation are most effective. This is because: 1) grasses resume active growth, 2) they have less chance to grow too large, 3) by waiting later, the alfalfa begins to canopy and interferes with spray coverage. Irrigations shortly (2 days) after a POAST application have been effective, but more consistent grass control is obtained when the irrigation was made before the application.

In large fields it may take several days for water to move across a field; and the grasses should not be allowed to grow too large on the side of the field irrigated first while waiting for the entire field to be irrigated. In these situations the field should be sprayed in segments, following the irrigation, to obtain best results.

Annual grass control

Apply POAST at the grass size and rate indicated in the following tables. If a grass has been cut, apply POAST after the regrowth reaches the minimum height (so there will be enough leaf area to absorb POAST) and before it exceeds the maximum height indicated in the table. Applications should be made before alfalfa canopies over the grasses and interferes with spray coverage. Also, applications made after an alfalfa cutting may need to be timed to follow an irrigation or a rainfall which will allow the grasses to regrow to a treatable size.

Some annual grasses are spring and summer germinating, while others are fall germinating, and the time they are actively growing and most susceptible to POAST may vary from area to area. Also some annuals germinate over a long period of time, and since control of small grasses is desired, applications after each weed flush may be needed. As a general guideline spray spring and summer germinating grasses as early in the season as possible. Optimum application timing may occur very early in the spring at initial green-up. Spray fall germinating weeds in the fall soon after they begin growing but before any killing frosts. This is because they are more susceptible when they begin growth in the fall and control is more complete. Late fall applications may be less effective due to environmental changes, such as frosts, or due to the onset of flowering.

Perennial grass control

POAST effectively controls or suppresses perennial grasses such as Bermudagrass, Johnsongrass, quackgrass, wirestem muhly and perennial ryegrass. However, their growth characteristics are such that they are more difficult to control than annual grasses especially in a perennial crop such as established alfalfa. A planned program consisting of repeated applications is usually necessary for best results.

The most economical way of controlling perennial grasses is to do so in the year of stand establishment, before rhizomes or stolons become large and difficult to kill. The field should be thoroughly disked before seeding to cut the rhizomes or stolons as small as possible.

In established stands it is important to begin applications in the spring when conditions favor active growth and before storage tissues have increased their reserves. Additional applications should be made on grass regrowth in later cuttings.

FORAGE CROPS—ANNUAL GRASSES—STANDARD RECOMMENDATIONS*
ALFALFA
HIGH AND ROLLING PLAINS OF TEXAS, OKLAHOMA AND EASTERN NEW MEXICO

				ADDITIVES (RATE/ACRE)	
				DASH OR OIL CONCENTRATE	UAN SOLUTION OR AMMONIUM SULFATE
GROUP	GRASS	TIME OF APPLICATION	RATE OF POAST PER ACRE	GROUND	
A	Goosegrass, Smooth Crabgrass	2-4″		2 pts.	—
	Large Crabgrass	2-4″		2 pts.	plus ½-1 gal. UAN or 2½ lbs. AMS is recommended
	Barnyardgrass, Broadleaf Signalgrass, Browntop Panicum, Fall Panicum Foxtails: Giant, Green, Yellow Johnsongrass, Seedling, Junglerice, Red Sprangletop, Texas Panicum, Witchgrass	3-8″	1½ pts. (5.3 acres/gal.)	2 pts.	—
	Shattercane/Wildcane If needed, retreat at the same rate and stage of growth.	6-18″		2 pts.	—
	Volunteer Corn, Maintain sufficient boom height above volunteer corn plants for best spray coverage.	6-20″		2 pts.	plus Always add ½-1 gal. UAN or 2½ lbs. AMS
B	Volunteer Cereals Barley, Oats, Rye, Wheat Not recommended for spring control of volunteer cereals that emerged the previous fall.	Before tillering, 2-4″ and prior to over-wintering.	2 pts. (4 acres/gal.)	2 pts.	plus ½-1 gal.UAN or 2½ lbs. AMS is recommended.

* For broad spectrum control of annual grasses in Group A (above), use 1½ pints of POAST per acre. When weed populations include grasses in Group B, increase the rate of POAST as indicated. If later flushes of annual grasses emerge after the first application, make additional applications at the same rate and at the same recommended stage of growth.

FORAGE CROPS—PERENNIAL GRASSES
ALFALFA
HIGH AND ROLLING PLAINS OF TEXAS, OKLAHOMA AND EASTERN NEW MEXICO

			ADDITIVE RATE (PER ACRE)
			DASH OR OIL CONCENTRATE RATE PER ACRE
GRASS	TIME OF APPLICATION	RATE OF POAST PER ACRE	GROUND
Bermudagrass • First Application	When stolon length is 3-6″	2½ pts.	2 pts.
• Second Application When regrowth occurs or new plants emerge.	1-4″ length of regrowth or new plants.	2½ pts.	2 pts.
Johnsongrass, Rhizome In seeding alfalfa make the first and second applications before the first cutting. In established alfalfa, control may be partial or inconsistent; johnsongrass regrowth will be suppressed. • First Application	6-10″	2½ pts.	2 pts.
• Second Application When regrowth occurs or new plants emerge.	4-8″	2½ pts.	2 pts.

WESTERN AND MOUNTAIN STATES—RECOMMENDATIONS FOR USE REGIONAL USE MAP

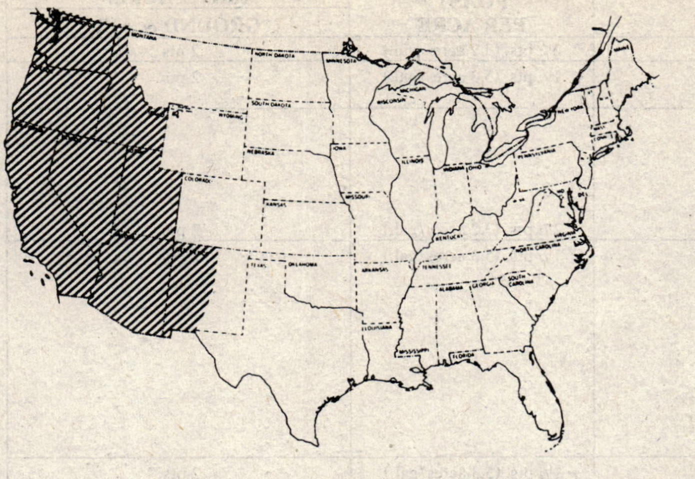

ROW CROPS—ANNUAL GRASSES—STANDARD RECOMMENDATIONS* COTTON, SUGAR BEETS AND SUNFLOWERS WESTERN AND MOUNTAIN STATES

GROUP	GRASS	TIME OF APPLICATION	RATE OF POAST PER ACRE	ADDITIVE (RATE/ACRE) DASH** OR OIL CONCENTRATE GROUND & AIR
A	Goosegrass, Smooth Crabgrass, Large Crabgrass Wild Oats (Idaho, Oregon, Washington only)	Up to 4"	1½ pts. (5.3 acres/gal.)	2 pts.
	Barnyardgrass, Small (for larger barnyardgrass see Group B below) Fall Panicum, Foxtails: Giant, Green, Yellow, Johnsongrass, Seedling, Junglerice, Ryegrass, Annual, Southwestern Cupgrass, Witchgrass	Up to 8"		2 pts.
	Shattercane/Wildcane If needed, retreat at the growth.	6-18"		2 pts.
	Volunteer Corn Maintain sufficient boom height above volunteer corn plants for best spray coverage.	Up to 12"		2 pts.
B	Barnyardgrass, Large (Apply before boot stage)	8-16"	2 pts. (4 acres/gal.)	2 pts.
	Volunteer Cereals Barley, Oats, Rye, Wheat Volunteer cereals which emerge from late spring through early summer (May through July) may be partially or incompletely controlled due to unfavorable conditions during this time.	Before tillering, (up to 4") and prior to over-wintering.		

* For broad spectrum control of annual grasses in Group A (above), use 1½ pints of POAST per acre. When weed populations include additional grasses in Group B, increase the rate of POAST as indicated. If later flushes of annual grasses emerge after the first application, making additional applications at the same rate and at the same recommended stage of growth.

** Dash® spray adjuvant is not to be used in California.

ROW CROPS—PERENNIAL GRASSES COTTON, SUGAR BEETS AND SUNFLOWERS WESTERN AND MOUNTAIN STATES

GRASS	TIME OF APPLICATION	RATE OF POAST PER ACRE	DASH OR OIL CONCENTRATE (RATE/ACRE) GROUND & AIR
Bermudagrass • First Application (In sunflowers, do not make a second application. Season-long control may not be obtained.)	Before plant stolon length exceeds 6".	2½ pts. (3.2 acres/gal.)	2 pts.
• Second Application	21 days after first application.	1½ pts. (5.3 acres/gal.)	2 pts.
• Third Application	1-4" length of regrowth, or new plants.	1½ pts.	2 pts.
Johnsongrass, Rhizome For best results, rhizomes should be thoroughly fragmented (less than 6"). (In sunflowers, do not make a second application. Season-long control may not be obtained.)			

GRASS	TIME OF APPLICATION	RATE OF POAST PER ACRE	DASH OR OIL CONCENTRATE (RATE/ACRE) GROUND & AIR
• First Application	6-10″	2½ pts. (3.2 acres/gal.)	2 pts.
• Subsequent Applications When regrowth occurs or new plants emerge.	4-8″	1½ pts. (5.3 acres/gal.)	2 pts.
Quackgrass For best results, rhizomes should be thoroughly fragmented (less than 6″) (In sunflowers, do not make a second application. Season-long control may not be obtained.) • First Application	6-8″	2½ pts. (3.2 acres/gal.)	2 pts.
• Subsequent Applications When regrowth occurs or new plants emerge. Depending upon environmental conditions and crop competition, season-long control may not always be obtained. However, competition of quackgrass with the crop will be reduced. **Note:** A cultivation no sooner than 14 days after application but within 21 days of application will aid in control.	6-8″	1½ pts. (5.3 acres/gal.)	2 pts.
Ryegrass, Perennial If regrowth occurs retreat at the same rate and stage of growth. (In sunflower retreat at 1 pt.)	3-8″	1½ pts. (5.3 acres/gal.)	2 pts.

POAST USE RECOMMENDATIONS IN ALFALFA—WESTERN AND MOUNTAIN STATES

POAST may be applied to seedling or established alfalfa grown for hay, silage, green chip, direct grazing or for seed. See the **Restrictions and limitations** on page 19 for the minimum length of time between application and harvest.

The effectiveness of POAST is dependent on the absorption and movement of POAST throughout the plant. For this to occur there must be enough leaf surface area to absorb the herbicide and the grass must be actively growing to move or translocate POAST to the roots and buds. Any stress conditions that slow the growth of the grass may decrease control or reduce the speed of control. These stress conditions may include mowing, lack of moisture, herbicide injury, mechanical injury or cold temperatures.

Mowing

Best control of annual grasses can be achieved if applications are made before grass weeds are mowed off. Once a grass is mowed it becomes tougher to control. This is because much of the leaf surface may be removed by the mowing and the grass is put under stress. Because of this, a lower rate is recommended for foxtail control (see following tables) before the second cutting. In southern areas without a killing frost some annuals can over-winter after having been mowed a number of times. These grasses can form large crowns which contain many viable buds. A large crown, even if it is an annual grass, may require repeated applications for control and depending on the species and size of the crown, control may not be complete.

IRRIGATION

Irrigation practices can be very critical to the successful use of POAST. This is because an irrigation may be necessary to start the grass weeds growing again. Generally, applications 2-4 days after an irrigation are most effective. This is because: 1) grasses resume active growth, 2) they have less of a chance to grow too large and, 3) by waiting later, the alfalfa begins to canopy and interferes with POAST spray coverage. Irrigation shortly after (2 days) a POAST application has been effective, but more consistent grass control is obtained when the irrigation was made before the application.

In large fields it may take several days for water to move across a field. And the grasses should not be allowed to grow too large on the side of the field irrigated first while waiting for the entire field to be irrigated. In these situations the field should be sprayed in segments, following the irrigation, to obtain best results.

Annual grass control

Apply POAST at the size and rate indicated in the following tables. If a grass has been cut, apply POAST after the regrowth reaches the minimum height (so there will be enough leaf area to absorb POAST) and before it exceeds the maximum height indicated in the table. Applications should be made before alfalfa canopies over the grasses and interferes with spray coverage. Also, applications made after an alfalfa cutting may need to be timed to follow an irrigation or a rainfall, which will allow the grasses to regrow to a treatable size.

Some annual grasses are spring and summer germinating and others are fall germinating, and the time they are actively growing and most susceptible to POAST may vary from area to area.

Since some annuals germinate over a long period of time, and since control of small grasses is desired, applications after each weed flush may be needed. As a general guideline, spray spring-germinating grasses in the spring and spray fall-germinating weeds in the fall soon after they begin growing. This is because they are more susceptible when they begin growth and control is more complete. Later applications may be less effective due to environmental changes, such as frosts, or due to the onset of flowering.

Perennial grass control

POAST effectively controls or suppresses perennial grasses such as Bermudagrass, Johnsongrass, quackgrass, and perennial ryegrass. However, their growth characteristics are such that they are more difficult to control than annual grasses, especially in a perennial crop such as established alfalfa. A planned program consisting of repeated applications is usually necessary for best results.

The most economical way of controlling perennial grasses is to do so in the year of stand establishment before rhizomes or stolons become large and difficult to kill. The field should be thoroughly disked before seeding to cut the rhizomes or stolons as small as possible.

In established stands it is important to begin applications in the spring when conditions favor active growth and before storage tissues have increased their reserves. Additional applications should be made on grass regrowth in later cuttings.

FORAGE CROPS—ANNUAL GRASSES—STANDARD RECOMMENDATIONS*
ALFALFA
WESTERN AND MOUNTAIN STATES

GROUP	GRASS		TIME OF APPLICATION	RATE OF POAST PER ACRE	ADDITIVE (RATE/ACRE) DASH** OR OIL CONCENTRATE GROUND
A	Goosegrass Smooth Crabgrass Large Crabgrass	Wild Oats (Idaho, Oregon, Washington only)	2-4″	1½ pts. (5.3 acres/gal.)	2 pts.
	Foxtails: Giant, Green, Yellow (Before the 2nd cutting; if application follows 2nd cutting, see Group B below). Barnyardgrass, Small	(For larger size barnyardgrass see Group B below) Fall Panicum, Johnsongrass, Seedling Junglerice, Ryegrass, Annual, Southwestern Cupgrass, Witchgrass	3-8″		2 pts.
	Shattercane/Wildcane		6-18″		2 pts.

GROUP	GRASS		TIME OF APPLICATION	RATE OF POAST PER ACRE	ADDITIVE (RATE/ACRE) DASH** OR OIL CONCENTRATE
					GROUND
	If needed, retreat at the same rate and stage of growth				
	Volunteer Corn Maintain sufficient boom	height above volunteer corn plants for best spray coverage	6-12"		2 pts.
B	Foxtails: Giant, Green, Yellow	(After the 2nd cutting)	3-8"	2 pts. (4 acres/gal.)	2 pts.
	Barnyardgrass, Large (in seed alfalfa only) Before boot stage.		8-16"		
	Volunteer Cereals Barley Oats Rye Wheat	Volunteer cereals which emerge from late spring through early summer (May through July) may be partially or incompletely controlled due to unfavorable conditions during this time.	Before tillering, 2-4" and prior to over-wintering.		

* For broad spectrum control of annual grasses in Group A (above), use $1\frac{1}{2}$ pints of POAST® herbicide per acre. When weed populations include additional grasses in Group B, increase the rate of POAST as indicated. If later flushes of annual grasses emerge after the first application, make additional applications at the same rate and at the same recommended stage of growth.

** Dash is not to be used in California.

FORAGE CROPS—PERENNIAL GRASSES
ALFALFA
WESTERN AND MOUNTAIN STATES

GRASS	TIME OF APPLICATION	RATE OF POAST PER ACRE	DASH OR OIL CONCENTRATE RATE/ACRE
			GROUND
Bermudagrass • First Application Bermudagrass growth will be suppressed.	When stolon length is 3-6".	$2\frac{1}{2}$ pts. (3.2 acres/gal.)	2 pts.
• Second Application When regrowth occurs or new plants emerge.	1-4" length of regrowth or new plants.	$2\frac{1}{2}$ pts.	2 pts.
Johnsongrass, Rhizome In seedling alfalfa make the first and second applications before the first cutting. In established alfalfa control may be partial or inconsistent. Johnsongrass growth will be suppressed. • First Application	6-10"	$2\frac{1}{2}$ pts.	2 pts.
• Second Application When regrowth occurs or new plants emerge.	4-8"	$2\frac{1}{2}$ pts.	2 pts.
Quackgrass (Idaho, Oregon and Washington only) • First Application	2-8"	$2\frac{1}{2}$ pts.	2 pts.
• Second Application When regrowth occurs or new plants emerge. In established alfalfa control may be partial or inconsistent. Quackgrass growth will be suppressed.	2-8"	$2\frac{1}{2}$ pts. (3.2 acres/gal.)	2 pts.
Ryegrass, Perennial If regrowth occurs retreat at the same rate and stage of growth.	3-8"	2 pts. (4 acres/gal.)	2 pts.

NONBEARING FOOD CROPS; ORNAMENTAL, NURSERY, AND OTHER NONFOOD CROPS
GENERAL RECOMMENDATIONS—ALL REGIONS

POAST should be applied when grasses are actively growing and before they reach the maximum size listed in **Recommendations for grass control** on pages 42-43. Many seedling, newly transplanted and established non-grassy ornamentals, trees, shrubs, and ground covers are tolerant to POAST. Very slight leaf speckling has been observed on a few species with no reduction in vigor or growth.

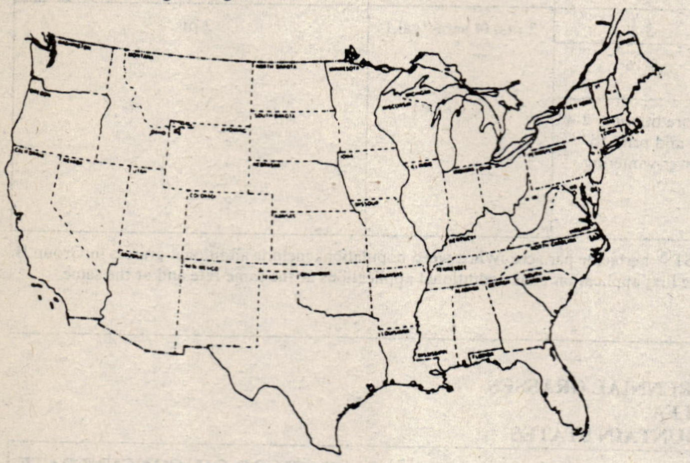

The following plants are tolerant to POAST:

Nonbearing food crops

Do not apply to nonbearing food crops within one year of harvest.

- Almonds
- Apples
- Apricots
- Asparagus
- Avocados
- Blackberries
- Blueberries
- Cherries
- Crabapples
- Cranberries
- Dates
- Figs
- Filberts
- Grapefruit
- Grapes
- Lemons
- Limes
- Macadamia
- Nectarines
- Olives
- Oranges
- Peaches
- Pears
- Pecans
- Pistachios
- Plums
- Pomegranates
- Prunes
- Raspberries
- Tangelos
- Tangerines
- Walnuts

Ornamentals Trees

Common Name	Scientific Name
Arborvitae (var.: Techny, Globe, Pyramidalis)	*Thuja occidentalis*
Ash, Green	*Fraxinus pennsylvanica*
Birch, Paper	*Betula papyrifera*
Dogwood, Flowering	*Cornus florida*
Dogwood, Red Osier	*Cornus sericea*
Fir, Douglas	*Pseudotsuga menziesii*
Fir, Fraser	*Abies fraseri*
Hemlock, Canada	*Tsuga canadensis*
Holly, Chinese (var.: Burfordii, Rotunda)	*Ilex cornuta*
Holly, Japanese (var.: Convexa, Compacta, Hellri)	*Ilex crenata*
Honeylocust	*Gleditsia triacanthos*
Magnolia, Southern	*Magnolia grandiflora*
Maple, Red	*Acer rubrum*
Maple, Silver	*Acer saccharinum*
Oak, Water	*Quercus nigra*
Oak, Willow	*Quercus phello*
Olive, Russian	*Elaeagnus angustifolia*
Pine, Austrian	*Pinus nigra*
Pine, Eastern White or White	*Pinus strobus*
Pine, Jack	*Pinus banksiana*
Pine, Jap. Black	*Pinus thunbergi*
Pine, Jap. White	*Pinus parviflora*
Pine, Loblolly	*Pinus taeda*
Pine, Mugho	*Pinus mugho*
Pine, Red	*Pinus resinosa*
Pine, Scotch	*Pinus sylvestris*
Pine, Shore	*Pinus contorta*
Pine, Slash	*Pinus elliottii*
Pine, Southern or Longleaf	*Pinus palustris*
Pine, Virginia	*Pinus virginiana*
Pine, Western Yellow	*Pinus ponderosa*
Poplar, Hybrid	*Populus alba*
Spruce, Black Halls (var: Densata)	*Picea glauca*
Spruce, Colorado Blue*	*Picea pungens glauca*
Spruce, Norway	*Picea abies*
Spruce, White	*Picea glauca*
Sweet Gum	*Liquidambar styraciflua*
Sycamore	*Platanus occidentalis*

* Certain blue-colored species are turned green for a season by POAST and oil concentrate; to avoid this, direct spray away from foliage.

Shrubs

Common Name	Scientific Name
Alpine Current	*Ribes alpinum*
Arrowwood, Southern	*Biburnum dentatum*
Autumn Olive	*Elaeagnus umbellata*
Boxwood	*Buxus sempervirens*
Camellia	*Camella japonica*
Cotoneaster, Bearberry	*Cotoneaster dammeri*
Cotoneaster, Cranberry	*Cotoneaster apiculata*
Cranberry Bush, European (var.: Aureum)	*Viburnum opulus*
Cranberry Bush, American	*Viburnum trilobum*
Euonymus, Spreading	*Euonymus kiautschovica*
Forsythia (var. Bronxensis)	*Forsythia viridissima*
Honeysuckle, Amur	*Lonicera maackii*
Honeysuckle, Fly	*Lonicera xylosteum*
Honeysuckle, Japanese	*Lonicera japonica*
Honeysuckle, Tatarian	*Lonicera tatarica*
Jojoba	*Simmondsia chinensis*
Juniper, Chinese (var.: Maney, Hetzii)*	*Juniperus chinesis*
Juniper, Creeping (var.: Wiltonii, Bar Harbor)*	*Juniperus horizontalis*
Juniper, Pfitzer*	*Juniperus* spp.
Juniper, Rocky Mountain (var.: Blue Heaven, Welchii)*	*Juniperus scopulorum*
Juniper, Shore (var.: Compacta)*	*Juniperus conferta*
Lilac, Common	*Syringa vulgaris*
Nandina	*Nandina domestica*
Nannyberry	*Viburnum lentago*
Ninebark (var.: Nanus)	*Physocarpus opulifolius*
Rhododendron, Azalea (var.: Hinocrimson, Hershey Red, Formosa Flame)	*Phododendron* spp.
Sandcherry, Purpleleaf	*Prunus cistena*
Snowball Viburnum (var.: Sterile)	*Viburnum opulus*
Spindle Tree	*Euonymus kiautschovica*
Spirea (var.: Anthony Waterer)	*Spiraea bumalda*
Wayfaring Tree, Twistwood	*Viburnum lantana*
Yew, Japanese	*Taxus cuspidata*

* Certain blue-colored species are turned green for a season by POAST and oil concentrate; to avoid this, direct spray away from foliage.

Garden Flowers

Common Name	Scientific Name
Begonia	*Begonia semperflorens*
Chrysanthemum	*Chrysanthemum indicum*
Coleus	*Coleus* spp.
Geranium	*Pelargonium hortorum*
Gladiolus	*Gladiolus* spp.
Impatiens	*Impatiens* spp.
Iris	*Iris* spp.
Maiden Pink	*Dianthus deltoides*
Marigold	*Tagetes* spp.
Periwinkle, Dwarf	*Vinca minor*
Petunia*	*Petunia hybrida*
Snapdragon	*Antirrhinum majus*

Common Name	Scientific Name
Sprenger, Asparagus Fern	*Asparagus densiflorus*
Sweet William	*Dianthus barbatus*
Zinnia*	*Zinnia elegans*

* Application of POAST may damage open flowers.

Notice to user
Due to variability within species and in application techniques neither the manufacturer nor the seller has determined whether or not POAST can be safely used on all nonbearing food crops, ornamentals, nursery and other nonfood crops under all conditions. It is therefore recommended that the professional user should determine if POAST can be used safely prior to broad use.

Recomendations for grass control

- Apply to actively growing grasses before tillering and/or seed head formation.
- Follow water volume and spray pressure recommendations.
- Apply to grasses at the sizes indicated below.
- In irrigated areas it may be necessary to irrigate prior to treatment to ensure weeds are growing actively.
- **Always add 2 pints per acre of oil concentrate.**

Ground Covers

Common Name	Scientific Name
Ivy, Algerian	*Hedera canaiensis*
Ivy, English	*Hedera helix*
Ice Plant*	*Lampranthus aureus*
Ice Plant*	*Lampranthus spectabilis*
Ice Plant*	*Mesembryanthemum* spp.
Ice Plant*, Hottentot Fig*	*Carpobrotus edulis*
Pachysandra	*Pachysandra terminalis*

* Application of POAST may damage open flowers.

RECOMMENDATIONS FOR GRASS CONTROL
NONBEARING FOOD AND NONFOOD CROPS—ANNUAL GRASSES

Grass	Rate of POAST per Acre*		Oil Concentrate (Rate/Acre)
	Grass Up to 6" Height	Grass Up to 12" Height	
Barnyardgrass Broadleaf Signalgrass Fall Panicum Foxtails, Giant, Green, Yellow Goosegrass Johnsongrass, Seedling Junglerice Large Crabgrass Lovegrass Orchardgrass, Seedling Red Sprangletop** Shattercane/Wildcane Smooth Crabgrass Tall Fescue, Seedling Texas Panicum Wild Proso Millet Witchgrass Woolly Cupgrass	1½ pts.	2½ pts.	2 pts.

Nonbearing Food and Nonfood Crops—Perennial Grasses

Grass	Maximum Size Range	Rate of POAST per Acre*	Oil Concentrate (Rate/Acre)
Bermudagrass (Wiregrass)	Up to 6" runners	2½ pts.	2 pts.
Johnsongrass, Rhizome	15-20" height		
Quackgrass	6-8" height		
Wirestem Muhly	Up to 6" height	1½ pts.	

* Repeat applications as needed.
** Not recommended in AZ, CA, NM, OR.

Spot treatment application
For control of grasses when using knapsack sprayers or high volume equipment utilizing handguns or other suitable nozzle arrangements, prepare a solution of POAST plus oil concentrate in water according to the table below. Shake sprayer occasionally to maintain a uniform mix of POAST and oil concentrate. Apply to actively growing grasses before tillering and/or seed head formation. Apply to the foliage of grasses on a spray-to-wet basis. Spray coverage should be uniform and complete. Do not spray to the point of runoff.

RECOMMENDATIONS FOR GRASS CONTROL
SPOT TREATMENT APPLICATION TABLE
NONBEARING FOOD AND NONFOOD CROPS—ANNUAL GRASSES

Grass	Concentration in Spray Solution**		
	POAST*		Oil Concentrate
	Grass Up To 6"Height	Grass Up To 12"Height	
See Annual grasses listed in **Broadcast Application Table above.**	1%	1½%	1%

PERENNIAL GRASSES

Grasses	Maximum Size Range	Concentration in Spray Solution**	
		POAST*	Oil Concentrate
Bermudagrass (Wiregrass)	Up to 6" runners	1½%	1%
Johnsongrass, Rhizome	15-20" height	1½%	1%
Quackgrass	6-8" height	1½%	1%
Wirestem Muhly	Up to 6" height	1%	1%

* Repeat applications as needed.
** Refer to **Solution Table** below for preparation of desired spray solution volume.

Solution Table

Desired Spray Solution Volume	Amount of POAST or Oil Concentrate to be Added for Solution	
	1%	1½%
1 gallon	1¼ fl. oz.	2 fl. oz.
3 gallons	3¾ fl. oz.	6 fl. oz.
5 gallons	6¼ fl. oz.	10 fl. oz.

1 tablespoon = ½ fl. oz.

POAST PLUS GOAL 1.6E TANK MIX

Tank mix of POAST with Goal 1.6E for use on conifers grown for Christmas trees for postemergence broadleaf and grass weed control. For use only in the states of AL, GA, KY, NC, SC, TN, VA and WV.

Grasses controlled: See species listed on page 42.

Broadleaf weeds controlled: Goal® herbicide will control certain broadleaf weeds; see Goal 1.6E label.

POAST and Goal rates: A maximum of 2½ pints per acre of POAST may be tank mixed with Goal 1.6E. A maximum of 2½ pints of Goal 1.6E may be tank mixed with POAST. See prior pages for minimum recommended rates of POAST and see Goal 1.6E label for minimum recommended rates of Goal 1.6E. Two or three applications of the tank mix may be needed for season-long control.
In some cases reduced grass control with POAST may be experienced when tank mixed with Goal 1.6E.

Water volume and spray pressure: Use 20 gallons of spray volume pressure and 40 psi with this tank mixture.

The following plants are tolerant to a tank mix of POAST and Goal 1.6E:

Common Name	Scientific Name
Fir, Fraser	*Abies fraseri*
Hemlock, Canada*	*Tsuga canadensis**
Pine, Virginia	*Pinus virginiana*
Pine, White or Eastern White	*Pinus strobus*
Spruce, Norway	*Picea abies*

* Canada Hemlock has a prolonged period of bud break and new growth; thus, directed applications are recommended during this period.

Timing: Applications should be made when weeds are actively growing and before conifer bud break or after conifer foliage has had an opportunity to harden-off. Broadleaf weeds must be within the height indicated on the Goal 1.6E label.

Mixing: Add components in the following sequence, gently mixing between component additions:

1) Oil concentrate
2) POAST
3) Goal 1.6E

Restrictions and limitations for POAST with Goal 1.6E tank mix

Follow all conifer specific and general use restrictions on Goal 1.6E label.

Do not apply the tank mix to conifer seedlings less than 10 months old.

Do not apply the tank mix when temperatures exceed 90 degrees F.

Do not apply this tank mixture with air equipment.

Do not substitute Dash for oil concentrate.

TALL FESCUE GROWTH SUPPRESSION WITH POAST OR POAST PLUS GOAL 1.6E ON CONIFERS GROWN FOR CHRISTMAS TREES

For use only in the states of AL, GA, KY, NC, SC, TN, VA and WV.

General information: POAST can control tall fescue *(Festuca arundinacea)* vegetative growth. Depending on the time of year and the developmental stage of tall fescue, the growth of seedheads of tall fescue can be suppressed or eliminated.

Application information

Apply POAST to actively growing tall fescue after it has had 4 to 6 inches of new growth, before the emergence of seedheads and before conifer bud break. Application from July 1 to mid-August may be less effective, especially if day temperatures reach 90°F. Tall fescue must be one year old before the first application of POAST.

Do not make applications to grasses under stress, such as stress due to lack of moisture, herbicide injury, or cold temperatures, since unsatisfactory suppression may result.

Adequate coverage of the leaf surface is necessary for absorption of this herbicide; thus, for optimum control, do not mow tall fescue turf for 30 days before or 14 days after application of POAST.

Rate: Apply POAST at a rate of 1 to 1¼ pints per acre. For greater tall fescue suppression, up to 2½ pints per acre of POAST can be used.

When POAST is tank mixed with Goal 1.6E (see preceding section) for tall fescue growth suppression, 1½ to 2½ pints of POAST should be used.

Users of POAST or POAST plus Goal 1.6E are advised to begin use of POAST at the minimum recommended rate and adjust rates as local conditions and experience dictates.

Note: Because of environmental differences at application and growth differences of tall fescue, control of tall fescue may exceed or fall short of that desired.

APPENDIX

The following are scientific names for the weeds listed in this section. For specific recommendations on control of these weeds, refer to the major crop and/or tank mix sections.

Grasses

Common Name	Scientific Name
Barnyardgrass	*Echinochloa crus-galli*
Bermudagrass	*Cynodon dactylon*
Broadleaf Signalgrass	*Brachiaria platyphylla*
Crabgrass, Large,	*Digitaria sanguinalis*
Smooth	*Digitaria ischaemum*
Cupgrass, Southwestern,	*Eriochloa gracilis*
Woolly	*Eriochloa villosa*
Field Sandbur	*Cenchrus incertus*
Foxtails, Giant,	*Setaria faberi*
Green,	*Setaria viridis*
Yellow	*Setaria glauca*
Goosegrass	*Eleusine indica*
Itchgrass	*Rottboellia exaltata*
Johnsongrass	*Sorghum halepense*
Junglerice	*Echinochloa colonum*
Lovegrass	*Eragrostis cilianensis*
Orchardgrass	*Dactylis glomerata*
Pigeon grass (see Foxtails)	
Panicum, Browntop,	*Panicum fasciculatum*
Fall,	*Panicum dichotomiflorum*
Texas	*Panicum texanum*
Quackgrass	*Agropyron repens*
Red Rice	*Oryza sativa*
Red Sprangletop	*Leptochloa filiformis*
Ryegrass, Annual,	*Lolium multiflorum*
Perennial	*Lolium perenee*
Tall Fescue	*Festuca arundinacea*
Tame Oats	*Avena sativa*
Volunteer Barley	*Hordeum vulgare*
Corn	*Zea mays*
Oats	*Avena sativa*
Rye	*Secale cereale*
Wheat	*Triticum aestivum*
Watergrass (see Barnyardgrass)	
Shattercane/Wildcane	*Sorghum bicolor*
Wild Oats	*Avena fatua*
Wild Proso Millet	*Panicum miliaceum*
Wiregrass (see Bermudagrass)	
Wirestem Muhly	*Muhlenbergia frondosa*
Witchgrass	*Panicum capillare*

Broadleaf Weeds

Common Name	Scientific Name
Balloonvine	*Cardiospermum halicacabum*
Beggarticks	*Bidens frondosa*
Black Nightshade	*Solanum nigrum*
Bristly Starbur	*Acanthospermum hispidum*
Buffalobur	*Solanum rostratum*
Burgherkin	*Cucumis anguria*
Canada Thistle	*Cirsium arvense*
Carpetweed	*Mollugo verticilla*
Citron	*Citrullus vulgaris*
Cocklebur	*Xanthium strumarium*
Coffee Senna	*Cassia occidentalis*
Common Lambsquarters	*Chenopodium album*
Common Purslane	*Portulaca oleracea*
Copperleaf, Hophornbeam	*Acalypha ostryaefolia*
Copperleaf, Virginia	*Acalypha virginica*
Crotalaria, Showy	*Crotalaria spectabilis*
Dayflower	*Commelina* spp.
Devilsclaw	*Probiscidea louisianica*
Florida Pusley	*Richardia scabra*
Galinsoga	*Galinsoga* spp.
Groundcherry, Cutleaf	*Physalis angulata*
Groundcherry, Lanceleaf	*Physalis lanceifolia*
Hairy Indigo	*Indigofera hirsuta*
Jimsonweed	*Datura stramonium*
Ladysthumb	*Polygonum persicaria*
Morningglories,	*Ipomoea* spp.
Smallflower	*Jacquemontia tamnifolia*
Nightshade, Black	*Solanum nigrum*
Nightshade, Eastern Black	*Solanum ptycanthum*
Palmer Amaranth	*Amaranthus palmeri*
Pennsylvania Smartweed	*Polygonum pensylvanicum*
Pigweed, Prostrate,	*Amaranthus blitoides*
Redroot,	*Amaranthus retroflexus*
Smooth	*Amaranthus hybridis*
Poorjoe	*Diodia teres*
Prickly Sida or Teaweed	*Sida spinosa*
Ragweed, Common,	*Ambrosia artemisiifolia*
Giant	*Ambrosia trifida*
Redweed	*Melochia corchorifolia*
Sesbania	*Sesbania exaltata*
Shepherdspurse	*Capsella bursa-pastoris*
Smellmelon	*Cucumis melo*
Spiny Amaranth	*Amaranthus spinosus*
Spurge, Prostrate	*Euphorbia supina*
Spurge, Spotted	*Euphorbia maculata*
Spurred Anoda	*Anoda cristata*
Tall Waterhemp	*Amaranthus tuberculatus*
Texas Gourd	*Cucurbita texana*
Tropic Croton	*Croton glandulosus*
Velvetleaf	*Abutilon theophrasti*
Venice Mallow	*Hibiscus trionum*
Volunteer Cowpea	*Vigna unguiculata*
Wild Buckwheat	*Polygonum convolvulus*
Wild Mustard	*Sinapis arvensis*
Wild Poinsettia	*Euphorbia heterophylla*
Wild Spiny Cucumber	*Cucumis dipsaceus*
Wild Sunflower	*Helianthus annuus*
Woolly Croton	*Croton capitatus*
Yellow Rocket	*Barbarea vulgaris*

Sedges

Common Name	Scientific Name
Yellow Nutsedge	*Cyperus esculentus*

CONDITIONS OF SALE AND WARRANTY

The Directions for use of this product reflect the opinion of experts based on field use and tests. The directions are believed to be reliable and should be followed carefully. However, it is impossible to eliminate all risks inherently associated with use of this product. Crop injury, ineffectiveness or other unintended consequences may result because of such factors as weather conditions, presence of other materials, or use of the product in a manner inconsistent with its labeling, all of which are beyond the control of BASF CORPORATION ("BASF") or the Seller. All such risks shall be assumed by the Buyer.

BASF warrants that this product conforms to the chemical description on the label and is reasonably fit for the purposes referred to in the Directions for use, subject to the inherent risks referred to above. BASF MAKES NO OTHER EXPRESS OR IMPLIED WARRANTY OF FITNESS OR MERCHANTABILITY OR ANY OTHER EXPRESS OR IMPLIED WARRANTY. IN NO CASE SHALL BASF OR THE SELLER BE LIABLE FOR CONSEQUENTIAL, SPECIAL OR INDIRECT DAMAGES RESULTING FROM THE USE OR HANDLING OF THIS PRODUCT. BASF and the Seller offer this product, and the Buyer and User accept it, subject to the foregoing Conditions of sale and warranty which may be varied only by agreement in writing signed by a duly authorized representative of BASF.

POAST and Basagran are registered trademarks of BASF AG. Blazer is a registered trademark and Dash is a trademark of BASF Corporation.

Goal is a registered trademark of Rohm & Hass Company.

Tackle and Butyrac are registered trademarks of Rhone-Poulenc Company.

Classic is a registered trademark of E.I. Dupont de Nemours and Company.

Butoxone is a registered trademark of Vertac Chemical Corporation.

PO 1 042 JJ6301 A0680888

SUPPLEMENTAL LABELING

POAST® HERBICIDE

POAST EPA Reg. No. 7969-58

Label approval pending in California.

FOR USE IN CABBAGE, BROCCOLI, AND CAULIFLOWER.

All applicable directions, restrictions, precautions and conditions of sale and warranty on the EPA-registered label are to be followed. This labeling must be in the possession of the user at the time of herbicide application.

Re-entry and workers' protection statements

Do not apply this product in such a manner as to directly or through drift expose workers or other persons, except those knowingly involved in the application. The area being treated must be vacated by unprotected persons.

Do not enter treated areas without protective clothing until sprays have dried.

Because certain states may require more restrictive re-entry intervals for various crops treated with this product, consult your State Department of Agriculture for further information.

Written or oral warnings must be given to workers who are expected to be in a treated area or in an area about to be treated with this product. Oral warnings must inform workers of areas or fields that may not be entered without specific protective clothing until sprays have dried, and appropriate actions to take in case of accidental exposure. When oral warnings are given, warnings shall be given in a language customarily understood by workers. Oral warnings must be given if there is reason to believe that written warnings cannot be understood by workers.

Written warnings must include the following information: "WARNING. Area treated with POAST® herbicide on (date of application). Do not enter without appropriate protective clothing until sprays have dried. If contacted, flush eyes immediately with water for 15 minutes. Get medical attention.

In case of contact with skin or clothing, remove contaminated clothing and wash skin thoroughly with soap and water. Call physician if irritation occurs. Wash contaminated clothing before re-use."

DIRECTIONS FOR USE

It is a violation of Federal law to use this product in a manner inconsistent with its labeling.

GENERAL INFORMATION

POAST is a selective broad spectrum postemergence herbicide for control of annual and perennial grass weeds. POAST does not control sedges or broadleaf weeds.

Since essentially all grass crops such as sorghum, corn, small grains and rice, as well as ornamental grasses such as turf, are susceptible to POAST, avoid all direct or indirect contact with any grass crop.

Control symptoms: POAST rapidly enters the plant through the foliage and translocates throughout the plant. Control symptoms exhibited by the grass plant progress from a slowing or stopping of growth (generally within two days), to reddening of the foliage to leaf tip burn. Subsequently, burn back of the foliage occurs. These symptoms will generally be observed within three weeks depending on environmental conditions.

Application information*

Apply POAST to actively growing grasses when they are at the proper growth stage as specified by the **Recommendations for grass control.**

Do not make applications to grasses under stress, such as stress due to lack of moisture, herbicide injury, mechanical injury or cold temperatures, since unsatisfactory control will probably result.

Thorough spray coverage of grass foliage is essential. For broadcast application use standard high pressure pesticide hollow cone or flat fan nozzles. Do not use flood or whirl chamber nozzles. Application of POAST with control drop applicator (CDA) nozzles is not recommended due to erratic coverage which causes inconsistent weed control. Use a minimum volume of 5 gallons and a maximum volume of 20 gallons of spray solution per acre. When using standard high pressure hollow cone or flat fan nozzles adjust pressure to a minimum of 40 psi and a maximum of 60 psi at the nozzle.

Always adjust spray pressure, spray volume and height of spray boom to ensure penetration of plant canopy and thorough coverage of grasses to be controlled.

Do not use selective application equipment such as recirculating sprayers, wiper applicators or shielded applicators.

*For use with ground equipment only.

Cultivation information

Do not cultivate within 5 days prior to application of POAST® herbicide or within 7 days following application.

A timely cultivation after 7 days may aid in providing season-long control.

Addition of oil concentrate

A nonphytotoxic oil concentrate (commonly referred to as oil concentrate) should always be added to the spray tank. The oil concentrate must contain either a petroleum or vegetable oil base and must meet the following criteria: 1) be nonphytotoxic, 2) contain only EPA-exempt ingredients, 3) provide good mixing quality in the jar test (see below), and 4) be successful in local experience.

The exact composition of suitable products will vary; however, vegetable and petroleum oil concentrates should contain emulsifiers which provide good mixing quality. For vegetable oil concentrates, it has been observed that highly refined vegetable oils are more satisfactory than unrefined vegetable oils. For additional information see **Jar test for estimating suitability of oil concentrates.**

Rate of oil concentrate:

2 pints/acre.

Jar test for estimating suitability of oil concentrates

1. **Water supply:** Use only water from intended source and at the source temperature.
2. **Amount of water in jar:**
   For 20 gal./A spray volume use 3⅓ cups (800 ml) of water.
   For 10 gal./A spray volume use 1⅔ cups (400 ml) of water.
   For 5 gal./A spray volume use ⅚ cups (200 ml) of water.
   For other spray volumes, adjust proportionately to above.
3. **Amount of herbicide and oil concentrate to add:** Add herbicide and oil concentrate at the rate of 1 teaspoon (5 ml) for each pint of recommended label rate.
4. **Add components in following sequence,** gently mixing between component additions:
   1) Oil concentrate.
   2) POAST.
5. **Cap jar,** invert 10 cycles, let stand for 15 minutes, evaluate.
6. **Evaluation:** An ideal tank mix combination will be uniform; thus, the suitability of the oil concentrate is questionable if any of the following are observed:
   Free oil at the surface-film or globules.
   Flocculation-fine particles which may be suspended in the liquid or found as a precipitated layer at the bottom of the jar.
   Clabbering-thickening texture (coagulated) resembling yogurt or a curd-like texture as with cottage cheese.

Mixing/spraying

Fill tank of a thoroughly clean sprayer half to two-thirds full with clean water. Start agitation and add oil concentrate; allow to mix thoroughly, Add POAST and remaining volume of water. Maintain constant agitation during application.

Restrictions and limitations

Do not apply POAST within 30 days of harvest.

Do not apply more than a total of 3 pints per acre in one season to cabbage, broccoli, and cauliflower.

Do not make applications to grasses under stress, such as stress due to lack of moisture or herbicide injury, as unsatisfactory control may result.

Do not apply if rainfall is expected within one hour following application as grass control will be unsatisfactory.

Do not mix or apply POAST with any other pesticide, additive, or fertilizer except as specifically recommended on this labeling.

Environmental hazards

Do not apply directly to water.

Do not contaminate water by cleaning of equipment or disposal of wastes.

Endangered species concerns

The use of any pesticide in a manner that may kill or otherwise harm an endangered or threatened species or adversely modify their habitat is a violation of Federal law. The use of this product is controlled to prevent death or harm to endangered or threatened species that occur in the following counties or elsewhere in their range.

Before using this pesticide in the following county you must obtain the appropriate EPA Endangered Species Bulletin. The use of this pesticide is prohibited in Solano County, California unless specified otherwise in the Bulletin. The EPA Bulletin is available from either your local pesticide distributor, your County Agricultural Extension Agent, the Endangered Species Specialist in your State Wildlife Agency Headquarters or the appropriate Regional Office of either the U.S. Fish and Wildlife Service or the U.S. Environmental Protection Agency. THIS BULLETIN MUST BE REVIEWED PRIOR TO PESTICIDE USE.

Regional use map

All weed control recommendations for a given crop and weed are on a regional basis. There are three regions delineated on the U.S. map. Best results are obtained by using recommendations for your region only.

—California, Arizona, Nevada, and Western New Mexico.

—High and Rolling Plains of Texas, Oklahoma and Eastern New Mexico.

—Midwest, South, Mountain (except Arizona, New Mexico and Nevada) and Pacific Northwest. (All states and areas except those listed above.)

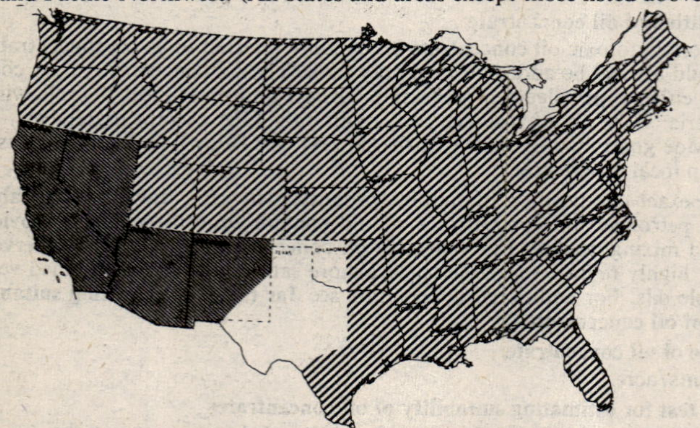

CALIFORNIA, ARIZONA, NEVADA AND WESTERN NEW MEXICO

Recommendations for grass control

- Apply to actively growing grasses at the sizes indicated in the tables of this regional section.
- Always follow recommendations given in the **Application information** section.
- Always adjust spray pressure, spray volume and height of spray boom to ensure penetration of plant canopy and thorough coverage of grasses to be controlled.
- Do not apply to drought-stressed grass or grass which has gone through an extended dry period.
- In irrigated areas it may be necessary to irrigate prior to treatment with POAST® herbicide to ensure weeds are growing actively.
- **Always add 2 pints of oil concentrate per acre.**

Cabbage, Broccoli, and Cauliflower
California, Arizona, Nevada and Western New Mexico

Group	Grass	Time of Application	Rate of POAST per Acre	Oil Concentrate Rate per Acre
A	Barnyardgrass Broadleaf Signalgrass Fall Panicum Foxtails: Giant, Green, Yellow Johnsongrass, Seedling Junglerice Ryegrass, Annual Southwestern Cupgrass Texas Panicum Witchgrass	3-8"	1½ pints	2 pints
	Goosegrass Large Crabgrass Smooth Crabgrass	Up to 4"		
	Shattercane/Wildcane If needed, retreat at the same rate and stage of growth.	6-18"		
	Volunteer Corn	8-12"		

* For broad spectrum control of annual grasses, use 1½ pints of POAST per acre. If subsequent flushes of annual grasses emerge after the first application, make an additional application at the same rate and at the recommended stage of growth.

HIGH AND ROLLING PLAINS OF TEXAS, OKLAHOMA AND EASTERN NEW MEXICO

Recommendations for grass control

- Apply to actively growing grasses at the sizes indicated in the tables of this regional section.
- Always follow recommendations given in the **Application information** section.
- Always adjust spray pressure, spray volume, and height of spray boom to ensure penetration of plant canopy and thorough coverage of grasses to be controlled.
- Do not apply to drought-stressed grass, or grass which has gone through an extended dry period.
- In irrigated areas it may be necessary to irrigate prior to treatment with POAST® herbicide to ensure weeds are growing actively.
- **Always add 2 pints of oil concentrate per acre.**

Cabbage, Broccoli, and Cauliflower
High and Rolling Plains of Texas, Oklahoma and Eastern New Mexico

Group	Grass	Time of Application	Rate of POAST per Acre	Oil Concentrate Rate per Acre
A	Barnyardgrass Broadleaf Signalgrass Browntop Panicum Fall Panicum Foxtails: Giant, Green, Yellow Johnsongrass, Seedling Junglerice Red Sprangletop Texas Panicum Witchgrass	2-6"	1½ pints	2 pints
	Goosegrass Large Crabgrass Smooth Crabgrass	Up to 4"		
	Shattercane/Wildcane If needed, retreat at the same rate and stage of growth.	6-10"		
	Volunteer Corn	8-10"		

* For broad spectrum control of annual grasses, use 1½ pints of POAST per acre. If subsequent flushes of annual grasses emerge after the first application, make an additional application at the same rate and at the recommended stage of growth.

Cabbage, Broccoli, and Cauliflower
High and Rolling Plains of Texas, Oklahoma and Eastern New Mexico

Grass	Time of Application	Rate of POAST per Acre	Oil Concentrate Rate per Acre
Bermudagrass • First Application Control may be partial or inconsistent, however bermudagrass growth will be suppressed.	Before plant diameter exceeds 6" or leaf height above ground exceeds 1".	1½ pints	2 pints
Second Application	21 days after first application	1½ pints	2 pints
Johnsongrass, Rhizome • First Application For best results, rhizomes should be thoroughly fragmented (less than 6"). Adjust volume of spray solution to a maximum of 10 gallons and a minimum of 5 gallons per acre while maintaining a ground speed of no more than 6 miles per hour.	6-10"	1½ pints	2 pints
• Subsequent Application When regrowth occurs or new plants emerge.	4-8"	1 pint	2 pints

MIDWEST, MOUNTAIN (EXCEPT ARIZONA, NEW MEXICO AND NEVADA), SOUTH, NORTHEAST AND PACIFIC NORTHWEST

Recommendations for grass control

- Apply to actively growing grasses at the sizes indicated in the tables of this regional section.
- Always follow recommendations given in the **Application information** section.
- Always adjust spray pressure, spray volume, and height of spray boom to ensure penetration of plant canopy and thorough coverage of grasses to be controlled.
- Do not apply to drought-stressed grass, or grass which has gone through an extended dry period.
- In irrigated areas it may be necessary to irrigate prior to treatment with POAST® herbicide to ensure weeds are growing actively.

- **Always add 2 pints of oil concentrate per acre.**

Cabbage, Broccoli, and Cauliflower
Annual Grasses*
Midwest, Mountain (except Arizona, New Mexico and Nevada)
South, Northeast and Pacific Northwest

Group	Grass	Time of Application	Rate of POAST per Acre	Oil Concentrate Rate per Acre
A	Wild Proso Millet	4-10"	½ pint	2 pints
B	Barnyardgrass Broadleaf Signalgrass Browntop Panicum Fall Panicum Foxtails: Giant, Green, Yellow Johnsongrass, Seedling Junglerice Red Sprangletop Ryegrass, Annual Texas Panicum Witchgrass Woolly Cupgrass	3-8"	1 pint	2 pints
	Goosegrass Large Crabgrass Smooth Crabgrass	Up to 4"		
	Shattercane/Wildcane If needed, retreat at the same rate and stage of growth.	6-18"		
	Volunteer Corn	6-20"		
C	Volunteer Cereals* Barley Rye Oats Wheat POAST is not recommended for spring control of volunteer cereals that emerged the previous fall.	Before tillering, up to 6" and prior to wintering.	1½ pints	2 pints
	Wild Oats	2-4"		

* For broad spectrum control of annual grasses in Groups A & B, use 1 pint of POAST per acre. When weed populations include additional grasses in Group C increase the rate of POAST to 1½ pints. If a second flush of annual grasses emerge after the first application, make an additional application at the same rate and at the recommended stage of growth.

Rescue treatment for controlling selected annual grasses

For best results, always apply POAST® herbicide to annual grasses at the growth stage and rate specified in the above table. However, if POAST cannot be applied at the recommended time, larger annual grasses can be controlled with a later application by increasing the rate of POAST. Apply to actively growing grasses at the rates and sizes indicated below.

Cabbage, Broccoli, and Cauliflower
Annual Grasses
Rescue Treatment
Midwest, South, Mountain (except Arizona, New Mexico and Nevada), Northeast and Pacific Northwest

Grass	Time of Application	Rate of POAST Per Acre	Oil Concentrate Rate per Acre
Wild Proso Millet	10-24"	1 pint	
Foxtails: Giant, Green, Yellow Johnsongrass, Seedling	8-16"	1½ pints	2 pints
Texas Panicum Fall Panicum Barnyardgrass Broadleaf Signalgrass	8-12"		
Large Crabgrass Smooth Crabgrass Goosegrass	6-8"		

Cabbage, Broccoli, and Cauliflower
Perennial Grasses
Midwest, South, Mountain (except Arizona, New Mexico and Nevada), Northeast and Pacific Northwest

Grass	Time of Application	Rate of POAST Per Acre	Oil Concentrate Rate per Acre
Bermudagrass			
• First Application	Before plant diameter exceeds 6" or leaf height above ground exceeds 1".	1½ pints	2 pints
• Subsequent Application If regrowth occurs or new plants emerge.	1-4" length of new plants.	1 pint	2 pints
Johnsongrass, Rhizome • First Application For best results, rhizomes should be thoroughly fragmented (less than 6").	15-25"	1½ pints	2 pints
• Subsequent Application If regrowth occurs or new plants emerge.	6-12"	1 pint	2 pints
Ryegrass, Perennial If regrowth occurs retreat at the same rate and stage of growth.	3-8"	1 pint	2 pints
Quackgrass • First Application For best results, rhizomes should be thoroughly fragmented (less than 6").	6-8"	1½ pints	2 pints
• Subsequent Application If regrowth occurs or new plants emerge. Control may be partial or inconsistent, however, quackgrass growth will be suppressed. **Note:** A cultivation no sooner than 14 days after application but within 21 days of application will aid in control.	6-8"	1½ pints	2 pints
Wiresten Muhly If regrowth occurs, retreat at the same rate and stage of growth.	Up to 6"	1¼ pints	2 pints

PROCEDURE FOR CLEANING EQUIPMENT

ATTENTION! Clean sprayer thoroughly before and after application of POAST® herbicide.

Clean sprayer thoroughly before application of POAST, particularly if a herbicide was used which has the potential to injure the crop to be sprayed with POAST.

Failure to clean sprayer thoroughly after the application of POAST may result in injury to any grass crop subsequently sprayed, such as corn, sorghum, small grains, rice and turf. Fill the sprayer with clean water and add a commercial sprayer cleaner or a surfactant/adjuvant at the recommended rate on its label. Circulate through entire sprayer system. Spray approximately half the tank solution through the hoses, booms and nozzles to clean these parts. Drain the tank and rinse the total system thoroughly several times with clean water.

STORAGE AND DISPOSAL

Do not contaminate water, food, or feed by storage or disposal.

Pesticide wastes are toxic. Improper disposal of excess pesticide, spray mixture, or rinsate is a violation of Federal law. If these wastes cannot be disposed of by use according to label instructions, contact your State Pesticide or Environmental Control Agency, or the Hazardous Waste representative at the nearest EPA Regional Office for guidance.

Triple rinse container (or equivalent). Then offer for recycling or reconditioning, or puncture and dispose of in a sanitary landfill, or by incineration, or, if allowed by state and local authorities, by burning. If burned, stay out of smoke.

CONDITIONS OF SALE AND WARRANTY

The Directions for use of this product reflect the opinion of experts based on field use and tests. The directions are believed to be reliable and should be followed carefully. However, it is impossible to eliminate all risk inherently associated with use of this product. Crop injury, ineffectiveness or other unintended consequences may result because of such factors as weather conditions, presence of other materials, or use of the product in a manner inconsistent with its labeling, all of which are beyond the control of BASF CORPORATION ("BASF") or the Seller. All such risks shall be assumed by the Buyer.

BASF warrants that this product conforms to the chemical description on the label and is reasonably fit for the purposes referred to in the Directions for use, subject to the inherent risks referred to above. BASF MAKES NO OTHER EXPRESS OR IMPLIED WARRANTY OF FITNESS OR

MERCHANTABILITY OR ANY OTHER EXPRESS OR IMPLIED WARRANTY. IN NO CASE SHALL BASF OR THE SELLER BE LIABLE FOR CONSEQUENTIAL, SPECIAL OR INDIRECT DAMAGES RESULTING FROM THE USE OR HANDLING OF THIS PRODUCT. BASF and the Seller offer this product, and the Buyer and User accept it, subject to the foregoing Conditions of sale and warranty which may be varied only by agreement in writing signed by a duly authorized representative of BASF.

POAST is a registered trademark of BASF AG.

PO I 025 JJ6201 A0170388

SUPPLEMENTAL LABELING

POAST® HERBICIDE

POAST EPA Reg. No. 7969-58

FOR USE ON CANTALOUPE, CUCUMBER, MUSK MELON, PUMPKIN, SUMMER SQUASH, WATERMELON AND WINTER SQUASH.

All applicable directions, restrictions, precautions and Conditions of sale and warranty on the EPA-registered label are to be followed. This labeling must be in the possession of the user at the time of herbicide application.

Re-entry and workers' protection statements

Do not apply this product in such a manner as to directly or through drift expose workers or other persons, except those knowingly involved in the application. The area being treated must be vacated by unprotected persons.

Do not enter treated areas without protective clothing until sprays have dried.

Because certain states may require more restrictive re-entry intervals for various crops treated with this product, consult your State Department of Agriculture for further information.

Written or oral warnings must be given to workers who are expected to be in a treated area or in an area about to be treated with this product. Oral warnings must inform workers of areas or fields that may not be entered without specific protective clothing until sprays have dried, and appropriate actions to take in case of accidental exposure. When oral warnings are given, warnings shall be given in a language customarily understood by workers. Oral warnings must be given if there is reason to believe that written warnings cannot be understood by workers. Written warnings must include the following information: "WARNING. Area treated with POAST® herbicide on (date of application). Do not enter without appropriate protective clothing until sprays have dried. If contacted, flush eyes immediately with water for 15 minutes. Get medical attention.

In case of contact with skin or clothing, remove contaminated clothing and wash skin thoroughly with soap and water. Call physician if irritation occurs. Wash contaminated clothing before re-use."

DIRECTIONS FOR USE

It is a violation of Federal law to use this product in a manner inconsistent with its labeling.

GENERAL INFORMATION

POAST is a selective broad spectrum postemergence herbicide for control of annual and perennial grass weeds. POAST does not control sedges or broadleaf weeds.

Since essentially all grass crops such as sorghum, corn, small grains and rice, as well as ornamental grasses such as turf, are susceptible to POAST, avoid all direct or indirect contact with any grass crop.

Control symptoms: POAST rapidly enters the plant through the foliage and translocates throughout the plant. Control symptoms exhibited by the grass plant progress from a slowing or stopping of growth (generally within two days), to reddening of the foliage to leaf tip burn. Subsequently, burn back of the foliage occurs. These symptoms will generally be observed within three weeks depending on environmental conditions.

Application information*

Apply POAST to actively growing grasses when they are at the proper growth stage as specified by the **Recommendations for grass control.**

Do not make applications to grasses under stress, such as stress due to lack of moisture, herbicide injury, mechanical injury or cold temperatures, since unsatisfactory control will probably result.

Thorough spray coverage of grass foliage is essential. For broadcast application use standard high pressure pesticide hollow cone or flat fan nozzles. Do not use flood or whirl chamber nozzles. Application of POAST with control drop applicator (CDA) nozzles is not recommended due to erratic coverage which causes inconsistent weed control. Use a minimum volume of 5 gallons and a maximum volume of 20 gallons of spray solution per acre. When using standard high pressure hollow cone or flat fan nozzles adjust pressure to a minimum of 40 psi and a maximum of 60 psi at the nozzle.

* Application may be made using ground equipment only

Always adjust spray pressure, spray volume and height of spray boom to ensure penetration of plant canopy and thorough coverage of grasses to be controlled.

Do not use selective application equipment such as recirculating sprayers, wiper applicators, or shielded applicators.

Cultivation information

Do not cultivate within 5 days prior to application of POAST® herbicide or within 7 days following application.

A timely cultivation after 7 days may aid in providing season-long control.

Addition of oil concentrate

A nonphytotoxic oil concentrate (commonly referred to as oil concentrate) should always be added to the spray tank. The oil concentrate must contain either a petroleum or vegetable oil base and must meet the following criteria: 1) be nonphytotoxic, 2) contain only EPA-exempt ingredients, 3) provide good mixing quality in the jar test (see below), and 4) be successful in local experience.

The exact composition of suitable products will vary; however, vegetable and petroleum oil concentrates should contain emulsifiers which provide good mixing quality. For vegetable oil concentrates, it has been observed that highly refined vegetable oils are more satisfactory than unrefined vegetable oils. For additional information see **Jar test for estimating suitability of oil concentrates.**

Rate of oil concentrate:

2 pints/acre.

Jar test for estimating suitability of oil concentrates

1. **Water supply:** Use only water from intended source and at the source temperature.
2. **Amount of water in jar:**
   For 20 gal./A spray volume use 3⅓ cups (800 ml) of water.
   For 10 gal./A spray volume use 1⅔ cups (400 ml) of water.
   For 5 gal./A spray volume use ⅚ cup (200 ml) of water.
   For other spray volumes, adjust proportionately to above.
3. **Amount of herbicide and oil concentrate to add:** Add herbicide and oil concentrate at the rate of 1 teaspoon (5 ml) for each pint of recommended label rate.
4. **Add components in following sequence,** gently mixing between component addition:
   1) Oil concentrate.
   2) POAST.
5. **Cap jar,** invert 10 cycles, let stand for 15 minutes, evaluate.
6. **Evaluation:** An ideal tank mix combination will be uniform; thus, the suitability of the oil concentrate is questionable if any of the following are observed:
   Free oil at the surface—film or globules.
   Flocculation—fine particles which may be suspended in the liquid or found as a precipitated layer at the bottom of the jar.
   Clabbering—thickening texture (coagulated) resembling yogurt or a curd-like texture as with cottage cheese.

Mixing/spraying

Fill tank of a thoroughly clean sprayer half to two-thirds full with clean water. Start agitation and add oil concentrate; allow to mix thoroughly. Add POAST and remaining volume of water. Maintain constant agitation during application.

Restrictions and limitations

Do not apply POAST within 14 days of harvest.

Do not apply more than a total of 3 pints per acre in one season to cucumber, melons, pumpkin or squash.

Under weather conditions of high humidity and/or temperature, some flowers may be damaged or dislodged by applications of POAST and oil concentrate.

Do not make applications to grasses under stress, such as stress due to lack of moisture or herbicide injury, as unsatisfactory control may result.

Do not apply if rainfall is expected within one hour following application as grass control will be unsatisfactory.

Do not mix or apply POAST with any other pesticide, additive, or fertilizer except as specifically recommended on this labeling.

Environmental hazards

Do not apply directly to water.

Do not contaminate water by cleaning of equipment or disposal of wastes.

Endangered species concerns

The use of any pesticide in a manner that may kill or otherwise harm an endangered or threatened species or adversely modify their habitat is a violation of Federal law.

The use of this product is controlled to prevent death or harm to Solano grass which occurs in Solano County, **California.** Before using this product in this county you must obtain the EPA Endangered Species Bulletin (EPA/ES-85-13) available from either your County Agricultural Extension Agent, the Endangered Species Specialist in the California Department of Fish and Game, or the Regional Offices of the U.S. Fish and Wildlife Service (Portland, Oregon) or the U.S. Environmental Protection Agency

(San Francisco, California). THIS BULLETIN MUST BE REVIEWED PRIOR TO PESTICIDE USE. THE USE OF THIS PRODUCT IS PROHIBITED IN THIS COUNTY UNLESS SPECIFIED OTHERWISE IN THE BULLETIN.

Regional use map

All weed control recommendations for a given crop and weed are on a regional basis. There are three regions delineated on the U.S. map. Best results are obtained by using recommendations for your region only.

—California, Arizona, Nevada and Western New Mexico.

—High and Rolling Plains of Texas, Oklahoma and Eastern New Mexico.

—Midwest, South, Mountain (except Arizona, New Mexico and Nevada), and Pacific Northwest. (All states and areas except those listed above).

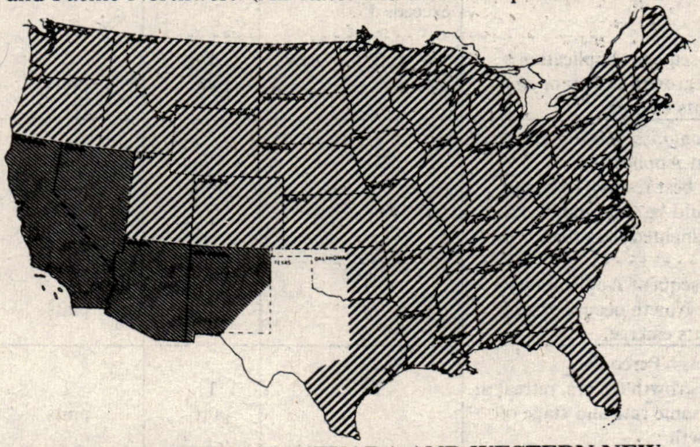

CALIFORNIA, ARIZONA, NEVADA AND WESTERN NEW MEXICO

Recommendations for grass control

- Apply to actively growing grasses at the sizes indicated in the table of this regional section.
- Cucumber, melons, pumpkin and squash at all stages of growth are tolerant to POAST® herbicide.
- Always follow recommendations given in the **Application information** section.
- Always adjust spray pressure, spray volume and height of spray boom to ensure penetration of plant canopy and thorough coverage of grasses to be controlled.
- Do not apply to drought-stressed grass or grass which has gone through an extended dry period.
- In irrigated areas it may be necessary to irrigate prior to treatment with POAST to ensure weeds are growing actively.
- **Always add 2 pints of oil concentrate per acre.**

Annual Grasses*
Cucumber, Melons, Pumpkin and Squash
California, Arizona, Nevada and Western New Mexico

Group	Grass	Time of Application	Rate of POAST per Acre	Oil Concentrate Rate per Acre
A	Barnyardgrass Broadleaf Signalgrass Fall Panicum Foxtails: Giant, Green, Yellow Johnsongrass, Seedling Junglerice Ryegrass, Annual Southwestern Cupgrass Texas Panicum Witchgrass Woolly Cupgrass	3-8"	1½ pints	2 pints
	Goosegrass Large Crabgrass Smooth Crabgrass	Up to 4"		
	Shattercane/Wildcane If needed, retreat at the same rate and stage of growth.	6-18"		
	Volunteer Corn	8-12"		

* For broad spectrum control of annual grasses, use 1½ pints of POAST per acre. If subsequent flushes of annual grasses emerge after the first application, make an additional application at the same rate and at the recommended stage of growth.

HIGH AND ROLLING PLAINS OF TEXAS, OKLAHOMA AND EASTERN NEW MEXICO

Recommendations for grass control

- Apply to actively growing grasses at the sizes indicated in the tables of this regional section.
- Cucumber, melons, pumpkin and squash at all stages of growth are tolerant to POAST® herbicide.
- Always follow recommendations given in the **Application information** section.
- Always adjust spray pressure, spray volume, and height of spray boom to ensure penetration of plant canopy and thorough coverage of grasses to be controlled.
- Do not apply to drought-stressed grass, or grass which has gone through an extended dry period.
- In irrigated areas it may be necessary to irrigate prior to treatment with POAST to ensure weeds are growing actively.
- **Always add 2 pints of oil concentrate per acre.**

Annual Grasses*
Cucumber, Melons, Pumpkin and Squash
High and Rolling Plains of Texas, Oklahoma and Eastern New Mexico

Group	Grass	Time of Application	Rate of POAST per Acre	Oil Concentrate Rate per Acre
A	Barnyardgrass Broadleaf Signalgrass Browntop Panicum Fall Panicum Foxtails: Giant, Green, Yellow Johnsongrass, Seedling Junglerice Red Sprangletop Texas Panicum Witchgrass	2-6"	1½ pints	2 pints
	Goosegrass Large Crabgrass Smooth Crabgrass	Up to 4"		
	Shattercane/Wildcane If needed, retreat at the same rate and stage of growth.	6-10"		
	Volunteer Corn	8-10"		

* For broad spectrum control of annual grasses, use 1½ pints of POAST per acre. If subsequent flushes of annual grasses emerge after first application, make an additional application at the same rate and at the recommended stage of growth.

Perennial Grasses
Cucumber, Melons, Pumpkin and Squash
High and Rolling Plains of Texas, Oklahoma and Eastern New Mexico

Grass	Time of Application	Rate of POAST per Acre	Oil Concentrate Rate per Acre
Bermudagrass • First Application Control may be partial or inconsistent, however bermudagrass growth will be suppressed.	Before plant diameter exceeds 6" or leaf height above ground exceeds 1".	1½ pints	2 pints
• Second Application	21 days after first application.	1½ pints	2 pints
Johnsongrass, Rhizome For best results, rhizomes should be thoroughly fragmented (less than 6"). Adjust volume of spray solution to a maximum of 10 gallons and a minimum of 5 gallons per acre while maintaining a ground speed of no more than 6 miles per hour.			
• First Application	6-10"	1½ pints	2 pints
• Subsequent Application When regrowth occurs or new plants emerge.	4-8"	1 pint	2 pints

MIDWEST, MOUNTAIN (EXCEPT ARIZONA, NEW MEXICO AND NEVADA) SOUTH, NORTHEAST AND PACIFIC NORTHWEST

Recommendations for grass control

- Apply to actively growing grasses at the sizes indicated in the tables of this regional section.
- Cucumber, melons, pumpkin and squash at all stages of growth are tolerant to POAST® herbicide.
- Always follow recommendations given in the **Application information** section.

- Always adjust spray pressure, spray volume and height of spray boom to ensure penetration of plant canopy and thorough coverage of grasses to be controlled.
- Do not apply to drought-stressed grass, or grass which has gone through an extended dry period.
- In irrigated areas, it may be necessary to irrigate prior to treatment with POAST to ensure weeds are growing actively.
- **Always add 2 pints of oil concentrate per acre.**

Annual Grasses*
Cucumber, Melons, Pumpkin and Squash
Midwest, Mountain (except Arizona, New Mexico and Nevada), South, Northeast and Pacific Northwest

Group	Grass	Time of Application	Rate of POAST per Acre	Oil Concentrate Rate per Acre
A	Wild Proso Millet	4-10"	½ pint	2 pints
B	Barnyardgrass Broadleaf Signalgrass Browntop Panicum Fall Panicum Foxtails: Giant, Green, Yellow Johnsongrass, Seedling Junglerice Red Sprangletop Ryegrass, Annual Texas Panicum Witchgrass Woolly Cupgrass	3-8"	1 pint	2 pints
	Goosegrass Large Crabgrass Smooth Crabgrass	Up to 4"		
	Shattercane/Wildcane If needed, retreat at the same rate and stage of growth.	6-18"		
	Volunteer Corn	6-20"		
C	Volunteer Cereals* Barley Rye Oats Wheat POAST is not recommended for spring control of volunteer cereals that emerged the previous fall.	Before tillering, up to 6" and prior to wintering.	1½ pints	2 pints
	Wild Oats	2-4"		

* For broad spectrum control of annual grasses in Groups A & B, use 1 pint of POAST per acre. When weed populations include additional grasses in Group C increase the rate of POAST to 1½ pints. If a second flush of annual grasses emerge after the first application, make an additional application at the same rate and at the recommended stage of growth.

Rescue treatment for controlling selected annual grasses

For best results, always apply POAST® herbicide to annual grasses at the growth stage and rate specified in the table on page 5. However, if POAST cannot be applied at the recommended time, larger annual grasses can be controlled with a later application by increasing the rate of POAST. Apply to actively growing grasses at the rates and sizes indicated at the right.

Annual Grasses
Cucumber, Melons, Pumpkin and Squash
Rescue Treatment
Midwest, South, Mountain (except Arizona, New Mexico and Nevada), Northeast and Pacific Northwest

Grass	Time of Application	Rate of POAST per Acre	Oil Concentrate Rate per Acre
Wild Proso Millet	10-24"	1 pint	2 pints
Foxtails: Giant, Green Yellow Johnsongrass, Seedling	8-16"	1½ pints	
Texas Panicum Fall Panicum Barnyardgrass Broadleaf Signalgrass	8-12"		
Large Crabgrass Smooth Crabgrass Goosegrass	6-8"		

Perennial Grasses
Cucumber, Melons, Pumpkin and Squash
Midwest, South, Mountain (except Arizona, New Mexico and Nevada), Northeast and Pacific Northwest

Grass	Time of Application	Rate of POAST per Acre	Oil Concentrate Rate per Acre
Bermudagrass • First Application	Before plant diameter exceeds 6" or leaf height above ground exceeds 1".	1½ pints	2 pints
• Subsequent Application If regrowth occurs or new plants emerge.	1-4" length of new plants.	1 pint	2 pints
Johnsongrass, Rhizome • First Application For best results, rhizomes should be thoroughly fragmented (less than 6").	15-25"	1½ pints	2 pints
• Subsequent Application If regrowth occurs or new plants emerge.	6-12"	1 pint	2 pints
Ryegrass, Perennial If regrowth occurs, retreat at the same rate and stage of growth.	3-8"	1 pint	2 pints
Quackgrass • First Application For best results, rhizomes should be thoroughly fragmented (less than 6").	6-8"	1½ pints	2 pints
• Subsequent Application If regrowth occurs or new plants emerge. Control may be partial or inconsistent, however, quackgrass growth will be suppressed. **Note: A cultivation no sooner than 14 days after application but within 21 days of application will aid in control.**	6-8"	1½ pints	2 pints
Wirestem Muhly If regrowth occurs, retreat at the same rate and stage of growth.	Up to 6"	1¼ pints	2 pints

PROCEDURE FOR CLEANING EQUIPMENT

ATTENTION! Clean sprayer thoroughly before and after application of POAST® herbicide.

Clean sprayer thoroughly before application of POAST, particularly if a herbicide was used which has the potential to injure the crop to be sprayed with POAST.

Failure to clean sprayer thoroughly after the application of POAST may result in injury to any grass crop subsequently sprayed, such as corn, sorghum, small grains, rice and turf. Fill the sprayer with clean water and add a commercial sprayer cleaner or a surfactant/adjuvant at the recommended rate on its label. Circulate through entire sprayer system. Spray approximately half the tank solution through the hoses, booms and nozzles to clean these parts. Drain the tank and rinse the total system thoroughly several times with clean water.

STORAGE AND DISPOSAL

Do not contaminate water, food, or feed by storage or disposal.

Pesticide wastes are toxic. Improper disposal of excess pesticide, spray mixture, or rinsate is a violation of Federal law. If these wastes cannot be disposed of by use according to label instructions, contact your State Pesticide or Environmental Control Agency, or the Hazardous Waste representative at the nearest EPA Regional Office for guidance.

Triple rinse container (or equivalent). Then offer for recycling or reconditioning, or puncture and dispose of in a sanitary landfill, or by incineration, or, if allowed by state and local authorities, by burning. If burned, stay out of smoke.

CONDITIONS OF SALE AND WARRANTY

The Directions for use of this product reflect the opinion of experts based on field use and tests. The directions are believed to be reliable and should be followed carefully. However, it is impossible to eliminate all risks inherently associated with use of this product. Crop injury, ineffectiveness or other unintended consequences may result from factors such as, but not limited to, weather conditions, presence of other materials, or use of the product in a manner inconsistent with its labeling, all of which are beyond the control of BASF CORPORATION ("BASF") or the Seller. All such risks shall be assumed by the Buyer.

BASF warrants that this product conforms to the chemical description on the label and is reasonably fit for the purposes referred to in the Directions for use, subject to the inherent risks referred to above. BASF MAKES NO OTHER EXPRESS OR IMPLIED WARRANTY OF FITNESS OR MERCHANTABILITY OR ANY OTHER EXPRESS OR IMPLIED WARRANTY. IN NO CASE SHALL BASF OR THE SELLER BE LIABLE FOR CONSEQUENTIAL, SPECIAL OR INDIRECT DAMAGES RESULTING FROM THE USE OR HANDLING OF THIS PRODUCT. BASF and the Seller offer this product, and the Buyer and User accept it, subject to the foregoing Conditions of sale and warranty which may be varied only by agreement in writing signed by a duly authorized representative of BASF.

POAST is a registered trademark of BASF AG.

PO 1 023 JJ6669 A0390789

SUPPLEMENTAL LABELING

POAST® HERBICIDE

POAST EPA Reg. No. 7969-58

Label approval pending in California

FOR USE IN CELERY, HEAD AND LEAF LETTUCE AND SPINACH.

All applicable directions, restrictions, precautions and Conditions of sale and warranty on the EPA-registered label are to be followed. This labeling must be in the possession of the user at the time of herbicide application.

Re-entry and workers' protection statements

Do not apply this product in such a manner as to directly or through drift expose workers or other persons, except those knowingly involved in the application. The area being treated must be vacated by unprotected persons.

Do not enter treated areas without protective clothing until sprays have dried.

Because certain states may require more restrictive re-entry intervals for various crops treated with this product, consult your State Department of Agriculture for further information.

Written or oral warnings must be given to workers who are expected to be in a treated area or in an area about to be treated with this product. Oral warnings must inform workers of areas or fields that may not be entered without specific protective clothing until sprays have dried, and appropriate actions to take in case of accidental exposure. When oral warnings are given, warnings shall be given in a language customarily understood by workers. Oral warnings must be given if there is reason to believe that written warnings cannot be understood by workers. Written warnings must include the following information: "WARNING. Area treated with POAST® herbicide on (date of application). Do not enter without appropriate protective clothing until sprays have dried. If contacted, flush eyes immediately with water for 15 minutes. Get medical attention.

In case of contact with skin or clothing, remove contaminated clothing, wash skin thoroughly with soap and water. Call physician if irritation occurs. Wash contaminated clothing before re-use."

DIRECTIONS FOR USE

It is a violation of Federal law to use this product in a manner inconsistent with its labeling.

GENERAL INFORMATION

POAST is a selective broad spectrum postemergent herbicide for control of annual and perennial grass weeds. POAST does not control sedges or broadleaf weeds.

Since essentially all grass crops such as sorghum, corn, and small grains, as well as ornamental grasses such as turf, are susceptible to POAST, avoid all direct or indirect contact with any grass crop.

Control symptoms: POAST rapidly enters the plant through the foliage and translocates throughout the plant. Control symptoms exhibited by the grass plant progress from a slowing or stopping of growth (generally within two days) to reddening of the foliage to leaf tip burn. Subsequently, burn back of the foliage occurs. These symptoms will generally be observed within three weeks depending on environmental conditions.

Under some conditions, or with certain oil concentrates, a very slight leaf speckling can occur. Spinach, head and leaf lettuce, and celery will outgrow these symptoms and later growth is not affected.

Application information*

Apply POAST to actively growing grasses when they are at the proper growth stage as specified by the **Recommendations for grass control.**

Thorough coverage of the grass foliage is essential. This is because the effectiveness of POAST is dependent on the absorption and movement of POAST throughout the plant.

Do not make applications to grasses under stress, such as stress due to lack of moisture, herbicide injury, mechanical injury or cold temperatures, since unsatisfactory control will probably result.

Broadcast application

Nozzle type: For broadcast application use standard high pressure pesticide hollow cone or flat fan nozzles. Do not use flood or whirl chamber nozzles.

Spray gallonage: Use a minimum volume of 5 gallons and a maximum volume of 20 gallons of spray solution per acre. In Western and Mountain States use a minimum of 10 gallons per acre. In the High and Rolling Plains of Texas, Oklahoma and Eastern New Mexico use a minimum of 5 gallons per acre and a maximum of 10 gallons per acre.

Spray pressure: When using standard high pressure hollow cone or flat fan nozzles adjust pressure to a minimum of 40 psi and a maximum of 60 psi at the nozzle.

* To be applied by ground equipment only.

Always adjust spray pressure, spray volume and height of spray boom to ensure penetration of plant canopy and thorough coverage of grasses to be controlled. Tall weeds, such as volunteer corn, may require the boom height to be as high as 20 inches above the corn height. Refer to the nozzle manufacturer's directions for recommended height.

Band applications: Banding of POAST® herbicide may be used to control annual grasses. Grasses which are not covered or only partly covered by a spray mixture of POAST will not be adequately controlled. All recommendations are on a broadcast basis unless otherwise stated. When banding, rates of POAST, oil concentrate and water should be reduced in proportion to the area sprayed.

Other spray equipment: Do not use selective application equipment such as recirculating sprayers, wiper applicators, or shielded applicators. Application of POAST with controlled drop applicator (CDA) nozzles is not recommended due to erratic coverage which causes inconsistent weed control.

Cultivation information

Do not cultivate within 5 days prior to application of POAST or within 7 days following application.

A timely cultivation after 7 days may aid in providing season-long control. For control of quackgrass, delay cultivation until 14-21 days after application of POAST.

Addition of oil concentrate

A nonphytotoxic oil concentrate (commonly referred to as oil concentrate) should always be added to the spray tank. The oil concentrate must contain either a petroleum or vegetable oil base and must meet the following criteria: 1) be nonphytotoxic, 2) contain only EPA-exempt ingredients, 3) provide good mixing quality in the jar test (see the following section), and 4) be successful in local experience.

The exact composition of suitable products will vary; however, vegetable and petroleum oil concentrates should contain emulsifiers which provide good mixing quality. For vegetable oil concentrates, it has been observed that highly refined vegetable oils are more satisfactory than unrefined vegetable oils. For additional information see **Jar test for estimating suitability of oil concentrates.**

Rate of oil concentrate: Ground application—2 pints/acre.

Jar test for estimating suitability of oil concentrates

1. **Water supply:** Use only water from intended source and at the source temperature.
2. **Amount of water in jar:**
   For 20 gal./A spray volume use 3⅓ cups (800 ml) of water.
   For 10 gal./A spray volume use 1⅔ cups (400 ml) of water.
   For 5 gal./A spray volume use ⅚ cups (200 ml) of water.
   For other spray volumes, adjust proportionately to above.
3. **Amount of herbicide and oil concentrate to add:** Add herbicide and oil concentrate at the rate of 1 teaspoon (5 ml) for each pint of recommended label rate.
4. **Add components in following sequence,** gently mixing between component additions:
   1) Oil concentrate.
   2) POAST.
5. **Cap jar,** invert 10 cycles, let stand for 15 minutes, evaluate.
6. **Evaluation:** An ideal tank mix combination will be uniform; thus, the suitability of the oil concentrate is questionable if any of the following are observed:
   Free oil at the surface—film or globules.
   Flocculation—fine particles which may be suspended in the liquid or found as a precipitated layer at the bottom of the jar.
   Clabbering—thickening texture (coagulated) resembling yogurt or a curd-like texture as with cottage cheese.

Mixing/spraying

Fill tank of a thoroughly clean sprayer half to two-thirds full with clean water. Start agitation and add oil concentrate; allow to mix thoroughly. Add POAST and remaining volume of water. Maintain constant agitation during application.

Desired Spray Solution Volume	Amount to be Added to Obtain a 1% Solution	
	POAST	Oil Concentrate
1 Gallon	1¼ Fl. Oz.	1¼ Fl. Oz.
25 Gallons	1 Qt.	1 Qt.
50 Gallons	2 Qts.	2 Qts.
100 Gallons	4 Qts.	4 Qts.

* 2 Tablespoons = 1 Fl. Oz.

Spot or small area treatment

When using knapsack sprayers or high volume spray equipment utilizing hand guns or other suitable nozzle arrangements, make 1% solution of POAST® herbicide in water. A recommended oil concentrate must also be used at a concentration of 1%.

Apply to foliage of grasses on a spray-to-wet basis. Spray coverage should be uniform and complete. Do not spray to point of runoff.

Prepare the desired volume of spray solution by mixing the amount of POAST and the amount of oil concentrate in water according to the table on Page 2.

Restrictions and limitations

Do not apply POAST within 30 days of harvest to celery and head lettuce.

Do not apply POAST within 15 days of harvest to leaf lettuce and spinach.

Do not apply more than a total of 3 pints per acre in one season to celery, head or leaf lettuce, or spinach.

Do not make applications to grasses under stress, such as stress due to lack of moisture or herbicide injury, as unsatisfactory control may result.

Do not apply if rainfall is expected within one hour following application as grass control will be unsatisfactory.

Do not mix or apply POAST with any other pesticide, additive, or fertilizer except as specifically recommended on this labeling.

Do not apply this product through any type of irrigation system.

Do not apply POAST when the air temperature exceeds 90°F **and** when the relative humidity exceeds 60%.

Environmental hazards

Do not apply directly to water.

Do not contaminate water by cleaning of equipment or disposal of wastes.

Endangered species concerns

The use of any pesticide in a manner that may kill or otherwise harm an endangered or threatened species or adversely modify their habitat is a violation of Federal law.

The use of this product is controlled to prevent death or harm to Solano grass which occurs in Solano County, **California.** Before using this product in this county you must obtain the EPA Endangered Species Bulletin (EPA/ES-85-13) available from either your County Agricultural Extension Agent, the Endangered Species Specialist in the California Department of Fish and Game, or the Regional Offices of the U.S. Fish and Wildlife Service (Portland, Oregon) or the U.S. Environmental Protection Agency (San Francisco, California). THIS BULLETIN MUST BE REVIEWED PRIOR TO PESTICIDE USE. THE USE OF THIS PRODUCT IS PROHIBITED IN THIS COUNTY UNLESS SPECIFIED OTHERWISE IN THE BULLETIN.

Regional use map

All weed control recommendations for a given crop and weed are on a regional basis. There are three regions delineated on the U.S. map. Best results are obtained by using recommendations for your region only.

—Midwest, South and Northeast (All states and areas except those listed below.)

—High and Rolling Plains of Texas, Oklahoma and Eastern New Mexico.

—Western and Mountain States.

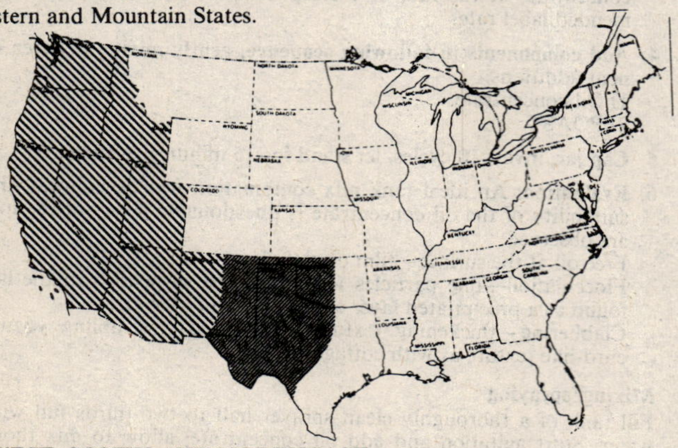

MIDWEST, SOUTH AND NORTHEAST

Recommendations for grass control

- Apply to actively growing grasses at the sizes indicated in the tables of this regional section.
- Celery, head and leaf lettuce and spinach are tolerant to POAST® herbicide at all stages of growth.
- Always follow recommendations given in the **Application information** section.
- Always adjust spray pressure, spray volume and height of spray boom to ensure penetration of plant canopy and thorough coverage of grasses to be controlled.
- Do not apply to drought-stressed grass, or grass which has gone through an extended dry period.
- In irrigated areas, it may be necessary to irrigate prior to treatment with POAST to ensure weeds are growing actively.
- **Always add 2 pints of oil concentrate per acre.**

Special Rate for Early Treatment*
Celery, Head and Leaf Lettuce and Spinach
Midwest, South and Northeast

Group	Grass	Time of Application	Rate of POAST per Acre	Additive Rate per Acre: Oil Concentrate: Ground
A	Wild Proso Millet	4-10"	½ pt. (16 acres per gallon)	2 pts.
B	Goosegrass	1-3"	¾ pt. (10.7 acres per gallon)	2 pts.
	Barnyardgrass** (Midwest Only) Broadleaf Signalgrass Fall Panicum Texas Panicum Foxtail: Giant, Green	1-4"		

* For broad spectrum control of annual grasses in Groups A & B, use ¾ pint of POAST per acre. If additional applications are needed, apply at the same rate and at the recommended stage of growth.

** In these states use 1 pint per acre as recommended in the next table: AL, AR, FL, GA, LA, MS, NC, SC, TN, TX, VA.

Annual Grasses—Standard Recommendations*
Celery, Head and Leaf Lettuce and Spinach
Midwest, South and Northeast

Group	Grass	Time of Application	Rate of POAST per Acre	Additive Rate per Acre: Oil Concentrate
A	Wild Proso Millet	4-10"	½ pt. (16 acres per gallon)	2 pts.
B	Wild Oats	Up to 4"	1 pt. (8 acres per gallon)	2 pts.
	Goosegrass Smooth Crabgrass	Up to 6"		2 pts.
	Large Crabgrass	Up to 6"		2 pts.
	Barnyardgrass Broadleaf Signalgrass Browntop Panicum Fall Panicum Foxtails: Giant, Green, Yellow Johnsongrass, Seedling Junglerice Red Sprangletop Ryegrass, Annual Texas Panicum Witchgrass Woolly Cupgrass	Up to 8"		2 pts.
	Shattercane/Wildcane If needed, retreat at the same rate and stage of growth.	6-18"		2 pts.
	Volunteer Corn Maintain sufficient boom height above volunteer corn plants for best spray coverage.	Up to 20"		2 pts.
C	Field Sandbur (Midwest only)	Up to 3"	1¼ pts. (6.4 acres per gallon)	2 pts.
D	Volunteer Cereals			

Group	Grass	Time of Application	Rate of POAST per Acre	Additive Rate per Acre
				Oil Concentrate
	Barley Oats Rye Wheat Not recommended for spring control of volunteer cereals that emerged the previous fall.	Before tillering, up to 4" and prior to overwintering	1½ pts. (5.3 acres per gallon)	2 pts.

* For broad spectrum control of annual grasses in Groups A and B (above), use 1 pint of POAST® herbicide per acre. When weed populations include additional grasses in Group C or D increase the rate of POAST as indicated. If later flushes of annual grasses emerge after the first application, make additional applications at the same rate and at the same recommended stage of growth. Do not exceed 3 pints/acre/season.

Rescue treatment for controlling selected annual grasses

For best results, always apply POAST® herbicide to annual grasses at the growth stage and rate specified in the table on page 5. However, if POAST cannot be applied at the recommended time, larger annual grasses can be controlled with a later application by increasing the rate of POAST. Apply to actively growing grasses at the rates and sizes indicated.

Annual Grasses—Rescue Treatment
Celery, Head and Leaf Lettuce and Spinach
Midwest, South and Northeast

Grass	Time of Application	Rate of POAST per Acre	Oil Concentrate Rate per Acre
Wild Proso Millet	10-24"	1 pt.	
Foxtails: Giant, Green, Yellow Johnsongrass, Seedling	8-16"	1½ pts. (5.3 acres per gallon)	2 pts. (4 acres per gallon)
Texas Panicum Fall Panicum Barnyardgrass Broadleaf Signalgrass	8-12"		
Large Crabgrass Smooth Crabgrass Goosegrass	6-8"		

Perennial Grasses
Celery, Head and Leaf Lettuce and Spinach
Midwest, South and Northeast

Grass	Time of Application	Rate of POAST per Acre	Additives Rate per Acre
			Oil Concentrate
Bermudagrass • First Application	Before plant diameter exceeds 6" or leaf height above ground exceeds 1".	1½ pts. (5.3 acres per gallon)	2 pts.
• Second Application If regrowth occurs or new plants emerge.	1-4" length of new plants.	1 pt. (8 acres per gallon)	2 pts.
Johnsongrass, Rhizome • First Application For best results, rhizomes should be thoroughly fragmented (less than 6"). (When using 11-20 gallons of spray volume per acre, use 1½ pints of POAST.)	15-25"	1 pt. (8 acres per gallon)	2 pts.
• Subsequent Application If regrowth occurs or new plants emerge.	6-12"	1 pt. (8 acres per gallon)	2 pts.
Ryegrass, Perennial If regrowth occurs or new plants emerge retreat at the same rate and stage of growth.	3-8"	1 pt. (8 acres per gallon)	2 pts.
Quackgrass • First Application For best results, rhizomes should be thoroughly fragmented (less than 6").	6-8"	1½ pts. (5.3 acres per gallon)	2 pts.
• Second Application If regrowth occurs or new plants emerge. Control may be partial or inconsistent, however, quackgrass growth will be suppressed. Note: A cultivation 14 to 21 days after application will aid in control.	6-8"	1½ pts. (5.3 acres per gallon)	2 pts.
Wirestem Muhly If regrowth occurs or new plants emerge retreat at the same rate and stage of growth.	Up to 6"	1¼ pts. (6.4 acres per gallon)	2 pts.

HIGH AND ROLLING PLAINS OF TEXAS, OKLAHOMA AND EASTERN NEW MEXICO

Recommendations for grass control

- Apply to actively growing grasses at the sizes indicated in the tables of this regional section.
- Celery, head and leaf lettuce and spinach are tolerant to POAST® herbicide at all stages of growth.
- Always follow recommendations given in the **Application information** section.
- Always adjust spray pressure, spray volume, and height of spray boom to ensure penetration of plant canopy and thorough coverage of grasses to be controlled.
- Do not apply to drought-stressed grass, or to grass which has gone through an extended dry period.
- In irrigated areas it may be necessary to irrigate prior to treatment with POAST to ensure weeds are growing actively.
- **Always add 2 pints of oil concentrate per acre.**

Annual Grasses—Standard Recommendations*
Celery, Head and Leaf Lettuce and Spinach
High and Rolling Plains of Texas, Oklahoma and Eastern New Mexico

Group	Grass	Time of Application	Rate of POAST per Acre	Additive Rate per Acre
				Oil Concentrate
A	Goosegrass Smooth Crabgrass	Up to 4"	1½ pts. (5.3 acres per gallon)	2 pts.
	Large Crabgrass	Up to 4"		2 pts.
	Barnyardgrass Broadleaf Signalgrass Browntop Panicum Fall Panicum Foxtails: Giant, Green, Yellow Johnsongrass, Seedling Junglerice Red Sprangletop Texas Panicum Witchgrass	Up to 8"		2 pts.
	Shattercane/Wildcane If needed, retreat at the same rate and stage of growth.	6-18"		2 pts.
	Volunteer Corn Maintain sufficient boom height above volunteer corn plants for best spray coverage.	Up to 20"		2 pts.

* For broad spectrum control of annual grasses in Group A (above), use 1½ pints of POAST per acre.

If later flushes of annual grasses emerge after the first application, make an additional application at the same rate and at the same recommended stage of growth.

Perennial Grasses
Celery, Head and Leaf Lettuce and Spinach
High and Rolling Plains of Texas, Oklahoma and Eastern New Mexico

Grass	Time of Application	Rate of POAST per Acre	Additive Rate per Acre
			Oil Concentrate
Bermudagrass			

Grass	Time of Application	Rate of POAST per Acre	Additive Rate per Acre
			Oil Concentrate
• First Application	Before plant diameter exceeds 6" or leaf height above ground exceeds 1".	$1^1/_2$ pts. (5.3 acres per gallon)	2 pts.
• Second Application If regrowth occurs or new plants emerge.	21 days after first application.	$1^1/_2$ pts. (5.3 acres per gallon)	2 pts.
Johnsongrass, Rhizome • First Application For best results, rhizomes should be thoroughly fragmented (less than 6"). Adjust volume of spray solution to a maximum of 10 gallons and a minimum of 5 gallons per acre while maintaining a speed of no more than 6 miles per hour.	6-10"	$1^1/_2$ pts. (5.3 acres per gallon)	2 pts.
• Subsequent Application If regrowth occurs or new plants emerge.	4-8"	1 pt. (8 acres per gallon)	2 pts.

WESTERN AND MOUNTAIN STATES

Recommendations for grass control

- Apply to actively growing grasses at the sizes indicated in the tables of this regional section.
- Celery, head and leaf lettuce and spinach are tolerant to POAST® herbicide at all stages of growth.
- Always follow recommendations given in the **Application information** section.
- Always adjust spray pressure, spray volume, and height of spray boom to ensure penetration of plant canopy and thorough coverage of grasses to be controlled.
- Do not apply to drought-stressed grass, or to grass which has gone through an extended dry period.
- In irrigated areas it may be necessary to irrigate prior to treatment with POAST to ensure weeds are growing actively.
- **Always add 2 pints of oil concentrate per acre.**

Annual Grasses—Standard Recommendations*
Celery, Head and Leaf Lettuce and Spinach
Western and Mountain States

Group	Grass	Time of Application	Rate of POAST per Acre	Additive Rate per Acre
				Oil Concentrate
A	Goosegrass Smooth Crabgrass Large Crabgrass Wild Oats (Idaho, Oregon, Washington only)	Up to 4"	$1^1/_2$ pts. (5.3 acres per gallon)	2 pts.
	Barnyardgrass Fall Panicum Foxtails: Giant, Green, Yellow Johnsongrass, Seedling Junglerice Ryegrass, Annual Southwestern Cupgrass Witchgrass	Up to 8"		2 pts.
	Shattercane/Wildcane If needed, retreat at the same rate and stage of growth.	6-18"		2 pts.
	Volunteer Corn Maintain sufficient boom height above volunteer corn plants for best spray coverage.	Up to 12"		2 pts.

* For broad spectrum control of annual grasses in Group A (above), use $1^1/_2$ pints of POAST per acre.

If later flushes of annual grasses emerge after the first application, make an additional application at the same rate and at the same recommended stage of growth.

Perennial Grasses
Celery, Head and Leaf Lettuce and Spinach
Western and Mountain States

Grass	Time of Application	Rate of POAST per Acre	Additive Rate per Acre
			Oil Concentrate
Ryegrass, Perennial If regrowth occurs or new plants emerge retreat at the same rate and stage of growth.	3-8"	$1^1/_2$ pts. (5.3 acres per gallon)	2 pts.

PROCEDURE FOR CLEANING EQUIPMENT

ATTENTION! Clean sprayer thoroughly before and after application of POAST® herbicide.

Clean sprayer thoroughly before application of POAST, particularly if a herbicide was used which has the potential to injure the crop to be sprayed with POAST.

Failure to clean sprayer thoroughly after the application of POAST may result in injury to any grass crop subsequently sprayed, such as corn, sorghum, small grains, rice and turf. Fill the sprayer with clean water and add a commercial sprayer cleaner or a surfactant/adjuvant at the recommended rate on its label. Circulate through entire sprayer system. Spray approximately half the tank solution through the hoses, booms and nozzles to clean these parts. Drain the tank and rinse the total system thoroughly several times with clean water.

CONDITIONS OF SALE AND WARRANTY

The Directions for use of this product reflect the opinion of experts based on field use and tests. The directions are believed to be reliable and should be followed carefully. However, it is impossible to eliminate all risks inherently associated with use of this product. Crop injury, ineffectiveness or other unintended consequences may result from factors such as, but not limited to, weather conditions, presence of other materials, or use of the product in a manner inconsistent with its labeling, all of which are beyond the control of BASF CORPORATION ("BASF") or the Seller. All such risks shall be assumed by the Buyer.

BASF warrants that this product conforms to the chemical description on the label and is reasonably fit for the purposes referred to in the Directions for use, subject to the inherent risks referred to above. BASF MAKES NO OTHER EXPRESS OR IMPLIED WARRANTY OF FITNESS OR MERCHANTABILITY OR ANY OTHER EXPRESS OR IMPLIED WARRANTY. IN NO CASE SHALL BASF OR THE SELLER BE LIABLE FOR CONSEQUENTIAL, SPECIAL OR INDIRECT DAMAGES RESULTING FROM THE USE OR HANDLING OF THIS PRODUCT. BASF and the Seller offer this product, and the Buyer and User accept it, subject to the foregoing Conditions of sale and warranty which may be varied only by agreement in writing signed by a duly authorized representative of BASF.

POAST is a registered trademark of BASF AG.

PO 1 029 JJ6203 A0190388

SUPPLEMENTAL LABELING

POAST® HERBICIDE

POAST EPA Reg. No. 7969-58

Label approval pending in California.

FOR USE ON POTATOES

All applicable directions, restrictions, precautions and Conditions of sale and warranty on the EPA-registered label are to be followed. This labeling must be in the possession of the user at the time of herbicide application.

Re-entry and workers' protection statements

Do not apply this product in such a manner as to directly or through drift expose workers or other persons, except those knowingly involved in the application. The area being treated must be vacated by unprotected persons.

Do not enter treated areas without protective clothing until sprays have dried.

Because certain states may require more restrictive re-entry intervals for various crops treated with this product, consult your State Department of Agriculture for further information.

Written or oral warnings must be given to workers who are expected to be in a treated area or in an area about to be treated with this product. Oral warnings must inform workers of areas or fields that may not be entered without specific protective clothing until sprays have dried, and appropriate actions to take in case of accidental exposure. When oral warnings are given, warnings shall be given in a language customarily understood by workers. Oral warnings must be given if there is reason to believe that written warnings cannot be understood by workers. Written warnings must include the following information: "WARNING. Area treated with POAST® herbicide on (date of application). Do not enter without appropriate protective clothing until sprays have dried. If contacted, flush eyes immediately with water for 15 minutes. Get medical attention.

In case of contact with skin or clothing, remove contaminated clothing, wash skin thoroughly with soap and water. Call physician if irritation occurs. Wash contaminated clothing before reuse."

DIRECTIONS FOR USE

It is a violation of Federal law to use this product in a manner inconsistent with its labeling.

General information

POAST is a selective broad spectrum postemergence herbicide for control of annual and perennial grass weeds. POAST does not control sedges or broadleaf weeds.

Since essentially all grass crops such as sorghum, corn, and small grains, as well as ornamental grasses such as turf, are susceptible to POAST, avoid all direct or indirect contact with any grass crop.

Control symptoms: POAST rapidly enters the plant through the foliage and translocates throughout the plant. Control symptoms exhibited by the grass plant progress from a slowing or stopping of growth (generally within two days), to reddening of the foliage to leaf tip burn. Subsequently, burn back of the foliage occurs. These symptoms will generally be observed within three weeks depending on environmental conditions.

Application information

Apply POAST to actively growing grasses when they are at the proper growth stage as specified by the **Recommendations for grass control.**

Do not make applications to grasses under stress, such as stress due to lack of moisture, herbicide injury, mechanical injury or cold temperatures, since unsatisfactory control will probably result.

Ground equipment: Thorough spray coverage of grass foliage is essential. For broadcast application use standard high pressure pesticide hollow cone or flat fan nozzles. Do not use flood or whirl chamber nozzles. Application of POAST with control drop applicator (CDA) nozzles is not recommended due to erratic coverage which causes inconsistent weed control. Use a minimum volume of 5 gallons and a maximum volume of 20 gallons of spray solution per acre. In Arizona, California, Idaho, Nevada, Oregon, Utah, Washington and Western New Mexico use a minimum of 10 gallons per acre. When using standard high pressure hollow cone or flat fan nozzles adjust pressure to a minimum of 40 psi and a maximum of 60 psi at the nozzle.

Always adjust spray pressure, spray volume and height of spray boom to ensure penetration of plant canopy and thorough coverage of grasses to be controlled.

Do not use selective application equipment such as recirculating sprayers, wiper applicators, or shielded applicators.

Air equipment: Thorough spray coverage of grass foliage is essential. Use a minimum of 5 gallons of water per acre. Increase water volume to 10 gallons per acre if grass foliage is dense.

Cultivation information

Do not cultivate within 5 days prior to application of POAST® herbicide or within 7 days following application.

A timely cultivation after 7 days may aid in providing season-long control. For control of quackgrass delay cultivation until 14-21 days after application of POAST.

Addition of oil concentrate

A nonphytotoxic oil concentrate (commonly referred to as oil concentrate) should always be added to the spray tank. The oil concentrate must contain either a petroleum or vegetable oil base and must meet the following criteria: 1) be nonphytotoxic, 2) contain only EPA-exempt ingredients, 3) provide good mixing quality in the jar test (see below), and 4) be successful in local experience.

The exact composition of suitable products will vary; however, vegetable and petroleum oil concentrates should contain emulsifiers which provide good mixing quality. For vegetable oil concentrates, it has been observed that highly refined vegetable oils are more satisfactory than unrefined vegetable oils. For additional information see **Jar test for estimating suitability of oil concentrates.**

Rate of oil concentrate: Ground and air application—2 pints/acre.

Jar test for estimating suitability of oil concentrates

1. **Water supply:** Use only water from intended source and at the source temperature.
2. **Amount of water in jar:** For 20 gal./A spray volume use $3\frac{1}{3}$ cup (800 ml) of water.
   For 10 gal./A spray volume use $1\frac{2}{3}$ cups (400 ml) of water.
   For 5 gal./A spray volume use $\frac{5}{6}$ cups (200 ml) of water.
   For other spray volumes, adjust proportionately to above.
3. **Amount of herbicide and oil concentrate to add:** Add herbicide and oil concentrate at the rate of 1 teaspoon (5 ml) for each pint of recommended label rate.
4. **Add components in following sequence,** gently mixing between component additions:
   1) Oil concentrate.
   2) POAST
5. **Cap jar,** invert 10 cycles, let stand for 15 minutes, evaluate.
6. **Evaluation:** An ideal tank mix combination will be uniform; thus, the suitability of the oil concentrate is questionable if any of the following are observed:
   Free oil at the surface—film or globules.
   Flocculation—fine particles which may be suspended in the liquid or found as a precipitated layer at the bottom of the jar.
   Clabbering—thickening texture (coagulated) resembling yogurt or a curd-like texture as with cottage cheese.

Mixing/spraying

Fill tank of a thoroughly clean sprayer half to two-thirds full with clean water. Start agitation and add oil concentrate; allow to mix thoroughly. Add POAST and remaining volume of water. Maintain constant agitation during application.

Restrictions and limitations

Do not make applications to grasses under stress, such as stress due to lack of moisture or herbicide injury, as unsatisfactory control may result.

Do not apply if rainfall is expected within one hour following application as grass control will be unsatisfactory.

Do not mix or apply POAST with any other pesticide, additive, or fertilizer except as specifically recommended on this labeling.

Do not apply POAST within 30 days of harvest.

Do not apply more than a total of 5 pints per acre in one season to potatoes.

Environmental hazards

Do not apply directly to water.

Do not contaminate water by cleaning of equipment or disposal of wastes.

Endangered species concerns

The use of any pesticide in a manner that may kill or otherwise harm an endangered or threatened species or adversely modify their habitat is a violation of Federal law.

The use of this product is controlled to prevent death or harm to Solano grass which occurs in Solano County, **California.** Before using this product in this county you must obtain the EPA Endangered Species Bulletin (EPA/ES-85-13) available from either your County Agricultural Farm Advisor in California, the Endangered Species Specialist in the California Department of Fish and Game, or the Regional Offices of the U.S. Fish and Wildlife Service (Portland, Oregon) or the U.S. Environmental Protection Agency (San Francisco, California). THIS BULLETIN MUST BE REVIEWED PRIOR TO PESTICIDE USE. THE USE OF THIS PRODUCT IS PROHIBITED IN THIS COUNTY UNLESS SPECIFIED OTHERWISE IN THE BULLETIN.

Regional use map

All weed control recommendations for a given crop and weed are on a regional basis. There are three regions delineated on the U.S. map. Follow recommendations for your region only.

—California, Utah, Idaho, Oregon, Washington, Arizona, Nevada and Western New Mexico.

—High and Rolling Plains of Texas, Oklahoma and Eastern New Mexico.

—Midwest, South, Northeast and Mountain States. (All states and areas except those listed above.)

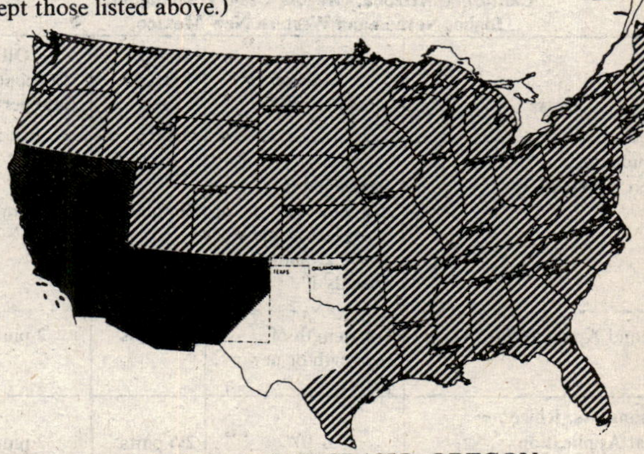

CALIFORNIA, ARIZONA, UTAH, IDAHO, OREGON, WASHINGTON, NEVADA AND WESTERN NEW MEXICO

Recommendations for grass control

- Apply to actively growing grasses at the sizes indicated in the tables of this regional section.
- Potatoes at all stages of growth are tolerant to POAST® herbicide.
- Always follow recommendations given in the **Application information** section.

- Always adjust spray pressure, spray volume and height of spray boom to ensure penetration of plant canopy and thorough coverage of grasses to be controlled.
- Do not apply to drought-stressed grass or grass which has gone through an extended dry period.
- In irrigated areas it may be necessary to irrigate prior to treatment with POAST to ensure weeds are growing actively.
- **Always add 2 pints of oil concentrate per acre.**

Annual Grasses*
Potato
California, Arizona, Utah, Idaho, Oregon, Washington, Nevada and Western New Mexico

Group	Grass	Time of Application	Rate of POAST per Acre	Oil Concentrate Rate per Acre
A	Barnyardgrass Fall Panicum Foxtails: Giant, Green, Yellow Johnsongrass, Seedling Junglerice Ryegrass, Annual Southwestern Cupgrass Witchgrass	3-8″	1½ pints	2 pints
	Goosegrass Large Crabgrass Smooth Crabgrass Wild Oats (Idaho, Oregon, and Washington only)	Up to 4″		
	Shattercane/Wildcane If needed, retreat at the same rate and stage of growth.	6-10″		
	Volunteer Corn**	8-12″		
B	Volunteer Cereals Barley Rye Oats Wheat Volunteer cereals which emerge from late spring through early summer (May through July) may be partially or incompletely controlled due to unfavorable conditions at time of application.	Before tillering, up to 4″	2 pints	2 pints

* For broad spectrum control of annual grasses, use 1½ pints of POAST per acre. When weed populations include additional grasses in Group B, increase the rate of POAST as indicated. If subsequent flushes of annual grasses emerge after the first application, make additional applications at the same rate and at the recommended stage of growth.

** Maintain sufficient boom height above volunteer corn plants for best spray coverage.

Perennial Grasses
Potato
California, Arizona, Oregon, Utah, Washington, Idaho, Nevada and Western New Mexico

Grass	Time of Application	Rate of POAST per Acre	Oil Concentrate Rate per Acre
Bermudagrass • First Application	Before plant diameter exceeds 6″ or leaf height above ground exceeds 1″.	2½ pints	2 pints
• Second Application	1-4″ length of regrowth or new plants.	1½ pints	2 pints
Johnsongrass, Rhizome • First Application	6-10″	2½ pints	2 pints
• Second Application When regrowth occurs or new plants emerge.	4-8″	1½ pints	2 pints
Ryegrass, Perennial If regrowth occurs, retreat at the same rate and stage of growth.	3-8″	1½ pints	2 pints
Quackgrass (Idaho, Oregon, Washington) For best results, rhizomes should be thoroughly fragmented (less than 6″). • First Application	6-8″	2½ pints	2 pints
• Subsequent Applications When regrowth occurs or new plants emerge. Depending upon environmental conditions and crop competition, season-long control may not be reduced. **Note:** A cultivation no sooner than 14 days after application but within 21 days of application will aid control.	6-8″	1½ pints	2 pints

HIGH AND ROLLING PLAINS OF TEXAS, OKLAHOMA AND EASTERN NEW MEXICO

Recommendations for grass control

- Apply to actively growing grasses at the sizes indicated in the tables of this regional section.
- Potatoes at all stages of growth are tolerant to POAST® herbicide.
- Always follow recommendations given in the **Application information** section.
- Always adjust spray pressure, spray volume, and height of spray boom to ensure penetration of plant canopy and thorough coverage of grasses to be controlled.
- Do not apply to drought-stressed grass, or grass which has gone through an extended dry period.
- In irrigated areas it may be necessary to irrigate prior to treatment with POAST to ensure weeds are growing actively.
- **Always add 2 pints of oil concentrate per acre.**

Annual Grasses*
Potato
High and Rolling Plains of Texas, Oklahoma and Eastern New Mexico

Group	Grass	Time of Application	Rate of POAST per Acre	Oil Concentrate Rate per Acre
A	Barnyardgrass Broadleaf Signalgrass Browntop Panicum Fall Panicum Foxtails: Giant, Green, Yellow Johnsongrass, Seedling Junglerice Red Sprangletop Texas Panicum Witchgrass	2-6″	1½ pints	2 pints
	Goosegrass Large Crabgrass Smooth Crabgrass	Up to 4″		
	Shattercane/Wildcane If needed, retreat at the same rate and stage of growth.	6-10″		
	Volunteer Corn**	8-10″		
B	Volunteer Cereals Barley Rye Oats Wheat POAST is not recommended for spring control of volunteer cereals that emerged the previous fall.	Before tillering, up to 4″ and prior to over-wintering	2 pints	2 pints

* For broad spectrum control of annual grasses, use 1½ pints of POAST per acre. When weed populations include additional grasses in Group B, increase the rate of POAST as indicated. If subsequent flushes of annual grasses emerge after the first application, make additional applications at the same rate and at the recommended stage of growth.

** Maintain sufficient boom height above corn plants for best spray coverage.

Perennial Grasses
Potato
High and Rolling Plains of Texas, Oklahoma and Eastern New Mexico

Grass	Time of Application	Rate of POAST per Acre	Oil Concentrate Rate per Acre
Bermudagrass • First Application	Before plant diameter exceeds 6" or leaf height above ground exceeds 1".	2 pints	2 pints
• Second Application	21 days after first application	1½ pints	2 pints
• Third Application	1-4" length of regrowth.	1½ pints	2 pints
Johnsongrass, Rhizome For best results, rhizomes should be thoroughly fragmented (less than 6"). Adjust volume of spray solution to a maximum of 10 gallons and a minimum of 5 gallons per acre while maintaining a ground speed of no more than 6 miles per hour.			
• First Application	6-10"	1½ pints	2 pints
• Subsequent Application When regrowth occurs or new plants emerge.	4-8"	1 pint	2 pints

MIDWEST, SOUTH, NORTHEAST AND MOUNTAIN STATES

Recommendations for grass control

- Apply to actively growing grasses at the sizes indicated in the tables of this regional section.
- Potatoes at all stages of growth are tolerant to POAST® herbicide.
- Always follow recommendations given in the **Application information** section.
- Always adjust spray pressure, spray volume and height of spray boom to ensure penetration of plant canopy and thorough coverage of grasses to be controlled.
- Do not apply to drought-stressed grass, or grass which has gone through an extended dry period.
- In irrigated areas, it may be necessary to irrigate prior to treatment with POAST to ensure weeds are growing actively.
- **Always add 2 pints of oil concentrate per acre.**

ANNUAL GRASSES*—SPECIAL RATE FOR EARLY TREATMENT
POTATO
MIDWEST, SOUTH, NORTHEAST AND MOUNTAIN STATES

GROUP	GRASS	TIME OF APPLICATION	RATE OF POAST PER ACRE	ADDITIVE RATE PER ACRE		
				OIL CONCENTRATE		UAN SOLUTION OR AMMONIUM SULFATE
A	Wild Proso Millet	4-10"	½ pint	2 pints		—
B	Goosegrass	1-3"	¾ pint	2 pints		—
	Barnyardgrass** (Midwest Only) Broadleaf Signalgrass Fall Panicum Texas Panicum Foxtail: Giant, Green	1-4"				
	Volunteer Corn	1-12"	¾ pint	2 pints	plus	1 gallon or 2½ lbs.

* For broad spectrum control of annual grasses in Groups A & B, use ¾ pint of POAST per acre. If additional applications are needed, apply at the same rate and at the recommended stage of growth.
** In these states use 1 pint per acre as recommended in the next table: AL, AR, FL, GA, LA, MS, NC, SC, TN, TX, VA.

ANNUAL GRASSES*
POTATO
MIDWEST, SOUTH, NORTHEAST AND MOUNTAIN STATES

GROUP	GRASS	TIME OF APPLICATION	RATE OF POAST PER ACRE	ADDITIVE RATE PER ACRE	
				OIL CONCENTRATE	UAN SOLUTION OR AMMONIUM SULFATE
A	Wild Proso Millet	4-10"	½ pint	2 pints	—
B	Barnyardgrass Broadleaf Signalgrass Fall Panicum Foxtail: Giant, Green, Yellow Johnsongrass, Seedling Junglerice Red Sprangletop Texas Panicum Witchgrass Woolly Cupgrass	Up to 8"	1 pint	2 pints	—
	Goosegrass Large Crabgrass Smooth Crabgrass	Up to 6"			
	Shattercane/Wildcane If needed, retreat at the same rate and stage of growth.	6-18"			
	Volunteer Corn**	Up to 20"	1 pint	2 pints	1 gallon or 2½ lbs.
	Wild Oats	Up to 4"			
C	Volunteer Cereals Barley Rye Oats Wheat	Before tillering, up to 6" and prior			

GROUP	GRASS	TIME OF APPLICATION	RATE OF POAST PER ACRE	ADDITIVE RATE PER ACRE	
				OIL CONCENTRATE	UAN SOLUTION OR AMMONIUM SULFATE
	POAST is not recommended for spring control of volunteer cereals that emerged the previous fall.	to overwintering.	1½ pints	2 pints	—
D	Itchgrass Red Rice	Up to 4"	2 pints	2 pints	—

* For broad spectrum control of annual grasses in Groups A & B, use 1 pint of POAST per acre. When weed populations include additional grasses in Groups C & D, increase the rate of POAST as indicated. If a second flush of annual grasses emerges after the first application, make additional applications at the same rate and at the recommended stage of growth.

* Maintain sufficient boom height above volunteer corn plants for best spray coverage.

Rescue treatment for controlling selected annual grasses

For best results, always apply POAST to annual grasses at the growth stage and rate specified in the above table. However, if POAST cannot be applied at the recommended time, larger annual grasses can be controlled with a later application by increasing the rate of POAST. Apply to actively growing grasses at the rates and sizes indicated below.

Potato—Annual Grasses—Rescue Treatment

Grass	Time of Application	Rate of POAST per Acre	Oil Concentrate Rate per Acre
Wild Proso Millet	10-24"	1 pint	
Foxtails: Giant, Green, Yellow Johnsongrass, Seedling	8-16"	1½ pints	2 pints
Texas Panicum Fall Panicum Barnyardgrass Broadleaf Signalgrass	8-12"		
Large Crabgrass Smooth Crabgrass Goosegrass	6-8"		

PERENNIAL GRASSES
POTATO
MIDWEST, SOUTH, NORTHEAST AND MOUNTAIN STATES

GRASS	TIME OF APPLICATION	RATE OF POAST PER ACRE	ADDITIVES RATE PER ACRE		
			OIL CONCENTRATE		UAN SOLUTION OR AMMONIUM SULFATE
Bermudagrass • First Application	Before plant diameter exceeds 6" or leaf height above ground exceeds 1".	1½ pints	2 pints		—
• Second Application If regrowth occurs or new plants emerge.	1-4" length of new plants.	1 pint	2 pints		—
Johnsongrass, Rhizome • First Application Use 5-10 gallons of spray solution per acre. Maintain a ground speed of no more than 6 miles per hour. For best results, rhizomes should be thoroughly fragmented (less than 6"). (When using 11-20 gallons of spray solution per acre, use 1½ pints of POAST.)	15-25"	1 pint	2 pints	plus	1 gallon or 2½ lbs.
• Subsequent Application When regrowth occurs or new plants emerge.	6-12"	1 pint	2 pints	plus	1 gallon or 2½ lbs.
Quackgrass • First Application For best results, rhizomes should be fragmented (less than 6").	6-8"	1½ pints	2 pints	plus	1 gallon or 2½ lbs.
• Subsequent Application If regrowth occurs of new plants emerge. Depending upon environmental conditions and crop cultural system, season-long control may not always be obtained. However, competition of quackgrass with the crop will be reduced. **Note:** A cultivation no sooner than 14 days after application but within 21 days of application will aid in control.	6-8"	1 pint	2 pints	plus	1 gallon or 2½ lbs.
Ryegrass, Perennial If regrowth occurs, retreat at same rate and stage of growth.	3-8"	1 pint	2 pints		—
Wirestem Muhly If regrowth occurs retreat at same rate and stage of growth.	Up to 6"	1¼ pints	2 pints		—

PROCEDURE FOR CLEANING EQUIPMENT

Attention! Clean sprayer thoroughly before and after application of POAST.

Clean sprayer thoroughly before application of POAST, particularly if a herbicide was used which has the potential to injure the crop to be sprayed with POAST.

Failure to clean sprayer thoroughly after application of POAST may result in injury to any grass crop subsequently sprayed, such as corn, sorghum, small grains, rice and turf. Fill the sprayer with clean water and add a commercial sprayer cleaner or a surfactant/adjuvant at the recommended rate on its label. Circulate through entire sprayer system. Spray approximately half the tank solution through the hoses, booms and nozzles to clean these parts. Drain the tank and rinse the total system thoroughly several times with clean water.

STORAGE AND DISPOSAL

Do not contaminate water, food, or feed by storage or disposal.

Pesticide wastes are toxic. Improper disposal of excess pesticide, spray mixture, or rinsate is a violation of Federal law. If these wastes cannot be disposed of by use according to label instructions, contact your State Pesticide or Environmental Control Agency, or the Hazardous Waste representative at the nearest EPA Regional Office for guidance.

Triple rinse container (or equivalent). Then offer for recycling or reconditioning, or puncture and dispose of in a sanitary landfill, or by incineration, or, if allowed by state and local authorities, by burning. If burned, stay out of smoke.

CONDITIONS OF SALE AND WARRANTY

The Directions for use of this product reflect the opinion of experts based on field use and tests. The directions are believed to be reliable and should be followed carefully. However, it is impossible to eliminate all risks inherently associated with use of this product. Crop injury, ineffectiveness or other unintended consequences may result from factors, such as, but not limited to, weather conditions, presence of other materials, or use of the product in a manner inconsistent with its labeling, all of which are beyond the control of BASF CORPORATION ("BASF") or the Seller. All such risks shall be assumed by the Buyer.

BASF warrants that this product conforms to the chemical description on the label and is reasonably fit for the purposes referred to in the Directions for use, subject to the inherent risks referred to above. BASF MAKES NO OTHER EXPRESS OR IMPLIED WARRANTY OF FITNESS OR MERCHANTABILITY OR ANY OTHER EXPRESS OR IMPLIED WARRANTY. IN NO CASE SHALL BASF OR THE SELLER BE LIABLE FOR CONSEQUENTIAL, SPECIAL OR INDIRECT DAMAGES RESULTING FROM THE USE OR HANDLING OF THIS PRODUCT. BASF and the Seller offer this product, and the Buyer and User accept it, subject to the foregoing Conditions of sale and warranty which may be varied only by agreement in writing signed by a duly authorized representative of BASF.

POAST is a registered trademark of BASF AG.

PO 1 027 JJ6291 A0580788

SUPPLEMENTAL LABELING

POAST® herbicide

POAST (EPA Reg. No. 7969-58)

FOR USE IN STRAWBERRIES AND RASPBERRIES

All applicable directions, restrictions, precautions and Conditions of sale and warranty on the EPA-registered label are to be followed. This labeling must be in the possession of the user at the time of herbicide application.

DIRECTIONS FOR USE

It is a violation of Federal law to use this product in a manner inconsistent with its labeling.

GENERAL INFORMATION

POAST® herbicide is a selective broad spectrum postemergence herbicide for control of annual and perennial grass weeds. POAST does not control sedges or broadleaf weeds.

Since essentially all grass crops such as sorghum, corn, small grains and rice, as well as ornamental grasses such as turf, are susceptible to POAST, avoid all direct or indirect contact with any grass crop.

Control symptoms: POAST rapidly enters the plant through the foliage and translocates throughout the plant. Control symptoms exhibited by the grass plant progress from a slowing or a stopping of growth (generally within two days), to reddening of the foliage and to leaf tip burn. Subsequently, burn back of the foliage occurs. These symptoms will generally be observed within three weeks depending on environmental conditions.

Application information*

Apply POAST to actively growing grasses by ground equipment when grasses are at the proper growth stage as specified by the **Recommendations for grass control** for each region. POAST herbicide can be applied before harvest or after harvest in a renovation program.

Do not make applications to grasses under stress, such as stress due to lack of moisture, herbicide injury, mechanical injury or cold temperatures, since unsatisfactory control may result.

Thorough spray coverage of grass foliage is essential. For broadcast application use standard high pressure pesticide hollow cone or flat fan nozzles. Do not use flood or whirl chamber nozzles. Application of POAST with control drop applicator (CDA) nozzles is not recommended due to erratic coverage which causes inconsistent weed control. Use a minimum volume of 5 gallons and a maximum volume of 20 gallons of spray solution per acre. When using standard high pressure hollow cone or flat fan nozzles adjust pressure to a minimum of 40 psi and a maximum of 60 psi at the nozzle.

Always adjust spray pressure, spray volume and height of spray boom to ensure penetration of plant canopy and thorough coverage of grasses to be controlled.

Do not use selective application equipment such as recirculating sprayers, wiper applicators, or shielded applicators.

Cultivation information

Do not cultivate 5 days prior to the application of POAST or within 7 days following application.

A timely cultivation after 7 days may aid in providing season-long control (exception-quackgrass, see individual recommendations).

Addition of oil concentrate

A nonphytotoxic oil concentrate (commonly referred to as oil concentrate) should always be added to the spray tank. The oil concentrate must contain either a petroleum or vegetable oil base and must meet the following criteria: 1) be nonphytotoxic, 2) contain only EPA exempt ingredients, 3) provide good mixing quality in the jar test (see page 2), and 4) be successful in local experience.

The exact composition of suitable products will vary; however, vegetable and pretroleum oil concentrates should contain emulsifiers which provide good mixing quality. For vegetable oil concentrates, it has been observed that highly refined vegetable oils are more satisfactory than unrefined vegetable oils. For additional information see **Jar test for estimating suitability of oil concentrates.**

Rate of oil concentrate:

Ground application—2 pints/acre.

*For use by ground equipment only.

Jar test for estimating suitability of oil concentrates

1. **Water supply:** Use only water from intended source and at the source temperature.
2. **Amount of water in jar:**
   For 20 gal./A spray volume uses 3 1/3 cups (800 ml) water.
   For 10 gal./A spray volume use 1 2/3 cups (400 ml) of water.
   For 5 gal./A spray volume use 5/6 cup (200 ml) of water.
   For other spray volumes, adjust proportionately to above.
3. **Amount of herbicide and oil concentrate to add:** Add herbicide and oil concentrate at the rate of 1 teaspoon (5 ml) for each pint of recommended label rate.
4. **Add components in following sequence,** gently mixing between component additions:
   a. Oil concentrate.
   b. POAST.
5. **Cap jar,** invert 10 cycles, let stand for 15 minutes, evaluate.
6. **Evaluation:** An ideal tank mix combination will be uniform; thus, the suitability of the oil concentrate is questionable if any of the following are observed:
   Free oil at the surface—film or globules.
   Flocculation—fine particles which may be suspended in the liquid or found as a precipitated layer at the bottom of the jar.
   Clabbering—thickening texture (coagulated) resembling yogurt or a curdlike texture as with cottage cheese.

Mixing/spraying

Fill tank of a thoroughly clean sprayer half to two-thirds full with clean water. Start agitation and add oil concentrate; allow to mix thoroughly. Add POAST and remaining volume of water. Maintain constant agitation during application.

Restrictions and limitations

Do not make applications to grasses under stress, such as stress due to lack of moisture or herbicide injury, as unsatisfactory control may result.

Do not apply if rainfall is expected within one hour following application as grass control will be unsatisfactory.

Do not mix or apply POAST with any other pesticide, additive, or fertilizer except as specifically recommended on this labeling.

Do not apply POAST within 30 days of harvest of strawberries. Do not apply POAST within 45 days of harvest of raspberries.

Do not apply more than a total of 2 1/2 pints per acre in one season to strawberries.

Do not apply more than a total of 5 pints per acre in one season to raspberries.

Do not apply this product through any type of irrigation system.

Note to strawberry growers:

Do not tank mix or sequentially apply POAST plus oil concentrate within one week before or after an application of Tenoran® herbicide as strawberry injury may occur.

Application of POAST plus oil concentrate applied up to 6 weeks after a Sinbar® herbicide application can occasionally cause strawberry leaf injury. It is believed to be variety related. Growers should determine injury potential by treating a small area first then waiting a week before treating the rest of the strawberry field with POAST plus oil concentrate.

Regional use map

All weed control recommendations for a given crop and weed are on a regional basis. There are three regions delineated on the U.S. map below. Best results are obtained by using recommendations for your region only.

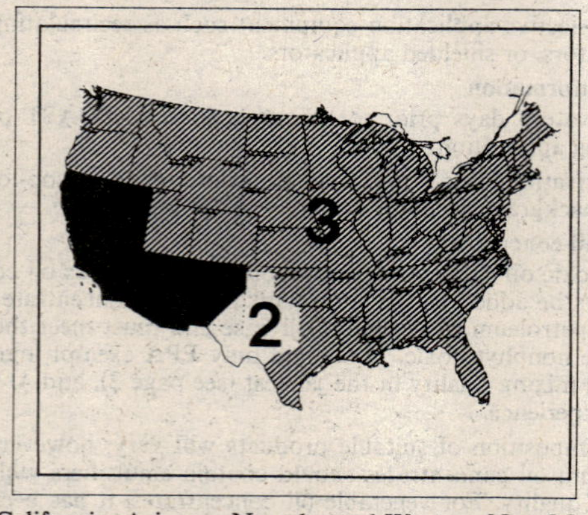

Region 1: California, Arizona, Nevada, and Western New Mexico.

Region 2. High and Rolling Plains of Texas, Oklahoma and Eastern New Mexico.

Region 3: Midwest, Mountain (except AZ and NM), South, Northeast and Pacific Northwest. (All states and areas except those listed in 1 and 2 above.)

REGION 1—CALIFORNIA, ARIZONA, NEVADA, AND WESTERN NEW MEXICO

Recommendations for grass control

- Apply to actively growing grasses as the sizes indicated in the tables of this strawberry and raspberry regional section.
- Always follow recommendations given in the **Application information** section.
- Always adjust spray pressure, spray volume and height of spray boom to ensure penetration of plant canopy and thorough coverage of grasses to be controlled.
- Do not apply to drought-stressed grass or grass which has gone through an extended dry period.
- In irrigated areas it may be necessary to irrigate prior to treatment with POAST to ensure that weeds are growing actively.
- **Always add 2 pints of oil concentrate per acre.**

Strawberry and Raspberry—Annual Grasses* (Region 1)
California, Arizona, Nevada, and Western New Mexico

Group	Grass	Time of Application	Rate of POAST per Acre	Oil Concentrate Rate per Acre
A	Barnyardgrass Broadleaf Signalgrass Fall Panicum Foxtails: Giant, Green, Yellow Johnsongrass, Seedling Junglerice Ryegrass, Annual Southwestern Cupgrass Texas Panicum Witchgrass	3-8"	2 pts.	2 pts.
	Goosegrass Large Crabgrass Smooth Crabgrass	Up to 4"		
	Shattercane/Wildcane If needed retreat at the same rate and stage of growth	6-18"		
	Volunteer Corn	8-12"		
B	Volunteer Cereals Barley Rye Oats Wheat Volunteer cereals which emerge from late spring through early summer (May through July) may be partially or incompletely controlled due to unfavorable conditions at time of application.	Before tillering, up to 4".	2½ pts.	2 pts.

* For broad spectrum control of annual grasses in Group A, use 2 pints of POAST per acre. When weed populations include additional grasses in Group B, increase the rate of POAST as indicated. In raspberries, if subsequent flushes of annual grasses emerge after the first application, make an additional application at the same rate and at the recommended stage of growth. Do not exceed a total of 2½ pts./A/season for strawberries and 5 pts./A/season for raspberries.

Strawberry and Raspberry—Perennial Grasses (Region 1)
(California, Arizona, Nevada, and Western New Mexico)

Grass	Time of Application	Rate of POAST per Acre	Oil Concentrate Rate per Acre
Bermudagrass • First application (One application may not give season-long control.)	Before plant diameter exceeds 6" or leaf height above ground exceeds 1".	2½ pts.	2 pts.
• Second application (Raspberry only)	21 days after first application.	2½ pts.	2 pts.
Johnsongrass, Rhizome For best results, rhizomes should be thoroughly fragmented (less than 6") • First application (One application may not give season-long control.)	6-10"	2½ pts.	2 pts.
• Second application (Raspberry only)	4-8"	2½ pts.	2 pts.

Do not exceed a total of 2½ pints/A/season for strawberries and 5 pints/A/season for raspberries.

REGION 2—HIGH AND ROLLING PLAINS OF TEXAS, OKLAHOMA AND EASTERN NEW MEXICO

Recommendations for grass control

- Apply to actively growing grasses as the sizes indicated in the tables of this strawberry and raspberry regional section.
- Always follow recommendations given in the **Application information** section.
- Always adjust spray pressure, spray volume and height of spray boom to ensure penetration of plant canopy and thorough coverage of grasses to be controlled.
- Do not apply to drought-stressed grass or grass which has gone through an extended dry period.
- In irrigated areas it may be necessary to irrigate prior to treatment with POAST to ensure that weeds are growing actively.
- **Always add 2 pints of oil concentrate per acre.**

Strawberry and Raspberry—Annual Grasses* (Region 2)
(High and Rolling Plains of Texas, Oklahoma and Eastern New Mexico)

Group	Grass	Time of Application	Rate of POAST per Acre	Oil Concentrate Rate per Acre
A	Barnyardgrass Broadleaf Signalgrass Browntop Panicum Fall Panicum Foxtails: Giant, Green, Yellow Johnsongrass, Seedling Junglerice Red Sprangletop Texas Panicum Witchgrass	2-6"	2 pts.	2 pts.
	Goosegrass Large Crabgrass Smooth Crabgrass	Up to 4"		
	Shattercane/Wildcane If needed, retreat at the same rate and stage of growth.	6-10"		
	Volunteer Corn	8-10"		
B	Volunteer Cereals Barley Rye Oats Wheat POAST is not recommended for spring control of volunteer cereals that emerged the previous fall.	Before tillering, up to 4" and prior to over-wintering.	2½ pts.	2 pts.

* For broad spectrum control of annual grasses in Group A, use 2 pints of POAST per acre. When weed populations include additional grasses in Group B, increase the rate of POAST as indicated. In raspberries, if subsequent flushes of annual grasses emerge after the first application, make an additional application at the same rate and at the recommended stage of growth. Do not exceed a total of 2½ pts./A/season for strawberries and 5 pts./A/season for raspberries.

Strawberry and Raspberry—Perennial Grasses* (Region 2)
(High and Rolling Plains of Texas, Oklahoma and Eastern New Mexico)

Grass	Time of Application	Rate of POAST per Acre	Oil Concentrate Rate per Acre
Bermudagrass • First application	Before plant diameter exceeds 6" or leaf height above ground exceeds 1".	1½ pts.	2 pts.
• Second application	21 days after first application.	1 pt.	2 pts.
• Subsequent applications. If regrowth occurs or new plants emerge. (raspberry only)	1-4" length of regrowth or new plants.	1 pt.	2 pts.
Johnsongrass, Rhizome For best results, rhizomes should be thoroughly fragmented (less than 6"). Adjust volume of spray solution to a maximum of 10 gallons and a minimum of 5 gallons per acre while maintaining a ground speed of no more than 6 miles per hour. • First application	6-10"	1½ pts.	2 pts.
• Second application When regrowth occurs or new plants emerge.	4-8"	1 pt.	2 pts.
• Subsequent applications. If regrowth occurs or new plants emerge. (raspberry only)	4-8"	1 pt.	2 pts.

Do not exceed a total of 2½ pints/A/season for strawberries and 5 pints/A/season for raspberries.

REGION 3—MIDWEST, MOUNTAIN (EXCEPT AZ AND NM), SOUTH, NORTHEAST AND PACIFIC NORTHWEST

Recommendations for grass control

- Apply to actively growing grasses at the sizes indicated in the tables of this strawberry and raspberry regional section.
- Always follow recommendations given in the **Application information** section.
- Always adjust spray pressure, spray volume and height of spray boom to ensure penetration of plant canopy and thorough coverage of grasses to be controlled.
- Do not apply to drought-stressed grass, or grass which has gone through an extended dry period.
- In irrigated areas it may be necessary to irrigate prior to treatment with POAST to ensure that weeds are growing actively.
- **Always add 2 pints of oil concentrate per acre.**

Strawberry and Raspberry—Annual Grasses* (Region 3)
(Midwest, Mountain (except AZ and NM), South, Northeast and Pacific Northwest)

Group	Grass	Time of Application	Rate of POAST per Acre	Oil Concentrate Rate per Acre
A	Wild Proso Millet	4-10"	¾ pt.	2 pts.
B	Barnyardgrass Broadleaf Signalgrass Browntop Panicum Fall Panicum Foxtails: Giant, Green, Yellow Johnsongrass, Seedling Junglerice Red Sprangletop Ryegrass, Annual Texas Panicum Witchgrass Woolly Cupgrass	3-8"	1½ pts.	2 pts.
	Goosegrass Large Crabgrass Smooth Crabgrass	Up to 4"		
	Shattercane/Wildcane If needed retreat at the same rate and stage of growth	6-18"		
	Volunteer Corn	6-20"		
C	Volunteer Cereals Barley Rye Oats Wheat POAST is not recommended for spring control of volunteer cereals that emerged the previous fall.	Before tillering, up to 6" and prior to wintering.	2 pts.	2 pts.
	Wild Oats	2-4"		
D	Itchgrass Red Rice	Up to 4"	2½ pts.	2 pts.

* For broad spectrum control of annual grasses in Groups A & B, use 1½ pints of POAST per acre. When weed populations include additional grasses in Groups C or D increase the rate of POAST to 2 pints. If a second flush of annual grasses emerge after the first application, make additional applications at the same recommended stage of growth. Do not exceed a total of 2½ pts./A/season for strawberries and 5 pts./A/season for raspberries.

Rescue treatment for controlling selected annual grasses in Region 3

For best results, always apply POAST to annual grasses at the growth stage and rate specified in the table on page 6. However, if POAST cannot be applied at the recommended time, larger annual grasses can be controlled with a later application by increasing the rate of POAST. Apply to actively growing grasses at the rates and sizes indicated.

Strawberry and Raspberry—Annual Grasses
Rescue Treatment (Region 3)
(Midwest, Mountain (ex. AZ and NM), South, Northeast and Pacific Northwest)

Grass	Time of Application	Rate of POAST per Acre	Oil Concentrate Rate per Acre
Wild Proso Millet	10-24"	1 pt.	
Foxtails: Giant, Green, Yellow Johnsongrass, Seedling	8-16"	2 pts.	2 pts.
Barnyardgrass Broadleaf Signalgrass Fall Panicum Texas Panicum	8-12"		
Goosegrass Large Crabgrass Smooth Crabgrass	6-8"		

Do not exceed a total of 2½ pts./A/season for strawberries and 5 pts./A/season for raspberries.

Strawberry and Raspberry—Perennial Grasses (Region 3)
(Midwest, Mountain (except AZ and NM), South, Northeast, and Pacific Northwest)

Grass	Time of Application	Rate of POAST per Acre	Oil Concentrate Rate per Acre
Bermudagrass • First application	Before plant diameter exceeds 6" or leaf height above ground exceeds 1".	1½ pts.	2 pts.
• Second application	21 days after first application.	1 pt.	2 pts.
• Subsequent applications if regrowth occurs or new plants emerge. (raspberry only)	1-4" length of new plants.	1 pt.	2 pts.
Johnsongrass, Rhizome For best results, rhizomes should be thoroughly fragmented (less than 6"). Adjust volume of spray solution to a maximum of 10 gallons and a minimum of 5 gallons per acre while maintaining a ground speed of no more than 6 miles per hour. • First application	6-10"	1½ pts.	2 pts.
• Second application When regrowth occurs or new plants emerge.	4-8"	1 pt.	2 pts.

Grass	Time of Application	Rate of POAST per Acre	Oil Concentrate Rate per Acre
• Subsequent applications if regrowth occurs or new plants emerge. (raspberry only)	4-8"	1 pt.	2 pts.
Ryegrass, Perennial • First application	3-8"	1½ pts.	2 pts.
• Second application Retreat if regrowth occurs or new plants emerge.	3-8"	1 pt.	2 pts.
Quackgrass For best results, rhizomes should be thoroughly fragmented (less than 6"). • First application	6-8"	2½ pts.	2 pts.
• Second application If regrowth occurs or new plants emerge. (raspberry only). Depending upon environmental conditions and crop cultural system, season-long control may not always be obtained. However, competition of quackgrass with the crop will be reduced. **Note:** A cultivation no sooner than 14 days after application but within 21 days of application will aid in control.	6-8"	2½ pts.	2 pts.
Wirestem Mulhy • First application	Up to 6"	1½ pts.	2 pts.
• Second application Retreat if regrowth occurs or new plants emerge.	Up to 6"	1 pt.	2 pts.

Do not exceed a total of 2½ pts./A/season for strawberries and 5 pts./A/season for raspberries.

PROCEDURE FOR CLEANING EQUIPMENT

ATTENTION: Clean sprayer thoroughly before and after application of POAST.

Clean sprayer thoroughly before application of POAST, particularly if a herbicide was used which has the potential to injure the crop to be sprayed with POAST.

Failure to clean sprayer thoroughly after the application of POAST may result in injury to any grass crop subsequently sprayed, such as corn, sorghum, small grains, rice and turf. Fill the sprayer with clean water and add a commercial sprayer cleaner or a surfactant/adjuvant at the recommended rate on its label. Circulate through entire sprayer system. Spray approximately half the tank solution through the hoses, booms and nozzles to clean these parts. Drain the tank and rinse the total system thoroughly several times with clean water.

STORAGE AND DISPOSAL

Do not contaminate water, food, or feed by storage or disposal.

Pesticide wastes are toxic. Improper disposal of excess pesticide, spray mixture, or rinsate is a violation of Federal law. If these wastes cannot be disposed of by use according to label instructions, contact your State Pesticide or Environmental Control Agency, or the Hazardous Waste representative at the nearest EPA Regional Office for guidance.

Triple rinse container (or equivalent). Then offer for recycling or reconditioning, or puncture and dispose of in a sanitary landfill, or by incineration, or, if allowed by state and local authorities, by burning. If burned, stay out of smoke.

CONDITIONS OF SALE AND WARRANTY

The Directions for use of this product reflect the opinion of experts based on field use and tests. The directions are believed to be reliable and should be followed carefully. However, it is impossible to eliminate all risk inherently associated with use of this product. Crop injury, ineffectiveness or other unintended consequences may result because of such factors as weather conditions, presence of other materials, or use of the product in a manner inconsistent with its labeling, all of which are beyond the control of BASF Corporation ("BASF") or the Seller. All such risks shall be assumed by the Buyer.

BASF warrants that this product conforms to the chemical description on the label and is reasonably fit for the purposes referred to in the Directions for use, subject to the inherent risks referred to above. BASF MAKES NO OTHER EXPRESS OR IMPLIED WARRANTY OF FITNESS OR MERCHANTABILITY OR ANY OTHER EXPRESS OR IMPLIED WARRANTY. IN NO CASE SHALL BASF OR THE SELLER BE LIABLE FOR CONSEQUENTIAL, SPECIAL OR INDIRECT DAMAGES RESULTING FROM THE USE OR HANDLING OF THIS PRODUCT. BASF and the Seller offer this product, and the Buyer and User accept it, subject to the foregoing Conditions of sale and warranty which may be varied only by agreement in writing signed by a duly authorized representative of BASF.

POAST is a registered trademark of BASF AG.

Tenoran is a registered trademark of Ciba-Geigy Corp.

Sinbar is a registered trademark of E.I. duPont de Nemours Co. Inc.

PO 1019 JJ6101 A0030188

SUPPLEMENTAL LABELING

POAST® HERBICIDE

POAST EPA Reg. No. 7969-58

FOR USE ON TOMATOES, PEPPERS AND EGGPLANT.

All applicable directions, restrictions, precautions and Conditions of sale and warranty on the EPA-registered label are to be followed. This labeling must be in the possession of the user at the time of herbicide application.

DIRECTIONS FOR USE

It is a violation of Federal law to use this product in a manner inconsistent with its labeling.

Re-entry and worker's protection statements

Do not apply this product in such a manner as to directly or through drift expose workers or other persons, except those knowingly involved in the application. The area being treated must be vacated by unprotected persons. Do not enter treated areas without protective clothing until sprays have dried.

Because certain states may require more restrictive re-entry intervals for various crops treated with this product, consult your State Department of Agriculture for futher information.

Written or oral warnings must be given to workers who are expected to be in a treated area or in an area about to be treated with this product. Oral warnings must inform workers of areas or fields that may not be entered without specific protective clothing until sprays have dried, and appropriate actions to take in case of accidental exposure. When oral warnings are given, warnings shall be given in a language customarily understood by workers. Oral warnings must be given if there is reason to believe that written warnings cannot be understood by workers. Written warnings must include the following information: "WARNING, Area treated with POAST® herbicide on (date of application). Do not enter without appropriate protective clothing until sprays have dried. Causes eye irritation. Do not get in eyes. Avoid contact with skin. Harmful if swallowed. If in eyes or on skin, immediately flush with plenty of water. Get medical attention if irritation persists."

GENERAL INFORMATION

POAST is a selective broad spectrum postemergence herbicide for control of annual and perennial grass weeds. POAST does not control sedges or broadleaf weeds.

POAST may be used on all types of tomatoes (examples: plum, cherry or other) peppers (bell, sweet, chili or other) and eggplants grown for fresh market or processing.

Since essentially all grass crops such as sorghum, corn, and small grains, as well as ornamental grasses such as turf, are susceptible to POAST, avoid all direct or indirect contact with any grass crop.

Control symptoms: POAST rapidly enters the plant through the foliage and translocates throughout the plant. Control symptoms exhibited by the grass plant progress from a slowing or stopping of growth (generally within two days), to reddening of the foliage to leaf tip burn. Subsequently, burn back of the foliage occurs. These symptoms will generally be observed within three weeks depending on environmental conditions.

Under some conditions, or with certain oil concentrates, a very slight leaf speckling can occur. Tomatoes, peppers and eggplant will outgrow these symptoms and later growth is not affected.

Application information*

Apply POAST to actively growing grasses when they are at the proper stage as specified by the **Recommendations for grass control.**

Thorough coverage of the grass foliage is essential. This is because the effectiveness of POAST is dependent on the absorption and movement of POAST throughout the plant.

Do not make applications to grasses under stress, such as stress due to lack of moisture, herbicide injury, mechanical injury or cold temperatures, since unsatisfactory control will probably result.

*To be applied by ground equipment only.

Broadcast application

Nozzle type: For broadcast application use standard high pressure pesticide hollow cone or flat fan nozzles. Do not use flood or whirl chamber nozzles.

Spray gallonage: Use a minimum volume of 5 gallons and a maximum volume of 20 gallons of spray solution per acre. In Arizona, California, Idaho, Nevada, Oregon, Utah, Washington and Western New Mexico use a minimum of 10 gallons per acre. In the High and Rolling Plains of Texas, Oklahoma and Eastern New Mexico use a minimum of 5 gallons per acre and a maximum of 10 gallons per acre.

Spray pressure: When using standard high pressure hollow cone or flat fan nozzles adjust the pressure to a minimum of 40 psi and maximum of 60 psi at the nozzle.

Always adjust spray pressure, spray volume and height of spray boom to ensure penetration of plant canopy and thorough coverage of grasses to be controlled. Tall weeds, such as volunteer corn may require the boom height to be as high as 20-inches above the corn height, refer to the nozzle manufacture's directions for recommended height.

Band applications: All recommendations are on a broadcast basis unless otherwise stated. Banding of POAST® herbicide may be used to control annual grasses. When banding, rates of POAST, oil concentrate and water should be reduced in proportion to the area sprayed. Grasses which are not covered or only partially covered by a spray mixture of POAST will not be adequately controlled. When treating taller weeds, such as volunteer corn, the sprayer must be high enough to spray the top leaves and into the whorl of the plant.

Other spray equipment: Do not use selective application equipment such as recirculating sprayers, wiper applicators, or shielded applicators. Application of POAST with control drop application (CDA) nozzles is not recommended due to erratic coverage which causes inconsistent weed control.

Cultivation information

Do not cultivate within 5 days prior to application of POAST or within 7 days following application. A timely cultivation after 7 days may aid in providing season-long control.

Addition of oil concentrate

A nonphytotoxic oil concentrate (commonly referred to as oil concentrate) should always be added to the spray tank. The oil concentrate must contain either a petroleum or vegetable oil base and must meet the following criteria: 1) be nonphytotoxic, 2) contain only EPA-exempt ingredients, 3) provide good mixing quality in the jar test (see below), and 4) be successful in local experience.

The exact composition of suitable products will vary; however, vegetable and petroleum oil concentrates should contain emulsifiers which provide good mixing quality. For vegetable oil concentrates, it has been observed that highly refined vegetable oils are more satisfactory than unrefined vegetable oils. For additional information see **Jar test for estimating suitability of oil concentrates.**

Rate of oil concentrate:

Ground application-2 pints/acre.

Jar test for estimating suitability of oil concentrates

1. **Water supply:** Use only water from intended source and at the source temperature.
2. **Amount of water in jar:**
   For 20 gals./A spray volume uses 3⅓ cups (800 ml) water.
   For 10 gals./A spray volume use 1⅔ cups (400 ml) of water.
   For 5 gals./A spray volume use ⅚ cup (200 ml) of water.
   For other spray volumes, adjust proportionately to above.
3. **Amount of herbicide and oil concentrate to add:** Add herbicide and oil concentrate at the rate of 1 teaspoon (5 ml) for each pint of recommended label rate.
4. **Add components in following sequence,** gently mixing between component additions:
   1. Oil concentrate.
   2. POAST.
5. **Cap jar,** invert 10 cycles, let stand for 15 minutes, evaluate.
6. **Evaluation:** An ideal tank mix combination will be uniform; thus, the suitability of the oil concentrate is questionable if any of the following are observed:
   Free oil at the surface–film or globules.
   Flocculation–fine particles which may be suspended in the liquid or found as a precipitated layer at the bottom of the jar.
   Clabbering–thickening texture (coagulated) resembling yogurt or a curd-like texture as with cottage cheese.

Mixing/spraying

Fill tank of a thoroughly clean sprayer half to two-thirds full with clean water. Start agitation and add oil concentrate; allow to mix thoroughly. Add POAST and remaining volume of water. Maintain constant agitation during application.

Spot or small area treatment

When using knapsack sprayers or high volume spray equipment utilizing handguns or other suitable nozzle arrangements, make 1% solution of POAST® herbicide in water. A recommended oil concentrate must also be used at a concentration of 1%.

Apply to foliage of grasses on a spray-to-wet basis. Spray coverage should be uniform and complete. Do not spray to the point of runoff.

Prepare the desired volume of spray solution by mixing the amount of POAST and the amount of oil concentrate in water according to the table below.

Desired Spray Solution Volume	Amount to Be Added to Obtain a 1% Solution	
	POAST	Oil Concentrate
1 gallon	1¼ fl. oz.	1¼ fl. oz.
25 gallons	1 qt.	1 qt.
50 gallons	2 qts.	2 qts.
100 gallons	4 qts.	4 qts.

* 2 tablespoons = 1 fl. oz.

Restrictions and limitations

Do not apply POAST within 20 days of harvest.

Do not apply more than a total of 4½ pints per acre in one season.

POAST plus oil concentrate should be used with caution under the conditions below. Under these environmental conditions the potential for leaf injury is greater.
– When the temperature exceeds 90°F and the relative humidity is 60% or greater, **or,**
– Anytime the temperature exceeds 100°F, regardless of the humidity.

Do not apply if rainfall is expected within one hour following application as grass control will be unsatisfactory.

Do not make applications to grasses under stress, such as stress due to lack of moisture or herbicide injury, as unsatisfactory control may result.

Do not mix or apply POAST with any other pesticide, additive, or fertilizer except as specifically recommended on this labeling.

Do not apply this product through any type of irrigation system.

Environmental hazards

Do not apply directly to water.

Do not contaminate water by cleaning of equipment or disposal of wastes.

Endangered species concerns

The use of any pesticide in a manner that may kill or otherwise harm an endangered or threatened species or adversely modify their habitat is a violation of Federal law.

The use of this product is controlled to prevent death or harm to Solano grass which occurs in Solano County, **California.** Before using this product in this county you must obtain the EPA Endangered Species Bulletin (EPA/ES-85-13) available from either your County Agricultural Extension Agent, the Endangered Species Specialists in the California Department of Fish and Game, or the Regional offices of the U.S. Fish and Wildlife Service (Portland, Oregon) or the U.S. Environmental Protection Agency (San Francisco, California). THIS BULLETIN MUST BE REVIEWED PRIOR TO PESTICIDE USE. THE USE OF THIS PRODUCT IS PROHIBITED IN THIS COUNTY UNLESS SPECIFIED OTHERWISE IN THE BULLETIN.

PROCEDURE FOR CLEANING EQUIPMENT

ATTENTION! Clean sprayer thoroughly before and after application of POAST.

Clean sprayer thoroughly before application of POAST, particularly if a herbicide was used which has the potential to injure the crop to be sprayed with POAST.

Failure to clean sprayer thoroughly after the application of POAST may result in injury to any grass crop subsequently sprayed, such as corn, sorghum, small grains, rice and turf. Fill the sprayer with clean water and add a commercial sprayer cleaner or a surfactant/adjuvant at the recommended rate on its label. Circulate through entire sprayer system. Spray approximately half the tank solution through the hoses, booms and nozzles to clean these parts. Drain the tank and rinse the total system thoroughly several times with clean water.

Regional use map

All weed control recommendations for a given crop and weed are on a regional basis. There are three regions delineated on the U.S. map. Best results are obtained by using recommendations for your region only.
–Midwest, South and Northeast (and all other regions not listed below)
–High and Rolling Plains of Texas, Oklahoma and Eastern New Mexico
–Western and Mountain States

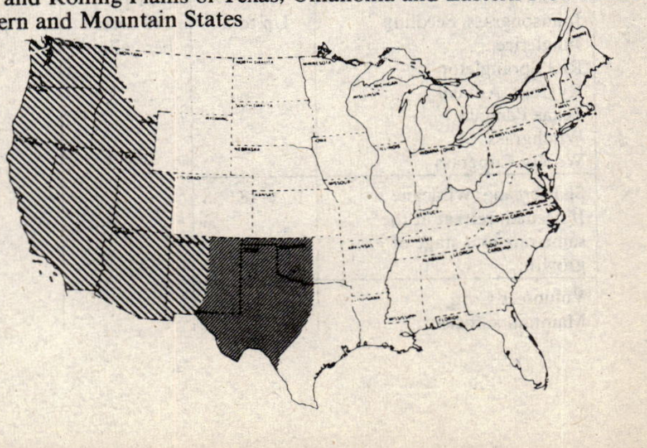

MIDWEST, SOUTH AND NORTHEAST

Recommendations for grass control

- Apply to actively growing grasses at the sizes indicated in the tables of this regional section.
- Always follow recommendations given in the **Application information** section.
- Always adjust spray pressure, spray volume and height of spray boom to ensure penetration of plant canopy and thorough coverage of grasses to be controlled.
- Do not apply to drought-stressed grass, or grass which has gone through an extended dry period.
- In irrigated areas, it may be necessary to irrigate prior to treatment with POAST® herbicide to ensure weeds are growing actively.
- Drop nozzles may be needed to obtain adequate coverage when crop height exceeds 24-inches.
- **Always add 2 pints of oil concentrate per acre.**

Special rate for early treatment

Tomatoes, Peppers and Eggplant
Midwest, South and Northeast

Group	Grass	Time of Application	Rate of POAST per Acre	Additive Rate per Acre: Oil Concentrate
A	Wild Proso Millet	4-10"	½ pt. (16 acres per gallon)	2 pts.
B	Goosegrass	1-3"	¾ pt. (10.7 acres per gallon)	2 pts.
	Barnyardgrass** (Midwest Only) Broadleaf Signalgrass Fall Panicum Texas Panicum Foxtail: Giant, Green	1-4"		

* For broad spectrum control of annual grasses in Groups A & B, use ¾ pint of POAST per acre. If additional applications are needed, apply at the same rate and at the recommended stage of growth.

** In these states use 1 pint per acre as recommended in the next table: AL, AR, FL, GA, LA, MS, NC, SC, TN, TX, VA.

Annual Grasses*

Tomatoes, Peppers and Eggplant
Midwest, South and Northeast

Group	Grass	Time of Application	Rate of POAST per Acre	Additive Rate per Acre: Oil Concentrate
A	Wild Proso Millet	4-10"	½ pt. (16 acres per gallon)	2 pts.
B	Wild Oats	Up to 4"	1 pt. (8 acres per gallon)	2 pts.
	Goosegrass Smooth Crabgrass Large Crabgrass	Up to 6"		
	Barnyardgrass Broadleaf Signalgrass Browntop Panicum Fall Panicum Foxtails: Giant, Green, Yellow Johnsongrass, Seedling Junglerice Red Sprangletop Rygrass, Annual Texas Panicum Witchgrass Woolly Cupgrass	Up to 8"		
	Shattercane/Wildcane If needed, retreat at the same rate and stage of growth.	6-18"		
	Volunteer Corn Maintain sufficient boom height above volunteer corn plants for best spray coverage.	Up to 20"		
C	Field Sandbur (Midwest only)	Up to 3"	1¼ pts. (6.4 acres per gallon)	2 pts.
D	Volunteer Cereals Barley Oats Rye Wheat Not recommended for spring control of volunteer cereals that emerged the previous fall.	Before tillering, up to 4" and prior to over-wintering.	1½ pts. (5.3 acres per gallon)	2 pts.

* For broad spectrum control of annual grasses in Group B (above), use 1 pint of POAST® herbicide per acre. When weed populations include additional grasses in Group C or D increase the rate of POAST as indicated. If later flushes of annual grasses emerge after the first application, make additional applications at the same rate and the same recommended stage of growth.

Perennial Grasses

Tomatoes, Peppers and Eggplant
Midwest, South and Northeast

Grass	Time of Application	Rate of POAST per Acre	Additive Rate per Acre: Oil Concentrate
Bermudagrass • First Application	Before stolon length exceeds 6"	1½ pts. (5.3 acres per gallon)	2 pts.
• Second Application If regrowth occurs or new plants emerge.	1-4" length of new plants.	1 pt. (8 acres per gallon)	2 pts.
Johnsongrass, Rhizome • First Application For best results, rhizomes should be thoroughly fragmented (less than 6").	15-25"	1 pt. (8 acres per gallon). (When using 11-20 gallons of spray volume per acre, use 1½ pints of POAST)	2 pts.
• Subsequent Application If regrowth occurs or new plants emerge.	6-12"	1 pt. (8 acres per gallon)	2 pts.
Ryegrass, Perennial If regrowth occurs or new plants emerge retreat at the same rate and stage of growth.	3-8"	1 pt. (8 acres per gallon)	2 pts.

Rescue treatment for controlling selected annual grasses

For best results, always apply POAST® herbicide to annual grasses at the growth stage and rate specified in the above table. However, if POAST cannot be applied at the recommended time, larger annual grasses can be controlled with a later application by increasing the rate of POAST. Apply to actively growing grasses at the rates and sizes indicated below.

Annual Grasses–Rescue Treatment

Tomatoes, Peppers and Eggplant

Grass	Time of Application	Rate of POAST per Acre	Oil Concentrate Rate per Acre
Wild Proso Millet	10-24"	1 pt.	2 pts. (4 acres per gallon)
Foxtails: Giant, Green Yellow Johnsongrass, Seedling	8-16"	1½ pts. (5.3 acres per gallon)	
Texas Panicum Fall Panicum Barnyardgrass Broadleaf Signalgrass	8-12"		
Large Crabgrass Smooth Crabgrass Goosegrass	6-8"		

HIGH AND ROLLING PLAINS OF TEXAS, OKLAHOMA AND EASTERN NEW MEXICO

Recommendations for grass control

- Apply to actively growing grasses at the sizes indicated in the tables of this regional section.

• Always follow recommendations given in the **Application information** section.

• Always adjust spray pressure, spray volume and height of spray boom to ensure penetration of plant canopy and thorough coverage of grasses to be controlled.

• Do not apply to drought-stressed grass, or grass which has gone through an extended dry period.

• In irrigated areas, it may be necessary to irrigate prior to treatment with POAST® herbicide to ensure weeds are growing actively.

• Drop nozzles may be needed to obtain adequate coverage when crop height exceeds 24-inches.

• **Always add 2 pints of oil concentrate per acre.**

Annual Grasses

Tomatoes, Peppers and Eggplant
High and Rolling Plains of Texas, Oklahoma and Eastern New Mexico

Grass	Time of Application	Rate of POAST per Acre	Additive Rate per Acre
			Oil Concentrate
Goosegrass Smooth Crabgrass Large Crabgrass	Up to 4"		
Barnyardgrass Broadleaf Signalgrass Browntop Panicum Fall Panicum Foxtails: Giant, Green, Yellow Johnsongrass, Seedling Junglerice Red Sprangletop Texas Panicum Witchgrass	Up to 8"	1½ pts. (5.3 acres per gallon)	2 pts.
Shattercane/Wildcane If needed, retreat at the same rate and stage of growth.	6-18"		
Volunteer Corn Maintain sufficient boom height above volunteer corn plants for best spray coverage.	Up to 20"		

* For broad spectrum control of annual grasses above, use 1½ pints of POAST per acre.
If later flushes of annual grasses emerge after the first application, make an additional application at the same rate and at the same recommended stage of growth.

Perennial Grasses

Tomatoes, Peppers and Eggplant
High and Rolling Plains of Texas, Oklahoma and Eastern New Mexico

Grass	Time of Application	Rate of POAST per Acre	Additive Rate per Acre
			Oil Concentrate
Bermudagrass • First Application	Before stolon length exceeds 6" above ground exceeds 1"	1½ pts. (5.3 acres per gallon)	2 pts.
• Second Application If regrowth occurs or new plants emerge.	21 days after first application.	1½ pts. (5.3 acres per gallon)	2 pts.
Johnsongrass, Rhizome • First Application For best results, rhizomes should be thoroughly fragmented (less than 6"). Adjust volume of spray solution to a maximum of 10 gallons and a minimum of 5 gallons per acre while maintaining a speed of no more than 6 miles per hour.	6-10"	1½ pts. (5.3 acres per gallon)	2 pts.
• Subsequent Application If regrowth occurs or new plants emerge.	4-8"	1 pt. (8 acres per gallon)	2 pts.

WESTERN AND MOUNTAIN STATES

Recommendations for grass control

• Apply to actively growing grasses at the sizes indicated in the tables of this regional section.

• Always follow recommendations given in the **Application information** section.

• Always adjust spray pressure, spray volume and height of spray boom to ensure penetration of plant canopy and thorough coverage of grasses to be controlled.

• Do not apply to drought-stressed grass, or grass which has gone through an extended dry period.

• In irrigated areas, it may be necessary to irrigate prior to treatment with POAST® herbicide to ensure weeds are growing actively.

• Drop nozzles may be needed to obtain adequate coverage when crop height exceeds 24-inches.

• **Always add 2 pints of oil concentrate per acre.**

Annual Grasses

Tomatoes, Peppers and Eggplant
Western and Mountain States

Grass	Time of Application	Rate of POAST per Acre	Additive Rate per Acre
			Oil Concentrate
Goosegrass Smooth Crabgrass Large Crabgrass Wild Oats (Idaho, Oregon, Washington only)	Up to 4"		
Barnyardgrass Fall Panicum Foxtails: Giant, Green, Yellow Johnsongrass, Seedling Junglerice Ryegrass, Annual Southwestern Cupgrass Witchgrass	Up to 8"	1½ pts. (5.3 acres per gallon)	2 pts.
Shattercane/Wildcane If needed, retreat at the same rate and stage of growth.	6-18"		
Volunteer Corn Maintain sufficient boom height above volunteer corn plants for best spray coverage.	Up to 12"		

* For broad spectrum control of annual grasses above, use 1½ pints of POAST per acre.
If later flushes of annual grasses emerge after the first application, make an additional application at the same rate and at the same recommended stage of growth.

Perennial Grasses

Tomatoes, Peppers and Eggplant
Western and Mountain States

Grass	Time of Application	Rate of POAST per Acre	Additive Rate per Acre
			Oil Concentrate
Ryegrass, Perennial If regrowth occurs or new plants emerge, retreat at the same rate and stage of growth.	3-8"	1½ pts. (5.3 acres per gallon)	2 pts.

CONDITIONS OF SALE AND WARRANTY

The Directions for use of this product reflect the opinion of experts based on field use and tests. The directions are believed to be reliable and should be followed carefully. However, it is impossible to eliminate all risks inherently associated with use of this product. Crop injury, ineffectiveness or other unintended consequences may result because of such factors as weather conditions, presence of other materials, or use of the product in a manner inconsistent with its labeling, all of which are beyond the control of BASF CORPORATION ("BASF") or the Seller. All such risks shall be assumed by the Buyer.

BASF warrants that this product conforms to the chemical description on the label and is reasonably fit for the purposes referred to in the Directions for use, subject to the inherent risks referred to above. BASF MAKES NO OTHER EXPRESS OR IMPLIED WARRANTY OF FITNESS OR MERCHANTABILITY OR ANY OTHER EXPRESS OR IMPLIED WARRANTY. IN NO CASE SHALL BASF OR THE SELLER BE LIABLE FOR CONSEQUENTIAL, SPECIAL OR INDIRECT DAMAGES RESULTING FROM THE USE OR HANDLING OF THIS PRODUCT. BASF and the Seller offer this product, and the Buyer and

User accept it, subject to the foregoing Conditions of sale and warranty which may be varied only by agreement in writing signed by a duly authorized representative of BASF.

POAST is a registered trademark of BASF AG.

PO 1 037 JJ6481 A0020189

POAST PLUS™

POSTEMERGENCE GRASS HERBICIDE

For use in cotton, peanuts and soybeans*

Active Ingredient

2-[1-(ethoxyimino)butyl]-5-[2-(ethylthio) propyl]-3-hydroxy-2-cyclohexen-1-one**	13%
Inert Ingredients	87%

** Equivalent to 1.0 pound per gallon

EPA Reg. No. 7969–88

KEEP OUT OF REACH OF CHILDREN

CAUTION

Causes moderate eye injury. Harmful if swallowed or absorbed through skin. Avoid contact with skin, eyes, or clothing. Wash thoroughly with soap and water after handling.

If in eyes: Flush with plenty of water. Call a physician if irritation persists. If on skin: Wash with plenty of soap and water. Get medical attention. If swallowed: Drink promptly a large quantity of milk, egg whites, gelatin solution or, if these are not available, large quantities of water. Avoid alcohol.

* Not Intended for use in California.

ENVIRONMENTAL HAZARDS

Do not apply directly to water. Do not contaminate water by cleaning of equipment or disposal of wastes.

Endangered species concerns

The use of any pesticide in a manner that may kill or otherwise harm an endangered or threatened species or adversely modify their habitat is a violation of Federal law.

DIRECTIONS FOR USE

cotton, peanuts and soybeans

It is a violation of Federal law to use this product in a manner inconsistent with its labeling.

GENERAL INFORMATION

POAST PLUS™ herbicide is an improved formulation for broad spectrum postemergence control of annual and perennial grasses. POAST PLUS does not control sedges or broadleaf weeds. Since all grass crops such as sorghum, corn, small grains and rice, as well as some ornamental grasses such as turf, are susceptible to POAST PLUS, avoid all direct or indirect contact with any desired grass plant.

Control symptoms: POAST PLUS rapidly enters the plant through the foliage and translocates throughout the plant. Control symptoms exhibited by the grass plant progress from a slowing and stopping of growth (generally within two days) to reddening of foliage, and to leaf tip burn. Subsequently, burn-back of the foliage occurs. These symptoms will generally be observed within three weeks, depending on environmental conditions.

Application Information

Apply POAST PLUS to actively growing grasses when they are at the proper growth stage as specified in the Recommendations for use tables.

Do not make applications to grasses under stress, such as stress due to a lack of moisture, herbicide injury, mechanical injury or cold temperatures, since unsatisfactory control will probably result.

Nozzle selection: Thorough spray coverage of grass foliage is essential. For broadcast application use standard high pressure pesticide hollow cone or flat fan nozzles. Do not use flood or whirl chamber nozzles. Application of POAST PLUS with control drop applicator (CDA) nozzles is not recommended due to erratic coverage which causes inconsistent weed control.

Ground equipment: A spray volume of 10 gallons per acre is optimum (5–20 GPA may be used.)

Spray pressure: 40-60 psi (measured at the nozzle).

Boom height: Use a broom height sufficient to cover entire grass plant. See nozzle manufacturer's recommendations.

Tall crop applications: When a crop, such as cotton, is 24 or more inches in height and the grasses may be below the crop canopy, drop nozzles should be used to insure good coverage of the grass species. Good coverage is essential for maximum control.

Band applications: Banding of POAST PLUS may be used to control annual grasses. Grasses which are not covered or only partly covered by the spray mixture will not be adequately controlled. All recommendations in the Rate and Time of Application tables are on a broadcast basis. When banding, rates of POAST PLUS, additives and water should be reduced in proportion to the areas sprayed.

Other spray equipment: Do not use selective application equipment such as recirculating sprayers, wiper applicators, or shielded applicators.

Air equipment: Use a minimum of 5 gallons per acre (except 10 gallons if foliage is dense) and a maximum of 40 psi pressure. Use only diaphragm-type nozzles production cone or fan spray patterns.

Cultivation Information

Do not cultivate within 5 days prior to application of POAST PLUS or within 7 days following application.

Mixing and spraying

Fill tank of a thoroughly clean sprayer one-half to two-thirds full with clean water. Start agitation and add Dash* spray adjuvant or oil concentrate; allow to mix thoroughly. Add POAST PLUS and remaining volume of water. Maintain constant agitation during application.

Addition of Dash* spray adjuvant or oil concentrate

A nonphytotoxic oil concentrate (commonly referred to as oil concentrate) should be added to the spray tank. The oil concentrate must contain either a petroleum or vegetable oil base and must meet the following criteria: 1) be nonphytotoxic, 2) contain only EPA-exempt ingredients, 3) provide good mixing quality in the jar test, and 4) be successful in local experience.

The exact composition of suitable products will vary, however, vegetable and petroleum oil concentrates should contain emulsifiers which provide good mixing quality. For vegetable oil concentrates, it has been observed that highly refined vegetable oils are more satisfactory than unrefined vegetable oils. For additional information see Jar test for estimating suitability of oil concentrates at the end of this section.

Dash may be used as a direct substitute for an oil concentrate with some exceptions. In some crops and tank mixes **Dash** is not recommended (see individual tables.)

Rate of Dash or oil concentrate

Ground and air application—2 pints/acre.

Jar test for estimating suitability of oil concentrates

1. Water supply: Use only water from intended source and at the source temperature.
2. Amount of water in jar:
   For 20 gals./A spray volume use 3⅓ cups (800 ml) of water.
   For 10 gals./A spray volume use 1 ⅔ (400 ml) of water.
   For 5 gals./A spray volume use ⅚ cup (200 ml) of water.
   For other spray volumes, adjust proportionately to above.
3. Amount of herbicide(s) and oil concentrate to add: Add herbicide(s) and oil concentrate at the rate of 1 teaspoon (5 ml) for each pint of recommended label rate.
4. Add components in following sequence, gently mixing between component additions:
   1) Water miscible or soluble products (such as Basagran* herbicide, Blazer* herbicide, ammonium sulfate, UAN solution) when applicable.
   2) Oil concentrate
   3) POAST PLUS™ herbicide (and other emulsifiable concentrates when applicable.)
5. Cap jar, invert 10 cycles, let stand for 15 minutes, evaluate.
6. Evaluation: An ideal tank mix combination will be uniform; thus the suitability of the oil concentrate is questionable if any of the following are observed:
   Free oil at the surface-film or globules.
   Flocculation-fine particles which may be suspended in the liquid or found as a precipitated layer at the bottom of the jar.
   Clabbering-thickening texture (coagulated) resembling yogurt or a curd-like texture as with cottage cheese.

Spot or small area treatment

Make a 1½% solution of POAST PLUS. Apply to grass foliage on a spray-to-wet basis.

Procedure for cleaning spray equipment

Attention! Clean sprayer thoroughly before and after application of POAST PLUS. Clean sprayer thoroughly prior to application of POAST PLUS, particularly if a herbicide was used which has the potential to injure crops.

The steps listed below are suggested for thorough cleaning of spray equipment prior to or following applications of POAST PLUS.

Step 1 Hose down thoroughly the inside as well as the outside of equipment while filling the spray tank half full of water. Flush by operating sprayer until the system is purged of this rinse water.

Step 2 Refill tank with water while adding 1 gallon household ammonia or 1 pint household dishwashing detergent per 100 gallons of water. Or add a commercial sprayer cleaner according to the manufacturer's directions.

Operate the pump to circulate the detergent solution through the sprayer system for 5 to 10 minutes and discharge a small amount of solution through the boom and nozzles. Let the solution stand for 24 hours.

Step 3 Flush the detergent solution out of the spray tank through the boom.

Step 4 Remove the nozzles and screens and flush the system with two tankfuls of water.

STORAGE AND DISPOSAL

Do not contaminate water, food, or feed by storage or disposal. Wastes resulting from the use of this product may be disposed of on site or at an approved waste disposal facility.

Triple rinse container (or equivalent). Then offer for recycling or reconditioning, or puncture and dispose of in a sanitary landfill, or by incineration, or, if allowed by state and local authorities, by burning. If burned, stay out of smoke.

General restrictions and limitations

Do not make applications to grass under stress, such as stress due to lack of moisture, herbicide injury, mechanical injury or cold temperatures, since unsatisfactory control will probably result.

Do not apply if rainfall is expected within one hour following application as grass control will probably be unsatisfactory.

Do not mix or apply POAST PLUS with any other pesticide, additive or fertilizer except as specifically recommended on this labeling, or EPA approved BASF supplemental labeling.

Do not apply POAST PLUS as a preplant or preemergent treatment prior to corn, milo, millet or sorghum.

Do not apply POAST PLUS through any type or irrigation system.

Restrictions and limitations for soybeans

Do not apply to soybeans within 90 days of harvest.

Do not apply more than a total of $7\frac{1}{2}$ pints (120 fl. oz.) of POAST PLUS per acre to soybeans in one season (including application before or after planting).

Do not graze treated soybean fields and do not feed treated soybean forage (green succulent) or ensilage to livestock. Treated soybean hay may be fed.

Classic* herbicide may cause antagonism of POAST PLUS when sprayed in a time period from 7 days prior to application to 1 day after application of POAST PLUS. This antagonism is more likely to occur during stress conditions.

Restrictions and limitations for cotton

Do not apply within 40 days of harvest.

Do not apply more than a total of $11\frac{1}{4}$ pints (180 fl. oz.) of POAST PLUS per acre in one season.

Do not graze treated cotton fields and do not feed treated forage to livestock.

Restrictions and limitations for peanuts

Do not apply POAST PLUS within 40 days of harvest.

Do not apply more than a total of $3\frac{3}{4}$ pints (60 fl. oz.) of POAST PLUS per acre in one season.

Do not feed treated peanut forage or hay to livestock.

Recommendations for grass control

Use of POAST PLUS™ herbicide is intended only for the states indicated on the map.

Apply to actively growing grasses at the sizes indicated on Table 5.

Soybeans, peanuts and cotton at all stages of growth are tolerant to POAST PLUS.

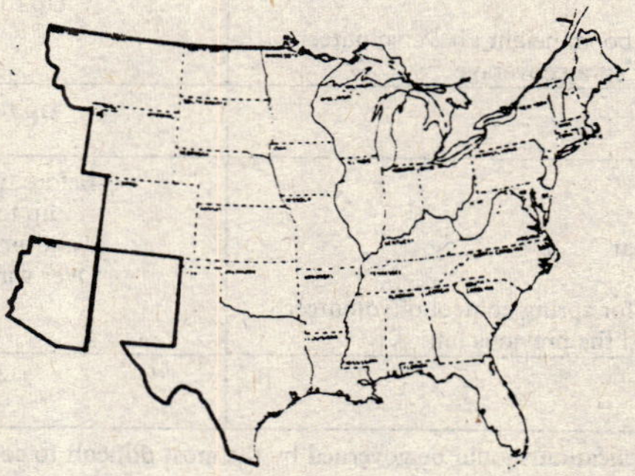

TABLE 1
ANNUAL GRASSES*—SPECIAL RATE FOR EARLY TREATMENT
POAST PLUS—SOYBEAN, COTTON AND PEANUTS
RATE AND TIME OF APPLICATION TABLE

GRASS	TIME OF APPLICATION	RATE OF POAST PLUS PER ACRE	ADDITIVE RATE PER ACRE: DASH OR OIL CONCENTRATE — GROUND OR AIR
Wild Proso Millet	4–10"	12 fl. oz. (10.7 acres/gal.)	2 pts.
Goosegrass	1–3"	18 fl. oz. (7.1 acres/gal.)	2 pts.
Barnyardgrass** (Midwest only) Broadleaf Signalgrass Fall Panicum Texas Panicum Foxtails: Giant Green	1–4"		
Volunteer Corn	1–12"		

* Broad spectrum application should be governed by the most difficult to control weeds.
** In the following states use 24 ounces per acre: AL, AR, AZ, FL, GA, LA, MS, NC, SC, TN, TX, VA.
If later flushes of annual grasses emerge after first application, make additional applications at the same recommended stage of growth.

TABLE 2
ANNUAL GRASSES—STANDARD RECOMMENDATIONS*
POAST PLUS—SOYBEAN, COTTON AND PEANUTS
RATE AND TIME OF APPLICATION

GRASS	TIME OF APPLICATION	RATE OF POAST PLUS PER ACRE	ADDITIVE RATE PER ACRE: DASH OR OIL CONCENTRATE — GROUND OR AIR
Wild Proso Millet	4–10"	12 fl. oz. (10.7 acres/gallon)	2 pts.
Wild Oats	Up to 4"		
Crabgrass, Large Crabgrass, Smooth Goosegrass	Up to 6"		

GRASS	TIME OF APPLICATION	RATE OF POAST PLUS PER ACRE	ADDITIVE RATE PER ACRE DASH OR OIL CONCENTRATE GROUND OR AIR
Barnyardgrass Broadleaf Signalgrass Browntop Panicum Fall Panicum Foxtails; Giant, Green, Yellow Johnsongrass, Seedling Junglerice Red Sprangletop Ryegrass, Annual Texas Panicum Witchgrass Woolly Cupgrass	Up to 8"	24 fl. oz. (5.3 acres/gallon)	2 pts.
Shattercane/Wildcane If needed, re-treat at the same rate and stage of growth.	6–18"		
Volunteer Corn Maintain sufficient boom height above volunteer corn plants for best spray coverage	Up to 20"		
Field Sandbur	Up to 3"	30 fl. oz. (4.3 acres/gallon)	2 pts.
Volunteer Cereals Barley Oats Rye Wheat Not recommended for spring control of volunteer cereals that emerged the previous fall.	Before tillering, up to 4" and prior to Over wintering	36. fl. oz. (3.5 acres/gallon)	2 pts.
Itchgrass Red Rice	2–4"	48 fl. oz. (2.7 acres/gallon)	2 pts.

* Broad spectrum application should be governed by the most difficult to control weed.

** If later flushes of annual grasses emerge after first application, make additional applications at the same rate and at the same recommended stage of growth.

TABLE 3
PERENNIAL GRASSES
POAST PLUS—SOYBEAN, COTTON AND PEANUTS
RATE AND TIME OF APPLICATION

GRASS	TIME OF APPLICATION	RATE OF POAST PER ACRE	ADDITIVES RATE PER ACRE DASH OR OIL CONCENTRATE GROUND OR AIR
Bermudagrass • First Application	Before stolon length exceeds 6".	36 fl. oz. (3.5 acres./gal.)	2 pts.
• Second Application If regrowth occurs or new plants emerge.	1-4" length of new plants or growth.	24 fl. oz. (5.3 acres/gal.)	2 pts.
Johnsongrass, Rhizome • First Application Use 5-10 gallons of spray solution per acre. Maintain a ground speed of no more than 6 miles per hour. For best results rhizomes should be thoroughly fragmented (less than 6"). (When using 11-20 gallons of spray solution per acre use 36 oz. pints of POAST PLUS)	15-25" (15-20" in no-till culture)	24 fl. oz. (5.3 acres/gal.)	2 pts.
• Second Application When regrowth occurs or new plants emerge.	6-12"	24 fl. oz. (5.3 acres/gal.)	2 pts.
Quackgrass • First Application For best results, rhizomes should be thoroughly fragmented (less than 6").	6-8"	36 fl. oz (3.5 acres/gal.)	2 pts.
• Second Application If regrowth occurs or new plants emerge.	6-8"	24 fl. oz. (5.3 acres/gal.)	2 pts.
Depending upon environmental conditions and crop cultural system, season-long control may not always be obtained. However, competition of quackgrass with the crop will be reduced. **Note:** In conventional wide row soybeans, a cultivation no sooner than 14 days after application but within 21 days of application will aid in control.			
Wirestem Muhly If regrowth occurs, retreat at the same rate and stage of growth.	Up to 6"	30 fl. oz. (4.3 acres/gal.)	2 pts.

POAST PLUS & Basagran tank mix—Soybeans
General and application information
Restrictions and limitations

GENERAL INFORMATION

POAST PLUS™ and Basagran* herbicides may be tank mixed for post-emergence control of the broadleaf and grass weeds shown in Table 4. Weeds must be actively growing and at the recommended growth stages.

Separate applications should be made if: a) all weeds to be controlled are not at the correct growth stage for treatment at the same time, or b) grasses to be controlled include rhizome johnsongrass, quackgrass, bermudagrass, wirestem muhly, shattercane, volunteer cereals, wild oats, red rice or itchgrass. See rate Tables 2 and 3 for POAST PLUS recommendations and Table 5, Separate Postemergence Application Systems.

Water volume and spray pressure

Ground equipment: Use 20 gallons of total spray solution per acre (broadcast basis) and a minimum pressure of 40 psi. Use standard high pressure hollow cone or flat fan nozzles spaced 20 inches apart. Do not use flood or whirl chamber nozzles.

Air equipment: Use a minimum of 5 gallons of spray solution per acre.

Mixing
Fill tank of a thoroughly clean sprayer half to two-thirds full with clean water. Start agitation and add Basagran, UAN or ammonium sulfate, oil concentrate or Dash* spray adjuvant, all to mix thoroughly. Add POAST PLUS and remaining volume of water. Maintain constant agitation during application.

Additives
At the low rate of POAST PLUS (24 fl. oz.) the additives, Dash plus UAN (or ammonium sulfate) must be used. At the higher rate of POAST PLUS, either Dash or oil concentrate must be used. UAN (or ammonium sulfate) may be added.

Coverage
Thorough coverage of actively growing weeds is essential. Large crop-and-weed leaf canopies shelter small weeds and can prevent adequate spray coverage. Soybeans at all stages of growth are tolerant to Basagran and POAST PLUS; however, under certain conditions soybeans may exhibit leaf burn, slight crinkles and bronzing of the beans.

Restrictions and limitations (partial list)
Read and follow the restrictions and limitations on the labels for POAST PLUS and Basagran. The most restrictive labeling applies in tank mixes. Do not use a tank mix of POAST PLUS + Basagran on cotton and peanuts.

TABLE 4
POAST PLUS + BASAGRAN TANK MIX—SOYBEANS
RATE AND TIME OF APPLICATION TABLE

PRODUCT	PRODUCT RATE PER ACRE	WEEDS CONTROLLED/WEED SIZE				ADDITIVE (RATE/ACRE) DASH OR OIL CONCENTRATE		ADDITIVE (RATE/ACRE) UAN SOLUTION OR AMMONIUM SULFATE
		ANNUAL GRASSES*						
POAST PLUS	24 fl. oz.	Wild Proso Millet** Fall Panicum Giant Foxtail	4-10″ 3-8″ 3-8″	Green Foxtail Witchgrass Woolly Cupgrass Volunteer Corn	3-8″ 3-8″ 3-8″ 1-12″	Dash only (2 pts.)	plus	½-1 gallon UAN or 2½ lbs. AMS
	36 fl. oz.	Barnyardgrass Broadleaf Signalgrass Yellow Foxtail Seedling Johnsongrass	3-8″ 3-8″ 3-8″ 3-8″	Junglerice Red Sprangletop Texas Panicum Goosegrass Large Crabgrass Smooth Crabgrass	3-8″ 3-8″ 3-8″ 3-6″ 3-6″ 3-6″	Dash or oil concentrate (2 pts.)	plus	½-1 gallon UAN or 2½ lbs. AMS may be added to this tank mixture.
plus	plus							
		BROADLEAVES AND SEDGE						
Basagran	1-2 pts./A according to weed species and size (see label for Basagran).	Balloonvine Beggarticks Bristly Starbur Canada Thistle*** Cocklebur Coffee Senna Common Lambsquarters Common Purslane Common Ragweed Cypressvine Morningglory Dayflower Devilsclaw Galinsoga Giant Ragweed Jimsonweed		Ladysthumb Pennsylvania Smartweed Prickly Sida or Teaweed Redweed Shepherdspurse Smallflower Morningglory Spurred Anoda Tropic Croton Velvetleaf Venice Mallow Wild Buckwheat Wild Mustard Wild Poinsettia Wild Sunflower Yellow Nutsedge				

* Tank mix does not control rhizome johnsongrass, quackgrass, bermudagrass, wirestem muhly, shattercane, volunteer cereals, wild oats, red rice, or itchgrass.
** For control of wild proso millet only, include POAST PLUS herbicide in the tank mix at 18 fl. ounces/A.
*** Requires two applications of Basagran in accordance with the label for control.

Separate application of POAST PLUS, preceded or followed by Basagran or Basagran + Blazer tank mix—Soybeans

Applications of POAST PLUS™ herbicide can be preceded or followed by Basagran[1] and/or Blazer[1] herbicides to obtain broad spectrum control of weeds listed on the respective product labels (refer to the labels for POAST PLUS, Basagran and Blazer). Also refer to these product labels for timing, rate and other iinformation for ground and aerial applications.

For best results when making separate applications, a minimum period of time is recommended between applications, depending upon their order according to Table 5.

TABLE 5
POSTEMERGENCE APPLICATION SYSTEMS
SEPARATE APPLICATIONS

Order of Application: First Product(s) Applied	Order of Application: Second Product(s) Applied	Minimum Time Between Applications
Basagran	POAST PLUS	24 hours
Basagran + Blazer*	POAST PLUS	7 days
POAST	Blazer or Basagran or Basagran + Blazer*	24 hours
Blazer	POAST PLUS	7 days

POAST PLUS + 2,4-D low volatile ester tank mix for use as a burndown prior to planting soybeans

GENERAL INFORMATION

For broad spectrum postemergence weed control a tank mix application of POAST PLUS with 2,4-D low volatile ester (LVE) may be made for control of emerged broadleaf and grass weeds **before planting soybeans.**

This tank mix does not control sedges or provide season-long control of hard-to-kill perennial weeds.

If grasses are larger than indicated in **Table 6** then use rate of POAST PLUS as recommended in **Annual Grasses—Standard Recommendations, Table 2.**

For application by ground equipment only

See equipment application section.

Additives

Dash* spray adjuvant or oil concentrate must be used with this tank mix.

Mixing

Fill tank of a thoroughly clean sprayer one-half to two-thirds full with clean water. Start agitation and add **Dash** or oil concentrate; allow to mix thoroughly. Add POAST PLUS then 2,4-D (LVE), then the remaining volume of water. Maintain constant agitation during application.

Selection of 2,4-D (LVE) formulation

Use only low volatile **ester** formulations of 2,4-D such as 2,4-D Isooctyl ester. Note that the recommended rate of 2,4-D is calculated on an acid equivalent basis. Make adjustments for the concentration of the 2,4-D formulation used. Since the exact composition of suitable products will vary, it is advised to conduct a compatibility test with each 2,4-D (LVE) formulation used.

Restrictions and limitations (partial list)

Do not apply this tank mix during or following planting or after soybeans emergence as severe soybeans injury will result.

Do not plant any crop until 3 months after treatment or until the 2,4-D (LVE) has disappeared from the soil.

Do not apply if rainfall is expected within 6 hours following application, as weed control will probably be unsatisfactory. Since all crops such as sorghum, corn, small grains, cotton, soybeans, rice, sugar beets, trees, shrubs, as well as turf, are extremely susceptible to POAST PLUS + 2,4-D (LVE) tank mix, avoid all direct or indirect postemergence contact with any desired plant.

Do not spray if the wind is blowing toward desired plants, or at anytime when the wind exceeds 6 miles per hour (refer to 2,4-D (LVE) label).

Observe all restrictions and limitations specified on labels for 2,4-D (LVE) and POAST PLUS. The most restrictive labeling applies in tank mixes.

TABLE 6
POAST PLUS + 2,4-D (LVE)—SOYBEANS
PREPLANT BURNDOWN RATE AND TIME OF APPLICATION

WEED SPECIES	TIME OF APPLICATION	RATE OF POAST PLUS PER ACRE	DASH OR OIL CONCENTRATE PER ACRE	2,4-D (A.E.) RATE PER ACRE**
GRASSES				
Wild Proso Millet	Up to 4"	12 fl. oz.	2 pts.	½ lb.
Barnyardgrass Broadleaf Signalgrass Fall Panicum Foxtails: Giant, Green, Yellow Johnsongrass, Seedling Witchgrass Woolly Cupgrass Large Crabgrass Smooth Crabgrass	Up to 3"			
BROADLEAVES				
Pennsylvania Smartweed	Up to 2"			
Field Bindweed* Wild Buckwheat*	Vine Length Up to 6"			
Canada Thistle* Common Cocklebur Common Dandelion Common Lambsquarters Common Ragweed Field Pennycress Giant Ragweed Marestail Horseweed Prickly Lettuce Redroot Pigweed Shepherdspurse Velvetleaf White Cockle* Wild Mustard Yellow Rocket	Up to 10"			

* Control may be partial or inconsistent.
** A.E. rate based on 2,4-D acid equivalent. See section entitled **Selection of 2,4-D (LVE) formulation.**

Appendix

The following are scientific names for the weeds listed in this section. For specific recommendations on control of these weeds, refer to the major crop and/or tank mix sections.

Grasses

Common Name	Scientific Name
Barnyardgrass	*Echinochloa crus-galli*
Bermudagrass	*Cynodon dactylon*
Broadleaf Signalgrass	*Brachiaria platyphylla*
Crabgrass, Large,	*Digitaria sanguinalis*
Smooth	*Digitaria ischaemum*
Cupgrass, Woolly	*Eriochloa villosa*
Foxtails, Giant,	*Setaria faberi*
Green,	*Setaria viridis*
Yellow	*Setaria glauca*
Goosegrass	*Eleusine indica*
Itchgrass	*Rottboellia exaltata*
Johnsongrass	*Sorghum halepense*
Junglerice	*Echinochloa colonum*
Pigeon grass (see Foxtails)	
Panicum, Browntop,	*Panicum fasciculatum*
Fall,	*Panicum dichotomiflorum*
Texas	*Panicum texanum*
Quackgrass	*Agropyron repens*
Red Rice	*Oryza sativa*
Red Sprangletop	*Leptochloa filiformis*
Ryegrass, Annual,	*Lolium multiflorum*
Perennial	*Lolium perenee*
Sandbur, Field	*Cenchrus incertus*
Volunteer Barley	*Hordeum vulgare*
Corn	*Zea mays*
Oats	*Avena sativa*
Rye	*Secale cereale*
Wheat	*Triticum aestivum*
Watergrass (see Barnyardgrass)	
Shattercane/Wildcane	*Sorghum bicolor*
Wild Oats	*Avena fatua*
Wild Proso Millet	*Panicum miliaceum*

Grasses	
Common Name	**Scientific Name**
Wiregrass (see Bermudagrass)	
Wirestem Muhly	*Muhlenbergia frondosa*
Witchgrass	*Panicum capillare*

Broadleaf Weeds

Common Name	Scientific Name
Balloonvine	*Cardiospermum halicacabum*
Beggarticks	*Bidens frondosa*
Bindweed, Field	*Convulvulus arvensis*
Bristly Starbur	*Acanthospermum hispidum*
Canada Thistle	*Cirsium arvense*
Cocklebur	*Xanthium strumarium*
Cockle, White	*Agrostemma githago*
Coffee Senna	*Cassia occidentalis*
Common Lambsquarters	*Chenopodium album*
Common Purslane	*Portulaca oleracea*
Crotalaria, Showy	*Crotalaria spectabilis*
Dandelion, Common	*Taraxacum officinale*
Dayflower	*Commelina* spp.
Devilsclaw	*Probiscidea louisianica*
Galinsoga	*Galinsoga* spp.
Horseweed (see Marestail)	
Lettuce, Prickly	*Lactuca serriola*
Jimsonweed	*Datura stramonium*
Ladysthumb	*Polygonum persicaria*
Marestail	*Hippuris vulgaris*
Morningglory, Cypressvine	*Ipomoea quamoollt*
Smallflower	*Jacquemontia tamnifolia*
Pennycress, Field	*Thiaspi aruense*
Pennsylvania Smartweed	*Polygonum pensylvanicum*
Pigweed, Redroot,	*Amaranthus retroflexus*
Smooth	*Amaranthus hybridis*
Prickly Sida or Teaweed	*Sida spinosa*
Ragweed, Common,	*Ambrosia artemisiifolia*
Giant	*Ambrosia trifida*
Redweed	*Melochia corchorifolia*
Shepherdspurse	*Capsella bursa-pastoris*
Spurred Anoda	*Anoda cristata*
Tropic Croton	*Croton glandulosus*
Velvetleaf	*Abutilon theophrasti*
Venice Mallow	*Hibiscus trionum*
Wild Buckwheat	*Polygonum convolvulus*
Wild Mustard	*Sinapis arvensis*
Wild Poinsettia	*Euphorbia heterophylla*
Wild Spiny Cucumber	*Cucumis dipsaceus*
Wild Sunflower	*Helianthus annuus*
Yellow Rocket	*Barbarea vulgaris*

Sedges

Common Name	Scientific Name
Yellow Nutsedge	*Cyperus esculentus*

CONDITIONS OF SALE AND WARRANTY

The Directions for use of this product reflect the opinion of experts based on field use and tests. The directions are believed to be reliable and should be followed carefully. However, it is impossible to eliminate all risks inherently associated with use of this product. Crop injury, ineffectiveness or other unintended consequences may result because of such factors as weather conditions, presence of other materials, or use of the product in a manner inconsistent with its labeling, all of which are beyond the control of BASF CORPORATION ("BASF") or the Seller. All such risks shall be assumed by the Buyer.

BASF warrants that this product conforms to the chemical description on the label and is reasonably fit for the purposes referred to in the Directions for use, subject to inherent risks referred to above. BASF MAKES NO OTHER EXPRESS OR IMPLIED WARRANTY OF FITNESS OR MERCHANTABILITY OR ANY OTHER EXPRESS OR IMPLIED WARRANTY. IN NO CASE SHALL BASF OR THE SELLER BE LIABLE FOR CONSEQUENTIAL, SPECIAL OR INDIRECT DAMAGES RESULTING FROM THE USE OR HANDLING OF THIS PRODUCT. BASF and the Seller offer this product and the Buyer and User accept it, subject to the foregoing Conditions of sale and warranty which may be varied only by agreement in writing signed by a duly authorized representative of BASF.

PYRAMIN® DF

HERBICIDE

For weed control in sugar beets and red table beets.

A dry flowable formulation containing:

Active ingredients*	67.7%
Pyrazon (5-amino-4-chloro-2-phenyl-3(2H)-pyridazinone)	65.0%
Related aminochloro and dichlorophenyl pyridazinones	2.7%
Inert ingredients	32.3%

*Pyrazon ANSI equals chloridazon BSI *et al.*

EPA Reg. No. 7969-81

KEEP OUT OF REACH OF CHILDREN.

CAUTION

Harmful if swallowed or absorbed through the skin. Avoid contact with skin, eyes and clothing. Wash thoroughly after handling. Remove and wash contaminated clothing before re-use.

STATEMENT OF PRACTICAL TREATMENT

In case of contact: for skin, wash with plenty of soap and water; for eyes flush with plenty of water. Get medical attention if irritation persists.

ENVIRONMENTAL HAZARDS

Do not apply directly to water or wetlands. Do not contaminate water when disposing of equipment washwaters.

DIRECTIONS FOR USE

It is a violation of Federal law to use this product in a manner inconsistent with its labeling.

STORAGE AND DISPOSAL

Do not contaminate water, food or feed by storage or disposal.

Wastes resulting from the use of this product may be disposed of on site or at an approved waste disposal facility. Triple rinse container (or equivalent). Then offer for recycling or reconditioning, or puncture and dispose of in a sanitary landfill, or by incineration, or, if allowed by state and local authorities, by burning. If burned, stay out of smoke.

GENERAL INFORMATION

Among the weeds controlled by PYRAMIN DF HERBICIDE are lambsquarters, pigweed, ragweed, shepherdspurse, fanweed, purslane, nightshade, mustard, henbit, smartweed, nettleleaf, goosefoot and velvetleaf (Michigan and Ohio only).

PYRAMIN DF will not consistently provide acceptable control of annual grasses. Where grasses are a serious problem, follow directions on this label for control of grasses with specific combinations featuring PYRAMIN DF or use a separate treatment for grasses labeled for use in sugar beets or red table beets.

The correct method of application for PYRAMIN DF varies depending on area and cultural practices. Always follow the label recommendations which fit your area and cultural system. Check with local experts to ensure that you are applying PYRAMIN DF according to label directions.

See sections entitled **Mixing directions and Restrictions and limitations.**

For band treatments see section entitled **Band treatment** in each crop section.

MIXING DIRECTIONS

Use 20 to 40 gallons of water per acre, broadcast basis, or proportionately less for banded applications. Fill sprayer tank halfway with water. Start agitation with a mechanical or good bypass agitator. Assure adequate hydraulic agitation. While agitating, add the required amount of PYRAMIN DF. Add water to fill tank to required volume. Continue agitation during spraying. Mix only the amount of spray that will be used immediately.

RESTRICTIONS AND LIMITATIONS

Do not use this product on soils classified as sands or loamy sands because this may result in crop injury.

Do not use preemergence on muck and peat.

Do not use on soils with organic matter higher than 5%, except in the Red River Valley (see special section), or less than 2½% in Michigan/Ohio for velvetleaf control (see special section), since this may result in unsatisfactory weed control or crop injury.

Do not use PYRAMIN DF as a broadcast treatment in the dry irrigated regions.

If the beet crop is lost due to climatic or soil conditions following application of PYRAMIN DF, do not plant other crops in the band treated with PYRAMIN DF during the same season. If fields are replanted to sugar beets, re-seed into the treated band. Do not use PYRAMIN DF again as a preemergence treatment on the replanted beets because crop injury may result.

Do not use the same spray equipment for other purposes unless thoroughly cleaned. Coarse sprays are less likely to drift. Do not store near fertilizers, seeds, insecticides or fungicides. Do not contaminate water used for irrigation or domestic purposes.

Where this material is used in combination with TCA, do not use treated beet tops for feed or forage.

Always follow the restrictions and limitations for all products used in a tank mixture. The most restrictive labeling applies in the case of a tank mixture.

Do not apply this product through any type of irrigation system.

SUGAR BEETS

Two applications of PYRAMIN DF can be used in the Central and Eastern states only. Do not apply more than a total of 11.7 pounds per acre per season as excessive chemical residue may result.

PREEMERGENCE APPLICATIONS

Central and Eastern states (MI, MN, ND, OH)

Preemergence without irrigation: Apply PYRAMIN DF at the rate of 4.8 to 5.7 pounds per acre in enough water for good coverage. Use the lower rate on sandy loam soils. Spray immediately after beet seeds are planted and before beets and weeds emerge. Where dry weather follows application and weeds emerge, use a shallow cultivation before weeds are 2 inches tall. Otherwise, do not disturb or cover the treated band.

Minnesota, North Dakota

For soils with organic matter of 5% to 7%: In the Red River Valley only, for the control of redroot and prostrate pigweeds, common lambsquarters and wild mustard in soils with organic matter content of 5% to 7%, apply PYRAMIN DF at the rate of 11.7 pounds per acre as a single preemergence application. Spray immediately after beet seeds are planted and before beets and weeds emerge. Do not follow with a postemergence treatment of PYRAMIN DF. Do not use the 11.7 pound rate on soils with organic matter higher than 7% or less than 5% since this may result in unsatisfactory weed control or crop injury, respectively. Where dry weather follows application or where annual grasses are also a problem, refer to section entitled **Preemergence without irrigation.**

Michigan and Ohio

For velvetleaf control preemergence: In Michigan and Ohio only, for the control of velvetleaf preemergence, apply PYRAMIN® DF HERBICIDE at the rate of 9.3 pounds per acre as a single preemergence application. Spray immediately after beet seeds are planted and before beets and weeds emerge. Do not use the 9.3 pound rate on soils with organic matter higher than 5% or less than 2½% since this may result in unsatisfactory weed control or crop injury, respectively. Where dry weather follows application or where annual grasses are also a problem, refer to section entitled **Preemergence without irrigation.**

Plains, Mountain and Western States (CA, CO, ID, KS, NE, OR, TX, UT, WA, WY)

Preemergence with sprinkler irrigation: If sprinkler irrigation is used, apply PYRAMIN DF as a banded surface preemergence application. Do not incorporate PYRAMIN DF where sprinkler irrigation is to be used, as crop injury may result. Do not use broadcast treatments. Prepare the seedbed and pre-irrigate to the field capacity of the soil. Plant seed ¾ to 1 inch deep.

Immediately, or within 3 days after planting, apply PYRAMIN DF at the rate of 4.8 pounds per acre on low organic matter loam soils or 5.7 pounds per acre on loam and clay soils. Follow directions for banded treatments (see section entitled **Band treatment**) and adjust application rates accordingly. Follow with no more than ¾ inch of sprinkler irrigation per set before beets and weeds emerge, as greater amounts may cause beet injury. Repeat as needed to get good beet emergence. Do not exceed ¾ inch of sprinkler irrigation per set until the beets have 2 true leaves. (Do not count the cotyledonary leaves, which are the first 2 leaves to appear.)

Preplant incorporation with furrow irrigation: On fields to receive furrow irrigation, use the following planting procedure. Prepare seedbed or form beds for planting. On low organic matter and sandy loam soils, apply and incorporate PYRAMIN DF at the rate of 4.8 pounds per acre. On loams, silt loams and clays, use the 5.7 pounds per acre rate. Follow directions for banded treatments (see section entitled **Band treatment**) and adjust application rates accordingly. Use a rotary tiller type of incorporator and incorporate not more than 2 inches deep. Plant beets and furrow irrigate. Since PYRAMIN DF must have moisture to control weeds effectively, irrigate until tops of beds are thoroughly wetted. Repeat furrow irrigation as often as necessary to ensure good beet emergence and growth. Sprinkler irrigation can be substituted for furrow irrigation after the beets have developed 2 true leaves. (Do not count the cotyledonary leaves, which are the first 2 leaves to appear.)

To assure planting into treated bands, do the following things all in one tractor operation: (1) spray PYRAMIN DF, (2) incorporate, and (3) plant beets. Treat a band 1 or 2 inches wider than the tiller head used. Do not incorporate PYRAMIN DF with a disk because unsatisfactory weed control and sugar beet injury may result.

TANK MIXES

PYRAMIN DF + TCA Herbicides (MI, MN, ND, OH)

Where annual grasses are also a problem, apply 4 pounds to 8 pounds per acre of TCA with the recommended rate of PYRAMIN DF. Follow label directions for both products.

PYRAMIN DF + Nortron EC Herbicides for winter-grown sugar beets in California

In fields where wild oats and volunteer cereals are expected to be a problem, use a tank mixture of PYRAMIN DF plus Nortron at recommended rates listed in the following rate table. When mixing PYRAMIN DF in the spray tank with Nortron, add PYRAMIN DF first and agitate spray solution thoroughly; then add Nortron.

Under sprinkler irrigation or where natural rainfall is adequate, apply this tank mix preemergence. See **Preemergence with sprinkler irrigation** for directions and precautions regarding application of sprinkler irrigation. Where furrow irrigation is to be used, apply this tank mix preplant. See **Preplant incorporation with furrow irrigation** for directions and precautions regarding application of furrow irrigation. Do not use the mixture under conditions where Nortron alone is not recommended. Before use, read label for Nortron for additional information and precautions.

See page 4 for **Tank Mix of PYRAMIN DF + Nortron EC Application Rate Table.**

PYRAMIN DF + Antor 4ES Herbicides (In California, Northwest and Intermountain States, Michigan and Ohio)

For control of a mixed population of weeds included on the labels for PYRAMIN DF and Antor 4ES, apply a tank mix according to the directions presented by region. Observe all cautions and limitations on labeling of both products.

Mixing directions: Add PYRAMIN DF to half the spray volume of water, agitate spray solution thoroughly, and then add Antor 4ES. Agitate while filling to final spray volume and during spraying.

California: For preplant incorporation with furrow irrigation, apply the tank mix to the soil surface prior to or at time of planting and incorporate to a depth of 1 to 2 inches using a rotary type tiller. Use the recommended rates in **Application Rate Table.** Irrigate until tops of beds are thoroughly wetted.

For preemergence application with sprinkler irrigation, apply the tank mix to the soil surface at time of planting or shortly after, but prior to weed germination, using the rates recommended in **Application Rate Table.** Irrigate prior to crop and weed germination. Follow directions and precautions on this label for use preemergence with sprinkler irrigation. Do not exceed ¾ inch of sprinkler irrigation per set until the sugar beets have 2 true leaves.

Northwest, Intermountain States: For use preemergence under sprinkler irrigation only, apply the tank mix to the soil surface at the time of planting or shortly after, but prior to weed germination, using the rates recommended in **Application Rate Table.** Irrigate prior to crop and weed germination. Follow directions and precautions on this label for use preemergence with sprinkler irrigation. Do not exceed ¾ inch of sprinkler irrigation per set until the sugar beets have 2 true leaves.

Michigan and Ohio: For use preemergence without irrigation, apply the tank mix to the soil surface at the time of planting or shortly after, but prior to weed germination using rates recommended in **Application Rate Table.**

Tank Mix of PYRAMIN DF + Nortron EC Application Rate Table

Soil Texture	PYRAMIN DF per Acre		Nortron EC(2) per Acre	
	Broadcast (For calibration purposes only)	10-inch Band Width(1) 30″ Row	Broadcast (For calibration purposes only)	10-inch Band Width(1) 30″ Row
Coarse Textured Soils: Sands, loamy sands and sandy loams	Not Recommended			
Medium Textured Soils: Silt loams, clay loams which contain less than 3% organic matter	4.8 pounds	1.6 Pounds	8 Pints	2⅔ Pints
Fine Textured Soils: Clay loams which contain more than 3% organic matter and clays	4.8 Pounds	1.6 Pounds	10 Pints	3½ Pints

(1) For other band or row widths, adjust rates in proportion to the area actually treated. Do not apply this mixture broadcast.

(2) Nortron EC (1.5 lbs. active ingredient per gallon).

Tank Mix of PYRAMIN DF + Antor 4ES
Application Rate Table

Area	Soil Texture	Broadcast Rate/Acre[1]	
		PYRAMIN DF	Antor 4ES
California, Northwest and Intermountain States, Michigan and Ohio	Coarse: Sandy loams only.	4.0 Pounds	2 Quarts
	Medium: Loams, silt loams and clay loams containing less than 3% organic matter.	4.8 Pounds	2 to 3 Quarts
	Fine: Clay loams and clays containing 3%-5% organic matter.	4.8 Pounds	3 Quarts

(1) See section of both labels on band application for calculation of proportional amounts.
(2) Use higher rate on clay loam soils.

POSTEMERGENCE APPLICATION (All States)

General information

Timing is very important. Observe weed growth daily. Treat after beets have 2 expanded true leaves (do not count the cotyledonary leaves, which are the first 2 leaves to appear) and before any weeds have more than 2 to 4 true leaves. Treatments on larger weeds will not be effective.

Good soil moisture prior to treatment is necessary. If soil is dry, irrigation is recommended prior to chemical application. For optimum performance, a postemergence application should be followed with moisture. Where furrow irrigation is possible, irrigate shortly after application to the point that top surface of beds has reached maximum water-holding capacity. If sprinkler irrigation is used, do not apply more than 3/4 inch of water on the first set after the herbicide is applied.

Do not apply PYRAMIN DF HERBICIDE if weeds are stressed by lack of moisture, excessive heat, high winds, frost or low temperature, as insufficient weed control may result. Wait until more favorable conditions prevail before applying.

In the Central and Eastern States, these combinations can follow preemergence use of PYRAMIN DF provided a total of 11.7 pounds PYRAMIN DF per acre per season is not exceeded.

The addition of a surfactant, or its equivalent, often improves results of postemergence treatments. However, certain surfactants will seriously injure sugar beets. Consult your Experiment Station or Extension Service Weed Specialist for the suggested surfactant in your area.

TANK MIXES

PYRAMIN DF + Crop Oil Concentrate or Nonphytotoxic Emulsifiable Oil: Addition of crop oil concentrate or a nonphytotoxic emulsifiable oil to PYRAMIN DF may improve broadleaf weed control and allow greater flexibility in timing of successful applications relative to size of weeds. Apply 4.8 to 5.7 pounds PYRAMIN DF per acre according to soil type and organic matter content with 1 quart of crop oil concentrate or 1 gallon of nonphytotoxic emulsifiable oil per acre (broadcast basis). Use sufficient spray solution for good coverage 40-100 gallons per acre treated.

A nonphytotoxic oil should contain 2% emulsifier and be characterized as having an unsulphonated residue of over 95%.

Add PYRAMIN DF first to half the spray volume of water and mix thoroughly in the spray tank: then add the emulsifiable oil with the agitator running. Finally, fill the tank with water to required volume.

Do not use oil when maximum temperatures are expected to exceed 90°F at any time during the five-day period following the application, as excessive crop injury may result.

For specific directions for using a nonphytotoxic oil with PYRAMIN DF in your locality and for recommended brands of oil to use, consult your local sugar company fieldman or BASF Representative.

Increased temporary sugar beet injury has been observed where PYRAMIN DF in combination treatments has followed preplant applications of Tillam or Ro-Neet herbicides.

PYRAMIN DF + Dalapon Herbicides: Where grass and broadleaf weeds are present, use 4.8 to 5.7 pounds of PYRAMIN DF per acre mixed with dalapon, according to state recommendations and directions on the dalapon product label.

Where broadleaf weeds are the principal problem, use PYRAMIN DF with 1 gallon per acre of nonphytotoxic, emulsifiable oil. These mixtures may burn beet leaves or cause temporary stunting under some conditions. Do not apply when sugar beet foliage is wet.

PYRAMIN DF + Betamix or Betanex EC Herbicides
Michigan and Ohio: For broad spectrum weed control, including pigweed, kochia, wild buckwheat, lambsquarters, smartweed, mustard, nightshade, ragweed and green and yellow foxtails, combine 4.8 pounds of PYRAMIN DF and 3 quarts of Betamix or Betanex per acre. This combination provides residual as well as quick postemergence weed control. Follow the directions on the labels of both products. Add the Betamix or Betanex to the half-full spray tank after the PYRAMIN DF is adequately suspended. Do not add additional surfactants to the mixture. Do not apply in more than 30 gallons of water per acre.

Band Treatment

When spraying a band over the row, the amount of PYRAMIN DF applied per banded acre is reduced in proportion to the area actually treated. Example: to spray a 7-inch band over beets in rows 28 inches apart, the amount of PYRAMIN DF applied per banded acre is 1/4 the amount recommended for broadcast applications.
4.8 pounds/acre
× 1/4 of area banded
= 1.2 pounds per banded acre.
See the table at the bottom of this page.

Using planter-mounted band sprayers ensures that all planted beets are treated, and that band width and rate of chemical are uniform.

Band and Row Width	Pounds PYRAMIN DF per Banded Acre for Broadcast Rate of:	
	4.8 Pounds/Acre	5.7 Pounds/Acre
7-inch band on 22-inch rows	1.6	2.0
7-inch band on 28-inch rows	1.2	1.6
10-inch band on 22-inch rows	2.0	2.8
10-inch band on 28-inch rows	1.6	2.0

RED TABLE BEETS

PREEMERGENCE APPLICATION

Apply 4.8 to 5.7 pounds per acre of PYRAMIN DF HERBICIDE in enough water for good coverage. Use lower rate on sandy loam soils. Spray after beet seeds are planted but before beets and weeds emerge.

If rain does not fall within 5 to 10 days after treatment, beets should be irrigated to activate PYRAMIN DF. If irrigation is not possible, use a shallow cultivation before weeds are 2 inches tall.

Early post emergence application

Apply 5.7 pounds per acre of PYRAMIN DF.

Timing is very important. Observe weed growth daily. Treat after beets have 2 expanded true leaves and before any weeds have more than 2 to 4 true leaves. This is usually within 2 weeks after planting. Treatment on larger weeds will not be effective.

Note: PYRAMIN DF may be used both pre- and postemergence in the same season. Follow directions given on this label for each type of application.

Band treatment

When spraying a band over the row, the amount of PYRAMIN DF applied per acre is in proportion to the area actually treated.

Beet Rows 24 Inches Apart and Band Width of:	Pounds PYRAMIN DF per Banded Acre for Broadcast Rate of:	
	4.8 Pounds/Acre	5.7 Pounds/Acre
6-inch band	1.2	1.6
12-inch band	2.4	2.8

CONDITIONS OF SALE AND WARRANTY

The Directions for use of this product reflect the opinion of experts based on field use and tests. The directions are believed to be reliable and should be followed carefully. However, it is impossible to eliminate all risks inherently associated with use of this product. Crop injury, ineffectiveness or other unintended consequences may result because of such factors as weather conditions, presence of other materials, or use of the product in a manner inconsistent with its labeling, all of which are beyond the control of BASF CORPORATION ("BASF") or the Seller. All such risks shall be assumed by the Buyer.

BASF warrants that this product conforms to the chemical description on the label and is reasonably fit for the purposes referred to in the Directions for use, subject to the inherent risks referred to above. BASF MAKES NO OTHER EXPRESS OR IMPLIED WARRANTY OF FITNESS OR MERCHANTABILITY OR ANY OTHER EXPRESS OR IMPLIED WARRANTY. IN NO CASE SHALL BASF OR THE SELLER BE LIABLE FOR CONSEQUENTIAL, SPECIAL OR INDIRECT DAMAGES RESULTING FROM THE USE OR HANDLING OF THIS PRODUCT. BASF and the Seller offer this product and the Buyer and User accept it, subject to the foregoing Conditions of sale and warranty which may be varied only by agreement in writing signed by a duly authorized representative of BASF.

Pyramin is a registered trademark of BASF AG.
Antor, Betanex, Betamix and Nortron are registered trademarks of NOR-AM Chemical Company.
Tillam and Ro-Neet are registered trademarks of ICI Americas Inc.
U.S. Patent 3,210,353

PY 1 004 JJ6445 AO981288

PYRAMIN® FL

HERBICIDE

For weed control in sugar beets and red table beets.

A flowable formulation containing:

ACTIVE INGREDIENTS:*	41.3%
Pyrazon (5-amino-4-chloro-2-phenyl-3(2H)- pyridazinone)	35.5%
Related aminochloro and dichlorophenyl pyridazinones	5.8%
INERT INGREDIENTS:	58.7%
TOTAL	100%

*Equivalent to 4.2 pounds per gallon of pyrazon and related aminochloro and dichlorophenyl pyridazinones. Pyrazon ANSI equals chloridazon BSI *et al.*

EPA Reg. No. 7969-50

KEEP OUT OF REACH OF CHILDREN

CAUTION

Harmful if swallowed or absorbed through the skin. Avoid contact with skin, eyes and clothing. Wash thoroughly after handling. Remove and wash contaminated clothing before re-use.

STATEMENT OF PRACTICAL TREATMENT

In case of contact: for skin, wash with plenty of soap and water; for eyes, flush with plenty of water. Get medical attention if irritation persists.

Shake well before using.

Made in West Germany.

DIRECTIONS FOR USE

It is a violation of Federal law to use this product in a manner inconsistent with its labeling.

GENERAL INFORMATION

Pyramin® FL herbicide is a flowable liquid for the control of certain broadleaf weeds in sugar beets and red table beets. Application methods include preplant incorporation, preemergence and postemergence applications.

Among the weeds conrolled by Pyramin® FL herbicide are lambsquarters, pigweed, ragweed, shepherdspurse, fanweed, purslane, nightshade, mustard, henbit, smartweed, nettleleaf goosefoot and velvetleaf (Michigan and Ohio only).

Pyramin FL will not consistently provide acceptable control of annual grasses. Where grasses are a serious problem, follow directions on this label for control of grasses with specific combinations featuring Pyramin FL or use a separate treatment for grasses labeled for use in sugar beets or red table beets.

The correct method of application for Pyramin FL varies depending on area and cultural practices. Always follow the label recommendations which fit your area and cultural system. Check with local experts to ensure that you are applying Pyramin FL according to label directions.

See sections entitled **Mixing directions** and **Precautions.**

For band treatments see section entitled **Band treatment** in each crop section.

MIXING DIRECTIONS

Use 20 to 40 gallons of water per acre, broadcast basis, or proportionately less for banded applications.

Fill sprayer tank halfway with water. Start agitation with a mechanical or good bypass agitator. Assure adequate hydraulic agitation. While agitating, add the required amount of Pyramin FL. Add water to fill tank to required volume.

Continue agitation during spraying. Mix only the amount of spray that will be used immediately.

RESTRICTIONS AND LIMITATIONS

Do not use this product on soils classified as sands or loamy sands because this may result in crop injury.

Do not use preemergence on muck and peat.

Do not use on soils with organic matter higher than 5%, except in the Red River Valley (see special section), or less than 2¹/₂% in Michigan/Ohio for velvetleaf control (see special section), since this may result in unsatisfactory weed control or crop injury.

Do not use Pyramin FL as a broadcast treatment in the dry irrigated regions.

If the beet crop is lost due to climatic or soil conditions following application of Pyramin FL, do not plant other crops in the band treated with Pyramin FL during the same season. If fields are replanted to sugar beets, re-seed into the treated band. Do not use Pyramin FL again as a preemergence treatment on the replanted beets because crop injury may result.

Do not use the same spray equipment for other purposes unless thoroughly cleaned. Coarse sprays are less likely to drift. Do not store near fertilizers, seeds, insecticides or fungicides. Do not contaminate water used for irrigation or domestic purposes.

Where this material is used in combination with TCA, do not use treated beet tops for feed or forage.

Always follow the restrictions and limitations for all products used in a tank mixture. The most restrictive labeling applies in the case of a tank mixture.

Do not apply this product through any type of irrigation system.

ENVIRONMENTAL HAZARDS

Keep out of lakes, ponds or streams.

Do not contaminate water by cleaning of equipment or disposal of wastes.

STORAGE AND DISPOSAL

Do not contaminate water, food, or feed by storage or disposal.

Wastes resulting from the use of this product may be disposed of on site or at an approved waste disposal facility.

Triple rinse container (or equivalent). Then offer for recycling or reconditioning, or puncture and dispose of in a sanitary landfill, or by incineration, or, if allowed by state and local authorities, by burning. If burned, stay out of smoke.

SUGAR BEETS

Two applications of Pyramin FL can be used in the Central and Eastern states only. Do not apply more than a total of 7¹/₄ quarts per acre per season as excessive chemical residue may result.

PREEMERGENCE APPLICATIONS

Central and Eastern states (MI, MN, ND, OH)

General information

Preemergence without irrigation. Apply Pyramin FL at the rate of 3 to 3¹/₂ quarts per acre in enough water for good coverage. Use the lower rate on sandy loam soils. Spray immediately after beet seeds are planted and before beets and weeds emerge.

Where dry weather follows application and weeds emerge, use a shallow cultivation before weeds are 2 inches tall. Otherwise, do not disturb or cover the treated band.

Minnesota, North Dakota

For soils with organic matter of 5% to 7%. In the Red River Valley only, for the control of redroot and prostrate pigweeds, common lambsquarters and wild mustard in soils with organic matter content of 5% to 7%, apply Pyramin FL at the rate of 7¹/₄ quarts per acre as a single preemergence application. Spray immediately after beet seeds are planted and before beets and weeds emerge. Do not follow with a postemergence treatment of Pyramin FL. Do not use the 7¹/₄ quarts rate on soils with organic matter higher than 7% or less than 5% since this may result in unsatisfactory weed control or crop injury, respectively.

Where dry weather follows application or where annual grasses are also a problem, refer to section entitled **Preemergence without irrigation.**

Michigan and Ohio

For velvetleaf control: In Michigan and Ohio only, for the control of velvetleaf, apply Pyramin FL at the rate of 5³/₄ quarts per acre as a single preemergence application. Spray immediately after beet seeds are planted and before beets and weeds emerge. Do not use the 5³/₄ quarts rate on soils with organic matter higher than 5% or less than 2¹/₂% since this may result in unsatisfactory weed control or crop injury, respectively.

Where dry weather follows application or where annual grasses are also a problem, refer to section entitled **Preemergence without irrigation.**

Plains, Mountain and Western States
(CA, CO, ID, KS, NE, OR, TX, UT, WA, WY)

Preemergence with sprinkler irrigation: If sprinkler irrigation is used, apply Pyramin FL as a banded surface preemergence application. Do not incorporate Pyramin FL where sprinkler irrigation is to be used, as crop injury may result. Do not use broadcast treatments. Prepare the seedbed and pre-irrigate to the field capacity of the soil. Plant seed ³/₄ to 1 inch deep.

Immediately, or within 3 days after planting, apply Pyramin FL at the rate of 3 quarts per acre on low organic matter loam soils or 3¹/₂ quarts per acre on loam and clay soils. Follow directions for banded treatments (see section entitled **Band treatment**) and adjust application rates accordingly. Follow with no more than ³/₄ inch of sprinkler irrigation per set before beets and weeds emerge, as greater amounts may cause beet injury. Repeat as needed to get good beet emergence. Do not exceed ³/₄ inch of sprinkler irrigation per set until the beets have 2 true leaves. (Do not count the cotyledonary leaves, which are the first 2 leaves to appear.)

Preplant incorporation with furrow irrigation: On fields to receive furrow irrigation, use the following planting procedure. Prepare seedbed or form beds for planting. On low organic matter and sandy loam soils, apply and incorporate Pyramin FL at the rate of 3 quarts per acre. On loams, silt loams and clays, use the 3½ quarts per acre rate. Follow directions for banded treatments (see section entitled **Band treatment**) and adjust application rates accordingly. Use a rotary tiller type of incorporator and incorporate not more than 2 inches deep. Plant beets and furrow irrigate. Since Pyramin FL must have moisture to control weeds effectively, irrigate until tops of beds are thoroughly wetted.

Repeat furrow irrigation as often as necessary to ensure good beet emergence and growth. Sprinkler irrigation can be substituted for furrow irrigation after the beets have developed 2 true leaves. (Do not count the cotyledonary leaves, which are the first 2 leaves to appear.)

To assure planting into treated bands, do the following things all in one tractor operation: (1) spray Pyramin FL, (2) incorporate, and (3) plant beets. Treat a band 1 or 2 inches wider than the tiller head used. Do not incorporate Pyramin FL with a disk because unsatisfactory weed control and sugar beet injury may result.

TANK MIXES

Pyramin FL + TCA herbicides (Michigan, Minnesota, N. Dakota, Ohio)
Where annual grasses are also a problem, apply 4 lbs. to 8 lbs. per acre of TCA with the recommended rate of Pyramin FL. Follow label directions for both products.

Pyramin FL + Nortron® EC herbicides for winter-grown sugar beets in California: In fields where wild oats and volunteer cereals are expected to be a problem, use a tank mixture of Pyramin FL plus Nortron at recommended rates listed in the following rate table. When mixing Pyramin FL in the spray tank with Nortron, add Pyramin FL first and agitate spray solution thoroughly; then add Nortron. See the **Application Rate Table** on page 4.

Under sprinkler irrigation or where natural rainfall is adequate, apply this tank mix preemergence. See **Preemergence with sprinkler irrigation** above for directions and precautions regarding application of sprinkler irrigation. Where furrow irrigation is to be used, apply this tank mix preplant. See **Preplant incorporation with furrow irrigation** above for directions and precautions regarding application of furrow irrigation. Do not use the mixture under conditions where Nortron alone is not recommended. Before use, read label for Nortron for additional information and precautions.

Pyramin FL + Antor® 4ES herbicides
(In California, Northwest, Intermountain States, Michigan and Ohio)
For control of a mixed population of weeds included on the labels for Pyramin® FL and Antor® 4ES herbicides, apply a tank mix according to the directions presented below by region. Observe all cautions and limitations on labeling of both products.

Mixing directions: Add Pyramin FL to half the spray volume of water, agitate spray solution thoroughly, and then add Antor 4ES. Agitate while filling to final spray volume and during spraying.

California: For preplant incorporation with furrow irrigation, apply the tank mix to the soil surface prior to, or at time of planting and incorporate to a depth of 1 to 2 inches using a rotary type tiller. Use the recommended rates in **Application Rate Table.** Irrigate until tops of beds are thoroughly wetted.

For preemergence application with sprinkler irrigation, apply the tank mix to the soil surface at time of planting or shortly after, but prior to weed germination, using the rates recommended in **Application Rate Table.** Irrigate prior to crop and weed germination. Follow directions and precautions on this label for use preemergence with sprinkler irrigation. Do not exceed ¾ inch of sprinkler irrigation per set until the sugar beets have two true leaves.

Northwest, Intermountain States: For use preemergence under sprinkler irrigation only, apply the tank mix to the soil surface at the time of planting or shortly after, but prior to weed germination, using the rates recommended in **Application Rate Table.** Irrigate prior to crop and weed germination. Follow directions and precautions on this label for use preemergence with sprinkler irrigation. Do not exceed ¾ inch of sprinkler irrigation per set until the sugar beets have two true leaves.

Michigan and Ohio. For use preemergence without irrigation, apply the tank mix to the soil surface at the time of planting or shortly after, but prior to weed germination using rates recommended in **Application Rate Table.**

Tank Mix or Pyramin FL + Nortron EC Application Rate Table

Soil Texture	Pyramin FL per Acre		Nortron EC(2) per Acre	
	Broadcast (For calibration purposes only)	10-inch Band Width(1) 30″ Row	Broadcast (For calibration purposes only)	10-inch Band Width(1) 30″ Row
Coarse Textured Soils: Sands, loamy sands and sandy loams	Not Recommended			
Medium Textured Soils: Silt loams, clay loams which contain less than 3% organic matter	3 Quarts	1 Quart	4 Quarts	2⅔ Pints
Fine Textured Soils: Clay loams which contain more than 3% organic matter and clays	3 Quarts	1 Quart	5 Quarts	3½ Pints

(1) For other band or row widths, adjust rates in proportion to the area actually treated. Do not apply this mixture broadcast.
(2) Nortron EC (1.5 lbs. active ingredient per gallon).

Tank Mix of Pyramin FL + Antor 4ES Application Rate Table

Area	Soil Texture	Broadcast Rate (Quarts/Acre)(1)	
		Pyramin FL	Antor 4ES
California, Northwest, Intermountain States, Michigan and Ohio	**Coarse:** Sandy loams only.	2½	2
	Medium: Loams, silt loams and clay loams containing less than 3% organic matter.	3	2 to 3(2)
	Fine: Clay loams and clays containing 3%-5% organic matter.	3	3

(1) See section of both labels on Band Application for calculation of proportional amounts.
(2) Use higher rate on clay loam soils.

POSTEMERGENCE APPLICATION
(All states)

GENERAL INFORMATION

Timing is very important. Observe weed growth daily. Treat after beets have 2 expanded true leaves (do not count the cotyledonary leaves, which are the first 2 leaves to appear) and before any weeds have more than 2 to 4 true leaves. Treatments on larger weeds will not be effective.

Good soil moisture **prior** to treatment is necessary. If soil is dry, irrigation is recommended prior to chemical application. For optimum performance, a postemergence application should be followed with moisture. Where furrow irrigation is possible, irrigate shortly after application to the point that top surface of beds has reached maximum water-holding capacity. If sprinkler irrigation is used, do not apply more than ¾ inch of water on the first set after the herbicide is applied.

Do not apply Pyramin FL if weeds are stressed by lack of moisture, excessive heat, high winds, frost or low temperature, as insufficient weed control may result. Wait until more favorable conditions prevail before applying.

In the Central and Eastern States, these combinations can follow preemergence use of Pyramin FL provided a total of 7¼ quarts Pyramin FL per acre per season is not exceeded.

The addition of a surfactant, or its equivalent, often improves results of postemergence treatments. However, certain surfactants will seriously injure sugar beets. Consult your Experiment Station or Extension Service Weed Specialist for the suggested surfactant in your area.

TANK MIXES

Pyramin FL + Nonphytotoxic Emulsifiable Oil: Addition of a nonphytotoxic emulsifiable oil to Pyramin FL may improve broadleaf weed control and allow greater flexibility in timing of successful applications relative to size of weeds. Apply 3 to 3½ quarts Pyramin FL per acre according to soil type and organic matter content with 1 gallon of nonphytotoxic, emulsifiable oil per acre (broadcast basis). Use sufficient spray solution for good coverage, 40-100 gallons per acre treated.

Use a nonphytotoxic oil containing 2% emulsifier and characterized as having an unsulphonated residue of over 95%.

Add Pyramin FL first to half the spray volume of water and mix thoroughly in the spray tank; then add the emulsifiable oil with the agitator running. Finally, fill the tank with water to required volume.

Do not use oil when maximum temperatures are expected to exceed 90°F at any time during the five-day period following the application, as excessive crop injury may result.

For specific directions for using a nonphytotoxic oil with Pyramin FL in your locality and for recommended brands of oil to use, consult your local sugar company fieldman or BASF representative.

Increased temporary sugar beet injury has been observed where Pyramin FL in combination treatments has followed preplant applications of Tillam® or Ro-Neet® herbicides.

Pyramin FL + Dalapon herbicides: Where grass and broadleaf weeds are present, use 3 to 3½ quarts of Pyramin FL per acre mixed with dalapon, according to state recommendations and directions on the dalapon product label.

Where broadleaf weeds are the principal problem, use Pyramin FL with 1 gallon per acre of nonphytotoxic, emulsifiable oil (for directions, see above). These mixtures may burn beet leaves or cause temporary stunting under some conditions. Do not apply when sugar beet foliage is wet.

Pyramin FL + Betanal® EC herbicides: For broad spectrum weed control including pigweed, kochia, wild buckwheat, lambsquarters, smartweed, mustard, nightshade, ragweed and green and yellow foxtails, combine 3 quarts of Pyramin FL and 3 quarts of Betanal per acre. This combination provides residual as well as quick postemergence weed control. Follow the directions on the labels of both products. Add the Betanal to the half-full spray tank after the Pyramin FL is adequately suspended. Do not add additional surfactants to the mixture. Do not apply in more than 30 gallons of water per acre.

Band treatment

When spraying a band over the row, the amount of Pyramin FL applied per banded acre is reduced in proportion to the area actually treated. Example: to spray a 7-inch band over beets in rows 28 inches apart, the amount of Pyramin FL applied per banded acre is 1/4 the amount recommended for broadcast applications.

Three quarts/acre × 1/4 of area banded = 3/4 quart per banded acre.

Using planter-mounted band sprayers ensures that all planted beets are treated, and that band width and rate of chemical are uniform.

Band and Row Width	Quarts Pyramin FL per Banded Acre for Broadcast Rate of: 3 Quarts/Acre	3½ Quarts/Acre
7-inch band on 22-inch rows	1	1¼
7-inch band on 28-inch rows	¾	1
10-inch band on 22-inch rows	1¼	1¾
10-inch band on 28-inch rows	1	1¼

RED TABLE BEETS

Preemergence application

Apply 3 to 3½ quarts per acre of Pyramin FL in enough water for good coverage. Use lower rate on sandy loam soils. Spray after beet seeds are planted but before beets and weeds emerge.

If rain does not fall within 5 to 10 days after treatment, beets should be irrigated to activate Pyramin FL. If irrigation is not possible, use a shallow cultivation before weeds are 2 inches tall.

Early postemergence application

Apply 3½ quarts per acre of Pyramin FL.

Beet Rows 24 inches apart and band width of:	Quarts Pyramin FL per Banded Acre for Broadcast Rate of: 3 Quarts/Acre	3½ Quarts/Acre
6-inch band	¾	1
12-inch band	1½	1¾

Timing is very important. Observe weed growth daily. Treat after beets have 2 expanded true leaves and before any weeds have more than 2 to 4 true leaves. This is usually within 2 weeks after planting. Treatment on larger weeds will not be effective.

Note: Pyramin FL may be used both pre and postemergence in the same season. Follow directions given on this label for each type of application.

Band treatment

When spraying a band over the row, the amount of Pyramin FL applied per acre is in proportion to the area actually treated.

CONDITIONS OF SALE AND WARRANTY

The Directions for Use of this product reflect the opinion of experts based on field use and tests. The directions are believed to be reliable and should be followed carefully. However, it is impossible to eliminate all risks inherently associated with use of this product. Crop injury, ineffectiveness or other unintended consequences may result because of such factors as weather conditions, presence of other materials, or use of the product in a manner inconsistent with its labeling, all of which are beyond the control of BASF CORPORATION ("BASF") or the Seller. All such risks shall be assumed by the Buyer.

BASF warrants that this product conforms to the chemical description on the label and is reasonably fit for the purposes referred to in the Directions for Use, subject to the inherent risks referred to above. BASF MAKES NO OTHER EXPRESS OR IMPLIED WARRANTY OF FITNESS OR MERCHANTABILITY OR ANY OTHER EXPRESS OR IMPLIED WARRANTY. IN NO CASE SHALL BASF OR THE SELLER BE LIABLE FOR CONSEQUENTIAL, SPECIAL OR INDIRECT DAMAGES RESULTING FROM THE USE OR HANDLING OF THIS PRODUCT. BASF and the Seller offer this product, and the Buyer and User accept it, subject to the foregoing Conditions of Sale and Warranty which may be varied only by agreement in writing signed by a duly authorized representative of BASF.

Pyramin is a registered trademark of BASF AG.
Antor, Betanal and Nortron are registered trademarks of NOR-AM Chemical Company.
Tillam and Ro-Neet are registered trademarks of Stauffer Chemical Company, U.S.A.
U.S. Patent 3,210,353

PY 1 002 JJ5963 A0410687

RONILAN® WP

FUNGICIDE

A wettable powder containing:

ACTIVE INGREDIENT:
3-(3, 5-Dichlorophenyl)-5-ethenyl-5-methyl-2,4-oxazolidinedione . . .50%
INERT INGREDIENTS: 50%
TOTAL 100%

EPA Reg. No. 7969-53

KEEP OUT OF REACH CHILDREN.

CAUTION

Causes eye irritation. Do not get on skin, in eyes or on clothing.

STATEMENT OF PRACTICAL TREATMENT

If contacted, flush eyes immediately with water for 15 minutes. Get medical attention. In case of contact with skin or clothing, remove contaminated clothing, wash skin thoroughly with soap and water. This product is a potential skin sensitizer. If irritation persists get medical attention.

ENVIRONMENTAL HAZARDS

Do not apply directly to water. Do not contaminate water by cleaning of equipment or disposal of wastes.

(2 × 1½ lb. bags. Do not remove bags from carton except for immediate use.)

Net contents 3 lbs.

RE-ENTRY AND WORKERS' PROTECTION STATEMENTS

Do not apply this product in such a manner as to directly or through drift expose workers or other persons, except those knowingly involved in the application. The area being treated must be vacated by unprotected persons.

Do not enter treated areas without protective clothing until sprays have dried.

Because certain states may require more restrictive re-entry intervals for various crops treated with this product, consult your State Department of Agriculture for further information.

Written or oral warnings must be given to workers who are expected to be in a treated area or in an area about to be treated with this product. Oral warnings must inform workers of areas or fields that may not be entered without specific protective clothing until sprays have dried, and appropriate actions to take in case of accidental exposure. When oral warnings are given, warnings shall be given in a language customarily understood by workers. Oral warnings must be given if there is reason to believe that written warnings cannot be understood by workers. Written warnings must include the following information: "WARNING. Area treated with Ronilan on (date of application). Do not enter without appropriate protective clothing until sprays have dried. If contacted, flush eyes immediately with water for 15 minutes. Get medical attention.

In case of contact with skin or clothing, remove contaminated clothing, wash skin thoroughly with soap and water. Call physician if irritation occurs. Wash contaminated clothing before re-use."

DIRECTIONS FOR USE

It is a violation of Federal law to use this product in a manner inconsistent with its labeling.

STORAGE AND DISPOSAL

Do not contaminate water, food, or feed by storage or disposal.

Wastes resulting from the use of this product may be disposed of on site or at an approved waste disposal facility.

Completely empty bag into application equipment. Then dispose of empty bag in a sanitary landfill or by incineration, or, if allowed by state and local authorities, by burning. If burned, stay out of smoke.

In case of emergency

In case of large-scale spillage regarding this product, call:
CHEMTREC 800-424-9300
BASF Corporation 201-316-3000

In case of medical emergency regarding this product, call:

1. Your local doctor for immediate treatment.
2. Your local poison control center (hospital).

3. BASF Corporation 201-316-3000
All applicable directions, restrictions, precautions and Conditions of sale and warranty are to be followed. This labeling must be in the possession of the user at the time of application.

GENERAL INFORMATION

Ronilan is a contact fungicide for the control of *Botrytis* fruit rot (gray mold) of strawberries, *Sclerotinia* "drop" (watery soft rot) of head lettuce, and brown rot blossom and twig blight, and fruit brown rot on stonefruit (see page 5 for list of stonefruit). Thorough coverage of plant parts to be protected is essential for effective disease control. If other diseases are a problem, an additional fungicide will be needed.

The repeated exclusive use of Ronilan, as is the case with the exclusive use of other fungicides, may result in the buildup of resistant strains of *Botrytis* and loss of disease control. A spray program alternating other fungicides with Ronilan may delay the buildup of resistant strains. If treatment becomes ineffective due to the presence of a Ronilan resistant strain of *Botrytis*, then prompt use of other fungicides is necessary to maintain disease control.

STRAWBERRIES

Time and rate of application for states other than California and Florida

Thorough spray coverage of the blossoms and developing fruit is essential. For full season control of *Botrytis* disease, the following spray program is recommended. The first application should be made no later than 10% primary bloom at rates indicated (see table). The interval between subsequent applications will vary according to weather conditions and resultant disease pressure. A rate of 1½ pounds product per acre is generally recommended. A one pound product per acre rate of Ronilan should be used only when low disease pressure can be predicted. A two pounds product per acre rate should be used when the foliage is dense and/or disease pressure is high. If a heavy rainfall occurs any time during this spray program or if a wet period (light rain, fog or dew) lasting more than 24 hours occurs, immediate retreatment is necessary at a rate of 1½ to 2 pounds product per acre as soon as conditions will allow the spray to dry on the plants.

Time and rate of application for California

Thorough spray coverage of blossoms and developing fruit is essential for good disease control. For full season control of *Botrytis* disease, fungicides may need to be applied at 7 to 14 day intervals throughout the production cycle. When using Ronilan, the first application should be made no later than 10% primary bloom. If conditions favorable for high disease pressure persist after the first application, a second application should be made 7 to 9 days after the first.

To reduce the hazard of resistance developing, further use of Ronilan should be reserved for periods of high disease pressure and/or highest economic return. A rate of 1½ pounds product per acre is generally recommended. A 2 pounds product per acre rate should be used when foliage is dense and disease pressure is high. **Make no more than a total of four applications of Ronilan per season.**

Time and rate of application for Florida

Thorough spray coverage of the developing fruit is essential. For full season control of *Botrytis* disease, fungicides should be applied at 7 day intervals throughout the production cycle. If a frost protection irrigation is planned, an application of Ronilan must be made immediately before and and after the irrigation. If a heavy rainfall occurs anytime during the spray program or if a wet period (light rain, fog, or dew) lasting more than 24 hours occurs, immediate retreatment is necessary as soon as conditions will allow the spray to dry on the plants.

Method of application

Ground equipment: Application of Ronilan should be made in not less than 100 gallons of spray solution per acre to obtain thorough coverage of the developing fruit. An operating pressure of 60-150 psi is recommended to obtain adequate penetration of the spray through the canopy. Cone-type nozzles are recommended. Spray booms with at least 3 nozzles per row (1 over row: 2 side drops) are recommended.

Application Rates For Strawberries In States Other Than California And Florida

Moisture Conditions	Spray* Interval (Days)	Rate (Pounds Product/A) 1st Year Plants or Sparse Foliage	Dense Foliage
Frequent natural moisture (intermittent rain, fog, dew) or when using sprinkler irrigation *(high disease pressure)*	7-9	1½	1½-2
Limited natural moisture or infrequent sprinkler irrigation *(low disease pressure)*	10-14	1	1½-2

* Use spray interval throughout the bearing cycle.

Application Rates For Strawberries In California

Moisture Conditions	Spray* Interval (Days)	Rate (Pounds Product/A) 1st Year Plants or Sparse Foliage	Dense Foliage
Frequent natural moisture (intermittent rain, fog, dew) or when using sprinkler irrigation *(high disease pressure)*	7-9	1½	2

* Make no more than four (4) applications per season.

Application Rates For Strawberries In Florida

Moisture Conditions	Spray Interval (Days)	Rate (Pounds Product/A)
Frequent natural moisture (intermittent rain, fog, dew) or when using sprinkler irrigation *(high disease pressure)*	7	2

Air equipment: Application of Ronilan should be made in not less than 20 gallons of spray solution per acre. Thorough spray coverage of the developing fruit is essential.

Restrictions and limitations for strawberries

Do not apply more than a total of 12 pounds of product per acre in one season, except in California where the total is restricted to 8 pounds in no more than four applications.

Ronilan does not control *Rhizopus* rot of strawberries in the field or in storage. There may be a competitive relationship between *Botrytis* and *Rhizopus* incidence such that control of *Botrytis* may result in an increase of *Rhizopus* rot in stored fruit. *Rhizopus* rot becomes a problem as a result of certain climatic conditions (such as prolonged warm, humid periods) and cultural practices (such as high nitrogen fertilization which may lead to the production of softer fruit). If conditions are conducive for *Rhizopus* development, including those described above, do not use Ronilan.

Do not use Ronilan as a plant dip, as injury will occur.

Do not apply this product through any type of irrigation system.

Rotational crop restrictions for states other than California

Lettuce (head types only) may be planted after strawberries.

Leafy vegetables may be planted 6 months after treatment that does not exceed 6 pounds active ingredient per acre (12 pounds product per acre).

Cucurbits may be planted 2 months after treatment that does not exceed 6 pounds active ingredient per acre (12 pounds product per acre).

Corn may be planted 2 months after treatment that does not exceed 6 pounds active ingredient per acre (12 pounds product per acre), provided only the corn grain is used for food and/or feed purposes.

Other grain crops may be planted 9 months after treatment that does not exceed 6 pounds active ingredient per acre (12 pounds product per acre).

Any rotational crops may be planted 20 days after treatment that does not exceed 3 pounds active ingredient per acre (6 pounds product per acre).

Rotational crop restrictions for California

Lettuce (head types only) may be planted after strawberries.

Leafy vegetables may be planted 6 months after treatment that does not exceed 4 pounds active ingredient per acre (8 pounds product per acre).

Cucurbits may be planted 2 months after treatment that does not exceed 4 pounds active ingredient per acre (8 pounds product per acre).

Corn may be planted 2 months after treatment that does not exceed 4 pounds active ingredient per acre (8 pounds product per acre), provided only the corn grain is used for food and/or feed purposes.

Other grain crops may be planted 9 months after treatment that does not exceed 4 pounds active ingredient per acre (8 pounds product per acre).

Any rotational crops may be planted 20 days after treatment that does not exceed 3 pounds active ingredient per acre (6 pounds product per acre).

Application Rates For Lettuce (Head Types Only)**

Application	Timing	Rate (Pounds Product/Acre) *Low Disease Pressure	*High Disease Pressure
	Direct Seeded Lettuce—Application should be made immediately (within 2 days) after thinning.		

Application	Timing	Rate (Pounds Product/Acre) *Low Disease Pressure	*High Disease Pressure
First	**Transplanted Lettuce**—Application should be made 7-10 days after transplanting.	1	1 ½-2
Second	Application should be made 14 days after first spray if cool, wet conditions (which favor disease) occur for periods lasting 48 hours or more.		
Third	**Direct Seeded Lettuce Only**—Application should be made 14 days after second spray if conditions favoring disease occur.		

* Based on previous history of disease infestation in the field or adjacent fields.

Leaf lettuce registration recently approved. See "Supplemental labeling" at end of this label.

LETTUCE (HEAD TYPES ONLY)

Time and rate of application

For control of *Sclerotinia* "drop," up to three applications may be made under certain conditions in one season as specified in the rate table. Thorough spray coverage of plant parts to be protected is essential for effective disease control. Do not disturb soil after application.

Method of application

Application of Ronilan should be made in not less than 100 gallons of spray solution per acre (broadcast basis) to obtain thorough coverage of the lower leaves, plant stem and bed surface. An operating pressure of 50-100 psi is recommended. Use cone or flat fan nozzles. Flood type nozzles should not be used. To apply Ronilan to one-row beds, spray booms should have at least 2 nozzles per row. For two-row beds use three nozzles or more per bed.

Restrictions and limitations for lettuce

Do not apply Ronilan within 28 days of harvest.

Do not use on leaf type lettuce.

Do not use Ronilan as a plant dip, as injury may occur.

Do not apply this product through any type of irrigation system.

STONEFRUIT

General Information

Ronilan is effective for the control of brown rot blossom and twig blight, and fruit brown rot of stonefruit including apricots, cherries, nectarines, peaches, plums, and prunes. Ronilan is a contact fungicide, therefore, thorough coverage of plant parts to be protected is essential for effective disease control.

Time and rate of application

Depending on conditions which favor disease development, one to three applications should be made during bloom for control of brown rot blossom and twig blight. One or two preharvest applications should be made for control of fruit brown rot (see table).

A rate of 1½ pounds product per acre is generally recommended. A one pound product per acre rate of Ronilan should be used only when low disease pressure can be predicted. Under high disease pressure or in conditions very favorable for disease development, use the higher treatment rates specified, multiple applications, and shorter application intervals. For large mature trees, use the higher treatment rates specified.

Consistency of disease control with low recommended rates can be enhanced by 8-16 oz./100 gal. of a nonionic surfactant in tank mix with Ronilan.

Method of application

Rates of Ronilan per 100 gallons are based on an application of 400 gallons per acre of dilute spray. Refer to the **Application Rates and Timing for Control of Brown Rot** if other than 400 gallons per acre is applied.

Apply Ronilan as a spray with ground equipment, using sufficient water to obtain **thorough coverage** (50-400 gallons per acre).

Mixing instructions: Thoroughly mix the required amount of Ronilan in a small volume of water. Partially fill the spray tank with clean water and begin agitation. Add the Ronilan mix to the tank and fill tank to the total volume required. Maintain agitation to keep the material in suspension and apply with properly calibrated spray equipment.

Ronilan is physically compatible with most pesticides used in a tank mix on stonefruit. To assure compatibility of Ronilan with other products, add correct proportions in a small container with water. Mix thoroughly and let the mixture stand for five minutes. The mixture is compatible if the combination remains mixed, or can be readily remixed.

Stonefruit
Application Rates And Timing For Control Of Brown Rot (*Monilinia* spp.)

Crop	Pounds of Product* per 100 Gallons	Acre	Timing
Apricots Cherries Nectarines Peaches Plums Prunes	0.25 to 0.5 Rate is based on 400 GPA	1.0 to 2.0	**Brown Rot Blossom & Twig Blight** Apply at **early bloom** Apricots—Red Bud Peaches & Nectarines—Pink Bud Cherries—Early Popcorn Plums and Prunes—Green Tip Apply again at **full bloom** if conditions persist which favor disease development.
	Do not exceed 2 lbs. product per acre per application.		Apply again at **petal fall** if conditions persist which favor disease development. Do not apply more than 3 bloom treatments. **Fruit Brown Rot** Apply when conditions favor disease development within 3 weeks of harvest. One additional application should be made 7 to 14 days later if these conditions persist. Do not apply more than 2 pre-harvest treatments. Do not apply within 3 days of harvest.

***See Time and rate of application.**

Restrictions and limitations for stonefruit

Do not apply Ronilan within 3 days of harvest.

Do not apply more than 2 pounds of Ronilan per acre per application.

Do not apply more than 10 pounds of Ronilan per acre per season (maximum of 6 pounds at bloom and 4 pounds pre-harvest).

Do not apply Ronilan during rain. Apply when conditions will permit spray to dry on the plants.

Do not apply this product through any type of irrigation system.

CONDITIONS OF SALE AND WARRANTY

The Directions for use of this product reflect the opinion of experts based on field use and tests. The directions are believed to be reliable and should be followed carefully. However, it is impossible to eliminate all risks inherently associated with use of this product. Crop injury, ineffectiveness or other unintended consequences may result because of such factors as weather conditions, presence of other materials, or use of the product in a manner inconsistent with its labeling, all of which are beyond the control of BASF CORPORATION ("BASF") or the Seller. All such risks shall be assumed by the buyer. BASF warrants that this product conforms to the chemical description on the label and is reasonably fit for the purposes referred to in the Directions for use, subject to the inherent risks referred to above. BASF MAKES NO OTHER EXPRESS OR IMPLIED WARRANTY OF FITNESS OR MERCHANTABILITY OR ANY OTHER EXPRESS OR IMPLIED WARRANTY. IN NO CASE SHALL BASF OR THE SELLER BE LIABLE FOR CONSEQUENTIAL, SPECIAL OR INDIRECT DAMAGES RESULTING FROM THE USE OR HANDLING OF THIS PRODUCT. BASF and the Seller offer this product, and the Buyer and User accept it, subject to the foregoing Conditions of sale and warranty which may be varied only by agreement in writing signed by a duly authorized representative of BASF.

EPA Establishment No. 39578-TX-1

Ronilan is a registered trademark of BASF AG.

Made in West Germany.

RO 2 013 JJ6329 A0751088

SUPPLEMENTAL LABELING

Ronilan® WP

EPA Reg. No. 7969-53

In lettuce (all types), raspberries, and dry bulb onions*

*Registration pending in California.

All applicable directions, restrictions, precautions and Conditions of sale and warranty on the EPA-registered label are to be followed. This labeling must be in the possession of the user at the time of application.

DIRECTIONS FOR USE

It is a violation of Federal law to use this product in a manner inconsistent with its labeling.

General Information

Ronilan® fungicide is a contact fungicide for control of Sclerotinia "drop" (watery soft rot) of lettuce, Botrytis fruit rot of raspberries and Sclerotinia watery soft rot, Sclerotium white rot, Botrytis neck rots, and Botrytis leaf blight of onions. A chemical barrier of **Ronilan** must be established and maintained to achieve effective disease control. For diseases which infect above ground plant parts (such as fruit rots and leaf blights caused by *Botrytis*), thorough spray coverage of the plant parts to be protected is essential. For diseases which infect plant parts at the soil surface of below ground (such as white rot on onions), thorough spray coverage of the soil and stem base is essential. Refer to the following sections for specific information.

LETTUCE

Time and rate of application

For control of Sclerotinia "drop", up to three applications may be made under certain conditions in one season as specified in the rate table. **Do not disturb soil after application.**

Method of application

Ground equipment: Application should be made with not less than 100 gallons of spray solution per acre (broadcast basis) to obtain thorough coverage of the lower leaves, plant stem and bed surface. An operating pressure of 50-100 PSI is recommended. Use cone or flat fan nozzles. Flood type nozzles should not be used. To apply **Ronilan** to one-row beds, spray booms should have at least 2 nozzles per row. For two-row beds, use 3 nozzles or more per bed.

Air equipment: Application of **Ronilan** should be made in not less than 20 gallons of spray solution per acre. Thorough spray coverage of the developing plant is essential.

Restrictions and limitations for lettuce

Do not apply **Ronilan** within 28 days of harvest.

Do not apply more than 6 lbs. of **Ronilan**/acre in one season on lettuce.

Do not apply **Ronilan** through any type of irrigation system.

Application Rate Table
Lettuce (All Types)

Application	Timing	Rate Lb. Product/Acre: Low* Disease Pressure	Rate Lb. Product/Acre: High* Disease Pressure
First	**Direct Seeded Lettuce** Application should be made immediately (within 2 days) after thinning. **Transplanted Lettuce** Application should be made 7-10 days after transplanting.	1	1½-2
Second	Application should be made 10-14 days after first spray if cool, wet conditions (which favor disease) occur for periods lasting 48 hours or more.		
Third	**Direct Seeded Lettuce Only** Application should be made 10-14 days after second spray if conditions favoring disease occur.		

* Based on previous history of disease infestation in the field or adjacent fields.

RASPBERRIES

Time and rate of application

For control of Botrytis fruit rot, the following spray program is recommended. The first application should be made not later than 10% primary bloom. Timing of subsequent applications should be made as indicated (see table) and will vary according to weather conditions and resultant disease pressure.

Apply **Ronilan** at rates ranging from 1 to 2 lbs. product/A. The 2 lb. rate will provide an added margin of control above lower rates and, therefore, should be used whenever disease pressure is predictably severe. Use the 1½ lb. rate when disease pressure is expected to be moderate. The 1 lb. rate should only be used in newly established areas when disease pressure is expected to be low.

If a heavy rainfall occurs anytime during this spray program or if a wet period (light rain, fog or dew) lasting more than 24 hours occurs, immediate retreatment is necessary at a rate of 2 lbs. product per acre as soon as conditions allow the spray to dry on the plants.

Method of application

Ground equipment: Applications of **Ronilan** should be made with not less than 100 gallons of spray solution per acre (broadcast basis) to obtain thorough coverage of the developing fruit. An operating pressure of 50-100 PSI is recommended. Direct spray nozzles toward the fruit bearing area of the canes.

Air equipment: Applications of **Ronilan** should be made in not less than 20 gallons of spray solution per acre. Thorough spray coverage of the developing fruit is essential.

Restrictions and limitations for raspberries

Do not apply more than a total of 8 lbs. of **Ronilan** per acre in one crop season.

Do not apply **Ronilan** within 9 days of harvest.

Do not apply **Ronilan** during rain. Wait until conditions are such that the spray will dry on the plants.

Do not apply **Ronilan** through any type of irrigation system.

Application Rate Table
Raspberries

Moisture Conditions	Spray Interval* (Days)	Rate Lbs. Product/A
Frequent natural moisture (intermittent rain, fog, dew) or when using sprinkler irrigation (high disease pressure)	7-9	1½-2
Limited natural moisture or infrequent sprinkler irrigation.	10-14	1-1½

*Use spray interval throughout the bearing cycle.

DRY BULB ONIONS

Time and rate of application

For control of Sclerotinia watery soft rot, Sclerotium white rot, and Botrytis neck rots, a combination of a soil treatment at planting (soil drench or spray band over the row) and foliar treatment at planting (soil drench or spray band over the row) and foliar treatments after crop emergence is needed. Botrytis leaf blight can be controlled only with foliar sprays. The following recommendations are made according to disease type.

Restrictions and limitations for onions

Do not apply more than a total of 10 lbs. of **Ronilan** per acre in one crop season.

Do not apply within 18 days of harvest.

Do not apply **Ronilan** during rain when controlling botrytis leaf blight. Wait until conditions are such that the spray will dry on the plants.

Do not apply **Botrytis** through any type of irrigation system.

Application Rate Table
Onions

For Sclerotinia watery soft rot, Sclerotium white rot, and Botrytis neck rots. A minimum program of 3 treatments is recommended.

Application	Timing	Rate, Volume and Site
First	At planting	Soil drench—2 lb. product/A in 400 gallons of water/A applied in a 4-6" band over the row **or** Spray band—2 lb. product/A in 100 gallons of water/A applied in a 2-4" band over the row.
Second	4-6 weeks after crop emergence.	Foliar—2 lbs. product/A in a minimum of 100 gallons of water/A directed at the stem base.
Third and Subsequent	Up to 3 additional sprays should be applied if disease pressure is high*. If a minimum spray program of 3 sprays is followed, the third application should be made 4-6 weeks before harvest.	Foliar—2 lbs. product/A in a minimum of 100 gallons of water/A directed at the stem base. Drop nozzles should be used to ensure spray reaches the stem and surrounding soil.

* High disease pressure based on previous history of disease infestation in the field or adjacent fields.

Application Rate Table
Onions
For Botrytis leaf blight. A minimum program of 2 foliar treatments* is recommended

Application	Timing	Rate Lbs. Product/A Low Disease Pressure	Rate Lbs. Product/A High Disease Pressure
First	4-6 weeks after crop emergence or as soon as disease starts to develop	$1\frac{1}{2}$	2
Second	2 weeks after first application	$1\frac{1}{2}$	2
Third and Subsequent	Up to 3 additional sprays should be applied if conditions occur which favor disease development (such as wet, humid, foggy conditions).	$1\frac{1}{2}$	2

* Applications should be made with not less than 100 gallons of spray solution per acre (broadcast basis) to obtain thorough coverage. Use cone or flat fan nozzles. Flood type nozzles should not be used.

"Refer to page 14 of this booklet for **Conditions of Sale and warranty**."

RONILAN® FL

FUNGICIDE

A flowable formulation containing:

ACTIVE INGREDIENT:
3-(3,5-Dichlorophenyl)-5-ethenyl-5-methyl-2,4-oxazolidinedione* 41.3%
INERT INGREDIENTS: 58.7%

TOTAL 100.0%
*Equivalent to 4.17 pounds active ingredient per gallon.
EPA Reg. No 7969-62

KEEP OUT OF REACH OF CHILDREN.
CAUTION

Causes eye irritation. Do not get on skin, in eyes or on clothing.

STATEMENT OF PRACTICAL TREATMENT

If contacted, flush eyes immediately with water for 15 minutes. Get medical attention. In case of contact with skin or clothing, remove contaminated clothing, wash skin thoroughly with soap and water. This product is a potential skin sensitizer. If irritation persists get medical attention.

ENVIRONMENTAL HAZARDS

Do not apply directly to water. Do not contaminate water by cleaning of equipment or disposal of wastes.

RE-ENTRY AND WORKERS' PROTECTION STATEMENTS

Do not apply this product in such a manner as to directly or through drift expose workers or other persons, except those knowingly involved in the application. The area being treated must be vacated by unprotected persons.

Do not enter treated areas without protective clothing until sprays have dried.

Because certain states may require more restrictive re-entry intervals for various crops treated with this product, consult your State Department of Agriculture for further information.

Written or oral warnings must be given to workers who are expected to be in a treated area or in an area about to be treated with this product. Oral warnings must inform workers of areas or fields that may not be entered without specified protective clothing until sprays have dried, and appropriate actions to take in case of accidental exposure. When oral warnings are given, warnings shall be given in a language customarily understood by workers. Oral warnings must be given if there is reason to believe that written warnings cannot be understood by workers. Written warnings must include the following information: "WARNING. Area treated with Ronilan on (date of application). Do not enter without appropriate protective clothing until sprays have dried. If contacted, flush eyes immediately with water for 15 minutes. Get medical attention.

In case of contact with skin or clothing, remove contaminated clothing, wash skin thoroughly with soap and water. Call physician if irritation occurs. Wash contaminated clothing before re-use."

DIRECTIONS FOR USE

It is a violation of Federal law to use this product in a manner inconsistent with its labeling.

STORAGE AND DISPOSAL

Do not contaminate water, food, or feed by storage or disposal.

Wastes resulting from the use of this product may be disposed of on site or at an approved waste disposal facility.

Triple rinse container (or equivalent). Then offer for recycling or reconditioning or puncture and dispose of in a sanitary landfill or by incineration, or, if allowed by state and local authorities, by burning. If burned, stay out of smoke.

In case of emergency

In case of large-scale spillage regarding this product, call:
CHEMTREC 800-424-9300
BASF Corporation 201-316-3000

In case of medical emergency regarding this product, call:

1. Your local doctor for immediate treatment.
2. Your local poison control center (hospital).
3. BASF Corporation 201-316-3000

All applicable directions, restrictions precautions and Conditions of sale and warranty are to be followed. This labeling must be in the possession of the user at the time of application.

GENERAL INFORMATION

Ronilan is a contact fungicide for the control of *Botrytis* fruit rot (gray mold) of strawberries, *Sclerotinia* "drop" (watery soft rot) of head lettuce, and brown rot blossom and twig blight, and fruit brown rot on stonefruit (see page 5 for list of stonefruit). Thorough coverage of plant parts to be protected is essential for effective disease control. If other diseases are a problem, an additional fungicide will be needed.

The repeated exclusive use of Ronilan, as is the case with the exclusive use of other fungicides, may result in the buildup of resistant strains of *Botrytis* and loss of disease control. A spray program alternating other fungicides with Ronilan may delay the buildup of resistant strains. If treatment becomes ineffective due to the presence of a Ronilan-resistant strain of *Botrytis*, then prompt use, of other fungicides is necessary to maintain disease control.

STRAWBERRIES

Time and rate of application for states other than California and Florida

Thorough spray coverage of the blossoms and developing fruit is essential. For full season control of *Botrytis* disease, the following spray program is recommended. The first application should be made no later than 10% primary bloom at rates indicated (see table). The interval between subsequent applications will vary according to weather conditions and resultant disease pressure. A rate of $1\frac{1}{2}$ pints product per acre is generally recommended. A one pint product per acre rate of Ronilan should be used only when low disease pressure can be predicted. A two pints product per acre rate should be used when the foliage is dense and/or disease pressure is high. If a heavy rainfall occurs any time during this spray program or if a wet period (light rain, fog or dew) lasting more than 24 hours occurs, immediate retreatment is necessary at a rate of $1\frac{1}{2}$ to 2 pints product per acre as soon as conditions will allow the spray to dry on the plants.

Time and rate of application for California

Thorough spray coverage of blossoms and developing fruit is essential for good disease control. For full season control of *Botrytis* disease, fungicides may need to be applied at 7 to 14 day intervals throughout the production cycle. When using Ronilan, the first application should be made no later than 10% primary bloom. If conditions favorable for high disease pressure persist after the first application, a second application should be made 7 to 9 days after the first.

To reduce the hazard of resistance developing, further use of Ronilan should be reserved for periods of high disease pressure and/or highest economic return. A rate of $1\frac{1}{2}$ pints product per acre is generally recommended. A 2 pints product per acre rate should be used when foliage is dense and disease pressure is high. **Make no more than a total of four applications of Ronilan per season.**

Time and rate of application for Florida

Thorough spray coverage of the developing fruit is essential. For full season control of *Botrytis* disease, fungicides should be applied at 7 day intervals throughout the production cycle. If a frost protection irrigation is planned, an application of Ronilan must be made immediately before and after the irrigation. If a heavy rainfall occurs any time during the spray program or if a wet period (light rain, fog, or dew) lasting more than 24 hours occurs, immediate retreatment is necessary as soon as conditions will allow the spray to dry on the plants.

Method of application

Ground equipment: Application of Ronilan should be made in not less than 100 gallons of spray solution per acre to obtain thorough coverage of the developing fruit. An operating pressure of 60-150 psi is recommended to obtain adequate penetration of the spray through the canopy. Cone-type nozzles are recommended. Spray booms with at least 3 nozzles per row (1 over row: 2 side drops) are recommended.

Application Rates for Strawberries in States other than California and Florida			
		Rate (Pints Product/A)	
Moisture Conditions	Spray* Interval (Days)	1st Year Plants of Sparse Foliage	Dense Foliage
Frequent natural moisture (intermittent rain, fog, dew) or when using sprinkler irrigation *(high disease pressure)*	7-9	1½	1½-2
Limited natural moisture or infrequent sprinkler irrigation *(low disease pressure)*	10-14	1	1½-2

*Use spray interval throughout the bearing cycle.

Application Rates for Strawberries in California			
		Rate (Pints Product/A)	
Moisture Conditions	Spray* Interval (Days)	1st Year Plants of Sparse Foliage	Dense Foliage
Frequent natural moisture (intermittent rain, fog, dew) or when using sprinkler irrigation *(high disease pressure)*	7-9	1½	2

*Make no more than four (4) applications per season.

Application Rates for Strawberries in Florida		
Moisture Conditions	Spray Interval (Days)	Rate (Pints Product/A)
Frequent natural moisture (intermittent rain, fog, dew) or when using sprinkler irrigation *(high disease pressure)*	7	2

Air equipment: Application of Ronilan should be made in not less than 20 gallons of spray solution per acre. Thorough spray coverage of the developing fruit is essential.

Restrictions and limitations for strawberries

Do not apply more than a total of 12 pints of product per acre in one season, except in California where the total is restricted to 8 pints in no more than four applications. Ronilan does not control *Rhizopus* rot of strawberries in the field or in storage. There may be a competitive relationship between *Botrytis* and *Rhizopus* incidence such that control of *Botrytis* may result in an increase of *Rhizopus* rot in stored fruit. *Rhizopus* rot becomes a problem as a result of certain climatic conditions (such as prolonged warm, humid periods) and cultural practices (such as high nitrogen fertilization which may lead to the production of softer fruit) if conditions are conducive for *Rhizopus* development, including those described above, do not use Ronilan.

Do not use Ronilan as a plant dip, as injury will occur.

Do not apply this product through any type of irrigation system.

Rotational crop restrictions for states other than California

Lettuce (head types only) may be planted after strawberries.

Leafy vegetables may be planted 6 months after treatment that does not exceed 6 pounds active ingredient per acre (12 pints product per acre).

Cucurbits may be planted 2 months after treatment that does not exceed 6 pounds active ingredient per acre (12 pints product per acre).

Corn may be planted 2 months after treatment that does not exceed 6 pounds active ingredient per acre (12 pints product per acre), provided only the corn grain is used for food and/or feed purposes.

Other grain crops may be planted 9 months after treatment that does not exceed 6 pounds active ingredient per acre (12 pints product per acre).

Any rotational crops may be planted 20 days after treatment that does not exceed 3 pounds active ingredient per acre (6 pints product per acre).

Rotational crop restrictions for California

Lettuce (head types only) may be planted after strawberries.

Leafy vegetables may be planted 6 months after treatment that does not exceed 4 pounds active ingredient per acre (8 pints product per acre).

Cucurbits may be planted 2 months after treatment that does not exceed 4 pounds active ingredient per acre (8 pints product per acre).

Corn may be planted 2 months after treatment that does not exceed 4 pounds active ingredient per acre (8 pints product per acre), provided only the corn grain is used for food and/or feed purposes.

Other grain crops may be planted 9 months after treatment that does not exceed 4 pounds active ingredient per acre (8 pints product per acre).

Any rotational crops may be planted 20 day after treatment that does not exceed 3 pounds active ingredient per acre (6 pints product per acre).

Application Rates for Lettuce (Head Types Only)			
		Rate (Pints Product/Acre)	
Application	Timing	*Low Disease Pressure	*High Disease Pressure
First	**Direct Seeded Lettuce**–Application should be made immediately (within 2 days) after thinning. **Transplanted Lettuce**–Application should be made 7-10 days after transplanting.	1	1½-2
Second	Application should be made 14 days after first spray if cool, wet conditions (which favor disease) occur for periods lasting 48 hours or more.		
Third	**Direct Seeded Lettuce Only**–Application should be made 14 days after second spray if conditions favoring disease occur.		

* Based on previous history of disease infestation in the field or adjacent fields.

LETTUCE (HEAD TYPES ONLY)

Time and rate of application

For control of *Sclerotinia* "drop," up to three applications may be made under certain conditions in one season as specified in the rate table. Thorough spray coverage of plant parts to be protected is essential for effective disease control. Do not disturb soil after application.

Method of application

Application of Ronilan should be made in not less than 100 gallons of spray solution per acre (broadcast basis) to obtain thorough coverage of the lower leaves, plant stem and bed surface. An operating pressure of 50-100 psi is recommended. Use cone or flat fan nozzles. Flood type should not be used. To apply Ronilan to one-row beds, spray booms should have at least 2 nozzles per row. For two-row beds use three nozzles or more per bed.

Restrictions and limitations for lettuce

Do not apply Ronilan within 28 days of harvest.

Do not use on leaf type lettuce.

Do not use Ronilan as a plant dip, as injury may occur.

Do not apply this product through any type of irrigation system.

STONEFRUIT

General Information

Ronilan is effective for the control of brown rot blossom and twig blight, and fruit brown rot of stonefruit including apricots, cherries, nectarines, peaches, plums, and prunes. Ronilan is a contact fungicide, therefore, thorough coverage of plant parts to be protected is essential for effective disease control.

Time and rate of application

Depending on conditions which favor disease development, one to three applications should be made during bloom for control of brown rot blossom and twig blight. One or two preharvest applications should be made for control of fruit brown rot (see table).

A rate of 1½ pints product per acre is generally recommended. A one pint product per acre rate of Ronilan should be in used only when low disease pressure can be predicted. Under high disease pressure or in conditions very favorable for disease development, use the higher treatment rates specified, multiple applications, and shorter application intervals. For large mature trees, use the higher treatment rates specified.

Consistency of disease control with low recommended rates can be enhanced by 8-16 oz./100 gal. of a non-ionic surfactant in tank mix with Ronilan.

Method of application

Rates of Ronilan per 100 gallons are based on an application of 400 gallons per acre of dilute spray. Refer to the **Application Rates and Timing for Control of Brown Rot** if other than 400 gallons per acre is applied.

Apply Ronilan as a spray with ground equipment, using sufficient water to obtain **thorough coverage** (50-400 gallons per acre).

Mixing instructions: Partially fill the spray tank with clean water and begin agitation. Measure the required amount of Ronilan and add to the tank. Fill the tank to the total volume required. Maintain agitation to keep the material in suspension and apply with properly calibrated spray equipment.

Ronilan is physically compatible with most pesticides used in a tank mix on stonefruit. To assure compatibility of Ronilan with other products, add correct proportions in a small container with water. Mix thoroughly and let the mixture stand for five minutes. The mixture is compatible if the combination remains mixed, or can be readily remixed.

Stonefruit
Application Rates and Timing for Control of Brown Rot (*Monilinia* spp.)

Crop	Pints of Product* per 100 Gallons	Pints of Product* per Acre	Timing
Apricots Cherries Nectarines Peaches Plums Prunes	0.25 to 0.5 Rate is based on 400 GPA.	1.0 to 2.0	**Brown Rot Blossom & Twig Blight** Apply at **early bloom** Apricots–Red Bud Peaches & Nectarines–Pink Bud Cherries–Early Popcorn Plums and Prunes–Green Tip Apply again at **full bloom** if conditions persist which favor disease development.
	Do not exceed 2 pints product per acre per application.		Apply again at **petal fall** if conditions persist which favor disease development. Do not apply more than 3 bloom treatments. **Fruit Brown Rot** Apply when conditions favor disease development within 3 weeks of harvest. One additional application should be made 7 to 14 days later if these conditions persist. Do not apply more than 2 preharvest treatments. Do not apply within 3 days of harvest.

*See Time and rate of application

Restrictions and limitations for stonefruit

Do not apply Ronilan within 3 days of harvest.

Do not apply more than 2 pints of Ronilan per acre per application.

Do not apply more than 10 pints of Ronilan per acre per season (maximum of 6 pints at bloom and 4 pints preharvest).

Do not apply Ronilan during rain.

Apply when conditions will permit spray to dry on the plants.

Do not apply this product through any type of irrigation system.

CONDITIONS OF SALE AND WARRANTY

The Directions for use of this product reflect the opinion of experts based on field use and tests. The directions are believed to be reliable and should be followed carefully. However, it is impossible to eliminate all risks inherently associated with use of this product. Crop injury, ineffectiveness or other unintended consequences may result because of such factors as weather conditions, presence of other materials, or use of the product in a manner inconsistent with its labeling, all of which are beyond the control of BASF CORPORATION ("BASF") or the Seller. All such risks shall be assumed by the Buyer.

BASF warrants that this product conforms to the chemical description on the label and is reasonably fit for the purposes referred to in the Directions for use, subject to the inherent risks referred to above. BASF MAKES NO OTHER EXPRESS OR IMPLIED WARRANTY OF FITNESS OR MERCHANTABILITY OR ANY OTHER EXPRESS OR IMPLIED WARRANTY. IN NO CASE SHALL BASF OR THE SELLER BE LIABLE FOR CONSEQUENTIAL, SPECIAL OR INDIRECT DAMAGES RESULTING FROM THE USE OR HANDLING OF THIS PRODUCT. BASF and the Seller offer this product, and the Buyer and User accept it, subject to the foregoing Conditions of sale and warranty which may be varied only by agreement in writing signed by a duly authorized representative of BASF.

EPA Establishment No. 7969-WG01

Ronilan is a registered trademark of BASF AG.

Made in West Germany

RO 1 012 JJ5968 A0470687

STORM®

POSTEMERGENCE HERBICIDE

For broad spectrum weed control in soybeans and peanuts

A soluble liquid formulation containing:

ACTIVE INGREDIENTS*:	
Sodium salt of bentazon	29.2%
Sodium salt of acifluorfen	13.4%
INERT INGREDIENTS:	57.4%
TOTAL	100.0%

*Equivalent to 2.67 pounds per gallon bentazon: 3-isopropyl-1 *H*-2,1,3-benzothiadiazin-4-(3*H*)-one 2, 2-dioxide, and 1.33 pounds per gallon of sodium acifluorfen: sodium 5-[2-chloro-4-(trifluoromethyl) phenoxy]-2-nitrobenzoate.

EPA Reg. No. 7969-76

KEEP OUT OF REACH OF CHILDREN.

DANGER/PELIGRO

PRECAUCIÓN AL USUARIO: Si usted no lee ingles, no use este producto hasta que la etiqueta haya sido explicada ampliamente.

PRECAUTIONARY STATEMENTS

HAZARDS TO HUMANS
DANGER

Causes irreversible eye damage. Harmful if swallowed, inhaled or absorbed through the skin. Do not get in eyes. Wear goggles or face shield when handling. Avoid breathing vapor or spray mist and contact with skin or clothing. In case of contact, immediately remove contaminated clothing and shoes. Wash thoroughly with soap and water after handling. Wash contaminated clothing with soap and hot water before re-use. This product may cause an allergic skin reaction.

STATEMENT OF PRACTICAL TREATMENT

If in eyes: Flush with large amounts of water for at least 15 minutes. Get medical attention.

If on skin: Wash with plenty of soap and water. Consult a physician if irritation persists.

If swallowed: Call a physician or Poison Control Center. Drink 1 or 2 glasses of water and induce vomiting by touching back of throat with finger. Do not induce vomiting or give anything by mouth to an unconscious person.

Note to physician: Emesis is recommended.

ENVIRONMENTAL HAZARDS

Do not apply directly to water or wetlands. Do not contaminate water when disposing of equipment washwaters. Do not apply when weather conditions favor drift from target area.

RE-ENTRY AND WORKERS' PROTECTION STATEMENTS

Do not apply this product in such a manner as to directly or through drift, expose workers or other persons. The area being treated must be vacated by unprotected persons. Do not enter treated areas without protective clothing until sprays have dried. Because certain states may require more restrictive re-entry intervals for various crops treated with this product, consult your State Department of Agriculture for further information.

DIRECTIONS FOR USE

It is a violation of Federal law to use this product in a manner inconsistent with its labeling.

STORAGE AND DISPOSAL

Keep from freezing. Store above 32°F. Do not contaminate water, food or feed by storage or disposal. Pesticide wastes are acutely hazardous. Improper disposal of excess pesticide spray mixture, or rinsate is a violation of Federal law. If these wastes cannot be disposed of by use according to label instructions, contact your State Pesticide or Environmental Control Agency, or the Hazardous Waste representative at the nearest EPA Regional Office for guidance.

Triple rinse container (or equivalent). Puncture and dispose of in a sanitary landfill, or by incineration, or if allowed by state and local authorities, by burning. If burned, stay out of smoke. Do not re-use empty container.

In case of emergency

In case of large scale spillage regarding this product, call:

CHEMTREC	800-424-9300
BASF CORPORATION	201-316-3000

In case of medical emergency regarding this product, call:

1. Your local doctor for immediate treatment.
2. Your local poison control center (hospital).
3. BASF Corporation 201-316-3000.

Steps to be taken in case material is released or spilled. Dike and contain spill with inert material (sand, earth, etc.) and transfer liquid and solid diking material to separate containers for disposal. Remove contaminated clothing and wash affected skin areas with water. Wash clothing before re-use. Keep spill out of all sewers and open bodies of water.

GENERAL INFORMATION

Storm® herbicide is intended for selective postemergence control of certain broadleaf weeds in soybeans and peanuts. Storm may provide partial control of some grasses. Storm is effective mainly through contact action; therefore, weeds must be thoroughly covered with spray. Storm may cause some soybean leaf-speckling and leaf-bronzing to occur under certain conditions. (See **Restrictions and limitations.**)

Timing of application

Apply postemergence applications of Storm early when weeds are small and actively growing and before weeds reach the maximum size listed in the **Application Rate Table.**

In peanuts, applications may be made from cracking through two expanded tetrafoliate leaves. Soybeans would generally be in the 2 to 3 trifoliate stage. Early application to weeds produces the most beneficial effect on weed control, and makes it easier to obtain through spray coverage. Delay in application which permits weeds to exceed the maximum size stated will result in inadequate control.

Cultivation before or during the application is not recommended. Cultivation may put weeds under stress, thus making control more difficult to obtain. Timely cultivation after application will usually assist in weed control.

Water volume and spray pressure

Apply recommended rates of Storm as follows:

Ground equipment: Use a minimum of 20 gallons of water per broadcast acre and a minimum of 40 psi pressure (measured at the boom not at the pump or in the line). When crop and weed foliage is dense use up to 50 gallons of water and up to 80 psi pressure. Use standard high pressure pesticide hollow cone or flat fan nozzles spaced 20 inches apart. Do not use flood, whirl chamber or controlled droplet applicator (CDA) nozzles.

Air equipment: Use a minimum of 10 gallons of water per acre and a maximum of 40 psi pressure. Use only diaphragm-type nozzles producing cone or fan spray patterns.

Aerial application—special directions.

To obtain uniform coverage and to avoid drift hazards, the following application equipment and practices should be used:

Nozzle height: 6 to 10 feet above crop.

Nozzle orientation: Nozzles must be oriented so as to discharge straight back with the air stream (opposite the direction of travel of the aircraft) and not more than 20 degrees downward. Nozzles must not be located further out than three-fourths the distance from the center of the aircraft to the end of the wing or rotor.

Water volume and spray pressure:

See **Air equipment.**

Do not apply Storm® herbicide by aircraft when wind is blowing at a velocity above 10 mph. Coarse sprays (larger droplets) are less likely to drift.

Do not apply Storm by air if ornamentals or sensitive non-target crops, such as cotton, sugar beets, sunflowers or okra are within 200 feet downwind.

Applicator must follow the most restrictive use cautions to avoid drift hazards, including those found in this labeling as well as applicable state and local regulations and ordinances.

Spray additives

Additives are needed with Storm to achieve consistent weed control.

Oil concentrate

A nonphytotoxic oil concentrate (commonly referred to as oil concentrate) should be added to the spray tank with Storm. The oil concentrate must contain either a petroleum or vegetable oil base and must meet the following criteria: 1) be nonphytotoxic, 2) contain only EPA-exempt ingredients, 3) provide good mixing quality in the jar test (see below), and 4) be successful in local experience.

The exact composition of suitable products will vary; however, vegetable and petroleum oil concentrates should contain emulsifiers which provide good mixing quality. For vegetable oil concentrates, it has been observed that highly refined vegetable oils are more satisfactory than unrefined vegetable oils. For additional information see **Jar test for estimating suitability of oil concentrates** at the end of this section.

With the addition of oil concentrate to Storm on soybeans and peanuts, a leaf burn may occur, but all new growth is normal and crop vigor is not reduced. The potential for leaf burn is increased when relative humidity and temperature are high. A few oil concentrates have exhibited excessive leaf burn. Refer to your supplier for information concerning successful local experience prior to purchasing any oil concentrate.

Rate of oil concentrate

Ground application—
2 pints/acre (maximum).
Air application—
1 pint/acre (maximum).

APPLICATION RATE TABLE—SOYBEANS AND PEANUTS

Apply Storm® herbicide when weeds are small and actively growing and before weeds reach the maximum size listed below. Such applications generally correspond to the soybean growth stages of unifoliate to two expanded trifoliate leaves and in peanuts to two expanded tetrafoliate leaves. Soybeans and peanuts may experience yellowing, bronzing, speckling, or burning of leaves under certain conditions but soybean and peanut plants generally outgrow this condition within 10 days.

APPLICATION FOR WEED GROWTH STAGES

WEEDS CONTROLLED	LEAF STAGE	MAX. HEIGHT	RATE OF STORM PER ACRE	OIL CONCENTRATE-RATE/A
Anoda, Spurred[a]	Up to 4	2"		
Carpetweed	3" diam.	2"		
Cocklebur	2-6	6"		
Crotalaria	Up to 6	6"		
Croton, Tropic,	1-2	<2"		
Woolly	1-2	<2"		
Jimsonweed	Up to 6	6"		
Ladysthumb	Up to 6	6"		
Lambsquarters[a]	4-6	2"		
Mallow, Venice	Up to 6	2"		
Morningglories	Up to 4	4"		

Spray additives

Additives are needed with Storm to achieve consistent weed control.

APPLICATION FOR WEED GROWTH STAGES

WEEDS CONTROLLED	LEAF STAGE	MAX. HEIGHT	RATE OF STORM PER ACRE	SPRAY ADDITIVE RATE/A
Mustard, Wild	Up to 6	4"		
Nightshade, Black	Up to 6	2"		See page ☐ for details.
Pigweed, Redroot,	Up to 6	2"		
Smooth	Up to 6	3"		
Ragweed, Common,	4-6	3"		
Giant	Up to 4	6		1-2 pts. oil concentrate
Redweed	2-4	3"	1½	or
Sesbania, Hemp	Up to 4	6"	pts.	½-1 gal. UAN
Sida, Prickly or Teaweed[a]	Up to 4	2"		

APPLICATION FOR WEED GROWTH STAGES				
WEEDS CONTROLLED	LEAF STAGE	MAX. HEIGHT	RATE OF STORM PER ACRE	SPRAY ADDITIVE RATE/A
Smartweed, Pennsylvania	Up to 6	6"		or
Starbur, Bristly	4-6	3"		1-2 pts./100 gals.
Texasweed	Up to 3	3"		surfactant
Velvetleaf[a]	Up to 4	2"		
Waterhemp, Tall	Up to 6	3"		

[a] Control may be inconsistent with this rate of Storm. A later application of Basagran® herbicide may be necessary (see label for Basagran).

Jar test for estimating suitability of oil concentrates

1. **Water supply:** Use only water from intended source and at the source temperature.
2. **Amount of water in jar:** Ground application—For 20 gals./A spray volume use 3⅓ cups or 800 ml of water.

   Air application—For 10 gals./A spray volume use 1⅔ cups or 400 ml of water. For other spray volumes, adjust proportionately to above.
3. **Amount of Storm and oil concentrate to add:** Add Storm and oil concentrate at the rate of 1 teaspoon or 5 ml for each pint of recommended label rate.
4. **Add components in following sequence,** gently mixing between component additions:
   a. Storm
   b. Oil concentrate
5. **Cap jar,** invert 10 cycles, let stand for 15 minutes, evaluate.
6. **Evaluation:** An ideal tank mix combination will be uniform, thus, the suitability of the oil concentrate is questionable if any of the following are observed:

   Free oil (film or globules) at the surface.

   Flocculation—fine particles which may be suspended in the liquid or found as a precipitated layer at the bottom of the jar.

   Clabbering—thickening texture (coagulated) resembling yogurt or a curd-like texture as with cottage cheese.

Urea ammonium nitrate (UAN)

Commonly referred to as 28%, 30% or 32% nitrogen solution, may be added in place of other spray additives for improved **pigweed** and **velvetleaf** control. The standard use rates are: ground application—½ to 1 gallon/acre and air application—½ gallon/acre

Nonionic spray adjuvants

The standard recommendation is 1-2 pints of an 80% active nonionic spray adjuvant per 100 gallons of water. Consult the spray adjuvant label for specific recommendations.

Mixing/spraying

Fill a thoroughly cleaned sprayer half to two-thirds full with clean water. Start agitation and add Storm; allow to mix thoroughly. Add oil concentrate and remaining volume of water. Maintain constant agitation during application.

Restrictions and limitations

Do not apply more than a total of 1½ pints of Storm per acre per season to soybeans or peanuts.

Do not apply Storm to soybeans or peanuts that have been subject to stress conditions such as hail damage, flooding, drought, injury from other herbicides or widely fluctuating temperatures, as crop injury may result.

Do not apply Storm to soybeans or peanuts that show injury (leaf phytotoxicity and/or plant stunting) produced by any other prior herbicide applications, because this injury may be enhanced and/or prolonged.

Do not apply Storm during prolonged periods of drought or during unseasonably cold weather, as unsatisfactory weed control may result.

Rainfall or overhead irrigation soon after application (within 8 hours) may nullify the effectiveness of Storm.

Do not mix or apply Storm with any other pesticide or any fertilizer except as specifically recommended on this labeling.

Clean sprayer thoroughly prior to application of Storm, particularly if a herbicide was used which has the potential to injure the crop to be sprayed.

In the case of crop failure, only soybeans or peanuts may be immediately replanted.

Crop rotation restriction: Root crops (such as carrots, turnips, sweet potatoes, etc.) must not be planted in fields treated with Storm for a period of 18 months following treatment.

Do not apply Storm within 50 days of soybean harvest and 75 days of peanut harvest.

Do not use treated plants for feed or forage.

Avoid drift to all other crops and nontarget areas.

Do not apply Storm through any type of irrigation system.

Storm + 2,4-DB tank mix in peanuts

General information

A tank mix of **Storm** plus 2,4-DB is recommended for control of larger morningglory, cocklebur, common ragweed, redroot pigweed, jimsonweed, burgherkin and citron in peanuts when the weed size exceeds that specified in **Table 1.** Control with this mix may decrease with increasing weed size or density of weed or crop canopy due to poor spray coverage. Do not apply the tank mix when peanuts are exhibiting injury from previously applied pesticides or are exhibiting stress symptoms from disease, nematodes, insects; excessive fertilizer or soil salts; wind injury; frost damage or high temperature stress or wilt; as increased crop response will result.

Timing of applications

For optimum control apply **Storm** plus 2,4-DB tank mix to actively growing weeds up to the 8-inch stage, usually 3 to 12 weeks after planting. Applications at later weed stages will result in partial control or suppression. Peanuts should be at least 2 weeks old when using a tank mix of **Storm** and 2,4-DB. Do not use after pod-filling stage begins.

Rate

Mix up to 1 pint of Butyrac 200, or Butoxene with 1½ pints of **Storm** for each acre being treated.

Spray additives

Add 1 pint of spray adjuvant per 100 gallons to increase control of weeds. Do not add crop oils to the tank mixture. The addition of adjuvants will increase the hormonal 2,4-DB crop response.

Water volume and spray pressure

For additional information refer to the section entitled **Directions for use,** page 2.

Ground equipment

For best results, the tank mix should be applied with ground equipment. For broadcast application and thorough coverage of weeds apply with flat fan or hollow cone nozzles spaced 20 inches apart in a minimum of 20 gallons of water per acre with a spray pressure of 40 psi.

Air equipment

Use a minimum of 10 gallons of total spray solution per acre. Aerial applicators should review **Restrictions and limitations** and **Drift hazards.**

Mixing

Fill half the spray tank with water and add the recommended amount of **Storm® herbicide,** 2,4-DB, spray adjuvant-while the agitator is running, than add the remaining quantity of water.

Drift hazards

Care must be taken when applying the tank mix to prevent drift to all non-target crops. Tobacco, ornamentals, mustards, sugar beets, potatoes, vegetables and cotton are a few of the crops known to be sensitive to this tank mix. Hormone type injury in non-target crops can result from trace amounts of 2,4-DB drift. The use of any cleared drift control agent may reduce this hazard; however, the drift control agent may also decrease the weed control activity.

Restrictions amd limitations

(partial list)

Read and follow all directions and use restrictions on **Storm** and 2,4-DB labels.

Do not apply the tank mixture within 75 days of harvest for peanuts.

Do not apply more than one application of the tank mixture to peanuts per growing season.

Do not use rates of **Storm** or 2,4 -DB in excess of those recommended on this label, or excessive injury and possible yield reduction could result.

Do not mix oils, liquid fertilizers or other pesticides with this tank mix except as specifically directed on this label or on other approved supplemental labeling.

Aerial applicators must be familiar with the EPA registered labels and follow the use precautions. In addition, aerial applicators should follow all applicable state and local regulations. In interpreting the label and the local regulations, the most restrictive limitations should apply to avoid hazards.

Table 2
Storm + 2,4-DB Tank Mix in Peanuts
Application Rate Table

Product	Product Rate	Weeds Controlled		Additive Information
Storm	**1½ pts. per Acre**	**See Table 1** for weeds controlled, leaf stage, maximum height.		Surfactant 1-2 pts./100 gal.[b]
plus	**plus**	**Weeds Controlled**		Do not add crop oil to the tank mixture
Butyrac 200[a] or Butoxone	8-16 oz./A	Burgherkin Citron Cocklebur Jimsonweed	Morningglory Pigweed, Redroot Ragweed, Common	

a When weed size exceeds those specified on **Table 1**
b The addition of a nonionic spray adjuvant will increase the hormonal 2,4-DB crop response.

CONDITION OF SALE AND WARRANTY

The Directions for use of this product reflect the opinion of experts based on field use and tests. The directions are believed to be reliable and should be followed carefully. However, it is impossible to eliminate all risks inherently associated with use of this product. Crop injury, ineffectiveness or other unintended consequences may result because of such factors as weather conditions, presence of other materials, or use of the product in a manner inconsistent with its labeling, all of which are beyond the control of BASF CORPORATION ("BASF") or the Seller. All such risks shall be assumed by the Buyer.

BASF warrants that this product conforms to the chemical description on the label and is reasonably fit for the purposes referred to in the Directions for use, subject to the inherent risks referred to above. BASF MAKES NO OTHER EXPRESS OR IMPLIED WARRANTY OF FITNESS OR MERCHANTABILITY OR ANY OTHER EXPRESS OR IMPLIED WARRANTY. IN NO CASE SHALL BASF OR THE SELLER BE LIABLE FOR CONSEQUENTIAL, SPECIAL OR INDIRECT DAMAGES RESULTING FROM THE USE OR HANDLING OF THIS PRODUCT. BASF and the Seller offer this product, and the Buyer and User accept, it, subject to the foregoing Conditions of sale and warranty which may be varied only by agreement in writing signed by a duly authorized representative of BASF.

ST 1001 JJ6611 A0340489

Appendix

The following are scientific names for the weeds listed in this section. For specific recommendations on control of these weeds refer to the **Application Rate Table.**

Broadleaf Weeds

Common Name	Scientific Name
Anoda, Spurred	*Anoda cristata*
Butterprint (see Velvetleaf)	
Buttonweed (see Velvetleaf)	
Carpetweed	*Mollugo verticillata*
Cocklebur	*Xanthium strumarium*
Crotolaria	*Crotalaria spectabilis*
Croton, Tropic,	*Croton glandulosus*
Woolly	*Croton capitatus*
Jimsonweed	*Datura stramonium*
Ladysthumb	*Polygonum persicaria*
Lambsquarters, Common	*Chenopodium album*
Mallow, Venice	*Hibiscus trionum*
Morningglory, Cypressvine,	*Ipomoea quamoclit*
Entireleaf	*Ipomoea hederacea* var. *intergruscula*
Ivyleaf	*Ipomoea hederacea*
Palmleaf	*Ipomoea wrightii*
Pitted	*Ipomoea lacunosa*
Purple Moonflower	*Ipomoea muricata*
Smallflower,	*Jacquemontia tamnifolia*
Tall (common)	*Ipomoea purpurea*
Mustard, Wild	*Sinapis arvensis*
Nightshade, Black	*Solanum nigrum*
Pigweed, Redroot	*Amaranthus retroflexus*
Smooth	*Amaranthus hybridis*
Ragweed, Common	*Ambrosia artemisiifolia*
Giant	*Ambrosia trifida*
Redweed	*Melochia corchorifolia*
Sesbania, Hemp	*Sesbania exaltata*
Sida, Prickly	*Sida spinosa*
Smartweed, Pennsylvania	*Polygonum pensylvanicum*
Starbur, Bristly	*Acanthospermum hispidum*
Texasweed	*Caperonia palustris*
Teaweed (see Prickly Sida)	
Velvetleaf	*Abutilon theophrasti*
Waterhemp, Tall	*Amaranthus tuberculatus*

CIBA-GEIGY Corporation

Agricultural Division
P.O. Box 18300
Greensboro, NC 27419
(919) 292-7100

AATREX® 4L

HERBICIDE

For season-long weed control in corn and sorghum

For weed control in certain other crops; in noncrop areas; and industrial sites

ACTIVE INGREDIENTS:	
Atrazine: 2-chloro-4-ethylamino-6-isopropylamino-*s*-triazine	40.8%
Related compounds	2.2%
INERT INGREDIENTS:	57.0%
TOTAL	100.0%

EPA Reg. No. 100-439 EPA Est. 100-LA-1

AATREX 4L contains 4 lbs. active ingredients per gallon.

Shake well before using.

Use entire contents at one time.

KEEP OUT OF REACH OF CHILDREN

CAUTION

See additional precautionary statements at end of label booklet.

See directions for use inside booklet.

AATREX® trademark of CIBA-GEIGY for atrazine

DIRECTIONS FOR USE AND CONDITIONS OF SALE AND WARRANTY

IMPORTANT: Read the entire **Directions for Use** and the **Conditions of Sale and Warranty** before using this product.

CONDITIONS OF SALE AND WARRANTY

The **Directions for Use** of this product reflect the opinion of experts based on field use and tests. The directions are believed to be reliable and should be followed carefully. However, it is impossible to eliminate all risks inherently associated with use of this product. Crop injury, ineffectiveness, or other unintended consequences may result because of such factors as weather conditions, presence of other materials, or the manner of use or application all of which are beyond the control of CIBA-GEIGY or the Seller. All such risks shall be assumed by the Buyer.

CIBA-GEIGY warrants that this product conforms to the chemical description on the label and is reasonably fit for the purposes referred to in the **Directions for Use** subject to the inherent risks referred to above. **CIBA-GEIGY makes no other express or implied warranty of Fitness or Merchantability or any other express or implied warranty. In no case shall CIBA-GEIGY or the Seller be liable for consequential, special, or indirect damages resulting from the use or handling of this product. CIBA-GEIGY** and the Seller offer this product, and the Buyer and user accept it, subject to the foregoing **Conditions of Sale and Warranty,** which may be varied only by agreement in writing signed by a duly authorized representative of CIBA-GEIGY.

DIRECTIONS FOR USE

It is a violation of federal law to use this product in a manner inconsistent with its labeling.

FAILURE TO FOLLOW THE DIRECTIONS FOR USE AND PRECAUTIONS ON THIS LABEL MAY RESULT IN POOR WEED CONTROL, CROP INJURY, OR ILLEGAL RESIDUES.

Do not apply this product in such a manner as to directly or through drift expose workers or other persons, except those knowingly involved in the application. The area being treated must be vacated by unprotected persons.

Do not enter treated areas without protective clothing until sprays have dried.

Because certain states may require more restrictive reentry intervals for various crops treated with this product, consult your State Department of Agriculture for further information.

Written or oral warnings must be given to workers who are expected to be in a treated area or in an area about to be treated with this product. Oral warnings must be given which inform workers of areas or fields that may not be entered without specific protective clothing until sprays have dried and appropriate actions to take in case of accidental exposure as described under **Precautionary Statements** on this label. When oral warnings are given, warnings shall be given in a language customarily understood by workers. Oral warnings must be given if there is reason to believe that written warnings cannot be understood by workers. Written warnings must include the following information: "CAUTION. Area treated with AATREX 4L on *(date of application)*. Do not enter without appropriate protective clothing until sprays have dried. In case of accidental exposure, flush eyes or skin with plenty of water. Call a physician if irritation persists. Remove and wash contaminated clothing before reuse."

GENERAL INFORMATION

This herbicide controls many annual broadleaf and grass weeds in corn, sorghum, sugarcane, pineapple, and certain other crops specified on this label. It is also effective in noncrop areas and industrial sites for control of most annual and many perennial broadleaf and grass weeds. This product may be applied before or after weeds emerge.

Following many years of continuous use of this product and chemically related products, biotypes of some of the weeds listed on this label have been reported which cannot be effectively controlled by this and related herbicides. Where this is known or suspected and weeds controlled by this product are expected to be present along with resistant biotypes, we recommend the use of this product in combinations or in sequence with other registered herbicides which are not triazines. If only resistant biotypes are expected to be present, use a registered non-triazine herbicide. Consult with your State Agricultural Extension Service for specific recommendations.

Since this product acts mainly through root absorption, its effectiveness depends on moisture to move it into the root zone. If weeds develop, a shallow cultivation or rotary hoeing will generally result in better weed control.

This product is nonflammable.

Avoid using near adjacent desirable plants or in greenhouses, or injury may occur.

To avoid spray drift, do not apply under windy conditions. Avoid spray overlap, as crop injury may result.

Where the use directions give a range of rates, use the lower rate on coarse-textured soil and soil low in organic matter; use the higher rate on fine-textured soil and soil high in organic matter.

Note: CIBA-GEIGY does not recommend applications in combination with other herbicides or oils, except as specifically described on the label or in literature published by CIBA-GEIGY.

APPLICATION PROCEDURES

Ground application: Use conventional ground sprayers equipped with nozzles that provide accurate and uniform application. Be certain that nozzles are uniformly spaced and are the same size. Calibrate sprayer before use and recalibrate at the start of each season and when changing carriers. Unless otherwise specified, use a minimum of 10 gals. of spray mixture/A for all preplant incorporated, preplant surface, preemergence, and postemergence applications (with or without oil or surfactant) with ground equipment.

Use a pump with capacity to (1) maintain 35-40 psi at nozzles, (2) provide sufficient agitation in tank to keep mixture in suspension, and (3) to provide a minimum of 20% bypass at all times. Use centrifugal pumps which provide propeller shear action for dispersing and mixing this product. The pump should provide a minimum of 10 gals./minute/100 gal. tank size circulated through a correctly positioned sparger tube or jets.

Use screens to protect the pump and to prevent nozzles from clogging. Screens placed on the suction side of the pump should be *16-mesh or coarser.* Do not place a screen in the recirculation line. Use 50-mesh or coarser screens between the pump and boom, and where required, at the nozzles. Check nozzle manufacturer's recommendations.

For band applications, calculate amount to be applied per acre as follows:

$$\frac{\text{band width in inches}}{\text{row width in inches}} \times \text{broadcast rate per acre} = \text{amount needed per acre of field}$$

Aerial application: Use aerial application only where broadcast applications are specified. Apply in a minimum of 1 qt. of water for each quart of this product applied per acre. For postemergence treatments on corn and sorghum, apply recommended rate in a minimum of 2 gals. of water/A. Avoid applications under conditions where uniform coverage cannot be obtained or where excessive spray drift may occur.

Avoid application to humans or animals. Flagmen and loaders should avoid inhalation of spray mist and prolonged contact with skin, and should wash thoroughly before eating and at the end of each day's operation.

Application in water or liquid fertilizer: Nitrogen solution or complete liquid fertilizer may replace all or part of the water as a carrier for preemergence, preplant incorporated, or preplant surface ground application on corn and sorghum. Check the compatibility of this product with liquid fertilizer and/or nitrogen solution as shown below before use. Do not apply in nitrogen solution or complete liquid fertilizer after corn or sorghum emerges, except as noted under **Lay-by treatment** for corn, or crop injury may occur.

Compatibility Test: Since liquid fertilizers can vary, even within the same analysis, always **check compatibility with herbicide(s) each time before use.** Be especially careful when using **complete** suspension or fluid fertilizers as serious compatibility problems are more likely to occur. Commercial application equipment may improve compatibility in some instances. The following test assumes a spray volume of 25 gals. per acre. For other spray volumes, make appropriate changes in the ingredients. Check compatibility using this procedure:

1. Add 1 pint of fertilizer to each of 2 one-quart jars with tight lids.
2. To **one** of the jars add 1/4 tsp. or 1.2 milliliters of a compatibility agent approved for this use, such as Compex® or Unite® (1/4 tsp. is equivalent to 2 pts. per 100 gals. spray). Shake or stir gently to mix.
3. To **both** jars add the appropriate amount of herbicide(s). If more than one herbicide is used, add them separately with dry herbicides first, flowables next, and emulsifiable concentrates last. After each addition, shake or stir gently to thoroughly mix. The appropriate amount of herbicides for this test follows:

   Dry herbicides: For each pound to be applied per acre, add 1.5 level teaspoons to each jar.

   Liquid herbicides: For each pint to be applied per acre, add 0.5 teaspoon or 2.5 milliliters to each jar.
4. After adding all ingredients, put lids on and tighten, and invert each jar ten times to mix. Let the mixtures stand 15 minutes and then look for separation, large flakes, precipitates, gels, heavy oily film on the jar, or other signs of incompatibility. Determine if the compatibility agent is needed in the spray mixture by comparing the two jars. If either mixture separates, but can be remixed readily, the mixture can be sprayed as long as good agitation is used. If the mixtures are incompatible, test the following methods of improving compatibility: (A) slurry the dry herbicides(s) in water before addition, or (B) add 1/2 of the compatibility agent to the fertilizer and the other 1/2 to the emulsifiable concentrate or flowable herbicide before addition to the mixture. If incompatibility is still observed, do not use the mixture.

Application in water plus emulsifiable oil or oil concentrate: Adding emulsifiable oil (petroleum-derived, petroleum-derived oil concentrate, or single or mixed crop-derived oil concentrate) to postemergence water-based sprays in corn and sorghum may improve weed control. However, under certain conditions, the use of either type of oil may seriously injure the crop. To minimize this possibility, observe the following directions:

Use one of the following properly emulsified:

1. A suitable oil concentrate containing at least 1% but not more than 20% suitable emulsifier or surfactant blend.
2. Petroleum-derived oil containing at least 1% suitable emulsifier.

Note: In the event of a compatibility problem when mixing oil with this product and water, a compatibility agent such as Compex or Unite should be used. Any of the above oils contaminated with water or other materials can cause compatibility problems and/or crop injury.

Mixing procedures—all uses: (1) Be sure sprayer is clean and not contaminated with any other materials, or crop injury or sprayer clogging may result. (2) Fill tank 1/4 full with clean water, nitrogen solution, or complete liquid fertilizer. (3) Start agitation. (4) Be certain that the agitation system is working properly and creates a rippling or rolling action on the liquid surface. (5) Pour product directly from container into tank. (6) Continue filling tank until 90% full. Increase agitation if necessary to maintain surface action. (7) Add emulsifiable oil, oil concentrate, or tank mix herbicide(s) after this product is thoroughly suspended. (8) Finish filling tank. (9) Empty tank as completely as possible before refilling to prevent buildup of oil or emulsifiable concentrate residue. Maintain agitation to avoid separation of materials. (10) If an oil or emulsifiable concentrate film starts to build up in tank, drain it and clean with strong detergent solution or solvent. (11) Clean sprayer thoroughly immediately after use by flushing system with water containing a detergent.

Center pivot sprinkler application

Preemergence or postemergence: Apply this product alone, or in tank mixtures with other herbicides on this label which are registered for center pivot application, with irrigation water. Apply either after planting before crop and weeds emerge, or after crop emergence, but before lay-by (20-30 inches) and before weeds exceed 1.5 inches in height. Apply this product only through a center pivot irrigation system. Do not apply this product through any other type of irrigation system. Crop injury, lack of effectiveness, or illegal pesticide residues in the crop can result from nonuniform distribution of treated water. If you have questions about calibration, you should contact State Extension Service specialists, equipment manufacturers or other experts, or refer to supplemental literature published by CIBA-GEIGY. Do not connect an irrigation system (including greenhouse systems) used for pesticide application to a public water system unless the pesticide label-prescribed safety devices for public water systems are in place. A person knowledgeable of the chemigation system and responsible for its operation, or under the supervision of the responsible person, shall shut the system down and make necessary adjustments should the need arise.

Operating Instructions

1. The system must contain a functional check valve, vacuum relief valve, and low pressure drain appropriately located on the irrigation pipeline to prevent water-source contamination from backflow.
2. The pesticide injection pipeline must contain a functional, automatic, quick-closing check valve to prevent the flow of fluid back toward the injection pump.
3. The pesticide injection pipeline must also contain a functional, normally closed, solenoid-operated valve located on the intake side of the injection pump and connected to the system interlock to prevent fluid from being withdrawn from the supply tank when the irrigation system is either automatically or manually shut down.
4. The system must contain functional interlocking controls to automatically shut off the pesticide injection pump when the water pump motor stops.
5. The irrigation line or water pump must include a functional pressure switch which will stop the water pump motor when the water pressure decreases to the point where pesticide distribution is adversely affected.
6. Systems must use a metering pump, such as a positive displacement injection pump (e.g., diaphragm pump) effectively designed and constructed of materials that are compatible with pesticides and capable of being fitted with a system interlock.
7. Do not apply when wind speed favors drift beyond the area intended for treatment.
8. Prepare a mixture with a minimum of 1 part of water to 1 part herbicide(s) and inject this mixture into the center pivot system. Injecting a larger volume of a more dilute mixture per hour will usually provide more accurate calibration of metering equipment. Maintain sufficient agitation to keep the herbicide in suspension.
9. Meter into irrigation water during entire period of water application.
10. Apply in 1/2-1 inch of water. Use the lower water volume (1/2 inch) on coarser-textured soils and the higher volume (1 inch) on finer-textured soils. More than 1 inch of water at application may reduce weed control by moving the herbicide below the effective zone in the soil.

Precautions for center pivot applications: Where sprinkler distribution patterns do not overlap sufficiently, unacceptable weed control may result. Where sprinkler distribution patterns overlap excessively, crop injury may result.

Within rate ranges in all tables on this label, use the lower rate on soil relatively coarse-textured or low in organic matter; use the higher rate on soil relatively fine-textured or high in organic matter.

AATREX 4L APPLIED ALONE—CORN OR GRAIN SORGHUM

Preplant Surface-Applied, Preplant Incorporated, or Preemergence (or Postemergence at 4 pts./A with Oil)

Broadleaf or Grass Weeds Controlled

barnyardgrass watergrass	cocklebur	nightshade
giant foxtail	groundcherry	pigweed
green foxtail	jimsonweed	purslane
large (hairy) crabgrass	kochia	ragweed
wild oats	lambsquarters	sicklepod
withchgrass	annual morningglory	velvetleaf
Panicum capillare)	mustards	(buttonweed)
yellow foxtail		

Postemergence with Emulsifiable Oil or Oil Concentrate in Water (2 + 2.4 pts./A)

Broadleaf Weeds Controlled

annual morningglory	lambsquarters	ragweed
cocklebur	mustards	smartweed
jimsonweed	pigweed	wild buckwheat
		velvetleaf

CORN

Preplant Surface-Applied (Broadleaf and grass control): Use on medium- and fine-textured soil with minimum-tillage or no-tillage systems only in CO, IL, IN, IA, KS, KY, MN, MO, MT, NE, ND, SD, WI, and WY. Apply 2/3 the recommended rate of AATREX shown in Table 1 for the soil texture as a split treatment 30-45 days prior to planting and the remainder at planting. Applications made less than 30 days prior to planting may be as either a split or single treatment. On *coarse-textured* soils, do not apply more than two weeks prior to planting.

If weeds are present at the time of treatment, apply in tank mix combination with a contact herbicide (for example, Gramoxone® or Roundup®).

Observe directions for use, precautions, and restrictions on the label of the contact herbicide.

Note: To the extent possible, do not move treated soil out of the row or move untreated soil to the surface during planting, or weed control will be diminished.

Preplant Incorporated (Broadleaf and grass control): Broadcast in spring after plowing at rate in Table 1. Apply to the soil and incorporate before, during, or after final seedbed preparation. Avoid deep incorporation. For best results, apply within two weeks prior to planting.

Preemergence (Broadleaf and grass control): Apply during or shortly after planting before weed emergence, at rate in Table 1.

Postemergence (Broadleaf and grass control): Apply before weeds exceed 1.5 inches in height, at rate in Table 1.

Table 1: Broadleaf and Grass Weed Control on Corn*

Soil Texture	Broadcast Rate/A
COARSE Sand, loamy sand, sandy loam	4 pts.
MEDIUM Loam, silt loam, silt, clay loam, sandy clay loam, silty clay loam, sandy clay, or silty clay with low organic matter	4.75 pts.
FINE Loam, silt loam, silt, clay loam, sandy clay loam, silty clay loam, sandy clay, or silty clay with medium to high organic matter, and clay (including the dark prairie soils of the Corn Belt)	6 pts.
Peat, muck, and high organic clay (apply postemergence only)	6 pts.

* *Broadleaf control (eastern CO, western KS, western NE, NM, OK Pan Handle, west TX, and eastern WY):* On sand, loamy sand, sandy loam, mild to strongly alkaline soil, and all recently leveled soil, apply no more than 2.4 pts./A, either preplant surface, preplant incorporated, or preemergence. On other soils in these areas, apply rate in Table 1 for broadleaf and grass control.

Lay-by treatment (Broadleaf and grass control): Broadcast 2-4 pts./A before weeds are 1.5 inches tall and before corn is 20-30 inches tall. When using nitrogen solutions, direct the spray to lower 3-4 inches of cornstalks to avoid corn foliage injury.

Postemergence with emulsifiable oil or oil concentrate in water: Add the following volume of one of the type oils indicated for aerial or ground application unless the oil label specifies otherwise:

Type Oil	Ground Application	Aerial Application
Oil concentrate (Crop or Petroleum-derived)	1 qt./A	½-1 qt./A
Petroleum-derived oil	1 gal./A	2 qts./A

Note: Crop-derived or petroleum-derived oil concentrates should contain at least 1% but not more than 20% suitable emulsifier or surfactant blend. Petroleum-derived oils should contain at least 1% suitable emulsifier.

Broadleaf and grass control: For postemergence control of those weeds listed under **Preplant Incorporated and Preemergence,** broadcast 4 pts./A plus emulsifiable oil or oil concentrate after weed emergence, but before weeds reach 1.5 inches in height.

Broadleaf control: For postemergence control of those weeds listed under **Postemergence with emulsifiable oil or oil concentrate in water,** broadcast 2.4 pts./A plus emulsifiable oil or oil concentrate before pigweed and lambsquarters reach 6 inches in height and before all other weeds reach 4 inches in height. A cultivation may be necessary if all weeds are not controlled or if weeds regrow.

Precautions for applications with emulsifiable oil or oil concentrate in water: (1) Inbred lines or any breeding stock may be severely injured by applications with emulsifiable oil or oil concentrate. (2) Adding other insecticides, herbicides, liquid fertilizers, or other materials is not recommended, because they may cause compatibility problems or crop injury. (3) Store and handle emulsifiable oil and oil concentrate carefully. Oil contaminated with even a small amount of water may not emulsify properly when added to the tank. To avoid crop injury, (4) Do not apply when crop is under stress from prolonged cold, wet weather, poor fertility, or other factors, or when crop is wet and succulent from recent rainfall. (5) Do not make more than one application per season except as recommended for control of yellow nutsedge, quackgrass, and Canada thistle.

Problem weeds

Yellow nutsedge and Canada thistle: This product will control yellow nutsedge *(Cyperus esculentus)* and Canada thistle *(Cirsium arvense)* when applied according to directions. For best results, apply each year until yellow nutsedge or Canada thistle is eliminated or reaches a level of infestation where neither weed is an economic problem. If yellow nutsedge or Canada thistle regrow following last application, cultivate once.

Four alternative methods of use for control of yellow nutsedge and Canada thistle are listed below in order of preference. If other weed species, such as annual grasses, are also expected, use alternative 2 or 4.

1. Broadcast 4 pts./A plus 1 gal. of emulsifiable oil or 1 qt. of oil concentrate. Apply after crop and yellow nutsedge or Canada thistle emerge, but before yellow nutsedge reaches a height of 3 inches or Canada thistle reaches a height of 6 inches. Repeat application before lay-by (20-30 inches), 10-20 days after the first application.
2. Broadcast 4 pts./A preplant incorporated for control of yellow nutsedge, or broadcast 4 pts./A during or shortly after planting, but prior to crop or weed emergence for control of Canada thistle. Follow with an application of 4 pts./A plus 1 gal. of emulsifiable oil or 1 qt. of oil concentrate after corn and weeds emerge, cut before yellow nutsedge or Canada thistle reaches a height of 6 inches.
3. Broadcast 8 pts./A plus 1 gal. of emulsifiable oil or 1 qt. of oil concentrate after crop emerges, but prior to lay-by (20-30 inches). Apply after yellow nutsedge and Canada thistle emerge, but before yellow nutsedge is 3 inches tall or Canada thistle is 6 inches tall.
4. Broadcast 8 pts./A before, during, or shortly after planting, but before crop or weed emergence (yellow nutsedge control only).

Note: Do not use emulsifiable oil or oil concentrate when corn is wet or under stress, especially when using 8 pts./A of this product, or crop injury may occur. See "Precautions for applications with emulsifiable oil or oil concentrate in water" for additional directions.

Quackgrass control on land going into corn production

Split application: Broadcast 4 pts./A in fall or spring. Plow 1-3 weeks later. Apply a second application at 4 pts./A in spring before, during, or after planting, but before weeds are 1.5 inches high. This split application will control quackgrass and most annual broadleaf and grass weeds.

Single application: Broadcast 6-8 pts./A in fall or spring. Plow 1-3 weeks later.

TANK MIXTURES FOR CORN

This product may be tank mixed with these herbicides for control of certain broadleaf and grass weeds in corn.

Dual® (metolachlor)	Lasso or Lasso EC + Roundup
Dual + Gramoxone	Lasso or Lasso EC + Gramoxone
Dual + Roundup (glyphosate)	Gramoxone
Dual + Princep®	Princep
Dual + Princep + Gramoxone	Princep + Gramoxone
Dual + Princep + Roundup	Princep + Roundup
Bexton® or Ramrod® (propachlor)	Roundup
Lasso® or Lasso EC (alachlor)	Sutan+®

Use tank mix directions appearing on the labels of the above herbicides when tank mixing with this product. Observe all precautions and limitations on labeling of products used in a particular tank mix.

Note: When the labels of the above herbicides refer to atrazine 80W, use equivalent rates of AATREX 4L. One lb. of 80W equals 1.6 pts. of 4L.

Princep 80W, Princep 4L, or Princep Caliber® 90

In addition to the weeds listed under **AATREX 4L Applied Alone—Corn and Grain Sorghum—Preplant Surface-Applied, Preplant Incorporated,** or **Preemergence,** this combination also controls crabgrass, fall panicum, and carpetweed.

Broadcast tank mix before planting, at planting, or after planting, but before crop and weeds emerge, at rates in Table 2. Use the 1:1 ratio for control of most weeds. Use the 1:2 ratio for expected heavy infestations of crabgrass and fall panicum. Cultivate shallowly if weeds develop.

Preplant Surface-Applied: Use on medium- and fine-textured soils with minimum-tillage or no-tillage systems only in CO, IL, IN, IA, KS, KY, MN, MO, MT, NE, ND, SD, WI, and WY. Apply ⅔ the recommended rate of AATREX and Princep shown in Table 2 for the soil texture as a split treatment 30-45 days prior to planting and the remainder at planting. Applications made less than 30 days prior to planting may be as either a split or single treatment. On *coarse-textured* soils, do not apply more than two weeks prior to planting. Refer to the AATREX Alone—Preplant Surface-Applied section of the corn label for additional details.

If weeds are present at time of treatment, apply in a tank mix combination with a contact herbicide (for example, Gramoxone or Roundup). Observe directions for use, precautions, and restrictions on the label of the contact herbicide.

Note: To the extent possible, do not move treated soil out of the row or move untreated soil to the surface during planting, or weed control will be diminished.

Preplant Incorporated: Apply to the soil and incorporate in the spring before, during, or after final seedbed preparation. Avoid deep incorporation. For best results, apply within two weeks prior to planting.

Preemergence: Apply during or shortly after planting, but before crop and weeds emerge.

Refer to Corn sections of this label and to Princep 80W, Princep Caliber 90, or Princep 4L labels for further directions, limitations, and precautions.

Table 2: Tank Mixtures with Princep on Corn

Soil Texture	Broadcast Rate/A			
	1:1 Ratio*		1:2 Ratio**	
	This Product	Princep 80W[1]	This Product	Princep 80W[1]
Sand, loamy sand, sandy loam	2 pts.	1.25 lbs.	1.32 pts.	1.67 lbs.
Loam, silt loam, silt, clay loam, sandy clay loam, silty clay loam, sandy clay, or silty clay with low organic matter	2.4 pts.	1.5 lbs.	1.6 pts.	2 lbs.
Loam, silt loam, silt, clay loam, sandy clay loam, silty clay loam, sandy clay, or silty clay with medium to high organic matter, and clay (including dark prairie soils of the Corn Belt)	3 pts.	1.8 lbs.	1.92 pts.	2.4 lbs.

* For control of most weeds.

** For control of expected heavy infestations of crabgrass and fall panicum.

[1] When using Princep Caliber 90 or Princep 4L use equivalent rates. One lb. of Princep 80W equals 0.9 lb. of Princep Caliber 90 or 1.6 pts. of Princep 4L.

Princep 80W, Princep 4L or Princep Caliber 90 plus Roundup: Use as tank mixture for preemergence and postemergence control of certain broadleaf and grass weeds where corn will be planted directly into a cover crop, established sod, or in previous crop residues. Refer to Roundup label for all directions, weeds controlled, precautions, and limitations.

Princep 80W, Princep 4L or Princep Caliber 90 plus Gramoxone: Use as tank mixture with Princep and Gramoxone to kill existing vegetation and for residual weed control where corn will be planted directly into a cover crop, established sod, or in previous crop residues. Add this product and Princep to water in spray tank, agitating until thoroughly mixed. Then add Gramoxone and a nonionic surfactant, such as X-77®. Continue agitation during application. Broadcast 2-4 pts. of this product plus 1.25-2.5 lbs. Princep 80W [or 2-4 pts. Princep 4L, or 1.1-2.2 lbs. Princep Caliber 90] plus 1.5-2.5 pts. Gramoxone in 20-60 gals. of water per sprayed acre. Apply before, during, or after planting, but before corn emerges. Add 0.5 pt. of a nonionic surfactant, such as X-77, per 100 gals. of spray mixture. Use the 2.5 pt. rate of Gramoxone if existing vegetation is 4-6 inches tall. This mixture will not control weeds taller than 6 inches.

Refer to further limitations and precautions on labels for this product, Princep, and Gramoxone.

Precautions for all applications to corn: (1) To avoid crop injury and illegal residues, do not apply more than 8 pts./A of this product per year. (2) Following harvest, plow (moldboard or disk-plow) and thoroughly till soil in fall or spring to minimize possible injury to spring-seeded rotational crops, regardless of rate used.

Note for all applications to corn: Do not graze or feed forage from treated areas for 21 days following application, or illegal residues may result.

Rotational crops: (1) Do not rotate to any crop except corn or sorghum until the following year, or injury may occur. (2) If applied after June 10, do not rotate with crops other than corn or sorghum the next year or crop injury may occur. (3) If the broadcast rate, or equivalent band application rate, exceeds 6 pts./A of AATREX 4L, or 6 pts./A of AATREX 4L plus Princep 4L (or equivalent of another Princep formulation), a crop of untreated corn or sorghum should precede the next rotational crop. (4) In the High Plains and Intermountain areas of the West where rainfall is sparse and erratic or where irrigation is required, use only when corn or sorghum is to follow corn or sorghum or when a crop of untreated corn or sorghum is to precede other rotational crops. (5) In eastern parts of the Dakotas, KS, western MN, and NE, do not rotate to soybeans if the rate applied to corn or sorghum was more than 4 pts./A or equivalent band application rate, or soybean injury may occur. (6) Injury may occur to soybeans planted the year following application on soils having a calcareous surface layer. (7) Do not plant sugar beets, tobacco, vegetables (including dry beans), spring-seeded small grains, or small-seeded legumes and grasses the year following application, or injury may occur.

SORGHUM AND SORGHUM-SUDAN HYBRIDS (GRAIN AND FORAGE TYPES)

Preplant Surface-Applied (Broadleaf and grass control): Use on medium- and fine-textured soil with minimum-tillage or no-tillage systems only in CO, IL, IN, IA, KS, KY, MN, MO, MT, NE, ND, SD, WI, and WY. Apply 2/3 the recommended rate of AATREX shown in Table 3 for the soil texture as a split treatment 30-45 days prior to planting and the remainder at planting. Applications made less than 30 days prior to planting may be as either a split or single treatment. Under dry conditions, irrigation after application is recommended to move AATREX into the soil.

If weeds are present at time of treatment, apply in a tank mix combination with a contact herbicide (for example, Gramoxone or Roundup). Observe directions for use, precautions, and restrictions on the label of the contact herbicide.

Note: To the extent possible, do not move treated soil out of the row or move untreated soil to the surface during planting, or weed control will be diminished.

Preplant Incorporated (Broadleaf and grass control): Broadcast in spring after plowing at rate shown in Table 3. Apply before, during, or after final seedbed preparation. If soil is tilled or worked after application, avoid deep incorporation. For best results, apply within two weeks prior to planting.

Preemergence (Broadleaf and grass control): Apply during or shortly after planting, but prior to weed or crop emergence at rate shown in Table 3.

Table 3: Preplant Surface, Preplant Incorporated, and Preemergence Broadleaf and Grass Weed Control in Sorghum*

Soil Texture	Organic Matter	Broadcast Rate/A
COARSE Sand, loamy sand, sandy loam	any level	DO NOT USE (except for preemergence use on bedded sorghum in AZ and CA)
MEDIUM AND FINE Loam, silt loam, silt, clay loam, sandy clay loam, silty clay loam, sandy clay, silty clay, clay	less than 1%	
	1-1.5%	3.2-4 pts.
	more than 1.5%	4-4.75 pts.

* Do not apply preplant surface or preplant incorporated in AL, AR, FL, GA, LA, MS, NC, NM, OK, SC, TN, or TX. Do not apply preemergence in NM, OK, or TX, except in northeast OK and the TX Gulf Coast.

In case of planting failure, sorghum or corn may be replanted. Do not make a second broadcast application, or injury may occur. If originally applied in a band and sorghum or corn is replanted in untreated row middles, this product may be applied in a band to the second planting.

Preemergence broadleaf weed control in furrow irrigated bedded sorghum (AZ and CA only): For preemergence control of many broadleaf weeds, broadcast 1.6-2.4 pts./A after bed preparation, during or after planting, but before sorghum and weeds emerge and before the first furrow irrigation. Follow with several regular irrigations, making sure to thoroughly wet all soil.

Precautions for preemergence application to furrow irrigated bedded sorghum in AZ and CA: To avoid possible sorghum injury, do not use on sand or loamy sand soil or on sorghum planted in furrows. Applications to sorghum growing on alkali soils or where cuts, fills, or erosion have exposed calcareous or alkali subsoils may cause crop injury. In case of crop failure, do not replant sorghum for 8 months following application. Corn may be planted immediately.

Postemergence (Broadleaf and grass control): Apply at rate shown in Table 4 before weeds exceed 1.5 inches in height. Apply up to close-in.

Table 4: Postemergence Broadleaf and Grass Weed Control in Sorghum

Soil texture	Minimum sorghum height at treatment	Broadcast rate/A
Sand or loamy sand	DO NOT USE	
Sandy loam	See directions for postemergence broadleaf weed control	
Silt loam to sandy clay loam	Completely emerged	4-4.75 pts.
Olton and Pullman clay soil	6 inches	4-4.75 pts.
Silty clay loam and finer textured soil	Completely emerged	6 pts.

Postemergence broadleaf weed control with emulsifiable oil or oil concentrate in water: Broadcast 2.4 pts./A for control of many broadleaf weeds. Apply before pigweed and lambsquarters reach 6 inches in height and before all other weeds reach 4 inches in height. In CO, western KS, NM, OK, TX, and desert regions of AZ and CA, apply when sorghum is 6-10 inches in height, but before it reaches boot stage. In all other areas, apply after sorghum reaches the 3-leaf stage. Add 1 gal. of emulsifiable oil/A for ground application and 0.5 gal./A for aerial application, or add 1 qt./A of oil concentrate for ground application. A cultivation may be necessary if all weeds are not controlled or if weeds regrow.

For the list of weeds controlled, see **AATREX 4L Applied Alone—Corn or Grain Sorghum—Postemergence with Emulsifiable Oil or Oil Concentrate in Water.**

Precautions for applications with emulsifiable oil or oil concentrate in water: See "Precautions for applications with emulsifiable oil or oil concentrate in water" in Corn section.

Postemergence broadleaf weed control with surfactant (CO, western KS, NM, OK, TX, and desert regions of AZ and CA only): Broadcast 2.4 pts./A plus 0.75-1.5 pts. of surfactant after sorghum reaches 6 inches in height, but before weeds exceed 1.5 inches in height. Apply only on sandy loam and finer textured soil.

Precautions for all applications to sorghum: (1) Heavy rain immediately following application tends to cause excessive concentrations of herbicide in seed furrow, resulting in possible crop injury. Do not apply to furrow-planted sorghum until furrows are leveled (plowed in). Level deep planter marks or seed furrows before application. (2) Application to sorghum growing under stress caused by minor element deficiency or to sorghum growing on highly calcareous soil may result in crop injury. (3) Following harvest, plow (moldboard or disk-plow) and thoroughly till soil in fall or spring to minimize possible injury to spring-seeded rotational crops, regardless of rate used. (4) Injury may occur if both this herbicide, preplant surface, preplant incorporated, or preemergence, and an at-planting systemic insecticide applied in-furrow are used.

Note: Do not graze or feed forage from treated areas for 21 days following application, or illegal residues may result.

Rotational crops: See "Rotational crops" in Corn section.

TANK MIXTURES FOR GRAIN SORGHUM

Igran® 80W: Use as tank mixture for control of certain broadleaf and grass weeds in grain sorghum. Refer to Igran 80W label for all directions, weeds controlled, precautions and limitations.

Dual 8E: Use as tank mixture with Dual 8E for control of those weeds listed on the Dual label, as well as on this label. Use this tank mixture only on sorghum seed treated with Concep®. Refer to the Dual 8E label for all directions, precautions, and limitations.

WINTER WEED CONTROL IN TEXAS

For postemergence control of winter weeds only, such as henbit, seedling dock, and annual thistle on fall bedded land in the Gulf Coast and Blacklands of Texas. Apply 0.8-1 qt./A postemergence to the weeds in November or December to land that will be planted to corn, grain sorghum, or forage sorghum the following spring. For best results, add a suitable surfactant, such as X-77, at the rate of 0.5% of the spray volume, an emulsifiable oil at the rate of 1.0% of the spray volume, or an oil concentrate at the rate of 1 qt. per acre.

Normal weed control programs may be used in the following corn, grain sorghum or forage sorghum crop.

Note: Do not plant any crops except corn, grain sorghum, or forage sorghum the spring following this treatment, or illegal residues may result.

AATREX 4L ALONE—CHEMICAL FALLOW

Wheat-Sorghum-Fallow: To control annual broadleaf and grass weeds following wheat harvest and in the following sorghum crop when grown under minimum tillage, broadcast 6 pts./A to wheat stubble immediately following wheat harvest. If weeds are present at application, remove them with a sweep plow or other suitable implement after application, or use an approved contact herbicide before or after the application of AATREX. Plant sorghum into wheat stubble the following spring with minimum soil disturbance. Use a surface planter or a planter leaving a shallow furrow. If weeds are present at planting, remove them with a sweep plow or other suitable implement before planting.

For the list of weeds controlled, see **AATREX 4L Applied Alone—Corn or Grain Sorghum—Preplant Surface-Applied, Preplant Incorporated, or Preemergence.**

Precautions: (1) use only on silt loam or finer textured soil, or crop injury may result. (2) Wheat-sorghum-fallow cropping sequence must be followed. (3) Do not apply following sorghum harvest.

Note: To avoid illegal residues, do not graze or feed forage from treated area to livestock. To avoid illegal residues and crop injury, do not plant any crop other than those on this label within 18 months following treatment.

Wheat-Corn-Fallow (KS, NE): This product controls cheatgrass (downy brome, chess), kochia, mustards, pigweed, Russian thistle, wild lettuce, wild sunflower, and volunteer wheat during period after wheat harvest. Weed control may extend into following corn crop grown under minimum tillage.

Follow directions for use, notes, and precautions in the "Wheat-Sorghum-Fallow" section above, substituting corn for references to sorghum.

Wheat-Fallow-Wheat (CO, KS, MT, NE, ND, SD, and WY): For preemergence control of cheatgrass (downy brome, chess), common lambsquarters, field pennycress, kochia, mustard, Russian thistle, wild lettuce, and suppression of volunteer wheat during fallow period of a wheat-fallow-wheat rotation, broadcast 1-2 pts./A on all soils except those listed under "Precautions." For control of pigweed and wild sunflower, use the higher rate. Apply to stubble ground. Treat only once during same fallow period.

TANK MIXTURES FOR CHEMICAL FALLOW

Wheat-Sorghum-Fallow or Wheat-Corn-Fallow (KS, NE)

Igran 80W: Use as tank mixture for control of certain broadleaf and grass weeds in wheat-sorghum-fallow or wheat-corn-fallow rotations. Refer to Igran 80W label for directions, weeds controlled, precautions, and limitations.

Gramoxone: If weeds are present at application, a tank mix with Gramoxone may be used. Broadcast 6 pts. of AATREX 4L plus 1.5-2.5 pts. of Gramoxone in 20-60 gals. of water/A by ground equipment. Add 0.5-1 pt. of a nonionic surfactant, such as X-77, per 100 gals. of spray mixture. Add AATREX to spray tank first and thoroughly mix with water. Then add Gramoxone, followed by surfactant. Use the 2.5 pt. rate of Gramoxone if weeds are 4-6 inches tall. This mixture will not control weeds taller than 6 inches. Apply to stubble ground. Treat only once during same fallow period. Refer to Gramoxone label for further directions, precautions, and limitations.

Wheat-Fallow-Wheat (CO, KS, MT, NE, ND, SD, and WY)

Igran 80W: Use as tank mixture for control of certain broadleaf and grass weeds in wheat-fallow-wheat rotations. Refer to Igran 80W label for directions, weeds controlled, precautions, and limitations.

Gramoxone: If weeds are present at application, a tank mix with Gramoxone may be used. Broadcast 1-2 pts. of AATREX 4L plus 1.5-2.5 pts. of Gramoxone in 20-60 gals. of water A by ground equipment. Add 0.5-1 pts. of a nonionic surfactant, such as X-77, per 100 gals. of spray mixture. Add AATREX to spray tank first and thoroughly mix with water. Then add Gramoxone, followed by surfactant. Use the 2.5 pt. rate of Gramoxone if weeds are 4-6 inches tall. This mixture will not control weeds taller than 6 inches. Apply to stubble ground. Treat only once during same fallow period. Refer to Gramoxone label for further directions, precautions, and limitations.

If weeds are present at application and this product is used alone, use either an approved contact herbicide before or after treatment, or tillage after treatment.

Use tillage to control weeds which escape during fallow period. Till before planting. For this product applied alone or in tank mixture with Gramoxone, plant at least 2 inches deep and 12 months or more after application.

Precautions: To avoid crop injury, (1) Do not use on sand soil. (2) Do not treat eroded hillsides, caliche and rocky outcroppings, or exposed calcareous subsoil. (3) Do not treat soils of the Rosebud and Canyon Series in western NE and adjoining counties in CO and WY. (4) Do not treat soils with calcareous surface layers. (5) Avoid spray overlap.

Note: Do not graze treated areas within 6 months after application, or illegal residues may result.

Aerial application: In order to assure that spray will be controllable within the target area when used according to label directions, make applications at a maximum height of 10 ft., using low drift nozzles at a maximum pressure of 40 psi, and restrict application to periods when wind speed does not exceed 10 mph. To assure that spray will not adversely affect adjacent sensitive nontarget plants, apply AATREX 4L alone by aircraft at a minimum upwind distance of 400 ft. from sensitive plants, or apply AATREX 4L + Igran 80W at a minimum upwind distance of 500 ft. from sensitive plants. When applying AATREX 4L + Igran 80W by aircraft, use a minimum of 5 gals. of water per acre.

PROSO MILLET

For preplant incorporated or preemergence control of foxtails, kochia, common lambsquarters, redroot pigweed, Russian thistle, slimleaf lambsquarters and spotted spurge, apply 1-2/pts. A on soils with less than 2% organic matter, or apply 2-4 pts./A on soils with 2% or higher organic matter.

Precaution: To avoid crop injury and illegal residues, make only one application per year.

Rotational crops: See "Rotational crops" in Corn section.

RANGELAND

To control certain weeds, including cheatgrass (downy brome, chess), common (annual) broomweed, little barley, medusahead, sagewort, and tumble mustard. Use lower rate where cheatgrass is a major problem weed; use higher rate for other weeds.

Poisonous plants: Tansymustard, pigweeds, lambsquarters, tumble mustard, cocklebur, and fiddleneck are controlled at 2 pts./A. Other mustards (*Brassica* spp.), groundcherry, (annual), broomweed, and threeawn or poverty grass (*Aristida* spp.) are partially controlled at 2 pts./A and controlled at 4 pts./A. Sneezeweed and dogbane are partially controlled at 4 pts./A.

Aerial application: In order to assure that spray will be controllable within the target area when used according to label directions, make applications at a maximum height of 10 ft. above vegetation, using low drift nozzles at a maximum pressure of 40 psi, and restrict application to periods when wind speed does not exceed 10 mph. To assure that spray will not adversely affect adjacent sensitive nontarget plants, apply AATREX 4L by aircraft at a minimum upwind distance of 400 ft. from sensitive plants.

Establishment of perennial range grasses in West (CA, ID, NV, OR, UT, WA): Broadcast 1.6-2 pts./A in fall after rains begin but before ground freezes. Perennial range grass may be seeded when rain begins the next fall following treatment. Plant in deep furrows such as those made by a deep furrow rangeland drill. Treated areas may be grazed 7 months after seeding. However, to provide sufficient time for establishment of desirable range grasses, treated areas should not be grazed until the fall following seeding.

Renovation of existing stands of perennial range grasses

Central Great Plains (CO, KS, MT, NE, ND, SD, WY): Broadcast 1-2 pts./A in fall before ground freezes.

Southwest (AR, KS, LA, MS, NM, OK, TX): Apply 2 pts./A AATREX 4L in early spring, before regrowth, to established rangeland or pasture species such as Asiatic (Old World) bluestems, little and big bluestems, buffalograss, switchgrass, and Indiangrass.

West (CA, ID, NV, OR, UT, WA): Do not graze area to be treated for 6 months prior to application. Broadcast 1-1.5 pts./A by ground equipment in fall before ground freezes.

Bluegrass and intermediate wheatgrass may be injured by this product at recommended rates.

Notes: To avoid illegal residues, (1) Do not apply more than once per year. (2) Apply only on neutral to acid soils. (3) Do not cut or feed range grass hay. (4) Do not graze treated areas within 3 months following spring application, or within 7 months following fall application. (5) Do not dump or spill product or dispose of container within reach of livestock.

Pure stands of Switchgrass and Big Bluestem (IA, IL, IN, KS, MN, MO, ND, NE, OH, SD, WI): For control of those weeds listed under **AATREX 4L Applied Alone—Corn or Grain Sorghum—Preplant Surface-Applied, Preplant Incorporated, or Preemergence.**

Use only on loam, silt loam, silty clay loam, clay loam, and silty clay soils with at least 1 percent organic matter.

Establishment: Broadcast 4 pts./A after planting but before weeds and crop emerge. Prepare a good, firm seedbed. Plant ½ inch deep with a grassland drill (preferred method) or a conventional drill. If a conventional drill is used on prepared seedbeds, remove all tension from the disk openers. For best results, cultipack or roll after planting. Clip weeds that escape in July or August. Avoid clipping switchgrass or big bluestem seedlings.

Renovation of existing stands: Broadcast 4 pts./A in early spring, before regrowth of the crop, and before weed emergence.

Notes: To avoid illegal residues, (1) Do not cut for hay. (2) Do not graze treated areas within 4 months following application at seeding or 3 months following application to established switchgrass or big bluestem. (3) Treat only once per year.

Rotational crops: See "Rotational crops" in Corn section. Do not plant small-seeded legumes the year following application, or injury may occur.

ROADSIDES

To control certain annual weeds in established perennial grasses along roadsides in CO, KS, MT, NE, ND, SD, and WY, including cheatgrass (downy brome, chess), common (annual) broomweed, little barley, medusahead, sagewort, and tumble mustard, broadcast 2 pts./A in a minimum of 10 gallons of water by ground equipment in the fall before ground freezes, or after thawing in the spring, but before the established grasses green-up and before weeds emerge. Examples of desirable established grasses include big bluestem, bluegrass, bromegrass, buffalograss, crested wheatgrass, Indiangrass, little bluestem, side-oats grama, switchgrass, and Western wheatgrass. Apply only once per year. Temporary discoloration or other form of injury to the desirable perennial grasses may occur following application.

Notes: To avoid illegal residues, (1) Do not cut or feed roadside grass hay. (2) Do not allow livestock to graze treated areas.

SUGARCANE

For control of many broadleaf and grass weeds, including amaranths, crabgrass, fireweed, Flora's paintbrush, foxtails, junglerice and wiregrass, broadcast 4-8 pts./A of AATREX 4L at time of planting or ratooning, but before sugarcane emerges. Broadcast aerially in a minimum of 5 gals. of spray per acre, or broadcast or band by ground equipment in a minimum of 20 gals. per acre, unless indicated otherwise. One additional application may be made over the sugarcane as it emerges, and two additional applications may be made interline after emergence as directed sprays. Repeat treatments, where needed, may be applied broadcast, band, or interline as suggested with the final application being prior to close-in. Do not exceed the rate of herbicide suggested for any one crop of sugarcane.

Note: Where high rates of AATREX are used alone, apply in a minimum of 1 qt. of water for each 1 qt. of AATREX 4L applied per acre.

Aerial application: In order to assure that spray will be controllable within the target area when used according to label directions, make applications at a maximum height of 10 ft., using low drift nozzles at a maximum pressure of 40 psi, and restrict application to periods when wind speed does not exceed 10 mph. To assure that spray will not adversely affect adjacent sensitive nontarget plants, apply AATREX 4L alone by aircraft at a minimum upwind distance of 400 ft. from sensitive plants.

Florida

For control of emerged pellitory weed: Apply 0.8-1.2 pts./A in at least 40 gals. of water as a directed spray by ground equipment prior to close-in. Add 4 qts. of surfactant for each 100 gals. of spray. Thoroughly cover weed foliage.

For control of alexandergrass, large crabgrass, pellitory (artillery) weed, and spiny amaranth, use one of the following methods at planting or ratooning:

1. Apply 8 pts./A preemergence. Follow with one or two applications, as needed, postemergence to sugarcane and weeds, at 4 pts./A. Treat before weeds exceed 1.5 inches in height.
2. Apply 1-3 times, as needed, at 4 pts./A postemergence to sugarcane and weeds. Treat before weeds exceed 1.5 inches in height.

Louisiana

For control of annual weeds during the summer fallow period, apply 4 pts./A to weed-free beds immediately after bed formation. Follow normal weed control after planting.

Precautions: To avoid crop injury, (1) Do not apply more than 10 qts./A to any one crop of sugarcane. (2) If making 4 pts./A application during summer fallow period, do not exceed 8 qts./A during the remainder of the growing season, or illegal residues may result.

Texas

Use AATREX 4L for control of barnyardgrass, pigweed, purslane and sunflower, in plant or ratoon sugarcane.

Apply 8 pts./A of AATREX 4L preemergence. Follow with one or two applications, as needed, at 6 pts./A postemergence to sugarcane and weeds.

For best results when weeds are emerged, add a nonionic surfactant at a concentration of 2 qts./100 gals. to the spray and apply before weeds exceed 1.5 inches in height.

Precautions: (1) Injury to sugarcane may occur when under moisture stress, when soil is of low adsorptive capacity, or when land is first cropped to sugarcane. (2) Do not apply after close-in. (3) Do not apply more than 20 pts./A to any one crop of sugarcane, or crop injury may result.

TURFGRASSES FOR SOD (FLORIDA ONLY)

St. Augustinegrass, Centipedegrass, and Zoysia Grass

Broadcast 4-8 pts./A according to soil texture to control those weeds listed under **AATREX 4L Applied Alone—Corn or Grain Sorghum—Preplant Surface-Applied, or Preplant Incorporated, or Preemergence.**

Muck or peat	8 pts.	Old beds: Within 2 days after lifting sod.
		New beds: 3-4 days after sprigging or plugging.
Sandy soil	4 pts.	Old beds: Within 2 days after lifting sod.
		New beds: 7-10 days after sprigging or plugging.

If weeds regrow, apply an additional 4 pts./A on muck or peat, or 2 pts./A on sandy soil.

Precautions: To avoid crop injury, (1) Do not apply within 30 days prior to cutting or lifting. (2) Do not apply in combination with surfactants or other spray additives. (3) Use only on turfgrass reasonably free of infestations of insects, nematodes, and diseases. (4) On newly sprigged turfgrass, temporary slowing of growth may follow application.

TURFGRASS FOR FAIRWAYS, LAWNS, SOD PRODUCTION* AND SIMILAR AREAS

*In states other than Florida. For use on turfgrass for sod in Florida, see "Turfgrasses for Sod (Florida only)" section above.

Bermudagrass, Centipedegrass, St. Augustinegrass, and Zoysia Grass

Apply AATREX 4L after October 1 before emergence of winter annual weeds for control of annual bluegrass, burclover, carpet burweed, chickweed, corn speedwell, henbit, hop clover, and spurweed. AATREX 4L will control annual bluegrass even if it is emerged at time of treatment. For control of summer annual weeds listed in the preemergence portion of the AATREX 4L Applied Alone—Corn or Grain Sorghum section of this label, also apply AATREX 4L in late winter before the weeds emerge. Apply in a minimum of 15 gals. of water per acre or 1 gal. per 1,000 sq. ft.

Where annual bluegrass is the major weed, use 1 qt./A (22 ml or 0.75 fl. oz. per 1,000 sq. ft.). Use 2 qts./A (44 ml or 1.5 fl. oz. per 1,000 sq. ft.) for control of the other weeds named above. Do not exceed 1 qt./A per treatment on newly sprigged turfgrass or on hybrid bermudagrass such as Tiflawn, Tifway, and Ormond.

For continued summer annual weed control, apply another 1 qt./A at least 30 days after the previous application, but not after April 15. However, do not make more than two applications of this product per year.

Precautions: On newly sprigged turfgrass and hybrid bermudagrass, temporary slowing of growth and yellowing may occur following application. To avoid turf injury, (1) Use only on turfgrass reasonably free of infestations of insects, nematodes, and diseases. (2) Do not use on golf greens. (3) Do not use north of NC (except may be used in Virginia Coastal Plains) or west of the high rainfall areas of eastern OK and eastern TX. (4) Do not use on muck or alkaline soils. (5) Do not apply over the rooting area of trees or ornamentals not listed on this label. (6) Do not overseed with desirable turfgrass within 4 months before or 6 months after treatment. (7) Do not apply this product to newly seeded bermudagrass until it has overwintered and has a well-developed rhizome system. Do not exceed 2 qts. product/A within 12 months of seeding bermudagrass.

Note: Do not graze or feed turf clippings to animals, or illegal residues may result.

MACADAMIA NUTS

For preemergence control of many broadleaf and grass weeds, including crabgrass, foxtail, wiregrass, Flora's paintbrush, spanishneedles, and fireweed, broadcast 4-8 pts./A before harvest and before weeds emerge. Repeat as necessary. Do not spray when nuts are on ground during harvest period. Do not apply by air.

PINEAPPLE

For control of purslane, spanishneedles, annual grasses, annual bindweed, ageratum, amaranth, rattlepod, Flora's paintbrush, fireweed, spurge, indigo, and papalo, apply up to 12.8 pts./A as a blanket spray immediately after planting or following harvest. Make additional blanket or interspace applications of up to 3.2 pts./A at 1-2 month intervals as needed, prior to differentiation. Apply in 20-40 gals. of water/A to assure thorough coverage.

Precautions: (1) Do not apply more than 60 pts. (7.5 gals.)/A per cycle, as crop injury or illegal residues may result. (2) Repeated monthly applications to plant foliage may slow plant growth and delay fruiting.

Note: Do not apply within 45 days of fruit harvest, or forage harvest if forage is to be fed to livestock, or illegal residues may result.

GUAVA

Use only on established plantings which are at least 18 months old. Apply as a directed spray at 4-8 pts./A of AATREX 4L in 20-50 gals. of spray mix preemergence or early postemergence to weeds. When applying postemergence, the use of a surfactant and greater spray volume (80-100 gals. of spray mix per acre) may enhance weed control. This product controls many annual broadleaf and grass weeds, including fireweed, purslane, scarlet pimpernel, spanishneedles, and sowthistle.

Notes: To avoid illegal residues, (1) Do not allow spray to contact foliage or fruit. (2) Do not apply more frequently than at 4 month intervals. (3) Do not apply more than 16 pts. of AATREX 4L per year.

CONIFERS

For control of annual broadleaf and grass weeds prior to transplanting, after transplanting or in established conifers [including Douglas fir, grand fir, noble fir, white fir, Austrian pine, bishop pine, Jeffrey pine, knobcone pine, loblolly pine, lodgepole pine (shore pine), monterey pine, ponderosa pine, Scotch pine, slash pine, blue spruce, and Sitka spruce]: Broadcast 4-8 pts. in a minimum of 5 gals. of water per acre by air or 10 gals. by ground before weeds are 1.5 inches tall. Apply to established trees between fall and early spring while trees are dormant. For new transplants, apply during or soon after transplanting. For applications prior to transplanting, allow sufficient precipitation to activate AATREX before transplanting. In areas where spring and summer rainfall is inadequate to activate AATREX, apply during fall prior to spring transplanting.

For the list of weeds controlled, see **AATREX 4L Applied Alone—Corn or Grain Sorghum—Preplant Surface-Applied, Preplant Incorporated, or Preemergence.**

Quackgrass control: Broadcast 8 pts. in a minimum of 5 gals. of water per acre by air or 10 gals. by ground between fall and early spring while trees are dormant and before quackgrass is more than 1.5 inches tall.

Precautions: (1) In areas west of the Rocky Mountains (except the Great Basin), grazing may begin 7 months after a fall application or 3 months after a winter or spring application. (2) To prevent illegal residues, do not graze treated areas of the Great Basin, or areas east of the Rocky Mountains. (3) Temporary injury to trees may occur following use of AATREX on coarse-textured soil. (4) To avoid crop injury, do not apply to seedbeds. (5) Also apply only once per year.

Aerial application: In order to assure that spray will be controllable within the target area when used according to label directions, make applications at a maximum height of 10 ft. above vegetation, using low drift nozzles at a maximum pressure of 40 psi, and restrict application to periods when wind speed does not exceed 10 mph. To assure that spray will not adversely affect adjacent sensitive nontarget plants, apply AATREX 4L by aircraft at a minimum upwind distance of 400 ft. from sensitive plants.

NONSELECTIVE WEED CONTROL ON NONCROP LAND

This product provides long-term weed control on industrial sites, highway medians and shoulders, railroad rights-of-way, lumberyards, petroleum tank farms, and in noncrop areas on farms, such as around buildings, equipment and fuel storage areas, along fences, and lanes. Do not use near adjacent desirable plants or in greenhouses, or injury may occur.

Apply before or soon after weeds begin growth. Make postemergence applications when weeds are young and actively growing. Use the higher rates on fine clay and muck soils. Use sufficient water to assure thorough coverage. Use at least 1 gal. of water for each qt. of product; use more if practical.

Note: Do not allow livestock to graze treated areas, or illegal residues may result.

Annual broadleaf and grass weeds (including barnyardgrass, cheatgrass, crabgrass, lambsquarters, foxtail, ragweed, puncturevine, and turkey mullein): Broadcast 4.8-10 qts./A.

Hard-to-kill annual and many perennial broadleaf and grass weeds (including bluegrass, burdock, Canada thistle, dogfennel, orchardgrass, plantain, quackgrass, purpletop, redtop, and smooth brome): Broadcast 10-20 qts./A.

Hard-to-kill biennial and perennial weeds (including bullthistle and sowthistle): Broadcast 20-40 qts./A.

For longer residual control in regions of high rainfall and a long growing season, broadcast 20-40 qts./A.

For small areas, 0.28 pt. per 1,000 sq. ft. is equivalent to 10 qts./A.

STORAGE AND DISPOSAL

Pesticide Disposal

Do not contaminate water, food, or feed by storage or disposal, or cleaning of equipment. Open dumping is prohibited. Wastes resulting from the use of this product are acutely toxic. Improper disposal of unused pesticide, spray mixture, or rinsate is a violation of federal law. Pesticide, spray mixture, or rinsate that cannot be used according to label instructions must be disposed of according to federal, state or local procedures. For guidance in proper disposal methods, contact your State Pesticide or Environmental Control Agency, or the Hazardous Waste representative at the nearest EPA Regional Office.

Container Disposal

Triple rinse (or equivalent) and offer for recycling or reconditioning, or puncture and dispose of in a sanitary landfill, or by incineration, or by open burning, if allowed by state and local authorities. If burned, keep out of smoke.

For minor spills, leaks, etc., follow all precautions indicated on this label and clean up immediately. Take special care to avoid contamination of equipment and facilities during cleanup procedures and disposal of wastes. In the event of a major spill, fire or other emergency, call (919) 292-7100 day or night.

PRECAUTIONARY STATEMENTS

HAZARDS TO HUMANS AND DOMESTIC ANIMALS

CAUTION

Harmful if swallowed, inhaled, or absorbed through skin. Do not breathe vapors or spray mist. Avoid contact with eyes, skin or clothing.

STATEMENT OF PRACTICAL TREATMENT

If swallowed: Call a physician or Poison Control Center immediately. Drink 1 or 2 glasses of water and induce vomiting by touching back of throat with finger, or, if available, by administering syrup of ipecac. Do not induce vomiting or give anything by mouth to an unconscious person.

If inhaled: Remove victim to fresh air. If not breathing, give artificial respiration, preferably mouth-to-mouth. Get medical attention.

If on skin: Wash with plenty of soap and water. Get medical attention.

If in eyes: Flush eyes with *plenty* of water. Call a physician if irritation persists.

Note to Physician: There is no specific antidote for atrazine. If this product is ingested, induce emesis or lavage stomach. The use of an aqueous slurry of activated charcoal may be considered.

ENVIRONMENTAL HAZARDS

Atrazine can travel (seep or leach) through soil and can enter ground water which may be used as drinking water. Atrazine has been found in ground water. Users are advised not to apply atrazine to sand and loamy sand soils where the water table (ground water) is close to the surface and where these soils are very permeable, i.e., well-drained. Your local agricultural agencies can provide further information on the type of soil in your area and the location of ground water.

This pesticide is toxic to aquatic invertebrates. Do not apply directly to water or wetlands. Do not apply where runoff is likely to occur. Do not apply when weather conditions favor drift from treated areas. Runoff and drift from treated areas may be hazardous to aquatic organisms in neighboring areas. Do not contaminate water by cleaning of equipment or disposal of wastes.

ENDANGERED SPECIES

Refer to booklet label for use restrictions to protect ENDANGERED SPECIES.

CHEMIGATION

Refer to booklet label for use directions for chemigation. Do not apply this product through any irrigation system unless the supplemental labeling for chemigation is followed.

AATREX® trademark of CIBA-GEIGY for atrazine
Bexton® trademark of Dow Chemical Co. for propachlor
Caliber® trademark of CIBA-GEIGY
Compex® trademark of KALO Agricultural Chemicals, Inc.
Concep® trademark of CIBA-GEIGY
Dual® trademark of CIBA-GEIGY for metolachlor
U.S. Patent No. 3,937,730
Gramoxone® trademark of ICI Americas for paraquat
Igran® trademark of CIBA-GEIGY for terbutryn
Lasso® trademark of Monsanto Company for alachlor
Princep® trademark of CIBA-GEIGY for simazine
Ramrod® trademark of Monsanto Company for propachlor
Roundup® trademark of Monsanto Company for glyphosate herbicide

Sutan +® trademark of Stauffer Chemical Co. for butylate
Unite® trademark of Hopkins Agricultural Chemical Co.
X-77® trademark of KALO Agricultural Chemicals, Inc.

CGA 7L38Y 078

This is a specimen label which was not prepared by CIBA-GEIGY. CIBA-GEIGY is not responsible for the accuracy of the information contained herein. As labels are subject to revision, always carefully read and follow the label on the product container.

AATREX® 80W

HERBICIDE

For season-long weed control in corn and sorghum

For weed control in certain other crops; in noncrop areas; and industrial sites

ACTIVE INGREDIENTS:	
Atrazine: 2-chloro-4-ethylamino-6-isopropylamino-*s*-triazine	76%
Related compounds	4%
INERT INGREDIENTS:	20%
TOTAL	100%

EPA Reg. No. 100-439 EPA Est. 100-LA-1

AATREX 80W is a wettable powder.

CGA 7L91K 078

KEEP OUT OF REACH OF CHILDREN

CAUTION

See additional precautionary statements at end of label booklet.

See directions for use inside booklet.

AATREX® trademark of CIBA-GEIGY for atrazine

DIRECTIONS FOR USE AND CONDITIONS OF SALE AND WARRANTY

IMPORTANT: Read the entire **Directions for Use** and the **Conditions of Sale and Warranty** before using this product.

CONDITIONS OF SALE AND WARRANTY

The **Directions for Use** of this product reflect the opinion of experts based on field use and tests. The directions are believed to be reliable and should be followed carefully. However, it is impossible to eliminate all risks inherently associated with use of this product. Crop injury, ineffectiveness, or other unintended consequences may result because of such factors as weather conditions, presence of other materials, or the manner of use or application all of which are beyond the control of CIBA-GEIGY or the Seller. All such risks shall be assumed by the Buyer.

CIBA-GEIGY warrants that this product conforms to the chemical description on the label and is reasonably fit for the purposes referred to in the **Directions for Use** subject to the inherent risks referred to above. **CIBA-GEIGY makes no other express or implied warranty of Fitness or Merchantability or any other express or implied warranty. In no case shall CIBA-GEIGY or the Seller be liable for consequential, special, or indirect damages resulting from the use or handling of this product.** CIBA-GEIGY and the Seller offer this product, and the Buyer and user accept it, subject to the foregoing **Conditions of Sale and Warranty**, which may be varied only by agreement in writing signed by a duly authorized representative of CIBA-GEIGY.

DIRECTIONS FOR USE

It is a violation of federal law to use this product in a manner inconsistent with its labeling.

FAILURE TO FOLLOW THE DIRECTIONS FOR USE AND PRECAUTIONS ON THIS LABEL MAY RESULT IN POOR WEED CONTROL, CROP INJURY, OR ILLEGAL RESIDUES.

Do not apply this product in such a manner as to directly or through drift expose workers or other persons, except those knowingly involved in the application. The area being treated must be vacated by unprotected persons.

Do not enter treated areas without protective clothing until sprays have dried.

Because certain states may require more restrictive reentry intervals for various crops treated with this product, consult your State Department of Agriculture for further information.

Written or oral warnings must be given to workers who are expected to be in a treated area or in an area about to be treated with this product. Oral warnings must be given which inform workers of areas or fields that may not be entered without specific protective clothing until sprays have dried, and appropriate actions to take in case of accidental exposure as described under **Precautionary Statements** on this label. When oral warnings are given, warnings shall be given in a language customarily understood by workers. Oral warnings must be given if there is reason to believe that written warnings cannot be understood by workers. Written warnings must include the following information: "CAUTION. Area treated with AATREX 80W on *(date of application)*. Do not enter without appropriate protective clothing until sprays have dried. In case of accidental exposure, flush eyes or skin with plenty of water. Call a physician if irritation persists. Remove and wash contaminated clothing before reuse."

GENERAL INFORMATION

This herbicide controls many annual broadleaf and grass weeds in corn, sorghum, sugarcane, pineapple, and certain other crops specified on this label. It is also effective in noncrop areas and industrial sites for control of most annual and many perennial broadleaf and grass weeds. This product may be applied before or after weeds emerge.

Following many years of continuous use of this product and chemically related products, biotypes of some of the weeds listed on this label have been reported which cannot be effectively controlled by this and related herbicides. Where this is known or suspected and weeds controlled by this product are expected to be present along with resistent biotypes, we recommend the use of this product in combinations or in sequence with other registered herbicides which are not triazines. If only resistant biotypes are expected to be present, use a registered non-triazine herbicide. Consult with your state Agricultural Extension Service for specific recommendations.

Since this product acts mainly through root absorption, its effectiveness depends on moisture to move it into the root zone. If weeds develop, a shallow cultivation or rotary hoeing will generally result in better weed control.

This product is noncorrosive to equipment and metal surfaces, nonflammable, and has low electrical conductivity.

Avoid using near adjacent desirable plants or in greenhouses, or injury may occur.

To avoid spray drift, do not apply under windy conditions. Avoid spray overlap, as crop injury may result.

Where the use directions give a range of rates, use the lower rate on coarse-textured soil and soil low in organic matter; use the higher rate on fine-textured soil and soil high in organic matter.

Note: CIBA-GEIGY does not recommend applications in combination with other herbicides or oils, except as specifically described on the label or in literature published by CIBA-GEIGY.

APPLICATION PROCEDURES

Ground application: Use conventional ground sprayers equipped with nozzles that provide accurate and uniform application. Be certain that nozzles are uniformly spaced and are the same size. Calibrate sprayer before use and recalibrate at the start of each season and when changing carriers. Unless otherwise specified, use a minimum of 10 gals. of spray mixture/A for all preplant incorporated, preplant surface, preemergence, and postemergence applications (with or without oil or surfactant) with ground equipment.

Use a pump with capacity to (1) maintain 35-40 psi at nozzles, (2) provide sufficient agitation in tank to keep mixture in suspension, and (3) to provide a minimum of 20% bypass at all times. Use centrifugal pumps which provide propeller shear action for dispersing and mixing this product. The pump should provide a minimum of 10 gals./minute/100 gal. tank size circulated through a correctly positioned sparger tube or jets.

Use screens to protect the pump and to prevent nozzles from clogging. Screens placed on the suction side of the pump should be *16-mesh or coarser.* Do not place a screen in the recirculation line. Use 50-mesh or coarser screens between the pump and boom, and where required, at the nozzles. Check nozzle manufacturer's recommendations.

For band applications, calculate amount to be applied per acre as follows:

$$\frac{\text{band width in inches}}{\text{row width in inches}} \times \text{broadcast rate per acre} = \text{amount needed per acre of field}$$

Aerial application: Use aerial application only where broadcast applications are specified. Apply in a minimum of 1 gal. of water for each 1 lb. of this product applied per acre. For postemergence treatments on corn and sorghum, apply recommended rate in a minimum of 2 gals. of water/A. Avoid applications under conditions where uniform coverage cannot be obtained or where excessive spray drift may occur.

Avoid application to humans or animals. Flagmen and loaders should avoid inhalation of spray mist and prolonged contact with skin, and should wash thoroughly before eating and at the end of each day's operation.

Application in water or liquid fertilizer: Nitrogen solution or complete liquid fertilizer may replace all or part of the water as a carrier for preemergence, preplant incorporated, or preplant surface ground application on corn and sorghum. Check the compatibility of this product with liquid fertilizer and/or nitrogen solution as shown below before use. Do not apply in nitrogen solution or complete liquid fertilizer after corn or sorghum emerges, except as noted under **Lay-by treatment** for corn, or crop injury may occur.

Compatibility Test: Since liquid fertilizers can vary, even within the same analysis, always **check compatibility with herbicide(s) each time before use.** Be especially careful when using **complete** suspension or fluid fertilizers as serious compatibility problems are more likely to occur. Commercial application equipment may improve compatibility in some instances. The following test assumes a spray volume of 25 gals. per acre. For other spray volumes, make appropriate changes in the ingredients. Check compatibility using this procedure:

1. Add 1 pint of fertilizer to each of 2 one-quart jars with tight lids.
2. To **one** of the jars add 1/4 tsp. or 1.2 milliliters of a compatibility agent approved for this use, such as Compex® or Unite® (1/4 tsp. is equivalent to 2 pts. per 100 gals. spray). Shake or stir gently to mix.
3. To **both** jars add the appropriate amount of herbicide(s). If more than one herbicide is used, add them separately with dry herbicides first, flowables next, and emulsifiable concentrates last. After each addition, shake or stir gently to thoroughly mix. The appropriate amount of herbicides for this test follows:
   Dry herbicides: For each pound to be applied per acre, add 1.5 level teaspoons to each jar.
   Liquid herbicides: For each pint to be applied per acre, add 0.5 teaspoon or 2.5 milliliters to each jar.
4. After adding all ingredients, put lids on and tighten, and invert each jar ten times to mix. Let the mixtures stand 15 minutes and then look for separation, large flakes, precipitates, gels, heavy oily film on the jar, or other signs of incompatibility. Determine if the compatibility agent is needed in the spray mixture by comparing the two jars. If either mixture separates, but can be remixed readily, the mixture can be sprayed as long as good agitation is used. If the mixtures are incompatible, test the following methods of improving compatibility: (A) slurry the dry herbicide(s) in water before addition, or (B) add 1/2 of the compatibility agent to the fertilizer and the other 1/2 to the emulsifiable concentrate or flowable herbicide before addition to the mixture. If incompatibility is still observed, do not use the mixture.

Application in water plus emulsifiable oil or oil concentrate: Adding emulsifiable oil (petroleum-derived, petroleum-derived oil concentrate, or single or mixed crop-derived oil concentrate) to postemergence water-based sprays in corn and sorghum may improve weed control. However, under certain conditions, the use of either type of oil may seriously injure the crop. To minimize this possibility, observe the following directions:

Use one of the following properly emulsified:

1. A suitable oil concentrate containing at least 1% but not more than 20% suitable emulsifier or surfactant blend.
2. Petroleum-derived oil containing at least 1% suitable emulsifier.

Note: In the event of a compatibility problem when mixing oil with this product and water, a compatibility agent such as Compex or Unite should be used. Any of the above oils contaminated with water or other materials can cause compatibility problems and/or crop injury.

Mixing procedures—all uses: (1) Be sure sprayer is clean and not contaminated with any other materials, or crop injury or sprayer clogging may result. (2) Fill tank 1/4 full with clean water, nitrogen solution, or complete liquid fertilizer. (3) Start agitation. (4) Be certain that the agitation system is working properly and creates a rippling or rolling action on the liquid surface. (5) Pour product directly from bag into tank. (6) Continue filling tank until 90% full. Increase agitation if necessary to maintain surface action. (7) Add emulsifiable oil, oil concentrate, or tank mix herbicide(s) after this product is thoroughly suspended. (8) Finish filling tank. (9) Empty tank as completely as possible before refilling to prevent buildup of oil or emulsifiable concentrate residue. Maintain agitation to avoid separation of materials. (10) If an oil or emulsifiable concentrate film starts to build up in tank, drain it and clean with strong detergent solution or solvent. (11) Clean sprayer thoroughly immediately after use by flushing system with water containing a detergent.

Center pivot sprinkler application

Preemergence or postemergence: Apply this product alone, or in tank mixtures with other herbicides on this label which are registered for center pivot application, with irrigation water. Apply either after planting before crop and weeds emerge, or after crop emergence, but before lay-by (20-30 inches) and before weeds exceed 1.5 inches in height. Apply this product only through a center pivot irrigation system. Do not apply this product through any other type of irrigation system. Crop injury, lack of effectiveness, or illegal pesticide residues in the crop can result from nonuniform distribution of treated water. If you have questions about calibration, you should contact State Extension Service specialists, equipment manufacturers or other experts, or refer to supplemental literature published by CIBA-GEIGY. Do not connect an irrigation system (including greenhouse systems) used for pesticide application to a public water system unless the pesticide label-prescribed safety devices for public water systems are in place. A person knowledgeable of the chemigation system and responsible for its operation, or under the supervision of the responsible person, shall shut the system down and make necessary adjustments should the need arise.

Operating Instructions

1. The system must contain a functional check valve, vacuum relief valve, and low pressure drain appropriately located on the irrigation pipeline to prevent water-source contamination from backflow.
2. The pesticide injection pipeline must contain a functional, automatic, quick-closing check valve to prevent the flow of fluid back toward the injection pump.
3. The pesticide injection pipeline must also contain a functional, normally closed, solenoid-operated valve located on the intake side of the injection pump and connected to the system interlock to prevent fluid from being withdrawn from the supply tank when the irrigation system is either automatically or manually shut down.
4. The system must contain functional interlocking controls to automatically shut off the pesticide injection pump when the water pump motor stops.
5. The irrigation line or water pump must include a functional pressure switch which will stop the water pump motor when the water pressure decreases to the point where pesticide distribution is adversely affected.
6. Systems must use a metering pump, such as a positive displacement injection pump (e.g., diaphragm pump) effectively designed and constructed of materials that are compatible with pesticides and capable of being fitted with a system interlock.
7. Do not apply when wind speed favors drift beyond the area intended for treatment.
8. Prepare a mixture with a minimum of 1 part of water to 1 part herbicide(s) and inject this mixture into the center pivot system. Injecting a larger volume of a more dilute mixture per hour will usually provide more accurate calibration of metering equipment. Maintain sufficient agitation to keep the herbicide in suspension.
9. Meter into irrigation water during entire period of water application.
10. Apply in 1/2-1 inch of water. Use the lower water volume (1/2 inch) on coarser-textured soils and the higher volume (1 inch) on finer-textured soils. More than 1 inch of water at application may reduce weed control by moving the herbicide below the effective zone in the soil.

Precautions for center pivot applications: Where sprinkler distribution patterns do not overlap sufficiently, unacceptable weed control may result. Where sprinkler distribution patterns overlap excessively, crop injury may result.

Within rate ranges in all tables on this label, use the lower rate on soil relatively coarse-textured or low in organic matter; use the higher rate on soil relatively fine-textured or high in organic matter.

AATREX 80W APPLIED ALONE—CORN OR GRAIN SORGHUM

Preplant Surface-Applied, Preplant Incorporated, or Preemergence (or Postemergence at 2.5 lbs./A with oil)

Broadleaf and Grass Weeds Controlled

barnyardgrass (watergrass)	witchgrass *(Panicum capillare)*	cocklebur	nighshade
giant foxtail	yellow foxtail	groundcherry	pigweed
green foxtail		jimsonweed	purslane
large (hairy) crabgrass		kochia	ragweed
wild oats		lambsquarters	sicklepod
		annual morningglory	velvetleaf (buttonweed)
		mustards	

Postemergence with Emulsifiable Oil or Oil Concentrate in Water (at 1.5 lbs./A)

Broadleaf Weeds Controlled

annual morningglory	lambsquarters	ragweed
cocklebur	mustards	smartweed
jimsonweed	pigweed	wild buckwheat
		velvetleaf

CORN

Preplant Surface-Applied (Broadleaf and grass control): Use on medium- and fine-textured soil with minimum-tillage or no-tillage systems only in CO, IL, IN, IA, KS, KY, MN, MO, MT, NE, ND, SD, WI, and WY. Apply 2/3 the recommended rate of AATREX shown in Table 1 for the soil texture as a split treatment 30-45 days prior to planting and the remainder at planting. Applications made less than 30 days prior to planting may be as either a split or single treatment. On *coarse-textured* soils, do not apply more than two weeks prior to planting.

If weeds are present at the time of treatment, apply in tank mix combination with a contact herbicide (for example, Gramoxone® or Roundup®). Observe directions for use, precautions, and restrictions on the label of the contact herbicide.

Note: To the extent possible, do not move treated soil out of the row or move untreated soil to the surface during planting, or weed control will be diminished.

Preplant Incorporated (Broadleaf and grass control): Broadcast in spring after plowing at rate in Table 1. Apply to the soil and incorporate before, during, or after final seedbed preparation. Avoid deep incorporation. For best results, apply within two weeks prior to planting.

Preemergence (Broadleaf and grass control): Apply during or shortly after planting before weed emergence, at rate in Table 1.

Postemergence (Broadleaf and grass control): Apply before weeds exceed 1.5 inches in height, at rate in Table 1.

Table 1: Broadleaf and Grass Weed Control on Corn*

Soil Texture	Broadcast rate/A
COARSE Sand, loamy sand, sandy loam	2.5 lbs.
MEDIUM Loam, silt loam, silt, clay loam, sandy clay loam, silty clay loam, sandy clay, or silty clay with low organic matter	3 lbs.
FINE Loam, silt loam, silt, clay loam, sandy clay loam, silty clay loam, sandy clay, or silty clay with medium to high organic matter, and clay (including the dark prairie soils of the Corn Belt)	3.75 lbs.
Peat, muck, and high organic clay (apply postemergence only)	3.75 lbs.

* *Broadleaf control (eastern CO, western KS, western NE, NM, OK Pan Handle, west TX, and eastern WY):* On sand, loamy sand, sandy loam, mild to strongly alkaline soil, and all recently leveled soil, apply no more than 1.5 lbs./A, either preplant surface, preplant incorporated, or preemergence. On other soils in these areas, apply rate in Table 1 for broadleaf and grass control.

Lay-by treatment (Broadleaf and grass control): Broadcast 1.25-2.5 lbs./A before weeds are 1.5 inches tall and before corn is 20-30 inches tall. When using nitrogen solutions, direct the spray to lower 3-4 inches of cornstalks to avoid corn foliage injury.

Postemergence with emulsifiable oil or oil concentrate in water: Add the following volume of one of the type oils indicated for aerial or ground application unless the oil label specifies otherwise:

Type Oil	Ground Application	Aerial Application
Oil concentrate (Crop or Petroleum-derived)	1 qt./A	½-1 qt./A
Petroleum-derived oil	1 gal./A	2 qts./A

Note: Crop-derived or petroleum-derived oil concentrates should contain at least 1% but not more than 20% suitable emulsifier or surfactant blend. Petroleum-derived oils should contain at least 1% suitable emulsifier.

Broadleaf and grass control: For postemergence control of those weeds listed under **Preplant Incorporated and Preemergence,** broadcast 2.5 lbs./A plus emulsifiable oil or oil concentrate after weed emergence, but before weeds reach 1.5 inches in height.

Broadleaf control: For postemergence control of those weeds listed under **Postemergence with emulsifiable oil or oil concentrate in water,** broadcast 1.5 lbs./A plus emulsifiable oil or oil concentrate before pigweed and lambsquarters reach 6 inches in height and before all other weeds reach 4 inches in height. A cultivation may be necessary if all weeds are not controlled or if weeds regrow.

Precautions for applications with emulsifiable oil or oil concentrate in water: (1) Inbred lines or any breeding stock may be severely injured by applications with emulsifiable oil or oil concentrate. (2) Adding other insecticides, herbicides, liquid fertilizers, or other materials is not recommended, because they may cause compatibility problems or crop injury. (3) Store and handle emulsifiable oil and oil concentrate carefully. Oil contaminated with even a small amount of water may not emulsify properly when added to the tank. To avoid crop injury, (4) Do not apply when crop is under stress from prolonged cold, wet weather, poor fertility, or other factors, or when crop is wet and succulent from recent rainfall. (5) Do not make more than one application per season except as recommended for control of yellow nutsedge, quackgrass, and Canada thistle.

Problem weeds

Yellow nutsedge and Canada thistle: This product will control yellow nutsedge *(Cyperus esculentus)* and Canada thistle *(Cirsium arvense)* when applied according to directions. For best results, apply each year until yellow nutsedge or Canada thistle is eliminated or reaches a level of infestation where neither weed is an economic problem. If yellow nutsedge or Canada thistle regrow following last application, cultivate once.

Four alternative methods of use for control of yellow nutsedge and Canada thistle are listed below in order of preference. If other weed species, such as annual grasses, are also expected, use alternative 2 or 4.

1. Broadcast 2.5 lbs./A plus 1 gal. of emulsifiable oil or 1 qt. of oil concentrate. Apply after crop and yellow nutsedge or Canada thistle emerge, but before yellow nutsedge reaches a height of 3 inches or Canada thistle reaches a height of 6 inches. Repeat application before lay-by (20-30 inches), 10-20 days after the first application.
2. Broadcast 2.5 lbs./A preplant incorporated for control of yellow nutsedge, or broadcast 2.5 lbs./A during or shortly after planting, but prior to crop or weed emergence for control of Canada thistle. Follow with an application of 2.5 lbs./A plus 1 gal. of emulsifiable oil or 1 qt. of oil concentrate after corn and weeds emerge, but before yellow nutsedge or Canada thistle reaches a height of 6 inches.
3. Broadcast 5 lbs./A plus 1 gal. of emulsifiable oil or 1 qt. of oil concentrate after crop emerges, but prior to lay-by (20-30 inches). Apply after yellow nutsedge and Canada thistle emerge, but before yellow nutsedge is 3 inches tall or Canada thistle is 6 inches tall.
4. Broadcast 5 lbs./A before, during, or shortly after planting, but before crop or weed emergence (yellow nutsedge control only).

Note: Do not use emulsifiable oil or oil concentrate when corn is wet or under stress, especially when using 5 lbs./A of this product, as crop injury may occur. See "Precautions for applications with emulsifiable oil or oil concentrate in water" for additional directions.

Quackgrass control on land going into corn production

Split application: Broadcast 2.5 lbs./A in fall or spring. Plow 1-3 weeks later. Apply a second application at 2.5 lbs./A in spring before, during, or after planting, but before weeds are 1.5 inches high. This split application will control quackgrass and most annual broadleaf and grass weeds.

Single application: Broadcast 3.75-5 lbs./A in fall or spring. Plow 1-3 weeks later.

TANK MIXTURES FOR CORN

This product may be tank mixed with these herbicides for control of certain broadleaf and grass weeds in corn:

Dual® (metolachlor)	Lasso or Lasso EC + Roundup
Dual + Gramoxone	Lasso or Lasso EC + Gramoxone
Dual + Roundup (glyphosate)	Gramoxone
Dual + Princep®	Princep
Dual + Princep + Gramoxone	Princep + Gramoxone
Dual + Princep + Roundup	Princep + Roundup
Bexton® or Ramrod® (propachlor)	Roundup
Lasso® or Lasso EC (alachlor)	Sutan + ®

Use tank mix directions appearing on the labels of the above herbicides when tank mixing with this product. Observe all precautions and limitations on labeling of products used in a particular tank mix.

Princep 80W, Princep 4L, or Princep Caliber® 90

In addition to the weeds listed under **AATREX 80W Applied Alone—Corn and Grain Sorghum—Preplant Surface-Applied, Preplant Incorporated, or Preemergence,** this combination also controls crabgrass, fall panicum, and carpetweed.

Broadcast tank mix before planting, at planting, or after planting, but before crop and weeds emerge, at rates in Table 2. Use the 1:1 ratio for control of most weeds. Use the 1:2 ratio for expected heavy infestations of crabgrass and fall panicum. Cultivate shallowly if weeds develop.

Preplant Surface-Applied: Use on medium- and fine-textured soils with minimum-tillage or no-tillage systems only in CO, IL, IN, IA, KS, KY, MN, MO, MT, NE, ND, SD, WI, and WY. Apply ⅔ the recommended rate of AATREX and Princep shown in Table 2 for the soil texture as a split treatment 30-45 days prior to planting and the remainder at planting. Applications made less than 30 days prior to planting may be as either a split or single treatment. On *coarse-textured* soils, do not apply more than two weeks prior to planting. Refer to the AATREX Alone—Preplant Surface-Applied section of the corn label for additional details.

If weeds are present at time of treatment, apply in a tank mix combination with a contact herbicide (for example, Gramoxone or Roundup). Observe directions for use, precautions, and restrictions on the label of the contact herbicide.

Note: To the extent possible, do not move treated soil out of the row or move untreated soil to the surface during planting, or weed control will be diminished.

Preplant Incorporated: Apply to the soil and incorporate in the spring before, during, or after final seedbed preparation. Avoid deep incorporation. For best results, apply within two weeks prior to planting.

Preemergence: Apply during or shortly after planting, but before crop and weeds emerge.

Refer to Corn sections of this label and to Princep 80W, Princep Caliber 90, or Princep 4L labels for further directions, limitations, and precautions.

Table 2: Tank Mixtures with Princep on Corn

Soil Texture	Broadcast rate/A			
	1:1 Ratio*		1:2 Ratio**	
	This product	Princep 80W[1]	This product	Princep 80W[1]
Sand, loamy sand, sandy loam	1.25 lbs.	1.25 lbs.	0.83 lb.	1.67 lbs.
Loam, silt loam, silt, clay loam, sandy clay loam, silty clay loam, sandy clay, or silty clay with low organic matter	1.5 lbs.	1.5 lbs.	1 lb.	2 lbs.

Soil Texture	Broadcast rate/A			
	1:1 Ratio*		1:2 Ratio**	
	This product	Princep 80W[1]	This product	Princep 80W[1]
Loam, silt loam, silt, clay loam, sandy clay loam, silty clay loam, sandy clay, or silty clay with medium to high organic matter, and clay (including dark prairie soils of the Corn Belt)	1.8 lbs.	1.8 lbs.	1.2 lbs.	2.4 lbs.

* For control of most weeds.
** For control of expected heavy infestations of crabgrass and fall panicum.
[1] When using Princep Caliber 90 or Princep 4L, use equivalent rates. One lb. of Princep 80W equals 0.9 lb. of Princep Caliber 90 or 1.6 pts. of Princep 4L.

Princep 80W, Princep 4L, or Princep Caliber 90 plus Roundup: Use as tank mixture for preemergence and postemergence control of certain broadleaf and grass weeds where corn will be planted directly into a cover crop, established sod, or in previous crop residues. Refer to Roundup label for all directions, weeds controlled, precautions, and limitations.

Princep 80W, Princep 4L, or Princep Caliber 90 plus Gramoxone: Use as tank mixture with Princep and Gramoxone to kill existing vegetation and for residual weed control where corn will be planted directly into a cover crop, established sod, or in previous crop residues. Add this product and Princep to water in spray tank, agitating until thoroughly mixed. Then add Gramoxone and a nonionic surfactant, such as X-77®. Continue agitation during application. Broadcast 1.25-2.5 lbs. of this product plus 1.25-2.5 lbs. Princep 80W [or 2-4 pts. Princep 4L, or 1.1-2.2 lbs. Princep Caliber 90] plus 1.5-2.5 pts. Gramoxone in 20-60 gals. of water per sprayed acre. Apply before, during, or after planting, but before corn emerges. Add 0.5 pt. of a nonionic surfactant, such as X-77, per 100 gals. of spray mixture. Use the 2.5 pt. rate of Gramoxone if existing vegetation is 4-6 inches tall. This mixture will not control weeds taller than 6 inches.

Refer to further limitations and precautions on labels for this product, Princep, and Gramoxone.

Precautions for all applications to corn: (1) To avoid crop injury and illegal residues, do not apply more than 5 lbs./A of this product per year. (2) Following harvest, plow (moldboard or disk-plow) and thoroughly till soil in fall or spring to minimize possible injury to spring-seeded rotational crops, regardless of rate used.

Note for all applications to corn: Do not graze or feed forage from treated areas for 21 days following application, or illegal residues may result.

Rotational crops: (1) Do not rotate to any crop except corn or sorghum until the following year, or injury may occur. (2) If applied after June 10, do not rotate with crops other than corn or sorghum the next year, or crop injury may occur. (3) If the broadcast rate, or equivalent band application rate, exceeds 3.75 lbs./A of AATREX 80W, or 3.75 lbs./A of AATREX 80W plus Princep 80W (or equivalent of another Princep formulation), a crop of untreated corn or sorghum should precede the next rotational crop. (4) In the High Plains and Intermountain areas of the West where rainfall is sparse and erratic or where irrigation is required, use only when corn or sorghum is to follow corn or sorghum or when a crop of untreated corn or sorghum is to precede other rotational crops. (5) In eastern parts of the Dakotas, KS, western MN, and NE, do not rotate to soybeans if the rate applied to corn or sorghum was more than 2.5 lbs./A or equivalent band application rate, or soybean injury may occur. (6) Injury may occur to soybeans planted the year following application on soils having a calcareous surface layer. (7) Do not plant sugar beets, tobacco, vegetables (including dry beans), spring-seeded small grains, or small-seeded legumes and grasses the year following application, or injury may occur.

SORGHUM AND SORGHUM-SUDAN HYBRIDS (GRAIN AND FORAGE TYPES)

Preplant Surface-Applied (Broadleaf and grass control): Use on medium- and fine-textured soil with minimum-tillage or no-tillage systems only in CO, IL, IN, IA, KS, KY, MN, MO, MT, NE, ND, SD, WI, and WY. Apply ⅔ the recommended rate of AATREX shown in Table 3 for the soil texture as a split treatment 30-45 days prior to planting and the remainder at planting. Applications made less than 30 days prior to planting may be as either a split or single treatment. Under dry conditions, irrigation after application is recommended to move AATREX into the soil.

If weeds are present at time of treatment, apply in a tank mix combination with a contact herbicide (for example, Gramoxone or Roundup). Observe directions for use, precautions, and restrictions on the label of the contact herbicide.

Note: To the extent possible, do not move treated soil out of the row or move untreated soil to the surface during planting, or weed control will be diminished.

Preplant Incorporated (Broadleaf and grass control): Broadcast in spring after plowing at rate shown in Table 3. Apply before, during, or after final seedbed preparation. If soil is tilled or worked after application, avoid deep incorporation. For best results, apply within two weeks prior to planting.

Preemergence (Broadleaf and grass control): Apply during or shortly after planting, but prior to weed or crop emergence at rate shown in Table 3.

Table 3: Preplant Surface, Preplant Incorporated, and Preemergence Broadleaf and Grass Weed Control in Sorghum*

Soil texture	Organic matter	Broadcast rate/A
COARSE Sand, loamy sand, sandy loam	any level	DO NOT USE (except for preemergence use on bedded sorghum in AZ and CA)
MEDIUM AND FINE Loam, silt loam, silt, clay loam, sandy clay loam, silty clay loam, sandy clay silty clay, clay	less than 1%	
	1-1.5%	2-2.5 lbs.
	more than 1.5%	2.5-3 lbs.

* Do not apply preplant surface or preplant incorporated in AL, AR, FL, GA, LA, MS, NC, NM, OK, SC, TN, or TX. Do not apply preemergence in NM, OK, or TX, except in northeast OK and the TX Gulf Coast.

In case of planting failure, sorghum or corn may be replanted. Do not make a second broadcast application, or injury may occur. If originally applied in a band and sorghum or corn is replanted in untreated row middles, this product may be applied in a band to the second planting.

Preemergence broadleaf weed control in furrow irrigated bedded sorghum (AZ and CA only): For preemergence control of many broadleaf weeds, broadcast 1-1.5 lbs./A after bed preparation, during or after planting, but before sorghum and weeds emerge and before the first furrow irrigation. Follow with several regular irrigations, making sure to thoroughly wet all soil.

Precautions for preemergence application to furrow irrigated bedded sorghum in AZ and CA: To avoid possible sorghum injury, do not use on sand or loamy sand soil or on sorghum planted in furrows. Applications to sorghum growing on alkali soils or where cuts, fills, or erosion have exposed calcareous or alkali subsoils may cause crop injury. In case of crop failure, do not replant sorghum for 8 months following application. Corn may be planted immediately.

Postemergence (Broadleaf and grass control): Apply at rate shown in Table 4 before weeds exceed 1.5 inches in height. Apply up to close-in.

Table 4: Postemergence Broadleaf and Grass Weed Control in Sorghum

Soil Texture	Minimum sorghum height at treatment	Broadcast rate/A
Sand or loamy sand	DO NOT USE	
Sandy loam	See directions for postemergence broadleaf weed control.	
Silty loam to sandy clay loam	Completely emerged	2.5-3 lbs.
Olton and Pullman clay soil	6 inches	2.5-3 lbs.
Silty clay loam and finer textured soil	Completely emerged	3.75 lbs.

Postemergence broadleaf weed control with emulsifiable oil or oil concentrate in water. Broadcast 1.5 lbs./A for control of many broadleaf weeds. Apply before pigweed and lambsquarters reach 6 inches in height and before all other weeds reach 4 inches in height. In CO, western KS, NM, OK, TX, and desert regins of AZ and CA, apply when sorghum is 6-10 inches in height, but before it reaches boot stage. In all other areas, apply after sorghum reaches the 3-leaf stage. Add 1 gal. of emulsifiable oil/A for ground application and 0.5 gal./A for aerial application, or add 1 qt./A of oil concentrate for ground application. A cultivation may be necessary if all weeds are not controlled or if weeds regrow.

For the list of weeds controlled, see **AATREX 80W Applied Alone—Corn or Grain Sorghum—Postemergence with Emulsifiable Oil or Oil Concentrate in Water.**

Precautions for applications with emulsifiable oil or oil concentrate in water: See "Precautions for applications with emulsifiable oil or oil concentrate in water" in Corn section.

Postemergence broadleaf weed control with surfactant (CO, western KS, NM, OK, TX, and desert regions of AZ and CA only): Broadcast 1.5 lbs./A plus 0.75-1.5 pts. of surfactant after sorghum reaches 6 inches in height, but before weeds exceed 1.5 inches in height. Apply only on sandy loam and finer textured soil.

Precautions for all applications to sorghum: (1) Heavy rain immediately followed application tends to cause excessive concentrations of herbicide in seed furrow, resulting in possible crop injury. Do not apply to furrow-planted sorghum until furrows are leveled (plowed in). Level deep planter marks or seed furrows before application. (2) Application to sorghum growing under stress caused by minor element deficiency to sorghum growing on highly calcareous soil may result in crop injury. (3) Following harvest, plow (moldboard or disk-plow) and thoroughly till soil in fall or spring to minimize pos-

sible injury to spring-seeded rotational crops, regardless of rate used. (4) Injury may occur if both this herbicide, preplant surface, preplant incorporated, or preemergence, and an at-planting systemic insecticide applied in-furrow are used.

Note: Do not graze or feed forage from treated areas for 21 days following application, or illegal residues may result.

Rotational crops: See "Rotational crops" in Corn section.

TANK MIXTURES FOR GRAIN SORGHUM

Igran® 80W: Use as tank mixture for control of certain broadleaf and grass weeds in grain sorghum. Refer to Igran 80W label for all directions, weeds controlled, precautions, and limitations.

Dual 8E: Use as tank mixture with Dual 8E for control of those weeds listed on the Dual label, as well as on this label. Use this tank mixture only on sorghum seed treated with Concep.®. Refer to the Dual 8E label for all directions, precautions, and limitations.

WINTER WEED CONTROL IN TEXAS

For postemergence control of winter weeds only, such as henbit, seedling dock, and annual thistle on fall bedded land in the Gulf Coast and Blacklands of Texas. Apply 1-1.25 lbs./A postemergence to the weeds in November or December to land that will be planted to corn, grain sorghum, or forage sorghum the following spring. For best results, add a suitable surfactant, such as X-77, at the rate of 0.5% of the spray volume, an emulsifiable oil at the rate of 1.0% of the spray volume, or an oil concentrate at the rate of 1 qt. per acre.

Normal weed control programs may be used in the following corn, grain sorghum, or forage sorghum crop.

Note: Do not plant any crops except corn, grain sorghum, or forage sorghum the spring following this treatment, or illegal residues may result.

AATREX ALONE—CHEMICAL FALLOW

Wheat-Sorghum-Fallow: To control annual broadleaf and grass weeds following wheat harvest and in the following sorghum crop when grown under minimum tillage, broadcast 3.75 lbs./A to wheat stubble immediately following wheat harvest. If weeds are present at application, remove them with a sweep plow or other suitable implement after application, or use an approved contact herbicide before or after the application of AATREX. Plant sorghum into wheat stubble the following spring with minimum soil disturbance. Use a surface planter or a planter leaving a shallow furrow. If weeds are present at planting, remove them with a sweep plow or other suitable implement before planting.

For the list of weeds controlled, see **AATREX 80W Applied Alone—Corn or Grain Sorghum—Preplant Surface-Applied, Preplant Incorporated, or Preemergence.**

Precautions: (1) Use only on silt loam or finer textured soil, or crop injury may result. (2) Wheat-sorghum-fallow cropping sequence must be followed. (3) Do not apply following sorghum harvest.

Note: to avoid illegal residues, do not graze or feed forage from treated area to livestock. To avoid illegal residues and crop injury, do not plant any crop other than those on this label within 18 months following treatment.

Wheat-Corn-Fallow (KS, NE): This product controls cheatgrass (downy brome, chess), kochia, mustards, pigweed, Russian thistle, wild lettuce, wild sunflower, and volunteer wheat during period after wheat harvest. Weed control may extend into following corn crop grown under minimum tillage.

Follow directions for use, notes, and precautions in the "Wheat-Sorghum-Fallow" section above, substituting corn for references to sorghum.

Wheat-Fallow-Wheat (CO, KS, MT, NE, ND, SD, and WY): For preemergence control of cheatgrass (downy brome, chess), common lambsquarters, field pennycress, kochia, mustard, Russian thistle, wild lettuce, and suppression of volunteer wheat during fallow period of a wheat-fallow-wheat rotation, broadcast 0.6-1.25 lbs./A on all soils except those listed under "Precautions." For control of pigweed and wild sunflower, use the higher rate. Apply to stubble ground. Treat only once during same fallow period.

TANK MIXTURES FOR CHEMICAL FALLOW

Wheat-Sorghum-Fallow or **Wheat-Corn-Fallow (KS, NE)**

Igran 80W: Use as tank mixture for control of certain broadleaf and grass weeds in wheat-sorghum-fallow or wheat-corn-fallow rotations. Refer to Igran 80W label for directions, weeds controlled, precautions, and limitations.

Gramoxone: If weeds are present at application, a tank mix with Gramoxone may be used. Broadcast 3.75 lbs. of AATREX 80W plus 1.5-2.5 pts. of Gramoxone in 20-60 gals. of water/A by ground equipment. Add 0.5-1 pt. of a nonionic surfactant, such as X-77, per 100 gals. of spray mixture. Add AATREX to spray tank first and thoroughly mix with water. Then add Gramoxone, followed by surfactant. Use the 2.5 pt. rate of Gramoxone if weeds are 4-6 inches tall. This mixture will not control weeds taller than 6 inches. Apply to stubble ground. Treat only once during same fallow period. Refer to Gramoxone label for further directions, precautions, and limitations.

Wheat-Fallow-Wheat (CO, KS, MT, NE, ND, SD, and WY)

Igran 80W: Use as tank mixture for control of certain broadleaf and grass weeds in wheat-fallow-wheat rotations. Refer to Igran 80W label directions, weeds controlled, precautions, and limitations.

Gramoxone: If weeds are present at application, a tank mix with Gramoxone may be used. Broadcast 0.6-1.25 lbs. of AATREX 80W plus 1.5-2.5 pts. of Gramoxone in 20-60 gals. of water/A by ground equipment. Add 0.5-1 pt. of a nonionic surfactant, such as X-77, per 100 gals. of spray mixture. Add AATREX to spray tank first and thoroughly mix with water. Then add Gramoxone, followed by surfactant. Use the 2.5 pt. rate of Gramoxone if weeds are 4-6 inches tall. This mixture will not control weeds taller than 6 inches. Apply to stubble ground. Treat only once during same fallow period. Refer to Gramoxone label for further directions, precautions, and limitations.

If weeds are present at application and this product is used alone, use either an approved contact herbicide before or after treatment, or tillage after treatment.

Use tillage to control weeds which escape during fallow period. Till before planting. For this product applied alone or in tank mixture with Gramoxone, plant at least 2 inches deep and 12 months or more after application.

Precautions: To avoid crop injury, (1) Do not use on sand soil. (2) Do not treat eroded hillsides, caliche and rocky outcroppings, or exposed calcareous subsoil. (3) Do not treat soils of the Rosebud and Canyon Series in western NE and adjoining counties in CO and WY. (4) Do not treat soils with calcareous surface layers. (5) Avoid spray overlap.

Note: Do not graze treated areas within 6 months after application, or illegal residues may result.

Aerial application: In order to assure that spray will be controllable within the target area when used according to label directions, make applications at a maximum height of 10 ft., using low drift nozzles at a maximum pressure of 40 psi, and restrict application to periods when wind speed does not exceed 10 mph. To assure that spray will not adversely affect adjacent sensitive nontarget plants, apply AATREX 80W alone by aircraft at a minimum upwind distance of 400 ft. from sensitive plants, or apply AATREX 80W + Igran 80W at a minimum upwind distance of 500 ft. from sensitive plants. When applying AATREX 80W +Igran 80W by aircraft, use a minimum of 5 gals. of water per acre.

PROSO MILLET

For preplant incorporated or preemergence control of foxtails, kochia, common lambsquarters, redroot pigweed, Russian thistle, slimleaf lambsquarters and spotted spurge, apply 0.6-1.25 lbs./A on soils with less than 2% organic matter, or apply 1.25-2.5 lbs./A on soils with 2% or higher organic matter.

Precaution: To avoid crop injury and illegal residues, make only one application per year.

Rotation crops: See "Rotational crops" in Corn section.

RANGELAND

To control certain weeds, including cheatgrass (downy brome, chess), common (annual) broomweed, little barley, medusahead, sagewort, and tumble mustard. Use lower rate where cheatgrass is a major problem weed; use higher rate for other weeds.

Poisonous plants: Tansymustard, pigweeds, lambsquarters, tumble mustard, cocklebur, and fiddleneck are controlled at 1.25 lbs./A. Other mustards (*Brassica* spp.), groundcherry (annual), broomweed, and threeawn or poverty grass (*Aristida* spp.) are partially controlled at 1.25 lbs./A and controlled at 2.5 lbs./A. Sneezeweed and dogbane are partially controlled at 2.5 lbs./A.

Aerial application: In order to assure that spray will be controllable within the target area when used according to label directions, make applications at a maximum height of 10 ft. above vegetation, using low drift nozzles at a maximum pressure of 40 psi, and restrict application to periods when wind speed does not exceed 10 mph. To assure that spray will not adversely affect adjacent sensitive nontarget plants, apply AATREX 80W by aircraft at a minimum upwind distance of 400 ft. from sensitive plants.

Establishment of perennial range grasses in West (CA, ID, NV, OR, UT, and WA): Broadcast 1-1.25 lbs./A in fall after rains begin but before ground freezes. Perennial range grass may be seeded when rain begins the next fall following treatment. Plant in deep furrows such as those made by a deep furrow rangeland drill. Treated area may be grazed 7 months after seeding. However, to provide sufficient time for establishment of desirable range grasses, treated areas should not be grazed until the fall following seeding.

Renovation of existing stands of perennial range grasses.

Central Great Plains (CO, KS, MT, NE, ND, SD, WY): Broadcast 0.6-1.25 lbs./A in fall before ground freezes.

Southwest (AR, KS, LA, MS, NM, OK, TX): Apply 1.25 lbs./A AATREX 80W in early spring, before regrowth to estabished rangeland or pasture species such as Asiatic (Old World) bluestems, little and big bluestems, buffalograss, switchgrass, and Indiangrass.

West (CA, ID, NV, OR, UT, WA): Do not graze area to be treated for 6 months prior to application. Broadcast 0.6-0.9 lb./A by ground equipment in fall before ground freezes.

Bluegrass and intermediate wheatgrass may be injured by this product at recommended rates.

Notes: To avoid illegal residues, (1) Do not apply more than once per year. (2) Apply only on neutral to acid soils. (3) Do not cut or feed range grass hay. (4) Do not graze treated areas within 3 months following spring application, or within 7 months following fall application. (5) Do not dump or spill product or dispose of container within reach of livestock.

Pure stands of Swithchgrass and Big Bluestem (IA, IL, IN, KS, MN, MO, ND, NE, OH, SD, and WI): For control of those weeds listed under **AATREX 80W Applied Alone—Corn or Grain Sorghum—Preplant Surface-Applied, Preplant Incorporated, or Preemergence.**

Use only on loam, silt loam, silty clay loam, clay loam, and silty clay soils with at least 1 percent organic matter.

Establishment: Broadcast 2.5 lbs./A after planting but before weeds and crop emerge. Prepare a good, firm seedbed. Plant 1/2 inch deep with a grassland drill (preferred method) or a conventional drill. If a conventional drill is used on prepared seedbeds, remove all tension from the disk openers. For best results, cultipack or roll after planting. Clip weeds that escape in July or August. Avoid clipping switchgrass or big bluestream seedlings.

Renovation of existing stands: Broadcast 2.5 lbs./A in early spring before regrowth of the crops, and before weed emergence.

Notes: To avoid illegal residues, (1) Do not cut for hay. (2) Do not graze treated areas within 4 months following application at seeding or 3 months following application to established switchgrass or big bluestem. (3) Treat only once per year.

Rotational crops: See "Rotational crops" in the Corn section. Do not plant small-seeded legumes the year following application, or injury may occur.

ROADSIDES

To control certain annual weeds in established perennial grasses along roadsides in CO, KS, MT, NE, ND, SD, and WY, including cheatgrass (downy brome, chess), common (annual) broomweed, little barley, medusahead, sagewort, and tumble mustard, broadcast 1.25 lbs./A in a minimum of 10 gallons of water by ground equipment in the fall before ground freezes, or after thawing in the spring, but before the established grasses green-up and before weeds emerge. Examples of desirable established grasses include big bluestem, bluegrama, bromegrass, buffalograss, crested wheatgrass, Indiangrass, little bluestem, side-oats grama, switchgrass, and Western wheatgrass. Apply only once a year. Temporary discoloration or other form of injury to the desirable perennial grasses may occur following application.

Notes: To avoid illegal residues, (1) Do not cut or feed roadside grass hay. (2) Do not allow livestock to graze treated areas.

SUGARCANE

For control of many broadleaf and grass weeds, including amaranths, crabgrass, fireweed, Flora's paintbrush, foxtails, junglerice and wiregrass, broadcast 2.5-5 lbs. of AATREX 80W at time of planting or ratooning, but before sugarcane emerges. Broadcast aerially in a minimum of 5 gals. of spray per acre, or broadcast or band by ground equipment in a minimum of 20 gals. per acre, unless indicated otherwise. One additional application may be made over the sugarcane as it emerges, and two additional applications may be made interline after emergence as directed sprays. Repeat treatments, where needed, may be applied broadcast, band, or interline as suggested with the final application being prior to close-in. Do not exceed the rate of herbicide suggested for any one crop of sugarcane.

Note: Where high rates of AATREX are used, apply in a minimum of 1 gal. of water for each 1 lb. of product applied per acre.

Aerial application: In order to assure that spray will be controllable within the target area when used according to label directions, make applications at a maximum height of 10 ft., using low drift nozzles at a maximum pressure of 40 psi, and restrict application to periods when wind speed does not exceed 10 mph. To assure tnat spray will not adversely affect adjacent sensitive nontarget plants, apply AATREX 80W alone by aircraft at a minimum upwind distance of 400 ft. from sensitive plants.

Florida

For control of emerged pellitory weed: Apply 0.5-0.75 lb./A in at least 40 gals. of water as a directed spray by ground equipment prior to close-in. Add 4 qts. of surfactant for each 100 gals. of spray. Thoroughly cover weed foliage.

For control of alexandergrass, large crabgrass, pellitory (artillery) weed, and spiny amaranth, use one of the following methods at planting or ratooning:

1. Apply 5 lbs./A preemergence. Follow with one or two applications, as needed, postemergence to sugarcane and weeds, at 2.5 lbs./A. Treat before weeds exceed 1.5 inches in height.
2. Apply 1-3 times, as needed, at 2.5 lbs./A postemergence to sugarcane and weeds. Treat before weeds exceed 1.5 inches in height.

Louisiana

For control of annual weeds during the summer fallow period, apply 2.5 lbs./A to weed-free beds immediately after bed formation. Follow normal weed control program after planting.

Precautions: To avoid crop injury, (1) Do not apply more than 12.5 lbs./A to any one crop of sugarcane. (2) If making 2.5 lbs./A application during the summer fallow period, do not exceed 10 lbs./A during the remainder of the growing season, or illegal residues may result.

Texas

Use AATREX 80W for control of barnyardgras, pigweed, purslane and sunflower, in plant or ratoon sugarcane.

Apply 5 lbs./A of AATREX 80W preemergence. Follow with one or two applications, as needed, at 3.75 lbs./A postemergence to sugarcane and weeds.

For best results when weeds are emerged, add a nonionic surfactant at a concentration of 2 qts./100 gals. to the spray and apply before weeds exceed 1.5 inches in height.

Precautions: (1) Injury to sugarcane may occur when under moisture stress, when soil is of low adsorptive capacity, or when land is first cropped to sugarcane. (2) Do not apply after close-in. (3) Do not apply more than 12.5 lbs./A to any one crop of sugarcane, or crop injury may result.

TURFGRASSES FOR SOD (FLORIDA ONLY)

St. Augustinegrass, Centipedegrass, and Zoysia Grass

Broadcast 2.5-5 lbs./A according to soil texture to control those weeds listed under **AATREX 80W Applied Alone—Corn or Grain Sorghum—Preplant Surface-Applied, Preplant Incorporated, or Preemergence.**

Muck or peat	5 lbs.	Old beds: Within 2 days after lifting sod
		New beds: 3-4 days after sprigging or plugging
Sandy soil	2.5 lbs.	Old beds: Within 2 days after lifting sod
		New beds: 7-10 days after sprigging or plugging

If weeds regrow, apply an additional 2.5 lbs./A on muck or peat, or 1.25 lbs./A on sandy soil.

Precautions: To avoid crop injury, (1) Do not apply within 30 days prior to cutting or lifting. (2) Do not apply in combination with surfactants or other spray additives. (3) Use only on turfgrass reasonably free of infestations of insects, nematodes, and diseases. (4) On newly sprigged turfgrass, temporary slowing of growth may follow application.

TURFGRASS FOR FAIRWAYS, LAWNS, SOD PRODUCTION* AND SIMILAR AREAS

* In states other than Florida. For use on turfgrass for sod in Florida, see "Turfgrasses for Sod (Florida only)" section above.

Bermudagrass, Centipedegrass, St. Augustinegrass, and Zoysia Grass

Apply AATREX 80W after October 1 before emergence of winter annual weeds for control of annual bluegrass, burclover, carpet burweed, chickweed, corn speedwell, henbit, hop clover, and spurweed. AATREX 80W will control annual bluegrass even if it is emerged at time of treatment. For control of summer annual weeds listed in the preemergence portion of the AATREX 80W Applied Alone—Corn or Grain Sorghum section of this label, also apply AATREX 80W in late winter before the weeds emerge. Apply in a minimum of 15 gals. of water per acre or 1 gal. per 1,000 sq. ft.

Where annual bluegrass is the major weed, use 1.25 lbs. of AATREX 80W/A (0.5 oz. per 1,000 sq. ft.). Use 2.5 lbs./A (1 oz. per 1,000 sq. ft.) for control of the other weeds named above. Do not exceed 1.25 lbs./A per treatment on newly sprigged turfgrass or on hybrid bermudagrass such as Tiflawn, Tifway, and Ormond.

For continued summer annual weed control, apply another 1.25 lbs./A at least 30 days after the previous application, but not after April 15. However, do not make more than two applications of this product per year.

Precautions: On newly sprigged turfgrass and hybrid bermudagrass, temporary slowing of growth and yellowing may occur following application. To avoid turf injury, (1) Use only on turfgrass reasonably free of infestations of insects, nematodes, and diseases. (2) Do not use on golf greens. (3) Do not use north of NC (except may be used in Virginia Coastal Plains) or west of the high rainfall areas of eastern OK and eastern TX. (4) Do not use on muck or alkaline soils. (5) Do not apply over the rooting area of trees or ornamentals not listed on this label. (6) Do not overseed with desirable turfgrass within 4 months before or 6 months after treatment. (7) Do not apply this product to newly seeded bermudagrass until it has overwintered and has a well-developed rhizome system. Do not exceed 2.2 lbs. product/A with 12 months of seeding bermudagrass.

Note: Do not graze or feed turf clippings to animals, or illegal residues may result.

MACADAMIA NUTS

For preemergence control of many broadleaf and grass weeds, including crabgrass, foxtail, wiregrass, Flora's paintbrush, spanishneedles, and fireweed, broadcast 2.5-5 lbs./A before harvest and before weeds emerge. Repeat as necessary. Do not spray when nuts are on ground during harvest period. Do not apply by air.

PINEAPPLE

For control of purslane, spanishneedles, annual grasses, annual bindweed, ageratum, amaranth, rattlepod, Flora's paintbrush, fireweed, spurge, indigo, and papalo, apply up to 8 lbs./A as a blanket spray immediately after planting or following harvest. Make additional blanket or interspace application of up to 2 lbs./A at 1-2 month intervals as needed, prior to differentiation. Apply in 20-40 gals. of water/A to assure thorough coverage.

Precautions: (1) Do not apply more than 37.5 lbs./A per cycle as crop injury or illegal residues may result. (2) Repeated monthly applications to plant foliage may slow plant growth and delay fruiting.

Note: Do not apply within 45 days of fruit harvest, or forage harvest if forage is to be fed to livestock, or illegal residues may result.

GUAVA

Use only on established plantings which are at least 18 months old. Apply as a directed spray at 2.5-5 lbs./A of AATREX 80W in 20-50 gals. of spray mix preemergence or early postemergence to weeds. When applying postemergence, the use of a surfactant and greater spray volume (80-100 gals. of spray mix per acre) may enhance weed control. This product controls many annual broadleaf and grass weeds, including fireweed, purslane, scarlet pimpernel, spanishneedles, and sowthistle.

Notes: To avoid illegal residues, (1) Do not allow spray to contact foliage or fruit. (2) Do not apply more frequently than at 4 month intervals. (3) Do not apply more than 10 lbs. of AATREX 80W per year.

CONIFERS

For control of annual broadleaf and grass weeds prior to transplanting, after transplanting or in established conifers [including Douglas fir, grand fir, noble fir, white fir, Austrian pine, bishop pine, Jeffrey pine, knobcone pine, loblolly pine, lodgepole pine (shore pine), monterey pine, ponderosa pine, Scotch pine, slash pine, blue spruce, and Sitka spruce]: Broadcast 2.5-5 lbs. in a minimum of 5 gals. of water per acre by air or 10 gals. by ground before weeds are 1.5 inches tall. Apply to established trees between fall and early spring while trees are dormant. For new transplants, apply during or soon after transplanting. For applications prior to transplanting, allow sufficient precipitation to activate AATREX before transplanting. In areas where spring and summer rainfall is inadequate to activate AATREX, apply during fall prior to spring transplanting.

For the list of weeds controlled, see **AATREX 80W Applied Alone—Corn or Grain Sorghum—Preplant Surface-Applied, Preplant Incorporated, or Preemergence.**

Quackgrass control: Broadcast 5 lbs. in a minimum of 5 gals. of water per acre by air or 10 gals. by ground between fall and early spring while trees are dormant and before quackgrass is more than 1.5 inches tall.

Precautions: (1) In areas west of the Rocky Mountains (except the Great Basin), grazing may begin 7 months after a fall application or 3 months after a winter or spring application. (2) To prevent illegal residues, do not graze treated areas of the Great Basin, or areas east of the Rocky Montains. (3) Temporary injury to trees may occur following use of AATREX on coarse-textured soil. (4) To avoid crop injury, do not apply to seedbeds. (5) Also apply only once per year.

Aerial application: In order to assure that spray will be controllable within the target area when used according to label directions, make applications at a maximum height of 10 ft. above vegetation, using low drift nozzles at a maximum pressure of 40 psi, and restrict application to periods when wind speed does not exceed 10 mph. To assure that spray will not adversely affect adjacent sensitive nontarget plants, apply AATREX 80W by aircraft at a minimum upwind distance of 400 ft. from sensitive plants.

NONSELECTIVE WEED CONTROL ON NONCROP LAND

This product provides long-term weed control on industrial sites, highway medians and shoulders, railroad rights-of-way, lumberyards, petroleum tank farms, and in noncrop areas on farms, such as around buildings, equipment and fuel storage areas, along fences, and lanes. Do not use near adjacent desirable plants or in greenhouses, or injury may occur.

Apply before or soon after weeds begin growth. Make postemergence applications when weeds are young and actively growing. Use the higher rates on fine clay and muck soils. Use sufficient water to assure thorough coverage. Use at least 1 gal. of water for each pound of product; use more if practical.

Note: Do not allow livestock to graze treated areas, or illegal residues may result.

Annual broadleaf and grass weeds (including barnyardgrass, cheatgrass, crabgrass, lambsquarters, foxtail, ragweed, puncturevine, and turkey mullein): Broadcast 6-12.5 lbs./A.

Hard-to-kill annual and many perennial broadleaf and grass weeds (including bluegrass, burdock, Canada thistle, dogfennel, orchardgrass, plantain, quackgrass, purpletop, redtop, and smooth brome): Broadcast 12.5-25 lbs./A.

Hard-to-kill biennial and perennial weeds (including bullthistle and sowthistle): Broadcast 25-50 lbs./A.

For longer residual control in regions of high rainfall and a long growing season, broadcast 25-50 lbs./A.

For small areas, 4.5 oz. per 1,000 sq. ft. is equivalent to 12.5 lbs./A.

STORAGE AND DISPOSAL

Pesticide Storage and Disposal

Store in a dry place. Do not contaminate water, food, or feed by storage, disposal, or cleaning of equipment. Wastes resulting from the use of this product may be disposed of on-site or at an approved waste disposal facility.

Container Disposal

Completely empty bag into application equipment. Dispose of empty bag in a sanitary landfill or by incineration, or by open burning, if allowed by state and local authorities. If burned, keep out of smoke.

For minor spills, leaks, etc., follow all precautions indicated on this label and clean up immediately. Take special care to avoid contamination of equipment and facilities during cleanup procedures and disposal of wastes. In the event of a major spill, fire or other emergency, call (919) 292-7100 day or night.

PRECAUTIONARY STATEMENTS

HAZARDS TO HUMANS AND DOMESTIC ANIMALS

CAUTION

Harmful if swallowed, inhaled, or absorbed through skin. Do not breathe dust or spray mist. Avoid contact with eyes, skin or clothing.

STATEMENT OF PRACTICAL TREATMENT

If swallowed: Call a physician or Poison Control Center immediately. Drink 1 or 2 glasses of water and induce vomiting by touching back of throat with finger, or, if available, by adminstering syrup of ipecac. Do not induce vomiting or give anything by mouth to an unconscious person.

If inhaled: Remove victim to fresh air. If not breathing, give artificial respiration, preferably mouth-to-mouth. Get medical attention.

If on skin: Wash with plenty of soap and water. Get medical attention.

If in eyes: Flush eyes with *plenty* of water. Call physician if irritation persists.

Note to Physician: There is no specific antidote for atrazine. If this product is ingested, induce emesis or lavage stomach. The use of an aqueous slurry of activated charcoal may be considered.

ENVIRONMENTAL HAZARDS

Atrazine can travel (seep or leach) through soil and can enter ground water which may be used as drinking water. Atrazine has been found in ground water. Users are advised not to apply atrazine to sand and loamy sand soils where the water table (ground water) is close to the surface and where these soils are very permeable, i.e., well-drained. Your local agricultural agencies can provide further information on the type of soil in your area and the location of ground water.

This pesticide is toxic to aquatic invertebrates. Do not apply directly to water or wetlands. Do not apply where runoff is likely to occur. Do not apply when weather conditions favor drift from treated areas. Runoff and drift from treated areas may be hazardous to aquatic organisms in neighboring areas. Do not contaminate water by cleaning of equipment or disposal of wastes.

AATREX® trademark of CIBA—GEIGY for atrazine
Bexton® trademark of Dow Chemical Co. for propachlor
Caliber® trademark of CIBA—GEIGY
Compex® trademark of KALO Agricultural Chemicals, Inc.
Concep® trademark of CIBA—GEIGY
Dual® trademark of CIBA—GEIGY for metolachlor
U.S. Patent No. 3,937,730
Gramoxone® trademark of ICI Americas for paraquat
Igran® trademark of CIBA—GEIGY for terbutryn
Lasso® trademark of Monsanto Company for alachlor
Princep® trademark of CIBA—GEIGY for simazine
Ramrod® trademark of Monsanto Company for propachlor
Roundup® trademark of Monsanto Company
for glyphosate herbicide
Sutan +® trademark of Stauffer Chemical Co. for butylate
Unite® trademark of Hopkins Agricultural Chemical Co.
X-77® trademark of KALO Agricultural Chemicals, Inc.

CGA 7L91J 067

This is a specimen label which was not prepared by CIBA-GEIGY. CIBA-GEIGY is not responsible for the accuracy of the information contained herein. As labels are subject to revision, always carefully read and follow the label on the product container.

AATREX® NINE-O®

HERBICIDE

For season-long weed control in corn and sorghum

For weed control in certain other crops; in noncrop areas; and industrial sites

ACTIVE INGREDIENTS:	
Atrazine: 2-chloro-4-ethylamino-6-isopropylamino-*s*-triazine. . . .	85.5%
Related compounds.	4.5%
INERT INGREDIENTS:	10.0%
TOTAL:	100.0%

EPA Reg. No. 100-585 EPA Est. 100-LA-1

AATREX NINE-O is a water dispersible granule.

KEEP OUT OF REACH OF CHILDREN

CAUTION

See additional precautionary statements at end of label booklet.

See directions for use inside booklet.

AATREX® trademark of CIBA-GEIGY for atrazine

CGA 7L101J 078

DIRECTIONS FOR USE AND CONDITIONS OF SALE AND WARRANTY

IMPORTANT: Read the entire **Directions for Use** and the **Conditions of Sale and Warranty** before using this product.

CONDITIONS OF SALE AND WARRANTY

The **Directions for Use** of this product reflect the opinion of experts based on field use and tests. The directions are believed to be reliable and should be followed carefully. However, it is impossible to eliminate all risks inherently associated with use of this product. Crop injury, ineffectiveness, or other unintended consequences may result because of such factors as weather conditions, presence of other materials, or the manner of use or application all of which are beyond the control of CIBA-GEIGY or the Seller. All such risks shall be assumed by the Buyer.

CIBA-GEIGY warrants that this product conforms to the chemical description on the label and is reasonably fit for the purposes referred to in the **Directions for Use** subject to the inherent risks referred to above. **CIBA-GEIGY makes no other express or implied warranty of Fitness or Merchantability or any other express or implied warranty. In no case shall CIBA-GEIGY or the Seller be liable for consequential, special, or indirect damages resulting from the use or handling of this product.** CIBA-GEIGY and the Seller offer this product, and the Buyer and user accept it, subject to the foregoing **Conditions of Sale and Warranty**, which may be varied only by agreement in writing signed by a duly authorized representative of CIBA-GEIGY.

DIRECTIONS FOR USE

It is a violation of federal law to use this product in a manner inconsistent with its labeling.

FAILURE TO FOLLOW THE DIRECTIONS FOR USE AND PRECAUTIONS ON THIS LABEL MAY RESULT IN POOR WEED CONTROL, CROP INJURY, OR ILLEGAL RESIDUES.

Do not apply this product in such a manner as to directly or through drift expose workers or other persons, except those knowingly involved in the application. The area being treated must be vacated by unprotected persons.

Do not enter treated areas without protective clothing until sprays have dried.

Because certain states may require more restrictive reentry intervals for various crops treated with this product, consult your State Department of Agriculture for further information.

Written or oral warnings must be given to workers who are expected to be in a treated area or in an area about to be treated with this product. Oral warnings must be given which inform workers of areas or fields that may not be entered without specific protective clothing until sprays have dried, and appropriate actions to take in case of accidental exposure as described under **Precautionary Statements** on this label. When oral warnings are given, warnings shall be given in a language customarily understood by workers. Oral warnings must be given if there is reason to believe that written warnings cannot be understood by workers. Written warnings must include the following information: "CAUTION. Area treated with AATREX NINE-O (*date of application*). Do not enter without appropriate protective clothing until sprays have dried. In case of accidental exposure, flush eyes or skin with plenty of water.

Call a physician if irritation persists. Remove and wash contaminated clothing before reuse."

GENERAL INFORMATION

This herbicide controls many annual broadleaf and grass weeds in corn, sorghum, sugarcane, pineapple, and certain other crops specified on this label. It is also effective in noncrop areas and industrial sites for control of most annual and many perennial broadleaf and grass weeds. This product may be applied before or after weeds emerge.

Following many years of continuous use of this product and chemically related products, biotypes of some of the weeds listed on this label have been reported which cannot be effectively controlled by this and related herbicides. Where this is known or suspected and weeds controlled by this product are expected to be present along with resistant biotypes, we recommend the use of this product in combinations or in sequence with other registered herbicides which are not triazines. If only resistant biotypes are expected to be present, use a registered non-triazine herbicide. Consult with your state Agricultural Extension Service for specific recommendations.

Since this product acts mainly through root absorption, its effectiveness depends on moisture to move it into the root zone. If weeds develop, a shallow cultivation or rotary hoeing will generally result in better weed control.

This product is noncorrosive to equipment and metal surfaces, nonflammable, and has low electrical conductivity.

Avoid using near adjacent desirable plants or in greenhouses, or injury may occur.

To avoid spray drift, do not apply under windy conditions. Avoid spray overlap, as crop injury may result.

Where the use directions give a range of rates, use the lower rate on coarse-textured soil and soil low in organic matter; use the higher rate on fine-textured soil and soil high in organic matter.

Note: CIBA-GEIGY does not recommend applications in combination with other herbicides or oils, except as specifically described on the label or in literature published by CIBA-GEIGY.

APPLICATION PROCEDURES

Ground application: Use conventional ground sprayers equipped with nozzles that provide accurate and uniform application. Be certain that nozzles are uniformly spaced and are the same size. Calibrate sprayer before use and recalibrate at the start of each season and when changing carriers. Unless otherwise specified, use a minimum of 10 gals. of spray mixture/A for all preplant incorporated, preplant surface, preemergence, and postemergence applications (with or without oil or surfactant) with ground equipment.

Use a pump with capacity to (1) maintain 35-40 psi at nozzles, (2) provide sufficient agitation in tank to keep mixture in suspension, and (3) to provide a minimum of 20% bypass at all times. Use centrifugal pumps which provide propeller shear action for dispersing and mixing this product. The pump should provide a minimum of 10 gals./minute/100 gal. tank size circulated through a correctly positioned sparger tube or jets.

Use screens to protect the pump and to prevent nozzles from clogging. Screens placed on the suction side of the pump should be *16-mesh or coarser.* Do not place a screen in the recirculation line. Use 50-mesh or coarser screens between the pump and boom, and where required, at the nozzles. Check nozzle manufacturer's recommendations.

For band applications, calculate amount to be applied per acre as follows:

$$\frac{\text{band width in inches}}{\text{row width in inches}} \times \text{broadcast rate per acre} = \text{amount needed per acre of field}$$

Aerial application: Use aerial application only where broadcast applications are specified. Apply in a minimum of 1 gal. of water for each 1 lb. of AATREX NINE-O applied per acre. For postemergence treatments on corn and sorghum, apply recommended rate in a minimum of 2 gals. of water/A. Avoid applications under conditions where uniform coverage cannot be obtained or where excessive spray drift may occur.

Avoid application to humans or animals. Flagmen and loaders should avoid inhalation of spray mist and prolonged contact with skin, and should wash thoroughly before eating and at the end of each day's operation.

Application in water or liquid fertilizer: Nitrogen solution or complete liquid fertilizer may replace all or part of the water as a carrier for preemergence, preplant incorporated, or preplant surface ground application on corn and sorghum. Check the compatibility of this product with liquid fertilizer and/or nitrogen solution as shown below before use. Do not apply in nitrogen solution or complete liquid fertilizer after corn or sorghum emerges, except as noted under **Lay-by treatment** for corn, or crop injury may occur.

Compatibility Test: Since liquid fertilizers can vary, even within the same analysis, always **check compatibility with herbicide(s) each time before use.** Be especially careful when using **complete** suspension or fluid fertilizers as serious compatibility problems are more likely to occur. Commercial application equipment may improve compatibility in some instances. The following test assumes a spray volume of 25 gals. per acre. For other spray volumes, make appropriate changes in the ingredients. Check compatibility using this procedure:

1. Add 1 pint of fertilizer to each of 2 one-quart jars with tight lids.

2. To **one** of the jars add 1/4 tsp. or 1.2 milliliters of a compatibility agent approved for this use, such as Compex® or Unite® (1/4 tsp. is equivalent to 2 pts. per 100 gals. spray). Shake or stir gently to mix.
3. To **both** jars add the appropriate amount of herbicide(s). If more than one herbicide is used, add them separately with dry herbicides first, flowables next, and emulsifiable concentrates last. After each addition, shake or stir gently to thoroughly mix. The appropriate amount of herbicides for this test follows:
   Dry herbicides: For each pound to be applied per acre, add 1.5 level teaspoons to each jar.
   Liquid herbicides: For each pint to be applied per acre, add 0.5 teaspoon or 2.5 milliliters to each jar.
4. After adding all ingredients, put lids on and tighten, and invert each jar ten times to mix. Let the mixtures stand 15 minutes and then look for separation, large flakes, precipitates, gels, heavy oily film on the jar, or other signs of incompatibility. Determine if the compatibility agent is needed in the spray mixture by comparing the two jars. If either mixture separates, but can be remixed readily, the mixture can be sprayed as long as good agitation is used. If the mixtures are incompatible, test the following methods of improving compatibility: (A) slurry the dry herbicide(s) in water before addition, or (B) add 1/2 of the compatibility agent to the fertilizer and the other 1/2 to the emulsifiable concentrate or flowable herbicide before addition to the mixture. If incompatibility is still observed, do not use the mixture.

Application in water plus emulsifiable oil or oil concentrate: Adding emulsifiable oil (petroleum-derived, petroleum-derived oil concentrate, or single or mixed crop-derived oil concentrate) to postemergence water-based sprays in corn and sorghum may improve weed control. However, under certain conditions, the use of either type of oil may seriously injure the crop. To minimize this possibility, observe the following directions:

Use one of the following properly emulsified:
1. A suitable oil concentrate containing at least 1% but not more than 20% suitable emulsifier or surfactant blend.
2. Petroleum-derived oil containing at least 1% suitable emulsifier.

Note: In the event of a compatibility problem when mixing oil with AATREX and water, a compatibility agent such as Compex or Unite should be used. Any of the above oils contaminated with water or other materials can cause compatibility problems and/or crop injury.

Mixing procedures—all uses: (1) Be sure sprayer is clean and not contaminated with any other materials, or crop injury or sprayer clogging may result. (2) Fill tank 1/4 full with clean water, nitrogen solution, or complete liquid fertilizer. (3) Start agitation. (4) Be certain that the agitation system is working properly and creates a rippling or rolling action on the liquid surface. (5) Pour product directly from bag into tank. (6) Continue filling tank until 90% full. Increase agitation if necessary to maintain surface action. (7) Add emulsifiable oil, oil concentrate, or tank mix herbicide(s) after this product is thoroughly suspended. (8) Finish filling tank. (9) Empty tank as completely as possible before refilling to prevent buildup of oil or emulsifiable concentrate residue.

Maintain agitation to avoid separation of materials. (10) If an oil or emulsifiable concentrate film starts to build up in tank, drain it and clean with strong detergent solution or solvent. (11) Clean sprayer thoroughly immediately after use by flushing system with water containing a detergent.

Center pivot sprinkler application

Preemergence or postemergence: Apply AATREX NINE-O alone, or in tank mixtures with other herbicides on this label which are registered for center pivot application, with irrigation water. Apply either after planting before crop and weeds emerge, or after crop emergence, but before lay-by (20-30 inches) and before weeds exceed 1.5 inches in height. Apply this product only through a center pivot irrigation system. Do not apply this product through any other type of irrigation system. Crop injury, lack of effectiveness, or illegal pesticide residues in the crop can result from nonuniform distribution of treated water. If you have questions about calibration, you should contact State Extension Service specialists, equipment manufacturers or other experts, or refer to supplemental literature published by CIBA-GEIGY. Do not connect an irrigation system (including greenhouse systems) used for pesticide application to a public water system unless the pesticide label-prescribed safety devices for public water systems are in place. A person knowledgeable of the chemigation system and responsible for its operation, or under the supervision of the responsible person, shall shut the system down and make necessary adjustments should the need arise.

Operating Instructions
1. The system must contain a functional check valve, vacuum relief valve, and low pressure drain appropriately located on the irrigation pipeline to prevent water-source contamination from backflow.
2. The pesticide injection pipeline must contain a functional, automatic quick-closing check valve to prevent the flow of fluid back toward the injection pump.
3. The pesticide injection pipeline must also contain a functional, normally closed, solenoid-operated valve located on the intake side of the injection pump and connected to the system interlock to prevent fluid from being withdrawn from the supply tank when the irrigation system is either automatically or manually shut down.
4. The system must contain functional interlocking controls to automatically shut off the pesticide injection pump when the water pump motor stops.
5. The irrigation line or water pump must include a functional pressure switch which will stop the water pump motor when the water pressure decreases to the point where pesticide distribution is adversely affected.
6. Systems must use a metering pump, such as a positive displacement injection pump (e.g., diaphragm pump) effectively designed and constructed of materials that are compatible with pesticides and capable of being fitted with a system interlock.
7. Do not apply when wind speed favors drift beyond the area intended for treatment.
8. Prepare a mixture with a minimum of 1 part of water to 1 part herbicide(s) and inject this mixture into the center pivot system. Injecting a larger volume of a more dilute mixture per hour will usually provide more accurate calibration of metering equipment. Maintain sufficient agitation to keep the herbicide in suspension.
9. Meter into irrigation water during entire period of water application.
10. Apply in 1/2-1 inch of water. Use the lower water volume (1/2 inch) on coarser-textured soils and the higher volume (1 inch) on finer-textured soils. More than 1 inch of water at application may reduce weed control by moving the herbicide below the effective zone in the soil.

Precautions for center pivot applications: Where sprinkler distribution patterns do not overlap sufficiently, unacceptable weed control may result. Where sprinkler distribution patterns overlap excessively, crop injury may result.

Within rate ranges in all tables on this label, use the lower rate on soil relatively coarse-textured or low in organic matter; use the higher rate on soil relatively fine-textured or high in organic matter.

AATREX NINE-O APPLIED ALONE—CORN OR GRAIN SORGHUM

Preplant Surface-Applied, Preplant Incorporated, or Preemergence (or Postemergence at 2.2 lbs./A with Oil)

Broadleaf and Grass Weeds Controlled

barnyardgrass (watergrass)	cocklebur	nightshade
giant foxtail	groundcherry	pigweed
green foxtail	jimsonweed	purslane
large (hairy) crabgrass	kochia	ragweed
wild oats	lambsquarters	sicklepod
witchgrass (*Panicum capillare*)	annual morningglory	velvetleaf (buttonweed)
yellow foxtail	mustards	

Postemergence with Emulsifiable Oil or Oil Concentrate in Water (at 1.3 lbs./A)

Broadleaf Weeds Controlled

annual morningglory	lambsquarters	ragweed
cocklebur	mustards	smartweed
jimsonweed	pigweed	wild buckwheat
		velvetleaf

CORN

Preplant Surface-Applied (Broadleaf and grass control): Use on medium- and fine-textured soils with minimum-tillage or no-tillage systems only in CO, IL, IN, IA, KS, KY, MN, MO, MT, NE, ND, SD, WI, and WY. Apply 2/3 the recommended rate of AATREX shown in Table 1 for the soil texture as a split treatment 30-45 days prior to planting and the remainder at planting. Applications made less than 30 days prior to planting may be as either a split or single treatment. On *coarse-textured* soils, do not apply more than two weeks prior to planting.

If weeds are present at the time of treatment, apply in tank mix combination with a contact herbicide (for example, Gramoxone® or Roundup®). Observe directions for use, precautions, and restrictions on the label of the contact herbicide.

Note: To the extent possible, do not move treated soil out of the row or move untreated soil to the surface during planting, or weed control will be diminished.

Preplant Incorporated (Broadleaf and grass control): Broadcast in spring after plowing at rate in Table 1. Apply to the soil and incorporate before, during, or after final seedbed preparation. Avoid deep incorporation. For best results, apply within two weeks prior to planting.

Preemergence (Broadleaf and grass control): Apply during or shortly after planting before weed emergence, at rate in Table 1.

Postemergence (Broadleaf and grass control): Apply before weeds exceed 1.5 inches in height, at rate in Table 1.

Table 1: Broadleaf and Grass Weed Control on Corn*

Soil texture	Broadcast rate/A
COARSE Sand, loamy sand, sandy loam	2.2 lbs.
MEDIUM Loam, silt loam, silt, clay loam, sandy clay loam, silty clay loam, sandy clay, or silty clay with low organic matter	2.6 lbs.
FINE Loam, silt loam, silt, clay loam, sandy clay loam, silty clay loam, sandy clay, or silty clay with medium to high organic matter, and clay (including the dark prairie soils of the Corn Belt)	3.3 lbs.
Peat, muck, and high organic clay (apply postemergence only)	3.3 lbs.

* *Broadleaf control (eastern CO, western KS, western NE, NM, OK Pan Handle, west TX, and eastern WY):* On sand, loamy sand, sandy loam, mild to strongly alkaline soil, and all recently leveled soil, apply no more than 1.3 lbs./A, either preplant surface, preplant incorporated, or preemergence. On other soils in these areas, apply rate in Table 1 for broadleaf and grass control.

Lay-by treatment (Broadleaf and grass control): Broadcast 1.1-2.2 lbs./A before weeds are 1.5 inches tall and before corn is 20-30 inches tall. When using nitrogen solutions, direct the spray to lower 3-4 inches of cornstalks to avoid corn foliage injury.

Postemergence with emulsifiable oil or oil concentrate in water: Add the following volume of one of the type oils indicated for aerial or ground application unless the oil label specifies otherwise:

Type Oil	Ground Application	Aerial Application
Oil concentrate (Crop or Petroleum-derived)	1 qt./A	½-1 qt./A
Petroleum-derived oil	1 gal./A	2 qts./A

Note: Crop-derived or petroleum-derived oil concentrates should contain at least 1% but not more than 20% suitable emulsifier or surfactant blend. Petroleum-derived oils should contain at least 1% suitable emulsifier.

Broadleaf and grass control: For postemergence control of those weeds listed under **Preplant Incorporated and Preemergence,** broadcast 2.2 lbs./A plus emulsifiable oil or oil concentrate after weed emergence, but before weeds reach 1.5 inches in height.

Broadleaf control: For postemergence control of those weeds listed under **Postemergence with emulsifiable oil or oil concentrate in water,** broadcast 1.3 lbs./A plus emulsifiable oil or oil concentrate before pigweed and lambsquarters reach 6 inches in height and before all other weeds reach 4 inches in height. A cultivation may be necessary if all weeds are not controlled or if weeds regrow.

Precautions for applications with emulsifiable oil or oil concentrate in water: (1) Inbred lines or any breeding stock may be severely injured by applications with emulsifiable oil or oil concentrate. (2) Adding other insecticides, herbicides, liquid fertilizers, or other materials is not recommended, because they may cause compatibility problems or crop injury. (3) Store and handle emulsifiable oil and oil concentrate carefully. Oil contaminated with even a small amount of water may not emulsify properly when added to the tank. To avoid crop injury, (4) Do not apply when crop is under stress from prolonged cold, wet weather, poor fertility, or other factors, or when crop is wet and succulent from recent rainfall. (5) Do not make more than one application per season except as recommended for control of yellow nutsedge, quackgrass, and Canada thistle.

Problem weeds

Yellow nutsedge and Canada thistle: This product will control yellow nutsedge (*Cyperus esculentus*) and Canada thistle (*Cirsium arvense*) when applied according to directions. For best results, apply each year until yellow nutsedge or Canada thistle is eliminated or reaches a level of infestation where neither weed is an economic problem. If yellow nutsedge or Canada thistle regrow following last application, cultivate once.

Four alternative methods of use for control of yellow nutsedge and Canada thistle are listed below in order of preference. If other weed species, such as annual grasses, are also expected, use alternative 2 or 4.

1. Broadcast 2.2 lbs./A plus 1 gal. of emulsifiable oil or 1 qt. of oil concentrate. Apply after crop and yellow nutsedge or Canada thistle emerge, but before yellow nutsedge reaches a height of 3 inches or Canada thistle reaches a height of 6 inches. Repeat application before lay-by (20-30 inches), 10-20 days after the first application.
2. Broadcast 2.2 lbs./A preplant incorporated for control of yellow nutsedge, or broadcast 2.2 lbs./A during or shortly after planting, but prior to crop or weed emergence for control of Canada thistle. Follow with an application of 2.2 lbs./A plus 1 gal. of emulsifiable oil or 1 qt. of oil concentrate after corn and weeds emerge, but before yellow nutsedge or Canada thistle reaches a height of 6 inches.
3. Broadcast 4.4 lbs./A plus 1 gal. of emulsifiable oil or 1 qt. of oil concentrate after crop emerges, but prior to lay-by (20-30 inches). Apply after yellow nutsedge and Canada thistle emerge, but before yellow nutsedge is 3 inches tall or Canada thistle is 6 inches tall.
4. Broadcast 4.4 lbs./A before, during, or shortly after planting, but before crop or weed emergence (yellow nutsedge control only).

Note: Do not use emulsifiable oil or oil concentrate when corn is wet or under stress, especially when using 4.4 lbs./A of this product, as crop injury may occur. See "Precautions for applications with emulsifiable oil or oil concentrate in water" for additional directions.

Quackgrass control on land going into corn production

Split application: Broadcast 2.2 lbs./A in fall or spring. Plow 1-3 weeks later. Apply a second application at 2.2 lbs./A in spring before, during, or after planting, but before weeds are 1.5 inches high. This split application will control quackgrass and most annual broadleaf and grass weeds.
Single application: Broadcast 3.3-4.4 lbs./A in fall or spring. Plow 1-3 weeks later.

TANK MIXTURES FOR CORN

This product may be tank mixed with these herbicides for control of certain broadleaf and grass weeds in corn:

Dual® (metolachlor)
Dual + Gramoxone
Dual + Roundup (glyphosate)
Dual + Princep®
Dual + Princep + Gramoxone
Dual + Princep + Roundup
Bexton® or Ramrod® (propachlor)
Lasso® or Lasso EC (alachlor)
Lasso or Lasso EC + Roundup
Lasso or Lasso EC + Gramoxone
Gramoxone
Princep
Princep + Gramoxone
Princep + Roundup
Roundup
Sutan+®

Use tank mix directions appearing on the labels of the above herbicides when tank mixing with this product. Observe all precautions and limitations on labeling of products used in a particular tank mix.

Note: When the labels of the above herbicides refer to atrazine 80W, use equivalent rate of AATREX NINE-O. One lb. of 80W equals 0.9 lb. NINE-O.

Princep 80W, Princep 4L, or Princep Caliber® 90

In addition to the weeds listed under **AATREX NINE-O Applied Alone—Corn and Grain Sorghum—Preplant Surface-Applied, Preplant Incorporated, or Preemergence,** this combination also controls crabgrass, fall panicum, and carpetweed.

Broadcast tank mix before planting, at planting, or after planting, but before crop and weeds emerge, at rates in Table 2. Use the 1:1 ratio for control of most weeds. Use the 1:2 ratio for expected heavy infestations of crabgrass and fall panicum. Cultivate shallowly if weeds develop.

Preplant Surface-Applied: Use on medium- and fine-textured soils with minimum-tillage or no-tillage systems only in CO, IL, IN, IA, KS, KY, MN, MO, MT, NE, ND, SD, WI, and WY. Apply ⅔ the recommended rate of AATREX and Princep shown in Table 2 for the soil textures as a split treatment 30-45 days prior to planting and the remainder at planting. Applications made less than 30 days prior to planting may be as either a split or single treatment. On *coarse-textured* soils, do not apply more than two weeks prior to planting. Refer to the AATREX Alone—Preplant Surface-Applied section of the corn label for additional details.

If weeds are present at time of treatment, apply in a tank mix combination with a contact herbicide (for example, Gramoxone or Roundup). Observe directions for use, precautions, and restrictions on the label of the contact herbicide.

Note: To the extent possible, do not move treated soil out of the row or move untreated soil to the surface during planting, or weed control will be diminished.

Preplant Incorporated: Apply to the soil and incorporate in the spring before, during, or after final seedbed preparation. Avoid deep incorporation. For best results, apply within two weeks prior to planting.

Preemergence: Apply during or shortly after planting, but before crop and weeds emerge.

Refer to Corn sections of this label and to Princep 80W, Princep Caliber 90, or Princep 4L labels for further directions, limitations, and precautions.

Table 2: Tank Mixtures with Princep on Corn

Soil texture	Broadcast rate/A			
	1:1 Ratio*		1:2 Ratio* *	
	This product	Princep 80W[1]	This product	Princep 80W[1]
Sand, loamy sand, sandy loam	1.1 lbs.	1.25 lbs.	0.73 lb.	1.67 lbs.
Loam, silt loam, silt, clay loam, sandy clay loam, silty clay loam, sandy clay, or silty clay with low organic matter	1.3 lbs.	1.5 lbs.	0.88 lb.	2 lbs.

Soil texture	Broadcast rate/A			
	1:1 Ratio*		1:2 Ratio* *	
	This product	Princep 80W[1]	This product	Princep 80W[1]
Loam, silt loam, silt, clay loam, sandy clay loam, silty clay loam, sandy clay, or silty clay with medium to high organic matter, and clay (including dark prairie soils of the Corn Belt)	1.6 lbs.	1.8 lbs.	1.07 lbs.	2.4 lbs.

* For control of most weeds.
* * For control of expected heavy infestations of crabgrass and fall panicum.
[1] When using Princep Caliber 90 or Princep 4L, use equivalent rates. One lb. of Princep 80W equals 0.9 lb. of Princep Caliber 90 or 1.6 pts. of Princep 4L.

Princep 80W, Princep 4L, or Princep Caliber 90 plus Roundup: Use as tank mixture for preemergence and postemergence control of certain broadleaf and grass weeds where corn will be planted directly into a cover crop, established sod, or in previous crop residues. Refer to Roundup label for all directions, weeds controlled, precautions, and limitations.

Princep 80W, Princep 4L, or Princep Caliber 90 plus Gramoxone: Use as tank mixture with Princep and Gramoxone to kill existing vegetation and for residual weed control where corn will be planted directly into a cover crop, established sod, or in previous crop residues. Add this product and Princep to water in spray tank, agitating until thoroughly mixed. Then add Gramoxone and a nonionic surfactant, such as X-77®. Continue agitation during application. Broadcast 1.1-2.2 lbs. of this product plus 1.25-2.5 lbs. Princep 80W [or 2-4 pts. Princep 4L, or 1.1-2.2 lbs. Princep Caliber 90] plus 1.5-2.5 pts. Gramoxone in 20-60 gals. of water per sprayed acre. Apply before, during, or after planting, but before corn emerges. Add 0.5 pt. of a nonionic surfactant, such as X-77, per 100 gals. of spray mixture. Use the 2.5 pt. rate of Gramoxone if existing vegetation is 4-6 inches tall. This mixture will not control weeds taller than 6 inches.

Refer to further limitations and precautions on labels for this product, Princep, and Gramoxone.

Precautions for all applications to corn: (1) To avoid crop injury and illegal residues, do not apply more than 4.4 lbs./A of this product per year. (2) Following harvest, plow (moldboard or disk-plow) and thoroughly till soil in fall or spring to minimize possible injury to spring-seeded rotational crops, regardless of rate used.

Note for all applications to corn: Do not graze or feed forage from treated areas for 21 days following application, or illegal residues may result.

Rotational crops: (1) Do not rotate to any crop except corn or sorghum until the following year, or injury may occur. (2) If applied after June 10, do not rotate with crops other than corn or sorghum the next year, or crop injury may occur. (3) If the broadcast rate, or equivalent band application rate, exceeds 3.3 lbs./A of AATREX NINE-O, or 3.3 lbs./A of AATREX NINE-O plus Princep Caliber 90 (or equivalent of another Princep formulation), a crop of untreated corn or sorghum should precede the next rotational crop. (4) In the High Plains and Intermountain areas of the West where rainfall is sparse and erratic or where irrigation is required, use only when corn or sorghum is to follow corn or sorghum or when a crop of untreated corn or sorghum is to precede other rotational crops. (5) In eastern parts of the Dakotas, KS, western MN, and NE, do not rotate to soybeans if the rate applied to corn or sorghum was more than 2.2 lbs./A or equivalent band application rate, or soybean injury may occur. (6) Injury may occur to soybeans planted the year following application on soils having a calcareous surface layer. (7) Do not plant sugar beets, tobacco, vegetables (including dry beans), spring-seeded small grains, or small-seeded legumes and grasses the year following application, or injury may occur.

SORGHUM AND SORGHUM-SUDAN HYBRIDS (GRAIN AND FORAGE TYPES)

Preplant Surface-Applied (Broadleaf and grass control): Use on medium- and fine-textured soils with minimum-tillage or no-tillage systems only in CO, IL, IN, IA, KS, KY, MN, MO, MT, NE, ND, SD, WI, and WY. Apply ⅔ the recommended rate of AATREX shown in Table 3 for the soil texture as a split treatment 30-45 days prior to planting and the remainder at planting. Applications made less than 30 days prior to planting may be as either a split or single treatment. Under dry conditions, irrigation after application is recommended to move AATREX into the soil.

If weeds are present at time of treatment, apply in a tank mix combination with a contact herbicide (for example, Gramoxone or Roundup). Observe directions for use, precautions, and restrictions on the label of the contact herbicide.

Note: To the extent possible, do not move treated soil out of the row or move untreated soil to the surface during planting, or weed control will be diminished.

Preplant Incorporated (Broadleaf and grass control): Broadcast in spring after plowing at rate shown in Table 3. Apply before, during, or after final seedbed preparation. If soil is tilled or worked after application, avoid deep incorporation. For best results, apply within two weeks prior to planting.

Preemergence (Broadleaf and grass control): Apply during or shortly after planting, but prior to weed or crop emergence at rate shown in Table 3.

Table 3: Preplant Surface, Preplant Incorporated, and Preemergence Broadleaf and Grass Weed Control in Sorghum*

Soil texture	Organic matter	Broadcast rate/A
COARSE Sand, loamy sand, sandy loam	any level	DO NOT USE (except for preemergence use on bedded sorghum in AZ and CA)
MEDIUM AND FINE Loam, silt loam, silt,	less than 1%	
clay loam, sandy clay loam, silty clay	1-1.5%	1.7-2.2 lbs.
loam, sandy clay, silty clay, clay	more than 1.5%	2.2-2.6 lbs.

* Do not apply preplant surface or preplant incorporated in AL, AR, FL, GA, LA, MS, NC, NM, OK, SC, TN, or TX. Do not apply preemergence in NM, OK, or TX, except in northeast OK and the TX Gulf Coast.

In case of planting failure, sorghum or corn may be replanted. Do not make a second broadcast application, or injury may occur. If originally applied in a band and sorghum or corn is replanted in untreated row middles, this product may be applied in a band to the second planting.

Preemergence broadleaf weed control in furrow irrigated bedded sorghum (AZ and CA only): For preemergence control of many broadleaf weeds, broadcast 0.9-1.3 lbs./A after bed preparation, during or after planting, but before sorghum and weeds emerge and before the first furrow irrigation. Follow with several regular irrigations, making sure to thoroughly wet all soil.

Precautions for preemergence application to furrow irrigated bedded sorghum in AZ and CA: To avoid possible sorghum injury, do not use on sand or loamy sand soil or on sorghum planted in furrows. Applications to sorghum growing on alkali soils or where cuts, fills, or erosion have exposed calcareous or alkali subsoils may cause crop injury. In case of crop failure, do not replant sorghum for 8 months following application. Corn may be planted immediately.

Postemergence (Broadleaf and grass control): Apply at rate shown in Table 4 before weeds exceed 1.5 inches in height. Apply up to close-in.

Table 4: Postemergence Broadleaf and Grass Weed Control in Sorghum

Soil texture	Minimum sorghum height at treatment	Broadcast rate/A
Sand or loamy sand	DO NOT USE	
Sandy loam	See directions for postemergence broadleaf weed control.	
Silt loam to sandy clay loam	Completely emerged	2.2-2.6 lbs.
Olton and Pullman clay soil	6 inches	2.2-2.6 lbs.
Silty clay loam and finer textured soil	Completely emerged	3.3 lbs.

Postemergence broadleaf weed control with emulsifiable oil or oil concentrate in water: Broadcast 1.3 lbs/A for control of many broadleaf weeds. Apply before pigweed and lambsquarters reach 6 inches in height and before all other weeds reach 4 inches in height. In CO, western KS, NM, OK, TX, and desert regions of AZ and CA, apply when sorghum is 6-10 inches in height, but before it reaches boot stage. In all other areas, apply after sorghum reaches the 3-leaf stage. Add 1 gal. of emulsifiable oil/A for ground application and 0.5 gal./A for aerial application, or add 1 qt. of oil concentrate for ground application. A cultivation may be necessary if all weeds are not controlled or if weeds regrow.

For the list of weeds controlled, see **AATREX NINE-O Applied Alone—Corn or Grain Sorghum—Postemergence with Emulsifiable Oil or Oil Concentrate in Water.**

Precautions for applications with emulsifiable oil or oil concentrate in water: See "Precautions for applications with emulsifiable oil or oil concentrate in water" in Corn section.

Postemergence broadleaf weed control with surfactant (CO, western KS, NM, OK, TX, and desert regions of AZ and CA only): Broadcast 1.3 lbs./A plus 0.75-1.5 pts. of surfactant after sorghum reaches 6 inches in height, but before weeds exceed 1.5 inches in height. Apply only on sandy loam and finer textured soil.

Precautions for all applications to sorghum: (1) Heavy rain immediately following application tends to cause excessive concentrations of herbicide in seed furrow, resulting in possible crop injury. Do not apply to furrow-planted sorghum until furrows are leveled (plowed in). Level deep planter marks or seed furrows before application. (2) Application to sorghum growing under stress caused by minor element deficiency or to sorghum growing on highly calcareous soil may result in crop injury. (3) Following harvest, plow (moldboard or disk-plow) and thoroughly till soil in fall or spring to minimize possible injury to spring-seeded rotational crops, regardless of rate used. (4) Injury may occur if both this herbicide, preplant surface, preplant incorporated, or preemergence, and an at-planting systemic insecticide applied in-furrow are used.

Note: Do not graze or feed forage from treated areas for 21 days following application, or illegal residues may result.

Rotational crops: See "Rotational crops" at end of Corn section.

TANK MIXTURES FOR GRAIN SORGHUM

Igran® 80W: Use as tank mixture for control of certain broadleaf and grass weeds in grain sorghum. Refer to Igran 80W label for all directions, weeds controlled, precautions, and limitations.

Dual 8E: Use as tank mixture with Dual 8E for control of those weeds listed on the Dual label, as well as on this label. Use this tank mixture only on sorghum seed treated with Concep®. Refer to the Dual 8E label for all directions, precautions, and limitations.

WINTER WEED CONTROL IN TEXAS

For postemergence control of winter weeds only, such as henbit, seedling dock, and annual thistle on fall bedded land in the Gulf Coast and Blacklands of Texas. Apply 0.9-1.1 lbs./A postemergence to the weeds in November or December to land that will be planted to corn, grain sorghum, or forage sorghum the following spring. For best results, add a suitable surfactant, such as X-77, at the rate of 0.5% of the spray volume, an emulsifiable oil at the rate of 1.0% of the spray volume, or an oil concentrate at the rate of 1 qt. per acre.

Normal weed control programs may be used in the following corn, grain sorghum, or forage sorghum crop.

Note: Do not plant any crops except corn, grain sorghum, or forage sorghum the spring following this treatment, or illegal residues may result.

AATREX ALONE—CHEMICAL FALLOW

Wheat-Sorghum-Fallow: To control annual broadleaf and grass weeds following wheat harvest and in the following sorghum crop when grown under minimum tillage, broadcast 3.3 lbs./A to wheat stubble immediately following wheat harvest. If weeds are present at application, remove them with a sweep plow or other suitable implement after application, or use an approved contact herbicide before or after the application of AATREX. Plant sorghum into wheat stubble the following spring with minimum soil disturbance. Use a surface planter or a planter leaving a shallow furrow. If weeds are present at planting, remove them with a sweep plow or other suitable implement before planting.

For the list of weeds controlled, see **AATREX NINE-O Applied Alone—Corn or Grain Sorghum—Preplant Surface-Applied, Preplant Incorporated, or Preemergence.**

Precautions: (1) Use only on silt loam or finer textured soil, or crop injury may result. (2) Wheat-sorghum-fallow cropping sequence must be followed. (3) Do not apply following sorghum harvest.

Notes: To avoid illegal residues, do not graze or feed forage from treated area to livestock. To avoid illegal residues and crop injury, do not plant any crop other than those on this label within 18 months following treatment.

Wheat-Corn-Fallow (KS, NE): This product controls cheatgrass (downy brome, chess), kochia, mustards, pigweed, Russian thistle, wild lettuce, wild sunflower, and volunteer wheat during period after wheat harvest. Weed control may extend into following corn crop grown under minimum tillage.

Follow directions for use, notes, and precautions in the "Wheat-Sorghum-Fallow" section above, substituting corn for references to sorghum.

Wheat-Fallow-Wheat (CO, KS, MT, NE, ND, SD, and WY): For preemergence control of cheatgrass (downy brome, chess), common lambsquarters, field pennycress, kochia, mustard, Russian thistle, wild lettuce, and suppression of volunteer wheat during fallow period of a wheat-fallow-wheat rotation, broadcast 0.5-1.1 lbs./A on all soils except those listed under "Precautions." For control of pigweed and wild sunflower, use the higher rate. Apply to stubble ground. Treat only once during same fallow period.

TANK MIXTURES FOR CHEMICAL FALLOW

Wheat-Sorghum-Fallow or Wheat-Corn-Fallow (KS, NE)

Igran 80W: Use as tank mixture for control of certain broadleaf and grass weeds in wheat-sorghum-fallow or wheat-corn-fallow rotations. Refer to Igran 80W label for directions, weeds controlled, precautions, and limitations.

Gramoxone: If weeds are present at application, a tank mix with Gramoxone may be used. Broadcast 3.3 lbs. of AATREX NINE-O plus 1.5-2.5 pts. of Gramoxone in 20-60 gals. of water/A by ground equipment. Add 0.5-1 pt. of a nonionic surfactant, such as X-77, per 100 gals. of spray mixture. Add AATREX to spray tank first and thoroughly mix with water. Then add Gramoxone, followed by surfactant. Use the 2.5 pt. rate of Gramoxone if weeds are 4-6 inches tall. This mixture will not control weeds taller than 6 inches. Apply to stubble ground. Treat only once during same fallow period. Refer to Gramoxone label for further directions, precautions, and limitations.

Wheat-Fallow-Wheat (CO, KS, MT, NE, ND, SD, and WY)

Igran 80W: Use as tank mixture for control of certain broadleaf and grass weeds in wheat-fallow-wheat rotations. Refer to Igran 80W label for directions, weeds controlled, precautions, and limitations.

Gramoxone: If weeds are present at application, a tank mix with Gramoxone may be used. Broadcast 0.5-1.1 lbs. of AATREX NINE-O plus 1.5-2.5 pts. of Gramoxone in 20-60 gals. of water/A by ground equipment. Add 0.5-1 pt. of a nonionic surfactant, such as X-77, per 100 gals. of spray mixture. Add AATREX to spray tank first and thoroughly mix with water. Then add Gramoxone, followed by surfactant. Use the 2.5 pt. rate of Gramoxone if weeds are 4-6 inches tall. This mixture will not control weeds taller than 6 inches. Apply to stubble ground. Treat only once during same fallow period. Refer to Gramoxone label for further directions, precautions, and limitations.

If weeds are present at application and this product is used alone, use either an approved contact herbicide before or after treatment, or tillage after treatment.

Use tillage to control weeds which escape during fallow period. Till before planting. For this product applied alone or in tank mixture with Gramoxone, plant at least 2 inches deep and 12 months or more after application.

Precautions: To avoid crop injury, (1) Do not use on sand soil. (2) Do not treat eroded hillsides, caliche and rocky out-croppings, or exposed calcareous subsoil. (3) Do not treat soils of the Rosebud and Canyon Series in western NE and adjoining counties in CO and WY. (4) Do not treat soils with calcareous surface layers. (5) Avoid spray overlap.

Note: Do not graze treated areas within 6 months after application, or illegal residues may result.

Aerial application: In order to assure that spray will be controllable within the target area when used according to label directions, make applications at a maximum height of 10 ft., using low drift nozzles at a maximum pressure of 40 psi, and restrict application to periods when wind speed does not exceed 10 mph. To assure that spray will not adversely affect adjacent sensitive nontarget plants, apply AATREX NINE-O alone by aircraft at a minimum upwind distance of 400 ft. from sensitive plants, or apply AATREX NINE-O + Igran 80W at a minimum upwind distance of 500 ft. from sensitive plants. When applying AATREX NINE-O + Igran 80W by aircraft, use a minimum of 5 gals. of water per acre.

PROSO MILLET

For preplant incorporated or preemergence control of foxtails, kochia, common lambsquarters, redroot pigweed, Russian thistle, slimleaf lambsquarters and spotted spurge, apply 0.6-1.1 lbs./A on soils with less than 2% organic matter, or apply 1.1-2.2 lbs./A on soils with 2% or higher organic matter.

Precaution: To avoid crop injury and illegal residues, make only one application per year.

Rotational crops: See "Rotational crops" in Corn section.

RANGELAND

To control certain weeds, including cheatgrass, (downy brome, chess), common (annual) broomweed, little barley, medusahead, sagewort, and tumble mustard: Use lower rate where cheatgrass is a major problem weed; use higher rate for other weeds.

Poisonous plants: Tansymustard, pigweeds, lambsquarters, tumble mustard, cocklebur, and fiddleneck are controlled at 1.1 lbs./A. Other mustards (*Brassica* spp.), groundcherry (annual), broomweed, and threeawn or poverty grass (*Aristida* spp.) are partially controlled at 1.1 lbs./A and controlled at 2.2 lbs./A. Sneezeweed and dogbane are partially controlled at 2.2 lbs./A.

Aerial application: In order to assure that spray will be controllable within the target area when used according to label directions, make applications at a maximum height of 10 ft. above vegetation, using low drift nozzles at a maximum pressure of 40 psi, and restrict application to periods when wind speed does not exceed 10 mph. To assure that spray will not adversely affect adjacent sensitive nontarget plants, apply AATREX NINE-O by aircraft at a minimum upwind distance of 400 ft. from sensitive plants.

Establishment of perennial range grasses in West (CA, ID, NV, OR, UT, WA): Broadcast 0.9-1.1 lbs./A in fall after rains begin but before ground freezes. Perennial range grass may be seeded when rain begins the next fall following treatment. Plant in deep furrows such as those made by a deep furrow rangeland drill. Treated areas may be grazed 7 months after seeding. However, to provide sufficient time for establishment of desirable range grasses, treated areas should not be grazed until the fall following seeding.

Renovation of existing stands of perennial range grasses

Central Great Plains (CO, KS, MT, NE, ND, SD, WY): Broadcast 0.6-1.1 lbs./A in fall before ground freezes.

Southwest (AR, KS, LA, MS, NM, OK, TX): Apply 1.1 lbs./A AATREX NINE-O in early spring, before regrowth, to established rangeland or pasture species such as Asiatic (Old World) bluestems, little and big bluestems, buffalograss, switchgrass, and Indiangrass.

West (CA, ID, NV, OR, UT, WA): Do not graze area to be treated for 6 months prior to application. Broadcast 0.6-0.8 lb./A in fall before ground freezes.

Bluegrass and intermediate wheatgrass may be injured by this product at recommended rates.

Notes: To avoid illegal residues, (1) Do not apply more than once per year. (2) Apply only on neutral to acid soils. (3) Do not cut or feed range grass hay. (4) Do not graze treated areas within 3 months following spring application, or within 7 months following fall application. (5) Do not dump or spill product or dispose of container within reach of livestock.

Pure stands of Switchgrass and Big Bluestem (IA, IL, IN, KS, MN, MO, ND, NE, OH, SD, WI): For control of those weeds listed under **AATREX NINE-O Applied Alone—Corn or Grain Sorghum—Preplant Surface-Applied, Preplant Incorporated, or Preemergence.**

Use only on loam, silt loam, silty clay loam, clay loam, and silty clay soils with at least 1 percent organic matter.

Establishment: Broadcast 2.2 lbs./A after planting but before weeds and crop emerge. Prepare a good, firm seedbed. Plant 1/2 inch deep with a grassland drill (preferred method) or a conventional drill. If a conventional drill is used on prepared seedbeds, remove all tension from the disk openers. For best results, cultipack or roll after planting. Clip weeds that escape in July or August. Avoid clipping switchgrass or big bluestem seedlings.

Renovation of existing stands: Broadcast 2.2 lbs./A in early spring before regrowth of the crop, and before weed emergence.

Notes: To avoid illegal residues, (1) Do not cut for hay. (2) Do not graze treated areas within 4 months following application at seeding or 3 months following application to established switchgrass or big bluestem. (3) Treat only once per year.

Rotational crops: See "Rotational crops" in Corn section. Do not plant small-seeded legumes the year following application, or injury may occur.

ROADSIDES

To control certain annual weeds in established perennial grasses along roadsides in CO, KS, MT, NE, ND, SD, and WY, including cheatgrass (downy brome, chess), common (annual) broomweed, little barley, medusahead, sagewort, and tumble mustard, broadcast 1.1 lbs./A in a minimum of 10 gallons of water by ground equipment in the fall before ground freezes, or after thawing in the spring, but before the established grasses green-up and before weeds emerge. Examples of desirable established grasses include big bluestem, bluegrama, bromegrass, buffalograss, crested wheatgrass, Indiangrass, little bluestem, side-oats grama, switchgrass, and Western wheatgrass. Apply only once per year. Temporary discoloration or other form of injury to the desirable perennial grasses may occur following application.

Notes: To avoid illegal residues, (1) Do not cut or feed roadside grass hay. (2) Do not allow livestock to graze treated areas.

SUGARCANE

For control of many broadleaf and grass weeds, including amaranths, crabgrass, fireweed, Flora's paintbrush, foxtails, junglerice and wiregrass, broadcast 2.2-4.4 lbs./A of AATREX NINE-O at time of planting or ratooning, but before sugarcane emerges. Broadcast aerially in a minimum of 5 gals. of spray per acre, or broadcast or band by ground equipment in a minimum of 20 gals. per acre, unless indicated otherwise. One additional application may be made over the sugarcane as it emerges, and two additional applications may be made interline after emergence as directed sprays. Repeat treatments, where needed, may be applied broadcast, band, or interline as suggested with the final application being prior to close-in. Do not exceed the rate of herbicide suggested for any one crop of sugarcane.

Note: Where high rates of AATREX are used, apply in a minimum of 1 gal. of water for each 1 lb. of product applied per acre.

Aerial application: In order to assure that spray will be controllable within the target area when used according to label directions, make applications at a maximum height of 10 ft., using low drift nozzles at a maximum pressure of 40 psi, and restrict application to periods when wind speed does not exceed 10 mph. To assure that spray will not adversely affect adjacent sensitive nontarget plants, apply AATREX NINE-O alone by aircraft at a minimum upwind distance of 400 ft. from sensitive plants.

Florida

For control of emerged pellitory weed: Apply 0.4-0.6 lb./A in at least 40 gals. of water as a directed spray by ground equipment prior to close-in. Add 4 qts. of surfactant for each 100 gals. of spray. Thoroughly cover weed foliage.

For control of alexandergrass, large crabgrass, pellitory (artillery) weed, and spiny amaranth, use one of the following methods at planting or ratooning:

1. Apply 4.4 lbs./A preemergence. Follow with one or two applications, as needed, postemergence to sugarcane and weeds, at 2.2 lbs./A. Treat before weeds exceed 1.5 inches in height.
2. Apply 1-3 times, as needed, at 2.2 lbs./A postemergence to sugarcane and weeds. Treat before weeds exceed 1.5 inches in height.

Louisiana

For control of annual weeds during the summer fallow period, apply 2.2 lbs./A to weed-free beds immediately after bed formation. Follow normal weed control program after planting.

Precautions: To avoid crop injury, (1) Do not apply more than 11 lbs./A to any one crop of sugarcane. (2) If making a 2.2 lbs./A application during the summer fallow period, do not exceed 8.8 lbs./A during the remainder of the growing season, or illegal residues may result.

Texas

Use AATREX NINE-O for control of barnyardgrass, pigweed, purslane and sunflower, in plant or ratoon sugarcane.

Apply 4.4 lbs./A of AATREX NINE-O preemergence. Follow with one or two applications, as needed, at 3.3 lbs./A postemergence to sugarcane and weeds.

For best results when weeds are emerged, add a nonionic surfactant at a concentration of 2 qts./100 gals. to the spray and apply before weeds exceed 1.5 inches in height.

Precautions: (1) Injury to sugarcane may occur when under moisture stress, when soil is of low absorptive capacity, or when land is first cropped to sugarcane. (2) Do not apply after close-in. (3) Do not apply more than 11 lbs./A to any one crop of sugarcane, or crop injury may result.

TURFGRASSES FOR SOD (FLORIDA ONLY)

St. Augustinegrass, Centipedegrass, and Zoysia Grass

Broadcast 2.2-4.4 lbs./A according to soil texture to control those weeds listed under **AATREX NINE-O Applied Alone—Corn or Grain Sorghum—Preplant Surface-Applied, Preplant Incorporated, or Preemergence.**

Muck or peat	4.4 lbs.	Old beds: Within 2 days after lifting sod
		New beds: 3-4 days after sprigging or plugging
Sandy soil	2.2 lbs.	Old beds: Within 2 days after lifting sod
		New beds: 7-10 days after sprigging or plugging

If weeds regrow, apply an additional 2.2 lbs./A on muck or peat, or 1.1 lbs./A on sandy soil.

Precautions: To avoid crop injury, (1) Do not apply within 30 days prior to cutting or lifting. (2) Do not apply in combination with surfactants or other spray additives. (3) Use only on turfgrass reasonably free of infestations of insects, nematodes, and diseases. (4) On newly sprigged turfgrass, temporary slowing of growth may follow application.

TURFGRASS FOR FAIRWAYS, LAWNS, SOD PRODUCTION* AND SIMILAR AREAS

*In states other than Florida. For use on turfgrass for sod in Florida, see "Turfgrass for Sod (Florida only)" section above.

Bermudagrass, Centipedegrass, St. Augustinegrass, and Zoysia Grass

Apply AATREX NINE-O after October 1 before emergence of winter annual weeds for control of annual bluegrass, burclover, carpet burweed, chickweed, corn speedwell, henbit, hop clover, and spurweed. AATREX NINE-O will control annual bluegrass even if it is emerged at time of treatment. For control of summer annual weeds listed in the preemergence portion of the AATREX NINE-O Applied Alone—Corn or Grain Sorghum section of this label, also apply AATREX NINE-O in late winter before the weeds emerge. Apply in a minimum of 15 gals. of water per acre or 1 gal. per 1,000 sq. ft.

Where annual bluegrass is the major weed, use 1.1 lbs. of AATREX NINE-O per acre (0.4 fl. oz. per 1,000 sq. ft.). Use 2.2 lbs./A (0.8 fl. oz. per 1,000 sq. ft.) for control of the other weeds named above. Do not exceed 1.1 lbs./A per treatment on newly sprigged turfgrass or on hybrid bermudagrass such as Triflawn, Wifway, and Ormond.

For continued summer annual weed control, apply another 1.1 lbs./A at least 30 days after the previous application, but not after April 15. However, do not make more than two applications of this product per year.

Precautions: On newly sprigged turfgrass and hybrid bermudagrass, temporary slowing of growth and yellowing may occur following application. To avoid turf injury, (1) Use only on turfgrass reasonably free of infestations of insects, nematodes, and diseases. (2) Do not use on golf greens. (3) Do not use north of NC (except may be used in Virginia Coastal Plains) or west of the high rainfall areas of eastern OK and eastern TX. (4) Do not use on muck or alkaline soils. (5) Do not apply over the rooting area of trees or ornamentals not listed on this label. (6) Do not overseed with desirable trufgrass within 4 months before or 6 months after treatment. (7) Do not apply this product to newly seeded bermudagrass until it has overwintered and has a well-developed rhizome system. Do not exceed 2.2 lbs. product/A within 12 months of seeding bermudagrass.

Note: Do not graze or feed turf clippings to animals, or illegal residues may result.

MACADAMIA NUTS

For preemergence control of many broadleaf and grass weeds, including crabgrass, foxtail, wiregrass, Flora's paintbrush, spanishneedles, and fireweed, broadcast 2.2-4.4 lbs./A before harvest and before weeds emerge. Repeat as necessary. Do not spray when nuts are on ground during harvest period. Do not apply by air.

PINEAPPLE

For control of purslane, spanishneedles, annual grasses, annual bindweed, ageratum, amaranth, rattlepod, Flora's paintbrush, fireweed, spurge, indigo, and papalo, apply up to 7.1 lbs./A as a blanket spray immediately after planting or following harvest. Make additional blanket or interspace applications of up to 1.7 lbs./A at 1-2 month intervals as needed, prior to differentiation. Apply in 20-40 gals. of water/A to assure thorough coverage.

Precautions: (1) Do not apply more than 33.3 lbs./A per cycle, as crop injury or illegal residues may result. (2) Repeated monthly applications to plant foliage may slow plant growth and delay fruiting.

Note: Do not apply within 45 days of fruit harvest, or forage harvest if forage is to be fed to livestock, or illegal residues may result.

GUAVA

Use only on established plantings which are at least 18 months old. Apply as a directed spray at 2.2-4.4 lbs./A of AATREX NINE-O in 20-50 gals. of spray mix preemergence or early postemergence to weeds. When applying postemergence, the use of a surfactant and greater spray volume (80-100 gals. of spray mix per acre) may enhance weed control. This product controls many annual broadleaf and grass weeds, including fireweed, purslane, scarlet pimpernel, spanishneedles, and sowthistle.

Notes: To avoid illegal residues, (1) Do not allow spray to contact foliage or fruit. (2) Do not apply more frequently than at 4 month intervals. (3) Do not apply more than 8.8 lbs. of AATREX NINE-O per year.

CONIFERS

For control of annual broadleaf and grass weeds prior to transplanting, after transplanting or in established conifers [including Douglas fir, grand fir, noble fir, white fir, Austrian pine, bishop pine, Jeffrey pine, knobcone pine, loblolly pine, lodgepole pine (shore pine), monterey pine, ponderosa pine, Scotch pine, slash pine, blue spruce, and Sitka spruce]: Broadcast 2.2-.4.4 lbs. in a minimum of 5 gals. of water per acre by air or 10 gals. by ground before weeds are 1.5 inches tall. Apply to established trees between fall and early spring while trees are dormant. For new transplants, apply during or soon after transplanting. For applications prior to transplanting, allow sufficient precipitation to activate AATREX before transplanting. In areas where spring and summer rainfall is inadequate to activate AATREX, apply during fall prior to spring transplanting.

For the list of weeds controlled, see **AATREX NINE-O Applied Alone—Corn or Grain Sorghum—Preplant Surface-Applied, Preplant Incorporated, or Preemergence.**

Quackgrass control: Broadcast 4.4 lbs. in a minimum of 5 gals. of water per acre by air or 10 gals. by ground between fall and early spring while trees are dormant and before quackgrass is more than 1.5 inches tall.

Precautions: (1) In areas west of the Rocky Mountains (except the Great Basin), grazing may begin 7 months after a fall application or 3 months after a winter or spring application. (2) To prevent illegal residues, do not graze treated areas of the Great Basin, or areas east of the Rocky Mountains. (3) Temporary injury to trees may occur following use of AATREX on coarse-textured soil. (4) To avoid crop injury, do not apply to seedbeds. (5) Also apply only once per year.

Aerial application: In order to assure that spray will be controllable within the target area when used according to label directions, make applications at a maximum height of 10 ft. above vegetation, using low drift nozzles at a maximum pressure of 40 psi. and restrict application to periods when wind speed does not exceed 10 mph. To assure that spray will not adversely affect adjacent sensitive nontarget plants, apply AATREX NINE-O by aircraft at a minimum upwind distance of 400 ft. from sensitive plants.

NONSELECTIVE WEED CONTROL ON NONCROP LAND

This product provides long-term weed control on industrial sites, highway, medians and shoulders, railroad rights-of-way, lumberyards, petroleum tank farms, and in noncrop areas on farms, such as around buildings, equipment and fuel storage areas, along fences, and lanes. Do not use near adjacent desirable plants or in greenhouses, or injury may occur.

Apply before or soon after weeds begin growth. Make postemergence applications when weeds are young and actively growing. Use the higher rates on fine clay and much soils. Use sufficient water to assure thorough coverage. Use at least 1 gal. of water for each pound of product; use more if practical.

Note: Do not allow livestock to graze treated areas, or illegal residues may result.

Annual broadleaf and grass weeds (including barnyardgrass, cheatgrass, crabgrass, lambsquarters, foxtail, ragweed, puncturevine, and turkey mullein): Broadcast 5.3-11.1 lbs./A.

Hard-to-kill annual and many perennial broadleaf and grass weeds (including bluegrass, burdock, Canada thistle, dogfennel, orchardgrass, plantain, quackgrass, purpletop, redtop, and smooth brome): Broadcast 11.1-22.2 lbs./A.

Hard-to-kill biennial and perennial weeds (including bullthistle and sowthistle): Broadcast 22.2-44.4 lbs./A.

For longer residual control in regions of high rainfall and a long growing season, broadcast 22.2-44.4 lbs./A.

For small areas, 4 oz. per 1,000 sq. ft. is equivalent to 11.1 lbs./A.

Tank Mixture with Princep 80W or Princep Caliber 90

Use only for long-term control of broadleaf and grass vegetation in areas listed in respective labels. Do not use on cropland, or near desirable plants, or in greenhouses, or injury may occur.

This tank mixture has similar activity and moisture requirements as described under above General information section, except that Princep provides preemergence control only.

For best results, apply to the soil shortly before weed growth begins or apply postemergence to young, actively growing plants not over 6 inches tall.

Use sufficient water for thorough soil and plant coverage. Use at least 1 gal. of water for each lb. of tank mixture herbicide. More water may be needed for heavy weed growth. Maintain agitation during mixing and application.

Broadcast tank mixtures at the following rates to control those weeds listed. Use the lower rates for light weed infestations and the higher rates for heavier infestations.

Use 2.9-5.6 lbs./A of this product plus 3.25-6.25 lbs./A of Princep 80W (or 2.9-5.6 lbs./A of Princep Caliber 90) to control weeds listed in the General Information and the several crop sections of the Princep 80W and Princep Caliber 90 labels.

The above rates will also control California burclover, *Orthocarpus purpuresces*, ripgut brome, smooth catsear, summer lupine,and turkey mullei.

Use 5.6 lbs./A of this product plus 6.25 lbs./A of Princep 80W (or 5.6 lbs./A of Princep Caliber 90) to control burdock, daisy fleabane, goldenrod, horseweed, little mallow, milkweed, orchardgrass, quackgrass, Russian thistle, shortpod mustard, wild barley, and wood sorrel.

Use 11.1 lbs. of this product plus 12.5 lbs./A of Princep 80W (or 11.1 lbs./A of Princep Caliber 90) to control eveningprimrose, shepherds-needle, and tall fescue.

For small areas, 1/2 cup (4 oz.) of this product per 1,000 sq. ft. is equivalent to 11.1 lbs./A; 2/3 cup (4.5 oz.) of Princep 80W or 1/2 cup (4 oz.) of Princep Caliber 90 per 1,000 sq. ft. is equivalent to 12.5 lbs./A of Princep 80W or 11.1 lbs./A of Princep Caliber 90, respectively.

Tank Mixture with Diuron 80W

Use only for long-term control of broadleaf and grass vegetation in areas listed in respective labels. Do not use on cropland, or near desirable plants, or in greenhouses, or injury may occur.

This tank mixture has activity and moisture requirements similar to that described in above General Information section and in the diuron 80W label. For best results, apply to soil shortly before weed growth begins or apply to young, actively growing plants not over 6 inches tall. For increased contact activity, add a surfactant, such as Surfactant WK, at 2 qts. per 100 gals. of spray volume. Add surfactant as last ingredient to nearly full spray tank.

Use sufficient water for thorough ground and plant coverage. Use at least 1 gal. of water for each lb. of tank mixture. More water may be needed for heavy weed growth. Maintain agitation during mixing and application.

Broadcast tank mixtures at the following rates to control those weeds listed. Use the lower rates for light weed infestations and the higher rates for heavier infestations.

Use 4.4-5.6 lbs./A of this product plus 5-6.25 lbs./A of diuron 80W, for example, Karmex® 80W, to control weeds listed in crop and noncrop Directions for Use on the diuron 80W label.

Use 5.6 lbs./A of this product plus 6.25 lbs./A of diuron 80W to control broomsedge, buckhorn plantain, Carolina geranium, chicory, downy brome, eveningprimrose, fleabane, goldenrod, horseweed, lespedeza, little mallow, orchardgrass, Russian thistle, shortpod mustard, tall fescue, wild barley, and wood sorrel.

Use 11.1 lbs./A of this product plus 12.5 lbs./A of diuron 80W to control alfalfa, black nightshade, leafy spurge, and shepherdsneedle.

To determine compatibility of tank mixtures, thoroughly mix products in a small container at the ratio of 1 qt. of water to 1.75 oz. of this product plus 2 oz. of diuron 80W. Let stand for 5 minutes. If the mixture remains mixed, or can be remixed readily, it is compatible and can be sprayed.

For small areas, 1/2 cup (4 oz.) of this product or 1.25 cups of diuron 80W, such as Karmex 80W, per 1,000 sq. ft. is equivalent to 11.1 lbs./A of this product or 12.5 lbs./A of diuron 80W.

STORAGE AND DISPOSAL

Pesticide Storage and Disposal

Store in a dry place. Do not contaminate water, food, or feed by storage, disposal, or cleaning of equipment. Wastes resulting from the use of this product may be disposed of on-site or at an approved waste disposal facility.

Container Disposal

Completely empty bag into application equipment. Dispose of empty bag in a sanitary landfill or by incineration, or by open burning, if allowed by state and local authorities. If burned, keep out of smoke.

For minor spills, leaks, etc., follow allow precautions indicated on this label and clean up immediately. Take special care to avoid contamination of equipment and facilities during cleanup procedures and disposal of wastes. In the event of a major spill, fire or other emergency, call (919) 292-7100 day or night.

PRECAUTIONARY STATEMENTS

HAZARDS TO HUMANS AND DOMESTIC ANIMALS

CAUTION

Harmful if swallowed, inhaled, or absorbed through skin. Do not breathe dust or spray mist. Avoid contact with eyes, skin or clothing.

STATEMENT OF PRACTICAL TREATMENT

If swallowed: Call a physician or Poison Control Center immediately. Drink 1 or 2 glasses of water and induce vomiting by touching back of throat with finger, or, if available, by administering syrup of ipecac. Do not induce vomiting or give anything by mouth to an unconscious person.

If inhaled: Remove victim to fresh air. If not breathing, give artificial respiration, preferably mouth-to-mouth. Get medical attention.

If on skin: Wash with plenty of soap and water. Get medical attention.

If in eyes: Flush eyes with *plenty* of water. Call a physician if irritation persists.

Note to Physician: There is no specific antidote for atrazine. If this product is ingested, induce emesis or lavage stomach. The use of an aqueous slurry of activated charcoal may be considered.

ENVIRONMENTAL HAZARDS

Atrazine can travel (seep or leach) through soil and can enter ground water which may be used as drinking water. Atrazine has been found in ground water. Users are advised not to apply atrazine to sand and loamy sand soils where the water table (ground water) is close to the surface and where these soils are very permeable, i.e., well-drained. Your local agricultural agencies can provide further information on the type of soil in your area and the location of ground water.

This pesticide is toxic to aquatic invertebrates. Do not apply directly to water or wetlands. Do not apply where runoff is likely to occur. Do not apply when weather conditions favor drift from treated areas. Runoff and drift from treated areas may be hazardous to aquatic organisms in neighboring areas. Do not contaminate water by cleaning of equipment or disposal of wastes.

AATREX® trademark of CIBA-GEIGY for atrazine
Bexton® trademark of Dow Chemical Co. for propachlor
Caliber® trademark of CIBA-GEIGY
Compex® trademark of KALO Agricultural Chemicals, Inc.
Concep® tradmark of CIBA-GEIGY
Dual® trademark of CIBA-GEIGY for metolachlor U.S. Patent No. 3,937,730
Gramoxone® trademark of ICI Americas for paraquat
Igran® trademark of CIBA-GEIGY for terbutryn
Karmex® trademark of E. I. duPont de Nemours & Co. for diuron
Lasso® trademark of Monsanto Company for alachlor
NINE-O® trademark of CIBA-GEIGY
Princep® trademark of CIBA-GEIGY for simazine
Ramrod® trademark of Monsanto Company for propachlor
Roundup® trademark of Monsanto Company for glyphosate herbicide
Sutan +® trademark of Stauffer Chemical Co. for butylate
Unite® trademark of Hopkins Agricultural Chemical Co.
X-77® trademark of KALO Agricultural Chemicals, Inc.

CGA 7L101J 078

This is a specimen label which was not prepared by CIBA-GEIGY. CIBA-GEIGY is not responsible for the accuracy of the information contained herein. As labels are subject to revision, always carefully read and follow the label on the product container.

APRON® 25W

FUNGICIDE

A seed treatment chemical for control of systemic downy mildew, seed rot and damping-off diseases of certain crops

For use only by commercial seed treaters

ACTIVE INGREDIENT:

Metalaxyl: *N*-(2,6-dimethylphenyl)-*N*-(methoxyacetyl) alanine methyl ester	25.0%
INERT INGREDIENTS:	75.0%
TOTAL:	100.0%

EPA Reg. No. 100-639

KEEP OUT OF REACH OF CHILDREN

Caution

See additional precautionary statements on panel at left.

DIRECTIONS FOR USE AND CONDITIONS OF SALE AND WARRANTY

IMPORTANT: Read the entire **Directions for Use** and the **Conditions of Sale and Warranty** before using this product.

CONDITIONS OF SALE AND WARRANTY

The **Directions for Use** of this product reflect the opinion of experts based on field use and tests. The directions are believed to be reliable and should be followed carefully. However, it is impossible to eliminate all risks inherently associated with use of this product. Crop injury, ineffectiveness, or other unintended consequences may result because of such factors as weather conditions, presence of other materials, or the manner of use or application all of which are beyond the control of CIBA-GEIGY or the Seller. All such risks shall be assumed by the Buyer.

CIBA-GEIGY warrants that this product conforms to the chemical description on the label and is reasonably fit for the purposes referred to in the **Directions for Use** subject to the inherent risks referred to above. **CIBA-GEIGY makes no other express or implied warranty of Fitness or Merchantability or any other express or implied warranty. In no case shall CIBA- GEIGY or the Seller be liable for consequential, special, or indirect damages resulting from the use or handling of this product.** CIBA-GEIGY and the Seller offer this product, and the Buyer and user accept it, subject to the foregoing **Conditions of Sale and Warranty,** which may varied only by agreement in writing signed by a duly authorized representative of CIBA-GEIGY.

GENERAL INFORMATION

APRON is a systemic fungicide seed dressing specifically for control of systemic downy mildews and diseases caused by soilborne *Pythium* and *Phytophthora* spp. For control of other soilborne diseases, such as *Rhizoctonia* spp., APRON 25W should be applied in combination with other registered seed dressing fungicides.

Note: APRON is a systemic fungicide having a specific mode of action. APRON could be subject to development of resistant strains of fungi. Development of resistance cannot be predicted. Therefore, CIBA-GEIGY cannot assume liability for crop damage resulting from resistant strains of fungi. Consult with your State Agricultural Experiment Station or Extension Service Specialist for guidance and ways to control any possible APRON resistant strains of fungi which may occur.

APRON 25W may be applied as a water-based slurry with other registered seed treatment insecticides and fungicides through standard slurry- or mist-type commercial seed treatment equipment.

DIRECTIONS FOR USE

It is a violation of federal law to use this product in a manner inconsistent with its labeling.

FAILURE TO FOLLOW THE DIRECTIONS FOR USE AND PRECAUTIONS ON THIS LABEL MAY RESULT IN CROP INJURY OR POOR DISEASE CONTROL.

Seed Treatment—Domestic Use	
Cotton	For Pythium seed rot and damping-off: Apply APRON 25W as a seed treatment to machine-delinted or acid- delinted cotton-seed at the rate of 1-2 ounces per 100 lbs. of seed.
Forage Grasses	Grasses grown for hay grazing or silage, corn (fodder or silage), and small grain grown for hay grazing or silage. For Pythium damping-off control: Apply APRON 25W as a seed treatment at the rate of 1-2 ounces per 100 lbs. of seed.
Forage Legumes	Alfalfa, clover, lespedeza, beans (for forage), soybeans (for forage), soybean hay, peas (for forage), pea vine hay, cowpeas (for forage), cowpea hay, trefoil, vetch and velvet beans (for forage). For Pythium damping-off and early season Phytophthora control: Apply APRON 25W as a seed treatment at the rate of 2

Seed Treatment—Domestic Use	
	ounces per 100 lbs. of seed. For suppression of early season downy mildew in alfalfa and soybeans: Apply APRON 25W as a seed treatment at the rate of 2 ounces per 100 lbs. of seed.
Garden Beets	For control of Pythium damping-off: Apply APRON 25W as a seed treatment at the rate of 2 ounces per 100 lbs. of seed.
Grain Crops	Wheat, barley, rye, oats, millet, rice, corn (field corn, sweet corn, and popcorn), buckwheat, and triticale. For Pythium damping-off control: Apply APRON 25W as a seed treatment at the rate of 1-2 ounces per 100 lbs. of seed. For control of systemic downy mildew of sweet corn: Apply APRON 25W as a seed treatment at the rate of 4 ounces per 100 lbs. of seed.
Turfgrasses (All)	For Pythium damping-off control on golf courses, home lawns, reclamation and erosion control projects: Apply APRON 25W as a seed treatment at the rate of 2-4 ounces per 100 lbs. of seed.
Seed and Pod Vegetables	Black-eyed peas, cowpeas, dill, edible soybeans, field beans, chick peas (garbanzo beans), field peas, garden peas, green beans, kidney beans, lima beans, navy beans, okra, peas, pole beans, snap beans, string beans, wax beans, lentils, and lupines. For Pythium damping-off and early season Phytophthora control: Apply APRON 25W as a seed treatment at the rate of 2 ounces per 100 lbs. of seed. For control of Pythium damping-off and systemic downy mildew of peas: Apply APRON 25W as a seed treatment at the rate of 5 ounces per 100 lbs. of seed.
Sorghum	For control of Pythium damping-off on all sorghum cultivars, use 2 ounces of APRON 25W per 100 lbs. of seed applied as a seed treatment. For control of systemic downy mildew, apply 2 ounces of APRON 25W per 100 lbs. of seed on cultivars which possess sorghum downy mildew resistance. To those cultivars which do not possess resistance to a particular race of sorghum downy mildew in heavily infected areas with recent severe history, apply 4 ounces of APRON 25W per 100 lbs. of seed as a seed treatment. Seed treatment with the 4 ounce rate should be planted and not carried over to the following year. Only high quality sorghum seed (90% germ and above) should be treated with APRON 25W.
Soybeans	For Pythium damping-off and early season Phytophthora control: Apply APRON 25W as a seed treatment at the rate of 1-2 ounces per 100 lbs. of seed.
Sugar Beets	For control of Pythium damping-off: Apply APRON 25W as a seed treatment at the rate of 2 ounces per 100 lbs. of seed.
Sunflowers	For control of systemic downy mildew: Apply APRON 25W as a seed treatment at the rate of 4 ounces per 100 lbs. of seed.

Seed Treatment - Export use

FOR TREATMENT OF THE FOLLOWING SEED FOR EXPORT ONLY TO COUNTRIES REQUIRING TREATMENT WITH METALAXYL BEFORE IMPORTATION.

Sorghum	For control of Pythium damping-off and systemic downy mildew of sorghum: Apply APRON 25W as a seed treatment at the rate of 4-14 ounces per 100 lbs. of seed.
Sunflower	For control of systemic downy mildew: Apply APRON 25W as a seed treatment at the rate of 8-14 ounces per 100 lbs. of seed.
Corn	For control of Pythium damping-off and systemic downy mildew of corn: Apply APRON 25W as a seed treatment at the rate of 4-14 ounces per 100 lbs. of seed.
Millet	For control of Pythium damping-off and systemic downy mildew of millet: Apply APRON 25W as a seed treatment at the rate of 4-14 ounces per 100 lbs. of seed.

Seed treated in accordance with the export use pattern must be labeled as follows: FOR EXPORT ONLY — NOT FOR DOMESTIC SALE OR USE.

Note: Federal law requires that bags containing treated seeds shall be labeled with the following information: "This seed has been treated with metalaxyl fungicide. Do not use for feed, food or oil purposes. Store away from feeds and foodstuffs." Use with an EPA-approved dye that imparts an unnatural color to the seed.

STORAGE AND DISPOSAL

Pesticide Storage and Disposal

Store in a dry place. Do not contaminate water, food or feed by storage, disposal, or cleaning of equipment. Open dumping is prohibited. Wastes resulting from the use of this product are acutely toxic. Improper disposal of unused pesticide, spray mixture, or rinsate is a violation of federal law. Pesticide, spray mixture, or rinsate that cannot be used according to label instructions must be disposed of according to federal, state or local procedures. For guidance in proper disposal methods, contact your State Pesticide or Environmental Control Agency, or the Hazardous Waste representative at the nearest EPA Regional Office.

Container Disposal

Completely empty bag into application equipment. Dispose of empty bag in a sanitary landfill or by incineration, or by open burning, if allowed by state and local authorities. If burned, keep out of smoke.

For minor spills, leaks, etc., follow all precautions indicated on this label and clean up immediately. Take special care to avoid contamination of equipment and facilities during cleanup procedures and disposal of wastes. In the event of a major spill, fire or other emergency, call (919) 292-7100 day or night.

PRECAUTIONARY STATEMENTS

HAZARDS TO HUMANS AND DOMESTIC ANIMALS

CAUTION

Harmful if swallowed, inhaled, or absorbed through the skin. Causes eye irritation. Avoid contact with eyes, skin, or clothing. Avoid breathing dust or spray mist. Wash thoroughly with soap and water after handling. Remove and wash contaminated clothing before reuse. Do not store near food or feed.

Practical Treatment: In case of contact, flush eyes or skin with plenty of water. Get medical attention if irritation persists. If swallowed, give large quantities of water to drink and induce vomiting. Never give anything by mouth to an unconscious person. Call a physician immediately. If inhaled, move to fresh air; if not breathing, give artifical respiration and get medical attention.

ENVIRONMENTAL HAZARDS

Do not apply directly to water. Apply only as specified on this label. Do not apply when weather conditions favor drift from treated areas. Do not contaminate water by cleaning of equipment or disposal of wastes.

PHYSICAL AND CHEMICAL HAZARDS

Do not use or store near heat or open flame.

APRON® trademark of CIBA-GEIGY for metalaxyl

CGA 66L7F 087

This is a specimen label which was not prepared by CIBA-GEIGY. CIBA-GEIGY is not responsible for the accuracy of the information contained herein. As labels are subject to revision, always carefully read and follow the label on the product container.

AQUAZINE® 90WDG

ALGICIDE

For control of algae and certain weeds in ponds

ACTIVE INGREDIENT:

Simazine: 2-chloro-4,6-bis(ethylamino)-*s*-traizine 90%

INERT INGREDIENTS: 10%

TOTAL 100%

EPA Reg. No. 100-650 EPA Est. 100-LA-1

KEEP OUT OF REACH OF CHILDREN

CAUTION

See additional precautionary statements on back of bag.

See directions for use on back of bag.

DIRECTIONS FOR USE AND CONDITIONS OF SALE AND WARRANTY

IMPORTANT: Read the entire **Directions for Use** and the **Conditions of Sale and Warranty** before using this product.

CONDITIONS OF SALE AND WARRANTY

The **Directions for Use** of this product reflect the opinion of experts based on field use and tests. The directions are believed to be reliable and should be followed carefully. However, it is impossible to eliminate all risks inherently associated with use of this product. Ineffectiveness or other unintended consequences may result because of such factors as weather conditions, presence of other materials, or the manner of use or application all of which are beyond the control of CIBA—GEIGY or the Seller. All such risks shall be assumed by the Buyer.

CIBA—GEIGY warrants that this product conforms to the chemical description on the label and is reasonably fit for the purposes referred to in the **Directions for Use** subject to the inherent risks referred to above. **CIBA—GEIGY makes no other express or implied warranty of Fitness or Merchantability or any other express or implied warranty. In no case shall CIBA—GEIGY or the Seller be liable for consequential, special, or indirect damages resulting from the use or handling of this product.** CIBA—GEIGY and the Seller offer this product, and the Buyer and user accept it, subject to the foregoing **Conditions of Sale and Warranty,** which may be varied only by agreement in writing signed by a duly authorized representative of CIBA—GEIGY.

DIRECTIONS FOR USE

It is a violation of federal law to use this product in a manner inconsistent with its labeling.

GENERAL INFORMATION

AQUAZINE 90WDG may be used in ponds containing water or it may be applied to the bottoms of drained ponds for control of several aquatic weeds and algae. Control of weeds is usually season-long. Algae may be controlled for one to three months or may require re-treatment later in the season. In ponds containing fish, applications should not be made to water during prolonged periods of hot weather because oxygen levels would normally be low at these times and treatment could induce additional oxygen stress to fish, resulting occasionally in fish kills. If fish stress is observed, aerate the water by any method to reduce the chances of a fish kill. Use only in ponds which will have little or no outflow after treatment. Do not use AQUAZINE 90WDG as a spot treatment.

Apply fertilizer 7-10 days after an AQUAZINE 90WDG application in ponds that have a regular fertilization program. A few days delay in bloom of planktonic algae and a reduction in the extent of bloom may occur when AQUAZINE 90WDG is used in ponds that receive fertilization.

Application to Water

When to Apply: Treat ponds after seasonal flow has ceased and early in the weed and algae growth period. Treat when 5-10% of the pond surface is covered with algae mats or scum and/or while submerged aquatic weeds are actively growing but before they reach the surface of the water. Usually, this will be between April 1 and May 15 in southern areas and between May 1 and June 15 in northern areas.

High water temperatures cause more rapid decay of dead weeds and algae. This leads to reduced oxygen content which can cause fish stress and/or death. Apply AQUAZINE 90WDG before water temperature exceeds 75°F (22°C).

Do not treat ponds with heavy infestations of weeds and/or algae, such as occurs in mid and late summer, since decomposition of heavy vegetative growth greatly reduces oxygen content of water which can cause fish stress and/or death.

How to Apply: After determining the amount of AQUAZINE 90WDG required to treat the entire pond, mix convenient portions such as 5 or 10 lbs. with enough water in a pail or bucket to form a thin paste or slurry. Toss the paste or slurry into the pond from the shoreline. Repeat this operation at several locations around the pond until the required amount has been applied. It is not necessary to use spray equipment or to evenly distribute AQUAZINE 90WDG over the surface since it will dissolve in water and evenly distribute itself throughout the entire pond. Do not spill paste or slurry on pond banks as injury to vegetation will occur.

Rates of Application: The appropriate rate of AQUAZINE 90WDG depends on the types of algae or weeds present and the extent of infestation. Rates are given in lbs. of AQUAZINE 90WDG per acre foot of water. Determine the number of acre feet of water in a pond by multiplying the surface acres by the average depth in feet.

Algae Control: Most unicellular algae (algae which impart a green, yellow, red, or brown turbid color to the water) and many floating mat (pond moss) algae may be controlled with 1.5 to 3 lbs. of AQUAZINE 90WDG per acre foot of water. If the algae infestation is light, use 1.5 lbs. If moderate, use 2.2 to 3 lbs. of AQUAZINE 90WDG per acre foot.

Cladophora, Chara, Hydrodictyon (water net algae), Pithophora, and most other algae which form on the pond bottom (benthic) or which produce underwater plumes that break loose and float on the surface, require 3.8 lbs. of AQUAZINE 90WDG per acre foot of water. Two applications at 3.8 lbs. per acre foot of water are usually required for control of Pithophora. If a second treatment is needed, apply at the first sign of algae buildup.

Algae control should occur within three to fourteen days after application, depending upon type of algae. Unicellular algae are usually controlled sooner than benthic filamentous algae.

Due to uncontrollable factors such as infestation with later appearing algae, some ponds may require re-treatment in mid or later summer. If re-treatment becomes necessary, apply at the first signs that algae (mats or moss) is building up. Do not treat in the fall or just before freezing is expected, since water oxygen levels may be lowered during ice cover.

Weed Control: Submerged weeds, including pond weeds (*Potamogeton* spp.), naiad (*Najas* spp.), watermilfoil (*Myriophyllum* spp.), and coontail (*Ceratophyllum* spp.) may be controlled with 3-6 lbs. of AQUAZINE 90WDG per acre foot of water. Use 3 lbs. for light infestations and 4.5-6 lbs. per acre foot of water for moderate infestations. For control of fanwort (*Cabomba* spp.) use 7.5 lbs. of AQUAZINE 90WDG per acre foot.

Control of submerged weeds, except coontail, occurs in 4-6 weeks. Coontail control may require 10 weeks.

Floating weeds, including duckweed (*Lemna* spp.) and watermeal (*Wolffia* spp.) may be controlled with 3-6 lbs. of AQUAZINE 90WDG per acre foot of water. Within this range use lower rates for light infestations and higher rates for moderate infestations.

Where watermeal occurs alone or with other weeds, control is most effective with a split application. Apply one-half the recommended rate initially, followed in 3-4 weeks with a repeat application if control is not satisfactory.

Control of duckweed occurs in 1-5 weeks, and watermeal in 5-9 weeks.

Application to Bottoms of Fish Hatchery, Fish Rearing and Other Ponds with Draining Capabilities

Treat bottoms of drained ponds having a minimum average depth of two feet before regrowth of existing submerged weeds begins, or before new weeds emerge.

Allow at least 72 hours for AQUAZINE 90WDG to become adsorbed to the soil before refilling the pond. Sprinkler irrigation or gentle rainfall after applying AQUAZINE 90WDG will help move AQUAZINE 90WDG into the surface soil, enhancing its effectiveness. Heavy rainfall during this period may erode treated soil, resulting in poor weed control.

Mixing Instructions: Make a slurry by adding AQUAZINE 90WDG to a small amount of water in a separate container. Then pour the slurry into the spray tank partially filled with clean water, and add the rest of the water. Apply AQUAZINE 90WDG with constant agitation in 20-40 gals. of water per acre. Do not spill the slurry on pond banks as injury to vegetation will occur.

Spray Equipment: Use conventional spray equipment with fan-type nozzles. Screens in nozzles as well as those in suction and in-line strainers, should be no finer than 50-mesh.

Use a pump with capacity to maintain 35-40 psi at the nozzles. If hydraulic agitation is used, the pump should also provide sufficient agitation in the tank to keep the mixture in suspension.

Wash sprayer thoroughly with clean water immediately after use.

Avoid application under conditions where uniform coverage cannot be obtained or where excessive spray drift may occur.

Algae Control: Apply 8.9-11.1 lbs. of AQUAZINE 90WDG per acre as a broadcast spray for control of the algae indicated under this section. Application to Water. Use 8.9 lbs. per acre for expected light infestations and 11.1 lbs. per acre for expected heavy infestations.

Weed Control: Apply 8.9-13.3 lbs. of AQUAZINE 90WDG per acre for control of American lotus, duckweeds, fanwort, naiad, pond weeds and watermilfoil.

Within the rate range, use the lower rate of AQUAZINE 90WDG for soils with large amounts of sand or where there have been light infestations of weeds, or where algae dominates. Use the higher rate for soils high in silt and clay and where there have been heavy infestations of weeds.

Note: Do not apply more than 8.9 lbs. of AQUAZINE 90WDG per acre where striped bass fry or fingerlings will be cultured immediately following the application of AQUAZINE 90WDG. Do not apply more than once each year to bottoms of drained ponds. Where heavy infestations of weeds and/or algae are treated, do not restock ponds with fish until adequate oxygen levels are restored.

Use of Fish and Water

Fish taken from treated ponds may be used for food.

Treated ponds may be used for swimming 4 hours after application.

Water from treated ponds may not be used for irrigation or spraying of agricultural crops, lawns, or ornamental plantings, or for watering cattle, goats, hogs, horses, poultry, or sheep or for human consumption until 12 months following treatment.

Precaution: Do not treat ponds which have bordering trees with roots visibly extended into the water, since injury to these trees may occur.

STORAGE AND DISPOSAL

Store in a dry place. Do not contaminate water, food, or feed by storage or disposal. Improper disposal of unused pesticide, pesticide mixture, or rinse water is a violation of federal law. If these wastes cannot be used according to label instructions, contact your State Pesticide or Environmental Control Agency, or the Hazardous Waste representative at the nearest EPA Regional Office for guidance in proper disposal methods.

Completely empty bag into application equipment. Dispose of empty bag in a sanitary landfill or by incineration, or if allowed by state and local authorities, by burning. Stay out of smoke from burning bags.

For minor spills, leaks, etc., follow all precautions indicated on this label and clean up immediately. Take special care to avoid contamination of equipment and facilities during cleanup procedures and disposal of wastes. In the event of a major spill, fire or other emergency, call (919) 292-7100 day or night.

PRECAUTIONARY STATEMENTS

HAZARDS TO HUMANS AND DOMESTIC ANIMALS

CAUTION

Harmful if swallowed, inhaled or absorbed through the skin. Avoid breathing dust or spray mist. Avoid contact with eyes, skin, or clothing.

First Aid: In case of contact, immediately flush eyes or skin with plenty of water. Get medical attention if irritation persists.

ENVIRONMENTAL HAZARDS

Do not contaminate domestic, livestock, or irrigation water supplies or lakes or streams. Do not contaminate water by cleaning of equipment or disposal of wastes. Do not apply when weather conditions favor drift from treated areas.

Aquazine® trademark of CIBA-GEIGY for simazine

CGA 47L5 048

This is a specimen label which was not prepared by CIBA-GEIGY. CIBA-GEIGY is not responsible for the accuracy of the information contained herein. As labels are subject to revision, always carefully read and follow the label on the product container.

ATRATOL® 8P

PELLETED HERBICIDE

For control of weeds on noncrop land

ACTIVE INGREDIENTS:

Atrazine: 2-chloro-4-ethylamino-6-isopropylamino-*s*-triazine	7.6%
Atrazine-related compounds	0.4%
Sodium chlorate ($NaCio_3$)	40.0%
Sodium metaborate* ($Na_2B_2O_4 5H_2O$)	47.0%
INERT INGREDIENTS:	5.0%
TOTAL	100.0%

*Equivalent to 4.58% boron expressed as elemental boron

EPA Reg. No. 100-475

KEEP OUT OF REACH OF CHILDREN

CAUTION

See additional caution statements on back of bag.

Control No. 47235

See directions for use on back of bag.

DIRECTIONS FOR USE AND CONDITIONS OF SALE AND WARRANTY

IMPORTANT: Read the entire **Directions for Use** and the **Conditions of Sale and Warranty** before using this product.

CONDITIONS OF SALE AND WARRANTY

The **Directions for Use** of this product reflect the opinion of experts based on field use and tests. The directions are believed to be reliable and should be followed carefully. However, it is impossible to eliminate all risks inherently associated with use of this product. Crop injury, ineffectiveness, or other unintended consequences may result because of such factors as weather conditions, presence of other materials, or the manner of use or application all of which are beyond the control of CIBA—GEIGY or the Seller. All such risks shall be assumed by the Buyer.

CIBA—GEIGY warrants that this product conforms to the chemical description on the label and is reasonably fit for the purposes referred to in the **Directions for Use** subject to the inherent risks referred to above. **CIBA—GEIGY makes no other express or implied warranty of Fitness or Merchantability or any other express or implied warranty. In no case shall CIBA—GEIGY or the Seller be liable for consequential, special, or indirect damages resulting from the use or handling of this product.** CIBA—GEIGY and the Seller offer this product, and the Buyer and user accept it, subject to the foregoing **Conditions of Sale and Warranty**, which may be varied only by agreement in writing signed by a duly authorized representative of CIBA—GEIGY.

DIRECTIONS FOR USE

It is a violation of federal law to use this product in a manner inconsistent with its labeling.

ATRATOL 8P is a nonselective herbicide that can be applied before or after plant growth begins. Since ATRATOL enters plants through their roots, its effectiveness is dependent on rainfall to move the chemical into the root zone. Very dry soil conditions and lack of rainfall may result in poor weed control.

Important: Use only in areas where complete control of all vegetation is desired, such as industrial sites, rights-of-way, lumberyards, petroleum tank farms, around farm buildings, along fence lines, and similar areas. When applied to the soil, this product usually inhibits all plant growth for a year or more. It should not be used on land to be cropped, or where adjacent desirable trees, shrubs, or plants might be injured.

To control annual broadleaf weeds and grasses: Apply 1/4-1/2 lb. per 100 sq. ft.

To control most perennial broadleaf weeds and grasses: Apply 1/2-1 lb. per 100 sq. ft. Where deep-rooted perennials are present, use the 1 lb. rate.

The higher rates listed above should be used in areas having a long growing season or high rainfall. The higher rates will also provide longer residual control.

STORAGE AND DISPOSAL

Store in a dry place. Do not contaminate water, food, or feed by storage or disposal. Pesticide wastes are acutely hazardous. Improper disposal of unused pesticide is a violation of federal law. If these wastes cannot be used according to label instructions, contact your State Pesticide or Environmental Control Agency, or the Hazardous Waste representative at the nearest EPA Regional Office for guidance in proper disposal methods.

Completely empty bag into application equipment. Dispose of empty bag in a sanitary landfill or by incineration, or if allowed by state and local authorities, by burning. Stay out of smoke from burning bags.

For minor spills, leaks, etc., follow all precautions indicated on this label and clean up immediately. Take special care to avoid contamination of equipment and facilities during cleanup procedures and disposal of wastes. In the event of a major spill, fire or other emergency, call (919) 292-7100 day or night.

CAUTION

Keep out of reach of children.

Harmful if swallowed. Avoid contamination of food and feed. Avoid inhalation of dust. Avoid contact with eyes, skin, and clothing. Wash clothing after application. Wash out equipment after application. Do not use equipment to apply insecticides, fungicides, or fertilizers. Do not contaminate water intended for domestic or irrigation use.

ATRATOL® trademark of CIBA—GEIGY

CGA 20L1N 106

This is a specimen label which was not prepared by CIBA-GEIGY. CIBA-GEIGY is not responsible for the accuracy of the information contained herein. As labels are subject to revision, always carefully read and follow the label on the product container.

ATRATOL® 90

HERBICIDE

Bare ground weed control with high suspensibility and easy wetting.
For nonselective weed control on industrial sites and noncrop areas
Do not use on agricultural lands

ACTIVE INGREDIENTS:

Atrazine: 2-chloro-4-ethylamino-6-isopropyl-amino-*s*-triazine	85.5%
Related compounds	4.5%
INERT INGREDIENTS:	10.0%
TOTAL	100.0%

EPA Est. 100-LA-1 EPA Reg. No. 100-622

Control No. 53160

Atratol 90 is a water dispersible granule

KEEP OUT OF REACH OF CHILDREN

CAUTION

See additional precautionary statements on panel below.

See directions for use on panel below.

DIRECTIONS FOR USE AND CONDITIONS OF SALE AND WARRANTY

IMPORTANT: Read the entire **Directions for Use** and the **Conditions of Sale and Warranty** before using this product.

CONDITIONS OF SALE AND WARRANTY

The **Directions for Use** of this product reflect the opinion of experts based on field use and tests. The directions are believed to be reliable and should be followed carefully. However, it is impossible to eliminate all risks inherently associated with use of this product. Crop injury, ineffectiveness, or other unintended consequences may result because of such factors as weather conditions, presence of other materials, or the manner of use or application all of which are beyond the control of CIBA–GEIGY or the Seller. All such risks shall be assumed by the Buyer.

CIBA–GEIGY warrants that this product conforms to the chemical description on the label and is reasonably fit for the purposes referred to in the **Directions for Use** subject to the inherent risks referred to above. **CIBA–GEIGY makes no other express or implied warranty of Fitness or Merchantability or any other express or implied warranty. In no case shall CIBA–GEIGY or the Seller be liable for consequential, special, or indirect damages resulting from the use or handling of this product.** CIBA–GEIGY and the Seller offer this product, and the Buyer and user accept it, subject to the foregoing **Conditions of Sale and Warranty**, which may be varied only by agreement in writing signed by a duly authorized representative of CIBA–GEIGY.

DIRECTIONS FOR USE

It is a violation of federal law to use this product in a manner inconsistent with its labeling.

FAILURE TO FOLLOW THE DIRECTIONS FOR USE AND PRECAUTIONS ON THIS LABEL MAY RESULT IN POOR WEED CONTROL AND/OR INJURY TO DESIRABLE PLANTS.

GENERAL INFORMATION

Atratol 90 is a nonselective herbicide for use on noncrop areas. Atratol 90 can be applied before or after plant growth begins. Although Atratol 90 has some foliage activity, most of its activity is through root absorption. Therefore, moisture is required to move it into the root zone. Very dry soil conditions and lack of sufficient rainfall may result in poor weed control. Postemergence applications should be made when weeds are young and growing actively.

Following many years of continuous use of this product and chemically related products, biotypes of some of the weeds listed on this label have been reported which cannot be effectively controlled by this and related herbicides. Where this is known or suspected and weeds controlled by this product are expected to be present along with resistant biotypes, we recommend the use of this product in registered combinations or in sequence with other registered herbicides which are not triazines. If only resistant biotypes are expected to be present, use a registered non-triazine herbicide. Consult with your state Agricultural Extension Service for specific recommendations.

Use only in areas where complete control of broadleaf and grass vegetation is desired, such as (1) railroad, highway, and utility rights-of-way, (2) petroleum tank farms, (3) industrial sites, (4) lumberyards, (5) around warehouses and chemical plants, and (6) along industrial fence lines. When applied to the soil, this product usually inhibits plant growth for one year or more. It should not be used on land to be cropped or near adjacent desirable trees, shrubs or plants, or in greenhouses, as injury may occur. Do not allow livestock to graze treated areas.

Atratol 90 is noncorrosive to equipment.

MIXING PROCEDURES

(1) Be sure sprayer is clean and not contaminated with any material, as sprayer clogging may result. (2) Fill tank 1/4 full with clean water, and start agitation. (3) Be certain that the agitation system is working properly and creates a rippling or rolling action on the liquid surface. (4) Pour product directly from bag into tank. (5) Continue filling tank until 90% full. Increase agitation if necessary to maintain surface action. (6) Add tank mix herbicides. (7) Complete filling tank, maintaining sufficient agitation at all times to ensure surface action. This applies to both spray and nurse tanks. (8) Empty tank as completely as possible before refilling to prevent buildup of emulsifiable concentrate residue. Always maintain agitation to avoid separation. (9) If an emulsifiable concentrate film starts to build up in tank, drain and clean with strong detergent solution or solvent. (10) Immediately after use, clean sprayer thoroughly by flushing system with water containing a detergent.

APPLICATION EQUIPMENT

Use conventional ground sprayers equipped with nozzles that provide accurate and uniform application. Be certain that nozzles are uniformly spaced and the same size. Calibrate sprayer before use and recalibrate at the start of each season.

Use a pump with capacity to (1) maintain 35-40 psi at nozzles, (2) provide sufficient agitation in tank to keep mixture in suspension, and (3) to provide a minimum of 20% bypass at all times. Use centrifugal pumps which provide propeller shear action for dispersing and mixing this product. The pump should provide a minimum of 10 gals./minute/100 gal. tank size circulated through a correctly positioned sparger tube or jets.

Use screens to protect the pump and to prevent nozzles from clogging. Screens placed on suction side of the pump should be *16-mesh or coarser.* Do not place a screen in the recirculation line. Use 50-mesh or coarser screens between the pump and boom, and where required, at the nozzles. Check nozzle manufacturer's recommendations.

APPLICATION RATES—ATRATOL 90 APPLIED ALONE

Use sufficient water to assure thorough ground and plant coverage. Use at *least* 1 gallon of water for each pound of Atratol 90; more water may be necessary where weed growth is heavy.

To control annual broadleaf and grass weeds (such as barnyardgrass, cheatgrass, crabgrass, foxtail, lambsquarters, puncturevine, and ragweed) and biennials (such as turkey mullein): Broadcast 5.3-11.1 lbs. of Atratol 90 per acre.

To control most annual and many perennial broadleaf and grass weeds (such as bluegrass, burdock, Canada thistle, dogfennel, orchardgrass, plantain, purpletop, quackgrass, redtop, and smooth brome): Broadcast 11.1-22.2 lbs. of Atratol 90 per acre.

To control hard-to-kill biennial and perennial weeds (such as bullthistle and sowthistle): Broadcast 22.2-44.4 lbs. of Atratol 90 per acre.

For longer residual control in regions of high rainfall and a long growing season, apply 22.2-44.4 lbs. of Atratol 90 per acre.

For small areas, 4 oz. per 1,000 sq. ft. is equivalent to 11.1 lbs. per acre.

TANK MIXTURES

Atratol 90 plus Princep® 80W or Princep Caliber® 90

Use only for long-term control of broadleaf and grass vegetation in areas listed in respective labels. Do not use on cropland, or near desirable trees, shrubs or other plants, or in greenhouses, as injury may occur.

This tank mixture has similar activity and moisture requirements as described in above General Information section, except that Princep provides preemergence control only.

For best results, apply to soil shortly before weed growth begins or apply postemergence to young, actively growing plants not over 6 inches tall.

Use sufficient water for thorough soil and plant coverage. Use at least 1 gal. of water for each lb. of Atratol 90 plus Princep 80W or Caliber 90 combination. More water may be needed for heavy weed growth. Maintain agitation during mixing and application.

Broadcast tank mixtures at the following rates to control weeds listed. Use the lower rates for light weed infestations and the higher rates for heavier infestations.

Use 2.9-5.6 lbs. of Atratol 90 plus 3.25-6.25 lbs. of Princep 80W or 2.9-5.6 lbs. of Princep Caliber 90 per acre to control weeds listed in the General Information and the several crop sections of the Princep 80W and Princep Caliber 90 labels.

The above rates will also control California burclover, *Orthocarpus purpurescens,* ripgut brome, smooth catsear, summer lupine, and turkey mullein.

Use 5.6 lbs. of Atratol 90 plus 6.25 lbs. of Princep 80W or 5.6 lbs. of Princep Caliber 90 per acre to control burdock, daisy fleabane, goldenrod, horseweed, little mallow, milkweed, orchardgrass, quackgrass, Russian thistle, shortpod mustard, wild barley, and wood sorrel.

Use 11.1 lbs. of Atratol 90 plus 12.5 lbs. of Princep 80W or 11.1 lbs. of Princep Caliber 90 per acre to control eveningprimrose, shepherds-needle, and tall fescue.

For small areas, 1/2 cup (4 oz.) of Atratol 90 or 2/3 cup (4.5 oz.) of Princep 80W or 1/2 cup (4 oz.) of Princep Caliber 90, per 1,000 sq. ft. is equivalent to 11.1 lbs. of Atratol 90 or 12.5 lbs. of Princep 80W or 11.1 lbs. of Princep Caliber 90 per acre, respectively.

Atratol 90 plus Diuron 80W

Use only for long-term control of broadleaf and grass vegetation in areas listed in respective labels. Do not use on cropland, or near desirable trees, shrubs or other plants, or in greenhouses, as injury may occur.

This tank mixture has activity and moisture requirements similar to that described in above **General Information** section and in the diuron 80W label. For best results, apply to soil shortly before weed growth begins or apply to young, actively growing plants not over 6 inches tall. For increased contact activity, add a surfactant, such as Surfactant WK, at 2 qts. per 100 gals. of spray volume. Add surfactant as last ingredient to nearly full spray tank.

Use sufficient water for thorough ground and plant coverage. Use at least 1 gal. of water for each pound of tank mixture. More water may be needed for heavy weed growth. Maintain agitation during mixing and application.

Broadcast tank mixtures at the following rates to control weeds listed. Use the lower rates for light weed infestations and the higher rates for heavier infestations.

Use 4.4-5.6 lbs. of Atratol 90 plus 5-6.25 lbs. of diuron 80W, for example, Karmex® 80W, per acre to control weeds listed in crop and noncrop Directions for Use in the diuron 80W label.

Use 5.6 lbs. of Atratol 90 plus 6.25 lbs. of diuron 80W per acre to control broomsedge, buckhorn, plantain, Carolina geranium, chicory, downy brome, eveningprimrose, fleabane, goldenrod, horseweed, lespedeza, little mallow, orchardgrass, Russian thistle, shortpod mustard, tall fescue, wild barley, and wood sorrel.

Use 11.1 lbs. of Atratol 90 plus 12.5 lbs. of diuron 80W per acre to control alfalfa, black nightshade, leafy spurge, and shepherds-needle.

To determine compatibility of tank mixtures, thoroughly mix products in a small container at the ratio of 1 qt. of water to 1.75 oz. of Atratol 90 plus 2 oz. of diuron 80W. Let stand for 5 minutes. If the mixture remains mixed, or can be remixed readily, it is compatible and can be sprayed.

For small areas, 1/2 cup (4 oz.) of Atratol 90 or 1.25 cups of diuron 80W, such as Karmex 80W, per 1,000 sq. ft. is equivalent to 11.1 lbs. of Atratol 90 or 12.5 lbs. of diuron 80W per acre.

Atratol 90 plus Oust®

Use this tank mixture to control the weeds listed below with either preemergence, early postemergence, or postemergence applications. Use the higher rates within the ranges given under conditions of heavy weed growth, on soil with over 2 1/2% organic matter, on areas with heavy decaying plant residues, or on high moisture areas such as along road edges or railroad shoulders. To improve wetting and/or contact activity on emerged weeds, add a nonionic surfactant such as duPont Surfactant WK or X-77® at 1 qt. per 100 gals. of spray. Do not apply by aircraft. Provide constant agitation during mixing and application to keep the mixture in suspension. Refer to the labels of both herbicides for specific noncrop sites and further directions, precautions, and limitations.

Preemergence to Early Postemergence: To control the weeds listed below apply 4.4-8.9 lbs. of Atratol 90 plus 3-5 oz. of Oust in a minimum of 25 gals. of water per acre.

Broadleaf Weeds (Up to 3-4" in Height or Diameter)

bouncingbet
buckhorn plantain
burclover
Carolina geranium
common chickweed
common dandelion
common speedwell
common yarrow
crimson clover
dogfennel
filaree
hoary cress
horsetail
kochia
lambsquarters
little mallow
mustard
ox-eye daisy
pigweed
prickly lettuce
puncturevine
purple starthistle
ragweed
Russian thistle
sowthistle
St. Johnswort
sunflower
sweet clover
tansymustard
tansy ragwort
tumble mustard
turkey mullein
vetch
wild carrot
yellow rocket

Grasses (Up to 6-12" in Height)

Alta fescue
annual bluegrass
annual ryegrass
bahia grass
barnyardgrass
cheatgrass
crabgrass
downy brome
fall panicum
fescue
foxtails
Indiangrass
Italian ryegrass
johnsongrass (short-term control)
little barley
red brome
red fescue
reed canarygrass
ripgut brome
ryegrass
smooth brome
sprangletop
witchgrass

Postemergence: To control actively growing weeds listed below apply 8.9-17.8 lbs. of Atratol 90 plus 6-12 oz. of Oust in a minimum of 25 gals. of water per acre.

Actively Growing Weeds

bedstraw
burdock
Canada thistle
curly dock
dewberry
fiddleneck
fleabane
goldenrod
hemlock
honeysuckle
Jerusalem artichoke
kudzu
mayweed
musk thistle
plantain
poison ivy
spanishneedles
Virginia pepperweed
wild blackberry
johnsongrass (season-long control)
orchardgrass
quackgrass
redtop
yellow nutsedge

Atratol 90 plus Arsenal®
(For Use Only in States Where Registered)

Use this tank mixture to provide postemergence and residual control of the weeds listed in the various categories below. This tank mixture provides quicker postemergence weed control over that of Arsenal applied alone. Use the higher rates within the rate ranges given where heavy or well established infestations of weeds occur. Apply uniformly to the noncrop land area to be treated in 30-60 gals. of water per acre with a spray pressure of 25 to 30 psi. Refer to the Arsenal label for the addition of surfactant when spray volumes over 60 gallons per acre are applied. Also refer to the Atratol 90 and Arsenal label for specific noncrop sites and further directions, precautions, and limitations.

Annual Weeds

Apply 4.4-11.1 lbs. Atratol 90 plus 2-3 pts. of Arsenal per acre.

barnyardgrass (*Echinochloa crusgalli*)
broadleaf signalgrass (*Brachiaria platyphylla*)
carpetweed (*Mollugo verticillata*)
cheat (*Bromus* spp.)
common ragweed (*Ambrosia artemisifolia*)
curly dock (*Rumex crispus*)
downy brome (*Bromus tectorum*)
fleabane (*Erigeron* spp.)
foxtails (*Setaria* spp.)
goldenrod (*Solidago* spp.)
hoary vervain (*Verbena stricta*)
jimsonweed (*Datura stramonium*)
kochia (*Kochia scoparia*)
lambsquarters (*Chenopodium alba*)
lespedeza (*Lespedeza* spp.)
marestail (*Erigeron canadensis*)
nightshade (*Solanum* spp.)
pigweed (*Amaranthus* spp.)
plaintain (*Plantago* spp.)
puncturevine (*Tribulus terrestris*)
purslane (*Portulaca* spp.)
ragweed (*Ambrosia* spp.)
sicklepod (*Cassia obtusifolia*)
smartweed (*Polygonum* spp.)
sorrel (*Rumex* spp.)
sowthistle (*Sonchrus* spp.)
sunflower (*Helianthus annuus*)
velvetleaf (*Abutilon theophrasti*)
wild buckwheat (*Polygonum convolvulus*)
wild mustard (*Brassica kaber*)
wild oats (*Avena fatua*)
yellow woodsorrel (*Oxalis stricta*)

Apply 4.4-11.1 lbs. Atratol 90 plus 3-4 pts. of Arsenal per acre.

cocklebur (*Xanthium pennsylvanicum*)
crabgrass (*Digitaria* spp.)
crotolaria (*Crotolaria* spp.)
goosegrass (*Eleusine indica*)
morningglory (*Ipomoea* spp.)

Biennial/Perennial Weeds

Apply 5.3-11.1 lbs. Atratol 90 plus 2-3 pts. of Arsenal per acre.

dandelion (*Taraxacum officinale*)
field bindweed (*Convolvulus arvensis*)
guineagrass (*Panicum maximum*)
honeylocust (*Gleditsia triacanthos*)
johnsongrass (*Sorghum halepense*)
multiflora rose (*Rosa multiflora*)
ox-eye daisy (*Chrysanthemum leucathemum*)
paragrass (*Brachiaria mutica*)
quackgrass (*Agropyron repens*)
snadspur (*Cenchrus* spp.)
tall fescue (*Festuca arundinacea*)
vaseygrass (*Paspalum urvillei*)
wild carrot (*Daucus carota*)

Apply 5.3-11.1 lbs. Atratol 90 plus 3-4 pts. of Arsenal per acre.

dewberry (*Rubus* spp.)
greenbriar (*Smilax* spp.)
honeysuckle (*Lonicera* spp.)
opossum grape (*Cissus sicyoides*)
poison ivy (*Rhus radicans*)
redvine (*Brunnichia cirrhosa*)
trumpetcreeper (*Campsis radicans*)
Virginia creeper (*Parthenocissus quinquefolia*)

Apply 5.3-11.1 lbs. Atratol 90 plus 4-6 pts. of Arsenal per acre.

bermudagrass (*Cynodon dactylon*)
blackberry (*Rubus* spp.)
Canada thistle (*Cirsium arvense*)
mulberry (*Morus* spp.)
sumac (*Rhus* spp.)

STORAGE AND DISPOSAL

Pesticide Storage and Disposal

Store in a dry place. Do not contaminate water, food, or feed by storage, disposal, or cleaning of equipment. Wastes resulting from the use of this product may be disposed of on-site or at an approved waste disposal facility.

Container Disposal

Completely empty bag into application equipment. Dispose of empty bag in a sanitary landfill or by incineration, or if allowed by state and local authorities, by burning. Stay out of smoke from burning bags.

For minor spills, leaks, etc., follow all precautions indicated on this label and clean up immediately. Take special care to avoid contamination of equipment and facilities during cleanup procedures and disposal of wastes. In the event of a major spill, fire or other emergency, call (919) 292-7100 day or night.

PRECAUTIONARY STATEMENTS

HAZARDS TO HUMANS AND DOMESTIC ANIMALS

CAUTION

Harmful if swallowed, inhaled, or absorbed through the skin. Avoid contact with eyes, prolonged contact with skin, and inhalation of dust. Wash thoroughly after handling and before eating or smoking. Remove and wash contaminated clothing before reuse.

STATEMENT OF PRACTICAL TREATMENT

If swallowed: Call a physician or Poison Control Center immediately. Drink 1 or 2 glasses of water and induce vomiting by touching back of throat with finger, or, if available, by administering syrup of ipecac. Do not induce vomiting or give anything by mouth to an unconscious person.

If inhaled: Remove victim to fresh air. If not breathing, give artificial respiration, preferably mouth-to-mouth. Get medical attention.

If on skin: Wash with plenty of soap and water. Get medical attention.

If in eyes: Flush eyes with *plenty* of water. Call a physician if irritation persists.

Note to Physician: There is no specific antidote for atrazine. If this product is ingested, induce emesis or lavage stomach. The use of an aqueous slurry of activated charcoal may be considered.

ENVIRONMENTAL HAZARDS

Atrazine can travel (seep or leach) through soil and can enter ground water which may be used as drinking water. Atrazine has been found in ground water. Users are advised not to apply atrazine to sand and loamy sand soils where the water table (ground water) is close to the surface and where these soils are very permeable, i.e., well-drained. Your local agricultural agencies can provide further information on the type of soil in your area and the location of ground water.

This pesticide is toxic to aquatic invertebrates. Do not apply directly to water or wetlands. Do not apply where runoff is likely to occur. Do not apply when weather conditions favor drift from treated areas. Runoff and drift from treated areas may be hazardous to aquatic organisms in neighboring areas. Do not contaminate water by cleaning of equipment or disposal of wastes.

Arsenal® trademark of American Cyanamid Company
Atratol® trademark of CIBA–GEIGY
Caliber® trademark of CIBA–GEIGY
Karmex® trademark of E.I. duPont de Nemours and Co., Inc. for diuron
Oust® trademark of E.I. duPont de Nemours and Co., Inc.
Princep® trademark of CIBA–GEIGY for simazine
X-77® trademark of KALO Agricultural Chemicals, Inc.

CGA 20L6F 086

This is a specimen label which was not prepared by CIBA-GEIGY. CIBA-GEIGY is not responsible for the accuracy of the information contained herein. As labels are subject to revision, always carefully read and follow the label on the product container.

BICEP® 6L

HERBICIDE

For weed control in corn and grain or forage sorghum

ACTIVE INGREDIENTS:

Atrazine: 2-chloro-4-ethyl-amino-6-isopropylamino-*s*-triazine	27.4%
Atrazine related compounds	1.5%
Metolachlor: 2-chloro-*N*-(2-ethyl-6-methylphenyl)-*N*-(2-methoxy-1-methylethyl) acetamide	36.1%
INERT INGREDIENTS:	35.0%
TOTAL:	100.0%

EPA Reg. No. 100-645 EPA Est. 100-LA-1

BICEP 6L contains 6 lbs. active ingredients per gallon.

KEEP OUT OF REACH OF CHILDREN

CAUTION

See additional precautionary statements at end of label booklet.

See directions for use inside booklet.

Bicep® trademark of CIBA–GEIGY

U.S. Patent No. 3,937,730
(metolachlor)

CGA 56L12K 069

DIRECTIONS FOR USE AND CONDITIONS OF SALE AND WARRANTY

IMPORTANT: Read the entire **Directions for Use** and the **Conditions of Sale and Warranty** before using this product.

CONDITIONS OF SALE AND WARRANTY

The **Directions for Use** of this product reflect the opinion of experts based on field use and tests. The directions are believed to be reliable and should be followed carefully. However, it is impossible to eliminate all risks inherently associated with use of this product. Crop injury, ineffectiveness, or other unintended consequences may result because of such factors as weather conditions, presence of other materials, or the manner of use or application all of which are beyond the control of CIBA-GEIGY or the Seller. All such risks shall be assumed by the Buyer.

CIBA–GEIGY warrants that this product conforms to the chemical description on the label and is reasonably fit for the purposes referred to in the **Directions for Use** subject to the inherent risks referred to above. **CIBA–GEIGY makes no other express or implied warranty of Fitness or Merchantability or any other express or implied warranty. In no case shall CIBA–GEIGY or the Seller be liable for consequential, special, or indirect damages resulting from the use of handling of this product.** CIBA–GEIGY and the Seller offer this product, and the Buyer and user accept it, subject to the foregoing **Conditions of Sale and Warranty,** which may be varied only by agreement in writing signed by a duly authorized representative of CIBA–GEIGY.

DIRECTIONS FOR USE

It is a violation of federal law to use this product in a manner inconsistent with its labeling.

FAILURE TO FOLLOW THE DIRECTIONS FOR USE AND PRECAUTIONS ON THIS LABEL MAY RESULT IN POOR WEED CONTROL, CROP INJURY, OR ILLEGAL RESIDUES.

Do not apply this product in such a manner as to directly or through drift expose workers or other persons, except those knowingly involved in the application. The area being treated must be vacated by unprotected persons.

REENTRY STATEMENT

Do not enter treated areas without protective clothing until sprays have dried.

Because certain states may require more restrictive reentry intervals for various crops treated with this product, consult your State Department of Agriculture for further information. Written or oral warnings must be given to workers who are expected to be in a treated area or in an area about to be treated with this product. Oral warnings must be given which inform workers of areas or fields that may not be entered without specific protective clothing until sprays have dried, and appropriate actions to take in case of accidental exposure as described under **Precautionary Statements** on this label. When oral warnings are given, warnings shall be given in a language customarily understood by workers. Oral warnings must be given if there is reason to believe that written warnings cannot be understood by workers. Written warnings must include the following information: "CAUTION. Area treated with BICEP 6L on (*date of application*). Do not enter without appropriate protective clothing until sprays have dried. In case of accidental exposure, flush eyes or skin with plenty of water. Call a physician if irritation persists. Remove and wash contaminated clothing before reuse."

GENERAL INFORMATION

BICEP 6L is a selective herbicide recommended as a preplant surface-applied, preplant incorporated, preemergence, early postemergence or lay-by treatment for control of most annual grasses and broadleaf weeds in corn (field corn, silage corn, sweet corn, or popcorn), and as a preplant surface-applied, preplant incorporated, or preemergence treatment for control of most annual grasses and broadleaf weeds in grain or forage sorghum provided the sorghum seed has been properly treated by the seed company with Concep®. This product may be tank mixed with Bladex® (4L, 80W, 90WDG), Dual® 8E, AAtrex® 4L (4LC, 80W, Nine-O®), or Princep® 4L (80W or Caliber® 90) for weed control in conventional tillage corn. This product may also be tank mixed with either Gramoxone® Super or Roundup® alone or in combination with AAtrex, Bladex, Dual or Princep, in minimum-tillage or no-tillage corn, or tank mixed with either Gramoxone or Roundup in minimum-tillage or no-tillage sorghum.

Following many years of continuous use of atrazine (one of the ingredients in BICEP 6L), and products chemically related to atrazine, biotypes of some of the weeds listed on this label which are controlled by the atrazine component have been reported to develop resistance to this and chemically related herbicides. Where this is known or suspected and weeds controlled by this product are expected to be present along with resistant biotypes, we recommend the use of BICEP 6L in combination or in sequence with registered herbicides which do not contain triazines. Consult with your State Agricultural Extension Service for specific recommendations.

Precautions: (1) If sorghum seed is not properly pretreated with Concep, BICEP 6L will severely injure the crop. (2) Injury may occur to corn or sorghum following the use of BICEP 6L under abnormally high soil moisture conditions during early development of the crop.

BICEP 6L alone or in tank mixture with AAtrex, Bladex, Dual, or Princep may be applied preplant incorporated, preplant surface, or preemergence on corn in water or fluid fertilizer. BICEP 6L may be applied in tank mix combination with Gramoxone Super or Roundup with or without the above herbicides preplant surface or preemergence to corn. Apply the early postemergence treatment on corn in water only. BICEP 6L alone may also be applied on sorghum preplant incorporated, preplant surface, or preemergence in water or in fluid fertilizer.

BICEP 6L may be applied in water by aircraft. Applications in fluid fertilizer should be only by ground equipment.

To avoid spray drift, do not apply under windy conditions. Avoid spray overlap, as crop injury may result.

This product may also be applied in irrigation water through center pivot irrigation systems.

Dry weather following preemergence application of BICEP 6L or a tank mixture may reduce effectiveness. Cultivate if weeds develop in conventional tillage corn or sorghum.

Observe all precautions and limitations on the label of each product used in tank mixtures.

Thoroughly clean sprayer prior to use. Do not use a sprayer contaminated with any other materials, or crop damage or sprayer clogging may result.

Mixing Instructions

Shake well before using. BICEP 6L is a liquid that may be mixed with water or fluid fertilizer and applied as a spray. BICEP 6L may also be sprayed onto dry bulk granular fertilizer and applied with the granular fertilizer.

Dry Bulk Granular Fertilizers

Many dry bulk granular fertilizers may be impregnated or coated with BICEP 6L and used to control weeds in corn or Concep-treated sorghum.

When applying BICEP 6L with dry bulk granular fertilizers, follow all directions for use and precautions on the BICEP 6L label regarding target crops, rates per acre, soil texture, application methods, and rotational crops.

All individual state regulations relating to dry bulk granular fertilizer blending, registration, labeling, and application are the responsibility of the individual and/or company selling the herbicide/fertilizer mixture.

Prepare the herbicide/fertilizer mixture by using any closed drum, belt, ribbon, or other commonly used dry bulk fertilizer blender. Nozzles used to spray BICEP 6L onto the fertilizer must be placed to provide uniform spray coverage.

If the herbicide/fertilizer mixture is too wet, use a highly absorptive material, such as Agsorb™ granules, Microcel E (Johns-Manville Products Corporation), diatomaceous earth or finely powdered clay, to obtain a dry free-flowing mixture. Add the absorptive material separately and uniformly to the herbicide/fertilizer mixture and blend to form a suitable free-flowing mixture. Generally, less than 2% by weight of absorptive material will be needed.

Calculate the amount of BICEP 6L to be used by the following:

$$\frac{2{,}000}{\text{lbs. of fertilizer per acre}} \times \text{qts. of BICEP 6L per acre} = \text{qts. of BICEP 6L per ton of fertilizer}$$

Pneumatic (Compressed Air) Application: High humidity, high urea concentrations, low fertilizer use rates, and dusty fertilizer may cause fertilizer mixture to build up or plug the distributor head, air tubes, or nozzle deflector plates. To minimize buildup, premix BICEP 6L with a noncombustible/nonflammable mineral oil at up to a 1:1 ratio of BICEP 6L to mineral oil per acre and spray mixture uniformly onto fertilizer. Mineral oil additive may be used either in a fertilizer blender or through direct injection systems. Drying agents should not be used when using oil additives. Use only those oils recommended by CIBA–GEIGY, i.e., KERMAC #600™ Solvent (CV-600 Mineral Seal Oil).

Notes: (1) Some separation of the BICEP 6L + mineral oil mixture may occur in temperatures below 40°F. To assure uniform application, use either in-tank agitation to keep the products mixed or keep products heated above 40°F before mixing. (2) When impregnating BICEP 6L in a blender before application, a drier mixture can be attained by substituting a drying agent for mineral oil. The use of Agsorb granules is recommended over other types of absorptive materials. (3) Drying agents are not recommended for use with On-The-Go impregnation equipment.

Precautions: To avoid potential for explosion, (1) Do not impregnate BICEP 6L on ammonium nitrate, potassium nitrate, or sodium nitrate either alone or in blends with other fertilizers. (2) Do not combine BICEP 6L with a single superphosphate (0-20-0) or treble superphosphate (0-46-0). (3) Do not use BICEP 6L on straight limestone, since absorption will not be achieved. Fertilizer blends containing limestone can be impregnated.

APPLICATION

Apply 200-700 pounds of the herbicide/fertilizer mixture per acre. For best results, apply the mixture uniformly to the soil with properly calibrated equipment immediately after blending. Uniform application of the herbicide/fertilizer mixture is essential in order to prevent possible crop injury or injury to subsequent rotational crops. Non-uniform application may also result in unsatisfactory weed control. In areas where conventional tillage is practiced, a shallow incorporation of the mixture into the soil is recommended to obtain satisfactory weed control. On fine or medium-textured soils in areas where soil incorporation is not planned, i.e., reduced tillage situations or in some conventional till situations, make applications approximately 30 days before planting to allow moisture to move the herbicide/fertilizer mixture into the soil. On coarse-textured soils, make applications approximately 14 days prior to planting.

Precautions: (1) To help avoid rotational crop injury, make applications as early as possible, since BICEP 6L impregnated onto dry bulk granular fertilizers can be expected to last longer in the soil than when BICEP 6L is applied as a spray in water or fluid fertilizer. (2) Do not use on crops where bedding occurs, or crop injury may occur.

Application in Water or Fluid Fertilizers

BICEP 6L Alone: Fill the spray tank one-half to three-fourths full with water or fluid fertilizer, add the proper amount of BICEP 6L, then add the rest of the water or fluid fertilizer. Provide sufficient agitation during mixing and application to maintain a uniform suspension.

Tank Mixtures: Fill the spray tank one-half to three-fourths full with water or fluid fertilizer, add the proper amount of BICEP 6L, then add AAtrex, Banvel®, linuron, or Princep; next add Dual; then add Gramoxone Super or Roundup, depending on the tank mix combination desired; and finally add the rest of the water or fluid fertilizer. When Bladex is used in the tank mixture, add it before BICEP 6L unless otherwise specified. (See Bladex section under **BICEP 6L Combinations—Corn** for further mixing instructions). Provide sufficient agitation during mixing and application to maintain a uniform suspension.

Compatibility Test: Check the compatability of BICEP 6L and tank mixtures in fluid fertilizer by mixing proportionate quantities in a small container, as described below, before mixing in the spray tank. Nitrogen solutions or complete fluid fertilizers may replace all or part of the water in the spray. Since liquid fertilizers can vary, even within the same analysis, **always check compatibility each time before reuse.** Be especially careful when using **complete** suspension or fluid fertilizers as serious compatibility problems are more apt to occur. Commercial application equipment may improve compatibility in some instances. The following test assumes a spray volume of 25 gals. per acre. For other spray volumes, make appropriate changes in the ingredients. Check compatibility using this procedure:

1. Add 1 pint of fertilizer to each of 2 one-quart jars with tight lids.
2. To **one** of the jars add 1/4 tsp. or 1.2 milliliters of a compatibility agent approved for this use, such as Compex® or Unite® (1/4 tsp. is equivalent to 2 pts. per 100 gals. spray). Shake or stir gently to mix.
3. To **both** jars add the appropriate amount of herbicide(s). If more than one herbicide is used, add them separately with dry herbicides first, flowables next, and emulsifiable concentrates last. After each addition, shake or stir gently to thoroughly mix. The appropriate amount of herbicides for this test follows:

   Dry herbicides: For each pound to be applied per acre, add 1.4 teaspoons to each jar.

   Liquid herbicides: For each pint to be applied per acre add 0.5 teaspoon or 2.5 milliliters to each jar.
4. After adding all ingredients, put lids on and tighten, and invert each jar ten times to mix. Let the mixtures stand 15 minutes and then look for separation, large flakes, precipitates, gels, heavy oily film on the jar, or other signs of incompatibility. Determine if the compatibility agent is needed in the spray mixture by comparing the two jars. If either mixture separates, but can be remixed readily, the mixture can be sprayed as long as good agitation is used. If the mixtures are incompatible, test the following methods of improving compatibility: (A) slurry the dry herbicide(s) in water before addition, or (B) add 1/2 of the compatibility agent to the fertilizer and the other 1/2 to the emulsifiable concentrate or flowable herbicide before addition to the mixture. If still incompatible, do not use the mixture.

Application Procedures

Ground Application: Use sprayers that provide accurate and uniform application. Screens in nozzles and in suction and in-line strainers should be no finer than 50-mesh. Use a pump with capacity to (1) maintain 35-40 psi at the nozzles, and (2) provide sufficient agitation in tank to keep mixture in suspension. Unless otherwise specified, use a minimum of 10 gals. of spray mixture per acre. Rinse sprayer thoroughly with clean water immediately after use.

For band applications, calculate amount to be applied per acre as follows:

$$\frac{\text{band width in inches}}{\text{row width in inches}} \times \text{broadcast rate per acre} = \text{amount needed per acre of field}$$

Aerial Application (for BICEP 6L alone): Use aerial application only where broadcast applications are specified. Apply a minimum of 1 gal. of water for each 1 gal. of this product applied per acre, but for rates below 1 gal./acre, use in sufficient water to equal 2 gals./acre of total spray. Avoid applications under conditions where uniform coverage cannot be obtained or where excessive spray drift may occur. In order to assure that spray will be controllable within the target area when used according to label directions, make applications at a maximum height of 10 ft., using low drift nozzles at a maximum pressure of 40 psi, and restrict application to periods when wind speed does not exceed 10 mph. To assure that spray will not adversely affect adjacent sensitive nontarget plants, apply BICEP 6L by aircraft at a minimum upwind distance of 400 ft. from sensitive plants.

Avoid application to humans or animals. Flagmen and loaders should avoid inhalation of spray mist and prolonged contact with skin, and should wash thoroughly before eating and at the end of each day's operation.

Center Pivot Irrigation Application: BICEP 6L alone or in tank mixture with other herbicides which are registered for center pivot application may be applied in irrigation water preemergence (after planting but before weeds or crop emerge) at rates recommended on this label. Apply this product only through a center pivot irrigation system. Do not apply this product through any other type of irrigation system. Crop injury, lack of effectiveness, or illegal pesticide residues in the crop can result from nonuniform distribution of treated water. If you have questions about calibration, you should contact State Extension Service specialists, equipment manufacturers or other experts. Do not connect an irrigation system (including greenhouse systems) used for pesticide application to a public water system unless the pesticide label-prescribed safety devices for public water systems are in place. A person knowledgeable of the chemigation system and responsible for its operation, or under the supervision of the responsible person, shall shut the system down and make necessary adjustments should the need arise.

Operating Instructions

1. The system must contain a functional check valve, vacuum relief valve, and low pressure drain appropriately located on the irrigation pipeline to prevent water-source contamination from backflow.
2. The pesticide injection pipeline must contain a functional, automatic, quick-closing check valve to prevent the flow of fluid back toward the injection pump.
3. The pesticide injection pipeline must also contain a functional, normally closed, solenoid-operated valve located on the intake side of the injection pump and connected to the system interlock to prevent fluid from being withdrawn from the supply tank when the irrigation system is either automatically or manually shut down.
4. The system must contain functional interlocking controls to automatically shut off the pesticide injection pump when the water pump motor stops.
5. The irrigation line or water pump must include a functional pressure switch which will stop the water pump motor when the water pressure decreases to the point where pesticide distribution is adversely affected.
6. Systems must use a metering pump, such as a positive displacement injection pump (e.g., diaphragm pump) effectively designed and constructed of materials that are compatible with pesticides and capable of being fitted with a system interlock.
7. Do not apply when wind speed favors drift beyond the area intended for treatment.
8. Prepare a mixture with a minimum of 1 part of water to 1 part herbicide(s) and inject this mixture into the center pivot system. Injecting a larger volume of a more dilute mixture per hour will usually provide more accurate calibration of metering equipment. Maintain sufficient agitation to keep the herbicide in suspension.
9. Meter into irrigation water during the entire period of water application.
10. Apply in 1/2-1 inch of water. Use the lower water volume (1/2 inch) on coarser-textured soils and the higher volume (1 inch) on finer-textured soils. More than 1 inch of water at application may reduce weed control by moving the herbicide below the effective zone in the soil.

Precautions for center pivot applications: Where sprinkler distribution patterns do not overlap sufficiently, unacceptable weed control may result. Where sprinkler distribution patterns overlap excessively, crop injury may result.

Within rate ranges in all tables on this label, use the lower rate on soil relatively coarse-textured or low in organic matter; use the higher rate on soil relatively fine-textured or high in organic matter.

BICEP 6L APPLIED ALONE—CORN, GRAIN SORGHUM, OR FORAGE SORGHUM

Preplant Surface-Applied, Preplant Incorporated, or Preemergence

Weeds Controlled		*Weeds Partially Controlled**
barnyardgrass (watergrass)	black nightshade	sandbur
browntop panicum	carpetweed	seedling johnsongrass
crabgrass	cocklebur	shattercane
crowfootgrass	common purslane	sicklepod
fall panicum	Florida pusley	volunteer sorghum
giant foxtail	galinsoga	
goosegrass	hairy nightshade	
green foxtail	jimsonweed	
red rice	lambsquarters	
signalgrass (*Brachiaria*)	morningglory	
southwestern cupgrass	pigweed	
witchgrass	ragweed	
yellow foxtail	smartweed	
yellow nutsedge	velvetleaf	

* Control of these weeds can be erratic due partially to variable weather conditions. Control may be improved by following these suggested procedures:

1. **Thoroughly till moist soil** to destroy germinating and emerged weeds. If BICEP 6L is to be applied preplant incorporated, this tillage may be used to incorporate BICEP 6L if uniform 2 inch incorporation is achieved as recommended under **Application Procedures.**
2. Plant crop into moist soil **immediately after tillage.** If BICEP 6L is to be used preemergence, apply at planting or immediately after planting.
3. If available, **sprinkler irrigate** within 2 days after application. Apply 1/2-1 inch of water. Use lower water volume (1/2 inch) on coarse-textured soils and higher volume (1 inch) on finer textured soils. Also, refer to the section on **Center Pivot Irrigation Application** for this method of applying BICEP 6L.
4. If irrigation is not possible and rain does not occur within 2 days after planting and application, weed control may be decreased. Under these conditions, a uniform, shallow cultivation is recommended as soon as weeds emerge.

Application Timings

Preplant Surface-Applied (Corn): Use on medium- and fine-textured soils with minimum-tillage or no-tillage systems in CO, IN, IL, IA, KS, KY, MN, MO, MT, NE, ND, SD, TN, WI, and WY. Apply 2/3 the recommended rate of BICEP 6L (2.4-3.0 qts./A on *medium* soils and 3.0-3.75 qts./A on *fine* soils) as a split treatment of 30-45 days before planting and the remainder at planting. Applications made less than 30 days prior to planting may be as either a split or single treatment. Use the lower rate for light expected weed infestations and the higher rate for heavy expected weed infestations. On *coarse-textured* soils, apply 2.4 qts./A not more than 2 weeks prior to planting. The above procedure may be followed if AAtrex or Dual 8E or Princep is used in tank mixtures with BICEP 6L. If weeds are present at the time of treatment, apply in a tank mixture combination with a contact herbicide (for example, Gramoxone Super or Roundup). Observe directions for use, precautions, and restrictions on the label of the contact herbicide.

On medium- and fine-textured soils with minimum- or no-tillage systems in DE, MD, MI, NY, OH, PA, VA, AND WV, preplant surface applications may be applied following the directions for use above. If the amount of rainfall results in unsatisfactory length of weed control following the earlier treatment, a postemergence application of an appropriately labeled broadleaf and/or grass weed herbicide may be used, i.e., AAtrex, Basagran® 2,4-D, Banvel®, bromoxynil (Brominal® or Buctril®). If the postemergence treatment includes the herbicide used preplant surface-applied, do not exceed the labeled rate for corn on a given soil texture. Observe all directions for use, precautions, and limitations on the label of the postemergent herbicide.

BICEP 6L may be used according to the above directions to control winter wheat planted as a cover crop in IN, KY, and OH, in addition to providing residual weed control. The wheat must be less than 6 inches tall (preferably still in a dormant or semi-dormant state coming out of winter) at the time of application. Depending on rainfall, 10-20 days may be required to completely kill the wheat. In the event that adequate rainfall does not occur, control of the winter wheat may be unsatisfactory and the application of a contact herbicide (i.e., Gramoxone Super or Roundup) may be required before planting the crop.

On medium- and fine-textured soils following final seed bed preparation in the Blackland and Gulf Coast areas of Texas, a preplant surface application of BICEP 6L at 1.8 to 2.2 qts./A may be made 30-45 days before planting. Grass suppression of 2-3 weeks after planting can be expected as a result of this application. Do not incorporate or disturb the soil before planting and avoid moving the soil during the planting operation. A follow-up application of Dual 8E may be needed in fields with a history of heavy grass pressure. Apply after planting, but before corn and grass weeds emerge.

Note: If a follow-up application of Dual 8E is needed, do not exceed a total of 2.5 lbs. ai of metolachlor/A including the preplant BICEP 6L application on medium- or fine-textured soils. On fine-textured soils with more than 3% organic matter, do not exceed 3 lbs. ai or metolachlor.

Note: To the extent possible, do not move treated soil out of the row or move untreated soil to the surface during planting, or weed control will be diminished.

Preplant Surface-Applied (Sorghum-Seed Treated with Concep): For minimum-tillage and no-tillage systems only, BICEP 6L may be applied up to 45 days before planting grain sorghum in IL, IA, eastern KS, MO, NE, and SD. Use only split applications for treatments made 30 to 45 days before planting with 2/3 the recommended rate applied initially and the remaining 1/3 at planting. On medium soils with greater than 1.5% organic matter, and on fine soils with less than 1.5% organic matter, apply 2.4-2.7 qts./A of BICEP 6L; on fine soils with greater than 1.5% organic matter, apply 2.7-3.0 qts./A of BICEP 6L. Applications made less than 30 days prior to planting may be made as either a split or single application.

If weeds are present at the time of treatment, apply in a tank mixture combination with a contact herbicide (for example, Gramoxone Super or Roundup). Observe directions for use, precautions, and restrictions on the label of the contact herbicide. Under dry conditions, irrigation after application is recommended to move BICEP 6L into the soil.

Note: To the extent possible, do not move treated soil out of the row or move untreated soil to the surface during planting, or weed control will be diminished. Do not use on coarse soils. Do not use on medium soils with less than 1.5% organic matter.

On medium- and fine-textured soils following final seed bed preparation in the Blackland and Gulf Coast areas of Texas, a preplant surface application of BICEP 6L at 1.8 to 2.2 qts./A may be made 30-45 days before planting. Grass suppression of 2-3 weeks after planting can be expected as a result of this application. Do not incorporate or disturb the soil before planting and avoid moving the soil during the planting operation. A follow-up application of Dual 8E may be needed in fields with a history of heavy grass pressure. Apply after planting, but before sorghum and grass weeds emerge.

Note: (1) Do not use on soils with a pH greater than 8.0 if grain sorghum is to be planted. (2) If a follow-up application of Dual 8E is needed, do not exceed a total of 2.25 lbs. ai of metolachlor/A, including the preplant BICEP 6L application on medium-textured soils. On fine-textured soils, do not exceed 2.5 lbs. ai of metolachlor/A.

Preplant Incorporated or Preemergence (Corn or Sorghum-Seed Treated with Concep): Apply BICEP 6L preplant incorporated or preemergence using the appropriate rates from Table 1 for corn, or from Table 2 for sorghum.

Preplant Incorporated: Apply to the soil and incorporate into the top 2 inches of the soil within 14 days prior to planting using a finishing disk harrow, rolling cultivator, or similar implement capable of providing uniform 2-inch incorporation. Use preplant incorporated if furrow irrigation is used or when a period of dry weather after application is expected. If crop is to be planted on beds, apply and incorporate after bed formation.

Preemergence: Apply to the soil surface of planting (behind the planter), or after planting but before weeds or crop emerge.

Table 1: BICEP 6L Preplant Incorporated, or Preemergence—Corn (Field corn, silage corn, sweet corn, popcorn)

Soil texture	Broadcast rate per acre	
	Less than 3% organic matter	3% organic matter or greater
COARSE: Sand, loamy sand, sandy loam	1.5 qts.	1.8 qts.
MEDIUM: Loam, silt loam, silt	1.8 qts.	2.4 qts.
FINE: Silty clay loam, sandy clay loam, silty clay, sandy clay, clay loam, clay	2.4 qts.	2.4-3 qts.*
Muck or peat soils (more than 20% organic matter)	DO NOT USE	

* For cocklebur, yellow nutsedge, and velvetleaf control on fine-textured soils above 3% organic matter: Apply 3 qts. of BICEP 6L per acre.

Notes: (1) In the event of escape of annual broadleaf weeds following a preplant surface, preplant incorporated, or preemergence treatment of Bicep applied alone or in combination, follow with a postemergence application of an appropriately labeled broadleaf and/or grass weed herbicide, i.e., AAtrex, Basagran, Brominal, Buctril, Banvel or 2,4-D. If the postemergence treatment includes the herbicide used in the earlier treatment, do not exceed the labeled rate for corn on a given soil texture. (2) Brominal or Buctril may be applied postemergence alone or in tank mix combination with AAtrex. Do not exceed 1.2 lbs. ai/A of AAtrex in tank mix combination with Brominal or Buctril postemergence. Refer to the Brominal, Buctril and AAtrex labels for specific rates and precautions.

Table 2: BICEP 6L—Grain or Forage Sorghum† (Seed treated with Concep)

Soil texture	Organic matter	Broadcast rate per acre
COARSE: Sand, loamy sand, sandy loam	any level	DO NOT USE
MEDIUM AND FINE: Loam, silt loam, silt,	less than 1%	DO NOT USE
sandy clay loam, clay loam, silty clay loam,	1-1.5%	1.8 qts.
sandy clay, silty clay, clay	more than 1.5%	1.8-2.4 qts.

† Do not use in NM, OK, or TX except in northeast OK and Texas Gulf Coast and Blacklands areas. Do not apply preplant incorporated in AZ or the Imperial Valley of CA.

Precautions: To avoid possible crop injury, (1) Do not apply BICEP 6L on highly alkaline soils or on eroded areas where calcareous subsoils are exposed. (2) Do not apply BICEP 6L when sorghum is planted in deep furrows because heavy rains following application can cause excessive concentrations of herbicide in the furrow. (3) Do not apply to sorghum grown under dry mulch tillage. (4) Injury may occur if both BICEP 6L applied preplant surface, preplant incorporated, or preemergence and an at-plant systemic insecticide applied in-furrow are used. (5) In addition, sorghum growing under stress caused by minor element deficiency may be injured by BICEP 6L.

Rotational Crops: (1) If treated crop is lost due to poor germination, hail, flood, insects, etc., corn may be replanted immediately or sorghum may be replanted immediately provided the seed has been properly treated with Concep. Do not make a second broadcast application. If the original application was banded and the second crop is planted in the untreated row middles, a second banded treatment may be applied. (2) Corn, sorghum, soybeans, cotton, or peanuts may be planted the spring following treatment. Do not graze or feed forage or fodder from cotton to livestock or illegal residues may result. Injury may occur to soybeans planted the year following application on soils having a calcareous surface layer, including, but not limited to those found in north-central and northwest Iowa; south-central, western and southwest Minnesota; northeast Nebraska; North Dakota and eastern South Dakota. (3) If applied after June 10, do not rotate with crops other than corn or sorghum the next year, or crop injury may occur. (4) In the High Plains and Intermountain areas of the West where rainfall is sparse and erratic or where irrigation is required, use only when corn or sorghum is to be planted the following year, or a crop of untreated corn or sorghum is to precede other rotational crops. (5) Small grains may be planted 15 months following treatment. (6) All other crops may be planted 18 months after application.

Postemergence—Corn

Weeds Controlled		Weeds Partially Controlled
barnyardgrass (watergrass)	jimsonweed	cocklebur
crabgrass	kochia	morningglory
crowfootgrass	lambsquarters	yellow nutsedge
fall panicum	mustard	
giant foxtail	pigweed	
green foxtail	prickly sida	
yellow foxtail	purslane	
	ragweed	
	smartweed	
	velvetleaf	

Application: Apply early postemergence using the appropriate rate from Table 3. Apply this treatment before grass and broadleaf weeds pass the 2-leaf stage and before corn exceeds 5 inches in height. Application to weeds larger than the 2-leaf stage will generally result in unsatisfactory control. Occasional corn leaf burn may result, but this should not affect later growth or yield. Do not apply postemergence in fluid fertilizer or severe crop injury may occur.

Table 3: Postemergence — Corn

Soil texture	Broadcast rate per acre
COARSE: Sand, loamy sand, sandy loam	1.8 qts.
MEDIUM: Loam, silt loam, silt	2.4 qts.
FINE: Sandy clay loam, silty clay loam, clay loam, sandy clay, silty clay, clay	2.4-3 qts.*

* For better control of cocklebur, morningglory, velvetleaf, and yellow nutsedge on fine-textured soils above 3% organic matter, apply 3 qts. of BICEP 6L per acre.

Rotational Crops: Follow the preceding crop rotation instructions for BICEP 6L Preplant Surface-Applied, Preplant Incorporated, or Preemergence.

Lay-by— Corn

BICEP 6L may be applied at 1.5-3 qts./A in a minimum of 15 gals. of water as a lay-by directed treatment to corn to extend control of weeds listed in the **Preplant Surface-Applied, Preplant Incorporated, Preemergence,** or **Postemergence** section of the corn label. Apply using the appropriate rate from Table 4.

For best results, apply BICEP 6L to weed free soil following use of a preplant incorporated or preemergence herbicide, or following a lay-by cultivation. If weeds have emerged at the time of BICEP 6L application, apply before grass and broadleaf weeds exceed the 2-leaf stage. Application to weeds larger than the 2-leaf stage will generally give unsatisfactory control. Do not apply to corn exceeding 40 inches in height. Minimize contact with corn leaves. Do not apply postemergence in fluid fertilizer, or severe crop injury may occur.

Notes: (1) If BICEP 6L has been applied preplant surface, preplant incorporated or preemergence, do not exceed a total of 6 qts./A of BICEP 6L on a corn crop or illegal residues may result. If AAtrex (atrazine) or AAtrex plus Dual tank mixtures have been applied preplant incorporated or preemergence, limit the BICEP 6L lay-by application not to exceed a total of 4 lbs. of active AAtrex or 6 lbs. of active Dual per acre on a corn crop, or illegal residues may result.

Table 4: Lay-by—Corn

Soil texture	Broadcast rate per acre
COARSE: Sand, loamy sand, sandy loam	1.5 qts.
MEDIUM: Loam, silt loam, silt	2.4 qts.

Soil texture	Broadcast rate per acre
FINE: Sandy clay loam, silty clay loam, clay loam, sandy clay, silty clay, clay	2.4-3 qts.*

* For better control of cocklebur, morningglory, velvetleaf, and yellow nutsedge on fine-textured soils above 3% organic matter, apply 3 qts. of BICEP 6L per acre.

Rotational Crops: (1) Do not rotate to any crop except corn or sorghum the following year, or injury may occur. (2) If the total amount of AAtrex, applied from preplant incorporated or preemergence applications of AAtrex or BICEP 6L plus postemergence applications of BICEP 6L, exceeds 3 lbs. active ingredient, a crop of untreated corn or sorghum should precede the next rotational crop.

BICEP 6L COMBINATIONS — CORN

Tank Mixture with AAtrex, Princep, or Dual 8E — Conventional Tillage

AAtrex (4L, 4LC, 80W, or Nine-O): Add up to 1 qt. of AAtrex 4L or 4LC (1.25 lbs. 80W or 1.1 lbs. Nine-O) per acre to the rate of BICEP 6L recommended in Table 1 in the southeastern U.S. where high rainfall can shorten the duration of control of broadleaf weeds, and in all areas where heavy infestations of cocklebur, morningglory, velvetleaf, or other broadleaf weeds claimed are expected.

Dual 8E: Add up to 1/2 pt. of Dual 8E per acre to the rate of BICEP 6L recommended in Table 1 when heavy infestations of yellow nutsedge, sandbur, or seedling johnsongrass are expected.

Princep (4L, 80W, or Caliber 90): Add up to 1 qt. of Princep 4L (1.25 lbs. 80W or 1.1 lbs. Caliber 90) per acre to the rate of BICEP 6L recommended in Table 1 in the northeastern U.S. where heavy infestations of crabgrass or fall panicum are expected.

Bladex (4L, 80W, or 90WDG) [Field Corn and Silage Corn Only]: Add up to 2 qts. of Bladex 4L (2.5 lbs. 80W or 2.2 lbs. 90WDG) per acre. When Bladex is added, the rate of BICEP 6L suggested in Table 1 should be reduced by not more than 25% for a given soil texture in the Clarion-Nicollet-Webster soil association in northern Iowa and southern Minnesota, or in other areas where soybean rotational concerns exist.

Note: Check the compatibility of BICEP 6L tank mixtures with Bladex before mixing in spray tank by using the procedure described under **Application in Water or Fluid Fertilizers.** Always use Unite (or an equivalent compatibility agent) at 2 pts./100 gals. when using Bladex 80W or Bladex 90WDG in tank mixtures with BICEP 6L. Compatibility testing is especially critical when using Bladex 80W or Bladex 90WDG combinations. Always add Bladex to the spray tank before BICEP 6L, regardless of which Bladex formulation is used.

Tank Mixtures of BICEP 6L Alone or BICEP 6L plus AAtrex, Bladex, Dual 8E, or Princep, with Gramoxone Super or Roundup for Minimum-Tillage or No-Tillage Systems

In minimum-tillage or no-tillage systems where corn is planted directly into a cover crop, stale seedbed, established sod, or previous crop residues, the contact herbicides Gramoxone Super or Roundup should be tank mixed with BICEP 6L alone or with BICEP 6L plus AAtrex, Bladex, Dual 8E, or Princep. When used as directed, the Gramoxone Super portion of the tank mixture controls most emerged annual weeds and suppresses many perennial weeds. Roundup combinations will control emerged annual and perennial weeds when applied as directed on the Roundup label. The BICEP 6L portion of the tank mixture provides preemergence control of the weeds listed on this label in the **BICEP 6L Alone** section for corn. The addition of AAtrex, Bladex, Dual 8E, or Princep offers the advantage indicated for each under "Conventional Tillage" above.

Application: Apply before, during, or after planting but before corn emerges, at the appropriate rate in Table 5. Up to 1 qt. of AAtrex 4L/4LC (1.25 lbs. 80W or 1.1 lbs. Nine-O), or 2 qts. of Bladex 4L (2.5 lbs. 80W or 2.2 lbs. 90WDG), or 1/2 pt. of Dual 8E, or 1 qt. of Princep 4L (1.25 lbs. 80W or 1.1 lbs. Caliber 90) per acre may be added to the rate of BICEP 6L recommended in Table 5. Add Gramoxone or Roundup at the following broadcast rates:

Gramoxone: 1.5-2.5 pts. per acre plus 8 oz. of X-77® Spreader per 100 gals. of spray mixture. Use the lower rate for control of annual weeds less than 4 inches tall and the higher rate for weeds 4-6 inches tall. This treatment will not consistently control weeds taller than 6 inches.

Roundup: 1.5 qts. per acre for control of existing annual weeds, or 2-4 qts. per acre for existing perennial weeds. See the Roundup label for weeds controlled and recommended rates for specific weeds.

Apply in 20-60 gals. of water per acre with conventional spray equipment.

Tank Mixtures of BICEP 6L Alone or BICEP 6L plus AAtrex or Bladex, with 2,4-D or 2,4-D plus Banvel for Minimum-Tillage or No-Tillage Systems

In minimum-tillage or no-tillage systems where corn is planted directly into a cover crop, stale seedbed, established sod, or previous crop residues, BICEP 6L may be applied in combination with AAtrex or Bladex. When used as directed, the BICEP 6L portion of the tank mixture provides preemergence control of the weeds listed on this label in the **BICEP 6L Alone** section for corn. The addition of AAtrex or Bladex offers the advantage indicated for each under "Conventional Tillage" above.

Application: Apply BICEP 6L before, during, or after planting but before corn emerges, at the appropriate rate in Table 5. Up to 1 qt. of AAtrex 4L/4LC (1.25 lbs. 80W or 1.1 lbs. Nine-O) or 2 qts. of Bladex 4L (2.5 lbs. 80W or 2.2 lbs. 90WDG) per acre may be added to the rate of BICEP 6L recommended in Table 5.

Where heavy crop residues exist, add an appropriately-labeled 2,4-D amine or low volatile ester to the spray tank last and apply in a minimum of 25 gals. of carrier/A.

As carriers, nitrogen solutions and complete liquid fertilizers, applied before corn emergence, enhance burndown of existing weeds, and therefore are recommended instead of water. Add X-77 surfactant at 1.0-2.0 qts./100 gals. of diluted spray, or another surfactant cleared for use on growing crops at its recommended rate. Apply before weeds exceed 3 inches in height. If alfalfa is present, add Banvel to the spray mixture at 0.33-0.5 pt./A and apply before alfalfa exceeds 6 inches in height.

For fields with existing sod grasses (e.g., bromegrass, orchardgrass, rye or timothy), when existing weeds exceed 3 inches in height or when very dry conditions exist, add Gramoxone Super at the rate of 2.5 pts./A in place of or in addition to 2,4-D as indicated above. Do not apply Gramoxone Super in suspension-type liquid fertilizer. Observe all directions for use, precautions and limitations on the respective product labels when applying these products in tank mix combination.

Note: When applying BICEP 6L and Bladex in tank mix combination, follow directions under "Conventional Tillage" above to ensure compatibility of these products in the tank mixture.

Table 5: BICEP 6L for Minimum-Tillage or No-Tillage Corn

Soil texture	Broadcast rate per acre
COARSE: Sand, loamy sand, sandy loam	1.8 qts.
MEDIUM: Loam, silt loam, silt	2.4 qts.
FINE: Sandy clay loam, silty clay loam, clay loam, sandy clay, silty clay, clay	2.4-3 qts.*
Muck or peat soils	DO NOT USE

* For cocklebur, yellow nutsedge, and velvetleaf control on fine-textured soils above 3% organic matter, apply 3 qts. of BICEP 6L per acre.

Tank Mixture with Linuron for Control of Lambsquarters and Pigweed

For prolonged control of lambsquarters and pigweed in DE, MD, NJ, NY, PA, VA and WV, BICEP 6L may be applied preemergence in combination with linuron. Apply BICEP 6L according to the rates in Table 1 and linuron according to the rates below.

Soil texture	Broadcast rate per acre
Sandy loam (1-3% organic matter)	0.67 lb. linuron 50W*
Sandy loam (3-6% organic mater)	1.0 lb. linuron 50W*
Medium and fine-textured soils (1-6% organic matter)	1.0 lb. linuron 50W*

* When using linuron L (4 lbs. a.i. per gal.), use equivalent rates. One pt. of linuron L equals one lb. of linuron wettable powder.

Follow instructions and precautions on the BICEP 6L and linuron labels when tank mixing these products.

Rotational Crops: Follow the crop rotation instructions in the **BICEP 6L Alone** section for corn.

BICEP 6L COMBINATIONS — GRAIN SORGHUM (SEED TREATED WITH CONCEP)

Tank Mixtures of BICEP 6L with Gramoxone Super or Roundup for Minimum-Tillage or No-Tillage Systems

In minimum-tillage or no-tillage systems where grain sorghum is planted directly into a cover crop, stale seedbed, established sod, or previous crop residues, the contact herbicides Gramoxone Super or Roundup may be tank mixed with BICEP 6L. When used as directed, the Gramoxone Super portion of the tank mixture controls most emerged annual weeds and suppresses many perennial weeds. Roundup combinations will control emerged annual and perennial weeds when applied as directed on the Roundup label. The BICEP 6L portion of the tank mixture provides preemergence control of the weeds listed on this label in the **BICEP 6L Applied Alone** section.

Refer to the label of each product used in combination and observe the planting details, restrictions, and all other precautions and limitations.

Application: Apply before, during, or after planting but before grain sorghum emerges, at the appropriate rate in Table 6. Add Gramoxone Super or Roundup at the following broadcast rates:

Gramoxone Super: 1.5-2.5 pts. per acre plus 8 oz. of X-77 Spreader per 100 gals. of spray mixture. Use the lower rate for control of annual weeds less than 4 inches tall and the higher rate for weeds 4-6 inches tall. This treatment will not consistently control weeds taller than 6 inches.

Roundup: 1.5 qts. per acre for control of existing annual weeds, or 2-4 qts. per acre for existing perennial weeds. See the Roundup label for weeds controlled and recommended rates for specific weeds.

Apply in a minimum of 20 gals. of water per acre with conventional spray equipment.

Table 6: BICEP 6L for Minimum-Tillage or No-Tillage Grain Sorghum† (Seed treated with Concep)

Soil texture	Organic matter	Broadcast rate per acre
COARSE: Sand, loamy sand, sandy loam	any level	DO NOT USE
MEDIUM AND FINE: Loam, silt loam, silt, sandy clay loam, clay loam, silty clay loam, sandy clay, silty clay, clay	less than 1%	DO NOT USE
	1-1.5%	1.8 qts.
	more than 1.5%	2.1-1.4 qts.

† Do not use in NM, OK, or TX except in northeast OK and Texas Gulf Coast and Blacklands areas. Do not apply preplant incorporated in AZ or the Imperial Valley of CA.

Rotational Crops: Follow the crop rotation instructions in the **BICEP 6L Alone** section.

ROADSIDES

To control certain annual weeds in established perennial grasses along roadsides in CO, KS, MT, NE, ND, SD, and WY, including barnyardgrass, cheatgrass (downy brome, chess), common (annual) broomweed, crabgrass, fall panicum, giant foxtail, goosegrass, green foxtail, little barley, medusahead, sagewort, tumble mustard, witchgrass, and yellow foxtail, broadcast 1.5 qts./A in a minimum of 10 gallons of water by ground equipment in the fall before ground freezes, or after thawing in the spring, but before the established grasses green-up and before weeds emerge.

Examples of desirable established grasses include big bluestem, bluegrama, bromegrass, buffalograss, crested wheatgrass, Indiangrass, little bluestem, side-oats grama, switchgrass, and western wheatgrass. Apply only once a year. Temporary discoloration or other form of injury to the desirable perennial grasses may occur following application.

Notes: (1) Keep off desirable flowers, ornamentals and shrubs. (2) Do not attempt to reseed treated roadsides with desirable perennial grasses for 12 months after application. To avoid illegal residues, (3) Do not cut or feed roadside grass, hay and (4) Do not allow livestock to graze treated areas.

STORAGE AND DISPOSAL

Pesticide Disposal

Do not contaminate water, food, or feed by storage, disposal, or cleaning of equipment. Open dumping is prohibited. Improper disposal of unused pesticide, spray mixture, or rinsate is a violation of federal law. Pesticide, spray mixture, or rinsate that cannot be used according to label instructions must be disposed of according to federal, state or local procedures. For guidance in proper disposal methods, contact your State Pesticide or Environmental Control Agency, or the Hazardous Waste representative at the nearest EPA Regional Office.

Container Disposal

Do not reuse empty container. Triple rinse (or equivalent), puncture and dispose of in a sanitary landfill, or by incineration, or by open burning if allowed by state and local authorities. If burned, keep out of smoke.

For minor spills, leaks, etc., follow all precautions indicated on this label and clean up immediately. Take special care to avoid contamination of equipment and facilities during cleanup procedures and disposal of wastes. In the event of a major spill, fire or other emergency, call 1-800-888-8372 day or night.

PRECAUTIONARY STATEMENTS

HAZARDS TO HUMANS AND DOMESTIC ANIMALS

CAUTION

Harmful is swallowed, inhaled or absorbed through the skin. Avoid breathing vapors or spray mists. Causes eye and skin irritation. May be a skin sensitizer. Avoid contact with eyes, skin or clothing. Wash thoroughly with soap and water after handling. Remove contaminated clothing and wash before reuse.

STATEMENT OF PRACTICAL TREATMENT

If swallowed: Call a physician or Poison Control Center. Drink 1 or 2 glasses of water and induce vomiting by touching back of throat with finger. Do not induce vomiting or give anything by mouth to an unconscious person.

If in eyes: Flush eyes with plenty of water. Call a physician if irritation persists.

If on skin: Wash with plenty of soap and water. Get medical attention if irritation persists.

If inhaled: Remove victim to fresh air.

Note to Physician: There is no specific antidote. If swallowed, induce emesis or lavage stomach. The use of an aqueous slurry of activated charcoal should be considered.

ENVIRONMENTAL HAZARDS

Do not apply directly to water or wetlands (swamps, bogs, marshes, and potholes). Do not contaminate water by cleaning of equipment or disposal of wastes. This pesticide contains atrazine, which has been shown to be toxic to aquatic invertebrates. Runoff and drift from treated areas may be hazardous to aquatic organisms in neighboring areas.

Ground Water and Surface Water Advisory

BICEP 6L contains both the active ingredients atrazine and metolachlor.

Atrazine can travel (seep or leach) through soil and can enter ground water which may be used as drinking water. Atrazine has been found in ground water. Users are advised not to apply atrazine to sand and loamy sand soils where the water table (ground water) is close to the surface and where these soils are very permeable, i.e., well-drained. Your local agricultural agencies can provide further information on the type of soil in your area and the location of ground water.

Metolachlor has been identified in limited sampling of ground water and there is the possibility that it may leach through the soil to ground water, especially where soils are coarse and ground water is near the surface. Following application and during rainfall events that cause runoff, metolachlor may reach surface water bodies including streams, rivers, and reservoirs.

Care must be taken when using this product to prevent back siphoning into wells, spills, or improper disposal of excess pesticide, spray mixtures, or rinsates.

Check valves or antisiphoning devices must be used on all mixing and/or irrigation equipment.

Chemigation

Refer to supplemental labeling in attached booklet for use directions on chemigation. Do not apply this product through any irrigation system unless the supplemental labeling on chemigation is followed.

AAtrex® trademark of CIBA-GEIGY for atrazine
Agsorb™ trademark of Oil-Dri Corporation
Banvel® trademark of Velsicol Chemical Corporation
Basagran® trademark of BASF Company for bentazon
Bicep® trademark of CIBA-GEIGY U.S. Patent No. 3,937,730 (metolachlor)
Bladex® trademark of Shell Chemical Company for cyanazine
Brominal® trademark of Union Carbide Corporation
Buctril® trademark of Rhone-Poulenc Chemical Company
Caliber® trademark of CIBA-GEIGY
Compex® trademark of KALO Agricultural Chemicals, Inc.
Concep® trademark of CIBA-GEIGY U.S. Patent No. 4,070,389
Dual® trademark of CIBA-GEIGY for metolachlor U.S. Patent No. 3,937,730
Gramoxone® trademark of ICI Americas for paraquat
KERMAC #600™ trademark of Kerr McGee Refining Corporation
Nine-O® trademark of CIBA-GEIGY
Princep® trademark of CIBA-GEIGY for simazine
Roundup® trademark of Monsanto Company for glyphosate herbicide
Unite® trademark of Hopkins Agricultural Chemical Company
X-77® trademark of KALO Agricultural Chemicals, Inc.

CGA 56L12K 169

This is a specimen label which was not prepared by CIBA-GEIGY. CIBA-GEIGY is not responsible for the accuracy of the information contained herein. As labels are subject to revision, always carefully read and follow the label on the product container.

CAPAROL® 4L

HERBICIDE

For selective weed control in cotton, celery, and pigeon peas

ACTIVE INGREDIENT:

Prometryn: 2, 4-bis (isopropylamino)-6-(methylthio)-*s*-triazine	44.4%
INERT INGREDIENTS:	55.6%
TOTAL:	100.0%

EPA Reg. No. 100-620 EPA Est. 100-LA-1

Caparol 4L contains 4 lbs. active ingredient per gallon.

Shake well before using.

KEEP OUT OF REACH OF CHILDREN

CAUTION

See additional precautionary statements at end of label booklet.

See directions for use inside booklet.

Caparol® trademark of CIBA-GEIGY for prometryn

CGA 17L7E 038

DIRECTIONS FOR USE AND CONDITIONS OF SALE AND WARRANTY

IMPORTANT: Read the entire **Directions for Use** and the **Conditions of Sale and Warranty** before using this product.

CONDITIONS OF SALE AND WARRANTY

The **Directions for Use** of this product reflect the opinion of experts based on field use and tests. The directions are believed to be reliable and should be followed carefully. However, it is impossible to eliminate all risks inherently associated with use of this product. Crop injury, ineffectiveness, or other unintended consequences may result because of such factors as weather conditions, presence of other materials, or the manner of use or application all of which are beyond the control of CIBA-GEIGY or the Seller. All such risks shall be assumed by the Buyer.

CIBA-GEIGY warrants that this product conforms to the chemical description on the label and is reasonably fit for the purposes referred to in the **Directions for Use** subject to the inherent risks referred to above. **CIBA-GEIGY makes no other express or implied warranty of Fitness or Merchantability or any other express or implied warranty. In no case shall CIBA-GEIGY or the Seller be liable for consequential, special, or indirect damages resulting from the use or handling of this product.** CIBA-GEIGY and the Seller offer this product, and the Buyer and user accept it, subject to the foregoing **Conditions of Sale and Warranty,** which may be varied only by agreement in writing signed by a duly authorized representative of CIBA-GEIGY.

DIRECTIONS FOR USE

It is a violation of federal law to use this product in a manner inconsistent with its labeling.

FAILURE TO FOLLOW THE DIRECTIONS FOR USE AND PRECAUTIONS ON THIS LABEL MAY RESULT IN CROP INJURY, POOR PERFORMANCE, AND/OR ILLEGAL RESIDUES.

Do not apply this product through any type of irrigation system.

Do not apply this product in such a manner as to directly or through drift expose workers or other persons except those knowingly involved in the application. The area being treated must be vacated by unprotected persons.

REENTRY STATEMENT

Do not enter treated areas without protective clothing until sprays have dried.

Because certain states may require more restrictive reentry intervals for various crops treated with this product, consult your State Department of Agriculture for further information.

Written or oral warnings must be given to workers who are expected to be in a treated area or in an area about to be treated with this product. Oral warnings must inform workers of areas or fields that may not be entered without specific protective clothing until sprays have dried, and appropriate actions to take in case of accidental exposure, as described under **Precautionary Statements** on this label. When oral warnings are given, warnings shall be given in a language customarily understood by workers. Oral warnings must be given if there is reason to believe that written warnings cannot be understood by workers. Written warnings must include the following information: CAUTION. Area treated with Caparol 4L on (date of application). Do not enter without appropriate protective clothing until sprays have dried. In case of accidental exposure, flush eyes or skin with plenty of water. Call physician if irritation persists. Remove and wash contaminated clothing before reuse.

GENERAL INFORMATION

Caparol 4L is a selective herbicide that may be applied either before or after weeds emerge for control of most annual broadleaf weeds and grasses, including groundcherry, lambsquarters, annual morningglory, malva, mustard, black nightshade, pigweed (carelessweed), purslane, Florida pusley, ragweed, smartweed, teaweed (prickly sida), barnyardgrass (watergrass), crabgrass, foxtail, goosegrass, junglerice, *Panicum* spp., signalgrass (and other *Brachiaria* spp.), and wild oats. Caparol also controls shallow-germinating seedlings of cocklebur, coffeeweed, and sandbur. Caparol will also provide partial control of spurred anoda (cottonweed), rough blackfoot (ironweed, cluster flaveria), and prairie sunflower in New Mexico and western Texas. Caparol does not control johnsongrass, bermudagrass, other established perennials, or sprangletop at selective rates.

When applied before weeds emerge, Caparol enters weeds through their roots. Thus, its effectiveness depends on moisture to move it into the soil. Under very dry soil conditions after application, a shallow cultivation or rotary hoeing will generally result in better weed control.

When applied to emerged weeds, Caparol provides foliar knock-down and/or residual control of later germinating weeds, depending on the rate applied.

Following many years of continuous use of this product and chemically related products, biotypes of some of the weeds listed on this label have been reported which cannot be effectively controlled by this and related herbicides. Where this is known or suspected and weeds controlled by this product are expected to be present along with resistant biotypes, we recommend the use of this product in registered combinations or in sequence with other registered herbicides which are not triazines. If only resistant biotypes are expected to be present, use a registered non-triazine herbicide. Consult with your State Agricultural Extension Service for specific recommendations.

To avoid spray drift, do not apply under windy conditions. Avoid spray overlap, as crop injury may result.

APPLICATION PROCEDURES

Ground application: (All uses) Use conventional ground sprayers equipped with nozzles that provide accurate and uniform application.

Calibrate sprayer before use and recalibrate at the start of each season and when changing carriers. Unless otherwise specified, use a minimum of 20 gals. of spray mixture/A for all preplant incorporated, preemergence, and postemergence applications (with or without surfactant) with ground equipment.

Use a pump with capacity to (1) maintain 35-40 psi at nozzles and (2) provide sufficient agitation in tank to keep mixture in suspension. A centrifugal pump which provides propeller shear action is recommended for dispersing and mixing this product. The pump should provide a minimum of 20 gals./minute/100 gal. tank size circulated through a correctly positioned sparger tube or jets.

For preplant incorporated or preemergence application, use flat fan nozzle tips. For postemergence band application, use off-center nozzle tips. For postemergence broadcast application, use flat fan or off-center nozzle tips. Use flood nozzle tips only in Arizona and California for lay-by treatment in cotton at least 18 inches tall.

Use screens to protect pump and to prevent nozzles from clogging. Screens placed on the suction side of the pump should be *16-mesh or coarser.* Do not place a screen in the recirculation line. Use 50-mesh or coarser screens between the pump and boom, and where required, at the nozzles. Check nozzle manufacturer's recommendations.

For band applications, calculate amount to be applied per acre as follows:

$$\frac{\text{band width in inches}}{\text{row spacing in inches}} \times \begin{matrix}\text{broadcast}\\ \text{rate}\\ \text{per acre}\end{matrix} = \begin{matrix}\text{amount}\\ \text{needed per}\\ \text{acre of field}\end{matrix}$$

Aerial Application: (cotton and pigeon peas only) Use aerial application only where broadcast applications are specified. Use a minimum of 5 gals. of spray mixture per acre. Avoid applications under conditions where uniform coverage cannnot be obtained or where excessive spray drift may occur.

Avoid application to humans or animals. Flagmen and loaders should avoid inhalation of spray mist and prolonged contact with skin, and should wash thoroughly before eating and at the end of each day's operation.

To assure that spray will be controllable within the target area when used according to label directions, make applications at a maximum height of 10 ft. above vegetation, using low drift nozzles at a maximum pressure of 40 psi, and restrict application to periods when wind speed does not exceed 10 mph. To assure that spray will not adversely affect adjacent sensitive nontarget plants, apply Caparol 4L by aircraft at a minimum upwind distance of 400 ft. from sensitive plants.

Mixing procedures—all uses: (1) Be sure sprayer is clean and not contaminated with any other materials, or crop injury or sprayer clogging may result. (2) Fill tank 1/4 full with clean water. (3) Start agitation. (4) Be certain that the agitation system is working properly and creates a rippling or rolling action on the liquid surface. (5) Pour product directly from container into tank. (6) Continue filling tank until 90% full. Increase agitation if necessary to maintain surface action. (7) Add tank mix herbicide(s).

Cleaning: Wash sprayer thoroughly with clean water immediately after use. Do not use the same sprayer without thoroughly cleaning on sensitive crops, as even small residues of Caparol in the tank may cause injury to these crops.

Seedbed Preparation

To insure proper placement of Caparol 4L, seedbeds must be well prepared and as free as possible from trash and clods. A firm seedbed is best for obtaining effective weed control. Uniformity in height and width of seedbeds is essential for proper postemergence applications of Caparol 4L. Beds should be low and flat. Take care to avoid planter marks. Wide planter packing wheels or rollers are recommended. Wheel furrows should be uniform in depth. Mount the sprayer so that it follows the same rows as the planter.

COTTON

Caparol 4L may be applied preplant incorporated or preemergence and/or postemergence as recommended in the following tables. The postemergence applications may follow preplant incorporated or preemergence treatments of Caparol 4L.

Do not use on glandless cotton varieties or crop injury will occur.

Note: Do not feed treated forage to livestock, or graze treated areas, or illegal residues may result.

Preplant incorporation (Arizona, California, and New Mexico)

Apply Caparol 4L at the appropriate rate in Table 1 as a broadcast or band treatment. If broadcast, treat the flat soil surface prior to listing. If banded, apply over partially finished or finished beds. Incorporate up to 4 inches deep immediately after application with PTO-driven equipment, double disk, rolling cultivator, rolling cultivators in tandem, or bed conditioner.

Table 1: Preplant Incorporation

Region	Soil texture	Broadcast rate per acre
Arizona, California, and New Mexico	sand, loamy sand	DO NOT USE
	sandy loam (AZ and CA only)	2.4-3.2 pts.
	sandy loam, loams (NM only)	3.2 pts.
	silt loam, clay	4.8 pts.

Precautions: To avoid crop injury, (1) Do not use Caparol 4L in cut areas of newly leveled fields, in areas of excess salt, or in areas where flooding over the beds is likely to occur. (2) Do not plant cotton in tractor wheel depressions. (3) On mulch planted cotton, water back only after cotton seedlings are well established. (4) In New Mexico, apply either preplant incorporated or preemergence (not both)—see Preemergence section.

Preemergence

Apply at planting or shortly after planting at the appropriate rate in Table 2. Caparol 4L may be used on cotton planted flat, on beds, or in furrow. To avoid concentration of Caparol 4L in the seed furrow, do not make broadcast applications to cotton planted in furrows deeper than 2 inches. Band applications may be made to cotton planted in furrows deeper than 2 inches, but band width should not exceed the width of the bottom of the furrow. If banded, do not cover treated bands with soil while cultivating untreated row middles. To avoid crop injury, do not use on sand or loamy sand, on shallow soils with caliche subsoils, or in areas with caliche outcroppings.

Cotton may be replanted in soil previously treated with Caparol 4L. Do not apply a second preemergence application of Caparol or injury may occur.

Table 2: Preemergence

Region	Soil texture	Broadcast rate per acre
Mid-South and Southeast other than Mississippi River Delta in Mississippi	sandy loam	3.2-4 pts.
	silt and clay loam	4.8 pts.
	Sharkey clay (Arkansas only)	5.6 pts.
Mississippi River Delta in Mississippi	sandy loam	4-4.8 pts.
	silt and clay loam	5.6 pts.
	Sharkey clay	DO NOT USE
Blacklands of Texas and Oklahoma, Texas Gulf Coast, and Texas Coastal Bend	loam	2.4 pts.
	clay	4.8 pts.
Rio Grande Valley of Texas*	loam	3.2 pts.
	clay	4.8 pts.
High Plains, Rolling Plains and Edwards Plateau of Texas, Southwest Texas and New Mexico**	sand, loamy sand	DO NOT USE
	sandy loam	1.6 pts.
	loam, sandy clay loam	2.4 pts.
	other clay soils	3.2 pts.
Arizona and California	DO NOT USE	

***Rio Grande Valley of Texas**—Furrow irrigation cotton—if adequate rain does not fall soon after application, a shallow cultivation will insure good weed control.

****New Mexico**—Apply either preplant incorporated or preemergence (not both)—See Preplant Incorporation section.

Winter Weed Control in Texas

For control of winter weeds **only**, such as henbit (purpletop) and seedling dock on fall bedded cotton land in the Texas Gulf Coast and Blacklands of Texas, apply 1.2-1.6 pts. of Caparol 4L per acre in the fall or winter to land that will be planted to cotton the following spring. For best results, apply before weeds emerge. Caparol 4L will give effective control of emerged henbit if applied before it reaches 4-6 inches tall. For postemergence henbit control, add a suitable surfactant,such as X-77®, at 0.5% of spray volume or an emulsifiable oil at 1.0% of spray volume.

Winter Weed Control in California

For control of winter weeds on fall bedded cotton land, apply Caparol 4L after bedding either preemergence or postemergence to weeds less than 2 inches tall. Winter weeds controlled include:

chickweed	London rocket	redmaids
fiddleneck	mustards	shepherdspurse
filarees	pineappleweed	sowthistle, annual

On sandy loam soil, apply 3.2 pts. per acre; on medium or fine soil, apply 4 pts. per acre. To avoid crop injury, do not use on sandy or loamy sand. For postemergence weed control, add a suitable surfactant, such as X-77®, at 0.5% of spray volume or an emulsifiable oil at 1.0% of spray volume. Rainfall or sprinkler irrigation is necessary to activate the preemergence activity of Caparol 4L.

After pre-irrigation and before planting in the spring, knock off the top 1/3 to 1/2 of the seedbed. Then make a preplant application of Caparol 4L over the surface of the seedbed using a power-tiller, rolling cultivator, or similar implement that will provide uniform incorporation. Refer to Table 1, for preplant incorporation rates of Caparol 4L in California. To avoid crop injury, do not cultivate treated soil back toward the cotton until after cotton emergence and just before the first irrigation.

Precautions: To avoid crop injury, do not use Caparol 4L for winter weed control in areas of excess salt or calcareous soil.

Note: To avoid illegal residues, do not use more than 10.3 pts. of Caparol 4L on sandy loam soil or 11.9 pts. Caparol 4L on medium or fine soil per acre per year including winter weed control, preplant incorporation, chemical hoe, and lay-by applications.

Postemergence Directed

Be especially careful in applying Caparol 4L postemergence to prevent contact of the spray with cotton leaves, or injury may occur. Use precision application equipment so the spray is accurately directed to the base of the cotton plants and still thoroughly covers the soil and weeds beneath the cotton plants. Apply during calm periods to prevent drift. Use leaf lifters or shields if leaf contact cannot be avoided merely by directing the spray. Apply only when all plants have exceeded the minimum recommended height. Apply to level, well prepared surfaces such as relatively clod-free beds made with bed-shapers.

To avoid crop injury, do not apply to furrow-planted cotton until furrows are leveled (plowed in). Do not treat cotton under stress from drought, cultivator damage, or fertilizer application.

When applying to emerged weeds, add 2 qts. of surfactant per 100 gals. of spray mixture. Use a surfactant that is compatible with Caparol 4L when applied in cotton and is approved by EPA for use on food and feed crops. Examples include: X-77, Tronic, and Triton.

Chemical Hoe (Emerged Weeds only): Apply Caparol 4L at the appropriate rate in Table 3, two or three times if necessary. In cotton 3-6 inches tall, be extremely careful to avoid spray contact with cotton leaves by applying Caparol 4L with a precision applicator equipped with fenders or shields such as Bell Row Shield, Dickey Fenders, or W&A Fenders. In cotton less than 10 inches tall, apply only if cotton is bed or flat-planted.

Table 3: Chemical Hoe

Height of cotton and area of use	Height of weeds	Broadcast rate per acre
3 to 6 inches (Arkansas, Louisiana, Mississippi, Missouri, Tennessee, and Texas)	less than 1 inch	1 pt.
6 or more inches (all regions)	less than 2 inches	1-1.3 pts.

Lay-by (Emerged Weeds and Germinating Weeds): Apply Caparol 4L at the appropriate rate in Table 4, once per season when cotton is at least 12 inches tall (18 inches where flood nozzles are used in Arizona and California). Apply before weeds are 2 inches tall.

Table 4: Lay-by

Region	Soil texture	Broadcast rate per acre
Mid-South and Southeast	sandy	2.4 pts.
	loam	2.8 pts.
	clay	3.2 pts.
Blacklands of Texas and Oklahoma	loam	1.6 pts.
	clay	3.2 pts.
High Plains of Texas and New Mexico	sandy	1.6 pts.
	loam and clay	2.4 pts.
Southwest Texas	loam	2.4 pts.
	clay	3.2 pts.
Rio Grande Valley of Texas	DO NOT USE	
Arizona and California (Do not use in the Coachella Valley)	sand and loamy sand	DO NOT USE
	sandy loam	2.4-2.3 pts.
	loam	3.2 pts.

Rotational Crops

The following vegetable and cover crops may be planted in the fall when Caparol 4L was applied on cotton by no more than one of these methods that year: preplant incorporated, preemergence, or only one chemical hoe treatment. Where lay-by or multiple applications are made, do not plant rotational crops until the following year as indicated below.

Vegetables

Cabbage, okra, onions, peas, red beets, sweet corn

Cover Crops (must be plowed down and not used for food or feed)
Oats, sorghum, winter barley, winter rye, winter wheat

Spring-seeded crops in California and Arizona and spring-seeded vegetables in the Rio Grande Valley of Texas should not be planted until after April 1.

CAPAROL 4L COMBINATIONS FOR COTTON

Prowl® (Arizona, California, New Mexico, and the upper and lower El Paso Valley of Texas)

This preplant incorporated tank mixture controls all weeds listed on this label and on the Prowl cotton label. Apply prior to listing or over partially finished or finished beds and incorporate immediately. Refer to the Prowl label for specific mixing, spraying, and incorporation methods. Continuous agitation in the spray tank is required to keep the material in suspension.

Apply the tank mixture at the appropriate rates from Table 5.

Table 5: Tank Mixture with Prowl

Soil Texture	Broadcast rate per acre	
	Prowl	Caparol 4L
sand, loamy sand	DO NOT USE	
sandy loam	1-1.5 pts.	2.4-3.2 pts.
loam	1.5-2 pts.	3.2 pts.
silt loam, silt, sandy clay loam	1.5-2 pts.	3.2-4.8 pts.
clay loam, silty clay loam, clay	1.5-3 pts.	3.2-4.8 pts.

Use the high rate for each soil texture above if heavy weed populations are anticipated. Use the 3 pt. rate of Prowl for heavy clay soils.

Precautions: To avoid crop injury, (1) Do not use in cut areas of newly leveled fields, in areas of excess salt, or in areas where flooding over the bed is likely to occur. (2) Do not plant cotton in tractor wheel depressions. (3) Do not use this tank mixture when cotton is irrigated up. (4) On mulch planted cotton, water back only after cotton seedlings are well established.

Note: Do not feed treated forage to livestock, or graze treated areas, or illegal residues may result.

Rotational Crops: If treated crop is lost, cotton may be replanted. Do not rework the soil. Refer to the Prowl label and the Cotton section of this label for rotational crop restrictions.

Treflan® EC Tank Mixture (Arizona, California, New Mexico and the upper and lower El Paso Valley of Texas)

This combination controls weeds listed on this label and on the Treflan label. This combination also controls shallow-germinating seedlings of cocklebur and coffeeweed.

Follow procedures on the Treflan label for soil preparation and incorporation. Apply the tank mix combination to the flat soil before disking.

Pour Caparol 4L directly into spray tank one-half to three-fourths full of water, allow it to disperse with agitation, add Treflan EC, and then add the rest of the water. Under conditions of very soft water and low spray volume (5-10 gals./A), compatibility of Caparol 4L + Treflan EC may be improved by adding the Treflan EC first, agitate, and then add the Caparol 4L. Continuous agitation in the spray tank is required to keep the material in suspension. Apply the tank mixture at the appropriate rates from Table 6.

Table 6: Tank Mixture with Treflan EC

Soil Texture	Broadcast rate per acre	
	Treflan EC	Caparol 4L
sand, loamy sand	DO NOT USE	
sandy loam	1 pt.	2.4-3.2 pts.*
medium soils	1.5 pts.	4 pts.
fine soils	2 pts.	4 pts.
muck or peat	DO NOT USE	

* Use less than 3.2 pts. per acre only in Arizona and California.

Precautions: To avoid crop injury, (1) Do not use in cut areas of newly leveled fields, in areas of excess salt, or in areas where flooding over the bed is likely to occur. (2) Do not plant cotton in tractor wheel depressions. (3) On mulch planted cotton, water back only after cotton seedlings are well established.

Note: Do not feed treated forage to livestock, or graze treated areas, or illegal residues may result.

Rotational Crops: Cabbage, celery, okra, onions, and peas may be planted in the fall after a spring application of Treflan + Caparol. Winter barley, rye, and wheat can be planted in the fall if they are plowed down and not used for food or feed. Refer to the Treflan EC label for other directions and precautions.

Treflan EC Split Application (Arizona and California)

Apply a preplant incorporated application of Treflan EC as directed on that label, except use the appropriate rate from Table 6. Do not apply Treflan EC before January 1. Follow at planting or just before planting with a preplant incorporated treatment of Caparol 4L as directed in the Cotton section of this label, except use the appropriate rate from Table 6.

DSMA or MSMA

For faster knockdown of the weeds controlled by Caparol 4L alone, apply 1-1.3 pts. of Caparol 4L plus 3 lbs. active ingredient of DSMA or 2 lbs. active ingredient of MSMA per acre, following the same directions, precautions, and limitations as given on this label for Caparol 4L applied alone postemergence directed (chemical hoe). If needed, make a second application 1-3 weeks after the first application. Do not apply after first bloom.

Several formulations of DSMA and MSMA are available under various trade names from several manufacturers. Observe the directions, limitations, and precautions on the label of the product used.

CELERY

Seedbeds (Florida)

Broadcast 1.2-1.6 pts. in a minimum of 20 gals. of water per acre after celery has 2-5 true leaves. Application may be made over the celery. Apply only after seedbed covers have been removed from seedbeds for at least one week. Apply only once per year to seedbeds.

Direct-seeded Celery (California only)

Apply Caparol 4L at rates given below in a minimum of 20 gals. of water per acre. Within the rate ranges given, use the lower rates on coarse-textured soils and soils low in organic matter; use the higher rates on fine-textured soils and soils high in organic matter.

Preemergence: Broadcast 2.4-3.2 pts. per acre at planting or shortly after planting before celery emerges.

Postemergence: Broadcast 1.6-2 pts. pere acre after celery has 2-5 true leaves. Application may be made over the celery. Apply before weeds are 2 inches tall.

To Avoid Injury to Direct-seeded Celery: (1) Make either one preemergence or one postemergence application (not both) per celery crop. (2) Do not use on sand or loamy sand. (3) Do not apply if celery is under water stress. (4) Do not apply postemergence treatments of Caparol 4L with other pesticides. Apply only after foliar applications of other pesticides are dry. (5) Do not apply within two weeks after an application of a herbicidal oil, such as "carrot" oil.

Transplants

Apply one application at the appropriate rate from Table 7 in a minimum of 20 gals. of water per acre during the 2-6 week period after transplanting. Within the rate ranges given, use the lower rate on relatively coarse-textures soils and soils low in organic matter; use the higher rate on relatively fine-textured soils and soils high in organic matter. Application may be made over the celery. Apply before weeds are 2 inches tall.

Table 7: Transplanted Celery

State	Soil	Broadcast rate per acre
Florida	sandy or muck	1.6 -3.2 pts.
California	coarse-textured	2-3.2 pts.
	fine-textured	3.2-4 pts.
Michigan, Ohio	fine-textured or muck	2-4 pts.
Wisconsin	fine-textured	3.2-4 pts.

Rotational Crops

The following crops may be seeded 5 months after applying no more than 4 pts. Caparol 4L per acre on celery: cabbage, celery, corn, onions, peas, and red beets.

PIGEON PEAS (PUERTO RICO ONLY)

For preemergence control of annual weeds, such as horse purslane, junglerice, wild spider flower, jimsonweed, spurge, pigweed and Florida pusley, apply 4 pts. of Caparol 4L per acre on loam soils, or 6 pts. per acre on clay soils. Apply at planting or immediately after planting before the crop or weeds emerge.

Precautions: To avoid crop injury, (1) Make only one application per year. (2) Do not use on sandy or loamy sand soils.

Note: Do not graze or feed forage or hay to livestock, or illegal residues may result.

Rotational Crops

The following crops may be seeded 5 months after applying no more than 4 pts. Caparol 4L per acre on pigeon peas: cabbage, celery, corn, onions, peas, and red beets.

STORAGE AND DISPOSAL

Do not contaminate water, food, or feed by storage or disposal. Pesticide wastes are toxic. Improper disposal of unused pesticide, spray mixture, or rinse water is a violation of federal law. If these wastes cannot be used according to label instructions, contact your State Pesticide or Environmental Control Agency, or the Hazardous Waste representative at the nearest EPA Regional Office for guidance in proper disposal methods.

Triple rinse container (or equivalent). Then offer for recycling or reconditioning, or puncture and dispose of in a sanitary landfill or by other procedures approved by state and local authorities.

For minor spills, leaks, etc., follow all precautions indicated on this label and clean up immediately. Take special care to avoid contamination of equipment and facilities during cleanup procedures and disposal of wastes. In the event of a major spill, fire or other emergency, call (919) 292-7100 day or night.

PRECAUTIONARY STATEMENTS

HAZARDS TO HUMANS AND DOMESTIC ANIMALS

CAUTION

Harmful if absorbed through skin or inhaled. Causes moderate eye irritation. Avoid contact with eyes, skin, or clothing. Avoid repeated skin contact, as this product may cause skin sensitization reactions in certain individuals. Avoid breathing vapor or spray mist. Wash thoroughly with soap and water after handling. Remove contaminated clothing and wash before reuse.

First Aid: If in eyes, flush with plenty of water. If on skin, wash with plenty of soap and water. Get medical attention if eye or skin irritation persists.

ENVIRONMENTAL HAZARDS

Do not apply directly to water or wetlands (swamps, bogs, marshes, and potholes). Do not contaminate water when disposing of equipment wash waters. Do not apply when weather conditions favor drift from areas treated.

Do not apply this product through any type of irrigation system.

Caparol® trademark of CIBA-GEIGY for prometryn
Prowl® trademark of American Cyanamid for pendimethalin
Treflan® trademark of Elanco Products for trifluralin
X-77® trademark of KALO Agricultural Chemicals, Inc.

CGA 17L7E 038

This is a specimen label which was not prepared by CIBA-GEIGY. CIBA-GEIGY is not responsible for the accuracy of the information contained herein. As labels are subject to revision, always carefully read and follow the label on the product container.

CONCEP® II 70W

HERBICIDE ANTIDOTE

For use as a seed dressing to protect grain or forage sorghum from the phytotoxic effects of metolachlor herbicide

ACTIVE INGREDIENT:
CGA-92194: α-[(1,3-dioxolan-2-ylmethoxy)imino]
benzene-acetonitrile 70%
INERT INGREDIENTS: 30%
TOTAL. 100%

KEEP OUT OF REACH OF CHILDREN

CAUTION

See additional precautionary statements on back of bag.

See directions for use on side of bag.

DIRECTIONS FOR USE AND CONDITIONS OF SALE AND WARRANTY

IMPORTANT: Read the entire **Directions for Use** and the **Conditions of Sale and Warranty** before using this product.

CONDITIONS OF SALE AND WARRANTY

The **Directions for Use** of this product reflect the opinion of experts based on field use and tests. The directions are believed to be reliable and should be followed carefully. However, it is impossible to eliminate all risks inherently associated with use of this product. Crop injury, ineffectiveness, or other unintended consequences may result because of such factors as weather conditions, presence of other materials, or the manner of use or application all of which are beyond the control of CIBA–GEIGY or the Seller. All such risks shall be assumed by the Buyer.

CIBA–GEIGY warrants that this product conforms to the chemical description on the label and is reasonably fit for the purposes referred to in the **Directions for Use** subject to the inherent risks referred to above. **CIBA–GEIGY makes no other express or implied warranty of Fitness or Merchantability or any other express or implied warranty. In no case shall CIBA–GEIGY or the Seller be liable for consequential, special, or indirect damages resulting from the use or handling of this product.** CIBA–GEIGY and the Seller offer this product, and the Buyer and user accept it, subject to the foregoing **Conditions of Sale and Warranty,** which may be varied only by agreement in writing signed by a duly authorized representative of CIBA–GEIGY.

DIRECTIONS FOR USE

It is a violation of federal law to use this product in a manner inconsistent with its labeling.

USE IN WELL VENTILATED AREA. WEAR AN APPROVED RESPIRATOR SUCH AS 3M 8710 WHEN HANDLING THIS MATERIAL.

GENERAL INFORMATION

Concep II is a herbicide antidote which, when used as a seed treatment, protects grain and forage sorghum from herbicidal injury due to metolachlor (contained in the herbicides Bicep® and Dual®). Without a Concep II seed treatment, Bicep, Dual, and other products containing metolachlor will generally cause severe sorghum injury. Concep II protects sorghum from the phytoxic effects of only metolachlor and not from other herbicide components including the triazine component of Bicep.

Application Information

Apply Concep II to high quality sorghum seed using seed treatment equipment that will apply a water-based seed treatment slurry **accurately** and **uniformly** to assure treatment within the acceptable rate range of 1.0-1.5 gm active ingredient (a.i.) of safener per kg of seed.

Application Rate

Sorghum Seed (Units*)	Concep II to Add to Slurry	Total Slurry Volume	
		Concep II Alone	Combination with Fungicide/ Insecticide
1	1.2-1.7 oz.	4-8 fl. oz.	7-11 fl. oz.
2	2.4-3.4 oz.	8-16 fl. oz.	14-22 fl. oz.
100	7.5-10.6 lbs.	3.1-6.3 gals.	5.5-8.6 gals.

* 1 Unit = 50 lbs. seed.

Mixing

Concep II does not contain a binder/sticker. A binder such as Ortho Seed Protectant Binder or Gustafson Phase II Binder will help prevent dusting-off below required level of safener.

1. Fill the premix tank with one-half to two-thirds of the water required and provide agitation.
2. Add dye as needed, then slowly add the recommended amount of Concep II to form a smooth slurry.
3. Add binder as recommended on that label and then add the remaining amount of water needed to bring the slurry up to the desired volume.
4. If Concep II is to be mixed with a fungicide/insecticide, add these after Concep II has been thoroughly mixed into a smooth slurry, and then add water to bring the slurry mixture to the desired volume.

Notes: (1) Do not allow the mixture to remain in equipment without agitation. (2) Do not allow the mixture to freeze.

Compatibility with Other Seed Treatments

Concep II may be tank mixed with labeled wettable powder or water dispersible granule seed treatment formulations of captan, captan + methoxychlor, captan + heptachlor or Lorsban®. Concep II may also be tank mixed with labeled flowable seed treatment formulations of methoxychlor and captan that do not contain dye. Flowable formulations containing dye may be incompatible under certain circumstances. Compatibility should be checked before starting seed treatment operation. Do not mix with organic solvent formulations of heptachlor. Refer to the respective fungicide/insecticide label(s) for directions for use of those products on sorghum.

Gas Chromatography Determinations

Analyze seed treated with Concep II using gas chromatography methods supplied by the Biochemistry Department, Agricultural Division, CIBA–GEIGY Corporation or your CIBA–GEIGY representative to assure that the entire seed lot has been properly treated with a minimum rate of 1.0 gm a.i. Concep II per kg of seed. Collect a representative sample for analysis from each seed lot by taking several subsamples during treatment of the lot.

Seed Quality

Seed treated with Concep II may exhibit a slight germination reduction due to differences inherent to the sorghum line or seed quality as affected by production, harvest and/or storage conditions. To assure adequate germination, treat a sample of each seed lot with Concep II combined with all other intended seed treatment chemicals at proper use rates and check germination prior to treating the entire lot. Only those seed lots meeting acceptable standards from these tests should be treated in large quantities.

Retreatment of Carried Over Seed

Provided germination is acceptable, Concep II-treated seed may be carried over and planted the year following treatment and still provide adequate safening. If a seed lot treated with Concep II the previous year does not analyze at least 1.0 gm a.i. per kg of seed, the seed lot may be retreated with an adequate amount of Concep II to raise the amount on the seed to within the acceptable rate range of 1.0-1.5 gm a.i. per kg of seed provided a pretreated sample, representative of the lot, meets acceptable germination standards.

Tagging of Carried Over Seed

Grain sorghum seed safened with Concep II may be tagged and sold as unsafened seed provided the following statement appears on the tag:

"CAUTION: TREATED SEED.—DO NOT USE FOR FOOD, FEED OR OIL PURPOSES."

Seed treatments such as fungicides and insecticides have specific requirements in seed laws and regulations which define the type of labeling required on bags of the treated seed. Concep II is not in the same regulatory category as fungicides or insecticides, and there are no federal requirements for labeling the safened seed. However, since certain state seed laws may be more restrictive than the federal law, the above CAUTION statement should be used.

This is the most restrictive requirement that applies to seed treated with Concep II. If you object to the above statement it is recommended that the seed laws of all states in which Concep II-treated seed is to be sold be reviewed, and that clarification be sought from a qualified attorney or representative of the state with respective to seed safeners and labeling requirements because some states may have lesser requirements under their labeling laws.

Other Information

Do not feed seed treated with Concep II to livestock.

Refer to the Bicep, and Dual labels for directions for use of these herbicides on sorghum.

STORAGE AND DISPOSAL

Store in a dry place where temperatures do not exceed 105°F. Do not contaminate water, food, or feed by storage or disposal. Open dumping is prohibited. This chemical, spray mixture, or rinsate that cannot be used according to label instructions must be disposed of according to applicable federal, state, or local procedures.

Completely empty bag by shaking, and tapping sides and bottom to loosen clinging particles. Empty residue into application equipment. Then dispose of bags in a sanitary landfill or by incineration if allowed by state and local authorities.

PRECAUTIONARY STATEMENTS

HAZARDS TO HUMANS AND DOMESTIC ANIMALS

CAUTION

Harmful if inhaled or absorbed through the skin. Avoid contact with skin, eyes or clothing. Avoid breathing dust. Wear an approved respirator such as 3M 8710 when handling this material. Do not contaminate food or feed. Wash thoroughly with soap and water after handling. Remove contaminated clothing and wash before reuse.

STATEMENT OF PRACTICAL TREATMENT

If in eyes: Flush eyes with plenty of water. Call a physician if irritation persists.

If inhaled: Remove victim to fresh air.

If on skin: Wash with plenty of soap and water. Get medical attention if irritation persists.

If swallowed: Give one or two glasses of water and induce vomiting. Never give anything by mouth to an unconscious person. Contact a local Poison Control Center, hospital or physician immediately.

Note to Physician

There is no specific antidote if Concep II 70W is ingested. If a large amount has been ingested and emesis has been inadequate, lavage stomach. An aqueous suspension of 30 to 50 gm of activated charcoal can be given. Treat symptomatically.

ENVIRONMENTAL HAZARDS

Keep out of any body of water. Do not contaminate water when disposing of equipment wash waters.

PHYSICAL AND CHEMICAL HAZARDS

Do not store near ignition sources such as electrical sparks, flames, or heated surfaces. Do not smoke while using this product. Follow good housekeeping practices to avoid dust accumulations. Ground all metallic surfaces.

Bicep® trademark of CIBA–GEIGY for metolachlor + atrazine
U.S. Patent No. 3,937,730 (metolachlor)
Concep® trademark of CIBA–GEIGY
U.S. Patent No. 4,269,775
Dual® trademark of CIBA–GEIGY for metolachlor
U.S. Patent No. 3,937,730
Lorsban® trademark of the Dow Chemical Company

CGA 74L4C 118

This is a specimen label which was not prepared by CIBA-GEIGY. CIBA-GEIGY is not responsible for the accuracy of the information contained herein. As labels are subject to revision, always carefully read and follow the label on the product container.

COTORAN® 80W

HERBICIDE

For preemergence and postemergence weed control in cotton

ACTIVE INGREDIENT:	
Fluometuron: 1,1-dimethyl-3-(α, α, α-trifluoro-*m* -tolyl) urea	80%
INERT INGREDIENTS:	20%
TOTAL	100%

EPA Reg. No. 100-609

Cotoran 80W is a wettable powder.

KEEP OUT OF REACH OF CHILDREN
WARNING

See additional precautionary statements on side of bag.

See directions for use on panel at right.

DIRECTIONS FOR USE AND CONDITIONS OF SALE AND WARRANTY

IMPORTANT: Read the entire **Directions for Use** and the **Conditions of Sale and Warranty** before using this product.

CONDITIONS OF SALE AND WARRANTY

The **Directions for Use** of this product reflect the opinion of experts based on field use and tests. The directions are believed to be reliable and should be followed carefully. However, it is impossible to eliminate all risks inherently associated with use of this product. Crop injury, ineffectiveness, or other unintended consequences may result because of such factors as weather conditions, presence of other materials, or the manner of use or application all of which are beyond the control of CIBA–GEIGY or the Seller. All such risks shall be assumed by the Buyer.

CIBA–GEIGY warrants that this product conforms to the chemical description on the label and is reasonably fit for the purposes referred to in the **Directions for Use** subject to the inherent risks referred to above. **CIBA–GEIGY makes no other express or implied warranty of Fitness or Merchantability or any other express or implied warranty. In no case shall CIBA–GEIGY or the Seller be liable for consequential, special, or indirect damages resulting from the use or handling of this product.** CIBA–GEIGY and the Seller offer this product, and the Buyer and user accept it, subject to the foregoing **Conditions of Sale and Warranty,** which may be varied only by agreement in writing signed by a duly authorized representative of CIBA–GEIGY.

GENERAL INFORMATION

Cotoran 80W is effective for both annual grass and annual broadleaf weed control in cotton.

Do not apply this product through any type of irrigation system.

Weed Control

Cotoran 80W applied according to directions will selectively control annual weeds in cotton, such as:

Annual Grass Weeds		
Barnyardgrass	Crowfootgrass	Goosegrass
Brachiaria	Fall panicum	Ryegrass
Crabgrass	Foxtail	
Annual Broadleaf Weeds		
Buttonweed	Jimsonweed	Purslane
Cocklebur	Lambsquarters	Ragweed
Florida pusley	Morningglory	Sesbania
Goathead	Pigweed	Sicklepod
Groundcherry (Wright)	Prickly sida (Ironweed or teaweed)	Smartweed
		Tumbleweed

Cotoran 80W at rates up to 2.5 lbs. per acre in cotton is not effective for the control of established perennial weeds, such as johnsongrass from rhizomes.

APPLICATION PROCEDURES

Mixing Instructions

Cotoran 80W is highly concentrated and should be thoroughly mixed as a paste or thick slurry in water in a separate container before adding to the spray tank. The spray tank should contain 1/3-1/2 the total amount of water or nitrogen solution to be used before adding the slurry.

To determine compatibility, add one tablespoon of Cotoran 80W to one pint of nitrogen solution (equals 2.5 lbs. in 25 gals; if using other application rates, adjust compatibility test amounts proportionally). Stir or shake thoroughly. Let stand 5 minutes. If it remains mixed or can be remixed readily, the mixture is compatible and can be used. If Cotoran 80W settles out when mixed with nitrogen solution, a suspensibility agent, such as Compex® at 1-2 pts. per 100 gals. fertilizer, may improve compatibility. Add 1/8-1/4 teaspoon of Compex to the compatibility test to determine its effectiveness.

Use bypass or mechanical agitation in the spray tank. Avoid leaving spray mixture in tank without constant agitation. If bypass agitation is used, it should terminate at the bottom of the tank to minimize foaming. Wash sprayer thoroughly with clean water immediately after use.

Equipment

Thoroughly clean sprayer prior to use. Do not use a sprayer contaminated with 2,4-D or other materials as crop damage or sprayer clogging may result.

For uniform distribution use fan-type nozzles. Use 50-mesh or coarser screens in nozzles and strainers. Use a suitable pump with capacity to deliver (1) the necessary volume to the nozzles at 35-40 psi, and (2) an additional volume to maintain the mixture in suspension in the spray tank. Apply in 20-40 gals. of spray mixture per acre.

Band Treatment

For band treatment, calculate the amount of Cotoran 80W needed by the formula:

$$\frac{\text{band width in inches}}{\text{row width in inches}} \times \text{broadcast rate per acre} = \text{amount needed per acre of field}$$

DIRECTIONS FOR USE

It is a violation of federal law to use this product in a manner inconsistent with its labeling.

COTTON

Cotoran 80W may be applied as a preemergence treatment to the surface of the soil, or lightly incorporated immediately after planting, or as a directed, semi-directed, or over-the-top postemergence application. For preemergence treatment only, Cotoran may be tank-mixed with most liquid nitrogen fertilizers. Cotoran may be combined with other herbicides in cotton as directed in the section, "Combinations."

Cotoran 80W is most effective when used as follows:

1st Application: Preemergence broadcast. See Preemergence Treatment below.

2nd Application: Postemergence, directed, semi-directed, or over-the-top when cotton is 3-6 inches high. See Postemergence Treatments below.

3rd Application: If needed, postemergence, directed, or semi-directed, at lay-by (last normal cultivation), using no more than 1.25 lbs. per acre.

Preemergence Treatment

Apply Cotoran 80W at a rate of 1.25-2.5 lbs. per acre. On light silt and sandy soils, low in organic matter, use the lower rate.

Cotoran may be applied to the surface of the soil, or lightly incorporated into the soil, after planting cotton.

Preplant Incorporated Use (Arizona, California, and New Mexico)

Cotoran 80W may be applied preplant as a broadcast spray at 2 lbs. per acre after the beds for preplant irrigation have been formed. Apply before the beds are dragged off (using a bed shaper set to remove 1-2 inches from beds) in preparation for planting. Furrow irrigation should precede the application and planting should follow as soon as possible. Plant into untreated soil after drag-off and return treated soil to the beds when furrows are re-formed. Do not allow furrow irrigation water to pond over the beds.

Postemergence Treatments

Cotoran 80W may be applied postemergence to cotton but preemergence to weeds, or postemergence to both cotton and weeds. Cotoran 80W should be applied at rates of 1.25-2.5 lbs. per acre. Apply as a directed, semi-directed, or over-the-top spray. Cotoran may be applied to cotton plants at any stage of growth from 3 inches high to lay-by. Use the higher rate when applying postemergence to weeds; weeds should be 2 inches high or less. Use of a surfactant that is approved by FDA for use on food and feed crops and is compatible with Cotoran 80W when applied to cotton may enhance the postemergence herbicidal activity of Cotoran. The surfactant rate should not exceed 0.5% of the carrier volume (1 pt. per 25 gals. of water).

Precautions: Rainfall that is sufficient to germinate seeds or for growing crops is adequate to activate the herbicidal activity of Cotoran; where dry weather conditions prevail, the herbicidal activity of Cotoran may be delayed or reduced.

Do not make more than three applications of Cotoran to the same crops or field in any one year. Do not plant crops other than cotton within 6 months of the last application of Cotoran or injury may result.

The use of Cotoran 80W as a preemergence or early postemergence application, following the use of a systemic insecticide at planting, may result in injury to cotton.

West Texas only: Do not use Cotoran on sandy, loamy sand, or fine sandy loam soils. Do not use Cotoran on cotton planted in furrows.

Arkansas, Louisiana, and Mississippi only: Use 1 lb. of Cotoran 80W per acre on sandy loam soils low in organic matter.

Arizona, California, and New Mexico only: Cotton, corn, and grain sorghum can be planted the next spring. Do not plant treated areas to other crops until one year after last application. Do not use on sandy or coarse-textured soil of less than 1% organic matter.

Note: Do not apply within 60 days of harvest. Do not feed foliage from treated cotton plants or gin trash to livestock.

Combinations

Treflan® EC: Use in a tank mixture with Cotoran 80W (all areas except Arizona and California) or before an overlay with Cotoran 80W in cotton for control of all weeds listed on both labels. For the **tank mixture,** refer to the Treflan EC label for directions, precautions, and limitations. For the **overlay,** apply Cotoran 80W as a preemergence surface treatment as recommended on this label at 1.25-2.5 lbs. per acre after a preplant soil-incorporated treatment of Treflan EC. Apply Treflan EC as recommended on its label at 1-2.5 pts. per acre.

Prowl®: Apply Cotoran 80W as a preemergence surface treatment as recommended on this label at 1.25-2.5 lbs. per acre after a preplant soil-incorporated treatment of Prowl for control of all weeds listed on both labels. Apply Prowl as recommended on its label at 1-3 pts. per acre.

DSMA or MSMA: Cotoran may be tank-mixed with DSMA or MSMA and applied as a directed spray to cotton from three inches high to first bloom. Use the broadcast rate of 1.25-2.5 lbs. of Cotoran per acre and the necessary amount of DSMA or MSMA plus 0.5% (1 pt. per 25 gals.) of a surfactant that is approved by FDA for use on food and feed crops and is compatible with Cotoran 80W + DSMA or Cotoran 80W + MSMA when applied to cotton. When using this mixture, do not allow spray to contact cotton foliage. Do not graze treated areas or feed treated forage or gin trash to livestock. Adhere strictly to the dosage and limitations on the labels of the products.

STORAGE AND DISPOSAL

Store in a dry place. Do not contaminate water, food, or feed by storage or disposal. Improper disposal of unused spray mixture, or rinse water is a violation of federal law. If these wastes cannot be used according to label instructions, contact your State Pesticide or Environmental Control Agency, or the Hazardous Waste representative at the nearest EPA Regional Office for guidance in proper disposal methods.

Completely empty bag into application equipment. Dispose of empty bag in a sanitary landfill or by incineration, or if allowed by state and local authorities, by burning. Stay out of smoke from burning bags.

For minor spills, leaks, etc., follow all precautions indicated on this label and clean up immediately. Take special care to avoid contamination of equipment and facilities during cleanup procedures and disposal of wastes. In the event of a major spill, fire or other emergency, call (919) 292-7100 day or night.

PRECAUTIONARY STATEMENTS

HAZARDS TO HUMANS AND DOMESTIC ANIMALS

WARNING

Causes eye irritation. Do not get in eyes. Harmful if swallowed, inhaled, or absorbed through the skin. Wear protective clothing, as this product may cause skin rash after repeated exposure. Avoid contact with skin or clothing. Avoid breathing dust or spray mist.

First Aid: In case of contact with the eyes, flush with water for at least 15 minutes and get medical attention if irritation persists. In case of contact with skin, wash immediately with plenty of soap and water. If swallowed, drink promptly a large quantity of milk, egg whites, or gelatin solution; if these are not available, drink large quantities of water. Induce vomiting. Avoid alcohol. Call a physician immediately. In case of inhalation exposure, move from the contaminated area.

Note to Physician: There is no specific antidote for Cotoran 80W. Induce emesis and lavage stomach. Give a saline laxative and supportive therapy.

ENVIRONMENTAL HAZARDS

Do not contaminate water used for irrigation or domestic purposes. Do not contaminate water by cleaning of equipment or disposal of wastes.

Cotoran® trademark of CIBA-GEIGY for fluometuron
Compex® trademark of KALO Agricultural Chemicals, Inc.
Prowl® trademark of American Cyanamid for pendimethalin
Treflan® trademark of Elanco Products for trifluralin

CGA 33L11D 038

This is a specimen label which was not prepared by CIBA-GEIGY. CIBA-GEIGY is not responsible for the accuracy of the information contained herein. As labels are subject to revision, always carefully read and follow the label on the product container.

COTORAN® 4L
HERBICIDE

For preemergence and postemergence weed control in cotton

ACTIVE INGREDIENT:
Fluometuron: 1,1-dimethyl-3-(α,α,α-trifluoro-*m*-tolyl)urea 41.7%
INERT INGREDIENTS: 58.3%
TOTAL 100.0%

EPA Reg. No. 100-642

EPA Est. 32761-MO-3

COTORAN 4L contains 4 lbs. active ingredient per gallon.
Use entire contents at one time.

KEEP OUT OF REACH OF CHILDREN
CAUTION

See additional precautionary statements at end of label booklet.

See directions for use inside booklet.

COTORAN® trademark of CIBA-GEIGY for fluometuron

CGA 33L14C 097

DIRECTIONS FOR USE AND CONDITIONS OF SALE AND WARRANTY

IMPORTANT: Read the entire **Directions for Use** and the **Conditions of Sale and Warranty** before using this product.

CONDITIONS OF SALE AND WARRANTY

The **Directions for Use** of this product reflect the opinion of experts based on field use and tests. The directions are believed to be reliable and should be followed carefully. However, it is impossible to eliminate all risks inherently associated with use of this product. Crop injury, ineffectiveness, or other unintended consequences may result because of such factors as weather conditions, presence of other materials, or the manner of use or application all of which are beyond the control of CIBA–GEIGY or the Seller. All such risks shall be assumed by the Buyer.

CIBA–GEIGY warrants that this product conforms to the chemical description on the label and is reasonably fit for the purposes referred to in the **Directions for Use** subject to the inherent risks referred to above. **CIBA–GEIGY makes no other express or implied warranty of Fitness or Merchantability or any other express or implied warranty. In no case shall CIBA–GEIGY or the Seller be liable for consequential, special, or indirect damages resulting from the use or handling of this product.** CIBA–GEIGY and the Seller offer this product, and the Buyer and user accept it, subject to the foregoing **Conditions of Sale and Warranty,** which may be varied only by agreement in writing signed by a duly authorized representative of CIBA–GEIGY.

DIRECTIONS FOR USE

It is a violation of federal law to use this product in a manner inconsistent with its labeling.

FAILURE TO FOLLOW THE DIRECTIONS FOR USE AND PRECAUTIONS ON THIS LABEL MAY RESULT IN CROP INJURY, POOR WEED CONTROL, AND/OR ILLEGAL RESIDUES.

Do not apply this product through any type of irrigation system.

GENERAL INFORMATION

COTORAN 4L is effective for both annual grass and annual broadleaf weed control in cotton.

Weed Control

COTORAN 4L applied according to directions will selectively control annual weeds in cotton, such as:

	Annual Grass Weeds	
Barnyardgrass	Fall panicum	Ryegrass
Crabgrass	Foxtail	Signalgrass
Crowfootgrass	Goosegrass	*(Brachiaria)*
	Annual Broadleaf Weeds	
Buttonweed	Lambsquarters	Ragweed
Cocklebur	Morningglory	Sesbania
Florida pusley	Pigweed	Sicklepod
Groundcherry (Wright)	Prickly sida (teaweed)	Smartweed
		Tumbleweed
Jimsonweed	Purslane	

COTORAN 4L at rates up to 4 pts. per acre in cotton is not effective for the control of established perennial weeds, such as johnsongrass from rhizomes.

APPLICATION PROCEDURES

Mixing Instructions

(1) Fill spray tank 1/2 to 3/4 full with clean water or nitrogen solution. (2) Start agitation. (3) Pour COTORAN 4L directly from the container into the partially filled spray tank. (4) If COTORAN 4L is to be tank mixed with another herbicide, add that product after the COTORAN 4L is thoroughly mixed in the partially filled spray tank. (5) Add the rest of the water or nitrogen solution. (6) Provide agitation during mixing and application to maintain a uniform suspension.

To determine compatibility, add two teaspoons of COTORAN 4L to one pint of nitrogen solution (equals 4 pts. in 25 gals.; if using other application rates, adjust compatibility test amounts proportionally). Stir or shake thoroughly. Let stand 5 minutes. If it remains mixed or can be remixed readily, the mixture is compatible and can be used. If COTORAN 4L settles out when mixed with a nitrogen solution, a suspensibility agent, such as Unite®, Compex®, or equivalent at 1-3 pts. per 100 gals. fertilizer, may improve compatibility. Add 1/8-3/8 teaspoon of Unite to the compatibility test to determine its effectiveness.

Equipment

Thoroughly clean sprayer prior to use. Do not use a sprayer contaminated with 2,4-D or other materials as crop damage or sprayer clogging may result.

Use properly calibrated conventional spray equipment that will apply spray uniformly and has hydraulic or mechanical agitation except in California or Arizona where only mechanical agitators are recommended. Screens in nozzles, in suction, and in-line strainers should be no finer than 50-mesh. Use a pump with capacity to maintain 35-40 psi at the nozzles. If hydraulic agitation is used, the pump should also provide sufficient agitation in the tank to keep the mixture in suspension. Avoid leaving spray mixture in tank without constant agitation. If bypass agitation is used, it should terminate at the bottom of the tank to minimize foaming. Wash sprayer thoroughly with clean water immediately after use.

Apply in a minimum of 10 gals. per acre for preplant incorporated or preemergence applications, or apply in a minimum of 20 gals. per acre for postemergence application.

Band Treatment

For band treatment, calculate the amount of COTORAN 4L needed by the formula:

$$\frac{\text{band width in inches}}{\text{row width in inches}} \times \text{broadcast rate per acre} = \text{amount neeed per acre of field}$$

COTTON

COTORAN 4L may be applied as a preemergence treatment to the surface of the soil, or lightly incorporated immediately after planting, or as a directed, semidirected, or over-the-top postemergence application. For preemergence treatment only. COTORAN 4L may be tank-mixed with most liquid nitrogen fertilizers. COTORAN 4L may be combined with other herbicides in cotton as directed in the section, "Combinations."

COTORAN 4L is most effective when used as follows:

1st Application: Preemergence broadcast. See Preemergence Treatment below.

2nd Application: Post emergence, directed, semidirected, or over-the-top when cotton is 3-6 inches high. See Postemergence Treatments below.

3rd Application: If needed, postemergence, directed, or semi-directed, at lay-by (last normal cultivation), using no more than 2 pts. per acre.

Preemergence Treatment

Apply COTORAN 4L at a rate of 2-4 pts. per acre. On light silt and sandy soils, low in organic matter, use the lower rate.

COTORAN 4L may be applied to the surface of the soil, or lightly incorporated into the soil, after planting cotton.

Preplant Incorporated Use (Arizona, California, and New Mexico)

COTORAN 4L may be applied preplant as a broadcast spray at 3.2 pts. per acre after the beds for preplant irrigation have been formed. Apply before the beds are dragged off (using a bed shaper set to remove 1-2 inches from beds) in preparation for planting. Furrow irrigation should precede the application and planting should follow as soon as possible. Plant into untreated soil after drag-off and return treated soil to the beds when furrows are reformed. Do not allow furrow irrigation water to pond over the beds.

Postemergence Treatments

COTORAN 4L may be applied postemergence to cotton but preemergence to weeds, or postemergence to both cotton and weeds. COTORAN 4L should be applied to rates of 2-4 pts. per acre. Apply as directed, semidirected, or over-the-top spray. COTORAN 4L may be applied to cotton plants at any stage of growth from 3 inches high to lay-by. Use the higher rate when applying postemergence to weeds; weeds should be 2 inches high or less. Use of a surfactant that is approved by FDA for use on food and feed crops and is compatible with COTORAN 4L when applied to cotton may enhance the postemergence herbicidal activity of COTORAN 4L. The surfactant rate should not exceed 0.5% of the carrier volume (1 pt. per 25 gals. of water).

Precaution: Rainfall that is sufficient to germinate seeds or for growing crops is adequate to activate the herbicidal activity of COTORAN 4L; where dry weather conditions prevail, the herbicidal activity of COTORAN 4L may be delayed or reduced. Cultivate if weeds develop.

Do not make more than three applications of COTORAN 4L to the same crops or field in any one year. Do not plant crops other than cotton within 6 months of the last application of COTORAN 4L or injury may result.

The use of COTORAN 4L as a preemergence or early postemergence application, following the use of a systemic insecticide at planting, may result in injury to cotton.

West Texas only: Do not use COTORAN 4L on sandy, loamy sand, or fine sandy loam soils. Do not use COTORAN 4L on cotton planted in furrows.

Arkansas, Louisiana, and Mississippi only: use 1.6 pts. of COTORAN 4L per acre on sandy loam soils low in organic matter.

Arizona, California, and New Mexico only: Cotton, corn, and grain sorghum can be planted the next spring. Do not plant treated areas to other crops until one year after last application. Do not use on sandy or coarse-textured soil of less than 1% organic matter.

Note: Do not apply within 60 days of harvest. Do not feed foliage from treated cotton plants or gin trash to livestock.

Combinations

Dual® 8E: Use in a tank mixture with COTORAN 4L preemergence in cotton for control of all weeds as listed on both labels. This combination will also control spotted spurge, hyssop spurge, nodding spurge, and prostrate spurge. Apply to the soil surface at planting or after planting, but before weeds or crop emerge, using the appropriate rates from the table below.

COTORAN 4L + Dual 8E—Cotton (AR, LA, MS, Eastern OK, TN, Bootheel of MO, and Gulf Coast, Rio Grande Valley and Eastern TX)

	Broadcast rates per acre	
Soil texture	**COTORAN 4L**	**Dual 8E**
Sand, loamy sand	DO NOT USE	
Sandy loam*	2 pts	1.25-1.5 pts.
Loam, silt, silt loam	2-3.2 pts.	1.5-2 pts.
Silty clay loam, sandy clay loam, silty clay, sandy clay, clay loam, clay	3.2-4 pts.	2 pts.

*Use on sandy loam soils only in eastern OK and Gulf Coast, Rio Grande Valley, and eastern TX.

Precautions: (1) Do not apply COTORAN 4L + Dual 8E on sand or loamy sand soils or in areas where water is likely to "pond" over the bed, or crop injury may occur. (2) Do not apply broadcast to furrow-planted cotton. (3) The use of COTORAN 4L following the use of a systemic insecticide at planting may result in crop injury. (4) Do not use on Taloka silt loam, or crop injury may occur.

Refer to the Dual 8E label for further instructions, precautions, and limitations.

Note: Do not feed treated forage or gin trash to livestock, or graze treated areas.

Rotational Crops: Refer to the crop rotation instructions for COTORAN 4L alone on this label and for Dual 8E alone on the Dual 8E label.

Treflan® EC: Use in a tank mixture with COTORAN 4L (all areas except Arizona and California) or before an overlay with COTORAN 4L in cotton for control of all weeds listed on both labels.

For the **tank mixture,** follow soil preparation and incorporation procedures for Treflan EC. Apply COTORAN 4L + Treflan EC tank mix at rates specified below in a minimum of 15 gals. of water per acre.

COTORAN 4L + Treflan EC Tank Mixture—Cotton (All areas except Arizona and California)

	Broadcast rates per acre	
Soil texture	**COTORAN 4L**	**Treflan EC**
Sand, loamy sand, sandy loam	2 pts.	1 pt.
Loam, silty clay loam*, silt loam, silt, sandy clay loam*	3.2 pts.	1.5 pts.
Clay, clay loam, silty clay loam*, silty clay, sandy clay, sandy clay loam*	4 pts.	2 pts.

*Silty clay loam and sandy loam are transitional soils. If these soils are predominantly sand or silt, they are usually classified as medium-textured soils. If predominantly clay, they are classified as fine-textured soils.

For the **overlay,** apply COTORAN 4L as a preemergence soil surface treatment as recommended on this label at 2-4 pts. per acre after a preplant soil incorporated treatment of Treflan EC. Apply Treflan EC as recommended on its label at 1-2.5 pts. per acre.

Precautions: Do not use the tank mix in Arizona and California. Do not apply the tank mix of COTORAN 4L + Treflan EC in liquid fertilizer. Refer to the Treflan EC label for directions, precautions, and limitations for use of that product, as well as the directions, precautions, and limitations on this COTORAN 4L label.

Prowl®: Apply COTORAN 4L as a preemergence surface treatment as recommended on this label at 2-4 pts. per acre after a preplant soil-incorporated treatment of Prowl for control of all weeds listed on both labels. Apply Prowl as recommended on its label at 1-3 pts. per acre. Refer to the Prowl label for directions, precautions, and limitations for use of that product.

DSMA or MSMA: COTORAN 4L may be tank-mixed with DSMA or MSMA and applied as a directed spray to cotton from three inches high to first bloom. Use the broadcast rate of 2-4 pts. of COTORAN 4L per acre and the necessary amount of DSMA or MSMA plus 0.5% (1 pt. per 25 gals.) of a surfactant that is approved by FDA for use on food and feed crops and is compatible with COTORAN 4L + DSMA or COTORAN 4L + MSMA when applied to cotton. When using this mixture, do not allow spray to contact cotton foliage. Do not graze treated areas or feed treated forage or gin trash to livestock. Adhere strictly to the dosage, precautions and limitations on the labels of the products.

STORAGE AND DISPOSAL

Do not contaminate water, food, or feed by storage or disposal. Pesticide wastes may be toxic. Improper disposal of unused pesticide, spray mixture, or rinse water is a violation of federal law. If these wastes cannot be used according to label instructions, contact your State Pesticide or Environmental Control Agency, or the Hazardous Waste representative at the nearest EPA Regional Office for guidance in proper disposal methods.

Triple rinse container (or equivalent). Then offer for recycling or reconditioning, or puncture and dispose of in a sanitary landfill or by incineration, or if allowed by state and local authorities, by burning. Stay out of smoke from burning jugs.

For minor spills, leaks, etc., follow all precautions indicated on this label and clean up immediately. Take special care to avoid contamination of equipment and facilities during cleanup procedures and disposal of wastes. In the event of a major spill, fire or other emergency, call (919) 292-7100 day or night.

PRECAUTIONARY STATEMENTS

HAZARDS TO HUMANS AND DOMESTIC ANIMALS

CAUTION

Harmful if swallowed, inhaled, or absorbed through the skin. Avoid breathing spray mist. Wash thoroughly with soap and water after handling. Remove and wash contaminated clothing before reuse.

First Aid: In case of contact with eyes, immediately flush with plenty of water. Get medical attention if irritation occurs. In case of contact with skin, wash thoroughly with soap and water. Get medical attention if irritation occurs. If swallowed, drink water and induce vomiting by placing fingers at back of throat. Immediately contact a physician, hospital, or local poison control center. In case of inhalation exposure, move from the contaminated area.

Notes to Physician: If COTORAN 4L is ingested, lavage stomach or induce vomiting. A slurry of activated charcoal in water can be left in the stomach. Give a saline laxative and supportive therapy.

ENVIRONMENTAL HAZARDS

Do not apply directly to water or wetlands. Do not contaminate water by cleaning of equipment or disposal of wastes.

Do not apply this product through any type of irrigation system.

COTORAN® trademark of CIBA-GEIGY for fluometuron
Compex® trademark of KALO Agricultural Chemicals, Inc.
Dual® trademark of CIBA-GEIGY for metolachlor
Prowl® trademark of American Cyanamid for pendimethalin
Treflan® trademark of Elanco Products for trifluralin
Unite® trademark of Hopkins-Agricultural Chemical Co.

CGA 33L14C 097

This is a specimen label which was not prepared by CIBA-GEIGY. CIBA-GEIGY is not responsible for the accuracy of the information contained herein. As labels are subject to revision, always carefully read and follow the label on the product container.

COTORAN® DF

HERBICIDE

For preemergence and postemergence weed control in cotton

ACTIVE INGREDIENT:

Fluometuron: 1,1-dimethyl-3-(α,α,α-trifluoro-*m*-tolyl) urea	85%
INERT INGREDIENTS:	15%
TOTAL:	100%

EPA Reg. No. 100-704 EPA Est. 9779-AR-13

10 Pounds Net Weight

Cotoran DF is a water dispersible granule.

KEEP OUT OF REACH OF CHILDREN

WARNING

See additional precautionary statements at end of label booklet.

See directions for use inside booklet.

Cotoran® trademark of CIBA-GEIGY for fluometuron

DIRECTIONS FOR USE AND CONDITIONS OF SALE AND WARRANTY

IMPORTANT: Read the entire **Directions for Use** and the **Conditions of Sale and Warranty** before using this product.

CONDITIONS OF SALE AND WARRANTY

The **Directions for Use** of this product reflect the opinion of experts based on field use and tests. The directions are believed to be reliable and should be followed carefully. However, it is impossible to eliminate all risks inherently associated with use of this product. Crop injury, ineffectiveness, or other unintended consequences may result because of such factors as weather conditions, presence of other materials, or the manner of use or application all of which are beyond the control of CIBA-GEIGY or the Seller. All such risks shall be assumed by the Buyer.

CIBA-GEIGY warrants that this product conforms to the chemical description on the label and is reasonably fit for the purposes referred to in the **Directions for Use** subject to the inherent risks referred to above. **CIBA-GEIGY makes no other express or implied warranty of Fitness or Merchantability or any other express or implied warranty. In no case shall CIBA-GEIGY or the Seller be liable for consequential, special, or indirect damages resulting from the use or handling of this product.** CIBA-GEIGY and the Seller offer this product, and the Buyer and user accept it, subject to the foregoing **Conditions of Sale and Warranty**, which may be varied only by agreement in writing signed by a duly authorized representative of CIBA-GEIGY.

DIRECTIONS FOR USE

It is a violation of federal law to use this product in a manner inconsistent with its labeling.

FAILURE TO FOLLOW THE DIRECTIONS FOR USE AND PRECAUTIONS ON THIS LABEL MAY RESULT IN CROP INJURY, POOR WEED CONTROL, AND/OR ILLEGAL RESIDUES.

Do not apply this product through any type of irrigation system.

GENERAL INFORMATION

Cotoran DF is effective for both annual grass and annual broadleaf weed control in cotton.

Weed Control

Cotoran DF applied according to directions will selectively control annual weeds in cotton, such as:

ANNUAL GRASS WEEDS		
Barnyardgrass	Fall panicum	Ryegrass
Crabgrass	Foxtail	Signalgrass
Crowfootgrass	Goosegrass	*(Brachiaria)*

ANNUAL BROADLEAF WEEDS		
Spurred anoda	Morningglory	Sicklepod
Cocklebur	Pigweed	Smartweed
Florida pusley	Prickly sida (teaweed)	Russian Thistle (tumbleweed)
Groundcherry (Wright)	Purslane	
Jimsonweed	Ragweed	
Lambsquarters	Sesbania	

Cotoran DF at rates up to 2.4 lbs. per acre in cotton is not effective for the control of established perennial weeds, such as johnsongrass from rhizomes.

Application Procedures

Mixing Instructions

(1) Fill spray tank ¾ full with clean water or nitrogen solution. (2) Start agitation. (3) Pour Cotoran DF directly from the container into the partially filled spray tank. (4) If Cotoran DF is to be tank mixed with another herbicide, add that product after the Cotoran DF is thoroughly mixed in the partially filled spray tank. (5) Add the rest of the water or nitrogen solution. (6) Provide agitation during mixing and application to maintain a uniform suspension.

To determine compatibility, add three teaspoons of Cotoran DF to one pint of nitrogen solution (equals 2.4 lbs. in 25 gals.; if using other application rates, adjust compatibility test amounts proportionally). Stir or shake thoroughly. Let stand 5 minutes. If it remains mixed or can be remixed readily, the mixture is compatible and can be used. If Cotoran DF settles out when mixed with a nitrogen solution, a suspensibility agent, such as Unite®, Compex®, or equivalent at 1–3 pts. per 100 gals. fertilizer, may improve compatibility. Add ⅛–⅜ teaspoon of Unite to the compatibility test to determine its effectiveness.

Equipment

Thoroughly clean sprayer prior to use. Do not use a sprayer contaminated with 2,4-D or other materials as crop damage or sprayer clogging may result.

Use properly calibrated conventional spray equipment that will apply spray uniformly and has hydraulic or mechanical agitation, except in California or Arizona where only mechanical agitators are recommended. Screens in nozzles, in suction, and in-line strainers should be no finer than 50-mesh. Use a pump with capacity to maintain 35–40 psi at the nozzles. If hydraulic agitation is used, the pump should also provide sufficient agitation in the tank to keep the mixture in suspension. Avoid leaving spray mixture in tank without constant agitation. If bypass agitation is used, it should terminate at the bottom of the tank to minimize foaming. Wash sprayer thoroughly with clean water immediately after use.

Apply in a minimum of 10 gals. per acre for preplant incorporated or preemergence applications, or apply in a minimum of 20 gals. per acre for postemergence application.

Band Treatment

For band treatment, calculate the amount of Cotoran DF needed by the formula:

$$\frac{\text{band width in inches}}{\text{row width in inches}} \times \text{broadcast rate per acre} = \text{amount needed per acre of field}$$

Cotton

Cotoran DF may be applied as a preemergence treatment to the surface of the soil, or lightly incorporated immediately after planting, or as a directed, semi-directed, or over-the-top postemergence application. For preemergence treatment only, Cotoran DF may be tank-mixed with most liquid nitrogen fertilizers. Cotoran DF may be combined with other herbicides in cotton as directed in the section, "Combinations."

Cotoran DF is most effective when used as follows:

1st Application: Preemergence broadcast. See Preemergence Treatment below.

2nd Application: Postemergence, directed, semi-directed, or over-the-top when cotton is 3–6 inches high. See Postemergence Treatments below.

3rd Application: If needed, postemergence, directed, or semi-directed, at lay-by (last normal cultivation), using no more than 1.2 lbs. per acre.

Preemergence Treatment

Apply Cotoran DF at a rate of 1.2–2.4 lbs. per acre. On light silt and sandy soils, low in organic matter, use the lower rate.

Cotoran DF may be applied to the surface of the soil, or lightly incorporated into the soil, after planting cotton.

Preplant Incorporated Use (Arizona, California, and New Mexico)

Cotoran DF may be applied preplant as a broadcast spray at 1.9 lbs. per acre after the beds for preplant irrigation have been formed. Apply before the beds are dragged off (using a bed shaper set to remove 1–2 inches from beds) in preparation for planting. Furrow irrigation should precede the application and planting should follow as soon as possible. Plant into untreated soil after drag-off and return treated soil to the beds when furrows are reformed. Do not allow furrow irrigation water to pond over the beds.

Postemergence Treatments

Cotoran DF may be applied postemergence to cotton but preemergence to weeds, or postemergence to both cotton and weeds. Cotoran DF should be applied at rates of 1.2–2.4 lbs. per acre. Apply as a directed, semi-directed, or over-the-top spray. Cotoran DF may be applied to cotton plants at any stage of growth from 3 inches high to lay-by. Use the higher rate when applying post-emergence to weeds; weeds should be 2 inches high or less. Use of a surfactant that is approved by FDA for use on food and feed crops and is compatible with Cotoran DF when applied to cotton may enhance the post-emergence herbicidal activity of Cotoran DF. The surfactant rate should not exceed 0.5% of the carrier volume (1 pt. per 25 gals. of water).

Precautions: Rainfall that is sufficient to germinate seeds or for growing crops is adequate to activate the herbicidal activity of Cotoran DF; where dry weather conditions prevail, the herbicidal activity of Cotoran DF may be delayed or reduced. Cultivate if weeds develop.

Do not make more than three applications of Cotoran DF to the same crops or field in any one year. Do not plant crops other than cotton or sugarcane within 6 months of the last application of Cotoran DF or injury may result.

The use of Cotoran DF as a preemergence or early postemergence application, following the use of a systemic insecticide at planting, may result in injury to cotton.

West Texas only: Do not use Cotoran DF on sandy, loamy sand, or fine sandy loam soils. Do not use Cotoran DF on cotton planted in furrows.

Arkansas, Louisiana, and Mississippi only: Use 1.0 lb. of Cotoran DF per acre on sandy loam soils low in organic matter.

Arizona, California, and New Mexico only: Cotton, corn, and grain sorghum can be planted the next spring. Do not plant treated areas to other crops until one year after last application. Do not use on sandy or coarse-textured soil of less than 1% organic matter.

Note: Do not apply within 60 days of harvest. Do not feed foliage from treated cotton plants or gin trash to livestock.

Combinations

Dual® 8E: Use in a tank mixture with Cotoran DF preemergence in cotton for control of all weeds as listed on both labels. This combination will also control spotted spurge, hyssop spurge, nodding spurge, and prostrate spurge. Apply to the soil surface at planting or after planting, but before weeds or crop emerge, using the appropriate rates from the table below. *Note:* Incompatibility may occur, add X-77 at 0.5% volume/volume spray (4 pints/100 gal.) to help overcome this condition.

Cotoran DF + Dual 8E — Cotton (AR, LA, MS, Eastern OK, TN, Bootheel of MO, and Gulf Coast, Rio Grande Valley and Eastern TX)

Soil texture	Broadcast Rates Per Acre	
	Cotoran DF	Dual 8E
Sand, loamy sand	DO NOT USE	
Sandy loam*	1.2 lbs.	1.25–1.5 pts.
Loam, silt, silt loam	1.2–1.9 lbs.	1.5–2 pts.
Silty clay loam, sandy clay loam, silty clay, sandy clay, clay loam, clay	1.9–2.4. lbs.	2 pts.

* Use on sandy loam soils only in eastern OK and Gulf Coast, Rio Grande Valley, and eastern TX.

Precautions: (1) Do not apply Cotoran DF + Dual 8E on sand or loamy sand soils or in areas where water is likely to "pond" over the bed, or crop injury may occur. (2) Do not apply broadcast to furrow-planted cotton. (3) The use of Cotoran DF following the use of a systemic insecticide at planting may result in crop injury. (4) Do not use on Taloka silt loam, or crop injury may occur.

Refer to the Dual 8E label for further instructions, precautions, and limitations.

Note: Do not feed treated forage or gin trash to livestock, or graze trated areas.

Rotational Crops: Refer to the crop rotation instructions for Cotoran DF alone on this label and for Dual 8E alone on the Dual 8E label.

Treflan EC

Apply Cotoran DF as a preplant incorporated or preemergence soil surface treatment as recommended on this label after a preplant soil incorporated treatement of Treflan EC. Refer to the Treflan EC label for directions, precautions, and limitations for use of that product, as well as the directions, precautions, and limitations on this Cotoran DF label.

Prowl®: Apply Cotoran DF as a preplant incorporated or preemergence soil surface treatment as recommended on this Cotoran DF label after a preplant soil-incorporated treatment of Prowl. Refer to the Prowl label for directions, precautions, and limitations for use of that product.

Zorial® Rapid 80: Apply Cotoran DF according to directions on this Cotoran DF label as a preemergence soil surface treatment following a preplant incorporated application of Zorial Rapid 80 or tank mix with Zorial Rapid 80 and apply as a preemergence treatment. Refer to the Zorial Rapid 80 label for directions for use, precautions, and limitations for use of that product.

DSMA or MSMA: Cotoran DF may be tank-mixed with DSMA or MSMA and applied as a directed spray to cotton from three inches high to first bloom. Use the broadcast rate of 1.2–2.4 lbs. of Cotoran DF per acre and the necessary amount of DSMA or MSMA plus 0.5% (1 pt. per 25 gals.) of a surfactant that is approved by FDA for use on food and feed crops and is compatible with Cotoran DF + DSMA or Cotoran DF + MSMA when applied to cotton. When using this mixture, do not allow spray to contact cotton foliage. Do not graze treated areas or feed treated forage or gin trash to livestock. Adhere strictly to the dosage and limitations on the labels of the products.

STORAGE AND DISPOSAL

Do not contaminate water, food, or feed by storage or disposal. Pesticide wastes may be toxic. Improper disposal of unused pesticide, spray mixture, or rinse water is a violation of federal law. If these wastes cannot be used according to label instructions, contact your State Pesticide or Environmental Control Agency, or the Hazardous Waste representative at the nearest EPA Regional Office for guidance in proper disposal methods.

Triple rinse container (or equivalent). Then offer for recycling or reconditioning, or puncture and dispose of in a sanitary landfill or by incineration, or if allowed by state and local authorities, by burning. Stay out of smoke from burning jugs.

For minor spills, leaks, etc., follow all precautions indicated on this label and clean up immediately. Take special care to avoid contamination of equipment and facilities during cleanup procedures and disposal of wastes. In the event of a major spill, fire or other emergency, call 1-800-888-8372 day or night.

PRECAUTIONARY STATEMENTS

HAZARDS TO HUMANS AND DOMESTIC ANIMALS

WARNING

Causes eye irritation. Do not get in eyes. Harmful if swallowed, inhaled, or absorbed through the skin. Wear protective clothing, as this product may cause skin rash after repeated exposure. Avoid contact with skin or clothing. Avoid breathing dust or spray mist.

First Aid: In case of contact with eyes, immediately flush with plenty of water. Get medical attention if irritation occurs. In case of contact with skin, wash thoroughly with soap and water. Get medical attention if irritation occurs. If swallowed, drink water and induce vomiting by placing fingers at back of throat. Immediately contact a physician, hospital, or local poison control center. In case of inhalation exposure, move from the contaminated area.

Notes to Physician: If Cotoran DF is ingested, lavage stomach or induce vomiting. A slurry of activated charcoal in water can be left in the stomach. Give a saline laxative and supportive therapy.

ENVIRONMENTAL HAZARDS

Do not apply directly to water or wetlands. Do not contaminate water by cleaning of equipment or disposal of wastes.

Cotoran® trademark of CIBA-GEIGY for fluometuron
Compex® trademark of KALO Agricultural Chemicals, Inc.
Prowl® trademark of American Cyanamid for pendimethalin
Treflan® trademark of Elanco Products for trifluralin
Unite® trademark of Hopkins Agricultural Chemical Co.

CGA 33L25A 049

This is a specimen label which was not prepared by CIBA-GEIGY. CIBA-GEIGY is not responsible for the accuracy of the information contained herein. As labels are subject to revision, always carefully read and follow the label on the product container.

CURACRON® 6E

INSECTICIDE MITICIDE

RESTRICTED USE PESTICIDE

FOR RETAIL SALE TO AND USE ONLY BY CERTIFIED APPLICATORS OR PERSONS UNDER THEIR DIRECT SUPERVISION. AND ONLY FOR THOSE USES COVERED BY THE CERTIFIED APPLICATOR'S CERTIFICATION.

For insect and mite control in cotton

ACTIVE INGREDIENT:
Profenofos: *O*-(4-bromo-2-chlorophenyl) *O*-ethyl *S*-propyl phosphorothioate 59.4%
INERT INGREDIENTS: 40.6%
TOTAL: 100.0%

EPA Reg. No. 100-599 EPA Est. 100-AL-1

CURACRON 6E contains 6 lbs. active ingredient per gallon.

KEEP OUT OF REACH OF CHILDREN

WARNING

See additional precautionary statements at end of label booklet.

See directions for use inside booklet.

CURACRON® trademark of CIBA—GEIGY for profenofos

U.S. Patent No. 3,992,533

CGA 55L6G 068

DIRECTIONS FOR USE AND CONDITIONS OF SALE AND WARRANTY

IMPORTANT: Read the entire **Directions for Use** and the **Conditions of Sale and Warranty** before using this product.

CONDITIONS OF SALE AND WARRANTY

The **Directions for Use** of this product reflect the opinion of experts based on field use and tests. The directions are believed to be reliable and should be followed carefully. However, it is impossible to eliminate all risks inherently associated with use of this product. Crop injury, ineffectiveness, or other unintended consequences may result because of such factors as weather conditions, presence of other materials, or the manner of use or application all of which are beyond the control of CIBA—GEIGY or the Seller. All such risks shall be assumed by the Buyer. CIBA—GEIGY warrants that this product conforms to the chemical description on the label and is reasonably fit for the purposes referred to in the **Directions for Use** subject to the inherent risks referred to above. **CIBA—GEIGY makes no other express or implied warranty of Fitness or Merchantability or any other express or implied warranty. In no case shall CIBA—GEIGY or the Seller be liable for consequential, special, or indirect damages resulting from the use or handling of this product.** CIBA—GEIGY and the Seller offer this product, and the Buyer and user accept it, subject to the foregoing **Conditions of Sale and Warranty,** which may be varied only by agreement in writing signed by a duly authorized representative of CIBA—GEIGY.

GENERAL INFORMATION

CURACRON 6E is an emulsifiable insecticide-miticide which is effective for control of cotton bollworm, tobacco budworm, certain other insects, and mites on cotton.

Best results can be expected from fields under Pest Management/Scouting programs.

Note: When using CURACRON 6E as a larvicide, best results will be obtained when application is timed to control first to third instar larvae.

Mixing Instructions

Prepare no more spray mixture than is needed for the immediate operation. Thoroughly clean spray equipment before using this product. Agitate thoroughly before and during application. Good bypass agitation is sufficient to keep CURACRON 6E in emulsion.

CURACRON 6E Alone: Add ²/₃ of the required amount of water to the spray tank with the agitator running, then add the desired amount of CURACRON 6E. Continue agitation while adding the remainder of the water.

CURACRON 6E + Tank Mixtures: Add ²/₃ of the required amount of water to the spray tank. With the agitator running if possible, add the desired amount of the other products recommended on this label. Then add the desired amount of CURACRON 6E. Continue agitation while adding the remainder of water.

Note: Do not let spray mixture stand over-night in the spray tank. Rinse spray tank at the end of the day.

Chemigation: Do not apply this product through any type of irrigation system.

DIRECTIONS FOR USE

It is a violation of federal law to use this product in a manner inconsistent with its labeling.

FAILURE TO FOLLOW ALL DIRECTIONS AND PRECAUTIONS ON THIS LABEL MAY RESULT IN CROP INJURY, POOR CONTROL, AND/OR ILLEGAL RESIDUES.

Cotton

Apply CURACRON 6E at the specified rates below in a minimum of 5 gallons of water per acre with ground equipment or in a minimum of 1 gallon of water per acre by air. For better coverage of late season "rank" cotton increase gallonage. To avoid spray drift, do not apply under windy conditions. Avoid spray overlap because crop injury may result.

If the need to control boll weevil becomes apparent while using CURACRON, Guthion® or methyl parathion may be tank mixed with CURACRON. Refer to the Guthion or methyl parathion labels for directions, limitations, and precautions.

Aphids, plant bugs/lygus bugs* (All cotton growing states): Apply ²/₃ pt. of CURACRON 6E per acre at 5-7 day intervals as needed starting when economic thresholds are reached. Economic thresholds will vary by locality. Consult your pest control advisor or cotton consultant to determine economic thresholds in your area.

* In Arizona and California, CURACRON 6E offers suppression of lygus bugs.

Cotton bollworm, tobacco budworm (All cotton growing states except Arizona and California):

Egg and Larvae Control

Early to mid-season cotton, light to moderate pest pressure: If careful scouting of early to mid-season cotton indicates pest populations are light to moderate, eggs and larvae may be controlled by tank mixing ¹/₃-²/₃ pints of CURACRON 6E with recommended rates of Galecron® 4E or SP. Refer to Galecron 4E or SP labels for directions, limitations, or precautions.

Later season cotton, light to heavy pest pressure: For light to moderate infestations, tank mix ²/₃ to 1 pint of CURACRON 6E with recommended rates of Galecron 4E or SP. For moderate to heavy infestations, tank mix 1 to 1¹/₃ pints CURACRON 6E with recommended rates of Galecron 4E or SP. Apply at 5-7 day intervals as needed beginning when infestations first appear. Refer to the Galecron 4E or SP label for directions, limitations, or precautions.

Contact Ovicide Activity

For contact ovicidal activity, apply ¹/₃-²/₃ pint of CURACRON 6E per acre when egg lay exceeds thresholds. Continue at 5-7 day intervals as needed. If larvae survive, use higher rates listed below.

Larvae Control Only

CURACRON alone: To control larvae only, apply ²/₃ to 1 pint of CURACRON 6E for light to moderate infestations. Apply 1 to 1¹/₃ pints CURACRON 6E for moderate to heavy infestations. Apply at 5-7 day intervals as needed beginning when infestations first appear.

Cotton bollworm, tobacco budworm (Arizona and California): Apply CURACRON 6E at 5-7 day intervals as needed beginning when infestations first appear. Use ²/₃ to 1 pt. per acre for light to moderate infestations and 1-1¹/₃ pts. per acre for moderate to heavy infestations. Eggs may be controlled by tank mixing CURACRON 6E with Galecron SP or 4E. Refer to the Galecron SP or 4E label for directions, limitations or precautions.

Cotton leafperforator, beet armyworm (All cotton growing states): Apply CURACRON 6E at 5-7 day intervals as needed starting when larvae first appear. Use ²/₃ 1 pt. per acre for light to moderate infestations and 1-1¹/₃ pts. per acre for moderate to heavy infestations.

Fall armyworm (All cotton growing states): Apply 1-1¹/₃ pints of CURACRON 6E per acre at 5-7 day intervals as needed starting when larvae first appear. For best results, application should be made to young larvae still feeding on leaf surfaces and before migration to blooms and bolls has occurred.

Mites (All cotton growing states): Apply ²/₃-1¹/₃ pts. of CURACRON 6E per acre at 5-7 day intervals as needed starting when mites first appear.

Whitefly suppression (All cotton growing states): To aid in the reduction of whitefly populations, apply 1-1¹/₃ pints of CURACRON 6E at a 5-7 day intervals as needed starting when whiteflies first appear.

CURACRON 6E + Tank Mixtures (All cotton growing states): To broaden the spectrum of insect control and make a more useful spray application, if multiple pests occur on cotton, it may be advantageous to tank mix CURACRON 6E with other EPA-registered insecticides.

CURACRON 6E has been found to be compatible in tank mixtures with the products listed below. When using CURACRON 6E in tank mixtures, observe all limitations, precautions, and rate recommendations which appear on the labels for these products: Ambush®, Ammo®, Azodrin®, Cymbush®, Galecron, Guthion, methomyl, methyl parathion, Pay-Off®, Pounce® and Pydrin®.

Note: Applications to cotton may cause spotting or reddening of leaves.

Note: To avoid illegal residues: (1) Do not apply more than 4 qts. of CURACRON 6E per acre per season. (2) Do not apply CURACRON 6E within 14 days of harvest. (3) Do not allow livestock to graze treated cotton. (4) Do not feed gin trash or foliage from treated cotton plants to livestock.

Fields treated with CURACRON may be rotated to other crops.

STORAGE AND DISPOSAL

Do not contaminate water, food, or feed by storage or disposal.

Pesticide disposal: Pesticide wastes are toxic. Improper disposal of excess pesticide, spray mixture, or rinsate is a violation of Federal law. If these wastes cannot be disposed of by use according to label instructions, contact your State Pesticide or Environment Control Agency or the Hazardous Waste Representative at the nearest EPA Regional Office for guidance.

Container disposal: Triple rinse (or equivalent). Then puncture container and dispose of in a sanitary landfill, or incinerate. The container may be disposed of by open burning if allowed by state and local authorities. Stay out of smoke from burning container.

Emergency Response

In the event of a major spill, fire, or other emergency, call (919) 292-7100

PRECAUTIONARY STATEMENTS

HAZARDS TO HUMANS AND DOMESTIC ANIMALS

WARNING

May be fatal if absorbed through skin. Harmful if swallowed or inhaled. Causes moderate eye irritation. Do not get in eyes, on skin, or on clothing. Avoid breathing vapors or spray mist. Wear protective clothing and rubber gloves when handling. Wash thoroughly with soap and water before eating or smoking. Remove contaminated clothing and wash before reuse.

STATEMENT OF PRACTICAL TREATMENT

IF IN EYES immediately flush with plenty of water. Get medical attention.

IF ON SKIN wash with plenty of soap and water. Remove contaminated clothing. Get medical attention.

IF SWALLOWED call a physician or poison control center. Give water or milk if conscious. May present aspiration hazard. DO NOT induce vomiting. Get medical attention immediately.

IF INHALED remove to fresh air.

Note to Physician: If symptoms of cholinergic poisoning occur, atropine plus 2-PAM may be helpful. If swallowed, lavage stomach, taking care to avoid aspiration of stomach contents into the lungs.

ENVIRONMENTAL HAZARDS

Due to the hazard to endangered fish, the application of this product is prohibited in Reeves County, Texas and within one mile of the Dexter National Fish Hatchery in New Mexico. This pesticide is toxic to fish and wildlife. Use with care when applying to areas frequented by wildlife or adjacent to any body of water. Do not apply directly to water. Do not apply with aircraft within 300 feet upwind of impounded water. Do not apply when weather conditions favor drift from treated areas. Do not apply with aircraft when wind speed is greater than 10 mph. Apply this pesticide only as specified on this label. Do not contaminate water by cleaning of equipment or disposal of wastes.

This product is highly toxic to bees exposed to direct treatment or residues on crops or weeds. Do not apply this product or allow it to drift to crops or weeds on which bees are foraging. Additional information may be obtained from your Cooperative Agricultural Extension Service.

Do not apply this product through any type of irrigation system.

Ambush® trademark of ICI Americas, Inc. for permethrin
Ammo® trademark of FMC Corporation for cypermethrin
Azodrin® trademark of Shell Chemical Company for monocrotophos
CURACRON® trademark of CIBA—GEIGY for profenofos
U.S. Patent No. 3,992,533
Cymbush® trademark of ICI Americas, Inc.
Galecron® trademark of CIBA—GEIGY for chlordimeform
U.S. Patent Nos. 3,378,437; 3,502,720; 3,478,156; 3,629,460; 3,462,537
Guthion® trademark of the Parent Company of Farbenfabriken Bayer GmbH, Leverkusen for azinphos-methyl
Pay-Off® trademark of American Cyanamid for flucythrinate
Pounce® trademark of FMC Corporation for permethrin
Pydrin® trademark of Shell Chemical Company for fenvalerate

CGA 55L6G 068

This is a specimen label which was not prepared by CIBA-GEIGY. CIBA-GEIGY is not responsible for the accuracy of the information contained herein. As labels are subject to revision, always carefully read and follow the label on the product container.

CURACRON® 8E

INSECTICIDE-MITICIDE

RESTRICTED USE PESTICIDE

FOR RETAIL SALE TO AND USE ONLY BY CERTIFIED APPLICATORS OR PERSONS UNDER THEIR DIRECT SUPERVISION. AND ONLY FOR THOSE USES COVERED BY THE CERTIFIED APPLICATOR'S CERTIFICATION.

For insect and mite control in cotton

ACTIVE INGREDIENT:	
Profenofos: *O*-(4-bromo-2-chlorophenyl) *O*-ethyl *S*-propyl phosphorothioate	72.7%
INERT INGREDIENTS*:	27.3%
TOTAL	100.0%

U.S. Patent No. 3,992,533

*Contains petroleum distillates

EPA Reg. No. 100-669 EPA Est. 100-AL-1

CURACRON 8E contains 8 lbs. active ingredient per gallon.

KEEP OUT OF REACH OF CHILDREN

WARNING

See additional precautionary statements at end of label booklet.

See directions for use inside booklet.

Curacron ® trademark of CIBA—GEIGY for profenofos

CGA 55L7C 068

DIRECTIONS FOR USE AND CONDITIONS OF SALE AND WARRANTY

IMPORTANT: Read the entire **Directions for Use** and the **Conditions of Sale and Warranty** before using this product.

CONDITIONS OF SALE AND WARRANTY

The **Directions for Use** of this product reflect the opinion of experts based on field use and tests. The directions are believed to be reliable and should be followed carefully. However, it is impossible to eliminate all risks inherently associated with use of this product. Crop injury, ineffectiveness, or other unintended consequences may result because of such factors as weather conditions, presence of other materials, or the manner of use or application all of which are beyond the control of CIBA—GEIGY or the Seller. All such risks shall be assumed by the Buyer.

CIBA—GEIGY warrants that this product conforms to the chemical description on the label and is reasonably fit for the purposes referred to in the **Directions for Use** subject to the inherent risks referred to above. **CIBA—GEIGY makes no other express or implied warranty of Fitness or Merchantability or any other express or implied warranty. In no case shall CIBA—GEIGY or the Seller be liable for consequential, special, or indirect damages resulting from the use or handling of this product.** CIBA—GEIGY and the Seller offer this product, and the Buyer and user accept it, subject to the foregoing **Conditions of Sale and Warranty,** which may be varied only by agreement in writing signed by a duly authorized representative of CIBA—GEIGY.

GENERAL INFORMATION

Curacron 8E is an emulsifiable insecticide-miticide which is effective for control of cotton bollworm, tobacco budworm, certain other insects, and mites on cotton.

Best results can be expected from fields under Pest Management/Scouting programs.

Note: When using Curacron 8E as a larvicide, best results will be obtained when application is timed to control first to third instar larvae.

Mixing Instructions

Prepare no more spray mixture than is needed for the immediate operation. Thoroughly clean spray equipment before using this product. Agitate thoroughly before and during application. Good bypass agitation is sufficient to keep Curacron 8E in emulsion.

Curacron 8E Alone: Add 2/3 of the required amount of water to the spray tank with the agitator running, then add the desired amount of Curacron 8E. Continue agitation while adding the remainder of the water.

Curacron 8E + Tank Mixtures: Add 2/3 of the required amount of water to the spray tank. With the agitator running if possible, add the desired amount of the other products recommended on this label. Then add the desired amount of Curacron 8E. Continue agitation while adding the remainder of water.

Note: Do not let spray mixture stand overnight in the spray tank. Rinse spray tank at the end of the day.

Chemigation: Do not apply this product through any type of irrigation system.

DIRECTIONS FOR USE

It is a violation of federal law to use this product in a manner inconsistent with its labeling.

FAILURE TO FOLLOW ALL DIRECTIONS AND PRECAUTIONS ON THIS LABEL MAY RESULT IN CROP INJURY, POOR CONTROL, AND/OR ILLEGAL RESIDUES.

Use Curacron 8E in the following states only: Alabama, Arkansas, Florida, Georgia, Louisiana, Kentucky, Mississippi, Missouri, North Carolina, New Mexico, Oklahoma, South Carolina, Tennessee, Texas, and Virginia.

Cotton

Apply Curacron 8E at the specified rates below in a minimum of 5 gallons of water per acre with ground equipment or in a minimum of 1 gallon of water per acre by air. For better coverage of late season "rank" cotton increase gallonage. To avoid spray drift, do not apply under windy conditions. Avoid spray overlap because crop injury may result.

If the need to control boll weevil becomes apparent while using Curacron, Guthion® or methyl parathion may be tank mixed with Curacron. Refer to the Guthion or methyl parathion labels for directions, limitations, and precautions.

Aphids, plant bugs/lygus bugs: Apply 1/2 pt. of Curacron 8E per acre at 5-7 day intervals as needed starting when economic thresholds are reached. Economic thresholds will vary by locality. Consult your pest control advisor or cotton consultant to determine economic thresholds in your area.

Cotton bollworm and tobacco budworm

Egg and Larvae Control

Early to mid-season cotton, light to moderate pest pressure: If careful scouting of early to mid-season cotton indicates pest populations are light to moderate, eggs and larvae may be controlled by tank mixing 1/4 - 1/2 pint of Curacron 8E with recommended rates of Galecron® 4E or SP. Refer to Galecron 4E or SP labels for directions, limitations, or precautions.

Later season cotton, light to heavy pest pressure: For light to moderate infestations, tank mix 1/2 to 3/4 pint of Curacron 8E with recommended rates of Galecron 4E or SP. For moderate to heavy infestations, tank mix 3/4 to 1 pint Curacron 8E with recommended rates of Galecron 4E or SP. Apply at 5-7 day intervals as needed beginning when infestations first appear. Refer to the Galecron 4E or SP label for directions, limitations, or precautions.

Contact Ovicide Activity

For contact ovicidal activity, apply 1/4-1/2 pint of Curacron 8E per acre when egg lay exceeds thresholds. Continue at 5-7 day intervals as needed. If larvae survive, use higher rates listed below.

Larvae Control Only

Curacron alone: To control larvae only, apply 1/2 to 3/4 pint of Curacron 8E for light to moderate infestations. Apply 3/4 to 1 pint Curacron 8E for moderate to heavy infestations. Apply at 5-7 day intervals as needed beginning when infestations first appear.

Cotton leafperforator, beet armyworm: Apply Curacron 8E at 5-7 day intervals as needed starting when larvae first appear. Use 1/2 to 3/4 pt. per acre for light to moderate infestations and 3/4 to 1 pt. per acre for moderate to heavy infestations.

Fall armyworm: Apply 3/4 to 1 pt. of Curacron 8E per acre at 5-7 day intervals as needed starting when larvae first appear. For best results, application should be made to young larvae still feeding on leaf surfaces and before migration to blooms and bolls has occurred.

Mites: Apply 1/2 to 1 pt. of Curacron 8E per acre at 5-7 day intervals as needed starting when mites first appear.

Whitefly suppression: To aid in the reduction of whitefly populations, apply 3/4 to 1 pt. of Curacron 8E at a 5-7 day intervals as needed starting when whiteflies first appear.

Curacron 8E + Tank Mixtures: To broaden the spectrum of insect control and make a more useful spray application, if multiple pests occur on cotton, it may be advantageous to tank mix Curacron 8E with other EPA-registered insecticides.

Curacron 8E has been found to be compatible in tank mixtures with the products listed below. When using Curacron 8E in tank mixtures, observe all limitations, precautions, and rate recommendations which appear on the labels for these products: Ambush®, Ammo®, Azodrin®, Cymbush®, Galecron, Guthion, methomyl, methyl parathion, Pay-Off®, Pounce® and Pydrin®.

Note: Applications to cotton may cause spotting or reddening of leaves.

Note: To avoid illegal residues: (1) Do not apply more than 3 qts. of Curacron 8E per acre per season. (2) Do not apply Curacron 8E within 14 days of harvest. (3) Do not allow livestock to graze treated cotton. (4) Do not feed gin trash or foliage from treated cotton plants to livestock.

Fields treated with Curacron may be rotated to other crops.

STORAGE AND DISPOSAL

Do not contaminate water, food, or feed by storage or disposal.

Pesticide disposal: Pesticide wastes are toxic. Improper disposal of excess pesticide, spray mixture, or rinsate is a violation of Federal law. If these wastes cannot be disposed of by use according to label instructions, contact your State Pesticide or Environmental Control Agency or the Hazardous Waste Representative at the nearest EPA Regional Office for guidance.

Container disposal: Triple rinse (or equivalent). Then puncture container and dispose of in a sanitary landfill, or incinerate. The container may be disposed of by open burning if allowed by state and local authorities. Stay out of smoke from burning container.

Emergency Response

In the event of a major spill, fire, or other emergency, call (919) 292-7100.

PRECAUTIONARY STATEMENTS

HAZARDS TO HUMANS AND DOMESTIC ANIMALS

WARNING

May be fatal if absorbed through skin. Causes eye irritation. Do not get in eyes, on skin, or on clothing. Avoid breathing vapors or spray mist. Wear protective clothing, goggles, face shield or safety glasses, and rubber gloves when handling. Harmful if swallowed or inhaled. Wash thoroughly with soap and water before eating, drinking or using tobacco. Remove contaminated clothing and wash before reuse.

STATEMENT OF PRACTICAL TREATMENT

IF SWALLOWED call a physician or poison control center. May present aspiration hazard. DO NOT induce vomiting. Drink large quantities of water. Get medical attention immediately.

IF ON SKIN wash with plenty of soap and water. Remove contaminated clothing. Get medical attention.

IF IN EYES hold eyelids open and flush with plenty of water. Call a physician if irritation persists.

IF INHALED remove victim to fresh air.

Note to Physician: If symptoms of cholinergic poisoning occur, atropine plus 2-PAM may be helpful. If swallowed, lavage stomach, taking care to avoid aspiration of stomach contents into the lungs.

ENVIRONMENTAL HAZARDS

Due to the hazard to endangered fish, the application of this product is prohibited in Reeves County, Texas and within one mile of the Dexter National Fish Hatchery in New Mexico. This pesticide is toxic to fish and wildlife. Use with care when applying to areas frequented by wildlife or adjacent to any body of water. Do not apply directly to water. Do not apply with aircraft within 300 feet upwind of impounded water. Do not apply when weather conditions favor drift from treated areas. Do not apply with aircraft when wind speed is greater than 10 mph. Apply this pesticide only as specified on this label. Do not contaminate water by cleaning of equipment or disposal of wastes.

This product is highly toxic to bees exposed to direct treatment or residues on crops or weeds. Do not apply this product or allow it to drift to crops or weeds on which bees are foraging. Additional information may be obtained from your Cooperative Agricultural Extension Service.

PHYSICAL OR CHEMICAL HAZARDS

Do not use or store near heat or open flame.

Do not apply this product through any type of irrigation system.

Ambush® trademark of ICI Americas, Inc. for permethrin
Ammo® trademark of FMC Corporation for cypermethrin
Azodrin® trademark of Shell Chemical Company for monocrotophos
Curacron® trademark of CIBA—GEIGY for profenofos
U.S. Patent No. 3,992,533
Cymbush® trademark of ICI Americas, Inc.
Galecron® trademark of CIBA—GEIGY for chlordimeform
U.S. Patent Nos. 3,378,437; 3,502,720; 3,478,156; 3,629,460; 3,462,537
Guthion® trademark of the Parent Company of Farbenfabriken Bayer GmbH, Leverkusen for azinphos-methyl
Pay-Off® trademark of American Cyanamid for flucythrinate
Pounce® trademark of FMC Corporation for permethrin
Pydrin® trademark of Shell Chemical Company for fenvalerate

CGA 55L7C 068

This is a specimen label which was not prepared by CIBA-GEIGY. CIBA-GEIGY is not responsible for the accuracy of the information contained herein. As labels are subject to revision, always carefully read and follow the label on the product container.

D.Z.N® diazinon 4E

INSECTICIDE

For control of cockroaches, and ornamental pests, and certain other household pests (as listed in directions for use)

Recommended for commercial/industrial use only

D.z.n diazinon 4E contains 4 lbs. diazinon per gallon

This product must not be used on golf courses and sod farms.

ACTIVE INGREDIENT:

Diazinon: [O,O-diethyl O-(2-isopropyl-6-methyl-4-pyrimidinyl) phosphorothioate]	47.5%
INERT INGREDIENTS*:	52.5%
TOTAL:	100.0%

* Contains xylene

EPA Reg. No. 100-463 EPA Est. 100-AL-1

KEEP OUT OF REACH OF CHILDREN

WARNING

STATEMENT OF PRACTICAL TREATMENT

If swallowed, call a physician immediately. DO NOT INDUCE VOMITING UNLESS NO OTHER TREATMENT IS AVAILABLE. Vomiting may cause aspiration pneumonia as product contains petroleum distillates. If it is necessary to induce vomiting, give victim one or two glasses of water and insert finger in back of throat. Do not induce vomiting or give anything by mouth if person is unconscious or convulsing.

If inhaled, remove victim to fresh air and apply artificial respiration if indicated.

If on skin, wash promptly with soap and water. Rinse thoroughly.

If in eyes, rinse eyes with plenty of water and call a physician immediately.

Note to Physician: Solvent may present aspiration hazard. Gastric lavage may be indicated if product was taken internally. Diazinon is an organophosphate insecticide. If symptoms of cholinesterase inhibition are present, atropine sulfate by injection is antidotal. 2-PAM is also antidotal and may be administered but only in conjuction with atropine.

See directions for use on panel at right.

DIRECTIONS FOR USE AND CONDITIONS OF SALE AND WARRANTY

IMPORTANT: Read the entire **Directions for Use** and the **Conditions of Sale and Warranty** before using this product.

CONDITIONS OF SALE AND WARRANTY

The **Directions for Use** of this product reflect the opinion of experts based on field use and tests. The directions are believed to be reliable and should be followed carefully. However, it is impossible to eliminate all risks inherently associated with use of this product. Crop injury, ineffectiveness, or other unintended consequences may result because of such factors as weather conditions, presence of other materials, or the manner of use or application all of which are beyond the control of CIBA-GEIGY or the Seller. All such risks shall be assumed by the Buyer.

CIBA-GEIGY warrants that this product conforms to the chemical description on the label and is reasonably fit for the purposes referred to in the **Directions for Use** subject to the inherent risks referred to above. **CIBA-GEIGY makes no other express or implied warranty of Fitness or Merchantability or any other express or implied warranty. In no case shall CIBA-GEIGY or the Seller be liable for consequential, special, or indirect damages resulting from the use or handling of this product.** CIBA-GEIGY and the Seller offer this product, and the Buyer and user accept it, subject to the foregoing **Conditions of Sale and Warranty,** which may be varied only by agreement in writing signed by a duly authorized representative of CIBA-GEIGY.

PRECAUTIONARY STATEMENTS

HAZARDS TO HUMANS AND DOMESTIC ANIMALS

WARNING

May be fatal if swallowed. May be absorbed through skin. Do not breathe spray mist. Do not get in eyes, on skin, or on clothing. Wash thoroughly after handling. Avoid contamination of food and feed. Food utensils such as tablespoons or measuring cups must not be used for food purposes after use in measuring pesticides. Keep out of reach of domestic animals. Do not use on humans, household pets, or livestock. Do not permit children or pets to go onto sprayed grass until spray has completely dried. Do not contaminate ornamental fish ponds.

STATEMENT OF PRACTICAL TREATMENT

If swallowed, call a physician immediately. DO NOT INDUCE VOMITING UNLESS NO OTHER TREATMENT IS AVAILABLE. Vomiting may cause aspiration pneumonia as product contains petroleum distillates. If it is necessary to induce vomiting, give victim one or two glasses or water and insert finger in back of throat. Do not induce vomiting or give anything by mouth if person is unconscious or convulsing.

If inhaled, remove victim to fresh air and apply artificial respiration if indicated.

If on skin, wash promptly with soap and water. Rinse thoroughly.

If in eyes, rinse eyes with plenty of water and call a physician immediately.

Note to Physician: Solvent may present aspiration hazard. Gastric lavage may be indicated if product was taken internally. Diazinon is an organophosphate insecticide. If symptoms of cholinesterase inhibition are present, atropine sulfate by injection is antidotal. 2-PAM is also antidotal and may be administered but only in conjunction with atropine.

ENVIRONMENTAL HAZARDS

This product is toxic to birds, fish and wildlife including waterfowl. Birds feeding on treated areas may be killed. Keep out of any body of water. Apply this product only as specified on this label.

This pesticide is highly toxic to bees exposed to direct treatment or to residues remaining on the treated area. Do not apply when bees are actively visiting the crop, cover crop, or weeds blooming in the treatment area. Applications should be timed to provide the maximum possible interval between treatment and the next period of bee activity.

PHYSICAL OR CHEMICAL HAZARDS

Do not use, pour, spill, or store near heat or open flame.

DIRECTIONS FOR USE

It is a violation of federal law to use this product in a manner inconsistent with its labeling.

GENERAL INFORMATION

D·z·n diazinon 4E is an emulsifiable solution containing 4 lbs. of diazinon per gallon. It is intended for use by pest control operators after dilution with water for the control of those pests specified on this label.

D·z·n diazinon 4E is not intended for use in preparing oil base sprays.

Notes: (1) The solvent used in this formulation is a petroleum distillate which may stain certain plastic, rubber, and asphalt materials such as tiles and floor coverings. Do not treat such materials. (2) Care should also be taken to avoid spotting of wallpaper and fabrics. (3) When used in dwellings, care should be taken to avoid deposits which could be frequently contacted by children. (4) Do not allow children in treated areas until surfaces are dry. (5) Do not contaminate food, food containers, or cooking utensils. (6) Do not apply this product through any type of irrigation system.

Household Pests

Dilute D·z·n diazinon 4E with water to make 1/2% or 1% spray and apply as indicated below:

Gals. of finished spray	Amount D·z·n diazinon 4E to use for	
	1/2% spray	1% spray
1	1 1/4* fl. oz.	2 1/2 fl. oz.
2	2 1/2 fl. oz.	5 fl. oz.
3	3 3/4 fl. oz.	7 1/2 fl. oz.
10	12 1/2 fl. oz.	25 fl. oz. (1 3/5 pts.)
25	32 fl. oz. (2 pts.)	64 fl. oz. (4 pts.)

* 2 1/2 tablespoons = 1 1/4 fl. oz.

Cockroaches (Waterbugs), Ants, Silverfish: Make spot applications of 1/2% coarse spray or apply with a paintbrush to baseboards, areas around water pipes, cracks and crevices, surfaces behind and beneath cabinets, refrigerators, sinks, stoves, and similar areas where these insects hide.

For heavy cockroach infestations, make the first application at a strength of 1%, followed by monthly applications of 1/2% spray.

Carpet Beetles: Apply a 1/2% spray in spot applications along baseboards and edges of carpeting, under carpeting and rugs and furniture, in closets and on shelving, and wherever these insects are seen or suspected.

Spiders and Scorpions: Apply a 1/2% spray around windows, doors, and along baseboards. Make spot applications to other areas over which these pests may crawl.

Pantry Pests—Exposed Stages (Sawtoothed grain beetle, Flour beetle, Rice weevil, Cigarette beetle, Drugstore beetle, Indian meal moth). Apply a 1/2% spray to shelves and cupboards where food containers suspected of being infested are stored. Remove all food and shelving paper before spraying. Do not replace the food packages until spray has dried. Cover shelves with shelving paper. This treatment is supplemental to good housekeeping, since it will have no effect on insects breeding within food packages. Infested food packages should be destroyed.

Brown Dog Ticks in Buildings: Make thorough application of a 1/2% spray to infested areas around baseboards, window and door frames, wall cracks, sleeping quarters of household pets, and localized areas of floors and floor coverings. Fresh bedding should be placed in animal quarters following treatment. Do not treat animals with this formulation.

Food and Feed Handling Establishments (including places other than private residences in which exposed food or feed is held, processed, prepared or served.)

Apply D·z·n diazinon 4E as either a crack and crevice or spot treatment as directed below. Limit spot treatments to floor surfaces where pests have been seen or where they are suspected of hiding. Limit individual spot treatments to an area no larger than 20 percent of the floor surface. Any individual spot treatment shall not exceed two square feet. Do not apply to areas, surfaces, or utensils which will come in contact with food. Take extreme care to avoid contamination of food or food contact surfaces or introducing the material into the air.

Food/Feed Areas (Application limited to Crack and Crevice and Spot Treatments Only): Food/Feed areas include areas for receiving, storage, packing (canning, bottling, wrapping, boxing), preparing, edible waste storage, and enclosed processing systems (mills, dairies, edible oils, syrups). Serving area (when food is exposed and facility is in operation) is considered a food area and includes facilities where prepared foods are served, such as dining rooms, but excluding areas where foods may be prepared or held.

Crack and Crevice Treatment—As a crack and crevice treatment, use equipment capable of delivering a pin stream of insecticide and apply 1/2-1% in small amounts directly into such spaces as expansion joints, in points between different elements of construction, between equipment and floors, wall boards, baseboards or openings leading to voids and hollow spaces in walls, hollow equipment legs and bases where cockroaches (waterbugs), ants, spiders, and silverfish hide.

Spot Treatment—As a spot treatment, apply 1/2-1% as a coarse, low pressure fan spray to selective floor surfaces, areas around water pipes, beneath cabinets, refrigerators, sinks, stoves, storage areas, baseboards, under elements of construction and similar areas where cockroaches, ants, spiders, and silverfish hide.

Limit spot treatment in food areas to floor surfaces where pests have been seen or where they are suspected of hiding. Limit individual spot treatments to an area no larger than 20 percent of the floor surface. Any individual spot treatment must not exceed two square feet. Avoid treating surfaces likely to be contacted by food. (Do not apply when facility is in operation or foods are exposed).

These treatments also aid in the control of pantry pests (exposed stages of sawtoothed grain beetle, flour beetle, rice weevil, cigarette beetle, drugstore beetle, and Indian meal moth) which come into contact with treated surfaces. DO NOT use this product in conduits, motor housings, and electrical switch boxes. Repeat applications as necessary.

Applications of this product in the food/feed areas of food or feed handling establishments other than as crack and crevice and spot treatments are not permitted.

Nonfood/Nonfeed Areas: Nonfood/Nonfeed areas include garbage rooms, lavatories, floor drains (to sewers), entries and vestibules, offices, locker rooms, machine rooms, boiler rooms, garages, mop closets, and storage (after canning or bottling).

In nonfood/nonfeed areas, apply 1/2-1% as a spot treatment using a coarse, low pressure for spray to baseboard areas, around water pipes, surfaces behind and beneath sinks, lockers, tables, pallets and similar areas where insects hide or through which they may enter.

Ornamental Pests

To control certain pests on ornamentals, such as Arborvitae, Azalea, Birch, Boxwood, Camellia, Carnation, Chrysanthemum, Douglas fir, Elm, Gladioli, Hawthorn, Holly, Juniper, Lilac, Locust, Maple, Oak, Pine, Plum, Poplar, Rhododendron, Rose, Spruce and Willow, apply the recommended rates indicated in the following chart.

Do not use on Ferns, Poinsettia, Hibiscus, Papaya, Pilea, and Gardenia, since injury to the plants may occur.

Pests	Amount D·z·n diazinon 4E to use	
	per 3 gals. water	per 100 gals. water
Aphids, Bagworms, Carnation bud mites, Carnation shoot mites, Clover mites, Cyclamen mites, Dipterous leafminers, European pine shoot moths, European red mites, Flea beetles, Holly bud months, Leafhoppers, Obscure root weevils, Omnivorous leaftiers, Privet mites, Scale crawlers: Cottonycushion, Lecanium, Pine needle, San Jose scale, Thrips, Twospotted mites, Webworms, Whiteflies	1/2 fl. oz.*	1 pt.*
Apple-and-thorn skeletonizers, Cotoneaster webworms, Fall webworms, Hemlock chermes, Oak loopers, Obliquebanded leafrollers, Pear slugs, Tent caterpillars	1 1/2 fl. oz.*	3 pts.*

* 1 fl. oz. = 2 tbsp.; 16 fl. oz. = 1 pt.

Mimosa Webworm: Mix 1/2 fl. oz. (1 tablespoon) in 3 gals. of water (or 1 pt. in 100 gals. of water) and apply as a thorough foliar spray when adults first appear. Make a second and third application at 4-5 week intervals.

Lawn Pests: (excluding golf courses and sod farms)

To control pests listed below, apply recommended rates. For pests such as Ants, Brown dog ticks, Clover mites, Crickets, Earwigs, Fleas, Millipedes, Sowbugs, and Springtails (Collembola), which are commonly found near the house and may enter the house, spray a 5-ft band of soil around the house, as well as the house foundation wall to a height of 2-3 ft. Repeat application if necessary.

Precautions: Birds, especially waterfowl, feeding or drinking on treated areas may be killed. Apply this product only as specified on this label. Because of the migratory habits of certain waterfowl, do not apply in Nassau County, New York between November 1 and May 20. Do not exceed maximum permitted label rates since rates above those recommended significantly increase potential hazards to birds. Avoid overlapping of sprays. If waterfowl, i.e., ducks, geese can be expected to enter the treated area after treatment (except spot treatments), water lawn with at least 1/4 inch water immediately after application, however, stop watering before puddling occurs.

Boxelder Bugs: Spray bugs outside the home with 1 1/4-2 1/2 fl. oz. (2 1/2-5 tablespoons) in 1 gal. of water. At the same time, spray thoroughly around foundation, windows, doors, tree trunks, and into cracks and other places where the insects hide.

Pests	Rate per 1,000 sq. ft.		Remarks
	D·z·n diazinon 4E	Water	
Lawn chinch bugs	2-3 fl. oz.	3 gals.	Use higher rate for longer residual control and in lawns of dense growth, such as St. Augustine lawns.
Ants, Armyworms, Clover mites, Springtails (Collembola). Crickets, Cutworms. Digger wasps, Earwigs, Frit flies, Lawn billbugs, Sod webworms (Lawn moth), Sowbugs	2-3 fl. oz.	3 gals.	Spot spray anthills and wasp ground nest openings. For frit flies, mow grass and water well before treatment but delay application until grass is dry. For digger wasps, apply at dusk when the wasps are not active.
Brown dog ticks, Bermudagrass mites, Chiggers, Fleas, Leafhoppers	1 1/4 fl. oz.	3 gals.	Note: Do not apply to animals. For brown dog ticks, spray grass and under shrubbery, particularly near house.
Millipedes	3 fl. ox.	3 gals.	
Rhodesgrass scales	3 fl. oz.	3 gals.	Apply when crawlers first emerge.

STORAGE AND DISPOSAL

Do not contaminate water, food, or feed by storage or disposal. Open dumping is prohibited.

Pesticide: Pesticide wastes are toxic. Improper disposal of excess pesticide, spray mixture, or rinsate is a violation of federal law. If these wastes cannot be disposed of by use according to label instructions, contact your State Pesticide or Environmental Control Agency or the Hazardous Waste Representative at the nearest EPA Regional Office for guidance.

Container: Triple rinse or equivalent. Then offer for recycling, reconditioning, or puncture and dispose of in a sanitary landfill, or by other procedures approved by state and local authorities.

For minor spills, leaks, etc., follow all precautions indicated on this label and clean up immediately. Take special care to avoid contamination of equipment and facilities during cleanup procedures and disposal of wastes. In the event of a major spill, fire or other emergency, call 1-800-888-8372 day or night.

D·z·n® trademark of CIBA-GEIGY for diazinon

CGA 1L47Y

This is a specimen label which was not prepared by CIBA-GEIGY. CIBA-GEIGY is not responsible for the accuracy of the information contained herein. As labels are subject to revision, always carefully read and follow the label on the product container.

D.Z.N® diazinon 14G

GRANULAR INSECTICIDE

For Agricultural or Commercial Use Only

For control of certain soil insects including wireworms, cutworms, and other crop pests

This product must not be used on golf courses and sod farms.

ACTIVE INGREDIENT:

Diazinon: O,O-Diethyl 0-(2-isopropyl-6-methyl-4-pyrimidinyl) phosphorothioate 14.3%

INERT INGREDIENTS: 85.7%

TOTAL: 100.0%

EPA Reg. No. 100-469 EPA Est. 39578-TX-1

D·z·n diazinon 14G is a granular insecticide.

KEEP OUT OF REACH OF CHILDREN

CAUTION

See additional precautionary statements on back of bag.

Control No. 06850

See directions for use on back of bag.

DIRECTIONS FOR USE AND CONDITIONS OF SALE AND WARRANTY

IMPORTANT: Read the entire **Directions for Use** and the **Conditions of Sale and Warranty** before using this product.

CONDITIONS OF SALE AND WARRANTY

The **Directions for Use** of this product reflect the opinion of experts based on field use and tests. The directions are believed to be reliable and should be followed carefully. However, it is impossible to eliminate all risks inherently associated with use of this product. Crop injury, ineffectiveness, or other unintended consequences may result because of such factors as weather conditions, presence of other materials, or the manner of use or application all of which are beyond the control of CIBA–GEIGY or the Seller. All such risks shall be assumed by the Buyer.

CIBA–GEIGY warrants that this product conforms to the chemical description on the label and is reasonably fit for the purposes referred to in the **Directions for Use** subject to the inherent risks referred to above. **CIBA–GEIGY makes no other express or implied warranty of Fitness or Merchantability or any other express or implied warranty. In no case shall CIBA–GEIGY or the Seller be liable for consequential, special, or indirect damages resulting from the use or handling of this product.** CIBA–GEIGY and the Seller offer this product, and the Buyer and user accept it, subject to the foregoing **Conditions of Sale and Warranty,** which may be varied only by agreement in writing signed by a duly authorized representative of CIBA–GEIGY.

DIRECTIONS FOR USE

It is a violation of federal law to use this product in a manner inconsistent with its labeling.

REENTRY STATEMENT FOR AGRICULTURAL USES

No user shall permit the application of this product in such a manner as to directly or indirectly expose workers or other persons except those knowingly involved in the application. The area being treated must be vacated by unprotected persons.

Do not enter treated areas without protective clothing until dusts have settled. Protective clothing includes at least a hat or other suitable head covering, a long-sleeved shirt and long-legged trousers or a coverall type garment (all of a closely woven fabric) covering the body including arms and legs, socks and shoes.

Because certain states may require more restrictive reentry intervals for various crops treated with this product, consult your State Department of Agriculture for further information.

Written or oral warnings must be given to workers who are expected to be in a treated area or in an area about to be treated with this product. Oral warnings must be given which inform workers of areas or fields that may not be entered without specific protective clothing until dusts have settled and appropriate actions to take as described under **Caution** statements at the end of this label. When oral warnings are given, warnings shall be given in a language customarily understood by workers. Oral warnings must be given if there is reason to believe that written warnings cannot be understood by workers. Written warnings must include the following information: "CAUTION. Area treated with Diazinon 14G on (date of application). Do not enter without appropriate protective clothing until dusts have settled. In case of accidental exposure, flush eyes or skin with plenty of water. Call a physician if irritation persists. Remove and wash contaminated clothing before reuse."

Applications to the soil should be worked into the top few inches immediately after application with a rotary hoe, cultivator, disk, harrow, or other suitable means for greatest effectiveness.

FAILURE TO FOLLOW ALL DIRECTIONS FOR USE AND PRECAUTIONS ON THIS LABEL MAY RESULT IN CROP INJURY, POOR INSECT CONTROL, AND/OR ILLEGAL RESIDUES.

FIELD AND FORAGE INSECTS

Beans (Pinto)

Lesser Cornstalk Borers: Apply 7-14 lbs. per acre in a 6-inch band centered over the row when beans are $\frac{1}{2}$-1 inch high.

Corn

Corn Rootworm Larvae: (Postemergence Basal Treatment)

Apply $4\frac{1}{2}$-$8\frac{1}{2}$ oz. per 1,000 linear feet of row in a band over the row at first sign of rootworms, but not later than June 24 in the Middle Corn Belt or July 1 in the Northern Corn belt. Cover treated band with 2-3 inches of soil.

Apply amount per acre recommended according to row spacing chart for basal treatment.

Lesser Cornstalk Borers: Apply $8\frac{1}{2}$-17 oz. per 1,000 linear feet of row in a 10-inch band centered over the row. Make application at planting time or when crop begins to emerge. Cover the treated band with 1-2 inches of soil.

Apply amount per acre recommended according to row spacing chart for basal treatment.

Suggested Rates According to Corn Row Spacing

Row Spacing	Linear feet of corn row per acre	Amount of D·z·n diazinon 14G Per Acre — Postemergence Basal Treatment
48"	10,890	3-6 lbs.
42"	12,455	$3\frac{1}{4}$-$6\frac{1}{2}$ lbs.
40"	13,068	$3\frac{1}{2}$-7 lbs.
38"	13,939	$3\frac{3}{4}$-$7\frac{1}{2}$ lbs.
36"	14,520	4-8 lbs.
30"	17,424	$4\frac{3}{4}$-$9\frac{1}{2}$ lbs.
24"	21,780	$5\frac{3}{4}$-$11\frac{1}{2}$ lbs.
20"	26,136	7-14 lbs.
15"	34,848	$9\frac{1}{4}$-$18\frac{1}{2}$ lbs.

Note: For row spacings not listed, calculate the appropriate dosage rates based on the 40-in. row figures.

European Corn Borers: Apply 7-14 lbs. per acre over the corn so granules fall into whorls, when 75% of plants show first-generation borers feeding. Make a second application 7 days later if needed. To control second-generation corn borers, apply 7-14 lbs. per acre, as above, when field count shows 100 egg masses per 100 plants.

D·z·n diazinon 14G may be applied by aircraft for the above use. This material should be uniformly applied to the area to be treated at the recommended rates given above.

Fall Armyworms: Apply 7-14 lbs. per acre over the corn so granules fall into whorls when armyworms first appear. Repeat application if necessary.

Southwestern Corn Borers: Apply 7-14 lbs. per acre over the corn so granules fall into whorls when second-generation borers first appear. Repeat application one week later.

Note: Do not feed treated fodder to dairy or beef cattle or keep for 10 days following application. Corn may be picked immediately. Temporary spotting of foliage sometimes occurs following application.

Cutworms: Broadcast 14-28 lbs. per acre. For surface cutworms apply just prior to planting or when larvae first appear and work into soil 1-2 inches. Do not apply after lay-by. For subterranean cutworms, apply just prior to planting and work into soil 3-6 inches.

Seed Corn Maggots: Broadcast 14-28 lbs. per acre just prior to planting. Work into soil 2 inches.

Garden Symphylans (Centipedes): Broadcast 70 lbs. per acre just prior to planting and work into soil 4-8 inches. This preplant treatment will also control other soil insects such as cutworms (surface and subterranean) and wireworms.

Wireworms: Broadcast 21-28 lbs. per acre before planting. Work into soil 4-6 inches.

Peanuts

Southern Corn Rootworms: Broadcast 28 lbs. per acre or the equivalent in a band over the fruiting zone, just prior to pegging. Work into top few inches of soil.

Cutworms: Apply 14 lbs. per acre in a 12-inch band over the row when cutworms first appear. Up to three additional applications at three-week intervals may be necessary. Do not apply within 21 days of peanut harvest.

Lesser Cornstalk Borers: Apply 14-21 lbs. per acre in a band over the fruiting zone just prior to pegging. Cover the band with 1-2 inches of soil.

Note: Livestock (1) may be fed peanut hulls, (2) may be fed peanut forage 7 days following treatment and (3) may be fed peanut hay 21 days following treatment.

Sorghum

Cutworms: Broadcast 14-28 lbs. per acre just prior to planting. Work into soil 2-3 inches to control surface cutworms; 3-6 inches to control subterranean cutworms.

Lesser Cornstalk Borers: Apply 7-14 lbs. per acre in a 10-inch band over the row at planting time or as crop emerges. Cover the band with 1-2 inches of soil.

Note: Do not apply directly to the seed furrow as reduction of sorghum seed germination may occur.

Soybeans

Cutworms: Broadcast 14-28 lbs. per acre just prior to planting. Work into soil 2-3 inches to control surface cutworms; 3-6 inches to control subterranean cutworms.

Lesser Cornstalk Borers: Apply 7-14 lbs. per acre in a 6-inch band over the row just prior to planting or when soybeans are 1-2 inches high. Cover the band with 1-2 inches of soil.

Sugar Beets

Sugar Beet Root Maggots: Apply 7-14 lbs. per acre in a 5-7 inch band (based on row width of 22 in.) or 4.7-9.4 oz. per 1,000 linear feet of row in a 5-7 inch band either incorporated 2-3 inches at planting or applied above the seed after the seed is covered. Apply ahead of the press wheel. If a heavy maggot infestation is anticipated or a wet spring occurs, a second application may be made, early postemergence, when the adult maggots (flies) first start to appear. If a second application is necessary, apply 7-$10\frac{1}{2}$ lbs. per acre in a 5-7 inch band (based on row width of 22 in.) or 4.7-7.1 oz. per 1,000 linear feet of row in a 5-7 inch band directly over the row.

Note: D·z·n diazinon 14G should not come in direct contact with the seed or injury will occur.

Wireworms: Broadcast 21-28 lbs. per acre just prior to planting. Work into soil 4-8 inches.

Sugarcane

Surgarcane Borers: Broadcast 10½-17½ lbs. per acre after joints have begun to form and when at least 5% of plants show borers feeding in or under the leaf sheath, but which have not bored into the stalks. Additional applications may be necessary to assure continuing control.

Note: Do not make more than four applications per growing season. Allow 28 days between last application and harvest. At that time bagasse and forage may be fed to livestock.

Wireworms: Apply 42 lbs. per acre in a 12-inch band over the open furrow, directly to seed pieces.

Tobacco

Cutworms: Broadcast 14-21 lbs. per acre before transplanting. Work into soil 2-4 inches.

Mole Crickets: Apply 7 lbs. per acre when insects are present. Make application to plant beds or fields 1-2 days prior to sowing of seed or transplanting and work into soil 1-2 inches.

Seed Corn Maggots: Broadcast 10½ lbs. per acre 3-4 weeks before transplanting. Work into soil 2 inches.

Wireworms: Broadcast 14-21 lbs. per acre 3-4 weeks before transplanting. Work into soil 6-9 inches.

Forage Crops (Alfalfa, Clover, Cowpeas, and Lespedeza)

Cutworms: Broadcast 14-28 lbs. per acre just prior to planting. Work into soil 2-3 inches to control surface cutworms, 3-6 inches to control subterranean cutworms.

VEGETABLE CROP INSECTS

Broccoli, Brussels Sprouts, Cabbage, and Cauliflower

Root Maggots: Broadcast 14-21 lbs. per acre before planting or transplanting. Work into soil 3-4 inches.

Onions

Onion Maggots: Apply 14-28 lbs. per acre as broadcast soil treatment just prior to planting. Thoroughly mix the chemical into the top 3-4 inches of soil by disking, harrowing, or other suitable means.

Radishes

Root Maggots: Apply 10½-14 lbs. per acre (2½-2⅘ oz. per 1,000 linear feet of row) in the seed furrow at planting time.

Sweet Corn

See Corn under Field and Forage Insects.

Tomatoes

Vinegar Flies (*Drosophila spp.*): Broadcast 7 lbs. per acre when flies first appear. Repeat in 7-10 day intervals or as required.

OTHER VEGETABLE INSECTS

Cutworms (Surface and Subterranean)

Broccoli	Endive (escarole)	Pole Beans
Brussels Sprouts	Kale	Snap Beans
Cabbage	Lettuce	Spinach
Cantaloupes	Lima Beans	Summer Squash
Cauliflower	Muskmelons	Swiss Chard
Celery	Parsley	Tomatoes
Collards	Peas	Watermelons
Cucumbers	Peppers	Winter Squash

Broadcast 14-28 lbs. per acre just prior to planting. Work into soil 2-3 inches to control surface cutworms, 3-6 inches to control subterranean cutworms.

Potatoes: Broadcast 14-28 lbs. per acre before planting. Work into soil 4-6 inches.

Garden Symphylans (Centipedes)

Cabbage	Peas	Red Beets
Carrots	Pole Beans	Snap beans
Lettuce	Radishes	Tomatoes
		Turnips

Broadcast 1.6 lbs. per 1,000 square feet just prior to planting. Work into soil 4-8 inches. This preplant treatment will also control other soil insects such as cutworms (surface and subterranean), mole crickets, root maggots, and wireworms.

Mole Crickets

Broccoli	Endive (escarole)	Radishes
Brussels Sprouts	Kale	Tomatoes
Cabbage	Lettuce	Turnips
Cauliflower	Mustard	
Collards	Peppers	

Apply 7 lbs. per acre when insects are present. Make application to plant beds or fields 1-2 days prior to sowing of seed or transplanting and work into soil 1-2 inches.

Lesser Cornstalk Borers

Beans (Lima, Pole, Snap): Apply 7-14 lbs. per acre in a 6-inch band centered over the row. Apply when plants are ½-1 inch tall.

Wireworms

Broccoli	Endive (escarole)	Pole Beans
Brussels Sprouts	Kale	Snap Beans
Cabbage	Lettuce	Spinach
Cantaloupes	Lima Beans	Summer Squash
Carrots	Muskmelons	Sweet Potatoes
Cauliflower	Onions	Swiss Chard
Celery	Parsley	Tomatoes
Collards	Peas	Watermelons
Cucumbers	Peppers	Winter Squash

Broadcast 21-28 lbs. per acre just prior to planting. Work into soil 4-8 inches.

Potatoes: Broadcast 21-42 lbs. per acre before planting. Work into soil 4-8 inches or apply 7-21 lbs. per acre at planting, placing D·z·n diazinon 14G in bands 3-6 inches deep along both sides of row.

Sweet Potatoes (Southeastern U.S.): Broadcast 21 lbs. per acre over the tops of vines when roots begin to enlarge.

LAWN PEST CONTROL (EXCLUDING GOLF COURSES AND SOD FARMS)

Lawn Chinch Bugs: Apply 0.3-0.6 lb. (5-10 oz.) per 1,000 sq. ft. (1.5-3.0 lbs. per 5,000 sq. ft.) and brush lawn to work granules down to the soil surface. Repeat application in 7-10 days if necessary. For longer residual control and on dense growing lawns such as St. Augustine grass, use the higher rate.

Billbugs; White Grubs such as Japanese Beetle Larva, European Chafer, and Southern Chafer: Apply 0.6 lbs. (10 oz.) per 1,000 sq. ft. of lawn. For grub control make application any time between late July and early October. For billbug control apply when activity is first observed or when chewed or brown grass is present, indicating damage from this insect.

Hyperodes Weevil (New York only): Apply 0.6 lb. (10 oz.) per 1,000 sq. ft. of grass. Treat problem areas in mid-April and again in mid-May.

Precautions: Birds, especially waterfowl, feeding or drinking on treated areas may be killed. Apply this product only as specified on this label. Because of the migratory habits of certain waterfowl, do not apply in Nassau County, New York between November 1 and May 20. Do not exceed maximum permitted label rates, since rates above those recommended significantly increase potential hazards to birds. Avoid overlapping of sprays. If waterfowl, i.e., ducks, geese can be expected to enter the treated area after treatment (except spot treatments), water lawn with at least 0.25 inch water immediately after application, however stop watering before puddling occurs.

STORAGE AND DISPOSAL

Do not contaminate water, food, or feed by storage or disposal. Store in a cool, dry, well-ventilated area away from foodstuffs.

Pesticide: Pesticide or equipment rinsate that cannot be used according to label instructions may be disposed of on site or at an approved waste disposal facility.

Container: Completely empty bag by shaking and tapping sides and bottom to loosen clinging particles into application equipment. Then dispose of empty bag in a sanitary landfill or incinerate, or burn if allowed by state and local authorities. Stay out of smoke during burning. Do not reuse container.

CAUTION

Keep out of reach of children.

Harmful if swallowed. Avoid contamination of food and feed. Do not store near food or feed products. Avoid inhalation of dust. Avoid contact with skin, eyes, and clothing. Wash thoroughly after handling. May be absorbed through skin.

Fish and Wildlife Caution

This product is toxic to birds, fish and wildlife including waterfowl. Birds and waterfowl feeding on treated areas may be killed. Keep out of lakes, streams, ponds, tidal marshes and estuaries (except for effluence from treated watercress beds). Do not apply to water that will be used for recreational purposes and human or livestock consumption. Shrimp and crab may be killed at application rates recommended on this label. Do not apply where fish, shrimp, crab, and other aquatic life are important resources.

This pesticide is highly toxic to bees exposed to direct treatment or to residues remaining on the treated area. Do not apply when bees are actively visiting the crop, cover crop, or weeds blooming in the treatment area. Application should be timed to provide the maximum possible interval between treatment and the next period of bee activity.

Apply this product only as specified on this label.

CGA 1L60Z 128

This is a specimen label which was not prepared by CIBA-GEIGY. CIBA-GEIGY is not responsible for the accuracy of the information contained herein. As labels are subject to revision, always carefully read and follow the label on the product container.

D.Z.N® diazinon 50W

INSECTICIDE

For control of certain insects and mites on fruits, nuts, vegetables, field crops, lawns, and ornamentals. For specific pests, refer to use directions. This product must not be used on golf courses and sod farms.

ACTIVE INGREDIENT:
Diazinon: O,O-Diethyl O-(2-isopropyl-6-methyl-4-pyrimidinyl) phosphorothioate 50%
INERT INGREDIENTS: 50%
TOTAL: 100%

D·z·n diazinon 50W is a wettable powder.

KEEP OUT OF REACH OF CHILDREN

WARNING/AVISO

PRECAUCIÓN AL USUARIO: Si usted no lee inglés, no use este producto hasta que la etiqueta haya sido explicado ampliamente.

See additional precautionary statements at end of booklet.

See directions for use inside booklet.

DIRECTIONS FOR USE AND CONDITIONS OF SALE AND WARRANTY

IMPORTANT: Read the entire **Directions for Use** and the **Conditions of Sale and Warranty** before using this product.

CONDITIONS OF SALE AND WARRANTY

The **Directions for Use** of this product reflect the opinion of experts based on field use and tests. The directions are believed to be reliable and should be followed carefully. However, it is impossible to eliminate all risks inherently associated with use of this product. Crop injury, ineffectiveness, or other unintended consequences may result because of such factors as weather conditions, presence of other materials, or the manner of use or application all of which are beyond the control of CIBA-GEIGY or the Seller. All such risks shall be assumed by the Buyer.

CIBA-GEIGY warrants that this product conforms to the chemical description on the label and is reasonably fit for the purposes referred to in the **Directions for Use** subject to the inherent risks referred to above. **CIBA-GEIGY makes no other express or implied warranty of Fitness or Merchantability or any other express or implied warranty. In no case shall CIBA-GEIGY or the Seller be liable for consequential, special, or indirect damages resulting from the use or handling of this product.** CIBA-GEIGY and the Seller offer this product, and the Buyer and user accept it, subject to the foregoing **Conditions of Sale and Warranty,** which may be varied only by agreement in writing signed by a duly authorized representative of CIBA-GEIGY.

GENERAL INFORMATION

D·z·n diazinon 50W should be mixed with sufficient water to assure thorough coverage unless otherwise noted in the Directions for Use. Best control is obtained when application is made at first sign of pests. Applications should be repeated as necessary to maintain control.

Since D·z·n diazinon 50W is a wettable powder formulation, constant agitation is necessary during application.

DIRECTIONS FOR USE

It is a violation of federal law to use this product in a manner inconsistent with its labeling.

Do not apply this product in such a manner as to directly or through drift expose workers or other persons except those knowingly involved in the application. The area being treated must be vacated by unprotected persons.

Do not apply this product through any type of irrigation system.

FAILURE TO FOLLOW ALL PRECAUTIONS ON THIS LABEL MAY RESULT IN POOR INSECT CONTROL, CROP INJURY, OR ILLEGAL RESIDUES.

REENTRY STATEMENT

Do not enter treated areas without protective clothing until sprays have dried. Protective clothing includes at least a hat or other suitable head covering, a long-sleeved shirt and long-legged trousers or a coverall type garment (all of a closely woven fabric covering the body, including arms and legs), socks and shoes.

Because certain states may require more restrictive reentry intervals for various crops treated with this product, consult your State Department of Agriculture for further information. (See **Note** below.)

Written or oral warnings must be given to workers who are expected to be in a treated area or in an area about to be treated with this product. Oral warnings must inform workers of areas or fields that may not be entered without specific protective clothing until sprays have dried, and appropriate actions to take in case of accidental exposure, as described under Precautionary Statements on this label. When oral warnings are given, warnings shall be given in a language customarily understood by workers. Oral warnings must be given if there is reason to believe that written warnings cannot be understood by workers. Written warnings must include the following information: WARNING. Area treated with D·z·n diazinon 50W on (date of application). Do not enter without appropriate protective clothing until sprays have dried. In case of accidental exposure, flush eyes or skin with plenty of water. Call physician if irritation persists. Remove and wash contaminated clothing before reuse.

Note: The California Department of Agriculture has set a 5-day reentry for diazinon on citrus, grapes, peaches, and nectarines. Five days should elapse between the time of application and the time a worker may enter the field to engage in any activity requiring substantial contact with treated foliage. When a mixture of two or more organic phosphate pesticides are applied in combination, the interval should be prolonged by adding to the largest applicable interval an additional 50% of that interval.

FRUIT AND NUT INSECTS

D·z·n diazinon 50W may be applied using ground or aerial application equipment. For dilute ground sprays, mix specified rate of D·z·n diazinon 50W and apply at 300 to 600 gals. of water per acre or 7 to 14 gals. per 1,000 square feet. For concentrate ground sprays, use equivalent rates* of D·z·n diazinon 50W in minimums of 20-50 gals. of water per acre or 1/2-1 1/4 gals. per 1,000 square feet. For aerial applications, use equivalent ground rates* of D·z·n diazinon 50W in minimums of 2-5 gals. of water per acre. Observe stated time intervals between last application and harvest.

***Note:** The rate of D·z·n diazinon 50W for concentrate ground and aerial sprays is based upon a dilute ground spray with conventional ground equipment applied at 300 to 600 gals. of water per acre. If directions call for 1 lb. of product per 100 gals. of water, this amounts to 3 to 6 lbs. of D·z·n diazinon 50W per acre of 1.1 to 2.2 oz. per 1,000 square feet when applied at dilute ground spray volumes.

Almonds

Mites, Twig borers: Apply 1-1 1/2 lbs. D·z·n diazinon 50W per 100 gals. of water as infestations develop.

Scales (including San Jose, parlatoria, black, brown, apricot; Mite eggs (including European red mite, brown mite); Twig borers; Aphid eggs (including apple, black cherry, mealy plum); Mealybugs: Apply as a dormant spray 1-1 1/2 lbs. D·z·n diazinon 50W plus 2-3 gals. dormant oil or 1-1 1/2 gals. superior type oil per 100 gals. of water.

Almond hulls may be fed to livestock.

Apples

Leafhoppers: Apply 1 lb. per 100 gals. of water in 3-5 applications beginning at petal fall and continuing through the growing season, or apply as infestations develop.

Scales (including San Jose, parlatoria, black, brown, apricot); Mite eggs (including European red mite, brown mite); Twig borers; Aphid eggs (including apple, black cherry, mealy plum); Mealy bugs: Apply as a dormant or delayed-dormant spray 1 lb. of D·z·n diazinon 50W plus 2 gals. dormant oil or 1-1/2 gals. superior type oil per 100 gals. of water.

Eyespotted bud moth, Fruittree leafroller, Mealybug: Mix 1 lb. per 100 gals. of water. Make 3-8 cover spray applications, beginning in the pink stage.

Do not apply within 14 days of harvest.

Apples and Pears

Codling moth, Rosy apple aphid, Green apple aphid, Wooly apple aphid, Mites: Mix 1 lb. D·z·n diazinon 50W per 100 gals. of water. Make 3-8 cover spray applications, beginning with the first cover. Applications should be timed according to insect infestations.

Apple maggot: Mix 1 lb. of D·z·n diazinon 50W per 100 gals. of water. Make at least 3 cover spray applications when flies are active and laying eggs.

San Jose scale crawlers, Forbes scale crawlers: Mix 1 lb. per 100 gals. of water. Apply early in the season when crawlers first appear. If infestation warrants, make a section application 7-10 days later.

Note: Mites will be suppressed if D·z·n diazinon 50W is used in the regular cover spray program. Slight russeting may occur on some varieties of apples, such as Golden Delicious.

Do not apply to apples or pears within 14 days of harvest.

Apricots

Scales (including San Jose, parlatoria, black, brown, apricot, walnut); Mite eggs (including European red mite, brown mite); Twig borers; Aphid eggs (including apple, black cherry, mealy plum); Mealybugs: Apply as a dormant spray 1 lb. of D·z·n diazinon 50W plus 2 gals. of dormant oil or 1-1 1/2 gals. superior type oil per 100 gals. of water.

Aphids, Brown mite (Clover mite), Twospotted spider mite, Olive scale crawlers, San Jose scale crawlers: Mix 1 lb. per 100 gals. of water. Apply when aphids, mites, and scale crawlers first appear.

Olive scale crawlers: Mix 1/2 lb. of D·z·n diazinon 50W plus 1 1/2 gals. of light medium horticultural oil per 100 gals. of water. Apply when scale crawlers are present.

Apricot mealybug: Mix 1 lb. per 100 gals. of water. Apply as a cover spray from petal fall to June.

Do not apply within 10 days of harvest.

Blueberries

Cranberry fruitworm, Cherry fruitworm, Blueberry maggot, Aphids, Thrips, Twospotted mite: Mix 1 lb. per 100 gals. of water. Apply up to 200 gals. of spray per acre, or apply 2 lbs. per acre in sufficient water for thorough coverage. Do not apply within 7 days of harvest.

Caneberries (Blackberries, Boysenberries, Dewberries, Loganberries, Raspberries)

Leafhoppers, Thrips, Aphids, Twospotted mite, Raspberry sawfly: Mix 1 lb. per 100 gals. of water and apply up to 200 gals. of spray per acre, or apply 2 lbs. per acre in sufficient water for thorough coverage.

Raspberry fruitworm: Apply at the above rates when blossom buds separate and repeat application immediately before blossoms open.

Dryberry mite: Mix 2 lbs. per 100 gals. of water. Apply as a dormant spray in early spring, when canes are still on the ground and buds are closed. Make a second application when the buds are well developed, but still closed.

Raspberry crown borer (Raspberry root borer): Mix 4 lbs. per 100-200 gals. of water and apply as a drench to the crown area and to the lower canes. Apply either in the spring before berries begin to form, or in the fall after harvest.

Do not apply within 7 days of harvest.

Cherries

Scales (including San Jose, parlatoria, black, brown, apricot); Mite eggs (including European red mite, brown mite); Twig borers; Aphid eggs (including apple, black cherry, mealy plum); Mealybugs: Apply as a dormant spray 1 1/4 lbs. D·z·n diazinon 50W plus 3 gals. dormant oil or 1-1 1/2 superior type oil per 100 gals. of water.

Eyespotted bud moth, Fruittree leafroller: Mix 1 lb. per 100 gals. of water. Apply in cover sprays as necessary, beginning with petal fall.

Leafhoppers: Apply 1 lb. per 100 gals. of water in 3-5 applications beginning at petal fall and continuing through the growing season, or apply as infestations develop.

Cherry fruit fly: Mix 1/2-1 lb. per 100 gals. of water. Make 3 or 4 applications, beginning when adult flies begin to emerge and repeating at 10-day intervals.

Black cherry aphid: Mix 1 lb. per 100 gals. of water. Apply as a cover spray when aphids appear and repeat after 10 days.

San Jose scale crawler: Mix 1 lb. per 100 gals. of water. Apply during the pre-pink stage or when crawlers first appear.

Cherry rust mite: Mix 1/2-1 lb. per 100 gals. of water. Apply as a supplemental spray after harvest.

Do not apply within 10 days of harvest.

Cranberries

Blackheaded fireworm, *Rhopobota naevana:* Apply 4 lbs. in 400 gals. of water per acre.

Cranberry fruitworm: Apply 6 lbs. in 400 gals. of water per acre or 2 1/4 oz. in 9 1/4 gals. of water per 1,000 square feet.

Do not apply within 7 days of harvest.

Figs

Vinegar fly (*Drosophila* spp.), Dried fruit beetle: Apply 1 lb. per 100 gals. of water. Do not apply within 5 days of harvest.

Filberts

Filbert leafroller, Aphids: Apply 2-4 lbs. in 250-400 gals. of water per acre, as a thorough cover spray.

Citrus Fruits (Grapefruit, Lemons, Limes, Oranges, Tangelos, Tangerines, Citrus Citron, Kumquats, and hybrids of these)

Citrus aphid, Cottonycushion scale crawlers, and Soft scale crawlers: Apply 1/2-1 lb. per 100 gals. of water (maximum 1,000 gals. per acre) as a thorough cover spray when aphids and crawler stages of scale insects first appear.

Fruittree leafroller: Mix 1/2-1 lb. per 100 gals. of spray (maximum 1,000 gals. per acre) when an average of 50 percent of eggs in egg masses have hatched in the spring.

Citrus thrip: Apply 1 lb. in 100 gals. of water (maximum 1,000 gals. per acre) when thrips first appear.

Citrus snow scale: Apply 1-2 lbs. per 100 gals. of water (maximum 1,000 gals. per acre) as a postbloom spray and again as a summer spray. For heavy snow scale infestations, use the higher dosage rate. Thorough coverage of tree trunk, limbs, and twigs is essential.

Do not apply during bloom. Do not apply to citrus within 21 days of harvest.

Grapes

Pacific spider mite, Leafhoppers, Grape berry moth, Grape leaffolder: Mix 1 lb. per 100 gals. of water and apply as a thorough cover spray, or apply 2 lbs. per acre in sufficient water for thorough coverage. Do not apply within 18 days of harvest.

Drosophilia spp.: Mix 6 oz. per 100 gals. of water and apply as a thorough cover spray, or apply a maximum of 18 oz. per acre in sufficient water for thorough coverage when insects first appear. Make additional applications at weekly intervals as necessary. Do not apply within 7 days of harvest. If a spreader-sticker is used, do not apply within 18 days of harvest.

Olives

Olive scale: Mix 3/4-1 lb. of D·z·n diazinon 50W plus 1 1/2 gals. of light medium summer horticultural oil per 100 gals. of water. Apply as a cover spray in June or July.

Do not apply during bloom to avoid injury to pollinating insects or within 75 days of harvest.

Peaches and Nectarines

Scales (including San Jose, parlatoria, black, brown, apricot); Mite eggs (including European red mite, brown mite); Twig borers; Aphid eggs (including apple, black cherry, mealy plum); Mealybugs: Apply as a dormant spray 1 lb. D·z·n diazinon 50W plus 2-3 gals. dormant oil or 1-1 1/2 superior type oil per 100 gals. of water.

Peach twig borer, Parlatoria scale: Mix 1 lb. D·z·n diazinon 50w plus 1 gal. of light medium summer horticultural oil per 100 gals. of water. Apply at petal fall.

Aphids, Brown mite (Clover mite), Twospotted spider mite, Olive scale crawlers, San Jose scale crawlers: Mix 1 lb. per 100 gals. of water. Apply when aphids, mites, and scale crawlers first appear.

Olive scale crawlers: Mix 1/2 lb. of D·z·n diazinon 50W plus 1 1/2 gals. light medium horticultural oil per 100 gals. of water. Apply when scale crawlers are present.

Oriental fruit moth: Mix 1 lb. per 100 gals. of water and apply at 100% petal fall. Make 2 additional applications at 10-15 day intervals. Time the last application to peak adult Oriental fruit month flight, if possible.

Peach twig borer: Mix 1 lb. per 100 gals. of water and apply to peaches in May and June.

Leafhoppers (Peaches only): Apply 1 lb. per 100 gals. of water in 3-5 applications beginning at petal fall and continuing through the growing season, or apply as infestations develop.

White peach scale (Peaches only): As a postharvest spray apply 1 1/2-2 lbs. per 100 gals. of water timed to coincide with peak crawler and immature scale activity. Under conditions of heavy infestations, 2 applications may be necessary.

Do not apply to peaches within 20 days, or to nectarines within 10 days, of harvest.

Pears

Scales (including San Jose, parlatoria, black, brown, apricot); Mite eggs (including European red mite, brown mite, pearleaf blister mite); Twig borers; Aphid eggs (including apple, black cherry, mealy plum); Mealybugs: Apply as a dormant spray 1 lb. of D·z·n diazinon 50W plus 2 gals. dormant oil or 1-1 1/2 gals. superior type oil per 100 gals. of water.

Pear psylla: Mix 1/2-1 lb. per 100 gals. of water. Apply as a cover spray at pre-pink stage. Make 1-2 additional applications during the first and second cover sprays if necessary.

Tentiform leafminer, Mealybug: Mix 1 lb. per 100 gals. of water. Apply in 3 cover sprays, beginning in April.

Refer also to the Apples section.

Do not apply within 14 days of harvest.

Pecans

Aphids, Mites, Pecan nut casebearer, Walnut caterpillar: Apply 2-6 lbs. per acre as a foliage spray or 8/10 to 2 1/4 oz. per 1,000 square feet. Do not apply after husks crack.

Pineapple

Pineapple scale: Apply 10 lbs. D·z·n diazinon 50W plus 2 gals. of light or medium horticultural oil in 500 gals. of water per acre or 1/4 lb. plus 6 4/10 oz. in 11 1/2 gals. of water per 1,000 square feet to the vegetative slips on the plant following harvest of the pineapple.

Pineapple mealybug: Apply 4 lbs. per acre. May be applied on day of harvest.

Pineapple forage and refuse may be fed to livestock 7 days following last application.

Plums and Prunes

Scales (including San Jose, parlatoria, black, brown, apricot); Mite eggs (including European red mite, brown mite, eriophyid mite); Twig borers; Aphid eggs (including apple, black cherry, mealy plum); Mealybugs: Apply as a dormant spray 1 lb. of D·z·n diazinon 50W plus 2-3 gals. dormant oil or 1-1 1/2 gals. superior type oil per 100 gals. of water. For late April or

May applications, mix 1 lb. of D·z·n diazinon 50W plus 1 gal. of light medium summer horticultural oil per 100 gals. of water.

Brown mite (Clover mite), European red mite: Apply 1 lb. in 100 gals. of water.

Leafcurl plum aphid, Mealy plum aphid, Thistle aphid: Apply $\frac{1}{2}$-1 lb. per 100 gals. of water.

Do not apply within 10 days of harvest.

Strawberries

Aphids, Twospotted spider mite: Apply 1 lb. in 100 gals. of water per acre.

Cyclamen mite: Apply 2 lbs. in 100 gals. of water per acre, directing spray to plant crowns. Plants should be sufficiently agitated to assure thorough coverage of the foliage. Make additional applications, if necessary, after harvest.

Mole cricket: Broadcast 2 lbs. per acre when insects are present. Apply 1-2 days before transplanting and mix into the top 1-2 inches of soil immediately.

Strawberry leafroller: Mix $\frac{3}{4}$-1 lb. per 100 gals. of water. Apply 200 gals. spray per acre when blossoms show color.

Do not apply within 5 days of harvest.

Walnuts

Aphids, Mites: Apply 2 lbs. per acre as a foliage spray.

Codling moth, Scale crawlers, Walnut caterpillar: Apply 6 lbs. per acre.

Do not apply after the husks open.

VEGETABLE FOLIAR INSECTS

D·z·n diazinon 50W should always be applied as a spray in sufficient water to assure thorough coverage of the foliage.

Beans (Pole Beans, Snap Beans, Lima Beans)

Black bean aphid, Mexican bean beetle, Leafminers: Apply 1-1$\frac{1}{2}$ lbs. per acre.

Cucumber beetle: Apply $\frac{1}{2}$-$\frac{3}{4}$ lb. in 100 gals. of water per acre.

Do not apply within 7 days of bean harvest. Bean forage or bean hay may be fed to dairy and beef cattle and sheep. Wait 4 days following treatment if forage is to be cut for hay.

Lima Beans

Dipterous leafminer, Leafhoppers, Mites: Apply 1 lb. per acre. Do not apply within 7 days of bean harvest. Bean forage or bean hay may be fed to dairy and beef cattle and sheep. Wait 4 days following treatment if forage is to be cut for hay.

Broccoli, Cabbage, Cauliflower

Aphids, Diamondback moth, Imported cabbageworm, Dipterous leafminer: Apply $\frac{1}{2}$-1 lb. per acre. Applications at 7-10 day intervals will usually be required. Do not apply to broccoli or cauliflower within 5 days, or to cabbage with 7 days of harvest.

Brussels Sprouts

Aphids, Diamondback moth, Imported cabbageworm: Apply 1 lb. per acre. Applications at 7-10 day intervals will usually be required. Do not apply to brussels sprouts within 7 days of harvest.

Carrots

Aphids: Apply 1 lb. per acre. Do not apply within 10 days of harvest.

Celery

Aphids, Flea beetles: Apply 1 lb. in 100 gals. of water per acre.

Dipterous leafminer: Apply $\frac{1}{2}$-1 lb. per acre in sufficient water to achieve thorough coverage when infestations first appear. Repeat as necessary.

Do not apply within 10 days of harvest. Plants should be market trimmed (tops removed) before being shipped or used. Tops should not be used for feed or food.

Collards, Kale, Swiss Chard, Turnip Tops

Aphids, Imported cabbageworm, Diamondback moth, Harlequin cabbage bug, Corn earworm: Apply $\frac{1}{2}$-1 lb. per acre. Do not apply to collards, kale, or turnip tops within 10 days, or to Swiss chard within 12 days of harvest.

Cucumbers

Aphids, Serpentine leafminer, Thrips: Apply 1 lb. per acre. Do not apply within 7 days of harvest.

Lettuce and Endive

Aphids, Dipterous leafminer: Apply $\frac{1}{2}$-1 lb. per acre. Do not apply within 10 days of harvest.

Melons and Squash (Cantaloupes, Casabas, Crenshaws, Honeydew Melons, Muskmelons, Persian Melons, and hybrids of these; Watermelons and their hybrids; Summer and Winter Squash)

Aphids, Leafhoppers, Mites: Apply 1-1$\frac{1}{2}$ lbs. per acre.

Serpentine leafminer, Thrips: Apply $\frac{1}{2}$-1 lb. per acre.

Do not apply to melons or winter squash within 3 days, or to summer squash within 7 days, of harvest.

Mushroom Houses

Mushroom flies (Phorids and Sciarids): Mix 4-8 lbs. in 50 gals. of water (0.5% to 1.0% actual diazinon) and spray on outside and inside walls and ventilators of mushroom houses, after compost in beds has been pasteurized by beating. Do not spray compost in beds or after spawning.

To treat wooden sides of beds, posts, doors, and door frames after pasteurizing, mix $\frac{1}{2}$ lb. in 6 gals. of water and apply with a brush.

Do not treat compost.

Onions

Onion thrips: Apply 1 lb. per acre. Do not apply within 10 days of harvest.

Peas

Aphids, Leafminers: Apply $\frac{3}{4}$-1 lb. per acre. Peas may be picked immediately. Pea vines and pea vine hay may be fed to dairy and beef cattle and sheep. Wait 4 days following treatment if forage is to be cut for hay.

Peppers

Aphids, Serpentine leafminer: Apply $\frac{1}{2}$ lb. per acre. Do not apply within 5 days of harvest.

Potatoes

Aphids, Flea beetles, Colorado potato beetle, Dipterous leafminer: Apply $\frac{1}{2}$-$\frac{3}{4}$ lb. per acre.

Banded cucumber beetle, Leafhoppers, Southern armyworm: Apply $\frac{3}{4}$-1 lb. per acre.

Do not apply within 35 days of harvest.

Radishes, Parsnips, Turnips

Aphids, Flea beetles, Dipterous leafminer: Apply $\frac{1}{2}$-1 lb. per acre. Do not apply within 10 days of harvest.

Spinach and Beets

Aphids, Dipterous leafminer: Apply $\frac{1}{2}$-1 lb. per acre. Do not apply to spinach within 10 days, or to beets within 14 days of harvest.

Squash (Summer and Winter)

See Melons and Squash above.

Sweet Corn

Corn earworm: Apply 2-2$\frac{1}{2}$ lbs. per acre when silks first appear. Two or three repeat applications at 2-3 day intervals will usually be necessary. Corn may be picked immediately following last application. Corn forage may be fed to dairy and beef cattle and sheep.

Refer also to Corn under Field and Forage Insects.

Tomatoes

Aphids, Dipterous leafminer: Apply $\frac{1}{2}$ lb. per acre. To control leafminers weekly applications are usually required.

Banded cucumber beetle, Fall armyworm, Southern armyworm: Apply $\frac{3}{4}$-1 lb. per acre.

Vinegar fly (*Drosophila* spp.): Apply $\frac{1}{2}$-1$\frac{1}{2}$ lbs. per acre. Weekly applications are usually required.

Do not apply within 1 day of harvest.

Watercress (Hawaii Only)

Cyclamen mite: Apply 1 lb. in 100 gals. of water per acre. Direct spray to crowns and growing tips of plants. If infestation is heavy, make a second application no later than 10 days after the first application. Repeat applications as necessary. Do not apply within 5 days of harvest.

VEGETABLE SOIL INSECTS

D·z·n diazinon 50W should always be applied as a spray in sufficient water to assure thorough coverage of the soil.

Carrot rust fly

Carrots: Apply 4 lbs. per acre as a drench to the seed furrow at planting time.

Cutworms (Surface and Subterranean)

Broccoli	Kale	Snap Beans
Brussels Sprouts	Lettuce	Spinach
Cabbage	Lima Beans	Summer Squash
Cantaloupes	Muskmelons	Sweet Corn
Cauliflower	Parsley	Swiss Chard
Celery	Peas	Tomatoes
Collards	Peppers	Watermelons
Cucumbers	Potatoes	Winter Squash
Endive (escarole)		

Broadcast 4-8 lbs. per acre or 1$\frac{1}{2}$ to 3 oz. per 1,000 square feet just prior to planting. Work into the soil immediately, 2-3 inches for surface cutworms, 3-6 inches for subterranean cutworms.

Garden Symphylans (Centipedes)

Cabbage	Lettuce	Radishes	Tomatoes
Carrots	Peas	Red Beets	Turnips
Corn	Pole Beans	Snap Beans	

Broadcast 20 lbs. per acre or $\frac{1}{2}$ lb. per 1,000 square feet just prior to planting. Work into soil 4-8 inches immediately. This preplant treatment will also control other soil insects: cutworms (surface and subterranean), mole crickets, root maggots, and wireworms.

Mole Crickets

Broccoli	Cauliflower	Kale	Radishes
Brussels Sprouts	Collards	Lettuce	Tomatoes
	Endive (escarole)	Mustard	Turnips
Cabbage		Peppers	

Broadcast 2 lbs. per acre when insects are present. Make application to plant beds or fields 1-2 days prior to sowing of seed or transplanting and work into soil 1-2 inches immediately.

Onion maggots

Onions: Furrow Application: Apply 2 lbs. per acre in sufficient water to drench the seed furrow at planting time.

Broadcast Application: Broadcast 4-8 lbs. per acre or 1½ to 3 oz. per 1,000 square feet just prior to planting. Work into soil 3-4 inches immediately.

Root maggots

Broccoli, Brussels sprouts, Cabbage, and Cauliflower: (1) Broadcast Application: Broadcast 4-6 lbs. per acre or 1½ to 2¼ oz. per 1,000 square feet and immediately work into the soil 3-4 inches before seeding, either in the seedbed or in the field or before transplanting. (2) Transplant Water (Drench) Application: Mix ¼-½ lb. in 50 gals. of transplant water, when 200-300 gals. of water are used per acre. May be applied at the rate of ½-1 cupful per plant, either by hand or by tractor-mounted sprayer equipped with drop nozzles to direct spray to the base of the plant. Transplant water treatments may result in stand reduction due to plant stress at time of transplanting.

Radishes: Mix 2 lbs. in 50 gals. of water. Apply 80-100 gals. per acre (1-1⅓ gals. per 1,000 linear feet of row) in the seed furrow at planting time.

Wireworms

Broccoli	Endive (escarole)	Snap Beans
Brussels Sprouts	Kale	Spinach
Cabbage	Lettuce	Summer Squash
Cantaloupes	Lima Beans	Sweet Corn
Carrots	Muskmelons	Sweet Potatoes
Cauliflower	Onions	Swiss Chard
Celery	Parsley	Tomatoes
Collards	Peas	Watermelons
Cucumbers	Peppers	Winter Squash

Broadcast 6-8 lbs. per acre or 2¼ to 3 oz. per 1,000 square feet prior to planting, and work into soil 4-8 inches immediately.

Potatoes: Broadcast 6-12 lbs. per acre or 2¼ to 4½ oz. per 1,000 square feet prior to planting. Immediately disk or harrow into soil 4-8 inches.

Southern potato wireworm

Potatoes (Southeastern U.S.): For potatoes on mineral soils, broadcast 4-8 lbs. per acre or 1½ to 3 oz. per 1,000 square feet prior to planting, and immediately work into soil 3-6 inches depending on location of wireworms.

FIELD AND FORAGE INSECTS

D·z·n diazinon 50W should always be applied as a spray in sufficient water to assure thorough coverage of the foliage or soil, depending on the type of application.

Alfalfa, Clover and Trefoil

Spotted alfalfa aphid, Yellow clover aphid. Pea aphid, Leafhoppers, Plant bug nymphs, Mites: Apply ¾-1 lb. per acre.

Grasshoppers: Apply 1 lb. per acre.

Cutworms (Surface and Subterranean): Broadcast 4-8 lbs. per acre or 1½ to 3 oz. per 1,000 square feet just prior to planting. Work into the soil immediately, 2-3 inches for surface cutworms, 3-6 inches for subterranean cutworms.

Dairy and beef cattle and sheep (1) may be grazed immediately following application, (2) may be fed green alfalfa or clover immediately following cutting, and (3) may be fed hay provided alfalfa or clover is not cut for hay for 7 days following treatment.

Alfalfa weevil larvae: For control in the spring, apply 2-3 lbs. per acre when feeding damage to growing tips becomes evident. Repeat application, if necessary. If damage is already extensive, cut the hay and make application to the stubble. Do not graze dairy and beef animals or sheep for 2 days. Do not cut for hay for 10 days following treatment.

Do not spray livestock directly except as indicated below under Livestock Insects, with either air or ground equipment.

Do not treat during bloom to avoid injury to pollinating insects.

Corn

Corn rootworm larvae—(Postemergence Basal Treatment)

Apply 1¼-2½ oz. per 1,000 linear feet of corn row in sufficient water to wet base of corn plant and a narrow band of soil next to the plant. Refer to row spacing chart for amounts recommended per acre. Cover treated band with 2-3 inches of soil using a disk hiller or cultivator immediately after spraying. Application should be made at the first sign of corn rootworm larvae. Application should not be made later than June 24 in the Middle Corn Belt or July in the Northern Corn Belt.

SUGGESTED RATES ACCORDING TO ROW SPACING FOR CORN ROOTWORM LARVAE CONTROL

Spacing	Linear Feet of Corn Row/Acre	Amount of D·z·n diazinon 50W Per Acre: Postemergence Basal Treatment
48″	10,890	14-28 oz.
42″	12,455	15½-31 oz.
40″	13,068	16-32 oz. (1-2 lbs.)
38″	13,939	17½-35 oz.
36″	14,250	18-36 oz.
30″	17,424	22-44 oz. (1½-2¾ lbs.)
24″	21,780	27½-55 oz.
20″	26,136	32-64 oz. (2-4 lbs.)
15″	34,848	44-88 oz.

Note: For row spacings not listed, calculate the appropriate dosage rates based on the 40-inch row figures.

Seedcorn maggot: Broadcast 4-8 lbs. per acre or 1½ to 3 oz. per 1,000 square feet just prior to planting. Work into 2 inches of soil immediately.

Cutworms (Surface and Subterranean): Broadcast 4-8 lbs. per acre or 1½ to 3 oz. per 1,000 square feet. For surface cutworms apply just prior to planting or when larvae first appear and work into 1-2 inches of soil immediately. Do not apply later than lay-by. For subterranean cut-worms, apply just prior to planting and work into 3-6 inches of soil immediately.

Wireworms: Apply 6-8 lbs. per acre or 2¼ to 3 oz. per 1,000 square feet prior to planting and work into 4-8 inches of soil immediately.

Corn leaf aphid: Apply 1-2 lbs. per acre in sufficient water to provide complete coverage of the upper portions of plants. Repeat application as necessary.

Grasshoppers: Apply 1 lb. per acre.

Sap beetle: Apply 2-2½ lbs. per acre. Add 1-2 gals. of soluble mineral oil per acre to the spray mixture. Consult local agricultural authorities for proper timing of sprays.

Corn may be picked immediately following last application. Corn forage may be fed to dairy and beef cattle, and sheep.

Cotton

Spider mites, Aphids, Leafhoppers, Lygus bug, Cotton leafworm, Cotton leafperforator: Apply 1-2 lbs. in 5-25 gals. of water per acre. Repeat as infestation warrants.

Do not apply within 14 days of harvest. Do not graze treated cotton plants, or feed gin wastes to livestock.

Cowpeas

Cutworms (Surface and Subterranean): Broadcast 4-8 lbs. per acre or 1½ to 3 oz. per 1,000 square feet just prior to planting. Work into 2-3 inches of soil for surface cutworms, 3-6 inches for subterranean cutworms.

Hops

Aphids, Mites: Apply 2 lbs. per acre. Do not apply within 14 days of harvest.

Lespedeza

Cutworms (Surface and Subterranean): Broadcast 4-8 lbs. per acre or 1½ to 3 oz. per 1,000 square feet just prior to planting. Work into 2-3 inches of soil immediately for surface cutworms, 3-6 inches for subterranean cutworms.

Peanuts

Cutworms: Apply 4 lbs. per acre in a 12-inch band over the row. Use sufficient water to achieve thorough coverage. Begin application when small larvae appear. Two or three additional applications at 3-week intervals may be necessary. Do not apply within 21 days of peanut harvest.

Note: Livestock (1) may be fed peanut hulls, (2) may be fed peanut forage 7 days following treatment, and (3) may be fed peanut hay 21 days following treatment.

Sorghum

Sorghum midge: Apply ½-1 lb. per acre when 90% of the heads have emerged from the boot. Repeat application within 4 days.

Crop may be fed to livestock immediately following treatment. Do not apply within 7 days of harvesting for grain.

Cutworms (Surface and Subterranean): Broadcast 4-8 lbs. per acre or 1½ to 3 oz. per 1,000 square feet just prior to planting. Work into 2-3 inches of soil immediately for surface cutworms, 3-6 inches for subterranean cutworms.

Soybeans

Cutworms (Surface and Subterranean): Broadcast 4-8 lbs. per acre or 1½ to 3 oz. per 1,000 square feet just prior to planting. Work into 2-3 inches of soil immediately for surface cutworms, 3-6 inches for subterranean cutworms.

Sugar Beets

Beet leafhopper, Dipterous leafminer, Aphids: Apply ¾-1 lb. per acre.

Grasshoppers: Apply 1 lb. per acre.

Wireworms: Broadcast 6-8 lbs. per acre or 2¼ to 3 oz. per 1,000 square feet just prior to planting, and work into 4-8 inches of soil immediately. Tops may be fed to beef and dairy animals.

Sugarcane

Aphids: Apply ½-1 lb. per acre.

Note: Allow 1 day between last application and harvest. Bagasse and forage may be fed to livestock.

Wireworms: Apply 8 lbs. per acre on a 12-inch band centered over the furrow in sufficient water to obtain thorough coverage. Make application directly to seed pieces in open furrow at covering time.

Tobacco

Aphids, Flea beetles: Apply ½-¾ lb. per acre. Do not apply within 3 days of harvest.

Cutworms: Broadcast 4-6 lbs. per acre or 1½ to 2¼ oz. per 1,000 square feet before transplanting and work into 2-4 inches of soil immediately.

Green June beetle larvae: Mix ¼ lb. in 50 gals. of water and apply as a drench (100 gals. per 100 sq. yds.) to uprooted areas of plant beds with sprinkling can.

Mole cricket: Broadcast 2 lbs. per acre, when insects are present. Make application to beds or fields 1-2 days prior to sowing of seed or transplanting, and work into 1-2 inches of soil immediately.

Seedcorn maggot: Broadcast Application: Broadcast 3 lbs. per acre, 3-4 weeks before transplanting. Work into 2 inches of soil immediately.

Transplant Water Application: Mix 1½-3 oz. in 50 gals. of transplant water.

Wireworms: Broadcast Application: Broadcast 4-6 lbs. per acre or 1½ to 2¼ oz. per 1,000 square feet 3-4 weeks before transplanting. Work into 6-9 inches of soil immediately.

Transplant Water Application: Mix 1½-3 oz. in 50 gals. of transplant water.

RANGE, PASTURE, AND GRASSLAND INSECTS

Bermudagrass

Thrips, Whiteflies, Grasshoppers: Apply 1 lb. per acre.

Rangeland, Pasture, Grass Forage, Ditch Banks, Roadsides, Wasteland, Noncrop Areas, Barrier Strips

Grasshoppers: Apply ¾-1 lb. per acre in sufficient water or oil to obtain thorough coverage when insects first appear, preferably in the nymphal stage.

Dairy and beef cattle, and sheep (1) may be grazed immediately following application, (2) may be fed green forage immediately following cutting, and (3) may be fed hay, provided grass is not cut for hay for 21 days if the spray is a water solution, or for 30 days if the spray is an oil solution. Do not spray livestock directly with either air or ground equipment except as indicated under Livestock Insects. Do not repeat application for 30 days.

LIVESTOCK INSECTS

Sheep "Ticks" (Keds), Lice

High gallonage, high pressure spray: Mix ½ lb. in 100 gals. of water, and apply at the rate of 1 gal. per animal.

Low gallonage, low pressure spray: Mix 1 lb. in 100 gals. of water. Apply at the rate of 1 qt. per animal.

Sprinkler Can: Mix 1 oz. in 6 gals. of water. Crowd sheep into a small pen and sprinkle 1 qt. of this mixture over the head, neck, top, and side of each sheep. Stir or agitate mixture frequently.

Note: Do not slaughter sheep within 14 days after treatment. Repeat applications as necessary.

ORNAMENTAL INSECTS

To control certain insects on ornamentals, such as Arborvitae, Azalea, Birch, Boxwood, Camellia, Carnation, Chrysanthemum, Douglas fir, Elm, Gladioli, Hawthorn, Holly, Juniper, Lilac, Locust, Maple, Oak, Pine, Plum, Poplar, Rhododendron, Rose, Spruce, and Willow, apply the recommended rates indicated below.

Insects	Amount D·z·n diazinon 50W to Use: per 3 gals. water	per 100 gals. water
Aphids, Bagworms, Carnation bud mites, Carnation shoot mites, Clover mites, Cyclamen mites, Dipterous leafminers, European pine shoot moths, European red mites, Flea beetles, Holly bud moths, Leafhoppers, Obscure root weevils, Omnivorous leaftiers, Privet mites, Scale crawlers (Cottonycushion, Lecanium, Pine needle, San Jose, Soft scale), Thrips, Twospotted mites, Webworms, Whiteflies	½ oz	1 lb.
Apple-and-thorn skeletonizers, Cotoneaster webworms, Fall webworms, Hemlock chermes, Oak loopers, Obliquebanded leafrollers, Pear slugs, Tent caterpillars	1½ oz.	3 lbs.

Do not use on Ferns, Poinsettia, Hibiscus, Papaya, Pilea, and Gardenia, since injury to the plants may occur.

Mimosa webworms: Mix ½ oz. in 3 gals. of water (or 1 lb. in 100 gals. of water) and apply as a thorough foliar spray when adults first appear. Make a second and third application at 4-5 week intervals.

LAWN PEST CONTROL (EXCLUDING SOD FARMS AND GOLF COURSES)

To control pests listed below, apply recommended rates. For pests, such as Ants, Brown dog ticks, Clover mites, Crickets, Earwigs, Fleas, Millipedes, Sowbugs, and Springtails (Collembola), which are commonly found near the house and may enter the house, spray a 5 ft. band of soil around the house as well as the house foundation wall to a height of 2-3 ft. Repeat application if necessary.

Precautions: Birds, especially waterfowl feeding or drinking on treated areas may be killed. Apply this product only as specified on this label. Because of the migratory habits of certain waterfowl, do not apply in Nassau County, New York between November 1, and May 20. Do not exceed maximum permitted label rates, since rates above those recommended significantly increase potential hazards to birds. Avoid overlapping sprays. If waterfowl, i.e., ducks, geese can be expected to enter treated area after treatment (except spot treatments), water lawn with at least 0.25 inch water immediately after application, however, stop irrigation before puddling occurs.

Pests	Rate/1,000 sq. ft.: D·z·n diazinon 50W	Water	Remarks
Lawn chinch bugs	2-3 oz.	3 gals.	Use higher rate for longer residual control and in turf-grass of dense growth, such as St. Augustine.
Ants, Armyworms, Clover mites, Springtails (Collembola), Crickets, Cutworms, Digger wasps, Earwigs, Frit flies, Lawn billbugs, Sod webworms (Lawn moths), Sowbugs	3 oz.	3 gals.	Spot spray ant hills and wasp ground next openings. Armyworms, For frit flies, mow grass and water well before treatment, but delay application until grass is dry.
Brown dog ticks, Bermudagrass mites, Chiggers, Fleas, Leafhoppers	1¼ oz.	3 gals.	Do not apply to animals. For brown dog ticks. Spray grass and under shrubbery.
Millipedes	3 oz.	3 gals.	
Rhodesgrass scale	3 oz.	3 gals.	Apply when crawlers first emerge.

Boxelder bugs: Spray bugs outside the home with 1½-2½ oz. in 1 gal. of water. At the same time, spray thoroughly around foundation, windows, doors, tree trunks, and into cracks and other places where the insects hide.

SEED TREATMENT—PLANTER-BOX TREATMENT

Corn

Seedcorn beetle: Mix 3 oz. per bushel of seed as a dry mixture in the planter box as a seed treatment prior to planting.

Corn, Peas, Soybeans, and Beans (Lima, Snap, White, and Kidney) Seedcorn maggot: Mix ½ oz. per bushel of seed as a dry mixture in the planter box as a seed treatment prior to planting.

Planter-Box Mixing Instructions: Stir seed slightly using a mixing stick. Additional mixing will result from the operation of the planter. DO NOT MIX WITH BARE HANDS.

Note: If seed has not been treated with a fungicide, use Captan or Thiram in combination with D·z·n diazinon 50W at rates indicated on fungicide package. Failure to use a fungicide with D·z·n diazinon 50W may result in injury to the seed. If the seed treatment increases the friction between seeds sufficiently to reduce the rate of delivery through the planter, 1/2 oz. of graphite per bushel of seed should be added as a drill lubricant. Adjust the planter to give the required rate of seed delivery with the treated seed.

Caution: D·z·n diazinon 50W treated grain cannot be used for feed for livestock, or for the preparation of products for human consumption.

FLY CONTROL

Fly control in barns and animal sleeping quarters (except dairy barns, milk rooms, and poultry houses).

Residual Sprays: Remove animals from buildings prior to treatment and keep animals out for at least four hours. Mix 2 lbs. of D·z·n diazinon 50W in 25 gals. of water and apply as an overall spray to ceilings and walls of livestock sheds, calf barns, hog barns, loafing sheds, and other farm buildings except poultry houses, dairy barns, and milk rooms. Areas where flies congregate, such as garbage dumps and corrals, should be sprayed thoroughly. Longer residual control may be obtained by increasing the rate to 4 lbs. in 25 gals. of water. Applications should be made to point of runoff. One gal. of the above spray should cover from 350-750 sq. ft. of area, depending on the surface treated. Repeat applications as fly populations warrant.

Bait Sprays: Mix the appropriate amount of D·z·n diazinon 50W with 1 lb. of sugar (or 2 cups of syrup or molasses) for the desired application.

For knapsack or similar type sprayer, mix 1/2 lb. with the sugar in 2 1/2 gals. of water and spot treat areas frequented by flies including cracks, crevices, doorways, around windows, or other areas where flies congregate. Repeat applications as fly populations warrant.

For sprinkling can application, mix 2 oz. with the sugar in 5 gals. of water and sprinkle lightly over the floor in areas frequented by flies. Applications every day or two will generally be required.

Do not contaminate milk, feed, drinking water, or apply to animals or poultry. Do not use in edible product areas of food processing plants or other areas where food is commercially prepared or processed. Do not use in dwellings.

STORAGE AND DISPOSAL

Do not contaminate water, food, or feed by storage or disposal. Open dumping is prohibited.

Pesticide: Pesticide wastes are toxic. Improper disposal of excess pesticide, spray mixture, or rinsate is a violation of Federal law. If pesticide, spray mixture or rinsates cannot be disposed of by use according to label instructions, contact your State Pesticide or Environmental Control Agency, or the Hazardous Waste representative at the nearest EPA Regional Office for guidance.

Container: Completely empty bag into application equipment. Then dispose of empty bag in a sanitary landfill or by incineration or, if allowed by state and local authorities, by burning. If burned, stay out of smoke.

PRECAUTIONARY STATEMENTS

HAZARDS TO HUMANS AND DOMESTIC ANIMALS

WARNING

May be fatal if swallowed. May be absorbed through skin. Do not breathe dust or spray mist. Avoid contact with skin, eyes, and clothing. Wash thoroughly after handling and before eating or smoking. Avoid contamination of food and feed. Keep out of reach of domestic animals. Do not use on humans or household pets. Do not permit children or pets to go onto sprayed grass until spray has completely dried.

STATEMENT OF PRACTICAL TREATMENT

If swallowed: Call a physician or Poison Control Center immediately. Drink one or two glasses of water and induce vomiting by touching back of throat with finger. Repeat until vomit fluid is clear. Do not induce vomiting or give anything by mouth to an unconscious person.

If inhaled: Remove victim to fresh air. Apply artifical respiration if indicated.

If on skin: Remove contaminated clothing and wash affected areas thoroughly with soap and water.

If in eyes: Flush eyes with plenty of water. Call a physician if irritation develops or persists.

Note to Physician: Gastric lavage is indicated. Atropine sulfate and 2-PAM and specific antidotes.

ENVIRONMENTAL HAZARDS

This product is toxic to birds, fish and wildlife including waterfowl. Birds feeding on treated areas may be killed. Keep out of lakes, streams, ponds, tidal marshes and estuaries (except for effluence from treated watercress beds). Do not apply to water that will be used for recreational purposes and human and livestock consumption. Shrimp and crab may be killed at application rates recommended on this label. Do not apply where fish, shrimp, crab, and other aquatic life are important resources.

This pesticide is highly toxic to bees exposed to direct treatment or to residues remaining on the treated area. Do not apply when bees are actively visiting the crop, cover crop, or weeds blooming in the treatment area. Applications should be timed to provide the maximum possible interval between treatment and the next period of bee activity.

Apply this product only as specified on this label.

PHYSICAL OR CHEMICAL HAZARDS

Do not use, pour, spill, or store near heat or open flame.

D·z·n® trademark of CIBA-GEIGY for diazinon

CGA 1L66K 048

This is a specimen label which was not prepared by CIBA-GEIGY. CIBA-GEIGY is not responsible for the accuracy of the information contained herein. As labels are subject to revision, always carefully read and follow the label on the product container.

D.Z.N® diazinon AG 500

INSECTICIDE

For Agricultural or Commercial Use Only

For control of certain insects on fruits, nuts, vegetables, field crops, lawns and ornamentals

This product must not be used on golf courses and sod farms.

ACTIVE INGREDIENT:	
Diazinon: O,O-Diethyl O-(2-isopropyl-6-methyl-4-pyrimidinyl) phosphorothioate	48.0%
INERT INGREDIENTS:	52.0%
TOTAL:	100.0%

EPA Reg. No. 100-461 EPA Est. 100-AL-1

D·z·n diazinon AG500 contains 4 lbs. diazinon per gallon.

KEEP OUT OF REACH OF CHILDREN

WARNING/AVISO

PRECAUCIÓN AL USUARIO: Si usted no lee inglés, no use este producto hasta que la etiqueta haya sido explicado ampliamente.

See additional precautionary statements at end of booklet.

See directions for use inside booklet.

D·z·n® trademark of CIBA-GEIGY for diazinon

CGA 1L128C 048

DIRECTIONS FOR USE AND CONDITIONS OF SALE AND WARRANTY

IMPORTANT: Read the entire **Directions for Use** and the **Conditions of Sale and Warranty** before using this product.

CONDITIONS OF SALE AND WARRANTY

The Directions for Use of this product reflect the opinion of experts based on field use and tests. The directions are believed to be reliable and should be followed carefully. However, it is impossible to eliminate all risks inherently associated with use of this product. Crop injury, ineffectiveness, or other unintended consequences may result because of such factors as weather conditions, presence of other materials, or the manner of use or application all of which are beyond the control of CIBA-GEIGY or the Seller. All such risks shall be assumed by the Buyer.

CIBA-GEIGY warrants that this product conforms to the chemical description on the label and is reasonably fit for the purposes referred to in the **Directions for Use** subject to the inherent risks referred to above. **CIBA-GEIGY makes no other express or implied warranty of Fitness or Merchantability or any other express or implied warranty. In no case shall CIBA-GEIGY or the Seller be liable for consequential, special, or indirect damages resulting from the use or handling of this product.** CIBA-GEIGY and the Seller offer this product, and the Buyer and user accept it, subject to the foregoing **Conditions of Sale and Warranty**, which may be varied only by agreement in writing signed by a duly authorized representative of CIBA-GEIGY.

GENERAL INFORMATION

D·z·n diazinon AG500 is an emulsifiable solution which, when diluted with water according to the directions, is intended for the control of insect pests of agricultural crops and other plant pests. Do not use in barns or dwellings. Do not use on livestock.

In the directions for use which follow, best control is obtained when application is made at first sign of pests. Application should be repeated as necessary to maintain control.

To avoid spray drift, do not apply under windy conditions. Avoid spray overlap, as crop or plant injury may result.

DIRECTIONS FOR USE

It is a violation of federal law to use this product in a manner inconsistent with its labeling.

Do not apply this product in such a manner as to directly or through drift expose workers or other persons except those knowingly involved in the application. The area being treated must be vacated by unprotected persons.

Do not apply this product through any type of irrigation system.

FAILURE TO FOLLOW THE DIRECTIONS FOR USE AND PRECAUTIONS ON THIS LABEL MAY RESULT IN POOR INSECT CONTROL, CROP INJURY, OR ILLEGAL RESIDUES.

REENTRY STATEMENT

Do not enter treated areas without protective clothing until sprays have dried. Protective clothing includes at least a hat or other suitable head covering, a long-sleeved shirt and long-legged trousers or a coverall type garment (all of closely woven fabric covering the body including arms and legs), socks and shoes.

Because certain states may require more restrictive reentry intervals for various crops treated with this product, consult your State Department of Agriculture for further information. (See *Note* below.)

Written or oral warnings must be given to workers who are expected to be in a treated area or in an area about to be treated with this product. Oral warnings must inform workers of areas or fields that may not be entered without specific protective clothing until sprays have dried, and appropriate actions to take in case of accidental exposure, as described under **Precautionary Statements** on this label. When oral warnings are given, warnings shall be given in a language customarily understood by workers. Oral warnings must be given if there is reason to believe that written warnings cannot be understood by workers. Written warnings must include the following information: WARNING. Area treated with D·z·n diazinon AG500 on (date of application). Do not enter without appropriate protective clothing until sprays have dried. In case of accidental exposure, flush eyes or skin with plenty of water. Call physician if irritation persists. Remove and wash contaminated clothing before reuse.

Note: The California Department of Agriculture has set a 5-day reentry for diazinon on citrus, grapes, peaches, and nectarines. Five days should elapse between the time of application and the time a worker may enter the field to engage in any activity requiring substantial contact with treated foliage. When a mixture of two or more organic phosphate pesticides is applied, the interval shall be prolonged by adding to the longest applicable interval an additional 50% of that interval.

FRUIT AND NUT CROPS

Note: Unless otherwise stated, the following rates should be applied as a dilute full cover spray.

Almonds

Mites, Twig Borers: Apply 1-1½ pts. per 100 gals. of water.

Parlatoria Scales, San Jose Scales: Mix 1-1½ pts. of D·z·n diazinon AG500 plus 2-3 gals. dormant oil per 100 gals. of water and apply as a dormant spray.

Almond hulls may be used for feed for livestock.

Apricots

Aphids, Brown Mites (Clover Mites), Twospotted Spider Mites, Olive Scale Crawlers, San Jose Scale Crawlers: Mix 1 pt. per 100 gals. of water. Apply when aphids, mites, and scale crawlers first appear.

Olive Scale Crawlers: Mix ½ pt. D·z·n diazinon AG500 plus 1½ gals. of light medium horticultural oil per 100 gals. of water. Apply when scale crawlers are present.

Apricot Mealybugs: Mix 1 pt. per 100 gals. of water. Apply as a cover spray from petal fall to June.

Do not apply within 10 days of harvest or illegal residues may result.

Blueberries

Cranberry Fruitworms, Cherry Fruitworms, Blueberry Maggots, Aphids, Thrips, Twospotted Mites: Mix 1 pt. per 100 gals. of water. Apply up to 200 gals. of spray per acre. Do not apply within 7 days of harvest or illegal residues may result.

Caneberries

Blackberries, Boysenberries, Dewberries, Loganberries, Raspberries

Leafhoppers, Thrips, Aphids, Twospotted Mites, Raspberry Sawflies: Apply 2 pts. per acre in a minimum of 100 gals. of water.

Raspberry Fruitworms: Apply at the above rates when blossom buds separate and repeat application immediately before blossoms open.

Dryberry Mites: Apply 1-2 qts. per acre in a minimum of 100 gals. of water. Apply as a dormant spray in early spring, when canes are still on the ground and buds are closed. Make a second application when the buds are well developed, but still closed.

Do not apply within 7 days of harvest or illegal residues may result.

Raspberry Crown Borers (Raspberry Root Borers): Apply 2 qts. per acre in a minimum of 100 gals. of water as a drench to the crown area and to the lower canes. Apply either in the spring before berries begin to form, or in the fall after harvest.

Cherries

Eyespotted Bud Moths, Fruittree Leafrollers: Mix 1 pt. per 100 gals. of water. Apply in cover sprays as necessary beginning with petal fall.

Cherry Fruit Flies: Mix ½-1 pt. per 100 gals. of water. Make 3 or 4 applications, beginning when adult cherry fruit flies begin to emerge and repeat at 10-day intervals.

Leafhoppers: Apply 1 pt. per 100 gals. of water in 3-5 applications beginning at petal fall and continuing through the growing season, or apply as infestations develop.

Black Cherry Aphids: Mix 1 pt. per 100 gals. of water. Apply as a cover spray when aphids appear and repeat after 10 days.

San Jose Scale Crawlers: Mix 1 pt. per 100 gals. of water. Apply during the pre-pink stage or when crawlers first appear.

Do not apply within 10 days of harvest or illegal residues may result.

Cherry Rust Mites: Mix ½-1 pt. per 100 gals. of water. Apply as a supplemental spray after harvest.

Citrus Fruits

Grapefruit, Lemons, Limes, Oranges, Tangelos, Tangerines, Citrus Citron, Kumquats, and hybrids of these

Citrus Aphids, Fruittree Leafrollers, Cottonycushion Scale Crawlers, Soft Scale Crawlers: Apply ½-pt. per 100 gals. of water (maximum 1,000 gals. per acre) as a thorough cover spray when aphids and crawler stages of scale insects first appear.

Fruittreee Leafrollers: Use ½-1 pt. per 100 gals. of spray (maximum 1,000 gals. per acre) when an average of 50 percent of eggs in egg masses have hatched in the spring.

Citrus Thrips: Apply 1 pt. in 100 gals. of water (maximum 1,000 gals. per acre) when thrips first appear.

Citrus Snow Scales: Apply 1-2 pts. per 100 gals. of water (maximum 1,000 gals. per acre) as a postbloom spray and again as a summer spray. For heavy snow scale infestations use the higher dosage rate. Thorough coverage of tree trunks, limbs, and twigs is essential.

Precaution: Do not apply during bloom. Do not apply to citrus within 21 days of harvest or illegal residues may result. Shrimp and crab may be killed at application rates recommended on this label. Do not apply where these are important resources.

Cranberries

Blackheaded Fireworms *(Rhopobota naevana)*: Apply 2 qts. in 400 gals. of water per acre.

Cranberry Fruitworms: Apply 3 qts. in 400 gals. of water per acre.

Do not apply within 7 days of harvest or illegal residues may result.

Figs

Vinegar Flies (*Drosophila* spp.), Dried Fruit Beetles: Apply 1 pt. per 100 gals. of water. Do not apply within 5 days of harvest or illegal residues may result.

Filberts

Filbert Leafrollers, Aphids: Apply 1-2 qts. in 250-400 gals. of water per acre as a thorough cover spray.

Grapes

Pacific Spider Mites, Leafhoppers, Grape Berry Moths, Grape Leaffolders: Apply 1-2 pts. per acre in a minimum of 100 gals. of water as a thorough cover spray. Do not apply within 18 days of harvest or illegal residues may result.

Drosophila spp.: Apply 12-18 fl. oz. per acre in a minimum of 100 gals. of water as a thorough cover spray, or apply when insects first appear. Make additional applications at weekly intervals as necessary.

Do not apply within 7 days of harvest or illegal residues may result.

Olives

Olive Scales: Mix ¾-1 pt. of D·z·n diazinon AG500 plus 1½ gals. of light medium summer horticultural oil per 100 gals. of water. Apply as a cover spray in June or July.

Precaution: Do not apply during bloom to avoid injury to pollinating insects or within 75 days of harvest or illegal residues may result.

Peaches, Nectarines

Aphids, Brown Mites (Clover Mites), Twospotted Spider Mites, Olive Scale Crawlers, San Jose Scale Crawlers: Mix 1 pt. per 100 gals. of water. Apply when aphids, mites, and scale crawlers first appear.

Olive Scale Crawlers: Mix ½ pt. of D·z·n diazinon AG500 plus 1½ gals. light medium horticultural oil per 100 gals. of water. Apply when scale crawlers are present.

Oriental Fruit Moths: Mix 1 pt. per 100 gals. of water and apply at 100% petal fall. Make 2 additional applications at 10-15 day intervals. Time the last application to peak adult oriental fruit moth flight, if possible.

Peach Twig Borers: Mix 1 pt. per 100 gals. of water and apply to peaches in May and June.

Leafhoppers (Peaches only): Apply 1 pt. per 100 gals. of water in 3-5 applications beginning at petal fall and continuing through the growing season, or apply as infestations develop.

White Peach Scale (Peaches only): As a post-harvest spray apply 1½-2 pts. per 100 gals. of water timed to coincide with peak crawler and immature scale activity. Under conditions of heavy infestations, two applications may be necessary.

Do not apply to peaches within 20 days, or to nectarines within 10 days of harvest or illegal residues may result.

Pecans

Aphids, Mites, Pecan Nut Casebearers, Walnut Caterpillars: Apply 1-3 qts. per acre as a foliage spray in a minimum of 100 gals. of water. Do not apply after husks crack or illegal residues may result.

Pineapple

Scales: Apply 5 qts. of D·z·n diazinon AG500 plus 2 gals. of light or medium horticultural oil in 500 gals. of water per acre to the vegetative slips on the plants following harvest of pineapple.

Mealybugs: Apply 2 qts. per acre in a minimum of 100 gals. of water. May be applied on day of harvest.

Pineapple forage and refuse may be fed to livestock 7 days following last application. Illegal residues may result if fed sooner than indicated.

Plums, Prunes

Brown Mites (Clover Mites), European Red Mites: Apply 1 pt. per 100 gals. of water.

Leafcurl Plum Aphids, Mealy Plum Aphids, Thistle Aphids: Apply ½-1 pt. per 100 gals. of water.

Do not apply within 10 days of harvest or illegal residues may result.

Strawberries

Aphids, Twospotted Spider Mites: Apply 1 pt. per 100 gals. of water per acre.

Cyclamen Mites: Apply 1 qt. per 100 gals. of water per acre, directing spray to plant crowns. Plants should be sufficiently agitated to assure thorough coverage of the foliage. Make additional applications, if necessary, after harvest.

Mole Crickets: Broadcast 1 qt. per acre when insects are present. Apply 1-2 days before transplanting and immediately mix into the top 1-2 inches of soil.

Strawberry Leafrollers: Mix ¾-1 pt. per 100 gals. of water. Apply 200 gals. of spray per acre when blossoms show color.

Do not apply within 5 days of harvest or illegal residues may result.

Walnuts

Aphids, Mites: Apply 1 qt. per acre as a foliage spray in a minimum of 100 gals. of water.

Codling Moths, Scale Crawlers, Walnut Caterpillars: Apply 3 qts. per acre as a foliage spray in a minimum of 100 gals. of water.

Do not apply after the husks open or illegal residues may result.

VEGETABLE FOLIAR INSECTS

Unless otherwise stated, foliar applications of D·z·n diazinon AG500 should always be applied to vegetable crops as a spray in a minimum of 2 gals. of water per acre for aerial applications and a minimum of 10 gals. of water per acre for ground applications.

Beans

Pole Beans, Snap Beans, Lima Beans

Black Bean Aphids, Mexican Bean Beetles, Leafminers: Apply 1-1½ pts. per acre.

Cucumber Beetles: Apply ¾ pt. in 100 gals. of water per acre.

Do not apply within 7 days of bean harvest or illegal residues may result. Bean forage or bean hay may be fed to dairy and beef cattle and sheep. Wait 4 days following treatment if forage is to be cut for hay or illegal residues may result.

Lima Beans

Dipterous Leafminers, Leafhoppers, Mites: Apply 1 pt. per acre. Do not apply within 7 days of harvest or illegal residues may result. Bean forage or bean hay may be fed to dairy and beef cattle and sheep. Wait 4 days following treatment if forage is to be cut for hay or illegal residues may result.

Broccoli, Cabbage, Cauliflower

Aphids, Diamondback Moths, Imported Cabbageworms: Apply ½-1 pt. per acre. Applications at 7-10 day intervals will usually be required. Do not apply to broccoli or cauliflower within 5 days, or to cabbage within 7 days of harvest or illegal residues may result.

Brussels Sprouts

Aphids, Diamondback Moths, Imported Cabbageworms: Apply 1 pt. per acre. Generally, applications at 7-10 day intervals will be required. Do not apply to brussels sprouts within 7 days of harvest or illegal residues may result.

Carrots

Aphids: Apply 1 qt. per acre. Do not apply within 10 days of harvest or illegal residues may result.

Celery

Aphids, Flea Beetles: Apply 1 pt. in 100 gals. of water per acre. Do not apply within 10 days of harvest or illegal residues may result. Plants should be market trimmed (tops removed) before being shipped or used or illegal residues may result. Tops should not be used for feed or food or illegal residues may result.

Collards, Kale, Swiss Chard, Turnip Tops

Aphids, Imported Cabbageworms, Diamondback Moths, Harlequin Cabbage Bugs: Apply ½-1 pt. per acre. Do not apply to collards, kale, or turnip tops within 10 days, or to Swiss Chard within 12 days of harvest or illegal residues may result.

Cucumbers

Aphids, Serpentine Leafminers, Thrips: Apply 1 pt. per acre. Do not apply within 7 days of harvest or illegal residues may result.

Lettuce, Endive

Aphids, Dipterous Leafminers: Apply ½-1 pt. per acre. Do not apply within 10 days of harvest or illegal residues may result.

Melons, Squash

Cantaloupes, Casabas, Crenshaws, Honeydew Melons, Muskmelons, Persian Melons, and hybrids of these; Watermelons and their hybrids; Summer and Winter Squash

Aphids, Leafhoppers, Mites: Apply 1-1½ pts. per acre.

Serpentine Leafminers, Thrips: Apply ½-1 pt. per acre.

Do not apply to melons and winter squash within 3 days, or to summer squash within 7 days of harvest or illegal residues may result.

Mushroom Houses

Mushroom Flies (Phorids and Sciarids): Mix ½-1 gal. in 50 gals. of water (0.5% to 1.0% actual diazinon) and spray on outside and inside walls and ventilators of mushroom houses, after compost in beds has been pasteurized by heating. Do not spray compost in beds or after spawning or illegal residues may result.

To treat wooden sides of beds, posts, doors, and door frames after pasteurizing, mix ½ pt. in 6 gals. of water and apply with a brush.

Do not treat compost or illegal residues may result.

Onions

Onion Thrips: Apply 1 pt. per acre. Do not apply within 10 days of harvest or illegal residues may result.

Peas

Aphids, Leafminers: Apply ¾-1 pt. per acre. Peas may be picked immediately. Pea vines and pea vine hay may be fed to dairy and beef cattle and sheep. Wait 4 days following treatment if forage is to be cut for hay or illegal residues may result.

Peppers

Aphids, Serpentine Leafminers: Apply ½ pt. per acre. Do not apply within 5 days of harvest or illegal residues may result.

Potatoes

Aphids, Flea Beetles, Colorado Potato Beetles, Dipterous Leafminers: Apply at the rate of ½-¾ pt. per acre.

Banded Cucumber Beetles, Leafhoppers, Southern Armyworms: Apply ¾-1 pt. per acre.

Do not apply within 35 days of harvest or illegal residues may result.

Radishes, Parsnips, Turnips

Aphids, Flea Beetles, Dipterous Leafminers: Apply ½-1 pt. per acre. Do not apply within 10 days of harvest or illegal residues may result.

Spinach, Beets

Aphids, Dipterous Leafminers: Apply ½-1 pt. per acre. Do not apply to spinach within 10 days, or to beets within 14 days of harvest or illegal residues may result.

Squash (Summer and Winter)

See Melons and Squash section.

Sweet Corn

Corn Earworms: Apply 1-1¼ qts. per acre when silks first appear. Two or three repeat applications at 2-3 day intervals will usually be necessary. Corn may be picked immediately following last application. Corn forage may be fed to dairy and beef cattle and sheep.

Refer also to Corn under Field and Forage Foliar Insects and Field and Forage Soil Insects.

Tomatoes

Aphids, Dipterous Leafminers: Apply ½ pt. per acre. To control leafminers, weekly applications are usually required.

Banded Cucumber Beetles, Fall Armyworms, Southern Armyworms: Apply ¾-1 pt. per acre.

Vinegar Flies (*Drosophila* spp.): Apply ½-1½ pts. per acre. Weekly applications are usually required.

Do not apply within 1 day of harvest or illegal residues may result.

Watercress (Hawaii only)

Cyclamen Mites: Mix 1 pt. per 100 gals. of water and apply to one acre as a thorough cover spray. Spray should be directed into the crowns and growing tips of plants. If infestation is heavy, make a second application no later than 10 days after the first application. Repeat applications as necessary. Do not apply within 5 days of harvest or illegal residues may result.

VEGETABLE SOIL INSECTS

Unless otherwise stated, soil applications of D·z·n diazinon AG500 should always be applied to vegetable crops as a spray in a minimum of 5 gals. of water per acre for ground applications.

Cutworms (Surface and Subterranean)

Broccoli	Endive (Escarole)	Potatoes
Brussels Sprouts	Kale	Snap Beans
Cabbage	Lettuce	Spinach
Cantaloupes	Lima Beans	Summer Squash
Cauliflower	Muskmelons	Sweet Corn
Celery	Parsley	Swiss Chard
Collards	Peas	Tomatoes
Cucumbers	Peppers	Watermelons
		Winter Squash

Broadcast 2-4 qts. per acre just prior to planting. Work into the soil immediately, 2-3 inches for surface cutworms, 3-6 inches for subterranean cutworms.

Garden Symphylans (Centipedes)

Cabbage	Lettuce	Radishes	Tomatoes
Carrots	Peas	Red Beets	Turnips
Corn	Pole Beans	Snap Beans	

Broadcast 10 qts. per acre just prior to planting. Immediately work into soil 4-8 inches. This pre-plant treatment will also control other soil insects: cutworms (surface and subterranean), mole crickets, root maggots, and wireworms.

Mole Crickets

Broccoli	Cauliflower	Kale	Radishes
Brussels Sprouts	Collards	Lettuce	Tomatoes
Cabbage	Endive (Escarole)	Mustard	Turnips
		Peppers	

Broadcast 1 qt. per acre when insects are present. Make application to plant beds or field 1-2 days prior to sowing of seed or transplanting and immediately work into soil 1-2 inches.

Onion Maggots

Onions:

Furrow Application: Apply 1 qt. per acre in sufficient water to drench the seed furrow at planting time.

Broadcast Application: Broadcast 2-4 qts. per acre just prior to planting. Immediately work into soil 3-4 inches.

Root Maggots

Broccoli, Brussels Sprouts, Cabbage, Cauliflower:

Broadcast Application: 2-3 qts. per acre and immediately work into the soil 3-4 inches, before seeding, either in the seedbed or in the field, or before transplanting.

Transplant water (drench) application: Mix 1/4-1/2 pt. in 50 gals. of transplant water, when 200-300 gals. of water are used per acre. May be applied at the rate of 1/2-1 cupful per plant, either by hand or by tractor-mounted sprayer equipped with drop nozzles to direct spray to the base of the plant.

Precaution: Transplant water treatments may result in stand reduction due to plant stress at time of transplanting.

Radishes: Mix 1 qt. in 50 gals. of water. Apply 80-100 gals. per acre (1-1 1/3 gals. per 1,000 linear feet of row) in the seed furrow at planting time.

Wireworms

Broccoli	Endive (Escarole)	Potatoes
Brussels Sprouts	Kale	Snap Beans
Cabbage	Lettuce	Spinach
Cantaloupes	Lima Beans	Summer Squash
Carrots	Muskmelons	Sweet Corn
Cauliflower	Onions	Sweet Potatoes
Celery	Parsley	Swiss Chard
Collards	Peas	Tomatoes
Cucumbers	Peppers	Watermelons
		Winter Squash

Broadcast 3-4 qts. per acre prior to planting, and immediately work into soil 4-8 inches.

Potatoes: Broadcast 3-6 qts. per acre prior to planting. Immediately disk or harrow into soil 4-8 inches.

Southern Potato Wireworms (Southeastern U.S.)

Potatoes: For use to control southern potato wireworms in potatoes grown on mineral soils in the Southeast, apply 2-4 qts. per acre as a soil treatment prior to planting, and immediately mix thoroughly by disking, or other suitable means into the top 3-6 inches of soil, depending on location of wireworms.

FIELD AND FORAGE FOLIAR INSECTS

Unless otherwise stated, foliar applications of D·z·n diazinon AG500 should always be applied to field and forage crops as a spray in a minimum of 2 gals. of water per acre for aerial applications and a minimum of 10 gals. of water per acre for ground applications.

Alfalfa, Clover

Spotted Alfalfa Aphids, Yellow Clover Aphids, Pea Aphids, Leafhoppers, Plant Bug Nymphs, Mites: Apply 3/4-1 pt. per acre.

Grasshoppers: Apply 1 pt. per acre.

Following the above applications, dairy and beef cattle and sheep (1) may be grazed immediately following application, (2) may be fed green alfalfa or clover immediately following cutting, and (3) may be fed hay provided alfalfa or clover is not cut for hay for 7 days following treatment or illegal residues may result.

Alfalfa Weevil Larvae: For control in the spring, apply 2-3 pts. per acre when feeding damage to growing tips become evident. Repeat application, if necessary. If damage is already extensive, cut the hay and make application to the stubble. Do not graze dairy and beef animals or sheep for 2 days. Do not cut for hay for 10 days following treatment or illegal residues may result.

Precaution: Do not spray livestock directly with either air or ground equipment or illegal residues may result. Do not treat during bloom to avoid injury to pollinating insects.

Clover

Striped Ground Crickets: Broadcast 1-2 pts. per acre following seeding in the fall but prior to clover emergence.

Corn

Use a minimum of 1 gal. of water for aerial applications and 5 gals. of water for ground applications.

Corn Rootworm Adults: Apply 1/2-1 pt. per acre.

Corn Leaf Aphids: Apply 1-2 pts. per acre. Complete coverage of the upper portions of the plant is essential. Repeat applications as necessary.

Mites: Apply 1 pt. per acre when infestations first appear. Repeat applications as necessary.

Flea Beetles: Apply 1 pt. per acre when insects first appear. Repeat applications as necessary.

Grasshoppers: Apply 1 pt. per acre.

Sap Beetles: Apply 2-2 1/2 pts. per acre. Add 1-2 gals. of soluble mineral oil per acre to the spray mixture. Consult State Agricultural Experiment Stations or State Extension Service Specialists for proper timing of sprays.

Corn may be picked immediately following last application. Corn forage may be fed to dairy and beef cattle and sheep.

Cotton

Spider Mites, Aphids, Leafhoppers, Lygus Bug, Cotton Leafworms, Cotton Leafperforators: Apply 1-2 pts. in 5-25 gals. of water per acre. Repeat as infestation warrants.

Do not apply within 14 days of harvest or illegal residues may result. Do not graze treated cotton plants, or feed gin waste to livestock or illegal residues may result.

Guar (Texas and Oklahoma only)

Midge: Apply 1/2-1 pt. per acre in a minimum of 10 gals. of water for ground application or 2 gals. of water for aerial application when 15% of the blooms are infested. Should the infestation warrant, repeat applications should be made in 5-6 days.

Do not make more than 3 applications to the same crop. Do not apply within 12 weeks of harvest or illegal residues may result.

Do not graze or feed treated forage to livestock within 12 weeks of harvest or illegal residues may result.

Hops

Aphids, Mites: Apply 1 qt. per acre. Do not apply within 14 days of harvest or illegal residues may result.

Sorghum

Use a minimum of 1 gal. of water for aerial applications and 5 gals. of water for ground applications.

Sorghum Midge: Apply 1/2-1 pt. per acre when 90% of heads have emerged from the boot. Repeat application within 4 days.

Mites: Apply 1 pt. per acre when infestation first appears and repeat applications as necessary.

Greenbug: Apply at the rates indicated below when infestation first appears and repeat as necessary.

Texas and Oklahoma—Apply 1/2-1 pt. per acre.

All Other Areas —Apply 1 pt. per acre.

Crop may be fed to livestock immediately following treatment. Do not apply within 7 days of harvesting for grain or illegal residues may result.

Sugar Beets

Beet Leafhoppers, Dipterous Leafminers, Aphids: Apply 3/4-1 pt. per acre.

Grasshoppers: Apply 1 pt. per acre.

Tops may be fed to beef and dairy animals.

Sugarcane

Aphids: Apply 1/2-1 pt. per acre. Allow 1 day between last application and harvest or illegal residues may result. Bagasse and forage may be fed to livestock.

Tobacco

Aphids, Flea Beetles: Apply 1/2-3/4 pt. per acre. Do not apply within 3 days of harvest or illegal residues may result.

Grasshoppers: Apply 1 pt. per acre. Do not apply within 14 days of harvest or illegal residues may result.

FIELD AND FORAGE SOIL INSECTS

Unless otherwise stated, soil applications of D·z·n diazinon AG500 should always be applied to field and forage crops as a spray in a minimum of 5 gals. of water per acre for ground applications.

Alfalfa, Clover

Cutworms (Surface and Subterranean): Broadcast 2-4 qts. per acre just prior to planting. Work into the soil immediately, 2-3 inches for surface cutworms, 3-6 inches for subterranean cutworms.

Corn

Corn Rootworm Larvae: Postemergence Basal Treatment: Apply 1 1/4-2 1/2 oz. per 1,000 linear feet of corn row in sufficient water to thoroughly wet base of corn plant and a narrow band of soil next to the plant. Refer to the row spacing chart for amounts recommended per acre. Cover treated band with 2-3 inches of soil using a disk hiller or cultivator immediately after spraying. Application should be made at the first sign of corn rootworm larvae, but not later than June 24 in the Middle Corn Belt, or July 1 in the Northern Corn Belt.

Suggested Rates According To Row Spacing For Corn Rootworm Larvae Control

Row Spacing	Linear Feet of Corn Row/Acre	Postemergence Basal Treatment (Fl. oz.)
48"	10,890	14-28
42"	12,455	15.5-31
40"	13,068	16-32 (1/2-1 qt.)
38"	13,939	17.5-35
36"	14,250	18-36
30"	17,424	22-44
24"	21,780	27.5-55
20"	26,136	32-64 (1-2 qts.)
15"	34,848	44-88

Note: For row spacings not listed, calculate the appropriate dosage based on the 40-inch row figure.

Seedcorn Maggots: Broadcast 2-4 qts. per acre just prior to planting. Work immediately into soil 2 inches.

Cutworms (Surface and Subterranean): Broadcast 2-4 qts. per acre. For surface cutworms apply just prior to planting or when larvae first appear and work into soil 1-2 inches immediately. Do not apply later than lay-by.

For subterranean cutworms apply just prior to planting and immediately work into soil 3-6 inches.

Wireworms: Apply 3-4 qts. per acre prior to planting and immediately work into the soil 4-8 inches.

Corn may be picked immediately following last application. Corn forage may be fed to dairy and beef cattle and sheep.

Cowpeas

Cutworms (Surface and Subterranean): Broadcast 2-4 qts. per acre, just prior to planting. Work into the soil immediately, 2-3 inches for surface cutworms, 3-6 inches for subterranean cutworms.

Lespedeza

Cutworms (Surface and Subterranean): Broadcast 2-4 qts. per acre just prior to planting. Work into the soil immediately, 2-3 inches for surface cutworms, 3-6 inches for subterranean cutworms.

Peanuts

Cutworms: Apply 2 qts. per acre in a 12-inch band over the row. Thorough coverage of the foliage is essential. Begin application when small larvae appear. Two or three additional applications at three-week intervals may be necessary. Do not apply within 21 days of peanut harvest or illegal residues may result.

Livestock (1) may be fed peanut hulls, (2) may be fed peanut forage 7 days following treatment, and (3) may be fed peanut hay 21 days following treatment. Illegal residues may result if fed sooner than indicated.

Sorghum

Cutworms (Surface and Subterranean): Broadcast 2-4 qts. per acre just prior to planting. Work into the soil immediately, 2-3 inches for surface cutworms, 3-6 inches for subterranean cutworms.

White Grubs: Apply 3 pints per acre prior to listing. Broadcast spray on the soil surface, then incorporate immediately with a disk or other suitable equipment to a 4-5 inch depth. Shallow incorporation will not give as effective control. Listing may be done following incorporation.

Soybeans

Cutworms (Surface and Subterranean): Broadcast 2-4 qts. per acre just prior to planting. Work into the soil immediately, 2-3 inches for surface cutworms, 3-6 inches for subterranean cutworms.

Sugar Beets

Wireworms: Broadcast 3-4 qts. per acre just prior to planting, and immediately work into soil 4-8 inches.

Tops may be fed to beef and dairy animals.

Sugarcane

Wireworms: Apply 1 gal. per acre in a 12-inch band centered over the furrow. Make application directly to seed pieces in open furrow at covering time.

Tobacco

Cutworms: Broadcast 2-3 qts. per acre before transplanting and immediately work into the soil 2-4 inches.

Green June Beetle Larvae: Mix 1/4 pt. in 50 gals. of water and apply as a drench (100 gals. per 100 sq. yds.) to uprooted areas of plant beds with sprinkling can.

Mole Crickets: Broadcast 1 qt. per acre, when insects are present. Make application to beds or fields 1-2 days prior to sowing of seed or transplanting, and immediately work into soil 1-2 inches.

Seedcorn Maggots: Broadcast 3 pts. per acre, 3-4 weeks before transplanting. Immediately work into soil 2 inches.

Wireworms: Broadcast 2-3 qts. per acre, 3-4 weeks before transplanting. Immediately work into soil 6-9 inches.

RANGE, PASTURE, AND GRASSLAND INSECTS

Unless otherwise stated, D·z·n diazinon AG500 should always be applied as a spray in a minimum of 1 gal. of water per acre for aerial applications and a minimum of 5 gals. of water per acre for ground applications.

Bermudagrass and Forage Grasses (other than Range and Pasture Grasses)

Thrips, Whiteflies, Grasshoppers: Apply 1 pt. per acre.

Dairy and beef cattle and sheep (1) may be grazed immediately following application, (2) may be fed green forage immediately following cutting. Forage grasses may be cut for hay 7 days following treatment. Do not repeat application for 30 days or illegal residues may result.

Rangeland, Pasture, Ditch Banks, Roadsides, Wasteland, Noncrop Areas, Barrier Strips

Grasshoppers: Apply 3/4-1 pt. per acre in water or oil (minimum of 1 gal. of oil per acre) when insects first appear, preferably in the nymphal stage. Thorough coverage of the foliage is essential.

Dairy and beef cattle and sheep (1) may be grazed immediately following application, (2) may be fed green forage imemdiately following cutting, and (3) may be fed hay provided grass is not cut for hay for 21 days if the spray is a water solution, or for 30 days if the spray is an oil solution or illegal residues may result. Do not spray livestock directly with either air or ground equipment or illegal residues may result. Do not repeat application for 30 days or illegal residues may result.

ORNAMENTAL INSECT CONTROL

To control certain insects on Arborvitae, Azalea, Birch, Boxwood, Camellia, Carnation, Chrysanthemum, Douglas Fir, Elm, Gladioli, Hawthorn, Holly, Juniper, Lilac, Locust, Maple, Oak, Pine, Plum, Poplar, Rhododendron, Rose, Spruce, and Willow, apply the recommended rates indicated below:

Insects	Rate per 3 gals. water	Rate per 100 gals. water
Aphids, Bagworms, Carnation Bud Mites, Carnation Shoot Mites, Clover Mites, Cyclamen Mites, Dipterous Leafminers, European Pine Shoot Moths, European Red Mites, Flea Beetles, Holly Bud Moths, Leafhoppers, Obscure Root Weevils, Omnivorous Leaftiers, Privet Mites, Scale Crawlers (Cottonycushion, Lecanium, Pine Needle, San Jose, Soft Scale), Thrips, Twospotted Mites, Webworms, Whiteflies	1/2 fl. oz.*	1 pt.**
Apple-and-Thorn Skeletonizers, Cotoneaster Webworms, Fall Webworms, Hemlock Chermes, Oak Loopers, Obliquebanded Leafrollers, Pear Slugs, Tent Caterpillars	1 1/2 fl. oz.*	3 pts.**

* 1 fl. oz. = 2 tablespoons

** 16 fl. oz. = 1 pint

Precaution: Do not use on Ferns, Poinsettia, Hibiscus, Papaya, Pilea, and Gardenia, since injury to the plants may occur.

Mimosa Webworms: Mix 1/2 fl. oz. in 3 gals. of water (or 1 pt. in 100 gals. of water) and apply as a thorough foliar spray when adults first appear. Make a second and third application at 4-5 week intervals.

LAWN PEST CONTROL (EXCLUDING GOLF COURSES AND SOD FARMS)

To control pests listed below, apply recommended rates. For control of Ants, Brown Dog Ticks, Clover Mites, Crickets, Earwigs, Fleas, Millipedes, Sowbugs, and Springtails (Collembola), which are commonly found near the house and may enter the house, spray a 5-ft. band of soil around the house, as well as the house foundation wall to a height of 2-3 ft. Repeat application if necessary.

Precautions: Birds, especially waterfowl, feeding or drinking on treated areas may be killed. Apply this product only as specified on this label. Because of the migratory habits of certain waterfowl, do not apply in Nassau County, New York between November 1 and May 20. Do not exceed maximum permitted label rates, since rates above those recommended significantly increase potential hazards to birds. Avoid overlapping of sprays. If waterfowl, i.e. ducks, geese can be expected to enter the treated area after treatment (except spot treatments), water lawn with at least 0.25 inch water immediately after application, however stop watering before puddling occurs.

	Rate/1,000 sq. ft.		
Pests	**D·z·n diazinon AG500**	**Water**	**Remarks**
Lawn Chinch Bugs	2-3 fl. oz.	3 gals.	Use higher rate for longer residual control and in lawns of dense growth, such as St. Augustine lawns.
Ants, Armyworms, Clover Mites, Springtails (Collembola), Crickets, Cutworms, Digger Wasps, Earwigs, Frit Flies, Lawn Billbugs, Sod Webworms (Lawn Moths), Sowbugs	2-3 fl. oz.	3 gals.	Spot spray ant hills and wasp ground nest openings. For frit flies, mow grass and water well before treatment but delay application until grass is dry.
Brown Dog Ticks, Bermudagrass Mites, Chiggers, Fleas, Leafhoppers	1 1/4 fl. oz.	3 gals.	*Note:* Do not apply to animals. For brown dog ticks, spray grass and under shrubbery, particularly near house.
Millipedes	3 fl. oz.	3 gals.	
Rhodesgrass Scales	3 fl. oz.	3 gals.	Apply when crawlers first emerge.

Billbugs, White Grubs (such as Japanese Beetle Larvae, European Chafer, Southern Chafer): Apply 3 fl. oz. per 1,000 sq. ft. of lawn. For grub control make application any time between late July and early October. For billbug control apply when activity is first observed or when chewed or brown grass is present, indicating damage from this insect.

Hyperodes Weevil (New York only): Apply 3 fl. oz. per 1,000 sq. ft. of grass. Treat problem areas in mid-April and again in mid-May.

Boxelder bugs: Spray bugs outside the home with 1 1/4-2 1/2 fl. oz. (2 1/2-5 tablespoons) in 1 gal. water. At the same time, thoroughly spray around foundation, windows, doors, tree trunks, and into cracks and other places where insects hide.

STORAGE AND DISPOSAL

Do not contaminate water, food, or feed by storage or disposal. Open dumping is prohibited.

Pesticide

Pesticide wastes are toxic. Improper disposal of excess pesticide, spray mixture or rinsate is a violation of Federal law. If pesticide, spray mixture or rinsate cannot be disposed of by use according to label instructions, contact your State Pesticide or Environmental Control Agency, or the Hazardous Waste representative at the nearest EPA Regional Office for guidance.

Container

Triple rinse (or equivalent). Then puncture and dispose of in a sanitary landfill, or incinerate, or, burn if allowed by state and local authorities. Stay out of smoke from burning container.

PRECAUTIONARY STATEMENTS

HAZARDS TO HUMANS AND DOMESTIC ANIMALS

WARNING

May be fatal if swallowed. May be absorbed through skin. Do not breathe spray mist. Do not get in eyes, on skin, or on clothing. Wash thoroughly after handling. Avoid contamination of food. Do not use on humans or permit children to go onto sprayed grass until spray has completely dried.

Keep out of reach of domestic animals. Do not contaminate feed or use on household pets or livestock. Do not permit pets to go onto sprayed grass until spray has completely dried.

STATEMENT OF PRACTICAL TREATMENT

If swallowed, call a physician immediately. DO NOT INDUCE VOMITING UNLESS NO OTHER TREATMENT IS AVAILABLE. Vomiting may cause aspiration pneumonia. If it is necessary to induce vomiting, give victim one or two glasses of water and insert finger in back of throat. Repeat until vomit fluid is clear. Do not induce vomiting or give anything by mouth to an unconscious person.

If inhaled, remove victim to fresh air and apply artificial respiration if indicated.

If on skin, wash promptly with soap and water. Rinse thoroughly.

If in eyes, rinse eyes with plenty of water and call a physician immediately.

Note to Physician: Gastric lavage is indicated. Contains a petroleum-based solvent which presents aspiration hazard. Atropine sulfate and 2-PAM are specific antidotes.

ENVIRONMENTAL HAZARDS

This product is toxic to birds, fish, and wildlife including waterfowl. Birds feeding on treated areas may be killed. Keep out of lakes, streams, ponds, tidal marshes and estuaries (except for effluence from treated watercress beds). Do not apply to water that will be used for recreational purposes and human and livestock consumption. Shrimp and crab may be killed at application rates recommended on this label. Do not apply where fish, shrimp, crab, and other aquatic life are important resources.

This pesticide is highly toxic to bees exposed to direct treatment or to residues remaining on the treated area. Do not apply when bees are actively visiting the crop, cover crop, or weeds blooming in the treatment area. Applications should be timed to provide the maximum possible interval between treatment and the next period of bee activity.

Apply this product only as specified on this label.

PHYSICAL OR CHEMICAL HAZARDS

Do not use, pour, spill, or store near hear or open flame.

D·z·n® trademark of CIBA-GEIGY for diazinon

CGA 1L128C 048

This is a specimen label which was not prepared by CIBA-GEIGY. CIBA-GEIGY is not responsible for the accuracy of the information contained herein. As labels are subject to revision, always carefully read and follow the label on the product container.

DUAL®

HERBICIDE

For weed control in corn, cotton, nonbearing citrus, nonbearing grapes, peanuts, pod crops, potatoes, safflowers, grain or forage sorghum, and soybeans

ACTIVE INGREDIENT:

Metolachlor: 2-chloro-*N*-(2-ethyl-6-methylphenyl)-*N*-(2-methoxy-1-methylethyl) acetamide 86.4%

INERT INGREDIENTS: 13.6%

Total 100.0%

EPA Reg. No. 100-673

Dual® trademark of CIBA-GEIGY for metolachlor

U.S. Patent No. 3,937,730

2 1/2 Gallons U.S. Standard Measure

Dual contains 8 lbs. active ingredient per gallon.

KEEP OUT OF REACH OF CHILDREN

WARNING/AVISO

PRECAUCIÓN AL USUARIO: Si usted no lee inglés, no use este producto hasta que la etiqueta haya sido explicado ampliamente.

See additional precautionary statements inside booklet.

See directions for use inside booklet.

DIRECTIONS FOR USE AND CONDITIONS OF SALE AND WARRANTY

IMPORTANT: Read the entire Directions for Use and the Conditions of Sale and Warranty before using this product.

Conditions of Sale and Warranty

The **Directions for Use** of this product reflect the opinion of experts based on field use and tests. The directions are believed to be reliable and should be followed carefully. However, it is impossible to eliminate all risks inherently associated with use of this product. Crop injury, ineffectiveness, or other unintended consequences may result because of such factors as weather conditions, presence of other materials, or the manner of use or application all of which are beyond the control of CIBA-GEIGY or the Seller. All such risks shall be assumed by the Buyer.

CIBA-GEIGY warrants that this product conforms to the chemical description on the label and is reasonably fit for the purposes referred to in the **Directions for Use** subject to the inherent risks referred to above. **CIBA-GEIGY makes no other express or implied warranty of Fitness or Merchantability or any other express or implied warranty. In no case shall CIBA-GEIGY or the Seller be liable for consequential, special, or indirect damages resulting from the use or handling of this product.** CIBA-GEIGY and the Seller offer this product, and the Buyer and user accept it, subject to the foregoing **Conditions of Sale and Warranty,** which may be varied only by agreement in writing signed by a duly authorized representative of CIBA-GEIGY.

DIRECTIONS FOR USE

It is a violation of federal law to use this product in a manner inconsistent with its labeling.

FAILURE TO FOLLOW THE DIRECTIONS FOR USE AND PRECAUTIONS ON THIS LABEL MAY RESULT IN POOR WEED CONTROL, CROP INJURY, OR ILLEGAL RESIDUES.

Do not apply this product in such a manner as to directly or through drift expose workers or other persons except those knowingly involved in the application. The area being treated must be vacated by unprotected persons.

Reentry Statement

Do not enter treated areas without protective clothing until sprays have dried.

Because certain states may require more restrictive reentry intervals for various crops treated with this product, consult your State Department of Agriculture for further information.

Written or oral warnings must be given to workers who are expected to be in a treated area or in an area about to be treated with this product. Oral warnings must inform workers of areas or fields that may not be entered without specific protective clothing until sprays have dried, and appropriate actions to take in case of accidental exposure, as described under **Precautionary Statements** on this label. When oral warnings are given, warnings shall be given in a language customarily understood by workers. Oral warnings must be given if there is reason to believe that written warnings cannot be understood by workers. Written warnings must include the following information: "WARNING. Area treated with Dual on (date of application). Do not enter without appropriate protective clothing until sprays have dried. In case of accidental exposure, flush eyes or skin with plenty of water. Call physician if irritation persists. Remove and wash contaminated clothing before reuse."

To avoid spray drift, do not apply under windy conditions. Avoid spray overlap, as crop injury may result.

A. GENERAL INFORMATION

Observe all precautions and limitations on the labels of each product used in tank mixtures. Tank mixtures are permitted only in those states where the tank mix partner is registered.

Dual is a selective herbicide recommended as a preplant surface-applied, preplant incorporated, or preemergence treatment in water or fluid fertilizer for control of most annual grasses and certain broadleaf weeds in nonbearing citrus, corn, cotton, nonbearing grapes, peanuts, pod crops, potatoes, safflowers, grain or forage sorghum and soybeans.

Note: Do not use in nurseries, turf, or landscape plantings.

Where directions specify a Dual tank mixture with AAtrex formulations, other brands of atrazine may be used. Follow the rates, recommendations, and limitations on the respective product label, if other brands of atrazine are used.

If Dual is incorporated, any supplemental tillage before planting must not exceed the depth of incorporation.

Dry weather following preemergence application of Dual or a tank mixture may reduce effectiveness. Cultivate if weeds develop.

Where reference is made to weeds partially controlled, partial control can either mean erratic control from good to poor, *or* consistent control at a level below that generally considered acceptable for commercial weed control.

Precaution: Injury may occur following the use of Dual under abnormally high soil moisture conditions during early development of the crop.

1. Soil Textures and Herbicide Rates

Where rates are based on coarse-, medium-, or fine-textured soils, it is understood that soil textural classes are generally categorized as follows:

Coarse	Medium	Fine	
Sand	Loam	Silty clay loam	Sandy clay
Loamy sand	Silt	Sandy clay loam	Clay loam
Sandy loam	Silt loam	Silty clay	Clay

Within rate ranges in the rate tables and elsewhere on this label, use the lower rate on soils relatively coarse-textured or low in organic matter; use the higher rate on soils relatively fine-textured or high in organic matter.

Note: Dual may be applied preemergence alone, or in combination with tank mix partners specified on this label, following preplant incorporated herbicides when used according to their label recommendations, provided that such use is not prohibited on the respective labels.

Thoroughly clean sprayer prior to use. Do not use a sprayer contaminated with any other materials, or crop damage or sprayer clogging may result.

2. Mixing Instructions

Dual Alone: Mix Dual with water or fluid fertilizer and apply as a spray. Fill the spray tank one-half to three-fourths full with water or fluid fertilizer, add the proper amount of Dual, then add the rest of the water or fluid fertilizer. Provide sufficient agitation during mixing and application to maintain a uniform emulsion.

Tank Mixtures: Fill the spray tank one-fourth full with water, and start agitation; add 2,4-D, AAtrex®, Amiben®, Balan®, Banvel®, Bladex®*, Canopy®, Caparol® 4L, Command®, Cotoran®, Eptam®, Gemini®, Lexone®, Lorox®, Lorox Plus®, Preview®, Princep®, Prowl®, Pursuit®, AAtrex + Princep, Scepter®, Sencor®, Sonalan®, or Treflan®, and allow it to become dispersed, then add Dual, then add Gramoxone® Super or Roundup® if these products are being used, and finally the rest of the water. For tank mixtures with AAtrex, Amiben, Banvel, Bladex, Canopy, Caparol 4L, Command, Cotoran*, Eptam, Gemini, Lexone, Lorox, Lorox Plus, Preview, Princep, Prowl*, Pursuit, AAtrex + Princep, Scepter, Sencor, Sonalan or Treflan, fluid fertilizers may replace all or part of the water as carrier, except in the AAtrex postemergence and the Banvel postemergence tank mixes. For tank mixtures with AAtrex, see additional mixing instructions on the AAtrex label. For each mixture check compatibility with fluid fertilizer, as described below, before mixing in spray tank. For all tank mixtures, agitate during mixing and application to maintain a uniform suspension.

***See Special Mixing Instructions** for tank mixtures with Bladex, Cotoran, and with AAtrex or Princep plus Prowl under the appropriate tank mixture section.

Compatibility Tests: Since liquid fertilizers can vary, even within the same analysis, **always check compatibility with herbicide(s) each time before use.** Be especially careful when using **complete** suspension or fluid fertilizers as serious compatibility problems are more likely to occur. Commercial application equipment may improve compatibility in some instances. The following test assumes a spray volume of 25 gals. per acre. For other spray volumes, make appropriate changes in the ingredients. Check compatibility using this procedure:

1. Add 1 pint of fertilizer to each of 2 one-quart jars with tight lids.
2. To **one** of the jars add 1/4 tsp. or 1.2 milliliters of a compatibility agent approved for this use, such as Compex® or Unite® (1/4 tsp. is equivalent to 2 pts. per 100 gals. spray). Shake or stir gently to mix.
3. To **both** jars add the appropriate amount of herbicide(s). If more than one herbicide is used, add them separately with dry herbicides first, flowables next, and emulsifiable concentrates last. After each addition, shake or stir gently to thoroughly mix. The appropriate amount of herbicides for this test follows:

Dry herbicides: For each pound to be applied per acre, add 1.5 level teaspoons to each jar.

Liquid herbicides: For each pint to be applied per acre, add 0.5 teaspoon or 2.5 milliliters to each jar.

Note: For Dual tank mixtures with AAtrex + Princep, use 1/3- 1/2 the amount of AAtrex specified above and the remainder as Princep, depending on whether the 1:2 or 1:1 ratio of AAtrex to Princep is to be applied.

4. After adding all ingredients, put lids on and tighten, and invert each jar ten times to mix. Let the mixtures stand 15 minutes and then look for separation, large flakes, precipitates, gels, heavy oily film on the jar, or other signs of incompatibility. Determine if the compatibility agent is needed in the spray mixture by comparing the two jars. If either mixture separates, but can be remixed readily, the mixture can be sprayed as long as good agitation is used. If the mixtures are incompatible, test the following methods of improving compatibility: (A) slurry the dry herbicide(s) in water before addition, or (B) add 1/2 of the compatibility agent to the fertilizer and the other 1/2 to the emulsifiable concentrate or flowable herbicide before addition to the mixture. If incompatibility is still observed, do not use the mixture.

3. Application Procedures

Application Timing

Dual alone or in some tank mixtures with other labeled herbicides may be applied for weed control in certain crops at various times. Refer to the given crop section of the label to determine if application timings listed below are recommended.

1. Preplant Surface Applied: For minimum-tillage or no-tillage systems only. Dual alone and some Dual tank mixtures may be applied up to 45 days before planting certain crops. Use only split applications for treatments made 30-45 days before planting, with 2/3 the recommended broadcast rate for the crop and soil texture applied initially and the remaining 1/3 at planting. Treatments less than 30 days before planting may be made either as a split or a single application. Refer to individual crop to determine if early preplant surface application is recommended. If weeds are present at the time of treatment, apply in a tank mixture combination with a contact herbicide (for example, Gramoxone Super or Roundup). Observe directions for use, precautions, and restrictions on the label of the contact herbicide. To the extent possible, do not move treated soil out of the row or move untreated soil to the surface during planting, or weed control will be diminished.

2. Preplant Incorporated: Apply Dual to the soil and incorporate into the top 2 inches of soil within 14 days before planting using a finishing disk, harrow, rolling cultivator, or similar implement capable of providing uniform 2 inch incorporation. Use a preplant incorporated application if furrow irrigation is used or when a period of dry weather after application is expected. If crop will be planted on beds, apply and incorporate Dual after bed formation unless specified otherwise.

3. Preemergence: Apply Dual during planting (behind the planter) or after planting, but before weeds or crops emerge.

Special Application Procedures - California Only (Corn, Safflower, Pod Crops): Preplant Incorporated: Broadcast Dual to the soil and thoroughly incorporate with a disk or similar implement set to till 4-6" deep. For more thorough incorporation, till the soil in two different directions (cross-till). Crops may be planted on flat surface or on beds. Caution should be used when forming the beds that only soil from the Dual treated zone is used (i.e., untreated soil should not be brought to soil surface). If the application is made to preformed beds, incorporate Dual with a tillage implement set to till 2- 4" deep. Care should be taken during tilling to keep the tilled (Dual treated) soil on the beds. *Preemergence:* Apply Dual after planting. Water with sprinkler or flood irrigation within 7-10 days.

Ground Application: Apply Dual alone or in tank mixtures by ground equipment in a minimum of 10 gals. of spray mixture per acre unless otherwise specified.

Use sprayers that provide accurate and uniform application. For Dual tank mixtures with wettable powder or dry flowable formulations, screens and strainers should be no finer than 50-mesh. Rinse sprayer thoroughly with clean water immediately after use.

Calculate the amount of herbicide needed for band treatment by the formula:

$$\frac{\text{band width in inches}}{\text{row width in inches}} \times \text{broadcast rate per acre} = \text{amount needed per acre of field}$$

Aerial Application: Apply Dual in water alone or in tank mixtures with AAtrex, Lorox, Lexone, or Sencor in a minimum total volume of 2 gals. per acre by aircraft. Avoid application under conditions where uniform coverage cannot be obtained or where excessive spray drift may occur. In order to assure that spray will be controllable within the target area when used according to label directions, make applications at a maximum height of 10 ft., using low drift nozzles at a maximum pressure of 40 psi. and restrict application to periods when wind speed does not exceed 10 mph. To assure that spray will not adversely affect adjacent sensitive nontarget plants, apply Dual alone or Dual + AAtrex by aircraft at a minimum upwind distance of 400 ft. from sensitive plants, or apply Dual + Lexone, Sencor, or Lorox at a minimum upwind distance of 300 ft. from sensitive plants.

Avoid application to humans or animals. Flagmen and loaders should avoid inhalation of spray mist and prolonged contact with skin and should wash thoroughly before eating and at the end of each day's operation.

Center Pivot Irrigation Application: Dual alone or in tank mixture with AAtrex or other herbicides on this label which are registered for center pivot application may be applied in irrigation water preemergence (after planting but before weeds or crop emerge) at rates recommended on this label. Apply this product only through a center pivot irrigation system. Do not apply this product through any other type of irrigation system. Crop injury, lack of effectiveness, or illegal pesticide residues in the crop can result from nonuniform distribution of treated water. If you have questions about calibration, you should contact State Extension specialists, equipment manufacturers or other experts. Do not connect an irrigation system (including greenhouse systems) used for pesticide application to a public water system unless the pesticide label-prescribed safety devices for public water systems are in place. A person knowledgeable of the chemigation system and responsible for its operation, or under the supervision of the responsible person, shall shut the system down and make necessary adjustments should the need arise.

Operating Instructions

1. The system must contain a functional check valve, vacuum relief valve, and low pressure drain appropriately located on the irrigation pipeline to prevent water-source contamination from backflow.
2. The pesticide injection pipeline must contain a functional, automatic, quick-closing check valve to prevent the flow of fluid back toward the injection pump.
3. The pesticide injection pipeline must also contain a functional, normally closed, solenoid-operated valve located on the intake side of the injection pump and connected to the system interlock to prevent fluid from being withdrawn from the supply tank when the irrigation system is either automatically or manually shut down.
4. The system must contain functional interlocking controls to automatically shut off the pesticide injection pump when the water pump motor stops.
5. The irrigation line or water pump must include a functional pressure switch which will stop the water pump motor when the water pressure decreases to the point where pesticide distribution is adversely affected.
6. Systems must use a metering pump, such as a positive displacement injection pump (e.g., diaphragm pump) effectively designed and constructed of materials that are compatible with pesticides and capable of being fitted with a system interlock.
7. Do not apply when wind speed favors drift beyond the area intended for treatment.
8. Prepare a mixture with a minimum of 1 part of water to 1 part herbicide(s) and inject this mixture into the center pivot system. Injecting a larger volume of a more dilute mixture per hour will usually provide more accurate calibration of metering equipment. Maintain sufficient agitation to keep the herbicide in suspension.
9. Meter into irrigation water during entire period of water application.
10. Apply in ½-1 inch of water. Use the lower water volume (½ inch) on coarser-textured soils and the higher volume (1 inch) on finer-textured soils. More than 1 inch of water at application may reduce weed control by moving the herbicide below the effective zone in the soil.

Precaution for center pivot applications: Where sprinkler distribution patterns do not overlap sufficiently, unacceptable weed control may result. Where sprinkler distributions patterns overlap excessively, crop injury may result.

4. Dry Bulk Granular Fertilizers

Many dry bulk granular fertilizers may be impregnated or coated with Dual alone or selected Dual tank mixtures which are registered for preplant incorporated or preplant-surface application which are used to control weeds in crops on the Dual label and are not prohibited from use on dry bulk granular fertilizers.

When applying Dual or Dual mixtures with dry bulk granular fertilizers, follow all directions for use and precautions on the respective product labels regarding target crops, rates per acre, soil texture, application methods (including timing of application), and rotational crops.

All individual state regulations relating to dry bulk granular fertilizer blending, registration, labeling, and application are the responsibility of the individual and/or company selling the herbicide/fertilizer mixture.

Prepare the granular herbicide/fertilizer mixtures by using any closed drum, belt, ribbon, or other commonly used dry bulk fertilizer blender. Nozzles used to spray Dual and Dual mixtures onto the fertilizer must be placed to provide uniform spray coverage.

If the herbicide/fertilizer mixture is too wet, use a highly absorptive material, such as Agsorb™ granules, Microcel E (Johns-Manville Products Corporation), diatomaceous earth or finely powdered clay, to obtain a dry free-flowing mixture. Add the absorptive material separately and uniformly to the herbicide/fertilizer mixture and blend to form a suitable free-flowing mixture. Generally, less than 2% by weight of absorptive material will be needed.

Calculate amounts of Dual, AAtrex, AAtrex + Princep, Bladex, Princep, Sencor, Lexone, or Sonalan by the following formula:

$$\frac{2{,}000}{\text{lbs. of fertilizer per acre}} \times \text{pts./acre of liquid or flowable product} = \text{pts. of liquid or flowable product per ton of fertilizer}$$

$$\frac{2{,}000}{\text{lbs. of fertilizer per acre}} \times \text{lbs./acre of dry product} = \text{lbs. of dry product per ton of fertilizer}$$

Pneumatic (Compressed Air) Application (Dual Alone): High humidity, high urea concentrations, low fertilizer use rates, and dusty fertilizer may cause fertilizer mixture to build up or plug the distributor head, air tubes, or nozzle deflector plates. To minimize buildup, premix Dual with a noncombustible/nonflammable mineral oil, at 1 to 2 pints of mineral oil per acre and spray mixture uniformly onto fertilizer. Mineral oil additive may be used either in a fertilizer blender or through direct injection systems. Use only those oils recommended by CIBA-GEIGY, i.e., KERMAC #600™ Solvent (CV-Mineral Seal Oil). Drying agents should not be used when using oil additives.

Notes: (1) Some separation of the Dual + mineral oil mixture may occur in temperatures below 40°F. To assure uniform application, use either in-tank agitation to keep the products mixed or keep the products heated above 40°F before mixing. (2) When impregnating Dual in a blender before application, a drier mixture can be attained by substituting a drying agent for mineral oil. The use of Agsorb is recommended over other types of absorptive materials. (3) Drying agents are not recommended for use with On-The-Go impregnation equipment.

Precautions: To avoid potential for explosion. (1) Do not impregnate Dual or Dual mixtures on ammonium nitrate, potassium nitrate or sodium nitrate either alone or in blends with other fertilizers. (2) Do not combine mixtures of Dual plus any other herbicide with single superphosphate (0-20-0) or treble superphosphate (0-46-0). (3) Do not use Dual or Dual mixtures on straight limestone, since absorption will not be achieved. Fertilizer blends containing limestone can be impregnated.

Application

Apply 200-700 pounds of the herbicide/fertilizer mixture per acre. For best results, apply the mixture uniformly to the soil with properly calibrated equipment immediately after blending. Uniform application of the herbicide/fertilizer mixture is essential to prevent possible crop injury. Non-uniform application may also result in unsatisfactory weed control. In areas where conventional tillage is practiced, a shallow incorporation of the mixture into the soil may improve weed control. On fine or medium-textured soils in areas where soil incorporation is not planned, i.e., reduced tillage situations or in some conventional till situations, make applications approximately 30 days before planting to allow moisture to move the herbicide/fertilizer mixture into the soil. On coarse-textured soils, make applications approximately 14 days prior to planting.

Precaution: To avoid crop injury, do not use the herbicide/fertilizer mixture on crops where bedding occurs.

B. DUAL APPLIED ALONE

1. *Weeds Controlled*

barnyardgrass (watergrass)	green foxtail	yellow foxtail
crabgrass	prairie cupgrass	yellow nutsedge
crowfootgrass	red rice	
fall panicum	signalgrass (*Brachiaria*)	black nightshade
foxtail millet	southwestern cupgrass	carpetweed
giant foxtail	witchgrass	Florida pusley
goosegrass		galinsoga
		pigweed

Weeds Partially Controlled:* common purslane, Florida beggarweed**, hairy nightshade, sandbur, seedling johnsongrass, shattercane, Texas panicum***, volunteer sorghum, wild proso millet.

* See General Information section. Control of these weeds can be erratic due partially to variable weather conditions. Control may be improved by following these suggested procedures:

1. **Thoroughly till moist soil** to destroy germinating and emerged weed. If Dual is to be applied preplant incorporated, this tillage may be used to incorporate Dual if uniform 2 inch incorporation is achieved as recommended under **Application Procedures.**
2. Plant crop into moist soil **immediately after tillage.** If Dual is to be used preemergence, apply at planting or immediately after planting.
3. If available, **sprinkler irrigate** within 2 days after application. Apply ½-1 inch of water. Use lower water volume (½ inch) on coarse textured soils and higher volume (1 inch) on finer textured soils. Also, refer to the section on **Center Pivot Irrigation Application** for this method of applying Dual.
4. If irrigation is not possible and rain does not occur within 2 days after planting and application, weed control may be decreased. Under these conditions, a uniform, shallow cultivation is recommended as soon as weeds emerge.

** For partial control of this weed, use a minimum of 2 pts./A and apply preemergence.

*** For partial control of this weed, use a minimum of 2 pts./A and apply through a center pivot irigation system.

2. Rotational Crops: Dual Alone: (1) If crop treated with Dual alone is lost, any crop on this label may be replanted immediately. Do not make a second broadcast application of Dual. If the original application was banded and the second crop is planted in the untreated row middles, a second banded treatment may be applied. (2) Barley, oats, rye, or wheat may be planted 4½ months following treatment: alfalfa may be planted 4 months following application. (3) Any crop on this label in addition to root crops, barley, buckwheat, milo, oats, rice, rye, or wheat may be planted in the spring following treatment. Clover may be seeded 9 months following application. Do not graze or feed forage or fodder from cotton to livestock. All other rotational crops may be planted 18 months after application. (4) Following a lay-by treatment or multiple treatments applied the previous season, any crop on this label may be planted in the spring. All other rotational crops may be planted 18 months after a lay-by application. **Dual Tank Mixtures:** For Rotational Crops restrictions for Dual used in tank mixtures, refer to the statements/restrictions above for Dual and to the respective product labels of any mixing partner(s) for additional statements/restrictions.

Important Notes: To avoid injury to rotational alfalfa or clover, (1) Do not apply more than 3 lbs. ai of metolachlor (3 pts. Dual) preemergence (including preplant-surface, preplant incorporated, postplant incorporated, etc.). (2) Do not make lay-by or other postemergent applications of Dual.

C. CITRUS - NONBEARING - DUAL ALONE AND COMBINATIONS

Citrus including grapefuits, lemons, limes, oranges, tangelos, citrus citron, kumquats and hybrids of these.

Broadcast 2-4 pts./A of Dual by ground equipment to weed-free soil in a minimum of 10 gals. of water to control weeds listed in the **Dual Applied Alone** section of this label. Apply the lower rates on coarse-textured soils with low organic matter and where the weed infestation is expected to be low. Apply the higher rates on fine-textured soils, soils high in organic matter, and where a heavy weed infestation is expected. Use caution when applying to keep the spray off the foliage to avoid possible foliar injury.

When weeds are present at the time of treatment, Dual may be applied in tank mix combination with Gramoxone Super or Roundup. Gramoxone Super may be applied at rates of 3-5 pts./A. Roundup may be applied at rates ranging from 1-5 qts./A, depending on the weed species. Refer to the Roundup label for the rate to be applied for various weed species. Direct applications containing Gramoxone Super or Roundup over the weeds, making certain to keep spray off foliage and stems of trees. Refer to the Gramoxone Super and Roundup labels for further directions, precautions and limitations relative to their use.

Precaution: Do not apply to trees transplanted in the grove less than 30 days and only after depressions around the trees have been filled in, or crop injury may occur.

Note: (1) Do not apply to trees that will bear harvestable fruit within 12 months of application, and (2) Do not graze livestock in treated areas or illegal residues may result.

D. CORN - DUAL ALONE

Apply Dual either preplant surface, preplant incorporated, or preemergence using the appropriate rate specified below.

Preplant Surface Applied: Refer to instructions for use of Dual alone under **Application Procedures.** Use on medium- and fine-textured soils with minimum-tillage or no-tillage systems in CO, IL, IN, IA, KS, KY, MN, MO, MT, NE, ND, SD, TN, WI, and WY. Apply ⅔ the recommended rate of Dual (2.5 pts./A on *medium soils* and 3 pts./A on *fine soils*) as a split treatment 30-45 days before planting and the remainder at planting. Applications made less than 30 days prior to planting may be as either a split or single treatment. Apply 2 pts./A on *coarse soils* not more than 2 weeks prior to planting.

On medium- and fine-textured soils with minimum- or no-tillage systems in DE, MD, MI, NY, OH, PA, VA, and WV, preplant surface applications may be applied following the directions for use above. If the amount of rainfall results in unsatisfactory length of weed control following the earlier treatment, a post-emergence application of an appropriately labeled broadleaf and/or grass weed herbicide may be used, i.e., AAtrex, Bicep 2, 4-D, Banvel, Basagran® or bromoxynil (Brominal® or Buctril®). If the postemergence treatment includes the herbicide used preplant surface applied, do not exceed the total labeled rate for corn on a given soil texture. Observe all directions for use, precautions and limitations on the label of the postemergent herbicide.

Preplant Incorporated or Preemergence: Follow instructions for use of Dual alone under **Application Procedures.** On *coarse soils,* apply 1.5-2 pts./A of Dual if organic matter content is less than 3%, or 2 pts./A if organic matter content is 3% or greater. On *medium soils,* apply 2-2.5 pts./A of Dual. On *fine soils,* apply 2-2.5 pts./A of Dual if organic matter content is less than 3%, or 2.5-3 pts./A if organic matter content is 3% or greater.

Lay-by: To extend the duration of weed control in corn, a maximum rate of 3 pts./A of Dual may be applied after corn emergence until the corn plants reach 40 inches in height following any preplant surface-applied, preplant incorporated or preemergence herbicide application including Dual. For best results, applications should be made to soil free of emerged weeds and directed towards the base of corn plants in excess of 5 inches tall. The total Dual rate applied on corn during any one crop year should not exceed 6 pts./A, depending on soil texture.

Shattercane and Wild Proso Millet - Partial Control: For more consistent partial control of shattercane or wild proso millet, apply 1.5-2 pts./A of Dual preplant incorporated followed by 1.5-2 pts./A of Dual preemergence. Make the preemergence application during or after planting, but before weeds and corn emerge. Apply the 2 pts./A rate of Dual when a heavy infestation of shattercane or wild proso millet is expected. A shallow cultivation may be needed following the preemergence application to control any late emerging shattercane or wild proso millet plants.

Note: (1) Do not apply more than the labeled application rate for a given soil texture per year either as a single or split treatment or illegal residues may result. (2) In corn, Dual may be used up to 4 pts./A as either a preplant surface, preplant incorporated, or preemergence treatment on soils having an organic matter content between 6% and 20%. (3) In the event of escape of annual broadleaf weeds following a preplant surface, preplant incorporated, or preemergence treatment of Dual, follow with a postemergence application of an appropriately labeled broadleaf and/or grass weed herbicide, i.e., AAtrex, Basagran, Bicep, Brominal, Buctril, Banvel or 2,4-D. If the postemergence treatment includes the herbicide used in the earlier treatment, do not exceed the total labeled rate for corn on a given soil texture. (4) Brominal or Buctril may be applied postemergence alone or in tank-mix combination with AAtrex. Do not exceed 1.2 lbs. ai/A of AAtrex in tank-mix combination with Brominal or Buctril postemergence. Refer to the Brominal, Buctril and AAtrex labels for specific rates and precautions. (5) Do not use Dual on peat or muck soils.

E. CORN - DUAL COMBINATIONS

Dual in any tank mixture for corn (except Dual + AAtrex postemergence and Dual + Banvel postemergence) may be applied in water or fluid fertilizer. Use only water in the Dual + AAtrex or the Dual + Banvel postemergence tank mixes.

1. Tank Mixture with AAtrex or Princep, or AAtrex plus Princep - Preplant Surface, Preplant Incorporated, or Preemergence

In addition to the weeds controlled by Dual alone, Dual + AAtrex or Princep, or Dual + AAtrex + Princep, applied preplant surface, preplant incorporated or preemergence, also controls the following weeds: browntop panicum, cocklebur, common purslane, hairy nightshade, lambsquarters, morningglory, ragweed, smartweed, and velvetleaf.

Apply Dual + AAtrex or Princep, or Dual + AAtrex + Princep either preplant surface, preplant incorporated, or preemergence.

Preplant Surface Applied: Follow instructions for use of Dual alone under **Application Procedures** and under application instructions for Dual alone on corn. Apply Dual + AAtrex or Princep, or Dual + AAtrex + Princep on *medium soils* (2.5 pts./A Dual + 3.2-4 pts./A of AAtrex 4L or Princep 4L, or AAtrex 4L + Princep 4L combined) and on *fine soils* (2.5-3 pts./A Dual + 4-5 pts./A of AAtrex 4L or Princep 4L, or AAtrex 4L + Princep 4L combined in minimum-tillage and no-tillage systems in CO, IL, IN, IA, KS, KY, MN, MO, MT, NE, ND, SD, TN, WI, and WY. Apply the tank mixtures as a split or single treatment in those states and as indicated in the Dual alone Preplant Surface Applied section of the label for corn. On *coarse soils*, apply 2 pts./A of Dual and 3.2 pts./A of AAtrex 4L or Princep 4L, or AAtrex 4L + Princep 4L combined.

Preplant Incorporated or Preemergence: Follow instructions for use of Dual alone under **Application Procedures.** Apply Dual + AAtrex or Princep, or Dual + AAtrex + Princep using the appropriate rates from Table 1.

Note: Do not apply more than the labeled rate for a given soil texture per year either as a split or single treatment, or illegal residues may result.

Shatterane and Wild Proso Millet - Partial Control

For more consistent partial control of shatterecane or wild proso millet where Dual is applied in tank mixture or sequentially with other registered corn herbicides, the following applications may be made:

1. Apply 1.5-2 pts./A of Dual plus 2 lbs. ai/A of AAtrex or Princep preplant incorporated followed by 1.5-2 pts./A of Dual preemergence. Make the preemergence application during or after planting, but before weeds and corn emerge.
2. Apply Dual at 2 pts./A alone or in tank mix combination with up to 2 lbs. ai/A of AAtrex, Bladex or Princep preplant incorporated followed by a post-directed application of Evik® 80W at 2.5 lbs./A. Do not exceed the total rate of triazine herbicide recommended for corn grown on a given soil texture. Refer to the Evik 80W label for specific directions for the post-directed application.
3. Apply Eradicane® or Sutan® (or equivalent EPTC or butylate formulations) at labeled rates preplant incorporated followed by a preemergence application of Dual at 1.5-2 pts./A. Do not use Eradicane or Sutan on soils where rapid degradation has been shown to occur. Make the preemergence application during or after planting, but before weeds and corn emerge.
4. For partial control of wild proso millet only, apply Dual preemergence at 1.5-2 pts./A followed by a postemergence application of Prowl plus Bladex at labeled rates. Refer to the Prowl and Bladex labels for application instructions. Postemergence application of Prowl and Bladex may cause corn injury.

Precaution: When following the application regimes in numbers 1-4 above, a shallow cultivation may be needed after the preemergence or postemergence application to help control any late emerging shattercane or wild proso millet plants.

Note: Do not exceed a total of 3 pts. ai/A of Dual in the preplant incorporated plus preemergence application on soils with less than 6% organic matter, or crop injury may occur.

Table 1: DUAL + AAtrex or Princep, or DUAL + AAtrex + Princep, Preplant Incorporated or Preemergence—Corn

Soil Texture	Broadcast rates per acre: Less than 3% organic matter: DUAL + AAtrex Nine-O®* or Princep Caliber® 90*	OR DUAL + AAtrex Nine-O** + Princep Caliber 90**	3% organic matter or greater: DUAL + AAtrex Nine-O* or Princep Caliber 90*	OR DUAL + AAtrex Nine-O** + Princep Caliber 90**
COARSE	1.25-1.5 pts. + 1.1-2.2 lbs.	1.25-1.5 pts. + .6-1.1 lbs. + .6-1.1 lbs.	1.5 pts. + 1.3-2.2 lbs.	1.5 pts. + .7-1.1 lbs. + .7-1.1 lbs.
MEDIUM	1.5-2 pts. + 1.3-2.2 lbs.	1.5-2 pts. + .7-1.1 lbs. + .7-1.1 lbs.	2 pts. + 1.8-2.2 lbs.	2 pts. + .9-1.1 lbs. + .9-1.1 lbs.
FINE	2 pts. + 1.8-2.2 lbs.	2 pts. + .9-1.1 lbs. + .9-1.1 lbs.	2-2.5 pts. + 1.8-2.2 lbs.***	2-2.5 pts. + .9-1.1 lbs.*** + .9-1.1 lbs.***
Muck or Peat (soils with more than 20% organic matter)	DO NOT USE			

*When using AAtrex 4L (or Princep 4L), AAtrex 4LC, or AAtrex 80W (or Princep 80W), use equivalent rates. One lb. of Nine-O or Caliber 90 equals 1.8 pts. of 4L (4LC) or 1.1 lb. 80W. Use Princep in preference to AAtrex when heavy infestations of crabgrass or fall panicum are expected. On soils having between 6% and 20% organic matter. Dual may be used up to 3.5 pts./A in tank mix combination with 3.3 lbs./A of AAtrex Nine-O, or equivalent rates of AAtrex 80W or AAtrex 4L (4LC).

**When using the tank mixture of Dual + AAtrex Nine-O + Princep Caliber 90, use equal rates of each as shown when heavy broadleaf weed infestations are expected. When heavy infestations of crabgrass or fall panicum are expected, use a 1:2 ratio of AAtrex + Princep instead of the 1:1 ratio given in Table 1 above. (*Example:* Total AAtrex Nine-O + Princep Caliber 90 = 1.2 lbs./A, use 0.4 lb. of AAtrex plus 0.8 lb. of Princep, respectively.) Refer to the asterisk (*) above for AAtrex 4L, 4LC and 80W, and Princep 4L and 80W conversions.

***For cocklebur, yellow nutsedge, and velvetleaf control on fine-textured soils above 3% organic matter, apply 2.25 lbs./A of AAtrex Nine-O, or equivalent rates of AAtrex 4L, 4LC or 80W, or the same total amount of AAtrex + Princep with 2.-2.5 pts./A of Dual.

2. Tank Mixture with AAtrex Postemergence

Weeds Controlled

barnyardgrass (watergrass)	giant foxtail	kochia	purslane
crabgrass	green foxtail	lambsquarters	ragweed
crowfootgrass	yellow foxtail	mustard	smartweed
fall panicum	jimsonweed	pigweed	velvetleaf
		prickly sida	

Weeds Partially Controlled: cocklebur, morningglory, yellow nutsedge.

Apply 1.5 pts./A of Dual + 1.3 lbs./A of AAtrex Nine-O* on *coarse soils.* 2 pts./A of Dual + 1.8 lbs./A of AAtrex Nine-O on *medium soils.* or 2-2.5 pts./A of Dual + 1.8-2.2 lbs./A** of AAtrex Nine-O on *fine soils.* Apply this tank mixture before grass and broadleaf weeds pass the 2-leaf stage and before corn exceeds 5 inches in height. Application to weeds larger than the 2-leaf stage will generally result in unsatisfactory control.

Lay-by: Apply to corn plants not more than 40 inches tall. Applications to corn in excess of 5 inches should be directed to the base of the corn plants: whereas applications to corn plants less than 5 inches tall may be made over the top. Occassionally some corn leaf burn may result, but this should not affect later growth or yield. Do not apply this postemergence tank mixture in fluid fertilizer, or severe crop injury may occur.

*When using AAtrex 4L, AAtrex 4LC, or AAtrex 80W, use equivalent rates. One lb. of Nine-O equals 1.8 pts. of 4L (4LC) or 1.1 lb. 80W.

**For better control of cocklebur, morningglory, velvetleaf, and yellow nutsedge on fine-textured soils above 3% organic matter, apply 2.2 lbs./A of AAtrex Nine-O, or equivalent rate of AAtrex 80W or AAtrex 4L (4LC), with 2-2.5 pts./A of Dual.

Tank mixtures of Dual plus AAtrex may be applied following use of any registered preplant surface-applied, preplant incorporated or preemergence corn herbicide, including Dual plus AAtrex.

Note: The total Dual rate should not exceed 6 pts., nor the AAtrex rate more than 4 lbs. active ingredient per acre during any one crop year, or illegal residues may result. Refer to the AAtrex label for geographic, soil-texture and rotational restrictions.

3. Tank Mixture with Bladex (Field Corn and Silage Corn Only)

In addition to the weeds controlled by Dual alone, Dual + Bladex, applied preplant incorporated or preemergence, also controls the following broadleaf weeds: jimsonweed, lambsquarters, mustard, ragweed, smartweed, cocklebur*, morningglory*, and velvetleaf*.

*Partially controlled.

Special Mixing Instructions: Fill the spray tank one-fourth full with water and start agitation. When using Bladex in tank mixtures, first add a compatibility agent, such as Unite or Compex at 2-3 pts./100 gals. of spray mixture. Then add the Bladex and allow it to become dispersed. Then add other herbicides registered for tank mix with Bladex, allowing each to fully disperse before adding the rest of the water or fluid fertilizer. Add the other herbicides in the following order: dry flowables and wettable powders first, then flowables, followed by emulsifiable concentrates. Add contact herbicides last; i.e., Gramoxone Super or Roundup. Agitate during mixing and application to maintain a uniform suspension. Tank mixtures with Bladex should not be allowed to stand without agitation. Empty the spray tank during the spraying operation. If fluid fertilizers are used as carriers, check the compatability of tank mixtures containing Bladex by using the **Compatibility Test** in the **General Information** section of this label.

Preplant Incorporated or Preemergence: Apply Dual + Bladex preplant incorporated or preemergence using the appropriate rate from Table 2. Follow instructions for use of Dual alone under **Application Procedures.**

Table 2: DUAL + Bladex—Field Corn and Silage Corn —Broadcast Rates Per Acre

Soil Texture	Percent organic matter in soil: Less than 1%	1-2.5%	2.5-4%	Over 4%
	Pts. DUAL + Qts. Bladex 4L*	Pts. DUAL + Qts. Bladex 4L*	Pts. DUAL + Qts. Bladex 4L*	Pts. DUAL + Qts. Bladex 4L*
COARSE Sand, loamy sand	DO NOT USE	1.25-1.5 + .8-1.4	1.5-1.75 + 1.4-1.6	2 + 1.75
COARSE Sandy loam	1.25 + .8	1.25-1.5 + 1.2-1.4	1.5-1.75 + 1.4-1.75	2 + 2
MEDIUM	1.5 + 1.2	1.5-1.75 + 1.4-1.75	1.75-2 + 1.75-2	2.25 + 2.25
FINE	1.5 + 1.4	1.75-2 + 1.6-2	2-2.25 + 2-2.25	2.5 + 2.5
Muck or Peat soils (soils with more than 20% organic matter)	DO NOT USE			

*When using Bladex 80W or Bladex 90DF, use equivalent rates. One qt. of 4L equals 1.25 lbs. of 80W or 1.1 lbs. of Bladex 90DF.

4. Tank Mixtures with AAtrex plus Bladex Formulations (Field Corn and Silage Corn Only)

Note: See **Tank Mixtures with Bladex** section of this label for specific mixing instructions.

Apply Dual + AAtrex + Bladex preplant incorporated or preemergence to corn using the rates from Table 3. Refer to the AAtrex and Bladex labels, as well as the **General Information** section of this label, for all directions, precautions, and limitations.

Rotational Crops: Refer to the crop rotation statements for Dual alone on this label and for AAtrex and Bladex alone on their respective labels.

Table 3: DUAL + AAtrex + Bladex—Field Corn and Silage Corn Only Broadcast Rate Preplant Incorporated or Preemergence

Soil Texture	Percent organic matter in soil: Less than 1%	1-2.5%	2.5-4%	Over 4%
	Pts. DUAL + Lbs. AAtrex Nine-O* + Qts. Bladex 4L*	Pts. DUAL + Lbs. AAtrex Nine-O* + Qts. Bladex 4L*	Pts. DUAL + Lbs. AAtrex Nine-O* + Qts. Bladex 4L*	Pts. DUAL + Lbs. AAtrex Nine-O* + Qts. Bladex 4L*
COARSE Sand, loamy sand	DO NOT USE	1.25-1.5 + .4-1.1 + .6-1	1.5-1.75 + .5-1.3 + 1-1.2	2 + .7-1.8 + 1.4-1.6
COARSE Sandy loam	1.25 + .5-1.1 + .6-1	1.25-1.5 + .7-1.3 + 1-1.2	1.5-1.75 + .9-1.8 + 1.2-1.6	2 + 1.1-2 + 1.6-2
MEDIUM Loam, silt loam, silt	1.5 + .7-1.3 + 1-1.2	1.5-1.75 + .7-1.8 + 1.2-1.6	1.75-2 + .9-1.8 + 1.4-1.6	2.25 + 1.1-2 + 1.8
FINE Sandy clay loam, silty clay loam, clay loam, sandy clay, silty clay	1.5 + .7-1.5 + 1.2-1.4	1.75-2 + .9-1.8 + 1.4-1.6	2-2.25 + .9-2 + 1.6-1.8	2.5 + 1.1-2.2 + 2
Muck or Peat soils (soils with more than 20% organic matter)	DO NOT USE			

* When using AAtrex 4L (4LC), AAtrex 80W, Bladex 80W, or Bladex 90DF, use equivalent rates. One lb. of AAtrex Nine-O equals 0.9 qts. (1.8 pts.) of AAtrex 4L or 1.1 lbs. of AAtrex 80W. One qt. of Bladex 4L equals 1.25 lbs. of Bladex 80W or 1.1 lbs. of Bladex 90DF. Use lower rates of AAtrex in the mixture where broadleaf weeds are not a major problem. Use higher rates of AAtrex where heavy infestations of such broadleaf weeds as cocklebur, morningglory, smartweed and velvetleaf are expected to be the dominant weeds present.

5. Tank Mixture with Banvel

Preemergence: Use this tank mixture only on field corn which is flat-planted (no furrows) in IL, IN, IA, MN, NE, OH, SD, and WI.

In addition to the weeds controlled by Dual alone, Dual + Banvel, applied preemergence, also controls lambsquarters, ragweed, smartweed, cocklebur*, jimsonweed*, morningglory*, and velvetleaf*.

*Partially controlled.

Apply Dual + Banvel preemergence. Broadcast 1 pt./A of Banvel with 2 pts./A of Dual on *medium soils,* or with 2-2.5 pts./A of Dual on *fine soils.* Do not apply on *coarse soils* or on soils with less than 2.5% organic matter. Apply this tank mixture to the soil surface at planting or after planting but before corn emerges. Plant corn at least 1.5 inches deep and apply behind planting equipment avoiding incorporation by the planter wheel or other seed covering device. Do not incorporate before corn emergence. If it is necessary to rotary hoe to break the soil crust, do not disturb the soil more than one-half inch deep.

Postemergence for Control of Pigweed (Mid-Atlantic states including DE, MD, PA, VA, and WV): Apply 1.5-2.25 pts. of Dual plus 0.5-1 pt./A of Banvel or 1-2 pt./A of Banvel II by ground equipment when pigweed plants are less than 3 inches tall and before corn exceeds 5 inches in height in a minimum of 20 gallons of spray per acre. Use the lower rate on coarse-textured and low organic matter soils. Use the higher rate on fine-textured and high organic matter soils.

Precautions: (1) Avoid drift to sensitive nontarget plants such as soybeans during application, or injury may occur. (2) Do not apply with aircraft.

6. Tank Mixture with AAtrex and Lorox for Control of Lambsquarters and Pigweed

For prolonged control of lambsquarters and pigweed in DE, MD, NJ, NY, PA, VA and WV, Dual may be applied preemergence in tank mix combination with AAtrex + Lorox. Apply Dual and AAtrex according to the rates in Table 1 and Lorox according to the rates below.

Soil Texture	Broadcast Rates per Acre
Sandy loam (1-3% organic matter)	.67 lb. Lorox*
Sandy loam (3-6% organic matter)	1.0 lb. Lorox*
Medium and fine-textured soils (1-6% organic matter)	1.0 lb. Lorox*

* When using Lorox L or Lorox DF, use equivalent rates. One pt. of Lorox L equals one lb. of Lorox DF.

Observe all directions for use, precautions and limitations on the Dual, AAtrex and Lorox labels when applying these products in tank mix combinations.

7. Tank Mixture with AAtrex or Princep Plus Prowl for Prolonged Control of Lambsquarters and Pigweed in Field Corn Only (Northeast U.S., including MI, IN, KY and States East of These)

For prolonged control of lambsquarters and pigweed, in addition to a broad spectrum of annual broadleaf and grass weeds. Dual in tank mix combination with AAtrex* or Princep plus Prowl 4E may be applied after planting but before corn or weeds emerge. Apply by ground equipment in a minimum of 10 gallons of water or 20 gallons of liquid fertilizer. Apply by air in a minimum of 5 gallons of water. Refer to Table 1 of this label for rates of Dual, AAtrex or Princep to be applied. Apply Prowl 4E according to the rates below in Table 4.

*Do not apply Dual in tank mix combination with AAtrex 80W plus Prowl, as this combination is not compatible. Other AAtrex formulations may be used.

Mixing Instructions: Fill the spray tank one-fourth full with water or fluid fertilizer and start agitation. To aid in compatibility, add a compatibility agent, such as Unite or X-77®, at 4 pts./100 gals. of spray mixture to the spray tank first, then add the AAtrex or Princep and allow it to become dispersed, then add Dual and Prowl 4E, and finally the rest of the water.

Table 4. Prowl 4E—Broadcast Rates Per Acre

Soil Texture	Percent Organic Matter in Soil: Less than 1.5%	1.5-3%	Over 3%
COARSE	1.5-2.0 pts.	2.0 pts.	3.0 pts.
MEDIUM	2.0 pts.	3.0 pts.	3.0 pts.
FINE	2.0 pts.	3.0 pts.	3.0 pts.

Observe all directions for use, precautions, and limitations on the respective produce labels when applying these products in tank mix combination. Refer to the Prowl 4E label for replanting instructions in the event of crop loss.

8. Tank Mixture with AAtrex, Bladex*, or Princep, AAtrex plus Bladex or AAtrex plus Princep, with Gramoxone Super or Roundup for Minimum-Tillage or No-Tillage Systems

*Field corn and silage corn only. See **Tank Mixture with Bladex** section of this label for special mixing instructions.

In minimum-tillage or no-tillage systems where corn is planted directly into a cover crop, stale seedbed, established sod, or previous crop residues, the contact herbicides Gramoxone Super or Roundup may be added to a tank mix of Dual + AAtrex, Bladex or Princep, Dual + AAtrex + Bladex or Dual + AAtrex + Princep. When used as directed, the Gramoxone Super portion of the tank mixture controls most emerged annual weeds and suppresses many perennial weeds. Roundup combinations will control emerged annual and perennial weeds when applied as directed on the Roundup label. The Dual + AAtrex or Princep, or Dual + AAtrex + Princep portion of the tank mixture provides preemergence control of the weeds listed on this label in the tank mixture section for Dual + AAtrex or Princep, or Dual + AAtrex + Princep Preplant Surface. Preplant Incorporated, or Preemergence. The Dual + Bladex or Dual + AAtrex + Bladex tank mixture provides preemergence control of weeds listed on this label in the tank mixture sections with AAtrex and Bladex.

Application: Apply before, during, or after planting, but before the corn emerges, at the rates specified below. Add Gramoxone Super or Roundup at the following broadcast rates:

Gramoxone Super: 1.5-2.5 pts./A plus 8 fl. oz. of X-77 Spreader per 100 gals. of spray mixture. Use the lower rate for control of annual weeds less than 4 inches tall and the higher rate for weeds 4-6 inches tall. This treatment will not consistently control weeds taller than 6 inches.

Note: Do not apply combinations containing Gramoxone Super in a suspension type liquid fetilizers because the activity of paraquat will be reduced.

Roundup: 1.5 qts./A for existing annual weeds, or 2-4 qts. per acre for existing perennial weeds. See the Roundup label for weeds controlled and recommended rates for specific weeds.

Apply in 20-60 gals. of water or fluid fetilizer per acre with ground equipment.

On *coarse soils*, apply 1.5 pts./A of Dual with 1.3 lbs. AAtrex Nine-O* or Princep Caliber 90*, or with 0.7 lb. Nine-O** plus 0.7 lb. Caliber 90**. On *medium soils*, apply 2 pts./A of Dual with 1.8 lbs. AAtrex Nine-O or Princep Caliber 90, or with 0.9 lb. Nine-O plus 0.9 lbs. Caliber 90 On *fine soils* ***, apply 2-2.5 pts./A of Dual with 1.8-2.2 lbs. AAtrex Nine-O or Princep Caliber 90, or with 0.9-1.1 lbs. of Nine-O plus 0.9-1.1 lbs. Caliber 90.

*When using AAtrex 4L (or Princep 4L), AAtrex 4LC, or AAtrex 80W (or Princep 80W), use equivalent rates. One lb. of Nine-O or Caliber 90 equals 1.8 pts. of 4L (4LC) or 1.1 lbs. of 80W. Use Princep in preference to AAtrex when heavy infestations of crabgrass or fall panicum are expected.

**When using the tank mixture of Dual + AAtrex Nine-O + Princep Caliber 90, use equal rates of AAtrex and Princep as shown when heavy broadleaf weed infestations are expected. When heavy infestations of crabgrass or fall panicum are expected, use a 1:2 ratio of AAtrex + Princep instead of the 1:1 ratio given above. (*Example:* Total AAtrex Nine-O + Princep Caliber 90 = 1.8 lbs./A, use 0.6 lb. of AAtrex plus 1.2 lbs. of Princep, respectively.) Refer to the asterisk (*) above for AAtrex 4L, 4LC and 80W, and Princep 4L and Princep 80W conversions.

***For cocklebur, yellow nutsedge, and velvetleaf control on fine-textured soils above 3% organic matter, apply 2.25 lbs./A of AAtrex Nine-O, or equivalent rates of AAtrex 80W or AAtrex 4L (4LC), or the same total amount of AAtrex + Princep, with 2-2.25 pts./A of Dual.

9. Tank Mixture with AAtrex or AAtrex plus 2,4-D or AAtrex Plus 2,4-D Banvel for Minimum-Tillage or No-Tillage Systems

In minimum-tillage or no-tillage systems where corn is planted directly into a cover crop, stale seedbed, established sod, or previous crop residues, Dual applied in combination with AAtrex will kill most emerged small annual weeds. Apply Dual + AAtrex before, during, or after planting, but before corn emerges according to the rates in Table 1.

Where heavy crop residues exist, add 0.8-1.6 pts./A of an appropriately labeled 3.8 lb. ai/gal. 2,4-D amine (such as Weedar 64, Weedar 64A, DMA-4 Herbicide, or Formula 40) to the spray tank last and apply in a minimum of 25 gals. of carrier/A.

As carriers, nitrogen solutions and complete liquid fertilizers, applied before corn emergence, enhance burndown of existing weeds, and therefore are recommended instead of water. Add X-77 surfactant at 1.0-2.0 qts/100 gals. of diluted spray, or another appropriate surfactant at its recommended rate. Apply before weeds exceed 3 inches in height. If alfalfa is present, add Banvel to the spray mixture at 0.33-0.5 pt./A and apply before alfalfa exceeds 6 inches in height.

For fields with existing sod grasses (e.g., bromegrass, orchardgrass, rye or timothy), when existing weeds exceed 3 inches in height or when very dry conditions exist, add Gramoxone Super at the rate of 2.5 pts./A in place of or in addition to 2,4-D as indicated above. Do not apply Gramoxone Super in suspension-type liquid fertilizer. Observe all directions for use, precautions and limitations on the respective product labels when applying these products in tank mix combination.

10. Tank Mixture with Bladex* or Bladex plus AAtrex plus 2,4-D or Bladex plus AAtrex plus 2,4-D plus Banvel for Minimum-Tillage or No-Tillage Systems (Field corn and silage corn only)

*See **Tank Mixture with Bladex** section of this label for special mixing instructions.

In minimum-tillage or no-tillage systems where corn is planted directly into a cover crop, stale seedbed, established sod, or previous crop residues. Dual applied in combination with Bladex or Bladex + AAtrex will provide control of emerged small annual weeds, and extended control of those species on the Dual and Bladex labels.

Apply Dual in tank mix combination with Bladex according to the "**Tank Mixtures with Bladex**" and "**Tank Mixtures with AAtrex. . .**" sections of this label or the preemergence section of the Bladex and AAtrex labels. Where heavy corn residues exist at the time of treatment, follow directions for use in the "**Bladex Plus Dual**" tank mixture in the **Minimum- or No-Till Weed Control** section of the Bladex label. Using a minimum of 25 gals./A of carrier, add an appropriately labeled 2,4-D amine or low volatile ester to the spray tank last.

As carriers, nitrogen solutions and complete liquid fertilizers, applied before corn emergence, enhance burndown of existing weeds, and therefore are recommended instead of water. Add X-77 surfactant at 1.0 to 2.0 qts./100 gals. of diluted spray, or another appropriate surfactant at its recommended rate. Apply before weeds exceed 3 inches in height. If alfalfa is present, add Banvel to the spray mixture at 0.33-0.5 pt./A and apply before alfalfa exceeds 6 inches in height.

For fields with existing sod grasses (e.g., bromegrass, orchardgrass, rye or timothy), when existing weeds exceed 3 inches in height or when very dry conditions exist, add Gramoxone Super at the rate of 2.5 pts./A in place of or in addition to 2,4-D as indicated above. Do not apply Gramoxone Super in suspension-type liquid fertilizer. Observe all directions for use, precautions and limitations on the respective product labels when applying these products in tank mix combination.

F. COTTON - DUAL ALONE - AR, LA, MS, NM, OK, TN, TX, AND BOOTHEEL OF MO

Application: Apply Dual either preplant incorporated or preemergence at the rate of 1.5 pts./A on sandy loams*, 1.5-2 pts./A on *medium soils*, or 2 pts. /A on *fine soils. Do not use on sands and loamy sands.*

*Use on sandy loam soils only in NM, OK, and TX.

Preplant Incorporated (NM, OK and TX Only): Apply to the soil and incorporate into the top inch of soil immediately before planting, at planting, or after planting but before crop or weeds emerge. Use a rolling cultivator or similar implement to uniformly incorporate not more than one inch deep. Use a preplant incorporated application if furrow irrigation is used or when a period of dry weather after application is expected. Where furrow irrigation is used, wet the top of the bed for best results. If the crop is to be planted on beds, apply and incorporate after bed formation. Cotton should be planted below the zone of incorporation; i.e., at least 1 inch on fine soils and 1.5 inches on coarse and medium soils. If incorporated prior to planting, use a planter that will result in a minimum of soil disturbance.

Preemergence: Apply to the soil surface at planting, or after planting but before weeds or crop emerge.

Note: For best control of yellow nutsedge and suppression of seedling johnsongrass, apply Dual preplant incorporated at the maximum rate for the soil texture, whether applied alone or mixed with Caparol 4L.

Precautions: To avoid crop injury, (1) Do not apply Dual on sand or loamy sand soils or in areas where water is likely to "pond" over the bed; (2) To avoid concentration in the seed furrow, do not make broadcast applications of Dual to cotton planted in furrows more than two inches deep. Band applications may be made to cotton planted in furrows deeper than two inches, but band width should not exceed the width of the bottom of the furrow; and (3) Do not apply to Taloka silt loam.

G. COTTON - DUAL COMBINATIONS

1. Tank Mixture with Caparol 4L

Dual tank mixtures with Caparol 4L may be applied in water or fluid fertilizer. When fluid fertilizer is used as a carrier for Dual, either alone or in combination with Caparol 4L, mix only the amount that will be sprayed in one operation. These mixtures should not be allowed to stand without agitation.

In addition to those weeds controlled by Dual alone, Dual + Caparol 4L, applied preplant incorporated or preemergence, also controls the following weeds: junglerice, wild oats, annual morningglory, groundcherry, hairy nightshade, lambsquarters, malva, mustard, prickly sida (teaweed), purslane, ragweed, and shallow-germinating seedlings of cocklebur and coffeeweed.

Preplant Incorporated or Preemergence: Apply Dual + Caparol 4L either preplant incorporated or preemergence using the appropriate rate from Table 5. Cotton should be planted below the zone of incorporation; i.e., at least 1.0 inch on fine soils and 1.5 inches on coarse and medium soils. If incorporated before planting, use a planter that will result in a minimum of soil disturbance.

Table 5: DUAL + Caparol 4L—Cotton (NM, OK, TX)

Use areas	Soil texture	Broadcast rates per acre	
		DUAL	Caparol 4L
ALL	sand, loamy sand	DO NOT USE	
Oklahoma, and Blacklands and Gulf Coast of Texas	loams	1.25-2 pts.	2.4 pts.
	clays	2 pts.	4.8 pts.
Rio Grande Valley of Texas	loams	1.25-2 pts.	3.2 pts.

Use areas	Soil texture	Broadcast rates per acre DUAL	Caparol 4L
New Mexico and High Plains, Rolling Plains, Edwards Plateau of Texas and Southwest Texas	clays	2 pts.	4.8 pts.
	sandy loam	1.25-1.5 pts.	1.6 pts.
	loams	1.25-2 pts.	2.4 pts.
	sandy clay loams	2 pts.	2.4 pts.
	other clay soils	2 pts.	3.2 pts.

Precautions: (1) To avoid concentration in the seed furrow, do not make broadcast applications of DUAL + Caparol 4L to cotton planted in furrows more than two inches deep. Band applications may be made to cotton planted in furrows deeper than two inches, but band width should not exceed the width of the bottom of the furrow. To avoid crop injury, (2) Do not apply on sand or loamy sand soils, or in areas where water is likely to "pond" over the bed; (3) Do not apply in cut areas of newly leveled fields, or in areas of excess salt; (4) Do not apply to glandless cotton varieties; and (5) Do not apply on Taloka silt loam.

Note: Do not feed treated forage to livestock, or graze treated areas, or illegal residues may result.

Refer to the Caparol 4L label for further instructions and limitations.

2. Tank Mixture with Cotoran

Dual may be applied in tank mixture with Cotoran DF preemergence for control of those weeds controlled by Dual alone and those as listed on the Cotoran DF label. This combination will also control spotted spurge, hyssop spurge, nodding spurge, and prostrate spurge. Apply to the soil surface at planting or after planting, but before weeds or crop emerge, using the appropriate rates from Table 6 below.

Mixing Instructions: Incompatibility may occur when tank mixing Dual and Cotoran DF. To help overcome this condition, fill the spray tank one-fourth full with water or fluid fertilizer and start agitation, add the Cotoran DF and allow to become dispensed. Add X-77 at 0.5% volume/volume final spray (4 pts./100 gals.), then add the Dual and finally the rest of the water or fluid fertilizer. Agitate during mixing and application to maintain a uniform suspension.

Table 6: DUAL + Cotoran DF—Cotton (AR, LA, MS, Eastern OK, TN, Bootheel of MO, and Gulf Coast, Rio Grande Valley and Eastern TX)

Soil Texture	Broadcast rates per acre DUAL	Cotoran DF
Sand, loamy sand	DO NOT USE	
Sandy loam*	1.25-1.5 pts.	1.2 lbs.
Loam, silt, silt loam	1.5-2 pts.	1.2-1.9 lbs.
Silty clay loam, sandy clay loam, silty clay, sandy clay, clay loam, clay	2 pts.	1.9-2.4 lbs.

*Use on sandy loam soils only in eastern OK and Gulf Coast, Rio Grande Valley, and eastern TX.

**When using Cotoran 4L use equivalent rates. Multiply lbs. of Cotoran DF by 1.7 to get pts. of Cotoran 4L.

Precautions: (1) Do not apply Dual + Cotoran on sand or loamy sand soils or in areas where water is likely to "pond" over the bed, or crop injury may occur. (2) To avoid concentration in the seed furrow, do not make broadcast applications of Dual + Cotoran to cotton planted in furrows more than two inches deep. Band applications may be made to cotton planted in furrows deeper than two inches, but band width should not exceed the width of the bottom of the furrow. (3) The use of Cotoran following the use of a systemic insecticide at planting may result in crop injury. (4) Do not use on Taloka silt loam, or crop injury may occur.

Refer to the Cotoran labels for further instructions, precautions, and limitations.

Note: To avoid possible illegal residues, do not feed treated forage or gin trash to livestock, or graze treated areas.

Rotational Crops: Refer to the crop rotation instructions for Dual alone on this label and for Cotoran alone on the Cotoran labels.

H. Grapes - Nonbearing - Dual Alone and Combinations

Broadcast 2-4 pts./A of Dual by ground equipment to weed-free soil in a minimum of 10 gals. of water to control weeds listed in the **Dual Applied Alone** section of this label. Apply the lower rates on coarse-textured soils with low organic matter and where the weed infestation is expected to be low. Apply the higher rates on fine-textured soils, soils high in organic matter, and where a heavy weed infestation is expected. Use caution when applying to keep the spray off the foliage to avoid possible foliar injury.

When weeds are present at the time of treatment. Dual may be applied in tank mix combination with Gramoxone Super or Roundup. Gramoxone Super may be applied at rates of 3-5 pts./A. Roundup may be applied at rates ranging from 1-5 qts./A, depending on the weed species. Direct applications containing Gramoxone Super or Roundup over the weeds, making certain to keep spray off grape foliage and vines. Refer to the Gramoxone Super and Roundup labels for further directions, precautions and limitations relative to their use.

Precaution: Do not apply to vines established in the vineyard less than 30 days and only after depressions around the vines have been filled in, or crop injury may occur.

Note: Do not apply to vines that will bear harvestable fruit within 12 months of application, or illegal residues may result.

I. Peanuts - Dual Alone

Apply Dual either preplant incorporated, postplant incorporated, preemergence or lay-by using the appropriate rate specified below. *Preplant Incorporated or Preemergence:* Follow instructions for use of Dual alone under **Application Procedures** *Postplant Incorporated:* Apply and shallowly incorporate Dual into the soil after planting, but before peanut germination. Incorporation depth and incorporating implements must be kept above the seed or seed will be damaged. *Lay-by:* Apply Dual to the soil immediately after the last normal cultivation.

Apply Dual alone preplant incorporated, postplant incorporated, preemergence or lay-by at a broadcast rate of 1.5-2 pts./A in the Southeast* and 1.25-2 pts./A in OK, TX, and NM.

*In the Southeast, use 2-3 pts./A and apply preemergence for partial control of Florida beggarweed.

Note: Dual alone may be applied as directed after any of the following preplant incorporated herbicides when used according to their label recommendations: Balan at 3-4 qts./A; Treflan E.C. at 1 pt./A; or Vernam® at 2.33-3 pts./A.

J. Peanuts - Dual Combinations

1. Tank Mixture with Balan L.C.

Dual + Balan tank mixture applied preplant incorporated controls those weeds listed under **Dual Applied Alone** and those weeds as listed on the Balan label.

Apply 1.5-2 pts./A of Dual + 3-4 qts./A of Balan in a minimum of 10 gals. of spray volume per acre for ground application or in a minimum of 5 gals. of spray volume per acre for aerial application. Follow the recommended procedures for Balan on the Balan label for soil preparation and incorporation of this tank mix. Apply and incorporate Dual + Balan up to 14 days prior to planting.

Note: Follow all restrictions and precautions on the Balan label.

2. Multiple Applications (Southeast Only: AL, FL, GA, NC, SC or VA)

Where weed pressure is heavy or where species difficult to control are expected, Dual is most effective when used as follows:

1st Application: Apply Dual preplant incorporated as directed under **Peanuts-Dual Alone** or apply Dual + Balan preplant incorporated as directed previously in this section. Refer to the respective section for weeds controlled.

2nd Application: Apply Dual any time from preemergence up to "ground cracking" at 1.5-3 pts./A for extended control of weeds not yet emerged. Refer to the **Dual Applied Alone** section for a list of weeds controlled.

3rd Application: Apply Dual at lay-by as directed under **Peanuts-Dual Alone.** Use only when late germinating weeds are expected to be a problem. Refer to the **Dual Applied Alone** section for a list of weeds controlled.

Notes: (1) Do not apply more than the equivalent of 6 pts. of Dual/A during any one year or illegal residues may result. This includes both Dual and Dual 25G (4 lbs. of Dual 25G equals 1 pt. of Dual). (2) Do not graze or feed peanut forage or fodder to livestock for 30 days following application and (3) Do not apply within 90 days of harvest, or illegal residues may result.

Rotational Crops: For a single application, refer to the rotational crop information under **Dual Applied Alone.** For multiple applications of Dual, corn, cotton, soybeans, peanuts, potatoes, safflowers, sorghum, or the pod crops listed on this label may be planted the spring following treatment. All other rotational crops may be planted 18 months after application.

K. Pod Crops - Dual Alone

Pod crops including: garbanzo, great northern beans, kidney beans, lima beans, mung beans, navy beans, peas (English*; southern peas, such as blackeye, pinkeye, crowder, etc.), pinto beans, snap beans (green, wax, string), and lupines (sweet, white, white sweet and grain).

Apply Dual either preplant incorporated or preemergence using the appropriate rate specified below. *Preplant Incorporated or Preemergence:* Follow instructions for use of Dual alone under **Application Procedures.** On *coarse soils* with less than 3% organic matter, apply 1.5-2 pts./A of Dual or 2 pts./A if organic matter is 3% or greater. On *medium soils,* apply 2-2.5 pts./A of Dual. On *fine soils,* apply 2-2.5 pts./A of Dual if organic matter content is less than 3%, or 2.5-3 pts./A if organic matter content is 3% or greater.

*On English peas, use only preemergence applications. Do not use on English peas in northeastern U.S., or injury may occur.

Note: Do not cut for hay within 120 days following a Dual application, or illegal residues may result.

L. Pod Crops - Dual Combinations

Note: When applying Dual in combination on pod crops, do not cut for hay within 120 days following application or illegal residues may result.

1. Tank Mixture and Sequential Applications with Eptam - Beans (Green or Dry)

This mixture controls all weeds controlled by Dual alone and by Eptam alone. Refer to the **Dual Applied Alone** section of this label for weeds controlled by Dual alone and to the Eptam label for weeds controlled by Eptam.

Preplant Incorporated: Follow instructions for use of Dual alone under **Application Procedures.** *Sequential:* Apply Eptam alone Preplant Incorporated, as specified on that label. Follow with a preemergence application of Dual, at rates specified for Dual alone, during planting (behind the planter) or after planting, but before the weeds or crop emerge.

Refer to the **General Information** section of this label and to the Eptam label for weather, cultural practices, and all other precautions and limitations that affect performance of these products.

Apply 3.5-4.5 pts./A of Eptam 7E* with Dual as specified. On *coarse soils,* apply 1.25 pts./A of Dual if organic matter content is less than 3%, or 1.5 pts./A if organic matter content is 3% or greater. On *medium soils,* apply 1.5 pts./A of Dual if organic matter content is less than 3%, or 2 pts./A if organic matter content is 3% or greater. On *fine soils,* apply 2 pts./A of Dual per acre if organic matter is less than 3%, or 2-2.5 pts./A if organic matter is 3% or greater.

*Refer to the Eptam label for rate limits depending on geographical area, and for species and varietal restrictions.

Precaution: Do not exceed 3.5 pts./A Eptam 7E on small white beans or green beans grown on coarse-textured soils.

2. Tank Mixture with Treflan - Beans (Dry - Kidney, Navy, Pinto, etc., Lima and Snap)

Dual + Treflan tank mix applied preplant incorporated controls those weeds listed under **Dual Applied Alone** and those weeds listed for Treflan alone on the Treflan E.C. label. Dual + Treflan E.C. may be applied by ground or aerial equipment and incorporated up to 14 days prior to planting. Follow the recommended procedures on this label and on the Treflan E.C. label using equipment that provides uniform 2 inch incorporation.

Apply Dual + Treflan tank mix using the appropriate Dual rate specified for Dual alone, and the Treflan rate from the **Dry Beans,** and the **Lima and Snap Beans** sections of the Treflan E.C. label. Choose the product rate for the specific soil texture/organic matter classification and weed species expected.

Note: Follow all restrictions and precautions on the Treflan E.C. label and on the Dual alone portion of this label.

3. Tank Mixture with Amiben - Beans (White, Navy, Kidney, Pinto, and Lima)

Dual + Amiben tank mix applied either preplant incorporated or preemergence controls those weeds listed under **Dual Applied Alone** and those weeds listed for Amiben alone on the Amiben/Amiben DS labels.

Preplant Incorporated or Preemergence: Apply Dual and Amiben either preplant incorporated or preemergence using the Dual rate as specified for Dual alone, and the Amiben rate as specified on the Amiben/Amiben DS labels according to the soil texture/organic matter classification and weed species expected. Follow instructions for use of Dual alone under **Application Procedures.**

Refer to the **General Information** section of this label and to the Amiben labels for weather and cultural practices that affect performance of these products.

M. Potatoes - Dual Alone

Apply Dual either incorporated, preemergence, or after hilling/lay-by according to directions specified below for control of weeds listed under the **General Information** section. Within a rate range, use the lower rate on soils relatively coarse-textured or low in organic matter; use the higher rate on soils relatively fine-textured or high in organic matter. Effectiveness will be reduced if later cultural practices expose untreated soil.

Incorporated: Apply Dual at 1.5-3 pts./A to the soil and incorporate into the top 3 inches before planting using a finishing disk, harrow, rolling cultivator, or similar implement. Planting and later cultural practices should not bring untreated soil to the surface. Postplant incorporated application may be made any time after planting to drag-off, but before potatoe emergence. Use an implement that evenly distributes Dual in the top 2 inches of soil. Do not damage potato seed pieces or sprouts with incorporation equipment.

Preemergence: Apply Dual at 1.5-3 pts./A either after planting as a preemergence, delayed preemergence, after drag-off or hilling treatment, but before weeds emerge. Up to 4 pts./A of Dual alone may be used where soil organic matter is between 6% and 20%.

After Hilling/Lay-by: Apply 2.5 pts./A Dual after hilling/at lay-by to control Dual-sensitive species for remainder of the growing season. This application will not control emerged weeds. It may be applied over a previous Dual application, but do not apply more than 5.5 pts./A of Dual in a single crop season.

Precaution: (1) Do not use on muck or peat soils. If cool, wet soil conditions occur after application. Dual may delay maturity and/or reduce yield of Superior and other early maturing potato varieties. To avoid crop injury, (2) Do not use on sweet potatoes or yams; (3) Do not apply both as a preemergence and an incorporated treatment; and (4) Do not use in Kern County, California.

Note: Potatoes treated with Dual should not be harvested within 60 days after the at planting to drag-off application, or within 40 days after a lay-by application, or illegal residues may result.

N. Potatoes - Dual Combinations

1. Tank Mixture with Sencor or Lexone

In addition to those weeds controlled by Dual alone, Dual applied in tank mix combination with or sequentially with any of the registered Sencor or Lexone formulations, also controls the following broadleaf weeds; cocklebur*, hairy nightshade*, hemp sesbania, jimsonweed*, lambsquarters, prickly sida, ragweed, smartweed, velvetleaf, Venice mallow, and wild mustard.

*Partially controlled.

Dual at 1.5-3 pts./A plus the labeled Sencor/Lexone use rate may be used preemergence through after last hilling. Apply 1.5-2 pts./A of Dual on *coarse soils* and 2-3 pts./A on other soil textures. Within this rate range, use the lower rate on soils relatively coarse-textured or low in organic matter; use the higher rate on soils relatively fine-textured or high in organic matter. Effectiveness will be reduced if later cultural practices expose untreated soil. Dual will not control emerged weeds.

Refer to the Sencor or Lexone labels for precautionary statements, restrictions, application information and weeds controlled.

Precautions: (1) Postemergence applications to potatoes should be made only as a directed or semi-directed spray to avoid chlorosis, minor necrosis or leaf distortion. To avoid crop injury, (2) Do not use Dual + Sencor or Lexone on potatoes in Kern County, California, and (3) Do not apply to sweet potatoes or yams. (4) Do not use this tank mixture on muck or peat soils.

Notes: (1) Potatoes treated with Dual in tank mixture with Sencor or Lexone cannot be harvested within 60 days after application or illegal residues may result. (2) Potatoes may not be harvested within 40 days after a lay-by application of Dual, or illegal residues may result.

2. Dual + Lorox Tank Mixture (East of Rocky Mountains)

Dual may be applied in a tank mix combination with any of the registered Lorox formulations as a preemergence broadcast application to potatoes. Apply to the soil surface after planting and before emergence of the crop or after final drag-off according to the rates specified in Table 7.

Table 7: DUAL + Lorox—Potatoes (East of Rocky Mountains)

	Broadcast rates per acre			
	1 to less than 3% organic matter		3 to 5% organic matter	
Soil Texture	DUAL	Lorox*	DUAL	Lorox*
COARSE Sandy loam	1.5 pts.	1-1.5 lbs.	2 pts.	1.5-2 lbs.
MEDIUM Loam, silt loam, silt	2 pts.	1.5-2 lbs.	2.5-3 pts.	2-2.5 lbs.

*When using Lorox L or Lorox DF, use equivalent rates. One pt. of Lorox L equals one lb. of Lorox DF.

Precautions: To avoid crop injury, (1) Do not use on sands or loamy sands, and (2) Do not incorporate or spray over the top of emerged potatoes.

Refer to the **General Information** section of this label and to the Lorox label for precautionary statements, restrictions, application information and weed controlled.

O. Safflowers - Dual Alone

Preplant Incorporated or Preemergence: Follow instructions for use of Dual alone under **Application Procedures.**

On *coarse soils,* apply 1.5-2 pts./A of Dual if organic matter content is less than 3%, or 2 pts./A if organic matter is 3% or greater. On *medium soils,* apply 2-2.5 pts./A of Dual. On *fine soils,* apply 2-2.5 pts./A of Dual if organic matter content is less than 3%, or 2.5-3 pts./A if organic matter content is 3% or greater.

P. GRAIN OR FORAGE SORGHUM (SEED TREATED WITH CONCEP®) - DUAL ALONE

Apply Dual either preplant surface, preplant incorporated, or preemergence using the appropriate rate specified below.

Preplant Surface Applied: Refr to instructions for use of Dual under **Application Procedures.** For minimum-tillage or no-tillage systems only. Dual may be applied up to 45 days before planting in CO, IL, IA, KS, MO, NE, and SD. Use only split applications for treatments made 30-45 days prior to planting, with ⅔ of the broadcast rate applied initially and the remaining ⅓ at planting. Apply 2.25 pts./A of Dual on medium soils or 2.5 pts./A on fine soils. Treatments less than 30 days prior to planting may be made either as a split or single application. Apply 2 pts./A of Dual on coarse soils not more than 2 weeks prior to planting.

Under dry conditions, irrigation after application is recommended to move Dual into the soil.

Preplant Incorporated or Preemergence: Refer to instructions for use of Dual under **Application Procedures.** Broadcast 1.5-2 pts./A of Dual on coarse soils, 2-2.25 pts./A on medium soils, or 2.25 pts./A on fine soils.

Precautions: (1) If sorghum seed is not properly treated with Concep, Dual will severly injure the crop. (2) Under high soil moisture conditions prior to sorghum emergence, injury may occur following the use of Dual. The crop will normally outgrow this effect. (3) Do not use Dual on sorghum grown under dry mulch tillage or injury may occur. (4) Do not make more than one application per year, or illegal residues may result.

Q. GRAIN OR FORAGE SORGHUM (SEED TREATED WITH CONCEP) - DUAL COMBINATIONS

Dual tank mixtures with AAtrex may be applied in water or fluid fertilizer. Apply Dual alone or in the tank mixtures only when the sorghum seed has been properly treated by the seed company with Concep.

Precautions: (1) Applications of Dual + AAtrex on highly alkaline soils or on eroded areas where calcareous subsoils are exposed may cause sorghum injury. (2) If sorghum seed is not properly treated with Concep, Dual + AAtrex may severly injure the crop. (3) Under high soil moisture conditions prior to sorghum emergence, injury may occur following the use of Dual + AAtrex. The crop will normally outgrow this effect. (4) Do not use Dual + AAtrex on sorghum grown under dry mulch tillage or injury may occur. (5) Do not make more than one application per year, or illegal residues may result.

1. Tank Mixture with AAtrex

In addition to the weeds controlled by Dual alone, Dual + AAtrex also controls the following broadleaf weeds when applied either preplant surface, preplant incorporated, or preemergence: cocklebur, common purslane, hairy nightshade, lambsquarters, morningglory, ragweed, smartweed, and velvetleaf.

Preplant Surface Applied: Refer to instructions for use of Dual under **Application Procedures.** For minimum-tillage or no-tillage systems only, Dual + AAtrex may be applied up to 45 days prior to planting in IL, IA, eastern KS, MO, NE and SD. Use only split applications for treatments made 30-45 days prior to planting, with 2/3 of the broadcast rate applied initially and the remaining 1/3 at planting. Apply 2.25 pts./A of Dual + 1.7-2 lbs./A of AAtrex Nine-O* on *medium soils* with 1.5% organic matter or greater. Apply 2.25 pts./A of Dual + 1.7-2 lbs./A of AAtrex Nine-O on *fine soils* with less than 1.5 organic matter, or apply 2.5 pts./A of Dual + 2-2.2 lbs./A AAttrex Nine-O on *fine soils* with 1.5% organic matter or greater. Treatments less than 30 days prior to planting may be made either as a split or single application. Under dry conditions, irrigation after application is recommended to move Dual + AAtrex into the soil.

Precautions: To avoid crop injury, (1) Do not use on coarse soils, and (2) Do not use on medium soils with less than 1.5% organic matter.

Preplant Incorporated or Preemergence: Refer to instructions for use of Dual under **Application Procedures.** On *medium soils* with 1.5% organic matter or greater, apply 1.5 pts./A of Dual + 1.3 lbs./A of AAtrex Nine-O*. On *fine soils* with less than 1.5% organic matter, apply 1.5 pts./A of Dual plus 1.3 lbs./A of AAtrex Nine-O; on *fine soils* with 1.5% organic matter or greater, apply 1.75-2 pts./A. of Dual plus 1.6-1.8 lbs./A of AAtrex Nine-O.

*When using AAtrex 80W, AAtrex 4L, or AAtrex 4LC, use equivalent rates. One lb. of AAtrex Nine-O equals 1.1 lb. of AAtrex 80W or 1.8 pts. of AAtrex 4L (4LC).

Precautions: To avoid crop injury, (1) Do not use on coarse soils; (2) Do not use on medium soils with less than 1.5% organic matter; (3) Do not use in NM, OK, or TX except in northeast OK and the Texas Gulf Coast and Blacklands areas; and (4) Do not apply preplant incorporated in AZ or the Imperial Valley of CA.

2. Tank Mixture of Dual or Dual + AAtrex, with Gramoxone Super or Roundup for Minimum-Tillage or No-Tillage Systems

In minimum-tillage or no-tillage systems where sorghum (seed treated with Concep) is planted directly into a cover crop, stale seedbed, established sod or previous crop residues, the contact herbicides Gramoxone Super or Roundup may be tank mixed with Dual or Dual + AAtrex. When used as directed, the Gramoxone Super portion of the tank mixture controls most emerged annual weeds and suppresses many perennial weeds. Roundup combinations will control emerged annual and perenenial weeds when applied as directed on the Roundup label. The Dual or Dual + AAtrex portion of the tank mixture provides preemergence control of the weeds listed on this label under the respective sections.

Refer to the label of each product used in combination and observe the planting details, restrictions, and all other precautions and limitations.

Applications: Apply before, during, or after planting but before sorghum emerges, at the appropriate rates listed under **Grain or Forage Sorghum - Dual Alone** or **- Dual Combinations - Dual + AAtrex,** respectively. Add Gramoxone Super or Roundup at the following broadcast rates:

Gramoxone Super: 1.5-2.5 pts. per acre plus 8 fl. oz. of X-77 Spreader per 100 gals. of spray mixture. Use the lower rate for control of annual weeds less than 4 inches tall and the higher rate for weeds 4-6 inches tall. This treatment will not consistently control weeds taller than 6 inches.

Roundup: 1.5 qts. per acre for control of existing annual weeds, or 2-4 qts. per acre for existing perennial weeds. See the Roundup label for weeds controlled and recommended rates for specific weeds.

Apply in a minimum of 20 gals. of water per acre with conventional spray equipment.

R. Soybeans - Dual Alone

Apply Dual either preplant surface-applied, preplant incorporated or preemergence using the appropriate rate specified below. *Preplant Surface-Applied, Preplant Incorporated or Preemergence:* Follow instructions for use of Dual alone under **Application Procedures.**

Preplant Surface-Applied: Use on *medium* and *fine soils* with minimum-tillage or no-tillage systems in CO, DE, IL, IN, IA, KS, KY, MD, MI, MN, MO, MT, NE, ND, NY, OH, PA, SD, TN, VA, WI, WV and WY. Apply 2/3 the recommended rate of Dual (2.25 pts./A on *medium soils* and 3 pts./A on *fine soils*) as a split treatment 30-45 days prior to planting and the remainder at planting. Applications made less than 30 days before planting may be as either a split or single treatment. Apply 2 pts./A on *coarse soils* not more than 2 weeks prior to planting.

Preplant Incorporated or Preemergence: On *coarse soils,* apply 1.5-2 pts./A of Dual if organic matter content is less than 3%, or 2 pts./A if organic matter content is 3% or greater. On *medium soils,* apply 2-2.5 pts./A of Dual. On *fine soils,* apply 2-2.5 pts./A of Dual if organic matter content is less than 3%, or 2.5-3 pts./A if organic matter content is 3% or greater.

Note: On soybeans, Dual may be used up to 4 pts./A as a preplant surface-applied, preplant incorporated or preemergence treatment on soils having an organic matter content between 6% and 20%. The total Dual rate applied to soybeans during any one crop should not exceed 4 pts./A.

S. Soybeans - Dual Combinations

Water or fluid fertilizer may be used as carrier for Dual in combination with Amiben, Sencor, Lexone, Lorox, Lorox Plus, Gemini, Canopy, Preview, Pursuit, Scepter, Sonalan or Command.

Note: For all of the following combinations, Dual may be used up to 3.5 pts./A on soils having an organic matter content between 6% and 20%. The total Dual rate applied to soybeans during any one crop year should not exceed 4 pts./A.

1. Tank Mixture with Sencor or Lexone

In addition to those weeds controlled by Dual alone, Dual + Sencor or Lexone, when applied as directed, also controls the following broadleaf weeds: cocklebur*, hairy nightshade, hemp sesbania, jimsonweed*, lambsquarters, prickly sida, ragweed, smartweed, velvetleaf, Venice mallow, and wild mustard.

*Partially controlled.

Apply Dual and Sencor or Lexone preplant incorporated or preemergence using the appropriate rates from Table 8. *Preplant Incorporated or Preemergence:* Follow instructions for use of Dual alone under **Application Procedures.** *Sequential:* Apply Dual alone *Preplant Incorporated,* as specified in Table 8 for this tank mixture. Follow with a preemergence application of Sencor or Lexone during planting (behind the planter) or after planting before weeds or soybeans emerge.

Refer to the Sencor or Lexone labels for planting details and soybean variety restrictions.

Table 8: DUAL + Sencor or Lexone—Soybeans

	Broadcast Rates Per Acre					
	0.5% to less than 3% organic matter			3% organic matter or greater		
Soil Texture**	DUAL	+	Sencor/ Lexone DF*	DUAL	+	Sencor/ Lexone DF*
COARSE Loamy sand (over 2% organic matter), sandy loam	1.25-1.5 pts.	+	.33 lb.	1.5 pts.	+	.5 lb.
MEDIUM	1.5-2 pts.	+	.5 lb.	2 pts.	+	.67 lb.***
FINE	2 pts.	+	.67 lb.	2-2.5 pts.	+	.67 lb.
Mississippi Delta only: Silty clay, clay	2 pts.	+	1.0 lbs.	2-2.5 pts.	+	1.0 lbs.
Muck or Peat soils (soils with more than 20% organic matter)	DO NOT USE					

*When using Sencor 4 or Lexone 4L, multiply lbs. of DF by 1.5 to get pts./A.

**On all sand and on loamy sand with less than 2% organic matter, do not use this tank mixture preemergence, or the sequential treatment. Do not use the tank mixture preplant incorporated on any sand, loamy sand, or sandy loam, or crop injury may occur.

***Use 0.5 lb./A if applied preplant incorporated.

Precautions: (1) Do not use the tank mix or sequential application on soil with less than 0.5% organic matter or on alkaline soil with a pH over 7.4, or crop injury may occur. (2) If heavy rain occurs soon after application, crop injury may result, especially in poorly drained areas where water stands for several days.

2. Tank Mixture with Lorox

In addition to those weeds controlled by Dual alone, Dual + Lorox, applied preemergence, also controls the following broadleaf weeds: cocklebur*, jimsonweed*, lambsquarters, morningglory*, prickly sida, ragweed, smartweed, velvetleaf*, Venice mallow, and wild mustard.

*Partially controlled.

Preemergence: Apply during planting (behind planter) or after planting, but before weeds or soybeans emerge. Refer to the Lorox label for planting details. Apply the appropriate rates from Table 9.

Table 9: DUAL + Lorox—Soybeans

	Broadcast rates per acre					
	0.5% to less than 3% organic matter			3% organic matter or greater		
Soil Texture*	DUAL	+	Lorox DF***	DUAL	+	Lorox DF***
COARSE**	1.25 pts.	+	1 lb.	1.5 pts.	+	1-1.5 lbs.
MEDIUM	1.5 pts.	+	1-1.5 lbs.	2 pts.	+	1.5-2 lbs.
FINE	2 pts.	+	2 lbs.	2-2.5 pts.	+	2.5-3 lbs.
Muck, or Peat soils (soils with more than 20% organic matter)	DO NOT USE					

*Do not use on sand, gravelly soils, or exposed subsoils.

**Do not use on loamy sand, except in the northeastern U.S. on loamy sand with over 1% organic matter.

***When using Lorox L or Lorox DF, use equivalent rates. One pt. of Lorox L equals one lb. of Lorox DF.

Precaution: Do not use on soil with less than 0.5% organic matter or crop injury may occur.

8. Tank Mixture with Amiben

This tank mixture controls all weeds controlled by Dual alone and by Amiben alone. Refer to the **Dual Applied Alone** section of this label for weeds controlled by Dual alone and to the Amiben labels for weeds controlled by Amiben.

Preplant Incorporated or Preemergence: Apply 1 gal./A of Amiben* plus either 1.5 pts./A of Dual on *coarse soils,* 1.5-2 pts./A of Dual on *medium soils,* or 2-2.5 pts./A of Dual on *fine soils.* Use 1.5 gals. of Amiben/A on *fine-textured high organic matter soils* or where a dense stand of difficult to control broadleaf weeds, such as black nightshade, velvetleaf, or wild mustard is expected to occur. *Preplant Incorporated:* Apply the tank mixture to the soil and incorporate into the top 2 inches of soil just before planting. Follow instructions for use, other than the preplant interval, of Dual alone under **Application Procedures.** *Preemergence:* Apply the tank mixture during planting (behind the planter) or after planting before weeds or soybeans emerge. *Sequential:* Apply Dual alone *Preplant Incorporated,* as specified above for this tank mixture. Follow with a preemergence application of Amiben during planting (behind the planter) or after planting before weeds or soybeans emerge.

*When using Amiben DS, multiply gallons of Amiben by 2.4 for lbs. Amiben DS/A.

Refer to the **General Information** section of this label and to the Amiben labels for weather and cultural practices that affect performance of these products.

4. Tank Mixture with Amiben and Sencor or Lexone

This tank mixture effectively controls those weeds as listed on the current Dual, Amiben/Amiben DS, and Lexone 4L/DF or Sencor 4/DF labels.

Apply Dual + Amiben + Lexone or Sencor preplant incorporated or preemergence using the appropriate rates from Table 10. *Preplant Incorporated or Preemergence:* Follow instructions for use of Dual alone under **Application Procedures.**

Table 10: DUAL + Amiben + Sencor or Lexone—Soybeans

	Broadcast rates per acre*				
Soil Texture	DUAL	+	Amiben	+	Sencor** or Lexone DF
COARSE Loamy sand (over 2% organic matter), sandy loam	1.5-2 pts.	+	3-4 qts.	+	.33 lb.
MEDIUM	2-2.5 pts.	+	4-5 qts.	+	.33-5 lb.***
FINE	2.5-3 pts.	+	5-6 qts.	+	.5-.67 lb.***
Muck or Peat soils (soils with more than 20% organic matter)	DO NOT USE				

*Refer to the Amiben/Amiben DS + Dual + Sencor/Lexone section of the Amiben labels for determining product use rate within rate ranges.

When using Amiben DS, multiply quarts of Amiben by 0.6 for lbs. Amiben DS/A.

**When using Sencor 4 or Lexone 4L, multiply lbs. of DF by 1.5 to get pts./A.

***On Clarion-Webster or similar alkaline (calcareous) soils with a pH of 7.5 or above, apply the DF at 0.33 lb./A on medium soils and 0.33-0.50 lb./A on fine soils.

Special Precautions: Refer to the Dual + Sencor/Lexone section of this label, the Amiben + Dual + Sencor/Lexone section of the Amiben labels and the Sencor/Lexone labels for planting details, variety limitations, and all cautions and restrictions concerning use of these products on soybeans.

5. Tank Mixture with Treflan

Dual + Treflan tank mix applied preplant incorporated controls those weeds listed under **Dual Applied Alone** and those weeds listed for Treflan alone on the Treflan E.C. label. Dual + Treflan may be applied by ground or aerial equipment and incorporated up to 14 days prior to planting. Follow the recommended procedures on the Treflan E.C. and Dual labels using equipment that provides uniform 2 inch incorporation.

Apply Dual + Treflan tank mix using the appropriate rate from the Dual alone section of this label and the Treflan alone section of the Treflan E.C. label for the specific soil texture/organic matter classification and weed species expected.

Note: Follow all restrictions and precautions on the Treflan E.C. label and on the Dual alone portion of this label.

6. Tank Mixture with Scepter

This tank mixture controls all weeds controlled by Dual alone and by Scepter alone. Refer to the **Dual Applied Alone** section for weeds controlled by Dual and to the Scepter label for weeds controlled by Scepter. Refer to the Scepter label for geographical locations where this tank mixture may be applied.

Apply Dual + Scepter preplant incorporated or preemergence using rates in the table below. Follow use directions under **Application Instructions** on the Scepter label. For preplant incorporated applications apply and incorporate within 30 days before planting. Observe all other precautions and limitations on the Scepter labels.

Table 11: DUAL + Scepter—Soybeans

	Broadcast Rates Per Acre			
	Less than 3% Organic Matter		3% or More Organic Matter	
Soil Texture	DUAL	Scepter	DUAL	Scepter
Coarse	1.25 pts.	.67 pt.	1.5 pts.	.67 pt.
Medium	1.5 pts.	.67 pt.	2.0 pts.	.67 pt.
Fine	2.0 pts.	.67 pt.	2-2.5* pts.	.67 pt.
Muck or Peat Soils (soils with more than 20% organic matter)	DO NOT USE			

*Use the higher rate of Dual if heavy weed infestations are expected.

Notes: (1) Do not apply within 90 days of harvest, and (2) Do not graze or feed treated soybean forage, hay or straw to livestock, or illegal residues may result.

7. Tank Mixture with Lorox Plus

This tank mixture controls all weeds controlled by Dual alone and by Lorox Plus alone. Refer to the **Dual Applied Alone** section for weeds controlled by Dual and to the Lorox Plus label for weeds controlled by Lorox Plus.

Apply Dual + Lorox Plus preemergence after planting but before soybeans emerge using rates in Table 12.

Note: Follow all use directions, limitations, precautions, information regarding application to soybeans and rotational restrictions on the Dual and Lorox Plus labels.

Table 12: DUAL + Lorox Plus—Soybeans

	Broadcast Rates Per Acre	
	0.5-3% Organic Matter	
Soil Texture	DUAL	Lorox Plus (60DF)
Coarse	1.25 pts.	12-14 oz.
Medium	1.5 pts.	14-16 oz.
Fine	2.0 pts.	16-18 oz.

Precaution: Do not apply to sand or to any soil with less than 0.5% organic matter or to any soil with pH greater than 6.8.

8. Tank Mixture with Gemini

This tank mixture controls all weeds controlled by Dual alone and by Gemini alone. Refer to the **Dual Applied Alone** section for weeds controlled by Dual and to the Gemini label for weeds controlled by Gemini.

Apply Dual + Gemini preemergence after planting but before soybeans emerge using rates in Table 13.

Note: Follow all use directions, limitations, precautions, information regarding application to soybeans and rotational restrictions on the Dual and Gemini labels.

Table 13: DUAL + Gemini—Soybeans

Soil Texture	Broadcast Rates Per Acre 0.5-3% Organic Matter	
	DUAL	Gemini (60DF)
Coarse (Sandy Loam Only)	1.25 pts.	12-16 oz.
Medium	1.5 pts.	16-20 oz.
Fine	2.0 pts.	20-24 oz.

Precaution: Do not apply to sand or loamy sand or to any soil with less than 0.5% organic matter or to any soil with pH greater than 7.0 except as noted on the Gemini label.

9. Tank Mixture with Canopy

This tank mixture controls all weeds controlled by Dual alone and by Canopy alone. Refer to the **Dual Applied Alone** section for weeds controlled by Dual and to the Canopy label for weeds controlled by Canopy.

Apply preplant incorporated or preemergence using the appropriate rates from Table 14. *Preplant Incorporated:* Apply within 2 weeks of planting. Uniformly incorporate into the top 1-2" of soil before planting soybeans. *Preemergence:* Apply after planting but before soybeans emerge.

Note: Follow all use directions, varietal restrictions, limitations, precautions, information regarding application to soybeans and rotational restrictions on the Dual and Canopy labels.

Table 14: DUAL + Canopy—Soybeans

Soil Texture	Broadcast Rates Per Acre 0.5%-Less Than 3% Organic Matter		3-5% Organic Matter	
	DUAL	Canopy 75DF	DUAL	Canopy 75DF
Coarse	1.25 pts.	6-10 oz.	1.5 pts.	6-10 oz.
Medium	1.5 pts.	8-12 oz.	2.0 pts.	8-12 oz.
Fine	2.0 pts.	10-14 oz.	2.0-2.5 pts.	10-14 oz.

Precaution: Do not apply to sand or to any soil with less than 0.5% organic matter or to any soil with pH greater than 7.0 except as noted on the Canopy label.

10. Tank Mixture with Preview

This tank mixture controls all weeds controlled by Dual alone and by Preview alone. Refer to the **Dual Applied Alone** section for weeds controlled by Dual and to the Preview label for weeds controlled by Preview.

Apply preplant incorporated or preemergence using the appropriate rates from Table 15. *Preplant Incorporated:* Apply within 2 weeks of planting. Uniformly incorporate in to the top 1-2"of soil before planting soybeans. *Preemergence:* Apply after planting but before soybeans emerge.

Note: Follow all use directions, varietal restrictions, limitations, precautions, information regarding application to soybeans and rotational restrictions on the Dual and Preview labels.

Table 15: DUAL + Preview—Soybeans

Soil Texture	Broadcast Rates Per Acre 0.5%-Less Than 3% Organic Matter		3-5% Organic Matter	
	DUAL	Preview 75DF	DUAL	Preview 75DF
Coarse	1.25 pts.	6 oz.	1.5 pts.	7 oz.
Medium	1.5 pts.	7 oz.	2.0 pts.	8 oz.
Fine	2.0 pts.	8 oz.	2.0-2.5 pts.	9-10 oz.

Precaution: Do not apply to sand or to any soil with less than 0.5% organic matter or to any soil with pH greater than 6.8.

11. Tank Mixture with Command*

This tank mixture controls all weeds controlled by Dual alone and by Command alone. Refer to the **Dual Applied Alone** section for weeds controlled by Dual and to the Command label for weeds controlled by Command.

Apply Dual + Command preplant incorporated using rates in Table 16. Follow all Command application instructions as to incorporation interval, geographical location, equipment operation, soil moisture conditions, etc.*

**Note:* Before making applications, read and strictly follow all use directions, limitations, precautions, information regarding application to soybeans and rotational restrictions on the Dual and Command labels.

Table 16: DUAL + Command—Soybeans

Soil Texture	Broadcast Rates Per Acre DUAL 0.5%-3% Organic Matter	DUAL Greater than 3% Organic Matter	Command 4E Northern Area	Command 4E Southern Area
Coarse	1.25 pts.	1.5 pts.	1.5-2 pts.	2-2.5 pts.
Medium	1.5 pts.	2.0 pts.	1.5-2 pts.	2-2.5 pts.
Fine	2.0 pts.	2.0-2.5 pts.	1.5-2 pts.	2-2.5 pts.

12. Tank Mixture with Sonalan

This tank mixture controls all weeds controlled by Dual alone and by Sonalan alone. Refer to the **Dual Applied Alone** section for weeds controlled by Dual and to the Sonalan label for weeds controlled by Sonalan.

Apply Dual and Sonalan preplant incorporated using the appropriate rates from Table 17.

Preplant Incorporated: Follow recommended soil preparation procedures for Sonalan. Refer to the **Sonalan/Dual Tank Mixture** label for incorporation specifications.

Sequential: Apply Sonalan alone preplant incorporated as specified on the Sonalan label. Follow with a preemergence application of Dual during planting (behind the planter) or after planting before weeds or soybeans emerge.

Table 17: DUAL + Sonalan—Soybeans

Soil Texture	Broadcast Rates Per Acre Less Than 3% Organic Matter		3% or More Organic Matter	
	DUAL	Sonalan	DUAL	Sonalan
Coarse	1.5-2 pts.	1.25-2 pts.	2 pts.	1.25-2 pts.
Medium*	2-2.5 pts.	1.75-2.5 pts.	2-2.5 pts.	1.75-2.5 pts.
Fine*	2-2.5 pts.	2.25-3 pts.	2.5-3 pts.	2.25-3 pts.
Muck or Peat Soils (soils with more than 20% organic matter)	DO NOT USE			

*For eastern black nightshade on these soils, apply Sonalan at 3 pts./A on medium and 3.5 pts./A on fine-textured soils and follow with two incorporation passes.

Note: Follow all use directions, limitations, precautions, and information regarding application to soybeans on the Dual and Sonalan labels.

13. Tank Mixture with Pursuit

This tank mixture controls all weeds controlled by Dual alone and by Pursuit alone. Refer to the **Dual Applied Alone** section for weeds controlled by Dual and to the Pursuit label for weeds controlled by Pursuit. Refer to the Pursuit label for geographical locations where this tank mixture may be applied.

Apply Dual + Pursuit early preplant, preplant incorporated, or preemergence after planting using rates in Table 18. Application can be made in water or liquid fertilizer. Follow all use directions under **Soil Applications** on the Pursuit label. For early preplant and preplant incorporated applications, apply within 30 days before planting.

Note: Follow all use directions, limitations, precautions, and information regarding application to soybeans and rotational restrictions on the Dual and Pursuit labels.

Table 18: DUAL + Pursuit—Soybeans

Soil Texture	Broadcast Rates Per Acre Less than 3% Organic Matter	3% or More Organic Matter	
	DUAL	DUAL	Pursuit
Coarse	1.25 pts.	1.5 pts.	0.25 pts.
Medium	1.5 pts.	2.0 pts.	0.25 pts.
Fine	2.0 pts.	2.0-2.5 pts.	0.25 pts.

14. Tank Mixture with Lexone, Sencor, Scepter, Lorox, Lorox Plus, Gemini, Canopy, Preview or Pursuit plus Gramoxone Super or Roundup for Minimum-Tillage or No-Tillage Systems

In minimum-tillage or no-tillage systems where soybeans are planted directly into a cover crop, stale seedbed, established sod, or previous crop residues, the contact herbicides Gramoxone Super or Roundup may be added to a tank mix of either Dual + Sencor or Lexone, Dual + Scepter, Dual + Lorox, Dual + Lorox Plus, Dual + Gemini, Dual + Canopy, Dual + Preview or Dual + Pursuit. When used as directed, the Gramoxone Super portion of the tank mixture controls most emerged weeds and suppresses many perennial weeds. Roundup combinations will control emerged annual and perennial weeds when applied as directed on the Roundup label. The Dual + Sencor/Lexone, Scepter, Lorox, Lorox Plus, Gemini, Canopy, Preview, or Pursuit portion or the tank mixture provides preemergence control of the weeds listed on this label in the tank mixture section for Dual + Sencor/Lexone, Dual + Scepter, Dual + Lorox, Dual + Lorox Plus, Dual + Gemini, Dual + Canopy, Dual + Preview and Dual + Pursuit, respectively.

Refer to the label of each product used in combination and observe the planting details, soybean variety restrictions, information regarding application to soybeans, geographical restrictions and all other precautions and limitations.

Refer below for rates of Gramoxone Super or Roundup, Dual + Sencor/Lexone, Dual + Scepter, Dual + Lorox, Dual + Lorox Plus, Dual + Gemini, Dual + Canopy, Dual + Preview, and Dual + Pursuit, respectively.

Application: Apply before, during, or after planting, but before the soybeans emerge, at the rates specified below. Add Gramoxone Super or Roundup at the following broadcast rates:

Gramoxone Super: 1.5-2.5 pts./A plus 8 fl. oz. of X-77 Spreader per 100 gals. of spray mixture. Use the lower rate for control of annual weeds less than 4 inches tall and the higher rate for weeds 4-6 inches tall. This treatment will not consistently control weeds taller than 6 inches.

Note: Do not apply combinations containing Gramoxone Super in suspension type liquid fertilizers as the activity of paraquat will be reduced.

Roundup: 1.5 qts./A for existing annual weeds, or 2-4 qts./A for existing perennial weeds. See the Roundup label for weeds controlled and recommended rates for specific weeds.

Apply in 20-60 gals. of water or fluid fertilizer per acre with ground equipment.

Dual + Lexone/Sencor + Gramoxone Super or Roundup

On *loamy sand* with over 2% organic matter, apply 1.5 pts./A of Dual + 0.33-0.5 lb./A of Sencor or Lexone DF. On *medium soils*, apply 2 pts./A Dual + 0.5-0.67 lb./A of Sencor or Lexone DF. On *fine soils*, apply 2-2.5 pts./A of Dual + 0.67 lb./A of Sencor or Lexone DF.

*When using Sencor 4 or Lexone 4L, multiply lbs. of DF by 1.5 to get pts./A.

Precautions: To avoid crop injury, (1) Do not use this tank mixture on soil with less than 0.5% organic matter, on alkaline soil with a pH over 7.4, or on all sand and on loamy sand with less than 2% organic matter. (2) If heavy rain occurs soon after application, crop injury may result, especially in poorly drained areas where water stands for several days or where the seeding slit has not been properly closed.

Dual + Scepter + Gramoxone Super or Roundup

On *coarse soils*, apply 1.5 pts./A of Dual + .67 pt./A of Scepter. On *medium soils*, apply 2 pts./A of Dual + .67 pt./A of Scepter. On *fine soils*, apply 2.5 pts./A of Dual + .67 pt./A of Scepter.

Notes: (1) Do not apply within 90 days of harvest, and (2) Do not graze or feed treated soybean forage, hay, or straw to livestock, or illegal residues may result.

Dual + Lorox + Gramoxone Super or Roundup

On *coarse soils**, apply 1.5 pts./A of Dual + 1-1.5 lbs./A of Lorox DF**. On *medium soils*, apply 2 pts./A of Dual + 1-2 lbs../A of Lorox DF. On *fine soils*, apply 2-2.5 pts./A of Dual + 2-3 lbs./A of Lorox DF.

*Do not use on loamy sand, except in the northeastern U.S. on loamy sand with over 1% organic matter, or injury may occur. Do not use on sand, gravelly soils, or exposed subsoils, or injury may occur.

**When using Lorox or Lorox DF, use equivalent rates. One pt. of Lorox L equals one lb. of Lorox DF.

Precaution: Do not use on soil with less than 0.5% organic matter or crop injury may ocur.

Dual + Lorox Plus + Gramoxone Super or Roundup

Use only where soils have 0.5-3% organic matter. On *coarse soils*, apply 1.5 pts./A of Dual + 12-14 oz./A of Lorox Plus 60DF. On *medium soils*, apply 2 pts./A of Dual + 14-16 oz./A of Lorox Plus. On *fine soils*, apply 2-2.5 pts./A of Dual + 16-18 oz./A of Lorox Plus.

Precaution: Do not apply to sand or to any soil with pH greater than 6.8.

Dual + Gemini + Gramoxone Super or Roundup

Use only where soils have 0.5-3% organic matter. On *coarse soils* (sandy loam only) apply 1.5 pts./A of Dual + 12-16 oz./A of Gemini 60DF. On *medium soils*, apply 2 pts./A of Dual + 16-20 oz./A of Gemini. On *fine soils*, apply 2-2.5 pts./A of Dual + 20-24 oz./A of Gemini.

Precaution: Do not apply to sand or loamy sand or to any soil with less than 0.5% organic matter or to any soil with pH greater than 7.0 except as noted on the Gemini label.

Dual + Canopy + Gramoxone Super or Roundup

Use only where soils have 0.5-5% organic matter. On *coarse soils* (except sand), apply 1.5 pts./A of Dual + 6-10 oz./A of Canopy 75DF. On *medium soils*, apply 2 pts./A of Dual + 8-12 oz./A of Canopy. On *fine soils*, apply 2-2.5 pts./A of Dual + 10-14 oz./A of Canopy.

Precaution: Do not apply to sand or to any soil with less than 0.5% organic matter or to any soil with pH greater than 7.0 exept as noted on the Canopy label.

Dual + Preview + Gramoxone Super or Roundup

Use only where soils have 0.5-5% organic matter. On *coarse soils* (except sand), apply 1.5 pts./A of Dual + 6*-7 oz./A of Preview 75DF. On *medium soils*, apply 2 pts./A of Dual + 7*-8 oz./A of Preview. On *fine soils*, apply 2-2.5 pts./A of Dual + 8*to 9-10 oz./A of Preview.

*Use this rate where the soil organic matter is in the 0.5 to less than 3.0% range.

Precaution: Do not apply to sand or to any soil with less than 0.5% or to any soil with pH greater than 6.8.

Dual + Pursuit + Gramoxone Super or Roundup

On *coarse soils*, apply 1.5 pts./A of Dual + 0.25 pt./A of Pursuit. On *medium soils*, apply 2 pts./A of Dual + 0.25 pt./A of Pursuit. On *fine soils*, apply 2.5 pts./A of Dual + 0.25 pt./A of Pursuit.

T. Storage and Disposal

Pesticide Disposal

Do not contaminate water, food or feed by storage or disposal. Open dumping is prohibited. Wastes resulting from the use of this product are toxic. Improper disposal of unused pesticide, spray mixture, or rinsate is a violation of federal law. Pesticide, spray mixture, or rinsate that cannot be used according to label instructions must be disposed of according to federal, state or local procedures. For guidance in proper disposal methods, contact your State Pesticide or Environmental Control Agency, or the Hazardous Waste representative at the nearest EPA Regional Office.

Container Disposal

Do not reuse empty container. Triple rinse (or equivalent), puncture and dispose of in a sanitary landfill, or by incineration, or by open burning, if allowed by state and local authorities. Keep out of smoke from burning containers.

This product may be stored at temperatures down to 30 degrees below 0°F.

For minor spills, leaks, etc., follow all precautions indicated on this label and clean up immediately. Take special care to avoid contamination of equipment and facilities during cleanup procedures and disposal of wastes. In the event of a major spill, fire or other emergency, call 1-800-888-8372 day or night.

U. Precautionary Statements

HAZARDS TO HUMANS AND DOMESTIC ANIMALS

WARNING

Causes substantial but temporary eye injury. Do not get in eyes, on skin, or on clothing. Harmful if swallowed, inhaled, or absorbed through skin. Avoid breathing spray mist. Wash thoroughly with soap and water after handling. The active ingredient, metolachlor, may cause skin sensitization reactions in certain individuals. Wear protective clothing (coveralls, gloves and safety glasses) when handling or using this product. Remove contaminated clothing and wash before reuse.

STATEMENT OF PRACTICAL TREATMENT

If in eyes: Hold eyelids open and flush with a steady, gentle stream of water for 15 minutes.

If swallowed: Drink promptly a large quantity of milk, egg whites, gelation solution, or if these are not available, large quantities of water. Avoid alcohol. Call a physician or Poison Control Center. Induce vomiting by touching back of throat with finger, or if available, by administering syrup of ipecac. Do not induce vomiting or give anything by mouth to an unconscious person.

If on skin: Wash with plenty of soap and water. Get medical attention if irritation persists.

If inhaled: Remove victim to fresh air.

Note to Physician: If swallowed, there is no specific antidote. Induce emesis and lavage stomach. Treat symptomatically! The use of an aqueous slurry of activated charcoal (such as Norit A) and a saline cathartic should be considered.

ENVIRONMENTAL HAZARDS

Do not apply directly to water or wetlands (swamps, bogs, marshes, and potholes). Do not contaminate water when disposing of equipment wash waters.

Ground Water and Surface Water Advisory

Metolachlor has been identified in limited sampling of ground water and there is the possibility that it may leach through the soil to ground water, especially where soils are coarse and ground water is near the surface. Following application and during rainfall events that cause runoff, metolachlor may reach surface water bodies including streams, rivers and reservoirs.

Care must be taken when using this product to prevent back siphoning into wells, spills or improper disposal of excess pesticide, spray mixtures or rinsates.

Check valves or antisiphoning devices must be used on all mixing and/or irrigation equipment.

Chemigation

Refer to supplemental labeling in attached booklet for use directions for chemigation. Do not apply this product through any irrigation system unless the supplemental labeling on chemigation is followed.

AAtrex®, Bicep®, Caliber®, Caparol®, Concep®, Cotoran®, Dual®, Evik®, Nine-O®, and Princep® are trademarks of CIBA-GEIGY Corporation

Agsorb™ trademark of Oil-Dri Corporation

Amiben®, Brominal®, and Buctril® are trademarks of Rhone-Poulenc Chemical Company

Balan®, Sonalan®, and Treflan® are trademarks of Elanco Products Company
Banvel® trademark of Sandoz Chemical Corporation
Basagran® trademark of BASF Company
Bladex®, Canopy®, Gemini®, Lexone®, Lorox®, and Preview® are trademarks of E.I. duPont de Nemours and Company
Command® trademark of FMC Corporation
Compex® and X-77® are trademarks of KALO Agricultural Chemicals, Inc.
Eptam®, Eradicane®, Sutan®, Vernam®, and Gramoxone® are trademarks of ICI Americas
KERMAC #600™ trademark of Kerr McGee Refining Corporation
Prowl®, Pursuit®, and Scepter® are trademarks of American Cyanamid
Roundup® trademark of Monsanto Company
Sencor® trademark of the Parent Company of Farbenfabriken Bayer GmbH, Leverkusen
Unite® trademark of Hopkins Agricultural Chemical Company

CGA 49L57 069

This is a specimen label which was not prepared by CIBA-GEIGY. CIBA-GEIGY is not responsible for the accuracy of the information contained herein. As labels are subject to revision, always carefully read and follow the label on the product container.

DUAL® 25G

GRANULAR HERBICIDE

For weed control in corn, peanuts, potatoes, and soybeans

ACTIVE INGREDIENT:
Metolachlor: 2-chloro-*N*-(2-ethyl-6-methyl-phenyl)-*N*-(2-methoxy-1-methylethyl) acetamide 25.0%
INERT INGREDIENTS: 75.0%
TOTAL: 100.0%

EPA Reg. No. 100-638 EPA Est. 32761-MO-3

DUAL 25G is a granular herbicide

KEEP OUT OF REACH OF CHILDREN

CAUTION

See additional precautionary statements on back of bag.

See directions for use on back of bag.

DIRECTIONS FOR USE AND CONDITIONS OF SALE AND WARRANTY

IMPORTANT: Read the entire **Directions for Use** and the **Conditions of Sale and Warranty** before using this product.

CONDITIONS OF SALE AND WARRANTY

The **Directions for Use** of this product reflect the opinion of experts based on field use and tests. The directions are believed to be reliable and should be followed carefully. However, it is impossible to eliminate all risks inherently associated with use of this product. Crop injury, ineffectiveness, or other unintended consequences may result because of such factors as weather conditions, presence of other materials, or the manner of use or application all of which are beyond the control of CIBA–GEIGY or the Seller. All such risks shall be assumed by the Buyer.

CIBA–GEIGY warrants that this product conforms to the chemical description on the label and is reasonably fit for the purposes referred to in the **Directions for Use** subject to the inherent risks referred to above. **CIBA–GEIGY makes no other express or implied warranty of Fitness or Merchantability or any other express or implied warranty. In no case shall CIBA–GIEGY or the Seller be liable for consequential, special, or indirect damages resulting from the use or handling of this product.** CIBA–GEIGY and the Seller offer this product, and the Buyer and user accept it, subject to the foregoing **Conditions of Sale and Warranty,** which may be varied only by agreement in writing signed by a duly authorized representative of CIBA–GEIGY.

DIRECTIONS FOR USE

It is a violation of federal law to use this product in a manner inconsistent with its labeling.

No user shall permit the application of this product in such a manner as to directly or indirectly expose workers or other persons except those knowingly involved in the application. The area being treated must be vacated by unprotected persons.

REENTRY STATEMENT

Do not enter treated areas without protective clothing until dust has settled.

Because certain states may require more restrictive reentry intervals for various crops treated with this product, consult your State Department of Agriculture for further information.

Written or oral warnings must be given to workers who are expected to be in a treated area or in an area about to be treated with this product. Oral warnings must be given which inform workers of areas or fields that may not be entered without specific protective clothing until dust has settled, and appropriate actions to take as described under **Precautionary Statements** at the end of this label. When oral warnings are given, warnings shall be given in a language customarily understood by workers. Oral warnings must be given if there is reason to believe that written warnings cannot be understood by workers. Written warnings must include the following information: "CAUTION. Area treated with DUAL 25G on (date of application). Do not enter without appropriate protective clothing until dust has settled. In case of accidental exposure, flush eyes or skin with plenty of water. Call a physician if irritation persists. Remove and wash contaminated clothing before reuse."

FAILURE TO FOLLOW THE DIRECTIONS FOR USE AND PRECAUTIONS ON THIS LABEL MAY RESULT IN POOR WEED CONTROL, CROP INJURY, AND/OR ILLEGAL RESIDUES.

GENERAL INFORMATION

DUAL 25G will control many annual grasses and certain broadleaf weeds in corn, peanuts, potatoes, and soybeans. It should be applied prior to the emergence of weeds. It may be preplant surface-applied (corn and soybeans only), or incorporated into the surface 1-2 inches of soil prior to planting or left on the soil surface during or after planting. Dry weather following application of DUAL 25G may reduce its effectiveness. Shallow cultivate if weeds develop or if tillage is necessary to improve the physical condition of the soil.

Precaution: Injury may occur following the use of DUAL 25G under abnormally high moisture conditions during early development of the crop.

Weeds Controlled

barnyardgrass (watergrass)	green foxtail	yellow foxtail
crabgrass	prairie cupgrass	yellow nutsedge
crowfootgrass	red rice	
fall panicum	signalgrass *(Brachiaria)*	black nightshade
foxtail millet	southwestern cupgrass	carpetweed
giant foxtail	witchgrass	Florida pusley
goosegrass		galinsoga
		pigweed

Weeds Partially Controlled*: common purslane, Florida beggarweed, hairy nightshade, sandbur, seedling johnsongrass, shattercane and volunteer sorghum.

* Control of these weeds can be erratic due partially to variable weather conditions. Control may be improved by following these suggested procedures:

1. **Thoroughly till moist soil** to destroy germinating and emerged weeds. If DUAL 25G is to be applied preplant incorporated, this tillage may be used to incorporate DUAL 25G if uniform 2 inch incorporation is achieved as recommended under **Application Procedures.**
2. Plant crop into moist soil **immediately after tillage.** If DUAL 25G is to be used preemergence, apply at planting or immediately after planting.
3. If available, **sprinkler irrigate** within 2 days after application. Apply ½-1 inch of water. Use lower water volume (½ inch) on coarse-textured soils and higher volume (1 inch) on finer textured soils.
4. If irrigation is not possible and rain does not occur within 2 days after planting and application, weed control may be decreased. Under these conditions, a uniform, shallow cultivation is recommended as soon as weeds emerge.

Application Procedures

DUAL 25G may be applied either broadcast or banded.

Broadcast Treatment—DUAL 25G may be applied with a broadcast type granular applicator or with a cyclone type spreader. Use applicators and techniques that will apply the granules uniformly to the soil.

Band Treatment—DUAL 25G may be applied with any standard granular applicator capable of uniformly distributing a predetermined amount of granules to a desired area. Band treatment made at seeding should only be applied behind the press wheel.

Note: Caution should be taken with seeding equipment so that any track (soil depression) made by the planter press wheel is minimal. This will reduce concentration of granules in the press wheel track by heavy rain which can result in crop injury. Soil tillage should not follow the use of band applications unless it is adapted so as not to displace the band and hence reduce weed control in the band.

Applicator settings are given for band applications in Chart 1. These settings were determined under laboratory conditions to $\pm 5\%$ accuracy. Note that the settings may differ depending on speed of application, as well as rate per acre. **It is important that the user calibrate each individual granular applicator prior to its use. Since the actual setting for each individual granular applicator may vary, the settings shown should serve only as guidelines for calibration.**

Calculate the amount of DUAL 25G needed for band treatment as follows:

$$\frac{\text{Band width in inches}}{\text{Row spacing in inches}} \times \text{Broadcast rate per acre} = \text{Band rate per acre}$$

Example: If 16 lbs. per acre is recommended for broadcast application and you wish to treat a band 14 inches wide in crop rows spaced 30 inches apart:

$$\frac{14}{30} \times 16 = \text{7.5 lbs. of granules needed for each acre that is to be treated.}$$

Do not apply DUAL 25G during periods of high winds. Wind may disrupt placement of the granules and reduce weed control. Locate granular diffusers close to the ground to reduce granular displacement. Wind shield attachments are advisable.

Chart 1: Applicator Settings For 14-inch Band Width*

Applicator Brand	Speed MPH	Equivalent Broadcast Rate Per Acre: 6 lbs.	8 lbs.	10 lbs.	12 lbs.	16 lbs.
Gandy Two-Opening (For 14-inch band) Model No. S-902	4	8.5	10	11	12	14
	5	9.5	11	12	13	15.5
	6	10.5	12	13	15	17
International Harvester 1980 & Before Allis Chalmers 1979 & After	4	.80	1.0	1.1	1.2	1.35
	5	.92	1.1	1.15	1.38	1.6
	6	1.0	1.13	1.25	1.45	1.9
International Harvester 1981 & After	4	1.65	1.75	1.85	1.9	2.1
	5	1.75	1.85	1.95	2.0	2.25
	6	1.85	1.95	2.0	2.1	2.45
John Deere (Metal Box)	4	1/19	1/21	1/23	1/25	1/29
	5	1/21	1/23	1/25	1/28	2/1
	6	1/22	1/24	1/27	1/30	2/5
John Deere 1978 & Before Meter Scale 0.. 10..20..etc.	4	7	8	10	11	14
	5	8	9	11	13	17
	6	9	10	12	14	18
John Deere 1979 & Later Meter Scale 0..5..15..etc.	4	5	6	7	9	12
	5	6	7	9	10	13
	6	7	8	10	11	15
Noble 1972 & Before	4	4	5	6	8	10
	5	5	6	8	9	12
	6	6	7	9	11	13
Noble 1973 & Later	4	3	4	5	6	7
	5	4	5	6	7	9
	6	5	6	7	8	10

* Settings are for application of the equivalent broadcast rates per acre indicated in a 14-inch band at speed (miles per hour) shown. If other than a 14-inch band is to be used, these settings do not apply and additional calibration will be necessary.

Calibration of Band Applicators

To calibrate a granular applicator for a standard 14-inch band, a sample has to be collected while the applicator is either running or simulating a run over a distance of 500 ft.

Procedure:

1. Fill applicator box with DUAL 25G.
2. Refer to Chart 1 and adjust the setting on your applicator to correspond to the desired rate per acre.
3. Attach a plastic bag or similar container to the applicator spout after detaching the delivery tube.
4. Make the 500 ft. run at planting speed.
   Speed/Time required to cover 500 feet:
   4 mph = 1 min. 25 sec.
   5 mph = 1 min. 8 sec.
   6 mph = 57 sec.
5. Weigh granules collected (or measure volume) and, if necessary, adjust and make successive runs until the setting delivers the correct number of ounces per 500 ft. of row, as shown in Chart 2.
6. Repeat this procedure for each granular applicator box.

Chart 2: The Equivalent Broadcast Rate Per Acre in Volume and Weight for 500 Feet on the Basis of a 14-inch Band

Broadcast Rate Per Acre	Volume Collected In 14-inch Band 500 Linear Feet	Weight Collected In 14-inch Band 500 Linear Feet
6 lbs.	1.6 fl. oz.	1.3 oz.
8 lbs.	2.2 fl. oz.	1.7 oz.
10 lbs.	2.8 fl. oz.	2.2 oz.
12 lbs.	3.2 fl. oz.	2.6 oz.
16 lbs.	4.5 fl. oz.	3.5 oz.

For band widths other than 14 inches, the calibration distance will differ from 500 ft.

For other band widths, calculate the relative calibration distance in which to collect the corresponding weight or volume of DUAL 25G in Chart 2 as follows:

$$\frac{\text{14 (Inches)}}{\text{Desired band width (Inches)}} \times 500 = \text{Calibration distance in feet}$$

Example: A band width of 10 inches is to be used, instead of 14 inches, to apply DUAL 25G at an equivalent rate of 10 lbs./broadcast acre.

A. $$\frac{\text{14 Inches}}{\text{10 Inches}} \times \text{500 ft.} = \text{700 ft.}$$

B. Collect 2.8 fl. oz. by volume or 2.15 oz. by weight for the newly calculated calibration distance of 700 ft. for a 10-inch band.

CORN

For control of annual grasses and broadleaf weeds listed under the **General Information** section, apply DUAL 25G either preplant surface, preplant incorporated or preemergence to corn before weeds emerge. For extended control of annual weeds, DUAL 25G may be applied to weed free soil at the time of the last cultivation (lay-by application) either as a surface or shallow incorporated treatment.

Conventional Tillage

Preplant Incorporated: Broadcast DUAL 25G to the soil and incorporate within the surface 1-2 inches of soil using the appropriate rate from Table 1 within 14 days before planting using appropriate surface incorporating equipment such as a finishing disk, harrow, rolling cultivator, or similar implement capable of providing uniform incorporation.

Preemergence: Band or broadcast DUAL 25G to the soil surface during or after planting and before weeds emerge using the appropriate rate from Table 1. For band treatment, apply granules behind the press wheel.

Lay-by (Last Cultivation) Application: Apply the appropriate rate of DUAL 25G from Table 1 on a broadcast basis at lay-by. Applications should be made before the corn plants exceed 40 inches in height. The soil should be free of weeds at the time of application, and the granules either left on the surface or shallowly incorporated into the soil. Use an application method that will not allow granules to get into the corn whorl.

This application of DUAL 25G may follow any preplant or preemergence herbicide application including DUAL 8E or DUAL 25G.

Table 1: DUAL 25G—Corn**

Soil texture	Broadcast rate per acre: Less than 3% organic matter	3-6%* organic matter
COARSE: Sand, loamy sand, sandy loam	6-8 lbs.	8 lbs.
MEDIUM: Loam, silt loam, silt	8-10 lbs.	8-10 lbs.
FINE: Silty clay loam, sandy clay loam, silty clay, sandy clay, clay loam, clay	8-10 lbs.	10-12 lbs.
Muck or peat soils	DO NOT USE	

* DUAL 25G may be used up to 16 lbs. per acre, broadcast basis, on soils having an organic matter content greater than 6%.

** DUAL 25G may be used sequentially in place of DUAL 8E with Bladex® or Banvel® applied at the tank mix rate specified on the respective product labels.

Reduced-Tillage and No Tillage

Preplant Surface-Applied: Broadcast DUAL 25G up to 45 days before planting corn on medium- and fine-textured soils in CO, IL, IN, IA, KS, KY, MN, MO, MT, NE, ND, SD, TN, WI, and WY. Apply ⅔ the recommended broadcast rate of DUAL 25G from Table 1 as a split treatment 30-45 days before planting and the remainder immediately before planting, or at planting. Applications less than 30 days before planting may be made either as a split or as a single treatment. Apply 8 lbs./A on coarse soils not more than 2 weeks before planting. If weeds are present at the time of treatment, apply a labeled contact herbicide (for example, Gramoxone® or Roundup®). Observe the directions for use, precautions, and restrictions on the label of the contact herbicide. To the extent possible, do not move treated soil out of the row or move untreated soil to the surface during planting, or weed control will be diminished. To avoid reduced weed control, do not apply to frozen soil as rainfall may change the pattern of the DUAL granules. Injury may occur from movement of DUAL 25G into the planter slit if the slit is not properly closed during planting.

On medium- and fine-textured soils with minimum- or no-tillage systems, in DE, MD, MI, NY, OH, PA, VA, and WV, preplant surface applications may be applied following the directions for use above. If the amount of rainfall results in unsatisfactory length of weed control following the earlier treatment, a postemergence application of an appropriately labeled broadleaf and/or grass weed herbicide could be used, i.e., AAtrex, 2,4-D, Banvel, Basagran®, bromoxynil (Brominal® or Buctril®). Observe all directions for use, precautions and limitations on the label of the postemergence herbicide.

Preemergence: Broadcast DUAL 25G at the appropriate rate from Table 1 during or after planting, before weeds emerge, or after weeds present have been killed with a labeled contact. Injury may occur from movement of DUAL 25G into the planter slit if the slit is not properly closed during planting.

Note: Do not apply more than the equivalent of 25 lbs. of DUAL 25G/A during any one year, or illegal residues may result. This includes both DUAL 25G and DUAL 8E (1 pt. of DUAL 8E equals 4 lbs. of DUAL 25G).

PEANUTS

DUAL 25G Single Applications—All Peanut Growing Areas

Apply DUAL 25G either preplant incorporated, postplant incorporated or preemergence or lay-by, using the appropriate rate specified below. **Preplant Incorporated:** Broadcast DUAL 25G to the soil surface and incorporate into the top 1-2 inches of soil before planting using a finishing disk, harrow, rolling cultivator, or similar implement capable of providing uniform 1-2 inch incorporation. Use a preplant incorporated application if furrow irrigation is to be used, or when a period of dry weather after application is expected. If crop will be planted on beds, apply and incorporate DUAL 25G after bed formation. **Postplant Incorporated:** Broadcast or band DUAL 25G to the soil surface and shallowly incorporate into the soil after planting, but before peanut germination. Incorporation depth and incorporating implements must be kept above the seed or seed will be damaged. **Preemergence:** Broadcast or band DUAL 25G to the soil surface during planting (behind the planter) or after planting, but before weeds or crop emerge. **Lay-by:** Broadcast or band DUAL 25G to the soil surface after the last normal cultivation. Do not apply when peanuts are wet with dew or rain.

Apply DUAL 25G preplant incorporated, postplant incorporated, preemergence or lay-by at a broadcast rate of 6-12 lbs. per acre in the Southeast* or 5-8 lbs. per acre in OK, TX and NM.

* In the southeast, use 8-12 lbs. per acre and apply preemergence for partial control of Florida beggarweed.

Note: DUAL 25G may be applied as directed after any of the following preplant incorporated herbicides when used according to their label recommendations: Balan® at 3-4 qts. per acre; Treflan® at 1 pt. per acre; or Vernam® at 2.3-3 pts. per acre.

DUAL 25G Multiple Applications—Southeast Only (AL, FL, GA, NC, SC or VA):

Where weed pressure is heavy or where species difficult to control are expected, DUAL 25G is most effective when used in multiple applications.

At application timings listed below, apply DUAL 25G at a broadcast rate of 6-12 lbs. per acre.

1st Application: Apply DUAL 25G preplant incorporated as directed under **DUAL 25G Single Applications.** DUAL 8E may be substituted as directed on that label.

2nd Application: Broadcast or band DUAL 25G anytime from preemergence up to "ground cracking" for control of weeds not yet emerged. Refer to **General Information** for a list of weeds controlled.

3rd Application: Apply DUAL 25G lay-by as directed under **DUAL 25G Single Applications.** Use only when late germinating weeds are expected to be a problem.

Notes: (1) Do not apply more than the equivalent of 24 lbs. of DUAL 25G per acre during any one year or illegal residues may result. This includes both DUAL 25G and DUAL 8E (1 pts. of DUAL 8E equals 4 lbs. of DUAL 25G). (2) Do not graze or feed peanut forage or fodder to livestock for 30 days following application and (3) Do not apply within 90 days of harvest, or illegal residues may result.

SOYBEANS

For control of annual grasses and broadleaf weeds in the General Information section, apply DUAL 25G either preplant surface, preplant incorporated, or preemergence to soybeans before weeds emerge.

Conventional Tillage

Preplant Incorporated: Broadcast DUAL 25G to the soil surface and incorporate in the top 2 inches of soil within 14 days before planting, using a finishing disk, harrow, or other appropriate tillage equipment at the appropriate rate from Table 2.

Preemergence: Band or broadcast DUAL 25G to the soil surface during or after planting and before weeds emerge using the appropriate rate from Table 2. For band treatment, apply granules behind the press wheel.

Table 2: DUAL 25G—Soybeans*

Soil texture	Broadcast rate per acre: Less than 3% organic matter	Broadcast rate per acre: 3-6%** organic matter
COARSE: Sand, loamy sand, sandy loam	6-8 lbs.	8 lbs.
MEDIUM: Loam, silt loam, silt	8-10 lbs.	8-10 lbs.
FINE: Silty clay loam, sandy clay loam, silty clay, sandy clay, clay loam, clay	8-10 lbs.	10-12 lbs.
Muck or peat soils	DO NOT USE	

* DUAL 25G may be used sequentially in place of DUAL 8E with Sencor®/Lexone®, Lorox®, or Amiben® applied at the tank mix rate specified on the respective product labels.

** DUAL 25G may be used up to 16 lbs./A on soils having a high organic matter content (greater than 6%).

Reduced-Tillage and No Tillage

Preplant Surface-Applied: Broadcast DUAL 25G up to 45 days before planting soybeans on medium- and fine-textured soils in CO, IL, IN, IA, KS, KY, MN, MO, MT, NE, ND, SD, TN, WI, and WY. Apply 2/3 the recommended broadcast rate of DUAL 25G as a split treatment 30-45 days before planting and the remainder immediately before planting, or at planting. Applications less than 30 days before planting may be made either as a split or as a single treatment. Apply 8 lbs./A on coarse soils not more than 2 weeks before planting. If weeds are present at the time of treatment, apply a labeled contact herbicide (for example, Gramoxone or Roundup). Observe the directions for use, precautions, and restrictions on the label of the contact herbicide. To the extent possible, do not move treated soil out of the row or move untreated soil to the surface during planting, or weed control will be diminished. To avoid reduced weed control, do not apply to frozen soil as rainfall may change the pattern of the DUAL granules. Injury may occur from movement of DUAL 25G into the planter slit if the slit is not properly closed during planting.

Preemergence: Broadcast DUAL 25G at the appropriate rate from Table 2 during or after planting, before weeds emerge, or after weeds present have been killed with a labeled contact. Injury may occur from movement of DUAL 25G into the planter slit if the slit is not properly closed during planting.

POTATOES

Apply DUAL 25G either incorporated, preemergence, or after hilling/lay-by according to the directions specified below for control of weeds listed under the General Information section. Within a rate range, use the lower rate on soils relatively coarse-textured or low in organic matter; use the higher rate on soils relatively fine-textured or high in organic matter. Effectiveness will be reduced if later cultural practices expose untreated soil.

Incorporated: Apply DUAL 25G at 6-12 lbs. per acre to the soil and incorporate into the surface of the soil 1-2 inches using appropriate tillage equipment. Planting and later cultural practices should not bring untreated soil to the surface. Postplant application may be made any time after planting to drag-off, but before potato emergence. Incorporate using an implement that distributes DUAL within the top 1-2 inches of soil. Do not damage potato seed pieces or sprouts with incorporation equipment.

Preemergence: Apply DUAL 25G at 6-12 lbs. per acre either after planting as a preemergence, delayed preemergence, after drag-off or hilling treatment, but before weeds emerge. Up to 16 lbs. per acre of DUAL 25G alone may be used where soil organic matter is greater than 6%.

After Hilling/Lay-by: Apply 10 lbs. per acre of DUAL 25G after hilling/at lay-by to control DUAL sensitive weeds for remainder of the growing season. This application will not control emerged weeds. It may be applied over a previous DUAL 25G application, but do not apply more than 22 lbs. of DUAL 25G per acre in a single crop season.

Precautions: (1) Do not use on muck or peat soils. (2) If cool, wet soil conditions occur after application, DUAL may delay maturity and/or reduce yield of Superior and other early maturing potato varieties. (3) Do not use on sweet potatoes or yams. (4) Do not apply both preemergence and incorporated treatments.

Notes: (1) To avoid illegal residues, potatoes treated with DUAL should not be harvested within 60 days after a planting to drag-off application, or within 40 days after a lay-by application. (2) Do not use in Kern County, California.

Rotational Crops

(1) If crop treated with DUAL 25G is lost any crop on this label may be replanted immediately. Do not make a second broadcast application of DUAL 25G. If the original application was banded and the second crop is planted in the untreated row middles, a second banded treatment may be applied. (2) Barley, oats, rye, or wheat may be planted 4½ months following application; alfalfa may be planted 4 months following application. (3) Any crop on this label in addition to cotton, pod crops, safflower, grain sorghum*, stone fruits, tree nuts, root crops, barley, buckwheat, milo,

oats, rice, rye and wheat may be planted in the spring following treatment. Do not graze or feed forage or fodder from cotton to livestock. All other rotational crops may be planted 18 months after application. (4) Any crop on this label in addition to cotton, pod crops, safflower, sorghum, stone fruits and tree nuts may be planted in the spring following a lay-by treatment or multiple treatments applied the previous season. All other rotational crops may be planted 18 months after a lay-by application.

* Provided the seed has been properly treated with Concep® seed protectant.

Important Notes: To avoid injury to rotational alfalfa or clover, (1) Do not apply more than 3 lbs. ai of metolachlor (12 lbs. Dual 25G) to corn, peanuts, potatoes, or soybeans either preplant surface, preplant incorporated, preemergence or postplant incorporated and (2) Do not make a lay-by or after-hilling application of Dual 25G.

STORAGE AND DISPOSAL

Pesticide Storage and Disposal

Store in a dry place. Do not contaminate water, food, or feed by storage, disposal, or cleaning of equipment. Open dumping is prohibited. Improper disposal of unused pesticide, spray mixture, or rinsate is a violation of federal law. If these wastes cannot be used according to label instructions, contact your State Pesticide or Environmental Control Agency, or the Hazardous Waste representative at the nearest EPA Regional Office for guidance in proper disposal methods.

Container Disposal

Completely empty bag into application equipment. Dispose of empty bag in a sanitary landfill or by incineration or, if allowed by state and local authorities, by burning. Stay out of smoke from burning bags.

For minor spills, follow all precautions indicated on this label and clean up immediately. Take special care to avoid contamination of equipment and facilities during cleanup procedures and disposal of wastes. In the event of a major spill, fire or other emergency, call (919) 292-7100 day or night.

PRECAUTIONARY STATEMENTS

HAZARDS TO HUMANS AND DOMESTIC ANIMALS

CAUTION

The active ingredient, metolachlor, may cause skin sensitization reactions in certain individuals. Wear protective clothing (coveralls and gloves) while handling or using this product. Harmful if swallowed, inhaled or absorbed through the skin. Avoid breathing dust. Causes eye irritation. Avoid contact with eyes, skin or clothing. Do not contaminate food or feed.

STATEMENT OF PRACTICAL TREATMENT

If in eyes: Flush eyes with plenty of water. Call a physician if irritation persists.

If on skin: Wash with plenty of soap and water. Get medical attention if irritation persists.

If inhaled: Move to fresh air.

If swallowed: Call a physician or Poison Control Center. Drink 1 or 2 glasses of water and induce vomiting by touching back of throat with finger. Do not induce vomiting or give anything by mouth to an unconscious person.

ENVIRONMENTAL HAZARDS

Cover or incorporate granules that are spilled during loading or are visible on soil surface in turn areas. Do not contaminate water by cleaning of equipment or disposal of wastes.

Ground Water and Surface Water Advisory

Metolachlor has been identified in limited sampling of ground water and there is a possibility that it may leach through the soil to ground water, especially where soils are coarse and ground water is near the surface. Following application and during rainfall events that cause runoff, metolachlor may reach surface water bodies including streams, rivers and reservoirs.

Care must be taken using this product to prevent spills or improper disposal of excess pesticide granules.

DUAL® trademark of CIBA–GEIGY for metolachlor U.S. Patent No. 3,937,730
Amiben® trademark of Union Carbide Agricultural Products, Inc.
AAtrex® trademark of CIBA–GEIGY for atrazine
Balan® trademark of Elanco Products for benefin
Banvel® trademark of Velsicol Chemical Corp. for dicamba
Basagran® trademark of BASF Company for bentazon
Bladex® trademark of Shell Chemical Company for cyanazine
Brominal® trademark of Union Carbide Corporation for bromoxynil
Buctril® trademark of Rhone-Poulenc Chemical Company for bromoxynil
Concep® trademark of CIBA–GEIGY U.S. Patent No. 4,070,389
Gramoxone® trademark of ICI Americas for paraquat
Lexone® trademark of E.I. duPont de Nemours and Co. for metribuzin
Lorox® trademark of E.I. duPont de Nemours and Co. for linuron
Princep® trademark of CIBA–GEIGY for simazine
Roundup® trademark of Monsanto Company for glyphosate herbicide
Sencor® trademark of Parent Company of Farbenfabriken Bayer GmbH, Leverkusen, for metribuzin
Treflan® trademark of Elanco Products for trifluralin
Vernam® trademark of Stauffer Chemical Company for vernolate

CGA 49L34F 098

This is a specimen label which was not prepared by CIBA-GEIGY. CIBA-GEIGY is not responsible for the accuracy of the information contained herein. As labels are subject to revision, always carefully read and follow the label on the product container.

EVIK® 80W

HERBICIDE

For weed control in bananas, corn, grapefruit, oranges, pineapple, sugarcane, and noncrop areas

For use as a potato vine desiccant

ACTIVE INGREDIENTS: Ametryn: 2-ethylamino-4-isopropylamino-6-methylthio-*s*-triazine	76%
Related compounds	4%
INERT INGREDIENTS:	20%
TOTAL	100%

EPA Reg. No. 100-473 EPA Est. 100-LA-1

EVIK 80W is a wettable powder.

KEEP OUT OF REACH OF CHILDREN

CAUTION

See additional precautionary statements on left panel.

See directions for use on panel at right.

DIRECTIONS FOR USE AND CONDITIONS OF SALE AND WARRANTY

IMPORTANT: Read the entire **Directions for Use** and the **Conditions of Sale and Warranty** before using this product.

CONDITIONS OF SALE AND WARRANTY

The **Directions for Use** of this product reflect the opinion of experts based on field use and tests. The directions are believed to be reliable and should be followed carefully. However, it is impossible to eliminate all risks inherently associated with use of this product. Crop injury, ineffectiveness, or other unintended consequences may result because of such factors as weather conditions, presence of other materials, or the manner of use or application all of which are beyond the control of CIBA-GEIGY or the Seller. All such risks shall be assumed by the Buyer.

CIBA-GEIGY warrants that this product conforms to the chemical description on the label and is reasonably fit for the purposes referred to in the **Directions for Use** subject to the inherent risks referred to above. **CIBA-GEIGY makes no other express or implied warranty of Fitness or Merchantability or any other express or implied warranty. In no case shall CIBA-GEIGY or the Seller be liable for consequential, special, or indirect damages resulting from the use or handling of this product.** CIBA-GEIGY and the Seller offer this product, and the Buyer and user accept it, subject to the foregoing **Conditions of Sale and Warranty**, which may be varied only by agreement in writing signed by a duly authorized representative of CIBA-GEIGY.

DIRECTIONS FOR USE

It is a violation of federal law to use this product in a manner inconsistent with its labeling.

FAILURE TO FOLLOW ALL PRECAUTIONS ON THIS LABEL MAY RESULT IN POOR WEED CONTROL, CROP INJURY, AND/OR ILLEGAL RESIDUES.

Do not apply this product in such a manner as to directly or through drift expose workers or other persons except those knowingly involved in the application. The area being treated must be vacated by unprotected persons.

Do not apply this product through any type of irrigation system.

REENTRY STATEMENT

Do not enter treated areas without protective clothing until sprays have dried.

Because certain states may require more restrictive reentry intervals for various crops treated with this product, consult your State Department of Agriculture for further information.

Written or oral warnings must be given to workers who are expected to be in a treated area or in an area about to be treated with this product. Oral warnings must inform workers of areas or fields that may not be entered without specific protective clothing until sprays have dried, and appropriate actions to take in case of accidental exposure, as described under **Precautionary Statements** on this label. When oral warnings are given, warnings shall be given in a language customarily understood by workers. Oral warnings must be given if there is reason to believe that written warnings cannot be understood by workers. Written warnings must include the following information: CAUTION. Area treated with EVIK 80W on (date of application). Do not enter without appropriate protective clothing until sprays have dried. In case of accidental exposure, flush eyes or skin with plenty of water. Call physician if irritation persists. Remove and wash contaminated clothing before reuse.

GENERAL INFORMATION

EVIK 80W controls most annual broadleaf and grass weeds (see list under each crop). When applied before weed emergence, EVIK 80W kills weeds as they germinate by entering roots. On existing weeds, it is effective through leaf contact.

Following many years of continuous use of this product and chemically related products, biotypes of some of the weeds listed on this label have been reported which cannot be effectively controlled by this and related herbicides. Where this is known or suspected, we recommend the use of this product in combination with other registered herbicides which are not triazines. Consult with your state Agricultural Extension Service for specific recommendations.

Avoid using EVIK 80W where adjacent desirable plants may be injured.

EVIK 80W is noncorrosive to equipment and metal surfaces, non-flammable, and has low electrical conductivity.

Application Procedures

Ground application: Use conventional ground sprayers equipped with nozzles that provide accurate and uniform application. Be certain that nozzles are uniformly spaced and the same size. Calibrate sprayer before use and recalibrate at the start of each season and when changing carriers.

Use a pump with capacity to (1) maintain 35-40 psi at nozzles, (2) provide sufficient agitation in tank to keep mixture in suspension, and (3) to provide a minimum of 20% bypass at all times. Use certrifugal pumps which provide propeller shear action for dispersing and mixing this product. The pump should provide a minimum of 10 gals./minute/100 gal. tank size circulated through a correctly positioned sparger tube or jets.

Use screens to protect the pump and to prevent nozzles from clogging. Screens placed on the suction side of the pump should be *16-mesh or coarser.* Do not place a screen in the recirculation line. Use 50-mesh or coarser screens between the pump and boom, and where required, at the nozzles. Check nozzle manufacturer's recommendations.

For band applications, calculate amount to be applied per acre as follows:

$$\frac{\text{band width in inches}}{\text{row width in inches}} \times \text{broadcast rate per acre} = \text{amount needed per acre of field}$$

Aerial application: Use aerial application only where specified in the use directions. Avoid applications under conditions where uniform coverage cannot be obtained or where excessive spray drift may occur.

Avoid application to humans or animals. Flagmen and loaders should avoid inhalation of spray mist and prolonged contact with skin, and should wash thoroughly before eating and at the end of each day's operation.

Mixing procedures—all uses: (1) Be sure sprayer is clean and not contaminated with any other materials, or crop injury or sprayer clogging may result. (2) Fill tank 1/4 full with clean water or nitrogen solution (may be used in corn). (3) Start agitation. (4) Be certain that the agitation system is working properly and creates a rippling or rolling action on the liquid surface. (5) Make a slurry by adding EVIK 80W to a small amount of water in a separate container and pour slurry into tank. (6) Continue filling tank until 90% full. Increase agitation if necessary to maintain surface action. (7) Add tank mix herbicide(s) after this product is thoroughly suspended. (8) Finish filling tank. Maintain agitation to avoid separation of materials. (9) Empty tank as completely as possible before refilling to prevent buildup of emulsifiable concentrate residue from possible tank mix herbicides. (10) If an emulsifiable concentrate film starts to build up in tank, drain it and clean with strong detergent solution or solvent. (11) Clean sprayer thoroughly immediately after use by flushing system with water containing a detergent.

Compatibility test: Nitrogen solutions may replace all or part of the water in the spray in corn. Since nitrogen solutions can vary, even within the same analysis, check compatibility each time before use. Commercial application equipment may improve compatibility in some instances. The following test assumes a spray volume of 25 gals. per acre. For other spray volumes, make appropriate changes in the ingredients. Check compatibility using this procedure.

1. Add 1 pint of fertilizer to each of 2 one-quart jars with tight lids.
2. To **one** of the jars add 1/4 tsp. or 1.2 milliliters of a compatibility agent approved for this use (1/4 tsp. is equivalent to 2 pts. per 100 gals. spray). Shake or stir gently to mix. Examples of compatibility agents include Compex® and Unite®.
3. To both jars add 1.4 teaspoons of EVIK 80W for each pound per acre to be applied.
4. After adding all ingredients, put lids on and tighten, and invert each jar ten times to mix. Let the mixtures stand 15 minutes and then look for separation, large flakes, precipitates, gels, heavy oily film on the jar, or other signs of incompatibility. Determine if the compatibility agent is needed in the spray mixture by comparing the two jars. If either mixture separates, but can be remixed readily, the mixture can be sprayed as long as good agitation is used. If the mixtures are incompatible, try slurrying the EVIK 80W in water before addition. If still incompatible, do not use the nitrogen solution and EVIK 80W in the same spray tank.

Bananas and Plantains

In Hawaii, EVIK 80W controls broadleaf and grass weeds including rattlebox (*Crotalaria* spp.), dallisgrass, goosegrass (*Eleusine indica*), Japanese tea, kukaipuaa and other crabgrass species (*Digitaria* spp.), pualele (sowthistle), common purslane, *Richardia* spp., spanishneedles, wild pea bean, *Amaranthus* spp., Flora's paintbrush, foxtail, junglerice, fireweed, and *Panicum* spp.

In Puerto Rico, EVIK 80W controls broadleaf and grass weeds including *Commelina diffusa, Paspalum* spp., *Axonopus compressus, Panicum* spp., *Digitaria* spp., and *Setaria* spp.

Apply EVIK 80W in Hawaii or Puerto Rico as a directed basal spray immediately after setting banana or plantain plants, or any time thereafter. Apply before weeds emerge or to emerged, growing weeds. Use the lower rates for applications made before weed emergence and for young, easy-to-kill weeds. Use the higher rates for more residual control and for controlling hard-to-kill weeds. Repeat as needed at 3-4 month intervals.

	Rates of application per acre	
Area	Per application (minimum of 30 gals. water/A)	Maximum per year
Hawaii	5-10 lbs.	30 lbs.
Puerto Rico	4-6 lbs.	16 lbs.

In Central America, apply 2-4 lbs. of EVIK 80W in a minimum of 20 gals. of water/A covering all vegetation thoroughly. Use the lower rate for application to young weeds, and the higher rate for application to heavier growth of older, hard-to-kill weeds. Repeat application 30 days later, covering all vegetation. Apply the third and subsequent applications as needed, covering only living vegetation. Use a maximum of 16 lbs. of EVIK 80W/A per year.

Corn (Field corn, sweet corn, popcorn)

EVIK 80W controls annual broadleaf and grass weeds including Texas panicum (Texas millet) fall panicum, signalgrass (*Brachiaria* spp.), goosegrass, crabgrass, barnyardgrss, giant foxtail, yellow and green foxtails, nutsedge, shattercane, wild proso millet, cocklebur, lambsquarters, Florida pusley, morningglory, pigweed, wild mustard, ragweed, velvetleaf, and smartweed. Weeds taller than specified in the rate table will not be controlled.

Apply EVIK 80W as a postemergence directed spray to weeds after the smallest corn is at least 12 inches tall (measured to the highest leaf surface on free standing plants).

Precaution: Do not spray over top of corn or injury will occur. Do not apply within 3 weeks of tasseling.

Apply in a minimum of 20 gals. of water/A to assure uniform coverage (nonpressure nitrogen solution may be substituted for all or part of the water). The entire weed must receive spray to be killed. Add a surfactant such as X-77®, duPont WK, or Tronic at the rate of 2 qts./100 gals. of spray mixture (0.5% of spray volume).

It is recommended that gauge wheels and/or leaf lifter equipment be used to prevent leaf contact with the spray. Drop nozzles may be used, but extreme care must be taken to keep the spray or drift from contacting the leaves and especially the whorl of the corn plant. Be sure the entire spray pattern is directed downward. Apply at a spray pressure of 30 psi or less to prevent "bounce back"—spray bouncing off soil or weeds and settling on corn leaves.

Use rates of application according to the following geographical areas:

Area 1

Arizona	Michigan	Ohio
California	Minnesota	Oregon
Colorado	Missouri	Pennsylvania
Connecticut	Montana	Rhode Island
Idaho	Nebraska	South Dakota
Illinois	Nevada	Utah
Indiana	New Hampshire	Vermont
Iowa	New Jersey	Washington
Kansas	New Mexico	Wisconsin
Maine	New York	Wyoming
Massachusetts	North Dakota	

Area 2

Alabama	Louisiana	Tennessee
Arkansas	Maryland	Texas
Delaware	Mississippi	Virginia
Florida	North Carolina	West Virginia
Georgia	Oklahoma	
Kentucky	South Carolina	

Where a rate range is indicated, use the higher rate for relatively dense weed infestations.

Rates of Application for Corn

Weed height	Weeds controlled*	Broadcast rate per acre Area 2	Broadcast rate per acre Area 1
Up to 2 inches	*Brachiaria* broadleafs	.75 lb.	2 lbs.
	Texas panicum fall panicum barnyardgrass goosegrass	1.25 lbs.	
	crabgrass	2-2.5 lbs.	
	foxtail	2-2.5 lbs.	2.5 lbs.
	shattercane nutsedge	Do not apply	
Up to 4 inches	wild proso millet		
2-4 inches	*Brachiaria* broadleafs	1.25 lbs.	2.5 lbs.
	Texas panicum fall panicum barnyardgrass goosegrass	2-2.5 lbs.	
	crabgrass	Do not apply	
	foxtail	2-2.5 lbs.	
	shattercane nutsedge	Do not apply	
4-6 inches	*Brachiaria* broadleafs	2-2.5 lbs.	2.5 lbs.
	fall panicum nutsedge	Do not apply	

* A mechanical cultivation may be required if weeds regrow.

Suggestions For Crop Rotation: Small grains, such as wheat, oats, and rye, may be planted in the fall following the recommended application. Do not plant any rotational crop other than small grains until the following year.

Note: Allow 30 days after the application before harvesting, grazing, or feeding forage to livestock.

Grapefruit and Oranges

Florida: Apply EVIK 80W twice per year (see following table) in a minimum of 40 gals./A as a directed spray to thoroughly cover existing weeds. Weeds controlled include common bermudagrass (*Cynodon dactylon*), spanishneedles (*Bidens bipinnata*), and Florida pusley (*Richardia scabra*). Mow 7-10 days before application to obtain most effective control of dense stands of common bermudagrass. EVIK is most effective when good moisture conditions exist.

Make the first application in the spring when weeds have resumed active growth, but before they exceed 4 inches in height. Make the second application 90-100 days later to obtain season-long control. For heavy bermudagrass infestations, use the higher rates in the first application.

Florida application	Broadcast rate per acre Light weed infestation	Broadcast rate per acre Heavy weed infestation
First (spring)	5-6 lbs.	7-8 lbs.
Second (summer)	5-6 lbs.	7 lbs.

Texas: Apply EVIK 80W in a minimum of 40 gals./A as a directed spray to thoroughly cover existing weeds. Use 3 lbs./A for control of heavy weed infestations and for top-kill of johnsongrass and marinevine (ivy treebine). Weeds controlled include barnyardgrass, buffelgrass, sandbur, Texas panicum (Colorado-grass), silver leaf nightshade, pigweed, common purslane, and sunflower. Use 2 lbs./A to control light weed infestations. For both rates, add a nonionic surfactant, such as duPont WK, X-77, or LOC, at the rate of 2 qts./100 gals. of spray mixture (0.5% of spray volume).

Make the first application in the spring when weeds have resumed active growth, but before they exceed 4 inches in height. Make additional applications as needed up to 15 lbs./A per year. EVIK is most effective when weeds are actively growing and 1-4 inches in height.

Precautions (Florida and Texas): (1) Do not apply where trees have been established in the grove less than two years. (2) Apply no more than 15 lbs. of EVIK 80W/A per year. (3) Do not allow spray to contact stems, fruit, or foliage on trees. (4) Do not apply where uniform coverage cannot be obtained or where excessive spray drift may occur.

Notes (Florida and Texas): (1) Do not apply within 30 days of harvest. (2) Do not graze treated areas.

Pineapple

For control of broadleaf and grass weeds including rattlebox (*Crotalaria* spp.), dallisgrass, goosegrass (*Eleusine indica*), Japanese tea, kukaipuaa and other crabgrass species (*Digitaria* spp.), pualele (sowthistle), common purslane, *Richardia* spp., spanishneedles, wild pea bean, *Amaranthus* spp., Flora's paintbrush, foxtail, junglerice, fireweed, and *Panicum* spp., apply up to 10 lbs. of EVIK 80W/A as a blanket spray immediately after planting, or after plant crop harvest is completed and before weeds emerge. Additional blanket or interspace applications may be made at 1-2 month intervals prior to differentiation as needed, but should not exceed 2½ lbs./A per application. Apply in a minimum of 30 gals. of water/A.

Note: (1) Do not apply more than 37½ lbs. EVIK 80W/A per crop cycle. (2) Do not make the last application within 160 days of harvest.

Sugarcane

To control weeds specified in the various states, apply EVIK 80W alone or in tank mix combinations. Broadcast aerially in a minimum volume of 5 gals. of spray/A, or broadcast or band by ground in a minimum of 20 gals./A, unless indicated otherwise. Repeat treatments, where needed, may be applied broadcast, band, or interline as recommended, with final application prior to close-in.

Aerial Application: Use aerial application only where broadcast applications are specified. Unless specified otherwise, apply a minimum of 1 gal. of water for each 1 lb. of product applied/A, but not less than 5 gals. total volume/A. Avoid applications under conditions where uniform coverage cannot be obtained or where excessive spray drift may occur. In order to assure that spray will be controllable within the target area when used according to label directions, make applications at a maximum height of 10 ft., using low drift nozzles at a maximum pressure of 40 psi, and restrict application to periods when wind speed does not exceed 10 mph. To assure that spray will not adversely affect adjacent sensitive nontarget plants, apply EVIK 80W alone by aircraft at a minimum upwind distance of 800 ft. from sensitive plants, or apply EVIK 80W + Karmex® at a minimum upwind distance of 500 ft. from sensitive plants.

Florida: Apply 0.5-1.5 lbs. of EVIK 80W in a minimum of 20 gals. of water/A directed to the base of plant or ratoon sugarcane to emerged weeds. Avoid wetting sugarcane foliage or injury may occur. Use nozzle tips which will minimize atomization or spray drift. Use the higher rate for high grass populations. Apply up to 2 repeat applications, if needed, at 30-day intervals before close-in. To control alexandergrass (*Brachiaria plantaginea*), apply at the 3-4 leaf stage or before 3 inches tall. For mixed weed infestations, use 1.5 lbs. of EVIK 80W plus 0.5 lb. acid equivalent of 2,4-D amine/A and/or 2 qts. of surfactant, such as X-114 or ACL 209, for each 100 gals. of spray to improve weed control. Observe all precautions and limitations on labeling of all products used in mixtures.

Hawaii: Use one of the following methods in plant or ratoon sugarcane for control of ageratum, rattlebox (*Crotalaria* spp.), dallisgrass, fireweed, goosegrass, (*Eleusine indica*), guineagrass, Japanese tea, kukaipuaa, and other crabgrass species (*Digitaria* spp.), morningglory, pualele (sowthistle), common purslane, *Richardia* spp., spanishneedles, wild pea bean, *Amaranthus* spp., Flora's paintbrush, foxtail, junglerice, and swollen fingergrass.

1. Apply 5-10 lbs./A before weeds or sugarcane emerge. A second and third application at 2.5-5 lbs./A may be made, as needed, to emerged weeds prior to close-in.
2. Apply 2.5-5 lbs. of EVIK 80W plus 2.5-5 lbs. of Karmex 80W before sugarcane and weeds emerge.
   A second application at 2.5-5 lbs. of EVIK 80W plus 2.5-5 lbs. of Karmex 80W may be made, as needed, postemergence to sugarcane and weeds. A third application at 2.5-3 lbs. of EVIK 80W plus 2.5-3 lbs. of Karmex 80W may be applied prior to close-in.
   For best results when Karmex 80W is used on emerged weeds, add a nonionic surfactant to the spray at the rate of 1-2 qts. per 100 gals. and apply as a directed spray.

Note: Use the minimum preemergence rates on nonirrigated sugarcane (high rainfall areas), on land first cropped to sugarcane, and for light weed infestations.

Precautions: (1) Sugarcane growing in areas of exposed subsoil, in rocky areas, or in soils of low adsorptive capacity may show temporary chlorosis following treatment. (2) Injury to sugarcane may occur when under moisture stress. (3) Certain sugarcane varieties may show a temporary chlorosis or stunting as a result of over-the-top application.

Louisiana: Use directions below for control of these weeds:

Weed height	Weeds controlled
Up to 3 inches	*itchgrass (raoulgrass)
Up to 4 inches	barnyardgrass, crabgrass, fall panicum, foxtail, goosegrass, Texas panicum
Up to 5 inches	annual sowthistle, common chickweed, henbit, paleseed plantain, swinecress
Up to 6 inches	*Brachiaria* spp., browntop panicum, cocklebur, Florida pusley, common lambsquarters, morningglory, pigweed, ragweed, smartweed, velvetleaf, wild mustard

* Controls emerged itchgrass. May not control itchgrass germinating after treatment.

Broadcast or band by ground equipment over the top of plant or ratoon sugarcane 2.5 lbs. of EVIK 80W plus 0.5 lb. acid equivalent of 2,4-D amine plus 1 qt. of crop oil concentrate, such as Agri-Dex, Amoco, or Unico (or 1 pt. of nonionic surfactant, such as duPont WK, X-77, or LOC) in a minimum of 20 gals. of water/A.

Follow with one or two repeat over-the-top or directed applications, as needed, using 3 lbs. of EVIK 80W plus 0.5 lb. of 2,4-D amine plus 1 qt. of crop oil concentrate (or 1 pt. of nonionic surfactant) in a minimum of 20 gals. of water/A. To avoid injury, do not apply over the top of sugarcane after April 10 or after sugarcane exceeds 20 inches in height. If needed, follow with one or two additional applications directed to the base of sugarcane (using same rates as in second or third application) before close-in. Observe all precautions and limitations on labeling of all products used in mixtures.

Precaution: Temporary yellowing of sugarcane leaves may follow over-the-top applications.

Puerto Rico: Use one of the following methods for control of *Amaranthus* spp., crabgrass, dallisgrass, foxtail, goosegrass *(Eleusine indica)*, itchgrass (raoulgrass), junglerice, milkweed, morningglory, *Panicum* spp., pigweed, purpletop, purslane, *Richardia* spp., spanishneedles, sandbur, and sowthistle in plant and/or ratoon sugarcane as specified:

1. **Plant or Ratoon Sugarcane:** Apply 5-10 lbs. of EVIK 80W/A before sugarcane or weeds emerge. Apply one or two additional applications, as needed, at 2.5-5 lbs./A before close-in, directed to the base of sugarcane.
2. **Plant Sugarcane:** Apply 5 lbs./A before sugarcane emerges and before weeds exceed 5 inches in height. A second application at 4 lbs./A may be made, as needed, over-the-top of sugarcane after the sugarcane and weeds emerge but prior to close-in.
3. **Ratoon Sugarcane:** Apply 4 lbs./A interline or over-the-top of sugarcane and weeds prior to close-in.
   In plant or ratoon sugarcane, add 1 pt. of a nonionic surfactant to the spray volume to be applied/A where weeds have emerged at application. Treat before weeds exceed 5 inches in height.

Precautions: (1) Sugarcane growing in areas of exposed subsoil, in rocky areas or during time of water stress may show a temporary chlorosis. (2) Do not apply EVIK 80W in combination with other herbicides when making over-the-top applications. (3) Temporary chlorosis or stunting may occur after over-the-top applications.

Texas: Use directions below for control of these weeds.

Weed height	Weeds controlled
Up to 2 inches	fall panicum, Texas panicum
Up to 4 inches	barnyardgrass, *Brachiaria* spp., cocklebur, Florida pusley, common lambsquarters, morningglory, ragweed, smartweed, velvetleaf, wild mustard
Up to 6 inches	pigweed, sunflower

Broadcast 1.5-2.5 lbs./A preemergence or postemergence to sugarcane or weeds. Add a nonionic surfactant at the rate of 2 qts./100 gals. of spray mixture.

Follow with one or two repeat applications, as needed. Make the final application before close-in. Use the higher rates of EVIK 80W for heavier weed infestations.

Precaution: Temporary yellowing of sugarcane leaves may follow over-the-top applications.

Potato Vine Desiccant

To kill potato vines and weed growth, apply EVIK 80W 10-14 days before harvest at the rates in the following table. Apply in a minimum of 100 gals. of water/A. Thorough coverage must be obtained in order to achieve a satisfactory kill. Complete vine kill is normally obtained in 10-14 days in the Southeast and in 14-17 days elsewhere. Vine kill may be delayed by unseasonably cold weather.

Note: Do not graze livestock on treated areas.

Area	Rate EVIK 80W per acre
Florida	1.25-2.5 lbs.
Southeast (NC, TN, and south except FL)	1.25-2.5 lbs. or 1 lb. *plus* 2 gals. nonphytotoxic oil
Northeast (VA, KY, and north), Midwest, North Central, CO, and WY	2.5-3 lbs.

Noncrop Land

EVIK 80W provides weed control on industrial sites, railroad rights-of-way, around highway guard rails, lumberyards, petroleum tank farms, utility substations and airport runways. Use only where control of all vegetation is desired. Do not use near adjacent desirable trees, shrubs or plants, or in greenhouses, as injury may occur.

EVIK 80W controls barnyardgrass, crabgrass, fall panicum, foxtails (giant, green, yellow), goosegrass, nutsedge, shattercane, signalgrass (*Brachiaria* spp.), Texas panicum (Texas millet), cocklebur, Florida pusley, lambsquarters, morningglory, wild mustard, pigweed, ragweed, smartweed, and velvetleaf.

Apply to emerged actively growing weeds at the rates recommended in the rate chart "Rates of Application for Corn" on this label. Apply in at least 20 gals. of water/A and add a surfactant such as X-77, duPont WK, or Tronic at the rate of 2 qts./100 gals. of spray mixture (0.5% of spray volume). Cover weeds thoroughly with spray for good control.

Do not plant desirable plants in treated areas for one year.

STORAGE AND DISPOSAL

Store in a cool, dry place. Do not contaminate water, food, or feed by storage or disposal. Improper disposal of unused pesticide, spray mixture, or rinse water is a violation of federal law. If these wastes cannot be used according to label instructions, contact your State Pesticide or Environmental Control Agency, or the Hazardous Waste representative at the nearest EPA Regional Office for guidance in proper disposal methods.

Completely empty bag into application equipment. Dispose of empty bag and box in a sanitary landfill or by incineration, or if allowed by state and local authorities, by burning. Stay out of smoke from burning containers.

For minor spills, leaks, etc., follow all precautions indicated on this label and clean up immediately. Take special care to avoid contamination of equipment and facilities during cleanup procedures and disposal of wastes. In the event of a major spill, fire or other emergency, call (919) 292-7100 day or night.

PRECAUTIONARY STATEMENTS

HAZARDS TO HUMANS AND DOMESTIC ANIMALS

CAUTION

Harmful if swallowed, inhaled, or absorbed through the skin. Causes eye irritation. Avoid contact with skin, eyes, or clothing. Avoid breathing vapors, dust, or spray mist. Remove contaminated clothing and wash before reuse. Wash thoroughly with soap and water after handling.

First Aid: If swallowed, contact your local poison control center, hospital, or physician. If the patient is unconscious, maintain breathing and heartbeat (cardiopulmonary resuscitation). If the patient is conscious and alert, induce vomiting (syrup of ipecac or stimulate the back of the throat with a finger). Never give anything by mouth to an unconscious person!

If inhaled, move to clean air. If in eyes, flush with plenty of water. If on skin, wash with plenty of soap and water. If irritation persists, seek medical attention.

Note to Physician: If ingested, there is no specific antidote. Induce emesis or lavage stomach. Give saline laxative and supportive therapy.

ENVIRONMENTAL HAZARDS

Do not apply directly to any body of water.
Do not contaminate water by cleaning of equipment or disposal of wastes.

EVIK® trademark of CIBA-GEIGY for ametryn
Karmex® trademark of E. I. duPont de Nemours and Co. for diuron
X-77® trademark of KALO Laboratories, Inc.

CGA 15L5Q 097

This is a specimen label which was not prepared by CIBA-GEIGY. CIBA-GEIGY is not responsible for the accuracy of the information contained herein. As labels are subject to revision, always carefully read and follow the label on the product container.

ORBIT™

FUNGICIDE

For control of certain diseases in pecans

Orbit contains 3.6 lbs. active ingredient per gallon.

ACTIVE INGREDIENT:	
1-[[2-(2,4-dichlorophenyl)-4-propyl-1,3-dioxolan-2-yl]methyl]-1*H*-1,2,4-triazole	41.8%
INERT INGREDIENTS:	58.2%
TOTAL	100.0%

EPA Reg. No. 100-702 EPA Est. 100-AL-1

KEEP OUT OF REACH OF CHILDREN

WARNING

See additional precautionary statements on back of container.

See directions for use in attached booklet.

PRECAUTIONARY STATEMENTS

HAZARDS TO HUMANS AND DOMESTIC ANIMALS

WARNING

This product may cause skin sensitization reactions in certain individuals. Wear rubber gloves and long-sleeved shirt when mixing or handling the undiluted product. Remove contaminated clothing and wash before reuse.

Causes eye irritation. Do not get in eyes, on skin, or on clothing. Harmful if swallowed, inhaled, or absorbed through skin. Wash thoroughly with soap and water after handling. Avoid breathing spray mist.

First Aid: If in eyes, flush with plenty of water and get medical attention if irritation persists. If on skin, wash with plenty of soap and water. Get medical attention if irritation persists. If swallowed, drink promptly a large quantity of milk, egg whites, or gelatin solution; if these are not available, drink large quantities of water. Never give anything by mouth to an unconscious person. Do not induce vomiting. Call a physician or Poison Control Center immediately. In case of inhalation exposure, move to fresh air.

Note to Physician: There is no specific antidote for this product. Induce emesis or lavage stomach, taking care to avoid aspiration of stomach contents into lungs. Give a saline laxative and supportive therapy.

ENVIRONMENTAL HAZARDS

This pesticide is toxic to fish. Do not apply directly to water. Do not contaminate water by cleaning of equipment or disposal of wastes. Do not contaminate water when disposing of equipment wash waters. Refer to product labeling for use restrictions to protect endangered species.

PHYSICAL OR CHEMICAL HAZARDS

Do not use, pour, spill, or store near heat or open flame. In event of spill, fire, or other emergency, call (919) 292-7100 day or night.

Store at temperatures above 0°F.

Chemigation: Do not apply this product through any type of irrigation system.

ORBIT™ trademark of CIBA–GEIGY
U.S. Patent No. 4,079,062

CGA 84L1B 029

DIRECTIONS FOR USE AND CONDITIONS OF SALE AND WARRANTY

IMPORTANT: Read the entire **Directions for Use** and the **Conditions of Sale and Warranty** before using this product.

CONDITIONS OF SALE AND WARRANTY

The **Directions for Use** of this product reflect the opinion of experts based on field use and tests. The directions are believed to be reliable and should be followed carefully. However, it is impossible to eliminate all risks inherently associated with use of this product. Crop injury, ineffectiveness, or other unintended consequences may result because of such factors as weather conditions, presence of other materials, or the manner of use or application all of which are beyond the control of CIBA–GEIGY or the Seller. All such risks shall be assumed by the Buyer.

CIBA–GEIGY warrants that this product conforms to the chemical description on the label and is reasonably fit for the purposes referred to in the **Directions for Use** subject to the inherent risks referred to above. **CIBA–GEIGY makes no other express or implied warranty of Fitness or Merchantability or any other express or implied warranty. In no case shall CIBA–GEIGY or the Seller be liable for consequential, special, or indirect damages resulting from the use or handling of this product.** CIBA–GEIGY and the Seller offer this product, and the Buyer and user accept it, subject to the foregoing **Conditions of Sale and Warranty,** which may be varied only by agreement in writing signed by a duly authorized representative of CIBA–GEIGY.

DIRECTIONS FOR USE

It is a violation of federal law to use this product in a manner inconsistent with its labeling.

GENERAL INFORMATION

ORBIT is a broad spectrum fungicide for the control of certain diseases in pecans.

FAILURE TO FOLLOW DIRECTIONS AND PRECAUTIONS ON THIS LABEL MAY RESULT IN CROP INJURY, POOR DISEASE CONTROL, OR ILLEGAL RESIDUES.

Spray Equipment

Thorough coverage is necessary to provide good disease control.

To avoid spray drift, do not spray when conditions favor drift beyond the target area. Avoid spray overlap as crop injury may occur.

Equip sprayers with nozzles that provide accurate and uniform application. Be certain that nozzles are the same size and uniformly spaced across the boom. Calibrate sprayer before use.

Use a pump with capacity to (1) maintain 35-40 psi at nozzles, and (2) provide sufficient agitation in tank to keep mixture in suspension—this requires recirculation of 10 percent of tank volume per minute. Use a jet agitator or liquid sparge tube for agitation. Do not air sparge.

Although ORBIT is an emulsifiable concentrate, it is suggested screens be used to protect the pump and to prevent nozzles from clogging. Screens placed on suction side of pump should be *16-mesh or coarser.* Do not place a screen in the recirculation line. Use 50-mesh or coarser screens between the pump and boom, and where required, at the nozzles. Check nozzle manufacturer's recommendations.

For more information on spray equipment and calibration, consult sprayer manufacturers and state recommendations. For specific local directions and spray schedules, consult the current agricultural experiment station recommendations.

To determine the physical compatibility of ORBIT with other products, use a jar test. Using a quart jar, add the proportionate amounts of the products to 1 qt. of water. Add wettable powders and water dispersible granular products first, then liquid flowables, and emulsifiable concentrates last. After thoroughly mixing, let stand for at least 5 minutes. If the combination remains mixed or can be remixed readily, it is physically compatible. Once compatibility has been proven, use the same procedure for adding required ingredients to the spray tank.

Mixing Procedures

(1) Wear rubber gloves and long-sleeved shirt when mixing and loading this product. (2) Be sure sprayer is clean; if it is contaminated with any material, crop injury or sprayer clogging may result. (3) Fill tank 1/4-1/2 full with clean water, and start agitation. (4) Be certain that the agitation system is working properly and creates a rippling or rolling action on the liquid surface. (5) Add product to the tank. (6) Complete filling tank, maintaining sufficient agitation at all times to ensure surface action. This applies to both spray and nurse tanks. (7) Do not leave tank mixtures in the tank without adequate agitation. Always maintain agitation to avoid separation and buildup of residues. If a residue starts to build up in tank, drain and clean with a strong detergent and ammonia solution and rinse thoroughly with clean water. (8) Immediately after use, clean sprayer thoroughly by flushing system with water containing a detergent and rinse with clean water.

Aerial Application: Avoid application under conditions where uniform coverage cannot be obtained or where excessive spray drift may occur. Do not apply directly to humans or animals. Flagmen and loaders should wash thoroughly before eating, drinking, or smoking, and at the end of each day's operation.

Chemigation: Do not apply this product through any type of irrigation system.

Pecans (bearing)

Apply ORBIT for control of pecan scab, downy spot, liver spot, vein spot, Zonate leaf spot and fungal leaf scorch in a regular spray program beginning at bud break when young leaves are unfolding. Continue applications while small nuts are forming, repeating at two-week intervals as needed, not exceeding four applications per growing season. Do not apply after shuck-split. Do not graze livestock in treated areas or cut treated cover crops for feed.

Apply ORBIT at the following rates in sufficient water to obtain good coverage, using a minimum spray volume of 50 gals./A by ground. For ground applications, the applicator must be in a closed cab during application.

Pecan Scab—South (AL, AR, FL, GA, KY, LA, MO, MS, NC, SC and TN): Apply 4 fl. oz./A on a 14-day schedule during bud break and pre-pollination sprays. Apply 8 fl. oz./A during nut formation and cover sprays.

Pecan Scab—Southwest (AZ, NM, OK and TX): Apply 4 fl. oz./A on a 14-day schedule during bud break and pre-pollination sprays. Apply 6 fl. oz./A in case bearer and cover sprays.

Foliar Diseases: ORBIT may be applied for control of mid-to late-season foliar diseases (such as Zonate leaf spot, *fungal leaf scorch, vein spot,* and downy spot) at 4 fl. oz./A with the recommended rate of Du-Ter® or Super Tin® in the cover sprays. Observe all directions, precautions, and limitations on the labeling of these products.

For pecans, ORBIT is compatible with most insecticides recommended, such as diazinon 50W, diazinon AG500, Zolone® 3E, dimethoate, Supracide® 2E and carbaryl 50W. Do not mix ORBIT with Cyprex® on pecans, or crop injury may occur.

Endangered Species Restrictions: The use of any pesticide in a manner that may kill or otherwise harm an endangered or threatened species or adversely modify their habitat is a violation of federal laws. The use of this product in pecans is controlled to prevent death or harm to endangered species. Do not use ORBIT in the following counties:

State	Counties			
Alabama	Colbert Greene Jackson	Lamar Lauderdale Limestone	Madison Marshall Morgan	Pickens Sumter
Arkansas	Clark Clay	Cross Dallas	Hot Springs Randolph	Sharp St. Francis
Kentucky	Ballard Butler	Green Hart	Livingston McCracken	Taylor Warren

State	Counties			
Tennessee	Bedford Decatur	Hardin Lincoln	Marshall Maury	Perry Sequatchie
Missouri	Gasconade Jefferson	Miller Osage	Ripley St. Louis	Wayne
Mississippi	Lowndes	Monroe	Noxubee	

Pecans (nonbearing)

Apply 4.0 to 5.5 fl. oz. of ORBIT /A in sufficient water to obtain good coverage for control of pecan scab, downy spot, brown leaf spot, liver spot, vein spot, Zonate leaf spot, and fungal leaf scorch. Begin applications at bud break and repeat at 2-4 week intervals as needed, not exceeding four applications per growing season. For ground applications, the applicator must be in a closed cab during application. Do not graze livestock in treated areas. Observe endangered species restrictions noted above in the bearing pecan section.

STORAGE AND DISPOSAL

Store at temperatures above 0°F. Do not contaminate water, food or feed by storage or disposal. Pesticide wastes may be toxic. Improper disposal of unused pesticide, spray mixture, or rinse water is a violation of federal law. If these wastes cannot be used according to label instructions, contact your State Pesticide or Environmental Control Agency, or the Hazardous Waste representative at the nearest EPA Regional Office for guidance in proper disposal methods.

Triple rinse container (or equivalent). Then offer for recycling, or reconditioning, or puncture and dispose of in a sanitary landfill, by incineration, or, if allowed by state and local authorities, by burning. Stay out of smoke from burning containers.

For minor spills, leaks, etc., follow all precautions indicated on this label and clean up immediately. Take special care to avoid contamination of equipment and facilities during cleanup procedures and disposal of wastes. In the event of a major spill, fire, or other emergency, call (919) 292-7100 day or night.

PRECAUTIONARY STATEMENTS

HAZARDS TO HUMANS AND DOMESTIC ANIMALS

WARNING

This product may cause skin sensitization reactions in certain individuals. Wear rubber gloves and long-sleeved shirt when mixing or handling the undiluted product. Remove contaminated clothing and wash before reuse.

Causes eye irritation. Do not get in eyes, on skin, or on clothing. Harmful if swallowed, inhaled, or absorbed through skin. Wash thoroughly with soap and water after handling. Avoid breathing spray mist.

First Aid: If in eyes, flush with plenty of water and get medical attention if irritation persists. If on skin, wash with plenty of soap and water. Get medical attention if irritation persists. If swallowed, drink promptly a large quantity of milk, egg whites, or gelatin solution; if these are not available, drink large quantities of water. Never give anything by mouth to an unconscious person. Do not induce vomiting. Call a physician or Poison Control Center immediately. In case of inhalation exposure, move to fresh air.

Note to Physician: There is no specific antidote for this product. Induce emesis or lavage stomach, taking care to avoid aspiration of stomach contents into lungs. Give a saline laxative and supportive therapy.

ENVIRONMENTAL HAZARDS

This pesticide is toxic to fish. Do not apply directly to water. Do not contaminate water by cleaning of equipment or disposal of wastes. Do not contaminate water when disposing of equipment wash waters. Refer to product labeling for use restrictions to protect endangered species.

PHYSICAL OR CHEMICAL HAZARDS

Do not use, pour, spill, or store near heat or open flame.

Cyprex® trademark of American Cyanamid for dodine
Du-Ter® trademark of Griffin Ag Products Co., Inc. for triphenyltin hydroxide
ORBIT™ trademark of CIBA-GEIGY U.S. Patent No. 4,079,062
Super Tin® trademark for Griffin Ag Products Co., Inc. for triphenyltin hydroxide
Supracide® trademark of CIBA-GEIGY for methidathion
Zolone® trademark of Rhône-Poulenc for phosalone

EPA Reg. No. 100-617

CGA 84L1B 029

PRAMITOL® 25E

HERBICIDE

Controls johnsongrass, bindweed, and other hard-to-kill weeds

Bare ground weed control around buildings, storage areas, fences, pumps, machinery, fuel tanks, drainage ditch banks, recreational areas, roadways, guard rails, airports, military installations, highway medians, pipelines, railroads, lumberyards, industrial sites, rights-of-way . . . or similar areas

See directions for use on panel at right.

For weed control on industrial sites, noncrop areas around-the-farm, and for use in and under asphalt.

PRAMITOL 25E contains 2 lbs. prometon per gallon.

ACTIVE INGREDIENT:	
Prometon: 2,4-bis(isopropyl-amino)-6-methoxy-*s*-triazine	25%
INERT INGREDIENTS:	75%
TOTAL	100.0%

EPA Est. 100-AL-1 EPA Reg. No. 100-443

KEEP OUT OF REACH OF CHILDREN

DANGER

See additional precautionary statements left of container seam.

Control No. 06550

DIRECTIONS FOR USE AND CONDITIONS OF SALE AND WARRANTY

IMPORTANT: Read the entire **Directions for Use** and the **Conditions of Sale and Warranty** before using this product.

CONDITIONS OF SALE AND WARRANTY

The **Directions for Use** of this product reflect the opinion of experts based on field use and tests. The directions are believed to be reliable and should be followed carefully. However, it is impossible to eliminate all risks inherently associated with use of this product. Crop injury, ineffectiveness, or other unintended consequences may result because of such factors as weather conditions, presence of other materials or the manner of use or application all of which are beyond the control of CIBA–GEIGY or the Seller. All such risks shall be assumed by the Buyer.

CIBA–GEIGY warrants that this product conforms to the chemical description on the label and is reasonably fit for the purposes referred to in the **Directions for Use** subject to the inherent risks referred to above. **CIBA–GEIGY makes no other express or implied warranty of Fitness or Merchantability or any other express or implied warranty. In no case shall CIBA–GEIGY or the Seller be liable for consequential, special, or indirect damages resulting from the use or handling of this product.** CIBA–GEIGY and the Seller offer this product, and the Buyer and user accept it, subject to the foregoing **Conditions of Sale and Warranty,** which may be varied only by agreement in writing signed by a duly authorized representative of CIBA–GEIGY.

DIRECTIONS FOR USE

It is a violation of federal law to use this product in a manner inconsistent with its labeling.

GENERAL INFORMATION

PRAMITOL 25E is a nonselective herbicide that can be applied in water or oil before or after plant growth begins. Although PRAMITOL 25E has considerable activity through foliar contact, much of its activity is through roots; therefore, its effectiveness is dependent on rainfall to move it into the root zone. Very dry soil conditions and lack of sufficient rainfall may result in poor weed control.

Use only in areas where complete control of all vegetation is desired, such as industrial sites, rights-of-way, lumberyards, petroleum tank farms, around farm buildings, along fence lines, etc. When applied to the soil, this product usually inhibits plant growth for a year or more. It should not be used on land to be cropped, or near adjacent desirable trees, shrubs or plants, or injury may occur.

Do not use for weed control in greenhouses.

Following many years of continuous use of this product and chemically related products, biotypes of some of the weeds listed on this label have been reported which cannot be effectively controlled by this and related herbicides. Where this is known or suspected, we recommend the use of this product in combination with other registered herbicides which are not triazines. Consult with your State Agricultural Extension Service for specific recommendations.

It is generally noncorrosive to equipment, but may cause swelling or more rapid deterioration of hoses and fittings containing natural rubber.

Industrial Sites and Noncrop Areas

PRAMITOL 25E can be applied prior to or up to 3 months after weed emergence. Application rates vary from 5-30 gals. of PRAMITOL 25E per acre, depending on climatic conditions, soil type, the weeds present, and the stage of growth of the weeds. In the folowing rate recommendations, the higher rates are intended for use on heavier soils where weed growth is heavy, and where rainfall is expected to be relatively high. The higher rates should also be used where longer residual control is desired in regions with a long growing season.

For best results, apply PRAMITOL 25E prior to weed emergence or when weeds are young and actively growing.

For control of annual and susceptible perennial weeds (such as downy bromegrass, oatgrass, goosegrass, quackgrass, puncturevine, goldenrod, and plantain): Broadcast 5-7½ gals. PRAMITOL 25E in 50-100 gals. of water per acre. For small areas, apply 1-1½ pts. PRAMITOL 25E per 1,000 sq. ft. in sufficient water to give thorough and uniform coverage.

For control of hard-to-kill perennial weeds and grasses (such as johnsongrass, bindweed, and wild carrot): Broadcast 20-30 gals. PRAMITOL 25E in 50-100 gals. of water per acre. For small areas, use 4-5½ pts. PRAMITOL 25E per 1,000 sq. ft. in sufficient water to give thorough and uniform coverage.

For faster top-kill of existing vegetation, apply PRAMITOL 25E in diesel oil, fuel oil, or weed oil. Use oils at rates of 100-200 gals. per acre. Tall dense vegetation such as johnsongrass will generally require 150 or more gals. of oil per acre for thorough coverage.

Asphalt Pavement

PRAMITOL 25E may be used to extend the useful life of asphalt pavement by preventing weeds from emerging through it. PRAMITOL 25E may be applied to the ground before laying asphalt, or it may be mixed with cutback asphalts.

PRAMITOL 25E should be used only where the area to be treated has been prepared according to good construction practices. If rhizomes, tubers, or other vegetative parts are present in the site, they should be removed by scalping with a grader blade. Recompaction of the site following disking or plowing will allow good penetration of PRAMITOL 25E, and will help to ensure proper asphalt curing and good weed control.

PRAMITOL 25E will not control woody vegetation such as small trees, brush, or woody vines. Roots of such species should be grubbed and removed from the site prior to application.

Under Asphalt: PRAMITOL 25E may be applied to the ground before laying rapid, medium, or slow curing asphalt coatings such as those used on parking lots, highway shoulders and median strips, roadways, and other industrial sites. PRAMITOL 25E should be applied just prior to the laying of the asphalt coating to prevent possible lateral movement of the herbicide by rainfall or other mechanical means. Sprayers should be equipped to provide continuous agitation of the spray mixture during application to ensure application of a uniform spray emulsion.

Apply PRAMITOL 25E uniformly at the rate of 20-25 gals. per acre (0.53-0.66 fl. oz. per sq. yd.) in a minimum of 100 gals. of water. Use the lower rate to control annual and susceptible perennial weeds such as downy bromegrass, oatgrass, goosegrass, quackgrass, puncturevine, goldenrod, and plantain. Use the higher rate to control hard-to-kill perennial weeds and grasses such as johnsongrass, bindweed, and wild carrot.

Precaution: Do not use PRAMITOL 25E under asphalt coatings less than 2 inches thick.

Mixed with Cutback Asphalt: PRAMITOL 25E may be mixed with standard and special cutback asphalts (such as RC, MC, and SC) for weed control in areas being stabilized, such as highway shoulders, highway median strips, under highway fences and guard rails, along airport runways and lights, soil and small stone aggregates on shoulders, and cracked asphalt surfaces being resurfaced.

PRAMITOL 25E should be simultaneously mixed with the cutback asphalt as it is being pumped into the applicator or transport truck to assure a homogeneous mix. Temperatures in the applicator or transport truck should not exceed 170°F.

PRAMITOL 25E should be used at the rate of 0.53-0.80 fl. oz. per sq. yd. (equivalent to 20-30 gals./A of PRAMITOL 25E). Use the lower rate on ordinary weed and grass complexes such as dallisgrass, King Ranch bluestem, knotroot bristlegrass, and the higher rate where johnsongrass, bermudagrass, or other deep-rooted perennials are present.

In arid areas where deep-rooted perennials are a problem, best results can be obtained by making the application from midsummer through late fall or prior to the rainy season.

Precautions: 1) Do not apply cutback asphalt mixtures containing PRAMITOL 25E to soil saturated with water. The condition of the soil should be favorable for penetration or proper asphalt curing may be affected. 2) Do not apply PRAMITOL 25E with asphalt emulsions such as SS.

STORAGE AND DISPOSAL

Do not contaminate water, food, or feed by storage or disposal. Pesticide wastes may be acutely hazardous. Improper disposal of unused pesticide, spray mixture, or rinse water is a violation of federal law. If these wastes cannot be used according to label instructions, contact your State Pesticide or Environmental Control Agency, or the Hazardous Waste representative at the nearest EPA Regional Office for guidance in proper disposal methods.

Triple rinse container (or equivalent). Then offer for recycling or reconditioning, or puncture and dispose of in a sanitary landfill or by other procedures approved by state and local authorities.

For minor spills, leaks, etc., follow all precautions indicated on this label and clean up immediately. Take special care to avoid contamination of equipment and facilities during cleanup procedures and disposal of wastes. In the event of a major spill, fire or other emergency, call (919) 292-7100 day or night.

PRECAUTIONARY STATEMENTS

HAZARDS TO HUMANS AND DOMESTIC ANIMALS

DANGER

Corrosive—causes eye damage. Wear goggles or face shield and rubber gloves when handling. Causes skin irritation. Harmful if swallowed, inhaled or absorbed through the skin. Avoid breathing spray mists. Do not get in eyes, on skin, or on clothing. Avoid contamination of food.

First Aid: In case of contact with eyes, immediately flush with plenty of water for 15 minutes and get medical attention; with skin, wash immediately with plenty of soap and water. If swallowed, drink *promptly* a large quantity of milk, egg whites, or gelatin solution; or if these are not available, drink large quantities of water. Avoid alcohol. Call a physician immediately. In case of inhalation exposure, remove from contaminated areas. Wash thoroughly after handling and before eating and smoking. Remove and wash contaminated clothing before reuse.

Note to Physician: Probable mucosal damage may contraindicate the use of gastric lavage. Treat for circulatory shock, respiratory depression, and convulsions if needed.

ENVIRONMENTAL HAZARDS

Do not contaminate water used for irrigation or domestic purposes.

Do not contaminate water by cleaning of equipment, or disposal of wastes. Apply this product only as specified on this label.

PHYSICAL OR CHEMICAL HAZARDS

Do not use, pour, spill, or store near heat or open flame. Store at temperatures above 32°F.

PRAMITOL® trademark of CIBA–GEIGY for prometon

CGA 8L5V 054

This is a specimen label which was not prepared by CIBA-GEIGY. CIBA-GEIGY is not responsible for the accuracy of the information contained herein. As labels are subject to revision, always carefully read and follow the label on the product container.

PRAMITOL® 5PS

PELLETED HERBICIDE

For control of weeds in noncrop areas

Bare ground weed control around . . . buildings, storage areas, fences, pumps, machinery, fuel tanks, recreational areas, roadways, guard rails, airports, military installations, highway medians, pipelines, railroads, lumberyards, industrial sites, rights-of-way . . . or similar areas

ACTIVE INGREDIENTS:

Prometon: 2,4-bis (isopropylamino)-6-methoxy-*s*-triazine	5.00%
Simazine: 2-chloro-4,6-bis (ethylamino)-*s*-triazine	0.75%
Sodium chlorate ($NaClO_3$)	40.00%
Sodium metaborate* ($Na_2B_2O_4$-$5H_2O$)	50.00%
INERT INGREDIENTS:	4.25%

* Equivalent to 4.87% boron expressed as elemental boron

EPA Reg. No. 100-479 EPA Est. 33560-TN-1ⓒ

EPA Est. 10659-MS-1 S (Superscript is first letter of lot number on bag)

KEEP OUT OF REACH OF CHILDREN

DANGER

IMPORTANT: Read the entire **Directions for Use** and the **Conditions of Sale and Warranty** before using this product.

CONDITIONS OF SALE AND WARRANTY

The **Directions for Use** of this product reflect the opinion of experts based on field use and tests. The directions are believed to be reliable and should be followed carefully. However, it is impossible to eliminate all risks inherently associated with use of this product. Crop injury, ineffectiveness, or other unintended consequences may result because of such factors as weather conditions, presence of other materials, or the manner of use or application all of which are beyond the control of CIBA–GEIGY or the Seller. All such risks shall be assumed by the Buyer.

CIBA–GEIGY warrants that this product conforms to the chemical description on the label and is reasonably fit for the purposes referred to in the **Directions for Use** subject to the inherent risks referred to above. **CIBA–GEIGY makes no other express or implied warranty of Fitness or Merchantability or any other express or implied warranty. In no case shall CIBA–GEIGY or the Seller be liable for consequential, special, or indirect damages resulting from the use or handling of this product.** CIBA–GEIGY and the Seller offer this product, and the Buyer and user accept it, subject to the foregoing **Conditions of Sale and Warranty**, which may be varied only by agreement in writing signed by a duly authorized representative of CIBA–GEIGY.

DIRECTIONS FOR USE

It is a violation of federal law to use this product in a manner inconsistent with its labeling.

GENERAL INFORMATION

PRAMITOL 5PS is a nonselective herbicide that can be applied before or after plant growth begins. Since PRAMITOL enters plants through their roots, its effectiveness is dependent on rainfall to move the chemical into the root zone. Very dry soil conditions and lack of rainfall may result in poor weed control.

Important: Use only in areas where complete control of all vegetation is desired, such as industrial sites, rights-of-way, lumberyards, petroleum tank farms, around farm buildings, along fence lines, and similar areas. When applied to the soil, this product usually inhibits all plant growth for a year or more. It should not be used on land to be cropped, or near adjacent desirable trees, shrubs, plants, or in greenhouses or injury may occur.

To control annual broadleaf weeds and grasses: Apply ½-1 lb. per 100 sq. ft.

To control most perennial broadleaf weeds and grasses: Apply 1-2 lbs. per 100 sq. ft. Where deep-rooted perennials are present use the 2 lb. rate.

The higher rates listed above should be used in areas having a long growing season or high rainfall. The higher rates will also provide longer residual control.

Wash out equipment after application. Do not use equipment to apply insecticides, fungicides, or fertilizers.

STORAGE AND DISPOSAL

Store in a dry place. Do not contaminate water, food, or feed by storage or disposal. Pesticide wastes may be acutely hazardous. Improper disposal of unused pesticide is a violation of federal law. If these wastes cannot be used according to label instructions, contact your State Pesticide or Environmental Control Agency, or the Hazardous Waste representative at the nearest EPA Regional Office for guidance in proper disposal methods.

Completely empty bag into application equipment. Dispose of empty bag in a sanitary landfill or by incineration, or if allowed by state and local authorities, by burning. Stay out of smoke from burning bags.

For minor spills, leaks, etc., follow all precautions indicated on this label and clean up immediately. Take special care to avoid contamination of equipment and facilities during cleanup procedures and disposal of wastes. In the event of a major spill, fire or other emergency, call (919) 292-7100 day or night.

PRECAUTIONARY STATEMENTS

HAZARDS TO HUMANS AND DOMESTIC ANIMALS

DANGER

Corrosive—causes eye and skin damage. Do not get in eyes, on skin or clothing. Wash clothing after handling. Wear goggles or face shield and rubber gloves when handling. Harmful if swallowed, inhaled or absorbed through the skin. Do not breathe dusts. Do not contaminate food of feed.

First Aid: In case of contact with skin, wash immediately with plenty of soap and water. In case of contact with the eyes, flush with water for at least 15 minutes and get medical attention. If swallowed drink *promptly* a large quantity of milk, egg whites, gelatin solution or if these are not available, drink large quantities of water. Avoid alcohol. Call a physician immediately. In case of inhalation exposure, move from the contaminated area.

Note to Physician: Probable mucosal damage may contraindicate the use of gastric lavage. Treat for circulatory shock, respiratory depression, and convulsion if needed.

ENVIRONMENTAL HAZARDS

Do not apply directly to water or wetlands. Do not contaminate water by cleaning of equipment or disposal of wastes. Simazine is a chemical which can travel (seep or leach) through soil and enter ground water which may be used as drinking water. Simazine has been found in ground water as a result of its use as a herbicide. Users of this product are advised not to apply simazine where the water table (ground water) is close to the surface and where the soils are very permeable, i.e., well-drained soils such as loamy sands. Users are advised to consult with their local agricultural agencies to obtain information on the location of ground water and the type of soil in their area.

PRAMITOL® trademark of CIBA–GEIGY for prometon

CGA 27L3N 047

This is a specimen label which was not prepared by CIBA-GEIGY. CIBA-GEIGY is not responsible for the accuracy of the information contained herein. As labels are subject to revision, always carefully read and follow the label on the product container.

PRIME+®

A Plant Growth Regulator for Control of Tobacco Suckers

ACTIVE INGREDIENT:

2-chloro-*N*-[2,6-dinitro-4-(trifluoro-methyl) phenyl]-*N*-ethyl-6-fluoro-benzene-methanamine	15%
INERT INGREDIENTS:	85%
TOTAL	100%

EPA Reg. No.100-640 EPA Est. 100-LA-1

PRIME+ contains 1.2 lbs. active ingredient per gallon.

KEEP OUT OF REACH OF CHILDREN

DANGER/PELIGRO

PRECAUCIÓN AL USUARIO: Si usted no lee inglés, no use este producto hasta que la etiqueta haya sido explicado ampliamente.

See additional precautionary statements inside booklet.

See directions for use inside booklet.

PRIME+® trademark of CIBA–GEIGY

CGA 73L9D 077

DIRECTIONS FOR USE AND CONDITIONS OF SALE AND WARRANTY

IMPORTANT: Read the entire **Directions for Use** and the **Conditions of Sale and Warranty** before using this product.

CONDITIONS OF SALE AND WARRANTY

The **Directions for Use** of this product reflect the opinion of experts based on field use and tests. The directions are believed to be reliable and should be followed carefully. However, it is impossible to eliminate all risks inherently associated with the use of this product. Crop injury, ineffectiveness, or other unintended consequences may result because of such factors as weather conditions, presence of other materials, or the manner of use or application all of which are beyond the control of CIBA–GEIGY or the Seller. All such risks shall be assumed by the Buyer.

CIBA–GEIGY warrants that this product conforms to the chemical description on the label and is reasonably fit for the purposes referred to in the **Directions for Use** subject to the inherent risks referred to above. **CIBA–GEIGY makes no other express or implied warranty of Fitness or Merchantability or any other express or implied warranty. In no case shall CIBA–GEIGY or the Seller be liable for consequential, special, or indirect damages resulting from the use or handling of this product.** CIBA–GEIGY and the Seller offer this product, and the Buyer and user accept it, subject to the foregoing **Conditions of Sale and Warranty**, which may be varied only by agreement in writing signed by a duly authorized representative of CIBA–GEIGY.

DIRECTIONS FOR USE

It is a violation of federal law to use this product in a manner inconsistent with its labeling.

FAILURE TO FOLLOW THE DIRECTIONS FOR USE AND PRECAUTIONS ON THIS LABEL MAY RESULT IN POOR SUCKER CONTROL AND/OR CROP INJURY.

Do not apply this product through any type of irrigation system.

Do not apply this product in such a manner as to directly or through drift expose workers or other persons except those knowingly involved in the application. The area being treated must be vacated by unprotected persons.

REENTRY STATEMENT

Do not enter treated areas without protective clothing until sprays have dried.

Because certain states may require more restrictive reentry intervals for various crops treated with this product, consult your State Department of Agriculture for further information.

Written or oral warnings must be given to workers who are expected to be in a treated area or in an area about to be treated with this product. Oral warnings must inform workers of areas or fields that may not be entered without specific protective clothing until sprays have dried, and appropriate actions to take in case of accidental exposure, as described under **Precautionary Statements** on this label. When oral warnings are given, warnings shall be given in a language customarily understood by workers. Oral warnings must be given if there is reason to believe that written warnings cannot be understood by workers. Written warnings must include the following information: "DANGER. Area treated with PRIME+ on (date of application). Do not enter without appropriate protective clothing until sprays have dried. In case of accidental eye exposure, flush eyes with plenty of water. Get medical attention. In case of accidental skin contact, wash skin with soap and water. Call physician if irritation persists. Remove and wash contaminated clothing before reuse."

GENERAL INFORMATION

PRIME+ is a plant growth regulator for the control of sucker growth on flue-cured, burley, dark, Maryland, and cigar tobacco.

One application of PRIME+ will provide full season control of excessive sucker development in tobacco. Read this label carefully and observe all directions and precautions.

MIXING INSTRUCTIONS

Mix 1 gal. of PRIME+ in 49 gals. of water or 2.5 fl. oz. PRIME+ per one gallon of water. Provide sufficient agitation during mixing and application to maintain a uniform mixture.

Note: Do not mix PRIME+ with insecticides or fungicides.

APPLICATION INSTRUCTIONS

Time of Application

Apply PRIME+ when most tobacco plants are in the elongated button to early flower stage, either before or after topping. If PRIME+ is applied after topping, apply within 24 hours of topping. During topping, remove all suckers one inch or longer from all plants. (See notes listed below.)

Precision Hand Application

Mix 1 gal. of PRIME+ in 49 gals. of water and apply as a coarse spray or drench to the top of the tobacco stalk using a handheld drop line, or mix 2.5 fl. oz. of PRIME+ per gallon of water when using the jug method or backpack sprayer. **Apply only once per plant per season.** Apply the spray mixture at a rate sufficient to run down the stalk contacting each leaf axil and just reach the ground line. **Excessive spray volume may result in carryover injury to rotational crops.** Spray volume per plant will generally be 1/3 - 2/3 fl. oz. depending on plant height. **Do not exceed 30 gals. spray volume per acre when using hand application methods.** (See notes listed below.)

Application with Powered Spray Equipment (3 nozzles per row)

Mix 1 gal. of PRIME+ in 49 gals. of water and apply at a rate of 30 to 50 gals. spray volume per acre. Use the high spray volume in tall types of tobacco, such as flue-cured, burley, Maryland, and cigar. Use the low spray volume only in short types of tobacco, such as dark. **Apply only once per season. Avoid application to plants treated by precision hand application method.** Spray volume should be sufficient to run down the stalk contacting each leaf axil and just reach the ground line. **Excessive spray volume beyond the point of runoff to the soil may result in carryover injury to rotational crops.** (See notes listed below.)

Apply PRIME+ as a coarse spray directed downward to the top of the tobacco stalk. This is best accomplished with three full cone nozzles per row; one nozzle directed downward over the center of the row and one 7-9 inches to each side of the center, directed at the top of the stalk. Maintain nozzles at a height of 12-16 inches above the top of the stalk during application. Maintain boom pressure at 20-25 psi, not exceeding 30 psi.

Sequential Application with Maleic Hydrazide

Mix 2 qts. of PRIME+ in 49.5 gals. of water and apply either by Precision Hand Application or Application with Powered Spray Equipment as described in those sections of this label. Timing of PRIME+ should be during the elongated button to early flower stage of tobacco.

Approximately one week after the PRIME+ Application, apply maleic hydrazide according to the directions, precautions, and restrictions on that label.

Notes

1. Timing is important; applications should be made at timing described above so that number and size of suckers will be small. Applications made before the elongated button stage of growth may result in chemical topping or distortion of leaves that were too immature at time of application. Spraying too late may result in poor sucker control.
2. PRIME+ will not control large suckers. All suckers one inch or longer on all plants must be removed before application of PRIME+. Suckers that escape treatment should be removed by hand about 10-14 days after application.
3. Where a contact (fatty alcohol) is commonly used to control early suckers and provide a more uniform crop, the normal sequence would be: (a) apply contact material according to its label, and (b) about 5 days after application of contact material, apply PRIME+. Timing of PRIME+ application should be during the elongated button to early flower stage.
4. Where a contact (fatty alcohol) is not used and a wide range of flowering exists, early flowering plants should be topped and suckers removed by hand. Then, when most plants are in the elongated button to early flower stage, remove all suckers one inch or longer, and apply PRIME+. If PRIME+ is applied using precision hand application, early flowering plants can be topped and treated selectively; later flowering plants can be topped and treated as they reach the proper treatment stage; and a contact may not be needed.
5. For proper sucker control, PRIME+ must contact the small suckers on the axil where the leaf and stalk connect. PRIME+ will not control suckers on plants that are leaning at time of application. Applications with powered spray equipment during windy periods are not recommended. High wind will disrupt the spray pattern of nozzles on boom sprayers and turn leaves causing improper contact and flow of the spray solution, resulting in poor sucker control. Application to wet plants, or drought stressed plants may result in poor sucker control.
6. Carefully follow the use rates and spray volume recommended, and spray field uniformly. Volume should be sufficient to run down the stalk, contact each leaf axil, and just reach the ground line.
7. Rain occurring within 2 hours after spraying may reduce effectiveness.

Precautions: (1) To avoid illegal residues and possible injury to crops planted in the fall or following spring, do not apply excessive spray volume to the point of runoff onto the soil. PRIME+ residues in the soil may interact with dinitroaniline herbicides such as Balan®, Basalin®, Paarlan®, Prowl®, or Treflan® to increase early season crop stunting. Thorough disking and deep tillage in the fall are recommended where excessive amounts of PRIME+ may have reached the soil. (2) When tobacco is in a state of very rapid, succulent growth due to high nitrogen and rainfall, PRIME+ may cause injury. Under these conditions, use only the precision hand application method and avoid contacting leaves with the spray solution.

STORAGE AND DISPOSAL

Pesticide Disposal

Do not contaminate water, food, or feed by storage or disposal. Open dumping is prohibited. Wastes resulting from the use of this product are acutely hazardous. Improper disposal of unused pesticide, spray mixture, or rinsate that cannot be used according to label instructions must be disposed of according to federal, state or local procedures. For guidance in proper disposal methods, contact your State Pesticide or Environmental Control Agency, or the Hazardous Waste representative at the nearest EPA Regional Office.

Container Disposal

Do not reuse empty container. Triple rinse (or equivalent) and offer for recycling or reconditioning, or puncture and dispose of in a sanitary landfill, or by incineration or by open burning, if allowed by state and local authorities. If burned, keep out of smoke.

For minor spills, leaks, etc., follow all precautions indicated on this label and clean up immediately. Take special care to avoid contamination of equipment and facilities during cleanup procedures and disposal of wastes. In the event of a major spill, fire, or other emergency, call (919) 292-7100 day or night.

Store at temperatures above 32°F.

PRECAUTIONARY STATEMENTS

HAZARDS TO HUMANS AND DOMESTIC ANIMALS

DANGER

Corrosive—causes eye damage. Causes skin irritation or skin sensitization on repeated contact. Do not get in eyes, on skin, or on clothing. Wear goggles or face shield and protective clothing when handling. Harmful if swallowed, inhaled, or absorbed through the skin. Avoid breathing spray mist.

First Aid: In case of contact with the eyes, immediately flush with plenty of water and get medical attention. In case of contact with skin, wash immediately with soap and water. If swallowed, drink plenty of water. Do not induce vomiting. Avoid alcohol. Call a phisician immediately. In case of inhalation exposure, move from the contaminated area.

Note to Physician: There is no specific antidote for PRIME+. Lavage stomach. Give a saline laxative or slurry of activated carbon in water and supportive therapy.

ENVIRONMENTAL HAZARDS

This pesticide is toxic to fish. Do not apply directly to lakes, streams, or ponds. Do not apply when weather conditions favor drift from treated areas. Do not contaminate water by cleaning of equipment or disposal of wastes.

Do not apply this product through any type of irrigation system.

Balan® trademark of Elanco Products for benefin
Basalin® trademark of BASF/Wyandotte Corp. for fluchloralin
Paarlan® trademark of Elanco Products for isopropalin
PRIME+® trademark of CIBA-GEIGY
Prowl® trademark of American Cyanamid for pendimethalin
Treflan® trademark of Elanco Products for trifluralin

CGA 73L9D 077

This is a specimen label which was not prepared by CIBA-GEIGY. CIBA-GEIGY is not responsible for the accuracy of the information contained herein. As labels are subject to revision, always carefully read and follow the label on the product container.

PRINCEP® 4G

GRANULAR HERBICIDE

For weed control in certain crops, ornamental plantings, on industrial sites, and around-the-farm

Princep 4G is a granular herbicide. See directions for use inside booklet.

ACTIVE INGREDIENT:	
Simazine: 2-chloro-4,6-bis (ethylamino)-*s*-triazine.	4%
INERT INGREDIENTS:	96%
TOTAL.	100%

EPA Reg. No. 100-435

KEEP OUT OF REACH OF CHILDREN CAUTION

See additional precautionary statements at end of label booklet.

DIRECTIONS FOR USE AND CONDITIONS OF SALE AND WARRANTY

IMPORTANT: Read the entire **Directions for Use** and the **Conditions of Sale and Warranty** before using this product.

CONDITIONS OF SALE AND WARRANTY

The **Directions for Use** of this product reflect the opinion of experts based on field use and tests. The directions are believed to be reliable and should be followed carefully. However, it is impossible to eliminate all risks inherently associated with use of this product. Crop injury, ineffectiveness, or other unintended consequences may result because of such factors as weather conditions, presence of other materials, or the manner of use or application all of which are beyond the control of CIBA-GEIGY or the Seller. All such risks shall be assumed by the Buyer.

CIBA-GEIGY warrants that this product conforms to the chemical description on the label and is reasonably fit for the purposes referred to in the **Directions for Use** subject to the inherent risks referred to above. **CIBA-GEIGY makes no other express or implied warranty of Fitness or Merchantability or any other express or implied warranty. In no case shall CIBA-GEIGY or the Seller be liable for consequential, special, or indirect damages resulting from the use or handling of this product.** CIBA-GEIGY and the Seller offer this product, and the Buyer and user accept it, subject to the foregoing **Conditions of Sale and Warranty**, which may be varied only by agreement in writing signed by a duly authorized representative of CIBA-GEIGY.

DIRECTIONS FOR USE

It is a violation of federal law to use this product in a manner inconsistent with its labeling.

FAILURE TO FOLLOW THE DIRECTIONS FOR USE AND PRECAUTIONS ON THIS LABEL MAY RESULT IN CROP INJURY, POOR WEED CONTROL, AND/OR ILLEGAL RESIDUES.

Do not apply this product through any type of irrigation system.

Do not apply this product in such a manner as to directly expose workers or other persons except those knowingly involved in the application. The area being treated must be vacated by unprotected persons.

Written or oral warnings must be given to workers who are expected to be in an area about to be treated with this product. Oral warnings must inform workers of areas or fields that may not be entered without specific protective clothing during application and appropriate actions to take in case of accidental exposure, as described under **Precautionary Statements** on this label. When oral warnings are given, warnings shall be given in a language customarily understood by workers. Oral warnings must be given if there is reason to believe that written warnings cannot be understood by workers. Written warnings must include the following information: CAUTION. Area will be treated with PRINCEP 4G. Do not enter without appropriate protective clothing during application. In case of accidental exposure, flush eyes or skin with plenty of water. Call physician if irritation persists. Remove and wash contaminated clothing before reuse.

GENERAL INFORMATION

Apply PRINCEP 4G herbicide before weeds emerge or following removal of weed growth. PRINCEP controls a wide variety of annual broadleaf and grass weeds when used at selective rates in agricultural crops and ornamental plantings. When used at higher, nonselective rates in noncrop areas, it also controls many perennial broadleaf and grass weeds.

Since PRINCEP enters weeds mainly through their roots, moisture is needed to move it into the root zone. Very dry soil conditions and lack of rainfall following application may necessitate shallow cultivation.

PRINCEP 4G is noncorrosive to equipment, nonflammable, and has low electrical conductivity.

Annual Weeds Controlled

alyssum	knawel (German moss)
annual bluegrass	common lambsquarters
annual morningglory	nightshade
annual ryegrass	pepperweed
barnyardgrass (watergrass)	pigweed
burclover	pineappleweed
carelessweed	prickly lettuce
carpetweed	common purslane
common chickweed	quackgrass
crabgrass (*Digitaria* spp.)	ragweed
downy brome (cheatgrass)	rattail fescue
fall panicum	redmaids
fiddleneck	Russian thistle
filaree	shepherdspurse
fireweed	signalgrass (*Brachiaria* spp.)
fivehook bassia	silver hairgrass
Flora's paintbrush	smartweed
Florida pusley	spanishneedles
foxtails	speedwell
goosegrass	tansymustard
groundsel	wild mustard
henbit	wild oats
junglerice	witchgrass
	yellow flower pepperweed

Following many years of continuous use of this product and chemically related products, biotypes of some of the weeds listed on this label have been reported which cannot be effectively controlled by this and related herbicides. Where this is known or suspected and weeds controlled by this product are expected to be present along with resistant biotypes, we recommend the use of this product in sequential combination with other registered herbicides which are not triazines. If only resistant biotypes are expected to be present, use a registered non-triazine herbicide. Consult with your State Agricultural Extension Service for specific recommendations.

SELECTIVE WEED CONTROL

Where a range of dosage rates is given, use the low rate on coarser textured soil and soil lower in organic matter; use the high rate on finer textured soil and soil higher in organic matter. For band applications, reduce the rate of PRINCEP 4G in proportion to the area treated. For example: when applying a 2-foot band to a crop planted in rows 6 feet apart, apply one-third the recommended rate of PRINCEP 4G per acre of field.

Distribute PRINCEP 4G uniformly over the soil surface by means of a cyclone-type spreader or other suitable application equipment. For treatment of small areas, application by hand may be adequate.

After treatment with PRINCEP 4G, do not plant any crop other than those specified on this label until the following year or injury may occur. Do not rotate with oats, sugar beets, or vegetables.

FRUIT CROPS

Blueberries

Apply 50-100 lbs. per acre in the spring before bud break and weed emergence, **or** apply a split application of 50 lbs. per acre in the spring before bud break and weed emergence, and 50 lbs. per acre in the fall. On plantings less than 6 months old, use 1/2 the above rate.

To control quackgrass, apply 100 lbs. per acre in the fall **or** split the application, applying 50 lbs. per acre in the fall and 50 lbs. per acre in the early spring, when quackgrass is actively growing and before bud break.

Note: Do not apply when fruit is present, or illegal residues may result.

Caneberries; i.e., blackberries, boysenberries, loganberries, raspberries

Apply 50-100 lbs. per acre in the spring before bud break and weed emergence. On plantings less than 6 months old, use 1/2 the above rate.

To control quackgrass, apply 100 lbs. per acre in the fall **or** split the application, applying 50 lbs. per acre in the fall and 50 lbs. per acre in the early spring when quackgrass is actively growing.

Note: Do not apply when fruit is present, or illegal residues may result.

Cranberries

Massachusetts: Apply up to 100 lbs. per acre either in the spring before cranberry growth or in the fall after harvest.

Other areas: Apply 50 lbs. per acre in the spring before new cranberry growth.

Grapefruit and Oranges

Florida: Apply 160-240 lbs. per acre in the spring (March or April) prior to weed emergence. Repeat annually.

Texas: Apply 100-125 lbs. per acre in the spring prior to weed emergence. Repeat annually.

Precautions: To avoid crop injury, (1) Do not use in nurseries. (2) Do not apply to bedded citrus. (3) Use only where trees have been established in the grove for one year or more. (4) Do not apply to trees under stress from freeze damage for one year after the freeze. (5) Avoid contact with fruit or foliage.

NURSERIES, CHRISTMAS TREE PLANTINGS, SHELTERBELTS

Nurseries (see list below)

Apply 50-75 lbs. per acre in fall or spring.

Precaution: To avoid plant injury, do not apply for at least one year after transplanting.

Christmas Tree Plantings and Shelterbelts (see list below)

Remove weed growth before application. Apply 50-100 lbs. per acre after transplanting. Use the same rate for annual maintenance applications.

For quackgrass control, apply 100 lbs. per acre in the fall or apply a split application of 50 lbs. per acre in the fall plus 50 lbs. per acre in early spring, after quackgrass begins growth.

Precautions: To avoid tree injury, (1) Do not use PRINCEP 4G on seedbeds or cutting beds. (2) In CA, OR, and WA, do not apply to Christmas trees or shelterbelts sooner than one year after transplanting. In other areas, do not apply to Christmas tree or shelterbelt transplants less than three years of age. (3) Do not use until soil is firmly settled around roots. (4) Do not apply more than once a year, except as directed for quackgrass control.

Apply PRINCEP 4G to these species of trees and shrubs as recommended above:

Conifers

arborvitae	Scotch pine	Fraser fir
Austrian pine	white pine	white fir
* Bishop pine	blue spruce	hemlock
* knobcone pine	Norway spruce	juniper
lodgepole pine	red spruce	red cedar
(shore pine)	white spruce	white cedar
* Monterey pine	balsam fir	yew
Mugho pine	Douglas fir	(*Taxus* spp.)
red pine		
(Norway pine)		* For California only

Deciduous Trees and Woody Ornamentals

American elm	Oregon grape
Siberian elm	(*Mahonia* spp.)
barberry	*Pieris* spp.
boxelder	(max. 80 lbs./A)
bush honeysuckle	red oak
caragana	Russian olive
cotoneaster	* oleander
dogwood	* palm (50 lbs./A)
eucalyptus	* bottle brush
holly (*Ilex* spp.)	* carob
(max. 80 lbs./A)	
honey locust	* For California only.

TURFGRASS FOR FAIRWAYS, LAWNS, AND SIMILAR AREAS

Bermudagrass, Centipedegrass, St. Augustinegrass, and Zoysia Grass

Apply PRINCEP 4G after September 1 (after October 1 for annual bluegrass) before emergence of winter annual weeds. PRINCEP will control annual bluegrass, burclover, lawn burweed, common and mouseear chickweed, corn speedwell, henbit, hop clover, spurweed, and parsley-piert. PRINCEP will also control these weeds soon after emergence. For control of summer annual weeds listed in the **General Information** section of this label, apply PRINCEP in late winter before the weeds emerge.

Irrigate with 1/2 inch of water if rainfall does not occur within 10 days after preemergence treatment.

Where annual bluegrass is the major weed, use 25 lbs. of PRINCEP 4G per acre (9 oz. per 1,000 sq. ft.). Use 25-50 lbs. per acre (9-18 oz. per 1,000 sq. ft.) for control of other weeds. However, do not exceed 25 lbs. per acre per treatment on newly sprigged turfgrass or on hybrid bermudagrass such as Tiflawn, Tifway, and Ormond.

For control of summer annuals which emerge after the initial application, apply an additional 25 lbs. per acre at least 30 days after the initial application, but not after June 1. However, do not make more than two applications per year.

Precautions: On newly sprigged turfgrass, hybrid bermudagrass, nondormant bermudagrass or nondormant zoysia grass, temporary slowing of growth and yellowing may occur following application. To avoid turf injury, (1) Use only on turfgrass reasonably free of infestations of insects, nematodes, and diseases. (2) Do not use on golf greens. (3) Do not use north of NC (except may be used in the Virginia Coastal Plains) or west of Highway 281 in Texas and Oklahoma. (4) Do not use on muck or alkaline soils. (5) Do not apply over the rooting area of trees or ornamentals not listed on this label. (6) Do not seed or overseed with desirable turfgrass within 4 months before or 6 months after treatment. (7) Do not apply this product to newly seeded grasses until they have overwintered and have a well-developed rhizome system. (8) Do not exceed 48 lbs. per acre within 12 months of seeding grasses.

Note: Do not graze or feed turf clippings to animals, or illegal residues may result.

NONSELECTIVE WEED CONTROL ON NONCROP LAND

Apply PRINCEP 4G at rates recommended below. Use the higher rates on heavy clay and muck soils. Best results are obtained when PRINCEP 4G is applied before weeds emerge.

Do not contaminate domestic or irrigation water supplies, or lakes, streams, or ponds.

PRINCEP 4G provides long-term weed control on industrial sites, highway medians and shoulders, railroad rights-of-way, lumberyards, petroleum tank farms, and in noncrop areas on farms, such as around buildings, equipment and fuel storage areas, along fences, roadsides, and lanes.

Precaution: Do not use this product for nonselective weed control on land to be cropped, near adjacent desirable trees, shrubs or plants, or in greenhouses, as injury may occur.

To control most annual broadleaf and grass weeds (such as barnyardgrass, cheatgrass, crabgrass, lambsquarters, foxtail, ragweed, puncturevine, and turkey mullein): Broadcast 250 lbs. per acre.

To control most annual and many perennial broadleaf and grass weeds (including quackgrass, bluegrass, redtop, burdock, Canada thistle, orchardgrass, purple top, smooth brome, dogfennel, and plantain): Broadcast 250-450 lbs. per acre.

For small areas, 5 3/4 lbs. per 1,000 sq. ft. is equivalent to 250 lbs. per acre.

STORAGE AND DISPOSAL

Store in a dry place. Do not contaminate water, food, or feed by storage or disposal. Improper disposal of unused pesticide is a violation of federal law. If wastes cannot be used according to label instructions, contact your State Pesticide or Environmental Control Agency, or the Hazardous Waste representative at the nearest EPA Regional Office for guidance in proper disposal methods.

Completely empty bag into application equipment. Dispose of empty bag and box in a sanitary landfill or by incineration, or if allowed by state and local authorities, by burning. Stay out of smoke from burning containers.

For minor spills, follow all precautions indicated on this label and clean up immediately. Take special care to avoid contamination of equipment and facilities during cleanup procedures and disposal of wastes. In the event of a major spill, fire or other emergency, call (919) 292-7100 day or night.

PRECAUTIONARY STATEMENTS

HAZARDS TO HUMANS AND DOMESTIC ANIMALS

CAUTION

Causes eye irritation. Harmful if swallowed, inhaled, or absorbed through the skin. Avoid breathing dust. Avoid contact with skin, eyes, or clothing. Wash thoroughly with soap and water after handling. Remove and wash contaminated clothing before reuse.

First Aid: If in eyes, flush with plenty of water. Get medical attention if irritation persists. If on skin, wash with plenty of soap and water. If inhaled, remove victim to fresh air. If not breathing, give artificial respiration, preferably mouth-to-mouth; get medical attention.

ENVIRONMENTAL HAZARDS

Do not apply directly to water or wetlands. Do not contaminate water by cleaning of equipment or disposal of wastes. Simazine is a chemical which can travel (seep or leach) through soil and enter ground water which may be used as drinking water. Simazine has been found in ground water as a result of its use as a herbicide. Users of this product are advised not to apply simazine where the water table (ground water) is close to the surface and where the soils are very permeable, i.e., well drained soils such as loamy sands. Users are advised to consult with their local agricultural agencies to obtain information on the location of ground water and the type of soil in their area.

CGA 6L13Z 028

PRINCEP® 4L

HERBICIDE

For weed control in certain crops, ornamental plantings, on industrial sites, and around-the-farm

ACTIVE INGREDIENT:	
Simazine: 2-chloro-4,6-bis (ethylamino)-*s*-triazine	41.9%
INERT INGREDIENTS:	58.1%
TOTAL	100.0%

EPA Reg. No. 100-526 EPA Est. 100-LA-1

PRINCEP 4L contains 4 lbs. active ingredient per gallon.
Use entire contents at one time.

KEEP OUT OF REACH OF CHILDREN

CAUTION

See additional precautionary statements at end of label booklet.

See directions for use inside booklet.

PRINCEP® trademark of CIBA-GEIGY for simazine

CGA 6L32L 028

DIRECTIONS FOR USE AND CONDITIONS OF SALE AND WARRANTY

IMPORTANT: Read the entire **Directions for Use** and the **Conditions of Sale and Warranty** before using this product.

CONDITIONS OF SALE AND WARRANTY

The **Directions for Use** of this product reflect the opinion of experts based on field use and tests. The directions are believed to be reliable and should be followed carefully. However, it is impossible to eliminate all risks inherently associated with use of this product. Crop injury, ineffectiveness, or other unintended consequences may result because of such factors as weather conditions, presence of other materials, or the manner of use or application all of which are beyond the control of CIBA-GEIGY or the Seller. All such risks shall be assumed by the Buyer.

CIBA-GEIGY warrants that this product conforms to the chemical description on the label and is reasonably fit for the purposes referred to in the **Directions for Use** subject to the inherent risks referred to above. **CIBA-GEIGY makes no other express or implied warranty of Fitness or Merchantability or any other express or implied warranty. In no case shall CIBA-GEIGY or the Seller be liable for consequential, special, or indirect damages resulting from the use or handling of this product.** CIBA-GEIGY and the Seller offer this product, and the Buyer and user accept it, subject to the foregoing **Conditions of Sale and Warranty**, which may be varied only by agreement in writing signed by a duly authorized representative of CIBA-GEIGY.

DIRECTIONS FOR USE

It is a violation of federal law to use this product in a manner inconsistent with its labeling.

FAILURE TO FOLLOW THE DIRECTIONS FOR USE AND PRECAUTIONS ON THIS LABEL MAY RESULT IN CROP INJURY, POOR WEED CONTROL, AND/OR ILLEGAL RESIDUES.

Do not apply this product through any type of irrigation system.

Do not apply this product in such a manner as to directly or through drift expose workers or other persons except those knowingly involved in the application. The area being treated must be vacated by unprotected persons.

REENTRY STATEMENT

Do not enter treated areas without protective clothing until sprays have dried.

Because certain states may require more restrictive reentry intervals for various crops treated with this product, consult your State Department of Agriculture for further information.

Written or oral warnings must be given to workers who are expected to be in a treated area or in an area about to be treated with this product. Oral warnings must inform workers of areas or fields that may not be entered without specific protective clothing until sprays have dried, and appropriate actions to take in case of accidental exposure, as described under **Precautionary Statements** on this label. When oral warnings are given, warnings shall be given in a language customarily understood by workers. Oral warnings must be given if there is reason to believe that written warnings cannot be understood by workers. Written warnings must include the following information: "CAUTION. Area treated with PRINCEP 4L on (*date of application*). Do not enter without appropriate protective clothing until sprays have dried. In case of accidental exposure, flush eyes or skin with plenty of water. Call physician if irritation persists. Remove and wash contaminated clothing before reuse."

GENERAL INFORMATION

Apply PRINCEP 4L Herbicide before weeds emerge or after removal of weed growth. PRINCEP controls a wide variety of annual broadleaf and grass weeds when used at selective rates in agricultural crops and ornamental plantings. When used at higher, nonselective rates in noncrop areas, it also controls many perennial broadleaf and grass weeds.

Where a range of application rates is given, use the low rate on coarser textured soil and soil lower in organic matter; use the high rate on finer textured soil and soil higher in organic matter.

To avoid spray drift, do not apply under windy conditions. Avoid spray overlap, as crop injury may result.

Since PRINCEP enters weeds mainly through their roots, moisture is needed to move it into the root zone. Very dry soil conditions following application may necessitate shallow cultivation or rotary hoeing.

PRINCEP 4L is nonflammable.

Annual Weeds Controlled

alyssum	fireweed	common purslane
annual bluegrass	fivehook bassia	ragweed
annual morningglory	Flora's paintbrush	rattail fescue
	Florida pusley	redmaids
annual ryegrass	foxtails	Russian thistle
barnyardgrass (watergrass)	goosegrass	shepherdspurse
	groundsel	signalgrass (*Brachiaria* spp.)
burclover	henbit	
carelessweed	junglerice	silver hairgrass
carpetweed	knawel (German moss)	smartweed
common chickweed		spanishneedles
crabgrass (*Digitaria* spp.)	common lambsquarters	speedwell
		tansymustard
downy brome (cheatgrass)	nightshade	wild mustard
	pepperweed	wild oats
fall panicum	pigweed	witchgrass
fiddleneck	pineappleweed	yellow flower
filaree	prickly lettuce	pepperweed

Following many years of continuous use of this product and chemically related products, biotypes of some of the weeds listed on this label have been reported which cannot be effectively controlled by this and related herbicides. Where this is known or suspected and weeds controlled by this product are expected to be present along with resistant biotypes, we recommend the use of this product in registered combinations or in sequence with other registered herbicides which are not triazines. If only resistant biotypes are expected to be present, use a registered non-triazine herbicide. Consult with your State Agricultural Extension Service for specific recommendations.

APPLICATION PROCEDURES

Ground application: Use conventional ground sprayers equipped with nozzles that provide accurate and uniform application. Be certain that nozzles are uniformly spaced and the same size. Calibrate sprayer before use and recalibrate at the start of each season and when changing carriers. Unless otherwise specified, use a minimum of 20 gals. of spray mixture per acre.

Use a pump with capacity to (1) maintain 35-40 psi at nozzles, (2) provide sufficient agitation in tank to keep mixture in suspension, and (3) to provide a minimum of 20% bypass at all times. Use centrifugal pumps which provide propeller shear action for dispersing and mixing this product. The pump should provide a minimum of 10 gals./minute/100 gal. tank size circulated through a correctly positioned sparger tube or jets.

Use screens to protect the pump and to prevent nozzles from clogging. Screens placed on the suction side of the pump should be *16-mesh or coarser*. Do not place a screen in the recirculation line. Use 50-mesh or coarser screens between the pump and boom, and where required, at the nozzles. Check nozzle manufacturer's recommendations.

For band applications, calculate amount to be applied per acre as follows:

$$\frac{\text{band width in inches}}{\text{row width in inches}} \times \text{broadcast rate per acre} = \text{amount needed per acre of field}$$

Aerial application: Use aerial application only where specified in the use directions. Apply in a minimum of 1 gal. of water for each quart of PRINCEP 4L applied per acre. Avoid applications under conditions where uniform coverage cannot be obtained or where excessive spray drift may occur.

Avoid application to humans or animals. Flagmen and loaders should avoid inhalation of spray mist and prolonged contact with skin, and should wash thoroughly before eating and at the end of each day's operation.

Mixing procedures—all uses: (1) Be sure sprayer is clean and not contaminated with any other materials, or crop injury or sprayer clogging may result. (2) Fill tank 1/4 full with clean water, nitrogen solution, or complete fluid fertilizer. (3) Start agitation. (4) Be certain that the agitation system is working properly and creates a rippling or rolling action on the liquid surface. (5) If any dry herbicides are being tank mixed, add them to tank and allow them to become thoroughly suspended. (6) Then pour PRINCEP 4L directly from jug into tank. (7) Continue filling tank until 90% full. Increase agitation if necessary to maintain surface action. (8) Add other liquid tank mix herbicide(s) after this product is thoroughly suspended. (9) Finish filling tank. Maintain agitation to avoid separation of materials. (10) Empty tank as completely as possible before refilling to prevent buildup of emulsifiable concentrate residue from possible tank mix herbicides. (11) If an emulsifiable concentrate film starts to build up in tank, drain it and clean with strong detergent solution or solvent. (12) Clean sprayer thoroughly immediately after use by flushing system with water containing a detergent.

Compatibility test: To determine the tank mix compatibility of PRINCEP 4L with liquid fertilizer, crop oil, spreaders, or other recommended pesticides, use this test method. Nitrogen solutions or complete fluid fertilizers may replace all or part of the water in the spray in corn. Since liquid fertilizers can vary, even within the same analysis, check compatibility each time before use. Be especially careful when using **complete** suspension or fluid fertilizers as serious compatibility problems are more apt to occur. Commercial application equipment may improve compatibility in some instances. Check compatibility using this procedure:

1. Add 1 pint of fertilizer to each of 2 one-quart glass jars with tight lids.
2. To **one** of the jars add 1/4 tsp. of a compatibility agent approved for this use (1/4 tsp. is equivalent to 2 pts. per 100 gals. spray). Cap and shake until mixed. Examples of compatibility agents include Compex® and Unite®.
3. To **both** jars add the appropriate amount of herbicide(s). If more than one herbicide is used, add them separately with dry herbicides first, flowables next, and emulsifiable concentrates last. After each addition, cap and shake until thoroughly mixed. The appropriate amount of herbicides for this test follows (assuming a spray volume of 25 gals. per acre):

   Dry herbicides: **For each pound per acre** add 1.5 teaspoons to each jar. Fluff up wettable powder products before measuring.

   Liquid herbicides: **For each pint per acre,** add 0.5 teaspoon to each jar.

   For a spray volume other than 25 gals. per acre, change the teaspoons added to each jar as follows:

   $$\frac{\text{25 gals.} \times \text{no. teaspoons given above}}{\text{desired gals. spray volume per acre}} = \frac{\text{teaspoons to add}}{\text{to each jar}}$$

4. After adding all ingredients, put lids on and tighten. Shake jars vigorously one minute. Let the mixtures stand 15 minutes and then look for separation, large flakes, precipitates, grease, gels, medium to heavy oily film on the jar, or other signs of incompatibility. Determine if the compatibility agent is needed in the spray mixture by comparing the two jars. If either mixture separates, but can be remixed readily, the mixture can probably be sprayed as long as good agitation is used. If the mixtures are incompatible, test the following methods of improving compatibility: (A) slurry the dry herbicide(s) in water before addition, or (B) add 1/2 of the compatibility agent to the fertilizer and the other 1/2 to the emulsifiable concentrate or flowable herbicide before addition to the mixture. If still incompatible, do not use the materials mixed in the same spray tank.

Rotational Crops and Perennial Crop Replanting

To avoid crop injury, observe the following precautions: (1) If rotating treated land the year following application, plant only corn, unless otherwise stated in this label. (2) If replanting perennial crops or if rotating land to crops other than corn, do not apply PRINCEP in the year preceding planting these crops.

FRUIT AND NUT CROPS

Apply the spray to the orchard or vineyard floor avoiding contact with fruit, foliage, or stems. Recommended rates are based on broadcast treatment. For band applications or spot applications around trees in fruit or nut plantings, reduce the broadcast rate of PRINCEP 4L and water per acre in proportion to the area actually sprayed.

Precautions (All fruit and nut crops): To avoid crop injury, (1) Apply only to orchards or groves where trees have been established 12 months or more unless specified differently. (2) Make only one application per year, except as noted otherwise. (3) Do not use on gravelly, sand, or loamy sand soil. (4) Immediately following application, limit overhead sprinkler irrigation to 1/2 inch.

Almonds and Peaches (CA only)

Apply 1-2 qts. of PRINCEP 4L per treated acre in a 2-4 ft. band on each side of the tree row. Apply before weeds emerge in late fall or early winter. Weeds controlled by 1 qt. of PRINCEP 4L include burclover, common chickweed, wild mustard, and shepherdspurse. Apply only once per year.

Precautions: To avoid crop injury, (1) Do not treat trees established in the grove less than 3 years. (2) Do not treat the Mission (Texas) variety of almonds. (3) Do not apply to almond trees propagated on plum rootstocks. (4) Do not replant almonds or peaches in treated soil for 12 months after treatment. (5) Do not apply on soil with less than 1% organic matter. (6) Do not treat areas where water will accumulate.

Apples, Pears, Sour Cherries

Apply 2-4 qts. per acre.

Avocados

CA and FL only: Apply 2-4 qts. per acre after final preparation of grove.

Blueberries and Caneberries (blackberries, boysenberries, loganberries, raspberries)

Apply 2-4 qts. per acre in the spring or apply a split application of 2 qts. per acre in the spring plus 2 qts. per acre in the fall. Apply in a minimum of 40 gals. of water per acre.

On plantings less than 6 months old, use 1/2 the above rate.

To control quackgrass, apply 4 qts. per acre in the fall or split the application applying 2 qts. per acre in the fall plus 2 qts. per acre in the spring, when quackgrass is actively growing.

Note: Do not apply when fruit is present, or illegal residues may result.

Cranberries

MA: Apply up to 4 qts. per acre either before spring growth begins or in the fall after harvest.

Other areas: Apply 2 qts. per acre before spring growth begins.

Filberts

Apply 2-4 qts. per acre in the fall or apply a split application of 2 qts. per acre in the fall plus 2 qts. per acre in the spring.

Precaution: If trees are planted on a hillside, excessive soil erosion may result from the elimination of weeds.

Note: Do not apply when nuts are on the ground during the harvest period, or illegal residues may result.

Grapes

Apply 2-4.8 qts. per acre any time between harvest and early spring.

Precaution: Do not use in vineyards established less than three years, or crop injury may occur.

Grapefruit, Lemons, Oranges

AZ (Lemons and Oranges only): Apply a split application of 1.6 qts. per acre in the spring plus 1.6 qts. per acre in the fall.

CA: Apply 2-4 qts. per acre in a single application; or apply 2 qts. per acre in the fall and 2 qts. per acre in the spring.

Precaution: Do not use in the Imperial, Coachella, or Palo Verde Valleys, or crop injury may occur.

FL (Grapefruit and Oranges only): (Ridge Areas) Apply only once per year. For control of weeds listed under General Information, apply 6.4-9.6 qts. per acre in the spring; to control milkweed vine, use 8-9.6 qts. per acre either preemergence or before vines are 12 inches tall. Complete control of milkweed vine may require repeat application over a period of 2-3 years. (Bedded Areas) Apply 3.2 qts. per acre in bedded areas to control balsam apple vine and other weeds. Apply prior to periods of major weed seed germination in soil where moisture conditions are good. Apply only once per year.

TX (Grapefruit and Oranges only): Apply 4-4.8 qts. per acre.

Precautions (All areas): To avoid crop injury, (1) Do not use in nurseries. (2) Do not apply to bedded grapefruit, lemons, or oranges (except for FL grapefruit and oranges). (3) Do not apply to trees under stress from freeze damage for one year after the freeze.

Macadamia Nuts

Apply 2-4 qts. in 50 gals. of water per acre before harvest and just prior to weed emergence. Repeat application as necessary.

Note: Do not apply when nuts are on the ground during the harvest period, as illegal residues may result.

Olives

Apply 2-4 qts. per acre following grove preparation in the fall. Repeat annually in midwinter.

Peaches, Plums, Sweet Cherries

Apply 1.6-4 qts. per acre. Apply in late fall to early spring prior to weed emergence.

Precautions: To avoid crop injury, (1) Peaches: use only in AR, LA, MO, OK, TX, and states east of the Miss. River. For California, see specific directions in the section "Almonds and Peaches (CA only)." (2) Plums and sweet cherries: use only in MO and states east of the Miss. River except TN.

Pecans

Apply 2-4 qts. per acre before weeds emerge in the spring.

Precautions: To avoid crop injury, (1) Do not use west of the Pecos River in TX or in NM, AZ, or CA. (2) Do not make applications to transplanted trees that have been established less than two years in the grove.

Note: Do not apply when nuts are on the ground, or illegal residues may result. Do not allow animals to graze treated areas.

Strawberries

Oregon and Washington: For control of chickweed, groundsel, mustard, and shepherdspurse, apply broadcast 1 qt. per acre. In fields where overhead irrigation is used to activate this product, apply after harvest at time of bed renovation. In fields where overhead irrigation is not available, apply during early October through November.

Precautions: To avoid crop injury, (1) Make only one application per growing season. (2) Do not apply within 4 months after transplanting.

Walnuts

Apply 2-4 qts. per acre. Leveling and furrowing operations after application will lessen effectiveness of weed control.

Note: Do not apply when nuts are on the ground, or illegal residues may result.

PRINCEP 4L Plus Evik® 80W for Grapefruit and Oranges

Use in grapefruit and oranges in FL and TX for control of weeds listed in the following directions and under General Information. Apply the rates given below in a minimum of 40 gals. of water per acre thoroughly covering existing vegetation, but avoiding contact with citrus fruit, foliage, or stems. Apply when weeds have resumed active growth in the spring, but before they exceed 4 inches in height (except as noted for vines). Observe all precautions and limitations on the labeling of PRINCEP 4L and Evik 80W. When mixing, add Evik 80W slurry to water in spray tank, agitate thoroughly, then add PRINCEP 4L and agitate thoroughly again.

FL (Ridge Area): Apply 4 qts. of PRINCEP 4L plus 5 lbs. of Evik 80W per acre; controls milkweed vine when applied either preemergence or on vines up to 24 inches tall.

FL (Bedded Areas): Apply 3.2 qts. of PRINCEP 4L plus 6 lbs. of Evik 80W per acre; controls balsamapple when applied either preemergence or on vines up to 12 inches tall.

TX: Apply 3.2-4 qts. of PRINCEP 4L plus 2-3 lbs. of Evik 80W per acre plus 2 qts. nonionic surfactant, such as X-77®, per 100 gals. of spray mixture (add surfactant to spray tank last); controls buffelgrass, marinevine (ivy treebine), Texas panicum, sandbur, and sunflower, and gives top-kill of johnsongrass. Use 4 qts. of PRINCEP 4L per acre for longer residual control; use 3 lbs. of Evik 80W per acre for control of heavy weed growth and for top-kill of johnsongrass; use the higher rate of both products for control of marinevine.

Precautions (All areas): To avoid injury, (1) Do not use in nurseries, or around trees that have been established in the grove less than two years, except in FL. (2) Do not use in nurseries, or where trees are under stress from freeze damage for one year after the freeze. (3) Apply only once per year.

Note (All areas): Do not apply with 30 days before harvest, or illegal residues may result. Do not graze treated areas.

PRINCEP 4L Plus Bromacil 80W for Grapefruit and Oranges (Florida Only)

Use in grapefruit and oranges in FL for control of balsamapple, black nightshade, carpetweed, crabgrass, cudweed, dayweed, Florida pusley, horseweed, pepperweed, pigweed, poorjoe, ragweed, rattlebox, spanishneedles and sandbur, and for partial control of bermudagrass, bahiagrass, pangolagrass, paragrass, and torpedograss. Apply 3.2-6.4 qts. of PRINCEP 4L plus 3-4 lbs. of bromacil 80W per acre beneath trees in a minimum of 40 gals. of water per acre before or soon after weed growth begins. When mixing, add bromacil 80W slurry to water in spray tank, agitate thoroughly, then add PRINCEP 4L and agitate thoroughly again. Use the lower rates for light weed infestations or all applications in bedded citrus areas. Use the higher rates for heavy weed infestations only in ridge grown citrus areas. Temporary yellowing of citrus leaves may occur following treatment.

Precautions: To avoid crop injury, (1) Do not use in nurseries, or where trees are under stress from freeze damage for one year after the freeze. (2) Do not use on soil with less than 1% organic matter or on poorly drained soil. (3) Do not treat trees planted in irrigation furrows. (4) Do not treat diseased trees such as those with foot rot. (5) Do not use in groves interplanted with other trees or desirable plants, nor in home grapefruit or orange plantings, or in areas where roots of other valuable plants or trees may extend. (6) Treated areas may be planted to citrus trees one year after application. (7) Do not rotate to other crops within two years after application.

Note: Apply only once per year and avoid contact with foliage and fruit with spray or mist, or illegal residues may result.

PRINCEP 4L Plus Gramoxone® Super

This tank mix is effective in the following fruit and nut crops for kill of existing vegetation and for residual control of the annual broadleaf and grass weeds claimed for PRINCEP 4L applied alone. This combination is also effective for top-kill and suppression of perennial weeds.

Use this tank mix on these crops:

almonds (CA only)	macadamia nuts
apples	olives
avocados (CA and FL[1] only)	oranges[2]
cherries (sour and sweet[4])	peaches[3]
filberts	pears
grapefruit (CA, FL[2] and TX only)	pecans
grapes	plums[4]
lemons (AZ and CA only)	walnuts

Apply the PRINCEP 4L rate given under the appropriate crop on this label plus 1.5-2.5 qts. Gramoxone Super* in 50-200 gals. (30-50 gals. for pecans) of water per acre to the orchard floor avoiding contact with fruit, foliage, or stems. Add a nonionic surfactant, such as X-77, at 0.5 pt. per 100 gals. of spray. Apply when weeds are succulent and new growth is 1-6 inches tall. For mature woody weeds or difficult to control perennial weeds, re-treat or spot treat with Gramoxone Super if regrowth occurs.

Add PRINCEP 4L to the spray tank first (refer to Mixing Procedures section of this label), then add Gramoxone Super and add the surfactant last. Provide constant agitation during mixing and application to keep the mixture in suspension.

Precautions: To avoid crop injury, (1) Apply the tank mix only once per year. (2) Use a shield for young trees or vines.

Refer to the PRINCEP 4L and Gramoxone Super labels for further directions, specific weeds controlled, and precautions and limitations on each crop.

[1] In avocados in Florida, this tank mix also controls balsamapple, rattail amaranth, and at the higher rate of each herbicide, it suppresses coral vine.

[2] In grapefruit and oranges in Florida, apply 3.2 qts. PRINCEP 4L per acre in bedded areas and 6.4-9.6 qts. PRINCEP 4L per acre in ridge areas.

[3] Limited to CA, AR, LA, MO, OK, TX, and states east of the Miss. River. As appropriate, refer to PRINCEP 4L sections "Almonds and Peaches (CA only)" or "Peaches, Plums, Sweet Cherries" for rate of PRINCEP 4L and other information.

[4] Limited to MO and states east of the Miss. River except TN.

* Based on a product containing 1.5 lbs. Gramoxone Super cation per gallon.

Tank Mixture with Roundup®

This tank mixture is effective in grape vineyards and in the following bearing or nonbearing tree crops for control of existing vegetation and for residual control of the annual broadleaf and grass weeds claimed for this product applied alone. This combination is also effective for partial control of perennial weeds contacted by the spray mixture during application. Use this tank mixture on these crops:

almonds** (CA only)	macadamia nuts**
apples*	oranges* (AZ, CA, FL[1] and TX)
avocados* (CA and FL only)	peaches*[2]
cherries (sour and sweet[3])	pears*
filberts**	pecans**
grapefruits* (CA, FL[1] and TX)	plums*[3]
grapes*	walnuts**
lemons* (AZ and CA)	

Use the appropriate rate given elsewhere on this label for PRINCEP applied alone to the crop being treated. Add to the spray tank 1-5 qts. of Roundup per acre depending on weeds present and their growth stage. Also add an agriculturally approved nonionic surfactant at 0.5% by volume of spray solution. Apply the mixture in 10-40 gals. of water per acre as a postemergence spray to the weeds at the appropriate weed growth stage given on the Roundup label.

Add PRINCEP to the spray tank first, then add Roundup. Provide constant agitation during mixing and application to keep the mixture in suspension. Refer to Application Procedures section of this label for further directions.

Precautions: To avoid crop injury, take extreme care to avoid contact of herbicide solution, spray, drift, or mist with foliage or green bark of trunk, branches, suckers, fruit, or other parts of trees or vines. Observe precautions on both the PRINCEP and Roundup labels for each crop involved.

Refer to the labels of both herbicides for further directions, specific weeds controlled, precautions and limitations on each crop.

* Allow a minimum of 14 days between last application and harvest.

** Allow a minimum of 21 days between last application and harvest of these crops.

[1] In grapefruit and oranges in Florida, apply 3.2 qts. of this product per acre in bedded areas and 6.4-9.6 qts. per acre in ridge areas.

[2] Limited to CA, AR, LA, MO, OK, TX, and states east of the Mississippi River. For CA, see specific directions in the section "Almonds and Peaches (CA only)."

[3] Plums and sweet cherries: limited to MO and states east of the Mississippi River except TN.

PRINCEP 4L Plus Surflan®

Use this tank mix in the following crops for preemergence control of all weeds claimed on both labels:

almonds	cherries	lemons	pecans
apples	filberts	oranges	plums
avocados	grapes	peaches	English

caneberries	grapefruit	pears	walnuts

Apply the PRINCEP 4L rate given under the appropriate crop on this label plus 2.67-5.33 lbs. of Surflan 75W or 2-4 qts. of Surflan A.S. in 20-40 gals. of water per acre. Refer to the Surflan 75W or Surflan A.S. label for complete tank mix directions, and use 0.8 qt. of PRINCEP 4L for each pound of PRINCEP 80W. Observe all precautions and limitations on the PRINCEP and Surflan labels.

FIELD CROPS

Corn

Nitrogen solution or complete liquid fertilizer may replace all or part of the water in the spray. Determine the physical compatibility of PRINCEP 4L with fertilizer before use. Refer to the General Information section of this label for a compatibility test procedure. Do not apply after corn has emerged, as there is danger of liquid fertilizer causing crop injury. Use a minimum of 10 gals. of spray mixture per acre.

Preemergence: Apply before weeds and corn emerge. Use the appropriate rate in the following table.

Preplant: Broadcast in the spring after plowing at the appropriate rate in the following table. Apply before, during, or after final seedbed preparation. If soil is tilled or worked after application, avoid deep incorporation of PRINCEP 4L. Best results will be obtained when PRINCEP is applied within 2 weeks before planting. Under dry weather conditions, preplant applications may give better weed control than preemergence. If weeds develop, particularly under relatively dry conditions, a shallow cultivation will generally result in better weed control.

Soil Texture	Broadcast* Rate Per Acre
Sand, silt, and loam that is low in organic matter	4 pts.
Soil containing moderate amounts of clay and organic matter	4.8 pts.
Loam that is high in organic matter and clay (including dark prairie soils of Corn Belt)	6 pts.
Peat, muck and high-organic clay	8 pts.

* For calculation of band treatment rate, see the General Information section.

Quackgrass Control: Broadcast 6-8 pts. per acre in the fall. Plow 2-3 weeks later, or if erosion is a problem, delay plowing until spring.

Precautions: (1) Do not apply more than 8 pts. of PRINCEP 4L per acre of corn per year. (2) Do not plant any crop except corn until the following year or injury may occur. (3) Following harvest, plow and thoroughly till the soil in fall or spring to minimize possible injury to spring-seeded rotational crops, regardless of the rate used. (4) If PRINCEP 4L is used at a rate higher than 6 pts. per acre or equivalent band rate, a crop of untreated corn should precede the next rotational crop. (5) Do not apply PRINCEP 4L preplant incorporated for weed control in corn in the High Plains and Intermountain areas of the West (including central and western KS, western NE, western OK, and the Panhandle of TX) where rainfall is sparse and erratic or where irrigation is required. (6) In the High Plains and Intermountain areas of the West where rainfall is sparse and erratic or where irrigation is required, use PRINCEP 4L only when corn is to follow corn or when a crop of untreated corn is to precede other rotational crops. (7) In western MN and eastern parts of the Dakotas, NE and KS, do not rotate to soybeans if the rate applied was more than 4 pts. per acre of equivalent band rate or injury may occur. (8) Injury may occur to soybeans planted in north-central and northwest IA, south-central and southwest MN, northeast NE, southeast SD, and other areas the year following application on soils having a calcareous surface layer. (9) Do not plant sugar beets, tobacco, vegetables (including dry beans), spring-seeded small grains, or small-seeded legumes and grasses the year following PRINCEP 4L application or injury may occur.

Note: Do not graze treated areas, or illegal residues may result.

Tank Mixtures on Corn

AAtrex®: Use PRINCEP 4L in a preplant or preemergence tank mixture with AAtrex Nine-O®, AAtrex 80W, or AAtrex 4L for control of many annual weeds, including carpetweed, crabgrass, fall panicum, foxtail, lambsquarters, morningglory, pigweed, ragweed, and velvetleaf.

Apply at the rates given below. Use the 1:1 ratio for most weeds and the 2:1 ratio for expected heavy infestations of crabgrass and fall panicum.

Soil Texture	Broadcast Rate Per Acre			
	1:1 Ratio		2:1 Ratio	
	PRINCEP 4L	AAtrex 80W*	PRINCEP 4L	AAtrex 80W*
Sand, loamy sand, sandy loam	1 qt.	1.25 lbs.	1.33 qts.	0.83 lb.
Loam, silt loam, silt, clay loam, sandy clay loam, silty clay loam, sandy clay, or silty clay with low organic matter	1.2 qts.	1.5 lbs.	1.6 qts.	1 lb.
Loam, silt loam, silt, clay loam, sandy clay loam, silty clay loam, sandy clay, or silty clay with medium to high organic matter, and clay (including dark prairie soils of the Corn Belt)	1.44 qts.	1.8 lbs.	1.92 qts.	1.2 lbs.

* When using AAtrex 4L or AAtrex Nine-O, use equivalent rates. One lb. of AAtrex 80W equals 0.8 qt. of AAtrex 4L or 0.9 lb. of AAtrex Nine-O.

Refer to the AAtrex label for complete directions. Also refer to the AAtrex label and the Corn section of this label for precautions and limitations.

Eradicane®: Use in a preplant incorporated tank mixture for control of all weeds claimed on both the PRINCEP and Eradicane labels including partial control (suppression) of shattercane (wild cane). Fluid ferilizer may replace all or part of the water in the spray. Check the physical compatibility of mixture with fertilizer before use. Refer to the General Information section of this label for a compatibility test procedure. Use a minimum of 20 gals. of spray volume per acre. Refer to the Eradicane label for incorporation directions. Use the higher rate of PRINCEP on fine-textured soil and where heavy broadleaf infestations are expected. Use the higher rate of Eradicane for heavy nutsedge and bermudagrass infestations. For partial control (suppression) of shattercane, broadcast and incorporate **immediately** before planting 2-4 pts. of PRINCEP 4L plus 7.33 pts. of Eradicane 6.7E (or 8 pts. Eradicane Extra) per acre. For control of the other weeds claimed on both labels, broadcast and incorporate 2-6 pts. of PRINCEP 4L plus 4.75-7.33 pts. of Eradicane 6.7E (or 5.33 pts. of Eradicane Extra) per acre. Observe all precautions and limitations on the PRINCEP 4L and Eradicane labels.

Gramoxone Super: Use in a tank mixture where corn will be planted directly in a cover crop, established sod, or previous crop residues. This combination controls existing vegetation and provides residual control of the annual broadleaf and grass weeds listed under General Information.

Add PRINCEP 4L to the spray tank, mix thoroughly with water, and then add Gramoxone Super and a nonionic surfactant. Provide constant agitation during mixing and application to keep the mixture in suspension.

Apply 4-6 pts. of PRINCEP 4L plus 1.5-2.5 pts. of Gramoxone Super* in 20-60 gals. of water per acre as a broadcast spray either before or after planting but before corn emerges. Add a nonionic surfactant at the rate of 0.5 pt. per 100 gals. of spray volume.

For further information, see general information, caution and warning statements, precautions, and notes on the PRINCEP 4L and the Gramoxone Super labels.

* Based on a product containing 1.5 lbs. of Gramoxone Super cation per gallon.

Sutan+®: Use in a preplant incorporated tank mixture for control of all weeds claimed on both the PRINCEP and Sutan+ labels including partial control (suppression) of shattercane (wild cane). Fluid fertilizer may replace all or part of the water in the spray. Check the physical compatibility of mixture with fertilizer before use. Refer to the General Information section of this label for a compatibility test procedure. Use a minimum of 20 gals. of spray volume per acre. Refer to the Sutan+ label for incorporation directions. Use the higher rate of PRINCEP on fine-textured soil and where heavy broadleaf infestations are expected. For partial control (suppression) of shattercane, broadcast and incorporate **immediately** before planting 2-4 pts. of PRINCEP 4L plus 7.33 pts. of Sutan+ 6.7E per acre. For control of the other weeds claimed on both labels, broadcast and incorporate 2-6 pts. of PRINCEP 4L plus 4.75 pts. of Sutan+ 6.7E per acre. Observe all precautions and limitations on the PRINCEP 4L and Sutan+ labels.

Sugarcane

Apply 2-4 qts. per acre at planting or after harvest if ratoon crop is to be grown, but before weeds or sugarcane emerge. PRINCEP 4L may also be applied interline-directed or over the sugarcane to provide continuing control of germinating weeds.

Precautions: To avoid crop injury or illegal residues, (1) Do not apply more than one 2-4 qt. application over the plants or more than two interline-directed applications. (2) Do not apply after close-in. (3) Total application should not exceed 10 qts. per acre per crop cycle.

Note: Do not feed or graze treated forage to livestock.

NURSERIES, CHRISTMAS TREE PLANTINGS, SHELTERBELTS

Nurseries (see list below)

Apply 2-3 qts. in at least 25 gals. of water per acre in fall or spring.

Precaution: To avoid plant injury, do not apply for at least one year after transplanting.

Christmas Tree Plantings and Shelterbelts (see list below)

Remove weed growth before application. Apply 2-4 qts. in at least 25 gals. of water per acre after transplanting. Use the same rate for annual maintenance applications.

For quackgrass control apply 4 qts. per acre in the fall or apply a split application of 2 qts. per acre in the fall plus 2 qts. per acre in early spring, after quackgrass begins growth.

Precautions: To avoid tree injury, (1) Do not use PRINCEP 4L on seedbeds or cutting beds. (2) In CA, OR, and WA, do not apply to Christmas trees or shelterbelts sooner than one year after transplanting. In other areas, do not apply to Christmas trees or shelterbelt transplants less than three years of age. (3) Do not use until soil is firmly settled around roots. (4) Do not apply more than once a year, except as directed for quackgrass control.

Apply PRINCEP 4L to these species of trees and shrubs, as recommended above:

Conifers

arborvitae	Norway spruce
Austrian pine	red spruce
*Bishop pine	white spruce
*knobcone pine	balsam fir
lodgepole pine (shore pine)	Douglas fir
	Fraser fir
*Monterey pine	white fir
Mugho pine	hemlock
red pine (Norway pine)	juniper
	red cedar
Scotch pine	white cedar
white pine	yew (*Taxus* spp.)
blue spruce	

Deciduous Trees and Woody Ornamentals

American elm	honey locust
Siberian elm	Oregon grape (*Mahonia* spp.)
barberry	
boxelder	
bush honeysuckle	*Pieris* spp. (max. 3.2 qts./A)
caragana	
cotoneaster	red oak
dogwood	Russian olive
eucalyptus	*oleander
holly (*Ilex* spp.) (max. 3.2 qts./A)	*palm (2 qts./A)
	*bottle brush
	*carob

* For California only

Surflan Tank Mix: On Christmas tree plantings, use this tank mix for preemergence control of weeds listed on the Surflan 75W (or Surflan A.S.) and PRINCEP 4L labels. Use on field grown conifer species listed on the labels for each herbicide plus grand fir, alpine fir, Engelmann spruce, black spruce, Colorado blue spruce, Coulter pine, giant redwood, and Veitchi fir. Broadcast the mixture as a directed spray to the soil surface as an overtop spray using 2-4 qts. of PRINCEP 4L and 2.67-5.33 lbs. of Surflan 75W (or 2-4 qts. of Surflan A.S.). Apply in sufficient water per acre to uniformly treat the area. Follow overtop sprays with sprinkler irrigation to move the herbicide from leaf surfaces to the soil. Remove weed growth before application. Mix weed residues, prunings or trash into the soil, or remove them before treatment. Soil should be in good tilth and free of clods at time of application. Shallow cultivation (1-2 inches) after treatment will not reduce weed control. Observe all precautions and limitations on the PRINCEP 4L and Surflan labels.

Dual® 8E Tank Mix: On field and liner* grown woody ornamentals, use this tank mix for preemergence control of yellow nutsedge, barnyardgrass, crowfootgrass, fall panicum, giant foxtail, green foxtail, yellow foxtail, goosegrass, large crabgrass, smooth crabgrass, redroot pigweed, galinsoga, Venice mallow, lambsquarters, purslane, ragweed, smartweed and other weeds listed on the Dual 8E alone label. Control should last about 60 days. Apply 1.6-2 pts. of PRINCEP 4L plus 2-3 pts. of Dual 8E in a minimum of 15 gals. of water per acre. For band applications use a proportional amount of herbicide. Direct the spray toward the base of established ornamentals or ornamentals transplanted at least 10 days. Apply before weeds emerge or after removing existing weeds. Use the higher rates on fine-textured soil and where broadleaf weeds are expected to form a significant part of the infestation. Use the lower rates on coarse-textured soil and where light infestations of broadleaf weeds are expected. Apply post-transplant to the following field and liner grown ornamentals:

Aucuba (*Aucuba* spp.)	Leucothoe (*Leucothoe* spp.)
Arborvitae (*Thuja* spp.)	Liriope (*Liriope* spp.) Silver Sunproof
Cotoneaster (*Cotoneaster* spp.)	
Crabapple (*Malus* spp.)	Maple Whips (*Acer* spp.)
Holly (*Ilex* spp.) Boxleaf Japanese, Japanese	Osmanthus (*Osmanthus* spp.)
	Red Oak (*Quercus* spp.)
Japanese Andromeda (*Pieris* spp.)	Rhododendron (*Rhododendron* spp.)
Juniper (*Juniperus* spp.) Andorra, Creeping, Shore, Bar Harbor, Blue Rug	
	Weigela (*Weigela* spp.)
	Yew (*Taxus* spp.) Dwarf Japanese

Note: Length of weed control may be reduced when continuous wet soil conditions follow herbicide application.

Precautions: To avoid plant injury, (1) Do not use on seedbeds or on unrooted cuttings. (2) Do not use in greenhouses or other enclosed areas.

* Plants transplanted normally in rows in a nursery or similar area for further growth prior to transplanting to final growing place.

TURFGRASSES FOR SOD (FLORIDA ONLY)

St. Augustinegrass, Centipedegrass, and Zoysia Grass

Apply 2-4 qts. per acre, according to soil texture as indicated below.

Muck or peat	4 qts.	Old beds	within 2 days after lifting of sod
		New beds	3-4 days after sprigging or plugging.
Sandy soil	2 qts.	Old beds	within 2 days after lifting of sod
		New beds	7-10 days after sprigging or plugging

If weeds regrow, apply an additional 2 qts. on muck or peat, or 1 qt. on sandy soil.

Precautions: To avoid crop injury, (1) Do not apply within 30 days prior to cutting or lifting. (2) Do not apply in combination with surfactants or other spray additives. (3) Use only on turfgrass reasonably free of infestations of insects, nematodes, and diseases. (4) On newly sprigged turfgrass, temporary slowing of growth may follow application.

TURFGRASS FOR FAIRWAYS, LAWNS, SOD PRODUCTION* AND SIMILAR AREAS

* In states other than Florida. For use on turfgrass for sod in Florida, see "Turfgrasses for Sod (Florida only)" section above.

Bermudagrass, Centipedegrass, St. Augustinegrass, and Zoysia Grass

Apply PRINCEP 4L after September 1 (after October 1 for annual bluegrass) before emergence of winter annual weeds. PRINCEP will control annual bluegrass, burclover, lawn burweed, common and mouseear chickweed, corn speedwell, henbit, hop clover, spurweed, and parsley-piert. PRINCEP will also control these weeds soon after emergence. For control of summer annual weeds listed in the **General Information** section of this label, apply PRINCEP in late winter before the weeds emerge. Apply in a minimum of 15 gals. of water per acre.

Irrigate with 1/2 inch of water if rainfall does not occur within 10 days after preemergence treatment.

Where annual bluegrass is the major weed, use 1 qt. of princep 4L per acre (22 ml or 0.75 fl. oz. per 1,000 sq. ft.). Use 2 qts. per acre (22-44 ml. or 0.75-1.5 fl. oz. per 1,000 sq. ft.) for control of other weeds. However, do not exceed 1 qt. per acre per treatment on newly sprigged turfgrass or on hybrid bermudagrass such as Tiflawn, Tifway, and Ormond.

For control of summer annuals which emerge after the initial application, apply an additional 1 qt. per acre at least 30 days after the initial application, but not after June 1. However, do not make more than two applications per year.

Precautions: On newly sprigged turfgrass, hybrid bermudagrass, nondormant bermudagrass or nondormant zoysia grass, temporary slowing of growth and yellowing may occur following application. To avoid turf injury, (1) Use only on turfgrass reasonably free of infestations of insects, nematodes, and diseases. (2) Do not use on golf greens. (3) Do not use north of NC (except may be used in the Virginia Coastal Plains) or west of Highway 281 in Texas and Oklahoma. (4) Do not use on muck or alkaline soils. (5) Do not apply over the rooting area of trees or ornamentals not listed on this label. (6) Do not seed or overseed with desirable turfgrass within 4 months before or 6 months after treatment. (7) Do not apply this product on newly seeded grasses until they have overwintered and have a well-developed rhizome system. (8) Do not exceed 2 qts. product per acre within 12 months of seeding grasses.

Note: Do not graze or feed turf clippings to animals, or illegal residues may result.

VEGETABLE CROPS

Asparagus (Established)

Apply 2-4 qts. per acre to beds established one year or more. Apply once per year either after disking in the spring before weeds emerge, or following harvest after removal of existing weed growth. Spring application must be made at least 3 days before first cutting.

Precautions (WA and OR only): To avoid crop injury, (1) Do not apply on gravelly, sand, or loamy sand soil. (2) Do not allow irrigation water to stand or to form pools in the field.

Artichokes

Apply 2-4 qts. per acre as a directed spray after the last fall tillage.

Precautions: To avoid crop injury, (1) Do not apply on gravelly, sand, or loamy sand soil. (2) Apply only once per year.

NONSELECTIVE WEED CONTROL ON NONCROP LAND

PRINCEP 4L provides long-term weed control on industrial sites, highway medians and shoulders, railroad rights-of-way, lumberyards, petroleum tank farms, and in noncrop areas on farms, such as around buildings, equipment and fuel storage areas, along fences, roadsides, and lanes.

Best results are obtained when PRINCEP 4L is applied before weeds emerge. If weed growth has begun and removal is not feasible, include a contact or translocated herbicide in the spray. Aerial applications may be made where it is feasible. Use at least 1 gal. of water for each 1 qt. of PRINCEP 4L for both ground and aerial applications. Use more water if practical.

Precautions: (1) Do not use this product for nonselective weed control on land to be cropped, near adjacent desirable trees, shrubs, or plants, or in greenhouses, or injury may occur. (2) Do not use in areas accessible to livestock or allow livestock to graze treated weed foliage.

Annual broadleaf and grass weeds (including barnyardgrass, cheat, crabgrass, lambsquarters, foxtail, ragweed, puncturevine, and turkey mullein): Apply 4.8-10 qts. per acre.

Annual and perennial broadleaf and grass weeds (including quackgrass, bluegrass, redtop, burdock, Canada thistle, orchardgrass, purpletop, smooth brome, dogfennel, and plantain): Apply 2.5-4.4 gals. per acre.

Tank Mixture with Roundup

This tank mixture will provide control of emerged annual weeds and residual control of weeds listed on the PRINCEP label. The combination also will partially control emerged perennial weeds listed on the Roundup label. Add to the spray tank the appropriate rate of PRINCEP noted in the above section for PRINCEP alone. Then add 1-5 qts. of Roundup per acre depending on weeds present and their growth stage. Also add an agriculturally approved nonionic surfactant at 0.5% by volume of spray solution. Apply the mixture in 10-40 gals. of water per acre as a postemergence spray at the appropriate weed growth stage given on the Roundup label. Provide constant agitation during mixing and application to keep the mixture in suspension. Refer to the Application Procedures section of this label for further directions. Refer to the labels of both herbicides for specific noncrop sites, rates, weeds controlled, and further directions, precautions, and limitations.

Tank Mixture with Oust®

Use this tank mixture to control the weeds listed below with either preemergence, early postemergence, or postemergence applications. Use the higher rates within the ranges given under conditions of heavy weed growth, on soil with over 2½% organic matter, on areas with heavy decaying plant residues, or on high moisture areas such as along road edges or railroad shoulders. To improve wetting and/or contact activity on emerged weeds, add a nonionic surfactant such as duPont Surfactant WK or X-77 at 1 qt. per 100 gals. of spray. Do not apply by aircraft. Provide constant agitation during mixing and application to keep the mixture in suspension. Refer to the labels of both herbicides for specific noncrop sites and further directions, precautions, and limitations.

Preemergence to Early Postemergence: To control the weeds listed below apply 1-2 gals. of PRINCEP 4L plus 3-5 ounces of Oust in a minimum of 25 gals. of water per acre.

Broadleaf Weeds (Up to 3-4" in Height or Diameter)

bouncingbet	horsetail	sowthistle
buckhorn plantain	kochia	St. Johnswort
burclover	lambsquarters	sunflower
Carolina geranium	little mallow	sweet clover
common chickweed	mustard	tansymustard
common dandelion	ox-eye daisy	tansy ragwort
common speedwell	pigweed	tumble mustard
common yarrow	prickly lettuce	turkey mullein
crimson clover	puncturevine	vetch
dogfennel	purple starthistle	wild carrot
filaree	ragweed	yellow rocket
hoary cress	Russian thistle	

Grasses (Up to 6-12" in Height)

Alta fescue	fall panicum	red brome
annual bluegrass	fescue	red fescue
annual ryegrass	foxtails	reed canarygrass
bahiagrass	Indiangrass	ripgut brome
barnyardgrass	Italian ryegrass	ryegrass
cheatgrass	johnsongrass (short-term control)	smooth brome
crabgrass		sprangletop
downy brome	little barley	witchgrass

Postemergence: To control actively growing weeds listed below apply 2-4 gals. of PRINCEP 4L plus 6-12 ounces of Oust in a minimum of 25 gals. of water per acre.

Actively Growing Weeds

bedstraw	Jerusalem artichoke	johnsongrass (season-long control)
Canada thistle	kudzu	
curly dock	mayweed	
dewberry	musk thistle	yellow nutsedge
fiddleneck	poison ivy	
fleabane	spanishneedles	
goldenrod	turkey mullein	
hemlock	Virginia pepperweed	
honeysuckle	wild blackberry	

STORAGE AND DISPOSAL

Do not contaminate water, food, or feed by storage or disposal. Pesticide wastes may be toxic. Improper disposal of unused pesticide, spray mixture, or rinse water is a violation of federal law. If these wastes cannot be used according to label instructions, contact your State Pesticide or Environmental Control Agency, or the Hazardous Waste representative at the nearest EPA Regional Office for guidance in proper disposal methods.

Triple rinse container (or equivalent). Then offer for recycling or reconditioning, or puncture and dispose of in a sanitary landfill or by other procedures approved by state and local authorities.

For minor spills, leaks, etc., follow all precautions indicated on this label and clean up immediately. Take special care to avoid contamination of equipment and facilities during cleanup procedures and disposal of wastes. In the event of a major spill, fire or other emergency, call (919) 292-7100 day or night.

PRECAUTIONARY STATEMENTS

HAZARDS TO HUMANS AND DOMESTIC ANIMALS

CAUTION

Causes eye irritation. Harmful if inhaled or absorbed through the skin. Avoid breathing vapors or spray mist. Avoid contact with eyes, skin, or clothing. Wash thoroughly with soap and water after handling. Remove and wash contaminated clothing before reuse.

First Aid: If in eyes, flush with plenty of water. Get medical attention if irritation persists. If on skin, wash with plenty of soap and water. If inhaled, remove victim to fresh air. If not breathing, give artificial respiration, preferably mouth-to-mouth; get medical attention.

ENVIRONMENTAL HAZARDS

Do not apply directly to water or wetlands. Do not contaminate water by cleaning of equipment or disposal of wastes. Simazine is a chemical which can travel (seep or leach) through soil and enter ground water which may be used as drinking water. Simazine has been found in ground water as a result of its use as a herbicide. Users of this product are advised not to apply simazine where the water table (ground water) is close to the surface and where the soils are very permeable, i.e., well drained soils such as loamy sands. Users are advised to consult with their local agricultural agencies to obtain information on the location of ground water and the type of soil in their area.

Do not apply this product through any type of irrigation system.

AAtrex® trademark of CIBA-GEIGY for atrazine
Compex® trademark of KALO Agricultural Chemicals, Inc.
Dual® trademark of CIBA-GEIGY for metolachlor
U.S. Patent No. 3,937,730
Eradicane® trademark of Stauffer Chemical Co.
Evik® trademark of CIBA-GEIGY for ametryn
Gramoxone® trademark of ICI Americas for paraquat
Nine-O® trademark of CIBA-GEIGY
Oust® trademark of E.I. duPont de Nemours and Co., Inc.
PRINCEP® trademark of CIBA-GEIGY for simazine
Roundup® trademark of Monsanto Company for glyphosate herbicide
Surflan® trademark of Elanco Products Co. for oryzalin
Sutan+® trademark of Stauffer Chemical Co.
Unite® trademark of Hopkins Agricultural Chemical Co.
X-77® trademark of KALO Agricultural Chemicals, Inc.

CGA 6L32L 028

This is a specimen label which was not prepared by CIBA-GEIGY. CIBA-GEIGY is not responsible for the accuracy of the information contained herein. As labels are subject to revision, always carefully read and follow the label on the product container.

PRINCEP® 80W

HERBICIDE

For weed control in certain crops, ornamental plantings, on industrial sites, and around-the-farm

PRINCEP 80W is a wettable powder.

ACTIVE INGREDIENT:	
Simazine: 2-chloro-4,6-bis (ethylamino)-*s*-triazine	80%
INERT INGREDIENTS:	20%
TOTAL	100%

EPA Reg No. 100-437 EPA Est. 100-LA-1

KEEP OUT OF REACH OF CHILDREN

CAUTION

See additional precautionary statements at end of booklet.

See directions for use inside booklet.

DIRECTIONS FOR USE AND CONDITIONS OF SALE AND WARRANTY

IMPORTANT: Read the entire **Directions for Use** and the **Conditions of Sale and Warranty** before using this product.

CONDITIONS OF SALE AND WARRANTY

The **Directions for Use** of this product reflect the opinion of experts based on field use and tests. The directions are believed to be reliable and should be followed carefully. However, it is impossible to eliminate all risks inherently associated with use of this product. Crop injury, ineffectiveness, or other unintended consequences may result because of such factors as weather conditions, presence of other materials, or the manner of use or application all of which are beyond the control of CIBA-GEIGY or the Seller. All such risks shall be assumed by the Buyer.

CIBA-GEIGY warrants that this product conforms to the chemical description on the label and is reasonably fit for the purposes referred to in the **Directions for Use** subject to the inherent risks referred to above. **CIBA-GEIGY makes no other express or implied warranty of Fitness or Merchantability or any other express or implied warranty. In no case shall CIBA-GEIGY or the Seller be liable for consequential, special, or indirect damages resulting from the use or handling of this product.** CIBA-GEIGY and the Seller offer this product, and the Buyer and user accept it, subject to the foregoing **Conditions of Sale and Warranty**, which may be varied only by agreement in writing signed by a duly authorized representative of CIBA-GEIGY.

DIRECTIONS FOR USE

It is a violation of federal law to use this product in a manner inconsistent with its labeling.

FAILURE TO FOLLOW THE DIRECTIONS FOR USE AND PRECAUTIONS ON THIS LABEL MAY RESULT IN POOR WEED CONTROL, CROP INJURY, AND/OR ILLEGAL RESIDUES.

Do not apply this product through any type of irrigation system.

Do not apply this product in such a manner as to directly or through drift expose workers or other persons except those knowingly involved in the application. The area being treated must be vacated by unprotected persons.

REENTRY STATEMENT

Do not enter treated areas without protective clothing until sprays have dried.

Because certain states may require more restrictive reentry intervals for various crops treated with this product, consult your State Department of Agriculture for further information.

Written or oral warnings must be given to workers who are expected to be in a treated area or in an area about to be treated with this product. Oral warnings must inform workers of areas or fields that may not be entered without specific protective clothing until sprays have dried, and appropriate actions to take in case of accidental exposure, as described under **Precautionary Statements** on this label. When oral warnings are given, warnings shall be given in a language customarily understood by workers. Oral warnings must be given if there is reason to believe that written warnings cannot be understood by workers. Written warnings must include the following information: CAUTION. Area treated with PRINCEP 80W on (*date of application*). Do not enter without appropriate protective clothing until sprays have dried. In case of accidental exposure, flush eyes or skin with plenty of water. Call physician if irritation persists. Remove and wash contaminated clothing before reuse.

GENERAL INFORMATION

Apply PRINCEP 80W Herbicide before weeds emerge or after removal of weed growth. PRINCEP controls a wide variety of annual broadleaf and grass weeds when used at selective rates in agricultural crops and ornamental plantings. When used at higher, nonselective rates in noncrop areas, it also controls many perennial broadleaf and grass weeds.

Where a range of application rates is given, use the low rate on coarser texturted soil and soil lower in organic matter; use the high rate on finer textured soil and soil higher in organic matter.

To avoid spray drift, do not apply under windy conditions. Avoid spray overlap, as crop injury may result.

Since PRINCEP enters weeds mainly through their roots, moisture is needed to move it into the root zone. Very dry soil conditions and lack of rainfall following application may necessitate shallow cultivation.

PRINCEP 80W is noncorrosive to equipment, nonflammable, and has low electrical conductivity.

Annual Weeds Controlled

alyssum	fivehook bassia	ragweed
annual bluegrass	Flora's paintbrush	rattail fescue
annual morningglory	Florida pusley	redmaids
annual ryegrass	foxtails	Russian thistle
barnyardgrass	goosegrass	shepherdspurse
(watergrass)	groundsel	signalgrass
burclover	henbit	(*Brachiaria* spp.)
carelessweed	junglerice	silver hairgrass
carpetweed	knawel	smartweed
common chickweed	(German moss)	spanishneedles
crabgrass	common	speedwell
(*Digitaria* spp.)	lambsquarters	tansymustard
downy brome	nightshade	wild mustard
(cheatgrass)	pepperweed	wild oats
fall panicum	pigweed	witchgrass
fiddleneck	pineappleweed	yellow flower
filaree	prickly lettuce	pepperweed
fireweed	common purslane	

Following many years of continuous use of this product and chemically related products, biotypes of some of the weeds listed on this label have been reported which cannot be effectively controlled by this and related herbicides. Where this is known or suspected and weeds controlled by this product are expected to be present along with resistant biotypes, we recommend the use of this product in registered combinations or in sequence with other registered herbicides which are not triazines. If only resistant biotypes are expected to be present, use a registered nontriazine herbicide. Consult with your State Agricultural Extension Service for specific recommendations.

APPLICATION PROCEDURES

Ground application: Use conventional ground sprayers equipped with nozzles that provide accurate and uniform application. Be certain that nozzles are uniformly spaced and the same size. Calibrate sprayer before use and recalibrate at the start of each season and when changing carriers. Unless otherwise specified, use a minimum of 20 gals. of spray mixture per acre.

Use a pump with capacity to (1) maintain 35-40 psi at nozzles, (2) provide sufficient agitation in tank to keep mixture in suspension, and (3) to provide a minimum of 20% bypass at all times. Use centrifugal pumps which provide propeller shear action for dispersing and mixing this product. The pump should provide a minimum of 10 gals./minute/100 gal. tank size circulated through a correctly positioned sparger tube or jets.

Use screens to protect the pump and to prevent nozzles from clogging. Screens placed on the suction side of pump should be *16-mesh or coarser.* Do not place a screen in the recirculation line. Use 50-mesh or coarser screens between the pump and boom, and where required, at the nozzles. Check nozzles manufacturer's recommendations.

For band applications, calculate amount to be applied per acre as follows:

$$\frac{\text{band width in inches}}{\text{row width in inches}} + \text{broadcast rate per acre} = \text{amount needed per acre of field}$$

Aerial application: Use aerial application only where specified in the use directions. Apply in a minimum of 1 gal. of water for each 1-1.5 lbs. of wettable powder applied per acre. Avoid applications under conditions where uniform coverage cannot be obtained or where excessive spray drift may occur.

Avoid application to humans or animals. Flagmen and loaders should avoid inhalation of spray mist and prolonged contact with skin, and should wash thoroughly before eating and at the end of each day's operation.

Mixing procedures—all uses: (1) Be sure sprayer is clean and not contaminated with any other materials, or crop injury or sprayer clogging may result. (2) Fill tank 1/4 full with clean water, nitrogen solution, or complete fluid fertilizer. (3) Start agitation. (4) Be certain that the agitation system is working properly and creates a rippling or rolling action on the liquid surface. (5) Pour product directly from bag to tank. (6) Continue filling tank until 90% full. Increase agitation if necessary to maintain surface action. (7) Add tank mix herbicide(s) after this product is thoroughly suspended. (8) Finish filling tank. Maintain agitation to avoid separation of materials. (9) Empty tank as completely as possible before refilling to prevent buildup of emulsifiable concentrate residue from possible tank mix herbicides. (10) If an emulsifiable concentrate film starts to build up in tank, drain it and clean with strong detergent solution or solvent. (11) Clean sprayer thoroughly immediately after use by flushing system with water containing a detergent.

Compatibility test: To determine the tank mix compatibility of PRINCEP 80W with liquid fertilizer, crop oil, spreaders, or recommended pesticides, use this test method. Nitrogen solutions or complete fluid fertilizers may replace all or part of the water in the spray in alfalfa and corn. Since liquid fertilizers can vary, even within the same analysis, check compatibility each time before use. Be especially careful when using **complete** suspension or fluid fertilizers as serious compatibility problems are more apt to occur. Commercial application equipment may improve compatability in some instances. Check compatibility using this procedure:

1. Add 1 pint of fertilizer to each of 2 one-quart glass jars with tight lids.
2. To **one** of the jars add 1/4 tsp. of a compatibility agent approved for this use (1/4 tsp. is equivalent to 2 pts. per 100 gals. spray). Cap and shake

until mixed. Examples of compatibility agents include Compex® and Unite®.

3. To **both** jars add the appropriate amount of herbicide(s). If more than one herbicide is used, add them separately with dry herbicides first, flowables next, and emulsifiable concentrates last. After each addition, cap and shake until thoroughly mixed. The appropriate amount of herbicides for this test follows (assuming a spray volume of 25 gals. per acre):

Dry herbicides:	**For each pound per acre** add 1.5 teaspoons to each jar. Fluff up wettable powder products before measuring.
Liquid herbicides:	**For each pint per acre** add 0.5 teaspoon to each jar.

For a spray volume other than 25 gals. per acre, change the teaspoons added to each jar as follows:

$$\frac{\text{25 gals.} \times \text{no. teaspoons given above}}{\text{desired gals. spray volume per acre}} = \text{teaspoons to add to each jar}$$

4. After adding all ingredients, put lids on and tighten. Shake jars vigorously one minute. Let the mixtures stand 15 minutes and then look for separation, large flakes, precipitates, grease, gels, medium to heavy oily film on the jar, or other signs of incompatibility. Determine if the compatibility agent is needed in the spray mixture by comparing the two jars. If either mixture separates, but can be remixed readily, the mixture can probably be sprayed as long as good agitation is used. If the mixtures are incompatible, test the following methods of improving compatibility: (A) slurry the dry herbicide(s) in water before addition, or (B) add 1/2 of the compatibility agent to the fertilizer and the other 1/2 to the emulsifiable concentrate or flowable herbicide before addition to the mixture. If still incompatible, do not use the materials mixed in the same spray tank.

Rotational Crops and Perennial Crop Replanting

To avoid crop injury, observe the following precautions: (1) If rotating treated land the year following application, plant only corn, unless otherwise stated in this label. (2) If replanting perennial crops or if rotating land to crops other than corn, do not apply PRINCEP in the year preceding planting these crops.

FRUIT AND NUT CROPS

Apply the spray to the orchard or vineyard floor avoiding contact with fruit, foliage, or stems. Recommended rates are based on broadcast treatment. For band applications or spot applications around trees in fruit or nut plantings, reduce the broadcast rate of PRINCEP 80W and water per acre in proportion to the area actually sprayed.

Precautions (All fruit and nut crops): To avoid crop injury, (1) Apply only to orchards or groves where trees have been established 12 months or more unless specified differently. (2) Make only one application per year, except as noted otherwise. (3) Do not use on gravelly, sand, or loamy sand soil. (4) Immediately following application, limit overhead sprinkler irrigation to 1/2 inch.

Almonds and Peaches (CA only)

Apply 1.25-2.5 lbs. of PRINCEP 80W per treated acre in a 2-4 ft. band on each side of the tree row. Apply before weeds emerge in late fall or early winter. Weeds controlled by 1.25 lbs. of PRINCEP 80W include burclover, common chickweed, wild mustard, and shepherdspurse. Apply only once per year.

Precautions: To avoid crop injury, (1) Do not treat trees established in the grove less than 3 years. (2) Do not treat the Mission (Texas) variety of almonds. (3) Do not apply to almond trees propagated on plum rootstocks. (4) Do not replant almonds or peaches in treated soil for 12 months after treatment. (5) Do not apply on soil with less than 1% organic matter. (6) Do not treat areas where water will accumulate.

Apples, Pears, Sour Cherries

Apply 2.5-5 lbs. per acre.

Avocados

CA and FL only: Apply 2.5-5 lbs. per acre after final preparation of grove.

Blueberries and Caneberries (blackberries, boysenberries, loganberries, raspberries)

Apply 2.5-5 lbs. per acre in the spring or apply a split application of 2.5 lbs. per acre in the spring plus 2.5 lbs. per acre in the fall. Apply in a minimum of 40 gals. of water per acre.

On plantings less than 6 months old, use 1/2 the above rate.

To control quackgrass, apply 5 lbs. per acre in the fall or split the application applying 2.5 lbs. per acre in the fall plus 2.5 lbs. per acre in the spring, when quackgrass is actively growing.

Note: Do not apply when fruit is present, or illegal residues may result.

Cranberries

MA: Apply up to 5 lbs. per acre either before spring growth begins or in the fall after harvest.

Other areas: Apply 2.5 lbs. per acre before spring growth begins.

Filberts

Apply 2.5-5 lbs. per acre in the fall or apply a split application of 2.5 lbs. per acre in the fall plus 2.5 lbs. per acre in the spring.

Precaution: If trees are planted on a hillside, excessive soil erosion may result from the elimination of weeds.

Note: Do not apply when nuts are on the ground during the harvest period, or illegal residues may result.

Grapes

Apply 2.5-6 lbs. per acre any time between harvest and early spring.

Precaution: Do not use in vineyards established less than three years, or crop injury may occur.

Grapefruit, Lemons, Oranges

AZ (Lemons and Oranges only): Apply a split application of 2 lbs. per acre in the spring plus 2 lbs. per acre in the fall.

CA: In grapefruit, lemons, and oranges, apply 2.5-5 lbs. per acre in a single application; or apply 2.5 lbs. per acre in the fall and 2.5 lbs. per acre in the spring.

Precaution: Do not use in the Imperial, Coachella, or Palo Verde valleys, or crop injury may occur.

FL (Grapefruit and Oranges only): (Ridge Areas) Apply only once per year. For control of weeds listed under General Information, apply 8-12 lbs. per acre in the spring; to also control milkweed vine, use 10-12 lbs. per acre either preemergence or before vines are 12 inches tall. Complete control of milkweed vine may require repeat application over a period 2-3 years. (Bedded Areas) Apply 4 lbs. per acre in bedded areas to control balsam apple vine and other weeds. Apply prior to periods of major weed seed germination in soil where moisture conditions are good. Apply only once per year.

TX (Grapefruit and Oranges only): Apply 5-6 lbs. per acre.

Precautions (All areas): To avoid crop injury (1) Do not use in nurseries. (2) Do not apply to bedded grapefruit, lemons, or oranges except for FL grapefruit and oranges. (3) Do not apply to trees under stress from freeze damage for one year after the freeze.

Macadamia Nuts

Apply 2.5 lbs. in 50 gals. of water per acre before harvest and just prior to weed emergence. Repeat application as necessary.

Note: Do not apply when nuts are on the ground during the harvest period, as illegal residues may result.

Olives

Apply 2.5-5 lbs. per acre following grove preparation in the fall. Repeat annually in midwinter.

Peaches, Plums, Sweet Cherries

Apply 2-5 lbs. per acre. Apply in late fall to early spring prior to weed emergence.

Precautions: To avoid crop injury, (1) Peaches: use only in AR, LA, MO, OK, TX, and states east of the Miss. River. For CA, see specific directions in the section "Almonds and Peaches (CA only)." (2) Plums and sweet cherries: use only in MO and states east of the Miss. River except TN.

Pecans

Apply 2.5-5 lbs. per acre before weeds emerge in the spring.

Precautions: To avoid crop injury, (1) Do not use west of the Pecos River in TX or NM, AZ, or CA. (2) Do not make applications to transplanted trees that have been established less than 2 years in the grove.

Note: Do not apply when nuts are on the ground, or illegal residues may result. Do not allow animals to graze treated areas.

Strawberries

Oregon and Washington: For control of chickweed, groundsel, mustard, and shepherdspurse, apply broadcast 1.25 lbs. per acre. In fields where overhead irrigation is used to activate this product, apply after harvest at time of bed renovation. In fields where overhead irrigation is not available, apply during early October through November.

Precautions: To avoid crop injury, (1) Make only one application per growing season. (2) Do not apply within 4 months after transplanting.

Walnuts

Apply 2.5-5 lbs. per acre. Leveling and furrowing operations after application will lessen effectiveness of weed control.

Note: Do not apply when nuts are on the ground, or illegal residues may result.

PRINCEP 80W Plus Evik® 80W for Grapefruit and Oranges

Use in grapefruit and oranges in FL and TX for control of weeds listed in the following directions and under General Information. Apply the rates given below in a minimum of 40 gals. of water per acre thoroughly covering existing vegetation, but avoiding contact with citrus fruit, foliage, or stems. Apply when weeds have resumed active growth in the spring, but before they exceed 4 inches in height (except as noted for vines). Observe all precautions and limitations on the labeling of this product and Evik 80W.

FL (Ridge area): Apply 5 lbs. of PRINCEP 80W plus 5 lbs. of Evik 80W per acre; controls milkweed vine when applied either preemergence or on vines up to 24 inches tall.

FL (Bedded areas): Apply 4 lbs. of PRINCEP 80W plus 6 lbs. of Evik 80W per acre; controls balsamapple when applied either preemergence or on vines up to 12 inches tall.

TX: Apply 4-5 lbs. of PRINCEP 80W plus 2-3 lbs. Evik 80W per acre plus 2 qts. nonionic surfactant, such as X-77®, per 100 gals. of spray mixture (add surfactant to spray tank last); controls buffelgrass, marinevine (ivy treebine), Texas panicum, sandbur, and sunflower, and gives top-kill of johnsongrass. Use 5 lbs of PRINCEP 80W per acre for longer residual control; use 3 lbs. of Evik 80W per acre for control of heavy weed growth and for top-kill of johnsongrass; use the higher rate of both products for control of marinevine.

Precautions (All areas): To avoid injury, (1) Do not use in nurseries, or around trees that have been established in the grove less than two years except in FL, (2) Do not use in nurseries, or where trees are under stress from freeze damage for one year after the freeze. (3) Apply only once per year.

Note (All areas): Do not apply within 30 days before harvest, or illegal residues may result. Do not graze treated areas.

PRINCEP 80W Plus Bromacil 80W for Grapefruit and Oranges (Florida Only)

Use in grapefruit and oranges in FL for control of balsamapple, black nightshade, carpetweed, crabgrass, cudweed, dayweed, Florida pusley, horseweed, pepperweed, pigweed, poorjoe, ragweed, rattlebox, spanishneedles and sandbur, and for partial control of bermudagrass, bahiagrass, pangolagrass, paragrass, and torpedograss. Apply 4-8 lbs. of PRINCEP 80W plus 3-4 lbs. of bromacil 80W per acre beneath trees in a minimum of 40 gals. of water per acre before or soon after weed growth begins. Use the lower rates for light weed infestations or all applications in bedded citrus areas. Use the higher rates for heavy weed infestations only in ridge grown citrus areas. Temporary yellowing of citrus leaves may occur following treatment.

Precautions: To avoid crop injury, (1) Do not use in nurseries, or where trees are under stress from freeze damage for one year after the freeze. (2) Do not use on soil with less than 1% organic matter or on poorly drained soil. (3) Do not treat trees planted in irrigation furrows. (4) Do not treat diseased trees such as those with foot rot. (5) Do not use in groves interplanted with other trees or desirable plants, nor in home grapefruit or orange plantings, or in areas where roots of other valuable plants or trees may extend. (6) Treated areas may be planted to citrus trees one year after application. (7) Do not rotate to other crops within two years after application.

Note: Apply only once per year and avoid contact with foliage and fruit with spray or mist, or illegal residues may result.

PRINCEP 80W Plus Gramoxone® Super

This tank mix is effective in the following fruit and nut crops for kill of existing vegetation and for residual control of the annual broadleaf and grass weeds claimed for PRINCEP 80W applied alone. This combination is also effective for top-kill and suppression of perennial weeds.

Use this tank mix on these crops:

almonds (CA only)	grapefruit	olives
apples	(CA, FL[2] and TX	oranges[2]
avocados	only)	peaches[3]
(CA and FL[1] only)	grapes	pears
cherries	lemons	pecans
(sour and sweet[4])	(AZ and CA only)	plums[4]
filberts	macadamia nuts	walnuts

Apply the PRINCEP 80W rate given under the appropriate crop on this label plus 1.5-2.5 qts. Gramoxone Super in 50-200 gals. (30-50 gals. for pecans) of water per acre to the orchard floor avoiding contact with fruit, foliage, or stems. Add a nonionic surfactant, such as X-77, at 0.5 pt. per 100 gals. of spray. Apply when weeds are succulent and new growth is 1-6 inches tall. For mature woody weeds or difficult to control perennial weeds, re-treat or spot treat with Gramoxone Super if regrowth occurs.

Add PRINCEP 80W to the spray tank first (refer to Mixing instructions section of this label), then add Gramoxone Super, and add the surfactant last. Provide constant agitation during mixing and application to keep the mixture in suspension.

Precautions: To avoid crop injury, (1) Apply the tank mix only once per year. (2) Use a shield for young trees or vines.

Refer to the PRINCEP 80W and Gramoxone Super labels for further directions, specific weeds controlled, and precautions and limitations on each crop.

[1] In avocados in Florida, this tank mix also controls balsamapple vine, rattail amaranth, and at the higher rate of each herbicide, it suppresses coral vine.

[2] In grapefruit and oranges in Florida, apply 4 lbs. PRINCEP 80W per acre in bedded areas and 8-12 lbs. PRINCEP 80W per acre in ridge areas.

[3] Limited to CA, AR, LA, MO, OK, TX, and states east of the Miss. River. As appropriate, refer to PRINCEP 80W sections "Almonds and Peaches (CA only)" or "Peaches, Plums, Sweet Cherries" for rate of PRINCEP 80W and other information.

[4] Limited to MO and states east of the Miss. River except TN.

* Based on a product containing 1.5 lbs. Gramoxone Super cation per gallon.

Tank Mixture with Roundup®

This tank mixture is effective in grape vineyards and in the following bearing or nonbearing tree crops for control of existing vegetation and for residual control of the annual broadleaf and grass weeds claimed for this product applied alone. This combination is also effective for partial control of perennial weeds contacted by the spray mixture during application. Use this tank mixture on these crops:

almonds**(CA only)	grapefruits*	peaches*[2]
apples*	(CA, FL[1] and TX)	pears*
avacados*	grapes*	pecans**
(CA and FL only)	lemons* (AZ and CA)	plums*[3]
cherries	macadamia nuts**	walnuts**
(sour and sweet[3])	oranges*	
filberts**	(AZ, CA, FL[1] and TX)	

Use the appropriate rate given elsewhere on this label for PRINCEP applied alone to the crop being treated. Add to the spray tank 1-5 qts. of Roundup per acre depending on weeds present and their growth stage. Also add an agriculturally approved nonionic surfactant at 0.5% by volume of spray solution. Apply the mixture in 10-40 gals. of water per acre as a postemergence spray to the weeds at the appropriate weed growth stage given on the Roundup label.

Add PRINCEP to the spray tank first, then add Roundup. Provide constant agitation during mixing and application to keep the mixture in suspension. Refer to Application Procedures section of this label for further directions.

Precaution: To avoid crop injury, take extreme care to avoid contact of herbicide solution, spray, drift, or mist with foliage or green bark of trunk, branches, suckers, fruit, or other parts of trees or vines. Observe precautions on both the PRINCEP and Roundup labels for each crop involved.

Refer to the labels of both herbicides for further directions, specific weeds controlled, precautions and limitations on each crop.

* Allow a minimum of 14 days between last application and harvest.

** Allow a minimum of 21 days between last application and harvest of these crops.

[1] In grapefruit and oranges in Florida, apply 4 lbs. of this product per acre in bedded areas and 8-12 lbs. per acre in ridge areas.

[2] Limited to CA, AR, LA, MO, OK, TX, and states east of the Mississippi River. For CA, see specific directions in the section "Almonds and Peaches (CA only)."

[3] Plums and sweet cherries: limited to MO and states east of the Mississippi River except TN.

PRINCEP 80W Plus Surflan®

Use this tank mix in the following crops for preemergence control of all weeds claimed on both labels:

almonds	cherries	lemons	pecans
apples	filberts	oranges	plums
avocados	grapes	peaches	English
caneberries	grapefruit	pears	walnuts

Apply the PRINCEP 80W rate given under the appropriate crop on this label plus 2.67-5.33 lbs. of Surflan 75W or 2-4 qts. of Surflan A.S. in 20-40 gals. of water per acre. Refer to the Surflan 75W or Surflan A.S. label for complete tank mix directions. Observe all precautions and limitations on the PRINCEP and Surflan labels.

FIELD AND FORAGE CROPS

Corn

Nitrogen solutions or complete fluid fertilizers may replace all or part of the water as a carrier in the spray. Determine the physical compatibility of PRINCEP 80W with fertilizer before use. Refer to the General Information section of this label for a compatibility test. Do not apply after corn has emerged as there is danger of liquid fertilizers causing crop injury.

Preemergence: Apply before weeds and corn emerge. Use the appropriate rate in the table below.

Preplant: Apply in the spring after plowing at the appropriate rate in the table below. Apply before, during, or after final seedbed preparation. If soil is tilled or worked after application, avoid deep incorporation of PRINCEP 80W. Best results will be obtained when PRINCEP 80W is applied within 2 weeks before planting. Under dry weather conditions, preplant applications may give better weed control. If weeds develop, particularly under relatively dry conditions, a shallow cultivation will generally result in better weed control.

Soil texture	Broadcast* rate per acre
Coarse-textured soil: Sand, silt, and loam that is low in organic matter	2.5 lbs.
Medium-textured soil: Soil containing a moderate amount of clay and organic matter	3 lbs.

Soil texture	Broadcast* rate per acre
Fine-textured soil: Loam that is high in organic matter and clay (including dark prairie soils of Corn Belt)	3.75 lbs.
Organic soil: Peat, muck, and high-organic clay	5 lbs.

* For calculation of band treatment rate, see the General Information section.

To control quackgrass: Apply 3.75-5 lbs. per acre in the fall. Plow two to three weeks later, or if erosion is a problem, delay plowing until spring.

Precautions: (1) Do not apply more than 5 lbs. of PRINCEP 80W per acre to corn per year. (2) Do not rotate to any crop except corn until the following year or injury may occur. (3) After harvest, plow and thoroughly till the soil in fall or spring to minimize possible injury to spring-seeded rotational crops, regardless of the rate used. (4) If more than 3.75 lbs. of PRINCEP 80W is used per acre (or equivalent rate in a band), a crop of untreated corn should precede the next rotational crop. (5) Do not apply PRINCEP 80W preplant incorporated in corn in the High Plains and Intermountain areas of the West (including central and western KS, western NE, western OK, and the Panhandle of TX) where rainfall is sparse and erratic or where irrigation is required. (6) In the High Plains and Intermountain areas of the West where rainfall is sparse and erratic or where irrigation is required, use PRINCEP 80W to control weeds in corn only when corn is to follow corn or when a crop of untreated corn is to precede another rotational crop. (7) In western MN and eastern parts of the Dakotas, NE, and KS, do not plant soybeans following corn treated with PRINCEP if more than 2.5 lbs. of PRINCEP 80W per acre (or equivalent rate in a band) was applied or injury may occur. (8) Injury may occur to soybeans planted in north-central and northwest IA, south-central and southwest MN, northeast NE, southeast SD, and other areas the year following application on soils having a calcareous surface layer. (9) Do not plant sugar beets, tobacco, vegetables (including dry beans), spring-seeded small grains, or small-seeded legumes and grasses the year after PRINCEP 80W application, or injury may occur.

Note: Do not graze treated areas, or illegal residues may result.

Tank Mixtures on Corn

AAtrex®: Use PRINCEP 80W in a preplant or preemergence tank mixture with AAtrex Nine-O®, AAtrex 80W, or AAtrex 4L for control of many annual weeds, including carpetweed, crabgrass, fall panicum, foxtail, lambsquarters, morningglory, pigweed, ragweed, and velvetleaf.

Apply at the rates given below. Use the 1:1 ratio for most weeds and the 2:1 ratio for expected heavy infestations of crabgrass and fall panicum.

Soil texture	Broadcast rate per acre			
	1:1 Ratio		2:1 Ratio	
	PRINCEP 80W	AAtrex 80W*	PRINCEP 80W	AAtrex 80W*
Sand, loamy sand, sandy loam	1.25 lbs.	1.25 lbs.	1.67 lbs.	0.83 lb.
Loam, silt loam, silt, clay loam, sandy clay loam, silty caly loam, sandy clay, or silty clay with low organic matter	1.5 lbs.	1.5 lbs.	2 lbs.	1 lb.
Loam, silt loam, silt, clay loam, sandy clay loam, silty clay loam, sandy clay, or silty clay with medium to high organic matter, and clay (including dark prairie soils of the Corn Belt)	1.8 lbs.	1.8 lbs.	2.4 lbs.	1.2 lbs.

* When using AAtrex 4L or AAtrex Nine-O, use equivalent rates. One lb. of AAtrex 80W equals 0.8 qt. of AAtrex 4L or 0.9 lb. of AAtrex Nine-O.

Refer to the AAtrex label for complete directions. Also refer to the AAtrex label and the Corn section of this label for precautions and limitations.

Eradicane®: Use in a preplant incorporated tank mixture for control of all weeds claimed on both the PRINCEP and Eradicane labels including partial control (suppression) of shattercane (wild cane). Fluid fertilizer may replace all or part of the water in the spray. Check the physical compatibility of mixture with fertilizer before use. Refer to the General Information section of this label for a compatibility test procedure. Use a minimum of 20 gals. of spray volume per acre. Refer to the Eradicane label for incorporation directions. Use the higher rate of PRINCEP on fine-textured soil and where heavy infestations are expected. Use the higher rate of Eradicane for heavy nutsedge and bermudagrass infestations. For partial control (suppression) of shattercane, broadcast and incorporate **immediately** before planting 1.25-2.5 lbs. of PRINCEP 80W plus 7.33 pts. of Eradicane 6.7E (or 8 pts. Eradicane Extra) per acre. For control of the other weeds claimed on both labels, broadcast and incorporate 1.25-3.75 lbs. of PRINCEP 80W plus 4.75-7.33 pts. of Eradicane 6.7E (or 5.33-8 pts. of Eradicane Extra) per acre. Observe all precautions and limitations on the PRINCEP 80W and Eradicane labels.

Gramoxone Super: Use in a tank mixture where corn will be planted directly in a cover crop, established sod, or previous crop residues. This combination controls existing vegetation and provides residual control of the annual broadleaf and grass weeds listed under General Information.

Add PRINCEP 80W to the spray tank, mix thoroughly with water, and then add Gramoxone Super and a nonionic surfactant. Provide constant agitation during mixing and application to keep the mixture in suspension.

Apply 2.5-3.75 lbs. of PRINCEP 80W plus 1.5-2.5 pts. of Gramoxone Super in 20-60 gals. of water per acre as a broadcast spray either before or after planting but before corn emerges. Add a nonionic surfactant at the rate of 0.5 pt. per 100 gals. of spray volume.

For further information, see general information, caution and warning statements, precautions, and notes on the PRINCEP 80W and the Gramoxone Super labels.

* Based on a product containing 1.5 lbs. Gramoxone Super cation per gallon.

Sutan+®: Use in a preplant incorporated tank mixture for control of all weeds claimed on both the PRINCEP and Sutan+ labels including partial control (suppression) of shattercane (wild cane). Fluid fertilizer may replace all or part of the water in the spray. Check the physical compatibility of mixture with fertilizer before use. Refer to the General Information section of this label for a compatibility test procedure. Use a minimum of 20 gals. of spray volume per acre. Refer to the Sutan+ label for incorporation directions. Use the higher rate of PRINCEP on fine-textured soil and where heavy broadleaf infestations are expected. For partial control (suppression) of shattercane, broadcast and incorporate **immediately** before planting 1.25-2.5 lbs. of PRINCEP 80W plus 7.33 pts. of Sutan+ 6.7E per acre. For control of the other weeds claimed on both labels, broadcast and incorporate 1.25-3.75 lbs. of PRINCEP 80W plus 4.75 pts. of Sutan+ 6.7E per acre. Observe all precautions and limitations on the PRINCEP 80W and Sutan+ labels.

Sugarcane

Apply 2.5-5 lbs. per acre at planting or after harvest if ratoon crop is to be grown, but before weeds or sugarcane emerge. PRINCEP 80W may also be applied interline-directed or over the sugarcane to provide continuing control of germinating weeds.

Precautions: To avoid crop injury or illegal residues, (1) Do not apply more than one 2.5-5 lbs. application over the plants or more than two interline-directed applications. (2) Do not apply after close-in. (3) Total application should not exceed 12.5 lbs. per acre crop cycle.

Note: Do not feed or graze treated forage to livestock.

NURSERIES, CHRISTMAS TREE PLANTINGS, AND SHELTERBELTS

Nurseries (see list below)

Apply 2.5-3.75 lbs. in at least 25 gals. of water per acre in fall or spring.

Precaution: To avoid plant injury, do not apply for at least one year after transplanting.

Christmas Tree Plantings and Shelterbelts (see list below)

Remove weed growth before application. Apply 2.5-5 lbs. in at least 25 gals. of water per acre after transplanting. Use the same rate for annual maintenance applications.

For quackgrass control apply 5 lbs. per acre in the fall or apply a split application of 2.5 lbs. per acre in the fall plus 2.5 lbs. per acre in early spring, after quackgrass begins growth.

Precautions: To avoid tree injury, (1) Do not use PRINCEP 80W on seedbeds or cutting beds. (2) In CA, OR, and WA do not apply to Christmas trees or shelterbelts sooner than one year after transplanting. In other areas, do not apply to Christmas trees or shelterbelt transplants less than three years of age. (3) Do not use until soil is firmly settled around roots. (4) Do not apply more than once a year, except as directed for quackgrass control.

Apply PRINCEP 80W to these species of trees and shrubs, as recommended above:

Conifers

arborvitae	Scotch pine	hemlock
Austrian pine	white pine	juniper
*Bishop pine	blue spruce	red cedar
*knobcone pine	Norway spruce	white cedar
lodgepole pine (shore pine)	red spruce	yew (*Taxus* spp.)
	white spruce	
*Monterey pine	balsam fir	
Mugho pine	Douglas fir	*For CA only
red pine (Norway pine)	Fraser fir	
	white fir	

Deciduous Trees and Woody Ornamentals

American elm	eucalyptus	Russian olive
Siberian elm	holly (*Ilex* spp.)	*oleander

barberry	(max. 4 lbs./A)	*palm
boxelder	honey locust	(2.5 lbs./A)
bush honeysuckle	Oregon grape (*Mahonia* spp.)	*bottle brush
caragana	*Pieris* spp.	*carob
cotoneaster	(max. 4lbs./A)	*For CA only
dogwood	red oak	

Surflan Tank Mix: On Christmas tree plantings, use this tank mix for preemergence control of weeds listed on this label and the Surflan 75W (or Surflan A.S.) label. Use on field grown conifer species listed on the labels for each herbicide plus grand fir, alpine fir, Engelmann spruce, black spruce, Colorado blue spruce, Coulter pine, giant redwood, and Veitchi fir. Broadcast the mixture as a directed spray to the soil surface or an an overtop spray using 2.5-5 lbs. of PRINCEP 80W and 2.67-5.33 lbs. of Surflan 75W (or 2-4 qts. of Surflan A.S.). Apply in sufficient water per acre to uniformly treat the area. Follow overtop sprays with sprinkler irrigation to move the herbicide from leaf surfaces to the soil. Remove weed growth before application. Mix weed residues, prunings or trash into the soil, or remove them before treatment. Soil should be in good tilth and free of clods at time of application. Shallow cultivation (1-2 inches) after treatment will not reduce weed control. Observe all precautions and limitations on the PRINCEP 80W and Surflan labels.

Dual® 8E Tank Mix: On field and liner* grown woody ornamentals, use this tank mix for preemergence control of yellow nutsedge, barnyardgrass, crowfootgrass, fall panicum, giant foxtail, green foxtail, goosegrass, large crabgrass, smooth crabgrass, redroot pigweed, galinsoga, Venice mallow, lambsquarters, purslane, ragweed, smartweed and other weeds listed on the Dual 8E alone label. Control should last about 60 days. Apply 1-1.25 lbs. of PRINCEP 80W plus 2-3 pts. of Dual 8E in a minimum of 15 gals. of water per acre. For band applications use a proportional amount of herbicide. Direct the spray toward the base of established ornamentals or ornamentals transplanted at least 10 days. Apply before weeds emerge or after removing existing weeds. Use the higher rates on fine-textured soil and where broadleaf weeds are expected to form a significant part of the infestation. Use the lower rates on coarse-textured soil and where light infestations of broadleaf weeds are expected. Apply post-transplant to the following field and liner grown ornamentals:

Aucuba (*Aucuba* spp.)	Liriope (*Liriope* spp.) Silver Sunproof
Arborvitae (*Thuja* spp.)	Maple Whips (*Acer* spp.)
Cotoneaster (*Cotoneaster* spp.)	Osmanthus (*Osmanthus* spp.)
Crabapple (*Malus* spp.)	Red Oak (*Quercus* spp.)
Holly (*Ilex* spp.), Boxleaf Japanese, Japanese	Rhododendron (*Rhododendron* spp.)
Japanese Andromeda (*Pieris* spp.)	Weigela (*Weigela* spp.)
Juniper (*Juniperus* spp.) Andorra, Creeping, Shore, Bar Harbor, Blue Rug	Yew (*Taxus* spp.) Dwarf Japanese
Leucothoe (*Leucothoe* spp.)	

Note: Length of weed control may be reduced when continuous wet soil conditions follow herbicide application.

Precautions: To avoid plant injury, (1) Do not use on seedbeds or on unrooted cuttings. (2) Do not use in greenhouses or other enclosed areas.

* Plants transplanted normally in rows in a nursery or similar area for further growth prior to transplanting to final growing place.

TURFGRASSES FOR SOD (FLORIDA ONLY)

St. Augustinegrass, Centipedegrass, and Zoysia Grass

Apply 2.5-5 lbs. per acre, according to soil texture as indicated below.

Muck or peat	5 lbs.	Old beds	within 2 days after lifting of sod
		New beds	3-4 days after sprigging or plugging
Sandy soil	2.5 lbs.	Old beds	within 2 days after lifting of sod
		New beds	7-10 days after sprigging or plugging

If weeds regrow, apply an additional 2.5 lbs. on muck or peat, or 1.25 lbs. on sandy soil.

Precautions: To avoid crop injury, (1) Do not apply within 30 days prior to cutting or lifting. (2) Do not apply in combination with surfactants or other spray additives. (3) Use only on turfgrass reasonably free of infestations of insects, nematodes, and diseases. (4) On newly sprigged turfgrass, temporary slowing of growth may follow application.

TURFGRASS FOR FAIRWAYS, LAWNS, SOD PRODUCTION* AND SIMILAR AREAS

* In states other than Florida. For use on turfgrass for sod in Florida see "Turfgrasses for Sod (Florida only)" section above.

Bermudagrass, Centipedegrass, St. Augustinegrass, and Zoysia Grass

Apply PRINCEP 80W after September 1 (after October 1 for annual bluegrass) before emergence of winter annual weeds. PRINCEP will control annual bluegrass, burclover, lawn burweed, common and mouseear chickweed, corn speedwell, henbit, hop clover, spurweed, and parsley-piert. PRINCEP will also control these weeds soon after emergence. For control of summer annual weeds listed in the **General Information** section of this label, apply PRINCEP in late winter before the weeds emerge. Apply in a minimum of 15 gals. of water per acre.

Irrigate with 1/2 inch of water if rainfall does not occur within 10 days after preemergence treatment.

Where annual bluegrass is the major weed, use 1.25 lbs. of PRINCEP 80W per acre (0.5 oz. per 1,000 sq. ft.). Use 1.25-2.5 lbs. per acre (0.5-1 oz. per 1,000 sq. ft) for control of other weeds. However, do not exceed 1.25 lbs. per acre per treatment on newly sprigged turfgrass or on hybrid bermudagrass such as Tiflawn, Tifway, and Ormond.

For control of summer annuals which emerge after the initial application, apply an additional 1.25 lbs. per acre at least 30 days after the initial application, but not after June

1. However, do not make more than two applications per year.

Precautions: On newly sprigged turfgrass, hybrid bermudagrass, nondormant bermudagrass or nondormant zoysia grass, temporary slowing of growth and yellowing may occur following application. To avoid turf injury, (1) Use only on turfgrass reasonably free of infestations of insects, nematodes, and disease. (2) Do not use on golf greens. (3) Do not use north of NC (except may be used in the Virginia Coastal Plains) or west of Highway 281 in Texas and Oklahoma. (4) Do not use on muck or alkaline soils. (5) Do not apply over the rooting area of trees or ornamentals not listed on this label. (6) Do not seed or overseed with desirable turfgrass within 4 months before or 6 months after treatment. (7) Do not apply this product to newly seeded grasses until they have overwintered and have a well-developed rhizome system. (8) Do not exceed 2.5 lbs. per acre within 12 months of seeding grasses.

Note: Do not graze or feed turf clippings to animals, or illegal residues may result.

VEGETABLE CROPS

Asparagus (Established)

Apply 2.5-5 lbs. per acre to beds established one year or more. Apply once per year either after disking in the spring before weeds emerge, or following harvest after removal of existing weed growth. Spring application must be made at least 3 days before first cutting.

Precautions: (WA and OR only): To avoid crop injury, (1) Do not apply on gravelly, sand, or loamy sand soil. (2) Do not allow irrigation water to stand or to form pools in the field.

Artichokes

Apply 2.5-5 lbs. per acre as a directed spray after the last fall tillage.

Precautions: To avoid crop injury, (1) Do not apply on gravelly, sand, or loamy sand soil. (2) Apply only once per year.

NONSELECTIVE WEED CONTROL ON NONCROP LAND

Best results are obtained when PRINCEP 80W is applied before weeds emerge. If weed growth has begun and removal is not feasible, include a contact or translocated herbicide in the spray.

PRINCEP 80W provides long-term weed control on industrial sites, highway medians and shoulders, railroad rights-of-way, lumberyards, petroleum tank farms, and in noncrop areas on farms, such as around buildings, equipment and fuel storage areas, along fences, roadsides, and lanes.

Aerial applications may be made where it is feasible. Use at least 1 gal. of water for each 1 lb. of PRINCEP 80W for both ground and aerial applications. Use more water if practical.

Precautions: (1) Do not use this product for nonselective weed control on land to be cropped, near adjacent desirable trees, shrubs, or plants, or in greenhouses, as injury may occur. (2) Do not use in areas accessible to livestock or allow livestock to graze treated weed foliage.

To control annual broadleaf and grass weeds (including barnyardgrass, cheat, crabgrass, common lambsquarters, foxtail, ragweed, puncturevine, and turkey mullein) apply 6-12.5 lbs. per acre.

To control most annual and many perennial broadleaf and grass weeds (including quackgrass, bluegrass, redtop, burdock, Canada thistle, orchardgrass, purpletop, smooth brome, dogfennel, and plantain) apply 12.5-22.5 lbs. per acre.

Tank Mixture with Roundup

This tank mixture will provide control of emerged annual weeds and residual control of weeds listed on the PRINCEP label. The combination also will partially control emerged perennial weeds listed on the Roundup label. Add to the spray tank the appropriate rate of PRINCEP noted in the above section for PRINCEP alone. Then add 1-5 qts. of Roundup per acre depending on weeds present and their growth stage. Also add an agriculturally approved nonionic surfactant at 0.5% by volume of spray solution. Apply the mixture in 10-40 gals. of water per acre as a postemergence spray at the appropriate weed growth stage given on the Roundup label. Provide constant agitation during mixing and application to keep the mixture in suspension. Refer to the Application Procedures section of this label for further directions. Refer to the labels of both herbicides for specific noncrop sites, rates, weeds controlled, and further directions, precautions, and limitations.

Tank Mixture with Oust®

Use this tank mixture to control the weeds listed below with either preemergence, early postemergence, or postemergence applications. Use the higher rates within the ranges given under conditions of heavy weed growth, on soil with over 2½% organic matter, on areas with heavy decaying plant residues, or on high moisture areas such as along road edges or railroad shoulders. To improve wetting and/or contact activity on emerged weeds, add a nonionic surfactant such as duPont Surfactant WK or X-77 at 1 qt. per 100 gals. of spray. Do not apply by aircraft. Provide constant agitation during mixing and application to keep the mixture in suspension. Refer to the labels on both herbicides for specific noncrop sites and further directions, precautions, and limitations.

Preemergence to Early Postemergence: To control the weeds listed below apply 5-10 lbs. of PRINCEP 80W plus 3-5 ounces of Oust in a minimum of 25 gals. of water per acre.

Broadleaf Weeds (Up to 3-4" In Height or Diameter)

bouncingbet	horsetail	sowthistle
buckhorn plantain	kochia	St. Johnswort
burclover	lambsquarters	sunflower
Carolina geranium	little mallow	sweet clover
common chickweed	mustard	tansymustard
common dandelion	ox-eye daisy	tansy ragwort
common speedwell	pigweed	tumble mustard
common yarrow	prickly lettuce	turkey mullein
crimson clover	puncturevine	vetch
dogfennel	purple starthistle	wild carrot
filaree	ragweed	yellow rocket
hoary cress	Russian thistle	

Grasses (Up to 6-12" In Height)

Alta fescue	fescue	red brome
annual bluegrass	foxtails	red fescue
annual ryegrass	Indiangrass	reed canarygrass
bahiagrass	Italian ryegrass	ripgut brome
barnyardgrass	johnsongrass	ryegrass
cheatgrass	(short-term	smooth brome
crabgrass	control)	sprangletop
downy brome	little barley	witchgrass
fall panicum		

Postemergence: To control actively growing weeds listed below apply 10-20 lbs. of PRINCEP 80W plus 6-12 ounces of Oust in a minimum of 25 gals. of water per acre.

Actively Growing Weeds

bedstraw	Jerusalem artichoke	johnsongrass
Canada thistle	kudzu	(season-long
curly dock	mayweed	control)
dewberry	musk thistle	yellow nutsedge
fiddleneck	poison ivy	
fleabane	spanishneedles	
goldenrod	turkey mullein	
hemlock	Virginia pepperweed	
honeysuckle	wild blackberry	

STORAGE AND DISPOSAL

Store in a dry place. Do not contaminate water, food, or feed by storage or disposal. Improper disposal of unused spray mixture, or rinse water is a violation of federal law. If these wastes cannot be used according to label instructions, contact your State Pesticide or Environmental Control Agency, or the Hazardous Waste representative at the nearest EPA Regional Office for guidance in proper disposal methods.

Completely empty bag into application equipment. Dispose of empty bag in a sanitary landfill or by incineration, or if allowed by state and local authorities, by burning. Stay out of smoke from burning bags.

For minor spills, leaks, etc., follow all precautions indicated on this label and clean up immediately. Take special care to avoid contamination of equipment and facilities during cleanup procedures and disposal of wastes. In the event of a major spill, fire or other emergency, call (919) 292-7100 day or night.

PRECAUTIONARY STATEMENTS

HAZARDS TO HUMANS AND DOMESTIC ANIMALS

CAUTION

Causes eye irritation. Harmful if inhaled or absorbed through the skin. Avoid breathing dust or spray mist. Avoid contact with eyes, skin, or clothing. Wash thoroughly with soap and water after handling. Remove and wash contaminated clothing before reuse.

First Aid: If in eyes, flush with plenty of water. Get medical attention if irritation persists. If on skin, wash with soap and water. If inhaled, remove victim to fresh air. If not breathing, give artificial respiration, preferably mouth-to-mouth; get medical attention.

ENVIRONMENTAL HAZARDS

Do not apply directly to water or wetlands. Do not contaminate water by cleaning of equipment or disposal of wastes. Simazine is a chemical which can travel (seep or leach) through soil and enter ground water which may be used as drinking water. Simazine has been found in ground water as a result of its use as a herbicide. Users of this product are advised not to apply simazine where the water table (ground water) is close to the surface and where the soils are very permeable, i.e., well drained soils such as loamy sands. Users are advised to consult with their local agricultural agencies to obtain information on the location of ground water and the type of soil in their area.

AAtrex® trademark of CIBA-GEIGY for atrazine
Compex® trademark of KALO Agricultural Chemicals, Inc.
Dual® trademark of CIBA-GEIGY for metolachlor
Eradicane® trademark of Stauffer Chemical Co.
Evik® trademark of CIBA-GEIGY for ametryn
Gramoxone® trademark of ICI Americas for paraquat
Nine-O® trademark of CIBA-GEIGY
Oust® trademark of E.I. duPont de Nemours & Co., Inc.
PRINCEP® trademark of CIBA-GEIGY for simazine
Roundup® trademark of Monsanto Company for glyphosate herbicide
Surflan® trademark of Elanco Products Co. for oryzalin
Sutan+® trademark of Stauffer Chemical Co.
Unite® trademark of Hopkins Agricultural Chemical Co.
X-77® trademark of KALO Agricultural Chemicals, Inc.

CGA 6L46H 028

This is a specimen label which was not prepared by CIBA-GEIGY. CIBA-GEIGY is not responsible for the accuracy of the information contained herein. As labels are subject to revision, always carefully read and follow the label on the product container.

PRINCEP® CALIBER® 90

HERBICIDE

For weed control in certain crops, ornamental plantings, on industrial sites, and around-the-farm

ACTIVE INGREDIENT:

Simazine: 2-chloro-4,6-bis (ethylamino)-*s*-trazine	90%
INERT INGREDIENTS:	10%
TOTAL	100%

EPA Reg. No. 100-603/EPA Est. 100-LA-1

PRINCEP CALIBER 90 is a water dispersible granule.

KEEP OUT OF REACH OF CHILDREN

CAUTION

See additional precautionary statements at end of booklet.

See directions for use inside booklet.

PRINCEP® trademark of CIBA-GEIGY for simazine

CGA 6L48E 028

DIRECTIONS FOR USE AND CONDITIONS OF SALE AND WARRANTY

IMPORTANT: Read the entire **Directions for Use** and the **Conditions of Sale and Warranty** before using this product.

CONDITIONS OF SALE AND WARRANTY

The **Directions for Use** of this product reflect the opinion of experts based on field use and tests. The directions are believed to be reliable and should be followed carefully. However, it is impossible to eliminate all risks inherently associated with use of this product. Crop injury, ineffectiveness, or other unintended consequences may result because of such factors as weather conditions, presence of other materials, or the manner of use or application all of which are beyond the control of CIBA-GEIGY or the Seller. All such risks shall be assumed by the Buyer.

CIBA-GEIGY warrants that this product conforms to the chemical description on the label and is reasonably fit for the purposes referred to in the **Directions for Use** subject to the inherent risks referred to above. **CIBA-GEIGY makes no other express or implied warranty of Fitness or Merchantability or any other express or implied warranty. In no case shall CIBA-GEIGY or the Seller be liable for consequential, special, or indirect damages resulting from the use or handling of this product.** CIBA-GEIGY and the Seller offer this product, and the Buyer and user accept it, subject to the foregoing **Conditions of Sale and Warranty,** which may be varied only by agreement in writing signed by a duly authorized representative of CIBA-GEIGY.

DIRECTIONS FOR USE

It is a violation of federal law to use this product in a manner inconsistent with its labeling.

FAILURE TO FOLLOW THE DIRECTIONS FOR USE AND PRECAUTIONS ON THIS LABEL MAY RESULT IN CROP INJURY, POOR WEED CONTROL, AND/OR ILLEGAL RESIDUES.

Do not apply this product through any type of irrigation system.

Do not apply this product in such a manner as to directly or through drift expose workers or other persons except those knowingly involved in the application. The area being treated must be vacated by unprotected persons.

REENTRY STATEMENT

Do not enter treated areas without protective clothing until sprays have dried.

Because certain states may require more restrictive reentry intervals for various crops treated with this product, consult your State Department of Agriculture for further information.

Written or oral warnings must be given to workers who are expected to be in a treated area or in an area about to be treated with this product. Oral warnings must inform workers of areas or fields that may not be entered without specific protective clothing until sprays have dried, and appropriate actions to take in case of accidental exposure, as described under **Precautionary Statements** on this label. When oral warnings are given, warnings shall be given in a language customarily understood by workers. Oral warnings must be given if there is reason to believe that written warnings cannot be understood by workers. Written warnings must include the following information: "CAUTION. Area treated with PRINCEP CALIBER 90 on (*date of application*). Do not enter without appropriate protective clothing until sprays have dried. In case of accidental exposure, flush eyes or skin with plenty of water. Call physician if irritation persists. Remove and wash contaminated clothing before reuse."

GENERAL INFORMATION

Apply this herbicide before weeds emerge or after removal of weed growth. PRINCEP controls a wide variety of annual broadleaf and grass weeds when used at selective rates in agricultural crops and ornamental plantings. When used at higher, nonselective rates in noncrop areas, it also control many perennial broadleaf and grass weeds.

Where a range of application rates is given, use the low rate on coarser textured soil and soil lower in organic matter; use the high rate on finer textured soil and soil higher in organic matter.

To avoid spray drift, do not apply under windy conditions. Avoid spray overlap, as crop injury may result.

Since this product enters weeds mainly through their roots, moisture is needed to move it into the root zone. Very dry soil conditions and lack of rainfall following application may necessitate shallow cultivation.

This product is noncorrosive to equipment, nonflammable, and has low electrical conductivity.

Annual Weeds Controlled

alyssum	fivehook bassia	ragweed
annual bluegrass	Flora's paintbrush	rattail fescue
annual morningglory	Florida pusley	redmaids
annual ryegrass	foxtails	Russian thistle
barnyardgrass (watergrass)	goosegrass	shepherdspurse
burclover	groundsel	signalgrass (*Brachiaria* spp.)
carelessweed	henbit	silver hairgrass
carpetweed	junglerice	smartweed
common chickweed	knawel (German moss)	spanishneedles
crabgrass (*Digitaria* spp.)	common lambsquarters	speedwell
downy brome (cheatgrass)	nightshade	tansymustard
fall panicum	pepperweed	wild mustard
fiddleneck	pigweed	wild oats
filaree	pineappleweed	witchgrass
fireweed	prickly lettuce	yellow flower pepperweed
	common purslane	

Following many years of continuous use of this product and chemically related products, biotypes of some of the weeds listed on this label have been reported which cannot be effectively controlled by this and related herbicides. Where this is known or suspected and weeds controlled by this product are expected to be present along with resistant biotypes, we recommend the use of this product in registered combinations or in sequence with other registered herbicides which are not triazines. If only resistant biotypes are expected to be present, use a registered non-triazine herbicide. Consult with your State Agricultural Extension Service for specific recommendations.

APPLICATION PROCEDURES

Ground application: Use conventional ground sprayers equipped with nozzles that provide accurate and uniform application. Be certain that nozzles are uniformly spaced and the same size. Calibrate sprayer before use and recalibrate at the start of each season and when changing carriers. Unless otherwise specified, use a minimum of 20 gals. of spray mixture per acre.

Use a pump with capacity to (1) maintain 35-40 psi at nozzles, (2) provide sufficient agitation in tank to keep mixture in suspension, and (3) to provide a minimum of 20% bypass at all times. Use centrifugal pumps which provide propeller shear action for dispersing and mixing this product. The pump should provide a minimum of 10 gals./minute/100 gal. tank size circulated through a correctly positioned sparger tube or jets.

Use screens to protect the pump and to prevent nozzles from clogging. Screens placed on suction side of pump should be *16-mesh or coarser.* Do not place a screen in the recirculation line. Use 50-mesh or coarser screens between the pump and boom, and where required, at the nozzles. Check nozzle manufacturer's recommendations.

For band applications, calculate amount to be applied per acre as follows:

$$\frac{\text{band width in inches}}{\text{row width in inches}} \times \text{broadcast rate per acre} = \text{amount needed per acre of field}$$

Aerial application: Use aerial application only where specified in the use directions. Apply in a minimum of 1 gal. of water for each 1-1.5 lbs. of herbicide applied per acre. Avoid application under conditions where uniform coverage cannot be obtained or where excessive spray drift may occur.

Avoid application to animals or humans. Flagmen and loaders should avoid inhalation of spray mist and prolonged contact with skin, and should wash thoroughly before eating and at the end of each day's operation.

Mixing procedures—all uses: (1) Be sure sprayer is clean and not contaminated with any other materials, as crop injury or sprayer clogging may result. (2) Fill tank 1/4 full with clean water, nitrogen solution or complete fluid fertilizer. (3) Start agitation. (4) Be certain that the agitation system is working properly and creates a rippling or rolling action on the liquid surface. (5) Pour product directly from bag into tank. (6) Continue filling tank until 90% full. Increase agitation if necessary to maintain surface action. (7) Add tank mix herbicide(s) after this product is thoroughly suspended. (8) Finish filling tank. Maintain agitation to avoid separation of materials. (9) Empty tank as completely as possible before refilling to prevent buildup of emulsifiable concentrate residue from possible tank mix herbicides. (10) If an emulsifiable concentrate film starts to build up in tank, drain it and clean with strong detergent solution or solvent. (11) Clean sprayer thoroughly immediately after use by flushing system with water containing a detergent.

Compatibility test: To determine the tank mix compatibility of PRINCEP CALIBER 90 with liquid fertilizer, crop oil, spreaders, or recommended pesticides, use this test method. Nitrogen solutions or complete fluid fertilizers may replace all or part of the water in the spray in alfalfa and corn. Since liquid fertilizers can vary, even within the same analysis, check compatibility each time before use. Be especially careful when using **complete** suspension or fluid fertilizers as serious compatibility problems are more apt to occur. Commercial application equipment may improve compatibility in some instances. Check compatibility using this procedure:

1. Add 1 pint of fertilizer to each of 2 one-quart glass jars with tight lids.
2. To **one** of the jars add 1/4 tsp. of compatibility agent approved for this use (1/4 tsp. is equivalent to 2 pts. per 100 gals. spray). Cap and shake until mixed. Examples of compatibility agents include Compex® and Unite®.
3. To **both** jars add the appropriate amount of herbicide(s). If more than one herbicide is used, add them separately with dry herbicides first, flowables next, and emulsifiable concentrates last. After each addition, cap and shake until thoroughly mixed. The appropriate amount of herbicides for this test follows (assuming a spray volume of 25 gals. per acre):

Dry herbicides:	**For each pound per acre** add 1.5 teaspoons to each jar. Fluff up wettable powder products before measuring.
Liquid herbicides:	**For each pint per acre** add 0.5 teaspoon to each jar.

For a spray volume other than 25 gals. per acre, change the teaspoons added to each jar as follows:

25 gals. × no. teaspoons given above / desired gals. spray volume per acre = teaspoons to add to each jar

4. After adding all ingredients, put lids on and tighten. Shake jars vigorously one minute. Let the mixtures stand 15 minutes and then look for separation, large flakes, precipitates, grease, gels, medium to heavy oily film on the jar, or other signs of incompatibility. Determine if the compatibility agent is needed in the spray mixture by comparing the two jars. If either mixture separates, but can be remixed readily, the mixture probably can be sprayed as long as good agitation is used. If the mixtures are incompatible, test the following methods of improving compatibility: (A) slurry the dry herbicide(s) in water before addition, or (B) add 1/2 of the compatibility agent to the fertilizer and the other 1/2 to the emulsifiable concentrate or flowable herbicide before addition to the mixture. If still incompatible, do not use the materials mixed in the same spray tank.

Rotational Crops and Perennial Crop Replanting

To avoid crop injury, observe the following precautions: (1) If rotating treated land the year following application, plant only corn, unless otherwise stated in this label. (2) If replanting perennial crops or if rotating land to crops other than corn, do not apply this product in the year preceding planting these crops.

FRUIT AND NUT CROPS

Apply the spray to the orchard or vineyard floor avoiding contact with fruit, foliage, or stems. Recommended rates are based on broadcast treatment. For band applications or spot applications around fruit or nut trees, reduce the broadcast rate of this product and water per acre in proportion to the area actually sprayed.

Precautions (All fruit and nut crops): To avoid crop injury, (1) Apply only to orchards or groves where trees have been established 12 months or more unless specified differently. (2) Make only one application per year, except as noted otherwise. (3) Do not use on gravelly, sand, or loamy sand soil. (4) Immediately following application, limit overhead sprinkler irrigation to 1/2 inch.

Almonds and Peaches (CA only)

Apply 1.1-2.2 lbs. per treated acre in a 2-4 ft. band on each side of the tree row. Apply before weeds emerge in late fall or early winter. Weeds controlled by 1.1 lbs. include burclover, common chickweed, wild mustard, and shepherdspurse. Apply only once per year.

Precautions: To avoid crop injury, (1) Do not treat trees established in the grove less than 3 years. (2) Do not treat the Mission (Texas) variety of almonds. (3) Do not apply to almond trees propagated on plum rootstocks. (4) Do not replant almonds or peaches in treated soil for 12 months after treatment. (5) Do not apply on soil with less than 1% organic matter. (6) Do not treat areas where water will accumulate.

Apples, Pears, Sour Cherries

Apply 2.2-4.4 lbs. per acre.

Avocados

CA and FL only: Apply 2.2-4.4 lbs. per acre after final preparation of grove.

Blueberries, and Caneberries (blackberries, boysenberries, loganberries, raspberries)

Apply 2.2-4.4 lbs. per acre in the spring or apply a split application of 2.2 lbs. per acre in the spring plus 2.2 lbs. per acre in the fall. Apply a minimum of 40 gals. of water per acre.

On plantings less than 6 months old, use 1/2 the above rate.

To control quackgrass, apply 4.4 lbs. per acre in the fall or split the application applying 2.2 lbs. per acre in the fall plus 2.2 lbs. per acre in the spring, when quackgrass is actively growing.

Note: Do not apply when fruit is present, or illegal residues may result.

Cranberries

MA: Apply up to 4.4 lbs. per acre either before spring growth begins or in the fall after harvest.

Other areas: Apply 2.2 lbs. per acre before spring growth begins.

Filberts

Apply 2.2-4.4 lbs. per acre in the fall or apply a split application of 2.2 lbs. per acre in the fall plus 2.2 lbs. per acre in the spring.

Precaution: If trees are planted on a hillside, excessive soil erosion may result from the elimination of weeds.

Note: Do not apply when nuts are on the ground during the harvest period, or illegal residues may result.

Grapes

Apply 2.2-5.3 lbs. per acre any time between harvest and early spring.

Precaution: Do not use in vineyards established less than three years, or crop injury may occur.

Grapefruit, Lemons, Oranges

AZ (Lemond and Oranges only): Apply a split application of 1.75 lbs. per acre in the spring plus 1.75 lbs. per acre in the fall.

CA: In grapefruit, lemons, and oranges, apply 2.2-4.4 lbs. per acre in a single application; or apply 2.2 lbs. per acre in the fall and 2.2 lbs. per acre in the spring.

Precaution: Do not use in the Imperial. Coachella, or Palo Verde valleys, or crop injury may occur.

FL (Grapefruit and Oranges only): (Ridge Areas) Apply only once per year. For control of weeds listed under General Information, apply 7.1-10.6 lbs. per acre in the spring; to also control milkweed vine, use 8.9-10.6 lbs. per acre either preemergence or before vines are 12 inches tall. Complete control of milkweed vine may require repeat application over a period of 2-3 years. (Bedded Areas) Apply 3.5 lbs. per acre in bedded areas to control balsam apple vine and other weeds. Apply prior to periods of major weed seed germination in soil where moisture conditions are good. Apply only once per year.

TX (Grapefruit and Oranges only): Apply 4.4-5.3 lbs. per acre.

Precautions (All areas): To avoid crop injury, (1) Do not use in nurseries. (2) Do not apply to bedded grapefruit, lemons, or oranges (except for FL grapefruit and oranges). (3) Do not apply to trees under stress from freeze damage for one year after the freeze.

Macadamia Nuts

Apply 2.2-4.4 lbs. in 50 gals. of water par acre before harvest and just prior to weed emergence. Repeat application as necessary.

Note: Do not apply when nuts are on the ground during the harvest period, as illegal residues may result.

Olives

Apply 2.2-4.4 lbs. per acre following grove preparation in the fall. Repeat annually in midwinter.

Peaches, Plums, Sweet Cherries

Apply 1.75-4.4 lbs. per acre. Apply in late fall to early spring prior to weed emergence.

Precautions: To avoid crop injury, (1) Peaches: use only in AR, LA, MO, OK, TX, and states east of the Miss. River. For CA, see specific directions in the section "Almonds and Peaches (CA only)." (2) Plums and sweet cherries: use only in MO and states east of the Miss. River except TN.

Pecans

Apply 2.2-4.4 lbs. per acre before weeds emerge in the spring.

Precautions: To avoid crop injury, (1) Do not use west of the Pecos River in TX or in NM, AZ, or CA. (2) Do not make applications to transplanted trees that have been established less than two years in the grove.

Note: Do not apply when nuts are on the ground, or illegal residues may result. Do not allow animals to graze treated areas.

Strawberries

Oregon and Washington: For control of chickweed, groundsel, mustard, and shepherdspurse, apply broadcast 1.1 lbs. per acre. In fields where overhead irrigation is used to activate this product, apply after harvest at time of bed renovation. In fields where overhead irrigation is not available, apply during early October through November.

Precautions: To avoid crop injury, (1) Make only one application per growing season. (2) Do not apply within 4 months after transplanting.

Walnuts

Apply 2.2-4.4 lbs. per acre. Leveling and furrowing operations after application will lessen effectiveness of weed control.

Note: Do not apply when nuts are on the ground, or illegal residues may result.

Tank Mixture with Evik® 80W for Grapefruit and Oranges

Use in grapefruit and oranges in FL and TX for control of weeds listed in the following directions and under General Information. Apply the rates given below in a minimum of 40 gals. of water per acre thoroughly covering existing vegetation, but avoiding contact with citrus fruit, foliage, or stems. Apply when weeds have resumed active growth in the spring, but before they exceed 4 inches in height (except as noted for vines). Observe all precautions and limitations on the labeling of this product and Evik 80W.

FL (Ridge Area): Apply 4.4 lbs. of this product plus 5 lbs. of Evik 80W per acre; controls milkweed vine when applied either preemergence or on vines up to 24 inches tall.

FL (Bedded Areas): Apply 3.5 lbs. of this product plus 6 lbs. of Evik 80W per acre; controls balsamapple when applied either preemergence or on vines up to 12 inches tall.

TX: Apply 3.5-4.4 lbs. of this product plus 2-3 lbs. of Evik 80w per acre plus 2 qts. nonionic surfactant, such as X-77®, per 100 gals. of spray mixture (add surfactant to spray tank last); controls buffelgrass, marinevine (ivy treebine), Texas panicum, sandbur, and sunflower, and gives top-kill of johnsongrass. Use 4.4 lbs. of this product per acre for longer residual control; use 3 lbs. of Evik 80W per acre for control of heavy weed growth and for top-kill of johnsongrass; use the higher rate of both products for control of marinevine.

Precautions (All Areas): To avoid injury, (1) Do not use in nurseries, or around trees that have been established in the grove less than two years, except in FL. (2) Do not use in nurseries, or where trees are under stress from freeze damage for one year after the freeze. (3) Apply only once per year.

Note (All Areas): Do not apply within 30 days before harvest, or illegal residues may result. Do not graze treated areas.

Tank Mixture with Bromacil 80W for Grapefruit and Oranges (Florida Only)

Use in grapefruit and oranges in FL for control of balsamapple, black nightshade, carpetweed, crabgrass, cudweed, dayweed, Florida pusley, horseweed, pepperweed, pigweed, poorjoe, ragweed, rattlebox, spanishneedles and sandbur, and for partial control of bermudagrass, bahiagrass, pangolagrass, paragrass, and torpedograss. Apply 3.5-7.1 lbs. of this product plus 3-4 lbs. of bromacil 80W per acre beneath trees in a minimum of 40 gals. of water per acre before or soon after weed growth begins. Use the lower rates for light weed infestations or all applications in bedded citrus areas. Use the higher rates for heavy weed infestations only in ridge grown citrus areas. Temporary yellowing of citrus leaves may occur following treatment.

Precautions: To avoid crop injury, (1) Do not use in nurseries or where trees are under stress from freeze damage for one year after the freeze. (2) Do not use on soil with less than 1% organic matter or on poorly drained soil. (3) Do not treat trees planted in irrigation furrows. (4) do not treat diseased trees such as those with foot rot. (5) Do not use in groves interplanted with other trees or desirable plants, not in home grapefruit or orange plantings, or in areas where roots of other valuable plants or trees may extend. (6) Treated areas may be planted to citrus trees one year after application. (7) Do not rotate to other crops within two years after application.

Note: Apply oly once per year and avoid contact with foliage and fruit with spray or mist, or illegal residues may result.

Tank Mixture with Gramoxone® Super

This tank mix is effective in the following fruit and nut crops for kill of existing vegetation and for residual control of the annual broadleaf and grass weeds claimed for this product applied alone. This combination is also effective for top-kill and suppression of perennial weeds.

Use this tank mix on these crops:

almonds (CA only)	grapefruit (CA, FL[2] & TX only)	oranges[2]
apples	grapes	peaches[3]
avocados (CA & FL[1] only)	lemons (AZ & CA only)	pears
cherries (sour and sweet[4])	macadamia nuts	pecans
filberts	olives	plums[4]
		walnuts

Apply the rate given under the appropriate crop on this label plus 1.5-2.5 qts. Gramoxone Super* in 50-200 gals. (30-50 gals for pecans) of water per acre to the orchard floor avoiding contact with fruit, foliage, or stems. Add a nonionic surfactant, such as X-77, at 0.5 pt. per 100 gals. of spray. Apply when weeds are succulent and new growth is 1-6 inches tall. For mature woody weeds or difficult to control perennial weeds, re-treat or spot treat with Gramoxone Super if regrowth occurs.

Add this product to the spray tank first (refer to Mixing procedures section of this label), then add Gramoxone Super and add the surfactant last. Provide constant agitation during mixing and application to keep the mixture in suspension.

Precautions: To avoid crop injury, (1) Apply the tank mix only once per year. (2) Use a shield for young trees or vines.

Refer to the labels of both products for further directions, specific weeds controlled, and precautions and limitations on each crop.

[1] In avocados in Florida, this tank mix also controls balsamapple vine, rattail amaranth, and at the higher rate of each herbicide, it suppresses coral vine.

[2] In grapefruit and oranges in Florida, apply 3.5 lbs. of this product per acre in bedded areas and 7.1-10.6 lbs. per acre in ridge areas.

[3] Limited to CA, AR, LA, MO, OK, TX, and states east of the Miss. River. As appropriate, refer to the sections "Almonds and Peaches (CA only)" or "Peaches, Plums, Sweet Cherries" for rate of this product and other information.

[4] Limited to MO and states east of the Miss. River except TN.

* Based on a product containing 1.5 lbs. Gramoxone Super cation per gallon.

Tank Mixture with Roundup®

This tank mixture is effective in grape vineyards and the following bearing or nonbearing tree crops for control of existing vegetation and for residual control of the annual broadleaf and grass weeds claimed for this product applied alone. This combination is also effective for partial control of perennial weeds contacted by the spray mixture during application. Use this tank mixture on these crops:

almonds**(CA only)	grapefruits* (CA, FL[1] and TX)	peaches*[2]
apples*	grapes*	pears*
avocados* (CA and FL only)	lemons* (AZ and CA)	pecans**
cherries (sour and sweet[3])	macadamia nuts**	plums*[3]
filberts**	oranges* (AZ, CA, FL[1] TX)	walnuts**

Use the appropriate rate given elsewhere on this label for PRINCEP applied alone to the crop being treated. Add to the spray tank 1.5 qts. of Roundup per acre depending on weeds present and their growth stage. Also add an agriculturally approved nonionic surfactant at 0.5% by volume of spray solution. Apply the mixture in 10-40 gals of water per acre as a postemergence spray to the weeds at the appropriate weed growth stage given on the Roundup label.

Add PRINCEP to the spray tank first, then add Roundup. Provide constant agitation during mixing and application to keep the mixture in suspension. Refer to Application Procedures section of this label for further directions.

Precaution: To avoid crop injury, take extreme care to avoid contact of herbicide solution, spray drift, or mist with foliage or green bark of trunk, branches, suckers, fruit, or other parts of trees or vines. Observe precautions on both the PRINCEP and Roundup labels for each crop involved.

Refer to the labels of both herbicides for further directions specific weeds controlled, precautions and limitations on each crop.

* Allow a minimum of 14 days between last application and harvest.

** Allow a minimum of 21 days between last application and harvest of these crops.

[1] In grapefruit and oranges in Florida, apply 3.5 lbs. of this product per acre in bedded areas and 7.1-10.6 lbs. per acre in ridge areas.

[2] Limited to CA, AR, LA, MO, OK, TX, and states east of the Mississippi River. For CA, see specific directions in the section "Almonds and Peaches (CA only)."

[3] Plums and sweet cherries: limited to MO and states east of the Mississippi River except TN.

Tank Mixture with Surflan®

Use this tank mix in the following crops for preemergence control of all weeds claimed on both labels:

almonds	cherries	lemons	pecans
apples	filberts	oranges	plums
avocados	grapes	peaches	English
caneberries	grapefruit	pears	walnuts

Apply the PRINCEP CALIBER 90 rate given under the appropriate crop on this label plus 2.67-5.33 lbs. of Surflan 75W or 22-4 qts. of Surflan A.S. in 20-40 gals. of water per acre. Refer to the Surflan 75W or Surflan A.S. label for complete tank mix directions. Observe all precautions and limitations on the PRINCEP and Surflan labels.

FIELD AND FORAGE CROPS

Corn

Nitrogen solutions or complete fluid fertilizers may replace all or part of the water as a carrier in the spray. Determine the physical compatibility of this product with fertilizer before use. Refer to the General Information section of this label for a compatibility test. Do not apply after corn has emerged as there is danger of liquid fertilizers causing crop injury.

Preemergence: Apply before weeds and corn emerge. Use the appropriate rate in the table below.

Preplant: Apply in the spring after plowing at the appropriate rate in the table below. Apply before, during, or after final seedbed preparation. If soil is tilled or worked after application, avoid deep incorporation. Best results will be obtained when this product is applied within 2 weeks before planting. Under dry weather conditions, preplant applications may give better weed control. If weeds develop, particularly under relatively dry conditions, a shallow cultivation will generally result in better weed control.

Soil texture	Broadcast* rate per acre
Coarse-textured soil: Sand, silt, and loam that is low in organic matter	2.2 lbs.
Medium-textured soil: Soil containing a moderate amount of clay and organic matter	2.6 lbs.
Fine-textured soil: Loam that is high in organic matter and clay (including dark prairie soils of Corn Belt)	3.3 lbs.
Organic soil: Peat, muck, and high-organic clay	4.4 lbs.

* For calculation of band treatment rate, see the General Information section.

To control quackgrass: Apply 3.3-4.4 lbs. per acre in the fall. Plow two to three weeks later, or if erosion is a problem, delay plowing until spring.

Precautions: (1) Do not apply more than 4.4 lbs. per acre to corn per year. (2) Do not rotate to any crop except corn until the following year or injury may occur. (3) After harvest, plow and thoroughly till the soil in fall or spring to minimize possible injury to spring-seeded rotational crops, regardless of the rate used. (4) If more than 3.3 lbs. is used per acre (or equivalent rate in a band), a crop of untreated corn should precede the next rotational crop. (5) Do not apply preplant incorporated in corn in the High Plains and Intermountain areas of the West (including central and western KS, western NE, western OK, and the Panhandle of TX) where rainfall is sparse and erratic or where irrigation is required. (6) In the High Plains and Intermountain areas of the West where rainfall is sparse and erratic or where irrigation is required, use this product to control weeds in corn only when corn is to follow corn or when a crop of untreated corn is to precede another rotational crop. (7) In western MN and eastern parts of the Dakotas, NE, and KS, do not plant soybeans following corn treated with this product if more than 2.2 lbs. per acre (or equivalent rate in a band) was applied or injury may occur.

(8) Injury may occur to soybeans planted in north-central and northwest IA, south-central and southwest MN, northeast NE, southeast SD, and other areas the year following application on soils having a calcareous surface layer. (9) Do not plant sugar beets, tobacco, vegetables (including dry beans), spring-seeded small grains, or small-seeded legumes and grasses the year after an application, or injury may occur.

Note: Do not graze treated areas, or illegal residues may result.

Tank Mixtures on Corn

AAtrex®: Use this product in a preplant or preemergence tank mixture with AAtrex Nine-O®, AAtrex 80W, or AAtrex 4L for control of many annual weeds, including carpetweed, crabgrass, fall panicum, foxtail, lambsquarters, morningglory, pigweed, ragweed, and velvetleaf.

Apply at the rates given below. Use the 1:1 ratio for most weeds and the 2:1 ratio for expected heavy infestations of crabgrass and fall panicum.

	Broadcast rate per acre			
	1:1 Ratio		2:1 Ratio	
Soil texture	PRINCEP Caliber 90	AAtrex 80W*	PRINCEP Caliber 90	AAtrex 80W*
Sand, loamy sand, sandy loam	1.1 lbs.	1.25 lbs.	1.5 lbs.	0.83 lb.
Loam, silt loam, silt, clay loam, sandy clay loam, silty clay loam, sandy clay, or silty clay with low organic matter	1.33 lbs.	1.5 lbs.	1.8 lbs.	1 lb.
Loam, silt loam, silt, clay loam, sandy clay loam, silty clay loam, sandy clay, or silty clay with medium to high organic matter, and clay (including dark prairie soils of the Corn Belt)	1.6 lbs.	1.8 lbs.	2.1 lbs.	1.2 lbs.

* When using AAtrex 4L or AAtrex Nine-O, use equivalent rates. One lb. of AAtrex 80W equals 0.8 qt. of AAtrex 4L or 0.9 lb. of AAtrex Nine-O.

Refer to the AAtrex label for complete directions, and use 0.9 lb. of this product for each pound of PRINCEP 80W. Also refer to the AAtrex label and the Corn section of this label for precautions and limitations.

Eradicane®: Use in preplant incorporated tank mixture for control of all weeds claimed on both the PRINCEP and Eradicane labels including partial control (suppression) of shattercane (wild cane). Fluid fertilizer may replace all or part of the water in the spray. Check the physical compatibility of mixture with fertilizer before use. Refer to the General Information section of this label for a compatibility test procedure. Use a minimum of 20 gals. of spray volume per acre. Refer to the Eradicane label for incorporation directions. Use the higher rate of PRINCEP on fine-textured soil and where heavy broadleaf infestations are expected. Use the higher rate of Eradicane for heavy nutsedge and bermudagrass infestations. For partial control (suppression) of shattercane, broadcast and incorporate **Immediately** before planting 1.1-2.2 lbs. of this product plus 7.33 pts. of Eradicane 6.7E (or 8 pts. Eradicane Extra) per acre. For control of the other weeds claimed on both labels, broadcast and incorporate 1.1-3.3 lbs. of this product plus 4.75-7.33 pts. of Eradicane 6.7E (or 5.33-8 pts. of Eradicane Extra) per acre. Observe all precautions and limitations on the PRINCEP and Eradicane labels.

Gramoxone Super: Use in a tank mixture where corn will be planted directly in a cover crop, established sod, or previous crop residues. This combination controls existing vegetation and provides residual control of the annual broadleaf and grass weeds listed under General Information.

Add this product to the spray tank, mix thoroughly with water, and then add Gramoxone Super and a nonionic surfactant. Provide constant agitation during mixing and application to keep the mixture in suspension.

Apply 2.2-3.3 lbs. of this product plus 1.5-2.5 pts. of Gramoxone Super* in 20-60 gals. of water per acre as a broadcast spray either before or after planting but before corn emerges. Add a nonionic surfactant at the rate of 0.5 pt. per 100 gals. of spray volume.

For further information, see general information, caution and warning statements, precautions, and notes on the PRINCEP and the Gramoxone Super labels.

* Based on a product containing 1.5 lbs. Gramoxone Super cation per gallon.

Sutan +®: Use in a preplant incorporated tank mixture for control of all weeds claimed on both the PRINCEP and Sutan + labels including partial control (suppression) of shattercane (wild cane). Fluid fertilizer may replace all or part of the water in the spray. Check the physical compatibility of mixture with fertilizer before use. Refer to the General Information section of this label for a compatibility test procedure. Use a minimum of 20 gals. of spray volume per acre. Refer to the Sutan + label for incorporation directions. Use the higher rate of PRINCEP on fine-textured soil and where heavy broadleaf infestations are expected. For partial control (suppression) of shattercane, broadcast and incorporate **immediately** before planting 1.1-2.2 lbs. of this product plus 7.33 pts. of Sutan + 6.7E per acre. For control of the other weeds claimed on both labels, broadcast and incorporate 1.1-3.3 lbs. of this product plus 4.75 pts. of Sutan + 6.7E per acre Observe all precautions and limitations on the PRINCEP and Sutan + labels.

Sugarcane

Apply 2.2-4.4 lbs. per acre at planting or after harvest if ratoon crop is to be grown, but before weeds or sugarcane emerge. This product may also be applied interline-directed or over the sugarcane to provide continuing control of germinating weeds.

Precautions: To avoid crop injury or illegal residues, (1) Do not apply more than one 2.2-4.4 lbs. application over the plants or more than two interline-directed applications. (2) Do not apply after close-in. (3) Total application should not exceed 11.1 lbs. per acre crop cycle.

Note: Do not feed or graze treated forage to livestock.

NURSERIES, CHRISTMAS TREE PLANTINGS, SHELTERBELTS

Nurseries (see list below)

Apply 2.2-3.4 lbs. in at least 25 gals. of water per acre in fall or spring.

Precaution: To avoid plant injury, do not apply for at least one year after transplanting.

Christmas Tree Plantings and Shelterbelts (see list below)

Remove weed growth before application. Apply 2.2-4.4 lbs. in at least 25 gals. of water per acre after transplanting. Use the same rate for annual maintenance applications.

For quackgrass control apply 4.4 lbs. per acre in the fall or apply a split application of 2.2 lbs. per acre in the fall plus 2.2 lbs. per acre in early spring, after quackgrass begins growth.

Precautions: To avoid tree injury, (1) Do not use on seedbeds or cutting beds. (2) In CA, OR, and WA do not apply to Christmas trees or shelterbelts sooner than one year after transplanting. In other areas, do not apply to Christmas trees or shelterbelt transplants less than three years of age. (3) Do not use until soil is firmly settled around roots. (4) Do not apply more than once a year, except as directed for quackgrass control.

Apply to thest species of trees and shrubs, as recommended above:

Conifers		
arborvitae	Scotch pine	hemlock
Austrian pine	white pine	juniper
*Bishop pine	blue spruce	red cedar
*knobcone pine	Norway spruce	white cedar
lodgepole pine	red spruce	yew
(shore pine)	white spruce	(*Taxus* spp.)
*Monterey pine	balsam fir	
Mugho pine	Douglas fir	
red pine	Fraser fir	
(Norway pine)	white fir	*For CA only

Deciduous Trees and Woody Ornamentals		
American elm	eucalyptus	red oak
Siberian elm	holly (*Ilex* spp.)	Russian olive
barberry	(max. 3.5 lbs./A)	*oleander
boxelder	honey locust	*palm (2.2 lbs./A)
bush honeysuckle	Oregon grape	*bottle brush
caragana	(*Mahonia* spp.)	*carob
cotoneaster	*Pieris* spp.	
dogwood	(max. 3.5 lbs./A)	*For CA only

Surflan Tank Mix: On Christmas tree plantings, use this tank mix for preemergence control of weeds listed on this label and the Surflan 75W (or Surflan A.S.) label. Use on field grown conifer species listed on the labels for each herbicide plus grand fir, alpine fir, Engelmann spruce, black spruce, Colorado blue spruce, Coulter pine, giant redwood, and Veitchi fir. Broadcast the mixture as a directed spray to the soil surface or as an over-top spray using 2.2-4.4 lbs. of PRINCEP CALIBER 90 and 2.67-5.33 lbs. of Surflan 75W (or 2-4 qts. of Surflan A.S.). Apply in sufficient water per acre to uniformly treat the area. Follow overtop sprays with sprinkler irrigation to move the herbicide from leaf surfaces to the soil. Remove weed growth before application. Mix weed residues, prunings or trash into the soil, or remove them before treatment. Soil should be in good tilth and free of clods at time of application. Shallow cultivation (1-2 inches) after treatment will not reduce weed control. Observe all precautions and limitations on the PRINCEP and Surflan labels.

Dual® 8E Tank Mix: On field and liner* grown woody ornamentals, use this tank mix for preemergence control of yellow nutsedge, barnyardgrass, crowfootgrass, fall panicum, giant foxtail, green foxtail, yellow foxtail, goosegrass large crabgrass, smooth crabgrass, redroot pigweed, galinsoga, Venice mallow, lambsquarters, purslane, ragweed, smartweed and other weeds listed on the Dual 8E alone label. Control should last about 60 days. Apply 0.9-1.1 lbs. of PRINCEP CALIBER 90 plus 2-3 pts. of Dual 8E in a minimum of 15 gals. of water per acre. For band applications use a proportional amount of herbicide. Direct the spray toward the base of established ornamentals or ornamentals transplanted at least 10 days. Apply before weeds emerge or after removing existing weeds. Use the higher rates on fine-textured soil and where broadleaf weeds are expected to form a significant part of the infestation. Use the lower rates on coarse-textured soil and where light infestations of broadleaf weeds are expected. Apply post-transplant to the following field and liner grown ornamentals.

Aucuba (*Aucuba* spp.)
Arborvitae (*Thuja* spp.)
Cotoneaster (*Cotoneaster* spp.)
Crabapple (*Malus* spp.)
Holly (*Ilex* spp.)
  Boxleaf Japanese, Japanese
Japanese Andromeda (*Pieris* spp.)
Juniper (*Juniperus* spp.)
  Andorra, Creeping Shore,
  Bar Harbor, Blue Rug
Leucothoe (*Leucothoe* spp.)
Liriope (*Liriope* spp.)
  Silver Sunproof
Maple Whips (*Acer* spp.)
Osmanthus
  (*Osmanthus* spp.)
Red Oak (*Quercus* spp.)
Rhododendron
  (*Rhododendron* spp.)
Weigela (*Weigela* spp.)
Yew (*Taxus* spp.)
  Dwarf Japanese

Note: Length of weed control may be reduced when continuous wet soil conditions follow herbicide application.

Precautions: To avoid plant injury, (1) Do not use on seedbeds or on unrooted cuttings. (2) Do not use in greenhouses or other enclosed areas.

* Plants transplanted normally in rows in a nursery or similar area for further growth prior to transplanting to final growing place.

TURFGRASSES FOR SOD (FLORIDA ONLY)

St. Augustinegrass, Centipedegrass, and Zoysia Grass

Apply 2.2-4.4 lbs. per acre, according to soil texture as indicated below.

Muck or peat	4.4 lbs.	Old beds	within 2 days after lifting of sod
		New beds	3-4 days after sprigging or plugging
Sandy soil	2.2 lbs.	Old beds	within 2 days after lifting of sod
		New beds	7-10 days after sprigging or plugging

If weeds regrow, apply an additional 2.2 lbs. on muck or peat, or 1.1 lbs. on sandy soil.

Precautions: To avoid crop injury, (1) Do not apply within 30 days prior to cutting or lifting. (2) Do not apply in combination with surfactants or other spray additives. (3) Use only on turfgrass reasonably free of infestations of insects, nematodes and diseases. (4) On newly sprigged turfgrass, temporary slowing of growth may follow application.

TURFGRASS FOR FAIRWAYS, LAWNS, SOD PRODUCTION*AND SIMILAR AREAS

*In states other than Florida. For use on turfgrass for sod in Florida, see "Turfgrasses for Sod (Florida only)" section above.

Bermudagrass, Centipedegrass, St. Augustinegrass, and Zoysia Grass

Apply PRINCEP CALIBER 90 after September 1 (after October 1 for annual bluegrass) before emergence of winter annual weeds. PRINCEP will control annual bluegrass, burclover, lawn burweed, common and mouseear chickweed, corn speedwell, henbit, hop clover, spurweed, and parsley-piert. PRINCEP will also control these weeds soon after emergence. For control of summer annual weeds listed in the **General Information** section of this label, apply PRINCEP in late winter before the weeds emerge. Apply in a minimum of 15 gals. of water per acre.

Irrigate with 1/2 inch of water if rainfall does not occur within 10 days after preemergence treatment.

Where annual bluegrass is the major weed, use 1.1 lbs. of PRINCEP CALIBER 90 per acre (0.4 oz. per 1,000 sq. ft.). Use 1.1- 2.2 lbs. per acre (0.4-0.8 oz. per 1,000 sq. ft.) for control of other weeds. However, do not exceed 1.1 lbs. per acre per treatment on newly sprigged turfgrass or on hybrid bermudagrass such as Tiflawn, Tifway, and Ormond.

For control of summer annuals which emerge after the initial application, apply an additional 1.1 lbs. per acre at least 30 days after the initial application, but not after June 1. However, do not make more than two applications per year.

Precautions: On newly sprigged turfgrass, hybrid bermudagrass, nondormant bermudagrass or nondormant zoysia grass, temporary slowing of growth and yellowing may occur following application. To avoid turf injury, (1) Use only on turfgrass reasonably free of infestations of insects, nematodes, and diseases. (2) Do not use on golf greens. (3) Do not use north of NC (except may be used in the Virginia Coastal Plains) or west of Highway 281 in Texas and Oklahoma. (4) Do not use on muck or alkaline soils. (5) Do not apply over the rooting area of trees or ornamentals not listed on this label. (6) Do not seed or overseed with desirable turfgrass within 4 months before or 6 months after treatment. (7) Do not apply this product to newly seeded grasses until they have overwintered and have a well-developed rhizome system. (8) Do not exceed 2.2 lbs. per acre within 12 months of seeding grasses.

Note: Do not graze or feed turf clippings to animals, or illegal residues may result.

VEGETABLE CROPS

Asparagus (Established)

Apply 2.2-4.4 lbs. per acre to beds established one year or more. Apply once per year either after disking in the spring before weeds emerge, or following harvest after removal of existing weed growth. Spring application must be made at least 3 days before first cutting.

Precautions: (WA and OR only): To avoid crop injury, (1) Do not apply on gravelly, sand, or loamy sand soil. (2) Do not allow irrigation water to stand or to form pools in the field.

Artichokes

Apply 2.2-4.4 lbs. per acre as a directed spray after the last fall tillage.

Precautions: To avoid crop injury, (1) Do not apply on gravelly, sand, or loamy sand soil. (2) Apply only once per year.

NONSELECTIVE WEED CONTROL ON NONCROP LAND

Best results are obtained when this product is applied before weeds emerge. If weed growth has begun and removal is not feasible, include a contact or translocated herbicide in the spray.

This product provides long-term weed control on industrial sites, highway medians and shoulders, railroad rights-of-way, lumberyards, petroleum tank farms, and in noncrop areas on farms, such as areound buildings, equipment and fuel storage areas, along fences, roadsides, and lanes.

Aerial applications may be made where it is feasible. Use at lest 1 gal. of water for each 0.9 lb. of this product for both ground and aerial applications. Use more water if practical.

Precautions: (1) Do not use this product for nonselective weed control on land to be cropped, near adjacent desirable trees, shrubs, or plants, or in greenhouses, or injury may occur. (2) Do not use in areas accessible to livestock or allow livestock to graze treated weed foliage.

To control annual broadleaf and grass weeds (including barnyardgrass, cheat, crabgrass, common lambsquarters, foxtail, ragweed, puncturevine, and turkey mullein) apply 5.3-11.1 lbs. per acre.

To control most annual and many perennial broadleaf and grass weeds (including quackgrass, bluegrass, redtop, burdock, Canada thistle, orchardgrass, purpletop, smooth brome, dogfennel, and plantain) apply 11.1-20 lbs. per acre.

Tank Mixture with Roundup

This tank mixture will provide control of emerged annual weeds and residual control of weeds listed on the PRINCEP label. The combination also will partially control emerged perennial weeds listed on the Roundup label. Add to the spray tank the appropriate rate of PRINCEP noted in the above section for PRINCEP alone. Then add 1-5 qts. of Roundup per acre depending on weeds present and their growth stage. Also add an agriculturally approved nonionic surfactant at 0.5% by volume of spray solution. Apply the mixture in 10-40 gals. of water per acre as a postemergence spray at the appropriate weed growth stage given on the Roundup label. Provide constant agitation during mixing and application to keep the mixture in suspension. Refer to the Application Procedures section of this label for further directions. Refer to the labels of both herbicides for specific noncrop sites, rates, weeds controlled, and further directions, precautions, and limitations.

Tank Mixture with Oust®

Use this tank mixture to control the weeds listed below with either preemergence, early postemergence, or postemergence applications. Use the higher rates within the ranges given under conditions of heavy weed growth, on soil with over 2 1/2% organic matter, on areas with heavy decaying plant residues, or on high moisture areas such as along road edges or railroad shoulders. To improve wetting and/or contact activity on emerged weeds, add a nonionic surfactant such as duPont Surfactant WK or X-77 at 1 qt. per 100 gals. of spray. Do not apply by aircraft. Provide constant agitation during mixing and application to keep the mixture in suspension. Refer to the labels of both herbicides for specific noncrop sites and further directions, precautions, and limitations.

Preemergence to Early Postemergence: To control the weeds listed below apply 4.4-8.9 lbs. of PRINCEP CALIBER 90 plus 3-5 ounces of Oust in a minimum of 25 gals. of water per acre.

Broadleaf Weeds (Up to 3-4" in Height or Diameter)

Bouncingbet	horsetail	sowthistle
buckhorn plantain	kochia	St. Johnswort
burclover	lambsquarters	sunflower
Carolina geranium	little mallow	sweet clover
common chickweed	mustard	tansymustard
common dandelion	ox-eye daisy	tansy ragwort
common speedwell	pigweed	tumble mustard
common yarrow	prickly lettuce	turkey mullein
crinson clover	puncturevine	vetch
dogfennel	purple starthistle	wild carrot
filaree	ragweed	yellow rocket
hoary cress	Russian thistle	

Grasses (Up to 6-12″0 In Height)

Alta fescue	fall panicum	red brome
annual bluegrass	fescue	red fescue
annual ryegrass	foxtails	reed canarygrass
bahiagrass	Indiangrass	ripgut brome
barnyardgrass	Italian ryegrass	ryegrass
cheatgrass	johnsongrass	smooth brome
crabgrass	(short-term control)	sprangletop
downy brome	little barley	witchgrass

Postemergence: To control actively growing weeds listed below apply 8.9-17.8 lbs. of PRINCEP CALIBER 90 plus 6-12 ounces of Oust in a minimum of 25 gals. of water per acre.

Actively Growing Weeds

bedstraw	Jerusalem artichoke	johnsongrass
Canada thistle	kudzu	(season-long
curly dock	mayweed	control)
dewberry	musk thistle	yellow nutsedge
fiddleneck	poison ivy	
fleabane	spanishneedles	
goldenrod	turkey mullein	
hemlock	Virginia pepperweed	
honeysuckle	wild blackberry	

STORAGE AND DISPOSAL

Store in a dry place. Do not contaminate water, food, or feed by storage or disposal. Pesticide wastes may be toxic. Improper disposal of unused pesticide, spray mixture, or rinse water is a violation of federal law. If these wastes cannot be used according to label instructions, contact your State Pesticide or Environmental Control Agency, or the Hazardous Waste representative at the nearest EPA Regional Office for guidance in proper disposal methods.

Completely empty bag into application equipment. Dispose of empty bag in a sanitary landfill or by incineration, or if allowed by state and local authorities, by burning. Stay out of smoke from burning bags.

For minor spills, leaks, etc., follow all precautions indicated on this label and clean up immediately. Take special care to avoid contamination of equipment and facilities during cleanup procedures and disposal of wastes. In the event of a major spill, fire or other emergency, call (919) 292-7100 day or night.

PRECAUTIONARY STATEMENTS

HAZARDS TO HUMANS AND DOMESTIC ANIMALS

CAUTION

Harmful if swallowed, inhaled or absorbed through the skin. Avoid breathing dust or spray mist. Avoid contact with eyes, skin, or clothing.

First Aid: In case of contact, immediately flush eyes or skin with plenty of water. Get medical attention if irritation persists.

ENVIRONMENTAL HAZARDS

Do not apply directly to water or wetlands. Do not contaminate water by cleaning of equipment or disposal of wastes. Simazine is a chemical which can travel (seep or leach) through soil and enter ground water which may be used as drinking water. Simazine has been found in ground water as a result of its use as a herbicide. Users of this product are advised not to apply simazine where the water table (ground water) is close to the surface and where the soils are very permeable, i.e., well drained soils such as loamy sands. Users are advised to consult with their local agricultural agencies to obtain information on the location of ground water and the type of soil in their area.

AAtrex® trademark of CIBA-GEIGY for atrazine.
Caliber® trademark of CIBA-GEIGY
Compex® trademark of KALO Agricultural Chemicals, Inc.
Dual® trademark of CIBA-GEIGY for metolachlor
Eradicane® trademark of Stauffer Chemical Co.
Evik® trademark of CIBA-GEIGY for ametryn
Gramoxone® trademark of ICI Americas for paraquat
Nine-O® trademark of CIBA-GEIGY
Oust® trademark of E.I. duPont de Nemours and Co., Inc.
PRINCEP® trademark of CIBA-GEIGY for simazine
Roundup® trademark of Monsanto Company for glyphosate herbicide
Surflan® trademark of Elanco Products Co. for oryzalin
Sutan+® trademark of Stauffer Chemical Co.
Unite® trademark of Hopkins Agricultural Chemical Co.
X-77® trademark of KALO Agricultural Chemicals, Inc.

CGA 6L48E 028

This is a specimen label which was not prepared by CIBA-GEIGY. CIBA-GEIGY is not responsible for the accuracy of the information contained herein. As labels are subject to revision, always carefully read and follow the label on the product container.

SEE SUPPLEMENT FOR NEW LABEL INFORMATION

RIDOMIL® 2E

FUNGICIDE

For the control of certain diseases in various crops caused by the Oomycete class of fungi

ACTIVE INGREDIENT:
Metalaxyl: *N*-(2,6-dimethyl-phenyl)-*N*-(methloxyacetyl) alanine methyl ester 25.1%
INERT INGREDIENTS: 74.9%
TOTAL 100.0%

EPA Reg. No. 100-607 EPA Est. 100-AL-1

RIDOMIL 2E contains 2 lbs. active ingredient per gallon.

KEEP OUT OF REACH OF CHILDREN

WARNING/AVISO

PRECAUCIÓN AL USUARIO: Si usted no lee inglés, no use este producto hasta que la etiqueta haya sido explicado ampliamente.

See additional precautionary statements inside booklet.

See directions for use inside booklet.

RIDOMIL® trademark of CIBA-GEIGY for metalaxyl
U.S. Patent No. 4,151,299

DIRECTIONS FOR USE AND CONDITIONS OF SALE AND WARRANTY

IMPORTANT: Read the entire **Directions for Use** and the **Conditions of Sale and Warranty** before using this product.

CONDITIONS OF SALE AND WARRANTY

The **Directions for Use** of this product reflect the opinion of experts based on field use and tests. The directions are believed to be reliable and should be followed carefully. However, it is impossible to eliminate all risks inherently associated with the use of this product. Crop injury, ineffectiveness, or other unintended consequences may result because of such factors as weather conditions, presence of other materials, or the manner of use or application all of which are beyond the control of CIBA-GEIGY or the Seller. All such risks shall be assumed by the Buyer.

CIBA-GEIGY warrants that this product conforms to the chemical description on the label and is reasonably fit for the purposes referred to in the **Directions for Use** subject to the inherernt risks referred to above. **CIBA-GEIGY makes no other express or implied warranty of Fitness or Merchantability or any other express or implied warranty. In no case shall CIBA-GEIGY or the Seller be liable for consequential, special, or indirect damages resulting from the use or handling of this product.** CIBA-GEIGY and the Seller offer the product, and the Buyer and user accept it, subject to the foregoing **Conditions of Sale and Warranty,** which may be varied only by agreement in writing signed by a duly authorized representative of CIBA-GEIGY.

DIRECTIONS FOR USE

It is a violation of federal law to use this product in a manner inconsistent with its labeling.

FAILURE TO FOLLOW DIRECTIONS FOR USE AND PRECAUTIONS ON THIS LABEL MAY RESULT IN CROP INJURY OR POOR DISEASE CONTROL.

Do not apply this product in such manner as to directly, or through drift, expose workers or other persons, except those knowingly involved in the application. The area being treated must be vacated by unprotected persons.

REENTRY STATEMENTS

Do not enter treated areas without protective clothing until sprays have dried.

Because certain states may require more restrictive reentry intervals for various crops treated with this product, consult your State Department of Agriculture for further information.

Written or oral warnings must be given to workers who are expected to be in a treated area or in an area about to be treated with this product. Oral warnings must be given which inform workers of areas or fields that may not be entered without specific protective clothing until sprays have dried and appropriate actions to take in case of accidental exposure, as described under **Precautionary Statements** at the back of this label. When oral warnings are given, warnings shall be given in a language customarily understood by workers. Oral warnings must be given if there is reason to believe that written warnings cannot be understood by workers. Written warnings must include the following information: "WARNING. Area treated with RIDOMIL 2E on *(date of application).* Do not enter without appropriate protective clothing until sprays have dried. Flush eyes or skin with plenty of water. Call a physician if irritation persists. Remove contaminated clothing and wash before reuse."

GENERAL INFORMATION

RIDOMIL is a systemic fungicide for use on selected crops to control certain diseases caused by members of the Oomycete class of fungi. Other fungicides must be used to control diseases incited by other classes of fungi.

THIS PRODUCT IS NOT TO BE USED IN FOLIAR APPLICATIONS OR IN SOLUTIONS USED TO DIP PLANTS.

Note: RIDOMIL is a systemic fungicide having a specific mode of action and could be subject to development of resistant strains of fungi. Development of resistance cannot be predicted. Therefore, CIBA-GEIGY cannot assume liability for crop damage resulting from resistant strains of fungi. If treatment is not effective following the use of RIDOMIL as recommended, a resistant strain of fungi may be present. If the treatment is ineffective due to the presence of a RIDOMIL resistant strain of fungi, neither RIDOMIL nor any other fungicide with similar action will effectively control that disease. Consideration should then be given to the prompt use of other types of suitable fungicides. Consult with your State Agriculture Experiment Station or Extension Service Specialist for guidance on your particular crop and disease control situation.

Do not make foliar applications to field grown tobacco, or other crops since this practice may encourage more rapid development of resistance.

To avoid spray drift, do not apply under windy conditions. Avoid spray overlap, as crop injury may result.

Mixing Instructions

Add one-fourth to one-half of the required amount of water to the spray tank, add the proper amount of RIDOMIL 2E, then add the rest of the water. When tank mixing other products with RIDOMIL 2E, follow the proper sequence of adding products to the spray tank. Wettable powders or water dispersible granules should be added to the water in the tank first, followed by flowable products, with emulsifiable concentrates, such as RIDOMIL 2E, added last. Provide sufficient agitation during mixing and application to maintain a uniform emulsion.

RIDOMIL 2E is usually compatible with Balan®, Bravo®, Dasanit® + Di-Syston®, Dithane® M-22, Dithane M-45, Dasanit + Nemacur®, Dasanit, D-z-n®, Difolatan®, Enide®, Furadan®, Manzate®, Manzate 200, Mocap®, Mocap Plus 4-2EC, Paarlan®, and Tillam®.

To assure the compatibility of RIDOMIL 2E with these and other products, pour the products into a small container of water in the correct proportions. After thorough mixing, let stand for five minutes. If the combination remains mixed, or can be remixed readily, the mixture is compatible.

Calculate the amount of RIDOMIL 3E needed for band treatment by the formula:

$$\frac{\text{band width in inches}}{\text{row spacing in inches}} \times \frac{\text{broadcast rate}}{\text{per acre}} = \frac{\text{amount needed}}{\text{per acre of field}}$$

For banded applications, the area actually is the area covered by the band, not total cropland area planted. Some row-crop recommendations are based on the treating in-the-row and these rates generally are specified as amounts (ounces) of product per certain row length (often 1,000 feet). Other express rates as amount per treated acre which means the total area treated with the pesticide. If rates are expressed as amount per treated acre and banded applications are used, the amount of pesticide used per acre will be proportionately less. The formula above can be used to calculate the amount of RIDOMIL 2E needed per acre of crop when banded applications are made.

APPLICATION INSTRUCTIONS

Apply RIDOMIL 2E by ground or air in sufficient water or liquid fertilizer to provide uniform coverage of the soil surface. Apply in a minimum of 20 gallons/A for ground applications and 5 gallons/A by air. Refer to the specific crop directions for use for application recommendations.

BEFORE TANK MIXING RIDOMIL 2E WITH OTHER REGISTERED PRODUCTS FOR ANY USE ON THIS LABEL, READ THE LABEL OF THE TANK MIX PARTNER TO BE CERTAIN IT IS LABELED FOR USE ON THE PARTICULAR CROP AND THAT USE PATTERNS ARE COMPATIBLE WITH THOSE OF RIDOMIL 2E.

Application Through Irrigation Systems

RIDOMIL 2E, alone or in combination with other fungicides which are registered for application through irrigation systems, may be applied through irrigation systems. Apply this product only through center pivot, solid set, moving wheel, micro-sprinkler or drip irrigation systems. Do not apply this product through any other type of irrigation system. Crop injury, lack of effectiveness, or illegal pesticide residues in the crop can result from nonuniform distribution of treated water. If you have questions about calibrations, you should contact State Extension Service specialists, equipment manufacturers or other experts. Do not connect an irrigation system (including green-house systems) used for pesticide application to a public water system unless the pesticide label-prescribed safety devices for public water systems are in place. A person knowledgeable of the chemigation system and responsible for its operation, or under the supervision of the responsible person, shall shut the system down and make necessary adjustments should the need arise.

Operating Instructions

1. The system must contain a functional check valve, vacuum relief valve, and low pressure drain appropriately located on the irrigation pipeline to prevent water-source contamination from backflow.
2. The pesticide injection pipeline must contain a functional, automatic, quick-closing check valve to prevent the flow of fluid back toward the injection pump.
3. The pesticide injection pipeline must also contain a functional, normally closed, solenoid-operated valve located on the intake side of the injection pump and connected to the system interlock to prevent fluid from being withdrawn from the supply tank when the irrigation system is either automatically or manually shut down.
4. The system must contain functional interlocking controls to automatically shut off the pesticide injection pump when the water pump motor stops.
5. The irrigation line or water pump must include a functional pressure switch which will stop the water pump motor when the water pressure decreases to the point where pesticide distribution is adversely affected.
6. Systems must use a metering pump, such as a positive injection pump (e.g., diaphragm pump) effectively designed and constructed of materials that are compatible with pesticides and capable of being fitted with a system interlock.
7. Do not apply when wind speed favors drift beyond the area intended.

Application Instructions

RIDOMIL 2E must be applied on the schedule specified in the specific crop use recommendations, not according to the irrigation schedule.

With the exception of avocados and citrus, RIDOMIL 2E has not been sufficiently tested when applied through irrigation systems to assure consistent product performance for all labeled uses. The following calibration and application techniques are provided for user reference, but do not constitute a warranty of fitness for application through sprinkler or drip irrigation equipment. Users should check with state and local regulatory agencies for potential use restrictions before applying any agricultural chemical or drip irrigation equipment.

Note: Do not inject RIDOMIL 2E at full strength or deterioration of valves and seals may occur. Use a dilution ratio of at least 10 parts water to 1 part RIDOMIL 2E. RIDOMIL 2E is corrosive to many seal materials. Leather seals are best. EPDM or silicone rubber seals can be used, but should be replaced once a year. Do not use Viton, Buna-N, Neoprene or PVC seals.

Center Pivot Irrigation Equipment (Use only with drive systems which provide uniform water distribution)

- Determine the size of the area to be treated.
- Determine the time required to apply ½-1 inch of water over the area to be treated when the system and injection equipment are operated at normal pressures recommended by the equipment manufacturer. Run the system at 80-95% of the manufacturer's rated capacity.
- Using water, determine the injection pump output when operated at normal line pressure.
- Determine the amount of RIDOMIL 2E required to treat the area covered by the irrigation system.
- Add the required amount of RIDOMIL 2E and sufficient water to meet the injection time requirements to the solution tank.
- Maintain constant solution tank agitation during the injection period.
- Continue to operate the system until the RIDOMIL 2E solution has cleared the sprinkler head.

Solid Set and Moving Wheel Irrigation Equipment

- Determine the acreage covered by the sprinklers.
- Fill injector solution tank with water and adjust flow rate to use the contents over a 20- to 30-minute interval.
- Determine the amount of RIDOMIL 2E required to treat the area covered by the irrigation system.
- Add the required amount of RIDOMIL 2E into the same quantity of water used to calibrate the injection period.
- Operate the system at the same pressure and time interval established during the calibration.
- Inject RIDOMIL 2E at the end of the irrigation cycle in ½-1 inch of water or as a separate application to maximize the effectiveness of the fungicide.
- Stop injection equipment after treatment is completed. Continue to operate the system until the RIDOMIL 2E solution has cleared the last sprinkler head.

Micro Sprinkler or Drip Irrigation Systems

General Instructions

- Each run of the irrigation system must be calibrated separately to determine the time it takes water to move through the system and to make sure all emitters in the system are putting out the same amount of water.
- Only pressure injection or venturi equipment is recommended.
- Figure the area to be treated in each irrigation run.
- Measure the output of each of the emitters or drip tubes closest to and farthest from the injector site.

- For calibration, substituite a concentrated detergent (such as Wisk) or a soluble fertilizer for the RIDOMIL 2E in the injector tank. The detergent will bubble as it leaves the emitters. The time period over which bubbles occur should be checked for both the closest and farthest emitters. If these times are not within two minutes of each other, adjust the dilution ratio and/or the injection rate.
- If a soluble fertilizer is used, measure the time intervals with a salt bridge. If a drip system is being calibrated, substitute soluble fertilizer for the RIDOMIL 2E in the injector and measure the time intervals with a salt bridge.

Step-by-Step Instructions

- Before starting to calibrate, operate the system until all the emitters are putting out at equal flow rates or until the system is operating at full pressure.
- Make up an indicator solution of detergent or fertilizer, using the same ratio to be used when mixing RIDOMIL 2E.
- Set the injector to apply the indicator solution at the injection rate to be used in the actual RIDOMIL 2E application.
- Attach a 5-inch length of flexible tubing over the emitter closest to the injection point, another length over the emitter farthest away. Both emitters should be monitored to determine the time intervals that the indicator solutions are observed.
- Begin injecting the indicator solution. Direct the flow from the tubes at the emitters into a small container. Begin timing when the indicator solution is first detected, stop timing when the indicator solutions are no longer detected.
- If the period of detection of the indicator solution between the two emitters are within two minutes of each other, comparable coverage will be obtained. If they are not, make adjustments by increasing the dilution ratio, using more water per part of RIDOMIL 2E, or adjust the injector to a slower flow rate.
- Once the system is calibrated, dilute the needed amount of RIDOMIL 2E with water usiing a minimum of 10 parts water to 1 part of RIDOMIL 2E.
- Do not begin to inject RIDOMIL 2E into the system until all emitters are producing equal flow rates, or until the system is at full pressure.
- Inject the RIDOMIL 2E into the system at the end of the irrigation in ½-1 inch of irrigation water.

APPLES (BEARING AND NONBEARING TREES)

Use of RIDOMIL 2E will aid in the control of crown, collar, and root rot caused by *Phytophthora* spp. when used in conjunction with good cultural practices and rootstocks that are most tolerant to the disease. RIDOMIL 2E applications should be made before symptoms appear, especially in areas of the orchard favorable for disease development. RIDOMIL 2E will not revitalize trees showing moderate to severe disease symptoms.

Broadcast Spray or Banded Applications: Apply 2-4 gals./treated acre as a broadcast spray (6-12 fl. oz./1,000 sq. ft.) in sufficient water to obtain thorough coverage. The treated area is based on the area under the tree canopy or the area of the sprayed row (similar to herbicide strips). Soil surface sprays of RIDOMIL 2E will not be effective until the fungicide is moved into the root zone by rainfall or irrigation. Applications should be made in early spring before growth starts and in the fall after harvest but before the ground freezes up.

Drench: Mix 1 qt. of RIDOMIL 2E with 100 gals. of water. Apply the amount of diluted mixture indicated in the table below around the trunk of each tree. Applications should be made in early spring before growth starts and in the fall after harvest but before the ground freezes up. On new plantings, delay the first application until two weeks after planting.

Note: To determine trunk diameter, measure the trunk 12 inches above the soil line.

Trunk Diameter	Qts. of Diluted Mixture/Tree
< 1 inch	1 qt.
1-3 inches	2 qts.
3-5 inches	3 qts.
> 5 inches	4 qts.

Notes: (1) Do not dip roots of trees in or spray bare roots with solutions containing RIDOMIL 2E. (2) Do not graze or feed cover crops in treated orchards, or illegal residues may occur.

ASPARAGUS

RIDOMIL 2E will control crown rot and spear rot caused by *Phytophthora* spp.

Apply 2 qts./A in a minimum of 10 gals. of water over the beds.

Cutting Beds: Apply 30 to 60 days before the first cutting. For additional control, make another application just before harvest.

New Planting: Apply after planting seedlings or after covering one-year old crowns.

AVOCADOS

Root Rot—*Phytophthora cinnamomi*

Begin applications at the start of the growing season or at transplanting. Two additional applications should be made at three-month intervals. Applications are not needed during the winter months of November through February. RIDOMIL 2E may be applied as a sleeve drench at the time of transplanting, as a soil surface spray under sprinkler irrigation system, as a directed spray under drip emitters or injected into the irrigation water.

Sleeve Drench: Mix 1 fl. oz. of RIDOMIL 2E with 18 gals. of water. At time of transplanting, drench the roots inside the sleeve with 1 qt. of RIDOMIL 2E solution per tree.

Note: The sleeve drench will not replace the soil surface for long term control of root rot.

Sprinkler Irrigation: Apply as a soil surface spray under the canopy of the tree in sufficient water to obtain uniform coverage. See the following table for the amount of RIDOMIL 2E to use based on the diameter of the tree canopy. Start applications at the beginning of the growing season or at transplanting and continue at three-month intervals.

Drip Irrigation: Apply the recommended amount of RIDOMIL 2E (see table) to the soil directly under the emitter at each tree. If there is more than one emitter per tree, distribute the total amount of RIDOMIL 2E needed among the emitters.

Injection into Irrigation Water (Sprinkle or Drip Irrigation only): Inject RIDOMIL 2E into the irrigation water at a rate of 2-4 fl. oz./1,000 gals. (3¾-7½ ppm active ingredient) at each irrigation. If RIDOMIL 2E is not applied at each irrigation, use the table below to determine how much RIDOMIL 2E should be injected into the irrigation water. If RIDOMIL is to be used more frequently than every three months, adjust the rates so that no more than the specified amount is applied during each three-month period. See the **General Information** section of this label for further instructions and precautions when making applications through irrigation systems.

Diameter of Tree Canopy (Ft.)	Amount of RIDOMIL 2E Per Ten Trees/3 Months
2	½-1 fl. oz.
5	3-6 fl. oz.
10	13-26 fl. oz.
15	29-58 fl. oz.
20	51-103 fl. oz.

Notes: (1) For best results, use RIDOMIL 2E as soon as soil tests indicate the presence of Phytophthora. (2) For new planting, the use of Phytophthora resistant rootstocks with RIDOMIL 2E is recommended. Mature trees in moderate to advanced stages of decline cannot be cured with RIDOMIL 2E. (3) Do not apply more than 24 gals./A of RIDOMIL 2E/year or make an application within 28 days of harvest, or illegal residues may result.

CITRUS

Includes grapefruits, lemons, limes, oranges, tangelos, tangerines, citrus citron, kumquats and hybrids of these.

Use RIDOMIL 2E on citrus for control of citrus foot rot, root rot, and trunk cankers caused by *Phytophthora* spp. RIDOMIL 2E can be applied as a topical canker application and as a soil application, as a spray through sprinkler or drip irrigation systems. If trees are on a drip irrigation system, distribute the amount of RIDOMIL 2E needed per tree (see tables) to the soil directly under the drip emitters at each tree. If there is more than one emitter per tree, distribute the total amount of RIDOMIL 2E needed among the emitters.

Note: Where nematodes are a problem, best results can be achieved if effective EPA-registered nematicides are used. Nematicides can be used in combination or in sequence with RIDOMIL applications.

Citrus in Nurseries (Arizona, California, and Florida Only)

Make the first application of RIDOMIL 2E at the time of planting. Make repeat applications at three-month intervals during the period when trees are actively growing. For banded applications, use a band wide enough to cover the root system of the plants. Do not apply RIDOMIL 2E solutions to bare roots.

Soil Drench: Apply 4-6 fl. oz./100 gals. of water as a drench over the row at the rate of 100-250 gals./1,000 feet of row. The width of the drench treatment should be wide enough to cover thr root systems of the plants. Follow with a ½-1 inch irrigation over the treated area.

Soil Surface Spray: Apply 2 gals./treated acre in a broadcast or banded surface spray to seedbeds, liners, or bedded stock in sufficient water to obtain thorough coverage. If applications are banded, the treatment area should be wide enough to cover the root systems of the plants. Follow the applications with a ½-1 inch irrigation over the treated area.

Citrus Resets or New Plantings (Arizona, California, and Florida Only)

Make the first application of RIDOMIL 2E to citrus resets or new plantings at the time of transplanting. Make up to three repeat applications per year at three-month intervals or when rot growth flushes occur.

Water Ring Drench: Mix 4-6 fl. oz./100 gals. of water. Apply 5 gals. of mix around the base of each tree within the watering ring.

Soil Surface Spray: Apply 1-2 gals./treated acre (3-6 fl. oz./1,000 sq. ft.) in sufficient water to obtain uniform coverage of the soil surface. Apply spray to the soil surface beneath the tree canopy or apply through irrigation water. If natural rainfall is not expected within three days of a soil surface application, irrigate with 1/2-1 inch water over the treared area. See instructions below for application through irrigation water.

Established Plantings

Soil Applications (Florida Only): Apply 1-2 gals./treated acre (3-6 fl. oz./1,000 sq. ft.) as a soil surface spray under the canopy of the trees. Apply in sufficient water to provide uniform coverage or apply through irrigation water. See instructions below for application through irrigation water. Two additional applications may be made per year at three-month intervals. Use the table as a guide for treating small numbers of trees.

Diameter of Tree Canopy (Ft.)	Fl. Oz. RIDOMIL 2E per Ten Trees
5	1/2-1
10	2 1/2-5
15	5-10
20	10-20

Soil Surface Spray (Arizona and California): Make one application in March-April. One or two additional applications per year can be made at three-month intervals or to coincide with flushes of root growth. Use the following table to determine the proper rate based on tree size and the number of applications per year. For applications based on broadcast rates, use RIDOMIL 2E at 4 gal./A (12 fl. oz./1,000 sq. ft.) when three applications are planned and at 5.5 gals./A (16 fl. oz./1,000 sq. ft.) when two applications are planned. Apply in sufficient water to provide uniform coverage or apply through irrigation water. See instructions below for application through irrigation water.

Diameter of Tree Canopy (Ft.)	Fl. Oz. RIDOMIL 2E per Ten Trees	
	2 Applications/yr.	3 Applications/yr.
5	3 1/2	2 1/4
10	13	9
15	29	21
20	51	37

Trunk Spray for Control of Gummosis Caused by Phytophthora spp. (Arizona, California and Texas only): Add 1 gallon of RIDOMIL 2E to 15 gallons of water and spray the surface of the trunks using enough spray to thoroughly wet the cankers. Up to three applications per year may be made.

Notes: (1) In Arizona and California, do not make trunk spray applications if 3 soil applications are planned or illegal residues may result. (2) Do not apply more than 24 lbs. ai/A of metalaxyl per year (12 gals. of RIDOMIL 2E), or illegal residues may result.

Application Through Irrigation Water (Sprinkler or Drip Irrigation Only): See comments and precautions in the **General Information** section of this label. Inject RIDOMIL 2E into the irrigation water at rates specified in the directions.

COLE CROPS (BROCCOLI, CABBAGE, AND CAULIFLOWER)

Includes broccoli, Chinese broccoli (gai lon, white flowering broccoli), cabbage, Chinese cabbage (tight-heading varieties only), and cauliflower only.

RIDOMIL 2E applied as a soil application at planting will control damping-off caused by *Pythium* spp. and basal stem rot caused by *Phytophthora* spp. Applications may be made preplant incorporated or as a soil surface spray after planting.

Preplant Incorporated Application: Apply 4-8 pts./treated acre as a broadcast application in sufficient water or liquid fertilizer to provide uniform coverage and incorporate in the top two inches of soil. For control of Pythium damping-off use only 1-2 pts./A. For banded applications, a 7-inch band is recommended. Use the formula in the **General Information** section of this label to calculate the amount of RIDOMIL 2E needed per acre.

Surface Application: Apply 4-8 pts./treated acre at planting in sufficient water or liquid fertilizer to provide uniform coverage. For banded applications, a 7-inch band is recommended. Use the formula in the **General Information** section of this label to calculate the amount of RIDOMIL 2E needed per acre. If natural rainfall is not expected before the seeds start germinating, RIDOMIL 2E should be incorporated mechanically before planting or be moved into the seed zone after planting with 1/2-1 inch sprinkler irrigation.

Notes: (1) Do not use RIDOMIL 2E for disease control in greenhouse crops. (2) Do not dip plants in solutions containing RIDOMIL 2E, or crop injury may occur.

COTTON

See Rots and Seedling Diseases of Cotton Cause by Pythium spp.

Apply 1/4-1/2 pt.13,000 linear feet of row (0.3-0.6 fl. oz./1,000 linear feet) as an in-furrow spray in 5-15 gals. of water or liquid fertilizer at planting. Mount the spray nozzle so the spray is directed into the furrow over the seed just before the seeds are covered.

For control of Pythium and Rhizoctonia apply 2-4 qts. Terraclor® 2E or 1 1/3-2 3/4 lbs. Terraclor 75W 13,000 linear feet of row in tank mixture with RIDOMIL 2E or use labeled rates of RIDOMIL 2E PC® 11G.

Note: When RIDOMIL 2E is applied with Terraclor 2E or Terraclor 75W, observe all precautions and restrictions that appear on the Terraclor 2E or Terraclor 75W labels.

CUCURBIT VEGETABLES

Includes balsam pear (bitter melon), Chinese waxgourd (Chinese preserving melon), citron melon, cucumber, gherkin, edible gourds, cantalope, casaba, crenshaw, honeydew melon, honey balls, mango melon, muskmelon, Persian melon, pumpkin, summer squash, winter squash, watermelon, and cucurbit hybrids only.

RIDOMIL 2E applied at planting will provide control of damping-off and cottony leak caused by *Pythium* spp. Applications may be made preplant incorporated or as a soil surface spray after planting.

Preplant Incorporated Application: Apply 4-8 pts./treated acre as a broadcast soil application in sufficient water or liquid fertilizer to provide uniform coverage and incorporate in the top two inches of soil. For banded applications, a 7-inch band is recommended. Use the formula in the **General Information** section of this label to calculate the amount of RIDOMIL 2E needed per acre.

Surface Application: Apply 4-8 pts./treated acre at planting in sufficient water or liquid fertilizer to provide uniform coverage. For banded applications, a 7-inch band is recommended. Use the formula in the **General Information** section of this label to calculate the amount of RIDOMIL 2E needed per acre. If natural rainfall is not expected before the seeds start germinating, RIDOMIL 2E should be incorporated mechanically before planting or be moved into the seed zone after planting with 1/2-1 inch sprinkler irrigation.

Notes: (1) Do not Use RIDOMIL 2E for disease control in greenhouse crops. (2) Do not dip plants in solutions containing RIDOMIL 2E, or crop injury may occur.

DECIDUOUS FRUITS AND NUTS IN ORCHARDS (NONBEARING)

Phytophthora Diseases

Use of RIDOMIL 2E will aid in the control of crown, collar and root rot or deciduous fruit and nut trees caused by *Phytophthora* spp. when used in conjunction with good cultural practices and rootstocks that are most tolerant to the disease. Applications should be made before symptoms appear, especially in areas favorable for disease development. RIDOMIL 2E will not revitalize trees showing moderate to severe disease symptoms.

Make the first application at the time of planting (see Notes below.). Additional applications should be made at three-month intervals during the time when conditions are favorable for disease development. For established plantings, make the first application in the spring before the growth starts.

Apply 2-4 gals./treated acre (6-12 fl. oz./1,000 sq. ft.) in sufficient water to obtain thorough coverage of the soil under the canopy of the trees. For banded applications, use the formula in the **General Information** section of this label to calculate the amount of RIDOMIL 2E needed per acre. Use the high rate in areas known to be infested with Phytophthora. Soil surface sprays of RIDOMIL 2E will not be effective until the fungicide is moved into the root zone by rainfall or irrigation.

Notes: (1) Do not apply to plantings that will bear harvestable fruit within 12 months of the last application, or illegal residues may result. To avoid injury, (2) Do not dip the roots of trees in RIDOMIL 2E solutions, spray the roots, or concentrate it around the tree trunks; (3) Do not apply to trees under stress; (4) In California, do not apply RIDOMIL 2E to newly planted trees within 90 days of planting; and (5) When applying RIDOMIL 2E to apples, see the **Apples** section of this label for specific recommendations.

HOPS

Downy Mildew *(Pseudoperonospora humuli):* Apply 1 qt./A of hops in a minimum of 20 gals. of water or liquid fertilizer to the soil surface over the crowns after pruning, but before training. Early application before shoots are six inches long is preferable.

Notes: (1) Do not apply after training or make more than one application per year, and (2) Do not feed hop refuse to livestock, or illegal residues may result.

LEAFY VEGETABLES (HEAD LETTUCE AND SPINACH ONLY)

RIDOMIL 2E applied as a soil application will control damping-off caused by *Pythium* Spp. in head lettuce and spinach and white rust (*Albugo occidentalis*) in spinach. Applications may be made banded over the row, preplant incorporated, or injected with liquid fertilizer.

Preplant Incorporated Application: Apply 4-8 pts./treated acre as a broadcast soil application in sufficient water or liquid fertilizer to provide uniform coverage and incorporate in the top two inches of soil. For banded applications, a 7-inch band is recommended. Use the formula in the **General Information** section of this label to calculate the amount of RIDOMIL 2E needed per acre.

Surface Application: Apply 4-8 pts./treated acre at planting in sufficient water or liquid fertilizer to provide uniform coverage. For banded applications, a 7-inch band is recommemded. Use the formula in the **General Information** section of this label to calculate the amount of RIDOMIL 2E needed per acre. If natural rainfall is not expected before the seeds start germinating, RIDOMIL 2E should be incorporated mechanically before planting or be moved into the seed zone after planting with ½-1 inch sprinkler irrigation.

White Rust Control (Spinach Only): Apply 1 pt./A or crop as a shanked in application in water or liquied fertilizer. Make the first additional application 40-50 days after planting or after the first cutting. One other application may be made after the next cutting. Use sufficient mechanical or bypass agitation to keep the RIDOMIL 2E mixed with the water or fertilizer.

Notes: (1) The additional applications or RIDOMIL 2E noted above and made after each cutting by shanking the fungicide into the beds along with liquid fertilizer provide continuing control of white rust. However, white rust can only be controlled in a preventative disease control program that begins with an application RIDOMIL 2E to the soil at planting. **If RIDOMIL 2E is not used at planting, do not use RIDOMIL 2E at any other time throughout the season.** Do not apply RIDOMIL 2E in foliar applications or in situations where white rust infections are already established. The use of RIDOMIL 2E in curative applications greatly increases the risk of the fungus developing resistance to metalazyl. The development of resistance will destroy the effectiveness of RIDOMIL 2E in controlling white rust. (2) Do not harvest spinach within 21 days of a RIDOMIL 2E application. (3) Do not use RIDOMIL 2E for disease control in greenhouse crops.

LEGUME VEGETABLES (SUCCULENT OR DRIED)

Includes field beans, French beans, kidney beans, lima beans, mung beans, navy beans, pinto beans, runner beans, snap beans, wax beans, broad beans (fava beans), chickpeas (garbanzo beans), lentils, lupines (sweet, white sweet, white and grain), garden peas, field peas, sugar peas, southern peas (blackeyed peas, crowder peas, cowpeas, catjang), and edible soybeans, etc.

Pythium Damping-Off and Root Rot: RIDOMIL 2E applied at planting will control damping-off and root rot caused by *Pythium* spp. Applications may be made preplant incorporated, or as a soil surface spray after planting.

Preplant Incorporated Application: Apply 2-4 pts./treated acre as a broadcast soil application in sufficient water or liquid fertilizer to provide uniform coverage and incorporate in the top two inches of soil. For banded applications, a 7-inch band is recommended. Use the formula in the **General Information** section of this label to calculate the amount of RIDOMIL 2E needed per acre.

Surface Application: Apply 2-4 pts./treated acre at planting in sufficient water or liquid fertilizer to provide uniform coverage. For banded applications, a 7-inch band is recommended. Use the formula in the **General Information** section of this label to calculate the amount of RIDOMIL 2E needed per acre. If natural rainfall is not expected before the seeds start germinating, RIDOMIL 2E should be incorporated mechanically before planting or be moved into the seed zone after planting with ½-1 inch sprinkler irrigation.

Note: Do not use RIDOMIL 2E for disease control in greenhouse crops.

ONIONS—DRY BULB, GREEN, AND ONIONS GROWN FOR SEED

RIDOMIL 2E applied at planting will control damping-off caused by *Pythium* spp. Applications may be made preplant incorporated or as a soil surface spray after planting.

Preplant Incorporated Application: Apply 2-4 pts./treated acre as a broadcast soil application in sufficient water or liquid fertilizer to provide uniform coverage and incorporate in the top two inches of soil. For banded applications, a 7-inch band is recommended. Use the formula in the **General Information** section of this label to calculate the amount of RIDOMIL 2E needed per acre.

Surface Application: Apply 2-4 pts./treated acre at planting in sufficient water or liquid fertilizer to provide uniform coverage. For banded applications, a 7-inch band is recommended. Use the formula in the **General Information** section of this label to calculate the amount of RIDOMIL 2E needed per acre. If natural rainfall is not expected before the seeds start germinating RIDOMIL 2E should be incorporated mechanically before planting or be moved into the seed zone after planting with ½-1 inch sprinkler irrigation.

Note: Do not use RIDOMIL 2E for disease control in greenhouse crops.

PEANUTS

RIDOMIL 2E is a soil-applied systemic fungicide for use in control of seedling and pod diseases of peanuts incited by *Pythium* spp.

Seedling Diseases: Apply 1 pt./A of crop as an application to the seed in-furrow or in a 7-inch band at the time of planting. For in-furrow applications, position the spray so the fungicide is mixed with the soil covering the seed. Avoid spraying the seed directly with the spray solution or crop injury may occur.

Pod Rot: Apply 1-2 qts./A at early pod set or pegging through the overhead irrigation systems. See the **General Information** section of this label for further instructions and precautions when making applications through irrigation systems.

Notes: (1) Where pathogens other than *Pythium* spp. are present, use fungicides that control those diseases in combination with RIDOMIL 2E. (2) Where the predominant pod rot pathogens are *Pythium* spp. and *Rhizoctonia* spp.: use RIDOMIL PC 11G at labeled rates.

PEPPERS AND EGGPLANT

Soil applications of RIDOMIL 2E will provide control of Pythium damping-off and crown rot caused by *Phytophthora caspsici.* RIDOMIL 2E must be applied to the soil before the plants are infected with Phytophthora to obtain satisfactory disease control.

Apply 4-8 pts./treated acre (broadcast rate) at the time of planting in sufficient water or liquid fertilizer to provide uniform coverage. If rainfall is not expected before the plants begin growth, RIDOMIL 2E should be incorporated mechanically before planting or be moved into the root zone aftrer planting with ½-1 inch or sprinkler irrigation water. For banded applications, a 12-16 inch band is recommended. After the initial application, two supplemental post-directed applications at 4 pts./treated acre (broadcast rate) should be made at 30-day intervals. The spray should be directed at the base of the plants and cover 6-10 inches of soil on either side of the plants. Use the formula in the **General Information** section of this label to calculate the amount of RIDOMIL 2E needed per acre.

Precautions: (1) RIDOMIL 2E may cause some yellowing of the pepper leaves. (2) Plants already infected with Phytophthora cannot be cured with RIDOMIL 2E applications. (3) The foliar blight phase of Phytophthora cannot be controlled with foliar applications of RIDOMIL 2E .

Notes: (1) Do not apply within 7 days of harvest, and (2) Do not apply more than 12 pts. of RIDOMIL 2E per acre of crop per season, or illegal residues may result.

PINEAPPLE

RIDOMIL 2E , applied as a "seed piece" dip, provides effective control of heart rot disease of pineapple caused by *Phytophthora* spp.

Apply RIDOMIL 2E as a crown dip before planting at the rate of 1-2 qts./100 gals. of water. Use 75 to 100 gals. of sip solutions per planted acre, depending on crown size, plant density and dipping techniques.

Note: If there is a crop failure within one year of planting treated crowns, do not harvest plant material for animal feed.

RASPBERRIES

RIDOMIL 2E is a soil-applied systemic fungicide for use in the control of Phytophthora root rot.

Apply 1 pt./1,000 linear ft. or row to the soil surface in a three-foot band over the row (14.5 pts. treated acre on broadcast basis). Make one application in the spring and another in the fall after harvest.

Note: Do not apply RIDOMIL 2E within 45 days before harvest or illegal residues may result.

SOYBEANS

RIDOMIL 2E is a soil-applied systemic fungicide for use in the control of Phytophthora root and stem rot and Pythium damping-off. RIDOMIL 2E may be applied broadcast, banded or in the seed furrow before the seeds are covered. The seed furrow applications will provide more consistent results if rain is not expected before the seeds germinate.

For best results against Phytophthora root and stem rot, use RIDOMIL 2E with soybean varieties that have some tolerance to the races of Phytophthora present in the field. The higher rate of RIDOMIL 2E should be used in areas with a history of heavy Phytophthora damage. Under heavy late season Phytophthora pressure, RIDOMIL 2E may not provide complete control.

Surface Applications: For full season control, apply 5 pts./treated acre in sufficient water or liquid fertilizer to provide uniform coverage at the time of planting. For banded applications, a 7-inch band is recommended. Use the formula in the **General Information** section of this label to calculate the amount of RIDOMIL 2E needed per acre. For early to mid season control, apply 1½-3 pts./treated acre.

In-Furrow Applications: For full season control, apply 1¹/₁₀ fl. oz./1,000 linear feet of row as an in-furrow spray in 5-10 gals. of water or liquid fertilizer at planting. Position the spray so the fungicide is mixed with the soil covering the seed. Avoid spraying the seed directly with the spray solution or crop injury may occur. Use sufficient water or liquid fertilizer to provide uniform coverage. Use the table below to determine the amount of RIDOMIL 2E needed per acre based on row spacing. For early to mid season control, apply ³/₁₀-⁶/₁₀ fl. oz./1,000 linear feet of row.

Suggested Rates Per Acre According to Row Spacing

Row Spacing	Linear Feet of Soybean Row Per Acre	Fluid Ounces of RIDOMIL 2E Needed Per Acre — Rate Desired		
		³/₁₀ fl. oz.	⁶/₁₀ fl. oz.	1¹/₁₀ fl. oz.
38 in.	13,756	4	8	15

Suggested Rates Per Acre According to Row Spacing

Row Spacing	Linear Feet of Soybean Row Per Acre	Fluid Ounces of RIDOMIL 2E Needed Per Acre — Rate Desired: 3/10 fl. oz.	6/10 fl. oz.	1 1/10 fl. oz.
36 in.	14,520	4 1/2	9	16
30 in.	17,424	5 1/2	11	19
24 in.	21,780	6 1/2	13	24
20 in.	26,136	8	16	28

Note: RIDOMIL 2E is specific for Pythium and Phytophthora and will not control other diseases that may attack soybeans.

TOBACCO

RIDOMIL 2E is a soil-applied systemic fungicide for use in the field before transplanting for control of black shank (*Phytophthora parasitica*, var. *Nicotianae*) and blue mold (*Peronospora tabacina*) on all types of tobacco. RIDOMIL 2E may also be applied in tobacco plant beds for control of blue mold and damping-off caused by *Pythium* spp. For control of anthracnose and other tobacco diseases, use fungicides that control those diseases.

Tobacco Plant Bed

For control of blue mold and damping-off caused by *Pythium* spp., apply RIDOMIL 2E only as a preplant broadcast soil application at the rate of 1 qt./treated acre in 50 gals. of water or liquid fertilizer (1 fl. oz. or 2 tablespoons/150 sq. yds. of bed in 2 gals. of water or liquid fertilizer). Apply before or at the time of seeding to the surface of the plant bed and lightly incorporate or follow with 1/2 inch sprinkler irrigation water. In Pennsylvania, use 2 qts./treated acre (2 fl. oz. or 4 tablespoons/150 sq. yds. of bed in 2 gals. of water or liquid fertilizer).

Notes: (1) Do not use RIDOMIL 2E for disease control in greenhouse crops. (2) Do not dip plants in solutions containing RIDOMIL 2E , or crop injury may occur.

Field Planted Tobacco

Blue Mold: Apply RIDOMIL 2E as a broadcast soil application and incorporate in the top 2-4 inches of soil before forming beds. For flue-cured tobacco, use 1-2 qts./treated acre, depending on disease pressure and length of control desired. Under low disease pressure or for early season control, use 1 qt./treated acre. For burley and other tobacco types, use 2 qts./treated acre.

Note: For best control of blue mold, use RIDOMIL 2E in the plant bed and follow with an application in the field prior to transplanting. Do not use RIDOMIL 2E in transplant water or in foliar applications on field tobacco because of the potential for the development of resistance in the blue mold fungus, crop injury or potential excessive residues in the crop.

Black Shank: Use RIDOMIL 2E as a broadcast soil application and incorporate in the top 2-4 inches of soil before forming beds. Apply RIDOMIL 2E using conventioal ground application equipment in sufficient water or liquid/dry fertilizer to provdie uniform coverage. Use the table below to determine the amount of RIDOMIL 2E needed per acre depending on the black shank severity.

Type of Tobacco	Disease Level in Field	Rate of RIDOMIL 2E Per Acre
Flue-Cured	Low to Moderate (Less than 6% Disease)	2 qts.
	High (More than 6% Disease)	4 qts.*
Burley and Other**	Low to Moderate (Less than 6% Disease)	4 qts.
	High (More than 6% Disease)	6 qts.

* Florida and Georgia—Use 6 qts./treated acre of RIDOMIL 2E in fields with heavy black shank levels (greater than 6%).

** Pennsylvania—Do not use RIDOMIL 2E for black shank control.

Lay-by Application: For prolonged control of blue mold or black shank in field planted tobacco, make a supplemental application of 1 qt./acre of crop as a soil application at the last culitvation. Position the nozzles so that the spray is deposited under the plants and is covered with soil by the cultivator. Do not make this lay-by application if more than 2 qts./A of RIDOMIL 2E were applied at transplanting or if no RIDOMIL 2E was applied at transplanting.

Notes: (1) For best results against black shank, use RIDOMIL 2E with tobacco varieties that have some resistance to black shank and use crop rotation. In fields where there is a history of severe black shank, use the highest rate and plant a variety that is resistant to the race of Phythphthora present in the field. (Burley L8 hybrids are only resistant to Phytophthora Race O.) (2) RIDOMIL 2E is not recommended for use in high black shank areas on highly susceptible flue-cured varieties such Hicks, Virginia Gold or White Gold. (3) Failure to adequately control nematodes in fields treated with RIDOMIL 2E may result in poor control of black shank.

TOMATOES

Soil applications of RIDOMIL 2E at planting will provide control of damping-off caused by *Pythium* spp. and soil applications applied 4-12 weeks before harvest under the vines will reduce fruit and root rot caused by *Pythium* spp. and *Phytophthora* spp.

Damping-Off (Pythium spp.): Apply 4-8 pts./treated acre in sufficient water or liquid fertilizer to provide uniform coverage at the time of planting. If natural rainfall is not expected before the seeds start germinating. RIDOMIL 2E should be incorporated mechanically before planting or be moved into the seed zone after planting with 1/2-1 inch sprinkler irrigation. For banded applications, a 7-inch band is recommended. Use the formula in the **General Information** section of this label to calculate the amount of RIDOMIL 2E needed per acre.

Fruit and Root Rot (Pythium spp. and Phytophthora spp.): Additional applications can be made as soil surface applications under the vines to provide control of root and fruit rot. If only one application is planned, apply 4 pts./treated acre under the vines 4 weeks after the planting application to 4 weeks before harvest, depending on the expected occurrence of the diseases. If root and fruit rot are problems throughout the growing season or the occurrence of diseases cannot be predicted, make two additional applications at 2 pts./treated acre under the vines, one 4-6 weeks after the planting application and another 4 weeks before harvest. Use sufficient water to obtain thorough coverage. If rainfall is not expected within three days of the application, follow as soon as possible with 1/2-1 inch overhead irrigation. Use the formula shown in the **General Information** section to determine the amount of RIDOMIL 2E needed per acre.

RIDOMIL can be injected into the beds with liquid fertilizer. For calculating the amount of RIDOMIL 2E needed per acre, apply the same amount as would be applied on a 7-inch band at 4 pts./treated acre. Keep RIDOMIL 2E suspended in the fertilizer solution with mechanical or bypass agitation.

Notes: To avoid illegal residues. (1) Do not apply more than 12 pts./treated acre season and (2) do not apply within 7 days of harvest. (3) Do not use RIDOMIL 2E for disease control in greenhouse crops.

VEGETABLE BEDDING PLANTS

Includes broccoli, Chinese broccoli (gai lon, white flowering broccoli), cabbage, Chinese cabbage (bok choy and napa), cauliflower, cucurbit vegetables, head lettuce, spinach, legume vegetables, onion and tomatoes.

Pythium Damping-Off: Apply 2-4 pts./treated acre as a preplant broadcast spray in 50 gals. of water or liquid fertilizer (1-2 fl. oz. or 2-4 tablespoons 150 sq. yds. of bed in 2 gals. water or liquid fertilizer) before or at the time of seeding to the surface of the beds. If natural rainfall is not expected before the seeds start germinating. RIDOMIL 2E should be incorporated mechanically before planting or be moved in the seed zone after planting with 1/2-1 inch sprinkler irrigation.

Note: In Georgia and Florida, do not use RIDOMIL 2E for disease control in vegetable plants grown in an enclosed environment, i.e., greenhouse, glasshouse, etc.

REPLANTING

If replanting is necessary, additional applications of RIDOMIL 2E or RIDOMIL 5G can be made provided that the total amount of metalaxyl applied does not exceed the maximum allowed for the specific crop. The maximum amount of metalaxyl applied to the soil per crop cycle (combined plants, failed and replanted) is as follows.

To determine the rate per application, follow the directions for use on a particular crop.

Failed and Replanted Crops	Maximum Lbs. Metalaxyl	Maximum Amount of RIDOMIL 2E or 5G
Asparagus	2	4 qts. 2E
Cole Crops (Broccoli, Cabbage, Cauliflower)		
Cotton	2	4 qts. 2E or 40 lbs. 5G
Cucurbit Vegetables**	2	4 qts. 2E
Leafy Vegetables (Head Lettuce and Spinach)		
Legume Vegetables***		
Onions		
Peanuts	2	4 qts. 2E or 40 lbs. 5G
Peppers/Eggplant	2	4 qts. 2E
Pineapples		
Soybeans	2	4 qts. 2E or 40 lbs. 5G
Tobacco	3	6 qts. 2E
Tomatoes		

Rotation (Plantback) Restrictions

Rotation Crop	Planting Time From Last RIDOMIL 2E Application
Apples, Asparagus Avocados	0 days
Broccoli	
Cabbage, Cauliflower, Chinese Broccoli, (gallon, white flowering broccoli), Chinese Cabbage (tight-heading varieties only), Citrus, Cucurbit Vegetables**	
Deciduous Fruits and Nuts*	
Eggplant	
Hops, Head Lettuce	
Legume Vegetables***	
Onions	
Peanuts, Peppers, Pineapples, Potatoes	
Raspberries	
Soybeans, Spinach	
Tobacco, Tomatoes	
Wheat	14 days
Corn	9 months
Root Crops	12 months
Crops **Not Intended** for Food or Feed	0 days
All Other Crops **Intended** for Food or Feed	18 months

* These crops and other perennial crops may be planted immediately following last application of RIDOMIL 2E provided they will not bear harvestable fruit within 12 months.

** **Cucurbit Vegetables:** Includes balsam pear (bitter melon), Chinese waxgourd (Chinese preserving melon), citron melon, cucumber, gherkin, edible gourds, cantaloupe, casaba, crenshaw, honeydew melon, honey balls, mango melon, musk-melon, Persian melon, pumpkin, summer squash, winter squash, water-melon, and cucurbit hybrids.

*** **Legume Vegetables (Succulent or Dried):** Includes beans, French beans, kidney beans, lima beans, mung beans, navy beans pinto beans, runner beans, snap beans, wax beans, broad beans (fava beans), chickpeas (garbanzo beans), lentils, lupines, (sweet, white sweet, white and grain), garden peas, field peas, sugar peas, southern peas (blackeyed peas, crowder peas, cowpeas, catjang) and edible soy-beans.

STORAGE AND DISPOSAL

Store at temperatures above 40°F. Crystals may form at lower storage temperatures. If this occurs, place the product in a warm room (68°F or above) and roll or shake the container at frequent intervals until all crystals are dissolved.

Pesticide Disposal

Do not contaminate water, food or feed by storage or disposal. Open dumping is prohibited. Wastes resulting from the use of this product are acutely toxic. Improper disposal of unused pesticide, spray mixture, or rinsate is a violation of federal law. Pesticide, spray mixture, or rinsate that cannot be used according to label instructions must be disposed of according to federal, state or local procedures. For guidance in proper disposal methods, contact your State Pesticide or Environmental Control Agency, or the Hazardous Waste representative at the nearest EPA Regional Office.

Container Disposal

Do not reuse empty container. Triple rinse (or equivalent) and offer for recycling or reconditioning, or puncture and dispose of in a sanitary landfill, or by incineration, or by open burning, if allowed by state and local authorities. If burned, keep out of smoke.

PRECAUTIONARY STATEMENTS

HAZARDS TO HUMANS AND DOMESTIC ANIMALS

WARNING

Causes eye injury. Do not get in eyes. Avoid contact with skin or clothing. Harmful if swallowed, inhaled or absorbed through the skin. Avoid breathing vapors or spray mist.

Practical Treatment: In case of contact with eyes, immediately flush with plenty of water. Get medical attention. In case of contact with eyes, immediately flush with plenty of water. Get medical attention. In case of contact with skin, wash thoroughly with soap and water. Get medical attention if irritation occurs. If swallowed, immediately contact a physician, hospital or local Poison Control Center. Do not induce vomiting. If inhaled, move to fresh air. If not breathing, give artificial respiration (preferably mouth to mouth) and get medical attention. **Notes to Physician:** If RIDOMIL 2E is ingested, lavage stomach. A slurry of activated charcoal in water can be left in the stomach. Give a saline laxative and supportive therapy.

ENVIRONMENTAL HAZARDS

Do not apply directly to water. Apply only as specified on this label. Do not apply when weather conditions favor drift from treated areas. Do not contaminate water by cleaning of equipment or disposal of wastes.

PHYSICAL OR CHEMICAL HAZARDS

Do not use or store near heat or open flame.

Store at temperatures above 40°F.

CHEMIGATION

Refer to supplemental labeling in attached booklet for use directions on chemigation. Do not apply this product through any type of irrigation system unless the supplemental labeling on chemigation is followed.

Balan® trademark of Elanco Products for benefin
Bravo® trademark of Fermenta Plant Protection Company
Dasanit® trademark of the Parent Company of Farbenfabriken Bayer GmbH, Leverkusen for fensulfothion
Difolatan® trademark of Chevron Chemical Company for captafol
Di-Syston® trademark of the Parent Company of Barbenfabriken Bayer GmbH, Leverkusen for disulfoton
Dithane® trademark of Rohm and Haas Company for maneb mancozeb
D·z·n® trademark of CIBA-GEIGY for diazinon
Enide® trademark of Tuco Division of the Upjohn Company for diphenamid
Furadan® trademark of FMC Corporation for carbofuran
Manzate® trademark of E.I. duPont de Nemours and Company for maneb mancozeb
Mocap® trademark of Rhone Poulenc, Inc. for ethoprop
Nemacur® trademark for the Parent Company of Farbenfabriken Bayer GmbH, Leverkusen, for fenamiphos
Paarlan® trademark of Elanco Products for isopropalin
RIDOMIL® trademark of CIBA-GEIGY for metalaxyl U.S. Patent No. 4,151,299
Terractor® trademark of Uniroyal, Inc., for PCNB
Tillam® trademark of Stauffer Chemical Company for Pebulate

CGA 65L14J 028

SEE SUPPLEMENT SECTION FOR NEW LABEL INFORMATION

This is a specimen label which was not prepared by CIBA-GEIGY. CIBA-GEIGY is not responsible for the accuracy of the information contained herein. As labels are subject to revision, always carefully read and follow the label on the product container.

RIDOMIL MZ® 58

FUNGICIDE

For control of certain diseases of potatoes, tomatoes, onions, cucumbers, melons, and squash

ACTIVE INGREDIENTS:

Metalaxyl: *N*-(2,6-dimethylphenyl)-*N*-(methoxyacetyl) alanine methyl ester		10.0%
Zinc ion and manganese ethylene bisdithiocarbamate,*		48.0%
a coordination product of manganese++	9.6%	
zinc++	1.2%	
ethylene bisdithiocarbamate ion	37.2%	
INERT INGREDIENTS:		42.0%
TOTAL:		100.0%

*Same as the active ingredient found in Dithane® M-45 and Manzate® 200.

EPA Reg. No. 100-629 EPA Est. 32761-MO-3

RIDOMIL MZ58 is a wettable powder

KEEP OUT OF REACH OF CHILDREN
WARNING/AVISO

PRECAUCIÓN AL USUARIO: Si usted no lee inglés, no use este producto hasta que la etiqueta haya sido explicado ampliamente.

See additonal precautionary statements on side of bag.

See directions for use on panel at right.

DIRECTIONS FOR USE AND CONDITIONS OF SALE AND WARRANTY

IMPORTANT: Read the entire **Directions for Use** and the **Conditions of Sale and Warranty** before using this product.

CONDITIONS OF SALE AND WARRANTY

The **Directions for Use** of this product reflect the opinion of experts based on field use and tests. The directions are believed to be reliable and should be followed carefully. However, it is impossible to eliminate all risks inherently associated with use of this product. Crop injury, ineffectiveness, or other unintended consequences may result because of such factors as weather conditions, presence of other materials, or the manner of use or application all of which are beyond the control of CIBA-GEIGY or the Seller. All such risks shall be assumed by the Buyer.

CIBA-GEIGY warrants that this product conforms to the chemical description on the label and is reasonably fit for the purposes referred to in the **Directions for Use** subject to the inherent risks referred to above. **CIBA-GEIGY makes no other express or implied warranty of Fitness or Merchantability or any other express or implied warranty. In no case shall CIBA-GEIGY or the Seller be liable for consequential, special, or indirect damages resulting from the use or handling of this product.** CIBA-

GEIGY and the Seller offer this product, and the Buyer and user accept it, subject to the foregoing **Conditions of Sale and Warranty,** which may be varied only by agreement in writing signed by a duly authorized representative of CIBA-GEIGY.

DIRECTIONS FOR USE

It is a violation of federal law to use this product in a manner inconsistent with its labeling.

Do not apply this product in such a manner as to directly or through drift expose workers or other persons, except those knowingly involved in the application. The area being treated must be vacated by unprotected persons.

FAILURE TO FOLLOW THE DIRECTIONS FOR USE AND PRECAUTIONS ON THIS LABEL MAY RESULT IN POOR DISEASE CONTROL, CROP INJURY, OR ILLEGAL RESIDUES.

Do not enter treated areas without protective clothing until sprays have dried.

Because certain states may require more restrictive reentry intervals for various crops treated with this product, consult your State Department of Agriculture for further information.

Written or oral warnings must be given to workers who are expected to be in a treated area or in an area about to be treated with this product. Oral warnings must inform workers of areas or fields that may not be entered without specific protective clothing until sprays have dried, and appropriate actions to take in case of accidental exposure, as described under **Precautionary Statements** on this label. When oral warnings are given, warnings shall be given in a language customarily understood by workers. Oral warnings must be given if there is reason to believe that written warnings cannot be understood by workers. Written warnings must include the following information: "WARNING. Area treated with RIDOMIL MZ58 on (date of application). Do not enter without appropriate protective clothing until sprays have dried. In case of accidental eye exposure, flush eyes with plenty of water. Get medical attention. In case of contact with skin, wash thoroughly with soap and water. Call physician if irritation occurs. Remove and wash contaminated clothing before reuse."

GENERAL INFORMATION

RIDOMIL MZ58 contains 10% of the active ingredient metalaxyl, and 48% of the active ingredient mancozeb. Metalaxyl is a systemic fungicide for use on selected crops to control diseases caused by members of the Phycomycete family of fungi. Mancozeb is effective against a wide range of fungal pathogens.

Note: Metalaxyl is a systemic fungicide having a specific mode of action and could be subject to development of resistant strains of fungi. Development of resistance cannot be predicted. Therefore, CIBA-GEIGY cannot assume liability for crop damage resulting from resistant strains of fungi. If treatment is not effective following the use of RIDOMIL MZ58 as recommended, a resistant strain of fungi may be present. If the treatment is ineffective due to the presence of a metalaxyl resistant strain of fungi, neither RIDOMIL MZ58 nor any other fungicide with similar action will effectively control that disease. Consideration should then be given to the prompt use of other types of suitable fungicides. Consult with your State Agricultural Experiment Station or Extension Service Specialist for guidance in your particular crop and disease control situation. Do not use RIDOMIL MZ58 for disease control in greenhouse crops.

Do not make applications when weather conditions favor drift from target area. Avoid spray overlap as crop injury may result.

Mixing Instructions

Add one-half to three-fourths of the required amount of water to the spray tank, add the proper amount of RIDOMIL MZ58, then add the rest of the water. Provide sufficient agitation during mixing and application to maintain a uniform suspension.

Amount of water needed per acre will vary according to the amount of plant growth. For dilute ground applications, the water volume will vary from 20-150 gals./A. For aerial applications, 3-10 gals. of water/A are generally needed. To minimize the potential of drift from target area, applications at wind speeds greater than 10 mph are not recommended.

RIDOMIL MZ58 is usually compatible with Bravo®, Dithane M-22, Dithane M-45, D·z·n®, Difolatan®, Manzate, and Manzate 200.

To assure the compatibility of RIDOMIL MZ58 with these and other products, pour the products into a small container of water in the correct proportions. After thorough mixing, let stand for five minutes. If the combination remains mixed, or can be remixed readily, the mixture is compatible.

Application Through Irrigation Systems

RIDOMIL MZ58 alone or in combination with other fungicides which are registered for sprinkler irrigation may be applied through irrigation systems.

Apply this product only through center pivot, solid set or moving wheel irrigation systems. Do not apply this product through any other type of irrigation system. Crop injury, lack of effectiveness, or illegal pesticide residues in the crop can result from nonuniform distribution of treated water. If you have questions about calibration, you should contact State Extension Service specialists, equipment manufacturers or other experts.

Do not connect an irrigation system (including greenhouse systems) used for pesticide application to a public water system unless the pesticide label-prescribed safety devices for public water systems are in place. A person knowledgeable of the chemigation system and responsible for its operation, or under the supervision of the responsible person, shall shut the system down and make necessary adjustments should the need arise.

Operating Instructions

1. The system must contain a functional check valve, vacuum relief valve, and low pressure drain appropriately located on the irrigation pipeline to prevent water-source contamination from backflow.
2. The pesticide injection pipeline must contain a functional, automatic, quick-closing check valve to prevent the flow of fluid back toward the injection pump.
3. The pesticide injection pipeline must also contain a functional, normally closed, solenoid-operated valve located on the intake side of the injection pump and connected to the system interlock to prevent fluid from being withdrawn from the supply tank when the irrigation system is either automatically or manually shut down.
4. The system must contain functional interlocking controls to automatically shut off the pesticide injection pump when the water pump motor stops.
5. The irrigation line or water pump must include a functional pressure switch which will stop the water pump motor when the water pressure decreases to the point where pesticide distribution is adversely affected.
6. Systems must use a metering pump, such as a positive displacement injection pump (e.g., diaphragm pump) effectively designed and constructed of materials that are compatible with pesticides and capable of being fitted with a system interlock.
7. Do not apply when wind speed favors drift beyond the area intended for treatment.

Application Instructions

RIDOMIL MZ58 must be applied on the schedule specified in the specific crop use recommendations, not according to the irrigation schedule. If irrigation schedules are used, ground or aerial applications must supplement chemigation applications to achieve adequate disease control.

With the exception of potatoes, RIDOMIL MZ58 has not been sufficiently tested to assure consistent product performance for all labeled uses when applied through chemigation systems. The following calibration and application techniques are provided for user reference, but do not constitute a warranty of fitness for application through sprinkler irrigation equipment. Users should check with state and local regulatory agencies for potential use retrictions before applying any agricultural chemical through sprinkler irrigation equipment.

Center-Pivot Irrigation Equipment (Use only with drive systems which provide uniform water distribution)

1. Determine the size of the area to be treated.
2. Determine the time required to apply 1/4 inch water over the area to be treated when the system and injection equipment is operated at normal pressures recommended by the equipment manufacturer. Run the system at 80-95% of the manufacturer's rated capacity.
3. Using water, determine the injection pump output when operated at normal line pressure.
4. Determine the amount of RIDOMIL MZ58 required to treat the area covered by the irrigation system.
5. Add the required amount of RIDOMIL MZ58 and sufficient water to meet the injection time requirements to the solution tank.
6. Maintain constant solution tank agitation during the injection period.
7. Stop injection equipment after treatment is completed. Continue to operate the system until the RIDOMIL MZ58 solution has cleared the sprinkler head.

Solid Set and Moving Wheel Irrigation Equipment

1. Determine the acreage covered by the sprinklers.
2. Fill injector solution tank with water and adjust flow rate to use the contents over a 20 to 30 minute interval.
3. Determine the amount of RIDOMIL MZ58 required to treat the area covered by the irrigation system.
4. Add the required amount of RIDOMIL MZ58 into the same quantity of water used to calibrate the injection period.
5. Operate system at the same pressure and time interval established during the calibration.
6. Inject RIDOMIL MZ58 at the end of the irrigation cycle or as a separate application to maximize retention of the fungicide by the foliage.
7. Stop injection equipment after treatment is completed. Continue to operate the system until the RIDOMIL MZ58 solution has cleared the last sprinkler head.

IMPORTANT

Do not allow RIDOMIL MZ58 to become wet or overheated during storage. This may lead to chemical changes which will impair the fungicidal effectiveness of RIDOMIL MZ58 and may also generate flammable vapors. Keep bag closed when not in use.

CUCUMBERS, MELONS, AND SQUASH

RIDOMIL MZ58 provides effective control of downy mildew caused by *Pseudoperonospora cubensis* when used as a foliar spray in a preventative disease control program.

Apply 1½-2 lbs. RIDOMIL MZ58 per acre in sufficient water to obtain thorough coverage. Begin applications when conditions are favorable for disease, but before infection, and continue at 14-day intervals until the threat of disease is over. Under heavy disease pressure, use the higher rate of RIDOMIL MZ58.

Notes: (1) To minimize the potential for resistance, do not make more than 4 applications of RIDOMIL MZ58 per crop. (2) If downy mildew continues to be a problem after making 4 applications of RIDOMIL MZ58, use other fungicides registered for downy mildew control. (3) Do not make the last application within 5 days of harvest or illegal residues may result.

ONIONS (DRY BULB AND SEED ONIONS)

RIDOMIL MZ58 provides effective control of downy mildew caused by *Peronospora destructor* when used as a foliar spray in a preventative disease control program.

Apply 1½-2 lbs. RIDOMIL MZ58 per acre in sufficient water to obtain thorough coverage. Begin applications when conditions are favorable for disease, but before infection, and continue at 14-day intervals until the threat of disease is over. Use a suitable spreader-sticker at rates recommended on the product label. Under heavy disease pressure, use the higher rate of RIDOMIL MZ58.

Notes: (1) To minimize the potential for resistence, do not make more than 4 applications of RIDOMIL MZ58 per crop. (2) If downy mildew continues to be a problem after making 4 applications of RIDOMIL MZ58, use other fungicides registered for downy mildew control. (3) Do not make the last application within 7 days of harvest, or illegal residues may result.

POTATOES

RIDOMIL MZ58, when used as a foliar application in a preventative disease control program, will provide effective control of late blight and tuber rot caused by *Phytophthora infestans*, Pythium leak caused by *Pythium* spp. and pink rot caused by *Phytophthora ethroseptica*. Make applications of RIDOMIL MZ58 in sufficient water to obtain thorough coverage for ground applications and in a minimum of 3 gals. of water per acre for aerial applications.

Late Blight: Apply 1½-2 lbs. RIDOMIL MZ58 per acre. Begin applications when conditions are favorable for disease, but before infection, and continue at 14-day intervals until the threat of disease is over. Under heavy disease pressure, use the higher rate of RIDOMIL MZ58.

Tuber Rot, Pythium Leak and Pink Rot: Apply 2 lbs. RIDOMIL MZ58 per acre. Make the first application at flowering and another one 14 days later. If the field has a history of tuber disease problems, make a third application 14 days after the second application.

Note: If applications of RIDOMIL MZ58 are being made for late blight control that correspond to the timing of applications for tuber disease control, additional applications for tuber disease control are not needed.

For supplemental control of early blight, add 0.8 lb. of Manzate 200 80W, Dithane M-45 80W, Manzate 80W or Dithane M-22 80W to the RIDOMIL MZ58 spray or make an additonal application of any registered early blight fungicide between each RIDOMIL MZ58 spray. Where early blight disease pressure is heavy, use early blight fungicides according to product use directions, and use RIDOMIL MZ58 only when late blight threatens. Do not apply more than 0.8 lb. per acre per application of Manzate 200 80W, Dithane M-45 80W, Manzate 80W, or Dithane M-22 80W in combination with RIDOMIL MZ58.

Notes: (1) Do not apply RIDOMIL MZ58 within 7 days of harvest or illegal residues may result. (2) To minimize the potential for resistance, do not make more than 4 applications of RIDOMIL MZ58 per crop. (3) If late blight is still a problem after making 4 applications of RIDOMIL MZ58, use other fungicides registered for late blight control. (4) Do not tank mix with amobam or Polyram®.

TOMATOES

RIDOMIL MZ58, when used as a foliar application in a preventative disease control program, will provide effective control of late blight caused by *Phytophthora infestans.*

Apply 1½-2 lbs. RIDOMIL MZ58 per acre as a foliar spray in sufficient water to provide thorough coverage. Begin applications when conditions are favorable for disease, but before infection, and continue at 14-day intevals until the threat of disease is over. Under heavy disease pressure, use the higher rate of RIDOMIL MZ58.

Note: When conditions are favorable for development of diseases other than late blight, any registered fungicides which control those diseases may be applied according to their label directions. When making applications of Manzate 200, Dithane M-45, Manzate, or Dithane M-22, 1½ lbs./A of RIDOMIL MZ58 may be added to the spray mix every 14 days for control of late blight. Do not apply more than 2 lbs. per acre per application of Dithane M-45 80W, Manzate 200 80W, Dithane M-22 80W, or Manzate 80W in combination with RIDOMIL MZ58. Observe all precautions and restrictions that appear on the labels of other products applied with RIDOMIL MZ58.

Notes: (1) Do not apply RIDOMIL MZ58 within 5 days of harvest or illegal residues may result. (2) To minimize the potential for resistance, do not make more than 4 applications of RIDOMIL MZ58 per crop. (3) If late blight continues to be a problem after making 4 applications of RIDOMIL MZ58, use other fungicides registered for late blight control. (4) Do not tank mix with amobam, Polyram, nabam, thiram, zineb, or ziram.

Plant Back Restrictions

Crop	Planting Time From Last Application Of RIDOMIL MZ58
Apples, Asparagus, Avocados	0 Days
Broccoli	
Cabbage, Cauliflower, Chinese Broccoli (gai lon, white flowering broccoli), Chinese Cabbage (tight-heading varieties only), Citrus, Cotton, Cucurbit Vegetables*	
Deciduous Fruits and Nuts	
Hops, Head Lettuce	
Legume Vegetables**	
Onions	
Peanuts, Pineapples, Potatoes	
Raspberries	
Soybeans, Spinach	
Tobacco, Tomatoes	
Wheat	14 Days
Corn	270 Days
Root Crops	
Crops **Not Intended** for Food or Feed	0 Days
Other Crops **Intended** for Food or Feed	540 Days

* **Cucurbit Vegetables:** Includes balsam pear (bitter melon), Chinese waxground (Chinese preserving melon), citron melon, cucumber, gherkin, edible gourds, cantalope, casaba, crenshaw, honeydew melon, honey balls, mango melon, muskmelon, Persian melon, pumpkin, summer squash, winter squash, watermelon, and cucurbit hybrids.

** **Legume Vegetables (Succulent or Dried):** Includes field beans, French beans, kidney beans, lima beans, mung beans, navy beans, pinto beans, runner beans, snap beans, wax beans, broad beans (fava beans), chickpeas (gradbanzo beans), lentils, garden peas, field peas, sugar peas, southern peas (blackeyed peas, crowder peas, cowpeas, catjang) and edible soybeans.

STORAGE AND DISPOSAL

Pesticide and Storage Disposal: Store in a dry place. Do not contaminate water, food, or feed by storage, disposal, or cleaning of equipment. Open dumping is prohibited. Wastes resulting from the use of this product are acutely toxic. Improper disposal of unused pesticide, spray mixture, or rinsate is a violation of federal law. Pesticide, spray mixture, or rinsate that cannot be used according to label instructions must be disposed of according to federal, state, or local procedures. For guidance in proper disposal methods, contact your State Pesticide or Environmental Control Agency, or the Hazardous Waste representative at the nearest EPA Regional Office.

Container Disposal: Completely empty bag into application equipment. Dispose of empty bag in a sanitary landfill or by incineration, or by open burning, if allowed by state and local authorities. If burned, keep out of smoke.

For minor spills, etc., follow all precautions indicated on this label and clean up immediately. Take special care to avoid contamination of equipment and facilities during cleanup procedures and disposal of wastes. In the event of a major spill, fire or other emergency, call (919) 292-7100 day or night.

PRECAUTIONARY STATEMENTS

HAZARDS TO HUMANS AND DOMESTIC ANIMALS

WARNING

Causes eye irriation. Do not get in eyes.

Harmful if swallowed, inhaled or absorbed through the skin. Avoid breathing dust or spray mist. Avoid contact with skin or clothing. Protective clothing (long pants, long sleeve shirt, gloves, hat, and boots) must be worn during mixing and loading. This product may cause allergic skin reaction in some individuals.

First Aid: In case of contact, immediately flush eyes or skin with plenty of water. Get medical attention if irritation persists. If swallowed, drink large quantities of water and induce vomiting by placing finger in back of throat, and contact local poison control center. Avoid alcohol. Never give anything by mouth to an unconscious person. In case of inhalation exposure, move from the contaminated area.

Note to Physician: There is no specific antidote for RIDOMIL MZ58. Lavage stomach. Give a saline laxative and supportive therapy.

ENVIRONMENTAL HAZARDS

This product is toxic to fish. Do not apply directly to water. Apply only as specified on this label. Do not apply when weather conditions favor drift from treated areas. Do not contaminate water by cleaning of equipment or disposal of wastes.

RIDOMIL® trademark of CIBA-GEIGY for metalaxyl U.S. Patent No. 4,151, 299
RIDOMIL MZ® trademark of CIBA-GEIGY
Bravo® trademark of SDS BioTech Corporation for chlorothalonil
Difolatan® trademark of Chevron Chemical Company for captafol
Dithane® trademark of Rohm and Haas Company for maneb/mancozeb
D·z·n® trademark of CIBA-GEIGY for diazinon
Manzate® trademark of E.I. duPont de Nemours and Company for maneb/mancozeb
Polyram® trademark of BASF Corporation for metiram

CGA 65L23 128

This is a specimen label which was not prepared by CIBA-GEIGY. CIBA-GEIGY is not responsible for the accuracy of the information contained herein. As labels are subject to revision, always carefully read and follow the label on the product container.

RIDOMIL® 5G

FUNGICIDE

For control of Phytophthora and Pythium diseases of avocados, citrus, cotton, head lettuce, nonbearing deciduous fruits and nuts, peanuts, raspberries, soybeans and tomatoes

Ridomil 5G is a granular fungicide.

ACTIVE INGREDIENT:
Metalaxyl: *N*-(2,6-dimethylphenyl)-*N*-(methoxyacetyl) alanine methyl ester 5.0%
INERT INGREDIENTS: 95.0%
TOTAL 100.0%

KEEP OUT OF REACH OF CHILDREN

CAUTION

See additional precautionary statements on back of bag.

EPA Reg. No. 100-628 EPA Est. 100-AL-1

See directions for use on back of bag.

DIRECTIONS FOR USE AND CONDITIONS OF SALE AND WARRANTY

IMPORTANT: Read the entire **Directions for Use** and the **Conditions of Sale and Warranty** before using this product.

CONDITIONS OF SALE AND WARRANTY

The **Directions for Use** of this product reflect the opinion of experts based on field use and tests. The directions are believed to be reliable and should be followed carefully. However, it is impossible to eliminate all risks inherently associated with use of this product. Crop injury, ineffectiveness, or other unintended consequences may result because of such factors as weather conditions, presence of other materials, or the manner of use or application all of which are beyond the control of CIBA-GEIGY or the Seller. All such risks shall be assumed by the Buyer.

CIBA-GEIGY warrants that this product conforms to the chemical description on the label and is reasonably fit for the purposes referred to in the **Directions for Use** subject to the inherent risks referred to above. **CIBA-GEIGY makes no other express or implied warranty of Fitness or Merchantability or any other express or implied warranty. In no case shall CIBA-GEIGY or the Seller be liable for consequential, special, or indirect damages resulting from the use or handling of this product.** CIBA-GEIGY and the Seller offer this product, and the Buyer and user accept it, subject to the foregoing **Conditions of Sale and Warranty**, which may be varied only by agreement in writing signed by a duly authorized representative of CIBA-GEIGY.

DIRECTIONS FOR USE

It is a violation of federal law to use this product in a manner inconsistent with its labeling.

FAILURE TO FOLLOW THE DIRECTIONS FOR USE AND PRECAUTIONS ON THIS LABEL MAY RESULT IN POOR DISEASE CONTROL OR CROP INJURY.

Do not apply this product in such a manner as to directly or indirectly expose workers or other persons except those knowingly involved in the application. The area being treated must be vacated by unprotected persons.

REENTRY STATEMENT

Do not enter treated areas without protective clothing until dust has settled.

Because certain states may require restrictive reentry intervals for various crops treated with this product, consult your State Department of Agriculture for further information.

Written or oral warnings must be given to workers who are expected to be in a treated area or in an area about to be treated with this product. Oral warnings must inform workers of areas or fields that may not be entered without specific protective clothing until dust has settled and appropriate actions to take in case of accidental exposure, as described under **Precautionary Statements** on this label. When oral warnings are given, warnings shall be given in a language customarily understood by workers. Oral warnings must be given if there is reason to believe that written warnings cannot be understood by workers. Written warnings must include the following information: "CAUTION. Area treated with RIDOMIL 5G on (date of application). Do not enter without appropriate protective clothing until dust has settled. In case of accidental exposure, flush eyes or skin with plenty of water. Call physician if irritation persists. Remove and wash contaminated clothing before reuse."

GENERAL INFORMATION

RIDOMIL 5G is a systemic fungicide which provides control of certain diseases caused by members of the Oomycete class of fungi on selected crops.

Note: RIDOMIL is a systemic fungicide having a specific mode of action. RIDOMIL could be subject to development of resistant strains of fungi. Development of resistance cannot be predicted. Therefore, CIBA-GEIGY cannot assume liability for crop damage resulting from resistant strains of fungi. Consult with your State Agricultural Experiment Station or Extension Service Specialist for guidance or ways to control any possible RIDOMIL resistant strains of fungi which may occur. Do not use Ridomil 5G for disease control in greenhouse crops.

Calculate the amount of RIDOMIL 5G needed for band treatment by the formula:

$$\frac{\text{band width in inches}}{\text{row spacing in inches}} \times \text{broadcast rate per acre} = \text{amount needed per acre of field}$$

Applicator Settings

It is important the user calibrate each individual granular applicator before its use. The settings in the chart in the **Soybeans** section of the label SHOULD ONLY BE USED AS A STARTING POINT. Even after calibration, it is best to check the actual application rate under your operating conditions. Application rates can vary due to changing moisture conditions, equipment wear, normal variation in product with density and particle size distribution.

Calibration: To calibrate a granular applicator, follow the procedure as outlined below:

1. Fill the hopper with RIDOMIL 5G.
2. Attach a plastic bag to the spout opening.
3. Refer to the appropriate chart in the **Soybeans** section of this label and adjust the setting on the applicator to match the desired rate per acre.
4. Measure off 1,000 linear feet and drive at the speed selected.
5. Weigh or measure the volume of each bag, change settings as needed to obtain the desired rate and repeat steps 2-5 until the setting delivers the correct number of ounces per 1,000 ft.
6. Repeat steps 1-5 for each applicator box.

Conversions

1. Time to cover 1,000 feet at:
   4 mph = 2 mins. 50 secs.
   5 mph = 2 mins. 16. secs.
   6 mph = 1 min. 57 secs.
2. Ounces (wt.) of granules to **approximate** ounces (volume):
   6 oz. wt. = 8 fl. oz. (1 level cup)

When calibrating granular applicators, follow the recommendations below:

1. To reduce granule flow variability, adjust orifice openings by moving the controls in one direction only, such as from the closed position to open.
2. Granule flow through an orifice is not always proportional to ground speed, so make calibration checks at the same speed you plant at and maintain that ground speed as constant as possible when planting.
3. Granular pesticide products vary considerably in size, density and particle shape, so recalibrate applicators whenever you change brands and/or formulations. The same brand can even change from year to year.
4. Changing humidity and temperature during the day can change the flow rate of granules through the applicator's orifice. Recalibrate when conditions change and periodically readjust orifice settings to compensate.
5. Know the recommended rate. For banded applications, the area actually is the area covered by the band, not total cropland area planted. Some row-crop recommendations are based on treating in-the-row and these rates generally are specified as amounts (ounces) of granules per certain row length (often 1,000 feet). Others express rates as amount per treated acre which means the total area treated with the pesticide. If rates are expressed as amount per treated acre and banded applications

are used, the amount of pesticide used per acre will be proportionately less. The formula above can be used to calculate the amount of RIDOMIL 5G needed per acre of crop when banded applications are made.

6. Calibrate frequently. There are several acceptable calibration methods. Learn one. Do the arithmetic first, then complete the material collection phase under field conditions.

Preharvest Intervals

Unless otherwise stated in the directions for use for a specific crop, there is no preharvest interval for RIDOMIL 5G.

AVOCADOS

RIDOMIL 5G is a systemic fungicide which provides control of avocado root rot caused by *Phytophthora cinnamomi*. For best results, RIDOMIL should be used before or as soon as early symptoms of disease appear. For new plantings, the use of Phytophthora resistant root stocks with RIDOMIL 5G is recommended. Mature trees in advanced stages of decline cannot be cured with RIDOMIL. Applications are not needed during the winter months of November through February.

Sprinkler Irrigation: Apply ¾–1½ oz./sq. yd. evenly to the soil surface under the canopy of the tree. The table below shows the amount of RIDOMIL 5G to use based on the diameter of the tree canopy. Start applications at the beginning of the growing season or at transplanting and make two additional applications at three-month intervals.

Drip Irrigation: Place the recommended amount of RIDOMIL 5G (see table below) on the soil surface directly around the drip emitter at each tree. If there is more than one emitter per tree, distribute the total amount of RIDOMIL 5G among the emitters. Start applications at the beginning of the growing season or at transplanting and make two additional applications at three-month intervals.

Diameter of Tree Canopy in Feet	Amount of RIDOMIL 5G Per Tree
5	1½-3 oz.
10	6-12 oz.
15	14-28 oz.
20	25-50 oz.

Notes: (1) Do not make more than three applications per year and do not make the last application within 28 days of harvest or illegal residues may result. (2) Do not apply more than 960 lbs. RIDOMIL 5G per acre per year, or illegal residues may result.

CITRUS

Includes grapefruit, lemons, limes, oranges, tangelos, tangerines, citrus citron, kumquats and hybrids of these.

Use RIDOMIL 5G on citrus to control citrus foot rot, root rot and trunk canker caused by *Phytophthora* spp. Apply to the soil as a broadcast application under the canopy of the trees. If rain is not expected within three days of application, water in with ½-1 inch of irrigation. For banded applications, use the formula in the **General Information** section for calculating the amount of RIDOMIL 5G needed per acre of trees. If the trees are on a drip irrigation system, distribute the amount of RIDOMIL 5G needed per tree (see table) to the soil surface in the immediate vicinity of the drip emitter(s) at each tree. If there is more than one emitter at each tree, distribute the total amount of RIDOMIL 5G needed among the emitters.

Note: Where nematodes are a problem, best results can be achieved if effective EPA-registered nematicides are used. Nematicides can be used in combination or in sequence with RIDOMIL applications.

Citrus in Nurseries
(Arizona, California and Florida Only)

Make the first application of RIDOMIL 5G at the time of planting and make two additional applications at three-month intervals during the period when trees are actively growing. Apply 40-80 lbs./treated acre as a broadcast or banded application to the seedbeds, liners or bedded stock. For banded applications, use a band wide enough to cover the root systems of the plants.

Citrus Resets or New Plantings
(Arizona, California and Florida Only)

Make the first application of RIDOMIL 5G to citrus resets or new plantings at the time of transplanting. Make up to three repeat applications at three-month intervals to coincide with root growth flushes during the growing season.

Water Ring Treatment: Uniformly spread 1-1½ oz. within the watering ring, then drench with five gallons water per tree.

Soil Surface Treatment: Apply 40-80 lbs./treated acre (15-30 oz./1,000 sq. ft.) under the canopy of the trees. If natural rainfall is not expected within three days, lightly irrigate with ½-1 inch of water.

Established Plantings

Soil Surface Applications (Florida Only): Apply 40-80 lbs./treated acre (15-30 oz./1,000 sq. ft.) as a soil surface treatment under the canopy of the trees. Two additional applications may be made per year at three-month intervals. Use the table below as a guide for treating individual trees.

Diameter of Tree Canopy in Feet	Amount of RIDOMIL 5G Per Tree
5	¼-½ oz.
10	1¼-2½ oz.
15	2½-5 oz.
20	5-10 oz.

Soil Surface Applications (Arizona and California Only): Make one application in March-April. One or two additional applications can be made at three-month intervals to coincide with root flushes. Use the following table to determine the proper rate based on tree size and the number of applications per year. For applications based on broadcast rates, use RIDOMIL 5G at 160 lbs./treated acre (3⅔ lbs./1,000 sq. ft.) when three applications are planned and at 220 lbs./treated acre (5 lbs./1,000 sq. ft.) when two applications are planned. Immediately after application irrigate with ½-1 inch sprinkler irrigation.

Diameter of Tree Canopy in Feet	Ounces of RIDOMIL 5G per Tree	
	2 Applications Per Year	3 Applications Per Year
5	1½ oz.	1 oz.
10	6½ oz.	5 oz.
15	15 oz.	11 oz.
20	24 oz.	20 oz.

COTTON

RIDOMIL 5G provides control of seed rot and seedling diseases of cotton caused by *Pythium* spp. It does not control seedling diseases caused by other fungi.

Seed Rots and Seedling Diseases of Cotton Caused by Pythium spp.

Apply 1¼-2½ lbs./13,000 linear feet of row (1½-3 oz./1,000 linear ft.) as an in-furrow application at the time of planting. Mount the application tubes so the granules are mixed with the soil covering the seeds.

For broader spectrum disease control, RIDOMIL 5G can be applied with 10-20 lbs. Terraclor 10G/13,000 linear feet of row as an in-furrow application.

Note: Alternatively, RIDOMIL PC11G may be used if the predominant pathogens are *Pythium* spp. and *Rhizoctonia* spp.

DECIDUOUS FRUITS AND NUTS IN ORCHARDS (NONBEARING)

Phytophthora Diseases

Use of RIDOMIL 5G will aid in the control of crown, collar and root rot of deciduous fruit and nut trees caused by *Phytophthora* spp. when used in conjunction with good cultural practices and rootstocks that are most tolerant to the disease. RIDOMIL 5G applications should be made before symptoms appear, especially in areas favorable for disease development. RIDOMIL 5G will not revitalize trees showing moderate to severe disease symptoms.

Make the first application at the time of planting. (See **Notes** below.) Additional applications should be made at 2-3 month intervals during the time when conditions are favorable for disease development. For established plantings, make the first application in the spring before growth starts.

Apply 80-160 lbs./treated acre (2-4 lbs./1,000 sq. ft.) uniformly under the canopy of the tree. The treated area is based on the area under the tree canopy or the area of the row to be treated (similar to herbicide strips). Use the formula in the **General Information** section for calculating the amount of RIDOMIL 5G needed per acre of trees. Use the high rate in areas known to be infested with Phytophthora. Soil surface applications of RIDOMIL 5G will not be effective until the fungicide is moved into the root zone by rainfall or irrigation.

Notes: (1) Do not apply to plantings that will bear harvestable fruit within 12 months of the last application, or illegal residues may result. To avoid injury, (2) Do not treat the roots directly or concentrate the granules around the tree trunks. (3) Do not apply to trees under stress, and (4) In California, do not apply RIDOMIL 5G to newly planted trees within 90 days of planting.

Head Lettuce

Ridomil 5G applied as a soil application will control damping off in head lettuce caused by *Pythium* spp. Applications may be made preplant incorporated or preemergence.

Preplant Incorporated Application: Apply 20-40 lbs./treated acre as a broadcast soil application and incorporate in the top two inches of soil. For banded applications, a 7-inch band is recommended. Use the formula in the **General Information** section of this label to calculate the amount of Ridomil 5G needed per acre. When incorporated applications are made at the time of planting, position the drop tubes so the granules are mixed with the soil covering the seed at a rate of 10-20 oz. of granules/1,000 linear feet of row. Avoid placing the granules directly in contact with the seed or injury may occur.

Surface Application: Apply 20-40 lbs./treated acre uniformly to the soil surface at the time of planting. For banded applications, a 7-inch band is recommended. Use the formula in the **General Information** section of this label to calculate the amount of Ridomil 5G needed per acre. If natural rainfall is not expected before the seeds start germinating, Ridomil 5G should be incorporated mechanically before planting or be moved into the seed zone after planting $\frac{1}{2}$-1 inch sprinkler irrigation.

PEANUTS

RIDOMIL 5G is a systemic fungicide for use in control of seedling and pod diseases of peanuts caused by *Pythium* spp.

Seedling Diseases: Apply 6.5 oz./1,000 linear feet in-furrow or in a 7-inch band (5 lbs./A on 42" rows) at the time of planting. With in-furrow applications, mount the application tubes so the granules are mixed with the soil covering the seed.

Pod Rot: Apply 13 oz./1,000 linear feet of row in an 8 to 12 inch band (10 lbs./12,400 linear ft. on 42" rows) over the rows at early pegging. For best results, an irrigation should follow the application.

Notes: (1) Do not apply RIDOMIL 5G to wet foliage as foliar phytotoxicity and/or poor control may result. (2) Where pathogens other than *Pythium* spp. are present, use fungicides that control these diseases in combination with RIDOMIL. (3) Alternatively, RIDOMIL PC 11G may be used if the predominant pathogens are *Pythium* spp. and *Rhizoctonia* spp.

RASPBERRIES

RIDOMIL 5G is a systemic fungicide for use in the control of Phytophthora root rot.

Apply 5 lbs./1,000 ft. of row to the soil surface in a three foot band over the row (72.5 lbs./A on a broadcast basis). Make one application in the spring and another application in the fall after harvest.

Note: Do not apply RIDOMIL 5G within 45 days of harvest, or illegal residues may result.

SOYBEANS

RIDOMIL 5G is a systemic fungicide for use in the control of Phytophthora root and stem rot and Pythium damping-off. RIDOMIL 5G may be applied in a 7-inch band over the row at planting or in the seed furrow before the seeds are covered. The seed furrow application will provide more consistent results if rain is not expected before the seeds germinate.

For best results against Phytophthora root and stem rot, use RIDOMIL 5G with soybean varieties that have some tolerance to the races of Phytophthora present in the field. The high rate of RIDOMIL 5G should be used in areas with a history of severe Phytophthora damage. Under severe late season Phytophthora pressure, RIDOMIL 5G may not provide complete control.

Pythium and Phytophthora: For full season control apply 6 oz./1,000 linear feet of row in a 7-inch band over the row or as a seed furrow application at the time of planting. For early to mid-season control only, apply $1\frac{1}{2}$-3 oz./1,000 linear feet of row in the **seed furrow.** When making seed furrow applications, position the applicator tubes so the granules are mixed with the soil covering the seed or are applied in a "T-band." Use the table below to determine the amount of RIDOMIL 5G needed per acre based on the row spacing. Use the high rate in areas with a history of Phytophthora problems.

Suggested Rates Per Acre According to Row Spacing

Row Spacing	Linear Feet of Soybean Row Per Acre	Pounds of RIDOMIL 5G Needed Per acre — Rate Desired in Ounces		
		$1\frac{1}{2}$ oz.	3 oz.	6 oz.
38 in.	13,756	$1\frac{1}{3}$ lbs.	$2\frac{3}{5}$ lbs.	$5\frac{1}{6}$ lbs.
36 in.	14,520	$1\frac{2}{5}$ lbs.	$2\frac{3}{4}$ lbs.	$5\frac{1}{2}$ lbs.
30 in.	17,424	$1\frac{2}{3}$ lbs.	$3\frac{1}{4}$ lbs.	$6\frac{1}{2}$ lbs.
24 in.	21,780	2 lbs.	$4\frac{1}{10}$ lbs.	$8\frac{1}{6}$ lbs.
20 in.	26,136	$2\frac{1}{2}$ lbs.	$4\frac{9}{10}$ lbs.	$9\frac{8}{10}$ lbs.

Note: RIDOMIL 5G is specific for Pythium and Phytophthora and will not control other diseases that may attack soybeans.

Suggested Application Settings for a 7" Band Width or in the Seed Furrow

Applicator Brand	Speed MPH	Ounces/1,000 ft.				
		1.5	3	4	5	6
Gandy	4	8	13	15	17	19
	5	9	15	17	19	21
	6	10	16	19	21	23
	7	12	18	20	23	25
Allis Chalmers 1979 & After	4	1	1.3	1.4	1.5	1.8
	5	1.1	1.4	1.5	1.9	2
	6	1.2	1.5	1.8	2	2.3
	7	1.3	1.5	1.9	2.2	2.5
Noble 1972 & Before	4	2	4	6	7 or 8	9
	5	3	6	7 or 8	9	11
	6	3 or 4	7	9	11	13
	7	4	8	10	12	15
Noble 1973 & After	4	2	4	5	6	7
	5	2 or 3	5	6	7	8 or 9
	6	3	5	7	8 or 9	10
	7	4	6	8	10	12
John Deere Max Emerge II	4	8	12	14	16	18
	5	9	13	16	19	22
	6	10	15	18	22	26
	7	11	16	20	26	30
John Deere 1978 & Before Meter Scale 0.10..20..etc.	4	5	8	10	12	14
	5	6	10	12	14	16
	6	6 or 7	11	14	16	19
	7	7	12	15	19	22
John Deere 1979 & Later Meter Scale 0..5..15..etc.	4	4	7	8	9	10
	5	5	7 or 8	9	10	12
	6	5 or 6	8	10	12	14
	7	6	9	11	14	16
International Harvester 1981 & After	4	1	1.4	1.55	1.75	1.85
	5	1.1	1.5	1.75	1.85	2
	6	1.2	1.65	1.85	2	2.3
	7	1.3	1.8	1.9	2.3	2.5

TOMATOES

Soil applications of RIDOMIL 5G at planting will provide control of damping-off caused by *pythium* spp. and soil applications applied 4-12 weeks before harvest under the vines will reduce fruit and root rot caused by *Pythium* spp. and *Phytophthora* spp.

Damping-Off (*Pythium* spp.): Apply 20 lbs./treated acre uniformly over the surface of the soil at the time of planting. If natural rainfall is not expected before the seeds begin germinating, RIDOMIL 5G should be incorporated mechanically with $\frac{1}{2}$-1 inch sprinkler irrigation. For banded applications, a 7-inch band is recommended. Use the formula in the **General Information** section of this label to calculate the amount of RIDOMIL 5G needed per acre.

Fruit and Root Rot (*Pythium* spp. and *Phytophthora* spp.): Additional applications can be made as soil surface applications under the vines to provide control of root and fruit rot. If only one application is planned, apply 20 lbs./treated acre under the vines four weeks after the planting application to four weeks before harvest, depending on the expected occurrence of the diseases. If root and fruit rot is a problem throughout the growing season or the occurrence of diseases cannot be predicted, make two additional applications at 10 lbs./treated acre under the vines, one 4-6 weeks after the planting application and another four weeks before harvest. If rainfall is not expected within three days of the application, follow as soon as possible with $\frac{1}{2}$-1 inch overhead irrigation. Use the formula shown in the **General Information** section to determine the amount of RIDOMIL 5G needed per acre.

RIDOMIL 5G can be shanked into the beds with fertilizer. For calculating the amount of RIDOMIL 5G needed per acre, apply the same amount as would be applied on a 7-inch band at 20 lbs./treated acre. To obtain uniform disease control, RIDOMIL 5G must be mixed uniformly with the fertilizer.

Notes: To avoid illegal residues, (1) Do not apply more than 40 lbs./treated acre/season, and (2) Do not apply within 7 days of harvest.

REPLANTING

If replanting is necessary, additional applications of RIDOMIL 5G or 2E can be made provided that the total amount of metalaxyl applied does not exceed the maximum specified on the product label for the particular crop.

ROTATION (PLANTBACK) RESTRICTIONS

<table>
<tr><th>Rotation Crop</th><th>Planting Time From Last RIDOMIL 5G Application (Days)</th></tr>
<tr><td>Apples*, Asparagus, Avocados*</td><td rowspan="12">0 days</td></tr>
<tr><td>Broccoli</td></tr>
<tr><td>Cabbage, Cauliflower, Chinese Broccoli (gai lon, white flowering broccoli), Chinese Cabbage (tight-heading varieties only), Citrus*, Cotton, Cucurbit Vegetables**</td></tr>
<tr><td>Deciduous Fruits And Nuts*</td></tr>
<tr><td>Eggplant</td></tr>
<tr><td>Hops, Head Lettuce</td></tr>
<tr><td>Legume Vegetables***</td></tr>
<tr><td>Onions</td></tr>
<tr><td>Peanuts, Peppers, Pineapples, Potatoes</td></tr>
<tr><td>Raspberries*</td></tr>
<tr><td>Soybeans, Spinach</td></tr>
<tr><td>Tobacco, Tomatoes</td></tr>
<tr><td>Wheat</td><td>14 days</td></tr>
<tr><td>Corn</td><td>270 days</td></tr>
<tr><td>Root Crops</td><td>360 days</td></tr>
<tr><td>Crops Not Intended for Food or Feed</td><td>0 days</td></tr>
<tr><td>All Other Crops Intended for Food or Feed</td><td>540 days</td></tr>
</table>

* These crops and other perennial crops may be planted immediately following last application of RIDOMIL 5G provided they will not bear harvestable fruit within 12 months.

****Cucurbit Vegetables:** Includes balsam pear (bitter melon), Chinese waxgourd (Chinese preserving melon), citron melon, cucumber, gherkin, edible gourds, cantaloupe, casaba, crenshaw, honeydew melon, honey balls, mango melon, muskmelon, Persian melon, pumpkins, summer squash, winter squash, watermelon, and cucurbit hybrids.

*****Legume Vegetables (Succulent or Dried):** Includes field beans, french beans, kidney beans, lima beans, mung beans, navy beans, pinto beans, runner beans, snap beans, wax beans, broad beans (fava beans), chickpeas (garbanzo beans), lentils, lupines (sweet, white sweet, white and grain), garden peas, field peas, sugar peas, southern peas (blackeyed peas, crowder peas, cowpeas, catjang) and edible soybeans.

STORAGE AND DISPOSAL

Store in a dry place. Do not contaminate water, food, or feed by storage, disposal, or cleaning of equipment. Open dumping is prohibited.

Pesticide Disposal: Wastes resulting from the use of this product are acutely toxic. Improper disposal of unused pesticide is a violation of federal law. Pesticide that cannot be used according to label instructions must be disposed of according to federal, state, or local procedures. For guidance in proper disposal methods, contact your State Pesticide or Environmental Control Agency, or the Hazardous Waste representative at the nearest EPA Regional Office.

Container Disposal: Completely empty bag into application equipment. Dispose of empty bag in a sanitary landfill or by incineration, or by open burning, if allowed by state and local authorities. If burned, keep out of smoke.

For minor spills, leaks, etc., follow all precautions indicated on this label and clean up immediately. Take special care to avoid contamination of equipment and facilities during cleanup procedures and disposal of wastes. In the event of a major spill, fire or other emergency, call 1-800-888-8372 day or night.

PRECAUTIONARY STATEMENTS

HAZARDS TO HUMANS AND DOMESTIC ANIMALS

CAUTION

Harmful if inhaled or absorbed through the skin. Causes moderate eye irritation. Avoid breathing dust. Avoid contact with skin, eyes or clothing. Wash thoroughly with soap and water after handling. Remove contaminated clothing and wash before reuse.

First Aid: In case of contact with eyes, flush with plenty of water. If on skin, wash with plenty of soap and water. Get medical attention if irritation persists. If swallowed, drink large quantities of water and induce vomiting by placing finger in back of throat; call a physician or contact your local Poison Control Center. Never give anything by mouth to an unconscious person. Avoid alcohol. In case of inhalation exposure, remove victim to fresh air.

ENVIRONMENTAL HAZARDS

Do not apply to water bodies or wetland. Do not contaminate water by cleaning of equipment or disposal of waste.

Ridomil® trademark of CIBA-GEIGY
Terraclor® trademark of UniRoyal, Inc. for PCNB

CGA 65L7H 059

This is a specimen label which was not prepared by CIBA-GEIGY. CIBA-GEIGY is not responsible for the accuracy of the information contained herein. As labels are subject to revision, always carefully read and follow the label on the product container.

RIDOMIL PC® 11G

GRANULAR FUNGICIDE

For the control of diseases of beans, cotton and peanuts caused by Pythium and Rhizoctonia

ACTIVE INGREDIENTS:
PCNB: Pentachloronitrobenzene 10.0%
Metalaxyl: *N*-(2,6-dimethylphenyl)-*N*-methoxyacetyl) alanine methyl ester 1.0%
INERT INGREDIENTS: 89.0%
TOTAL.................... 100.0%

EPA Reg. No. 100-664 EPA Est. 8590-PA-1
Control No. 54810

RIDOMIL PC 11G is a granular fungicide.

KEEP OUT OF REACH OF CHILDREN

CAUTION

See additional precautionary statements on back of bag.

See directions for use on back of bag.

DIRECTIONS FOR USE AND CONDITIONS OF SALE AND WARRANTY

IMPORTANT: Read the entire **Directions for Use** and **Conditions of Sale and Warranty** before using this product.

CONDITIONS OF SALE AND WARRANTY

The **Directions for Use** of this product reflect the opinion of experts based on field use and tests. The directions are believed to be reliable and should be followed carefully. However, it is impossible to eliminate all risks inherently associated with use of this product. Crop injury, ineffectiveness, or other unintended consequences may result because of such factors as weather conditions, presence of other materials, or the manner of use or application all of which are beyond the control of CIBA-GEIGY or the Seller. All such risks shall be assumed by the Buyer.

CIBA-GEIGY warrants that this product conforms to the chemical description on the label and is reasonably fit for the purposes referred to in the **Directions for Use** subject to the inherent risks referred to above. **CIBA-GEIGY makes no other express or implied warranty of Fitness or Merchantability or any other express or implied warranty. In no case shall CIBA-GEIGY or the Seller be liable for consequential, special, or indirect damages resulting from the use or handling of this product.** CIBA-GEIGY and the Seller offer this product, and the Buyer and user accept it, subject to the foregoing **Conditions of Sale and Warranty**, which may be varied only by agreement in writing signed by a duly authorized representative of CIBA-GEIGY.

DIRECTIONS FOR USE

It is a violation of federal law to use this product in a manner inconsistent with its labeling.

FAILURE TO FOLLOW THE DIRECTIONS FOR USE AND PRECAUTIONS ON THIS LABEL MAY RESULT IN POOR DISEASE CONTROL OR CROP INJURY.

GENERAL INFORMATION

RIDOMIL PC 11G is a granular fungicide containing 1% metalaxyl, a systemic fungicide, and 10% PCNB (pentachloronitrobenzene). It provides control of damping–off and seed and seedling rot diseases caused by *Pythium* and *Rhizoctonia* in beans, cotton, and peanuts. RIDOMIL PC 11G will also provide suppression of pod rot in peanuts caused by *Sclerotium rolfsii*.

Note: Metalaxyl is a systemic fungicide having a specific mode of action. Metalaxyl could be subject to development of resistant strains of fungi. Development of resistance cannot be predicted. Therefore, CIBA-GEIGY cannot assume liability for crop damage resulting from resistant strains of fungi. Consult with your State Agricultural Experiment Station or Extension Service Specialist for guidance or ways to control any possible metalaxyl resistant strains of fungi which may occur.

Applicator Settings
It is important that the user calibrate each individual granular applicator before its use. The settings in the charts under the **Cotton** and **Peanuts** sections below SHOULD ONLY BE USED AS A STARTING POINT. Even after calibration, it is best to check the actual application rate under your operating conditions. Application rates can vary due to changing moisture conditions, equipment wear, normal variation in product bulk density and particle size distribution.

Calibration: To calibrate a granular applicator, follow the procedure as outlined below:
1. Fill the hopper with RIDOMIL PC 11G.
2. Attach a plastic bag to the spout opening.
3. Refer to the appropriate chart in either the **Cotton** or **Peanuts** section of this label and adjust the setting on the applicator to match the desired rate per acre.
4. Measure off 1,000 linear feet and drive at the speed selected.
5. Weigh or measure the volume of each bag, change settings as needed to obtain the desired rate and repeat steps 2-5 until the setting delivers the correct number of ounces per 1,000 feet.
6. Repeat steps 1-5 for each applicator box.

Conversions
1. Time to cover 1,000 feet at:
   2 mph = 5 mins. 40 secs.
   3 mph = 3 mins. 54 secs.
   4 mph = 2 mins. 50 secs.
   5 mph = 2 mins. 16 secs.
   6 mph = 1 min. 57 secs.
2. Ounces (wt.) of granules to *approximate* ounces (volume):
   6 oz. wt. = 8 fl. oz. (1 level cup)

When calibrating granular applicators, follow the recommendations below:
1. To reduce granule flow variability, adjust orifice openings by moving the controls in one direction only, such as from the closed position to open.
2. Granule flow through an orifice is not always proportional to ground speed, so make calibration checks at the same speed at which you plant and maintain that ground speed as constantly as possible when planting.
3. Granular pesticide products vary considerably in size, density, and particle shape, so recalibrate applicators whenever you change brands and/or formulations. The same brand can even change from year to year.
4. Changing humidity and temperature during the day can change the flow rate of granules through the applicator's orifice. Recalibrate when conditions change and periodically readjust orifice settings to compensate.
5. Know the recommended rate. For band applications, the area actually is the area covered by the band, not total cropland area planted. Some row-crop recommendations are based on treating in-the-row and these rates generally are specified as amounts (ounces) of granules per certain row length (often 1,000 feet).
6. Calibrate frequently. There are several acceptable calibration methods. Learn one. Do the arithmetic first, then complete the material collection phase under field conditions.

Calculate the amount of RIDOMIL PC 11G needed for band treatment as follows:

$$\frac{\text{Band Width in Inches}}{\text{Row Spacing in Inches}} \times \text{Broadcast Rate per Acre} = \text{Band Rate per Acre}$$

BEANS

RIDOMIL PC 11G will control damping-off and seed and seedling rots of dry and green beans (bush and pole types) caused by Pythium and Rhizoctonia when used at planting.

Apply 3/4 lb. (12 oz.) of RIDOMIL PC 11G per 1,000 linear feet of row at the time of planting. Adjust the application equipment so the granules are mixed with the soil surrounding the seed.

Note: Do not allow the feeding or grazing of cotton foliage by livestock, or illegal residues may result.

COTTON

RIDOMIL PC 11G will control damping-off, seed and seedling rot diseases caused by *Pythium* spp. and *Rhizoctonia solani*. For best results, use high quality seed and plant into a properly prepared seedbed.

Apply 7-10 lbs. of RIDOMIL PC 11G per 13,000 linear feet of row (8.6-12.3 oz./1,000 linear feet) at the time of planting. Adjust the application equipment so the granules are mixed with the soil surrounding the seed. See Chart 1 below for suggested applicator settings and the **General Information** section of this label for calibration instructions.

Note: Do not allow the feeding or grazing of cotton foliage by livestock, or illegal residues may result.

Chart 1: Cotton—Suggested Applicator Settings

Applicator Brand	Speed MPH	40" Rows In-Furrow Lbs./A: 10	9	8	7
		Equivalent Oz./1,000 Ft.: 12.3	11.1	9.8	8.6
Gandy	4	28	26.8	25	23.5
	5	31	30	28	26.3
	6	34.5	32.5	30.5	29
Allis Chalmers	4	2.4	2.1	1.9	1.6
	5	3	2.8	2.4	2.05
	6	3.6	3.2	2.9	2.5
Noble 1972 and Later	4	19	17	15	13
	5	23	21	19	17
	6	27	25	22	20
Noble 1973 and Later	4	14	13	12 or 11	10
	5	17 or 16	15	14	13 or 12
	6	19	18	16	15 or 14
John Deere Metal Box	4	2/12	2/10	2/8	2/5
	5	2/18	2/15	2/12	2/9
	6	2/22	2/19	2/16	2/13
John Deere 1978 and Before Meter Scale 0..10..20..etc.	4	28	27	24	22
	5	31	30	28	26
	6	33	32	31	29
John Deere 1979 and Later Meter Scale 0..5..15..etc.	4	18	17	16	15
	5	21	20	18	17
	6	23	22	21	19
International Harvester 1981 and Later	4	3.1	2.9	2.6	2.35
	5	3.6	3.4	3.1	2.8
	6	4.5	4	3.55	3.2

Rotation Crops
When RIDOMIL PC 11G is used for disease control in cotton, do not rotate to any root crop within 12 months following application, or illegal residues may result.

PEANUTS

RIDOMIL PC 11G applied at planting will provide control of seed and seedling rots of peanuts caused by Pythium and Rhizoctonia. Applications of RIDOMIL PC 11G during pegging will provide control of pod rot caused by Pythium and Rhizoctonia and will provide suppression of *Sclerotium rolfsii.*

Seed and Seedling Rots: Apply 12.5-25 lbs. of RIDOMIL PC 11G/A in a 9-inch band at the time of planting. Incorporate with mechanical equipment to a depth of 2-4 inches. For row spacings other than 36 inches and band widths other than 9 inches, calculate the amount of RIDOMIL PC 11G needed per acre based on the formula in the **General Information** section of this label, using a broadcast rate of 50-100 lbs./A. Use the higher rate in fields with a history of seedling disease problems or when weather conditions are expected to be unfavorable for rapid germination and growth of the seedlings.

Pod Rot: Apply 50-100 lbs./A (21.5-43 oz./1,000 linear feet) (broadcast basis) at early pegging. For banded applications, adjust the equipment so the granules are distributed over the row. The higher rate should be used in areas where pod rot is severe. See Chart 2 below for suggested applicator settings and the **General Information** section of this label for calibration instructions.

Sclerotium Rolfsii: When RIDOMIL PC 11G is used for pod rot control, some suppression of *Sclerotium rolfsii* will be obtained. In areas where *Sclerotium rolfsii* is a problem use other fungicides that are labeled for this disease.

Notes: (1) Do not allow the feeding or grazing of peanut foliage or harvest residues by livestock, or illegal residues may result. (2) Do not apply more than 100 lbs. of RIDOMIL PC 11G/A/season.

Chart 2: Peanuts—Suggested Applicator Settings for Pod Rot Control

Applicator Brand	Speed MPH	Rate for 14″ Band Width Oz./1,000 linear ft.* (one opening per row)					
		43	38.6	34.3	30	25.7	21.5
		Lbs./A for 40″ rows					
		35	31.5	28	24.5	21	17.5
Gandy	2	36	34	32	30	28	26
	3	42.5	40.5	39	36.5	33.5	31
	4	47.5	45.5	43	41.5	39	36
	5	53.5	50	47.5	45	42.5	40
	6	61	56	52	49	45	42.5
John Deere 1978 and Before Meter Scale 0..10..20..etc.	2	33	32	30	28	26	24
	3	39	37	36	34	32	30
	4	Max	43	40	38	36	33
	5	—	—	Max	43	39	36
	6	—	—	—	Max	43	39
John Deere 1979 and Later Meter Scale 0..5..15..etc.	2	24	22	21	20	18	16
	3	31	28	26	25	22	20
	4	37	34	32	30	27	24
	5	43	40	37	34	31	28
	6	—	—	41	38	34	31
Noble 1973 and Later	2	20	18	17	15	12	10
	3	28	26	24	22	18	16
	4	34	32	30	27	24	20
	5	38	36	34	31	28	24
	6	40	39	37	35	32	28
Noble 1972 and Before	2	28	26	23	20	17	13
	3	37	34	32	29	26	22
	4	41	40	38	35	32	28
	5	45	43	41	39	37	32
	6	50	46	44	42	40	37
John Deere Metal Box	2	2/25	2/21	2/19	2/16	2/11	2/06
	3	3/03	2/32	2/30	2/26	2/21	2/16
	4	3/12	3/08	3/04	3/00	2/29	2/24
	5	3/21	3/15	3/12	3/08	3/02	2/30
	6	3/33	3/23	3/20	3/14	3/08	3/02
International Harvester 1981 and Later	2	5/3	4/7	4/2	3/6	3/1	2/6
	3	7/8	7/0	6/2	5/4	4/7	3/9
	4	10/3	9/3	8/3	7/3	6/2	5/2
	5	—	11/6	10/3	9/0	7/7	6/4
	6	—	—	—	10/8	9/3	7/8
Allis Chalmers	2	5/1	4/6	4/2	3/8	3/3	2/9
	3	7/3	6/6	6/0	5/3	4/6	4/0
	4	9/6	8/6	7/7	6/8	5/9	5/0
	5	11/6	10/5	9/4	8/3	7/3	6/1
	6	—	—	11/2	9/9	8/5	7/2

* If an applicator has a dual metering unit (2 openings/row) set gauge at half indicated rate.

Note: To calculate the amount of RIDOMIL PC 11G needed for row spacings other than 40 inches and band widths other than 14 inches, refer to the formula in the **General Information** section.

Rotation Crops*

Crops	Rotational Interval
Asparagus	0 days
Beans (Dry and Snap), Broccoli	
Cabbage, Cauliflower, Cotton, Cucurbit Vegetables	
Head Lettuce, Hops	
Legume Vegetables Other than Beans	
Onions	
Peanuts, Peppers, Potatoes	
Soybeans, Spinach	
Tomatoes	
Brussels Sprouts	12 months
Corn	
Garlic	
Other Root Crops	
Wheat	
All Other Crops Intended for Food or Feed	

* These crops represent those which may be planted (rotated to) according to the interval listed following an application of RIDOMIL PC 11G to beans, cotton, or peanuts. Do not plant any crop not registered for use with metalaxyl in metalaxyl-treated soil for a period of 12 months.

STORAGE AND DISPOSAL

Pesticide

Store in a dry place. Do not contaminate water, food, or feed by storage, disposal, or cleaning of equipment. Open dumping is prohibited. Wastes resulting from the use of this product are toxic. Improper disposal of unused pesticide is a violation of federal law. Pesticide that cannot be used according to label instructions must be disposed of according to federal, state or local procedures. For guidance in proper disposal methods, contact your State Pesticide or Environmental Control Agency, or the Hazardous Waste representative at the nearest EPA Regional Office.

Container Disposal

Completely empty bag into application equipment. Disposal of empty bag in a sanitary landfill or by incineration, or by open burning, if allowed by state and local authorities. If burned, keep out of smoke.

For minor spills, etc., follow all precautions indicated on this label and clean up immediately. Take special care to avoid contamination of equipment and facilities during cleanup procedures and disposal of wastes. In the event of a major spill, fire or other emergency, call (919) 292-7100 day or night.

PRECAUTIONARY STATEMENTS

HAZARDS TO HUMANS AND DOMESTIC ANIMALS

CAUTION

Harmful if absorbed through the skin or inhaled. Causes moderate eye irritation. Avoid contact with skin, eyes or clothing. Avoid breathing dust. Wash thoroughly with soap and water after handling. Remove contaminated clothing and wash before reuse. Protective clothing including gloves, long-sleeved shirt, long pants, socks and shoes should be worn during mixing/loading procedures.

STATEMENT OF PRACTICAL TREATMENT

If in eyes: Flush eyes with plenty of water. Call a physician if irritation persists.

If on skin: Wash with plenty of soap and water. Get medical attention.

If inhaled: Move to fresh air.

ENVIRONMENTAL HAZARDS

This product is toxic to fish and aquatic organisms. Do not apply directly to water or wetlands. Cover or incorporate all granules. Do not contaminate water when disposing of equipment washwaters.

RIDOMIL PC® trademark of CIBA-GEIGY

CGA 81L3D 049

This is a specimen label which was not prepared by CIBA-GEIGY. CIBA-GEIGY is not responsible for the accuracy of the information contained herein. As labels are subject to revision, always carefully read and follow the label on the product container.

RIDOMIL® BRAVO® 81W

FUNGICIDE

For the control of certain diseases in broccoli, cabbage, cauliflower, cucurbit vegetables, onions, potatoes, and tomatoes

ACTIVE INGREDIENTS:

Metalaxyl: *N*-(2,6-dimethylphenyl)-*N*-(methoxyacetyl) alanine methyl ester	9.0%
Chlorothalonil: Tetrachloroisophthalonitrile	72.0%
INERT INGREDIENTS:	19.0%
TOTAL	100.0%

EPA Reg. No. 100-658

This outer protective container contains RIDOMIL/BRAVO 81W in an inner water-soluble bag. Entire inner bag and contents dissolve in water. After opening outer container, immediately dump entire unopened inner bag into the partially filled sprayer or mix tank. Do not handle the soluble bag or expose it to moisture, since this may cause rupturing.

RIDOMIL/BRAVO 81W is a wettable powder.

KEEP OUT OF REACH OF CHILDREN

DANGER/PELIGRO

PRECAUCIÓN AL USUARIO: Si usted no lee inglés, no use este producto hasta que la etiqueta haya sido explicado ampliamente.

See additional precautionary statements on left side of container.

STATEMENT OF PRACTICAL TREATMENT

If in eyes: Flush with plenty of water and get medical attention.

If swallowed: Induce vomiting by touching back of throat with finger. Drink promptly a large quantity of milk, egg whites, or gelatin solution. If none of these are available, drink large quantities of water. Avoid alcohol. Call a physician or Poison Control Center. Do not induce vomiting or give anything by mouth to an unconscious person.

If inhaled: Remove victim to fresh air. If not breathing give artificial respiration, preferably mouth to mouth. Get medical attention immediately.

If on skin: Wash promptly with plenty of soap and water. Rinse thoroughly. Get medical attention.

Note to Physician

If swallowed, probable mucosal damage may contraindicate the use of gastric lavage.

If in eyes, specialized ophthalmologic attention may be required.

See directions for use on right side of container.

DIRECTIONS FOR USE AND CONDITIONS OF SALE AND WARRANTY

IMPORTANT: Read the entire **Directions for Use** and the **Conditions of Sale and Warranty** before using this product.

CONDITIONS OF SALE AND WARRANTY

The **Directions for Use** of this product reflect the opinion of experts based on field use and tests. The directions are believed to be reliable and should be followed carefully. However, it is impossible to eliminate all risks inherently associated with use of this product. Crop injury, ineffectiveness, or other unintended consequences may result because of such factors as weather conditions, presence of other materials, or the manner of use or application all of which are beyond the control of CIBA-GEIGY or the Seller. All such risks shall be assumed by the Buyer.

CIBA-GEIGY warrants that this product conforms to the chemical description on the label and is reasonably fit for the purposes referred to in the **Directions for Use** subject to the inherent risks referred to above. **CIBA-GEIGY makes no other express or implied warranty of Fitness or Merchantability or any other express or implied warranty. In no case shall CIBA-GEIGY or the Seller be liable for consequential, special, or indirect damages resulting from the use or handling of this product.** CIBA-GEIGY and the Seller offer this product, and the Buyer and user accept it, subject to the foregoing **Conditions of Sale and Warranty,** which may be varied only by agreement in writing signed by a duly authorized representative of CIBA-GEIGY.

DIRECTIONS FOR USE

It is a violation of federal law to use this product in a manner inconsistent with its labeling.

FAILURE TO FOLLOW THE DIRECTIONS FOR USE AND PRECAUTIONS ON THIS LABEL MAY RESULT IN POOR DISEASE CONTROL, CROP INJURY, AND/OR ILLEGAL RESIDUES.

Do not apply this product in such a manner as to directly, or through drift, expose workers or other persons, except those knowingly involved in the application. The area being treated must be vacated by unprotected persons.

REENTRY STATEMENTS

Do not enter treated areas without protective clothing until sprays have dried.

Because certain states may require more restrictive reentry intervals for various crops treated with this product, consult your State Department of Agriculture for further information.

Written or oral warnings must be given to workers who are expected to be in the treated area or in an area about to be treated with this product. Oral warnings must be given which inform workers of areas or fields that may not be entered without specific protective clothing until sprays have dried and appropriate actions to take in case of accidental exposure, as described under **Precautionary Statements** on the back of this label. When oral warnings are given, warnings shall be given in a language customarily understood by workers. Oral warnings must be given if there is reason to believe that written warnings cannot be understood by workers. Written warnings must include the following information: "DANGER. Area treated with RIDOMIL/BRAVO 81W on (date of application). Do not enter without appropriate protective clothing until sprays have dried. Flush eyes or skin with plenty of water. Call a physician if irritation persists. Remove contaminated clothing and wash before reuse."

GENERAL INFORMATION

RIDOMIL/BRAVO 81W contains 9% of the active ingredient metalaxyl and 72% of the active ingredient chlorothalonil. Metalaxyl is a systemic fungicide that provides control of downy mildew and late blight diseases of selected crops. Chlorothalonil is a broad-spectrum protectant fungicide that controls many diseases of vegetables.

Note: Metalaxyl is a systemic fungicide having a specific mode of action and could be subject to development of resistant strains of fungi. Development of resistance cannot be predicted. Therefore, CIBA-GEIGY cannot assume liability for crop damage resulting from resistant strains of fungi. If treatment is not effective following the use of RIDOMIL/BRAVO 81W as recommended, a resistant strain of fungi may be present. If the treatment is ineffective due to the presence of a metalaxyl resistant strain of fungi, neither RIDOMIL/BRAVO 81W nor any other fungicide with similar action will effectively control that disease. Consideration should then be given to the prompt use of other types of suitable fungicides. Consult with your State Agricultural Experiment Station or Extension Service Specialist for guidance in your particular crop and disease control situation.

Mixing Instructions

RIDOMIL/BRAVO 81W is an effective fungicide when used according to label directions for control of a broad spectrum of plant diseases. Thorough, uniform coverage is essential for disease control. Prepare no more spray mixture than is needed for the immediate operation. Thoroughly clean spray equipment before using this product. Agitate thoroughly before and during application. Good bypass agitation is sufficient to keep RIDOMIL/BRAVO 81W thoroughly mixed. Flush spray tank thoroughly with water daily after use and dispose of pesticide rinsate by application to an already treated area.

Amount of water needed per acre will vary with crop and amount of plant growth. Apply in a minimum of 20 gals./A for dilute sprays and 5 gals. A for concentrate ground sprays and aerial applications.

To avoid spray drift, do not apply under windy conditions. Avoid spray overlap, as crop injury may result.

RIDOMIL/BRAVO 81W Alone: Add 2/3 of the required amount of water to the spray or mixing tank. With the agitator running, drop the required number of **unopened** soluble bags of

RIDOMIL/BRAVO 81W into the tank. Continue agitation while adding the remainder of water and during application to maintain a uniform suspension.

RIDOMIL/BRAVO 81W + Tank Mixtures: Add 2/3 of the required amount of water to the spray or mixing tank. With the agitator running, drop the required number of **unopened** soluble bags of RIDOMIL/BRAVO 81W into the tank. Then add the desired amount of the tank mix partner(s). Continue agitation while adding the remainder of the water and during application to maintain a uniform suspension.

Note: Do not let spray mixture stand overnight in the spray tank. Rinse spray tank at the end of the day.

Preharvest Intervals

Unless otherwise stated in the directions for use for a specific crop, there is no preharvest interval.

APPLICATION THROUGH IRRIGATION SYSTEMS

RIDOMIL/BRAVO 81W alone or in combination with other fungicides which are registered for sprinkler irrigation may be applied through irrigation systems.

Apply this product only through center pivot, solid set or moving wheel irrigation systems. Do not apply this product through any other type of irrigation system. Crop injury, lack of effectiveness, or illegal pesticide residues in the crop can result from nonuniform distribution of treated water. If you have questions about calibration, you should contact State Extension Service specialists, equipment manufacturers or other experts. Do not connect an irrigation system (including greenhouse systems) used for pesticide application to a public water system unless the pesticide label-prescribed safety devices for public water systems are in place. A person knowledgeable of the chemigation system and responsible for its operation, or under the supervision of the responsible person, shall shut the system down and make necessary adjustments should the need arise.

Posting Requirements

Posting of areas to be chemigated is required when (1) any part of a treated area is within 300 feet of sensitive areas such as residential areas, labor camps, businesses, day care centers, hospitals, in-patient clinics, nursing homes or any public areas such as schools, parks, playgrounds, or other public facilities not including public roads, or (2) when the chemigated area is open to the public such as golf courses or retail greenhouses.

Posting must conform to the following requirements. Treated areas shall be posted with signs at all usual points of entry and along likely routes of approach from the listed sensitive areas. When there are no usual points of entry, signs must be posted in the corners of the treated areas and in any other location affording maximum visibility to sensitive areas. The printed side of the sign should face away from the treated area towards the sensitive area. The signs shall be printed in English. Signs must be posted prior to application and must remain posted until foliage has dried and soil surface water has disappeared. Signs may remain in place indefinitely as long as they are composed of materials to prevent deterioration and maintain legibility for the duration of the posting period.

All words shall consist of letters at least 2 1/2 inches tall, and all letters and the symbol shall be a color which sharply contrasts with their immediate background. At the top of the sign shall be the words KEEP OUT, followed by an octagonal stop sign symbol at least 8 inches in diameter containing the word STOP. Below the symbol shall be the words PESTICIDES IN IRRIGATION WATER.

Operating Instructions

1. The system must contain a functional check valve, vacuum relief valve, and low pressure drain appropriately located on the irrigation pipeline to prevent water source contamination from backflow.
2. The pesticide injection pipeline must contain a functional, automatic, quick-closing check valve to prevent the flow of fluid back toward the injection pump.
3. The pesticide injection pipeline must also contain a functional, normally closed, solenoid-operated valve located on the intake side of the injection pump and connected to the system interlock to prevent fluid from being withdrawn from the supply tank when the irrigation system is either automatically or manually shut down.
4. The system must contain functional interlocking controls to automatically shut off the pesticide injection pump when the water pump motor stops.
5. The irrigation line or water pump must include a functional pressure switch which will stop the water pump motor when the water pressure decreases to the point where pesticide distribution is adversely affected.
6. Systems must use a metering pump, such as a positive displacement injection pump (e.g., diaphragm pump) effectively designed and constructed of materials that are compatible with pesticides and capable of being fitted with a system interlock.
7. Do not apply when wind speed favors drift beyond the area intended.

Application Instructions

RIDOMIL/BRAVO 81W must be applied on the schedule specified in the specific crop use recommendations, not according to the irrigation schedule. If irrigation schedules are used, ground or aerial applications must supplement chemigation applications to achieve adequate disease control.

With the exception of potatoes, RIDOMIL/BRAVO 81W has not been sufficiently tested when applied through irrigation systems to assure consistent product performance for all labeled uses. The following calibration and application techniques are provided for user reference, but do not constitute a warranty of fitness for application through sprinkler irrigation equipment. Users should check with state and local regulatory agencies for potential use restrictions before applying any agricultural chemical through sprinkler irrigation equipment.

Center-Pivot Irrigation Equipment (Use only with drive systems which provide uniform water distribution):

1. Determine the size of the area to be treated.
2. Determine the time required to apply 1/4 inch water over the area to be treated when the system and injection equipment is operated at normal pressures recommended by the equipment manufacturer. Run the system at 80-95% of the manufacturer's rated capacity.
3. Using water, determine the injection pump output when operated at normal line pressure.
4. Determine the amount of RIDOMIL/BRAVO 81W required to treat the area covered by the irrigation system.
5. Add the required number of RIDOMIL/BRAVO 81W soluble bags and sufficient water to meet the injection time requirements to the solution tank. (See **Mixing Instructions** section of this label.)
6. Maintain constant solution tank agitation during the injection period.
7. Stop injection equipment after treatment is completed. Continue to operate system until the RIDOMIL/BRAVO 81W solution has cleared the sprinkler head.

Solid Set and Moving Wheel Irrigation Equipment

1. Determine the acreage covered by the sprinklers.
2. Fill injector solution tank with water and adjust flow rate to use the contents over a 20 to 30 minute interval.
3. Determine the amount of RIDOMIL/BRAVO 81W required to treat the area covered by the irrigation system.
4. Add the required number of RIDOMIL/BRAVO 81W soluble bags into the same quantity of water used to calibrate the injection period. (See **Mixing Instructions** section of this label.)
5. Operate system at the same pressure and time interval established during the calibration.
6. Inject RIDOMIL/BRAVO 81W at the end of the irrigation cycle or as a separate application to maximize retention of the fungicide by the foliage.
7. Stop injection equipment after treatment is completed. Continue to operate the system until the RIDOMIL/BRAVO 81W solution has cleared the last sprinkler head.

COLE CROPS (BROCCOLI, CABBAGE, AND CAULIFLOWER)*

* Broccoli, Chinese broccoli (gai lon, white flowering broccoli), cabbage, Chinese cabbage (tight-heading varieties only), and cauliflower only.

RIDOMIL/BRAVO 81W, when used as a foliar application in a preventive disease control program, provides effective control of downy mildew and Alternaria leaf spot.

Downy Mildew: Apply 1½-2 lbs./A. Begin applications when conditions are favorable for disease, but before infection and continue at 14-day intervals until the threat of disease is over. Use the higher rate under severe disease pressure.

Alternaria Leaf Spot: Apply 1½-2 lbs./A. Begin applications when conditions are favorable for disease, but before infection, and continue at 14-day intervals until the threat of disease is over. Under severe disease pressure, use other effective EPA registered fungicides according to label directions between the RIDOMIL/BRAVO 81W sprays.

Notes: (1) Do not apply within 7 days of harvest, or illegal residues may result. (2) To minimize the potential for resistance, do not make more than 4 applications of RIDOMIL/BRAVO 81W per crop.

CUCURBIT VEGETABLES*

* Cucumber, cantaloupe, casaba, crenshaw, honeydew melon, honey balls, muskmelon, Persian melon, watermelon, pumpkin, summer squash, winter squash and curcurbit hybrids only.

RIDOMIL/BRAVO 81W, when used as a foliar spray in a preventive disease control program, provides effective control of many diseases of cucurbit vegetables.

Downy Mildew: Apply 1½-2 lbs./A. Begin applications when conditions are favorable for disease, but before infection, and continue at 14-day intervals until the threat of disease is over. Use the higher rate under severe disease pressure.

Anthracnose, Cercospora Leaf Spot, Gummy Stem Blight, Leaf Blight and Scab: Apply 2-3 lbs./A. Begin applications when conditions are favorable for disease, but before infection, and continue at 10-14 day intervals. Under severe disease pressure, alternate RIDOMIL/BRAVO 81W sprays with other effective EPA-registered fungicides every 7 days.

Note: To minimize the potential for resistance, do not make more than 4 applications of RIDOMIL/BRAVO 81W per crop.

ONIONS (DRY BULB, SEED AND GREEN ONIONS)

RIDOMIL/BRAVO 81W, when used as a foliar spray in a preventive disease control program, provides effective control of downy mildew, Botrytis leaf blight and purple blotch.

Downy Mildew: Apply 1½-2 lbs./A. Begin applications when conditions are favorable for disease, but before infection, and continue at 14-day intervals until the threat of disease is over. Use a compatible spreader-sticker at rates recommended on the product label. Use the higher rate under severe disease pressure.

Botrytis Leaf Blight (Blast) and Purple Blotch: Apply 2 lbs./A on a 14-day schedule **and** apply other effective EPA-registered fungicides between the RIDOMIL/BRAVO 81W sprays; **or** apply 2 lbs./A of RIDOMIL/BRAVO 81W on a 7-day schedule.

Notes: (1) Do not apply to sweet Spanish onions if they have received or will receive additional chlorothalonil applications and (2) Do not apply to dry bulb onions within 7 days of harvest or to green onions within 21 days of harvest or illegal residues may result. (3) Do not make more than 3 applications to green onions per season. If more than 14 days remain after 3 applications of RIDOMIL/BRAVO 81W, it is recommended that another EPA-registered fungicide which does not contain chlorothalonil be used. (4) To minimize the potential for resistance, do not make more than 4 applications of RIDOMIL/BRAVO 81W to dry bulb or seed onions.

POTATOES

RIDOMIL/BRAVO 81W, when used as a foliar spray in a preventative disease control program, provides effective control of late blight, early blight and Botrytis vine rot. Two to three applications will also provide control of tuber rot in the field caused by *Phytophthora* spp. and storage rot caused by *Pythium* and *Phytophthora* spp.

Late Blight: Apply 1½-2 lbs./A. Begin applications when conditions are favorable for disease, but before infection, and continue at 14-day intervals until threat of disease is over. Use the higher rate under severe disease pressure.

Tuber Rot and Storage Rot *(Phytophthora and Pythium spp.)*: Apply 2 lbs./A. Make one application at flowering and another 14 days later. If the field has a history of tuber rot or if foliar blight is present, make a third application 14 days after the second.

Early Blight and Botrytis Vine Rot: Apply 1½-2 lbs./A. Begin applications when plants are six inches high or when conditions are favorable for disease, but before infection, and continue at 14-day intervals until the threat of disease is over. Under severe disease pressure, use 1½ lbs./A on a 7-day schedule or use other effective EPA-registered fungicides between the RIDOMIL/BRAVO 81W sprays.

Notes: (1) Do not apply within 7 days of harvest, or illegal residues may result. (2) To minimize the potential for resistance, do not make more than 4 applications of RIDOMIL/BRAVO 81W per crop.

TOMATOES

RIDOMIL/BRAVO 81W, when used as a foliar spray in a preventive disease control program, provides effective control of several tomato diseases.

Late Blight and Fruit Rot Caused by Phytophthora spp: Apply 1½-2 lbs./A. Begin applications when conditions are favorable for disease, but before infection, and continue at 14-day intervals until threat of disease is over. Use the higher rate under severe disease pressure.

Early Blight, Gray Leaf Spot, Gray Leaf Mold, and Septoria Leaf Spot: Apply 2 lbs./A. Begin applications when conditions are favorable for disease, but before infection, and continue at 14-day intervals. Under severe disease pressure, use other effective EPA-registered fungicides between the RIDOMIL BRAVO 81W sprays.

Anthracnose, Alternaria Fruit Rot (Black Mold), Rhizoctonia Fruit Rot and Botrytis Gray Mold: Apply 2-3 lbs./A. Begin applications when conditions are favorable for disease, but before infection, and continue at 10-14 day intervals. Under severe disease pressure, use other effective EPA-registered fungicides between the RIDOMIL/BRAVO 81W sprays.

Note: To minimize the potential for resistance, do not make more than 4 applications of RIDOMIL/BRAVO 81W per crop.

Rotation (Plant Back) Restrictions

Rotation Crop	Planting Time From Last Application of RIDOMIL/BRAVO 81W
Apples*	0 Days
Asparagus	12 Months
Avocados*, Broccoli, Citrus*	0 Days
Corn	9 Months
Cabbage	0 Days
Celery	18 Months
Cauliflower	0 Days
Carrots	9 Months
Cranberry*	0 Days
Brussels Sprouts	18 Months
Eggplant	12 Months
Mint	18 Months
Grass Grown for Seed	0 Days
Shallots, Garlic	18 Months
Papaya*	0 Days
Peppers	12 Months
Potatoes, Cotton, Cucurbit Vegetables**	0 Days
Other Cucurbit Vegetables	12 Months
Deciduous Fruits and Nuts*	0 Days
Hops, Head Lettuce	12 Months
Dry Beans, Snap Beans	0 Days
Peas	12 Months
Onions (Bulb, Green and Seed), Peanuts	0 Days
Pineapples, Raspberries	12 Months
Soybeans	0 Days
Spinach, Tobacco	12 Months
Tomatoes	0 Days
Wheat, Root Crops	12 Months
Crops **Not Intended** for Food or Feed	0 Days
All Other Crops **Intended** for Food or Feed	18 Months

* These crops may be planted immediately following last application of RIDOMIL/BRAVO 81W, provided they will not bear harvestable fruit within 12 months.

** Cucurbit vegetables including cucumbers: cantaloupes, casabas, crenshaws, honeydew melons, honey balls, muskmelons, Persian melons, and their hybrids; watermelons, watermelon hybrids; pumpkins; summer and winter squash.

STORAGE AND DISPOSAL

Pesticide

Store in a dry place. Do not contaminate water, food, or feed by storage or disposal. Open dumping is prohibited. Wastes resulting from the use of this pesticide are acutely hazardous. Improper disposal of unused pesticide, spray mixture, or rinsate is a violation of federal law. Pesticide, spray mixture, or rinsate that cannot be used according to label instructions must be disposed of according to federal, state, or local procedures.

For guidance in proper disposal methods, contact your State Pesticide or Environmental Control Agency, or the Hazardous Waste representative at the nearest EPA Regional Office.

Container

Completely empty bag into application equipment. Dispose of empty bag in a sanitary landfill, by incineration or, if allowed by state and local authorities, by burning. Stay out of smoke from burning bags.

For minor spills, leaks, etc., follow all precautions indicated on this label and clean up immediately. Take special care to avoid contamination of equipment and facilities during cleanup procedures and disposal of wastes. In the event of a major spill, fire, or other emergency, call (919) 292-7100 day or night.

PRECAUTIONARY STATEMENTS

HAZARDS TO HUMANS AND DOMESTIC ANIMALS

DANGER

Corrosive. Causes irreversible eye damage. May be fatal if swallowed or inhaled. Harmful if absorbed through skin. Do not get in eyes, on skin, or on clothing. Wear goggles or face shield when handling this product. Do not breathe spray mist. Wear a mask or a jointly approved pesticide respirator (by the Mining Enforcement and Mining Administration and the National Institute for Occupational Safety and Health) when handling this product. Wash thoroughly with soap and water after handling and before eating, drinking, or using tobacco. Remove contaminated clothing and wash before reuse.

ENVIRONMENTAL HAZARDS

This product is toxic to fish. Do not apply directly to water. Apply only as specified on this label. Do not apply when weather conditions favor drift from treated area. Do not contaminate water when cleaning equipment or disposing of wastes.

RIDOMIL® trademark of CIBA-GEIGY Corporation for metalaxyl
U.S. Patent No. 4,151,299
Bravo® trademark of Fermenta Plant Protection Company for chlorothalonil

CGA 80L2C 108

This is a specimen label which was not prepared by CIBA-GEIGY. CIBA-GEIGY is not responsible for the accuracy of the information contained herein. As labels are subject to revision, always carefully read and follow the label on the product container.

SEQUESTRENE® 138 FE

IRON CHELATE* MICRONUTRIENT

For correction of iron deficiency in crops and ornamentals in alkaline and calcareous soils

* ACTIVE INGREDIENT:
Technical sodium ferric ethylenediamine di-(o-hydroxyphenylacetate)
GUARANTEED ANALYSIS:

Iron (Fe)	6%
100% Chelated Iron	
Moisture content not more than 13%	
Total Nitrogen	0%
Available	
Phosphoric acid	0%
Soluble Potash	0%

EPA Est. 100-AL-1

KEEP OUT OF REACH OF CHILDREN

CAUTION

See directions for use on panel at right.

DIRECTIONS FOR USE AND CONDITIONS OF SALE AND WARRANTY

IMPORTANT: Read the entire **Directions for Use** and the **Conditions of Sale and Warranty** before using this product.

CONDITIONS OF SALE AND WARRANTY

The **Directions for Use** of this product reflect the opinion of experts based on field use and tests. The directions are believed to be reliable and should be followed carefully. However, it is impossible to eliminate all risks inherently associated with use of this product. Crop injury, ineffectiveness, or other unintended consequences may result because of such factors as weather conditions, presence of other materials, or the manner of use or application all of which are beyond the control of CIBA-GEIGY or the Seller. All such risks shall be assumed by the Buyer.

CIBA-GEIGY warrants that this product conforms to the chemical description on the label and is reasonably fit for the purposes referred to in the **Directions for Use** subject to the inherent risks referred to above. **CIBA-GEIGY makes no other express or implied warranty of Fitness or Merchantability or any other express or implied warranty. In no case shall CIBA-GEIGY or the Seller be liable for consequential, special, or indirect damages resulting from the use or handling of this product.** CIBA-GEIGY and the Seller offer this product, and the Buyer and user accept it, subject to the foregoing **Conditions of Sale and Warranty,** which may be varied only by agreement in writing signed by a duly authorized representative of CIBA-GEIGY.

GENERAL INFORMATION

SEQUESTRENE 138 FE Iron Chelate is a source of iron in a form readily available to plants. It can be applied either as a soil application or as a foliar spray. SEQUESTRENE 138 FE is effective over a wide pH range, including soils where pH is 7.0 and above. Rates of application will depend upon the severity of deficiency, but should be kept within the recommended range for each crop.

To be absorbed from the soil, the SEQUESTRENE 138 FE must get into the root zone. The chelate can be placed in the root zone during planting or it can be moved into the soil by rainfall, irrigation, or tillage.

Soil applications of SEQUESTRENE 138 FE can be made as a drench, injected directly into the soil, or banded. For row crops, soil applications are most effective when made in a band placed in the soil at planting time, or as a sidedressing shortly after plants emerge or after transplanting. SEQUESTRENE 138 FE can be applied alone or in combination with dry or liquid fertilizers. Or, to get uniform distribution, it can be mixed with an inert material, such as sand or dry soil, and uniformly distributed over the soil surface around plants. When applied to the soil around trees, apply within the drip line and lightly incorporate into the soil or water in. SEQUESTRENE 138 FE can be used at any time, but application in the spring or around planting time is preferred because (1) spring rains will move SEQUESTRENE 138 FE into root zone, and (2) iron will be available during the early flush of plant growth. Injection into the irrigation water provides a good after-planting method of application.

Foliar spray applications in water should be made as thorough cover sprays carefully following the Directions for Use given below. Adding a small amount of wetting agent or detergent to the solution may be helpful to insure even distribution on the leaf surfaces and increase the effectiveness of foliar applications. SEQUESTRENE 138 FE is compatible with most commonly used insecticides and fungicides and can be applied in combination with them. However, because available pesticide combinations differ so widely, before applying chelate-pesticide combinations to large crop areas a small test area should be sprayed to determine that no phytotoxicity or undesirable effects result from the combination spray.

To facilitate the mixing of the product into solution it is suggested that the required amount be added to a small amount of water and mixed until completely dissolved. The pre-mix solution can then be added, using agitation, to the final water or liquid fertilizer solution.

DIRECTIONS FOR USE

Fruit and Nut Crops

Almonds, Apples, Apricots, Cherries, Nectarines, Pecans, Plums, Prunes, and Walnuts

Foliage: Mix 1-1½ lbs. per 100 gals. of water and apply as a thorough cover spray prior to bloom or after harvest. To avoid possible injury to plants do not use in combination with oils.

Apples, Apricots, Avocados, Cherries, Macadamias, Nectarines, Peaches, Pears, Plums, Prunes, and Walnuts

Soil: Apply ½-2 oz. per tree for each inch of trunk diameter, but do not apply more than 1 lb. per tree. Apply annually in the spring, either prior to or at the time of new spring growth.

Peaches and Pears

Foliage: Mix 1-3 lbs. per 100 gals. of water and apply as a thorough cover spray about 4-6 weeks after full bloom. If deficiency is severe, apply 1 or 2 additional sprays at 2-3 week intervals. To avoid possible injury to plants do not use in combination with oils.

Citrus

Foliage: Mix 1 lb. per 100 gals. of water and apply as a thorough cover spray any time after the harvest of the main crop and up until bloom of the succeeding main crop. Do not apply when the main fruit crop is on the tree. To avoid possible injury to plants do not use in combination with oils or miticides.

Soil: For nursery stock, apply ⅓-½ lb. per 300 sq. ft. For mature trees, apply ⅓-½ lb. per tree in light sandy soils and ½-1 lb. per tree in heavy soils. Broadcast evenly within the drip line prior to the winter flush of growth. For an annual maintenance program, apply ¼-⅓ lb. per tree.

Blackberries, Boysenberries, Dewberries, Loganberries, Raspberries, and Strawberries

Soil: Band or sidedress ½-1 lb. per 100 ft. of row in early spring or when deficiency symptoms first appear.

Grapes

Foliage: Mix 1½ lbs. per 100 gals. of water and apply as a thorough cover spray.

Soil: Band or sidedress ½-1 lb. per 100 ft. of row.

Ornamentals

Hardy Aster, Canna, Chrysanthemum, Daylily, Geranium, Iris, Nasturtium, Petunia, Verbena, and similar annuals and perennials

Soil: Apply 2 oz. per 100 sq. ft. and water in well.

Gladiolus

Soil: Band ½ lb. per 100 ft. of row.

Azalea, Crepe Myrtle, Euonymous, Forsythia, Gardenia, Honeysuckle, Rose, Spiraea, and similar shrubs and vines

Soil: Apply 2 oz. for small shrubs and 4 oz. for large shrubs. Apply directly beneath or immediately around the shrubs and water in well.

Dwarf Apple, Chinese Arborvitae, Chinese and American Elm, Juniper, Mimosa, Shortleaf Pine, Ponderosa Pine, Pin Oak, Sycamore, and similar trees

Soil: Apply up to 2 oz. per tree for each inch of trunk diameter. Broadcast evenly within the drip line and follow with thorough irrigation.

Field Crops

Alfalfa and Forage Grasses

Soil: Apply up to 5 lbs. per acre at planting or when deficiency symptoms first appear.

Cotton, Cowpeas, Field Peas, Rice, Safflower, Sugar Beets, and Sugarcane

Foliage: Apply 1½ lbs. per acre in sufficient water for thorough coverage. Apply 4-6 weeks after planting or when deficiency symptoms first appear. Repeat in 2-3 weeks if necessary.

Soil: Apply up to 5 lbs. per acre at planting or when deficiency symptoms first appear.

Corn, Peanuts, and Sorghum

Soil: Band or sidedress up to 5 lbs. per acre at planting or when deficiency symptoms first appear.

Flax

Soil: Apply up to 10 lbs. per acre at planting or within a few weeks after planting.

Corn, Flax, Peanuts, and Sorghum

Foliage: Apply 1½ lbs. per acre in sufficient water for thorough coverage. Apply 4-6 weeks after planting or when deficiency symptoms first appear. Repeat in 2-3 weeks if necessary.

Soybeans

Foliage: Apply 2-2½ lbs. per acre in a band over the soybean plants with sufficient water for thorough coverage. Apply within 3-7 days after chlorotic symptoms develop or when the beans are growing into the second trifoliolate stage.

Applications later than the second trifoliolate stage will not be as effective.

Vegetable Crops

Beans, Corn, Mustard, and Spinach

Soil: Band or sidedress up to 5 lbs. per acre at planting or when deficiency symptoms first appear.

Black-eyed Peas, Carrots, Cucumbers, Eggplants, Melons, Onions, Parsnips, Peas, Peppers, Potatoes, Radishes, Squash, Tomatoes, and Turnips

Foliage: Apply 1½ lbs. per acre in sufficient water for thorough coverage. Apply 4-6 weeks after planting or when deficiency symptoms first appear. Repeat in 2-3 weeks if necessary. Apply to cucumbers, eggplants, melons, peppers, squash, and tomatoes prior to bloom.

Soil: Band or sidedress up to 5 lbs. per acre at planting or when deficiency symptoms first appear.

Cabbage, Cauliflower, Celery, and Lettuce

Soil: Band or sidedress up to 5 lbs. per acre at planting or when deficiency symptoms first appear.

SEQUESTRENE® trademark of CIBA-GEIGY

Fla. Reg. No. 926-740

CGA 3L18R 079

This is a specimen label which was not prepared by CIBA-GEIGY. CIBA-GEIGY is not responsible for the accuracy of the information contained herein. As labels are subject to revision, always carefully read and follow the label on the product container.

SEQUESTRENE® 330 FE

IRON CHELATE* MICRONUTRIENT

For correction of iron deficiency in certain crops, ornamentals, and turf in slightly acid to slightly alkaline soils

GUARANTEED ANALYSIS:

Iron (Fe)	10%
100% Chelated Iron	
Moisture content not more than 7%	
Total Nitrogen	0%
Available Phosphoric acid	0%
Soluble Potash	0%

KEEP OUT OF REACH OF CHILDREN

DANGER

See additional precautionary statements on left side of bag.

***Active Ingredient: Technical sodium ferric diethylenetriamine pentaacetate**

See directions for use on panel at right.

DIRECTIONS FOR USE AND CONDITIONS OF SALE AND WARRANTY

IMPORTANT: Read the entire **Directions for Use** and the **Conditions of Sale and Warranty** before using this product.

CONDITIONS OF SALE AND WARRANTY

The **Directions for Use** of this product reflect the opinion of experts based on field use and tests. The directions are believed to be reliable and should be followed carefully. However, it is impossible to eliminate all risks inherently associated with use of this product. Crop injury, ineffectiveness, or other unintended consequences may result because of such factors as weather conditions, presence of other materials, or the manner of use or application all of which are beyond the control of CIBA-GEIGY or the Seller. All such risks shall be assumed by the Buyer.

CIBA-GEIGY warrants that this product conforms to the chemical description on the label and is reasonably fit for the purposes referred to in the **Directions for Use** subject to the inherent risks referred to above. **CIBA-GEIGY makes no other express or implied warranty of Fitness or Merchantability or any other express or implied warranty. In no case shall CIBA-GEIGY or the Seller be liable for consequential, special, or indirect damages resulting from the use or handling of this product.** CIBA-GEIGY and the Seller offer this product, and the Buyer and user accept it, subject to the foregoing **Conditions of Sale and Warranty,** which may be varied only by agreement in writing signed by a duly authorized representative of CIBA-GEIGY.

DIRECTIONS FOR USE

SEQUESTRENE 330 FE Iron Chelate is a source of iron in a form readily available to plants. It can be applied as a soil application or as a foliar spray. SEQUESTRENE 330 FE generally performs best on slightly acid to slightly alkaline and calcareous soils where pH is 6.0-7.5. Rates of application depend upon severity of deficiency, but should be kept within the recommended range for each crop. Optimum results, unless specified differently, occur when applications are made early in the growth cycle.

To be absorbed from the soil, the SEQUESTRENE 330 FE must get into the root zone. The chelate can be placed in the root zone during planting, or it can be moved into the soil by rainfall, irrigation, or tillage.

Soil applications of SEQUESTRENE 330 FE can be made in water or in combination with dry or liquid fertilizers. SEQUESTRENE 330 FE can also be mixed with an inert material, such as sand or dry soil, and uniformly distributed over the soil surface around the plants. For row crops, soil applications are most effective when made in a band placed in the soil at planting time, or as a sidedressing shortly after plants emerge or after transplanting. When treating soil around trees, apply within the drip line, and lightly incorporate into the soil or water in. For most ornamentals, shrubs and trees, dissolve SEQUESTRENE 330 FE in a convenient volume of water and sprinkle the solution on the soil around the plants, or pour the solution into 5 or 6 shallow holes around the plants.

Foliar spray applications in water should be made as thorough cover sprays, carefully following the directions given below. Adding up to 0.5% wetting agent (1 pint in 25 gals. spray) to the solution may be helpful to insure even distribution on the leaf surfaces and increase the effectiveness of foliar applications. SEQUESTRENE 330 FE is compatible with most commonly used insecticides and fungicides and can be applied in combination with them. However, because available pesticide combinations differ so widely, before applying chelate-pesticide combinations to large crop areas, a small test should be sprayed to determine that no phytotoxicity or undesirable effects result from the combination spray. Do no use in combination with Parathion.

To facilitate the mixing, add the required amount of powder to a small amount of water and mix until completely dissolved. The pre-mix solution can then be added, using agitation, to the final water or liquid fertilizer solution.

FRUIT AND NUT CROPS

Avocados

Soil: For 3-15 year old trees, apply 2-5 lbs. per tree. For 15-30 year old trees, apply 5-10 lbs. per tree.

Almonds, Apples, Apricots, Cherries, Grapes, Nectarines, Peaches, Pears, Pecans, Plums, Prunes, and Walnuts

Foliage: Mix 1 lb. per 100 gals. of water and apply as a thorough cover spray prior to bloom or after harvest. To avoid possible injury to plants, do not use in combination with oils.

Apples, Apricots, Cherries, Nectarines, Pears, Peaches, Plums, and Prunes

Soil: Apply 1-2 lbs. per tree, depending on size and age of tree.

Blackberries, Blueberries, Boysenberries, Dewberries, Grapes, Loganberries, Raspberries, and Strawberries

Soil: Band or sidedress 1/2-1 lb. per 100 ft. of row.

Citrus

Foliage: Mix 1 lb. per 100 gals. of water and apply as a thorough cover spray any time after the harvest of the main crop and up until bloom of the succeeding main crop. Do not apply when the main crop is on the tree. To avoid possible injury to plants, do not use in combination with oils or miticides.

Soil: Apply 1-2 lbs. per tree alone or mixed with fertilizer. SEQUESTRENE 330 FE may also be applied in irrigation water. For an annual preventive program, apply 1/4 lb. per tree.

FIELD CROPS

Alfalfa and Forage Grasses

Soil: Apply up to 10 lbs. per acre.

Beans, Cotton, Cowpeas, Field Peas, Rice, Safflower, Sorghum, Sugar Beets, and Sugarcane

Foliage: Apply 1 lb. per acre in sufficient water for thorough coverage. Apply 4-6 weeks after planting or when deficiency symptoms first appear. Repeat in 2-3 weeks if necessary.

Soil: Apply up to 10 lbs. per acre.

Corn and Peanuts

Soil: Band or sidedress up to 10 lbs. per acre at planting or when deficiency symptoms first appear.

Flax

Soil: Broadcast up to 10 lbs. per acre at planting or when deficiency symptoms first appear.

Soybeans

Foliage: Apply 1 lb. per acre in sufficient water for thorough coverage. Apply 4-6 weeks after planting or when deficiency symptoms first appear. Repeat in 2-3 weeks if necessary.

Soil: Band or sidedress up to 10 lbs. per acre at planting or when deficiency symptoms first appear.

VEGETABLE CROPS

Beans and Black-eyed Peas

Foliage: Apply 1 lb. per acre in sufficient water for thorough coverage. Apply 4-6 weeks after planting or when deficiency symptoms first appear. Repeat in 2-3 weeks if necessary.
Soil: Apply 10-20 lbs. per acre as a band or sidedress at planting or when deficiency symptoms first appear.

Cabbage, Cauliflower, Celery, and Lettuce

Soil: Band or sidedress up to 10 lbs. per acre.

Carrots, Cucumbers, Eggplants, Melons, Onions, Parsnips, Peas, Peppers, Potatoes, Radishes, Squash, Tomatoes, and Turnips

Foliage: Apply 1 lb. per acre in sufficient water for thorough coverage. Apply 4-6 weeks after planting or when deficiency symptoms first appear. Repeat in 2-3 weeks if necessary.

Soil: Band or sidedress up to 10 lbs. per acre.

Corn, Mustard, and Spinach

Foliage (corn only): Apply 1 lb. per acre in sufficient water for thorough coverage. Apply 4-6 weeks after planting or when deficiency symptoms first appear. Repeat in 2-3 weeks if necessary.

Soil: Band or sidedress up to 10 lbs. per acre at planting or when deficiency symptoms first appear.

ORNAMENTALS AND TURF

Azalea, Camellia, Cherry Laurel, Gardenia, Hibiscus, Holly, Hydrangea, Ixora, Ligustrum, Pieris Japonica, Primrose, Pyracantha, Rhododendron, Rose, and similar plants

Foliage: Mix 1 lb. per 100 gals of water (1-1 1/2 level tsp. per 1 gal.) and apply as a thorough cover spray.

Soil: For nurseries, greenhouses, or home gardens where plants are grown close together in beds or rows, apply 2-4 ozs. (6-12 tbsp.) per 100 sq. ft.

For individual garden plants, apply 1 tsp. per plant for plants up to 2 ft. in height; 2 tsp. per plant for plants 2-3 ft. in height; 1 tbsp. per plant for larger woody shrubs.

For potted plants, apply 1/4 tsp. per 8-inch pot, or 1/2 tsp. per 12-inch pot.

Chrysanthemum, Carnation, Gladiolus, Petunia, Snapdragon, and similar herbaceous plants

Foliage and Soil: Use one-half (1/2) the amounts suggested for azaleas, camellias, etc.

Camphor, Magnolia, Pin Oak, Russian Olive, Sandcherry, Spruce, Boxwood, Ornamental Citrus, and other shade trees and very large woody shrubs

Foliage: Mix 1 lb. per 100 gals. of water and apply as a thorough cover spray.

Soil: Apply 1-2 ozs. (3-6 tbsp.) per inch of trunk diameter at chest level. Broadcast around base of shrub or tree and follow with heavy watering, or apply as a soil drench.

Turf

Lawns: Apply 1 lb. per 1,000 sq. ft. either in combination with fertilizer or as a thorough cover spray. Follow with thorough watering.

Putting Greens: Apply 1/4 lb. per 1,000 sq. ft. as a thorough cover spray and water into the soil. Make 1-4 applications during the fall and winter months. Interval between applications should be no less than one month.

PRECAUTIONARY STATEMENTS

HAZARDS TO HUMANS AND DOMESTIC ANIMALS

DANGER

Corrosive—Causes eye and skin burns. Respiratory irritant. Harmful if swallowed. Do not get in eyes, on skin, or on clothing. Avoid breathing dust, vapor or mist. Wear impervious gloves and chemical goggles. Keep container closed. Contains nitrilotriacetic acid or its salt as a by-product.

First Aid: In case of contact,

Eyes—Immediately flush eyes with plenty of water.

Skin—Immediately flush skin with plenty of water. Remove contaminated clothing.

If inhaled, remove to fresh air. If not breathing, give artificial respiration. Preferably mouth-to-mouth. If breathing is difficult, give oxygen. If swallowed, do not induce vomiting. If conscious, give plenty of milk or water. Get medical attention!

Consult Material Safety Data Sheet prior to handling.

SEQUESTRENE® trademark of CIBA-GEIGY

Fla. Reg. No. 926-743

CGA 3L2T 016

This is a specimen label which was not prepared by CIBA-GEIGY. CIBA-GEIGY is not responsible for the accuracy of the information contained herein. As labels are subject to revision, always carefully read and follow the label on the product container.

SUPRACIDE® 2E

INSECTICIDE-MITICIDE

RESTRICTED USE PESTICIDE

FOR RETAIL SALE TO AND USE ONLY BY CERTIFIED APPLICATORS OR PERSONS UNDER THEIR DIRECT SUPERVISION, AND ONLY FOR THOSE USES COVERED BY THE CERTIFIED APPLICATOR'S CERTIFICATION.

For control of certain insects of alfalfa; artichokes; certain citrus, fruits and nuts; cotton; nursery stock; olives; safflowers; sunflowers; sorghum; and tobacco

ACTIVE INGREDIENT:

Methidathion: O,O-dimethyl phosphorodithioate, S-ester with 4-(mercaptomethyl)-2-methoxy-Δ^2-1,3,4-thiadiazolin-5-one	24.4%
INERT INGREDIENTS:	75.6%
TOTAL	100.0%

EPA Reg. No. 100-501 EPA Est. 100-AL-1

SUPRACIDE 2E contains 2 lbs. active ingredient per gallon.

KEEP OUT OF REACH OF CHILDREN

DANGER/ PELIGRO

PRECAUCIÓN AL USUARIO: Si usted no lee inglés, no use este producto hasta que la etiqueta haya sido explicado ampliamente.

See additional precautionary statements at end of label booklet.

See directions for use inside booklet.

SUPRACIDE® trademark of CIBA-GEIGY for methidathion

CGA 30L23F 059

DIRECTIONS FOR USE AND CONDITIONS OF SALE AND WARRANTY

IMPORTANT: Read the entire **Directions for Use** and the **Conditions of Sale and Warranty** before using this product.

CONDITIONS OF SALE AND WARRANTY

The **Directions for Use** of this product reflect the opinion of experts based on field use and tests. The directions are believed to be reliable and should be followed carefully. However, it is impossible to eliminate all risks inherently associated with use of this product. Crop injury, ineffectiveness, or other unintended consequences may result because of such factors as weather conditions, presence of other materials, or the manner of use or application all of which are beyond the control of CIBA-GEIGY or the Seller. All such risks shall be assumed by the Buyer.

CIBA-GEIGY warrants that this product conforms to the chemical description on the label and is reasonably fit for the purposes referred to in the **Directions for Use** subject to the inherent risks referred to above. **CIBA-GEIGY makes no other express or implied warranty of Fitness or Merchantability or any other express or implied warranty. In no case shall CIBA-GEIGY or the Seller be liable for consequential, special, or indirect damages resulting from the use or handling of this product.** CIBA-GEIGY and the Seller offer this product, and the Buyer and user accept it, subject to the foregoing **Conditions of Sale and Warranty**, which may be varied only by agreement in writing signed by a duly authorized representative of CIBA-GEIGY.

DIRECTIONS FOR USE

It is a violation of federal law to use this product in a manner inconsistent with its labeling.

FAILURE TO FOLLOW THE DIRECTIONS FOR USE AND PRECAUTIONS ON THIS LABEL MAY RESULT IN POOR INSECT AND MITE CONTROL, CROP INJURY, AND/OR ILLEGAL RESIDUES.

GENERAL INFORMATION

SUPRACIDE 2E is an emulsifiable concentrate which when applied according to the Directions for Use will control certain insect pests and mites on agricultural crops.

Wear protective clothing as described in the USDA/EPA Guide for Commercial Applicators during the preparation of the spray mixture and during application.

Do not apply this product in such a manner as to directly or through drift expose workers or other persons except those knowingly involved in the application. Do not treat while unprotected humans or domestic animals are present in the treatment area.

REENTRY STATEMENT

Two days must elapse between the time of application and the time a worker reenters the treated area when SUPRACIDE 2E is applied at rates up to 2 lbs. ai/A.

Fourteen days must elapse between the time of application and the time a worker reenters the treated area when SUPRACIDE 2E is applied at rates greater than 2 lbs. ai/A and up to the maximum label rate of 10 lbs. ai/A. Do not apply Supracide at a rate exceeding 10 lbs. ai/A.

If a worker chooses to reenter a treated field prior to the allowable time of reentry, protective clothing consisting of a protective suit worn over normal work attire, chemical resistant gloves, chemical resistant shoes, shoe covers or boots, a NIOSH or MSHA approved respiratory protection device, goggles or face shield must be worn. Protective clothing must be removed daily and laundry washed, keeping the clothing separate from regularly worn clothing. Fresh protective clothing must be worn each day that the worker enters a field prior to the allowable time.

Because certain states may require more restrictive reentry intervals for various crops treated with this product, consult your State Department of Agriculture for further information.

Written or oral warnings must be given to workers who are expected to be in a treated area or in an area about to be treated with this product. Oral warnings must inform workers of areas or fields that may not be entered without specific protective clothing until the intervals stated above have elapsed, and appropriate actions to take in case of accidental exposure, as described under **Precautionary Statements** on this label. When oral warnings are given, warnings shall be given in a language customarily understood by workers. Oral warnings must be given if there is reason to believe that written warnings cannot be understood by workers. Written warnings must include the following information: "DANGER. Area treated with SUPRACIDE 2E on (date of application). Do not enter without appropriate protective clothing until (date based on intervals stated above). In case of accidental exposure, flush eyes or skin with plenty of water. Call physician if irritation persists. Remove and wash contaminated clothing before reuse."

Mixing Procedures

(1) Be sure sprayer is clean; if it is contaminated with any material, crop injury or sprayer clogging may result. (2) Fill tank 1/4-1/2 full with clean water, and start agitation. (3) Be certain that the agitation system is working properly and creates a rippling or rolling action on the liquid surface. (4) Add SUPRACIDE to the tank. Then add tank mix ingredients. (5) Complete filling tank, maintaining sufficient agitation at all times to ensure surface action. This applies to both spray and nurse tanks. (6) Do not leave tank mixtures in the tank without adequate agitation. Always maintain agitation to avoid separation and buildup of residues. If a residue starts to build up in tank, drain and clean with strong detergent and ammonia solution and rinse thoroughly with clean water. (7) Immediately after use, clean sprayer thoroughly by flushing system with water containing a detergent and rinse with clean water.

Chemigation: Do not apply this product through any type of irrigation system.

Alfalfa (Pure Stands or Stands Containing Clover or Grass) and Pure Stands of Timothy Grass

Apply at the rates listed below in a minimum of 10 gals. of water per acre with ground equipment or 2 gals. of water per acre by air. For alfalfa weevil and Egyptian alfalfa weevil control, make foliar applications to standing alfalfa when 20-30% of the growing tips show insect feeding damage. Make stubble applications immediately following a cutting. For other insects listed on alfalfa, apply when insects first appear. Make no more than one stubble application and no more than one foliage application per cutting. Use the lower rate for light infestations and the higher rate for moderate to heavy infestations.

Table 1: Alfalfa—Rates Per Acre

Insects Controlled	Northeastern States	Southern States	California	Pacific Intermountain States Excluding CA
Alfalfa weevil	2–4 pts.	3–4 pts.	2 pts.	2–4 pts.
Aphids	2–4 pts.	3–4 pts.	2 pts.*	2–4 pts.
Egyptian alfalfa weevil	—	—	2 pts.*	2–4pts.
Grass scale *Eriococcus insignis*	—	—	—	4 pts. (Washington only)
Lygus	—	—	2 pts.	2–4 pts.
Leafhoppers	2–4 pts.	2–4 pts.	2 pts.	2–4 pts.
Spittlebugs	2–4 pts.	—	—	—

* Egyptian alfalfa weevil and pea aphid: (California only) Use 1–2 pts. where integrated control is practiced. Apply as above. When Egyptian alfalfa weevil larval counts exceed 20 per sweep, use the 2 pt. rate only.

Note: To avoid injury to pollinating insects, do not apply during bloom. To avoid illegal residues, do not apply within 14 days of cutting, grazing, or feeding to livestock.

Almonds

Dilute and Concentrate Sprays

Peach twig borer, San Jose scale, brown almond mite: Apply during the dormant or delayed dormant period, or as a cover spray. Mix a dilute spray at the rate of 1–2 pts. per 100 gals. of water and apply 300–600 gals. per acre. If applying as a concentrate spray*, apply 3–12 pts. per acre in a minimum of 50 gals. of water per acre. For rates less than 2 pts. of Supracide per 100 gals., tank mix with EPA-registered oil.

When concentrate applications are made, proper equipment must be used to obtain uniform coverage.

Aerial Sprays

Peach twig borer, San Jose scale: Apply during the dormant or delayed dormant period and use 6–12 pts. per acre as an aerial spray* in a minimum of 20 gals. of spray per acre. Spray swath should be no wider than the distance between the tree rows or the length of the spray boom.

EPA-registered oils may be added to the SUPRACIDE spray mixture in the dormant/delayed dormant period. Do not use more than 1.5 gals of oil per 100 gals. spray as a dilute spray or 6 gals. of oil per acre as a concentrate spray or 5 gals of oil per acre as an aerial spray.

* The rate of SUPRACIDE for concentrate and aerial sprays is based on a dilute spray with conventional ground equipment. A dilute spray is considered to be 300–600 gals. per acre. This amounts to a minimum of 6 pts. and a maximum of 12 pts. of SUPRACIDE per acre.

Note: To avoid illegal residues, (1) Do not make more than one application during the dormant or delayed dormant period nor more than one cover spray each season. (2) Do not apply within 80 days of harvest. (3) Do not graze treated areas or feed cover crops grown in treated areas to livestock.

Precaution: Apply the dormant/delayed dormant application before any blossoms open, or injury may occur. Do not combine with oils in the cover spray, or injury or excessive residues may occur. The cover spray of SUPRACIDE may be slightly phytotoxic on some almond varieties.

Artichokes

Apply in a minimum of 50 gals. of water with ground equipment or a minimum of 10 gals. of water by air. Begin applications when insects appear in newly planted fields or when shoots emerge following complete cutback of the plants. Continue applications at 14 day intervals until buds begin to appear. Do not plant crops except those listed on this label within one year after last application.

Artichoke plume moth, chrysanthemum leafminer: Apply 4 pts. per acre.

Note: To avoid illegal residues, (1) Do not apply after buds begin to form. (2) Avoid application under conditions which favor drift to adjacent artichoke fields where buds are present. (3) Do not exceed a total of 8 applications per crop cycle.

Cotton

Apply in a minimum of 10 gals. of water with ground equipment or in a minimum of 3 gals. of water by air unless otherwise specified.

Spider mites, boll weevil, bollworm: Apply 4 pts. per acre. Repeat applications at 5–7 day intervals if necessary.

Lygus bug: (Arizona, California only) Apply 2–4 pts. per acre. Use the low rate for light to moderate infestations and the high rate for moderate to heavy infestations. Repeat applications if necessary.

Pink bollworm: (Arizona, California only) Apply 2 pts. per acre. Repeat applications at 5–7 day intervals if necessary.

Bandedwing whitefly: (mid-South and southeast United States only) Apply 1–2 pts. per acre. Use the low rate for light to moderate infestations and the high rate for moderate to heavy infestations. Repeat applications if necessary.

Note: To avoid illegal residues, (1) Do not apply more than 16 pts. per acre during any one growing season. (2) Do not apply within 14 days of harvest. (3) Do not apply more than the 2 pt. per acre rate to **open** bolls. (4) Do not graze treated cotton plants or feed gin trash to livestock.

Deciduous Fruit

Dilute and Concentrate Sprays

See Table 2 for specific crops and insects. Apply during the dormant or delayed dormant period. Use a rate of 1–2 pts. per 100 gals. of water as a dilute spray (300–600 gals. per acre) or a rate of 3–12 pts. per acre as a concentrate spray* (a minimum of 50 gals. of water per acre). The lower rates may be used for light infestations and the higher rates for moderate to heavy infestations. For rates less than 2 pts. of SUPRACIDE per 100 gals., tank mix with an EPA-registered oil.

When concentrate applications are made, proper equipment must be used to obtain uniform coverage.

Table 2: Deciduous Fruits

Crop	Insects
Apples	San Jose scale, rosy apple aphid
Cherries, pears	San Jose scale
Nectarines, peaches, plums, prunes	Peach twig borer, San Jose scale, Green peach aphid[1]
Apricots	Peach twig borer
Pears†	Pear psylla[2]

† California only

[1] For green peach aphid, use 8 pts. SUPRACIDE + 4 gals. oil per acre or 12 pts. SUPRACIDE per acre alone.

[2] For pear psylla, use 8–12 pts. SUPRACIDE + 4–6 gals. oil in a minimum of 100 gals. total spray per acre.

Aerial Sprays

See Table 3 for specific crops and insects. Apply during the dormant or delayed dormant period and use 3–12 pts. per acre as an aerial spray* in a minimum of 20 gals. total spray per acre. Spray swath should be no wider than the distance between the tree rows or the length of the spray boom. The lower rates may be used for light infestations and the higher rates for moderate to heavy infestations. For rates less than 2 pts. of SUPRACIDE per 100 gals., tank mix with an EPA-registered oil.

Table 3: Deciduous Fruits (Continued)

Crop	Insects
Apples	San Jose scale, rosy apple aphid
Peaches, plums, prunes, nectarines, apricots	Peach twig borer, San Jose scale
Cherries, pears	San Jose scale

EPA-registered oils may be added to the SUPRACIDE spray mixture in the dormant/delayed dormant period. Do not use more than 1.5 gals. of oil per 100 gals. spray as a dilute spray or 6 gals. of oil per acre as a concentrate spray or 5 gals. of oil per acre as an aerial spray.

* The rate of SUPRACIDE for concentrate and aerial sprays is based on a dilute spray with conventional ground equipment. A dilute spray is considered to be 300-600 gals. per acre. This amounts to a minimum of 3 pts. and a maximum of 12 pts. of SUPRACIDE per acre.

Note: To avoid illegal residues, (1) Apply before any blossoms open or injury may occur. (2) Do not graze treated areas or feed cover crops grown in treated areas to livestock.

Precaution: Do not use SUPRACIDE on Red Beaut plums.

Grapefruits, Lemons, Oranges and Other Citrus Fruits

SUPRACIDE Alone

Apply the recommended rates anytime except during the bloom period, or within 14 days of harvest. Preferred timing is during the immediate post-bloom period, **before fruit becomes infested.** Use the low rate for light infestations and the high rate for moderate to heavy infestations. Thorough coverage is essential for good control.

Make no more than 2 applications per growing season and allow at least 45 days between applications. Severe infestations may require 2 treatments.

California red scale: (Arizona and California only) Apply 1-2 pts. per 100 gals. of water.

Black scale: (California only) Apply 1-2 pts. per 100 gals. of water.

Chaff, Glovers, purple, snow, yellow scale, and mealybug: (Florida only) Apply 1/2 -2 pts. per 100 gals. of water.

Black scale, whiteflies, and citrus blackfly: (Florida only) Apply 1-2 pts. per 100 gals. of water.

Brown soft, California red, chaff scale, and mealybug: (Texas only) Apply ½-2 pts. per 100 gals. of water.

The minimum amount of SUPRACIDE applied per acre is dependent upon the number of trees per acre and upon tree height. See example below.

Example: Mature trees of normal foliage require a minimum of 5 gals. of spray plus one additional gallon for each foot of tree height for adequate coverage. Eighty trees per acre 20 feet tall require $80 \times 25 = 2{,}000$ gals. of dilute spray per acre.

Note: To avoid illegal residues, (1) Do not apply more than 40 pts. per acre. (2) Do not graze treated areas. (3) Make no more than 2 applications per growing season nor apply within 14 days of harvest.

SUPRACIDE Plus Oil

In Florida and Texas, EPA-registered citrus spray oils at rates of 2-4 qts. per 100 gals. of water may be tank mixed with SUPRACIDE following the rates and timing listed above for SUPRACIDE alone.

In California, EPA-registered citrus spray oils at rates listed below may be tank mixed with SUPRACIDE. See the above directions for rates and timing. A rate of ½ pt. of SUPRACIDE per 100 gals. of water may be used if tank mixed with oil.

Type of oil	Amount per 100 gals. of water
Light-medium grade	1.6 - 1.8 gals.
Medium grade	1.6 - 1.8 gals.
Narrow-range 415 grade	1.2 - 1.4 gals.

For further details, consult the current *University of California Citrus Treatment Guide.*

Note: To avoid illegal residues, (1) Do not apply more than 40 pts. of SUPRACIDE per acre in any one season. (2) Do not graze treated areas. (3) Make no more than 2 applications with oil per growing season nor apply within 60 days of harvest. (4) Lemons only, do not exceed 10 pts. per acre or apply more than once per season when tank mixed with oil.

Mangoes

Apply during the postharvest to bloom stage with ground equipment in 100 gals. of spray mixture per acre. Spray foliage to thoroughly wet upper and lower leaf surfaces. Repeat as needed at 3 week intervals.

False oleander scale, *Pseudaulacaspis cockerelli* (cooley): Apply ½-1 pt. and 0.5% oil per acre.

Note: To avoid illegal residues, (1) Do not exceed 5 applications per season. (2) Do not graze or feed cover crops from treated groves.

Nursery Stock (woody ornamental or herbaceous plants commercially grown for transplanting)

Apply at the rate listed below in a thorough cover spray to runoff. Thorough coverage and timing of application are essential for effective control.

To control scale insects on nursery stock, including both potted and field grown plants, apply two pints per 100 gals. of water once per season after sale crawlers have hatched. Monitor populations closely to determine when crawlers are present.

Because of changing cultivars in the nursery industry, test each new cultivar for phytotoxicity by treating one plant with the recommended rate at least 2 days prior to treatment of the remaining plants.

Olives

Apply once during the postharvest or prebloom season at the rate listed below as a full cover spray in a minimum of 400 gals. of water per acre. Supreme or superior type oil may be added to the spray mixture at the rate of 1-1.5 gals. per 100 gals. of water.

Black scale: Apply 2 pts./100 gals.

Note: To avoid illegal residues, (1) Do not apply more than 12 pts. per acre per year. (2) Do not apply after buds break. (3) Do not graze treated areas or feed cover crops grown in treated areas to livestock.

Pecans (Southeastern U.S. Only)

Apply the appropriate rate listed below as a foliar or full cover spray in sufficient water for thorough coverage.

Yellow pecan aphid: Apply 2 pts. per 100 gals. and repeat as necessary at 14-21 day intervals.

Hickory shuckworm and pecan serpentine leafminer: Apply 2-4 pts. per 100 gals. when insects first appear or leafminer damage first appears and repeat as necessary at 14-21 day intervals. Use the 4 pt. per 100 gals. rate when infestations are heavy.

Precaution: Do not tank mix SUPRACIDE with Cyprex® as phytotoxicity may result.

Note: To avoid illegal residues, (1) Do not apply within 60 days of harvest. (2) Do not graze treated areas.

Safflowers

Apply in a minimum of 20 gals. of water by ground equipment or in a minimum of 5 gals. of water by air.

Lygus bugs and thrips: Apply 2 pts. per acre. Begin applications either at bud stage, 12-14 days before onset of bloom, or as infestations develop. Additional applications can be made on a 7-14 day schedule or as further infestations develop.

Note: To avoid illegal residues, (1) Do not apply within 28 days of harvest. (2) Do not make more than three applications per year. (3) Do not graze treated areas.

Sunflowers

Apply 2 pts. per acre. Make ground application in a minimum of 10 gals. of water per acre; aerial application in a minimum of 5 gals. of water per acre. Thorough coverage of the heads is necessary. One to three applications may be made on a 7-day schedule. The number of applications will be determined by the severity of the infestations.

Stem weevil*: Begin applications in early season at vegetative stage 2.3 (third alternate leaf developed) to vegetative stage 3.1 (inflorescence head visible but tightly surrounded by young leaves).

Sunflower moth, seed weevil**, banded sunflower moth, and sunflower maggot complex: Begin applications at the onset of bloom as the infestations develop.

Sunflower seed midge: Begin applications during early bud development.

* *Cylindrocopturus adspersus* and *Apion occidentalis.*

** *Smicronyx fulvus and Smicronyx sordidus.*

Note: To avoid illegal residues, (1) Do not apply within 50 days of harvest. (2) Do not graze treated areas or feed treated forage to livestock.

Grain Sorghum Grown for Forage and Grain (Colorado, Kansas, Oklahoma, and Texas Only)

Apply in a minimum of 10 gals. of water per acre with ground equipment or 2 gals. of water per acre by air. Up to three applications may be made on a 10-14 day schedule. The number of applications will be determined by the severity of the infestation.

Banks grass mite, sorghum greenbug: Apply 2 pts. per acre.

Precaution: Application may result in phytotoxicity to certain sorghum varieties. This phytotoxicity is a temporary spotting, reddening or chlorosis of the leaves.

Note: To avoid illegal residues, do not apply within 30 days of harvesting or feeding to livestock.

Tobacco

Apply at the rates listed below in a minimum of 25 gals. of water per acre when insects appear. Apply 1-3 repeat applications on a 7-14 day interval as needed.

Flea beetles, hornworms: Apply 2-3 pts. per acre.

Budworms: Apply 4 pts. per acre.

Precaution: When applying SUPRACIDE with hand sprayers, avoid concentrating the spray in the bud area since phytotoxicity may occur. Do not apply SUPRACIDE to tobacco plants which are under stress as injury may occur.

Note: Do not apply in mixture with other pesticides or within 3 days of tobacco harvest.

Walnuts

Dilute and Concentrate Sprays

Walnut scale, frosted scale, Italian pear scale: Apply during the dormant or delayed dormant period or as a cover spray when crawlers appear.

Codling moth, walnut aphid: Apply as a cover spray according to moth activity or when aphids appear.

Use a rate of 2 pts. per 100 gals. of water as a dilute spray (400 gals. per acre), or a rate of 8 pts. per acre as a concentrate spray* (a minimum of 100 gals. of water per acre).

When concentrate applications are made, proper equipment must be used to obtain uniform coverage. If scale infestations are severe, dilute applications should be used.

Aerial Sprays

Walnut scale, frosted scale: Apply during the dormant or delayed dormant period and use 8 pts. per acre as an aerial spray* in a minimum of 20 gals. of spray per acre. Spray swath should be no wider than the distance between the tree rows or the length of the spray boom.

EPA-registered oils may be added to the SUPRACIDE spray mixture in the dormant/delayed dormant period. Do not use more than 1 gal. of oil per acre in concentrate and aerial sprays or 4 gals. of oil per acre in dilute sprays.

*The rate of SUPRACIDE for concentrate and aerial sprays is based on a dilute spray with conventional ground equipment. A dilute spray is considered to be 400 gals. per acre. This amounts to 8 pts. of SUPRACIDE per acre.

Note: To avoid illegal residues, (1) Make no more than one application during the dormant or delayed dormant period nor more than two cover sprays each season. (2) Do not apply within 7 days of harvest. (3) Do not graze treated areas or feed cover crops grown in treated areas to livestock.

Precaution: Do not add oil to SUPRACIDE sprays to walnuts in southern or central California or to trees that may be subject to stress due to drought, drying winds, disease, or severe insect infestations.

STORAGE AND DISPOSAL

Do not contaminate water, food, or feed by storage or disposal.

Storage: Store at temperatures above 40°F. Crystals may form at lower storage temperatures. If this occurs, place the product in a warm room (68°F or above) and roll or shake the container at frequent intervals until all crystals are dissolved.

Pouring the product through a 50-mesh or finer screen will indicate adequacy of or need to repeat above procedure. Do not store near food, feed, or fertilizer. Do not use or store in or around the home.

Pesticide and Container Disposal: Pesticide wastes are acutely hazardous. Improper disposal of unused pesticide, spray mixture, or rinse water is a violation of federal law. If these wastes cannot be used according to label instructions, contact your State Pesticide or Environmental Control Agency, or the Hazardous Waste representative at the nearest EPA Regional Office for guidance in proper disposal methods.

Triple rinse container (or equivalent) and offer for recycling or reconditioning, or puncture and dispose of in a sanitary landfill, by incineration, or if allowed by state and local authorities, by burning. Stay out of smoke from burning containers.

For minor spills, leaks, etc., follow all precautions indicated on this label and clean up immediately. Take special care to avoid contamination of equipment and facilities during cleanup procedures and disposal of wastes. In the event of a major spill, fire or other emergency, call 1-800-888-8372 day or night.

PRECAUTIONARY STATEMENTS

HAZARDS TO HUMANS AND DOMESTIC ANIMALS

DANGER

Corrosive. Causes irreversible eye damage. May cause skin irritation. Do not get in eyes, on skin, or on clothing. Wear goggles or face shield and rubber gloves when mixing. May be harmful or fatal if swallowed or absorbed through skin. Do not breathe vapor or spray mist. Prolonged or repeated inhalation may be harmful. Wear a mask or pesticide respirator jointly approved by the Mining Enforcement Safety Administration (MESA) or the National Institute for Occupational Safety and Health (NIOSH) during periods of prolonged or repeated use. Wash thoroughly with soap and water after handling and before eating, drinking, or smoking. Remove contaminated clothing and wash before reuse or if not washable, discard. Keep out of reach of children and domestic animals.

STATEMENT OF PRACTICAL TREATMENT

If in eyes: Flush with plenty of water. Get medical attention immediately.

If swallowed: Call a physician or poison control center immediately. DO NOT INDUCE VOMITING UNLESS NO OTHER TREATMENT IS AVAILABLE. Vomiting may cause aspiration pneumonia. If it is necessary to induce vomiting, do so under medical supervision. Give victim one or two glasses of water and insert finger in back of throat. Repeat until vomit fluid is clear. Do not induce vomiting or give anything by mouth to an unconscious person.

If on skin: Remove contaminated clothing and wash skin with plenty of soap and water. Get medical attention.

If inhaled: Remove victim to fresh air. If not breathing, give artificial respiration, preferably mouth-to-mouth unless exposure was due to swallowing; then use cardiopulmonary resuscitation (CPR). Get medical attention.

Note to Physician: Lavage stomach with care to prevent aspiration into lung. After lavage, instill and leave 30 gm. activated charcoal in 3-4 oz. water. Atropine sulfate and 2-PAM are specific antidotes. Morphine, theophylline, aminophylline, and large amounts of intravenous fluids are contraindicated.

ENVIRONMENTAL HAZARDS

This pesticide is toxic to fish and wildlife. Do not apply directly to water or wetlands (swamps, bogs, marshes, and potholes). Drift and runoff from treated areas may be hazardous to aquatic organisms in neighboring areas. Do not contaminate water by cleaning of equipment or disposal of wastes.

This pesticide is highly toxic to bees exposed to direct treatment or to residues remaining on the treated area. Do not apply when bees are actively visiting the crop, cover crop, or weeds blooming in the treatment area. Applications should be timed to provide the maximum possible interval between treatment and the next period of bee activity.

PHYSICAL OR CHEMICAL HAZARDS

Do not use, pour, spill, or store near heat or open flame.

SUPRACIDE® trademark of CIBA-GEIGY for methidathion
Cyprex® trademark of American Cyanamid for dodine

CGA 30L23F 059

This is a specimen label which was not prepared by CIBA-GEIGY. CIBA-GEIGY is not responsible for the accuracy of the information contained herein. As labels are subject to revision, always carefully read and follow the label on the product container.

SWAT™
INSECTICIDE

RESTRICTED USE PESTICIDE

FOR RETAIL SALE TO AND USE ONLY BY CERTIFIED APPLICATORS OR PERSONS UNDER THEIR DIRECT SUPERVISION, AND ONLY FOR THOSE USES COVERED BY THE CERTIFIED APPLICATOR'S CERTIFICATION.

For Control of Insects in Cotton	By Wt.
ACTIVE INGREDIENTS	78.0%
*Phosphamidon	75.5%
Related Compounds	2.5%
INERT INGREDIENTS:	22.0%
TOTAL	100.0%

*2-chloro-2-diethylcarbamoyl-1-methylvinyl dimethyl phosphate

Contains 8 lbs. Phosphamidon insecticide per gallon. U.S. Patent No. 2,908,605. SWAT is a liquid systemic insecticide for the control of both sucking and chewing insects. It may be applied as a dilute spray or as a concentrate spray.

EPA Reg. No. 100-668 EPA Est. 100-AL-1

KEEP OUT OF REACH OF CHILDREN
DANGER/PELIGRO

DANGER–POISON

PRECAUCIÓN AL USUARIO: Si usted no lee inglés, no use este producto hasta que la etiqueta haya sido explicado ampliamente.

STATEMENT OF PRACTICAL TREATMENT

IF SWALLOWED: Call a physician immediately. Drink 1 or 2 glasses of water and induce vomiting by touching back of throat with finger. Repeat until vomit fluid is clear. Do not induce vomiting or give anything by mouth to an unconscious person. Have victim lie down and keep quiet.

IF INHALED: Remove victim to fresh air. Apply artificial respiration if indicated. Get medical attention.

IF IN EYES: Immediately flush eyes with plenty of clear water. Get medical attention if irritation persists.

IF ON SKIN: Remove contaminated clothing and immediately wash skin with soap and water.

NOTE TO PHYSICIAN: Atropine is antidotal. 2-PAM is also antidotal and may be administered in conjunction with Atropine.

DO NOT STORE OR USE AROUND THE HOME

SEE SIDE PANEL FOR ADDITIONAL PRECAUTIONS

DIRECTIONS FOR USE AND CONDITIONS OF SALE AND WARRANTY

IMPORTANT: Read the entire **Directions for Use** and the **Conditions of Sale and Warranty** before using this product.

CONDITIONS OF SALE AND WARRANTY

The **Directions for Use** of this product reflect the opinion of experts based on field use and tests. The directions are believed to be reliable and should be followed carefully. However, it is impossible to eliminate all risks inherently associated with use of this product. Crop injury, ineffectiveness, or other unintended consequences may result because of such factors as weather conditions, presence of other materials, or the manner of use or application all of which are beyond the control of CIBA-GEIGY or the Seller. All such risks shall be assumed by the Buyer.

CIBA-GEIGY warrants that this product conforms to the chemical description on the label and is reasonably fit for the purposes referred to in the **Directions for Use** subject to the inherent risks referred to above. **CIBA-GEIGY makes no other express or implied warranty of Fitness or Merchantability or any other express or implied warranty. In no case shall CIBA-GEIGY or the Seller be liable for consequential, special, or indirect damages resulting from the use or handling of this product.** CIBA-GEIGY and the Seller offer this product, and the Buyer and user accept it, subject to the foregoing **Conditions of Sale and Warranty**, which may be varied only by agreement in writing signed by a duly authorized representative of CIBA-GEIGY.

DIRECTIONS FOR USE

It is a violation of Federal Law to use this product in a manner inconsistent with its labeling.

FAILURE TO FOLLOW THE DIRECTIONS FOR USE AND PRECAUTIONS ON THIS LABEL MAY RESULT IN CROP INJURY, POOR PERFORMANCE, AND/OR ILLEGAL RESIDUES.

REENTRY STATEMENT

Do not enter treated areas without protective clothing until sprays have dried. Because certain states may require a more restrictive reentry interval for a crop treated with this product, consult your State Department of Agriculture for further information.

Written or oral warnings must be given to workers who are expected to be in a treated area or in an area about to be treated with this product. When oral warnings are given, warnings shall be given in a language customarily understood by workers. Oral warnings must be given if there is reason to believe that written warnings cannot be understood by workers. Written and oral warnings must include the following information: DANGER-Area treated with SWAT on (date of application). Do not enter without appropriate protective clothing until sprays have dried. In case of accidental exposure, follow instructions under Statement of Practical Treatment section.

GENERAL INFORMATION

SWAT is a liquid systemic insecticide.

Chemigation: Do not apply this product through any type of irrigation system.

To avoid spray drift, do not apply under windy conditions. Avoid spray overlap since illegal residues or crop injury may occur.

Cotton: Early Season (2 through 10 leaf stage)—For the control of aphids, thrips (excluding Western flower thrips), and fleahoppers, apply 2 or 3 applications of 3 fluid ounces per acre with ground equipment. Spray directly on young plants in sufficient water to completely cover, beginning when insects first appear.

Mid and Late Season: For the control of lygus (plant bugs), and leafhoppers, apply 8 fluid ounces per acre. For aphid control, apply 3 to 4 fluid ounces per acre. Use the higher rate when infestation is heavy. Applications may be made by ground in a minimum of 5 gallons of water per acre, or by air in a minimum of 1 gallon of water per acre. Begin application as insects appear, and repeat as necessary. Do not apply within 2 weeks of harvest.

STORAGE AND DISPOSAL

Flammable. Keep away from heat, open flame, sparks, or hot surfaces. Keep container closed. Store in a cool, dry place. (For more information, see Physical and Chemical Hazards section).

Do not contaminate water, food or feed by storage or disposal.

Pesticide Disposal: Pesticide wastes are acutely hazardous. Improper disposal of excess pesticide, spray mixture or rinsate is a violation of Federal Law. If these wastes cannot be disposed of by use according to label instructions, contact your State Pesticide or Environmental Control Agency, or the Hazardous Waste representative at the nearest EPA Regional Office for guidance.

Container Disposal: Triple rinse (or equivalent). Then offer for recycling or reconditioning, or puncture and dispose of in a sanitary landfill, or by other procedures approved by State and local authorities.

PRECAUTIONARY STATEMENTS

HAZARDS TO HUMANS AND DOMESTIC ANIMALS

DANGER

Poisonous if swallowed, inhaled or absorbed through skin. Wear a pesticide respirator jointly approved by the Mining Enforcement and Safety Administration and by the National Institute for Occupational Safety and Health under the provisions of 30 CFR Part 11. Rapidly absorbed through the skin. Causes eye irritation. Do not get in eyes, on skin or clothing. Wear natural rubber gloves, protective clothing and goggles. Wash hands, arms and face thoroughly with soap and water before eating or smoking. Wash all contaminated clothing with soap and water before reuse. Do not contaminate food or feed. Do not apply this product in such a manner as to directly or through drift expose workers or other persons. The area being treated must be vacated by unprotected persons.

ENVIRONMENTAL HAZARDS

This product is toxic to fish, birds, and other wildlife. Birds feeding on treated areas may be killed. Do not apply directly to water. Do not apply when weather conditions favor drift from areas treated. Do not apply where runoff is likely to occur. Do not contaminate water by cleaning of equipment or disposal of wastes. Apply this product only as specified on this label. This product is highly toxic to bees exposed to direct treatment or residues on crops. Protective information may be obtained from your Cooperative Agricultural Extension Service.

PHYSICAL OR CHEMICAL HAZARDS

Flammable. Keep away from heat, open flame, sparks, or hot surfaces. Use only in well-ventilated area. Keep container closed. Clean up spills immediately. This product presents a fire hazard. Liquid quickly evaporates and forms vapor (fumes) which can catch fire and burn with explosive violence. Invisible vapor spreads easily and can be set on fire by many sources such as pilot lights, welding equipment and electrical motors and switches. Fire hazard is greater as liquid temperatures rise above 76°.

Do not heat the container. Replace the cap or bung. Emptied container still contains hazardous or explosive vapor or liquid.

SWAT™ trademark of CIBA-GEIGY for phosphamidon U.S. Patent No. 2,908,605

CGA 39L6C 067

This is a specimen label which was not prepared by CIBA-GEIGY. CIBA-GEIGY is not responsible for the accuracy of the information contained herein. As labels are subject to revision, always carefully read and follow the label on the product container.

TILT®

FUNGICIDE

For control of certain diseases in cereals, grasses grown for seed, and sugarcane

ACTIVE INGREDIENT:	
1-[[2-(2,4-dichlorophenyl)-4-propyl-1,3-dioxolan-2-yl]methyl]-1*H*-1,2,4-triazole	41.8%
INERT INGREDIENTS:	58.2%
TOTAL	100.0%

EPA Reg. No. 100-617 EPA Est. 100-AL-1

KEEP OUT OF REACH OF CHILDREN.

WARNING

See additional precautionary statements and directions for use inside booklet.

TILT® trademark of CIBA-GEIGY

 038

DIRECTIONS FOR USE AND CONDITIONS OF SALE AND WARRANTY

IMPORTANT: Read the entire **Directions for Use** and the **Conditions of Sale and Warranty** before using this product.

CONDITIONS OF SALE AND WARRANTY

The **Directions for Use** of this product reflect the opinion of experts based on field use and tests. The directions are believed to be reliable and should be followed carefully. However, it is impossible to eliminate all risks inherently associated with use of this product. Crop injury, ineffectiveness, or other unintended consequences may result because of such factors as weather conditions, presence of other materials, or the manner of use or application all of which are beyond the control of CIBA-GEIGY or the Seller. All such risks shall be assumed by the Buyer.

CIBA-GEIGY warrants that this product conforms to the chemical description on the label and is reasonably fit for the purposes referred to in the **Directions for Use** subject to the inherent risks referred to above. **CIBA-GEIGY makes no other express or implied warranty of Fitness or Merchantability or any other express or implied warranty. In no case shall CIBA-GEIGY or the Seller by liable for consequential, special, or indirect damages resulting from the use or handling of this product.** CIBA-GEIGY and the Seller offer this product, and the Buyer and user accept it, subject to the foregoing **Conditions of Sale and Warranty,** which may be varied only by agreement in writing signed by a duly authorized representative of CIBA-GEIGY.

DIRECTIONS FOR USE

It is a violation of federal law to use this product in a manner inconsistent with its labeling.

GENERAL INFORMATION

TILT is a broad spectrum fungicide for the control of certain diseases in wheat, barley, rye, rice, sugarcane, pecans, and in grasses grown for seed.

FAILURE TO FOLLOW DIRECTIONS AND PRECAUTIONS ON THIS LABEL MAY RESULT IN CROP INJURY, POOR DISEASE CONTROL, OR ILLEGAL RESIDUES.

Spray Equipment

Thorough coverage is necessary to provide good disease control.

To avoid spray drift, do not apply when conditions favor drift beyond the target area. Avoid spray overlap, as crop injury may occur.

Equip sprayers with nozzles that provide accurate and uniform application. Be certain that nozzles are the same size and uniformly spaced across the boom. Calibrate sprayer before use.

Use a pump with capacity to (1) maintain 35-40 psi at nozzles, and (2) provide sufficient agitation in tank to keep mixture in suspension—this requires recirculation of 10 percent of tank volume per minute. Use a jet agitator or liquid sparge tube for agitation. Do not air sparge.

Although TILT is an emulsifiable concentrate, it is suggested screens be used to protect the pump and to prevent nozzles from clogging. Screens placed on suction side of pump should be *16-mesh or coarser.* Do not place a screen in the recirculation line. Use 50-mesh or coarser screens between the pump and boom, and where required, at the nozzles. Check nozzle manufacturer's recommendations.

For more information on spray equipment and calibration, consult sprayer manufacturers and state recommendations. For specific local directions on spray schedules, consult the current state agricultural experiment station recommendations.

To determine the physical compatibility of TILT with other products, use a jar test. Using a quart jar, add the proportionate amounts of the products to 1 qt. of water. Add wettable powders and water dispersible granular products first, then liquid flowables, and emulsifiable concentrates last. After thoroughly mixing, let stand for at least 5 minutes. If the combination remains mixed or can be remixed readily, it is physically compatible. Once compatibility has been proven, use the same procedure for adding required ingredients to the spray tank.

Aerial Application: For those crops where aerial applications are indicated, apply in a minimum of 5 gals. of water/A unless specified otherwise. Avoid applications under conditions when uniform coverage cannot be obtained or when excessive spray drift may occur. Do not apply directly to humans or animals. Flagmen and loaders should wash thoroughly before eating, drinking, or smoking, and at the end of each day's operation.

Ground Application: Apply TILT by ground equipment in a minimum of 15 gals. of water/A

Chemigation: Do not apply this product through any type of irrigation system.

Mixing Procedures

(1) Wear rubber gloves and long-sleeved shirt when mixing and loading this product. (2) Be sure sprayer is clean; if it is contaminated with any material, crop injury or sprayer clogging may result. (3) Fill tank 1/4-1/2 full with clean water, and start agitation. (4) Be certain that the agitation system is working properly and creates a rippling or rolling action on the liquid surface. (5) Add product to the tank. For wheat, barley, rye, and rice, custom applicators must use a closed system for mixing and loading this product. (6) Complete filling tank, maintaining sufficient agitation at all times to ensure surface action. This applies to both the spray and nurse tanks. (7) Do not leave tank mixtures in the tank without adequate agitation. Always maintain agitation to avoid separation and buildup of residues. If a residue starts to build up in tank, drain and clean with strong detergent and ammonia solution, and rinse thoroughly with clean water. (8) Immediately after use, clean sprayer thoroughly by flushing system with water containing a detergent and rinse with clean water.

Rotational Crops

To avoid possible illegal residues, (1) Do not double-crop treated acreage when TILT is applied to the first crop unless the second crop appears on this label. (2) Do not graze or feed forage, fodder, or straw from rotational crops planted in the fall or the spring following treatment.

CEREALS

Wheat, Barley, Rye

In wheat, barley, and rye, TILT controls rusts (*Puccinia* spp.), powdery mildew (*Erysiphe* spp.), leaf blight and glume blotch (*Septoria* spp.), tan spot (*Pyrenophora tritici-repentis*), Helminthosporium leaf blight, barley scald (*Rhynchosporium secalis*) and net blotch (*Pyrenophora teres*). Apply 4 fl. oz. of TILT/A by ground or aerial application. Highest yields are normally obtained when TILT is applied to the emerging flag leaf (Feekes growth stage 8). Do not apply after this growth stage to avoid possible illegal residues. TILT may be applied earlier if disease symptoms appear (especially applicable to barley). Make only one application per season.

For control of foot rot (*Pseudocercosporella* spp.), apply 4 fl. oz. of TILT/A plus half rates of other EPA-registered fungicides such as Benlate®, Mertect®, and Topsin® M. Apply at tillering but before elongation has occurred.

Do not graze or feed livestock treated forage or cut the green crop for hay or silage. After harvest the straw may be used for bedding.

Rice

In rice, TILT controls sheath blight (*Rhizoctonia solani*), brown leaf spot (*Helminthosporium oryzae*), narrow brown leaf spot and brown blotch (*Cereospora oryzae*), leaf smut (*Entyloma oryzae*) and sheath spot (*Rhizoctonia oryzae*). TILT also suppresses stem rot (*Sclerotium oryzae*). Apply TILT on either of the following schedules as an aerial spray in 5-10 gals. of water/A.

A. 6 fl. oz./A at first internode elongation (up to 2-inch panicle) and repeat at swollen boot. Make the second application 10-14 days after the first application, but before the boot splits and head emerges. TILT provides best control of sheath blight when the first application is applied at disease appearance in the field. The first application is recommended when 5 percent or fewer of the tillers are infected.

B. 10 fl. oz./A at first internode elongation (up to 2-inch panicle). The 10 oz. rate is recommended if greater than 10 percent of the tillers are infected with sheath blight. If disease reappears, use another registered fungicide for the second application.

Note: To avoid possible illegal residues (1) Do not apply to stubble or ratoon crop rice; (2) Do not use in rice fields where commercial farming of crayfish will be practiced; (3) Do not drain water from treated rice fields into ponds used for commercial catfish farming; (4) Do not use water drained from treated fields to irrigate other crops; (5) Do not use in California.

Endangered Species Restrictions: The use of TILT (propiconazole) on rice is restricted to protect the endangered fat pocketbook pearly mussel (*Potamilus capax*) and its habitat. Use is prohibited in the following areas of Arkansas:

Mississippi Country—Within the basin that drains directly into the Right Hand Chute of Little River, south of Big Lake National Wildlife Refuge.

Poinsett County—Between Crowley's Ridge and the levee east of the Right Hand Chute of Little River and the St. Francis Floodway. Use is also prohibited west of Rt. 140 and north of Rt. 63 at the siphon near Marked Tree. Except that the prohibited area does not include the area bounded by Arkansas highway 373 on the west, highway 63 on the east and highway 14 on the south.

Cross, St. Francis, and Lee Counties: Between Crowley's Ridge and the levee east of the Right Hand Chute of Little River and the St. Francis Floodway as far south as the confluence of L'Anguille River (Lee County).

Grasses Grown for Seed (Nebraska, Oregon, Washington, Idaho, and Minnesota only)

TILT controls rust (*Puccinia* spp.), powdery mildew, and Selenophoma stem eyespot in perennial ryegrass, fescues, bluegrass, orchardgrass, and/or wheatgrasses.

Mix 4 to 8 oz. of TILT (maximum 4 fl. oz. on bluegrass) in a minimum of 20 gals. of water per acre for ground equipment, or in a minimum of 10 gals. of water per acre for aerial application. Apply when powdery mildew and Selenophoma infections or rust pustules are noticeable and increasing in number in late spring or early summer. Repeat at 14 to 21 day intervals. Under severe rust pressure, use the higher rate of 8 fl. oz. (except on bluegrass), and/or make more frequent applications to maintain rust control until the seed is mature. Make the last application about two weeks before seed matures. For bluegrass, it is important to begin applications in early season.

Note: To avoid possible illegal residues do not:

(1) feed hay cut within 20 days of the last application or (2) graze treated areas within 140 daysof the last application.

Sugarcane (Hawaii only)

For control of pineapple disease of sugarcane (*Ceratocystis paradoxa*) apply TILT to cut seed pieces using 0.75 fl. oz. (22 ml) per 100 gals. water (1:17,000). Treatments can be applied in either a cold or hot water dip. Do not use treated seed pieces for food or feed purposes. Dispose of spent dip solution according to state and federal regulations.

Cold Water Dip—Immerse seed pieces to give thorough wetting, remove and allow to drain.

Hot Water Dip—Maintain water temperature at 52°C. Soak the seed pieces for 20-30 minutes, remove and allow to drain.

STORAGE AND DISPOSAL

Store at temperatures above 0° F. Do not contaminate water, food or feed by storage or disposal. Pesticide wastes may be toxic. Improper disposal of unused pesticide, spray mixture, or rinse water is a violation of federal law. If these wastes cannot be used according to label instructions, contact your State Pesticide or Environmental Control Agency, or the Hazardous Waste representative at the nearest EPA Regional Office for guidance in proper disposal methods.

Triple rinse container (or equivalent). Then offer for recycling or reconditioning, or puncture and dispose of in a sanitary landfill, by incineration, or, if allowed by state and local authorities, by burning. Stay out of smoke from burning containers.

For minor spills, leaks, etc., follow all precautions indicated on this label and clean up immediately. Take special care to avoid contamination of equipment and facilities during cleanup and disposal of wastes. In the event of a major spill, fire, or other emergency, call (919) 292-7100 day or night.

PRECAUTIONARY STATEMENTS

HAZARDS TO HUMANS AND DOMESTIC ANIMALS

WARNING

This product may cause skin sensitization reactions in certain individuals. Wear rubber gloves and long-sleeved shirt when mixing or handling this product. Remove contaminated clothing and wash before reuse.

Causes eye irritation. Do not get in eyes, on skin, or on clothing. Harmful if swallowed, inhaled, or absorbed through skin. Wash thoroughly with soap and water after handling. Avoid breathing spray mist.

First Aid: If in eyes, flush with plenty of water and get medical attention if irritation persists. If on skin, wash with plenty of soap and water. Get medical attention if irritation persists. If swallowed, drink promptly a large quantity of milk, egg whites, or gelatin solution; if these are not available, drink large quantities of water. Never give anything by mouth to an unconscious person. Do not induce vomiting. Call a physician or Poison Control Center immediately. In case of inhalation exposure, move to fresh air.

Note to Physician: There is no specific antidote for this product. Induce emesis or lavage stomach, taking care to avoid aspiration of stomach contents into lungs. Give a saline laxative and supportive therapy.

ENVIRONMENTAL HAZARDS

This pesticide is toxic to fish. Do not apply directly to water. Do not contaminate water by cleaning of equipment or disposal of wastes. Do not contaminate water when disposing of equipment wash waters. Refer to product labeling for use restrictions to protect endangered species.

PHYSICAL OR CHEMICAL HAZARDS

Do not use, pour, spill or store near heat or open flame.

Store at temperatures above 0°F.

Do not apply this product through any type of irrigation system.

Benlate® trademark of E. I. duPont de Nemours & Co. (Inc.) for benomyl

Mertect® trademark of Merck and Co., Inc. for thiabendazole

TILT® trademark of CIBA-GEIGY

U.S. Patent No. 4,079,062

Topsin® trademark of Pennwalt Corporation for thiophanate-methyl

CGA 72L1K 069

This is a specimen label which was not prepared by CIBA-GEIGY. CIBA-GEIGY is not responsible for the accuracy of the information contained herein. As labels are subject to revision, always carefully read and follow the label on the product container.

TRIGARD® 75W

INSECTICIDE

For leafminer control in celery and lettuce

For aerial application only

ACTIVE INGREDIENT:	
Cyromazine: N-cyclopropyl-1,3,5-triazine-2,4,6-triamine	75%
INERT INGREDIENTS:	25%
TOTAL	100%

EPA Reg. No. 100-654

This outer protective bag contains TRIGARD 75W in an inner water-soluble bag. Entire inner bag and contents dissolve in water. After opening outer bag, immediately dump *entire* unopened inner bag into the partially filled sprayer or mix tank. Do not handle the soluble bag or expose it to moisture, since this may cause rupturing.

TRIGARD 75W is a wettable powder.

KEEP OUT OF REACH OF CHILDREN

CAUTION

See additional precautionary statements on back of bag.

See directions for use on back of bag.

DIRECTIONS FOR USE AND CONDITIONS OF SALE AND WARRANTY

IMPORTANT: Read the entire **Directions for Use** and the **Conditions of Sale and Warranty** before using this product.

CONDITIONS OF SALE AND WARRANTY

The **Directions for Use** of this product reflect the opinion of experts based on field use and tests. The directions are believed to be reliable and should be followed carefully. However, it is impossible to eliminate all risks inherently associated with use of this product. Crop injury, ineffectiveness, or other unintended consequences may result because of such factors as weather conditions, presence of other materials, or the manner of use or application all of which are beyond the control of CIBA-GEIGY or the Seller. All such risks shall be assumed by the Buyer.

CIBA-GEIGY warrants that this product conforms to the chemical description on the label and is reasonably fit for the purposes referred to in the **Directions for Use** subject to the inherent risks referred to above. **CIBA-GEIGY makes no other express or implied warranty of Fitness or Merchantability or any other express or implied warranty. In no case shall CIBA-GEIGY or the Seller be liable for consequential, special, or indirect damages resulting from the use or handling of this product.** CIBA-GEIGY and the Seller offer this product, and the Buyer and user accept it, subject to the foregoing **Conditions of Sale and Warranty,** which may be varied only by agreement in writing signed by a duly authorized representative of CIBA-GEIGY.

DIRECTIONS FOR USE

It is a violation of federal law to use this product in a manner inconsistent with its labeling.

FAILURE TO FOLLOW THE DIRECTIONS FOR USE AND PRECAUTIONS ON THIS LABEL MAY RESULT IN CROP INJURY AND/OR POOR INSECT CONTROL.

TRIGARD is a 75% wettable powder which, when mixed with water according to the directions for use given below, will control certain insect species on designated plants.

Do not apply this product in such a manner as to directly or through drift expose workers or other persons, except those knowingly involved in the application. The area being treated must be vacated by unprotected persons.

REENTRY STATEMENT

Do not enter treated areas without protective clothing until seven days after last application.

Because certain states may require more restrictive reentry intervals for various crops treated with this product, consult your State Department of Agriculture for further information.

Written or oral warnings must be given to workers who are expected to be in a treated area or in an area about to be treated with this product. Oral warnings must be given which inform workers of areas or fields that may not be entered without specific protective clothing until seven days after last application and appropriate actions to take in case of accidental exposure as described under **Precautionary Statements** on this label. When oral warnings are given, warnings shall be given in a language customarily understood by workers. Oral warnings must be given if there is reason to believe that written warnings cannot be understood by workers. Written warnings must include the following information: "CAUTION. Area treated with TRIGARD on (date of application). Do not enter without appropriate protective clothing until seven days after last application. In case of accidental exposure, flush eyes or skin with plenty of water. Call a physician if irritation persists. Remove and wash contaminated clothing before reuse."

Mixing Instructions

Prepare no more spray mixture than is needed for the immediate operation. Thoroughly clean spray equipment before using this product. Agitate thoroughly before and during application. Good bypass agitation is sufficient to keep TRIGARD thoroughly mixed. Flush spray tank thoroughly with water daily after use and dispose of pesticide rinsate by application to an already treated area.

TRIGARD Alone: Add ²/₃ of the required amount of water to the spray or mixing tank. With the agitator running, drop the required number of **unopened** soluble bags of TRIGARD into the tank. (One 2-lb. water-soluble bag of TRIGARD will treat 12 acres of celery or lettuce.) Continue agitation while adding the remainder of the water.

Note: Do not let spray mixture stand overnight in the spray tank. Rinse spray tank at the end of the day.

To avoid spray drift, do not apply under windy conditions. Avoid spray overlap, as crop injury may result.

Chemigation: Do not apply this product through any type of irrigation system.

CELERY AND LETTUCE (HEAD ONLY)

For leafminer control, apply ¹/₆ lb. TRIGARD per acre as a foliar spray in a minimum of 5 gals. of water by air as leafminers first appear. Repeat applications at 7-day intervals or as necessary to maintain control.

Notes: (1) Do not make more than 12 applications to one crop of celery or 8 applications to one crop of lettuce, and (2) Do not make the last application within 7 days of harvest, or illegal residues may result.

Rotational Crops

All crops on this label may be planted following harvest of a TRIGARD-treated crop. Do not rotate to any other crop, except sweet corn or radishes. Do not plant sweet corn or radishes within 3 months after last application.

STORAGE AND DISPOSAL

Store in a cool, dry place. Do not contaminate water, food, or feed by storage or disposal. Do not use in animal feeds.

Storage—Do not store this product under wet conditions. Handle outer bag carefully when stored at temperatures less than 50°F to avoid breakage of inner soluble bag.

Pesticide Disposal—Improper disposal of unused pesticide, spray mixture, or rinsate is a violation of federal law. Pesticide, spray mixture, or rinsate that cannot be used according to label instructions, must be disposed of according to federal, state or local procedures. For guidance in proper disposal methods, contact your State Pesticide or Environmental Control Agency, or the Hazardous Waste representative at the nearest EPA Regional Office.

Container Disposal—Do not reuse outer bag. Dispose of empty outer bag in a sanitary landfill or by incineration or, if allowed by state and local authorities, by burning. If burned, stay out of smoke.

For minor spills, leaks, etc., follow all precautions indicated on this label and clean up immediately. Take special care to avoid contamination of equipment and facilities during cleanup procedures and disposal of wastes. In the event of a major spill, fire or other emergency, call (919) 292-7100 day or night.

PRECAUTIONARY STATEMENTS

HAZARDS TO HUMANS AND DOMESTIC ANIMALS

CAUTION

Causes moderate eye irritation. Harmful if swallowed, absorbed through the skin or inhaled. Avoid contact with eyes, skin or clothing. Wear long-sleeved shirt and pants when handling or applying this product. Avoid breathing dust. Wash thoroughly with soap and water after handling. Remove contaminated clothing and wash before reuse.

STATEMENT OF PRACTICAL TREATMENT

If swallowed: Call a physician or Poison Control Center. Drink 1 or 2 glasses of water and induce vomiting by touching back of throat with finger. Do not induce vomiting or give anything by mouth to an unconscious person.

If on skin: Wash with plenty of soap and water. If irritation develops or persists, get medical attention.

If in eyes: Flush with plenty of water. If irritation develops or persists, get medical attention.

If inhaled: Move to fresh air.

ENVIRONMENTAL HAZARDS

Do not apply directly to water or wetlands. Do not contaminate water by cleaning of equipment or disposal of wastes. Do not apply when weather conditions favor drift from target area.

TRIGARD® trademark of CIBA-GEIGY

CGA 69L30B 067

This is a specimen label which was not prepared by CIBA-GEIGY. CIBA-GEIGY is not responsible for the accuracy of the information contained herein. As labels are subject to revision, always carefully read and follow the label on the product container.

The Dow Chemical Company

Agricultural Products Department
P.O. Box 1706
Midland, MI 48640
1 (800) 258-2436

CONFRONT*

HERBICIDE

For the Control of Annual and Perennial Broadleaf Weeds in Ornamental Turf

To be used only under the direct supervision of commercial applicators responsible for turf weed control programs.

Active Ingredient(s):

Triclopyr (3,5,6-trichloro-2-pyridinyloxyacetic acid), as the triethylamine salt 33.0%

Clopyralid (3,6-dichloro-2-pyridinecarboxylic acid), as the triethylamine salt 12.1%

Inert Ingredient(s): 54.9%

ACID EQUIVALENT:

Triclopyr 23.7% - 2.25 lb/gal

Clopyralid 7.9% - 0.75 lb/gal

E.P.A. Registration No. 464-644

E.P.A. Est. 464-M-1

KEEP OUT OF REACH OF CHILDREN

DANGER

PELIGRO:

PRECAUCION AL USUARIO: Si usted no lee ingles, no use este producto hasta que la etiqueta le haya sido explicada ampliamente.

PRECAUTIONARY STATEMENTS

HAZARDS TO HUMANS AND DOMESTIC ANIMALS

CAUSES IRREVERSIBLE EYE DAMAGE • MAY CAUSE ALLERGIC SKIN REACTION (CONCENTRATE PRODUCT ONLY) • HARMFUL IF ABSORBED THROUGH SKIN, OR IF SWALLOWED OR INHALED

Do Not Get In Eyes, On Skin, Or Clothing • Wear Goggles Or Face Shield And Rubber Gloves When Handling • Avoid Breathing Vapors • Wash Thoroughly With Soap And Water After Handling And Before Eating Or Smoking • Remove Contaminated Clothing And Wash Before Reuse

Note To Applicator: In The Dilute Mix, No Skin Reaction Is Expected When Used As Directed

STATEMENT OF PRACTICAL TREATMENT

In case of contact, immediately flush eyes with plenty of water for at least 15 minutes and get medical attention promptly. Flush skin with water and get medical attention if signs of irritation develop.

If swallowed, promptly drink a large quantity of milk, egg whites, gelatin solution or, if these are not available, drink large quantities of water.

Avoid alcohol. Call a physician.

PHYSICAL OR CHEMICAL HAZARDS

Do Not Cut or Weld Container.

ENVIRONMENTAL HAZARDS

Keep out of lakes, streams or ponds. Do not contaminate water when disposing of equipment washwaters.

NOTICE

Read and understand the entire label before using. Use only according to label directions.

Before buying or using this product, read "WARRANTY LIMITATIONS AND DISCLAIMER" elsewhere on this label. If terms are not acceptable, return **unopened** package at once to seller for full refund of purchase price paid. Otherwise, use by the buyer or any other user constitutes acceptance of the terms under WARRANTY LIMITATIONS AND DISCLAIMER.

IN CASE OF AN EMERGENCY endangering life or property involving this product, call collect 517-636-4400

AGRICULTURAL CHEMICAL

Do Not Ship or Store with Food, Feeds, Drugs, or Clothing

See Side Panel for Important Use Precautions

GENERAL INFORMATION

CONFRONT Herbicide is a broad-spectrum non-phenoxy weed killer for control of broadleaf weeds in ornamental turf.

DIRECTIONS FOR USE

It is a violation of Federal law to use this product in a manner inconsistent with its labeling.

Do not apply this product through any type of irrigation system.

Do no use for manufacturing or formulating.

Weeds Controlled and Use Rate Recommendations

	Suggested Use Rate		
Weeds	**pt/A**	**fl.oz./1000 sq. ft.**	**tsp/1000 sq.ft.**
Black medic Red clover Hop clover White clover	1†	0.37 (11 ml)	2½
Common Cocklebur Common Vetch Creeping beggarweed Dwarf beggarweed False Dandelion Henbit Matchweed Spotted Catsear Spurweed	1.5	0.55 (16 ml)	3½
Broadleaf Plantain Burdock Coffeeweed Common Dandelion Common Ragweed Ground Ivy Lambsquarters Narrowleaf (Buckhorn) Plantain Shepherd's Purse Virginia pepperweed	1.5-2	0.55 to 0.74 (16 to 22 ml)	3½to 4½
Canada Thistle Curly Dock Germander Speedwell Goldenrod Lespedeza Musk Thistle Poison Ivy Smartweed Wild Buckwheat	2	0.74 (22 ml)	4½
(REPEAT TREATMENT MAY BE NECESSARY) Common Yellow Woodsorrel Creeping Woodsorrel Wild Violet	2	0.74 (22 ml)	4½

† For faster activity under good growing conditions, 1.5 pt/acre is recommended. For extended weed control, repeat applications are recommended.

Preparing the Spray

Add about one-half the desired amount of clean water to spray tank. Add CONFRONT and complete addition of water with agitation running. Mix thoroughly and continue moderate agitation while spraying.

Application Directions

Broad Treatment of Ornamental Turf: Foliage sprays should be applied during warm weather when weeds are actively growing. Application under drought conditions may provide less than desirable results. Use low spray pressures to minimize spray drift. (See Use Precautions) When hard-to-control species are prevalent and during applications made in late summer when the plants are mature and during drought conditions, use the higher rates.

Apply CONFRONT in a manner to avoid contacting nearby susceptible crops or other desirable plants. Very small amounts of spray drift may injure susceptible plants, including ornamental trees or shrubs. DO NOT apply to exposed roots of shallow rooted trees or shrubs.

Apply 1 to 2 pints of CONFRONT in enough water to make 35 to 175 gallons total spray per acre to control broadleaf weeds growing in perennial bluegrass, tall fescue, or perennial ryegrass. DO NOT use on other grass species, such as bentgrass or St. Augustine grass, unless injury can be tolerated. Apply from early spring through fall when weeds are actively growing. Broadleaf weed species germinate at different times. Only emerged weeds present at time of application are controlled. A maximum of 2 pt/acre per application is recommended. Do not use more than 5 pints per acre per year of treatment. Additional applications should not be made less than four weeks apart to minimize grass injury. Newly seeded turf should be mowed two or three times before being treated. Do not water for 24 hours after application. Do not reseed for 3 weeks after application.

Make application using equipment that will insure uniform coverage of the spray volumes applied. To improve spray coverage, the addition of an agricultural surfactant (such as ORTHO X-77 or TRONIC) at a rate of 1/4 to 1 pint per acre is suggested. Use the higher rates of surfactant for lower rates of product and lower spray volumes.

Spot Treatment of Ornamental Turf Using Portable Sprayers: Mix 0.5 to 1.0 fluid ounces of CONFRONT in enough water to make 3 gallons of spray (1 to 2 pints per 100 gallons of spray) and apply at any time broadleaf weeds are susceptible by wetting foliage of undesirable plants to point of runoff.

DO NOT apply to exposed roots of shallow rooted trees or shrubs.

USE PRECAUTIONS

Apply this product only as specified on this label.

Do not apply CONFRONT directly to, or otherwise permit it to come into direct contact with, cotton, grapes, tobacco, vegetable crops, flowers, fruit or ornamental trees, or other desirable broadleaf plants and do not permit spray mists containing it to drift onto them.

DO NOT apply to exposed roots of shallow rooted trees and shrubs.

Avoid injurious Spray Drift. Applications should be made only when hazards from spray drift are at a minimum. Very small quantities of spray, which may not be visible, may seriously injure susceptible plants. Do not spray when wind is blowing toward susceptible crops or ornamental plants near enough to be injured.

With ground broadcast equipment, drift can be reduced by keeping the spray boom as low as possible; by applying no less than 35 gallons of spray per acre; by using no more than 20 pounds spraying pressure with flat fan or flooding flat fan nozzle tips; and by spraying when wind velocity is low.

In handgun applications drift can be reduced by using low pressure Teejet flat fan, Raindrop or other drift controlling nozzles. Standard flat fan nozzles may also be used at no more than 20 psig. At this pressure the spray pattern may be narrower.

Do not apply on ditches used to transport irrigation water. Do not apply where runoff or irrigation water may flow onto susceptible crops as injury may result.

PRECAUTIONARY STATEMENTS

ENVIRONMENTAL HAZARDS

Clopyralid is a chemical which can travel (seep or leach) through soil and under certain conditions contaminate groundwater which may be used for irrigation or drinking purposes. Users are advised not to apply clopyralid where soils have a rapid to very rapid permeability throughout the profile (such as loamy sand to sand) and the water table of an underlying aquifer is shallow, or to soils containing sinkholes over limestone bedrock, severely fractured surfaces, and substrates which would allow direct introduction into an acquifer. Your local agricultural agencies can provide further information on the type of soil in your area and the location of groundwater.

STORAGE AND DISPOSAL

Do not contaminate water, food or feed by storage or disposal.

STORAGE: Store above 28 °F or agitate before use.

DISPOSAL:

Pesticide Disposal: Pesticide wastes are toxic. Improper disposal of excess pesticide, spray mixture, or rinstate is a violation of Federal Law. If these wastes cannot be disposed of by use according to label instructions, contact your State Pesticide or Environmental Control Agency, or the Hazardous Waste representative at the nearest EPA Regional Office for guidance.

Container Disposal: Triple rinse (or equivalent). Then offer for recycling or reconditioning, or puncture and dispose of in a sanitary landfill, or by other precedures approved by state and local authorities. Plastic containers, after triple rinsing, may be incinerated if allowed by state and local authorities. If burned, stay out of smoke.

General: Consult federal, state, or local disposal authorities for approved alternative procedures.

Be sure that use of this product conforms to all applicable regulations.

WARRANTY LIMITATIONS AND DISCLAIMER

The Dow Chemical Company warrants that this product conforms to the chemical description on the label and is reasonably fit for the purposes stated on the label when used in strict accordance with the directions therein under normal conditions of use. THIS IS THE ONLY WARRANTY MADE ON THIS PRODUCT. NO OTHER EXPRESS AND NO IMPLIED WARRANTY OF MERCHANTABILITY OR FITNESS FOR A PARTICULAR PURPOSE IS MADE OUTSIDE OF THIS LABEL. Therefore, neither this warranty nor any other warranty of merchantability or fitness for a particular purpose, express or implied, extends to the use of this product contrary to label instructions (including conditions noted on the label, such as unfavorable temperatures, soil conditions, etc.), under abnormal conditions (such as excessive rainfall, drought, tornadoes, hurricanes, etc.) or under conditions not reasonably foreseeable to or beyond the control of seller.

When buyer or user suffers losses or damages resulting from the use or handling of this product (including claims based on contract, negligence, strict liability, or other legal theories), buyer or user must promptly notify in writing The Dow Chemical Company of any claims to be eligible to receive either remedy given below. The EXCLUSIVE REMEDY OF THE BUYER OR USER and the LIMIT OF LIABILITY of The Dow Chemical Company or any other seller will be one of the following, at the election of The Dow Chemical Company:

(1) **Refund of purchase price** paid by buyer or user for product bought, **or**

(2) **Replacement** of amount of product used.

The seller **will not be liable for consequential or incidental damages or losses.**

The terms of this Warranty Limitations And Disclaimer cannot be varied by any written or verbal statements or agreements. Any employee or sales agent of the seller is not authorized to vary or exceed the terms of this Warranty Limitations And Disclaimer in any manner.

*Trademark of THE DOW CHEMICAL COMPANY

12169-L1 689

CROSSBOW*

Weed and Brush Killer

LOW VOLATILE BRUSH AND WEED HERBICIDE

For the Control of Most Kinds of Unwanted Trees and Brush, as Well as Annual and Perennial Broadleaf Weeds on Rangeland, Permanent Grass Pasture, Fence Rows, Roadsides, Non-Irrigation Ditchbank, Other Non-crop Areas and Industrial Sites

ACTIVE INGREDIENT(S):

2,4-Dichlorophenoxyacetic Acid, Butoxyethyl Ester	34.4%
Triclopyr (3,5,6-trichloro-2- pyridinyloxyacetic acid), Butoxyethyl Ester	16.5%
INERT INGREDIENTS:	49.1%
TOTAL	100.0%

Contains Petroleum Distillates

2,4-D Acid Equivalent 23.7% - 2 pounds per gallon

Triclopyr Acid Equivalent 11.9% - 1 pound per gallon

E.P.A. Registration No. 464-589 E.P.A. Est. 464-MI-1

KEEP OUT OF REACH OF CHILDREN

CAUTION

AVISO:

PRECAUCIÓN AL USUARIO:

Si usted no lee inglés, no use este producto hasta que la etiqueta le haya sido explicada ampliamente.

PRECAUTIONARY STATEMENTS

HAZARDS TO HUMANS AND DOMESTIC ANIMALS

HARMFUL IF SWALLOWED, INHALED, OR ABSORBED THROUGH SKIN • CAUSES EYE IRRITATION • THIS PRODUCT MAY CAUSE ALLERGIC SKIN REACTION

Avoid Contact With Skin, Eyes, Or Clothing • Avoid Breathing Vapor • Wash Thoroughly With Soap And Water After Handling And Before Eating Or Smoking • Remove And Wash Contaminated Clothing Before Reuse

STATEMENT OF PRACTICAL TREATMENT:

In case of contact, flush skin or eyes with plenty of water. Get medical attention if irritation persists.

If swallowed, DO NOT induce vomiting. CALL A PHYSICIAN OR POISON CONTROL CENTER.

PHYSICAL OR CHEMICAL HAZARDS

COMBUSTIBLE • Do Not Use or Store Near Heat or Open Flame. • Do Not Cut or Weld Container.

ENVIRONMENTAL HAZARDS

This pesticide is toxic to fish. Do not apply directly to bodies of water. Do not contaminate irrigation ditches or water used for irrigation or domestic purposes.

NOTICE

Read the entire label. Use only according to label directions.

Before buying or using this product, read "WARRANTY LIMITATIONS AND DISCLAIMER" elsewhere on this label. If terms are not acceptable, return **unopened** package at once to seller for full refund of purchase price paid. Otherwise, use by the buyer or any other user constitutes acceptance of the terms under "WARRANTY LIMITATIONS AND DISCLAIMER."

IN CASE OF AN EMERGENCY
endangering life or property involving this product, call collect 517-636-4400

AGRICULTURAL CHEMICAL
Do Not Ship or Store with Food, Feeds, Drugs, or Clothing

GENERAL INFORMATION

CROSSBOW Weed and Brush Killer is recommended for control of most species of unwanted woody plants, as well as annual and perennial broadleaf weeds, growing on rangeland, permanent grass pasture, fence rows, non-irrigation ditch banks, roadsides, other noncrop areas, and industrial sites.

Susceptible Woody Species Include: Alder, ash, aspen, birch, blackberry, black locust, boneset, Ceanothus, cherry, elderberry, elm, hazel, honeysuckle, maples, multiflora rose, pine, poison ivy, poison oak, oaks (except black, post and blackjack oaks), Russian olive, sassafras, scotchbroom, sumac, sweetgum, sycamore, tamarack, and willow.

Susceptible Broadleaf Weeds Include: annual fleabane, bedstraw, bullthistle, burdock, buttercup, Canada thistle, clover (including burclover), cocklebur, curly dock, dandelion, dogfennel (suppression), field pennycress, goldenrod, ground ivy, hemp dogbane, hemp sesbania, horseweed, horsenettle (suppression), kochia, lambsquarters, marshelder, mouseear chickweed, musk thistle, oxalis, plantains, pokeweed, ragweeds, redroot pigweed, sunflower, tall ironweed, tansy ragwort, vetch, wild carrot, and wild violet.

DIRECTIONS FOR USE

It is a violation of Federal law to use this product in a manner inconsistent with its labeling.

Do not apply this product through any type of irrigation system.

Foliar sprays should be applied during warm weather when brush and weeds are actively growing. Application under drought conditions may provide less than desirable results. Use low spray pressures to minimize spray drift. **Apply CROSSBOW Weed and Brush Killer in a manner to avoid contacting nearby susceptible crops or other desirable plants and to avoid contaminating water intended for irrigation or domestic use. Read and follow all use Precautions given on this label.**

Do not use on bentgrass. Do not use on newly seeded grasses until grass has established a good root system and is tillering.

Note: Do not spray pastures containing desirable broadleaf forbs, especially legumes such as clover, unless injury or loss of such plants can be tolerated. However, the stand and growth of established grasses usually is improved, particularly when rainfall is adequate and grazing is deferred.

PREPARING THE SPRAY

CROSSBOW Weed and Brush Killer in water forms an emulsion (not a solution), and separation may occur unless the spray mixture is agitated continuously.

Water Spray: Fill the spray tank about half full with clean water. Then add the CROSSBOW Weed and Brush Killer and complete filling the tank with agitation running. Mix thoroughly and continue moderate agitation while spraying.

Size of Sprayer	Amount of CROSSBOW for Spray Solution		
	1%	1.5%	4%
1 gal	1 1/3 fl oz	2 fl oz	5 1/3 fl oz
3 gal	4 fl oz	6 fl oz	1 pt
5 gal	6 2/3 fl oz	10 fl oz	1 2/3 pt
50 gal	2 qt	3 qt	2 gal
100 gal	1 gal	1.5 gal	4 gal

APPLICATION DIRECTIONS

Woody Plant Control (See Table 1)

High Volume Applications Through Handguns: Using a power or hand pressured spray-gun, apply a foliar wetting spray containing 1 to 1 1/2 gallons of this product in sufficient water to make 100 gallons of total spray mix. See attached chart for smaller amounts of this 1 to 1.5% spray mix.

Spray to give thorough coverage of the foliage, wetting all leaves and green stems to the drip point. Depending on the plant size and foliage density, the total amount of required spray is usually 100 to 200 gallons per sprayed acre.

For best results, applications should be made when woody plants are actively growing. This is most likely to occur for a period after full leaf in the spring to early summer when moisture and temperature are favorable. For multiflora rose control, the best time for treatment may be expected during the early to mid-flowering stage.

The required spray volume will increase substantially if the brush exceeds 5 feet in height. Brush over 8 feet tall is difficult to treat efficiently. Large brush or trees may be controlled better by basal or mechanical methods.

Foliar Broadcast Sprays (Ground Equipment and Helicopter): Apply 1.5 to 4 gallons of this product in enough water to deliver 15 to 30 gallons total spray per acre. Use a boom type or other broadcast spray equipment that provides uniform spray coverage over the top of the foliage and make applications when plants are growing well. The favorable period for treatment is most likely to occur after full leaf in the spring and continue into early summer, depending on soil moisture and other conditions. Follow-up treatment with foliar high-volume or basal type treatments may be needed, especially if treating under less favorable conditions.

Aerial Application (Helicopter only): On areas near susceptible crops or ornamental plants, use NALCO-TROL drift control additive as recommended by the manufacturer or the MICROFOIL boom, THRU-VALVE boom, or equivalent drift control system. Thickened sprays prepared by using high viscosity invert systems or other drift reducing systems may be utilized if they control spray drift as well as NALCO-TROL or the above mentioned booms. If a spray thickening agent is used, follow all use recommendations and precautions on the product label. Do not use a thickening agent with the MICRO-FOIL or THRU-VALVE booms or other systems that cannot accommodate thick sprays.

Dormant Stem Applications: To control susceptible woody species such as multiflora rose and blackberry, mix 1 to 4 gallons of this product in diesel oil, No. 1 or No. 2 fuel oil or kerosene to make 100 gallons of spray and apply to thoroughly wet upper and lower stems including the root collar and any ground sprouts. Treat at any time when the brush is dormant and the bark is dry. Best results have been obtained with late winter to early spring applications. Do not treat when snow or water prevent spraying to the groundline. For the most susceptible woody species such as blackberries substitute other diluents or oils only in accordance to manufacturer's recommendations. Apply mixture to thoroughly wet upper and lower stems as described above. The more tolerant species may require total oil carrier for better control. Brush over 8 feet in height is difficult to treat efficiently. Basal or mechanical methods may be better suited for control of large trees.

Basal Bark and Stump Applications: For control of susceptible woody plants and to prevent or control regrowth from cut stumps, mix 4 gallons of this product in diesel oil. No. 1 or No. 2 fuel oil or kerosene to make 100 gallons of spray mixture. Spray the basal parts of brush or trees to a height of 15 to 20 inches from the ground. Thoroughly wet all the basal bark area including crown buds and grounds sprouts. Spray runoff should visibly wet the ground at the base of the stems or trunks. Basal and cut stump applications can be made at any time of the year except when snow or water prevent spraying to the groundline. Best results have been obtained with winter to early spring applications. Basal treatments are less effective on trees with diameters larger than 6 to 8 inches. For better regrowth control, cut the larger trees and treat the stumps. Treat stumps the same as the trunks and also treat the freshly cut surface. The cambium layer just inside the bark is the most important area of the cut surface to treat.

Thinline Basal Applications: For the control of small multiflora rose, apply a horizontal thin line of undiluted herbicide across all the stems at a height where the stems are less than 1/2 inch in diameter and have thinner bark to penetrate. Treat when the bark is dry and rain is not forecasted. Apply approximately 20 ml undiluted product per bush. Wherever a stem over 1/2 inch in diameter is treated, it should be completely ringed with herbicide to obtain best results. Additional herbicide is likely to be needed for adequate coverage of these larger stems in a bush or clump. For bushes with large numbers of stems in a bush or clump. For bushes with large numbers of stems (over 3 or 4), coverage of each stem may be difficult. Basal bark or dormant stem applications may be more effective.

Old stems with thickened bark require more herbicide than young stems with thin bark. Where regrowth is treated, better root kill may result if resprouts are treated after they are one year old and the bark has lost its green color, but before sprouts reach one inch in diameter.

General Weed Control (See Table 2)

Broadcast Treatment (Ground Equipment and Helicopter): Use up to 1 1/2 gallons of CROSSBOW Weed and Brush Killer per acre in the amount of water needed for uniform coverage to control annual and perennial broadleaf weeds. Apply when weeds are actively growing. See recommendations regarding the use of drift control additives as listed in the "WOODY PLANT CONTROL" section under "Broadcast Sprays".

Spot Treatment: To control broadleaf weeds in small areas with a hand sprayer, use 4 to 6 fl. oz. of CROSSBOW Weed and Brush Killer in 3 gallons of water and spray to thoroughly wet all foliage.

Grazing Restrictions

Withdraw livestock from treated forage at least 3 days before slaughter during the year of treatment. Do not graze lactating dairy animals on treated areas for one year following treatment. Do not harvest grass for hay from treated areas for one year following treatment.

Do not graze areas treated with more than 1.5 galons per acre of product or harvest hay from treated areas for one year following treatment.

Conservation Reserve Program (CRP) For Seeding To Permanent Grasses
Use CROSSBOW Weed and Brush Killer on CRP acres only when the perennial grasses are established. Conditions that stress grasses, such as drought, will increase potential for injury to the grasses.

Restrictions: Grass grown on CRP acreas cannot be grazed or used for hay. Do not graze or crop treated acres for at least one year after treatment. Do not use CROSSBOW Weed and Brush Killer if legumes are a desired cover crop during CRP. Do not use on bentgrass or newly seeded grass.

Broadcast Application (Ground or Aerial): Apply 1 to 2 quarts of CROSSBOW for small weed control or up to 1.5 gallons of CROSSBOW for deep-rooted perennial and susceptible woody species control using enough water to deliver 10 or more gallons of total spray volume per acre.

Follow precautions and recommendations outlined under Foliar Low-Volume Broadcast Applications.

For basal and dormant brush treatments, follow application directions listed in "WOODY PLANT CONTROL".

TABLE 1
WOODY PLANT CONTROL

A—Easy-To-Control Species
(1.5 gal/acre broadcast application or 1-1.5% solutions for high-volume foliar applications).

Alder	Boneset	Maples	Sumac
Ash	Ceanothus spp	Multiflora rose	Sycamore
Aspen	Cherry[1]	Poison oak	Tamarack
Birch	Elderberry	Poison Ivy	Tanoak
Blackberry	Honeysuckle	Sassafrass	Willow
Black locust		Scotchbroom	White oak

[1] Except black

B—Harder-To-Control Species
(High-volume applications, 1.5% solution, conventional basal or dormant stem applications are recommended. Broadcast rates of 2 to 4 gal/acre will increase the degree of control of these species. See grazing restrictions when rates of application greater than 1.5 gal/acre are used).

Buckbrush (suppression) (*Symphoricarpos* sp)	Russian olive
Elm	Salmonberry (suppression)
Hazel	Sweetgum
Pine	Trumpetcreeper (top growth)

TABLE 2
BROADLEAF WEED CONTROL

Annual Weeds	Biennial Weeds	Perennial Weeds
2 qt/A or 1% solution	**2 qt/A or 1% solution**	**2-4 qt/A 1-1.5% solution**
Annual fleabane[1]	Bull thistle[1]	Bedstraw
Cocklebur[1]	Burdock	Broadleaf plantain
Common lambsquarter	Musk thistle[1]	Buttercup[2]
Common ragweed	Tansy ragwort	Canada thistle[2]
Field pennycress[1]	Wild carrot[1]	Clover
Hemp sesbania[1]		Curly dock
Horseweed (marestail)		Dandelion
Kochia[1] (suppression)		Dogfennel
Marshelder[1]		Goldenrod[3] (suppression)
Redroot pigweed[1]		Ground ivy
Sunflower		Hemp dogbane
Wild mustard		Horsenettle[2] (suppression)
		Milkweed[3] (suppression)
		Mouseear chickweed
		Oxalis
		Pokeweed[3]
		Tall ironweed[2]
		Vetch
		Wild violets

[1] 4 qt/acre broadcast or 1 to 1.5% solution for high-volume application
[2] Minimum of 4 qt/acre broadcast
[3] Only high-volume applications recommended, using 1.5% solution

USE PRECAUTIONS

Apply this product only as specified on this label.

Do not apply **CROSSBOW Weed and Brush Killer** directly to, or otherwise permit it to come into direct contact with cotton, grapes, tobacco, vegetable crops, flowers, fruit or ornamental trees, or other desirable broadleaf plants and do not permit spray mists containing it to drift onto them.

Avoid Injurious Spray Drift. Applications should be made only when hazards from spray drift are at a minimum. Very small quantities of spray, which may not be visible may seriously injure susceptible plants. Do not spray when wind is blowing toward susceptible crops or ornamental plants near enough to be injured. Spray drift can be reduced by adding a spray thickening agent such as Nalco-Trol or its equivalent to the spray mixture. If a spray thickening agent is used, follow all use recommendations and precautions on the product label.

With ground broadcast equipment, drift can be reduced by keeping the spray boom as low as possible; by applying no less than 20 gallons of spray per acre; by using no more than 20 pounds spraying pressure with flat fan or flooding flat fan nozzle tips; and by spraying only when wind velocity is less than 5 miles per hour. In handgun application, use gun pressure no higher than 50 psi and apply in such manner as to confine the spray to the target area.

With the helicopter, spray drift can be lessened by applying a coarse spray; by using no more than 30 pounds spray pressure at the nozzles; by using a spray boom no longer than 3/4 the rotor length; by spraying only when wind velocities are low; and by using an approved drift control system.

At high temperatures, vapors from this product may injure susceptible crops growing nearby. Excessive amounts of this herbicide in the soil may temporarily inhibit seed germination and plant growth.

STORAGE AND DISPOSAL

Do not contaminate water, food or feed by storage or disposal.

STORAGE: Store above 10°F or agitate before use.

DISPOSAL:

Pesticide Disposal: Pesticide wastes are toxic. Improper disposal of excess pesticide, spray mixture, or rinsate is a violation of Federal Law. If these wastes cannot be disposed of by use according to label instructions, contact your State Pesticide or Environmental Control Agency, or the Hazardous Waste representative at the nearest EPA Regional Office for guidance.

Container Disposal: Triple rinse (or equivalent). Then offer for recycling or reconditioning, or puncture and dispose of in a sanitary landfill, or by other procedures approved by state and local authorities. Plastic containers, after triple rinsing, may be incinerated or burned if allowed by state and local authorities. If burned, stay out of smoke.

General. Consult federal, state, or local disposal authorities for approved alternative procedures.

Be sure that use of this product conforms to all applicable regulations.

WARRANTY LIMITATIONS AND DISCLAIMER

The Dow Chemical Company warrants that this product conforms to the chemical description on the label and is reasonably fit for the purposes stated on the label when used in strict accordance with the directions therein under normal conditions of use. THIS IS THE ONLY WARRANTY MADE ON THIS PRODUCT. NO OTHER EXPRESS AND NO IMPLIED WARRANTY OF MERCHANTABILITY OR FITNESS FOR A PARTICULAR PURPOSE IS MADE OUTSIDE OF THIS LABEL. Therefore, neither this warranty nor any other warranty of merchantability or fitness for a particular purpose, express or implied, extends to the use of this product contrary to label instructions (including conditions noted on the label, such as unfavorable temperatures, soil conditions, etc.), under abnormal conditions (such as excessive rainfall, drought, tornadoes, hurricanes, etc.) or under conditions not reasonably foreseeable to or beyond the control of seller.

When buyer or user suffers losses or damages resulting from the use or handling of this product (including claims based on contract, negligence, strict liability, or other legal theories), buyer or user must promptly notify in writing The Dow Chemical Company of any claims to be eligible to receive either remedy given below. The EXCLUSIVE REMEDY OF THE BUYER OR USER and the LIMIT OF LIABILITY of The Dow Chemical Company or any other seller will be one of the following, at the election of The Dow Chemical Company:

(1) **Refund of purchase price** paid by buyer or user for product bought, **or**

(2) **Replacement** of amount of product used.

The seller **will not be liable for consequential or incidental damages or losses.**

The terms of this Warranty Limitations And Disclaimer cannot be varied by any written or verbal statements or agreements. Any employee or sales agent of the seller is not authorized to vary or exceed the terms of this Warranty Limitations And Disclaimer in any manner.

01662-L2 F588

*Trademark of THE DOW CHEMICAL COMPANY

CURTAIL*

HERBICIDE

For Selective Control of Broadleaf Weeds in Wheat and Barley Not Underseeded With a Legume, Fallow Cropland And Conservation Reserve Program (CRP) Acres

ACTIVE INGREDIENT(S):

Clopyralid (3,6-Dichloro-2-pyridinecarboxylic acid) as alkanolamine salts (of the ethanol and isopropanol series)	6.2%
2,4-Dichlorophenoxyacetic Acid as alkanolamine salts (of the ethanol and isopropanol series)	31.5%
INERT INGREDIENTS:	62.3%
TOTAL	100.0%

Acid Equivalents:
3,6-Dichloro-2-pyridinecarboxylic acid 3.9% - 0.38 lb/gal
2,4-Dichlorophenoxyacetic acid 20.8% - 2.0 lb/gal

E.P.A. Registration No. 464-563 E.P.A. Est. 464-MI-1

KEEP OUT OF REACH OF CHILDREN

DANGER
PELIGRO:

PRECAUCIÓN AL USUARIO:
Si usted no lee inglés, no use este producto hasta que la etiqueta le haya sido explicada ampliamente.

PRECAUTIONARY STATEMENTS

HAZARDS TO HUMANS AND DOMESTIC ANIMALS

CAUSES EYE DAMAGE AND SKIN IRRITATION • HARMFUL IF SWALLOWED

Do Not Get In Eyes, On Skin Or On Clothing • Wear Goggles Or Face Shield When Handling • Wash Thoroughly With Soap And Water After Handling • Remove Contaminated Clothing and Wash Before Reuse

STATEMENTS OF PRACTICAL TREATMENT:

If in eyes, flush with plenty of water for at least 15 minutes. Get medical attention.

If on skin, wash with plenty of soap and water. Get medical attention if irritation persists.

If swallowed, induce vomiting immediately by giving two glasses of water and sticking finger down throat. Call a physician. Do not induce vomiting or give anything by mouth to an unconscious person.

ENVIRONMENTAL HAZARDS

Keep out of lakes, streams or ponds. Do not contaminate water when disposing of equipment washwaters. Do not contaminate water used for irrigation or domestic purposes. Do not apply directly to any body of water or wetlands. See additional precautionary statements elsewhere on this label.

NOTICE

Read the entire label. Use only according to label directions.

Before buying or using this product, read "WARRANTY LIMITATIONS AND DISCLAIMER" elsewhere on this label. If terms are not acceptable, return **unopened** package at once to seller for full refund of purchase price paid. Otherwise, use by the buyer or any other user constitutes acceptance of the terms under "WARRANTY LIMITATIONS AND DISCLAIMER."

IN CASE OF AN EMERGENCY
endangering life or property involving this product, call collect 517-636-4400

AGRICULTURAL CHEMICAL
Do Not Ship or Store with Food, Feeds, Drugs, or Clothing

DIRECTIONS FOR USE

It is a violation of Federal law to use this product in a manner inconsistent with its labeling.

Do not apply this product through any type of irrigation system.

CURTAIL Herbicide is recommended for selective control of the following broadleaf weeds in wheat and barley not underseeded with a legume, fallow cropland (including summer fallow, post-harvest, and set-aside acres) and land in the Conservation Reserve Program (CRP).

alfalfa
artichoke, Jerusalem
buckwheat, wild
burdock, common
chamomile, false (scentless)
chamomile, mayweed (dogfennel)
clover, sweet
clover, red
nightshade, cutleaf
nightshade, hairy
pennycress, field (fanweed)
pigweed, redroot
pineappleweed
plantain
radish, wild
ragweed, common
ragweed, giant
cocklebur, common
coffeeweed
cornflower (bachelor button)
dandelion
dock, curly
flixweed[1]
groundsel, common
hawksbeard, narrowleaf
horseweed
jimsonweed
knapweed, diffuse
knapweed, Russian[1]
knapweed, spotted
kochia (2-4 leaf)[1]
ladysthumb
lambsquarters, common
lettuce, prickly
mustard, tumble (Jim Hill)
mustard, wild
salsify, meadow (goatsbeard)
shepherdspurse
sicklepod
smartweed, Pennsylvania
sorrel, red
sowthistle, annual
sowthistle, perennial[1]
starthistle, yellow
sunflower, common
tansymustard, pinnate[1]
thistle, Canada
thistle, musk
thistle, Russian (1-3 leaf)
velvetleaf
vetch
volunteer beans
volunteer lentils
volunteer peas

[1] These weeds may only be supressed. Suppression is a visual reduction in weed competition (reduced population or vigor) as compared to untreated areas. The degree of weed control and duration of effect will vary with weed size and density, spray rate and coverage, and growing conditions before, during and after the time of treatment.

For perennial weeds, CURTAIL Herbicide will control the initial topgrowth and inhibit regrowth during the season of application (season-long control). At higher use rates shown on this label, CURTAIL Herbicide may cause a reduction in shoot regrowth in the season following application, however, plant response may be inconsistent due to inherent variability in shoot regrowth from perennial root systems.

Timing: Apply to actively growing weeds. Extreme growing conditions such as drought or near freezing temperatures prior to, at and following time of application may reduce weed control and increase the risk of crop injury at all stages of growth. **Only weeds which are emerged at the time of application will be affected.** Wet foliage at the time of application may decrease control. The CURTAIL Herbicide treatment will be rainfast within 6-8 hours after application.

Rate: Generally, lower labeled application rates will be satisfactory for young, succulent growth of sensitive weed species. For less sensitive species, perennials and under conditions where control is more difficult (plant stress conditions such as drought or extreme temperatures, dense weed stands and/or larger weeds) the higher rates will be needed. Weeds in fallow or other areas where crop competition is not a factor will generally require higher rates to obtain control or suppression.

Coverage: Adequate spray coverage and drift control are important. Obtaining a balance between spray coverage and drift control may sometimes be difficult but can be achieved, provided the applicator understands the factors affecting coverage and drift. Factors affecting spray coverage include spray volume, crop canopy and weed density. As crop canopy and weed density increase, spray volume should be increased to obtain equivalent weed control. Refer to manufacturer's recommendations for information on the relationship between gallons per acre, spray pressure, sprayer speed, nozzle type and arrangement, nozzle height above the target canopy, droplet size and drift potential for respective application equipment. Do not apply less than 2 and not more than 40 gallons per acre total spray volume. For best results, apply 5 or more gallons per acre by air and 10 or more gallons per acre by ground. Reducing total spray volume may result in decreased coverage and weed control. Use enough total spray volume and a delivery system to provide thorough coverage and a uniform spray pattern. Do not apply where spray drift may be a problem due to proximity of susceptible crops or other desirable plants.

Use of Surfactants: Addition of wetting and/or penetration agents is not usually necessary when using CURTAIL Herbicide; however, if a surfactant will be added to the spray solution, use a nonionic surfactant approved for use on growing crops, such as X-77, of at least 80% active ingredient and **do not exceed** 4 pints per 100 gallons of spray solution (0.5% v/v). Adding a surfactant to the spray mixture may increase effectiveness on weeds but may reduce selectivity to the crop, particularly under conditions which promote plant stress.

Use With Sprayable Liquid Fertilizer Solutions: CURTAIL Herbicide is compatible with most non-pressurized liquid fertilizer solutions, however, if these solutions are to be sprayed with CURTAIL Herbicide, a compatibility test (jar test) should be made prior to mixing. Jar tests are particularly important when a new batch of fertilizer or pesticide is used, when water source changes, or when tank mixture ingredients or concentrations are changed. A compatibility test is performed by pouring the spray components (in the desired order and proportions) into a clear glass jar before mixing in the spray tank. Use of a compatability aid such as UNITE or COMPLEX may help obtain and maintain a uniform spray solution during mixing and application. Agitation in the spray tank must be vigorous to compare with jar test agitation. For best results, liquid fertilizer rates should not exceed 50% of the total spray volume. Premix CURTAIL Herbicide with water and add to the liquid fertilizer/water mixture while agitating contents of the spray tank. Apply the spray the same day it is prepared while maintaining continuous agitation. NOTE: Foliar-applied liquid fertilizers can cause yellowing or leaf burn of crop foliage.

WHEAT AND BARLEY

Timing: Apply CURTAIL Herbicide in the spring to actively growing wheat or barley once 4 leaves have unfolded on the main stem and tillering has begun up to the jointing stage (first node of main stem detectable). To control or suppress weeds listed above, make application after maximum emergence of the target weeds but before they exceed 3 inches in height or diameter (for rosettes). To obtain seasonlong control of perennial weeds such as Canada thistle, apply after the majority of the weed's basal leaves have emerged from the soil, but before bud state. A later timing of application (when the grain is between the jointing and boot stages) may be used to treat later-emerging weeds, however, this may increase the risk of crop injury. Do not apply after the boot stage. **NOTE:** Do not permit dairy animals or meat animals being finished for slaughter to forage or graze treated grain fields within 1 week after treatment. Do not harvest hay from treated grain fields.

Rate: Apply 2 to 2⅔ pints of CURTAIL Herbicide per acre. The higher rate may be used when the condition of the weeds and/or crop at the time of treatment may prevent optimum control. **Note: Higher rates of CURTAIL Herbicide or CURTAIL Herbicide applications following spring postemergence treatments with 2,4-D or MCPA may increase the risk of crop injury.**

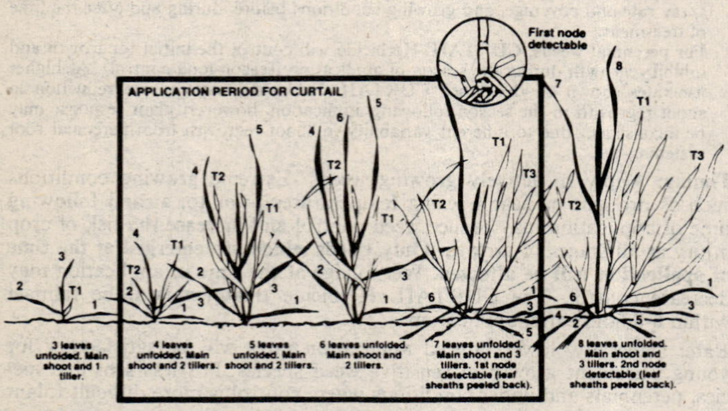

TANK MIXTURES FOR WHEAT AND BARLEY

This product may be tank mixed with the following herbicides for control of additional weeds in wheat and barley. Read and follow the label of each tank mix product used for precautionary statements, directions for use, weeds controlled and geographic and other restrictions. When tank mixing, use in accordance with the most restrictive of label limitations and precautions. No label dosages should be exceeded. This product cannot be mixed with any product containing a label prohibition against such mixing.

Active Ingredient	Product	Formulation	Amount of Product Per Acre
bromoxynil[1]	Buctril	2 lb/gal	¾ to 1 pt
	Buctril 4EC	4 lb/gal	⅜ to ½ pt
	Buctril ME4		
chlorsulfuron	Glean	75% DF	⅙ to ¼ wt oz
dicamba[1,2]	Banvel	4 lb/gal	⅛ to ¼ pt
difenzoquat	Avenge	2 lb/gal	2½ to 4 pt
diuron	Direx 4L	4 lb/gal	1 to 1½ pt
	Diuron 4L		
	Diruon 80 WDG	80% DF	½ to 1 lb
	Diruon DF	80% WP	
MCPA or 2,4-D[2]		4lb/gal	Up to ½ pt
metribuzin	Lexone DF	75% DF	2½ to 4 wt oz
	Sencor DF		
metsulfuron methyl	Ally	60% DF	⅒ wt oz
terbutyn	Igran 80 WR	80% WP	6 to 10 wt oz
thiameturon	Harmony	75% DF	⅓ to ½ wt oz

[1] Buctril or Banvel tank mixes with CURTAIL Herbicide may be useful in broadening the annual weed control spectrum but may reduce control of perennials, such as Canada thistle.

[2] Note: Do not tank mix CURTAIL Herbicide with 2,4-D or dicamba unless the risk of crop injury is acceptable.

FALLOW CROPLAND

Tming: CURTAIL Herbicide can be applied either postharvest, in the spring/summer (during fallow period) or to set-aside acres to control or suppress weeds listed above (refer to rotation restrictions). Apply to young, emerged weeds under conditions that promote active growth. For best results on tough perennial weeds such as Canada thistle, apply after the majority of the basal leaves have emerged, but before bud stage. Later applications may result in less consistent control. Extreme growing conditions (such as drought or near freezing temperatures) prior to, at, and following the time of application may reduce weed control.

For best results, wait 14 to 20 days after application before cultivating or fertilizing with shank-type applicators to allow for thorough translocation. To avoid potential phytotoxicity, allow at least 20 days after the application before seeding to wheat, barley or grasses.

Rate: Apply 2 to 4 pints of CURTAIL Herbicide per acre. Use the higher rate on perennial weeds or when the condition of the weeds at the time of treatment may prevent optimum control.

Tank Mixtures For Fallow Cropland: To improve control of certain broadleaf weeds, CURTAIL Herbicide at 2 pints per acre may be applied with up to 1.5 lb ae per acre additional 2,4-D or MCPA. When mixing, read and follow the label of each tank mix product used for precautionary statements, directions for use, weeds controlled and geographic and other restrictions. When tank mixing, use in accordance with the most restrictive of label limitations and precautions. No label doses should be exceeded. This product cannot be mixed with any product containing a label prohibition against such mixing.

GRASSES GROWN FOR SEED

Timing: Apply only to established grasses before the boot stage. Applications in the boot stage and beyond will result in increased injury. Do not apply to bentgrass unless injury can be tolerated.

Rate: Use 2 to 4 pints of CURTAIL Herbicide per acre for control of annual weeds and Canada thistle. The potential for crop injury exists due to the 2,4-D component of this product and must be balanced against the benefits of improved weed control.

Note: Do not graze (dairy) cattle in treated areas for 14 days after application. Remove meat animals from freshly treated areas for 7 days before slaughter. Withdrawal is not needed if 2 weeks or more have elapsed since application. Do not cut treated grass for hay within 30 days after application.

CONSERVATION RESERVE PROGRAM (CRP) FOR SEEDING TO PERMANENT GRASSES ONLY

Do not use CURTAIL Herbicide if legumes or bentgrass are a desired cover during CRP.

Grass grown on CRP acres cannot be grazed or used for hay.

Conditions that stress grasses, such as drought, will increase potential for injury to the grass at all stages of growth. Do not use in newly seeded areas until grass is established.

Timing; CURTAIL Herbicide can be applied when perennial grasses have become established (tillering, with at least 1.5 inches secondary roots, and good vigor). For control of weeds such as musk thistle (after rosette stage), Canada thistle and knapweed (diffuse, spotted and Russian), apply to actively growing weeds after the majority of the basal leaves have emerged, but before bud stage. Later applications may result in less consistent control.

Rate: Apply 2 to 5 quarts of CURTAIL Herbicide per acre.

USE PRECAUTIONS

Apply only once per 12-month period.

This product can affect susceptible broadleaf plants directly through foliage and indirectly by root uptake from treated soil. Therefore, do not apply CURTAIL Herbicide directly to or allow spray drift to come in contract with vegetables, flowers, grapes, tomatoes, potatoes, beans, lentils, peas, alfalfa, sunflowers, soybeans, safflower, or other desirable broadleaf crops and ornamental plants or soil where these sensitive crops will be planted the same season.

Do not contaminate irrigation ditches or water used for irrigation or domestic purposes.

Do not plant or rotate to any crop for one year after treatment, except for wheat, barley, oats, grass or sugar beets. To avoid potential phytotoxicity, do not plant peas, lentils, beans (except soybeans) or broadleaf seed crops for 18 months after treatment unless the risk of injury is acceptable.

Avoid Spray Drift: Applications should be made to avoid spray drift since very small quantities of the spray, which may not be visible, may severely injure susceptible crops during both growing and dormant periods. Use coarse sprays to minimize drift since, under adverse weather conditions, fine spray droplets may drift a mile or more. Fine droplets can present a drift hazard. A drift control or deposition agent such as NALCO-TROL may be used with this product to aid in reducing spray drift. If used, follow all use recommendations and precautions on the product label.

Ground Application: To minimize spray drift apply CURTAIL Herbicide in a total spray volume of 10 or more gallons per acre as a large-droplet, low pressure spray. Refer to manufacturer's recommendations for additional information on gallons per acre, spray pressure, sprayer speed, nozzle types and arrangements, nozzle heights above the target canopy, etc., for respective application equipment. Spot treatments should only be applied with a calibrated boom to prevent misapplication. With ground equipment, spray drift can be lessened by keeping the spray boom as low as possible; by applying no more than 20 gallons of spray per acre; by using no more than 30 pounds spraying pressure with large droplet-producing nozzle tips; by spraying when wind velocity is low; and by stopping all spraying when wind exceeds 6 to 7 miles per hour. Do not apply with hollow cone-type insecticide or other nozzles that produce a fine-droplet spray.

Aerial Application: With aircraft, drift can be lessened by applying a coarse spray; by using no more than 30 pounds spray pressure at the nozzles; by using straight-stream nozzles directed straight back; by using a spray boom no longer than 3/4 the wing span of the aircraft; and by spraying only when wind velocity is less than 6 mph.

Do Not Apply By Aircraft When An Air Temperature Inversion Exists. Such a condition is characterized by little or no wind and with lower air temperature near the ground than at higher levels. The use of a smoke device on the aircraft or continuous smoke column at or near site of application will indicate air direction and velocity, and whether a temperature inversion is present, which is shown by layering of the smoke.

Do not move treated soil and avoid situations where treated soil particles may blow into areas where susceptible crops are grown. Violent windstorms may move soil particles. If this product is on soil particles and they are blown onto susceptible plants, visible symptoms may appear. Serious injury is unlikely. The hazard of movement of this product on dust is reduced if treated fields are irrigated or if rain occurs shortly after application.

Straw from treated areas, or manure from animals that have grazed treated areas, cannot be used for composting or mulching on ground where susceptible crops may be grown the following season. To promote herbicide decomposition, plant material should be evenly incorporated or burned.

Do not use in a greenhouse. Excessive amounts of this herbicide in the soil may temporarily inhibit seed germination or plant growth.

Do not contaminate water when disposing of equipment washwaters. Apply this product only as specified on this label.

Do not transfer livestock from treated grazing areas onto sensitive broadleaf crop areas without first allowing 7 days of grazing on an untreated pasture. Otherwise, urine may contain enough clopyralid to cause injury to sensitive broadleaf plants.

PRECAUTIONARY STATEMENTS

ENVIRONMENTAL HAZARDS

Clopyralid is a chemical which can travel (seep or leach) through soil and under certain conditions contaminate groundwater which may be used for irrigation or drinking purposes. Users are advised not to apply clopyralid where soils have a rapid to very rapid permeability throughout the profile (such as loamy sand to sand) and the water table of an underlying aquifer is shallow, or to soils containing sinkholes over limestone bedrock, severely fractured surfaces, and substrates which would allow direct introduction into an aquifer. Your local agricultural agencies can provide further information on the type of soil in your area and the location of groundwater.

STORAGE AND DISPOSAL

Do not contaminate water, food or feed by storage or disposal.

Storage: Store above 10°F or warm and agitate before use.

Pesticide Disposal: Pesticide wastes are toxic. Improper disposal of excess pesticide, spray mixture, or rinsate is a violation of Federal Law. If these wastes cannot be disposed of by use according to label instructions, contact your State Pesticide or Environmental Control Agency, or the Hazardous Waste representative at the nearest EPA Regional Office for guidance.

Container Disposal: Metal Do not reuse container. Triple rinse (or equivalent). Puncture and dispose of in a sanitary landfill, or by other procedures approved by state and local authorities.

Container Disposal: Plastic Do not reuse container. Triple rinse (or equivalent). Puncture and dispose of in a sanitary landfill, or by incineration, or, if allowed by state and local authorities, by burning. If burned, stay out of smoke.

WARRANTY LIMITATIONS AND DISCLAIMER

The Dow Chemical Company warrants that this product conforms to the chemical description on the label and is reasonably fit for the purposes stated on the label when used in strict accordance with the directions therein under normal conditions of use. THIS IS THE ONLY WARRANTY MADE ON THIS PRODUCT. NO OTHER EXPRESS AND NO IMPLIED WARRANTY OF MERCHANTABILITY OR FITNESS FOR A PARTICULAR PURPOSE IS MADE OUTSIDE OF THIS LABEL. Therefore, neither this warranty nor any other warranty of merchantability or fitness for a particular purpose, express or implied, extends to the use of this product contrary to label instructions (including conditions noted on the label, such as unfavorable temperatures, soil conditions, etc.), under abnormal conditions (such as excessive rainfall, drought, tornadoes, hurricanes, etc.) or under conditions not reasonably foreseeable to or beyond the control of seller.

When buyer or user suffers losses or damages resulting from the use or handling of this product (including claims based on contract, negligence, strict liability, or other legal theories), buyer or user must promptly notify in writing The Dow Chemical Company of any claims to be eligible to receive either remedy given below. The EXCLUSIVE REMEDY OF THE BUYER OR USER and the LIMIT OF LIABILITY of The Dow Chemical Company or any other seller will be one of the following, at the election of The Dow Chemical Company:

(1) **Refund of purchase price** paid by buyer or user for product bought, **or**

(2) **Replacement** of amount of product used.

The seller **will not be liable for consequential or incidental damages or losses.**

The terms of this Warranty Limitations And Disclaimer cannot be varied by any written or verbal statements or agreements. Any employee or sales agent of the seller is not authorized to vary or exceed the terms of this Warranty Limitations And Disclaimer in any manner.

08103-L1 A1288

*Trademark of THE DOW CHEMICAL COMPANY

CURTAIL® M

HERBICIDE

For Selective Control of Broadleaf Weeds in Wheat, Barley and Oats Not Underseeded with a Legume and Conservation Reserve Program (CRP) Acres

Active Ingredient(s)
Clopyralid (3,6-Dichloro-2-pyridinecarboxylic acid) 5.0%
MCPA (2-Methyl-4-chlorophenoxyacetic acid)
as the isooctyl ester 43.4%
Inert Ingredient(s) 51.6%

Acid Equivalents:
3,6-Dichloro-2-pyridinecarboxylic acid, 5.0% - 0.42 lb/gal
2-Methyl-4-chlorophenoxyacetic acid, 27.8% - 2.35 lb/gal
Contains Petroleum Distillates

E.P.A. Registration No. 464-625
E.P.A. Est. 464-M1-1

KEEP OUT OF REACH OF CHILDREN

CAUTION
PRECAUTION

Precaucion al usuario: Si usted no lee inglés, no use este producto hasta que la etiqueta le haya sido explicada ampliamente.

PRECAUTIONARY STATEMENTS

Hazards to Humans and Domestic Animals

CAUSES EYE INJURY • HARMFUL IF SWALLOWED, INHALED OR ABSORBED THROUGH SKIN

Avoid Contact With Eyes, Skin, Or Clothing • Avoid Breathing Spray Mist • Wash Thoroughly With Soap And Water After Handling • Remove Contaminated Clothing And Wash Before Reuse

STATEMENT OF PRACTICAL TREATMENT

If in eyes, flush with plenty of water for at least 15 minutes. Get medical attention if irritation persists.

If on skin, wash with plenty of soap and water. Get medical attention if irritation persists.

If swallowed, do not induce vomiting. Call a physician immediately.

PHYSICAL OR CHEMICAL HAZARDS

COMBUSTIBLE • Do Not Use or Store Near Heat or Open Flame. • Do Not Cut or Weld Container.

ENVIRONMENTAL HAZARDS

Do not contaminate water when disposing of equipment washwaters. Do not contaminate water used for irrigation or domestic purposes. Do not apply directly to any body of water or wetlands. See additional precautionary statements elsewhere on this label.

NOTICE

Read the entire label. Use only according to label directions.

Before buying or using this product, read "WARRANTY LIMITATIONS AND DISCLAIMER" elsewhere on this label. If terms are not acceptable, return **unopened** package at once to seller for full refund of purchase price paid. Otherwise, use by the buyer or any other user constitutes acceptance of the terms under "WARRANTY LIMITATIONS AND DISCLAIMER."

IN CASE OF AN EMERGENCY endangering life or property involving this product, call collect 517-636-4000

AGRICULTURAL CHEMICAL

Do Not Ship or Store with Foods, Feeds, Drugs, or Clothing

DIRECTIONS FOR USE

It is a violation of Federal law to use this product in a manner inconsistent with its labeling.

Do not apply this product through any type of irrigation system.

CURTAIL M Herbicide is recommended for selective, postemergence control of the following broadleaf weeds in wheat, barley and oats not underseeded with a legume and land in the Conservation Reserve Program (CRP).

alfalfa	nightshade, cutleaf
artichoke, Jerusalem	nightshade, hairy
buckwheat, wild	pennycress, field (fanweed)
burdock, common	pigweed, redroot[1]
chamomile, false (scentless)	pineappleweed
chamomile, mayweed (dogfennel)	plantain
	radish, wild
clover, sweet	ragweed, common
clover, red	ragweed, giant
cocklebur, common	salsify, meadow (goatsbeard)
coffeeweed	
cornflower (bachelor button)	shepherdspurse
dandelion	sicklepod
dock, curly	smartweed, Pennsylvania
flixweed[1]	sorrel, red
groundsel, common	sowthistle, annual
hawksbeard, narrowleaf	sowthistle, perennial[1]
horseweed	starthistle, yellow
jimsonweed	sunflower, common
knapweed, diffuse	tansymustard, pinnate[1]
knapweed, Russian[1]	thistle, Canada
knapweed, spotted	thistle, musk
kochia (2-4 leaf)[1]	thistle, Russian (1-3 leaf)[1]
ladysthumb	velvetleaf
lambsquarters, common	vetch
lettuce, prickly	volunteer beans
mustard, tumble (Jim Hill)	volunteer lentils
mustard, wild	volunteer peas

[1] These weeds may only be suppressed. Suppression is a visual reduction in weed competition (reduced population of vigor) as compared to untreated areas. The degree of weed control and duration of effect will vary with weed size and density, spray rate and coverage, and growing conditions before, during and after the time of treatment.

For perennial weeds, CURTAIL M Herbicide will control the initial top-growth and inhibit regrowth during the season of application (season-long control). At higher use rates shown on this label, CURTAIL M Herbicide may cause a reduction in shoot regrowth in the season following application, however, plant response may be inconsistent due to inherent variability in shoot regrowth from perennial root systems.

Timing: Apply to actively growing weeds. Extreme growing conditions such as drought or near freezing temperatures prior to, at and following time of application may reduce weed control and increase the risk of crop injury at all stages of growth. **Only weeds which are emerged at the time of application will be affected.** Wet foliage at the time of application may decrease control. The CURTAIL M Herbicide treatment will be rainfast within 6-8 hours after application.

Rate: Generally, lower labeled application rates will be satisfactory for young, succulent growth of sensitive weed species. For less sensitive species, perennials and under conditions where control is more difficult (plant stress conditions such as drought or extreme temperatures, dense weed stands and/or larger weeds) the higher rates will be needed. Weeds in areas where crop competition is not a factor will generally require higher rates to obtain control or suppression.

Coverage: Adequate spray coverage and drift control are important. Obtaining a balance between spray coverage and drift control may sometimes be difficult but can be achieved, provided the applicator understands the factors affecting coverage and drift. Factors affecting spray coverage include spray volume, crop canopy and weed density. As crop canopy and weed density increase, spray volume should be increased to obtain equivalent weed control. Refer to manufacturer's recommendations for information on the relationship between gallons per acre, spray pressure, sprayer speed, nozzle type and arrangement, nozzle height above the target canopy, droplet size and drift potential for respective application equipment. Do not apply less than 2 and not more than 40 gallons per acre total spray volume. For best results, apply 5 or more gallons per acre by air and 10 or more gallons per acre by ground. Reducing total spray volume may result in decreased coverage and weed control. Use enough total spray volume and a delivery system to provide thorough coverage and a uniform spray pattern. Do not apply where spray drift may be a problem due to proximity of susceptible crops or other desirable plants.

Use of Surfactants: Addition of wetting and/or penetration agents is not usually necessary when using CURTAIL M Herbicide; however, if a surfactant will be added to the spray solution, use a nonionic surfactant approved for use on growing crops, such as X-77, of at least 80% active ingredient and **do not exceed** 4 pints per 100 gallons of spray solution (0.5% v/v). Adding a surfactant to the spray mixture may increase effectiveness on weeds but may reduce selectivity to the crop, particularly under conditions which promote plant stress.

Use With Sprayable Liquid Fertilizer Solutions: CURTAIL M Herbicide is compatible with most non-pressurized liquid fertilizer solutions, however, if these solutions are to be sprayed with CURTAIL M Herbicide, compatibility test (jar test) should be made prior to mixing. Jar tests are particularly important when a new batch of fertilizer or pesticide is used, when water source changes, or when tank mixture ingredients or concentrations are changed. A compatibility test is performed by pouring the spray components (in the desired order and proportions) into a clear glass jar before mixing in the spray tank. Use of a compatability aid such as UNITE or COMPEX may help obtain and maintain a uniform spray solution during mixing and application. Agitation in the spray tank must be vigorous to compare with jar test agitation. For best results, liquid fertilizer rates should not exceed 50% of the total spray volume. Premix CURTAIL M Herbicide with water and add to the liquid fertilizer/water mixture while agitating contents of the spray tank. Apply the spray the same day it is prepared while maintaining continuous agitation. NOTE: Foliar-applied liquid fertilizers can cause yellowing or leaf burn of crop foliage.

Wheat, Barley and Oats

Timing: Apply CURTAIL M Herbicide in the spring to actively growing wheat, barley or oats once 3 leaves have unfolded on the main stem and tillering has begun up to the jointing stage (first node of main stem detectable). To control or suppress weeds listed above, make application after maximum emergence of the target weeds but before they exceed 3 inches in height or diameter (for rosettes). To obtain season-long control of perennial weeds such as Canada thistle, apply after the majority of the weed's basal leaves have emerged from the soil, but before bud stage. A later timing of application (when the grain is between the jointing and boot stages) may be used to treat later-emerging weeds, however, this may increase the risk of crop injury. Do not apply after the boot stage. **Note:** Do not permit dairy animals or meat animals being finished for slaughter to forage or graze treated grain fields within 1 week after treatment. Do not harvest hay from treated grain fields.

Rate: Apply 1 3/4 to 2 1/3 pints of CURTAIL M Herbicide per acre. The higher rate may be used when the condition of the weeds and/or crop at the time of treatment may prevent optimum control. **Note: Higher rates of CURTAIL M Herbicide or CURTAIL M Herbicide applications following spring postemergence treatments with 2,4-D or MCPA may increase the risk of crop injury.**

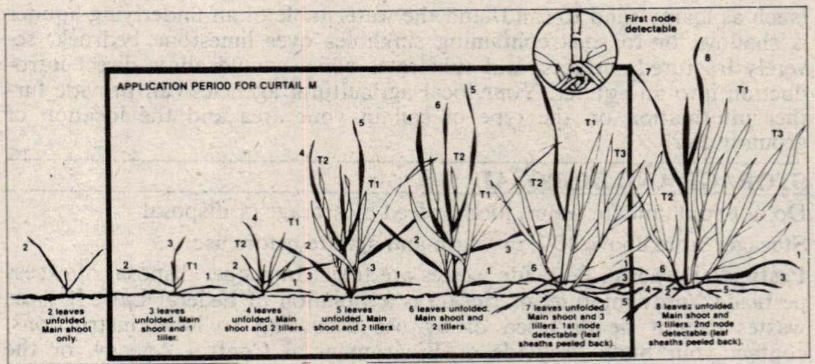

Tank Mixtures for Wheat, Barley and Oats: This product may be tank mixed with the following herbicides for control of additional weeds in wheat, barley and oats. Read and follow the label of each tank mix product used for precautionary statements, directions for use, weeds controlled and geographic and other restrictions. When tank mixing, use in accordance with the most restrictive of label limitations and precautions. No label dosages should be exceeded. This product cannot be mixed with any product containing a label prohibition against such mixing.

Active Ingredient	Product	Formulation	Amount of Product Per Acre
chlorsulfuron	Glean	75% DF	1/6 to 1/4 wt oz
dicamba[1,2]	Banvel	4 lb/gal	1/8 to 1/4 pt
difenzoquat[3]	Avenge	2 lb/gal	2 1/2 to 4 pt
diuron	Direx 4L	4 lb/gal	1 to 1 1/2 pt
	Diuron 4L		
	Diuron 80 WDG	80% DF	1/2 to 1 lb
	Diuron DF	80% WP	
2,4-D[2,3]		4 lb/gal	Up to 1/2 pt
metribuzin[3]	Lexone DF	75% DF	2 1/2 to 4 wt oz
	Sencor DF		
metsulfuron methyl	Ally	60% DF	1/10 wt oz
terbutryn[3]	Igran 80 WP	80% WP	6 to 10 wt oz
thiameturon	Harmony	75% DF	1/3 to 1/2 wt oz

[1] Banvel tank mixes with CURTAIL M Herbicide may be useful in broadening the annual weed control spectrum but may reduce control of perennials, such as Canada thistle.

[2] Note: Do not tank mix CURTAIL M Herbicide with 2,4-D or dicamba unless the risk of crop injury is acceptable.

[3] Tank mix for application on wheat and barley only.

Grasses Grown for Seed

Rates: Use 1 3/4 to 3 1/2 pints of CURTAIL M Herbicide per acre for control of annual weeds and Canada thistle. The potential for crop injury exists due to the MCPA component of this product and must be balanced against the benefits of improved weed control.

Timing: Apply only to established grasses before the boot stage. Applications in the boot stage and beyond will result in increased injury. Do not apply to bentgrass unless injury can be tolerated.

Conservation Reserve Program (CRP) For Seeding To Permanent Grasses Only

Do not use CURTAIL M Herbicide if legumes or bentgrass are a desired cover during CRP.

Grass grown on CRP acres cannot be grazed or used for hay.

Timing: CURTAIL M Herbicide can be applied when perennial grasses have become established (tillering, with at least 1.5 inches secondary roots, and good vigor). For control of weeds such as musk thistle (after rosette stage), Canada thistle and knapweed (diffuse, spotted and Russian), apply to actively growing weeds after the majority of the basal leaves have emerged, but before bud stage. Later applications may result in less consistent control.

After CRP, do not plant broadleaf crops in treated areas until an adequately sensitive bioassay shows that no detectable clopyralid is present in the soil.

RATE: Apply 1 3/4 to 4 3/4 quarts of CURTAIL M Herbicide per acre.

USE PRECAUTIONS

Apply only once per 12-month period.

This product can affect susceptible broadleaf plants directly through foliage and indirectly by root uptake from treated soil. Therefore, do not apply CURTAIL M Herbicide directly to or allow spray drift to come in contact with vegetables, flowers, grapes, tomatoes, potatoes, beans, lentils, peas, alfalfa, sunflowers, soybeans, safflower, or other desirable broadleaf crops and ornamental plants or soil where these sensitive crops will be planted the same season.

Do not plant or rotate to any crop for one year after treatment, except for wheat, barley, oats, grass or sugar beets. To avoid potential phytotoxicity, do not plant peas, lentils, beans (except soybeans) or broadleaf seed crops for 18 months after treatment unless the risk of injury is acceptable.

Avoid Spray Drift: Applications should be made to avoid spray drift since very small quantities of the spray, which may not be visible, may severely injure susceptible crops during both growing and dormant periods. Use coarse sprays to minimize drift since, under adverse weather conditions, fine spray droplets may drift a mile or more. Fine droplets can present a drift hazard. A drift control or deposition agent such as NALCO- TROL may be used with this product to aid in reducing spray drift. If used, follow all use recommendations and precautions on the product label.

Ground Application: To minimize spray drift, apply CURTAIL M Herbicide in a total spray volume of 10 or more gallons per acre as a large-droplet, low pressure spray. Refer to manufacturer's recommendations for additional information on gallons per acre, spray pressure, sprayer speed, nozzle types and arrangements, nozzle heights above the target canopy, etc., for respective application equipment. Spot treatments should only be applied with a calibrated boom to prevent misapplication. With ground equipment, spray drift can be lessened by keeping the spray boom as low as possible; by applying no more than 20 gallons of spray per acre; by using no more than 30 pounds spraying pressure with large droplet-producing nozzle tips; by spraying when wind velocity is low; and by stopping all spraying when wind exceeds 6 to 7 miles per hour. Do not apply with hollow cone-type insecticide or other nozzles that produce a fine-droplet spray.

Aerial Application: With aircraft, drift can be lessened by applying a coarse spray; by using no more than 30 pounds spray pressure at the nozzles; by using straight-stream nozzles directed straight back; by using a spray boom no longer than 3/4 the wing span of the aircraft; and by spraying only when wind velocity is less than 6 mph.

Do Not Apply By Aircraft When An Air Temperature Inversion Exists. Such a condition is characterized by little or no wind and with lower air temperature near the ground than at higher levels. The use of a smoke device on the aircraft or continuous smoke column at or near site of application will indicate air direction and velocity, and whether a temperature inversion is present, which is shown by layering of the smoke.

Do not move treated soil and avoid situations where treated soil particles may blow into areas where susceptible crops are grown. Violent windstorms may move soil particles. If this product is on soil particles and they are blown onto susceptible plants, visible symptoms may appear. Serious injury is unlikely. The hazard of movement of this product on dust is reduced if treated fields are irrigated or if rain occurs shortly after application.

Straw from treated areas, or manure from animals which have grazed treated areas, cannot be used for composting or mulching on ground where susceptible crops may be grown the following season. To promote herbicide decomposition, plant material should be evenly incorporated or burned.

Do not use in a greenhouse. Excessive amounts of this herbicide in the soil may temporarily inhibit seed germination or plant growth.

Do not contaminate water when disposing of equipment washwaters. Apply this product only as specified on this label.

Do not contaminate irrigation ditches or water used for domestic purposes.

Do not transfer livestock from treated grazing areas onto sensitive broadleaf crop areas without first allowing 7 days of grazing on an untreated pasture. Otherwise, urine may contain enough clopyralid to cause injury to sensitive broadleaf plants.

PRECAUTIONARY STATEMENTS

ENVIRONMENTAL HAZARDS

Clopyralid is a chemical which can travel (seep or leach) through soil and under certain conditions contaminate groundwater which may be used for irrigation or drinking purposes. Users are advised not to apply clopyralid where soils have a rapid to very rapid permeability throughout the profile (such as loamy sand to sand) and the water table of an underlying acquifer is shallow, or to soils containing sinkholes over limestone bedrock, severely fractured surfaces, and substrates which would allow direct introduction into an aquifer. Your local agricultural agencies can provide further information on the type of soil in your area and the location of groundwater.

STORAGE AND DISPOSAL

Do not contaminate water, food or feed by storage or disposal.

Storage: Store above 10°F or warm and agitate before use.

Pesticide Disposal. Pesticide wastes are toxic. Improper disposal of excess pesticide, spray mixture, or rinsate is a violation of Federal Law. If these wastes cannot be disposed of by use according to label instructions, contact your State Pesticide or Environmental Control Agency, or the Hazardous Waste representative at the nearest EPA Regional Office for guidance.

Container Disposal: Metal Do not reuse container. Triple rinse (or equivalent). Puncture and dispose of in a sanitary landfill, or by other procedures approved by state and local authorities.

Container Disposal: Plastic Do not reuse container. Triple rinse (or equivalent). Puncture and dispose of in a sanitary landfill, or by incineration, or, if allowed by state and local authorities, by burning. If burned, stay out of smoke.

WARRANTY LIMITATIONS AND DISCLAIMER

The Dow Chemical Company warrants that this product conforms to the chemical description on the label and is reasonably fit for the purposes stated on the label when used in strict accordance with the directions therein under normal conditions of use. THIS IS THE ONLY WARRANTY MADE ON THIS PRODUCT. NO OTHER EXPRESS AND NO IMPLIED WARRANTY OF MERCHANTABILITY OR FITNESS FOR A PARTICULAR PURPOSE IS MADE OUTSIDE OF THIS LABEL. Therefore, neither this warranty nor any other warranty of merchantability or fitness for a particular purpose, express or implied, extends to the use of this product contrary to label instructions (including conditions noted on the label, such as unfavorable temperatures, soil conditions, etc.), under abnormal conditions (such as excessive rainfall, drought, tornadoes, hurricanes, etc.) or under conditions not reasonably foreseeable to or beyond the control of seller.

When buyer or user suffers losses or damages resulting from the use or handling of this product (including claims based on contract, negligence, strict liability, or other legal theories), buyer or user must promptly notify in writing The Dow Chemical Company and any claims to be eligible to receive either remedy given below. The EXCLUSIVE REMEDY OF THE BUYER OR USER and the LIMIT OF LIABILITY of The Dow Chemical Company or any other seller will be one of the following, at the election of The Dow Chemical Company:

(1) **Refund of purchase price** paid by buyer or user for product bought, **or**

(2) **Replacement** of amount of product used.

The seller **will not be liable for consequential or incidental damages or losses.**

The terms of this Warranty Limitations And Disclaimer cannot be varied by any written or verbal statements or agreements. Any employee or sales agent of the seller is not authorized to vary or exceed the terms of this Warranty Limitations And Disclaimer in any manner.

12168-L1

*Trademark of THE DOW CHEMICAL COMPANY

A 389

LORSBAN* 4E

INSECTICIDE

For Control of Various Insects Infesting Certain Field, Fruit, Nut, and Vegetable Crops

ACTIVE INGREDIENT(S):

Chlorpyrifos [O,O-diethyl O-(3,5,6-trichloro-2-pyridinyl) phosphorothioate]	40.7%
INERT INGREDIENTS:	59.3%
TOTAL	100.0%

Contains 4 pounds of chlorpyrifos per gallon

E.P.A Registration No. 464-448 E.P.A. Est. 464-MI-1

KEEP OUT OF REACH OF CHILDREN

WARNING

AVISO:

PRECAUCIÓN AL USUARIO: Si usted no lee inglés, no use este producto hasta que la etiqueta le haya sido explicada ampliamente.

PRECAUTIONARY STATEMENTS

HAZARDS TO HUMANS AND DOMESTIC ANIMALS

MAY BE FATAL IF SWALLOWED • HARMFUL IF ABSORBED THROUGH SKIN • CAUSES SUBSTANTIAL BUT TEMPORARY EYE INJURY • CAUSES SKIN IRRITATION

Avoid Breathing Vapor Or Spray Mist • Do Not Get In Eyes, On Skin, Or On Clothing • Wear Protective Clothing, Rubber Gloves, And Goggles • Wash Thoroughly With Soap And Water After Handling And Before Eating Or Smoking • Immediately Remove Contaminated Clothing And Wash Before Reuse • Destroy Contaminated Shoes • Do Not Apply This Product In Such A Manner As To Directly Or Through Drift Expose Workers Or Other Persons • The Area Being Treated Must Be Vacated By Unprotected Persons • Keep Away From Food, Feedstuffs And Water Supplies

STATEMENTS OF PRACTICAL TREATMENT:

If Swallowed: Do not induce vomiting. Contains aromatic petroleum solvent. Call a physician or poison control center immediately.

If In Eyes: Flush with plenty of water for at least 15 minutes. Get medical attention.

If On Skin: Wash with plenty of soap and water. Get medical attention if irritation persists.

If Inhaled: Remove to fresh air if effects occur. Get medical attention.

NOTE TO PHYSICIANS: Chlorpyrifos is a cholinesterase inhibitor. Treat symptomatically. If exposed, plasma and red blood cell cholinesterase tests may indicate significance of exposure (baseline data are useful). Atropine, only by injection, is the preferable antidote. Oximes, such as 2-PAM/protopam, may be therapeutic if used early; however, use only in conjunction with atropine. In case of severe acute poisoning, use antidote immediately after establishing an open airway and respiration.

PHYSICAL OR CHEMICAL HAZARDS

COMBUSTIBLE • Do Not Use or Store Near Heat or Open Flame • Do Not Cut or Weld Container.

ENVIRONMENTAL HAZARDS

This pesticide is toxic to birds and wildlife, and extremely toxic to fish and aquatic organisms. Do not apply directly to water. Drift and runoff from treated areas may be hazardous to aquatic organisms in adjacent aquatic sites. Cover or incorporate spills. Do not contaminate water by cleaning of equipment or disposal of waste. This product is highly toxic to bees exposed to direct treatment or residues on blooming crops or weeds. Avoid use when bees are actively foraging. Protective information may be obtained from your Cooperative Agricultural Extension Service.

NOTICE

Read and understand the entire label before using. Use only according to label directions.

Before buying or using this product, read "WARRANTY LIMITATIONS AND DISCLAIMER" elsewhere on this label. If terms are not acceptable, return **unopened** package at once to seller for full refund of purchase price paid. Otherwise, use by the buyer or any other user constitutes acceptance of the terms under WARRANTY LIMITATIONS AND DISCLAIMER.

IN CASE OF AN EMERGENCY endangering life or property involving this product, call collect 517-636-4400

AGRICULTURAL CHEMICAL

Do Not Ship or Store with Food, Feeds, Drugs, or Clothing

REENTRY STATEMENT

Do not allow reentry into treated areas within 24 hours unless protective clothing is worn. Protective clothing consists of at least a hat or other suitable head covering, a long-sleeved shirt and long-legged trousers or a coverall type garment (all of closely woven fabric covering the body, including the arms and legs), shoes and socks. Because certain states may require more restrictive reentry intervals for various crops treated with this product, consult your State Department of Agriculture for further information.

Written or oral warnings must be given to workers who are expected to be in a treated area or in an area about to be treated with this product. These oral warnings shall inform workers of areas or fields that may not be entered without protective clothing and actions to take in case of accidental exposure as shown by the Statement of Practical Treatment section on this label. When oral warnings are given, warnings shall be given in a language customarily understood by workers. Written warnings must include the following information: WARNING. Area treated with LORSBAN 4E insecticide on (insert date of application). Do not enter without protective clothing for 24 hours. In case of accidental exposure (insert actions to take as shown by the Statement of Practical Treatment section on this label).

Do not use on turf or for structural pest control. Do not formulate this product into other end use products.

DIRECTIONS FOR USE

It is a violation of Federal law to use this product in a manner inconsistent with its labeling.

GENERAL INFORMATION

LORSBAN 4E insecticide forms an emulsion when diluted with water and is suitable for use in all conventional spray equipment.

MIXING INSTRUCTIONS

To prepare the spray, add a portion of the required amount of water to the spray tank and with agitation add the LORSBAN 4E. Complete filling the tank with the balance of water needed. Maintain sufficient agitation during both mixing and application to ensure uniformity of the spray mixture.

LORSBAN 4E insecticide may also be used in tank mixtures with certain herbicides and/or with non-pressure fertilizer solutions as recommended under specific crop use directions. Note that LORSBAN 4E may be inactivated by sulfur containing fertilizer solutions. Prepare tank mixtures in the same manner as recommended above for use of LORSBAN 4E alone. When tank mixtures of LORSBAN 4E and herbicides are involved, add wettable powders first, flowables second, and emulsifiable concentrates last. Where a fertilizer solution is involved, it is strongly recommended that a fertilizer pesticide compatibility agent such as Unite or Compex be used. Maintain constant agitation during both mixing and application to ensure uniformity of the spray mixture. Do not allow spray mixtures to stand overnight.

NOTE: Test compatibility of the intended tank mixture before adding LORSBAN 4E to the spray or mix tank. Add proportionate amounts of each ingredient to a pint or quart jar, cap, shake, and let set 15 minutes. Formation of precipitates that do not readily redisperse indicates an incompatible mixture that should not be used.

SPRINKLER IRRIGATION

LORSBAN 4E may be applied by sprinkler irrigation for the following crop uses:

Alfalfa, Citrus Orchard Floors, Field Corn, Popcorn,
Sweet Corn, Cotton, Cranberries, Sorghum, and
Soybeans

See the use sections for the individual crops for further application information. Do not apply this product to the above listed crops through any other type of irrigation system. Do not apply this product by chemigation to any other crop.

The following use directions are to be followed when LORSBAN 4E is applied through sprinkler irrigation systems. Thoroughly clean the injection system and tank of any fertilizer or chemical residues, and dispose of the residues according to state and federal laws. Flush the injector with soap and water. Determine the amount of insecticide needed to cover the desired acreage. Pump the required LORSBAN 4E insecticide into the steel tank, start mechanical or hydraulic agitation, and add in order the non-emulsifiable oil and/or water. Continually agitate the mixture containing LORSBAN 4E. Set the sprinkler system to deliver the desired inches of water per acre. Start the water pump and sprinkler, and let the system achieve the desired pressure and speed before starting the injector. Start the injector and calibrate the injector system according to section 14 (below). The mixture containing LORSBAN 4E must be injected continuously and uniformly into the irrigation water line as the sprinkler is moving. This procedure is necessary to deliver the desired rate per acre in a uniform manner. When the application is finished, allow the entire irrigation and injector system to be thoroughly flushed clean before stopping the system.

The following use precautions will result in a safe and successful application of mixtures containing LORSBAN 4E.

1. Apply this product only through sprinkler irrigation systems including center pivot, lateral move, end tow, side (wheel) roll, traveler, big gun, solid set, micro sprinkler, or hand move. Do not apply this product through any other type of irrigation system.
2. Crop injury, lack of effectiveness, or illegal pesticide residues in the crop can result from nonuniform distribution of treated water.

3. If you have questions about calibration, you should contact State Extension Service specialists, equipment manufacturers or other experts.
4. Do not connect an irrigation system (including greenhouse systems) used for pesticide application to a public water system.
5. A person knowledgeable of the chemigation system and responsible for its operation, or under the supervision of the responsible person, shall shut the system down and make necessary adjustments should the need arise.
6. The system must contain a functional check valve, vacuum relief valve, and low pressure drain appropriately located on the irrigation pipeline to prevent water source contamination from backflow. Refer to the American Society of Agricultural Engineer's Engineering Practice 409 for more information.
7. The pesticide injection pipeline must contain a functional, automatic, quick-closing check valve to prevent the flow of fluid back toward the injection pump.
8. The pesticide injection pipeline must also contain a functional, normally closed, solenoid-operated valve located on the intake side of the injection pump and connected to the system interlock to prevent fluid from being withdrawn from the supply tank when the irrigation system is either automatically or manually shut down.
9. The system must contain functional interlocking controls to automatically shut off the pesticide injection pump when the water pump motor stops.
10. The irrigation line or water pump must include a functional pressure switch which will stop the water pump motor when the water pressure decreases to the point where pesticide distribution is adversely affected.
11. Systems must use a metering pump, such as a positive displacement injection pump (e.g., diaphragm pump) effectively designed and constructed of materials that are compatible with pesticides and capable of being fitted with a system interlock. The metering pump must provide a greater pressure than that of the irrigation system at the point of injection. The pump must meet Section 675 for "Electrically Driven or Controlled Irrigation Machines" NEC 70 and must contain Viton or Teflon seals.
12. To insure uniform mixing of the insecticide into the water line, inject the mixture through a nozzle placed in the fertilizer injection port or just ahead of an elbow or tee in the irrigation line so that the turbulence created at those points will assist in mixing. It is suggested that the injection point be higher than the insecticide tank to prevent siphoning.
13. The steel tank holding the insecticide mixture should be large enough to allow the system to complete a revolution with one filling. It should be free of rust, fertilizer, sediment, and foreign material, and equipped with an in-line strainer situated between the tank and the injector pump.
14. In order to calibrate the irrigation system and injector to apply the mixture containing LORSBAN 4E, determine the following: 1) Calculate the number of acres irrigated by the system, 2) Set the irrigation rate and determine the number of minutes for the system to cover the intended treatment area, 3) Calculate the total gallons of insecticide mixture needed to cover the desired acreage. Divide the total gallons of insecticide mixture needed by the number of minutes to cover the treatment area. This value equals the gallons per minute output that the injector must deliver. Convert the gallons per minute to milliliters or ounces per minute. Calibrate the injector pump with the system in operation at the desired irrigation rate. It is suggested that the injector pump be calibrated at least twice before operation, and the system should be monitored during operation.
15. Do not apply when wind speed favors drift beyond the area intended for treatment. End guns must be turned off during the application, if they irrigate nontarget areas.
16. Do not allow irrigation water to collect or runoff and pose a hazard to livestock, wells, or adjoining crops.
17. Allow foliage to dry before reentering the field.
18. Do not apply through sprinkler systems which deliver a low coefficient of uniformity such as certain water drive units.

USE AND DOSAGE RECOMMENDATIONS

ALFALFA: Use LORSBAN 4E to control the following pests at the dosages indicated by application as a broadcast, foliar spray:

Aphids (suppression)	1/2 pint per acre;
Corn rootworm adults (spotted cucumber beetle), grasshoppers	1/2 to 1 pint per acre;
Alfalfa blotch leafminer, alfalfa looper, alfalfa weevil larvae and adults, armyworms, cutworms, Egyptian alfalfa weevil larvae and adults, plant bugs, leafhoppers, spittlebugs	1 to 2 pints per acre.

Note: Use higher rates to control spotted alfalfa aphid in California and Nevada. Stubble spray may be applied to control leafhopper in the Northeast.

Mix the required dosage with enough water to ensure thorough coverage of crop foliage and apply using aerial (fixed-wing or helicopter) or power-operated ground spray equipment. For aerial application use 2 to 5 gallons of water per acre. For best coverage when using ground application, a minimum of 20 gallons of water per acre with hollow cone nozzles is recommended. Control may be reduced at low spray volumes under high temperature and wind conditions. Treat when field counts or crop injury indicates that damaging pest populations are developing or present; however, do not apply more than once per crop cutting. Some reduction in insect control may be evident under excessively cool conditions. For Egyptian alfalfa weevil control in California, apply the specified dosage in a minimum of 5 gallons of water per acre when larvae are actively feeding and populations reach 15 to 20 larvae per 180 degree sweep with a 15-inch diameter net.

LORSBAN 4E may also be applied through **sprinkler irrigation systems** as a postemergence broadcast application to control the above listed foliar pests. For best results, use the recommended rate of LORSBAN 4E per acre. Maintain vigorous tank agitation to assure uniformity of the application throughout the injection period. See the Sprinkler Irrigation section of this label for further information.

LORSBAN 4E insecticide should not be tank-mixed with pesticides, surfactants, or fertilizer formulations unless prior use has shown the combination noninjurious under your current conditions of use. Some phytotoxic symptoms may be observed on young, tender rapidly growing alfalfa when treated with LORSBAN 4E. Alfalfa will outgrow the symptoms and no yield loss should be expected.

This product is highly toxic to bees exposed to direct treatment on alfalfa. Do not apply if nearby bees are clustered outside of hives and bees are foraging. Protective information may be obtained from your Agricultural Extension Service.

Restrictions: Do not cut or graze treated alfalfa within 7 days after application of 1/2 pint of LORSBAN 4E per acre, within 14 days after application of 1 pint per acre, or within 21 days after application of rates above 1 pint per acre. Do not make more than 4 applications per year or apply more than once per crop cutting.

ASPARAGUS: Use LORSBAN 4E to control cutworms, asparagus aphids, and asparagus beetles by application at the rate of 2 pints per acre. Mix the specified dosage in sufficient water to ensure thorough coverage of treated plants and apply as a broadcast, foliar spray. For cutworms, it is preferable to apply LORSBAN 4E when the soil is moist and worms are active on or near the soil surface. Applications may be made during the fern stage for control of asparagus beetles and asparagus aphids when field counts or crop injury indicates that damaging pest populations are developing or present.

Restrictions: Do not make more than one preharvest application per season or apply within one day of harvest. Do not make more than two postharvest applications during the fern stage. Based on available residue data, the use of LORSBAN 4E on asparagus is limited to the Midwest and Pacific Northwest.

CHERRIES: Use LORSBAN 4E for the control of lesser peachtree borer, greater peachtree borer, and American plum borer by application as a trunk spray. Mix 1 1/2 to 3 quarts of LORSBAN 4E with 100 gallons of water and apply as a coarse, low-pressure spray to give uniform coverage of tree trunks and lower limbs. Make a second application two weeks after the first one and a third application after harvest. Avoid contact with foliage in sweet cherries as premature leaf drop may result. Consult your State Agricultural Experiment Station or Extension Service Specialist for proper time to treat your area.

In addition, one of the three allowable applications per year may be applied as a dormant spray for control of San Jose scale, peach twig borer, and climbing cutworms. For control of these pests tank mix 1/2 to 1 pint of LORSBAN 4E with 1 to 2 gallons of a petroleum oil recommended for dormant use in 100 gallons of water and spray the entire tree by application to runoff using ground spray equipment. For low volume (concentrate) sprays (40 to 100 gallons of spray mixture per acre) use the same amounts of LORSBAN 4E and spray oil per acre required for application as a dilute spray and apply in a manner that will ensure thorough coverage of the trees. Use the higher dosage of LORSBAN 4E for severe infestations. Use oil as recommended by your State Agricultural Experiment Station or Extension Service Specialist.

Restrictions: Make only three applications per year. Do not apply within 6 days before harvest. Do not allow meat or dairy animals to graze in treated orchards.

CITRUS FRUITS: Use LORSBAN 4E insecticide at the rates indicated to control the following pests. Use the lower rates for light infestations and increase the dosage for heavier infestations.

CROP	PEST	DOSAGE LORSBAN 4E FOR SPRAY COVERAGE: DILUTE (PINTS/100 GAL. WATER)	DOSAGE LORSBAN 4E FOR SPRAY COVERAGE: DILUTE OR CONCENTRATE (PINTS/ACRE)	GALLONS OF SPRAY PER ACRE FOR DILUTE APPLICATIONS[3]	REMARKS
Grapefruit Lemons Oranges and Other Citrus Fruits	Aphids, Katydids Lepidopterous larvae Cutworms Fruittree leafroller Orange tortrix Western tussock moth	Outside Coverage Spray (OC) ½ to 1	2 to 7	200 to 700	Use directions apply in all states including California
	Citrus thrips[1], Mealy bugs[1] Scale insects[2] Black scale Brown soft scale California red scale Snow scale, Chaff scale	½ to 1	2 to 7	200 to 700	Use directions apply in all states except California (See footnotes 1 & 2)
	Citrus rust mite		4 to 7	100 to 700	Use directions apply in Florida, only.

[1] In California apply as Outside Coverage (OC) using 6 to 12 pints of LORSBAN 4E per acre in 100 to 300 gallons of finished spray per acre.
[2] In California apply as Thorough Coverage (TC) using 8 to 12 pints of LORSBAN 4E per acre. For dilute sprays apply 1200 to 2400 gallons of spray per acre but do not apply more than 12 pints of LORSBAN 4E per acre or use less than ½ pint of LORSBAN 4E per 100 gallons of finished spray per acre.
[3] Gallonage per acre based on average size of mature trees. Adjust to tree size per acre.

For dilute applications, use the specified dosage per 100 gallons of water for the spray type indicated for pest to be controlled. For **concentrate sprays,** use the same amount of LORSBAN 4E per acre that would be needed for application as a dilute spray, using appropriate application/coverage procedures. A petroleum spray oil recommended for use on citrus may be added to **dilute spray mixtures only** at a rate of up to 1.8 gallons per 100 gallons of water to improve control of aphids, mealybugs, scale insects, and thrips. Treat when insects become a problem or in accordance with the local spray schedule recommended by your State Extension Service Specialist. Do not apply when temperature exceeds 95°F. LORSBAN 4E is highly toxic to bees exposed to direct treatment and should not be applied when bees are actively visiting the area. Apply late in evening or early morning to groves where honey bees are expected to visit.

LORSBAN 4E may be applied in tank mixtures with ethion, dicofol, or Vendex. See the Mixing Instructions section of this label for further instructions. Read and carefully follow all applicable directions, restrictions, and precautions on labeling for the other products used in combination with LORSBAN 4E.

Precautions: Observe local use directions for tank mix combinations especially in regard to applications of LORSBAN 4E plus spray oil. Consult with a County Farm Advisor, County Agent, Extension Service personnel, or Agricultural Commissioner for such information regarding a given locality. In California LORSBAN 4E should not be used in combination with spray oil when temperatures are expected to exceed 95°F the day of application or for several consecutive days thereafter. Do not apply when trees are stressed by drought or high temperature. LORSBAN 4E should not be tank mixed with Difolatan 80 Sprills as crop injury may occur.

Restrictions: *Do not apply more than 2 applications per fruit year or more than 15 pints of LORSBAN 4E insecticide per acre per fruit year. Do not make second application within 30 days of the first application. Do not treat within 21 days before harvest for applications up to 7 pints of LORSBAN 4E per acre nor within 35 days for application of rates above 7 pints per acre. Do not pick fruit or do other work involving contact with trees within 2 days after treatment. Do not allow livestock to graze in treated areas.*

CITRUS ORCHARD FLOORS: Use LORSBAN 4E to control red imported fire ants and other ant species by applying the specified dose in 25 or more gallons of water with ground application equipment that will uniformly apply the spray to the orchard floor. To control foraging ants and suppress mounds, apply LORSBAN 4E to the orchard floor at the rate of ¾ to 1 quart per acre. Retreat as needed. For best insect control, uniform coverage of the orchard floor is necessary. Do not apply where weed growth or other obstructions would impede uniform coverage of the orchard floor. Do not apply in tank mixtures with Evik herbicide. Foliar applications of LORSBAN 4E may be made in addition to the orchard floor treatments.

LORSBAN 4E may also be applied to citrus orchard floors through sprinkler irrigation systems only if the system uniformly covers the soil surface at the base of the tree. For best results, use the recommended amount of LORSBAN 4E per acre. See the Sprinkler Irrigation section of this label for further information.

Restrictions: *Do not apply more than 10 quarts of LORSBAN 4E per acre per season. Do not apply last treatment within 28 days before harvest. Do not allow livestock to graze in treated areas. In Florida, do not make more than 3 applications per season.*

CRANBERRIES: Use LORSBAN 4E by application as a broadcast, foliar spray to control brown spanworm, cranberry fruitworm, cranberry weevil, cutworms, fireworms, and Sparganothis fruitworms at the rate of 3 pints per acre. Mix the specified dosage in enough water to ensure thorough coverage and apply no less than 5 gallons of spray per acre when using aerial equipment or no less than 100 gallons of spray per acre when using ground equipment. For weevil control, apply once at flower bud development (late May, early June) and, if weevils are present, once after 100% bloom (early to mid July). For other insects, treat when field counts indicate damaging insect populations are developing or present. Apply only after the winter flood has been removed. To avoid pesticide contamination of flood waters, make no applications while bogs are flooded.

LORSBAN 4E may also be applied through sprinkler irrigation systems to control the above listed pests. For best results, use the recommended rate of LORSBAN 4E per acre. Maintain vigorous tank agitation to assure uniformity of the application throughout the injection period. See the Sprinkler Irrigation section of this label for further information.

Restrictions: *Do not make more than two applications per year or apply within 60 days before harvest.*

FIELD CORN, POPCORN, SWEET CORN: For use to control cutworms, armyworms, corn earworm, corn rootworm adults, chinch bugs, grasshoppers, wireworms, flea beetle larvae and adults, aphids, billbugs, grubs, western bean cutworm, corn borers, symphylans, common stalk borer, and lesser cornstalk borer.

Preplant Incorporation Treatment: Use LORSBAN 4E at the following rates by application in sufficient water to the soil surface and incorporate into the soil:

Cutworms, symphylans	2 to 4 pints per acre;
Wireworms, billbugs, flea beetle larvae, grubs, seed corn maggots, seed corn beetle	4 pints per acre.
Lesser cornstalk borer, corn rootworm larvae	6 pints per acre.

NOTE: For suppression of nematodes in North Carolina, South Carolina, and Virginia, apply 4 to 6 pints of LORSBAN 4E per acre.

Use recommended rate in not less than 10 gallons of water per acre and apply as a broadcast spray to the soil surface using suitable power-operated ground spray equipment. On the same day of treatment, incorporate the insecticide into the top 2 to 4 inches of soil using a disc, field cultivator or equivalent equipment.

LORSBAN 4E insecticide may also be applied in tank mixtures with non-pressure fertilizer solutions and/or with Bladex, Eradicane, Sutan, Lasso, Dual, and atrazine herbicides. See the Mixing Instructions section of this label for further information. Read and carefully follow all applicable directions, restrictions, and precautions on labeling for the other products used in combination with LORSBAN 4E.

Preplant, At-Plant or Preemergence Treatment in Conservation Tillage: Use LORSBAN 4E at the following rates by application in sufficient water to surface trash and exposed soil:

Cutworms, armyworms	1 to 2 pints per acre.

Use recommended rate in not less than 20 gallons of water per acre and apply as a broadcast spray using suitable power-operated ground spray equipment. Use higher rates for residual control.

LORSBAN 4E may also be applied in tank mixtures with non-pressure fertilizer solutions and/or with paraquat and Roundup. See the Mixing Instructions section of this label for further information. Read and carefully follow all applicable directions, restrictions, and precautions on labeling for the other products used in combination with LORSBAN 4E.

Cultivation Time Treatment: Use LORSBAN 4E at the rate of 2 pints per acre to control corn rootworm larvae. Apply LORSBAN 4E as a water emulsion on both sides of the row at the base of the plants just ahead of the cultivator shovels. Cover the insecticide with soil around the brace roots. The best time to apply a basal treatment of a soil insecticide with cultivation is near the beginning of egg hatch. A cultivation application of LORSBAN 4E may be made in addition to an at planting application of LORSBAN 15G granular insecticide.

Postemergence Treatment: Use LORSBAN 4E at the following rate by application in sufficient water to ensure thorough coverage of treated plants:

Grasshoppers . . . . 1/2 to 1 pint per acre;
Armyworms, chinch bugs, aphids, corn rootworm adults, webworms, western bean cutworm, European corn borer (see note) . . . . 1 to 2 pints per acre;
Southwestern corn borer, corn earworm . . . . . . 1 1/2 to 2 pints per acre;
Cutworms, billbugs, lesser cornstalk borer, flea beetle adults, common stalk borer . . . . 2 to 3 pints per acre.

Note: The recommended dosage will control silk clipping by corn rootworm adults. For European corn borer control, use 1 1/2 to 2 pints per acre when application is made with power-operated ground and aerial equipment and 1 to 2 pints per acre when application is made through a sprinkler irrigation system.

Treat when field counts indicate that pests are or may become a problem. For best billbug, chinch bug, and flea beetle control, apply with sufficient water to ensure a minimum spray volume of 20 to 40 gallons per acre and 40 psi using **ground spray equipment.** On corn less than 6 inches tall, apply the insecticide spray in a 9 to 12 inch wide band over the row. On corn greater than 6 inches tall, apply the insecticide spray using drop nozzles directed to the base of the plant. Do not reduce the dosage for banded or directed applications. Concentrate the full labeled dosage rate in the treated zone. When chinch bugs continue to immigrate to corn over a prolonged period or under extreme pressure, a second application of LORSBAN 4E may be needed.

For cutworm, webworm, western bean cutworm, armyworm, aphid, European and southwestern corn borer, grasshopper, lesser cornstalk borer, corn rootworm adult, corn earworm, and common stalk borer control, apply as a broadcast spray using either aerial(fixed-wing or helicopter) or power-operated ground spray equipment. For aerial application use 2 to 5 gallons of spray per acre. Control may be reduced at low spray volumes under high temperature and wind conditions. For cutworms, it is preferable to apply LORSBAN 4E when soil is moist and worms are active on or near the soil surface. If ground is dry, cloddy or crusty at time of treatment, worms may be protected from the spray and effectiveness will be reduced. If such conditions exist, shallow incorporation using a rotary hoe or other suitable equipment immediately before or soon after treatment may improve control. Consult your Agricultural Experiment Station or Extension Service specialist for additional information concerning control practices in your area. For webworm control, shallow incorporation using a rotary hoe or other suitable equipment immediately before or soon after treatment is necessary. For first-generation European corn borer control, treat when 25 and 50 percent of the corn plants show pinhole feeding or leaf-feeding scars. Ground applications should be directed into the corn whorls. For second-generation European corn borer control, treat when field counts of egg masses indicate pests are or may become a problem. For southwestern corn borer control, treat when field counts of egg masses indicate pests are or may become a problem. A second application may be applied 10 to 14 days later, if needed due to reinfestation. For common stalk borer control, treat approximately 11 days after application of Roundup herbicide or after complete burndown with paraquat herbicide (3 to 5 days). Do not use LORSBAN 4E in combination with the burndown herbicide for control of common stalk borer.

LORSBAN 4E insecticide may also be applied through **sprinkler irrigation systems** as a postemergence broadcast application to control the above listed foliar insects. For best results, use the recommended rate of LORSBAN 4E in a tank mix with 2 pints per acre of non-emulsifiable oil. Maintain vigorous tank agitation to assure uniformity of the LORSBAN 4E plus oil mixture throughout the injection period. LORSBAN 4E may also be applied through sprinkler irrigation systems at the rate of 2 to 3 pints per acre to control corn rootworm larvae. Time application to coincide with the appearance of the second instar larvae. Maintain vigorous tank agitation to assure uniformity of the application throughout the injection period. Apply with enough water to wet the root zone to the depth control is needed. Under saturated soil conditions, allow enough soil drying to occur so that an application using a minimum water rate will not produce runoff. Consult University Extension personnel or other experienced consultants to determine the need to treat and to aid in application timing. See the Sprinkler Irrigation section of this label for further information.

Restrictions: Do not apply within 35 days before harvest of grain. Do not apply more than a total of 15 pints of LORSBAN 4E per acre per season. Do not allow livestock to graze in treated areas nor harvest treated corn silage as feed for meat or dairy animals within 14 days after last treatment. Do not feed treated corn fodder to meat or dairy animals within 35 days after last treatment.

SWEET CORN Grown Only in Florida and Georgia: Use LORSBAN 4E to control infestations of beet armyworm, fall armyworm and corn earworm by application as a broadcast, foliar spray at the rate of 1 to 2 pints per acre. Mix the specified dosage in enough water to ensure thorough coverage and apply using suitable aerial or ground spray equipment. For aerial application, use at least 2 gallons of spray per acre. Treat when field counts indicate damaging pest populations are developing or present. Retreat as necessary to maintain control but do not apply more than twenty-two 1-pint or eleven 2-pint treatments per season.

LORSBAN 4E insecticide may also be applied through **sprinkler irrigation systems** as a postemergence broadcast application to control the above listed foliar insects. For best results, use the recommended rate of LORSBAN 4E in a tank mix with 2 pints per acre of non-emulsifiable oil. Maintain vigorous tank agitation to assure uniformity of the LORSBAN 4E plus oil mixture throughout the injection period. See the Overhead Sprinkler Irrigation section of this label for further information.

Restrictions: Do not apply more than 22 pints of LORSBAN 4E per acre per season. Do not harvest corn ears, allow livestock to graze in treated areas, nor feed treated silage, fodder, or grain to meat or dairy animals within 21 days after treatment. Do not use in conjunction with postplant broadcast, foliar applications of LORSBAN 15G granular insecticide.

COTTON: Use LORSBAN 4E for control of the following pests in all states except Arizona and California at the dosages indicated.

Cotton fleahopper, plant bugs (*Lygus, Mirids*) . . 3/8 to 1 pint per acre;
Cotton aphid, fall armyworm, grasshoppers, thrips, yellow-striped armyworm . . . . 1/2 to 1 pint per acre;
Spider mites . . . . 1 pint per acre;
Beet armyworm, bollworms (*Heliothis spp.*), boll weevil, cutworms, pink bollworm, salt marsh caterpillar . . . . 1 1/2 to 2 pints per acre.

NOTE: The recommended dosage rate of 3/8 pint per acre will not achieve the high degree of control of the higher label rate, but will minimize the damage done by plant bugs and cotton fleahopper and allow the beneficial insects to survive, build up, and be available to aid in the control of bollworms infesting cotton.

Use LORSBAN 4E for control of the following pests in Arizona and California at the dosages indicated:

Armyworms, cotton aphid, cotton fleahopper, *Lygus*, salt marsh caterpillar, thrips . . . . 1 to 2 pints per acre;
Bollworms, (*Heliothis spp.*), boll weevil, cutworms, pink bollworm . . . . 2 pints per acre.

NOTE: The 2 pint rate will aid in the suppression of cotton leafperforator and spider mites.

Mix the required dosage with sufficient water to ensure thorough coverage of plants and apply using aerial or power-operated ground spray equipment. For aerial application, use at least 1 gallon of spray per acre. Treat when field counts indicate damaging insect populations are developing or present. Re-treat as necessary to maintain control.

LORSBAN 4E insecticide may also be applied through **sprinkler irrigation systems** as a postemergence broadcast application to control the above listed foliar pests. For best results, use the recommended rate of LORSBAN 4E per acre. Maintain vigorous tank agitation to assure uniformity of the application throughout the injection period. See the Sprinkler Irrigation section of this label for further information.

For effective control of spider mites when large numbers of eggs are present, apply a second spray 3 to 5 days after initial treatment to control newly hatched nymphs.

For best results on bollworms, it is suggested that fields be scouted twice per week and treatments made when worms are 1/4 inch or less in length. The following table illustrates the size of worms in relation to age and stage of development (instar) as a guide to timing of treatments for best control.

From the table it can be seen that a scouting schedule of only once per week will not be satisfactory since the worms may be too big to control effectively by the seventh or eighth day.

Timing for the best worm control

	Age (days)	Size	Instar
Get the worms at this stage	Hatch	1/16"	Hatch
	3	3/32"	I
	5	9/32"	II
	6	7/16"	III
	8	11/16"	IV

Proper application techniques help to ensure thorough spray coverage and correct dosage and are thus important in obtaining good control of pests. Consider these suggestions when applying LORSBAN 4E insecticide on cotton.

Aerial Application: Shorten boom length to avoid spray entering the vortices at the wing tips. Swath width should be reduced when wind direction is the same as direction of spraying.

The proper nozzle arrangement and swath width to avoid skips and vortices effect can be checked out by flying over a paper tape (adding machine paper) using water with or without soluble dye. (The dye gives a permanent record.)

Flying at a height of 5-15 feet above the target results in the best coverage.

Nozzle orientation of the boom is important. More break-up occurs when nozzles are pointed straight down versus the straight back position. Desired droplet size (100-200 microns) can be obtained by angling the nozzles somewhere in this range.

Marking of swath by flagging or permanent markers is essential.

Ground Application: Orient the boom and nozzles so that uniform coverage is obtained. The swath width should not be wider than the boom; drift spray is wasted spray so do not depend on it. Use flat fan or disc-core hollow cone nozzles with maximum spacing of 20 inches and spray pressure of 40-60 psi with a droplet size of 100-200 microns.

Restrictions: Do not apply within 14 days before harvest. Do not allow livestock to graze in treated areas. Do not feed gin trash or treated forage to livestock.

LORSBAN 4E insecticide may also be used in tank mixtures in all states to control armyworms, bollworms ***(Heliothis spp.),*** **boll weevil, cotton fleahopper, plant bugs** ***(Lygus, mirids),*** **and spider mites** as indicated in the following table.

Tank Mixture	Rate/Acre
LORSBAN 4E plus Ambush 2E	1 to 2 pints plus 1/5 to 2/5 pint
LORSBAN 4E plus Pounce 3.2E	1 to 2 pints plus 1/8 to 1/4 pint
LORSBAN 4E plus Pydrin 2.4E	1 to 2 pints plus 1/6 to 1/3 pint
LORSBAN 4E plus Bolstar 6E	1/2 to 1 pint plus 2/3 to 1 1/3 pints
LORSBAN 4E plus methyl parathion[1] Plus EPN[1]	1 pint plus 0.6 lb a.i plus 0.6 lb a.i.

[1] Use only emulsifiable formulations containing these active ingredients (a.i.).

Mix the tank mixture components in sufficient water to ensure thorough coverage of plants (at least 1 gallon of spray per acre) and apply using aerial spray equipment. Treat when field counts indicate damaging insect populations are developing or present. Re-treat as necessary to maintain control.

In California and Arizona, use LORSBAN 4E at 1 to 2 pints per acre plus Monitor 4E at 1 to 2 pints per acre or LORSBAN 4E at 1 to 2 pints per acre plus Orthene 75S at 0.33 to 1.3 pounds per acre for control of armyworms, bollworms, cabbage looper, cotton aphids, cotton fleahopper, cutworms, grasshopper, omnivorous leafroller, pink bollworm, plant bugs (*Lygus Mirids*), saltmarsh caterpillar, thrips, and whitefly (suppression). Apply in 10 to 25 gallons of water per acre with ground spray equipment or in 5 to 10 gallons of water per acre with aerial spray equipment. Treat when field counts indicate that damaging insect populations are developing or present. For tank mixtures which include Monitor 4E, apply before 65 percent of the bolls have opened. Re-treat as necessary to maintain control.

Restrictions: Make no applications of these tank mixtures closer to harvest than the longest pre-harvest interval shown for any of the products in the tank mixture. Do not allow livestock to graze in treated areas. Do not feed gin trash or treated forage to livestock. Workers entering the field within 48 hours after application should wear protective clothing.

FIGS: Use LORSBAN 4E at the rate of 2 quarts per acre for control of driedfruit beetle by application in sufficient water to the soil surface followed by incorporation into the top 3 inches of soil. Apply to fig orchard soil as a dormant application in late winter prior to beetle emergence and prior to leaf formation.

Restrictions: Make only one application per year. Do not apply within 7 months of harvest. Based on available residue data, use LORSBAN 4E on figs is restricted to California.

GRAPES: Use LORSBAN 4E insecticide for control of grape root borer by application just before the pest emerges from the soil. Mix 4 1/2 pints of LORSBAN 4E with 100 gallons of water and apply 2 quarts of the diluted spray mixture to the soil surface on a 15-square foot area around the base of each vine. Do not allow spray to contact fruit or foliage.

Restrictions: Do not make more than one application per season or apply within 35 days before harvest. Based upon available residue data, the use of LORSBAN 4E in grapes is restricted to states East of the Rocky Mountains.

MINT: Use LORSBAN 4E by application as a broadcast, foliar spray to control cutworms at the rate of 2 to 4 pints per acre and mint root borer at the rate of 4 pints per acre. Mix the specified dosage in water to give no less than 10 gallons of spray per acre and apply using ground spray equipment. For cutworm control, treat during May and June when field counts indicate damaging insect populations are developing or present. When larvae are less than 3/4 inch in length, use the 2 pint rate. When larvae are 3/4 inch or more in length, use the higher rate. **Make only one application during the growing season. Do not apply within 90 days before harvest.** For **mint root borer** control, apply post-harvest when field counts indicate damaging insect populations are developing or present. Follow treatment with approximately 1 acre inch of sprinkler irrigation immediately after application to incorporate the insecticide into the soil. **Make only one post-harvest application per season.**

NECTARINES, PEACHES: Use LORSBAN 4E insecticide for the control of peach tree borers by application as a trunk spray before newly hatched borers enter the trees. Mix 3 quarts of LORSBAN 4E with 100 gallons of water and apply as a coarse, low-pressure spray to give uniform coverage of tree trunks. Thoroughly wet all bark areas from ground level to scaffold limbs. Do not allow spray to contact fruit. Consult your State Agricultural Experiment Station's or Extension Service Specialist's written recommendations for proper time to treat in your area.

LORSBAN 4E may also be used as a preplant dip application for nonbearing peach trees at the equivalent application rate of 3 quarts per 100 gallons of water for control of peach tree borer. Dip trees several inches above the grafting bud scar and plant immediately or allow to dry before returning to storage. Do not allow peach trees to remain in contact with the dip solution.

Restrictions: Make only one application per season. Do not apply within 14 days before harvest. Do not allow meat or dairy animals to graze in treated orchards.

ONIONS (dry bulb): Use LORSBAN 4E insecticide to control onion maggot by application as an in-furrow drench. Apply LORSBAN 4E at the rate of 1.1 fluid ounce per 1000 linear feet of row at an 18-inch row spacing. Use a minimum of 40 gallons of total drench per acre. Incorporate to a depth of 1 to 2 inches.

Restrictions: Do not make more than one application per year.

PEANUTS: Use LORSBAN 4E for control of lesser cornstalk borer by application as a directed spray to the base of the plants in an 8 to 10 inch band. Apply LORSBAN 4E at the rate of 2-4 pints per acre when field counts indicate damaging insect populations are developing or present. Mix the required dosage in enough water to ensure thorough coverage. Do not apply as a foliage application.

Restrictions: Do not make more than one application per season. Do not harvest within 21 days after treatment. Do not feed treated peanut forage or hay to meat or dairy animals.

SORGHUM: Use LORSBAN 4E Insecticide for control of the following pests at the dosages indicated:

Sorghum midge . . . . (Apply when 30 to 50 percent of the seed heads are in bloom, repeat at 3-day intervals if necessary.);	1/2 pint per acre
Greenbugs, grasshoppers, yellow sugar cane aphid, and other aphids	1/2 to 1 pint per acre;
Chinch bugs, lesser cornstalk borer . . . . (Apply as a directed spray toward the base of the plant using power-operated ground spray equipment with sufficient water to ensure coverage of an 8 to 12 inch band centered on the row. On plants less than 6 inches high, apply an 8 to 12 inch band over the row. Do not reduce the dosage for banded or directed applications. Concentrate the full labeled dosage rate in the treated zone.);	1 to 2 pints per acre
Webworms	1 pint per acre;
Armyworms, corn earworm, cutworms	1 to 2 pints per acre;
European and southwestern corn borer	1 1/2 to 2 pints per acre.

Mix the specified dosage in enough water to ensure thorough coverage and apply using suitable aerial or ground spray equipment. To minimize chemical injury, do not apply LORSBAN 4E to drought stressed grain sorghum within 3 days following irrigation or rain except where the product is applied in irrigation water.

LORSBAN 4E insecticide may also be applied through **sprinkler irrigation systems** as a postemergence broadcast application to control the above listed foliar pests. For best results, use the recommended rate of LORSBAN 4E per acre. Maintain vigorous tank agitation to assure uniformity of the application throughout the injection period. See the Sprinkler Irrigation section of this label for further information.

Restrictions: The treated crop is not to be used for grain forage, fodder, hay, or silage within 30 days after application of 1 pint of LORSBAN 4E per acre or within 60 days after application of rates above 1 pint per acre. Do not treat sweet varieties of sorghum. Do not apply more than 3 pints of LORSBAN 4E per acre per season.

SOYBEANS: For use to control armyworms, bean leaf beetle, corn earworm, cutworms, European corn borer, grasshoppers, green cloverworm, lesser cornstalk borer, Mexican been beetle, saltmarsh caterpillar and other woollybears, southern green stink bug, spider mites, and velvetbean caterpillar.

Soil Treatment: Use LORSBAN 4E at the rate of 1 to 2 pints per acre to control cutworms and lesser cornstalk borer. Mix the specified dosage in a minimum of 10 gallons of spray per acre and apply to the soil surface using suitable ground spray equipment. Equivalent rates of insecticide spray required per 100 feet of row for various row spacing are given in the accompanying table. For at-plant treatments apply the insecticide over the row in a 4 to 6-inch band in front of the planter shoe or press wheel or after the press wheel followed by a drag chain for light incorporation. Do not apply as an in-furrow treatment. For postemergence rescue treatments, apply as a directed spray in a 9 to 12-inch band at the base of the plant. To plants under 6 inches high apply over the top in a 6- to 12-inch band. Treat when field counts or conditions indicate that pests are or may become a problem.

	Fluid Ounces of Spray Required Per 100 Feet of Row for Various Row Spacing			
Volume of Spray Per Acre	36″	32″	28″	24″
10 Gallons	8.8	7.9	6.9	5.9
15 Gallons	13.2	11.8	10.3	8.8
20 Gallons	17.6	15.7	13.7	11.8

Foliar Treatment: Use LORSBAN 4E at the following rate by application in sufficient water to ensure thorough coverage of treated plants:

European corn borer, southern green stink bug . . . 2 pints per acre
Bean leaf beetle, cutworms, corn earworm, saltmarsh caterpillar and other woollybears . . . . . . 1 to 2 pints per acre;
Mexican bean beetle, armyworms . . . . 1 to 1 1/2 pints per acre;
Velvetbean caterpillar, grasshoppers, green cloverworm, spider mites . . . . 1/2 to 1 pint per acre.

Apply as a broadcast spray using either aerial or ground equipment when field counts indicate damaging insect populations are developing or present; retreat as necessary to maintain control. For effective control of spider mites when large numbers of eggs are present, apply a second spray 3 to 5 days after initial treatment to control newly-hatched nymphs. **On determinate soybeans do not apply more than one application after pod set.**

LORSBAN 4E insecticide may also be applied through **sprinkler irrigation systems** as a postemergence broadcast application to control the above listed foliar pests. For best results, use the recommended rate of LORSBAN 4E per acre. Maintain vigorous tank agitation to assure uniformity of the application throughout the injection period. See the Sprinkler Irrigation section of this label for further information.

Restrictions: Do not apply more than 6 pints of LORSBAN 4E per acre or 3 pounds of chlorpyrifos (active ingredient) per acre per season. Do not apply last treatment within 28 days before harvest nor apply last two treatments closer than 14 days apart. Do not allow livestock to graze in treated areas or otherwise feed treated soybean forage, hay, and straw to meat or dairy animals.

STRAWBERRIES: Use LORSBAN 4E insecticide by application as a broadcast foliar spray to control strawberry bud weevil at the rate of 1 quart per acre. Apply in a minimum of 40 gallons of spray per acre when buds first appear and 10 to 14 days later. Do not apply after berries start to form or when berries are present. LORSBAN 4E should not be tank mixed with pesticides, surfactants, or fertilizer formulations unless prior use has shown the combination noninjurious under your current conditions of use. Phytotoxicity may occur when LORSBAN 4E is applied to strawberries experiencing high temperature and drought stress.

Restrictions: For pre-bloom use only. Do not make more than two applications per season or apply within 21 days before harvest.

SUNFLOWERS: For use to control cutworms, sunflower beetle larvae and adults, stem weevil, sunflower moth, banded sunflower moth, woollybears, seed weevil, and grasshoppers.

Preplant Incorporation Treatment: Use LORSBAN 4E insecticide at the following rates by application in sufficient water to the soil surface and incorporate into the soil:

Cutworms . . . . 2 to 4 pints per acre.

Use recommended rate in not less than 10 gallons of water per acre and apply as a broadcast spray to the soil surface using suitable power-operated ground spray equipment. On the same day of treatment, incorporate the insecticide into the top 2 to 4 inches of soil using a disc, field cultivator or equivalent equipment.

Postemergence Treatment: Use LORSBAN 4E for control of the following pests at the dosage indicated by application in sufficient water to ensure thorough coverage of treated plants:

Cutworms . . . . 2 to 3 pints per acre;
Sunflower beetle larvae and adults, stem weevil, sunflower moth, banded sunflower moth, woollybears, and seed weevil . . . . 1 to 1 1/2 pints per acre;
Grasshoppers . . . . 1 pint per acre.

Apply as a broadcast spray using either aerial (fixed-wing or helicopter) or power-operated ground spray equipment when field counts indicate that pests are or may become a problem. For cutworm control, a second treatment may be made 7 to 10 days later, if needed. For stem weevil control, optimal treatment time is within 5 to 7 days after adult weevils begin to appear. For sunflower moth control, make first application during early 1 to 5 percent bloom stage. A second treatment may be made 7 days later, if needed. For seed weevil control, treat when field counts indicate there are 10 to 12 adults per plant for oil crops and 1 to 3 adults per plant on confectionary crops. Additional treatments should be made at successive 7 to 10 day intervals if field counts indicate need to re-treat. For sunflower beetle larvae or adult control, treat when field counts indicate there are 10 larvae or 1 to 2 adults per seedling. Additional treatments may be made at successive 7 to 10 day intervals if field counts indicate need to retreat.

Restrictions: Do not apply more than 9 pints of LORSBAN 4E per acre per season. Do not apply within 42 days before harvest. Do not allow livestock to graze in treated areas.

SUGAR BEETS: Use LORSBAN 4E by application as a broadcast, foliar spray to control beet armyworms at a rate of 1 1/2 to 2 pints per acre and cutworms at a rate of 2 pints per acre. Treat when field counts indicate that damaging insect populations are or may be a problem. Mix the specified dosage with 2 to 5 gallons of water per acre and apply using suitable aerial spray equipment or with 10 to 30 gallons of water per acre when using power-operated ground spray equipment. Re-treat as necessary to maintain control.

Restrictions: Do not apply within 30 days before harvest of beet roots and tops. Do not apply more than a total of 8 pints of LORSBAN 4E per acre per season. Do not allow livestock to graze in treated areas nor harvest treated beet tops as feed for meat or dairy animals within 30 days after last treatment.

SWEET POTATOES: Use LORSBAN 4E to reduce the feeding damage caused by populations of *Conderus* wireworm, *Systena* flea beetle and the sweet potato flea beetle. Apply at the rate of 4 pints per acre as a broadcast (overall) spray to the soil surface followed by incorporation. Mix the specified dosage with enough water to obtain uniform coverage and apply as a coarse spray using suitable ground spray equipment. Incorporate the insecticide to a depth of 4 to 6 inches as soon as possible after application by using a rotary hoe, disc cultivator, or other suitable incorporation equipment. Plant the crop in the usual manner no later than 14 days after treatment (any delay in planting will reduce the length of time that LORSBAN 4E will protect against feeding damage). LORSBAN 4E will not control false wireworms or white fringe beetle or other grubs that attack sweet potatoes.

Restrictions: Do not make more than one application per season. Do not harvest within 125 days of treatment.

TOBACCO: Use LORSBAN 4E for preplant treatment to control larvae of cutworms, flea beetles, mole crickets, root maggots, and wireworms. Apply 2 to 3 quarts of LORSBAN 4E per acre in not less than 10 gallons of water as a broadcast (overall) spray to the soil surface one week before transplanting. Immediately following application, incorporate the insecticide into the soil to a depth of 2 to 4 inches using suitable equipment. The application of LORSBAN 4E will also suppress the movement of imported fire ants into treated fields.

To control the above insects and low to moderate populations of rootknot nematodes in North Carolina, South Carolina, and Virginia, use LORSBAN 4E at the rate of 5 quarts per acre. To control the above insects and moderate populations of rootknot nematodes in all tobacco growing regions, use LORSBAN 4E in a tank mix with Nemacur 3 at the rate of 2 quarts of LORSBAN 4E plus 4 quarts of Nemacur 3 per acre. Read and carefully follow all applicable directions, restrictions, and precautions on labeling for Nemacur 3 used in combination with LORSBAN 4E. Apply the specified dosage in not less than 10 gallons of water as a broadcast (overall) spray to the soil surface 24 to 48 hours before bedding and transplanting. Immediately following application, incorporate into the soil to a depth of at least 4 inches using suitable equipment. Where the nematode species *Meloidogyne arenaria* or *M. javanica* are present or high populations of *M. incognita*, apply TELONE* II soil fumigant at the recommended label rate.

Restrictions: Do not make more than one application per season.

TREE FRUITS: Use LORSBAN 4E as a dormant or delayed dormant spray at the rates indicated to control the following insects on the crops listed. While LORSBAN 4E may be used without oil, oil is recommended for additional pest control.

Crop	Insect	Dosage LORSBAN 4E
Apples	Rosy apple aphid San Jose scale Lygus Pandemis Leafroller Climbing cutworms	½ to 1 pint
Pears	San Jose scale Climbing cutworms Pear psylla adults	
Plums, Prunes	San Jose scale Mealy plum aphid Climbing cutworms Peach twig borer	
Almonds Peaches, Nectarines	San Jose scale Peach twig borer Climbing cutworms	

Tank mix the specified dosage with 1 to 2 gallons of a petroleum spray oil recommended for dormant use in 100 gallons of water and spray the entire tree by application to runoff using suitable ground spray equipment. (SEE PRECAUTIONS FOR USE IN CALIFORNIA.) For low volume (concentrate) sprays (40 to 100 gallons of spray mixture per acre), use the same amounts of LORSBAN 4E and spray oil per acre required for application as a dilute spray and apply in a manner that will ensure thorough coverage of the trees. Use the higher dosage of LORSBAN 4E for severe infestations. Use oil as recommended by your State Agricultural Experiment Station or Extension Service Specialist.

Precautions: Because cold or dry conditions may cause LORSBAN 4E plus oil sprays to infuse trees resulting in bud damage or drop, do not apply until winter rains or irrigation has replenished soil moisture such that bark and twigs are not desiccated.

Additional Precautions Specific to California: Use a minimum of 250 gallons of total spray volume per acre. Do not use more than 4 pints of LORSBAN 4E per acre. Do not use more than 4 gallons of spray oil per acre on almonds, peaches, or nectarines. Do not use any adjuvants or surfactants in addition to or as a substitute for a petroleum spray oil in a tank mix with LORSBAN 4E. Do not apply on almonds in the following counties in California: Butte, Colusa, Glenn, Solano, Sutter, Tehama, Yolo, and Yuba.

Restrictions: Make only one application during the dormant season. Do not allow meat or dairy animals to graze in treated orchards.

TREE NUTS: Use LORSBAN 4E at the dosages indicated by application as a foliar spray to control pests listed in the following table. Mix the required dosage in sufficient water to ensure thorough and complete coverage of the foliage and crop and apply as a concentrate or dilute spray using conventional, power-operated spray equipment. For dilute sprays applied to tree nut crops, mix the required dosage in sufficient water to allow for spray to runoff. For concentrate sprays, apply an equivalent amount of LORSBAN 4E per acre. Treat when pests appear or in accordance with local conditions. To suppress pecan leaf scorch mite on pecans, use in a preventative program. For aerial application on pecans, apply 2 to 4 pints of LORSBAN 4E in 5 to 15 gallons of spray volume per acre. Insect control by aerial application may be less than control by ground application because of less coverage. For ant control on pecans, do not apply where weed growth or other obstructions would impede uniform coverage of the orchard floor. For best pecan *Phylloxera* control, use 2 applications 7 to 10 days apart starting at bud swell and ending by early leaf expansion. Consult your State Agricultural Experiment Station, Certified Pest Control Advisor, or Extension Service specialist for specific use information in your area.

Crop	Insects Controlled	Dosage LORSBAN 4E	Restrictions
Almonds	Navel orangeworm Peach twig borer San Jose scale	4 pints per acre	Make no more than 3 applications per season on almonds and filberts and no more than 2 applications per season on walnuts. Do not apply within 14 days of harvest. Do not allow livestock to graze in treated orchards.
Filberts	Eyespotted bud moth Filbert aphid Filbert leafroller Filbertworm Obliquebanded leafroller Omniverous leaftier Winter moth	3 to 4 pints per acre	
Walnuts	Coding moth Walnut scale	4 pints per acre	
Pecans	Black pecan aphid Hickory shuckworm Pecan leaf scorch mite (suppression)	2 pints per 100 gallon	Make no more than 5 applications per season. Do not apply within 28 days of harvest. Do not allow livestock to graze in treated orchards. For dilute spray apply 200 to 600 gallons per acre. Make no applications of tank mixtures closer to harvest than the longest pre-harvest interval shown for any of the products in the tank mixture.
	Pecan nut casebearer Phylloxera spp. Fall webworm Spittlebug	1 pint per 100 gallon	
	Fire ants and other ant species (orchard floor)	2 pints per acre as a ground spray	
	Yellow pecan aphid complex	1 pint of LORSBAN 4E plus 2⅔ ounce of Pydrin 2.4E or 2.6 ounce of Cymbush 3E or 3 ounce of Ammo 2.5E per 100 gallon	

VEGETABLES: Use LORSBAN 4E to control root maggots on the following crops at the dosages indicated:

Cauliflower . . . . 1.6 to 2.4 fl oz per 1000 linear feet of row;

Broccoli, Brussels Sprouts, Cabbage, Chinese Cabbage, Collards, Kale, Kohlrabi, Turnips . . . . . . 1.6 to 2.75 fl oz per 1000 linear feet of row.

For **direct seeded crops** apply the specified dosage in a water-based spray as a 4-inch wide band over the row at planting time, behind the planter shoe and in front of the press wheel to achieve shallow incorporation. Apply in a minimum of 40 gallons of total spray per acre.

For **transplanted crops,** apply LORSBAN 4E as a water-based spray directed to the base of the plants immediately after setting. Use a minimum of 40 gallons of total spray per acre. Do not add any additional adjuvants, surfactants, or spreader stickers. **Do not apply as a foliage application.**

Restrictions: Do not apply more than 2 pints of LORSBAN 4E to cauliflower planted in 40-inch rows. Use proportional amounts for other row spacings not to exceed 4 pints of LORSBAN 4E per acre. Do not apply more than 2.6 pints of LORSBAN 4E per acre to broccoli, Brussels sprouts, cabbage, Chinese cabbage, collards, kale, kohlrabi, and turnips planted in 40-inch rows or more than 4½ pints of LORSBAN 4E per acre to these crops in 20-inch rows (or two rows per bed). Use proportional amounts for other row spacings not to exceed 4½ pints of LORSBAN 4E per acre. Do not make more than one application per season or apply within 30 days before harvest.

Radishes . . . . 1.0 fl oz per 1000 linear feet of row.

Apply the specified dosage as a water-based drench in the seed furrows with the seed at planting time. Use a minimum of 40 gallons of total drench per acre.

Restrictions: Do not apply more than 5½ pints of LORSBAN 4E per acre or make more than one application per season.

Rutabagas . . . . 1.6 to 3.3 fl oz per 1000 linear feet of row.

Apply the specified dosage in a water-based spray as a 4-inch wide band over the row at planting time, behind the planter shoe and in front of the press wheel to achieve shallow incorporation. Use a minimum of 40 gallons of total spray per acre.

Restrictions: Do not apply more than 4½ pints of LORSBAN 4E per acre or make more than one application per season. Do not use rutabaga tops for food or feed purposes.

To avoid phytotoxicity in vegetable crops, do not mix with other pesticide products or treat plants that are under extreme heat and drought stress.

STORAGE AND DISPOSAL

Do not contaminate water, food or feed by storage or disposal. Open dumping is prohibited.

Storage: Store in original container in secured dry storage area. Prevent cross-contamination with other pesticides and fertilizers. Do not store above 100°F for extended periods of time. Storage below 20°F may result in formation of crystals. If product crystallizes, store at 50 to 70°F and agitate to redissolve crystals. If container is damaged or spill occurs, use product immediately or dispose of product and damaged container as indicated below.

Pesticide Disposal: Pesticide wastes are toxic. Improper disposal of excess pesticide, spray mixture, or rinsate is a violation of Federal law. If these wastes cannot be disposed of by use according to label instructions, contact your State Pesticide or Environmental Control Agency, or the

Hazardous Waste representative at the nearest EPA Regional Office for guidance.

Container Disposal: Triple rinse (or equivalent). Then offer for recycling or reconditioning, or puncture and dispose of in a sanitary landfill, or by other procedures approved by state and local authorities. For reusable portable containers, do not rinse or put any other material into container. When returning reusable portable container, seal all openings and return to owner.

or

Container Disposal: Triple rinse (or equivalent). Then puncture and dispose of in a sanitary landfill, or by incineration, or, if allowed by state and local authorities, by burning. If burned, stay out of smoke.

ATTENTION
Do Not Cut or Weld Container

WARRANTY LIMITATIONS AND DISCLAIMER

The Dow Chemical Company warrants that this product conforms to the chemical description on the label and is reasonably fit for the purposes stated on the label when used in strict accordance with the directions therein under normal conditions of use. THIS IS THE ONLY WARRANTY MADE ON THIS PRODUCT. NO OTHER EXPRESS AND NO IMPLIED WARRANTY OF MERCHANTABILITY OR FITNESS FOR A PARTICULAR PURPOSE IS MADE OUTSIDE OF THIS LABEL. Therefore, neither this warranty nor any other warranty of merchantability or fitness for a particular purpose, express or implied, extends to the use of this product contrary to label instructions (including conditions noted on the label, such as unfavorable temperatures, soil conditions, etc.), under abnormal conditions (such as excessive rainfall, drought, tornadoes, hurricanes, etc.) or under conditions not reasonably foreseeable to or beyond the control of seller.

When buyer or user suffers losses or damages resulting from the use or handling of this product (including claims based on contract, negligence, strict liability, or other legal theories), buyer or user must promptly notify in writing The Dow Chemical Company of any claims to be eligible to receive either remedy given below. The EXCLUSIVE REMEDY OF THE BUYER OR USER and the LIMIT OF LIABILITY of The Dow Chemical Company or any other seller will be one of the following, at the election of The Dow Chemical Company: (1) **Refund of purchase price** paid by buyer or user for product bought, **or** (2) **Replacement** of amount of product used. The seller **will not be liable for consequential or incidental damages or losses.**

The terms of this Warranty Limitations And Disclaimer cannot be varied by any written or verbal statements or agreements. Any employee or sales agent of the seller is not authorized to vary or exceed the terms of this Warranty Limitations And Disclaimer in any manner.

SPECIAL LOCAL NEED REGISTRATIONS

General Information and Mixing Instructions. Use LORSBAN 4E by mixing with water to form an emulsion spray. To prepare the spray, start to fill the spray tank with the required amount of water, add the recommended amount of LORSBAN 4E, and complete filling the tank with the balance of the water needed. Maintain sufficient agitation during mixing and application to ensure uniformity of the spray mixture. Apply LORSBAN 4E with the amounts of water recommended in each SLN. LORSBAN 4E should not be tank mixed with pesticides, surfactants, or fertilizer formulations unless prior use has shown the combination noninjurous under local conditions of use. Consult your Cooperative Agricultural Extension Service for additional information concerning control practices in your area.

CARROTS GROWN FOR SEED: For use only within **Washington** (EPA SLN No. WA-850025) to control cutworms and lygus, apply LORSBAN 4E with the amounts of water recommended below. Consult your Cooperative Agricultural Extension Service for additional information concerning cutworms and lygus control practices in your area. Use LORSBAN 4E by application as a foliar, broadcast spray at the rate of 1 quart per acre. Mix the specified dosage in sufficient water to ensure thorough coverage of treated areas. For **aerial application** use at lest 5 gallons of water per acre. Make application after carrots have started to bolt and when field counts or crop injury indicate that damaging pest populations are developing or present.

Restrictions: Do not allow livestock to graze in treated areas. Do not feed treated carrot cuttings or seed screenings to livestock.

This product is highly toxic to bees exposed to direct treatment or residue on treated crops or weeds. Do not apply when bees are in the field. Application must be made after bees have ceased flying for the day. Consult your Cooperative Agricultural Extension Service for additional information concerning protection of pollinating bees.

CLOVER GROWN FOR SEED For use only within **Oregon** (EPA SLN No. OR-850032) to control garden symphylans, armyworms, cutworms, aphids, weevils, lygus and spittlebugs, apply LORSBAN 4E with the amounts of water recommended below. Consult your Cooperative Agricultural Extension Service for additional information concerning garden symphylan, armyworm, cutworm, aphid, weevil, lygus, and spittlebug control practices in your area. For **garden symphylan control** use LORSBAN 4E as a preplant, broadcast application to soil surfaces at the rate of 2 quarts per acre. Mix the specified dosage in at least 10 gallons of water per acre to ensure thorough coverage of the treated soil. On the same day of treatment, incorporate the insecticide into the top 2 to 4 inches of soil using a disc, field cultivator, or equivalent equipment. For **established planting,** apply LORSBAN 4E at the rate of 2 quarts per acre as a broadcast, foliar spray before spring growth. Apply in at least 40 gallons of water to ensure thorough coverage and penetration. Incorporate to a depth of 2 to 4 inches using a disc or other suitable equipment. For **armyworm, cutworm, weevil, lygus and spittlebug control,** use LORSBAN 4E as a broadcast, foliar application at the rate of 2 pints per acre. For aphid control, use LORSBAN 4E as a broadcast, foliar application at the rate of 1 to 2 pints per acre. Mix the specified dosage in sufficient water to ensure thorough coverage of treated areas. Treat when field counts or crop injury indicates that damaging pest populations are developing or present. For **best cutworm control** soil should be moist and worms active on or near the soil surface at the time of treatment.

Restrictions: Do not allow livestock to graze in treated areas. Do not feed treated clover cuttings or seed screenings to livestock or use hay for livestock bedding.

This product is highly toxic to bees exposed to direct treatment or residues on treated crops or weeds. Do not apply to clover in bloom. Evening or night time applications when forages are not present reduce exposure hazards. Consult your Cooperative Agricultural Extension Service for additional information concerning protection of pollinating bees.

FIELD AND SWEET CORN: For use only within **Florida** (sweet corn only) EPA SLN No. FL-780046) and Georgia (EPA SLN No. GA-790019) to control beet armyworms, fall armyworm, and corn earworm larvae and **Colorado** (sweet corn only) (EPA SLN No. CO-810017), **Maryland** (sweet corn only) (EPA SLN No. MD-840004), **New Jersey** (sweet corn only) (EPA SLN No. NJ-840008), **Pennsylvania** (sweet corn only) (EPA SLN No. PA-840008), and **Delaware** (sweet corn only) (EPA SLN No. DE-840003) to control armyworms and corn earworm larvae, use 1 pint of LORSBAN 4E in sufficient water to ensure thorough coverage of plants and apply using either aerial (fixed-wing or helicopter) or power-operated ground spray equipment. For aerial application use at least 3 gallons of water per acre. Treat when field counts indicate that worms are or may become a problem. Repeat the application as needed, but do not exceed 13 applications per crop season.

Restrictions: Do not treat within 7 days before harvest. Do not allow livestock to graze in treated areas. Do not feed treated corn silage, forage, or fodder to meat or dairy animals.

For use only within **Florida** (field and sweet corn grown only for planting seed) (EPA SLN No. FL-770025) to control beet armyworm, fall armyworm, and corn earworm, use 1 to 2 pints of LORSBAN 4E per acre in sufficient water to ensure thorough coverage of plants and apply using either aerial (fixed-wing or helicopter) or power-operated ground spray equipment. For aerial application use at least 3 gallons of water per acre. Treat when field counts indicate that worms are or may become a problem. Repeat the application as needed.

Restrictions: Apply only to corn to be harvested for seed. Do not allow livestock to graze in treated areas. Do not feed treated corn grain, silage, or forage to meat or dairy animals.

COTTON: For use only within **Alabama** (EPA SLN No. AL-850003) used alone or in tank mixtures with pyrethroid insecticides is recommended for control and suppression of aphids, fall armyworm, and spider mites infesting cotton. Use LORSBAN 4E at a rate of 1/2 to 1 pint per acre and pyrethroid insecticides at the dosage rate shown on their product labels. Mix the required dosage of LORSBAN 4E or LORSBAN 4E and pyrethroid insecticide in sufficient cottonseed oil or soybean oil to give a finished spray volume 1 of quart per acre. Apply a minimum of 1 quart of spray per acre with aircraft calibrated to give adequate coverage of plants. If the aircraft is calibrated to deliver higher volumes, use **emulsifiable** vegetable oil at a rate of 1 pint per acre with the above dosage of insecticide and add enough water for 2 gallons of finished spray per acre. Treat when field count indicate damaging insect populations are developing or present. However, for fall armyworm control, use only as part of a regularly scheduled program (5 to 7 days apart).

Restrictions: Make no applications of these tank mixtures closer to harvest than the longest pre-harvest interval shown for any of the products in the tank mixture. Do not allow livestock to graze in treated areas. Do not feed gin trash or treated forage to livestock.

For use only within **Arkansas** (EPA SLN No. AR-870001) to suppress infestations of plantbugs and fleahoppers and preserve populations of beneficial arthropods in cotton. Use LORSBAN 4E at the rate of 1/4 pint per acre. Mix with water to form an emulsion spray. Apply the specified dosage in sufficient water to ensure thorough coverage using aerial or ground spary equipment. Treat when field counts indicate damaging density levels of plantbugs or fieahoppers are present.

Restrictions: Do not apply within 14 days before harvest. Do not allow livestock to graze in treated areas or feed gin trash or treated forage to livestock.

For use only within **Mississippi** (EPA SLN No. MS-840004) to control larvae of wireworms and cutworms, apply 2 pints of LORSBAN 4E per acre for wireworm control and 1 to 2 pints of LORSBAN 4E per acre for cutworm control in not less than 10 gallons of water in a 16 to 18 inch band over the row prior to or at planting. Immediately following application incorporate the insecticide into the soil to a depth of 3 to 4 inches using suitable equipment.

Restrictions: Do not make more than one application per season.

For use only within **Mississippi** (EPA SLN No. MS-830025) to control and suppress spider mites, use 1/2 to 1 pint of LORSBAN 4E per acre in tank mixtures with pyrethroid insecticides at the dosage rate shown on their product label and sufficient cottenseed oil or soybean oil to give a finished spray volume of 1 quart per acre. Apply a minimum of 1 quart of spray per acre with aircraft calibrated to give adequate coverage of plants. Treat when field counts indicate damaging insect populations are developing or present.

Restrictions: Make no applications of these tank mixtures closer to harvest than the longest preharvest interval shown for any of the products in the tank mixture. Do not allow livestock to graze in treated areas. Do not feed gin trash or treated forage to livestock.

GRAPES: For use only within **Michigan** (EPA SLN No. MI-840002) and **Missouri** (EPA SLN No. MO-870004) to control climbing cutworms, use 1 quart of LORSBAN 4E per acre in a minimum of 25 gallons of water and apply as a prebloom, dilute, or concentrate spray. LORSBAN 4E should not be tank mixed with pesticides, surfactants, or fertilizer formulations unless prior use has shown the combination noninjurious under your conditions of use. Consult your State Agricultural Station or Extension Service Specialist for proper time to treat in your area.

Restrictions: Do not make more than one application per season. Do not use in conjunction with a soil surface spray application of LORSBAN 4E for grape root borer control.

For use only with **Tennessee** (EPA SLN No. TN-880002) and **Georgia** (EPA SLN No. GA-860001) to control grape root borer by split soil applications, mix 2 1/4 pints of LORSBAN 4E with 100 gallons of water and apply 2 quarts of the diluted spray mixture to the soil surface on a 15 square foot area around the base of each vine. Do not allow spray to contact fruit or foilage. Make the first application before the pest emerges (at least 35 days prior to harvest) followed by a second application after harvest. Consult with local extension entomologists for optimum application timing based on pest emergence.

Restrictions: Do not make more than 2 applications per season or apply within 35 days before harvest.

NON-BEARING GRAPES: For use only within **Idaho** (EPA SLN No. ID-820018), **Oregon** (EPA SLN No. OR-820001), and **Washington** (EPA SLN No. WA-810036) to control cutworms, apply 1 quart of LORSBAN 4E per acre as a broadcast spray using either aerial or power-operated ground spray equipment. For aerial application use at least three gallons of spray per acre. If sprinkler irrigation is available, apply 1/4 to 1/2 inch of water after application to help incorporate the material. Treat when cutworms first become active and when field counts indicate damaging insect populations are developing or present. Retreat as necessary to maintain control. Consult your State Agricultural Experiment Station or Extension Service specialist for additional information concerning cutworm control practices in your area.

Restrictions: Do not harvest grapes in the same growing season as application.

MINT: For use only within **Oregon** (EPA SLN No. OR-8500034) by application as a preplant, broadcast spray to soil surfaces at the rate of 2 quarts per acre to control garden symphylans infesting mint. Mix the specified dosage in at least 10 gallons of water per acre to ensure thorough coverage of treated soil. On the same day of treatment, incorporate the insecticide into the top 2 to 4 inches of soil using a disc, field cultivator, or equivalent equipment.

Restrictions: Do not make more than one preplant, incorporation application in the spring. Do not use in conjunction with a broadcast, foliar spray application of LORSBAN 4E for cutworm control.

ONIONS (Dry Bulb): For use only within **Michigan** (EPA SLN No. MI-850006) for early season application on onions (dry bulb), to control Onion maggot. Use LORSBAN 4E at the rate of 1 quart per acre (equivalent to one pound of chlorpyrifos per acre) and apply as a directed spray to the base of onion seedlings or transplants during peak egg laying by the onion maggot adult. Proper timing of the insecticide application is essential for satisfactory control. Use a minimum of 100 gallons of total spray per acre for thorough wetting of the soil at the base of the plants.

Restrictions: Do not make more than one directed spray application if LORSBAN 4E or LORSBAN 15G were used at planting. Do not use within 60 days of harvest or apply more than 2 pounds of chlorpyrifos per acre per growing season.

PEANUTS: For use only within **GEORIGA** (EPA SLN No. GA-870006) use LORSBAN 4E at the rate of 4 pints per acre to suppress wireworms. Apply the specified dosage in not less than 10 gallons of water as a preplant broadcast spray to the soil surface using suitable power-operated ground application equipment. Immediately following application, incorporate the insecticide into the soil to a depth of 3 to 4 iches using suitable equipment.

Restrictions: Do not make more than one application per season. Do not apply a postplant directed spray following a preplant broadcast spray.

PERENNIAL GRASS SEED CROPS: For use only within **Idaho** (EPA SLN No. ID-830005) and **Oregon** (EPA SLN No. OR-830007) to control billbugs, cutworms, and aphids and **Washington** (EPA SLN No. WA-830009) to control cutworms and aphids, use 1 quart of LORSBAN 4E per acre in sufficient water to ensure good coverage and apply as a broadcast spray using power-operated ground application equipment. For improved results, the thatch should be moist at the time of application and the insecticide washed into the thatch immediately after application. If sprinkler irrigation is available, apply 1/4 to 1/2 inch of water after application to help incorporate the material. Otherwise, apply to early morning dew-covered field or during or immediately following a light rain. For billbug control, apply in early April when adult billbugs are active. For cutworm control, apply in either spring or fall when larvae or larval damage first appear. Fall applications should follow sufficient irrigation and/or rainfall. Heavy carbon residue from field burning will prevent penetration of LORSBAN 4E into the plant crown where the insects feed. Retreat as necesary to maintain control. To control aphids, apply 1 to 2 pts of LORSBAN 4E per acre as a broadcast spray using power-operated ground application equipment. Treat as colonies begin forming before heading. Consult your Cooperative Agricultural Extension Service for additional information concerning billbug, cutworm, and aphid control practices in your area. This product is highly toxic to bees exposed to direct treatment or residues on treated crops or weeds. Protective information may be obtained from your Cooperative Agricultural Extension Service.

Restrictions: Apply only to perennial grass seed crops. Do not allow livestock to graze in treated areas. Do not feed treated grass cuttings or seed screenings to livestock nor use hay for livestock bedding.

For use only within **South Dakota** (EPA SLN No. SD-870006) to control billbugs and cutworms infesting perennial grass and seed crops. Apply 1 quarts of LORSBAN 4E per acre as a broadcast spray using power-operated ground application equipment. Apply LORSBAN 4E with enough water to ensure good coverage. For best results, use 20 gpa of spray mixture. For improved results, the thatch should be moist at the time of application and the insecticide washed into the thatch immediately after application. If sprinkler irrigation is available, apply 1/4 to 1/2 inches of water after application to help incorporate the material. Otherwise, apply to early morning dew-covered fields or during or immediately following a light rain. For **billbug control,** apply in early April when adult billbugs are active. For **cutworm control,** apply in either spring or fall when larvae or larval damage first appear. Fall applications should follow sufficient irrigation and/or rainfall. Heavy carbon residue from field burning will prevent penetration of LORSBAN 4E into the plant crown where the insects feed. Retreat as necessary to maintain control. To control aphids, apply 1 to 2 pts of LORSBAN 4E per acre as a broadcast spray using power-operated ground application equipment. Treat as colonies begin forming before heading. Consult your Cooperative Agricultural Extension Service for additional information concerning billbug, cutworm, and aphid control practices in your area.

For aerial application, follow above dosage recommendations and use two to five gallons of water per acre. Control may be reduced at low spray volumes under high temperature and wind conditions.

Restrictions: Apply only to perennial grass seed crops. Do not allow livestock to graze in treated areas. Do not feed treated grass cuttings or seed screenings to livestock nor use hay for livestock bedding.

RADISHES GROWN FOR SEED: For use only within **Oregon** (EPA SLN No. OR-850031) use LORSBAN 4E as a preplant, broadcast application to soil surface to control garden symphylan at the rate of 2 quarts per acre. Mix the specified dosage in at least 10 gallons of water per acre to ensure thorough coverage of the treated soil. On the same day of treatment, incorporate the herbicide into the top 2 to 4 inches of soil using a disc, field cultivator, or equivalent equipment.

For **cutworm control,** use LORSBAN 4E as a broadcast application to soil surface at the rate of 1 quart per acre. Mix the specified dosage in at least 10 gallons of water per acre to ensure thorough coverage of treated soil. Treat in the spring when cutworm larvae begin to appear. It is preferable to apply LORSBAN 4E when the soil is moist and worms are active on or near the soil surface.

Restrictions: Do not allow livestock to graze in treated areas. Do not feed treated radish cuttings or seed screenings to livestock.

This product is highly toxic to bees exposed to direct treatment or residue on treated crops or weeds. Do not apply to radishes in bloom. Evening or night time applications when foragers are not present in the fields reduce exposure hazards. Consult your Cooperative Agricultural Extension Service for additional information concerning protection of pollinating bees.

STRAWBERRIES: For use only within **Idaho** (EPA SLN No. ID-830028), **Oregon** (EPA SLN No. OR-830017), and **Washington** (EPA SLN No. WA-830033) to control garden symphylans, use 2 quarts of LORSBAN 4E per acre with enough water to ensure good coverage of the soil surface and apply pre-plant as a broadcast spray to the soil surface using suitable power-operated ground application equipment. On the same day of treatment, incorporate the insecticide into the top 2 to 4 inches of soil using a disc, field cultivator or equivalent equipment. LORSBAN 4E should not be tank mixed with pesticides, surfactants, or fertilizer formulations unless prior use has shown the combination noninjurious under local conditions of use. Consult your Cooperative Agricultural Extension Service for additional information concerning garden symphylan control practices in your area.

Restrictions: Apply only as a soil pre-plant incorporation application in the spring for protection of strawberries the following year.

For use only within **Oregon** (EPA SLN No. OR-840021) to control strawberry crown moth, apply 1 quart of LORSBAN 4E per acre as a directed spray to crowns of strawberry plants immediately after harvest and after plants are topped. Apply in a minimum of 100 gallons of spray per acre. **Do not** sprinkle irrigate for one weed following application. A second application may be required two to two and one half weeks later. Phytotoxicity may occur when LORSBAN 4E is applied to plants experiencing high temperature and water stress. LORSBAN 4E should not be tank mixed with pesticides, surfactants, or fertilizer formulations unless prior use has shown the combination noninjurious under local conditions. Consult your Cooperative Agricultural Extension Service for additional information concerning strawberry crown moth control practices in your area.

Restrictions: Use postharvest only. Do not make more than two applications per year.

SUGAR BEETS GROWN FOR SEED: For use only within **Idaho** (EPA SLN No. ID-830021) and **Oregon** (EPA SLN No. OR-830018) to control garden symphylans, use 2 quarts of LORSBAN 4E per acre with enough water to ensure good coverage of the soil surface and apply pre-plant as a broadcast spray to the soil surface using surface using suitable power-operated ground application equipment. LORSBAN 4E should not be tank mixed with pesticides, surfactants, or fertilizer formulations unless prior use has shown the combination noninjurious under local conditions of use. Consult your Cooperative Agricultural Extension Service for additional information concerning garden symphylan control practices in your area.

Restrictions: Apply only as a soil pre-plant incorporation application in the fall for protection of sugar beets the following year.

01545-L1 M689

*Trademark of THE DOW CHEMICAL COMPANY

LORSBAN* 15G

GRANULAR INSECTICIDE

For Control of Various Insects Infesting Certain Field and Vegetable Crops

ACTIVE INGREDIENT(S):	
Chlorpyrifos [O,O-diethyl O-(3,5,6-trichloro-2-pyridinyl) phosphorothioate]	15.0%
INERT INGREDIENTS:	85.0%
TOTAL	100.0%

E.P.A. Registration No. 464-523 E.P.A. Est. 464-MI-1

KEEP OUT OF REACH OF CHILDREN

CAUTION
AVISO:

PRECAUCIÓN AL USUARIO: Si usted no lee inglés, no use este producto hasta que la etiqueta le haya sido explicada ampliamente.

PRECAUTIONARY STATEMENTS

HAZARDS TO HUMANS AND DOMESTIC ANIMALS

HARMFUL IF SWALLOWED
• CAUSES EYE IRRITATION

Avoid Contact With Eyes, Skin, And Clothing • Avoid Breathing Dusts • Wash Thoroughly With Soap And Water After Handling • Remove Contaminated Clothing And Wash Before Reuse • Keep Away From Food, Feedstuffs And Water Supplies

STATEMENTS OF PRACTICAL TREATMENT:

If Swallowed: Call a physician or Poison Control Center. Drink 1 or 2 glasses of water and induce vomiting by touching back of throat with finger. Do not induce vomiting or give anything by mouth to an unconscious person.

If In Eyes: Flush with plenty of water for 5 minutes. Get medical attention if irritation persists.

If On Skin: Wash with plenty of soap and water. Get medical attention if irritation persists.

NOTE TO PHYSICIANS: Chlorpyrifos is a cholinesterase inhibitor. Treat symptomatically. If exposed, plasma and red blood cell cholinesterase tests may indicate significance of exposure (baseline data are useful). Atropine, only by injection, is the preferable antidote. Oximes, such as 2-PAM/protopam, may be therapeutic if used early; however, use only in conjunction with atropine. In case of severe acute poisoning, use antidote immediately after establishing an open airway and respiration.

ENVIRONMENTAL HAZARDS

This pesticide is toxic to birds and wildlife, and extremely toxic to fish and aquatic organisms. Do not apply directly to water. Drift and runoff from treated areas may be hazardous to aquatic organisms in adjacent aquatic sites. Cover or incorporate spills. Do not contaminate water by cleaning of equipment or disposal of waste.

NOTICE

Read and understand the entire label before using. Use only according to label directions.

Before buying or using this product, read "WARRANTY LIMITATIONS AND DISCLAIMER" elsewhere on this label. If terms are not acceptable, return **unopened** package at once to seller for full refund of purchase price paid. Otherwise, use by the buyer or any other user constitutes acceptance of the terms under WARRANTY LIMITATIONS AND DISCLAIMER.

IN CASE OF AN EMERGENCY endangering life or property involving this product, call collect 517-636-4400

AGRICULTURAL CHEMICAL

Do Not Ship or Store with Food, Feeds, Drugs, or Clothing

RQ CHLORPYRIFOS MIXTURE, NA 2783

DIRECTIONS FOR USE

It is a violation of Federal law to use this product in a manner inconsistent with its labeling.

Crop	Pest	Amount of LORSBAN 15G per 1,000 feet of row				Amount of LORSBAN 15G per acre for Preplant Broadcast Treatment	Use Directions	Restrictions
		Planting Time T-Band	Band	Treatment In-Furrow	Postplant Treatment			
Corn Field, Popcorn, Sweet Corn	Northern and Western Corn Rootworm Larvae	8 oz	8 oz	—	8 oz	—	**At Plant Band Treatment:** In conventional, minimum and no-till corn, place the granules in a 6 to 7 inch wide band over the row behind the planter shoe, **in front of the press wheel,** and incorporate into the top one inch of soil using tines or chains or other suitable equipment. For cutworm control under very dry soil conditions, best results will be obtained through proper incorporation. In soils over 10% organic matter, use the highest rate given. For suppression of nematodes in North Carolina, South Carolina, and Virginia, apply LORSBAN 15G as a 7-10 inch wide band at the rate of 16 oz per 1000 feet of row. Incorporate with drag chains, tines, or other suitable equipment to a depth of 1 to 2 inches. **At Plant T-Band Treatment:** In conventional, minimum and no-till corn, for control of rootworm, cutworm, and billbug, in addition to lesser cornstalk borer, symphylans, wireworm, seed corn beetle, seed corn maggot and grubs, apply as a T-band over an open seed furrow. For suppression of nematodes in the Southeast, use LORSBAN 15G as a T-band treatment at 16 ounces per 1,000 feet of row over an open seed furrow. LORSBAN 15G has demonstrated suppression of certain soil-borne pathogens which may result in certain physiological and agronomic advantages to corn under environmental stress conditions when compared to corn not treated with LORSBAN 15G. **At Plant In-Furrow Treatment:** Apply LORSBAN 15G at planting time as an in-furrow treatment in conventional, minimum and no-till corn. Direct the granules into the planter shoe with the seed or place the applicator tube directly behind the planter shoe so that the granules drop into the seed furrow or place the granular band applicator behind the planter shoe so that the granules fall on the soil surface and into the open seed furrow and are covered with soil. Equivalent rates of LORSBAN 15G per acre for various row spacings are given in the accompanying table. **Postplant Treatment:** Apply LORSBAN 15G insecticide at cultivation time to control corn rootworm larvae. Place the granules over the row and cultivate into the soil or sidedress both sides of the row and cultivate into the soil. For best rootworm control, place the granules at the base of the plants on both sides of the row just ahead of cultivation shovels and cover with soil. Consult your State Agricultural Experiment Station or extension service specialist for proper time to treat in your area. Equivalent rates of LORSBAN 15G per acre for various row spacings are given in the accompanying table Consult your State Agricultural Experiment Station or extension service specialist for further information concerning soil insect control in your area. Equivalent rates of LORSBAN 15G per acre for various row spacings are given in the accompanying table. **Postplant Broadcast Treatment:** Use LORSBAN 15G for control of European and southwestern corn borers by uniformly broadcasting the granules over the corn plants by aerial application or applying the granules into the corn whorls by ground application. Consult your State Agricultural Experiment Station or extension service specialist for local threshold information. Scouting for insect damage is strongly encouraged. **Preplant Broadcast Treatment:** Use equipment that will evenly distribute the granules over the treated area. Incorporate the granules into the soil using a S tine, C tine, disc cultivator, or other suitable incorporation equipment set to cut to a depth of 4 to 6 inches. Plant the crop in the usual manner as soon as possible after treatment.	For soil insect control, do not exceed the equivalent of 16 ounces of LORSBAN 15G per 1,000 feet of row or 13.5 pounds of LORSBAN 15G per acre per crop season. For foliar insect control, do not exceed the equivalent of 16 ounces of LORSBAN 15G per 1,000 feet of row or 13 pounds of LORSBAN 15G per acre per crop season. Do not apply within 35 cays before harvest of grain. Do not allow livestock to graze in treated areas nor harvest treated corn silage as feed for meat or dairy animals within 14 days after last treatment. Do not feed treated corn fodder to meat or dairy animals within 35 days after last treatment.
	Southern Corn Rootworm Larvae	8 oz.	8 oz	8 oz.	8 oz	—		
	Cutworms[1]	8 oz	8 oz	8 to 16 oz	—	6.75 to 13.5 lb		
	Wireworms[2]	8 to 16 oz	16 oz	8 to 16 oz	—	13.5 lb		
	Billbugs	8 to 16 oz	8 to 16 oz	—	—	13.5 lb		
	Seed Corn Maggots	8 oz	—	8 oz	—	—		
	Grubs	16 oz	—	8 to 16 oz	—	13.5 lb		
	Lesser Cornstalk Borer	8 to 16 oz	8 to 16 oz	—	—	—		
	Seed Corn Beetle	8 to 16 oz	—	8 to 16 oz	—	—		
	Flea Beetle Larvae	—	—	—	—	13.5 lb		
	Symphylans	8 to 12 oz	8 to 12 oz	—	—	—		
	European & Southwestern Corn Borer[3]	—	—	—	6 to 8 oz	—		
	Fire ants[2]	8 oz	—	—	—	—		
	Chinch Bug[2]	8 oz	—	8 oz	—	—		
Citrus Orchard Floors	Fire Ants, and other ant species	—	—	—	6.7lb/A	—	To control foraging ants and suppress mounds, apply granules with ground application equipment using a suitable granular applicator, such as a cyclone fertilizer spreader, that will uniformly broadcast the granules over the orchard floor. Retreat as needed. Do not apply where weed growth or other obstructions would impede uniform coverage of the orchard floor.	Do not apply more than 67 pounds of LORSBAN 15G per acre per season or apply last treatment within 28 days before harvest. Do not allow livestock to graze in treated areas.

[1] 8 oz rate-suppression only for in-furrow treatments.
[2] 8 oz rate-suppression only
[3] Use 5.0 to 6.5 lb. of LORSBAN 15G per acre applied broadcast by air.

Crop	Pest	Amount of LORSBAN 15G per 1,000 feet of row: Planting Time Treatment T-Band	Planting Time Treatment Band	Planting Time Treatment In-Furrow	Postplant Treatment	Amount of LORSBAN 15G per acre for Preplant Broadcast Treatment	Use Directions	Restrictions
Sorghum/ Milo	Lesser Cornstalk Borer	4 to 12 oz	4 to 12 oz	—	—	—	Apply in a 6 to 8 inch band over the row and lightly incorporate by using the press wheel or drag chain. Equivalent rates of LORSBAN 15G per acre for various row spacings are given in the accompanying table. Use the lowest rate for lesser cornstalk borer control when protection is desired for 2 to 3 weeks and higher rates for longer residual activity. It is absolutely necessary to incorporate the granules especially at lower rates.	Do not make more than one application per season.
	Corn Rootworm and Cutworms	8 oz	8 oz	—	—	—		
	Fire Ants[1]	8 oz	—	—	—	—		
	Chinch Bug[1]	8 oz	—	8 oz	—	—		
	[1] 8 oz rate —suppression only							
Onions (Dry Bulb)	Onion Maggot	—	—	3.7 oz	—	—	Apply the granules at the rate of 3.7 ounces per 1,000 feet of row at 18-inch row spacing by placing in the furrow.	Do not make more than one application per year.
Peanuts	Cutworms; Lesser Cornstalk Borer; Southern Corn Rootworm Larvae; Wireworms; Southern Blight (White Mold)	—	7.5 to 15 oz	—	7.5 to 15 oz	—	**At Plant Preventive Treatment:** For control of cutworms, lesser cornstalk borer, southern corn rootworm larvae, suppression of wireworms, and to inhibit the growth and development of white mold (southern blight) disease caused by *Sclerotium rolfsii*, apply LORSBAN 15G in a 6 to 12 inch band over the row behind the planter shoe and in front of the press wheel. Lightly incorporate with tines or chains or other suitable equipment. If the 7.5 oz rate is used at planting time, then another application of 7.5 oz per 1,000 feet of row should be made postplant to maintain control. **Postplant Preventive Treatment:** Apply LORSBAN 15G at early flowering in a 6 to 12 inch band over the row or during early pegging in a 10 to 18 inch band over the fruiting zone. Use the 7.5 oz rate only if a 7.5 oz rate was applied at plant. Irrigation or rain following application will enhance activity. Under conditions of heavy white mold pressure, a suitable fungicide may also be required and should be applied separately.	Do not apply more than 15 oz of LORSBAN 15G per 1,000 feet of row or 13.6 pounds of LORSBAN 15G per acre per crop season. Do not harvest within 21 days after treatment. Do not feed peanut forage or hay to meat or dairy animals.
	Lesser Cornstalk Borer[1]	—	—	—	7.5 to 15 oz	—	**Band Rescue Treatment:** Use LORSBAN 15G for the control of lesser cornstalk borer when the insect first appears, usually just prior to or at pegging. Apply in a 10 to 18-inch band over the fruiting zone. **Broadcast Rescue Treatment:** When lesser cornstalk borer first appears, uniformly broadcast the granules by aerial application. Care must be taken to ensure proper equipment calibration and delivery.	
	[1] Use 10 to 13 lb of LORSBAN 15G per acre applied broadcast by air.							
Soybeans	Lesser Cornstalk Borer, Cutworms	8 oz	8 oz	—	8 oz	—	Use LORSBAN 15G insecticide to control larvae of the lesser cornstalk borer and cutworms by application at planting time or postemergence as a band (row) treatment at the rate of 8 ounces per 1,000 linear feet of row. In the Southeast apply 4 to 8 ounces per 1,000 linear feet of row as an at-plant treatment. Equivalent rates of LORSBAN 15G per acre for various row spacing are given in the accompanying table. When applied at planting time incorporate the granules into the top 1 inch of soil by placing in a 4 to 10 inch band over the row behind the planter shoe and ahead of the press wheel. A drag chain can also be used for incorporation. For postemergence treatment when insects first appear incorporate the granules in a 4 to 10 inch band to a depth of ½ to 1 inch using a suitable cultivator. Apply LORSBAN 15G with equipment that will provide uniform distribution of the granules. Do not apply as an in-furrow treatment. For suppression of fire ants, use LORSBAN 15G at 8 ounces per 1,000 feet of row as an at plant T-band treatment.	Do not make more than one application per season.
Sugar Beets	Sugar Beet Root Maggot	6.5 to 9.0 oz	6.5 to 9.0 oz	4.5 to 9.0 oz	6.5 to 9.0 oz	—	**At Plant Band Treatment:** To control cutworms and sugar beet root maggot larvae at planting time, place LORSBAN 15G in a 5-inch wide band behind the planter shoe, over the drill row, and **in front** of the press wheel. Apply granules at the rate of 6.5 to 9 ounces per 1,000 feet of row (equivalent to 10 to 13.5 pounds per acre based on a 22-inch row spacing). Incorporate granules into the top 1 to 2 inches of soil using suitable equipment. **At Plant In-Furrow Treatment:** To control only sugar beet root maggot larvae, LORSBAN 15G may be placed in-furrow at planting time. Apply at the rate of 4.5 to 9 ounces per 1,000 feet of row (equivalent to 7 to 13.5 pounds per acre based on 22-inch row spacing). **Postemergence Band Treatment:** For postemergence control of sugar beet root maggot larvae, place LORSBAN 15G in a 5-inch wide band	Do not make more than one application per year.
	Cutworms	6.5 to 9.0 oz	6.5 to 9.0 oz	—	—	—		

Crop	Pest	Amount of LORSBAN 15G per 1,000 feet of row: Planting Time Treatment T-Band	Planting Time Treatment Band	Planting Time Treatment In-Furrow	Postplant Treatment	Amount of LORSBAN 15G per acre for Preplant Broadcast Treatment	Use Directions	Restrictions
							over the beets (up to 2 to 4 true leaf stage of the plant). Apply granules at the rate of 6.5 to 9 ounces per 1,000 feet of row (equivalent to 10 to 13.5 pounds per acre based on 22-inch row spacing). Incorporate the granules into the top ½ to 1 inch of soil using a suitable cultivator.	
Sunflowers	Cutworms	—	8 oz	—	—	—	Place the granules in a 7-inch wide band over the row behind the planter shoe **in front of the press wheel** and incorporate into the top one inch of soil using suitable equipment.	Do not make more than one application per season.
Sweet Potatoes	Wireworms *(Conoderus)*, Flea Beetles *(Systena)*, and Sweet Potato Beetle	—	—	—	—	13.5 lb	Use LORSBAN 15G to reduce the feeding damage caused by populations of the listed pests. Evenly broadcast the granules over the soil surface and then incorporate the granules into the soil to a depth of 4 to 6 inches using a rotary hoe, disc cultivator, or other suitable equipment. Plant the crop in the usual manner no later than 14 days after treatment (any delay in planting will reduce the length of time that LORSBAN 15G will protect against feeding damage). LORSBAN 15G will not control false wireworm or whitefringed beetle and other grubs that attack sweet potatoes.	Do not make more than one application per season. Do not harvest within 125 days of treatment.
Tobacco	Cutworms, Flea Beetle Larvae, Root Maggots, and Wireworms	—	—	—	—	13.5 to 20 lb	Apply the granules one week before transplanting, using equipment that will evenly distribute the granules over a treated area. Immediately following application, incorporate the granules into the soil to a depth of 2 to 4 inches using suitable equipment. The application of LORSBAN 15G will also suppress movement of imported fire ant into a treated field.	Do not make more than one application per season.
Vegetables Broccoli, Brussels Sprouts, Cabbage, Cauliflower, Chinese Cabbage, Collards, Kale, Kohlrabi, Rutabagas, Turnips	Root Maggots	4.6 to 9.2 oz	—	—	—	—	Place LORSBAN 15G across the seed row in a 4-inch wide band behind the planter shoe and **in front of the press wheel** to achieve shallow incorporation. This application requires a spreader or splitter on the end of the drop tube from the granule applicator.	Do not apply more than 7½ pounds of LORSBAN 15G per acre to crops planted in 40-inch rows or more than 15 pounds of LORSBAN 15G per acre to crops planted in 20-inch rows (or two rows per bed). Use proportional amounts for other row spacings not to exceed 15 pounds of LORSBAN 15G per acre. Do not make more than one application per season. Do not use rutabaga tops for food or feed purposes.
Radishes	Root Maggots	—	—	3.3 oz	—	—	Apply the granules in the seed furrow with the seed at planting time.	Do not apply more than 18.3 pounds of LORSBAN 15G per acre or make more than one application per season.

LORSBAN 15G Insecticide—Rate per 1,000 ft of row

Amount of LORSBAN 15G Per 1,000 Feet of Row	Pounds of LORSBAN 15G required Per Acre from Various Row Spacing: 40″	38″	36″	34″	32″	30″	22″	18″
3.7 ounces	3.0	3.2	3.4	3.6	3.8	4.0	5.5	6.7
4 ounces	3.3	3.4	3.6	3.8	4.1	4.4	5.9	7.3
6 ounces	4.9	5.2	5.4	5.8	6.1	6.5	8.9	10.9
6.5 ounces	5.3	5.6	5.9	6.2	6.6	7.1	9.7	11.8
7.5 ounces	6.1	6.4	6.8	7.2	7.7	8.2	11.1	13.6
8 ounces	6.5	6.9	7.3	7.7	8.2	8.7	11.9	17.5
9 ounces	7.4	7.7	8.2	8.6	9.2	9.8	13.4	16.3
12 ounces	9.8	10.3	10.9	11.5	12.3	13.1	17.8	21.8
15 ounces	12.3	12.9	13.9	14.4	15.3	16.3	22.3	27.2
16 ounces	13.1	13.8	14.5	15.4	16.3	17.4	23.8	29.0

GENERAL INSTRUCTIONS FOR EQUIPMENT CALIBRATION

Caution: The following chart lists suggested initial gauge settings for application of LORSBAN 15G Insecticide with one hopper opening per row. Be sure to check the actual application rate under your operating conditions.

1. Fill hopper.
2. Attach a plastic bag to tube opening.
3. Set your planter to the initial settings shown on chart.
4. Measure off 1,000 row feet and drive your planter the premeasured distance at your desired speed.
5. Each bag should contain 6 to 8 ounces (wt.) of granules depending on your desired rate.
6. If the result is over or under the desired rate—adjust the settings and repeat the calibration.

LORSBAN 15G Insecticide—Calibration

	Speed, mph: 4		5		6		7		8	
	Application Rate, oz. Per 1,000 row ft.: 8	16	8	16	8	16	8	16	8	16
Planter Type	Gauge Setting									
Gandy[1]	21.4	30.2	23.7	32.4	26.0	36.0	27.7	39.0	30.2	41.0
John Deere[1] Maxi-Emerge[2]	20	44	26	46	30	49	35	52	40	54
John Deere[1] 7000 Max-Emerge (Odd Nos. on Gate)	14	22	16	24	18	26	19	28	21	30
John Deere[1] 7000 Max-Emerge (Even Nos. on Gate)	17	30	20	33	24	35	26	36	28	38
John Deere[2] 71 Flexi-Planter and Older Planters	1/30	2/17	2/5	2/22	2/9	2/27	2/13	2/31	2/16	3/1
Allis Chalmers[3] 70 Series	8	13	8	13	8	13	8	13	8	13
Allis Chalmers[4] 78 & 79 Series	1/9.0	3/3.0	2/3.3	3/9.5	2/6.0	4/3.0	2/9.0	5/4.0	3/2.5	6/0
Noble[1] (New)	11	19	14	22	16	25	17	28	19	31
White Planter	11	19	14	22	16	25	17	28	19	31

LORSBAN 15G Insecticide—Calibration										
	Speed, mph									
	4		5		6		7		8	
	Application Rate, oz. Per 1,000 row ft.									
	8	16	8	16	8	16	8	16	8	16
Planter Type	Gauge Setting									
International Harvester[4]	1/9.0	3/3.0	2/3.3	3/9.5	2/6.0	4/3.0	2/9.0	5/4.0	3/2.5	6/0
Buffalo All-Flex[5] (Fleischer Mfg)	4⅞	10	4⅞	10	4⅞	10	—	—	—	—

[1] Gauge setting.

[2] Gauge setting with range 1 & 2—number is range/notch.

[3] Gauge setting is constant regardless of speed.

[4] Gauge setting shown with stem gates & dial settings—number shown is gate/dial.

[5] Number of turns open on the adjustment nut.

STORAGE AND DISPOSAL

Do not contaminate water, food or feed by storage or disposal. Open dumping is prohibited.

Pesticide Disposal: Wastes resulting from the use of this product may be disposed of on site or at an approved waste disposal facility.

Container Disposal: Completely empty bag into application equipment. Then dispose of empty bag in a sanitary landfill or by incineration or, if allowed by state and local authorities, by burning. If burned, stay out of smoke.

WARRANTY LIMITATIONS AND DISCLAIMER

The Dow Chemical Company warrants that this product conforms to the chemical description on the label and is reasonably fit for the purposes stated on the label when used in strict accordance with the directions therein under normal conditions of use. THIS IS THE ONLY WARRANTY MADE ON THIS PRODUCT. NO OTHER EXPRESS AND NO IMPLIED WARRANTY OF MERCHANTABILITY OR FITNESS FOR A PARTICULAR PURPOSE IS MADE OUTSIDE OF THIS LABEL. Therefore, neither this warranty nor any other warranty of merchantability or fitness for a particular purpose, express or implied, extends to the use of this product contrary to label instructions (including conditions noted on the label, such as unfavorable temperatures, soil conditions, etc.), under abnormal conditions (such as excessive rainfall, drought, tornadoes, hurricanes, etc.) or under conditions not reasonably foreseeable to or beyond the control of seller.

When buyer or user suffers losses or damages resulting from the use or handling of this product (including claims based on contract, negligence, strict liability, or other legal theories), buyer or user must promptly notify in writing The Dow Chemical Company of any claims to be eligible to receive either remedy given below. The EXCLUSIVE REMEDY OF THE BUYER OR USER and the LIMIT OF LIABILITY of The Dow Chemical Company or any other seller will be one of the following, at the election of The Dow Chemical Company:

(1) **Refund of purchase price** paid by buyer or user for product bought, **or**

(2) **Replacement** of amount of product used.

The seller **will not be liable for consequential or incidental damages or losses.**

The terms of this Warranty Limitations And Disclaimer cannot be varied by any written or verbal statements or agreements. Any employee or sales agent of the seller is not authorized to vary or exceed the terms of this Warranty Limitations And Disclaimer in any manner.

U.S. Patent No. 4,380,537

45405-L2 I389

* Trademark of THE DOW CHEMICAL COMPANY

LORSBAN* 50W

WETTABLE POWDER INSECTICIDE

For Control of Various Insects on Certain Fruit and Nut Crops

ACTIVE INGREGIENT(S):
Chlorpyrifos [O,O-diethyl O-(3,5,6-trichloro-2-pyridinyl) phosphorothioate] 50%
INERT INGREDIENTS: 50%
TOTAL 100%

E.P.A. Registration No. 464–552 E.P.A. Est. 447–TX–1

KEEP OUT OF REACH OF CHILDREN

WARNING

AVISO:

PRECAUCIÓN AL USUARIO: Si usted no lee inglés, no use este producto hasta que la etiqueta la haya sido explicada ampliamente.

PRECAUTIONARY STATEMENTS

HAZARDS TO HUMANS AND DOMESTIC ANIMALS

MAY BE FATAL IF SWALLOWED • HARMFUL IF ABSORBED THROUGH SKIN OR INHALED • CAUSES EYE IRRITATION

Avoid Contact With Eyes, Skin, Or Clothing • Avoid Breathing Dust Or Spray Mist • Wash Thoroughly With Soap And Water After Handling And Before Eating Or Smoking • When Loading The Spray Tank Or Handling The Powder, Wear A Pesticide Respirator Jointly Approved By The Mining Enforcement And Safety Administration (Formerly The U.S. Bureau Of Mines) And By The National Institute For Occupational Safety And Health Under The Provisions Of 30 CFR Part II • Wear Protective Clothing Such As Long Pants, Long Sleeve Shirt, Gloves, Hat, And Boots When Loading The Spray Tank Or Handling The Powder And During Application Of The Spray Mixture •Work To Windward To Stay Out Of Spray Drift Or Mist • Remove Contaminated Clothing And Wash Before Reuse • Do Not Apply This Product In Such A Manner As To Directly Or Through Drift Expose Workers Or Other Persons • The Area Being Treated Must Be Vacated By Unprotected Persons • Keep Away From Food, Feedstuffs And Water Supplies

STATEMENT OF PRACTICAL TREATMENT

If Swallowed: Call a physician or Poison Control Center. Drink 1 or 2 glasses of water and induce vomiting by touching back of throat with finger. Do not induce vomiting or give anything by mouth to an unconscious person.

If On Skin: Wash with plenty of soap and water. Get medical attention if irritation persists.

If in Eyes: Flush with plenty of water for 5 minutes. Get medical attention if irritation persists.

If Inhaled: Remove to fresh air if effects occur. Get medical attention.

NOTE TO PHYSICIANS: Chlorpyrifos is a cholinesterase inhibitor. Treat symptomatically. If exposed, plasma and red blood cell cholinesterase tests may indicate significance of exposure (baseline data are useful). Atropine, only by injection, is the prefereable antidote. Oximes, such as 2–PAM/protopam, may be therapeutic if used early; however, use only in conjunction with atropine. In case of severe acute poisoning, use antidote immediately after establishing an open airway and respiration.

ENVIRONMENTAL HAZARDS

This pesticide is toxic to birds and wildlife, and extremely toxic to fish and aquatic organisms. Do not apply directly to water. Drift and runoff from treated areas may be hazardous to aquatic organisms in adjacent aquatic sites. Cover or incorporate spills. Do not contaminate water when disposing of equipment washwaters. This product is highly toxic to bees exposed to direct treatment or residues on blooming crops or weeds. Avoid use when bees are actively foraging of if bees are likely to come in contact with wet or re-wetted treated surfaces. Protective information may be obtained from your Coorporative Extension Service. This product may pose a hazard to certain Federally endangered fish species. Refer to the Endangered Species section of this label for any geographic restrictions.

NOTICE

Read and understand the entire label before using. Use only according to label directions.

Before buying or using this product, read "WARRANTY LIMITATIONS AND DISCLAIMER" elsewhere on this label. If terms are not acceptable, return **unopened** package at once to seller for full refund of purchase price paid. Otherwise, use by the buyer or any other user constitutes acceptance of the terms under WARRANTY LIMITATIONS AND DISCLAIMER.

IN CASE OF AN EMERGENCY endangering life or property involving this product, call collect 517–636–4400

AGRICULTURAL CHEMICAL

Do Not Ship or Store with Food, Feeds, Drugs, or Clothing

ENDANGERED SPECIES RESTRICTIONS

The use of LORSBAN 50W may pose a hazard to certain Federally endangered fish species. This product **cannot** be used on sour cherry trees in the following countries:

COLORADO: Delta, Larimer and Mesa

OHIO: Washington

UTAH: Emery, Grand, Utah and Washington

WISCONSIN: Iowa, Pierce, Richland and Vernon

DIRECTIONS FOR USE

It is a violation of federal law to use this product in a manner inconsistent with its labeling.

GENERAL INFORMATION

LORSBAN 50W insecticide is a wettable powder formulation designed for use as a cover spray to control various pests. The pests controlled are listed in the accompanying table. LORSBAN 50W is compatible with insecticides, miticides, and fungicides commonly recommended in orchard spray schedules except for alkaline materials such as Bordeaux mixture and lime.

Test compatibility of the intended tank mixture before adding LORSBAN 50W to the spray or mix tank. Add proportionate amounts of each ingredient to a pint or quart jar, cap, shake, and let set 15 minutes. Formation of precipitates that do not readily redisperse indicates an incompatible mixture that should not be used.

MIXING INSTRUCTIONS

LORSBAN 50W mixes readily with water to form a suspension. Mix the required amount of LORSBAN 50W with a small quantity of water and add this premix while filling the spray tank. Alternatively, fill the tank with the required amount of water and then add the premix. Maintain sufficient agitation during both mixing and application to ensure uniformity of the spray mixture. LORSBAN 50W contains a wetting agent. Do not use additional wetting agents, spreaders or stickers.

USE AND DOSAGE RECOMMENDATIONS

Use LORSBAN 50W at the dosages indicated by application as a foliar spray to control pests listed in the following table. Where a range rate is listed in the use directions, use the higher rate for more severe infestations. Mix the required dosage in sufficient water to ensure thorough and complete coverage of the foliage and crop and apply as a concentrate or dilute spray using conventional, power-operated spray equipment. For dilute sprays applied to tree fruits and nuts, mix the required dosage in sufficient water to allow for spray to run off. For concentrate sprays, apply an equivalent amount of LORSBAN 50W per acre. Treat when pests appear or in accordance with local conditions. Mow or beat down orchard blooming weed cover crops before applying spray to tree fruit crops to reduce hazard to bees. This step is especially important in relation to early Spring cover sprays when morning conditions may be wet from dew. LORSBAN 50W will perform best in an almond or walnut pest control program that includes winter sanitation (removal of mummy nuts) and regular monitoring of target pests. To control apple aphid infesting apples, spray every 10 to 14 days on a preventative schedule. To suppress pecan leaf scorch mite on pecans, use in a preventative program. For best pecan *Phylloxera* control, use two applications 7 to 10 days apart starting at bud swell and ending by early leaf expansion. For ant control on pecans, do not apply where weed growth or other obstructions would impede uniform coverage of the orchard floor. Consult your State Agricultural Experiment Station, Certified Pest Control Advisor, or Extension Service specialist for specific use information in your area.

LORSBAN 50W may also be used on almonds and walnuts as a dormant/delayed dormant treatment applied at the dosages indicated in a dilute or concentrate spray. Proper equipment must be used to obtain complete coverage of the entire bark area. If scale infestations are severe, dilute sprays should be used to ensure complete coverage. Use oil as recommended by your state Agricultural Experiment Station, State Extension Service Specialist, or Certified Pest Control Advisor. Do not use oil with LORSBAN 50W on walnuts. Because dry or cold conditions may cause LORSBAN 50W plus oil sprays to infuse trees resulting in bud damage or drop, do not apply until winter rains or irrigation has replenished soil moisture such that bark and twigs are not dessicated.

DOSAGE AND USE RECOMMENDATIONS FOR LORSBAN 50W WETTABLE POWDER INSECTICIDE

CROP	INSECTS CONTROLLED	DOSAGE LORSBAN 50W	RESTRICTIONS
Apples	Apple aphid, Apple maggot, Codling moth, European apple sawfly, Eyespotted bud moth, European corn borer, Fruittree leafroller, Green fruitworm, Lesser appleworm, Obliquebanded leafroller, Orange tortrix, Pandemis leafroller, Oriental fruit moth, Periodical cicada, Plum curcullo, Redbanded leafroller, Rosy apple aphid, San Jose scale, Tarnished plant bug, Tufted apple budmoth, Variegated leafroller, Woolly apple aphid, Dogwood borer (trunk spray)	3 lb per acre	Make no more than 8 applications per season. Do not apply last treatment within 28 days before harvest or apply last two treatments closer than 21 days apart. Do not allow livestock to graze in treated orchards. In California do not do work involving contact with trees within 2 days after treatment.
Almonds	Navel orangeworm, Peach twig borer, San Jose scale	4 lb per acre	Make only one dormant/delayed dormant spray application and no more than three foliar spray applications on almonds per season, one dormant/delayed dormant spray application and no more than two foliar spray applications on walnuts per season, and no more than three foliar spray applications on filberts per season. Do not apply within 14 days of harvest. Do not allow livestock to graze in treated orchards.
Walnuts	Codling moth, Walnut scale, Walnut husk fly	4 lb per acre	
Filberts	Eyespotted bud moth, Filbert aphid, Filbert leafroller, Filbertworm Obliquebanded leafroller, Omniverous leaftier, Winter moth	3 to 4 lb per acre	
Pecans	Black pecan aphid, Hickory shuckworm, Pecan leaf scorch mite (suppression)	2 lb per 100 gallons	Make no more than 5 applications per season. Do not apply within 28 days of harvest. Do not allow livestock to graze in treated orchards.
	Pecan nut casebearer, Phylloxera spp., Fall webworm, Spittlebug	1 lb per 100 gallons	
	Fire ants and other ant species (orchard floor)	2 lb per acre	

CROP	INSECTS CONTROLLED	DOSAGE LORSBAN 50W	RESTRICTIONS
Brassica (cole) Broccoli Brussels sprouts Cabbage Cauliflower Collards Kale Kohlrabi	Cabbage aphid Cutworms Imported cabbageworm Striped flea beetle (adult)	2 lb per acre	Make no more than 6 applications per season. Do not apply within 21 days before harvest. To avoid phytotoxicity in vegetable crops, do not mix with other pesticide products or treat plants that are under extreme heat and drought stress.
Sour Cherries	Borers American plum Lesser peachtree Pacific flatheaded Peach twig Peachtree Shothole Climbing cutworm Green fruitworm Leafrollers Fruittree Obliquebanded Pandemis Redbanded Lesser appleworm Tarnished plant bug Moth Eyespotted bud Mineola Scale European fruit lecanium San Jose Stink bug	2 to 3 lb per acre	Make no more than 8 applications per season. Do not apply within 14 days of harvest. Do not allow livestock to graze in treated orchard. In California do not do work involving contact with trees within 2 days after treatment. Do not use on sweet cherries. Refer to the Endangered Species Section of the label for geographic restrictions on the use of LORSBAN 50W on sour cherries.
	Black cherry aphid Cherry fruitfly Plum curculio Rose chafer	3 lb per acre	

REENTRY STATEMENT

Do not allow reentry into treated areas within 24 hours unless protective clothing is worn. Protective clothing consists of at least a hat or other suitable head covering, a long-sleeved shirt and long-legged trousers or a coverall type garment (all of closely woven fabric covering the body, including the arms and legs), shoes and socks. Because certain states may require more restrictive reentry intervals for various crops treated with this product, consult your state Department of Agriculture for further information. Written or oral warnings must be given to workers who are expected to be in a treated area or in an area about to be treated with this product. These oral warnings shall inform workers of areas or fields that may not be entered without protective clothing and actions to take in case of accidental exposure as shown by the Statement of Practical Treatment section on this label. When oral warnings are given, warnings shall be given in a language customarily understood by workers. Written warnings must include the following information: WARNING. Area treated with LORSBAN 50W wettable powder insecticide on (insert date of application). Do not enter without protective clothing for 24 hours. In case of accidental exposure (insert actions to take as shown by the Statement of Practical Treatment section on this label).

STORAGE AND DISPOSAL

Do not contaminate water, food or feed by storage or disposal. Open dumping is prohibited.

Pesticide Disposal: Wastes resulting from the use of this product may be disposed of on site or at an approved waste disposal facility.

Container Disposal: Completely empty bag into application equipment. Then dispose of empty bag in a sanitary landfill or by incineration, or, if allowed by state and local authorities, by burning. If burned, stay out of smoke.

WARRANTY LIMITATIONS AND DISCLAIMER

The Dow Chemical Company warrants that this product conforms to the chemical description on the label and is reasonably fit for the purposes stated on the label when used in strict accordance with the directions therein under normal conditions of use. THIS IS THE ONLY WARRANTY MADE ON THIS PRODUCT. NO OTHER EXPRESS AND NO IMPLIED WARRANTY OF MERCHANTABILITY OR FITNESS FOR A PARTICULAR PURPOSE IS MADE OUTSIDE OF THIS LABEL. Therefore, neither this warranty nor any other warranty of merchantability, or fitness for a particular purpose, express or implied, extends to the use of this product contrary to label instructions (including conditions noted on the label, such as unfavorable temperatures, soil conditions, etc.) under abnormal conditions (such as excessive rainfall, drought, tornadoes, hurricanes, etc.) or under conditions not reasonably foreseeable to or beyond the control of seller.

When buyer or user suffers losses or damages resulting from the use or handling of this product (including claims based on contract, negligence, strict liability, or other legal theories), buyer or user must promptly notify in writing The Dow Chemical Company of any claims to be eligible to receive either remedy given below. The EXCLUSIVE REMEDY OF THE BUYER OR USER and the LIMIT OF LIABILITY of The Dow Chemical Company or any other seller will be one of the following, at the election of The Dow Chemical Company:

(1) **Refund of purchase price** paid by buyer or user for product bought, **or**

(2) **Replacement** of amount of product used.

The seller **will not be liable for consequential or incidental damages or losses**

The terms of this Warranty Limitations and Disclaimer cannot be varied by any written or verbal statements or agreements. Any employee or sales agent of the seller is not authorized to vary or exceed the terms of this Warranty Limitations And Disclaimer in any manner.

45386–L1 I189

* Trademark of THE DOW CHEMICAL COMPANY

N* SERVE* 24

FOR STABILIZED NITROGEN

Use to Delay Nitrification of Ammoniacal and Urea Nitrogen Fertilizer Compositions in the Soil by Controlling the Nitrification Process.

ACTIVE INGREDIENT(S):	
Nitrapyrin [2-chloro-6-(trichloromethyl) pyridine]	22.2%
Related chlorinated pyridines	2.5%
INERT INGREDIENTS:	75.3%
TOTAL	100.0%

Contains 2 pounds of active ingredients per gallon

E.P.A Registration No. 464-433 E.P.A. Est. 464-CA-1

KEEP OUT OF REACH OF CHILDREN

WARNING

AVISO:

PRECAUCIÓN AL USUARIO: Si usted no lee inglés, no use este producto hasta que la etiqueta le haya sido explicada ampliamente.

PRECAUTIONARY STATEMENTS

HAZARDS TO HUMANS AND DOMESTIC ANIMALS

HARMFUL IF SWALLOWED • HARMFUL IF ABSORBED THROUGH SKIN • CAUSES EYE IRRITATION • THE ACTIVE INGREDIENT IN THIS PRODUCT MAY CAUSE SKIN SENSITIZATION REACTIONS IN CERTAIN INDIVIDUALS

Avoid Contact With Skin, Eyes, Or Clothing • Avoid Breathing Vapor Or Spray Mist • Wear Protective Clothing And Rubber Gloves • Wash Thoroughly With Soap And Water After Handling • Remove Contaminated Clothing And Wash Before Reuse

STATEMENTS OF PRACTICAL TREATMENT:

If Swallowed: Do not induce vomiting. Contains aromatic petroleum solvent. Call a physician or Poison Control Center.

If in Eyes: Flush with plenty of water. Get medical attention if irritation persists.

If On Skin: Wash with plenty of soap and water. Get medical attention if irritation persists.

If Inhaled: Remove to fresh air if effects occur. Get medical attention.

PHYSICAL OR CHEMICAL HAZARDS

COMBUSTIBLE • Do Not Use, Pour, Spill or Store Near Heat or Open Flame • Do Not Cut or Weld Container.

ENVIRONMENTAL HAZARDS

Do not apply directly to water. Do not contaminate water by cleaning of equipment or disposal of wastes.

NOTICE

Read the entire label. Use only according to label directions.

Before buying or using this product, read "WARRANTY LIMITATIONS AND DISCLAIMER" elsewhere on this label. If terms are not acceptable, return **unopened** package at once to seller for full refund of purchase price paid. Otherwise, use by the buyer or any other user constitutes acceptance of the terms under "WARRANTY LIMITATIONS AND DISCLAIMER."

IN CASE OF AN EMERGENCY **endangering life or property involving this product, call collect 517–636–4400**

AGRICULTURAL CHEMICAL

Do Not Ship or Store with Food, Feeds, Drugs, or Clothing

DIRECTIONS FOR USE

It is a violation of Federal law to use this product in a manner inconsistent with its labeling.

GENERAL INFORMATION

N-SERVE 24 nitrogen stabilizer is an oil-soluble product that may be used with anhydrous ammonia, dry ammonium and urea fertilizers.

Do not apply this product through any type of irrigation system.

MIXING INSTRUCTIONS

Anhydrous Ammonia: N-SERVE 24, a product that is soluble in liquid anhydrous ammonia, may be pumped into the anhydrous tank through the liquid fill valve before, during or after loading. Pumping and metering systems are necessary for the injection of N-SERVE 24 into the tank. The ammonia containing N-SERVE 24 is then applied to cropland using conventional ammonia application equipment. Tank mixing with anhydrous ammonia is the preferred method of application; however, N-SERVE 24 may also be applied through downstream injection systems.

Granular Ammonium and Urea: N-SERVE 24 may be impregnated on most dry ammonia fertilizers or fertilizer blends containing ammoniacal fertilizers. Uniform application in the field and immediate incorporation are necessary to insure optimum results.

For impregnating N-SERVE 24 on dry fertilizers, use a closed rotary drum mixer equipped with suitable spraying equipment. The spray nozzle should be positioned inside the mixer to provide uniform spray coverage of the tumbling fertilizer. Check the crop section of this label to determine the per acre use rate of N-SERVE 24. The maximum amount of N-SERVE 24 to be added is one quart per 100 pounds of nitrogen fertilizer and minimum amount is one quart per 500 pounds of nitrogen fertilizer. At the higher concentrations of N-SERVE 24, the fertilizer may not readily absorb all of the liquid. For a suitable free-flowing mixture, an absorptive powder such as Micro-CelE should be added separately and uniformly to the fertilizer blend following addition of N-SERVE 24.

Bulk fertilizers impregnated with N-SERVE 24 should be applied immediately, NOT STORED. All bulk containers should be tightly covered while the products are being transported and applied to reduce the chance of loss of N-SERVE 24 via volatilization. All fertilizers impregnated with N-SERVE 24 should be **incorporated immediately** after application for optimal performance.

All individual state regulations including those related to flamable liquid handling and to dry bulk blending registration labeling and application are the responsibility of the individual and/or company selling N-SERVE 24 on fertilizer mixtures.

Do not use N-SERVE 24 with fertilizers containing nitrate such as ammonium nitrate, potassium nitrate or calcium nitrate. Do not mix seed with dry fertilizers impregnated with N-SERVE 24.

USE AND DOSAGE RECOMMENDATIONS

N-SERVE 24 nitrogen stabilizer must be injected or incorporated in a zone or band in the soil with the fertilizer at a minimum depth of 2 to 4 inches during or immediately after application.

CORN, SORGHUM AND WHEAT

Preplant or At-Plant Row or Band Application: Use N-SERVE 24 at the rate of 1/2 to 1 quart per acre. The higher rate is preferred with all fall, most spring applications, and high yield production programs.

Preplant or At-Plant Broadcast Application: Use N-SERVE 24 at the rate of 1 to 2 quarts per acre. The higher rate is preferred when fall applications are made to spring planted crops.

Split Application on Corn: In addition to a preplant or at-plant application, a second application may be made postplant at the rate of 1/2 to 1 quart of N-SERVE 24 per acre. For high yield production, use the 1 quart per acre rate. The total amount of N-SERVE 24 applied in a split application program involving a preplant or at-plant treatment followed by a postplant treatment must not exceed 2 quarts per acre per year.

COTTON

Preplant or At-Plant Band or Broadcast Application: Use N-SERVE 24 at the rate of 1/2 to 1 quart per acre. The higher rate should be used if N-SERVE 24 is appplied in the fall as a preplant treatment for spring planted cotton.

Postplant Row or Band Application: Use N-SERVE 24 at the rate of 1/2 quart per acre.

Do not allow livestock to graze in treated areas or feed forage to livestock.

STRAWBERRIES

Preplant Application: Use N-SERVE 24 at the rate of 1 quart per acre. Plant strawberries as soon as practical thereafter. Based on available residue data, use of N-SERVE 24 on strawberries is restricted to California.

RESTRICTIONS:

Do not apply more than 2 quarts of N-SERVE 24 per acre per year on corn, sorghum, and wheat and 1 quart of N-SERVE 24 per acre per year on cotton and strawberries. Make only one postplant application per year. Postplant applications can only be made up to 30 days postplant on corn and up to 45 days postplant on cotton. Crops listed on this label may be rotated in less than one year from the last application. All other crops are not to be rotated in less than one year from the last application.

STORAGE AND DISPOSAL

Do not contaminate water, food, or feed by storage or disposal. Open dumping is prohibited. Store above 18°F to keep product in solution. Below 14°F, crystals may form and settle to the bottom. To redissolve the crystals, allow the contents to warm up to about 70°F and to stand for several days at that temperature. Once the crystals have dissolved, a gentle rolling of the container will ensure a uniform mixture of the contents.

Pesticide Disposal: Pesticide wastes are toxic. Improper disposal of excess pesticide, spray mixture, or rinsate is a violation of Federal law. If these wastes cannot be disposed of by use according to label instructions, contact your State Pesticide or Environmental Control Agency, or the Hazardous Waste representative at the nearest EPA Regional Office for guidance.

Container Disposal: Triple rinse (or equivalent). Then offer for recycling or reconditioning, or puncture and dispose of in a sanitary landfill, or by other procedures approved by state and local authorities.

USE PRECAUTIONS

N-SERVE 24 nitrogen stabilizer is corrosive to aluminum components which may affect the operation of pumping, mixing and application equipment. Solutions of N-SERVE 24 and anhydrous ammonia may cause corrosion of submerged aluminum components. These parts should be replaced with mild steel or stainless steel materials of construction. Side mount or end mount liquid level float gauges containing aluminum should be replaced with stainless and mild steel guages. To minimize corrosion of equipment and for maximum effectiveness, do not store anhydrous ammonia containing N-SERVE 24 longer than three weeks.

Acme fitting gaskets should be replaced as needed with standard Buna-N or Buna-S gaskets. More resistant gasket materials such as Teflon may also be used.

WARRANTY LIMITATIONS AND DISCLAIMER

The Dow Chemical Company warrants that this product conforms to the chemical description on the label and is reasonably fit for the purposes stated on the label when used in strict accordance with the directions therein under normal conditions of use. THIS IS THE ONLY WARRANTY MADE ON THIS PRODUCT. NO OTHER EXPRESS AND NO IMPLIED WARRANTY OF MERCHANTABILITY OR FITNESS FOR A PARTICULAR PURPOSE IS MADE OUTSIDE OF THIS LABEL. Therefore, neither this warranty nor any other warranty of merchantability or fitness for a particular purpose, express or implied, extends to the use of this product contrary to label instructions (including conditions noted on the label, such as unfavorable temperatures, soil conditions, etc.), under abnormal conditions (such as excessive rainfall, drought, tornadoes, hurricanes, etc.), or under conditions not reasonably forseeable to or beyond the control of seller.

When buyer or user suffers losses or damages resulting from the use or handling of this product (including claims based on contract, negligence, strict liability, or other legal theories), buyer or user must promplty notify in writing The Dow Chemical Company of any claims to be eligible to receive either remedy given below. The EXCLUSIVE REMEDY OF THE

BUYER OR USER and the LIMIT OF LIABILITY of The Dow Chemical Company or any other seller will be one of the following, at the election of The Dow Chemical Company:

(1) **Refund of purchase price** paid by buyer or user for product bought, **or**

(2) **Replacement** of amount of product used.
The seller **will not be liable for consequential or incidental damages or losses.**

The terms of this Warranty Limitations And Disclaimer cannot be varied by any written or verbal statements or agreements. Any employee or sales agent of the seller is not authorized to vary or exceed the terms of this Warranty Limitations And Disclaimer in any manner.

56090–L3 D787

*Trademark of THE DOW CHEMICAL COMPANY

N* SERVE* 24E

FOR STABILIZED NITROGEN

Use to Delay Nitrification of Ammoniacal and Urea Nitrogen Fertilizer Compositions in the Soil by Controlling the Nitrification Process.

ACTIVE INGREDIENT(S):

Nitrapyrin [2-chloro-6-(trichloromethyl) pyridine]	21.9%
Related chlorinated pyridines	2.4%
INERT INGREDIENTS:	75.7%
TOTAL	100.0%

Contains 2 pounds of active ingredients per gallon

E.P.A Registration No. 464-432 E.P.A. Est. 464-CA-1(WP); 407-MN-1 (PK); 5905-AR-1(DM)

Superscript used corresponds to letter in Lot No.

KEEP OUT OF REACH OF CHILDREN

DANGER
PELIGRO:

PRECAUCIÓN AL USUARIO: Si usted no lee inglés, no use este producto hasta que la etiqueta le haya sido explicada ampliamente.

PRECAUTIONARY STATEMENTS

HAZARDS TO HUMANS AND DOMESTIC ANIMALS

CAUSES IRREVERSIBLE EYE DAMAGE • HARMFUL IF SWALLOWED •HARMFUL IF ABSORBED THROUGH SKIN • THE ACTIVE INGREDIENT IN THIS PRODUCT MAY CAUSE SKIN SENSITIZATION REACTIONS IN CERTAIN INDIVIDUALS

Do Not Get In Eyes Or On Clothing • Wear Goggles • Avoid Contact With Skin • Wear Protective Clothing And Rubber Gloves • Avoid Breathiing Vapor Or Spray Mist • Wash Thoroughly With Soap And Water After Handling • Remove Contaminated Clothing And Wash Before Reuse

STATEMENTS OF PRACTICAL TREATMENT:

If in Eyes: Flush with plenty of water for at least 15 minutes. Get medical attention. **If Swallowed:** Do not induce vomiting. Contains aromatic petroleum solvent. Call a physician or Poison Control Center immediately.

If On Skin: Wash with plenty of soap and water. Get medical attention if irritation persists.

If Inhaled: Remove to fresh air if effects occur. Get medical attention.

PHYSICAL OR CHEMICAL HAZARDS

COMBUSTIBLE • Do Not Use, Pour, Spill or Store Near Heat or Open Flame • Do Not Cut or Weld Container.

ENVIRONMENTAL HAZARDS

Do not apply directly to water. Do not contaminate water by cleaning of equipment or disposal of wastes.

NOTICE

Read the entire label. Use only according to label directions.

Before buying or using this product, read "WARRANTY LIMITATIONS AND DISCLAIMER" elsewhere on this label. If terms are not acceptable, return **unopened** package at once to seller for full refund of purchase price paid. Otherwise, use by the buyer or any other user constitutes acceptance of the terms under "WARRANTY LIMITATIONS AND DISCLAIMER."

IN CASE OF AN EMERGENCY endangering life or property involving this product, call collect 517–636–4400

AGRICULTURAL CHEMICAL

Do Not Ship or Store with Food, Feeds, Drugs, or Clothing

DIRECTIONS FOR USE

It is a violation of Federal law to use this product in a manner inconsistent with its labeling.

GENERAL INFORMATION

N-SERVE 24E nitrogen stabilizer is an emulsifiable product that may be used in formulation of and application with aqua ammonia, other liquid ammoniacal or urea nitrogen fertilizer compositions or liquid animal wastes. When liquid or dry ammonium and urea fertilizers and animal wastes are surface applied, N-SERVE 24E may be used with water and applied during the incorporation operation.

Do not apply this product through any type of irrigation system.

MIXING INSTRUCTIONS

To prepare the spray, add 1/2 of the required amount of water or fertilizer to the spray tank and, with agitation, add the N-SERVE 24E. Complete filling the tank with the balance of the water or fertilizer needed. When adding N-SERVE 24E through eductor systems on fertilizer application units, either premix one part of N-SERVE 24E to two parts water or be careful to open the inlet valve only about 1/3 of maximum. Either method will reduce the N-SERVE 24E to fertilizer ratio and thus reduce the possibility of formation of undesirable invert emulsions of N-SERVE 24E and fertilizer. When making applications of N-SERVE 24E and liquid nitrogen fertilizer compositions or liquid animal wastes, agitation must be provided before and during application to maintain a uniform mixture of N-SERVE 24E. The application system must be capable of keeping a suspended solid dispersed throughout the liquid.

TANK MIXING INSTRUCTIONS

N-SERVE 24E may also be applied in tank mixtures in water or in most urea-ammonium nitrate solutions and N-P-K solutions, slurries or suspensions with the following preplant incorporated herbicides Treflan, Prowl, or Tolban for use on cotton; atrazine or Bladex for use on sorghum; and Sutan +, atrazine, Bladex, Eradicane, Lasso, Dual, or combinations of Sutan + with atrazine or Bladex for use on corn. Physical compatibility of these tank mixtures should be determined before combining in the spray tank. Tank mixtures of N-SERVE 24E and these herbicides in water generally do not require a compatibility aid such as Compex or Unite while tank mixtures with fertilizers frequently will. Maintain constant agitation during both mixing and application to ensure uniformity of the spray mixture. Read and carefully follow all applicable directions including dosage rates, restrictions, and precautions for the specified crops on labeling for the other products used in combination with N-SERVE 24E.

The suggested mixing sequence for tank mixtures of N-SERVE 24E with herbicides in water or liquid nitrogen fertilizers is as follows:

1. Fill the spray tank 1/2 full of water or liquid fertilizer.

2. Begin agitation.

3. Add a compatibility aid, if needed. (If water slurried premixes of wettable powders or flowables are used as in Steps 4 and 5, a compatibility aid may be added to the premix for convenience.)

4. Add wettable powder formulations to the tank (make a water slurry with wettable powders first).

5. Add flowable formulations to the tank.

6. Add emulsifiable concentrate formulations including N-SERVE 24E to the tank (premix 1:2 with water first if introduced to the tank through eductor systems).

7. Add the balance of the water or liquid fertilizer and continue agitation until the tank is empty.

Note: Test compatibility of the intended tank mixture before adding N-SERVE 24E nitrogen stabilizer to the spray or mix tank. Add proportionate amounts of each ingredient to a pint or quart jar, cap, shake, and let set 15 minutes. Formation of precipitates or layers that do not readily re-disperse indicates an incompatible mixture that should not be used.

USE AND DOSAGE RECOMMENDATIONS

N-SERVE 24E nitrogen stabilizer must be injected or incorporated in a zone or band in the soil with the fertilizer at a minimum depth of 2 to 4 inches during or immediately after application.

CORN, SORGHUM AND WHEAT

Preplant or At-Plant Row or Band Application: Use N-SERVE 24E at the rate of 1/2 to 1 quart per acre. The higher rate is preferred with all fall, most spring applications, and high yield production programs.

Preplant or At-Plant Broadcast Application: Use N-SERVE 24E at the rate of 1 to 2 quarts per acre. The higher rate is preferred when fall applications are made to spring planted crops.

Application with Liquid Animal Wastes: Use N-SERVE 24E at the rate of 1 to 2 quarts per acre and apply preplant in the fall or spring. The 2 quart per acre rate is recommended for all fall applications with animal wastes.

Split Application on Corn: In addition to a preplant or at-plant application, a second application may be made postplant at the rate of 1/2 to 1 quart of N-SERVE 24E per acre. For high yield production, use the 1 quart per acre rate. The total amount of N-SERVE 24E applied in a split application program involving a preplant or at-plant treatment followed by a postplant treatment must not exceed 2 quarts per acre per year.

COTTON

Preplant or At-Plant Band or Broadcast Application: Use N-SERVE 24E at the rate of 1/2 to 1 quart per acre. the higher rate should be used if N-

SERVE 24E is applied in the fall as a preplant treatment for spring planted cotton.

Postplant Row or Band Application: Use N-SERVE 24E at the rate of 1/2 quart per acre.

Do not allow livestock to graze in treated areas or feed forage to livestock.

STRAWBERRIES

Preplant Application: Use N-SERVE 24E at the rate of 1 quart per acre. Premix N-SERVE 24E with the nitrogen containing fertilizer before application or mix N-SERVE 24E with water and overspray the nitorgen containing fertilizer following application. Plant strawberries as soon as practical thereafter. Based on available residue data, use of N-SERVE 24E on strawberries is restricted to California.

Restrictions: Do not apply more than 2 quarts of N-SERVE 24E per acre per year on corn, sorghum, and wheat and 1 quart of N-SERVE 24E per acre per year on cotton and strawberries. Make only one postplant application per year. Postplant application can only be made up to 30 days postplant on corn and up to 45 days on cotton. Crops listed on this label may be rotated in less than one year from the last application. All other crops are not to be rotated in less than one year from the last application.

STORAGE AND DISPOSAL

Do not contaminate water, food, or feed by storage or disposal. Open dumping is prohibited. Store above 18°F to keep product in solution. Below 14°F, crystals may form and settle to the bottom. To redissolve the crystals, allow the contents to warm up to about 70°F and to stand for several days at that temperature. Once the crystals have dissolved, a gentle rolling of the container will ensure a uniform mixture of the contents.

Pesticide Disposal: Pesticide wastes are acutely hazardous. Improper disposal of excess pesticide, spray mixture, or rinsate is a violation of Federal law. If these wastes cannot be disposed of by use according to label instructions, contact your State Pesticide or Environmental Control Agency, or the Hazardous Waste representative at the nearest EPA Regional Office for guidance.

Container Disposal: Triple rinse (or equivalent). Then offer for recycling or reconditioning, or puncture and dispose of in a sanitary landfill, or by other procedures approved by state and local authorities.

USE PRECAUTIONS

N-SERVE 24E nitrogen stabilizer is corrosive to aluminum components which may affect the operation of pumping, mixing and application equipment. To minimize corrosion of equipment and for maximum effectiveness, do not store liquid fertilizer solutions containing N-SERVE 24E. Use within 72 hours of mixing. Rinse application equipment used to apply liquid fertilizers and N-SERVE 24E with water after each use. Use with hot mix fertilizers is not recommended. High temperatures generated in production of these mixtures can cause rapid degradation of product and a reduction in effectiveness.

WARRANTY LIMITATIONS AND DISCLAIMER

The Dow Chemical Company warrants that this product conforms to the chemical description on the label and is reasonably fit for the purposes stated on the label when used in strict accordance with the directions therein under normal conditions of use. THIS IS THE ONLY WARRANTY MADE ON THIS PRODUCT. NO OTHER EXPRESS AND NO IMPLIED WARRANTY OF MERCHANTABILITY OR FITNESS FOR A PARTICULAR PURPOSE IS MADE OUTSIDE OF THIS LABEL. Therefore, neither this warranty nor any other warranty of merchantability or fitness for a particular purpose, express or implied, extends to the use of this product contrary to label instructions (including conditions noted on the label, such as unfavorable temperatures, soil conditions, etc.), under abnormal conditions (such as excessive rainfall, drought, tornadoes, hurricanes, etc.) or under conditions not reasonably forseeable to or beyond the control of seller.

When buyer or user suffers losses or damages resulting from the use or handling of this product (including claims based on contract, negligence, strict liability, or other legal theories), buyer or user must promplty notify in writing The Dow Chemical Company of any claims to be eligible to receive either remedy given below. The EXCLUSIVE REMEDY OF THE BUYER OR USER and the LIMIT OF LIABILITY of The Dow Chemical Company or any other seller will be one of the following, at the election of The Dow Chemical Company:

(1) **Refund of purchase price** paid by buyer or user for product bought, **or**

(2) **Replacement** of amount of product used.
The seller **will not be liable for consequential or incidental damages or losses.**

The terms of this Warranty Limitations And Disclaimer cannot be varied by any written or verbal statements or agreements. Any employee or sales agent of the seller is not authorized to vary or exceed the terms of this Warranty Limitations And Disclaimer in any manner.

56092–L3 B787

*Trademark of THE DOW CHEMICAL COMPANY

PATHWAY*

HERBICIDE

For Controlling Unwanted Trees Via Cut Surface Treatments

Active Ingredient(s):
Picloram (4-amino-3,5,6-trichloropicolinic acid), as the triisopropanolamine salt 5.4%
2,4-Dichlorophenoxyacetic Acid, as the triisopropanolamine salt . 20.9%
Inert Ingredient(s): 73.7%

Acid Equivalents: Picloram - 3.0% 2,4-Dichlorophenoxyacetic acid - 11.2%

E.P.A. Registration No. 464-510
E.P.A. Est. 464-MI-1

KEEP OUT OF REACH OF CHILDREN

WARNING

AVISO:

PRECAUCION AL USUARIO: Si usted no lee inglés, no use este producto hasta que la etiqueta le haya sido explicada ampliamente.

PRECAUTIONARY STATEMENTS

HAZARDS TO HUMANS AND DOMESTIC ANIMALS

HARMFUL OR FATAL IF SWALLOWED • CAUSES EYE INJURY • MAY CAUSE SKIN IRRITATION

Avoid Contact With Eyes, Skin And Clothing • Wash Thoroughly After Handling • Keep Container Closed • Do Not Store Near Food Or Feed • Do Not Drink Solution • Do Not Store In Open Or Unlabeled Containers

STATEMENT OF PRACTICAL TREATMENT

Contains ethylene glycol. **If swallowed,** induce vomiting immediately by giving two glasses of water and sticking finger down throat. Get medical attention. Never give anything by mouth to an unconscious person. **In case of eye contact,** promptly flush with plenty of water; and get medical attention. Remove contaminated clothing and wash before reuse.

PHYSICAL OR CHEMICAL HAZARDS

COMBUSTIBLE • Keep Away From Heat And Open Flame.

ENVIRONMENTAL HAZARDS

Do not apply directly to any body of water. Do not contaminate water used for irrigation or domestic purposes by cleaning of equipment or disposal of wastes. Do not allow run-off or spray to contaminate wells, irrigation ditches or any body of water used for irrigation or domestic purposes. Do not make application when circumstances favor movement from treatment site. See additional precautionary statements elsewhere on this label.

NOTICE

Read the entire label. Use only according to label directions.

Before buying or using this product, read "WARRANTY LIMITATIONS AND DISCLAIMER" elsewhere on this label. If terms are not acceptable, return **unopened** package at once to seller for full refund of purchase price paid. Otherwise, use by the buyer or any other user constitutes acceptance of the terms under "WARRANTY LIMITATIONS AND DISCLAIMER."

IN CASE OF AN EMERGENCY endangering life or property involving this product, call collect 517-636-4400

AGRICULTURAL CHEMICAL

Do Not Ship or Store with Food, Feeds, Drugs, or Clothing

GENERAL INFORMATION

PATHWAY Herbicide is a **ready-to-use product** that is effective in cut surface applications for killing unwanted trees and preventing undesirable sprouting of cut trees in forest and other non-crop areas such as fence rows and right-of-ways. Among the many species that can be controlled are: **ailanthus, alder, aspen, birch, cedar, cherry, dogwood, elm, firs, green ash, gum, hawthorn, hickory, hornbeam, maples, oaks, pecan, persimmon, serviceberry, sourwood, sweetbay.**

USE DIRECTIONS

It is a violation of Federal law to use this product in a manner inconsistent with its labeling.

Use undiluted as indicated below.

TREE INJECTION TREATMENT: Use 1 milliliter of undiluted PATHWAY Herbicide solution through the bark completely around the tree trunk at intervals of 2 to 3 inches between edges of the injector wounds. Make injections near ground level when using the tree injector or 2 to 4 feet above the ground when using a HYPOHATCHET Injector or similar device. Treatments can be made in any season. Maples should not be treated during the spring sap flow. With some difficult to control species such as dogwood, hickory, sugar maple, bigleaf maple, tanoak, and some firs, application to a continuous cut rather than to spaced cuts may be more uniformly effective.

FRILL OR GIRDLE TREATMENT: Make a single hack girdle or "frill" of overlapping axe cuts through the bark completely around the tree as close to the ground as feasible. Spray or paint the injured surface with undiluted PATHWAY Herbicide, using enough volume to wet treated areas.

STUMP TREATMENT: Spray or paint the cut surfaces of freshly cut stumps and stubs with undiluted PATHWAY Herbicide. The cambium area next to the bark is the most vital area to wet.

USE PRECAUTIONS

Use this product only as specified on this label. PATHWAY Herbicide is highly active as a herbicide, is water soluble, can move with surface runoff water, and can remain phytotoxic for a year or more if it gets into the soil. Untreated trees occasionally are affected by root uptake of herbicide leaches into the soil or excreted by the roots of nearby treated trees. If such injury cannot be tolerated, do not use PATHWAY close to desirable trees.

Do not contaminate cropland, water, or irrigation ditches.

Do Not Contaminate Water Intended for Irrigation or Domestic Purposes. To avoid injury to crops or other desirable plants, do not treat or allow spray drift or run-off to fall onto banks or bottoms of irrigation ditches, either dry or containing water, or other channels that carry water that may be used for irrigation or domestic purposes. Do not apply to snow or frozen ground.

Water contaminated with residues of picloram from drainage ditch bank uses shall not be used to irrigate crops which are not registered for use with this chemical.

Do not rotate food or feed crops on treated land if they are not registered for use with picloram until an adequately sensitive bioassay or chemical test shows that no detectable picloram is present in the soil.

Do not apply on or in the vicinity of sensitive crops of desirable broadleaf plants. Small amounts can cause injury to plants.

Do not allow careless application or spray drift. Do not permit any spray or spray drift to contact desirable plants. Use only gravity or very low spray pressure.

Do not remove product from original container except to apply as indicated under USE DIRECTIONS.

Do not store near food, feedstuff, fertilizer, seeds, insecticides, fungicides, or other pesticides.

PRECAUTIONARY STATEMENTS

ENVIRONMENTAL HAZARDS

Picloram is a chemical which can travel (seep or leach) through soil and under certain conditions contaminate groundwater which may be used for irrigation and drinking purposes. Users are advised not to apply picloram where soils have a rapid to very rapid permeability throughout the profile (such as loamy sand to sand) and the water table of an underlying aquifer is shallow or to soils containing sinkholes over limestone bedrock, severely fractured surfaces, and substrates which would allow direct introduction into an aquifer. Your local agricultural agencies can provide further information on the type of soil in your area and the location of groundwater.

An aquifer is defined as "an underground, saturated, permeable, geologic formation capable of producing significant quantities of water to a well or spring. It is the ability of the saturated zone, or portion of that zone, to yield water which makes it an aquifer" (American Chemical Society, 1983).

STORAGE AND DISPOSAL

Do not contaminate water, food or feed by storage or disposal.

Pesticide Disposal: Pesticide wastes are toxic. Improper use of excess pesticide, spray mixture, or rinsate is a violation of Federal law. If these wastes cannot be disposed of by use according to label instructions, contact your State Pesticide or Environmental Control Agency, or the Hazardous Waste representative at the nearest EPA Regional Office for guidance.

Rinse application equipment after use, at least three times with water, and dispose of rinse water in a non-cropland area away from water supplies.

Container Disposal: Do not reuse container. Triple rinse (or equivalent). Puncture and dispose of in a sanitary landfill, or by incineration, or, if allowed by state and local authorities, by burning. If burned, stay out of smoke.

General: Consult federal, state or local disposal authorities for approved alternative procedures.

Be sure that use of this product conforms to all application regulations.

WARRANTY LIMITATIONS AND DISCLAIMER

The Dow Chemical Company warrants that this product conforms to the chemical description on the label and is reasonably fit for the purposes stated on the label when used in strict accordance with the directions therein under normal conditions of use. THIS IS THE ONLY WARRANTY MADE ON THIS PRODUCT. NO OTHER EXPRESS AND NO IMPLIED WARRANTY OF MERCHANTABILITY OR FITNESS FOR A PARTICULAR PURPOSE IS MADE OUTSIDE OF THIS LABEL. Therefore, neither this warranty nor any other warranty of merchantability or fitness for a particular purpose, express or implied, extends to the use of this product contrary to label instructions (including conditions noted on the label, such as unfavorable temperatures, soil conditions, etc.), under abnormal conditions (such as excessive rainfall, drought, tornadoes, hurricanes, etc.) or under conditions not reasonably foreseeable to or beyond the control of seller.

When buyer or user suffers losses or damages resulting from the use or handling of this product (including claims based on contract, negligence, strict liability, or other legal theories), buyer or user must promptly notify in writing The Dow Chemical Company of any claims to be eligible to receive either remedy given below. THE EXCLUSIVE REMEDY OF THE BUYER OR USER and the LIMIT OF LIABILITY of The Dow Chemical Company or any other seller will be one of the following, at the election of The Dow Chemical Company:

(1) **Refund of purchase price** paid by buyer or user for product bought, **or**

(2) **Replacement** of amount of product used.

The seller **will not be liable for consequential or incidental damages or losses.**

The terms of this Warranty Limitations And Disclaimer cannot be varied by any written or verbal statements or agreements. Any employee or sales agent of the seller is not authorized to vary or exceed the terms of this Warranty Limitations And Disclaimer in any manner.

11387-L1 489

*Trademark of THE DOW CHEMICAL COMPANY

STINGER®

HERBICIDE

For Selective Postemergence Control of Broadleaf Weeds in Sugar Beets, Wheat, Barley and Oats Not Underseeded With a Legume, Fallow Cropland, Rangeland and Permanent Grass Pastures, Non-Cropland Areas and Conservation Reserve Program (CRP) Acres

Active Ingredient(s):

Clopyralid (3,6-dichloro-2-pyridinecarboxylic acid)
as the monoethanolamine salt 40.9%
Inert Ingredient(s): 59.1%

ACID EQUIVALENT: 3,6-dichloro-2-pyridinecarboxylic acid, 31%- 3lb/gal

E.P.A. Registration No. 464-600
E.P.A. Est. 464-MI-1

KEEP OUT OF REACH OF CHILDREN

CAUTION

PRECAUCION:

PRECAUCION AL USUARIO: Si usted no lee ingles, no use este producto hasta que la etiqueta le haya sido explicado ampliamente.

PRECAUTIONARY STATEMENTS

Hazards to Humans and Domestic Animals

CAUSES EYE INJURY • HAMFUL IF INHALED OR ABSORBED THROUGH SKIN

Avoid Contact With Eyes, Skin Or Clothing • Avoid Breathing Spray Mist • Wash Thoroughly With Soap And Water After Handling • Remove Contaminated Clothing And Wash Before Reuse

STATEMENT OF PRACTICAL TREATMENT

If in eyes, flush with plenty of water. Get medical attention if irritation persists. **If on skin,** wash with plenty of soap and water. Get medical attention.

PHYSICAL OR CHEMICAL HAZARDS

COMBUSTIBLE • Do Not Use or Store Near Heat or Open Flame. • Do Not Cut or Weld Container.

ENVIRONMENTAL HAZARDS

Do not contaminate water when disposing of equipment washwaters. Do not contaminate water used for irrigation or domestic purposes. Do not apply directly to any body of water or wetlands. See additional precautionary statements elsewhere on this label.

NOTICE

Read the entire label. Use only according to label directions.

Before buying or using this product, read "WARRANTY LIMITATIONS AND DISCLAIMER" elsewhere on this label. If terms are not acceptable, return **unopened** package at once to seller for full refund of purchase price paid. Otherwise, use by the buyer or any other user constitutes accptance of the terms under "WARRANTY LIMITATIONS AND DISCLAIMER."

IN CASE OF AN EMERGENCY endangering life or property involving this product, call collect 517-636-4400

AGRICULTURAL CHEMICAL

Do Not Ship Or Store with Food, Feeds, Drugs, or Clothing

DIRECTIONS FOR USE

It is a violation of Federal law to use this product in a manner inconsistent with its labeling.

Do not apply this product through any type of irrigation system.

STINGER Herbicide is recommended for the postemergence control of Canada thistle (rosette to pre-bud), wild buckwheat, ladysthumb, green smartweed, diffuse and spotted knapweed, mayweed (dogfennel), musk thistle (rossette to bud), scentless chamomile (false), sunflower (common and wild), sowthistle (annual and perennial), cocklebur, ragweed and common vetch infesting sugar beets, wheat, barley and oats not underseeded with a legume, fallow cropland, rangeland and permanent grass pastures, conservation reserve program (CRP) acres and non- cropland areas including fence rows, around farm buildings and equipment pathways.

For ground treatment, use enough total spray volume to provide uniform spray coverage. Apply 2 or more gallons total spray volume per acre when using a SPRA-COUPE and 5 or more gallons total spray volume per acre with other ground equipment. Do not apply where spray drift may be a problem due to proximity of susceptible crops or other desirable plants. Apply during warm weather when weeds are young and growing actively. Best results should be obtained when soil moisture is adequate for plant growth. This product can affect plants directly through foliage and indirectly by root uptake from treated soil.

When tank mixing, read and follow the label of each tank-mix product used for precautionary statements, directions for use, weeds controlled and georgraphic and other restrictions. Use in accordance with the most restrictive of label limitations and precautions. No label dosages should be exceeded. This product cannot be mixed with any product containing a label prohibition against such mixing.

Sugar Beets: STINGER Herbicide is recommended for the control of various annual and perennial broadleaf weeds infesting sugar beets. Apply 1/4 to 2/3 pint of STINGER Herbicide per acre with ground equipment as a broadcast foliar spray. Apply in 10 or more gallons total spray volume per acre when the sugar beets are in the 2-true leaf to 8 leaf stage of growth and the weeds are young and activly growing. Re-treat as necessary but do not exceed 2/3 pint of STINGER Herbicide per acre per season. Do not apply within 105 days before harvest of beet roots and tops. Do not apply to sugar beets grown for seed.

STINGER Herbicide may be applied as a band treatment. use the formulas below to determine the appropriate rate and volume per treated acre.

$$\frac{\text{Band width in inches}}{\text{Row width in inches}} \times \text{Broadcast rate per treated acre} = \text{Band rate per treated acre}$$

$$\frac{\text{Band width in inches}}{\text{Row width in inches}} \times \text{Broadcast volume per treated acre} = \text{Band volume per treated acre}$$

Weed Species Recommendations: Apply 1/4 to 2/3 pint of STINGER Herbicide per acre to control the following weeds:

Weed Species	Growth Stage	Amount of Stinger per acre[1]
Cocklebur wild sunflower volunteer sunflower giant ragweed volunteer soybeans volunteer alfalfa	up to 6 leaf	1/4 to 1/2 pt
common ragweed	up to 5 leaf	
wild buckwheat	3 to 5 leaf stage but before vining	
smartweeds (suppression)	3 to 4 leaf stage	
Canada thistle[2] sowthistle (suppression)	rosette to prebud	1/2 to 2/3 pt

[1] Use the lower rate for light to moderate infestations and good growing conditions and the higher rate for dense infestations or under poor growing conditions such as drought.

[2] For the most effective control of Canada thistle, apply as a broadcast treatment to the entire infested area.

To control additional weeds, tank mix 1/4 to 2/3 pint of STINGER Herbicide with 2 to 6.5 pints of Betamix or Betanex. When tank mixing, read and follow the label of each tank-mix product used for precautionary statements, directions for use, weeds controlled and geographic and other restrictions. Use in accordance with the most restrictive of the label limitations and precautions. No label dosages should be exceeded. This product cannot be tank mixed with any product containing a label prohibition against such mixing.

Fallow Cropland: Apply 1/4 to 1/2 pint of STINGER Herbicide per acre when weeds are 1 to 3 inches tall. Apply 1/2 pint per acre when weeds are 3 to 6 inches tall or under dry conditions. Where Canada thistle or knapweeds are the primary pest, best results are obtained by applying 1/2 pint of STINGER per acre when weeds are 3 to 6 inches tall.

Non-Cropland: Apply 1/4 to 1/2 pint of STINGER Herbicide per acre when weeds are 1 to 3 inches tall. Apply 1/2 pint per acre when weeds are 3 to 6 inches tall or under dry conditions. Where Canada thistle or knapweeds are the primary pest, best results are obtained by applying 1/2 pint of STINGER Herbicide per acre when weeds are 3 to 6 inches tall. To improve spectrum of activity or to increase activity against taller weeds, STINGER may be tank mixed with 0.5 to 2.0 lbs ae per acre of 2,4-D amine or low volatile ester.

Wheat, Barley and Oats: Apply 1/4 to 1/3 pint of STINGER Herbicide per acre from the 3 leaf stage up to early boot stage of growth. **NOTE:** Do not permit dairy animals or meat animals being finished for slaughter to forage or graze treated grain fields within 1 week after treatment. Do not harvest hay from treated grain fields.

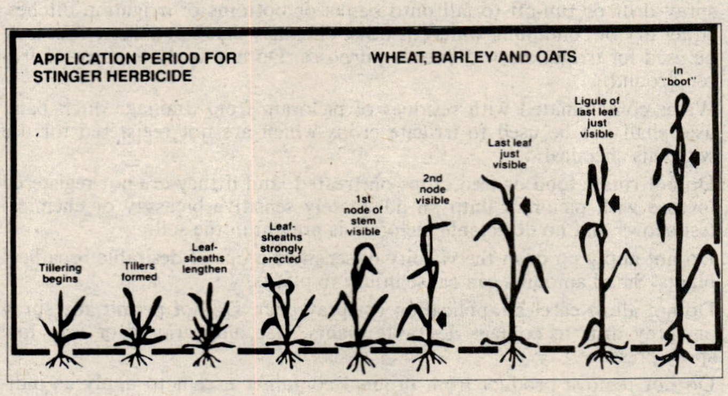

Tank Mixtures for Wheat, Barley and Oats: Tank mix 1/4 to 1/3 pint per acre of STINGER Herbicide with the herbicides listed below for the control of additional weeds.

Active Ingredient /Product Name		Amount of Product per Acre
2,4-D[1]		
2,4-D Amine		
or Ester	4 lb/gal	1/2-1 pt
MCPA		
MCPA Amine		
or Ester	4 lb/gal	1/2-1 pt
bromoxynil		
BROMINAL		
or BUCTRIL	2 lb/gal	3/4-1 pt
ME4BROMINAL	4 lb/gal	3/8 -1/2 pt
bromoxynil + MCPA		
BRONATE	2 + 2 lb/gal	1 pt
chlorsulfuron		
GLEAN 75% DF		1/6-1/4 wt oz
dicamba		
BANVEL	4 lb/gal	1/8-1/4 pt
terbutryn[1]		
IGRAN 80% WP		7.5-12.5 wt oz
metribuzin[1]		
SENCOR 75 DF		
or LEXONE DF		2.5-4 wt oz
LEXONE 4L		
or SENCOR 4	4 lb/gal	3.7-6 fl oz
LEXONE 50 WP		3.75-6 wt oz
diuron		
DIREX 4L		
or DIURON 4L	4 lb/gal	3/4 to 1 1/4 pt

[1] Tank mix for application on wheat and barley only.

Rangeland and Permanent Grass Pastures: Apply 2/3 to 1 1/3 pints of STINGER Herbicide per acre when weeds are young and actively growing. Grasses are tolerant, but new grass seedlings may be injured to varying degrees until the grass has become well established.

NOTE: Many forbs are susceptible to STINGER Herbicide. Do not spray pastures containing desirable forbs, especially legumes, unless injury can be tolerated. However, the stand and growth of established perennial grasses is usually improved after spraying, especially when rainfall is adequate and grazing is deferred.

Do not use hay or straw from treated areas for composting or mulching on susceptible broadleaf crops.

Conservation Reserve Program (CRP) For Seeding To Permanent Grasses Only

Do not use STINGER Herbicide if legumes or bentgrass are a desired cover during CRP.

Grass grown on CRP acres cannot be grazed or used for hay.

Conditions that stress grasses, such as drought, will increase potential for injury to the grass at all stages of growth. Do not use in newly seeded areas until grass is established.

After CRP, do not plant broadleaf crops in treated areas until an adequately sensitive bioassay shows that no detectable clopyralid is present in the soil.

Broadcast Applications (Ground): Applications of STINGER Herbicide should be made when perennial grasses have become established (has tillered, developed a good secondary root system and shows good vigor) since most perennial grasses have shown better tolerance to the herbicide at that stage.

For control of actively growing weeds such as musk thistle (after the rosette stage), Canada thistle and knapweed (spotted, diffuse and Russian), use 2/3 to 1 1/3 pints per acre of STINGER Herbicide after the grass is well established. For the control of wild buckwheat, volunteer sunflower and musk thistle rosettes, apply 2/3 pint per acre of STINGER Herbicide. STINGER Herbicide can also be tank mixed with 1/2 to 1 lb/acre of 2,4-D where species present are sensitive to 2,4-D. For best results, use in 5 or more gallons of water per acre by ground. Increasing the rate of application can increase the risk of injury. Application prior to the flowering stage is recommended (still in the bud stage).

USE PRECAUTIONS

Apply only once per 12 month period, except for sugar beets (see Directions for Use).

Do not plant or rotate to any crop for one year after treatment, except for wheat, barley, oats, grass or sugar beets. To avoid potential phytotoxicity, do not plant peas, lentils, beans (except soybeans) or broadleaf seed crops for 18 months after treatment unless the risk of injury is acceptable.

Do not apply STINGER Herbicide directly to or allow spray drift to come in contact with vegetables, flowers, grapes, tomatoes, potatoes, beans or other desirable broadleaf crops and ornamental plants.

Do not contaminate irrigation ditches or water used for irrigation or domestic purposes.

Avoid Spray Drift: Applications should be made to avoid spray drift since very small quantities of the spray, which may not be visible, may severely injure susceptible crops during both growing and dormant periods. Use coarse sprays to minimize drift since, under adverse weather conditions, fine spray droplets may drift a mile or more. Drift control additives such as NALCO-TROL may be used with this product to aid in reducing spray drift. All use recommendations and precautions on the drift control label should be followed completely to assure proper mixing and maximum effectiveness.

Ground Application: With ground equipment, spray drift can be lessened by keeping the spray boom as low as possible; by using no more than 45 pounds spraying pressure with nozzles designed for herbicide application that do not produce a fine droplet spray; by spraying when wind velocity is low; and by stopping all spraying when wind exceeds 10 miles per hour.

Do not apply by aircraft.

Do not transfer livestock from treated grazing areas onto sensitive broadleaf crop areas without fist allowing 7 days of grazing on an untreated pasture. Otherwise, urine may contain enough clopyralid to cause injury to sensitive broadleaf plants.

Do not use in a greenhouse. Excessive amounts of this herbicide in the soil may temporarily inhibit seed germination or plant growth.

PRECAUTIONARY STATEMENTS

ENVIRONMENTAL HAZARDS

Clopyralid is a chemical which can travel (seep or leach) through soil and under certain conditions contaminate groundwater which may be used for irrigation or drinking purposes. Users are advised not to apply clopyralid where soils have a rapid to very rapid permeability throughout the profit (such as loamy sand to sand) and the water table of an underlying aquifer is shallow, or to soils containing sinkholes over limestone bedrock, severely fractured surfaces, and substrates which would allow direct introduction into an aquifer. Your local agricultural agencies can provide further information on the type of soil in your area and the location of groundwater.

STORAGE AND DISPOSAL

Do not contaminate water, food or feed by storage or disposal.

Storage: Store above 28 °F or warm to 40 °F and agitate before use.

Pesticide Disposal: Wastes resulting from the use of this product may be disposed of on site or at an approved waste disposal facility.

Container Disposal; Metal Do not reuse container. Triple rinse (or equivalent). Puncture and dispose of in a sanitary landfill, or by other procedures approved by state and local authorities.

Container Disposal: Plastic Do not reuse container. Triple rinse (or equivalent). Puncture and dispose of in a sanitary landfill, or by incineration, or, if allowed by state and local authorities, by burning. If burned, stay out of smoke.

WARRANTY LIMITATIONS AND DISCLAIMER

The Dow Chemical Company warrants that this product conforms to the chemical description on the label and is reasonably fit for the purposes stated on the label when used in strict accordance with the directions therein under normal conditions of use. THIS IS THE ONLY WARRANTY MADE ON THIS PRODUCT. NO OTHER EXPRESS AND NO IMPLIED WARRANTY OF MERCHANTABILITY OR FITNESS FOR A PARTICULAR PURPOSE IS MADE OUTSIDE OF THIS LABEL. Therefore, neither this warranty nor any other warranty of merchantability or fitness for a particular purpose, express or implied, extends to the use of this product contrary to label instructions (including conditions noted on the label, such as unfavorable temperatures, soil conditions, etc.), under abnormal conditions (such as excessive rainfall, drought, tornadoes, hurricanes, etc.) or under conditions not reasonably foreseeable to or beyond the control of seller.

When buyer or user suffers losses or damages resulting from the use or handling of this product (including claims based on contract, negligence, strict liability, or other legal theories), buyer or user must promptly notify in writing The Dow Chemical Company or any claims to be eligible to receive either remedy given below. The EXCLUSIVE REMEDY OF THE BUYER OR USER and the LIMIT OF LIABILITY of The Dow Chemical Company or any other seller will be one of the following, at the election of The Dow Chemical Company:

(1) **Refund of purchase price** paid by buyer or user for product bought, **or**

(2) **Replacement** of amount of product used.

The seller **will not be liable for consequential or incidental damages or losses.**

The terms of this Warranty Limitations And Disclaimer cannot be varied by any written or verbal statements or agreements. Any employee or sales agent of the seller is not authorized to vary or exceed the terms of this Warranty Limitations And Disclaimer in any manner.

*Trademark of THE DOW CHEMICAL COMPANY

08505-L6 1188

TANDEM*

HERBICIDE

For Weed Control in Field Corn

ACTIVE INGREDIENT(S):

Tridiphane[2-(3,5-dichlorophenyl)-2-(2,2,2-trichloroethyl)oxirane]	43.6%
INERT INGREDIENTS:	56.4%
TOTAL	100.0%

Contains petroleum distillates
Contains 4.0 pounds of tridiphane per gallon

E.P.A. Registration No. 464-615 E.P.A. Est. 464-MI-1

KEEP OUT OF REACH OF CHILDREN

WARNING

AVISO:

PRECAUCIÓN AL USUARIO: Si usted no lee inglés, no use este producto hasta que la etiqueta le haya sido explicada ampliamente.

PRECAUTIONARY STATEMENTS

HAZARDS TO HUMANS AND DOMESTIC ANIMALS

CAUSES SUBSTANTIAL BUT TEMPORARY EYE INJURY • MAY BE FATAL IF SWALLOWED • HARMFUL IF INHALED • HARMFUL IF ABSORBED THROUGH SKIN • MAY CAUSE ALLERGIC SKIN REACTION

Do Not Get in Eyes, On Skin, Or On Clothing • Wear Goggles Or Face Shield And Rubber Gloves When Handling And Mixing • Do Not Take Internally • Wash Thoroughly With Soap And Water After Handling And Before Eating Or Smoking • Avoid Breathing Vapors Or Spray Mists • Remove And Wash Contaminated Clothing Before Reuse

STATEMENT OF PRACTICAL TREATMENT:

If In Eyes: Immediately flush eyes with plenty of water for at least 15 minutes. Promptly call a physician.

If On Skin: Wash with plenty of soap and water. Get medical attention if irritation persists.

If Swallowed: Call a physician or Poison Control Center. Do not induce vomiting. Contains petroleum distillates.

PHYSICAL OR CHEMICAL HAZARDS

COMBUSTIBLE • Do Not Use or Store Near Heat or Open Flame.

ENVIRONMENTAL HAZARDS

This product is toxic to fish. Do not apply directly to lakes, streams and ponds. Do not contaminate water by cleaning of equipment or disposal of wastes.

NOTICE

Read and understand the entire label before using. Use only according to label directions.

Before buying or using this product, read "WARRANTY LIMITATIONS AND DISCLAIMER" elsewhere on this label. If terms are not acceptable, return **unopened** package at once to seller for full refund of purchase price paid. Otherwise, use by the buyer or any other user constitutes acceptance of the terms under WARRANTY LIMITATIONS AND DISCLAIMER.

IN CASE OF AN EMERGENCY endangering life or property involving this product, call collect 517-636-4400

AGRICULTURAL CHEMICAL

Do Not Ship or Store with Food, Feeds, Drugs, or Clothing

DIRECTIONS FOR USE

It is a violation of Federal law to use this product in a manner inconsistent with its labeling.

POSTEMERGENCE APPLICATION

General Information

TANDEM herbicide is a selective postemergence herbicide to be applied as a **tank mixture** in combination with atrazine and/or Bladex 80W or Bladex 90DF for use to control **annual grasses and broadleaf weeds in field corn.** Residual preemergence activity will normally occur with these combinations provided soil coverage is achieved and adequate moisture is present. Weed species controlled by these combinations include the following:

Annual Grasses

Barnyardgrass *(Echinochloa crus-galli)*
Crabgrass *(Digitaria spp.)*
Foxtails
  Giant *(Setaria faberi)*
  Green *(Setaria viridis)*
  Yellow *(Setaria glauca)*
Goosegrass *(Eleusine indica)*
Millet, Wild proso *(Panicum miliaceum)* (1)
Panicum, Fall *(Panicum dichotomiflorum)* (1)
Panicum, Texas *(Panicum texanum)* (1) AL, FL, GA Only
Sandbur, Field *(Cenchrus incertus)* (1) CO, KS, NE, ONLY
Shattercane *(Sorghum bicolor)* (1)-Suppression
Signalgrass, Broadleaf *(Brachiaria platyphylla)*
Wild Oats *(Avena fatua)* (1)
Woolly Cupgrass *(Eriochloa villosa)* (1)

(1) Please see special use recommendation for control of these grasses.

Annual Broadleaves

Beggarweed, Florida *(Desmodium tortuosum)*
Buffalobur *(Solanum rostratum)*
Burcucumber *(Sicyos angulatus)*
Cocklebur, Common *(Xanthium strumarium)*
Jimsonweed *(Datura stramonium)*
Kochia *(Kochia scoparia)*
Ladysthumb *(Polygonum persicaria)*
Lambsquarters, Common *(Chenopodium album)*
Morningglory, Tall *(Ipomoea purpurea)*
Mustard *(Brassica spp.)*
Nightshade, Black *(Solanum nigrum)*
Pigweed *(Amaranthus spp.)*
Purslane, Common *(Portulaca oleracea)*
Ragweed, Common *(Ambrosia artemisiifolia)*
Ragweed, Giant *(Ambrosia trifida)*
Sicklepod *(Cassia obtusifolia)*
Smartweed, Pale *(Polygonum lapathifolium)*
Smartweed, Pennsylvania *(Polygonum pensylvanicum)*
Starbur, Bristly *(Acanthospermum hispidum)*
Sunflower, Common *(Helianthus annuus)*
Velvetleaf *(Abutilon theophrasti)*

Perennial Broadleaf

Thistle, Canada *(Cirsium arvense)* - Suppression

NOTE: Not effective on triazine-resistant species.

TANDEM herbicide works mainly through foliar uptake so weed control will be less than optimum when the herbicide is applied to weeds growing under drought-stressed conditions just prior to or after herbicide application. **For best weed control, apply only to weeds which are actively growing.**

Up to approximately 10 days must elapse after herbicide application for weed control to occur. First symptoms of weed control (observed 2-3 days after application) are burning of the leaf tips and margins. The leaf surface may also appear limp and almost transparent. As the symptoms progress, the entire leaf surface will turn brown and will eventually disintegrate.

Available data indicate that rainfall two to three hours following application will not reduce activity from a postemergence application of TANDEM and triazine herbicides. **Rainfall is needed after application in order to activate triazine in the soil, which in turn provides residual weed control.**

TIMING OF APPLICATION:

Apply tank mixtures of TANDEM herbicide plus atrazine and/or Bladex 80W or 90DF as a broadcast foliar spray to young **actively growing** weeds. **The number of leaves on grass weeds is the best indicator of proper timing of application.** For optimum control, apply after the first flush of annual grass has emerged and is in the 1 to 3 leaf stage as shown in the diagram below. Applications made to larger grasses or those not actively growing may result in reduced control. A cultivation 7 to 14 days after treatment can improve results.

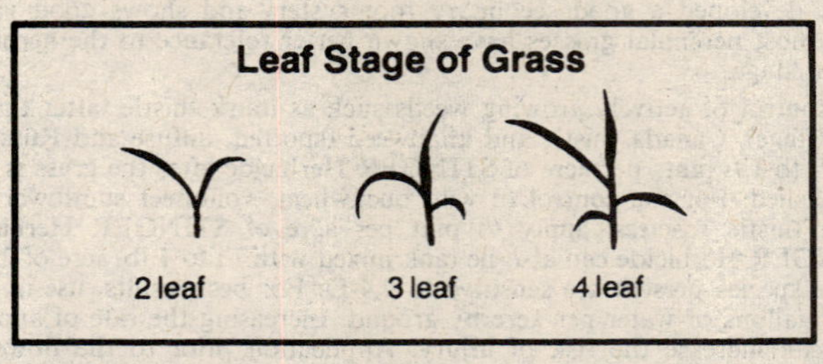

SPRAY ADDITIVES:

Emulsifiable crop oil concentrate or crop oil is required as an additive to all TANDEM plus atrazine combinations to improve wetting of weed foliage. **Do not use crop oil, crop oil concentrate or other spray adjuvants with mixtures containing Bladex 80W or 90DF.**

APPLICATION PROCEDURES

Apply TANDEM herbicide as follows to ensure uniform and thorough spray coverage of weed foliage.

GROUND APPLICATION

Use equipment which provides for uniform spray coverage. Use 20 to 30 gallons of spray volume per acre with nozzles which are operated at a pressure of 30-40 psi (at the boom). **The use of flat fan nozzles provides the best coverage and distribution for the volume and pressure range needed to apply a material like TANDEM. Use only spray nozzles designed for herbicide application. Use of spray nozzle tips which deliver coarse sprays is not recommended. Do not use flood jet nozzles. Screens in nozzles and line strainer should be no smaller than 50 mesh. Use higher spray pressures and volumes of 25 to 30 gallons per acre when weed foliage is dense and/or weeds are larger than the 2 leaf stage.**

Mixing Instructions (Ground):

1. Always use clean spray equipment to prevent contamination with other materials that may cause injury or clogging of nozzles or screens.
2. Fill the spray tank about one-half full with clean water.
3. With the spray tank agitation in operation, add the required amount of atrazine and/or Bladex 80W or 90DF. When wettable powder formulations of these herbicides are used, premixing with water to form a slurry before adding to the spray tank will help to ensure dispersion.
4. Continue filling the spray tank with water while adding TANDEM.
5. Add emulsifiable crop oil concentrate or crop oil after TANDEM but before filling of the spray tank is completed.
6. Maintain agitation during mixing and application. If agitation is stopped and/or the spray mixture is allowed to settle, thorough remixing is required. Before spraying is resumed, check tank bottom to ensure complete remixing has occurred.
7. In **hard water areas,** vigorous agitation of spray mixture may cause foaming to occur. If this happens, an **antifoam agent should be added** to the mixture.

High Flotation Equipment:

TANDEM herbicide plus triazine tank-mixtures can be applied through high flotation application equipment (such as Big-A's, TERRA-GATOR, etc.) as long as the following guidelines are followed. The guidelines are designed to maximize foliage deposition of the herbicide tank-mixtures.

1. Use 80° flat fan nozzles such as 8015.
2. Use a **minimum** of 30 gpa spray volume.
3. Use spray pressures of 35 to 40 psi—measured at the boom.
4. Application speed should not exceed 10 mph.
5. Boom set-up:
   (a) Arrange nozzles on a 30-inch spacing.
   (b) Nozzles should have a minimum 100% overlap.
   (c) Boom directly behind the wheels of the applicator should be adjusted to height (maximum) of 28 to 30 inches above the ground.
   (d) The side booms should be equipped with 80° nozzles, set at 90° or perpendicular to the ground. Portion of boom directly behind the wheels should have the nozzles set at 60° or turned back 30° to the ground.
   (e) Vertical movement of the boom should be minimized in order to keep a maximum boom height of 42 inches above the ground (excluding the boom section directly behind the wheels of the applicator).

6. Efforts should be made to reduce the amount of dust created behind the applicator. The use of mud flaps will help reduce the amount of dust and turbulence directly behind the applicator.

Spra-Coupes:

TANDEM herbicide plus triazine tank-mixtures can be applied through Spra-Coupe low volume ground application equipment as long as the following guidelines are followed. The guidelines are designed to maximize foliage deposition of the herbicide tank-mixtures.

1. Use 80° flat fan nozzles.
2. Use a **minimum** of 10 gpa spray volume.
3. Use minimum spray pressure of 60 psi measured at the boom.
4. Application speed of 10 mph or less.
5. Boom set-up.
   (a) Arrange nozzles on 20-inch centers.
   (b) Boom height should be 22 inches above the target.

AERIAL APPLICATION:

Apply TANDEM herbicide in combination with atrazine plus crop oil concentrate or crop oil, through fixed-wing aircraft. Apply TANDEM herbicide tank-mixtures when grass weeds are actively growing and are in the 1 to 3 leaf stage of growth. Apply at rates recommended on this label. Apply through fixed-wing aircraft which has been patterned and calibrated according to Operation Safe guidelines (contact your state Aerial Applicators Association or state Cooperative Extension Service for information on fixed-wing aircraft patterning and calibration). Apply herbicides in a **minimum** total spray volume of 5 gallons per acre. **Use of less than 5 gallons per acre will result in poor weed control.**

For maximum herbicidal activity of TANDEM herbicide, it is desired to have a particle spectrum within the spray pattern (measured within the corn canopy) with a volume median diameter of approximately 300 microns. In order to obtain the desired particle spectrum, the following nozzle orientations should be used. At low air speeds (90 to 100 mph) orient nozzles on the boom perpendicular to the ground. At fast air speeds (120 to 140 MPH) orient nozzles on the boom parallel to the ground.

Mixing Instructions (aerial):

1. Fill the spray tank about one-half full with clean water.
2. With agitation in operation, add the required amount of TANDEM herbicide plus crop oil concentrate or crop oil. Agitate tank until the TANDEM plus crop oil concentrate or crop oil are completely emulsified. A visual indication of complete emulsification will be the absence of any surface oil layer or pockets.
3. Continue filling the spray tank with water while adding the atrazine. With the wettable powder formulations, premixing with water to form a slurry before adding to the spray tank will help to ensure dispersion.
4. Maintain agitation during mixing and application. If agitation is stopped and/or the spray mixture is allowed to settle, thorough remixing is required. Before spraying is resumed, check tank-bottom to ensure complete re-mixing has occurred.
5. In hard water areas, vigorous agitation of spray mixture may cause foaming to occur. If this happens, an antifoam agent should be added to the mixture.

Restrictions: Do not apply where uniform coverage cannot be obtained or where excessive spray drift may occur. Avoid application directly to animals or humans. Flagmen or loaders should avoid inhalation, spray mist, or contact with skin and should wash thoroughly before eating and at the end of each day's operation. All other restrictions on this label should be adhered to.

CHEMIGATION

General Directions: Apply this product only through center-pivot or linear move sprinkler irrigation systems. Do not apply this product through any other type of irrigation system.

Crop injury, lack of effectiveness or illegal pesticide residues in the crop can result from non-uniform distribution of treated water.

If you have questions about calibration, you should contact State Extension Service specialists, equipment manufacturers or other experts.

Do not connect an irrigation system (including greenhouse systems) used for pesticide application to a public water system unless the pesticide label prescribed safety devices for public water systems are in place.

A person knowledgeable of the chemigation system and responsible for its operation, or under the supervision of the responsible person, shall shut the system down and make necessary adjustments should the need arise.

The system must contain a functional check valve, vacuum relief valve, and low pressure drain appropriately located on the irrigation pipeline to prevent water source contamination from backflow.

The pesticide injection pipeline must contain a functional, automatic, quick-closing check valve to prevent the flow of fluid back toward the injection pump.

The pesticide injection pipeline must also contain a functional, normally closed, solenoid-operated valve located on the intake side of the injection pump and connected to the system interlock to prevent fluid from being withdrawn from the supply tank when the irrigation system is either automatically or manually shut down.

The system must contain functional interlocking controls to automatically shut off the pesticide injection pump when the water pump motor stops.

The irrigation line or water pump must include a functional pressure switch which will stop the water pump motor when the water pressure decreases to the point where pesticide distribution is adversely affected.

Systems must use a metering pump, such as a positive displacement injection pump (e.g., diaphragm pump) effectively designed and constructed of materials that are compatible with pesticides and capable of being fitted with system interlock.

Do not apply when wind speed favors drift beyond the area intended for treatment.

Before mixing TANDEM and Atrazine herbicides in the chemigation mix tank, a tank mix compatibility test must first be conducted as outlined in the following section.

Tank Mix Compatibility Evaluation Procedure: To determine if the Atrazine 4L formulation is compatible with the TANDEM and non-emulsifiable crop oil to be used, conduct the following "jar test".

1. Add 1 cup water to clean one-quart jar.
2. Add 1 cup Atrazine 4L formulation and shake.
3. Add 2/3 cup of non-emulsifiable crop oil and shake vigorously until atrazine, oil and water form a uniform suspension.
4. Add 1/3 cup of TANDEM and shake.

After the final mixing, determine if any signs of incompatibility are present in the mixture. Signs of incompatibility are: clabbering or curdling; precipitate which does not resuspend; or precipitate which sticks tenaciously to the glass. If any of these occur, DO NOT proceed with mixing and application. Select an alternate Atrazine 4L formulation and repeat the jar test. Once a successful jar test has been completed, that mixture can be applied through the chemigation equipment.

Specific Directions: Apply TANDEM herbicide in combination with Atrazine 4L and **non-emulsifiable crop oil** to water applied through center-pivot or linear move sprinkler systems. Apply when grass weeds are actively growing and are in the 1 to 3 leaf stage of growth. Apply at rates recommended on this label. Apply non-emulsifiable crop oil at the rate of 1 qt per acre. Prepare mixture with a minimum of 1 part water to 1 part atrazine. Injecting a larger volume of a more dilute slurry per hour will assure more accurate calibration of metering equipment. Maintain sufficient agitation to keep herbicide in suspension. Meter slurry into irrigation water during entire period. Apply in 1/4 to 1/2 inch of water. Use the lower water volume if top 1 to 2 inches of soil are moist.

Use higher water volume if top 2 inches of soil are dry. Inject dilute slurry into the system through a positive displacement pump.

Tank Mix Directions:

1. Thoroughly clean the injection system and tank of any fertilizer or chemical residues, and dispose of the residues according to State and Federal Laws.
2. Flush the injector with soap and water.
3. Calibrate the injector system.
4. Prepare a mixture of Atrazine 4L, TANDEM, **non-emulsifiable crop oil** and water in a steel mix tank after determining the total amount of mixture that will be injected per hour at the desired rate and time of operation.
5. Fill tank one-half full with water, pump the required Atrazine 4L into the steel tank, and start mechanical agitation. Then add non-emulsifiable crop oil and vigorously agitate the oil until a uniform suspension with the Atrazine 4L and water mixture occurs. The required amount of TANDEM and additional water should then be added to obtain the desired total mixture volume.
6. Continually agitate the TANDEM herbicide plus Atrazine 4L plus oil mixture.
7. Set the sprinkler system to deliver the desired inches of water per acre (1/4 to 1/2 inch recommended).
8. Start the irrigation pump and system, and let the system achieve the desired pressure and speed before starting the injector.
9. The TANDEM herbicide plus Atrazine 4L plus non-emulsifiable crop oil mixture must be injected continuously and uniformly into the irrigation water line as the system is moving. This procedure is necessary to deliver the desired rate per acre in a uniform manner.
10. When application is finished, allow the entire irrigation and injector system to be thoroughly flushed clean before stopping the system. Flush for at least 15 minutes.

Precautions: The irrigation pumping plant and the chemical injection pump must be interlocked so that if the irrigation pumping plant stops, the chemical injection pump also will stop. This procedure will prevent the possibility of filling the entire irrigation pipeline with the chemical mixture from the supply tank.

A positive displacement pump is required to provide uniform injection into the water line. The injector pump must provide a greater pressure than that of the irrigation system at the point of injection.

To insure uniform mixing of the herbicide into the water line, inject the mixture into the fertilizer injection port or just ahead of an elbow or tee in the irrigation line so that the turbulence created at those points will assist in mixing. It is suggested that the injection point be higher than the chemical tank to prevent siphoning.

The steel tank holding the herbicidal mixture should be large enough to allow the system to complete a revolution with one filling. It should be free of rust, fertilizer, sediment, and foreign material, and equipped with an in-line strainer situated between the tank and the injector pump.

In order to calibrate the irrigation system and injector to apply the herbicidal mixture, determine the following: 1) calculate the number of acres covered by the system, 2) the number of minutes for the system to make one revolution, 3) the number of inches of water per acre delivered by the system, 4) the number of minutes required for the injector pump to inject one gallon of herbicidal mixture, and 5) the total gallons of herbicide mixture needed to cover the desired acreage. Divide the total gallons of mixture needed to cover the irrigated area by the number of minutes to make one circle. This value equals the gallons per minute output that the injector must deliver. It is suggested that the system be calibrated at least twice before operation, and the system should be monitored during operation.

End guns must be turned off during the application.

Do not apply with liquid fertilizer.

Do not allow irrigation runoff to collect or runoff and pose a hazard to livestock, wells, or adjoining crops.

Dosage Recommendations

Consult the following tables and text for dosage rates to be used for various size weeds and soil textures. See specific directions for: treatment of ESCAPES (1 to 3 leaf grasses); SALVAGE Programs (4-6 leaf grasses); soils with organic matter levels greater than 7%; wild proso millet; woolly cupgrass; wild oats; fall panicum; Texas panicum (AL, FL, GA Only); field sandbur (CO, KS, NE Only); shattercane suppression and no till/conservation tillage: Read the complete **Directions for Use** section of this label before mixing or applying tank mixtures of TANDEM herbicide plus atrazine and/or Bladex 80W or 90DF.

GENERAL RECOMMENDATIONS:

For control of actively growing annual grasses and broadleaf weeds, follow one of the programs listed in Table 1 or Table 2. Apply after the first flush of annual grass has emerged and is in the 1 to 3 leaf stage.

TABLE 1: Dosage Rates For Tank Mixtures Of TANDEM Plus Atrazine

	Broadcast Rate Per Acre		
	Timing Of Treatment—Weed Size		
	Postemergence—Grasses 1 To 3 Leaf		
Soil Texture[(a)]	TANDEM (pints)	Atrazine 4L[(b)] (pints)	Crop Oil[(c)] Concentrate (pints)
Coarse, Medium or Fine— less than 4% organic matter	1	3	2
Medium or Fine— 4% to 7% organic matter	1-1.5[(d)]	3-4[(d)]	2

(a) Soil texture, organic matter content and rainfall following application affect the length of residual activity. **Coarse** texture soils include sand, loamy sand and sandy loam. **Medium** includes loam, silt loam, silt and sandy clay loam. **Fine** includes silty clay loam, clay loam, sandy clay, silty clay and clay.

(b) Any atrazine product recommended for use on field corn may be used. Use the conversions listed below to determine the proper amount.

Atrazine 4L (pints)		Atrazine 80W (pounds)		Atrazine 90DF (pounds)
3	=	1.9	=	1.7
4	=	2.5	=	2.2

(c) Six to eight pints of crop oil can be substituted for 2 pints of crop oil concentrate.

(d) Use higher rate of TANDEM when weed pressures are heavy **(such as set aside acres)** and higher rate of atrazine to extend residual control.

TABLE 2: Dosage Rates For Tank Mixtures Of TANDEM Plus Bladex 90DF Or TANDEM Plus Atrazine And Bladex 90DF

	Broadcast Rate Per Acre		
	Timing Of Treatment—Weed Size		
	Postemergence—Grasses 1 To 3 Leaf		
Soil Texture[(a)]	TANDEM (pints)	Atrazine 4L[(b)] (pints)	Bladex 90DF[(c)(d)] (pounds)
Coarse—1% to 2% organic matter	1		1.3
Coarse— Medium or Fine— 2% to 7% organic matter	1-1.5[(d)]		1.7-2.2[(d)]
Coarse, Medium or Fine— 2% to 4% Organic matter	1-1.5[(d)]	1-1.5[(a)]	0.9-1.1[(d)]
Medium to Fine— 4% to 7% organic matter	1-1.5[(d)]	2	1.1

(a) Soil texture, organic matter content and rainfall following application affect the length of residual activity. **Coarse** texture includes sand, loamy sand and sandy loam. **Medium** includes loam, silt loam, silt and sandy clay loam. **Fine** includes silty clay loam, clay loam, sandy clay, silty clay and clay. **For high pH soils with potential atrazine carry-over problems, use lower rates on low organic matter, high pH soils. As organic matter increases, increase the rate of TANDEM and triazine within the guidelines of carry-over concern. Adjust the levels of atrazine and Bladex to maintain a total triazine level of 1.5 lb ai/A in these soils.**

(b) Any atrazine product recommended for use on field corn may be used. Use the conversions listed below to determine the proper amount.

Atrazine 4L (pints)		Atrazine 80W (pounds)		Atrazine 90DF (pounds)
1.5	=	0.94	=	0.83
2	=	1.25	=	1.1

(c) Bladex 80W may be used in place of Bladex 90DF. Use the conversions listed below to determine the proper amount.

Bladex 90DF (pounds)		Bladex 80W (pounds)
0.9	=	1.0
1.1	=	1.25
1.3	=	1.5
1.7	=	1.9
2.2	=	2.5

Do not use crop oil concentrate or crop oil in mixtures that contain Bladex 90DF or Bladex 80W. Do not apply mixtures that contain Bladex 90DF or Bladex 80W to corn larger than the 4 true leaf stage. Temporary yellowing may result following application of these treatments, particularly if cold, adverse growing conditions occur after application. However, recovery is rapid and yield is not reduced. Do not apply TANDEM plus Bladex 90DF or Bladex 80W when corn is under stress from frost, prolonged cold or wet weather or when plants are wet from recent rainfall. Extended or extreme cold and wet conditions after application may reduce stands. Do not use Bladex 90DF or Bladex 80W on sandy or loamy sand soil containing less than 1% organic matter. **Do not apply TANDEM plus Bladex 90DF or Bladex 80W on inbred lines or breeding stock.**

(d) Use higher rates of TANDEM and Bladex 80W or 90DF when weed pressures are heavy.

NO TILL/CONSERVATION TILLAGE

Tillage systems that retain a large amount of surface residue need special consideration. Weed growth at planting must be controlled by a burndown herbicide program. Follow manufacturers' label recommendations for these herbicides and apply sufficient rate and spray volume to assure complete control of vegetation at planting time. Combinations with TANDEM herbicide applied as recommended in the use tables are effective in a program to control the new flush of annual weeds that emerge after early preplant applications or at-plant applications with other herbicides.

Apply combinations with TANDEM herbicide listed in the use rate tables after the new flush of annual grass has emerged, is actively growing and is in the 1 to 3 leaf stage. At this time, broadleaf weeds may be larger and form a canopy over smaller grasses, resulting in poor coverage. Apply combinations with TANDEM earlier under these conditions, when good coverage of all emerged weeds can be accomplished.

SET ASIDE ACRES

When set aside acres are planted to corn, follow the recommendations listed below when applying TANDEM tank mixtures due to high weed densities.

1. Use high rates of TANDEM and Atrazine 4L and/or Bladex 80W or Bladex 90DF listed in the dosage tables.
2. Apply in 25 to 30 gallons per acre total spray volume.
3. Use higher spray pressures.

SPECIAL WEED CONTROL PROGRAMS

Escapes-Control of Escaped 1-3 Leaf Grasses

Use the following programs if preplant or preemergence weed control programs have failed to control grasses and the grass is in the 1 to 3 leaf stage.

Previous Triazine Herbicide Treatments: For control of 1-3 leaf annual grasses which break through preplant incorporated and preemergence herbicide treatments that contained triazine herbicides and were applied at label rates, follow one of the programs outlined in Table 1 or Table 2 under the "General Recommendations" section of the Dosage Recommendations. Use the rate of triazine appropriate for your soil type and do not exceed the rate of total triazine recommended for the soil type. For atrazine products, be sure to observe rotational crop suggestions to prevent possible injury to subsequent crops caused by atrazine carry-over. **Pro-**

grams which contain Bladex cannot be applied to corn larger than 4 true leaf stage.

Previous Non-Triazine Herbicide Treatments: For control of 1-3 leaf annual grasses which break through preplant incorporated or preemergence herbicide treatments which were applied at label rates, follow one of the two programs outlined in Table 3. **These programs cannot be applied to corn larger than 4 true leaf stage.**

TABLE 3: Control of Escapes (1-3 Leaf Grasses) from Previous Non-Triazine Herbicide Treatment.

Soil Texture[(a)]	Broadcast Rate Per Acre — Timing Of Treatment—Weed Size — Postemergence—Grasses 1 To 3 Leaf		
	TANDEM (pints)	Atrazine 4L[(b)] (pints)	Bladex 90DF[(c)] (pounds)
Coarse—	*OPTION 1*	—	2.2
Greater than 2% organic matter	1.5		
or			
Sandy Loam-2%	*OPTION 2*		
or more organic matter	1.5	2	1.8-2.2[(d)]
or			
Medium or Fine 1% or more organic matter			

(a) **Coarse** texture soils include sand and loamy sand. **Medium** includes loam, silt loam, silt and sandy clay loam. **Fine** includes silty clay loam, clay loam, sandy clay, silty clay and clay.

(b) Any atrazine product recommended for use on field corn may be used. Use the conversions listed below to determine the proper amount.

Atrazine 4L (pints)		Atrazine 80W (pounds)		Atrazine 90DF (pounds)
2	=	1.25	=	1.1

(c) Bladex 80W may be used in place of Bladex 90DF. Use the conversions listed below to determine the proper amount.

Bladex 90DF (pounds)		Bladex 80W (pounds)
1.8	=	2.0
2.2	=	2.5

Do not use crop oil concentrate or crop oil in mixtures that contain Bladex 90DF or Bladex 80W. Do not apply mixtures that contain Bladex 90DF or Bladex 80W to corn larger than the 4 true leaf stage. Temporary yellowing may result following application of these treatments, particularly if cold, adverse growing conditions occur after application. However, recovery is rapid and yield is not reduced.

Do not apply TANDEM plus Bladex 90DF or Bladex 80W when corn is under stress from frost, prolonged cold or wet weather or when plants are wet from recent rainfall. Extended or extreme cold and wet conditions after application may reduce stands. Do not use Bladex 90DF or Bladex 80W on sandy or loamy sand soil containing less than 1% organic matter. **Do not apply TANDEM plus Bladex 90DF or Bladex 80 W on inbred lines or breeding stock.**

(d) Use higher rates of TANDEM and Bladex 80W or 90DF when weed pressures are heavy.

Salvage Treatment of Larger Grasses (4 to 6 Leaf)

Use the programs in Table 4 or Table 5 if rain or other conditions delay normal TANDEM and triazine herbicide postemergence application, or if pre-plant incorporated, or preemergence weed control programs have failed to control grasses, and the grass is in the 4 to 6 leaf stage.

Previous Triazine Herbicide Application

Grasses are in the 4 to 6 leaf stage resulting from the failure of an earlier pre-plant incorporated, preemergence or early postemergence triazine herbicide application. Cultivation 7 to 14 days after TANDEM and triazine herbicide application may increase weed control.

TABLE 4: Treatment of 4 to 6 Leaf Grasses—Previous Triazine Application.

Soil Texture[(a)]	Broadcast Rate Per Acre — Timing Of Treatment—Weed Size — Postemergence—Grasses 4 To 6 Leaf		
	TANDEM (pints)	Atrazine 4L[(b)] (pints)	Crop Oil[(c)] Concentrate (pints)
Coarse, Medium or Fine— up to 7% organic matter	1.5	4	2

(a) Soil texture, organic matter content and rainfall following application affect the length of residual activity. **Coarse** texture soils include sand, loamy sand and sandy loam. **Medium** includes loam, silt loam, silt and sandy clay loam. **Fine** includes silty clay loam, clay loam, sandy clay, silty clay and clay.

(b) Any atrazine product recommended for use on field corn may be used. Use the conversion listed below to determine the proper amount.

Atrazine 4L (pints)		Atrazine 80W (pounds)		Atrazine 90DF (pounds)
4	=	2.5	=	2.2

(c) Six to eight pints of crop oil can be substituted for 2 pints of crop oil concentrate.

Previous Non-Triazine Herbicide Application

Grasses are in the 4 to 6 leaf stage resulting from the failure of earlier herbicide application NOT containing a triazine herbicide or Bladex, or because no herbicide was applied at all. Cultivation 5 to 10 days after initial TANDEM and atrazine application may improve weed control and may substitute for second atrazine treatment. However, when weed pressures are heavy, initial application of TANDEM plus Atrazine 4L plus crop oil concentrate followed 5 to 10 days later with Atrazine 4L plus crop oil concentrate, will be necessary for optimum weed control.

TABLE 5: Treatment of 4 to 6 leaf Grasses-Previous Non-Triazine Application or No Herbicide Application

Soil Texture[(a)]	Broadcast Rate Per Acre — Timing Of Treatment—Weed Size — Postemergence—Grasses 4 To 6 Leaf		
	TANDEM (pints)	Atrazine 4L[(b)] (pints)	Crop Oil[(c)] Concentrate (pints)
Medium—Silt and clay loams that are low in organic matter	INITIAL TREATMENT—GRASS 4 TO 6 LEAF		
	1.5	2.75	2
	SECOND TREATMENT 5 TO 10 DAYS AFTER INITIAL		
	—	2	2
Fine—Silt and clay loams with medium to high organic matter and clays.	INITIAL TREATMENT—GRASS 4 TO 6 LEAF		
	1.5	4	2
	SECOND TREATMENT 5 TO 10 DAYS AFTER INITIAL		
		2	2

(a) Soil texture, organic matter content and rainfall following application affect the length of residual activity.

(b) Any atrazine product recomended for use on field corn may be used. Use the conversions listed below to determine the proper amount.

Atrazine 4L (pints)		Atrazine 80W (pounds)		Atrazine 90DF (pounds)
2	=	1.3	=	1.1
2.75	=	1.72	=	1.53
4	=	3.75	=	3.3

(c) Six to eight pints of crop oil can be substituted for 2 pints crop oil concentrate.

Soils with Organic Matter Levels Greater Than 7%

For control of grasses and broadleaf weeds on soil containing organic matter levels greater than 7%, use tank mixtures combining TANDEM herbicide and triazine herbicides as outlined in the **three options** listed in table 6. Apply after the first flush of annual grass has emerged and is in the one to three leaf stage. Cultivation 7 to 14 days after TANDEM and triazine application may improve weed control

TABLE 6: Treatment Programs for Soils with Organic Matter Levels Greater Than 7%.

Preemergence	Broadcast Rate Per Acre — Timing Of Treatment—Weed Size — Postemergence—Grasses 1 To 3 Leaf		
Bladex 4L[(a)] (pints)	TANDEM (pints)	Atrazine 4L[(b)] (pints)	Crop Oil[(c)] Concentrate (pints)
	OPTION 1		
	INITIAL TREATMENT—GRASS 1 TO 3 LEAF		
—	1.5	3	2
	SECOND TREATMENT 5 TO 14 DAYS AFTER INITIAL		
—	—	3	2
	OPTION 2		
	INITIAL TREATMENT—GRASS 1 TO 3 LEAF		
—	0.75	3	2
	SECOND TREATMENT 5 TO 14 DAYS AFTER INITIAL		
—	0.75	3	2
	OPTION 3		
5	1.5	4	2

(a) Do not apply Bladex 4L preemergence to peat or muck soils. Do not apply Bladex 4L postemergence to corn.

(b) Any atrazine product recommended for use on field corn may be used. Use the conversions listed below to determine the proper amount.

Atrazine 4L (pints)		Atrazine 80W (pounds)		Atrazine 90DF (pounds)
3	=	1.9	=	1.7
4	=	2.5	=	2.2

Total amount of Atrazine 4L applied should not exceed 3 lb. active ingredient per acre.

(c) Six to eight pints of crop oil can be substituted for 2 pints of crop oil concentrate.

Wild Proso Millet

For control of wild proso millet, apply one of the tank mixtures listed in Table 7 when the wild proso millet is in the 1 to 3 leaf stage and corn is **no larger than the 4 true leaf stage.** Treatments listed below involve TANDEM and triazine herbicides applied either postemergence or split postemergence; or Bladex 4L applied preemergence followed by TANDEM and triazine applied postemergence. If the second split postemergence treatments are made when corn is beyond the 5 leaf stage. directing the spray with drop nozzles may provide better coverage and improve control. A layby cultivation following treatment may improve results. The combinations listed below can also be used to control a new flush of actively growing wild proso millet following: (1) preplant incorporated applications of Eradicane, Eradicane Extra or Sutan herbicides or (2) preemergence application of Lasso or Dual herbicides. Applications should be made when the grass is in the 1 to 3 leaf stage.

TABLE 7. TREATMENT PROGRAMS FOR WILD PROSO MILLET.

	BROADCAST RATE PER ACRE				
	TIMING OF TREATMENT—WEED SIZE				
SOIL TEXTURE[a]	PRE-EMERGENCE	POSTEMERGENCE—GRASSES 1 to 3 LEAF			
	BLADEX 4L[b] (PINTS)	TANDEM (PINTS)	ATRAZINE 4L[c] (PINTS)	BLADEX 90DF[d] (POUNDS)	CROP OIL[e] CONCENTRATE (PINTS)
Sand or loamy sand—Greater than 2% organic matter or Sandy Loam—2% or greater organic matter or Medium, Fine 1% or greater organic matter	*OPTION 1*				
		INITIAL TREATMENT—GRASS 1 T0 3 LEAF			
	—	1.5	—	2.2	—
		SECOND TREATMENT 7 TO 14 DAYS AFTER INITIAL			
	—	—	4	—	2
	OPTION 2				
		INITIAL TREATMENT—GRASS 1 TO 3 LEAF			
	—	1.5	2	1.1	—
		SECOND TREATMENT 7 TO 14 DAYS AFTER INITIAL			
	—	—	4	—	2
For the soil types listed above or Sandy loam-1% or greater organic matter		INITIAL TREATMENT—GRASS 1 TO 3 LEAF			
	—	0.75	—	1.8	—
		SECOND TREATMENT 7 TO 14 DAYS AFTER INITIAL			
	—	0.75	3	—	2
Sandy loam—greater than 2% organic matter or Medium or Fine—2% or greater organic matter	*OPTION 1*				
	4-6	1.5	—	2.2	—
	OPTION 2				
	4-6	1.5	2	1.1	—

(a) **Medium** texture soils include loam, silt loam, silt and sandy clay loam. **Fine** includes silty clay loam, clay loam, sandy clay, silty clay and clay.

(b) See Bladex 4L label for preemergence application rates in relation to soil texture and soil organic matter content.

(c) Any atrazine product recommended for use on field corn may be used. Use the conversions listed below to determine the proper amount.

Atrazine 4L (pints)		Atrazine 80W (pounds)		Atrazine 90DF (pounds)
2	=	1.25	=	1.1
3	=	1.9	=	1.7
4	=	2.5	=	2.2

(d) Bladex 80W may be used in place of Bladex 90DF. Use the conversions listed below to determine the proper amount.

Bladex 90DF (pounds)		Bladex 80W (pounds)
1.1	=	1.2
1.8	=	2.0
2.2	=	2.5

Do not use crop oil concentrate or crop oil in mixtures that contain Bladex 90DF or Bladex 80W. Do not apply mixtures that contain Bladex 90DF or Bladex 80W to corn larger than the 4 true leaf stage. Temporary yellowing may result following application of these treatments, particularly if cold, adverse growing conditions occur after application. However, recovery is rapid and yield is not reduced.

Do not apply TANDEM plus Bladex 90DF or Bladex 80W when corn is under stress from frost, prolonged cold or wet weather or when plants are wet from recent rainfall. Extended or extreme cold and wet conditions after application may reduce stands. Do not use Bladex 90DF or Bladex 80W on sandy or loamy sand soil containing less than 1% organic matter. **Do not apply TANDEM plus Bladex 90DF or Bladex 80W on inbred lines or breeding stock.**

(e) Six to eight pints of crop oil can be substituted for 2 pints of crop oil concentrate.

Woolly Cupgrass

For control of woolly cupgrass, apply one of the tank mixtures listed in Table 8 when the woolly cupgrass is in the 1 to 3 leaf stage and corn is **no larger than the 4 true leaf stage.** Treatments listed below involve TANDEM and triazine herbicides applied either postemergence or split postemergence; or Bladex 4L applied preemergence followed by TANDEM and triazine applied postemergence. If the second split postemergence treatments are made when corn is beyond the 5 leaf stage, directing the spray with drop nozzles may provide better coverage and improve control. A layby cultivation following treatment may improve results. The combinations listed below can also be used to control a new flush of actively growing woolly cupgrass following: (1) preplant incorporated applications of Eradicane, Eradicane Extra or Sutan herbicides or (2) preemergence applications of Lasso or Dual. Applications should be made when the grass is in the 1 to 3 leaf stage.

TABLE 8. TREATMENT PROGRAMS FOR WOOLLY CUPGRASS.

SOIL TEXTURE[a]	PRE-EMERGENCE BLADEX 4L[b] (PINTS)	BROADCAST RATE PER ACRE — TIMING OF TREATMENT—WEED SIZE — POSTEMERGENCE—GRASSES 1 to 3 LEAF: TANDEM (PINTS)	ATRAZINE 4L[c] (PINTS)	BLADEX 90DF[d] (POUNDS)	CROP OIL[e] CONCENTRATE (PINTS)
Sand or loamy sand—Greater than 2% organic matter or Sandy Loam—2% or greater organic matter or Medium, Fine 1% or greater organic matter	*OPTION 1*				
		INITIAL TREATMENT—GRASS 1 TO 3 LEAF			
	—	1.5	—	2.2	—
		SECOND TREATMENT 7 TO 14 DAYS AFTER INITIAL			
	—	—	4	—	2
	OPTION 2				
		INITIAL TREATMENT—GRASS 1 TO 3 LEAF			
	—	1.5	2	1.1	—
		SECOND TREATMENT 7 TO 14 DAYS AFTER INITIAL			
	—	—	4	—	2
	OPTION 3				
		INITIAL TREATMENT—GRASS 1 TO 3 LEAF			
	—	1.5	4	—	2
		SECOND TREATMENT 6 TO 14 DAYS AFTER INITIAL			
	—	—	2	—	2
For the soil types listed above or Sandy loam-1% or greater organic matter.		INITIAL TREATMENT—GRASS 1 TO 3 LEAF			
	—	0.75	—	1.8	—
		SECOND TREATMENT 7 TO 14 DAYS AFTER INITIAL			
	—	0.75	3	—	2
Sandy loam—greater than 2% organic matter or Medium or Fine—2% or greater organic matter	*OPTION 1* 4-6	1.5	—	2.2	—
	OPTION 2 4-6	1.5	2	1.1	—
	OPTION 3 4-6	1.5	4	—	2

(a) **Medium** texture soils include loam, silt loam, silt and sandy clay loam. **Fine** includes silty clay loam, clay loam, sandy clay, silty clay and clay.
(b) See Bladex 4L label for preemergence application rates in relation to soil texture and soil organic matter content.
(c) Any atrazine product recommended for use on field corn may be used. Use the conversions listed below to determine the proper amount.

Atrazine 4L (pint)		Atrazine 80W (pounds)		Atrazine 90DF (pounds)
2	=	1.25	=	1.1
3	=	1.9	=	1.7
4	=	2.5	=	2.2

(d) Bladex 80W may be used in place of Bladex 90DF. Use the conversions listed below to determine the proper amount.

Bladex 90DF (pounds)		Bladex 80W (pounds)
1.1	=	1.2
1.8	=	2.0
2.2	=	2.5

Do not use crop oil concentrate or crop oil in mixtures that contain Bladex 90DF or Bladex 80W. Do not apply mixtures that contain Bladex 90DF or Bladex 80W to corn larger than the 4 true leaf stage. Temporary yellowing may result following application of these treatments, particularly if cold, adverse growing conditions occur after application. However, recovery is rapid and yield is not reduced.
Do not apply TANDEM plus Bladex 90DF or Bladex 80W when corn is under stress from frost, prolonged cold or wet weather or when plants are wet from recent rainfall. Extended or extreme cold and wet conditions after application may reduce stands. Do not use Bladex 90DF or Bladex 80W on sandy or loamy sand soil containing less than 1% organic matter. **Do not apply TANDEM plus Bladex 90DF or Bladex 80W on inbred lines or breeding stock.**
(e) Six to eight pints of crop oil can be substituted for 2 pints of crop oil concentrate.

Wild Oats

For control of wild oats, follow the guideline listed in Table 9 when the wild oats are in the 1 to 3 leaf stage. A cultivation 7 to 14 days after the TANDEM plus Atrazine 4L plus Bladex 90DF application may improve weed control.

TABLE 9: Treatment of Wild Oats.

Soil Texture[a]	Broadcast Rate Per Acre — Timing of Treatment—Weed Size — Postemergence—Grasses 1 to 3 leaf: TANDEM (pints)	Atrazine 4L[b] (pints)	Bladex 90DF[c] (pounds)
Medium or Fine—2% or greater organic matter	1	1	1.1-1.7

(a) **Medium** texture soils include loam, silt loam, silt and sandy clay loam. **Fine** includes silty clay loam, sandy clay, silty clay and clay.
(b) Any atrazine product recommended for use on field corn may be used. If Atrazine 80W is used in place of Atrazine 4L, add 0.62 lb of product to the spray mixture per acre. If atrazine 90% dispersible granules are used, add 0.56 lb of product to the spray mixture per acre.
(c) Use higher rates of Bladex 90DF herbicide when weed pressures are heavy. Bladex 80W may be in used place of Bladex 90DF. Use the conversions listed below to determine the proper amount.

Bladex 90DF (pounds)		Bladex 80W (pounds)
1.1	=	1.2
1.7	=	1.9

Do not use oil concentrate or crop oil mixtures that contain Bladex 90DF or Bladex 80W. Do not apply mixtures that contain Bladex 90DF or Bladex 80W to corn larger than the 4 true leaf stage. Temporary yellowing may result following application of these treatments particularly if cold, adverse growing conditions occur after application. However, recovery is rapid and yield is not reduced.
Do not apply TANDEM plus Bladex 90DF or Bladex 80W when corn is under stress from frost, prolonged cold or wet weather or when plants are wet from recent rainfall. Extended or extreme cold and wet conditions after application may reduce stands. Do not use Bladex 90DF or Bladex 80W on sandy or loamy sand soil containing less than 1% organic matter. **Do not apply TANDEM plus Bladex 90DF or Bladex 80W on inbred lines or breeding stock.**

Fall Panicum

Following are two Tables for fall panicum control programs. For control of fall panicum in states other than North Carolina, South Carolina and Virginia, follow one of the three programs outlined in Table 10A. All programs require a second application of atrazine plus crop oil concentrate

7-12 days after the initial treatment. A cultivation may be substituted for the second application of atrazine plus crop oil concentrate, 7-12 days after the initial treatment.

TABLE 10A: Treatment programs for Fall Panicum in States other than North Carolina, South Carolina and Virginia.

Soil Texture[a]	Broadcast Rate Per Acre			
	Timing Of Treatment—Weed Size			
	Postemergence—Grasses 1 to 3 leaf			
	TANDEM (pints)	Atrazine 4L[b] (pints)	Bladex 90DF[c] (pounds)	Crop Oil[c] Concentrate (pints)
Coarse, Medium or Fine— 2-7% organic matter	*OPTION 1*			
	INITIAL TREATMENT			
	1-1.5[e]	3-4(e)	—	2
	SECOND TREATMENT 7 TO 12 DAYS LATER			
	—	2	—	2
	OPTION 2			
	INITIAL TREATMENT			
	1-1.5	—	1.8-2.2(e)	—
	SECOND TREATMENT 7 TO 12 DAYS LATER			
	—	2	—	2
	OPTION 3			
	INITIAL TREATMENT			
	1-1.5	1.5-2	0.9-1.1	—
	SECOND TREATMENT 7 TO 12 DAYS LATER			
	—	2	—	2

(a) **Coarse** texture soils include sand, loamy sand and sandy loam. **Medium** includes loam, silt and sandy clay loam. **Fine** includes silty clay loam, clay loam, sandy clay, silty clay and clay.

(b) Any atrazine product recommended for use on field corn may be used. Use the conversions listed below to determine the proper amount.

Atrazine 4L (pints)		Atrazine 80W (pounds)		Atrazine 90DF (pounds)
1.5	=	0.9	=	0.8
2	=	1.25	=	1.1
3	=	1.9	=	1.7
4	=	2.5	=	2.2

(c) Bladex 80W may be used in place of Bladex 90DF. Use the conversions listed below to determine the proper amount.

Bladex 90DF (pounds)		Bladex 80W (pounds)
0.9	=	1.0
1.1	=	1.25
1.8	=	2.0
2.2	=	2.5

Do not use crop oil concentrate or crop oil mixture that contain Bladex 90DF or Bladex 80W. Do not apply mixtures that contain Bladex 90DF or Bladex 80W to corn larger than the 4 true leaf stage. Temporary yellowing may result following application of these treatments, particularly if cold, adverse growing conditions occur after application. However, recovery is rapid and yield is not reduced.

Do not apply TANDEM plus Bladex 90DF or Bladex 80W when corn is under stress from frost, prolonged cold or wet weather or when plants are wet from recent rainfall. Extended or extreme cold and wet conditions after application may reduce stands. Do not use Bladex 90DF or Bladex 80W on sandy or loamy sand soil containing less than 1% organic matter. **Do not apply TANDEM plus Bladex 90DF or Bladex 80W on inbred lines or breeding stock.**

(d) Six to eight pints of crop oil can be substituted for 2 pints of crop oil concentrate.

(e) Use high application rate of TANDEM and atrazine and/or Bladex when weed pressures are heavy.

Fall Panicum—North Carolina, South Carolina and Virginia Only

Select one of the three treatment programs listed in Table 10B, which utilize TANDEM plus atrazine plus crop oil concentrate applied either postemergence or split postemergence; or atrazine applied preemergence followed by TANDEM plus atrazine plus crop oil concentrate applied postemergence.

TABLE 10B: Treatment programs for Fall Panicum in North Carolina, South Carolina and Virginia Only.

Soil Texture[a]	Broadcast Rate Per Acre			
	Timing Of Treatment—Weed Size			
	Pre-emergence	Postemergence—Grasses 1 to 3 leaf		
	Atrazine 4L[b] (pints)	TANDEM (pints)	Atrazine 4L[b] (pints)	Crop Oil[c] Concentrate (pints)
Coarse—less than 4% organic matter	INITIAL TREATMENT—GRASS 1 TO 3 LEAF			
	—	1	3	2
	SECOND TREATMENT 5 TO 10 DAYS AFTER INITIAL			
	—	—	1	2
Peat, muck or high organic matter	INITIAL TREATMENT—GRASS 1 TO 3 LEAF			
	—	1-1.5[d]	3	2
	SECOND TREATMENT 5 TO 10 DAYS AFTER INITIAL			
	—	—	3	2
Coarse—less than 4% organic matter	3	1	3	2

(a) **Coarse** texture soils include sand, loamy sand and sandy loam.

(b) Any atrazine product recommended for use on field corn may be used. Use the conversions listed below to determine the proper amount.

Atrazine 4L (pints)		Atrazine 80W (pounds)		Atrazine 90DF (pounds)
1	=	0.62	=	0.56
3	=	1.9	=	1.7

(c) Six to eight pints of crop oil can be substituted for 2 p ints of crop oil concentrate.

(d) Use higher rates of TANDEM when weed pressures are heavy.

Texas Panicum—Alabama, Florida and Georgia Only

Texas panicum is an atrazine-tolerant weed species. Season-long control of Texas panicum is seldom obtained because this species is a prolific seed producer. These relatively large grass seeds may remain viable in the soil for several years and may germinate from deep in the soil. These characteristics combined with the long growing season typical of the Southeast allow multiple grass flushes to occur throughout the growing season especially during the corn dry down period in mid-summer. Select one of the three programs listed below, which involves a postemergence application of TANDEM plus Atrazine 4L plus crop oil concentrate, or a preemergence application of Bladex 4L or Atrazine 4L followed by a postemergence application of TANDEM plus Atrazine 4L plus crop oil concentrate.

TABLE 11: Treatment of Texas Panicum in Alabama, Florida and Georgia Only.

Broadcast Rate Per Acre				
		Timing Of Treatment—Weed Size		
Preemergence		Postemergence—Grasses 1 To 3 Leaf		
Bladex 4L[a] (pints)	Atrazine 4L[b] (pints)	TANDEM (pints)	Atrazine 4L[b] (pints)	Crop Oil[c] Concentrate (pints)
		OPTION 1		
		INITIAL TREATMENT—GRASS 1 TO 3 LEAF		
		1-1.5[d]	3-4[d]	2
		SECOND TREATMENT 5 TO 10 DAYS AFTER INITIAL		
			2	2
		OPTION 2		
		INITIAL TREATMENT—GRASS 1 TO 3 LEAF		
		0.75	3	2
		SECOND TREATMENT 5 TO 10 DAYS AFTER INITIAL		
		0.75	3	2
		OPTION 3		
2 -OR-	2	1-1.5[d]	3-4[d]	2

(a) Do not apply Bladex 4L on sand or loamy sand soil with less than 1% organic matter content.

(b) Any atrazine product recommended for use on field corn may be used, including wettable powder and dispersible granule formulations. If Atrazine 80W is used, add 1.25 to 2.5 lb product to the spray mixture per acre. If atrazine 90% dispersible granules are used, add 1.1 to 2.2 lb of product to the spray mixture per acre.

(c) Six to eight pints of crop oil can be substituted for 2 pints of crop oil concentrate.

(d) Use higher rates of TANDEM and atrazine when weed pressures are heavy.

Field Sandbur—Colorado, Kansas and Nebraska Only

For the control of field sandbur, use one of the programs in Table 12 and follow the guidelines below. If rainfall does not occur within 5 to 10 days after application, a timely cultivation may be needed for supplemental control. Additionally, under droughty conditions, an overhead irrigation of 0.5 to 1 inch of water 5 to 10 days after application will improve control. Treatments listed below include a preemergence application of Atrazine 4L followed by a postemergence application of TANDEM plus Atrazine 4L plus crop oil concentrate, while the other treatment involves only a postemergence application of TANDEM plus Atrazine 4L plus crop oil concentrate. When the field to be treated has a previous history of heavy field sandbur infestations, it is **recommended** to use the preemergence Atrazine 4L treatment followed by the postemergence TANDEM plus Atrazine 4L plus crop oil concentrate treatment.

TABLE 12: Treatment of Field Sandbur in Colorado, Kansas and Nebraska Only

	Broadcast Rate Per Acre			
	Timing Of Treatment—Weed Size			
Soil Texture[(a)]	Pre-emergence	Postemergence—Grasses 1 To 3 Leaf		
	Atrazine 4L[(b)] (pints)	TANDEM (pints)	Atrazine 4L[(b)] (pints)	Crop Oil[(c)] Concentrate (pints)
Coarse—1.5% or less organic matter	—	*OPTION 1* 1-1.5[(d)]	3	2
Coarse—1.5% or less organic matter	2	*OPTION 2* Heavily Infested Farms 1-1.5[(d)]	3	2

(a) Coarse texture soils include sand, loamy sand and sandy loam.
(b) Any atrazine product recommended for use on field corn may be used. Use the conversions listed below to determine the proper amount.

Atrazine 4L (pints)		Atrazine 80W (pounds)		Atrazine 90DF (pounds)
2	=	1.25	=	1.1
3	=	1.9	=	1.7

(c) Six to eight pints of crop oil can be substituted for 2 pints of crop oil of concentrate.
(d) Use higher rate of TANDEM when weed pressures are heavy.

Shattercane

For enhanced suppression of shattercane following a labeled application of Eridicane Extra herbicide, follow the guidelines below. TANDEM plus triazine herbicide programs applied no later than 2 leaf stage of shattercane, following a previous Eradicane Extra PPI application will aid in the management of shattercane by reducing plant populations and by providing a stunting of remaining shattercane plants. This reduction in height usually lasts 7 to 14 days following application of a TANDEM plus triazine program, which provides an opportunity for the crop to obtain a height differential. Under droughty conditions, an overhead irrigation within 5 days after TANDEM plus triazine application will improve results. Treatment results will be reduced under adverse environmental conditions such as low rainfall and high temperatures and in fields where soils have a history of successive thiocarbamate use.

TABLE 13: SUPPRESSION PROGRAMS FOR SHATTERCANE FOLLOWING AN ERADICANE EXTRA PPI APPLICATION.

		BROADCAST RATE PER ACRE			
		TIMING OF TREATMENT—WEED SIZE			
PREEMERGENCE		POSTEMERGENCE—SHATTERCANE (SPIKE TO 2 LEAF)			
ERADICANE EXTRA[(a)] (PINTS)		TANDEM (PINTS)	ATRAZINE 4L[(b)] (PINTS)	BLADEX 90DF[(c)] (POUNDS)	CROP OIL CONCENTRATE (PINTS)
6.67-8	OPTION 1	1.5	4.0 CULTIVATE 7 TO 10 DAYS LATER	—	2
6.67-8	OPTION 2	1.5	2.0 CULTIVATE 7 TO 10 DAYS LATER	1.1	—
6.67-8	OPTION 3	1.5	— CULTIVATE 7 TO 10 DAYS LATER	2.2	—

(a) Follow all recommended cultural practices on the Eradican Extra label for suppression of shattercane and the label for recommended rate for your soil type.
(b) Any atrazine product recommended for use on field corn may be used. Use the conversions listed below to determine the proper amount.

Atrazine 4L (pints)		Atrazine 80W (pounds)		Atrazine 90DF (pounds)
2	=	1.25	=	1.1
4	=	2.5	=	2.2

(c) Bladex 80W may be used in place of Bladex 90DF. Use the conversions listed below to determine the proper amount.

Bladex 90DF (pounds)		Bladex 80WW (pounds)
1.1	=	1.2
2.2	=	2.5

Do not use crop oil concentrate or crop oil in mixtures that contain Bladex 90DF or Bladex 80W. Do not apply mixtures that contain Bladex 90DF or Bladex 80W to corn larger than the 4 true leaf stage. Temporary yellowing may result following application of these treatments, particularly if cold, adverse growing conditions occur after application. However, recovery is rapid and yield is not reduced.

Do not apply TANDEM plus Bladex 90DF or Bladex 80W when corn is under stress from frost, prolonged cold or wet weather or when plants are wet from recent rainfall. Extended or extreme cold and wet conditions after application may reduce stands. Do not use Bladex 90DF or Bladex 80W on sandy or loamy sand soil containing less than 1% organic matter. **Do not apply TANDEM plus Bladex 90DF or Bladex 80W on inbred lines or breeding stock.**

USE PRECAUTIONS AND RESTRICTIONS

Read and carefully follow all applicable directions, restrictions and cautions on labeling for other products used in combination with TANDEM. **For atrazine products, be sure to observe rotational crop suggestions to prevent possible injury to subsequent crops caused by atrazine carry-over.**

Do not apply more than a total of 2 quarts of TANDEM from any type of application per acre per growing season.

Do not apply TANDEM herbicide to sweet corn or pop corn.

Inbred Lines or Breeding Stock: Susceptibility of corn injury from tank mixtures of TANDEM herbicide plus atrazine is highly related to varietal response. Inbred lines or any breeding stock may be severely injured by tankmixed applications of TANDEM and atrazine plus emulsifiable crop oil concentrate or crop oil. Thus, use of TANDEM on inbred lines or breeding stock **IS AT THE END USER'S OWN RISK.** Contact your seed production agronomist for advice before tank mixtures of TANDEM plus atrazine are applied to inbred lines or breeding stock.

Do not apply TANDEM plus Bladex 80W or 90DF to inbred lines or breeding stock.

Do not apply TANDEM herbicide in combination with atrazine and/or Bladex Herbicides:

- When corn is under stress from prolonged cold, frost, wet weather, poor fertility, storm damage, or other weather stress.
- When crop leaf is wet and succulent from recent rainfall.
- In the State of Texas.

Do Not Apply: Bladex 4L, Bladex 80W, or Bladex 90DF by means of chemigation or aerial application.

Do Not Apply Bladex 4L, Bladex 80W, or Bladex 90DF to sand or loamy sand soil with less than 1% organic matter content.

On coarse soils, which are mild to strongly alkaline, the total amount of Bladex 4L, Bladex 80W or Bladex 90DF should not exceed 1.2 lb active ingredient per acre (Eastern CO, Western NE, and Eastern WY).

Do Not Apply Bladex 80W or Bladex 90DF postemergence to corn larger than 4 true leaf.

Tank-mixing TANDEM and triazine herbicides with other insecticides, herbicides, liquid fertilizers, or other materials is **not recommended** because of possible compatibility or crop injury problems.

Store and handle emulsifiable oil or crop oil concentrate carefully. Oil contaminated with even a small amount of water may not emulsify properly when added to the tank.

STORAGE AND DISPOSAL

Do not contaminate water, food or feed by storage or disposal.

Pesticide Disposal: Pesticide wastes are toxic. Improper disposal of excess pesticide, spray mixture, or rinsate is a violation of Federal Law. If these wastes cannot be disposed of by use according to label instructions, contact your State Pesticide or Environmental Control Agency, or the Hazardous Waste representative at the nearest EPA Regional Office for guidance.

Container Disposal: Triple rinse (or equivalent). Then dispose of in a sanitary landfill, or by incineration, or, if allowed by local authorities, by burning. If burned, stay out of smoke.

WARRANTY LIMITATIONS AND DISCLAIMER

The Dow Chemical Company warrants that this product conforms to the chemical description on the label and is reasonably fit for the purposes stated on the label when used in strict accordance with the directions therein under normal conditions of use. THIS IS THE ONLY WARRANTY MADE ON THIS PRODUCT. NO OTHER EXPRESS AND NO IMPLIED WARRANTY OF MERCHANTABILITY OR FITNESS FOR A PARTICULAR PURPOSE IS MADE OUTSIDE OF THIS LABEL. Therefore, neither this warranty nor any other warranty of merchantability or fitness for a particular purpose, express or implied, extends to the use of this product contrary to label instructions (including conditions noted on the label, such as unfavorable temperatures, soil conditions, etc.), under abnormal conditions (such as excessive rainfall, drought, tornadoes, hurricanes, etc.) or under conditions not reasonably foreseeable to or beyond the control of seller.

When buyer or user suffers losses or damages resulting from the use or handling of this product (including claims based on contract, negligence, strict liability, or other legal theories), buyer or user must promptly notify in writing The Dow Chemical Company of any claims to be eligible to receive either remedy given below. The EXCLUSIVE REMEDY OF THE BUYER OR USER and the LIMIT OF LIABILITY of The Dow Chemical Company or any other seller will be one of the following, at the election of The Dow Chemical Company:

(1) **Refund of purchase price** paid by buyer or user for product bought, **or**

(2) **Replacement** of amount of product used.

The seller **will not be liable for consequential or incidental damages or losses.**

The terms of this Warranty Limitations And Disclaimer cannot be varied by any written or verbal statements or agreements. Any employee or sales agent of the seller is not authorized to vary or exceed the terms of this Warranty Limitations And Disclaimer in any manner.

U.S. Patent No. 4,211,549

Other Patents Pending

85281-L2 C589

* Trademark of THE DOW CHEMICAL COMPANY

TELONE* II

SOIL FUMIGANT

RESTRICTED USE PESTICIDE

For retail sale to and use only by Certified Applicators or persons under their direct supervision and only for those uses covered by the Certified Applicator's certification.

A Liquid for Preplant Treatment of Soil to Control Plant Parasitic Nematodes and Certain Other Soil Pests

ACTIVE INGREDIENT(S):	(by weight)
1,3-Dichloropropene	94%
INERT INGREDIENTS:	6%
TOTAL	100%

1 gallon of TELONE II weighs 10.2 lb. @ 70°F

E.P.A. Registration No. 464-511 E.P.A. Est. 464-TX-1[(TB)]; 464-MI-1[(MI)]; 464-CA-1[(WP)]; 464-CA-2[(WL)]; 33776-WA-1[(WQ)]. Superscript used corresponds to letter in Lot No.

KEEP OUT OF REACH OF CHILDREN

WARNING

AVISO:

PRECAUCIÓN AL USUARIO: Si usted no lee inglés, no use este producto hasta que la etiqueta le haya sido explicada ampliamente.

PRECAUTIONARY STATEMENTS

HAZARDS TO HUMANS AND DOMESTIC ANIMALS

HAZARDOUS LIQUID AND VAPOR • MAY BE FATAL IF INHALED, ABSORBED THROUGH SKIN, OR SWALLOWED • CAUSES SUBSTANTIAL BUT TEMPORARY EYE DAMAGE • CAUSES SKIN IRRITATION AND, IF CONFINED, SKIN BURNS • MAY CAUSE ALLERGIC SKIN REACTION • MAY CAUSE LUNG, LIVER AND KIDNEY DAMAGE AND RESPIRATORY SYSTEM IRRITATION UPON PROLONGED CONTACT • THE USE OF THIS PRODUCT MAY BE HAZARDOUS TO YOUR HEALTH. TELONE II CONTAINS 1,3 DICHLOROPROPENE, WHICH HAS BEEN DETERMINED TO CAUSE TUMORS IN LABORATORY ANIMALS. RISKS CAN BE REDUCED BY CLOSELY FOLLOWING THE USE DIRECTIONS AND PRECAUTIONS, AND BY WEARING PROTECTIVE CLOTHING SPECIFIED ELSEWHERE ON THIS LABEL.

Do Not Breathe Vapor • Do Not Get In Eyes, On Skin, On Clothing • Do Not Take Internally • Do Not Use The Mouth To Siphon TELONE II From Containers Or To Blow Out Clogged Lines, Nozzles, Etc.• Use Only With Adequate Ventilation • Wear Chemical Goggles And Other Protective Equipment Required Under the Section So Designated To Avoid Contact When Handling TELONE II • Wash Thoroughly With Soap And Water After Handling And Before Eating Or Smoking • If protective gear, becomes contaminated, immediately wash with soap and water. Never wear protective gear having the odor of 1,3-dichloropropene. Aerate and wash all protective gear thoroughly after use until all odor is gone. • Render unusable and dispose of contaminated leather goods, including shoes. • READ AND FOLLOW FURTHER INSTRUCTIONS UNDER PROTECTIVE EQUIPMENT REQUIREMENTS SECTION

STATEMENT OF PRACTICAL TREATMENT:

If Inhaled: remove to fresh air. If not breathing, give artificial respiration, preferably mouth-to-mouth. If breathing is difficult, give oxygen. Call a physician.

If On Skin: immediately flush skin with plenty of water for at least 15 minutes while removing contaminated clothing and shoes. Call a physician. If water is not immediately available, remove excess chemical from skin with sorbent material such as towel or dry soil, then proceed at once to location where water is available and thoroughly wash contaminated skin with plenty of water. Call a physician.

If In Eyes: immediately flush eyes with plenty of water for at least 15 minutes. Call a physician.

If Swallowed: do not induce vomiting. Call a physician or Poison Control Center immediately. Never give anything by mouth to an unconscious person.

NOTE TO PHYSICIAN: Because rapid absorption may occur through lungs if product is aspirated and cause systemic effects, the decision to induce vomiting or not should be made by a physician. If lavage is performed, endotracheal and/or esophageal control is suggested. Danger from lung aspiration must be weighed against toxicity when considering emptying the stomach.

PHYSICAL OR CHEMICAL HAZARDS

FLAMMABLE • Do Not Use, Pour, Spill, or Store Near Heat or Open Flame • Do Not Cut or Weld Container.

ENVIRONMENTAL HAZARDS

Do not contaminate water by cleaning equipment or disposing of wastes. See section under Storage, Shipment, and Disposal. In case of spills properly dispose of contaminated materials.

NOTICE

Read and understand the entire label before using. Use only according to label directions.

Before buying or using this product, read "WARRANTY LIMITATIONS AND DISCLAIMER" elsewhere on this label. If terms are not acceptable, return **unopened** package at once to seller for full refund of purchase price paid. Otherwise, use by the buyer or any other user constitutes acceptance of the terms under "WARRANTY LIMITATIONS AND DISCLAIMER."

IN CASE OF AN EMERGENCY **endangering life or property involving this product, call collect 517-636-4400**

AGRICULTURAL CHEMICAL

Do Not Ship or Store with Food, Feeds, Drugs, or Clothing

DIRECTIONS FOR USE

It is a violation of Federal law to use this product in a manner inconsistent with its labeling.

PROTECTIVE EQUIPMENT REQUIREMENTS

1. The following clothing and equipment are required to be used by persons actually engaged in carrying out any operations that are likely to involve direct contact with TELONE II including equipment calibration, cleanup and repair; sampling; cleanup of spills; fumigant transfer; and rinsate disposal. This equipment must also be used for any operations that are done within 6 feet of unshielded, pressurized hoses containing TELONE II:
   a. NIOSH- MSHA-approved half-face organic vapor respirator and chemical goggles or a full-face respirator.
   b. Body covering that has long sleeves and long pants and is constructed of material resistant to TELONE II.
   c. Hat with a bill, such as a bump cap, resistant to TELONE II.
   d. Synthetic work gloves and heavy-duty, synthetic footgear resisant to TELONE II.
2. The following clothing and equipment must be readily available at all times for use by persons operating application and soil sealing equipment. It is recommended that such safety gear be carried on the equipment involved:
   a. NIOSH- MSHA-approved half-face or full-face organic vapor respirator. This equipment must be worn when the odor of TELONE II can be detected.

b. Heavy-duty, synthetic footgear resistant to TELONE II. This footgear must be worn when walking on treated soil within 72 hours after application.
c. Equipment operators who carry out any of the operations listed in Section 1, above, or who must work within 6 feet of unshielded, pressurized hoses containing TELONE II must have available and wear the protective equipment recommended for these situations

NOTE: There are no protective clothing materials that are completely impervious to penetration by liquid TELONE II. Thin layers of polyethylene (minimum 1 mil), rubber and vinyl protective gear give short-term protection and must be immediately discarded upon contamination. Heavy (5+mils) polyethylene, rubber and neoprene provide longer term protection. Leather gives no protection.

NIOSH- or MSHA-approved respiratory protection must be worn when TELONE II is exposed to the atmosphere or when conducting operations that vent to the atmosphere. A NIOSH- or MSHA-approved half-face organic vapor respirator or full-face respirator must be used. When in use canisters or cartridges must be replaced daily or sooner if specified by manufacturer or at first sign of odor breakthrough, whichever occurs first. NIOSH-approved organic vapor cartidges, such as the North N7500, will be adequate for short-term exposure situations. Where very high concentrations of vapors might be expected (such as large spills in poorly ventilated areas) a positive pressure self-contained or air-supplied respirator must be used.

FIELD REENTRY

Do not enter treated area within 72 hours after treatment unless heavy-duty synthetic footgear is worn. See section on Protective Equipment Requirements for further information. If the odor of TELONE II can be detected, an organic vapor repirator must be worn. Certain states may have more restrictive reentry intervals. Consult your State Department of Agriculture for further information.

Written or oral warnings must be given to workers who are expected to be in a treated area or in an area about to be treated with this product. These oral warnings shall inform workers of areas or fields that may not be entered without heavy-duty synthetic footgear until after 3 days post treatment. In case of accidental exposure follow directions as shown by the Statement of Practical Treatment section on this label. When oral warnings are given, warnings shall be given in a language customarily understood by workers. Written warnings must include the following: **WARNING.** Area treated with TELONE II soil fumigant on (insert date of application). Do not enter without heavy-duty synthetic footgear until after 3 days post treatment.

STORAGE, SHIPMENT AND DISPOSAL

Pesticide wastes are toxic. Improper disposal of excess pesticide is a violation of Federal Law. If these wastes cannot be disposed of by use according to label instructions, contact your State Pesticide or Environmental Control Agency, or the Hazardous Waste representative at the nearest EPA Regional Office for Guidance.

STORAGE: Store in tightly-closed original container in a cool place away from dwellings. Do not allow contamination of seeds, plants, fertilizers, or other pesticide chemicals. Do not contaminate food, feedstuffs, drugs, or domestic water supplies. In outside storage, store drums on sides to avoid accumulation of rain water in top or bottom recessed areas.

SHIPMENT: Do not ship or store with food, feeds, drugs, or clothing.

DISPOSAL: Triple rinse original container with fuel oil, kerosene or a similar type of petroleum solvent and dispose of rinsate by incorporation into field just treated or by other approved means. After aeration offer container to qualified reconditioner or dispose of as directed by state or local regulations.

Because TELONE II is corrosive under certain conditions, flush all application equipment with fuel oil, kerosene or a similar type of petroleum solvent immediately after use. Fill pumps and meters with new motor oil or a 50% motor oil/fuel oil mixture before storing. **Do not use water.** Dispose of rinsate by incorporation into field just treated or by other approved means. Never introduce rinsate or unused TELEONE II into surface or underground water supplies.

GENERAL INFORMATION

Read the entire label before using TELONE II.

Use TELONE II soil fumigant as a preplant soil treatment to control the following types of plant parasitic nematodes: burrowing, citrus, cyst (golden, sugar beet, soybean, carrot and wheat), dagger, lance, pin, reniform, ring, root knot, root lesion (meadow), spiral, sting and stubby root.

TELONE II can also be used to control garden centipedes (symphylans) and wireworms, suppress sugar beet *Rhizomania* disease, *Fusarium* wilt of cotton and *Verticillium* wilt of mint and potatoes and aid in the control of bacterial canker of peaches.

TELONE II will control pests that are present in the treatment zone at time of fumigation. It will not control pests that are introduced after fumigation from sources such as contaminated soil, equipment, irrigation water and planting material.

Fumigate soil to be planted to crops such as those listed below by applying TELONE II under the conditions and at the rates recommended under APPLICATION DIRECTIONS.

Note: Do not apply TELONE II through any type of irrigation system.

Vegetable Crops:

asparagus	egg plant	pimentoes
beans	endive	potatoes
beets	garlic	pumpkins
blackeyed peas	horseradish	radishes
broccoli	kale	rutabaga
brussels sprouts	kohlrabi	salsify
cabbage	leeks	shallots
cantaloupe	lettuce	spinach
carrots	melons	squash (summer)
cauliflower	mustard greens	squash (winter)
celery	okra	sweet potatoes
collards	onions	swiss chard
corn	parsnips	tomatoes
cowpeas	peas	turnips
cucumbers	peppers	watermelons.

Field Crops:

alfalfa	kenaf	rye
barley	lespedeza	safflower
birdsfoot trefoil	millet	sorghum
buckwheat	milo	soybeans
clover	mint	sugar beets
corn	oats	sugarcane
cotton	pasture grass	tobacco
flax	peanuts	vetch
grasses	popcorn	wheat.
hops		

Fruit and Nut Crops:

almonds	gooseberries	pecans
apples	grapefruit	persimmons
apricots	grapes	pineapple
bananas	hazelnuts (filberts)	plums
blackberries	hickory nuts	pomegranates
blueberries	huckleberries	prunes
boysenberries	kumquats	quince
cashew nuts	lemons	raspberries
cherries	limes	strawberries
chestnuts	loganberries	tangerines
cranberries	nectarines	tangelos
currants	olives	youngberries
dates	oranges	walnuts
dewberries	peaches	
figs	pears	

Nursery Crops including floral plants, ornamentals, shrubs and bushes; forest, shade, fruit and nut trees and vine and bramble fruits of all types. When used according to state nursery regulations, TELONE II soil fumigant may be used in the production of certified nursery stock.

APPLICATION DIRECTIONS

WHEN TO TREAT: TELONE II can be applied at any time of the year when soil conditions permit. Conditions that allow rapid diffusion of the fumigant as a gas through the soil normally give best results. Because TELONE II does not provide residual control of soil pests, it should be used before planting each crop. The following soil temperature and moisture conditions should exist at time of treatment. Failure to meet these conditions may result in unsatisfactory product performance:

Soil temperature at the depth of application must be between 40°F and 80°F. In areas where the soil temperature in the spring may not reach 40°F in time to allow application of TELONE II prior to planting, late summer or early fall treatment is recommended.

Soil moisture throughout the desired treatment zone generally should be at or near the permanent wilting point to allow optimum dispersion of the fumigant, which moves as a gas through the soil air spaces. The permanent wilting point varies with soil texture and organic matter content. Coarser textured soils can be fumigated under conditions of higher soil moisture than finer textured soils; however, if the soil moisture is too high, fumigant movement will be retarded and effectiveness of the treatment will be reduced. Previous and/or local experience with the soil to be treated or the crop to be planted can often serve as a guide to conditions that will be acceptable. If you do not know how to determine the soil moisture content of the area to be treated, consult your local Extension Service or Soil Conservation Service Specialist or Pest Control Advisor (Ag Consultant) for assistance.

In general, no irrigation should immediately precede subsoiling or fumigation; however, when surface soil moisture conditions are not likely to provide an adequate seal against fumigant loss, a very light sprinkler irrigation to wet the top 1 to 2 inches of soil may be useful.

SOIL PREPARATION: The soil should be free of clods. Plant residues should be incorporated into the soil and allowed to decompose prior to treatment. Very little crop residue should be present on the soil surface. Crop residue that is present should lie flat to permit the soil to be sealed effectively. Compacted soil layers within the desired treatment zone should be fractured before or during application of the fumigant. Deviation from the above conditions may result in unsatisfactory results.

PLACEMENT OF THE FUMIGANT: TELEONE II may be applied as either a broadcast (overall) or row treatment. It must be placed at least 10 inches below the final soil surface. Deeper placement is recommended when fumigating soil to be planted to deep-rooted plants, such as perennials fruit and nut crops, or to control deeply distributed pests.

APPLICATION RATES

NEMATODES:

Broadcast Application: Use chisel (shank), Nobel (sweep) plow or plow-sole application equipment with one or more fumigant outlets. For best results under most conditions, use chisel equipment with ripper-type, forward-swept shanks. Nobel plow equipment is particularly useful for fall fumigation when the soil may still contain some undecomposed plant material. Subsoiling may be necessary before application as described under **Soil Preparation.** The fumigant outlet spacing varies with the type of application equipment used.

With chisel equipment a fumigant outlet spacing of 12 to 24 inches is recommended. The outlet spacing for this equipment may be up to 1½ times the application depth but generally should be equal to the application depth and should not exceed the soil-shattering capability of the chisels. The maximum outlet spacing should not exceed 24 inches.

With plow-sole equipment a 12 inch outlet spacing is recommended.

With Nobel (sweep) plow equipment use an outlet spacing of 9 inches along the sweeps.

Broadcast application can be made in the same direction or at an angle to the direction of the planting row. Refer to **Table 1** for broadcast treatment rates for various crops.

Table 1. Broacast Treatment Rates for Nematode Control

Crops	Soil Texture	Recommended Rate: Gallons Per Acre	Recommended Rate: Fl. oz./ 1000 ft. of Row/Outlet(1)
Vegetable(2)	Mineral(3) Muck or Peat(5)	9-18(4) 24-36	26-53 71-106
Field	Mineral Muck or Peat	9-18(4) 24-36	26-53 71-106
Fruit, Nut and Nursery 6,7,8,9,10	Sand Sandy Loam Silt Loam Clay Loam	27-33 36-48 63-75 84-102	79-97 106-141 185-220 247-300

[1] Flow rates are based on 12-inch outlet spacing. Flow rates for alternate spacings can be calculated using the following formula: fl. oz/1000 ft. of row/outlet = 2.94 × rate in gallons/acre × outlet spacing in feet. For row treatment refer to Tables 2 and 3.

[2] Potatoes: Use 9 to 18 gallons of TELONE II per acre to control the Northern root knot nematode *Meloidogyne hapla* in mineral soil and 24 to 36 gallons per acre in muck soil.

For high populations of this species use the higher recommended rate. For more difficult-to-control root knot nematodes such as the Columbia root knot nematode *Meloidogyne chitwoodi*, qpply 20 gallons per acre (59 fl. oz/1000 ft of row/outlet based on 12-inch centers) in mineral soil. For best results apply the fumigant at least 18 inches below the final soil surface.

Before fumigation, soil sampling for the type and number of pests present is recommended. This can help to determine the need for additional treatment with a contact nematicide. Preharvest sampling for nematodes also is recommended. If the nematode population is high enough that the crop may be damaged, the potatoes can be harvested early.

[3] Mineral soil includes sand, sandy loam, silt, and clay loam. Use the higher rates for finer textured (heavier) soils.

[4] For cyst-forming nematodes use 18 gallons per acre (53 fl. oz./1000 ft./outlet).

[5] Greater than 20% organic matter content.

[6] Pineapple: Application may be made at the time of planting. For best results, seal the soil with polyethylene film, which acts as a gas permeability barrier. Dow High Barrier 419 polyethylene mulch film, with a minimum thickness of 1 mil, is recommended.

[7] Strawberries: For broadcast fumigation of mineral soils only, apply 24 to 36 gallons per acre.

[8] Tree Planting Sites: Use 24 fl. oz. (1.5 pints of TELONE II by application of the fumigant at a single point in the center of each planting site at a depth of 5 ft. below the final soil surface. Sites prepared by backhoeing to break up restrictive soil layers that may retard fumigant movement should be dug in the approximate dimensions of 10 × 10 × 10 ft. The hole should then be backfilled to a depth of 5 ft., the fumigant applied using a closed-system application tube and the remainder of the soil previously removed immediately added to the hole. For sites where no restrictive soil layers are present, the fumigant can be applied to a depth of 5 ft. using an injection auger. For best results, prepare and treat planting sites in the fall and plant in the spring.

[9] For shallow-rooted plants grown only one year, use 15 to 27 gallons per acre (44 to 79 fl. oz./1000 ft. of row/outlet).

[10] Citrus: For burrowing nematode control, inject TELONE II on 18-inch centers at least 12 inches deep. For buffers within existing groves or for tree planting sites within existing groves, do not apply within 5 feet of living trees. Keep the field free of plants susceptible to this nematode at least two years before planting to citrus.

Row Application: Use chisel equipament to treat a band of soil where the crop is to be planted, i.e. the plant row. One or two chisels per plant row is recommended. In general, when one chisel is used, apply TELONE II at the flow rates given in **Table 2.** When two chisels per plant row are used, space the chisels (fumigant outlets) 8 to 12 inches apart and divide the flow rates given in **Table 2** equally between the two outlets. Regardless of the number of chisels used, the amount of fumigant applied per 1000 feet of plant row should remain the same. With certain deeper rooted crops such as potatoes and sugar beets, higher flow rates may be necessary to ensure adequate treatment of the zone of soil where primary root growth occurs; however, in no case should the amount of fumigant applied per acre exceed the gallons per acre rates for broadcast treatment given in Table 1. To determine the amount (gallons) of TELONE II required per acre for various plant row spacings and flow rates, refer to **Table 3.** Note that as the distance between the plant row increases the amount of fumigant required decreases and vice versa.

To prevent seed germination problems caused by improper seed-to-soil contact or improper seedling depth, do not place the seed directly over the furrow left by the applicator chisel(s). When one chisel is used per plant row, place the seed about 4 inches to one side of the chisel furrow. When two chisels are used per plant row, plant the seed in the center of the area between the chisel furrows.

Table 2. Row Treatment Rates for Nematode Control Using a Single Chisel Per Row[1]

Crops	Soil Texture(2)	Recommended Rate Fl. oz./1000 ft. of Row(3)
Vegetable	Mineral Muck and Peat	56-106 142-212
Field(4)	Mineral Muck and Peat	52-106 142-212
Fruit, Nut and Nursery(5)	Mineral Muck and Peat	52-106 142-212

[1] For row spacing of 24 inches or less apply TELONE II as a broadcast treatment.

[2] For a description of soil texture see Table 1.

[3] To determine actual gallons per acre needed for various row spacings see Table 3.

[4] Sugar Beets: To control sugar beet cyst nematode, use 93 fl. oz/1000 ft. of row.

[5] Pineapples: To control reniform nematodes use 230 fl. oz/1000 ft. of row.

Table 3. Gallons of TELONE II Required Per Acre for Various Row Spacings and Fumigant Flow Rates(1)

Fl. oz./ 1000 ft. of Row	Plant Row Spacing (inches): 26	28	30	32	34	36	38	40	42	44	46	48	50
52	8.2	7.6	7.1	6.6	6.2	5.9	5.6	5.3	5.1	4.8	4.6	4.4	4.2
60	9.4	8.8	8.2	7.7	7.2	6.8	6.4	6.1	5.8	5.6	5.3	5.1	4.9
68	10.7	9.9	9.3	8.7	8.2	7.7	7.3	6.9	6.6	6.3	6.0	5.8	5.5
76	11.9	11.1	10.3	9.7	9.1	8.6	8.2	7.8	7.4	7.0	6.7	6.5	6.2
84	13.2	12.3	11.4	10.7	10.1	9.5	9.0	8.6	8.2	7.8	7.5	7.1	6.9
92	14.4	13.4	12.5	11.7	11.0	10.4	9.9	9.4	8.9	8.5	8.2	7.8	7.5
100	15.7	14.6	13.6	12.8	12.0	11.3	10.7	10.2	9.7	9.3	8.9	8.5	8.2
108	17.0	15.8	14.7	13.8	13.0	12.2	11.6	11.0	10.5	10.2	9.6	9.2	8.8
116	18.2	16.9	15.8	14.8	13.9	13.2	12.5	11.8	11.3	10.8	10.3	9.9	9.5
124	19.5	18.1	16.9	15.8	14.9	14.1	13.3	12.7	12.1	11.5	11.0	10.5	10.1
132	20.7	19.3	18.0	16.8	15.8	15.0	14.2	13.5	12 8	12.2	11.7	11.2	10.8
140	22.0	20.4	19.1	17.9	16.8	15.9	15.0	14.3	13.6	13.0	12.4	11.9	11.4
148	23.2	21.6	20.1	18.9	17.8	16.8	15.9	15.1	14.4	13.7	13.1	12.6	12.1
156	24.5	22.8	21.2	19.9	18.7	17.7	16.8	15.9	15.2	14.5	13.8	13.3	12.7
164	25.8	23.9	22.3	20.9	19.7	18.6	17.6	16.7	15.9	15.2	14.6	13.9	13.4
172	27.0	25.1	23.4	21.9	20.7	19.5	18.5	17.6	16.7	16. 0	15.3	14.6	14.0
180	28.3	26.3	24.5	23.0	21.6	20.4	19.3	18.4	17.5	16.7	16.0	15.3	14.7
188	29.5	27.4	25.6	24.0	22.6	21.3	20.2	19.2	18.3	17.4	16.7	16.0	15.3
196	30.8	28.6	26.7	25.0	23.5	22.2	21.1	20.0	19.1	18.2	17.4	16.7	16.0
204	32.0	29.8	27.8	26.0	24.5	23.1	21.9	20.8	19.8	18.9	18.1	17.4	16.7
212	33.3	30.9	28.9	27.0	25.5	24.0	22.8	21.6	20.6	19.7	18.8	18.0	17.3

[1] Refer to **Table 2** for the rate needed for a specific crop and/or soil texture. To obtain the gallons per acre used for a row spacing not shown in this table, use the following equation:

$$\frac{\text{fl. oz./1000 ft. of row}}{\text{row spacing (inches)}} \times 4.08^{(1)} = \text{gallons per acre}$$

(a) $4.08 = \frac{12 \text{ inches} \times 43.56 \text{ (no. 1000 ft./acre)}}{128 \text{ (fl. oz. per gallon)}}$

DISEASES:

Bacterial Canker of Peaches: TELONE II can be used as an aid in the control of this disease by application as a preplant, overall treatment of light (sandy) soils at the rate of 36 gallons per acre (106 fl. oz./1000 ft. row per outlet) preferably in the fall when the soil is warm (55-80°F at injection depth) and moist. Inject the fumigant at a depth of 10 to 12 inches with chisels mounted on 12-inch centers.

***Fusarium* Wilt of Cotton:** The effects of this disease can be suppressed by controlling the root knot nematodes associated with this disease/nematode complex. Use TELONE II as a row treatment at the rate of 46 to 106 fl. oz./1000 ft. of row. The lower rate is suitable for mineral soils whereas the higher rate should be used for heavier soils.

Sugar Beet *Rhizomania* Disease: Use TELONE II to suppress the effects of this disease by preplant application at the rate of at least 73.5 but not more than 132 fl. oz./1000 ft. of plant row. These flow rates are equivalent to 10 to 18 gallons per acre for sugar beets planted in 30-inch beds with one plant row per bed. For beets planted in 40-inch beds with two plant rows per bed the recommended flow rates are equivalent to 15 to 27 gallons per acre. Use the higher rates for heavier (finer textured) soils and/or for higher levels of disease infestation. TELONE II is believed to reduce the activity of *Polymyxa betae*, which has been identified as the vector of the *Rhizomania* disease virus. The fumigant should be placed at least 12 inches below the final soil surface. Immediately after application, mechanically compact (seal) the soil surface to prevent fumigant loss. Sealing can be accomplished by forming the beds during application or, when fumigating pre-formed beds, re-list the beds or use a ring roller, cultipacker, bed shaper, press sealer or similar device.

***Verticillium* Wilt of Mint and Potatoes:** To aid in the control of this disease, apply TELONE II as a broadcast treatment. For mint, use 59 gallons per acre (173 fl. oz./1000 ft. row/outlet) in the spring or, preferably, in the fall. For potatoes, use 17 to 25 gallons per acre (50 to 73 fl. oz./1000 ft. row/outlet) in the spring, or 25 to 34 gallons per acre (73 to 100 fl. oz./1000 ft. row/outlet) in the fall.

INSECTS:

Sympyhlans (Garden Centipedes): Use TELONE II for treatment of soil to be planted to crops where these pests have been shown to be a problem. Apply the fumigant only as a broadcast treatment at the rate of 18 to 36 gallons per acre (53 to 106 fl. oz./1000 ft. row/outlet) when soil temperature is warm (55 to 80°F) at the application depth.

Wireworms: Use TELONE II for treatment of soil to be planted to crops where these pests have been shown to be a problem. Apply the fumigant as a broadcast treatment at the rates recommended for nematode control (Table 1) by injection at least 14 inches below the final soil surface.

SEALING THE SOIL: Immediately after application, mechanically compact the soil to prevent fumigant loss. Use a ring roller, cultipacker, disc and roller, or similar device. If chisel or plow sole traces are likely to remain, disc the soil immediately before sealing. Note that little or no crop residue should be exposed through the soil surface following the sealing operation. After row application, sealing may be accomplished by forming the beds so that the fumigant is at least 12 inches below the final soil surface. When fumigating pre-formed beds, seal the soil with ring rollers, press sealers, or by reforming the beds.

Sealing can also be accomplished by applying plastic film, such as polyethylene, over the entire area or in strips. An adequate seal can sometimes be made by evenly moistening the upper 1 to 2 inches of soil over the entire treated area immediately after application of the fumigant.

For example, a light sprinkler irrigation to obtain a moisture level near field capacity in the top 2 inches of soil will aid in sealing a dry soil surface.

TREATMENT PERIOD: Leave the soil undistrubed and unplanted for at least 7 days after application of the fumigant. A longer treatment period is required if the soil becomes cold or wet, and the deep-rooted tree, shrub and vine planting sites.

After the treatment period, to prevent phytotoxicity, allow the fumigant to dissipate competely before planting the crop. Under optimum soil conditions for dissipation, 1 week for each 10 gallons is recommended. For fruit, nut, and nursery crops at least three months should elapse between treatment and planting. To hasten dissipation, especially if heavy rains or low temperature occur during the treatment period, till the soil to the depth of fumigant application. Use a knife-like chisel without turning the soil to reduce the possibility of recontaminating the treated soil. Dissipation is usually complete when the odor of TELONE II is no longer evident at the application depth. Lettuce seed can also be used as an indicator plant; contact Dow Chemical U.S.A. for test details. Do not plant if the odor of TELONE II is present at the depth of application.

USE PRECAUTIONS

Fertility Interactions: Fumigation may temporarily raise the level of ammonia nitrogen and soluble salts in the soil. This is most likely to occur when heavy rates of fertilizer and fumigant are applied to soils that are either cold, wet, acid, or high in organic matter. To avoid injury to certain crops, legumes (beans), lettuce, onions, and sugar beets, fertilize as indicated by soil tests made after fumigation. To avoid ammonia injury or nitrate starvation or both to crops on high organic soils do not use fertilizers containing ammonium salts and use only fertilizers containing nitrates, until after the crop is well established and the soil temperature is above 65°F.

When using high rates of TELONE II as required by certain State nursery regulations, liming of highly acid soils before fumigation may stimulate nitrification and reduce the possibility of ammonia toxicity. Certain nursery crops such as citrus seedlings, *Cornus sp.*, *Crataegus sp.*, spruce, and vegetable crops such as cauliflower have shown evidence of phosphorus deficiency following fumigation. To avoid this possible effect, it is suggested that additional phosphate fertilizer be used where experience indicates a deficiency may occur. Foliar applied phosphorus is recommended.

Recontamination Prevention: To avoid reinfestation of treated soil do not use irrigation water, transplants, tools, seed pieces, or crop remains that could carry soilborne pests from infested land. Clean equipment carefully before using.

WARRANTY LIMITATIONS AND DISCLAIMER

The Dow Chemical Company warrants that this product conforms to the chemical description on the label and is reasonably fit for the purposes stated on the label when used in strict accordance with the directions therein under normal conditions of use. THIS IS THE ONLY WARRANTY MADE ON THIS PRODUCT. NO OTHER EXPRESS AND NO IMPLIED WARRANTY OF MERCHANTABILITY OR FITNESS FOR A PARTICULAR PURPOSE IS MADE OUTSIDE OF THIS LABEL. Therefore, neither this warranty nor any other warranty of merchantability or fitness for a particular purpose, express or implied, extends to the use of this product contrary to label instructions (including conditions noted on the label, such as unfavorable temperatures, soil conditions, etc.), under abnormal conditions (such as excessive rainfall, drought, tornadoes, hurricanes, etc.) or under conditions not reasonably foreseeable to or beyond the control of seller.

When buyer or user suffers losses or damages resulting from the use or handling of this product (including claims based on contract, negligence, strict liability, or other legal theories), buyer or user must promptly notify in writing The Dow Chemical Company of any claims to be eligible to receive either remedy given below. The EXCLUSIVE REMEDY OF THE BUYER OR USER and the LIMIT OF LIABILITY of The Dow Chemical Company or any other seller will be one of the following, at the election of The Dow Chemical Company:

(1) **Refund of purchase price** paid by buyer or user for product bought, **or**

(2) **Replacement** of amount of product used.

The seller **will not be liable for consequential or incidental damages or losses.**

The terms of this Warranty Limitations And Disclaimer cannot be varied by any written or verbal statements or agreements. Any employee or sales agent of the seller is not authorized to vary or exceed the terms of this Warranty Limitations And Disclaimer in any manner.

85456-L6 J 989

* Trademark of THE DOW CHEMICAL COMPANY

TELONE* C-17

SOIL FUNGICIDE AND NEMATICIDE

RESTRICTED USE PESTICIDE

For retail sale to and use only by Certified Applicators or persons under their direct supervision and only for those uses covered by the Certified Applicator's certification.

A Multi-Purpose, Preplant Soil Fumigant for Use to Control Nematodes, Symphylans, Wireworms and Certain Soil Borne Diseases in Crop Lands

ACTIVE INGREDIENT(S):

1,3-Dichloropropene	74.0%
Chloropicrin	16.5%
INERT INGREDIENTS:	9.5%
TOTAL	100.0%

One gallon of TELONE C-17 weighs about 10.5 lb.
Contains 7.82 pounds of 1,3-dichloropropene and 1.79 pounds of chloropicrin per gallon.

E.P.A. Registration No. 464-379

E.P.A. Est. 464-CA-1$^{(WP)}$; 464-AR-1$^{(AM)}$; 464-MI-1$^{(MM)}$.

Superscript used corresponds to letter in Lot No.

KEEP OUT OF REACH OF CHILDREN

DANGER

PELIGRO:

PRECAUCIÓN AL USUARIO: Si usted no lee inglés, no use este producto hasta que la etiqueta le haya sido explicada ampliamente.

PRECAUTIONARY STATEMENTS

HAZARDS TO HUMANS AND DOMESTIC ANIMALS

HAZARDOUS LIQUID AND VAPOR • MAY BE FATAL IF INHALED, ABSORBED THROUGH SKIN, OR SWALLOWED • CAUSES SEVERE EYE DAMAGE • CAUSES BURNS OF SKIN • MAY CAUSE LUNG, LIVER AND KIDNEY DAMAGE AND RESPIRATORY SYSTEM IRRITATION UPON PROLONGED CONTACT • MAY CAUSE ALLERGIC SKIN REACTION

Do Not Breathe Vapor • Do Not Get In Eyes, On Skin, On Clothing • Do Not Take Internally • Use Only With Adequate Ventilation • Wear Eye and Skin Protection Necessary to Prevent Contact When Handling TELONE C-17 • Wash Thoroughly With Soap And Water After Handling And Before Eating Or Smoking • If protective gear, such as boots, or gloves, becomes contaminated, immediately wash with soap and water. Never wear protective gear having the odor of 1,3-dichloropropene. Aerate and wash all protective gear thoroughly after use until all odor is gone. • Render unusable and dispose of contaminated leather goods, including shoes. • Do not apply this product in such a manner as to directly or through drift expose workers or other personnel. The area being treated must be vacated by unprotected persons. • READ AND FOLLOW FURTHER INSTRUCTIONS UNDER HANDLING PRECAUTIONS SECTION

STATEMENT OF PRACTICAL TREATMENT:

If Inhaled: remove to fresh air. If not breathing, give artificial respiration, preferably mouth-to-mouth. If breathing is difficult, give oxygen. Call a physician.

If On Skin: immediately flush skin with plenty of water for at least 15 minutes while removing contaminated clothing and shoes. Call a physician. If water not immediately available, remove excess chemical from skin with sorbent material such as towel or dry soil, then proceed at once to location where water is available and thoroughly wash contaminated skin with plenty of water. Call a physician.

If In Eyes: immediately flush eyes with plenty of water for at least 15 minutes. Call a physician.

If Swallowed: do not induce vomiting. Call a physician or Poison Control Center immediately. Never give anything by mouth to an unconscious person.

NOTE TO PHYSICIAN: Because rapid absorption may occur through lungs if product is aspirated and cause systemic effects, the decision to induce vomiting or not should be made by a physician. If lavage is performed, endotracheal and/or esophageal control is suggested. Danger from lung aspiration must be weighed against toxicity when considering emptying the stomach.

PHYSICAL OR CHEMICAL HAZARDS

FLAMMABLE • Do Not Use, Pour, Spill, or Store Near Heat or Open Flame • Do Not Cut or Weld Container.

ENVIRONMENTAL HAZARDS

Do not contaminate water by cleaning of equipment or disposal of wastes. In case of spills properly dispose of contaminated materials.

Read the entire label. Use only according to label directions.

Before buying or using this product, read "WARRANTY LIMITATIONS AND DISCLAIMER" elsewhere on this label. If terms are not acceptable, return **unopened** package at once to seller for full refund of purchase price paid. Otherwise, use by the buyer or any other user constitutes acceptance of the terms under "WARRANTY LIMITATIONS AND DISCLAIMER."

IN CASE OF AN EMERGENCY endangering life or property involving this product, call collect 517-636-4400

AGRICULTURAL CHEMICAL

Do Not Ship or Store with Food, Feeds, Drugs, or Clothing

DIRECTIONS FOR USE

It is a violation of Federal law to use this product in a manner inconsistent with its labeling.

HANDLING PRECAUTIONS

CANCER HAZARD WARNING STATEMENT: The use of this product may be hazardous to your health. This product contains 1,3-dichloropropene which has been determined to cause tumors in laboratory animals. Risks can be reduced by closely following the use directions and precautions, and by wearing protective clothing specified elsewhere on this label.

DO NOT INHALE VAPORS: NIOSH- or MSHA-approved respiratory protection should be worn when TELONE C-17 liquid soil fumigant is exposed to the atmosphere or when conducting operations which vent to the atmosphere. A NIOSH- or MSHA-approved full-face respirator shall be used during small spills, repairs, calibrations, transfers, sampling and when working in poorly ventilated areas. When in use canisters or cartridges shall be replaced after each half hour of use or sooner if specified by manufacturer or at first sign of odor breakthrough, whichever comes first. NIOSH-approved organic vapor cartridges, such as a Welsh 7400-IL, will be adequate for short-term situations such as listed above.

Where very high concentrations of vapors might be expected (such as large spills in poorly ventilated areas) a self-contained or air-supplied respirator should be used.

DO NOT GET ON SKIN: The following clothing and equipment are required during the handling and application of TELONE C-17:

One-piece coveralls which have long sleeves and long pants constructed of laminated fabric as specified in the USDA/EPA Guide for Commercial Applicators.

Liquid-proof hat such as a plastic hard hat with a plastic sweat band.

Heavy-duty liquid proof (neoprene/synthetic) work gloves and boots.

Note: Any article worn while handling TELONE C-17 must be washed before reusing. Immediately remove all clothing which has been drenched or has otherwise absorbed TELONE C-17 from any spill. Dispose of contaminated clothing in a sanitary landfill, by incineration or if allowed by State and local authorities by burning. If burned, stay out of smoke.

There are no protective clothing materials that are *completely* impervious. Rubber, vinyl protective gear, thin layers of polyethylene (minimum 1 mil), give short-term protection and shall be immediately discarded upon contamination. Heavy (3+ mil) polyethylene, rubber, neoprene provide longer term protection. Leather gives no protection. If gear becomes contaminated, immediately wash with soap and water. Never wear protective gear having odor of TELONE C-17. Wash and aerate all protective gear thoroughly after use until all odor is gone.

DO NOT GET IN EYES: Wear eye protection such as full-face respirator when handling TELONE C-17.

DO NOT SWALLOW: Do not use the mouth to siphon TELONE C-17 from containers or to blow out clogged lines, nozzles, etc.

STORAGE, SHIPMENT AND DISPOSAL

Pesticide wastes are toxic. Improper disposal of excess pesticide is a violation of Federal Law. If these wastes cannot be disposed of by use according to label instructions, contact your State Pesticide or Environmental Control Agency, or the Hazardous Waste representative at the nearest EPA Regional Office for Guidance.

STORAGE: Store in tightly-closed original container in a cool place away from dwellings. Do not allow contamination of seeds, plants, fertilizers, or other pesticide chemicals. Do not contaminate food, feedstuffs, drugs, or domestic water supplies. In outside storage, store drums on sides to avoid accumulation of rain water in top or bottom recessed areas.

SHIPMENT: Do not ship or store with food, feeds, drugs, or clothing.

DISPOSAL: Triple rinse containers and dispose of rinsate in field just treated. After aeration offer containers to qualified reconditioner or destroy by puncturing or crushing and dispose of in an approved sanitary landfill or bury in the user's cropland or dispose of by other procedures approved by state and local authorities. Do not pollute surface or underground water supplies.

GENERAL INFORMATION

Apply TELONE C-17 soil fungicide and nematicide as a preplanting soil treatment to control or to aid in reducing the damaging effects of certain *soil borne diseases* [soil rot (soil pox) of sweet potatoes; Granville (bacterial) wilt, black root rot, black shank diseases of tobacco; Verticillium wilt of white potatoes and mint, pink root of onions, pod rot of peanuts]; *plant parasitic nematodes* [root-knot, meadow (lesion), citrus, cyst formers (golden, sugar beet, soybean), burrowing, lance, reniform, ring, spiral, sting, pin, stubby root, stylet, dagger and certain others]; *symphylans* (garden centipedes) and *wireworms.* Treat land to be planted to the crops listed below by applying TELONE C-17 under the conditions and at the rates recommended under DIRECTIONS FOR USE. Read this entire label before using TELONE C-17. **Note:** Do not apply TELONE C-17 through any type of irrigation system.

Vegetable Crops: asparagus, beans, beets, blackeyed peas, broccoli, brussels sprouts, cabbage, cantaloupe, carrots, cauliflower, celery, collards, corn, cowpeas, cucumbers, egg plant, endive, garlic, horseradish, kale, kohlrabi, leeks, lettuce, melons, mustard greens, okra, onions, parsnips, peas, peppers, pimentoes, potatoes, pumpkins, radishes, rutabaga, salsify, shallots, spinach, squash (summer), squash (winter), sweet potatoes, swiss chard, tomatoes, turnips, watermelons.

Field Crops: alfalfa, barley, birdsfoot trefoil, buckwheat, clover, corn, cotton, flax, grasses, hops, kenaf, lespedeza, millet, milo, mint, oats, peanuts, popcorn, rye, safflower, sorghum, soybeans, sugar beets, sugarcane, tobacco, vetch, wheat.

Citrus Fruit Tree Planting Sites: grapefruit, kumquats, lemons, limes, oranges, tangerines, tangelos.

Deciduous Fruit and Nut-Tree Planting Sites: almonds, apples, apricots, cashew nuts, cherries, chestnuts, dates, figs, filberts, hazelnuts, hickory nuts, nectarines, olives, peaches, pears, pecans, persimmons, pineapple, plums, pomegranates, prunes, quince, walnuts.

Bush and Vine Planting Sites: blackberries, boysenberries, cranberries, currants, dewberries, gooseberries, grapes, huckleberries, loganberries, raspberries, strawberries, youngberries.

Nursery Crops including floral plants, ornamentals, shrubs and bushes, forest shade, fruit and nut trees and vine and bramble fruits of all types.

APPLICATION DIRECTIONS

When to Treat: Apply TELONE C-17 whenever soil conditions are suitable but preferably in the late summer or fall or in the spring up to 2 or 3 weeks before planting the crop. Land to be planted to early spring crops, especially crops to be set out, such as celery, tomatoes, nursery and orchard stock, should be treated only in the late summer or early fall. For best results on annual crops use the treatment once each year.

Soil Preparation: At time of application the soil should be in good seed bed condition, free of clods and undecomposed plant material and with soil moisture at about one-half of field capacity. Soil temperature should be 40° to 80°F at the depth of injection. If undecomposed plant debris is present, it should be plowed down and allowed to decompose before applying TELONE C-17. Deep tillage, 12 to 18 inches, often improves results. **NOTE: Do not use TELONE C-17 in extremely heavy clay or muck soils.**

Application: TELONE C-17 may be applied as a **row** or as an **overall (broadcast) treatment** using a chisel-type applicator to inject the fumigant below the soil surface. **For row treatment** spring application is preferable. Fall treatment should be made only when land can be fumigated and bedded for early spring planting. Use a single chisel per row to treat the soil directly beneath the area where the crop will be planted. Inject the material at a depth of 10 to 12 inches below the soil surface. Immediately after application seal the soil by ridging or bedding the fumigated row area with enough soil to bring the soil surface 14 to 16 inches above the point of injection. **For overall (broadcast) treatment** apply whenever soil conditions are suitable. Space applicator chisels 12 inches apart and inject the fumigant 10 to 12 inches below the soil surface. Immediately after application seal the soil by compacting the surface with a roller, cultipacker or similar soil sealing device. When overall treatment is made in late summer or early fall, the field may be planted to a fall cover crop to be plowed under before planting the spring crop.

Rates to Use: Consult the following Dosage and Use Recommendations tables for dosage and application details to control nematodes, symphylans, wireworms and certain soil borne diseases in crops listed.

DOSAGE AND USE RECOMMENDATIONS TO CONTROL NEMATODES, SYMPHYLANS† AND WIREWORMS†

CROPS (CONSULT LIST OF INDIVIDUAL CROPS UNDER GENERAL INFORMATION)	TYPE OF TREATMENT	SOIL TYPE	DOSAGE	
			GALLONS PER ACRE[1]	FL. OZ./1000 FT. ROW PER CHISEL
Shallow Rooted plants: Field Crops, Floral Crops, Grasses and Turf, Small Fruits, Vegetables, Ornamentals	Row (42")[5]	Mineral	5.1 to 6.9	52.5 to 71
		Muck or peat	10.3 to 13.7	106 to 141
	Overall (Broadcast)	Mineral	10.3 to 17.1[2]	30.3 to 50.2
		Muck or Peat	27.4[3] to 41.1	80.5 to 120.8
Strawberries	Overall	Mineral	27.4 to 41.1	80.5 to 120.8
Sugar Beets Root Knot Nematode	Row (42")	Mineral	6	61.7
	Overall		12.8 to 17.1	37.6 to 50.2
Sugar Beet Nematode	Row (42")		10.3	106
	Overall		12.8 to 21.4	37.6 to 62.9
Pineapple	Row	Mineral	25.7 to 51.4	—
Citrus—Florida[4]	Overall	Mineral	34.2	100.5

Nursery and Field: Citrus Fruit Trees, Deciduous Fruit Trees, Forest Trees, Grapes, Nut Trees, Ornamentals (deep rooted)	Overall: gallons Per Acre to Penetrate Various Depths			
Mineral Soils	3 ft.	4 ft.	5 ft.	6 ft.
Sand	17.1	23.9	30.8	37.6
Sandy Loam	30.8	34.2	41.1	54.7
Silt Loam	47.9	58.1	71.8	85.5
Clay Loam	61.6	78.7	95.8	116.3

[1] Use the higher rates in heavier soil.
[2] For cyst-forming nematodes increase dosage to 21.4 gallons per acre (62.9 fl. oz./1000 ft. row per chisel).
[3] For muck soils containing less than 30% organic matter use 21.4 gallons per acre.
[4] For burrowing nematode in citrus inject on 18-inch centers, 12 inches deep. Keep free of plants susceptible to burrowing nematodes for 2 years before replanting to citrus.
[5] Row treatment is not recommended for potatoes in irrigated areas of western and northwestern states.
† **NOTE:** To control symphylans (garden centipedes) use only overall at 21.4 or more gallons per acre, and apply during late summer or early fall when the soil is warm. To control wireworms use dosages recommended for nematodes in overall or broadcast treatments.

DOSAGE AND USE RECOMMENDATIONS to Control Soil Borne Disease†

Crop	Type of Treatment	Dosage	
		Gallons Per Acre	Fl. Oz./1000 Ft. Row Per Chisel
Sweet Potatoes	Row (42") Overall	8.5 to 10.5 24 to 27.5	88[1] to 108[1] 70 to 81
Tobacco	Row (42") Overall	8.5 to 10.5 24 to 27.5	88[1] to 108[1] 70 to 81
Peanuts	Row (42")	2.5 to 5	26[1] to 52[1]
White Potatoes	Overall only	24 to 27.5	70 to 81
Mint	Overall only	24 to 27.5	70 to 81
Onions	Overall only	24 to 27.5	70 to 81

[1] Use the same flow rate per chisel for other single chisel row spacings. As the distance between rows increases, the gallons required per acre will decrease and vice versa.
† **NOTE:** Consult **General Information** to determine diseases controlled in crops listed. Nematodes also will be controlled.

Exposure Period: After application and sealing, leave the soil undisturbed for 7 to 14 days. Cold, wet soil retards diffusion of TELONE C-17, requiring a longer exposure period.

Aeration of Soil Before Planting: At the end of the exposure period allow the soil to aerate completely before planting the crop. Aeration is usually complete when the odor of TELONE C-17 is no longer evident. Under optimum soil and weather conditions, allow one week of aeration time for each 10 gallons of TELONE C-17 applied per acre. When TELONE C-17 is used for treating deep-rooted tree and shrub planting sites, a 3 to 6 months aeration period should be allowed. **To hasten aeration, especially if heavy rains or low temperatures occur during the exposure period, work the soil to the depth of the treatment zone.** After row treatment use a knife-like chisel in the bed without turning the soil, thus reducing possible recontamination of the treated soil. To hasten aeration after overall treatment, plow or deep cultivate to the depth of the treatment zone.

FIELD REENTRY: Workers entering treated areas within 72 hours after application of TELONE C-17 must wear protective clothing. Because certain States may require more restrictive reentry intervals for various crop areas treated with this product, consult your State Department of Agriculture for further information.

Written or oral warnings must be given to workers who are expected to be in a treated area or in an area about to be treated with this product. These oral warnings shall inform workers of areas or fields that may not be entered without protective clothing until after 3 days post treatment. In case of accidental exposure follow directions as shown by the Statement of Practical Treatment section on this label. When oral warnings are given, warnings shall be given in a language customarily understood by workers. Written warnings must include the following information: **DANGER.** Area treated with TELONE C-17 soil fungicide and nematicide on (insert date of application). Do not enter without protective clothing until 3 days post treatment. In case of accidental exposure take actions as shown by the Statement of Practical Treatment section on this label.

Attention: It is very important to avoid reinfestation of treated soil. Use only planting materials known to be free from disease and parasitic nematodes. It is good practice to cut sweet potato plants at or above the soil line and to use these "cut plants" for making field plantings. Do not use tools or crop residues that could carry soil borne pests from infested land to treated land.

Consult State Agricultural Experiment Station or Extension Service authorities for information on other practices such as post-harvest destruction of crop residues, weed control and cultural practices, and use of nematode and disease resistant crop varieties that also may aid in reducing crop losses caused by soil borne pests.

USE PRECAUTIONS

Important - Note Carefully: The treatment may temporarily raise the level of ammonia nitrogen and soluble salts in the soil. This is most likely to occur when heavy rates of fertilizer and TELONE C-17 are applied to soils that are cold, wet, acid, or high in organic matter. To avoid injury to plant roots, fertilize as indicated by soil tests made after treatment. To avoid ammonia injury or nitrate starvation, or both, to crops on high organic soils do not use fertilizers containing ammonium salts and use only fertilizers containing nitrates, until after the crop is well established and the soil temperature is above 65°F. In mineral soils do not apply more than $^2/_3$ of the nitrogen requirements from fertilizers containing ammonium salts until the crop is well established and soil temperature is above 65°F. Liming highly acid soils before treatment stimulates nitrification and reduces the possibility of ammonia and/or nitrite toxicity. Certain crops including cotton, sugarcane, and pineapple are tolerant to ammonia and the above rule does not apply to them. Citrus seedlings, *Cornus sp. Crataegus sp.*, spruce, vegetable crops such as cauliflower and certain other crops have shown evidence of phosphorus deficiency following fumigation. To avoid this possible effect, it is suggested that additional phosphate fertilizer be used on soils which tend to be deficient in this nutrient.

Since TELONE C-17 soil fungicide and nematicide is corrosive under certain conditions, flush all applicators with fuel oil or kerosene immediately after use. DO NOT USE WATER. Dispose of rinsate in field just treated or by other procedures approved by State and Local authorities. **Do not use containers, pumps, or other transfer equipment made of aluminum, magnesium or their alloys, as under certain conditions TELONE C-17 may be severely corrosive to such metals.**

WARRANTY LIMITATIONS AND DISCLAIMER

The Dow Chemical Company warrants that this product conforms to the chemical description on the label and is reasonably fit for the purposes stated on the label when used in strict accordance with the directions therein under normal conditions of use. THIS IS THE ONLY WARRANTY MADE ON THIS PRODUCT. NO OTHER EXPRESS AND NO IMPLIED WARRANTY OF MERCHANTABILITY OR FITNESS FOR A PARTICULAR PURPOSE IS MADE OUTSIDE OF THIS LABEL. Therefore, neither this warranty nor any other warranty of merchantability or fitness for a particular purpose, express or implied, extends to the use of this product contrary to label instructions (including conditions noted on the label, such as unfavorable temperatures, soil conditions, etc.), under abnormal conditions (such as excessive rainfall, drought, tornadoes, hurricanes, etc.) or under conditions not reasonably foreseeable to or beyond the control of seller.

When buyer or user suffers losses or damages resulting from the use or handling of this product (including claims based on contract, negligence, strict liability, or other legal theories), buyer or user must promptly notify in writing The Dow Chemical Company of any claims to be eligible to receive either remedy given below. The EXCLUSIVE REMEDY OF THE BUYER OR USER and the LIMIT OF LIABILITY of The Dow Chemical Company or any other seller will be one of the following, at the election of The Dow Chemical Company:

(1) **Refund of purchase price** paid by buyer or user for product bought, **or**

(2) **Replacement** of amount of product used.

The seller **will not be liable for consequential or incidental damages or losses.**

The terms of this Warranty Limitations And Disclaimer cannot be varied by any written or verbal statements or agreements. Any employee or sales agent of the seller is not authorized to vary or exceed the terms of this Warranty Limitations And Disclaimer in any manner.

85452-L6 J1087

* Trademark of THE DOW CHEMICAL COMPANY

TORDON* 22K

WEED KILLER

RESTRICTED USE PESTICIDE

For retail sale to and use only by Certified Applicators or persons under their direct supervision and only for those uses covered by the Certified Applicator's certification.

ACTIVE INGREDIENTS(S):	
Picloram (4-amino-3,5,6-tricholorpicolinic acid), as the potassium salt.	24.4%
INERT INGREDIENTS:	75.6%
TOTAL	100.0%

Acid Equivalent:
Picloram (4-amino-3,5,6-trichloropicolinic acid) - 21.1% - 2 lb/gal

E.P.A. Registration No. 464-323 E.P.A. Est. 464-MI-1

KEEP OUT OF REACH OF CHILDREN

CAUTION

AVISO:

PRECAUCIÓN AL USUARIO: Si usted no lee inglés, no use este producto hasta que la etiqueta le haya sido explicada ampliamente.

PRECAUTIONARY STATEMENTS

HAZARDS TO HUMANS AND DOMESTIC ANIMALS

MAY CAUSE IRRITATION

Avoid Contact With Eyes, Skin, Or Clothing • Avoid Breathing Spray Mist • Keep Container Closed • Wash Thoroughly After Handling

STATEMENTS OF PRACTICAL TREATMENT:

In case of contact, immediately flush eyes or skin with plenty of water. Get medical attention if irritation persists.

If swallowed, induce vomiting immediately by giving two glasses of water and sticking finger down throat. Call a physician. Do not induce vomiting or give anything by mouth to an unconscious person.

ENVIRONMENTAL HAZARDS

Do not apply directly to any body of water. Do not contaminate water used for irrigation or domestic purposes by cleaning of equipment or disposal of wastes. Do not allow run-off or spray to contaminate wells, irrigation ditches or any body of water used for irrigation or domestic purposes.

Do not make application when circumstances favor movement from treatment site. See additional precautionary statements elsewhere on this label.

NOTICE

Read the entire label. Use only according to label directions.

Before buying or using this product, read "WARRANTY LIMITATIONS AND DISCLAIMER" elsewhere on this label. If terms are not acceptable, return **unopened** package at once to seller for full refund of purchase price paid. Otherwise, use by the buyer or any other user constitutes acceptance of the terms under "WARRANTY LIMITATIONS AND DISCLAIMER."

IN CASE OF AN EMERGENCY endangering life or property involving this product, call collect 517-636-4400

AGRICULTURAL CHEMICAL

Do Not Ship or Store with Food, Feeds, Drugs, or Clothing

READ THE FEDERAL LABEL AND APPROPRIATE STATE LABELS BEFORE USING THIS PRODUCT

FEDERAL (E.P.A.) LABEL

GENERAL INFORMATION

Use TORDON 22K Weed Killer on non-cropland areas such as fence rows, around farm buildings, equipment pathways and roadsides, to control annual and deep rooted perennial weeds such as Absynth wormwood - Brackenfern - Buffalo bur - Bur ragweed - Burroweed - Cactus species - Camel thorn - Catclaw acacia - Chaparral species - Dock - Firebrush - Field bindweed (perennial morningglory) - Fringed sagebrush - Gambel oak - Goldenrod - Gorse - Guava - Hau - Horsenettles, Carolina, white - Java plum - Junipers - Knapweeds, diffuse, Russian, spotted squarrose - Lantana - Larkspurs - Leafy spurge - Live oak - Locoweeds - Lupines - Melastoma - Mesquite - Ox-eye daisy - Milkweed - Pamakani - Poison Oak - Pinyon - Pricklypear cactus - Rabbitbrush - Rush skeletonweed - Scotch broom - Snakeweeds - Sowthistle - Starthistles, lberian, purple, yellow - Tansy ragwort - Toadflaxes - Thistles, artichoke, Beaumont, Canada, distaff, golden, italian, musk, scotch, wavy leaf.

Picloram is highly potent, persistent and water soluble. Tiny amounts can kill or injure many broadleafed plants. To prevent damage to desirable crops and plants **follow all directions and precautions.**

DIRECTIONS FOR USE

Do not use for manufacturing or formulating.

It is a violation of Federal law to use this product in a manner inconsistent with its labeling.

Do not apply this product through any type of irrigation system.

Mix the required amount of TORDON 22K Weed Killer in water and apply as a coarse low pressure spray using ground equipment or aircraft. For best results treat when the weeds are growing actively in the spring before full bloom or late summer into fall. Treatments during full bloom or seed stage of some weeds may not give good control.

For General Use on Perennial Weeds on Non-Cropland, use 1 to 1½ gallons of TORDON 22K Weed Killer per acre in 50 to 100 gallons of water and spray to wet weed foliage and soil. **NOTE:** Local conditions may affect the use of herbicides. State agricultural experiment stations or extention service weed specialists in many states issue recommendations to fit local conditions. Be sure that use of this product conforms to all applicable regulations.

For Use As A Spot Treatment on Perennial Weeds. Mix at the rate of 1 gallon of TORDON 22K Weed Killer per 100 gallons of water. Apply at the rate of 100 gallons of spray mixture per acre. This will provide a rate of 2 pounds of picloram per acre. For small amounts use 2½ fluid ounces TORDON 22K per 2 gallons of water. For round patches apply as indicated in the table.

Feet across Round Patch to be treated (weed area plus 10 foot border)	Gallons of spray mixture to apply
25	1.0
50	4.5
75	10.0
100	18.0
235 or (1 acre)	100.0

Tank Mixture for Spot or Broadcast Treatment of Susceptible Annual and Perennial Broadleaf Weeds: TORDON 22K Weed Killer may also be tank mixed with 2,4-D products such as ESTERON* 99* Concentrate, FORMULA 40*, DMA* 4, or ESTERON 6E herbicides for use on areas having mixed species including those which respond well to 2,4-D. Read and follow all directions and use precautions on other product labels.

CONSERVATION RESERVE PROGRAM (CRP) FOR SEEDING TO PERMANENT GRASSES ONLY

Do not use TORDON 22K Weed Killer if legumes are a desired cover during CRP.

Grass grown on CRP acres cannot be grazed or used for hay.

Conditions that stress grasses, such as drought, will increase potential for injury to the grass at all stages of growth.

To reduce potential damage to subsequent small grain crops, use the lower rate or discontinue the use of TORDON 22K Weed Killer at least 2 years prior to the seeding of small grain crops. After CRP, do not plant broadleaf crops in treated acres until an adequately sensitive bioassay shows that no detectible picloram is present in the soil.

Broadcast Applications (Aerial and Ground): Applications of TORDON 22K Weed Killer should be made when perennial grasses have become established (has developed a good secondary root system and shows good vigor) since most perennial grasses have shown better tolerance to the herbicide at that stage.

For control of actively growing perennial weeds, use up to 1 quart per acre of TORDON 22K Weed Killer after the grass is established. For best results, use in 2 or more gallons of water per acre by air or in 5 or more gallons of water per acre by ground. Increasing the rate of application can increase the risk of injury.

For control of actively growing susceptible annual broadleaf weeds, (including Russian thistle) apply ¼ to ½ pint per acre of TORDON 22K Weed Killer. TORDON 22K can also be tank-mixed with ½ to 1 lb/acre 2,4-D where species present are sensitive to 2,4-D. Read and follow all directions for use and use precautions on other product labels.

Spot Application: For spot applications when perennial grasses are established, use 1 pint to 4 quarts per acre of TORDON 22K Weed Killer. Rates of 2 quarts per acre or more should only be used for control of deep-rooted perennial broadleaf weeds.

TORDON 22K at rates over 1 quart per acre may suppress certain established grasses such as bromegrass, bluegrama and buffalograss. However, subsequent grass growth should be improved by release from weed competition.

BE SURE YOU FOLLOW ALL USE PRECAUTIONS GIVEN ON THIS LABEL AND REMEMBER THESE KEY POINTS

1. Use only the recommended amounts.
2. Picloram is persistent. It will carry over in the soil.
3. TORDON 22K is water soluble. It can move with water in irrigation or drainage ditches.
4. Spray drift can damage crops.

USE PRECAUTIONS

Use this product only as specified on this label. Observe any special use and application restrictions and limitations, including method of application and permissible areas of use as promulgated by state or local authorities.

Do Not Make Application When Circumstances Favor Movement From Treatment Site.

Do Not Apply or Otherwise Permit TORDON 22K or Sprays Containing TORDON 22K to Contact Crops or Other Desirable Broadleaf Plants including but not limited to alfalfa, beans, grapes, melons, peas, potatoes, safflower, soybeans, sugarbeets, sunflower, tomatoes, and other vegetable crops, flowers, fruit plants, ornamentals, shade trees nor the soil containing roots of nearby valuable plants.

Avoid Injurious Spray Drift: Applications should be made only when there is no hazard from spray drift since very small quantities of the spray, which may not be visible, may severely injure susceptible crops during both growing and dormant periods. To minimize spray drift use low spray pressure, under 30 psi; spray when wind velocity is less than 10 mph; and apply as a coarse spray. To aid in further reducing spray drift, a drift control and deposition aid such as NALCO-TROL may be used with this product. If such a drift control aid is used, follow all use recommendations and precautions on the product label.

Ground Equipment: With ground equipment spray drift can be lessened by keeping the spray boom as low as possible; by applying 20 gallons or more of spray per acre; by using no more than 30 pounds spraying pressure with large droplet producing nozzle tips; by spraying when wind velocity is low; and by stopping all spraying when wind exceeds 10 miles per hour. Do not apply with hollow cone-type insecticide or other nozzles that produce fine-droplet spray.

Aerial Application: With aircraft, drift can be lessened by using straight stream nozzles directed straight back, and by using a spray boom no longer than ¾ the wing span of the aircraft.

Determine Air Movement and Direction Before Foliar Application: Do not spray when wind is blowing toward susceptible crops or ornamental plants near enough to be injured. It is suggested that a continuous smoke column at or near the spray site or a smoke generator on the spray equipment be used to detect air movements, lapse conditions, or temperature inversions (stable air). If the smoke layers or indicates a potential of hazardous spray drift, do not spray.

Do Not Contaminate Water Intended for Irrigation or Domestic Purposes. To avoid injury to crops or other desirable plants, do not treat or allow spray drift or run-off to fall onto banks or bottoms of irrigation ditches, either dry or containing water, or other channels that carry water that may be used for irrigation or domestic purposes. Do not apply to snow or frozen ground.

Water contaminated with residues of picloram from drainage ditch bank uses shall not be used to irrigate crops which are not registered for use with this chemical.

Do not rotate food or feed crops on treated land if they are not registered for use with picloram until an adequately sensitive bioassay or chemical test shows that no detectable picloram is present in the soil.

Do not move treated soil to other areas or use it to grow plants if they are not registered for use with picloram until an adequate sensitive bioassay or chemical test shows that no detectable picloram is present in the soil.

Do Not Mix with Dry Fertilizer.

PRECAUTIONARY STATEMENTS

ENVIRONMENTAL HAZARDS

Picloram is a chemical which can travel (seep or leach) through soil and under certain conditions contaminate groundwater which may be used for irrigation and drinking purposes. Users are advised not to apply picloram where soils have a rapid to very rapid permeability throughout the profile (such as loamy sand to sand) and the water table of an underlying aquifer is shallow or to soils containing sinkholes over limestone bedrock, severely fractured surfaces, and substrates which would allow direct introduction into an aquifer. Your local agricultural agencies can provide further information on the type of soil in your area and the location of groundwater.

An aquifer is defined as "an underground, saturated, permeable, geologic formation capable of producing significant quantities of water to a well or spring. It is the ability of the saturated zone, or portion of that zone, to yield water which makes it an aquifer" (American Chemical Society, 1983).

STORAGE AND DISPOSAL

Prohibitions: Do not contaminate water, food, or feed by storage or disposal. Open dumping is prohibited.

Pesticide Disposal: Wastes resulting from the use of this product may be disposed of on site or at an approved waste disposal facility.

Container Disposal: METAL Do not reuse container. Triple rinse (or equivalent). Puncture and dispose of in a sanitary landfill, or by other procedures approved by state and local authorities.

Container Disposal: PLASTIC Do not reuse container. Triple rinse (or equivalent). Puncture and dispose of in a sanitary landfill, or by incineration, or, if allowed by state and local authorities, by burning. If burned, stay out of smoke.

Rinse application equipment after use, at least three times with water, and dispose of rinse water in a non-cropland area away from water supplies.

General: Consult federal, state or local disposal authorities for approved alternative procedures.

SPECIAL LOCAL NEEDS

COLORADO . . . . EPA SLN NO. CO–850002

For The Control of Broadleaf Weeds in Rangelands (Private, Public and Forest) And Permanent Grass Pastures.

DIRECTIONS FOR USE

It is a violation of Federal law to use this product in a manner inconsistent with its labeling.

Do not apply this product through any type of irrigation system.

Read complete use directions and precautions on this label and container label before using.

RANGELANDS (PRIVATE, PUBLIC AND FOREST) AND PERMANENT GRASS PASTURES

Use TORDON 22K Weed Killer to control broadleaf annual and perennial weeds such as (but not limited to) Canada thistle and other thistles, field bindweed, leafy spurge, Russian knapweed, spotted and diffuse knapweed, plains or Geyer larkspur, Dalmation and yellow toadflax, fringed sagebrush, snakeweed, rabbit brush, pricklypear cactus, skeletonleaf bursage, locoweed, and lupines on rangeland and permanent grass pastures. Treat when weeds are actively growing. Retreatment may be necessary the following year.

Spot treatment: Use TORDON 22K rates as shown in the following table and apply in sufficient water to obtain adequate spray coverage. For perennial weeds, spray the weed infested area plus a 10-foot border around the infestation. Consult the product container label for guidance to prepare spray mixes to spot treat small patches.

Broadcast treatment: Apply as a broadcast spray on hard to control weeds at rates shown in the following table. Apply the selected rate of TORDON 22K in 2-4 gallons of total spray volume per acre (water diluent only) by air, or in 5 to 30 gallons of water per acre by ground equipment. For rates exceeding 1 quart of TORDON 22K per acre, the total area treated in a single season should not exceed 25% of a landowner's acreage found in any particular watershed.

Tank mixtures for spot or broadcast treatments: TORDON 22K may also be tank mixed with 2,4-D products for use on areas having mixed species including those which respond well to 2,4-D. In tank mix combinations, use 1 pint to 4 quarts of TORDON 22K with 1 to 2 quarts of low volatile ester or amine products containing 3.8 lb. 2,4-D acid equivalent per gallon or with 2/3 to 1 1/3 quarts of comparable products containing 5.6 lb. 2,4-D acid equivalent per gallon in spray volumes specified below. READ AND FOLLOW ALL DIRECTIONS AND USE PRECAUTIONS ON OTHER PRODUCT LABELS.

Note: TORDON 22K at rates over 1 quart may suppress certain grasses, such as bromegrass, bluegramma and buffalograss. Usually later grass growth will be improved by release from competition.

Grazing restrictions: When applying more than 1 quart of TORDON 22K per acre, do not cut grass for feed within two weeks after treatment. Meat animals grazing for up to two weeks after treatment should be removed from treated areas three days prior to slaughter. Do not graze lactating dairy animals on treated areas within two weeks after treatment.

Weed Species[4]	Rates of TORDON 22K Per Sprayed Area[3]
Diffuse Knapweed Spotted Knapweed Rabbitbrush Fringed Sagebrush Many seedling annual weeds	1 pint
Musk, Plumeless & Scotch Thistles Tall or Geyer Larkspur[1,2] Locoweed[1,2] Lupines[1] Snakeweed Pricklypear Cactus	1 to 2 pints
Russian Knapweed Skeletonleaf Bursage Canada Thistle Dalmation Toadflax Yellow Toadflax Field Bindweed	2 to 4 quarts
Leafy Spurge[1]	1 quart —Repeat annually for 3–4 years 2 quarts —Treat 2 consecutive years 4 quarts —Single application

[1]Denotes poisonous plants
[2] Use higher rate
[3] NOTE: Follow all above treatments with spot touch-up treatment as needed using TORDON 22K herbicide.
[4] For additional species or more specific rates consult the current CO Weed Control Guide.

USE PRECAUTIONS

Use this product only as specified on this label. Observe any special use and application restrictions and limitations, including method of application and permissible areas of use as promulgated by state or local authorities.

Avoid water contamination. Do not allow TORDON 22K Weed Killer to contaminate water used for drinking, irrigation or other domestic purposes. Do not apply on inner banks or bottom of irrigation ditches. Do not clean containers or application equipment on or near these areas.

Do not apply on or near susceptible crops or desirable plants including alfalfa, beans, grapes, melons, peas, potatoes, safflower, soybeans, sugar beets, sunflower, tomatoes, and other vegetable crops, flowers, fruit plants, ornamentals or shade trees.

Avoid Spray Drift. Applications should be made only when there is no hazard from spray drift since very small quantities of the spray, which may not be visible, may severely injure susceptible crops, such as those listed above, during both growing and dormant periods. Use coarse sprays to minimize drift since, under adverse weather conditions, fine spray droplets may drift a mile or more. A spray thickening agent, such as NALCO-TROL, may be used with this product to aid in reducing spray drift. If used, follow all use recommendations and precautions on the product label.

Ground Equipment: With ground equipment, spray drift can be lessened by keeping the spray boom to a height recommended for the nozzle tips in use and by spraying only when wind velocity is less than 10 miles per hour. Do not apply with hollow cone-type insecticide or other nozzles that produce a fine-droplet spray.

Aerial application: Drift can be lessened by applying a coarse spray, by using no more than 40 pounds spray pressure at nozzles; by using nozzles which do not produce hazardous amounts of fine spray droplets; by using a spray boom no longer than 3/4 of the rotor or wing length; and by spraying only when wind velocity is less than 10 mph.

Determine air movement and direction before foliar application: Use a smoke generator or other means at or near the application site or on aircraft for detection of air movement, air stability, or temperature inversions. Such a condition exists when there is little or no wind and air temperature is lower near the ground than at higher levels. Use appropriate drift control measures or stop application when smoke is moving toward nearby desirable susceptible plants or sensitive areas.

Do not rotate treated rangeland or pasture to other crop uses.

Do not spray pastures if the forage legume component is desired. TORDON 22K Weed Killer may injure or kill legumes. Also new legume seedlings may not be successful if made within 2 years following application of this herbicide.

Avoid movement of treated soil. Picloram may remain in treated soil for an extended period. Do not move treated soil to other areas and do not use such soil to grow plants until residues have reached a non-phytotoxic level.

Do not transfer livestock from treated grass areas onto broadleaf crop areas without first allowing 7 days of grazing on untreated grass pasture. Otherwise, urine may contain enough picloram to cause injury to sensitive broadleaf plants.

Rinse application equipment and container thoroughly, preferably at least three times with water, and dispose of rinse water in a non-cropland area away from water supplies.

Do not reuse containers for TORDON 22K Weed Killer for any purpose. Containers should be disposed of by punching holes in them and burying (in a non-crop area away from water) or by taking to an approved landfill. Where indicated, follow local container disposal regulations.

Be sure that use of this product conforms to all applicable regulations.

COLORADO . . . . EPA SLN NO. CO-880003

For Control of Annual and Perennial Broadleaf Weeds Such as Field Bindweed in Grainland Between Crops of Small Grains and in Non-Cropland Areas

GENERAL INFORMATION

Use TORDON 22K Weed Killer tank mixed with 2,4-D herbicides by ground application to control or reduce infestations of broadleaf annual and perennial weeds such as field bindweed on non-cropland and on grainland during the period following crop harvest and prior to planting winter wheat, barley, or oats. Picloram, the active ingredient in TORDON 22K Weed Killer, is highly active, water soluble and has moderate soil residual. To prevent damage to desirable crops and plants, follow all directions and precautions on this label.

DIRECTIONS FOR USE

It is a violation of Federal law to use this product in a manner inconsistent with its labeling.

Do not apply this product through any type of irrigation system.

Read complete use directions and precautions on this label and container label before using.

For reduction of perennial weeds such as field bindweed, use ground application equipment to broadcast treat infested areas using 0.5 to 1 pint TORDON 22K Weed Killer in 10 or more gallons of water per acre. The addition of 2,4-D herbicides at 1/2 to 2 lb ae/acre may significantly improve weed control. Read and follow all directions and use precautions on other product labels.

Between crops of small grains, treat annually to maintain control of perennial weed infestation. For best results with continuous small grain production, treat field bindweed infested areas in June or July following small grain harvest, when bindweed is actively growing, applying the 0.5 pint rate of TORDON 22K Weed Killer and 1/2 to 2 lb 2,4-D ae/acre (higher rate of TORDON 22K may be used but greater crop injury to fall planted grain may result compared to that which may occur with the lower rate).

For best results with small grain-fallow production, treat field bindweed infested areas in June or July following grain harvest, applying the 1 pint rate of TORDON 22K with 2,4-D; then around mid-May during the fallow season, treat the field bindweed infested area with the 0.5 pint rate of TORDON 22K and 1/2 to 2 lb 2,4-D ae/acre **(higher rate of TORDON 22K may be used but greater crop injury to fall planted grain may result compared to what may occur from the lower rate).**

Do not treat with TORDON 22K Weed Killer more than once each calendar year. Additional treatment with 2,4-D alone may be made any time weeds are present with adequate foliage and in good growing condition.

For long residual control of perennial weeds such as field bindweed, spot treat scattered infestations following use directions on the container label. When using these higher rate treatments of TORDON 22K, do not treat more than 10% of the area of an infested field in one calendar year. Wheat or other small grain crops may be planted in such treated fields, but crop injury in the spot treated areas may occur for one or more seasons after treatment.

To control volunteer grain or grass weeds that may be present, use cultivation or appropriate grass control herbicides alone or in combination with broadleaf weed control treatments. Consult the University Experiment Station or Extension Service for recommendations to fit local conditions.

Wheat or other small grain crops may be injured when planted at varying periods following application of TORDON 22K Weed Killer. Do not treat with TORDON 22K if the risk of injury to the small grain crop cannot be tolerated.

Roadsides and other non-cropland areas may be treated during bud to early bloom stage of field bindweed in the spring and again in the fall as needed. For low dose repeat treatments, use the 1 pint rate of TORDON 22K and 1/2 to 2 lb 2,4-D ae/acre in the fall. A follow-up spot application is suggested to maintain control. Yearly or twice yearly treatments over a 3 to 4 year period may be needed as part of an overall control program. For higher rate residual control of these and other perennial weeds, follow instruction on container label.

USE PRECAUTIONS

Use this product only as specified on this label. Observe any special use and application restrictions and limitations, including method of application and permissible areas of use as promulgated by state and local authorities.

Handle and Use Correctly. Small amounts of TORDON 22K Weed Killer can injure sensitive broadleaf plants. To avoid problems, observe the following precautions:

1. **Do Not Contaminate Water Intended for Irrigation or Domestic Purposes.** Do not treat or allow spray drift to fall onto innerbanks or bottom of irrigation ditches, either dry or containing water.
2. **Do not Apply to Sensitive Crops or Desirable Plants** including alfalfa, beans, cotton, grapes, melons, peanuts, peas, potatoes, sunflower, sugarbeets, soybeans, tobacco, tomatoes, other field or vegetable crops, flowers, fruit plants, ornamentals, or shade trees.
3. **Use a bioassay to determine that picloram residues are no longer detectable in the treated soil** before using treated areas to grow sensitive crops. Contact State Extension Specialist for further information on bioassay testing.
4. **Do not Move Treated Soil to Other Areas** where susceptible desirable plants may be exposed and damaged while phytotoxic residues are present.
5. **Avoid Spray Drift.** Applications should be made only when there is no hazard from spray drift since very small quantities of the spray, which may not be visible, may severely injure susceptible crops during both growing and dormant periods. To avoid spray drift, apply only when wind velocity is less than 10 mph and is not blowing toward susceptible crops; keep spray boom as low as possible; apply 20 gallons or more of spray per acre; use no more than 30 psi spraying pressure with large droplet producing tips.

   Do Not Apply With Nozzles That Produce Fine-Droplet Spray. A spray thickening agent such as NALCO-TROL may be used with this product to aid in reducing spray drift. If a drift reducing agent is used, follow use directions and precautions on the product label.
6. **Do Not Use Where a Porous Sandy Surface and Substrate Overlie Ground Water Ten Feet or Less Below the Surface.**
7. **Be Sure That Use of This Product Conforms to All Applicable Regulations.**
8. **Grazing Restrictions: None.**
9. **Follow Storage and Disposal Requirements on Container Label.**

HAWAII. . . . . EPA NO. HI-870010

For Use in the State of Hawaii to Control Certain Troublesome Woody Plants in Rangelands and Permanent Grass Pastures and Non-Cropland Areas.

RESTRICTED TO TORDON 22K PERMIT HOLDERS

USE DIRECTIONS

Use TORDON 22K Weed Killer to control troublesome woody plants such as lantana, melastoma, guava, Java plum, pamakani, cat's claw, gorse, firebush and hau as well as many other woody plants, broadleaf weeds and vines infesting rangelands and permanent grass pastures and non-cropland areas in the state of **Hawaii.** Treat anytime during the year when the plants are actively growing. Do not apply as a broadcast spray.

Do not apply this product through any type of irrigation system.

Individual Plant Foliar Treatment: Mix TORDON 22K at the rate of 2 quarts per 100 gallons of water. Add 2 quarts of surfactant to the spray mixture. Apply as a low pressure (10 to 30 psi) spray to thoroughly wet the leaves, stems and trunks of the brush. For smaller amounts of spray mixture, mix 2 1/2 fluid ounces of TORDON 22K and 2 1/2 fluid ounces of surfactant into 4 gallons of water and apply as above.

Re-treat areas in subsequent years as needed.

Do not graze dairy animals on treated areas within two weeks after application. Do not slaughter meat animals grazing on treated areas until two weeks have elapsed after application.

Cut Surface Treatments: For the control of firebush, hau, Java plum and guava with trunks of 3 inches in diameter or larger. Use TORDON 22K Weed Killer diluted to 1 to 4 in water as directed below.

Tree Injector: Application should be made by injecting 1 milliliter of the diluted solution through the bark at intervals of 3 inches between edges of the injector wound. The injections should completely surround the tree at any convenient height.

Frill or Girdle: Make a single girdle through the bark completely around the tree at a convenient height. Wet the cut surface with the diluted solution.

USE PRECAUTIONS

Do not allow spray drift. TORDON herbicides are highly active. Very small amounts may cause damage to plants if applied during either growing or dormant periods. Do not apply or otherwise permit TORDON 22K or sprays containing it to contact desirable plants such as pineapple, papaya, macadamia, coffee, vegetables, flowers, grapes, fruit trees, ornamentals, tomatoes, potatoes, beans of all types and other valuable broadleaf plants, nor the soil containing roots of nearby valuable plants. Apply TORDON 22K only when there is little or no wind or no hazard from spray drift.

Do not contaminate water. To avoid injury to crops or other desirable plants, do not contaminate irrigation ditches or water intended for irrigation or domestic purposes. Do not treat or allow spray drift to fall onto inner banks or bottom of irrigation ditches, either dry or containing water, or other channels that carry water that may be used for irrigation purposes.

Do not transfer livestock directly from treated areas onto broadleaf crop areas without allowing 7 days on untreated grass pastures. Urine may contain enough picloram to cause crop injury.

Do not use manure from animals grazing treated areas on land used for growing broadleaf crops, ornamentals, orchards or other valuable plants. Manure may contain enough picloram to cause crop injury.

Do not mix with other pesiticides except other herbicides in accordance with University of Hawaii suggestions.

Other precautions. Do not store near food, feedstuff, fertilizer, seeds, insecticides, fungicides or other pesticides.

Empty containers. Rinse with water, pour in spray tank. Repeat twice. Dispose of containers by punching holes in them and burying in non-cropland away from water supplies or in approved sanitary land fills.

Cleaning spray equipment. Flush spray tank with water and drain completely. Follow by thorough rinsing of tank, hoses and nozzles with soap and water. Dispose of all wastes by burying in non-cropland away from water supplies.

BE SURE THAT ALL USE OF TORDON 22K WEED KILLER CONFORMS TO LOCAL REGULATIONS

IDAHO . . . . **EPA SLN NO. ID-780009**
NEVADA . . . . **EPA SLN NO. NV-790007**
OREGON . . . . **EPA SLN NO. OR-780012**
UTAH . . . . **EPA SLN NO. UT-820011**
WASHINGTON . . . . **EPA SLN NO. WA-820068**

For Control of Yellow Starthistle, Scotch, Musk, and Canada Thistles, Spotted, Diffuse and Russian Knapweeds, Rush Skeletonweed, Larkspur, Leafy Spurge, Rabbitbrush spp., Tansy Ragwort, Field Bindweed, Poison Oak, Gorse, Dalmation Toadflax, Buffalo Bur, Henbane and other Susceptible Poisonous Plants, Broadleaf Weeds and Woody Plants on Rangeland and Permanent Grass Pastures.

DIRECTIONS FOR USE

It is a violation of Federal law to use this product in a manner inconsistent with its labeling.

Do not apply this product through any type of irrigation system.

Read complete use directions and precautions on this label and container label before using.

Mix the required amount of TORDON 22K Weed Killer in water and apply as a coarse, low pressure spray using ground equipment or aircraft.

For best results, treat when the weeds are growing actively in the spring before full bloom or in late summer or fall. Treatments during full bloom or seed stage of some weeds may not give good control.

BROADCAST TREATMENT: TORDON 22K Weed Killer can be applied as a broadcast treatment at rates of 1 quart or less per acre; higher rates can be used on patches but the total areas treated should be regulated so no more than 25 gals. of TORDON 22K Weed Killer is used on any 100 acre block in a single season. Suggested rates to control several broadleaf weeds are shown in the table below. Weeds requiring higher rates may be suppressed with rates of 1 quart per acre. Apply as a single broadcast spray during any one growing season. Retreat in subsequent years as needed using similar rates. TORDON 22K Weed Killer may be tank mixed with ESTERON* 99* Concentrate, FORMULA 40*, DMA* 4 or ESTERON 6E herbicides (2,4-D products) for use on areas having mixed species including those which respond well to 2,4-D, such as big sagebrush. In tank mix combinations, use 1/4 to 1 quart of TORDON 22K with 1 to 2 quarts ESTERON 99 Concentrate, FORMULA 40 or DMA 4 or with 2/3 to 1 1/3 quarts ESTERON 6E per acre.

Weed Species	Rates of TORDON 22K per Treated Acre
Yellow Starthistle†, Scotch Thistle, Musk Thistle, Ox-eye Daisy	1/4 to 1/2 quart
Diffuse Knapweed, Spotted Knapweed, Henbane, Buffalo Bur, Rabbitbrush spp., Tansy Ragwort	1/2 to 1 quart
Canada Thistle, Rush Skeletonweed Russian Knapweed, Dalamation Toadflax, Gorse	2-3 quarts‡
Leafy Spurge†, Larkspur†, Field Bindweed, Poison Oak†	4 quarts‡

†Denotes poisonous plants
‡Lower rates may be used for short-term suppression

GROUND APPLICATION: When applying TORDON 22K Weed Killer with ground equipment, use coarse low pressure spray (under 30 psi) and apply uniformly to provide good coverage of the weeds. Apply only when weather conditions are favorable for keeping spray on the target area. Do not allow spray drift to contact off-target susceptible plants, or areas to be planted to susceptible crops. Do not apply when wind velocity exceeds 10 mph. Where desirable susceptible plants, such as potatoes, beans, peas and other vegetable crops, ornamentals or legumes **are growing or may be planted within 1/2 mile,** apply TORDON 22K **only if air movement is continuously from a definite direction and away from these plants.**

AIR APPLICATION: When applying by aircraft, pilot must comply with all applicable state and local regulations. When applying TORDON 22K Weed Killer by air, use coarse, low pressure spray (under 25 psi) and apply uniformly to provide good coverage of the weeds. Apply only when weather conditions are favorable for keeping spray on target area. Do not allow spray drift to contact off-target susceptible plants. Do not apply when wind velocity exceeds 10 mph, or as required by state regulations. The distance between outermost nozzles should not exceed 3/4 of the wing or rotor length. Do not spray when air temperature exceeds 85°F. Where desirable susceptible plants, such as potatoes, beans, peas and other vegetable crops, ornamentals or legumes are growing or may be planted within 1 mile, apply TORDON 22K Weed Killer by helicopter only if air movement is continuously from a definite direction and away from these plants at the time of application. Fixed wing aircraft may be used only in areas east of the crest of the Cascade Mountains under the above conditions provided that fields or homesites where susceptible plants may be growing or be planted are over 3 miles **downwind** at the time of application.

SPOT TREATMENT: For spot treatment of small patches of broadleaf weeds, use 1/2 to 4 quarts of TORDON 22K in 100 gallons of water and spray weed foliage uniformly using 50 to 100 gallons of spray per treated acre.

NOTE: For treating small areas, 1 quart TORDON 22K Weed Killer in 100 gallons of water per acre is equivalent to 2 teaspoonsful per gallon of water applied to a 500 square foot area.

GRAZING RESTRICTIONS: (Where rates greater than 1 quart per acre are applied). Do not graze dairy animals on treated areas within 2 weeks after application. Other animals should be withdrawn from treated areas at least 3 days before slaughter. Observe grazing restrictions on other product labels when using tank mixtures.

USE PRECAUTIONS

Use this product only as specified on this label. Observe any special use and application restrictions and limitations, including method of application and permissible areas of use as promulgated by state or local authorities.

Do Not Contaminate Nontarget Land Areas, Cropland, Water, or Irrigation Ditches. Do not apply directly to standing or running water. Do not apply where surface water from treated areas is likely to run off to adjacent cropland, either planted or to be planted, or into streams, irrigation ditches, irrigation ponds, or wells. Do not clean containers nor application equipment on or near these areas. Do not apply on inner banks or bottom of irrigation ditches. Not apply to frozen ground.

Do not apply on or in the vicinity of susceptible crops or desirable plants including alfalfa, beans, clovers, grapes, melons, peas, potatoes, safflower, soybeans, sugar beets, sunflower, tomatoes and other vegetable crops, flowers, fruit plants, ornamentals or shade trees.

Avoid Spray Drift: Applications should be made only when there is no hazard from spray drift since very small quantities of the spray, which may not be visible, may severely injure susceptible crops during both growing and dormant periods. Use coarse sprays to minimize drift since, under adverse weather conditions, fine spray droplets may drift a mile or more. The spray thickening agent, NALCO-TROL, may be used with this product to aid in reducing spray drift. If used follow all use recommendations and precautions on the product label.

Ground Equipment: With ground equipment, spray drift can be lessened by keeping the spray boom as low as possible; by applying 20 gallons or more of spray per acre; by using no more than 30 pounds spraying pressure with large droplet-producing nozzle tips; by spraying when wind velocity is low; and by stopping all spraying when wind exceeds 10 miles per hour. Do not apply with hollow cone-type insecticide or other nozzles that produce a fine-droplet spray.

Aerial Application: Drift can be lessened by applying a coarse spray; by using no more than 25 pounds spray pressure at nozzles; by using nozzles which do not produce hazardous amounts of fine spray droplets; by using a spray boom no longer than 3/4 of the rotor or wing length; and by spraying only when wind velocity is less than 10 mph, or as required by state regulations.

Determine Air Movement and Direction Before Foliar Application: Use a smoke generator or other means at or near the application site or on aircraft for detection of air movement, air stability, or temperature inversions. Such a condition exists when there is little or no wind and air temperature is lower near the ground than at higher levels. Use appropriate drift control measures or avoid application when smoke is moving toward nearby desirable susceptible plants or sensitive areas.

Do not rotate treated rangeland or pasture to other crop uses.

Do not spray pastures if the forage legume component is desired. TORDON 22K Weed Killer may injure or kill legumes. Also, new legume seedings may not be successful if made within 2 years following application of this herbicide.

Do not move treated soil to other areas. Do not use it to grow plants, unless adquate sensitive bioassay or chemical tests show that no detectable picloram is present in the soil.

Do not transfer livestock from treated grass areas onto broadleaf crop areas without first allowing 7 days of grazing on untreated grass pasture. Otherwise, urine may contain enough picloram to cause injury to sensitive broadleaf plants.

Do not re-use containers for TORDON 22K Weed Killer for any purpose. Dispose by punching holes in them and burying with waste or by taking to an approved landfill. Where indicated, follow official local container disposal regulations.

Rinse application equipment after use, preferably at least three times with water, and dispose of rinse water in a non-cropland area away from water supplies.

Be sure that use of this product conforms to all applicable regulations.

READ AND FOLLOW MIXING AND USE INSTRUCTIONS AND PRECAUTIONS ON PRODUCT CONTAINER LABEL

IDAHO . . . . EPA SLN NO. ID-820030
OREGON. EPA SLN NO. OR-820061
WASHINGTON. . . . . EPA SLN NO. WA-840067

For Control of Susceptible and Annual Perennial Broadleaf Weeds in Non-Irrigated Grainland Not Rotated to Broadleaf Crops East of the Cascade Mountains

DIRECTIONS FOR USE

It is a violation of Federal law to use this product in a manner inconsistent with its labeling.

Do not apply this product through any type of irrigation system.

Read complete use directions and precautions on this label and container label before using.

To prepare the spray, add about half the desired amount of water to the spray tank. Then, with agitation, add the recommended amount of TORDON 22K and 2,4-D, MCPA or other approved herbicides or adjuvants. Finally, with continued agitation, add the rest of the water. **Addition of surfactants may aid control under dry conditions, but may cause injury to grain if used over the crop.**

ANNUAL BROADLEAF WEED CONTROL IN SPRING AND WINTER WHEAT, OATS AND BARLEY[1]

To control susceptible annual broadleaf weeds such as wild buckwheat, dogfennel, sunflower, wild mustard, pennycress, lambsquarters, pigweed and Russian thistle (Canada thistle may be suppressed), in wheat or barley use TORDON 22K Weed Killer as a tank mix with 2,4-D or MCPA or other registered herbicides. For use on oats, mix with MCPA or other registered herbicides. Read and follow directions and use precautions on other product labels.

Underseeded legumes may be injured seriously. **Use only on fields in continuous grain.**

Winter Wheat: Apply after resumption of active growth in the spring **until the early jointing stage.**

Spring Wheat, Barley and Oats: Apply from the **3-leaf stage to the early jointing stage** of growth, **as indicated in the chart below. Application at the 3 to 5-leaf stage** occasionally causes slight head malformations and straw shortening but normally **yields are not reduced.**

Durum Wheat: Do not treat durum since some varieites of durum wheat may be injured.

NOTE: Do not apply TORDON 22K Weed Killer within 50 days before harvest. Do not graze or feed forage from treated areas for 2 weeks after application.

NOTE: Spray mixtures may cause shorter straw on some varieties of cereals but grain yields are usually not affected.

USE RATES FOR WHEAT, BARLEY AND OATS[1]

WEEDS	WEEDS GROWTH STAGE[2]	GRAIN STAGE	AMOUNTS OF EACH PRODUCT PER ACRE 2,4-D OR MCPA		
			TORDON 22K	4 LB AE/GAL. OR 6 LB AE/GAL.	
More susceptible species, such as lambsquarters, pennycress, wild mustard & mayweed	1-3 inches	3-5 leaf	1 fl. oz.	1/2 pt.	1/3 pt.
	3-6 inches	To early jointing	1 1/2 fl. oz.	3/4 pt.	1/2 pt.
Less susceptible species, such as volunteer sunflower, wild buckwheat Russian thistle, pigweed and top growth suppression of Canada thistle	1-6 inches	To early jointing	1 1/2 fl. oz.	3/4 pt.	1/2 pt.

[1]For Oats, tankmix only with MCPA amine or MCP ester herbicides or other registered herbicides

[2]For best results, treat when weeds have 2 to 4 leaves and are actively growing.

PERENNIAL BROADLEAF WEED CONTROL IN FALLOW GRAINLAND

NOTE: TORDON 22K Weed Killer effectively controls perennial broadleaf weeds, however, use of TORDON 22K either as a spot or broadcast treatment may result in serious crop injury and/or complete yield loss in treated areas for one or more years. Growers should carefully evaluate the benefit of perennial weed control versus the risk of crop failure before using Tordon 22K on fallow grainland. **Treat only infested areas during fallow periods where substantial or complete yield loss can be tolerated for one or more years.**

To supress perennial broadleaf weeds such as field bindweed, Canada thistle, Rush skeletonweed and Russian knapweed in fallow grainland, apply TORDON 22K Weed Killer at 1/4 to 1 pint plus 1 to 2 lbs of 2,4-D per acre after crop harvest when weeds are actively growing. The potential for substantial or complete yield loss may be reduced by not planting treated areas for 12 months after application.

TORDON 22K at 1 pint or less per acre alone or plus 2,4-D or other registered herbicides may be applied by air in 3 to 10 gallons of spray under conditions that will prevent injury to any nearby susceptible crops or ornamentals.

Refer to Use Precautions for suggestions to reduce off-target drift hazard.

DO NOT APPLY MORE THAN 1 PINT OF TORDON 22K WEED KILLER PER ACRE BY AIR.

Spot Treatments of TORDON 22K at rates from 1 to 8 pints per acre can be made if the treated areas comprise less than 10% of the immediate field in any one year. Do not exceed more than 1 gallon of TORDON 22K per acre per year. The addition of 1 to 2 lbs of 2,4-D per acre will improve control when using rates less than 4 pints of TORDON 22K per acre.

The treated area should extend at least 10 feet beyond the visible weeds to control regrowth from perennial roots and seed germination. A permanent record of rates of TORDON 22K and dates of treatment should be made to prevent inadvertent planting of susceptible crops before disappearance of injurious residues.

Livestock may graze treated fields with the following restrictions when applying more than 1 quart TORDON 22K per acre: (1) Meat animals grazing for up to two weeks after treatment should be removed from treated areas 3 days prior to slaughter. (2) Do not graze dairy animals on treated areas within 2 weeks after treatment. (3) Observe restrictions imposed by other herbicides in the spray.

USE PRECAUTIONS

Use this product only as specified on this label and product label. Observe any special use and application restrictions and limitations, including method of application and permissible areas of use as promulgated by state and local authorities.

Do Not Contaminate Water Intended for Irrigation or Domestic Purposes. Do not treat or allow spray drift to fall onto innerbanks or bottom of irrigation ditches, either dry or containing water or other channels that carry water that may be used for irrigation purposes.

Do not apply on or in the vicinity of susceptible crops or desirable plants including alfalfa, beans, grapes, melons, peas, potatoes, safflower, sugar beets, sunflower, tomatoes and other vegetable crops, flowers, fruit plants, ornamentals or shade trees.

Avoid Spray Drift: Applications should be made only when there is no hazard from spray drift since very small quantities of the spray, which may not be visible, may severely injure susceptible crops during both growing and dormant periods. Use coarse sprays to minimize drift since, under adverse weather conditions such as temperature inversions, fine spray droplets may drift a mile or more. Spray thickening agents, such as NALCO-TROL, may be used with this product to aid in reducing spray drift. If used follow all use recommendations and precautions on the product label.

Ground Equipment: With ground equipment, spray drift can be lessened by keeping the spray boom as low as possible; by applying 20 gallons or more of spray per acre; by using no more than 30 pounds spraying pressure with large droplet producing nozzle tips; by spraying when wind velocity is low; and by stopping all spraying when wind exceeds 10 miles per hour. Do not apply with hollow cone-type insecticide or other nozzles that produce fine-droplet spray.

Aerial Application: Drift can be lessened by applying a coarse spray; by using no more than 25 pounds spray pressure at nozzles; by using nozzles which do not produce hazardous amounts of fine spray droplets; by using a spray boom no longer than 3/4 of the rotor or wing length; and by spraying only when wind velocity is less than 10 mph, or as required by state regulations.

Determine Air Movement and Direction Before Application: Use a smoke generator or other suitable means at or near the application site or on aircraft for detection of air movement, air stability, or temperature inverstions (stable air). Use appropriate drift control measures or avoid application when smoke layers or is moving toward nearby desirable susceptible plants or sensitive areas.

Do not move treated soil to other areas. Do not use it to grow plants, unless adequately sensitive bioassays show that no detectable picloram is present in the soil.

To not transfer livestock from treated grazing areas onto broadleaf crop areas during first full season after treatment without allowing 7 days of grazing on an untreated area. Otherwise, urine may contain enough picloram to cause injury to sensitive broadleaf plants.

Do not re-use containers for TORDON 22K Weed Killer for any purpose. Dispose by punching holes in them and burying with waste or by taking to an approved landfill. Where indicated, follow official local container disposal regulations.

Rinse application equipment after use, preferably at least three times with water, and dispose of rinse water in a noncropland area away from water supplies.

Be sure that use of this product conforms to all applicable regulations.

Do not use where a sandy porous surface and substrate overlie groundwater 10 feet or less below the surface.

READ AND FOLLOW MIXING AND USE INSTRUCTIONS AND PRECAUTIONS ON PRODUCT CONTAINER LABEL

IOWA . . . . EPA SLN NO. IA-790007

For the Control of Musk Thistle in Rangeland and Permanent Grass Pastures.

GENERAL INFORMATION

Use TORDON 22K Weed Killer to control musk thistle growing in rangeland or permanent grass pastures. Picloram, the active ingredient in TORDON 22K Weed Killer, is highly active, persistent, and water soluble. To prevent damage to desirable crops and plants follow all directions on this label.

DIRECTIONS FOR USE

It is a violation of Federal law to use this product in a manner inconsistent with its labeling.

Do not apply this product through any type of irrigation system.

Mix the required amount of TORDON 22K Weed Killer in the specified amount of water and apply broadcast with conventional ground equipment. Use enough spray volume to provide uniform coverage of the weeds, usually 10 to 20 gallons per acre. Do not apply by air or with a mist blower.

Apply when musk thistle plants are in the rosette stage of growth or before flower stock elongates over 2 inches above leaves in the spring or in the fall prior to soil freeze-up. Treat no more than once in any 12-month period and no more than twice in any three-year period.

Use 6 to 8 fl oz (3/8 to 1/2 pt) of TORDON 22K Weed Killer in 10 to 20 gallons of water per acre and apply as a coarse, low-pressure (20 to 30 psi) spray. Apply only when weather conditions are favorable for keeping spray on target area. Do not apply when wind velocity exceeds 10 mph. Where desirable susceptible plants (see **Use Precautions**) are growing within 1/2 mile of the application site, apply TORDON 22K only if air movement is continuous from a definite direction away from these plants.

USE PRECAUTIONS

Use this product only as specified on this label. Observe any special use and application restrictions and limitations including method of application and permissible areas of use, promulgated by state or local authorities.

Handle and Use Correctly. Small amounts of TORDON 22K Weed Killer can injure sensitive broadleaf plants. To avoid problems, observe the following precautions:

1. Do Not Contaminate Nontarget Land Areas, Cropland, Water, or Irrigation Ditches. Do not apply directly to standing or running water. Do not apply where surface water from treated areas can run off to adjacent croplands, either planted or to be planted, or into drainage channels, streams, irrigation ditches, ponds, or wells. Do not clean containers or application equipment on or near these areas. Do not apply on inner banks or bottoms of irrigation ditches.

2. Do Not Apply to or in the Vicinity of Sensitive Crops or Desirable Plants including alfalfa, beans, cotton, grapes, melons, peanuts, peas, potatoes, safflower, sugar beets, soybeans, sunflower, tobacco, tomatoes, other field or vegetable crops, flowers, fruit plants, ornamentals, or shade trees. Do not apply on areas which may be planted at a later time to such crops or plants.

3. Do Not Apply to Cropland Used for Production of Desirable Crops Other Than Forage Grasses. Do not rotate treated rangeland or pasture to other crop uses.

4. Do Not Spray Pastures if the Forage Legume Component is Substantial and Desired. TORDON 22K Weed Killer may injure or kill forage legumes. Also, new legume seedling establishment may not be successful if seeded within 2 years following application of this herbicide.

5. Do Not Move Treated Soil to Other Areas. Do not use it to grow plants, unless adequately sensitive bioassay or chemical tests show that no detectable picloram is present in the soil.

6. Do Not Transfer Livestock from treated grass area onto a broadleaf crop area without first allowing 7 days of grazing on untreated grass pasture. Otherwise, urine may contain enough picloram to cause injury to sensitive broadleaf plants.

7. Do Not Store Near Food, feedstuff, fertilizer, seed, insecticides, fungicides, or other pesticides.

8. Do Not Reuse Containers for TORDON 22K Weed Killer for Any Purpose. Dispose by punching holes in them and burying with waste or by taking to an approved sanitary landfill. Rinse application and handling equipment thoroughly after use, preferably a minimum of three times with clean water, and dispose of rinse water in a non-cropland area away from water supplies. Where indicated, follow official local container disposal regulations.

9. Do Not Allow Spray Drift. Do not permit any spray or spray drift to contact desirable plants off the area treated or the soil containing roots of such plants.

To avoid spray drift, use low spray pressure, under 30 psi and apply as a coarse spray. Do not use pressures nor spray nozzles that produce fine droplets which would present a drift hazard. Spray TORDON 22K Weed Killer only when wind velocity is less than 10 mph and **when there is no hazard from spray drift off the target area. Do not spray when the wind is blowing toward susceptible crops or ornamental plants that are one mile from the target area.**

10. Do Not Use Where a Sandy Porous Surface and Substrate Overlie Ground Water Ten Feet or Less Below the Surface.

11. Be Sure That Use of This Product Conforms to All Applicable Regulations.

KANSAS . . . . EPA SLN NO. KS-850007

For Control of Annual and Perennial Broadleaf Weeds Such As Field Bindweed in Grainland Between Crops of Small Grains and in Non-Cropland Areas

GENERAL INFORMATION

Use TORDON 22K Weed Killer tank mixed with 2,4-D herbicides by ground application to control or reduce infestations of broadleaf annual and perennial weeds such as field bindweed or non-cropland and on grainland during the period following crop harvest and prior to planting winter wheat, barley, or oats. Picloram, the active ingredient in TORDON 22K Weed Killer, is highly active, water soluble and has moderate soil residual. To prevent damage to desirable crops and plants, follow all directions and precautions on this label.

DIRECTIONS FOR USE

It is a violation of Federal law to use this product in a manner inconsistent with its labeling.

Do not apply this product through any type of irrigation system.

Read complete use directions and precautions on this label and container label before using.

For reduction of perennial weeds such as field bindweed, use ground application equipment to broadcast treat infested areas using 0.5 to 1 pint TORDON 22K Weed Killer in 10 or more gallons of water per acre. The addition of 2,4-D herbicides at 1/2 to 2 lb ae/acre may significantly improve weed control. Read and follow all directions and use precautions on other product labels.

Between crops of small grains, treat annually to maintain control of perennial weed infestation. For best results with continuous small grain production, treat field bindweed infested areas in June or July following small grain harvest, when bindweed is actively growing, applying the 0.5 pint rate of TORDON 22K Weed Killer and 1/2 to 2 lb 2,4-D ae/acre (higher rate of TORDON 22K may be used but greater crop injury to fall planted grain may result compared to that which may occur with the lower rate).

For best results with small-grain fallow production, treat field bindweed infested areas in June or July following grain harvest, applying the 1 pint rate of TORDON 22K with 2,4-D; then around mid-May during the fallow season, treat the field bindweed infested area with the 0.5 pint rate of TORDON 22K and 1/2 to 2 lb 2,4,-D ae/acre **(higher rate of TORDON 22K may be used but greater crop injury to fall planted grain may result compared to what may occur from the lower rate).**

Do not treat with TORDON 22K Weed Killer more than once each calendar year. Additional treatment with 2,4-D alone may be made any time weeds are present with adequate foliage and in good growing condition.

For long residual control of perennial weeds such as field bindweed, spot treat scattered infestations following use directions on the container label. When using these higher rate treatments of TORDON 22K, do not treat more than 10% of the area of an infested field in one calendar year. Wheat or other small grain crops may be planted in such treated fields, but crop injury in the spot treated areas may occur for one or more seasons after treatment.

To control volunteer grain or grass weeds that may be present, use cultivation or appropriate grass control herbicides alone or in combination with broadleaf weed control treatments. Consult the University Experiment Station or Extension Service for recommendations to fit local conditions.

Wheat or other small grain crops may be injured when planted at varying periods following application of TORDON 22K Weed Killer. Do not treat with TORDON 22K if the risk of injury to the small grain crop cannot be tolerated.

Roadsides and other non-cropland areas may be treated during bud to early bloom stage of field bindweed in the spring and again in the fall as needed. For low dose repeat treatments, use the 1 pint rate of TORDON 22K and 1/2 to 2 lb 2,4-D ae/acre in the fall. A follow-up spot application is suggested to maintain control. Yearly or twice yearly treatments over a 3 to 4 year period may be needed as part of an overall control program. For higher rate residual control of these and other perennial weeds, follow instructions on container label.

USE PRECAUTIONS

Use this product only as specified on this label. Observe any special use and application restrictions and limitations, including method of application and permissible areas of use as promulgated by state and local authorities.

Handle and Use Correctly. Small amounts of TORDON 22K Weed Killer can injure sensitive broadleaf plants. To avoid problems, observe the following precautions:

1. Do Not Contaminate Water Intended for Irrigation or Domestic Purposes. Do not treat or allow spray drift to fall onto innerbanks or bottom of irrigation ditches, either dry or containing water.

2. Do Not Apply to Sensitive Crops or Desirable Plants including alfalfa, beans, cotton, grapes, melons, peanuts, peas, potatoes, sunflower, sugarbeets, soybeans, tobacco, tomatoes, other field or vegetable crops, flowers, fruit plants, ornamentals, or shade trees.

3. Use a bioassay to determine that picloram residues are no longer detectable in the treated soil before using treated areas to grow sensitive crops. Contact State Extension Specialists for further information on bioassay testing.

4. Do Not Move Treated Soil to Other Areas where susceptible desirable plants may be exposed and damaged while phytotoxic residues are present.

5. Avoid Spray Drift. Applications should be made only when there is no hazard from spray drift since very small quantities of the spray, which may not be visible, may severely injure susceptible crops during both growing and dormant periods. To avoid spray drift, apply only when wind velocity is less than 10 mph and is not blowing toward susceptible crops; keep spray boom as low as possible; apply 20 gallons or more of spray per acre; use no more than 30 psi spraying pressure with large droplet producing tips.

Do Not Apply With Nozzles That Produce Fine-Droplet Spray. A spray thickening agent such as NALCO-TROL may be used with this product to aid in reducing spray drift. If a drift reducing agent is used, follow use directions and precautions on the product label.

6. Do Not Use Where a Porous Sandy Surface and Substrate Overlie Ground Water Ten Feet or Less Below the Surface.

7. Be Sure That Use of This Product Conforms to All Applicable Regulations.

8. Grazing Restrictions: None.

9. Follow Storage and Disposal Requirements on Container Label.

KANSAS . . . . **EPA SLN NO. KS-850003**

For The Control of Musk Thistle, Pricklypear Cactus and Broom Snakeweed in Rangeland, Permanent Grass Pastures and Non-Crop Areas.

GENERAL INFORMATION

Use TORDON 22K Weed Killer to control musk thistle, pricklypear cactus and broom snakeweed growing in rangeland, permanent grass pasture, on roadsides and non-crop areas. Picloram, the active ingredient in TORDON 22K Weed Killer, is highly active, persistent and water soluble. To prevent damage to desirable crops and plants, follow all directions on this label.

DIRECTIONS FOR USE

It is a violation of Federal law to use this product in a manner inconsistent with its labeling.

Do not apply this product through any type of irrigation system.

Mix the required amount of TORDON 22K Weed Killer in the specified amount of water and apply broadcast by air or ground equipment. Apply as a coarse, low-pressure spray using enough spray volume to provide uniform coverage of weeds. Apply the recommended rate of TORDON 22K in 2 or more gallons of water per acre by air or in 10 or more gallons of water per acre with ground application. Do not apply with a mist blower.

MUSK THISTLE

Apply in the fall when musk thistle plants are in the rosette stage of growth. Treat from October 1 until soil freezes. **For spring application apply with ground equipment only.** Treat when musk thistle plants are in the rosette stage and before the flower stalk elongates over two inches above the rosette leaves. Spraying after this time is likely to result in unsatisfactory control. Treat no more than once in any 12-month period, and no more than twice in any 3-year period.

Use 6 - 8 fluid ounces (3/8 - 1/2 pint) of TORDON 22K Weed Killer per acre. The higher rate is preferred if flower stalks are nearing the two-inch height.

Apply only when weather conditions are favorable for keeping spray on the target area. Do not treat by aerial application any musk thistle infestation in the understory of areas designated as forest (land supporting timber stands 120 feet wide), wooded strips or wooded pasture where the tree component is desirable. These areas should be treated by spot application only. Where wooded areas or other desirable susceptible plants are growing within 1/2 mile of the application site, apply TORDON 22K only if air movement is continuous in a direction away from these plants. (See Use Precautions.)

PRICKLYPEAR CACTUS

Treat during the spring or early summer when pricklypear is actively growing, but before full bloom, usually May-June. Apply 1 - 2 pints of TORDON 22K as a broadcast spray or as a spot treatment at 1/4 - 1/2 teaspoon/100 square feet. Use of a diesel oil-water-emulsifier spray mixture may provide better uptake of the chemical. Total effect of the spray application may not be evident for 2 to 3 years after application of the TORDON 22K.

BROOM SNAKEWEED (PERENNIAL BROOMWEED)

Treat broom snakeweed before bud stage when the plant is actively growing, using 1 - 2 pints of TORDON 22K. Apply as a broadcast spray or use as a spot treatment at 1/4 - 1/2 teaspoon per 100 square feet.

Apply the recommended rate of TORDON 22K Weed Killer in at least 2 or more gallons of water with added agricultural surfactant, or a diesel oil-water-emulsifier mixture, per acre by air. For broadcast or ground application, apply 10 or more gallons of carrier per acre.

Note: Chemical control is only a portion of an overall management program on rangeland and tame pasture. To maintain long-term weed control, a good grazing management program is necessary. For information on the management practices in your area, contact the State Extension range and pasture specialist or the State Soil Conservation Service range conservationist through the appropriate county office.

USE PRECAUTIONS

Use this product only as specified on this label. Observe any special use and application restrictions and limiations, including method of application and permissible areas of use as promulgated by state or local authorities.

Do Not Contaminate Water Intended for Irrigation or Domestic Purposes. Do not treat or allow spray drift to fall onto innerbanks or bottom of irrigation ditches, either dry or containing water or other channels that carry water that may be used for irrigation purposes.

Do not apply on or in the vicinity of susceptible crops or desirable plants including alfalfa, beans, grapes, melons, peas, potatoes, safflower, soybeans, sugar beets, sunflower, tomatoes and other vegetable crops, flowers, fruit plants, ornamentals, forest and shade trees or the soil containing roots of nearby valuable plants.

AVOID SPRAY DRIFT: Applications should be made only when there is no hazard from spray drift since very small quantities of the spray, which may not be visible, may severely injure susceptible crops during both growing and dormant periods. To avoid spray drift use low spray pressure, apply as a coarse spray, and spray when wind velocity is low. A spray thickening agent such as NALCO-TROL may be used with this product to aid in reducing spray drift. It a thickening agent is used, follow all use recommendations and precautions on the product label.

Ground Application: With ground equipment, spray drift can be lessened by keeping the spray boom as low as possible; by applying 20 gallons or more of spray per acre; by using no more than 30 pounds spraying pressure with large droplet producing nozzle tips; by spraying when wind velocity is low; and by stopping all spraying when wind exceeds 10 miles per hour. Do not apply with nozzles that produce a fine-droplet spray.

Aerial Application: Drift can be lessened by applying a coarse spray; by using no more than 25 pounds spray pressure at nozzles; by using nozzles which do not produce hazardous amounts of fine spray droplets; by using a spray boom no longer than 3/4 of the rotor or wing lenth; and by spraying only when wind velocity is less than 10 mph, or as required by state regulations.

Determine Air Movement and Direction Before Follar Application: Use a smoke generator or other means at or near the application site or on aircraft for detection of air movement, air stability, or temperature inversions. Such a condition exists when there is little or no wind and air temperature is lower near the ground than at higher levels. Use appropriate drift control measures or avoid application when smoke is moving toward nearby desirable susceptible plants or sensitive areas.

Do not rotate treated rangeland or pasture to other crop uses.

Do not spray pastures if the forage legume component is substantial and desired. TORDON 22K Weed Killer may injure or kill legumes. Also, new legume seedings may not be successful if made within two years following application of this herbicide.

Do not move treated soil to other areas. Do not use it to grow plants, unless adequately sensitive bioassay or chemical tests show that no detectable picloram is present in the soil.

Do not transfer livestock from treated grass areas onto broadleaf crop areas without first allowing 7 days of grazing on an untreated grass pasture. Otherwise, urine may contain enough picloram to cause injury to sensitive broadleaf plants.

Do not re-use containers for TORDON 22K Weed Killer for any purpose. Dispose by punching holes in them and burying with waste or by taking to an approved landfill. Where indicated, follow official local container disposal regulations.

Rinse application equipment after use, preferably at least three times with water, and dispose of rinse water in a non-cropland area away from water supplies.

Be sure that use of this product conforms to all applicable regulations.

Do not use where a sandy porous surface and substrate overlie ground water ten feet or less below the surface.

READ AND FOLLOW MIXING AND USE INSTRUCTIONS AND PRECAUTIONS ON PRODUCT CONTAINER LABEL

MISSOURI . . . . EPA SLN NO. MO-790004

For the Control of Musk Thistle, Goldenrod, Ironweed, and other Broadleaf Weeds in Rangeland and Permanent Grass Pastures.

GENERAL INFORMATION

Use TORDON 22K Weed Killer to control musk thistle growing in rangeland or permanent grass pastures. Picloram, the active ingredient in TORDON 22K Weed Killer, is highly active, water soluble and has moderate soil residual. Small amounts of TORDON 22K Weed Killer can injure sensitive broadleaf plants. To prevent damage to desirable crops and plants follow all directions on this label.

DIRECTIONS FOR USE

It is a violation of Federal law to use this product in a manner inconsistent with its labeling.

Do not apply this product through any type of irrigation system.

Broadcast Treatment: To control broadleaf weeds, treat when weeds are growing well and apply broadcast with conventional ground equipment. Do not apply by air or with mist blower.

For control of **musk thistle:** Apply when musk thistle plants are in the rosette stage or growth in the fall from October 1 to soil freeze-up or in the spring before flower stock elongates or "bolts" above the rosette leaves. Use 6 to 8 fluid ounces (3/8 to 1/2 pint) of TORDON 22K in 10 to 20 gallons of water. Generally, most consistent control is received from fall applications.

For control of **goldenrod:** Apply when plants are growing well at the rate of 6 to 8 fluid ounces (3/8 to 1/2 pint) of TORDON 22K in 10 to 20 gallons of water. Best results are obtained with applications made prior to early bud stage.

For control of **ironweed:** Apply when plants are growing well at the rate of 1 point of TORDON 22K in 10 to 20 gallons of water. For best results, apply prior to early bud stage.

Wick Applications: Mix 1 part of TORDON 22K with 2 parts of water to prepare a 33% solution. Apply when weeds are actively growing and are above most desirable plants. For musk thistle, apply before seed head develops into purple flower for best results. Later applications may not stop early season seed production. For ironweed and goldenrod, best results are obtained with applications made prior to early bud stage.

Wick applicator should be drained and cleaned after each use. Ropes should be changed when flow is reduced from wear, extended use, poor cleaning or intermittent use.

USE PRECAUTIONS

Use this product only as specified on this label. Observe any special use and application restrictions and limitations, including method of application and permissible areas of use as promulgated by state or local authorities.

Do not apply on or in the vicinity of susceptible crops or desirable plants including alfalfa, beans, grapes, melons, peas, potatoes, safflower, soybeans, sugar beets, sunflower, tomatoes and other vegetable crops, flowers, fruit plants, ornamentals, shade trees, or the soil containing roots of nearby valuable plants.

Avoid Spray Drift: Applications should be made only when there is no hazard from spray drift since very small quantities of the spray, which may not be visible, may severely injure susceptible crops during both growing and dormant periods. To avoid spray drift use low spray pressure, under 30 psi; spray when wind velocity is less than 10 mph; and apply as a coarse spray. Spray drift can be lessened by keeping the spray boom as low as possible; by applying 20 gallons or more of spray per acre; by using large droplet producing nozzle tips; and by stopping all spraying when wind exceeds 10 miles per hour. Do not apply with hollow cone-type insecticide or other nozzles that produce fine-droplet spray. Do not apply to cropland used for production of desirable crops other than forage grasses. No not rotate treated rangeland or pasture to other crop uses.

Do not treat pasture or rangeland if injury to the forage legume cannot be tolerated. TORDON 22K Weed Killer may injure or kill forage legumes when exposed to sprays or residues from wicked plants. Also, a new legume seedling establishment may not be successful if seeded within 2 years following application of herbicide.

Do Not Contaminate Water intended for irrigation, drinking or other domestic purposes. To avoid crop or other plant injury, do not treat or allow spray drift to fall onto inner banks or bottom of irrigation and drainage ditches.

Do not move treated soil to other areas or use it to grow plants unless adequately sensitive bioassay shows that no detectable picloram is present in the soil.

Do not transfer livestock from treated grass area onto a broadleaf crop area without first allowing 7 days of grazing on an untreated grass pasture. Otherwise, urine may contain enough picloram to cause injury to sensitive broadleaf plants.

Do not use where a sandy porous surface and substrate overlie ground water 10 feet or less below the surface.

STORAGE AND DISPOSAL

Do not contaiminate water, food or feed by storage or disposal.

Presticide Disposal: Pesticide, spray mixture or rinsate that cannot be used according to label instructions must be disposed of according to Federal, state or local procedures under the Resource Conservation and Recovery Act.

Container Disposal: Do not reuse containers. Dispose of them in a sanitary landfill or by other procedures approved by state and local authorities.

READ AND FOLLOW MIXING AND USE INSTRUCTIONS AND PRECAUTIONS ON PRODUCT CONTAINER LABEL

MONTANA . . . . EPA SLN NO. MT-780003

For the Control of Broadleaf Weeds in Rangelands, Permanent Grass Pastures, Spring Barley and Oats, and Spring and Winter Wheat.

DIRECTIONS FOR USE

It is a violation of Federal law to use this product in a manner inconsistent with its labeling.

Do not apply this product through any type of irrigation system.

Read complete use directions and precautions on this label and container label before using.

RANGELAND AND PASTURE

Use TORDON 22K Weed Killer to control broadleaf annual and perennial weeds such as Canada and other thistles, field bindweed, leafy spurge, Russian knapweed, spotted knapweed, tall larkspur, yellow toadflax, locoweed, snakeweed and lupines on rangeland and permanent grass pastures. Treat when weeds are growing well using low pressure sprays. Retreatment at the same rate may be necessary the following year.

Spot Treatment: (ground application only) Use TORDON 22K at rates of 2 to 4 quarts in 20 to 100 gallons of water per acre and apply as a spray to the foliage. Use the higher rates to control leafy spurge, larkspur, toadflax and knapweed. Use the lower rates for bindweed and thistles. For a 1,000 square foot infestation, apply about 2 1/4 fluid ounces of TORDON 22K in 1 gallon of water (equivalent to 3 quarts per acre of TORDON 22K).

NOTE: When spot treating, do not cut grass for feed within 2 weeks after treatment. Meat animals grazing for up to 2 weeks after treatment should be removed from treated areas 3 days prior to slaughter. Do not graze dairy animals on treated areas within 2 weeks after treatment.

Broadcast Treatment: (ground or aerial application) to suppress the growth of many perennial broadleaf weeds, apply as a broadcast spray using 1 quart of TORDON 22K in 1 to 4 gallons of water per acre by air or in 20 to 100 gallons of water per acre by ground equipment and apply as a broadcast spray during the growing season when weeds are growing well. (Many seedling annual weeds can be controlled using 1 pint per acre.)

Tank Mixture for Spot or Broadcast Treatments: TORDON 22K may also be tank mixed with 2,4-D products such as ESTERON*99* Concentrate, FORMULA 40*, DMA* 4 or ESTERON 6E herbicides for use on areas having mixed species including those which respond well to 2,4-D, such as big sagebrush. In tank mix combinations, use 1 pint to 1 quart of TORDON 22K with 1 to 2 quarts ESTERON 99 Concentrate, FORMULA 40 or DMA 4, or with 2/3 to 1 1/3 quarts of ESTERON 6E per acre, in spray volumes specified above. **Read and follow all directions and use precautions on other product lables.**

SPRING BARLEY AND OATS, SPRING AND WINTER WHEAT NOT UNDERSEEDED WITH A LEGUME

For the control of wild buckwheat and other annual broadleaf weeds normally controlled with 2,4-D or MCPA such as wild mustard, Russian thistle, pennycress, lambsquarters and pigweed, in spring wheat and barley and winter wheat, use TORDON 22K as a tank mix with a 2,4-D or MCPA formulation such as DMA 4, FORMULA 40, ESTERON 99 Concentrate, ESTERON 6E, or MCP Amine Herbicides. For use on spring oats, tank mix only with MCP Amine Herbicide. **Read and follow all directions and use precautions on other product labels.**

Spring Wheat, Barley and Oats: Apply during the 3 through 5 leaf stage of growth. Application of TORDON 22K occasionally causes slight head malformations and straw shortening but normally this does not affect yield.

Durum Wheat: Do not treat durum since at least some varieties appear to be more sensitive than other wheat.

Winter Wheat: Apply after resumption of active growth in the spring and before the early boot stage.

For aerial or ground treatment, use enough total spray volume to provide adequate spray coverage. Apply 1 to 4 gpa by air and 5 to 20 gpa by ground. Spray pressure should not exceed 30 psi. Use coarse spray to minimize spray drift.

To prepare the spray, mix only with water. Add about half the desired amount of water in the spray tank. Then with agitation, add the recommended amount of TORDON 22K and 2,4-D or MCPA as outlined in the table. Finally, with continued agitation, add the rest of the water.

The dosages recommended equate to 1/4 oz. picloram + 4 oz. 2,4-D or MCPA ae/acre when weeds are small or 3/8 + 6 oz/acre when weeds are more advanced or when dry soil conditions exist.

Use Rates for Barley, Oats, and Wheat

Weed Growth Stage	Amounts of Each Product per Acre†		
	TORDON 22k	*plus* DMA 4, FORMULA 40, ESTERON 99 Concentrate	*or* MCP Amine *or* ESTERON 6E
weeds 1-3 inches tall	1 fl. oz.	1/2 pt.	1/3 pt.
weeds 3-6 inches tall or under dry conditions	1 1/2 fl. oz.	3/4 pt.	1/2 pt.

†When measuring small amounts of TORDON 22K weed killer, special care should be taken not to exceed suggested rates.

NOTE: Use only on land that will be planted the following year to grass or grain crops such as small grains, corn, sorgum, or flax. Do not apply more than 3/8 ounce picloram per acre during any 12-month period.

USE PRECAUTIONS

Use this product only as specified on this label. Observe any special use and application restrictions and limitations, including method of application and permissible areas of use as promulgated by state or local authorities.

Do Not Contaminate Nontarget Land Areas, Cropland, Water, or Irrigation Ditches. Do not apply directly to standing or running water. Do not apply where surface water from treated areas can run off to adjacent cropland, either planted or to be planted, or into streams, irrigation ditches, irrigation ponds, or wells. Do not clean containers nor application equipment on or near these areas. Do not apply on inner banks or bottom of irrigation ditches.

Do not apply on or in the vicinity of susceptible crops or desirable plants including alfalfa, beans, grapes, melons, peas, potatoes, safflower, soybeans, sugar beets, sunflower, tomatoes and other vegetable crops, flowers, fruit plants, ornamentals or shade trees.

Avoid Spray Drift: Applications should be made only when there is no hazard from spray drift since very small quantities of the spray, which may not be visible, may severely injure susceptible crops during both growing and dormant periods. Use coarse sprays to minimize drift since, under adverse weather conditions, fine spray droplets may drift a mile or more. The spray thickening agent, NALCO-TROL, may be used with this product to aid in reducing spray drift. If used follow all use recommendations and precautions on the product label.

Ground Equipment: With ground equipment, spray drift can be lessened by keeping the spray boom as low as possible; by applying 20 gallons or more of spray per acre; by using no more than 30 pounds spraying pressures with large droplet producing nozzle tips; by spraying when wind velocity is low; and by stopping all spraying when wind exceeds 10 miles per hour. Do not apply with hollow cone-type insecticide or other nozzles that produce a fine-droplet spray.

Aerial Application: With aircraft, drift can be lessened by applying a coarse spray; by using no more than 30 pounds spray pressure at the nozzles; by using a straight stream nozzles directed straight back; by using a spray boom no longer than 3/4 the wing span of the aircraft; and by spraying only when wind velocity is less than 10 mph.

Do Not Apply By Aircraft When An Air Temperature Inversion Exists: Such a condition is characterized by little or no wind and with air temperature lower near the ground than at higher levels. The use of a continuous smoke column at or near the site of application or use of a smoke generating device on the aircraft is suggested to indicate direction and velocity of air movement, and to indicate a temperature inversion by layering of the smoke.

Do not rotate treated rangeland or pasture to other crop uses.

Do not spray pastures or grain if the forage legume component is desired. TORDON 22K Weed Killer may injure or kill legumes. Also new legume seedings may not be successful if made within 2 years following application of this herbicide.

Do not move treated soil to other areas. Do not use it to grow plants, unless adequate sensitive bioassay or chemical tests show that no detectable picloram is present in the soil.

Do not transfer livestock from treated grass areas onto broadleaf crop areas without first allowing 7 days of grazing on untreated grass pasture. Otherwise, urine may contain enough picloram to cause injury to sensitive broadleaf plants.

Do not re-use containers for TORDON 22K Weed Killer for any purpose. Dispose by punching holes in them and burying with waste or by taking to an approved landfill. Where indicated, follow official local container disposal regulations.

Rinse application equipment after use, preferably at least three times with water, and dispose of rinse water in a non-cropland area away from water supplies.

Be sure that use of this product conforms to all applicable regulations.

Do not use where a sandy porous surface and substrate overlie ground water 10 feet or less below the surface.

NEBRASKA . . . . EPA SLN NO. NE-840009

For the Control of Leafy Spurge, Musk Thistle and other Broadleaf Weeds in Rangeland and Permanent Grass Pastures.

DIRECTIONS FOR USE

It is a violation of Federal law to use this product in a manner inconsistent with its labeling.

Do not apply this product through any type of irrigation system.

Read complete use directions and precautions on this label and the container label before using.

GENERAL INFORMATION

Use TORDON 22K Weed Killer to control leafy spurge, musk thistle and other broadleaf weeds growing in rangeland or permanent grass pastures. Picloram, the active ingredient in TORDON 22K Weed Killer, is highly active, persistent, and water soluble. To prevent damage to desirable crops and plants follow all directions and precautions on this label.

Treat when weeds are growing well using low pressure sprays. Treat no more than once in any 12-month period and no more than twice in any 3-year period. Do not apply with a mist blower.

For applications at rates greater than 8 fl oz per acre, limit coverage to no more than 10% or 10 acres, whichever is greater, of an operator's acreage found in any particular watershed.

Spot Treatment (Ground Application Only): For control of broadleaf annual and perennial weeds such as leafy spurge, Canadian and other thistles, field bindweed, Russian knapweed, spotted knapweed, tall larkspur, locoweed, snakeweed and lupines use TORDON 22K at rates of 2 to 4 quarts in 5 to 100 gallons of water per acre and apply as a spray to the foliage. Use the higher rates to control leafy spurge, larkspur, toadflax and knapweed. Use the lower rates for bindweed and thistles. For a 1,000 square foot infestation, apply about 2 1/4 fluid ounces of TORDON 22K in 1 gallon of water (equivalent to 3 quarts per acre of TORDON 22K).

NOTE: When spot treating, do not cut grass for feed within 2 weeks after treatment. Meat animals grazing for up to 2 weeks after treatment should be removed from treated areas 3 days prior to slaughter. Do not graze dairy animals on treated areas within 2 weeks after treatment.

Broadcast Treatment - Ground or Aerial Application: To suppress the growth of many perennial broadleaf weeds apply as a broadcast spray using 1 quart of TORDON 22K in 2 to 4 gallons of water per acre by air or in 5 to 100 gallons of water per acre by ground equipment and apply as a broadcast spray during the growing season when weeds are growing well. (Many seedling annual weeds can be controlled using 1 pint per acre.)

Tank Mixture for Spot or Broadcast Treatments: TORDON 22K may also be tank mixed with 2,4-D herbicides for use on areas having mixed species including those which respond well to 2,4-D such as big sagebrush. **Read and follow all directions and use precautions on other product labels.**

For Control of Musk Thistle: Use 6 to 8 fluid ounces (3/8 to 1/2 pint) of TORDON 22K in 2 to 4 gallons of water per acre by air or in 10 to 20 gallons of water per acre by ground equipment. Apply in the spring before May 15, or when musk thistle plants are in the rosette stage of growth or before flower stock elongates over 2 inches above leaves, whichever comes first. Apply in the fall after October 1, but before soil freeze-up or December 1, whichever is earlier.

USE PRECAUTIONS

Use this product only as specified on this label. Observe any special use and application restrictions and limitations, including method of application and permissible areas of use as promulgated by state or local authorities.

Do Not Contaminate Nontarget Land Areas, Cropland, Water, or Irrigation Ditches. Do not apply directly to standing or running water. Do not apply where surface water from treated areas can run off to adjacent cropland, either planted or to be planted, or into streams, irrigation ditches, irrigation ponds, or wells. Do not clean containers nor application equipment on or near these areas. Do not apply on inner banks or bottom of irrigation ditches.

Do not apply on or in the vicinity of susceptible crops or desirable plants including alfalfa, beans, grapes, melons, peas, potatoes, safflower, soybeans, sugar beets, sunflower, tomatoes and other vegetable crops, flowers, fruit plants, ornamentals or shade trees.

Avoid Spray Drift: Applications should be made only when there is no hazard from spray drift since very small quantities of the spray, which may not be visible, may severely injure susceptible crops during both growing and dormant periods. Use coarse sprays to minimize drift since, under adverse weather conditions, fine spray droplets may drift a mile or more. The spray thickening agent, NALCO-TROL, may be used with this product to aid in reducing spray drift. If used follow all use recommendations and precautions on the product label.

Ground Equipment: With ground equipment, spray drift can be lessened by keeping the spray boom as low as possible; by applying 20 gallons or more of spray per acre; by using no more than 30 pounds spraying pressures with large droplet producing nozzle tips; by spraying when wind velocity is low; and by stopping all spraying when wind exceeds 10 miles per hour. Do not apply with nozzles that produce a fine-droplet spray.

Aerial Application: With aircraft, drift can be lessened by applying a coarse spray; by using no more than 30 pounds spray pressure at nozzles; by using straight stream nozzles directed straight back; by using a spray boom no longer than 3/4 the wing span of the aircraft; and by spraying only when wind velocity is less than 10 mph.

Do Not Apply By Aircraft When An Air Temperature Inversion Exists: Such a condition is characterized by little or no wind and with air temperature lower near the ground than at higher levels. The use of a continuous smoke column at or near the site of application or use of a smoke generating device on the aircraft is suggested to indicate direction and velocity of air movement, and to indicate a temperature inversion by layering of the smoke.

Do not rotate treated rangeland or pasture to other crop uses.

Do not spray pastures or gain if the forage legume component is desired. TORDON 22K Weed Killer may injure or kill legumes. Also new legume seedings may not be successful if made within 2 years following application of this herbicide.

Do not move treated soil to other areas. Do not use it to grow plants, unless adequate sensitive bioassay or chemical tests show that no detectable picloram is present in the soil.

Do not transfer livestock from treated grass areas onto broadleaf crop areas without first allowing 7 days of grazing on untreated grass pastures. Otherwise, urine may contain enough picloram to cause injury to sensitive broadleaf plants.

Do not re-use containers for TORDON 22K Weed Killer for any purpose. Dispose by punching holes in them and burying with waste or by taking to an approved landfill. Where indicated, follow official local container disposal regulations.

Rinse application equipment after use, preferably at least three times with water, and dispose of rinse water in a non-cropland area away from water supplies.

Be sure that use of this product conforms to all applicable regulations.

Do not use on flood, gravity, sprinkler, or sub-irrigated land.

Do not use where a sandy porous surface and substrate overlie ground water 15 feet or less below the surface.

Do not store near food, feedstuff, fertilizer, seeds, insecticides, fungicides or other pesticides.

FOLLOW STORAGE AND DISPOSAL REQUIREMENTS ON CONTAINER LABEL.

NEW MEXICO . . . . EPA SLN NO. NM-870003

For Control of Annual and Perennial Broadleaf Weeds Such as Field Bindweed in Fallow Grainland Between Crops of Small Grains.

DIRECTIONS FOR USE

It is a violation of Federal law to use this product in a manner inconsistent with its labeling.

Do not apply this product through any type of irrigation system.

Read complete use directions and precautions on this label and container label before using.

Use TORDON 22K Weed Killer by ground or aerially to control or reduce infestations of broadleaf annual and perennial weeds such as field bindweed on grainland during the period following crop harvest and prior to plating winter wheat, barley, or oats.

For **reduction of perennial** weeds such as field bindweed, broadcast treat infested areas using 0.5 to 1 pint TORDON 22K Weed Killer. The addition of 2,4-D herbicides at 1/2 to 2 lb ae/acre may significantly improve weed control. Read and follow all directions and use precautions on other product labels. To maintain or further reduce perennial weed infestations, treat annually between crops. For best results, treat in June or July following small grain harvest, applying the higher rates of TORDON 22K and 2,4-D. Initial treatment alternatively may be made soon after harvesting a fall maturing crop such as grain sorghum or corn. During the next calendar year, retreat the same area with the lower rate of TORDON 22K, prefereably in May or June during the fallow period prior to fall planting of the next small grain crop.

Do not treat with TORDON 22K Weed Killer more than once each calendar year. Additional treatment with 2,4-D alone may be made anytime weeds are present with adequate foliage and in good growing condition.

TORDON 22K at 1 pint or less per acre alone or plus 2,4-D or other compatible registered herbicides may be applied by air in 2 or more gallons of spray per acre under conditions that will prevent injury to any nearby susceptible crops or ornamentals. **Refer to Use Precautions for suggestions to reduce off-target drift hazard.**

DO NOT APPLY MORE THAN 1 PINT OF TORDON 22K WEED KILLER PER ACRE BY AIR.

To control volunteer grain or grass weeds that may be present, use cultivation or appropriate grass control herbicides alone or in combination with broadleaf weed control treatments. Consult the University Experiment Station or Extension Service for recommendations to fit local conditions.

Wheat or other small grain crops and seedling grasses may be injured when planted at varying periods following application of TORDON 22K Weed Killer. Do not treat with TORDON 22K if the risk of injury to the small grain crop cannot be tolerated.

For long residual control of perennial weeds such as field bindweed, spot treat scattered infestations following use directions on the container label. When using these higher rate treatments of TORDON 22K, do not treat more than 10% of the area of an infested field in one calendar year. Wheat or other small grain crops may be planted in such treated fields, but crop injury in the spot treated areas may occur for one or more seasons after treatment.

USE PRECAUTIONS

Use this product only as specified on this label. Observe any special use and application restrictions and limitations, including method of application and permissible areas of use as promulgated by state or local authorities.

Do not apply on or in the vicinity of susceptible crops or desirable plants including alfalfa, beans, grapes, melons, peas, potatoes, safflower, soybeans, sugar beets, sunflower, tomatoes and other vegetable crops, flowers, fruit plants, ornamentals, shade trees or the soil containing roots of nearby valuable plants.

Avoid Spray Drift: Applications should be made only when there is no hazard from spray drift since very small quantiies of the spray, which may not be visible, may severely injure susceptible crops during both growing and dormant periods. To avoid spray drift use low spray pressure, under 30 psi; spray when wind velocity is less than 10 mph; and apply as a coarse spray. The spray thickening agent, NALCO-TROL, may be used with this product to aid in reducing spray drift. If this thickening agent is used, follow all use recommendations and precautions on the product label.

Ground Equipment: With ground equipment, spray drift can be lessened by keeping the spray boom as low as possible; by applying 20 gallons or more of spray per acre; by using no more than 30 pounds spraying pressure with large droplet producing nozzle tips; by spraying when wind velocity is low; and by stopping all spraying when wind exceeds 10 miles per hour. Do not apply with hollow cone-type insecticide or other nozzles that produce a fine-droplet spray.

Aerial Equipment: Drift can be lessened by applying a coarse spray; by using no more than 25 pounds spray pressure at nozzles which do not produce hazardous amounts of fine spray droplets; by using a spray boom no longer than 3/4 of the rotor or wing length; and by spraying only when wind velocity is less than 10 mph, or as required by state regulations.

Do Not Contaminate Water Intended for Irrigation or Domestic Purposes. Do not treat or allow spray drift to fall onto innerbanks or bottom of irrigation ditches, either dry or containing water or other channels that carry water that may be used for irrigation purposes.

Do not move treated soil to other areas where susceptible desirable plants may be exposed and damaged while phytotoxic residues are present.

NORTH DAKOTA . . . . EPA SLN NO. ND-780002

For the Control of Broadleaf Weeds in Rangelands, Permanent Grass Pastures, Spring Barley and Oats, Spring and Winter Wheat, and Flax.

DIRECTIONS FOR USE

It is a violation of Federal law to use this product in a manner inconsistent with its labeling.

Do not apply this product through any type of irrigation system.

Read complete use directions and precautions on this label and container label before using.

SPRING BARLEY AND OATS, SPRING AND WINTER WHEAT NOT UNDERSEEDED WITH A LEGUME

For the control of wild buckwheat, volunteer sunflower and other annual broadleaf weeds normally controlled with 2,4-D or MCPA, such as wild mustard, Russian thistle, pennycress, lambsquarters and pigweed, in spring wheat and barley and winter wheat, use TORDON 22K Weed Killer as a tank mix with a 2,4-D or MCPA formulation such as FORMULA 40*, DMA* 4, ESTERON*99* Concentrate, ESTERON 6E, MCP Ester, or MCP Amine Herbicides. For use on spring oats, tank mix only with MCP Ester or MCP Amine Herbicide. **Read and follow all directions and use precautions on other product labels.**

Spring Wheat, Barley and Oats: Apply during the 3 through 5 leaf stage of growth. Application of TORDON 22K occasionally causes slight head malformations and straw shortening but normally this does not affect yield.

Durum Wheat: Do not treat durum since at least some varieties appear to be more sensitive than other wheat.

Winter Wheat: Apply after resumption of active growth in the spring and before the early boot stage.

For aerial or ground treatment, use enough total spray volume to provide adequate spray coverage. Apply 1 to 4 gpa by air and 5 to 20 gpa by ground.

To prepare the spray, mix only with water. Add about half the desired amount of water in the spray tank. Then with agitation, add the recommended amount of TORDON 22K and 2,4-D or MCPA as outlined in the table. Finally, with continued agitation, add the rest of the water.

USE RATES FOR WHEAT, BARLEY AND OATS†

WEEDS	WEED GROWTH STAGE‡	AMOUNTS OF EACH PRODUCT PER ACRE: TORDON 22k	*plus* FORMULA 40, DMA 4, ESTERON 99 CONCENTRATE, MCP ESTER, OR MCP Amine	*or* ESTERON 6E
More susceptible species such as lambsquarters, pennycress, and wild mustard	1-3 inches	1 fl. oz.	1/2 pt.	1/3 pt.
	3-6 inches	1 1/2 fl. oz.	3/4 pt.	1/2 pt.
Less susceptible species such as volunteer sunflower, wild buckwheat Russian thistle, pigweed, and top growth supression of Canada thistle	1-6 inches	1 1/2 fl. oz.	3/4 pt.	1/2 pt.

† For oats, tank mix only with MCP amine or MCP Ester Herbicides.

‡ For best results, treat when weeds have 2 to 4 leaves and are actively growing.

NOTE: Use only on land that will be planted the following year to grass or grain crops such as small grains, corn, sorghum, or flax. Do not spray a small grain crop more than once with TORDON 22K Weed Killer.

FLAX

For control or suppression of wild buckwheat, volunteer sunflower, pigweed and other susceptible broadleaf weeds, treat flax when 2 to 6 inches tall, before buds begin to form. Apply a tank mixture that delivers 1 fluid ounce TORDON 22K plus 1/2 pint MCP Amine herbicide per acre. For best results and crop safety, apply when flax is 2 to 3 inches tall and the weeds are young and growing well. Treat wild buckwheat before it begins to form a vine. For aerial or ground treatment, use enough water dilution to provide adequate spray coverage, usually 1 to 4 gpa by air or 5 to 20 gpa by ground equipment. **Read and follow all directions and use precautions on labels of all products contained in the tank mixture.**

NOTE: The use of some preplant herbicides such as Eptam or Treflan, or the presence of crop stress from drouth, disease or other elements, may increase risk of crop injury and poor weed control from postemergence herbicide treatment. Also, flax varieties may vary in tolerance to herbicides. Do not treat flax unless the possibility of crop injury can be tolerated. Consult local Extension or University Extension Specialists for further information.

RANGELAND AND PERMANENT GRASS PASTURES

Use TORDON 22K Weed Killer to control broadleaf annual and perennial weeds such as Canada Thistle and other thistles, field bindweed, leafy spurge, Russian knapweed, spotted and diffuse knapweed, plains or Geyer larkspur, Dalmation and yellow toadflax, locoweed, and lupines on rangeland and permanent grass pastures. Treat when weeds are growing well using low pressure sprays. Retreatment at the same rate may be necessary the following year.

SPOT TREATMENT: Use TORDON 22K at rates shown in the table for broadcast treatment and apply in water with sufficient dilution to obtain adequate spray coverage. For perennial weeds, spray the weed infested area plus a 10 foot border around the infestation. Consult the product container label for guidance to prepare spray mixes to spot treat small patches.

BROADCAST TREATMENT: Apply as a broadcast spray on hard to control weeds at rates shown in the following table. Apply the selected rate of TORDON 22K in 5 gallons of total spray volume per acre (water diluent only) by air, or in 20 to 100 gallons of water per acre by ground equipment. For rates exceeding 1 quart of TORDON 22K per acre, the total area treated in a single season should not exceed 25% of a landowner's acreage found in any particular watershed.

TANK MIXTURES FOR SPOT OR BROADCAST TREATMENTS: TORDON 22K may also be tank mixed with 2,4-D products for use on areas having mixed species, including those which respond well to 2,4-D, such as big sagebrush. READ AND FOLLOW ALL DIRECTIONS AND USE PRECAUTIONS ON OTHER PRODUCT LABELS.

Weed Species	Rates of TORDON 22K Per Sprayed Acre‡
Diffuse Knapweed Spotted Knapweed Many Seedling annual weeds	1 pint
Plains or Geyer Larkspur † Locoweed † Lupines †	1 quart
Russian Knapweed	2 quarts
Canada Thistle	2 quarts - Fall rosette & prebolt 3 quarts -Any other time
Dalmation Toadflax Yellow Toadflax Field Bindweed	3 quarts
Leafy Spurge †	1 quart - Repeat annually for 3 to 4 years 2 quarts -Treat 2 consecutive years. 4 quarts - Follow as needed with spot touchup treatment

† Denotes poisonous plants.

†† NOTE: When applying more than 1 quart of TORDON 22K per acre, do not cut grass for feed within 2 weeks after treatment. Meat animals grazing for up to two weeks after treatment should be removed from treated areas 3 days prior to slaughter. Do not graze dairy animals on treated areas within 2 weeks after treatment.

USE PRECAUTIONS

Use this product only as specified on this label. Observe any special use and application restrictions and limitations, including method of applica-

tion and permissible areas of use as promulgated by state or local authorities.

Do Not Contaminate Water Intended for Irrigation or Domestic Purposes. To avoid injury to crops or other desirable plants, do not treat or allow spray drift to fall onto innerbanks or bottom of irrigation ditches or other channels that carry water that may be used for irrigation purposes.

Do not apply on or in the vicinity of susceptible crops or desirable plants including alfalfa, beans, grapes, melons, peas, potatoes, safflower, soybeans, sugar beets, sunflower, tomatoes and other vegetable crops, flowers, fruit plants, ornamentals or shade trees.

Avoid Spray Drift: Applications should be made only when there is no hazard from spray drift since very small quantities of the spray, which may not be visible, may severely injure susceptible crops during both growing and dormant periods. Use coarse sprays to minimize drift since, under adverse weather conditions, fine spray droplets may drift a mile or more. A spray thickening agent, such as NALCO-TROL, may be used with this product to aid in reducing spray drift. If used, follow all use recommendations and precautions on the product label.

Ground Application: With ground equipment, spray drift can be lessened by keeping the spray boom as low as possible; by applying 20 gallons or more of spray per acre; by using no more than 30 pounds spraying pressures with large droplet producing nozzle tips; by spraying when wind velocity is low; and by stopping all spraying when wind exceeds 10 miles per hour. Do not apply with hollow cone-type insecticide or other nozzles that produce a fine-droplet spray.

Aerial Application: Drift can be lessened by applying a coarse spray; by using no more than 25 pounds spray pressure at the nozzles; by using nozzles which do not produce hazardous amounts of fine spray droplets; by using a spray boom no longer than 3/4 of the rotor or wing length; and by spraying only when wind velocity is less than 10 mph, or as required by state regulations.

Determine Air Movement and Direction Before Foliar Application: Use a smoke generator or other means at or near the application site or on aircraft for detection of air movement, air stability, or temperature inversions. Such a condition exists when there is little or no wind and air temperature is lower near the ground than at higher levels. Use appropriate drift control measures or avoid application when smoke is moving toward nearby desirable susceptible plants or sensitive areas.

Do not rotate treated rangeland or pasture to other crop uses.

Do not spray pastures if the forage legume component is desired. TORDON 22K Weed Killer may injure or kill legumes. Also new legume seedings may not be successful if made within 2 years following application of this herbicide.

Do not move treated soil to other areas. Do not use it to grow plants, unless adequate sensitive bioassay or chemical tests show that no detectable picloram is present in the soil.

Do not transfer livestock from treated grass areas onto sensitive broadleaf crop areas for 12 months after application or until picloram has disappeared from the soil, without first allowing 7 days of grazing on an untreated grass pasture. Otherwise, urine may contain enough picloram to cause injury to sensitive broadleaf plants.

Do not re-use containers for TORDON 22K Weed Killer for any purpose. Dispose by punching holes in them and burying with waste or by taking to an approved landfill. Where indicated, follow official local container disposal regulations.

Rinse application equipment after use, preferably at least three times with water, and dispose of rinse water in a non-cropland area away from water supplies.

Be sure that use of this product conforms to all applicable regulations.

Do not use where a sandy porous surface and substrate overlie ground water 10 feet or less below the surface.

This registration has been issued by the North Dakota State Laboratories Department pursuant to the provisions of section 24(c) of the Federal Insecticide, Fungicide, and Rodenticide Act, as amended.

OKLAHOMA . . . . EPA SLN NO. OK-840005

For Control of Annual and Perennial Broadleaf Weeds Such as Field Bindweed in Grainland Between Crops of Small Grains.

GENERAL INFORMATION

Use TORDON 22K Weed Killer tank mixed with 2,4-D herbicides by ground application to control or reduce infestations of broadleaf annual and perennial weeds such as field bindweed on grainland during the period following crop harvest and prior to planting winter wheat, barley, or oats. Picloram, the active ingredient in TORDON 22K Weed Killer, is highly active, water soluble and has moderate soil residual. To prevent damage to desirable crops and plants follow all directions on this label.

DIRECTIONS FOR USE

It is a violation of Federal law to use this product in a manner inconsistent with its labeling.

Do not apply this product through any type of irrigation system.

Read complete use directions and precautions on this label and container label before using.

For reduction of perennial weeds such as field bindweed, use ground application equipment to broadcast treat infested areas using 0.5 to 1 pint TORDON 22K Weed Killer tank mixed with 1 to 2 pints of 3.8 lb/gal. 2,4-D amine or low-volatile ester products or with 1/3 to 2/3 quarts of 5.8 lb/gallon comparable 2,4-D formulations, in 10 or more gallons of water per acre.

To maintain control of perennial weed infestations, treat annually between crops of small grains. For best results with continuous grain production treat field bindweed infested areas in June or July following small grain harvest, when bindweed is actively growing, applying the lower rate of TORDON 22K Weed Killer and either rate of 2,4-D **(higher rate of TORDON 22K Weed Killer may be used by greater crop injury to fall planted grain may result compared to that which may occur with the lower rate).** For best results with small grain-fallow production treat field bindweed infested areas in June or July following grain harvest, applying the higher rates of TORDON 22K Weed Killer and 2,4-D; then around mid-May during the fallow season treat the field bindweed infested area with the lower rate of TORDON 22K Weed Killer and either rate of 2,4-D **(higher rate of TORDON 22K Weed Killer may be used but greater crop injury to fall planted grain may result compared to what may occur from the lower rate).**

Do not treat with TORDON 22K Weed Killer more than once each calendar year. Additional treatment with 2,4-D alone may be made anytime weeds are present with adequate foliage and in good growing condition.

To control volunteer grain or grass weeds that may be present, use cultivation or appropriate grass control herbicides alone or in combination with broadleaf weed control treatments. Consult the University Experiment Station or Extension Service for recommendations to fit local conditions.

Wheat or other small grain crops may be injured when planted at varying periods following application of TORDON 22K Weed Killer. Do not treat with TORDON 22K if the risk of injury to the small grain crop cannot be tolerated.

For long residual control of perennial weeds such as field bindweed, spot treat scattered infestations following use directions on the container label. When using these higher rate treatments of TORDON 22K, do not treat more than 10% of the area of an infested field in any calendar year. Wheat or other small grain crops may be planted in such treated fields, but crop injury in the spot treated areas may occur for one or more seasons after treatment.

USE PRECAUTIONS

Use this product only as specified on this label. Observe any special use and application restrictions and limitations, including method of application, and permissible areas of use as promulgated by state and local authorities.

Handle and Use Correctly. Small amounts of TORDON 22K Weed Killer can injure sensitive broadleaf plants. To avoid problems, observe the following precautions:

1. Do Not Contaminate Water Intended for Irrigation or Domestic Purposes. Do not treat or allow spray drift to fall onto innerbanks or bottom of irrigation ditches, either dry or containing water.

2. Do Not Apply to or in the Vicinity of Sensitive Crops or Desirable Plants including alfalfa, beans, cotton, grapes, melons, peanuts, peas, potatoes, sunflower, sugarbeets, soybeans, tobacco, tomatoes, other field or vegetable crops, flowers, fruit plants, ornamentals, or shade trees.

3. Use a bioassay to determine that picloram residues are no longer detectable in the treated soil. Contact State Extension Specialists for further information on bioassay testing.

4. Do Not Move Treated Soil to Other Areas where susceptible desirable plants may be exposed and damaged while phytotoxic residues are present.

5. Avoid Spray Drift. Applications should be made only when there is no hazard from spray drift since very small quantities of the spray, which may not be visible, may severely injure susceptible crops during both growing and dormant periods. To avoid spray drift apply only when wind velocity is less than 10 mph and is not blowing toward susceptible crops, keep spray boom as low as possible; apply 20 gallons or more of spray per acre; use no more than 30 psi spraying pressure with large droplet producing tips.

Do Not Apply With Hollow Cone-Type Insecticide or Other Nozzles that Produce Fine-Droplet Spray. A spray thickening agent such as NALCO-TROL may be used with this product to aid in reducing spray drift. If a drift reducing agent is used, follow use directions and precautions on the product label.

6. Do Not Use Where a Sandy Porous Surface and Substrate Overlie Ground Water Ten Feet or Less Below the Surface.

7. Be Sure That Use of This Product Conforms to All Applicable Regulations.

8. Grazing Restrictions: None.

9. Follow Storage and Disposal Requirements on Container Label.

OKLAHOMA. . . . . EPA SLN No. OK-840013

For The Control of Broadleaf Weeds Such as Wild Buckwheat on Land Planted to Winter Wheat, Spring Barley and Oats, Not Underseeded With a Legume.

GENERAL INFORMATION

Use TORDON 22K Weed Killer by itself or in a tank mix with a 2,4-D or MCPA amine formulation to control wild buckwheat as well as many other annual broadleaf weeds. Application may be made either by air or ground application equipment. Picloram, the active ingredient in TORDON 22K Weed Killer, is highly active, water soluble and has moderate soil residual. To prevent damage to desirable crops and plants, follow all directions on this label.

DIRECTIONS FOR USE

It is a violation of Federal law to use this product in a manner inconsistent with its labeling.

Do not apply this product through any type of irrigation system.

Read complete directions and precautions on this label and container label before using.

Read complete use directions and precautions for all tank mixed herbicides and container labels before using.

For the control of wild buckwheat and other annual broadleaf weeds, use TORDON 22K by itself or as a tank mix with 2,4-D or MCPA amine formulations. For use on spring oats, tank mix only with MCPA Amine Herbicide formulations.

SPRING BARLEY AND OATS: Apply during the 3 through 5 leaf stage of growth. Application of TORDON 22K occasionally causes slight head malformations and straw shortening but normally does not affect yield.

WINTER WHEAT: Apply after resumption of active growth to early jointing.

For aerial or ground treatment use enough total spray volume to provide adequate spray coverage. Apply 2 to 4 gpa by air and 10 or more gallons of water per acre by ground. Spray pressure should not exceed 30 psi. Use spray nozzles designed for herbicide application to minimize spray drift.

To prepare the spray, mix only with water. Add about half the desired amount of water in the spray tank. Then, with agitation, add the recommended amount of TORDON 22K Weed Killer and either 2,4-D or MCPA Herbicide formulations as outlined in the table that follows. Finally, with continued agitation, add the rest of the water.

AMOUNTS OF EACH PRODUCT PER ACRE[1]

Growth Stage of Winter Wheat	TORDON* 22K plus	4 lb./gal. 2,4-D or	4 lb./gal. MCPA
To early jointing	1-1½ fl. oz.	—	—
	1-1½ fl. oz.	½-¾ pt.	—
	1-1½ fl. oz.	—	½-¾ pt.

AMOUNTS OF EACH PRODUCT PER ACRE[1]

Growth Stage of Spring Barley and Oats	TORDON* 22K plus	4 lb./gal. 2,4,-D or	4 lb./gal. MCPA
3 to 5 leaf stage.	1-1½ fl. oz.	—	—
	1-1½ fl. oz.	½-¾ pt.	—
	1-1½ fl. oz.	—	½-¾ pt.[2]

[1]When measuring small amounts of TORDON* 22K Weed Killer, special care should be taken not to exceed suggested rates.

[2]For use on spring oats, tank mix only with MCPA Amine Herbicide formulations.

The dosages recommended equate to ¼-⅜ oz. picloram (0.016 - 0.023 lbs. a.e.); 4 - 6 oz. 2,4-D (0.25 - 0.38 lbs. a.e.); 4 - 6 oz. MCPA (0.25 - 0.38 lbs. a.e.) per acre.

Note: Use only on land that will be planted the following fall to grass or small grain crops or fallowed. To not apply more than ⅜ ounce picloram (1½ fl. ounces of TORDON 22K Weed Killer; 0.023 lbs. a.e.) per acre for wild buckwheat control during the small grain growing season.

USE PRECAUTIONS

Use this product only as specified on this label. Observe any special use and application restrictions and limitations, including method of application, and permissible areas of use as promulgated by state and local authorities.

Handle and use correctly. Small amounts of TORDON 22K Weed Killer can injure sensitive broadleaf plants. To avoid problems, observe the following precautions:

1. DO NOT CONTAMINATE WATER INTENDED FOR IRRIGATION OR DOMESTIC PURPOSES. Do not treat or allow spray drift to fall onto innerbanks or bottom of irrigation ditches, either dry or containing water.

2. DO NOT APPLY TO OR IN THE VICINITY OF SENSITIVE CROPS OR DESIRABLE PLANTS including alfalfa, beans, cotton, grapes, melons, peanuts, peas, potatoes, sunflowers, sugarbeets, soybeans, tobacco, tomatoes, other field or vegetable crops, flowers, fruit plants, ornamentals, or shade trees.

3. DO NOT MOVE TREATED SOIL TO OTHER AREAS where susceptible desirable plants may be exposed and damaged while phytotoxic residues are present.

4. AVOID SPRAY DRIFT. Applications should be made only when there is no hazard from spray drift since very small quantities of the spray, which may not be visible, may severely injure susceptible crops during both growing and dormant periods. Use coarse sprays to minimize drift since, under adverse weather conditions, fine spray droplets may drift a mile or more. The spray thickening agent NALCO-TROL may be used with this product to aid in reducing spray drift. If used, follow all use recommendations and precautions on the product label.

GROUND EQUIPMENT: With ground equipment, spray drift can be lessened by keeping the spray boom as low as possible; by applying 10 gallons or more of spray per acre; by using no more than 30 p.s.i. spraying pressure with large droplet producing nozzle tips; by using spray nozzles designed for herbicide application; by spraying when wind velocity is low; and by stopping all spraying when wind exceeds 10 miles per hour. Do not apply with hollow cone-type insecticide or other nozzles that produce a fine-droplet spray.

AERIAL APPLICATION: With aircraft, drift can be lessened by applying a coarse spray; by using no more than 30 psi spray pressure at nozzles; by using straight stream nozzles directed straight back; by using a spray boom no longer than ¾ the wing span of the aircraft; and by spraying only when wind velocity is less than 10 mph.

DO NOT APPLY BY AIRCRAFT WHEN AN AIR TEMPERATURE INVERSION EXISTS. Such a condition is characterized by little or no wind and with air temperature lower near the ground than at higher levels. The use of a continuous smoke column at or near site of application or use of a smoke generating device on the aircraft is suggested to indicate direction and velocity of air movement, and to indicate a temperature inversion by layering of the smoke.

5. DO NOT SPRAY PASTURES OR GRAIN IF THE FORAGE LEGUME COMPONENT IS DESIRED. TORDON 22K Weed Killer may injure or kill legumes. Also, new legume seedlings may not be successful if made within 2 years following application of this herbicide.

6. DO NOT TRANSFER LIVESTOCK from treated grass areas onto broadleaf crop areas without first allowing 7 days of grazing on untreated grass pasture. Otherwise, urine may contain enough picloram to cause injury to sensitive broadleaf plants.

7. DO NOT USE WHERE A SANDY POROUS SURFACE AND SUBSTRATE OVERLIE GROUND WATER TEN FEET OR LESS BELOW THE SURFACE.

8. BE SURE THAT USE OF THIS PRODUCT CONFORMS TO ALL APPLICABLE REGULATIONS.

9. GRAZING RESTRICTIONS: NONE.

10. FOLLOW STORAGE AND DISPOSAL REQUIREMENTS ON CONTAINER LABEL.

SOUTH DAKOTA. . . . . EPA SLN NO. SD-870008

For the Control of Broadleaf Weeds in Rangelands, Permanent Grass Pastures, Spring Barley and Oats, and Spring and Winter Wheat.

DIRECTIONS FOR USE

It is a violation of Federal law to use this product in a manner inconsistent with its labeling.

Do not apply this product through any type of irrigation system.

Read complete use directions and precautions on this label and container label before using.

RANGELAND AND PASTURE

Use TORDON 22K Weed Killer to control broadleaf annual and perennial weeds such as Canadian and other thistles, field bindweed, leafy spurge, Russian knapweed, spotted knapweed, tall larkspur, yellow toadflax, locoweed, snakeweed and lupines on rangeland and permanent grass pastures. Treat when weeds are growing well using low pressure sprays. Retreatment at the same rate may be necessary the following year.

Spot Treatment: (ground application only) Use TORDON 22K at rates of 2 to 4 quarts in 20 to 100 gallons of water per acre and apply as a spray to the foliage. Use the higher rates to control leafy spurge, larkspur, toadflax and knapweed. Use the lower rates for bindweed and thistles. For a 1,000 square foot infestation, apply about 2¼ fluid ounces of TORDON 22K in 1 gallon of water (equivalent to 3 quarts per acre of TORDON 22K).

NOTE: When spot treating, do not cut grass for feed within 2 weeks after treatment. Meat animals grazing for up to 2 weeks after treatment should be removed from treated areas 3 days prior to slaughter. Do not graze dairy animals on treated areas within 2 weeks after treatment.

Broadcast Treatment: (ground or aerial application) To suppress the growth of many perennial broadleaf weeds, apply as a broadcast spray using 1 quart of TORDON 22K in 1 to 4 gallons of water per acre by air or in 20 to 100 gallons of water per acre by ground equipment and apply as a broadcast spray during the growing season when weeds are growing well. (Many seedling annual weeds can be controlled using 1 pint per acre.)

Tank Mixture for Spot or Broadcast Treatments: TORDON 22K may also be tank mixed with 2,4-D products such as ESTERON* 99* Concentrate, FORMULA 40*, DMA* 4 or ESTERON 6E herbicides for use on areas having mixed species including those which respond well to 2,4-D, such as big sagebrush. In tank mix combinations, use 1 pint to 1 quart of TORDON 22K with 1 to 2 quarts ESTERON 99 Concentrate, FOR-

MULA 40 or DMA 4, or with 2/3 to 1 1/3 quarts of ESTERON 6E per acre, in spray volumes specified above. **Read and follow all directions and use precautions on other product labels.**

For Control of Musk Thistle: Use 6 to 8 fluid ounces (3/8 to 1/2 pint) of TORDON 22K in 1 to 4 gallons of water per acre by air or in 10 to 20 gallons of water per acre by ground equipment, and apply in the spring or fall when musk thistle plants are in the rosette stage of growth.

SPRING BARLEY AND OATS, SPRING AND WINTER WHEAT NOT UNDERSEEDED WITH A LEGUME

For the control of wild buckwheat and other annual broadleaf weeds normally controlled with 2,4-D or MCPA such as wild mustard, Russian thistle, pennycress, lambsquarters and pigweed, in spring wheat and barley and winter wheat use TORDON 22K as a tank mix with a 2,4-D or MCPA formulation such as DMA 4, FORMULA 40, ESTERON 99 Concentrate, ESTERON 6E, or MCP Amine Herbicides. For use on spring oats, tank mix only with MCP Amine Herbicide. **Read and follow all directions and use precautions on other product labels.**

Spring Wheat, Barley and Oats: Apply during the 3 through 5 leaf stage of growth. Application of TORDON 22K occasionally causes slight head malformations and straw shortening but normally this does not affect yield.

Durum Wheat: Do not treat durum since at least some varieties appear to be more sensitive than other wheat.

Winter Wheat: Apply after resumption of active growth in the spring and before the early boot stage.

For aerial or ground treatment, use enough total spray volume to provide adequate spray coverage. Apply 1 to 4 gpa by air and 5 to 20 gpa by ground. Spray pressure should not exceed 30 psi. Use a coarse spray to minimize spray drift.

To prepare the spray, mix only with water. Add about half the desired amount of water in the spray tank. Then with agitation, add the recommended amount of TORDON 22K and 2,4-D or MCPA as outlined in the table. Finally, with continued agitation, add the rest of the water.

The dosages recommended equate to 1/4 oz. picloram + 4 oz. 2,4-D or MCPA ae/acre when weeds are small or 3/8 + 6 oz/acre when weeds are more advanced or when dry soil conditions exist.

Use Rates for Barley, Oats, and Wheat

Weed Growth Stage	Amounts of Each Product Per Acre†		
	TORDON 22K *plus*	DMA 4, FORMULA 40, ESTERON 99 Concentrate or MCP Amine *or*	ESTERON 6E
weeds 1-3 inches tall	1 fl. oz.	1/2 pt.	1/3 pt.
weeds 3-6 inches tall or under dry conditions	1 1/2 fl. oz.	3/4 pt.	1/2 pt.

† When measuring small amounts of TORDON 22K weed killer, special care should be taken not to exceed suggested rates.

NOTE: Use only on land that will be planted the following year to grass or grain crops such as small grains, corn, sorghum, or flax. Do not apply more than 3/8 ounce picloram per acre during any 12-month period.

USE PRECAUTIONS

Use this product only as specified on this label. Observe any special use and application restrictions and limitations, including method of application and permissible areas of use as promulgated by state or local authorities.

Do Not Contaminate Nontarget Land Areas, Cropland, Water, or Irrigation Ditches. Do not apply directly to standing or running water. Do not apply where surface water from treated areas can run off to adjacent cropland, either planted or to be planted, or into streams, irrigation ditches, irrigation ponds, or wells. Do not clean containers nor application equipment on or near these areas. Do not apply on inner banks or bottom of irrigation ditches.

Do not apply on or in the vicinity of susceptible crops or desirable plants including alfalfa, beans, grapes, melons, peas, potatoes, safflower, soybeans, sugar beets, sunflower, tomatoes and other vegetable crops, flowers, fruit plants, ornamentals or shade trees.

Avoid Spray Drift: Applications should be made only when there is no hazard from spray drift since very small quantities of the spray, which may not be visible, may severely injure susceptible crops during both growing and dormant periods. Use coarse sprays to minimize drift since, under adverse weather conditions, fine spray droplets may drift a mile or more. The spray thickening agent, NALCO-TROL, may be used with this product to aid in reducing spray drift. If used, follow all use recommendations and precautions on the product label.

Ground Equipment: With ground equipment, spray drift can be lessened by keeping the spray boom as low as possible; by applying 20 gallons or more of spray per acre; by using no more than 30 pounds spraying pressure with large droplet producing nozzle tips; by spraying when wind velocity is low; and by stopping all spraying when wind exceeds 10 miles per hour. Do not apply with hollow cone-type insecticide or other nozzles that produce a fine-droplet spray.

Aerial Application: With aircraft, drift can be lessened by applying a coarse spray; by using no more than 30 pounds spray pressure at nozzles; by using straight stream nozzles directed straight back, by using a spray boom no longer than 3/4 the wing span of the aircraft; and by spraying only when wind velocity is less than 10 mph.

Do Not Apply By Aircraft When An Air Temperature Inversion Exists: Such a condition is characterized by little or no wind and with air temperature lower near the ground than at higher levels. The use of a continuous smoke column at or near site of application or use of a smoke generating device on the aircraft is suggested to indicate direction and velocity of air movement, and to indicate a temperature inversion by layering of the smoke.

Do not rotate treated rangeland or pasture to other crop uses.

Do not spray pastures or grain if the forage legume component is desired. TORDON 22K Weed Killer may injure or kill legumes. Also new legume seedlings may not be successful if made within 2 years following application of this herbicide.

Do not move treated soil to other areas. Do not use it to grow plants, unless adequate sensitive bioassay or chemical tests show that no detectable picloram is present in the soil.

Do not transfer livestock from treated grass areas onto broadleaf crop areas without first allowing 7 days of grazing on untreated grass pasture. Otherwise, urine may contain enough picloram to cause injury to sensitive broadleaf plants.

Do not re-use containers for TORDON 22K Weed Killer for any purpose. Dispose by punching holes in them and burying with waste or by taking to an approved landfill. Where indicated, follow official local container disposal regulations.

Rinse application equipment after use, preferably at least three times with water, and dispose of rinse water in a non-cropland area away from water supplies.

Be sure that use of this product conforms to all applicable regulations.

Do not use where a sandy porous surface and substrate overlie ground water 10 feet or less below the surface.

TEXAS . . . . EPA SLN NO. TX-830003

For Control of Annual and Perennial Broadleaf Weeds Such as Field Bindweed in Grainland Between Crops of Small Grains.

DIRECTIONS FOR USE

It is a violation of Federal law to use this product in a manner inconsistent with its labeling.

Do not apply this product through any type of irrigation system.

Read complete use directions and precautions on this label and container label before using.

Use TORDON 22K Weed Killer by ground application to control or reduce infestations of broadleaf annual and perennial weeds such as field bindweed on grainland during the period following crop harvest and prior to planting winter wheat, barley, or oats.

For **reduction of perennial** weeds such as field bindweed, broadcast treat infested areas using 0.5 to 1 pint TORDON 22K tank mixed with 1 to 2 pints ESTERON* 99* Concentrate, FORMULA 40*, DMA* 4 herbicides, or with 1/3 to 2/3 quart ESTERON 6E herbicide in 10 or more gallons of water per acre. To maintain or further reduce perennial weed infestations, treat annually between crops. For best results treat in June or July following small grain harvest, applying the higher rates of TORDON 22K and 2,4-D. Initial treatment alternatively may be made soon after harvesting a fall maturing crop such as grain sorghum or corn. During the next calendar year, retreat the same area with the lower rate of TORDON 22K, preferably in May or June during the fallow period prior to fall planting of the next small grain crop.

Do not treat with TORDON 22K Weed Killer more than once each calendar year. Additional treatment with 2,4-D alone may be made anytime weeds are present with adequate foliage and in good growing condition.

To control volunteer grain or grass weeds that may be present, use cultivation or appropriate grass control herbicides alone or in combination with broadleaf weed control treatments. Consult the University Experiment Station or Extension Service for recommendations to fit local conditions.

Wheat or other small grain crops may be injured when planted at varying periods following application of TORDON 22K Weed Killer. Do not treat with TORDON 22K if the risk of injury to the small grain crop cannot be tolerated.

For long residual control of perennial weeds such as field bindweed, spot treat scattered infestations following use directions on the container label. When using these higher rate treatments of TORDON 22K, do not treat more than 10% of the area of an infested field in one calendar year. Wheat or other small grain crops may be planted in such treated fields, but crop injury in the spot treated areas may occur for one or more seasons after treatment.

USE PRECAUTIONS

Use this product only as specified on this label. Observe any special use and application restrictions and limitations, including method of application and permissible areas of use as promulgated by state and local authorities.

Do not apply on or in the vicinity of susceptible crops or desirable plants including alfalfa, beans, grapes, melons, peas, potatoes, safflower, soybeans, sugar beets, sunflower, tomatoes and other vegetable crops, flowers, fruit plants, ornamentals, shade treees, or the soil containing roots of nearby valuable plants.

AVOID SPRAY DRIFT: Applications should be made only when there is no hazard from spray drift since very small quantities of the spray, which may not be visible, may severely injure susceptible crops during both growing and dormant periods. To avoid spray drift use low spray pressure, under 30 psi; spray when wind velocity is less than 10 mph; and apply as a coarse spray. The spray thickening agent NALCO-TROL may be used with this product to aid in reducing spray drift. If this thickening agent is used, follow all use recommendations and precautions on the product label.

Ground Equipment: With ground equipment, spray drift can be lessened by keeping the spray boom as low as possible; by applying 20 gallons or more of spray per acre; by using no more than 30 pounds spraying pressure with large droplet producing nozzle tips; by spraying when wind velocity is low; and by stopping all spraying when wind exceeds 10 miles per hour. Do not apply with hollow cone-type insecticide or other nozzles that produce fine-droplet spray.

Do Not Contaminate Water Intended for Irrigation or Domestic Purposes. Do not treat or allow spray drift to fall onto innerbanks or bottom of irrigation ditches, either dry or containing water or other channels that carry water that may be used for irrigation purposes.

Do not move treated soil to other areas where susceptible desirable plants may be exposed and damaged while phytotoxic residues are present.

UTAH. . . . . EPA SLN NO. UT-840004

For the Control of Broadleaf Weeds in Rangeland, Forest and Permanent Grass Pastures, Spring Barley and Oats, and Spring and Winter Wheat.

DIRECTIONS FOR USE

It is a violation of Federal law to use this product in a manner inconsistent with its labeling.

Do not apply this product through any type of irrigation system.

Read complete use directions and precautions on this label and container label before using.

SPRING BARLEY AND OATS, SPRING AND WINTER WHEAT NOT UNDERSEEDED WITH A LEGUME

For the control of wild buckwheat, volunteer sunflower and other annual broadleaf weeds such as wild mustard, Russian thistle, pennycress, lambsquarters, pigweed and sowthistle in spring wheat and barley and winter wheat, use TORDON 22K weed killer as a tank mix with amine or ester formulations of 2,4-D or MCPA. For use on spring oats, tank mix only with MCP Amine Herbicide. READ AND FOLLOW ALL DIRECTIONS AND USE PRECAUTIONS AND USE PRECAUTIONS ON OTHER LABELS.

SPRING WHEAT, BARLEY AND OATS: Apply during the 3 through 5 leaf stage of growth. Application of TORDON 22K occasionally causes slight head malformations and straw shortening but normally this does not affect yield.

DURUM WHEAT: Do not treat durum since at least some varieties appear to be more sensitive than other wheat.

WINTER WHEAT: Apply after resumption of active growth in the spring and before the early boot stage.

For aerial or ground treatment, use enough total spray volume to provide adequate spray coverage. Apply 2 to 4 gpa by air and 5 to 10 gpa by ground. Spray pressure should not exceed 40 psi. Use coarse spray to minimize spray drift.

To prepare the spray, mix only with water. Add about half the desired amount of water in the spray tank. Then with agitation, add the recommended amount of TORDON 22K and 2,4-D or MCPA as outlined in the table. Finally, with continued agitation, add the rest of the water. Be sure to maintain agitation during application of tank mixes with 2,4-D or MCPA ester products.

USE RATES FOR WHEAT, BARLEY AND OATS[1]

WEEDS	WEED GROWTH STAGE[2]	AMOUNTS OF EACH PRODUCT PER ACRE: TORDON 22K PLUS	PRODUCTS CONTAINING 2,8 LB 2,4-D OR MCPA A.E. PER GAL. OR	5,6 LB 2,4-D A.E. PER GAL.
More susceptible species such as lambsquarters	1-3 inches	1 fl. oz.	1/2 pt.	1/3 pt.
	3-6 inches	1 1/2 fl. oz.	3/4 pt.	1/2 pt.
Less susceptible species such as volunteer sunflower, wild buckwheat, Russian thistle, annual and perennial sowthistle, pigweed, and top growth suppression of Canada thistle	1-6 inches	1 1/2 fl. oz.	3/4 pt.	1/2 pt.

[1] For oats, tank mix only with MCP Amine or MCP Ester Herbicides.

[2] For best results, treat when weeds have 2 to 4 leaves and are actively growing.

NOTE: Use only on land that will be planted the following year to grass or grain crops such as small grains, corn, sorghum, or flax.

Do not spray a small grain crop more than once with TORDON 22K Weed Killer, during a single season.

RANGELAND, FOREST AND PERMANENT GRASS PASTURES

Use TORDON 22K Weed Killer to control broadleaf annual and perennial weeds such as (but not limited to) Canada thistle and other thistles, field bindweed, leafy spurge, Russian knapweed, spotted and diffuse knapweed, plains or Geyer larkspur, Dalmation and yellow toadflax, fringed sagebrush, snakeweed, rabbit brush, pricklypear cactus, skeltonleaf bursage, locoweed, and lupines on rangeland and permanent grass pastures. Treat when weeds are actively growing. Retreatment may be necessary the following year.

SPOT TREATMENT: Use TORDON 22K rates shown in the following table and apply in sufficient water to obtain adequate spray coverage. For perennial weeds, spray the weed infested area plus a 10-foot border around the infestation. Consult the product container label for guidance to prepare spray mixes to spot treat small patches.

BROADCAST TREATMENT: Apply as a broadcast spray on hard to control weeds at rates shown in the following table. Apply the selected rate of TORDON 22K in 2-4 gallons of total spray volume per acre (water diluent only) by air, or in 5 to 30 gallons of water per acre by ground equipment. For rates exceeding 1 quart of TORDON 22K per acre, the total area treated in a single season should not exceed 25% of a landowner's acreage found in any particular watershed.

TANK MIXTURES FOR SPOT OR BROADCAST TREATMENTS: TORDON 22K may also be tank mixed with 2,4-D products for use on areas having mixed species including those which respond well to 2,4-D. In tank mix combinations, use 1 pint to 4 quarts of TORDON 22K with 1 to 2 quarts of low volatile ester or amine products containing 3.8 lb. 2,4-D acid equivalent per gallon or with 2/3 to 1 1/3 quarts of comparable products containing 5.6 lb. 2,4-D acid equivalent per gallon in spray volumes specified below. When tank mixing with 2,4-D ester products, agitation should be maintained throughout preparation and application. **READ AND FOLLOW ALL DIRECTIONS AND USE PRECAUTIONS ON OTHER PRODUCT LABELS.**

NOTE: TORDON 22K at rates over 1 quart may suppress certain grasses, such as bromegrass, bluegramma and buffalograss. Usually later grass growth will be improved by release from competition.

GRAZING RESTRICTIONS: When applying more than 1 quart of TORDON 22K per acre, do not cut grass for feed within two weeks after treatment. Meat animals grazing for up to two weeks after treatment should be removed from treated areas three days prior to slaughter. Do not graze lactating dairy animals on treated areas within two weeks after treatment.

Weed Species[1]	Rates Of TORDON 22K Per Sprayed Acre[2]
Diffuse Knapweed; Spotted Knapweed; Rabbitbrush; Fringed Sagebrush;	1 pint
Musk, Plumeless & Scotch Thistles; Tall or Geyer	1 to 2 pints

Weed Species[1]	Rates Of TORDON 22K Per Sprayed Acre[2]
Larkspur[3,4]; Locoweed[3,4] Lupines[4] Snakeweed; Pricklypear Cactus	
Russian Knapweed; Skeletonleaf Bursage	1 to 2 quarts
Canada Thistle; Dalmation Toadflax; Yellow Toadflax; Field Bindweed	2 to 4 quarts
Leafy Spurge[4]	1 quart Repeat annually for 3-4 years 2 quarts Treat 2 consecutive years 4 quarts Single application

[1] For additional species or more specific rates consult the current UT Weed Control Guide.

[2] NOTE: Follow all above treatments with spot touch-up treatment as needed using TORDON 22K herbicide.

[3] Use higher rates.

[4] Denotes poisonous plants.

USE PRECAUTIONS

Use this product only as specified on this label. Observe any special use and application restrictions and limitations, including method of application and permissible areas of use as promulgated by state or local authorities.

Avoid water contamination. Do not allow TORDON 22K Weed Killer to contaminate water used for drinking, irrigation or other domestic purposes. Do not apply on inner banks or bottoms of irrigation ditches. Do not clean containers or application equipment on or near these areas.

Do not apply on or near susceptible crops or desirable plants including alfalfa, beans, grapes, melons, peas, potatoes, safflower, soybeans, sugar beets, sunflower, tomatoes and other vegetable crops, flowers, fruit plants, ornamentals or shade trees.

Avoid Spray Drift: Applications should be made only when there is no hazard from spray drift since very small quantities of the spray, which may not be visible, may severely injure susceptible crops such as those listed above, during both growing and dormant periods. Use coarse sprays to minimize drift since, under adverse weather conditions, fine spray droplets may drift a mile or more. A spray thickening agent, such as NALCO-TROL may be used with this product to aid in reducing spray drift. If used, follow all use recommendations and precautions on the product label.

Ground Equipment: With ground equipment, spray drift can be lessened by keeping the spray boom to a height recommended for the nozzle tips in use and by spraying only when wind velocity is less than 5 miles per hour. Do not apply with nozzles that produce a fine-droplet spray.

Aerial Application: Drift can be lessened by applying a coarse spray; by using no more than 40 pounds spray pressure at nozzles; by using nozzles which do not produce hazardous amounts of fine spray droplets; by using a spray boom no longer than 3/4 of the rotor or wing length; and by spraying only when wind velocity is less than 5.0 mph.

Determine Air Movement and Direction Before Foliar Application: Use a smoke generator or other means at or near the application site or on aircraft for detection of air movement, air stability, or temperature inversions. Such a condition exists when there is little or no wind and air temperature is lower near the ground than at higher levels. Use appropriate drift control measures or stop application when smoke is moving toward nearby susceptible plants or sensitive areas.

DO NOT ROTATE TREATED RANGELAND OR PASTURE to susceptible crops until phytotoxic residues have disappeared from the soil.

DO NOT SPRAY PASTURES OR GRAIN IF THE FORAGE LEGUME COMPONENT IS DESIRED. TORDON 22K weed killer may injure or kill legumes. Also new legume seedlings may not be successful if made within 2 years following application of this herbicide.

Avoid movement of treated soil. Picloram may remain in treated soil for an extended period. Do not move treated soil to other areas and do not use such soil to grow susceptible plants until residues have reached a non-phytotoxic level.

LIVESTOCK TRANSFER RESTRICTIONS. If livestock are moved from a treated area within the first 12 months after application onto a sensitive broadleaved area, allow 7 days of grazing on untreated pasture. Otherwise, urine may contain enough picloram to cause injury to sensitive broadleaf plants.

Rinse application equipment and containers thoroughly, preferably at least three times with water, and dispose of rinse water in a non-cropland area away from water supplies.

DO NOT RE-USE CONTAINERS FOR TORDON 22K WEED KILLER for any purpose. Containers should be disposed of by punching holes in them and burying (in a non-crop area away from water) or by taking to an approved landfill. Where indicated, follow official container disposal regulations.

BE SURE THAT USE OF THIS PRODUCT CONFORMS TO ALL APPLICABLE REGULATIONS.

WYOMING . . . . EPA SLN NO. WY-840001

For the Control of Broadleaf Weeds in Rangeland, Forest and Permanent Grass Pastures, Spring Barley and Oats, and Spring and Winter Wheat.

DIRECTIONS FOR USE

It is a violation of Federal law to use this product in a manner inconsistent with its labeling.

Do not apply this product through any type of irrigation system.

Read complete use directions and precautions on this label and container label before using.

SPRING BARLEY AND OATS, SPRING AND WINTER WHEAT NOT UNDERSEEDED WITH A LEGUME

For the control of wildbuckwheat, volunteer sunflower and other annual broadleaf weeds such as wild mustard, Russian thistle, pennycress, lambsquarter and pigweed, in spring wheat and barley and winter wheat, use TORDON 22K weed killer as a tank mix with amine or ester formulations of 2,4-D or MCPA. For use on spring oats, tank mix only with MCP Amine Herbicide. READ AND FOLLOW ALL DIRECTIONS AND USE PRECAUTIONS AND USE PRECAUTIONS ON OTHER PRODUCT LABELS.

SPRING WHEAT, BARLEY AND OATS: Apply during the 3 through 5 leaf stage of growth. Application of TORDON 22K occasionally causes slight head malformations and straw shortening but normally this does not affect yield.

DURUM WHEAT: Do not treat durum since at least some varieties appear to be more sensitive than other wheat.

WINTER WHEAT: Apply after resumption of active growth in the spring and before the early boot stage.

For aerial or ground treatment, use enough total spray volume to provide adequate spray coverage. Apply 2 to 4 gpa by air and 5 to 10 gpa by ground. Spray pressure should not exceed 40 psi. Use coarse spray to minimize spray drift.

To prepare the spray, mix only with water. Add about half the desired amount of water in the spray tank. Then with agitation, add the recommended amount of TORDON 22K and 2,4-D or MCPA as outlined in the table. Finally, with continued agitation, add the rest of the water. Be sure to maintain agitation during application of tank mixes with 2,4-D ester products.

USE RATES FOR WHEAT, BARLEY AND OATS[1]

WEEDS	WEED GROWTH STAGE[2]	AMOUNTS OF EACH PRODUCT PER ACRE — PRODUCTS CONTAINING: TORDON 22K PLUS	3.8 LB 2,4-D OR MCPA A.E. PER GAL. OR	5.6 LB 2,4-D A.E. PER GAL.
More susceptible species such as lambsquarters	1-3 inches	1 fl. oz.	1/2 pt.	1/3 pt.
	3-6 inches	1 1/2 fl. oz.	3/4 pt.	1/2 pt.
Less susceptible species such as volunteer sunflower, wild buckwheat, Russian thistle, annual and perennial sowthistle, pigweed, and top growth suppression of Canada thistle	1-6 inches	1 1/2 fl. oz.	3/4 pt.	1/2 pt.

[1] For oats, tank mix only with MCP Amine or MCP Ester Herbicides.

[2] For best results, treat when weeds have 2 to 4 leaves and are actively growing.

NOTE: Use only on land that will be planted the following year to grass or grain crops such as small grains, corn, sorghum, or flax.

Do not spray a small grain crop more than once with TORDON 22K Weed Killer, during a single season.

RANGELAND, FOREST AND PERMANENT GRASS PASTURES

Use TORDON 22K Weed Killer to control broadleaf annual and perennial weeds such as (but not limited to) Canada thistle and other thistles, field bindweed, leafy spurge, Russian knapweed, spotted and diffuse knapweed, plains or Geyer larkspur, Dalmation and yellow toadflax, fringed sagebrush, snakeweed, rabbit brush, pricklypear cactus, skeltonleaf bursage, locoweed, and lupines on rangeland and permanent grass pastures. Treat when weeds are actively growing. Retreatment may be necessary the following year.

SPOT TREATMENT: Use TORDON 22K rates shown in the following table and apply in sufficient water to obtain adequate spray coverage. For perennial weeds, spray the weed infested area plus a 10-foot border around the infestation. Consult the product container label for guidance to prepare spray mixes to spot treat small patches.

BROADCAST TREATMENT: Apply as a broadcast spray on hard to control weeds at rates shown in the following table. Apply the selected rate of TORDON 22K in 2-4 gallons of total spray volume per acre (water diluent only) by air, or in 5 to 30 gallons of water per acre by ground equipment. For rates exceeding 1 quart of TORDON 22K per acre, the total area treated in a single season should not exceed 25% of a landowner's acreage found in any particular watershed.

TANK MIXTURES FOR SPOT OR BROADCAST TREATMENTS: TORDON 22K may also be tank mixed with 2,4-D products for use on areas having mixed species including those which respond well to 2,4-D. In tank mix combinations, use 1 pint to 4 quarts of TORDON 22K with 1 to 2 quarts of low volatile ester or amine products containing 3.8 lb. 2,4-D acid equivalent per gallon or with 2/3 to 1 1/3 quarts of comparable products containing 5.6 lb. 2,4-D acid equivalent per gallon in spray volumes specified below. When tank mixing with 2,4-D ester products, agitation should be maintained throughout preparation and application. READ AND FOLLOW ALL DIRECTIONS AND USE PRECAUTIONS ON OTHER PRODUCT LABELS.

NOTE: TORDON 22K at rates over 1 quart may suppress certain grasses, such as bromegrass, bluegramma and buffalograss. Usually later grass growth will be improved by release from competition.

GRAZING RESTRICTIONS: When applying more than 1 quart of TORDON 22K per acre, do not cut grass for feed within two weeks after treatment. Meat animals grazing for up to two weeks after treatment should be removed from treated areas three days prior to slaughter. Do not graze lactating dairy animals on treated areas within two weeks after treatment.

Weed Species[1]	Rates Of TORDON 22K Per Sprayed Acre[2]
Diffuse Knapweed; Spotted Knapweed; Rabbitbrush; Fringed Sagebrush;	1 pint
Musk, Plumeless & Scotch Thistles; Tall or Geyer Larkspur[3,4]; Locoweed[3,4] Lupines[4] Snakeweed; Pricklypear Cactus	1 to 2 pints
Russian Knapweed; Skeletonleaf Bursage	1 to 2 quarts
Canada Thistle; Dalmation Toadflax; Yellow Toadflax; Field Bindweed	2 to 4 quarts
Leafy Spurge[4]	1 quart Repeat annually for 3-4 years 2 quarts Treat 2 consecutive years 4 quarts Single application

[1] For additional species or more specific rates consult the current WY Weed Control Guide.

[2] NOTE: Follow all above treatments with spot touch-up treatment as needed using TORDON 22K herbicide.

[3] Use higher rates.

[4] Denotes poisonous plants.

USE PRECAUTIONS

Use this product only as specified on this label. Observe any special use and application restrictions and limitations, including method of application and permissible areas of use as promulgated by state or local authorities.

Avoid water contamination. Do not allow TORDON 22K Weed Killer to contaminate water used for drinking, irrigation or other domestic purposes. Do not apply on inner banks or bottoms of irrigation ditches. Do not clean containers or application equipment on or near these areas.

Do not apply on or near susceptible crops or desirable plants including alfalfa, beans, grapes, melons, peas, potatoes, safflower, soybeans, sugar beets, sunflower, tomatoes and other vegetable crops, flowers, fruit plants, ornamentals or shade trees.

Avoid Spray Drift: Applications should be made only when there is no hazard from spray drift since very small quantities of the spray, which may not be visible, may severely injure susceptible crops such as those listed above, during both growing and dormant periods. Use coarse sprays to minimize drift since, under adverse weather conditions, fine spray droplets may drift a mile or more. A spray thickening agent, such as NALCO-TROL may be used with this product to aid in reducing spray drift. If used follow all use recommendations and precautions on the product label.

Ground Equipment: With ground equipment, spray drift can be lessened by keeping the spray boom to a height recommended for the nozzle tips in use and by spraying only when wind velocity is less than 10 miles per hour. Do not apply with nozzles that produce a fine-droplet spray.

Aerial Application: Drift can be lessened by applying a coarse spray; by using no more than 40 pounds spray pressure at nozzles; by using nozzles which do not produce hazardous amounts of fine spray droplets; by using a spray boom no longer than 3/4 of the rotor or wing length; and by spraying only when wind velocity is less than 7.5 mph.

Determine Air Movement and Direction Before Foliar Application: Use a smoke generator or other means at or near the application site or on aircraft for detection of air movement, air stability, or temperature inversions. Such a condition exists when there is little or no wind and air temperature is lower near the ground than at higher levels. Use appropriate drift control measures or stop application when smoke is moving toward nearby susceptible plants or sensitive areas.

DO NOT ROTATE TREATED RANGELAND OR PASTURE to other sensitive crop uses.

DO NOT SPRAY PASTURES OR GRAIN IF THE FORAGE LEGUME COMPONENT IS DESIRED. TORDON 22K weed killer may injure or kill legumes. Also new legume seedlings may not be successful if made within 2 years following application of this herbicide.

Avoid movement of treated soil. Picloram may remain in treated soil for an extended period. Do not move treated soil to other areas and do not use such soil to grow susceptible plants until residues have reached a non-phytotoxic level

LIVESTOCK TRANSFER RESTRICTIONS. If livestock are moved from a treated area within the first 12 months after application onto a sensitive broadleaved area, allow 7 days of grazing on untreated pasture. Otherwise, urine may contain enough picloram to cause injury to sensitive broadleaf plants.

Rinse application equipment and containers thoroughly, preferably at least three times with water, and dispose of rinse water in a non-cropland area away from water supplies.

DO NOT RE-USE CONTAINERS FOR TORDON 22K WEED KILLER for any purpose. Containers should be disposed of by punching holes in them and burying (in a non-crop-land area away from water) or by taking to an approved landfill. Where indicated, follow official container disposal regulations.

BE SURE THAT USE OF THIS PRODUCT CONFORMS TO ALL APPLICABLE REGULATIONS.

SUPPLEMENTAL USE LABELS

NEBRASKA.................... EPA No. 464-07093-NE

To Control Broadleaf Weeds on Land Planted to Spring Barley and Spring and Winter Wheat.

USE DIRECTIONS

It is a violation of Federal law to use this product in a manner inconsistent with its labeling.

Do not apply this product through any type of irrigation system.

Do not apply where spray drift may be a problem due to proximity of susceptible crops or other desirable plants. Do not apply aerially when wind exceeds 5 mph or when picloram-susceptible crops or other plants are located within one mile downwind from the application site. Such crops and plants include field beans, soybeans, sunflowers, peas, potatoes, safflower, sugar beets, alfalfa, vegetables, fruit plants, flowers, and ornamentals.

FOR SPRING BARLEY, SPRING AND WINTER WHEAT NOT UNDER-SEEDED WITH A LEGUME: For the control of wild buckwheat as well as broadleaf weeds normally controlled with 2,4-D such as wild mustard, kochia, Russian thistle, fanweed, lambsquarter, and pigweed, use TORDON 22K Weed Killer as a tank mix only with a 2,4-D amine formulation containing 4 lb acid equivalent per gallon, such as DMA* 4 or FORMULA 40* herbicides.

For aerial or ground treatment, use enough total spray volume to provide adequate spray coverage. Apply 1 to 4 gpa by air and 5 to 20 gpa by ground. Spray pressure should not exceed 20 psi. Coarse sprays are less likely to drift.

NOTE: Do not apply aerially when wind exceeds 5 mph or when picloram-susceptible crops, including field beans, soybeans, peas, potatoes, sugar

beets, safflower or alfalfa, are located within one mile downwind from the application site.

To prepare the spray, mix only with water. Add about half the desired amount of water in the spray tank. Then with agitation, add the recommended amount of TORDON 22K and 2,4-D as outlined in the table. Finally with continued agitation, add the rest of the water.

Apply to spring barley and wheat during the period between the 4 leaf and early boot stages of growth. Apply to winter wheat after resumption of active growth in the spring and before the early boot stage.

Do not apply more than 3/8 ounce picloram per acre during any 12 month period.

Growth Stage	Amounts Of Products Per Acre†	
	TORDON 22K + 4 lb/gal 2,4-D Amine	
4 to 6 leaf (weeds small)	1 fl. oz	0.5 pt.
6 leaf to early boot (weeds more advanced or under dry conditions)	1.5 fl. oz.	0.75 pt.

†When measuring small amounts of TORDON 22K weed killer, special care should be taken not to exceed suggested rates.

The above table equates to 1/4 oz. picloram + 4 oz 2,4-D ae/acre for grains in the 4 to 6 leaf stage and 3/8 + 6 oz/acre for grain beyond the 6 leaf stage or when dry soil conditions exist.

NOTE: Since a specific growing season can be unusually dry and cool, to assure no carry-over problem, use only on land which will be fallowed or replanted to grass or grain crops the following year.

USE PRECAUTIONS

Use this product only as specified on this label. Observe any special use and application restrictions and limitations, including method of application and permissible areas of use as promulgated by state or local authorities.

Note: Do not apply on or in the vicinity of sensitive crops or desirable plants including alfalfa, beans, cotton, grapes, melons, peas, potatoes, safflower, sugar beets, sunflower, tobacco, tomatoes, other field and vegetable crops, flowers, fruit plants, ornamentals or shade trees. **Do not apply on areas which may be planted at a later time to such crops or plants.**

Do Not Contaminate Nontarget Land Areas, Cropland, Water, or Irrigation Ditches. Do not apply directly to standing or running water. Do not apply where surface water from treated areas can run off to adjacent cropland, either planted or to be planted, or into drainage channels, streams, irrigation ditches, ponds, or wells. Do not clean containers nor application equipment on or near these areas. Do not apply on inner banks or bottoms of irrigation ditches.

Do not allow spray drift. Do not permit any spray or spray drift to contact desirable plants off the area treated or the soil containing roots of such plants. To avoid spray drift, use low spray pressure, under 30 psi for ground equipment and under 10 psi for aircraft, and apply as a coarse spray. Do not use pressures nor spray nozzles that produce fine droplets which could present a drift hazard. Spray TORDON 22K Weed Killer only when wind velocity is less than 10 mph and **when there is no hazard from spray drift off the target area. Do not spray when the wind is blowing toward susceptible crops or ornamental plants that are closer than one mile from the target area.** Do not spray above an air inversion. Consult local authorities on possible inversion situation.

Do not rotate treated grain fields to other crop uses.

Do not spray grain if the forage legume component is desired. TORDON 22K Weed Killer may injure or kill legumes. Also, new legume seedlings may not be successful if made within 2 years following application of this herbicide.

Do not move treated soil to other areas. Do not use it to grow plants, unless adequately sensitive bioassay or chemical tests show that no detectable picloram is present in the soil.

Do not store near food, feedstuff, fertilizer, seed, insecticides, fungicides, or other pesticides.

Do not re-use containers for TORDON 22K Weed Killer for any purpose. Dispose by punching holes in them and burying with waste or by taking to an approved landfill. Rinse application equipment after use, preferably at least three times with water, and dispose of rinse water in a non-cropland area away from water supplies. Where indicated, follow official local container disposal regulations.

Be sure that use of this product conforms to all applicable regulations.

Do not use where a sandy porous surface and substrate overlie ground water ten feet or less below the surface.

WARRANTY LIMITATIONS AND DISCLAIMER

The Dow Chemical Company warrants that this product conforms to the chemical description on the label and is reasonably fit for the purposes stated on the label when used in strict accordance with the directions therein under normal conditions of use. THIS IS THE ONLY WARRANTY MADE ON THIS PRODUCT. NO OTHER EXPRESS AND NO IMPLIED WARRANTY OF MERCHANTABILITY OR FITNESS FOR A PARTICULAR PURPOSE IS MADE OUTSIDE OF THIS LABEL. Therefore, neither this warranty nor any other warranty of merchantability or fitness for a particular purpose, express or implied, extends to the use of this product contrary to label instructions (including conditions noted on the label, such as unfavorable temperatures, soil conditions, etc.), under abnormal conditions (such as excessive rainfall, drought, tornadoes, hurricanes, etc.) or under conditions not reasonably foreseeable to or beyond the control of seller.

When buyer or user suffers losses or damages resulting from the use or handling of this product (including claims based on contract, negligence, strict liability, or other legal theories), buyer or user must promptly notify in writing The Dow Chemical Company of any claims to be eligible to receive either remedy given below. The EXCLUSIVE REMEDY OF THE BUYER OR USER and the LIMIT OF LIABILITY of The Dow Chemical Company or any other seller will be one of the following, at the election of The Dow Chemical Company:

(1) **Refund of purchase price** paid by buyer or user for product bought, **or**

(2) **Replacement** of amount of product used.

The seller **will not be liable for consequential or incidental damages or losses.**

The terms of this Warranty Limitations And Disclaimer cannot be varied by any written or verbal statements or agreements. Any employee or sales agent of the seller is not authorized to vary or exceed the terms of this Warranty Limitations And Disclaimer in any manner.

87116-L12 D1188

* Trademark of THE DOW CHEMICAL COMPANY

TORDON* RTU

HERBICIDE

For Controlling Unwanted Trees Via Cut Surface Treatments

ACTIVE INGREDIENT(S):

Picloram (4-amino-3,5,6-trichloropicolinic acid), as the triisopropanolamine salt 5.4%

2,4-Dichlorophenoxyacetic Acid, as the triisopropanol-amine salt . 20.9%

INERT INGREDIENTS: 73.7%

TOTAL 100.0%

Acid Equivalents:

Picloram - 3.0%

2,4-Dichlorophenoxyacetic acid - 11.2%

E.P.A Registration No. 464-510 E.P.A. Est. 464-MI-1

KEEP OUT OF REACH OF CHILDREN

WARNING

AVISO:

PRECAUCIÓN AL USUARIO: Si usted no lee inglés, no use este producto hasta que la etiqueta le haya sido explicada ampliamente.

PRECAUTIONARY STATEMENTS

HAZARDS TO HUMANS AND DOMESTIC ANIMALS

HARMFUL OR FATAL IF SWALLOWED • CAUSES EYE INJURY • MAY CAUSE SKIN IRRITATION

Avoid Contact With Eyes, Skin, And Clothing • Wash Thoroughly After Handling • Keep Container Closed • Do Not Store Near Food or Feed • Do Not Drink Solution • Do Not Store in Open or Unlabeled Containers

STATEMENTS OF PRACTICAL TREATMENT:

Contains ethylene glycol. If swallowed, induce vomiting immediately by giving two glasses of water and sticking finger down throat. Get medical attention. Never give anything by mouth to an unconscious person. In case of eye contact, prompty flush with plenty of water; and get medical attention. Remove contaminated clothing and wash before reuse.

PHYSICAL OR CHEMICAL HAZARDS

COMBUSTIBLE • Keep Away From Heat And Open Flame.

ENVIRONMENTAL HAZARDS

Do not apply directly to any body of water. Do not contaminate water used for irrigation or domestic purposes by cleaning of equipment or disposal of wastes. Do not allow run-off or spray to contaminate wells, irrigation ditches or any body of water used for irrigation or domestic purposes. Do not make application when circumstances favor movement from treatment site. See additional precautionary statements elsewhere on this label.

NOTICE

Read the entire label. Use only according to label directions.

Before buying or using this product, read "WARRANTY LIMITATIONS AND DISCLAIMER" elsewhere on this label. If terms are not acceptable, return **unopened** package at once to seller for full refund of purchase price paid. Otherwise, use by the buyer or any other user constitutes acceptance of the terms under "WARRANTY LIMITATIONS AND DISCLAIMER."

IN CASE OF AN EMERGENCY endangering life or property involving this product, call collect 517-636-4400

AGRICULTURAL CHEMICAL

Do Not Ship or Store with Food, Feeds, Drugs, or Clothing

GENERAL INFORMATION

TORDON RTU Herbicide is a **ready-to-use product** that is effective in cut surface applications for killing unwanted trees and preventing undesirable sprouting of cut trees in forest and other non-crop areas such as fence rows and right-of-ways. Among the many species that can be controlled are: **ailanthus, alder, aspen, birch, cedar, cherry, dogwood, elm, firs, green ash, gum, hawthorn, hickory, hornbeam, maples, oaks, pecan, persimmon, serviceberry, sourwood, sweetbay.**

USE DIRECTIONS

It is a violation of Federal law to use this product in a manner inconsistent with its labeling.

Use undiluted as indicated below.

TREE INJECTION TREATMENT: Use 1 milliliter of undiluted TORDON RTU Herbicide solution through the bark completely around the tree trunk at intervals of 2 to 3 inches between edges of the injector wounds. Make injections near ground level when using the tree injector or 2 to 4 feet above the ground when using a HYPOHATCHET Injector or similar device. Treatments can be made in any season. Maples should not be treated during the spring sap flow. With some difficult to control species such as dogwood, hickory, sugar maple, bigleaf maple, tanoak, and some firs, application to a continuous cut rather than to spaced cuts may be more uniformly effective.

FRILL OR GIRDLE TREATMENT: Make a single hack girdle or "frill" of overlapping axe cuts through the bark completely around the tree as close to the ground as feasible. Spray or paint the injured surface with undiluted TORDON RTU Herbicide, using enough volume to wet treated areas.

STUMP TREATMENT: Spray or paint the cut surfaces of freshly cut stumps and stubs with undiluted TORDON RTU Herbicide. The cambium area next to the bark is the most vital area to wet.

USE PRECAUTIONS

Use this product only as specified on this label. TORDON RTU Herbicide is highly active as a herbicide, is water soluble, can move with surface runoff water, and can remain phytotoxic for a year or more if it gets into the soil. Untreated trees occasionally are affected by root uptake of herbicide leaches into the soil or excreted by the roots of nearby treated trees. If such injury cannot be tolerated, do not use TORDON RTU close to desirable trees.

Do not contaminate cropland, water, or irrigation ditches.

Do Not Contaminate Water Intended for Irrigation or Domestic Purposes. To avoid injury to crops or other desirable plants, do not treat or allow spray drift or run-off to fall onto banks or bottoms of irrigation ditches, either dry or containing water, or other channels that carry water that may be used for irrigation or domestic purposes. Do not apply to snow or frozen ground.

Water contaminated with residues of picloram from drainage ditch bank uses shall not be used to irrigate crops which are not registered for use with this chemical.

Do not rotate food or feed crops on treated land if they are not registered for use with picloram until an adequately sensitive bioassay or chemical test shows that no detectable picloram is present in the soil.

Do not apply on or in the vicinity of sensitive crops of desirable broadleaf plants. Small amounts can cause injury to plants.

Do not allow careless application or spray drift. Do not permit any spray or spray drift to contact desirable plants. Use only gravity flow or very low spray pressure.

Do not remove product from original container except to apply as indicated under USE DIRECTIONS.

Do not store near food, feedstuff, fertilizer, seeds, insecticides, fungicides, or other pesticides.

PRECAUTIONARY STATEMENTS

ENVIRONMENTAL HAZARDS

Picloram is a chemical which can travel (seep or leach) through soil and under certain conditions contaminate groundwater which may be used for irrigation and drinking purposes. Users are advised not to apply picloram where soils have a rapid to very rapid permeability throughout the profile (such as loamy sand to sand) and the water table of an underlying aquifer is shallow or to soils containing sinkholes over limestone bedrock, severely fractured surfaces, and substrates which would allow direct introduction into an aquifer. Your local agricultural agencies can provide further information on the type of soil in your area and the location of groundwater.

An aquifer is defined as "an underground, saturated, permeable, geologic formation capable of producing significant quantities of water to a well or spring. It is the ability of the saturated zone, or portion of that zone, to yield water which makes it an aquifer" (American Chemical Society, 1983).

STORAGE AND DISPOSAL

Do not contaminate water, food or feed by storage or disposal.

Pesticide Disposal: Pesticide wastes are toxic. Improper use of excess pesticide, spray mixture, or rinsate is a violation of Federal law. If these wastes cannot be disposed of by use according to label instructions, contact your State Pesticide or Environmental Control Agency, or the Hazardous Waste representative at the nearest EPA Regional Office for guidance.

Rinse application equipment after use, at least three times with water, and dispose of rinse water in a non-cropland area away from water supplies.

Container Disposal: Do not reuse container. Triple rinse (or equivalent). Puncture and dispose of in a sanitary landfill, or by incineration, or, if allowed by state and local authorities, by burning. If burned, stay out of smoke.

General: Consult federal, state or local disposal authorities for approved alternative procedures.

Be sure that use of this product conforms to all application regulations.

WARRANTY LIMITATIONS AND DISCLAIMER

The Dow Chemical Company warrants that this product conforms to the chemical description on the label and is reasonably fit for the purposes stated on the label when used in strict accordance with the directions therein under normal conditions of use. THIS IS THE ONLY WARRANTY MADE ON THIS PRODUCT. NO OTHER EXPRESS AND NO IMPLIED WARRANTY OF MERCHANTABILITY OR FITNESS FOR A PARTICULAR PURPOSE IS MADE OUTSIDE OF THIS LABEL. Therefore, neither this warranty nor any other warranty of merchantability or fitness for a particular purpose, express or implied, extends to the use of this product contrary to label instructions (including conditions noted on the label, such as unfavorable temperatures, soil conditions, etc.), under abnormal conditions (such as excessive rainfall, drought, tornadoes, hurricanes, etc.) or under conditions not reasonably foreseeable to or beyond the control of seller.

When buyer or user suffers losses or damages resulting from the use or handling of this product (including claims based on contract, negligence, strict liability, or other legal theories), buyer or user must promplty notify in writing The Dow Chemical Company of any claims to be eligible to receive either remedy given below. The EXCLUSIVE REMEDY OF THE BUYER OR USER and the LIMIT OF LIABILITY of The Dow Chemical Company or any other seller will one of the following, at the election of The Dow Chemical Company: (1) **Refund of purchase price** paid by buyer or user for product bought, **or** (2) **Replacement** of amount of product used. The seller **will not be liable for consequential or incidental damages or losses.**

The terms of this Warranty Limitations And Disclaimer cannot be varied by any written or verbal statements or agreements. Any employee or sales agent of the seller is not authorized to vary or exceed the terms of this Warranty Limitations And Disclaimer in any manner.

87244-L2 B887

*Trademark of THE DOW CHEMICAL COMPANY

TURFLON* AMINE

HERBICIDE

For the control of Broadleaf Weeds in Ornamental Turf

To be used only under the direct supervision of commercial applicators responsible for turf weed control programs.

ACTIVE INGREDIENT(S):

Triclopyr (3,5,6-trichloro-2-pyridinyloxyacetic acid), as the triethylamine salt	44.4%
INERT INGREDIENTS:	55.6%
TOTAL	100.0%

Contains methanol

Acid equivalent: Triclopyr - 31.8% - 3 lb/gal

E.P.A. Registration No. 464-546 E.P.A. Est. 464-MI-1

KEEP OUT OF REACH OF CHILDREN

DANGER
PELIGRO:

PRECAUCIÓN AL USUARIO: Si usted no lee inglés, no use este producto hasta que la etiqueta le haya sido explicada ampliamente.

PRECAUTIONARY STATEMENTS

HAZARDS TO HUMANS AND DOMESTIC ANIMALS

CAUSES EYE DAMAGE AND SKIN IRRITATION • HARMFUL IF SWALLOWED

Do Not Get in Eyes, On Skin Or On Clothing • Wear Goggles Or Face Shield and Rubber Gloves When Handling • Avoid Contamination Of Food • Wash Thoroughly After Handling

STATEMENT OF PRACTICAL TREATMENT:

In case of contact, immediately flush eyes with plenty of water for at least 15 minutes and get medical attention promptly. Flush skin with water and get medical attention if signs of irritation develop. Remove and wash contaminated clothing before reuse.

If swallowed, promptly drink a large quantity of milk, egg whites, gelatin solution or if these are not available, drink large quantities of water.

Avoid alcohol. Call a physician.

PHYSICAL OR CHEMICAL HAZARDS

Do Not Use or Store Near Heat or Open Flame.

Do Not Cut or Weld Container

ENVIRONMENTAL HAZARDS

Keep out of lakes, streams or ponds. Do not apply directly to water or wetlands. Do not contaminate water when disposing of equipment washwaters.

NOTICE

Read and understand the entire label before using. Use only according to label directions.

Before buying or using this product, read "WARRANTY LIMITATIONS AND DISCLAIMER" elsewhere on this label. If terms are not acceptable, return **unopened** package at once to seller for full refund of purchase price paid. Otherwise, use by the buyer or any other user constitutes acceptance of the terms under WARRANTY LIMITATIONS AND DISCLAIMER.

IN CASE OF AN EMERGENCY endangering life or property involving this product, call collect 517-636-4400

AGRICULTURAL CHEMICAL

Do Not Ship or Store with Food, Feeds, Drugs, or Clothing See Side Panel for Important Use Precautions

GENERAL INFORMATION

TURFLON Amine Herbicide is recommended for the control of annual and perennial broadleaf weeds actively growing in tall fescue, perennial bluegrass, or perennial ryegrass ornamental turf.

Among the annual and perennial broadleaf weeds controlled are:

Bindweed	Field Bindweed	Smartweed
Burdock	Ground Ivy	Tansy Ragwort
Canada Thistle	Lambsquarters	Vetch
Chicory	Oxalis	Wild Lettuce
Curly Dock	Plantain	Wild Violet
Dandelion	Ragweed	

DIRECTIONS FOR USE

Do not use for manufacturing or formulating.

It is a violation of Federal law to use this product in a manner inconsistent with its labeling.

Do not apply this product through any type of irrigation system.

MIXING INSTRUCTIONS

Use TURFLON Amine Herbicide at rates of 1/2 to 2 quarts per acre to control broadleaf weeds. In all cases use the amount specified in enough water to give uniform and complete coverage of the plants to be controlled. Use only water suitable for spraying. The recommended order of addition to the spray tank is water, **Nalco-Trol** (if used), surfactant (if used), additional herbicide (if used), TURFLON Amine Herbicide. If combined with emulsifiable concentrate herbicides moderate continuous adequate agitation is required.

Before using any recommended tank mixtures, read the directions and all use precautions on both labels.

APPLICATIONS DIRECTIONS

Broadcast Treatment of Ornamental Turf: Foliage sprays should be applied during warm weather, early spring through fall, when weeds are actively growing. Appliction under drought conditions may provide less than desirable results. Use low spray pressures to minimize spray drift. (See Use Precautions.) When hard-to-control species are prevalent and during applictions made in late summer when the plants are mature and during drought conditions, use the higher rates.

Make application using equipment that will assure uniform coverage of the spray volumes applied. TO IMPROVE SPRAY COVERAGE, ADD AN AGRICULTURAL SURFACTANT (SUCH AS **ORTHO** X-77 or **TRONIC**) AT A RATE OF 1/4 TO 1 PINT PER ACRE. USE THE HIGHER RATES OF SURFACTANT FOR LOWER RATES OF PRODUCT AND LOWER SPRAY VOLUMES.

Apply 1/2 to 2 quarts of TURFLON Amine Herbicide in 35 to 100 gallons of water per acre to control broadleaf weeds actively growing in tall fescue, bluegrass or perennial ryegrass turf. **DO NOT** use on other species, such as bentgrass or St. Augustine grass, unless injury can be tolerated. To minimize grass injury, do not treat if grass is under heat or drought stress, and repeat applictions should be made at least four weeks apart.

Apply TURFLON Amine Herbicide in a manner to avoid contacting nearby susceptible crops or other desirable plants. Very small amounts of spray drift may injure susceptible plants, including ornamental trees and shrubs. **DO NOT** apply to exposed roots of shallow rooted trees or shrubs.

Broadleaf weed species germinate at different times before being treated. Only emerged weeds present at the time of application will be controlled. Newly seeded turf should be mowed two or three times before being treated. Do not water for 24 hours after application. Do not reseed for 3 weeks after application.

Spot Treatment of Ornamental Turf Using Portable Sprayers: Mix 1.0 to 1.25 fluid ounces of TURFLON Amine in enough water to make 3 gallons of spray (1 to 1.25 quarts per 100 gallons of spray) and apply at any time broadleaf weeds are susceptible by wetting foliage of undesirable plants to the point of runoff. Do not apply more than 2 1/2 quarts of TURFLON Amine on any one acre.

USE PRECAUTIONS

Apply this product only as specified on this label.

Before using any recommended tank mixtures, read the directions and all use precautions on both labels.

Do not apply TURFLON Amine Herbicide directly to, or otherwise permit it to come into direct contact with grapes, tobacco, vegetable crops, flowers or other desirable broadleaf plants and do not permit spray mists containing it to drift into them.

AVOID INJURIOUS SPRAY DRIFT: Applications should be made only when there is little or no hazard from spray drift. Very small quantities of spray, which may not be visible may seriously injure susceptible plants. Do not spray when wind is blowing toward susceptible crops or ornamental plants near enough to be injured.

In handgun applications, drift can be reduced by using low pressure Teejet flat fan, raindrop, or other drift controlling nozzles. Standard flat fan nozzles may also be used at no more than 20 psi. At this pressure the spray pattern may be narrower.

With ground broadcast equipment, drift can be reduced by keeping the spray boom as low as possible; by applying no less than 35 gallons of spray per acre; by using no more than 20 pounds of spraying pressure with flat fan or flooding fan tips; and by spraying when wind velocity is low.

Do not apply on ditches used to transport irrigation water. Do not apply where runoff or irrigation water may flow onto susceptible crops.

Do not graze treated areas or feed treated forage for one year following application.

Be sure that use of this product conforms to all applicable regulations.

STORAGE AND DISPOSAL

Do not contaminate water, food or feed by storage or disposal.

STORAGE: Store above 28°F or agitate before use.

DISPOSAL:

Prohibitions - Open dumping is prohibited.

Pesticide Disposal - Pesticide wastes are acutely hazardous. Improper disposal of excess pesticide, spray mixture, or rinsate is a violation of Federal Law. If these wastes cannot be disposed of by use according to label instructions, contact your State Pesticide or Environmental Control Agency, or the Hazardous Waste representative at the nearest EPA Regional Office for guidance.

Container Disposal - Triple rinse (or equivalent). Then offer for recycling or reconditioning, or puncture and dispose of in a sanitary landfill, or by other approved state and local procedures. Plastic containers may also be disposed of by incineration, or, if allowed by state and local authorities, by burning. If burned, stay out of smoke.

General - Consult federal, state, or local disposal authorities for approved alternative procedures.

WARRANTY LIMITATIONS AND DISCLAIMER

The Dow Chemical Company warrants that this product conforms to the chemical description on the label and is reasonably fit for the purposes stated on the label when used in strict accordance with the directions therein under normal conditions of use. THIS IS THE ONLY WARRANTY MADE ON THIS PRODUCT. NO OTHER EXPRESS AND NO IMPLIED WARRANTY OF MERCHANTABILITY OR FITNESS FOR A PARTICULAR PURPOSE IS MADE OUTSIDE OF THIS LABEL. Therefore, neither this warranty nor any other warranty of merchantability or fitness for a particular purpose, express or implied, extends to the use of this product contrary to label instructions (including conditions noted on the label, such as unfavorable temperatures, soil conditions, etc.), under abnormal conditions (such as excessive rainfall, drought, tornadoes, hurricanes, etc.) or under conditions not reasonable foreseeable to or beyond the control of seller.

When buyer or user suffers losses or damages resulting from the use or handling of this product (including claims based on contract, negligence, strict liability, or other legal theories), buyer or user must promptly notify in writing The Dow Chemical Company of any claims to be eligible to receive either remedy given below. The EXCLUSIVE REMEDY OF THE BUYER OR USER and the LIMIT OF LIABILITY of The Dow Chemical Company or any other seller will be one of the following, at the election of The Dow Chemical Company: (1) **Refund of purchase price** paid by buyer or user for product bought, **or** (2) **Replacement** of amount of product used. The seller **will not be liable for consequential or incidental damages or losses.**

The terms of this Warranty Limitations And Disclaimer cannot be varied by any written or verbal statements or agreements. Any employee or sales agent of the seller is not authorized to vary or exceed the terms of this Warranty Limitations And Disclaimer in any manner.

04163-L1 B1088

*Trademark of THE DOW CHEMICAL COMPANY

TURFLON* II AMINE

HERBICIDE

For the Control of Annual and Perennial Broadleaf Weeds in Ornamental Turf

To be applied only under the direct supervision of commercial applicators responsible for turf weed control programs.

ACTIVE INGREDIENT(S):

2,4-D (2,4-Dichlorophenoxyacetic acid) as the dimethylamine salt. 34.2%
Triclopyr (3,5,6-trichloro-2-pyridinyloxyacetic acid)
as the triethylamine salt. 15.2%
INERT INGREDIENTS: 50.6%
TOTAL 100.0%

2,4-D acid equivalent 28.4% - 2.78 lb/gal
Triclopyr acid equivalent 10.9% - 1.07 lb/gal

E.P.A. Registration No. 464-606 E.P.A. Est. 464-MI-1

KEEP OUT OF REACH OF CHILDREN

DANGER
PELIGRO:

PRECAUCIÓN AL USUARIO: Si usted no lee inglés, no use este producto hasta que la etiqueta le haya sido explicada ampliamente.

PRECAUTIONARY STATEMENTS

HAZARDS TO HUMANS AND DOMESTIC ANIMALS

CAUSES IRREVERSIBLE EYE DAMAGE • MAY BE FATAL IF ABSORBED THROUGH SKIN • HARMFUL IF SWALLOWED OR INHALED

Avoid Breathing Vapors • Do Not Get In Eyes, On Skin, Or Clothing • Wear Goggles Or Face Shield And Rubber Gloves When Handling • Wash Thoroughly With Soap And Water After Handling And Before Eating Or Smoking • Remove Contaminated Clothing And Wash Before Reuse

STATEMENT OF PRACTICAL TREATMENT:

In case of contact, immediately flush eyes with plenty of water for at least 15 minutes and get medical attention promptly. Flush skin with water and get medical attention if signs of irritation develop.

If swallowed, promptly drink a large quantity of milk, egg whites, gelatin solution or, if these are not available, drink large quantities of water.

Avoid alcohol. Call a physician.

NOTE TO PHYSICIAN: Probable mucosal damage may contraindicate use of gastric lavage.

PHYSICAL OR CHEMICAL HAZARDS

Do Not Cut or Weld Container.

ENVIRONMENTAL HAZARDS

Keep out of lakes, streams or ponds. Do not contaminate water by cleaning of equipment or disposal of wastes.

NOTICE

Read the entire label. Use only according to label directions.

Before buying or using this product, read "WARRANTY LIMITATIONS AND DISCLAIMER" elsewhere on this label. If terms are not acceptable, return **unopened** package at once to seller for full refund of purchase price paid. Otherwise, use by the buyer or any other user constitutes acceptance of the terms under "WARRANTY LIMITATIONS AND DISCLAIMER."

IN CASE OF AN EMERGENCY endangering life or property involving this product, call collect 517-636-4400

AGRICULTURAL CHEMICAL

Do Not Ship or Store with Food, Feeds, Drugs, or Clothing

See Side Panel for Important Use Precautions

GENERAL INFORMATION

TURFLON II Amine Herbicide is a broad-spectrum weed killer for control of broadleaf weeds in ornamental turf.

DIRECTIONS FOR USE

It is a violation of Federal law to use this product in a manner inconsistent with its labeling.

Do not apply this product through any type of irrigation system.

Do not use for manufacturing or formulating.

Application Rates

Weeds Controlled		Rate/Acre	Rate/1000 sq. ft.
Black medic Carpet weed Catnip Chamise Chickweeds Chicory Cinquefoil Clovers Cocklebur Coffeeweed Cornflower Creeping beggarweed Dandelion Dayflower Docks Dwarf beggarweed Field bindweed	Goldenrod Henbit Hop clover Knawel Lambsquarter Lespedeza Little starwart Mallow Matchweed Plantain Purslane speedwell Smartweed Sowthistle Spiderwort Spotted catsear Spurweed Vetch Yarrow	2-½ pints	1 oz or 2 Tbsp.
Burdock Buttercup Mustards	Poison ivy Poison oak Wild carrot	3 to 3-½ pints	1.3 oz or 2-½ Tbsp.
Cornspeedwell Ground ivy Oxalis (*stricta* and *corniculata*)	Parsley-piert Prostrate spurge Wild violet	3 pints, 2 applications 4 to 6 weeks apart	1.3 oz or 2-½ Tbsp. 2 applications 4 to 6 weeks apart

APPLICATION DIRECTIONS

Foliage sprays should be applied during warm weather when weeds are actively growing. Application under drought conditions may provide less than desirable results. Use low spray pressures to minimize spray drift. (See Use Precautions)

Apply TURFLON II Amine in a manner to avoid contacting nearby susceptible crops or other desirable plants. Very small amounts of spray drift may injure susceptible plants, including ornamental trees or shrubs. DO NOT apply to exposed roots of shallow rooted trees or shrubs.

Broadcast Treatment of Ornamental Turf: Apply 1 to 2 quarts of TURFLON II Amine in enough water to make 35 to 175 gallons total spray per acre to control broadleaf weeds growing in perennial bluegrass, tall fescue, or perennial ryegrass. DO NOT use on other grass species, such as bentgrass or St. Augustine grass, unless injury can be tolerated. Apply from early spring through fall when weeds are actively growing. Broadleaf weed species germinate at different times. Only emerged weeds present at time of application are controlled. Additional applications should be made four weeks apart to minimize grass injury. Newly seeded turf should be mowed two or three times before being treated. Do not water for 24 hours after application. Do not reseed for 3 weeks after application.

Spot Treatment of Ornamental Turf Using Portable Sprayers: Mix 1 to 2 fluid ounces of TURFLON II Amine in enough water to make 3 gallons of spray (1 to 2 quarts per 100 gallons of spray) and apply at any time broadleaf weeds are susceptible by wetting foliage of undesirable plants to point of runoff.

DO NOT apply to exposed roots of shallow rooted trees or shrubs. Do not apply more than 8 quarts of TURFLON II Amine Herbicide on any one acre at one time.

PREPARING THE SPRAY

Add about one-half the desired amount of clean water to spray tank. Add TURFLON II Amine and complete addition of water with agitation running. Mix thoroughly and continue moderate agitation while spraying.

USE PRECAUTIONS

Apply this product only as specified on this label.

Do not apply TURFON II Amine directly to, or otherwise permit it to come into direct contact with, cotton, grapes, tobacco, vegetable crops, flowers, fruit or ornamental trees, or other desirable broadleaf plants and do not permit spray mists containing it to drift onto them.

DO NOT apply to exposed roots of shallow rooted trees and shrubs.

Avoid Injurious Spray Drift. Applications should be made only when hazards from spray drift are at a minimum. Very small quantities of spray, which may not be visible, may seriously injure susceptible plants. Do not spray when wind is blowing toward susceptible crops or ornamental plants near enough to be injured.

With ground broadcast equipment, drift can be reduced by keeping the spray boom as low as possible; by applying no less than 32 gallons of spray per acre; by using no more than 20 pounds spraying pressure with flat fan or flooding flat fan nozzle tips; and by spraying when wind velocity is less than 5 miles per hour.

In handgun applications drift can be reduced by using low pressure Teejet flat fan, Raindrop or other drift controlling nozzles. Standard flat fan nozzles may also be used at no more than 20 psig. At this pressure the spray pattern may be narrower.

Do not apply on ditches used to transport irrigation water. Do not apply where runoff or irrigation water may flow onto susceptible crops as injury may result.

STORAGE AND DISPOSAL

Do not contaminate water, food or feed by storage or disposal.

STORAGE: Store above 28°F or agitate before use.

DISPOSAL:

Pesticide Disposal: Pesticide wastes are toxic. Improper disposal of excess pesticide, spray mixture, or rinsate is a violation of Federal Law. If these wastes cannot be disposed of by use according to label instructions, contact your State Pesticide or Environmental Control Agency, or the Hazardous Waste representative at the nearest EPA Regional Office for guidance.

Container Disposal: Triple rinse (or equivalent). Then offer for recycling or reconditioning, or puncture and dispose of in a sanitary landfill, or by other procedures approved by state and local authorities. Plastic containers, after triple rinsing, may be incinerated if allowed by state and local authorities. If burned, stay out of smoke.

General: Consult federal, state, or local disposal authorities for approved alternative procedures.

Be sure that use of this product conforms to all applicable regulations.

WARRANTY LIMITATIONS AND DISCLAIMER

The Dow Chemical Company warrants that this product conforms to the chemical description on the label and is reasonably fit for the purposes stated on the label when used in strict accordance with the directions therein under normal conditions of use. THIS IS THE ONLY WARRANTY MADE ON THIS PRODUCT. NO OTHER EXPRESS AND NO IMPLIED WARRANTY OF MERCHANTABILITY OR FITNESS FOR A PARTICULAR PURPOSE IS MADE OUTSIDE OF THIS LABEL. Therefore, neither this warranty nor any other warranty of merchantability or fitness for a particular purpose, express or implied, extends to the use of this product contrary to label instructions (including conditions noted on the label, such as unfavorable temperatures, soil conditions, etc.), under abnormal conditions (such as excessive rainfall, drought, tornadoes, hurricanes, etc.) or under conditions not reasonably foreseeable to or beyond the control of seller.

When buyer or user suffers losses or damages resulting from the use or handling of this product (including claims based on contract, negligence, strict liability, or other legal theories), buyer or user must promptly notify in writing The Dow Chemical Company of any claims to be eligible to receive either remedy given below. The EXCLUSIVE REMEDY OF THE BUYER OR USER and the LIMIT OF LIABILITY of The Dow Chemical Company or any other seller will be one of the following, at the election of The Dow Chemical Company: (1) **Refund of purchase price** paid by buyer or user for product bought, **or** (2) **Replacement** of amount of product used. The seller **will not be liable for consequential or incidental damages or losses.**

The terms of this Warranty Limitations And Disclaimer cannot be varied by any written or verbal statements or agreements. Any employee or sales agent of the seller is not authorized to vary or exceed the terms of this Warranty Limitations And Disclaimer in any manner.

04462-L1 B1188

*Trademark of THE DOW CHEMICAL COMPANY

TURFLON*D

HERBICIDE

Low Volatile Weed Killer

For the Control of Annual and Perennial Broadleaf Weeds in Ornamental Turf

To be applied only under the direct supervision of commercial applicators responsible for turf weed control programs.

ACTIVE INGREDIENT(S):	
2,4-Dichlorophenoxyacetic Acid, Butoxyethyl Ester	34.4%
Triclopyr (3,5,6-trichloro-2-pyridinyloxyacetic acid), Butoxyethyl Ester	16.5%
INERT INGREDIENTS:	49.1%
TOTAL	100.0%

Contains petroleum distillates
2,4-D Acid Equivalent 23.7% - 2 pounds per gallon
Triclopyr Acid Equivalent 11.9% - 1 pound per gallon
E.P.A. Registration No. 464-589 E.P.A. Est. 464-MI-1

KEEP OUT OF REACH OF CHILDREN

CAUTION

AVISO:

PRECAUCIÓN AL USUARIO: Si usted no lee inglés, no use este producto hasta que la etiqueta le haya sido explicada ampliamente.

PRECAUTIONARY STATEMENTS

HAZARDS TO HUMANS AND DOMESTIC ANIMALS

HARMFUL IF SWALLOWED, INHALED OR ABSORBED THROUGH SKIN • CAUSES EYE IRRITATION • THIS PRODUCT MAY CAUSE ALLERGIC SKIN REACTION

Avoid Contact With Eyes, Skin, Or Clothing • Avoid Breathing Vapor • Wash Thoroughly With Soap And Water After Handling And Before Eating Or Smoking • Remove And Wash Contaminated Clothing Before Reuse

STATEMENT OF PRACTICAL TREATMENT:

In case of contact, flush skin or eyes with plenty of water. Get medical attention if irritation persists.

If swallowed, DO NOT induce vomiting. CALL A PHYSICIAN OR POISON CONTROL CENTER.

PHYSICAL OR CHEMICAL HAZARDS

COMBUSTIBLE • Do Not Use or Store Near Heat or Open Flame. • Do Not Cut or Weld Container.

ENVIRONMENTAL HAZARDS

This pesticide is toxic to fish. Do not apply directly to bodies of water. Do not contaminate irrigation ditches or water used for irrigation or domestic purposes.

NOTICE

Read the entire label. Use only according to label directions.

Before buying or using this product, read "WRRANTY LIMITATIONS AND DISCLAIMER" elsewhere on this label. If terms are not acceptable, return **unopened** package at once to seller for full refund or purchase price paid. Otherwise, use by the buyer or any other user constitutes acceptance of the terms under "WARRANTY LIMITATIONS AND DISCLAIMER."

IN CASE OF AN EMERGENCY endangering life or property involving this product, call collect 517-636-4400

AGRICULTURAL CHEMICAL

Do Not Ship or Store with Food, Feeds, Drugs, or Clothing
See Back Panel for Important Use Precautions

Weeds Controlled:
Black medic
Bull thistle (a)
Burdock (a)
Buttercup (a)
Canada thistle (a)
Carpetweed
Catnip
Chamise
Chickweeds (common and mouseear)
Chicory
Cinquefoil
Clovers
Cocklebur
Coffeeweed
Cornflower
Cornspeedwell (b)
Creeping beggarweed
Dandelion
Dayflower
Docks
Dwarf beggarweed
Field bindweed
Goldenrold
Ground ivy (b)
Henbit
Hopclover
Knawel
Lambsquarter
Lespedeza
Little starwort
Mallow
Matchweed
Mustards (a)
Oxalis (*stricta* and *corniculata)* (b)
Parsley piert (b)
Plantain
Poison ivy
Poison oak
Prostrate spurge (b)
Purslane speedwell
Smartweed
Sowthistle
Spiderwort
Spotted catsear
Spurweed
Vetch
Wild carrot (a)
Wild violet (b)
Yarrow

DIRECTIONS FOR USE

It is a violation of Federal law to use this product in a manner inconsistent with its labeling.

Do not use for manufacturing or formulating.

Do not apply this product through any type of irrigation system.

Foliage sprays should be applied during warm weather when weeds are actively growing. Application under drought conditions may provide less than desirable results. Use low spray pressures to minimize spray drift. **Apply TURFLON D in a manner to avoid contacting nearby susceptible crops or other desirable plants and to avoid contaminating water intended for irrigation or domestic use. Read and follow all Use Precautions given on this label.**

PREPARING THE SPRAY

TURFLON D in water forms an emulsion (not a solution), and separation may occur unless the spray mixture is agitated continuously.

Water Spray: Fill the spray tank about half full with clean water. Then add the **TURFLON D** and complete filling the tank with agitation running. Mix thoroughly and continue moderate agitation while spraying.

APPLICATION DIRECTIONS

Broadcast Treatment of Ornamental Turf: Apply 1 to 2 quarts of TURFLON D in enough water to make 35 to 200 gallons total spray per acre to control broadleaf weeds growing in tall fescue, bluegrass, or perennial ryegrass turf. **Do not** use on other grass species unless injury can be tolerated. TURFLON D at 3 pints per acre (1.1 ounce per 1,000 square feet) will provide control of most weeds listed. The use of 4 pints per acre (1.5 ounces per 1,000 square feet) is suggested for those followed by (a). Optimum control of those species followed by (b) has been obtained when two applications of 3 pints per acre have been made 4 weeks apart. Apply from early spring through early fall when weeds are growing. To control a variety of broadleaf weed species which may germinate at different periods, three or four applications of TURFLON D may be needed during a year. Applications should be made four weeks apart to minimize grass injury.

Newly seeded turf should be mowed two or three times before being treated. Do not water for 24 hours after application. Do not reseed for 3 weeks after application.

Spot Treatment of Ornamental Turf Using Portable Sprayers: Mix one to two fluid ounces of TURFLON D in enough water to make three gallons of spray (one to two quarts per 100 gallons of spray) and apply at any time broadleaf weeds are susceptible by wetting foliage of undesirable plants to point of runoff.

DO NOT apply to exposed roots of shallow rooted trees or shrubs. Do not apply more than 8 quarts of TURFLON D Herbicide on any one acre at one time.

STORAGE AND DISPOSAL

Do not contaminate water, food or feed by storage or disposal.

STORAGE: Store above 10°F or agitate before use.

DISPOSAL:

Pesticide Disposal: Pesticide wastes are toxic. Improper disposal of excess pesticide, spray mixture, or rinsate is a violation of Federal Law. If these wastes cannot be disposed of by use according to label instructions, contact your State Pesticide or Environmental Control Agency, or the Hazardous Waste representative at the nearest EPA Regional Office for guidance.

Container Disposal: Triple rinse (or equivalent). Then offer for recycling or reconditioning, or puncture and dispose of in a sanitary landfill, or by other procedures approved by state and local authorities. Plastic containers after triple rinsing, may be incinerated if allowed by state and local authorities. If burned, stay out of smoke.

General: Consult federal, state, or local disposal authorities for approved alternative procedures.

Be sure that use of this product conforms to all applicable regulations.

USE PRECAUTIONS

Apply this product only as specified on this label.

Do not apply **TURFLON D** directly to, or otherwise permit it to come into contact with cotton, grapes, tobacco, vegetable crops, flowers, fruit or ornamental trees, or other desirable broadleaf plants and do not permit spray mists containing it to drift onto them.

Avoid Injurious Spray Drift. Applications should be made only when hazards from spray drift are at a minimum. Very small quantities of spray, which may not be visible may seriously injure susceptible plants. Do not spray when wind is blowing toward susceptible crops or ornamental plants near enough to be injured.

With ground broadcast equipment, drift can be reduced by keeping the spray boom as low as possible; by applying no less than 32 gallons of spray per acre; by using no more than 20 pounds spraying pressure with flat fan or flooding flat fan nozzle tips; and by spraying only when wind velocity is low.

In handgun applications drift can be reduced by using low pressure Teejet flat fan, raindrop, or other drift controlling nozzles. Standard flat fan nozzles may also be used at no more than 20 psi. At this pressure the spray pattern may be narrower.

At high temperatures, vapors from this product may injure susceptible crops growing nearby. Excessive amounts of this herbicide in the soil may temporarily inhibit seed germination and plant growth.

WARRANTY LIMITATIONS AND DISCLAIMER

The Dow Chemical Company warrants that this product conforms to the chemical description on the label and is reasonably fit for the purposes stated on the label when used in strict accordance with the directions therein under normal conditions of use. THIS IS THE ONLY WARRANTY MADE ON THIS PRODUCT. NO OTHER EXPRESS AND NO IMPLIED WARRANTY OF MERCHANTABILITY OR FITNESS FOR A PARTICULAR PURPOSE IS MADE OUTSIDE OF THIS LABEL. Therefore, neither this warranty nor any other warranty of merchantability or fitness for a particular purpose, express or implied, extends to the use of this product contrary to label instructions (including conditions noted on the label, such as unfavorable temperatures, soil conditions, etc.), under abnormal conditions (such as excessive rainfall, drought, tornadoes, hurricanes, etc.) or under conditions not reasonably foreseeable to or beyond the control of seller.

When buyer or user suffers losses or damages resulting from the use or handling of this product (including claims based on contract, negligence, strict liability, or other legal theories), buyer or user must promptly notify in writing The Dow Chemical Company of any claims to be eligible to receive either remedy given below. The EXCLUSIVE REMEDY OF THE BUYER OR USER and the LIMIT OF LIABILITY of The Dow Chemical Company or any other seller will be one of the following, at the election of The Dow Chemical Company: (1) **Refund of purchase price** paid by buyer or user for product bought, **or** (2) **Replacement** of amount of product used. The seller **will not be liable for consequential or incidental damages or losses.**

The terms of this Warranty Limitations And Disclaimer cannot be varied by any written or verbal statements or agreements. Any employee or sales agent of the seller is not authorized to vary or exceed the terms of this Warranty Limitations And Disclaimer in any manner.

01091-L1 E988

*Trademark of THE DOW CHEMICAL COMPANY

TURFLON * ESTER

HERBICIDE

For Control of Certain Broadleaf Weeds in Ornamental Turf

To be used only under the direct supervision of commercial applicators responsible for turf weed control programs.

ACTIVE INGREDIENT(S):

Triclopyr (3,5,6-trichloro-2-pyridinyloxyacetic acid), Butoxyethyl Ester	61.6%
INERT INGREDIENTS:	38.4%
TOTAL	100.0%

Acid Equivalent:
Triclopyr - 44.3% - 4 lb/gal
Contains petroleum distillates

E.P.A. Registration No. 464-554 E.P.A. Est. 464-MI-1

KEEP OUT OF REACH OF CHILDREN

CAUTION

AVISO:

PRECAUCIÓN AL USUARIO: Si usted no lee inglés, no use este producto hasta que la etiqueta le haya sido explicada ampliamente.

PRECAUTIONARY STATEMENTS

HAZARDS TO HUMANS AND DOMESTIC ANIMALS

HARMFUL IF SWALLOWED, INHALED OR ABSORBED THROUGH SKIN

Avoid Contact With Eyes, Skin, Or Clothing • Avoid Breathing Mists Or Vapors • Avoid Contamination Of Food • Wash Thoroughly After Handling • Remove And Wash Contaminated Clothing Before Reuse

STATEMENT OF PRACTICAL TREATMENT:

In case of skin contact, flush skin with plenty of water. Get medical attention if irritation persists.

If swallowed, do not induce vomiting. **Call a physician.**

PHYSICAL OR CHEMICAL HAZARDS

COMBUSTIBLE • Do Not Use or Store Near Heat or Open Flame. • Do Not Cut or Weld Container.

ENVIRONMENTAL HAZARDS

This pesticide is toxic to fish. Keep out of lakes, ponds, or streams. Do not contaminate water by cleaning of equipment or disposal of wastes.

NOTICE

Read the entire label. Use only according to label directions.

Before buying or using this product, read "WARRANTY LIMITATIONS AND DISCLAIMER" elsewhere on this label. If terms are not acceptable, return **unopened** package at once to seller for full refund of purchase price paid. Otherwise, use by the buyer or any other user constitutes acceptance of the terms under "WARRANTY LIMITATIONS AND DISCLAIMER."

IN CASE OF AN EMERGENCY endangering life or property involving this product, call collect 517-636-4400

AGRICULTURAL CHEMICAL

Do Not Ship or Store with Food, Feeds, Drugs, or Clothing

GENERAL INFORMATION

TURFLON*Ester Herbicide is recommended for the control of annual and perennial broadleaf weeds actively growing in tall fescue, perennial bluegrass, or perennial ryegrass turf.

Among the annual and perennial broadleaf weeds controlled are:

Black Medic	**Matchweed**
Bull Thistle	**Mustard**
Burdock	**Oxalis**
Canada Thistle	**Plantain**
Chicory	**Ragweed**
Clover	**Smartweed**
Creeping Beggarweed	**Sweet Clover**
Curly Dock	**Vetch**
Dandelion	**Wild Carrot (Queen Annes Lace)**
Field Bindweed	
Goldenrod	**Wild Lettuce**
Ground Ivy	**Wild Violet**
Lambsquarters	**Yarrow**
Lespedeza	

USE DIRECTIONS

Do not use for manufacturing or formulating.

It is a violation of Federal law to use this product in a manner inconsistent with its labeling.

Do not apply this product through any type of irrigation system.

MIXING INSTRUCTIONS

TURFLON Ester Herbicide in water forms an emulsion (not a solution) and separation may occur unless the spray mixture is agitated continuously.

Add about one-half the desired amount of clean water to spray tank. Add TURFLON Ester and complete addition of water with agitation running. Mix thoroughly and continue moderate agitation while spraying.

APPLICATION DIRECTIONS

Broadcast Treatment of Ornamental Turf: Foliage sprays should be applied during warm weather, early spring through fall, when weeds are actively growing. Application under drought conditions may provide less than desirable results. Use low spray pressures to minimize spray drift. (See Use Precautions.) When hard-to-control species are prevalent and during applications made in late summer when the plants are mature and during drought conditions, use the higher rates.

Make application using equipment that will assure uniform coverage of the spray volumes applied.

Apply 1/2 to 1 quart of TURFLON Ester in 35 to 100 gallons of water per acre to control broadleaf weeds actively growing in tall fescue, bluegrass, or perennial ryegrass turf. DO NOT use on other species, such as bentgrass or St. Augustine grass, unless injury can be tolerated. To minimize grass injury, do not treat if grass is under heat or drought stress and repeat applications should be made at least four weeks apart.

Apply TURFLON Ester Herbicide in a manner to avoid contacting nearby susceptible crops or other desirable plants. Very small amounts of spray drift may injure susceptible plants, including ornamental trees and shrubs. DO NOT apply to exposed roots of shallow rooted trees or shrubs.

Broadleaf weed species germinate at different times before being treated. Only emerged weeds present at the time of application will be controlled. Newly seeded turf should be mowed two or three times before being treated. Do not water for 24 hours after application. Do not reseed for 3 weeks after application.

To improve spectrum of activity, low volatile amine salt or ester formulations of 2,4-D may be tank mixed with TURFLON Ester Herbicide.

Spot Treatment of Ornamental Turf: Mix 1/4 to 1/2 fluid ounces of TURFLON Ester Herbicide in enough water to make three gallons of spray (1/2 to 1 pint per 100 gallons spray) and apply at any time broadleaf weeds are susceptible by wetting foliage of undesirable plants to point of runoff. Do not apply more than 2 quarts of TURFLON Ester Herbicide (400 gallons spray mixture) on any one acre.

USE PRECAUTIONS

Read and follow the general directions, the precautionary statements, and all use precautions on this product label and on labels for any products used in tank mixtures.

This labeling must be in the possession of the user at the time of application.

Local conditions may affect the use of herbicides. Consult your local specialist for advice in selecting treatments from this label to best fit local conditions.

Do not apply TURFLON Ester Herbicide directly to, or otherwise permit it to come into direct contact with, grapes, tobacco, vegetable crops, flowers or other desirable broadleaf plants and do not permit spray mists containing TURFLON Ester Herbicide to drift onto them.

AVOID INJURIOUS SPRAY DRIFT: Applications should be made only when there is little or no hazard from spray drift. Very small quantities of spray, which may not be visible, may seriously injure susceptible plants. Do not spray when wind is blowing toward susceptible crops or ornamental plants near enough to be injured.

In handgun applications, drift can be reduced by using low pressure Teejet flat fan, raindrop, or other drift controlling nozzles. Standard flat fan nozzles may also be used at no more than 20 psi. At this pressure, the spray pattern may be narrower.

With ground broadcast equipment, drift can be reduced by keeping the spray boom as low as possible; by applying no less than 35 gallons of spray per acre; by using no more than 20 pounds of spraying pressure with flat fan or flooding fan tips; and by spraying when wind velocity is low.

Do not apply on ditches used to transport irrigation water. Do not apply where runoff or irrigation water may flow onto susceptible crops.

STORAGE AND DISPOSAL

Do not contaminate water, food, or feed by storage or disposal.

STORAGE: Store above 28°F or agitate before use.

DISPOSAL:

Prohibitions: Open dumping is prohibited.

Pesticide Disposal - Pesticide spray mixture or rinse water that cannot be used according to label instructions must be disposed of according to applicable Federal, state, or local procedures.

Container Disposal - Triple rinse (or equivalent). Then offer for recycling or reconditioning or puncture and dispose of in a sanitary landfill, or by other approved state and local procedures. Plastic containers may also be disposed of by incineration or, if allowed by state and local authorities, by burning. If burned, stay out of smoke.

General - Consult Federal, state, or local disposal authorities for approved alternative procedures.

Be sure that use of this product conforms to all applicable regulations.

WARRANTY LIMITATIONS AND DISCLAIMER

The Dow Chemical Company warrants that this product conforms to the chemical description on the label and is reasonably fit for the purposes stated on the label when used in strict accordance with the directions therein under normal conditions of use. THIS IS THE ONLY WARRANTY MADE ON THIS PRODUCT. NO OTHER EXPRESS AND NO IMPLIED WARRANTY OF MERCHANTABILITY OR FITNESS FOR A PARTICULAR PURPOSE IS MADE OUTSIDE OF THIS LABEL. Therefore, neither this warranty nor any other warranty of merchantability or fitness for a particular purpose, express or implied, extends to the use of this product contrary to label instructions (including conditions noted on the label, such as unfavorable temperatures, soil conditions, etc.), under abnormal conditions (such as excessive rainfall, drought, tornadoes, hurricanes, etc.) or under conditions not reasonably foreseeable to or beyond the control of seller.

When buyer or user suffers losses or damages resulting from the use or handling of this product (including claims based on contract, negligence, strict liability, or other legal theories), buyer or user must promptly notify in writing The Dow Chemical Company of any claims to be eligible to receive either remedy given below. The EXCLUSIVE REMEDY OF THE BUYER OR USER and the LIMIT OF LIABILITY of The Dow Chemical Company or any other seller will be one of the following, at the election of The Dow Chemical Company: (1) **Refund of purchase price** paid by buyer or user for product bought, **or** (2) **Replacement** of amount of product used. The seller **will not be liable for consequential or incidental damages or losses.**

The terms of this Warranty Limitations And Disclaimer cannot be varied by any written or verbal statements or agreements. Any employee or sales agent of the seller is not authorized to vary or exceed the terms of this Warranty Limitations And Disclaimer in any manner.

88946-L2 F888

*Trademark of THE DOW CHEMICAL COMPANY

E. I. du Pont de Nemours & Co. (Inc.)

Agricultural Products Department
Wilmington, DE 19898
(800) 441-7515

ALLY®

HERBICIDE
DRY FLOWABLE GRANULE

ACTIVE INGREDIENT	BY WEIGHT:
Metsulfuron Methyl Methyl 2-[[[[(4-methoxy-6-methyl-1,3,5-triazin-2-yl)-amino]carbonyl]-amino]sulfonyl]benzoate	60%
INERT INGREDIENTS:	40%
TOTAL	100%

EPA Reg. No. 352-435
U.S. Patent 4,383,113

KEEP OUT OF REACH OF CHILDREN
CAUTION

PRECAUTIONARY STATEMENTS

HAZARDS TO HUMANS AND DOMESTIC ANIMALS

CAUTION!

Harmful if absorbed through skin. Causes eye irritation. Avoid contact with skin, eyes or clothing. Avoid breathing dust or spray mist. Wash thoroughly with soap and water after handling. Remove and wash contaminated clothing before reuse.

STATEMENT OF PRACTICAL TREATMENT

In case of contact with eyes, immediately flush with plenty of water. If on skin, wash with plenty of soap and water. Get medical attention if irritation persists.
For medical emergencies involving this product, call toll free 1-800-441-3637.

ENVIRONMENTAL HAZARDS

Do not apply directly to water or wetlands. Do not contaminate water by cleaning of equipment or disposal of wastes.

IMPORTANT INFORMATION—(Read Before Using)

Injury to or loss of desirable trees or vegetation may result from failure to observe the following: Do not apply, drain or flush equipment on or near desirable trees or other plants, or on areas where their roots may extend, or in locations where the chemical may be washed or moved into contact with their roots. Do not use on lawns, walks, driveways, tennis courts or similar areas. Prevent drift of spray to desirable plants. Do not contaminate any body of water, including irrigation water that may be used on other crops. Carefully observe sprayer cleanup instructions, as spray tank residue may damage crops other than wheat or barley.

INFORMATION ON RESISTANT WEEDS:

Naturally-occurring weed biotypes* resistant to this product are known to exist. To delay the development of resistant biotypes, only spray "Ally" in tank mixtures with broadleaf herbicides having a different mode of action**, such as: 2,4-D, Banvel[1]***, Buctril[2], Bronate[2], Curtail[3], MCPA, Du Pont "Karmex" Herbicide or Du Pont "Lexone" Herbicide—as specified in the "Tank Mixtures For Resistant Weed Management" section of this label.

*Biotypes are naturally-occurring individuals of the species which have a slightly different genetic makeup. Resistant biotypes may look exactly the same as susceptible biotypes. Herbicide resistant biotypes are able to survive a use rate several times higher than needed to control susceptible biotypes.

**Mode of action is the chemical interaction that interupts a biological process necessary for plant growth and development.

***Tank mixes with "Banvel" may result in reduced control of some broadleaf weeds.

GENERAL INFORMATION

Du Pont "Ally" Herbicide is recommended for use in CO, ID, KS, MN, MT, NE, NM, ND, OK, OR, SD, TX, UT WA, and WY.

"Ally" is recommended for use on land primarily dedicated to the production of wheat (including durum) and barley. Rotation options are provided for certain other crops such as oats, proso millet, dryland grain sorghum, dryland corn, soybeans, flax, sunflower, safflower, alfalfa, hay, and dry beans. In areas having a short growing season, prolonged periods of low soil temperature and low annual rainfall, "Ally" can remain in the soil for 34 months or more and cause severe injury to certain crops other than those listed in the "Crop Rotation Guidelines" section of the label. Read and follow the "Crop Rotation Guidelines" section for the specific rotation intervals for all following crops. Before using "Ally", carefully consider your crop rotation plans and options. For maximum rotational flexibility, do not treat all your wheat or barley acres with "Ally".

"Ally" is a 60% active ingredient herbicide formulated as a dry flowable granule to be mixed in water and applied for use as a uniform broadcast spray for selective weed control in wheat (including durum), barley and in grasses on acreage enrolled in the Conservation Reserve Program (CRP). "Ally" may be applied by air or with ground spray equipment. It is noncorrosive, nonflammable, nonvolatile and does not freeze.

For application to wheat or barley, "Ally" should be applied postemergence to actively growing broadleaf weeds. Herbicide combinations may be required for certain weeds as indicated under "Tank Mixtures" in the "Weed Control in Wheat and Barley" sections.

"Ally" rapidly inhibits growth of susceptible weeds; however, typical symptoms (discoloration) of dying weeds may not be noticeable for 1 to 3 weeks after application depending on growing conditions and weed susceptibility. Warm, moist conditions following treatment enhance the activity of "Ally", while cold, dry conditions delay activity. Weeds hardened-off by cold weather or drought stress may not be fully controlled or suppressed and regrowth may occur. Snow or rainfall received within 4 hours after application can reduce the level of weed control.

Degree of control and duration of effect depend on: Weed spectrum and density; weed size and variability; growing conditions prior to, at and following time of application; amount of precipitation, and spray coverage. With adequate rainfall for soil activation, short-term residual control of the more sensitive species may be obtained for a few weeks after application.

READ AND FOLLOW ALL APPROPRIATE SECTIONS OF LABEL INCLUDING PRECAUTIONS BEFORE USING THIS PRODUCT.

GRAZING

"Ally" has no grazing restrictions.

DIRECTIONS FOR USE

It is a violation of federal law to use this product in a manner inconsistent with its labeling.

"Ally" should be used only in accordance with recommendations on this label or in separate published Du Pont recommendations available through local dealers.

Du Pont will not be responsible for losses or damages resulting from the use of this product in any manner not specifically recommended by Du Pont. User assumes all risks associated with such nonrecommended use.

MAXIMUM USE RATE AND SOIL pH LIMITATION

In CO, ID, Western KS and Western NE (west of highway 183), MN, MT, NM, ND, OK Panhandle, OR, TX Panhandle, SD, UT, WA and WY, the maximum rate is 1/10 oz/A in a 22 month period.

In Central KS, Central NE, Central OK and North Central TX, the maximum crop use rate is 1/10 oz/A in a 10 month period.

Do not use "Ally" on soils with a pH greater than 7.9 as the extended soil residual activity could adversely affect crop rotation options beyond normal intervals and under certain conditions cause injury to wheat and barley.

NOTE: Prior to use of "Ally", take soil samples at 0-4" depth and determine soil pH by laboratory analysis using a 1:1 (soil:water) suspension. Samples should be representative of the different conditions in the field (for example, slope, soil texture, low areas, eroded areas, etc.) Consult local extension publications for recommended soil sampling procedures.

WEED CONTROL IN WHEAT AND BARLEY

For best weed control or weed suppression, apply postemergence when environmental conditions favor active growth of broadleaf weeds and when crop canopy will allow thorough coverage of target weeds. Unless otherwise directed, always include a surfactant.

Determine crop rotation plans according to "Crop Rotation Guidelines."

TIMING OF CROP APPLICATION

For winter wheat and winter barley, apply "Ally" (1/10 oz/A) postemergence after crop is in the 2-leaf stage—**but before the boot stage.** Do not apply during boot stage or early heading as crop injury may occur.

For spring barley and spring wheat (except durum or variety Wampum), apply "Ally" (1/10 oz/A) postemergence after crop is in the 2-leaf stage—**but before the boot stage.** Do not apply during boot stage or early heading as crop injury may occur.

For durum spring wheat and Wampum variety of spring wheat, apply "Ally" (1/10 oz/A) postemergence only after crop is tillering (refer to 2,4-D or MCPA manufacturers' labels)—**but before the boot stage** and only in combination with either 2,4-D or MCPA. Do not apply during boot stage or early heading as crop injury may occur.

Irrigated Cerals (wheat/barley)—on land dedicated to cereal production which includes supplemental irrigation, delay first post treatment irrigation for at least 3 days after treatment. The first post treatment irrigation should not exceed 1". Apply "Ally" after crop tillering has begun. Do not apply "Ally" to stressed plants.

WEEDS CONTROLLED BY "ALLY"

1/10 Ounce Per Acre

(80 acres treated per 8 ounce container)

For best results, always use "Ally" in a tank mix with 2,4-D or MCPA. Unless otherwise directed, treat when weeds are less than 4" tall or in diameter and are actively growing. See "Specific Weed Problems", "Tank Mixtures For Resistant Weed Management" and "Tank Mixtures" sections for additional information.

Bur buttercup (testiculate)	Pigweed (redroot, smooth, tumble)
Chickweed (common)	Plains coreopsis
Common purslane	†Prickly lettuce
Conical catchfly	Shepherdspurse
Cow cockle	Smallseed falseflax
False chamomile	Smartweed (green, ladysthumb, pale)
Field pennycress (fanweed)	Snow speedwell
Filaree	Tumble mustard (Jim Hill)
Groundsel (common)	Volunteer sunflower
Henbit	Waterpod
Mayweed	
Miners lettuce	

WEEDS SUPPRESSED[1] BY "ALLY"

Annual Ryegrass**	Knotweed (prostrate)**
Canada thistle**	Sowthistle**
Common sunflower**	Wild buckwheat**
Cron gromwell**	

[1] Weed suppression is a visual reduction in weed competition (reduced population and/or vigor) as compared to an untreated area. Degree of suppression will vary with rate used, size of weeds and environmental conditions following treatment.

† Naturally-occurring resistant biotypes of these weeds are known to occur in the Central Plains and the Pacific Northwest. See "Tank Mixtures For Resistant Weed Management" section of label for additional information.

**See "Specific Weed Problems".

WEEDS CONTROLLED WITH A TANK MIX OF "ALLY" PLUS 2,4-D OR MCPA

1/10 Ounce Per Acre

(80 acres treated per 8 ounce container)

For best results, add a surfactant to the "Ally" tank mix and apply when weeds are both actively growing and no larger than 4" tall or in diameter, and crop canopy will allow thorough coverage of target weeds. See "Specific Weed Problems" and "Tank Mixtures" sections for additional information.

Blue mustard (purple)**	Lambsquarters (common, slimleaf)
Coast fiddleneck (tarweed)	†Russian thistle
Flixweed**	Tansymustard**
†Kochia	Treacle mustard

† Naturally-occurring resistant biotypes of these weeds are known to occur in the Central Plains and the Pacific Northwest. See "Tank Mixtures For Resistant Weed Management" section of label for additional information.

SPECIFIC WEED PROBLEMS

Annual Ryegrass (in OK, TX): To obtain the best results, a sequential treatment of Du Pont "Glean" Herbicide in the fall followed by "Ally" in the spring is recommended. Apply "Glean" at 1/2 oz/A preemergence to ryegrass. 1/2 to 1" of rainfalls is needed to move "Glean" into the weed root zone prior to ryegrass emergence. Remove grazing cattle during wet (muddy) field conditions to avoid disturbing the herbicide barrier. After completion of wheat grazing but prior to boot stage of the wheat, apply "Ally" with a surfactant or with a liquid nitrogen fertilizer topdressing application. For fields not grazed, apply the sequential application of "Ally" as soon as ryegrass starts to grow after winter dormancy. Do not add a surfactant to liquid nitrogen fertilizer plus "Ally" combinations. In mixing "Ally" with liquid fertilizer, slurry "Ally" in water then thoroughly mix the slurry into the fertilizer. Run a tank mix compatibility test before mixing "Ally" in fertilizer solution. DO NOT use with fertilizers having a pH of 3.0 or less as rapid product degradation can occur. The addition of 2,4-D is not recommended for annual ryegrass suppression.

Blue Mustard, Flixweed and Tansymustard: In ID, MN, MT, ND, OR, SD, UT, WA and WY, for best results, apply "Ally" tank mixtures with 2,4-D or MCPA postemergence to mustards but before bloom.

Canada Thistle and Sowthistle - Apply either "Ally" plus surfactant or "Ally" plus 2,4-D or MCPA in the spring after the majority of thistles have emerged and are small (rosette stage to 6" elongating stems) and actively growing. An application will inhibit the ability of emerged thistles to compete with the crop.

Sunflower (common/volunteer) - Apply either "Ally" plus surfactant or "Ally" plus 2,4-D or MCPA after the majority of sunflowers have emerged, are 2" to 4" tall and are actively growing. Thorough coverage is important. Use minimum spray volumes of 3 gal. by air and 5 gal. by ground.

Corn Gromwell and Prostrate Knotweed - Apply "Ally" plus surfactant when weeds are actively growing, no larger than 2" tall and crop canopy will allow thorough coverage. The addition of 2,4-D or MCPA may or may not improve the results.

Wild Buckwheat - For best results, apply "Ally" plus 2,4-D or "Ally" plus MCPA when plants have no more than 3 true leaves (not counting the cotyledons). If plants are not actively growing, delay treatment until environmental conditions favoring active weed growth are present. Thorough coverage is important.

TANK MIXTURES FOR RESISTANT WEED MANAGEMENT

Central Kansas, Central Nebraska, Central Oklahoma and North Central TX:

- Only apply "Ally" as a tank mix treatment with either 2,4-D (amine or ester) or MCPA (amine or ester). Use 1/10 oz/A of "Ally" plus 1/4 to 1/2 lb. active ingredient 2,4-D or MCPA (ester fomulations of 2,4-D or MCPA have provided best results). Surfactant may be added at one to two pints per 100 gallons of spray volume; however, the addition of surfactant may increase the chance for crop injury. Apply "Ally" plus MCPA from 3-5 leaf but prior to boot stage. Apply "Ally" plus 2,4-D after tillering (refer to 2,4-D manufacturer's label) but prior to boot stage.
- "Ally" tank mixes can be applied annually in this area.
- If resistant weeds are known to be present, consider using another herbicide treatment or adjust the use rate of the "Ally" tank mix partner so that it alone will control the resistant species.
- Read and follow all use instructions, warnings and precautions for the companion herbicide.
- Do not apply "Ally" during fallow.

Idaho, Minnesota, Montana, North Dakota, Oregon, South Dakota, Utah, Washington and Nothern Wyoming:

- Only apply "Ally" as a tank mix treatment with either 2,4-D (amine or eater) or MCPA (amine or ester). Use 1/10 oz/A of "Ally" plus 1/4 to 1/2 lb. active ingredient 2,4-D or MCPA (ester formulations of 2,4-D or MCPA have provided best results). Surfactant may be added at one to two pint per 100 gallons of spray volume; however, the addition of surfactant may increase the chance for crop injury. Apply "Ally" plus MCPA from 3-5 leaf but prior to boot stage. Apply "Ally" plus 2,4-D after tillering (refer to 2,4-D manufacturer's label) but prior to boot stage.
- Do not apply "Ally" during fallow.
- Do not apply "Ally" more often than once in a 22 month period for a given field.
- Do not apply "Ally" for 22 months before, or after a "Glean" treatment.
- If resistant weeds are known to be present, consider using another herbicide treatment or adjust the use rate of the "Ally" tank mix partner so that it alone will control the resistant species.
- Read and follow all use instructions, warnings and precautions for the companion herbicide.

Colorado, Western Kansas and Western Nebraska (west of highway 183), Eastern New Mexico, Oklahoma Panhandle, Texas Panhandle and Southeastern Wyoming:

- Only apply "Ally" as a tank mix treatment with either 2,4-D (amine or eater) or MCPA (amine or ester). Use 1/10 oz/A of "Ally" plus 1/4 to 1/2 lb. active ingredient 2,4-D or MCPA (ester formulations of 2,4-D or MCPA have provided best results). Surfactant may be added at one to two pints per 100 gallons of spray volume; however, the addition of surfactant may increase the chance for crop injury. Apply "Ally" plus MCPA from 3-5 leaf but prior to boot stage. Apply "Ally" plus 2,4-D after tillering (refer to 2,4-D manufacturer's label) but prior to boot stage.
- Do not apply "Ally" more often than once in a 22 month period on a given field.
- Do not apply "Ally" during fallow.
- If resistant weeds are known to be present, consider using another herbicide treatment or adjust the use rate of the "Ally" tank mix partner so that it alone will control the resistant species.
- Read and follow all use instructions, warnings and precautions for the companion herbicide.

TANK MIXTURES

2,4-D (amine or ester) or MCPA (amine or ester) - Use "Ally" at 1/10 oz/A plus 1/4 to 1/2 lb active ingredient 2,4-D or MCPA (ester formulations have provided best results). Surfactant may be added at 1 to 2 pints per 100 gallons of spray volume 1/2; however, the addition of surfactant may increase the chance of crop injury.

Always mix "Ally" in water prior to adding 2,4-D or MCPA and surfactant. Read and follow all label instructions on timing, precautions and warning for these herbicides prior to using these tank mixtures.

Other Tank Mixtures - "Ally" must be in suspension before adding the companion herbicide(s) or spray adjuvants. "Ally" will not control wild oats or other grasses. If broadleaf weeds plus wild oats and/or grasses are present, apply "Ally" with a suitable registered product either as a tank mix or sequential treatment.

Follow all use instructions, label rates, warnings, precautions and surfactant recommendations of companion herbicide(s).

DO NOT tank mix with "Hoelon" 3EC[4] as wild oat or green foxtail control may be reduced.

Tank mixes of "Ally" + "Banvel" may result in reduced control of some broadleaf weeds.

"Ally" may be tank mixed with insecticides registered for use on cereal grains. However, under certain conditions, (drought stress, crop in 2-4 leaf stage) tank mixes of "Ally" plus organophosphate insecticides (such as methyl or ethyl, parathion, "Di-Syston" [5] etc.) may produce temporary crop yellowing or, in severe cases, crop injury. The potential for crop injury is greatest when there are fluctuations in day/night temperatures just prior to or soon after application. Limit first use to a small area before treating large acreage.

Do not apply "Ally" within 60 days of crop emergence where an organophosphate insecticide (such as "Di-Syston") has been applied as an in-furrow treatment, as crop injury may result.

DO NOT USE "ALLY" PLUS MALATHION AS CROP INJURY MAY RESULT.

SPRAY PREPARATION, ADDITIVES, PRODUCT MEASUREMENT, SURFACTANT AND LIQUID FERTILIZER

Spray Preparation: Pour the proper amount of "Ally" into the necessary volume of water in the spray tank with the agitator running. Continuous agitation is required for a uniform suspension and application. "Ally" must be added first to the spray tank followed by any other tank mix chemicals and surfactant.

Use spray preparation of "Ally" within 24 hours or product degradation may occur. If spray preparation is left standing without agitation, thoroughly agitate before reusing.

Additives: Do not use with spray additives that lower the pH of the spray solution below pH 3.0, as rapid product degradation can occur.

Product Measurement: The "Ally" volumetric measuring cylinder is to be used as a guide, since the degree of accuracy varies by plus or minus 7.5%. For more precise measurement, use scales calibrated in ounces.

Surfactant - Unless directed otherwise, use a surfactant of at least 80% active ingredient and add it as the last ingredient at the rate of 1 to 2 quarts per 100 gallons of spray volume on winter wheat or 1/2 to 1 quart on spring wheat, spring or winter barley, durum spring wheat and Wampum variety of spring wheat. Antifoaming agents may be needed. DO NOT use liquid fertilizer in addition to or as a substitute for a surfactant.

Liquid Fertilizer - Slurry "Ally" in water; then thoroughly mix the slurry into the liquid fertilizer. DO NOT add a surfactant. Run a tank mix compatibility test before mixing "Ally" in fertilizer solution. DO NOT use with fertilizers having a pH of 3.0 or less as rapid product degradation can occur. If 2,4-D is included in "Ally" and liquid fertilizer mixture, the ester formulations are generally more compatible.

WEED CONTROL FOR THE CONSERVATION RESERVE PROGRAM (CRP)

"Ally" is registered for CRP use in CO, Southern ID, KS, MT, NE, NM, ND, OK, SD, TX and UT and WY. Consult the "Ally" supplemental label for CRP use instructions.

WEED CONTROL IN REDUCED TILLAGE FALLOW

DO NOT USE "ALLY" IN FALLOW

EQUIPMENT - SPRAY VOLUMES

Apply using properly calibrated air or ground equipment. Select a spray volume and delivery system that will insure thorough coverage and a uniform spray pattern. Avoid overlapping, and shut off spray booms while starting, turning, slowing or stopping, or injury to the crop or following crops may result.

Agitation - Continuous agitation is required to keep "Ally" in suspension.

Spray Equipment - Refer to specific manufacturer's recommendations for additional information on GPA, pressure, speed, nozzle types and arrangements, nozzle heights above the target canopy, etc., for respective application equipment.

Ground Application - For optimum spray distribution and thorough coverage, use flat fan or low volume flood nozzles. For flat fan nozzles, do not use less than 3-gallon spray volume per acre (GPA).

For flood nozzles on 30-inch nozzle spacings, use not less than 10 GPA and no larger than TK10 or equivalent and not less than 30 psi. On 40-inch nozzle spacings, use not less than 13 GPA. 100% overlapping of nozzle spray pattern is recommended for 30, and 40-inch spacings.

With "Raindrop"[6] nozzles, do not use less than 30 GPA and insure for 100% overlap of nozzle spray patterns.

Use 50-mesh screens or larger.

Aerial Application - Use nozzle types and arrangements that will provide for optimum spray distribution and maximum coverage at 1 to 5 GPA. Do not apply during inversion conditions, when winds are gusty, or when other conditions will favor poor coverage and/or drift.

Caution - Avoid Spray Drift

Follow these practices to minimize drift.

Do not allow spray from either ground or aerial equipment to drift onto adjacent crops or land, as even small amounts will injure other plants. When spraying near adjacent, sensitive crops or plants, do everything possible to reduce spray drift. This includes:

- Stop spraying if wind speed becomes excessive. DO NOT SPRAY IF WIND SPEED IS 10 MPH OR GREATER. Spray drift can occur at wind speeds less than 10 MPH. If sensitive crops or plants are downwind, extreme caution must be used even in relatively low wind conditions! DO NOT SPRAY IF WINDS ARE GUSTY.
- High temperatures, drought, and low relative humidity increase the possibility of spray drift. EXTREME CAUTION MUST BE USED WHEN THESE CONDITIONS ARE PRESENT AND SENSITIVE CROPS OR PLANTS ARE NEARBY, REGARDLESS OF WIND SPEED.
- Do not apply when an inversion exists. An inversion is characterized by little or no air movement and an increase in air temperature with an increase in altitude. In humid regions, a fog or mist may form. An inversion may be detected by producing a smoke column and checking for a layering effect. Smoke-producing devices on aircraft are recommended. If not sure whether inversion conditions are present, consult with local weather services before making an application.
- Drift from aerial or ground equipment may be further reduced by:
  1. Using coarse sprays to minimize drift. DO NOT APPLY WITH HOLLOW-CONE INSECTICIDE NOZZLES ON GROUND EQUIPMENT. Do not use nozzles that produce fine droplets, such as Sprayfoil[7] or airblast-type nozzles. Nozzles should be oriented at an angle between straight down and straight back for ground applications. For aerial applications orient nozzles straight back along the windstream. If using flood-type nozzles on aircraft, orient them so spray is produced in direction of the airstream.
  2. Increasing volume of spray mix per acre (for example, minimum 5 GPA by air, 10 GPA by ground) by using higher flow rate nozzles.
  3. Reducing pressure (PSI). DO NOT EXCEED 40 PSI when applying "Ally". (Vehicle speed must also be reduced to maintain spray mix volume per acre.) Consult manufacturer's catalogs for details on correct calibration.
  4. Apply as close to target plants as possible while still maintaining a good spray pattern.

NOTE: Do not allow spray to drift onto adjacent crops, or onto agricultural land scheduled to be planted to crops other than wheat, as injury to the crop may occur. Extreme care must be taken to prevent drift to desirable plants or nontarget agricultural land.

CROP ROTATION GUIDELINES

Crop rotation plans are determined by the crop to be planted and a minimum rotation interval. Minimum rotation interval is the time from the last application of "Ally" to the anticipated date of planting. For maximum rotational flexibility, do not use "Ally" on all your wheat or barley.

Unless a Crop Rotation Interval is specified, a bioassay must be completed before rotating to any crop other than those listed below. See "Bioassay" section. Do not use on soils with a pH greater than 7.9.

Wherever "Ally" is used on land previously treated with "Glean" read the rotational guidelines on both labels and follow the one with the longest interval stated for your situation.

These crops can be planted on nonirrigated land following the use of "Ally" at 1/10 oz/A:

CROP TO BE PLANTED

Winter and spring wheat

Area: all[1]
Soil pH*: 7.9 or less
Cumulative Precipitation** (inches): none
Minimum Rotation Interval (months): 1

CRP grasses[2]

Area: all
Soil pH*: 7.9 or less
Cumulative Precipitation** (inches): none
Minimum Rotation Interval (months): 4

Durum wheat, barley, spring/winter oats

Area: all
Soil pH*: 7.9 or less
Cumulative Precipitation** (inches): none
Minimum Rotation Interval (months): 10

Grain sorghum, proso millet

Area: CO, KS, NE, NM, OK, TX, Southern WY
Soil pH*: 7.9 or less
Cumulative Precipitation** (inches): none
Minimum Rotation Interval (months): 10

Area: SD[3]
Soil pH*: 7.9 or less

Cumulative Precipitation** (inches): 13
Minimum Rotation Interval (months): 12

Area: MT, ND (west of State Hwy 1), Northern WY
Soil pH*: 7.9 or less
Cumulative Precipitation** (inches): none
Minimum Rotation Interval (months): 22

Area: ND (east of State Hwy 1)
Soil pH*: 7.9 or less
Cumulative Precipitation** (inches): none
Minimum Rotation Interval (months): 34

Field corn

Area: Central KS[4]
Soil pH*: 7.9 or less
Cumulative Precipitation** (inches): 25
Minimum Rotation Interval (months): 14

Area: CO, KS, NE, Southern WY[5]
Soil pH*: 7.5 or less
Cumulative Precipitation** (inches): 15
Minimum Rotation Interval (months): 12

Area: CO, KS, NE, Southern WY[5]
Soil pH*: 7.6 to 7.9
Cumulative Precipitation** (inches): 22
Minimum Rotation Interval (months): 22

Area: SD[6]
Soil pH*: 7.9 or less
Cumulative Precipitation** (inches): 15
Minimum Rotation Interval (months): 12

Area: MT, ND (west of State Hwy 1), Northern WY
Soil pH*: 7.9 or less
Cumulative Precipitation** (inches): 22
Minimum Rotation Interval (months): 22

Area: ND (east of State Hwy 1)
Soil pH*: 7.9 or less
Cumulative Precipitation** (inches): none
Minimum Rotation Interval (months): 34

Soybeans

Area: Central KS[7]
Soil pH*: 7.9 or less
Cumulative Precipitation** (inches): 25
Minimum Rotation Interval (months): 14

Area: KS, NE[8]
Soil pH*: 7.5 or less
Cumulative Precipitation** (inches): 22
Minimum Rotation Interval (months): 22

Area: KS, NE[8]
Soil pH*: 7.6 to 7.9
Cumulative Precipitation** (inches): 33
Minimum Rotation Interval (months): 34

Cotton (dryland only)

Area: OK (east of the Panhandle), North Central TX[9]
Soil pH*: 7.9 or less
Cumulative Precipitation** (inches): 25
Minimum Rotation Interval (months): 14

Area: OK Panhandle and TX Panhandle, Eastern NM
Soil pH*: 7.9 or less
Cumulative Precipitation** (inches): 30
Minimum Rotation Interval (months): 22

Alfalfa (hay only)

Area: MT Soil
pH*: 7.9 or less
Cumulative Precipitation** (inches): none
Minimum Rotation Interval (months): 34

Dry beans

Area: ND (west of State Hwy 1)
Soil pH*: 7.9 or less
Cumulative Precipitation** (inches): none
Minimum Rotation Interval (months): 22

Area: ND (east of State Hwy 1)
Soil pH*: 7.9 or less
Cumulative Precipitation** (inches): none
Minimum Rotation Interval (months): 34

Flax, Safflower, Sunflower

Area: CO, Southern ID, KS, MT, NE, NM, ND (west of State Hwy 1), OK, SD, TX, UT, WY
Soil pH*: 7.9 or less
Cumulative Precipitation** (inches): none
Minimum Rotation Interval (months): 22

Area: ND (east of State Hwy 1)
Soil pH*: 7.9 or less
Cumulative Precipitation** (inches): none
Minimum Rotation Interval (months): 34

All other crops

Area: all
Soil pH*: 7.9 or less
Cumulative Precipitation** (inches): none
Minimum Rotation Interval (months): Field Bioassay/LRBsm Bioassay[10]

*Soil pH is to be determined by laboratory analysis using the 1:1; soil:water suspension method on representative soil samples taken at 0–4" depth. Consult local extension publications for recommended soil sampling procedures.

**Cumulative Precipitation equals the total amount received from the date of "Ally" application to the date of planting. Should accumulated precipitation not be sufficient to meet the indicated amounts, do not rotate to the indicated crops until the following growing season.

[1] All—CO, Southern ID, KS, MT, NE, NM, ND, OK, SD, TX, UT, WY

[2] CRP grasses—

Blue Grama
Bluestems - Big, Little, Plains, Sand, WW Spar
Buffalograss
Green Sprangletop
Indiangrass
Kleingrass
Lovegrasses - Atherstone, Sand, Weeping, Wilman
Orchardgrass
Sideoats Grama
Switchgrass - Blackwell
Wheatgrasses - Bluebunch, Crested, Intermediate, Pubescent, Siberian, Slender, Streambank, Tall, Thickspike, Western
Wildrye grass - Russian

The planting of grass and legume mixtures is not recommended as injury to the legume may occur.

[3] **SD**—South of state highway 212 and east of the Missouri River, and south of state highway 34 and west of the Missouri River.

[4] **Central KS**—Generally east of state highway 183 and west of the Flinthills.

[5] **CO**—Generally north of I-70.

[5] **KS**—Generally north of I-70 and west of state highway 183.

[5] **NE**—Generally west of state highway 183.

[5] **WY**—Counties of Goshen, Laramie, Platte.

[6] **SD**—East of the Missouri River and south of state highway 14 west of the Missouri River.

[7] **Central KS**—Generally east of state highway 183 and west of the Flinthills.

[8] **KS**—Generally north of I-70 and west of state highway 183.

[8] **NE**—Generally west of state highway 183 and east of the Panhandle.

[9] Counties of:

Archer	Baylor	Bell	Bosque	Bowie
Callahan	Camp	Cass	Clay	Collin
Cooke	Coryell	Dallas	Delta	Denton
Eastland	Ellis	Falls	Fannin	Foard
Franklin	Grayson	Hardeman	Haskell	Hill
Hood	Hopkins	Hunt	Jack	Johnson
Kaufman	Knox	Lamar	Limestone	McLennan
Milam	Montague	Morris	Navarro	Palo Pinto
Parker	Rains	Red River	Robertson	Rockwall
Shackelford	Somervell	Stephens	Tarrent	Throckmorton
Titus	Upshur	Van Zandt	Wilbarger	Wichita
Williamson	Wise	Wood	Young	

[10] **IMPORTANT**—Land previously treated with "Ally" at 1/10 oz/acre cannot be rotated to crops other than those listed until either a field bioassay or LRBsm confirms that residues of "Ally" that could cause crop injury are not present. See "Bioassay" section for details.

SPRAYER CLEANUP

To avoid subsequent injury to crops other than wheat or barley, immediately after spraying and prior to spraying other crops, thoroughly remove all traces of "Ally" from mixing and spray equipment as follows:

1. Drain tank, then flush tank, boom and hoses with clean water for a minimum of 10 minutes.
2. Fill the tank with clean water then add 1/2 gallon chlorine bleach (containing 5 1/4% sodium hypochlorite) per 100 gallons of water. Flush through boom and hoses, then allow to sit for 15 minutes with agitation, then drain.
3. Repeat Step 2.
4. Nozzles and screens should be removed and cleaned separately. To remove traces of chlorine bleach, rinse the tank thoroughly with clean water and flush through hoses and boom.

NOTE: To reduce the amount of water required in the above procedure, see separate Du Pont bulletin, "Reduced Volume Cleanout Procedure for Large Sprayers."

CAUTION: Do not use chlorine bleach with ammonia. All traces of liquid fertilizer containing ammonia, ammonium nitrate or ammonium sulphate must be rinsed with water from the mixing and application equipment before adding chlorine bleach solution. Failure to do so will release a gas with a musty chlorine odor which can cause eye, nose, throat and lung irritation. Do not clean equipment in an enclosed area.

PRECAUTIONS

Do not apply an "Ally" plus "Glean" tank mixture.

In CO, ID, Western KS and Western NE (west of highway 183), MN, NM, ND, OK Panhandle, OR, TX Panhandle, SD, UT, WA and WY, the maximum use rate is 1/10 oz/A in a 22 month period.

In central KS, Central NE, Central OK and North central TX, the maximum use rate is 1/10 oz/A in a 10 month period.

Do not use on soils with pH greater than 7.9 (for example highly calcareous soils) as extended soil residual activity could adversely affect minimum rotation intervals for all crops.

Wherever "Ally" is used on land previously treated with Du Pont "Glean" read the rotational guidelines on both labels and follow the one with the longest interval stated for your situation.

Wherever land has been or will be treated with "Assert" [8] Herbicide and "Ally" Herbicide, plant only wheat or barley until a bioassay (see "Bioassay" section of label) demonstrates that other crops can be successfully grown. On land that is frequently rotated to crops other than wheat or barley, do not use "Ally" wherever "Assert" has been or will be used. The additive effect of soil residues from these treatments has not been determined and crop rotation guidelines and minimum rotation intervals are not known; injury to rotational crops may occur.

Do not apply to irrigated land where tail water will be used to irrigate crops other than wheat and barley.

Do not apply to frozen ground where surface runoff may occur.

Do not apply to snow covered ground.

Varieties of wheat and barley differ in their tolerance to herbicides. When using "Ally" for the first time on a particular variety, limit initial use to one 8 oz jug. If no symptoms of crop injury occur within 14 days after treatment, balance of acreage can be treated.

Do not apply "Ally" to wheat or barley that is stressed by severe weather conditions, drought, low fertility, water saturated soil, disease or insect damage as crop injury may result. Severe winter stress, drought, disease or insect damage following application also may result in crop injury.

Under certain conditions such as heavy rainfall, prolonged cool weather (daily high temperatures less than 50°F) or wide fluctuations in day/night temperatures just prior to or soon after treatment, temporary discoloration and/or crop injury may occur. Risk of injury is greatest when crop is in the 2 to 5 leaf stage.

Tank mixtures of "Ally" and organophosphate insecticides (such as methyl or ethyl parathion or "Di-Syston", etc.) may cause temporary discoloration or crop injury. The potential for crop injury is greatest when there are wide fluctuations in day/night temperatures just prior to or soon after treatment.

The combined treatment effects of "Ally" postemergence preceded by preemergence wild oat herbicides may cause crop injury to spring wheat when crop stress (soil crusting, planting too deep, prolonged cold weather or drought) causes poor seedling vigor.

To prevent cold weather-related crop injury, avoid making applications during winter months when weather conditions are unpredictable and can be severe.

Do not apply to wheat or barley undersown with legumes or grasses as injury to the forage may result.

To reduce the potential for movement of treated soil due to wind erosion, do not apply to powdery dry or light sandy soils until they have been stabilized by rainfall, trashy mulch, reduced tillage or other such cultural practices. Injury to adjacent crops may occur when treated soil is blown onto land used to produce crops other than cereal grains.

For ground applications applied to weeds when dry, dusty field conditions exist, control of weeds in wheel track areas may be reduced. The addition of 2,4-D or MCPA should improve weed control under these conditions.

Preplant or preemergence applications of 2,4-D or herbicides containing 2,4-D made within two weeks of planting spring cereals may cause crop injury when used in conjunction with early postemergence applications of "Ally". Under these conditions, delay "Ally" treatment until crop tillering has begun.

BIOASSAY

A bioassay (field or LRB℠) must be completed before rotating to crops not listed on this label, or rotating at intervals shorter than those listed in the "Crop Rotation Recommendations (Noncereal Crops)" section.

FIELD BIOASSAY

"Ally" is a useful tool for weed control in wheat or barley; however, under some conditions small amounts of "Ally" can remain in the soil and injure crops other than those listed on the "Ally" label under "Crop Rotation Guidelines" for 34 months or more after application; therefore, before you use "Ally", you should carefully consider your crop rotation plans during the three (or more) year period following treatment.

A field bioassay involves growing test strips of the crop or crops you plan to grow the following year in fields previously treated with "Ally". Crop response will indicate whether or not to rotate to the crop(s) grown in the test strips.

"Ally" breaks down most rapidly in soils that have high microbial populations. Factors that favor microbial activity include having annual rainfall of 10" or more and having long growing seasons with warm soil temperatures. Factors that reduce microbial activity, hence slow the disappearance of "Ally" in soils, are low rainfall and prolonged periods of soil temperatures less than 40°F.

Microbial activity, soil temperature, and to a large degree soil moisture, can vary greatly from year-to-year, and from area-to-area. Consequently, it is not always possible to accurately predict when areas treated with "Ally" can be rotated to crops other than those listed on the label.

A biological assay of your "Ally" treated field is the only sure way of determining when crops other than those listed on the label can be grown and is conducted as follows:

1. The accuracy and reliability of any field bioassay is largely dependent on the location and number of strips planted. Be sure to select areas of the field previously treated with "Ally" that are representative of the various field conditions. Be sure to consider factors such as field size, soil texture, drainage, turn-around areas, eroded knolls or alkaline spots when selecting the sites that are most representative of the soil conditions in the field. Even in small fields, more than one test strip is required to accurately determine whether it is safe to rotate to a crop not listed on the label. On large fields, several test strips will be needed in order to obtain reliable results based on the field variables mentioned above.
2. Plant the test strips perpendicular to the direction in which the field was sprayed. Each strip should be long enough to cross the width of several spray swaths. A large test strip area is more reliable than a small one. Suggested size is 1/4 to 1/2 acre per test strip.
3. Use standard tillage and seeding equipment to plant the bioassay.
4. Prepare a seed bed and plant the crops and varieties you want the option of growing the following year. **IT IS IMPORTANT TO USE THE SAME PLANTING TIME, CONDITIONS, TECHNIQUES AND CULTURAL PRACTICES YOU NORMALLY USE TO PLANT AND GROW THE BIOASSAY CROP(S).** If possible, plant into an adjacent area not treated with "Ally" to use as a comparison.
5. Do not overspray the test strips with herbicides that may damage the bioassay crop(s).
6. If the crop(s) in the test strip(s) grow to maturity with a normal harvest, the assay is positive and you may now rotate to the new crop. However, if crop(s) in the test strips die, are stunted, or fail to yield a normal harvest, the assay is negative and you should not rotate to the new crop(s). Run the assay until positive results are obtained before rotating to the new crop(s).
7. If the bioassay indicates that "Ally" residues are still present, do not rotate to crop(s) other than those listed on the label until bioassay results indicate that the assay crops are growing normally.

DU PONT LRB℠ BIOASSAY SERVICE

In the states of ID, MT, ND, SD, OR, and WA, the Du Pont LRB℠ bioassay service is available through certain dealers and/or consultants. This service uses soil samples taken by Du Pont certified individuals for laboratory bioassay analysis. LRB℠ results will serve as a crop rotation recommendation.

Check with your local Du Pont representative or call toll free 1-800-782-3557 for information regarding the LRB℠ bioassay service.

With any chemical, follow labeling instructions and warnings carefully.

1 Registered trademark of Sandoz Crop Protection Corporation.
2 Registered trademark of Rhone-Poulenc Ag Company.
3 Registered trademark of The Dow Chemical Company.
4 Registered trademark of Hoechst-Roussel Agri-Vet Company.
5 Registered trademark of Bayer AG, Leverkusen.
6 Registered trademark of Delavan Corporation.
7 Registered trademark of D. and W. Corporation.
8 Registered trademark of American Cyanamid Company.

NOTICE TO BUYER: Purchase of this material does not confer any rights under patents of countries outside of the United States.

STORAGE AND DISPOSAL

Storage: Store product in original container only, away from other pesticides, fertilizer, food or feed.

Product Disposal: Do not contaminate water, food or feed by storage, disposal or cleaning of equipment. Wastes resulting from the use of this product may be disposed of on site or at an approved waste disposal facility.

Container Disposal: Triple rinse (or equivalent) and then offer for recycling or reconditioning, or puncture and dispose of in a sanitary landfill, or by incineration, or, if allowed by state and local authorities, by burning. If burned, stay out of smoke.

NOTICE OF WARRANTY

Du Pont warrants that this product conforms to the chemical description on the label thereof and is reasonably fit for purposes stated on such label only when used in accordance with directions under normal use conditions. It is impossible to eliminate all risks inherently associated with the use of this product. Crop injury, ineffectiveness, or other unintended consequences may result because of such factors as weather conditions, presence of other materials, or the manner of use or application, all of which are beyond the control of Du Pont. In no case shall Du Pont be liable for consequential, special or indirect damages resulting from the use or handling of this product. All such risks shall be assumed by the buyer. DU PONT MAKES NO WARRANTIES OF MERCHANTABILITY OR FITNESS FOR A PARTICULAR PURPOSE NOR ANY OTHER EXPRESS OR IMPLIED WARRANTY EXCEPT AS STATED ABOVE.

AG-3261 8049 4/13/89

This is a specimen label. DuPont is not responsible for the accuracy of the information contained herein. As labels are subject to revision, always carefully read and follow the label on the product container.

ASANA®

INSECTICIDE
1.9 EMULSIBLE CONCENTRATE

RESTRICTED USE PESTICIDE

For retail sale to and use only by Certified Applicators or persons under their direct supervision and only for those uses covered by the Certified Applicator's certification.

This product contains 1.9 lbs. Active Ingredient per gallon.

ACTIVE INGREDIENT:	BY WEIGHT
(S)-cyano(3-phenoxyphenyl) methyl-(S)-4-chloro-alpha-(1-methylethyl)benzeneacetate	24%
INERT INGREDIENTS:*	76%
TOTAL	100%

*CONTAINS XYLENE RANGE AROMATIC SOLVENT

EPA Reg. No. 352-502 EPA Est. No. 5905-AR-1

KEEP OUT OF REACH OF CHILDREN

DANGER
PELIGRO

PRECAUCION AL USUARIO: Si usted no lee ingles, no use este producto hasta que le etiqueta haya sido explicado ampliamente.

STATEMENT OF PRACTICAL TREATMENT

If less than 2 ml is swallowed per kg of body weight, do not induce vomiting. If vomiting occurs spontaneously, keep head below hips to prevent aspiration of liquid into lungs. If more than 2 ml per kg body weight is swallowed and vomiting has not occurred, emesis should be induced with supervision by a physician or professional staff. If symptoms such as loss of gag reflex, convulsions or unconsciousness occur before emesis gastric lavage should be considered following intubation with a cuffed endotracheal tube.

For eye contact flush eyes with plenty of water for at least 15 minutes while holding eyelids open. Get medical attention.

For skin contact wash with soap and water immediately. After drying, apply vitamin E cream or oil if available. If not available, apply vegetable oil liberally over painful areas. The oil or cream may be used repeatedly until relief is achieved. Get medical help if irritation persists.

If inhaled remove victim to fresh air and provide oxygen if breathing is difficult. Give artificial respiration if not breathing. Get medical attention.

Remove and wash clothing before reuse.

For medical emergencies involving this product, call toll free 1-800-441-3637.

PRECAUTIONARY STATEMENTS

HAZARDS TO HUMANS

DANGER

Corrosive. Causes eye damage and skin burns. May be fatal if swallowed or inhaled. Do not get in eyes, on skin, or on clothing. Wear goggles or face shield, protective clothing, and rubber gloves when handling. Wash thoroughly with soap and water after handling. Remove contaminated clothing and wash before reuse. May cause allergic reactions.

For medical emergencies involving this product, call toll free 1-800-441-3637.

ENVIRONMENTAL HAZARDS

This pesticide is toxic to wildlife and extremely toxic to fish. Use with care when applying in areas adjacent to any body of water. Do not apply directly to water. Do not apply when weather conditions favor drift from treated areas. Do not contaminate water by cleaning of equipment or disposal of wastes. Apply this product only as specified on this label.

This product is highly toxic to bees exposed to direct treatment on blooming crops or weeds. Do not apply ASANA or allow it to drift to blooming crops or weeds while bees are actively visiting the treatment area. Additional information may be obtained by consulting your Cooperative Extension Service.

WILDLIFE PROTECTION IS EVERYONE'S RESPONSIBILITY

DIRECTIONS FOR USE

It is a violation of Federal Law to use this product in a manner inconsistent with the labeling.

REENTRY STATEMENTS

Do not apply this product in such a manner as to directly or through drift expose workers or other persons. The area being treated must be vacated by unprotected persons.

Do not enter treated areas without protective clothing until sprays have dried.

Because certain states may require more restrictive reentry intervals for various crops treated with this product, consult your State Department of Agriculture for further information.

Written or oral warnings must be given to workers who are expected to be in a treated area or in an area about to be treated with this product. (Indicate specific oral warnings which inform workers of areas or fields that may not be entered without specific protective clothing, period of time field must be vacated and appropriate actions to take in case of accidental exposure.) When oral warnings are given, warnings shall be given in a language customarily understood by workers. Oral warnings must be given if there is reason to believe that written warnings cannot be understood by workers. Written warnings must include the following information: "WARNING! Area treated with ASANA Insecticide on (date of application). Do not enter without appropriate protective clothing until sprays have dried. (Insert here action to take in case of accidental exposure. See front panel STATEMENT OF PRACTICAL TREATMENT for details.)"

GENERAL DIRECTIONS

ASANA Insecticide 1.9 Emulsible Concentrate contains 1.9 pounds of active ingredient per gallon. For the applications given below, mix the required amount of ASANA in sufficient diluent to provide uniform coverage (Refer to Use Tables).

Chemigation. Apply this product only through sprinkler including center pivot, lateral move, end tow, side (wheel) row, traveler, big gun, solid set, or hand move irrigation systems. Do not apply this product through any other type of irrigation system. ASANA 1.9 may be premixed in a supply tank with water, oil, fertilizer, or other appropriate tank mixed agricultural chemicals. Agitation may be necessary. Application should be in sufficient water and of sufficient duration to apply the recommended rate evenly to the entire treated area. No run-off can be permitted during chemigation. Crop injury, lack of effectiveness, or illegal pesticide residues in the crop can result from nonuniform distribution of treated water. Do not apply when wind speed favors drift beyond the area intended for treatment.

If you have questions about calibration, you should contact State Extension Service specialists, equipment manufacturers or other experts. A person knowledgeable of the chemigation system and responsible for its operation, or under the supervision of the responsible person, shall shut the system down and make necessary adjustments should the need arise.

Do not connect an irrigation system (including greenhouse systems) used for ASANA 1.9 application to a public water system.

The system must contain a functional check valve, vacuum relief valve, and low pressure drain appropriately located on the irrigation pipeline to prevent water source contamination from backflow. The pesticide injection pipeline must also contain a functional, automatic, quick-closing check valve to prevent the flow of fluid back toward the injection pump. The pesticoid injection pipeline must also contain a functional, normally closed, solenoid-operated valve located on the intake side of the injection pump and connected to the system interlock to prevent fluid from being withdrawn from the supply tank when the irrigation system is either automatically or manually shut down. The system must contain functional interlocking controls to automatically shut-off the pesticide injection pump when the water pump motor stops. The irrigation line or water pump must include a functional pressure switch which will stop the water pump motor when the water pressure decreases to the point where pesticide distribution is adversely affected. Systems must use a metering pump, such as a positive displacement injection pump (e.g., diaphragm pump) effectively designed and constructed of materials that are compatible with pesticides and capable of being fitted with a system interlock.

Posting of areas to be chemigated is required when 1) any part of a treated area is within 300 feet of sensitive areas such as residential areas, labor camps, businesses, day care centers, hospitals, in-patient clinics, nursing homes or any public areas such as schools, parks, playgrounds, or other public facilities not including public roads, or 2) when the chemigated areas is open to the public as golf courses or retail greenhouses.

Posting must conform to the following requirements. Treated areas shall be posted with signs at all usual points of entry and along likely routes of approach from the listed sensitive areas. When there are no usual points of entry, signs must be posted in the corners of the treated areas and in any other location affording maximum visibility to sensitive areas. The printed side of the sign should face away from the treated area towards the sensitive area. The signs shall be printed in English. Signs must be posted prior to application and must remain posted until foliage has dried and soil surface water has disappeared. Signs may remain in place indefinitely as long as they are composed of materials to prevent deterioration and maintain legibility for the duration of the posting period. All words shall consist of letters at least 2½ inches tall, and all letters and the symbol shall be be a color which sharply contrasts with their immediate background. At the top of the sign shall be the words KEEP OUT, followed by an octagonal stop sign symbol at least 8 inches in diameter containing the word STOP. Below the symbol shall be words "PESTICIDES IN IRRIGATION WATER."

Posting required for chemigation does not replace other posting and reentry interval requirements for farmworker safety.

Do not plant root crops other than those in this label within 12 months after last application. Root crops in this label and all other rotation crops may be planted immediately following last applications.

COTTON

CROP	INSECT	DOSAGE: LB AI/ ACRE	DOSAGE: FL OZ/ ACRE	NUMBER ACRES TREATED WITH GALLON	FURTHER USE INSTRUCTIONS	DAYS TO HARVEST
Cotton	Cotton Bollworm Tobacco Budworm Boll Weevil Beet Armyworm Cabbage Looper Saltmarsh Caterpillar Cotton Leafworm Cutworms Lypus Bugs, Plant Bugs Thrips (on seedling cotton) Leafhoppers, Fleahoppers, Grasshoppers, Whitefly*	0.03–0.05	2.0–3.4	63–38	May be applied in water or nonvolatile vegetable oils. When using oil, apply a total spray volume of 1 or more quarts/acre. When diluted with water, apply no less than one gallon per acre by air (min. of 3 gal/a in Arizona and 5 gallons per acre in California) or 4 gallons per acre by ground. Repeat as necessary throughout season to maintain control. For the control of Boll Weevil, a 3-5 day interval between sprays may be required. Apply no more than 0.5 lbs/a to a crop in growing season. Do not graze livestock on treated fields or feed treated trash.	21
	NOTE: For light infestations of the above insects.	0.02	1.4	95		
	Pink Bollworm	0.015–0.03	1.0–2.0	128–63		
	Cotton Leaf Perforator	0.03	2.0	63		

* Aids in the suppression.

COTTON—TANK MIX APPLICATIONS
READ AND FOLLOW ALL PRECAUTIONS AND RESTRICTIONS ON THE PRODUCT LABELING.

TANK MIX COMBINATION	INSECT	DOSAGE	FURTHER USE INSTRUCTIONS	DAYS TO HARVEST
LANNATE L NUDRIN 1.8 Methomyl Insecticide[1]Solution	Cotton Bollworm and Tobacco Budworm eggs	ASANA (from cotton table) with ½ pt/acre LANNATE L/NUDRIN 1.8		21
LANNATE NUDRIN/90 Methomyl Insecticide[1]Water Soluble Powder (With adjuvant)	Cotton Bollworm and Tobacco Budworm eggs	ASANA (from cotton table) with ⅛ lb/acre LANNATE/NUDRIN 90	Add adjuvant (X-77, Bio-88, Unite, Triton B-1956 or equivalent†) to spray tank, then LANNATE/NUDRIN 90; then ASANA, agitate until thoroughly mixed.	
AZODRIN 5 Water Miscible Insecticide[1]	Thrips	ASANA (from cotton table) with 3-5 oz/acre AZODRIN 5	Use the higher rates when insect populations are high.	
	Whiteflies	ASANA (from cotton table) with ⅜-1⅝ pt/acre AZODRIN 5		
	Mites	ASANA (from cotton table) with ⅞-1⅝ pt/acre AZODRIN 5		

† Compatibility of alternate should be determined before use.

FIELD CROPS

CROP	INSECT	DOSAGE: LB AI/ ACRE	DOSAGE: FL OZ/ ACRE	NUMBER ACRES TREATED WITH GALLON	FURTHER USE INSTRUCTIONS	DAYS TO HARVEST
Field Corn	Cutworm Grasshoppers	0.025–0.05	1.7–3.4	76–38	Apply as necessary to maintain control.	21
	Western Bean Cutworm	0.0125–0.025	0.85–1.7	152–76	Apply as necessary to maintain control. Application must be made before larvae enter the ear.	
	Corn Earworm	0.025–0.05	1.7–3.4	76–38	First application should be at or before silking. Subsequent applications should be made at 3-5 day intervals until silking is completed. Repeat applications may be applied if economically damaging populations exist.	
	Stalk Borer Flea Beetles	0.025–0.05	1.7–3.4	76–38	Application must be made early in migration from grassy areas to corn, before borers enter the plant. More than one application may be necessary.	

CROP	INSECT	DOSAGE LB AI/ ACRE	FL OZ/ ACRE	NUMBER ACRES TREATED WITH GALLON	FURTHER USE INSTRUCTIONS	DAYS TO HARVEST
	Corn Rootworm (adult control)	0.025-0.05	1.7-3.4	76-38	Spray at first sign of silk feeding. Reinfestation may occur. Should reinfestation occur, repeat as necessary to maintain control.	
	Chinch Bugs	0.025–0.05	1.7–3.4	76-38	Over the top sprays, as used for control of ear and stalk infesting corn pests, are not adequate for chinch bug control. It is very important that the spray be directed at the base of the plant through the use of drop nozzles or some other mechanism. Repeat as necessary to control migrating chinch bugs.	
	European Corn Borer	0.04-0.05	2.7-3.4	48-38	First brood: spray while eggs are in the blackhead stage or before the larvae enter the whorl. Application by ground equipment is suggested. Good coverage of both upper and lower leaf surfaces is essential. This can be accomplished with drop nozzles over the row and on each side of the corn plant. Multiple applications may be required when egg laying is prolonged or where moderate to heavy populations are present. A higher rate is recommended for moderate to heavy populations. Proper coverage by ground equipment usually requires 20-30 gallons of carrier. Once larvae enter the whorl, foliar sprays will not provide adequate control. Second brood: make applications when sufficient egg masses are found. Spray when eggs are in the blackhead stage or starting to hatch. When egg laying is prolonged or a third generation is present, additional sprays may be required. A higher rate is recommended for moderate to heavy populations. Good coverage above, below, and in the ear zone is essential. This usually requires 2-3 gallons of carrier by air. If ground equipment is used, drop nozzles on each side of the plant will provide best coverage.	
	Southwestern Corn Borer	0.025-0.05	1.7-3.4	76-38	Apply as necessary to maintain control. Two or more applications may be required. Higher rates (0.036-0.05 lb a.i.) are recommended for moderate to heavy populations.	
	Armyworm (True Armyworm)	0.025-0.05	1.7-3.4	76-38	Apply as necessary to maintain control.	
	Adult Corn Rootworm European Corn Borer	ASANA and Parathions from respective corn labels			ASANA insecticide may be tank-mixed with Parathion (methyl and/or ethyl) and applied on field corn (and seed corn) where supplemental control of Adult Corn Rootworm and/or European Corn Borer is desired in conjunction with insects controlled by ASANA when used alone. For supplemental control of second brood European Corn Borer, make applications when sufficient egg masses are found. Spray when eggs are in the blackhead stage or starting to hatch. When egg laying is prolonged or a third generation is present, additional sprays may be required. A higher rate is recommended for moderate to heavy populations. Good coverage above, below, and in the ear zone is essential. This usually requires 2-3 gallons of carrier by air. If ground equipment is used, drop nozzles on each side of the plant will provide best coverage. For supplemental control of Adult Corn Rootworm, spray at first sign of silk feeding. Reinfestation may occur. Should reinfestation occur, repeat as necessary to maintain control. Refer to the ASANA and Parathion (methyl or ethyl) labels for appropriate rates of the individual products for controlling the respective insects.	
	Corn Leaf Aphid Oat Bird-Cherry Aphid	0.025-0.05	1.7-3.4	76-38	For optimum results, direct the spray at the aphid population so as to achieve maximum coverage of the exposed insects. Aphids not contacted by the spray, such as in whorls and leaf axis, may not be adequately controlled.	
					Do not exceed 0.25 lb a.i. per acre per season.	

ASANA 1.9 EC may be tank-mixed with the following herbicides when insect populations require control concurrent with the need for weed control. Follow all herbicidal label directions regarding proper herbicide usage. Jar-test compatibility should be conducted before field mixtures are made and applications initiated.

BLADEX 80 WP[1]	ATRAZINE 80 WP	BICEP 4.5L[2]	DUAL 8E[2]
BLADEX 4L[1]	ATRAZINE 4L	BANVEL 4EC[3]	ROUNDUP[4]
LASSO EC[4]	GRAMOXONE SUPER[5]		

CROP	INSECT	DOSAGE LB AI/ ACRE	FL OZ/ ACRE	NUMBER ACRES TREATED WITH GALLON	FURTHER USE INSTRUCTIONS	DAYS TO HARVEST
Seed Corn	For specific insect control recommendations refer to Field Corn (above). Follow directions carefully.				Multiple applications and/or shortened intervals between sprays must be used to insure proper insect control. Do not exceed 2.5 lb al per acre per season.	21
Pop Corn	For specific insect control recommendations refer to Field Corn (above). Follow directions carefully.				Multiple applications and/or shortened intervals between sprays must be used to insure proper insect control. Do not exceed 0.5 lbs ai per acre per season.	1
Peanuts	Corn Earworm, Red-necked Peanut Worm, Potato Leaf-hopper, Velvetbean Cater-pillar	0.0125–0.025	0.85-1.7	152-76	Repeat as necessary to maintain control. Do not feed or graze livestock on treated vines. Do not exceed 0.15 lb. ai per acre per season.	21
	Beet Armyworm, Granulate Cutworm, Grashoppers, Cutworms	0.025-0.05	1.7-3.4	76-38	**Aids in control.	
	Lesser Cornstalk Borer** Fall Armyworm***	0.05	3.4	38	***Aids in control. For heavy infestations, use of LANNATE/NUDRIN is recommended. Please refer to the LANNATE/NUDRIN label for use instructions.	
Soybean	Mexican Bean Beetle, Green Cloverworm, Velvetbean Caterpillar, Potato Leafhopper, Saltmarsh Caterpillar, Woolybear Caterpillar	0.0125-0.025	0.85-1.7	152-76	Repeat as necessary to maintain control. Do not feed or graze livestock on treated plants. Do not exceed 0.2 lb ai per acre per season. When applying in nonvolatile vegetable oils use a total spray volume of 1 or more quarts.	21
	Cabbage Looper, Corn Earworm, Southern Green Stinkbug, Bean Leaf Beetle, Cutworms, Three-cornered Alfalfa Hopper, Beet Armyworm, Grasshoppers, Japanese Beetle (adult)	0.025-0.05	1.7-3.4	76-38		

FRUITS AND NUTS

DILUTE SPRAY: Apply specified dosage per 100 gallons of water in a uniform spray applied to the point of drip with conventional ground equipment. Do not exceed maximum number of gallons per acre indicated.

NOTE: In order to apply the correct amount of ASANA Insecticide to your orchard you must know the number of gallons of water needed to spray one acre of your trees to the point of drip. If you do not already know this gallonage, you should conduct a test to determine it. If you do not know how to conduct such a test with your equipment, you should ask assistance from your equipment dealer or State Extension Specialist.

CONCENTRATE SPRAY: Apply specified dosage per acre in no less than 30 gals. of water per acre by ground. When applying ASANA by air, consult your Cooperative Extension Service for application guidelines.

CROP	INSECT	DOSAGE: CONCENTRATE SPRAY LB AI/ ACRE	CONCENTRATE SPRAY FL OZ./ ACRE	DILUTE SPRAY FL OZ/ 100 GALS	NUMBER ACRES TREATED WITH GALLON	FURTHER USE INSTRUCTIONS	DAYS TO HARVEST
Apples	Oblique Banded Leafroller, Codling Moth, Variegated Leafroller, White Apple Leafhopper, Tentiform Leaf Miner, San Jose Scale (fruit infestations only). Apple Aphid, Tufted Apple Budmoth, Plum Curculio, Oriental Fruit Moth, Apple Maggot, Red-Banded Leafroller, Lesser Appleworm, Green Fruitworm, Rosy Apple Aphid, Plant Bugs (Tarnished Plant Bug, Stink Bugs), Periodical Cicada	0.025-0.075	1.7-5.0	0.7-2.0	76-25	Repeat as necessary to maintain control. Do not feed or graze livestock on treated orchard floors. Do not exceed 0.525 lbs. a.i. per acre per season. For dilute spray apply 200-600 gals per acre, but do not exceed 5 fl. oz. of ASANA 1.9 EC per acre per treatment. Plant Bug, Rosy Apple Aphid Control —Time of application is critical in achieving control. Use prebloom and postbloom spray timings recommended by State Extension Services.	21
Pecans	Pecan Weevil, Hickory Shuckworm, Pecan Nut Casebearer, Pecan Aphids, Pecan Spittlebug Pecan Stem Phylloxera, Pecan Leaf Phylloxera	0.025-0.075	1.7-5.0	0.7-2.0	76-25	Repeat as necessary to maintain control. Do not feed or graze livestock on treated orchard floors. Do not exceed 0.3 lbs. ai per acre per season. For dilute spray apply 200-600 gals. per acre, but do not exceed 5 fl. oz. of ASANA 1.9 EC per acre per treatment. Correct timing of spray applications is critical in achieving optimum control of leaf and stem phylloxera. Consult local spray recommendations for correct times of application.	21
Walnuts	Walnut Aphid, Walnut Husk Fly, Codling Moth, Navel Orangeworm	0.05-0.1	3.4-6.7	1.4	38-19	Repeat as necessary to maintain control. Do not exceed 0.2 lb. ai per acre per season. For dilute sprays, apply 200-400 gals per acre. Do not feed or graze livestock on treated crop	21

CROP	INSECT	DOSAGE: CONCENTRATE SPRAY: LB AI/ ACRE	DOSAGE: CONCENTRATE SPRAY: FL OZ./ ACRE	DOSAGE: DILUTE SPRAY: FL OZ/ 100 GALS	NUMBER ACRES TREATED WITH GALLON	FURTHER USE INSTRUCTIONS	DAYS TO HARVEST
						floors. NOTE: Use of baits in Walnut Husk fly sprays is recommended where endorsed by local Agricultural Extension Service. CONCENTRATE SPRAY: Apply specified dosage per acre in no less than 30 gals. of water per acre by ground or 5 gals. of water per acre by air.	

VEGETABLE CROPS

CROP	INSECT	DOSAGE: LB AI/ ACRE	DOSAGE: FL OZ ACRE	NUMBER ACRES TREATED WITH GALLON	FURTHER USE INSTRUCTIONS	DAYS TO HARVEST
Artichokes	Artichoke Plume Moth	0.025-0.05	1.7-3.4	76-38	Apply no more than three times per crop. Begin application when larvae first appear with subsequent applications at 14 day intervals. Apply in a minimum of 10 gallons per acre by ground (use sufficient water to obtain coverage without excessive runoff).	7
Broccoli, Cabbage,	Diamondback Moth	0.0125-0.025	0.85-1.7	152-76	Repeat as necessary to maintain control. Do not exceed 0.4 lbs. ai per acre per season.	3
Cauliflower	Imported Cabbageworm, Cabbage Looper, Beet Armyworm, Alfalfa Looper, Flea Beetle, Grasshopper, Cutworm	0.025-0.05	1.7-3.4	76-38		
Cucumber, Melons (Cantaloupe, Honeydew Melons, Watermelon, Muskmelon), Pumpkin, Squash (Summer, Winter)	Pickleworm, Rindworms, Leafhoppers, Corn Earworm, Cabbage Looper, Squash Bug, Cutworms (seedling spray), Cucumber Beetle (adult control). Grasshoppers, Plant Bugs (Lygus Bugs, Stink Bugs), Squash Vine Borer	0.025-0.05	1.7-3.4	76-38	Repeat as necessary to maintain control. Do not exceed 0.25 lb ai per acre per season.	3
Dry Beans Dry Peas	Mexican Bean Beetle, Saltmarsh Caterpillar	0.0125-0.025	0.85-1.7	152-76	Repeat as necessary to maintain control. Do not exceed 0.2 lb. ai per acre per season. Do not feed or graze livestock on treated vines.	21
	Corn Earworm, Potato Leafhopper, Pea Aphid, Grasshoppers, Leafhoppers, Cowpea Curculio, Cutworms, Cabbage Looper, Western Bean Cutworm, Velvetbean Caterpillar, Beet Armyworm, Cucumber Beetle, Corn Rootworm (Adults), Painted Lady Butterfly (larvae), Green Cloverworm	0.025-0.05	1.7-3.4	76-38		
Eggplant	Colorado Potato Beetle, Flea Beetles, Corn Earworm, Loopers, European Corn Borer	0.025-0.05	1.7-3.4	76-38	Begin application when insects or their damage is observed and repeat at 7 to 10 day intervals to achieve control. Do not exceed 0.35 lbs. ai per acre per season.	7
Green Peas	Pea Aphid, Green Cloverworm	0.0125-0.025	0.85-1.7	152-76	Repeat as necessary to maintain control but do not exceed 0.1 lb ai/acre per season. Do not feed treated pea vines to livestock.	3
	Cabbage Looper, Celery Looper, Alfalfa Looper, Cowpea Curculio, Cutworms, Pea Leaf Weevil, Armyworm, Corn Earworm, Imported Cabbageworm, Alfalfa Caterpillar	0.025-0.05	1.7-3.4	76-38		
Peppers	European Corn Borer, Beet Armyworm, Flea Beetles, Corn Earworm, Loopers, Colorado Potato Beetle, Pepper Weevil*, Cucumber Beetle (adults), Southern Armyworm	0.025-0.05	1.7-3.4	76-38	Begin application when insects or their damage is observed and repeat at 7 to 10 day intervals to achieve control. Do not exceed 0.35 lbs. ai per acre per season.	7
Potato	Potato Psyllid, Potato Tuberworm	0.0125-0.025	0.85-1.7	152-76	Repeat as necessary to maintain control. Do not graze livestock on treated vines. Do not exceed 0.35 lbs. ai per acre per season.	7
	Fleabeetles, Tarnished Plant Bug, Cabbage Looper, Potato Aphid, Potato Leafhopper, Cucumber	0.025-0.05	1.7-3.4	76-38		

CROP	INSECT	DOSAGE LB AI/ ACRE	DOSAGE FL OZ ACRE	NUMBER ACRES TREATED WITH GALLON	FURTHER USE INSTRUCTIONS	DAYS TO HARVEST
	Beetle (adult control), Grasshoppers, European Corn Borer, Beet Armyworm, Western Yellow-striped Armyworm, Buckthorn Aphid, Cutworms					
	Colorado Potato Bettle					
	Long Island, NY	0.05	3.4	38		
Sweet Corn	Corn Earworm, Southwestern Corn Borer, European Corn Borer, Beet Armyworm, Flea Beetles, Sap Beetles (adults), Banded Cucumber Beetle, Tarnished Plant Bug	0.025-0.05	1.7-3.4	76-38	Do not exceed 0.5 lb ai per acre per season. Begin applications either before or at time of silking and continue as necessary to maintain control. For protection against corn borers other than at silking time, treat when necessary to maintain control. For additional information consult directions for use under field corn.	1
	Stalk Borer, Cutworms, Corn Rootworms (adult control), Grasshoppers, Chinch Bugs, Armyworm	0.025-0.05	1.7-3.4	76-38	For additional information consult directions for use under field corn.	
	Western Bean Cutworm	0.0125-0.025	0.85-1.7	152-76		
	Corn Leaf Aphid Oat Bird-Cherry Aphid	0.025-0.05	1.7-3.4	76-38	For optimum results, direct the spray at the aphid population so as to achieve maximum coverage of the exposed insects. Aphids not contacted by the spray, such as in whorls and leaf axils, may not be adequately controlled.	
Tomato	Tomato Hornworm Tobacco Hornworm	0.0125-0.025	0.85-1.7	152-76	Repeat as necessary to maintain control. Do not feed or graze livestock on treated vines.	1
	Tomato Pinworm, Flea Beetle, Tomato Fruitworm, Colorado Potato Beetle, Cutworms, Cabbage Looper, Southern Armyworm, Sugarbeet Armyworm, Western Yellow-Striped Armyworm, Grasshopper, Whitefly, Potato Aphid	0.025-0.05	1.7-3.4	76-38	Do not exceed 0.5 lbs. ai per acre per season.	
	Vegetable Leafminer**	0.05	3.4	38		

* Not Recommended in Florida.

SPECIALTY USES

CROP	INSECT	DOSAGE LB AI/ ACRE	DOSAGE FL OZ/ ACRE	NUMBER ACRES TREATED WITH GALLON	FURTHER USE INSTRUCTIONS
Non-crop Land (excluding public land such as forests, parks, or recreational areas.)	Grasshoppers, Saltmarsh Caterpillar	0.0125-0.025	0.85-1.7	152-76	Spray non-crop land adjacent to tilled areas to control migrating insects (Grasshoppers, Armyworms) which are a threat to crops. Repeat as necessary to maintain control but do not exceed 0.5 lbs. active ingredient per acre per year. Do not feed treated crop to livestock. Do not spray ditch banks or areas adjacent to water.
	Armyworms, Army Cutworms, Chinch Bugs	0.025-0.05	1.7-3.4	76-38	

CROP	INSECT	DOSAGE HIGH VOLUME SPRAYS LB AI/ 100 GALS	DOSAGE HIGH VOLUME SPRAYS FL OZ/ 100 GALS.	DOSAGE LOW VOLUME SPRAYS (MIST-BLOWER)	FURTHER USE DIRECTIONS
Christmas Tree Plantings, Pine Seed Orchards, Forest Tree Nurseries	Nantucket Pine Tip Moth, Spittlebugs, Pineleaf Chermid, Redheaded Pine Sawfly, Balsam Woolly Adelgid, Pine Chafer, Red Pine Sawfly, Pine Conelet Bug, European Pine Sawfly, Cranberry Girdler	0.025-0.05	1.7-3.4	0.025-0.05	Apply as needed to maintain control. Spray sufficient gallonage to obtain good coverage of entire tree.

CROP	INSECT	DOSAGE HIGH VOLUME SPRAYS LB AI/ 100 GALS	FL OZ/ 100 GALS.	LOW VOLUME SPRAYS (MIST-BLOWER)	FURTHER USE DIRECTIONS
	(adult control), Pine Needle Midge, Balsam Twig Aphid				
	Pales Weevil, Northern Pine Weevil	0.025-0.05	1.7-3.4		Apply as needed to maintain control. Dilute formulation in kerosene, instead of water, and spray tree stumps to control developing larvae. Use sufficient spray volume to thoroughly cover stump.
	Coneworms, Seedbugs, Seed Chalcid	3.4 fl oz/100 gals water for high volume sprayers. 18 fl oz/100 gals water for low volume sprayers. 0.19 lb ai/acre application in not less than 10 gals of water for aerial applications.			Apply first application within 1 week of female flower closure or peak pollen flight for Webbing Coneworm control. For other Coneworms and Seedbugs, apply first application within 30 days following female flower closure. Repeat application at intervals of 4 weeks but do not exceed 1.6 lbs ai/ac/yr. For Seed Chalcid control, apply when all cones are pendant, and repeat at 1-2 week intervals for 2 or more sprays, or as needed. Apply approximately 5-10 gals of the 3.4 fl oz/100 gal dilution per tree with high volume sprayers. With low volume sprayers apply 100 gals of the 18 fl oz/100 gal dilution per acre. Do not graze or harvest cover crop. Avoid contact with open water.

[1]Registered trademark of E. I. Du Pont de Nemours & Co. Inc.
[2]Registered trademark of Ciba-Geigy Corp.
[3]Registered trademark of Sandoz Crop Protection Corp.
[4]Registered trademark of Monsanto Agricultural Co.
[5]Registered trademark of ICI Americas, Inc.

STORAGE AND DISPOSAL

Do not contaminate water, food, or feed by storage or disposal.

STORAGE:
Store in a secure, dry and temperate area. Store in original container. Keep container closed when not in use. Do not store near food or feed. Do not use or store around the home. Avoid contact with water. In case of spill or leak, soak up with sand, earth or synthetic absorbent (do not use alkaline absorbents) and dispose of wastes in compliance with local, State and Federal regulations.

PESTICIDE DISPOSAL:
Pesticide wastes are toxic. Improper disposal of excess pesticide, spray mixture, or rinsate is a violation of Federal Law. If these wastes cannot be disposed of by use according to label instructions contact your State Pesticide or Environmental Control Agency, or the Hazardous Waste representative at the nearest EPA Regional Office for guidance.

CONTAINER DISPOSAL:
Triple rinse (or equivalent). Then offer for recycling or reconditioning, or puncture and dispose of in a sanitary landfill, or incineration, or, if allowed by state and local authorities, by burning. If burned, stay out of smoke.

In case of a significant spill, call CHEMTREC (800) 424-9300.

NOTICE OF WARRANTY

DU PONT MAKES NO WARRANTY OF MERCHANTABILITY, FITNESS FOR ANY PURPOSE, OR OTHERWISE, EXPRESS OR IMPLIED, concerning this product or its uses which extend beyond the use of the product under normal conditions in accord with the statements made on this label.

Sumitomo Chemical Company, Ltd., holds U.S. Patents for the active ingredient. Sales by E. I. Du Pont de Nemours Co. (Inc.) of products containing the active ingredient are made under agreements with Sumitomo Chemical Company, Ltd.

AG-3214 8098

12/29/86

This is a specimen label. DuPont is not responsible for the accuracy of the information contained herein. As labels are subject to revision, always carefully read and follow the label on the product container.

ASANA® XL

INSECTICIDE

0.66 EMULSIBLE CONCENTRATE

RESTRICTED USE PESTICIDE

For retail sale to and use only by Certified Applicators or persons under their direct supervision and only for those uses covered by the Certified Applicator's certification.

ACTIVE INGREDIENT:	BY WEIGHT
(S)-cyano(3-phenoxyphenyl)methyl-(S)-4-chloro-alpha-(1-methylethyl)benzeneacetate	8.4%
INERT INGREDIENTS:*	91.6%
TOTAL	100.0%

*CONTAINS XYLENE RANGE AROMATIC SOLVENT
This product contains 0.66 lbs. Active Ingredient per gallon.
EPA Reg. No. 352-515 EPA Est. No. 5905-AR-1

KEEP OUT OF REACH OF CHILDREN

WARNING

AVISO PRECAUCION AL USUARIO: Si usted no lee ingles, no use este producto hasta que la etiqueta haya sido explicada ampliamente.

STATEMENT OF PRACTICAL TREATMENT

If swallowed: Do not induce vomiting. If vomiting occurs spontaneously, keep head below hips to prevent aspiration of liquid into lungs. If vomiting has not occurred, emesis should be induced with supervision by a physician or professional staff. If symptoms such as loss of gag reflex, convulsions or unconsciousness occur before emesis, gastric lavage should be considered following intubation with a cuffed endotracheal tube.

If on eyes: Flush eyes with plenty of water for at least 15 minutes while holding eyelids open. Get medical attention.

If on skin: Wash with soap and water immediately. After drying, apply vitamin E cream or oil if available. If not available, apply vegetable oil liberally over painful areas. The oil or cream may be used repeatedly until relief is achieved. Get medical help if irritation persists.

If inhaled: Remove victim to fresh air and provide oxygen if breathing is difficult. Give artificial respiration if not breathing. Get medical attention.

Remove and wash clothing before reuse.

For medical emergencies involving this product, call toll free 1-800-441-3637.

PRECAUTIONARY STATEMENTS

HAZARDS TO HUMANS AND DOMESTIC ANIMALS

WARNING May be fatal if swallowed. Harmful if inhaled. Do not get in eyes, on skin, or on clothing. Wear goggles or face shield, protective clothing, and rubber gloves when handling. Wash thoroughly with soap and water after handling. Remove contaminated clothing and wash before reuse.

For medical emergencies involving this product, call toll free 1-800-441-3637.

ENVIRONMENTAL HAZARDS

This pesticide is toxic to wildlife and extremely toxic to fish. Use with care when applying in areas adjacent to any body of water. Do not apply directly to water. Do not apply when weather conditions favor drift from treated areas. Do not contaminate water when disposing of equipment washwaters. Apply this product only as specified on this label.

This product is highly toxic to bees exposed to direct treatment on blooming crops or weeds. Do not apply ASANA XL or allow it to drift to blooming crops or weeds while bees are actively visiting the treatment area. Additional information may be obtained by consulting your Cooperative Extension Service.

WILDLIFE PROTECTION IS EVERYONE'S RESPONSIBILITY

DIRECTIONS FOR USE

It is a violation of Federal Law to use this product in a manner inconsistent with the labeling.

REENTRY STATEMENTS

Do not apply this product in such a manner as to directly or through drift expose workers or other persons. The area being treated must be vacated by unprotected persons.

Do not enter treated areas without protective clothing until sprays have dried.

Because certain states may require more restrictive reentry intervals for various crops treated with this product, consult your State Department of Agriculture for further information.

Written or oral warnings must be given to workers who are expected to be in a treated area or in an area about to be treated with this product. (Indicate specific oral warnings which inform workers of areas or fields that may not be entered without specific protective clothing, period of time field must be vacated and appropriate actions to take in case of accidental exposure.) When oral warnings are given, warnings shall be given in a language customarily understood by workers. Oral warnings must be given if there is reason to believe that written warnings cannot be understood by workers. Written warnings must include the following information: "WARNING! Area treated with ASANA XL Insecticide on (date of application). Do not enter without appropriate protective clothing until sprays have dried. (Insert here action to take in case of accidental exposure. See front panel STATEMENT OF PRACTICAL TREATMENT for details.)"

GENERAL DIRECTIONS

ASANA XL Insecticide 0.66 Emulsible Concentrate contains 0.66 pounds of active ingredient per gallon. For the applications given below, mix the required amount of ASANA XL in sufficient diluent to provide uniform coverage (Refer to Use Tables). ASANA XL may be applied by ground or aerial application equipment. Do not enter treated areas before spray is dry.

Resistance: If treatment with ASANA XL Insecticide is not effective after recommended use, a resistant (tolerant) population of insects may be present. Poor insect control may result. Promptly consult your local State Agricultural Extension Service for advice on the use of alternative insecticides.

Herbicide Tank Mixing: ASANA XL may be tank mixed with the herbicide products listed below when insect populations require control concurrent with the need for weed control. Follow all herbicide and ASANA XL label directions regarding proper usage. A "jar test" for physical compatibility should be conducted before field mixtures are made and applications initiated. For best results, use of spray equipment having continuous agitation is recommended.

ASANA XL may be used in combination with 2,4-D herbicides providing that the following mixing directions are followed: 1) Do not apply the combination in a volume of water less than 2 gallons per acre total spray. 2) Always mix the ASANA XL thoroughly in the total volume of spray water first, followed by the addition of the 2,4-D herbicide. Because of the availability of a great variety of 2,4-D herbicide products, a jar test for physical compatibility should be conducted before field mixtures of a particular combination are made.

Bladex 80WP (1)	Lasso 4EC (4)	Bladex 4L (1)
Atrazine 4L	Classic (1)	Atrazine 80WP
Preview (1)	Gramoxone Super (5)	Velpar L (1)
Bicep 4.5L (2)	Velpar (1)	Banvel 4EC (3)
Lorox Plus (1)	Dual 8E (2)	Extrazine 90DF (1)
Roundup (4)	Conquest 90DF (1)	

Chemigation. Apply this product only through sprinkler including center pivot, lateral move, end tow, side (wheel) row, traveler, big gun, solid set, or hand move irrigation systems. Do not apply this product through any other type of irrigation system. ASANA XL may be premixed in a supply tank with water, oil, fertilizer, or other appropriate tank mixed agricultural chemicals. A pretest of physical compatibility for untried tank mixes is advised. Agitation may be necessary. Application should be in sufficient water and of sufficient duration to apply the recommended rate evenly to the entire treated area. No run-off can be permitted during chemigation. Crop injury, lack of effectiveness, or illegal pesticide residues in the crop can result from nonuniform distribution of treated water. Do not apply when wind speed favors drift beyond the area intended for treatment.

If you have questions about calibration, you should contact State Extension Service specialists, equipment manufacturers or other experts. A person knowledgeable of the chemigation system and responsible for its operation, or under the supervision of the responsible person, shall shut the system down and make necessary adjustments should the need arise.

Do not connect an irrigation system (including greenhouse systems) used for ASANA XL application to a public water system.

The system must contain a functional check valve, vacuum relief valve, and low pressure drain appropriately located on the irrigation pipeline to prevent water source contamination from backflow. The pesticide injection pipeline must contain a functional, automatic, quick-closing check valve to prevent the flow of fluid back toward the injection pump. The pesticide injection pipeline must also contain a functional, normally closed, solenoid-operated valve located on the intake side of the injection pump and connected to the system interlock to prevent fluid from being withdrawn from the supply tank when the irrigation system is either automatically or manually shut down. The system must contain functional interlocking controls to automatically shut-off the pesticide injection pump when the water pump motor stops. The irrigation line or water pump must include a functional pressure switch which will stop the water pump motor when the water pressure decreases to the point where pesticide distribution is adversely affected. Systems must use a metering pump, such as a positive displacement injection pump (e.g., diaphragm pump) effectively designed and constructed of materials that are compatible with pesticides and capable of being fitted with a system interlock. Do not enter treated area before crop is dry.

Crop Rotation: Do not plant root crops other than those in this label within 12 months after last application. Root crops in this label and all other rotation crops may be planted immediately following last application.

[1] Registered trademark of E. I. Du Pont de Nemours & Co. Inc.
[2] Registered trademark of Ciba-Geigy Corp.
[3] Registered trademark of Sandoz Crop Protection Corp.
[4] Registered trademark of Monsanto Agricultural Co.
[5] Registered trademark of ICI Americas Inc.

COTTON

CROP	INSECT	DOSAGE LB ai/ ACRE	DOSAGE FL OZ/ ACRE	Number Acres Treated with One Gallon of ASANA XL	FURTHER USE INSTRUCTIONS	DAYS TO HARVEST
Cotton	Cotton Bollworm Tobacco Budworm Boll Weevil Beet Armyworm* Cabbage Looper Saltmarsh Caterpillar Cotton Leafworm Cutworms Lygus Bugs, Plant Bugs Thrips (on seedling cotton) Leafhoppers, Fleahoppers, Grasshoppers, Whitefly**	0.03-0.05	5.8-9.6	22-13	May be applied in water or nonvolatile vegetable oils. When using oil, apply a total spray volume of 1 or more quarts/acre. When diluted with water, apply no less than one gallon per acre by air (min. of 3 gal/a in Arizona and 5 gal/a in California) or 4 gal/a by ground. Repeat as necessary throughout season to maintain control. For the control of Boll Weevil, a 3-5-day interval between sprays may be required. Apply no more than 0.5 lbs ai/a to a crop in a growing season. Do not graze livestock on treated fields or feed treated trash.	21
	NOTE: For light infestations of the above insects	0.02	3.9	33		
	Pink Bollworm	0.015-0.03	2.9-5.8	44-22		
	Cotton Leaf Perforator	0.03	5.8	22	*Aids in control. **Aids in the suppression	

COTTON—TANK MIX APPLICATIONS
READ AND FOLLOW ALL PRECAUTIONS AND RESTRICTIONS ON THE PRODUCT LABELING.

TANK MIX COMBINATION	INSECT	DOSAGE	FURTHER USE INSTRUCTIONS	DAYS TO HARVEST
LANNATE L/ NUDRIN 1.8 Methomyl Insecticide[1] Solution	Cotton Bollworm and Tobacco Budworm eggs	ASANA XL (from cotton table) with 1/2 pt/acre LANNATE L/ NUDRIN 1.8		21
LANNATE/ NUDRIN 90 Methomyl Insecticide[1] Water Soluble Powder (with adjuvant)	Cotton Bollworm and Tobacco Budworm eggs	ASANA XL (from cotton table) with 1/8 lb/acre LANNATE/ NUDRIN 90	Add adjuvant (X-77, Bio-88, Unite, Triton B-1956 or equivalent*) to spray tank, then ASANA XL, then LANNATE/NUDRIN 90; agitate until thoroughly mixed. *Compatibility of alternate should be determined before use.	

FIELD CROPS

CROP	INSECT	DOSAGE LB ai/ ACRE	DOSAGE FL OZ/ ACRE	Number of Acres Treated with One Gallon of ASANA XL	FURTHER USE INSTRUCTIONS	DAYS TO HARVEST
CORN Field Corn	Cutworm Grasshoppers	0.03-0.05	5.8-9.6	22-13	Apply as necessary to maintain control. Applications for cutworm control may be applied before, during, or after planting as required to protect emerging or emerged corn seedlings. Physical compatibility with herbicides and/or fertilizers should be determined prior to tank mixing ASANA XL with these products. Follow all herbicide/fertilizer recommendations and directions regarding proper usage.	21
	Western Bean Cutworm	0.015-0.03	2.9-5.8	44-22	Apply as necessary to maintain control. Application must be made before larvae enter the ear.	
	Corn Earworm	0.03-0.05	5.8-9.6	22-13	First application should be at or before silking. Subsequent applications should be made at 3-5 day intervals until silking is completed. Repeat applications may be applied if economically damaging populations exist.	

CROP	INSECT	DOSAGE LB ai/ ACRE	FL OZ/ ACRE	Number of Acres Treated with One Gallon of ASANA XL	FURTHER USE INSTRUCTIONS	DAYS TO HARVEST
	Stalk Borer Flea Beetles	0.03-0.05	5.8-9.6	22-13	Application must be made early in migration from grassy areas to corn, before borers enter the plant. More than one application may be necessary.	
	Corn Rootworm (adult control)	0.03-0.05	5.8-9.6	22-13	Spray at first sign of silk feeding. Reinfestation may occur. Should reinfestation occur, repeat as necessary to maintain control.	
	Chinch Bugs	0.03-0.05	5.8-9.6	22-13	Over the top sprays, as used for control of ear and stalk infesting corn pests, are not adequate for chinch bug control. It is very important that the spray be directed at the base of the plant through the use of drop nozzles or some other mechanism. Repeat as necessary to control migrating chinch bugs.	21
	European Corn Borer	0.04-0.05	7.8-9.6	16-13	First brood: spray while eggs are in the blackhead stage or before the larvae enter the whorl. Application by ground equipment is suggested. Good coverage of both upper and lower leaf surfaces is essential. This can be accomplished with drop nozzles over the row and on each side of the corn plant. Multiple applications may be required when egg laying is prolonged or where moderate to heavy populations are present. A higher rate is recommended for moderate to heavy populations. Proper coverage by ground equipment usually requires 20-30 gallons of carrier. Once larvae enter the whorl, foliar sprays will not provide adequate control. Second brood: make applications when sufficient egg masses are found. Spray when eggs are in the blackhead stage or starting to hatch. When egg laying is prolonged or a third generation is present, additional sprays may be required. A higher rate is recommended for moderate to heavy populations. Good coverage above, below, and in the ear zone is essential. This usually requires 2-3 gallons of carrier by air. If ground equipment is used, drop nozzles on each side of the plant will provide best coverage.	
	Southwestern Corn Borer	0.03-0.05	5.8-9.6	22-13	Apply as necessary to maintain control. Two or more applications may be required. Higher rates (0.036-0.05 lb. a.i.) are recommended for moderate to heavy populations.	
	Armyworm (True Armyworm)	0.03-0.05	5.8-9.6	22-13	Apply as necessary to maintain control.	
	Adult Corn Rootworm, European Corn Borer	ASANA XL and Parathions from respective corn labels			ASANA XL Insecticide may be tank-mixed with Parathion (methyl and/or ethyl) and applied on field corn (and seed corn) where supplemental control of Adult Corn Rootworm and/or European Corn Borer is desired in conjunction with insects controlled by ASANA XL when used alone. For supplemental control of second brood European Corn Borer, make applications when sufficient egg masses are found. Spray when eggs are in the blackhead stage or starting to hatch. When egg laying is prolonged or a third generation is present, additional sprays may be required. A higher rate is recommended for moderate to heavy populations. Good coverage above, below, and in the ear zone is essential. This usually requires 2-3 gallons of carrier by air. If ground equipment is used, drop nozzles on each side of the plant will provide best coverage. For supplemental control of Adult Corn Rootworm, spray at first sign of silk feeding. Reinfestation may occur. Should reinfestation occur, repeat as necessary to maintain control.	21

CROP	INSECT	DOSAGE LB ai/ ACRE	DOSAGE FL OZ/ ACRE	Number of Acres Treated with One Gallon of ASANA XL	FURTHER USE INSTRUCTIONS	DAYS TO HARVEST
					Refer to the ASANA XL and Parathion (methyl or ethyl) labels for appropriate rates of the individual products for controlling the respective insects.	
	Corn Leaf Aphid Oat Bird-Cherry Aphid	0.03-0.05	5.8-9.6	22-13	For optimum results, direct the spray at the aphid population so as to achieve maximum coverage of the exposed insects. Aphids not contacted by the spray, such as in whorls and leaf axis, may not be adequately controlled.	21
	Do not exceed 0.25 lb. a.i. per acre per season.					
Seed Corn	For specific insect control recommendations refer to Field Corn (above). Follow directions carefully.				Multiple applications and/or shortened intervals between sprays must be used to insure proper insect control. Do not exceed 0.25 lb a.i. per acre per season.	21
Pop Corn	For specific insect control recommendations refer to Field Corn (above). Follow directions carefully.				Multiple applications and/or shortened intervals between sprays must be used to insure proper insect control. Do not exceed 0.5 lbs. a.i. per acre per season.	1
Peanuts	Corn Earworm, Red-necked Peanut Worm, Potato Leafhopper, Velvetbean Caterpillar	0.015-0.03	2.9-5.8	44-22	Repeat as necessary to maintain control. Do not feed or graze livestock on treated vines. Do not exceed 0.15 lb a.i. per acre per season.	21
	Beet Armyworm**, Granulate Cutworm, Grasshoppers, Cutworms	0.03-0.05	5.8-9.6	22-13	** Aids in control.	
	Lesser Cornstalk Borer** Fall Armyworm***	0.05	9.6	13	*** Aids in control. For heavy infestations use of LANNATE/NUDRIN Insecticide is recommended. Please refer to the LANNATE/NUDRIN label for use instructions.	
Soybean	Mexican Bean Beetle, Green Cloverworm, Velvetbean Caterpillar, Potato Leafhopper, Saltmarsh Caterpillar Woolybear Caterpillar	0.015-0.03	2.9-5.8	44-22	Repeat as necessary to maintain control. Do not feed or graze livestock on treated plants. Do not exceed 0.2 lb a.i. per acre per season. When applying in nonvolatile vegetable oils use a total spray volume of 1 or more quarts.	21
	Cabbage Looper, Corn Earworm, Southern Green Stinkbug, Bean Leaf Beetle, Cutworms, Three -cornered Alfalfa Hopper, Beet Armyworm*, Grasshoppers, Japanese Beetle (adult)	0.03-0.05	5.8-9.6	22-13	*Aids in control	
Sugarcane	Sugarcane Borer	0.03-0.05	5.8-9.6	22-13	Repeat as necessary to maintain control. Do not exceed 0.2 lb ai/acre per season	21
Sunflower	Sunflower Beetle	0.015-0.03	2.9-5.8	44-22	Repeat as necessary to maintain control. Do not exceed 0.2 lb. ai/acre per season	28
	Grasshoppers Cutworms Heliothis (complex) Sunflower Moth Sunflower Seed Weevil Banded Sunflower Moth Beet Armyworm* Sunflower Maggot Leafhoppers Sunflower Stem Weevil	0.03-0.05	5.8-9.6	22-13	*Aids in control.	

FRUITS AND NUTS

DILUTE SPRAY: Apply specified dosage per 100 gallons of water in a uniform spray applied to the point of drip with conventional ground equipment. Do not exceed maximum number of gallons per acre indicated.

NOTE: In order to apply the correct amount of ASANA XL insecticide to your orchard you must know the number of gallons of water needed to spray one acre of your trees to the point of drip. If you do not already know this gallonage, you should conduct a test to determine it. If you do not know how to conduct such a test with your equipment, you should ask assistance from your equipment dealer or State Extension specialist.

CONCENTRATE SPRAY: Apply specified dosage per acre in no less than 30 gals. of water per acre by ground. When applying ASANA XL by air, consult your Cooperative Extension Service for application guidelines.

FRUIT AND NUTS

CROP	INSECT	DOSAGE: CONCENTRATE SPRAY LB ai/ ACRE	DOSAGE: CONCENTRATE SPRAY FL OZ/ ACRE	DOSAGE: DILUTE SPRAY FL OZ/ 100 GALS	Number of Acres Treated with One Gallon of ASANA XL	FURTHER USE INSTRUCTIONS	DAYS TO HARVEST
Apple	Oblique Banded Leafroller, Codling Moth, Variegated Leafroller, White Apple Leafhopper, Tentiform Leaf Miner, San Jose Scale (fruit infestations only), Apple Aphid, Tufted Apple Budmoth, Plum Curculio, Oriental Fruit Moth, Apple Maggot, Red-Banded Leafroller, Lesser Appleworm, Green Fruitworm, Rosy Apple Aphid, Plant Bugs (Tarnished Plant Bug, Stink Bugs), Periodical Cicada	0.025-0.075	4.8-14.5	2.0-5.8	26-9	Repeat as necessary to maintain control. Do not feed or graze livestock on treated orchard floors. Do not exceed 0.525 lbs. a.i. per acre per season. For dilute spray apply 200-600 gals per acre, but do not exceed 14.5 fl. oz. of ASANA XL per acre per treatment. Plant Bug, Rosy Apple Aphid Control—Time of application is critical in achieving control. Use prebloom and post bloom spray timings recommended by State Extension Services.	21
Stone Fruits (including peaches, apricots, nectarines, plums, prune plums, cherries)	Peach Twig Borer, Oriental Fruit Moth, Peach Tree Borer, Plum Curculio, Lesser Peach Tree Borer, Periodical Cicada, Cherry-Fruit Fly, Western Cherry Fruit Fly, American Plum Borer, Black Cherry Aphid, Leafrollers, Leafhoppers, Green Fruit worm, Plant Bugs (Tarnished Plant Bug, Stink Bugs)	0.025-0.075	4.8-14.5	2.0-5.8	26-9	Repeat as necessary to maintain control. Do not exceed 0.375 lbs a.i. per acre per season with no more than 0.3 lbs. a.i. per acre per season between bloom and harvest. For dilute spray apply 200-400 gals per acre, but do not exceed 14.5 fl. oz. of ASANA XL per acre per treatment. Do not graze livestock on treated orchard floors. Peach Tree Borer, American Plum Borer Lesser Peach Tree Borer Control—Apply as directed trunk and scaffold limb spray. Thorough coverage of trunk and scaffold limbs is required. Peach Twig Borer, Plum Curculio, Oriental Fruit Moth, Cherry Fruit Fly, Western Cherry Fruit Fly, Leafrollers, Black Cherry Aphid, Periodica Cicada Control—Apply by ground sprayer to achieve thorough coverage of all aerial portions of the tree. Plant Bug Control— Time of application is critical in achieving control. Use prebloom and post bloom spray timings recommended by State Extension Services.	14
Pear (Dormant)	Pear Psylla	0.05-0.1	9.6-19.2	7.3-12.8	13.2-6.6	Apply during dormant to prebloom (white bud) stage only. Do not apply more than 0.2 lb. a.i. per acre per season. Do not graze orchard floor. For dilute spray apply 150-250 gals per acre, but do not exceed 19.2 fl. oz of ASANA XL per acre per treatment.	
Pear	Pear Psylla Codling Moth Green Fruitworm Plum Curculio Leafrollers Pear Slug Periodical Cicada	.025-0.075	4.8-14.5	2.0-5.8	26-9	Repeat as necessary to maintain control. Do not exceed 0.375 lbs. a.i. per acre per season, nor more than 0.225 lb. a.i. per acre between bloom and harvest. Do not feed or graze livestock on treated orchard floors. For dilute spray apply 200-600 gals per acre, but do not exceed 14.5 fl ozs of ASANA XL per acre per treatment.	28
Almonds	Naval Orangeworm Peach Twig Borer	0.05-0.1	9.6-19.2	7.3-12.8	13.2-6.6	Repeat as needed to maintain control but do not exceed 0.2 lb. a.i. per acre per season. Do not graze livestock on treated orchard floors.	21
Filberts	Filbert Worm Oblique Banded Leafroller	0.05-0.1	9.6-19.2	7.3-12.8	13.2-6.6	Make first application after emergence of filbert worm moths in early summer. Do not apply a second treatment earlier than three weeks after the first. Do not apply	21

CROP	INSECT	DOSAGE: CONCENTRATE SPRAY LB ai/ ACRE	DOSAGE: CONCENTRATE SPRAY FL OZ/ ACRE	DOSAGE: DILUTE SPRAY FL OZ/ 100 GALS	Number of Acres Treated with One Gallon of ASANA XL	FURTHER USE INSTRUCTIONS	DAYS TO HARVEST
						more than 0.2 lb. a.i. per acre per season. For dilute spray, apply 200-400 gals. per acre, but do not exceed 19.2 fl. oz of ASANA XL per acre per treatment. Do not graze livestock in treated orchards	
Pecans	Pecan Weevil, Hickory Shuckworm, Pecan Nut Casebearer, Pecan Aphids, Pecan Spittlebug	0.025-0.075	4.8-14.5	2.0-5.8	26-9	Repeat as necessary to maintain control. Do not feed or graze livestock on treated orchard floors. Do not exceed 0.3 lbs. a.i. per acre per season. For dilute spray apply 200-600 gals per acre, but do not exceed 14.5 fl. oz. of ASANA XL. per acre per treatment.	21
	Pecan Stem Phylloxera Pecan Leaf Phylloxera	0.025-0.075	4.8-14.5	2.0-5.8	26-9	Correct timing of spray applications is critical in achieving optimum control of leaf and stem phylloxera. Consult local spray recommendations for correct times of application.	21
Walnuts	Walnut Aphid, Walnut Husk Fly, Codling Moth, Navel Orangeworm	0.05-0.1	9.6-19.2	4.0	13-6	Repeat as necessary to maintain control. Do not exceed 0.2 lb. a.i. per acre per season. For dilute sprays, apply 200-400 gals per acre. Do not feed or graze livestock on treated crop floors. NOTE: Use of baits in Walnut Husk fly sprays is recommended where endorsed by local Agricultural Extension Service. CONCENTRATE SPRAY: Apply specified dosage per acre in no less than 30 gals of water per acre by ground or 5 gals. or water per acre by air.	21

VEGETABLE CROPS

CROP	INSECT	DOSAGE LB ai/ ACRE	DOSAGE FL OZ/ ACRE	Number of Acres Treated with One Gallon of ASANA XL	FURTHER USE INSTRUCTIONS	DAYS TO HARVEST
Artichokes	Artichoke Plume Moth	0.03-0.05	5.8-9.6	22-13	Repeat as necessary to maintain control, but not more often than each 14 days. Applyno more than 0.075 lb. ai/acre between bud formation and harvest of an individual fruit. Apply in a minimum of 10 gallons per acre by air and 50-200 gallons per acre by ground (use sufficient water to obtain coverage without excessive runoff)	7
Broccoli, Cabbage, Cauliflower, Chinese Cabbage* (tight-headed varieties only)	Imported Cabbageworm	0.015-0.03	2.9-5.8	44-22	Repeat as necessary to maintain control. Do not exceed 0.4 lbs. a.i. per acre per season.	3
	Cabbage Looper, Beet Armyworm*, Alfalfa Looper, Flea Beetle, Grasshopper, Cutworm	0.03-0.05	5.8-9.6	22-13	*Aids in control.	
Carrots	Aster Leafhopper, Cutworms Leafhoppers	0.03-0.05	5.8-9.6	22-13	Apply as needed for control, but do not exceed 0.5 lb ai/acre per season. Use ground application only.	7
	Carrot Weevil*	0.05	9.6	13	*Begin treatment when weevils become active.	
Collards	Imported Cabbageworm Cabbage Looper Beet Armyworm* Alfalfa Looper Flea Beetle Grasshopper Cutworm	0.03-0.05	5.8-9.6	22-13	Repeat as necessary to maintain control. Do not feed livestock on treated plant parts. Do not exceed 0.2 lbs. a.i. per acre per season. Use ground application equipment only *Aids in control.	7

CROP	INSECT	DOSAGE LB ai/ ACRE	FL OZ/ ACRE	Number of Acres Treated with One Gallon of ASANA XL	FURTHER USE INSTRUCTIONS	DAYS TO HARVEST
Cucumber, Melons (Cantaloupe, Honeydew Melons, Watermelon, Muskmelon), Pumpkin, Squash (Summer, Winter)	Pickleworm, Rindworms, Leafhoppers, Corn Earworm, Cabbage Looper, Squash Bug, Cutworms (seedling spray), Cucumber Beetle (adults), Grasshoppers, Plant Bugs (Lygus Bugs, Stink Bugs), Squash Vine Borer	0.03-0.05	5.8-9.6	22-13	Repeat as necessary to maintain control. Do not exceed 0.25 lb a.i. per acre per season.	3
Beans, Snap	Mexican Bean Beetle Saltmarsh Caterpillar	0.015-0.03	2.9-5.8	44-22	Repeat as needed to maintain control but do not exceed 0.2 lb/ai acre per season. Do not allow livestock to graze treated bean fields. Do not harvest treated bean vines for livestock forage, fodder, or hay	3
	Potato Leafhopper European Cornborer Corn Earworm Cucumber Beetle (adults) Flea Beetle Green Cloverworm Cutworm (seedling spray) Grasshopper Pea Aphid Leafhopper Cabbage Looper, Western Bean Cutworm Velvet Bean Caterpillar Beet Armyworm* Cucumber Beetle Corn Rootworm (adults) Leafminer (Guam only)	0.03-0.05	5.8-9.6	22-13	*Aids in control.	
Dry Beans, Dry Peas, Lentils	Mexican Bean Beetle, Saltmarsh Caterpillar	0.015-0.03	2.9-5.8	44-22	Repeat as necessary to maintain control. Do not exceed 0.2 lb a.i. per acre per season. Do not feed or graze livestock on treated vines.	21
	Corn Earworm, Potato Leafhopper, Pea Aphid, Grasshoppers, Leafhoppers, Cowpea Curculio, Cutworms, Cabbage Looper, Western Bean Cutworm, Velvetbean Caterpillar, Beet Armyworm*, Cucumber Beetle, Corn Rootworm (Adults), Painted Lady Butterfly (larvae), Green Cloverworm	0.03-0.05	5.8-9.6	22-13		
Eggplant	Colorado Potato Beetle, Flea Beetles, Corn Earworm, Loopers, European Corn Borer	0.03-0.05	5.8-9.6	22-13	Begin application when insects or their damage is observed and repeat at 7 to 10 day intervals to achieve control. Do not exceed 0.35 lbs. a.i. per acre per season.	7
Green Peas	Pea Aphid, Green Cloverworm	0.015-0.03	2.9-5.8	44-22	Repeat as necessary to maintain control but do not exceed 0.1 lb ai/acre per season. Do not feed treated pea vines to livestock.	3
	Cabbage Looper, Celery Looper, Alfalfa Looper, Cutworms, Armyworm, Corn Earworm, Imported Cabbageworm, Alfalfa Caterpillar	0.03-0.05	5.8-9.6	22-13		
Peppers	European Corn Borer, Beet Armyworm*, Flea Beetles, Corn Earworm, Loopers, Colorado Potato Beetle, Pepper Weevil*, Cucumber Beetle (adults), Southern Armyworm	0.03-0.05	5.8-9.6	22-13	Begin application when insects or their damage is observed and repeat at 7 to 10 day intervals to achieve control. Do not exceed 0.35 lbs. a.i. per acre per season. *Aids in control.	7
Potato	Potato Psyllid, Potato Tuberworm	0.015-0.03	2.9-5.8	44-22	Repeat as necessary to maintain control. Do not graze livestock on treated vines. Do not exceed 0.35 lbs. a.i. per acre per season	7
	Fleabeetles, Tarnished Plant Bug, Cabbage Looper, Potato Aphid, Potato Leafhopper,	0.03-0.05	5.8-9.6	22-13		

CROP	INSECT	DOSAGE LB ai/ ACRE	DOSAGE FL OZ/ ACRE	Number of Acres Treated with One Gallon of ASANA XL	FURTHER USE INSTRUCTIONS	DAYS TO HARVEST
	Cucumber Beetle (adult control), Grasshoppers, European Corn Borer, Beet Armyworm*, Western Yellow-striped Armyworm, Buckthorn Aphid, Cutworms					
	Colorado Potato Beetle	0.03-0.05	5.8-9.6	22-13		
	(Long Island, NY)	0.05	9.6	13	*Aids in control.	
Radishes	Armyworms Beetles	0.03-0.05	5.8-9.6	22-13	Repeat as necessary to maintain control. Do not apply more than 0.1 lb. ai/acre per season.	7
Sweet Corn	Corn Earworm, Southwestern Corn Borer, European Corn Borer, Beet Armyworm*, Flea Beetles, Sap Beetles (adults), Banded Cucumber Beetle, Tarnished Plant Bug	0.03-0.05	5.8-9.6	22-13	Do not exceed 0.5 lb a.i. per acre per season. Begin applications either before or at time of silking and continue as necessary to maintain control. For protection against corn borers other than at silking time, treat when necessary to maintain control. For additional information consult directions for use under field corn. *Aids in control.	1
	Stalk Borer, Cutworms, Corn Rootworms (adults), Grasshoppers, Chinch Bugs, Armyworm	0.03-0.05	5.8-9.6	22-13	For additional information consult directions for use under field corn.	
	Western Bean Cutworm	0.015-0.03	2.9-5.8	44-22		
	Corn Leaf Aphid, Oat Bird-Cherry Aphid	0.03-0.05	5.8-9.6	22-13	For optimum results, direct the spray at the aphid population so as to achieve maximum coverage of the exposed insects. Aphids not contacted by the spray, such as in whorls and leaf axils, may not be adequately controlled.	
Tomato	Tomato Hornworm Tobacco Hornworm	0.015-0.03	2.9-5.8	44-22	Repeat as necessary to maintain control. Do not feed or graze livestock on treated vines. Do not exceed 0.5 lbs. a.i. per acre per season.	1
	Tomato Pinworm, Flea Beetle, Tomato Fruitworm, Colorado Potato Beetle, Cutworms, Cabbage Looper, Southern Armyworm, Beet Armyworm*, Western Yellow-Striped Armyworm, Grasshopper, Whitefly, Potato Aphid	0.03-0.05	5.8-9.6	22-13	*Aids in control.	
	Vegetable Leafminer**	0.05	9.6	13	**Not recommended in Florida	

SPECIALTY USES

CROP	INSECT	DOSAGE LB ai/ ACRE	DOSAGE FL OZ/ ACRE	Number of Acres Treated with One Gallon of ASANA XL	FURTHER USE INSTRUCTIONS
Non-Crop Land (excluding public land such as forests, parks, or recreational)	Grasshoppers, Saltmarsh Caterpillar	0.015-0.03	2.9-5.8	44-22	Spray non-crop land adjacent to tilled areas to control migrating insects (Grasshoppers, Armyworms) which are a threat to crops. Repeat as necessary to maintain control but do not exceed 0.5 lbs. active ingredient per acre per year. Do not feed treated crop to livestock. Do not spray ditch banks or areas adjacent to water.
	Armyworms, Army Cutworms Chinch Bugs	0.03–0.05	5.8-9.6	22-13	

CROP	INSECT	DOSAGE: HIGH VOLUME SPRAYS LB ai/ 100 GALS.	FL OZ/ 100 GALS.	LOW VOLUME SPRAYS (MISTBLOWER) LB ai/A	FURTHER USE DIRECTIONS
Christmas Tree Plantings, Conifer Seed Orchards, Forest Tree Nurseries	Nantucket Pine Tip Moth, Spittlebugs, Pineleaf Chermid, Redheaded Pine Sawfly, Balsam Woolly Adelgid, Pine Chafer, Red Pine Sawfly, Pine Conelet Bug, European Pine Sawfly, Cranberry Girdler (adult control), Pine Needle Midge, Balsam Twig Aphid, Spruce budworm	0.03-0.05	5.8-9.6	0.03-0.05	Apply as needed to maintain control. Spray sufficient gallonage to obtain good coverage of entire tree.
	Pales Weevil, Northern Pine Weevil	0.03-0.05	5.8-9.6	Apply as needed to maintain control. Dilute formulation in kerosene, instead of water, and spray tree stumps to control developing larvae. Use sufficient spray volume to thoroughly cover stump.	
	Coneworms, Seedbugs, Seed Chalcid	9.6 fl oz/100 gals water for high volume sprayers. 52 fl oz/100 gals water for low volume sprayers. 0.19 lb. ai/acre application in not less than 10 gals of water for aerial applications.		Apply first application within 1 week of female flower closure or peak pollen flight for Webbing Coneworm control. For other Coneworms and Seedbugs, apply first application within 30 days following female flower closure. Repeat application at intervals of 4 weeks but do not exceed 1.6 lbs. ai/ac/yr. For Seed Chalcid control, apply when all cones are pendant, and repeat at 1-2 week intervals for 2 or more sprays, or as needed. Apply approximately 5-10 gals of the 9.6 fl oz/100 gal dilution per tree with high volume sprayers. With low volume sprayers apply 100 gals of the 52 fl oz/100 gal dilution per acre. Do not graze or harvest cover crop. Avoid contact with open water.	

STORAGE AND DISPOSAL

Do not contaminate water, food, or feed by storage or disposal.

STORAGE: Store in a secure, dry and temperate area. Store in original container. Keep container closed when not in use. Do not store near food or feed. Do not use or store around the home. Avoid contact with water. In case of spill or leak, soak up with sand, earth or synthetic absorbent (do not use alkaline absorbents) and dispose of wastes in compliance with local, State and Federal regulations.

PESTICIDE DISPOSAL: Pesticide wastes are toxic. Improper disposal of excess pesticide, spray mixture, or rinsate is a violation of Federal Law. If these wastes cannot be disposed of by use according to label instructions contact your State Pesticide or Environmental Control Agency, or the Hazardous Waste representative at the nearest EPA Regional Office for guidance.

CONTAINER DISPOSAL: Triple rinse (or equivalent). Then offer for recycling or reconditioning if container reuse is permitted. If container reuse is prohibited, puncture metal containers and dispose of in a sanitary landfill, or by other approved State and local procedures. Dispose of plastic containers in a sanitary landfill, or by incineration if allowed by State and and local authorities.

In case of a significant spill, call CHEMTREC (800) 424-9300.

NOTICE OF WARRANTY

DU PONT MAKES NO WARRANTY OF MERCHANTABILITY, FITNESS FOR ANY PURPOSE, OR OTHERWISE, EXPRESS OR IMPLIED, concerning this product or its uses which extend beyond the use of the product under normal conditions in accord with the statements made on this label.

Sumitomo Chemical Company, Ltd., holds U.S. Patents for the active ingredient. Sales by E. I. Du Pont de Nemours & Co. (Inc.) of products containing the active ingredient are made under agreements with Sumitomo Chemical Company, Ltd.

AG-3244 8029 2/7/89

This is a specimen label. DuPont is not responsible for the accuracy of the information contained herein. As labels are subject to revision, always carefully read and follow the label on the product container.

ASSURE®

HERBICIDE EMULSIFIABLE CONCENTRATE

ACTIVE INGREDIENT	BY WEIGHT
2-[4-[(6-chloro-2-quinoxalinyl)oxy]phenoxyl]propionic acid, ethyl ester	9.5%*
INERT INGREDIENTS:	90.5%
TOTAL	100.0%

Contains petroleum distillates

* Equivalent to 0.8 pounds ai per gallon

U.S. Pat. 4,629,493 EPA Reg. No. 352-441

KEEP OUT OF REACH OF CHILDREN

PRECAUTIONARY STATEMENTS

HAZARDS TO HUMANS AND DOMESTIC ANIMALS

CAUTION!

MAY IRRITATE EYES, NOSE, THROAT AND SKIN

Avoid breathing vapors or spray mist. Avoid contact with skin, eyes and clothing. In case of contact, immediately flush eyes or skin with plenty of water. Get medical attention if skin irritation persists. Wash thoroughly with soap and water after handling. Remove and wash contaminated clothing before reuse.

For medical emergencies involving this product, call toll free 1-800-441-3637.

ENVIRONMENTAL HAZARDS

Do not apply directly to water or wetlands. Do not contaminate water by cleaning of equipment or disposal of waste.

PHYSICAL OR CHEMICAL HAZARDS

Combustible. Keep away from heat, sparks, and open flames. Keep container closed.

IMPORTANT

Injury to or loss of desirable trees or vegetation may result from failure to observe the following: Do not apply or drain or flush equipment on or near desirable trees or plants, or on areas where their roots may extend, or in locations where the chemical may be washed or moved into contact with their roots. Do not use on lawns, walks, driveways, tennis courts, or similar areas. Prevent drift of spray to desirable plants. Do not contaminate any body of water. Keep from contact with fertilizers, insecticides, fungicides, and seeds during storage.

Use caution when handling "ASSURE" Herbicide as accidental spills or splashes may harm the finish on painted equipment.

Thoroughly clean all traces of "ASSURE" from application equipment immediately after use. Flush tank, pump, hoses and boom with several changes of water after removing nozzle tips and screens (clean these parts separately). Failure to thoroughly clean the equipment may result in injury to subsequently sprayed grass crops.

GENERAL INFORMATION

"ASSURE" is a selective postemergence herbicide recommended for the control of annual and perennial grasses in soybeans.

"ASSURE" does not control sedges or broadleaf weeds.

Recommended rates and timing for "ASSURE" Herbicide application will provide control of the emerged grasses present. Subsequent flushes of grass weeds require a separate treatment.

Most grass crops, including wheat, barley, rye, oats, sorghum, rice and corn are highly sensitive to "ASSURE" Herbicide and all direct or indirect contact should be avoided.

When not in use, keep container closed to prevent evaporation or loss through spillage.

Under certain environmental conditions, minor leaf spotting may occur on soybeans. New growth and crop yields are unaffected.

BIOLOGICAL ACTIVITY

"ASSURE" is a systemic herbicide which is rapidly absorbed and readily translocated from the treated foliage to the root system and growing points of the plant.

Control Symptoms—Treated plants show a reduction in growth and a loss of competitiveness. An early chlorosis/necrosis of the younger plant tissues is followed by a progressive collapse of the remaining foliage. These symptoms will generally be observed in one to three weeks depending on the grass species treated and the environmental conditions.

DIRECTIONS FOR USE

It is a violation of federal law to use this product in a manner inconsistent with its labeling.

Do not apply this product through any type of irrigation system.

TIMING OF APPLICATION

Apply "ASSURE" Herbicide postemergence to actively growing grasses according to the accompanying rate table recommendations. Provide thorough and uniform spray coverage.

Applications made to grass plants stressed due to insufficient moisture or cold temperatures, or to grass plants exceeding the stage of growth recommended in the rate table may result in unsatisfactory control.

In arid regions where irrigation is used to supplement limited rainfall, "ASSURE" Herbicide should be applied as soon as possible after an irrigation.

Under no-till crop production, increase spray pressure and volume to ensure thorough coverage.

APPLICATION INFORMATION

SPRAY ADJUVANTS

Always include a spray adjuvant with postemergence applications of "ASSURE". Petroleum oil concentrates should be of good quality and contain a minimum of 15% emulsifier/surfactant. Nonionic surfactants should contain at least 80% active ingredient. DO NOT use vegetable (crop origin) oils as spray adjuvants with "ASSURE" herbicide. (USE ONLY EPA APPROVED OIL CONCENTRATES AND SURFACTANTS).

Ground Application

Nozzles: Use standard fan or hollow cone nozzles to apply "ASSURE" herbicide. Do not use flood nozzles.

Spray Volume: Use a minimum of 10 gallons of water per acre in Area I (adequate rainfall[1]). Use a minimum of 15 gallons of water per acre in Area II (arid conditions[1]). Do not exceed 40 gallons of water per acre in Area I or Area II.

Spray Adjuvant: Always include a nonphytotoxic petroleum oil concentrate at 1% v/v (4 Qts/100 Gal) or a nonionic surfactant at 0.25% v/v (1 Qt/100 Gal).

Aerial Application

Spray Volume: Use a minimum of 3 gallons of water per acre in Area I (adequate rainfull[1]). Use a minimum of 5 gallons of water in Area II (arid conditions[1]).

Spray Adjuvant: Always include a nonphytotoxic petroleum oil concentrate at 0.5 % v/v or (2 Qts/100 Gal) a nonionic surfactant at 0.25% v/v (1 Qt/100 Gal).

[1] A description of the area boundaries is included.

Areas I and II

Area I—encompasses most of the eastern half of the United States. It is characterized as having adequate rainfall to support crop production and generally does not require irrigation during the growing season. States included in area I are; Alabama, Arkansas, Connecticut, Delaware, Florida, Georgia, Illinois, Indiana, Iowa, Kansas (broadly defined as east of Interstate 135), Kentucky, Louisiana, Maine, Maryland, Massachusetts, Michigan, Minnesota, Mississippi, Missouri, Nebraska (broadly defined as east of State Highway 183), New Hampshire, New Jersey, New York, North Carolina, North Dakota, Ohio, Oklahoma (broadly defined as east of Interstate 35), Pennsylvania, Rhode Island, South Carolina, South Dakota, Tennessee, Texas (broadly defined as east of Interstate 35), Vermont, Virginia, West Virginia, and Wisconsin.

Area II—encompasses the western portion of four states west of the Mississippi river. It is characterized as having arid conditions and in many locations requires irrigation during the growing season.[1] States included in area II are; Kansas (broadly defined as west of Interstate 135), Nebraska (broadly defined as west of State Highway 183), Oklahoma (broadly defined as west of Interstate 35) and Texas (broadly defined as west of Interstate 35).

[1] Note: Applications of "ASSURE" herbicide should be made to actively growing grasses. If a field is to receive an irrigation, wait until after the irrigation to apply "ASSURE" herbicide.

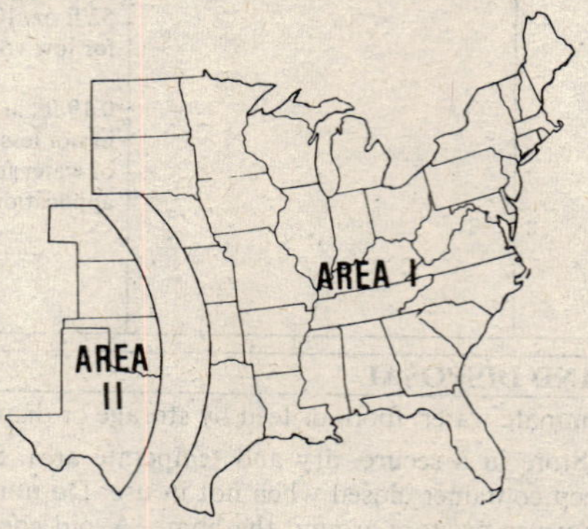

RESTRICTIONS

Do not graze treated fields or harvest for forage or hay.

Do not use vegetable (crop origin) oils as spray adjuvant or carrier.

Do not apply "ASSURE" Herbicide to plants stressed due to insufficient moisture. Grass weeds under this condition frequently become less sensitive to herbicide activity.

Do not cultivate within 7 days before or 7 days after an application of "ASSURE" as control may be unsatisfactory.

Do not apply if rain is expected within 1 hour of application as control may be unsatisfactory.

Do not apply "ASSURE" Herbicide when environmental conditions are such that drift to other crops and non-target areas is possible. Grass crops such as corn, sorghum, wheat, rice, etc. are highly sensitive to "ASSURE" Herbicide.

Do not apply more than 40 ounces of product per acre per season.

Do not apply "ASSURE" within 80 days of harvest.

Do not apply after pod set.

Do not tank mix "ASSURE" Herbicide with any pesticide, fertilizer, or additive except as directed in this label. Application of a broadleaf herbicide within one day following an application of "ASSURE" or within seven days preceeding an application of "ASSURE" may result in reduced grass control.

Do not rotate to crops other than soybeans within 120 days after application.

Users in Texas must confer with personnel of regional U.S. Fish and Wildlife Service offices to insure that federally-protected plants are not exposed to this product.

RECOMMENDATIONS FOR GRASS CONTROL IN SOYBEANS

Area I—(Areas with Adequate Rainfall)

- Apply only to actively growing grasses at the recommended weed stages.
- Always add a nonphytotoxic petroleum oil concentrate or a nonionic surfactant according to label recommendations.
- Follow spray volume recommendations and field calibrate equipment to assure uniform plant coverage.

ANNUAL GRASSES	SPECIES	SIZE AT APPLICATION (INCHES)	"ASSURE" OZ PRODUCT/A	—SPRAY ADJUVANT (% CONCENTRATION)— OIL CONCENTRATE GROUND	AIR	OR	NONIONIC SURFACTANT GROUND/AIR
Corn (volunteer)	Zea mays	6-18	10	1.0% v/v (4 Qt/100 Gal)	0.5% v/v (2 Qt/100 Gal)	or	0.25% v/v (1 Qt/100 Gal)
Shattercane	Sorghum bicolor	6-12	10				
Johnsongrass (seedling)	Sorghum halepense	2-8	10				
Wild Proso Millet	Panicum miliaceum	2-6	10				
Crowfootgrass	Dactyloctenium aegyptium	2-6	14	1.0% v/v (4 Qt/100 Gal)	0.5% v/v (2 Qt/100 Gal)	or	0.25% v/v (1 Qt/100 Gal)
Fall Panicum	Panicum dichotomiflorum	2-6	14				
Foxtails							
Giant	Setaria faberi	2-8	14				
Green	Setaria viridis	2-4	14				
Yellow	Setaria lutescens	2-4	14				
Bristly	Setaria verticillata	2-4	14				
Goosegrass	Eleusine indica	2-6[a]	14				
Itchgrass	Rottboellia exaltata	2-8	14				
Sandbur (field)	Cenchrus incertus	2-6	14				
Sprangletop	Leptochloa filiformis	2-6	14				
Volunteer Cereals							
Wheat	Triticum aestivum	2-6	14				
Barley	Hordeum vulgare	2-6	14				
Oats	Avena sativa	2-6	14				
Rye	Secale cereale	2-6	14				
Wild Oat	Avena fatua	2-6	14				
Witchgrass	Panicum capillare	2-6	14				
Woolycupgrass	Eriochloa villosa	2-4	14				
Barnyardgrass	Echinochloa crus-galli	2-6	16	1.0% v/v (4 Qt/100 Gal)	0.5% v/v (2 Qt/100 Gal)	or	0.25% v/v (1 Qt/100 Gal)
Broadleaf signalgrass	Brachiaria platyphylla	2-6	16				
Crabgrass							
Large	Digitaria sanguinalis	2-6[a]	16				
Smooth	Digitaria ischaemum	2-6[a]	16				
Junglerice	Echinochloa colonum	2-6	16				
Red Rice	Oryza sativa	1-4	16				
Texas panicum	Panicum texanum	2-4	16				

PERENNIAL GRASSES	SPECIES	SIZE AT APPLICATION (INCHES)	"ASSURE" OZ PRODUCT/A	—SPRAY ADJUVANT (% CONCENTRATION)— OIL CONCENTRATE GROUND	AIR	OR	NONIONIC SURFACTANT GROUND/AIR
Bermudagrass	Cynodon dactylon						
First Application	. . . .	3 tall (or up to 6" runners)	20	1.0% v/v (4 Qt/100 Gal)	0.5% v/v (2 Qt/100 Gal)	or	0.25% v/v (1 Qt/100 Gal)
Rhizome Johnsongrass	Sorghum halepense						
First Application	. . . .	10-24	20				
Quackgrass	Agropyron repens						
First Application	. . . .	6-10	20				
Wirestem Muhly	Muhlenbergia frondosa						
First Application	. . . .	4-8	16				

If regrowth of perennial grasses occurs, make a second application of "ASSURE" using 14 ounces product/A timed to grass weed sizes as follows: Bermudagrass (3" tall or up to 6" runners), Rhizome Johnsongrass (6"-10"), Quackgrass (4"-8"), Wirestern Muhly (4"-8").

Area II—(Arid conditions)

- Apply only to actively growing grasses at the recommended weed stages.
- Always add a nonphytotoxic petroleum oil concentrate or a nonionic surfactant according to the label recommendations.
- Make applications as soon as possible after a rainfall or irrigation.
- Follow spray volume recommendations and field calibrate equipment to assure uniform plant coverage.

ANNUAL GRASSES	SPECIES	SIZE AT APPLICATION (INCHES)	"ASSURE" OZ PRODUCT/A	—SPRAY ADJUVANT (% CONCENTRATION)— OIL CONCENTRATE GROUND	AIR	OR	NONIONIC SURFACTANT GROUND/AIR
Corn (volunteer)	Zea mays	6-10	14	1.0% v/v (4 Qt/100 Gal)	0.5% v/v (2 Qt/100 Gal)	or	0.25% v/v (1 Qt/100 Gal)
Shattercane	Sorghum bicolor	6-10	14				
Johnsongrass (seedling)	Sorghum halepense	2-6	14				
Sandbur (field)	Cenchrus insertus	2-6	14				
Barnyardgrass	Echinochloa crus-galli	2-4	16	1.0% v/v (4 Qt/100 Gal)	0.5% v/v (2Qt/100 Gal)	or	0.25% v/v (1 Qt/100 Gal)
Broadleaf Signalgrass	Brachiaria platyphylla	2-4	16				
Crabgrass (large)	Digitaria sanguinalis	2-4	16				
Foxtails							
Green	Setaria viridis	2-4	16				
Yellow	Setaria lutescens	2-4	16				
Junglerice	Echinochloa colonum	2-4	16				
Texas Panicum	Panicum texanum	2-4	20				
Volunteer cereals							
Wheat	Triticum aestivum	2-4	16				
Barley	Hordeum vulgare	2-4	16				
Oats	Avena sativa	2-4	16				
Rye	Secale cereale	2-4	16				

PERENNIAL GRASSES	SPECIES	SIZE AT APPLICATION (INCHES)	"ASSURE" OZ PRODUCT/A	—SPRAY ADJUVANT (% CONCENTRATION)— OIL CONCENTRATE GROUND	AIR	OR	NONIONIC SURFACTANT GROUND/AIR
Bermudagrass	Cynodon dactylon						
First Application	. . . .	3 tall (or up to 6" runners)	24	1.0% v/v (4 Qt/100 Gal)	0.5% v/v 2 Qt/100 Gal)	or	0.25% v/v (1 Qt/100 Gal)
Rhizome Johnsongrass	Sorghum halepense						
First Application	. . . .	10-16	24				

If regrowth of perennial grasses occurs, make a second application of "ASSURE" using 16 ounces product/A timed to grass weed sizes as follows: Bermudagrass (3" tall or up to 6" runners). Rhizome Johnsongrass (6"-10").

RECOMMENDATIONS FOR GRASS CONTROL IN SOYBEANS (cont'd)

Spot/Small Area Spray Recommendations (Areas I and II)

Make a 3/4 percent solution of "ASSURE" in water for spot treating annual species such as volunteer corn or perennial species such as rhizome johnsongrass. The spray solution should include a nonphytotoxic oil concentrate at 1% v/v or a nonionic surfactant at 0.25% v/v. Treat plants on a spray-to-wet basis insuring a good coverage.

DESIRED SPRAY VOLUME	"ASSURE"	PETROLEUM OIL CONCENTRATE	NONIONIC SURFACTANT
1 gallon	1.0 fl. oz. (2 tbsp)	1.25 fl. oz. (2.5 tbsp)	0.3 fl. oz. (2 tsp)
25 gallons	24.0 fl. oz	1.0 quart	8.0 fl oz (1 cup)
50 gallons	1.5 quarts	2.0 quarts	0.5 qt.
100 gallons	3.0 quarts	4.0 quarts	1.0 qt.

Tank-mix Applications of "ASSURE" with "Basagran"[1] or "Classic"[2]

- Tank mixes of "ASSURE" herbicide with post broadleaf herbicides have typically shown some reduction in grass control over applications of "ASSURE" alone. If "ASSURE" is to be tank-mixed with "Basagran"[1], or "Classic"[2] the label rate of "ASSURE" should be increased by 4 ounces product/A.
  - — Do not use "ASSURE"/"Basagran" or "ASSURE"/"Classic" tank-mixes for infestations of annual grasses including; barnyardgrass, broadleaf signalgrass, crabgrass, red rice, texas panicum, or for infestations of perennial grasses including; bermudagrass, quackgrass, or wirestem muhly.
  - — Do not use "ASSURE"/"Basagran" or "ASSURE"/"Classic" tank-mixes in Area II (as defined in this label).

Sequential Applications with Post Broadleaf Herbicides

- Some reduction in grass control is possible when "ASSURE" is applied immediately prior to or sequentially after an application of a post broadleaf herbicide.
  - — After an application of "ASSURE", wait a minimum of 24 hours before applying a post broadleaf herbicide.
  - — In fields treated with a post broadleaf herbicide, do not apply "ASSURE" until grass plants are again developing new leaves (generally 7 days after the post broadleaf herbicide application).

STORAGE AND DISPOSAL

STORAGE: Do not subject to temperatures below 32° F. Do not contaminate water, food or feed by storage or disposal. Keep from contact with fertilizers, insecticides, fungicides, and seeds during storage.

PESTICIDE DISPOSAL: Pesticides, spray mixture, or rinsate that cannot be used according to label instructions must be disposed of according to applicable Federal, State or local procedures.

CONTAINER DISPOSAL: Triple rinse (or equivalent) and dispose of in a sanitary landfill or by incineration if allowed by State and local authorities.

NOTICE TO BUYER: Purchase of this material does not confer any rights under patents of countries outside of the United States.

[1] Basagran™ - Trademark of BASF Corporation
[2] Classic® - Trademark of E. I. du Pont De Nemours & Co. (Inc.)

NOTICE OF WARRANTY

Du Pont warrants that this product conforms to the chemical description on the label thereof and is reasonably fit for the purposes stated on such label only when used in accordance with the directions under normal use conditions. It is impossible to eliminate all risks inherently associated with the use of this product. Crop injury, ineffectiveness, or other unintended consequences may result because of such factors as weather conditions, presence of other materials, or the manner of use or application, all of which are beyond the control of Du Pont. In no case shall Du Pont be liable for consequential, special or indirect damages resulting from the use or handling of this product. All such risks are assumed by the buyer. DU PONT MAKES NO WARRANTIES OF MERCHANTABILITY OR FITNESS FOR A PARTICULAR PURPOSE NOR ANY OTHER EXPRESS OR IMPLIED WARRANTY EXCEPT AS STATED ABOVE.

AG-3764 8058 6/10/88

This is a specimen label. DuPont is not responsible for the accuracy of the information contained herein. As labels are subject to revision, always carefully read and follow the label on the product container.

ATRAZINE 4L

HERBICIDE

ACTIVE INGREDIENTS:	BY WEIGHT
Atrazine (2-chloro-4-ethylamino-6-isopropylamino-s-triazine) . . .	40.8%
Related compounds.	2.2%
INERT INGREDIENTS:	57.0%
TOTAL.	100.0%

Contains 4 lbs. Active Ingredients per Gallon.

This product contains the toxic inert ingredient formaldehyde.

EPA Reg. No. 352-490 EPA Est. No. 352-1L-1

KEEP OUT OF REACH OF CHILDREN

CAUTION

STATEMENT OF PRACTICAL TREATMENT

IF SWALLOWED, drink one or two glasses of water immediately and induce vomiting by sticking finger down back of throat.

IN CASE OF CONTACT, immediately flush eyes or skin with plenty of water. Get medical attention if irritation persists.

PRECAUTIONARY STATEMENTS

HAZARDS TO HUMANS AND DOMESTIC ANIMALS

CAUTION—May be fatal if swallowed. Harmful if absorbed through the skin. Avoid breathing vapors or spray mist. Avoid contact with eyes, skin or clothing. Wash thoroughly after handling. Remove and wash contaminated clothing before reuse.

For medical emergencies involving this product, call toll free 1-800-441-3637.

Do not apply this product in such a manner as to directly or through drift expose workers or other persons. The area being treated must be vacated by unprotected persons.

ENVIRONMENTAL HAZARDS

ATRAZINE can travel (seep or leach) through soil and can enter groundwater which may be used as drinking water. ATRAZINE has been found in groundwater. Users are advised not to apply ATRAZINE to sand and loamy sand soils where the water table (groundwater) is close to the surface and where the soils are very permeable. Your local agriculture agencies can provide further information on the type of soil in your area and the location of groundwater.

ATRAZINE is toxic to aquatic invertebrates. Do not apply directly to water or wetlands. Do not apply where runoff is likely to occur. Do not apply when weather conditions favor drift from treated areas. Runoff and drift from treated areas may be hazardous to aquatic organisms in neighboring areas. Do not contaminate water by cleaning of equipment or disposal of wastes. In case of significant spill, call CHEMTREC (800) 424-9300.

DIRECTIONS FOR USE

It is a violation of Federal law to use this product in a manner inconsistent with its labeling.

REENTRY STATEMENTS

Do not enter treated areas without protective clothing until sprays have dried. Because certain states may require more restrictive reentry intervals for various crops treated in this product, consult your State Department of Agriculture for further information.

Written or oral warnings must be given to workers who are expected to be in a treated area or in an area about to be treated with this product. (Indicate specific oral warnings which inform workers of areas or fields that may not be entered without specific protective clothing, period of time field must be vacated and appropriate actions to take in case of accidental exposure.) When oral warnings are given warnings shall be given in a language customarily understood by workers. Oral warnings must be given if there is reason to believe that written warnings cannot be understood by workers. Written warnings must include the following information: "CAUTION! Area treated with ATRAZINE Herbicide on (date of application). Do not enter without appropriate protective clothing until sprays have dried. (Insert here actions to take in case of accidental exposure. See STATEMENT OF PRACTICAL TREATMENT for details.)"

GENERAL INFORMATION

"ATRAZINE" 4L Herbicide controls many annual broadleaf and grass weeds in corn and sorghum including barnyardgrass, witchgrass, (Panicum capillare), yellow foxtail, green foxtail, wild oats, large (hairy) crabgrass, giant foxtail, velvetleaf, morningglory, lambsquarters, pigweed, ragweed, nightshade, purslane, and mustard. "ATRAZINE" 4L may be applied before or after weeds emerge.

Apply at the rates given under each crop before or after weeds emerge. Use lower rates on light soils and low organic matter soils. Use higher rates on heavy soils and high organic matter soils.

"ATRAZINE" 4L acts mainly through root absorption so rainfall or irrigation is needed to move it into the root zone. Should weeds develop, a shallow cultivation or rotary hoeing will generally result in better weed control.

Avoid using "ATRAZINE" 4L near adjacent desirable trees, shrubs, or plants or injury may occur.

ATRAZINE 4L is non-flammable.

APPLICATION PROCEDURES

General: "ATRAZINE" 4L, a liquid formulation, should be mixed with water or liquid fertilizer and applied as a spray. Pour into the tank during or after filling. Hydraulic (jet) or mechanical agitation is recommended during mixing and application to keep the material in suspension. All return lines to the tank must discharge below liquid level and agitation should not be so violent as to cause air bubbles to form in the liquid. Wash sprayer thoroughly after use.

Ground Application: Apply "ATRAZINE" 4L in 5-40 gallons of water per acre to assure even coverage unless otherwise specified. For uniform distribution, use 80° flat fan-type nozzles. Screens in nozzles, as well as those in suction and in-line strainers, should be no finer than 50-mesh. Use a minimum of 10 gal. of water per acre for all postemergence applications combined with oil or surfactant.

Use a suitable pump with capacity to deliver (1) the necessary volume to the nozzles at 35-40 p.s.i. and (2) an additional volume to maintain the mixture in suspension in the spray tank.

For band applications, calculate amount to be applied per acre as follows:

$$\frac{\textbf{Band Width in Inches}}{\textbf{Row Width in Inches}} \times \textbf{Broadcast rate per acre} = \textbf{Amount needed Per acre of field}$$

Aerial Application

Use only where broadcast applications are specified. Apply a minimum of 1 gallon of water for each 2 pints of "ATRAZINE" 4L to be applied per acre. For postemergence treatments, apply recommended rate in a minimum of 2 gal. water per acre. Do not apply where uniform coverage cannot be obtained or where excessive spray drift may occur. Avoid application directly to animals or humans. Flagmen or loaders should avoid inhalation of spray mist or contact with skin and should wash thoroughly before eating and at the end of each day's operation.

Application in Liquid Fertilizer

Nitrogen solutions or complete liquid fertilizers may replace all or part of the water as a carrier for preemergence and preplant applications. Mix as described under Application Procedures. Do not apply after corn or sorghum emerges as there is danger of liquid fertilizers causing crop injury.

Check to ensure the compatibility of ATRAZINE 4L with liquid fertilizer prior to mixing in the spray tank. A convenient method for testing the compatibility of ATRAZINE 4L with liquid fertilizer follows:

A. Add one pint of liquid fertilizer or other carrier to be used to two quart jars. Mark one quart jar "with" and the other jar "without."

B. To the jar marked "with" add 1/4 teaspoon of a suitable compatibility agent. Shake gently for 5 to 10 seconds to mix. (1/4 teaspoon/pint = 2 pints/100 gallons.)

C. To each jar mix the number of teaspoons of herbicide shown on the following chart times the number of qts. per acre of ATRAZINE 4L to be applied corresponding to the gallons per acre of liquid fertilizer to be used. (For example, if 4 qts. per acre of ATRAZINE 4L are to be applied in 20 gallons of liquid fertilizer per acre, then 4 1 1/4 teaspoons, or 5 teaspoons should be mixed with one pint of liquid fertilizer.) If more than one herbicide is used, add them separately; wettable powders first, flowables second and emulsifiable concentrates last. Shake gently for 5 to 10 seconds after each addition.

D. Let stand one-half hour and make observations. If any separation, agglomeration or precipitation has occurred, again shake the jar 10 to 15 seconds.

1. If mixtures appear to be compatible in both jars, the herbicide(s) can probably be used without a compatibility agent.
2. If the mixture in the jar marked "with" is compatible, but the one in the jar marked "without" is incompatible, a compatibility agent should be used.
3. If both mixtures are incompatible, the liquid fertilizer and the herbicide(s) should not be used in the same spray tank. If a test such as outlined indicates that components of a proposed mix are compatible, materials to be combined in the spray tank should still be added in the sequence according to directions prescribed on the label of the herbicides or pesticides involved.

JAR TEST FOR ATRAZINE 4L COMPATIBILITY WITH LIQUID FERTILIZER

Gal. Liquid Fert/Acre	10	15	20	25	30	40
Teaspoons ATRAZINE 4L /Pint of Fertilizer	2-1/2	1-1/2	1-1/4	1	3/4	1/2

Application in Water Plus Emulsifiable Oil or Oil Concentrate

The addition of oil to ATRAZINE 4L in water sprays may speed the activity of ATRAZINE 4L and provide quicker kill of weeds. However, under certain conditions, the use of oil may seriously damage corn or sorghum. To minimize the possibility of damage, follow the directions, procedures and precautions given below.

Use a crop oil containing at least 1% suitable emulsifier or a crop oil concentrate containing not more than 20% emulsifier or surfactant blend designated for use with ATRAZINE. Emulsifiable oil or oil concentrate contaminated with water or other materials can cause compatibility problems and/or crop injury.

Mixing Procedures—all uses: (1) Thoroughly clean sprayer prior to use. Do not use sprayer contaminated with 2,4-D or other materials as crop damage or sprayer clogging may result. (2) Fill spray tank with water, allowing for the addition of the "ATRAZINE" 4L. (3) With the agitation system in operation, add the "ATRAZINE" 4L. (4) Add the emulsifiable oil or oil concentrate last. (5) Empty the tank as completely as possible before refilling to prevent a buildup of oil in the tank. Keep the agitation system in operation to avoid a separation of the oil from the water in the spray mixture still remaining in the tank. (6) If an oil film starts to build up in the tank, drain the tank and clean out with a strong detergent solution or solvent. (7) Clean the sprayer thoroughly immediately after use by flushing the system with water containing a detergent. (8) All return lines to the tank must discharge below liquid level.

Precautions for Applications with Emulsifiable Oil or Oil Concentrate in Water: (1) Do not use oil in "ATRAZINE" 4L sprays when corn or sorghum is under stress from prolonged cold, wet weather, poor fertility, or other factors or when corn or sorghum is wet and succulent from recent rainfall as crop injury may occur. (2) Do not use oils in sprays when treating inbred lines or any breeding stock as injury may occur. (3) Adding other insecticides, herbicides, liquid fertilizers, or other materials is not recommended because they may cause compatibility problems or crop injury. (4) Store and handle emulsifiable oil or oil concentrate carefully. Oil contaminated with even a small amount of water may not emulsify properly when added to the tank. (5) Do not make more than one application per season except as recommended for control of yellow nutsedge and Canada thistle on this label.

CORN

ATRAZINE 4L may be applied either before planting, at planting or after planting at the rate indicated below.

Preplant

Broadcast in the spring after plowing at the rate indicated in Table 1. Apply before, during or after final seedbed preparation. If soil is tilled or worked after application, avoid deep incorporation of ATRAZINE 4L. Best results have been obtained when ATRAZINE 4L is applied within two weeks prior to planting.

Preemergence

Apply during or shortly after planting prior to weed emergence at the rate indicated in Table 1.

Postemergence

Apply before weeds exceed 1.5 inches in height at the rate indicated in Table 1.

TABLE 1 BROADLEAF AND GRASS WEED CONTROL IN CORN*	
Soil Texture	Broadcast Rate Per Acre of ATRAZINE 4L
COARSE soils: Sands, loamy sands, and sandy loams	4 pints
MEDIUM soils: Silt and clay loams that are low inorganic matter	4.75 pints
FINE soils: Silt and clay loams with medium to high organic matter and clays	6 pints
Peats, mucks, and high organic clays (Apply postemergence only)	6 pints

* Broadleaf weed control (eastern CO, western KS, western NE, NM, OK Panhandle, west TX, and eastern WY). On sand, loamy sand, sandy loam, mild to strongly alkaline soil and all recently leveled soil apply 2.4 pt. per acre, either preplant or preemergence. Controls many broadleaf weeds including pigweed, lambsquarters, nightshade, purslane, and kochia. On other soil types in these areas, apply rate in Table 1 for broadleaf and grass control.

Lay-by Treatment (Broadleaf and grass control)

Broadcast 2-4 pt. per acre in a minimum of 5 gals. of water or nitrogen solution, before weeds are 1.5 inches high and corn is 20-30 inches high. When using nitrogen solution, direct the spray to the lower 3-4 inches of cornstalks to avoid corn foliage injury. Maintain agitation in spray tank during application.

Postemergence with Emulsifiable Oil or Oil Concentrate in Water Broadleaf and Grass Control

Ground Application

Broadcast 4 pts. of ATRAZINE 4L and 1 gal. of emulsifiable oil or 1 qt. of oil concentrate in 20-40 gals. of water per acre after emergence but before weeds reach 1.5 inches in height.

Aerial Application

Broadcast 4 pts. of ATRAZINE 4L and 0.5 gal. of emulsifiable oil in no less than 5 gals. of water per acre after weed emergence but before weeds reach 1.5 inches in height.

Broadleaf Control

Broadcast 2.4 pt. per acre for control of broadleaf weeds, such as annual morningglory, cocklebur, lambsquarters, mustards, pigweed, ragweed, smartweed, and wild buckwheat. Add emulsifiable oil at the rate of 1 gal. per acre for ground application and 0.5 gal. per acre for aerial applications. Apply before pigweed and lambsquarters reach 6 inches in height and before all other weeds reach 4 inches in height. A cultivation may be necessary if all weeds are not controlled or if weeds regrow.

CHEMIGATION

Application in Center Pivot Sprinkler Irrigation Water (For use only in the states of CO, KS, NE, SD, WY)

Apply this product only through a center pivot sprinkler irrigation system. Do not apply this product through any other type of irrigation system. Do not use this product in an irrigation system connected to a public water system.

Crop injury, lack of effectiveness or illegal pesticide residues in the crop can result from non-uniform distribution of treated water.

If you have questions about calibration, you should contact State Extension Service specialists, equipment manufacturers or other experts.

A person knowledgeable of the chemigation system and responsible for its operation, or under the supervision of the responsible person, shall shut the system down and make necessary adjustments should the need arise.

Irrigation System Requirements:

1. The system must contain a functional check valve, vacuum relief valve, and a low pressure drain appropriately located on the irrigation pipeline to prevent water source contamination from backflow.
2. The pesticide injection pipeline must contain a functional, automatic, quick-closing check valve to prevent the flow of fluid back toward the injection pump.
3. The pesticide injection pipeline must also contain a functional, normally closed, solenoid-operated valve located on the intake side of the injection pump and connected to the system interlock to prevent fluid from being withdrawn from the supply tank when the irrigation system is either automatically or manually shut down.
4. The system must contain functional interlocking controls to automatically shut off the pesticide injection pump when the water pump motor stops.
5. The irrigation line or water pump must include a functional pressure switch which will stop the water pump motor when the water pressure decreases to the point where pesticide distribution is adversely affected.
6. Systems must use a metering pump, such as a positive displacement injection pump (e.g., diaphragm pump) effectively designed and constructed of materials that are compatible with pesticides and capable of being fitted with a system interlock.
7. The system must be designed to inject the herbicide mix with a positive displacement pump into the main line ahead of a right angle turn to ensure adequate mixing.

Use Specifications

1. ATRAZINE 4L alone or in tank-mix combinations with other herbicides on this label which are registered for center pivot application, may be applied in center pivot sprinkler irrigation water. Apply any time after planting but prior to crop emergence. Pre-mix the herbicide(s) in water to form a slurry and inject at a carefully calibrated rate into the irrigation water during the entire period of the run. Mix as directed for spray application. Provide good agitation for the herbicide mixture throughout the application. Select the correct application rate of ATRAZINE 4L or the ATRAZINE tank-mix combinations from the rate tables listed on this label.

   Example:

   A center pivot sprinkler that is 1320 feet long (1/4 mile) will irrigate 125.7 acres. A sandy loam soil will require 4 pints of ATRAZINE 4L per acre. (See Table 1)

   125.7 A x 4.0 pts. = 502.8 pts. or 62.9 gal. ATRAZINE 4L.

   In the case of a mixing tank size of 600 gallons and where the sprinkler takes 20 hours to apply 1/2 inch of water to the complete circle;

   Injection rate = 600/20 = 30 gal./hr. or 0.5 gal./minute.

   Calibrate the injection pump to deliver herbicide mix into the irrigation system at the rate of 30 gallons per hour or 0.5 gallons per minute. Add 502.8 pints or 62.9 gals. of ATRAZINE 4L to the tank.
2. For tank-mix combinations select the correct dosage rate and follow the same procedures outlined above for each herbicide component.
3. Apply in 1/2 to 1 acre-inch of water. Use this method of application only with irrigation systems designed for uniform application of water.
4. Application of more than label recommended quantities of irrigation water per acre may result in decreased product performance by removing the chemical from the zone of effectiveness.
5. Do not apply when wind speed favors drift beyond the area intended for treatment.
6. Do not apply when system connectors leak, when nozzles do not provide uniform distribution or when lines containing the product must be drained and dismantled.
7. Greater accuracy in calibration and distribution will be achieved by injecting a large volume or more dilute slurry per hour.

Problem Weeds

ATRAZINE 4L will control yellow nutsedge (Cyperus esculentus) and Canada thistle (Cirsium arvense) when applied according to directions. For best results, apply each year until yellow nutsedge or Canada thistle is eliminated or reaches a level of infestation where neither weed is a problem. If yellow nutsedge or Canada thistle regrow following last application, cultivate once. When ATRAZINE 4L is applied postemergence to the weeds, add 1 gal. of emulsifiable crop oil or 1 qt. of oil concentrate per acre.

Six alternative methods of use for control of yellow nutsedge and Canada thistle are listed below in order of preference. If other weed species, such as annual grasses, are also expected, use alternative 2, 3, 5, or 6.

1. Broadcast 4 pt. plus 1 gal. of emulsifiable oil or 1 qt. of oil concentrate per acre after crop and yellow nutsedge or Canada thistle emerge, but before yellow nutsedge reaches a height of 3 inches or Canada thistle reaches a height of 6 inches. Repeat application before lay-by (20-30 inches), 10-20 days after the first application has been made.
2. Broadcast 4 pt. per acre preplant. Follow with an application of 4 pt. plus 1 gal. of emulsifiable oil or 1 qt. of oil concentrate per acre, after corn and weeds emerge, but before yellow nutsedge reaches a height of 3 inches (yellow nutsedge control only).
3. Broadcast 4 pt. per acre during or shortly after planting, but prior to crop or weed emergence. Follow with an application of 4 pt. plus 1 gal. of emulsifiable oil or 1 qt. of oil concentrate per acre, after corn and weeds emerge, but before yellow nutsedge reaches a height of 3 inches or Canada thistle reaches a height of 6 inches.
4. Broadcast 8 pt. plus 1 gal. of emulsifiable oil or 1 qt. of oil concentrate per acre after crop emerges, but prior to lay-by (20-30 inches) and after yellow nutsedge and Canada thistle emerge, but before yellow nutsedge is 3 inches tall or Canada thistle is 6 inches tall.
5. Broadcast 8 pt. per acre preplant (yellow nutsedge control only).
6. Broadcast 8 pt. per acre during or shortly after planting, but prior to crop or weed emergence (yellow nutsedge control only).

NOTE: Do not use emulsifiable oil or oil concentrate when corn is wet or under stress, especially when using 8 pt. of this product per acre. See "Precautions for Applications with Emulsifiable Oil or Oil Concentrate in Water" for additional directions.

Quackgrass Control on Land Going Into Corn Production

Split application: Broadcast 4 pt. per acre in the fall or spring. Plow 1-3 weeks later. Broadcast a second application at 4 pt. per acre in the spring before, during or after planting but before weeds are 1.5 inches high. This split application will control both quackgrass and most annual broadleaf and grass weeds.

Single application: Broadcast 6-8 pt. per acre in the fall or spring. Plow 1-3 weeks later.

TANK MIXTURES ON CORN

"DUAL" 8E

Use as tank mixture for control of certain broadleaf and grass weeds in corn. Refer to "Dual" 8E label for all directions, weeds controlled, precautions and limitations.

"LASSO" or "LASSO" EC (alachlor)

Use as tank mixture for control of many annual broadleaf and grass weeds in corn (field, hybrid seed, silage, sweet, popcorn) including barnyardgrass, carpetweed, crabgrass, fall panicum, Florida pusley, foxtails (giant, green, yellow), goosegrass, jimsonweed, kochia, lambsquarters, mustard, nightshade, pigweed, purslane, ragweed, signalgrass (Brachiaria), smartweed, and witchgrass. This tank mix will reduce competition from the hard to control annual weeds—annual morningglory, cocklebur, and velvetleaf (buttonweed).

For preemergence or early postemergence broadcast application, use the rates in Table 2. Apply tank mixture any time from immediately after planting until weeds reach 2-leaf state and corn in no more than 5 inches tall. On sweet corn, apply before crop and weeds emerge. Application after the 2-leaf stage will not give satisfactory control. With postemergence application, occasional corn leaf burn should not affect growth or yield. If applied with liquid fertilizers, spray before crop emerges.

For preplant incorporated applications, use 0.4 pt. more of ATRAZINE 4L plus 0.5 qt. of "Lasso" per acre than indicated in Table 2. Use no more than 3.2 pts. of ATRAZINE 4L plus 2.5 qt. of "Lasso" per acre. Broadcast within 7 days before planting and incorporate 2 inches deep.

Apply tank mixture in a minimum of 20 gal. of water per acre. Liquid fertilizer may replace all or part of the water. Add ATRAZINE 4L to spray tank, thoroughly mix with water, then add Lasso. Follow mixing procedures in the section "Application Procedures" of this label.

TABLE 2
TANK MIXTURES WITH "LASSO" ON CORN

	Broadcast Rate/Acre			
	Less than 3% Organic Matter		3% or more Organic Matter	
Soil Texture	ATRAZINE 4L	"Lasso"	ATRAZINE 4L	"Lasso"
COARSE* Sand, loamy sand, sandy loam	2 pt.	1.5 qt.	2 pt.	1.5 qt.
MEDIUM Loam, silt loam, silt	2-2.4 pt.	1.75 qt.	2.4-2.8 pt.	2 qt.
FINE Silty clay loam, sandy clay loam, silty clay, sandy clay, clay loam, clay	2.4-3.2 pt.	2.25 qt.	2.4-3.2 pt.	2.5 qt.

* When applied via center pivot irrigation on coarse soils, apply 2 pt. of Atrazine 4L and 2 qt. of Lasso per acre.

Refer to Lasso label for information concerning center pivot irrigation applications and other directions, limitations, and cautions.

"PRINCEP" 80W or "PRINCEP" 4L (Simazine)

Use as a tank mixture for control of many annual weeds including carpetweed, crabgrass, fall panicum, foxtail, lambsquarters, morningglory, pigweed, ragweed, and velvetleaf. Apply before planting, at planting, or after planting, but before crop and weeds emerge, at rates in Table 3. Use 1:1 ratio for control of above weeds. Use 1:2 ratio for expected heavy infestations of crabgrass and fall panicum. Apply in a minimum of 20 gal. water per acre. Cultivate shallowly if weeds develop.

Preplant

Apply in spring during or after final seedbed preparation. If soil is tilled or worked after application, avoid deep incorporation. For best results, apply within two weeks before planting.

Preemergence

Apply during or shortly after planting, but before crop and weeds emerge. Refer to Corn sections of this label and to "Princep" 80W or "Princep" 4L label for further directions, limitations, and cautions.

TABLE 3
TANK MIXTURES WITH "PRINCEP" ON CORN

	Broadcast Rate/Acre			
Soil Texture	ATRAZINE 4L	1:1 Ratio* "PRINCEP" 80W or "PRINCEP" 4L	ATRAZINE 4L	1:2 Ratio** "PRINCEP" 80W or "PRINCEP" 4L
COARSE: Sand, loamy sand, sandy loam	2 pt.	1.25 lb. or 2 pt.	1.32 pt.	1.67 lb. or 2.67 pt.
MEDIUM: Loam, silt loam, silt, clay loam, sandy clay loam, silty clay loam, sandy clay, or silty clay with low organic matter	2.4 pt.	1.5 lb. or 2.4 pt.	1.6 pt.	2 lb. or 3.2 pt.
FINE: Loam, silt loam, silt, clay loam, sandy clay loam, silty clay loam, sandy clay or silty clay with medium to high organic matter and clay (including dark prairie soils of the Corn Belt)	3 pt.	1.8 lb. or 3 pt.	1.92 pt	2.4 lb. or 3.84 pt.

* For control of above weeds.

** For control of expected heavy infestations of crabgrass and fall panicum

PROPACHLOR 65WP (except in California)

Use a tank mixture for control of many annual broadleaf and grass weeds in corn (field, hybrid seed, silage and sweet corn only), including annual morningglory, annual ryegrass, barnyardgrass (watergrass), buttonweed (velvetleaf), carpetweed, cocklebur, crabgrass, fall panicum, Florida pusley, giant foxtail, green foxtail, yellow foxtail, goosegrass, groundsel, jimsonweed, lambsquarters, mustard, nightshade, pigweed, purslane, ragweed, smartweed and sunflower. Broadcast 2.1-3.2 pt. of ATRAZINE 4L plus 3.8-6 lb. of propachlor 65WP per acre on soil surface any time after planting until broadleaf and grass weeds reach the 2-leaf stage. Use lower rates on coarse-textured soil low in organic matter. Use higher rates on fine-textured soil high in organic matter.

Use a minimum of 2.8 pt. per acre of ATRAZINE 4L in tank mixture for better control of annual morningglory, buttonweed (velvetleaf), cocklebur, and sunflower.

Apply in a minimum of 20 gal. per acre. Non-pressure fluid fertilizer may replace all or part of the water used as a carrier for applications before crop and weeds emerge. Add ATRAZINE 4L to spray tank and thoroughly mix with water. Then, cut the propachlor 65WP bag and fill with water to fill line. Grasp neck of bag firmly, shake vigorously, and add contents to spray tank.

Refer to propachlor 65WP label for further directions, limitations, and cautions.

PARAQUAT

For control of existing vegetation and residual control where corn will be planted directly into cover crop, established sod, or in previous crop residues, broadcast 4-6 pt. of ATRAZINE 4L and 1-2 pts./A (2 lbs./gal.) or 1.3-2.7 pts./A (1.5 lbs./gal.) paraquat in 20-60 gal. of water per acre. Add ATRAZINE 4L to water in spray tank, agitate until thoroughly mixed with water, and then add paraquat and surfactant.

"PRINCEP" 80W plus Paraquat

Use as a tank mixture with Princep 80W and paraquat for kill of existing vegetation and residual weed control where corn will be planted directly into a cover crop, established sod, or in previous crop residues. Add ATRAZINE 4L and Princep 80W to water in spray tank, agitating until thoroughly mixed. Then add paraquat and a nonionic surfactant such as Ortho X-77. Continue agitation during application. Apply 2-4 pt. of ATRAZINE 4L plus 1.25-2.5 lb. of Princep 80W plus 1-2 pts./A (2 lbs./gal.) or 1.3-2.7 pts./A (1.5 lbs./gal.) paraquat in 20-60 gal. of water per sprayed acre within 3 weeks before, during, or after planting, but before corn crop emerges. Add 8 fl. oz. of a nonionic surfactant, such as Ortho X-77, per 100 gal. diluted spray. Use lower rates of ATRAZINE 4L and of Princep 80W on coarse soil, and higher rates on fine-textured soil. Use the 2 pts./A (2 lbs./gal.) or 2.7 pts./A (1.5 lbs.)gal.) rate of paraquat if existing vegetation is 4-6 inches tall. This mixture will not control weeds taller than 6 inches.

Refer to further limitations and precautions on labels for this product, Princep 80W, and paraquat.

PRECAUTIONS FOR ALL APPLICATIONS TO CORN

(1) Do not apply more than 8 pts. ATRAZINE 4L to corn in any one year. (2) Land treated with ATRAZINE 4L should not be planted to any crop except corn or sorghum until the following year or injury may occur. (3) Injury may occur in any crop other than corn or sorghum planted the year following treatment. (4) Following harvest of a treated crop, plow (moldboard or disk-plow) and thoroughly till the soil in the fall or spring to minimize possible injury to rotational spring-seeded crops, regardless of the rate used. (5) To avoid the possibility of injury to following fall or spring seeded crops, the use of BLADEX Herbicide in place of ATRA-

ZINE 4L is recommended. Follow directions on the BLADEX label. (6) The use of ATRAZINE 4L on land that is to be rotated the following year to soybeans that will be treated with metribuzin (SENCOR or LEXONE) may increase the chance of injury. (7) If ATRAZINE 4L is applied after June 10, do not rotate with crops other than corn or sorghum the next year or injury may occur. (8) In the High Plains and Intermountain areas of the West where rainfall is sparse and erratic or where irrigation is required, use ATRAZINE 4L to control weeds in corn or sorghum only where corn or sorghum is to follow corn or a crop of untreated corn or sorghum is to precede other rotational crops. (9) Do not graze treated area or feed treated forage to livestock for 21 days following application.

SORGHUM AND SORGHUM-SUDAN HYBRIDS
(Grain and Forage Types)

Apply before planting, at planting, or after planting. See Corn section for weeds controlled.

Preplant (Broadleaf and Grass Control)

Broadcast in spring after plowing at rate in Table 4. Apply before, during, or after final seedbed preparation. If soil is tilled or worked after application, avoid deep incorporation. For best results, apply within two weeks before planting.

Preemergence (Broadleaf and Grass Control)

Apply during or shortly after planting before weed or crop emergence, at rate in Table 4.

In case of planting failure, sorghum may be replanted. Do not make a second broadcast application, as injury may occur. If originally applied in a band and sorghum is replanted in untreated row middles, ATRAZINE 4L may be applied in a band to the second planting.

Preemergence Broadleaf Control in Furrow Irrigated Bedded Sorghum (AZ and CA only):

For preemergence control of many broadleaf weeds, including groundcherry, lambsquarters, morningglory, mustard, pigweed, and purslane, broadcast 1.6-2.4 pt. per acre. Apply after bed preparation, during or after planting, but before sorghum and weeds emerge and before the first furrow irrigation. Follow with several regular irrigations to thoroughly wet all soil. Use lower rate on coarse-textured soil and soil low in organic matter; use higher rate on fine-textured soil and soil high in organic matter.

Precautions for Preemergence Application to Furrow Irrigated Bedded Sorghum Grown in AZ and CA

To avoid possible sorghum injury, do not use on sand or loamy sand soil or on sorghum planted in the furrow. Applications to sorghum growing on alkali soils or where cuts, fills, or erosions have exposed calcareous or alkali subsoils may cause crop injury. Do not replant sorghum for 8 months following application. Corn may be planted immediately.

TABLE 4
PREPLANT AND PREEMERGENCE BROADLEAF AND GRASS WEED CONTROL IN SORGHUM*

Soil Texture	Organic Matter	Broadcast Rate Acre
COARSE: Sand, loamy sand, sandy loam, sandy clay loam	any level	Do not use (except for preemergent use on bedded sorghum in AZ and CA as indicated below)
MEDIUM and FINE: Silt loam, clay loam, clay	less than 1%	
	1-1.5%	3.2-4 pt.
	more than 1.5%	4-4.75 pt.

* Do not apply preplant in AL, AR, FL, GA, LA, MS, NC, NM, SC, TN, or TX. Do not apply preemergence in NM, OK, or TX, except in northeast OK and the TX Gulf Coast.

TABLE 5
POSTEMERGENCE BROADLEAF AND GRASS WEED CONTROL IN SORGHUM

Soil Texture	Minimum Sorghum Height at Treatment	Broadcast Rate/Acre
Sand or loamy sand	Do not use	
Sandy loam	See directions for postemergence broadleaf weed control below	
Silt loam to sandy clay loam	Completely Emerged	4-4.75 pt.
Olton and Pullman clay soil	At least 6 inches	4-4.75 pt.
Silty clay loam and finer textured soil	Completely Emerged	6 pt.

Postemergence (Broadleaf and Grass Control)

Apply at rate in Table 5 before weeds exceed 1.5 inches in height. Apply up to "close-in".

Postemergence Broadleaf Control with Emulsifiable Oil or Oil Concentrate in Water

Broadcast 2.4 pt. per acre for control of many broadleaf weeds, including annual morningglory, cocklebur, lambsquarters, mustard, pigweed, ragweed, smartweed, and wild buckwheat. Apply before pigweed and lambsquarters reach 6 inches in height and before all other weeds reach 4 inches in height. In CO, Western KS, NM, OK, TX and desert regions of AZ and CA, apply when sorghum is 6-10 inches in height, but before it reaches boot stage. In all other areas, apply after sorghum reaches the 3-leaf stage. Add emulsifiable oil at 1 gal. per acre for ground application and 0.5 gal. per acre for aerial applications. A cultivation may be necessary if all weeds are not controlled or if weeds regrow.

Precautions for Applications with Emulsifiable Oil or Oil Concentrate in Water

See "Precautions for Applications with Emulsifiable Oil or Oil Concentrate in Water" in Application Procedures section of this label.

Postemergence Broadleaf Control with Surfactant

(CO, Western KS, NM, OK, TX and Desert Regions of AZ and CA only) Broadcast 2.4 pt. plus 0.75-1.5 pt. of surfactant per acre after sorghum reaches 6 inches in height, but before weeds reach 1.5 inches in height. Apply only on sandy loam and finer textured soil.

Precautions for All Applications to Sorghum

(1) Heavy rain immediately following application tends to cause excessive concentrations of herbicide in seed furrow, resulting in possible crop injury. Do not apply to furrow-planted sorghum until furrows are leveled (plowed-in). Level deep planter marks or seed furrows before application. (2) Application to sorghum growing under stress caused by minor element deficiency or to sorghum growing on highly calcareous soil may result in crop injury. (3) Injury may occur in any crop other than corn or sorghum planted the year following treatment. (4) Following harvest, plow (moldboard or disk-plow) and thoroughly till soil in fall or spring to minimize possible injury to rotational spring-seeded crops, regardless of rate used. (5) Injury may occur if both this herbicide, preplant or preemergence, and an at-planting systemic insecticide are used.

NOTE: Do not graze or feed forage from treated areas for 21 days following application.

Additional Precautions

See "Precautions for All Applications to Corn" in Corn section of this label.

CHEMICAL FALLOW

Wheat-Sorghum-Fallow

This treatment controls annual broadleaf and grass weeds following wheat harvest and in the following sorghum crop when grown under minimum tillage.

Apply 6 pt. to wheat stubble immediately following wheat harvest. If weeds are present, remove them with a sweep plow or other suitable implement after application. Plant sorghum into wheat stubble the following spring with minimum disturbance of the soil. Use a surface planter or a planter leaving a shallow furrow. If weeds are present at planting, remove them with a sweep plow or other suitable implement before planting.

Precautions

(1) Use only on a silt loam or finer textured soil. (2) Wheat-sorghum-fallow cropping sequence must be followed. (3) Do not apply following sorghum harvest.

NOTE: Do not graze or feed forage from treated area to livestock. Do not plant any crop other than those on this label within 18 months following treatment.

Wheat-Corn-Fallow (KS, NE)

ATRAZINE 4L controls cheatgrass (downy brome, chess), kochia, mustards, pigweed, Russian thistle, wild lettuce, wild sunflower, and volunteer wheat during period after wheat harvest. Weed control may extend into following corn crop grown under minimum tillage.

Apply with ground equipment only.

Follow directions for use, notes, and precautions in the "Wheat-Sorghum-Fallow" section above, substituting corn for references to sorghum.

Wheat-Fallow-Wheat (CO, KS, MT, NE, ND, SD, and WY)

For preemergence control of cheatgrass (downy brome, chess), common lambsquarters, field pennycress, kochia, mustard, Russian thistle, wild lettuce, and suppression of volunteer wheat during fallow period of a wheat-fallow-wheat rotation, apply 1-2 pts. in 10-40 gal. of water per acre for ground application, or in a minimum of 5 gal. of water per acre for aerial application, on all soils except those listed under "Precautions." For control of pigweed and wild sunflower, use the higher rate. Apply to stubble ground. Treat only once during same fallow period.

If weeds are present at application, a tank mix with paraquat may be used. Broadcast 1-2 pts. of ATRAZINE 4L plus 1-2 pts./A. (2 lbs./gal.) or 1.3-2.7 pts./A. (1.5 lbs./gal) paraquat. Add 8 fl.oz. of a nonionic surfactant, such as Ortho X-77, per 100 gal. of spray mixture. Add ATRAZINE 4L to spray tank first and thoroughly mix with water. Then add paraquat, followed by surfactant. Use the 2 pts./A (2 lbs./gal.) or 2.7 pts./A (1.5 lbs./gal.) rate of paraquat if weeds are 4-6 inches tall. This mixture will not control weeds taller than 6 inches. Apply to stubble ground. Treat only once during same fallow period. Refer to paraquat label for further directions, precautions, and limitations.

If weeds are present at application and ATRAZINE 4L is used alone, use either an approved contact herbicide before or after treatment, or tillage after treatment.

Use tillage to control weeds which escape during fallow period. Till before planting. For ATRAZINE 4L applied alone or in tank mixture with paraquat, plant at least 2 inches deep and 12 months or more after application.

Precautions

(1) Do not use on sand soil. (2) Do not treat eroded hillsides, caliche and rocky outcroppings, or exposed calcareous subsoil. (3) Do not treat soils of the Rosebud and Canyon Series in Western NE and adjoining counties in CO and WY. (4) Do not treat soils with calcareous surface layers. (5) Avoid spray overlap.

NOTE: Do not graze treated areas within 6 months after application.

RANGELAND AND PASTURE

To control certain weeds, including cheatgrass (downy brome, chess), tumble mustard, sagewort, annual or common broomweed, use lower rate on areas where cheatgrass is the major problem weed; use higher rate for other weeds.

For ground applications, use a minimum of 10 gals. of water per acre.

For aerial applications, use a minimum of 5 gals. of water per acre.

See "Application Procedures" for additional information.

Establishment of Perennial Range Grasses (CA, ID, NV, OR, UT, WA)

Apply 1.6-2 pts. per acre in fall after rains begin, but before ground freezes. Perennial range grass may be seeded when rain begins the next fall following treatment. Plant in deep furrows such as those made by a deep furrow rangeland drill. Treated areas may be grazed 7 months after seeding. However, to provide sufficient time for establishment of desirable range grasses, treated areas should not be grazed until the fall following seeding.

Renovation of Existing Stands of Perennial Range Grasses

Central Great Plains (CO, KS, MT, ND, NE, SD, WY): Apply 1-2 pts. per acre in fall before ground freezes.

Southern Great Plains (OK, TX): Apply 1 qt./A of ATRAZINE 4L in early spring prior to April 30 to established rangeland or pasture, including such species as Asiatic (Old World) bluestems, little and big bluestems, buffalograss, switchgrass, and Indiangrass. Apply 1-2 qts./A of "ATRAZINE" 4L on common or improved bermudagrass selections. Use the lower rate on coarse-textured soil, soil low in organic matter, and/or where cheatgrass is the major problem weed; use the higher rate on fine-textured soil, soil high in organic matter, and/or where weeds other than cheatgrass are the major problem.

West (CA, ID, NV, OR, UT, WA): Do not graze area to be treated for six months prior to application. Apply 1-1.5 pt. per acre by ground equipment in fall before the ground freezes. Bluegrass and intermediate wheatgrass may be injured by this product at recommended rates.

Do not apply to crested and intermediate wheatgrasses. Do not cut or feed range grass hay. Do not cut or feed areas within seven months following fall application or three months following spring application. Apply only once per year.

Warm Season Perennial Range Grasses—(Midwest):

Switchgrass: For control of many annual broadleaf and grass weeds including cheatgrass, foxtails (giant, green, and yellow), lambsquarters, pigweed, ragweed, smartweed, and velvetleaf.

For ground application use a minimum of 15 gal. of water per acre.

Use only on loam, silt loam, silty clay loam, clay loam and silty clay soils with at least 1 percent organic matter.

New seedling: Apply 4 pt. per acre after planting but before weeds emerge. Prepare a good, firm seedbed. Plant 0.5 inch deep with a grassland drill (preferred method) or a conventional drill. If a conventional drill is used on prepared seedbeds, remove all tension from the disk openers. For best results, cultipack or roll after planting. Clip weeds that escape in July or August. Avoid clipping switchgrass seedlings.

Established Stands: Apply 4 pt. per acre in April or early May, before weed emergence.

Precautions: See "Precautions for All Applications To Corn" in Corn section of this label. NOTE: Do not cut for hay. Do not graze treated areas within 4 months following application at seeding or 3 months following application to established switchgrass. Treat only once per year.

SUGARCANE

For control of many annual broadleaf and grass weeds, including amaranths, crabgrass, fireweed, Flora's paintbrush, foxtails, junglerice, and wiregrass: Broadcast 4-8 pts. in 20-50 gals. of water for adequate coverage of the soil surface at time of planting or ratooning, but before the cane emerges. One additional application may be made over the cane as it emerges, and two additional applications may be made interline after emergence as directed sprays.

For control of emerged pellitory weed (Florida only): Apply 0.8-1.2 pt. per acre in at least 40 gals. of water per acre as a directed spray. Add 2 qts. of surfactant for each 50 gals. of spray. Thoroughly cover weed foliage.

Precautions:

(1) Do not apply after "close-in." (2) Do not apply more than 20 pt. per acre to any one crop of cane.

For control of alexandergrass, large crabgrass, pellitory (artillery) weed, and spiny amaranth, use one of the following methods at planting or ratooning (Florida only):

1. Apply 8 pt. in 20-50 gal. of water per acre, preemergence, broadcast or banded. Follow with one or two broadcast or banded over-the-top applications as needed, postemergence to sugarcane and weeds, at 4 pt. in 20-50 gal. of water per acre. Treat before weeds exceed 1.5 inches in height.
2. Apply 1-3 times as needed, at 4 pt. in 20-50 gal. of water per acre either broadcast or banded over-the-top, postemergence to sugarcane and weeds. Treat before weeds exceed 1.5 inches in height.

FORESTS AND CHRISTMAS TREE PLANTATIONS

For control of annual broadleaf and grass weeds prior to transplanting, after transplanting or in established conifers including Douglas fir, Grand fir, Noble fir, White fir, Austrian pine, Bishop pine, Jeffrey pine, Knobcone pine, Loblolly pine, Lodgepole pine, Monterey pine, Ponderosa pine, Scotch pine, Slash pine, Blue spruce and Sitka spruce.

Annual broadleaf and grass weed control: Broadcast 4 to 8 pints in a minimum of 5 gallons of water per acre by air or 10 gallons by ground to assure thorough coverage. Apply before weed seedlings are more than 1.5 inches high. Apply to established trees between fall and early spring while trees are dormant. For new transplants, apply during or soon after transplanting. For applications prior to transplanting, allow sufficient precipitation to activate ATRAZINE 4L before transplanting. In areas where spring and summer rainfall is inadequate to activate "ATRAZINE", apply during fall prior to spring transplanting.

Quackgrass Control: Broadcast 8 pints in a minimum of 5 gallons of water per acre by air or 10 gallons by ground between fall and early spring while trees are dormant and before Quackgrass is more than 1.5 inches tall.

Precautions:

(1) Do not graze treated areas. (2) Do not apply to seedbeds. (3) Do not make more than one application per year.

Aerial Application:

In order to assure that spray will be controllable within the target area when used according to label directions, make applications at a maximum height of 10 ft. above vegetation, using low drift nozzles at a maximum pressure of 40 psi, and restrict application to periods when wind speed does not exceed 10 mph. To assure that spray will not adversely affect adjacent sensitive non-target plants, apply ATRAZINE 4L by aircraft at a minimum upwind distance of 400 ft. from sensitive plants.

NON-SELECTIVE WEED CONTROL IN NON-CROP LAND

Apply ATRAZINE 4L at rates recommended below before or soon after weeds begin growth. Postemergence applications should be made when weeds are young and actively growing. Use the higher rates on heavy clay and muck soils.

ATRAZINE 4L can be used to provide long term weed control on industrial sites, highway medians and shoulders, railroad rights-of-way, lumberyards, petroleum tank farms, and in non-crop areas on farms, such as around buildings, equipment and fuel storage areas, along fences and lanes.

Use sufficient water to assure thorough coverage. Use at least 1 gal. of water for each quart of ATRAZINE 4L, more if practical. To control many annual broadleaf and grass weeds (such as barnyardgrass, plantain, quackgrass, purpletop, redtop and smooth brome): Broadcast 10-20 qts. per acre. To control hard-to-kill biennial and perennial weed (such as bull thistle and sowthistle): Broadcast 20-40 qts. per acre.

For longer residual control in regions of high rainfall and a long growing season: Broadcast 20-40 qts. per acre.

For small areas 7.2 ozs. per 1,000 sq. ft. is equivalent to 10 qts. per acre

STORAGE AND DISPOSAL

STORAGE: Do not contaminate water, food, or feed by storage or disposal. Do not use or store around the home environment. Avoid contact with water. In case of spill or leak, soak up with sand, earth or synthetic absorbent (do not use alkaline absorbents) and dispose of wastes in compliance with local, state and Federal regulations.

PESTICIDE DISPOSAL: Wastes resulting from the use of this product may be disposed of on site or at an approved waste disposal facility.

CONTAINER DISPOSAL: Triple rinse (or equivalent). Then offer for recycling or reconditioning, or puncture and dispose of in a sanitary landfill, or by incineration, or, if allowed by state and local authorities, by burning. If burned, stay out of smoke.

NOTICE OF WARRANTY

Du Pont warrants that this product conforms to the chemical description on the label thereof and is reasonably fit for the purposes stated on such label only when used in accordance with the directions under normal use conditions. It is impossible to eliminate all risks inherently associated with the use of this product. Crop injury, ineffectiveness or other unintended consequences may result because of such factors as weather conditions, presence of other materials, or the manner of use or application, all of which are beyond the control of Du Pont. In no case shall Du Pont be liable for consequential, special or indirect damages resulting from the use or handling of this product. All such risks shall be assumed by the buyer. DU PONT MAKES NO WARRANTIES OF MERCHANTABILITY OR FITNESS FOR A PARTICULAR PURPOSE NOR ANY OTHER EXPRESS OR IMPLIED WARRANTY EXCEPT AS STATED ABOVE.

BLADEX is a registered trademark of E.I. DuPont de Nemours and Co.
LASSO is a registered trademark of Monsanto Company.
PRINCEP is a registered trademark of Ciba-Geigy.
ORTHO X-77 is a registered trademark of Chevron Chemical Co. , Inc.
SENCOR is a registered trademark of Mobay Chemical Company.
LEXONE is a registered trademark of E.I. DuPont de Nemours and Co.

AG-3268 8069 5/30/89

This is a specimen label. DuPont is not responsible for the accuracy of the information contained herein. As labels are subject to revision, always carefully read and follow the label on the product container.

BENLATE®

FUNGICIDE
WETTABLE POWDER

ACTIVE INGREDIENT:

Benomyl [Methyl 1-(butylcarbamoyl)-2-benzimidazolecarbamate]	50%
INERT INGREDIENTS:	50%
TOTAL	100%

U.S. Pat. 3,631,176

EPA Reg. No. 352-354

KEEP OUT OF REACH OF CHILDREN

PRECAUTIONARY STATEMENTS

HAZARDS TO HUMANS AND DOMESTIC ANIMALS

CAUTION!

MAY IRRITATE EYES, NOSE, THROAT AND SKIN.

Avoid breathing dust or spray mist. Avoid contact with skin, eyes and clothing. Wash thoroughly after using.

Do not apply this product in such a manner as to directly or through drift expose workers, or other persons. The area being treated must be vacated by unprotected persons.

Protective clothing must be worn during periods of exposure. See Directions for Use.

This product may cause a temporary allergic skin reaction in a few susceptible persons. This condition should be treated as an allergic dermatitis. There is no evidence of aftereffects or permanent injury.

First Aid: In case of contact, flush skin or eyes with plenty of water; for eyes, get medical attention.

Aerial Mixer-Loaders: Wear a dust mask or respirator during handling and mixing as this product may be harmful if inhaled. See Directions for Use.

For medical emergencies involving this product, call toll free 1-800-441-3637.

ENVIRONMENTAL HAZARDS

This pesticide is toxic to fish. Do not apply directly to water. Do not apply where runoff is likely to occur. Drift and runoff from treated areas may be hazardous to fish in adjacent areas. Do not contaminate water by cleaning of equipment or disposal of wastes. Do not apply when weather conditions favor drift from areas treated.

For registered aquatic uses: Aquatic organisms may be killed at recommended application rates.

PHYSICAL OR CHEMICAL HAZARDS

Keep away from fire or sparks.

DIRECTIONS FOR USE

It is a violation of federal law to use this product in a manner inconsistent with its labeling.

Du Pont "Benlate" should be used only in accordance with recommendations on this label, or in separate published Du Pont recommendations available through local dealers.

Du Pont will not be responsible for losses or damages resulting from use of this product in any manner not specifically recommended by Du Pont. User assumes all risk associated with such nonrecommended use.

"Benlate" is a systemic fungicide recommended for the control of many important plant diseases. If treatment is not effective following use of "Benlate" as recommended, a resistant strain of the fungus may be present. If treatment is ineffective due to the presence of a benomyl resistant strain, then neither "Benlate", nor any other benzimidazole or thiophanate type fungicide will effectively control that disease; consideration should be given to prompt use of other types of suitable fungicides.

The repeated exclusive use of "Benlate" may lead to buildup of resistant strains of fungi and loss of disease control. A spray program using other fungicides may delay resistant strain buildup. Consult your state extension specialist or official state recommendations for guidance on your particular crop and disease control situation.

Note: Do not tank mix or alternate "Benlate" with benzimidazole or thiophanate products such as "Mertect"[1] or "Topsin"[2].

Do not enter or allow entry into treated areas for 24 hours after application unless wearing the protective clothing specified for this product or unless entering to perform post application activities on strawberries or bulbs.

WORK SAFETY RULES

Keep all unprotected persons, children, livestock and pets away from treated area or where there is danger of drift.

During mixing and loading of the concentrated product wear the following protective clothing and equipment:

Long sleeve shirt and long legged pants; chemical resistant gloves, chemical resistant apron; chemical resistant shoes, shoe coverings, or boots. Wear a dust mask or a respirator approved by the National Institutes of Occupational Safety and Health (30 CFR Part 11) when mixing or loading for aerial application.

Wear the following protective clothing during application, equipment repair, equipment cleaning, and during reentry to treated areas before expiration of the 24 hour reentry interval:

Long sleeve shirt and long legged pants; chemical resistant gloves; chemical resistant shoes, shoe coverings, or boots. During aerial application from enclosed cockpits wear long sleeve shirt and long legged pants. A clean pair of chemical resistant gloves must be carried in the cockpit and worn during exiting from the aircraft.

Important! Before removing gloves, wash them with soap and water. Always wash hands, face, and arms with soap and water before smoking, eating, drinking, or toileting.

After work, take off all clothes and shoes. Shower using soap and water. Wear only clean clothes. Do not use contaminated clothing. Wash protective clothing and protective equipment with soap/detergent and water after each use. Personal and protective clothing worn during use must be laundered separately from household articles. Clothing or protective equipment heavily contaminated or drenched with benomyl must be destroyed according to state or local regulations. Heavily contaminated or drenched clothing cannot be adequately decontaminated.

Flaggers must be in totally enclosed vehicles.

Apply as a spray with ground equipment (except as otherwise directed), using sufficient water to obtain thorough coverage of the plants. Under severe disease conditions use the higher rate and shorter interval specified for each crop; also, for tree crops, use the higher rate for large mature trees. For aerial application (listed crops only) use the following gals. per acre: Rice and Soybeans, 3 to 10; Cabbage (seed crop), Celery, Cucurbits, Peanuts and Sugar Beets, 5 to 10; Almonds, Avocados, Beans, Pecans, Stone Fruits and Strawberries, 10 to 20; Grapes, 15 to 20; Roses, Flowers, Ornamentals and Shade Trees, 20 gals. per acre minimum.

For use in small gardens and orchards (less than 1 acre), application rates may be converted to lbs. per 100 gals. by dividing the lbs. per acre rate in half, and applying the resulting spray mixture at the rate of 4.5 gals. per 1000 sq. ft. (1 lb. "Benlate" per 100 gals. equals 1 tablespoonful per gal).

Add required amount of "Benlate" to necessary volume of water in spray tank agitated by hydraulic or mechanical means; continuous agitation is required to keep the material in suspension. Do not tank mix "Benlate" with lime or alkaline pesticides such as Bordeaux mixture or lime sulfur.

Where use of spray oil is recommended (apples, peanuts, pecans, stone fruits), use a nonphytotoxic superior-type (60 to 70 second viscosity) spray oil; add as last ingredient to spray tank. Before applying other pesticides in conjunction with spray oil or immediately before or after oil application, consult product labels. Observe all cautions and limitations on labeling of all products used in mixtures.

Grazing or feeding livestock is prohibited on forage and hay of benomyl treated wheat, and on forage and hay of barley, oats, rye and wheat following seed treatment.

CHEMIGATION: Apply "Benlate" Fungicide only through sprinkler including center pivot, lateral move, end tow, side (wheel) roll, traveler, big gun, solid set or hand move irrigation systems only on crops specifically designated in the "Directions For Use" for that particular crop. Do not apply "Benlate" through any other type of irrigation system or on any other crops.

Crop injury, lack of effectiveness, or illegal pesticide residues in the crop can result from nonuniform distribution of treated water.

If you have questions about calibration, you should contact State Extension Service specialists, equipment manufacturers or other experts.

Do not connect an irrigation system (including greenhouse systems) used for pesticide application to a public water system unless the pesticide label-prescribed safety devices for the public water systems are in place.

A person knowledgeable of the chemigation system and responsible for its operation, or under the supervision of the responsible person, shall shut the system down and make necessary adjustments should the need arise.

Specific Instructions for Public Water Systems:

1. Public water system means a system for the provision to the public of piped water for human consumption if such system has at least 15 service connections or regularly serves an average of at least 25 individuals daily at least 60 days out of the year.
2. Chemigation systems connected to public water systems must contain a functional, reduced-pressure zone, backflow preventer (RPZ) or the funtional equivalent in the water supply line upstream from the point of pesticide introduction. As an option to the RPZ, the water from the public water system should be discharged into a reservoir tank prior to pesticide introduction. There shall be a complete physical break (air gap) between the outlet end of the fill pipe and the top or overflow rim of the reservoir tank of at least twice the inside diameter of the fill pipe.
3. The pesticide injection pipeline must contain a functional, automatic, quick-closing check valve to prevent the flow of fluid back toward the injection pump.
4. The pesticide injection pipeline must contain a functional, normally closed, solenoid-operated valve located on the intake side of the injection pump and connected to the system interlock to prevent fluid from being withdrawn from the supply tank when the irrigation system is either automatically or manually shut down.
5. The system must contain functional interlocking controls to automatically shut off the pesticide injection pump when the water pump motor stops, or in cases where there is no water pump, when the water pressure decreases to the point where pesticide distribution is adversely affected.
6. Systems must use a metering pump, such as a positive displacement injection pump (e.g., diaphragm pump) effectively designed and constructed of materials that are compatible with pesticides and capable of being fitted with a system interlock.
7. Do not apply when wind speed favors drift beyond the area intended for treatment.

Specific Instructions for Sprinkler Irrigation Systems:

1. The system must contain a functional check valve, vacuum relief valve, and low pressure drain appropriately located on the irrigation pipeline to prevent water source contamination from backflow.
2. The pesticide injection pipeline must contain a functional, automatic, quick-closing check valve to prevent the flow of fluid back toward the injection pump.
3. The pesticide injection pipeline must also contain a functional, normally closed, solenoid-operated valve located on the intake side of the injection pump and connected to the system interlock to prevent fluid from being withdrawn from the supply tank when the irrigation system is either automatically or manually shut down.
4. The system must contain functional interlocking controls to automatically shut off the pesticide injection pump motor when the water pump motor stops.
5. The irrigation line or water pump must include a functional pressure switch which will stop the water pump motor when the water pressure decreases to the point where pesticide distribution is adversely affected.
6. Systems must use a metering pump, such as a positive displacement injection pump (e.g., diaphragm pump) effectively designed and constructed of materials that are compatible with pesticides and capable of being fitted with a system interlock.
7. Do not apply when wind speed favors drift beyond the area intended for treatment.
8. Good agitation is required in the injection tank.
9. In moving systems, apply specified dosage of "Benlate" as a continuous injection. In nonmoving systems inject "Benlate" for 15 to 30 minutes at end of cycle. Use the least amount of water possible consistent with uniform coverage.
10. Mix the amount of "Benlate" needed for acreage to be treated into the quantity of water determined during prior calibration. For moving systems inject into the system continuously for one complete revolution of the field. For nonmoving systems inject into system for the time established during calibration.
11. Stop injection equipment after treatment is completed and continue to operate irrigation equipment until all "Benlate" is flushed from system.

ALMONDS: Brown Rot Blossom Blight—Apply 1 to 1½ lbs per acre at pink bud. Under severe disease conditions and on highly susceptible varieties, make a second application during half- to full-bloom.

APPLES: For applications through cover sprays, use "Benlate" as a tank mixture as detailed below. Apply 200 to 500 gals of spray per acre with hydraulic ground equipment or equivalent amount of products per acre with concentrate sprayers. Do not graze livesock in treated orchards.

"Benlate" + Manzate® 200 Fungicide: Scab, Powdery Mildew, Sooty Blotch, Flyspeck, Cedar Apple Rust, Quince Rust, Bitter Rot, Black Rot, Brown Rot—Use 2 to 3 ozs "Benlate" plus 12 ozs "Manzate" 200 per 100 gals of water; 1 qt spray oil may be added per 100 gals. Apply at ½" green tip and repeat at 7- to 14-day intervals (or as needed) through the cover sprays.

Use the 3 oz rate of "Benlate" and add spray oil to the spray mixture for varieties more susceptible to powdery mildew, and for scab if an application is missed during an infection period (apply as soon as possible after infection period in order to deactivate scab and to prevent further infection). Do not apply within 30 days of harvest.

"Benlate" + Captan: Scab, Powdery Mildew, Sooty Blotch, Flyspeck, Bitter Rot, Black Rot—Use 2 to 3 ozs "Benlate" plus 12 to 16 ozs Captan 50 WP Fungicide (or 7½ to 10 ozs Captan 80WP) per 100 gals of water. Apply at ½" green tip and repeat at 7- to 14-day intervals (or as needed) through the cover sprays. Use the 3 oz rate of "Benlate" for varieties more susceptible to powdery mildew. If an application is missed during an infection period, apply the higher rates as soon as possible after the infection period in order to deactivate scab and to prevent further infection. **Note:** Spray injury may result if Captan is used with, immediately before, or closely following an oil spray.

Postharvest Fruit Rots (Botrytis spp., Penicillium spp., Gloeosporium spp.)—Make a single application of 6 ozs "Benlate" per 100 gals anytime from 3 weeks before harvest up to day of harvest. For additional protection of fruit to be held in storage, thoroughly wet harvested fruit by dipping or spraying at 8 ozs per 100 gals.

AVOCADOS (Florida): Scab, Cercospora Spot, Anthracnose—Apply 1 to 2 lbs per acre; begin when buds swell and repeat at 3- to 4-week intervals. Do not apply within 30 days of harvest.

BEANS: White Mold (Sclerotinia), Gray Mold (Botrytis)—Use on beans grown as fresh vegetables, for processing, or for the dry bean market. Apply 1½ to 2 lbs per acre at 25% to 50% bloom; repeat at peak bloom. For narrow-row (20-24") irrigated dry beans in Montana, Nebraska, Colorado and Wyoming, apply at initial bloom and repeat 7 to 10 days later; only partial control of white mold may result. Do not apply within 14 days of harvest (28 days for lima beans); do not use where crop is grown only for forage purposes.

BLUEBERRIES: Apply 1 lb per acre. Do not make more than 4 applications before harvest; do not apply within 21 days of the harvest. **Mummy Berry, Botrytis Blossom Blight**—Apply at green tip and repeat at 7- to 10-day intervals through petal fall. **Anthracnose Leafspot**—Apply when disease first appears and make one additional application 14 days later. After harvest, make up to 4 applications to the bushes at 14-day intervals as needed.

CABBAGE (Seed Crop, Pacific Northwest); White Blight (Sclerotinia Stalk Rot)—Apply 2 lbs per acre by aircraft in 5 to 10 gals of water; add a spreader-sticker to aid in wetting plants. Make first application at first petal fall; make two additional applications at 14-day intervals if conditions favor development of disease. **Note:** Do not graze treated areas; do not use seed or plant parts for food or feed purposes.

CANEBERRIES—RASPBERRIES, BLACKBERRIES, BOYSENBERRIES, LOGANBERRIES, DEWBERRIES: Botrytis, Powdery Mildew, Penicillium Rots—Apply ¾ lb per acre at early bloom (5 to 10%) and at full bloom; make up to 3 additional applications at 14-day intervals as needed. Do not apply within 3 days of harvest.

CELERY: Early Blight (Cercospora), Late Blight (Septoria)—Apply ¼ to ½ lb per acre; begin when disease first appears and repeat at 7- 10-day intervals. Do not apply within 7 days of harvest.

CITRUS: Scab—Apply 1½ to 3 lbs per acre. Under conditions of severe disease pressure, apply at pinhead stage (just prior to first flush) and repeat at ⅔ petal fall; otherwise, make a single application at ⅔ petal fall.

Greasy Spot—Make a single application of 1½ to 3 lbs per acre during the period mid-June to mid-July.

Fruit Decay (Green Mold, Blue Mold, Stem-end Rot)—Preharvest Spray—Make a single application of 1 to 2 lbs per acre anytime from 3 weeks prior to harvest up to day of harvest. **Postharvest**—Apply as a dip, flood, or spray using 1 to 2 lbs per 100 gals; do not immerse fruit for more than 5 min. When citrus wax is used, "Benlate" may be incorporated into the wax spray. Use the higher rate on more susceptible fruits and when excessive inoculum levels are present. For control of sporulation (*Penicillium spp.*), apply as a spray in citrus wax using 4 lbs "Benlate" per 100 gals.

Note: Do not graze livestock in treated groves.

CUCURBITS—CUCUMBERS, MELONS, PUMPKINS, SUMMER AND WINTER SQUASH: Target Spot (Cucumbers), Gummy Stem Blight, Powdery Mildew, Anthracnose—Apply ¼ to ½ lb per acre; for aerial application, use ½ lb per acre. Begin applications when plants begin to run or when disease first appears, and repeat at 7- to 14-day intervals as needed. For target spot, use 7-day intervals as needed.

GRAPES: Botrytis Bunch Rot—Apply 1 to 1½ lbs per acre at first bloom (no later than 5% bloom) and repeat 14 days later if severe disease conditions persist. Make an additional application 3 to 4 weeks before harvest or when sugar begins to build; repeat 14 days later if conditions favorable for disease persist. "Benlate" does not control bunch rots caused by other organisms such as *Rhizopus spp., Alternaria spp.,* and *Diplodia spp.;* these rots occur most frequently in high temperature areas such as the San Joaquin and Sacramento Valleys of California. **Powdery Moldew, Black Rot, Bitter Rot—East of Rockies**—Apply ¾ to 1½ lbs per acre when foliage first develops and repeat at 14- to 21-day intervals, or as needed, until berries are full size.

Note: Do not apply within 7 days of harvest.

MACADAMIA NUTS (Hawaii): Botrytis Blossom Blight—Apply 1¾ lbs per acre; a surfactant may be added to the spray to improve wetting of foliage. Begin applications 1 to 2 weeks prior to bloom, and repeat at 7- to 14-day intervals through the bloom period.

MANGOES: Anthracnose—Apply 1 to 2 lbs per acre. Begin applications at first appearance of panicles (approx. 2" long), and repeat at weekly intervals until all fruits are set. Continue at 3- to 4-week intervals. Do not apply within 14 days of harvest.

MUSHROOMS: Verticillium Spot (Dry Bubble)—Use 1 lb per 100 gals and apply to bed surface at the rate of 12½ gals per 1000 sq ft. Apply immediately after casing and repeat at pinning; alternatively, if disease has occurred, apply to beds after picking and repeat 10 days later. Do not apply within 2 days of harvest.

PEANUTS: "Benlate" + "Manzate" 200 Fungicide: Cercospora Leafspot, Rust, Ascochyta Web Blotch—Apply ¼ lb "Benlate" plus 1½ lbs "Manzate" 200 per acre; spray oil may be added at the rate of 1 pt to 1 qt per acre. Do not use "Benlate" alone. Begin applications 35 to 40 days after planting or when disease first appears. Repeat at following intervals: for Cercospora leafspot, 10 to 14 days; for rust, 7 to 10 days; for ascochyta web blotch, 7 to 14 days. Do not apply within 14 days of harvest; do not graze or feed treated vines, hay, or hulls to livestock.

PEARS: Scab, Powdery Mildew, Sooty Blotch, Flyspeck—Use 4 to 6 ozs per 100 gals of water; apply 200 to 500 gals of spray per acre with hydraulic ground equipment or equivalent amount of "Benlate" per acre with concentrate sprayers. Apply at ½" green tip and repeat at 7- to 14-day intervals (or as needed) through the cover sprays. If an application is missed during an infection period, use 6 ozs per 100 gals and apply as soon as possible after the infection period in order to deactivate scab and to prevent further infection. Do not graze livestock in treated orchards.

Postharvest Fruit Rots (Botrytis spp., Penicillium spp., Gloeosporium spp.)—Make a single application of 6 ozs per 100 gals anytime from 3 weeks before harvest up to day of harvest. For additional protection of fruit to be held in storage, thoroughly wet harvested fruit by dipping or spraying at 8 ozs per 100 gals.

Overwintering Scab—Apply 8 ozs per 100 gals after harvest but before leaf drop. Thorough wetting of foliage is necessary.

PECANS: Pecan Scab, Brown Leafspot, Downy Spot, Powdery Mildew, Liverspot, Zonate Leafspot, Fungal Leaf Scorch—Apply ½ to 1 lb per acre; use the higher rate on trees over 30' tall. For aerial application (Ark., La., Miss., Okla., Tex. only), use 1 lb per acre. Spray oil may be added at the rate of 1 to 2 gals per acre. Apply at prepollination when young leaves are unfolding, when small nuts are forming, and thereafter at 3- to 4-week intervals. Do not apply after shucks split.

PINEAPPLE: Thielaviopsis Rot (Fresh Fruit)—Use 2 to 4 lbs per 100 gals of water. Immediately after harvest, immerse or spray fruit to give thorough wetting and allow to drain; do not immerse for more than 5 min.

Pineapple Butt Rot (Thielaviopsis paradoxa)—Use 1¼ lbs per 100 gals of water as a preplant dip treatment. Immerse seedpieces to give thorough wetting; remove and allow to drain.

RICE (Except Calif.): Rice Blast, Stem Rot—Apply 1 to 2 lbs per acre at booting and repeat at heading. Do not apply within 21 days of harvest. Do not apply to stubble rice. Do not apply to fields where crayfish or catfish farming is practiced, nor drain water from treated areas into areas where such farming is practiced. Water drained from treated areas must not be used to irrigate other crops.

SOYBEANS: Diaporthe Pod-and-Stem Blight, Anthracnose, Septoria Brown Spot, Cercospora Frogeye Leafspot, Purple Seed Stain—Apply ½ to 1 lb per acre. For **determinate** varieties (generally grown in the South), apply at early pod set when majority of pods are ⅛ to ½" in length; for **indeterminate** varieties (generally grown in the North), apply when pods near the top of the plant are ½ to 1" in length. Make one additional application 14 to 21 days later. Do not apply within 35 days of harvest; do not graze or feed treated soybean vines or hay to livestock.

STONEFRUITS—APRICOTS, CHERRIES, NECTARINES, PEACHES, PLUMS, PRUNES: Treatment is most effective if applied just before rainfall; for aerial application, fly over every row or center.

EAST OF ROCKY MOUNTAINS—Use ¾ to 1½ lbs per acre on trees up to 12' tall; over 12', use 1½ to 2 lbs per acre.

Brown Rot Blossom Blight—Apply at early bloom stages (apricots—red bud; peaches, nectarines—pink bud; cherries—early popcorn, plums and prunes—green tip); for this application only, "Benlate" may be used in combination with spray oil. Make a second application at 75% to 100% bloom. If blossoming is prolonged or conditions favorable for disease persist, apply at petal fall.

Fruit Brown Rot—After blossom blight sprays, make two preharvest applications beginning 3 weeks before harvest up to day of harvest.

Peach Scab, Powdery Mildew—Use same schedule as for Brown Rot Blossom Blight plus applications at shuck split, shuck fall and 14 days later.

Cherry Leaf Spot—Use same schedule as for Brown Rot Blossom Blight and continue at 10- to 14-day intervals through harvest. Make an additional application 2 to 3 weeks after harvest.

WEST OF ROCKY MOUNTAINS—Use 1½ to 2 lbs per acre.

Brown Rot Blossom Blight—Apply at early bloom stages (apricots—red bud; peaches, nectarines—pink bud; cherries—early popcorn; plums and prunes—green tip); for this application only, "Benlate" may be used in combination with spray oil. If blossoming is prolonged or conditions favorable for disease persist, make a second application 14 days later.

Fruit Brown Rot—After blossom blight sprays, make a preharvest application (before rain) any time from 3 weeks before harvest to day of harvest. Make a second application if conditions favorable for disease persist or harvest is prolonged. Preharvest applications are most effective when applied with ground equipment, using sufficient volume to provide thorough and uniform coverage of fruit.

Powdery Mildew—Use same schedule as for Brown Rot Blossom Blight plus applications at shuck split, shuck fall, and 14 days later.

Cherry Leaf Spot—Use same schedule as for Brown Rot Blossom Blight and continue at 10- to 14-day intervals through harvest. Make an additional application 2 to 3 weeks after harvest.

POSTHARVEST FRUIT ROTS (U.S.)—Dip or spray fruit thoroughly as soon as possible after harvest; use ½ lb per 100 gals. When wax is used, "Benlate" may be incorporated into the wax spray.

NOTE: "Benlate" does not control peach leaf curl, shot hole (*Coryneum blight*) or bacterial blast, nor fruit rots caused by *Rhizopus spp.* and *Alternaria spp.* Do not graze livestock in treated orchards.

STRAWBERRIES: Gray Mold (Botrytis), Powdery Mildew, Leaf Scorch, Leaf Blight, Mycosphaerella Leaf Spot—Apply 1 lb per acre at 10% bloom and at full bloom; continue at 10- to 14-day intervals, using ½ lb per acre. **Anthracnose**—Apply 1 lb per acre when plants are established (plant bed or field) and repeat at 7-day intervals.

Transplants: Botrytis Crown Rot, Mycosphaerella Leaf Spot—Use ½ lb per 100 gals of water. Immerse plants to give thorough wetting; remove and allow to drain.

SUGARBEETS: Cercospora Leafspot—Apply ⅜ to ½ lb per acre. Begin application when disease first appears and repeat at 14- to 21-day intervals as needed. Do not apply within 21 days of harvest.

SUGARCANE (HAWAII): Pineapple Disease (Ceratocystis paradoxa)—Apply to cut seedpieces either as a cold dip or hot dip.

Cold Dip—Use ½ lb per 100 gals of water (1:1600). Immerse seedpieces to give thorough wetting; remove and allow to drain.

Hot Dip—Use ¼ lb per 100 gals of water (1:3200). Maintain temperature of the dip at 50°C. Soak seedpieces for 20 to 30 minutes; remove and allow to drain.

Note: Do not use treated seedpieces for food or feed purposes.

TOMATOES—Field and Greenhouses: Gray Mold (Botrytis), Leaf Mold (Cladosporium), White Mold (Sclerotinia), Cercospora Leafspot, Phoma Leafspot—For field tomatoes, apply ½ to 1 lb per acre; for greenhouse, use ½ to 1 lb per 100 gals of water. Begin applications when disease first appears and repeat at 7- to 14-day intervals as needed.

ROSES, FLOWERS, ORNAMENTALS, SHADE TREES—Field and Greenhouse: Foliar Spray—Begin applications when disease first appears and repeat at 10- to 14-day intervals throughout the growing season; shorten interval during humid, rainy weather. Use at the following rates: **½ lb per 100 gals** (1 tablespoonful per 2 gals)—for Powdery Mildew, Botrytis Gray Mold. **1 lb per 100 gals**—For Anthracnose (for shade trees and woody ornamentals, begin at bud break and make 2 or 3 additional applications at 10- to 14-day intervals); Black Spot of roses; Cercospora, Entomosporium, Ramularia, and Septoria Leafspots; Ascochyta and Phomopsis blights; Didymellina Leafspot of iris; Corynespora leafspot of Ligustrum; Ovulinia Blight of azalea and rhododendron (begin as flowers open); Scab of pyracantha and flowering crab. Addition of a surfactant to the spray mixture improves distribution of the spray on hard-to-wet plants such as roses. For aerial application, use ½ to 1 lb per acre.

Drench Treatment—Botrytis, Fusarium, Rhizoctonia and Sclerotinia stem, crown and root rots on herbaceous annuals, perennials and bedding plants: Cylindrocladium and Thielaviopsis rots on woody ornamentals such as azaleas, rhododendrons, conifers, and poinsettias—Use 1 lb per 100 gals; apply as a drench or heavy spray (1 to 2 pts per sq ft) after transplanting into propagation beds or containers. Repeat at 2- to 4-week intervals during periods favorable for disease. "Benlate" does not control *Pythium spp.* or *Phytophthora spp.*

Preplant Dip Treatment—For disease listed under Drench Treatment, use 1 lb per 100 gals of water; immerse plants or cuttings for 10 to 15 min.; remove and allow to drain.

BULBS (Easter Lily, Tulip, Gladiolus, Daffodil, Iris): Fusarium and Penicillium Rots—Use 1⅔ lbs per 100 gals of water (2 tablespoonfuls per gal). Soak cleaned bulbs for 15 to 30 minutes in warm dip (80 to 85°F.), preferably within 48 hours after digging. Dry bulbs after treatment. If bulbs are for forcing, treat after bulbs have been heat-cured.

NOTICE TO BUYER: Purchase of this material does not confer any rights under patents of countries outside of the United States.

[1]Registered trademark of Merck & Co.
[2]Registered trademark of Nippon Soda Co., Japan

STORAGE AND DISPOSAL

STORAGE: Never allow "Benlate" to become wet during storage. This may lead to certain chemical changes which will reduce the effectiveness of "Benlate" as a fungicide. Keep container tightly closed when not in use. Store product in original container only, away from other pesticides, fertilizer, food or feed.

DISPOSAL: Do not contaminate water, food or feed by storage or disposal. Wastes resulting from the use of this product may be disposed of on site or at an approved waste disposal facility. Completely empty bag into application equipment. Then dispose of empty bag in a sanitary landfill or by incineration, or, if allowed by state and local authorities, by burning. If burned, stay out of smoke.

NOTICE

Do not enter treated areas without protective clothing until sprays have dried. Because certain states may require more restrictive reentry intervals for various crops treated with this product, consult your State Department of Agriculture for further information.

Written or oral warnings must be given to workers who are expected to be in a treated area or in an area about to be treated with this product. When oral warnings are given, warnings shall be given in a language customarily understood by workers. Oral warnings must be given if there is reason to believe that written warnings cannot be understood by workers. Warnings must include the following information: "CAUTION. Area treated with 'Benlate' on (date of application). Do not enter without appropriate protective clothing until sprays have dried. In case of contact, flush skin or eyes with plenty of water; for eyes, get medical attention."

NOTICE OF WARRANTY

Du Pont warrants that this product conforms to the chemical description on the label thereof and is reasonably fit for purposes stated on such label only when used in accordance with the directions under normal use conditions. It is impossible to eliminate all risks inherently associated with the use of this product. Crop injury, ineffectiveness or other unintended consequences may result because of such factors as weather conditions, presence of other materials, or the manner of use or application, all of which are beyond the control of Du Pont. In no case shall Du Pont be liable for consequential, special or indirect damages resulting from the use or handling of this product. All such risks shall be assumed by the Buyer. DU PONT MAKES NO WARRANTIES OF MERCHANTABILITY OR FITNESS FOR A PARTICULAR PURPOSE NOR ANY OTHER EXPRESS OR IMPLIED WARRANTY EXCEPT AS STATED ABOVE.

AG-3729 8058 4/28/88

This is a specimen label. DuPont is not responsible for the accuracy of the information contained herein. As labels are subject to revision, always carefully read and follow the label on the product container.

BENLATE® 50 DF

FUNGICIDE
DISPERSIBLE GRANULES

ACTIVE INGREDIENT:	BY WEIGHT
Benomyl [Methyl 1-(butylcarbamoyl)-2-benzimidazolecarbamate]	50%
INERT INGREDIENTS:	50%
TOTAL	100%

EPA Reg. No. 352-447
U.S. Pat. 3,631,176

KEEP OUT OF REACH OF CHILDREN

PRECAUTIONARY STATEMENTS

HAZARDS TO HUMANS AND DOMESTIC ANIMALS

CAUTION!

MAY IRRITATE EYES, NOSE, THROAT AND SKIN.

Avoid breathing dust or spray mist. Avoid contact with skin, eyes and clothing. Wash thoroughly after using.

Do not apply this product in such a manner as to directly or through drift expose workers, or other persons. The area being treated must be vacated by unprotected persons.

Protective clothing must be worn during periods of exposure. See Directions for Use.

This product may cause a temporary allergic skin reaction in a few susceptible persons. This condition should be treated as an allergic dermatitis. There is no evidence of aftereffects or permanent injury.

First Aid: In case of contact, flush skin or eyes with plenty of water; for eyes, get medical attention.

Aerial Mixer-Loaders: Wear a dust mask or respirator during handling and mixing as this product may be harmful if inhaled. See Directions for Use.

For medical emergencies involving this product, call toll free 1-800-441-3637.

ENVIRONMENTAL HAZARDS

This pesticide is toxic to fish. Do not apply directly to water. Do not apply where runoff is likely to occur. Drift and runoff from treated areas may be hazardous to fish in adjacent areas. Do not contaminate water when disposing of equipment washwaters. Do not apply when weather conditions favor drift from areas treated.

For registered aquatic uses: Aquatic organisms may be killed at recommended application rates.

PHYSICAL OR CHEMICAL HAZARDS

Keep away from fire or sparks.

NOTICE

Do not enter treated areas without protective clothing until sprays have dried. Because certain states may require more restrictive reentry intervals for various crops treated with this product, consult your State Department of Agriculture for further information.

Written or oral warnings must be given to workers who are expected to be in a treated area or in an area about to be treated with this product. When oral warnings are given, warnings shall be given in a language customarily understood by workers. Oral warnings must be given if there is reason to believe that written warnings cannot be understood by workers. Warnings must include the following information: "CAUTION. Area treated with 'BENLATE' 50 DF on (date of application). Do not enter without appropriate protective clothing until sprays have dried. In case of contact, flush skin or eyes with plenty of water; for eyes, get medical attention."

DIRECTIONS FOR USE

It is a violation of federal law to use this product in a manner inconsistent with its labeling.

Du Pont "BENLATE" 50 DF should be used only in accordance with recommendations on this label, or in separate published Du Pont recommendations available through local dealers.

Du Pont will not be responsible for losses or damages resulting from use of this product in any manner not specifically recommended by Du Pont. User assumes all risk associated with such nonrecommended use.

"BENLATE" 50 DF is a systemic fungicide recommended for the control of many important plant diseases. If treatment is not effective following use of "BENLATE" 50 DF as recommended, a resistant strain of the fungus may be present. If treatment is ineffective due to the presence of a benomyl resistant strain, then neither "BENLATE" 50 DF, nor any other benzimidazole or thiophanate type fungicide will effectively control that disease; consideration should be given to prompt use of other types of suitable fungicides.

The repeated exclusive use of "BENLATE" 50 DF may lead to buildup of resistant strains of fungi and loss of disease control. A spray program using other fungicides may delay resistant strain buildup. Consult your state extension specialist or official state recommendations for guidance on your particular crop and disease control situation.

Note: Do not tank mix or alternate "BENLATE" 50 DF with benzimidazole or thiophanate products such as "Mertect"[1] or "Topsin"[2].

Do not enter or allow entry into treated areas for 24 hours after application unless wearing the protective clothing specified for this product or unless entering to perform post application activities on strawberries or bulbs.

WORK SAFETY RULES

Keep all unprotected persons, children, livestock and pets away from treated area or where there is danger of drift.

During mixing and loading of the concentrated product wear the following protective clothing and equipment:

> Long sleeve shirt and long legged pants; chemical resistant gloves, chemical resistant apron; chemical resistant shoes, shoe coverings, or boots. Wear a dust mask or a respirator approved by the National Institute of Occupational Safety and Health (30 CFR Part 11) when mixing or loading for aerial application.

Wear the following protective clothing during application, equipment repair, equipment cleaning, and during reentry to treated areas before expiration of the 24 hour reentry interval:

> Long sleeve shirt and long legged pants; chemical resistant gloves; chemical resistant shoes, shoe coverings, or boots. During aerial appli-

cation from enclosed cockpits wear long sleeve shirt and long legged pants. A clean pair of chemical resistant gloves must be carried in the cockpit and worn during exiting from the aircraft.

Important! Before removing gloves, wash them with soap and water. Always wash hands, face, and arms with soap and water before smoking, eating, drinking, or toileting.

After work, take off all clothes and shoes. Shower using soap and water. Wear only clean clothes. Do not use contaminated clothing. Wash protective clothing and protective equipment with soap/detergent and water after each use. Personal and protective clothing worn during use must be laundered separately from household articles. Clothing or protective equipment heavily contaminated or drenched with benomyl must be destroyed according to state or local regulations. Heavily contaminated or drenched clothing cannot be adequately decontaminated.

Flaggers must be in totally enclosed vehicles.

Apply as a spray with ground equipment (except as otherwise directed), using sufficient water to obtain thorough coverage of plants. Under severe disease conditions use the higher rate and shorter interval specified for each crop; also, for tree crops, use the higher rate for large mature trees. For aerial application (listed crops only) use the following gals per acre: Rice and Soybeans, 3 to 10; Cabbage (seed crop), Celery, Cucurbits, Peanuts and Sugar Beets, 5 to 10; Almonds, Avocados, Beans, Pecans, Stone Fruits and Strawberries, 10 to 20; Grapes, 15 to 20; Roses, Flowers, Ornamentals and Shade Trees, 20 gals per acre minimum.

For use in small gardens and orchards (less than 1 acre), application rates may be converted to lbs per 100 gals by dividing the lbs per acre rate in half, and applying the resulting spray mixture at the rate of 4.5 gals per 1000 sq ft (1 lb "BENLATE" 50 DF per 100 gals equals 1 tablespoonful per gal).

Add required amount of "BENLATE" 50 DF to necessary volume of water in spray tank agitated by hydraulic or mechanical means; continuous agitation is required to keep the material in suspension. Do not tank mix "BENLATE" 50 DF with lime or alkaline pesticides such as Bordeaux mixture or lime sulfur.

Where use of spray oil is recommended (apples, peanuts, pecans, stone fruits), use a nonphytotoxic superior-type (60 to 70 second viscosity) spray oil; add as last ingredient to spray tank. Before applying other pesticides in conjunction with spray oil or immediately before or after oil application, consult product labels. Observe all cautions and limitations on labeling of all products used in mixtures.

Grazing or feeding livestock is prohibited on forage and hay of benomyl treated wheat, and on forage and hay of barley, oats, rye and wheat following seed treatment.

CHEMIGATION: Apply "BENLATE" 50 DF only through sprinkler including center pivot, lateral move, end tow, side (wheel) roll, traveler, big gun, solid set or hand move irrigation systems only on crops specifically designated in the "Directions For Use" for that particular crop. Do not apply "BENLATE" 50 DF through any other type of irrigation system or on any other crops.

Crop injury, lack of effectiveness, or illegal pesticide residues in the crop can result from nonuniform distribution of treated water.

If you have questions about calibration, you should contact State Extension Service specialists, equipment manufacturers or other experts.

Do not connect an irrigation system (including greenhouse systems) used for pesticide application to a public water system unless the pesticide label-prescribed safety devices for the public water systems are in place.

A person knowledgeable of the chemigation system and responsible for its operation, or under the supervision of the responsible person, shall shut the system down and make necessary adjustments should the need arise.

Specific Instructions for Public Water Systems:

1. Public water system means a system for the provision to the public of piped water for human consumption if such system has at least 15 service connections or regularly serves an average of at least 25 individuals daily at least 60 days out of the year.
2. Chemigation systems connected to public water systems must contain a functional, reduced-pressure zone, backflow preventer (RPZ) or the functional equivalent in the water supply line upstream from the point of pesticide introduction. As an option to the RPZ, the water from the public water system should be discharged into a reservoir tank prior to pesticide introduction. There shall be a complete physical break (air gap) between the outlet end of the fill pipe and the top or overflow rim of the reservoir tank of at least twice the inside diameter of the fill pipe.
3. The pesticide injection pipeline must contain a functional, automatic, quick-closing check valve to prevent the flow of fluid back toward the injection pump.
4. The pesticide injection pipeline must contain a functional, normally closed, solenoid-operated valve located on the intake side of the injection pump and connected to the system interlock to prevent fluid from being withdrawn from the supply tank when the irrigation system is either automatically or manually shut down.
5. The system must contain functional interlocking controls to automatically shut off the pesticide injection pump when the water pump motor stops, or in cases where there is no water pump, when the water pressure decreases to the point where pesticide distribution is adversely affected.
6. Systems must use a metering pump, such as a positive displacement injection pump (e.g., diaphragm pump) effectively designed and constructed of materials that are compatible with pesticides and capable of being fitted with a system interlock.
7. Do not apply when wind speed favors drift beyond the area intended for treatment.

Specific Instructions for Sprinkler Irrigation Systems:

1. The system must contain a functional check valve, vacuum relief valve, and low pressure drain appropriately located on the irrigation pipeline to prevent water source contamination from backflow.
2. The pesticide injection pipeline must contain a functional, automatic, quick-closing check valve to prevent the flow of fluid back toward the injection pump.
3. The pesticide injection pipeline must also contain a functional, normally closed, solenoid-operated valve located on the intake side of the injection pump and connected to the system interlock to prevent fluid from being withdrawn from the supply tank when the irrigation system is either automatically or manually shut down.
4. The system must contain functional interlocking controls to automatically shut off the pesticide injection pump when the water pump motor stops.
5. The irrigation line or water pump must include a functional pressure switch which will stop the water pump motor when the water pressure decreases to the point where pesticide distribution is adversely affected.
6. Systems must use a metering pump, such as a positive displacement injection pump (e.g., diaphragm pump) effectively designed and constructed of materials that are compatible with pesticides and capable of being fitted with a system interlock.
7. Do not apply when wind speed favors drift beyond the area intended for treatment.
8. Good agitation is required in the injection tank.
9. In moving systems, apply specified dosage of "BENLATE" 50 DF as a continuous injection. In nonmoving systems inject "BENLATE" 50 DF for 15 to 30 minutes at end of cycle. Use the least amount of water possible consistent with uniform coverage.
10. Mix the amount of "BENLATE" 50 DF needed for acreage to be treated into the quantity of water determined during prior calibration. For moving systems inject into the system continuously for one complete revolution of the field. For nonmoving systems inject into system for the time established during calibration.
11. Stop injection equipment after treatment is completed and continue to operate irrigation equipment until all "BENLATE" 50 DF is flushed from system.

ALMONDS: Brown Rot Blossom Blight—Apply 1 to 1½ lbs. per acre at pink bud. Under severe disease conditions and on highly susceptible varieties, make a second application during half- to full-bloom.

APPLES: For applications through cover sprays, use "BENLATE" 50 DF as a tank mixture as detailed below. Apply 200 to 500 gals of spray per acre with hydraulic ground equipment or equivalent amount of products per acre with concentrate sprayers. Do not graze livestock in treated orchards.

"BENLATE" 50 DF + "Manzate" 200 Fungicide: Scab, Powdery Mildew, Sooty Blotch, Flyspeck, Cedar Apple Rust, Quince Rust, Bitter Rot, Black Rot, Brown Rot—Use 2 to 3 ozs "BENLATE" 50 DF plus 12 ozs "Manzate" 200 per 100 gals of water; 1 qt spray oil may be added per 100 gals. Apply at ½" green tip and repeat at 7- to 14-day intervals (or as needed) through the cover sprays.

Use the 3 oz rate of "BENLATE" 50 DF and add spray oil to the spray mixture for varieties more susceptible to powdery mildew, and for scab if an application is missed during an infection period (apply as soon as possible after infection period in order to deactivate scab and to prevent further infection). Do not apply within 30 days of harvest.

"BENLATE" 50 DF + Captan: Scab, Powdery Mildew, Sooty Blotch, Flyspeck, Bitter Rot, Black Rot—Use 2 to 3 ozs "BENLATE" 50 DF plus 12 to 16 ozs Captan 50WP Fungicide (or 7½ to 10 ozs Captan 80WP) per 100 gals of water. Apply at ½" green tip and repeat at 7- to 14-day intervals (or as needed) through the cover sprays. Use the 3 oz rate of "BENLATE" 50 DF for varieties more susceptible to powdery mildew. If an application is missed during an infection period, apply the higher rates as soon as possible after the infection period in order to deactivate scab and to prevent further infection. **Note:** Spray injury may result if Captan is used with, immediately before, or closely following an oil spray.

Postharvest Fruit Rots (Botrytis spp., Penicillium spp., Gloeosporium spp.)—Make a single application of 6 ozs "BENLATE" 50 DF per 100 gals anytime from 3 weeks before harvest up to day of harvest. For additional protection of fruit to be held in storage, thoroughly wet harvested fruit by dipping or spraying at 8 ozs per 100 gals.

AVOCADOS (Florida): Scab, Cercospora Spot, Anthracnose—Apply 1 to 2 lbs per acre; begin when buds swell and repeat at 3- to 4-week intervals. Do not apply within 30 days of harvest.

BEANS: White Mold (Sclerotinia), Gray Mold (Botrytis)—Use on beans grown as fresh vegetables, for processing, or for the dry bean market. Apply 1½ to 2 lbs per acre at 25% to 50% bloom; repeat at peak bloom. For narrow-row (20-24") irrigated dry beans in Montana, Nebraska, Colorado and Wyoming, apply at initial bloom and repeat 7 to 10 days later; only partial control of white mold may result. Do not apply within 14 days of harvest (28 days for lima beans); do not use where crop is grown only for forage purposes.

BLUEBERRIES: Apply 1 lb per acre. Do not make more than 4 applications before harvest; do not apply within 21 days of the harvest. **Mummy Berry, Botrytis Blossom Blight**—Apply at green tip and repeat at 7- to 10-day intervals through petal fall. **Anthracnose Leafspot**—Apply when disease first appears and make one additional application 14 days later. After harvest, make up to 4 applications to the bushes at 14-day intervals as needed.

CABBAGE (Seed Crop—Pacific Northwest): White Blight (Sclerotinia Stalk Rot)—Apply 2 lbs per acre by aircraft in 5 to 10 gals of water; add a spreader-sticker to aid in wetting plants. Make first application at first petal fall; make two additional applications at 14-day intervals if conditions favor development of disease. **Note:** Do not graze treated areas; do not use seed or plant parts for food or feed purposes.

CANEBERRIES—RASPBERRIES, BLACKBERRIES, BOYSENBERRIES, LOGANBERRIES, DEWBERRIES: Botrytis, Powdery Mildew, Penicillium Rots—Apply ¾ lb per acre at early bloom (5 to 10%) and at full bloom; make up to 3 additional applications at 14-day intervals as needed. Do not apply within 3 days of harvest.

CELERY: Early Blight (Cercospora), Late Blight (Septoria)—Apply ¼ to ½ lb per acre; begin when disease first appears and repeat at 7- to 10-day intervals. Do not apply within 7 days of harvest.

CITRUS: Scab—Apply 1½ to 3 lbs per acre. Under conditions of severe disease pressure, apply at pinhead stage (just prior to first flush) and repeat at ⅔ petal fall; otherwise, make a single application at ⅔ petal fall.

Greasy Spot—Make a single application of 1½ to 3 lbs per acre during the period mid-June to mid-July.

Fruit Decay (Green Mold, Blue Mold, Stem-end Rot)—Preharvest Spray—Make a single application of 1 to 2 lbs per acre anytime from 3 weeks prior to harvest up to day of harvest. **Postharvest**—Apply as a dip, flood, or spray using 1 to 2 lbs per 100 gals; do not immerse fruit for more than 5 min. When citrus wax is used, "BENLATE" 50 DF may be incorporated into the wax spray. Use the higher rate on more susceptible fruits and when excessive inoculum levels are present. For control of sporulation (*Penicillium spp.*), apply as a spray in citrus wax using 4 lbs "BENLATE" 50 DF per 100 gals.

Note: Do not graze livestock in treated groves.

CUCURBITS—CUCUMBERS, MELONS, PUMPKINS, SUMMER AND WINTER SQUASH: Target Spot (Cucumbers), Gummy Stem Blight, Powdery Mildew, Anthracnose—Apply ¼ to ½ lb per acre; for aerial application, use ½ lb per acre. Begin applications when plants begin to run or when disease first appears, and repeat at 7- to 14-day intervals as needed. For target spot, use 7-day intervals as needed.

GRAPES: Botrytis Bunch Rot—Apply 1 to 1½ lbs per acre at first bloom (no later than 5% bloom) and repeat 14 days later if severe disease conditions persist. Make an additional application 3 to 4 weeks before harvest or when sugar begins to build; repeat 14 days later if conditions favorable for disease persist. "BENLATE" 50 DF does not control bunch rots caused by other organisms such as *Rhizopus spp., Alternaria spp.,* and *Diplodia spp.;* these rots occur most frequently in high temperature areas such as the San Joaquin and Sacramento Valleys of California. **Powdery Mildew, Black Rot, Bitter Rot—East of Rockies**—Apply ¾ to 1½ lbs per acre when foliage first develops and repeat at 14- to 21-day intervals, or as needed, until berries are full size.

Note: Do not apply within 7 days of harvest.

MACADAMIA NUTS (Hawaii): Botrytis Blossom Blight—Apply 1¾ lbs per acre; a surfactant may be added to the spray to improve wetting of foliage. Begin applications 1 to 2 weeks prior to bloom, and repeat at 7- to 14-day intervals through the bloom period.

MANGOES: Anthracnose—Apply 1 to 2 lbs per acre. Begin applications at first appearance of panicles (approx. 2" long), and repeat at weekly intervals until all fruits are set. Continue at 3- to 4-week intervals. Do not apply within 14 days of harvest.

MUSHROOMS: Verticillium Spot (Dry Bubble)—Use 1 lb per 100 gals and apply to bed surface at the rate of 12½ gals per 1000 sq ft. Apply immediately after casing and repeat at pinning; alternatively, if disease has occurred, apply to beds after picking and repeat 10 days later. Do not apply within 2 days of harvest.

PEANUTS: "BENLATE" 50 DF + "Manzate" 200 Fungicide: Cercospora Leafspot, Rust, Ascochyta Web Blotch—Apply ¼ lb "BENLATE" 50 DF plus 1½ lbs "Manzate" 200 per acre; spray oil may be added at the rate of 1 pt to 1 qt per acre. Do not use "BENLATE" 50 DF alone. Begin applications 35 to 40 days after planting or when disease first appears. Repeat at following intervals: for Cercospora leafspot, 10 to 14 days; for rust, 7 to 10 days; for Ascochyta web blotch, 7 to 14 days. Do not apply within 14 days of harvest; do not graze or feed treated vines, hay, or hulls to livestock.

PEARS: Scab, Powdery Mildew, Sooty Blotch, Flyspeck—Use 4 to 6 ozs per 100 gals of water, apply 200 to 500 gals of spray per acre with hydraulic ground equipment or equivalent amount of "BENLATE" 50 DF per acre with concentrate sprayers. Apply at ½" green tip and repeat at 7- to 14-day intervals (or as needed) through the cover sprays. If an application is missed during an infection period, use 6 ozs per 100 gals and apply as soon as possible after the infection period in order to deactivate scab and to prevent further infection. Do not graze livestock in treated orchards.

Postharvest Fruit Rots (Botrytis spp., Penicillium spp., Gloeosporium spp.)—Make a single application of 6 ozs per 100 gals anytime from 3 weeks before harvest up to day of harvest. For additional protection of fruit to be held in storage, thoroughly wet harvested fruit by dipping or spraying at 8 ozs per 100 gals.

Overwintering Scab—Apply 8 ozs per 100 gals after harvest but before leaf drop. Thorough wetting of foliage is necessary.

PECANS: Pecan Scab, Brown Leafspot, Downy Spot, Powdery Mildew, Liverspot, Zonate Leafspot, Fungal Leaf Scorch—Apply ½ to 1 lb per acre; use the higher rate on trees over 30' tall. For aerial application (Ark., La., Miss., Okla., Tex. only), use 1 lb per acre. Spray oil may be added at the rate of 1 to 2 gals per acre. Apply at prepollination when young leaves are unfolding, when small nuts are forming, and thereafter at 3- to 4-week intervals. Do not apply after shucks split.

PINEAPPLE: Thielaviopsis Rot (Fresh Fruit)—Use 2 to 4 lbs per 100 gals of water. Immediately after harvest, immerse or spray fruit to give thorough wetting and allow to drain; do not immerse for more than 5 min.

Pineapple Butt Rot (Thielaviopsis paradoxa)—Use 1¼ lbs per 100 gals of water as a preplant dip treatment. Immerse seedpieces to give thorough wetting; remove and allow to drain.

RICE (Except Calif.): Rice Blast, Stem Rot—Apply 1 to 2 lbs per acre at booting and repeat at heading. Do not apply within 21 days of harvest. Do not apply to stubble rice. Do not apply to fields where crayfish or catfish farming is practiced, nor drain water from treated areas into areas where such farming is practiced. Water drained from treated areas must not be used to irrigate other crops.

SOYBEANS: Diaporthe Pod-and-Stem Blight, Anthracnose, Septoria Brown Spot, Cercospora Frogeye Leafspot, Purple Seed Stain—Apply ½ to 1 lb per acre. For **determinate** varieties (generally grown in the South), apply at early pod set when majority of pods are ⅛ to ½" in length; for **indeterminate** varieties (generally grown in the North), apply when pods near the top of the plant are ½ to 1" in length. Make one additional application 14 to 21 days later. Do not apply within 35 days of harvest; do not graze or feed treated soybean vines or hay to livestock.

STONEFRUITS—APRICOTS, CHERRIES, NECTARINES, PEACHES, PLUMS, PRUNES: Treatment is most effective if applied just before rainfall; for aerial application, fly over every row or center.

EAST OF ROCKY MOUNTAINS—Use ¾ to 1½ lbs per acre on trees up to 12' tall; over 12', use 1½ to 2 lbs per acre.

Brown Rot Blossom Blight—Apply at early bloom stages (apricots—red bud; peaches, nectarines—pink bud; cherries—early popcorn; plums and prunes—green tip); for this application only, "BENLATE" 50 DF may be used in combination with spray oil. Make a second application at 75% to 100% bloom. If blossoming is prolonged or conditions favorable for disease persist, apply at petal fall.

Fruit Brown Rot—After blossom blight sprays, make two preharvest applications beginning 3 weeks before harvest up to day of harvest.

Peach Scab, Powdery Mildew—Use same schedule as for Brown Rot Blossom Blight plus applications at shuck split, shuck fall and 14 days later.

Cherry Leaf Spot—Use same schedule as for Brown Rot Blossom Blight and continue at 10- to 14-day intervals through harvest. Make an additional application 2 to 3 weeks after harvest.

WEST OF ROCKY MOUNTAINS—Use 1½ to 2 lbs per acre.

Brown Rot Blossom Blight—Apply at early bloom stages (apricots—red bud; peaches, nectarines—pink bud; cherries—early popcorn; plums and prunes—green tip); for this application only, "BENLATE" 50 DF may be used in combination with spray oil. If blossoming is prolonged or conditions favorable for disease persist, make a second application 14 days later.

Fruit Brown Rot—After blossom blight sprays, make a preharvest application (before rain) any time from 3 weeks before harvest to day of harvest. Make a second application if conditions favorable for disease persist or harvest is prolonged. Preharvest applications are most effective when applied with ground equipment, using sufficient volume to provide thorough and uniform coverage of fruit.

Powdery Mildew—Use same schedule as for Brown Rot Blossom Blight plus applications at shuck split, shuck fall, and 14 days later.

Cherry Leaf Spot—Use same schedule as for Brown Rot Blossom Blight and continue at 10- to 14-day intervals through harvest. Make an additional application 2 to 3 weeks after harvest.

POSTHARVEST FRUIT ROTS (U.S.)—Dip or spray fruit thoroughly as soon as possible after harvest; use 1/2 lb per 100 gals. When wax is used, "BENLATE" 50 DF may be incorporated into the wax spray.

NOTE: "BENLATE" 50 DF does not control peach leaf curl, shot hole (*Coryneum blight*) or bacterial blast, nor fruit rots caused by *Rhizopus spp.* and *Alternaria spp.* Do not graze livestock in treated orchards.

STRAWBERRIES: Gray Mold (Botrytis), Powdery Mildew, Leaf Scorch, Leaf Blight, Mycosphaerella Leafspot—Apply 1 lb per acre at 10% bloom and at full bloom; continue at 10- to 14-day intervals, using 1/2 lb per acre. **Anthracnose**—Apply 1 lb per acre when plants are established (plant bed or field) and repeat at 7-day intervals.

Transplants: Botrytis Crown Rot, Mycosphaerella Leaf Spot—Use 1/2 lb per 100 gals of water. Immerse plants to give thorough wetting; remove and allow to drain.

SUGAR BEETS: Cercospora Leafspot—Apply 3/8 to 1/2 lb per acre. Begin application when disease first appears and repeat at 14- to 21-day intervals as needed. Do not apply within 21 days of harvest.

SUGARCANE (HAWAII): Pineapple Disease (Ceratocystis paradoxa)—Apply to cut seedpieces either as a cold dip or hot dip.

Cold Dip—Use 1/2 lb per 100 gals of water (1:1600). Immerse seedpieces to give thorough wetting; remove and allow to drain.

Hot Dip—Use 1/4 lb per 100 gals of water (1:3200). Maintain temperature of the dip at 50°C. Soak seedpieces for 20 to 30 minutes; remove and allow to drain.

Note: Do not use treated seedpieces for food or feed purposes.

TOMATOES—Field and Greenhouse: Gray Mold (Botrytis), Leaf Mold (Cladosporium), White Mold (Sclerotinia), Cercospora Leafspot, Phoma Leafspot—For field tomatoes, apply 1/2 to 1 lb per acre; for greenhouse, use 1/2 to 1 lb per 100 gals of water. Begin applications when disease first appears and repeat at 7- to 14-day intervals as needed.

ROSES, FLOWERS, ORNAMENTALS, SHADE TREES—Field and Greenhouse: Foliar Spray—Begin applications when disease first appears and repeat at 10- to 14-day intervals throughout the growing season; shorten interval during humid, rainy weather. Use at the following rates: **1/2 lb per 100 gals** (1 tablespoonful per 2 gals)—for Powdery Mildew, Botrytis Gray Mold. **1 lb per 100 gals**—for Anthracnose (for shade trees and woody ornamentals, begin at bud break and make 2 or 3 additional applications at 10- to 14-day intervals); Black Spot of roses; Cercospora, Entomosporium, Ramularia, and Septoria Leafspots, Ascochyta and Phomopsis blights; Didymellina Leafspot of iris; Cornespora Leafspot of Ligustrum; Ovulinia Blight of azalea and rhododendron (begin as flowers open); Scab of pyracantha and flowering crab. Addition of a surfactant to the spray mixture improves distribution of the spray on hard-to-wet plants such as roses. For aerial application, use 1/2 to 1 lb per acre.

Drench Treatment—Botrytis, Fusarium, Rhizoctonia and Sclerotinia stem, crown and root rots on herbaceous annuals, perennials and bedding plants: Cylindrocladium and Thielaviopsis rots on woody ornamentals such as azaleas, rhododendrons, conifers, and poinsettias—Use 1 lb per 100 gals; apply as a drench or heavy spray (1 to 2 pts per sq ft) after transplanting into propagation beds or containers. Repeat at 2- to 4-week intervals during periods favorable for disease. "BENLATE" 50 DF does not control *Pythium spp.* or *Phytophthora spp.*

Preplant Dip Treatment—For diseases listed under Drench Treatment, use 1 lb per 100 gals of water; immerse plants or cuttings for 10 to 15 min.; remove and allow to drain.

BULBS (Easter Lily, Tulip, Gladiolus, Daffodil, Iris): Fusarium and **Penicillium Rots**—Use 1 2/3 lbs per 100 gals of water (2 tablespoonfuls per gal). Soak cleaned bulbs for 15 to 30 minutes in warm dip (80 to 85°F), preferably within 48 hours after digging. Dry bulbs after treatment. If bulbs are for forcing, treat after bulbs have been heat-cured.

STORAGE AND DISPOSAL

STORAGE: Never allow "BENLATE" 50 DF to become wet during storage. This may lead to certain chemical changes which will reduce the effectiveness of "BENLATE" 50 DF as a fungicide. Keep container tightly closed when not in use. Store product in original container only, away from other pesticides, fertilizer, food or feed.

PESTICIDE DISPOSAL: Do not contaminate water, food or feed by storage or disposal. Wastes resulting from the use of this product may be disposed of on site or at an approved waste disposal facility.

CONTAINER DISPOSAL: Completely empty bag into application equipment. Then dispose of empty bag in a sanitary landfill or by incineration, or, if allowed by state and local authorities, by burning. If burned, stay out of smoke.

NOTICE TO BUYER: Purchase of this material does not confer any rights under patents of countries outside of the United States.

NOTICE OF WARRANTY

Du Pont warrants that this product conforms to the chemical description on the label thereof and is reasonably fit for purposes stated on such label only when used in accordance with the directions under normal use conditions. It is impossible to eliminate all risks inherently associated with the use of this product. Crop injury, ineffectiveness or other unintended consequences may result because of such factors as weather conditions, presence of other materials, or the manner of use or application, all of which are beyond the control of Du Pont. In no case shall Du Pont be liable for consequential, special or indirect damages resulting from the use or handling of this product. All such risks shall be assumed by the Buyer. DU PONT MAKES NO WARRANTIES OF MERCHANTABILITY OR FITNESS FOR A PARTICULAR PURPOSE NOR ANY OTHER EXPRESS OR IMPLIED WARRANTY EXCEPT AS STATED ABOVE.

[1]Registered trademark of Merck & Co.
[2]Registered trademark of Nippon Soda Co., Japan

AG-3727 8058 4/28/88

This is a specimen label. DuPont is not responsible for the accuracy of the information contained herein. As labels are subject to revision, always carefully read and follow the label on the product container.

BIDRIN® 8

WATER MISCIBLE INSECTICIDE

RESTRICTED USE PESTICIDE

For retail sale to and use only by Certified Applicators or persons under their direct supervision and only for those uses covered by the Certified Applicator's certification.

ACTIVE INGREDIENT	BY WEIGHT:
Dicrotophos (Dimethyl phosphate of 3-hydroxy N, N-dimethyl-cis-crotonamide)	82%
INERT INGREDIENTS:	18%
TOTAL	100%

Contains 8 lb. active ingredient per gallon.

EPA Reg. No. 352-466

U.S. Patent Number 2,802,855

KEEP OUT OF REACH OF CHILDREN

DANGER–POISON

PELIGRO

PRECAUCION AL USUARIO: Si usted no lee ingles, no use este producto hasta que la etiqueta haya sido explicada ampliamente.

STATEMENT OF PRACTICAL TREATMENT

ATROPINE IS THE EMERGENCY ANTIDOTE FOR BIDRIN INSECTICIDE POISONING. 2-PAM is also antidotal and may be used in conjunction with atropine. CALL A PHYSICIAN IN ALL CASES OF SUSPECTED POISONING.

IF SWALLOWED, induce vomiting immediately by giving two glasses of water and sticking a finger down throat. Call a physician. NEVER INDUCE VOMITING OR GIVE ANYTHING BY MOUTH TO AN UNCONSCIOUS PERSON.

IF ON SKIN, immediately remove patient from the vicinity of the insecticide, remove all contaminated clothing, and wash skin with soap and running water.

IF IN EYES, wash immediately with running water for fifteen minutes.

IF WARNING SYMPTOMS APPEAR, (See "Warning Symptoms") keep patient prone and quiet. Start artificial respiration immediately if patient is not breathing. Transport the patient immediately to the nearest hospital.

For medical emergencies involving this product, call toll free 1-800-441-3637.

PRECAUTIONARY STATEMENTS

HAZARDS TO HUMANS AND DOMESTIC ANIMALS

DANGER! Poisonous if swallowed, inhaled, or absorbed through skin. Rapidly absorbed through skin. Inhalation or skin contact may, without symptoms, progressively increase susceptibility to "BIDRIN" Insecticide poisoning. This product has a strong skin sensitizing potential. Do not swallow or get in eyes, on skin, or on clothing. Do not breathe vapors. Keep out of reach of children. Do not contaminate food or feed products.

PRECAUTIONS IN USING

Wear clean natural rubber gloves, clean protective clothing and goggles. Replace gloves frequently and destroy used gloves. During commercial or prolonged exposure in spray mixing and loading operations, wear a pesticide respirator approved by the National Institute for Occupational Safety and Health. Wash thoroughly with soap and water after handling and before eating or smoking.

Decontaminate areas and equipment by washing with a dilute solution of alkali (less than 5% in water) and detergent. Absorb waste solution. Dispose of collected waste materials according to pesticide disposal instructions.

NOTE TO PHYSICIAN

Warning Symptoms: Symptoms include weakness, headache, tightness in chest, blurred vision, non-reactive pin-point pupils, salivation, sweating, nausea, vomiting, diarrhea, and abdominal cramps.

Treatment: Atropine is the specific therapeutic antagonist of choice against parasympathetic nervous stimulation. If there are signs of parasympathetic stimulation, atropine sulfate should be injected at 10-minute intervals, in doses of 1 to 2 milligrams, until complete atropinization has occurred.

Pralidoxime chloride (2-PAM chloride) may also be used as an effective antidote in addition to and while maintaining full atropinization. In adults, an initial dose of 1 gram of 2-PAM should be injected, preferably as an infusion in 250 cc of saline over a 15- to 30-minute period. If this is not practical, 2-PAM may be administered slowly by intravenous injection as a 5 percent solution in water over not less than two minutes. After about an hour, a second dose of 1 gram of 2-PAM will be indicated, if muscle weakness has not been relieved. For infants and children the dose of 2-PAM is 0.25 grams.

Morphine is an improper treatment.

Clear chest by postural drainage. Oxygen administration may be necessary. Observe patient continuously for 48 hours. Repeated exposure to cholinesterase inhibitors may, without warning, cause prolonged susceptibility to very small doses of any cholinesterase inhibitor. Allow no further exposure until cholinesterase regeneration has been attained as determined by blood test.

For medical emergencies involving this product, call toll free 1-800-441-3637.

ENVIRONMENTAL HAZARDS

WILDLIFE PRECAUTIONS

This pesticide is extremely toxic to fish, birds, and other wildlife. DO NOT SPRAY BORDER AREAS. Keep border areas within 25 feet of field free of bird feed and shelter. Birds feeding on treated area may be killed. Do NOT use whenever large flocks of birds are in the vicinity or when they are noted feeding in or around fields. Avoid spray drift. To minimize spray drift, do NOT spray when windy. Do not apply directly to lakes, streams, ponds, or wetlands. Runoff from treated areas may be hazardous to aquatic organisms in neighboring areas. Do not contaminate water by cleaning of equipment or disposal of wastes.

This product is highly toxic to bees exposed to direct treatment or residues on blooming crops or weeds. Do not apply this product or allow it to drift to blooming crops or weeds if bees are visiting the treatment area. Protective information may be obtained from your Cooperative Agricultural Extension Service.

WILDLIFE PROTECTION IS EVERYONE'S RESPONSIBILITY

PHYSICAL OR CHEMICAL HAZARDS

Keep away from heat and open flame.

DIRECTIONS FOR USE

It is a violation of Federal law to use this product in a manner inconsistent with its labeling.

Du Pont "BIDRIN" 8 Insecticide should be used only in accordance with recommendations on this label.

Du Pont will not be responsible for losses or damages resulting from use of this product in any manner not specifically recommended by Du Pont. User assumes all risk associated with such non-recommended use.

REENTRY STATEMENTS

Do not apply this product in such a manner as to directly or through drift expose workers or other persons. The area being treated must be vacated by unprotected persons.

Do not enter treated areas for forty-eight (48) hours unless appropriate protective clothing is worn. Because certain States may require more restrictive reentry intervals for various crops treated with this product, consult your State Department of Agriculture for further information.

Written or oral warnings must be given to workers who are expected to be in a treated area or in an area about to be treated with this product. (Indicate specific oral warnings which inform workers of areas or fields that may not be entered without specific protective clothing, period of time field must be vacated and appropriate actions to take in case of accidental exposure.)

When oral warnings are given, warnings shall be given in a language customarily understood by workers. Oral warnings must be given if there is reason to believe that written warnings cannot be understood by workers.

Written warnings must include the following information:

"DANGER! Area treated with "BIDRIN" Insecticide on (date of application). Do not enter without appropriate protective clothing until sprays have dried. (Insert here actions to take in case of accidental exposure. See STATEMENT OF PRACTICAL TREATMENT for details.)"

GENERAL DIRECTIONS

"BIDRIN" 8 Insecticide is a water-miscible product which contains 2 pounds of active ingredient per quart. For foliage applications, mix the required amount of BIDRIN in sufficient water to provide uniform coverage. If container is emptied, triple rinse adding rinsate to the spray tank.

Do not apply this product through any type of irrigation system.

Use and Directions for Application on Cotton

Insect	Dosage Fluid Ounces Per Acre	Directions For Application
Aphids, Thrips, Spider Mites, Cotton Fleahopper	For Early Season Control 1.6-3.2	Apply to give uniform coverage. Repeat as necessary, but do not apply more than three times per season. Do not apply within 30 days of harvest. Do not graze livestock on treated fields, or feed treated gin trash.
Grasshoppers, Boll Weevil	4.0	
Aphids, Stinkbugs, Cotton Fleahopper Spider Mites, Grasshoppers, Boll Weevil, Black Fleahopper, Plant Bugs (Lygus), Saltmarsh Caterpillars, Leaf Perforators	For Mid or Late Season Control 4.0-8.0	

STORAGE AND DISPOSAL

Do not contaminate water, food, or feed by storage or disposal. Do not reuse empty container. Open dumping is prohibited.

STORAGE:

Do not store in or around the home. Store in a secure, dry and temperate area. Store in original container. Keep away from heat and open flame. If spilled, avoid exposure. If significant spill, do not attempt to clean up; call CHEMTREC (800) 424-9300. Spill can be contained by covering with a sufficient amount of absorbent (sand, soil, sawdust, etc.) to collect all liquid. Collected waste must be disposed of according to pesticide disposal below. CAUTION: Wear a pesticide respirator and protective clothing.

PESTICIDE DISPOSAL: Pesticide wastes are acutely hazardous. Improper disposal of excess pesticide, spray mixture or rinsate is a violation of Federal Law. If these wastes cannot be disposed of by use according to label instructions, contact your State Pesticide or Environmental Control Agency, or the Hazardous Waste representative at the nearest EPA Regional Office for guidance.

CONTAINER DISPOSAL: Triple rinse (or equivalent). Then dispose of in a sanitary landfill or by other approved State and local procedures.

NOTICE OF WARRANTY

Du Pont warrants that this product conforms to the chemical description on the label thereof and is reasonably fit for the purposes stated on such label only when used in accordance with the directions under normal use conditions. It is impossible to eliminate all risks inherently associated with the use of this product. Crop injury, ineffectiveness or other unintended consequences may result because of such factors as weather conditions, presence of other materials, or the manner of use or application, all of which are beyond the control of Du Pont. In no case shall Du Pont be liable for consequential, special or indirect damages resulting from the use or handling of this product. All such risks shall be assumed by the buyer. DU PONT MAKES NO WARRANTIES OF MERCHANTABILITY OR FITNESS FOR A PARTICULAR PURPOSE NOR ANY OTHER EXPRESS OR IMPLIED WARRANTY EXCEPT AS STATED ABOVE.

AG-3247 8029 4/20/87

This is a specimen label. DuPont is not responsible for the accuracy of the information contained herein. As labels are subject to revision, always carefully read and follow the label on the product container.

BLADEX® 4L

HERBICIDE LIQUID

RESTRICTED USE PESTICIDE

For retail sale to and use only by Certified Applicators or persons under their direct supervision and only for those uses covered by the Certified Applicator's certification. See Precautionary Statements for reasons this product is classified Restricted Use.

Contains 4 lbs. active ingredient per gallon.

ACTIVE INGREDIENTS:

Cyanazine: 2[[4-chloro-6-(ethylamino)-s-triazin- 2-yl]amino]-2-methylpropionitrile	43%
INERT INGREDIENTS:	57%
TOTAL	100%

EPA Reg. No. 352-470
Covered by U.S. Patents Nos. 3,505,325 and 3,620,710

KEEP OUT OF REACH OF CHILDREN

WARNING

AVISO PRECAUCION AL USUARIO: Si usted no lee ingles, no use este producto hasta que la etiqueta haya sido explicada ampliamente.

STATEMENT OF PRACTICAL TREATMENT

IF SWALLOWED, call a physician or poison control center. Drink 1 or 2 glasses of water and induce vomiting by touching back of throat with finger. Do not induce vomiting or give anything by mouth to an unconscious person.

IF IN EYES, flush with plenty of water. If irritation persists, call a physician.

IF ON SKIN, wash immediately with plenty of soap and water.

IF INHALED, remove victim to fresh air. If not breathing, give artificial respiration, preferably mouth-to-mouth and get medical attention.

For medical emergencies involving this product, call toll free 1-800-441-3637.

PRECAUTIONARY STATEMENTS

HAZARDS TO HUMANS AND DOMESTIC ANIMALS

WARNING! May be fatal if swallowed. Harmful if inhaled or absorbed through the skin. Causes temporary eye injury.

This product may be hazardous to your health. This product is classified "Restricted Use" because, at doses which caused serious maternal illness in laboratory animals, birth defects were present. Use of protective clothing and equipment and following the precautions below can reduce risk. Avoid breathing spray mist. Avoid contact with skin, eyes, or clothing. Do not get in eyes or on clothing. Wear a face shield when mixing and loading. Wash thoroughly with soap and water after handling and before eating or smoking.

Wear long trousers and long-sleeved clothing when applying this product. Wear chemical resistant gloves extending above the wrist, a chemical resistant apron, long trousers and long-sleeved clothing when mixing or loading or when adjusting, repairing or cleaning equipment. Protective gloves must be washed with soap and water after use and before removing from the hands. Remove contaminated clothing and wash before reuse. Contaminated clothing should be laundered separately from household laundry to prevent cross-contamination of other laundry. Heavily contaminated or drenched clothing and protective equipment must be discarded or destroyed in accordance with State and local regulations.

Do not apply this product in such a manner as to directly, or through drift, expose workers or other persons. The area being treated must be vacated by unprotected persons.

Keep out of reach of domestic animals, particularly cattle. Consumption of this product, spray solutions, or water contaminated with product can result in serious illness or possible death of bovines.

ENVIRONMENTAL HAZARDS

Do not apply directly to water or wetlands. Do not contaminate water by cleaning equipment or disposal of wastes.

BLADEX is a pesticide which can move (seep or travel) through soil and can contaminate groundwater which may be used as drinking water. BLADEX has been found in groundwater as a result of agricultural use. Users are advised not to apply BLADEX where the water table (groundwater) is close to the surface and where the soils are very permeable (i.e., well drained soils such as loamy sands). Your local agricultural agencies can provide further information on the type of soil in your area and the location of groundwater.

In case of significant spill, call CHEMTREC 1-800-424-9300.

DIRECTIONS FOR USE

It is a violation of Federal law to use this product in a manner inconsistent with its labeling.

GENERAL INFORMATION

BLADEX 4L Herbicide is a selective herbicide for the control of annual grasses and broadleaved weeds in field corn, popcorn, sweet corn, cotton and grain sorghum.

Consult your local Agricultural Extension Agent for help in determining soil texture, organic matter content, and the most appropriate herbicide rate for local conditions.

Do not apply this product in irrigation water with any kind of irrigation system.

Do not apply this product with aerial application equipment.

BLADEX 4L is not effective when used preemergence on peat or muck soils. DO not use BLADEX 4L on sands or loamy sands (soils consisting of more than 70% sand) containing less than 1% organic matter.

Weather Effects: As a preemergence herbicide, BLADEX 4L is active mainly through the roots, and therefore, its effect on weeds is dependent on adequate rainfall or sprinkler irrigation to move the herbicide into the root zone. Moisture should be sufficient to thoroughly wet the soil throughout the zone where weed seeds may germinate and enough to make soil too wet to cultivate. Rotary hoeing or shallow cultivation is recommended for those applications which are not incorporated at the time of treatment, if adequate rainfall or sprinkler irrigation has not occurred within about ten days after application of BLADEX 4L.

Heavy rainfall between planting and crop emergence may cause crop injury or stand loss. Rainfall tends to cause excessive concentrations of herbicide in the seed furrow, resulting in possible crop injury. Level deep planter marks or seed furrrows before application.

Under conditions which delay weed germination, such as low temperatures, lack of soil surface moisture, or when germination extends over a long period, the effectiveness of the herbicide may be impaired. Rotary hoeing, shallow cultivation or a postemergence herbicide treatment may be of benefit under these circumstances.

If the crop is cultivated, tillage should be shallow to minimize herbicide dilution in the soil. Should the crop stand be lost due to adverse weather conditions, insects, etc., the field can be replanted the same season to corn or sorghum.

To enhance weed control in areas of less than 25 inches of rainfall or where long dry periods are common, these treatments may require shallow incorporation with a tool such as a field cultivator operated from 5-7 mph. Incorporation should not be more than three inches deep to keep from burying the herbicide. A spike-toothed harrow, deep tillage disk or rolling basket device is not recommended for incorporating BLADEX.

When applied as a post-emergence herbicide, BLADEX 4L is also active through foliage as well as through the roots. Yellowing and/or stunting of the crop may result from this treatment, particularly if cold, adverse growing conditions occur after application. Extended or extreme cold and wet conditions may reduce stands. Do not apply BLADEX postemergence to a crop that is damaged or growing under stress.

OBSERVE ALL CAUTIONS AND LIMITATIONS ON LABELING OF ALL PRODUCTS USED IN MIXTURES.

Triazine Resistant Weeds: In fields where triazine resistant biotypes of weeds have been identified, BLADEX should be used in combination with or in sequence with other registered non-triazine herbicides. (Triazine resistant biotypes of Kochia and Pigweed have been identified in some fields in the Western Great Plains and triazine resistant biotypes of Pigweed and Lambquarters have been identified in some fields in various states.) Consult with appropriate state agricultural extension service representatives for specific recommendations.

APPLICATION DIRECTIONS

GENERAL MIXING AND SPRAYING INSTRUCTIONS

This product may not be applied by means of chemigation or aerial application.

Use sufficient agitation to ensure that the BLADEX 4L is completely dispersed and in uniform suspension prior to application or tank mixing with other formulations.

The following general mixing instructions are recommended when using this or any other liquid suspension formulation.

I. General

1. Unless otherwise specified, use at least 7.5 gallons of water per acre for soil applications and at least 15 gallons of water per acre for foliar applications for all applications with ground equipment. NOTE: Sufficient carrier must be used to assure uniform application. Follow label requirement of all products used in tank mix combinations.
2. A nitrogen solution or complete liquid fertilizer may replace all or part of the water as a carrier for preemergence or preplant application. Do not apply fertilizer mixtures after crop emerges, because injury may occur.
3. Always check the tank mix compatibility (TMC) of this or any other formulation before mixing with liquid fertilizer carriers or other formulations. A simple but generally reliable TMC evaluation procedure has been provided for your use in step II of these mixing instructions.
4. Start with thoroughly clean equipment. (See the label of previous compound used for cleaning instructions.)

5. Fill tank 1/2 full with carrier. Start and maintain consistent agitation through all mixing and spraying procedures. Make sure that the agitation system is working properly and creates a rippling or rolling action on the liquid surface.
6. Add the BLADEX 4L to the tank with agitation.
7. Fill tank to 75 percent capacity with carrier. Filling and bypass lines should be kept below liquid surface. Increase tank agitation if necessary to maintain surface action.
8. When desired, appropriate emulsible oil, oil concentrate, or other tank mix formulations should be added at this time. Pre-slurry these added ingredients before addition, if the compatibility test shows it to be necessary.
9. Complete filling tank maintaining sufficient agitation at all times to ensure surface action. This applies to both spray and nurse tanks.
10. Tank mixtures should always be applied immediately after preparation. If for any reason this is not possible, assure that sufficient agitation has been provided to re-mix all products and check for complete resuspension prior to application.
11. Empty tank as completely as possible before refilling to prevent buildup of oil or emulsible concentrate residues when tank mixing with these formulations. Always maintain agitation to avoid separation.
12. If an oil or emulsible concentrate film starts to build up after using these formulations, drain and clean the tank with strong detergent solution or appropriate solvent.
13. It is recommended that the sprayer be thoroughly cleaned by flushing with a detergent solution at the end of each work day when any emulsible oil, oil concentrate, or other emulsible formulation has been used either alone or in tank mix combinations with other pesticide formulations, even if no obvious problems have been encountered. This precaution will ensure a clean sprayer and continued trouble-free operation.

II. Tank Mix Compatibility Evaluation Procedure

1. Add one pint of carrier liquid to each of two one-quart jars. Mark one quart jar "with" and the other "without".
2. Add 1/4 teaspoon of a suitable tank mix compatibility agent (1/4 teaspoon/pint = 2 pints/100 gallons of carrier) to the jar marked "with," cap the jar, and shake gently for five to ten seconds to mix.
3. Add the appropriate amount of herbicide to both jars, cap each jar, and shake gently for five to ten seconds to mix. If problems are encountered in mixing wettable powder or dry flowable formulations into a liquid fertilizer, then pre-slurry these formulations in water prior to their addition to the liquid fertilizer and proceed with the test. The following chart has been provided to assist you in selecting the appropriate BLADEX 4L use rate for this evaluation. If more than one herbicide is to be used in the tank mixture, each should be added separately as follows: water solubles first, wettable powders or dry flowables second, liquid flowables third, and emulsible concentrate or oil formulations last with each jar capped and gently shaken for five to ten seconds for each addition.

Jar Test for BLADEX 4L Compatibility

Gallons of Liquid Carrier per Acre	4	7.5	15	20	25	30	40
Teaspoons of BLADEX 4L per Pint of Liquid Carrier	6.0	3.2	1.6	1.2	1.0	0.8	0.6

This chart is based on one quart of BLADEX 4L (1 pound active ingredients) per acre in the indicated carrier volumes. Intended field use rates are achieved by varying the amount of BLADEX 4L; i.e., for a field use rate of 2.5 quarts of BLADEX 4L in 15 gallons of carrier per acre, add 4 teaspoons of BLADEX 4L to the quart jars containing one pint of carrier. Calculation: For 2.5 quarts of BLADEX 4L/15 gallons of carrier per acre, multiply 2.5 × 1.6 to get 4 teaspoons of BLADEX 4L per pint of carrier.

4. Let each jar stand one-half hour and make observations. If any separation, agglomeration, or precipitation has occurred, shake the jar again for 10 to 15 seconds, and note whether any of the following occur: a. Separated phases do not re-mix uniformly. b. Screen/nozzle plugging lumps do not disperse. c. Precipitate does not re-suspend readily. d. Precipitate sticks tenaciously to the glass.
5. If none of the above problems occur in either jar, then the herbicides can, in most cases, be safely used without a compatibility agent.
6. If problems 4.a or 4.b occur in the jar marked "without" but do not occur in the jar marked "with," the compatibility agent should be used.
7. If problems 4.a or 4.b are seen in both jars, then the herbicides and carrier mixture are incompatible and should not be used in the same spray tank. Alternatively, a different tank mix compatibility agent can be evaluated.
8. If problems 4.c or 4.d occur in the jar marked "without" but do not occur in the jar marked "with," the compatibility agent should be used unless constant, thorough agitation can be maintained and immediate clean-out of spray system is performed.
9. If problems 4.c and 4.d are seen in the jar marked "with," the user proceeds with mixing and application at his own risk should agitation in the system be insufficient or curtailed.
10. Those mixtures defined as compatible in this test should then be mixed for use as indicated in steps 1-12 of the general mixing instructions listed above.

If a test such as outlined indicates that components of a proposed mix are compatible, the applicator still has the responsibility of combining materials in sequence to the spray tank in accordance with directions prescribed on the label of the herbicides or pesticides involved.

Tests have indicated that compatibility agents, noted below by the various tank mix combinations, may give improved compatibility in liquid fertilizers.

Tank Mix Combination	Compatibility Agents
BLADEX 4L/"Lasso" (Liquid Fertilizer Grade)	Probably not needed in 28-0-0, 10-34-0. "Compex" may help in others.
BLADEX 4L/"Genate Plus", "Sutan+" 6.7E	Probably not needed in 28-0-0. Incompatible in 10-34-0. "Unite", "Spray-Mate", "Kem-Link", may help in others.
BLADEX 4L/"Dual" 8E	Probably not needed in 28-0-0. "Unite", "Spraymate", "Ivory Liquid"

III. Application Equipment

1. Use application equipment fitted with nozzles that provide accurate and uniform coverage. Be certain that nozzles are uniformly spaced and the same size. Calibrate sprayer before use and recheck frequently during use whenever possible.
2. Use a pump with capacity to: a. Maintain 35-40 psi at nozzles. b. Provide sufficient agitation in tank to keep mixture in suspension. c. Provide a minimum of 20 percent bypass at all times.
3. Use centrifugal pumps which provide sufficient shear action for dispersing and mixing this product. The pump should provide a minimum of 10 gallons/minute/100-gallon tank size circulated through the jets of a correctly-positioned sparger tube.
4. Use screens to protect the pump and to prevent nozzles from clogging. Screens placed on suction side of pump should be 10 to 16 mesh. Do not place a screen in the recirculation line. Use a 40- to 50-mesh screen between the pump and boom and, where required, 50-mesh screens at the nozzles. Check our equipment manufacturer's literature for specific recommendations.

FERTILIZER IMPREGNATION

BLADEX 4L Herbicide may be applied when coated on or impregnated in dry granular fertilizer for early preplant, preemergence or preplant incorporated weed control in field corn. All recommendations, cautions and special precautions on this label must be followed along with state regulations relating to dry bulk fertilizer blending, impregnating and labeling. Also, follow the precautions on the label of any product mixed with this product.

General Blending Directions

BLADEX 4L may be coated on or impregnated in dry bulk fertilizers using tower blenders, rotary drum blenders or blending augurs or conveyors. DO NOT impregnate BLADEX or tank mixes containing BLADEX on or in fertilizers containing Ammonium Nitrate, Potassium Nitrate, or Sodium Nitrate. Do not use on straight limestone since absorption will not be achieved. Fertilizer blends containing limestone can be impregnated when using BLADEX alone. Use a minimum of 200 lbs. and a maximum of 450 lbs. per acre of dry fertilizer.

Use equipment that will give a uniform distribution of the herbicide throughout each batch of impregnated fertilizer. Non-uniform impregnation can cause crop injury or unsatisfactory performance.

A. BLADEX 4L may be used as the only herbicide for impregnation.

1. Add BLADEX 4L when at least 1/2 the total fertilizer volume required is in the mixer. A minimum of 200 lbs per acre of an approved fertilizer should be used.
2. BLADEX 4L can be moved from the chemical holding tank to the mixer by using an air system or a liquid pump with a hose at least 1 inch in diameter.
3. Position the spray nozzles to achieve uniform coverage of the BLADEX 4L on the dry fertilizer without spraying the walls of the mixer.
4. Flush the spray lines with water and spray onto the dry fertilizer.
5. Add remaining fertilizer, plus drying agents when necessary, and blend thoroughly for at least three minutes.
6. Add 2-5% of a suitable drying agent to insure a herbicide/fertilizer mixture that will spread through air spreaders. The need for a drying agent is determined by the wetness of the fertilizer batch. Wetness can change with humidity, nitrogen content, fertilizer rates and herbicide rates.

B. BLADEX 4L may be used in tank mixes with other herbicides.

1. While fertilizer is blending, add the BLADEX 4L first and the tank mix fertilizer last.
2. Add any necessary drying agent to insure a spreadable herbicide/fertilizer mixture and blend thoroughly for at least three minutes.
3. Follow the other appropriate use instructions found in Section A for use with BLADEX 4L tank mixes.

(For more information on Drying Agents, Application Equipment, Calibration Guide and variations of these methods, see Du Pont Bulletin on "Fertilizer Impregnation").

CLEAN OUT:

Equipment used to impregnate or apply fertilizer impregnated with BLADEX or combinations including BLADEX must be cleaned out by running at least 1,000 lbs. of fertilizer not impregnated with BLADEX through the impregnation equipment and application equipment, if the next batch of material is to be applied to a crop for which BLADEX or a combination herbicide is not registered.

APPLICATION:

Uniform application of BLADEX 4L which has been impregnated in or coated on dry fertilizer is essential for satisfactory weed control and crop safety. Accurate calibration of the fertilizer applicator is necessary. Applying while turning at the ends of the fields may result in excessive application rates causing crop injury. Do not double apply across the ends or sides of the field.

Crop injury and/or poor weed control may occur where the impregnated fertilizer is not uniformly applied. Air flow or augur metered application equipment is preferred (one pass application). If other equipment is used, the recommended method of application is to apply 1/2 the recommended rate and overlap 50 percent to double apply by splitting the middles to obtain the best distribution pattern.

Apply immediately after impregnation. Impregnated fertilizer may become lumpy and difficult to spread if stored.

RATES AND TIMING:

Use the application rates and timing shown in the appropriate sections of this label. Follow the precautions on the labels of all products used.

CORN

WEEDS CONTROLLED BY "BLADEX" 4L ALONE AND IN COMBINATION WITH OTHER HERBICIDES ON CORN

Grasses

Annual bluegrass
Annual fescues
Annual (Italian) ryegrass
Barnyardgrass(1)
Bullgrass
Crabgrass
Fall panicum
Giant foxtail
Goosegrass
Green foxtail
Junglerice
Stinkgrass (Indian lovegrass)
Witchgrass
Yellow foxtail

Broadleaves

Annual groundcherry
Annual morningglory
Annual sedge
Black mustard
Florida pusley (Florida purslane)
Hedge mustard
Jimsonweed(1)
Ragweed (Common)
Russian thistle
Shepherdspurse
Smallflower galinsoga
Buffalobur
Buttercup (annual)
Carpetweed
Cocklebur(2)
Common chickweed
Common groundsel
Common mallow
Common purslane
Corn spurry
Curly dock (seedling)
Fiddleneck
Kochia
Ladysthumb
Lambsquarters
Mayweed
Nightshade (annual)
Pigweed(1)
Pineappleweed
Plantain
Poorjoe
Prickly sida (teaweed)
Prostrate knotweed
Prostrate spurge
Smartweed (Pennsylvania)
Sunflower(2) (wild, annual, common)
Tarweed cuphea (Gumweed)
Velvetleaf(1)
Wild buckwheat
Wild mustard
Wild radish
Wild turnip

(1) Under conditions such as low temperatures, lack of soil surface moisture or other factors that may cause delay in germination of the seeds, the effectiveness of BLADEX may be impaired against these weeds.

(2) The degree of control will be reduced is soil moisture and temperature conditions cause deep germination of the seed.

CORN

PREEMERGENCE-PREPLANT INCORPORATED

Apply BLADEX treatments just before, at or after planting but before crop has emerged. Avoid removal of treated soil from seedrow prior to or during the planting operation.

BLADEX may also be applied early prior to planting or in a split application if pre-season weed control is desired. For split applications, do not exceed the total amount of BLADEX for the soil texture and organic matter shown in Table 1. If BLADEX is applied early, more than 15 days before planting, a split application of BLADEX or some other herbicide treatment may be necessary at or after planting to provide additional length of weed control. For further information see "Early Preplant" recommendations in the Conservation Tillage section of this label.

Rotary hoeing is recommended for preemergence applications which do not receive adequate rainfall or sprinkler irrigation to wet the top 2 inches of soil or depth of germinating weeds within about 10 days after application.

BLADEX alone or in tank mix combinations should not be incorporated more than three inches deep to keep from burying the herbicide. Single or two pass incorporation with a tool such as a field cultivator operated at 5-7 mph is acceptable. A spike-toothed harrow, deep tillage disk or rolling basket device is not recommended for incorporating BLADEX.

BLADEX Applied Alone

Use the proper rate for the soil texture and organic matter indicated in Table 1. Any rotational crop may be planted in the fall or spring following this treatment.

TABLE 1
PREEMERGENCE BROADCAST APPLICATION RATES PER ACRE FOR "BLADEX" 4L APPLIED ALONE ON CORN

SOIL TEXTURE DESCRIPTION	QUARTS OF "BLADEX" 4L — PERCENT ORGANIC MATTER IN SOIL*					
	LESS THAN 1%	1%	2%	3%	4%	5% & OVER
Sand, Loamy sand	DO NOT USE	1.25	1.5	2.25	2.75	3.25
Sandy loam	1.25	1.75	2.0	2.5	3.0	3.5
Loam, Silt loam, Silt	1.5	2.0	2.5	3.0	3.5	4.0
Sandy clay loam, Clay loam, Silty clay loam	2.0	2.5	3.0	3.5	4.0	4.5
Sandy clay, Silty clay, Clay	2.75	3.0	3.5	4.0	4.5	4.75
Peat or Muck	NOT RECOMMENDED					

* For organic matter content between those listed, adjust the rate proportionately.

"BLADEX" COMBINATIONS

BLADEX plus Atrazine

Use BLADEX 4L plus atrazine at the proper rate for soil texture and organic matter indicated in Tables 2 and 3. These tables provide rates for generally weedy conditions. The ratio of the amounts of each herbicide may be adjusted as necessary for particular weed conditions as long as the combined rate of the two products does not exceed the combined rate for the soil shown in Table 2 or 3. For grassier conditions use a ratio that contains higher levels of BLADEX (3:1). For fields with more broadleaves use a ratio that contains higher levels of atrazine (1:1).

Rotational Crops: (1) Plant only corn, peanuts, sorghum, or soybeans the year following the use of this mixture. (2) If soybeans are to be planted, injury may occur due to the carryover of atrazine. (3) If applied after June 10, do not rotate with crops other than corn or sorghum the next year or injury may occur. (4) In the high plains and intermountain areas of the West where rainfall is sparse and erratic or where irrigation is required, use only when corn or sorghum is to be planted the following year, or a crop of corn or sorghum not treated with atrazine is to precede other rotational crops. (5) Small grains may be planted 15 months following treatment. (6) All other crops may be planted 18 months after application.

For all states except Kentucky, Missouri, Tennessee and Kansas east of Highway 99 use Table 2.

In Kentucky, Missouri, Tennessee and Kansas east of Highway 99 use Table 3.

TABLE 2
PREEMERGENCE BROADCAST APPLICATION RATES PER ACRE FOR
TANK-MIX COMBINATIONS OF "BLADEX" 4L PLUS ATRAZINE 4L ON CORN
FOR USE IN ALL STATES EXCEPT KENTUCKY, MISSOURI, TENNESSEE AND KANSAS EAST OF HIGHWAY 99

	QUARTS OF "BLADEX" 4L + QUARTS OF ATRAZINE 4L**					
	PERCENT ORGANIC MATTER IN SOIL*					
SOIL TEXTURE DESCRIPTION	LESS THAN 1%	1%	2%	3%	4%	5% & OVER
Sand, Loamy sand	DO NOT USE	0.8 + 0.5	1.0 + 0.5	1.5 + 0.5	1.8 + 0.8	2.25 + 1.0
Sandy loam	0.8 + 0.5	1.0 + 0.5	1.5 + 0.5	1.8 + 0.8	2.25 + 1.0	2.75 + 1.25
Loam, Silt loam, Silt	1.0 + 0.5	1.5 + 0.5	2.0 + 0.8	2.25 + 1.0	2.75 + 1.25	3.25 + 1.25
Sandy clay loam, Clay loam, Silty clay loam	1.5 + 0.5	2.0 + 0.8	2.25 + 1.0	2.75 + 1.25	3.25 + 1.25	3.50 + 1.25
Sandy clay, Silty clay, Clay	2.0 + 0.8	2.25 + 1.0	2.75 + 1.25	3.25 + 1.25	3.5 + 1.25	3.75 + 1.5
Peat or Muck	NOT RECOMMENDED					

* For organic matter content between those listed, adjust the rate proportionately.
** If Atrazine 80W is used, multiply rates shown by 1.25 to equal pounds of Atrazine 80W.
If Atrazine 90% is used, multiply rates shown by 1.11 to equal pounds of Atrazine 90%.

TABLE 3
PREEMERGENCE BROADCAST APPLICATION RATES PER ACRE FOR
TANK-MIX COMBINATIONS OF "BLADEX" 4L PLUS ATRAZINE 4L ON CORN
FOR USE ONLY IN ALL KENTUCKY, MISSOURI, TENNESSEE AND KANSAS EAST OF HIGHWAY 99

	QUARTS OF "BLADEX" 4L + QUARTS OF ATRAZINE 4L**					
	PERCENT ORGANIC MATTER IN SOIL*					
SOIL TEXTURE DESCRIPTION	LESS THAN 1%	1%	2%	3%	4%	5 & OVER
Sand, Loamy sand	DO NOT USE	0.8 + 0.5	1.0 + 0.5	1.5 + 0.6	1.8 + 0.8	2.25 + 1.0
Sandy loam	0.8 + 0.5	1.4 + 0.7	1.6 + 0.8	1.8 + 0.9	2.2 + 1.1	2.8 + 1.3
Loam, Silt loam, Silt	1.4 + 0.7	2.0 + 1.0	2.2 + 1.1	2.4 + 1.2	2.8 + 1.3	3.25 + 1.3
Sandy clay loam, Clay loam, Silty clay loam	1.6 + 0.8	2.2 + 1.1	2.4 + 1.2	2.8 + 1.3	3.25 + 1.3	3.5 + 1.4
Sandy clay, Silty clay, Clay	2.0 + 1.0	2.4 + 1.2	2.8 + 1.3	3.25 + 1.3	3.5 + 1.4	3.75 + 1.5
Peat or Muck	NOT RECOMMENDED					

* For organic matter content between those listed, adjust the rate proportionately.
** If Atrazine 80W is used, multiply rates shown by 1.25 to equal pounds of Atrazine 80W.
If Atrazine 90% is used, multiply rates shown by 1.11 to equal pounds of Atrazine 90%.

BLADEX plus "Lasso" 4EC

Use BLADEX 4L at the proper rate for the soil texture and organic matter shown in Table 4 plus 2 quarts per acre of "Lasso". (Use 2.5 quarts of "Lasso" on clay soils containing 5 percent organic matter and over.) Any rotational crop may be planted the fall or spring following this treatment.

BLADEX plus "Sutan +" 6.7E, "Genate Plus" 6.7EC or "Eradicane" 6.7E

Use BLADEX 4L at the proper rate for the soil texture and organic matter shown in Table 4 plus 1.8 quarts per acre of "Sutan+", "Genate Plus" or "Eradicane" for control of many annual grasses and broadleaf weeds. (Use 2.4 quarts of "Sutan+", "Genate Plus" or "Eradicane" on loam soils containing 5 percent or more organic matter, and on clay loams and clays containing 4 percent or more organic matter.) Do not use on sands and loamy sands having less than 1 percent organic matter in the light sandy soils of eastern coastal states. Do not use on corn seed stock.

Apply before planting. Incorporate the mixture immediately after application using power-driven cultivation equipment set for 2-3 inches in depth, or a tandem disc set to cut to a depth of about 4 inches while operating at 4-6 mph. For thorough mixing, disc in two directions (cross disc), and follow with a harrow, drag, or other leveling device. Prior to the second discing, readjust the disc to prevent cutting deeper than 4 inches. BLADEX 4L may be applied preemergence as an overlay over previously incorporated "Sutan+", "Genate Plus" or "Eradicane". Any rotation crop may be planted in the fall or spring following these treatments.

Existing stands of quackgrass, purple and yellow nutsedge must be turned under and thoroughly chopped up prior to chemical treatments. Additional weeds controlled by "Sutan+", "Genate Plus" or "Eradicane" combinations:

Grasses:

Sandbur	Shattercane (Wild Cane)*
Texas Panicum	Quackgrass ("Eradicane" only)
	Wild Proso Millet* ("Eradicane" only)

Perennial Weeds:

Yellow Nutsedge (Nutgrass) Purple Nutsedge Nutgrass)

* Suppression only - refer to "Sutan+", "Genate Plus" or "Eradicane" label for appropriate supplemental cultural and tillage practices.

For fields with moderate to heavy infestations of these weeds refer to the "Sutan+", "Genate Plus" or "Eradicane" label for appropriate higher rates.

BLADEX plus "Dual" 8E

Use BLADEX 4L at the proper rate for soil texture and organic matter shown in Table 4. Use "Dual" as follows:

Soil Texture	Broadcast Rate Per Acre For "Dual"
Coarse Sand, Loamy sand, Sandy loam	1.25 - 1.5 pints
Medium Loam, Silt loam, Silt	1.5 - 2.0 pints
Fine Sandy clay loam, Silty clay loam, Clay loam, Sandy clay, Silty clay, Clay	1.5 - 2.5 pints

The low end of the rate range should be used for lowest organic matter soils and the rate increased as organic matter increases to a point that soils containing 4 percent organic matter or more require the highest rate shown for that soil texture. Refer to the Dual label for precautions on rotational crops.

BLADEX plus Atrazine plus "Lasso", "Sutan+", "Genate Plus", "Eradicane", or "Dual"

Use BLADEX 4L plus atrazine at the proper rate for soil texture and organic matter shown in Table 5. Use "Lasso", "Sutan+", "Genate Plus", "Eradicane", or "Dual" according to rates shown in "BLADEX Combinations" in this section of the label. Rotational Crops: Refer to Rotational Crops section of "BLADEX plus Atrazine" in this section of the label.

TABLE 4
PREEMERGENCE BROADCAST APPLICATION RATES PER ACRE FOR "BLADEX" 4L USED IN TANK-MIX COMBINATIONS WITH "LASSO", "SUTAN+", "GENATE PLUS", "ERADICANE", OR "DUAL" ON CORN

SOIL TEXTURE DESCRIPTION	QUARTS OF "BLADEX" 4L					
	PERCENT ORGANIC MATTER IN SOIL*					
	LESS THAN 1%	1%	2%	3%	4%	5% & OVER
Sand, Loamy sand	0.6**	0.75	1.25	1.5	1.75	2.0
Sandy loam	0.75	1.25	1.5	1.75	2.0	2.25
Loam, Silt loam, Silt, Sandy clay loam,	1.25	1.5	1.75	2.0	2.25	2.5
Clay loam, Silty clay loam	1.5	1.75	2.0	2.25	2.5	2.75
Sandy clay, Silty clay, Clay	1.75	2.0	2.25	2.5	2.75	3.0
Peat or Muck	NOT RECOMMENDED					

* For organic matter content between those listed, adjust the rate proportionately.
** Do not use in the light sandy soils of the Atlantic Coastal Plain.

TABLE 5
PREEMERGENCE BROADCAST APPLICATION RATES PER ACRE FOR "BLADEX" 4L PLUS ATRAZINE 4L USED IN TANK-MIX COMBINATIONS WITH "LASSO", "SUTAN+", "GENATE PLUS", "ERADICANE", OR "DUAL" ON CORN

SOIL TEXTURE DESCRIPTION	QUARTS OF "BLADEX" 4L + QUARTS OF ATRAZINE 4L***					
	PERCENT ORGANIC MATTER IN SOIL*					
	LESS THAN 1%	1%	2%	3%	4%	5% & OVER
Sand, Loamy sand	0.4+0.2**	0.5+0.25	0.75+0.5	1.0+0.5	1.25+0.5	1.25+0.75
Sandy loam	0.5+0.25	0.75+0.5	1.0+0.5	1.25+0.5	1.25+0.75	1.5+0.75
Loam, Silt loam, Silt	0.75+0.5	1.0+0.5	1.25+0.5	1.25+0.75	1.5+0.75	1.75+0.75
Sandy clay loam, Clay loam, Silty clay loam	1.0+0.5	1.25+0.5	1.25+0.75	1.5+0.75	1.75+0.75	2.0+0.75
Sandy clay, Silty clay, Clay	1.25+0.5	1.25+0.75	1.5+0.75	1.75+0.75	2.0+0.75	2.0+1.0
Peat or Muck	NOT RECOMMENDED					

* For organic matter content between those listed, adjust the rate proportionately.
** Do not use in the light sandy soils of the Atlantic Coastal Plain.
*** If Atrazine 90% is used, multiply rates shown by 1.11 to equal quarts of Atrazine 4L. If Atrazine 80W is used, multiply rates shown by 1.25 to equal pounds of Atrazine 80W.

CONVERSATION TILLAGE WEED CONTROL

Early Preplant (EPP)

BLADEX 4L may be used for Early Preplant or Preemergence weed control for land going into production of corn under conservation tillage programs. Complete any planned early spring tillage prior to application. Apply herbicide treatment before weeds germinate or before weed seedlings are more than 3 inches tall. Tillage after application may reduce the effectiveness of the herbicide treatment. A nitrogen solution or complete fertilizer solution may replace all or part of the water as a carrier. The spray gallonage and spray boom design must be adequate to give thorough uniform coverage of the weed foliage. Follow label requirements of all products used in tank mix combinations.

BLADEX alone or in combination with atrazine: Apply 15 to 30 days prior to planting. Use the proper rate for soil texture and organic matter indicated in Table 1, 2 or 3. Where heavy crop residues exist, the rates shown in Tables 1, 2 or 3 should be increased by 25%.

BLADEX 4L plus "Princep" 4L or "Princep" "Caliber" 90 or BLADEX 4L plus Atrazine plus "Princep" 4L or "Princep" "Caliber" 90: Apply 30 days or more prior to planting. Use the proper rate of BLADEX 4L or of BLADEX 4L plus Atrazine for the soil texture and organic matter as shown in Table 1, 2 or 3 and add 1 quart/acre of "Princep" 4L or 1.11 pounds/acre of "Princep" "Caliber" 90.

Burndown of Existing Weeds:

Combinations with 2,4-D: Where broadleaf weeds are present at the time of application, add 2,4-D LV Ester at 1-1/3 - 2 pts./A. (6 lb./gal.) or 2-3 pts./A. (4 lbs./gal.) (or 2,4-D Amine at recommended rates) plus "Ortho X-77" surfactant at 1 qt./100 gals. of diluted spray, or other suitable surfactant at its recommended rate.

Combinations with Paraquat Products: When grasses are present and/or when existing weeds exceed 3 inches in height add 1-2 pts./A. (2 lbs./gal.) or 1.3-2.7 pts./A. (1.5 lbs./gal.) of paraquat. Well established weeds 6 inches tall or taller may not be well controlled.

Apply BLADEX 4L in at least 20 gal./A. of carrier by ground sprayer. (The volume of carrier and the application equipment must be adequate to give a uniform application.) Add Ortho surfactant at 1-2 qts./100 gals. of diluted spray (or other suitable surfactant at recommended rates) where paraquat is used. Crop oil concentrate or emulsible vegetable oil are not needed where paraquat is used.

Depending upon weather conditions following the EPP application, a postemergence treatment of BLADEX or some other herbicide treatment may be necessary at or after planting to provide additional length of weed control. If desired, 1-1/2-2 pts./A. of "Dual" 8E or 2 qts./A. of "Lasso" 4EC may be tank-mixed with the BLADEX EPP treatment or applied preemergence at planting.

Rotational Corps: Refer to Rotational Crops section for each treatment in the Preemergence section of the label.

At Planting

BLADEX 4L applied alone or in combination with atrazine and/or "Lasso" or "Dual" according to the following directions will kill most existing small weeds and suppress many emerged perennial weeds when corn is planted into no-till stalk ground (corn, sorghum), stubble ground (soybean, small grains), and any minimum-till situation. This treatment then provides residual control of annual weeds as in conventional tillage.

Apply BLADEX 4L alone or with other products according to the directions for those treatments in the Preemergence section of the label. Where heavy crop residues exist, the BLADEX rate shown in Tables 1 through 5 should be increased by 25%. Add 0.5-1.0 pt./A. of 2,4-D LV (6 lbs./gal.) or 0.75 to 1.5 pts./A. (4 lbs./gal.) (or 2,4-D Amine at recommended rates). Add the 2,4-D to the spray tank last.

Use a minimum of 15 gals./A. of carrier. Complete spray coverage of the weeds is essential for best performance. Nitrogen solutions and complete liquid fertilizers are the preferred carriers for this treatment because they aid in the burndown of existing weeds. Add "Ortho X-77" surfactant at 1-2 qts. per 100 gals. of diluted spray, or other suitable surfactant at its recommended rate. If water is used as a carrier, crop oil concentrate may be used as a surfactant. Apply before weeds exceed 3 inches in height. For control of existing alfalfa add 1/3 - 1/2 pt./A. of "Banvel" to the spray mixture. Apply before the alfalfa exceeds 6 inches in height.

For fields with existing sod grasses such as orchardgrass, bromegrass, rye or timothy, or when very dry conditions exist, or when existing weeds exceed 3 inches in height, add paraquat to the tank mix. Use 2 pts../A. (2 lbs./gal.) or 1.3 pts./A. (1.5 lbs./gal.) of paraquat in combination with BLADEX as described above in this section, except the 2,4-D may be ommited, if desired, and the gallonage should be increased to 20-40 gals./acre to give thorough coverage of the weed growth. Do not apply paraquat in suspension type liquid fertilizer.

POSTEMERGENCE

Under warm, sunny, dry conditions of low humidity and the absence of dew formation at night, BLADEX 4L may be applied preemergence on field corn only. Do not use a surfactant, crop oil, or any other adjuvant. Addition of a surfactant, EV oil, or fertilizer solution is not recommended because moderate to severe injury, including stand loss, may occur.

Do not apply this treatment under cold, wet weather conditions or to corn growing under stress caused by weather, insects, disease, etc. Yellowing of the corn, stunting or stand loss may result from this treatment, particularly if high rainfall or cold, adverse growing conditions occur after application. Do not apply postemergence on popcorn, sweet corn or corn grown for seed.

BLADEX Applied Alone

Use BLADEX 4L at the proper rate for the soil texture and organic matter shown in Table 6 or 7. Use rates shown in Table 6 if BLADEX, CONQUEST, EXTRAZINE or EXTRAZINE II have not been applied to the soil this season. Use rates shown in Table 7 if BLADEX, CONQUEST EXTRAZINE or EXTRAZINE II have been applied to the soil this season. This treatment may be used on peat or muck soils for burndown and suppression of existing weeds but will not provide residual control. Apply from crop emergence through the four-leaf stage of corn growth but before weeds exceed about 1 1/2 inches in height. Do not apply over the top of corn if the fifth leaf is visible. Apply in water only. Do not spray emerged corn plants with BLADEX in a liquid fertilizer carrier or in tank mix combinations with EC formulation herbicides.

Any rotational crop may be planted the fall or spring following this treatment.

TABLE 6
POSTEMERGENCE BROADCAST APPLICATION RATES PER ACRE FOR "BLADEX" 4L ON FIELD CORN
NO PRIOR APPLICATION OF "BLADEX", "CONQUEST", "EXTRAZINE" OR "EXTRAZINE" II

	QUARTS OF "BLADEX" 4L			
	PERCENT ORGANIC MATTER IN SOIL*			
SOIL TEXTURE	LESS THAN 1%	1%	2%	OVER 2%
Sand, Loamy sand	DO NOT USE	1.1	1.5	2.0
Sandy loam	1.1	1.5	2.0	2.0
Loam, Silt loam, Silt	1.5	2.0	2.0	2.0
All other textures	2.0	2.0	2.0	2.0

*For organic matter content between those listed, adjust the rate proportionately.

TABLE 7
POSTEMERGENCE BROADCAST APPLICATION RATES PER ACRE FOR "BLADEX" 4L ON FIELD CORN
"BLADEX", "CONQUEST", "EXTRAZINE" OR "EXTRAZINE" II USED IN PRIOR APPLICATION

	QUARTS OF "BLADEX" 4L			
	PERCENT ORGANIC MATTER IN SOIL*			
SOIL TEXTURE	LESS THAN 1%	1%	2%	OVER 2%
Sand, Loamy sand	DO NOT USE	DO NOT USE	1.3	1.3
Sandy Loam	DO NOT USE	DO NOT USE	1.5	1.8
Loam, Silt loam, Silt	DO NOT USE	1.3	1.8	1.8
All other textures	DO NOT USE	1.8	1.8	2.0

*For organic matter content between those listed, adjust the rate proportionately.

BLADEX COMBINATIONS

BLADEX plus Atrazine

Apply as directed in "Postemergence-BLADEX Applied Alone" section of this label. Use an amount of BLADEX 4L plus Atrazine equal to the rate shown in Table 6 or 7 for the proper soil texture and organic matter. To determine the amount of BLADEX 4L to use, multiply the rate in Table 6 or 7 by 0.6. To determine the amount of Atrazine 4L to use, multiply the rate indicated in Table 6 or 7 by 0.4. (Multiply by 0.5 for pounds of Atrazine 80W.) (For the 2.0 qts/A. rate shown in Table 6 or 7, use 1.2 qts/A. of BLADEX 4L plus 0.8 qts/A. of Atrazine 4L or 1.0 lbs/A. of Atrazine 80W.)

Rotational Crops: See Rotational section in the "Preemergence BLADEX plus Atrazine" section of this label.

SWEET CORN

BLADEX 4L may be applied preemergence or preplant incorporated for the control of annual grasses and broadleaved weeds in sweet corn.

NOTE: BLADEX may cause injury or stand loss on new or "super sweet" varieties of sweet corn. Consult with Agricultural Extension Agencies and sweetcorn seed suppliers about the sensitivity of new varieties to potential injury.

Apply BLADEX treatments just before, at or after planting but before crop has emerged. Avoid removal of treated soil from seedrow prior to or during the planting operation. Do not apply postemergence to sweet corn. Rotary hoeing is recommended for preemergence application which do not receive adequate rainfall or sprinkler irrigation to wet the top 1 1/2 to 2 inches of soil within about 10 days after application. If a BLADEX mixture is to be incorporated, except as noted, single or two pass incorporation is acceptable. Care should be taken to incorporate the BLADEX mixture no deeper than the top two inches of soil.

TABLE 8
PREEMERGENCE BROADCAST APPLICATION RATES PER ACRE FOR
TANK-MIX COMBINATIONS OF "BLADEX" 4L PLUS ATRAZINE 4L ON SWEET CORN

	QUARTS OF "BLADEX" 4L + QUARTS OF ATRAZINE 4L**					
	PERCENT ORGANIC MATTER IN SOIL*					
SOIL TEXTURE DESCRIPTION	LESS THAN 1%	1%	2%	3%	4%	5% & OVER
Sand, Loamy sand	DO NOT USE	0.8+0.4	1.1+0.4	1.3+0.7	1.5+0.9	2.2+1.1
Sandy loam	DO NOT USE	1.1+0.4	1.3+0.7	1.5+0.9	2.0+1.1	2.8+1.3
Loam, Silt loam, Silt	DO NOT USE	1.3+0.7	1.5+0.9	2.0+1.1	2.5+1.3	3.2+1.3
Sandy clay loam, Clay loam, Silty clay loam	DO NOT USE	1.5+0.9	1.8+1.1	2.5+1.3	3.2+1.3	3.4+1.6
Sandy clay, Silty clay loam	DO NOT USE	1.8+1.1	2.8+1.3	3.2+1.3	3.4+1.6	3.6+1.8
Peat or Muck	NOT RECOMMENDED					

* For organic matter content between those listed, adjust the rate proportionately.
** If Atrazine 90% is used, multiply rates shown by 1.11 to equal lbs. of 90% Atrazine products. If Atrazine 80W is used, multiply rates shown by 1.25 equal lbs. of Atrazine 80W.

BLADEX PLUS ATRAZINE

Apply BLADEX 4L at the proper rate for soil texture and organic matter indicated in Table 8. Table 8 provides rates for generally weedy conditions. The ratio of the amounts of each herbicide may be adjusted as necessary for particular weed conditions as long as the combined rate of the two products does not exceed the combined rate for the soil shown in Table 8. For grassier conditions use a ratio that contains higher levels of BLADEX (3:1). For fields with more broadleaves use a ratio that contains higher levels of atrazine (1:1).

Rotational Crops: (1) Plant only corn, sorghum or soybeans the year following the use of this mixture. (2) If soybeans are to be planted, injury may occur due to the carryover of Atrazine. (3) If applied after June 10, do not rotate with crops other than corn or sorghum the next year or injury may occur. (4) Small grains may be planted 15 months following treatments. (5) All other crops may be planted 18 months after application.

"BLADEX" PLUS "LASSO" 4EC

Use BLADEX 4L at the proper rate for the soil texture and organic matter shown in Table 9 plus 2 quarts per acre of "Lasso". (Use 2.5 quarts

"Lasso" on clay soils containing 5 percent organic matter and over.) Any rotational crop may be planted the fall or spring following this treatment.

"BLADEX" PLUS "SUTAN +" 6.7E, "GENATE PLUS" 6.7EC OR "ERADICANE" 6.7E

Use BLADEX 4L at the proper rate for the soil texture and organic matter shown in Table 9 plus 1.8 quarts per acre of "Sutan+", "Genate Plus" or "Eradicane" for control of many annual grasses and broadleaf weeds. (Use 2.4 quarts of "Sutan+", "Genate Plus" or "Eradicane" on loam soils containing 5 percent or more organic matter, and clay loams and clays containing 4 percent or more organic matter.) Do not use on sands and loamy sands having less than 1 percent organic matter nor on the light sandy soils of eastern coastal states. Do not use on corn grown for seed.

Apply before planting. Incorporate the mixture immediately upon application using power-driven cultivation equipment set for 2-3 inch depth, or tandem disc set to cut about 4 inches deep while operating at 4-6 mph. For thorough mixing, disc in two directions (cross disc), and follow with a harrow, drag, or other leveling device. Prior to the second discing, readjust the disc to prevent cutting deeper than 4 inches. BLADEX 4L may be applied preemergence as an overlay over previously incorporated "Sutan+", "Genate Plus" or "Eradicane", if desired. Any rotation crop may be planted in the fall or spring following these treatments.

Existing stands of quackgrass, purple and yellow nutsedge must be turned under and thoroughly chopped up prior to chemical treatments.

Additional weeds controlled by "Sutan+", "Genate Plus" or "Eradicane" combinations:

Grasses:

Sandbur
Texas Panicum
Shattercane (Wild Cane)*
Quackgrass ("Eradicane" Only)
Wild Proso Millet* ("Eradicane" only)

Perennial Weeds

Yellow Nutsedge (Nutgrass)
Purple Nutsedge (Nutgrass)

* Suppression only - refer to "Sutan+", "Genate Plus" or "Eradicane" label for appropriate supplemental cultural and tillage practices.

For fields with moderate to heavy infestations of these weeds refer to the "Sutan+", "Genate Plus" or "Eradicane" labels for appropriate higher rates.

BLADEX PLUS "DUAL" 8E

Use BLADEX 4L at the proper rate for soil texture and organic matter shown in Table 9. Use "Dual" as follows:

Soil Texture	Broadcast Rate Per Acre for "Dual"
Coarse Sand, Loamy sand, Sandy loam	1.25 - 1.5 pints
Medium Loam, Silt loam, Silt	1.5 - 2.0 pints
Fine Sandy clay loam, Silty clay loam, Clay loam, Sandy clay, Silty clay, Clay	1.5 - 2.5 pints

The low end of the rate range should be used for lowest organic matter soils and the rate increased proportionately as the organic matter increases. Soils containing 4 percent organic matter or more require the highest rate shown for that soil texture. Refer to the "Dual" label for precautions or rotational crops.

TABLE 9
PREEMERGENCE BROADCAST APPLICATION RATES PER ACRE FOR "BLADEX" 4L USED IN TANK MIX COMBINATIONS WITH "LASSO", "SUTAN+", "GENATE PLUS", "ERADICANE", OR "DUAL" ON SWEET CORN

SOIL TEXTURE DESCRIPTION	QUARTS OF "BLADEX" 4L — PERCENT ORGANIC MATTER IN SOIL*					
	LESS THAN 1%	1%	2%	3%	4%	5% & OVER
Sand, Loamy sand	DO NOT USE	0.8	1.2	1.4	1.6	2.0
Sandy loam	DO NOT USE	1.2	1.4	1.6	2.0	2.2
Loam, Silt loam, Silt	DO NOT USE	1.4	1.6	2.0	2.2	2.6
Sandy clay loam, Clay loam, Silty clay loam	DO NOT USE	1.8	2.0	2.2	2.6	2.8
Sandy clay, Silty clay loam	DO NOT USE	2.0	2.2	2.6	2.8	3.0
Peat or Muck	NOT RECOMMENDED					

* For organic matter content between those listed, adjust the rate proportionately.
** Do not use in the light sandy soils of the Atlantic Coastal Plain.

TABLE 10
PREEMERGENCE BROADCAST APPLICATION RATES PER ACRE FOR "BLADEX" 4L PLUS ATRAZINE 4L USED IN TANK-MIX COMBINATIONS WITH "LASSO", "SUTAN+", "GENATE PLUS", "ERADICANE", OR "DUAL" ON SWEET CORN

SOIL TEXTURE DESCRIPTION	QUARTS OF "BLADEX" 4L + QUARTS OF ATRAZINE 4L** — PERCENT ORGANIC MATTER IN SOIL*					
	LESS THAN 1%	1%	2%	3%	4%	5% & OVER
Sand, Loamy sand	DO NOT USE	0.6+0.2	0.9+0.4	1.0+0.5	1.0+0.6	1.4+0.6
Sandy loam	DO NOT USE	0.8+0.4	1.0+0.5	1.1+0.6	1.4+0.6	1.6+0.6
Loam, Silt loam, Silt	DO NOT USE	1.0+0.5	1.2+0.6	1.4+0.6	1.6+0.6	1.8+0.9
Sandy clay loam, Clay loam, Silty clay loam	DO NOT USE	1.2+0.6	1.4+0.6	1.6+0.6	1.8+0.9	2.0+0.9
Sandy clay, Silty clay loam	DO NOT USE	1.4+0.6	1.6+0.9	1.8+0.9	2.0+0.9	2.0+1.1
Peat or Muck	NOT RECOMMENDED					

* For organic matter content between those listed, adjust the rate proportionately.
** If a 90% Atrazine is used multiply Atrazine rates shown by 1.11 to equal lbs. of 90% Atrazine products. If Atrazine 80W is used, multiply rates shown by 1.25 to equal lbs. of Atrazine 80W.

"BLADEX" PLUS ATRAZINE PLUS "LASSO", "SUTAN+", "GENATE PLUS", "ERADICANE" OR "DUAL"

NOTE: Do not use combinations with "Sutan+", "Genate Plus" or "Eradicane" in New Jersey. Use BLADEX 4L plus Atrazine at the proper rate of soil texture and organic matter shown in Table 10. Use "Lasso", "Sutan+", "Genate Plus", "Eradicane" or "Dual" according to rates shown in BLADEX combinations in this section of the label.

Rotational Crops: Refer to Rotational Crops section of BLADEX plus Atrazine in this section of the label.

COTTON

IDLE SEASON EARLY PREPLANT WEED CONTROL (CALIFORNIA ONLY)

Weeds Controlled

Annual Bluegrass
Annual Ryegrass
Barnyardgrass*
Bristly Foxtail
Rabbitsfoot Grass
Volunteer Small Grains (suppression)
Wild Oat*
Yellow Foxtail

Broadleaves

Annual henbit
Black nightshade
Burclover
Cheeseweed*
Chickweed
Fiddleneck
Lambsquarters
London rocket
Pineapple weed
Prickly lettuce
Shepherdspurse
Sowthistle

Groundsel
Knotweed
Marestail
Miners lettuce
Wild mustard
Wild radish

* Under soil moisture conditions favoring deep germination, these species may not be completely controlled.

BLADEX may be used for burndown of small existing annual weeds and residual control of weeds during the winter and early spring season prior to planting cotton in California only. Complete any planned tillage prior to application. Apply herbicide treatment before weeds germinate or before weed seedlings are more than 3 inches tall. Tillage after application may reduce the effectiveness of the herbicide treatment.

Apply BLADEX at least 30 days prior to planting. Apply the proper rate for the soil texture, organic matter and time interval between application and planting indicated in the Table 11. Where existing weeds are present, add crop oil concentrate, surfactant, or emulsible vegetable oil at its recommended rate to aid in the burndown of small weeds.

Where existing weeds are greater than 3 inches in height, when very dry conditions exist or where volunteer grains are a major problem, tank-mix BLADEX with 1-2 pts./A. (2 lbs./gal.) or 1.3-2.7 pts./A. (1.5 lbs./gal.) of paraquat. Well established weeds 6 inches tall or taller may not be well controlled.

Apply BLADEX 4L in at least 20 gal./A. of carrier by ground sprayer. (The volume of carrier and the application equipment must be adequate to give a uniform application.) Add "Ortho X-77" surfactant at 1-2 qts./100 gals. of diluted spray (or other suitable surfactant at recommended rates) where paraquat is used. Crop oil concentrate or emulsible vegetable oil are not needed where paraquat is used.

BLADEX can also be tank-mixed with "Treflan" or "Prowl" and incorporated for fall listed cotton beds instead of surface applied as described above. Precautions: 1). Failure to wait the recommended time interval between application and planting may result in crop injury. 2). At least one inch of rainfall or an equivalent irrigation that waters the surface of the soil after application must precede planting. 3). The use of this treatment on calcareous or caliche soil outcroppings may result in crop injury. 4). Do not graze or feed foliage from treated areas to livestock. 5). Do not apply BLADEX to cotton land in irrigation water.

TABLE 11
BROADCAST APPLICATION RATES PER ACRE OF "BLADEX" 4L FOR IDLE SEASON OR EARLY PREPLANT TREATMENT ON COTTON

SOIL TEXTURE DESCRIPTION	DAYS PRIOR TO PLANTING*					
	30 DAYS		60 DAYS		90 DAYS	
	QTS. OF "BLADEX" 4L		QTS. OF "BLADEX" 4L		QTS. OF "BLADEX" 4L	
	ORGANIC MATTER		ORGANIC MATTER		ORGANIC MATTER	
	UNDER 2%	OVER 2%	UNDER 2%	OVER 2%	UNDER 2%	OVER 2%
Sands, Loamy sands	1.5	2.0	2.5	3.0	3.0	3.5
All other Soils	2.0	2.5	3.0	3.5	3.5	4.0

* For time intervals between those listed, adjust the rates proportionately

PREEMERGENCE

Apply preemergence on cotton only in the States of Alabama, Arkansas, Louisiana and Mississippi.

Weeds Controlled

Annual morningglory
Cocklebur
Prickly sida (Teaweed)
Spurge

BLADEX 4L is a selective preemergence herbicide for early season weed control in cotton. Supplemental practices (such as BLADEX applied directed postemergence) may be necessary to control late season weeds. BLADEX 4L can be used in a tank-mix combination with Zorial Rapid 80.

Carefully match the BLADEX rate with the soil texture. Do not use on fields where the soil texture changes from coarse to fine. Avoid overlapping the spray pattern or overdosing the field with BLADEX 4L. Application rates above those recommended for the soil texture can result in yellowing or stunting of the crop and may result in stand reduction.

While cotton exhibits tolerance to BLADEX 4L, adverse growing conditions such as excessive rains, standing water or cold weather may result in stand reduction.

DO NOT GRAZE OR FEED FOLIAGE FROM TREATED AREAS TO LIVESTOCK.

BLADEX plus "Zorial Rapid 80".

Apply BLADEX 4L plus "Zorial Rapid 80" at the proper rate for the soil texture shown in Table 12. The soil must contain at least 1.0 percent organic matter. Seed placement should be 1/2-3/4 inch from the soil surface. Plant only cotton within six months after the last application of "Zorial Rapid 80" or injury may occur.

DIRECTED POSTEMERGENCE - LAYBY

Weeds Controlled

Annual morningglory*
Bristly starbur
Cocklebur
Crotalaria
Jimsonweed
Lambsquarters
Nightshade (annual)
Palmer amaranth
Pigweed (redroot and spiny)
Prickly sida (Teaweed)
Sicklepod
Spurge
Tropic croton
Wright groundcherry

* The degree of preemergence control from a layby treatment will be reduced if soil moisture and temperature conditions cause deep germination of the seed.

TABLE 12
PREEMERGENCE BROADCAST APPLICATION RATES PER ACRE FOR "BLADEX" 4L + "ZORIAL RAPID 80" ON COTTON

Soil Texture*	Quarts/Acre "Bladex" 90DF	+	Quarts/Acre "Zorial Rapid 80"
Coarse Soils** Sandy loam	0.5	+	0.8
Medium Soils Silt and Silt loam,	0.6	+	1.3
Loam, Clay loam, Sandy clay loam, Sandy clay	0.9	+	1.3
Fine Soils Silty clay loam, Silty clay, Clay	1.2	+	1.6

* The soil must contain at least 1% organic matter.
** Do not use on coarse soils (Sands and Loamy sands) containing more than 70% sand.

TABLE 13
DIRECTED POSTEMERGENCE
APPLICATION RATES PER ACRE FOR "BLADEX" 4L ON COTTON

BROADCAST	BANDED 38" ROW	
	12" BAND	19" BAND
0.6 - 1.0 qt.	0.2 - 0.3 qt.	0.3 - 0.5 qt.

Use the maximum rate when dry or arid conditions exist.

TABLE 14
LAYBY APPLICATION RATES PER ACRE FOR "BLADEX" 4L ON COTTON

HEIGHT OF COTTON	SOIL TEXTURE		BROADCAST RATES
12 inches or more	Coarse	Sandy loam	0.8 qt.
	Medium	Silt, Silt loam, Loam, Clay loam, Sandy clay loam, Sandy clay	1.2 qts.

HEIGHT OF COTTON	SOIL TEXTURE		BROADCAST RATES
	Fine	Silty clay loam, Silty clay and Clay	1.6 qts.

TABLE 15
DIRECTED POSTEMERGENCE
APPLICATION RATES PER ACRE "BLADEX" 4L + MSMA ON COTTON

PRODUCT	BROADCAST	BANDED 38" ROW	
		12" BAND	19" BAND
BLADEX 4L + MSMA (4 lb/gal) or MSMA (6.6 lb/gal)	0.6 - 1.0 qt. + 4 pints or 2.4 pints	0.2 - 0.3 qt. + 1.3 pints or 0.8 pint	0.3 - 0.5 qt. + 2 pints or 1.2 pints

BLADEX 4L and tank-mix combinations may be applied directed postemergence or layby to cotton and either preemergence to weeds or postemergence to weeds in all cotton growing States.

Apply before weeds are more than 2 inches tall. Apply the directed postemergence treatment after the cotton has attained the minimum height of 6 inches. For layby treatment, apply BLADEX 4L after the cotton has attained the height of 12 inches or more.

The spray mixture should be directed to the soil around the base of the cotton plants. Care should be taken to prevent the spray from striking the cotton leaves as injury will occur. The use of leaf lifters or shields on application equipment is recommended to avoid spraying the cotton foliage.

BLADEX may be applied directed postemergence and/or layby following a preemergence application of BLADEX. Apply no more than two directed postemergence and one premergence application to the same crop in any one year. If BLADEX is not used preemergence, apply no more than three directed postemergence applications including layby to the same crop in any one year. (In California, apply no more than two directed postemergence applications including layby.)

When applied as a layby treatment before weeds emerge, the effectiveness of BLADEX 4L depends on rainfall or irrigation to move it into the soil. When irrigation water activation is used, every row must be watered and for skip row cotton all treated soil must be irrigated.

Any rotational crop may be planted the fall or spring following any of the treatments in this section providing the soil is plowed or deep-disced prior to planting the rotation crop.

Do not apply BLADEX to cotton in irrigation water.

DO NOT GRAZE OR FEED FOLIAGE FROM TREATED AREAS TO LIVESTOCK.

BLADEX Applied Alone

Apply BLADEX 4L directed-postemergence at the rate shown in Table 13. Apply at layby at the rates for the soil texture indicated in Table 14. Add an agricultural surfactant such as "Ortho X-77" or other surfactant suitable for use on growing cotton at the rate of 2 quarts per 100 gallons of spray mixture (or as directed by the manufacturer).

BLADEX plus MSMA

Apply a tank-mix combination of BLADEX 4L plus MSMA plus surfactant after the cotton is 6 inches tall but before it reaches the bloom stage. Apply no more than two applications of this mixture before the first bloom stage. Tank-mix BLADEX plus MSMA at the rates indicated in Table 15. Add a surfactant, such as "Ortho X-77", at the rate of 2 qts./100 gals of spray mixture (or as directed by the manufacturer).

GRAIN SORGHUM (MILO)
(See the GENERAL INFORMATION section of this label)

Tank-mix combinations of "BLADEX" 4L plus atrazine, metolachlor (Dual), alachlor (Lasso), propazine (Milogard) or propachlor (Ramrod) may be used for selective preemergence weed control in grain sorghum. In addition, BLADEX 4L may be used in tank-mix combinations with atrazine, "Dual" or "Lasso" for the control of weeds in early spring, early preplant, 14 to 35 days or more prior to planting grain sorghum. BLADEX 4L in tank mix combinations may be applied preemergence or early preplant on grain sorghum grown under conventional or conservation tillage systems.

Do not use on forage sorghums. See the appropriate sections of the label for geographic distribution restrictions.

WEEDS CONTROLLED BY "BLADEX" 4L IN TANK-MIX OR SEQUENTIAL COMBINATIONS WITH OTHER HERBICIDES ON GRAIN SORGHUM (MILO)

Grasses

Cheatgrass	Stinkgrass (Indian lovegrass)
Green foxtail	Volunteer Wheat (2)
Crabgrass	Yellow foxtail
Downy brome	

Broadleaves

Annual Moringglory	Lambsquarters	Russian thistle
Carpetweed	Pennsylvania smartweed	Shepherdspurse
Cocklebur (1)	Prostrate pigweed	Sunflower (1)
Common purslane	Prickly lettuce	Tansy and other mustards
Horseweed (marestail)	Ragweed (Common)	Velvetleaf
Kochia	Redroot pigweed	

(1) Under soil moisture and temperature conditions favoring deep germination or other factors that may cause delayed germination, these species may not be completely controlled.

(2) When the herbicide treatment is applied two weeks or more before planting, weed control of these species may break early if heavy rainfall occurs between application and planting.

General Directions For Use On Grain Sorghum

BLADEX used in tank-mixes with the products listed above should be applied only once per crop season or in split dosage treatments. If replanting of grain sorghum is necessary, it may be planted in soil previously treated with these mixtures. Apply BLADEX 4L in these tank-mix combinations before the crop has emerged.

Precautions: Do not make an additional application of BLADEX, EXTRAZINE, or any product containing cyanazine or crop injury may occur. Heavy rainfall beteen planting and crop emergence may cause crop injury or stand loss. Rainfall tends to cause excessive concentrations of herbicide in seed furrow, resulting in possible crop injury. Level deep planter marks or seed furrows before application. Do not apply to furrow-planted sorghum.

Sorghum growing under stress caused by minor element deficiency, cold, wet weather or sorghum growing on highly calcareous soil (high pH) may suffer in injury including stand reduction. Sorghum subjected to high winds, sand cutting, hail damage, or cold temperatures may be more susceptible to injury from the chemical treatment with possible stand loss. Where crop residues are pressed into the planter slot or any other factor keeps the slot from closing, crop injury or stand reduction may occur, caused by herbicides coming into direct contact with the seed from the spray or by being washed into the slot after a heavy rain.

When using "Dual" on sorghum, only "Concep II" safened seed may be used. When using "Lasso" on sorghum, only "Screen" safened seed may be used.

Early preplant treatments will require a preemergence herbicide treatment other than BLADEX 90DF, 80W or 4L, or EXTRAZINE 90DF or 4L at planting, and/or postemergence herbicide in the growing crop to provide required weed control, if the early preplant application is made more than 35 days prior to planting, or if weeds are present at planting time.

WHEAT/SORGHUM/FALLOW ROTATION

For sorghum grown under a wheat/sorghum/fallow rotation, BLADEX 4L plus atrazine may be applied early preplant or preemergence. If an early preplant application with BLADEX 4L plus atrazine is used, it should be used in conjunction with either a residual herbicide after wheat harvest the previous year, a preemergence herbicide treatment at planting, and/or a postemergence herbicide treatment in the growing crop. The total rate of atrazine permitted in conjunction with a sorghum crop (postharvest plus early preplant plus preemergence) is limited to 3.0 lbs. active ingredient per acre. Even this or lower rates may carry over to injure rotational crops.

Rotational Crops: (1) Plant only corn, sorghum, or soybeans the year following the use of this mixture. (2) If soybeans are to be planted, injury may occur. (3) If applied after June 10, do not rotate with crops other than corn or sorghum the next year or injury may occur. (4) In the high plains and intermountain areas of the West where rainfall is sparse and erratic or where irrigation is required, use only when corn or sorghum is to be planted the following year, or a crop of corn or sorghum not treated with this mixture or atrazine is to precede other rotational crops. (5) Small grains may be planted 15 months following treatment. (6) All other crops may be planted 18 months after application.

Conservation Tillage

For grain sorghum grown under conservation tillage, any of the herbicide treatments listed above may be tank-mixed with paraquat, and/or 2,4-D where weeds and grasses are present at the time of application exceeding two inches in height.

If volunteer wheat or cheatgrass is over 2 inches tall, heavily tillered, and/or growing in a dense mat, or if the wheat stubble has been tilled (under-cut or disc) or grazed and seed is buried, complete control may not be achieved. Control of volunteer and other weeds is usually better in complete no-till situations than were prior tillage "planted" the seed and allowed extensive root development.

Complete spray coverage of the weeds is essential for best performance. Apply the desired treatment in 15 to 30 gallons of spray mixture per acre by ground rig. When using paraquat in a tank-mix, apply the desired rates in 20 to 40 gallons of spray mixture per acre. Use the higher volumes where there are heavy crop residues on the soil surface. Nitrogen solutions are the preferred carriers for these treatments as they aid in the burndown of existing weeds. In addition, crop oil or a nonionic surfactant may be added to the tank-mix as they aid in the burndown of existing weeds.

When tank-mixing with 2,4-D to control broadleaf weeds 14-35 days prior to planting, use 1 to 3 pints per acre of 4 pounds ai per gallon, 2,4-D LV (2/3 to 2 pints of 6 pounds ai per gallon 2,4-D LV) (or 2,4-D Amine at recommended rates). Use the higher rates where overwintering weeds are present or when directed on the 2,4-D label for the control of specific hard-to-kill weed species, such as perennials.

CAUTION: Use only these 2,4-D product with properly registered labels that permit such use and application rates.

When tank-mixing with paraquat to control grass and broadleaf weeds at the time of planting, use an approved nonionic adjuvant such as "Ortho X-77" at the rate of 1 quart per 100 gallons of dilute spray. Use 1-2 pints of paraquat (2 lbs./gal) or 1.3-2.7 pints (1.5 lbs/gal) per acre. Use the higher rate when weed growth is heavy or over 4 inches tall or when dry weather conditions prevail. Established weeds 6 inches tall or taller may not be completely controlled with paraquat.

OBSERVE ALL CAUTIONS AND LIMITATIONS ON LABELING OF ALL PRODUCTS USED IN MIXTURES.

EARLY PREPLANT

BLADEX 4L in tank-mix combinations with other herbicides may be applied Early Preplant on grain sorghum only in the states of Kansas, Nebraska and South Dakota. The tank-mixes may be applied on grain sorghum grown under conventional or conservation tillage systems. Do not use on forage sorghum.

TABLE 16
EARLY PREPLANT BROADCAST APPLICATION RATES OF "BLADEX" 4L PLUS ATRAZINE 4L ON GRAIN SORGHUM (0.8% TO 3% ORGANIC MATTER SOILS)

	QUARTS OF "BLADEX" 4L + QUARTS OF ATRAZINE 4L		
	DAYS PRIOR TO PLANTING (b,c)		
SOIL TEXTURE DESCRIPTION	14 DAYS	28 DAYS	35 DAYS
Sand, Loamy sand	---------------- DO NOT USE ----------------		
Sandy loam	1.2 + 0.6(a,d)	1.6 + 0.8	2.0 + 1.0
Loam, Silty loam, silt	1.5 + 0.7	2.0 + 1.0	2.3 + 1.2
Sandy clay loam, Clay loam, Silty clay loam	2.0 + 1.0	2.3 + 1.2	2.6 + 1.4
Sandy clay, Silty clay, Clay	2.2 + 1.1	2.3 + 1.2	3.0 + 1.5
Peat or Muck	NOT RECOMMENDED		
Eroded Slopes or Knobs Soils with pH greater than 8.0	NOT RECOMMENDED		

(a) The first number is quarts/acre of BLADEX 4L. The second number is quarts/acre of atrazine 4L. For pounds/acre of atrazine 80W multiply by 1.25
(b) For intervals longer than 35 days due to delayed planting, etc., another herbicide treatment is needed before, at, or after planting.
(c) For sandy loam soils containing less than 1.5% O.M., increase the treatment interval from 14 to 21 days. Use the 14 day interval for sandy soils containing more than 1.5%.
(d) For rates between those listed at 14 and 35 days, adjust the rate proportionately.

TABLE 17
EARLY PREPLANT BROADCAST APPLICATION RATES PER ACRE IN QUARTS OF "BLADEX" 4L, QUARTS OF ATRAZINE 4L AND PINTS OF "DUAL" 8E ON GRAIN SORGHUM (b)

	DAYS PRIOR TO PLANTING (c) (0.8% TO 3% ORGANIC MATTER)					
SOIL TEXTURE DESCRIPTION	14 DAYS* QTS "BLADEX"/QTS ATRAZ/PTS DUAL			28 DAYS QTS "BLADEX"/QTS ATRAZ/PTS DUAL		
Sand, Loamy sand	DO NOT USE			1.0(a)	0.5	1.8
Sandy loam	0.9(d)	0.4	1.35	1.1	0.5	1.8
Loam, Silt loam, Silt	1.1	0.5	1.35	1.35	0.7	1.8
Sandy clay loam, Clay loam, Silty clay loam	1.35	0.7	1.6	1.7	0.8	1.8
Sandy clay, Silty clay, Clay	1.5	0.75	1.6	1.8	0.9	1.8
Peat or Muck	NOT RECOMMENDED					
Eroded Slopes or Knobs Soils with pH greater than 8.0	NOT RECOMMENDED					

(a) The first number is quarts/acre of BLADEX 4L. The second number is quarts/acre of atrazine 4L. The third number is pints/acre of Dual 8E.
(b) If using Dual 8E on sorghum, Concep II safened seed should be planted.
(c) For intervals longer than 28 days due to delayed planting, etc., another herbicide treatment is needed before, at, or after planting.
(d) For sandy loam soils containing less than 1.5% O.M., increase the treatment interval from 14 to 21 days. Use the 14 day interval for sandy soils containing more than 1.5%.
* For rates between those listed at 14 and 28 days, adjust the rate proportionately.

TABLE 18
EARLY PREPLANT BROADCAST APPLICATION RATES PER ACRE IN QUARTS OF "BLADEX" 4L, QUARTS OF ATRAZINE 4L AND QUARTS OF "LASSO" 4EC ON GRAIN SORGHUM (b)

	DAYS PRIOR TO PLANTING (c) (0.8% TO 3% ORGANIC MATTER)					
SOIL TEXTURE DESCRIPTION	14 DAYS* QTS "BLADEX"/QTS ATRAZ/QTS "LASSO"			28 DAYS QTS "BLADEX"/QTS ATRAZ/QTS "LASSO"		
Sand, Loamy sand	DO NOT USE			0.9(a)	0.45	2.25
Sandy loam	0.7(d)	0.4	1.8	0.9	0.5	2.25
Loam, Silt loam, Silt	0.9	0.5	1.8	1.2	0.6	2.25
Sandy clay loam, Clay loam, Silty clay loam	1.2	0.6	2.0	1.5	0.75	2.25

SOIL TEXTURE DESCRIPTION	DAYS PRIOR TO PLANTING (c) (0.8% TO 3% ORGANIC MATTER)					
	14 DAYS* QTS "BLADEX"/QTS ATRAZ/QTS "LASSO"			28 DAYS QTS "BLADEX"/QTS ATRAZ/QTS "LASSO"		
Sandy clay, Silty clay, Clay	1.35	0.7	2.0	1.7	0.8	2.25
Peat or Muck	NOT RECOMMENDED					
Eroded Slopes or Knobs Soils with pH greater than 8.0	NOT RECOMMENDED					

(a) The first number is quarts/acre of BLADEX 4L. The second number is quarts/acre of atrazine 4L. The third number is quarts/acre of "Lasso" 4EC.
(b) If using "Lasso" 4EC on sorghum, Screen safened seed should be planted.
(c) For intervals longer than 28 days due to delayed planting, etc., another herbicide treatment is needed before, at, or after planting.
(d) For sandy loam soils containing less than 1.5% O.M., increase the treatment interval from 14 to 21 days. Use the 14 day interval for sandy soils containing more than 1.5%.
* For rates between those listed at 14 and 28 days, adjust the rate proportionately.

BLADEX PLUS ATRAZINE

BLADEX 4L plus atrazine provides control of weeds listed on this label in the Weeds Controlled section for BLADEX 4L on grain sorghum. Precautions: Crop injury can occur if the soil stays dry between application and planting. Delay planting until at least 10 days after a soaking rain after treatment (soaks the soil to a depth of 4 inches or more). Heavy rains between planting and emergence can cause crop injury or stand loss. Rotational Crops: Refer to the General Information section of the label. Fields treated with this tank-mix may be planted only to corn or grain sorghum within 12 months after this treatment. Other crops should not be planted for 18 months following this treatment.

Use BLADEX 4L plus atrazine 4L at the proper rate for soil texture and time interval indicated in Table 16.

Split Applications: BLADEX 4L plus Atrazine 4L dosage rates in Table 16 may be applied in a split application of 80% of the dosage 21 days or more before planting and the remaining 20% at planting time. If the season has been dry without sufficient rainfall to activate the herbicide after the initial application, the remaining 20% at planting time may be omitted to reduce the chances of crop injury.

BLADEX IN THREE-WAY COMBINATIONS

BLADEX 4L plus atrazine in combination with "Dual" or "Lasso" applied early preplant provides control of weeds listed on this label in the Weeds Controlled section for BLADEX 4L on grain sorghum. Additional weeds controlled by one of these combinations include:

Grasses

Barnyardgrass (1)	Stinkgrass (2)
Fall Panicum	Witchgrass (2)
Giant Foxtail	

(1) (2) Refer to Weeds Controlled section on grain sorghum for explanation.

Rotational Crops: If the crop treated with any of these combinations is lost, corn or grain sorghum may be replanted immediately without retreatment. Refer to the General Information section of the label for additional information regarding crop rotation.

BLADEX PLUS ATRAZINE PLUS "DUAL"

Dual may be used on grain sorghum only with seed treated with "Concep II" seed safener.

Use BLADEX 4L plus atrazine 4L plus "Dual" 8E at the proper rate for soil texture and time interval indicated in Table 17.

BLADEX plus atrazine plus "Lasso"

"Lasso" may be used on grain sorghum only with seed treated with "Screen" seed safener.

Use BLADEX 4L plus Atrazine 4L plus "Lasso" 4EC at the proper rate for soil texture and time interval indicated in Table 18.

PREEMERGENCE OR SHALLOW PREPLANT INCORPORATION

Tank-mix combinations of BLADEX 4L plus atrazine, metolachlor (Dual), alachlor (Lasso), propachlor (Ramrod) or propazine (Milogard) may be used for selective preemergence weed control in grain sorghum. BLADEX 4L and its tank mixes may be applied preemergence on grain sorghum grown under conventional or conservation tillage systems. Do not use on forage sorghum. BLADEX 4L in two-way tank-mix combinations with propachlor or propazine or in three-way combinations with atrazine and propachlor, "Lasso" or "Dual" should be applied only once per crop season or in split dosage treatments. (See the Early Preplant section.) If replanting of grain sorghum is necessary, it may be planted in soil previously treated with these mixtures. Do not make a second application of BLADEX EXTRAZINE or any product containing cyanazine or crop injury may occur.

Apply these tank- mix combinations at planting or after planting, but before the crop and weeds have emerged. Heavy rain immediately following application tends to cause excessive concentrations of herbicide in seed furrow and can cause crop injury and stand reduction. Do not apply to furrow-planted sorghum until furrows are leveled (plowed-in). Level deep planter marks or seed furrows before application. Sorghum growing under stress caused by minor element deficiency or cold, wet weather or sorghum growing on highly calcareous soil may suffer injury and stand loss.

BLADEX 4L plus Propachlor (Ramrod)

Weeds Controlled

Apply this tank-mix to grain sorghum only in States East of the Rocky Mountains.

Grasses	Giant foxtail
Barnyardgrass	Green foxtail
Crabgrass	Yellow foxtail
Fall panicum	
Broadleaves	Pigweed
Annual morningglory	Ragweed (Common)
Carpetweed	Smartweed (Pennsylvania)
Cocklebur*	Velvetleaf*
Common purslane	
Lambsquarters	

*Under conditions such as low temperatures, lack of soil surface moisture or other factors causing a delay in germination of the seeds, the degree of control may be impaired against these weeds.

Apply BLADEX 4L plus propachlor at the proper rate for soil texture and organic matter shown in Table 19.

Any rotational crop may be planted the fall or spring following this treatment.

BLADEX plus Propazine (Milogard)

Weeds Controlled

Apply to grain sorghum only in the States of Texas, Oklahoma and Kansas.

Weeds Controlled

Grasses	
Crabgrass	Green foxtail
Fall panicum	Yellow foxtail
Giant foxtail	
Broadleaves	
Annual morningglory	Pigweed
Carpetweed	Puncturevine
Cocklebur	Ragweed (Common)
Common purslane	Smartweed
Lambsquarters	Velvetleaf

TABLE 19
PREEMERGENCE BROADCAST APPLICATION RATES IN QUARTS OF "BLADEX" 4L PLUS QUARTS OF PROPACHLOR (RAMROD) 4L PER ACRE ON GRAIN SORGHUM

SOIL TEXTURE DESCRIPTION	PERCENT ORGANIC MATTER IN THE SOIL* 2% QTS "BLADEX" +	QTS "RAMROD"	3% QTS "BLADEX" +	QTS "RAMROD"
Sand, Loamy sand	DO NOT USE			
Sandy loam	1.0(a)	2.5	1.2	3.0
Loam, Silt loam, Silt	1.2	3.0	1.35	3.5

SOIL TEXTURE DESCRIPTION	PERCENT ORGANIC MATTER IN THE SOIL* 2% QTS "BLADEX" +	QTS "RAMROD"	3% QTS "BLADEX" +	QTS "RAMROD"
Sandy clay loam, Clay loam, Silty clay loam	1.35	3.5	1.6	4.0
Sandy clay, Silty clay, Clay	1.6	4.0	1.6	4.0
Peat or Muck	NOT RECOMMENDED			
Eroded Slopes or Knobs Soils with pH greater than 8.0	NOT RECOMMENDED			

(a) The first number is quarts/acre of BLADEX 4L. The second number is quarts/acre of Ramrod 4L.
* For organic matter content between those listed, adjust the rate proportionately.

TABLE 20
PREEMERGENCE BROADCAST APPLICATION RATES IN QUARTS OF BLADEX 4L PLUS QUARTS OF PROPAZINE (MILOGARD) 4L PER ACRE ON GRAIN SORGHUM

SOIL TEXTURE DESCRIPTION	"BLADEX" 4L QTS/ACRE	PROPAZINE 4L* LBS/ACRE
Sand, Loamy sand, Sandy loam	DO NOT USE	
Loam, Silt loam, Silt	0.7	0.75
Sandy clay loam, Clay loam, Silty clay loam	1.0	1.0
Sandy clay, Silty clay, Clay	1.2	1.25
Peat or Muck	NOT RECOMMENDED	
Eroded Slopes or Knobs Soils with pH greater than 8.0	NOT RECOMMENDED	

* If Propazine 80W is used to multiply rate shown by 1.25 to equal pounds of Propazine 80W.
If Propazine 90DF is used multiply rate shown by 1.11 to equal pounds of Propazine 90DF.

TABLE 21
PREEMERGENCE OR SHALLOW PREPLANT INCORPORATION BROADCAST APPLICATION RATES IN QUARTS OF "BLADEX" 4L, QUARTS OF ATRAZINE 4L PLUS PINTS OF "DUAL" 8E PER ACRE ON GRAIN SORGHUM (a,b)

SOIL TEXTURE DESCRIPTION	PRODUCT		PERCENT ORGANIC MATTER IN SOIL* 1%	2%	3%
Sand, Loamy sand, Sandy loam			DO NOT USE		
Loam Silty loam Silt	BLADEX 4L Atrazine 4L "Dual" 8E	Quarts Quarts Pints	0.66 0.3 1.5	0.7 0.4 1.5	0.8 0.4 1.5
Sandy clay loam Clay loam Silty clay loam	BLADEX 4L Atrazine 4L "Dual" 8E	Quarts Quarts Pints	0.7 0.4 1.5	0.8 0.4 1.75	0.9 0.5 1.75
Sandy clay Silty clay Clay	BLADEX 4L Atrazine 4L "Dual" 8E	Quarts Quarts Pints	0.8 0.4 1.5	0.9 0.5 1.75	0.9 0.5 2.0
Peat or Muck	NOT RECOMMENDED				
Eroded Slopes or Knobs Soils with pH greater than 8.0	NOT RECOMMENDED				

(a) When using "Dual" 8E on sorghum, only "Concep II" safened seed should be planted.
(b) To enhance weed control in areas of less than 25 inches of annual precipitation or where long dry periods are common, this treatment may require shallow incorporation with a tool such as a field cultivator operated at 5-7 mph.
* For organic matter content between those listed, adjust the rate proportionately.

TABLE 22
PREEMERGENCE OR SHALLOW PREPLANT INCORPORATION BROADCAST APPLICATION RATES IN QUARTS OF "BLADEX" 4L, QUARTS OF ATRAZINE 4L AND QUARTS OF "LASSO" EC PER ACRE ON GRAIN SORGHUM (a,b)

SOIL TEXTURE DESCRIPTION	PRODUCT		PERCENT ORGANIC MATTER IN SOIL* 1%	2%	3%
Sand, Loamy sand			DO NOT USE		
Sandy loam	BLADEX 4L Atrazine 4L "Lasso" 4EC	Quarts Quarts Quarts	none none none	0.66 0.3 2.0	0.7 0.4 2.0
Loam Silt loam Silt	BLADEX 4L Atrazine 4L "Lasso" 4EC	Quarts Quarts Quarts	0.66 0.3 2.0	0.7 0.4 2.0	0.8 0.4 2.0
Sandy clay loam Clay loam Silty clay loam	BLADEX 4L Atrazine 4L "Lasso" 4EC	Quarts Quarts Quarts	0.7 0.4 2.25	0.8 0.4 2.25	0.9 0.5 2.25
Sandy clay Silty clay Clay	BLADEX 4L Atrazine 4L "Lasso" 4EC	Quarts Quarts Quarts	0.8 0.4 2.25	0.9 0.5 2.25	0.9 0.5 2.5
Peat or Muck	NOT RECOMMENDED				
Eroded Slopes or Knobs Soils with pH greater than 8.0	NOT RECOMMENDED				

(a) When using "Lasso" 4EC on sorghum, only "Screen" safened seed should be planted.
(b) To enhance weed control in areas of less than 25 inches of annual precipitation or where long dry periods are common, this treatment may require shallow incorporation with a tool such as a field cultivator operated at 5-7 mph.
* For organic matter content between those listed, adjust the rate proportionately.

Use BLADEX 4L plus propazine (Milogard) 4L at the proper rate for soil texture and organic matter indicated in Table 20.

Rotational Crops: In the Texas Gulf Coast and Texas Blacklands, fields treated with this tank-mix may be planted to cotton, soybeans, or corn 12 months after treatment. In Oklahoma and west Texas, fields treated with this tank-mix may be planted to cotton or corn 12 months after application. In Kansas, corn may be planted in rotation 12 months after treatment. Other crops should not be planted for 18 months following treatment.

BLADEX IN THREE-WAY COMBINATIONS

BLADEX 4L plus plus atrazine plus "Dual", "Lasso" or "Ramrod" may be applied preemergence or with shallow incorporation for weed control in grain sorghum only in the states of Kansas, Nebraska and South Dakota. These treatments provide control of weeds listed on this label in the Weeds Controlled section for BLADEX 4L on grain sorghum. Additional weeds controlled by these combinations include:

Grasses

Barnyardgrass (1)	Witchgrass (2)
Stinkgrass (2)	Giant Foxtail
Fall Panicum	

(1) (2) Refer to Weeds Controlled section on grain sorghum for explanation.

Rotational Crops: If the crop treated with this combination is lost, corn or grain sorghum may be replanted immediately without retreatment. Refer to the "General Information" section of the BLADEX label and to the tank mix product labels for additional information regarding crop rotation.

"BLADEX" PLUS ATRAZINE PLUS "DUAL"

"Dual" can be used on grain sorghum only with seed treated with "Concep II" seed safener.

Use BLADEX 4L plus "Dual" 8E at the proper rate for soil texture and organic matter indicated in Table 21. Use only on soils having at least 1% organic matter.

"BLADEX" PLUS ATRAZINE PLUS "LASSO"

"Lasso" can be used on grain sorghum only with seed treated with Screen seed safener.

Use BLADEX 4L plus Atrazine 4L plus "Lasso" 4EC at the proper rate for soil texture and organic matter indicated in Table 22. Use only on soils having at least 1% organic matter.

"BLADEX" PLUS ATRAZINE PLUS "RAMROD"

Use BLADEX 4L plus Atrazine 4L plus "Ramrod" at the proper rate for soil texture and organic matter indicated in Table 23.

TABLE 23

PREEMERGENCE BROADCAST APPLICATION RATES IN QUARTS OF "BLADEX" 4L, QUARTS OF ATRAZINE 4L AND QUARTS OF PROPACHLOR (RAMROD) 4L PER ACRE ON GRAIN SORGHUM

	PERCENT ORGANIC MATTER IN THE SOIL*					
	2% QTS "BLADEX"/QTS ATRAZ/QTS "RAMROD"			3% QTS "BLADEX"/QTS ATRAZ/QTS "RAMROD"		
SOIL TEXTURE DESCRIPTION						
Sand, Loamy sand	DO NOT USE					
Sandy loam	0.6(a)	0.3	2.25	0.7	0.4	2.7
Loam, Silt loam, Silt	0.7	0.4	2.7	0.8	0.4	3.15
Sandy clay loam, Clay loam, Silty clay loam	0.8	0.4	3.15	0.9	0.5	3.6
Sandy clay, Silty clay, Clay	0.9	0.5	3.6	0.9	0.5	3.6
Peat or Muck	NOT RECOMMENDED					
Eroded Slopes or Knobs Soils with pH greater than 8.0	NOT RECOMMENDED					

(a) The first number is quarts/acre of BLADEX 4L. The second number is quarts/acre of atrazine 4L. The third number is quarts/acre of Ramrod 4L.

* For organic matter content between those listed, adjust the rate proportionately.

STORAGE AND DISPOSAL

STORAGE: Do not contaminate water, food or feed by storage or disposal. Do not use or store around the home environment. Avoid contact with water. In case of spill or leak, soak up with sand, earth or synthetic absorbant (do not use alkaline absorbants) and dispose of wastes in compliance with local, State or Federal regulations.

PESTICIDE DISPOSAL: Pesticide, spray mixture or rinsate that cannot be used according to label instructions must be disposed of according to applicable Federal, State or local procedures.

CONTAINER DISPOSAL: Triple rinse (or equivalent). Then offer for recycling or reconditioning, or puncture and dispose of in a sanitary landfill or by incineration, or, if allowed by state and local authorities, by burning. If burned, stay out of smoke.

NOTICE OF WARRANTY

Du Pont warrants that this product conforms to the chemical description on the label thereof and is reasonably fit for the purposes stated on such label only when used in accordance with the directions under normal use conditions. It is impossible to eliminate all risks inherently associated with the use of this product. Crop injury, ineffectiveness or other unintended consequences may result because of such factors as weather conditions, presence of other materials, or the manner of use or application, all of which are beyond the control of Du Pont. In no case shall Du Pont be liable for consequential, special or indirect damages resulting from the use or handling of this product. All such risks shall be assumed by the buyer. DU PONT MAKES NO WARRANTIES OF MERCHANTABILITY OR FITNESS FOR A PARTICULAR PURPOSE NOR ANY OTHER EXPRESS OR IMPLIED WARRANTY EXCEPT AS STATED ABOVE.

BLADEX., CONQUEST, EXTRAZINE-Trademark of E.I. Du Pont de Nemours & Co. (Inc)
"Banvel" - Trademark of Sandoz Crop Protection Corp.
"Dual", "Milogard", "Concep II", - Trademarks of CIBA-Geigy Corp.
"Eradicane", "Sutan+"- Trademarks of ICI Americas, Inc
"Genate Plus" - Trademark of PPG Industries, Inc.
"Lasso", "Ramrod", "Screen" - Trademarks of Monsanto Co.
"Ortho X-77" - Trademark of Chevron Chemical Co.
"Princep" "Caliber", - Trademarks of CIBA-Geigy Corp.
"Prowl" Trademark ofAmerican Cyanamid.
"Treflan" Trademark of Elanco Products Company.
"Zorial Rapid 80" - Trademark of Zoecon Corporation.

AG-3276 8089 8/2/89

This is a specimen label. DuPont is not responsible for the accuracy of the information contained herein. As labels are subject to revision, always carefully read and follow the label on the product container.

BLADEX® 90DF
HERBICIDE

RESTRICTED USE PESTICIDE

For retail sale to and use only by Certified Applicators or persons under their direct supervision and only for those uses covered by the Certified Applicator's certification. See Precautionary Statements for reasons this product is classified Restricted Use.

Bladex® 90DF Herbicide is a water dispersible granule.

ACTIVE INGREDIENTS	BY WEIGHT
Cyanazine: 2[[4-chloro-6-(ethylamino)-s-triazin-2-yl]amino]-2-methylpropionitrile	90%
INERT INGREDIENTS	10%
TOTAL	100%

Covered by U.S. Patents Nos. 3,505,325 and 3,620,710

EPA Reg. No. 352-495

KEEP OUT OF REACH OF CHILDREN
WARNING

AVISO PRECAUCION AL USUARIO: Si usted no lee ingles, no use este producto hasta que la etiqueta haya sido explicada ampliamente.

STATEMENT OF PRACTICAL TREATMENT

IF SWALLOWED, call a physician or poison control center. Drink 1 or 2 glasses of water and induce vomiting by touching back of throat with finger. Do not induce vomiting or give anything by mouth to an unconscious person.

IF IN EYES, flush with plenty of water. Get medical attention if irritation persists.

IF ON SKIN, wash immediately with plenty of soap and water.

IF INHALED, remove victim to fresh air. If not breathing, give artificial respiration, preferably mouth-to-mouth. Get medical attention.

For medical emergencies involving this product, call 1-800-441-3637.

PRECAUTIONARY STATEMENTS

HAZARDS TO HUMANS AND DOMESTIC ANIMALS

WARNING: May be fatal if swallowed. Harmful if inhaled or absorbed through the skin. Causes temporary eye injury.

This product may be hazardous to your health. It is classified "Restricted Use" because, at doses which caused serious maternal illness in laboratory animals, birth defects were present. Use of protective clothing and equipment and following the precautions below can reduce risk.

Avoid breathing spray mist. Avoid contact with skin, eyes, or clothing. Do not get in eyes or on clothing. Wear a face shield when mixing and loading. Wash thoroughly with soap and water after handling and before eating or smoking.

Wear long trousers and long-sleeved clothing when applying this product. Wear chemical resistant gloves extending above the wrist, a chemical resistant apron, long trousers and long-sleeved clothing when mixing or loading or when adjusting, repairing or cleaning equipment. Protective gloves must be washed with soap and water after use and before removing from the hands. Remove contaminated clothing and wash before reuse. Contaminated clothing should be laundered separately from household laundry to prevent cross-contamination of other laundry. Heavily contaminated or drenched clothing and protective equipment must be discarded or destroyed in accordance with State and local regulations.

Do not apply this product in such a manner as to directly, or through drift, expose workers or other persons. The area being treated must be vacated by unprotected persons.

Keep out of reach of domestic animals, particularly cattle. Consumption of this product, spray solutions, or water contaminated with product can result in serious illness or possible death of bovines.

ENVIRONMENTAL HAZARDS

Do not apply directly to water or wetlands. Keep out of lakes, streams or ponds. Do not contaminate water by cleaning equipment or disposal of wastes.

BLADEX is a pesticide which can move (seep or travel) through soil and can contaminate groundwater which may be used as drinking water. BLADEX has been found in groundwater as a result of agricultural use. Users are advised not to apply BLADEX where the water table (groundwater) is close to the surface and where the soils are very permeable (i.e., well drained soils such as loamy sands). Your local agricultural agencies can provide further information on the type of soil in your area and the location of groundwater.

In case of significant spill, call CHEMTREC 1-800-424-9300.

DIRECTIONS FOR USE

It is a violation of Federal law to use this product in a manner inconsistent with its labeling.

GENERAL INFORMATION

BLADEX 90DF Herbicide is a selective herbicide for the control of annual grasses and broadleaved weeds in field corn, popcorn, sweet corn, cotton, grain sorghum, wheat and fallow cropland.

Consult your local Agricultural Extension Agent for help in determining soil texture, organic matter content, and the most appropriate herbicide rate for local conditions.

Where surfactants or emulsible vegetable oils are added to BLADEX 90DF for over the top of corn postemergence applications, use on field corn only.

Do not apply this product in irrigation water with any kind of irrigation system.

Do not apply this product with aerial application equipment.

BLADEX 90DF is not effective when used preemergence on peat or muck soils. Do not use BLADEX 90DF on sands or loamy sands (soils consisting of more than 70% sand) containing less than 1% organic matter.

Weather Effects: As a preemergence herbicide, BLADEX 90DF is active mainly through the roots, and therefore, its effect on weeds is dependent on adequate rainfall or sprinkler irrigation to move the herbicide into the root zone. Moisture should be sufficient to thoroughly wet the soil throughout the zone where weed seeds may germinate and enough to make the soil too wet to cultivate. Rotary hoeing or shallow cultivation is recommended for those applications which are not incorporated at the time of treatment, if adequate rainfall or sprinkler irrigation has not occurred within about ten days after application of BLADEX 90DF.

Heavy rainfall between planting and crop emergence may cause crop injury or stand loss. Rainfall tends to cause execessive concentrations of herbicide in the seed furrow, resulting in possible crop injury. Level deep planter marks or seed furrows before application.

Under conditions which delay weed germination, such as low temperatures, lack of soil surface moisture, or when germination extends over a long period, the effectiveness of the herbicide may be impaired. Rotary hoeing, shallow cultivation or a postemergence herbicide treatment may be of benefit under these circumstances.

If the crop is cultivated, tillage should be shallow to minimize herbicide dilution in the soil. Should the crop stand be lost due to adverse weather conditions, insects, etc., the field can be replanted the same season to corn or sorghum.

To enhance weed control in areas of less than 25 inches of rainfall or where long dry periods are common, these treatments may require shallow incorporation with a tool such as a field cultivator operated from 5-7 mph. Incorporation should not be more than three inches deep to keep from burying the herbicide. A spike-toothed harrow, deep tillage disk or rolling basket device is not recommended for incorporating BLADEX.

When applied as a post-emergence herbicide, BLADEX 90DF is also active through foliage as well as through the roots. Yellowing and/or stunting of the crop may result from this treatment, particularly if cold, adverse growing conditions occur after application. Extended or extreme cold and wet conditions may reduce stands. Do not apply BLADEX postemergence to a crop that is damaged or growing under stress.

OBSERVE ALL CAUTIONS AND LIMITATIONS ON LABELING OF ALL PRODUCTS USED IN MIXTURES.

Triazine Resistant Weeds: In fields where triazine resistant biotypes of weeds have been identified, BLADEX should be used in combination with or in sequence with other registered non-triazine herbicides. (Triazine resistant biotypes of Kochia and Pigweed have been identified in some fields in the Western Great Plains and triazine resistant biotypes of Pigweed and Lambsquarters have been identified in some fields in various states.) Consult with appropriate state agricultural extension service representatives for specific recommendations.

APPLICATION DIRECTIONS

GENERAL MIXING AND SPRAYING INSTRUCTIONS

This product may not be applied by means of chemigation or aerial application.

Use sufficient agitation to ensure that the BLADEX 90DF is completely dispersed and in uniform suspension prior to application or tank mixing with other formulations.

The following general mixing instructions are recommended when using this or any other liquid suspension formulation.

I. General

1. Unless otherwise specified, use at least 7.5 gallons of water per acre for soil applications and at least 15 gallons of water per acre for foliar applications for all applications with ground equipment. NOTE: Sufficient carrier must be used to assure uniform application. Follow label requirement of all products used in tank mix combinations.
2. A nitrogen solution or complete liquid fertilizer may replace all or part of the water as a carrier for preemergence or preplant application. Do not apply fertilizer mixtures after crop emerges, because injury may occur.
3. Always check the tank mix compatibility (TMC) of this or any other formulation before mixing with liquid fertilizer carriers or other formulations. A simple but generally reliable TMC evaluation procedure has been provided for your use in step II of these mixing instructions.
4. Start with thoroughly clean equipment. (See the label of previous compound for cleaning instructions.)
5. Fill tank 1/2 full with carrier. Start and maintain consistent agitation through all mixing and spraying procedures. Make sure that the agitation system is working properly and creates a rippling or rolling action on the liquid surface.
6. Slowly add the dry flowable (DF) to the tank or inductor.
7. Fill tank to 75 percent capacity with carrier. Filling and bypass lines should be kept below liquid surface. Increase tank agitation if necessary to maintain surface action.
8. When desired, appropriate emulsible oil, oil concentrate, or other tank mix formulations should be added at this time. Pre-slurry these added ingredients before addition, if the compatibility test shows it to be necessary.
9. Complete filling tank maintaining sufficient agitation at all times to ensure surface action. This applies to both spray and nurse tanks.
10. Tank mixtures should always be applied immediately after preparation. If for any reason this is not possible, assure that sufficient agitation has been provided to re-mix all products and check for complete resuspension prior to application.
11. Empty tank as completely as possible before refilling to prevent buildup of oil or emulsible concentrate residues when tank mixing with these formulations. Always maintain agitation to avoid separation.
12. If an oil or emulsible concentrate film starts to build up after using these formulations, drain and clean the tank with strong detergent solution or appropriate solvent.
13. It is recommended that the sprayer be thoroughly cleaned by flushing with a detergent solution at the end of each work day when any emulsible oil, oil concentrate, or other emulsible formulation has been used either alone or in tank mix combinations with other pesticide formulations, even if no obvious problems have been encountered. This precaution will ensure a clean sprayer and continued trouble-free operation.

II. Tank Mix Compatibility Evaluation Procedure

1. Add one pint of carrier liquid to each of two one-quart jars. Mark one quart jar "with" and the other "without."

2. Add 1/4 teaspoon of a suitable tank mix compatibility agent (1/4 teaspoon/pint = 2 pints/100 gallons of carrier) to the jar marked "with," cap the jar, and shake gently for five to ten seconds to mix.
3. Add the appropriate amount of herbicide to both jars, cap each jar, and shake gently for five to ten seconds to mix. If problems are encountered in mixing wettable powder or dry flowable formulations into a liquid fertilizer, then pre-slurry these formulations in water prior to their addition to the liquid fertilizer and proceed with the test. The following chart has been provided to assist you in selecting the appropriate BLADEX 90DF use rate for this evaluation. If more than one herbicide is to be used in the tank mixture, each should be added separately as follows: water solubles first, wettable powders or dry flowables second, liquid flowables third, and emulsible concentrate or oil formulations last, with each jar capped and gently shaken for five to ten seconds for each addition.

Jar Test for BLADEX 90DF Compatibility

Gallons of Liquid Carrier per Acre	4	7.5	15	20	25	30	40
Teaspoons of BLADEX 90DF per Pint of Liquid Carrier	6.0	3.2	1.6	1.2	1.0	0.8	0.6

This chart is based on one pound of BLADEX 90DF (0.9 pound active ingredients) per acre in the indicated carrier volumes. Intended field use rates are achieved by varying the amount of BLADEX 90DF; i.e., for a field use rate of 2.5 pounds of BLADEX 90DF in 15 gallons of carrier per acre, add 4.0 teaspoons of BLADEX 90DF to the quart jars containing one pint of carrier. Calculation: For 2.5 pounds of BLADEX 90DF/15 gallons of carrier per acre, multiply 2.5 × 1.6 to get 4.0 teaspoons of BLADEX 90DF per pint of carrier.

4. Let each jar stand one-half hour and make observations. If any separation, agglomeration, or precipitation has occurred, shake the jar again for 10 to 15 seconds, and note whether any of the following occur:
   a. Separated phases do not re-mix uniformly.
   b. Screen/nozzle plugging lumps do not disperse.
   c. Precipitate does not re-suspend readily.
   d. Precipitate sticks tenaciously to the glass.
5. If none of the above problems occur in either jar, then the herbicides can, in most cases, be safely used without a compatibility agent.
6. If problems 4.a or 4.b occur in the jar marked "without" but do not occur in the jar marked "with," the compatibility agent should be used.
7. If problems 4.a or 4.b are seen in both jars, then the herbicides and carrier mixture are incompatible and should not be used in the same spray tank. Alternatively, a different tank mix compatibility agent can be evaluated.
8. If problems 4.c or 4.d occur in the jar marked "without" but do not occur in the jar marked "with," the compatibility agent should be used unless constant, thorough agitation can be maintained and immediate clean-out of spray system is performed.
9. If problems 4.c or 4.d are seen in the jar marked "with," the user proceeds with mixing and application at his own risk should agitation in the system be insufficient or curtailed.
10. Those mixtures defined as compatible in this test should then be mixed for use as indicated in steps 1-12 of the general mixing instructions listed above.

If a test such as outlined indicates that components of a proposed mix are compatible, the applicator still has the responsibility of combining materials in sequence to the spray tank in accordance with directions prescribed on the label of the herbicides or pesticides involved.

Tests have indicated that compatibility agents, noted below by the various tank mix combinations, may give improved compatibility in liquid fertilizers.

Tank Mix Combination	Compatibility Agents
BLADEX 90DF/"Lasso". (Liquid Fertilizer Grade)	Probably not needed in 28-0-0, 10-34-0. "Compex" may help in others.
BLADEX 90DF/"Genate Plus", "Sutan+" 6.7E	Probably not needed in 28-0-0. Incompatible in 10-34-0. "Unite", "Spray-Mate", "Kem-Link", may help in others.
BLADEX 90DF/"Dual" 8E	Probably not needed in 28-0-0. "Unite", "Spray-Mate", "Ivory Liquid" may help in others.

III. Application Equipment

1. Use application equipment fitted with nozzles that provide accurate and uniform coverage. Be certain that nozzles are uniformly spaced and the same size. Calibrate sprayer before use and recheck frequently during use whenever possible.
2. Use a pump with capacity to:
   a. Maintain 35-40 psi at nozzles.
   b. Provide sufficient agitation in tank to keep mixture in suspension.
   c. Provide a minimum of 20 percent bypass at all times.
3. Use centrifugal pumps which provide sufficient shear action for dispersing and mixing this product. The pump should provide a minimum of 10 gallons/minute/100-gallon tank size circulated through the jets of a correctly-positioned sparger tube.
4. Use screens to protect the pump and to prevent nozzles from clogging. Screens placed on suction side of pump should be 10 to 16 mesh. Do not place a screen in the recirculation line. Use a 40- to 50-mesh screen between the pump and boom and, where required, 50-mesh screens at the nozzles. Check your equipment manufacturer's literature for specific recommendations.

FERTILIZER IMPREGNATION

BLADEX 90DF Herbicide may be applied when coated on or impregnated in dry granular fertilizer for early preplant, preemergence or preplant incorporated weed control in field corn. All recommendations, cautions and special precautions on this label must be followed along with state regulations relating to dry bulk fertilizer blending, impregnating and labeling. Also, follow the precautions on the label of any product mixed with this product.

General Blending Directions

BLADEX 90DF may be coated on or impregnated in dry bulk fertilizers using tower blenders, rotary drum blenders or blending augurs or conveyors. DO NOT impregnate BLADEX or tank mixes containing BLADEX on or in fertilizers containing Ammonium Nitrate, Potassium Nitrate, or Sodium Nitrate. Do not use on straight limestone since absorption will not be achieved. Fertilizer blends containing limestone can be impregnated when using BLADEX alone. Use a minimum of 200 lbs. and a maximum of 450 lbs. per acre of dry fertilizer. Use equipment that will give a uniform distribution of the herbicide throughout each batch of impregnated fertilizer. Non-uniform impregnation can cause crop injury or unsatisfactory performance.

A. BLADEX 90DF may be used as the only herbicide for impregnation.
   1. Add BLADEX 90DF to 1/2 the total fertilizer volume required.
   2. Spray one gallon of water (to break down the 90DF) and one gallon of diesel fuel (to prevent evaporation and crusting) per ton of fertilizer and allow to mix thoroughly.
      NOTE: If the fertilizer is dusty, add the diesel fuel to the fertilizer before adding the herbicide.
   3. Add remaining fertilizer and mix thoroughly. (3 minutes or more for rotary blenders.)
   4. Use 2-3% Ag-Sorb or 1-2% MP-79 drying agent (or a suitable amount of another effective drying agent) to insure a spreadable herbicide/fertilizer mixture. The need for a drying agent is determined by the wetness of the fertilizer batch. Wetness can change with humidity, nitrogen content, fertilizer types, fertilizer rates and herbicide rates.

B. BLADEX 90DF may be used in tank mixes with other dry herbicides including "Princep"™ "Caliber"™ 90. Follow the procedure as above in "A".

C. BLADEX 90DF may be used in tank mixes where an EC or other liquid herbicide acts as the sticking agent. This may eliminate the need for water and/or diesel fuel.
   1. While fertilizer is blending, add the BLADEX 90DF. Experience has shown this will provide the most consistent performance due to the grinding action of the fertilizer on the 90DF.
   2. Spray in the EC herbicide and mix thoroughly. (3 minutes or more for rotary drum blenders.)
   3. Add drying agent to insure a spreadable herbicide/fertilizer mixture. Usually less drying agent is required when using BLADEX 90DF.

D. Pre-slurried BLADEX 90DF can be used alone or in a tank mix for impregnation. For rotary-drum mixers, the liquids can be moved into the drum using an air system or a liquid pump. Do not add extra water. Add drying agent to ensure a spreadable herbicide/fertilizer mixture.

(For more information on Drying Agents, Application Equipment, Calibration Guide and variations of these methods, see Du Pont Bulletin on "Fertilizer Impregnation").

CLEAN OUT:

Equipment used to impregnate or apply fertilizer impregnated with BLADEX or combinations including BLADEX must be cleaned out by running at least 1,000 lbs. of fertilizer not impregnated with BLADEX through the impregnation equipment and application equipment, if the next batch of material is to be applied to a crop for which BLADEX or a combination herbicide is not registered.

APPLICATION:

Uniform application of BLADEX 90DF which has been impregnated in or coated on dry fertilizer is essential for satisfactory weed control and crop safety. Accurate calibration of the fertilizer applicator is necessary. Applying while turning at the ends of the fields may result in excessive application rates causing crop injury. Do not double apply across the ends or sides of the field.

Crop injury and/or poor weed control may occur where the impregnated fertilizer is not uniformly applied. Air flow or augur metered application equipment is preferred (one pass application). If other equipment is used, the recommended method of application is to apply 1/2 the recommended rate and overlap 50 percent to double apply by splitting the middles to obtain the best distribution pattern.

Apply immediately after impregnation. Impregnated fertilizer may become lumpy and difficult to spread if stored.

RATES AND TIMING:

Use the application rates and timing shown in the appropriate sections of this label. Follow the precautions on the labels of all products used.

CORN

WEEDS CONTROLLED BY "BLADEX" 90DF ALONE AND IN COMBINATION WITH OTHER HERBICIDES ON CORN

Grasses

Annual bluegrass
Annual fescues
Annual (Italian) ryegrass
Barnyardgrass(1)
Bullgrass
Crabgrass
Fall panicum
Giant foxtail
Goosegrass
Green foxtail
Junglerice
Stinkgrass (Indian lovegrass)
Witchgrass
Yellow foxtail

Broadleaves

Annual groundcherry
Annual morningglory
Annual sedge
Black mustard
Buffalobur
Florida pusley (Florida purslane)
Hedge mustard
Jimsonweed(1)
Kochia
Ragweed (Common)
Russian thistle
Shepherdspurse
Smallflower galinsoga
Smartweed
Buttercup (annual)
Carpetweed
Cocklebur(2)
Common chickweed
Common groundsel
Common mallow
Common purslane
Corn spurry
Curly dock (seedling)
Fiddleneck
Ladysthumb
Lambsquarters
Mayweed
Nightshade (annual)
Pigweed(1)
Pineappleweed
Plantain
Poorjoe
Prickly sida (teaweed)
Prostrate knotweed
Prostrate spurge
(Pennsylvania)
Sunflower(2) (wild, annual, common)
Tarweed cuphea (Gumweed)
Velvetleaf(1)
Wild buckwheat
Wild mustard
Wild radish
Wild turnip

(1) Under conditions such as low temperatures, lack of soil surface moisture or other factors that may cause delay in germination of the seeds, the effectiveness of BLADEX may be impaired against these weeds.

(2) The degree of control will be reduced if soil moisture and temperature conditions cause deep germination of the seed.

CORN

PREEMERGENCE-PREPLANT INCORPORATED

Apply BLADEX treatments just before, at or after planting but before crop has emerged. Avoid removal of treated soil from seedrow prior to or during the planting operation.

BLADEX may also be applied early prior to planting or in a split application if pre-season weed control is desired. For split applications, do not exceed the total amount of BLADEX for the soil texture and organic matter shown in Table 1. If BLADEX is applied early, more than 15 days before planting, a split application of BLADEX or some other herbicide treatment may be necessary at or after planting to provide additional length of weed control. For further information see "Early Preplant" recommendations in the Conservation Tillage section of this label.

TABLE 1
PREEMERGENCE BROADCAST APPLICATION RATES PER ACRE FOR "BLADEX" 90DF APPLIED ALONE ON CORN

SOIL TEXTURE DESCRIPTION	POUNDS OF "BLADEX" 90DF — PERCENT ORGANIC MATTER IN SOIL*					
	LESS THAN 1%	1%	2%	3%	4%	5% & OVER
Sand, Loamy sand	DO NOT USE	1.3	1.8	2.4	3.1	3.6
Sandy loam	1.3	2.0	2.2	2.7	3.3	4.0
Loam, Silt loam, Silt	1.8	2.2	2.7	3.3	4.0	4.4
Sandy clay loam, Clay loam, Silty clay loam	2.2	2.7	3.3	4.0	4.4	4.9
Sandy clay, Silty clay, Clay	3.1	3.3	4.0	4.4	4.9	5.3
Peat or Muck	NOT RECOMMENDED					

* For organic matter content between those listed, adjust the rate proportionately.

Rotary hoeing is recommended for preemergence applications which do not receive adequate rainfall or sprinkler irrigation to wet the top 2 inches of soil or depth of germinating weeds within about 10 days after application.

BLADEX alone or in tank mix combinations should not be incorporated more than three inches deep to keep from burying the herbicide. Single or two pass incorporation with a tool such as a field cultivator operated at 5-7 mph is acceptable. A spike-toothed harrow, deep tillage disk or rolling basket device is not recommended for incorporating BLADEX.

TABLE 2
PREEMERGENCE BROADCAST APPLICATION RATES PER ACRE FOR TANK-MIX COMBINATIONS OF "BLADEX" 90DF PLUS ATRAZINE 4L ON CORN FOR USE IN ALL STATES EXCEPT KENTUCKY, MISSOURI, TENNESSEE AND KANSAS EAST OF HIGHWAY 99

SOIL TEXTURE DESCRIPTION	POUNDS OF "BLADEX" 90DF + QUARTS OF ATRAZINE 4L** — PERCENT ORGANIC MATTER IN SOIL*					
	LESS THAN 1%	1%	2%	3%	4%	5% & OVER
Sand, Loamy sand	DO NOT USE	0.9 + 0.3	1.3 + 0.4	1.6 + 0.6	1.8 + 0.8	2.4 + 1.0
Sandy loam	0.9 + 0.3	1.3 + 0.4	1.6 + 0.6	1.8 + 0.8	2.4 + 1.0	3.1 + 1.2
Loam, Silt loam, Silt	1.3 + 0.4	1.6 + 0.6	2.2 + 0.8	2.4 + 1.0	3.1 + 1.2	3.6 + 1.3
Sandy clay loam, Clay loam, Silty clay loam	1.6 + 0.6	2.2 + 0.8	2.4 + 1.0	3.1 + 1.2	3.6 + 1.3	3.8 + 1.4
Sandy clay, Silty clay, Clay	2.2 + 0.8	2.4 + 1.0	3.1 + 1.2	3.6 + 1.3	3.8 + 1.4	4.0 + 1.6
Peat or Muck	NOT RECOMMENDED					

* For organic matter content between those listed, adjust the rate proportionately.
** If Atrazine 80W is used, multiply rates shown by 1.25 to equal pounds of Atrazine 80W.
If Atrazine 90% is used, multiply rates shown by 1.11 to equal pounds of Atrazine 90%.

BLADEX Applied Alone

Use the proper rate for the soil texture and organic matter indicated in Table 1. Any rotational crop may be planted in the fall or spring following this treatment.

BLADEX COMBINATIONS

BLADEX plus Atrazine

Use BLADEX 90DF plus atrazine at the proper rate for soil texture and organic matter indicated in Tables 2 and 3. These tables provide rates for generally weedy conditions. The ratio of the amounts of each herbicide may be adjusted as necessary for particular weed conditions as long as the combined rate of the two products does not exceed the combined rate for the soil shown in Table 2 or 3. For grassier conditions use a ratio that contains higher levels of BLADEX (3:1). For fields with more broadleaves use a ratio that contains higher levels of atrazine (1:1).

TABLE 3
PREEMERGENCE BROADCAST APPLICATION RATES PER ACRE FOR TANK-MIX COMBINATIONS OF "BLADEX" 90DF PLUS ATRAZINE 4L ON CORN FOR USE ONLY IN ALL KENTUCKY, MISSOURI, TENNESSEE AND KANSAS EAST OF HIGHWAY 99

SOIL TEXTURE DESCRIPTION	POUNDS OF "BLADEX" 90DF + QUARTS OF ATRAZINE 4L** PERCENT ORGANIC MATTER IN SOIL*					
	LESS THAN 1%	1%	2%	3%	4%	5 & OVER
Sand, Loamy sand	DO NOT USE	0.9 + 0.5	1.1 + 0.5	1.6 + 0.6	1.9 + 0.8	2.5 + 1.0
Sandy loam	0.9 + 0.5	1.5 + 0.7	1.8 + 0.8	2.0 + 0.9	2.4 + 1.1	3.1 + 1.3
Loam, Silt loam, Silt	1.5 + 0.7	2.2 + 1.0	2.4 + 1.1	2.7 + 1.2	3.1 + 1.3	3.6 + 1.3
Sandy clay loam, Clay loam, Silty clay loam	1.8 + 0.8	2.4 + 1.1	2.7 + 1.2	3.1 + 1.3	3.6 + 1.3	3.9 + 1.4
Sandy clay, Silty clay, Clay	2.2 + 1.0	2.7 + 1.2	3.1 + 1.3	3.6 + 1.3	3.9 + 1.4	4.2 + 1.5
Peat or Muck	NOT RECOMMENDED					

* For organic matter content between those listed, adjust the rate proportionately.
** If Atrazine 80W is used, multiply rates shown by 1.25 to equal pounds of Atrazine 80W.
If Atrazine 90% is used, multiply rates shown by 1.11 to equal pounds of Atrazine 90%.

Rotational Crops: (1) Plant only corn, peanuts, sorghum, or soybeans the year following the use of this mixture. (2) If soybeans are to be planted, injury may occur due to the carryover of atrazine. (3) If applied after June 10, do not rotate with crops other than corn or sorghum the next year or injury may occur. (4) In the high plains and intermountain areas of the West where rainfall is sparse and erratic or where irrigation is required, use only when corn or sorghum is to be planted the following year, or a crop of corn or sorghum not treated with atrazine is to precede other rotational crops. (5) Small grains may be planted 15 months following treatment. (6) All other crops may be planted 18 months after application.

For all states except Kentucky, Missouri, Tennesee and Kansas east of Highway 99 use Table 2.

In Kentucky, Missouri, Tennessee and Kansas east of Highway 99 use Table 3.

BLADEX plus "Lasso".4EC

Use BLADEX 90DF at the proper rate for the soil texture and organic matter shown in Table 4 plus 2 quarts per acre of "Lasso" (Use 2.5 quarts of "Lasso" on clay soils containing 5 percent organic matter and over.) Any rotational crop may be planted the fall or spring following this treatment.

BLADEX plus "Sutan+" 6.7E, "Genate Plus" or "Eradicane" 6.7E

Use BLADEX 90DF at the proper rate for the soil texture and organic matter shown in Table 4 plus 1.8 quarts per acre of "Sutan+", "Genate Plus" or "Eradicane" for control of many annual grasses and broadleaf weeds. (Use 2.4 quarts of "Sutan+", "Genate Plus" or "Eradicane" on loam soils containing 5 percent or more organic matter, and on clay loams and clays containing 4 percent or more organic matter.) Do not use on sands and loamy sands having less than 1 percent organic matter in the light sandy soils of eastern coastal states. Do not use on corn seed stock.

Apply before planting. Incorporate the mixture immediately after application using power-driven cultivation equipment set for 2-3 inches in depth, or a tandem disc set to cut to a depth of about 4 inches while operating at 4-6 mph. For thorough mixing, disc in two directions (cross disc), and follow with a harrow, drag, or other leveling device. Prior to the second discing, readjust the disc to prevent cutting deeper than 4 inches. BLADEX 90DF may be applied preemergence as an overlay over previously incorporated "Sutan+", "Genate Plus" or "Eradicane". Any rotation crop may be planted in the fall or spring following these treatments.

Existing stands of quackgrass, purple and yellow nutsedge must be turned under and thoroughly chopped up prior to chemical treatments.

Additional weeds controlled by "Sutan+", "Genate Plus" or "Eradicane" combinations:

Grasses:	Sandbur Shattercane (Wild Cane)* Texas Panicum Quackgrass (Eradicane only) Wild Proso Millet* (Eradicane only)
Perennial Weeds:	Yellow Nutsedge (Nutgrass) Purple Nutsedge (Nutgrass)

*Suppression only - refer to "Sutan+", "Genate Plus" or "Eradicane" label for appropriate supplemental cultural and tillage practices.

For fields with moderate to heavy infestations of these weeds refer to the "Sutan+", "Genate Plus" or "Eradicane" label for appropriate higher rates.

BLADEX plus "Dual" 8E

Use BLADEX 90DF at the proper rate for soil texture and organic matter shown in Table 4. Use "Dual" as follows:

Soil Texture	Broadcast Rate Per Acre For "Dual"
Coarse Sand, Loamy sand, Sandy loam	1.25 - 1.5 pints
Medium Loam, Silt loam, Silt	1.5 - 2.0 pints
Fine Sandy clay loam, Silty clay loam, Clay loam, Sandy clay, Silty clay, Clay	1.5 - 2.5 pints

The low end of the rate range should be used for lowest organic matter soils and the rate increased as organic matter increases to a point that soils containing 4 percent organic matter or more require the highest rate shown for that soil texture. Refer to the "Dual" label for precautions on rotational crops.

TABLE 4
PREEMERGENCE BROADCAST APPLICATION RATES PER ACRE FOR "BLADEX" 90DF USED IN TANK-MIX COMBINATIONS WITH "LASSO", "SUTAN+", "GENATE PLUS", "ERADICANE", OR "DUAL" ON CORN

SOIL TEXTURE DESCRIPTION	POUNDS OF "BLADEX" 90DF PERCENT ORGANIC MATTER IN SOIL*					
	LESS THAN 1%	1%	2%	3%	4%	5% & OVER
Sand, Loamy sand	0.7**	0.9	1.3	1.6	1.8	2.2
Sandy loam	0.9	1.3	1.6	1.8	2.2	2.4
Loam, Silt loam, Silt	1.3	1.6	1.8	2.2	2.4	2.9
Sandy clay loam, Clay loam, Silty clay loam	1.6	2.0	2.2	2.4	2.9	3.1
Sandy clay, Silty clay, Clay	2.0	2.2	2.7	2.9	3.1	3.3
Peat or Muck	NOT RECOMMENDED					

* For organic matter content between those listed, adjust the rate proportionately.
** Do not use in the light sandy soils of the Atlantic Coastal Plain.

BLADEX plus Atrazine plus "Lasso", "Sutan+", "Genate Plus", "Eradicane", or "Dual"

Use BLADEX 90DF plus atrazine at the proper rate for soil texture and organic matter shown in Table 5. Use "Lasso", "Sutan+", "Genate Plus", "Eradicane", or "Dual" according to rates shown in "BLADEX Combinations" in this section of the label. Rotational Crops: Refer to Rotational Crops section of "BLADEX plus Atrazine" in this section of the label.

CONSERVATION TILLAGE WEED CONTROL

Early Preplant (EPP)

BLADEX 90DF may be used for Early Preplant or Preemergence weed control for land going into production of corn under conservation tillage programs.

TABLE 5
PREEMERGENCE BROADCAST APPLICATION RATES PER ACRE FOR "BLADEX" 90DF PLUS ATRAZINE 4L USED IN TANK-MIX COMBINATIONS WITH "LASSO", "SUTAN+", "GENATE PLUS", "ERADICANE", OR "DUAL" ON CORN

SOIL TEXTURE DESCRIPTION	POUNDS OF "BLADEX" 90DF + QUARTS OF ATRAZINE 4L***					
	PERCENT ORGANIC MATTER IN SOIL*					
	LESS THAN 1%	1%	2%	3%	4%	5% & OVER
Sand, Loamy sand	0.5+0.2**	0.7+0.25	0.9+0.5	1.2+0.5	1.1+0.5	1.6+0.75
Sandy loam	0.7+0.25	0.9+0.5	1.2+0.5	1.1+0.5	1.6+0.75	1.8+0.75
Loam, Silt loam, Silt	0.9+0.5	1.2+0.5	1.3+0.5	1.6+0.75	1.8+0.75	2.0+0.75
Sandy clay loam, Clay loam, Silty clay loam	1.2+0.5	1.3+0.5	1.6+0.75	1.8+0.75	2.0+0.75	2.2+0.75
Sandy clay, Silty clay, Clay	1.3+0.5	1.6+0.75	1.8+0.75	2.0+0.75	2.2+0.75	2.2+1.0
Peat or Muck	NOT RECOMMENDED					

* For organic matter content between those listed, adjust the rate proportionately.
** Do not use in the light sandy soils of the Atlantic Coastal Plain.
*** If Atrazine 90% is used, multiply rates shown by 1.11 to equal pounds of Atrazine 90%. If Atrazine 80W is used, multiply rates shown by 1.25 to equal pounds of Atrazine 80W.

Complete any planned early spring tillage prior to application. Apply herbicide treatment before weeds germinate or before weed seedlings are more than 3 inches tall. Tillage after application may reduce the effectiveness of the herbicide treatment. A nitrogen solution or complete fertilizer solution may replace all or part of the water as a carrier. The spray gallonage and spray boom design must be adequate to give thorough uniform coverage of the weed foliage. Follow label requirements of all products used in tank mix combinations.

BLADEX 90DF alone or in combination with atrazine: Apply 15 to 30 days prior to planting. Use the proper rate for soil texture and organic matter indicated in Table 1, 2 or 3. In no-till corn where heavy crop residues exist, the rates shown in Tables 1,2 or 3 should be increased by 25%.

BLADEX 90DF plus "Princep" 4L or "Princep" "Caliber" 90 or BLADEX 90DF plus Atrazine plus "Princep" 4L or "Princep" "Caliber" 90: Apply 30 days or more prior to planting. Use the proper rate of BLADEX 90DF or of BLADEX 90DF plus Atrazine for the soil texture and organic matter as shown in Table 1, 2 or 3 and add 1 quart/acre of "Princep" 4L or 1.11 pounds/acre of "Princep" "Caliber" 90.

Burndown of Existing Weeds:

Combinations with 2,4-D: Where broadleaf weeds are present at the time of application, add 2,4-D LV Ester at 1-1/3-2 pts./A. (6 lb./gal.) or 2-3 pts./A. (4 lbs./gal.) (or 2,4-D Amine at recommended rates) plus "Ortho X-77" surfactant at 1 qt./100 gals. of diluted spray, or other suitable surfactant at its recommended rate.

Combinations with Paraquat Products: When grasses are present and/or when existing weeds exceed 3 inches in height add 1-2 pts./A. (2 lbs./gal.) or 1.3-2.7 pts./A. (1.5 lbs./gal.) of paraquat. Well established weeds 6 inches tall or taller may not be well controlled.

Apply BLADEX 90DF plus paraquat in at least 20 gal./A. of carrier by ground sprayer. (The volume of carrier and the application equipment must be adequate to give a uniform application.) Add "X-77" surfactant at 1-2 qts./100 gals. of diluted spray (or other suitable surfactant at recommended rates) where paraquat is used. Crop oil concentrate or emulsible vegetable oil are not needed where paraquat is used.

Depending upon weather conditions following the EPP application, a postemergence treatment of BLADEX or some other herbicide treatment may be necessary at or after planting to provide additional length of weed control. If desired, 1-1/2 to 2 pts./A. of "Dual" 8E or 2 qts./A. of "Lasso" 4EC may be tank-mixed with the BLADEX EPP treatment or applied preemergence at planting.

Rotational Crops: Refer to Rotational Crops section for each treatment in the Preemergence section of the label.

At Planting

BLADEX 90DF applied alone or in combination with atrazine and/or "Lasso" or "Dual" according to the following directions will kill most existing small weeds and suppress many emerged perennial weeds when corn is planted into no-till stalk ground (corn, sorghum), stubble ground (soybean, small grains), and any minimum-till situation. This treatment then provides residual control of annual weeds as in conventional tillage.

Apply BLADEX 90DF alone or with other products according to the directions for those treatments in the Preemergence section of the label. Where heavy crop residues exist, the BLADEX rate shown in Tables 1 through 5 should be increased by 25%. Add 0.5-1.0 pt./A. of 2,4-D LV (6 lbs./gal.) or 0.75 to 1.5 pts./A. (4 lbs./gal.) (or 2,4-D Amine at recommended rates). Add the 2,4-D to the spray tank last.

TABLE 6
POSTEMERGENCE BROADCAST APPLICATION RATES PER ACRE FOR "BLADEX" 90DF ON FIELD CORN NO PRIOR APPLICATION OF "BLADEX", "CONQUEST", "EXTRAZINE" OR "EXTRAZINE" II

SOIL TEXTURE	POUNDS OF "BLADEX" 90DF			
	PERCENT ORGANIC MATTER IN SOIL*			
	LESS THAN 1%	1%	2%	OVER 2%
Sand, Loamy sand	DO NOT USE	1.3	1.8	2.2
Sandy loam	1.3	1.8	2.2	2.2
Loam, Silt loam, Silt	1.8	2.2	2.2	2.2
All other textures	2.2	2.2	2.2	2.2

* For organic matter content between those listed, adjust the rate proportionately.

Use a minimum of 15 gals./A. of carrier. Complete spray coverage of the weeds is essential for best performance. Nitrogen solutions and complete liquid fertilizers are the preferred carriers for this treatment because they aid in the burndown of existing weeds. Add "Ortho X-77". surfactant at

1-2 qts. per 100 gals. of diluted spray, or other suitable surfactant at its recommended rate. If water is used as a carrier, crop oil concentrate may be used as a surfactant. Apply before weeds exceed 3 inches in height. For control of existing alfalfa add 1/3 - 1/2 pt./A. of "Banvel", to the spray mixture. Apply before the alfalfa exceeds 6 inches in height.

For fields with existing sod grasses such as orchardgrass, bromegrass, rye or timothy, or when very dry conditions exist, or when existing weeds exceed 3 inches in height, add paraquat to the tank mix. Use 1 to 2 pts./A. (2 lbs./gal.) or 1.3 to 2.7 pts./A. (1.5 lbs./gal.) of paraquat in combination with BLADEX as described above in this section, except the 2,4-D may be omitted, if desired, and the gallonage should be increased to 20-40 gals./acre to give thorough coverage of the weed growth. Do not apply paraquat in suspension type liquid fertilizer.

POSTEMERGENCE

Under dry, arid conditions of low humidity and the absence of dew formation at night, add a surfactant such as "Ortho X-77"., or an emulsible vegetable (EV) oil suitable for use on growing corn, at its recommended rate. Do not use petroleum-based crop oils. Addition of a surfactant or EV oil is not recommended under moist, rainy conditions and when dew forms at night because injury may occur.

Do not apply this treatment under cold, wet weather conditions or to corn growing under stress caused by weather, insects, disease, etc. Yellowing of the corn may result from this treatment, particularly if cold, adverse growing conditions occur after application. Extended or extreme cold and wet conditions may reduce stands.

Do not apply postemergence on popcorn, sweet corn or corn grown for seed.

BLADEX Applied Alone

Use BLADEX 90DF at the proper rate for the soil texture and organic matter shown in Table 6 or 7. Use rates shown in Table 6 if BLADEX, CONQUEST, EXTRAZINE or EXTRAZINE II have not been applied to the soil this season. Use rates shown in Table 7 if BLADEX, CONQUEST, EXTRAZINE or EXTRAZINE II have been applied to the soil this season. This treatment may be used on peat or muck soils for burndown and suppression of existing weeds but will not provide residual control. Apply from crop emergence through the four-leaf stage of corn growth but before weeds exceed about 1-1/2 inches in height. Do not apply over the top of corn if the fifth leaf is visible. Apply in water only. Do not spray emerged corn plants with BLADEX in a liquid fertilizer carrier.

Any rotational crop may be planted the fall or spring following this treatment.

TABLE 7

POSTEMERGENCE BROADCAST APPLICATION RATES PER ACRE FOR "BLADEX" 90DF ON FIELD CORN "BLADEX", "CONQUEST", "EXTRAZINE" OR "EXTRAZINE" II USED IN PRIOR APPLICATION

	POUNDS OF "BLADEX" 90DF			
	PERCENT ORGANIC MATTER IN SOIL*			
SOIL TEXTURE	LESS THAN 1%	1%	2%	OVER 2%
Sand, Loamy sand	DO NOT USE	DO NOT USE	1.5	1.5
Sandy Loam	DO NOT USE	DO NOT USE	1.8	2.2
Loam, Silt loam, Silt	DO NOT USE	1.5	2.2	2.2
All other textures	DO NOT USE	2.0	2.2	2.2

* For organic matter content between those listed, adjust the rate proportionately.

BLADEX COMBINATIONS

BLADEX plus Atrazine

Apply as directed in "Postemergence-BLADEX Applied Alone" section of this label. Use an amount of BLADEX 90DF plus Atrazine equal to the rate shown in Table 6 or 7 for the proper soil texture and organic matter. To determine the amount of BLADEX 90DF to use, multiply the rate in Table 6 or 7 by 0.7. To determine the amount of Atrazine 4L to use, multiply the rate indicated in Table 6 or 7 by 0.24. (Multiply by 0.3 for pounds of Atrazine 80W.) For the 2.5 lbs./A. rate shown in Table 6 or 7, use 1.75 lbs/A. of BLADEX 90DF plus 0.6 qts/A. of Atrazine 4L or 0.75 lbs/A. of Atrazine 80W.

Rotational Crops: See Rotational Crops section in the "Preemergence - BLADEX plus Atrazine" section of this label.

BLADEX plus "Banvel"

Apply as directed under "Postemergence - BLADEX Applied Alone" section of this label. Add 1/2 to 2/3 pt./A. of "Banvel", to the mixture. Do not use with a surfactant or emulsible vegetable oil.

SWEET CORN

BLADEX 90DF may be applied preemergence or preplant incorporated for the control of annual grasses and broadleaf weeds in sweet corn.

NOTE: BLADEX may cause injury or stand loss on new or "super sweet" varieties of sweet corn. Consult with Agricultural Extension Agencies and sweet cornseed suppliers about the sensitivity of new varieties to potential injury.

Apply BLADEX treatments just before, at or after planting but before crop has emerged. Avoid removal of treated soil from seedrow prior to or during the planting operation. Do not apply postemergence to sweet corn.

Rotary hoeing is recommended for preemergence applications which do not receive adequate rainfall or sprinkler irrigation to wet the top 1 1/2 to 2 inches of soil within about 10 days after application.

If a BLADEX mixture is to be incorporated, except as noted, single or two pass incorporation is acceptable. Care should be taken to incorporate the BLADEX mixture no deeper than the top two inches of soil.

BLADEX PLUS ATRAZINE

Apply BLADEX 90DF at the proper rate for soil texture and organic matter indicated in Table 8. Table 8 provides rates for generally weedy conditions. The ratio of the amounts of each herbicide may be adjusted as necessary for particular weed conditions as long as the combined rate of the two products does not exceed the combined rate for the soil shown in Table 8. For grassier conditions use a ratio that contains higher levels of BLADEX (3:1). For fields with more broadleaves use a ratio that contains higher levels of atrazine (1:1).

Rotational Crops: (1) Plant only corn, sorghum or soybeans the year following the use of this mixture. (2) If soybeans are to be planted, injury may occur due to the carryover of Atrazine. (3) If applied after June 10, do not rotate with crops other than corn or sorghum the next year or injury may occur. (4) Small grains may be planted 15 months following treatments. (5) All other crops may be planted 18 months after application.

"BLADEX" PLUS "LASSO" 4EC

Use BLADEX 90DF at the proper rate for the soil texture and organic matter shown in Table 9 plus 2 quarts per acre of "Lasso". (Use 2.5 quarts "Lasso", on clay soils containing 5 percent organic matter and over.) Any rotational crop may be planted the fall or spring following this treatment.

"BLADEX" PLUS "SUTAN+" 6.7E, "GENATE PLUS" OR "ERADICANE" 6.7E

Use BLADEX 90DF at the proper rate for the soil texture and organic matter shown in Table 9 plus 1.8 quarts per acre of "Sutan+", "Genate Plus" or "Eradicane" for control of many annual grasses and broadleaf weeds. (Use 2.4 quarts of "Sutan+" or "Eradicane" on loam soils containing 5 percent or more organic matter, and clay loams and clays containing 4 percent or more organic matter.) Do not use on sands and loamy sands having less than 1 percent organic matter nor on the light sandy soils of eastern coastal states. Do not use on corn grown for seed.

Apply before planting. Incorporate the mixture immediately upon application using power-driven cultivation equipment set for 2-3 inch depth, or tandem disc set to cut about 4 inches deep while operating at 4-6 mph. For thorough mixing, disc in two directions (cross disc), and follow with a harrow, drag, or other leveling device. Prior to the second discing, readjust the disc to prevent cutting deeper than 4 inches. BLADEX 90DF may be applied preemergence as an overlay over previously incorporated "Sutan+", "Genate Plus" or "Eradicane", if desired. Any rotation crop may be planted in the fall or spring following these treatments.

Existing stands of quackgrass, purple and yellow nutsedge must be turned under and thoroughly chopped up prior to chemical treatments.

TABLE 8
PREEMERGENCE BROADCAST APPLICATION RATES PER ACRE FOR TANK-MIX COMBINATIONS OF "BLADEX" 90DF PLUS ATRAZINE 4L ON SWEET CORN

SOIL TEXTURE DESCRIPTION	POUNDS OF "BLADEX" 90DF + QUARTS OF ATRAZINE 4L** — PERCENT ORGANIC MATTER IN SOIL*					
	LESS THAN 1%	1%	2%	3%	4%	5% & OVER
Sand, Loamy sand	DO NOT USE	0.9+0.4	1.2+0.4	1.4+0.7	1.7+0.9	2.4+1.1
Sandy loam	DO NOT USE	1.2+0.4	1.4+0.7	1.7+0.9	2.2+1.1	3.1+1.3
Loam, Silt loam, Silt	DO NOT USE	1.4+0.7	1.7+0.9	2.2+1.1	2.8+1.3	3.6+1.3
Sandy clay loam, Clay loam, Silty clay loam	DO NOT USE	1.7+0.9	2.0+1.1	2.8+1.3	3.6+1.3	3.8+1.6
Sandy clay, Silty clay, Clay	DO NOT USE	2.0+1.1	3.1+1.3	3.6+1.3	3.8+1.6	4.0+1.8
Peat or Muck	NOT RECOMMENDED					

* For organic matter content between those listed, adjust the rate proportionately.
** If 90% Atrazine is used, multiply rates shown by 1.11 to equal lbs. of 90% Atrazine products. If Atrazine 80W is used, multiply rates shown by 1.25 to equal lbs. of Atrazine 80W.

TABLE 9
PREEMERGENCE BROADCAST APPLICATION RATES PER ACRE FOR "BLADEX" 90DF USED IN TANK MIX COMBINATIONS WITH "LASSO", "SUTAN+", "GENATE PLUS", "ERADICANE", OR "DUAL" ON SWEET CORN

SOIL TEXTURE DESCRIPTION	POUNDS OF "BLADEX" 90DF — PERCENT ORGANIC MATTER IN SOIL*					
	LESS THAN 1%	1%	2%	3%	4%	5% & OVER
Sand, Loamy sand	DO NOT USE	0.9	1.3	1.6	1.8	2.2
Sandy loam	DO NOT USE	1.3	1.6	1.8	2.2	2.5
Loam, Silt loam, Silt	DO NOT USE	1.6	1.8	2.2	2.5	2.9
Sandy clay loam, Clay loam, Silty clay loam	DO NOT USE	2.0	2.2	2.5	2.9	3.1
Sandy clay, Silty clay, Clay	DO NOT USE	2.2	2.5	2.9	3.1	3.3
Peat or Muck	NOT RECOMMENDED					

* For organic matter content between those listed, adjust the rate proportionately.
** Do not use in the light sandy soils of the Atlantic Coastal Plain.

Additional weeds controlled by "Sutan+", "Genate Plus" or "Eradicane" combinations:

Grasses:	Sandbur Shattercane (Wild Cane)* Texas Panicum Quackgrass ("Eradicane" only) Wild Proso Millet* ("Eradicane" only)
Perennial Weeds:	Yellow Nutsedge (Nutgrass) Purple Nutsedge (Nutgrass)

*Suppression only - refer to "Sutan+", "Genate Plus" or "Eradicane" label for appropriate supplemental cultural and tillage practices.

For fields with moderate to heavy infestations of these weeds refer to the "Sutan+", "Genate Plus" or "Eradicane" labels for appropriate higher rates.

"BLADEX" PLUS "DUAL" 8E

Use BLADEX 90DF at the proper rate for soil texture and organic matter shown in Table 9. Use "Dual" as follows:

Soil Texture	Broadcast Rate Per Acre for "Dual" 8E
Coarse Sand, Loamy sand, Sandy loam	1.25 - 1.5 pints
Medium Loam, Silt loam, Silt	1.5 - 2.0 pints
Fine Sandy clay loam, Silty clay loam, Clay loam, Sandy clay, Silty clay, Clay	1.5 - 2.5 pints

The low end of the rate range should be used for lowest organic matter soils and the rate increased proportionately as the organic matter increases. Soils containing 4 percent organic matter or more require the highest rate shown for that soil texture. Refer to the Dual label for precautions or rotational crops.

TABLE 10
PREEMERGENCE BROADCAST APPLICATION RATES PER ACRE FOR "BLADEX" 90DF PLUS ATRAZINE 4L USED IN TANK-MIX COMBINATIONS WITH "LASSO", "SUTAN+" "GENATE PLUS", "ERADICANE", OR "DUAL" ON SWEET CORN

SOIL TEXTURE DESCRIPTION	POUNDS OF "BLADEX" 90DF + QUARTS OF ATRAZINE 4L** — PERCENT ORGANIC MATTER IN SOIL*					
	LESS THAN 1%	1%	2%	3%	4%	5% & OVER
Sand, Loamy sand	DO NOT USE	0.7+0.2	0.9+0.4	1.1+0.5	1.2+0.6	1.6+0.6
Sandy loam	DO NOT USE	0.9+0.4	1.1+0.5	1.2+0.6	1.6+0.6	1.8+0.6
Loam, Silt loam, Silt	DO NOT USE	1.1+0.5	1.3+0.6	1.6+0.6	1.8+0.6	2.0+0.9
Sandy clay loam, Clay loam, Silty clay loam	DO NOT USE	1.3+0.6	1.6+0.6	1.8+0.6	2.0+0.9	2.2+0.9
Sandy clay, Silty clay, Clay	DO NOT USE	1.6+0.6	1.8+0.9	2.0+0.9	2.2+0.9	2.2+1.1
Peat or Muck	NOT RECOMMENDED					

* For organic matter content between those listed, adjust the rate proportionately.
** Do not use in the light sandy soils of the Atlantic Coastal Plain.
*** If a 90% Atrazine is used multiply Atrazine rates shown by 1.11 to equal lbs. of 90% Atrazine products. If Atrazine 80W is used, multiply rates shown by 1.25 to equal lbs. of Atrazine 80W.

COTTON

IDLE SEASON EARLY PREPLANT WEED CONTROL (CALIFORNIA ONLY)

Weeds Controlled

Grasses

Annual Bluegrass	Bristly Foxtail	Wild Oat*
Annual Ryegrass	Rabbitsfoot Grass	Yellow Foxtail
Barnyardgrass*	Volunteer Small Grains (suppression)	

Broadleaves

Annual henbit	Groundsel	Pineapple weed
Black nightshade	Knotweed	Prickly lettuce
Burclover	Lambsquarters	Shepherdspurse
Cheeseweed*	London rocket	Sowthistle
Chickweed	Marestail	Wild mustard
Fiddleneck	Miners lettuce	Wild radish

* Under soil moisture conditions favoring deep germination, these species may not be completely controlled.

BLADEX may be used for burndown of small existing annual weeds and residual control of weeds during the winter and early spring season prior to planting cotton in California only. Complete any planned tillage prior to application. Apply herbicide treatment before weeds germinate or before weed seedlings are more than 3 inches tall. Tillage after application may reduce the effectiveness of the herbicide treatment.

"BLADEX" PLUS ATRAZINE PLUS "LASSO", "SUTAN +", "GENATE PLUS", "ERADICANE" OR "DUAL"

NOTE: Do not use combinations with "Sutan+", "Genate Plus", "Eradicane" in New Jersey. Use BLADEX 90DF plus Atrazine at the proper rate of soil texture and organic matter shown in Table 10. Use "Lasso", "Sutan+", "Genate Plus", "Eradicane" or "Dual" according to rates shown in BLADEX combinations in this section of the label.

Rotational Crops: Refer to Rotational Crops section of BLADEX plus Atrazine in this section of the label.

Apply BLADEX at least 30 days prior to planting. Apply the proper rate for the soil texture, organic matter and time interval between application and planting indicated in the Table 11. Where existing weeds are present, add crop oil concentrate, surfactant, or emulsible vegetable oil at its recommended rate to aid in the burndown of small weeds.

Where existing weeds are greater than 3 inches in height, when very dry conditions exist or where volunteer grains are a major problem, tank-mix BLADEX with 1-2 pts./A. (2 lbs./gal.) or 1.3-2.7 pts./A. (1.5 lbs./gal.) of paraquat. Well established weeds 6 inches tall or taller may not be well controlled.

Apply BLADEX 90DF plus paraquat in at least 20 gal./A. of carrier by ground sprayer. (The volume of carrier and the application equipment must be adequate to give a uniform application.) Add "Ortho X-77" surfactant at 1-2 qts./100 gals. of diluted spray (or other suitable surfactant at recommended rates) where paraquat is used. Crop oil concentrate or emulsible vegetable oil are not needed where paraquat is used. Do not apply paraquat combinations in suspension type fertilizer.

BLADEX can also be tank-mixed with "Treflan", or "Prowl", and incorporated for fall listed cotton beds instead of surface applied as described above. Precautions: 1). Failure to wait the recommended time interval between application and planting may result in crop injury. 2). At least one inch of rainfall or an equivalent irrigation that waters the surface of the soil after application must precede planting. 3). The use of this treatment on calcareous or caliche soil outcroppings may result in crop injury. 4). Do not graze or feed foliage from treated areas to livestock. 5). Do not apply BLADEX to cotton land in irrigation water.

TABLE 11
BROADCAST APPLICATION RATES PER ACRE OF "BLADEX" 90DF FOR IDLE SEASON OR EARLY PREPLANT TREATMENT ON COTTON

SOIL TEXTURE DESCRIPTION	DAYS PRIOR TO PLANTING*					
	30 DAYS		60 DAYS		90 DAYS	
	LBS. OF "BLADEX" 90DF		LBS. OF "BLADEX" 90DF		LBS. OF "BLADEX" 90DF	
	ORGANIC MATTER		ORGANIC MATTER		ORGANIC MATTER	
	UNDER 2%	OVER 2%	UNDER 2%	OVER 2%	UNDER 2%	OVER 2%
Sands, Loamy sands	1.75	2.25	2.75	3.25	3.25	4.0
All other Soils	2.25	2.75	3.25	4.0	4.0	4.5

* For time intervals between those listed, adjust the rates proportionately

PREEMERGENCE

Apply preemergence on cotton only in the states of Alabama, Arkansas, Louisiana and Mississippi.

Weeds Controlled

Annual morningglory	Prickly sida (Teaweed)
Cocklebur	Spurge

BLADEX 90DF is a selective preemergence herbicide for early season weed control in cotton. Supplemental practices (such as BLADEX applied directed postemergence) may be necessary to control late season weeds. BLADEX 90DF can be used in a tank-mix combination with Zorial Rapid 80.

Carefully match the BLADEX rate with the soil texture. Do not use on fields where the soil texture changes from coarse to fine. Avoid overlapping the spray pattern or overdosing the field with BLADEX 90DF. Application rates above those recommended for the soil texture can result in yellowing or stunting of the crop and may result in stand reduction.

While cotton exhibits tolerance to BLADEX 90DF, adverse growing conditions such as excessive rains, standing water or cold weather may result in stand reduction.

DO NOT GRAZE OR FEED FOLIAGE FROM TREATED AREAS TO LIVESTOCK.

BLADEX plus "Zorial Rapid 80"

Apply BLADEX 90DF plus Zorial Rapid 80 at the proper rate for the soil texture shown in Table 12. The soil must contain at least 1.0 percent organic matter. Seed placement should be 1/2-3/4 inch from the soil surface. Plant only cotton within six months after the last application of Zorial Rapid 80 or injury may occur.

DIRECTED POSTEMERGENCE - LAYBY

BLADEX 90DF and tank-mix combinations may be applied directed postemergence or layby to cotton and either preemergence to weeds or postemergence to weeds in all cotton growing States.

Weeds Controlled

Annual morningglory*	Palmer amaranth
Bristly starbur	Pigweed (redroot and spiny)
Cocklebur	
Crotalaria	Prickly sida (Teaweed)
Jimsonweed	Sicklepod
Lambsquarters	Spurge
Nightshade (annual)	Tropic croton
	Wright groundcherry

*The degree of preemergence control from a layby treatment will be reduced if soil moisture and temperature conditions cause deep germination of the seed.

Apply before weeds are more than 2 inches tall. Apply the directed postemergence treatment after the cotton has attained the minimum height of 6 inches. For layby treatment, apply BLADEX 90DF after the cotton has attained the height of 12 inches or more.

TABLE 12
PREEMERGENCE BROADCAST APPLICATION RATES PER ACRE FOR "BLADEX" 90DF + "ZORIAL RAPID 80" ON COTTON

SOIL TEXTURE*		POUNDS/ACRE BLADEX 90DF	+	POUNDS/ACRE ZORIAL RAPID 80
Coarse Soils**	Sandy loam	0.6		0.8
Medium Soils	Silt and Silt loam,	0.7		1.3
	Loam, Clay loam, Sandy clay loam, Sandy clay	1.0		1.3
Fine Soils	Silty clay loam, Silty clay, Clay	1.3		1.6

* The soil must contain at least 1% organic matter.
** Do not use on coarse soils (Sands and Loamy sands) containing more than 70% sand.

TABLE 13
DIRECTED POSTEMERGENCE APPLICATION RATES PER ACRE FOR "BLADEX" 90DF ON COTTON

BROADCAST	BANDED 38" ROW 12" BAND	BANDED 38" ROW 19" BAND
0.7 - 1.1 lb.	0.2 - 0.4 lb.	0.4 - 0.5 lb.

Use the maximum rate when dry or arid conditions exist.

The spray mixture should be directed to the soil around the base of the cotton plants. Care should be taken to prevent the spray from striking the cotton leaves as injury will occur. The use of leaf lifters or shields on application equipment is recommended to avoid spraying the cotton foliage.

BLADEX may be applied directed postemergence and/or layby following a preemergence application of BLADEX. Apply no more than two directed postemergence and one preemergence application to the same crop in any one year. If BLADEX is not used preemergence, apply no more than three directed postemergence applications including layby to the same crop in any one year. (In California, apply no more than two directed postemergence applications including layby.)

When applied as a layby treatment before weeds emerge, the effectiveness of BLADEX 90DF depends on rainfall or irrigation to move it into the soil. When irrigation water activation is used, every row must be watered and for skip row cotton all treated soil must be irrigated.

Any rotational crop may be planted the fall or spring following any of the treatments in this section providing the soil is plowed or deep-disced prior to planting the rotation crop.

Do not apply BLADEX to cotton in irrigation water.

DO NOT GRAZE OR FEED FOLIAGE FROM TREATED AREAS TO LIVESTOCK.

TABLE 14
LAYBY APPLICATION RATES PER ACRE FOR "BLADEX" 90DF ON COTTON

HEIGHT OF COTTON	SOIL TEXTURE		BROADCAST RATES
12 inches or more	Coarse	Sandy loam	0.9 lb.
	Medium	Silt, Silt loam, Loam, Clay loam, Sandy clay loam, Sandy clay	1.3 lbs.
	Fine	Silty clay loam, Silty clay and Clay	1.8 lbs.

TABLE 15
DIRECTED POSTEMERGENCE APPLICATION RATES PER ACRE "BLADEX" 90DF + MSMA ON COTTON

PRODUCT	BROADCAST	BANDED 38" ROW 12" BAND	BANDED 38" ROW 19" BAND
BLADEX 90DF	0.7 - 1.1 lb.	0.2 - 0.4 lb.	0.4 - 0.5 lb.
+	+	+	+
MSMA (4 lb/gal)	4 pints	1.3 pints	2 pints
or	or	or	or
MSMA (6.6 lb/gal)	2.4 pints	0.8 pint	1.2 pints

BLADEX Applied Alone

Apply BLADEX 90DF directed-postemergence at the rate shown in Table 13. Apply at layby at the rates for the soil texture indicated in Table 14. Add an agricultural surfactant such as "Ortho X-77" or other surfactant suitable for use on growing cotton at the rate of 2 quarts per 100 gallons of spray mixture (or as directed by the manufacturer).

BLADEX plus MSMA

Apply a tank-mix combination of BLADEX 90DF plus MSMA plus surfactant after the cotton is 6 inches tall but before it reaches the bloom stage. Apply no more than two applications of this mixture before the first bloom stage. Tank-mix BLADEX plus MSMA at the rates indicated in Table 15. Add a surfactant, such as Ortho X-77, at the rate of 2 qts./100 gals of spray mixture (or as directed by the manufacturer).

GRAIN SORGHUM (MILO)
(See the GENERAL INFORMATION section of this label)

Tank-mix combinations of BLADEX 90DF plus atrazine, metolachlor "Dual", alachlor "Lasso", propazine "Milogard" or propachlor "Ramrod" may be used for selective preemergence weed control in grain sorghum. In addition, BLADEX 90DF may be used in tank-mix combinations with atrazine, "Dual" or "Lasso" for the control of weeds in early spring, early preplant, 14 to 35 days or more prior to planting grain sorghum. BLADEX 90DF in tank mix combinations may be applied preemergence or early preplant on grain sorghum grown under conventional or conservation tillage systems.

Do not use on forage sorghums. See the appropriate sections of the label for geographic distribution restrictions.

WEEDS CONTROLLED BY "BLADEX" 90DF IN TANK-MIX OR SEQUENTIAL COMBINATIONS WITH OTHER HERBICIDES ON GRAIN SORGHUM (MILO)

Grasses

Cheatgrass	Stinkgrass (Indian lovegrass)
Green foxtail	Volunteer Wheat (2)
Crabgrass	Yellow foxtail
Downy brome	

Broadleaves

Annual Morningglory	Lambsquarters	Russian thistle
Carpetweed	Pennsylvania smartweed	Shepherdspurse
Cocklebur (1)	Prostrate pigweed	Sunflower (1)
Common purslane	Prickly lettuce	Tansy and other mustards
Horseweed (marestail)	Ragweed (Common)	Velvetleaf
Kochia	Redroot pigweed	

(1) Under soil moisture and temperature conditions favoring deep germination or other factors that may cause delayed germination, these species may not be completely controlled.
(2) When the herbicide treatment is applied two weeks or more before planting, weed control of these species may break early if heavy rainfall occurs between application and planting.

General Directions For Use On Grain Sorghum

BLADEX used in tank-mixes with the products listed above should be applied only once per crop season or in split dosage treatments. If replanting of grain sorghum is necessary, it may be planted in soil previously treated with these mixtures. Apply BLADEX 90DF in these tank-mix combinations before the crop has emerged.

Precautions: Do not make an additional application of BLADEX, EXTRAZINE, or any product containing cyanazine or crop injury may occur. Heavy rainfall between planting and crop emergence may cause crop injury or stand loss. Rainfall tends to cause excessive concentrations of herbicide in seed furrow, resulting in possible crop injury. Level deep planter marks or seed furrows before application. Do not apply to furrow-planted sorghum.

Sorghum growing under stress caused by minor element deficiency, cold, wet weather or sorghum growing on highly calcareous soil (high pH) may suffer injury including stand reduction. Sorghum subjected to high winds, sand cutting, hail damage, or cold temperatures may be more susceptible to injury from the chemical treatment with possible stand loss. Where crop residues are pressed into the planter slot or any other factor keeps the slot from closing, crop injury or stand reduction may occur, caused by herbicides coming into direct contact with the seed from the spray or by being washed into the slot after a heavy rain.

When using "Dual" on sorghum, only "Concep" II safened seed may be used. When using "Lasso" on sorghum, only "Screen" safened seed may be used.

Early preplant treatments will require a preemergence herbicide treatment other than BLADEX 90DF, 80W or 4L, or "EXTRAZINE" 90DF or 4L at planting, and/or postemergence herbicide in the growing crop to provide required weed control, if the early preplant application is made more than 35 days prior to planting, or if weeds are present at planting time.

Wheat/Sorghum/Fallow Rotation

For sorghum grown under a wheat/sorghum/fallow rotation, BLADEX 90DF plus atrazine may be applied early preplant or preemergence. If an early preplant application with BLADEX 90DF plus atrazine is used, it should be used in conjunction with either a residual herbicide after wheat harvest the previous year, a preemergence herbicide treatment at planting, and/or a postemergence herbicide treatment in the growing crop. The total rate of atrazine permitted in conjunction with a sorghum crop (postharvest plus early preplant plus preemergence) is limited to 3.0 lbs. active ingredient per acre. Even this or lower rates may carry over to injure rotational crops.

Rotational Crops: (1) Plant only corn, sorghum, or soybeans the year following the use of this mixture. (2) If soybeans are to be planted, injury may occur. (3) If applied after June 10, do not rotate with crops other than corn or sorghum the next year or injury may occur. (4) In the high plains and intermountain areas of the West where rainfall is sparse and erratic or where irrigation is required, use only when corn or sorghum is to be planted the following year, or a crop of corn or sorghum not treated with this mixture or atrazine is to precede other rotational crops. (5) Small grains may be planted 15 months following treatment. (6) All other crops may be planted 18 months after application.

Conservation Tillage

For grain sorghum grown under conservation tillage, any of the herbicide treatments listed above may be tank-mixed with paraquat, and/or 2,4-D where weeds and grasses are present at the time of application exceeding two inches in height.

If volunteer wheat or cheatgrass is over 2 inches tall, heavily tillered, and/or growing in a dense mat, or if the wheat stubble has been tilled (undercut or disc) or grazed and seed is buried, complete control may not be achieved. Control of volunteer and other weeds is usually better in complete no-till situations than where prior tillage "planted" the seed and allowed extensive root development.

Complete spray coverage of the weeds is essential for best performance. Apply the desired treatment in 15 to 30 gallons of spray mixture per acre by ground rig. When using paraquat in a tank-mix, apply the desired rates in 20 to 40 gallons of spray mixture per acre. Use the higher volumes where there are heavy crop residues on the soil surface. Nitrogen solutions are the preferred carriers for these treatments as they aid in the burndown of existing weeds. In addition, crop oil or a nonionic surfactant may be added to the tank-mix as they aid in the burndown of existing weeds.

When tank-mixing with 2,4-D to control broadleaf weeds 14-35 days prior to planting, use 1 to 3 pints per acre of 4 pounds ai per gallon, 2,4-D LV (2/3 to 2 pints of 6 pounds ai per gallon 2,4-D LV) (or 2,4-D Amine at recommended rates). Use the higher rates where overwintering weeds are present or when directed on the 2,4-D label for the control of specific hard-to-kill weed species, such as perennials.

CAUTION: Use only these 2,4-D products with properly registered labels that permit such use and application rates.

When tank-mixing with paraquat to control grass and broadleaf weeds at the time of planting, use an approved nonionic adjuvant such as "Ortho X-77" at the rate of 1 quart per 100 gallons of dilute spray. Use 1-2 pints of paraquat (2 lbs./gal) or 1.3-2.7 pints (1.5 lbs/gal) per acre. Use the higher rate when weed growth is heavy or over 4 inches tall or when dry weather conditions prevail. Established weeds 6 inches tall or taller may not be completely controlled with paraquat.

OBSERVE ALL CAUTIONS AND LIMITATIONS ON LABELING OF ALL PRODUCTS USED IN MIXTURES.

EARLY PREPLANT

BLADEX 90DF in tank-mix combinations with other herbicides may be applied Early Preplant on grain sorghum only in the states of Kansas, Nebraska and South Dakota. The tank-mixes may be applied on grain sorghum grown under conventional or conservation tillage systems. Do not use on forage sorghum.

BLADEX PLUS ATRAZINE

BLADEX 90DF plus atrazine provides control of weeds listed on this label in the Weeds Controlled section for BLADEX 90DF on grain sorghum.

Precautions: Crop injury can occur if the soil stays dry between application and planting. Delay planting until at least 10 days after a soaking rain after treatment (soaks the soil to a depth of 4 inches or more). Heavy rains between planting and emergence can cause crop injury or stand loss.

Rotational Crops: Refer to the General Information section of the label. Fields treated with this tank-mix may be planted only to corn or grain sorghum within 12 months after this treatment. Other crops should not be planted for 18 months following this treatment.

Use BLADEX 90DF plus atrazine 4L at the proper rate for soil texture and time interval indicated in Table 16.

Split Applications: BLADEX 90DF plus Atrazine 4L dosage rates in Table 16 may be applied in a split application of 80% of the dosage 21 days or more before planting and the remaining 20% at planting time. If the season has been dry without sufficient rainfall to activate the herbicide after the initial application, the remaining 20% at planting time may be omitted to reduce the chances of crop injury.

"BLADEX" IN THREE-WAY COMBINATIONS

BLADEX 90DF plus atrazine in combination with "Dual" or "Lasso" applied early preplant provides control of weeds listed on this label in the Weeds Controlled section for BLADEX 90DF on grain sorghum. Additional weeds controlled by one of these combinations include:

Grasses

Barnyardgrass (1)	Stinkgrass (2)
Fall Panicum	Witchgrass (2)
Giant Foxtail	

(1) (2) Refer to Weeds Controlled section on grain sorghum for explanation.

Rotational Crops: If the crop treated with any of these combinations is lost, corn or grain sorghum may be replanted immediately without retreatment. Refer to the General Information section of the label for additional information regarding crop rotation.

BLADEX PLUS ATRAZINE PLUS DUAL

Dual may be used on grain sorghum only with seed treated with "Concep" II seed safener.

Use BLADEX 90DF plus atrazine 4L plus "Dual" 8E at the proper rate for soil texture and time interval indicated in Table 17.

"BLADEX" PLUS ATRAZINE PLUS "LASSO"

"Lasso" may be used on grain sorghum only with seed treated with "Screen" seed safener.

Use BLADEX 90DF plus Atrazine 4L plus "Lasso" 4EC at the proper rate for soil texture and time interval indicated in Table 18.

PREEMERGENCE OR SHALLOW PREPLANT INCORPORATION

Tank-mix combinations of BLADEX 90DF plus atrazine, metolachlor (Dual), alachlor (Lasso), propachlor (Ramrod) or propazine (Milogard) may be used for selective preemergence weed control in grain sorghum. BLADEX 90DF and its tank mixes may be applied preemergence on grain sorghum grown under conventional or conservation tillage systems. Do not use on forage sorghum.

TABLE 16
EARLY PREPLANT BROADCAST APPLICATION RATES OF "BLADEX" 90DF PLUS ATRAZINE 4L ON GRAIN SORGHUM (0.8% TO 3% ORGANIC MATTER SOILS)

SOIL TEXTURE DESCRIPTION	POUNDS OF "BLADEX" 90DF + QUARTS OF ATRAZINE 4L DAYS PRIOR TO PLANTING (b,c) 14 DAYS	28 DAYS	35 DAYS
Sand, Loamy sand	DO NOT USE		
Sandy loam	1.3 + 0.6(a,d)	1.8 + 0.8	2.2 + 1.0
Loam, Silt loam, Silt	1.7 + 0.7	2.2 + 1.0	2.6 + 1.2
Sandy clay loam, Clay loam, Silty clay loam	2.2 + 1.0	2.6 + 1.2	2.9 + 1.4
Sandy clay, Silty clay, Clay	2.4 + 1.1	2.6 + 1.2	3.3 + 1.5
Peat or Muck	NOT RECOMMENDED		
Eroded Slopes or Knobs Soils with pH greater than 8.0	NOT RECOMMENDED		

(a) The first number is pounds/acre of BLADEX 90DF. The second number is quarts/acre of atrazine 4L. For pounds/acre of atrazine 80W multiply by 1.25
(b) For intervals longer than 35 days due to delayed planting, etc., another herbicide treatment is needed before, at, or after planting.
(c) For sandy loam soils containing less than 1.5% O.M., increase the treatment interval from 14 to 21 days. Use the 14 day interval for sandy soils containing more than 1.5%.
(d) For rates between those listed at 14 and 35 days, adjust the rate proportionately.

BLADEX 90DF in two-way tank-mix combinations with propachlor or propazine or in three-way combinations with atrazine and propachlor, "Lasso" or "Dual" should be applied only once per crop season or in split dosage treatments. (See the Early Preplant section.) If replanting of grain sorghum is necessary, it may be planted in soil previously treated with these mixtures. Do not make a second application of BLADEX, EXTRAZINE or any product containing cyanazine or crop injury may occur. Apply these tank-mix combinations at planting or after planting, but before the crop and weeds have emerged. Heavy rain immediately following application tends to cause excessive concentrations of herbicide in seed furrow and can cause crop injury and stand reduction. Do not apply to furrow-planted sorghum until furrows are leveled (plowed-in). Level deep planter marks or seed furrows before application. Sorghum growing under stress caused by minor element deficiency or cold, wet weather or sorghum growing on highly calcareous soil may suffer injury and stand loss.

TABLE 17
EARLY PREPLANT BROADCAST APPLICATION RATES PER ACRE IN POUNDS OF "BLADEX" 90DF, QUARTS OF ATRAZINE 4L AND PINTS OF "DUAL" 8E ON GRAIN SORGHUM (b)

SOIL TEXTURE DESCRIPTION	DAYS PRIOR TO PLANTING (c) (0.8% TO 3% ORGANIC MATTER) 14 DAYS* LBS "BLADEX"/QTS ATRAZ/PTS "DUAL"			28 DAYS LBS "BLADEX"/QTS ATRAZ/PTS "DUAL"		
Sand, Loamy sand	DO NOT USE			1.1(a)	0.5	1.8
Sandy loam	1.0(d)	0.4	1.35	1.2	0.5	1.8
Loam, Silt loam, Silt	1.2	0.5	1.35	1.5	0.7	1.8
Sandy clay loam, Clay loam, Silty clay loam	1.5	0.7	1.6	1.9	0.8	1.8
Sandy clay, Silty clay, Clay	1.7	0.75	1.6	2.0	0.9	1.8
Peat or Muck	NOT RECOMMENDED					
Eroded Slopes or Knobs Soils with pH greater than 8.0	NOT RECOMMENDED					

(a) The first number is pounds/acre of BLADEX 90DF. The second number is quarts/acre of atrazine 4L. The third number is pints/acre of "Dual" 8E.
(b) If using "Dual" 8E on sorghum, "Concep" II safened seed should be planted.
(c) For intervals longer than 28 days due to delayed planting, etc., another herbicide treatment is needed before, at, or after planting.
(d) For sandy loam soils containing less than 1.5% O.M., increase the treatment interval from 14 to 21 days. Use the 14 day interval for sandy soils containing more than 1.5%.
* For rates between those listed at 14 and 28 days, adjust the rate proportionately.

TABLE 18
EARLY PREPLANT BROADCAST APPLICATION RATES PER ACRE IN POUNDS OF "BLADEX" 90DF, QUARTS OF ATRAZINE 4L AND QUARTS OF "LASSO" 4EC ON GRAIN SORGHUM (b)

SOIL TEXTURE DESCRIPTION	DAYS PRIOR TO PLANTING (c) (0.8% TO 3% ORGANIC MATTER) 14 DAYS* LBS "BLADEX"/QTS ATRAZ/QTS "LASSO"			28 DAYS LBS "BLADEX"/QTS ATRAZ/QTS "LASSO"		
Sand, Loamy sand	DO NOT USE			1.0(a)	0.45	2.25
Sandy loam	0.8(d)	0.4	1.8	1.0	0.5	2.25
Loam, Silt loam, Silt	1.0	0.5	1.8	1.3	0.6	2.25
Sandy clay loam, Clay loam, Silty clay loam	1.3	0.6	2.0	1.7	0.75	2.25
Sandy clay, Silty clay, Clay	1.5	0.7	2.0	1.9	0.8	2.25
Peat or Muck	NOT RECOMMENDED					
Eroded Slopes or Knobs Soils with pH greater than 8.0	NOT RECOMMENDED					

(a) The first number is pounds/acre of BLADEX 90DF. The second number is quarts/acre of atrazine 4L. The third number is quarts/acre of "Lasso" 4EC.
(b) If using "Lasso" 4EC on sorghum, Screen safened seed should be planted.
(c) For intervals longer than 28 days due to delayed planting, etc., another herbicide treatment is needed before, at, or after planting.
(d) For sandy loam soils containing less than 1.5% O.M., increase the treatment interval from 14 to 21 days. Use the 14 day interval for sandy soils containing more than 1.5%.
* For rates between those listed at 14 and 28 days, adjust the rate proportionately.

BLADEX 90DF plus Propachlor (Ramrod)

Weeds Controlled

Apply this tank-mix to grain sorghum only in States East of the Rocky Mountains.

Grasses	Giant foxtail
Barnyardgrass	Green foxtail
Crabgrass	Yellow foxtail
Fall panicum	
Broadleaves	Pigweed
Annual morningglory	Ragweed (Common)
Carpetweed	Smartweed (Pennsylvania)
Cocklebur*	Velvetleaf*
Common purslane	
Lambsquarters	

* Under conditions such as low temperatures, lack of soil surface moisture or other factors causing a delay in germination of the seeds, the degree of control may be impaired against these weeds.

Apply BLADEX 90DF plus propachlor at the proper rate for soil texture and organic matter shown in Table 19.

Any rotational crop may be planted the fall or spring following this treatment.

BLADEX plus Propazine (Milogard)

Apply to grain sorghum only in the States of Texas, Oklahoma and Kansas.

Weeds Controlled

Grasses	
Crabgrass	Green foxtail
Fall panicum	Yellow foxtail
Giant foxtail	
Broadleaves	
Annual morningglory	Pigweed
Carpetweed	Puncturevine
Cocklebur	Ragweed (Common)
Common purslane	Smartweed
Lambsquarters	Velvetleaf

Use BLADEX 90DF plus propazine (Milogard) 4L at the proper rate for soil texture and organic matter indicated in Table 20.

Rotational Crops: In the Texas Gulf Coast and Texas Blacklands, fields treated with this tank-mix may be planted to cotton, soybeans, or corn 12 months after treatment. In Oklahoma and west Texas, fields treated with this tank-mix may be planted to cotton or corn 12 months after application. In Kansas, corn may be planted in rotation 12 months after treatment. Other crops should not be planted for 18 months following treatment.

TABLE 19
PREEMERGENCE BROADCAST APPLICATION RATES IN POUNDS OF "BLADEX" 90DF PLUS QUARTS OF PROPACHLOR (RAMROD) 4L PER ACRE ON GRAIN SORGHUM

SOIL TEXTURE DESCRIPTION	PERCENT ORGANIC MATTER IN THE SOIL* 2% LBS "BLADEX" +	2% QTS "RAMROD"	3% LBS "BLADEX" +	3% QTS "RAMROD"
Sand, Loamy sand	DO NOT USE			
Sandy loam	1.1(a)	2.5	1.3	3.0
Loam, Silt loam, Silt	1.3	3.0	1.5	3.5
Sandy clay loam, Clay loam, Silty clay loam	1.5	3.5	1.8	4.0
Sandy clay, Silty clay, Clay	1.8	4.0	1.8	4.0
Peat or Muck	NOT RECOMMENDED			
Eroded Slopes or Knobs Soils with pH greater than 8.0	NOT RECOMMENDED			

(a) The first number is pounds/acre of BLADEX 90DF. The second number is quarts/acre of "Ramrod" 4L.
* For organic matter content between those listed, adjust the rate proportionately.

BLADEX IN THREE-WAY COMBINATIONS

BLADEX 90DF plus plus atrazine plus "Dual", "Lasso" or "Ramrod" may be applied preemergence or with shallow incorporation for weed control in grain sorghum only in the states of Kansas, Nebraska and South Dakota. These treatments provide control of weeds listed on this label in the Weeds Controlled section for BLADEX 90DF on grain sorghum. Additional weeds controlled by these combinations include:

Grasses	
Barnyardgrass (1)	Witchgrass (2)
Stinkgrass (2)	Giant Foxtail
Fall Panicum	

(1) (2) Refer to Weeds Controlled section on grain sorghum for explanation.

Rotational Crops If the crop treated with this combination is lost, corn or grain sorghum may be replanted immediately without retreatment. Refer to the "General Information" section of the BLADEX label and to the tank mix product labels for additional information regarding crop rotation.

"BLADEX" PLUS ATRAZINE PLUS "DUAL"

"Dual" can be used on grain sorghum only with seed treated with "Concep" II seed safener.

Use BLADEX 90DF plus "Dual" 8E at the proper rate for soil texture and organic matter indicated in Table 21. Use only on soils having at least 1% organic matter.

TABLE 20
PREEMERGENCE BROADCAST APPLICATION RATES IN POUNDS OF "BLADEX" 90DF PLUS QUARTS OF PROPAZINE (MILOGARD) 4L PER ACRE ON GRAIN SORGHUM

SOIL TEXTURE DESCRIPTION	BLADEX 4L LBS/ACRE	PROPAZINE 4L* QTS/ACRE
Sand, Loamy sand, Sandy loam	DO NOT USE	
Loam, Silt loam, Silt	0.9	0.75
Sandy clay loam, Clay loam, Silty clay loam	1.1	1.0
Sandy clay, Silty clay, Clay	1.3	1.25
Peat or Muck	NOT RECOMMENDED	
Eroded Slopes or Knobs Soils with pH greater than 8.0	NOT RECOMMENDED	

* If Propazine 80W is used to multiply rate shown by 1.25 to equal pounds of Propazine 80W.
If Propazine 90DF is used multiply rate shown by 1.11 to equal pounds of Propazine 90DF.

TABLE 21
PREEMERGENCE OR SHALLOW PREPLANT INCORPORATION BROADCAST APPLICATION RATES IN POUNDS OF "BLADEX" 90DF, QUARTS OF ATRAZINE 4L PLUS PINTS OF "DUAL" 8E PER ACRE ON GRAIN SORGHUM (a,b)

SOIL TEXTURE DESCRIPTION	PRODUCT	PERCENT ORGANIC MATTER IN SOIL* 1%	2%	3%
Sand, Loamy sand, Sandy loam		DO NOT USE		

SOIL TEXTURE DESCRIPTION	PRODUCT		PERCENT ORGANIC MATTER IN SOIL* 1%	2%	3%
Loam Silty loam Silt	BLADEX 90DF Atrazine 4L "Dual" 8E	Pounds Quarts Pints	0.75 0.3 1.5	0.8 0.4 1.5	0.9 0.4 1.5
Sandy clay loam Clay loam Silty clay loam	BLADEX 90DF Atrazine 4L "Dual" 8E	Pounds Quarts Pints	0.8 0.4 1.5	0.9 0.4 1.75	1.0 0.5 1.75
Sandy clay Silty clay Clay	BLADEX 90DF Atrazine 4L "Dual" 8E	Pounds Quarts Pints	0.9 0.4 1.5	1.0 0.5 1.75	1.0 0.5 2.0
Peat or Muck	NOT RECOMMENDED				
Eroded Slopes or Knobs Soils with pH greater than 8.0	NOT RECOMMENDED				

(a) When using "Dual" 8E on sorghum, only "Concep" II safened seed should be planted.
(b) To enhance weed control in areas of less than 25 inches of annual precipitation or where long dry periods are common, this treatment may require shallow incorporation with a tool such as a field cultivator operated at 5-7 mph.
* For organic matter content between those listed, adjust the rate proportionately.

"BLADEX" PLUS ATRAZINE PLUS "LASSO"

"Lasso" can be used on grain sorghum only with seed treated with "Screen" seed safener.

Use BLADEX 90DF plus Atrazine 4L plus "Lasso" 4EC at the proper rate for soil texture and organic matter indicated in Table 22. Use only on soils having at least 1% organic matter.

TABLE 22

PREEMERGENCE OR SHALLOW PREPLANT INCORPORATION BROADCAST APPLICATION RATES IN POUNDS OF "BLADEX" 90DF, QUARTS OF ATRAZINE 4L AND QUARTS OF "LASSO" EC PER ACRE ON GRAIN SORGHUM (a,b)

SOIL TEXTURE DESCRIPTION	PRODUCT		PERCENT ORGANIC MATTER IN SOIL* 1%	2%	3%
Sand, Loamy sand			DO NOT USE		
Sandy loam	BLADEX 90DF Atrazine 4L "Lasso" 4EC	Pounds Quarts Quarts	none none none	0.75 0.3 2.0	0.8 0.4 2.0
Loam Silt loam Silt	BLADEX 90DF Atrazine 4L "Lasso" 4EC	Pounds Quarts Quarts	0.75 0.3 2.0	0.8 0.4 2.0	0.9 0.4 2.0
Sandy clay loam Clay loam Silty clay loam	BLADEX 90DF Atrazine 4L "Lasso" 4EC	Pounds Quarts Quarts	0.8 0.4 2.25	0.9 0.4 2.25	1.0 0.5 2.25
Sandy clay Silty clay Clay	BLADEX 90DF Atrazine 4L "Lasso" 4EC	Pounds Quarts Quarts	0.9 0.4 2.25	1.0 0.5 2.25	1.0 0.5 2.5
Peat or Muck	NOT RECOMMENDED				
Eroded Slopes or Knobs Soils with pH greater than 8.0	NOT RECOMMENDED				

(a) When using "Lasso" 4EC on sorghum, only Screen safened seed should be planted.
(b) To enhance weed control in areas of less than 25 inches of annual precipitation or where long dry periods are common, this treatment may require shallow incorporation with a tool such as a field cultivator operated at 5-7 mph.
* For organic matter content between those listed, adjust the rate proportionately.

TABLE 23

PREEMERGENCE BROADCAST APPLICATION RATES IN POUNDS OF "BLADEX" 90DF, QUARTS OF ATRAZINE 4L AND QUARTS OF PROPACHLOR (RAMROD) 4L PER ACRE ON GRAIN SORGHUM

SOIL TEXTURE DESCRIPTION	PERCENT ORGANIC MATTER IN THE SOIL* 2% LBS "BLADEX"/QTS ATRAZ/QTS "RAMROD"			3% LBS "BLADEX"/QTS ATRAZ/QTS "RAMROD"		
Sand, Loamy sand	DO NOT USE					
Sandy loam	0.7(a)	0.3	2.25	0.8	0.4	2.7
Loam, Silt loam, Silt	0.8	0.4	2.7	0.9	0.4	3.15
Sandy clay loam, Clay loam, Silty clay loam	0.9	0.4	3.15	1.0	0.5	3.6
Sandy clay, Silty clay, Clay	1.0	0.5	3.6	1.0	0.5	3.6
Peat or Muck	NOT RECOMMENDED					
Eroded Slopes or Knobs Soils with pH greater than 8.0	NOT RECOMMENDED					

(a) The first number is pounds/acre of BLADEX 90DF. The second number is quarts/acre of atrazine 4L. The third number is quarts/acre of "Ramrod" 4L.
* For organic matter content between those listed, adjust the rate proportionately.

"BLADEX" PLUS ATRAZINE PLUS "RAMROD"

Use BLADEX 90DF plus Atrazine 4L plus Ramrod at the proper rate for soil texture and organic matter indicated in Table 23.

DO NOT USE IN CALIFORNIA.

TABLE 24
PREPLANT BROADCAST APPLICATION RATES PER ACRE FOR "BLADEX" 90DF FOR WINTER WHEAT 60 DAYS BEFORE PLANTING*

SOIL TEXTURE DESCRIPTION	POUNDS OF "BLADEX" 90DF — PERCENT ORGANIC MATTER IN SOIL	
	LESS THAN 2%	MORE THAN 2%
SOIL TEXTURE DESCRIPTION	LESS THAN 2%	MORE THAN 2%
Sand, Loamy sand	2.2	2.7
All other textures	2.7	3.1

* For applications made 45 days prior to planting, decrease the BLADEX rate indicated by 0.4 pound/acre.

WHEAT

EARLY PREPLANT

BLADEX 90DF may be used for Early Preplant or Post-harvest weed control in continuous winter wheat.

Weeds controlled: See FALLOW CROPLAND section of this label.

Apply after wheat harvest but 45-60 days prior to planting winter wheat. Use at the proper rate for soil texture and organic matter indicated in Table 24. Add "Ortho X-77" at 1-2 qts./100 gals. of diluted spray. Use the higher rate of surfactant when dry weather conditions prevail. Failure to wait the recommended time interval between application and planting may result in crop injury and stand reduction. Use of this treatment on calcareous or caliche subsoil outcroppings may result in crop injury and stand reduction.

For best results, during harvest use a straw chopper/spreader on the combine that distributes the straw uniformly over the soil surface unless the straw is to be baled and removed prior to treatment. Avoid tillage after application as it may reduce the effectiveness of the herbicide treatment.

Where broadleaf weeds are present at time of application, tank-mix 1-1/3 - 2 pts./A. of 2,4-D low volatile 6 lb./gal. ester (2-3 pts./A. of 2,4-D LV 4 lb./gal.) (or 2,4-D Amine at recommended rates). Use the high rate when weeds are over 4 inches tall or when directed on the 2,4-D label for control of hard-to-kill weed species, such as perennials.

Where grasses and broadleaf weeds are present and exceed 2 inches in height, tank-mix with 1-2 pts./A. (2 lbs./gal.) or 1.3-2.7 pts./A. (1.5 lbs./gal.) of paraquat. Well established weeds over 6 inches tall may not be well controlled. Apply at least 25 gals./A. of spray mixture by ground sprayer where paraquat is included in the tank mixture.

No more than two applications of BLADEX may be made prior to planting wheat including applications made under the FALLOW CROPLAND section of this label.

FALLOW CROPLAND

Weeds Controlled by BLADEX 90DF Alone and in Tank-Mix Combinations

Grasses

Annual ryegrass (Italian)	Indian lovegrass (Stinkgrass)
Barnyardgrass*	Volunteer wheat
Crabgrass	Wild oat*
Cheatgrass	Witchgrass
Downy brome	Yellow foxtail
Green foxtail	

Broadleaves

Cocklebur*	Prostrate spurge
Common chickweed	Purslane
Dog fennel	Russian thistle
False flax	Shepherdspurse
Henbit	Smartweed (Pennsylvania)
Horseweed (marestail)	Sunflower* (wild)
Kochia	Purple mustard
Lambsquarters	Tansy mustard
Pennycress	Tumble mustard
Pigweed*	Wild radish
Prickly lettuce	Wild buckwheat*
Prostrate knotweed	

*Under soil moisture and temperature conditions favoring deep germination or other factors that may cause delayed germination, these species may not be completely controlled.

Additional weeds controlled are listed in the CORN section of this label. BLADEX 90DF may be used alone or in tank-mix combination with atrazine for the control of certain annual weeds during a fallow program.

BLADEX 90DF or BLADEX 90DF plus atrazine should be used in tank-mix combination with paraquat and/or 2,4-D as described below if growing vegetation is present. Should weeds become established before adequate rainfall for herbicide activation occurs, sweep tillage may be employed to destroy them.

Application Directions

Apply fallow cropland herbicide treatments uniformly to the soil surface. Adjust boom height on ground rigs to obtain the correct spray pattern at the top of the stubble rather than the ground. At sprayer speeds over 8 mph and when crop residues are heavy, use flood type nozzles and at least 20 gal./A. of carrier.

BLADEX 90DF Applied Alone

Select the appropriate rate of BLADEX 90DF from Table 25. Add "Ortho X-77" surfactant at the rate of 1 qt./100 gals. of diluted spray or other suitable surfactant at recommended rates. Winter wheat may be planted four months or more after treatment. Spring wheat or durum wheat and other rotational crops may be planted nine months after treatment. Grain sorghum or field corn may be planted in the spring following a late fall application.

BLADEX 90DF Plus Atrazine

BLADEX 90DF may be used in tank-mix combination with atrazine where a maximum period of weed control is desired in a fallow cropland program. Treatments must be applied before November 15 of the year before planting winter wheat or at least 11 months before planting spring wheat or durum wheat. Select the appropriate rates of BLADEX 90DF plus atrazine for a particular location from Table 26. Add Ortho X-77 surfactant at 1 qt./100 gals. of diluted spray or other appropriate surfactant at recommended rates.

TABLE 25
BROADCAST APPLICATION RATES PER ACRE FOR "BLADEX" 90DF ON FALLOW CROPLAND*

SOIL TEXTURE DESCRIPTION	POUNDS PER ACRE OF "BLADEX" 90DF		
	POST HARVEST TREATMENT FOR FALL WEED CONTROL	LATE FALL TREATMENT FOR SPRING WEED CONTROL	SPRING TREATMENT FOR SPRING WEED CONTROL
Sand, Loamy sand, Sandy loam, Loam, Silt loam, Silt, Sandy clay loam, Clay loam	2.7	3.6	2.7
Silty clay loam, Sandy clay Silty clay, Clay	3.1	3.6	2.7

* For soils containing more than 2% organic matter, use an additional 0.4 lb/acre of BLADEX 90DF.
For soils containing more than 3% organic matter, use an additional 1.0 lb/acre BLADEX 90DF.

TABLE 26
BROADCAST APPLICATION RATES PER ACRE FOR TANK MIX COMBINATIONS OF "BLADEX" 90DF PLUS ATRAZINE ON FALLOW CROPLAND*

LOCATION/TREATMENT			LBS/ACRE "BLADEX" 90DF	+	QTS/ACRE ATRAZINE 4L**
FALL APPLICATION FOR WINTER WHEAT Kansas, southern Nebraska, Oklahoma, Colorado East Slope between Hwy. I-76 & State Hwy. 96			1.8	+	0.8
FALL APPLICATION FOR WINTER OR SPRING WHEAT Nebraska Panhandle, Wyoming, Utah, Colorado West Slope & East Slope north of Hwy. I-76 & south of Hwy. 96, Idaho, Montana, North Dakota, South Dakota			1.8	+	0.6
SPRING APPLICATION FOR SPRING WHEAT Idaho, Montana, North Dakota, South Dakota, northern Utah, northern Wyoming			2.2	+	0.4
FALL APPLICATION FOR WINTER WHEAT Columbia Basin areas of Washington & Oregon	Average Annual Rainfall	over 15"	2.2	+	0.4
		10-15"	2.2	+	0.27
		under 10"	2.2	+	0.2

* Use an additional 0.5 lb./A. of BLADEX 90DF for soils with 2.0 to 3.0 percent organic matter. Use an additional 1.0 lb./A. of BLADEX 90DF for soils containing more than 3.0 percent organic matter.

** If Atrazine 80W is used multiply rate shown by 1.25 to equal pounds of Atrazine 80W. If Atrazine 90DF is used, multiply rate shown by 1.1 to equal pounds of Atrazine 90DF.

Do not use this treatment on sands or on Rosebud or Canyon series soils, or on calcareous or caliche subsoil outcroppings because of possible atrazine carryover damage to the succeeding crop.

Do not graze or feed foliage from treated areas to livestock within six months after application.

NOTE: If conditions cause weeds to begin to germinate in the spring or summer following a post-harvest treatment of BLADEX or BLADEX plus atrazine, an application of BLADEX 90DF may be made for additional weed control in the spring or summer prior to planting winter wheat. Apply as directed under "BLADEX 90DF Applied Alone" in this section of the label. Do not make more than two applications of BLADEX or any compound containing cyanazine prior to planting.

BLADEX Combinations with Paraquat

On fallow crop land having an existing or established weed population, paraquat may be tank mixed at 1-2 pts./A. (2 lbs./gal.) or 1.3-2.7 pts./A. (1.5 lbs./gal.) with either BLADEX 90DF or BLADEX 90DF/atrazine tank-mix combination as previously described in this section. Apply the recommended rates in at least 25 gallons of spray mixture per acre by ground rig. Use higher volumes and the high rates of paraquat when weed growth is heavy or when dry weather conditions prevail. Add "Ortho X-77" surfactant at 1 qt./100 gals. of diluted spray or other suitable surfactant at recommended rates. Established weeds over 6 inches tall may not be completely controlled.

BLADEX Combinations with 2,4-D

2,4-D LV Ester may be added to any treatment in this section to help control broadleaf weeds growing at the time of application. Use 1-1/3 to 2 pts./A. of 2,4-D LV 6 lb. Ester (2-3 pts./A. of 2,4-D LV 4 lb. Ester) (or 2,4-D Amine at recommended rates). Use the high rate when weeds are over 4 inches tall or when directed on the 2,4-D label for the control of hard-to-kill weed species, such as perennials. When 2,4-D is used, it should be added to the spray tank last.

Use of Supplemental Tillage

In fields where established weeds are too large to be effectively controlled with paraquat or 2,4-D, sweep tillage should be employed. Till before applying the herbicide treatment. This type of tillage will preserve a maximum amount of existing stubble on the surface for soil protection. Similar tillage may also be used if weeds become established prior to receiving adequate rainfall for activation of the herbicide treatment. At some point prior to seeding wheat, the herbicide will degrade and no longer be effective. Limited tillage should be employed at this time. This tillage should be kept shallow to preserve as much moisture as possible for the crops.

STORAGE AND DISPOSAL

STORAGE: Do not contaminate water, food or feed by storage or disposal. Do not use or store around the home environment. Avoid contact with water. In case of spill or leak, avoid breathing dust or vapors, clean up and dispose of wastes in compliance with local, State and Federal regulations.

PESTICIDE DISPOSAL: Pesticide, spray mixture or rinsate that cannot be used according to label instructions must be disposed of according to applicable Federal, State or local procedures.

CONTAINER DISPOSAL: Completely empty bag into application equipment. Then dispose of empty bag in a sanitary landfill or by incineration, or, if allowed by state and local authorities, by burning. If burned, stay out of smoke.

BLADEX., CONQUEST., EXTRAZINE. -
Trademarks of E.I. DuPont de Nemours & Co. (Inc.)
"Banvel" - Trademark of Sandoz Crop Protection Corp.
"Dual", "Milogard", "Concep" II, - Trademarks of CIBA-Geigy Corp.
"Eradicane", "Sutan+" - Trademarks of ICI Americas, Inc.
"Genate Plus" -Trademark of PPG Industries, Inc.
"Lasso", "Ramrod", "Screen" - Trademarks of Monsanto Co.
"Ortho X-77"- Trademark of Chevron Chemical Co.
"Princep" "Caliber", - Trademarks of CIBA-Geigy Corp.
"Prowl" - Trademark of American Cyanamid.
"Treflan" - Trademark of Elanco Products Company.
"Zorial" Rapid 80 - Trademark of Zoecon Corporation.

AG - 3201 8078 5/27/88

NOTICE OF WARRANTY

Du Pont warrants that this product conforms to the chemical description on the label thereof and is reasonably fit for the purposes stated on such label only when used in accordance with the directions under normal use conditions. It is impossible to eliminate all risks inherently associated with the use of this product. Crop injury, ineffectiveness or other unintended consequences may result because of such factors as weather conditions, presence of other materials, or the manner of use or application, all of which are beyond the control of Du Pont. In no case shall Du Pont be liable for consequential, special or indirect damages resulting from the use or handling of this product. All such risks shall be assumed by the buyer. DU PONT MAKES NO WARRANTIES OF MERCHANTABILITY OR FITNESS FOR A PARTICULAR PURPOSE NOR ANY OTHER EXPRESS OR IMPLIED WARRANTY EXCEPT AS STATED ABOVE.

This is a specimen label. DuPont is not responsible for the accuracy of the information contained herein. As labels are subject to revision, always carefully read and follow the label on the product container.

CANOPY®

HERBICIDE DISPERSIBLE GRANULES

	BY WEIGHT
ACTIVE INGREDIENTS:	75.0%
Metribuzin: [4-Amino-6-1(1,1-dimethylethyl)-3-(methylthio)-1,2,4-triazin-5(4H)-one]	64.3%
Chlorimuron Ethyl: Ethyl 2-[[[[(4-chloro-6-methoxypyrimidin-2-yl) amino]carbonyl]amino]sulfonyl]benzoate	10.7%
INERT INGREDIENTS:	25.0%
TOTAL	100.0%

EPA Reg. No. 352-444

U.S. Pats. 4,346,220, 4,394,506, 4,547,215 & 4,645,530

KEEP OUT OF REACH OF CHILDREN

PRECAUTIONARY STATEMENTS

HAZARDS TO HUMANS AND DOMESTIC ANIMALS

CAUTION! HARMFUL IF SWALLOWED. CAUSES MODERATE EYE INJURY.

Avoid breathing dust or spray mist. Avoid contact with skin, eyes and clothing.

In case of contact with eyes, immediately flush with plenty of water. Get medical attention if irritation persists. Wash thoroughly with soap and water after handling. Remove and wash contaminated clothing before reuse.

STATEMENT OF PRACTICAL TREATMENT

IF IN EYES: Flush eyes with plenty of water. Call a physician if irritation persists.

IF SWALLOWED: Call a physician or Poison Control Center. Drink 1 or 2 glasses of water and induce vomiting by touching back of throat with finger. Do not induce vomiting or give anything by mouth to an unconscious person.

For medical emergencies involving this product, call toll free 1-800-441-3637.

ENVIRONMENTAL HAZARDS

Do not apply directly to water or wetlands. Do not contaminate water by cleaning of equipment or disposal of wastes.

IMPORTANT

Injury to or loss of desirable trees or vegetation may result from failure to observe the following: Do not apply or drain or flush equipment on or near desirable trees or other plants, or on areas where their roots may extend, or in locations where the chemical may be washed or moved into contact with their roots. Do not use on lawns, walks, driveways, tennis courts or similar areas. Prevent drift of spray to desirable plants. Do not contaminate any body of water. Keep from contact with fertilizers, insecticides, fungicides and seeds during storage.

PRIOR TO USING "CANOPY" HERBICIDE, CONSIDERATION SHOULD BE GIVEN TO CROP ROTATION PLANS. Crops other than soybeans may be extremely sensitive to low concentrations of "CANOPY" remaining in the soil the next planting season. Choice of rotation crop is restricted following application of "CANOPY". (See "ROTATIONAL CROP GUIDELINES".) "CANOPY" should not be applied on soils above pH 7.0 except in those states listed in Group III of the "ROTATIONAL CROP GUIDELINES" where application can be made to soils with a pH as high as 7.5.

Thoroughly clean all traces of "CANOPY" from application equipment immediately after use and prior to spraying crops other than soybeans. Failure to remove even small amounts of "CANOPY" from application equipment may result in injury to subsequently sprayed crops.

Metribuzin is a chemical which can travel (seep or leach) through soil and can contaminate groundwater which may be used as drinking water. Metribuzin has been found in groundwater as a result of agricultural use. Users are advised not to apply metribuzin where the water table (groundwater) is close to the surface and where the soils are very permeable, i.e., well drained soils such as loamy sands. Your local agricultural agencies can provide further information on the type of soil in your area and the location of groundwater.

GENERAL INFORMATION

Du Pont "CANOPY" Herbicide is a dispersible granule formulation to be mixed with water and sprayed for selective preemergence and preplant incorporated weed control in soybeans. When applied according to the instructions on this label, it will control many broadleaf weeds and provide partial control of nutsedge.

Preemergence and preplant incorporated applications of "CANOPY" require rainfall or sprinkler irrigation to activate the herbicide. Degree of control and duration of effect depend on: rate used, weed spectrum, growing conditions at and following time of treatment, soil pH, texture, organic matter, moisture and precipitation.

Because most crops are highly sensitive to "CANOPY", all direct or indirect contact (such as spray drift) to crops or to land scheduled to be planted to crops other than soybeans should be avoided.

BIOLOGICAL ACTIVITY

"CANOPY" rapidly inhibits the growth of susceptible weeds. Following application of preplant incorporation or preemergence treatment, susceptible weeds may germinate and emerge, but growth then ceases and leaves become yellow 3-5 days after emergence. Death of leaf tissue and growing point will follow in some species while others will remain green but stunted and noncompetitive. "CANOPY" provides partial control of some annual grasses when used preplant or preemergence but other products may be needed to ensure adequate grass control.

DIRECTIONS FOR USE

It is a violation of federal law to use this product in a manner inconsistent with its labeling. Do not apply this product through any type of irrigation system.

EQUIPMENT—SPRAY VOLUMES

Apply uniformly by ground equipment with a properly calibrated low pressure (20 to 40 psi) fixed-boom sprayer equipped with fan-type nozzles and screens no finer than 50 mesh. Use 10 to 40 gals of water per acre. Continuous agitation in the spray tank is required to keep the material in suspension. Avoid overlap and shut off spray booms while starting, turning, slowing or stopping, as injury to the crop may result.

EARLY PREPLANT OR PREPLANT INCORPORATED (See Rate Chart)

"CANOPY" can be applied at planting or up to 30 days prior to planting either preemergence or preplant incorporated. If incorporated, uniformly incorporate the treatment no deeper than the top 1-2" of soil prior to planting soybeans. Use equipment suited to proper incorporation into the top 1-2 inches of soil, eg: Do-all, field cultivator, roterra. "CANOPY" can be tank mixed with "Command"[10] and applied preplant incorporated. Refer to the "Command" label for specific use instructions. If tank-mixed with a grass herbicide such as a DNA like "Treflan"[1], "Sonalan[1]", or acetanilide such as "Dual"[2], or "Lasso"[3], follow label instructions for proper incorporation of the grass herbicide into the top 1 to 2" of soil. Improper soil incorporation can result in erratic weed control or crop injury. Rainfall* (~1") is required to activate preplant incorporated herbicides such as "CANOPY" before weed emergence.

"CANOPY" + "Treflan," "Prowl,"[4] "Dual" or "Sonalan" — Apply within 2 weeks of planting.
"CANOPY" + "Treflan" - (Early PPI) — Apply within 30 days of planting.
"CANOPY" + "Lasso" — Apply within 7 days of planting.
"CANOPY" + "Command" — Apply within 30 days of planting.

PREEMERGENCE (See Rate Chart)

Preemergence application: Apply before soybeans emerge. Do not spray overtop of emerged soybeans, as severe injury or death of crop will occur.

Preemergence activity of "CANOPY" requires rainfall* to activate. Best results are obtained if treatment is made to moist soil and followed by rainfall or irrigation (~1") before weeds germinate. On dry soil, more moisture is required for activation (1 to 2") before weed emergence. If moisture is insufficient to activate the herbicide, a rotary hoeing or shallow cultivation should be made after emergence of crop while weeds are small enough to be controlled by mechanical means. "CANOPY" can be applied preemergence in combination with "Lasso," "Dual," "Prowl" "Surflan" or following the use of a preplant grass material such as "Treflan" or "Sonalan". Consult labels for rates and use instructions.

*Rainfall required to activate ppi or pretreatments depends on soil moisture when activating moisture is received. Several small rainfalls of less than 1/4" are not as beneficial as one large rainfall of 1/2 inch or more for activation.

MINIMUM TILLAGE OR NO-TILL (See Rate Chart)

"CANOPY" applied preemergence to soybeans but postemergence to broadleaf weeds and small annual grasses will provide some burndown of existing weeds as well as normal preemergence broadleaf weed control. When used for burndown, add 0.25% v/v (1 qt/100 gal) surfactant such as Surfactant WK or crop oil concentrate at 0.25% v/v (1 qt./100 gal.). For added postemergence control, especially on grasses, "CANOPY" can be applied with "Roundup"[3], paraquat, or "Bronco"[3]. For added preemergence grass control, "CANOPY" can be mixed with "Dual", "Surflan" or "Lasso". For added postemergence broadleaf weed control, "CANOPY" may be tank mixed with 2,4-D low volatile ester. Do not apply 2,4-D low volatile ester within 30 days of planting. Consult labels for rates and use instructions. Maintain constant agitation of spray mixture.

RATE CHART

Apply as follows:

Soil Texture		Broadcast Rate (Ounces Per Acre) — Organic Matter: Percent in Soil 1/2-3%	Percent in Soil 3-5%
Coarse:	Loamy sand Sandy loam	6	8
Medium:[a]	Loam Silt loam* Silt Sandy clay loam	8	10
Fine:	Silty clay loam Clay loam Clay	10	12

[a] Medium soils in southeast KS, southwest MO and northeast OK with 1/2-3% O.M., use 5-6 oz/A.

* Silt loam soils in Tennessee and Kentucky use 6-8 oz/A.

Do not apply to soil with less than 1/2% O.M.

On soils with higher organic matter or heavy pressure from large, deep-germinating weed seeds, an additional 15% more material should be added to the recommended rate for each soil type.

For band applications, use proportionately less than those listed above.

Plant seed 1 1/2" to 2" deep on a flat or raised seedbed only or crop injury may occur.

Soybean stunting may occur if excessive rainfall occurs after application but before soybeans germinate. Injury is more prevalent under poor drainage or compacted conditions or when soil is saturated for long periods of time. Soybeans rapidly outgrow stunting once favorable growing conditions return.

Do not apply "CANOPY" to soils with pH greater than 7.0 except in those states as directed in the "ROTATIONAL CROP GUIDELINES" section. Soybean varieties such as Altona, Coker 102 and 156, Govan, Semmes, Tracy, Vansoy, Terra Vig 505 and 606, AP 71, NKS 1884, Agropro 55, Asgrow 6520, Maple Amber, Portage and Vinton 81 are sensitive to "CANOPY." Injury may occur if "CANOPY" is used on these varieties. Before use on any soybean variety, tolerance to "CANOPY" must first be determined. Injury to soybeans may occur if "CANOPY" is used on soils having a calcareous surface layer or pH greater than 7.5, or if used in conjunction with soil-applied organic phosphate pesticides such as "Dasanit"[5], "Disyston"[5], "Mocap"[6], "Nemacur"[5], "Thimet"[4], parathion, or "Lorsban"[7].

NOTE: Injury may occur if atrazine was applied on the soil the year before use of "CANOPY". Seedling disease, cold weather, deep planting (more than 2"), excessive moisture, high salt concentration, or drought may weaken soybean seedlings and increase possibility of crop injury.

Do not apply to Black Belt Soils of Alabama and Mississippi with a soil pH greater than 7.0 or history of nutrient deficiency such as iron chlorosis, as injury may occur.

WEEDS CONTROLLED

The following weeds will be controlled at the above rates both preplant incorporated and preemergence:

*Cocklebur	Pigweed
Florida beggarweed	Palmer
Hemp sesbania	Redroot
Hophornbeam copperleaf	Smooth
Jimsonweed	Spiny amaranth
Lambsquarters	Tall waterhemp
Mustard, wild	Poinsettia (wild)
*Morningglory, annual	Prickly sida (teaweed)
Ivyleaf	Purslane, common
Entireleaf	Ragweed, common
Pitted	*Ragweed, giant
Smallflower	*Sicklepod
Tall	Smartweed (annual)
	Spotted spurge
	Sunflower
	Velvetleaf

PARTIAL CONTROL:

Barnyardgrass	Johnsongrass, seedling
Yellow nutsedge	Mexicanweed
Purple nutsedge	Burcucumber
Crabgrass	Panicum, Texas and fall
Foxtail, species	Signalgrass, broadleaf

* Large-seeded weeds, germinating deep in the soil such as morningglory, sicklepod, cocklebur and giant ragweed or other weeds which may emerge at various times during the growing season may require a cultivation or an application of Du Pont "Classic" Herbicide for season-long control.

NOTE: Consult "Lasso", "Dual", "Prowl", "Surflan", "Treflan", "Sonalan", or "Command" labels for additional weeds controlled when "CANOPY" is used in conjunction with these herbicides.

SPRAY TANK PREPARATION

Fill tank 1/4 full with water. Start agitation system, add "CANOPY" and continue adding water. Add separately each additional component of any tank mix while adding water. Continue agitation throughout. If poor mixing should occur with any component, premix the component with two parts water before adding to the spray tank.

A fertilizer solution may be used in the spray mixture. Small quantities should be tested for compatibility by the following procedures before full-scale mixing.

1. Put 1 pint of fertilizer solution in a quart jar.
2. Mix 2 teaspoons "CANOPY" with 2 tablespoons of water; mix thoroughly and add to fertilizer solution.
3. Close jar and shake well.
4. If other herbicides are to be used in the mixture, premix 2 teaspoons of wettable powder or 1 teaspoon of liquid with 2 tablespoons of water; add to "CANOPY"/fertilizer solution mixture.
5. Close jar and shake well.
6. Watch mixture for several seconds; check again in 30 minutes.
7. If mixture does not separate, foam, gel, or become lumpy, it may be used.
8. Mixing ability may be improved by adding compatibility agents such as Kalo Laboratories' "Compex"[8] or Witco Chemicals' "Sponto"[9] 168D. Follow directions on container.

Provided the above procedure shows the mixture to be compatible, prepare the tank mixture as follows: Add the fertilizer solution to the spray tank first, with the agitator running, add the required amount of "CANOPY" and thoroughly mix. For tank mixtures with other herbicides, follow directions above. For tank mixtures with other herbicides, all applicable directions, restrictions and precautions for the additional herbicides are also to be followed.

Use "CANOPY" spray preparations the same day as mixed or product degradation may occur. Thoroughly reagitate and remix before using, if allowed to settle. When tank mixing with other herbicides, all applicable directions, restrictions and precautions for the additional herbicides are also to be followed.

ROTATIONAL CROP GUIDELINES

The minimum length of time in months before "CANOPY" treated soil can be planted with major crops is given below. Choice of rotational crop is determined by geographical location (state), rate, soil pH and time interval elapsed since last application (either band or broadcast). Guidelines are divided into three groups based on these criteria.

Group I

- States of AL, AR, FL, GA, LA, MS or TX - Soil pH 7.0 or less
- In southeast KS, southwest MO and northeast OK - Soil pH 7.0 or less and use rates of 5.0-6.0 ounces per acre

Group II

- States of DE, KY, MD, MO Bootheel, NJ, NC, PA, SC, TN VA or WV - Soil pH 6.8 or less

AND

—Use rate 8-10 oz./A

Group III

- All other states not listed in Groups I or II with soil pH 7.0 or less OR
- States of AL, AR, FL, GA, LA, MS or TX with soil pH 7.1-7.5 OR
- States of DE, KY, MD, MO Bootheel, NJ, NC, PA, SC, TN, VA or WV

—Soil pH 6.9-7.5 OR
—Fields not meeting rate requirements in Group II.

Minimum length of time in months before planting recrop:

Crops	Group I	Group II	Group III
Soybeans	Anytime	Anytime	Anytime
Barley	4	4	4
Ryegrass	4	4	4
Wheat	4	4	4
Alfalfa	10	10	18
Clover	12	12	18
Field Corn	9	10	18
Cotton	10	10	18
Peanuts	12	12	18
Rice	10	10	18
Sorghum	10	10	18
Tobacco (Transplant)	10	10	18
Tomato (Transplant)	10	10	18

All other crops require a minimum 18 month rotational interval and a successful field bioassay.**

The suggested interval before initiating a field bioassay is 10 months for Group I and 15 months for Groups II and III.

**The season before planting any crop not listed above, a successful field bioassay must be completed. The field bioassay will detect small quantities of "CANOPY" which can remain in the soil and injure rotational crops. A successful field bioassay means growing to maturity a test strip of the crop(s) intended for production the following year. The test strip should cross the whole field including knolls and low areas.

Combination Treatment: When recommended tank mix combinations are used, consult the product labels for replanting and rotational crop instructions and observe the directions that are the most restrictive.

SPRAYER CLEANUP

To avoid subsequent injury to crops other than soybeans, immediately after spraying and prior to spraying other crops, thoroughly remove all traces of "CANOPY" from mixing and spray equipment. For more information, consult supplemental bulletin on sprayer cleanout following use of Du Pont "Classic", "CANOPY" or "Gemini" Herbicides.

PRECAUTIONS

Do not apply "CANOPY" Herbicide to soils with a pH greater than 7.0 except in those states in Group III of the "ROTATIONAL CROP GUIDELINES" where applications can be made to soils with a pH as high as 7.5.

CAUTION: Soil pH varies greatly even within the same field; pH variations as much as 2 pH units are common. Determine soil pH by laboratory analysis using a 1:1 soil:water suspension.

Do not graze treated fields or harvest for forage or hay.

Use of "CANOPY" may affect rotational crops. Follow "ROTATIONAL CROP GUIDELINES".

Do not use on sand, muck, or peat soils, nor on soils with less than 1/2% organic matter, as crop injury may occur.

Do not apply "CANOPY" after the soybean crop has emerged.

Do not apply to Black Belt Soils of Alabama and Mississippi with soil pH greater than 7.0 or on soils with a history of nutrient deficiency such as iron chlorosis or injury may occur.

If "Scepter"[4], "Pursuit"[3] or any product containing the active ingredients, imazaquin or imazethapyr (such as "Squadron"[4], "Partner"[3]) is applied the same year as "CANOPY", do not plant crops other than soybeans for at least 15 months from the last application. Crops listed in the "'CANOPY' Rotational Crop Guidelines" may be planted after 15 months; all other crops require a successful field bioassay.

Do not apply in land that has been or will be treated with "Ally", "Glean" or "Finesse" herbicide in the states of Nebraska, Kansas, North Dakota and South Dakota.

[1] Trademark of Elanco Products Co., Indianapolis, Indiana
[2] Trademark of Ciba-Geigy Corp., Greensboro, North Carolina
[3] Trademark of Monsanto Agricultural Products Co., St. Louis, Missouri
[4] Trademark of American Cyanamid Corp., Princeton, New Jersey
[5] Trademark of Mobay Chemical Corp., Kansas City, Missouri
[6] Trademark of Rhone-Poulenc, Inc., Monmouth Junction, New Jersey
[7] Trademark of Dow Chemical Co., Midland, Michigan
[8] Trademark of Kalo Agricultural Chemicals, Inc., Overland Park, Kansas
[9] Trademark of Witco Chemical Corp., Houston, Texas
[10] Trademark of FMC Corp., Philadelphia, Pennsylvania

NOTICE TO BUYER: Purchase of this material does not confer any rights under patents of countries outside of the United States.

STORAGE AND DISPOSAL

Storage: Store product in original container only, away from other pesticides, fertilizer, food or feed.

Pesticide Disposal: Do not contaminate water, food or feed by storage or disposal. Wastes resulting from the use of this product may be disposed of on site or at an approved waste disposal facility.

Container Disposal: Triple rinse (or equivalent), then offer for recycling or reconditioning, or puncture and dispose of in a sanitary landfill, or by incineration, or, if allowed by state and local authorities, by burning. If burned, stay out of smoke.

NOTICE OF WARRANTY

Du Pont warrants that this product conforms to the chemical description on the label thereof and is reasonably fit for the purpose stated on such label only when used in accordance with the directions under normal use conditions. It is impossible to eliminate all risks inherently associated with the use of this product. Crop injury, ineffectiveness, or other unintended consequences may result because of such factors as weather conditions, presence of other materials, or the manner of use or application, all of which are beyond the control of Du Pont. In no case shall Du Pont be liable for consequential, special or indirect damages resulting from the use or handling of this product. All such risks are assumed by the buyer. DU PONT MAKES NO WARRANTIES OF MERCHANTABILITY OR FITNESS FOR A PARTICULAR PURPOSE NOR ANY OTHER EXPRESS OR IMPLIED WARRANTY EXCEPT AS STATED ABOVE.

AG-3225 8098 4/11/88

This is a specimen label. DuPont is not responsible for the accuracy of the information contained herein. As labels are subject to revision, always carefully read and follow the label on the product container.

CLASSIC®

HERBICIDE
DISPERSIBLE GRANULES

ACTIVE INGREDIENT	BY WEIGHT
Chlorimuron Ethyl Ethyl 2-[[[[(4-chloro-6-methoxypyrimidin-2-yl)amino]-carbonyl]amino]sulfonyl] benzoate	25%
INERT INGREDIENTS:	75%
TOTAL	100%

EPA Reg. No. 352-436

U.S. Pat. No. 4,394,506 & 4,547,215

KEEP OUT OF REACH OF CHILDREN

PRECAUTIONARY STATEMENTS

HAZARDS TO HUMANS AND DOMESTIC ANIMALS

CAUTION!

MAY IRRITATE EYES, NOSE, THROAT AND SKIN.

May be harmful if absorbed through skin. Avoid breathing dust or spray mist. Avoid contact with skin, eyes and clothing. Get medical attention if irritation persists. Wash thoroughly with soap and water after handling. Remove and wash contaminated clothing before reuse.

STATEMENT OF PRACTICAL TREATMENT

IF ON SKIN: Wash with plenty of soap and water. Get medical attention if irritation persists.

IF IN EYES: Flush eyes with plenty of water. Call a physician if irritation persists.

For medical emergencies involving this product, call toll free 1-800-441-3637.

ENVIRONMENTAL HAZARDS

Do not apply directly to water or wetlands. Do not contaminate water by cleaning of equipment or disposal of wastes.

IMPORTANT

Injury to or loss of desirable trees or vegetation may result from failure to observe the following: Do not apply or drain or flush equipment on or near desirable trees or other plants, or on areas where their roots may extend, or in locations where the chemical may be washed or moved into contact with their roots. Do not use on lawns, walks, driveways, tennis courts or similar areas. Prevent drift of spray to desirable plants. Do not contaminate any body of water. Keep from contact with fertilizers, insecticides, fungicides and seeds during storage.

Thoroughly clean all traces of "CLASSIC" Herbicide from application equipment immediately after use and prior to spraying crops other than soybeans.

GENERAL INFORMATION

Du Pont "CLASSIC" Herbicide is a dispersible granule formulation to be mixed with water and sprayed for selective postemergence weed control in soybeans. When applied according to the instructions on this label, it will control many broadleaf weeds and yellow nutsedge.

"CLASSIC" will provide best postemergence results when applied to young actively growing weeds. Degree of control, and duration of effect depend on: rate used; weed spectrum; weed size; growing conditions at and following time of treatment; soil moisture; precipitation; and spray adjuvants. Some weeds, such as pigweed, are more affected by stress than others. Delay application until stress passes and weeds start to grow again. If weeds get large, use higher rate and spray volume. Treating weeds under stress or large weeds may result in only partial control.

Because most crops, other than soybeans, are highly sensitive to "CLASSIC", all direct or indirect contact (such as spray drift) to crops or land scheduled to be planted during the current growing season to crops other than soybeans should be avoided.

IMPORTANT: Prior to using "CLASSIC", consideration should be given to crop rotation plans. Crops other than soybeans can be extremely sensitive to low concentrations of "CLASSIC" remaining in the soil the next planting season. Choice of rotation crop is restricted following application of "CLASSIC". Crop rotation may be further restricted if "CLASSIC" is applied over Du Pont "Canopy®", "Gemini®", "Preview®" or "Lorox Plus™" Herbicides (See "'CLASSIC' ROTATIONAL CROP GUIDELINES"). Do not apply to soil with a pH greater than 7.0 except in those states as directed in "'CLASSIC' ROTATIONAL CROP GUIDELINES", as extended soil residual activity could adversely effect crop rotation options beyond indicated intervals.

BIOLOGICAL ACTIVITY

"CLASSIC" rapidly inhibits the growth of susceptible weeds. Following postemergence application, leaves of susceptible plants appear yellow in 3-5 days followed by death of the growing point. Susceptible plants are controlled in 7-21 days. Death of leaf tissue and growing point will follow in some species while others will remain green but stunted and noncompetitive.

DIRECTIONS FOR USE

It is a violation of federal law to use this product in a manner inconsistent with its labeling.

SPRAY TANK PREPARATION

"CLASSIC" should be thoroughly mixed with water in the spray tank before adding any other material (e.g., surfactant). Agitation is required for uniform mixing and application. Apply "CLASSIC" spray preparation within 24 hours of product mixing or product degradation may occur. Thoroughly reagitate before using if allowed to settle.

APPLICATION RECOMMENDATIONS

Applications should be made by ground or air to young actively growing weeds. Applications of "CLASSIC" must include a nonionic surfactant of 0.25% v/v (1 qt/100 gal) or a crop oil concentrate at 1% v/v (1 gal/100 gal). Use a good quality surfactant with at least 80% active ingredient to improve wetting and/or contact activity of "CLASSIC".

Under hot, dry conditions, weed control may be improved by replacing surfactant with crop oil concentrate at 1% v/v (1 gal/100 gal). Use a petroleum based crop oil concentrate of good quality with at least 15% emulsifiers/surfactant. Crop oil concentrate is especially helpful in control of giant ragweed and pigweed. Crop oil concentrate may increase phytotoxicity to soybeans.

AERIAL APPLICATION: —minimum 3 gallons water per acre
—use appropriate spray nozzle types,

cores, and orifice disks in proper arrangement to deliver optimum spray distribution and maximum coverage
—under heavy weed pressure or dense crop foliage, increase minimum spray volume to 5 gallons per acre
—do not apply by air during air inversion conditions, gusty winds, or any condition which favors drift or poor coverage

GROUND APPLICATION:
—minimum 10 gallons water per acre
—use flat fan nozzle at 25-40 psi or hollow cone nozzle at 40-60 psi
—use proper boom height for good weed coverage
—under heavy weed pressure or dense crop foliage, increase minimum spray volume to 15 gallons per acre
—do not use low pressure spray nozzle tips (flood type)

Do not apply "CLASSIC" if rain is expected within 1 hour, or weed control may be decreased.

Do not apply this product through any type of irrigation system.

For band application, use proportionately less. Carefully calibrate band applicator to not exceed labeled rates or injury may occur.

A second application of "CLASSIC" may be made 2-3 weeks after the initial application if conditions warrant. Do not exceed 1 1/2 oz total "CLASSIC" per season.

POSTEMERGENCE APPLICATION GUIDE

"CLASSIC" can be applied after first trifoliate until 60 days before soybean harvest.

APPLICATION RATE

	½ oz/acre		⅔ oz/acre		¾ oz./acre	
Weeds Controlled	**Leaf Stage†**	**Maximum Height (Inches)**	**Leaf Stage†**	**Maximum Height (Inches)**	**Leaf Stage†**	**Maximum Height (Inches)**
Beggarticks (Bidens sp.)	4	4	6	6	8	8
Bristly Starbur	2	2	3	3	4	4
Cocklebur	6	6	8	8	10	12
Florida Beggarweed	3	4	4	5	6	6
Hemp Sesbania	3	4	4	5	5	6
Jerusalem Artichoke (above ground portion)	—	—	—	—	8	8
Jimsonweed	4	4	5	5	6	6
*Morningglory (annual)						
Entireleaf	2	2	3	3	4	4
Ivyleaf	2	2	3	3	3	4
Pitted	2	2	3	3	4	4
Smallflower	2	2	3	3	3	4
Tall	2	2	3	3	3	4
Mustard	4	4**	5	5**	6	6**
Pigweed						
Redroot	4	2	5	3	6	4
Ragweed						
Common	—	—	5	3	6	4
*Giant	—	—	—	—	6	6
*Sicklepod	1	2	2	3	3	4
Smartweed						
Ladysthumb	4	2	5	3	6	4
Pennsylvania	4	2	5	3	6	4
Sunflower	6	5	7	6	8	8
Wild Poinsettia	—	—	2	2	4	4
Yellow Nutsedge	4	3	5	3	6	4
*Velvetleaf††	—	—	6	4	8	6
Weeds Suppressed						
*Burcucumber	—	—	3	3	4	6
*Cowpea	—	—	3	5	4	6
Marestail	4	3	5	5	6	6
Purple Nutsedge	3	3	4	4	6	5
Smooth Pigweed	3	2	4	3	6	4
Tall Waterhemp	—	—	—	—	4	3

* See " 'CLASSIC' SPLIT APPLICATION" section.
** Diameter
† When counting leaves for correct application time, do not count the cotyledons. Count the first true leaves to determine correct stage to spray.
†† The addition of ammonium nitrogen fertilizers such as 1 gallon 28% nitrogen per acre or 1-2 qts. 10-34-0 per acre to the spray solution is required for control of velvetleaf. A surfactant or crop oil concentrate must also be used with the fertilizer. A cultivation at approximately 14 days after application will improve velvetleaf control.

"CLASSIC" sprayed at stage indicated will either control or suppress (stunt) the weeds, delaying growth for 2-3 weeks. Weeds larger than on the label will be suppressed but may recover after 2-3 weeks. Spraying weeds in cotyledon stage is not recommended. A cultivation approximately 14 days after application but before regrowth will greatly aid control.

Weed suppression is a visual reduction in weed competition (reduced population, size, and/or vigor) as compared to an untreated area. Degree of control can be increased by treating weeds when actively growing (not under stress) and by using maximum use rate.

Temporary leaf yellowing and/or growth retardation of soybeans may occur following application of "CLASSIC". These effects will generally be most evident 5-7 days after application to soybeans under stress. Under favorable soybean growing conditions, the crop will quickly recover.

Cultivation before, during or within 7 days after the application is not recommended. Cultivation may put weeds under stress by pruning roots, thus making control more difficult to obtain. The best time to cultivate is approximately 14 days after application.

Under stress conditions, delay application until more favorable growing conditions prevail and weeds resume growth.

"CLASSIC" SPLIT APPLICATIONS

Some weeds with multiple germination times or suppressed (stunted) weeds, may require two applications of "CLASSIC". Applications should be timed 14-21 days apart or when sufficient regrowth has occurred to require a second treatment. A cultivation approximately 14 days after treatment may replace the need for the second application. Weeds such as burcucumber, cocklebur, cowpea, giant ragweed, morningglory, pigweed, sicklepod and velvetleaf may require a second application.

TANK MIX APPLICATIONS

"CLASSIC" Plus 2,4-DB:

"CLASSIC" may be tank mixed with 1-2 fluid ounces per acre of 2,4-DB for improved control of annual morningglory and other broadleaf weeds. Apply by ground to young actively growing weeds, and before annual morningglory is 4 inches in height.

Temporary yellowing, leaf crinkling, and/or soybean growth retardation may occur following application of "CLASSIC" plus 2,4-DB. These effects will generally be most evident 5-7 days after application to soybeans which are under stress. Under favorable growing conditions, the crop will quickly recover. Consult 2,4-DB label for use precautions.

"CLASSIC" Plus Post Grass Herbicides:
"CLASSIC" may be tank mixed with post grass herbicides such as "Poast"[2]. The activity of "CLASSIC" will not be affected. However, under certain conditions, "CLASSIC" may be antagonistic to the grass herbicide. Refer to the grass herbicide label for specific use information and precautions.

"CLASSIC" ROTATIONAL CROP GUIDELINES

TIME INTERVAL BEFORE REPLANTING IN MONTHS (from last "CLASSIC" application)

SECTION I: "CLASSIC"

Group A

All states with:
■ Soil pH 7.0 or less

—OR—

■ States of AL, AR, FL, GA, LA, NC, MS, SC, TN, TX and MO Bootheel with soil pH 7.1-7.5

Group B

■ States of: DE, MD, NJ, PA, NC, SC, WV, VA, and KY, with soil pH 7.1-7.5

Minimum Time in Months Before Planting Rotational Crop[a]

Crops	Group A	Group B
Soybeans	Anytime	Anytime
Cereal grains	3	3
Ryegrass	3	3
Field Corn	9†	15
Cotton	9	15
Alfalfa	9	15
Clover	9	15
Dry beans	9	15
Sorghum	9	15
Peanut	9	15
Rice	9	15
Tobacco	9	15
Tomato (transplant)	9	15

All other crops require a successful field bioassay[b]. Suggested interval before initiating bioassay is 9 months from Group A and 15 months from Group B.
† In states of AL, AR, FL, GA, LA, MS, and TX with soil pH 7.0 or less, field corn can be recropped after 8 months.

SECTION II: "CLASSIC" FOLLOWING "CANOPY" OR "GEMINI" HERBICIDE

Group C

■ States of: AL, AR, FL, GA, LA, MS, TX, with soil pH 7.0 or less

—OR—

■ States of: DE, MD, NJ, PA, NC, SC, WV, VA, KY, TN and MO Bootheel with soil pH 7.0 or less

AND

—"Gemini": use rate 16-22 oz/A OR
—"Canopy": use rate 7-10 oz/A

AND

—Only one "CLASSIC" application per season or rotational crop injury may occur.

Group D

■ All other states with soil pH 7.0 or less

—OR—

■ States listed in Section II Group C not meeting pH or rate requirement or receiving more than one "CLASSIC" application.

Minimum Time in Months Before Planting Rotational Crop[a]

Crops*	Group C	Group D
Soybeans	Anytime	Anytime
Cereal grains	3	3
Ryegrass	3	3
Field Corn	9	15
Cotton	9	15
Alfalfa	9	15
Clover	9	15
Dry beans	9	15
Sorghum	9	15
Peanut	9	15
Rice	9	15
Tobacco (transplant)	9	15
Tomato (transplant)	9	15

All other crops require a successful field bioassay[b]. Suggested interval before initiating bioassay is 9 months from Group C and 15 months from Group D.
* Consult "Canopy" and "Gemini" labels on rotational crop intervals. Some crop rotations may be longer than indicated here for one or the other product because of metribuzin or linuron restrictions. Use the longest rotational crop interval.

SECTION III: "CLASSIC" FOLLOWING "PREVIEW" OR "LOROX PLUS" HERBICIDE

Group E

■ "Preview" 8 oz/A or less or "Lorox Plus" 16 oz/A or less

AND

■ One "CLASSIC" application per season

Group F

■ "Preview" or "Lorox Plus" rates greater than listed in Group E

—OR—

■ More than one "CLASSIC" application

Minimum Time in Months Before Planting Rotational Crop[a]

Crops*	Group E	Group F
Soybeans	Anytime	Anytime
Cereal grains	3	3
Field Corn	10	15
Alfalfa	11	11
Clover	11	11
Dry beans	11	11
Sorghum	11	15
Tobacco	11	15
Tomato (transplant)	11	15

All other crops require a successful field bioassay[b].
Suggested interval before initiating bioassay is 11 months from Group E and 15 months from Group F.

* Consult "Preview" and "Lorox Plus" labels on rotational crop intervals. Some crop rotations may be longer than indicated here for one or the other product because of metribuzin or linuron restriction. Use the longest rotational crop interval.
[a] If "CLASSIC" is applied after August 1, extend recrop interval on corn, cotton, rice and sorghum 2 months or rotational crop may be injured.
[b] Field bioassay—The season before planting any crop not listed above, a successful field bioassay must be completed. The field bioassay will detect small quantities of "CLASSIC" which can remain in the soil and injure rotational crops. A successful field bioassay means growing to maturity a test strip of crop(s) intended for production the following year. The test strip should cross the whole field including knolls and low areas. Suggested interval before planting bioassay crop is listed on "Recrop Table."

SPRAYER CLEANUP

IMPORTANT—To avoid subsequent injury to crops other than soybeans, immediately after spraying and prior to spraying other crops, thoroughly remove all traces of "CLASSIC" from mixing and spray equipment. For more information, consult supplemental bulletin on sprayer cleanout following use of Du Pont "CLASSIC", "Canopy", "Preview", "Gemini" or "Lorox Plus" Herbicides.

PRECAUTIONS

- Do not apply "CLASSIC" before first trifoliate opens or later than 60 days before soybean maturity.
- Do not apply "CLASSIC" on soil with a pH greater than 7.0 except in those states as directed in the "'CLASSIC' ROTATIONAL CROP GUIDELINES" section.
- CAUTION: soil pH varies greatly even within the same field. pH variations as much as 2 pH units are common. Determine soil pH by laboratory analysis using a 1:1 soil:water suspension.
- Do not graze treated fields or harvest for forage or hay.
- Do not apply "CLASSIC" to plants under stress from abnormal weather (hot or cold) or growing conditions, drought, water saturated soil, disease, insect or prior herbicide injury, as crop injury or poor weed control may result. Severe stress, drought, disease, insect damage or nutrient deficiency following application may also result in crop injury. Delay application until stress passes and weeds and soybeans resume growth.
- Do not apply "CLASSIC" to Black Belt Soils of Alabama and Mississippi with a soil pH greater than 7.0 or history of nutrient deficiency such as iron chlorosis, or injury may occur.
- Do not apply "CLASSIC" if rain is expected within 1 hour of application as control may be unsatisfactory.
- Follow extended crop rotation interval as directed on this label when "CLASSIC" is applied over "Canopy", "Gemini", "Preview" or "Lorox Plus" Herbicides.
- If "Scepter"[1], "Pursuit"[1] or any product containing the active ingredients imazaquin or imazethapyr (such as "Squadron"[1], "Partner"[3]) is applied the same year as "CLASSIC", do not plant crops other than soybeans for at least 15 months from the last application. Crops listed in the " 'CLASSIC' ROTATIONAL CROP GUIDELINES" may be planted after 15 months; all other crops require a successful field bioassay.
- "CLASSIC" may cause antagonism to post grass herbicides such as "Poast"[2] when tank mixed or sprayed within 7 days prior to or 1 day after the grass herbicide. Antagonism is more apt to occur under stress conditions.
- Do not apply to land that has been or will be treated with Du Pont "Glean®", "Ally®" or "Finesse®" Herbicides in the states of Nebraska, Kansas, South Dakota and North Dakota.

[1] Trademark of American Cyanamid Corp., Princeton, NJ
[2] Trademark of BASF Wyandotte Corporation
[3] Trademark of Monsanto Company

NOTICE TO BUYER: Purchase of this material does not confer any rights under patents of countries outside of the United States.

STORAGE AND DISPOSAL

STORAGE: Store product in original container only, away from other pesticides, fertilizer, food or feed.

PESTICIDE DISPOSAL: Do not contaminate water, food or feed by storage or disposal. Wastes resulting from the use of this product may be disposed of on site or at an approved waste disposal facility.

CONTAINER DISPOSAL: Triple rinse (or equivalent). Then offer for recycling or reconditioning, or puncture and dispose of in a sanitary landfill, or incineration, or, if allowed by state and local authorities, by burning. If burned, stay out of smoke.

NOTICE OF WARRANTY

Du Pont warrants that this product conforms to the chemical description on the label thereof and is reasonably fit for the purpose stated on such label only when used in accordance with the directions under normal use conditions. It is impossible to eliminate all risks inherently associated with the use of this product. Crop injury, ineffectiveness, or other unintended consequences may result because of such factors as weather conditions, presence of other materials, or the manner of use or application, all of which are beyond the control of Du Pont. In no case shall Du Pont be liable for consequential, special or indirect damages resulting from the use or handling of this product. All such risks are assumed by the buyer. DU PONT MAKES NO WARRANTIES OF MERCHANTABILITY OR FITNESS FOR A PARTICULAR PURPOSE NOR ANY OTHER EXPRESS OR IMPLIED WARRANTY EXCEPT AS STATED ABOVE.

AG-3226 8098 2/10/88

This is a specimen label. DuPont is not responsible for the accuracy of the information contained herein. As labels are subject to revision, always carefully read and follow the label on the product container.

ESCORT®

HERBICIDE DRY FLOWABLE

ACTIVE INGREDIENT	BY WEIGHT
Metsulfuron Methyl Methyl 2-[[[[(4-methoxy-6-methyl-1,3,5-triazin-2-yl)-amino]carbonyl]amino]sulfonyl]benzoate	60%
INERT INGREDIENTS:	40%
TOTAL	100%

PACKAGE SIZE 8 - 8 OUNCE JARS

EPA Reg. No. 352-439

U.S. Patent 4,383,113

KEEP OUT OF REACH OF CHILDREN

PRECAUTIONARY STATEMENTS

HAZARDS TO HUMANS AND DOMESTIC ANIMALS

CAUTION!

Harmful if absorbed through skin. Causes eye irritation. Avoid contact with skin, eyes or clothing. Avoid breathing dust or spray mist. Wash thoroughly with soap and water after handling. Remove and wash contaminated clothing before reuse.

STATEMENT OF PRACTICAL TREATMENT

In case of contact with eyes, immediately flush with plenty of water. If on skin, wash with plenty of soap and water. Get medical attention if irritation persists.

For medical emergencies involving this product, call toll free 1-800-441-3637.

ENVIRONMENTAL HAZARDS

Keep out of any body of water. Do not contaminate water by cleaning of equipment or disposal of waste.

IMPORTANT

DO NOT USE ON FOOD OR FEED CROPS. Injury to or loss of desirable trees or other plants may result from failure to observe the following: Do not apply "ESCORT" Herbicide (except as recommended), or drain or flush equipment on or near desirable trees or other plants, or on areas where their roots may extend or in locations where the chemical may be washed or moved into contact with their roots.

Do not use on lawns, walks, driveways, tennis courts or similar areas. Keep from contact with fertilizers, insecticides, fungicides and seeds.

Do not apply to any body of water, including streams, irrigation water or wells. Do not apply where runoff water may flow onto agricultural land, as injury to crops may result. Do not apply "ESCORT" during periods of intense rainfall or to water-saturated soils, or to frozen ground as off-target movement may occur.

Do not apply this product through any type of irrigation system.

Do not allow spray to drift onto adjacent crops or other desirable plants or trees as injury may occur.

In areas where sensitive crops are grown, make applications to soils whose surface has been settled by rain. Do not treat powdery dry soil or light sandy soils, when there is little likelihood of rainfall soon after treatment, as wind may cause off-target movement.

Following an "ESCORT" application, do not use sprayer for application to crops. This is extremely important, as low rates of "ESCORT" can kill or severely injure most crops.

GENERAL INFORMATION

Du Pont "ESCORT" Herbicide is a dispersible granule to be mixed in water and applied as a foliar spray for control of unwanted woody plants and annual and perennial broadleaf weeds, basal soil treatment of multiflora rose, for preemergence and postemergence industrial weed control, and for the growth suppression of certain desirable grasses in noncrop areas such as airports, highways, roadsides, utility and pipeline rights-of-ways, petroleum tank farms, storage areas, pumping installations, plant sites, industrial turf, roadside turf, railroad rights-of-way and fencerows. "ESCORT" rapidly stops growth of susceptible plant species; however, typical symptoms (discoloration) may not be noticeable for several weeks after application, depending on growing conditions and susceptibility.

Weeds absorb "ESCORT" through both the foliage and roots resulting in visual effects progressing from growth inhibition to reddish-purplish coloration, chlorosis, necrosis, vein discoloration and death of terminal. Initial effects are usually seen 2 or 3 weeks following application; however, the final effects on weeds are evident at about 4 to 6 weeks after application. On brush, ultimate effects occur during the season after treatment.

Warm, moist conditions following treatment promote the activity of "ESCORT", while cold, dry conditions may reduce or delay activity. Weeds and brush hardened off by cold weather or drought stress may not be controlled. Degree of control and duration of effect are dependent on the rate used, sensitivity and size of target species, as well as soil moisture and soil temperature. "ESCORT" is noncorrosive, nonflammable, nonvolatile and does not freeze.

DIRECTIONS FOR USE

It is a violation of federal law to use this product in a manner inconsistent with its labeling.

"ESCORT" Herbicide should be used only in accordance with recommendations on this label or in separate published Du Pont recommendations available through local dealers.

Du Pont will not be responsible for losses or damages resulting from the use of this product in any manner not specifically recommended by Du Pont. User assumes all risks associated with such nonrecommended use.

Du Pont "ESCORT" Herbicide is recommended as a postemergence application for control of brush and certain perennial broadleaf weeds and preemergence and postemergence for weed control on noncropland. "ESCORT" is also recommended for use on established, unimproved, industrial turf for weed control. On fescue/bluegrass turf, "ESCORT" is recommended for growth suppression and seedhead inhibition.

BRUSH CONTROL

Foliar applications should be made after brush is fully leafed until the beginning of fall leaf coloration. Complete coverage of all foliage and stems is required for brush control. Effectiveness may be reduced if rainfall occurs within 24 hours after application.

High Volume Foliar Ground Applications:

Apply as a full coverage spray to foliage, stems and limbs at a volume of 50 to 400 gallons of total spray per acre depending on plant species, height and density of growth and on the type of spray equipment used. Spray to runoff for thorough coverage. On tall, dense brush, it is often necessary to spray from both sides to obtain adequate coverage. Use the following amounts of "ESCORT" per 100 gallons of water (include surfactant at 1 quart/100 gallons) and spray to runoff for control of:

½ to 1 OUNCE*	½ to 1 OUNCE*	1 to 2 OUNCES*	3 to 4 OUNCES
Ash	Elm	Elder	Kudzu
Aspen	Multiflora rose	Oaks	
Blackberry	Salmonberry	Ocean Spray (Holodiscus)	
Cherry	Willow	Red Maple	
		Cottonwood	

* Where mixed stands occur, use the higher rate for the most difficult-to-control species.

For broader spectrum of control, "ESCORT" may be tank mixed with "Arsenal"[1] or "Garlon"[2] or "Roundup"[3] using labeled use rates for each product.

Low Volume Foliar Ground Applications:

Prepare spray solution of "ESCORT" Herbicide at a rate of ½ to 1 ounce product per 5 to 25 gallons of water using a spray volume of no more than 25 gallons of water per acre for control of ash, cherry and elm; add surfactant at 0.25% by volume.

Apply as a full coverage (runoff) spray to foliage, stems and limbs using equipment which will assure particle breakup and uniform coverage of the low spray volumes being applied. Total spray volume per acre will depend upon plant species, height and density of growth and the type of equipment used.

Spotgun Basal Soil Treatment:

For control of multiflora rose, prepare a spray suspension of "ESCORT" using 1 ounce product per gallon of water. Mix vigorously until "ESCORT" is dispersed and agitate periodically while applying the spray suspension.

Apply the spray preparation with an exact-delivery hand gun applicator. This equipment delivers a thin stream of a predetermined volume when triggered. Apply at the rate of 4 milliliters for each 2 feet of rose canopy diameter. Direct the treatment to the soil within 2 feet of the stem union of the rose to be controlled. When treating large plants and more than one delivery of "ESCORT" dispersion is required, make applications on opposite sides of the plant.

Applications should be made from early spring to summer, when the ground is not frozen.

WEED CONTROL

"ESCORT" may be applied preemergence or postemergence to the weeds, but for best results, apply postemergence to young, actively growing weeds (less than 2" tall or 2" across) at any time of the year, except when ground is frozen.

"ESCORT" should be used during seasons when rainfall normally occurs as moisture is required to move "ESCORT" into the root zone of weeds before they develop an established root system. Preemergence treatments control or suppress weeds through root uptake, while postemergence works through root uptake and foliar uptake. Under limited rainfall conditions, "ESCORT" may not provide satisfactory control of hard-to-kill perennials.

Treated soils should be left undisturbed to reduce the potential for "ESCORT" movement by soil erosion due to wind or water. Injury to crops may occur when treated soil is blown or moved onto land used to produce crops.

Rate selection is based on weed species, weed size and soil texture. Use the higher rates on established plants and on fine textured soils, and lower rates on smaller weeds and coarse textured soils.

Since the degree of control and duration of effect will vary with the amount of chemical applied, soil texture, soil pH, soil organic matter, weed size, rainfall and other conditions, it is suggested that users limit their first use to small areas.

NONSELECTIVE WEED CONTROL

Apply "ESCORT" in a minimum of 10 gallons of water per acre. Spray volumes of 20 to 40 gallons per acre and pressures of 25 to 35 psi are preferred. Use the following rates of "ESCORT" (including surfactant at 0.25% to 0.5% by volume) to control weeds listed:

⅓ to 1 ounce "ESCORT"/acre*

Annual sowthistle	Common Groundsel	Dogfennel	Multiflora rose	Smooth pigweed
Aster	Common purslane	False chamomile	Pennsylvania smartweed	Sweet clover
Bittercress	Common sunflower	Fiddleneck tarweed	Prickly lettuce	Tansymustard
Blue Mustard	Common yarrow	Field pennycress	Prostrate knotweed	Treacle mustard
Bur buttercup	Conical catchfly	Flixweed	Redroot pigweed	Tumble Mustard
Chicory	Corn Cockle	Henbit	Redstem filaree	Wild carrot
Clover (white)	Cow Cockle	Lambquarters	Shepherdspurse	Wild lettuce
Goldenrod	Crown vetch	Miners lettuce	Smallseed falseflax	Wild mustard
Common chickweed				

* Where mixed stands occur, use the higher recommended rate for the most difficult-to-control species.

1 to 2 ounces "ESCORT"/acre*

Bull thistle	Common mullein	Hoary cress (whitetop)	Russian thistle
Buckhorn plantain	Curly dock	Kochia	Salsify
Canada thistle**	Gumweed	Musk thistle	Scouringrush

* Where mixed stands occur, use the higher recommended rate for the most difficult-to-control species.
** Suppression

SELECTIVE WEED CONTROL

"ESCORT" is recommended for use on unimproved industrial turf where certain grasses are well-established and desired as ground cover. An application of "ESCORT" may cause temporary discoloration of the grasses.

Applications can be made anytime, except when ground is frozen. Apply uniformly with properly calibrated ground equipment, using 10 to 100 gallons of spray solution per acre. Sprayer pressures of 25 to 35 psi are preferred.

NOTE: Do not apply "ESCORT" to turf that is under stress from drought, insects, disease, cold temperatures or poor fertility, as injury may result.

Do not apply to turf less than 1 year old.

Do not use on bahiagrass, as severe injury will result.

Fescue/Bluegrass: Apply 1/4 to 1/2 ounce "ESCORT" per acre, but do not exceed 1/2 ounce per acre within a nine-month period.

Crested Wheatgrass and Smooth Brome: Apply 1/4 to 1 ounce "ESCORT" per acre for weed control in these grasses. Some chlorosis or stunting may occur following the application.

Bermudagrass: Apply 1/4 to 2 ounces "ESCORT" per acre for weed control.

Some chlorosis or stunting may occur following the application.

These treatments will control the following weeds:

1/4 to 1/3 ounce "ESCORT"/acre	1/3 to 1/2 ounce "ESCORT"/acre	1/3 to 1/2 ounce "ESCORT"/acre	1/2 to 1 ounce "ESCORT"/acre	1 to 2 ounces "ESCORT"/acre
Bittercress	Annual sowthistle	Henbit	Buckhorn plantain	Bull thistle
Bur buttercup	Aster	Lambquarters	Canada Thistle**	Russian thistle
Blue mustard	Bahiagrass	Miners lettuce	Curly dock	Scouringrush
Chickweed	Clover	Plantain	Kochia	Gumweed
Chicory	Common mullein	Prickly lettuce	Common sunflower	Salsify
Field pennycress	Common yarrow	Redroot pigweed	Prostrate knotweed	Musk thistle
Redstem filaree	Conical catchfly	Shepherdspurse	Hoary cress (whitetop)	
Wild carrot	Cow cockle	Smooth Pigweed		
	Crown vetch	Smallseed falseflax		
	Dandelion	Sweet clover		
	False chamomile	Tansymustard		
	Fiddleneck tarweed	Treacle mustard		
	Flixweed	Tumble mustard		
	Goldenrod	Wild lettuce		
		Wood sorrell		

** Weeds suppressed involving a visual reduction in competition compared to an untreated area.

GROWTH SUPPRESSION AND SEEDHEAD INHIBITION

Fescue/Bluegrass: "ESCORT" is recommended for use on unimproved turf in noncropland areas for growth suppression and seedhead inhibition of well-established turf. Apply 1/4 to 1/2 ounce "ESCORT" anytime from after green-up until seedhead appearance. For improved performance, apply 1/4 to 1/3 ounce "ESCORT" plus 1/8 to 1/4 pint "Embark"[4] per acre.

Apply uniformly with properly calibrated ground equipment using at least 10 gallons of spray per acre. Addition of a nonionic surfactant at 0.25% by volume (1 qt per 100 gals) provides maximum performance, but may temporarily increase chlorosis of turf.

NOTE: Use the lower rates for minimum chlorosis of turf (particularly fescue).

Do not apply "ESCORT" to turf that is under stress from drought, insects, disease, cold temperatures or poor fertility, as injury may result.

Do not apply to turf less than 1 year old.

Do not use on bahiagrass, as severe injury will result.

Do not exceed 1/2 ounce "ESCORT" per acre within nine months.

Weed Control Application Technique: Before spraying, calibrate the equipment to determine the quantity of water necessary to uniformly cover the vegetation and soil in a measured area to be treated. Use a fixed boom, off center nozzle or boomless straight stream nozzles properly calibrated to a constant speed of travel and rate of delivery. Avoid overlapping and shut off spray booms while starting, turning, slowing or stopping when spraying industrial turf as injury to the desired species may result. Extreme care must be taken to prevent drift.

SPRAY PREPARATION

Fill the spray tank half full of water and, with agitator running, add the proper amount of "ESCORT". Finish adding the required amount of water. Continuous agitation is required to keep the product in suspension. For postemergence applications, a nonionic surfactant of at least 80% active ingredient may be added to improve wetting and/or contact activity. To minimize drift, a drift control agent may be added at the recommended label rate. "ESCORT" spray preparations may degrade in acid solutions if not used in 24 hours; they are stable in alkaline solutions. Thoroughly reagitate before using. Tank mixes with other registered herbicides should be tested for compatibility before full scale mixing.

Sprayer Agitation: Use mechanical or bypass agitation to thoroughly mix the spray solution. After initial mixing, do not use excessive agitation when using a surfactant as foaming problems may result. To avoid foaming problems, use antifoam agents.

Sprayer Cleanup: Immediately after spraying thoroughly remove all traces of "ESCORT" from mixing and spray equipment as follows:

1. Drain tank, then flush tank, boom and hoses with clean water for a minimum of 10 minutes.
2. Fill the tank with clean water, then add 1/2 gallon chlorine bleach (containing 5 1/4% sodium hypochlorite) per 100 gallons of water. Turn on sprayer long enough to flush through boom and hoses. Stop spraying, but keep agitator working in the tank for 15 minutes, then drain.
3. Repeat Step 2.
4. Nozzles and screens should be removed and cleaned separately. To remove traces of chlorine bleach, rinse the tank thoroughly with clean water and flush through hoses and boom.

NOTE: This sprayer cleanup procedure is only effective for "ESCORT" and for general uses specified under "Directions for Use"; sprayer should not be used on plants such as fine turf, ornamentals or crops.

[1] Registered trademark of American Cyanamid Company.
[2] Registered trademark of Dow Chemical Company.
[3] Registered trademark of Monsanto Company.
[4] Registered trademark of 3M Agricultural Products.

NOTICE TO BUYER: Purchase of this material does not confer any rights under patents of countries outside of the United States.

STORAGE AND DISPOSAL

Storage: Store product in original container, away from other pesticides, fertilizer, food or feed.

Pesticide Disposal: Do not contaminate water, food or feed by storage, disposal or cleaning of equipment. Wastes resulting from the use of this product may be disposed of on site or at an approved waste disposal facility.

Container disposal: Triple rinse (or equivalent) the container and then offer for recycling or reconditioning, or puncture and dispose of in a sanitary landfill, or by incineration, or, if allowed by state and local authorities, by burning. If burned, stay out of smoke.

NOTICE OF WARRANTY

Du Pont warrants that this product conforms to the chemical description on the label thereof and is reasonably fit for purposes stated on such label only when used in accordance with directions under normal use conditions. It is impossible to eliminate all risks inherently associated with the use of this product. Injury to adjacent crops, ineffectiveness or other unintended consequences may result because of such factors as weather conditions, presence of other material, or the manner of use or application, all of which are beyond the control of Du Pont. In no case shall Du Pont be liable for consequential, special or indirect damages resulting from the use or handling of this product. All such risks shall be assumed by the buyer. DU PONT MAKES NO WARRANTIES OF MERCHANTABILITY OR FITNESS FOR A PARTICULAR PURPOSE NOR ANY OTHER EXPRESS OR IMPLIED WARRANTY EXCEPT AS STATED ABOVE.

AG-2978 8048 4/11/88

This is a specimen label. DuPont is not responsible for the accuracy of the information contained herein. As labels are subject to revision, always carefully read and follow the label on the product container.

EXPRESS®

HERBICIDE DISPERSIBLE GRANULES

ACTIVE INGREDIENT:	BY WEIGHT
Methyl 2-[[[[N-(4-methoxy-6-methyl-1, 3,5-triazin-2-yl)-methylamino] carbonyl]amino]sulfonyl]benzoate	75%
INERT INGREDIENTS	25%
TOTAL	100%

EPA Reg. No. 352-509 U.S. Patent 4,383,113

KEEP OUT OF REACH OF CHILDREN

CAUTION

PRECAUTIONARY STATEMENTS

HAZARDS TO HUMANS AND DOMESTIC ANIMALS

CAUTION!

Avoid contact with skin, eyes and clothing. In case of contact with eyes, immediately flush with plenty of water. Get medical attention if irritation persists. Wash thoroughly after handling. Remove and wash contaminated clothing before reuse.

For medical emergencies involving this product, call toll free 1-800-441-3637.

ENVIRONMENTAL HAZARDS

Do not apply directly to water or wetlands (swamps, bogs, marshes, potholes). Do not contaminate water by disposal of equipment washwaters.

DIRECTIONS FOR USE

It is a violation of federal law to use this product in a manner inconsistent with its labeling.

Do not apply this product through any type of irrigation system.

"EXPRESS" Herbicide should be used only in accordance with recommendations on this label or in separately published Du Pont recommendations.

Du Pont will not be responsible for losses or damages resulting from the use of this product in any manner not specifically recommended by Du Pont.

IMPORTANT INFORMATION

Injury to or loss of desirable trees or vegetation may result from failure to observe the following: Do not apply or drain or flush equipment on or near desirable trees or other plants, or on areas where their roots may extend, or in locations where the chemical may be washed or moved into contact with their roots. Do not use on lawns, walks, driveways, tennis courts or similar areas. Prevent drift of spray to desirable plants. Do not contaminate any body of water. Thoroughly clean all traces of Du Pont "EXPRESS" Herbicide from application equipment immediately after use. (Refer to "SPRAYER CLEANUP" section of this label.) Failure to follow these procedures may result in injury to subsequently sprayed crops.

GENERAL INFORMATION

Du Pont "EXPRESS" Herbicide is recommended for selective postemergence control of certain broadleaf weeds in wheat (including durum) and barley. Do not use "EXPRESS" on any other crop. "EXPRESS" is a water-dispersible granule to be mixed in water or other recommended carrier and applied as a uniform broadcast spray. It is noncorrosive, nonflammable, nonvolatile and does not freeze.

Best results are obtained when "EXPRESS" is applied to young, actively-growing weeds. The use rate will depend on weed spectrum and size of weed at time of application. The degree of control and duration of effect are dependent on rate used, sensitivity and size of target weed and environmental conditions at the time of and following application.

"EXPRESS" stops growth of susceptible weeds rapidly. However, typical symptoms of dying weeds (discoloration) may not be noticeable for 1-3 weeks after application (2-5 weeks for wild garlic) depending on the environmental conditions and weed susceptibility. Warm, moist conditions following treatment promote the activity of "EXPRESS", while cold, dry conditions delay the activity. Weeds hardened off by cold weather or drought stress will be less susceptible.

A vigorous growing crop will aid weed control by shading and providing competition for weeds. However, a dense crop canopy at time of application can intercept spray and result in reduced weed control. Weeds may not be adequately controlled in areas of thin crop stand or seeding skips.

Naturally occurring weed biotypes* that are resistant to "Glean", "Finesse" or "Ally" will also be resistent to "EXPRESS". In areas where these weed biotypes are known to exist, only spray "EXPRESS" in tank mixtures with other broadleaf herbicides having a different mode of action** such as: 2,4-D, "Banvel"[1]/"Banvel" II***, "Buctril"[2], "Bronate"[2], "Curtail"[3], MCPA, "Karmex" or "Lexone". Refer to TANK MIXTURES.

* Biotypes are naturally-occurring individuals of the species which have a slightly different genetic makeup. Resistant biotypes may look exactly the same as susceptible biotypes. Herbicide resistant biotypes are able to survive a use rate several times higher than needed to control susceptible biotypes.

** Mode of action is the chemical interaction that interrupts a biological process necessary for plant growth and development.

*** Tank mixes with "Banvel" or "Banvel" II may result in reduced control of some broadleaf weeds.

Read and follow all appropriate sections of label, including precautions, before using this product.

WEED CONTROL IN WHEAT AND BARLEY

CROP STAGE AT APPLICATION

Make applications after the crop is in the 2-leaf stage, but before the flag leaf is visible. **Do not** Harvest sooner than 45 days after the last application of "EXPRESS".

APPLICATION RATES

Apply 1/6 to 1/3 ounce "EXPRESS" per acre to wheat (including durum) or barley for control or partial control of the weeds listed below.

Use 1/3 ounce "EXPRESS" per acre when weed infestation is heavy and predominantly consists of those weeds listed under partial control, or when application timing and environmental conditions are marginal (refer to APPLICATION AND ENVIRONMENTAL CONDITIONS FOR OPTIMUM PERFORMANCE).

Use 1/6 to 1/4 ounce "EXPRESS" per acre when weed infestation is light and predominantly consists of those weeds listed under weeds controlled, and when optimum application conditions occur.

Sequential treatments of "EXPRESS" may be made provided the total amount of "EXPRESS" applied to the crop does not exceed 1/3 ounce per acre; for example, 1/6 ounce followed by 1/6 ounce. The final application must be made before the flag leaf is visible. **Do not** harvest sooner than 45 days after the last application of "EXPRESS".

WEEDS CONTROLLED

Black mustard	Kochia
Blue/purple mustard	Mayweed chamomile
Bushy wallflower/Treacle mustard	Miners lettuce
Coast fiddleneck	Pineappleweed
Common chickweed	Prickly lettuce
Common groundsel	Russian thistle
Common lambsquarters	Slimleaf lambsquarters
Corn spurry	Tarweed fiddleneck
False chamomile	Wild chamomile
Field pennycress	Wild mustard
Flixweed	

PARTIAL CONTROL*

Annual sowthistle	Redroot pigweed
Canada thistle**	Shepherdspurse
Common sunflower	Tansymustard
Hairy nightshade	Tumble/Jim Hill mustard
Henbit	Wild buckwheat
Pennsylvania smartweed	Wild garlic
Prostrate knotweed	Wild radish
Redmaids	

*Partial Control: A visual reduction of weed population as well as a significant loss of vigor for individual weed plants. For best results, use 1/3 ounce "EXPRESS" per acre.

**See SPECIFIC WEED PROBLEMS for more information.

SPECIFIC WEED PROBLEMS

Canada thistle: For best results apply 1/3 oz. per acre plus surfactant in the spring when all thistles have emerged above the soil, are actively growing, and are 4 to 8 inches tall with 2 to 6 inches of new growth.

APPLICATION AND ENVIRONMENTAL CONDITIONS FOR OPTIMUM PERFORMANCE

Crop Stage: Make applications after the crop is in the 2-leaf stage, but before the flag leaf is visible. Do not harvest sooner than 45 days after the last application of "EXPRESS".

Since thorough coverage is required, avoid crop canopy obstruction of the spray contacting the weed foliage.

Crop Competition: A vigorous growing crop will aid weed control by shading and providing competition for weeds. However, a dense crop canopy at time of application can intercept spray and result in reduced weed control. Weeds may not be adequately controlled in areas of thin crop stand or seeding skips.

Pest Stage: Since "EXPRESS" has very little or no soil activity, only those weeds that have germinated above the soil surface will be controlled. Consequently, application of "EXPRESS" should be made when all or the majority of weeds have germinated. Annual broadleaf weeds should be past the cotyledon stage, actively growing, and less than 4 inches tall or across. Wild garlic plants should be 6 to 12 inches tall with 4 to 6 inches of new growth. See SPECIFIC WEED PROBLEMS for more information on Canada thistle.

Application: Foliar absorption is the primary means of "EXPRESS" uptake by plants; therefore, thorough spray coverage of weeds is essential.

Weather: Conditions which are conducive to healthy, actively growing plants optimize "EXPRESS" weed control performance. Ideal conditions include warm temperatures (70 Deg.F. or more) and adequate soil moisture before, during and immediately after application.

Avoid making applications of "EXPRESS" to weeds when rainfall is threatening. Rainfall immediately after treatment can wash "EXPRESS" off weed foliage and result in reduced weed control effectiveness. Several hours of dry weather are needed to allow "EXPRESS" to be absorbed by weed foliage.

Surfactant: Use only EPA-approved surfactants authorized for use on food crops.

Water Spray Carrier - Unless specified otherwise, add a nonionic surfactant of at least 80% active ingredient strength at 0.25% vol/vol (1 quart per 100 gallons of spray solution).

Liquid N Carrier - Always use surfactant when applying "EXPRESS" in liquid nitrogen fertilizer. When "EXPRESS" is applied using liquid nitrogen fertilizer solution as a spray carrier, early, temporary, crop yellowing and stunting may occur. Use 0.125 - 0.25% vol/vol surfactant (1 pint - 1 quart per 100 gallons of spray solution) when applying "EXPRESS" in liquid nitrogen fertilizer solution. Refer to TANK MIXTURES for surfactant information when tank mixing with other products.

EQUIPMENT-SPRAY VOLUMES

Prior to spraying "EXPRESS", thoroughly clean sprayer (tanks, booms, hoses, nozzles, mixing cones, tanks, etc.).

Apply using properly calibrated air or ground equipment. Select a spray volume and delivery system that will insure thorough coverage and a uniform spray pattern. Avoid overlapping, and shut off spray booms while starting, turning, slowing or stopping, or injury to the crop may result.

Do not apply "EXPRESS" through any type of irrigation system.

Spray Equipment: Refer to specific manufacturer's recommendations for additional information on spray volume (GPA), pressure, speed, nozzle types and arrangements, nozzle heights above the target area, etc., for respective application equipment.

Ground Application: For optimum spray distribution and thorough coverage, use flat fan or low volume flood nozzles. For flat fan nozzles, do not use less than 5-gallon spray volume per acre (GPA).

For flood nozzles on 30-inch spacing, use not less than 10 GPA and no larger than TK10 or equivalent and not less than 30 psi. On 40-inch nozzle spacings, use not less than 13 GPA or not less than 20 GPA when nozzles are on a 60-inch spacing. 100% overlapping of nozzle spray pattern is essential for 30, 40 and 60-inch spacings.

When using liquid nitrogen solution as a spray carrier, for flood nozzles use not less than 30 GPA and no larger than TK20. See statements on liquid N solutions below for additional information.

"Raindrop"[4] nozzles are not recommended for "EXPRESS" applications as weed control performance may be reduced.

Use 50-mesh screens or larger.

Aerial Application: Use orifice discs, cores and nozzle types and arrangements that will provide for optimum spray distribution and maximum coverage at 3 to 5 GPA. Do not apply during inversion conditions, when winds are gusty, or when other conditions will favor poor coverage and/or off target spray movement.

Liquid Nitrogen Fertilizer Solutions: "EXPRESS" can be added directly to liquid N solutions (for example 28-0-0, 32-0-0). Use 0.125 - 0.25% vol/vol surfactant (1 pint - 1 quart per 100 gallons of spray solution) when applying "EXPRESS" in liquid nitrogen fertilizer solution. The addition of liquid nitrogen fertilizer to the spray solution, or the use of liquid nitrogen fertilizer as the total carrier of the spray solution will increase the weight of the spray solution as compared to water (see table below). Consequently, appropriate conversions of spray nozzles and/or pressure must be made in order to maintain proper spray volumes. Refer to the conversion chart below for guidance (taken from Spray Systems Company Catalog 39, page 2).

Influence of adding liquid 28% N fertilizer to spray solution weight per gallon:

% of Spray Solution	Weight of Spray Solution Water	28% N (Lbs/Gallon)
100	0	8.34
50	50	9.5
0	100	10.6

SPRAYING SYSTEMS CONVERSION CHART

SPRAY SOLUTIONS OTHER THAN WATER--NEW METHOD

Since all the tabulations are based on spraying water, which weighs 8.34 lbs per USA gallon, conversion factors must be used when spraying solutions which are heavier or lighter than water. To determine the proper size nozzle for the solution to be sprayed, first multiply the desired GPM or GPA of solution by the water rate conversion factor. Then use the new converted GPM or GPA rate to select the proper size nozzle.

Example: 10 GPA (28%N) × 1.13 = 11.3 GPA (water)

Weight of Solution	Specific Gravity	Conversion Factors
7.0 lbs per gallon	0.84	0.92
8.0 lbs per gallon	0.96	0.98
8.34 lbs per gallon	1.00	1.00
9.0 lbs per gallon	1.08	1.04
10.0 lbs per gallon	1.20	1.13
10.65 lbs per gallon-28% nitrogen	1.28	1.13
11.0 lbs per gallon	1.32	1.15
12.0 lbs per gallon	1.44	1.20
14.0 lbs per gallon	1.68	1.30
Published in Spraying Systems Company Catalog 39, page 2.		

Agitation: Continuous agitation is required to keep "EXPRESS" in suspension.

CAUTION - AVOID SPRAY DRIFT

Follow these practices to minimize drift.

Do not allow spray from either ground or aerial equipment to drift onto adjacent crops or land, as even small amounts will injure other plants. When spraying near adjacent, sensitive crops or plants, do everything possible to reduce spray drift. This includes:

- Stop spraying wind speed becomes excessive. DO NOT SPRAY IF WIND SPEED IS 10 MPH OR GREATER. Spray drift can occur at wind speeds less than 10 MPH. If sensitive crops or plants are downwind, extreme caution must be used even in relatively low wind conditions! DO NOT SPRAY IF WINDS ARE GUSTY.
- High temperatures, drought, and low relative humidity increase the possibility of harmful spray drift. EXTREME CAUTION MUST BE USED WHEN THESE CONDITIONS ARE PRESENT AND SENSITIVE CROPS OR PLANTS ARE NEARBY, REGARDLESS OF WIND SPEED.
- Do not apply when an inversion exists. An inversion is characterized by little or no air movement and an increase in air temperature with an increase in altitude. In humid regions, a fog or mist may form. An inversion may be detected by producing a smoke column and checking for a layering effect. Smoke-producing devices on aircraft are recommended. If not sure whether inversion conditions are present, consult with local weather services before making an application.
- Postemergence grass herbicides (such as "Avenge"[5] and "Assert"[5]) are often applied using high pressure. When "EXPRESS" is tank mixed with these products, do not exceed 40 psi.
- Drift from aerial or ground equipment may be further reduced by:

  1. Using coarse sprays to minimize drift. DO NOT APPLY WITH HOLLOW-CONE INSECTICIDE NOZZLES ON GROUND EQUIPMENT. Do not use nozzles that produce fine droplets, such as Sprayfoil[6] or airblast-type nozzles. Nozzles should be oriented at an angle between straight down and straight back for ground applications. For aerial applications orient nozzles straight back along the windstream. If using flood-type nozzles on aircraft, orient them so spray is produced in direction of the airstream.
  2. Increasing volume of spray mix per acre (for example, minimum GPA by air, 10 GPA by ground) by using higher flow rate nozzles.
  3. Reducing pressure (PSI). DO NOT EXCEED 40 PSI when applying "EXPRESS". [Vehicle speed must also be reduced to maintain spray mix volume per acre.] Consult manufacturer's catalogs for details on correct calibration.
  4. Apply as close to target plants as possible while still maintaining a good spray pattern.

TANK MIXTURES

2,4-D (amine or ester) or MCPA (amine or ester) - Use "EXPRESS" plus 1/8 to 3/8 lb active ingredient 2,4-D or MCPA (ester formulations have provided best results). Surfactant may be added at 0.125 - 0.25% vol/vol (1 to 2 pints per 100 gallons of spray volume); however, the addition of surfactant may increase the chance of crop injury. Use the 1 to 2 pint rate of surfactant with 1/8 lb active ingredient of 2,4-D or MCPA. Use the 1 pint rate of surfactant with 1/4 to 3/8 lb active ingredint of 2,4-D or MCPA. Higher rates of 2,4-D or MCPA may be used, but do not exceed highest rate allowed on the label.

Always mix "EXPRESS" in water prior to adding 2,4-D or MCPA and surfactant. Read and follow all label instructions on timing, precautions and warnings for these herbicides prior to using these tank mixtures.

"EXPRESS" may be tank mixed with other suitable registered herbicides (for example "Karmex" and "Lexone") to control weeds other than those listed. Read and follow all manufacturer's label recommendations for the companion herbicide. If these recommendations conflict with this "EXPRESS" label , do not use as a tank mix with "EXPRESS".

"EXPRESS" will not control wild oats or other grasses. For wild oat control, "EXPRESS" can be tank mixed with "Avenge" or "Assert". When tank mixing "EXPRESS" and "Assert", ALWAYS include another broadleaf weed herbicide with a different mode of action, for example: MCPA ester, "Bronate". DO NOT tank mix with "Hoelon"[7] 3EC as grass weed control may be reduced.

Always mix "EXPRESS" in water PRIOR to adding other products, including surfactants. When mixing "EXPRESS" in liquid nitrogen fertilizer solutions (for example 28-0-0, 32-0-0), simply add the "EXPRESS" directly to the liquid fertilizer with the agitator running.

Refer to SPRAY PREPARATION for further information.

Tank mixes of "EXPRESS" plus "Banvel" or "Banvel" II may result in reduced control of some broadleaf weeds.

Tank mixes of "EXPRESS" plus metribuzin may result in reduced control of wild garlic.

"EXPRESS" may be tank mixed or used sequentially with insecticides registered for use on cereal grains. However, under certain conditions, (drought stress, crop in 2-4 leaf stage) tank mixes or sequential applications of "EXPRESS" plus organophosphate insecticides (such as parathion) may produce temporary crop yellowing or, in severe cases, crop injury. Limit first use to a small area before treating large areas.

DO NOT USE "EXPRESS" PLUS MALATHION.

SPRAY PREPARATION

Mix the proper amount of "EXPRESS" into the necessary volume of water in the spray tank with the agitator running, then add the companion herbicide to the tank after all the "EXPRESS" is in suspension. ALWAYS MIX "EXPRESS" IN WATER FIRST, PRIOR TO ADDING OTHER PRODUCTS IN THE SAME SPRAY TANK.

Use the spray preparation of "EXPRESS" within 24 hours as product degradation may occur. If spray preparation is left standing, thoroughly agitate before reusing.

Surfactant: Unless specified otherwise, add a nonionic surfactant of at least 80% active ingredient strength at 0.25% vol/vol (1 quart per 100 gallons of spray solution). Use only EPA-approved surfactants authorized for use on food crops.

Liquid Nitrogen Fertilizer: Liquid nitrogen fertilizer solutions such as 28-0-0 or 32-0-0, may be used as a carrier in place of water. To apply "EXPRESS" with liquid nitrogen fertilizer, simply add the "EXPRESS" directly to the liquid fertilizer with the agitator running (for at least 3 to 5 minutes). For fertilizer solutions that contain more than just nitrogen, for example 10-34-0, slurry the desired amount of "EXPRESS" in a clean bucket using water until a flowable mixture is produced. Add this slurry to the agitating spray tank of liquid nitrogen fertilizer solution. Thoroughly rinse all of the "EXPRESS" slurry into the spray tank. Do not use "EXPRESS" in liquid fertilizer solutions of less than pH 3. Run a tank mix compatibility test before mixing "EXPRESS" in fertilizer solution.

Use 0.125 - .025% vol/vol surfactant (1 pint - 1 quart per 100 gallons of spray solution) when applying "EXPRESS" in liquid nitrogen fertilizer solution.

Use of "EXPRESS" and a surfactant in liquid nitrogen fertilizer solutions may cause early, temporary crop injury (discoloration and stunting).

When the spray solution contains liquid nitrogen fertilizer, the weight per gallon of solution varied significantly from the weight of water (8.34 lbs per USA gallon). Consequently, liquid nitrogen fertilizer spray solutions must use the appropriate conversion in order to insure proper spray volume. See EQUIPMENT-SPRAY VOLUMES for further information.

CROP ROTATION

Any crop may be planted 60 days after the application of "EXPRESS".

SPRAYER CLEANUP

To avoid subsequent injury to crops other than wheat or barley, immediately after spraying thoroughly remove all traces of "EXPRESS" from mixing and spray equipment as follows:

1. Drain tank; rinse interior surfaces of tank; then flush tank, boom and hoses with clean water for a minimum of 5 minutes.
2. Fill the tank full with clean water, then add cleaning solution*. Flush solution through boom, hoses and nozzles, then allow to sit for 15 minutes with agitation running; then drain.
3. Repeat Step 2.
4. Repeat Step 1.
5. Nozzles and screens should then be removed and cleaned separately. To remove traces of cleaning solution, rinse the tank thoroughly with clean water and flush through hoses and boom.
6. Flush boom and hoses with clean water for 5 minutes just prior to using the sprayer for the first time after the "EXPRESS" application.

*Use any one of the following cleaning solutions:

1. Nutra-Sol (carefully read and follow Nutra-Sol label directions).
2. Loveland Spray Tank Cleaner (carefully read and follow Loveland Spray Tank Cleaner label directions).
3. One-half gallon chlorine bleach (containing 5 1/4% sodium hypochlorite) per 100 gallons of water.
4. One gallon ammonia (containing 3% active) per 100 gallons of water.

CAUTION: Do not use chlorine bleach with ammonia. All traces of liquid fertilizer containing ammonia, ammonium nitrate or ammonium sulfate must be rinsed with water from the mixing and application equipment before adding chlorine bleach solution. Failure to do so will release a gas with a musty chlorine odor which can cause eye, nose, throat and lung irritation. Do not clean equipment in an enclosed area.

PRECAUTIONS

The total rate of "EXPRESS" cannot exceed 1/3 ounce product per acre applied to any one crop during one growing season.

Do not harvest sooner than 45 days after the last application of "EXPRESS".

Do not graze or feed forage or hay from treated areas to livestock (dry-harvested straw may be used for bedding and/or feed).

Do not plant to any crop other than wheat or barley for 60 days after application of "EXPRESS".

Do not apply to wheat or barley that is stressed by severe winter, drought, water-saturated soil, disease or insect damage as crop injury may result. Under certain conditions such as prolonged cool weather (daily high temperature less than 50 Deg.F.) or wide fluctuations in day/night temperatures just prior to or soon after treatment, temporary yellowing and/or crop stunting may occur. Risk of injury is greatest when crop is in the 2 to 5-leaf stage.

Do not apply to wheat or barley crops underseeded with another crop.

"EXPRESS" is only registered on wheat and barley. Do not use on any other crop.

For ground applications applied to weeds when dry, dusty field conditions exist, control of weeds in wheel track areas may be reduced.

The "EXPRESS" volumetric measuring cylinder is to be used only as a guide, since the degree of accuracy varies by plus or minus 7.5%. For more precise measurement, use scales calibrated in ounces.

STORAGE AND DISPOSAL

STORAGE: Store product in original container only, away from other pesticides, fertilizer, food or feed.

PRODUCT DISPOSAL: Do not contaminate water, food or feed by storage or dispcsal. Waste resulting from the use of this product may be disposed of on site or at an approved waste disposal facility.

CONTAINER DISPOSAL: Triple-rinse (or equivalent) the container. Then offer for recycling or reconditioning, or puncture and dispose of in a sanitary landfill, or by incineration, or, if allowed by state and local authorities, by burning. If burned, stay out of smoke.

1 Registered trademark of Sandoz Crop Protection Corporation.
2 Registered trademark of Rhone - Poulenc Ag Company.
3 Registered trademark of the Dow Chemical Company.
4 Registered trademark of Delavan Corporation.
5 Registered trademark of American Cyanamid Company.
6 Registered trademark of D. AND W. INDUSTRIES.
7 Registered trademark of Hoechst-Roussel Agri-Vet Company.

NOTICE TO BUYER: Purchase of this material does not confer any rights under patents of countries outside of the United States.

NOTICE OF WARRANTY

Du Pont warrants that this product conforms to the chemcial description on the label thereof and is reasonably fit for purposes stated on such label only when used in accordance with directions under normal use conditions. It is impossible to eliminate all risks inherently associated with the use of this product. Crop injury, ineffectiveness, or other unintended consquences may result because of such factors as weather conditions, presence of other materials, or the manner of use or application, all of which are beyond the control of Du Pont. In no case shall Du Pont be liable for consequential, special or indirect damages resulting from the use of handling of this product. All such risks shall be assumed by the buyer. DU PONT MAKES NO WARRANTIES OF MERCHANTABILITY OR FITNESS FOR A PARTICULAR PURPOSE NOR ANY OTHER EXPRESSED OR IMPLIED WARRANTY EXCEPT AS STATED ABOVE.

AG - 3270 8079 6/30/89

This is a specimen label. DuPont is not responsible for the accuracy of the information contained herein. As labels are subject to revision, always carefully read and follow the label on the product container.

EXTRAZINE® II 4L

HERBICIDE
LIQUID

RESTRICTED USE PESTICIDE

For retail sale to and use only by Certified Applicators or persons under their direct supervision and only for those uses covered by the Certified Applicator's certification. See Precautionary Statements for reasons this product is classified Restricted Use.

ACTIVE INGREDIENTS:	BY WEIGHT
Cyanazine: 2-[[4-chloro-6-(ethylamino)-s-triazin-2-yl]amino]-2-methylpropionitrile	32.0%
Atrazine: 2-chloro-4-(ethylamino)-6-(isopropylamino)-s-triazine	10.4%
Related compounds	0.6%
INERT INGREDIENTS:	57.0%
TOTAL	100.0%

This product contains 4 pounds active ingredients per gallon.

EPA Reg. No. 352-500

KEEP OUT OF REACH OF CHILDREN
WARNING

AVISO

PRECAUCION AL USUARIO: Si usted no lee ingles, no use este producto hasta que la etiqueta haya sido explicada ampliamente.

STATEMENT OF PRACTICAL TREATMENT

IF SWALLOWED: Call a physician or poison control center. Drink 1 or 2 glasses of water and induce vomiting by touching back of throat with finger. Do not induce vomiting or give anything by mouth to an unconscious person.

IF IN EYES: Flush with plenty of water. Call a physician.

IF ON SKIN: Wash immediately with plenty of soap and water.

IF INHALED: Remove victim to fresh air. If not breathing, give artificial respiration, preferably mouth-to-mouth. Get medical attention.

For medical emergencies involving this product, call 1-800-441-3637.

PRECAUTIONARY STATEMENTS

HAZARDS TO HUMANS AND DOMESTIC ANIMALS

WARNING:

May be fatal if swallowed. Harmful if inhaled or absorbed through the skin. Causes temporary eye injury.

This product may be hazardous to your health. This product has been classified "Restricted Use" because, at doses which caused serious maternal illness in laboratory animals, birth defects were present. Use of protective clothing and equipment and following the precautions below can reduce risk.

Avoid breathing spray mist. Avoid contact with skin, eyes, or clothing. Do not get in eyes or on clothing. Wear a face shield when mixing and loading. Wash thoroughly with soap and water after handling and before eating or smoking.

Wear long trousers and long-sleeved clothing when applying this product. Wear chemical resistant gloves extending above the wrist, long trousers, a chemical resistant apron, and long-sleeved clothing when mixing or loading or when adjusting, repairing or cleaning equipment. Protective gloves must be washed with soap and water after use and before removing from the hands. Remove contaminated clothing and wash before reuse. Contaminated clothing should be laundered separately from household laundry to prevent cross-contamination of other laundry. Heavily contaminated or drenched clothing and protective equipment must be discarded or destroyed in accordance with State and local regulations.

Do not apply this product in such a manner as to directly, or through drift, expose workers or other persons. The area being treated must be vacated by unprotected persons.

Keep out of reach of domestic animals, particularly cattle. Consumption of this product, spray solutions, or water contaminated with product can result in serious illness or possible death of bovines.

ENVIRONMENTAL HAZARDS

Cyanazine and Atrazine, the active ingredients of EXTRAZINE II, are pesticides which can move (seep or travel) through soil and can contaminate groundwater which may be used as drinking water. Cyanazine and Atrazine have been found in groundwater as a result of agricultural use. Users are advised not to apply EXTRAZINE II where the water table (groundwater) is close to the surface and where the soils are very permeable (i.e. well drained soils such as loamy sands). Your local agricultural agencies can provide further information on the type of soil in your area and the location of groundwater.

Atrazine is toxic to aquatic invertebrates. Do not apply directly to water or wetlands. Do not apply where runoff is likely to occur. Do not apply when weather conditions favor drift from treated areas. Runoff and drift from treated areas may be hazardous to aquatic organisms in neighboring areas. Do not contaminate water by cleaning of equipment or disposal of wastes.

In case of significant spill, call CHEMTREC 1-800-424-9300.

DIRECTIONS FOR USE

It is a violation of Federal law to use this product in a manner inconsistent with its labeling.

GENERAL INFORMATION

EXTRAZINE II 4L Herbicide is a selective herbicide for the control of annual grasses and broadleaf weeds in field corn, popcorn, and sweet corn.

Consult your local Agricultural Extension Agent for help in determining soil type, organic matter content, and the most appropriate herbicide rate for local conditions.

Do not apply this product in irrigation water with any kind of irrigation system.

Do not apply this product with aerial application equipment.

EXTRAZINE II 4L is not effective when used preemergence on peat or muck soils. Do not use EXTRAZINE II 4L on sands or loamy sands (soils consisting of more than 70% sand) containing less than 1% organic matter.

Weather Effects: As a preemergence herbicide, EXTRAZINE II 4L is active mainly through the roots, and, therefore, its effect on weeds is dependent on adequate rainfall to move the herbicide into the root zone. Moisture should be sufficient to thoroughly wet the soil throughout the zone where weed seeds may germinate and enough to make the soil too wet to cultivate. Rotary hoeing or shallow cultivation is recommended for those applications which are not incorporated at the time of treatment, if adequate rainfall or sprinkler irrigation has not occurred within about ten days after application of EXTRAZINE II 4L.

Heavy rainfall between planting and crop emergence may cause crop injury or stand loss. Rainfall tends to cause excessive concentrations of herbicide in the seed furrow, resulting in possible crop injury. Level deep planter marks or seed furrows before application.

Under conditions which delay weed germination, such as low temperatures, lack of soil surface moisture, or when germination extends over a long period, the effectiveness of the herbicide may be impaired. Rotary hoeing, shallow cultivation or a postemergence herbicide treatment may be of benefit under these circumstances.

If the crop is cultivated, tillage should be shallow to minimize herbicide dilution in the soil. Should the crop stand be lost due to adverse weather conditions, insects, etc., the field can be replanted the same season to corn or sorghum.

To enhance weed control in areas of less than 25 inches of rainfall or where long dry periods are common, these treatments may require shallow incorporation with a tool such as a field cultivator operated from 5-7 mph. Incorporation should not be more than three inches deep to keep from burying the herbicide. A spike-toothed harrow, deep tillage disk or rolling basket device is not recommended for incorporating EXTRAZINE II 4L.

When applied as a post-emergence herbicide, EXTRAZINE II 4L is also active through foliage as well as through the roots. Yellowing and/or stunting of the crop may result from this treatment, particularly if cold, adverse growing conditions occur after application. Extended or extreme cold and wet conditions may reduce stands. Do not apply EXTRAZINE II 4L postemergence to a crop that is damaged or growing under stress.

OBSERVE ALL CAUTIONS AND LIMITATIONS ON LABELING OF ALL PRODUCTS USED IN MIXTURES.

Rotational crops: (1) Plant only corn, sorghum, or soybeans the year following the use of this mixture. (2) If soybeans are to be planted, injury may occur. (3) If applied after June 10, do not rotate with crops other than corn or sorghum the next year or injury may occur. (4) In the high plains and inter-mountain areas of the West where rainfall is sparse and erratic or where irrigation is required, use only when corn or sorghum is to be planted the following year, or a crop of corn or sorghum not treated with this mixture or atrazine is to precede other rotational crops. (5) Small grains may be planted 15 months following treatment. (6) All other crops may be planted 18 months after application.

Triazine Resistant Weeds: In fields where triazine resistant biotypes of weeds have been identified, EXTRAZINE II 4L should be used in combination with or in sequence with other registered non-triazine herbicides. (Triazine resistant biotypes of Kochia and Pigweed have been identified in some fields in the Western Great Plains and triazine resistant biotypes of Pigweed and Lambsquarters have been identified in some fields in various states.) Consult with appropriate state agricultural extension service representatives for specific recommendations.

APPLICATION DIRECTIONS

GENERAL MIXING AND SPRAYING DIRECTIONS

This product may not be applied by means of chemigation or aerial application.

Use sufficient agitation to ensure that the EXTRAZINE II 4L is completely dispersed and in uniform suspension prior to application or tank mixing with other formulations.

The following general mixing instructions are recommended:

I. General

1. Unless otherwise specified, use at least 7.5 gallons of water per acre for soil applications and at least 15 gallons of water per acre for foliar applications for all applications with ground equipment.
   NOTE: Sufficient carrier must be used to assure uniform application. Follow label requirements of all products used in tank mix combinations.
2. A nitrogen solution or complete liquid fertilizer may replace all or part of the water as a carrier for preemergence or preplant application on corn. Do not apply fertilizer mixtures after crop emerges, as injury may occur.
3. Always check the tank mix compatibility (TMC) of this or any other formulation before mixing with liquid fertilizer carriers or other formulations. A simple but generally reliable TMC evaluation procedure has been provided for your use in Step II of these mixing instructions.
4. Start with thoroughly clean equipment. (See the labels of the previous compounds used for cleaning instructions.)
5. Fill the tank at least 1/2 full with carrier. Start and maintain consistent agitation through all mixing and spraying procedures. Make sure the agitation system is working properly and creates a rippling or rolling action on the liquid surface.
6. Add the recommended amount of EXTRAZINE II 4L to the tank.
7. Fill the tank to 75 percent capacity with carrier. Filling and bypass lines should be kept below liquid surface. Increase tank agitation if necessary to maintain surface action.
8. When desired, appropriate emulsible crop oil, crop oil concentrate, or other tank mix formulations should be added at this time. Pre-slurry these added ingredients before addition, if the compatibility test shows it to be necessary.
9. Complete filling tank, maintaining sufficient agitation at all times to ensure surface action. This applies to both spray and nurse tanks.
10. Tank mixtures should always be applied immediately after preparation. If, for any reason, this is not possible, assure that sufficient agitation has been provided to re-mix all products and check for complete re-suspension prior to application.
11. Empty tank as completely as possible before refilling to prevent buildup of oil or emulsible concentrate residues when tank mixing with these formulations. Always maintain agitation to avoid separation.
12. If an oil or emulsible concentrate film starts to build up after using these formulations, drain and clean the tank with strong detergent solution or appropriate solvent.
13. It is recommended that the sprayer be thoroughly cleaned by flushing with a detergent solution at the end of each work day when any emulsible crop oil, crop oil concentrate, or other emulsible formulation has been used either alone or in tank mix combinations with other pesticide formulations. This precaution will ensure a clean sprayer and continued trouble-free operation.

II. Tank Mix Compatibility Evaluation Procedure

1. Add 1 pint of carrier liquid to each of 2 one-quart jars. Mark 1 quart jar "with" and the other "without."
2. Add 1/4 teaspoon of a suitable compatibility agent (1/4 teaspoon/pint = 2 pints/100 gallons of carrier) to the jar marked "with", cap the jar and shake gently for 5 to 10 seconds to mix.
3. Add the appropriate amount of herbicide to both jars, cap each jar and shake gently for 5 to 10 seconds to mix. If problems are encountered in mixing wettable powder or dry flowable formulations into a liquid fertilizer, pre-slurry these formulations in water prior to their addition to the liquid fertilizer and proceed with the test. The following chart has been provided to assist in selecting the approximate EXTRAZINE II 4L use rate for this evaluation.

Jar Test for EXTRAZINE II 4L Compatibility

Gallons of Liquid Carrier per acre	4	7.5	15	20	25	30
Teaspoons of EXTRAZINE II 4L per pint of Liquid Carrier	6.0	3.2	1.6	1.2	1.0	0.8

This chart is based on 1 quart of EXTRAZINE II 4L (1.0 pound active ingredient) per acre in the indicated carrier volumes. Intended field use rates are achieved by varying the amount of EXTRAZINE II 4L. (e.g., for a field use rate of 3 quarts of EXTRAZINE II 4L in 15 gallons of carrier per acre, add 4.8 teaspoons of EXTRAZINE II 4L to the quart jars containing 1 pint of carrier. Calculation: 3 quarts of EXTRAZINE II 4L/15 gallons of carrier per acre = 3 × 1.6 = 4.8 teaspoons of EXTRAZINE II 4L per pint of carrier.)

4. Let each jar stand one-half hour and make observations. If any separation, agglomeration, or precipitation has occurred, shake the jar again for 10 to 15 seconds, and note whether any of the following occur:
   a. Separated phases do not re-mix uniformly.
   b. Screen/nozzle plugging lumps do not disperse.
   c. Precipitate does not re-suspend readily.
   d. Precipitate sticks tenaciously to the glass.
5. If none of the above problems occur in either jar, the herbicides can, in most cases, be safely used in that carrier without a compatibility agent.
6. If problem 4.a or 4.b occurs in the jar marked "without" but does not occur in the jar marked "with", the compatibility agent should be used.
7. If problem 4.a or 4.b is seen in both jars, then the herbicides and carrier are incompatible and should not be used in the same spray tank. Alternatively, a different tank mix compatibility agent can be evaluated.
8. If problem 4.c or 4.d occurs in the jar marked "without" but does not occur in the jar marked "with", the compatibility agent should be used unless constant, thorough agitation can be maintained and immediate clean-out of the spray system is performed.
9. If problem 4.c or 4.d is seen in the jar marked "with," the user proceeds with mixing and application at his own risk should the agitation in the system be insufficient or curtailed.
10. Those mixtures defined as compatible in this test should then be mixed for use as indicated in Steps 1 through 12 of the general mixing instructions listed above.

If a test such as outlined indicates that components of a proposed mix are compatible, the applicator still has the responsibility of combining materials in sequence to the spray tank in accordance with directions prescribed on the label of the herbicides or pesticides involved.

Tests have indicated that compatibility agents, noted below by the various tank mix combinations, may give improved compatibility in liquid fertilizers.

Tank Mix Combination	Compatibility Agents
EXTRAZINE II 4L/"Lasso" (Liquid Fertilizer Grade)	Probably not needed in 28-0-0, 10-34-0. "Compex" may help in others.
EXTRAZINE II 4L/"Genate Plus" "Sutan+" 6.7E	Probably not needed in 28-0-0. Incompatible in 10-34-0. "Unite", "Spray-Mate", "Kem-Link", may help in others.
EXTRAZINE II 4L/"Dual" 8E	Probably not needed in 28-0-0. "Unite", "Spray-Mate", "Ivory Liquid" may help in others.

III. Application Equipment

1. Use application equipment fitted with nozzles that provide accurate and uniform coverage. Be certain that nozzles are uniformly spaced and the same size. Calibrate sprayer before use and re-check frequently during use whenever possible.
2. Use a pump with capacity to:
   a. Maintain 35 to 40 psi at the nozzles.
   b. Provide sufficient agitation in tank to keep mixture in suspension.
   c. Provide a minimum of 20 percent bypass at all times.
3. Use centrifugal pumps which provide sufficient shear action for dispersing and mixing this product. The pump should provide a minimum of 10 gallons/minute/100-gallon tank size circulated through the jets of a correctly-positioned sparger tube.
4. Use screens to protect the pump and to prevent nozzles from clogging. Screens placed on suction side of pump should be 10 to 16 mesh. Do not place a screen in the recirculation line. Use a 40- to 50-mesh screen between the pump and boom and, where required, 50-mesh screens at the nozzles. Check your equipment manufacturer's literature for specific recommendations.

FERTILIZER IMPREGNATION

EXTRAZINE II 4L Herbicide may be applied when coated on or impregnated in dry granular fertilizer for early preplant, preemergence or preplant incorporated weed control in field corn. All recommendations, cautions and special precautions on this label must be followed along with state regulations relating to dry bulk fertilizer blending, impregnating and labeling.

General Blending Directions

EXTRAZINE II 4L may be coated on or impregnated in dry bulk fertilizers using tower blenders, rotary drum blenders or blending augurs or conveyors. DO NOT impregnate EXTRAZINE II 4L or tank mixes containing EXTRAZINE II 4L on or in fertilizers containing Ammonium Nitrate, Potassium Nitrate, or Sodium Nitrate. Do not use on straight limestone since absorption will not be achieved. Fertilizer blends containing limestone can be impregnated when using EXTRAZINE II 4L alone. Use a minimum of 200 lbs. and a maximum of 450 lbs. per acre of dry fertilizer.

Use equipment that will give a uniform distribution of the herbicide throughout each batch of impregnated fertilizer. Non-uniform impregnation can cause crop injury or unsatisfactory performance.

A. EXTRAZINE II 4L may be used as the only herbicide for impregnation.
   1. Add EXTRAZINE II 4L to the total fertilizer volume required and mix thoroughly.
   2. Add "Ag-Sorb" or "MP-79" drying agent (or a suitable amount of another effective drying agent) as needed to insure a spreadable herbicide/fertilizer mixture. The need for a drying agent is determined by the wetness of the fertilizer batch. Wetness can change with humidity, nitrogen content, fertilizer types, fertilizer rates and herbicide rates.
B. EXTRAZINE II 4L may be used in tank mixes with other herbicides registered for this use. The amount of drying agent needed will vary depending on whether the product is a dry or liquid formulation. EXTRAZINE II 4L and "Sutan+", "Eradicane", "Dual" or "Lasso" can be mixed together in the proper use ratio for simultaneous impregnation. If there is a mixing compatibility problem then add the EXTRAZINE II 4L first as follows:
   1. While fertilizer is blending, add the EXTRAZINE II 4L.
   2. Spray in the EC herbicide and mix thoroughly for at least 3 minutes.
   3. Add drying agent to insure a spreadable herbicide/fertilizer mixture. Generally less than 2% by weight of drying agent will be needed.

(For more information on Drying Agents, Application Equipment, Calibration Guide and variations of these methods, see Du Pont Bulletin on "Fertilizer Impregnation").

CLEAN OUT:

Equipment used to impregnate or apply fertilizer impregnated with EXTRAZINE II 4L or combinations including EXTRAZINE II 4L must be cleaned out by running at least 1,000 lbs. of fertilizer not impregnated with EXTRAZINE II 4L through the impregnation equipment and application equipment, if the next batch of material is to be applied to a crop for which EXTRAZINE II 4L or a combination herbicide is not registered.

APPLICATION:

Uniform application of EXTRAZINE II 4L which has been impregnated in or coated on dry fertilizer is essential for satisfactory weed control and crop safety. Accurate calibration of the fertilizer applicator is necessary. Applying while turning at the ends of the fields may result in excessive application rates causing crop injury. Do not double apply across the ends or sides of the field.

Crop injury and/or poor weed control may occur where the impregnated fertilizer is not uniformly applied. Air flow or augur metered application equipment is preferred (one pass application). If other equipment is used, the recommended method of application is to apply ½ the recommended rate and overlap 50 percent to double apply by splitting the middles to obtain the best distribution pattern.

Apply immediately after impregnation. Impregnated fertilizer may become lumpy and difficult to spread if stored.

RATES AND TIMING:

Use the application rates and timing shown in the appropriate sections of this label. Follow the precautions on the labels of all products used.

CORN

Weeds Controlled by EXTRAZINE II 4L Alone and in Combination with Other Herbicides on Corn

Grasses

Annual bluegrass	Crabgrass	Junglerice
Annual fescues	Fall panicum	Stinkgrass (Indian lovegrass)
Annual (Italian) ryegrass	Giant foxtail	Witchgrass
Barnyardgrass(1)	Goosegrass	Yellow foxtail
Bullgrass	Green foxtail	

Broadleaves

Annual groundcherry	Florida pusley (Florida purslane)	Ragweed (Common)
Annual morningglory	Hedge mustard	Russian Thistle
Annual sedge	Jimsonweed(1)	Shepherdspurse
Black mustard	Kochia	Smallflower galinsoga
Buffalobur	Ladysthumb	Smartweed (Pennsylvania)
Buttercup (annual)	Lambsquarters	Spiny sida
Carpetweed	Mayweed	Sunflower(2) (wild, annual, common)
Common chickweed	Nightshade (annual)	Tarweed cuphea (Gumweed)
Cocklebur(2)	Pigweed(1)	Velvetleaf(1)
Common groundsel	Pineappleweed	Wild buckwheat
Common mallow	Plantain	Wild mustard
Common purslane	Poorjoe	Wild radish
Corn spurry	Prickly sida (teaweed)	Wild turnip
Curly dock (seedling)	Prostrate knotweed	
Fiddleneck	Prostrate spurge	

(1) Under conditions such as low temperatures, lack of soil surface moisture, or other factors that may cause delay in germination of the seeds, the effectiveness of EXTRAZINE II 4L may be impaired against these weeds.
(2) The degree of control will be reduced if soil moisture and temperature conditions cause deep germination of the seed.

PREEMERGENCE-PREPLANT INCORPORATED

Apply EXTRAZINE II 4L treatments just before, at, or after planting but before crop has emerged. Avoid removal of treated soil from seedrow prior to or during the planting operation.

EXTRAZINE II 4L may also be applied early prior to planting or in a split application if pre-season weed control is desired. For split applications, do not exceed the total amount of EXTRAZINE II 4L for the soil texture and organic matter shown in Table 1. If EXTRAZINE II 4L is applied early, more than 15 days before planting, a split application of EXTRAZINE II 4L or some other herbicide treatment may be necessary at or after planting to provide additional length of weed control. For further information, see "Early Preplant" recommendations in the Conservation Tillage section of label.

Rotary hoeing is recommended for preemergence applications which do not receive adequate rainfall or sprinkler irrigation to wet the top 2 inches of soil or depth of germinating weeds within about 10 days after application.

EXTRAZINE II 4L alone or in tank mix combinations should not be incorporated more than three inches deep to keep from burying the herbicide. Single or two pass incorporation with a tool such as a field cultivator operated at 5-7 mph is acceptable. A spike-toothed harrow, deep tillage disk or rolling basket device is not recommended for incorporating EXTRAZINE II 4L.

See the Rotational Crops statement in the General Information section of the label.

TABLE 1
PREEMERGENCE BROADCAST APPLICATION RATES PER ACRE FOR "EXTRAZINE" II 4L APPLIED ALONE ON CORN

	QUARTS OF EXTRAZINE II 4L					
	PERCENT ORGANIC MATTER IN SOIL*					
SOIL TEXTURE DESCRIPTION	LESS THAN 1%	1%	2%	3%	4%	5% & OVER
Sand, Loamy sand	DO NOT USE	1.25	1.5	2.0	2.5	3.25
Sandy loam	1.25	1.5	2.0	2.5	3.25	4.0
Loam, Silt loam, Silt	1.5	2.0	2.75	3.25	4.0	4.25
Sandy Clay loam, Clay loam, Silty clay loam	2.0	2.75	3.25	4.0	4.5	4.75
Sandy clay, Silty clay, Clay	2.75	3.25	4.0	4.5	4.75	5.25
Peat or Muck	NOT RECOMMENDED					

* For organic matter content between those listed, adjust the rate proportionately.

EXTRAZINE II 4L Applied Alone

Use at the proper rate for soil texture and organic matter indicated in Table 1.

EXTRAZINE II 4L COMBINATIONS

EXTRAZINE II 4L plus "Lasso" 4EC

Use EXTRAZINE II 4L at the proper rate for the soil texture and organic matter shown in Table 2 plus 2 quarts per acre of "Lasso". (Use 2.5 quarts "Lasso" on clay soils containing 5 percent organic matter and over).

EXTRAZINE II 4L plus "Sutan+" 67E/"Genate Plus" or "Eradicane" 67E

Use EXTRAZINE II 4L at the proper rate for the soil texture and organic matter shown in Table 2 plus 1.8 quarts per acre of "Sutan+"/"Genate Plus" or "Eradicane" for control of many annual grasses and broadleaf weeds. (Use 2.4 quarts of "Sutan+"/"Genate Plus" or "Eradicane" on loam soils containing 5 percent or more organic matter.) Do not use on sands and loamy sands of less than 1 percent organic matter in the light sandy soils of eastern coastal states. Do not use on corn seed stock.

Apply before planting. Incorporate the mixture immediately upon application using power-driven cultivation equipment set for 2 to 3 inch depth, or a tandem disc set to cut to a depth of about 4 inches while operating at 4 to 6 mph. For thorough mixing, disc in two directions (cross disc), and follow with a harrow, drag, or other leveling device. Prior to the second discing, readjust the disc to prevent cutting deeper than 4 inches. EXTRAZINE II 4L may be applied preemergence as an overlay over previous incorporated "Sutan+"/"Genate Plus" or "Eradicane" if desired.

Existing stands of quackgrass, purple and yellow nutsedge must be turned under and thoroughly chopped up prior to chemical treatments.

Additional weeds controlled by "Sutan+"/"Genate Plus" or "Eradicane" combinations:

Grasses	
Sandbur	Shattercane (Wild Cane)*
Texas Panicum	Quackgrass (Eradicane only)
	Wild Proso Millet* (Eradicane only)

Perennial Weeds	
Yellow Nutsedge (nutgrass)	Purple Nutsedge (nutgrass)

* Suppression only—refer to "Sutan+"/"Genate Plus" or "Eradicane" label for appropriate supplement cultural and tillage practices.

For fields with moderate to heavy infestations of these weeds, refer to the "Sutan+"/"Genate Plus" or "Eradicane" labels for appropriate higher rates.

EXTRAZINE II 4L plus "Dual" 8E

Use EXTRAZINE II 4L at the proper rate for soil texture and organic matter shown in Table 2. Use "Dual" as follows:

Soil Texture	Broadcast Rate Per Acre For "Dual" 8E
Coarse Sand, Loamy sand, Sandy loam	1.25 - 1.5 pints
Medium Loam, Silt loam, Silt	1.5 - 2.0 pints
Fine Sandy clay loam, Silty clay loam, Clay loam, Sandy clay, Silty clay, Clay	1.5 - 2.5 pints

The low end of the rate range should be used for lowest organic matter soils and the rate increased as organic matter increases to a point that soils containing 4 percent organic matter or more require the highest rate shown for that soil texture. Refer to the "Dual" label for additional precautions on rotational crops.

CONSERVATION TILLAGE WEED CONTROL

Early Preplant (EPP)

EXTRAZINE II 4L may be used for Early Preplant or Preemergence weed control for land going into the production of corn under conservation tillage programs. Complete any planned early spring tillage prior to application. Apply herbicide treatment before weeds germinate or before weed seedlings are more than 3 inches tall. Tillage after application may reduce the effectiveness of the herbicide treatment. A nitrogen solution or complete fertilizer solution may replace all or part of the water as a carrier. The spray gallonage and spray boom design must be adequate to give thorough uniform coverage of the weed foliage. Follow label requirements of all products used in tank mix combinations.

EXTRAZINE II 4L alone: Apply 15 to 30 days prior to planting. Use the proper rate for soil texture and organic matter indicated in Table 1. Where heavy crop residues exist, the rates shown in Table 1 should be increased by 25 percent.

TABLE 2
PREEMERGENCE BROADCAST APPLICATION RATES PER ACRE FOR "EXTRAZINE" II 4L USED IN TANK MIX COMBINATIONS WITH "LASSO", "SUTAN+", "GENATE PLUS", "ERADICANE", OR "DUAL" ON CORN

SOIL TEXTURE DESCRIPTION	QUARTS OF EXTRAZINE II 4L					
	PERCENT ORGANIC MATTER IN SOIL*					
	LESS THAN 1%	1%	2%	3%	4%	5% & OVER
Sand, Loamy sand	0.6**	0.75	1.25	1.5	1.75	2.0
Sandy loam	0.75	1.25	1.5	1.75	2.0	2.25
Loam, Silt loam, Silt, Sandy clay loam	1.25	1.5	1.75	2.0	2.25	2.5
Clay loam, Silty clay loam	1.5	1.75	2.0	2.25	2.5	2.75
Sandy clay, Silty clay, Clay	1.75	2.0	2.25	2.5	2.75	3.0
Peat or Muck	NOT RECOMMENDED					

* For organic matter content between those listed, adjust the rate proportionately.
** Do not use on the light sandy soils of the Atlantic Coastal Plain.

EXTRAZINE II 4L plus "Princep" 4L or EXTRAZINE II 4L plus "Princep Caliber" 90: Apply 30 days or more prior to planting. Use the proper rate of EXTRAZINE II 4L for the soil texture and organic matter indicated in Table 1 and add 1 quart/acre of "Princep" 4L or 1.11 pounds/acre of "Princep" Caliber 90.

Burndown of Existing Weeds:

Combinations with 2,4-D: Where broadleaf weeds are present at the time of application, add 2,4-D LV Ester at 1⅓ - 2 pts./A. (6 lbs./gal.) or 2-3 pts./A. (4 lbs./gal.) (or 2,4-D Amine at recommended rates) plus Ortho. X-77 surfactant at 1 qt./100 gals. of diluted spray, or other suitable nonionic surfactant at its recommended rate. When 2,4-D Ester is tank mixed with EXTRAZINE II 4L as described, these additional weeds will be controlled: Buckwheat, Dandelion, Dock, Marestail, Pennycress, Prickly lettuce and Tansy Mustard.

Combinations with Paraquat Products or "Roundup": When grasses are present and/or when existing weeds exceed 3 inches in height, add paraquat at 1-2 pts./A. (2 lb/gal) or 1.3-2.7 pts./A. (1.5 lb/gal) or 2 pts./A. of "Roundup" to the above mixture. Well established weeds over 6 inches tall may not be well controlled. With paraquat use 20-40 gal./A. of spray mixture by ground rig. (Use the higher gallonages for heavy infestations of weeds.) Add 1 to 2 pints of a non-ionic surfactant per 100 gallons of spray. Using fertilizer solutions containing phosphates as a carrier will reduce the activity of paraquat products.

Sequential Treatments: Depending upon the application timing and the weather conditions following the EPP application, a sequential herbicide treatment may be necessary to provide additional length of weed control. This may be a postemergence treatment with BLADEX, CONQUEST, EXTRAZINE, EXTRAZINE II or some other herbicide treatment applied at or after planting. If desired, 1½ - 2 pts./A. of "Dual" or 2 qts./A. of "Lasso" may be tank mixed with the EXTRAZINE II 4L EPP treatment or applied preemergence at planting.

Rotational Crops: Refer to the General Information section of this label and to the labels of all products used in combination or in sequence.

At Planting

EXTRAZINE II 4L applied alone or in combination with Lasso or Dual according to the following directions will kill most existing small weeds and suppress many emerged perennial weeds when corn is planted into no-till stalk ground (corn, sorghum), stubble ground (soybean, small grains), and any minimum-till situation. This treatment then provides residual control of annual weeds as in conventional tillage.

Apply EXTRAZINE II 4L alone or with other products according to the directions for those treatments in the Preemergence section of the label. Where heavy crop residues exist, the EXTRAZINE II 4L rate shown in Table 1 should be increased by 25 percent. Add 2,4-D LV Ester at ½ - 1 pt./A. (6 lbs./gal.) or 0.75 - 1.5 pt./A (4 lbs./gal.) (or 2,4-D Amine at recommended rates). Add the 2,4-D to the spray tank last. Use a minimum of 25 gals./A. of carrier. Complete spray coverage of the weeds is essential for best performance.

Nitrogen solutions and complete liquid fertilizers are the preferred carriers for this treatment because they aid in the burndown of existing weeds. Add Ortho X-77, surfactant at 1-2 qts. per 100 gals. of diluted spray, or other suitable surfactant at its recommended rate. If water is used as a carrier, crop oil concentrate may be used as an adjuvant. Apply before weeds exceed 3 inches in height. For control of existing alfalfa, add ⅓-½ pt./A. of "Banvel" to the spray mixture. Apply before the alfalfa exceeds 6 inches in height.

For fields with existing sod grasses such as orchardgrass, bromegrass, rye or timothy, or when very dry conditions exist, or when existing weeds exceed 3 inches in height add paraquat to the tank-mix. Use 2 pints/A. (2 lbs/gal) or 2.7 pints/A. (1.5 lbs/gal) paraquat in combination with EXTRAZINE II 4L as described above in this section, except the 2,4-D may be omitted, if desired. Do not apply paraquat in suspension type liquid fertilizer.

POSTEMERGENCE

Under warm, sunny, dry conditions of low humidity and the absence of dew formation at night, EXTRAZINE II 4L may be applied postemergence on field corn only. Do not use a surfactant, crop oil, or any other adjuvant. Addition of a surfactant, EV oil, or fertilizer solution is not recommended because moderate to severe injury, including stand loss, may occur.

Do not apply this treatment under cold, wet weather conditions or to corn growing under stress caused by weather, insects, disease, etc. Yellowing of the corn, stunting or stand loss may result from this treatment, particularly if high rainfall or cold, adverse growing conditions occur after application.

Do not apply postemergence on popcorn, sweet corn or corn grown for seed.

EXTRAZINE II 4L Applied Alone

Use EXTRAZINE II 4L at the proper rate for the soil texture and organic matter shown in Table 3 or 4. Use rates shown in Table 3 if BLADEX, CONQUEST, EXTRAZINE or EXTRAZINE II has not been applied to the soil this season. Use rates shown in Table 4 if BLADEX, CONQUEST, EXTRAZINE or EXTRAZINE II has been applied to the soil this season. This treatment may be used on peat or muck soils for burndown and suppression of existing weeds but will not provide residual control. Apply from crop emergence through the four-leaf stage of corn growth before weeds exceed about 1½ inches in height. Do not apply over the top of corn if the fifth leaf is visible. Apply in water only. Do not spray emerged corn plants in a liquid fertilizer carrier or in tank mix combinations with EC formulation herbicides.

Rotational Crops: Refer to the General Information section of the label.

SWEET CORN

EXTRAZINE II 4L may be applied preemergence or preplant incorporated for the control of annual grasses and broadleaf weeds in sweet corn.

NOTE: EXTRAZINE II 4L may cause injury or stand loss on new or "super sweet" varieties of sweet corn. Consult with Agricultural Extension Agencies and sweetcorn seed suppliers about the sensitivity of new varieties to potential injury.

Apply EXTRAZINE II 4L treatments just before, at or after planting but before crop has emerged. Avoid removal of treated soil from seedrow prior to or during the planting operation. Do not apply postemergence to sweet corn.

Rotary hoeing is recommended for preemergence applications which do not receive adequate rainfall or sprinkler irrigation to wet the top 1½ to 2 inches of soil within about 10 days after application.

If a EXTRAZINE II 4L mixture is to be incorporated, except as noted, single or two pass incorporation is acceptable. Care should be taken to incorporate the EXTRAZINE II 4L mixture no deeper than the top two inches of soil.

Rotational Crops: (1) Plant only corn, sorghum or soybeans the year following the use of this mixture. (2) If soybeans are to be planted, injury may occur due to the carryover of Atrazine. (3) If applied after June 10, do not rotate with crops other than corn or sorghum the next year or injury may occur. (4) Small grains may be planted 15 months following treatments. (5) All other crops may be planted 18 months after application.

TABLE 3
POSTEMERGENCE BROADCAST APPLICATION RATES PER ACRE FOR "EXTRAZINE" II 4L ON CORN
NO PRIOR APPLICATION OF "BLADEX", "CONQUEST", "EXTRAZINE" OR "EXTRAZINE" II

SOIL TEXTURE DESCRIPTION	QUARTS OF EXTRAZINE II 4L — PERCENT ORGANIC MATTER IN SOIL*: LESS THAN 1%	1%	2%	OVER 2%
Sand, Loamy sand	DO NOT USE	1.25	1.5	2.0
Sandy loam	1.25	1.5	2.0	2.0
Loam, Silt loam, Silt	1.5	2.0	2.0	2.0
All other textures	2.0	2.0	2.0	2.0

* For organic matter content between those listed, adjust the rate proportionately.

TABLE 4
POSTEMERGENCE BROADCAST APPLICATION RATES PER ACRE FOR "EXTRAZINE" II 4L ON CORN
"BLADEX", "CONQUEST", "EXTRAZINE" OR "EXTRAZINE" II USED IN PRIOR APPLICATION

SOIL TEXTURE DESCRIPTION	QUARTS OF EXTRAZINE II 4L — PERCENT ORGANIC MATTER IN SOIL*: LESS THAN 1%	1%	2%	OVER 2%
Sand, Loamy sand	DO NOT USE	DO NOT USE	1.4	1.4
Sandy loam	DO NOT USE	DO NOT USE	1.6	2.0
Loam, Silt loam, Silt	DO NOT USE	1.4	2.0	2.0
All other textures	DO NOT USE	1.8	2.0	2.0

* For organic matter content between those listed, adjust the rate proportionately.

TABLE 5
PREEMERGENCE BROADCAST APPLICATION RATES PER ACRE FOR "EXTRAZINE" II 4L APPLIED ALONE ON SWEET CORN

SOIL TEXTURE DESCRIPTION	QUARTS OF EXTRAZINE II 4L — PERCENT ORGANIC MATTER IN SOIL*: LESS THAN 1%	1%	2%	3%	4%	5% & OVER
Sand, Loamy sand	DO NOT USE	1.2	1.5	2.0	2.4	3.3
Sandy loam	DO NOT USE	1.5	2.0	2.4	3.1	4.1
Loam, Silt loam, Silt	DO NOT USE	2.0	2.4	3.1	3.8	4.5
Sandy clay loam, Clay loam, Silty clay loam	DO NOT USE	2.4	2.9	3.8	4.5	5.0
Sandy clay, Silty clay, Clay	DO NOT USE	2.9	4.1	4.5	5.0	5.25
Peat or Muck	NOT RECOMMENDED					

* For organic matter content between those listed, adjust the rate proportionately.

EXTRAZINE II 4L Alone

Apply EXTRAZINE II 4L at the proper rate for soil texture and organic matter indicated in Table 5.

EXTRAZINE II 4L PLUS "LASSO" 4EC

Use EXTRAZINE II 4L at the proper rate for the soil texture and organic matter shown in Table 6 plus 2 quarts per acre of "Lasso". (Use 2.5 quarts "Lasso" on clay soils containing 5 percent organic matter and over.)

EXTRAZINE II 4L PLUS "SUTAN+" 6.7E, "GENATE PLUS" OR "ERADICANE" 6.7E

Do not use combinations with "Sutan+", "Genate Plus" or "Eradicane" in New Jersey.

Use EXTRAZINE II 4L at the proper rate for the soil texture and organic matter shown in Table 6 plus 1.8 quarts per acre of "Sutan+", "Genate Plus" or "Eradicane" for control of many annual grasses and broadleaf weeds. (Use 2.4 quarts of "Sutan+" or "Eradicane" on loam soils containing 5 percent or more organic matter, and clay loams and clays containing 4 percent or more organic matter.) Do not use on sands and loamy sands having less than 1 percent organic matter nor on the light sandy soils of eastern coastal states. Do not use on corn grown for seed.

Apply before planting. Incorporate the mixture immediately upon application using power-driven cultivation equipment set for 2-3 inch depth, or tandem disc set to cut about 4 inches deep while operating at 4-6 mph. For thorough mixing, disc in two directions (cross disc), and follow with a harrow, drag, or other leveling device. Prior to the second discing, readjust the disc to prevent cutting deeper than 4 inches. EXTRAZINE II 4L may be applied preemergence as an overlay over previously incorporated "Sutan+", "Genate Plus" or "Eradicane", if desired.

Existing stands of quackgrass, purple and yellow nutsedge must be turned under and thoroughly chopped up prior to chemical treatments.

TABLE 6
PREEMERGENCE BROADCAST APPLICATION RATES PER ACRE FOR "EXTRAZINE" II 4L USED IN TANK MIX COMBINATIONS WITH "LASSO", "SUTAN+", "GENATE PLUS", "ERADICANE" OR "DUAL" ON SWEET CORN

SOIL TEXTURE DESCRIPTION	QUARTS OF EXTRAZINE II 4L					
	PERCENT ORGANIC MATTER IN SOIL*					
	LESS THAN 1%	1%	2%	3%	4%	5% & OVER
Sand, Loamy sand	DO NOT USE	0.8	1.3	1.5	1.6	2.0
Sandy loam	DO NOT USE	1.2	1.5	1.7	2.0	2.2
Loam, Silt loam, Silt	DO NOT USE	1.5	1.8	2.0	2.2	2.7
Sandy clay loam, Clay loam, Silty clay loam	DO NOT USE	1.8	2.0	2.2	2.7	2.9
Sandy clay, Silty clay loam	DO NOT USE	2.0	2.5	2.7	2.9	3.1
Peat or Muck	NOT RECOMMENDED					

* For organic matter content between those listed, adjust the rate proportionately.
** Do not use in the light sandy soils of the Atlantic Coastal Plain.

Additional weeds controlled by "Sutan+", "Genate Plus" or "Eradicane" combinations:

Grasses:

Sandbur	Shattercane (Wild Cane)*
Texas Panicum	Quackgrass (Eradicane. Only)
	Wild Proso Millet* (Eradicane. only)

Perennial Weeds

Yellow Nutsedge (Nutgrass)	Purple Nutsedge (Nutgrass)

* Suppression only—refer to "Sutan+", "Genate Plus" or "Eradicane" label for appropriate supplemental cultural and tillage practices.

For fields with moderate to heavy infestations of these weeds refer to the "Sutan+", "Genate Plus" or "Eradicane" labels for appropriate higher rates.

"EXTRAZINE" II 4L PLUS "DUAL" 8E

Use EXTRAZINE II 4L at the proper rate for soil texture and organic matter shown in Table 6. Use "Dual" as follows:

Soil Texture	Broadcast Rate Per Acre for "Dual" 8E
Coarse Sand, Loamy sand, Sandy loam	1.25 - 1.5 pints
Medium Loam, Silt loam, Silt	1.5 - 2.0 pints
Fine Sandy clay loam, Silty clay loam, Clay loam, Sandy clay, Silty clay, Clay	1.5 - 2.5 pints

The low end of the rate range should be used for lowest organic matter soils and the rate increased proportionately as the organic matter increases. Soils containing 4 percent organic matter or more require the highest rate shown for that soil texture. Refer to the "Dual" label for precautions or additional information on rotational crops.

STORAGE AND DISPOSAL

Storage: Do not contaminate water, food, or feed by storage or disposal. Do not use or store around the home environment. Avoid contact with water. In case of spill or leak, soak up with sand, earth or synthetic absorbent. Do not use alkaline absorbents, and dispose of wastes in compliance with local, state and Federal regulations.

Pesticide Disposal: Pesticide, spray mixture, or rinsate that cannot be used according to label instructions must be disposed of according to applicable Federal, state, or local procedures.

Container Disposal: Triple rinse (or equivalent). Then offer for recycling or reconditioning, or puncture and dispose of in a sanitary landfill, or by incineration, or, if allowed by state and local authorities, by burning. If burned, stay out of smoke.

NOTICE OF WARRANTY

Du Pont warrants that this product conforms to the chemical description on the label thereof and is reasonably fit for the purposes stated on such label only when used in accordance with the directions under normal use conditions. It is impossible to eliminate all risks inherently associated with the use of this product. Crop injury, ineffectiveness or other unintended consequences may result because of such factors as weather conditions, presence of other materials, or the manner of use or application, all of which are beyond the control of Du Pont. In no case shall Du Pont be liable for consequential, special or indirect damages resulting from the use or handling of this product. All such risks shall be assumed by the buyer. DU PONT MAKES NO WARRANTIES OF MERCHANTABILITY OR FITNESS FOR A PARTICULAR PURPOSE NOR ANY OTHER EXPRESS OR IMPLIED WARRANTY EXCEPT AS STATED ABOVE.

BLADEX, CONQUEST, EXTRAZINE.
Trademarks of E.I. Du Pont de Nemours and Co. (Inc.)
"Banvel". Trademark of Sandoz Crop Protection Corp.
"Dual". Trademark of CIBA-Geigy Corp.
"Eradicane". Trademark of ICI Americas, Inc.
"Genate Plus". Trademark of PPG Industries, Inc.
"Lasso", "Roundup". Trademarks of Monsanto Co.
"Ortho X-77". Trademark of Chevron Chemical Co.
"Princep", "Caliber". Trademarks of CIBA-Geigy Corp.
"Sutan+". Trademark of ICI Americas, Inc.

AG-3194 8068 5/27/88

This is a specimen label. DuPont is not responsible for the accuracy of the information contained herein. As labels are subject to revision, always carefully read and follow the label on the product container.

EXTRAZINE® II DF

HERBICIDE
DISPERSIBLE GRANULE

RESTRICTED USE PESTICIDE

For retail sale to and use only by Certified Applicators or persons under their direct supervision and only for those uses covered by the Certified Applicator's certification. See Precautionary Statements for reasons this product is classified Restricted Use.

EXTRAZINE II DF is a water dispersible granule.

ACTIVE INGREDIENTS	BY WEIGHT
Cyanazine: 2-[[4-chloro-6-(ethylamino)-s-triazin-2-yl]amino]-2-methylpropionitrile	67.5%
Atrazine: 2-chloro-4-(ethylamino)-6-(isopropylamino)-s-triazine	21.4%
Related compounds	1.1%
INERT INGREDIENTS	10.0%
TOTAL	100.0%

This product contains 90 percent by weight active ingredients.

EPA Reg. No. 352-497

KEEP OUT OF REACH OF CHILDREN
WARNING
AVISO

PRECAUCION AL USUARIO: Si usted no lee ingles, no use este producto hasta que la etiqueta haya sido explicada ampliamente.

STATEMENT OF PRACTICAL TREATMENT

IF SWALLOWED: Call a physician or poison control center. Drink 1 or 2 glasses of water and induce vomiting by touching back of throat with finger. Do not induce vomiting or give anything by mouth to an unconscious person.

IF IN EYES: Flush with plenty of water. Get medical attention if irritation persists.

IF ON SKIN: Wash with plenty of soap and water.

IF INHALED: Remove victim to fresh air. If not breathing, give artificial respiration, preferably mouth-to-mouth. Get medical attention.

For medical emergencies involving this product, call 1-800-441-3637.

PRECAUTIONARY STATEMENTS

HAZARDS TO HUMANS AND DOMESTIC ANIMALS

WARNING: May be fatal if swallowed. Harmful if inhaled or absorbed through the skin. Causes temporary eye injury.

This product may be hazardous to your health. This product has been classified "Restricted Use" because, at doses which caused serious maternal illness in laboratory animals, birth defects were present. Use of protective clothing and equipment and following the precautions below can reduce risk.

Avoid breathing spray mist. Avoid contact with skin, eyes, or clothing. Do not get in eyes or on clothing. Wear a face shield when mixing and loading. Wash thoroughly with soap and water after handling and before eating or smoking.

Wear long trousers and long-sleeved clothing when applying this product. Wear chemical resistant gloves extending above the wrist, a chemical resistant apron, long trousers, and long-sleeved clothing when mixing or loading or when adjusting, repairing or cleaning equipment. Protective gloves must be washed with soap and water after use and before removing from the hands. Remove contaminated clothing and wash before reuse. Contaminated clothing should be laundered separately from household laundry to prevent cross-contamination of other laundry. Heavily contaminated or drenched clothing and protective equipment must be discarded or destroyed in accordance with State and local regulations.

Do not apply this product in such a manner as to directly, or through drift, expose workers or other persons. The area being treated must be vacated by unprotected persons.

Keep out of reach of domestic animals, particularly cattle. Consumption of this product, spray solutions, or water contaminated with product can result in serious illness or possible death of bovines.

ENVIRONMENTAL HAZARDS

Cyanazine and Atrazine, the active ingredients of EXTRAZINE II, are pesticides which can move (seep or travel) through soil and can contaminate groundwater which may be used as drinking water. Cyanazine and Atrazine have been found in groundwater as a result of agricultural use. Users are advised not to apply EXTRAZINE II where the water table (groundwater) is close to the surface and where the soils are very permeable (i.e. well drained soils such as loamy sands). Your local agricultural agencies can provide further information on the type of soil in your area and the location of groundwater.

Atrazine is toxic to aquatic invertebrates. Do not apply directly to water or wetlands. Do not apply where runoff is likely to occur. Do not apply when weather conditions favor drift from treated areas. Runoff and drift from treated areas may be hazardous to aquatic organisms in neighboring areas. Do not contaminate water by cleaning of equipment or disposal of wastes.

In case of significant spill, call CHEMTREC 1-800-424-9300.

DIRECTIONS FOR USE

It is a violation of Federal law to use this product in a manner inconsistent with its labeling.

GENERAL INFORMATION

EXTRAZINE II DF Herbicide is a selective herbicide for the control of annual grasses and broadleaf weeds in field corn, popcorn, and sweet corn.

Consult your local Agricultural Extension Agent for help in determining soil type, organic matter content, and the most appropriate herbicide rate for local conditions.

Where surfactants or emulsible vegetable oils are added to EXTRAZINE II DF for over the top of corn postemergence applications, use on field corn only.

Do not apply this product in irrigation water with any kind of irrigation system.

Do not apply this product with aerial application equipment.

EXTRAZINE II DF is not effective when used preemergence on peat or muck soils. Do not use EXTRAZINE II DF on sands or loamy sands (soils consisting of more than 70% sand) containing less than 1% organic matter.

Weather Effects: As a preemergence herbicide, EXTRAZINE II DF is active mainly through the roots, and, therefore, its effect on weeds is dependent on adequate rainfall to move the herbicide into the root zone. Moisture should be sufficient to thoroughly wet the soil throughout the zone where weed seeds may germinate and enough to make the soil too wet to cultivate. Rotary hoeing or shallow cultivation is recommended for those applications which are not incorporated at the time of treatment, if adequate rainfall or sprinkler irrigation has not occurred within about ten days after application of EXTRAZINE II DF.

Heavy rainfall between planting and crop emergence may cause crop injury or stand loss. Rainfall tends to cause excessive concentrations of herbicide in the seed furrow, resulting in possible crop injury. Level deep planter marks or seed furrows before application.

Under conditions which delay weed germination, such as low temperatures, lack of soil surface moisture, or when germination extends over a long period, the effectiveness of the herbicide may be impaired. Rotary hoeing, shallow cultivation or a postemergence herbicide treatment may be of benefit under these circumstances.

If the crop is cultivated, tillage should be shallow to minimize herbicide dilution in the soil. Should the crop stand be lost due to adverse weather conditions, insects, etc., the field can be replanted the same season to corn or sorghum.

To enhance weed control in areas of less than 25 inches of rainfall or where long dry periods are common, these treatments may require shallow incorporation with a tool such as a field cultivator operated from 5-7 mph. Incorporation should not be more than three inches deep to keep from burying the herbicide. A spike-toothed harrow, deep tillage disk or rolling basket device is not recommended for incorporating EXTRAZINE II.

When applied as a post-emergence herbicide, EXTRAZINE II DF is also active through foliage as well as through the roots. Yellowing and/or stunting of the crop may result from this treatment, particularly if cold, adverse growing conditions occur after application. Extended or extreme cold and wet conditions may reduce stands. Do not apply EXTRAZINE II postemergence to a crop that is damaged or growing under stress.

OBSERVE ALL CAUTIONS AND LIMITATIONS ON LABELING OF ALL PRODUCTS USED IN MIXTURES.

Rotational crops: (1) Plant only corn, sorghum, or soybeans the year following the use of this mixture. (2) If soybeans are to be planted, injury may occur. (3) If applied after June 10, do not rotate with crops other than corn or sorghum the next year or injury may occur. (4) In the high plains and inter-mountain areas of the West where rainfall is sparse and erratic or where irrigation is required, use only when corn or sorghum is to be planted the following year, or a crop of corn or sorghum not treated with this mixture or atrazine is to precede other rotational crops. (5) Small grains may be planted 15 months following treatment. (6) All other crops may be planted 18 months after application.

Triazine Resistant Weeds: In fields where triazine resistant biotypes of weeds have been identified, EXTRAZINE II should be used in combination with or in sequence with other registered non-triazine herbicides. (Triazine resistant biotypes of Kochia and Pigweed have been identified in some fields in the Western Great Plains and triazine resistant biotypes of Pigweed and Lambsquarters have been identified in some fields in various states.) Consult with appropriate state agricultural extension service representatives for specific recommendations.

APPLICATION DIRECTIONS

GENERAL MIXING AND SPRAYING DIRECTIONS

This product may not be applied by means of chemigation or aerial application.

Use sufficient agitation to ensure that the EXTRAZINE II DF is completely dispersed and in uniform suspension prior to application or tank mixing with other formulations.

The following general mixing instructions are recommended:

I. General

1. Unless otherwise specified, use at least 7.5 gallons of water per acre for soil applications and at least 15 gallons of water per acre for foliar applications for all applications with ground equipment. NOTE: Sufficient carrier must be used to assure uniform application. Follow label requirements of all products used in tank mix combinations.
2. A nitrogen solution or complete liquid fertilizer may replace all or part of the water as a carrier for preemergence or preplant application on corn. Do not apply fertilizer mixtures after crop emerges, as injury may occur.
3. Always check the tank mix compatibility (TMC) of this or any other formulation before mixing with liquid fertilizer carriers or other formulations. A simple but generally reliable TMC evaluation procedure has been provided for your use in Step II of these mixing instructions.
4. Start with thoroughly clean equipment. (See the labels of the previous compounds used for cleaning instructions.)
5. Fill the tank at least 1/2 full with carrier. Start and maintain consistent agitation through all mixing and spraying procedures. Make sure the agitation system is working properly and creates a rippling or rolling action on the liquid surface.
6. Slowly add the recommended amount of EXTRAZINE II DF to the tank or inductor.
7. Fill the tank to 75 percent capacity with carrier. Filling and bypass lines should be kept below liquid surface. Increase tank agitation if necessary to maintain surface action.
8. When desired, appropriate emulsible crop oil, crop oil concentrate, or other tank mix formulations should be added at this time. Pre-slurry these added ingredients before addition, if the compatibility test shows it to be necessary.
9. Complete filling tank, maintaining sufficient agitation at all times to ensure surface action. This applies to both spray and nurse tanks.

10. Tank mixtures should always be applied immediately after preparation. If, for any reason, this is not possible, assure that sufficient agitation has been provided to re-mix all products and check for complete re-suspension prior to application.
11. Empty tank as completely as possible before refilling to prevent buildup of oil or emulsible concentrate residues when tank mixing with these formulations. Always maintain agitation to avoid separation.
12. If an oil or emulsible concentrate film starts to build up after using these formulations, drain and clean the tank with strong detergent solution or appropriate solvent.
13. It is recommended that the sprayer be thoroughly cleaned by flushing with a detergent solution at the end of each work day when any emulsible crop oil, crop oil concentrate, or other emulsible formulation has been used either alone or in tank mix combinations with other pesticide formulations. This precaution will ensure a clean sprayer and continued trouble-free operation.

II. Tank Mix Compatibility Evaluation Procedure

1. Add 1 pint of carrier liquid to each of 2 one-quart jars. Mark 1 quart jar "with" and the other "without."
2. Add 1/4 teaspoon of a suitable compatibility agent (1/4 teaspoon/pint = 2 pints/100 gallons of carrier) to the jar marked "with", cap the jar and shake gently for 5 to 10 seconds to mix.
3. Add the appropriate amount of herbicide to both jars, cap each jar and shake gently for 5 to 10 seconds to mix. If problems are encountered in mixing wettable powder or dry flowable formulations into a liquid fertilizer, pre-slurry these formulations in water prior to their addition to the liquid fertilizer and proceed with the test. The following chart has been provided to assist in selecting the approximate EXTRAZINE II DF use rate for this evaluation.

JAR TEST FOR EXTRAZINE II DF COMPATIBILITY

Gallons of Liquid Carrier per acre	4	7.5	15	20	25	30
Teaspoons of EXTRAZINE II DF per pint of Liquid Carrier	6.0	3.2	1.6	1.2	1.0	0.8

This chart is based on 1 pound of EXTRAZINE II DF (0.9 pounds active ingredient) per acre in the indicated carrier volumes. Intended field use rates are achieved by varying the amount of EXTRAZINE II DF. (e.g., for a field use rate of 3 pounds of EXTRAZINE II DF in 15 gallons of carrier per acre, add 4.8 teaspoons of EXTRAZINE II DF to the quart jars containing 1 pint of carrier. Calculation: 3 pounds of EXTRAZINE II DF/15 gallons of carrier per acre = 3 × 1.6 = 4.8 teaspoons of EXTRAZINE II DF per pint of carrier.)

4. Let each jar stand one-half hour and make observations. If any separation, agglomeration, or precipitation has occurred, shake the jar again for 10 to 15 seconds, and note whether any of the following occur:
   a. Separated phases do not re-mix uniformly.
   b. Screen/nozzle plugging lumps do not disperse.
   c. Precipitate does not re-suspend readily.
   d. Precipitate sticks tenaciously to the glass.
5. If none of the above problems occur in either jar, the herbicides can, in most cases, be safely used in that carrier without a compatibility agent.
6. If problem 4.a or 4.b occurs in the jar marked "without" but does not occur in the jar marked "with", the compatibility agent should be used.
7. If problem 4.a or 4.b is seen in both jars, then the herbicides and carrier are incompatible and should not be used in the same spray tank. Alternatively, a different tank mix compatibility agent can be evaluated.
8. If problem 4.c or 4.d occurs in the jar marked "without" but does not occur in the jar marked "with", the compatibility agent should be used unless constant, thorough agitation can be maintained and immediate clean-out of the spray system is performed.
9. If problem 4.c or 4.d is seen in the jar marked "with," the user proceeds with mixing and application at his own risk should the agitation in the system be insufficient or curtailed.
10. Those mixtures defined as compatible in this test should then be mixed for use as indicated in Steps 1 through 12 of the general mixing instructions listed above.

If a test such as outlined indicates that components of a proposed mix are compatible, the applicator still has the responsibility of combining materials in sequence to the spray tank in accordance with directions prescribed on the label of the herbicides or pesticides involved.

Tests have indicated that compatibility agents, noted below by the various tank mix combinations, may give improved compatibility in liquid fertilizers.

Tank Mix Combination	Compatibility Agents
EXTRAZINE II DF/Lasso. (Liquid Fertilizer Grade)	Probably not needed in 28-0-0, 10-34-0. Compex may help in others.
EXTRAZINE II DF/Genate Plus. Sutan+. 6.7E	Probably not needed in 28-0-0. Incompatible in 10-34-0. Unite, Spray-Mate, Kem-Link, may help in others.
EXTRAZINE II DF/Dual. 8E	Probably not needed in 28-0-0. Unite, Spray-Mate, Ivory Liquid may help in others.

III. Application Equipment

1. Use application equipment fitted with nozzles that provide accurate and uniform coverage. Be certain that nozzles are uniformly spaced and the same size. Calibrate sprayer before use and re-check frequently during use whenever possible.
2. Use a pump with capacity to:
   a. Maintain 35 to 40 psi at the nozzles.
   b. Provide sufficient agitation in tank to keep mixture in suspension.
   c. Provide a minimum of 20 percent bypass at all times.
3. Use centrifugal pumps which provide sufficient shear action for dispersing and mixing this product. The pump should provide a minimum of 10 gallons/minute/100-gallon tank size circulated through the jets of a correctly-positioned sparger tube.
4. Use screens to protect the pump and to prevent nozzles from clogging. Screens placed on suction side of pump should be 10 to 16 mesh. Do not place a screen in the recirculation line. Use a 40- to 50-mesh screen between the pump and boom and, where required, 50-mesh screens at the nozzles. Check your equipment manufacturer's literature for specific recommendations.

FERTILIZER IMPREGNATION

EXTRAZINE II DF Herbicide may be applied when coated on or impregnated in dry granular fertilizer for early preplant, preemergence or preplant incorporated weed control in field corn. All recommendations, cautions and special precautions on this label must be followed along with state regulations relating to dry bulk fertilizer blending, impregnating and labeling.

General Blending Directions

EXTRAZINE II DF may be coated on or impregnated in dry bulk fertilizers using tower blenders, rotary drum blenders or blending augurs or conveyors. DO NOT impregnate EXTRAZINE II or tank mixes containing EXTRAZINE II on or in fertilizers containing Ammonium Nitrate, Potassium Nitrate, or Sodium Nitrate. Do not use on straight limestone since absorption will not be achieved. Fertilizer blends containing limestone can be impregnated when using EXTRAZINE II alone. Use a minimum of 200 lbs. and a maximum of 450 lbs. per acre of dry fertilizer.

Use equipment that will give a uniform distribution of the herbicide throughout each batch of impreganted fertilizer. Non-uniform impregnation can cause crop injury or unsatisfactory performance.

A. EXTRAZINE II DF may be used as the only herbicide for impregnation.
1. Add EXTRAZINE II DF to 1/2 the total fertilizer volume required.
2. Spray one gallon of water (to break down 90DF) and one gallon of diesel fuel (to prevent evaporation and crusting) per ton of fertilizer and allow to mix thoroughly. NOTE: If the fertilizer is dusty, add the diesel fuel before adding the herbicide.
3. Add remaining fertilizer and mix thoroughly. (3 minutes or more for rotary drum blenders.)
4. Add 2-3% Ag-Sorb or 1-2% MP-79 drying agent (or a suitable amount of another effective drying agent) to insure a spreadable herbicide/fertilizer mixture. The need for a drying agent is determined by the wetness of the fertilizer batch. Wetness can change with humidity, nitrogen content, fertilizer types, fertilizer rates and herbicide rates.

B. EXTRAZINE II DF may be used in tank mixes with other dry herbicides including Princep Caliber 90. Follow the procedure as above in "A".

C. EXTRAZINE II DF may be used in tank mixes where an EC or other liquid herbicide acts as the sticking agent. This may eliminate the need for water and/or diesel fuel.
1. While fertilizer is blending, add the EXTRAZINE II DF. Experience has shown that this will provide the most consistent performance due to the grinding action of the fertilizer on the 90DF.
2. Spray in the EC herbicide and mix thoroughly. (3 minutes or more for rotary drum blenders).
3. Add drying agent to insure a spreadable herbicide/fertilizer mixture. Usually less drying agent is required when using EXTRAZINE II DF.

D. Pre-slurried EXTRAZINE II DF can be used alone or in a tank mix for impregnation. For rotary-drum mixers, the liquids can be moved into the drum using an air system or liquid pump. Do not add extra water. Add drying agent to insure a spreadable herbicide/fertilizer mixture.

(For more information on Drying Agents, Application Equipment, Calibration Guide and variations of these methods, see Du Pont Bulletin on "Fertilizer Impregnation").

CLEAN OUT

Equipment used to impregnate or apply fertilizer impregnated with EXTRAZINE II or combinations including EXTRAZINE II must be cleaned out by running at least 1,000 lbs. of fertilizer not impregnated with

EXTRAZINE II through the impregnation equipment and application equipment, if the next batch of material is to be applied to a crop for which EXTRAZINE II or a combination herbicide is not registered.

APPLICATION

Uniform application of EXTRAZINE II DF which has been impregnated in or coated on dry fertilizer is essential for satisfactory weed control and crop safety. Accurate calibration of the fertilizer applicator is necessary. Applying while turning at the ends of the fields may result in excessive application rates causing crop injury. Do not double apply across the ends or sides of the field.

Crop injury and/or poor weed control may occur where the impregnated fertilizer is not uniformly applied. Air flow or augur metered application equipment is preferred (one pass application). If other equipment is used, the recommended method of application is to apply ½ the recommended rate and overlap 50 percent to double apply by splitting the middles to obtain the best distribution pattern.

Apply immediately after impregnation. Impregnated fertilizer may become lumpy and difficult to spread if stored.

RATES AND TIMING

Use the application rates and timing shown in the appropriate sections of this label. Follow the precautions on the labels of all products used.

CORN

Weeds Controlled by EXTRAZINE II DF Alone and in Combination with Other Herbicides on Corn

Grasses

Annual bluegrass	Crabgrass	Junglerice
Annual fescues	Fall panicum	Stinkgrass (Indian lovegrass)
Annual (Italian) ryegrass	Giant foxtail	Witchgrass
Barnyardgrass(1)	Goosegrass	Yellow foxtail
Bullgrass	Green foxtail	

Broadleaves

Annual groundcherry	Florida pusley (Florida purslane)	Ragweed (Common)
Annual morningglory	Hedge mustard	Russian Thistle
Annual sedge	Jimsonweed(1)	Shepherdspurse
Black mustard	Kochia	Smallflower galinsoga
Buffalobur	Ladysthumb	Smartweed (Pennsylvania)
Buttercup (annual)	Lambsquarters	Spiny sida
Carpetweed	Mayweed	Sunflower(2) (wild, annual, common)
Cocklebur(2)	Nightshade (annual)	Tarweed cuphea (Gumweed)
Common chickweed	Pigweed(1)	Velvetleaf(1)
Common groundsel	Pineappleweed	Wild buckwheat
Common mallow	Plantain	Wild mustard
Common purslane	Poorjoe	Wild radish
Corn spurry	Prickly sida (teaweed)	Wild turnip
Curly dock (seedling)	Prostrate knotweed	
Fiddleneck	Prostrate spurge	

(1) Under conditions such as low temperatures, lack of soil surface moisture, or other factors that may cause delay in germination of the seeds, the effectiveness of EXTRAZINE II may be impaired against these weeds.

(2) The degree of control will be reduced if soil moisture and temperature conditions cause deep germination of the seed.

TABLE 1
PREEMERGENCE BROADCAST APPLICATION RATES PER ACRE FOR "EXTRAZINE" II DF APPLIED ALONE ON CORN

SOIL TEXTURE DESCRIPTION	POUNDS OF EXTRAZINE II DF — PERCENT ORGANIC MATTER IN SOIL*					
	LESS THAN 1%	1%	2%	3%	4%	5% & OVER
Sand, Loamy sand	DO NOT USE	1.4	1.7	2.2	2.8	3.6
Sandy loam	1.4	1.7	2.2	2.8	3.6	4.4
Loam, Silt loam, Silt	1.7	2.2	3.1	3.6	4.4	5.0
Sandy clay loam, Clay loam, Silty clay loam	2.2	3.1	3.6	4.4	5.0	5.3
Sandy clay, Silty clay, Clay	3.1	3.6	4.4	5.0	5.3	5.8
Peat or Muck	NOT RECOMMENDED					

*For organic matter content between those listed, adjust the rate proportionately.

PREEMERGENCE-PREPLANT INCORPORATED

Apply EXTRAZINE II treatments just before, at, or after planting but before crop has emerged. Avoid removal of treated soil from seedrow prior to or during the planting operation.

EXTRAZINE II may also be applied early prior to planting or in a split application if pre-season weed control is desired. For split applications, do not exceed the total amount of EXTRAZINE II for the soil texture and organic matter shown in Table 1. If EXTRAZINE II is applied early, more than 15 days before planting, a split application of EXTRAZINE II or some other herbicide treatment may be necessary at or after planting to provide additional length of weed control. For further information, see "Early Preplant" recommendations in the Conservation Tillage section of this label.

Rotary hoeing is recommended for preemergence applications which do not receive adequate rainfall or sprinkler irrigation to wet the top 2 inches of soil or depth of germinating weeds within about 10 days after application.

EXTRAZINE II alone or in tank mix combinations should not be incorporated more than three inches deep to keep from burying the herbicide. Single or two pass incorporation with a tool such as a field cultivator operated at 5-7 mph is acceptable. A spike-toothed harrow, deep tillage disk or rolling basket device is not recommended for incorporating EXTRAZINE II.

See the Rotational Crops statement in the General Information section of the label.

EXTRAZINE II Applied Alone

Use at the proper rate for soil texture and organic matter indicated in Table 1.

EXTRAZINE II COMBINATIONS

EXTRAZINE II plus Lasso. 4EC

Use EXTRAZINE II DF at the proper rate for the soil texture and organic matter shown in Table 2 plus 2 quarts per acre of Lasso. (Use 2.5 quarts Lasso on clay soils containing 5 percent organic matter and over).

EXTRAZINE II plus Sutan+. 6.7E/Genate Plus. or Eradicane. 6.7E

Use EXTRAZINE II DF at the proper rate for the soil texture and organic matter shown in Table 2 plus 1.8 quarts per acre of Sutan+/Genate Plus or Eradicane for control of many annual grasses and broadleaf weeds. (Use 2.4 quarts of Sutan+/Genate Plus or Eradicane on loam soils containing 5 percent or more organic matter.) Do not use on sands and loamy sands of less than 1 percent organic matter in the light sandy soils of eastern coastal states. Do not use on corn seed stock.

Apply before planting. Incorporate the mixture immediately upon application using power-driven cultivation equipment set for 2 to 3 inch depth, or a tandem disc set to cut to a depth of about 4 inches while operating at 4 to 6 mph. For thorough mixing, disc in two directions (cross disc), and follow with a harrow, drag, or other leveling device. Prior to the second discing, readjust the disc to prevent cutting deeper than 4 inches. EXTRAZINE II DF may be applied preemergence as an overlay over previous incorporated Sutan+/Genate Plus or Eradicane if desired.

Existing stands of quackgrass, purple and yellow nutsedge must be turned under and thoroughly chopped up prior to chemical treatments.

Additional weeds controlled by Sutan+/Genate Plus or Eradicane combinations:

Grasses

Sandbur	Shattercane (Wild Cane)*
Texas Panicum	Quackgrass (Eradicane only)
	Wild Proso Millet*(Eradicane only)

Perennial Weeds

Yellow Nutsedge (nutgrass)	Purple Nutsedge (nutgrass)

*Suppression only—refer to Sutan+/Genate Plus or Eradicane label for appropriate supplement cultural and tillage practices.

For fields with moderate to heavy infestations of these weeds, refer to the Sutan+/Genate Plus or Eradicane labels for appropriate higher rates.

TABLE 2
PREEMERGENCE BROADCAST APPLICATION RATES PER ACRE FOR EXTRAZINE II DF USED IN TANK-MIX COMBINATIONS WITH LASSO, SUTAN+, GENATE PLUS, ERADICANE, OR DUAL ON CORN

SOIL TEXTURE DESCRIPTION	POUNDS OF EXTRAZINE II DF — PERCENT ORGANIC MATTER IN SOIL*					
	LESS THAN 1%	1%	2%	3%	4%	5% & OVER
Sand, Loamy sand	0.7**	0.8	1.4	1.7	1.9	2.2
Sandy loam	0.8	1.4	1.7	1.9	2.2	2.5
Loam, Silt loam, Silt	1.4	1.7	1.9	2.2	2.5	2.8
Sandy clay loam, Clay loam, Silty clay loam	1.7	1.9	2.2	2.5	2.8	3.1
Sandy clay, Silty clay, Clay	1.9	2.2	2.5	2.8	3.1	3.3
Peat or Muck	NOT RECOMMENDED					

*For organic matter content between those listed, adjust the rate proportionately.
**Do not use on the light sandy soils of the Atlantic Coastal Plain.

EXTRAZINE II plus Dual 8E

Use EXTRAZINE II DF at the proper rate for soil texture and organic matter shown in Table 2. Use Dual as follows:

SOIL TEXTURE	BROADCAST RATE PER ACRE FOR DUAL 8E
Coarse Sand, Loamy sand, Sandy loam	1.25—1.5 pints
Medium Loam, Silt loam, Silt	1.5—2.0 pints
Fine Sandy clay loam, Silty clay loam, Clay loam, Sandy clay, Silty clay, Clay	1.5—2.5 pints

The low end of the rate range should be used for lowest organic matter soils and the rate increased as organic matter increases to a point that soils containing 4 percent organic matter or more require the highest rate shown for that soil texture. Refer to the Dual label for additional precautions on rotational crops.

CONSERVATION TILLAGE WEED CONTROL

Early Preplant (EPP)

EXTRAZINE II DF may be used for Early Preplant or Preemergence weed control for land going into the production of corn under conservation tillage programs. Complete any planned early spring tillage prior to application. Apply herbicide treatment before weeds germinate or before weed seedlings are more than 3 inches tall. Tillage after application may reduce the effectiveness of the herbicide treatment. A nitrogen solution or complete fertilizer solution may replace all or part of the water as a carrier. The spray gallonage and spray boom design must be adequate to give thorough uniform coverage of the weed foliage. Follow label requirements of all products used in tank mix combinations.

EXTRAZINE II DF alone: Apply 15 to 30 days prior to planting. Use the proper rate for soil texture and organic matter indicated in Table 1. Where heavy crop residues exist, the rates shown in Table 1 should be increased by 25 percent.

EXTRAZINE II DF plus Princep 4L or EXTRAZINE II DF plus Princep Caliber 90: Apply 30 days or more prior to planting. Use the proper rate of EXTRAZINE II DF for the soil texture and organic matter indicated in Table 1 and add 1 quart/acre of Princep 4L or 1.11 pounds/acre of Princep Caliber 90.

Burndown of Existing Weeds

Combinations with 2,4-D: Where broadleaf weeds are present at the time of application, add 2,4-D LV Ester at 1-1/3-2 pts./A. (6 lbs./gal.) or 2-3 pts./A. (4 lbs./gal.) (or 2,4-D Amine at recommended rates) plus Ortho X-77. surfactant at 1 qt./100 gals. of diluted spray, or other suitable nonionic surfactant at its recommended rate. When 2,4-D Ester is tank mixed with EXTRAZINE II as described, these additional weeds will be controlled: Buckwheat, Dandelion, Dock, Marestail, Pennycress, Prickly lettuce, and Tansy mustard.

Combinations with Paraquat Products and Roundup: When grasses are present and/or when existing weeds exceed 3 inches in height, add paraquat 1-2 pts./A. (2 lb/gal) or 1.3-2.7 pts./A. (1.5 lb/gal) or 2 pts./A. of Roundup to the above mixture. Well established weeds over 6 inches tall may not be well controlled. With paraquat products use 20-40 gal./A. of spray mixture by ground rig. (Use the higher gallonages for heavy infestations of weeds.) Add 1 to 2 pints of a non-ionic surfactant per 100 gallons of spray. Using fertilizer solutions containing phosphates as a carrier will reduce the activity of paraquat products.

Sequential Treatments: Depending upon the application timing and the weather conditions following the EPP application, a sequential herbicide treatment may be necessary to provide additional length of weed control. This may be a postemergence treatment with BLADEX, CONQUEST, EXTRAZINE, EXTRAZINE II or some other herbicide treatment applied at or after planting. If desired, 1-1/2-2 pts./A. of Dual or 2 qts./A. of Lasso may be tank mixed with the EXTRAZINE II EPP treatment or applied preemergence at planting.

Rotational Crops: Refer to the General Information section of this label and to the labels of all products used in combination or in sequence.

At Planting

EXTRAZINE II DF applied alone or in combination with Lasso or Dual according to the following directions will kill most existing small weeds and suppress many emerged perennial weeds when corn is planted into no-till stalk ground (corn, sorghum), stubble ground (soybean, small grains), and any minimum-till situation. This treatment then provides residual control of annual weeds as in conventional tillage.

Apply EXTRAZINE II DF alone or with other products according to the directions for those treatments in the Preemergence section of the label. Where heavy crop residues exist, the EXTRAZINE II rate shown in Table 1 should be increased by 25 percent. Add 2,4-D LV Ester at 1/2-1 pt./A.(6 lbs./gal.) or 0.75-1.5 pt./A.(4 lbs./gal.) (or 2,4-D Amine at recommended rates). Add the 2,4-D to the spray tank last. Use a minimum of 25 gals./A. of carrier. Complete spray coverage of the weeds is essential for best performance.

Nitrogen solutions and complete liquid fertilizers are the preferred carriers for this treatment because they aid in the burndown of existing weeds. Add Ortho X-77. surfactant at 1-2 qts. per 100 gals. of diluted spray, or other suitable surfactant at its recommended rate. If water is used as a carrier, crop oil concentrate may be used as an adjuvant. Apply before weeds exceed 3 inches in height. For control of existing alfalfa, add 1/3-1/2 pt./A. of Banvel. to the spray mixture. Apply before the alfalfa exceeds 6 inches in height.

For fields with existing sod grasses such as orchardgrass, bromegrass, rye or timothy, or when very dry conditions exist, or when existing weeds exceed 3 inches in height add paraquat to the tank-mix. Use 2 pints/A. (2 lbs/gal) or 2.7 pints/A. (1.5 lbs/gal) paraquat in combination with EXTRAZINE II as described above in this section, except the 2,4-D may be omitted, if desired. Do not apply paraquat in suspension type liquid fertilizer.

POSTEMERGENCE

Under dry, arid conditions of low humidity and the absence of dew formation at night, add a surfactant, such as Ortho X-77, or an emulsible vegetable (EV) oil suitable for use on growing corn at its recommended rate. Do not use petroleum-based crop oils. Addition of a surfactant or EV oil is not recommended under moist, rainy conditions and when dew forms at night as injury may occur.

Do not apply this treatment under cold, wet, weather conditions or to corn growing under stress caused by weather, insects, disease, etc. Yellowing of the corn may result from this treatment, particularly if cold, adverse growing conditions occur after application. Extended or extreme cold and wet conditions may reduce stands.

Do not apply postemergence on popcorn, sweet corn or corn grown for seed.

EXTRAZINE II Applied Alone

Use EXTRAZINE II DF at the proper rate for the soil texture and organic matter shown in Table 3 or 4. Use rates shown in Table 3 if BLADEX, CONQUEST, EXTRAZINE or EXTRAZINE II has not been applied to the soil this season. Use rates shown in Table 4 if BLADEX, CONQUEST, EXTRAZINE or EXTRAZINE II has been applied to the soil this season. This treatment may be used on peat or muck soils for burndown and suppression of existing weeds but will not provide residual control. Apply from crop emergence through the four-leaf stage of corn growth before weeds exceed about 1 1/2 inches in height. Do not apply over the top of corn if the fifth leaf is visible. Apply in water only. Do not spray emerged corn plants in a liquid fertilizer carrier or in tank mix combinations with EC formulation herbicides.

Rotational Crops: Refer to the General Information section of the label.

TABLE 3 POSTEMERGENCE BROADCAST APPLICATION RATES PER ACRE FOR EXTRAZINE II DF ON CORN NO PRIOR APPLICATION OF BLADEX, CONQUEST, EXTRAZINE OR EXTRAZINE II				
	POUNDS OF EXTRAZINE II DF			
	PERCENT ORGANIC MATTER IN SOIL*			
SOIL TEXTURE	LESS THAN 1%	1%	2%	OVER 2%
Sand, Loamy sand	DO NOT USE	1.3	1.8	2.2
Sandy loam	1.3	1.8	2.2	2.2
Loam, Silt loam, Silt	1.8	2.2	2.2	2.2
All other textures	2.2	2.2	2.2	2.2

*For organic matter content between those listed, adjust the rate proportionately.

TABLE 4 POSTEMERGENCE BROADCAST APPLICATION RATES PER ACRE FOR EXTRAZINE II DF ON CORN BLADEX, CONQUEST, EXTRAZINE OR EXTRAZINE II USED IN PRIOR APPLICATION				
	POUNDS OF EXTRAZINE II DF			
	PERCENT ORGANIC MATTER IN SOIL*			
SOIL TEXTURE	LESS THAN 1%	1%	2%	OVER 2%
Sand, Loamy sand	DO NOT USE	DO NOT USE	1.5	1.5
Sandy loam	DO NOT USE	DO NOT USE	1.75	2.2
Loam, Silt loam, Silt	DO NOT USE	1.5	2.2	2.2
All other textures	DO NOT USE	2.0	2.2	2.2

*For organic matter content between those listed, adjust the rate proportionately.

SWEET CORN

EXTRAZINE II DF may be applied preemergence or preplant incorporated for the control of annual grasses and broadleaf weeds in sweet corn.

NOTE: EXTRAZINE II may cause injury or stand loss on new or "super sweet" varieties of sweet corn. Consult with Agricultural Extension Agencies and sweetcorn seed suppliers about the sensitivity of new varieties to potential injury.

Apply EXTRAZINE II treatments just before, at or after planting but before crop has emerged. Avoid removal of treated soil from seedrow prior to or during the planting operation. Do not apply postemergence to sweet corn. Rotary hoeing is recommended for preemergence applications which do not receive adequate rainfall or sprinkler irrigation to wet the top 1½ to 2 inches of soil within about 10 days after application.

If an EXTRAZINE II mixture is to be incorporated, except as noted, single, or two pass incorporation is acceptable. Care should be taken to incorporate the EXTRAZINE II mixture no deeper than the top two inches of soil.

Rotational Crops: (1) Plant only corn, sorghum or soybeans the year following the use of this mixture. (2) If soybeans are to be planted, injury may occur due to the carryover of Atrazine. (3) If applied after June 10, do not rotate with crops other than corn or sorghum the next year or injury may occur. (4) Small grains may be planted 15 months following treatments. (5) All other crops may be planted 18 months after application.

EXTRAZINE II Alone

Apply EXTRAZINE II DF at the proper rate for soil texture and organic matter indicated in Table 5.

EXTRAZINE II plus LASSO 4EC

Use EXTRAZINE II DF at the proper rate for the soil texture and organic matter shown in Table 6 plus 2 quarts per acre of Lasso. (Use 2.5 quarts Lasso. on clay soils containing 5 percent organic matter and over.)

TABLE 5 PREEMERGENCE BROADCAST APPLICATION RATES PER ACRE FOR "EXTRAZINE" II DF APPLIED ALONE ON SWEET CORN						
	POUNDS OF EXTRAZINE II DF					
	PERCENT ORGANIC MATTER IN SOIL*					
SOIL TEXTURE DESCRIPTION	LESS THAN 1%	1%	2%	3%	4%	5% & OVER
Sand, Loamy sand	DO NOT USE	1.3	1.6	2.1	2.6	3.5
Sandy loam	DO NOT USE	1.6	2.1	2.6	3.3	4.4
Loam, Silt loam, Silt	DO NOT USE	2.1	2.6	3.3	4.1	4.9
Sandy clay loam, Clay loam, Silty clay loam	DO NOT USE	2.6	3.1	4.1	4.9	5.4
Sandy clay, Silty clay, Clay	DO NOT USE	3.1	4.4	4.9	5.4	5.8
Peat or Muck	NOT RECOMMENDED					

*For organic matter content between those listed, adjust the rate proportionately.

TABLE 6 PREEMERGENCE BROADCAST APPLICATION RATES PER ACRE FOR "EXTRAZINE" II DF USED IN TANK MIX COMBINATIONS WITH "LASSO", "SUTAN+", "GENATE PLUS", "ERADICANE" OR "DUAL" ON SWEET CORN						
	POUNDS OF EXTRAZINE II DF					
	PERCENT ORGANIC MATTER IN SOIL*					
SOIL TEXTURE DESCRIPTION	LESS THAN 1%	1%	2%	3%	4%	5% & OVER
Sand, Loamy sand	DO NOT USE	0.9	1.3	1.6	1.8	2.2
Sandy loam	DO NOT USE	1.3	1.6	1.8	2.2	2.4
Loam, Silt loam, Silt	DO NOT USE	1.6	1.8	2.2	2.4	2.9
Sandy clay loam, Clay loam, Silty clay loam	DO NOT USE	2.0	2.2	2.4	2.9	3.1
Sandy clay, Silty clay loam	DO NOT USE	2.2	2.7	2.9	3.1	3.3
Peat or Muck	NOT RECOMMENDED					

*For organic matter content between those listed, adjust the rate proportionately.

EXTRAZINE II plus SUTAN+6.7E, GENATE PLUS OR ERADICANE 6.7E

Do not use combinations with "Sutan+", "Genate Plus" or "Eradicane" in New Jersey.

Use EXTRAZINE II DF at the proper rate for the soil texture and organic matter shown in Table 6 plus 1.8 quarts per acre of "Sutan+", "Genate Plus" or "Eradicane" for control of many annual grasses and broadleaf weeds. (Use 2.4 quarts of "Sutan+" or "Eradicane" on loam soils containing 5 percent or more organic matter, and clay loams and clays containing 4 percent or more organic matter.) Do not use on sands and loamy sands having less than 1 percent organic matter nor on the light sandy soils of eastern coastal states. Do not use on corn grown for seed.

Apply before planting. Incorporate the mixture immediately upon application using power-driven cultivation equipment set for 2-3 inch depth, or tandem disc set to cut about 4 inches deep while operating at 4-6 mph. For thorough mixing, disc in two directions (cross disc), and follow with a harrow, drag, or other leveling device. Prior to the second discing, readjust the disc to prevent cutting deeper than 4 inches. EXTRAZINE II DF may be applied preemergence as an overlay over previously incorporated "Sutan+", "Genate Plus" or "Eradicane", if desired.

Existing stands of quackgrass, purple and yellow nutsedge must be turned under and thoroughly chopped up prior to chemical treatments.

Additional weeds controlled by "Sutan+", "Genate Plus" or "Eradicane" combinations:

Grasses:

Sandbur	Shattercane (Wild Cane)*
Texas Panicum	Quackgrass ("Eradicane" Only)
Wild Proso Millet*("Eradicane" only)	

Perennial Weeds

Yellow Nutsedge (Nutgrass)	Purple Nutsedge (Nutgrass)

*Suppression only—refer to "Sutan+", "Genate Plus" or "Eradicane" label for appropriate supplemental cultural and tillage practices.

For fields with moderate to heavy infestations of these weeds refer to the "Sutan+", "Genate Plus" or "Eradicane" labels for appropriate higher rates.

EXTRAZINE II plus DUAL 8E

Use EXTRAZINE II DF at the proper rate for soil texture and organic matter shown in Table 6. Use "Dual" as follows:

SOIL TEXTURE	BROADCAST RATE PER ACRE FOR "DUAL" 8E
Coarse Sand, Loamy sand, Sandy loam	1.25—1.5 pints
Medium Loam, Silt loam, silt	1.5—2.0 pints
Fine Sandy clay loam, Silty clay loam, Clay loam, Sandy clay, Silty clay, Clay	1.5—2.5 pints

The low end of the rate range should be used for lowest organic matter soils and the rate increased proportionately as the organic matter increases. Soils containing 4 percent organic matter or more require the highest rate shown for that soil texture. Refer to the "Dual" label for precautions or additional information on rotational crops.

STORAGE AND DISPOSAL

Storage: Do not contaminate water, food, or feed by storage or disposal. Do not use or store around the home environment. Avoid contact with water. In case of spill or leak, soak up with sand, earth, or synthetic absorbent. Do use alkaline absorbents and dispose of wastes in compliance with local, state and Federal regulations.

Pesticide Disposal: Pesticide, spray mixture, or rinsate that cannot be used according to label instructions must be disposed of according to applicable Federal, state, or local procedures.

Container Disposal: Completely empty bag into application equipment. Then dispose of empty bag in a sanitary landfill, or by incineration, or, if allowed by state and local authorities, by burning. If burned, stay out of smoke.

NOTICE OF WARRANTY

Du Pont warrants that this product conforms to the chemical description on the label thereof and is reasonably fit for the purposes stated on such label only when used in accordance with the directions under normal use conditions. It is impossible to eliminate all risks inherently associated with the use of this product. Crop injury, ineffectiveness or other unintended consequences may result because of such factors as weather conditions, presence of other materials, or the manner of use or application, all of which are beyond the control of Du Pont. In no case shall Du Pont be liable for consequential, special or indirect damages resulting from the use or handling of this product. All such risks shall be assumed by the buyer. DU PONT MAKES NO WARRANTIES OF MERCHANTABILITY OR FITNESS FOR A PARTICULAR PURPOSE NOR ANY OTHER EXPRESS OR IMPLIED WARRANTY EXCEPT AS STATED ABOVE.

BLADEX., CONQUEST, EXTRAZINE
—Trademarks of E.I. Du Pont de Nemours and Co. (Inc.)
"Banvel". Trademark of Sandoz Crop Protection Corp.
"Dual". Trademark of CIBA-Geigy Corp.
"Eradicane". Trademark of ICI Americas, Inc.
"Genate Plus". Trademark of PPG Industries, Inc.
"Lasso"., "Roundup". Trademarks of Monsanto Co.
"Ortho X-77". Trademark of Chevron Chemical Co.
"Princep", "Caliber" Trademarks of CIBA-Geigy Corp.
"Sutan+" Trademark of ICI Americas, Inc.

AG-3195 8068 5/27/88

This is a specimen label. DuPont is not responsible for the accuracy of the information contained herein. As labels are subject to revision, always carefully read and follow the label on the product container.

FINESSE®

HERBICIDE
DRY FLOWABLE

ACTIVE INGREDIENTS	BY WEIGHT
Chlorsulfuron	
2-Chloro-N-[(4-methoxy-6-methyl-1,3,5-triazin-2-yl) aminocarbonyl]benzenesulfonamide	62.5%
Metsulfuron Methyl	
Methyl 2-[[[[(4-methoxy-6-methyl-1,3,5-triazin-2-yl) amino]carbonyl]amino]sulfonyl]benzoate	12.5%
INERT INGREDIENTS	25.0%
TOTAL	100.0%

EPA Reg. No. 352-445

U.S. Pats. 4,127,405 4,383,113

KEEP OUT OF REACH OF CHILDREN
CAUTION

PRECAUTIONARY STATEMENTS

HAZARDS TO HUMANS AND DOMESTIC ANIMALS

CAUTION!

MAY IRRITATE EYES, NOSE, THROAT OR SKIN

Harmful if absorbed through skin or inhaled. Causes eye irritation. Avoid contact with skin, eyes or clothing. Avoid breathing dust or spray mist. Wash thoroughly with soap and water after handling. Remove and wash contaminated clothing before reuse.

STATEMENT OF PRACTICAL TREATMENT

In case of contact with eyes, immediately flush with plenty of water. Get medical attention if irritation persists.

For medical emergencies involving this product, call toll free 1-800-441-3637.

ENVIRONMENTAL HAZARDS

Do not apply directly to water or wetlands. Do not contaminate water by cleaning of equipment or disposal of wastes.

IMPORTANT INFORMATION

Du Pont "FINESSE" Herbicide is recommended for use in Northern Idaho, Oregon, and Washington on land primarily used for production of wheat and barley. "FINESSE" should not be used in areas where annual crop rotations are frequently practiced as "FINESSE" can remain in the soil for 2 to 3 years and cause severe injury to crops other than those listed in the "Cereal Recropping Intervals and Crop Rotation Recommendations (Noncereal Crops)" section of this label.

Before using "FINESSE", carefully consider your crop rotation plans and options. For maximum rotational flexibility, do not treat all your wheat or barley acres.

Injury to or loss of desirable trees or vegetation may result from failure to observe the following: Do not apply, drain or flush equipment on or near desirable trees or other plants, or on areas where their roots may extend, or in locations where the chemical may be washed or moved into contact with their roots. Do not use on lawns, walks, driveways, tennis courts or similar areas. Prevent drift of spray to desirable plants. Do not contaminate any body of water, including irrigation water that may be used on other crops. Carefully observe sprayer cleanup instructions, as spray tank residue may damage crops other than wheat or barley.

INFORMATION ON RESISTANT WEEDS:

Following the use of "FINESSE" in monoculture cereals production (continuous cereals or cereal-fallow-cereal), some naturally-occurring biotypes* of certain weeds listed on this label may not be effectively controlled by this product.

If weeds listed on this label are not satisfactorily controlled, respray problem areas in a timely and effective manner using a broadleaf herbicide having a different mode of action**, such as: 2,4-D, Banvel(1)***, Buctril[2], Bronate[2], Curtail[3], MCPA, "Karmex" or "Lexone".

To delay the occurrence of resistant biotypes, use FINESSE only in tank mixes with other herbicides effective on the same broadleaf weed species. Do not let weed escapes go to seed.

Consult your Ag chemical dealer, applicator, consultant, appropriate state agricultural extension service representative or your local Du Pont representative for specific recommendations.

*Biotypes are naturally-occurring individuals of the species which have a slightly different genetic makeup. Resistant biotypes may look exactly the same as susceptible biotypes. Herbicide resistant biotypes are able to survive a use rate several times higher than needed to control susceptible biotypes.

**Mode of action is the chemical interaction that interrupts a biological process necessary for plant growth and development.

***Tank mixes with "Banvel" may result in reduced control of some broadleaf weeds.

GENERAL INFORMATION

"FINESSE" is a dry flowable granule to be mixed in water and applied as a uniform broadcast spray for selective weed control in winter and spring wheat, and winter and spring barley.

"FINESSE" is not registered for use in rye or triticale. "FINESSE" is also recommended for selective weed control in wheatgrass and sheep fescue on acreage enrolled in the Conservation Reserve Program. It is noncorrosive, nonflammable, nonvolatile and does not freeze.

"FINESSE" rapidly inhibits growth of susceptible weeds. However, typical symptoms (discoloration) of dying weeds may not be noticeable for 1 to 3 weeks after application, depending on growing conditions and weed susceptibility.

Degree of control and duration of effect depend on: a) rate used, b) weed spectrum and density, c) weed size and variability, d) growing conditions at and following time of treatment, e) soil pH, f) soil organic matter, g) length of growing season, h) amount of precipitation, and i) spray coverage.

IMPORTANT: UNLESS OTHERWISE SPECIFIED IN THE "CROP ROTATION RECOMMENDATIONS (NONCEREAL CROPS)" SECTION OF THIS LABEL, land previously treated with "FINESSE" cannot be rotated to crops other than wheat, oats, barley, rye or triticale until a bioassay confirms that residues of "FINESSE" that could cause crop injury are not present. See "Bioassay" section of this label for details. Failure to follow these instructions could result in injury to subsequent crops.

For crop rotation flexibility do not use "FINESSE" on all your wheat or barley.

READ AND FOLLOW ALL APPROPRIATE SECTIONS OF LABEL INCLUDING PRECAUTIONS BEFORE USING THIS PRODUCT.

GRAZING

"FINESSE" has no grazing restriction.

DIRECTIONS FOR USE

It is a violation of federal law to use this product in a manner inconsistent with its labeling.

"FINESSE" should be used only in accordance with recommendations on this label or in separate published Du Pont recommendations available through local dealers.

Du Pont will not be responsible for losses or damages resulting from the use of this product in any manner not specifically recommended by Du Pont. User assumes all risks associated with such nonrecommended use.

SOIL RESIDUAL ACTIVITY

In the states of Northern ID, OR and WA, "FINESSE" is recommended for use on land having a soil pH of 7.9 or lower and dedicated to the long-term production of wheat or barley. THE SOIL RESIDUAL ACTIVITY OF "FINESSE" CAN INJURE CROPS OTHER THAN WHEAT, BARLEY, OATS, RYE OR TRITICALE FOR 2 TO 3 YEARS OR MORE. "FINESSE" should not be used on soils above pH 7.9, as extended soil residual activity could adversely affect crop rotation options beyond normal intervals.

Rainfall, soil temperature and soil pH are important factors affecting "FINESSE" breakdown in soil. "FINESSE" breakdown is more rapid under conditions of low soil pH, high soil temperature and moist soil. The breakdown process is slow under conditions of high soil pH, low soil temperature and dry soil.

MAXIMUM USE RATES AND SOIL pH LIMITATIONS

IN THE STATES OF NORTHERN ID, OR AND WA, the maximum use rate is 0.5 oz/A per crop period on soils having a pH of 7.9 or lower. Do not use more than 0.5 oz/A in an 18 month period. Do not use on soils having a pH greater than 7.9.

Wherever "FINESSE" and "Glean" are used on the same land having a pH of 7.9 or lower, the combined total use rate per crop period should not exceed 0.5 oz/A in an 18 month period.

NOTE: Prior to use of "FINESSE", take representative soil samples at 0-4" depth and determine soil pH by laboratory analysis using 1:1 soil:water suspension. Samples should be representative of the different conditions in the field (for example, slope, soil texture, low areas, eroded areas, etc.) Consult local extension publications for recommended soil sampling procedures.

TIMING OF APPLICATION

Make a single or split application of "FINESSE", but do not exceed 0.5 oz/A per crop period. For best results with postemergence application, apply to actively growing weeds and add a surfactant of at least 80% active ingredient at the rate of 1 to 2 qts/100 gals of spray solution. For best results, make postemergence treatments while the crop canopy is open to ensure maximum spray coverage. Use the higher rates for heavy weed infestations and large weeds, and for fall applications in high rainfall areas. To control weeds that germinate after treatment, rainfall must move "FINESSE" into the weed root zone before weed seeds germinate and develop an established root system. Spring applications in low rainfall areas may not receive enough rainfall after treatment, resulting in poor residual weed control.

See following instructions for details on postemergence treatments, single and split applications.

NOTE: Varieties of wheat and barley differ in their tolerance to herbicides. When using "FINESSE" for the first time on a particular variety, limit initial use to one 18 oz jug. If no symptoms of crop injury occur 14 days after treatment, balance of acreage can be treated.

Postemergence To Winter/Spring Wheat And Winter/Spring Barley

o WINTER WHEAT AND WINTER BARLEY

Apply "FINESSE" 0.3 to 0.5 oz/A in the fall or spring anytime after crop is in the 2-leaf stage - but before boot stage. Do not apply during boot stage or early heading as crop injury may occur. Postemergence applications should be made only as tank mixtures with another broadleaf herbicide. Refer to "TANK MIXTURES FOR RESISTANT WEED MANAGEMENT".

Do not make an early postemergence treatment to late seeded wheat or barley as the combined effect of herbicide stress plus cold weather and/or moisture stress could cause crop injury. Delay making a postemergence treatment to late seeded wheat or barley until crop has started to tiller.

Do not apply "FINESSE" within 60 days of crop emergence where an organophosphate insecticide (such as "Di-Syston", etc.) has been used as an in-furrow treatment, as crop injury may occur.

o SPRING WHEAT AND SPRING BARLEY - East of Cascades Only

Apply "FINESSE" 0.3 to 0.5 oz/A anytime after crop is in the 2-leaf stage through 2nd joint stage. Postemergence applications should be made only as tank mixtures with another broadleaf herbicide. Refer to "TANK MIXTURES FOR RESISTANT WEED MANAGEMENT". DO NOT apply once the flag leaf is visible, as crop injury may occur.

To avoid the risk of cold weather related crop injury, apply "FINESSE" when good growing conditions (adequate soil moisture, daily high temp. of 50 Deg.F or more) are expected to continue until crop has started to tiller.

Do not apply prior to tillering when cold and/or dry weather can reduce seedling vigor making crop more vulnerable to the combination of herbicide and weather stress. The combined effect of herbicide plus stress from cold and/or dry weather can result in temporary yellowing or crop injury (yield reduction).

Do not make a late fall, winter or early spring application to wheat or barley until crop is well established and has started to tiller.

Split Applications to Winter Wheat and Winter Barley

o WINTER WHEAT AND WINTER BARLEY

"FINESSE" can be applied fall postemergence plus spring postemergence provided that each application is made with another broadleaf herbicide. Refer to "TANK MIXTURES FOR RESISTANT WEED MANAGEMENT".

When using a split application, do not use less than 0.2 oz/A per treatment or more than 0.3 oz/A per treatment and do not exceed 0.5 oz/A per crop. Do not use more than 0.5 oz/A in an 18 month period. Follow all instructions in the "POSTEMERGENCE" directions for use.

Postemergence To Wheatgrass Or Sheep Fescue On Acreage Enrolled In The Conservation Reserve Program

Apply 0.3 to 0.5 oz/A in the fall or spring anytime after wheatgrass or sheep fescue is in the 3 to 4 leaf stage. For best results, apply preemergence to weeds or early postemergence when weeds are actively growing. Add surfactant at the rate of 1 to 2 quarts per 100 gals for applications made postemergence to weeds.

2,4-D can be tank mixed with "FINESSE" at 1/4 lb. AI/A for all labeled grasses larger than the 5-leaf stage. For fully tillered stands, up to 1/2 lb. AI/A of 2,4-D may be used. Surfactant may be added at 1 to 2 pts. per 100 gallons of spray. However, the addition of surfactant may increase the chance of grass injury.

To control weeds that germinate after treatment, rainfall must move "FINESSE" into the weed root zone before weed seeds germinate and develop an established root system. Spring applications in low rainfall areas may not receive enough rainfall after treatment resulting in poor residual weed control.

Note: Wherever wheatgrass or sheep fescue are planted in a mix with other grasses, limit first use of "FINESSE" to a small area. If injury to other grasses is not evident in 14 days after treatment, balance of acreage can be treated.

"FINESSE" is not recommended for mixed plantings that include legumes as injury to the legumes may occur.

WEED CONTROL IN REDUCED TILLAGE FALLOW

DO NOT USE "FINESSE" IN FALLOW.

WEED CONTROL/USE RATE TABLE

NOTE: Read and follow all instructions under "Specific Weed Problems" for all weeds marked with "†".

WEEDS CONTROLLED OR SUPPRESSED*					RATE–OUNCES PER ACRE	ACRES TREATED PER 18 OUNCE CONTAINER
Annual bluegrass*†	Coast fiddleneck	Groundsel (common)	Prickly lettuce**	Tumble mustard (Jim	0.3 to 0.5	At the 0.3 oz/A rate, an
Annual ryegrass*†	(tarweed)	Hempnettle	Prickly poppy	Hill)		18 oz container will treat
Annual sowthistle	Conical catchfly	Henbit	(pinnate)	Vetch†		60 acres
Bedstraw*†	Corn gromwell*†	Knotweed (prostrate)†	Prostrate pigweed	White cockle		
Blue mustard	Corn spurry	Ladysthumb	Purslane (common)	Wild buckwheat		At the 0.4 oz/A rate, an
Broadleaf dock	Cow cockle	Lambsquarters	Redroot pigweed	Wild carrot		18 oz container will treat
Bur beakchervil	Dovefoot geranium	Little bittercress	Redstem filaree	Wild mustard		45 acres
Bur buttercup	False chamomile	Mayweed	Shepherdspurse	Wild radish†		
(testiculate)	Flixweed*†	Miners lettuce	Smooth pigweed			At the 0.5 oz/A rate, an
Canada thistle*†	Field pennycress	Pennsylvania	Speedwell			18 oz container will treat
Chickweed (common,	Green foxtail	smartweed	(common, ivyleaf)*			36 acres
jagged, moussear)	(pigeongrass)*	Pineappleweed	Tansymustard*†			

* When used as directed, weeds marked with an asterisk are suppressed and/or controlled. Weed suppression is a visual reduction in weed competition (reduced population and/or vigor) as compared to an untreated area. Degree of suppression will vary with rate used, size of weeds and environmental conditions following treatment.

† See "Specific Weed Problems" for more information regarding control or suppression of these weeds.

** Naturally-occurring resistant biotypes of prickly lettuce are known to occur. See "TANK MIXTURES FOR RESISTANT WEED MANAGEMENT" section of this label for additional details.

SPECIFIC WEED PROBLEMS

Note: Always use "FINESSE" in tank mixtures with another suitable herbicide. Refer to "TANK MIXTURES FOR RESISTANT WEED MANAGEMENT".

Annual Bluegrass/Annual Ryegrass: For best results, apply "FINESSE" preemergence to ryegrass. 1/2 to 1 inch of rainfall is needed to move "FINESSE" into the weed root zone prior to ryegrass emergence. Tank mix with "Karmex" Herbicide at 1 1/2 lbs/A per acre to increase control. See "Specific Tank Mixtures" for additional details.

Bedstraw: Use the higher rate of "FINESSE". For postemergence treatments, apply before bedstraw is greater than 2 inches long and use the 2 qts/100 gals rate of surfactant.

Canada Thistle: Apply "FINESSE" plus surfactant (2 qts/100 gals) after the majority of thistles have emerged and while they are small (rosette stage to 4 to 6" tall), but actively growing. A single application will effectively inhibit the ability of Canada thistle to compete with the crop. For maximum long-term effect, yearly treatment may be required.

Corn Gromwell: Where corn gromwell is a major weed problem, use the higher rate of "FINESSE" or tank mix "FINESSE" with "Karmex". See "Specific Tank Mixtures" for additional details.

Flixweed/Tansymustard: For best results with postemergence applications, apply "FINESSE" at 0.3 to 0.5 oz/A in a tank mixture with 2,4-D (ester or amine) when weeds are actively growing. See "TANK MIXTURES FOR SPECIFIC WEED COMPLEXES" section of label. If weeds are inactive due to adverse weather conditions (cold, dry weather), delay application until active weed growth resumes. For best results with a fall application, "FINESSE" should be applied at the highest recommended rate to provide adequate residual activity.

Prostrate Knotweed: For best results, apply preemergence to knotweed in the fall. For postemergence treatments, apply "FINESSE" plus surfactant to small (no more than 4 true leaves) actively growing plants. For maximum postemergence control, knotweed plants should remain actively growing for 3 to 4 days following application.

Vetch: For control, apply 0.4 oz/A "FINESSE" plus 4 oz active ingredient/A MCPA (amine or ester).

Wild Radish: Postemergence applications will provide best results.

TANK MIXTURES FOR RESISTANT WEED MANAGEMENT

Northern Idaho, Oregon and Washington (west of the Cascades only):

- Split applications (fall postemergence plus spring postemergence) can be made providing the maximum use rate per 18 month (0.5 oz/A) is not exceeded and all applications are in a tank mix with another broadleaf herbicide. See below for tank mix partners. For "FINESSE" do not use less than 0.2 oz/A per treatment or more than 0.3 oz/A per treatment. Note: Du Pont "Harmony" Hebicide in tank mixes can be used as a sequential treatment after a fall application of "FINESSE".
- Single postemergence application in combination with another broadleaf herbicide can be made in the fall or spring anytime the majority of broadleaf weeds have emerged and after the crop is in the 2-leaf stage. See below for tank mix partners. Use rate range for "FINESSE" is 0.3 to 0.5 oz/A.
- Tank mix partners' include:

2,4-D (amine or ester)	4 to 8 oz active ingredient/acre
MCPA (amine or ester)	4 to 8 oz active ingredient/acre
"Buctril" 4EC	1/4 pt to 1 pt/acre
"Bronate"	1/2 pt to 2 pt/acre
"Karmex" DF or diuron DF	1/2 lb to 1 1/2 lb/acre
"Lexone" DF	1/8 to 2/3 lb/acre
"Banvel"	1/16 to 1/4 pt/acre
"Curtail"	1 to 2 pt/acre

* Follow all instructions, warnings and precautions on the companion product label.

** Tank mixes with "Banvel" may result in reduced weed control of some broadleaf weeds.

- "FINESSE" is no longer recommended for the control of Russian thistle or kochia.
- Do not use more than 0.5 oz/A in an 18 month period.
- Do not apply "FINESSE" during fallow period.
- If resistant weeds are known to be present, consider using another herbicide treatment or adjust the use rate of the "FINESSE" tank mix partner so that it alone will control the resistant species.

TANK MIXTURES FOR SPECIFIC WEED COMPLEXES

"Lexone" DF Herbicide Plus "FINESSE": Where broadleaf weeds and cheatgrass (downy brome) are the main problems, a fall application of "Lexone" DF at 1/3 to 1/2 lb/A with "FINESSE" at 0.3 to 0.4 oz/A is recommended for best results. Apply after wheat or barley is well tillered and has a 2" secondary root system established throughout the field. 1/2 to 1" of rainfall is needed within 1 to 2 weeks of application. Follow all restrictions on the "Lexone" DF label.

"Karmex" DF or Diuron DF Herbicide Plus "FINESSE": Where annual bluegrass, annual ryegrass, corn gromwell, green foxtail (pigeongrass), and wild buckwheat are the main weed problems, apply 1 to 1 1/2 lbs/A of diuron DF plus 0.3 to 0.4 oz/A "FINESSE". Apply preemergence or early postemergence to actively growing weeds less than 2 inches tall or across. 1/2 to 1 inch of rainfall is needed within 1 to 2 weeks after application. Follow all restrictions on the "Karmex" or diuron labels.

2,4-D (amine or ester) OR MCPA (AMINE OR ESTER) Plus "FINESSE": Tank mixtures of "FINESSE" plus 2,4-D/MCPA are recommended when weeds are large and/or stressed due to adverse conditions (cold temperature, low soil mosture, dry, dusty field conditions) or when dense crop canopy makes it difficult to obtain thorough spray coverage. Use "FINESSE" at 0.3 to 0.5 oz/A plus 1/4 to 1/2 lb active ingredient/A. of 2,4-D/MCPA (ester formulations of 2,4-D have provided best results). Surfactant may be added at 1 qt/100 gals of spray; however, the addition of surfactant may increase the chance of crop injury. Do not use surfactant with rates higher than 0.3 oz/A. "FINESSE" should be mixed in water with the agitator running prior to adding 2,4-D/MCPA. Read and follow all label instructions on timing, precautions and warnings for these herbicides prior to using these tank mixtures.

IMPORTANCE OF APPLICATION TIMING RELATIVE TO GROWING CONDITIONS AND RAINFALL

How Growing Conditions Affect Weed Control:

Applications made postemergence to weeds are most effective when "FINESSE" is applied to young, actively growing weeds and an open crop canopy allows thorough spray coverage. Warm, moist growing conditions promote active weed growth and enhance the activity of "FINESSE" by allowing maximum foliar uptake. If cold, dry conditions exist, delay postemergence treatment until weather conditions promote active weed growth. For best results with treatments made postemergence to weeds during periods of cold temperatures, apply "FINESSE" when minimum day/night temperatures are 40 degrees F or higher. Avoid postemergence applications to weeds which are inactive due to adverse weather conditions. Weeds hardened off by cold weather or drought stress may not be controlled.

How Growing Conditions Affect Crop Safety:

Prolonged cold weather (daily maximum temperatures below 50°F) while crop is in seedling stage (1 to 5 leaf) can cause crop injury following either a preemergence or postemergence treatment.

To avoid the risk of cold weather related injury, apply "FINESSE" when good growing conditions are expected to continue until crop has started to tiller.

Rainfall After Treatment:

Rainfall after treatment will affect "FINESSE" performance when applied postemergence or preemergence to weeds.

Postemergence treatments control or suppress weeds primarily through foliar uptake. Rainfall within two weeks after application may increase the control by "FINESSE" on some hard to control weeds (such as corn gromwell, flixweed, tansymustard).

Rainfall immediately after treatment can wash "FINESSE" off weed foliage and result in reduced weed control effectiveness. Do not make postemergence applications to weeds when rainfall is occurring or threatening.

For best results preemergence to weeds, it is important to apply "FINESSE" when you can expect at least 1 to 2" (clay soils may require more) of rain or sprinkler irrigation to move "FINESSE" into the soil profile before weed seeds germinate or develop an established root system. Weeds that germinate after treatment and develop an established root system before rainfall moves "FINESSE" into the weed root zone may not be controlled.

When weed emergence is uneven, control of weeds that germinate after treatment will be dependent on the timing and amount of rainfall following application. Sufficient rainfall is needed to move "FINESSE" into the weed root zone before weeds that germinate after treatment can develop an established root system. When favorable growing conditions exist, rainfall may be needed within a few days after treatment.

SPRAY PREPARATION, ADDITIVES, PRODUCT MEASUREMENT AND SURFACTANT

SPRAY PREPARATION: Mix the proper amount of "FINESSE" into the necessary volume of water in the spray tank with the agitator running. Continuous agitation is required for uniform mixing and application.

Use "FINESSE" spray preparations soon after mixing. If spraying is delayed, thoroughly reagitate before using.

ADDITIVES: Do not use with spray additives that lower the pH of the spray solution below pH 3.0 as rapid product degradation can occur.

PRODUCT MEASUREMENT: The "FINESSE" volumetric measuring cylinder is to be used only as a guide, as the degree of accuracy varies by plus or minus 10%. For more precise measurement, use scales calibrated in ounces.

SURFACTANT: Use a surfactant of at least 80% active ingredient in postemergence application to weeds to improve wetting and/or foliar activity of "FINESSE". Add surfactant at 1 to 2 qts/100 gals of spray volume as the last ingredient. The higher rate of surfactant is particularly useful with spray volumes of 5 gpa or less, when using low rates of "FINESSE", and when spraying some weed species (see "Specific Weed Problems" section). Antifoaming agents may be needed.

LIQUID FERTILIZER TANK MIXTURES

LIQUID FERTILIZER: To apply "FINESSE" with liquid fertilizer, slurry the "FINESSE" in water; then thoroughly mix the slurry into the liquid fertilizer. The addition of surfactant to tank mixtures of "FINESSE" plus liquid fertilizer increases the risk of crop injury.

Run a tank mix compatibility test before mixing "FINESSE" in fertilizer solutions.

Do not use with liquid fertilizers having a pH of 3.0 or less as rapid product degradation can occur.

DO NOT USE LIQUID FERTILIZER AS A SUBSTITUTE FOR SURFACTANT.

TANK MIXTURES WITH OTHER HERBICIDES, INSECTICIDES AND FUNGICIDES

When using a tank mixture for the first time, run a standard compatibility test and use on a small portion of a field to be sure there is adequate crop safety and satisfactory performance before adopting large scale use. "FINESSE" must be in suspension before adding the companion pesticides. Follow all instructions, warnings and precautions on the companion product label.

Other Herbicides: Use a suitable registered companion herbicide if weeds and grasses other than those listed on this label are present. "FINESSE" must be in suspension in the spray tank before adding the companion herbicide. Follow the surfactant recommendation on the companion herbicide label. Tank mixtures with "Hoelon"[5] 3EC may result in reduced wild oat control.

Insecticides: "FINESSE" may be tank mixed with insecticides registered for use on cereal grains. However, under certain conditions, (drought or cold stress while crop is in 2-4 leaf stage) tank mixtures of sequential treatments of "FINESSE" and organophosphate insecticides (such as methyl or ethyl parathion, "Di-Syston", etc.) may produce temporary crop yellowing or, in severe cases, crop injury. The potential for crop injury is greatest when there are wide fluctuations in day/night temperatures just prior to or soon after treatment. Limit first use to a small area. If no symptoms of crop injury occur 14 days after treatment, balance of acreage can be treated.

Do not apply "FINESSE" within 60 days of crop emergence where an organophosphate insecticide (such as "Di-Syston") has been applied as an in-furrow treatment, as crop injury may result.

DO NOT USE "FINESSE" PLUS MALATHION AS CROP INJURY MAY RESULT.

FUNGICIDES: "FINESSE" may be tank mixed with Du Pont "Benlate" Fungicide or Du Pont "Manzate" 200 Fungicide/Du Pont "Manzate"200 DF Fungicide or other fungicides whenever the proper timing for herbicide and fungicide treatments coincide.

SPRAY EQUIPMENT AND SPRAY VOLUME

Apply using properly calibrated air or ground equipment. Select a spray volume and delivery system that will ensure thorough coverage and a uniform spray pattern. Avoid overlapping, and shut off spray booms while starting, turning, slowing or stopping, or injury to the crop or following crops may result.

Do not apply this product through any type of irrigation system.

SPRAY EQUIPMENT: Refer to specific manufacturer's recommendations for additional information on gpa, pressure, speed, nozzle types and arrangements, nozzle heights above the target canopy, etc., for respective application equipment.

GROUND APPLICATION: For optimum spray distribution and thorough coverage, use flat fan or low volume flood nozzles. For flat fan nozzles, do not use less than 3-gallon spray volume per acre (gpa).

For flood nozzles on 30-inch nozzle spacings, use not less than 10 gpa and no larger than "TK" 10 or equivalent and not less than 30 psi. On 60-inch nozzle spacings, use not less than 20 gpa. 100% overlapping of nozzle spray pattern is recommended for both 30-inch and 60-inch spacings.

With "Raindrop"[6] nozzles, do not use less than 30 gpa and insure for 100% overlap of nozzle spray pattern.

Use 50-mesh screens or larger.

Use higher spray volumes to obtain better coverage when crop canopy is dense.

AERIAL APPLICATION: Use nozzle types and arrangements that will provide for optimum spray distribution and maximum coverage at 3 to 5 gpa. Do not apply during inversion conditions, when winds are gusty, or when other conditions will favor poor coverage and/or drift.

Use higher spray volumes to obtain better coverage when crop canopy is dense.

CAUTION-AVOID SPRAY DRIFT

Follow these practices to minimize drift.

Do not allow spray from either ground or aerial equipment to drift onto adjacent crops or land, as even small amounts will injure other plants. When spraying near adjacent, sensitive crops or plants, do everything possible to reduce spray drift. This includes:

- Stop spraying if wind speed becomes excessive. DO NOT SPRAY IF WIND SPEED IS 10 MPH OR GREATER. Spray drift can occur at wind speeds less than 10 MPH. If sensitive crops or plants are downwind, extreme caution must be used even in relatively low wind conditions! DO NOT SPRAY IF WINDS ARE GUSTY.
- High temperatures, drought, and low relative humidity increase the possibility of spray drift. EXTREME CAUTION MUST BE USED WHEN THESE CONDITIONS ARE PRESENT AND SENSITIVE CROPS OR PLANTS ARE NEARBY, REGARDLESS OF WIND SPEED.
- Do not apply when an inversion exists. An inversion is characterized by little or no air movement and an increase in air temperature with an increase in altitude. In humid regions, a fog or mist may form. An inversion may be detected by producing a smoke column and checking for a layering effect. Smoke-producing devices on aircraft are recommended. If not sure whether inversion conditions are present, consult with local weather services before making an application.
- Drift from aerial or ground equipment may be further reduced by:

  1. Using coarse sprays to minimize drift. DO NOT APPLY WITH HOLLOW-CONE INSECTICIDE NOZZLES ON GROUND EQUIPMENT. Do not use nozzles that produce fine droplets, such as Sprayfoil [7] or airblast-type nozzles. Nozzles should be oriented at an angle between straight back for ground applications. For aerial applications orient nozzles straight back along the windstream. If using flood-type nozzles on aircraft, orient them so spray is produced in direction of the airstream.

  2. Increasing volume of spray mix per acre (for example minimum 5

GPA by air, 10 GPA by ground) by using higher flow rate nozzles.

3. Reducing pressure (PSI). DO NOT EXCEED 40 PSI when applying "FINESSE". (Vehicle speed must also be reduced to maintain spray mix volume per acre.) Consult manufacturer's catalogs for details on correct calibration.

4. Apply as close to target plants as possible while still maintaining a good spray pattern.

NOTE: Do not allow spray tc drift onto adjacent crops, or onto agricultural land scheduled to be planted to crops other than wheat as injury to the crop may occur. Extreme care must be taken to prevent drift to desirable plants or nontarget agricultural land.

CEREAL RECROPPING INTERVALS

RECROPPING TO WHEAT, OATS, BARLEY, RYE AND TRITICALE IN NORTHERN ID, OR, AND WA:

Recropping plans are determined by soil pH, rate of "FINESSE" applied and a minimum recropping interval. The minimum recropping interval is from the time from the last application of "FINESSE" to the anticipated date of planting.

Soil pH*	Use Rate Oz/Acre	Minimum Recropping Interval (Months) Wheat/Rye/Triticale	Oats	Barley
6.5 or lower	0.2 to 0.4	0	10	10
6.5 or lower	0.5	4	10	10
6.6 to 7.5	0.2 to 0.4	0	10	16
6.6 to 7.5	0.5	4	16	24
7.6 to 7.9	0.2 to 0.4	4	16	24
above 7.9	Do not use	------------Not Applicable------------		

* Soil pH is to be determined by laboratory analysis using 1:1 soil:water suspension method on representative soil samples taken at 0-4" depth. Consult local extension publications for recommended soil sampling procedures.

ROTATION INTERVAL FOR PLANTING GRASSES ON CONSERVATION RESERVE PROGRAM ACRES

Wherever "Finesse"has been previously used in wheat, barley or fallow, the following grasses may be planted after the intervals specified in the tables below. The planting of grass and legume mixtures is not recommended as injury to the legume may occur.

Bentgrasses
Blue Grama
Bluestems - Big, Little, Plains, Sand, WW Spar
Buffalograss
Galleta
Green needlegrass
Green sprangletop
Indiangrass
Indian ricegrass
Lovegrass - Sand, Weeping
Orchardgrass (excluding Piaute)
Praire sandreed
Sand dropseed
Sheep fescue
Sideoats Grama
Switchgrass
Wheatgrasses - Crested, Intermediate, Pubescent, Slender, Streambank, Tall, Thickspike, Western
Wild-rye grasses - Beardless, Russian

ROTATION INTERVALS IN STATES OF:
NORTHERN ID, OR AND WA

Soil pH*	Use Rate Oz/A	Minimum Interval for Planting Grasses
7.9 or lower	0.2 to 0.4	2 months
7.5 or lower	0.5	4 months

* Soil pH is to be determined by laboratory analysis using the 1:1; soil:water suspension method on representative soil samples taken at 0–4"depth. Consult local extension publications for recommended soil sampling procedures.

CROP ROTATION RECOMMENDATIONS (NONCEREAL CROPS)

Note: The crop rotation intervals specified in this section of the label must be followed unless a field or LRB℠ bioassay indicates a shorter planting interval.

NORTHERN ID, NORTHEASTERN OR, EASTERN WA

Unless a Crop Rotation Interval is specified, a field or LRB℠ bioassay must be completed before rotating to any crop other than those listed below. See "Bioassay" section.

NOTE: Successful rotation to peas and lentils can be expected in fields of even terrain having well drained soils with a uniform pH of 6.5 or less. Localized areas of crop injury may occur in fields that have highly variable terrain with areas of poor drainage and/or areas of high soil pH (eroded knolls, exposed calcareous subsoil where pH is above 6.5

PEAS AND LENTILS:

Northern Idaho, Northeastern Oregon, Eastern Washington Counties:

ID	OR	WA
Benewah	Baker	Asotin
Bonner	Umatilla	Columbia
Boundary	Union	Garfield
Clearwater	Wallowa	Pend Oreille
Idaho		Spokane
Koontenai		Stevens
Letah		Walla Walla
Lewis		Whitman
Nez Perce		

In the above counties on nonirrigated land, the intervals are:

Crop	Soil pH*	Use Rate (Oz/Acre)	Cumulative Precipitation** (in)	Rotation Interval (months)
Peas	6.5 or lower	0.2 to 0.4	35	24
(Alaskan,	6.5 or lower	0.5	—	Field or LRB℠ Bioassay
Columbian)	6.6 to 7.5	0.2 to 0.4	—	Field or LRB℠ Bioassay
	7.6 to 7.9	0.2 to 0.4	—	Field or LRB℠ Bioassay
Lentils	6.5 or lower	0.2 to 0.4	50	36
(Chilean)	6.5 or lower	0.5	—	Field or LRB℠ Bioassay
	6.6 to 7.5	0.2 to 0.4	—	Field or LRB℠ Bioassay
	7.6 to 7.9	0.2 to 0.4	—	Field or LRB℠ Bioassay

* Soil pH is to be determined by laboratory analysis using the 1:1; soil:water suspension method on representative soil samples taken at 0–4" depth. Consult local extension publications for recommended soil sampling procedures.

** Cumulative Precipitation equals the total amount received from the date of "FINESSE" application to the date of planting. Should accumulated precipitation not be sufficient to meet the indicated amounts, do not rotate to the indicated crops until the following growing season.

For recropping flexibility, do not use "FINESSE" on all your wheat or barley acreage.

WESTERN OREGON

Unless a crop rotation is specified, a field or LRB℠ bioassay must be completed before rotating crops other than those listed on this or other "FINESSE" labels. See Oregon 24(c) label for rotation intervals for annual and perennial ryegrasses and crimson and red clovers.

SPRAYER CLEANUP

To avoid subsequent injury to crops other than wheat or barley, immediately after spraying thoroughly remove all traces of "FINESSE" from mixing and spray equipment as follows:

1) Drain tank; then flush tank, boom and hoses with clean water for a minimum of 10 minutes.
2) Fill the tank with clean water then add 1/2 gallon chlorine bleach (containing 5 1/4% sodium hypochlorite) per 100 gallons of water. Flush solution through boom and hoses, then allow to sit for 15 minutes with agitation; then drain.
3) Repeat Step 2.
4) Nozzles and screens should be removed and cleaned separately. To remove traces of chlorine bleach, rinse the tank thoroughly with clean water and flush through hoses and boom.

NOTE: To reduce the amount of water required in the above procedure, see separate Du Pont Bulletin, "Reduced Volume Cleanout Procedure for Large Sprayers".

CAUTION: Do not use chlorine bleach with ammonia. All traces of liquid fertilizer containing ammonia, ammonium nitrate or ammonium sulphate must be rinsed with water from the mixing and application equipment before adding chlorine bleach solution. Failure to do so will release a gas with a musty chlorine odor which can cause eye, nose, throat and lung irritation. Do not clean equipment in an enclosed area.

PRECAUTIONS

Varieties of wheat and barley differ in their tolerance to herbicides. When using "FINESSE" for the first time on a particular variety, limit initial use to one 18 oz jug. If no symptoms of crop injury occur 14 days after treatment, balance of acreage can be treated.

Do not apply "FINESSE" to wheat and barley that is stressed by severe weather conditions, drought, low fertility, water saturated soil, disease or insect damage as crop injury may result. Severe winter stress, drought, disease, or insect damage following application may also result in crop injury.

"FINESSE" is not registered for use in rye or triticale.

IN THE STATES OF NORTHERN ID, OR AND WA, the maximum use rate is 0.5 oz/A per crop period on soils having a pH of 7.9 or lower. Do not use more than 0.5 oz/A in an 18 month period. Do not use on soils having a pH greater than 7.9.

Whenever "FINESSE" and "GLEAN" are used on the same land having a pH of 7.9, or lower, the combined total use rate per crop period should not exceed 0.5 oz/A in an 18 month period.

Do not apply to wheat or barley undersown with legumes and/or grasses as injury to the forage may result.

Do not apply to frozen ground where surface runoff may occur.

Do not apply to snow covered ground.

Do not apply to irrigated land where tail water will be used to irrigate other crop land.

Do not use on fields that have variable soil conditions where large areas are gravelly or sandy, have eroded knolls, calcium deposits, or widely variable pH readings or organic matter content. Use of "FINESSE" on fields with these conditions may result in injury to the crop or following crops.

Under certain conditions such as heavy rainfall, prolonged cool weather (daily high temperatures less than 50°F) or wide fluctuations in day/night temperatures just prior to or soon after treatment, temporary discoloration and/or crop injury may occur. Risk of injury is greatest when crop is in 1 to 5 leaf stage.

Tank mixtures of "FINESSE" and organophosphate insecticides (such as methyl or ethyl parathion, "Di–Syston", etc.) may cause temporary discoloration or crop injury. The potential for crop injury is greatest when there are wide fluctuations in day/night temperatures just prior to or soon after treatment.

Wherever land has been or will be treated with "Assert"[8] herbicide and "FINESSE", plant only wheat or barley until a bioassay (see "Bioassay" section of label) demonstrates that other crops can be successfully grown. On land that is frequently rotated to crops other than wheat or barley, do not use "FINESSE" wherever "Assert" has been or will be used. The additive effect of soil residues from these treatments has not been determined and crop rotation guidelines and minimum rotation intervals are not known; injury to rotational crops may occur.

Fall applications on coarse textured soils (expecially those having a pH of greater than 7.0) may not provide adequate control or suppression of spring germinating weeds.

To reduce the potential for movement of treated soil due to wind erosion, do not apply to powdery dry or light sandy soils until they have been stabilized by rainfall, trashy mulch, reduced tillage or other cultural practices. Injury to adjacent crops may occur when treated soil is blown onto land used to produce crops other than cereal grains.

For ground applications applied postemergence to weeds when dry, dusty field conditions exist, control of weeds in wheel track areas may be reduced. The addition of 2,4-D should improve weed control under these conditions.

To prevent cold weather related crop injury, avoid making preemergence or early postemergence (2-4 leaf) applications during late fall, winter and early spring months when cold weather conditions are unpredictable and can be severe. The combined affects of herbicide stress plus winter stress can result in greater crop injury than either stress factor alone.

Weed control or suppression may be unsatisfactory on soils containing 5% or more organic matter.

The "FINESSE" volumetric measuring cylinder is to be used only as a guide as the degree of accuracy varies by plus or minus 10%. For more precise measurement, use scales calibrated in ounces.

BIOASSAY

A bioassay (field or LRB℠) must be completed before rotating to crops not listed on this label or rotating at intervals shorter than those listed in the "Crop Rotation Recommendations (Noncereal Crops)" section.

• FIELD BIOASSAY

"FINESSE" is a useful tool for weed control in wheat or barley; however, under some conditions small amounts of "FINESSE" can remain in the soil and injure crops other than wheat, barley, rye or triticale for 2-3 years after application; therefore, before you use "FINESSE", you should carefully consider your crop rotation plans during the 2-3 year period following treatment.

A field bioassay will be necessary if crops other than wheat, barley, rye or triticale are to be planted on land previously treated with "FINESSE".

A bioassay involves growing test strips of the crop or crops you plan to grow in fields previously treated with "FINESSE". Crop response will indicate whether or not to rotate to the crop(s) grown in the test strips.

"FINESSE" breaks down more rapidly in soils having a pH less than 7.0 and in areas having 20" or more of annual rainfall and a long growing season with warm soil temperatures. "FINESSE" residues break down more slowly as soil pH increases above 7.0. Other contributing factors that slow the disappearance of "FINESSE" are low rainfall and prolonged periods of soil temperatures less than 40°F.

Of the key factors that influence the rate of disappearance, only soil pH remains relatively constant from year to year. Soil temperature, and to a larger degree soil moisture, can vary greatly from year to year, and from area to area. Consequently, it is not always possible to accurately predict when areas treated with "FINESSE" can be rotated to crops other than those listed on label.

A bioassay of your "FINESSE" treated field is the only sure way of determining when crops other than those listed on the label can be grown and is conducted as follows:

1. The accuracy and reliability of any field bioassay is largely dependent on the location and number of strips planted. Be sure to select areas of the field previously treated with "FINESSE" that are representative of the various field conditions. Be sure to consider factors such as field size, soil texture, drainage, turnaround areas, eroded knolls or alkaline spots when selecting the sites that are most representative of the soil conditions in the field.

   Even in small fields, more than one test strip is required to accurately determine whether it is safe to rotate to a noncereal crop. On large fields, several test strips will be needed in order to obtain reliable results based on the field variables mentioned above.
2. Plant the test strips perpendicular to the direction in which the field was sprayed. Each strip should be long enough to cross the width of several spray swaths. A large test strip area is more reliable than a small one. Suggested size is 1/4 to 1/2 acre per site.
3. Use standard tillage and seeding equipment to plant the bioassay crop(s).
4. Prepare a seed bed and plant the crops and varieties you want the option of growing the following year. It is important to use the same planting time, conditions, techniques and cultural practices you normally use to plant and grow the bioassay crop(s). If possible, plant into an adjacent area not treated with "FINESSE" to use as a comparison.
5. Do not overspray the test strips with herbicides that may damage the bioassay crop(s).
6. If the crop(s) in the test strips grow to maturity with a normal harvest, the assay is positive and you may now rotate to the new crop. However, if crop(s) in the test strips die, are stunted, or fail to yield a normal harvest, the assay is negative and you should not rotate to the new crop(s). Run the assay until positive results are obtained before rotating to the new crop(s).
7. If the bioassay indicates that "FINESSE" residues are still present, do not rotate to crop(s) other than those listed on the label until bioassay results indicate that the assay crops are growing normally.

• DU PONT LRB℠ BIOASSAY SERVICE

In the states of Northern ID, OR and WA, the Du Pont LRB℠ bioassay service is available through certain dealers and/or consultants. This service uses soil samples taken by Du Pont certified individuals for laboratory bioassay analysis. LRB℠ results will serve as a crop rotation recommendation.

Check with your local Du Pont representative or call toll free 1-800-782-3557 for information regarding the LRB℠ bioassay service.

With Any Chemical, Follow Labeling Instructions And Warnings Carefully.

STORAGE AND DISPOSAL

STORAGE: Store product only in original container, away from other pesticides, fertilizer, food or feed.

PRODUCT DISPOSAL: Do not contaminate water, food or feed by storage or disposal. Wastes resulting from the use of this product may be disposed of on site or at an approved waste disposal facility.

CONTAINER DISPOSAL: Triple rinse (or equivalent) the container. Then offer for recycling or reconditioning, or puncture and dispose of in a sanitary landfill, or by incineration, or, if allowed by state and local authorities, by burning. If burned, stay out of smoke.

NOTICE TO BUYER: Purchase of this material does not confer any rights under patents of countries outside of the United States.

(1) Registered trademark of Sandoz Crop Protection Corporation.
(2) Registered trademark of Rhône-Poulene Ag Company.
(3) Registered trademark of Dow Chemical Company.
(4) Registered trademark of Bayer AG, Leverkusen.
(5) Registered trademark of Hoechst-Roussel Agri-Vet Company.
(6) Registered trademark of Delavan Corporation.
(7) Registered trademark of D. and W. Industries.
(8) Registered trademark of American Cyanamid Company.

NOTICE OF WARRANTY

Du Pont warrants that this product conforms to the chemical description on the label thereof and is reasonably fit for purposes stated on such label only when used in accordance with directions under normal use conditions. It is impossible to eliminate all risks inherently associated with the use of this product. Crop injury, ineffectiveness, or other unintended consequences may result because of such factors as weather conditions, presence of other materials, or the manner of use or application, all of which are beyond the control of Du Pont. In no case shall Du Pont be liable for consequential, special or indirect damages resulting from the use or handling of this product. All such risks shall be assumed by the buyer. DU PONT MAKES NO WARRANTIES OF MERCHANTABILITY OR FITNESS FOR A PARTICULAR PURPOSE NOR ANY OTHER EXPRESS OR IMPLIED WARRANTY EXCEPT AS STATED ABOVE.

AG-3262 8049 4/19/89

This is a specimen label. DuPont is not responsible for the accuracy of the information contained herein. As labels are subject to revision, always carefully read and follow the label on the product container.

GEMINI®

HERBICIDE
DISPERSIBLE GRANULES

ACTIVE INGREDIENTS: 60%
Linuron: [3-(3,4-dichlorophenyl)-1-methoxy-1-methylurea] . . 55.4%
Chlorimuron Ethyl: Ethyl 2-[[[[(4-chloro-6-methoxy-pyrimidin-2-yl)amino]- carbonyl]amino]sulfonyl]benzoate . . . 4.6%
INERT INGREDIENTS: 40%
TOTAL 100%

EPA Reg. No. 352-443

U.S. Pat. 4,394,506 4,547,215

KEEP OUT OF REACH OF CHILDREN

PRECAUTIONARY STATEMENTS

HAZARDS TO HUMANS AND DOMESTIC ANIMALS

WARNING!

CAUSES SUBSTANTIAL BUT TEMPORARY EYE INJURY AND CAUSES SKIN IRRITATION. AVOID CONTACT WITH SKIN, EYES AND CLOTHING. HARMFUL IF SWALLOWED.

In case of eye contact, flush with plenty of water and get medical attention.

In case of skin contact, wash with soap and water and get medical attention. Wash thoroughly with soap and water after handling. Remove and wash contaminated clothing before reuse.

If swallowed, call a physician or Poison Control Center. Drink 1 or 2 glasses of water and induce vomiting by touching back of throat with finger. Do not induce vomiting or give anything by mouth to an unconscious person.

For medical emergencies involving this product, call toll free 1-800-441-3637.

ENVIRONMENTAL HAZARDS

Do not apply directly to water or wetlands. Do not contaminate water by cleaning of equipment or disposal of wastes.

IMPORTANT

When mixing/loading and spraying, wear a wide-brimmed hat, mid-forearm to elbow length chemical resistant gloves, and one-piece overalls which have long sleeves and long pants. Overalls should be constructed of finely woven fabric as specified in the USDA/EPA Guide for Commercial Applicators.

Any article of clothing worn while applying product must be cleaned before reusing. Clothing which has been drenched or has otherwise absorbed concentrated pesticide must be disposed of in a sanitary landfill, or incinerated, or, if allowed by state and local authorities, by burning.

Instead of clothing and equipment specified above, the applicator can use an enclosed tractor cab which provides a filtered air supply (as specified by Taschenberg and Bourke, 1975).

Injury to or loss of desirable trees or vegetation may result from failure to observe the following: Do not apply or drain or flush equipment on or near desirable trees or other plants, or on areas where their roots may extend, or in locations where the chemical may be washed or moved into contact with their roots. Do not use on lawns, walks, driveways, tennis courts, or similar areas. Prevent drift of spray to desirable plants. Do not contaminate any body of water. Keep from contact with fertilizers, insecticides, fungicides, and seeds during storage.

PRIOR TO USING "GEMINI", CONSIDERATION SHOULD BE GIVEN TO CROP ROTATION PLANS. Crops other than soybeans can be extremely sensitive to low concentrations of "Gemini" remaining in the soil the next planting season. Choice of rotational crop is restricted following application of "Gemini". (See "ROTATIONAL CROP GUIDELINES".)

Thoroughly clean all traces of "Gemini" from application equipment immediately after use and prior to spraying crops other than soybeans. Failure to remove even small amounts of "Gemini" from application equipment may result in injury to subsequently sprayed crops.

GENERAL INFORMATION

Du Pont "Gemini" Herbicide is a dispersible granule formulation to be mixed with water and sprayed for selective preemergence weed control in soybeans. When applied according to the instructions on this label, it will control many broadleaf weeds and provide partial control of nutsedge and annual grasses.

Preemergence and preplant Incorporated applications of "Gemini" require rainfall or sprinkler irrigation to activate the herbicide. Degree of control and duration of effect depend on: rate used, weed spectrum, growing conditions at and following time of treatment, soil pH, texture, organic matter, moisture, and precipitation.

Because most crops are highly sensitive to "Gemini", all direct or indirect contact (such as spray drift) to crops or to land scheduled to be planted to crops other than soybeans should be avoided.

BIOLOGICAL ACTIVITY

"Gemini" rapidly inhibits the growth of susceptible weeds. Following activation of preplant incorporated or preemergence treatment, susceptible weeds may germinate and emerge, but growth then ceases and leaves become yellow 3-5 days after emergence. Death of leaf tissue and growing point will follow in some species while others remain green but stunted and noncompetitive. "Gemini" provides partial control of some annual grasses when used preplant or preemergence but other products may be needed to ensure adequate grass control.

DIRECTIONS FOR USE

It is a violation of federal law to use this product in a manner inconsistent with its labeling.

Do not apply this product through any type of irrigation system.

EQUIPMENT - SPRAY VOLUMES

Apply uniformly by ground equipment with a properly calibrated low pressure (20 to 40 psi) fixed-boom sprayer equipped with fan-type nozzles and screens no finer than 50 mesh. Use 10 to 40 gals. of water per acre.

Continuous agitation in the spray tank is required to keep the material in suspension. Avoid overlap, and shut off spray booms while starting, turning, slowing or stopping, or injury to the crop may result.

EARLY PREPLANT OR PREPLANT INCORPORATED (See rate chart)

"Gemini" may be applied at planting or up to 30 days prior to planting either preemergence or preplant incorporated. If incorporated, "Gemini" should be uniformly incorporated no deeper than the top 1-2" of soil prior to planting soybeans. Use equipment suited for proper incorporation into the top 1-2 inches of soil e.g.: Do-all, field cultivator, Roterra. If tank-mixed with a grass herbicide, such as a DNA like "Treflan"[4], "Sonalan"[4], or acetanilide such as "Dual"[2], or "Lasso"[1], follow label instructions for proper incorporation of the grass herbicide into the top 1 to 2" of soil. Improper soil incorporation can result in erratic weed control or crop injury. Rainfall* (—1") is required to activate preplant incorporated herbicides such as "Gemini" before weed emergence.

Use of a disk for incorporation is not recommended.

"Gemini" + "Treflan", "Prowl", "Dual" or "Sonalan" - Apply within 2 weeks of planting.

"Gemini" + "Lasso" - Apply within 7 days of planting.

PREEMERGENCE (See rate chart)

Preemergence application: Apply before soybeans emerge. Do not spray over top of emerged soybeans as severe injury or death of crop will occur.

Preemergence activity of "Gemini" requires rainfall* to activate. Best results are obtained if treatment is made to moist soil and followed by rainfall or irrigation (—1") before weeds germinate. On dry soil more moisture is required for activation (1 to 2") before weed emergence. If moisture is insufficient to activate the herbicide, a rotary hoeing or shallow cultivation should be made after emergence of crop while weeds are small enough to be controlled by mechanical means. "Gemini" can be applied preemergence in combination with "Lasso", "Dual", "Prowl"[3], "Surflan"[4], or following the use of a preplant grass material such as "Treflan" or "Sonalan." Consult labels for rates and use instructions.

* Rainfall required to activate ppi or pre treatments depends on soil moisture when activating moisture is received. Several small rainfalls of less than 1/4" are not as beneficial as one large rain of 1/2 inch or more for activation.

MINIMUM TILLAGE OR NO-TILL (See rate chart)

"Gemini" applied preemergence to soybeans but postemergence to broadleaf weeds and small annual grasses will provide some burndown of existing weeds as well as normal preemergence broadleaf control. When used for burndown add 0.25% (1 qt./100 gal) surfactant such as "Surfactant WK" or crop oil concentrate at 0.25% v/v (1 qt./100 gal). For added postemergence broadleaf weed control, "Gemini" may be tank mixed with 2,4-D low volatile ester. Do not apply 2,4-D ester within 30 days of planting. For added postemergence control especially on grasses, "Gemini" can be applied with "Roundup"[1], paraquat, or "Bronco"[1]. For added preemergence grass control "Gemini" can be mixed with "Dual", "Surflan" or "Lasso." Consult labels for rates and use instructions. Maintain constant agitation of spray mixture.

RATE CHART

Apply as follows:

Soil Texture (1/2 - 3% Organic Matter)	Rate (Ounces Per Acre)
Coarse: Loamy sand, Sandy loam	12-16
Medium: Loam, Silty loam, Silt	16-20

Soil Texture (1/2 - 3% Organic Matter)	Rate (Ounces Per Acre)
Fine: Silty Clay loam Clay loam Clay	20-24

Do not use on sand or loamy sand nor on any soil with less than 1/2 percent organic matter.

Do not apply "Gemini" on soils with pH greater than 7.0 except in those states as directed in the "ROTATIONAL CROP GUIDELINES" where application can be made to soils with a pH up to 7.5.

For band applications use proportionately less than listed in rate chart.

Plant seed 1 1/2 to 2" deep on a flat or raised seedbed only, or crop injury may occur.

Soybean stunting may occur if excessive rainfall occurs after application. Injury is more prevalent under poor drainage or compacted conditions or when soil is saturated for long periods of time. Soybeans rapidly outgrow stunting once favorable growing conditions return.

Injury to soybeans may occur if "Gemini" is used on soils having a calcareous surface layer or a pH of 7.0 or higher, or if used in conjunction with soil-applied insecticides. Seedling disease, cold or hot weather, deep planting (more than 2"), excessive moisture, high salt concentration, or drought may weaken soybean seedlings and increase possibility of crop injury. Do not apply to Black Belt Soils of Alabama and Mississippi with a soil pH greater than 7.0 or a history of nutrient deficiency such as iron chlorosis, as injury may occur.

Replanting Soybeans

If initial seeding fails to produce a stand, treated fields may be replanted to soybeans. Do not rework soil; do not retreat field with a second application as injury to the crop may result. Do not replant treated areas to any crop other than soybeans within four months after treatment as injury to subsequent crops may result.

WEEDS CONTROLLED

The following weeds will be controlled at the above rates:

Carpetweed	Mustards	Prickly sida (teaweed)
Cocklebur	Pigweed species	Purslane
Florida beggarweed	Palmer	Ragweed, common
Jimsonweed	Redroot	*Ragweed, giant
Lambsquarters	Smooth	Smartweed
*Morningglory, annual	Spiny amaranth	Sunflower
Ivyleaf	Tall waterhemp	Velvetleaf
Entireleaf		
Pitted		
Smallflower		
Tall		

Partial Control:

Annual grasses: Burcucumber, Hemp sesbania, Purple nutsedge, Sicklepod*, Wild poinsettia, Yellow nutsedge

* Large seeded weeds germinating deep in the soil such as morningglory and giant ragweed may require a cultivation or an application of Du Pont "Classic" Herbicide postemergence for season long control. Consult "Classic" label for application directions, recropping intervals and use restrictions.

NOTE: Consult "Lasso", "Dual", "Prowl", "Sonalan", "Surflan", or "Treflan" labels for additional weeds controlled when "Gemini" is used in conjunction with these herbicides.

SPRAY TANK PREPARATION

Fill tank 1/4 full with water. Start agitation system, add "Gemini" and continue adding water. Add separately each additional component of any tank mix while adding water. Continue agitation throughout. If poor mixing should occur with any component, premix the component with two parts water before adding to the spray tank.

A fertilizer solution may be used in the spray mixture. Small quantities should be tested for compatibility by the following procedures before full scale mixing.

1. Put 1 pint of fertilizer solution in a quart jar.
2. Mix 2 teaspoons "Gemini" with 2 tablespoons of water; mix thoroughly and add to fertilizer solution.
3. Close jar and shake well.
4. If other herbicides are to be used in the mixture, premix 2 teaspoons of wettable powder or 1 teaspoon of liquid with 2 tablespoons of water; add to "Gemini"/fertilizer solution mixture.
5. Close jar and shake well.
6. Watch mixture for several seconds; check again in 30 minutes.
7. If mixture does not separate, foam, gel, or become lumpy, it may be used.
8. Mixing ability may be improved by adding compatibility agents such as Kalo Laboratories' "Compex"[5] or Witco Chemicals' "Sponto"[6] 168D. Follow directions on container.

Provided the above procedure shows the mixture to be compatible, prepare the tank mixture as follows: Add the fertilizer solution to the spray tank first, with the agitator running, add the required amount of "Gemini" and thoroughly mix. For tank mixtures with other herbicides, follow directions above.

When tank mixing with other herbicides, all applicable directions, restrictions and precautions for the additional herbicides are also to be followed. Use "Gemini" spray preparations the same day as mixed or product degradation may occur. Thoroughly reagitate before using if allowed to settle.

RECROP INTERVAL FOR "GEMINI" HERBICIDE

Crops that can be planted after using "Gemini" are listed below. Choice of rotational crop is determined by geographical location (state), soil pH, rate and time interval elapsed since last application.

ROTATIONAL CROP GUIDELINES

GROUP I

States of: AL, AR, FL, GA, LA, MS, TX

■ Soil pH 7.0 or less

GROUP II

States of: DE, KY, MD, MO Bootheel, NJ, NC, PA, SC, TN, VA, WV

■ Soil pH 6.8 or less

AND

■ Use rate 16-22 oz/A

GROUP III

■ All other states not listed in Group I or II

■ Soil pH 7.0 or less OR

States of: AL, AR, FL, GA, LA, MS, TX

■ Soil pH 7.1-7.5 OR

States of: DE, KY, MD, MO Bootheel, NJ, NC, PA, SC, TN, VA, WV

■ Soil pH 6.9-7.5 OR Fields not meeting rate requirement in Group II

TIME INTERVAL IN MONTHS BEFORE PLANTING ROTATIONAL CROP

Crops	Group I	Group II	Group III
Soybeans	Anytime	Anytime	Anytime
Cereal grains	4	4	4
Field Corn	9	10	18
Clover	10	12	18
Cotton	10	10	18
Sorghum	10	10	18
Tobacco	10	10	18
Tomato (transplant)	10	10	18

All other crops require a minimum 18 month rotational interval and a successful field bioassay[a].

Suggested interval (months) before initiating bioassay is 10 months for Group I, 12 months for Group II and 15 months for Group III bioassay interval.

Do not apply "Gemini" Herbicide in areas where the average annual precipitation is less than 25 inches.

Extended dry or drought conditions following application of "Gemini" may affect rotational intervals. If such conditions are encountered, consideration should be given to recropping only to soybeans or extending the recommended rotational interval of other crops.

[a] Field Bioassay - The field bioassay will detect small quantities of "Gemini" which can remain in the soil and injure rotational crops. A successful field bioassay must be completed the season before planting any crop not listed above. A successful field bioassay means growing to maturity a test strip of the crop(s) intended for production the following year. The test strip should cross the whole field, including knolls and low areas.

SPRAYER CLEANUP

To avoid subsequent injury to crops other than soybeans, immediately after spraying and prior to spraying other crops, thoroughly remove all traces of "Gemini" Herbicide from mixing and spray equipment. For more information, consult supplemental bulletin on sprayer cleanout following use of Du Pont "Classic", "Canopy" or "Gemini" Herbicides.

PRECAUTIONS

Do not apply "Gemini" to soils with a pH greater than 7.0, except for those states listed in Group III of the "ROTATIONAL CROP GUIDELINES" where "Gemini" can be applied to soils up to 7.5 pH.

CAUTION: Soil pH varies greatly even within the same field. pH variations as much as 2 pH units are common. Determine soil pH by laboratory analysis using a 1:1 soil:water suspension.

Do not graze treated fields or harvest for forage or hay.

Do not use on sand, muck, or peat soils, nor on soils with less than 1/2% organic matter as crop injury may result.

Do not apply in areas where the average annual rainfall is less than 25 inches.

Do not apply "Gemini" after the soybean crop has emerged.

Use of "Gemini" may affect rotational crops. Follow "ROTATIONAL CROP GUIDELINES".

Do not apply to Black Belt Soils of Alabama and Mississippi with soil pH greater than 7.0, or on soils with a history of nutrient deficiencies such as iron chlorosis or injury may occur.

When recommended tank mix combinations are used, consult the product label for replanting and recropping instructions and observe the directions that are the most restrictive.

If "Scepter"[3], "Pursuit"[3] or any product containing the active ingredients, imazaquin or imazethapyr (such as "Squadron"[1], "Partner"[5]) is applied the same year as "Gemini", do not plant crops other than soybeans for at least 15 months from the last application. Crops listed in the "Gemini" Rotational Crop Guidelines" may be planted after 15 months; all other crops require a successful field bioassay.

Do not apply to land that has been or will be treated with "Ally", "Finesse" or "Glean" herbicides in the states of Nebraska, Kansas, North Dakota and South Dakota.

[1] Trademark of Monsanto Agricultural Products Co., St. Louis, MO
[2] Trademark of Ciba-Geigy Corp., Greensboro, NC
[3] Trademark of American Cyanamid Corp., Princeton, NJ
[4] Trademark of Elanco Products Co., Indianapolis, IN
[5] Trademark of Kalo Agricultural Chemicals, Inc., Overland Park, KS
[6] Trademark of Witco Chemical Corp., Houston, TX

NOTICE TO BUYER: Purchase of this material does not confer any rights under patents of countries outside of the United States.

STORAGE AND DISPOSAL

Storage: Store product in original container only, away from other pesticides, fertilizer, food, or feed.

Disposal: Do not contaminate water, food, or feed by storage or disposal. Wastes resulting from the use of this product may be disposed of on site or at an approved waste disposal facility.

Container Disposal: Triple rinse (or equivalent). Then offer for recycling or reconditioning, or puncture and dispose of in a sanitary landfill, or incineration, or, if allowed by state and local authorities, by burning. If burned, stay out of smoke.

NOTICE OF WARRANTY

Du Pont warrants that this product conforms to the chemical description on the label thereof and is reasonably fit for the purposes stated on such label only when used in accordance with the directions under normal use conditions. It is impossible to eliminate all risks inherently associated with the use of this product. Crop injury, ineffectiveness, or other unintended consequences may result because of such factors such as weather conditions, presence of other materials, or the manner of use or application, all of which are beyond the control of Du Pont. In no case shall Du Pont be liable for consequential, special or indirect damages resulting from the use or handling of this product. All such risks are assumed by the buyer. DU PONT MAKES NO WARRANTIES OF MERCHANTABILITY OR FITNESS FOR A PARTICULAR PURPOSE NOR ANY OTHER EXPRESS OR IMPLIED WARRANTY EXCEPT AS STATED ABOVE.

AG-3707 8048 4/11/88

This is a specimen label. DuPont is not responsible for the accuracy of the information contained herein. As labels are subject to revision, always carefully read and follow the label on the product container.

GLEAN®

HERBICIDE
DRY FLOWABLE

ACTIVE INGREDIENT:	BY WEIGHT
Chlorsulfuron 2-Chloro-N-[(4-methoxy-6-methyl-1,3,5-triazin-2-yl) aminocarbonyl]benzenesulonamide	75%
INERT INGREDIENTS:	25%
TOTAL	100%

EPA Reg. No. 352-404

U.S. Pat. 4,127,405

KEEP OUT OF REACH OF CHILDREN

PRECAUTIONARY STATEMENTS

HAZARDS TO HUMANS AND DOMESTIC ANIMALS

CAUTION!

MAY IRRITATE EYES, NOSE, THROAT OR SKIN.

Harmful if swallowed or absorbed through skin. Avoid contact with skin, eyes or clothing. Wash thoroughly with soap and water after handling. Remove contaminated clothing and wash before reuse.

STATEMENT OF PRACTICAL TREATMENT

If swallowed: Call a physician or Poison Control Center. Drink 1 or 2 glasses of water and induce vomiting by touching back of throat with finger. Do not induce vomiting or give anything by mouth to an unconscious person.

If in eyes: Flush eyes with plenty of water. Call a physician if irritation persists.

If on skin: Wash with plenty of soap and water. Call a physician if irritation persists.

For medical emergencies involving this product, call toll-free 1-800-441-3637.

ENVIRONMENTAL HAZARDS

Do not apply directly to water or wetlands. Do not contaminate water by cleaning of equipment or disposal of waste.

IMPORTANT INFORMATION—(READ BEFORE USING)

"GLEAN" is recommended for use on land primarily dedicated to the long-term production of wheat, barley or oats. "GLEAN" should not be used in any area where annual crop rotation is frequently practiced except as indicated for the states of AR, KS, LA, OK and TX. See "Cereal Recropping Intervals" and "Crop Rotation Recommendations (Noncereal Crops)" sections for details.

In areas having soil pH greater than 7.0, a short growing season, prolonged periods of low soil temperature and low annual rainfall, "GLEAN" can remain in the soil for 3 to 4 years or more and cause severe injury to crops other than wheat, barley or oats.

Before using "GLEAN", carefully consider your crop rotation plans and options. For rotational flexibility, do not treat all your wheat, barley, oats or fallow acres.

Injury to or loss of desirable trees or vegetation may result from failure to observe the following: Do not apply, drain or flush equipment on or near desirable trees or other plants, or on areas where their roots may extend, or in locations where the chemical may be washed or moved into contact with their roots. Do not use on lawns, walks, driveways, tennis courts or similar areas. Prevent drift of spray to desirable plants. Do not contaminate any body of water, including irrigation water that may be used on other crops. Carefully observe sprayer cleanup instructions, as spray tank residue may damage crops other than wheat, oats or barley.

INFORMATION ON RESISTANT WEEDS:

Following the use of "GLEAN" in monoculture cereals production (continuous cereals or cereal-fallow-cereal), some naturally-occurring biotypes* of certain weeds listed on this label may not be effectively controlled by this product.

If weeds listed on this label are not satisfactorily controlled, respray problem areas in a timely and effective manner using a broadleaf herbicide having a different mode of action**, such as: 2,4-D, Banvel[1]***, Buctril[2], Bronate[2], Curtail[3], MCPA, Du Pont "Karmex" Herbicide or Du Pont "Lexone" Herbicide.

To delay the occurrence of resistant biotypes, use "GLEAN" in tank mixes and/or sequential treatments with other herbicides effective on the same broadleaf weed species. Do not let weed escapes go to seed.

Consult your Ag chemical dealer, applicator, consultant, appropriate state agricultural extension service representative or your local Du Pont representatives for specific recommendations.

*Biotypes are naturally-occurring individuals of the species which have a slightly different genetic makeup. Resistant biotypes may look exactly the same as susceptible biotypes. Herbicide resistant biotypes are able to survive a use rate several times higher than needed to control susceptible biotypes.

**Mode of action is the chemical interaction that interrupts a biological process necessary for plant growth and development.

***Tank mixes with "Banvel" may result in reduced control of some broadleaf weeds.

GENERAL INFORMATION

"GLEAN" Herbicide is a dry flowable granule containing 75% active ingredient, to be mixed in water and applied as a uniform broadcast spray for selective weed control in wheat (including durum), barley and spring oats. In the states of Texas, western Oregon and western Washington, "GLEAN" may also be used on winter oats. "GLEAN" is also recommended for selective weed control in grasses on acreage enrolled in the Conservation Reserve Program. It is noncorrosive, nonflammable, nonvolatile and does not freeze.

Prior to using "GLEAN", careful consideration should be given to crop rotation plans. Crops other than wheat, barley, oats, and triticale can be extremely sensitive to low concentrations of "GLEAN" in the soil.

READ AND FOLLOW ALL APPROPRIATE SECTIONS OF LABEL INCLUDING PRECAUTIONS BEFORE USING THIS PRODUCT.

GRAZING

"GLEAN" has no grazing restrictions.

SOIL RESIDUAL ACTIVITY

In the states of CA, Northern ID, OR and WA, "GLEAN" is recommended for use on land having a soil pH of 7.9 or lower and dedicated to the long-term production of cereal grains. **The soil residual activity of "GLEAN" can injure crops other than wheat, barley or oats for 2 to 3 years or more.** "GLEAN" should not be used on soils above pH 7.9, as extended soil residual activity could adversely affect crop rotation options beyond normal intervals.

In the states of Southern ID, MN, MT, ND, SD, UT and Northern WY, "GLEAN" is recommended for use on land having a soil pH of 7.9 or lower and dedicated to the long-term production of cereal grains. **The soil residual activity of "GLEAN" can injure crops other than wheat, barley or oats for 3 to 4 years or more.** "GLEAN" should not be used on soils above pH 7.9, as the extended soil residual activity could adversely affect crop rotation options beyond normal intervals and under certain conditions cause injury to wheat, barley or oats.

In the states of AR, Central KS, LA, Central NE, Central OK and North Central TX, "GLEAN" is recommended for use on land having a soil pH of 7.9 or lower. **The soil residual activity of "GLEAN" can injure crops other than wheat, barley or oats for 2 to 4 years or more.** "GLEAN" should not be used on soils higher than pH 7.9, as extended soil residual activity could adversely affect crop rotation options beyond normal intervals.

Rainfall, soil temperature and soil pH are important factors affecting "GLEAN" breakdown in soil. "GLEAN" breakdown is more rapid under conditions of low soil pH, high soil temperature and moist soil. The breakdown process is slow under conditions of high soil pH, low soil temperature and dry soil.

IMPORTANT: UNLESS OTHERWISE SPECIFIED IN THE "CROP ROTATION RECOMMENDATIONS (NONCEREAL CROPS)" SECTION OF THIS LABEL, land previously treated with "GLEAN" cannot be rotated to crops other than wheat, oats, barley or triticale until a bioassay confirms that residues of "GLEAN" that could cause crop injury are not present. See "Bioassay" section of this label for details. Failure to follow these instructions could result in injury to subsequent crops.

For crop rotation flexibility do not use "GLEAN" on all your wheat, barley, or spring oats.

HOW APPLICATION TIMING AND ENVIRONMENTAL CONDITIONS AFFECT WEED CONTROL AND CROP SAFETY

How Growing Conditions and Crop Density Affect Weed Control—Postemergence application to weeds are most effective when "GLEAN" is applied to young, actively growing weeds which are less than 2" tall or 2" across. Warm, moist growing conditions promote active weed growth and enhance the activity of "GLEAN" by allowing maximum foliar uptake. If cold, dry conditions exist, delay postemergence treatment until weather conditions promote active weed growth. For best results with treatments made postemergence to weeds during periods of cold temperatures, apply "GLEAN" when minimum day/night temperatures are 40 Deg. F or higher.

Avoid postemergence applications to weeds which are inactive due to adverse weather conditions. Weeds hardened off by cold weather or drought stress may not be controlled.

A vigorous, growing crop will aid weed control by shading and providing competition for weeds. However, a dense crop canopy at time of application can intercept spray and result in reduced weed control. Weeds may not be adequately controlled in areas of thin crop stand or seeding skips.

How Rainfall After Treatment Affects Weed Control—Rainfall after treatment will affect "GLEAN" performance when applied postemergence or preemergence to weeds. Without sufficient rainfall to move "GLEAN" into the weed root zone, weeds that germinate after treatment will not be controlled.

Postemergence treatments control or suppress weeds through both foliar and root uptake. To maximize "GLEAN" activity on existing weeds, sufficient rainfall is needed soon after treatment to move "GLEAN" into the weed root zone, **before** weeds develop an established root system and grow beyond the seedling stage.

Avoid making applications postemergence to weeds when rainfall is threatening. Rainfall immediately after treatment can wash "GLEAN" off weed foliage and result in reduced weed control effectiveness. Several hours of dry weather is needed to allow "GLEAN" to be absorbed by weed foliage.

When weed emergence is uneven, control of weeds that germinate after treatment will be dependent on the timing and amount of rainfall following application. Sufficient rainfall is needed to move "GLEAN" 2 to 3" deep into the weed root zone before weeds that germinate after treatment can develop an established root system. When favorable growing conditions exist, rainfall may be needed within a few days after treatment.

For best preemergence results, it is important to apply "GLEAN" when you can expect at least 1 to 2" (clay soils require more than sandy soils) of rain or sprinkler irrigation to move "GLEAN" 2 to 3" deep into the soil profile before weed seeds germinate and develop an established root system. Weeds that germinate after treatment and develop an established root system before rainfall moves "GLEAN" into the soil profile may not be controlled.

How Growing Conditions Affect Crop Safety—Prolonged cold weather (daily maximum temperature below 50°F) and/or drought, and/or low fertility while crop is in seedling stage (1 to 5 leaf) can cause crop injury following either a preemergence or postemergence treatment.

To avoid the risk of cold weather related injury, apply "GLEAN" when good growing conditions are expected to continue until crop has started to tiller.

DIRECTIONS FOR USE

It is a violation of federal law to use this product in a manner inconsistent with its labeling.

"GLEAN" Herbicide should be used only in accordance with recommendations on this label or in separate published Du Pont recommendations available through local dealers.

Du Pont will not be responsible for losses or damages resulting from the use of this product in any manner not specifically recommended by Du Pont. User assumes all risks associated with such nonrecommended use.

MAXIMUM USE RATES AND SOIL PH LIMITATIONS

In the states of CA, Northern ID, OR and WA, the maximum use rate is 1/3 oz/A per crop on soils having a pH of 7.9 or lower. Do not apply more than 1/3 oz/A in a 18 month period. Do not use on soils having a pH greater than 7.9.

In the states of Southern ID, MN, MT, ND, SD, UT and Northern WY, the maximum use rate is 1/3 oz/A in a 48 month period on soils having a pH of 7.9 or lower. Do not use on soils having a pH greater than 7.9.

In the states of Central KS, Central NE and Central OK, the maximum use rate is 1/3 oz/A per crop period on soils having a pH of 7.9 or lower. Do not use on soils with a pH greater than 7.9.

In the states of AR, LA, Central and North Central TX and Southern OK, the maximum use rate is 1/2 oz/A per crop period on soils having a pH of 7.9 or lower. Do not use on soils with a pH greater than 7.9.

NOTE: Prior to use of "GLEAN", take soil samples at 0-4" depth and determine soil pH by laboratory analysis using a 1:1 (soil:water) suspension. Samples should be representative of the different conditions in the field (for example, slope, soil texture, low areas, eroded areas, etc.). Consult local extension publications for recommended soil sampling procedures.

APPLICATION TECHNIQUES AND TIMING

Preemergence (After Planting) To Wheat (Including Durum)

Preemergence applications of "GLEAN" are only recommended where foxtail or annual ryegrass are the target weeds.

Apply "GLEAN" (1/6 - 1/3 oz/A for foxtail or 1/6 to 1/2 oz/A for ryegrass) after planting, but before crop emergence. Rainfall or sprinkler irrigation following treatment is necessary to activate "GLEAN" before weed seeds germinate and develop an established root system. **Wheat must be planted at least 1" deep.** For best results apply "GLEAN" uniformly to a smooth seedbed.

Do not apply preemergence to late fall seedlings when cold and/or dry weather can delay seedling emergence and reduce seedling vigor. If these conditions exist, delay treatment until crop has emerged and weather conditions allow active wheat growth and wheat is showing good vigor.

Preemergence applications of "GLEAN" are not recommended where organophosphate insecticides (such as "Di-Syston" [4], etc.) have been used as an in-furrow treatment as crop injury may occur.

When environmental conditions cause delayed seedling emergence and/or poor seedling vigor, delay post treatment irrigation until after the wheat is actively growing and is showing good vigor.

NOTE: DO NOT APPLY PREEMERGENCE (FALL OR SPRING) TO IRRIGATED DURUM WHEAT.

NOTE: DO NOT APPLY PREEMERGENCE (FALL OR SPRING) TO BARLEY, SPRING OATS OR WAMPUM VARIETY OF SPRING WHEAT AS CROP INJURY MAY RESULT.

Split-Treatment To Wheat

"GLEAN" can be applied fall postemergence plus spring postemergence provided that each application is made with another broadleaf herbicide. Refer to "TANK MIXTURES FOR RESISTANT WEED MANAGEMENT". Allow at least 30 days between treatments. Do not apply more than the maximum use rate, per crop year, as indicated in the MAXIMUM USE RATES AND SOIL PH LIMITATIONS section of label. Do not make more than 2 treatments per crop. Apply last application before boot stage. Base recropping interval on date of last application and total amount of "GLEAN" used.

Preplant Incorporation (PPI) and Preplant Surface (PPS) Applications To Early Seeded Winter Wheat in the States of MT, ND, SD and Northern Wyoming Where Foxtail is the Target Weed

NOTE: Preplant incorporation and preplant surface treatments are only recommended for early seeded winter wheat where growing conditions are favorable (good soil moisture, moderate temperatures) for good stand establishment prior to winter dormancy.

Apply "GLEAN" as a uniform broadcast spray not more than 3 weeks prior to the anticipated planting of early seeded winter wheat.

Use 1/3 oz/acre.

Winter wheat may be planted anytime after treatment.

- **INCORPORATION (PPI)**—Follow spray application with shallow (not deeper than 3 to 4") mechanical incorporation. Use either single pass or double pass incorporation (second pass at right angle to first pass) with sweeps (duckfoot cultivator), spring tooth or field cultivator. Incorporation may be improved if a harrow is pulled behind the primary incorporation implement.

- **SEEDING AFTER EITHER PPI OR PPS TREATMENT**—The use of disc type drills is recommended because of minimal soil disturbance. A hoe type drill may be used if drill spacing is not more than 10" wide and tractor speed is at least 5 mph. If a hoe drill is used, weed control results may be variable depending on amount of soil displacement in seed row. Where practical, a harrow pulled behind a hoe drill should increase the effectiveness of this treatment.
- **PRECAUTIONS**

Because of variations in incorporation equipment and seeding techniques, it is recommended that growers limit first use of either PPI or PPS to a small area to be sure weed control results are satisfactory.

Excessive displacement of treated soil may result in poor weed control in the seed row.

In high rainfall situations or on low pH soils (pH less than 6.5) a second application may be needed in the spring. Refer to instructions for split application.

Do not apply prior to late fall plantings as cold/dry weather can delay seedling emergence and reduce seedling vigor, making crop more vulnerable to the combination of herbicide and weather stress, resulting in crop injury.

DO NOT MAKE A PREPLANT INCORPORATED OR PREPLANT SURFACE APPLICATION PRIOR TO PLANTING BARLEY OR SPRING OATS.

Dry Fertilizer Impregnation And Application To Winter/Spring Wheat And Durum In Montana And The Dakotas

- **IMPREGNATION**—The herbicide/fertilizer impregnation process must be done at commercial fertilizer or chemical dealerships that are properly equipped for this procedure.

NOTE: The practice of impregnating "GLEAN" on dry fertilizer is recommended only for dealers whose primary crop business is wheat, barley and oats. Failure to thoroughly clean all traces of "GLEAN" from equipment used to mix or apply dry fertilizer for use on other crops may result in crop injury.

Not more than 1/4 to 1/3 oz of "GLEAN" should be impregnated on a minimum of 150 lbs of dry fertilizer per acre.

Slurry the "GLEAN" in water using 1 part "GLEAN" to at least 5 parts water (1-18 oz jug in 3 qts water). Do not exceed slurry volume of 1 pt per 100 lbs of fertilizer. Continuous agitation (mechanical or recirculating) is required to keep "GLEAN" in suspension during the impregnation process.

To impregnate, mix and blend the dry fertilizer and herbicide in a closed rotary drum-type mixer allowing sufficient time to ensure uniform coverage. The delivery nozzle(s) must be placed inside the mixer and positioned to provide uniform spray coverage of the tumbling fertilizer. Use "GLEAN" impregnated dry fertilizer as soon as possible after blending.

Before using blending and/or application equipment to subsequently mix or apply fertilizer to crops other than wheat, barley or oats, thoroughly clean all traces of "GLEAN" and "GLEAN" impregnated fertilizer from equipment.

NOTE: All state regulations relating to dry bulk fertilizer blending, registration, labeling and application are the responsibility of the individual and/or company selling the fertilizer/herbicide mixture.

- APPLICATION—Apply "GLEAN" impregnated dry fertilizer as an early fall treatment before planting or after crop emergence. Spring applications should only be made before planting. Spring wheat and durum may be planted into fall applications in accordance to directions provided in the "Fall Application Prior to Planting Spring Wheat (including Durum)" section of this label. Spread the herbicide treated fertilizer uniformly with a properly calibrated applicator. When using fan spreaders, a 100% overlap is recommended. Fan spreaders should be calibrated to apply 1/2 the desired rate per acre. Application pattern should be overlapped to cover 1/2 of the previous swath.
- INCORPORATION

  FALL—Mechanical incorporation is not needed for early fall applications as fall and winter rain and snow is usually sufficient to move "GLEAN" into the weed root zone. Weed control may not be satisfactory in dry years or from late fall applications.

  If mechanical incorporation is desired prior to planting, use single pass, double incorporation with sweeps (duckfoot cultivator) followed by springtooth (flextine) harrow, or use double-pass incorporation (second pass at right angle to first pass) with a culti-harrow, spike tooth or springtooth harrow or sweeps (duckfoot cultivator). Shallow incorporation not deeper than 3-4" is recommended.

  SPRING—Because spring rainfall is often undependable, mechanical incorporation is recommended prior to planting. Use same incorporation procedures described for fall mechanical incorporation. Best results are obtained when rainfall (1-2") follows mechanical incorporation prior to weed emergence. In dry conditions, weed control may not be satisfactory.

Fall Application Prior To Planting Spring Wheat (Including Durum) in the States of MT, ND, SD and Northern Wyoming Where Foxtail is the Target Weed

Apply "GLEAN" (1/3 oz/A) in the fall to undisturbed stubble where straw is spread evenly or after cultivation to a uniform soil surface. Shallow tillage, not more than 4" deep, may be done after application. In the spring use shallow tillage to prepare a seedbed. Do not moldboard plow. Fall application is not effective for Canada thistle emerging the following spring. See Canada thistle under, SPECIFIC WEED PROBLEMS.

NOTE: DO NOT PLANT IRRIGATED DURUM WHEAT AFTER A FALL APPLICATION OF "GLEAN".

NOTE: DO NOT PLANT SPRING BARLEY, WAMPUM SPRING WHEAT OR SPRING OATS AFTER A FALL APPLICATION OF "GLEAN".

Postemergence To Winter/Spring Wheat, Durum*, Winter/Spring Barley and Spring Oats

- WINTER WHEAT/WINTER BARLEY

  Apply "GLEAN" (1/6 - 1/3 oz/A) in the fall or spring anytime after the crop is in the 2-leaf stage—**but before boot stage.**

  Do not make an early postemergence treatment to late seeded wheat or barley as the combined effect of herbicide stress plus cold weather and/or moisture stress could cause crop injury. Delay making a postemergence treatment to late seeded wheat or barley until crop has started to tiller.

  In areas where cold weather conditions can be severe (ID, OR, UT, WA, WY, MT, ND, SD, MN) do not make a late fall, winter or early spring application to wheat or barley until crop is well established and has started to tiller.

  Do not apply "GLEAN" within 60 days of crop emergence where organophosphate insecticides (such as "Di-Syston", etc.) have been used as an in-furrow treatment as crop injury may result.

  DO NOT apply during boot stage or early heading as crop injury may occur.
- SPRING WHEAT, DURUM*, SPRING BARLEY AND SPRING OATS

  In the Pacific Northwest, apply "GLEAN" (1/6 to 1/3 oz/A) anytime after crop is in the 2-leaf stage through the 2nd joint stage. DO NOT apply once the flag leaf is visible as crop injury may occur.

  In all other areas, apply "GLEAN" (1/6 to 1/3 oz/A) anytime after crop is in the 2-leaf stage—but before boot stage. Do not apply during boot stage as crop injury may occur.

For irrigated cereal crops, delay first post treatment irrigation for at least 3 days after treatment.

To avoid the risk of cold weather related crop injury, apply "GLEAN" when good growing conditions (adequate soil moisture, daily high temp. of 50°F or more) are expected to continue until crop has started to tiller.

Do not apply prior to tillering when cold and/or dry weather can reduce seedling vigor making crop more vulnerable to the combination of herbicide and weather stress. The combined effect of herbicide plus stress from cold and/or dry weather can result in temporary yellowing or crop injury (yield reduction).

* NOTE: APPLY TO VIC DURUM AFTER EARLY TILLERING, BUT BEFORE BOOT STAGE.

Weed Control For The Conservation Reserve Program (CRP)

"GLEAN" is registered for CRP use in CO, IA, ID, KS, MN, MT, NE, NM, ND, OK, OR, SD, TX, UT, WA and WY. Consult "GLEAN" supplemental label for CRP use instructions.

Preemergence Or Postemergence Application To Winter Oats - Texas, Western Oregon And Western Washington Only

Preemergence to Oats: Apply "GLEAN" at 1/3 oz/A as a post planting preemergence treatment to early seeded winter oats. Use the 1/2 oz/A (Central and Northeast TX) rate where annual ryegrass is the primary weed problem.

Do not make a preemergence treatment to late fall plantings (after November 1) as herbicide stress plus cold weather stress can cause crop injury.

Remove grazing cattle during wet (muddy) field conditions to avoid disturbing the herbicide barrier.

Heavy rainfall between the time of treatment and the 2-leaf crop stage can result in temporary yellowing and stunting and may result in crop injury.

Postemergence to Oats (broadleaf control only): Apply "GLEAN" at 1/6 to 1/3 oz/A when crop is in 2-leaf to boot stage. When weeds are present at the time of application, add a surfactant (80% active ingredient or more) at the rate of 1 to 2 quarts per 100 gallons of spray solution. Fall applications of less than 1/3 oz/A may not provide adequate control of spring germinating broadleaf weeds.

NOTE: Under abnormally wet conditions, especially on coarse textured soils, fall applications may not provide adequate control of ryegrass and/or spring control of ryegrass and/or spring germinating broadleaf weeds.

Remove grazing cattle during wet (muddy) field conditions to avoid disturbing the herbicide barrier.

Postemergence applications generally do not provide adequate suppression of annual ryegrass.

WEED CONTROL IN REDUCED TILLAGE FALLOW

DO NOT USE "GLEAN" IN FALLOW EXCEPT FOR FALL APPLICATIONS PRIOR TO PLANTING SPRING WHEAT FOR THE CONTROL OF FOXTAIL IN THE STATES OF MT, ND, SD AND NORTHERN WYOMING.

WEED CONTROL

GENERAL INFORMATION

"GLEAN" rapidly inhibits growth of susceptible weeds. However, typical symptoms (discoloration) of dying weeds may not be noticeable for 1 to 3 weeks after application depending on growing conditions and weed susceptibility.

Degree of control and duration of effect depend on: a) rate used, b) weed spectrum, c) weed size, d) degree of weed infestation, e) growing conditions at and following time of treatment, f) length of growing season, g) soil pH, h) soil organic matter and i) precipitation.

For maximum weed control or suppression, always use the highest recommended rate for your area, soil pH and weed problem. Do not use less than 1/6 oz/A.

Because of rapid breakdown in soil, "GLEAN" may not provide season long weed control on soils below pH 6.5.

For best results with application made postemergence to weeds, apply "GLEAN" to small (less than 2"tall or 2"across), actively growing weeds. Add a surfactant of at least 80% active ingredient at the rate of 1-2 quarts/100 gallons of spray solution. The higher rate of surfactant is particularly useful with spray volumes of 5 GPA or less and when using low rates of "GLEAN". The use of surfactants having less than 80% active ingredient may reduce weed control.

Sufficient rainfall after preemergence or postemergence treatment is necessary to move "GLEAN" 2-3" into the weed root zone **before** weed seeds germinate and develop an established root system or existing weeds grow beyond the seedling stage. In most areas, fall treatments provide the best opportunity for rainfall activation and most consistent residual weed control. Late spring applications may not receive enough rainfall after treatment resulting in poor weed control. Without sufficient rainfall to move "GLEAN" into the weed root zone, weeds that germinate after treatment will not be controlled.

Weed Control/Use Rate Table

NOTE: Read and follow all instructions under "Specific Weed Problems" for all weeds marked with "†".

The 1/6 oz/A use rate is recommended only for short-term control or suppression. Use 1/3 oz/A where soil residual weed control is important.

Where soil pH is 6.5 or lower, use the 1/3 oz/A rate where maximum soil residual weed control is important.

The 1/2 oz/A use rate is recommended only for the control/suppression of annual ryegrass in AR, LA, OK and TX.

Weeds Controlled at 1/6 thru 1/4 Ounce Per Acre		
Blue mustard	Henbit	Smooth pigweed
Conical catchfly	Mayweed	Tansymustard† (except ID, OR, UT, WA)
Curly dock	Miners Lettuce	Treacle mustard
Field pennycress	Pineappleweed	Tumble mustard (Jim Hill)
Flixweed† (except ID, OR, UT, WA)	Prostrate pigweed	Waterpod
Hempnettle	Redroot pigweed	Wild mustard
	Shepherdspurse	

Weeds Controlled at 1/3 Ounce Per Acre			Weeds Suppressed* at 1/3 Ounce Per Acre		
Bur beakchervil	False flax	White cockle	Annual ryegrass†	+ Kochia†	+ Russian thistle†
Buttercup	+ Kochia† (in MT and ND only)	Wild carrot	Bedstraw	Pennsylvania smartweed	Sunflower† (in TX partial control only)
Coast Fiddleneck (tarweed)	Ladysthumb	Wild turnip	Canada thistle†	+ Prickly lettuce	Speedwell
Common chickweed	Lambsquarters†		Corn gromwell	Prostrate knotweed†	Tansymustard†
Common groundsel	Mouseear chickweed		Flixweed†		Wild buckwheat
Corn spurry	Purslane (common)		Giant foxtail†		Wild garlic/Wild onion†
Cow cockle	Redstem filaree		Green foxtail† (pigeongrass)		Wild radish†
False chamomile	+ Russian thistle† in MT and ND only)				Yellow foxtail†

* Weed suppression is a visual reduction in weed competition (reduced population and/or vigor) as compared to an untreated area. Degree of suppression will vary with rate used, size of weeds and environmental conditions following treatment.

+ Naturally-occurring resistant biotypes of these weeds are known to occur in the Central Plains and the Pacific Northwest. See "Tank Mixtures For Resistant Weed Management" section of label for additional information.

† See "Specific Weed Problems".

SPECIFIC WEED PROBLEMS

Canada Thistle: Apply "GLEAN" plus surfactant after majority of thistles have emerged and while they are small (rosette stage to 4"-6" tall), but actively growing. A single application will effectively inhibit the ability of Canada thistle to compete with the crop. For maximum long-term effect, yearly treatment may be required.

Annual Ryegrass: (Southwest Arkansas/Northwest Louisiana): Apply "GLEAN" at 1/2 oz/A preemergence to ryegrass. 1/2 to 1" of rainfall is needed to move "GLEAN" into the weed root zone prior to ryegrass emergence. Remove grazing cattle during wet (muddy) field conditions to avoid disturbing the herbicide barrier. Under abnormally wet conditions, especially on coarse textured soils, fall applications may not adequately control ryegrass and/or spring germinating broadleaf weeds.

Annual Ryegrass: (Southeast Oklahoma, Central and North Central TX): Apply "GLEAN" at 1/2 oz/A preemergence to ryegrass. 1/2 to 1" of rainfall is needed to move "GLEAN" into the weed root zone prior to ryegrass emergence. Remove grazing cattle during wet (muddy) field conditions to avoid disturbing the herbicide barrier. Under abnormally wet conditions, especially on coarse textured soils, fall applications may not adequately control ryegrass and/or spring germinating broadleaf weeds.

For best results, a sequential treatment of "GLEAN" followed by Du Pont "Ally" Herbicide is recommended. Apply "GLEAN" as stated above then follow with 1/10 oz/A of "Ally" after completion of wheat grazing, but prior to boot stage of the wheat. For fields not grazed, apply the sequential application of "Ally" as soon as ryegrass starts to grow after winter dormancy. "Ally" may be applied with a surfactant or with a liquid nitrogen fertilizer topdressing application. Do not add a surfactant to liquid nitrogen fertilizer plus "Ally" combinations. In mixing "Ally" with liquid fertilizer, slurry "Ally" in water then thoroughly mix the slurry into the fertilizer. Run a tank mix compatibility test before mixing "Ally" in fertilizer solution. DO NOT use with fertilizers having a pH of 3.0 or less as rapid product degradation can occur. The addition of 2,4-D is not recommended for annual ryegrass suppression.

Foxtail/Pigeongrass (green, yellow and giant): In MN, MT, ND and SD, fall applications at 1/3 oz/A in winter wheat or prior to planting spring wheat will control these foxtail species. Applications made in the spring (only on land that has been in fallow the previous year), also give suppression. For best results in the spring, apply preemergence to foxtail at 1/3 oz/A. Postemergence applications at 1/3 oz/A should be made with surfactant before the foxtail is more than 1" tall or beyond the 1-2 leaf stage. 1/2 to 1" of rainfall is needed (after either a pre or postemergence treatment) to move "GLEAN" into the weed root zone before the foxtail is beyond the 2-3 leaf stage. Without adequate rainfall incorporation before the foxtail is beyond the 2-3 leaf stage, foxtail suppression may not be adequate.

Fall application may only provide short-term suppression.

Excessive rainfall after treatment may result in unsatisfactory weed control performance.

Wild Buckwheat: In MN, MT, ND and SD, fall applications at 1/3 oz/A will control wild buckwheat. Applications made in the spring give suppression. For best results in the spring, apply "GLEAN" with surfactant just after seedlings have emerged and are actively growing. Late spring applications may not receive enough rainfall after treatment, resulting in poor residual weed suppression.

In KS, OK and TX, best results are from fall or early spring applications made preemergence to wild buckwheat. Postemergence applications should be made with surfactant just after seedlings have emerged and are actively growing.

In all other states, fall applications should provide best results. If treatment is delayed until spring, apply "GLEAN" with surfactant just after seedlings have emerged and are actively growing.

Lambsquarters: For best results, use not less than 1/3 oz/A applied in the fall. For spring postemergence application, apply when lambsquarters are less than 2" tall or 2" across and are actively growing. Use not less than the 1/3 oz/A rate of "GLEAN" plus 2 qt surfactant/100 gal of spray solution.

Russian Thistle and Kochia:

"GLEAN" is not recommended for the control of these two weeds in ID, OR and WA.

Postemergence Suppression: Apply "GLEAN" plus either 2,4-D (ester or amine) or MCPA (ester or amine) after majority of weeds have emerged. For best results, weeds must be actively growing at time of application (adequate soil moisture and daily temperatures above 60 Deg. F). Add surfactant at 1/2 but not more than 1 qt/100 gal of spray solution. Thorough coverage is important. See "Tank Mixtures for Resistant Weed Management" and "Tank Mixtures" section of label.

Sunflower—For best results in NM, OK (Panhandle) and TX, apply "GLEAN" after majority of sunflowers have emerged and are small (not more than 2" tall) and actively growing. Add surfactant at 2 qt/100 gal of water. If "GLEAN" is applied preemergence, make application in early spring to allow for timely and adequate rainfall to move "GLEAN" into the weed root zone before weeds germinate or develop an established root system.

In all other areas, fall applications provide best results. If treatment is delayed until spring, best results are from a postemergence application made after most of the sunflowers have emerged and are not more than 2" tall. The addition of 2 qt of surfactant/100 gal of water will improve postemergence activity. Sunflowers that emerge after a spring treatment may not be controlled.

NOTE: In high rainfall areas, fall applications may not provide adequate residual control of sunflowers. Deep germinating sunflowers that emerge after a spring treatment may not be controlled.

Flixweed, Tansymustard:

STATES OF ID, OR, UT, WA

POSTEMERGENCE TREATMENTS

For best results, tank mix "GLEAN" at 1/3 oz/A with another herbicide effective on flixweed and tansymustard such as 2,4-D (ester or amine) to improve control of these weeds. See "Tank Mixtures for Resistant Weed Management" and "Tank Mixtures" section of label.

ALL OTHER AREAS

Rates of 1/6 to 1/3 oz/A applied when weeds are small and actively growing will provide control. If weeds are inactive due to adverse weather conditions (cold, dry weather before and/or after treatment), delay application until moisture and temperature conditions are favorable for active weed growth, or use a tank mix treatment with 2,4-D or MCPA.

Wild Garlic/Wild Onion: "GLEAN" will provide aerial bulblet control only.

Wild Radish: Postemergence application will provide best results.

Prostrate Knotweed: For best results apply in the fall.

TANK MIXTURES FOR RESISTANT WEED MANAGEMENT

Central Kansas, Central Nebraska, Central Oklahoma and North Central Texas:

- "GLEAN" may be used annually as a tank mix treatment after weeds have emerged. Use 1/6 to 1/3 oz/A of "GLEAN" plus 1/4 to 1/2 lb. active ingredient MCPA or 2,4-D (ester formulations of 2,4-D or MCPA have provided best results). Surfactant may be added at 1/2, but not more than 1 qt/100 gal. of spray; however, the addition of surfactant may increase the chance of crop injury. Apply "GLEAN" plus MCPA from 3-5 leaf stage but prior to boot stage. Apply "GLEAN" plus 2,4-D after tillering (refer to appropriate 2,4-D manufacturer's label) but prior to boot stage.
- Make only one application per crop.
- Do not apply "GLEAN" during fallow.
- Do not make an early season treatment where a tank mix cannot be used.
- If resistant weeds are known to be present, consider using another herbicide treatment or adjust the use rate of the "GLEAN" tank mix partner so that it alone will control the resistant species.
- Read and follow all use instructions, warnings and precautions for the companion herbicide.

Colorado, Western Kansas and Western Nebraska (west of highway 183), Eastern New Mexico, Oklahoma Panhandle, Texas Panhandle and Southeastern Wyoming:

"GLEAN" is not recommended for use in this area. Use "Ally" or Du Pont "Harmony" Herbicide in tank mix with other broadleaf herbicides. See "Ally" and "Harmony" labels for details.

Southern Idaho, Montana, North Dakota, South Dakota and Northern Wyoming:

- Apply "GLEAN" not more than once every 48 months.
- Make only one application per crop period.
- Do not use "GLEAN" during fallow.
- Discontinue the use of "GLEAN" for at least 48 months on any fields where "GLEAN" has been previously applied 3 times or more.
- For all applications made after weeds have emerged, use "GLEAN" in a tank mix with another broadleaf herbicide such as 2,4-D, MCPA etc.
- If resistant weeds are known to be present, consider using another herbicide treatment or adjust the use rate of the "GLEAN" tank mix partner so that it alone will control the resistant species.
- Read and follow all use instructions, warnings and precautions for the companion herbicide.

Northern Idaho, Oregon and Washington:

- Do not use more than 1/3 oz/A in an 18 month period.
- Do not make an early season treatment where a tank mix cannot be made.
- Do not use "GLEAN" for the control of kochia or Russian thistle.
- Do not apply "GLEAN" during fallow.
- Apply as a postemergence tank mix or split[1] treatment in the fall and/or spring anytime after the majority of weeds have emerged and after crops is in the 2 leaf stage. Use 1/6 to 1/3 oz/A of "GLEAN" plus one of the products* listed below:

2,4-D (amine or ester)	4 to 8 oz active ingredient/acre
MCPA (amine or ester)	4 to 8 oz active ingredient/acre
"Buctril" 4EC	1/4 pt to 1 pt/acre
"Bronate"	1/2 pt to 2 pt/acre
"Karmex" DF or diuron DF	1/2 lb to 1-1/2 lb/acre
"Lexone" DF	1/8 to 2/3 lb/acre
"Banvel"**	1/16 to 1/4 pt/acre
"Curtail"	1 to 2 pt/acre

*Read and follow all instructions, warnings and precautions on the companion product label.

**Tank mixes with "Banvel" may result in reduced control of some broadleaf weeds.

1 Split treatments (fall postemergence plus spring postemergence) can be made providing the maximum use rate per 18 month period (1/3 oz/A) is not exceeded and all applications are made in a tank mix with another broadleaf herbicide.

Note: "Harmony" Herbicide in tank mixes can be use as a sequential treatment after a fall application of "GLEAN".

If resistant weeds are known to be present, consider using another herbicide treatment or adjust the use rate of the "GLEAN" tank mix partner so that it alone will control the resistant species.

TANK MIXTURES WITH MCPA, 2,4-D, "KARMEX" AND "LEXONE" DF HERBICIDES

MCPA (amine or ester) or 2,4-D (amine or ester) plus "GLEAN": Tank mixtures of "GLEAN" plus either MCPA or 2,4-D are recommended when weeds are large and/or stressed due to adverse environmental conditions (cold temperature, low soil moisture, dry, dusty field conditions) or when dense crop canopy makes it difficult to obtain thorough spray coverage.

Use "GLEAN" at 1/6 to 1/3 oz/A plus 1/4 to 1/2 lb active ingredient MCPA or 2,4-D (ester formulations of 2,4-D and MCPA have provided best results). Surfactant may be added at 1/2, but not more than 1 qt/100 gal of spray; however, the addition of surfactant may increase the chance of crop injury.

"GLEAN" should be mixed in water with the agitator running prior to adding 2,4-D or MCPA. Read and follow all label instructions on timing, precautions and warnings for these herbicides prior to using these tank mixtures.

"Karmex" plus "GLEAN": Where prickly lettuce, corn gromwell, wild buckwheat, annual ryegrass and annual bluegrass are the main weed problems, apply "Karmex" at 1 to 1 1/2 lb/A with "GLEAN" at 1/6 to 1/3 oz/A to improve weed control. Apply preemergence or postemergence to actively growing weeds less than 2" tall or 2" across. 1/2 to 1" of rainfall is needed within 1 to 2 weeks after application. Follow all restrictions on the "Karmex" label.

Do not use "Karmex" plus "GLEAN" in CA.

"Lexone" DF plus "GLEAN": In the states of ID, OR, UT and WA, where downy brome, gromwell, speedwell and jagged chickweed are the main problems in winter wheat and barley, an application of "Lexone" DF at 1/3 to 1/2 lb/A with "GLEAN" at 1/6 to 1/3 oz/A is recommended. Apply after winter wheat or barley is well tillered and has a 2" secondary root system established throughout the field. 1/2 to 1" of rainfall is needed within 1 to 2 weeks of application. Do not use "Lexone" DF plus "GLEAN" on spring wheat. Follow all restrictions on the "Lexone" DF label.

In the states of KS, OK and TX, "Lexone" DF at 1/3 to 1/2 lb/A is recommended for downy brome and cheatgrass control and may be mixed with "GLEAN" at 1/6 to 1/3 oz/A to broaden the spectrum of weeds controlled. Apply after winter wheat is well tillered and has a 2" secondary root system throughout the field. 1/2 to 1" of rainfall is needed within 1 to 2 weeks of application.

Do not use "Lexone" DF plus "GLEAN" on barley in the states of KS, OK and TX.

Do not use "Lexone" DF plus "GLEAN" in CA.

SPRAY PREPARATION, PRODUCT MEASUREMENT, SURFACTANT AND LIQUID FERTILIZER

Spray Preparation:

Mix the proper amount of "GLEAN" into the necessary volume of water in the spray tank with the agitator running. Agitation is required for uniform mixing and application. If spray preparation is left standing, thoroughly reagitate before using.

Additives: Do not use with spray tank additives that lower the pH of the spray solution below pH 3.0, as rapid product degradation can occur.

Product Measurement: The "GLEAN" volumetric measuring cylinder is to be used as a guide as the degree of accuracy is plus or minus 10%. For more precise measurement, use scales calibrated in ounces.

Surfactant: Use a surfactant of at least 80% active ingredient in postemergence applications to weeds to improve wetting and/or foliar activity of "GLEAN". Add surfactant at 1 to 2 qt/100 gal of spray as the last ingredient. The higher rate of surfactant is particularly useful with spray volumes of 5 GPA or less and when using low rates of "GLEAN". Antifoaming agents may be needed.

Do not use liquid fertilizer as a substitute for surfactant.

Liquid fertilizer: To apply "GLEAN" with liquid fertilizer, slurry the "GLEAN" in water; then thoroughly mix the slurry into the liquid fertilizer. The addition of surfactant to tank mixtures of "GLEAN" plus liquid fertilizer increases the risk of crop injury.

Run a tank mix compatibility test before mixing "GLEAN" in fertilizer.

Do not use with liquid fertilizers having a pH of 3.0 or less as rapid product degradation can occur.

Do not use liquid fertilizer as a substitute for surfactant.

TANK MIXTURES WITH OTHER HERBICIDES, INSECTICIDES AND FUNGICIDES

When using a tank mixture for the first time, run a standard compatibility test and use on a small portion of the field to be sure there is adequate crop safety and satisfactory performance before adopting large scale use. "GLEAN" must be in suspension before adding companion pesticides. Follow all instructions, warnings and precautions on the companion product label.

Other Herbicides: Use a suitable registered companion herbicide if weeds greater than 2" tall or 2" across are present or if weeds and grasses other than those listed on this label are present. "GLEAN" must be in suspension in the spray tank before adding the companion herbicide. Follow the surfactant recommendation on the companion herbicide label.

Insecticides: "GLEAN" may be tank mixed with insecticides registered for use on cereal grains. However, under certain conditions, (drought or cold stress, while crop is in 2-4 leaf stage) tank mixtures or sequential treatments of "GLEAN" and organophosphate insecticides (such as methyl or ethyl, parathion, "Di-Syston", etc.) may produce temporary crop yellowing or, in severe cases, crop injury. Limit first use to a small area. If no symptoms of crop injury occur 14 days after treatment, balance of acreage can be treated.

DO NOT USE "GLEAN" AND MALATHION AS CROP INJURY MAY RESULT.

Do not apply "GLEAN" within 60 days of crop emergence where an organophosphate insecticide (such as "Di-Syston") has been applied as an in-furrow treatment as crop injury may occur.

Fungicides: "GLEAN" may be tank mixed with Du Pont "Benlate" Fungicide or Du Pont "Manzate" 200 Fungicide/Du Pont "Manzate" 200 DF Fungicide or other fungicides whenever the proper timing for herbicide and fungicide treatments coincide.

EQUIPMENT—SPRAY VOLUMES

Apply using properly calibrated air or ground equipment. Select a spray volume and delivery system that will insure thorough coverage and a uniform spray pattern. For ground application, flat fan nozzles are recommended (minimum 3 GPA). When using flood jet or "Raindrop"[5] nozzles, use higher spray volume (minimum 20 GPA) to ensure thorough coverage. However, "GLEAN" may be applied at not less than 10 GPA when using small orifice flooding nozzles such as flood jet TK 5 to TK 7.5 or equivalent, providing these nozzles are on a 30-inch spacing or not less than 13 GPA when these flooding nozzles are on a 40-inch spacing.

Do not apply this product through any type of irrigation system.

Use 50-mesh screens or larger.

Unless otherwise stated, use at least 1 gallon spray volume (GPA) per acre by air. Use higher spray volumes to obtain better coverage when either the crop canopy or stubble is dense.

Continuous agitation is required to keep "GLEAN" in suspension. Avoid overlapping, and shut off spray booms while starting, turning, slowing or stopping, or injury to the crop may result.

CAUTION - AVOID SPRAY DRIFT

Follow these practices to minimize drift.

Do not allow spray from either ground or aerial equipment to drift onto adjacent crops or land, as even small amounts will injure other plants. When spraying near adjacent, sensitive crops or plants, do everything possible to reduce spray drift. This includes:

- Stop spraying if wind speed becomes excessive. DO NOT SPRAY IF WIND SPEED IS 10 MPH OR GREATER. Spray drift can occur at wind speeds less than 10 MPH. If sensitive crops or plants are downwind, extreme caution must be used even in relatively low wind conditions! DO NOT SPRAY IF WINDS ARE GUSTY.
- High temperatures, drought, and low relative humidity increase the possibility of spray drift. EXTREME CAUTION MUST BE USED WHEN THESE CONDITIONS ARE PRESENT AND SENSITIVE CROPS OR PLANTS ARE NEARBY, REGARDLESS OF WIND SPEED.
- Do not apply when an inversion exists. An inversion is characterized by little or no air movement and an increase in air temperature with an increase in altitude. In humid regions, a fog or mist may form. An inversion may be detected by producing a smoke column and checking for a layering effect. Smoke-producing devices on aircraft are recommended. If not sure whether inversion conditions are present, consult with local weather services before making an application.
- Drift from aerial or ground equipment may be further reduced by:
  1. Using coarse sprays to minimize drift. DO NOT APPLY WITH HOLLOW-CONE INSECTICIDE NOZZLES ON GROUND EQUIPMENT. Do not use nozzles that produce fine droplets, such as Sprayfoil[6] or airblast-type nozzles. Nozzles should be oriented at an angle between straight down and straight back for ground applications. For aerial applications orient nozzles straight back along the windstream. If using flood-type nozzles on aircraft, orient them so spray is produced in direction of the airstream.
  2. Increasing volume of spray mix per acre (for example, minimum 5 GPA by air, 10 GPA by ground) by using higher flow rate nozzles.
  3. Reducing pressure (PSI). DO NOT EXCEED 40 PSI when applying "GLEAN". (Vehicle speed must also be reduced to maintain spray mix volume per acre.) Consult manufactuer's catalogs for details on correct calibration.
  4. Apply as close to target plants as possible while still maintaining a good spray pattern.

NOTE: Do not allow spray to drift onto adjacent crops, or onto agricultural land scheduled to be planted to crops other than wheat, as injury to the crop may occur. Extreme care must be taken to prevent drift to desirable plants or nontarget agricultural land.

CEREAL RECROPPING INTERVALS

RECROPPING TO WHEAT, OATS, BARLEY, AND TRITICALE IN AR, CO, KS, LA, NE, NM, OK, TX AND SOUTHEASTERN WY:

Recropping plans are determined by soil pH, rate of "GLEAN" applied and a minimum recropping interval. The minimum recropping interval is from time of last application to the anticipated date of planting.

Soil pH*	Use Rate (oz/acre)	Minimum Recropping Interval (Months)		
		Wheat/Triticale	Oats	Barley
7.9 or lower	1/6 to 1/3	0	10	10
7.9 or lower	1/2	4	10	16
above 7.9	Do Not Use	Not applicable		

* Soil pH is to be determined by laboratory analysis using the 1:1; soil:water suspension method on representative soil samples taken at 0-4" depth. Consult local extension publications for recommended soil sampling procedures.

RECROPPING TO WHEAT, OATS, BARLEY, AND TRITICALE IN MN, MT, ND, SD AND NORTHERN WY:

Recropping plans are determined by soil pH, rate of "GLEAN" applied and a minimum recropping interval. The minimum recropping interval is from the time of last application to the anticipated date of planting.

Soil pH*	Use Rate (oz/acre)	Minimum Recropping Interval (Months)		
		Wheat/Triticale	Oats	Barley
6.5 or lower	1/6 to 1/3	0	10	10
6.5 or lower	1/2	4	10	10
6.6 to 7.9	1/6 to 1/3	0	10	16
above 7.9	Do Not Use	Not Applicable		

* Soil pH is to be determined by laboratory analysis using the 1:1; soil:water suspension method on representative soil samples taken at 0-4" depth. Consult local extension publications for recommended soil sampling procedures.

RECROPPING TO WHEAT, OATS, BARLEY AND, TRITICALE IN SOUTHERN ID AND UT:

Recropping plans are determined by soil pH, rate of "GLEAN" applied and a minimum recropping interval. The minimum recropping interval is from time of last application to the anticipated date of planting.

Soil pH*	Use Rate (oz/acre)	Minimum Recropping Interval (Months)		
		Wheat/Triticale	Oats	Barley
6.5 or lower	1/6 to 1/3	0	10	10
6.5 or lower	1/2	4	10	10
6.6 to 7.9	1/6 to 1/3	0	10	24
above 7.9	Do Not Use	Not Applicable		

* Soil pH is to be determined by laboratory analysis using the 1:1; soil:water suspension method on representative soil samples taken at 0-4" depth. Consult local extension publications for recommended soil sampling procedures.

RECROPPING TO WHEAT, OATS, BARLEY, AND TRITICALE IN CA, NORTHERN ID, OR AND WA:

Recropping plans are determined by soil pH, rate of "GLEAN" applied and a minimum recropping interval. The minimum recropping intervals is from time of last application to the anticipated date of planting.

Soil pH*	Use Rate (oz/acre)	Minimum Recropping Interval (Months)		
		Wheat/Triticale	Oats	Barley
6.5 or lower	1/6 to 1/3	0	10	10
6.5 or lower	1/2	4	10	10
6.6 to 7.5	1/6 to 1/3	0	10	16
6.6 to 7.5	1/2	4	16	24
7.6 to 7.9	1/6 to 1/3	4	16	24
above 7.9	Do Not Use	———Not Applicable———		

* Soil pH is to be determined by laboratory analysis using the 1:1; soil:water suspension method on representative soil samples taken at 0-4" depth. Consult local extension publications for recommended soil sampling procedures.

ROTATION INTERVAL FOR PLANTING GRASSES ON CONSERVATION RESERVE PROGRAM ACRES

Wherever "GLEAN" has been previously used in wheat, barley, oats or fallow, the following grasses may be planted after the intervals specified in the tables below. The planting of grass and legume mixtures is not recommended as injury to the legume may occur.

- Bentgrasses
- Blue Grama
- Bluestems—Big, Little, Plains, Sand, WW Spar
- Buffalograss
- Galleta
- Green needlegrass
- Green sprangletop
- Indiangrass
- Indian ricegrass
- Lovegrasses—Sand, Weeping
- Orchardgrass (excluding Piaute)
- Praire sandreed
- Sand dropseed
- Sheep fescue
- Sideoats Grama
- Switchgrass
- Wheatgrasses—Crested, Intermediate, Pubescent, Slender, Streambank, Tall, Thickspike, Western
- Wild-rye grasses—Beardless, Russian

ROTATION INTERVALS IN STATES OF:

Soil pH*	Use Rate oz/acre	Minimum Interval for Planting Grasses
Southern ID, MN, MT ND, SD, UT, AND NORTHERN WY		
6.5 or lower	1/6 to 1/2	2 months (all grasses)
6.6 to 7.5	1/6 to 1/3	4 months (all grasses)
7.6 to 7.9	1/6 to 1/3	4 months (Wheatgrasses only)
AR, CO, KS, LA, NE, NM, OK, TX, and SOUTHEASTERN WY		
7.9 or lower	1/6 to 1/3	2 months (all grasses)
7.9 or lower	1/2	4 months (all grasses)
CA, NORTHERN ID, OR AND WA		
7.9 or lower	1/6 to 1/2	2 months (all grasses)
7.5 or lower	1/2	4 months (all grasses)

* Soil pH is to be determined by laboratory analysis using the 1:1; soil water suspension method on representative soil samples taken at 0-4" depth. Consult local extension publications for recommended soil sampling procedures.

CROP ROTATION RECOMMENDATIONS (NONCEREAL CROPS)

Note: The crop rotation intervals specified in this section of the label must be followed unless a field or LRB℠ bioassay indicates a shorter planting interval.

ARKANSAS/LOUISIANA

Unless a Crop Rotation Interval is specified, a field bioassay must be completed before rotating to any crop other than those listed below. See "Bioassay" section. DO NOT USE ON SOILS WITH A pH GREATER THAN 7.9.

Cotton, Grain Sorghum, Soybeans:

In Southwest Arkansas and Northwest Louisiana on nonirrigated land, the interval for these crops is:

Crop	Soil pH*	Use Rate (oz/acre)	Cumulative Precipitation** (in.)	Rotation Interval (months)
Cotton Grain Sorghum Soybeans	7.9 or lower	1/6 to 1/2	25	14

* Soil pH is to be determined by laboratory analysis using the 1:1; soil:water suspension method on representative soil samples taken at 0-4" depth. Consult local extension publications for recommended soil sampling procedures.

** Cumulative Precipitation equals the total amount received from the date of "GLEAN" application to the date of planting. Should accumulated precipitation not be sufficient to meet the indicated amounts, do not rotate to the indicated crops until the following growing season.

COLORADO

Unless a Crop Rotation Interval is specified, a field bioassay must be completed before rotating to any crop other than those listed below. See "Bioassay" section. DO NOT USE ON SOILS WITH A pH GREATER THAN 7.9.

Field Corn, Proso and Setaria (Hay) Millets, Grain Sorghum:

In the counties of Adams, Arapahoe, Logan, Morgan, Phillips, Sedgwick, Washington and Yuma on nonirrigated land, the intervals for field corn and millets are:

Crop	Soil pH*	Use Rate (oz/acre)	Cumulative Precipitation** (in.)	Rotation Interval (months)
Field Corn Millets	7.5 or lower	1/6 to 1/3	30	24
	7.5 or lower	1/2	45	36
	7.6 to 7.9	1/6 to 1/3	45	36
	7.6 to 7.9	1/2	60	48

In Eastern Colorado on nonirrigated land, the intervals for grain sorghum are:

Grain Sorghum	7.5 or lower	1/6 to 1/2	45	36
	7.6 to 7.9	1/6 to 1/2	60	48

* Soil pH is to be determined by laboratory analysis using the 1:1; soil:water suspension method on representative soil samples taken at 0-4" depth. Consult local extension publications for recommended soil sampling procedures.

** Cumulative Precipitation equals the total amount received from the date of "GLEAN" application to the date of planting. Should accumulated precipitation not be sufficient to meet the indicated amounts, do not rotate to the indicated crops until the following growing season.

KANSAS

Unless a Crop Rotation interval is specified, a field bioassay must be completed before rotating to any crop other than those listed below. See "Bioassay" section. DO NOT USE ON SOILS WITH A pH GREATER THAN 7.9.

Grain Sorghum and Soybeans:

In Central Kansas (generally east of highway 183 and west of the Flinthills) on nonirrigated land, the intervals for grain sorghum and soybeans are:

Crop	Soil pH*	Use Rate (oz/acre)	Cumulative Precipitation** (in.)	Rotation Interval (months)
Grain Sorghum	7.9 or lower	1/6 to 1/2	25	14
	7.5 or lower	1/6to 1/3	25	14
	7.5 or lower	1/2	46	26
Soybean	7.6 to 7.9	1/6 to 1/3	46	26
	7.6 to 7.9	1/2	64	36

In West Central and Western Kansas (generally west of highway 183 to the western edge of these counties—Grant, Kearny, Logan, Rawlings, Stevens, Thomas, Wichita) on nonirrigated land, the intervals are:

Grain Sorghum	7.5 or lower	1/6 to 1/3	21	14
	7.5 or lower	1/2	42	26
	7.6 to 7.9	1/6 to 1/3	42	26
	7.6 to 7.9	1/2	54	36

Far Western Kansas

In the last tier of counties along the Kansas/Colorado border (Cheyenne, Greeley, Hamilton, Morton, Sherman, Stanton, Wallace) on nonirrigated land, the intervals are:

Crop	Soil pH*	Use Rate (oz/acre)	Cumulative Precipitation** (in.)	Rotation Interval (months)
Grain Sorghum	7.9 or lower	1/6 to 1/3	54	36
	7.9 or lower	1/2	72	48

* Soil pH is to be determined by laboratory analysis using the 1:1; soil:water suspension method on representative soil samples taken at 0-4" depth. Consult local extension publications for recommended soil sampling procedures.

** Cumulative Precipitation equals the total amount received from the date of "GLEAN" application to the date of planting. Should accumulated precipitation not be sufficient to meet the indicated amounts, do not rotate to the indicated crops until the following growing season.

MINNESOTA

A field or LRB^{SM} bioassay must be completed before rotating to crops other than the cereal grains or Conservation Reserve Program grasses listed on this label.

MONTANA

Unless a Crop Rotation Interval is specified, a field or LRB^{SM} bioassay must be completed before rotating to any crop other than those listed below. See "Bioassay" section. DO NOT USE ON SOILS WITH A pH GREATER THAN 7.9.

Safflower:
In Montana on nonirrigated land, the interval is:

Crop	Soil pH*	Use Rate (oz/acre)	Cumulative Precipitation** (in.)	Rotation Interval (months)
Safflower	7.9 or lower	1/6 to 1/3	39	34
	6.5 or lower	1/2	—	Field or LRB^{SM} Bioassay

Note: Safflower may be planted sooner than 34 months upon the successful completion of a field bioassay or when recommended by the LRB^{SM} bioassay.

* Soil pH is to be determined by laboratory analysis using the 1:1; soil:water suspension method on representative soil samples taken at 0-4" depth. Consult local extension publications for recommended soil sampling procedures.

** Cumulative Precipitation equals the total amount received from the date of "GLEAN" application to the date of planting. Should accumulated precipitation not be sufficient to meet the indicated amounts, do not rotate to the indicated crops until the following growing season.

NEBRASKA

Unless a Crop Rotation Interval is specified, a field bioassay must be completed before rotating to any crop other than those listed below. See "Bioassay" section. DO NOT USE ON SOILS WITH A pH GREATER THAN 7.9.

Grain Sorghum and Soybeans:
In the South Central Nebraska counties of Franklin, Nuckools, Thayer and Webster on nonirrigated land, the intervals for grain sorghum and soybeans are:

Crop	Soil pH*	Use Rate (oz/acre)	Cumulative Precipitation** (in.)	Rotation Interval (months)
Grain Sorghum	7.9 or lower	1/6 to 1/2	25	14
	7.5 or lower	1/6 to 1/3	25	14
	7.5 or lower	1/2	46	26
Soybeans	7.6 to 7.9	1/6 to 1/3	46	26
	7.6 to 7.9	1/2	64	36

Field Corn, Proso and Setaria (Hay) Millets, Grain Sorghum, Soybeans:
In Western Nebraska (generally west of highway 183 to the Wyoming border) on nonirrigated land, the intervals are:

Crop	Soil pH*	Use Rate (oz/acre)	Cumulative Precipitation** (in.)	Rotation Interval (months)
Field Corn	7.5 or lower	1/6 to 1/3	40	24
Millets	7.5 or lower	1/2	60	36
Grain Sorghum	7.6 to 7.9	1/6 to 1/3	60	36
Soybeans	7.6 to 7.9	1/2	80	48

* Soil pH is to be determined by laboratory analysis using the 1:1; soil:water suspension method on representative soil samples taken at 0-4" depth. Consult local extension publications for recommended soil sampling procedures.

** Cumulative Precipitation equals the total amount received from the date of "GLEAN" application to the date of planting. Should accumulated precipitation not be sufficient to meet the indicated amounts, do not rotate to the indicated crops until the following growing season.

NEW MEXICO

Unless a Crop Rotation Interval is specified, a field bioassay must be completed before rotating to any crop other than those listed below. See "Bioassay" section. DO NOT USE ON SOILS WITH A pH GREATER THAN 7.9.

Grain Sorghum
In the counties of Curry and Quay, the interval for grain sorghum on nonirrigated land is:

Crop	Soil pH*	Use Rate (oz/acre)	Cumulative Precipitation** (in.)	Rotation Interval (months)
Grain Sorghum	7.9 or lower	1/6 to 1/3	30	25

* Soil pH is to be determined by laboratory analysis using the 1:1; soil:water suspension method on representative soil samples taken at 0-4" depth. Consult local extension publications for recommended soil sampling procedures.

** Cumulative Precipitation equals the total amount received from the date of "GLEAN" application to the date of planting. Should accumulated precipitation not be sufficient to meet the indicated amounts, do not rotate to the indicated crops until the following growing season.

NORTH DAKOTA

Unless a Crop Rotation Interval is specified, a field or LRB^{SM} bioassay must be completed before rotating to any crop other than those listed below. See "Bioassay" section. DO NOT USE ON SOILS WITH A pH GREATER THAN 7.9.

Safflower:
In North Dakota on nonirrigated land, the interval is:

Crop	Soil pH*	Use Rate (oz/acre)	Cumulative Precipitation** (in.)	Rotation Interval (months)
Safflower	7.9 or lower	1/6 to 1/3	45	34
	6.5 or lower	1/2	—	Field or LRB^{SM} Bioassay

Note: Safflower may be planted sooner than 34 months upon the successful completion of a field bioassay or when recommended by the LRB^{SM} bioassay.

* Soil pH is to be determined by laboratory analysis using the 1:1; soil:water suspension method on representative soil samples taken at 0-4" depth. Consult local extension publications for recommended soil sampling procedures.

** Cumulative Precipitation equals the total amount received from the date of "GLEAN" application to the date of planting. Should accumulated precipitation not be sufficient to meet the indicated amounts, do not rotate to the indicated crops until the following growing season.

OKLAHOMA

Unless a Crop Rotation Interval is specified, a field bioassay must be completed before rotating to any crop other than those listed below. See "Bioassay" section. DO NOT USE ON SOILS WITH A pH GREATER THAN 7.9.

Cotton, Mungbeans, Grain Sorghum, Soybeans:
In Central and Eastern Oklahoma (generally east of highway 183) on nonirrigated land, the intervals for these crops are:

Crop	Soil pH*	Use Rate (oz/acre)	Cumulative Precipitation** (in.)	Rotation Interval (months)
Grain Sorghum Cotton Mungbeans Soybeans	7.9 or lower	1/6 to 1/2	25	14

In Western Oklahoma (generally west of highway 183 and east of the Panhandle) on nonirrigated land, the interval for cotton and grain sorghum is:

Cotton	7.9 or lower	1/6 to 1/3	25	14
Grain Sorghum	7.9 or lower	1/2	46	26

In the Oklahoma Panhandle, on nonirrigated land, the interval for grain sorghum is:

Grain Sorghum	7.9 or lower	1/6 to 1/3	30	25

* Soil pH is to be determined by laboratory analysis using the 1:1; soil:water suspension method on representative soil samples taken at 0-4" depth. Consult local extension publications for recommended soil sampling procedures.

** Cumulative Precipitation equals the total amount received from the date of "GLEAN" application to the date of planting. Should accumulated precipitation not be sufficient to meet the indicated amounts, do not rotate to the indicated crops until the following growing season.

OREGON

Unless a Crop Rotation Interval is specified, a field or LRB bioassay must be completed before rotating to crops other than those listed on this or other "GLEAN" labels.

See Oregon 24(c) label for rotation intervals for annual and perennial ryegrasses, crimson and red clovers, snap beans and corn.

PACIFIC NORTHWEST (NORTHERN ID, NORTHEASTERN OR, EASTERN WA)

Unless a Crop Rotation Interval is specified, a field or LRB^{SM} bioassay must be completed before rotating to any crop other than those listed below. See "Bioassay" section.

NOTE: Successful rotation to peas and lentils can be expected in fields of even terrain having well drained soils with a uniform pH of 6.5 or less. Localized areas of crop injury may occur in fields that have highly variable terrain with areas of poor drainage and/or areas of high soil pH (eroded knolls, exposed calcereous subsoil where pH is above 6.5).

Peas and Lentils:
Northern Idaho, Northeastern Oregon, Eastern Washington Counties:

ID		OR	WA	
Benewah	Koontenai	Baker	Asotin	Spokane
Bonner	Letah	Umatilla	Columbia	Stevens
Boundary	Lewis	Union	Garfield	Walla Walla
Clearwater	Nez Perce	Wallowa	Pend Oreille	Whitman
Idaho				

In the above counties on nonirrigated land, the intervals are:

Crop	Soil pH*	Use Rate (oz/acre)	Cumulative Precipitation** (in.)	Rotation Interval (months)
Peas	6.5 or lower	1/6 to 1/3	35	24
(Alaska,	6.5 or lower	1/2	—	***
Columbian)	6.6 to 7.5	1/6 to 1/2	—	***
	7.6 to 7.9	1/6 to 1/3	—	***
	6.5 or lower	1/6 to 1/3	50	36
Lentils	6.5 or lower	1/2	—	***
(Chilean)	6.6 to 7.5	1/6 to 1/2	—	***
	7.6 to 7.9	1/6 to 1/3	—	***

* Soil pH is to be determined by laboratory analysis using the 1:1; soil:water suspension method on representative soil samples taken at 0-4" depth. Consult local extension publications for recommended soil sampling procedures.

** Cumulative Precipitation equals the total amount received from the date of "GLEAN" application to the date of planting. Should accumulated precipitation not be sufficient to meet the indicated amounts, do not rotate to the indicated crops until the following growing season.

***Field or LRB[sm] Bioassay

SOUTH DAKOTA

A field or LRB[SM] bioassay must be completed before rotating to crops other than the cereal grains of Conservation Reserve Program grasses listed on this label.

TEXAS

Unless a Crop Rotation Interval is specified, a field bioassay must be completed before rotating to any crop other than those listed below. See "Bioassay" section. DO NOT USE ON SOILS WITH A pH GREATER THAN 7.9.

Cotton, Mungbeans, Grain Sorghum, Soybeans:

Eastern Texas Counties:

Archer	Kaufman
Bell	Lamar
Bosque	Limestone
Bowie	McLennan
Camp	Milam
Cass	Montague
Clay	Morris
Colin	Navarro
Cooke	Palo Pinto
Coryell	Parker
Dallas	Rains
Delta	Red River
Denton	Robertson
Ellis	Rockwall
Falls	Somervell
Fannin	Tarrent
Franklin	Titus
Grayson	Upshur
Hill	Van Zandt
Hood	Wichita
Hopkins	Williamson
Hunt	Wise
Jack	Wood
Johnson	Young

In the above counties of Eastern Texas on nonirrigated land, the interval for these crops is:

Crop	Soil pH*	Use Rate (oz/acre)	Cumulative Precipitation** (in.)	Rotation Interval (months)
Grain Sorghum Cotton Mungbeans Soybeans	7.9 or lower	1/6 to 1/2	25	14

Central Texas Counties:

Baylor	Knox
Callahan	Shackelford
Eastland	Stephens
Foard	Throckmorton
Hardeman	Wilbarger
Haskell	

In the above counties of Central Texas on nonirrigated land, the interval for cotton and grain sorghum is:

Crop	Soil pH*	Use Rate (oz/acre)	Cumulative Precipitation** (in.)	Rotation Interval (months)
Cotton	7.9 or lower	1/6 to 1/3	25	14
Grain Sorghum	7.9 or lower	1/2	46	26

In the Texas Panhandle, on nonirrigated land, the interval for commercial grain sorghum is:

Grain Sorghum	7.9 or lower	1/6 to 1/3	30	25

NOTE: Do not plant sorghum grown for hybrid seed production.

* Soil pH is to be determined by laboratory analysis using the 1:1; soil:water suspension method on representative soil samples taken at 0-4" depth. Consult local extension publications for recommended soil sampling procedures.

** Cumulative Precipitation equals the total amount received from the date of "GLEAN" application to the date of planting. Should accumulated precipitation not be sufficient to meet the indicated amounts, do not rotate to the indicated crops until the following growing season.

SOUTHEASTERN WYOMING

Unless a Crop Rotation Interval is specified, a field bioassay must be completed before rotating to any crop other than those listed below. See "Bioassay" section. DO NOT USE ON SOILS WITH A pH GREATER THAN 7.9.

Proso and Setaria (Hay) Millets:

In the counties of Goshen, Laramie and Platte on nonirrigated land, the intervals are:

Crop	Soil pH*	Use Rate (oz/acre)	Cumulative Precipitation** (in.)	Rotation Interval (months)
	7.5 or lower	1/6 to 1/3	30	24
Millets	7.5 or lower	1/2	45	36
	7.6 to 7.9	1/6 to 1/3	45	36
	7.6 to 7.9	1/2	60	48

* Soil pH is to be determined by laboratory analysis using the 1:1; soil:water suspension method on representative soil samples taken at 0-4" depth. Consult local extension publications for recommended soil sampling procedures.

** Cumulative Precipitation equals the total amount received from the date of "GLEAN" application to the date of planting. Should accumulated precipitation not be sufficient to meet the indicated amounts, do not rotate to the indicated crops until the following growing season.

SPRAYER CLEANUP

To avoid subsequent injury to crops other than wheat, oats or barley, immediately after spraying thoroughly remove all traces of "GLEAN" from mixing and spray equipment as follows:

1) Drain tank, then flush tank, boom and hoses with clean water for a minimum of 10 minutes.

2) Fill the tank with clean water, then add 1/2 gallon chlorine bleach (containing 5 1/4% sodium hypochlorite) per 100 gallons of water. Flush through boom and hoses, allow to sit for 15 minutes with agitation, then drain.

3) Repeat Step 2.

4) Nozzles and screens should be removed and cleaned separately. To remove traces of chlorine bleach, rinse the tank thoroughly with clean water and flush through hoses and boom.

NOTE: To reduce the amount of water required in the above procedure, see separate Du Pont bulletin, "Reduced Volume Cleanout Procedure for Large Sprayers".

CAUTION: Do not use chlorine bleach with ammonia. All traces of liquid fertilizer containing ammonia, ammonium nitrate or ammonium sulphate must be rinsed with water from the mixing and application equipment before adding chlorine bleach solution. Failure to do so will release a gas with a musty chlorine odor which can cause eye, nose, throat and lung irritation. Do not clean equipment in an enclosed area.

PRECAUTIONS

Varieties of wheat, oats and barley differ in their tolerance to herbicides. When using "GLEAN" for the first time on a particular variety, limit initial use to one 18 oz jug. If no symptoms of crop injury occur 14 days after treatment, balance of acreage can be treated.

Do not apply "GLEAN" to wheat, oats or barley that are stressed by severe weather conditions, drought, low fertility, water saturated soil, disease, or insect damage, as crop injury may result. Severe winter stress, drought, disease, or insect damage following application may also result in crop injury.

In the states of CA, Northern ID, OR and WA, the maximum use rate is 1/3 oz/A per crop or fallow period on soils having a pH of 7.9 or lower. Do not apply more than 1/3 oz/A in an 18 month period. Do not use on soils having a pH greater than 7.9.

In the states of Southern ID, MN, MT, ND, SD, UT and Northern WY, the maximum use rate is 1/3 oz/A in a 48 month period on soils having a pH of 7.9. or lower. Do not use on soils having a pH greater than 7.9.

In the states of Central KS, Central NE and Central OK, the maximum use rate is 1/3 oz/A per crop period on soils having a pH of 7.9 or lower. Do not use on soils with a pH greater than 7.9.

In the states of AR, LA, Central and North Central TX and Southern OK, the maximum use rate is 1/2 oz/A per crop period on soils having a pH of 7.9 or lower. Do not use on soils with a pH greater than 7.9.

Do not apply to wheat, barley or oats undersown with legumes and grasses, as injury to the forages will result.

Do not apply to frozen ground where surface runoff may occur.

Do not apply to snow covered ground.

Do not apply to irrigated land where tailwater will be used to irrigate other cropland.

Do not use on fields that have variable soil conditions where large areas are gravelly or sandy, have eroded knolls, calcium deposits or widely variable pH readings or organic matter content. Use of "GLEAN" on fields with these conditions may result in crop injury or adversely affect crop rotation intervals.

The combined effects of preemergence "GLEAN" plus preemergence wild oat herbicides may cause crop injury to spring wheat when crop stress (soil crusting, planting too deep, prolonged cold wet weather, or drought) causes poor seedling vigor.

Tank mixtures or sequential treatments of "GLEAN" and organophosphate insecticides (such as methyl or ethyl parathion, "Di-Syston", etc.) may cause temporary discoloration or crop injury.

Wherever land has been or will be treated with "Assert"[7] herbicide and "GLEAN", plant only wheat or barley until a bioassay (see "Bioassay" section of label) demonstrates that other crops can be successfully grown. On land that is frequently rotated to crops other than wheat or barley, do not use "GLEAN" wherever "Assert" has been or will be used. The additive effect of soil residues from these treatments has not been determined and crop rotation guidelines and minimum rotation intervals are not known; injury to rotational crops may occur.

Under certain conditions such as hot, dry weather, heavy rainfall, prolonged cool weather (daily high temperature less than 50°F) or wide fluctuations in day/night temperatures just prior to or soon after treatment, temporary discoloration and/or crop injury may occur. Risk of injury is greatest when crop is in the 1-5 leaf stage.

To prevent cold weather related crop injury, avoid making preemergence applications or early postemergence applications (2-4 leaf stage) to wheat or barley during late fall, winter or early spring when cold weather conditions are unpredictable and can be severe. The combined effects of herbicide stress plus cold weather stress can result in greater crop injury than either stress factor alone.

Preemergence weed control or suppression may be unsatisfactory on soils containing 5% or more organic matter.

Fall applications on coarse textured soils (especially those having a pH of greater than 7.0) may not provide adequate control or suppression of spring germinating weeds.

To reduce the potential for movement of treated soil due to wind erosion, do not apply to powdery dry or light sandy soils until they have been stabilized by rainfall, trashy mulch, reduced tillage or other cultural practices. Injury to adjacent crops may occur when treated soil is blown onto land used to produce crops other than cereal grains.

For ground applications applied postemergence to weeds when dry, dusty field conditions exist, control of weeds in wheel track areas may be reduced. The addition of 2,4-D or MCPA should improve weed control under these conditions.

Preemergence applications of 2,4-D or herbicides containing 2,4-D made within two weeks of planting spring cereals may cause crop injury when used in conjunction with preemergence or early postemergence applications of "GLEAN".

BIOASSAY

A bioassay (field or LRB℠) must be completed before rotating to crops not listed on this label or rotating at intervals shorter than those listed in the "Crop Rotation Recommendations (Noncereal Crops)" section.

- FIELD BIOASSAY

"GLEAN" herbicide is a useful tool for weed control in wheat, barley, or spring oats. However, under some conditions small amounts of "GLEAN" can remain in the soil and injure crops other than wheat, barley, oats, or triticale for 2 to 4 years or more after application. Therefore, before you use "GLEAN", you should carefully consider your crop rotation plans during the 2 to 4 year period following treatment.

A field bioassay will be necessary if crops other than wheat, barley, spring oats, triticale or those listed on the label are to be planted on land previously treated with "GLEAN". Crop response will indicate whether or not to rotate to the crop(s) grown in the test strips.

A field bioassay involves growing test strips of the crop or crops you plan to grow the following year in fields previously treated with "GLEAN". Crop response will indicate whether or not to rotate to the crop(s) grown in the test strips.

"GLEAN" breaks down most rapidly in soils having a pH less than 7.0, in areas having 20" or more of annual rainfall, and a long growing season with warm soil temperatures. "GLEAN" residues breakdown more slowly as soil pH increases above 7.0. Other contributing factors that slow the disappearance of "GLEAN" are low rainfall and prolonged periods of soil temperatures less than 40°F.

Of the key factors that influence the rate of disappearance, only soil pH remains relatively constant from year to year. Soil temperature, and to a larger degree soil moisture, can vary greatly from year to year and from area to area. Consequently, it is not always possible to accurately predict when areas treated with "GLEAN" can be rotated to crops other than those listed on label.

A bioassay of your "GLEAN" treated field is the only sure way of determining when crops other than those listed on label can be grown.

1. The accuracy and reliability of any field bioassay is largely dependent on the location and number of strips planted. Be sure to select areas of the field previously treated with "GLEAN" that are representative of the various field conditions. Be sure to consider factors such as field size, soil texture, drainage, turnaround areas, eroded knolls or alkaline spots when selecting the sites that are most representative of the soil conditions in the field.
   Even in small fields, more than one test strip is required to accurately determine whether it is safe to rotate to a crop not listed on the label. On large fields, several test strips will be needed in order to obtain reliable results based on the field variables mentioned above.
2. Plant the test strips perpendicular to the direction in which the field was sprayed. Each strip should be long enough to cross the width of several spray swaths. A large test strip area is more reliable than a small one. Suggested size is 1/4 to 1/2 acre per test strip.
3. Use standard tillage and seeding equipment to plant the bioassay.
4. Prepare a seed bed and plant the crops and varieties you want the option of growing the following year. IT IS IMPORTANT TO USE THE SAME PLANTING TIME, CONDITIONS, TECHNIQUES AND CULTURAL PRACTICES YOU NORMALLY USE TO PLANT AND GROW THE BIOASSAY CROP(S). If possible, plant into an adjacent area not treated with "GLEAN" to use as a comparison.
5. Do not overspray the test strips with herbicides that may damage the bioassay crop(s).
6. If the crop(s) in the test strip(s) grow to maturity with a normal harvest, the assay is positive and you may now rotate to the new crop. However, if crop(s) in the test strips die, are stunted, or fail to yield a normal harvest, the assay is negative and you should not rotate to the new crop(s). Run the assay until positive results are obtained before rotating to the new crop(s).
7. If the bioassay indicates that "GLEAN" residues are still present, do not rotate to crops other than wheat, barley, oats or triticale or those listed on label until bioassay results indicate that the assay crops are growing normally.

DU PONT LRB℠ BIOASSAY SERVICE

In the states of ID, MN, MT, ND, OR, SD and WA, the Du Pont LRB℠ bioassay service is available through certain dealers and/or consultants. This service uses soil samples taken by Du Pont certified individuals for laboratory bioassay analysis. LRB℠ results will serve as a crop rotation recommendation.

Check with your local Du Pont representative or call toll free 1-800-782-3557 for information regarding the LRB℠ bioassay service.

With any chemical, follow labeling instructions and warnings carefully.

STORAGE AND DISPOSAL

STORAGE: Store product in original container only, away from other pesticides, fertilizer, food or feed.

PRODUCT DISPOSAL: Do not contaminate water, food or feed by storage or disposal. Wastes resulting from the use of this product may be disposed of on site or at an approved waste disposal facility.

CONTAINER DISPOSAL: Triple rinse (or equivalent). Then offer for recycling or reconditioning, or puncture and dispose of in a sanitary landfill or incineration, or, if allowed by state and local authorities, by burning. If burned, stay out of smoke.

NOTICE TO BUYER: Purchase of this material does not confer any rights under patents of countries outside of the United States.

NOTICE OF WARRANTY

Du Pont warrants that this product conforms to the chemical description on the label thereof and is reasonably fit for purposes stated on such label only when used in accordance with directions under normal use conditions. It is impossible to eliminate all risks inherently associated with the use of this product. Crop injury, ineffectiveness, or other unintended consequences may result because of such factors as weather conditions, presence of other materials, or the manner of use or application, all of which are beyond the control of Du Pont. In no case shall Du Pont be liable for consequential, special or indirect damages resulting from the use or handling of this product. All such risks shall be assumed by the buyer. DU PONT MAKES NO WARRANTIES OF MERCHANTABILITY OR FITNESS FOR A PARTICULAR PURPOSE NOR ANY OTHER EXPRESSED OR IMPLIED WARRANTY EXCEPT AS STATED ABOVE.

[1] Registered trademark of Sandoz Crop Protection Corporation.
[2] Registered trademark of Rhone-Poulenc Ag Company.
[3] Registered trademark of The Dow Chemical Company.
[4] Registered trademark of Bayer Ag, Leverkusen.
[5] Registered trademark of Delavan Corporation.
[6] Registered trademark of D. and W. Corporation.
[7] Registered trademark of American Cyanamid Company

AG-3260 8049 4/13/89

This is a specimen label. DuPont is not responsible for the accuracy of the information contained herein. As labels are subject to revision, always carefully read and follow the label on the product container.

HARMONY®

HERBICIDE

ACTIVE INGREDIENT	BY WEIGHT
Methyl 3-[[[[(4-methoxy-6-methyl-1,3,5-triazin-2-yl)-amino] carbonyl]amino]sulfonyl]-2-thiophenecarboxylate	75%
INERT INGREDIENTS:	25%
TOTAL	100%

EPA Reg. No. 352-446

U.S. Patent 4,481,029

KEEP OUT OF REACH OF CHILDREN

PRECAUTIONARY STATEMENTS

HAZARDS TO HUMANS AND DOMESTIC ANIMALS

CAUTION!

Avoid contact with skin, eyes, and clothing. In case of contact with eyes, immediately flush with plenty of water. Get medical attention if irritation persists. Wash thoroughly after handling. Remove and wash contaminated clothing before reuse.

For medical emergencies involving this product, call toll free 1-800-441-3637.

ENVIRONMENTAL HAZARDS

Do not apply directly to water or wetlands. Do not contaminate water by cleaning of equipment or disposal of wastes.

DIRECTIONS FOR USE

It is a violation of federal law to use this product in a manner inconsistent with its labeling.

"HARMONY" Herbicide should be used only in accordance with recommendations on this label or in separately published Du Pont recommendations.

Du Pont will not be responsible for losses or damages resulting from the use of this product in any manner not specifically recommended by Du Pont.

IMPORTANT INFORMATION

Injury to or loss of desirable trees or vegetation may result from failure to observe the following: Do not apply or drain or flush equipment on or near desirable trees or other plants, or on areas where their roots may extend, or in locations where the chemical may be washed or moved into contact with their roots. Do not use on lawns, walks, driveways, tennis courts or similar areas. Prevent drift of spray to desirable plants. Do not contaminate any body of water. Thoroughly clean all traces of Du Pont "HARMONY" Herbicide from application equipment immediately after use. (Refer to SPRAYER CLEANUP SECTION of this label.) Failure to follow these procedures may result in injury to subsequently sprayed crops.

GENERAL INFORMATION

Du Pont "HARMONY" herbicide is recommended for selective postemergence control of certain broadleaf weeds in wheat (including durum) and barley. Do not use "HARMONY" on any other crop. "HARMONY" is a water dispersible granule to be mixed in water or other recommended carrier and applied as a uniform broadcast spray. It is noncorrosive, nonflammable, nonvolatile, and does not freeze.

Best results are obtained when "HARMONY" is applied to young, actively growing weeds. The use rate will depend on weed spectrum and size of weed at time of application. The degree of control and duration of effect are dependent on rate used, sensitivity and size of target weed, and environmental conditions at the time of and following application. "HARMONY" stops growth of susceptible weeds rapidly. However, typical symptoms of dying weeds (discoloration) may not be noticeable for 1-3 weeks after application (2-5 weeks for wild garlic) depending on the environmental conditions and weed susceptibility. Warm, moist conditions following treatment promote the activity of "HARMONY", while cold, dry conditions delay the activity. Weeds hardened off by cold weather or drought stress will be less susceptible.

A vigorous growing crop will aid weed control by shading and providing competition for weeds. However, a dense crop canopy at time of application can intercept spray and result in reduced weed control. Weeds may not be adequately controlled in areas of thin crop stand or seeding skips.

Naturally-occurring weed biotypes* that are resistant to "Glean", "Finesse" or "Ally" will also be resistant to "HARMONY". In areas where these weed biotypes are known to exist, only spray "HARMONY" in tank mixtures, with other broadleaf herbicides having a different mode of action** such as: 2,4-D, Banvel[1]***, Buctril[2], Bronate[2], Curtail[3], MCPA, "Karmex" or "Lexone". Refer to TANK MIXTURES.

* Biotypes are naturally-occurring individuals of the species which have a slightly different genetic makeup. Resistant biotypes may look exactly the same as susceptible biotypes. Herbicide resistant biotypes are able to survive a use rate several times higher than needed to control susceptible biotypes.

** Mode of action is the chemical interaction that interrupts a biological process necessary for plant growth and development.

*** Tank mixes with "Banvel" may result in reduced control of some broadleaf weeds.

Read and follow all appropriate sections of label, including Precautions, before using this product.

WEED CONTROL IN WHEAT AND BARLEY

CROP STAGE AT APPLICATION

For winter wheat, make applications after the crop is in the 2-leaf stage, but before the 3rd node is detectable (Zadoks 12 through 32).

For spring wheat and barley, make applications after the crop is in the 2-leaf stage, but before the 1st node is detectable (Zadoks 12 through 30).

APPLICATION RATES

Apply 1/2 ounce "HARMONY" per acre to wheat (including durum) or barley for control or partial control of the weeds listed below.

Use 2/3 ounce "HARMONY" per acre when weed infestation is heavy and predominately consists of those weeds listed under partial control, or when application timing and environmental conditions are marginal (refer to APPLICATION AND ENVIRONMENTAL CONDITIONS FOR OPTIMUM PERFORMANCE).

Use 1/3 ounce "HARMONY" per acre when weed infestation is light and predominately consists of those weeds listed under weeds controlled, and when optimum application conditions occur.

Sequential treatments of "HARMONY" may be made provided the total amount of "HARMONY" applied to the crop does not exceed 2/3 ounce per acre; for example, 1/3 ounce followed by 1/3 ounce. The final application must be made before the 3rd node is detectable for winter wheat, or before the 1st node is detectable for spring wheat and barley.

Weeds Controlled		Partial Control*
Annual knawel	Mouseear chickweed	Common cocklebur
Annual sowthistle	Mouseear cress	Common mallow
Black mustard	Pennsylvania smartweed	Common sunflower
Bushy wallflower/ Treacle mustard	Redroot pigweed	Cutleaf eveningprimrose
Carolina geranium	Redmaids	Henbit
Coast fiddleneck	Prostrate knotweed	Prickly lettuce**
Common buckwheat	Russian thistle—in ID, OR and WA partial control only.	Tansymustard**
Common chickweed**	Scentless mayweed	Wild radish**
Common groundsel	Shepherdspurse	
Common lambsquarters	Smallflower buttercup	
Common mallow	Stinking mayweed/ Dogfennel	
Corn chamomile	Swamp smartweed	
Corn spurry	Swinecress	
Curly dock	Tarweed fiddleneck	
False chamomile	Tumble mustard	
Field pennycress	Volunteer lentils	
Flixweed	Volunteer peas	
Green smartweed	Volunteer sunflower	
Kochia	Wild buckwheat**	
Ladysthumb	Wild chamomile	
London rocket	Wild garlic**	
Malva/little mallow	Wild mustard	
Marshelder		
Miners lettuce		

* Partial control: A visual reduction of weed population as well as a significant loss of vigor for individual weed plants. For best results, use 1/2 or 2/3 ounce "HARMONY" per acre.

** See SPECIFIC WEED PROBLEMS for more information.

SPECIFIC WEED PROBLEMS

Common chickweed and wild buckwheat: For best results, apply a minimum of 1/2 ounce "HARMONY" per acre plus surfactant when all or the majority of weeds have germinated and are past the cotyledon stage. Weeds should be less than 3 inches tall or across at the time of "HARMONY" application.

Prickly lettuce, tansymustard and wild radish: For best results, use 1/2 to 2/3 ounce "HARMONY" per acre plus 2,4-D or MCPA. Refer to TANK MIXTURES for more information.

Wild garlic: For best results, apply 1/2 to 2/3 ounce "HARMONY" per acre plus surfactant when wild garlic plants are less than 12 inches tall with 4 to 6 inches of new growth. For severe infestations, use the 2/3 ounce per acre rate of "HARMONY". Control may be reduced when plants are hardened-off by cold weather and/or drought stress. Control is enhanced when applications are made during warm temperatures (60°F. or more) to actively growing wild garlic plants. Although aerial bulblet formation will be controlled, typical symptoms of dying wild garlic plants (discoloration and collapse) may not be noticeable for 2-5 weeks.

APPLICATION AND ENVIRONMENTAL CONDITIONS FOR OPTIMUM PERFORMANCE

Crop Stage: For winter wheat, make applications after the crop is in the 2-leaf stage, but before the 3rd node is detectable (Zadoks 12 through 32).

For spring wheat and barley, make applications after the crop is in the 2-leaf stage, but before the 1st node is detectable (Zadoks 12 through 30). Since thorough coverage is required, avoid crop canopy obstruction of the spray contacting the weed foliage.

Crop Competition: A vigorous growing crop will aid weed control by shading and providing competition for weeds. However, a dense crop canopy at time of application can intercept spray and result in reduced weed control. Weeds may not be adequately controlled in areas of thin crop stand or seeding skips.

Pest Stage: Since "HARMONY" has very little or no soil activity, only those weeds that have germinated above the soil surface will be controlled. Consequently, application of "HARMONY" should be made when all or the majority of weeds have germinated. Annual broadleaf weeds should be past the cotyledon stage, actively growing, and less than 4 inches tall or across. Wild garlic plants should be 6 to 12 inches tall with 4 to 6 inches of new growth. See SPECIFIC WEED PROBLEMS for more information on common chickweed, wild buckwheat, prickly lettuce, tansymustard, wild radish and wild garlic.

Application: Foliar absorption is the primary means of "HARMONY" uptake by plants; therefore, thorough spray coverage of weeds is essential.

Weather: Conditions which are conducive to healthy, actively growing plants optimize "HARMONY" weed control performance. Ideal conditions include warm temperatures (70°F. or more) and adequate soil moisture before, during and immediately after application.

Avoid making applications of "HARMONY" to weeds when rainfall is threatening. Rainfall immediately after treatment can wash "HARMONY" off weed foliage and result in reduced weed control effectiveness. Several hours of dry weather are needed to allow "HARMONY" to be absorbed by weed foliage.

Surfactant: Use only EPA approved surfactants authorized for use on food crops.

Water Spray Carrier: Add a nonionic surfactant of at least 80% active ingredient strength at 0.25% vol/vol (1 quart per 100 gallons of spray solution). Do not use a surfactant in the low desert valleys of Imperial and Riverside counties of California or in Arizona as unacceptable crop injury may occur.

Liquid N Carrier: Always use surfactant when applying "HARMONY" in liquid nitrogen fertilizer. When "HARMONY" is applied using liquid nitrogen fertilizer solution as a spray carrier, early, temporary, crop yellowing and stunting may occur. Use 0.125 - 0.25% vol/vol surfactant (1 pint - 1 quart per 100 gallons of spray solution) when applying "HARMONY" in liquid nitrogen fertilizer solution. Refer to TANK MIXTURES for surfactant information when tank mixing with other products.

EQUIPMENT-SPRAY VOLUMES

Prior to spraying "HARMONY", thoroughly clean sprayer (tanks, booms, hoses, nozzles, mixing cones, tanks, etc.).

Apply using properly calibrated air or ground equipment. Select a spray volume and delivery system that will insure thorough coverage and a uniform spray pattern. Avoid overlapping, and shut off spray booms while starting, turning, slowing or stopping, or injury to the crop may result.

Do not apply "HARMONY" through any type of irrigation system.

Spray Equipment: Refer to specific manufacturer's recommendations for additional information on spray volume (GPA), pressure, speed, nozzle types and arrangements, nozzle heights above the target area, etc., for respective application equipment.

Ground Application: For optimum spray distribution and thorough coverage, use flat fan or low volume flood nozzles. For flat fan nozzles, do not use less than 5 gallon spray volume per acre (GPA).

For flood nozzles on 30-inch spacing, use not less than 10 GPA and no larger than TK10 or equivalent and not less than 30 psi. On 40-inch nozzle spacings, use not less than 13 GPA or not less than 20 GPA when nozzles are on a 60-inch spacing. 100% overlapping of nozzle spray pattern is essential for 30, 40 and 60-inch spacings.

When using liquid nitrogen solution as a spray carrier, for flood nozzles use not less than 30 GPA and no larger than TK20. See statements on liquid N solutions below for additional information.

"Raindrop"[4] nozzles are not recommended for "HARMONY" applications as weed control performance may be reduced.

Use 50-mesh screens or larger.

Aerial Application: Use orifice discs, cores and nozzle types and arrangements that will provide for optimum spray distribution and maximum coverage at 3 to 5 GPA. Do not apply during inversion conditions, when winds are gusty, or when other conditions will favor poor coverage and/or off target spray movement.

Liquid Nitrogen Fertilizer Solutions: Always slurry "HARMONY" in clean water PRIOR to mixing in liquid N solutions. Use 0.125 - 0.25% vol/vol surfactant (1 pint - 1 quart per 100 gallons of spray solution) when applying "HARMONY" in liquid nitrogen fertilizer solution. The addition of liquid nitrogen fertilizer to the spray solution, or the use of liquid nitrogen fertilizer as the total carrier of the spray solution will increase the weight of the spray solution as compared to water (see table below). Consequently, appropriate conversions of spray nozzles and/or pressure must be made in order to maintain proper spray volumes. Refer to the conversion chart below for guidance (taken from Spray Systems Company Catalog 39, page 2).

Influence of adding liquid 28% N fertilizer to spray solution weight per gallon:

% of Spray Solution		Weight of Spray Solution
Water	28% N	(Lbs/Gallon)
100	0	8.34
50	50	9.5
0	100	10.6

SPRAYING SYSTEMS CONVERSION CHART
SPRAYING SOLUTIONS OTHER THAN WATER—NEW METHOD

Since all the tabulations are based on spraying water, which weighs 8.34 lbs per USA gallon, conversion factors must be used when spraying solutions which are heavier or lighter than water. To determine the proper size nozzle for the solution to be sprayed, first multiply the desired GPM or GPA of solution by the water rate conversion factor. Then use the new converted GPM or GPA rate to select the proper size nozzle.

Example: 10 GPA (28%N) × 1.13 = 11.3 GPA (water)

Weight of Solution	Specific Gravity	Conversion Factors
7.0 lbs per gallon	.84	.92
8.0 lbs per gallon	.96	.98
8.34 lbs per gallon-Water	1.00	1.00
9.0 lbs per gallon	1.08	1.04

Weight of Solution	Specific Gravity	Conversion Factors
10.0 lbs per gallon	1.20	1.10
10.65 lbs per gallon-28% nitrogen	1.28	1.13
11.0 lbs per gallon	1.32	1.15
12.0 lbs per gallon	1.44	1.20
14.0 lbs per gallon	1.68	1.30

Published in Spraying Systems Company Catalog 38, page 2.

Agitation: Continuous agitation is required to keep "HARMONY" in suspension.

CAUTION-AVOID SPRAY DRIFT

Do not allow spray from either ground or aerial equipment to drift onto adjacent crops or land, as even small amounts will injure other plants. When spraying near adjacent, sensitive crops or plants, do everything possible to reduce spray drift. This includes:

- Stop spraying if wind speed becomes excessive. DO NOT SPRAY IF WIND SPEED IS 10 MPH OR GREATER. Spray drift can occur at wind speeds less than 10 MPH. If sensitive crops or plants are downwind, extreme caution must be used even in relatively low wind conditions! DO NOT SPRAY IF WINDS ARE GUSTY.
- High temperatures, drought, and low relative humidity increase the possibility of harmful spray drift. EXTREME CAUTION MUST BE USED WHEN THESE CONDITIONS ARE PRESENT AND SENSITIVE CROPS OR PLANTS ARE NEARBY, REGARDLESS OF WIND SPEED.
- Do not apply when an inversion exists. An inversion is characterized by little or no air movement and an increase in air temperature with an increase in altitude. In humid regions, a fog or mist may form. An inversion may be detected by producing a smoke column and checking for a layering effect. Smoke-producing devices on aircraft are recommended. If not sure whether inversion conditions are present, consult with local weather services before making an application.
- Postemergence grass herbicides (such as Hoelon[5], Avenge[6] and Assert[6]) are often applied using high pressure. When "HARMONY" is tank mixed with these products, do not exceed 40 psi.
- Drift from aerial or ground equipment may be further reduced by:
    1. Using coarse sprays to minimize drift. DO NOT APPLY WITH HOLLOW-CONE INSECTICIDE NOZZLES ON GROUND EQUIPMENT. Do not use nozzles that produce fine droplets, such as Sprayfoil [7] or airblast-type nozzles. Nozzles should be oriented at an angle between straight down and straight back for ground applications. For aerial applications orient nozzles straight back into the windstream. If using flood-type nozzles on aircraft orient them so spray is produced in direction of the airstream.
    2. Increasing volume of spray mix per acre (minimum 5 GPA by air, 10 GPA by ground) by using higher flow rate nozzles.
    3. Reducing pressure (PSI). DO NOT EXCEED 40 PSI. [Vehicle speed must also be reduced to maintain spray mix volume per acre.] Consult manufacturer's catalogs for details on correct calibration.
    4. Apply as close to target plants as possible while still maintaining a good spray pattern.

TANK MIXTURES

2,4-D (amine or ester) or MCPA (amine or ester)-Use "HARMONY" plus 1/8 to 3/8 lb active ingredient 2,4-D or MCPA (ester formulations have provided best results). Surfactant may be added at 1 to 2 pints per 100 gallons of spray volume; however, the addition of surfactant may increase the chance of crop injury. Use the 1 to 2 pint rate of surfactant with 1/8 lb active ingredient rate of 2,4-D or MCPA. Use the 1 pint rate of surfactant with 1/4 to 3/8 lb active ingredient of 2,4-D or MCPA. Higher rates of 2,4-D or MCPA may be used, but do not exceed highest rate allowed on the label.

Always mix "HARMONY" in water prior to adding 2,4-D or MCPA and surfactant. Read and follow all label instructions on timing, precautions and warnings for these herbicides prior to using these tank mixtures.

"HARMONY" may be tank mixed with suitable registered herbicides (for example, "Karmex" and "Lexone") to control weeds other than those listed. Read and follow all manufacturer's label recommendations for the companion herbicide. If these recommendations conflict with this "HARMONY" label, do not use as a tank mix with "HARMONY".

"HARMONY" will not control wild oats or other grasses. For wild oat control, "HARMONY" can be tank mixed with "Hoelon", "Avenge" or "Assert". For green or yellow foxtail (pigeongrass) suppression, "HARMONY" can be tank mixed with "Hoelon". When tank mixing "HARMONY" and "Hoelon", the addition of a surfactant is not required. When tank mixing "HARMONY" and "Assert", ALWAYS include another broadleaf weed herbicide with a different mode of action, for example: MCPA ester, Bronate.

Always mix "HARMONY" in water PRIOR to adding other products, including surfactants. When mixing "HARMONY" in liquid nitrogen fertilizer solutions, first mix the "HARMONY" in clean water and then add the "HARMONY" and water slurry to the liquid nitrogen solution. Refer to SPRAY PREPARATION for further information.

Tank mixes of "HARMONY" plus "Banvel" may result in reduced control of some broadleaf weeds.

Tank mixes of "HARMONY" plus metribuzin may result in reduced control of wild garlic.

HARMONY may be tank mixed or used sequentially with insecticides registered for use on cereal grains. However, under certain conditions, drought stress, crop in 2-4 leaf stage) tank mixes or sequential applications of "HARMONY" plus organophosphate insecticides (such as parathion) may produce temporary crop yellowing or, in severe cases, crop injury. Limit first use to a small area before treating large areas.

DO NOT USE "HARMONY" PLUS MALATHION.

SPRAY PREPARATION

Mix the proper amount of "HARMONY" into the necessary volume of water in the spray tank with the agitator running, then add the companion herbicide to the tank after all the "HARMONY" is in suspension. ALWAYS MIX "HARMONY" IN WATER FIRST, PRIOR TO ADDING OTHER PRODUCTS IN THE SAME SPRAY TANK.

Use the spray preparation of "HARMONY" within 24 hours as product degradation may occur. If spray preparation is left standing, thoroughly agitate before reusing.

Surfactant: Add a nonionic surfactant of at least 80% active ingredient strength at 0.25% vol/vol (1 quart per 100 gallons spray solution). **Do not use a surfactant in the low desert valleys of Imperial and Riverside counties of California or in Arizona as unacceptable crop injury may occur.** Use only EPA approved surfactants authorized for use on food crops.

Liquid Nitrogen Fertilizer: Liquid nitrogen fertilizer solutions may be used as a carrier in place of water. Slurry the desired amount of "HARMONY" in a clean bucket using water until a flowable mixture is produced. Add this slurry to the agitating spray tank of liquid nitrogen fertilizer solution. Thoroughly rinse all of the "HARMONY" slurry into the spray tank. Do not use "HARMONY" in liquid fertilizer solutions of less than pH 3. Run a tank mix compatibility test before mixing "HARMONY" in fertilizer solution.

Use 0.125 - 0.25% vol/vol surfactant (1 pint - 1 quart per 100 gallons of spray solution) when applying "HARMONY" in liquid nitrogen fertilizer solution.

Use of "HARMONY" and a surfactant in liquid nitrogen fertilizer solutions may cause early, temporary crop injury (discoloration and stunting).

When the spray solution contains liquid nitrogen fertilizer, the weight per gallon of solution varies significantly from the weight of water (8.34 lbs per USA gallon). Consequently, liquid nitrogen fertilizer spray solutions must use the appropriate conversion in order to insure proper spray volume. See EQUIPMENT-SPRAY VOLUMES for further information.

CROP ROTATION

Any crop may be planted 60 days after the application of "HARMONY".

SPRAYER CLEANUP

To avoid subsequent injury to crops other than wheat or barley, immediately after spraying thoroughly remove all traces of "HARMONY" from mixing and spray equipment as follows:

1. Drain tank; rinse interior surfaces of tank; then flush tank, boom and hoses with clean water for a minimum of 5 minutes.
2. Fill the tank full with clean water, then add cleaning solution*. Flush solution through boom, hoses and nozzles, then allow to sit for 15 minutes with agitation running; then drain.
3. Repeat Step 2.
4. Repeat Step 1.
5. Nozzles and screens should then be removed and cleaned separately. To remove traces of cleaning solution, rinse the tank thoroughly with clean water and flush through hoses and boom.
6. Flush boom and hoses with clean water for 5 minutes just prior to using the sprayer for the first time after the "HARMONY" application.

* Use any one of the following cleaning solutions:

1. Nutra-Sol (carefully read and follow Nutra-Sol label directions).
2. Loveland Spray Tank Cleaner (carefully read and follow Loveland Spray Tank Cleaner label directions).
3. One-half gallon chlorine bleach (containing 5 1/4% sodium hypochlorite) per 100 gallons of water.
4. One gallon ammonia (containing 3% active) per 100 gallons of water.

CAUTION: Do not use chlorine bleach with ammonia. All traces of liquid fertilizer containing ammonia, ammonium nitrate or ammonium sulfate must be rinsed with water from the mixing and application equipment before adding chlorine bleach solution. Failure to do so will release a gas with a musty chlorine odor which can cause eye, nose, throat and lung irritation. Do not clean equipment in an enclosed area.

PRECAUTIONS

The total rate of "HARMONY" cannot exceed 2/3 ounce product per acre applied to any one crop during one growing season.

Do not graze or feed forage or hay from treated areas to livestock (harvested straw may be used for bedding and/or feed).

Do not plant to any crop other than wheat or barley for 60 days after application of "HARMONY".

Do not apply to wheat or barley that is stressed by severe winter, drought, water-saturated soil, disease or insect damage as crop injury may result. Under certain conditions such as prolonged cool weather (daily high temperature less than 50°F.) or wide fluctuations in day/night temperatures just prior to or soon after treatment, temporary yellowing and/or crop stunting may occur. Risk of injury is greatest when crop is in the 2 to 5-leaf stage.

Do not apply to wheat or barley crops underseeded with another crop.

"HARMONY" is only registered on wheat and barley. Do not use on any other crop.

For ground applications applied to weeds when dry, dusty field conditions exist, control of weeds in wheel track areas may be reduced.

The "HARMONY" volumetric measuring cylinder is to be used only as a guide, since the degree of accuracy varies by plus or minus 7.5%. For more precise measurement, use scales calibrated in ounces.

NOTICE TO BUYER: Purchase of this material does not confer any rights under patents of countries outside of the United States.

STORAGE AND DISPOSAL

STORAGE: Store product in original container only, away from other pesticides, fertilizer, food or feed.

PESTICIDE DISPOSAL: Do not contaminate water, food or feed by storage or disposal. Waste resulting from the use of this product may be disposed of on site or at an approved waste disposal facility.

CONTAINER DISPOSAL: Triple-rinse (or equivalent) the container. Then offer for recycling or reconditioning, or puncture and dispose of in a sanitary landfill, or by incineration, or, if allowed by state and local authorities, by burning. If burned, stay out of smoke.

1 Registered trade mark of Sandoz Crop Protection Corporation.
2 Registered trademark of Rhone - Poulenc Ag Company.
3 Registered trademark of the Dow Chemical Company.
4 Registered trademark of Delavan Corporation.
5 Registered trademark of Hoechst-Roussel Agri-Vet Company.
6 Registered trademark of American Cyanamid Company.
7 Registered trademark of D. AND W. INDUSTRIES.

NOTICE OF WARRANTY

Du Pont warrants that this product conforms to the chemical description on the label thereof and is reasonably fit for purposes stated on such label only when used in accordance with directions under normal use conditions. It is impossible to eliminate all risks inherently associated with the use of this product. Crop injury, ineffectiveness, or other unintended consequences may result because of such factors as weather conditions, presence of other materials, or the manner of use or application, all of which are beyond the control of Du Pont. In no case shall Du Pont be liable for consequential, special or indirect damages resulting from the use or handling of this product. All such risks shall be assumed by the buyer. DU PONT MAKES NO WARRANTIES OF MERCHANTABILITY OR FITNESS FOR A PARTICULAR PURPOSE NOR ANY OTHER EXPRESSED OR IMPLIED WARRANTY EXCEPT AS STATED ABOVE.

AG-3238 8019 1/3/89

This is a specimen label. DuPont is not responsible for the accuracy of the information contained herein. As labels are subject to revision, always carefully read and follow the label on the product container.

HARMONY® EXTRA

HERBICIDE
DISPERSIBLE GRANULE

ACTIVE INGREDIENTS	BY WEIGHT
Methyl-3-[[[[(4-methoxy-6-methyl-1,3,5-triazin-2-yl) amino] carbonyl]amino]sulfonyl]-2-thiophenecarboxylate	50%
Methyl 2-[[[[N-(4-methoxy-6-methyl-1,3,5-triazin-2-yl) methylamino] carbonyl]amino]sulfonyl]benzoate	25%
INERT INGREDIENTS	25%
TOTAL	100%

EPA Reg. No. 352 – 538 U.S. Patent 4,481,029 & 4,383,113

KEEP OUT OF REACH OF CHILDREN
CAUTION

PRECAUTIONARY STATEMENTS

HAZARDS TO HUMANS AND DOMESTIC ANIMALS

CAUTION! Avoid contact with skin, eyes and clothing. In case of contact with eyes, immediately flush with plenty of water. Get medical attention if irritation persists. Wash thoroughly after handling. Remove and wash contaminated clothing before reuse.

For medical emergencies involving this product, call toll free 1-800-441-3637.

ENVIRONMENTAL HAZARDS

Do not apply directly to water or wetlands (bogs, marshes, potholes and swamps). Do not contaminate water by disposal of equipment washwaters.

DIRECTIONS FOR USE

It is a violation of federal law to use this product in a manner inconsistent with its labeling.

Do not apply this product through any type of irrigation system.

"HARMONY" EXTRA HERBICIDE should be used only in accordance with recommendations on this label or in separately published Du Pont recommendations.

Du Pont will not be responsible for losses or damages resulting from the use of this product in any manner not specifically recommended by Du Pont.

IMPORTANT INFORMATION

Injury to or loss of desirable trees or vegetation may result from failure to observe the following: Do not apply or drain or flush equipment on or near desirable trees or other plants, or on areas where their roots may extend, or in locations where the chemical may be washed or moved into contact with their roots. Do not use on lawns, walks, driveways, tennis courts or similar areas. Prevent drift of spray to desirable plants. Do not contaminate any body of water. Thoroughly clean all traces of Du Pont "Harmony" Extra Herbicide from application equipment immediately after use. (Refer to "SPRAYER CLEANUP" section of this label.) Failure to follow these procedures may result in injury to subsequently sprayed crops.

GENERAL INFORMATION

Du Pont "HARMONY EXTRA herbicide is recommended for selective postemergence control of certain broadleaf weeds in wheat (including durum) and barley. Do not use "HARMONY" EXTRA is a water dispersible granule to be mixed in water or other recommended carrier and applied as a uniform broadcast spray. It is noncorrosive, nonflammable, nonvolatile and does not freeze.

Best results are obtained when "HARMONY" EXTRA is applied to young, actively growing weeds. The use rate will depend on weed spectrum and size of weed at time of application. The degree of control and duration of effect are dependent on rate used, sensitivity and size of target weed and environmental conditions at the time of and following application.

"HARMONY" EXTRA stops growth of susceptible weeds rapidly. However, typical symptoms of dying weeds (discoloration) may not be noticeable for 1–3 weeks after application (2–5 weeks for wild garlic) depending on the environmental conditions and weed susceptibility. Warm, moist conditions following treatment promote the activity of "HARMONY" EXTRA, while cold, dry conditions delay the activity. Weeds hardened off by cold weather or drought stress will be less susceptible.

A vigorous growing crop will aid weed control by shading and providing competition for weeds. However, a dense crop canopy at time of application can intercept spray and result in reduced weed control. Weeds may not be adequately controlled in areas of thin crop stand or seeding skips.

Naturally-occurring weed biotypes* that are resistant to Du Pont "Glean" Herbicide, Du Pont "Finesse" Herbicide or Du Pont "Ally" Herbicide will also be resistant to "Harmony" Extra. In areas where these weed biotypes are known to exist, only spray "HARMONY" EXTRA in tank mixtures with other broadleaf herbicides having a different mode of action** such as: 2,4-D, Banvel[1]***, Buctril[2], Bronate[2], Curtail[3], MCPA or Du Pont "Lexone" DF Herbicide. Refer to TANK MIXTURES.

* Biotypes are naturally-occurring individuals of the species which have a slightly different genetic makeup. Resistant biotypes may look exactly the same as susceptible biotypes. Herbicide resistant biotypes are able to survive a use rate several times higher than needed to control susceptible biotypes.

** Mode of action is the chemical interaction that interrupts a biological process necessary for plant growth and development.

*** Tank mixes with "Banvel" may result in reduced control of some broadleaf weeds.

Read and follow all appropriate sections of label, including precautions, before using this product.

WEED CONTROL IN WHEAT AND BARLEY

CROP STAGE AT APPLICATION

For winter wheat, make applications after the crop is in the 2-leaf stage, but before the 3rd node is detectable.

For spring wheat and barley, make applications after the crop is in the 2-leaf stage, but before the 1st node is detectable.

APPLICATION RATES

Apply 0.5 ounce "HARMONY" EXTRA per acre to wheat (including durum) or barley for control or partial control of the weeds listed below.

Use 0.6 ounce "HARMONY" EXTRA per acre when weed infestation is heavy and predominately consists of those weeds listed under partial control, or when application timing and environmental conditions are marginal (refer to APPLICATION AND ENVIRONMENTAL CONDITIONS FOR OPTIMUM PERFORMANCE).

Use 0.3 to 0.4 ounce "HARMONY" EXTRA per acre when weed infestation is light and predominately consists of those weeds listed under weeds controlled, and when optimum application conditions occur.

Do not use less than 0.3 ounce "HARMONY" EXTRA per acre.

Sequential treatments of "HARMONY" EXTRA may be made provided the total amount of "HARMONY" EXTRA applied to the crop does not exceed 1.0 ounce per acre; for example, 0.5 ounce followed by 0.5 ounce. Do not make the first application until after winter wheat is in the 2-leaf stage and make the last application before the 3rd is detectable. Do not make the first application until after spring wheat or barley is in the 2-leaf stage and make the last application before the 1st node is detectable.

WEEDS CONTROLLED	
Annual knawel	Mouseear chickweed
Annual sowthistle	Mouseearcress
Black mustard	Narrowleaf lambsquarters
Blue/Purple mustard	Nightflowering catchfly
Broadleaf dock	Pennsylvania smartweed
Bur buttercup	Pineappleweed
Bushy wallflower/Treacle mustard	Prickly lettuce
Clasping pepperweed	Prostrate knotweed
Coast fiddleneck	Prostrate pigweed
Common buckwheat	Redmaids
Common chickweed	Redroot pigweed
Common groundsel	Russian thistle
Common lambsquarters	Scentless chamomile/mayweed
Common radish	Shepherdspurse
Common sunflower	Slimleaf lambsquarters
Corn chamomile	Smallflower buttercup
Corn gromwell**	Smallseed falseflax
Corn spurry	Stinking chickweed
Cowcockle	Stinking mayweed/Dogfennel
Curly dock	Swinecress
False chamomile	Tansymustard
Field chickweed	Tarweed fiddleneck
Field pennycress	Tumble/Jim Hill mustard
Flixweed	Volunteer lentils
Green smartweed	Volunteer peas
Henbit	Volunteer sunflower
Kochia	Wild buckwheat**
Ladysthumb	Wild chamomile
London rocket	Wild garlic**
Marshelder	Wild mustard
Mayweed chamomile	Wild radish**
Miners lettuce	

PARTIAL CONTROL*	
Canada thistle**	Common vetch
Carolina geranium	Cutleaf eveningprimrose
Catchweed bedstraw	Hairy nightshade
Common cocklebur	Little mallow
Common mallow	

*Partial control: A visual reduction of weed population as well as a significant loss of vigor for individual weed plants. For best results, use 0.5 or 0.6 ounce "HARMONY" EXTRA per acre.

**See SPECIFIC WEED PROBLEMS for more information.

SPECIFIC WEED PROBLEMS

Canada thistle: For best results apply 0.6 ounce "HARMONY" EXTRA per acre plus surfactant in the spring when all thistles have emerged above the soil, are actively growing, and are 4 to 8 inches tall with 2 to 6 inches of new growth. Canada thistle control will be improved by using "HARMONY" EXTRA in combination with 2,4–D (refer to TANK MIXTURES).

Corn gromwell, Wild buckwheat: For best results, use 0.5 to 0.6 ounce "HARMONY" EXTRA per acre plus surfactant.

Wild garlic: For best results, apply 0.5 to 0.6 ounce "HARMONY" EXTRA per acre plus surfactant when wild garlic plants are less than 12 inches tall with 2 to 4 inches of new growth. For severe infestations, use the 0.6 ounce per acre rate of "HARMONY" EXTRA. Control may be reduced when plants are hardened-off by cold weather and/or drought stress. Control is enhanced when applications are made during warm temperatures to actively growing wild garlic plants. Typical symptoms of dying wild garlic plants (discoloration and collapse) may not be noticeable for 2–5 weeks.

Thorough spray coverage of all garlic plants is essential.

Wild radish: For best results, apply 0.5 to 0.6 ounce "HARMONY" EXTRA per acre plus surfactant either in the fall or spring to wild radish rosettes less than 6" in diameter. Applications made later than 30 days after weed emergence will result in partial control. Fall applications should be made prior to hardening-off of plants.

APPLICATION AND ENVIRONMENTAL CONDITIONS FOR OPTIMUM PEFORMANCE

Crop Stage: For winter wheat, make applications after the crop is in the 2-leaf stage, but before the 3rd node is detectable.

For spring wheat and barley, make applications after the crop is in the 2-leaf stage, but before the 1st node is detectable.

Since thorough coverage is required, avoid crop canopy obstruction of the spray contacting the weed foliage.

Crop Competition: A virgorous growing crop will aid weed control by shading and providing competition to the weeds. However, a dense crop canopy at time of application can intercept spray and result in reduced weed control. Weeds may not be adequately controlled in areas of thin crop stand or seeding skips.

Pest Stage: Since "HARMONY" EXTRA has very little or no soil activity, only those weeds that have germinated above the soil surface will be controlled. Consequently, application of "HARMONY" EXTRA should be made when all or the majority of weeds have germinated. Annual broadleaf weeds should be past the cotyledon stage, actively growing, and less than 4 inches tall or across. Wild garlic plants should be less than 12 inches tall with 2 to 4 inches of new growth. See SPECIFIC WEED PROBLEMS for more information on Canada thistle, corn gromwell, wild buckwheat, wild garlic, and wild radish.

Application: Foliar absorption is the primary means of "HARMONY" EXTRA uptake by plants; therefore, thorough spray coverage of alal target weeds is essential.

Weather: Conditions which are conducive to healthy, actively growing plants optimize "HARMONY" EXTRA weed control performance. Ideal conditions include warm temperatures and adequate soil moisture before, during and immediately after application.

Avoid making applications of "HARMONY" EXTRA to weeds when rainfall is threatening. Rainfall immediately after treatment can wash "HARMONY" EXTRA off weed foliage and result in reduced weed control effectiveness. Several hours of dry weather are needed to allow "HARMONY" EXTRA to be absorbed by weed foliage.

Surfactant: Use only EPA approved surfactants authorized for use on food crops.

Water Spray Carrier - Unless specified otherwise, always add a nonionic surfactant of at least 80% active ingredient strength at 0.25% vol/vol (1 quart per 100 gallons of spray solution).

Liquid N Carrier - Surfactant use is recommended when applying "HARMONY" EXTRA in liquid nitrogen fertilizer. When "HARMONY" EXTRA is applied using liquid nitrogen fertilizer solution as a spray carrier, early, temporary, crop yellowing and stunting may occur. These symptoms will be more noticeable when surfactant is used. Use 1/16– 1/4% vol/vol surfactant (1/2 pint - 1 quart per 100 gallons of spray solution) when applying "HARMONY" EXTRA in liquid nitrogen fertilizer solution. Refer to TANK MIXTURES for surfactant information when tank mixing with other products. If surfactant is not used when applying "HARMONY" EXTRA in liquid nitrogen fertilizer, weed control may be significantly reduced.

EQUIPMENT-SPRAY VOLUMES

Prior to spraying "HARMONY" EXTRA, thoroughly clean sprayer (tanks, booms, hoses, nozzles, mixing cones, all mixing equipment, etc.)

Apply using properly calibrated air or ground equipment. Select a spray volume and delivery system that will insure thorough coverage and a uniform spray pattern. Avoid overlapping, and shut off spray booms while starting, turning, slowing or stopping, or injury to the crop may result.

Do not use equipment and/or spray volumes that will cause spray to drift onto non-target sites. Do not make applications during weather conditions which cause spray to drift onto non-target sites. Refer to CAUTION-AVOID SPRAY DRIFT.

Do not apply "HARMONY" EXTRA through any type of irrigation system.

Spray Equipment: Refer to specific manufacturer's recommendations for additional information on spray volume (GPA), pressure, speed, nozzle types and arrangements, nozzle heights above the target area, etc., for respective application equipment.

Ground Application: For optimum spray distribution and thorough coverage, use flat fan or low volume flood nozzles. For flat fan nozzles, do not use less than 5 gallon spray volume per acre (GPA).

For flood nozzles on 30-inch spacing, use not less than 10 GPA and no larger than TK10 or equivalent and not less than 30 psi. On 40-inch nozzle spacings, use not less than 13 GPA or not less than 20 GPA when nozzles are on a 60-inch spacing. 100% overlapping of nozzle spray pattern is essential for 30, 40 and 60-inch spacings.

When using liquid nitrogen solution as a spray carrier, for flood nozzles use not less than 30 GPA and no larger than TK20. See statements on liquid N solutions below for additional information.

"Raindrop"[4] nozzles are not recocmmended for "HARMONY" EXTRA applications as weed control performance may be reduced.

Use 50-mesh screens or larger.

Aerial Application: Use nozzle types and arrangements that will provide for optimum spray distribution and maximum coverage at 3 to 5 GPA. Do not apply during inversion conditions, when winds are gusty, or when other conditions will favor poor coverage and/or off target spray movement.

Liquid Nitrogen Fertilizer Solutions: Always slurry "HARMONY" EXTRA in clean water PRIOR to mixing in liquid N solutions. Surfactant use is recommended when applying "HARMONY" EXTRA in liquid nitrogen fertilizer.

If surfactant is not used, weed control may be significantly reduced. When "HARMONY" EXTRA is applied using liquid nitrogen fertilizer solution as a carrier, early, temproary crop yellowing and stunting may occur. The symptoms will be more noticeable when surfactant is used. Use 1/16 — 1/4% vol/vol surfactant (1/2 pint - 1 quart per 100 gallons of spray solution) when applying "HARMONY" EXTRA in liquid nitrogen fertilizer solution.

The addition of liquid nitrogen fertilizer to the spray solution, or the use of liquid nitrogen fertilizer as the total carrier of the spray solution will increase the weight of the spray solution as compared to water (see table below). Consequently, appropriate conversions of spray nozzles and/or pressure must be made in order to maintain proper spray volumes. Refer to the conversion chart below for guidance (taken from Spray Systems Company Catalog 39, page 2).

Influence of adding liquid 28% N fertilizer to spray solution weight per gallon:

% of Spray Solution		Weight of Spray Solution (Lbs/Gallon)
Water	28% N	
100	0	8.34
50	50	9.5
0	100	10.6

SPRAYING SYSTEMS CONVERSION CHART

SPRAYING SOLUTIONS OTHER THAN WATER—NEW METHOD

Since all the tabulations in the Spray Systems Catalog are based on spraying water, which weighs 8.34 lbs. per USA gallon, conversion factors must be used when spraying solutions which are heavier or lighter than water. To determine the proper size nozzle for the solution to be sprayed, first multiply the desired GPM or GPA of solution by the water rate conversion factor. Then use the new converted GPM or GPA rate to select the proper size nozzle.

Example: 10 GPA (28%N) X 1.13 = 11.3 GPA (water)

Weight of Solution		Specific Gravity	Conversion Factors
7.0	lbs per gallon	0.84	0.92
8.0	lbs per gallon	0.96	0.98
8.34	lbs per gallon-Water	1.00	1.00
9.0	lbs per gallon	1.08	1.04
10.0	lbs per gallon	1.20	1.10
10.65	lbs per gallon-28% nitrogen	1.28	1.13
11.0	lbs per gallon	1.32	1.15
12.0	lbs per gallon	1.44	1.20
14.0	lbs per gallon	1.68	1.30

Published in Spraying Systems Company Catalog 39, page 2.

Agitation: Continuous agitation is required to keep "HARMONY" EXTRA in suspension.

CAUTION - AVOID SPRAY DRIFT

Follow these practices to minimize drift.

Do not allow spray from either ground or aerial equipment to drift onto adjacent crops or land, as even small amounts will injur other plants. When spraying, do everything possible to reduce spray drift. This includes:

- Stop spraying if wind speed becomes excessive. DO NOT SPRAY IF WIND SPEED IS 10 MPH OR GREATER. Spray drift can occur at wind speeds less than 10 MPH. If sensitive crops or plants are downwind, extreme caution must be used even in relatively low wind conditions! DO NOT SPRAY IF WINDS ARE GUSTY.
- High temperatures, drought, and low relative humidity increase the possibility of spray drift. EXTREME CAUTION MUST BE USED WHEN THESE CONDITIONS ARE PRESENT AND SENSITIVE CROPS OR PLANTS ARE NEARBY, REGARDLESS OF WIND SPEED.
- Do not apply when an inversion exists. An inversion is characterized by little or no air movement and an increase in air temperature with an increase in altitude. In humid regions, a fog or mist may form. An inversion may be detected by producing a smoke column and checking for a layering effect. Smoke-producing devices on aircraft are recommended. If not sure whether inversion conditions are present, consult with local weather serevices before making an application.
- Postemergence grass herbicides (such as Avenge[5] and Assert[5]) are often applied using high pressure. When "HARMONY" EXTRA is tank mixed with these products, do not exceed 40 psi.
- Drift from aerial or ground equipment may be further reduced by:
  1. Using coarse sprays to minimize drift. DO NOT APPLY WITH HOLLOW-CONE INSECTICIDE NOZZLES ON GROUND EQUIPMENT. Do not use nozzles that produce fine droplets, such as Sprayfoil[6] or airblast-type nozzles. Nozzles should be oriented at an angle between straight down and straight back for ground applications. For aerial applications orient nozzles straight back along the windstream. If using flood-type nozzles on aircraft, orient them so spray is produced in direction of the airstream.
  2. Increasing volume of spray mix per acre (for example, minimum 5 GPA by air, 10 GPA by ground) by using higher flow rate nozzles.
  3. Reducing pressure (PSI). DO NOT EXCEED 40 PSI when applying "HARMONY" EXTRA. [Vehicle speed must also be reduced to maintain spray mix volume per acre.] Consult manufacturer's catalogs for details on correct calibration.
  4. Apply as close to target plants as possible while still maintaining a good spray pattern.

TANK MIXTURES

2,4–D (amine or ester) or MCPA (amine or ester) - Use "HARMONY" EXTRA plus 1/8 to 3/8 lb active ingredient 2,4–D or MCPA (ester formulations have provided best results). Surfactant may be added at 0.125–0.25% vol/vol (1 pint to 1 quart per 100 gallons of spray volume); however, the addition of surfactant may increase the chance of crop injury. Use the 1 pint to 1 quart rate of surfactant with 1/8 lb active ingredient rate of 2,4–D or MCPA. Use the 1 pint rate of surfactant with 1/4 to 3/8 lb active ingredient of 2,4–D or MCPA. Higher rates of 2,4–D or MCPA may be used, but do not exceed highest rate allowed on the label.

Always mix "HARMONY" EXTRA in water prior to adding 2,4–D or MCPA. Always add surfactant last. Read and follow all label instructions on timing, precautions and warnings for these herbicides prior to using these tank mixtures.

"HARMONY" EXTRA may be tank mixed with other suitable registered herbicides (for example "Lexone") to control weeds other than those listed. Read and follow all manufacturer's label recommendations for the companion herbicide. If these recommendations conflict with this "HARMONY" EXTRA label or supplemental "HARMONY" EXTRA labels, do not use as a tank mix with "HARMONY" EXTRA.

"HARMONY" EXTRA will not control wild oats or other grasses. For wild oat control, "HARMONY" EXTRA can be tank mixed with "Avenge" or "Assert". When tank mixing "HARMONY" EXTRA and "Assert", ALWAYS include another broadleaf weed herbicide with a different mode of action, for example: 2,4–D ester, MCPA ester, "Buctril" or "Bronate". **DO NOT tank mix with Hoelon[7] 3EC as grass weed control may be reduced.**

Always mix "HARMONY" EXTRA in water PRIOR to adding other products, including surfactants. When mixing "HARMONY" EXTRA in liquid nitrogen fertilizer solutions, first mix the "HARMONY" EXTRA in clean water and then add the "HARMONY" EXTRA and water slurry to the liquid nitrogen solution. Refer to SPRAY PREPARATION for further information.

Tank mixes of "HARMONY" EXTRA plus "Banvel" may result in reduced control of some broadleaf weeds.

Tank mixes of "HARMONY" EXTRA plus metribuzin may result in reduced control of wild garlic.

HARMONY" EXTRA may be tank mixed or used sequentially with fungicides or insecticides registered for use on cereal grains. However, under certain conditions, (drought stress, crop in 2–4 leaf stage) tank mixes or sequential applications of "HARMONY" EXTRA plus organophosphate insecticides (such as parathion) may produce temporary crop yellowing or, in severe cases, crop injury. Limit first use to a small area before treating large areas.

DO NOT USE "HARMONY" EXTRA PLUS MALATHION OR "LORSBAN"[3] as crop injury will result.

SPRAY PREPARATION, ADDITIVES, PRODUCT MEASUREMENT, SURFACTANT AND LIQUID NITROGEN FERTILIZER

Spray Preparation: Mix the proper amount of "HARMONY" EXTRA into the necessary volume of water in the spray tank with the agitator running, then add the companion herbicide to the tank after all the "HARMONY" EXTRA is in suspension. ALWAYS MIX "HARMONY" EXTRA IN WATER FIRST, PRIOR TO ADDING OTHER PRODUCTS IN THE SAME SPRAY TANK.

Use the spray preparation of "HARMONY" EXTRA within 24 hours as product degradation may occur. If spray preparation is left standing, thoroughly agitate before reusing.

Additives: Do not use with spray additives that alter the pH of the spray solution below pH 5.0 or above pH 9.0, as rapid product degradation can occur. Spray solutions of pH 6.0 - 8.0 allow for optimum stability of "HARMONY" EXTRA.

Product Measurement: The "HARMONY" EXTRA volumetric measuring cylinder is to be used as a guide, since the degree of accuracy varies by plus or minus 7.5%. For more precise measurement, use scales calibrated in ounces.

Surfactant: Unless specified otherwise, add a nonionic surfactant of at least 80% active ingredient strength at 0.25% vol/vol (1 quart per 100 gallons of spray solution). Use only EPA approved surfactants authorized for use on food crops. Refer to TANK MIXTURES and EQUIPMENT - SPRAY VOLUMES for variations on surfactant rate.

Liquid Nitrogen Fertilizer: Liquid nitrogen fertilizer solutions may be used as a carrier in place of water. Slurry the desired amount of "HARMONY" EXTRA in a clean bucket using water until a flowable mixture is produced. Add this slurry to the agitating spray tank of liquid nitrogen fertilizer solution. Thoroughly rinse all of the "HARMONY" EXTRA slurry into the spray tank. Do not use "HARMONY" EXTRA in liquid fertilizer solutions of less than pH 5.0. Run a tank mix compatibility test before mixing "HARMONY" EXTRA in fertilizer solution.

Use 1/16 - 1/4% vol/vol surfactant (1/2 pint - 1 quart per 100 gallons of spray solution) when applying "HARMONY" EXTRA in liquid nitrogen fertilizer solution.

Use of "HARMONY" EXTRA and a surfactant in liquid nitrogen fertilizer solutions may cause early, temporary crop injury (discoloration and stunting).

When the spray solution contains liquid nitrogen fertilizer, the weight per gallon of solution varies significantly from the weight of water (8.34 lbs per USA gallon). Consequently, liquid nitrogen fertilizer spray solutions must use the appropriate conversion in order to insure proper spray volume. See EQUIPMENT-SPRAY VOLUMES for further information.

CROP ROTATION

Any crop may be planted 60 days after the application of "HARMONY" EXTRA.

SPRAYER CLEANUP

To avoid subsequent injury to crops other than wheat or barley, immediately after spraying thoroughly remove all traces of "HARMONY" EXTRA from mixing and spray equipment as follows:

1. Drain tank; rinse interior surfaces of tank; then flush tank, boom and hoses with clean water for a minimum of 5 minutes.
2. Fill the tank full with clean water, then add cleaning solution*. Flush solution through boom, hoses and nozzles, then allow to sit for 15 minutes with agitation running; then drain.
3. Repeat Step 2.
4. Repeat Step 1.
5. Nozzles and screens should then be removed and cleaned separately. To remove traces of cleaning solution, rinse the tank thoroughly with clean water and flush through hoses and boom.
6. Flush boom and hoses with clean water for 5 minutes just prior to using the sprayer for the first time after the "HARMONY" EXTRA application.

*Use any one of the following cleaning solutions:

1. Nutra-Sol[8] (carefully read and follow Nutra-Sol label directions).
2. Loveland Spray Tank Cleaner[9] (carefully read and follow Loveland Spray Tank Cleaner label directions).
3. Tank-Aid[10] (carefully read and follow Tank-Aid label directions).
4. One-half gallon chlorine bleach (containing 5 1/4% sodium hypochlorite) per 100 gallons of water.
5. One gallon ammonia (containing 3% active) per 100 gallons of water.

CAUTION: Do not use chlorine bleach with ammonia. All traces of liquid fertilizer containing ammonia, ammonium nitrate or ammonium sulfate must be rinsed with water from the mixing and application equipment before adding chlorine bleach solution. Failure to do so will release a gas with a musty chlorine odor which can cause eye, nose, throat and lung irritation. Do not clean equipment in an enclosed area.

PRECAUTIONS

The total rate of "HARMONY" EXTRA cannot exceed 1.0 ounce product per acre applied to any one crop during one growing season.

Do not graze or feed forage or hay from treated areas to livestock (dry-harvested straw may be used for bedding and/or feed).

Do not plant to any crop other than wheat or barley for 60 days after application of "HARMONY" EXTRA.

Do not apply to wheat or barley that is stressed by severe winter, drought, water-saturated soil, disease or insect damage as crop injury may result. Under certain conditions such as prolonged cool weather (daily high temperature less than 50 Deg.F.) or wide fluctuations in day/night temperatures just prior to, during or soon after treatment, temporary yellowing and/or crop stunting may occur. Risk of injury is greatest when crop is in the 2 to 5-leaf stage.

Do not apply to wheat or barley crops underseeded with another crop.

"HARMONY" EXTRA is only registered on wheat and barley. Do not use on any other crop.

For ground applications applied to weeds when dry, dusty field conditions exist, control of weeds in wheel track areas may be reduced.

STORAGE AND DISPOSAL

STORAGE: Store product in original container only, away from other pesticides, fertilizer, food or feed.

PRODUCT DISPOSAL: Do not contaminate water, food or feed by storage or disposal. Waste resulting from the use of this product may be disposed of on site or at an approved waste disposal facility.

CONTAINER DISPOSAL: Triple-rinse (or equivalent) the container. Then offer for recycling or reconditioning, or puncture and dispose of in a sanitary landfill, or by incineration, or, if allowed by state and local authorities, by burning. If burned, stay out of smoke.

NOTICE TO BUYER: Purchase of this material does not confer any rights under patents of countries outside of the United States.

NOTICE OF WARRANTY

Du Pont warrants that this product conforms to the chemical description on the label thereof and is reasonably fit for purposes stated on such label only when used in accordance with directions under normal use conditions. It is impossible to eliminate all risks inherently associated with the use of this product. Crop injury, ineffectiveness, or other unintended consequences may result because of such factors as weather conditions, presence of other materials, or the manner of use or application, all of which are beyond the control of Du Pont. In no case shall Du Pont be liable for consequential, special or indirect damages resulting from the use or handling of this product. All such risks shall be assumed by the buyer. DU PONT MAKES NO WARRANTIES OF MERCHANTABILITY OR FITNESS FOR A PARTICULAR PURPOSE NOR ANY OTHER EXPRESSED OR IMPLIED WARRANTY EXCEPT AS STATED ABOVE.

[1] Registered trademark of Sandoz Crop Protection Corporation.
[2] Registered trademark of Rhone-Poulenc Ag Company.
[3] Registered trademark of the Dow Chemical Company.
[4] Registered trademark of Delavan Corporation.
[5] Registered trademark of American Cyanamid Company.
[6] Registered trademark of D. AND W. INDUSTRIES.
[7] Registered trademark of Hoechst-Roussel Agri-Vet Company.
[8] Sold by Thomas G. Kilfoil, San Bruno, CA.
[9] Registered trademark of Loveland Industries, Inc.
[10] Manufactured for Cornbelt Chemical Company.

AG-3277 8089 8/3/89

This is a specimen label. DuPont is not responsible for the accuracy of the information contained herein. As labels are subject to revision, always carefully read and follow the label on the product container.

HYVAR® X

HERBICIDE
WETTABLE POWDER

ACTIVE INGREDIENT:	BY WEIGHT
Bromacil (5-bromo-3-sec-butyl-6-methyluracil)	80%
INERT INGREDIENTS:	20%
TOTAL	100%

EPA Reg. No. 352-287

KEEP OUT OF REACH OF CHILDREN
CAUTION

PRECAUTIONARY STATEMENTS

HAZARDS TO HUMANS AND DOMESTIC ANIMALS

CAUTION! MAY IRRITATE EYES, NOSE, THROAT, AND SKIN.

Avoid breathing dust. Avoid contact with skin, eyes or clothing. Wash thoroughly with soap and water after handling.

For medical emergencies involving this product, call toll free 1-800-441-3637.

STATEMENT OF PRACTICAL TREATMENT

If in eyes: Flush with plenty of water. Get medical attention if irritation persists.

If on skin: Wash with plenty of soap and water. Get medical attention if irritation persists.

ENVIRONMENTAL HAZARDS

Do not apply directly to water or wetlands. Do not contaminate water by cleaning of equipment or disposal of wastes.

Bromacil is a chemical which can travel (seep or leach) through soil and can enter ground water which may be used as drinking water. Bromacil has been found in ground water as a result of agricultural use. The correct use rates by geographical area, specified herein, and proper mixing-loading site precautions and procedures must be followed to minimize potential bromacil movement into groundwater.

IMPORTANT

Injury to or loss of desirable trees or other plants may result from failure to observe the following: Do not apply (except as recommended for crop use), or drain or flush equipment on or near desirable trees or other plants, or on areas where their roots may extend, or in locations where the chemical may be washed or moved into contact with their roots. Do not use on lawns, walks, driveways, tennis courts, or similar areas. Prevent drift of dry powder or spray to desirable plants. Do not contaminate domestic waters. Keep from contact with fertilizers, insecticides, fungicides, and seeds. Thoroughly clean all traces of "HYVAR" X from application equipment immediately after use. Flush tank, pump, hoses, and boom with several changes of water after removing nozzle tips and screens (clean these parts separately).

PESTICIDE HANDLING

- Calibrate sprayers only with clean water away from the well site.
- Make scheduled checks of spray equipment.
- Assure accurate measurement of pesticides by all operation employees.
- Mix only enough product for the job at hand.
- Avoid over-filling of spray tank.
- Do not discharge excess material on the soil at a single spot in the field/grove or mixing/loading station.
- Dilute and agitate excess solution and apply at labeled rates/uses.
- Avoid storage of pesticides near well sites.

GENERAL INFORMATION

Du Pont "HYVAR" X Herbicide is a wettable powder to be mixed in water and applied as a spray for control of weeds and brush. It is non-corrosive to equipment, non-flammable and non-volatile.

"HYVAR" X is an effective general herbicide for the control of many annual weeds at low rates and perennial weeds at higher rates and is particularly useful for control of perennial grasses. It may be used on non-cropland for non-selective weed and brush control and for selective weed control in certain crops.

Effects are slow to appear and may not become apparent until the chemical has been carried into the root zone of the weeds by moisture. The degree of control and duration of effect will vary with the amount of herbicide applied, soil texture, rainfall, and other conditions.

DIRECTIONS FOR USE

It is a violation of Federal Law to use this product in a manner inconsistent with its labeling.

Du Pont "HYVAR" X Herbicide should be used only in accordance with recommendations on this label, or in separate published Du Pont recommendations available through local dealers.

Du Pont will not be responsible for losses or damages resulting from use of this product in any manner not specifically recommended by Du Pont. User assumes all risks associated with such nonrecommended use.

Do not apply this product through any type of irrigation system.

Do not apply this product in such a manner as to directly or through drift expose workers or other persons. The area being treated must be vacated by unprotected persons.

Farm workers performing hand labor operations should not enter treated areas without protective clothing until sprays have dried. Because certain states may require more restrictive reentry intervals for various crops treated with this product, consult your State Department of Agriculture for further information.

Written or oral warnings must be given to farm workers performing hand labor operations who are expected to be in a treated cropland area or in a cropland area about to be treated with this product. When oral warnings are given, warnings shall be given in a language customarily understood by workers. Oral warnings must be given if there is reason to believe that written warnings cannot be understood by workers. Warnings must include the following information: "CAUTION. Area treated with "HYVAR" X on (date of application). Do not enter without appropriate protective clothing until sprays have dried. In case of contact, flush eyes with plenty of water; wash skin with plenty of soap and water. Get medical attention if irritation persists."

Apply "HYVAR" X as a spray just before or during the period of active growth of plants to be controlled and when rainfall can be expected for soil activation. If dense growth is present, results will be improved if vegetation is removed before treatment. Do not apply when ground is frozen.

Before spraying, calibrate equipment to determine quantity of water necessary to uniformly cover measured area to be treated. Weigh the proper amount of "HYVAR" X and mix into necessary volume of water.

For crop use, apply with a fixed-boom power sprayer properly calibrated to a constant speed and rate of delivery. Use sufficient water (min. 40 gal. per acre) to provide thorough and uniform coverage of the ground. Spray booms must be shut off while starting, turning, slowing, or stopping, or injury to the crop or successive crops may result.

For non-crop use, application also may be made with a handgun sprayer using at least 200 gals. spray per acre to insure uniform coverage. For small areas, a hand sprayer or sprinkling can may be used.

Nozzle screens should be 50 mesh or larger. Continuous agitation in the spray tank is required to keep the material in suspension. Agitate by mechanical or hydraulic means in the spray tank. If by-pass or return line is used, it should terminate at bottom of tank to minimize foaming. Do not use air agitation.

NON-CROP USE

WEED CONTROL

To control most weeds for an extended period of time on non-cropland areas such as RAILROAD, HIGHWAY and PIPELINE RIGHT-OF-WAYS, PETROLEUM TANK FARMS, LUMBERYARDS, STORAGE AREAS and INDUSTRIAL PLANT SITES:

Apply 3 to 6 lbs. per acre for control of annuals such as cheat, crabgrass, downy brome, foxtail, lambsquarters, puncturevine, ragweed, ryegrass, turkey mullein, and wild oats. When applied just prior to or after emergence of annuals, rates as low as 2 lbs. per acre control many annual weeds and grasses in low rainfall areas and give short term control in higher rainfall areas.

Apply 7 to 15 lbs. per acre for control of perennials such as bahiagrass, bermudagrass, bluegrass, broomsedge, dandelion, dog fennel, goldenrod, johnsongrass, nutsedge, plantain, purpletop, natalgrass (redtop), quackgrass, smooth brome, and wild carrot. In areas with low or seasonal rainfall, rates as low as 5 lbs. per acre control many perennial weeds and grasses.

Where limited rainfall (usually less than 4 inches) occurs during the active growth period, such as some areas of the West, "HYVAR" X usually will not provide satisfactory control of hard-to-kill, deep-rooted perennial weeds such as johnsongrass. Note—Use the higher levels of the dosage ranges on adsorptive soils (those high in organic matter or carbon).

Retreatment—Apply 2 to 6 lbs. per acre when annual weeds and grasses reappear on sites where weed growth has been controlled.

For Small Areas— 1/4 cupful of "HYVAR" X per 250 sq. ft. is approximately 15 lbs. per acre.

Combinations—For broader spectrum control of emerged weeds, apply "HYVAR" X (as listed above) plus surfactant with "Banvel" D (dicamba) or 2,4-D amine, or dalapon, or MSMA or paraquat or "Phytar" 560 (Na salt of cacodylic acid); use the higher rates of "HYVAR" X on heavier soils or soils high in organic matter; use the companion herbicides as recommended on package labels using higher rates on heavy, dense vegetation.

BRUSH CONTROL

To control undesirable woody plants on non-cropland areas such as RAILROAD RIGHT-OF-WAYS, STORAGE AREAS, INDUSTRIAL PLANT SITES, and DRAINAGE DITCHES:

Apply in spring or summer as a broadcast or basal (spot) treatment; for use on drainage ditches, apply as a basal (spot) treatment only. **Note:** For effective brush control and prevention of damage to desirable vegetation: do not apply to brush standing in water; do not use water from treated ditches for irrigation; do not use in irrigation ditches nor on right-of-ways or other sites where marketable timber or other desirable trees or shrubs are immediately adjacent to the treated area.

Broadcast Treatment—Apply 7 to 15 lbs. per acre to control brush such as oak, pine, sweet gum, and willow. Use the higher rates on adsorptive soils (those high in organic matter or carbon).

Basal (Spot) Treatment—Mix 2½ lbs. in 5 gals. of water and apply at the rate of 1 to 2 fl. oz. per stem 2" to 4" in basal diameter; wet base of stem to run-off. Treatment controls woody plants such as cottonwood, hackberry, maple, oak, poplar, red bud, sweet gum, wild cherry, willow, and winged elm.

SELECTIVE USE IN CROPS

All dosages of "HYVAR" X are expressed as broadcast rates. For band treatment, use proportionately less.

Moisture is necessary to activate the chemical; best results are obtained if moisture is supplied by rainfall or irrigation within two weeks after application.

CITRUS

Apply as a band or broadcast treatment beneath and/or between trees. Avoid contact of foliage and fruit with spray or mist. Temporary yellowing of citrus leaves may occur following treatment. Do not use on citrus in Kern County, California.

Dosage rates are expressed as broadcast rates; for band treatment use proportionately less.

Because injury to citrus trees may result: do not use on soils low in organic matter (less than 1%) nor on poorly drained soils; do not apply more than 8 lbs. per acre per year; do not treat trees planted in irrigation furrows; do not treat diseased trees such as those with foot rot. Do not use in citrus orchards interplanted with other trees or desirable plants, nor in home citrus plantings or in areas where roots of other valuable plants or trees may extend as plant injury may result. Do not graze cattle in treated areas. Treated areas may be planted to citrus trees one year after last application; do not replant to other crops within 2 years after last application as plant injury may result.

FLORIDA ONLY

East Coast/Flatwoods Areas—(low permeable soils)

— The maximum allowable is 6.4 lbs active (bromacil) per acre per year. The allowable amount is inclusive of all bromacil formulations used within one year.

— Under the above guidelines, do not apply more than 8 lbs per acre per year of "Hyvar" which contains 6.4 lbs active (bromacil).

— Multiple applications may improve control of "hard-to-kill" species.

Trees Established Less Than One Year: For control of annuals, apply 2–3 lbs of "HYVAR" X per acre as needed to maintain weed control. Do not apply at less than 60 day intervals; do not apply more than 6 pounds per acre within any 6 month period, or more than 8 pounds during the first year.

Trees Established One to Three Years: For control of annuals, apply 2–3 lbs of "HYVAR" X per acre as needed to maintain weed control. Do not apply more than 6 pounds per acre within any 6 month period, or more than 8 pounds per acre per year.

Trees Established Three or More Years: Apply 2–4 lbs per acre as needed to maintain weed control. Do not use more than 4 pounds per acre in any one application, nor more than 8 pounds per acre per year. Use the higher rate for control of perennials listed below. Repeat annual treatments are required to control perennials.

Ridge areas—(highly permeable soils; sands, loamy sands)

— The maximum allowable is 4 lbs active (bromacil) per acre per year. The allowable amount is inclusive of all bromacil formulations used within one year.

— Under the above guidelines, do not apply more than 5 lbs per acre per year of "HYVAR" X which contains 4 lbs active (bromacil).

— Multiple applications may improve control of "hard-to-kill" species.

Trees Established Less Than One Year: For control of annual weeds, apply 2–3 lbs of "HYVAR" X per acre as needed to maintain weed control. Do not apply at less than 60 day intervals; do not use more than 5 lbs per acre per year.

Trees Established One to Three Years: For control of annual weeds, apply 2–3 lbs of "HYVAR" X per acre as needed to maintain weed control. Do not use more than 5 lbs per acre per year.

Trees Established Three or More Years: For control of annual weeds, apply 2–4 lbs of "HYVAR" X per acre as needed to maintain weed control. Do not use more than 4 pounds per acre in any one application, nor more than 5 lbs per acre per year. To control established perennials (listed below) use the seasonal use rate of 5 lbs "HYVAR" X per acre. Repeat annual treatments are required to control perennials.

WEEDS CONTROLLED (annuals)	
barnyardgrass	puncturevine
crabgrass	purslane
crowfootgrass	sandspur
Florida pusley	sedge (annual)
henbit	sprangletop
lambsquarter	Texas panicum
mustard	turkey mullein
natalgrass (red top)	

WEEDS CONTROLLED (perennials)	
bermudagrass	Bahiagrass
torpedograss	Johnsongrass
paragrass	nutsedge
pangolagrass	

Note: Partial control of perennials usually occurs with a single treatement; repeat applications are required to control perennials. Control of perennials may be improved by cultivation prior to treatment; otherwise, avoid working the soil as long as weed control continues or else effectiveness of the treatment may be reduced.

TEXAS ONLY

Trees Established for at Least One Year to Four Years

Apply 2 to 4 lbs "HYVAR"X per acre. Use the lower rate for 1 to 2 year old trees and the higher rate for trees more than 2 years old. A second application may be made if needed, but do not apply more than 4 lbs "HYVAR" X per acre per year as injury to the trees may result.

U.S. (Except Florida)

Do not use on citrus in Kern County, California.

Trees Established for Four Years or More

Annual Weeds—Including barnyardgrass (watergrass), crabgrass, crowfootgrass, Florida pusley, henbit, lambsquarters, mustard, natalgrass (red top), puncturevine, purslane, sandspur, sedge (annual), sprangletop, Texas panicum (Colorado grass), and turkey mullein, apply 2 to 4 lbs. per acre. Apply anytime of the year, preferably shortly before or after weed growth begins when adequate moisture is available.

Perennial Weeds—Best results are obtained if application is made shortly before or shortly after weed growth begins; if dense growth is present, remove tops and spray the ground. Effects on perennial weeds are slow to appear, usually progressing over a period of several months.

Make a single application per year during the period from winter to early summer; use at the following rates:

Soil Texture	Lbs. "HYVAR" X Per Acre
Sand, loamy sand	4 to 5
Sandy loam	5 to 6
Silt loam, clay loam	6 to 8

Alternatively, make two applications of 3 to 4 lbs. "HYVAR" X per acre per year. In Texas, and Louisiana, apply in spring and summer; in California and Arizona, apply in fall and spring.

Note: Partial control usually occurs with a single treatment; repeat applications are required to control perennials.

"HYVAR" X controls the following:

Bermudagrass	All areas U.S.
Johnsongrass	Texas
Nutsedge	Texas, California

Control of perennials may be improved by cultivation prior to treatment; otherwise, avoid working the soil as long as weed control continues or else effectiveness of the treatment may be reduced.

PINEAPPLE

Do not replant treated areas to any crop other than pineapple within 2 years after last application as injury to subsequent crops may result.

Hawaii and Florida—For control of seedling weeds such as amaranthus, balsamapple, chloris, crabgrass, Floras paintbrush, foxtail, goosegrass (wiregrass), and Hialoa, apply 2 to 6 lbs. per acre broadcast before the planting material begins to grow. Use the lower rates in low rainfall areas (5 to 10 inches annually) and on clean-culture fields; use the higher rates in high rainfall areas (above 10 inches annually) and for trashmulch fields. Follow with 2 to 4 lbs. per acre as directed interline spray as needed prior to differentiation, or apply 2 lbs. per acre broadcast as needed after plants are 8 months old but before differentiation. Do not apply more than 10 lbs. total per acre per plant crop.

For ratoon crop, apply 1 to 4 lbs. per acre broadcast after harvesting plant crop but before differentiation. Do not apply more than 4 lbs. total per acre per ratoon crop.

Puerto Rico—For control of seedling weeds such as crabgrass, goosegrass, jungle rice, pigweed, and purslane, apply 2 to 4 lbs. per acre broadcast immediately after planting and before planting material begins to grow.

SPRAY PREPARATION

Mixing in water — Fill tank 1/2 full with water. Start agitation system, add "HYVAR" X and continue adding water. Add separately each additional component of any tank-mix while adding water. Continue agitation throughout.

Mixing in liquid fertilizer (citrus only) — A fertilizer solution may be used in the spray mixture. Small quantities should be tested for compatibility by the following procedures before full scale mixing.

1. Put 1 pint fertilizer solution in a quart jar.
2. Mix 2 teaspoons "HYVAR" X with 2 tablespoonfuls of water; mix thoroughly and add to fertilizer solution.
3. Close jar and shake well.
4. If other herbicides are used in the mixture, premix 2 teaspoonfuls of dry materials to 1 teaspoonful of liquids with 2 tablespoons of water; add to "HYVAR" X/fertilizer solution mixture.
5. Close jar and shake well.
6. Watch mixture for several seconds; check again in 30 minutes.
7. If mixtures does not separate, form, gel or become lumpy, it may be used.

Provided the above procedure shows the mixture to be compatible, prepare the tank mixture as follows: Add the fertilizer solution to the spray tank first; mix the required amount of "HYVAR" X with water to form a slurry, with the agitator running, slowly add the slurry and mix thoroughly. For other herbicides, follow directions above. When tank mixing with other herbicides, all applicable directions, restrictions, and precautions for the additional herbicides are also to be followed.

Thoroughly reagitate before using if allowed to settle.

NOTICE TO BUYER Purchase of this material does not confer any rights under patents of countries outside of the United States.

STORAGE AND DISPOSAL

STORAGE: Store product in original container only, away from other pesticides, fertilizer, food, or feed.

PESTICIDE DISPOSAL: Do not contaminate water, food, or feed by storage or disposal. Wastes resulting from the use of this product may be disposed of on site or at an approved waste disposal facility.

CONTAINER DISPOSAL: Completely empty bag into application equipment. Then dispose of empty bag in a sanitary landfill or by incineration, or, if allowed by State and local authorities, by burning. If burned, stay out of smoke.

NOTICE OF WARRANTY

Du Pont warrants that this product conforms to the chemical description on the label thereof and is reasonably fit for purposes stated on such label only when used in accordance with directions under normal use conditions. It is impossible to eliminate all risks inherently associated with use of this product. Crop injury, ineffectiveness or other unintended consequences may result because of such factors as weather conditions, presence of other materials, or the manner of use or application, all of which are beyond the control of Du Pont. In no case shall Du Pont be liable for consequential, special or indirect damages resulting from the use or handling of this product. All such risks shall be assumed by the Buyer. DU PONT MAKES NO WARRANTIES OF MERCHANTABILITY OR FITNESS FOR A PARTICULAR PURPOSE NOR ANY OTHER EXPRESS OR IMPLIED WARRANTY EXCEPT AS STATED ABOVE.

AG-3258 8049/8059 4/13/89

This is a specimen label. DuPont is not responsible for the accuracy of the information contained herein. As labels are subject to revision, always carefully read and follow the label on the product container.

HYVAR® X-L

HERBICIDE

WATER SOLUBLE LIQUID

ACTIVE INGREDIENT:	BY WEIGHT
Lithium salt of bromacil (5-bromo-3-sec-butyl-6-methyluracil)	21.9%
INERT INGREDIENTS:	78.1%
TOTAL	100%

1 Gallon Contains 2 Pounds Bromacil

EPA Reg. No. 352-346

U.S. Pats. 3,352,862, 3,909,236

KEEP OUT OF REACH OF CHILDREN

WARNING

PRECAUTIONARY STATEMENTS

HAZARDS TO HUMANS AND DOMESTIC ANIMALS

WARNING!

HARMFUL OR FATAL IF SWALLOWED. CAUSES EYE IRRITATION. MAY IRRITATE NOSE, THROAT, AND SKIN.

Do not get in eyes. Avoid contact with skin and clothing. Avoid breathing spray mist. Wash thoroughly with soap and water after handling.

STATEMENT OF PRACTICAL TREATMENT

If in eyes: Flush with plenty of water. Get medical attention if irritation persists.

If on skin: Wash with plenty of soap and water. Get medical attention if irritation persists.

For medical emergencies involving this product, call toll free 1-800-441-3637.

ENVIRONMENTAL HAZARDS

Do not apply directly to water or wetlands. Do not contaminate water by cleaning of equipment or disposal of wastes.

PHYSICAL AND CHEMICAL HAZARDS

Combustible. Do not use or store near heat or open flame.
Keep container closed when not in use.

IMPORTANT—Injury to or loss of desirable trees or other plants may result from failure to observe the following: Do not apply, or drain or flush equipment on or near desirable trees or other plants or on areas where their roots may extend, or in locations where the chemical may be washed or moved into contact with their roots. Do not use on lawns, walks, driveways, tennis courts, or similar areas. Prevent drift of spray to desirable plants. Do not contaminate domestic waters. Do not use water from treated ditches for irrigation. Keep from contact with fertilizers, insecticides, fungicides, and seeds. Thoroughly clean all traces of "Hyvar" X-L from application equipment immediately after use. Flush tank, pump, hoses, and boom with several changes of water after removing nozzle tips and screens (Clean these parts separately).

GENERAL INFORMATION

Du Pont "Hyvar" X-L Herbicide is an effective general herbicide for use on non-cropland areas for control of brush and many annual and perennial weeds, and is particularly useful for control of perennial grasses. It is a water soluble liquid to be mixed in water and applied as a spray; "Hyvar" X-L may also be applied undiluted for basal (soil) treatment of brush. Effects are slow to appear and may not become apparent until the chemical has been carried into the root zone of the weeds by moisture. The degree of control and duration of effect will vary with the amount of chemical applied, soil type, rainfall, and other conditions.

"Hyvar" X-L, after mixing in water, is non-volatile, non-flammable, and non-corrosive to metals except aluminum. Do not use with aluminum spray nozzles.

DIRECTIONS FOR USE

It is a violation of federal law to use this product in a manner inconsistent with its labeling.

"Hyvar" X-L Herbicide should be used only in accordance with recommendations on this label, or in separate published Du Pont recommendations available through local dealers. Do not use on any crop in Kern County, California.

Du Pont will not be responsible for losses or damages resulting from use of this product in any manner not specifically recommended by Du Pont. User assumes all risks associated with such nonrecommended use.

Apply "Hyvar" X-L before or during the period of active growth of plants to be controlled when rainfall can be expected for soil activation. If dense growth is present, results will be improved if vegetation is removed before treatment. Do not apply when ground is frozen.

Before spraying, calibrate equipment to determine quantity of water necessary to uniformly cover measured area to be treated. Measure the proper amount of "Hyvar" X-L and mix into the necessary volume of water. After "Hyvar" X-L has been thoroughly mixed in the spray tank, agitation of the spray solution is not required.

Use a fixed-boom power sprayer properly calibrated to a constant speed and rate of delivery. For application with a hand-gun sprayer, use at least 200 gals. spray per acre. For treating small areas a tank type hand sprayer or sprinkling can may be used. For basal (soil) treatment of brush, "Hyvar" X-L may be applied undiluted with an exact delivery hand gun applicator.

Do not apply this product through any type of irrigation system.

WEED CONTROL: To control most weeds for an extended period of time on non-cropland areas such as RAILROAD, HIGHWAY and PIPELINE RIGHT OF WAYS, PETROLEUM TANK FARMS, LUMBERYARDS, STORAGE AREAS, and INDUSTRIAL PLANT SITES:

Apply 1½ to 3 gals. "Hyvar" X-L per acre to control ANNUAL WEEDS and GRASSES such as foxtail, ryegrass, wild oats, crabgrass, cheatgrass, bromegrass, ragweed, lambsquarters, puncturevine, orchardgrass and turkey mullein. When applied just prior to or after emergence of annuals, rates as low as ¾ gal. per acre control many annual weeds and grasses in low rainfall areas and give short term control in higher rainfall areas.

Apply 3 to 6 gals. "Hyvar" X-L per acre to control PERENNIAL WEEDS and GRASSES such as smooth brome, bahiagrass, bluegrass, redtop, purpletop, quackgrass, broomsedge, aster, dandelion, dog fennel, goldenrod, plantain and wild carrot. In areas with low or seasonal rainfall, rates as low as 2¼ gals. per acre control many perennial weeds and grasses.

Apply 6 to 12 gals. "Hyvar" X-L per acre to control JOHNSONGRASS; use at the same rate for OTHER HARD-TO-KILL PERENNIAL WEEDS and GRASSES such as bermudagrass, dallisgrass, nutsedge, vaseygrass, saltgrass, bouncingbet, dogbane, bracken fern and horsetail. Where limited rainfall (usually less than 4 inches) occurs during the active growth period, such as some areas of the West, "Hyvar" X-L will not provide satisfactory control of hard-to-kill, deep-rooted perennial weeds such as johnsongrass.

Use the higher levels of the dosage ranges on adsorptive soils (usually those high in organic matter or carbon).

Retreatment—Apply 3/4 to 3 gals. "Hyvar" X-L per acre when annual weeds and grasses reappear on sites where weed growth has been controlled.

For Small Areas—Three fl. oz. of "Hyvar" X-L per 250 sq. ft. is approximately 4 gals. per acre.

BRUSH CONTROL

To control undesirable woody plants on non-cropland areas such as RAILROAD RIGHT OF WAYS, STORAGE AREAS, INDUSTRIAL PLANT SITES and DRAINAGE DITCHES:

Apply as a broadcast or basal (soil) treatment; for use on drainage ditches, apply as a basal (soil) treatment only. **Note:** For effective brush control and prevention of damage to desirable vegetation: do not apply to brush standing in water; do not use water from treated ditches for irrigation; do not use in irrigation ditches nor on right of ways or other sites where marketable timber or other desirable trees or shrubs are immediately adjacent to the treated areas.

Broadcast Treatment—Apply 2¼ to 5 gals. per acre to control oak, willow, sweet gum, and pine; apply 6 to 12 gals. per acre to control brush such as American elm, winged elm, hackberry, sumac, and cottonwood. Use the higher levels of the dosage ranges on adsorptive soils (those high in organic matter or carbon).

Basal (Soil) Treatment—Controls woody plants such as cottonwood, hackberry, maple, oak, poplar, red bud, sweet gum, wild cherry, willow, winged elm, American elm and sumac.

Diluted—Mix 1 gal. "Hyvar" X-L in 5 gals. of water and apply at the rate of 1 to 2 fl. oz. per stem 2" to 4" in basal diameter; wet base of stem to run-off.

Undiluted—Apply "Hyvar" X-L undiluted with an exact delivery hand gun applicator; this equipment delivers a thin stream of a predetermined volume when triggered. Apply at the rate of 5 to 10 milliliters per stem 2" to 4" in basal diameter; direct the treatment to a spot at the base of the brush (root collar area). When treating large stems and more then one delivery of solution is needed per stem, apply on the opposite side of the stem. **Note:** "Hyvar" X-L is a combustible product. Do not smoke while applying the undiluted product, and keep away from heat and open flame.

NOTICE TO BUYER: Purchase of this material does not confer any rights under patents of countries outside of the United States.

STORAGE AND DISPOSAL

STORAGE: Store product in original container only, away from other pesticides, fertilizer, food, or feed. Keep container closed when not in use.

PESTICIDE DISPOSAL: Do not contaminate water, food, or feed by storage or disposal. Wastes resulting from the use of this product may be disposed of on site or at approved waste disposal facility.

CONTAINER DISPOSAL: Triple rinse (or equivalent). Then offer for recycling or reconditioning, or puncture and dispose of in a sanitary landfill, or by incineration, or, if allowed by State and local authorities, by burning. If burned, stay out of smoke.

NOTICE OF WARRANTY

Du Pont warrants that this product conforms to the chemical description on the label thereof and is reasonably fit for purposes stated on such label only when used in accordance with directions under normal use conditions. It is impossible to eliminate all risks inherently associated with use of this product. Ineffectiveness or other unintended consequences may result because of such factors as weather conditions, presence of other materials, or the manner of use or application, all of which are beyond the control of Du Pont. In no case shall Du Pont be liable for consequential, special or indirect damages resulting from the use or handling of this product. All such risks shall be assumed by the Buyer. DU PONT MAKES NO WARRANTIES OF MERCHANTABILITY OR FITNESS FOR A PARTICULAR PURPOSE NOR ANY OTHER EXPRESS OR IMPLIED WARRANTY EXCEPT AS STATED ABOVE.

AG-3206 8088 10/4/85

This is a specimen label. DuPont is not responsible for the accuracy of the information contained herein. As labels are subject to revision, always carefully read and follow the label on the product container.

HYVAR® L

CITRUS HERBICIDE
WATER SOLUBLE LIQUID

For Use In The States of Florida and Texas Only

ACTIVE INGREDIENT	BY WEIGHT
Lithium salt of bromacil (5-bromo-3-sec-butyl-6-methyluracil). . .	33.8%
INERT INGREDIENTS:	66.2%
TOTAL	100.0%

1 Gallon Contains 3.2 Pounds Bromacil

EPA Reg. No. 352-425

U.S. Pats. 3,352,862 & 3,909,236

KEEP OUT OF REACH OF CHILDREN
WARNING

PRECAUTIONARY STATEMENTS

HAZARDS TO HUMANS AND DOMESTIC ANIMALS

WARNING!

HARMFUL OR FATAL IF SWALLOWED. CAUSES EYE IRRITATION. MAY IRRITATE NOSE, THROAT, AND SKIN.

Do not get in eyes. Avoid contact with skin and clothing. Avoid breathing vapors. Wash thoroughly with soap and water after handling and before eating or smoking.

For medical emergencies involving this product, call toll free 1-800-441-3637.

STATEMENT OF PRACTICAL TREATMENT

If in eyes: Flush with plenty of water. Get medical attention if irritation persists.

If on skin: Wash with plenty of soap and water. Get medical attention if irritation persists.

ENVIRONMENTAL HAZARDS

Do not apply directly to any body of water. Do not contaminate water by cleaning of equipment or disposal of wastes.

Bromacil is a chemical which can travel (seep or leach) through soil and can enter ground water which may be used as drinking water. Bromacil has been found in ground water as a result of agricultural use. The correct use rates by geographical area, specified herein, and proper mixing-loading site precautions and procedures must be followed to minimize potential bromacil movement into groundwater.

PHYSICAL OR CHEMICAL HAZARDS

Combustible. Keep container closed when not in use.

IMPORTANT—Injury to or loss of desirable trees or other plants may result from failure to observe the following: Do not apply, drain, or flush equipment on or near desirable trees or other plants or on areas where their roots may extend, or in locations where the chemical may be washed or moved into contact with their roots. Do not use on lawns, walks, driveways, tennis courts, or similar areas. Prevent drift or spray to desirable plants. Do not contaminate domestic waters. Thoroughly clean all traces of "HYVAR" L Citrus Herbicide from application equipment immediately after use. Flush tank, pump, hoses, and boom with several changes of water after removing nozzle tips and screens (clean these parts separately).

PESTICIDE HANDLING

- Calibrate sprayers only with clean water away from the well site.
- Make scheduled checks with spray equipment.
- Assure accurate measurement of pesticides by all operation employees.
- Mix only enough product for the job at hand.
- Avoid over-filling of spray tank.
- Do not discharge excess material on the soil at a single spot in the field/grove or mixing/loading station.
- Dilute and agitate excess solution and apply at labeled rates/uses.
- Avoid storage of pesticides near well sites.

GENERAL INFORMATION

Du Pont "HYVAR" L Citrus Herbicide is an effective general herbicide for use on citrus for control of many annual and perennial weeds, and is particularly useful for control of perennial grasses. It is a water soluble liquid to be mixed in water and applied as a spray.

"HYVAR" L, after mixing in water, is non-volatile, non-flammable, and non-corrosive to metals except aluminum.

"HYVAR" L controls many annual and certain perennial weeds. For annuals, apply any time of the year, preferably shortly before or after weed growth begins when adequate moisture is available. For perennials, best results are obtained if application is made shortly before or shortly after weed growth begins; if dense growth is present, remove tops and spray the ground. Effects on perennial weeds are slow to appear, usually progressing over a period of several months.

Effects are slow to appear and may not become apparent until the chemical has been carried into the root zone of the weeds by moisture. Moisture is necessary to activate the herbicide; best results are obtained if treatment is made to moist soil, and moisture is supplied by rainfall or sprinkler irrigation within two weeks after application. The degree of control and duration of effect will vary on the application rate, soil texture, rainfall and other conditions.

The degree of control and duration of effect will vary with the amount of chemical applied, soil type, rainfall, and other conditions.

DIRECTIONS FOR USE

It is a violation of federal law to use this product in a manner inconsistent with its labeling.

"HYVAR" L should be used only in accordance with recommendations on this label, or in separate published Du Pont recommendations available through local dealers.

Du Pont will not be responsible for losses or damages resulting from use of this product in any manner not specifically recommended by Du Pont. User assumes all risks associated with such nonrecommended use.

Do not apply this product through any type of irrigation system.

Do not apply this product in such a manner as to directly or through drift expose workers or other persons. The area being treated must be vacated by unprotected persons.

Do not enter treated areas without protective clothing until sprays have dried. Because certain states may require more restrictive reentry intervals for various crops treated with this product, consult your State Department of Agriculture for further information.

Written or oral warnings must be given to workers who are expected to be in a treated area or in an area about to be treated with this product. When oral warnings are given, warnings shall be given in a language customarily understood by workers. Oral warnings must be given if there is reason to believe that written warnings cannot be understood by workers. Warnings must include the following information: "WARNING. Area treated with "HYVAR" L on (date of application). Do not enter without appropriate protective clothing until sprays have dried. In case of contact, flush eyes with plenty of water; wash skin with plenty of soap and water. Get medical attention if irritation persists."

Apply "HYVAR" L before or during the period of active growth of plants to be controlled when rainfall can be expected for soil activation. If dense growth is present, results will be improved if vegetation is removed before treatment.

Before spraying, calibrate equipment to determine quantity of water necessary to uniformly cover measured area to be treated. Measure the proper amount of "HYVAR" L and mix into the necessary volume of water. After "HYVAR" L has been thoroughly mixed in the spray tank, agitation of the spray solution is not required. However, if "HYVAR" L is tank mixed with wettable powders, adequate, continuous agitation is required.

Use a fixed-boom power sprayer properly calibrated to a constant speed and rate of delivery.

NOTE: Do not tank mix "HYVAR®" L with acidic herbicides such as Roundup[1] or with liquid fertilizers.

Best results are obtained if "HYVAR" L is applied to bare ground. If dense populations of hard-to-kill species are present, control of these weeds prior to application of "HYVAR" L is recommended.

Apply as a band or broadcast treatment beneath and/or between trees. Avoid contact of foilage and fruit with spray or mist.

Dosages of "HYVAR" L are expressed as broadcast rates; for band treatment use proportionately less.

Do not graze cattle in treated areas.

Temporary yellowing or vein clearing of citrus leaves may occur following treatment. Because injury to citrus trees may result; do not use on soils with less than 1% organic matter, poorly drained soils, gravelly soils, nor thinly covered or exposed subsoils; do not treat plants in irrigation furrows; do not treat diseased or stressed trees.

Do not use in citrus groves interplanted with other trees or desirable plants, or in home citrus plantings or in areas where roots of valuable plants or trees may extend as plant injury may result.

Treated areas may be planted to citrus trees one year after last application; do not replant to other crops within 2 years after last application as plant injury may result.

Control of perennials may be improved by cultivation, otherwise, avoid working the soil as long as weed control effectiveness of the treatment may be reduced.

CITRUS

Apply with a properly calibrated sprayer as a band or broadcast treatment beneath and/or between trees. Use sufficient water (minimum 40 gallons per acre) to obtain uniform coverage.

Do not use with aluminum nozzles.

Avoid contact of foliage and fruit with spray or mist.

Avoid overlapping of the herbicide swath, and shut off spray boom while starting, turning, slowing or stopping, or injury to trees may occur.

WEEDS CONTROLLED

Annuals

barnyardgrass	puncturevine
crabgrass	purslane
crowfootgrass	sandspur
Florida pusley	sedge (annual)
henbit	sprangletop
lambsquarter	Texas panicum
mustard	turkey mullein
natalgrass (red top)	

Perennials

bermudagrass	Bahiagrass
torpedograss	Johnsongrass
paragrass	nutsedge
pangolagrass	

FLORIDA

*** East Coast/Flatwoods Areas - (low permeable soils):**

- The maximum allowable is 6.4 lbs active (bromacil) per acre per year. The allowable amount is inclusive of all bromacil formulations used within one year.
- Under the above guidelines, do not apply more than 8 qts. per acre per year of "Hyvar" L which contains 6.4 lbs active (bromacil).
- Multiple applications may improve control of "hard-to-kill" species.

Trees Established Less Than One Year

For control of annuals, apply 2 to 3 quarts of "Hyvar" L per acre as needed to maintain weed control. Do not apply at less than 60 day intervals; do not apply more than 6 quarts per acre within any 6 month period, or more than 8 quarts during the first year.

Trees Established One to Three Years

For control of annuals, apply 2 to 3 quarts of "Hyvar" L per acre as needed to maintain weed control. Do not apply more than 6 quarts per acre within any 6 month period, or more than 8 quarts per acre per year.

Trees Established Three or More Years

Apply 2 to 4 quarts per acre as needed to maintain weed control. Do not use more than 5 quarts per acre in any one application, nor more than 8 quarts per acre per year. For control of perennials, use the high rate. Partial control usually occurs with a single treatment; repeat applications are required to control perennials.

*** Ridge Areas -** (highly permeable soils)

- The maximum allowable is 4.2 lbs active (bromacil) per acre per year. The allowable amount is inclusive of all bromacil formulations used within one year.
- Under the above guidelines, do not apply more than 5.25 qts per acre per year of "Hyvar" L which contains 4.2 lbs active (bromacil).
- Multiple applications may improve control of "hard-to-kill" species.

Trees Established Less Than One Year

For control of annuals, apply 2 to 3 quarts of "Hyvar" L per acre as needed to maintain weed control. Do not apply at less than 60 day intervals; do not use more than 5.25 qts per acre per year.

Trees Established One to Three Years

For control of annuals, apply 2 to 3 quarts per of "Hyvar" L per acre as needed to maintain weed control. Do not use more than 5.25 qts per acre per year.

Trees Established Three or More Years

Apply 2 to 4 quarts per acre as needed to maintain weed control. Do not use more than 4 quarts per acre in any one application, nor more than 5.25 qts per acre per year. For control of perennials, use the high rate. Partial control usually occurs with a single treatment; repeat applications are required to control perennials.

TEXAS

Trees Less Than One Year Old

In newly developed groves or in groves that have been heavily reset, apply 2 to 3 quarts "HYVAR" L per acre as needed to maintain weed control. Do not apply at less than 60 day intervals; do not apply more than 6 quarts per acre within any 6 month period, or more than 8 quarts per acre during the first year.

Trees Established One to Three Years

Apply 2 to 3 quarts per acre as needed to maintain weed control. Do not apply more than 6 quarts per acre within any 6 month period, or more than 8 quarts per acre per year.

Trees Established for Four Years or More

For annual weed control apply 2 to 4 quarts per acre.

For perennial weed control make a single application per year during the period from winter to early summer; use at the following rates:

Soil Texture	Qts. "HYVAR" L Per Acre
Sand, loamy sand	4 to 5
Sandy loam	5 to 6
Silt loam, clay loam	6 to 8

Alternatively, make two applications of 3 to 4 qts. "HYVAR" L per acre per year in spring and summer. Partial control of perennials usually occurs with a single treatment; repeat applications are required for control.

[1]Registered trademark of Monsanto Company

NOTICE TO BUYER: Purchase of this material does not confer any rights under patents of countries outside of the United States.

STORAGE AND DISPOSAL

STORAGE: This product should not be stored at temperatures below 32°F. Store product in original container only, away from other pesticides, fertilizer, food, or feed. Keep away from heat and flame.

PESTICIDE DISPOSAL: Do not contaminate water, food, or feed by storage or disposal. Wastes resulting from the use of this product may be disposed of on site or at approved waste disposal facility.

CONTAINER DISPOSAL Triple rinse (or equivalent) the container. Then offer for recycling or reconditioning, or puncture and dispose of in a sanitary landfill, or by incineration, or, if allowed by state and local authorities, by burning. If burned, stay out of smoke.

NOTICE OF WARRANTY

Du Pont warrants that this product conforms to the chemical description on the label thereof and is reasonably fit for purposes stated on such label only when used in accordance with directions under normal use conditions. It is impossible to eliminate all risks inherently associated with use of this product. Ineffectiveness or other unintended consequences may result because of such factors as weather conditions, presence of other materials, or the manner of use or application, all of which are beyond the control of Du Pont. In no case shall the seller be liable for consequential, special, or indirect damages resulting from the use or handling of this product. All such risks shall be assumed by the buyer. DU PONT MAKES NO WARRANTIES OF MERCHANTABILITY OR FITNESS FOR A PARTICULAR PURPOSE NOR ANY OTHER EXPRESS OR IMPLIED WARRANTY EXCEPT AS STATED ABOVE.

AG 3257 8049 4/13/89

This is a specimen label. DuPont is not responsible for the accuracy of the information contained herein. As labels are subject to revision, always carefully read and follow the label on the product container.

KARMEX® DF

HERBICIDE
DISPERSIBLE GRANULE

ACTIVE INGREDIENT:	BY WEIGHT
Diuron [3-(3,4-dichlorophenyl)-1,1-dimethylurea]	80%
INERT INGREDIENTS:	20%
TOTAL	100%

EPA Reg. No. 352-508

KEEP OUT OF REACH OF CHILDREN
WARNING

PRECAUTIONARY STATEMENTS

HAZARDS TO HUMANS AND DOMESTIC ANIMALS

WARNING!

AVISO

PRECAUCION AL USUARIO Si usted no lee ingles, no use este producto hasta que la etiquets haya sido explicada ampliamente.

CAUSES EYE IRRITATION. MAY IRRITATE NOSE, THROAT, AND SKIN.

Avoid breathing dust or spray mist. Avoid contact with skin, eyes, and clothing. Wash thoroughly with soap and water after handling.

STATEMENT OF PRACTICAL TREATMENT

If on skin: Wash with plenty of soap and water; get medical attention if irritation persists.

If in eyes: Flush with plenty of water; get medical attention if irritation persists.

For medical emergencies involving this product, call toll free 1-800-441-3637.

ENVIRONMENTAL HAZARDS

Do not apply directly to water. Do not contaminate water by cleaning of equipment or disposal of wastes. Cover or incorporate spills.

IMPORTANT

Injury to or loss of desirable trees or other plants may result from failure to observe the following:

Do not apply (except as recommended for crop use), or drain or flush equipment on or near desirable trees or other plants, or on areas where their roots may extend, or in locations where the chemical may be washed or moved into contact with their roots. Do not use on home plantings of trees, shrubs or herbaceous plants, nor on lawns, walks, driveways, tennis courts or similar areas. Prevent drift of dry powder or spray to desirable plants. Do not contaminate any body of water. Keep from contact with fertilizers, insecticides, fungicides and seeds. Do not apply this product through any type of irrigation sysytem.

Thoroughly clean all traces of "KARMEX" DF from application equipment immediately after use. Flush tank, pump, hoses, and boom with several changes of water after removing nozzle tips and screens (clean these parts separately).

GENERAL INFORMATION

Du Pont "KARMEX" DF Herbicide is a dispersible granule to be mixed in water and applied as a spray for selective control of weeds in certain crops and for nonselective weed control on non-cropland areas. It is noncorrosive to equipment, nonflammable and nonvolatile.

"KARMEX" DF may be applied to soil prior to emergence of weeds to control susceptible weed seedlings for an extended period of time; the degree of control and duration of effect will vary with the amount of chemical applied, soil texture, rainfall and other conditions. Soils high in clay or organic matter require higher dosages than soil low in clay or organic matter to obtain equivalent herbicide performance. Moisture is required to activate the chemical; best results occur if rainfall (or sprinkler irrigation) occurs within 2 weeks of application.

"KARMEX" DF applied preemergence, before emergence of crop and weeds, is an effective procedure because susceptible weeds are controlled in an early, vulnerable seedling state before they compete with the crop. With favorable moisture conditions, "KARMEX" DF continues to control weeds for some time as the crop becomes better able to compete. Should weed seedlings begin to break through the preemergence treatment in significant numbers, secondary weed control procedures should be implemented; these include cultivation and postemergence herbicide application.

"KARMEX" DF may also be used to control emerged weeds. Results vary with rate applied and environmental conditions; best results are obtained on succulent weeds growing under conditions of high humidity and temperatures of 70°F or higher. Addition of a surfactant such as Du Pont Surfactant WK to the spray (where recommended) increases contact effects of "KARMEX" DF.

"KARMEX" DF may be used as a directed postemergence application, where spray nozzles are adjusted so the weeds are sprayed but the crop is not, on the following crops: artichoke, corn (field), cotton, sorghum (grain), sugarcane and established plantings of apples, bananas, plantains, blueberries, caneberries, gooseberries, citrus, grapes, macadamia nuts, olives, papayas, peaches, pears, pecans, walnuts and certain tree plantings.

Under specified conditions (see Directions for Use), "KARMEX" DF without surfactant may be applied over the top of alfalfa (established, dormant or semidormant), asparagus (established), birdsfoot trefoil (established, dormant), grass seed crops (established), oats, red clover (established, dormant), surgarcane, wheat, pineapple and plumosus fern (established, mowed).

Weed species vary in susceptibility to "KARMEX" DF and they may be more difficult to control when under stress. Combinations of "KARMEX" DF with other herbicides (as registered) increase the number of weed species controlled; consult labels of the companion product for this and other information.

Since the effect of "KARMEX" DF varies with soils, uniformity of application, and environmental conditions, it is suggested that growers limit their first use to small areas. Observe all cautions and limitations on labeling of all products used in mixtures.

DIRECTIONS FOR USE

It is a violation of federal law to use this product in a manner inconsistent with its labeling.

Du Pont "KARMEX" DF Herbicide should be used only in accordance with recommendations on this label, or in separate published Du Pont recommendations available through local dealers.

Du Pont will not be responsible for losses or damages resulting from use of this product in any manner not specifically recommended by Du Pont. User assumes all risk associated with such non-recommended use.

Do not apply this product in such a manner as to directly or through drift expose workers or other persons. The area being treated must be vacated by unprotected persons.

Farm workers performing hand labor operations should not enter treated cropland areas without protective clothing until sprays have dried. Because certain states may require more restrictive reentry intervals for various crops treated with this product, consult your State Department of Agriculture for further information.

Written or oral warnings must be given to farm workers performing hand operations who are expected to be in a treated cropland area or in a cropland area about to be treated with this product. When oral warnings are given, warnings shall be given in a language customarily understood by workers. Oral warnings must be given if there is reason to believe that written warnings cannot be understood by workers. Warnings must include the following information: "CAUTION. Area treated with "KARMEX" DF on (date of application). Do not enter without appropriate protective clothing until sprays have dried. In case of contact, wash skin with plenty of soap and water; flush eyes with plenty of water. Get medical attention if irritation to skin or eyes persist".

Do not use in Kern County, California, except for non-cropland and citrus weed control.

SELECTIVE USE IN CROPS

PREEMERGENCE USE (Germinating Weeds): "KARMEX" DF, at recommended rates, controls annual weeds such as:

Broadleaves		
¾ to 1 Lb/Acre	1½ to 2 Lbs/Acre	2 to 6 Lbs/Acre
Lambsquarters Pigweed Purslane Ragweed	Annual groundcherry Annual morningglory Chickweed Corn spurry Dogfennel Fiddleneck (amsinckia) Gromwell Knawel Pennycress Sheperdspurse Tansy-mustard Wild buckwheat Wild lettuce Wild mustard	Ageratum Annual smartweed Annual sowthistle Corn speedwell Dayflower Flora's paintbrush Hawksbeard Horseweed Kochia Marigold Mexican clover Pineappleweed Pokeweed Rabbit tobacco Spanishneedles Velvetleaf (buttonweed) Wild radish

Partial control of the following weeds usually occurs at rates stated:

Broadleaves	
1 Lb/Acre	4 Lbs/Acre
Annual morningglory Cocklebur Prickly sida (teaweed) Sesbania Sicklepod	Horsenettle

Grasses		
¾ to 1 Lb/Acre	1½ to 2 Lbs/Acre	2 to 6 Lbs/Acre
Barnyardgrass (watergrass) Crabgrass	Annual bluegrass Annual sweet vernalgrass Foxtail Rattail fescue Red sprangletop Velvetgrass	Annual lovegrass Annual ryegrass Kyllinga Orchardgrass Peppergrass Ricegrass Sanbur Seedling johnsongrass

Partial control of the following weeds usually occurs at rates stated:

Grasses	
4 Lb/Acre	8 to 10 Lbs/Acre
Quackgrass	Guineagrass Maidengrass Pangolagrass

POSTEMERGENCE USE (Emerged Seedling Weeds)

"KARMEX" DF at recommended rates, controls annual weeds such as annual morningglory, barnyardgrass (watergrass), crabgrass, crowfoot, goosegrass, pigweed and purslane. Addition of Surfactant WK to the spray (where recommended) increases contact effects of "KARMEX" DF. Best results are obtained on succulent weeds growing under conditions of high humidity and temperatures of 70°F or higher.

EQUIPMENT—SPRAY VOLUMES AND PRESSURES

Use a boom power sprayer properly calibrated to a constant speed and rate of delivery. Openings in screens should be equal to or larger than 50 mesh. Continuous agitation in the spray tank is required to keep the material in suspension. Agitate by mechanical or hydraulic means; if by-pass or return line is used, it should terminate at bottom of tank to minimize foaming. Avoid overlapping, and shut off spray booms while starting, turning, slowing or stopping, or injury to the crop may result.

For preemergence application, use 25 to 40 gals. per acre and spray pressure of 30 to 40 psi. For postemergence application, use sufficient volume (min. 25 gals. per acre) for thorough coverage of weed foliage; use spray pressure of 20 to 25 psi to keep spray drift to a minimum.

Aerial: For alfalfa, asparagus, barley (winter), cotton (preplant or preemergence only), grass seed crops, pineapple, sugarcane and wheat (winter), application may be made by aircraft (5 to 10 gals. per acre); avoid overlapping of spray swath and avoid application under conditions where excessive drift may occur. Where land is bedded, make application parallel to rows.

SPRAY PREPARATION

Mix proper amount of "KARMEX" DF into necessary volume of water; where use of Du Pont Surfactant WK is recommended, dilute with 10 parts of water and add as last ingredient to nearly full tank.

USE RATES

All dosages of "KARMEX" DF are expressed as broadcast rates; for band treatment, use proportionately less. For example, use 1/3 of the broadcast rate when treating a 14" band where row spacing is 42". Where a range of dosages is given, use the lower rate on coarse textured soils (low in clay or organic matter) and the higher rate on the fine textured soils (high in clay or organic matter); for postemergence application, use the lower rate on smaller weeds and the higher rate on larger weeds.

SOIL LIMITATIONS

Crop injury may result from failure to observe the following:

Unless otherwise directed, do not use on sand, loamy sand, gravelly soils or exposed subsoils; nor on pecans where organic matter is less than 1/2 %; nor on alfalfa, apples, artichoke, barley (winter), bermudagrass pasture, citrus, cotton, grapes, oats, olives, papayas, peaches, pears, plumosus fern, sorghum, sugarcane, walnuts and wheat (winter) where organic matter is less than 1%; nor on blueberries, birdsfoot trefoil, caneberries, gooseberries, macadamia nuts and peppermint where organic matter is less than 2%.

Preemergence weed control will be reduced on high organic matter soils (greater than 5%, such as peat or muck).

REPLANTING

Unless otherwise directed, do not replant treated areas to any crop within 2 years after last application as injury to subsequent crops may result.

FIELD CROPS (See Soil Limitations)

A good seedbed must be prepared before preemergence use of "KARMEX" DF as crop injury may result if application is made to ground which is cloddy or compacted resulting in improperly planted seed. Plant seed to depth specified. Unless otherwise directed, surface of the soil should not be cultivated or disturbed after application of "KARMEX" DF and before emergence of the crop as weed control may be reduced and crop injury may result. However, if moisture is insufficient to activate the herbicide, a shallow cultivation (rotary hoe preferred) should be made after emergence of crops while weeds are small enough to be controlled by mechanical means.

ALFALFA

Treat only stands established for 1 year or more. Do not apply to seedling alfalfa nor to alfalfagrass mixtures; do not apply to alfalfa under stress from disease, insect damage, shallow root penetration (such as on shallow hard pans), alkali spots; nor to flooded fields as crop injury may result. Do not spray on snow-covered or frozen ground.

Idaho, Oregon, Washington: Use 1½ to 3 lbs. per acre; for control of volunteer alfalfa, use 4 lbs. per acre. Apply in fall after alfalfa becomes dormant but no later than mid-December.

California (Dormant and Semi-Dormant Varieties): Use 1½ to 3 lbs. per acre; for control of volunteer alfalfa, use 4 lbs. per acre. Apply in fall or winter after alfalfa becomes dormant or semi-dormant, but before growth begins in the spring. Crop injury may result if application is made to actively growing alfalfa. For best results, apply before weeds have emerged or become established (2" in height or diameter). Control of established weeds is improved by applying "KARMEX" DF with a suitable contact herbicide registered for such use. Sufficient rainfall for soil activation of "KARMEX" DF is unlikely in California after February 1. Treated areas may be replanted to any crop after one year from last application if rate does not exceed 2 lbs. per acre.

Arizona, Nevada: Use 1½ to 3 lbs. per acre; apply in fall after alfalfa becomes dormant but no later than January.

Eastern Colorado, Kansas: For control of tansymustard, apply 1 lb. per acre shortly after emergence of mustard in the fall or winter; use 2 lbs. per acre if weeds are 2" to 4" in height. Alternatively, if other annual weeds are present, apply 2 to 3 lbs. per acre in February or March.

Other Areas Where Alfalfa Becomes Winter Dormant: Use 1½ to 3 lbs. per acre (1½ to 2 lbs. per acre East of Appalachian Mountains). Apply in March or early April, but before spring growth begins.

ARTICHOKE—California

Apply 2 to 4 lbs. per acre in late fall or early winter after the last cultivation. Apply before weeds germinate or to emerging seedlings. Direct spray to cover the area between the rows and at the base of artichoke plants, keeping contact with crop plants at a minimum.

ASPARAGUS

Apply as a band or broadcast treatment. Do not apply to young plants during the first growing season (except as noted below), nor to newly seeded asparagus, nor on plants with exposed roots as severe injury may result. Preemergence weed control will be reduced on high organic matter soils (greater than 5%).

Established Plantings: On light sandy soils and other soils low in clay or organic matter, apply 1 to 2 lbs. per acre. On soils high in clay or organic matter, use 2 to 4 lbs. per acre. Two applications may be used; the first application should be made before weeds become established but no earlier than 4 weeks before spear emergence and no later than the early cutting period (if weeds are controlled into the cutting period by cultural practices, application may be delayed until immediately after the last cultivation); a second application may be made immediately following completion of harvest provided rainfall is expected. When two applications are used in one season, do not exceed 3 lbs. per acre per application. In Washington (irrigated crop), apply a single treatment of 4 lbs. per acre. If treatment is delayed until late winter or early spring, incorporation of the chemical in the top 1" to 2" of soil may substitute for lack of rain to activate the herbicide.

Newly Planted Crowns—California (San Joaquin Delta): Make a single application of 2 to 4 lbs. per acre on soils high in clay or organic matter; use the lower rate on clay loams and the higher rate on peat soils. Do not use on soils containing less than 2% organic matter. Soil must be settled by rainfall or irrigation prior to treatment. Do not treat crowns planted to a depth of less than 2".

BARLEY, WINTER (Drill-Planted)—Western Oregon and Western Washington

Make a single application of 1½ to 2 lbs. per acre as soon as possible after planting but before emergence of barley. Do not replant treated areas to any crop within 1 year after last application as injury to subsequent crops may result.

BERMUDAGRASS PASTURES (Newly Sprigged)

Apply 1 to 3 lbs. after planting and before emergence of bermudagrass or weeds. Alternatively, for control of emerged annual weeds up to 4" in height, apply ½ to 1 lb. per acre; add 1 pt. Surfactant WK per 25 gals. of spray. If bermudagrass has emerged at time of treatment, temporary burn of exposed plant parts may occur.

Plant sprigs (stolons) 2" deep in a well-prepared seedbed; do not treat areas where sprigs are planted less than 2" deep as crop injury may result. Do not graze or feed foliage from treated areas to livestock within 70 days after application.

BIRDSFOOT TREFOIL (Lotus)—Western Oregon

Treat only stands established for a least 1 year; do not apply to seedling trefoil as injury may result. Make a single application of 2 lbs. per acre when trefoil is dormant (October 15 to December 15). Do not replant treated areas to any crop within 1 year after last application as injury to subsequent crops may result.

CORN (FIELD)

Postemergence—Make a single application of 3/4 lb. per acre in combination with non-pressure nitrogen solution. If nitrogen solution is not used, apply 1 lb. per acre; add 1 pt. Surfactant WK per 25 gals. of spray. Apply as a directed spray when corn is a least 20" high and weeds are no taller than 3". DO NOT APPLY OVER TOP OF CORN. Do not replant to any crop within 1 year, except that cotton, corn and grain sorghum may be planted the spring following treatment.

Preemergence—Arkansas, Louisiana, Mississippi and Tennessee: Make a single application of 2/3 to 1 lb. per acre as a broadcast or band treatment after planting but before corn emerges. Plant corn at least 1½" deep. Do not replant treated areas to crops other than corn or cotton within 4 months following band treatment and 6 months following broadcast treatment as crop injury may result.

COTTON

During a single crop season, do not exceed the following amounts of "KARMEX" DF per acre as injury to subsequent crops may result: 1 lb. on loamy sand; 1½ lbs. on sandy loam; 2 lbs. on clay loam; 2¾ lbs. on clay. Injury may occur if "KARMEX" DF is used in conjunction with soil-applied organic phosphate pesticides. Do not allow livestock to graze treated cotton.

Preplant—Arizona and California: Use "KARMEX" DF alone, or apply as a separate operation following preplant broadcast treatment with "Treflan"[1] (incorporated according to directions on "Treflan" label). Apply "KARMEX" DF as a broadcast spray after beds are formed, preirrigated, and final seedbeds prepared. Prior to planting, drag-off the tops of the beds and plant in moist soil not treated with "KARMEX" DF. Treated soil is returned to the bed after planting when irrigation furrows are reformed after cotton has emerged. If more than two furrowing-out operations are made prior to lay-by, or deep furrows are made early, weed control may be reduced in furrow bottoms. Use at the following rates:

"KARMEX" DF Alone: 1 to 2½ lbs. per acre.

"KARMEX" DF Following "Treflan":

Soil Texture	Product Per Acre—Preplant "Treflan"	Product Per Acre—Preplant "KARMEX" DF
Sandy loam, loam, silt loam, silt	1 pt.	⅔ to 1 lb.
Sandy clay loam, clay loam, silty clay loam, sandy clay, clay	1½ pts.	1 to 1¼ lbs.

Note: Seedling disease may weaken plants and increase the possibility of injury from the use of "Treflan" followed by "KARMEX" DF. These treatments should be used only in conjunction with a standard fungicide seed treatment plus a good supplemental soil fungicide program such as captan-PCNB mixture.

Preemergence—U.S., except Arizona, California, and areas west of Interstate 35 or 35W in Texas and Oklahoma: Use "KARMEX" DF alone or apply as a separate operation following preplant treatment with "Treflan". Apply "KARMEX" DF after planting but before cotton emerges. Do not treat cotton in deep furrows as crop injury may result; use only where cotton is planted on flat or raised seedbeds. Shallow incorporation (no deeper than ¼") with a rotary hoe or similar equipment following planting usually improves results especially during dry weather. A wide press wheel should be used on the planter to provide a level seedbed for subsequent early season postemergence treatments. If moisture is insufficient to activate "KARMEX" DF or if soil becomes crusted before crop emerges, a shallow rotary hoeing (no deeper than ¼") should be made before weeds become established.

"KARMEX" DF Alone: Make a single application as a broadcast or band spray, using the following broadcast rates; for band treatment, use proportionately less.

Soil Texture**	Lbs. "KARMEX" DF Per Acre
Loamy sand	⅔
Sandy loam, loam, silt loam, silt	1
Sandy clay loam, clay loam, silty clay loam, sandy clay	1¼
Silty clay, clay	2

** Do not use on soils with less than 1% organic matter as crop injury may result.

"KARMEX" DF Following "Treflan" Preplant: Apply "Treflan" prior to planting as a broadcast or band treatment; incorporate according to directions on "Treflan" label. As a separate operation, apply "KARMEX" DF as a band treatment (14" to 20" wide) after planting but before cotton emerges. Use at the following broadcast rates; for band treatment, use proportionately less. See "Note" under Preplant above.

Soil Texture**	Product Per Acre: Preplant "Treflan"	Product Per Acre: Preemergence "KARMEX" DF
Loamy sand	½ pt.	⅔ lb
Sandy loam, loam, silt loam, silt	1 pt	1 lb
Sandy clay loam, clay loam, silty clay loam, sandy clay, silty clay, clay	1½ pts	1¼ - 2 lbs.

** Do not use on soils with less than 1% organic matter as crop injury may result.

Postemergence—U.S.: Apply only as a directed spray to cover weed foliage; adjust nozzles to minimize contact of cotton leaves with spray or drift or crop injury may result. DO NOT SPRAY OVER TOP OF COTTON.

Early Season—Apply when cotton is a least 6" tall and when weeds are actively growing and do not exceed 2" in height. Apply as a band treatment at following rates; for each 25 gals. of spray, add 1 pt. Surfactant WK. Two applications may be made if needed.

Weed Problem (Up to 2" Tall)	Lbs. "KARMEX" DF Per Acre (Broadcast Basis)
Annual grasses	½
Pigweed	¼

For control of seedling perennial grasses such as johnsongrass and partial control of nutsedge or when weed growth is under drought stress or as high as 4", add 2 to 3½ lbs. disodium methylarsonate (DSMA; 63% anhydrous or equivalent) to above spray mixture. If DSMA is used, do not apply after first bloom.

Late Season (Lay-By)—Apply 1 to 1½ lbs. per acre (1 to 2 lbs. in Arizona and California) when cotton is at least 12" tall (at least 20" tall for Pima S-2). For control of germinating weed seedlings, apply to soil beneath cotton plants and between rows immediately after last cultivation. In irrigated cotton, best weed control is obtained if the field is irrigated within 3 to 4 days after application; thoroughly wet the surface of the ground over the row to carry the herbicide into the root zone of germinating weeds. Alternatively, for control of emerged annual weeds (up to 4" in height) at lay-by time, make a single application in combination with Surfactant WK (1 pt.

per 25 gals. spray), or use 1/2 to 3/4 lb. "KARMEX" DF (plus surfactant) per acre and repeat later if needed.

Replanting: If initial seeding fails to produce a stand, cotton may be replanted in soil treated preplant or preemergence with "KARMEX" DF, alone or following "Treflan". Where ever possible, avoid disturbing original bed. If necessary to rework soil before replanting, use shallow cultivation such as discing; do not relist nor move soil into the original drill area. Plant seed at least 1" deep. Do not retreat field with a second preplant or preemergence application during the same crop year as injury to the crop may result.

Subsequent Crops:

"KARMEX" DF — Type of Application	Crops That May Follow Treated Cotton
Band preemergence or postemergence	Any crop 4 months after last application.
Band preemergence plus postemergence —or— Broadcast preemergence (and preplant) —or— Broadcast preemergence plus band postemergence	Cotton, soybeans, corn or grain sorghums (not sorgos or forage sorghums nor grass sorghums) the next spring. Do not replant treated areas to any other crop within one year after last application as injury to subsequent crops may result.
Broadcast postemergence (lay-by)	Cotton, corn, grain sorghums (not sorgos or forage sorghums nor grass sorghums) the next spring. Do not replant treated areas to any other crop within one year after last application as injury to subsequent crops may result.

For subsequent crops in fields where "Treflan" is used, follow instructions on "Treflan" label.

GRASS SEED CROPS (Perennial) Except as noted, apply only to established plantings at least 1 year old.

Colorado, Kansas, New Mexico and Oklahoma: On sand bluestem, side oats grama and switchgrass, apply 2 to 3 lbs. per acre during the dormant period shortly before weed seedlings emerge. Do not apply after crop begins growth in the spring as crop injury may result. In fields where ash residues have accumulated from burning straw, use 3 lbs. per acre; spread unburned chaff or straw with a harrow or chopper before application.

Western Oregon: On alto fescue, Astoria bentgrass, Highland bentgrass, Kentucky bluegrass (Merion bluegrass) and orchardgrass, apply 2 to 4 lbs. per acre between October 1 and November 15. In fields where ash residues have accumulated from burning straw, use 3 to 4 lbs. per acre; spread unburned chaff or straw with a harrow or chopper before application. If perennial velvetgrass (Holcus lanatus) is a problem, use 4 lbs. per acre. For best results, apply as soon as possible after fall rains start. Established weeds (beyond 2 to 4 leaf stage) should be removed prior to treatment.

Well established vigorous stands of spring-planted alta fescue, Kentucky bluegrass and orchardgrass may be treated the following fall provided the crop is planted before April 1 and treatment is not applied before October 15; use 2 lbs. per acre.

Oregon: For use in newly planted bentgrass, chewing fescue, Kentucky bluegrass, perennial ryegrass, orchardgrass and tall fescue. During planting operation, spray Aqua Nu-Char[2] or Gro-Safe[3] or other suitable brands of activated charcoal as a 1" band on soil surface at rate of 300 lbs. per acre (broadcast basis; equivalent to 15 lbs. per acre of crop where row spacing is 20"). Mount nozzles to apply directly over seed rows to prevent crop injury. Follow with "KARMEX" DF as a single broadcast spray at rate of 2 1/2 to 3 lbs. per acre; apply as soon as possible after planting but before crops or weeds emerge and before rains or sprinkler irrigation. Fall or spring plantings may be treated; best results usually occur with early fall plantings. Treatment will not control downy brome or wild oats.

OATS (Drill-Planted)

Do not replant treated areas to any crop within one year after last application as injury to subsequent crops may result.

Spring oats—Idaho, Eastern Oregon, Eastern Washington: Use in areas where average annual rainfall exceeds 16". Make a single application of 1 to 1 1/2 lbs. per acre after planting, either before or after oats emerge but within 6 weeks of planting. Best results are usually obtained when application is made 3 to 4 weeks after planting. Apply before weeds are 3" to 4" tall.

Winter Oats and Mixtures with Peas or Vetch—Western Oregon and Western Washington: Make a single application of 1 1/2 to 2 lbs. per acre as soon as possible after planting but before emergence of the crop.

PEPPERMINT—Pacific Northwest

Apply 3 lbs. per acre just after the last cultivation in the spring prior to emergence of peppermint. Do not apply to newly planted (less than 1 year) nor to emerged peppermint as injury may result.

RED CLOVER—Western Oregon

Make a single application of two pounds per acre on established red clover stands (at least 9 months). Apply "KARMEX" DF when red clover is dormant (October 15 to December 15). Do not apply to seedling red clover, and do not replant treated area to any crop within one year after last application.

Treatment will control annual weeds such as bluegrass, chickweed, hawksbeard, rattail fescue, rye grass, and velvet grass.

SORGHUM (GRAIN)—Southwestern States

Apply 1/4 to 1/2 lb. per acre; add 1 pt. Surfactant WK per 25 gals. of spray. Apply as a directed postemergence broadcast or band spray after sorghum is 15" tall to control weeds 2" to 4" in height. DO NOT SPRAY OVER TOP OF SORGHUM. Use the lower rate on broadleaved weeds up to 2" tall; use the higher rate on grasses up to 2" and broadleaved weeds up to 4" tall. When the lower rate is used, a second application may be made if needed provided the amount applied in one crop year does not exceed 1/2 lb. per acre. Treatment of weeds under drought stress is usually ineffective.

Do not replant treated areas to crops other than cotton or corn within 4 months following band treatment and 6 months following broadcast treatment as crop injury may result.

SUGARCANE

To prevent possible crop injury on new cane varieties, tolerance to "KARMEX" DF should be determined prior to adoption as field practice. Do not treat surgarcane growing on thinly covered subsoils or rocky areas as crop injury may result. Temporary chlorosis of the crop may result from application over emerged cane; to minimize chlorosis, use directed postemergence sprays.

Florida: Preemergence—For high organic soils, apply 2 to 4 lbs. per acre as a broadcast or band spray prior to weed emergence after planting or after harvesting plant crop (for ratoon crop). **Postemergence**—Make 1 or 2 applications of 2 lbs. per acre as needed by directed spray inter-row. Alternatively, for panicum control, make up to 3 applications of 1/2 to 1 lb. per acre as a directed spray after cane has emerged but before panicum exceeds 2" in height; add 1 qt. Surfactant WK per 100 gals. of spray. Adjust nozzles to spray beneath cane plants and between rows to cover weed foliage and to minimize contact of cane leaves with spray or drift. Do not apply more than 6 lbs. total per acre between planting (or ratooning) and harvest.

Hawaii and Puerto Rico: Apply 4 to 8 lbs. per acre as a broadcast spray prior to weed emergence after planting or after harvesting plant crop (for ratoon crop). A second and third application of 2 to 4 lbs. per acre may be made as a broadcast spray over emerged cane or by directed spray inter-row.

If weeds are emerged, add a surfactant (such as Surfactant WK, "Osamul" 95 or "Sterox" SK) to the spray at the rate of 1 to 2 qts. per 100 gals. and apply as a directed spray. DO NOT SPRAY OVER TOP OF CANE.

Do not apply more than 3 treatments nor more than 10 lbs. (Puerto Rico) or 12 lbs. (Hawaii) total per acre between planting (or ratooning) and harvest. Treated areas may be planted to sugarcane or pineapple one year after last application.

Louisiana: Use on plant cane seeded on fallowed ground. Make a single application of 3 to 3 3/4 lbs. per acre at either of the following times. **Fall Treatment (August through October)**—Treat a 2 ft. band over the row after planting of cane, but before weeds or cane emerge. **Spring Treatment (January through April)**—if shaving and off-barring are practiced, treat a 2 ft. band over the row before weeds or cane emerge.

WHEAT, WINTER (Drill-Planted)

Crop injury may result where severe winter stress, disease or insect damage follows application; winter-sensitive varieties such as McDermid and Hyslop may be less tolerant of "KARMEX" DF than winter-hardy varieties such as Gaines and Nugaines. Crop injury may also result from failure to observe the following: Do not use on sand or loamy sand soils, nor on gravelly or sandy loams low in organic matter (less than 1%), nor on thinly covered or exposed subsoil areas (clay knobs); do not treat wheat planted less than 1" deep; do not treat wheat where winter climatic conditions have caused "heaving" of plants; do not treat wheat plants lacking in vigor due to poor emergence, insect damage, disease, high alkalinity or other causes; do not apply after wheat has reached the "boot" stage of maturity; do not use with surfactants, or nitrogen solution. Do not replant treated areas to any other crop within 1 year after last treatment (except as noted) as injury to subsequent crops may result.

Idaho, Oregon and Washington—East of Cascade Range: Areas Where Average Annual Rainfall Exceeds 16 Inches: Make a single application of 1 to 1 1/2 lbs. per acre. **Fall Treatment:** For early fall-planted wheat (seeded before September 10), apply 3 to 6 weeks after planting but before weeds are 3" to 4" tall. Treatment after October 1 has generally given best results. Application should not be made after soil freezes in the fall. Wheat planted in late October should not be treated until the following spring. **Spring Treatment:** Apply as soon as wheat starts to grow in the spring. Treatment made prior to April 10 will usually give good results provided weed growth is less than 4" tall. Application later than May 1 may give poor results.

Alternatively, make a single application of 1/2 to 1 lb. "KARMEX" DF plus 1/4 lb. bromoxynil per acre as a tank mixture, either in the fall after wheat has emerged but before soil freezes or in the spring as soon as soil thaws; apply before weeds are 2" tall or across.

Areas Where Average Annual Rainfall is 10 to 16 Inches: After wheat is planted in the fall, make a single application of 1 to 1½ lbs. per acre when sufficient moisture is available to germinate wheat seed. Apply before soil freezes and before weeds are 2" tall. Application later than March 1 may give poor results.

Note—If fall-planted wheat fails to grow due to winter kill or adverse growing conditions after fall treatment, only fields treated before November 1 may be replanted to spring wheat. Spring wheat should not be planted before April 1, and only after deep discing and plowing to a depth of 4" to 6" prior to planting. Do not retreat field with a second application during the same crop year as injury to the crop may result.

Oregon and Washington—West of Cascade Range: Make a single application of 1½ to 2 lbs. per acre as soon as possible after planting; if wheat and weeds have emerged, apply before weeds are 3" to 4" tall. Alternatively, apply a tank mixture of "KARMEX" DF plus bromoxynil as detailed above for "East of Cascade Range".

Other Areas: Make a single application in the spring as soon as wheat (fall-planted) starts to grow and before weeds are 2" tall. Application later than May 1 may give poor results.

Central Plains and Midwest: Use 1 to 2 lbs. per acre.

Kansas, Oklahoma and Texas: Do not use on sand or sandy loam soils. Use 1 lb. per acre on silt and silt loam soils and 1½-2 lbs. per acre on clay, clay loam, and silty clay loam soils.

Northeast: Use 1 to 1½ lbs. per acre.

FRUIT AND NUT CROPS (See Soil Limitations)

Unless otherwise directed, make a single application per year as a directed spray, avoiding contact of foliage and fruit with spray or drift. Do not graze livestock in treated orchards or groves.

APPLES:

Use "KARMEX" DF alone, or apply as a tank mixture with Du Pont "Sinbar" Herbicide.

"KARMEX" DF Alone—Use only under trees established in the orchard for at least 1 year; do not treat varieties grafted on full-dwarf root stocks. Apply 4 lbs. per acre in the spring (March through May). In the Far West, treatment may be made in winter (December through February), or apply 2 lbs. per acre as a postharvest treatment followed by 2 lbs. in the spring.

"KARMEX" DF + "Sinbar"—Use only under trees established in the orchard for at least 2 years. Apply either in the spring or after harvest in the fall before weeds emerge or during early seedling stage of weed growth.

Soil Texture	"Karmex" DF + "Sinbar" Lbs. Product Per Acre	
	1 to 2% Organic Matter	More than 2% Organic Matter
Sandy loam	1 + 1	1½ + 1½
Loam, silt loam, silt	1½ + 1½	2 + 2
Clay loam, clay	2 + 2	2 + 2

Where crop is grown under furrow irrigation or under raised-berm flood irrigation (trees 4" to 6" above waterline), apply only as a band treatment. Do not treat trees planted in the bottom of irrigation furrows, nor trees grown under flat flood or basin irrigation, as injury to trees may result. Where complete weed control to harvest is desired, additional weed control measures may be required during the growing season.

Georgia—Apply 2 to 3 lbs. per acre in the spring. Repeat application in the fall but do not use more than 4 lbs. per acre per year. Add Du Pont Surfactant WK at 1 pint per 25 gals. spray mixture to improve control of small, emerged weeds.

BANANAS AND PLANTAINS—New Plantings

To control annual weeds, apply 1½ to 3 lbs. per acre after planting but before weeds emerge. Do not apply to loose soil directly over the planting material.

Established Plantings: For control of annuals and for top-kill of perennials such as bermudagrass, birdseed grass and guineagrass, apply 3 to 6 lbs. per acre plus 1 pt. Surfactant WK (or suitable equivalent) per 25 gals. of spray; avoid contact of plants with spray or drift as injury may result. When tall, dense weed growth is present, remove weed growth before application. If application is made to soil free of weeds, omit the surfactant from the spray. Repeat treatment as needed, but do not apply more often than 6-week intervals nor more than a total of 12 lbs. per acre (broadcast basis) in a 12-month period.

Note: Do not replant treated areas to any crop within 2 years after last application as injury to subsequent crops may result, except that sugarcane or pineapple may be planted one year after last application.

BLUEBERRIES, CANEBERRIES AND GOOSEBERRIES

Use only in fields which have been established for a least 1 year. Do not apply to berries interplanted with fruit trees; do not apply to plants whose roots are exposed as injury may result. Apply as a band treatment at a base of canes or bushes; for spring application, apply before germination and growth of annual weeds.

Georgia—Blueberries:
Apply 1½ to 2 lbs. per acre in the spring and repeat treatment after harvest in the fall. Add Du Pont Surfactant WK at 1 pint per 25 gals. spray mixture to improve control of small, emerged weeds.

Indiana, Michigan and Ohio—Blueberries: Apply 2 to 4 lbs. per acre in late spring; alternatively, apply 2 lbs. per acre in the fall and repeat at same rate in the spring. Raspberries: Apply 3 lbs. per acre in the spring.

Massachusetts—Blueberries: Apply 2 lbs. per acre in late spring.

New Jersey—Blueberries: For control of winter annuals, apply 2 lbs. per acre in October, November or December, or a single application of 2½ lbs. per acre may be applied in early to mid spring.

California—Raspberries, Blackberries, Boysenberries, Dewberries and Loganberries: For control of winter annuals, apply 2 lbs. per acre in October or November; repeat at same rate in late spring to control summer annuals. A single application of 3 lbs. per acre in January or February will control both winter and summer annuals in some areas, but the separate fall and spring schedule is preferred.

Western Oregon and Western Washington—Blueberries, Caneberries and Gooseberries: Use same schedule as recommended for California.

CITRUS

Use only under trees established in the grove for at least 1 year. Time application as indicated for specific areas, except application may be made any time of the year where sprinkler or flood irrigation can be timed to activate the herbicide. Established perennial weeds require other special control procedures. Do not apply under citrus trees that have been subjected to freezing within 6 months.

Arizona (except Yuma area) and California (except Imperial and Coachella Valleys): Apply 3 to 4 lbs. per acre shortly after grove has been laid-up in final form (nontillage program) in late fall or early winter. Alternatively, apply 2 lbs. per acre in October or November and repeat at the same rate in March or April. Subsequent annual applications of 2 to 3 lbs. per acre will usually give adequate weed control.

Florida and Puerto Rico: Make a single application of 4 to 8 lbs. per acre, or apply 3 to 4 lbs. per acre followed by the same rate 4 to 6 months later. On bearing citrus, apply any time when seasonal rains are expected; on nonbearing trees, apply when winter banks are pulled down.

For control of guineagrass, loosetrife, maidencane, paragrass, primrose willow and seamyrtle in ditches adjacent to citrus groves, use 1 lb. per 1000 sq. ft. (40 lbs. per acre) in sufficient water (min. 4 gals. per 1000 sq. ft.) to provide thorough and uniform coverage. Apply in the spring before weed growth starts or after removal of vegetation. Repeat treatment on a spot basis to control hard-to-kill species such as guineagrass. In bedded groves, do not treat water furrows between the beds as injury to the trees may result.

Texas: Apply 2 to 4 lbs. per acre for annual weeds; use 4 to 6 lbs. per acre for control of johnsongrass seedlings. Best results accompany application in the spring; well established weeds should be eliminated by cultivation prior to treatment.

GRAPES

Apply only to established vineyards (at least 3 years old) as a band treatment to grape rows. On soils low in clay or organic matter (1 to 2%), severe plant injury may result if heavy rainfall or more than one inch of irrigation occurs soon after treatment. This risk must be assumed by the user.

East of the Rocky Mountains: On soils low in clay or organic matter (1 to 2%), apply 2 to 3 lbs. per acre on soils high in clay or organic matter, apply 3 to 6 lbs. per acre. Apply in the spring just prior to germination and growth of annual weeds.

West of the Rocky Mountains: Apply in November, December or January. For initial treatment, apply 3 to 4 lbs. per acre; subsequent annual applications of 2 lbs. per acre will usually give adequate weed control. Do not apply to vines with trunks less than 1½" in diameter as injury may result.

New York and Pennsylvania—Perennial Grasses: Use only in established vineyards (at least 4 years old) for spot control of perennial grasses such as orchardgrass, quackgrass and ryegrass. Apply in the spring as a band treatment to ridged soil (2" to 4" high) under the trellis at the rate of 8 to 12 lbs. per acre. Band width should not exceed 30". Do not apply more than once every 4 years. Use only on heavy soils such as loams, silt loams, clay loams. Do not use in areas where grape roots are shallow or exposed because of high bedrock, poor drainage, or erosion as injury to grapes may result.

MACADAMIA NUTS—Hawaii

Use only under trees established in the orchard for at least 1 year. Apply 2 to 6 lbs. per acre immediately after harvest, preferably before weeds emerge; if weeds have emerged, add 1 pt. Surfactant WK per 25 gals. of spray. Retreat as needed but do not exceed 10 lbs. per acre per year.

OLIVES—California

Use only under trees established in the grove for at least 1 year. Apply 2 lbs. per acre after grove has been laid-up in final form in late October or November; repeat at same rate in March or April. Remove weed growth prior to treatment.

PAPAYAS

Use only under trees established in the orchard for at least 1 year. Apply 2½ to 5 lbs. per acre, preferably before weeds emerge; if weeds have emerged, add 1 pt. Surfactant WK per 25 gals. of spray.

PEACHES

Use "KARMEX" DF alone, or apply as a tank mixture with "Sinbar".

"KARMEX" DF Alone—Use only under trees established in the orchard for at least 3 years. Apply 2 to 5 lbs. per acre in the early spring before weeds emerge or during the early seedling stage of weed growth. Do not apply within 3 months of harvest; in the Far West, do not apply within 8 months of harvest.

"KARMEX" DF + "Sinbar"—Use only under trees established in the orchard for at least 2 years. Apply either in the spring or after harvest in the fall before weeds emerge or during early seedling stage of weed growth.

Soil Texture	"Karmex" DF + "Sinbar" Lbs. Product Per Acre	
	1 to 2% Organic Matter	More than 2% Organic Matter
Sandy loam	1 + 1	1½ + 1½
Loam, silt loam, silt	1½ + 1½	2 + 2
Clay loam, clay	2 + 2	2 + 2

Where crop is grown under furrow irrigation or under raised-berm flood irrigation (trees 4" to 6" above waterline), apply only as a band treatment. Do not treat trees planted in the bottom of irrigation furrows, nor trees grown under flat flood or basin irrigation, as injury to trees may result. Where complete weed control to harvest is desired, additional weed control measures may be required during the growing season.

Georgia—On trees established for at least 2 years, apply 2 to 3 lbs. per acre in the spring. Repeat application in the fall but do not exceed 5 lbs. per acre per year. Add Du Pont Surfactant WK at 1 pint per 25 gals. spray mixture to improve control of small, emerged weeds.

Where crop is grown under furrow irrigation or under raised-berm flood irrigation (trees 4" to 6" above waterline), apply only as a band treatment. Do not treat trees planted in the bottom of irrigation furrows, nor trees grown under flat flood or basin irrigation, as injury to trees may result. Where complete weed control to harvest is desired, additional weed control measures may be required during the growing season.

PEARS

Use only under trees established in the orchard for at least 1 year; do not treat varieties grafted on full-dwarf root stocks. Apply 4 lbs. per acre in the spring (March through May). In the Far West, treatment may be made in winter (December through February), or apply 2 lbs. per acre as a post-harvest treatment followed by 2 lbs. in the spring.

PECANS

Use "KARMEX" DF alone or apply as a tank mixture with "Sinbar". Make a single band or broadcast application as a directed spray using a minimum of 30 gals. of water per acre. Apply in the spring before weeds emerge or during the early seedling stage of growth.

	Lbs Product Per Acre		
Soil Texture	"KARMEX" DF Alone*	-OR-	Tank Mixture "KARMEX" DF + "Sinbar"**
Sandy loam	2	1½	+ 1½
Loam, silt loam, silt	3	1¾	+ 1¾
Clay loam, clay	4	2	+ 2

* Use only on trees established in grove for at least 3 yrs. and on soils with at least ½% organic matter.

** Use on trees established in the grove for at least 1 yr. and on soils with at least 1% organic matter.

Note: Do not use on eroded areas where subsoil or roots are exposed, nor on trees that are diseased or lacking in vigor or on trees planted in irrigation furrows as injury to the trees may result.

PINEAPPLE—Hawaii and Florida

Apply 4 to 8 lbs. per acre as a broadcast spray just before or immediately after planting but prior to weed emergence. Use 4 lbs. per acre after harvesting plant crop (for ratoon crop). For plant crop only, a second and third broadcast or interspace application may be made prior to differentiation at the rate of 2 lbs. per acre at intervals of not less than 2 months. Additional applications to plant crop may be made as needed to interspace only using 2 lbs. per acre. Do not apply more than 3 broadcast sprays (maximum 12 lbs. per acre) prior to differentiation nor more than 16 lbs. total per acre per plant crop. Treated areas may be planted to pineapple or sugarcane 1 year after last application.

WALNUTS (English)—California

Use only under trees established in the orchard for at least 1 year. As an initial treatment, apply 3 to 5 lbs. per acre after the orchard has been laid up in final from (nontillage program) in late fall or early winter; retreat annually with 2 to 3 lbs. per acre. Alternatively, apply 2 lbs. per acre in October or November and repeat at same rate in March or April.

ORNAMENTAL CROPS
(See Soil Limitations)

ORNAMENTAL BULB CROPS (Bulbous Iris, Narcissus)—Western Washington

Make a single application of 4 lbs. per acre. Apply after planting but no later than 4 weeks prior to bulb emergence (usually late September or October). Do not replant treated areas to any crop within 1 year after last application as injury to subsequent crops may result.

PLUMOSUS FERN—Florida

Hand weed and mow fern; then make a single application of 3 lbs. per acre within 3 to 5 days. Do not cultivate or disturb soil after application as crop injury may result. Treat only established stands at least 1 year old.

TREE PLANTINGS—Colorado, Montana, Nebraska, North Dakota, South Dakota, Wyoming

Use only under established plantings (1 year or older) of American elm, caragana, cottonwood, Douglas fir, green ash, honeysuckle, Ponderosa pine, redcedar, Russian olive and Siberian elm. Use 2½ to 5 lbs. per acre; apply as a band 4 ft. wide in the tree row (2 ft. on each side of row). For example, 1 oz. "KARMEX" DF (4 level tablespoonfuls) treats 135 ft. of tree row (2 ft. on each side of row) at the rate of 5 lbs. per acre. Apply as a directed spray in early spring before weeds emerge and before trees leaf out. Do not apply to foliage of trees, nor under trees growing in low areas as injury to the trees may result.

NON-CROP WEED CONTROL

"KARMEX" DF is an effective herbicide for the control of many weeds on non-cropland areas where bare ground is desired. The degree of control and duration of effect will vary with the amount of chemical applied, soil texture, rainfall and other conditions.

"KARMEX" DF may be used as a preemergence treatment at any time of the year except when ground is frozen, provided adequate moisture is supplied by rainfall or artificial means to activate the herbicide. Best results are obtained if application is made to the soil shortly before weed growth begins. If dense growth is present, remove tops and spray the ground.

Increased contact activity on established weeds may be obtained by the addition of a non-ionic surfactant of at least 80% active ingredient at 0.25 to 0.5% by volume of water (1-2 qts/100 gals water). Apply as a drenching spray to actively growing weeds during warm weather when daily temperature will exceed 70°F.

Except for small areas, use a fixed-boom power sprayer properly calibrated to insure a constant rate of application. Mix proper amount of "KARMEX" DF into volume of water necessary to obtain uniform coverage; if Surfactant WK is used, dilute with 10 parts of water and add as last ingredient to nearly full tank. Material must be kept in suspension at all times. Agitate by mechanical or hydraulic means in the spray tank; if bypass or return line is used, it should terminate at bottom of tank to minimize foaming. Openings in screens should be equal to or larger than 50 mesh.

General Weed Control: To control most weeds for an extended period of time on non-cropland such as utility, highway, pipeline and railroad right of ways, petroleum tank farms, lumberyards, storage areas, industrial plant sites, and around farm buildings—apply 5 to 15 lbs. per acre to control most annual weeds.

Irrigation and Drainage Ditches: Apply 5 to 15 lbs. per acre to control most annual weeds. Apply only when water is not in the ditch. For irrigation ditches, apply during the non-crop season, and when ditch is not in use. To minimize movement of "KARMEX" DF with irrigation water (to avoid possible crop injury), it is essential that the herbicide be fixed in the soil by moisture. Apply before expected seasonal rainfall (if possible when soil in the ditch is still moist). Following treatment, if rainfall has not totaled at least 4 inches, fill ditch with water and allow to stand for 72 hours; drain off and waste remaining water before using ditch. Do not treat any ditch into which roots of trees or other desirable plants may extend as injury may result.

1 Reg. trademark of Elanco Products Co.
2 Reg. trademark of Westvaco Corp.
3 Reg. trademark of ICI United States Inc.

STORAGE AND DISPOSAL

STORAGE: Store product in original container only, away from other pesticides, fertilizer, food, or feed.

PESTICIDE DISPOSAL: Do not contaminate water, food, or feed by storage or disposal. Wastes resulting from the use of this product may be disposed of on site or at an approved waste disposal facility.

CONTAINER DISPOSAL: Completely empty bag into application equipment. Then dispose of empty bag in a sanitary landfill or by incineration, or, if allowed by State and local authorities, by burning. If burned, stay out of smoke.

NOTICE OF WARRANTY

Du Pont warrants that this product conforms to the chemical description on the label thereof and is reasonably fit for purposes stated on such label only when used in accordance with the directions under normal use conditions. It is impossible to eliminate all risks inherently associated with the use of this product. Crop injury, ineffectiveness or other unintended consequences may result because of such factors as weather conditions, presence of other materials, or the manner of use or application, all of which are beyond the control of Du Pont. In no case shall Du Pont be liable for consequential, special or indirect damages resulting from the use or handling of this product. All such risks shall be assumed by the Buyer. DU PONT MAKES NO WARRANTIES OF MERCHANTABILITY OR FITNESS FOR A PARTICULAR PURPOSE NOR ANY OTHER EXPRESS OR IMPLIED WARRANTY EXCEPT AS STATED ABOVE.

AG-3243 8029 1/3/89

This is a specimen label. DuPont is not responsible for the accuracy of the information contained herein. As labels are subject to revision, always carefully read and follow the label on the product container.

KRENITE® S

BRUSH CONTROL AGENT WATER SOLUBLE LIQUID

Contains 4 Lbs. Active Per Gallon

ACTIVE INGREDIENT	BY WEIGHT
Ammonium salt of fosamine [ethyl hydrogen (aminocarbonyl) phosphonate]	41.5%
INERT INGREDIENTS:	58.5%
TOTAL	100.0%

EPA Reg. No. 352-395

U.S. Pats. 3,627,507 & 3,846,512

KEEP OUT OF REACH OF CHILDREN WARNING

PRECAUTIONARY STATEMENTS

HAZARDS TO HUMANS AND DOMESTIC ANIMALS

WARNING! CAUSES EYE IRRITATION.

Do not get in eyes, on skin, or on clothing. Harmful if swallowed. Avoid breathing spray or mist.

STATEMENT OF PRACTICAL TREATMENT

If in eyes: flush with plenty of water for 15 minutes. Get medical attention if irritation persists.

If on skin: wash with plenty of soap and water.

For medical emergencies involving this product, call toll free 1-800-441-3637.

ENVIRONMENTAL HAZARDS

Do not contaminate water by cleaning of equipment or disposal of wastes. Do not apply directly to water.

IMPORTANT

Do not use on food crops. Do not allow drift or spray mist to contact desirable trees, shrubs, or other plants, as injury may result. Keep from contact with fertilizers, insecticides, fungicides, and seeds.

Thoroughly clean all traces of "KRENITE" S from application equipment after use. Flush tank, pump, hoses and boom with several changes of water after removing the nozzle tips and screens (clean these parts separately).

GENERAL INFORMATION

Du Pont "KRENITE" S Brush Control Agent is a water soluble liquid to be diluted with water and applied as a foliar spray for control and/or growth suppression of many woody species. A spray directed to only part of susceptible brush species will provide control of the portion sprayed, resulting in a trimming effect. It is applied to brush during the period from July to first fall coloration.

It is recommended for use on noncropland adjacent to and surrounding domestic water supply reservoirs, supply streams, lakes and ponds, as well as on drainage ditchbanks, highway rights-of-way, industrial plant sites, railroad rights-of-way, reforestation areas prior to planting, storage areas, utility and pipeline rights-of-way and other similar areas. "KRENITE" S is non-flammable and non-volatile.

Not registered for sale or use in California or Arizona.

DIRECTIONS FOR USE

It is a violation of federal law to use this product in a manner inconsistent with its labeling.

"KRENITE" S Brush Control Agent should be used only in accordance with recommendations on this label, or in separate published Du Pont recommendations available through local dealers.

Du Pont will not be responsible for losses or damages resulting from use of this product in any manner not specifically recommended by Du Pont. User assumes all risk associated with such non-recommended use.

Do not apply this product through any type of irrigation system.

General Brush Control — Non-Cropland Areas—"KRENITE" S at the following rates controls or suppresses undesirable brush/herbaceous plant species such as:

1½ to 6 gal "KRENITE" S per acre†		2 to 6 gal "KRENITE" S per acre†
Bigleaf maple*	Red alder	American elder
Birch	Red maple*	Basswood*
Blackberry	Red oak	Eastern cottonwood
Black cherry*	Salmonberry	Eastern white pine
Blackgum*	Sassafras*	Field bindweed**
Black locust	Sourwood*	Multiflora rose
Chinese tallow	Sumac	Slippery elm
Bracken (fern)	Sweetgum	Sycamore
Chokecherry*	Thimbleberry	Tree-of-heaven
Elm*	Tuliptree (yellow popular)*	Wild grape
Hawthorn*		Wild plum
Hickory*	Vine maple	Winged elm*
Leafy spurge**	Virginia pine	
Loblolly pine	Water oak	
Persimmon*	White ash*	
Pin cherry	White oak	
Quaking aspen	Willow*	

* Partial control and growth suppression.

** Apply "KRENITE" S after plants begin to bloom.

† Do not use more than 3 gal "KRENITE" S per acre for aerial treatment.

Spray Preparation

Measure the proper amount of "KRENITE" S and use agitation to thoroughly mix into the necessary volume of water. To enhance the activity of the "KRENITE" S treatment, the addition of ¼% (by volume) of a non-phytotoxic superior-type (60-70 second viscosity) oil as the last ingredient may be desirable in some areas. (Consult local sales personnel). After "KRENITE" S has been thoroughly mixed in the spray tank, agitation of the spray solution is not required.

Spray Application Methods

Before applying, properly calibrate spray equipment. For woody plant control, make a single foliar application of the recommended rate of "KRENITE" S during the period from July to first fall coloration. For control of only a portion of a plant, as in trimming, direct the spray to thoroughly cover only the section of the plant to be controlled. Use the highest recommended rate on stands in which species partially controlled and suppressed are dominant.

Prepare a spray solution using 1½ to 3 gallons "KRENITE" S per 100 gallons of water. Do not prepare less than a 1½% solution or exceed 6 gals. "KRENITE" S per acre. Apply a quantity of spray solution which will thoroughly and uniformly cover the target brush without causing unnecessary run-off. Spray pressure at nozzles should be 100 to 300 psi. Total rate and volume per acre will depend on plant species, height and density of growth and on type of equipment used. On tall dense brush, it is often necessary to spray from both sides to obtain adequate coverage.

For aerial treatment, apply 1½ to 3 gals of "KRENITE" S per acre using 10 to 40 gallons of water to thoroughly and uniformly cover the target brush without causing unnecessary run-off. Total rate and volume per acre will depend on plant species, height and density of growth, and on type of equipment used. If foaming is encountered, use an anti-foam agent.

Note: Effectiveness may be reduced if, following treatment, rainfall occurs on the same day. Do not cut treated brush until stems are dead, or sprouting may occur.

NOTICE TO BUYER: Purchase of this material does not confer any rights under patents of countries outside of the United States.

STORAGE AND DISPOSAL

STORAGE: Store product in original container only, away from other pesticides, fertilizer, food, or feed.

PESTICIDE DISPOSAL: Do not contaminate water, food, or feed by storage or disposal. Wastes resulting from the use of this product may be disposed of on site or at an approved waste disposal facility.

CONTAINER DISPOSAL: Triple rinse (or equivalent) the container. Then offer for recycling or reconditioning, or puncture and dispose of in a sanitary landfill, or incineration, or, if allowed by state and local authorities, by burning. If burned, stay out of smoke.

NOTICE OF WARRANTY

Du Pont warrants that this product conforms to the chemical description on the label thereof and is reasonably fit for purposes stated on such label only when used in accordance with directions under normal use conditions. It is impossible to eliminate all risks inherently associated with the use of this product. Ineffectiveness or other unintended consequences may result because of such factors as weather conditions, presence of other materials, or the manner of use or application, all of which are beyond the control of Du Pont. In no case shall Du Pont be liable for consequential, special or indirect damages resulting from the use or handling of this product. All such risks shall be assumed by the Buyer. DU PONT MAKES NO WARRANTIES OF MERCHANTABILITY OR FITNESS FOR A PARTICULAR PURPOSE NOR ANY OTHER EXPRESS OR IMPLIED WARRANTY EXCEPT AS STATED ABOVE.

AG-3210 8088 10/1/87

This is a specimen label. DuPont is not responsible for the accuracy of the information contained herein. As labels are subject to revision, always carefully read and follow the label on the product container.

KROVAR® I DF

HERBICIDE
DISPERSIBLE GRANULE

ACTIVE INGREDIENTS:		80%
Bromacil [5-bromo-3-sec-butyl-6-methyluracil]	40%	
Diuron [3-(3,4-dichlorophenyl)-1,1-dimethylurea]	40%	
INERT INGREDIENTS:		20%
TOTAL		100%

EPA Reg. No. 352-505

U.S. Pat. 3,914,122 and 3,914,230

KEEP OUT OF REACH OF CHILDREN
DANGER

PRECAUTIONARY STATEMENTS

HAZARDS TO HUMANS AND DOMESTIC ANIMALS

DANGER! CAUSES SKIN DAMAGE.

Do not get in eyes, on skin or on clothing. Harmful if swallowed. Avoid breathing dust or spray mist. Wash thoroughly with soap and water after handling.

STATEMENT OF PRACTICAL TREATMENT

If on skin: Wash with plenty of soap and water. Get medical attention.

If in eyes: Flush with plenty of water. Get medical attention if irritation persists.

If swallowed: Call a physician or Poison Control Center. Drink 1 or 2 glasses of water and induce vomiting by touching back of throat with finger. Do not induce vomiting or give anything by mouth to an unconscious person.

For medical emergencies involving this product, call toll free 1-800-441-3637.

ENVIRONMENTAL HAZARDS

Do not apply directly to water or wetlands. Do not contaminate water by cleaning of equipment or disposal of wastes.

IMPORTANT—Injury to or loss of desirable trees or other plants may result from failure to observe the following:

Do not apply (except as recommended for crop use), or drain or flush equipment on or near desirable trees or other plants, or on areas where their roots may extend, or in locations where the chemical may be washed or moved into contact with their roots. Do not use on lawns, walks, driveways, tennis courts or similar areas. Do not use in home fruit plantings nor in citrus orchards interplanted to other trees or desirable plants. Prevent drift of dry powder or spray to desirable plants. Do not contaminate any body of water. Keep from contact with fertilizers, insecticides, fungicides and seeds.

Thoroughly clean all traces of "KROVAR" I DF from application equipment immediately after use. Flush tank, pump, hoses and "boom" with several changes of water after removing nozzle tips and screens (clean these parts separately).

GENERAL INFORMATION

Du Pont "KROVAR" I DF Herbicide is dispersible granule to be mixed in water and applied as a spray for selective control of weeds in citrus and for non-crop weed control.

"KROVAR" I DF controls annual weeds such as barnyardgrass (watergrass), chickweed, crabgrass, filaree, fleabane, Florida pusley, foxtail, groundsel, horseweed, johnsongrass seedlings, junglerice, lambsquarters, natalgrass (red top), nightshade (annual), pigweed, pineappleweed, puncturevine, purslane, ragweed, sandbur (sandspur), shepherdspurse, sowthistle (annual), spanishneedles, wild lettuce, and wild mustard. At higher use rates, treatment controls certain perennial weeds such as balsamapple vine seedlings, bermudagrass, heart leaf, drymary, guineagrass and milkweed (strangler) vine seedlings; and provides suppression of established balsamapple vine and milkweed (strangler) vine. Effects on perennial weeds are slow to appear, usually progressing over a period of several months.

Moisture is necessary to activate the herbicide; best results are obtained if treatment is made to moist soil, and moisture is supplied by rainfall or sprinkler irrigation within two weeks after application. The degree and duration of control will vary with the amount applied, soil texture, rainfall, and other conditions.

DIRECTIONS FOR USE

It is a violation of federal law to use this product in a manner inconsistent with its labeling.

Du Pont "KROVAR" I DF should be used only in accordance with recommendations on this label, or in separate published Du Pont recommendations available through local dealers.

Du Pont will not be responsible for losses or damages resulting from use of this product in any manner not specifically recommended by Du Pont. User assumes all risk associated with such non-recommended use.

Do not apply this product through any type of irrigation system.

Farm workers performing hand labor operations should not enter treated areas without protective clothing until sprays have dried. Because certain states may require more restrictive reentry intervals for various crops treated with this product, consult your State Department of Agriculture for further information.

Written or oral warnings must be given to farm workers performing hand labor operations who are expected to be in a treated cropland area or in a cropland area about to be treated with this product. When oral warnings are given, warnings shall be given in a language customarily understood by workers. Oral warnings must be given if there is reason to believe that written warnings cannot be understood by workers. Warnings must include the following information: "DANGER. Area treated with "KROVAR" I DF on (date of application). Do not enter without appropriate protective clothing until sprays have dried. In case of contact, flush eyes with plenty of water; wash skin with plenty of soap and water. Get medical attention if irritation persists."

CITRUS

Apply any time of year provided overhead moisture (rainfall or sprinkler irrigation) is available to activate the herbicide, preferably just before or just after weeds have germinated.

Use only where trees have been established in the orchard for at least three years (except in Florida). Apply with a properly calibrated fixed-boom power sprayer as a band or broadcast treatment beneath and/or between trees. Use sufficient water (min. 40 gals. per acre) to obtain uniform coverage; avoid contact of foliage and fruit with spray or mist. Avoid overlapping, and shut off spray boom while starting, turning, slowing or stopping, or injury to trees may result. Continuous agitation in the spray tank is required to keep the material in suspension. Best results are obtained if application is made shortly before or shortly after weed growth begins; if dense growth is present, remove tops and spray ground. Do not graze cattle in treated areas. Dosages are expressed as broadcast rates; for band treatment, use proportionately less.

ANNUAL WEEDS

California, Arizona—Best results occur when applied in late fall or early winter, but before winter annuals become well established. Application should be made after the first fall or early winter rains have settled the soil.

For the initial treatment, apply 4-5 lbs. per acre on coarse soils containing 1-2% organic matter and 5-6 lbs. per acre on fine soils, or soils with organic matter of 2½% or more and repeat annually as needed. Alternatively, apply 3-4 lbs. per acre in the fall and repeat at 2-4 lbs. per acre in the spring. Do not exceed 6 lbs. per acre per year. Use the higher range of rates where groundsel or puncturevine are known to be a problem. The higher rates will also suppress low density stands of bermudagrass and yellow nutsedge. Repeat annually.

Louisiana—Make a single application of 2 to 4 lbs. per acre on coarser soils (sands, loamy sands, sandy loams) and 4 to 6 lbs. per acre on finer soils (silt loams, clay loams, or soils with organic matter of 2½% or more); use the higher rates for maximum suppression of perennials. Alternatively, make two applications per year at rates of 2 lbs. per acre on coarser soils and 3 lbs. per acre on finer soils; make the second application when needed to maintain weed control.

FLORIDA AND TEXAS

Trees Established Less Than One Year:

Apply 2 to 4 lbs. "KROVAR" I DF per acre as needed to maintain weed control. Do not apply at less than 60-day intervals; do not apply more than 6 lbs. per acre during any 6-month period or more than 8 lbs. per acre during the first year.

Trees Established One or Two Years:

Apply 2 to 4 lbs. "KROVAR" I DF per acre. A second application may be made when needed to maintain weed control, but do not exceed 8 lbs. per acre per year.

Trees Established Three or More Years:*

Make one or two applications of 4 to 8 lbs. per acre. Make the second application when needed to maintain weed control. Do not apply more than 16 lbs. per acre per year.

Tank Mixtures:

Tank mixtures of "KROVAR" I DF and paraquat are recommended for weed control in citrus. For rates of individual products, refer to the respective product label. Do not apply tank mixture to foliage or green wood of citrus as injury may result.

* Texas—Use 2-4 lbs per acre on coarser soils (sands, loamy sands, sandy loams) and 4-6 lbs per acre on finer soils (silt loams, clay loams, or soils with organic matter of 2½% or more). Use the higher rates for maximum suppression of perennials.

Note: Temporary yellowing of citrus leaves may occur following treatment. Because injury to citrus trees may result: do not use on soils with less than 1% organic matter (1/2% in Florida), poorly drained soils, gravelly soils, nor thinly covered or exposed subsoils; do not treat trees planted in irrigation furrows; do not treat diseased trees such as those with foot rot.

Do not replant treated areas to any crop within two years after last application as injury to subsequent crops may result, except that citrus trees may be planted one year after the last application.

NON-CROP WEED CONTROL

For short-term control of annual weeds on non-cropland areas such as roadsides, storage areas, and plant sites, apply 4 to 6 lbs. per acre in 40 to 100 gals. of water. For extended control of annuals and partial control of perennials such as bermudagrass and nutsedge, apply 7 to 18 lbs. per acre. For control of hard-to-kill perennials such as bermudagrass, bouncingbet, dogbane, johnsongrass, nutsedge, and saltgrass, apply 19 to 40 lbs. per acre.

Use the higher levels of dosage rates on adsorptive soils (high in organic matter or carbon). Best results occur when application is made just before weed emergence or in early stages of weed growth.

Retreatment: Apply 4 to 6 lbs. per acre when annual weeds and grasses reappear on sites where weed growth has been controlled.

Small Areas: ¼ cupful of "KROVAR" I DF per 200 sq. ft. is approximately 15 lbs. per acre.

STORAGE AND DISPOSAL

STORAGE: Store product in original container only, away from other pesticides, fertilizer, food, or feed.

PESTCIDE DISPOSAL: Do not contaminate water, food, or feed by storage or disposal. Wastes resulting from the use of this product may be disposed of on site or at an approved waste disposal facility.

CONTAINER DISPOSAL: Completely empty bag into application equipment. Then dispose of empty bag in a sanitary landfill or by incineration, or, if allowed by State and local authorities, by burning. If burned, stay out of smoke.

NOTICE OF WARRANTY

Du Pont warrants that this product conforms to the chemical description on the label thereof and is reasonably fit for the purposes stated on such label only when used in accordance with directions under normal use conditions. It is impossible to eliminate all risks inherently associated with the use of this product. Crop injury, ineffectiveness, or other unintended consequences may result because of such factors as weather conditions, presence of other materials, or the manner of use or application, all of which are beyond the control of Du Pont. In no case shall Du Pont be liable for consequential, special or indirect damages resulting from the use or handling of this product. All such risks shall be assumed by the Buyer. DU PONT MAKES NO WARRANTIES OF MERCHANTABILITY OR FITNESS FOR A PARTICULAR PURPOSE NOR ANY OTHER EXPRESS OR IMPLIED WARRANTY EXCEPT AS STATED ABOVE.

AG-3211 8088 12/19/86

This is a specimen label. DuPont is not responsible for the accuracy of the information contained herein. As labels are subject to revision, always carefully read and follow the label on the product container.

KROVAR® II DF

HERBICIDE DISPERSIBLE GRANULE

		BY WEIGHT
ACTIVE INGREDIENTS:		80%
Bromacil [5-bromo-3-sec-butyl-6-methyluracil]	53%	
Diuron [3-(3,4-dichlorophenyl)-1,1-dimethylurea]	27%	
INERT INGREDIENTS:		20%
TOTAL		100%

EPA Reg. No. 352-440 U.S. Pat. 3,914,122 & 3,914,230

KEEP OUT OF REACH OF CHILDREN

CAUTION!

PRECAUTIONARY STATEMENTS

HAZARDS TO HUMANS AND DOMESTIC ANIMALS

CAUTION!

HARMFUL IF SWALLOWED. CAUSES MODERATE EYE INJURY.

Avoid breathing dust or spray mist. Avoid contact with skin, eyes or clothing. In case of contact with eyes, immediately flush with plenty of water. Get medical attention if irritation persists. Wash thoroughly with soap and water after handling. Remove and wash contaminated clothing before reuse.

STATEMENT OF PRACTICAL TREATMENT

If in eyes: Flush with plenty of water. Call a physician if irritation persists.

If swallowed: Call a physician or Poision Control Center. Drink 1 or 2 glasses of water and induce vomiting by touching back of throat with finger. Do not induce vomiting or give anything by mouth to an unconscious person.

ENVIRONMENTAL HAZARDS

Do not apply directly to any body of water. Do not contaminate water by cleaning of equipment or disposal of wastes.

Bromacil is a chemical which can travel (seep or leach) through soil and can enter ground water which may be used as drinking water. Bromacil has been found in ground water as a result of agricultural use. The correct use rates by geographical area, specified herein, and proper mining-loading site precautions and procedures must be followed to minimize potential bromacil movement into groundwater.

IMPORTANT—Injury to or loss of desirable trees or other plants may result from failure to observe the following:

Do not apply (except as recommended for crop use), or drain or flush equipment on or near desirable trees or other plants, or on areas where their roots may extend, or in locations where the chemical may be washed or moved into contact with their roots. Do not use on lawns, walks, driveways, tennis courts or similar areas. Do not use in home fruit plantings nor in citrus orchards interplanted to other trees or desirable plants. Prevent drift of dry powder or spray to desirable plants. Do not contaminate any body of water. Keep from contact with fertilizers, insecticides, fungicides and seeds.

Thoroughly clean all traces of "KROVAR" II DF from application equipment immediately after use. Flush tank, pump, hoses and "boom" with several changes of water after removing nozzle tips and screens (clean these parts separately).

PESTICIDE HANDLING-Loading site precautions and procedures:

- Calibrate sprayers only with clean water away from the well site.
- Make scheduled checks of spray equipment.
- Assure accurate measurement of pesticides by all operation employees.
- Mix only enough product for the job at hand.
- Avoid over-filling of spray tank.
- Do not discharge excess material on the soil at a single spot in the field/grove or mixing/loading station.
- Dilute and agitate excess solution and apply at labeled rates/uses.
- Avoid storage of pesticides near well sites.

GENERAL INFORMATION

Du Pont "KROVAR" II DF Herbicide is a dry flowable granule to be mixed in water and applied as a spray for selective control of weeds in citrus and for non-crop weed control.

"KROVAR" II DF controls certain perennial grasses and annual weeds.

Moisture is necessary to activate the herbicide; best results are obtained if treatment is made to moist soil, and moisture is supplied by rainfall or sprinkler irrigation within two weeks after application. The degree of control and duration of effect will vary with the amount applied, soil texture, rainfall and other conditions.

DIRECTIONS FOR USE

It is a violation of federal law to use this product in a manner inconsistent with its labeling.

Du Pont "KROVAR" II DF should be used only in accordance with recommendations on this label, or in separate published Du Pont recommendations available through local dealers.

Du Pont will not be responsible for losses or damages resulting from use of this product in any manner not specifically recommended by Du Pont. User assumes all risk associated with such non-recommended use.

Do not apply this product through any type of irrigation system.

Farm workers performing hand labor operations should not enter treated areas without protective clothing until sprays have dried. Because certain states may require more restrictive reentry intervals for various crops treated with this product, consult your State Department of Agriculture for further information.

Written or oral warnings must be given to workers who are expected to be in a treated area or in an area about to be treated with this product. When oral warnings are given, warnings shall be given in a language customarily understood by workers. Oral warnings must be given if there is reason to believe that written warnings cannot be understood by workers. Warnings must include the following information: "CAUTION. Area treated with "KROVAR" II DF on (date of application). Do not enter without appropriate protective clothing until sprays have dried. In case of contact, flush eyes with plenty of water; wash skin with plenty of soap and water. Get medical attention if irritation persists."

Do not apply this product in such a manner as to directly or through drift expose workers, other persons or animals. The area being treated must be vacated by unprotected persons.

CITRUS

"KROVAR" II DF Controls:

Annual Weeds

barnyard grass	nightshade (annual)
crabgrass	pigweed
Florida pusley	purslane
foxtail	ragweed
goosegrass	sandbur (sandspur)
lambsquarter	sowthistle
natalgrass (red top)	spanishneedles

Perennial Weeds (At maximum rates and repeat treatments)

balsamapple vine (seedlings)	milkweed vine (seedlings)
bermudagrass	guineagrass (partial control)

Apply any time of year provided overhead moisture (rainfall or sprinkler irrigation) is available to activate the herbicide, preferably just before or just after weeds have germinated.

Apply with a properly calibrated fixed-boom power sprayer as a band or broadcast treatment beneath and/or between trees. Use sufficient water (min. 40 gals. per acre) to obtain uniform coverage; avoid contact of foliage and fruit with spray or mist. Avoid overlapping, and shut off spray boom while starting, turning, slowing or stopping, or injury to trees may result. Continuous agitation in the spray tank is required to keep the material in suspension.

Best results are obtained if "KROVAR" II DF is applied to bare ground. If vegetation is present at application, tank mixes with foliar active herbicides are recommended to control existing weeds (see tank-mixes below).

If dense populations of hard-to-kill species are present, control of these weeds prior to application to "KROVAR" II DF is recommended.

Dosages are expressed as broadcast rates; for band treatment, use proportionately less.

Do not graze cattle in treated areas.

Temporary yellowing or vein cleaning of citrus leaves may occur following treatment. Because injury to citrus trees may result; do not use on soils with less than 1% organic matter, poorly drained soils, gravely soils, nor thinly covered or exposed subsoils; do not treat plants in irrigation furrows; do not treat diseased or stressed trees.

Do not use in citrus groves interplanted with other trees or desirable plants, or in home citrus plantings or in areas where roots of valuable plants or trees may extend as plant injury may result.

Do not replant treated areas to any crop within two years after last application as injury to subsequent crops may result, except that citrus trees may be planted one year after the last application.

TANK MIXTURES

"KROVAR" II DF is recommended for use on citrus for weed control in tank mixtures with paraquat. For rates of the individual products, refer to the respective product label. Do not apply tank mixture to foliage or green wood as injury may result.

Arizona, California, Lousiana, Texas (Do not use in Kern County, CA)

Trees Established Four Years or More

Perennial Weeds: For bermudagrass, nutsedge make a single application per year during the period from winter to early summer at the following rates:

Soil Texture	Lbs. Per Acre
Sand, loamy sand	4 to 6
Sandy loam	5 to 7
Silt loam, clay loam	6 to 8

Alternatively, make two applications of 3 to 4 lbs. per acre per year; apply in spring and summer.

Partial control of perennials usually occurs with a single treatment; repeat applications are required to control perennials. Control may be improved by cultivation prior to treatment; otherwise, avoid working the soil as long as weed control continues or else effectiveness of the treatment may be reduced.

Annual Weeds: Apply 2 to 5 lbs. per acre using the lower rates on coarse soils; apply any time of year provided overhead moisture (rainfall or sprinkler irrigation) is available to activate the herbicide, preferably just before or just after weed germinate.

TEXAS

Trees Established One to Four Years

Apply 3 to 6 lbs "KROVAR" II DF per acre; for details on timing, equipment, spray volume and other use directions, refer to product label. Use the lower rate for 1 to 2 year old trees and the higher rate for trees more than 2 years old. A second application may be made if needed, but do not apply more than 8 lbs "KROVAR" II DF per acre per year as injury to the trees may result.

FLORIDA

Trees Established Less Than One Year

Apply 2 to 4 lbs "KROVAR" II DF per acre as needed to maintain weed control. Do not apply at less than 60-day intervals; do not apply more than 6 lbs per acre during any 6-month period or more than 8 lbs per acre during the first year.

Trees Established One to Three Years

Apply 2 to 4 lbs "KROVAR" II DF per acre as needed to maintain weed control. Do not exceed 8 lbs per acre per year.

Trees Established Three or More Years

East Coast/Flatwoods Areas - (low permeable soils)

- Apply 3 to 6 lbs "KROVAR" II DF per acre, but do not apply more than 12 lbs per acre per year; 12 lbs "KROVAR" II DF contains 6.4 lbs active bromacil.
- The maximum allowable is 6.4 lbs active (bromacil) per acre per year. The allowable amount is inclusive of all bromacil formulations use within one year.
- Multiple applications may improve control of "hard-to-kill" species.

Ridge Areas - (highly permeable soils)

- Apply 2 to 4 lbs "KROVAR" II DF per acre, but do not apply more than 8 lbs per acre per year; 8 lbs "KROVAR" II DF contains 4.2 lbs active (bromacil).
- The maximum allowable is 4.2 lbs active (bromacil) per acre per year. The allowable amount is inclusive of all bromacil formulations used within one year.
- Multiple applications may improve control of "hard-to-kill" species.

Tank Mixtures

Tank mixtures of "KROVAR" II DF and paraquat are recommended for control of emerged weeds in citrus. For rates of individual products, refer to the respective product label. Do not apply tank mixture to foliage or green wood as injury may result.

Seedling Vines (balsam apple and milkweed)—Florida: Remove established vines from trees prior to treatment. Apply 6 lbs. per acre after vine growth begins in the spring (or after cultivation or chopping), but before vines reach tree limbs. For band treatment, the treated area should extend at least 1 foot beyond the dripline of the tree. Add Du Pont Surfactant WK at a rate of 1 qt. per 50 gals. of spray mixture.

A repeat application may be made in late summer or early fall to control seedlings which germinate late in the season.

When used in conjunction with good mechanical cultivation practices, repeated annual applications aid in the suppression of established vines.

SPRAY PREPARATION

Mixing in Water

Fill tank 1/2 full with water. Start agitation system, add "KROVAR" II DF and continue adding water. Add separately each additional component of any tank-mix while adding water.

Mixing in liquid fertilizer

A fertilizer solution may be used in the spray mixture. Small quantities should be tested for compatibility by the following procedures before full scale mixing.

1. Put 1 pint fertilizer solution in a quart jar.
2. Mix 2 teaspoons "KROVAR" II DF with 2 tablespoons of water; mix thoroughly and add to fertilizer solution.
3. Close jar and shake well.
4. If other herbicides are used in the mixture, premix 2 teaspoons of dry materials to 1 teaspoon of liquids with 2 tablespoons of water; add to "KROVAR" II DF/fertilizer solution mixture.
5. Close jar and shake well.
6. Watch mixture for several seconds; check again in 30 minutes.
7. If mixture does not separate, foam, gel or become lumpy, it may be used.

Provided the above procedure shows the mixture to compatible, prepare the tank mixture as follows: Add the fertilizer solution to the spray tank first; mix the required amount of "KROVAR" II DF with water to form a slurry, with the agitator running, slowly add the slurry and mix thoroughly. For other herbicides, follow directions above. When tank mixing with other herbicides, all applicable directions, restrictions and precautions for the additional herbicides are also to be followed.

Thoroughly reagitate before using if allowed to settle.

NON-CROP WEED CONTROL

For short-term control of annual weeds on non-cropland areas such as roadsides, storage areas, and plant sites, apply 2 to 5 lbs. per acre in 40 to 100 gals. water. For extended control of annuals and partial control of perennials such as bermudagrass and nutsedge, apply 6 to 14 lbs. per acre. For control of hard-to-kill perennials such as bermudagrass, bouncingbet, dogbane, johnsongrass, nutsedge, and saltgrass, apply 15 to 23 lbs. per acre.

Use the higher levels of dosage rates on adsorptive soils (high in organic matter or carbon). Best results occur when application is made just before weed emergence or in early stages of weed growth.

Retreatment: Apply 2 to 5 lbs. per acre when annual weeds and grasses reappear on sites where weed growth has been controlled.

Small Areas: 1/4 cupful of "KROVAR" II DF per 200 sq. ft. is approximately 15 lbs. per acre.

NOTICE TO BUYER: Purchase of this material does not confer any rights under patents of countries outside of the United States.

STORAGE AND DISPOSAL

STORAGE: Store product in original container only, away from other pesticides, fertilizer, food or feed.

PRODUCT DISPOSAL: Do not contaminate water, food, or feed by storage or disposal. Wastes resulting from the use of this product may be disposed of on site or at an approved waste disposal facility.

CONTAINER DISPOSAL: Completely empty box into application equipment. Then dispose of empty box in a sanitary landfill or by incineration, or, if allowed by state and local authorities, by burning. If burned, stay out of smoke.

NOTICE OF WARRANTY

Du Pont warrants that this product conforms to the chemical description on the label thereof and is reasonably fit for purposes stated on such label only when used in accordance with directions under normal use conditions. It is impossible to eliminate all risks inherently associated with the use of this product. Crop injury, ineffectiveness, or other unintended consequences may result because of such factors as weather conditions, presence of other materials, or the manner of use or application, all of which are beyond the control of Du Pont. In no case shall Du Pont be liable for consequential, special or indirect damages resulting from the use or handling of this product. All such risks shall be assumed by the buyer. DU PONT MAKES NO WARRANTIES OF MERCHANTABILITY OR FITNESS FOR A PARTICULAR PURPOSE NOR ANY OTHER EXPRESS OR IMPLIED WARRANTY EXCEPT AS STATED ABOVE.

AG-3259 8049 4/13/89

This is a specimen label. DuPont is not responsible for the accuracy of the information contained herein. As labels are subject to revision, always carefully read and follow the label on the product container.

LANNATE®

INSECTICIDE
WATER SOLUBLE POWDER

ACTIVE INGREDIENT:	BY WEIGHT
Methomyl {S-methyl N-[(methylcarbamoyl)oxy]thioacetimidate}	90%
INERT INGREDIENTS:	10%
TOTAL	100%

EPA Reg. No. 352-342

U.S. Pat. 3,639,633

KEEP OUT OF REACH OF CHILDREN

DANGER–POISON

PELIGRO

PRECAUCION AL USUARIO: Si usted no lee ingles, no use este producto hasta que la etiqueta haya sido explicada ampliamente.

STATEMENT OF PRACTICAL TREATMENT

If swallowed: Call a physician or Poison Control Center. Drink 1 or 2 glasses of water and induce vomiting by touching back of throat with finger. Do not induce vomiting or give anything by mouth to an unconscious person.

If inhaled: Remove from exposure and have patient lie down and keep quiet. If patient is not breathing, start artificial respiration immediately. Never give anything by mouth to an unconscious person.

In case of contact, wash skin with plenty of soap and water; for eyes, flush with water for 15 minutes and get medical attention; remove and wash contaminated clothing before re-use.

ATROPINE IS AN ANTIDOTE — SEEK MEDICAL ATTENTION AT ONCE IN ALL CASES OF SUSPECTED POISONING.

If warning symptons appear (see WARNING SYMPTOMS), get medical attention.

For medical emergencies involving this product, call toll free 1-800-441-3637.

PRECAUTIONARY STATEMENTS

HAZARDS TO HUMANS AND DOMESTIC ANIMALS

DANGER–POISON

DANGER–MAY BE FATAL IF SWALLOWED. POISONOUS IF INHALED.

Do not breathe vapors or spray mist. Do not get in eyes, on skin, or on clothing. When mixing spray, loading, applying, or otherwise handling, wear protective clothing, goggles, and a mask or respirator jointly approved by the Mine Safety and Health Administration (formerly the U.S. Bureau of Mines) and by the National Institute for Occupational Safety and Health. Wear clean clothes daily. Wash thoroughly after handling and before eating or smoking. Pilot should not assist in the mixing and loading operation.

Do not apply this product in such a manner as to directly or through drift expose workers, other persons or animals. The area being treated must be vacated by unprotected persons.

WARNING SYMPTOMS — Methomyl poisoning produces effects associated with anticholinesterase activity which may include weakness, blurred vision, headache, nausea, abdominal cramps, discomfort in the chest, constriction of pupils, sweating, slow pulse, muscle tremors.

NOTE TO PHYSICIAN

TREATMENT — Atropine sulfate should be used for treatment. Administer repeated doses, 1.2 to 2.0 mg. intravenously every 10 to 30 minutes until full atropinization is achieved. Maintain atropinization until the patient recovers. Artificial respiration or oxygen may be necessary. Allow no further exposure to any cholinesterase inhibitor until recovery is assured.

Do not use 2-PAM for exposure to "LANNATE" alone. However, for exposure to combinations of "LANNATE" and organophosphorous insecticides, 2-PAM may be used as required to supplement the atropine sulfate treatment. Do not use morphine.

For medical emergencies involving this product, call toll free 1-800-441-3637.

ENVIRONMENTAL HAZARDS

This product is toxic to fish, birds, and other wildlife. Keep out of any body of water. Do not apply where runoff is likely to occur. Do not apply when weather conditions favor drift from areas treated. Do not contaminate water by cleaning of equipment or disposal of wastes.

This product is toxic to bees exposed to direct application. Do not apply this product while bees are actively visiting the treatment area. Time applications to coincide with periods of minimum bee activity.

Refer to ENDANGERED SPECIES section of this manual for use restrictions on this product.

PHYSICAL OR CHEMICAL HAZARDS

Keep container closed. Use with adequate ventilation.

GENERAL INFORMATION

Scouting — Monitoring insect populations to determine whether or not there is a need for application of "LANNATE" is recommended. More than one treatment of "LANNATE" may be required to control a population of pests.

Insect Predators — "LANNATE" at rates of 1/8 to 1/4 lb. per acre helps conserve certain beneficials, including big-eyed bugs, damsel bugs, flower bugs and spiders in cotton and soybeans. While these beneficials cannot be relied upon to control pests, they are of potential value and should be monitored along with pests in pest management programs on these crops.

Compatibility — Since formulations may be changed and new ones introduced, it is recommended that users premix a small quantity of a desired tank mix and observe for possible adverse changes (settling out, flocculation, etc.). Avoid mixtures of several materials and very concentrated spray mixtures.

Do not use "LANNATE" with Bordeaux mixture, "Du Ter" (triphenyltin hydroxide), lime sulfur, "Rayplex" iron nor in highly alkaline solutions. Use mildly alkaline mixtures immediately after mixing to prevent loss of insecticidal activity.

Tank-mix solutions containing Boron may affect solubility of the water soluble film. When using Boron containing solutions in a tank-mix, follow these procedures:

-Add the correct amount of "LANNATE" water soluble packets first

-Be sure the soluble packets are completely dissolved

-Introduce Boron containing solutions last

If the above procedure cannot be followed, and Boron is an essential part of the spray mix, substitute "LANNATE" L formulation for the water soluble packets.

DIRECTIONS FOR USE

It is a violation of federal law to use this product in a manner inconsistent with its labeling.

Du Pont "LANNATE" Insecticide should be used only in accordance with recommendations on this label, or in separate published Du Pont recommendations available through local dealers.

Du Pont will not be responsible for losses or damages resulting from use of this product in any manner not specifically recommended by Du Pont. User assumes all risks associated with such non-recommended use. "LANNATE" is a dry powder to be dissolved in water for application by ground or air equipment. Use only in commercial and farm plantings; do not use in home plantings.

Do not enter treated areas without appropriate protective clothing until sprays have dried. Because certain states may require more restrictive reentry intervals for various crops treated with this product, consult your State Department of Agriculture for further information.

Written or oral warnings must be given to workers who are expected to be in a treated area or in an area about to be treated with this product. When oral warnings are given, warnings shall be given in a language customarily understood by workers. Oral warnings must be given if there is reason to believe that written warnings cannot be understood by workers. Written (or oral) warnings must include the following information: "DANGER—POISON. Area treated with 'LANNATE' on (date of application)." Do not enter treated areas without appropriate protective clothing until sprays have dried. If warnings symptoms appear, refer to Statement of Practical Treatment on front of 'LANNATE' label and seek medical attention at once."

SPRAY PREPARATION—Fill spray tank 1/4 to 1/2 full of water. Add "LANNATE" directly to spray tank. Mix thoroughly to fully dissolve the insecticide; once dissolved continued agitation is not required. Use mechanical or hydraulic means; do not use air agitation. Be sure the soluble packet is completely dissolved before adding any other ingredients. Spray mix should not be stored overnight in spray tank.

Do not apply this product through any type of irrigation system.

APPLICATION—Apply at the recommended rates when insects first appear. Unless otherwise noted, continue application at 5- to 7-day intervals or as needed. Use sufficient water to obtain thorough coverage. Since "LANNATE" is a fast acting contact insecticide, best results follow direct spraying of the target insect. This can be accomplished with fine spray droplets at low volume.

For aerial, use a minimum of 2 gals. per acre (gpa) except 10 gpa for grapes, nectarines and strawberries; 15 gpa for oranges, lemons, grapefruit, tangelos and tangerines.

For certain crops a minimum of 1 gallon per acre may be used providing the following conditions are met:

- equipment is capable of delivering small spray droplets to insure thorough coverage,
- equipment is adjusted to distribute spray uniformly over the spray swath
- wind conditions and other factors such as temperature and humidity are such that the spray is delivered to the target area,
- local regulations do not prohibit low-volume aerial sprays.
- use rates are applied as directed on the package label or supplemental labeling for the following crops:

Alfalfa	Celery	Peas (succulent)
Anise	Collards	Peppermint
Asparagus	Corn	Peppers
Barley	Cotton	Potato
Beans	Cucumber	Rye
Broccoli	Lettuce	Sorghum
Brussels sprouts	Melons	Soybean
Cabbage	Mint	Spinach
Carrot	Oats	Sugar beet
Cauliflower	Peanuts	Summer Squash
		Wheat

Apply the low rates on small plants, small insects and light infestations of insects. Use intermediate rates on large insects and heavier infestations of insects. Use 1 to 3 applications of the highest recommended rate for controlling severe infestations. Thereafter, use the lowest rate possible to maintain control.

NOTICE TO BUYER — Purchase of this material does not confer any rights under patents of countries outside of the United States.

STORAGE AND DISPOSAL

STORAGE: Handle this package carefully to prevent breakage of inner bag when stored at low temperatures. Allow to warm above 50°F for normal handling. Store product in original container only, away from other pesticides, fertilizer, food or feed. Not for use or storage in or around the home.

PESTICIDE DISPOSAL: Do not contaminate food or feed by storage or disposal. Pesticide Wastes are acutely hazardous. Improper disposal of excess pesticide, spray mixture, or rinsate is a violation of Federal Law. If these wastes cannot be disposed of by use according to label instructions contact your State Pesticide or Environmental Control Agency, or the Hazardous Waste representative at the nearest EPA regional office for guidance.

CONTAINER DISPOSAL: Completely empty bag into application equipment. Then dispose of empty package in a sanitary landfill, or if allowed by State and local authorities, by burning. If burned, stay out of smoke.

NOTICE OF WARRANTY

Du Pont warrants that this product conforms to the chemical description on the label thereof and is reasonably fit for the purposes stated on such label only when used in accordance with the directions under normal use conditions. It is impossible to eliminate all risks inherently associated with the use of the product. Crop injury, ineffectiveness, or other unintended consequences may result because of such factors as weather conditions, presence of other materials, or the manner of use or application, all of which are beyond the control of Du Pont. In no case shall Du Pont be liable for consequential, special or indirect damages resulting from the use or handling of this product. All such risks shall be assumed by the buyer. DU PONT MAKES NO WARRANTIES OF MERCHANTABILITY OR FITNESS FOR A PARTICULAR PURPOSE NOR ANY OTHER EXPRESSED OR IMPLIED WARRANTY EXCEPT AS STATED ABOVE.

CROPS	INSECTS	RATES: "LANNATE" LBS. PER ACRE	LAST APPLICATION—DAYS: TO HARVEST	LAST APPLICATION—DAYS: TO LIVESTOCK GRAZING/ FEEDING
Alfalfa (Do not apply to dormant or semi-dormant alfalfa when min. daily temp. is 50°F. or lower.)	Pea Aphid, Lygus Bugs, Blotch Leafminer, Aphids, Egyptian Alfalfa Weevil Larvae	1/2–1		7
	Fall Armyworm	1/2		
	Alfalfa Weevil Larvae	1		
	Loopers, Yellowstriped Armyworm, Western Yellowstriped Armyworm, Armyworn	1/4–1/2		
	Variegated Cutworm, Beet Armyworm	1/4–1		
	Alfalfa Caterpillar	1/4–1/2		
Anise (Fennel)	Cabbage Looper	1	7	
	Beet Armyworm	1/2–1		
Apples **G (Do not use on Early MacIntosh & Wealthy varieties)	Apple Aphid, Rosy Apple Aphid, Tufted Apple Budmoth, Green Fruitworm, Tarnished Plant Bug	1/4–1/2/100 gals.; up to 400 gals. per acre	8	10
	Codling Moth (10–12 day intervals)			
	Leafrollers (Fruittree, Obliquebanded, Redbanded, Variegated), Lesser Appleworm, White Apple Leafhopper, Tentiform Leafminer, Cutworm	1/2/100 gals.; up to 400 gals. per acre		

CROPS	INSECTS	RATES "LANNATE" LBS. PER ACRE	LAST APPLICATION—DAYS TO HARVEST	LAST APPLICATION—DAYS TO LIVESTOCK GRAZING/FEEDING
Asparagus	Beet Armyworm, Western Yellowstriped Armyworm, Asparagus Beetle, Spotted Asparagus Beetle, White Cutworm, Redbacked Cutworm	½–1	1	
	Variegated Cutworm	½		
Avocado	Avocado Looper, Avocado Leafroller, Omniverous Leafroller	½–1	1	10
Barley	Armyworm, Cereal Leaf Beetle, Aphids,	¼–½	7	10
Beans (Includes Blackeyed Peas & Cowpeas)	Leafhopper, Mexican Bean Beetle	¼–1	Succulent Beans ¼–½ lb.—1; over ½ lb.—3; Dry Beans—25	3 Vines 7 Hay
	Fall Armyworm, Variegated Cutworm	½		
	Beet Armyworm, Loopers, Corn Earworm, Saltmarsh Caterpillar, Yellowstriped Armyworm, Western Yellowstriped Armyworm, Lygus Bugs, Thrips, Aphids	½–1		
	European Corn Borer***—Initiate when moth flights first appear and continue preventive treatments at 3–4 day intervals to control eggs and larvae			
	Spotted Cucumber Beetle	¼–½		
Beets (Table)	Imported Cabbageworm, Beet Armyworm, Cabbage Looper, Diamondback Moth	¼–1	0–Roots	
	Cucumber Beetle, Variegated Cutworm	½	14–Tops	
Bermudagrass	Fall Armyworm, Armyworm	¼–1		7 Forage 3 Dehydrated Hay
Blueberries****	Blueberry Leafhopper, Aphids, Tussock Moth, Weevil, Sharp-Nosed Leafhopper	½	3	
	Cranberry Fruitworm, Cherry Fruitworm **G	½–1		
	Flea Beetle (larvae), Sawfly (larvae), Blueberry Leafroller	1		
	Blueberry Maggot	¼–½		
Broccoli	Loopers, Imp. Cabbageworm, Diamondback Moth	¼–1*	3	
Brussels Sprouts	Loopers, Imp. Cabbageworm, Diamondback Moth	½–1*	3	
	Variegated Cutworm	½*		
Cabbage	Loopers, Imp. Cabbageworm, Diamondback Moth	¼–1*	1	
	Fall Armyworm, Variegated Cutworm	½*		
Carrot	Beet Armyworm	⅓–1	1	
	Aster Leafhopper	½–1		
	Armyworms	¼–½		
	Variegated Cutworm	¼–½		
Cauliflower	Imp. Cabbageworm, Loopers, Diamondback Moth	¼–1*	3	
	Variegated Cutworm	½*		
Celery	Beet Armyworm, Aster Leafhopper	½–1	7	
	Loopers	1		
	Variegated Cutworm	½		
	Armyworm	¼–½		
Chicory	Beet Armyworm, Variegated Cutworm, Leafhoppers	½–1	30	
Chinese Cabbage **G	Loopers, Beet Armyworm	½–1	10	
Collards (Fresh market only—do not apply when temp. is less than 50°F. Do not apply when crop is less than 10" tall.)	Loopers, Diamondback Moth, Variegated Cutworm	½	10	
	Imported Cabbageworm, Beet Armyworm	½–1		
Corn (Sweet) Note: Certain hybrid varieties of sweet corn are susceptible to methomyl injury. Treat a small area to determine crop safety before full scale spraying	Earworm—Whorl as needed	⅓–½	0 Ears	3 Forage
	Fall Armyworm, Armyworm, Earworm (Ovicide/Larvicide), European Corn Borer—Ears 1–3 days or as needed, Corn Rootworm (adult beetles), Flea Beetles, Picnic Beetles, Aphids	¼–½		
	Variegated Cutworm, Beet Armyworm	½		
Corn (Field & Popcorn)	Earworm, Armyworm, Fall Armyworm, Corn Rootworm (adult beetles), Flea Beetles, Picnic Beetles, Aphids	¼–½	0 Ears	3 Forage
	Variegated Cutworm, Beet Armyworm	½		
Cotton (Use may redden cotton—If excessive, stop or alternate with other insecticides.) U.S.	**Ovicide/Larvicide** Bollworm, Tobacco Budworm (Initiate schedule when significant numbers of eggs are present. Continue at 5– to 7–day intervals while eggs are present and larval control is adequate. If significant larvae survive, use higher rates below.) Lygus Bugs (adults and nymphs) start treatment on low level population for suppression.	⅛–¼ (See Predators)		
	Cotton Leafworm	¼–½		
	Aphids, Thrips	¼		

CROPS	INSECTS	RATES "LANNATE" LBS. PER ACRE	LAST APPLICATION—DAYS TO HARVEST	LAST APPLICATION—DAYS TO LIVESTOCK GRAZING/FEEDING
East of Rockies only	Cotton Fleahopper (as needed)	1/8–1/4	15	Do not Graze or Feed
	Larvicide for worms Bollworm, Tobacco Budworm, Beet Armyworm, Cotton Leafperforator, Fall Armyworm, Lygus Bugs (adults and nymphs) use as occasional spray in regular schedule but not more often than every 10 days.	1/2		
	Bollworm, Tobacco Budworm, Beet Armyworm, Cotton Leafperforator, Fall Armyworm, Lygus Bugs (adult and nymphs) up to 3 applications at 3–5 day intervals after desired boll load set on plants.	1/2–3/4		
West of Rockies only	**Larvicide for worms** Bollworm, Fall Armyworm, Tobacco Budworm, Lygus Bugs	1/2–3/4		
	Beet Armyworm	1/3–1/2		
	Cotton Leafperforator	1/3–3/4		
Southwest Only	Cotton Fleahopper (as needed)	1/8–1/4		
Cucumber	Loopers, Tobacco Budworm, Beet Armyworm, Yellowstriped Armyworm, Granulate Cutworm, Flea Beetles, Cucumber Beetles, Melon Aphid, Melonworm, Pickleworm	1/2–1	1/2 lb.—1 Over 1/2 lb.—3	
	Fall Armyworm Variegated Cutworm	1/2		
Eggplant	Green Peach Aphid	1/4–1	5	
	Tomato Pinworm **G, (Beet Armyworm)	1/2–1		
Endive, Escarole	Beet Armyworm	1/2–1	10	
Forest Woodlots, Plantations and Nurseries **G	Spruce Budworm (treat when larvae are feeding on buds, retreat after 3–4 days if needed)	1/4–1/2		
	Forest Tent Caterpillar, Western Tent Caterpillar	1/2–1		
Garlic	Beet Armyworm	1/2*	7	
Grapefruit CA & AZ only	Thrips, Fruittree Leafroller, Orange Tortrix, Western Tussock Moth	1/2–1	1	10
U.S.	Beet Armyworm			
FL only	Spirea Aphid			
	Whiteflies	1		
CA, AZ, & FL only	Yellow Scale, Chaff Scale, Purple Scale, Soft Brown Scale, Glover Scale	1/2/100 gals. up to 800 gals. per acre		
Grapes East of Rockies only	Grape Berry Moth (apply pre- & post-bloom & repeat at 10–14 day intervals or as needed), Leafhoppers **G	1/2–1	1 Fresh & Raisin 14 Wine Grapes	
U.S.	Omniverous Leafroller, Grape Leaffolder, Orange Tortrix	1/2–1		
West of Rockies only	Saltmarsh Caterpillar			
	Leafhoppers **G	1/2–1		
	Climbing Cutworm (Treat infestation and repeat at 7–14 day intervals as needed)	1		
CA only	Grape Leaf Skeletonizer **G	1/2–1		
	Thrips	1/8–1/4		
Horseradish **G	Aphids, Thrips	1/2	65	
Leafy Green Vegetables: Beet (Tops), Dandelions, Kale, Mustard Greens, Parsley, Swiss Chard, Turnip Greens	Beet Armyworm, Cabbage Looper, Diamondback Moth, Imported Cabbageworm	1/2–1	10	
Lemon CA & AZ only	Thrips, Western Tussock Moth, Orange Tortrix	1/2–1	1	10
U.S.	Beet Armyworm			
FL only	Spirea Aphid			
	Whiteflies	1		
CA, AZ, & FL only	Yellow Scale, Chaff Scale, Purple Scale, Soft Brown Scale, Glover Scale	1/2/100 gals. up to 800 gals. per acre		
Lentils	Western Yellowstriped Armyworm	1/2–1	21	
Lettuce, (head varieties U.S. and Leaf varieties in CA and AZ)	Beet Armyworm, Loopers	1/4–1	1/4–1/2 lb.—7 over 1/2 lb.—10	
	Thrips, Aphids, Corn Earworm, Aster Leafhopper	1/2–1		
	Variegated Cutworm	1/2		
Melons	Loopers, Tobacco Budworm, Beet Armyworm, Yellowstriped Armyworm, Granulate Cutworm, Flea Beetles, Cucumber Beetles, Melon Aphid, Melonworm, Pickleworm	1/2–1	1/2 lb.—1 day over 1/2 lb.—3 days	
	Fall Armyworm, Variegated Cutworm	1/2		
Mint	Variegated Cutworm, Alfalfa Looper	1	14	
	Flea Beetles	3/4–1		
Nectarine CA & AZ only	Thrips (up to 3 applications)	1/2–1	1	10
Oats	Armyworms, Cereal Leaf Beetle, Aphids	1/4–1/2	7	10
Onions	Beet Armyworm	1/2–1*	7 Green and	

CROPS		INSECTS		RATES "LANNATE" LBS. PER ACRE	LAST APPLICATION—DAYS TO HARVEST	LAST APPLICATION—DAYS TO LIVESTOCK GRAZING/FEEDING
		Thrips, Variegated Cutworm		1	Dry Onions	
Oranges	CA & AZ only	Thrips, Western Tussock Moth, Orange Tortrix, Fruittree Leafroller, Citrus Cutworm		1/2–1	1	10
	U.S.	Beet Armyworm				
	FL only	Spirea Aphid				
		Whiteflies		1		
	CA, AZ, & FL only	Yellow Scale, Chaff Scale, Purple Scale, Soft Brown Scale, Glover Scale		1/2/100 gals. up to 800 gals. per acre		
Ornamentals **G (See Suppl. Labeling)		Bulletin E-40477 11/87 obtainable on request		1/4–1/2/ 100 gals.		
Peaches		Catfacing Insects (Plant Bugs and Stink Bugs)—begin at petal fall and continue in cover sprays at 7- to 10-day intervals		1–2 lb. (or 1/4–1/2/100 gals. up to 400 gals. per acre)	4	10
		Oriental Fruit Moth—begin at petal fall, use trapping devices and frequent field inspection to determine need for treatment. Continue treatment in cover sprays and alternate with residual-type insecticides registered for this use.**G				
		Green Peach Aphid				
Peanuts		Corn Earworm***, Beet Armyworm, Potato Leafhopper—Up to 3 applications		1/4–1	21	Do not feed treated vines
		Loopers, Green Cloverworm, Velvetbean Caterpillar, Thrips, Granulate Cutworm—Up to 3 applications		1/2–1		
		Fall Armyworm		1/4–1/2		
Pears	Northeast only	Green Fruitworm, Obliquebanded Leafroller—One application only		1/2/100 gals. up to 400 gals./acre	7	10
Peas (Succulent)		Armyworm, Green Cloverworm		1/4–1/2	1	5 Forage 14 Hay
		Loopers, Pea Aphid, Beet Armyworm, Saltmarsh Caterpillar, Variegated Cutworm		1/2–1		
		Alfalfa Caterpillar		1/4–1		
Pecans	Southeast only	Aphids		1/2–1	30	10
Peppers		Green Peach Aphid, Fall Armyworm, Loopers, Beet Armyworm		1/2	3	
		Armyworm, Variegated Cutworm		1/4–1/2		
Pomegranates		Omniverous Leafroller		1	14	
Potato		Leafhoppers, Tuberworm, Loopers, Aphids, Beet Armyworm		1/2–1	6	
		Variegated Cutworm, Fall Armyworm, Flea Beetles		1/2		
Rye		Armyworms, Cereal Leaf Beetle, Aphids		1/4–1/2	7	10
Sorghum		Sorghum Webworm		1/2	14	14
		Sorghum Midge—Apply when 50% bloom and 3-5 days later if needed		1/4–1/2		
		Fall Armyworm (Budworm), Beet Armyworm, Corn Earworm, Armyworm				
Soybean	U.S.	Green Cloverworm, Velvetbean Caterpiller, Mexican Bean Beetle	Light to moderate infestations	1/8–1/4 (See Predators)	14	under 1/2 lb. 3 Forage 7 Hay
			Moderate to severe infestations	1/4–1/2		
		Corn Earworm	Moderate to severe infestations			
			Light to moderate infestations	1/8–1/4 (See Predators)		
		Beet Armyworm, Salt Marsh Caterpiller, Bean Leaf Beetle, Thrips, Fall Armyworm, Silver Spotted Skipper	Light to moderate infestations			1/2–1 lb. 10 Forage 12 Hay
			Moderate to severe infestations	1/4–3/8		
				3/8–1/2		
	AR, AL, LA, MS, TN, & TX only	Soybean Looper (worms up to 1/2" long)	Use higher rate for severe infestations	1/2–1		
Spinach (Do not apply when min. daily temp. is 32°F. or lower. Do not apply to seedlings less than 3" diameter.)		Loopers, Beet Armyworm		1/2–1	7	
		Fall Armyworm, Variegated Cutworm		1/2		
Strawberry		Beet Armyworm, Thrips, Aphids		1/2–1	3-Fresh 10-Processing Fruit	
		Lygus Bugs, Omniverous Leaf Tiers		1		
		Armyworm		3/4–1		
Sugar Beet		Beet Webworm, Flea Beetles, Carrion Beetles, Beet Armyworm, Aphids, Western Yellowstriped Armyworm—Up to 4 applications		1/4–1	7	30 Tops
		Variegated Cutworm		1/2		

CROPS		INSECTS	RATES "LANNATE" LBS. PER ACRE	LAST APPLICATION—DAYS TO HARVEST	LAST APPLICATION—DAYS TO LIVESTOCK GRAZING/FEEDING
Summer Squash		Loopers, Tobacco Budworm, Beet Armyworm, Yellowstriped Armyworm, Granulate Cutworm, Flea Beetles, Cucumber Beetles, Melon Aphid, Melonworm, Pickleworm, Fall Armyworm	1/2–1	1/2 lb.—1 day over 1/2 lb. 3 days	
Tangelo	CA & AZ only	Thrips, Western Tussock Moth, Orange Tortrix			
	U.S.	Beet Armyworm	1/2–1		
	FL only	Spirea Aphid		1	10
		Whiteflies	1		
	CA, AZ, & FL only	Yellow Scale, Chaff Scale, Purple Scale, Soft Brown Scale, Glover Scale	1/2/100 gals up to 800 gals. per acre		
Tangerine	CA & AZ only	Thrips, Western Tussock Moth, Orange Tortrix			
	U.S.	Beet Armyworm	1/2–1		
	FL only	Spirea Aphid		1	10
		Whiteflies	1		
	CA, AZ, & FL only	Yellow Scale, Chaff Scale, Purple Scale, Soft Brown Scale, Glover Scale	1/2/100 gals. up to 800 gals. per acre		
Tobacco (Except shade)		Flea Beetle, Hornworm	1/4–1/2	5 Flue cured	
		Loopers, Aphids, Tobacco Budworm, Fall Armyworm	1/2	14 Air or fire cured	
Tomato		Tomato Fruitworm, Aphids, Hornworm, Loopers, Beet Armyworm, Southern Armyworm	1/2–1		
		Pinworm		1	
		Fall Armyworm, Variegated Cutworm	1/2		
		Armyworm	1/4–1/2		
Turf **G		Sod Webworm (after application, sprinkle irrigate for 15 minutes)	1–2 (2/5–4/5 oz. per 1000 sq. ft.)		Do not graze or feed
Watercress **G		Beet Armyworm, Cabbage Looper, Diamondback Moth, Imported Cabbageworm (Use in minimum of 30 gpa of water. Water should be turned off 24 hours prior to application.)	1/2–1	10	
Wheat		Armyworm, Cereal Leaf Beetle, Aphids	1/4–1/2	7	10

* Add wetting agent
** All registrations for aerial and ground use unless marked G for ground use only.
*** Ovicidal and larvicidal control.

AG-572 0884 8097 10/12/84

This is a specimen label. DuPont is not responsible for the accuracy of the information contained herein. As labels are subject to revision, always carefully read and follow the label on the product container.

LANNATE® L

INSECTICIDE
WATER SOLUBLE LIQUID

RESTRICTED USE PESTICIDE

For retail sale to and use only by Certified Applicators or persons under their direct supervision and only for those uses covered by the Certified Applicator's certification.

CONTAINS 1.8 LBS. ACTIVE INGREDIENT PER GALLON

ACTIVE INGREDIENT	BY WEIGHT
Methomyl {S-methyl N-[(methylcarbamoyl)oxy]thioacetimidate}	24%
INERT INGREDIENTS:	76%
TOTAL	100%

EPA Reg. No. 352-370
U. S. Pats. 3,639,633 & 3,862,316

KEEP OUT OF REACH OF CHILDREN

DANGER–POISON

PELIGRO

PRECAUCION AL USUARIO: Si usted no lee ingles, no use este producto hasta que la etiqueta haya sido explicada ampliamente.

STATEMENT OF PRACTICAL TREATMENT

If swallowed: Call a physician or Poison Control Center. Drink 1 or 2 glasses of water and induce vomiting by touching back of throat with finger. Do not induce vomiting or give anything by mouth to an unconscious person.

If inhaled: Remove from exposure and have patient lie down and keep quiet. If patient is not breathing, start artificial respiration immediately. Never give anything by mouth to an unconscious person.

In case of contact, wash skin with plenty of soap and water; for eyes, flush with water for 15 minutes and get medical attention; remove and wash contaminated clothing before re-use.

ATROPINE IS AN ANTIDOTE — SEEK MEDICAL ATTENTION AT ONCE IN ALL CASES OF SUSPECTED POISONING.

If warning symptoms appear (see WARNING SYMPTOMS), get medical attention.

For medical emergencies involving this product, call toll free 1-800-441-3637.

PRECAUTIONARY STATEMENTS

HAZARDS TO HUMANS AND DOMESTIC ANIMALS

DANGER–POISON

CONTAINS METHANOL. MAY BE FATAL OR CAUSE BLINDNESS IF SWALLOWED. POISONOUS IF INHALED. CAUSES EYE DAMAGE.

Do not breathe vapors or spray mist. Do not get in eyes, on skin, or on clothing. When mixing spray, loading, applying, or otherwise handling, wear protective clothing, goggles, and a mask or respirator jointly approved by the Mine Safety and Health Administration (formerly the U. S. Bureau of Mines) and by the National Institute for Occupational Safety and Health. Wear clean clothes daily. Wash thoroughly after handling and before eating or smoking. Pilot should not assist in the mixing and loading operation.

Do not apply this product in such a manner as to directly or through drift expose workers, other persons or animals. The area being treated must be vacated by unprotected persons.

WARNING SYMPTOMS — Methomyl poisoning produces effects associated with anticholinesterase activity which may include weakness, blurred vision, headache, nausea, abdominal cramps, discomfort in the chest, constriction of pupils, sweating, slow pulse, muscle tremors.

NOTE TO PHYSICIAN

TREATMENT — Atropine sulfate should be used for treatment. Administer repeated doses, 1.2 to 2.0 mg. intravenously every 10 to 30 minutes until full atropinization is achieved. Maintain atropinization until the patient recovers. Artificial respiration or oxygen may be necessary. Allow no further exposure to any cholinesterase inhibitor until recovery is assured.

Do not use 2-PAM for exposure to "LANNATE" L alone. However, for exposure to combinations of "LANNATE" L and organophosphorous insecticides, 2-PAM may be used as required to supplement the atropine sulfate treatment.

For medical emergencies involving this product, call toll free 1-800-441-3637.

ENVIRONMENTAL HAZARDS

This product is toxic to fish, birds, and other wildlife. Keep out of any body of water. Do not apply where runoff is likely to occur. Do not apply when weather conditions favor drift from areas treated. Do not contaminate water by cleaning of equipment or disposal of wastes.

This product is toxic to bees exposed to direct application. Do not apply this product while bees are actively visiting the treatment area. Time applications to coincide with periods of minimum bee activity.

PHYSICAL OR CHEMICAL HAZARDS

Flammable. Keep away from heat, sparks, and open flame. Keep container closed. Use with adequate ventilation.

GENERAL INFORMATION

Scouting — Monitoring insect populations to determine whether or not there is a need for application of "LANNATE" L is recommended. More than one treatment of "LANNATE" L may be required to control a population of pests.

Insect Predators —"LANNATE" L at rates of 1/2 to 1 pt. per acre helps conserve certain beneficials, including big-eyed bugs, damsel bugs, flower bugs and spiders in cotton and soybeans. While these beneficials cannot be relied upon to control pests, they are of potential value and should be monitored along with pests in pest management programs on these crops.

Compatibility — Since formulations may be changed and new ones introduced, it is recommended that users premix a small quantity of a desired tank mix and observe for possible adverse changes (settling out, flocculation, etc.). Avoid mixtures of several materials and very concentrated spray mixtures.

Do not use "LANNATE" L with Bordeaux mixture, "Du Ter" (triphenyltin hydroxide), lime sulfur, "Rayplex" iron nor in highly alkaline solutions. Use mildly alkaline mixtures immediately after mixing to prevent loss of insecticidal activity.

DIRECTIONS FOR USE

It is a violation of federal law to use this product in a manner inconsistent with its labeling.

Du Pont "LANNATE" L Insecticide should be used only in accordance with recommendations on this label, or in separate published Du Pont recommendations available through local dealers.

Du Pont will not be responsible for losses or damages resulting from use of this product in any manner not specifically recommended by Du Pont. User assumes all risks associated with such non-recommended use.

Do not enter treated areas without appropriate protective clothing until sprays have dried. Because certain states may require more restrictive reentry intervals for various crops treated with this product, consult your State Department of Agriculture for further information.

Written or oral warnings must be given to workers who are expected to be in a treated area or in an area about to be treated with this product. When oral warnings are given, warnings shall be given in a language customarily understood by workers. Oral warnings must be given if there is reason to believe that written warnings cannot be understood by workers. Written (or oral) warnings must include the following information: "DANGER — POISON. Area treated with 'LANNATE' L on (date of application). Do not enter treated areas without appropriate protective clothing until sprays have dried. If warning symptoms appear, refer to Statement of Practical Treatment on front panel of 'LANNATE' L label and seek medical attention at once."

"LANNATE" L is a water soluble liquid to be diluted with water for application by ground or air equipment. Use only in commercial plantings; do not use in home plantings.

SPRAY PREPARATION — Fill spray tank 1/4 to 1/2 full of water. Add "LANNATE" L directly to spray tank and mix thoroughly, using mechanical or hydraulic means; do not use air agitation. Spray mix should not be stored overnight in spray tank.

Do not apply this product through any type of irrigation system.

APPLICATION — Apply at the recommended rates when insects first appear. Unless otherwise noted, continue application at 5- to 7-day intervals or as needed. Use sufficient water to obtain thorough, uniform coverage. Since "LANNATE" L is a fast acting contact insecticide, best results follow direct spraying of the target insect. This can be accomplished with fine spray droplets at low volume.

For aerial, use a minimum of 2 gals. per acre (gpa) except 10 gpa for grapes, nectarines and strawberries; 15 gpa for oranges, lemons, grapefruit, tangelos and tangerines.

For certain crops a minimum of 1 gallon per acre may be used providing the following conditions are met.

- equipment is capable of delivering small spray droplets to insure thorough coverage,
- equipment is adjusted to distribute spray uniformly over the spray swath,
- wind conditions and other factors such as temperature and humidity are such that the spray is delivered to the target area,
- local regulations do not prohibit low-volume aerial sprays,
- use rates are applied as directed on the package label or supplemental labeling for the following crops:

Alfalfa	Carrot	Melons	Rye
Anise	Cauliflower	Mint	Sorghum
Asparagus	Celery	Oats	Soybean
Barley	Collards	Peanuts	Spinach
Beans	Corn	Peas (succulent)	Sugar beet
Broccoli	Cotton	Peppermint	Summer squash
Brussels sprouts	Cucumber	Peppers	Wheat
Cabbage	Lettuce	Potato	

Apply the low rates on small plants, small insects and light infestations of insects. Use intermediate rates on large insects and heavier infestations of insects. Use 1 to 3 applications of the highest recommended rate for controlling severe infestations. Thereafter, use the lowest rate possible to maintain control.

NOTICE TO BUYER — Purchase of this material does not confer any rights under patents of countries outside of the United States.

STORAGE AND DISPOSAL

STORAGE: Do not subject to temperatures below 32°F. Store product in original container only, away from other pesticides, fertilizer, food or feed. Not for use or storage in or around the home.

PESTICIDE DISPOSAL: Do not contaminate water, food, or feed by storage or disposal. Pesticide wastes are toxic. Improper disposal of excess pesticide, spray mixture, or rinsate is a violation of Federal Law. If these wastes cannot be disposed of by use according to label instructions, contact your State Pesticide or Environmental Control Agency, or the Hazardous Waste representative at the nearest EPA Regional Office for guidance.

CONTAINER DISPOSAL: Triple rinse (or equivalent) the container. Then offer for recycling or reconditioning, or puncture and dispose of in a sanitary landfill, or, if allowed by state and local authorities, by burning. If burned, stay out of smoke.

NOTICE OF WARRANTY

Du Pont warrants that this product conforms to the chemical description on the label thereof and is reasonably fit for the purposes stated on such label only when used in accordance with the directions under normal use conditions. It is impossible to eliminate all risks inherently associated with the use of this product. Crop injury, ineffectiveness, or other unintended consequences may result because of such factors as weather conditions, presence of other materials, or the manner of use or application, all of which are beyond the control of Du Pont. In no case shall Du Pont be liable for consequential, special or indirect damages resulting from the use or handling of this product. All such risks shall be assumed by the buyer. DU PONT MAKES NO WARRANTIES OF MERCHANTABILITY OR FITNESS FOR A PARTICULAR PURPOSE NOR ANY OTHER EXPRESSED OR IMPLIED WARRANTY EXCEPT AS STATED ABOVE.

		RATES	LAST APPLICATION—DAYS	
CROPS	INSECTS	"LANNATE" L PINTS PER ACRE	TO HARVEST	TO LIVESTOCK GRAZING/ FEEDING
Alfalfa (Do not apply to dormant or semi-dormant alfalfa when min. daily temp is 50°F. or lower.)	Pea Aphid, Lygus Bugs, Blotch Leafminer, Aphids, Egyptian Alfalfa Weevil Larvae	2-4		7
	Fall Armyworm	2		
	Alfalfa Weevil Larvae	4		
	Loopers, Yellowstriped Armyworm, Western Yellowstriped Armyworm, Armyworm	1-2		
	Variegated Cutworm, Beet Armyworm	1-4		
	Alfalfa Caterpillar	1-2		
Anise (Fennel)	Cabbage Looper	4	7	
	Beet Armyworm	2-4		
Apple (Do not use on Early MacIntosh & Wealthy varieties) **G	Apple Aphid, Rosy Apple Aphid, Tufted Apple Budmoth, Green Fruitworm, Tarnished Plant Bug	1-2/100 gals.; up to 400 gals. per acre	8	10
	Codling Moth (10-12 day intervals)			
	Leafrollers (Fruittree, Obliquebanded, Redbanded, Variegated), Lesser Appleworm, White Apple Leafhopper, Tentiform Leafminer, Cutworm	2/100 gals. up to 400 gals. per acre		
Asparagus	Beet Armyworm, Western Yellowstriped Armyworm, Asparagus Beetle, Spotted Asparagus Beetle, White Cutworm, Redbacked Cutworm	2-4	1	
	Variegated Cutworm	2		
Avocado	Avocado Looper, Avocado Leafroller, Omniverous Leafroller	2-4	1	10
Barley	Armyworms, Cereal Leaf Beetle, Aphids	1-2	7	10
Beans (Includes Blackeyed Peas & Cowpeas)	Leafhopper, Mexican Bean Beetle	1-4	Succulent Beans 1-2 pts.—1, over 2 pts.—3; Dry Beans—25	3 Vines 7 Hay
	Fall Armyworm, Variegated Cutworm	2		
	Beet Armyworm, Loopers, Corn Earworm, Saltmarsh Caterpillar, Yellowstriped Armyworm, Western Yellowstriped Armyworm, Lygus Bugs, Thrips, Aphids	2-4		
	European Corn Borer***—Initiate when moth flights first appear and continue preventive treatments at 3-4 day intervals to control eggs and larvae			
	Spotted Cucumber Beetle	1-2		
Beets (Table)	Imported Cabbageworm, Beet Armyworm, Cabbage Looper, Diamondback Moth	1-4	0-Roots	
	Cucumber Beetle, Variegated Cutworm	2	14-Tops	
Bermudagrass	Fall Armyworm, Armyworm	1-4		7 Forage 3 Dehydrated Hay
Blueberries****	Blueberry Leafhopper, Aphids, Tussock Moth, Weevil, Sharp-Nosed Leafhopper	2	3	
	Cranberry Fruitworm, Cherry Fruitworm **G	2-4		
	Flea Beetle (larvae), Sawfly (larvae), Blueberry Leafroller	4		
	Blueberry Maggot	1-2		
Broccoli	Loopers, Imp. Cabbageworm, Diamondback Moth	1-4*	3	
Brussels Sprouts	Loopers, Imp. Cabbageworm, Diamondback Moth	2-4*	3	
	Variegated Cutworm	2*		
Cabbage	Loopers, Imp. Cabbageworm, Diamondback Moth	1-4*	1	
	Fall Armyworm, Variegated Cutworm	2*		
Carrot	Beet Armyworm	1⅓-4	1	
	Aster Leafhopper	2-4		
	Armyworms	1-2		
	Variegated Cutworm	1-2		
Cauliflower	Imp. Cabbageworm, Loopers, Diamondback Moth	1-4*	3	
	Variegated Cutworm	2*		
Celery	Beet Armyworm, Aster Leafhopper	2-4	7	
	Loopers	4		
	Variegated Cutworm	2		
	Armyworms	1-2		
Chicory	Beet Armyworm, Variegated Cutworm, Leafhoppers	2-4	80	
Chinese Cabbage **G	Loopers, Beet Armyworm	2-4	10	

CROPS	INSECTS	RATES "LANNATE" L PINTS PER ACRE	LAST APPLICATION—DAYS TO HARVEST	LAST APPLICATION—DAYS TO LIVESTOCK GRAZING/ FEEDING
Collards (Fresh market only—do not apply when temp. is less than 50°F. Do not apply when crop is less than 10" tall.)	Loopers, Diamondback Moth, Variegated Cutworm	2	10	
	Imported Cabbageworm, Beet Armyworm	2-4		
Corn (Sweet) Note: Certain hybrid varieties of sweet corn are susceptible to methomyl injury. Treat a small area to determine crop safety before full scale spraying.	Earworm—Whorl as needed	1⅓-2	0 Ears	3 Forage
	Fall Armyworm, Armyworm, Earworm (Ovicide/Larvicide), European Corn Borer–Ears 1-3 days or as needed, Corn Rootworm (adult beetles), Flea Beetles, Picnic Beetles, Aphids	1-2		
	Variegated Cutworm, Beet Armyworm	2		
Corn (Field & Popcorn)	Earworm, Armyworm, Fall Armyworm, Corn Rootworm (adult beetles), Flea Beetles, Picnic Beetles, Aphids	1-2	0 Ears	3 Forage
	Variegated Cutworm, Beet Armyworm	2		
Cotton (Use may redden cotton — If excessive, stop or alternate with other insecticides U.S.	**Ovicide/Larvicide** Bollworm, Tobacco Budworm (Initiate schedule when significant numbers of eggs are present. Continue at 5- to 7-day intervals while eggs are present and larval control adequate. If significant larvae survive, use higher rates below.) Lygus Bugs (adults and nymphs) start treatment on low level population for suppression.	½-1 (See Predators)	15	Do not Graze or Feed
	Cotton Leafworm	1-2		
	Aphids, Thrips	1		
	Cotton Fleahopper (as needed)	½-1		
East of Rockies only.	**Larvicide for worms** Bollworm, Tobacco Budworm, Beet Armyworm, Cotton Leafperforator, Fall Armyworm, Lygus Bugs (adults and nymphs) use as occasional spray in regular schedule but not more often than every 10 days.	2		
	Bollworm, Tobacco Budworm, Beet Armyworm, Cotton Leafperforator, Fall Armyworm, Lygus Bugs (adults and nymphs) up to 3 applications at 3-5 day intervals after desired boll load set on plants.	2-3		
West of Rockies only	**Larvicide for worms** Bollworm, Fall Armyworm, Tobacco Budworm, Lygus bugs	2-3		
	Beet Armyworm	1⅓-2		
	Cotton Leafperforator	1⅓-3		
Southwest only	Cotton Fleahopper (as needed)	½-2		
Cucumber	Loopers, Tobacco Budworm, Beet Armyworm, Yellowstriped Armyworm, Granulate Cutworm, Flea Beetles, Cucumber Beetles, Melon Aphid, Melonworm, Pickleworm	2-4	2 pts.—1 Over 2 pts.—3	
	Fall Armyworm, Variegated Cutworm	2		
Eggplant	Green Peach Aphid	1-4	5	
	Tomato Pinworm **G. (Beet Armyworm)	2-4		
Endive, Escarole	Beet Armyworm	2-4	10	
Forest Woodlots, Plantations, and Nurseries **G	Spruce Budworm (treat when larvae are feeding on buds, retreat after 3-4 days if needed)	1-2		
	Forest Tent Caterpillar, Western Tent Caterpillar	2-4		
Garlic	Beet Armyworm	2*	7	
Grapefruit CA & AZ only	Thrips, Fruittree Leafroller, Orange Tortix, Western Tussock Moth	2-4	1	10
U.S.	Beet Armyworm			
Fl only	Spirea Aphid			
	Whiteflies	4		
CA, AZ, & FL only	Yellow Scale, Chaff Scale, Purple Scale, Soft Brown Scale, Glover Scale	2/100 gals. up to 800 gals. per acre		
Grapes East of Rockies only	Grape Berry Moth (apply pre- & post-bloom & repeat at 10-14 day intervals or as needed), Leafhoppers **G	2-4	1 Fresh & Raisin 14 Wine Grapes	
U.S.	Omniverous Leafroller, Grape Leaffolder, Orange Tortrix	2-4		
West of Rockies only	Saltmarsh Caterpillar	2-4		
	Leafhoppers **G			
	Climbing Cutworm (Treat infestation and repeat at 7-14 day intervals as needed)	4		
CA only	Grape Leaf Skeletonizer **G	2-4		
	Thrips	½-1		
Horseradish **G	Aphids, Thrips	2	65	

CROPS	INSECTS	RATES "LANNATE" L PINTS PER ACRE	LAST APPLICATION—DAYS TO HARVEST	LAST APPLICATION—DAYS TO LIVESTOCK GRAZING/ FEEDING
Leafy Green Vegetables: Beet (Tops), Dandelions, Kale, Mustard Greens, Parsley, Swiss Chard, Turnip Greens	Beet Armyworm, Cabbage Looper, Diamondback Moth, Imported Cabbageworm	2-4	10	
Lemon CA & AZ only	Thrips, Western Tussock Moth, Orange Tortrix		1	10
U.S.	Beet Armyworm	2-4		
FL only	Spirea Aphid			
	Whiteflies	4		
CA, AZ, & FL only	Yellow Scale, Chaff Scale, Purple Scale, Soft Brown Scale, Glover Scale	2/100 gals. up to 800 gals. per acre		
Lentils	Western Yellowstriped Armyworm	2-4	21	
Lettuce, (head varieties in U.S. and Leaf varieties in CA and AZ)	Beet Armyworm, Loopers	1-4	1-2 pts.—7 over 2 pts.—10	
	Thrips, Aphids, Corn Earworm, Aster Leafhopper	2-4		
	Variegated Cutworm	2		
Melons	Loopers, Tobacco Budworm, Beet Armyworm, Yellowstriped Armyworm, Granulate Cutworm, Flea Beetles, Cucumber Beetles, Melon Aphid, Melonworm, Pickleworm	2-4	2 pts.—1 day over 2 pts.—3 days	
	Fall Armyworm, Variegated Cutworm	2		
Mint	Variegated Cutworm, Alfalfa Looper	4	14	
	Flea Beetles	3-4		
Nectarine CA & AZ only	Thrips (up to 3 applications)	2-4	1	10
Oats	Armyworms, Cereal Leaf Beetle, Aphids	1-2	7	10
Onions	Beet Armyworm	2-4*	7 Green and Dry Onions	
	Thrips, Variegated Cutworm	2*		
Oranges CA & AZ only	Thrips, Western Tussock Moth, Orange Tortrix, Fruitree Leafroller, Citrus Cutworm		1	10
U.S.	Beet Armyworm	2-4		
FL only	Spirea Aphid			
	Whiteflies	4		
CA, AZ, & FL only	Yellow Scale, Chaff Scale, Purple Scale, Soft Brown Scale, Glover Scale	2/100 gals. up to 800 gals. per acre		
Ornamentals **G (See Suppl. Labeling)	Bulletin E-40477 11/87 obtainable on request	1-2 per 100 gals.		
Peaches	Catfacing Insects (Plant Bugs and Stink Bugs)—begin at petal fall and continue in cover sprays at 7- to 10-day intervals	4-8 (or 1-2 pts./100 gals. up to 400 gals. per acre)	4	10
	Oriental Fruit Moth—begin at petal fall; use trapping devices and frequent field inspection to determine need for treatment. Continue treatment in cover sprays and alternate with residual-type insecticides registered for this use. **G			
	Green Peach Aphid			
Peanuts	Corn Earworm***, Beet Armyworm, Potato Leafhopper—Up to 3 applications	1-4	21	Do not feed treated vines
	Loopers, Green Cloverworm, Velvetbean Caterpillar, Thrips, Granulate Cutworm—Up to 3 applications.	2-4		
	Fall Armyworm	1-2		
Pears Northeast only	Green Fruitworm, Obliquebanded Leafroller—One application only	2/100 gals. up to 400 gals./acre	7	10
Peas (Succulent)	Armyworm, Green Cloverworm	1-2	1	5 forage 14 Hay
	Loopers, Pea Aphid, Beet Armyworm, Saltmarsh Caterpillar, Variegated Cutworm	2-4		
	Alfalfa Caterpillar	1-4		
Pecans Southeast only	Aphids	2-4	30	10
Peppers	Green Peach Aphid, Fall Armyworm, Loopers, Beet Armyworm	2	3	
	Armyworm, Variegated Cutworm	1-2		
Pomegranates	Omniverous Leafroller	4	14	
Potato	Tuberworm, Loopers, Aphids, Beet Armyworm Leafhoppers	2-4	6	
	Variegated Cutworm, Fall Armyworm, Flea Beetles	2		
Rye	Armyworms, Cereal Leaf Beetle, Aphids	1-2	7	10
Sorghum	Sorghum Webworm	2	14	14
	Sorghum Midge—Apply when 50% bloom and 3-5 days later if needed	1-2		
	Fall Armyworm (Budworm), Beet Armyworm, Corn Earworm, Armyworm			

CROPS		INSECTS		RATES: "LANNATE" L PINTS PER ACRE	LAST APPLICATION—DAYS: TO HARVEST	LAST APPLICATION—DAYS: TO LIVESTOCK GRAZING/ FEEDING
Soybean	U.S.	Green Cloverworm, Velvetbean Caterpillar, Mexican Bean Beetle	Light to moderate infestations	½-1 (See Predators)		
			Moderate to severe infestations	1-2		under 2 pts. 3 Forage 7 Hay
		Corn Earworm (worms up to ¼" long)	Light to moderate infestations			
		(worms up to ½" long)	Moderate to severe infestations	½-1 (See Predators)	14	2-4 pts.
		Beet Armyworm, Salt Marsh Caterpillar, Bean Leaf Beetle, Thrips, Fall Armyworm, Silver Spotted Skipper	Light to moderate infestations	1-1½		10 Forage 12 Hay
			Moderate to severe infestations	1½-2		
	AR, AL, LA, MS TN, & TX only	Soybean Looper (worms up to ½" long)	Use higher rate for severe infestations	2-4		
Spinach (Do not apply when min. daily temp. is 32°F. or lower. Do not apply to seedlings less than 3" diameter.)		Loopers, Beet Armyworm		2-4	7	
		Fall Armyworm, Variegated Cutworm		2		
Strawberry		Beet Armyworm, Thrips, Aphids		2-4	3-Fresh	
		Lygus Bugs, Omniverous Leaf Tiers		4	10-Processing	
		Armyworms		3-4	Fruit	
Sugar Beet		Beet Webworm, Flea Beetles, Carrion Beetles, Beet Armyworm, Aphids, Western Yellowstriped Armyworm—Up to 4 applications		1-4	7	30 Tops
		Variegated Cutworm		2		
Summer Squash		Loopers, Tobacco Budworm, Beet Armyworm, Yellowstriped Armyworm, Granulate Cutworm, Flea Beetles, Cucumber Beetles, Melon Aphid, Melonworm, Pickleworm, Fall Armyworm		2-4	2 pts.—1 day over 2 pts. 3 days	
Tangelo	CA & AZ only	Thrips, Western Tussock Moth, Orange Tortrix		2-4	1	10
	U.S.	Beet Armyworm				
	FL only	Spirea Aphid				
		Whiteflies		4		
	CA, AZ, & FL only	Yellow Scale, Chaff Scale, Purple Scale, Soft Brown Scale, Glover Scale		2/100 gals. up to 800 gals. per acre		
Tangerine	CA & AZ only	Thrips, Western Tussock Moth, Orange Tortrix		2-4	1	10
	U.S.	Beet Armyworm				
	FL only	Spirea Aphid				
		Whiteflies		4		
	CA, AZ, & FL only	Yellow Scale, Chaff Scale, Purple Scale, Soft Brown Scale, Glover Scale		2/100 gals up to 800 gals. per acre		
Tobacco (Except shade)		Flea Beetle, Hornworm		1-2	5 Flue cured 14 Air or Fire cured	
		Loopers, Aphids, Tobacco Budworm, Fall Armyworm		2		
Tomato		Tomato Fruitworm, Aphids, Hornworm, Loopers, Beet Armyworm, Southern Armyworm, Pinworm		2-4	1	
		Fall Armyworm, Variegated Cutworm		2		
		Armyworm		1-2		
Turf **G		Sod Webworm (after application, sprinkle irrigate for 15 minutes)		4-8 (1.5-3 fl. ozs. per 1000 sq. ft.)		Do not graze or feed
Watercress **G		Beet Armyworm, Cabbage Looper, Diamondback Moth, Imported Cabbageworm (Use in minimum of 30 gpa of water. Water should be turned off 24 hours prior to application.)		2-4	10	

CROPS	INSECTS	RATES	LAST APPLICATION—DAYS	
		"LANNATE" L PINTS PER ACRE	TO HARVEST	TO LIVESTOCK GRAZING/ FEEDING
Wheat	Armyworm, Cereal Leaf Beetle, Aphids	1-2	7	10

* Add wetting agent

** All registrations for aerial and ground use unless marked G for ground use only.

*** Ovicidal and larvicidal control.

**** Do not apply during bloom.

AG-195 0884 10/12/84

This is a specimen label. DuPont is not responsible for the accuracy of the information contained herein. As labels are subject to revision, always carefully read and follow the label on the product container.

LANNATE® LV

INSECTICIDE
WATER SOLUBLE LIQUID

RESTRICTED USE PESTICIDE

For retail sale to and use only by Certified Applicators or persons under their direct supervision and only for those uses covered by the Certified Applicator's certification.

CONTAINS 2.4 lbs. ACTIVE INGREDIENT PER GALLON

ACTIVE INGREDIENT	BY WEIGHT
Methomyl (S-methyl N-[(methylcarbamoyl)oxy[thioacetimidate)	29%
INERT INGREDIENTS:	71%
TOTAL	100%

EPA Reg. No. 352-384

U.S. Pats. 3,639,633 & 3,975,518

KEEP OUT OF REACH OF CHILDREN

DANGER–POISON

PELIGRO

PRECAUCION AL USUARIO: Si usted no lee ingles, no use este producto hasta que la etiqueta haya sido explicada ampliamente.

STATEMENT OF PRACTICAL TREATMENT

If swallowed: Call a physician or Poison Control Center. Drink 1 or 2 glasses of water and induce vomiting by touching back of throat with finger. Do not induce vomiting or give anything by mouth to an unconcious person.

If inhaled: Remove from exposure and have patient lie down and keep quiet. If patient is not breathing, start artificial respiration immediately. Never give anything by mouth to an unconscious person.

If in eyes: Flush eyes with plenty of water. Get medical attention.

If on skin: Wash skin with plenty of soap and water, remove and wash contaminated clothing before re-use. Get medical attention.

ATROPINE IS AN ANTIDOTE—SEEK MEDICAL ATTENTION AT ONCE IN ALL CASES OF SUSPECTED POISONING.

If warning symptoms appear (see **WARNING SYMPTOMS**), get medical attention.

For medical emergencies involving this product, call toll free 1-800-441-3637.

PRECAUTIONARY STATEMENTS

HAZARDS TO HUMANS AND DOMESTIC ANIMALS

DANGER–POISON

CONTAINS METHANOL. MAY BE FATAL OR CAUSE BLINDNESS IF SWALLOWED. POISONOUS IF INHALED. CAUSES EYE DAMAGE.

Do not breathe vapors or spray mist. Do not get in eyes, on skin, or on clothing. When mixing spray, loading, applying or otherwise handling, wear protective clothing, goggles, and a mask or respirator jointly approved by the Mine Safety and Health Administration (formerly the U.S. Bureau of Mines) and by the National Institute for Occupational Safety and Health. Wear clean clothes daily. Wash thoroughly after handling and before eating or smoking. Pilot should not assist in the mixing and loading operation. Do not apply this product in such a manner as to directly or through drift expose workers, other persons or animals. The area being treated must be vacated by unprotected persons.

WARNING SYMPTOMS—Methomyl poisoning produces effects associated with anticholinesterase activity which may include weakness, blurred vision, headache, nausea, abdominal cramps, discomfort in the chest, constriction of pupils, sweating, slow pulse, muscle tremors.

NOTE TO PHYSICIAN

TREATMENT—Atropine sulfate should be used for treatment. Administer repeated doses, 1.2 to 2.0 mg. intravenously every 10 to 30 minutes until full atropinization is achieved. Maintain atropinization until the patient recovers. Artificial respiration or oxygen may be necessary. Allow no further exposure to any cholinesterase inhibitor until recovery is assured.

Do not use 2-PAM for exposure to "LANNATE" LV alone. However, for exposure to combinations of "LANNATE" LV and organophosphorous insecticides, 2-PAM may be used as required to supplement the atropine sulfate treatment. Do not use morphine.

For medical emergencies involving this product, call toll free 1-800-441-3637.

ENVIRONMENTAL HAZARDS

This product is toxic to fish, birds, and other wildlife. Keep out of any body of water. Do not apply where runoff is likely to occur. Do not apply when weather conditions favor drift from areas treated. Do not contaminate water by cleaning of equipment or disposal of wastes.

This product is toxic to bees exposed to direct application. Do not apply this product while bees are actively visiting the treatment area. Time applications to coincide with periods of minimum bee activity.

PHYSICAL OR CHEMICAL HAZARDS

Combustible. Do not use or store near heat or open flame. Keep container closed. Use with adequate ventilation.

GENERAL INFORMATION

Scouting — Monitoring insect populations to determine whether or not there is a need for application of "LANNATE" LV is recommended. More than one treatment of "LANNATE" LV may be required to control a population of pests.

Insect Predators — "LANNATE" LV at rates of 1/2 to 3/4 pt. per acre helps conserve certain beneficials, including big-eyed bugs, damsel bugs, flower bugs, and spiders in cotton and soybeans. While these beneficials cannot be relied upon to control pests, they are of potential value and should be monitored along with pests in pest management programs on these crops.

Compatibility — Since formulations may be changed and new ones introduced, it is recommended that users premix a small quantity of a desired tank mix and observe for possible adverse changes (settling out, flocculation, etc.). Avoid mixtures of several materials and very concentrated spray mixtures.

Do not use "LANNATE" LV with Bordeaux mixture, "Du Ter" (triphenyltin hydroxide), lime sulfur, "Rayplex" iron nor in highly alkaline solutions. Use mildly alkaline mixtures immediately after mixing to prevent loss of insecticidal activity.

DIRECTIONS FOR USE

It is a violation of Federal Law to use this product in a manner inconsistent with its labeling.

Du Pont "LANNATE" LV Insecticide should be used only in accordance with recommendations on this label, or in separate published Du Pont recommendations available through local dealers.

Du Pont will not be responsible for losses or damages resulting from use of this product in any manner not specifically recommended by Du Pont. User assumes all risks associated with such non-recommended use.

"LANNATE" LV is a water soluble liquid to be diluted with water for application by ground or air equipment. Use only in commercial plantings; do not use in home plantings.

Do not enter treated areas without appropriate protective clothing until sprays have dried. Because certain states may require more restrictive reentry intervals for various crops treated with this product, consult your State Department of Agriculture for further information.

Written or oral warnings must be given to workers who are expected to be in a treated area or in an area about to be treated with this product. When oral warnings are given, warnings shall be given in a language customarily understood by workers. Oral warnings must be given if there is reason to believe that written warnings cannot be understood by workers.

Written (or oral) warnings must include the following information: "DANGER — POISON. Area treated with "LANNATE" LV on (date of application). Do not enter treated areas without appropriate protective clothing until sprays have dried. If warning symptoms appear, refer to Statement of Practical Treatment on front panel of "LANNATE" LV label and seek medical attention at once."

SPRAY PREPARATION—Fill spray tank 1/4 to 1/2 full of water. Add "LANNATE" LV directly to spray tank and mix thoroughly, using mechanical or hydraulic means; do not use air agitation. Spray mix should not be stored overnight in spray tank.

Do not apply this product through any type of irrigation system.

APPLICATION—Apply at the recommended rates when insects first appear. Unless otherwise noted, continue application at 5-7 day intervals or as needed. Use sufficient water to obtain thorough, uniform coverage.

For aerial, use a minimum of 2 gallons per acre (gpa) except 10 gpa for grapes, nectarines and strawberries; 15 gpa for oranges, lemons, grapefruit, tangelos and tangerines.

"LANNATE" LV is recommended for use as a low volume aerial spray 0.53 gpa (2L) for cotton* and soybeans* and 1 gpa for the crops listed below providing the following conditions are met:

- equipment is capable of delivering small spray droplets to insure thorough coverage,
- equipment is adjusted to distribute spray uniformly over the spray swath,
- wind conditions and other factors such as temperature and humidity are such that the spray is delivered to the target area,
- local regulations do not prohibit low-volume aerial sprays,
- use rates are applied as directed on the package label or supplemental labeling for the following crops:

Alfalfa	Carrot	Melons	Rye
Anise	Cauliflower	Mint	Sorghum
Asparagus	Celery	Oats	Soybean
Barley	Collards	Peanuts	Spinach
Beans	Corn	Peas (succulent)	Sugar beet
Broccoli	Cotton	Peppermint	Summer
Brussels sprouts	Cucumber	Peppers	squash
Cabbage	Lettuce	Potato	Wheat

Apply the low rates on small plants, small insects and light infestations of insects. Use intermediate rates on large insects and heavier infestations of insects. Use 1 to 3 applications of the highest recommended rate for controlling severe infestations. Thereafter, use the lowest rate possible to maintain control.

*Not Registered for aerial application in a diluted volume of less than 1 gal in CA.

NOTICE TO BUYER — Purchase of this material does not confer any rights under patents of countries outside of the United States.

STORAGE AND DISPOSAL

STORAGE: Do not subject to temperatures below 32°F. Store product in original container only, away from other pesticides, fertilizer, food or feed. Not for use or storage in or around the home.

PESTICIDE DISPOSAL: Do not contaminate water, food, or feed by storage or disposal. Pesticide wastes are acutely hazardous. Improper disposal of excess pesticide, spray mixture, or rinsate is a violation of Federal Law. If these wastes cannot be disposed of by use according to label instructions, contact your State Pesticide or Environmental Control Agency, or the Hazardous Waste representative at the nearest EPA Regional Office for guidance.

CONTAINER DISPOSAL: Triple rinse (or equivalent) the container. Then offer for recycling or reconditioning, or puncture and dispose of in a sanitary landfill, or by other procedures approved by state and local authorities.

NOTICE TO BUYER—Purchase of this material does not confer any rights under patents of countries outside of the United States.

NOTICE OF WARRANTY

Du Pont warrants that this product conforms to the chemical description on the label thereof and is reasonably fit for the purposes stated on such label only when used in accordance with the directions under normal use conditions. It is impossible to eliminate all risks inherently associated with the use of this product. Crop injury, ineffectiveness, or other unintended consequences may result because of such factors as weather conditions, presence of other materials, or the manner of use or application, all of which are beyond the control of Du Pont. In no case shall Du Pont be liable for consequential, special or indirect damages resulting from the use or handling of this product. All such risks shall be assumed by the buyer. DU PONT MAKES NO WARRANTIES OF MERCHANTABILITY OR FITNESS FOR A PARTICULAR PURPOSE NOR ANY OTHER EXPRESSED OR IMPLIED WARRANTY EXCEPT AS STATED ABOVE.

CROPS	INSECTS	RATES "LANNATE" LV PINTS PER ACRE	LAST APPLICATION—DAYS TO HARVEST	TO LIVESTOCK GRAZING/ FEEDING
Alfalfa (Do not apply to dormant or semi-dormant alfalfa when min. daily temp is 50°F. or lower.)	Pea Aphid, Lygus Bugs, Blotch Leafminer, Aphids, Egyptian Alfalfa Weevil Larvae	1½-3		7
	Fall Armyworm	1½		
	Alfalfa Weevil Larvae	3		
	Loopers, Yellowstriped Armyworm, Western Yellowstriped Armyworm, Armyworm	3/4-1½		
	Variegated Cutworm, Beet Armyworm	3/4-3		
	Alfalfa Caterpillar	1-1½		
Anise (Fennel)	Cabbage Looper	3	7	
	Beet Armyworm	1½-3		
Apple (Do not use on Early MacIntosh & Wealthy varieties) **G	Apple Aphid	3/8-1½/100 gals. up to 400 gals. per acre	8	10
	Rosy Apple Aphid, Tufted Apple Budmoth, Green Fruitworm, Tarnished Plant Bug			
	Codling Moth (10-12 day intervals)	3/4-1½/100 gals. up to 400 gals. per acre		
	Leafrollers (Fruittree, Obliquebanded, Redbanded, Variegated), Lesser Appleworm, White Apple Leafhopper, Tentiform Leafminer	1½/100 gals. up to 400 gals. per acre		
	Cutworm	1½/100 gals.		
Asparagus	Beet Armyworm, Western Yellowstriped Armyworm, Asparagus Beetle, Spotted Asparagus Beetle, White Cutworm, Redbacked Cutworm	1½-3	1	

CROPS	INSECTS	RATES	LAST APPLICATION—DAYS	
		"LANNATE" LV PINTS PER ACRE	TO HARVEST	TO LIVESTOCK GRAZING/ FEEDING
	Variegated Cutworm	1½		
Avocado	Avocado Looper, Avocado Leafroller, Omniverous Leafroller	1½-3	1	10
Barley	Armyworm, Cereal Leaf Beetle, Aphids	¾-1½	7	10
Beans (Includes Blackeyed Peas & Cowpeas)	Leafhopper, Mexican Bean Beetle	¾-3	Succulent Beans ¾-1½ pts.—1, over 1½ pts.—3; Dry Beans—25	3 Vines 7 Hay
	Fall Armyworm, Variegated Cutworm	1½		
	Beet Armyworm, Loopers, Corn Earworm, Saltmarsh Caterpillar, Yellowstriped Armyworm, Western Yellowstriped Armyworm, Lygus Bugs, Thrips, Aphids	1½-3		
	European Corn Borer***—Initiate when moth flights first appear and continue preventive treatments at 3-4 day intervals to control egs and larvae			
	Spotted Cucumber Beetle	¾-1½		
Beets (Table)	Imported Cabbageworm, Beet Armyworm, Cabbage Looper, Diamondback Moth	¾-3	0-Roots 14-Tops	
	Cucumber Beetle, Variegated Cutworm	1½		
Bermudagrass	Fall Armyworm, Armyworm	¾-3		7 Forage 3 Dehydrated Hay
Blueberries****	Blueberry Leafhopper, Aphids, Tussock Moth, Weevil, Sharp–Nosed Leafhopper	1½	3	
	Cranberry Fruitworm, Cherry Fruitworm **G	1½-3		
	Flea Beetle (larvae), Sawfly (larvae), Blueberry Leafroller	3		
	Blueberry Maggot	¾-1½		
Broccoli	Loopers, Imp. Cabbageworm, Diamondback Moth	¾-3*	3	
Brussels Sprouts	Loopers, Imp. Cabbageworm, Diamondback Moth	1½-3*	3	
	Variegated Cutworm	1½*		
Cabbage	Loopers, Imp. Cabbageworm, Diamondback Moth	¾-3*	1	
	Fall Armyworm, Variegated Cutworm	1½*		
Carrot	Beet Armyworm	1-3	1	
	Aster Leafhopper	1½-3		
	Armyworms	¾-1½		
	Variegated Cutworm	¾-1½		
Cauliflower	Imp. Cabbageworm, Loopers, Diamondback Moth	¾-3*	3	
	Variegated Cutworm	1½*		
Celery	Beet Armyworm, Aster Leafhopper	1½-3	7	
	Loopers	3		
	Variegated Cutworm	1½		
	Armyworms	¾-1½		
Chicory	Beet Armyworm, Variegated Cutworm, Leafhoppers	1½-3	80	
Chinese Cabbage **G	Loopers, Beet Armyworm	1½-3	10	
Collards (Fresh market only—do not apply when temp. is less than 50°F. Do not apply when crop is less than 10" tall.)	Loopers, Diamondback Moth, Variegated Cutworm	1½	10	
	Imported Cabbageworm, Beet Armyworm	1½-3		
Corn (Sweet) Note: Certain hybrid varieties of sweet corn are susceptible to methomyl injury. Treat a small area to determine crop safety before full scale spraying.	Earworm—Whorl as needed	1-1½	0 Ears	3 Forage
	Fall Armyworm, Armyworm, Earworm (Ovicide/Larvicide), European Corn Borer—Ears 1-3 days or as needed, Corn Rootworm (adult beetles), Flea Beetles, Picnic Beetles, Aphids	¾-1½		
	Variegated Cutworm, Beet Armyworm	1½		
Corn (Field & Popcorn)	Earworm, Armyworm, Fall Armyworm, Corn Rootworm (adult beetles), Flea Beetles, Picnic Beetles, Aphids	¾-1½	0 Ears	3 Forage
	Variegated Cutworm, Beet Armyworm	1½		
Cotton (Use may redden cotton—If excessive, stop or alternate with other insecticides.) U.S.	**Ovicide/Larvicide** Bollworm, Tobacco Budworm (Initiate schedule when significant numbers of eggs are present. Continue at 5- to 7-day intervals while eggs are present and larval control adequate. If significant larvae survive, use higher rates below) Lygus Bugs (adults and nymphs) start treatment on low level population for suppression	⅖-¾ (See Predators)		
	Cotton Leafworm	¾-1½		
	Aphids, Thrips	¾		

CROPS	INSECTS	RATES: "LANNATE" LV PINTS PER ACRE	LAST APPLICATION—DAYS: TO HARVEST	LAST APPLICATION—DAYS: TO LIVESTOCK GRAZING/ FEEDING
East of Rockies only.	Cotton Fleahopper (as needed)	2/5-3/4		
	Larvicide for worms Bollworm, Tobacco Budworm, Beet Armyworm, Cotton Leafperforator, Fall Armyworm, Lygus Bugs (adults and nymphs) use as occasional spray in regular schedule but not more often than every 10 days.	1 1/2	15	Do not Graze or Feed
	Bollworm, Tobacco Budworm, Beet Armyworm, Cotton Leafperforator, Fall Armyworm, Lygus Bugs (adults and nymphs) up to 3 applications at 3-5 day intervals after desired boll load set on plants.	1 1/2-2 1/4		
West of Rockies only	**Larvicide for worms** Bollworm, Fall Armyworm, Tobacco Budworm, Lygus bugs	1 1/2-2 1/4		
	Beet Armyworm	7/8-1 1/2		
	Cotton Leafperforator	7/8-2 1/4		
Southwest only	Cotton Fleahopper (as needed)	2/5-3/4		
Cucumber	Loopers, Tobacco Budworm, Beet Armyworm, Yellowstriped Armyworm, Granulate Cutworm, Flea Beetles, Cucumber Beetles, Melon Aphid, Melonworm, Pickleworm	1 1/2-3	1 1/2 pts.—1 Over 1 1/2 pts.—3	
	Fall Armyworm, Variegated Cutworm	1 1/2		
Eggplant	Green Peach Aphid	3/4-3	5	
	Tomato Pinworm **G, Beet Armyworm	1 1/2-3		
Endive, Escarole	Beet Armyworm	1 1/2-3	10	
Forest Woodlots, Plantations, and Nurseries **G	Spruce Budworm (treat when larvae are feeding on buds, retreat after 3-4 days if needed)	3/4-1 1/2		
	Forest Tent Caterpillar, Western Tent Caterpillar	1 1/2-3		
Garlic	Beet Armyworm	1 1/2*	7	
Grapefruit — CA & AZ only	Thrips, Fruittree Leafroller, Orange Tortix, Western Tussock Moth	1 1/2-3	1	10
U.S.	Beet Armyworm			
Fl only	Spirea Aphid			
	Whiteflies	3		
CA, AZ, & FL only	Yellow Scale, Chaff Scale, Purple Scale, Soft Brown Scale, Glover Scale	1 1/2/100 gals. up to 800 gals. per acre		
Grapes — East of Rockies only	Grape Berry Moth (apply pre- & post-bloom & repeat at 10-14 day intervals or as needed), Leafhoppers **G	1 1/2-3	1 Fresh & Raisin 14 Wine Grapes	
U.S.	Omniverous Leafroller, Grape Leaffolder, Orange Tortrix	1 1/2-3		
West of Rockies only	Saltmarsh Caterpillar Leafhoppers **G	1 1/2-3		
	Climbing Cutworm (Treat infestation and repeat at 7-14 day intervals as needed)	3		
CA only	Grape Leaf Skeletonizer **G	1 1/2-3		
	Thrips	3/8-3/4		
Horseradish **G	Aphids, Thrips	1 1/2	65	
Leafy Green Vegetables: Beet (Tops), Dandelions, Kale, Mustard Greens, Parsley, Swiss Chard, Turnip Greens	Beet Armyworm, Cabbage Looper, Diamondback Moth, Imported Cabbageworm	1 1/2-3	10	
Lemon — CA & AZ only	Thrips, Western Tussock Moth, Orange Tortrix	1 1/2-3	1	10
U.S.	Beet Armyworm			
FL only	Spirea Aphid			
	Whiteflies	3		
CA, AZ, & FL only	Yellow Scale, Chaff Scale, Purple Scale, Soft Brown Scale, Glover Scale	1 1/2/100 gals. up to 800 gals. per acre		
Lentils	Western Yellowstriped Armyworm	1 1/2-3	21	
Lettuce, (head varieties in U.S. and Leaf varieties in CA and AZ)	Beet Armyworm, Loopers	3/4-3	3/4-1 1/2 pts.—7 over 1 1/2 pts.—10	
	Thrips, Aphids, Corn Earworm, Aster Leafhopper	1 1/2-3		
	Variegated Cutworm	1 1/2		
Melons	Loopers, Tobacco Budworm, Beet Armyworm, Yellowstriped Armyworm, Granulate Cutworm, Flea Beetles, Cucumber Beetles, Melon Aphid, Melonworm, Pickleworm	1 1/2-3	1 1/2 pts.—1 day over 1 1/2 pts.—3 days	
	Fall Armyworm, Variegated Cutworm	1 1/2-3		
Mint	Variegated Cutworm, Alfalfa Looper	3	14	
	Flea Beetles	2 1/4-3		
Nectarine — CA & AZ only	Thrips (up to 3 applications)	1 1/2-3	1	10
Oats	Armyworms, Cereal Leaf Beetle, Aphids	3/4-1 1/2	7	10

CROPS	INSECTS		RATES "LANNATE" LV PINTS PER ACRE	LAST APPLICATION—DAYS TO HARVEST	LAST APPLICATION—DAYS TO LIVESTOCK GRAZING/ FEEDING
Onions	Beet Armyworm		1½-3*	7 Green and Dry Onions	
	Thrips, Variegated Cutworm		3*		
Oranges CA & AZ only	Thrips, Western Tussock Moth, Orange Tortrix, Fruitree Leafroller, Citrus Cutworm		1½-3	1	10
U.S.	Beet Armyworm				
FL only	Spirea Aphid				
	Whiteflies		3		
CA, AZ, & FL only	Yellow Scale, Chaff Scale, Purple Scale, Soft Brown Scale, Glover Scale		1½/100 gals. up to 800 gals. per acre		
Peaches	Catfacing Insects (Plant Bugs and Stink Bugs)—begin at petal fall and continue in cover sprays at 7- to 10-day intervals		3-6 (or ¾-1½/100 gals. up to 400 gals. per acre)	4	10
	Oriental Fruit Moth—begin at petal fall, use trapping devices and frequent field inspection to determine need for treatment. Continue treatment in cover sprays and alternate with residual-type insecticides registered for this use. **G				
	Green Peach Aphid		2¼-6		
Peanuts	Corn Earworm***, Beet Armyworm, Potato Leafhopper—Up to 3 applications		¾-3	21	Do not feed treated vines
	Loopers, Green Cloverworm, Velvetbean Caterpillar, Thrips, Granulate Cutworm—Up to 3 applications.		1½-3		
	Fall Armyworm		¾-1½		
Pears Northeast only	Green Fruitworm, Obliquebanded Leafroller—One application only		1½/100 gals. up to 400 gals./acre	7	10
Peas (Succulent)	Armyworm, Green Cloverworm		¾-1½	1	5 forage 14 Hay
	Loopers, Pea Aphid, Beet Armyworm, Saltmarsh Caterpillar, Variegated Cutworm		1½-3		
	Alfalfa Caterpillar		¾-3		
Pecans Southeast only	Aphids		1½-3	30	10
Peppers	Green Peach Aphid, Fall Armyworm, Loopers, Beet Armyworm		1½	3	
	Armyworm, Variegated Cutworm		¾-1½		
Pomegranates	Omniverous Leafroller		3	14	
Potato	Tuberworm, Loopers, Aphids, Beet Armyworm		1½-3	6	
	Leafhoppers Variegated Cutworm, Fall Armyworm, Flea Beetles		1½		
Rye	Armyworms, Cereal Leaf Beetle, Aphids		¾-1½	7	10
Sorghum	Sorghum Webworm		1½	14	14
	Sorghum Midge—Apply when 50% bloom and 3-5 days later if needed		¾-1½		
	Fall Armyworm (Budworm), Beet Armyworm, Corn Earworm, Armyworm				
Soybean† U.S.	Green Cloverworm, Velvetbean Caterpillar, Mexican Bean Beetle	Light to moderate infestations	2/5-3/4 (see Predators)	14	
		Moderate to severe infestations	¾-1½		under 1½ pts. 3 Forage 7 Hay
	Corn Earworm (worms up to ½" long)	Moderate to severe infestations			
	(worms up to ¼" long)	Light to moderate infestations	2/5-3/4		1½-3 pts. 10 Forage 12 Hay
	Beet Armyworm, Salt Marsh Caterpillar, Fall Armyworm, Bean Leaf Beetle, Thrips	Light to moderate infestations	¾-1		
		Moderate to severe infestations	1-1½		
AR, AL, LA, MS, TN, & TX only	Soybean Looper (worms up to ½" long)	Use higher rate for severe infestations	1½-3		
Spinach (Do not apply when min. daily temp. is 32°F. or lower. Do not apply to seedlings less than 3" diameter.)	Loopers, Beet Armyworm		1½-3	7	
	Fall Armyworm, Variegated Cutworm		1½		
Strawberry	Beet Armyworm, Thrips, Aphids		1½-3	3-Fresh	
	Lygus Bugs, Omniverous Leaf Tiers		3	10-Fruit	
	Armyworms		2¼-3	Processing	
Sugar Beet	Beet Webworm, Flea Beetles, Carrion Beetles, Beet Armyworm, Aphids, Western Yellowstriped Armyworm—Up to 4 applications		¾-3	7	30 Tops
	Variegated Cutworm		1½		

CROPS		INSECTS	RATES	LAST APPLICATION—DAYS	
			"LANNATE" LV PINTS PER ACRE	TO HARVEST	TO LIVESTOCK GRAZING/ FEEDING
Summer Squash		Loopers, Tobacco Budworm, Beet Armyworm, Yellowstriped Armyworm, Granulate Cutworm, Flea Beetles, Cucumber Beetles, Melon Aphid, Melonworm, Pickleworm, Fall Armyworm	1½-3	1½ pts.—1 day over 1½ pts. 3 days	
Tangelo	CA & AZ only	Thrips, Western Tussock Moth, Orange Tortrix	1½-3	1	10
	U.S.	Beet Armyworm			
	FL only	Spirea Aphid			
	FL only	Whiteflies	3		
	CA, AZ, & FL only	Yellow Scale, Chaff Scale, Purple Scale, Soft Brown Scale, Glover Scale	1½/100 gals. up to 800 gals. per acre		
Tangerine	CA & AZ only	Thrips, Western Tussock Moth, Orange Tortrix	1½–3	1	10
	U.S.	Beet Armyworm			
	FL only	Spirea Aphid			
	FL only	Whiteflies	3		
	CA, AZ, & FL only	Yellow Scale, Chaff Scale, Purple Scale, Soft Brown Scale, Glover Scale	1½/100 gals up to 800 gals. per acre		
Tobacco (Except shade)		Flea Beetle, Hornworm	¾-1½	5 Flue cured	
		Loopers, Aphids, Tobacco Budworm, Fall Armyworm	1½	14 Air or Fire cured	
Tomato		Tomato Fruitworm, Aphids, Hornworm, Loopers, Beet Armyworm, Southern Armyworm	1½-3	1	
		Pinworm			
		Fall Armyworm, Variegated Cutworm	1½		
		Armyworm	¾-1½		
Turf **G		Sod Webworm (after application, sprinkle irrigate for 15 minutes)	3-6 (1.1-2.3 fl. ozs. per 1000 sq. ft.)		Do not graze or feed
Watercress **G		Beet Armyworm, Cabbage Looper, Diamondback Moth, Imported Cabbageworm (Use in minimum of 30 gpa of water. Water should be turned off 24 hours prior to application.)	1½-3	10	
Wheat		Armyworm, Cereal Leaf Beetle, Aphids	¾-1½	7	10

† Not registered for aerial application in a diluted volume of less than 1 gallon per acre in California.
* Add wetting agent.
** All registrations for aerial and ground use unless marked G for ground use only.
*** Ovicidal and larvicidal control.
**** Do not apply during bloom.

AG 100884 10/12/84

This is a specimen label. DuPont is not responsible for the accuracy of the information contained herein. As labels are subject to revision, always carefully read and follow the label on the product container.

LEXONE® DF

HERBICIDE DISPERSIBLE GRANULES

ACTIVE INGREDIENT:	BY WEIGHT
Metribuzin[4-Amino-6-(1,1-dimethylethyl)-3-(methylthio)-1,2,4-triazin-5(4H)-one]	75%
INERT INGREDIENTS:	25%
TOTAL	100%

EPA Reg. No. 352-390
U.S. Pat. 4,346,220

KEEP OUT OF REACH OF CHILDREN

CAUTION

PRECAUTIONARY STATEMENTS

HAZARDS TO HUMANS AND DOMESTIC ANIMALS

CAUTION! MAY BE HARMFUL IF INHALED OR SWALLOWED

Avoid breathing dust or spray mist. Avoid contact with skin, eyes and clothing.

Wash thoroughly with soap and warm water after handling. Wash contaminated clothing with soap and hot water before reuse.

STATEMENT OF PRACTICAL TREATMENT

If in eyes: Flush with plenty of water. Get medical attention if irritation persists.

If on skin: Wash with plenty of soap and water. Get medical attention if irritation persists.

For medical emergencies involving this product, call toll free 1-800-441-3637.

ENVIRONMENTAL HAZARDS

Do not apply directly to water or wetlands. Do not contaminate water by cleaning of equipment or disposal of wastes.

GENERAL INFORMATION

Du Pont "LEXONE" DF Herbicide is a dispersible granule to be mixed in water or liquid fertilizer and applied as a spray for selective control of certain broadleaf weeds and grasses in alfalfa, asparagus, bermudagrass turf, corn, fallow potatoes, sainfoin, soybeans, sugarcane and tomatoes. It is noncorrosive to equipment, nonflammable, and nonvolatile. For soybeans, "LEXONE" DF may be used in conjunction with "Treflan"[1], "Sonalan"[1], "Surflan"[1], "Lasso"[2], "Roundup"[2], "Bronco"[2], "Dual"[3], "Prowl"[4], Paraquat[5] or "Amiben"[6]. Observe all cautions and limitations on labeling of all products used in mixtures.

The degree of control and duration of effect will vary with:

Soil Organic Matter Content—High organic matter requires higher dosages than low organic matter for equivalent herbicide performance.

Soil Texture—High clay content requires higher dosage rates than those with less clay.

Soil pH—Activity of "LEXONE" DF increases as the pH increases.

Rainfall—Moisture is necessary to activate the herbicide; best results are obtained if treatment is made to moist soil and moisture is supplied by rainfall or sprinkler irrigation (¼" to ½") within one week after application. If moisture is insufficient to activate the herbicide, a shallow cultivation (rotary hoe preferred) should be made after emergence of crop while weeds are small enough to be controlled by mechanical means. If heavy rains occur within 3 weeks of planting, injury to the crop may occur; after 3 weeks, any effects are likely to be negligible.

Amount Applied—The higher recommended rate is more active than the lower rate on a given soil.

Other Conditions—Factors such as incorporation depth of "LEXONE" DF, competition of the crop with weeds, weed species, drought stress of weeds, high salt content of soil, crop seed quality, soil residues of herbicides previously applied, soil insecticides and combination herbicide treatments all can affect performance.

PREEMERGENCE USE (Germinating Weeds):
"LEXONE" DF, when used as directed, controls for an extended time annual weeds such as:

BROADLEAVES*		GRASSES	
Beggarweeds	Pepperweed	"LEXONE" DF at rates above ⅔ lb per acre:	
Wild buckwheat	Palmer pigweed	Foxtails	Cheatgrass
Common chickweed	Prostrate pigweed	Crabgrass	Downy brome
Carpetweed	Redroot pigweed* (careless weed)	Crowfootgrass	
White cockle	Smooth pigweed*	Combinations of "LEXONE" DF with "Treflan", "Sonalan", "Surflan", "Lasso", "Dual", or "Prowl" control annual grasses listed above and:	
Hophornbeam copperleaf	Common ragweed	Barnyardgrass	Crabgrass
Flixweed	Hemp sesbania (coffeebean)	Foxtails	Signalgrass
Henbit	Shepherdspurse	Fall panicum	
Jimsonweed	Sicklepod (cassia, coffeeweed)		
Kochia*	Prickly sida (spiny sida, teaweed)		
Lambsquarters*	Pennsylvania smartweed		
Venice mallow	Spotted spurge		
Meadow salsify	Spurred anoda		
Blue mustard	Yellow rocket		
Tansy mustard	Velvetleaf		
Wild mustard			
* except triazine-resistant weeds			
PARTIAL CONTROL		Combinations of "LEXONE" DF with "Treflan", "Sonalan", or "Surflan" controls annual grasses listed above and:	
Cocklebur	Florida purslane (Florida pusley)	Seedling johnsongrass	
Common purslane			
Sunflower			

Note: Consult "Treflan", "Dual", "Surflan", "Sonalan", "Lasso", "Prowl", Paraquat, "Amiben", "Bronco" or "Roundup" labels for additional weeds controlled when "LEXONE" DF is used in conjunction with these herbicides.

"LEXONE" DF applied before emergence of asparagus, corn, potatoes, soybeans, sugarcane and weeds is an effective procedure because susceptible weeds are controlled in an early, vulnerable seedling stage before they compete with the crop. With favorable moisture conditions, "LEXONE" DF continues to control weeds for some time as the crop becomes better able to compete. Should weed seedlings begin to break through the preemergence treatment in significant numbers, secondary weed control procedures should be implemented; these include cultivation and postemergence herbicide application.

POSTEMERGENCE USE (EMERGED SEEDLING OR PERENNIAL WEEDS):
"LEXONE" DF, when used as directed, controls weeds such as:

BROADLEAVES*		GRASSES
Carpetweed	Mexican weed	Bluegrass
Cocklebur	Mustards	Crabgrass (large, smooth)
Common purslane	Pigweeds (prostrate, redroot, smooth)*	Goosegrass
Common ragweed	Sicklepod (cassia)	
Dayflower	Prickly sida (spiny sida, teaweed)	
Florida beggarweed	Pennsylvania smartweed	
Fumitory	Toadflax	
Hemp sesbania	Russian thistle	
Jimsonweed	Velvetleaf	
Ladysthumb		
Lambsquarters*		
PARTIAL CONTROL		**PARTIAL CONTROL OR SUPPRESSION**
Curly dock	Kochia*	Barnyardgrass (watergrass)
Dandelion	Morningglory	Broadleaf signalgrass
Florida purslane (Florida pusley)	Spotted spurge	Foxtails
Horsenettle	Sunflower	Quackgrass
	Wild poinsettia	

* Except triazine resistant weeds

"LEXONE" DF may be sprayed over the top of certain emerged crops; they are alfalfa (established, dormant), potatoes, sainfoin (dormant), sugarcane and tomatoes (seeded and transplants) to effectively control susceptible emerged seedling weeds.

Directed postemergence applications of "LEXONE" DF, where spray nozzles are adjusted so that weeds are sprayed, but the crop is not, may be used in emerged soybeans (southern and southeastern states) and established tomatoes to control susceptible emerged seedling weeds.

Results of postemergence treatment of emerged weeds vary with rate applied and environmental conditions; best results are obtained on succulent weeds growing under conditions of high humidity and temperatures of 70°F or higher. Addition of a surfactant to the spray (where recommended) increases contact effects of "LEXONE" DF. Application will also provide control of emerging susceptible weed seedlings for an extended period of time.

IMPORTANT—Injury to or loss of desirable trees or other plants may result from failure to observe the following:

Do not apply (except as recommended), or drain or flush equipment on or near desirable trees or other plants, or on areas where their roots may extend, or in locations where the chemical may be washed or moved into contact with their roots. Do not use on lawns, walks, driveways, tennis courts or similar areas. Prevent drift of spray to desirable plants. Do not contaminate any body of water. Keep from contact with fertilizers, insecticides, fungicides and seeds during storage.

Thoroughly clean all traces of "LEXONE" DF from application equipment immediately after use. Flush tank, pump, hoses and boom with several changes of water after removing nozzle tips and screens (clean these parts separately).

Varietal sensitivity to "LEXONE" DF is known in several crops for which "LEXONE" DF is registered. All new crop varieties should be evaluated on a limited basis for their tolerance prior to using "LEXONE" DF.

DIRECTIONS FOR USE

It is a violation of federal law to use this product in a manner inconsistent with its labeling.

Du Pont "LEXONE" DF Herbicide should be used only in accordance with recommendations on this label, or in separate published Du Pont recommendations available through local dealers.

Du Pont will not be responsible for losses or damages resulting from use of this product in any manner not specifically recommended by Du Pont. User assumes all risks associated with such nonrecommended use.

Do not use in Kern County, California.

Metribuzin is a chemical which can travel (seep or leach) through soil and can contaminate ground water which may be used as drinking water. Metribuzin has been found in ground water as a result of agricultural use. Users are advised not to apply metribuzin where the water table (ground water) is close to the surface and where the soils are very permeable, i.e., well drained soils such as loamy sands. Your local agricultural agencies can provide further information on the type of soil in your area and the location of ground water.

SPRAY PREPARATION:

Mixing in Water -

Fill tank 1/4 full with water. Start agitation system, add "LEXONE" DF and continue adding water. Add separately each additional component of any tank mix while adding water. If a surfactant is used, add it last when the tank is nearly full. Continue agitation throughout. If poor mixing should occur with any component, premix the component with two parts water before adding to the spray tank; follow this procedure when mixing "Treflan" or "Prowl". Use spray mixture promptly. "LEXONE" DF plus "Treflan" tank mixtures should not be stored overnight; flush sprayer with water after use and dispose of mix properly.

Mixing in Liquid Fertilizers -

For preemergence application to potatoes or soybeans, nonpressure nitrogen or fertilizer solution may be used in the spray mixture. Small quantities should be tested for compatibility by the following procedures before full scale mixing:

1. Put 1 pint of fertilizer solution in a quart jar.
2. Mix 2 teaspoons "LEXONE" DF with 2 tablespoons of water; mix thoroughly and add to fertilizer solution.
3. Close jar and shake well.
4. If other herbicides are to be used in the mixture, premix 2 teaspoons of wettable powders or 1 teaspoon of liquids with 2 tablespoons of water; add to "LEXONE" DF/fertilizer solution mixture.
5. Close jar and shake well.
6. Watch mixture for several seconds; check again in 30 minutes.
7. If mixture does not separate, foam, gel, or become lumpy, it may be used.
8. Mixing ability may be improved by adding compatibility agents such as Kalo Laboratories "Compex" or Witco Chemicals "Sponto" 168D. Follow directions on container.

Provided the above procedure shows the mixture to be compatible, prepare the tank mixture as follows: add the fertilizer solution to the spray tank first, with the agitator running, add the required amount of "LEXONE" DF and thoroughly mix. For tank mixtures with other herbicides, follow directions above.

EQUIPMENT—SPRAY VOLUMES:

Apply uniformly by air (except as noted) or with a properly calibrated low pressure (20 to 40 psi) boom sprayer equipped with fan-type nozzles and screens no finer than 50 mesh. Use 10 to 40 gals of water per acre with ground equipment (20 to 60 gals with mixture containing paraquat and 20 to 30 gals with mixture containing "Roundup"). Use 2 to 10 gals per acre for aerial application, except use 7 gals per acre for sugarcane (irrigated only) in Hawaii, and 5 to 10 gals per acre for potatoes west of Rocky Mountains and for sugarcane in Louisiana and Texas.

Continuous agitation in the spray tank is required to keep the material in suspension. Avoid overlapping, and shut off spray booms while starting, turning, slowing or stopping, or injury to the crop may result. With aerial equipment, avoid overlapping of spray swath and avoid application under conditions where excessive drift may occur.

For Use In Chemigation Systems (Potatoes Only)

Apply this product only through sprinkler (including center pivot and wheel roll) irrigation systems. Do not apply this product through any other type of irrigation system. Crop injury, lack of effectiveness, or illegal pesticide residues in the crop can result from nonuniform distribution of treated water. If you have questions about calibration, you should contact State Extension Service specialists, equipment manufacturers or other experts. Do not connect an irrigation system (including greenhouse systems) used for pesticide application to a public water system unless the pesticide label prescribed safety devices for public water systems are in place. A person knowledgeable of the chemigation system and responsible for its operation, or under the supervision of the responsible person, shall shut the system down and make necessary adjustments should the need arise.

Chemigation Systems Connected to Public Water Systems

Public water systems means a system for the provision to the public of piped water for human consumption if such system has at least 15 service connections or regularly serves an average of at least 25 individuals daily at least 60 days out of the year. Chemigation systems connected to public water systems must contain a functional, reduced-pressure zone, backflow preventer (RPZ) or the functional equivalent in the water supply line upstream from the point of pesticide introduction. As an option to the RPZ, the water from the public water system should be discharged into a reservoir tank prior to pesticide introduction. There shall be a complete physical break (air gap) between the outlet end of the fill pipe and the top or overflow rim of the reservoir tank of at least twice the inside diameter of the fill pipe. The pesticide injection pipeline must contain a functional, automatic, quick-closing check valve to prevent the flow of fluid back toward the injection pump. The pesticide injection pipeline must contain a functional, normally closed, solenoid-operated valve located on the intake side of the injection pump and connected to the system interlock to prevent fluid from being withdrawn from the supply tank when the irrigation system is either automatically or manually shut down. The system must contain functional interlocking controls to automatically shut off the pesticide injection pump when the water pump motor stops, or in cases where there is no water pump, when the water pressure decreases to the point where pesticide distribution is adversely affected. Systems must use a metering pump, such as a positive displacement injection pump (e.g., diaphragm pump) effectively designed and constructed of materials that are compatible with pesticides and capable of being fitted with a system interlock. Do not apply when wind speed favors drift beyond the area intended for treatment.

Sprinkler Chemigation

The sprinkler chemigation system must contain a functional check valve, vacuum relief valve, and low pressure drain appropriately located on the irrigation pipeline to prevent water source contamination from backflow. The pesticide injection pipeline must contain a functional, automatic, quick-closing check valve to prevent the flow of fluid back toward the injection pump. The pesticide injection pipeline must also contain a functional, normally closed, solenoid-operated valve located on the intake side of the injection pump and connected to the system interlock to prevent fluid from being withdrawn from the supply tank when the irrigation system is either automatically or manually shut down. The system must contain functional interlocking controls to automatically shut off the pesticide injection pump when the water pump motor stops. The irrigation line or water pump must include a functional pressure switch which will stop the water pump motor when the water pressure decreases to the point where pesticide distribution is adversely affected. Systems must use a metering pump, such as a positive displacement injection pump (e.g., diaphragm pump) effectively designed and constructed of materials that are compatible with pesticides and capable of being fitted with a system interlock. Do not apply when wind speed favors drift beyond the area intended for treatment.

Agitation is recommended in the pesticide spray tank. See "SPRAY PREPARATION" for specific details on agitation and spray preparation.

Directions regarding application of metribuzin into the water application and of quantity of water to be applied per acre are described in the crop section "Potatoes".

USE RATES:

All rates are expressed as broadcast rates per acre; for band treatment, use proportionately less. For example, for 14" band on 42" row, use 1/3 of broadcast rate. Where a range of dosages is given, use the lower rate on coarser soils (low in clay or organic matter) and the higher rate on finer soils (high in clay or organic matter).

REPLANTING RECOMMENDATIONS:

If initial seeding fails to produce a stand, crops registered for the rate of "LEXONE" DF that has been applied may be replanted into the treated area. Do not retreat during the same crop year as injury to the crop may result.

CROP ROTATION RECOMMENDATIONS:

Unless otherwise noted, observe the following:

Areas in the Northwest where up to 1/2 lb "LEXONE" DF per acre has been used on lentils and peas may be recropped to wheat and barley after 4 months.

Areas treated with up to 2/3 lb "LEXONE" DF per acre may be recropped to wheat and barley 8 months after treatment.

Following preemergence treatment of soybeans, winter wheat may be planted 4 months after application. Areas in the Midsouth where soybeans have been treated preemergence with up to 1 lb "LEXONE" DF per acre may be recropped to cotton and rice after 8 months.

Areas treated with up to 1 1/3 lbs "LEXONE" DF per acre may be recropped to alfalfa, asparagus, corn, forage grasses, potatoes, sainfoin, soybeans, sugarcane and tomatoes after 4 months.

Areas treated with up to 1 1/3 lbs "LEXONE" DF may be recropped to wheat and barley after 10 months.

Areas treated with 1 1/3 to 2 2/3 lbs "LEXONE" DF per acre may be recropped to soybeans after 4 months; 10 months after application, alfalfa, asparagus, corn, forage grasses, potatoes, sainfoin and tomatoes may be planted.

Areas treated with 2²/₃ to 5¹/₃ lbs "LEXONE" DF per acre may be recropped to sugarcane 4 months after application; 13 months after application, alfalfa, corn, forage grasses, sainfoin, soybeans, and tomatoes may be planted.

Other Crops: For other crops, do not plant within 12 months; do not plant root crops within 18 months.

Cover crops may be planted any time, but stand reduction may occur; do not graze or harvest these cover crops for feed or food.

Special Conditions: For winter or drought periods, add the number of weeks of winter or drought to the above listed recropping intervals.

For soil pH 7.5 or above, add 2 months to the recropping intervals listed.

Combination Treatments: When recommended tank mix combinations are used, consult the product label for replanting and recropping instructions and observe the directions that are the most restrictive.

CROP USES

ALFALFA AND SAINFOIN (Alone or with Grasses)

Use on alfalfa produced for forage or hay; do not use on alfalfa grown for seed.

Make a single application in the fall after plants become dormant or in the spring before new growth starts. Do not apply during the first growing season after seeding. Do not graze or harvest within 28 days after application. Stress which may be caused in part by low fertility, disease, insects, winter kill, overcutting, drought or frost may increase the chance of crop injury. In Washington, Oregon, Idaho and Utah, partial control of downy brome, flixweed and tansy mustard will occur at the lower use rate; for improved control of these weeds in dormant pure stands of alfalfa only, add ²/₅ to ¹/₂ lb active paraquat per acre.

In California, use only north of U.S. Interstate 80.

Pure Stands of Alfalfa or Sainfoin: Select the rate from the table below:

Soil Texture	Lbs "LEXONE" DF Per Acre: ½ to 2% Organic Matter	Lbs "LEXONE" DF Per Acre: More Than 2% Organic Matter
Coarse Loamy sand, Sandy loam	¹/₂ to ²/₃	²/₃ to ³/₄
Medium Loam, Silt Loam, Silt, Sandy clay, Sandy clay loam	²/₃ to ³/₄	³/₄ to 1
Fine Silty clay, Silty clay loam, Clay, Clay Loam	³/₄ to 1	1 to 1¹/₃

Do not use on sand nor on soils with less than ¹/₂% organic matter as crop injury may result.

Mixed Stands with Grasses: The higher range of rates for a given soil texture and organic matter classification in the table above may provide partial reduction of forage grass stands. Rates of 1 to 1¹/₃ lbs per acre may severely reduce stands of forage grass.

ASPARAGUS

Make a single preemergence application of 1¹/₃ to 2²/₃ lbs per acre in early spring prior to spear emergence. Use the lower rate on coarse textured soils (low in clay or organic matter) and the higher rate on fine textured soils (high in clay or organic matter). If field is to be disked, apply "LEXONE" DF after disking, but before crop emerges.

In the Northwest, apply as above; or make a single preemergence application of ²/₃ to 1¹/₃ lbs "LEXONE" DF per acre in early spring prior to spear emergence. Following harvest but before fern emerges, apply ²/₃ to 2 lbs "LEXONE" DF per acre. Do not exceed 2²/₃ lbs "LEXONE" DF per acre per growing season; or make a single preemergence application of ²/₃ to 1¹/₃ lbs. "LEXONE" DF per acre in early spring prior to spear emergence. Following harvest but before fern emerges, apply a tank-mix combination of ²/₃ to 1 lb "LEXONE" DF plus 1 to 2 lbs. "KARMEX" per acre. For irrigated asparagus, after the initial application of "LEXONE" DF, do not exceed more than ¹/₂" of water for the first irrigation. Additional irrigations may be normal programs.

Do not apply within 14 days of harvest; do not apply after spear emergence. Do not use on newly seeded asparagus nor on young plants during the first growing season after setting crowns.

BERMUDAGRASS

Hawaii (only) - For control of goosegrass in established turf, apply ²/₃ to 1¹/₃ lbs "LEXONE" DF at 3 month intervals, as needed, but do not apply more than 5¹/₃ lbs per acre per year. Use the lower rate where soil pH exceeds 7.5 and the higher on fine-textured soils.

Apply when temperature is between 80 and 90 degrees F, when bermudagrass is growing vigorously and not under stress conditions. Do not water bermudagrass for at least 24 hours after application. Temporary discoloration of bermudagrass may occur.

Avoid spray overlaps.

Do not apply to areas of close-mowed turf nor within the root zone of desirable shrubs, trees, or other ornamentals, as injury may occur.

Do not mow for at least 3 days.

Do not add surfactant.

Do not graze or use grass clippings for animal feed.

Do not use on residual lawns unless under the supervision of a commercial applicator.

CORN (FIELD)

ILLINOIS, INDIANA, IOWA, KANSAS, KENTUCKY, MICHIGAN, MINNESOTA, MISSOURI, NEBRASKA, OHIO, SOUTH DAKOTA AND WISCONSIN only — Apply "LEXONE" DF plus atrazine plus "Lasso" or "Dual" tank mixture after planting but before corn emerges. Plant seed at least 1¹/₂" deep on flat or raised seedbeds only or injury to the crop may result. Soil should be well prepared and as free as possible from trash and clods before planting.

Treatment controls many annual weeds such as velvetleaf, common lambsquarters*, common ragweed, Pennsylvania smartweed, redroot pigweed*, jimsonweed, giant foxtail, yellow foxtail, green foxtail, smooth crabgrass, large crabgrass, barnyardgrass, fall panicum and witchgrass. "LEXONE" DF in the combination improves control of velvetleaf, common lambsquarters* and Pennsylvania smartweed. For additional weeds controlled, see atrazine and "Lasso" or "Dual" labeling.

* Except triazine resistant weeds

SOIL TEXTURE	2% to 6% ORGANIC MATTER ONLY: "LEXONE" DF (LBS)		ATRAZINE (LBS)		"LASSO" (QTS)		OR "DUAL" 8E (PTS)
Medium Loam, Silt loam, Silt, Sandy clay, Sandy clay loam	¹/₄–¹/₃	+	³/₄ to 1¹/₂ (Active)	+	1¹/₂ to 2	or	1¹/₂ to 2
Fine Silty clay, Silty clay loam, Clay, Clay loam	¹/₃	+	³/₄ to 1¹/₂ (Active)	+	2 to 2¹/₂	or	2 to 2¹/₂

Do not use this mixture on coarse soils (sand, loamy sand, sandy loam) as crop injury may occur. Do not use on muck soils.

Do not use on soils having a calcareous surface layer or a pH of 7.5 or higher.

Before use on any inbred parent or variety of corn, tolerance to "LEXONE" DF must first be determined; use only on inbred parents or varieties that are tolerant to "LEXONE" DF.

If initial seeding fails to produce a stand, treated fields may be replanted to corn; do not rework soil. Do not retreat field with a second application as injury to the crop may result.

For other crops that may follow corn, see atrazine and "Lasso" or "Dual" labeling.

FALLOW

Corn - Fall Bedded Fields In California (Only)

Apply ²/₃ to 1 lb "LEXONE" DF per acre in the fall using the lower rate on ¹/₂ to 2% organic matter soils and the higher rates on soils over 2% organic matter. Do not plant corn within 120 days of application to prevent crop injury. At least 2 inches of moisture is needed to activate "LEXONE" DF for weed control and to begin breakdown of the herbicide in soil for crop safety. Do not disturb soil surface after application as this reduces weed control. Do not use in Kern County, CA.

Tomatoes - Fall Bedded Fields In California (Only)

Apply ²/₃ to 1 lb "LEXONE" DF per acre in the fall using the lower rate on ¹/₂ to 2% organic matter soils and the higher rates on soils over 2% organic matter. Do not plant tomatoes within 90 days of application to prevent crop injury. At least 2 inches of moisture is needed to activate "LEXONE" DF for weed control and to begin breakdown of the herbicide in soil for crop safety. Do not disturb soil surface after application as this reduces weed control. Do not use in Kern County, CA.

WHEAT (Reduced Tillage Fallow)

Apply ²/₃ to 1 lb "LEXONE" DF per acre in the fall or apply ¹/₂ to ²/₃ lb in the spring to fallow ground that will be followed by winter wheat. Use the lower rate on coarse textured soils and the higher rates on medium to fine textured (heavier) soils.

Time treatment to new weed growth to control:

Blue mustard*	Fanweed	Miners lettuce
Bur buttercup	Filared	Pigweed (prostrate, redroot)**
Burcherval*	Flixweed*	Pineappleweed
Chickweed (common, mouse-ear, jagged)	Gromwell*	Prickly lettuce
	Henbit	Russian thistle*
Conical catchfly*	Jacobs ladder	Speedwell (ivyleaf)
Corn cockle	Jim Hill mustard	Tansy mustard*
Dog fennel (Mayweed)*	Knotweed	Tarweed (fiddleneck)
False Flax	Kochia**	Treacle mustard
Lambsquarters**	Wild buckwheat*	Wild mustard

Annual ryegrass*	Japanese brome
Cheatgrass	Pacific meadow foxtail (Blackgrass)*
Downy Brome*	Windgrass*
Green foxtail*	Yellow foxtail*

* Weeds and grasses partially controlled. Partial control is an observable reduction in weed competition (reduced weed population and/or vigor) as compared to an untreated area. Degree of partial control can vary with soil type, environmental factors, and size of weeds at time of application.

**Except triazine resistant weeds.

If weeds larger than 2" tall or across are present, use a suitable registered companion herbicide such as "Glean", paraquat, "Round-up" or "Chem Hoe" 135 as a tank mix. Refer to labeling for these products for additional use instructions and precautions.

Where spring application is appropriate in Colorado, Kansas, Montana, Nebraska, North Dakota, Oklahoma, South Dakota, Texas, and Wyoming for improved control of downy brome, Russian thistle, and tansymustard, tank mix 1/3 to 2/3 lbs "LEXONE" DF with 2 to 2.5 lbs "Igran" 80 W per acre.

Do not plant spring seeded cereals following "LEXONE" DF fall fallow application.

Do not apply "LEXONE" DF in the spring if "LEXONE" DF was used the preceding fall or winter.

Use supplemental control procedures such as cultivation or other herbicides if seasonal weed control is desired.

Recrop Intervals - See "Crop Rotation Recommendations".

POTATOES

Preemergence: Make a single application of 2/3 to 1 1/3 lbs per acre after planting or after drag-off but before crop emerges. Use the lower rates on coarse textured soils and the higher rates on fine textured soils (high in clay or organic matter).

(Northeast U.S.): For improved control of annual grasses, apply 2/3 to 1 1/3 lbs "LEXONE" DF plus 1.5 to 3 pints "Dual" 8E per acre after planting or after drag-off but before crop emerges. Use the lower rates on coarse textured soils and the higher rates in fine textured soils (high in clay and organic matter).

Postemergence: Do not use on early maturing smooth-skinned white or red-skinned varieties. Apply only if there have been at least 3 successive days of sunny weather prior to application. Treat before weeds are 1" tall. Make a single application of 2/3 lb per acre. For redroot pigweed and lambsquarters only, use 1/2 to 2/3 lb per acre. Treatment may cause some chlorosis or minor necrosis. Do not apply within 60 days of harvest.

Application by Sprinkler Irrigation: Refer to "Chemigation" in "Directions for Use" before applying product. Apply specified dosage of "LEXONE" DF in 1/4 to 3/4 inch of water per acre (1/4 to 1/2 inch of water per acre on sandy soil) as a continuous injection in center pivot and self-propelled wheel-move systems or in the last 15-30 minutes of set-in permanent solid-set sprinkler systems.

Irrigation systems must be equipped with automatic shut-off devices which prevent back-flow to the water source. Maintain continuous agitation in injection nurse tanks during application. Treatment may be made as a single postemergence or split postemergence application. Check irrigation system to insure uniform application of water to all areas. Failure to apply "LEXONE" DF uniformly may result in crop injury and/or poor weed control.

Preemergence plus Postemergence: Apply as directed above. Do not use more than 1 1/3 lbs total per acre per season. Do not apply within 60 days of harvest.

Postemergence plus Postemergence—(Oregon, Washington, Idaho): Use only if "LEXONE" DF was not applied preemergence. Allow at least 14 days between applications. For best results, apply before weeds are 1" in height or diameter. Do not apply within 24 hours of treatment with other pesticides.

Oregon and Washington—Make 2 applications of 1/3 to 2/3 lb per acre. On coarse (sandy) soils containing 1/2 to 1% organic matter do not exceed 1/3 lb per acre per application.

Idaho—Make 2 applications of 1/6 to 2/3 lb per acre. On coarse (sandy) soils containing 1/2 to 1% organic matter, do not exceed 1/3 lb per acre per application.

CROP ROTATION RECOMMENDATIONS: In the Northwest, wheat and barley may be planted 8 months after an application of "LEXONE" DF made no later than June 30; crops other than wheat, barley, and potatoes cannot be planted before one year after treatment. Cover crops planted to prevent soil erosion or for soil building purposes may be planted anytime after application of "LEXONE" DF but do not graze or harvest for feed or food.

See "Replanting" and "Crop Rotation Recommendations" for further details.

Note: Because potato varieties vary in their resistance to herbicides, determine tolerance to "LEXONE" DF prior to adoption as a field practice to prevent possible crop injury. Crop injury may occur if Acadia Russett, Atlantic, Batoche, Belchip, Campbell 12 and 13, Caribe, Centennial, Cobbler, Gold Rus, Hampton, Jemseg, Monona, MN7973, Norchip, Norgold, NY 64, Ontario, Sebago and Shepody varieties are under stress from lack of moisture, fertility, sunlight or affected by insects or disease when treated postemergence. Use of "LEXONE" DF on sands or loamy sands containing 1/2 to 1 1/2% organic matter may result in crop injury. Do not use on any soil containing less than 1/2% organic matter, as crop injury may result. Preemergence use on soils containing more than 10% organic matter may result in only partial weed control. Postemergence use on rill irrigated potatoes may not provide adequate weed control in absence of rainfall. Do not apply to sweet potatoes or yams.

SOYBEANS

Select one of the following herbicide treatments and apply as directed. The combination treatments control weeds that are susceptible to "LEXONE" DF plus those controlled by the companion products (refer to product labeling). Dosage rates listed are broadcast rates; for band applications use proportionately less.

Note: Soybean varieties such as Altona, Coker 102 and 156, Govan, NKS 1884, Semmes, Tracy, Vansoy, Terra Vig 505 and 606, Agripro 55, Asgrow 6520, Maple Amber, Portage, Vinton 81 and AP 71 are sensitive to "LEXONE" DF. Injury may occur if "LEXONE" DF is used on these varieties.

Varieties showing above average tolerance to "Lexone" are American Revere, Asgrow 1937, Asgrow 3659, Asgrow 3860, DSR 171, DSR 207, Essex, Fayette, Hisoy 170, Lakota, Lawrence, LOL 4207, NKS 1492, Pride B216, Pride B242, SRF 250, SRF 350 P, Union, Wayne, Wells II and Williams 82; for maximum weed control on these varieties, use the higher rate (where a range of rates is listed) for the appropriate soil type and organic matter as shown in the following tables.

Injury to soybeans may occur under the following conditions:

- Improper sprayer calibration
- Over application or uneven application
- Use on very coarse, porous soils
- Soils with pH of 7.5 or higher or those with a calcareous surface layer
- Use with soil-applied organic phosphate insecticides
- Heavy rains within 3 weeks of planting

Seedling disease, cold weather, deep planting (more than 2") excessive moisture, high salt concentration or drought may weaken soybean seedlings and increase possibility of crop injury.

Preemergence To Soybeans

Conventional Tillage

Preplant Incorporated: Incorporate herbicide treatment into soil before planting of soybeans according to instructions on label of product used in conjunction with "LEXONE" DF; improper soil incorporation can result in erratic weed control or crop injury. Do not plant seed more than 2" deep. Use combination as tank mixtures in water. Select the proper rate from the following chart and use it with a "companion" herbicide listed below:

Soil Texture	Pounds "LEXONE" DF Per Acre	
	1/2 to 3% Organic Matter	More Than 3% Organic Matter
Coarse Loamy sand, Sandy loam	1/3	1/2
Medium Loam, Silt Loam, Silt, Sandy clay, Sandy clay loam	1/2	2/3
Fine Silty clay, Silty clay loam, Clay, Clay Loam	2/3	2/3
Mississippi Delta Only— Silty Clay to Heavy clay	1	1

Do not use on sand. Do not use on loamy sand or sandy loam with less than 1% organic matter, nor on other soils with less than 1/2% organic matter as crop injury may result. Do not use on peat. Treated vines may be grazed or fed to livestock 40 days after application unless specified otherwise on the companion product.

Any one of the following "Companion" products may be tank mixed with "LEXONE" DF to improve weed control:

"Treflan", "Dual", "Sonalan"	Apply within 14 days of planting
"Lasso", "Prowl"	Apply within 7 days of planting

Note: Read product labels for weeds controlled, rates to use (etc.), and observe the most limiting use restrictions for the combination selected.

Preplant Non-Incorporated:
"LEXONE" DF + "Surflan" AS or "Surflan" 75W

Preplant Surface Application: Apply within 4 weeks of planting. Treatment also controls dayflower and goosegrass.

Soil Texture Description	Organic Matter	RATE PER ACRE		
		"Lexone" DF Pounds	+ "Surflan" AS Quarts	or "Surflan" 75W Pounds
Coarse: Sandy loam	1 to 5%	1/3	+ 1/2 to 3/4	or 2/3 to 1
Medium: Loam, Silt Loam, Silt, Sandy clay, Sandy clam loam	1/2 to 5%	1/2	+ 3/4 to 1	or 1 to 1 1/3
Fine: Silty clay, Silty clay loam, Clay, Clay loam	1/2 to 5%	2/3 to 1	+ 1 to 1 1/4	or 1 1/3 to 1 2/3
Mississippi Delta Only—Silty clay to Heavy clay	1/2 to 5%	2/3 to 1	+ 1 to 1 1/4	or 1 1/3 to 1 2/3

* For sicklepod control on Coastal Plains soils of Southeast and Midsouth, use 1/3 to 1/2 lbs "LEXONE" DF.

Do not use on sand or loamy sand. Do not use on sandy loam with less than 1% organic matter. Do not use on other soils with less than 1/2% organic matter. Do not use preemergence application on soils with more than 3% organic matter nor the preplant surface application on soils with more than 5% organic matter as poor weed control and/or crop injury may result. Do not use treated vines for feed or forage.

Preplant Incorporation of "Treflan", "Sonalan", or "Prowl", "Lasso" or "Dual" Followed by "LEXONE" DF Preemergence: Apply the preplant incorporated treatment according to directions on product label. As a separate operation, follow with "LEXONE" DF (alone) using rates given under "Preemergence Application" (below):

Preemergence Application: Apply after planting but before soybeans emerge; do not spray over top of emerged soybeans. Plant seed 1 1/2" to 2" deep on flat or raised seedbeds only or injury to the crop may result; soil should be well prepared and as free as possible from trash and clods before planting. Use combinations as tank mixtures in water.

Where a range of "LEXONE" DF dosage rate is listed for a specific soil type and organic matter, use the higher rate for difficult control problems, such as cocklebur, sicklepod, jimsonweed, hemp sesbania, prickly sida or velvetleaf. Use the higher rate also for minimum or no-tillage soybeans, and in the states of Georgia and South Carolina.

Select the appropriate rate of "LEXONE" DF and apply it in combination with products listed below for improved weed control:

Soil Texture	Lbs "LEXONE" DF Per Acre	
	Organic Matter	
	1/2 to 3%	More Than 3%
Coarse Sandy loam	1/6 to 1/3*	1/2
Medium Loam, Silt Loam, Silt, Sandy clay, Sandy clay loam	1/3 to 1/2	2/3
Fine Silty clay, Silty clay loam, Clay, Clay Loam	1/2 to 2/3	2/3
Mississippi Delta Only — Silty clay to heavy clay	2/3 to 1	1

* For sicklepod control on Coastal Plains soils of Southeast and Midsouth, use 1/3 to 1/2 lb "LEXONE" DF.

Do not use on soils with less than 1/2% organic matter, nor on sand, loamy sand or muck soils as crop injury may result. Do not use on peat.

Treated vines may be grazed or fed to livestock 40 days after application.

Any one of the following products may be mixed with "Lexone" to improve weed control: "Lasso", "Dual", "Prowl", "Surflan", and "Amiben".

NOTE: Read the product labels for weeds controlled, rates to use and observe the most limiting use restrictions for the combination selected.

Preplant Incorporated (PPI) Combinations plus Preemergence - Use combinations as tank mixes in water. Promptly incorporate PPI treatment into soil before planting of soybeans. Improper soil incorporation can result in erratic weed control or crop injury. Do not plant seed more than 2" deep. After planting of soybeans and before emergence, apply a preemergence treatment of "LEXONE" DF.

Select a combination as specified in the section "Preplant Incorporated" and apply as directed.

After planting but before emergence of the soybeans, apply a second application of "LEXONE" DF (without a companion product) as follows:

Soil Texture	Lbs "LEXONE" DF Per Acre		
	Percent Organic Matter in Soil		
	Less than 2%	2 to 4%	Over 4%
Coarse Sandy loam	1/6	1/6	1/3
Medium Loam, Silt Loam, Silt, Sandy clay, Sandy clay loam	1/6	1/6 to 1/3	1/3 to 1/2
Fine Silty clay, Silty clay loam, Clay, Clay loam	1/6	1/6 to 1/3	1/3 to 1/2

Do not use on sand. Do not use on loamy sand or sandy loam with less than 1% organic matter, nor on other soils with less than 1/2% organic matter as crop injury may result. Do not use on peat. Treated vines may be grazed or fed to livestock 40 days after application unless specified otherwise on the companion product(s).

Minimum or No-Tillage Soybeans

Preemergence to Crop and Weeds

"LEXONE" DF plus paraquat, "Roundup", "Lasso", "Dual", or "Surflan" may be used as a tank mixture for preemergence and postemergence control of many broadleaved weeds and grasses and for top kill or suppression of established annuals and some perennials where soybeans will be planted directly into a performed bed (stale seedbed), cover crop, or in previous crop residues such as corn or small grain stubble. For improved grass and volunteer small grain control, the addition of "Lasso" (or "Lasso" EC) or "Surflan" to the "LEXONE" DF plus paraquat or "Roundup" tank mixture is recommended. For all combinations, except "Surflan", treated vines may be grazed or fed to livestock 40 days after application.

Apply with ground equipment, during or after planting but before crop emerges. Use 20 to 60 gals of water per acre for paraquat and use the higher gallonage for dense stubble or vegetation. Use 20 gals of water per acre for "Roundup" combinations.

Preemergence to Crop and Postemergence to Weeds:

"LEXONE" DF - Paraquat—Select the rate from the table for preemergence application of "LEXONE" DF and thoroughly mix in spray tank first, then add paraquat as directed under "Paraquat Combinations".

"LEXONE" DF + "Lasso" (or "Lasso" EC) + Paraquat—Select the rate from the table for preemergence application of "LEXONE" DF + "Lasso" tank mixture and thoroughly mix in spray tank first according to directions; then add paraquat as directed under "Paraquat Combinations".

"LEXONE" DF + "Surflan" + Paraquat—Select the rate from the table for preemergence application of "LEXONE" DF + "Surflan" tank mixture preemergence and thoroughly mix in spray tank first according to directions; then add paraquat as directed under "Paraquat Combinations".

Do not use treated vines for feed or forage.

Paraquat Combinations—Select the "LEXONE" DF combination treatment above and mix in the spray tank according to directions; then add paraquat at a rate of 1/4 to 1/2 lb active ingredient per acre for control of emerged weeds. Use the higher rate for weeds 4" to 6" tall. As the last ingredient, add 8 oz "Ortho" X-77 Spreader per 100 gals of spray mixture. Maintain constant agitation.

"Roundup" Combinations ("LEXONE" DF + "Lasso" + "Roundup"): Select the rate for preemergence application of "LEXONE" DF + "Lasso" tank mixture and thoroughly mix in spray tank first according to directions; then add 1 to 2 qts "Roundup" per acre to the mixture and maintain constant agitation. Use the higher rates for maximum top kill of vegetation.

"Bronco" Combinations—Select the rate from the table for preemergence application of "LEXONE" DF and thoroughly mix in spray tank first; then add "Bronco" at rate of 3 1/4 to 5 qts for coarse soils, and 4 to 5 qts for medium or fine soils. Use the higher rate of "Bronco" for dense stubble, heavy crop residues, or heavy weed populations.

Postemergence to Soybeans

Since the degree of weed control with "LEXONE" DF (combinations) preemergence treatment is influenced by weather, weed species, etc. (see General Information), provision should be made for follow-up weed control procedures including cultivation and postemergence herbicides. Where appropriate, follow preemergence application with an early season post-emergence treatment of Du Pont "Classic" Herbicide over-the-top of soybeans for improved control of emerged seedling cocklebur, pigweeds, common ragweed, giant ragweed, sicklepod, hemp sesbania, Florida beggarweed, sunflower, annual morningglories, jimsonweed, annual smartweed, bristly starbur and yellow nutsedge. (See "Classic" package label for use instructions). Follow with cultivation, as needed.

"LEXONE" DF (alone) may also be applied as directed postemergence treatment in Southern and Southeastern states. **DO NOT SPRAY OVER THE TOP OF SOYBEANS.** Apply when soybeans are at least 12 inches tall and when broadleaf weeds do not exceed 3 inches and grasses and common ragweed do not exceed 1 inch in height. Direct spray to cover weeds with minimal or no contact to soybean plant. Crop injury may result if spray is higher than 2 inches on soybean stem. Apply 1/3 to 2/3 lb per acre in 25 to 40 gallons water. Use the lower rate for seedling weeds and the higher rate for larger weeds. Application of 1/3 lb per acre controls Florida beggarweed, carpetweed, cocklebur, dayflower, Mexican weed, prostrate and smooth pigweed, redroot pigweed (carelessweed)*, common purslane, sicklepod (Cassia), velvetleaf and crabgrass (large, smooth); 1/3 to 2/3 lb per acre controls hemp sesbania, prickly sida (spiny sida, teaweed); 2/3 lb per acre controls jimsonweed, Pennsylvania smartweed, common ragweed and gives partial control of morningglory species, horsenettle, Florida purslane (Florida pusley), spotted spurge and wild poinsettia when less than 3 inches tall and broadleaf signalgrass up to 1 inch tall. Do not use on sand, nor on soils with less than 1/2% organic matter as crop injury may result. Add a nonionic surfactant such as "Ortho" X-77 Spreader to the spray mixture to obtain better wetting of weed leaf surfaces. If needed, a second application may be made after 7 days. Apply only with ground equipment designed for precise application of directed sprays such as oiling shoes or gauge wheels. Do not exceed nozzle pressure of 25 psi or use nozzle tips smaller than 8002 T-Jet (or equivalent). For broadcast application, use a single flood-type spray nozzle ("K" series or equivalent) per middle, mounted on an oiling shoe or gauge wheel. For band treatment, use two nozzles per row mounted on oiling shoes or gauge wheels, one on each side of a row. Do not apply under weather conditions which favor drift.

* Except triazine resistant weeds

Treated vines may be grazed or fed to livestock 40 days after application.

SUGARCANE

Hawaii—Make a single broadcast application as follows: for aerial application (irrigated cane only), use at least 7 gals water per acre. Spot treatment may also be made at the rate of 3 1/3 to 6 2/3 lbs per acre, but do not exceed 10 2/3 lbs total per acre per crop cycle. Do not apply within 17 months of harvest. Do not use treated foliage for feed or forage.

	Lbs "LEXONE" DF Per Acre	
Time of Application	**Irrigated**	**Non-Irrigated**
Preemergence—apply within 2 weeks after planting	5 1/3 to 8	2 2/3 to 5 1/3
	—or—	
Early Postemergence—apply over cane before weeds are 3" tall	5 1/3 to 8	2 2/3 to 5 1/3
	—or—	
Postemergence—apply prior to "close in" time and before weeds are 3" tall	2 2/3 to 5 1/3	2 2/3 to 5 1/3

In addition to the weeds listed elsewhere on this label, treatment (non-irrigated cane) controls ageratum, spiny amaranth, wild euphoria, fireweed, Flora's paintbrush, Richardia, garden spurge, graceful spurge, tarweed, guineagrass, plushgrass, ricegrass and wiregrass (goosegrass); in irrigated cane, the foregoing weeds plus spleen amaranth, haole koa, hioloa, hila hila, rattlepod and Alexander grass.

Louisiana and Texas—Apply 2 to 4 lbs per acre during the fall after planting but before cane emerges, or to the stubble after harvest. Make a second application in early spring prior to new cane growth. If necessary, a third application may be made in late spring at layby. Do not apply within 60 days of harvest. Do not use treated foliage for feed or forage.

In addition to the weeds listed elsewhere on this label, treatment controls curly dock, annual morningglory, marestail, sowthistle and winter oats.

Louisiana—For use on fallow ground to be followed by sugarcane, apply 1 1/3 to 2 lbs "LEXONE" DF per acre. Use the lower rate if applied less than 30 days before planting. Application should be made after the rows are built and ground should not be disturbed until planting or loss of control will result. Treatment provides short term control of seedling johnsongrass and other annual weeds (see Preemergence Use). To aid in control of johnsongrass, stubble sugarcane should be destroyed after harvest or in early spring. Land should be disked frequently in the spring to control rhizome johnsongrass. In late May to early June, rows should be built. The soil surface should be free of trash and clods. Application of "LEXONE" DF should be made at this time. Rhizome johnsongrass will not be controlled by "LEXONE" DF; therefore, spot treatment with an appropriate postemergence herbicide will be necessary.

TOMATOES

East of Rocky Mountains—"LEXONE" DF may be applied as a preplant incorporated treatment alone or in a tank mixture with "Treflan" EC in transplanted tomatoes or "LEXONE" DF may be applied alone as a postemergence broadcast or directed spray to established tomatoes including seeded established or transplanted tomatoes. Apply with a properly calibrated low-pressure (20 to 40 psi) fixed-boom sprayer wtih fan type nozzles. Use 10 to 40 gallons of water per acre with ground equipment for preplant incorporated treatment and 20 to 75 gallons for postemergence broadcast or directed spray application. Do not use air blast or other high pressure spray equipment for postemergence spray applications.

Preemergence To Transplanted Tomatoes: Preplant Incorporated "LEXONE" DF alone or "LEXONE" DF Plus "Treflan" EC Tank Mix (Transplant Tomatoes: Except Florida): Apply 1/3 to 2/3 lbs per acre "LEXONE" DF alone or as a tank mix combination with "Treflan" EC as a broadcast spray immediately before transplanting. Use the higher rate for heavy weed pressure or for difficult-to-control weeds. Incorporate uniformly to a depth of 2 to 4 inches. Apply "Treflan" EC at rates of 1 to 2 pts per acre on coarse soils, 1 1/4 to 2 pts per acre on medium soils, and 1 1/2 to 2 pts per acre on fine soils. Consult "Treflan" EC label for specific use instructions and observe all cautions and limitations. When "LEXONE" DF is tank mixed with "Treflan" EC, apply immediately after mixing; do not allow tank mixture to stand for prolonged periods of time.

Postemergence To Established Tomatoes: Broadcast Application (Established Tomatoes; Except Florida): Apply 1/3 to 2/3 lbs "LEXONE" DF per acre over-the-top of established tomato plants. Use the higher rate for heavy weed pressure or for difficult to control weeds. Treat before weeds are 1 inch tall. One or more applications may be made per crop season; allow at least 14 days between applications. Do not apply more than 1 1/3 lbs "LEXONE" DF per acre total within a 35-day period per crop season. Do not treat seeded or transplanted tomatoes until plants have reached the 5 to 6 leaf stage or until transplants have recovered from transplant shock and new growth has started. Apply only if there have been at least 3 successive days of sunny weather prior to application, or crop injury may occur. Do not apply within 24 hours of treatment with other pesticides. Do not tank mix with other pesticides.

Directed Application (Established Tomatoes; East of Rocky Mountains): Apply 2/3 to 1 1/3 lbs "LEXONE" DF per acre as a directed spray according to directions under "Broadcast-Postemergence Application" but do not allow spray to contact tomato foliage. This method of application is recommended for fields with heavy weed pressure of difficult-to-control weed species.

CROP ROTATION RECOMMENDATIONS: For rotation crops in fields treated with "Treflan", follow instructions on product label.

See "Replanting" and "Crop Rotation Recommendations" for further details.

Note: Because tomato varieties vary in their resistance to "LEXONE" DF, determine tolerance prior to adoption as a field practice to prevent possible crop injury. Do not use hot caps on tomatoes within 7 days before or at any time after application of "LEXONE" DF. Crop injury or delayed maturity may result from postemergence or directed spray applications if tomatoes are growing under stress conditions caused by low fertility, disease, insects, drought, cool, wet or cloudy weather prior to application. Do not apply within 7 days of harvest. Do not apply more than 1 1/3 lbs "LEXONE" DF per acre per crop season.

NONCROP USES

For nonselective use, apply 1 1/3 to 8 lbs "LEXONE" DF per acre in 20 to 60 gallons of water. For best results, apply shortly before weed growth begins. Do not use on rights-of-way.

For control of established seedling weeds add:

Paraquat at a rate of 1/4 to 1/2 lb active ingredient per gallon per acre for control of emerged weeds. Apply 20 to 60 gallons of water per acre using the higher spray volume for weeds 4" to 6" tall. As the last ingredient, add 8 ounces "Ortho" X-77 Spreader (or equivalent) per 100 gallons of spray mixture or add 1 to 2 quarts "Roundup" in 20 to 30 gallons of water per acre. Use the higher rates for maximum top kill of vegetation.

1 Registered trademarks of Elanco Products Co.; "Treflan" contains 4 lbs trifluralin per gal, "Surflan" contains 75% oryzalin; "Sonalan" contains 3 lbs ethafluralin per gallon.

1 Registered trademarks of Monsanto Co.; "Lasso" contains 4 lbs alachlor per gal, "Roundup" contains 4 lbs glyphosate per gal; "Bronco" contains 2.6 lbs alachlor and 1.04 lbs glyphosate per gallon.

3 Registered trademark of Ciba-Geigy Corp.; "Dual" 8E contains 8 lbs metolachlor per gal.

4 Registered trademark of American Cyanamid Co.; "Prowl" contains 4 lbs pendimethalin per gal.

5 Paraquat is a restricted use pesticide.

6 Registered trademark of Union Carbide Co.; "Amiben" contains 2 lbs chloramben per gal.

NOTICE TO BUYER: Purchase of this material does not confer any rights under patents of countries outside of the United States.

STORAGE AND DISPOSAL

STORAGE: Store product in original container only, away from other pesticides, fertilizer, food or feed.

PRODUCT DISPOSAL: Do not contaminate water, food or feed by storage or disposal. Wastes resulting from the use of this product may be disposed of on site or at approved waste disposal facility.

CONTAINER DISPOSAL: Triple rinse (or equivalent) the container. Then offer for recycling or reconditioning, or puncture and dispose of in a sanitary landfill, or by incineration, or if allowed by state and local authorities, by burning. If burned, stay out of smoke.

NOTICE OF WARRANTY

Du Pont warrants that this product conforms to the chemical description on the label thereof and is reasonably fit for the purposes stated on such label only when used in accordance with the directions under normal use conditions. It is impossible to eliminate all risks inherently associated with the use of this product. Crop injury, ineffectiveness or other unintended consequences may result because of such factors as weather conditions, presence of other materials, or the manner of use or application, all of which are beyond the control of Du Pont. In no case shall Du Pont be liable for consequential, special or indirect damages resulting from the use or handling of this product. All such risks shall be assumed by the buyer. DU PONT MAKES NO WARRANTIES OF MERCHANTABILITY OR FITNESS FOR A PARTICULAR PURPOSE NOR ANY OTHER EXPRESS OR IMPLIED WARRANTY EXCEPT AS STATED ABOVE.

AG-3253 8049 4/13/89

This is a specimen label. DuPont is not responsible for the accuracy of the information contained herein. As labels are subject to revision, always carefully read and follow the label on the product container.

LEXONE® 4L

HERBICIDE
LIQUID

CONTAINS 4 LBS ACTIVE INGREDIENT PER GALLON

ACTIVE INGREDIENT:	BY WEIGHT
Metribuzin [4-Amino-6-(1,1-dimethylethyl)-3- (methylthio)-1,2,4-triazin-5(4H)-one]	42.1%
INERT INGREDIENTS:	57.9%
TOTAL	100%

EPA Reg. No. 352-382
U.S. Pat. 4,346,220

KEEP OUT OF REACH OF CHILDREN
CAUTION

PRECAUTIONARY STATEMENTS

HAZARDS TO HUMANS AND DOMESTIC ANIMALS

CAUTION

MAY BE HARMFUL IF INHALED OR SWALLOWED

Avoid breathing spray mist. Avoid contact with skin, eyes and clothing. Wash thoroughly with soap and warm water after handling. Wash contaminated clothing with soap and hot water before reuse.

STATEMENT OF PRACTICAL TREATMENT

If in eyes: Flush with plenty of water. Get medical attention if irritation persists.

If on skin: Wash with plenty of soap and water. Get medical attention if irritation persists.

For medical emergencies involving this product, call toll free 1-800-441-3637.

ENVIRONMENTAL HAZARDS

Do not apply directly to water or wetlands. Do not contaminate water by cleaning of equipment or disposal of wastes.

GENERAL INFORMATION

Du Pont "LEXONE" 4L Herbicide is a liquid to be mixed in water or liquid fertilizer and applied as a spray for selective control of certain broadleaf weeds and grasses in alfalfa, asparagus, bermudagrass turf, corn, fallow, potatoes, sainfoin, soybeans, sugarcane and tomatoes. It is noncorrosive to equipment, nonflammable, and nonvolatile. For soybeans, "LEXONE" 4L may be used in conjunction with "Treflan"[1], "Sonalan"[1], "Surflan"[1], "Lasso"[2], "Roundup"[2], "Bronco"[2], "Dual"[3], "Prowl"[4], Paraquat [5] or "Amiben"[6]. Observe all cautions and limitations on labeling of all products used in mixtures.

The degree of control and duration of effect will vary with:

Soil Organic Matter Content—High organic matter requires higher dosages than low organic matter for equivalent herbicide performance.

Soil Texture—High clay content requires higher dosage rates than those with less clay.

Soil pH—Activity of "LEXONE" 4L increases as the pH increases.

Rainfall—Moisture is necessary to activate the herbicide; best results are obtained if treatment is made to moist soil and moisture is supplied by rainfall or sprinkler irrigation (1/4" to 1/2") within one week after application. If moisture is insufficient to activate the herbicide, a shallow cultivation (rotary hoe preferred) should be made after emergence of crop while weeds are small enough to be controlled by mechanical means. If heavy rains occur within 3 weeks of planting, injury to the crop may occur; after 3 weeks, any effects are likely to be negligible.

Amount Applied—The higher recommended rate is more active than the lower rate on a given soil.

Other Conditions—Factors such as incorporation depth of "LEXONE" 4L, competition of the crop with weeds, weed species, drought stress of weeds, high salt content of soil, crop seed quality, soil residues of herbicides previously applied, soil insecticides and combination herbicide treatments all can affect performance.

PREEMERGENCE USE (Germinating Weeds):

"LEXONE"4L, when used as directed, controls for an extended time annual weeds such as:

BROADLEAVES*		GRASSES
Beggarweeds Wild buckwheat Common chickweed Carpetweed White cockle Hophornbeam copperleaf Flixweed Henbit Jimsonweed Kochia* Lambsquarters* Venice mallow Meadow salsify Blue mustard Tansy mustard Wild mustard	Pepperweed Palmer pigweed Prostrate pigweed Redroot pigweed* (careless weed) Smooth pigweed* Common ragweed Hemp sesbania (coffeebean) Shepherdspurse Sicklepod (cassia, coffeeweed) Prickly sida (spiny sida, teaweed) Pennsylvania smartweed Spotted spurge Spurred anoda Yellow rocket Velvetleaf	"LEXONE" 4L at rates above 1 pt per acre: Foxtails, Crabgrass, Crowfootgrass, Cheatgrass, Downy brome Combinations of "LEXONE" 4L with "Treflan", "Sonalan", "Surflan", "Lasso", "Dual", or "Prowl" control annual grasses listed above and: Barnyardgrass, Crabgrass, Fall panicum, Foxtails, Signalgrass
* except triazine-resistant weeds		
Partial Control		Combinations of "LEXONE" 4L with "Treflan", "Sonalan", or "Surflan" controls annual grasses listed above and:
Cocklebur Common purslane	Florida purslane (Florida pusley) Sunflower	Seedling johnsongrass

Note: Consult "Treflan", "Dual", "Surflan", "Sonalan", "Lasso", "Prowl", Paraquat, "Amiben", "Bronco" or "Roundup" labels for additional weeds controlled when "LEXONE" 4L is used in conjunction with these herbicides.

"LEXONE" 4L applied before emergence of asparagus, corn, potatoes, soybeans, sugarcane and weeds is an effective procedure because susceptible weeds are controlled in an early, vulnerable seedling stage before they compete with the crop. With favorable moisture conditions, "LEXONE" 4L continues to control weeds for some time as the crop becomes better able to compete. Should weed seedlings begin to break through the preemergence treatment in significant numbers, secondary weed control procedures should be implemented; these include cultivation and postemergence herbicide application.

POSTEMERGENCE USE (EMERGED SEEDLING OR PERENNIAL WEEDS):
"LEXONE" 4L, when used as directed, controls weeds such as:

BROADLEAVES*			GRASSES	
Carpetweed	Hemp sesbania	Prickly sida (spiny sida, teaweed)	Bluegrass	Goosegrass
Cocklebur	Jimsonweed	Pennsylvania smartweed	Crabgrass (large, smooth)	
Common purslane	Ladysthumb	Toadflax		
Common ragweed	Lambsquarters*	Russian thistle		
Dayflower	Mexican weed	Velvetleaf		
Florida beggarweed	Mustards			
Fumitory	Pigweeds (prostrate, redroot, smooth)*			
	Sicklepod (cassia)			
	Partial Control		**Partial Control or Suppression**	
Curly dock	Horsenettle	Spotted spurge	Barnyardgrass (watergrass)	Foxtails
Dandelion	Kochia*	Sunflower	Broadleaf signalgrass	Quackgrass
Florida purslane (Florida pusley)	Morningglory	Wild poinsettia		
* except triazine resistant weeds				

"LEXONE" 4L may be sprayed over the top of certain emerged crops; they are alfalfa (established, dormant), potatoes, sainfoin (dormant), sugarcane and tomatoes (seeded and transplants) to effectively control susceptible emerged seedling weeds.

Directed postemergence applications of "LEXONE" 4L, where spray nozzles are adjusted so that weeds are sprayed, but the crop is not, may be used in emerged soybeans (southern and southeastern states) and established tomatoes to control susceptible emerged seedling weeds.

Results of postemergence treatment of emerged weeds vary with rate applied and environmental conditions; best results are obtained on succulent weeds growing under conditions of high humidity and temperatures of 70°F or higher. Addition of a surfactant to the spray (where recommended) increases contact effects of "LEXONE" 4L. Application will also provide control of emerging susceptible weed seedlings for an extended period of time.

IMPORTANT—Injury to or loss of desirable trees or other plants may result from failure to observe the following:

Do not apply (except as recommended), or drain or flush equipment on or near desirable trees or other plants, or on areas where their roots may extend, or in locations where the chemical may be washed or moved into contact with their roots. Do not use on lawns, walks, driveways, tennis courts or similar areas. Prevent drift of spray to desirable plants. Do not contaminate any body of water. Keep from contact with fertilizers, insecticides, fungicides and seeds during storage.

Thoroughly clean all traces of "LEXONE" 4L from application equipment immediately after use. Flush tank, pump, hoses and boom with several changes of water after removing nozzle tips and screens (clean these parts separately).

Varietal sensitivity to "LEXONE" 4L is known in several crops for which "LEXONE" 4L is registered. All new crop varieties should be evaluated on a limited basis for their tolerance prior to using "LEXONE" 4L.

DIRECTIONS FOR USE

It is a violation of federal law to use this product in a manner inconsistent with its labeling.

Du Pont "LEXONE" 4L Herbicide should be used only in accordance with recommendations on this label, or in separate published Du Pont recommendations available through local dealers.

Du Pont will not be responsible for losses or damages resulting from use of this product in any manner not specifically recommended by Du Pont. User assumes all risks associated with such nonrecommended use.

Do not use in Kern County, California.

Metribuzin is a chemical which can travel (seep or leach) through soil and can contaminate ground water which may be used as drinking water. Metribuzin has been found in ground water as a result of agricultural use. Users are advised not to apply metribuzin where the water table (ground water) is close to the surface and where the soils are very permeable, i.e., well drained soils such as loamy sands. Your local agricultural agencies can provide further information on the type of soil in your area and the location of ground water.

SPRAY PREPARATION:

Mixing in Water -

Fill tank 1/4 full with water. Start agitation system, add "LEXONE" 4L and continue adding water. Add separately each additional component of any tank mix while adding water. If a surfactant is used, add it last when the tank is nearly full. Continue agitation throughout. If poor mixing should occur with any component, premix the component with two parts water before adding to the spray tank; follow this procedure when mixing "Treflan" or "Prowl". Use spray mixture promptly. "LEXONE" 4L plus "Treflan" tank mixtures should not be stored overnight; flush sprayer with water after use and dispose of mix properly.

Mixing in Liquid Fertilizers -

For preemergence application to potatoes or soybeans, nonpressure nitrogen or fertilizer solution may be used in the spray mixture. Small quantities should be tested for compatibility by the following procedures before full scale mixing:

1. Put 1 pint of fertilizer solution in a quart jar.
2. Mix 2 teaspoons "LEXONE" 4L with 2 tablespoons of water; mix thoroughly and add to fertilizer solution.
3. Close jar and shake well.
4. If other herbicides are to be used in the mixture, premix 2 teaspoons of wettable powders or 1 teaspoon of liquids with 2 tablespoons of water; add to "LEXONE" 4L/fertilizer solution mixture.
5. Close jar and shake well.
6. Watch mixture for several seconds; check again in 30 minutes.
7. If mixture does not separate, foam, gel, or become lumpy, it may be used.
8. Mixing ability may be improved by adding compatibility agents such as Kalo Laboratories "Compex" or Witco Chemicals "Sponto" 168D. Follow directions on container.

Provided the above procedure shows the mixture to be compatible, prepare the tank mixture as follows: add the fertilizer solution to the spray tank first, with the agitator running, add the required amount of "LEXONE" 4L and thoroughly mix. For tank mixtures with other herbicides, follow directions above.

EQUIPMENT—SPRAY VOLUMES:

Apply uniformly by air (except as noted) or with a properly calibrated low pressure (20 to 40 psi) boom sprayer equipped with fan-type nozzles and screens no finer than 50 mesh. Use 10 to 40 gals of water per acre with ground equipment (20 to 60 gals with mixture containing paraquat and 20 to 30 gals with mixture containing "Roundup"). Use 2 to 10 gals per acre for aerial application, except use 7 gals per acre for sugarcane (irrigated only) in Hawaii, and 5 to 10 gals per acre for potatoes west of Rocky Mountains and for sugarcane in Louisiana and Texas.

Continuous agitation in the spray tank is required to keep the material in suspension. Avoid overlapping, and shut off spray booms while starting, turning, slowing or stopping, or injury to the crop may result. With aerial equipment, avoid overlapping of spray swath and avoid application under conditions where excessive drift may occur.

For Use In Chemigation Systems (Potatoes Only)

Apply this product only through sprinkler (including center pivot and wheel roll) irrigation systems. Do not apply this product through any other type of irrigation system. Crop injury, lack of effectiveness, or illegal pesticide residues in the crop can result from nonuniform distribution of treated water. If you have questions about calibration, you should contact State Extension Service specialists, equipment manufacturers or other experts. Do not connect an irrigation system (including greenhouse systems) used for pesticide application to a public water system unless the pesticide label prescribed safety devices for public water systems are in place. A person knowledgeable of the chemigation system and responsible for its operation, or under the supervision of the responsible person, shall shut the system down and make necessary adjustments should the need arise.

Chemigation Systems Connected to Public Water Systems

Public water systems means a system for the provision to the public of piped water for human consumption if such system has at least 15 service connections or regularly serves an average of at least 25 individuals daily at least 60 days out of the year. Chemigation systems connected to public water systems must contain a functional, reduced-pressure zone, backflow preventer (RPZ) or the functional equivalent in the water supply line upstream from the point of pesticide introduction. As an option to the RPZ, the water from the public water system should be discharged into a reservoir tank prior to pesticide introduction. There shall be a complete physical break (air gap) between the outlet end of the fill pipe and the top or overflow rim of the reservoir tank of at least twice the inside diameter of the fill pipe. The pesticide injection pipeline must contain a functional, automatic, quick-closing check valve to prevent the flow of fluid back toward the injection pump. The pesticide injection pipeline must contain a functional, normally closed, solenoid-operated valve located on the intake side of the injection pump and connected to the system interlock to prevent fluid from being withdrawn from the supply tank when the irrigation system is either automatically or manually shut down. The system must contain functional interlocking controls to automatically shut off the pesticide injection pump when the water pump motor stops, or in cases where there is no water pump, when the water pressure decreases to the point where pesticide distribution is adversely affected. Systems must use a metering pump, such as a positive displacement injection pump (e.g., diaphragm pump) effectively designed and constructed of materials that are compatible with pesticides and capable of being fitted with a system interlock. Do not apply when wind speed favors drift beyond the area intended for treatment.

Sprinkler Chemigation

The sprinkler chemigation system must contain a functional check valve, vacuum relief valve, and low pressure drain appropriately located on the irrigation pipeline to prevent water source contamination from backflow. The pesticide injection pipeline must contain a functional, automatic, quick-closing check valve to prevent the flow of fluid back toward the injection pump. The pesticide injection pipeline must also contain a functional, normally closed, solenoid-operated valve located on the intake side of the injection pump and connected to the system interlock to prevent fluid from being withdrawn from the supply tank when the irrigation system is either automatically or manually shut down. The system must contain functional interlocking controls to automatically shut off the pesticide injection pump when the water pump motor stops. The irrigation line or water pump must include a functional pressure switch which will stop the water pump motor when the water pressure decreases to the point where pesticide distribution is adversely affected. Systems must use a metering pump, such as a positive displacement injection pump (e.g., diaphragm pump) effectively designed and constructed of materials that are compatible with pesticides and capable of being fitted with a system interlock. Do not apply when wind speed favors drift beyond the area intended for treatment.

Agitation is recommended in the pesticide spray tank. See "SPRAY PREPARATION" for specific details on agitation and spray preparation.

Directions regarding application of metribuzin into the water application and of quantity of water to be applied per acre are described in the crop section "Potatoes".

USE RATES:

All rates are expressed as broadcast rates per acre; for band treatment, use proportionately less. For example, for 14" band on 42" row, use 1/3 of broadcast rate. Where a range of dosages is given, use the lower rate on coarser soils (low in clay or organic matter) and the higher rate on finer soils (high in clay or organic matter).

REPLANTING RECOMMENDATIONS:

If initial seeding fails to produce a stand, crops registered for the rate of "LEXONE" 4L that has been applied may be replanted into the treated area. Do not retreat during the same crop year as injury to the crop may result.

CROP ROTATION RECOMMENDATIONS:

Unless otherwise noted, observe the following:

Areas in the Northwest where up to 3/4 pint "LEXONE" 4L per acre have been used on lentils and peas may be recropped to wheat and barley after 4 months.

Areas treated with up to 1 pt "LEXONE" 4I per acre may be recropped to wheat and barley 8 months after treatment.

Following preemergence treatment of soybeans, winter wheat may be planted 4 months after application. Areas in the Midsouth where soybeans have been treated preemergence with up to 1 1/2 pts "LEXONE" 4L per acre may be recropped to cotton and rice after 8 months.

Areas treated with up to 2 pts "LEXONE" 4L per acre may be recropped to alfalfa, asparagus, corn, forage grasses, potatoes, sainfoin, soybeans, sugarcane and tomatoes after 4 months.

Areas treated with up to 2 pts "LEXONE" 4L may be recropped to wheat and barley after 10 months.

Areas treated with 2 to 4 pts "LEXONE" 4L per acre may be recropped to soybeans after 4 months; 10 months after application, alfalfa, asparagus, corn, forage grasses, potatoes, sainfoin and tomatoes may be planted.

Areas treated with 4 to 8 pts "LEXONE" 4L per acre may be recropped to sugarcane 4 months after application; 13 months after application, alfalfa, corn, forage grasses, potatoes, sainfoin, soybeans, and tomatoes may be planted.

Other Crops: For other crops, do not plant within 12 months; do not plant root crops within 18 months.

Cover crops may be planted any time, but stand reduction may occur; do not graze or harvest these cover crops for feed or food.

Special Conditions: For winter or drought periods, add the number of weeks of winter or drought to the above listed recropping intervals.

For soil pH 7.5 or above, add 2 months to the recropping intervals listed.

Combination Treatments: When recommended tank mix combinations are used, consult the product label for replanting and recropping instructions and observe the directions that are the most restrictive.

CROP USES

ALFALFA AND SAINFOIN (Alone or with Grasses)

Use on alfalfa produced for forage or hay; do not use on alfalfa grown for seed.

Make a single application in the fall after plants become dormant or in the spring before new growth starts. Do not apply during the first growing season after seeding. Do not graze or harvest within 28 days after application. Stress which may be caused in part by low fertility, disease, insects, winter kill, overcutting, drought or frost may increase the chance of crop injury. In Washington, Oregon, Idaho and Utah, partial control of downy brome, flixweed and tansy mustard will occur at the lower use rate; for improved control of these weeds in dormant pure stands of alfalfa only, add 2/5 to 1/2 lb active paraquat per acre.

In California, use only north of U.S. Interstate 80.

Pure Stands of Alfalfa or Sainfoin: Select the rate from the table below:

Soil Texture	Pints "LEXONE" 4L Per Acre	
	1/2 to 2% Organic Matter	More than 2% Organic Matter
Coarse Loamy sand, Sandy loam	3/4 to 1	1 to 1 1/4
Medium Loam, Silt Loam, Silt, Sandy clay, Sandy clay loam	1 to 1 1/4	1 1/4 to 1 1/2
Fine Silty clay, Silty clay loam, Clay, Clay Loam	1 1/4 to 1 1/2	1 1/2 to 2

Do not use on sand nor on soils with less than 1/2% organic matter as crop injury may result.

Mixed Stands with Grasses: The higher range of rates for a given soil texture and organic matter classification in the table above may provide partial reduction of forage grass stands. Rates of 1 1/2 to 2 pints per acre may severely reduce stands of forage grass.

ASPARAGUS

Make a single preemergence application of 2 to 4 pints per acre in early spring prior to spear emergence. Use the lower rate on coarse textured soils (low in clay or organic matter) and the higher rate on fine textured soils (high in clay or organic matter). If field is to be disked, apply "LEXONE" 4L after disking, but before crop emerges.

In the Northwest, apply as above; or make a single preemergence application of 1 to 2 pts "LEXONE" 4L per acre in early spring prior to spear emergence. Following harvest but before fern emerges, apply 1 to 3 pts "LEXONE" 4L per acre. Do not exceed 4 pts "LEXONE" 4L per acre per growing season; or make a single preemergence application of 1 to 2 pts "LEXONE" 4L per acre in early spring prior to spear emergence. Following harvest but before fern emerges, apply a tank-mix combination of 1 to 1 1/2 pts "LEXONE" 4L plus 1 to 2 lbs "Karmex" per acre. For irrigated asparagus, after the initial application of "LEXONE" 4L, do not exceed more than 1/2" of water for the first irrigation. Additional irrigations may be normal programs.

Do not apply within 14 days of harvest; do not apply after spear emergence. Do not use on newly seeded asparagus nor on young plants during the first growing season after setting crowns.

BERMUDAGRASS

Hawaii - for control of goosegrass in established turf, apply 1 to 2 pts "LEXONE" 4L at 3 month intervals, as needed, but do not apply more than 8 pints per acre per year. Use the lower rate where soil pH exceeds 7.5 and the higher on fine-textured soils.

Apply when temperature is between 80 and 90 degrees F, when bermudagrass is growing vigorously and not under stress conditions. Do not water bermudagrass for at least 24 hours after application. Temporary discoloration of bermudagrass may occur.

Avoid spray overlaps.

Do not apply to areas of close-mowed turf nor within the root zone of desirable shrubs, trees, or other ornamentals, as injury may occur.

Do not mow for at least 3 days.

Do not add surfactant.

Do not graze or use grass clippings for animal feed.

Do not use on residual lawns unless under the supervision of a commercial applicator.

CORN (FIELD)

ILLINOIS, INDIANA, IOWA, KANSAS, KENTUCKY, MICHIGAN, MINNESOTA, MISSOURI, NEBRASKA, OHIO, SOUTH DAKOTA AND WISCONSIN only — Apply "LEXONE" 4L plus atrazine plus "Lasso" or "Dual" tank mixture after planting but before corn emerges. Plant seed at least 1½" deep on flat or raised seedbeds only or injury to the crop may result. Soil should be well prepared and as free as possible from trash and clods before planting.

Treatment controls many annual weeds such as velvetleaf, common lambsquarters*, common ragweed, Pennsylvania smartweed, redroot pigweed*, jimsonweed, giant foxtail, yellow foxtail, green foxtail, smooth crabgrass, large crabgrass, barnyardgrass, fall panicum and witchgrass. "LEXONE" 4L in the combination improves control of velvetleaf, common lambsquarters* and Pennsylvania smartweed. For additional weeds controlled, see atrazine and "Lasso" or "Dual" labeling.

* Except triazine resistant weeds

SOIL TEXTURE	2% to 6% ORGANIC MATTER ONLY							
	"LEXONE" 4L (Pts)		ATRAZINE (Lbs)		"LASSO" (Qts)	or	"DUAL" 8E (Pts)
Medium Loam, Silt loam, Silt, Sandy clay, Sandy clay loam	⅔ to ½	+	¾ to 1½ (Active)	+	1½ to 2	or	1½ to 2
Fine Silty clay, Silty clay loam, Clay, Clay loam	½	+	¾ to 1½ (Active)	+	2 to 2½	or	2 to 2½

Do not use this mixture on coarse soils (sand, loamy sand, sandy loam) as crop injury may occur. Do not use on muck soils.

Do not use on soils having a calcareous surface layer or a pH of 7.5 or higher.

Before use on any inbred parent or variety of corn, tolerance to "LEXONE" 4L must first be determined; use only on inbred parents or varieties that are tolerant to "LEXONE" 4L.

If initial seeding fails to produce a stand, treated fields may be replanted to corn; do not rework soil. Do not retreat field with a second application as injury to the crop may result.

For other crops that may follow corn, see atrazine and "Lasso" or "Dual" labeling.

FALLOW

Corn - Fall Bedded Fields In California

Apply 0.9 to 1.5 pts "LEXONE" 4L per acre in the fall using the lower rate on ½ to 2% organic matter soils and the higher rates on soils over 2% organic matter. Do not plant corn within 120 days of application to prevent crop injury. At least 2 inches of moisture is needed to activate "LEXONE" 4L for weed control and to begin breakdown of the herbicide in soil for crop safety. Do not disturb soil surface after application as this reduces weed control. Do not use in Kern County, CA.

Tomatoes - Fall Bedded Fields In California

Apply 1 to 1.5 pts of "LEXONE" 4L per acre in the fall using the lower rate on ½ to 2% organic matter soils and the higher rates on soils over 2% organic matter. Do not plant tomatoes within 90 days of application to prevent crop injury. At least 2 inches of moisture is needed to activate "LEXONE" 4L for weed control and to begin breakdown of the herbicide in soil for crop safety. Do not disturb soil surface after application as this reduces weed control. Do not use in Kern County, CA.

WHEAT (REDUCED TILLAGE FALLOW)

Apply 0.67 to 1.5 pts "LEXONE" 4L per acre in the fall or apply ⅔ to 1 pts in the spring to fallow ground that will be followed by winter wheat. Use the lower rate on coarse textured soils and the higher rates on medium to fine textured (heavier) soils.

Time treatment to new weed growth to control:

Blue mustard*	Fanweed	Miners lettuce
Bur buttercup	Filared	Pigweed (prostrate, redroot)**
Burcherval*	Flixweed*	Pineappleweed
Chickweed (common, mouse-ear, jagged)	Gromwell*	Prickly lettuce
Conical catchfly*	Henbit	Russian thistle*
Corn cockle	Jacobs ladder	Speedwell (ivyleaf)
Dog fennel (May-weed)*	Jim Hill mustard	Tansy mustard*
False Flax	Knotweed	Tarweed (fiddleneck)
Lambsquarters**	Kochia**	Treacle mustard
	Wild buckwheat*	Wild mustard
Annual ryegrass*	Japanese brome	
Cheatgrass	Pacific meadow foxtail (Blackgrass)*	
Downy Brome*	Windgrass*	
Green foxtail*	Yellow foxtail*	

* Weeds and grasses partially controlled. Partial control is an observable reduction in weed competition (reduced weed population and/or vigor) as compared to an untreated area. Degree of partial control can vary with soil type, environmental factors, and size of weeds at time of application.

** Except triazine resistant weeds.

If weeds larger than 2" tall or across are present, use a suitable registered companion herbicide such as "Glean", paraquat, "Round-up" or "Chem Hoe" 135 as a tank mix. Refer to labeling for these products for additional use instructions and precautions.

Where spring application is appropriate in Colorado, Kansas, Montana, Nebraska, North Dakota, Oklahoma, South Dakota, Texas, and Wyoming for improved control of downy brome, Russian thistle, and tansymustard, tank mix ½ to 1 pt "LEXONE" 4L with 2 to 2.5 lbs "Igran" 80 W per acre.

Do not plant spring seeded cereals following "Lexone" DF fall fallow application.

Do not apply "LEXONE" 4L in the spring if "LEXONE" 4L was used the preceding fall or winter.

Use supplemental control procedures such as cultivation or other herbicides if seasonal weed control is desired.

Recrop Intervals-See "Crop Rotation Recommendations"

POTATOES

Preemergence: Make a single application of 1 to 2 pints per acre after planting or after drag-off but before crop emerges. Use the lower rates on coarse textured soils and the higher rates on fine textured soils (high in clay or organic matter).

Postemergence: Do not use on early maturing smooth-skinned white or red-skinned varieties. Apply only if there have been at least 3 successive days of sunny weather prior to application. Treat before weeds are 1" tall. Make a single application of 1 pint per acre. For redroot pigweed and lambsquarters only, use ½ to 1 pint per acre. Treatment may cause some chlorosis or minor necrosis. Do not apply within 60 days of harvest.

Application by Sprinkler Irrigation: Refer to "Chemigation" in "Directions for Use" before applying product. Apply specified dosage of "LEXONE" 4L in ¼ to ¾ inch of water per acre (¼ to ½ inch of water per acre on sandy soil) as a continuous injection in center pivot and self-propelled wheel-move systems or in the last 15-30 minutes of set-in permanent solid-set sprinkler systems.

Irrigation systems must be equipped with automatic shut-off devices which prevent back-flow to the water source. Maintain continuous agitation in injection nurse tanks during application. Treatment may be made as a single postemergence or split postemergence application. Check irrigation system to insure uniform application of water to all areas. Failure to apply "LEXONE" 4L uniformly may result in crop injury and/or poor weed control.

Preemergence plus Postemergence: Apply as directed above. Do not use more than 2 pints total per acre per season. Do not apply within 60 days of harvest.

Postemergence plus Postemergence—(Oregon, Washington, Idaho): Use only if "LEXONE" 4L was not applied preemergence. Allow at least 14 days between applications. For best results, apply before weeds are 1" in height or diameter. Do not apply within 24 hours of treatment with other pesticides.

Oregon and Washington—Make 2 applications of ½ to 1 pint per acre. On coarse (sandy) soils containing ½ to 1% organic matter do not exceed ½ pint per acre per application.

Idaho—Make 2 applications of ¼ to 1 pint per acre. On coarse (sandy) soils containing ½ to 1% organic matter, do not exceed ¾ pint per acre per application.

CROP ROTATION RECOMMENDATIONS: In the Northwest, wheat and barley may be planted 8 months after an application of "LEXONE" 4L made no later than June 30; crops other than wheat, barley, and potatoes cannot be planted before one year after treatment. Cover crops planted to prevent soil erosion or for soil building purposes may be planted anytime after application of "LEXONE" 4L but do not graze or harvest for feed or food.

See "Replanting" and "Crop Rotation Recommendations" for further details.

Note: Because potato varieties vary in their resistance to herbicides, determine tolerance to "LEXONE" 4L prior to adoption as a field practice to prevent possible crop injury. Crop injury may occur if Acadia Russett, Atlantic, Batoche, Belchip, Campbell 12 and 13, Caribe, Centennial, Cobbler, Gold Rus, Hampton, Jemseg, Monona, MN7973, Norchip, Norgold, NY 64, Ontario, Sebago and Shepody varieties are under stress from lack of moisture, fertility, sunlight or affected by insects or disease when treated postemergence. Use of "LEXONE" 4L on sands or loamy sands containing 1/2 to 1 1/2% organic matter may result in crop injury. Do not use on any soil containing less than 1/2% organic matter, as crop injury may result. Preemergence use on soils containing more than 10% organic matter may result in only partial weed control. Postemergence use on rill irrigated potatoes may not provide adequate weed control in absence of rainfall. Do not apply to sweet potatoes or yams.

SOYBEANS

Select one of the following herbicide treatments and apply as directed. The combination treatments control weeds that are susceptible to "LEXONE" 4L plus those controlled by the companion products (refer to product labeling). Dosage rates listed are broadcast rates; for band applications use proportionately less.

Note: Soybean varieties such as Altona, Coker 102 and 156, Govan, NKS 1884, Semmes, Tracy, Vansoy, Terra Vig 505 and 606, Agripro 55, Asgrow 6520, Maple Amber, Portage, Vinton 81 and AP 71 are sensitive to "LEXONE" 4L. Injury may occur if "LEXONE" 4L is used on these varieties.

Varieties showing above average tolerance to "Lexone" are American Revere, Asgrow 1937, Asgrow 3659, Asgrow, 3860, DSR 171, DSR 207, Essex, Fayette, Hisoy 170, Lakota, Lawrence, LOL 4207, NKS 1492, Pride B216, Pride B242, SRF 250, SRF 350 P, Union, Wayne, Wells II and Williams 82; for maximum weed control on these varieties, use the higher rate (where a range of rates is listed) for the appropriate soil type and organic matter as shown in the following tables.

Injury to soybeans may occur under the following conditions:

- Improper sprayer calibration
- Over application or uneven application
- Use on very coarse, porous soils
- Soils with pH of 7.5 or higher or those with a calcareous surface layer
- Use with soil-applied organic phosphate insecticides
- Heavy rains within 3 weeks of planting

Seedling disease, cold weather, deep planting (more than 2") excessive moisture, high salt concentration or drought may weaken soybean seedlings and increase possibility of crop injury.

Preemergence To Soybeans

Conventional Tillage

Preplant Incorporated: Incorporate herbicide treatment into soil before planting of soybeans according to instructions on label of product used in conjunction with "LEXONE" 4L; improper soil incorporation can result in erratic weed control or crop injury. Do not plant seed more than 2" deep. Use combination as tank mixtures in water. Select the proper rate from the following chart and use it with a "companion" herbicide listed below:

	Pints "LEXONE" 4L Per Acre	
Soil Texture	**1/2 to 3% Organic Matter**	**More than 3% Organic Matter**
Coarse Loamy sand, Sandy loam	1/2	3/4
Medium Loam, Silt Loam, Silt, Sandy clay, Sandy clay loam	3/4	1
Fine Silty clay, Silty clay loam, Clay, Clay Loam	1	1
Mississippi Delta Only— Silty Clay to Heavy clay	1 1/2	1 1/2

Do not use on sand. Do not use on loamy sand or sandy loam with less than 1% organic matter, nor on other soils with less than 1/2% organic matter as crop injury may result. Do not use on peat. Treated vines may be grazed or fed to livestock 40 days after application unless specified otherwise on the companion product.

Any one of the following "Companion" products may be tank mixed with "LEXONE" 4L to improve weed control:

"Treflan", "Dual", "Sonalan"	Apply within 14 days of planting
"Lasso", "Prowl"	Apply within 7 days of planting

Note: Read product labels for weeds controlled, rates to use (etc.), and observe the most limiting use restrictions for the combination selected.

Preplant Non-Incorporated:
"LEXONE" 4L + "Surflan" AS or "Surflan" 75W

Preplant Surface Application: Apply within 4 weeks of planting. Treatment also controls dayflower and goosegrass.

Soil Texture	Organic Matter	"LEXONE" 4L (Pts)	+	Rate Per Acre "Surflan" AS or "Surflan" 75W
Coarse Sandy loam	1-5%	1/2*	+	1/2 to 3/4 (qt) or 2/3 to 1 (lb)
Medium Loam, Silt loam, Silt, Sandy clay, Sandy clay loam	1/2-5%	3/4	+	3/4 to 1 (qt) or 1 to 1 1/3 (lbs)
Fine Silty clay, Silty clay loam, Clay, Clay loam	1/2-5%	1	+	1 to 1 1/4 (qts) or 1 1/3 to 1 2/3 (lbs)
Mississippi Delta Only Silty clay to heavy clay	1/2-5%	1 to 1 1/4	+	1 to 1 1/4 (qts) or 1 1/3 to 1 2/3 (lbs)

* For sicklepod control on Coastal Plains soils of Southeast and Midsouth, use 1/2 to 3/4 pts "LEXONE" 4L.

Do not use on sand or loamy sand. Do not use on sandy loam with less than 1% organic matter. Do not use on other soils with less than 1/2% organic matter. Do not use preemergence application on soils with more than 3% organic matter nor the preplant surface application on soils with more than 5% organic matter as poor weed control and/or crop injury may result. Do not use treated vines for feed or forage.

Preplant Incorporation of "Treflan", "Sonalan", or "Prowl", "Lasso" or "Dual" Followed by "LEXONE" 4L Preemergence: Apply the preplant incorporated treatment according to directions on product label. As a separate operation, follow with "LEXONE" 4L (alone) using rates given under "Preemergence Application" (below):

Preemergence Application: Apply after planting but before soybeans emerge; do not spray over top of emerged soybeans. Plant seed 1 1/2" to 2" deep on flat or raised seedbeds only or injury to the crop may result; soil should be well prepared and as free as possible from trash and clods before planting. Use combinations as tank mixtures in water.

Where a range of "LEXONE" 4L dosage rate is listed for a specific soil type and organic matter, use the higher rate for difficult control problems, such as cocklebur, sicklepod, jimsonweed, hemp sesbania, prickly sida or velvetleaf. Use the higher rate also for minimum or no-tillage soybeans, and in the states of Georgia and South Carolina.

Select the appropriate rate of "LEXONE" 4L and apply it in combination with products listed below for improved weed control:

	Pints "LEXONE" 4L Per Acre	
	Organic Matter	
Soil Texture	**1/2 to 3%**	**More than 3%**
Coarse Sandy loam	1/4 to 1/2*	3/4
Medium Loam, Silt Loam, Silt, Sandy clay, Sandy clay loam	1/2 to 3/4	1
Fine Silty clay, Silty clay loam, Clay, Clay Loam	3/4 to 1	1
Mississippi Delta Only— Silty clay to heavy clay	1 to 1 1/2	1 1/2

* For sicklepod control on Coastal Plains soils of Southeast and Midsouth, use 1/2 to 3/4 pts "LEXONE" 4L.

Do not use on soils with less than 1/2% organic matter, nor on sand, loamy sand or muck soils as crop injury may result. Do not use on peat.

Treated vines may be grazed or fed to livestock 40 days after application.

Any one of the following products may be mixed with "Lexone" to improve weed control:

"Lasso", "Dual", "Prowl", "Surflan", and "Amiben".

NOTE: Read the product labels for weeds controlled, rates to use and observe the most limiting use restrictions for the combination selected.

Preplant Incorporated (PPI) Combinations plus Preemergence - Use combinations as tank mixes in water. Promptly incorporate PPI treatment into soil before planting of soybeans. Improper soil incorporation can result in erratic weed control or crop injury. Do not plant seed more than 2" deep. After planting of soybeans and before emergence, apply a preemergence treatment of "LEXONE" 4L.

Select a combination as specified in the section "Preplant Incorporated" and apply as directed.

After planting but before emergence of the soybeans, apply a second application of "LEXONE" 4L (without a companion product) as follows:

	Pints "LEXONE" 4L Per Acre		
	Organic Matter		
Soil Texture	Less than 2%	2 to 4%	Over 4%
Coarse Sandy loam	1/4	1/4	1/2
Medium Loam, Silt loam, Silt, Sandy clay, Sandy clay loam	1/4	1/4 to 1/2	1/2 to 3/4
Fine Silt clay, Silty clay loam, Clay, Clay loam	1/4	1/4 to 1/2	1/2 to 3/4

Do not use on sand. Do not use on loamy sand or sandy loam with less than 1% organic matter, nor on other soils with less than 1/2% organic matter as crop injury may result. Do not use on peat. Treated vines may be grazed or fed to livestock 40 days after application unless specified otherwise on the companion product(s).

Minimum Or No-Tillage Soybeans
Preemergence to Crop and Weeds
"LEXONE" 4L plus paraquat, "Roundup", "Lasso", "Dual", or "Surflan" may be used as a tank mixture for preemergence and postemergence control of many broadleaved weeds and grasses and for top kill or suppression of established annuals and some perennials where soybeans will be planted directly into a performed bed (stale seedbed), cover crop, or in previous crop residues such as corn or small grain stubble. For improved grass and volunteer small grain control, the addition of "Lasso" (or "Lasso" EC) or "Surflan" to the "LEXONE" 4L plus paraquat or "Roundup" tank mixture is recommended. For all combinations, except "Surflan", treated vines may be grazed or fed to livestock 40 days after application.

Apply with ground equipment, during or after planting but before crop emerges. Use 20 to 60 gals of water per acre for paraquat and use the higher gallonage for dense stubble or vegetation. Use 20 gals of water per acre for "Roundup" combinations.

Preemergence to Crop and Postemergence to Weeds:
"LEXONE" 4L - Paraquat—Select the rate from the table for preemergence application of "LEXONE" 4L and thoroughly mix in spray tank first, then add paraquat as directed under "Paraquat Combinations".

"LEXONE" 4L + "Lasso" (or "Lasso" EC) + Paraquat—Select the rate from the table for preemergence application of "LEXONE" 4L + "Lasso" tank mixture and thoroughly mix in spray tank first according to directions; then add paraquat as directed under "Paraquat Combinations".

"LEXONE" 4L + "Surflan" + Paraquat—Select the rate from the table for preemergence application of "LEXONE" 4L + "Surflan" tank mixture preemergence and thoroughly mix in spray tank first according to directions; then add paraquat as directed under "Paraquat Combinations".

Do not use treated vines for feed or forage.

Paraquat Combinations—Select the "LEXONE" 4L combination treatment above and mix in the spray tank according to directions; then add paraquat at a rate of 1/4 to 1/2 lb active ingredient per acre for control of emerged weeds. Use the higher rate for weeds 4" to 6" tall. As the last ingredient, add 8 oz "Ortho" X-77 Spreader per 100 gals of spray mixture. Maintain constant agitation.

Roundup Combinations ("LEXONE" 4L + "Lasso" + "Roundup"): Select the rate for preemergence application of "LEXONE" 4L + "Lasso" tank mixture and thoroughly mix in spray tank first according to directions; then add 1 to 2 qts "Roundup" per acre to the mixture and maintain constant agitation. Use the higher rates for maximum top kill of vegetation.

"Bronco" Combinations—Select the rate from the table for preemergence application of "LEXONE" 4L and thoroughly mix in spray tank first; then add "Bronco" at rate of 3 1/4 to 5 qts for coarse soils, and 4 to 5 qts for medium or fine soils. Use the higher rate of "Bronco" for dense stubble, heavy crop residues, or heavy weed populations.

Postemergence to Soybeans
Since the degree of weed control with "LEXONE" 4L (combinations) preemergence treatment is influenced by weather, weed species, etc. (see General Information), provision should be made for follow-up weed control procedures including cultivation and postemergence herbicides. Where appropriate, follow preemergence application with an early season postemergence treatment of Du Pont "Classic" Herbicide over-the-top of soybeans for improved control of emerged seedling cocklebur, pigweeds, common ragweed, giant ragweed, sicklepod, hemp sesbania, Florida beggarweed, sunflower, annual morningglories, jimsonweed, annual smartweed, bristly starbur and yellow nutsedge. (See "Classic" package label for use intructions). Follow with cultivation, as needed.

"LEXONE" 4L (alone) may also be applied as directed postemergence treatment in Southern and Southeastern states. **DO NOT SPRAY OVER THE TOP OF SOYBEANS.** Apply when soybeans are at least 12 inches tall and when broadleaf weeds do not exceed 3 inches and grasses and common ragweed do not exceed 1 inch in height. Direct spray to cover weeds with minimal or no contact to soybean plant. Crop injury may result if spray is higher than 2 inches on soybean stem. Apply 1/2 to 1 pt per acre in 25 to 40 gallons water. Use the lower rate for seedling weeds and the higher rate for larger weeds. Application of 1/2 pt per acre controls Florida beggarweed, carpetweed, cocklebur, dayflower, Mexican weed, prostrate and smooth pigweed, redroot pigweed (carelessweed)*, common purslane, sicklepod (Cassia), velvetleaf and crabgrass (large, smooth); 3/4 to 1 pt per acre controls hemp sesbania, prickly sida (spiny sida, teaweed); 1 pt per acre controls jimsonweed, Pennsylvania smartweed, common ragweed and gives partial control of morningglory species, horsenettle, Florida purslane (Florida pusley), spotted spurge and wild poinsettia when less than 3 inches tall and broadleaf signalgrass up to 1 inch tall. Do not use on sand, nor on soils with less than 1/2% organic matter as crop injury may result. Add a nonionic surfactant such as "Ortho" X-77 Spreader to the spray mixture to obtain better wetting of weed leaf surfaces. If needed, a second application may be made after 7 days. Apply only with ground equipment designed for precise application of directed sprays such as oiling shoes or gauge wheels. Do not exceed nozzle pressure of 25 psi or use nozzle tips smaller than 8002 T-Jet (or equivalent). For broadcast application, use a single flood-type spray nozzle ("K" series or equivalent) per middle, mounted on an oiling shoe or gauge wheel. For band treatment, use two nozzles per row mounted on oiling shoes or gauge wheels, one on each side of a row. Do not apply under weather conditions which favor drift.

* Except triazine resistant weeds

Treated vines may be grazed or fed to livestock 40 days after application.

SUGARCANE
Hawaii—Make a single broadcast application as follows: for aerial application (irrigated cane only), use at least 7 gals water per acre. Spot treatment may also be made at the rate of 5 to 10 pints per acre, but do not exceed 16 pts. total per acre per crop cycle. Do not apply within 17 months of harvest. Do not use treated foliage for feed or forage.

	Pints "LEXONE" 4L Per Acre	
Time of Application	Irrigated	Non-Irrigated
Preemergence—apply within 2 weeks after planting	8 to 12	4 to 8
—or—		
Early Postemergence—apply over cane before weeds are 3" tall	8 to 12	4 to 8
—or—		
Postemergence—apply prior to "close in" time and before weeds are 3" tall	4 to 8	4 to 8

In addition to the weeds listed elsewhere on this label, treatment (non-irrigated cane) controls ageratum, spiny amaranth, wild euphoria, fireweed, Flora's paintbrush, Richardia, garden spurge, graceful spurge, tarweed, guineagrass, plushgrass, ricegrass and wiregrass (goosegrass); in irrigated cane, the foregoing weeds plus spleen amaranth, haole koa, hioloa, hila hila, rattlepod and Alexander grass.

Louisiana and Texas—Apply 3 to 6 pints per acre during the fall after planting but before cane emerges, or to the stubble after harvest. Make a second application in early spring prior to new cane growth. If necessary, a third application may be made in late spring at layby. Do not apply within 60 days of harvest. Do not use treated foliage for feed or forage.

In addition to the weeds listed elsewhere on this label, treatment controls curly dock, annual morningglory, marestail, sowthistle and winter oats.

Louisiana—For use on fallow ground to be followed by sugarcane, apply 2 to 3 pints "LEXONE" 4L per acre. Use the lower rate if applied less than 30 days before planting. Application should be made after the rows are built and ground should not be disturbed until planting or loss of control will result. Treatment provides short term control of seedling johnsongrass and other annual weeds (see Preemergence Use). To aid in control of johnsongrass, stubble sugarcane should be destroyed after harvest or in early spring. Land should be disked frequently in the spring to control rhizome johnsongrass. In late May to early June, rows should be built. The soil surface should be free of trash and clods. Application of "LEXONE" 4L should be made at this time. Rhizome johnsongrass will not be controlled by "LEXONE" 4L; therefore, spot treatment with an appropriate postemergence herbicide will be necessary.

TOMATOES
East of Rocky Mountains—"LEXONE" 4L may be applied as a preplant incorporated treatment alone or in a tank mixture with "Treflan" EC in transplanted tomatoes or "LEXONE" 4L may be applied alone as a postemergence broadcast or directed spray to established tomatoes including seeded established or transplanted tomatoes. Apply with a properly calibrated low-pressure (20 to 40 psi) fixed-boom sprayer wtih fan type nozzles. Use 10 to 40 gallons of water per acre with ground equipment for preplant incorporated treatment and 20 to 75 gallons for postemergence broadcast or directed spray application. Do not use air blast or other high pressure spray equipment for postemergence spray applications.

Preemergence To Transplanted Tomatoes:

Preplant Incorporated "LEXONE" 4L alone or "LEXONE" 4L Plus "Treflan" EC Tank Mix (Transplant Tomatoes: Except Florida): Apply 1/2 to 1 pint per acre "LEXONE" 4L alone or as a tank mix combination with "Treflan" EC as a broadcast spray immediately before transplanting. Use the higher rate for heavy weed pressure or for difficult-to-control weeds. Incorporate uniformly to a depth of 2 to 4 inches. Apply "Treflan" EC at rates of 1 to 2 pts per acre on coarse soils, 1 1/4 to 2 pts per acre on medium soils, and 1 1/2 to 2 pts per acre on fine soils. Consult "Treflan" EC label for specific use instructions and observe all cautions and limitations. When "LEXONE" 4L is tank mixed with "Treflan" EC, apply immediately after mixing; do not allow tank mixture to stand for prolonged periods of time.

Postemergence To Established Tomatoes:

Broadcast Application (Established Tomatoes; Except Florida): Apply 1/2 to 1 pint "LEXONE" 4L per acre over-the-top of established tomato plants. Use the higher rate for heavy weed pressure or for difficult to control weeds. Treat before weeds are 1 inch tall. One or more applications may be made per crop season; allow at least 14 days between applications. Do not apply more than 2 pts. "LEXONE" 4L per acre total within a 35-day period per crop season. Do not treat seeded or transplanted tomatoes until plants have reached the 5 to 6 leaf stage or until transplants have recovered from transplant shock and new growth has started. Apply only if there have been at least 3 successive days of sunny weather prior to application, or crop injury may occur. Do not apply within 24 hours of treatment with other pesticides. Do not tank mix with other pesticides.

Directed Application (Established Tomatoes; East of Rocky Mountains): Apply 1 to 2 pints "LEXONE" 4L per acre as a directed spray according to directions under "Broadcast-Postemergence Application" but do not allow spray to contact tomato foliage. This method of application is recommended for fields with heavy weed pressure of difficult-to-control weed species.

CROP ROTATION RECOMMENDATIONS: For rotation crops in fields treated with "Treflan", follow instructions on product label.

See "Replanting" and "Crop Rotation Recommendations" for further details.

Note: Because tomato varieties vary in their resistance to "LEXONE" 4L, determine tolerance prior to adoption as a field practice to prevent possible crop injury. Do not use hot caps on tomatoes within 7 days before or at any time after application of "LEXONE" 4L. Crop injury or delayed maturity may result from postemergence or directed spray applications if tomatoes are growing under stress conditions caused by low fertility, disease, insects, drought, cool, wet or cloudy weather prior to application. Do not apply within 7 days of harvest. Do not apply more than 2 pts "LEXONE" 4L per acre per crop season.

NONCROP USES

For nonselective use, apply 2 to 12 pints "LEXONE" 4L per acre in 20 to 60 gallons of water. For best results, apply shortly before weed growth begins. Do not use on rights-of-way.

For control of established seedling weeds add:

Paraquat at a rate of 1/4 to 1/2 lb active ingredient per gallon per acre for control of emerged weeds. Apply 20 to 60 gallons of water per acre using the higher spray volume for weeds 4" to 6" tall. As the last ingredient, add 8 ounces "Ortho" X-77 Spreader (or equivalent) per 100 gallons of spray mixture or add 1 to 2 quarts "Roundup" in 20 to 30 gallons of water per acre. Use the higher rates for maximum top kill of vegetation.

1 Registered trademarks of Elanco Products Co.; "Treflan" contains 4 lbs trifluralin per gal, "Surflan" contains 75% oryzalin; "Sonalan" contains 3 lbs ethafluralin per gallon.
2 Registered trademarks of Monsanto Co.; "Lasso" contains 4 lbs alachlor per gal, "Roundup" contains 4 lbs glyphosate per gal; "Bronco" contains 2.6 lbs alachlor and 1.04 lbs glyphosate per gallon.
3 Registered trademark of Ciba-Geigy Corp.; "Dual" 8E contains 8 lbs metolachlor per gal.
4 Registered trademark of American Cyanamid Co.; "Prowl" contains 4 lbs pendimethalin per gal.
5 Paraquat is a restricted use pesticide.
6 Registered trademark of Union Carbide Co.; "Amiben" contains 2 lbs chloramben per gal.

NOTICE TO BUYER: Purchase of this material does not confer any rights under patents of countries outside of the United States.

STORAGE AND DISPOSAL

STORAGE: Store product in original container only, away from other pesticides, fertilizer, food or feed.

PRODUCT DISPOSAL: Do not contaminate water, food or feed by storage or disposal. Wastes resulting from the use of this product may be disposed of on site or at approved waste disposal facility.

CONTAINER DISPOSAL: Triple rinse (or equivalent) the container. Then offer for recycling or reconditioning, or puncture and dispose of in a sanitary landfill, or by incineration, or if allowed by state and local authorities, by burning. If burned, stay out of smoke.

NOTICE OF WARRANTY

Du Pont warrants that this product conforms to the chemical description on the label thereof and is reasonably fit for the purposes stated on such label only when used in accordance with the directions under normal use conditions. It is impossible to eliminate all risks inherently associated with the use of this product. Crop injury, ineffectiveness or other unintended consequences may result because of such factors as weather conditions, presence of other materials, or the manner of use or application, all of which are beyond the control of Du Pont. In no case shall Du Pont be liable for consequential, special or indirect damages resulting from the use or handling of this product. All such risks shall be assumed by the buyer. DU PONT MAKES NO WARRANTIES OF MERCHANTABILITY OR FITNESS FOR A PARTICULAR PURPOSE NOR ANY OTHER EXPRESS OR IMPLIED WARRANTY EXCEPT AS STATED ABOVE.

AG-3252 8049 4/13/89

This is a specimen label. DuPont is not responsible for the accuracy of the information contained herein. As labels are subject to revision, always carefully read and follow the label on the product container.

LONDAX®

HERBICIDE

FOR USE ON RICE IN THE STATES OF ARKANSAS, LOUISIANA, MISSISSIPPI, MISSOURI AND TEXAS

DRY FLOWABLE

ACTIVE INGREDIENTS	BY WEIGHT
Methyl-2-[[[[(4,6-dimethoxypyrimidin-2-yl)amino]carbonyl amino]sulfonyl]methyl]benzoate	60%
INERT INGREDIENTS	40%
TOTAL	100%

EPA Reg. No. 352-506 U.S. Pat. 4,420,325

KEEP OUT OF REACH OF CHILDREN

WARNING

PRECAUTIONARY STATEMENTS

HAZARDS TO HUMANS AND DOMESTIC ANIMALS

WARNING! CAUSES EYE IRRITATION.

Do not get in eyes, on skin or clothing. Wear goggles or face shield when handling. Avoid breathing spray mist. Harmful if swallowed. Wash contaminated clothing before reuse.

STATEMENT OF PRACTICAL TREATMENT

If in eyes, immediately flush with plenty of water. Get medical attention.

If on skin, wash with plenty of soap and water. Get medical attention if irritation persists.

If swallowed, call a physician or poison control center. Drink 1 or 2 glasses of water and induce vomiting by touching back of throat with fingers. Do not induce vomiting or give anything by mouth to an unconscious person.

For medical emergencies involving this product, call toll free: 1-800-441-3637.

ENVIRONMENTAL HAZARDS

Do not apply directly to water or wetlands (swamps, bogs, marshes and potholes) except as specified on this label for use in rice. Do not contaminate water by cleaning of equipment or disposal of wastes.

IMPORTANT

Injury to or loss of desirable trees or vegetation may result from failure to observe the following:

Do not apply or drain or flush equipment on or near desirable trees or other plants, or on areas where their roots may extend, or in locations where the chemical may be washed or moved into contact with their roots.

Do not use on lawns, walks, driveways, tennis courts or similar areas. Prevent drift of spray to desirable plants.

Do not contaminate any body of water.

Keep from contact with fertilizers, insecticides, fungicides and seeds during storage.

Thoroughly clean all traces of "LONDAX" Herbicide from application equipment immediately after use and prior to spraying crops other than rice. Cleanup procedures are described in the "SPRAYER CLEANUP" section of this label. Failure to follow these procedures may result in injury to subsequently sprayed crops.

GENERAL INFORMATION

"LONDAX" Herbicide is a dry flowable formulation for selective preemergence and early postemergence weed control after establishment of the permanent flood. When applied according to the inst.uctons on the label, it will control many broadleaf and sedge weeds. Best control will be achieved when "LONDAX" is applied to very young emerging and actively growing weeds (less than 3 leaves). Degree and duration of control may depend on: use rate, weed spectrum and infestation intensity, weed size at application time, growing conditions at and following time of treatment, soil pH, texture and organic matter, and water management.

Because most crops other than rice are highly sensitive to "LONDAX", all direct or indirect (such as spray drift) contact to crops or land scheduled to be planted to crops other than rice should be avoided as injury may result.

BIOLOGICAL ACTIVITY

"LONDAX" rapidly inhibits growth of susceptible weeds. Following preemergence to early postemergence applications (to weeds), leaves of susceptible plants will appear chlorotic in 3 to 5 days and necrosis of the growing point will subsequently occur. Susceptible plants are controlled in 7 to 21 days, depending on species, as evidenced by complete necrosis of the leaf tissue and growing point. In some cases, affected plants will be stunted and remain green, but are not competitive with the crop.

The herbicidal action of "LONDAX" may be influenced by temperature. Expression of herbicide symptoms is accelerated at warmer temperatures and may be delayed when air or water temperatures are below 65 degrees F. Under cool conditions, the appearance of chlorosis and eventual necrosis in susceptible plants may be delayed beyond 5 days. Occasional temporary chlorosis and/or growth retardation of rice may occur following "LONDAX" application. These symptoms are accentuated at high ambient temperatures and/or cold water temperatures. These symptoms are normally transient and disappear within 2 to 3 weeks after application.

DIRECTIONS FOR USE

It is a violation of Federal law to use this product in a manner inconsistent with its labeling. This labeling must be in possession of the user at the time of pesticide application.

SPRAY TANK PREPARATION

"LONDAX" should be thoroughly mixed with water in a spray tank before adding any other material (i.e., antidrift agent). Agitation is required for uniform mixing and application. Apply "LONDAX" spray preparations within 24 hours of product mixing or product degradation may occur. Thoroughly reagitate before using.

"LONDAX" is compatible with many commercially available antidrift agents, for example (polyvinyl alcohol or polyacrylamide based materials). Antidrift agents are recommended with applications of "LONDAX" in accordance with the labeling directions on the specific antidrift agent. Only antidrift agents approved by EPA can be used.

APPLICATION RECOMMENDATION

"LONDAX" must be applied after establishment of the permanent flood directly into standing water.

WATER SEEDED RICE

When applied as a water mixed spray, to water seeded rice, use a minimum of 5 gallons of water per acre. For best results, "LONDAX" should be applied preemergence to early postemergence to submerged weeds. Best control will be achieved when "LONDAX" is applied to very young emerging and actively growing weeds (less than 3 leaves) after establishment of the permanent flood.

DRY SEEDED RICE

When applied as a water mixed spray, to dry seeded rice, use a minimum of 10 gallons of water per acre to assist in the penetration of the spray mixture through the rice canopy. For best results, "LONDAX" should be applied pre-emergence to early postemergence to submerged weeds. Best control will be achieved when "LONDAX" is applied to very young actively growing weeds (less than 3 leaves) after establishment of the permanent flood.

WATER MANAGEMENT FOLLOWING APPLICATION

Good water management at the time of "LONDAX" application and for at least 7 days following will ensure maximum product performance. Hold water static on the field for a minimum of 7 days after application. Holding water on the field for more than 7 days will enhance product performance. Failure to hold water for the 7-day minimum will decrease product performance. Rainfall sufficient to cause water flow off the field may reduce performance if it occurs within 7 days of application.

At the time of application and for the following 7 days, water on the field should be held static. Introduction of water may reduce herbicidal efficacy in areas adjacent to water inlet sites. Water level at the time of application and within 7 days of application should be high enough so that the soil is completely covered with water (preferably at least 3 inches with no exposed soil). Product performance will be reduced in areas where soil is not covered with water. Target weeds' foilage must remain covered with water at time of application and during the water holding period.

Note: Static water is defined as not allowing water to exit the field. Water may be introduced into the field to maintain flood level, but this may result in reduced control in areas around the water inlet(s).

WEEDS

"LONDAX" effectively controls the following weeds when used according to instructions on the label:

Annual arrowhead spp.: Sagittaria spp.	Pickerelweed: Pontederia cordata
Blunt Spikerush: Eleocharis obtusa	Purple ammannia: Ammanaia coccinea
Dayflower: Commelina communis	Redstem: Ammannia auriculata
Ducksalad: Heteranthera limosa	Rice Flatsedge: Cyperus iria
Eclipta: Eclipta alba	Roughseed bulrush: Scirpus mucronatus
Eisen waterhyssop: Bacaopa eisenii	Southern naiad: Najas quadalupensis
Falsepimpernel: Lindernia spp.	Texasweed: Caperonia palustris
Gooseweed: Sphenoclea zeylanica	Water plantain (seedling) Alisma spp.
Mexicanweed: Caperonia castanaefolia	Waterwort: Elatine spp.
Mud plantain: Heteranthera reniformis	Yellow nutsedge: Cyperus esculentus

"LONDAX" will result in marked reduction in growth and vigor (stunting):

Alligatorweed: Alternanthera philoxeroides

USE RATE: A maximum of 1 oz. active ingredient (1 2/3 oz. product) "LONDAX" Herbicide per acre per season is recommended.

A recommended rate range in Arkansas, Louisiana, Mississippi, Missouri and Texas is 1 to 1 2/3 oz. product per acre. For best results "LONDAX" should be applied to static water immediately after the establishment of permanent flood.

For Alligator weed (runners under 12 inches) 1 2/3 oz. "LONDAX" will result in marked reduction in growth and vigor (stunting).

For all other weeds on label, 1 to 1 2/3 oz. "LONDAX" will provide control if the labeled application and water management recommendations are followed. Use the higher rate where weed density is high or where the rice canopy reduces penetration of the spray mixture to the water.

If two applications of "LONDAX" are made per year, the total amount of "LONDAX" can not exceed 1 2/3 oz. of product per acre (1.0 oz. active ingredient).

COMBINATION RECOMMENDATIONS

"LONDAX" may be used in conjunction with rice herbicides used for grass control.

"LONDAX" can be applied following propanil applications. Do not apply "LONDAX" until a permanent flood has been established following propanil application.

Observe all applicable directions, restrictions (including water holding requirements) and precautions on the Ordram[2], Bolero[3] and propanil labels.

Note: Hold water in "LONDAX" treated fields for a minimum of 7 days.

SPRAYER CLEANUP

IMPORTANT - When "LONDAX" is applied as a spray, tank and spray equipment clean out is important and should be done immediately after spraying to avoid subsequent injury to crops other than rice. Use the following procedure for spray equipment clean out before using sprayer to treat rice:

1) Drain tank, then flush tank, boom and hoses with clean water for 10 minutes.
2) Repeat Step 1.
3) Fill the tank with clean water, then add Nutra-Sol[4] at a rate of 32 ounces of product per 100 gallons of water. Flush through the boom and hoses, then allow to sit for 15 minutes with agitation (and recirculation, if possible); drain equipment making sure to flush the boom and all hoses thoroughly.
4) Nozzles and screens should be removed and cleaned separately.
5) Thoroughly rinse sprayer, boom and hoses with clean water to remove Nutra-sol.
6) The rinsate may be disposed of on site or at an approved waste disposal facility.

Use this clean-out procedure if the sprayer will be used to treat crops other than rice:

1) Drain tank, then flush tank, boom and hoses with clean water for 10 minutes.
2) Fill the tank with clean water, then add 1/2 gallon of chlorine bleach (containing 5 1/4% sodium hypochlorite) per 100 gallons of water. Flush through boom and hoses, allow to sit for 15 minutes with agitation, then drain.
3) Repeat step 2.
4) Nozzles and screens should be removed and cleaned separately. To remove traces of chlorine bleach, rinse the tank thoroughly with clean water and flush through hoses and boom.
5) Dispose of rinsates only at an approved waste disposal facility.

CAUTION: Do not use chlorine bleach with ammonia. All traces of liquid fertilizer containing ammonia, ammonium nitrate or ammonium sulphate must be rinsed with water from the mixing and application equipment before adding chlorine bleach solution. Failure to do so will release a gas with a musty odor which can cause eye, nose, throat and lung irritation. Do not clean equipment in an enclosed area.

RESTRICTIONS

Do not apply more than 1 oz. "LONDAX" active ingredient per acre per season (1²/₃ oz. product).

Do not graze treated fields or feed treated forage within 80 days of last application.

Do not apply "LONDAX" within 80 days of harvest.

Do not apply "LONDAX" to rice under stress from abnormal weather or growing conditions, drought, disease, insect or prior herbicide injury, as crop injury may occur. Severe stress, drought, disease or insect damage following application may also result in crop injury.

Water drained directly from treated fields must not be used to irrigate other crops.

Do not mix "LONDAX" with any additives except as directed by this label.

Do not use "LONDAX" on wild rice (Zizania aquatica).

Do not rotate to crops other than rice for 120 days following application.

Do not farm crawfish in treated fields.

LONDAX - Registered trademark of E.I. Du Pont de Nemours & Co. (Inc.).
[2] Registered trademark of ICI Americas, Inc., Wilmington, Delaware.
[3] Registered trademark of Chevron Chemical Company, San Francisco, California.
[4] Nutra-Sol is a product of Thomas G. Kilfoil Company, Inc., San Bruno, California.

STORAGE AND DISPOSAL

STORAGE: Store product in original container only, away from other pesticides, fertilizer, food or feed. Not for use or storage in or around the home. Keep container closed.

PESTICIDE DISPOSAL: Do not contaminate water, food or feed by disposal. Waste resulting from the use of this product may be disposed of on site or at an approved waste disposal facility.

CONTAINER DISPOSAL: Emptied paper (or cardboard) containers may be burned if allowed by state and local authorities. If burned, stay out of smoke.

NOTICE TO BUYER Purchase of this material does not confer any rights under patents of countries outside the United States.

NOTICE OF WARRANTY

Du Pont warrants that this product conforms to the chemical description on the label thereof and is reasonably fit for the purposes stated on such label only when used in accordance with the directions under normal use conditions. It is impossible to eliminate all risks inherently associated with the use of this product. Crop injury, ineffectiveness, or other unintended consequences may result because of such factors as weather conditions, presence of other materials, or the manner of use or application, all of which are beyond the control of Du Pont. In no case shall Du Pont be liable for consequential, special or indirect damages resulting from the use or handling of this product. All such risks shall be assumed by the buyer. DU PONT MAKES NO WARRANTIES OF MERCHANTABILITY OR FITNESS FOR A PARTICULAR PURPOSE NOR ANY OTHER EXPRESS OR IMPLIED WARRANTY EXCEPT AS STATED ABOVE.

AG-3246 8029 2/17/89

This is a specimen label. DuPont is not responsible for the accuracy of the information contained herein. As labels are subject to revision, always carefully read and follow the label on the product container.

LONDAX®

HERBICIDE
FOR USE ON RICE IN THE STATE OF CALIFORNIA

DRY FLOWABLE

ACTIVE INGREDIENTS	BY WEIGHT
Methyl 2-[[[[(4,6-dimethoxypyrimidin-2-yl)amino]carbonyl]amino]sulfonyl]methyl]benzoate	60%
INERT INGREDIENTS	40%
TOTAL	100%

EPA Reg. No. 352-506 U.S. Pat. 4,420,325

KEEP OUT OF REACH OF CHILDREN
WARNING

PRECAUTIONARY STATEMENTS

HAZARDS TO HUMANS AND DOMESTIC ANIMALS

WARNING! CAUSES EYE IRRITATION

Do not get in eyes, on skin or clothing. Wear goggles or face shield when handling. Avoid breathing spray mist. Harmful if swallowed. Wash contaminated clothing before reuse.

STATEMENT OF PRACTICAL TREATMENT

If in eyes, immediately flush with plenty of water. Get medical attention.

If on skin, wash with plenty of soap and water. Get medical attention if irritation persists.

If swallowed, call a physician or poison control center. Drink 1 or 2 glasses of water and induce vomiting by touching back of throat with fingers. Do not induce vomiting or give anything by mouth to an unconscious person.

For medical emergencies involving this product, call toll free: 1-800-441-3637.

ENVIRONMENTAL HAZARDS

Do not apply directly to water or wetlands (swamps, bogs, marshes and potholes) except as specified on this label for use in rice. Do not contaminate water by cleaning of equipment or disposal of wastes.

IMPORTANT

Injury to or loss of desirable trees or vegetation may result from failure to observe the following:

Do not apply or drain or flush equipment on or near desirable trees or other plants, or on areas where their roots may extend, or in locations where the chemical may be washed or moved into contact with their roots.

Do not use on lawns, walks, driveways, tennis courts or similar areas. Prevent drift of spray to desirable plants.

Do not contaminate any body of water.

Keep from contact with fertilizers, insecticides, fungicides and seeds during storage.

Thoroughly clean all traces of "LONDAX" Herbicide from application equipment immediately after use and prior to spraying crops other than rice. Cleanup procedures are described in the "SPRAYER CLEANUP" section of this label. Failure to follow these procedures may result in injury to subsequently sprayed crops.

GENERAL INFORMATION

"LONDAX" Herbicide is a dry flowable formulation for selective preemergence and early postemergence weed control after establishment of the permanent flood. When applied according to the instructions on the label, it will control many broadleaf and sedge weeds. Best control will be achieved when "LONDAX" is applied to very young emerging and actively growing weeds (less than 3 leaves). Degree and duration of control may depend on: use rate, weed spectrum and infestation intensity, weed size at application time, growing conditions at and following time of treatment, soil pH, texture and organic matter, and water management.

Because most crops other than rice are highly sensitive to "LONDAX", all direct or indirect (such as spray drift) contact to crops or land scheduled to be planted to crops other than rice should be avoided as injury may result.

BIOLOGICAL ACTIVITY

"LONDAX" rapidly inhibits growth of susceptible weeds. Following preemergence to early postemergence applications (to weeds), leaves of susceptible plants will appear chlorotic in 3 to 5 days and necrosis of the growing point will subsequently occur. Susceptible plants are controlled in 7 to 21 days, depending on species, as evidenced by complete necrosis of the leaf tissue and growing point. In some cases, affected plants will be stunted and remain green, but are not competitive with the crop.

The herbicidal action of "LONDAX" may be influenced by temperature. Expression of herbicide symptoms is accelerated at warmer temperatures and may be delayed when air or water temperatures are below 65 degrees F. Under cool conditions, the appearance of chlorosis and eventual necrosis in susceptible plants may be delayed beyond 5 days. Occasional temporary chlorosis and/or growth retardation of rice may occur following "LONDAX" application. These symptoms are accentuated at high ambient temperatures and/or cold water temperatures. These symptoms are normally transient and disappear within 2 to 3 weeks after application.

DIRECTIONS FOR USE

It is a violation of Federal law to use this product in a manner inconsistent with its labeling. This labeling must be in possession of the user at the time of pesticide application.

SPRAY TANK PREPARATION

"LONDAX" should be thoroughly mixed with water in a spray tank before adding any other material (i.e., antidrift agent). Agitation is required for uniform mixing and application. Apply "LONDAX" spray preparations within 24 hours of product mixing or product degradation may occur. Thoroughly reagitate before using.

"LONDAX" is compatible with many commercially available antidrift agents including, Nalcotrol[2], Unite[3], Poly Control[4], Air Drop[5], for example (polyvinyl alcohol or polyacrylamide based materials). Antidrift agents are recommended with applications of "LONDAX" in accordance with the labeling directions of the specific antidrift agent. Only antidrift agents approved by EPA can be used.

APPLICATION RECOMMENDATION

"LONDAX" must be applied after establishment of the permanent flood directly into standing water.

WATER SEEDED RICE

When applied as a water mixed spray, to water seeded rice, use a minimum of 5 gallons of water per acre. For best results, "LONDAX" should be applied preemergence to early postemergence to submerged weeds and post emergence to rice from the 1- to 3-leaf stage. Best control will be achieved when "LONDAX" is applied to very young emerging and actively growing weeds (less than 3 leaves) after establishment of the permanent flood. "LONDAX" can be applied to rice beyond the 3-leaf stage, but late applications should be targeted to the preemergence to early postemergence stage of weeds.

WATER MANAGEMENT FOLLOWING APPLICATION

Good water management at the time of "LONDAX" application and for at least 5 days following will ensure maximum product performance. Hold water static on the field for a minimum of 5 days after application. Holding water on the field for more than 5 days will enhance product performance. Failure to hold water for the 5-day minimum will decrease product performance. Rainfall sufficient to cause water flow off the field may reduce performance if it occurs within 5 days of application.

At the time of application and for the following 5 days, water on the field should be held static. Introduction of water may reduce herbicidal efficacy in areas adjacent to water inlet sites. Water level at the time of application and within 5 days of application should be high enough so that the soil is completely covered with water (preferably at least 3 inches with no exposed soil). Product performance will be reduced in areas where soil is not covered with water. Target weeds' foliage must remain covered with water at time of application and during the water holding period.

Note: Static water is defined as not allowing water to exit the field. Water may be introduced into the field to maintain flood level, but this may result in reduced control in areas around the water inlet(s).

WEEDS

"LONDAX" effectively controls the following weeds when used according to instructions on the label:

Blunt Spikerush: Eleocharis obtusa
California Arrowhead: Sagittaria montevidensis calycina
Ducksalad: Heteranthera limosa
Eisen waterhyssop: Bacaopa eisenii
Purple ammannia: Ammanaia coccinea
Redstem: Ammannia auriculata
Roughseed bulrush: Scirpus mucronatus
Southern naiad: Najas quadalupensis
Smallflower umbrellaplant: Cyperus difformis
Water plantain (seedling) Alisma spp
Waterwort: Elatine spp

USE RATE: A maximum of 1 oz. active ingredient "LONDAX" Herbicide per acre per season is recommended.

The recommended rate in California is $1^2/_3$ oz. product per acre.

COMBINATION RECOMMENDATIONS

"LONDAX" may be used in conjunction with rice herbicides used for grass control. Sequential treatment of "LONDAX" with "Ordram"[6] or "Bolero"[7] will aid in control of barnyardgrass and watergrass, in addition to the above listed weeds. Applications of "LONDAX" should be made on the same day as "Ordram" and "Bolero" applications, if possible, or as soon as possible prior to or after application of these herbicides.

"LONDAX" can be applied following propanil applications. Do not apply "LONDAX" until a permanent flood has been established following propanil application.

"LONDAX" can be applied prior to propanil applications providing that all water management practices described by this label with respect to "LONDAX" use are followed.

Observe all applicable directions, restrictions (including water holding requirements) and precautions on the "Ordram", "Bolero" and propanil labels.

"LONDAX" has been successfully tank mixed with Red Top Methyl Parathion 5 Spray[8] in California. Tank mixes containing "LONDAX" and this brand of parathion should be used immediately after mixing. Consult your local Du Pont representative if tank mixes with other brands of parathion are planned.

SPRAYER CLEANUP

IMPORTANT - When "LONDAX" is applied as a spray, tank and spray equipment clean out is important and should be done immediately after spraying to avoid subsequent injury to crops other than rice. Use the following procedure for spray equipment clean out before using sprayer to treat rice:

1) Drain tank, then flush tank, boom and hoses with clean water for 10 minutes.

2) Repeat Step 1.

3) Fill the tank with clean water, then add Nutra-Sol[9] at a rate of 32 ounces of product per 100 gallons of water. Flush through the boom and hoses, then allow to sit for 15 minutes with agitation (and recirculation, if possible); drain equipment making sure to flush the boom and all hoses thoroughly.

4) Nozzles and screens should be removed and cleaned separately.

5) Thoroughly rinse sprayer, boom and hoses with clean water to remove Nutra-sol.

6) The rinsate may be disposed of on site or at an approved water disposal facility.

Use this clean-out procedure if the sprayer will be used to treat crops other than rice:

1) Drain tank, then flush tank, boom and hoses with clean water for 10 minutes.

2) Fill the tank with clean water, then add $^1/_2$ gallon of chlorine bleach (containing $5^1/_4$% sodium hypochlorite) per 100 gallons of water. Flush through boom and hoses, allow to sit for 15 minutes with agitation, then drain.

3) Repeat step 2.

4) Nozzles and screens should be removed and cleaned separately. To remove traces of chlorine bleach, rinse the tank thoroughly with clean water and flush through hoses and boom.

5) Dispose of rinsates only at an approved waste disposal facility.

CAUTION: Do not use chlorine bleach with ammonia. All traces of liquid fertilizer containing ammonia, ammonium nitrate or ammonium sulphate must be rinsed with water from the mixing and application equipment before adding chlorine bleach solution. Failure to do so will release a gas with a musty odor which can cause eye, nose, throat and lung irritation. Do not clean equipment in an enclosed area.

RESTRICTIONS

Do not apply more than 1 oz. active ingredient "LONDAX" per acre per season ($1^2/_3$ oz. product).

Do not graze treated fields or feed treated forage within 80 days of last application.

Do not apply "LONDAX" within 80 days of harvest.

Do not apply "LONDAX" to rice under stress from abnormal weather or growing conditions, drought, disease, insect or prior herbicide injury, as crop injury may occur. Severe stress, drought, disease or insect damage following application may also result in crop injury.

Water drained directly from treated fields must not be used to irrigate other crops.

Do not mix "LONDAX" with any additives except as directed by this label.

Do not use "LONDAX" on wild rice (Zizania aquatica).

Do not rotate to crops other than rice for 120 days following application.

LONDAX - Registered trademark of E. I. Du Pont de Nemours & Co. (Inc.).
[2] Nalcotrol is a product of Nalco Chemical Company, Carson, California.
[3] Unite is a product of Hopkins Agricultural Chemical Company, Madison, Wisconsin.
[4] Poly Control is a product of JLB International Chemical Company, Inc., Vero Beach, Florida.
[5] Air Drop is a product of Knapp Manufacturing Co., Fresno, California.
[6] Registered trademark of ICI Americas, Inc., Wilmington, Delaware.
[7] Registered trademark of Chevron Chemical Company, San Francisco, California.
[8] Red Top Methyl Parathion 5 Spray is a product of Wilbur Ellis Company, Fresno, California.
[9] Nutra-Sol is a product of Thomas G. Kilfoil Company, Inc., San Bruno, California.

STORAGE AND DISPOSAL

STORAGE: Store product in original container only, away from other pesticides, fertilizer, food or feed. Not for use or storage in or around the home. Keep container closed.

PESTICIDE DISPOSAL: Do not contaminate water, food or feed by disposal. Waste resulting from the use of this product may be disposed of on site or at an approved waste disposal facility.

CONTAINER DISPOSAL: Emptied paper (or cardboard) containers may be burned if allowed by state and local authorities. If burned, stay out of smoke.

NOTICE TO BUYER Purchase of this material does not confer any rights under patents of countries outside the United States.

NOTICE OF WARRANTY

Du Pont warrants that this product conforms to the chemical description on the label thereof and is reasonably fit for the purposes stated on such label only when used in accordance with the directions under normal use conditions. It is impossible to eliminate all risks inherently associated with the use of this product. Crop injury, ineffectiveness, or other unintended consequences may result because of such factors as weather conditions, presence of other materials, or the manner of use or application, all of which are beyond the control of Du Pont. In no case shall Du Pont be liable for consequential, special or indirect damages resulting from the use or handling of this product. All such risks shall be assumed by the buyer. DU PONT MAKES NO WARRANTIES OF MERCHANTABILITY OR FITNESS FOR A PARTICULAR PURPOSE NOR ANY OTHER

EXPRESS OR IMPLIED WARRANTY EXCEPT AS STATED ABOVE.

AG - 3245 8029 2/17/89

This is a specimen label. DuPont is not responsible for the accuracy of the information contained herein. As labels are subject to revision, always carefully read and follow the label on the product container.

LOROX® DF

HERBICIDE
DISPERSIBLE GRANULES

ACTIVE INGREDIENT:	BY WEIGHT
Linuron [3-(3,4-dichlorophenyl)-1-methoxy-1-methylurea]	50%
INERT INGREDIENTS:	50%
TOTAL	100%

EPA Reg. No. 352-394

KEEP OUT OF REACH OF CHILDREN

PRECAUTIONARY STATEMENTS

HAZARDS TO HUMANS AND DOMESTIC ANIMALS

CAUTION!

HARMFUL IF SWALLOWED. CAUSES EYE IRRITATION. MAY IRRITATE NOSE, THROAT AND SKIN.

Avoid breathing dust or spray mist. Avoid contact with skin, eyes and clothing. Wash thoroughly with soap and water after handling.

STATEMENT OF PRACTICAL TREATMENT

If in eyes: Flush with plenty of water. Get medical attention if irritation persists.

If on skin: Wash with plenty of soap and water. Get medical attention if irritation persists.

If swallowed: Call a physician or Poison Control Center. Drink 1 or 2 glasses of water and induce vomiting by touching back of throat with finger. Do not induce vomiting or give anything by mouth to an unconscious person.

For medical emergencies involving this product, call toll free 1-800-441-3637.

ENVIRONMENTAL HAZARDS

Do not apply directly to water or wetlands. Do not contaminate water by cleaning of equipment or disposal of wastes.

IMPORTANT—When mixing/loading and spraying wear a wide-brimmed hat, mid-forearm to elbow length chemical resistant gloves, and one-piece overalls which have long sleeves and long pants. Overalls should be constructed of finely woven fabric as specified in the USDA/EPA Guide for Commercial Applicators.

Any article of clothing worn while applying product must be cleaned before reusing. Clothing which has been drenched or has otherwise absorbed concentrated pesticide must be disposed in a sanitary landfill, or incinerated, or if allowed by State and local authorities, by burning.

Instead of clothing and equipment specified above, the applicator can use an enclosed tractor cab which provides a filtered air supply (as described by Taschenberg and Bourke, 1975).

Injury to or loss of desirable trees or other plants may result from failure to observe the following:

Do not apply (except as recommended), or drain or flush equipment on or near desirable trees or other plants, or on areas where their roots may extend, or in locations where the chemical may be washed or moved into contact with their roots. Do not use on home plantings of trees, shrubs or herbaceous plants, nor on lawns, walks, driveways, tennis courts, or similar areas. Prevent drift of dry powder or spray to desirable plants. Do not contaminate any body of water. Keep from contact with fertilizers, insecticides, fungicides and seeds.

Do not apply aerially.

Thoroughly clean all traces of "LOROX" DF from application equipment immediately after use. Flush tank, pump, hoses, and boom with several changes of water after removing nozzle tips and screens (clean these parts separately).

GENERAL INFORMATION

Du Pont "LOROX" DF Herbicide is a dispersible granule to be mixed in water and applied as a spray for selective control of weeds in certain crops and for non-selective weed control on non-cropland areas. It is non-corrosive to equipment, non-flammable and non-volatile.

"LOROX" DF is an effective preemergence treatment of soybeans, asparagus, bulbs, carrots, corn, hybrid poplar, parsnips, potatoes and sorghum for control of weeds listed below. "LOROX" DF is particularly effective in non-tillage soybean production. Treatment is made after planting but before emergence of soybeans. Weeds have usually emerged and the treatment controls susceptible weeds. Treatment also controls susceptible emerging weeds for an extended time.

"LOROX" DF may be sprayed postemergence over-the-top of emerged asparagus (before or after cutting and to small ferns), carrots (at least 3" tall) and celery (transplants) to effectively control susceptible emerged seedling weeds.

Directed postemergence applications of "LOROX" DF, where spray nozzles are adjusted so that weeds are sprayed but the crop is not, may be used in emerged asparagus (larger ferns), corn (field and sweet), cotton, sorghum and soybeans to effectively control susceptible emerged seedling weeds.

"LOROX" DF may be applied to soil prior to emergence of weeds to control susceptible weed seedlings for an extended period of time; the degree of control and duration of effect will vary with:

Soil Organic Matter Content—High organic matter requires higher dosages than low organic matter for equivalent herbicide performance. Best results accompany preemergence use of "LOROX" DF on soils of 1/2% to 3% organic matter. The combination of "LOROX" DF plus Du Pont LEXONE DF Herbicide (or Du Pont LEXONE 4L Herbicide) provides most effective preemergence weed control on soils from 1% to 4% organic matter content.

Soil Texture—High clay content requires higher dosage rates than those with less clay. Best results accompany preemergence use of "LOROX" DF on coarse to medium soils.

Rainfall—Moisture is required to activate the chemical following preemergence application to soil. Best results accompany rainfall (or irrigation) that occurs within 2 weeks of application. In the Colombia River Basin, use "LOROX" DF only if the crop is sprinkler irrigated.

Weed Species And Conditions of Growth—Species vary in susceptibility and may be more difficult to control when under stress.

"LOROX" DF is recommended for use in conjunction with certain other herbicides for treatment of field corn, potatoes, sorghum, and soybeans; observe all cautions and limitations on labeling of all products used in mixtures.

Since the effect of "LOROX" DF varies with soils, crop varieties, uniformity of application, and environmental conditions, it is suggested that growers limit their first use to small areas.

PREEMERGENCE USE (Germinating Weeds): "LOROX" DF, at recommended rates, controls annual weeds such as:

Broadleaves		Grasses
Florida beggarweed	Mustards	Barnyardgrass (watergrass)
Carpetweed	Nettleleaf goosefoot	Canarygrass
Chickweed	Pigweeds	Crabgrasses
Common dayflower	Purslane (common)	Foxtails (including giant)
Florida purslane (Florida pusley)	Wild radish	Goosegrass
Galinsoga	Ragweed (common)	Fall panicum
Lambsquarters	Smartweed (Pennsylvania)	

Partial Control

Cocklebur (common)
Annual morningglory
Prickly sida (teaweed)
Sicklepod
Velvetleaf (buttonweed)

"LOROX" DF will not control established perennials such as bermudagrass, Canada thistle, field bindweed, johnsongrass and purple nutsedge. The lower dosage rates are effective on coarser soils and the higher rates on finer soils and on the more resistant seedling weeds. Sufficient moisture (1/2" to 1" on moist soils; 1" to 2" on dry soils) in the form of rainfall or sprinkler irrigation is necessary after treatment to carry the chemical into the root zone of germinating weeds; best results are obtained when this occurs within two weeks after application. If heavy rainfall occurs soon after application, injury to crop may result.

"LOROX" DF applied preemergence, before emergence of soybeans, asparagus, bulbs, carrots, corn (field), parsnips, potatoes, and sorghum, and weeds, is an effective procedure because susceptible weeds are controlled in an early, vulnerable seedling stage before they compete with the crop. With favorable moisture conditions, "LOROX" DF continues to control weeds for some time as the crop becomes better able to compete. Should weed seedlings begin to break through the preemergence treatment in significant numbers, secondary weed control procedures should be implemented; these include cultivation and postemergence herbicide application.

A good seed bed must be prepared before application of "LOROX" DF as crop injury may result if application is made to ground which is cloddy or compacted resulting in improperly planted seed. Plant seed to depth specified. Surface of the soil should not be cultivated or disturbed after application of "LOROX" DF and before emergence of the crop as weed control may be reduced and crop injury may result. However, if moisture is insufficient to activate the herbicide, a shallow cultivation (rotary hoe preferred) should be made after emergence of row crops while weeds are small enough to be controlled by mechanical means. Deep cultivation reduces the effectiveness of "LOROX" DF.

Note: Consult "Amiben"[1], atrazine, "Bronco"[3], "Dual"[2] 8E, "Lasso"[3], paraquat[4], propazine, "Prowl"[5], "Ramrod"[3], "Roundup"[3], "Surflan"[6], and "Treflan"[6] labels for additional weeds controlled when "LOROX" DF is used in conjunction with these herbicides.

POSTEMERGENCE USE (Emerged Seedling Weeds): "LOROX" DF, at recommended rates, controls weeds such as:

Broadleaves		Grasses
Annual morningglory	Lambsquarters	Annual ryegrass
Carpetweed	Mustards	Barnyardgrass (watergrass)
Cocklebur (common)	Nettleleaf goosefoot	Broadleaf signalgrass
Common dayflower	Pigweeds	
Common ragweed	Prickly sida (teaweed)	Canarygrass
Dog fennel		
Fiddleneck (amsinckia)	Purslane (common)	Crabgrasses
	Sesbania	Fall panicum
Florida beggarweed	Sicklepod	Foxtails (including giant)
Florida purslane (Florida pusley)	Smartweed, (Pennsylvania)	
Groundsel	Velvetleaf (buttonweed)	Goosegrass
Knawel		Rattail fescue
Lambsquarters	Wild buckwheat	Texas panicum

Results of postemergence treatment of emerged weeds vary with rate applied and environmental conditions; best results are obtained on succulent weeds growing under conditions of high humidity and temperatures of 70°F or higher. Addition of a surfactant to the spray (where recommended) increases contact effects of "LOROX" DF. Application will also provide control of emerging susceptible weed seedlings for an extended period of time.

Note: Consult bromoxynil, "Butoxone"[7], and "Butyrac" 200[1] labels for additional weeds controlled when "LOROX" DF is used in conjunction with these herbicides. Best results are obtained under conditions of high humidity and temperatures over 70°F. Control of emerged weeds under drought stress is usually impractical.

DIRECTIONS FOR USE

It is a violation of federal law to use this product in a manner inconsistent with its labeling.

"LOROX" DF Herbicide should be used only in accordance with recommendations on this label, or in separate published Du Pont recommendations available through local dealers. Do not use on any crop in Kern County, California, except carrots and asparagus when a permit has been obtained from the Agricultural Commission.

Do not apply this product through any type of irrigation system.

Du Pont will not be responsible for losses or damages resulting from use of this product in any manner not specifically recommended by Du Pont. User assumes all risk associated with such non-recommended use.

Fill tank 1/4 full with water. Start agitation system; add "LOROX" DF, and continue adding water. Add separately each additional component of any tank mix while adding water. If a surfactant is used, add it last when the tank is nearly full. Continue agitation throughout. If poor mixing should occur with any component, premix the component with two parts water before adding to the spray tank.

FERTILIZER SPRAY MIXTURES: For preemergence application, nonpressure nitrogen or fertilizer solution may be used in the spray mixture unless otherwise directed. Small quantities should be tested for compatibility by the following procedure before full scale mixing:

1. Put 1 pint of fertilizer solution in a quart jar.
2. Mix 2 teaspoons "LOROX" DF with 2 tablespoons of water; mix thoroughly and add to fertilizer solution.
3. Close jar and shake well.
4. If other herbicides are to be used in the mixture, premix 2 teaspoons of wettable powders or 1 teaspoon of liquids with 2 tablespoons of water; add to "LOROX" DF/fertilizer mixture.
5. Close jar and shake well.
6. Watch mixture for several seconds; check again in 30 minutes.
7. If mixture does not separate, foam, gel, or become lumpy, it may be used.
8. Mixing ability may be improved by adding compatibility agents such as Kalo Laboratories "Complex" or Witco Chemicals "Sponto" 168D. Follow directions on container.

If the mixture is compatible, prepare spray by adding the fertilizer solution to the spray tank first; then dilute "LOROX" DF with 2 parts of water, thoroughly mix, and add slowly to spray tank with agitator running. For tank mixtures with other herbicides, follow directions above.

EQUIPMENT—SPRAY VOLUMES AND PRESSURES: Application is limited to a ground boom sprayer properly calibrated to a constant speed and rate of delivery. Openings in screens should be equal to or larger than 50 mesh. Continuous agitation in the spray tank is required to keep the material in suspension. Agitate by mechanical or hydraulic means; if bypass or return line is used, it should terminate at bottom of tank to minimize foaming. Avoid overlapping and shut off spray booms while starting, turning, slowing or stopping, or injury to the crop may result.

For preemergence application, use 20 to 60 gals per acre and spray pressure of 30 to 40 psi.

For postemergence application, use sufficient volume (min 20 gals per acre) for thorough coverage of weed foliage; use spray pressure of 20 to 25 psi to keep spray drift to a minimum.

USE RATES: All rates of "LOROX" DF (and tank mixtures) are expressed as broadcast rates; for band treatment, use proportionately less. For example, use 1/3 of the broadcast rate when treating a 14" band where row spacing is 42". Where a range of dosages is given, use the lower rate on coarser soils (low in clay or organic matter) and the higher rate on finer soils (high in clay or organic matter); for postemergence application, use the lower rate on smaller weeds and the higher rate on larger weeds.

SOIL LIMITATIONS: Unless otherwise directed, do not use on sand, loamy sand, gravelly soils or exposed subsoils nor on soils containing less than 1% organic matter as injury to the treated or subsequent crops may result.

REPLANTING: If initial seeding fails to produce a stand, crops registered for the rate of "LOROX" DF that has been applied may be replanted into the treated area. Thoroughly rework soil before replanting. Do not retreat during the same crop year as injury to the crop may result.

CROP ROTATION RECOMMENDATIONS: Unless otherwise directed, any crop may be planted after 4 months except for cereals where only barley, oats, rye and wheat may be planted.

West of Rocky Mountains: Carrots or celery may be planted 4 months after last application; do not plant any other crop until 1 year after the last application as injury may result.

CROP USES

ASPARAGUS

(California, Michigan, Minnesota, North Carolina, Oregon, Washington only)

Direct Seeded or Newly Planted Crowns—Do not exceed 4 lbs total per acre per season; do not use surfactant or fertilizer solution in spray mixture.

Preemergence Application:* Make a single application of 2 to 4 lbs per acre after planting seed 1 1/2" deep in coarse soil and 1" deep in fine soils. During planting operation, spray activated charcoal ("Aqua Nu-Char"[8] or "Gro-Safe"[8]) as a 1" band on soil surface directly over seed rows at the rate of 300 lbs per acre (equivalent to 15 lbs per acre of crop with 20" row spacing).

Postemergence Application: Make 1 or 2 applications of 1 to 2 lbs per acre when ferns are in 6" to 18" stage and weeds are not over 4" tall.

Established Beds: Do not apply within 1 day of harvest; do not exceed 8 lbs total per acre per season; do not use surfactant or fertilizer solution in spray mixture. Do not reenter treated areas for 24 hours following application unless protective clothing is worn.

Preemergence Application:* Make a single application of 2 to 4 lbs per acre.

Postemergence Application: Make 1 to 4 applications of 1 to 2 lbs per acre before weeds exceed 4" in height. Apply before cutting season or immediately after cutting. Do not reenter treated areas for 24 hours following application unless protective clothing is worn.

Directed Postemergence Application (Fern Stage): Make a single application of 4 to 8 lbs per acre as a directed. Spray to base of plants for control of dudain melon.

Note: If more than 4 lbs "LOROX" DF per acre is applied per season, do not plant any other crop until 1 year after last application. See "Replanting". Do not enter treated postemergence for 24 hours following application unless protective clothing is worn.

In California, dudain melon and annual nightshade are controlled.

*Preemergence weed control will be reduced in soils with high organic matter (greater than 5% and peat or muck.)

BULBS (Tulip, Calla Lily, Daffodil and Dutch Iris Only)

California

After planting of bulbs, settle the soil with sprinkler irrigation (rainfall will serve the same purpose); then before emergence of plants (bulbs), apply 2 pounds "LOROX" DF per acre in a minimum of 20 gallons of water per acre. Treat only once during the growing season.

Note: Unless other directed, do not use on sand, loamy sand, gravelly soils, or exposed subsoils nor on soils containing less than 1% organic matter as injury to the crop may result. Do not reenter treated areas for 24 hours following application unless protective clothing is worn.

CARROTS

Preemergence Application—Florida, Michigan, Ohio, and Wisconsin: Make a single application of 1 to 2 lbs per acre in Florida and 1 to 3 lbs per acre in Michigan, Ohio, and Wisconsin; apply after planting but before carrots emerge; plant seed at least 1/2" deep. Use the lower rate on lighter soils (low in clay or organic matter) and higher rate on heavier soils (high in clay or organic matter). Subsequent postemergence application may be made provided the total does not exceed 4 lbs "LOROX" DF per acre per season.

Postemergence Application—U.S.: Apply 1 1/2 to 3 lbs per acre as a non-directed spray after carrots are at least 3" tall. Apply before annual grasses exceed 2" in height and before broadleaf weeds exceed 6" in height. Repeat application may be made but do not exceed 4 lbs "LOROX" DF per acre (West of Rocky Mountains, do not exceed 3 lbs "LOROX" DF per acre). Do not exceed 40 psi spray nozzle pressure as crop injury may result.

"LOROX" DF may be applied following an application of Stoddard solvent provided treatments are at least one day apart. Stoddard solvent may be applied following application of "LOROX" DF provided treatments are at least 2 weeks apart. Shorter time intervals between applications may result in injury to the crop. Do not apply "LOROX" DF as a tank mixture with Stoddard solvent, surfactants, nitrogen or fertilizer solution, other pesticides, nor when temperature exceeds 85°F. as crop injury may result.

Note: Because carrot varieties vary in their resistance, determine tolerance to "LOROX" DF prior to adoption as a field practice to prevent possible crop injury; do not treat susceptible varieties which show an initial burning of foliage following postemergence treatment with "LOROX" DF. Do not reenter treated areas treated postemergence for 24 hours following application unless protective clothing is worn. See "Replanting".

CELERY

East of Rocky Mountains—Post-transplant Application: Make a single application of 1 1/2 to 3 lbs per acre. Apply as a non-directed spray after celery is transplanted and established, but before celery is 8" tall. Apply before annual grasses exceed 2" in height and before broadleaf weeds exceed 6" in height. In the Northeast, use only on celery grown on muck soils.

Do not exceed 40 psi spray nozzle pressure, and do not apply when temperature exceeds 85°F. nor as a tank mixture with surfactants, nitrogen or fertilizer solution, or other pesticides as injury to the crop may result. Do not replant to crops other than celery or carrots within 4 months after application as injury to subsequent crops may result. See "Replanting". Do not reenter treated areas for 24 hours following application unless protective clothing is worn.

CORN (FIELD)

Preemergence Application—East of Rocky Mountains: Select one of the following herbicide treatments for application as a tank mixture. Make a single application after planting but before crop emerges. Plant seed at least 1 3/4" deep on flat or raised seedbeds only or injury to the crop may result. Do not spray over top of emerged corn.

"LOROX" DF at 3/4 to 1 lb per acre (tank mixed with standard registered herbicides for corn) controls triazine-resistant lambsquarters on soils of 1 to 4% organic matter east of the Rocky Mts.

"LOROX" DF + "Lasso"

Soil Texture	Product Per Acre					
	1 to 3% Organic Matter			3 to 6% Organic Matter		
	"LOROX" DF (lbs)	+	"Lasso" (qts)	"LOROX" DF (lbs)	+	"Lasso" (qts)
Coarse Sandy loam	2/3 to 1 1/4	+	3/4 to 1	1 1/4 to 2	+	1 to 1 1/2
Medium Loam, Silt loam, Silt, Sandy clay, Sandy clay loam	1 to 1 2/3	+	1 to 1 1/2	1 2/3 to 2 1/2	+	1 1/2 to 2
Fine Silty Clay, Silty clay loam, Clay, Clay loam	1 1/4 to 2	+	1 1/2 to 2	2 to 3	+	2 to 2 1/2

Replanting: Corn or soybeans may be replanted within 4 months. See "Replanting".

"LOROX" DF + "Ramrod"

Soil Texture	Product Per Acre					
	1 to 3% Organic Matter			3 to 6% Organic Matter		
	"LOROX" DF (lbs)	+	"Ramrod" (4 lbs/gal)	"LOROX" DF (lbs)	+	"Ramrod" (4 lbs/gal)
Coarse Sandy loam	2/3 to 1 1/3	+	1 1/3 to 2 3/5	1 1/3 to 2	+	2 3/5 to 3 9/10
Medium Loam, Silt loam, Silt, Sandy clay, Sandy clay loam	1 to 1 2/3	+	2 to 3 1/3	1 2/3 to 2 1/2	+	3 1/3 to 5 1/5
Fine Silty clay, Silty clay loam, Clay, Clay loam	1 1/4 to 2	+	2 3/5 to 3 9/10	2 to 3	+	3 9/10 to 6 1/5

Replanting: Corn may be replanted within 4 months; after 4 months, see "Replanting"

"LOROX" DF + atrazine (Tank Mixture)—

For improved control of annual grasses, add "Lasso" as recommended on "Lasso" labeling for "Lasso"/atrazine combinations.

Soil Texture	Product Per Acre					
	1 to 2% Organic Matter			2 to 5% Organic Matter		
	"LOROX" DF (lbs)	+	atrazine 80% (lbs)	"LOROX" DF (lbs)	+	atrazine 80% (lbs)
Coarse Sandy loam	2/3 to 1	+	1/2 to 2/3	1 to 2	+	2/3 to 1 1/4
Medium Loam, Silt loam, Silt, Sandy clay, Sandy clay loam	1 to 1 1/2	+	2/3 to 1	1 1/2 to 2 1/2	+	1 to 1 1/2
Fine Silty clay, Silty clay loam, Clay, Clay loam	1 1/3 to 1 2/3	+	3/4 to 1	1 2/3 to 3	+	1 to 2

Replanting: Corn may be replanted within 6 months; after 6 months, see "Replanting"

CORN (FIELD AND SWEET)

Directed Postemergence Application: Make a single application as a directed spray after corn is at least 15" high (measured to the highest leaf surface on free standing plants). Do not spray over top of corn. Apply only when there is sufficient differential between height of corn and weeds so that the directed spray thoroughly covers all weed foliage without contact of upper leaves or whorl of corn by spray or drift, as such contact may cause crop injury. Early cultivation (rotary hoe or other suitable equipment) will aid in achieving proper differential between height of corn and weeds.

Use 1 1/4 to 3 lbs per acre; add 1 pint surfactant for each 25 gals spray mixture. For field corn, non-pressure nitrogen solution may be substituted for all or part of the water. Use the lower rate on lighter soils (low in clay or organic matter) and when weeds do not exceed 2" in height; use the higher rate on heavier soils (high in clay or organic matter) for weeds up to 5" in height. Do not reenter treated areas for 24 hours following application unless protective clothing is worn. See "Replanting".

COTTON

Directed Postemergence Application—East of Rocky Mountains: Apply as a directed spray; adjust nozzles to minimize contact to cotton leaves with spray or drift or crop injury may result. Do not spray over top of cotton.

Apply 1 to 1 1/2 lbs per acre when cotton is at least 8" tall and emerged weeds do not exceed 2" in height; add 1 pint surfactant for each 25 gals spray mixture. If needed, a second application at same rate may be made one week or later after initial treatment. Alternatively, after cotton is 20" tall, make a single application of 2 to 3 lbs per acre following last cultivation; if emerged weeds are present, add surfactant as directed above. **Note:** Do not use on Pima varieties of cotton. Do not graze treated fields or feed forage from treated areas to livestock. Do not feed gin trash to livestock. Do not reenter treated areas for 24 hours following application unless protective clothing is worn. See "Replanting".

HYBRID POPLAR (MIDWEST)

Apply to 2 to 4 lbs "LOROX" DF per acre before bud break in the spring.

For application after bud break, apply to 2 to 4 lbs "LOROX" DF per acre as a directed spray. Spray should be directed weed growth and to avoid contact with the poplar plant; do not spray over the top of the poplar as injury to the plant will result.

Use the lower rate on light soils and higher rate on heavier soils. For best results on emerged weeds, treat at the seedling stage.

More than one treatment may be made but no more than 8 lbs "LOROX" DF per acre should be applied per year.

PARSNIPS

Preemergence Application: Make a single application of 1 1/2 to 3 lbs per acre. Apply after planting but before crop emerges. Plant seed at least 1/2" deep. See "Replanting".

POTATOES

Preemergence Application: Make a single application as a broadcast spray after planting but before crop emerges. Plant seed at least 2" deep. Do not spray over top of emerged potatoes. If beds are to be "dragged" and/or "hilled", apply after the final "dragging" or "hilling" operation. Apply before grasses are 2" tall and before broadleaf weeds are 6" tall, preferably just before or when weed seedlings emerge. If emerged weeds are present, add 1 pint surfactant for each 25 gals spray mixture. In irrigated areas, best results are obtained when application is made to moist soil, followed within 2 weeks by 1" to 2" of sprinkler irrigation (or rainfall). On powder dry soils, irrigate prior to herbicide application and follow with sprinkler irrigation to activate the herbicide.

East of Rocky Mountains: Apply 1½ to 2½ lbs per acre on the coarser soils (sandy loams, silt loams; 1 to 2% organic matter) and 2½ to 4 lbs per acre on finer soils (silts, clay loams; 2 to 5% organic matter); on soils over 5% organic matter, use 4 lbs per acre and apply to emerged weeds (before potatoes emerge).

Wisconsin—Central Sands Area: Apply 1 lb per acre on sands and 2 lbs per acre on loamy sands.

Northeast: For improved control of annual grasses (see "Dual" labeling), apply:

Soil Texture	Product Per Acre					
	1 to 3% Organic Matter			3 to 5% Organic Matter		
	"LOROX" DF (lbs)	+	"Dual" 8E (pts)	"LOROX" DF (lbs)	+	"Dual" 8E (pts)
Coarse Sandy loam	1 to 1½	+	1½	1½ to 2	+	2
Medium Loam, Silt loam, Silt, Sandy clay, Sandy clay loam	1½ to 2	+	2	2 to 2½	+	2½ to 3
See Replanting.						

SORGHUM

Preemergence Application: Select a registered herbicide treatment for application as a tank mixture with "LOROX" DF. Make a single application after planting but before crop emerges. Plant seed at least 1" deep on flat or raised seedbeds only or injury to the crop may result. Do not spray over top of emerged sorghum.

"LOROX" DF-Southwest	Pounds of "LOROX" DF per acre	
	Percent Organic Matter in Soil	
Soil Texture Description	1 to 2%	2 to 4%
Coarse: Sandy Loam	⅝ to 1	1 to 1½
Medium: Loam, Silt loam, Silt, Sandy clay, Sandy clay loam	1 o 1½	1 to 2
Replanting: See "Replanting" and replanting on the companion product labeling. Prior to replanting, thorough seedbed preparation including fall or spring plowing is recommended.		

Directed Postemergence Application: Make a single application of "LOROX" DF as a directed spray; add 1 pint surfactant for each 25 gals spray mixture. If sprayer is equipped with skids, shoes or shields, apply 1 lb per acre when sorghum is 12" tall (free standing plants) and weeds are up to 2" in height; Use 1 to 2 lbs per acre when sorghum is 15" tall and weeds are 2" to 4" in height. If boom drops are used, apply 1 to 2 lbs per acre when sorghum is at least 18" tall and weeds are 2" to 4" in height. Apply only when there is sufficient differential between height of sorghum and weeds so that the directed spray thoroughly covers all weed foliage without contact of upper leaves or whorl of sorghum by spray or drift as such contact may cause crop injury.

Note: Do not graze or feed plants to livestock within 3 months after postemergence application. See "Replanting". Do not reenter treated areas for 24 hours following application unless protective clothing is worn.

SOYBEANS—CONVENTIONAL TILLAGE

Preemergence Application: For broad spectrum weed control, select one of the following herbicide combination treatments and make a single application after planting but before crop emerges. Plant seed at least 1¾" deep on flat or raised seedbeds only or injury to the crop may result. Injury to soybeans may result if application is made to fields with standing water or fields too wet to cultivate. Do not spray over top of emerged soybeans. Do not use on sand or loamy sand nor any soil containing less organic matter than listed below.

"LOROX" DF + "Lexone" DF*

For control of common ragweed, hemp sesbania, jimsonweed, lambsquarters, pigweeds, prickly sida, sicklepod, velvetleaf, spotted spurge, Venice mallow, purslane, carpetweed, Pennsylvania smartweed, and partial control of cocklebur*, use the following:

Soil Texture	Product Per Acre					
	½ to 3% Organic Matter			3 to 6% Organic Matter		
	"LOROX" DF (lbs)	+	"Lexone" DF (lbs)	"LOROX" DF (lbs)	+	"Lexone" DF (lbs)
Coarse Sandy loam	⅓ to ½	+	⅙ to ¼	½ to ¾	+	¼ to ⅓
	or ("Lexone" 4L—¼ to ⅖ pt)			or ("Lexone" 4L—⅖ to ½ pt)		
Medium Loam, Silt loam, Silt, Sandy clay, Sandy clay loam	½ to ¾	+	¼ to ⅓	¾ to 1½	+	⅓ to ½
	or ("Lexone" 4L—⅖ to ½ pt)			or ("Lexone" 4L—½ to ¾ pt)		
Fine Silty clay, Silty clay loam, Clay, Clay loam	¾ to 1½	+	⅓ to ½	1½ to 2	+	½ to ⅔
	or ("Lexone" 4L—½ to ¾ pt)			or ("Lexone" 4L—¾ to 1 pt)		

*For improved control of annual grasses, tank mix with "Lasso" or "Dual" (per labeling):

Herbicide	Recropping/Restrictions
"Lasso"	**Soybeans or field corn may be replanted within 4 months; after 4 months, see "Replanting". Treated vines may be grazed or fed to livestock 40 days after application.**
or	
"Dual"	**Soybeans or field corn may be replanted within 4 months, for rotation crops, see "Replanting" and follow instructions on "Dual" label. Do not graze or feed forage from treated areas to livestock.**

Note: Soybean varieties such as Altona, Coker 102 and 156, Govan, NKS 1884, Semmes, Tracy, Vansoy, Terra Vig 505 and 606, Agripro 55, Asgrow 6520, Maple Amber, Portage, Vinton 81, and AP 71 are sensitive to "Lexone". Injury may occur if "Lexone" is used on these varieties. Before use on any other soybean variety, tolerance to "Lexone" must first be determined. Varieties showing above average tolerance to "Lexone" are Americana Revere, Asgrow 1937, Asgrow 3659, Asgrow 3860, DSR 171, DSR 207, Essex, Fayette, Hisoy 170, Lakota, Lawrence, LOL 4207, NKS 1492, Pride B216, Pride B242, SRF 250, SRF 350 P, Union, Wayne, Wells II and Williams 82; for maximum weed control, use the higher rate (where a range of rates is listed) for the appropriate soil type and organic matter as shown in the following tables. Injury to soybeans may occur if "Lexone" is used on soils having a calcareous surface layer or pH of 7.5 or higher, or if used in conjunction with soil applied organic phosphate pesticides such as "Dasanit", "Disyston", "Mocap", "Nemacur", "Thimet", "Parathion", "Lorsban" 15G or Counter. Injury may occur if atrazine was applied on the soil the year before use of "Lexone". Seedling disease, cold weather, deep planting (more than 2"), excessive moisture, high soil pH (pH 7.5 or higher), high salt concentration, or drought may weaken soybean seedlings and increase possibility of crop injury.

Do not use on sand nor loamy sand.

Cultivation after planting: Treated soil may be shallow-cultivated, rotary-hoed, or hand-hoed without reducing the weed control activity of the tank mixture. Do not cultivate deeper than the treated layer of soil since this may bring untreated soil to the surface and poor weed control may result.

Replanting: If initial seeding fails to produce a stand, treated fields may be replanted to soybeans; do not rework soil; do not retreat field with a second application as injury to the crop may result. Do not replant treated areas to any crop other than soybeans within four months after treatment as injury to subsequent crops may result. See "Replanting".

"LOROX" DF*

Soil Texture	Product Per Acre	
	½ to 3% Organic Matter "LOROX" DF (lbs)	3 to 6% Organic Matter "LOROX" DF (lbs)
Coarse Sandy loam	⅔ to 1¼	1¼ to 2
Medium Loam, Silt loam, Silt, Sandy clay, Sandy clay loam	1 to 1⅔	1⅔ to 2½
Fine Silty clay, Silty clay loam, Clay, Clay loam	1¼ to 2	2 to 3

* For improved control of annual grasses, add as per the following labeling:

Herbicide	Recropping/Restrictions
"Lasso"	**Soybeans or field corn may be replanted within 4 months; after 4 months, see "Replanting.**
or	
"Dual"	**Soybeans or field corn may be replanted within 4 months; for rotation crops, see "Replanting" and follow instructions on "Dual" label. Do not graze or feed forage from treated areas to livestock.**
or	

"Prowl"	**Soybeans or field corn may be replanted within 4 months; crops listed in "Replanting" may be planted the following year.**
or	
"Surflan"	**Soybeans may be replanted within 4 months; after 4 months, see "Replanting" but do not plant potatoes within 12 months. Do not use treated vines for feed or forage.**
or	
"Amiben"	**Soybeans or field corn may be replanted within 4 months; after 4 months, see "Replanting."**

For control of black nightshade, apply:

Soil Texture	Product Per Acre 1 to 3% Organic Matter "LOROX" DF (lbs)	"Lasso" (qts)
Coarse Sandy loam	¾ to 1½	2
Medium Loam, Silt loam, Silt, Sandy clay, Sandy clay loam	1 to 1⅔	2½ to 3
Fine Silty clay, Silty clay loam, Clay, Clay loam	1¼ to 2	3

Do not use on sand or loamy sands.

Replanting: Soybeans or field corn may be replanted within 4 months; after 4 months, see "Replanting".

For control of black nightshade in Kentucky, Illinois, Indiana and Ohio:

Soil Texture	Product Per Acre 1 to 3% Organic Matter "LOROX" DF (lbs)	"Dual" 8E (pts)
Coarse Sandy loam	¾ to 1½	1½
Medium Loam, Silt loam, Silt, Sandy clay, Sandy clay loam	1 to 1⅔	2
Fine Silty clay, Silty clay loam, Clay, Clay loam	1¼ to 2	2 to 2½

Do not use on sand or loamy sands.

Replanting: Soybeans or field corn may be replanted within 4 months; after 4 months, see "Replanting".

Preemergence Following "Treflan" or "Prowl" Preplant: Where "Treflan" or "Prowl" has been used as a preplant incorporated treatment (according to directions on product label), apply "LOROX" DF preemergence (after planting and before emergence of soybeans) as a separate operation using rates recommended below for "LOROX" DF alone. For rotation crops, follow instructions on "Treflan" or "Prowl" labels and see "Replanting". Plant seed at least 1¾" deep on flat or raised seedbeds only or injury to the crop may result. Injury to soybeans may result if application is made to fields with standing water or fields too wet to cultivate. Do not spray over top of emerged soybeans. Do not use on sand or loamy sand nor any soil containing less organic matter than listed below.

"LOROX" DF Alone—If weeds have emerged, add 1 pint surfactant for each 25 gals spray mixture.

Soil Texture	Pounds "LOROX" DF Per Acre ½ to 2% Organic Matter	2 to 5% Organic Matter
Coarse Sandy loam	1 to 1⅔	1⅔ to 3
Medium Loam, Silt loam, Silt, Sandy clay, Sandy clay loam	1¼ to 2⅓	2⅓ to 4
Fine Silty clay, Silty clay loam, Clay, Clay loam	1⅓ to 2⅔	2⅔ to 5 (over 5% organic matter, use 6 lbs)

Replanting: Soybeans or field corn may be replanted within 4 months; after 4 months, see "Replanting".

Preemergence Followed by Postemergence (Directed) Application: Both preemergence and postemergence (directed to weeds with minimum contact with soybean plants) treatment may be required for control of:

Annual morningglory—"LOROX" DF preemergence (any combination listed) followed by a directed postemergence spray of "LOROX" DF + 2,4-DB.

Broadleaf signalgrass—"LOROX" DF + "Lexone" DF preemergence followed by a directed postemergence spray of "LOROX" DF, if needed.

Cocklebur (common)—"LOROX" DF + "Lexone" DF preemergence followed by a directed postemergence spray of "LOROX" DF + 2,4-DB.

Hemp sesbania—"LOROX" DF + "Lexone" DF preemergence followed by a directed postemergence application of "LOROX" DF, if needed.

Prickly sida—"LOROX" DF + "Lexone" DF preemergence followed by a directed postemergence spray of "LOROX" DF, if needed.

Velvetleaf—"LOROX" DF + "Lexone" DF preemergence followed by a directed postemergence application of "LOROX" DF, if needed.

Other weeds—Directed postemergence application of "LOROX" DF should be used as a backup treatment for listed preemergence treatments and will extend the duration and degree of weed control obtained with preemergence treatment. Postemergence treatment should be used only as needed.

For dosage rates, refer to "Preemergence" and "Directed Postemergence" sections under "Soybeans".

SOYBEANS-MINIMUM OR NO-TILLAGE

Preemergence to Soybeans Postemergence to Weeds: "LOROX" DF preemergence to soybeans may be used for preemergence and postemergence control of many broadleaf weeds and grasses where soybeans will be planted directly into a preformed bed (stale seed bed), cover crop or in previous crop residues such as corn or small grain stubble. Apply with ground equipment immediately before, during or after planting but before crop emerges; maintain constant agitation of spray mixture.

"LOROX" DF Alone

Soil Texture	Pounds "LOROX" DF Per Acre ½ to 2% Organic Matter	2 to 5% Organic Matter
Coarse Sandy loam	1 to 1⅔	1⅔ to 3
Medium Loam, Silt loam, Silt, Sandy clay, Sandy clay loam	1¼ to 2⅓	2⅓ to 4
Fine Silty clay, Silty clay loam, Clay, Clay loam	1⅓ to 2⅔	2⅔ to 5 (over 5% organic matter, use 6 lbs)

Replanting: Soybeans or field corn may be replanted within 4 months; after 4 months, see "Replanting".

If small seedling weeds are present, add 1 pint surfactant for each 25 gals spray mixture for improved contact activity. On larger weeds, add "Paraquat" or "Roundup" or "Bronco" as described under these combinations; these treatments will also suppress some perennial weeds.

"LOROX" DF + "Lasso" or "Dual" 8E or "Surflan"—will improve control of grasses and volunteer small grains.

"LOROX" DF + "Lasso"* (or "Dual"* or "Surflan"*) + Paraquat or "Roundup"—Thoroughly mix "LOROX" DF and companion herbicide in spray tank first according to directions; then add paraquat, or "Roundup" as directed under Paraquat Combinations or "Roundup" Combinations below.

Soil Texture	Product Per Acre ½ to 3% Organic Matter "LOROX" DF (lbs)	+	"Lasso"* (qts)	3 to 6% Organic Matter "LOROX" DF (lbs)	+	"Lasso"* (qts)
Coarse Sandy loam	¾ to 1½	+	2	1½ to 2⅖	+	2½
Medium Loam, Silt loam, Silt, Sandy clay, Sandy clay loam	1⅛ to 2	+	2½	2 to 3	+	3
Fine Silty clay, Silty clay loam, Clay, Clay loam	1¼ to 2¼	+	2½	2¼ to 3½	+	3

*For improved control of annual grasses and volunteer small grains, add as per labeling:

Herbicide	**Recropping/Restrictions**
"Lasso"	**Soybeans or field corn may be replanted within 4 months; after 4 months, see "Replanting".**
or	
"Dual"	**Soybeans or field corn may be replanted within 4 months;**

for rotation crops see "Replanting" and follow instructions on "Dual" label. Do not graze or feed forage from treated areas to livestock.

or

"Surflan" **Do not use treated vines for feed or forage. Soybeans may be replanted within 4 months; after 4 months, see "Replanting" and do not plant potatoes within 12 months.**

For control of black nightshade apply with paraquat or "Roundup" as shown below:

Soil Texture	Product Per Acre 1 to 3% Organic Matter "LOROX" DF (lbs)	"Lasso" (qts)
Coarse Sandy loam	3/4 to 1 1/2	2
Medium Loam, Silt loam, Silt, Sandy clay, Sandy clay loam	1 to 2	2 1/2 to 3
Fine Silty clay, Silty clay loam, Clay, Clay loam	1 1/4 to 2 1/4	3

Replanting: Soybeans or field corn may be replanted within 4 months; after 4 months, see "Replanting".

Paraquat Combinations—Select and tank mix in water one of the above treatments; then add 1/4 to 1/2 lb. active ingredient paraquat per acre for control of emerged weeds. Use the higher rate for weeds 4" to 6" tall. As the last ingredient, add 1/2 pint "Ortho" X-77 Spreader per 100 gals of spray mixture. Maintain constant agitation. Use 20 to 60 gals of water per acre. Use the higher gallonage for dense stubble or vegetation.

"Roundup" Combinations—Select and tank mix one of the above treatments. As last ingredient, add 1 1/2 quarts of "Roundup" per acre for control of emerged annual weeds or 2 to 4 quarts per acre for control of emerged perennial and annual weeds. Use 20 to 30 gals of water per acre.

"LOROX" DF + "Bronco"—Do not add paraquat or "Roundup" to this combination.

Soil Texture	Product Per Acre ½ to 5% Organic Matter "LOROX" DF (lbs)		"Bronco" (b) (qts)
Coarse Sandy loam	3/4 to 2	+	3 1/4 to 5
Medium Loam, Silt loam, Silt, Sandy clay, Sandy clay loam	1 1/8 to 2 1/2	+	4 to 5
Fine Silty clay, Silty clay loam, Clay, Clay loam	1 1/4 to 3	+	4 to 5

(a) Do not use on soils with less than 1/2% organic matter nor on sand, loamy sand, or muck soils as crop injury may result.

(b) Use the higher rate for dense stubble, heavy crop residue, or heavy weed population.

Replanting Soybeans: If initial seeding fails to produce a stand, treated fields may be replanted to soybeans; do not rework soil; do not retreat field with a second application as injury to the crop may result. Do not replant treated areas to any crop other than soybeans within four months after treatment as injury to subsequent crops may result. See "Replanting".

SOYBEANS—CONVENTIONAL OR MINIMUM OR NO-TILLAGE

Directed Postemergence Application: Apply "LOROX" DF alone or as a tank mixture with 2,4-DB, as a directed spray to cover weed foliage with minimum contact of the soybean plant. Do not spray higher than 3" on the soybean stem or crop injury may result. Do not spray over top of soybean plants. For broadcast application, use a single flood-type spray nozzle ("K" series or equivalent) per middle mounted on an oiling shoe or gauge wheel. For band treatment, use two nozzles per row mounted on an oiling shoe or gauge wheels, one of each side of row. To avoid spray drift, which may cause crop injury, do not exceed nozzle pressure of 25 psi nor use nozzle tips smaller than 8002 T-Jet (or equivalent) and do not spray under windy conditions. Add 1 pint surfactant for each 25 gals spray mixture. For best results, use a preemergence treatment (such as "LOROX" DF) or cultivation to control early weed growth and to increase the differential between height of soybeans and weeds.

NOTE: Do not use on soils with less than 1/2% organic matter. Do not apply more than 2 lbs "LOROX" DF per acre per season for postemergence treatments. Do not apply within 60 days of harvest. Do not feed soybean forage or hay to livestock from fields treated postemergence. Harvested soybeans may be used for food, feed or oil purposes. See "Replanting". Do not reenter treated areas for 24 hours following application unless protective clothing is worn.

"LOROX" DF Alone—For soybeans at least 8" high and when weeds do not exceed 2" in height, apply 1/2 to 1 lb per acre. Make a second application at same rate if new flush of weeds occur.

For soybeans at least 12" high and when weeds do not exceed 4" in height, make a single application of 1 to 2 lbs per acre. Alternatively, make a split application of 1 lb per acre followed by a second application at the same rate after 1 week or later.

"LOROX" DF + 2,4-DB—For soybeans at least 8" high and when weeds do not exceed 4" in height, apply 1 lb "LOROX" DF plus 1/5 lb 2,4-DB (13 fl oz "Butyrac" 200 or 1 pint "Butoxone") per acre. A second application may be made if needed, but do not make more than 2 applications per season.

NON-CROP WEED CONTROL

For short-term control of annual weeds on non-cropland areas such as roadsides and fence rows—apply 2 to 6 lbs "LOROX" DF per acre in 40 to 100 gals of water. For best results, apply shortly before weed growth begins or at early seedling stage of growth. For control of established annual weeds, add surfactant at rate of 2 qts per 100 gals of spray mixture and apply as a thorough coverage spray during periods when daily temperatures exceed 70°F. and before weed growth exceeds 8" in height.

Do not reenter treated areas for 24 hours following application unless protective clothing is worn.

[1] Registered trademark of Union Carbide Agricultural Products Company, Inc. "Amiben" contains 2 lbs chloramben per gal; "Butyrac" 200 contains 2 lbs 2,4-DB per gal.
[2] Registered trademark of Ciba-Geigy Corporation, "Dual" 8E contains 8 lbs metolachlor per gal.
[3] Registered trademark of Monsanto Company, "Lasso" contains 4 lbs alachlor per gal. "Roundup" contains 4 lbs glyphosate per gal. "Bronco" contains 2.6 lbs alachlor and 1.04 lbs glyphosate per gal. "Ramrod" contains 4 lbs propachlor per gal.
[4] Paraquat contains 2 lbs active per gal, paraquat is a restricted use pesticide.
[5] Registered trademark of American Cyanamid Company, "Prowl" contains 4 lbs pendimethalin per gal.
[6] Registered trademark of Elanco Products Company, "Surflan" 75W contains 75% W.P. oryazalin, "Surflan" AS contains 4 lbs oryazalin per gal. "Treflan" contains 4 lbs trifluralin per gal.
[7] Registered trademark of Rhone-Poulenc Inc., "Butoxone" contains 1.75 lbs dimethylamine salt and 2 lbs isooctyl ester of 2,4-DB per gal.
[8] Registered trademark of Westvaco Corporation.
[9] Registered trademark of ICI United States, Inc.

STORAGE AND DISPOSAL

STORAGE: Keep from freezing. Store product in original container only, away from other pesticides, fertilizer, food, or feed.

PRODUCT DISPOSAL: Do not contaminate water, food, or feed by storage or disposal. Wastes resulting from the use of this product may be disposed of on site or at approved waste disposal facility.

CONTAINER DISPOSAL: Triple rinse (or equivalent) the container. Then offer for recycling or reconditioning, or puncture and dispose of in a sanitary landfill, or by incineration, or, if allowed by state and local authorities, by burning. If burned, stay out of smoke.

NOTICE OF WARRANTY

Du Pont warrants that this product conforms to the chemical description on the label thereof and is reasonably fit for purposes stated on such label only when used in accordance with the use directions on this product. Crop injury, ineffectiveness or other unintended consequences may result because of such factors as weather conditions, presence of other materials, or the manner of use or application, all of which are beyond the control of Du Pont. In no case shall Du Pont be liable for consequential, special or indirect damages resulting from the use of handling of this product. All such risks shall be assumed by the Buyer. DU PONT MAKES NO WARRANTIES OF MERCHANTABILITY OR FITNESS FOR A PARTICULAR PURPOSE NOR ANY OTHER EXPRESS OR IMPLIED WARRANTY EXCEPT AS STATED ABOVE.

AG-3254 8049 4/13/89

LOROX PLUS®

HERBICIDE
DISPERSIBLE GRANULES

FOR USE ON SOYBEANS ONLY

	BY WEIGHT
ACTIVE INGREDIENTS:	60%
Linuron: [3-(3,4-dichlorophenyl)-1-methoxy-1-methylurea]	56.9%
Chlorimuron Ethyl: Ethyl 2-[[[[(4-chloro-6-methoxypyrimidin-2-yl)amino]carbonyl]-amino]sulfonyl]benzoate	3.1%
INERT INGREDIENTS:	40%
TOTAL	100%

EPA Reg. No. 352-451
U.S. Pats. 4,394,506 & 4,547,215

KEEP OUT OF REACH OF CHILDREN

PRECAUTIONARY STATEMENTS

HAZARDS TO HUMANS AND DOMESTIC ANIMALS

WARNING!

CAUSES SUBSTANTIAL BUT TEMPORARY EYE INJURY AND CAUSES SKIN IRRITATION.

AVOID CONTACT WITH SKIN, EYES AND CLOTHING. AVOID BREATHING DUST OR SPRAY MIST. WASH THOROUGHLY WITH SOAP AND WATER AFTER HANDLING. REMOVE AND WASH CONTAMINATED CLOTHING BEFORE REUSE.

STATEMENT OF PRACTICAL TREATMENT

If in eyes: Flush with plenty of water and get medical attention.

If on skin: Wash with soap and water and get medical attention.

If swallowed: Call a physician or Poison Control Center. Drink 1 or 2 glasses of water and induce vomiting by touching back of throat with finger. Do not induce vomiting or give anything by mouth to an unconscious person.

For medical emergencies involving this product, call toll free 1-800-441-3637.

ENVIRONMENTAL HAZARDS

Do not apply directly to water or wetlands. Do not contaminate water by cleaning of equipment or disposal of wastes.

IMPORTANT

When mixing/loading and spraying, wear a wide-brimmed hat, mid-forearm to elbow length chemical resistant gloves, and one-piece overalls which have long sleeves and long pants. Overalls should be constructed of finely woven fabric as specified in the USDA/EPA Guide for Commercial Applicators.

Any article of clothing worn while applying product must be cleaned before reusing. Clothing which has been drenched or has otherwise absorbed concentrated pesticide must be disposed of in a sanitary landfill, or incinerated, or, if allowed by state and local authorities, by burning.

Instead of clothing and equipment specified above, the applicator can use an enclosed tractor cab which provides a filtered air supply (as specified by Taschenberg and Bourke, 1975).

Injury to or loss of desirable trees or vegetation may result from failure to observe the following: Do not apply or drain or flush equipment on or near desirable trees or other plants, or on areas where their roots may extend, or in locations where the chemical may be washed or moved into contact with their roots. Do not use on lawns, walks, driveways, tennis courts, or similar areas. Prevent drift of spray to desirable plants. Do not contaminate any body of water. Keep from contact with fertilizers, insecticides, fungicides, and seeds during storage.

Prior to using "Lorox Plus", consideration should be given to crop rotation plans. Crops other than soybeans can be extremely sensitive to low concentrations of "Lorox Plus" remaining in the soil the next planting season. Choice of rotational crop is restricted following application of "Lorox Plus". (See "ROTATIONAL CROP GUIDELINES").

Do not apply if soil pH is greater than 6.8.

"Lorox Plus" is registered for use on soybeans only. Use on crops other than soybeans will result in severe crop injury or crop death.

GENERAL INFORMATION

Du Pont "Lorox Plus" Herbicide is a dispersible granule formulation to be mixed with water and sprayed for selective preemergence weed control in soybeans. When applied according to the instructions on this label, it will control many broadleaf weeds and provide partial control of nutsedge.

Preemergence applications of "Lorox Plus" require rainfall or sprinkler irrigation to activate the herbicide. Degree of control and duration of effect depend on: rate used, weed spectrum, growing conditions at and following time of treatment, soil pH, texture, organic matter, moisture and precipitation.

Because most crops are highly sensitive to "Lorox Plus", all direct or indirect contact (such as spray drift) to crops or to land scheduled to be planted to crops other than soybeans should be avoided.

BIOLOGICAL ACTIVITY

"Lorox Plus" rapidly inhibits the growth of susceptible weeds. Following preemergence treatment, susceptible weeds may germinate and emerge, but growth then ceases and leaves become yellow 3-5 days after emergence. Death of leaf tissue and growing point will follow in some species while others will remain green but stunted and noncompetitive. "Lorox Plus" provides partial control of some annual grasses, but other products may be needed to ensure adequate grass control.

DIRECTIONS FOR USE

It is a violation of federal law to use this product in a manner inconsistent with the terms of the label.

Do not apply this product through any type of irrigation system.

EQUIPMENT-SPRAY VOLUMES

Apply uniformly by ground equipment with a properly calibrated low pressure (20 to 40 psi) fixed-boom sprayer equipped with fan-type nozzles or other carefully calibrated nozzles and screens no finer than 50 mesh. Use 10 to 40 gals of water per acre.

Continuous agitation in the spray tank is required to keep the material in suspension. Avoid overlap and shut off spray booms while starting, turning, slowing or stopping, as injury to the crop may result.

EARLY PREPLANT OR PREPLANT INCORPORATED (See Rate Chart)

"Lorox Plus" can be applied at planting or up to 30 days prior to planting either preemergence or preplant incorporated. If incorporated, "Lorox Plus" should be uniformly incorporated no deeper than the top 1-2" of soil prior to planting soybeans. Use equipment suited for proper incorporation into the top 1-2 inches of soil e.g.: Do-all, field cultivator or Roterra. If tank-mixed with a grass herbicide, such as a DNA like "Treflan"[4], "Sonalan"[4], or acetanilide such as "Dual"[2], or "Lasso"[1], follow label instructions for proper incorporation of the grass herbicide into the top 1 to 2" of soil. Improper soil incorporation can result in erratic weed control or crop injury. Rainfall* (~1") is required to activate preplant incorporated herbicides such as "Lorox Plus" before weed emergence.

Use of a disk for incorporation is not recommended.

PREEMERGENCE USE ON SOYBEANS (See Rate Chart)

Apply before soybeans emerge. Do not spray over top of emerged soybeans or severe crop injury or crop death will occur.

Preemergence activity of "Lorox Plus" requires rainfall or irrigation to activate. Best results are obtained if treatment is made to moist soil and followed by rainfall or irrigation (1/2" to 1") before weeds germinate. On dry soil, chemical activation requires more moisture (1-2"). If moisture is insufficient to activate the herbicide, a rotary hoeing or shallow cultivation should be made after emergence of crop while weeds are small enough to be controlled by mechanical means. "Lorox Plus" can be applied preemergence in combination with "Lasso", "Dual", "Prowl"[3], "Surflan"[4] or following the use of a preplant grass material such as "Treflan" "Prowl" or "Sonalan". Consult labels for rates and use instructions.

SOYBEANS — MINIMUM TILLAGE OR NO-TILL (See Rate Chart)

"Lorox Plus" applied preemergence to soybeans but postemergence to broadleaf weeds and small annual grasses will provide limited burndown of existing weeds. When used for burndown add 0.25% V/V surfactant (1 qt/100 gal) such as Surfactant WK or crop oil concentrate at 0.25% V/V (1 qt/100 gal). For added burndown of broadleaf weeds, "Lorox Plus" may be tank mixed with 2,4-D low volatile ester. Do not apply 2,4-D ester within 30 days of planting. For added postemergence control, especially on grasses, "Lorox Plus" can be applied with "Roundup"[1], paraquat or "Bronco"[1]. For added preemergence control, especially on grasses, "Lorox Plus" can be mixed with "Dual", "Lasso", or "Surflan". Consult labels for rates and use instructions. Maintain constant agitation of spray mixture.

RATE CHART

Apply within 2 weeks of planting as follows:

Soil Texture (1/2-3% Organic Matter*)	Rate (Ounces Per Acre)
Coarse: Loamy sand, Sandy loam	12-14
Medium: Loam, Silty loam, Silt	14-16
Fine: Silty clay loam, Clay loam, Clay	16-18

* Do not apply to soil with less than 1/2% O.M.

Do not apply "Lorox Plus" to soils with pH greater than 6.8.

Dosage rates listed are broadcast rates. For band applications use proportionally less.

Plant seed 1 1/2" to 2" deep on a flat or raised seedbed only, or crop injury may occur.

Soybean stunting may occur if excessive rainfall occurs after application but before soybeans germinate. Injury is more prevalent under poor drainage or compacted conditions or when soil is saturated for long periods of time. Soybeans rapidly outgrow stunting once favorable growing conditions return.

NOTE: Seedling disease, cold weather, deep planting (more than 2"), excessive moisture, high salt concentration or drought may weaken soybean seedlings and increase the possibility of crop injury.

WEEDS CONTROLLED

The following weeds will be controlled at the above rates preemergence:

Carpetweed	Pigweeds species	Purslane
Cocklebur*	Palmer amaranth	Ragweed, common
Eastern black nightshade†	Redroot	Smartweeds, annual
	Smooth	Spotted spurge
Jimsonweed	Spiny amaranth	Velvetleaf
Lambsquarters†	Tall waterhemp	
Mustards	Prickly sida (Teaweed)	

Partial Control:

Burcucumber	Morningglory, annual*	Nutsedge species
Crabgrass species		Panicum species
Eastern black nightshade	Tall	Ragweed, giant*
	Ivyleaf	Sunflower*
Foxtail species	Pitted	
Johnsongrass, seedling	Entireleaf	

† Requires the addition of "Lasso".

* Large seeded weeds, germinating deep in the soil, such as morningglory, sunflower, cocklebur and giant ragweed or weeds with subsequent flushes may require a cultivation or an application of Du Pont "Classic" Herbicide postemergent for season-long control. Consult "Classic" label for application directions, recropping intervals and use restrictions. NOTE: Consult "Lasso", "Dual", "Prowl", "Treflan", "Sonalan" or "Surflan" labels for additional weeds controlled when "Lorox Plus" is used in conjunction with these herbicides.

SPRAY TANK PREPARATION

Fill tank 1/4 full with water. Start agitation system, add "Lorox Plus" and continue adding water. Add separately each additional component of any tank mix while adding water. Continue agitation throughout. If poor mixing should occur with any component, premix the component with two parts water before adding to the spray tank. A fertilizer solution may be used in the spray mixture. Small quantities should be tested for compatibility by the following procedures before full scale mixing.

1. Put 1 pint of fertilizer solution in quart jar.
2. Mix 2 teaspoons "Lorox Plus" with 2 tablespoons of water; mix thoroughly and add to fertilizer solution.
3. Close jar and shake well.
4. If other herbicides are to be used in the mixture, premix 2 teaspoons of wettable powder or 1 teaspoon of liquid with 2 tablespoons of water; add to "Lorox Plus"/fertilizer solution mixture.
5. Close jar and shake well.
6. Watch mixture for several seconds; check again in 30 minutes.
7. If mixture does not separate, foam, gel or become lumpy, it may be used.
8. Mixing ability may be improved by adding compatibility agents such as Kalo Laboratories "Compex"[5] or Witco Chemicals "Sponto"[6] 168D. Follow directions on container.

Provided the above procedure shows the mixture to be compatible, prepare the tank mixture as follows: Add the fertilizer solution to the spray tank first; with the agitator running, add the required amount of "Lorox Plus" and thoroughly mix. For tank mixtures with other herbicides, follow directions above.

When tank mixing with other herbicides, all applicable directions, restrictions and precautions for the additional herbicides are also to be followed. Use "Lorox Plus" spray preparations the same day as mixed or product degradation may occur. Thoroughly reagitate before using if allowed to settle.

ROTATIONAL CROP GUIDELINES

Crops that can be planted after using "Lorox Plus" are listed below. Choice of rotational crop is determined by time interval elapsed since last application. Prior to 18 months following application, do not rotate crops used for food or feed, which are not registered for use with "Lorox Plus" or Du Pont "Lorox" Herbicide onto areas previously treated with these chemicals.

Crops	Time Interval in Months Before Replanting
Soybeans	anytime
Wheat, Barley, Oats, Rye	4
Field Corn	10**
Clover	12
Sorghum	10**
Tobacco (transplant)	10**
Tomato (transplant)	10**

All Other Crops require a successful Field Bioassay***

** For applications of "Lorox Plus" after June 15, extend recrop interval.

***Field Bioassay - The season before planting any crop not listed above, a successful field bioassay must be completed. The field bioassay will detect small quantities of "Lorox Plus" which can remain in the soil and injure rotational crops. A successful field bioassay means growing to maturity a test strip of the crop(s) intended for production the following year. The test strip should cross the whole field including knolls and low areas.

Combination Treatment: When recommended tank mix combinations are used, consult the product label for replanting and recropping instructions and observe the directions that are the most restrictive.

SPRAYER CLEANUP

To avoid subsequent injury to crops other than soybeans, immediately after spraying and prior to spraying other crops, thoroughly remove all traces of "Lorox Plus" from mixing and spray equipment. Be careful to rinse both the spray tank, boom and clean nozzles and screens. For more information, consult supplemental bulletin on sprayer cleanout following use of Du Pont "Classic", "Canopy", "Gemini", "Preview" or "Lorox Plus" Herbicides.

PRECAUTIONS

Do not apply "Lorox Plus" Herbicide to soils with a pH greater than 6.8.

CAUTION: Soil pH varies greatly even within the same field. pH variations as much as 2 pH units are common when the soil has a calcareous sub soil. Determine soil pH by laboratory analysis using a 1:1 soil:water suspension.

Do not graze treated fields or harvest for forage or hay.

Use of "Lorox Plus" may affect rotational crops. Follow "ROTATIONAL CROP GUIDELINES."

Do not use on sand, muck, or peat soils, nor on soils with less than 1/2% organic matter, as crop injury may occur.

Do not apply "Lorox Plus" after the soybean crop has emerged.

If "Scepter"[3], "Pursuit"[3] or any product containing the active ingredients, imazaquin or imazethapyr (such as "Squadron"[3], "Partner"[3]) is applied the same year as "Lorox Plus", do not plant crops other than soybeans for at least 15 months from the last application. Crops listed in the "'Lorox Plus' Rotational Crop Guidelines" may be planted after 15 months; all other crops require a successful field bioassay.

Do not apply to land that has been or will be treated with "Glean", "Ally" or "Finesse" herbicides in the States of Nebraska, Kansas, North Dakota and South Dakota.

[1]Registered Trademark of Monsanto Agricultural Products Co., St. Louis, MO
[2]Registered Trademark of Ciba-Geigy Corp., Greensboro, NC
[3]Registered Trademark of American Cyanamid Corp., Princeton, NJ
[4]Registered Trademark of Elanco Products Co. Indianapolis, IN
[5]Registered Trademark of Kalo Agricultural Chemicals Inc., Overland Park, KS
[6]Registered Trademark of Witco Chemical Corp., Houston, TX

NOTICE TO BUYER: Purchase of this material does not confer any rights under patents of countries outside of the United States.

STORAGE AND DISPOSAL

Storage: Store product in original container only, away from other pesticides, fertilizer, food or feed.

Pesticide Disposal: Do not contaminate water, food or feed by storage or disposal. Wastes resulting from the use of this product may be disposed of on site or at an approved waste disposal facility.

Container Disposal: Triple rinse (or equivalent). Then offer for recycling or reconditioning, or puncture and dispose of in a sanitary landfill, or incineration, or, if allowed by state and local authorities, by burning. If burned, stay out of smoke.

NOTICE OF WARRANTY

Du Pont warrants that this product conforms to the chemical description on the label thereof and is reasonably fit for the purposes stated on such label only when used in accordance with the directions under normal conditions. It is impossible to eliminate all risks inherently associated with the use of this product. Crop injury, ineffectiveness or other unintended consequences may result because of such factors as weather conditions, presence of other materials, or the manner of use or application, all of which are beyond the control of Du Pont. In no case shall Du Pont be liable for consequential, special or indirect damages resulting from the use or handling of this product. All such risks are assumed by the buyer. DU PONT MAKES NO WARRANTIES OF MERCHANTABILITY OR FITNESS FOR A PARTICULAR PURPOSE NOR ANY OTHER EXPRESS OR IMPLIED WARRANTY EXCEPT AS STATED ABOVE.

AG-3228 8098 4/11/88

This is a specimen label. DuPont is not responsible for the accuracy of the information contained herein. As labels are subject to revision, always carefully read and follow the label on the product container.

LOROX® L

HERBICIDE
LIQUID SUSPENSION

ACTIVE INGREDIENT:	BY WEIGHT
Linuron [3-(3,4-dichlorophenyl)-1-methoxy-1-methylurea]	40.7%
INERT INGREDIENTS:	59.3%
TOTAL	100.0%

SHAKE WELL BEFORE USING • KEEP FROM FREEZING

Contains 4 lbs Linuron Per Gallon

EPA Reg. No. 352-391

KEEP OUT OF REACH OF CHILDREN

PRECAUTIONARY STATEMENTS

HAZARDS TO HUMANS AND DOMESTIC ANIMALS

CAUTION!

HARMFUL IF SWALLOWED.

Avoid breathing spray mist. Avoid contact with skin, eyes and clothing.

STATEMENT OF PRACTICAL TREATMENT

If in eyes: Flush with plenty of water. Get medical attention if irritation persists.

If on skin: Wash with plenty of soap and water. Get medical attention if irritation persists.

If swallowed: Call a physician or Poison Control Center. Drink 1 or 2 glasses of water and induce vomiting by touching back of throat with finger. Do not induce vomiting or give anything by mouth to an unconscious person.

For medical emergencies involving this product, call toll free 1-800-441-3637.

ENVIRONMENTAL HAZARDS

Do not apply directly to water or wetlands. Do not contaminate water by cleaning of equipment or disposal of wastes.

IMPORTANT—When mixing/loading and spraying wear a wide-brimmed hat, mid-forearm to elbow length chemical resistant gloves, and one-piece overalls which have long sleeves and long pants. Overalls should be constructed of finely woven fabric as specified in the USDA/EPA Guide for Commercial Applicators.

Any article of clothing worn while applying product must be cleaned before reusing. Clothing which has been drenched or has otherwise absorbed concentrated pesticide must be disposed in a sanitary landfill, or incinerated, or if allowed by State and local authorities, by burning.

Instead of clothing and equipment specified above, the applicator can use an enclosed tractor cab which provides a filtered air supply (as described by Taschenberg and Bourke, 1975).

Injury to or loss of desirable trees or other plants may result from failure to observe the following:

Do not apply (except as recommended), or drain or flush equipment on or near desirable trees or other plants, or on areas where their roots may extend, or in locations where the chemical may be washed or moved into contact with their roots. Do not use on home plantings of trees, shrubs or herbaceous plants, nor on lawns, walks, driveways, tennis courts, or similar areas. Prevent drift of spray to desirable plants. Do not contaminate any body of water. Keep from contact with fertilizers, insecticides, fungicides and seeds.

Do not apply aerially.

Thoroughly clean all traces of "LOROX" L from application equipment immediately after use. Flush tank, pump, hoses, and boom with several changes of water after removing nozzle tips and screens (clean these parts separately).

GENERAL INFORMATION

Du Pont "LOROX" L Herbicide is a liquid suspension to be mixed in water and applied as a spray for selective control of weeds in certain crops and for nonselective weed control on noncropland areas. It is noncorrosive to equipment, nonflammable and nonvolatile.

"LOROX" L is an effective preemergence treatment of soybeans, asparagus, bulbs, carrots, corn, hybrid poplar, parsnips, potatoes and sorghum for control of weeds listed below. "LOROX" L is particularly effective in non-tillage soybean production. Treatment is made after planting but before emergence of soybeans. Weeds have usually emerged and the treatment controls susceptible weeds. Treatment also controls susceptible emerging weeds for an extended time.

"LOROX" L may be sprayed postemergence over-the-top of emerged asparagus (before or after cutting and to small ferns), carrots (at least 3" tall) and celery (transplants) to effectively control susceptible emerged seedling weeds.

Directed postemergence applications of "LOROX" L, where spray nozzles are adjusted so that weeds are sprayed but the crop is not, may be used in emerged asparagus (larger ferns), corn (field and sweet), cotton, sorghum and soybeans to effectively control susceptible emerged seedling weeds.

"LOROX" L may be applied to soil prior to emergence of weeds to control susceptible weed seedlings for an extended period of time, the degree of control and duration of effect will vary with:

Soil Organic Matter Content—High organic matter requires higher dosages than low organic matter for equivalent herbicide performance. Best results accompany preemergence use of "LOROX" L on soils of 1/2% to 3% organic matter. The combination of "LOROX" L plus Du Pont Lexone® L Herbicide (or Du Pont Lexone® 4L Herbicide) provides most effective preemergence weed control on soils from 1% to 4% organic matter content.

Soil Texture—High clay content requires higher dosage rates than those with less clay. Best results accompany preemergence use of "LOROX" L on coarse to medium soils.

Rainfall—Moisture is required to activate the chemical following preemergence application to soil. Best results accompany rainfall (or irrigation) that occurs within 2 weeks of application. In the Colombia River Basin, use "LOROX" L only if the crop is sprinkler irrigated.

Weed Species And Conditions of Growth—Species vary in susceptibility and may be more difficult to control when under stress.

"LOROX" L is recommended for use in conjunction with certain other herbicides for treatment of field corn, potatoes, sorghum, and soybeans; observe all cautions and limitations on labeling of all products used in mixtures.

Since the effect of "LOROX" L varies with soils, crop varieties, uniformity of application, and environmental conditions, it is suggested that growers limit their first use to small areas.

PREEMERGENCE USE (Germinating Weeds): "LOROX" L, at recommended rates, controls annual weeds such as:

Broadleaves		Grasses
Florida beggarweed	Mustards	Barnyardgrass (watergrass)
Carpetweed	Nettleleaf goosefoot	Canarygrass
Chickweed	Pigweeds	Crabgrasses
Common dayflower	Purslane (common)	Foxtails (including giant)
Florida purslane (Florida pusley)	Wild radish	Goosegrass
Galinsoga	Ragweed (common)	Fall panicum
Lambsquarters	Smartweed (Pennsylvania)	

Partial Control

- Cocklebur (common)
- Annual morningglory
- Prickly sida (teaweed)
- Sicklepod
- Velvetleaf (buttonweed)

"LOROX" L will not control established perennials such as bermudagrass, Canada thistle, field bindweed, johnsongrass and purple nutsedge. The lower dosage rates are effective on coarser soils and the higher rates on finer soils and on the more resistant seedling weeds. Sufficient moisture (1/2" to 1" on moist soils; 1" to 2" on dry soils) in the form of rainfall or sprinkler irrigation is necessary after treatment to carry the chemical into the root zone of germinating weeds; best results are obtained when this occurs within two weeks after application. If heavy rainfall occurs soon after application, injury to crop may result.

"LOROX" L applied preemergence, before emergence of soybeans, asparagus, bulbs, carrots, corn (field), parsnips, potatoes, and sorghum, and weeds, is an effective procedure because susceptible weeds are controlled in an early, vulnerable seedling stage before they compete with the crop. With favorable moisture conditions, "LOROX" L continues to control weeds for some time as the crop becomes better able to compete. Should weed seedlings begin to break through the preemergence treatment in significant numbers, secondary weed control procedures should be implemented; these include cultivation and postemergence herbicide application.

A good seed bed must be prepared before application of "LOROX" L as crop injury may result if application is made to ground which is cloddy or compacted resulting in improperly planted seed. Plant seed to depth specified. Surface of the soil should not be cultivated or disturbed after application of "LOROX" L and before emergence of the crop as weed control may be reduced and crop injury may result. However, if moisture is insufficient to activate the herbicide, a shallow cultivation (rotary hoe preferred) should be made after emergence of row crops while weeds are small enough to be controlled by mechanical means. Deep cultivation reduces the effectiveness of "LOROX" L.

Note: Consult "Amiben"[1], atrazine, "Bronco"[3], "Dual"[2] 8E, "Lasso"[3], paraquat[4], propazine, "Prowl"[5], "Ramrod"[3], "Roundup"[3], "Surflan"[6], and "Treflan"[6] labels for additional weeds controlled when "LOROX" L is used in conjunction with these herbicides.

POSTEMERGENCE USE (Emerged Seedling Weeds): "LOROX" L, at recommended rates, controls weeds such as:

Broadleaves		Grasses
Annual morningglory	Lambsquarters	Annual ryegrass
Carpetweed	Mustards	Barnyardgrass (watergrass)
Cocklebur (common)	Nettleleaf goosefoot	Broadleaf signalgrass
Common dayflower	Pigweeds	Canarygrass
Common ragweed	Prickly sida (teaweed)	Crabgrasses
Dog fennel	Purslane (common)	Fall panicum
Fiddleneck (amsinckia)	Sesbania	Foxtails (including giant)
Florida beggarweed	Sicklepod	Goosegrass
Florida purslane (Florida pusley)	Smartweed (Pennsylvania)	Rattail fescue
Groundsel	Velvetleaf (buttonweed)	Texas panicum
Knawel	Wild buckwheat	

Results of postemergence treatment of emerged weeds vary with rate applied and environmental conditions; best results are obtained on succulent weeds growing under conditions of high humidity and temperatures of 70°F or higher. Addition of a surfactant to the spray (where recommended) increases contact effects of "LOROX" L. Application will also provide control of emerging susceptible weed seedlings for an extended period of time.

Note: Consult bromoxynil, "Butoxone"[7], and "Butyrac"[1] 200[1] labels for additional weeds controlled when "LOROX" L is used in conjunction with these herbicides. Best results are obtained under conditions of high humidity and temperatures over 70°F. Control of emerged weeds under drought stress is usually impractical.

DIRECTIONS FOR USE

It is a violation of federal law to use this product in a manner inconsistent with its labeling.

"LOROX" L Herbicide should be used only in accordance with recommendations on this label, or in separate published Du Pont recommendations available through local dealers. Do not use on any crop in Kern County, California, except carrots and asparagus when a permit has been obtained from the Agricultural Commission.

Do not apply this product through any type of irrigation system.

Du Pont will not be responsible for losses or damages resulting from use of this product in any manner not specifically recommended by Du Pont. User assumes all risk associated with such nonrecommended use.

Fill tank 1/4 full with water. Start agitation system; add "LOROX" L, and continue adding water. Add separately each additional component of any tank mix while adding water. If a surfactant is used, add it last when the tank is nearly full. Continue agitation throughout. If poor mixing should occur with any component, premix the component with two parts water before adding to the spray tank.

FERTILIZER SPRAY MIXTURES: For preemergence application, nonpressure nitrogen or fertilizer solution may be used in the spray mixture unless otherwise directed. Small quantities should be tested for compatibility by the following procedure before full scale mixing:

1. Put 1 pint of fertilizer solution in a quart jar.
2. Mix 2 teaspoons "LOROX" L with 2 tablespoons of water; mix thoroughly and add to fertilizer solution.
3. Close jar and shake well.
4. If other herbicides are to be used in the mixture, premix 2 teaspoons of wettable powders or 1 teaspoon of liquids with 2 tablespoons of water; add to "LOROX" L/fertilizer mixture.
5. Close jar and shake well.
6. Watch mixture for several seconds; check again in 30 minutes.
7. If mixture does not separate, foam, gel, or become lumpy, it may be used.
8. Mixing ability may be improved by adding compatibility agents such as Kalo Laboratories "Complex" or Witco Chemicals "Sponto" 168D. Follow directions on container.

If the mixture is compatible, prepare spray by adding the fertilizer solution to the spray tank first; then dilute "LOROX" L with 2 parts of water, thoroughly mix, and add slowly to spray tank with agitator running. For tank mixtures with other herbicides, follow directions above.

EQUIPMENT—SPRAY VOLUMES AND PRESSURES: Application is limited to a ground boom sprayer properly calibrated to a constant speed and rate of delivery. Openings in screens should be equal to or larger than 50 mesh. Continuous agitation in the spray tank is required to keep the material in suspension. Agitate by mechanical or hydraulic means; if bypass or return line is used, it should terminate at bottom of tank to minimize foaming. Avoid overlapping and shut off spray booms while starting, turning, slowing or stopping, or injury to the crop may result.

For preemergence application, use 20 to 60 gals per acre and spray pressure of 30 to 40 psi.

For postemergence application, use sufficient volume (min 20 gals per acre) for thorough coverage of weed foliage; use spray pressure of 20 to 25 psi to keep spray drift to a minimum.

USE RATES: All rates of "LOROX" L (and tank mixtures) are expressed as broadcast rates; for band treatment, use proportionately less. For example, use 1/3 of the broadcast rate when treating a 14" band where row spacing is 42". Where a range of dosages is given, use the lower rate on coarser soils (low in clay or organic matter) and the higher rate on finer soils (high in clay or organic matter); for postemergence application, use the lower rate on smaller weeds and the higher rate on larger weeds.

SOIL LIMITATIONS: Unless otherwise directed, do not use on sand, loamy sand, gravelly soils or exposed subsoils nor on soils containing less than 1% organic matter as injury to the treated or subsequent crops may result.

REPLANTING: If initial seeding fails to produce a stand, crops registered for the rate of "LOROX" L that has been applied may be replanted into the treated area. Thoroughly rework soil before replanting. Do not retreat during the same crop year as injury to the crop may result.

CROP ROTATION RECOMMENDATIONS: Unless otherwise directed, any crop may be planted after 4 months except for cereals where only barley, oats, rye and wheat may be planted.

West of Rocky Mountains: Carrots or celery may be planted 4 months after last application; do not plant any other crop until 1 year after the last application as injury may result.

CROP USES

ASPARAGUS

(California, Michigan, Minnesota, North Carolina, Oregon, Washington only)

Direct Seeded or Newly Planted Crowns—Do not exceed 4 pints total per acre per season; do not use surfactant or fertilizer solution in spray mixture.

Preemergence Application:* Make a single application of 2 to 4 pints per acre after planting see 1½" deep in coarse soil and 1" deep in fine soils. During planting operation, spray activated charcoal ("Aqua Nu-Char"[8] or "Gro-Safe"[8]) as a 1" band on soil surface directly over seed rows at the rate of 300 lbs per acre (equivalent to 15 lbs per acre of crop with 20" row spacing).

Postemergence Application: Make 1 or 2 applications of 1 to 2 pints per acre when ferns are in 6" to 18" stage and weeds are not over 4" tall.

Established Beds: Do not apply within 1 day of harvest; do not exceed 8 pints total per acre per season; do not use surfactant or fertilizer solution in spray mixture. Do not reenter treated areas for 24 hours following application unless protective clothing is worn.

Preemergence Application:* Make a single application of 2 to 4 pints per acre.

Postemergence Application: Make 1 to 4 applications of 1 to 2 pints per acre before weeds exceed 4" in height. Apply before cutting season or immediately after cutting. Do not reenter treated areas for 24 hours following application unless protective clothing is worn.

Directed Postemergence Application (Fern Stage): Make a single application of 4 to 8 pints per acre as a directed spray to base of plants for control of dudain melon.

Note: If more than 4 pts "LOROX" L per acre is applied per season, do not plant any other crop until 1 year after last application. See "Replanting". Do not enter areas treated postemergence for 24 hours following application unless protective clothing is worn.

*Preemergence weed control will be reduced in soils with high organic matter (greater than 5% and peat or muck).

BULBS (Tulip, Calla Lily, Daffodil, and Dutch Iris Only)
California

After planting of bulbs, settle the soil with sprinkler irrigation (rainfall will serve the same purpose); then before emergence of plants (bulbs), apply 2 pints "LOROX" L per acre in a minimum of 20 gallons of water per acre. Treat only once during the growing season.

Note: Unless other directed, do not use on sand, loamy sand, gravelly soils, or exposed subsoils nor on soils containing less than 1% organic matter as injury to the crop may result. Do not reenter treated areas for 24 hours following application unless protective clothing is worn.

CARROTS

Preemergence Application—Florida, Michigan, Ohio, and Wisconsin: Make a single application of 1 to 2 pints per acre in Florida and 1 to 3 pints per acre in Michigan, Ohio, and Wisconsin; apply after planting but before carrots emerge; plant seed at least ½" deep. Use the lower rate on lighter soils (low in clay or organic matter) and higher rate on heavier soils (high in clay or organic matter). Subsequent postemergence application may be made provided the total does not exceed 4 pints "LOROX" L per acre per season.

Postemergence Application—U.S.: Apply 1½ to 3 pints per acre as a nondirected spray after carrots are at least 3" tall. Apply before annual grasses exceed 2" in height and before broadleaf weeds exceed 6" in height. Repeat application may be made but do not exceed 4 pints "LOROX" L per acre (West of Rocky Mountains, do not exceed 3 pints "LOROX" L per acre). Do not exceed 40 psi spray nozzle pressure as crop injury may result.

"LOROX" L may be applied following an application of Stoddard solvent provided treatments are at least one day apart. Stoddard solvent may be applied following application of "LOROX" L provided treatments are at least 2 weeks apart. Shorter time intervals between applications may result in injury to the crop. Do not apply "LOROX" L as a tank mixture with Stoddard solvent, surfactants, nitrogen or fertilizer solution, other pesticides, nor when temperature exceeds 85°F. as crop injury may result.

Note: Because carrot varieties vary in their resistance, determine tolerance to "LOROX" L prior to adoption as a field practice to prevent possible crop injury; do not treat susceptible varieties which show an initial burning of foliage following postemergence treatment with "LOROX" L. Do not reenter treated areas treated postemergence for 24 hours following application unless protective clothing is worn. See "Replanting".

CELERY

East of Rocky Mountains—Post-transplant Application: Make a single application of 1 1/2 to 3 pints per acre. Apply as a non-directed spray after celery is transplanted and established, but before celery is 8" tall. Apply before annual grasses exceed 2" in height and before broadleaf weeds exceed 6" in height. In the Northeast, use only on celery grown on muck soils.

Do not exceed 40 psi spray nozzle pressure, and do not apply when temperature exceeds 85°F. nor as a tank mixture with surfactants, nitrogen or fertilizer solution, or other pesticides as injury to the crop may result. Do not replant to crops other than celery or carrots within 4 months after application as injury to subsequent crops may result. See "Replanting". Do not reenter treated areas for 24 hours following application unless protective clothing is worn.

CORN (FIELD)

Preemergence Application—East of Rocky Mountains: Select one of the following herbicide treatments for application as a tank mixture. Make a single application after planting but before crop emerges. Plant seed at least 1 3/4" deep on flat or raised seedbeds only or injury to the crop may result. Do not spray over top of emerged corn.

"LOROX" L at 3/4 to 1 pint per acre (tank mixed with standard registered herbicide for corn) controls triazine-resistant lambsquarters on soils of 1 to 4% organic matter east of the Rocky Mountains.

"LOROX" L + "Lasso"

Soil Texture	Product Per Acre					
	1 to 3% Organic Matter			3 to 6% Organic Matter		
	"LOROX" L (pts)	+	"Lasso" (qts)	"LOROX" L (pts)	+	"Lasso" (qts)
Coarse Sandy loam	2/3 to 1 1/4	+	3/4 to 1	1 1/4 to 2	+	1 to 1 1/2
Medium Loam, Silt loam, Silt, Sandy clay, Sandy clay loam	1 to 1 2/3	+	1 to 1 1/2	1 2/3 to 2 1/2	+	1 1/2 to 2
Fine Silty clay, Silty clay loam, Clay, Clay loam	1 1/4 to 2	+	1 1/2 to 2	2 to 3	+	2 to 2 1/2

Replanting: Corn or soybeans may be replanted within 4 months, see "Replanting".

"LOROX" L + "Ramrod"

Soil Texture	Pints Per Acre					
	1 to 3% Organic Matter			3 to 6% Organic Matter		
	"LOROX" L (pts)	+	"Ramrod" (4 lbs/gal)	"LOROX" L (pts)	+	"Ramrod" (4 lbs/gal)
Coarse Sandy loam	2/3 to 1 1/3	+	1 1/3 to 2 3/5	1 1/3 to 2	+	2 3/5 to 3 9/10
Medium Loam, Silt loam, Silt, Sandy clay, Sandy clay loam	1 to 1 2/3	+	2 to 3 1/3	1 2/3 to 2 1/2	+	3 1/3 to 5 1/5
Fine Silty clay, Silty clay loam, Clay, Clay loam	1 1/4 to 2	+	2 3/5 to 3 9/10	2 to 3	+	3 9/10 to 6 1/5

Replanting: Corn may be replanted within 4 months; after 4 months, see "Replanting".

"LOROX" L + atrazine (Tank Mixture)—

For improved control of annual grasses, add "Lasso" as recommended on "Lasso" labeling for "Lasso"/atrazine combinations.

Soil Texture	Product Per Acre					
	1 to 2% Organic Matter			2 to 5% Organic Matter		
	"LOROX" L (pts)	+	atrazine 80% (lbs)	"LOROX" L (pts)	+	atrazine 80% (lbs)
Coarse Sandy loam	2/3 to 1	+	1/2 to 2/3	1 to 2	+	2/3 to 1 1/4
Medium Loam, Silt loam, Silt, Sandy clay, Sandy clay loam	1 to 1 1/2	+	2/3 to 1	1 1/2 to 2 1/2	+	1 to 1 1/2
Fine Silty clay, Silty clay loam, Clay, Clay loam	1 1/3 to 1 2/3	+	3/4 to 1	1 2/3 to 3	+	1 to 2

Replanting: Corn may be replanted within 6 months; after 6 months, see "Replanting".

CORN (FIELD AND SWEET)

Directed Postemergence Application: Make a single application as a directed spray after corn is at least 15" high (measured to the highest leaf surface on free standing plants). Do not spray over top of corn. Apply only when there is sufficient differential between height of corn and weeds so that the directed spray thoroughly covers all weed foliage without contact of upper leaves or whorl of corn by spray or drift, as such contact may cause crop injury. Early cultivation (rotary hoe or other suitable equipment) will aid in achieving proper differential between height of corn and weeds.

Use 1 1/4 to 3 pints per acre; add 1 pint surfactant for each 25 gals spray mixture. For field corn, non-pressure nitrogen solution may be substituted for all or part of the water. Use the lower rate on lighter soils (low in clay or organic matter) and when weeds do not exceed 2" in height; use the higher rate on heavier soils (high in clay or organic matter) for weeds up to 5" in height. Do not reenter treated areas for 24 hours following application unless protective clothing is worn. See "Replanting".

COTTON

Directed Postemergence Application—East of Rocky Mountains: Apply as a directed spray; adjust nozzles to minimize contact to cotton leaves with spray or drift or crop injury may result. Do not spray over top of cotton.

Apply 1 to 1 1/2 pints per acre when cotton is at least 8" tall and emerged weeds do not exceed 2" in height; add 1 pint surfactant for each 25 gals spray mixture. If needed, a second application at same rate may be made one week or later after initial treatment. Alternatively, after cotton is 20" tall, make a single application of 2 to 3 pints per acre following last cultivation; if emerged weeds are present, add surfactant as directed above. **Note:** Do not use on Pima varieties of cotton. Do not graze treated fields or feed forage from treated areas to livestock. Do not feed gin trash to livestock. Do not reenter treated areas for 24 hours following application unless protective clothing is worn. See "Replanting".

HYBRID POPLAR (MIDWEST)

Apply 2 to 4 pints "LOROX" L per acre before bud break in the spring.

For application after bud break, apply 2 to 4 pints "LOROX" L per acre as a directed spray. Spray should be directed to weed growth and to avoid contact with the poplar plant; do not spray over the top of the poplar as injury to the plant will result.

Use the lower rate on light soils and higher rate on heavier soils. For best results on emerged weeds, treat at the seedling stage.

More than one treatment may be made but no more than 8 pints "LOROX" L per acre should be applied per year.

PARSNIPS

Preemergence Application: Make a single application of 1 1/2 to 3 pints per acre. Apply after planting but before crop emerges. Plant seed at least 1/2" deep. See "Replanting".

POTATOES

Preemergence Application: Make a single application as a broadcast spray after planting but before crop emerges. Plant seed at least 2" deep. Do not spray over top of emerged potatoes. If beds are to be "dragged" and/or "hilled", apply after the final "dragging" or "hilling" operation. Apply before grasses are 2" tall and before broadleaf weeds are 6" tall, preferably just before or when weed seedlings emerge. If emerged weeds are present, add 1 pint surfactant for each 25 gals spray mixture. In irrigated areas, best results are obtained when application is made to moist soil, followed within 2 weeks by 1" to 2" of sprinkler irrigation (or rainfall). On powder dry soils, irrigate prior to herbicide application and follow with sprinkler irrigation to activate the herbicide.

East of Rocky Mountains: Apply 1 1/2 to 2 1/2 pints per acre on the coarser soils (sandy loams, silt loams; 1 to 2% organic matter) and 2 1/2 to 4 pints per acre on finer soils (silts, clay loams; 2 to 5% organic matter); on soils over 5% organic matter, use 4 pints per acre and apply to emerged weeds (before potatoes emerge).

Wisconsin—Central Sands Area: Apply 1 pint per acre on sands and 2 pints per acre on loamy sands.

Northeast: For improved control of annual grasses (see "Dual" labeling), apply:

Soil Texture	Product Per Acre					
	1 to 3% Organic Matter			3 to 5% Organic Matter		
	"LOROX" L (pts)	+	"Dual" 8E (pts)	"LOROX" L (pts)	+	"Dual" 8E (pts)
Coarse Sandy loam	1 to 1 1/2	+	1 1/2	1 1/2 to 2	+	2
Medium Loam, Silt loam, Silt, Sandy clay, Sandy clay loam	1 1/2 to 2	+	2	2 to 2 1/2	+	2 1/2 to 3
See Replanting.						

SORGHUM

Preemergence Application: Select a registered herbicide treatment for application as a tank mixture with "LOROX" L. Make a single application after planting but before crop emerges. Plant seed at least 1" deep on flat or raised seedbeds only or injury to the crop may result. Do not spray over top of emerged sorghum.

"LOROX" L –Southwest	Pints of "LOROX" L per acre	
	Percent Organic Matter in Soil	
Soil Texture Description	**1 to 2%**	**2 to 4%**
Coarse: Sandy Loam	5/8 to 1	1 to 1 1/2
Medium: Loam, Silt loam, Silt, Sandy clay, Sandy clay loam	1 to 1 1/2	1 to 2
Replanting: See "Replanting on the companion product labeling. Prior to replanting, through seeded preparation including fall or spring plowing is recommended.		

Directed Postemergence Application: Make a single application of "LOROX" L as a directed spray; add 1 pint surfactant for each 25 gals spray mixture. If sprayer is equipped with skids, shoes or shields, apply 1 pint per acre when sorghum is 12" tall (free standing plants) and weeds are up to 2" in height. Use 1 to 2 pints per acre when sorghum is 15" tall and weeds are 2" to 4" in height. If boom drops are used, apply 1 to 2 pints per acre when sorghum is at least 18" tall and weeds are 2" to 4" in height. Apply only when there is sufficient differential between height of sorghum and weeds so that the directed spray thoroughly covers all weed foliage without contact of upper leaves or whorl of sorghum by spray or drift as such contact may cause crop injury.

Note: Do not graze or feed plants to livestock within 3 months after postemergence application. See "Replanting". Do not reenter treated areas for 24 hours following application unless protective clothing is worn.

SOYBEANS—CONVENTIONAL TILLAGE

Preemergence Application: For broad spectrum weed control, select one of the following herbicide combination treatments and make a single application after planting but before crop emerges. Plant seed at least 1 3/4" deep on flat or raised seedbeds only or injury to the crop may result. Injury to soybeans may result if application is made to fields with standing water or fields too wet to cultivate. Do not spray over top of emerged soybeans. Do not use on sand or loamy sand nor any soil containing less organic matter than listed below.

"LOROX" L + "Lexone" DF*

For control of common ragweed, hemp sesbania, jimsonweed, lambsquarters, pigweeds, prickly sida, sicklepod, velvetleaf, spotted spurge, Venice mallow, purslane, carpetweed, Pennsylvania smartweed, and partial control of cocklebur*, use the following:

Soil Texture	Product Per Acre					
	1/2 to 3% Organic Matter in Soil			3 to 6% Organic Matter in Soil		
	"LOROX" L (pts)	+	"Lexone" DF (lbs)	"LOROX" L (pts)	+	"Lexone" DF (lbs)
Coarse Sandy loam	1/3 to 1/2	+	1/6 to 1/4	1/2 to 3/4	+	1/4 to 1/3
		or ("Lexone" 4L—1/4 to 2/5 pt)			or ("Lexone" 4L—2/5 to 1/2 pt)	
Medium Loam, Silt loam, Silt, Sandy clay, Sandy clay loam	1/2 to 3/4	+	1/4 to 1/3	3/4 to 1 1/2	+	1/3 to 1/2
		or ("Lexone" 4L—2/5 to 1/2 pt)			or ("Lexone" 4L—1/2 to 3/4 pt)	
Fine Silty clay, Silty clay loam, Clay, Clay loam	3/4 to 1 1/2	+	1/3 to 1/2	1 1/2 to 2	+	1/2 to 2/3
		or ("Lexone" 4L—1/2 to 3/4 pt)			or ("Lexone" 4L—3/4 to 1 pt)	

*For improved control of annual grasses, tank mix with "Lasso" or "Dual" (per labeling):

Herbicide	Recropping/Restrictions
"Lasso"	**Soybeans or field corn may be replanted within 4 months; after 4 months, see "Replanting". Treated vines may be grazed or fed to livestock 40 days after application.**
or	
"Dual"	**Soybeans or field corn may be replanted within 4 months, for rotation crops, see "Replanting" and follow instructions on "Dual" label. Do not graze or feed forage from treated areas to livestock.**

Note: Soybean varieties such as Altona, Coker 102 and 156, Govan, NKS 1884, Semmes, Tracy, Vansoy, Terra Vig 505 and 606, Agripro 55, Asgrow 6520, Maple Amber, Portage, Vinton 81, and AP 71 are sensitive to "Lexone". Injury may occur if "Lexone" is used on these varieties. Before use on any other soybean variety, tolerance to "Lexone" must first be determined. Varieties showing above average tolerance to "Lexone" are Americana Revere, Asgrow 1937, Asgrow 3659, Asgrow 3860, DSR 171, DSR 207, Essex, Fayette, Hisoy 170, Lakota, Lawrence, LOL 4207, NKS 1492, Pride B216, Pride B242, SRF 250, SRF 350 P, Union, Wayne, Wells II and Williams 82; for maximum weed control, use the higher rate (where a range of rates is listed) for the appropriate soil type and organic matter as shown in the following tables. Injury to soybeans may occur if "Lexone" is used on soils having a calcareous surface layer or pH of 7.5 or higher, or if used in conjunction with soil applied organic phosphate pesticides such as "Dasanit", "Disyston", "Mocap", "Nemacur", "Thimet", "Parathion", "Lorsban" 15G or Counter. Injury may occur if atrazine was applied on the soil the year before use of "Lexone". Seedling disease, cold weather, deep planting (more than 2"), excessive moisture, high soil pH (pH 7.5 or higher), high salt concentration, or drought may weaken soybean seedlings and increase possibility of crop injury.

Do not use on sand nor loamy sand.

Cultivation after planting: Treated soil may be shallow-cultivated, rotary-hoed or hand-hoed without reducing the weed control activity of the tank mixture. Do not cultivate deeper than the treated layer of soil since this may bring untreated soil to the surface and poor weed control may result.

Replanting: If initial seeding fails to produce a stand, treated fields may be replanted to soybeans; do not rework soil; do not retreat field with a second application as injury to the crop may result. Do not replant treated areas to any crop other than soybeans within four months after treatment as injury to subsequent crops may result. See "Replanting".

"LOROX" L*

Soil Texture	Product Per Acre	
	1/2 to 3% Organic Matter "LOROX" L (pts)	3 to 6% Organic Matter "LOROX" L (pts)
Coarse Sandy loam	2/3 to 1 1/4	1 1/4 to 2
Medium Loam, Silt loam, Silt, Sandy, clay, Sandy clay loam	1 to 1 2/3	1 2/3 to 2 1/2
Fine Silty clay, Silty clay loam, Clay, Clay loam	1 1/4 to 2	2 to 3

*For improved control of annual grasses, add as per the following labeling:

Herbicide	Recropping/Restrictions
"Lasso"	**Soybeans or field corn may be replanted within 4 months; after 4 months, see "Replanting."**
or	
"Dual"	**Soybeans or field corn may be replanted within 4 months; for rotation crops, see "Replanting" and follow instructions on "Dual" label. Do not graze or feed forage from treated areas to livestock.**
or	
"Prowl"	**Soybeans or field corn may be replanted within 4 months; crops listed in "Replanting" may be planted the following year.**
or	
"Surflan"	**Soybeans may be replanted within 4 months; after 4 months, see "Replanting" but do not plant potatoes within 12 months. Do not use treated vines for feed or forage.**
or	
"Amiben"	**Soybeans or field corn may be replanted within 4 months; after 4 months, see "Replanting."**

For control of black nightshade, apply:

Soil Texture	Product Per Acre	
	1 to 3% Organic Matter	
	"LOROX" L (pts)	"Lasso" (qts)

For control of black nightshade, apply:

Soil Texture	Product Per Acre: 1 to 3% Organic Matter	
	"LOROX" L (pts)	"Lasso" (qts)
Coarse Sandy loam	¾ to 1½	2
Medium Loam, Silt loam, Silt, Sandy clay, Sandy clay loam	1 to 1⅔	2½ to 3
Fine Silty clay, Silty clay loam, Clay, Clay loam	1¼ to 2	3

Do not use on sand or loamy sands.

Replanting: Soybeans or field corn may be replanted within 4 months; after 4 months, see "Replanting".

For control of black nightshade in Kentucky, Illinois, Indiana and Ohio:

Soil Texture	Product Per Acre: 1 to 3% Organic Matter	
	"LOROX" L (pts)	"Dual" 8E (pts)
Coarse Sandy loam	¾ to 1½	1½
Medium Loam, Silt loam, Silt, Sandy clay, Sandy clay loam	1 to 1⅔	2
Fine Silty clay, Silty clay loam, Clay, Clay loam	1¼ to 2	2 to 2½

Do not use on sand or loamy sands.

Replanting: Soybeans or field corn may be replanted within 4; after 4 months, see "Replanting".

Preemergence Following "Treflan" or "Prowl" Preplant: Where "Treflan" or "Prowl" has been used as a preplant incorporated treatment (according to directions on product label), apply "LOROX" L preemergence (after planting and before emergence of soybeans) as a separate operation using rates recommended below for "LOROX" L alone. For rotation crops, follow instructions on "Treflan" or "Prowl" labels and see "Replanting". Plant seed at least 1¾" deep on flat or raised seedbeds only or injury to the crop may result. Injury to soybeans may result if application is made to fields with standing water or fields too wet to cultivate. Do not spray over top of emerged soybeans. Do not use on sand or loamy sand nor any soil containing less organic matter than listed below.

"LOROX" L Alone—If weeds have emerged, add 1 pint surfactant for each 25 gals spray mixture.

Soil Texture	Pints "LOROX" L Per Acre	
	½ to 2% Organic Matter	2 to 5% Organic Matter
Coarse Sandy loam	1 to 1⅔	1⅔ to 3
Medium Loam, Silt loam, Silt, Sandy clay, Sandy clay loam	1¼ to 2⅓	2⅓ to 4
Fine Silty clay, Silty clay loam, Clay, Clay loam	1⅓ to 2⅔	2⅔ to 5 (over 5% organic matter, use 6 pts)

Replanting: Soybeans or field corn may be replanted within 4 months; after 4 months, see "Replanting".

Preemergence Followed by Postemergence (Directed) Application: Both preemergence and postemergence (directed to weeds with minimum contact with soybean plants) treatment may be required for control of:

Annual morningglory—"LOROX" L preemergence (any combination listed) followed by a directed postemergence spray of "LOROX" L + 2,4-DB.

Broadleaf signalgrass—"LOROX" L + "Lexone" DF preemergence followed by a directed postemergence spray of "LOROX" L, if needed.

Cocklebur (common)—"LOROX" L + "Lexone" DF preemergence followed by a directed postemergence spray of "LOROX" L + 2,4-DB.

Hemp sesbania—"LOROX" L + "Lexone" DF preemergence followed by a directed postemergence application of "LOROX" L, if needed.

Prickly sida—"LOROX" L + "Lexone" DF preemergence followed by a directed postemergence spray of "LOROX" L, if needed.

Velvetleaf—"LOROX" L + "Lexone" DF preemergence followed by a directed postemergence application of "LOROX" L, if needed.

Other weeds—Directed postemergence application of "LOROX" L should be used as a backup treatment for listed preemergence treatments and will extend the duration and degree of weed control obtained with preemergence treatment. Postemergence treatment should be used only as needed.

For dosage rates, refer to "Preemergence" and "Directed Postemergence" sections under "Soybeans".

SOYBEANS—MINIMUM OR NO-TILLAGE

Preemergence to Soybeans Postemergence to Weeds: "LOROX" L preemergence to soybeans may be used for preemergence and postemergence control of many broadleaf weeds and grasses where soybeans will be planted directly into a preformed bed (stale seed bed), cover crop or in previous crop residues such as corn or small grain stubble. Apply with ground equipment immediately before, during or after planting but before crop emerges; maintain constant agitation of spray mixture.

"LOROX" L Alone

	Pints "LOROX" L Per Acre	
	½ to 2% Organic Matter	2 to 5% Organic Matter
Coarse Sandy loam	1 to 1⅔	1⅔ to 3
Medium Loam, Silt loam, Silt, Sandy clay, Sandy clay loam	1¼ to 2⅓	2⅓ to 4
Fine Silty clay, Silty clay loam, Clay, Clay loam	1⅓ to 2⅔	2⅔ to 5 (over 5% organic matter, use 6 pts)

Replanting: Soybeans or field corn may be replanted within 4 months; after 4 months, see "Replanting".

If small seedling weeds are present, add 1 pint surfactant for each 25 gals spray mixture for improved contact activity. On larger weeds, add "Paraquat" or "Roundup" or "Bronco" as described under these combinations; these treatments will also suppress some perennial weeds.

"LOROX" L + "Lasso" or "Dual" 8E or "Surflan"—will improve control of grasses and volunteer small grains.

"LOROX" L + "Lasso"* (or "Dual"* or "Surflan"*) + Paraquat or "Roundup"—Thoroughly mix "LOROX" L and companion herbicide in spray tank first according to directions; then add paraquat, or "Roundup" as directed under Paraquat Combinations or "Roundup" Combinations below.

Soil Texture	Product Per Acre: ½ to 3% Organic Matter			Product Per Acre: 3 to 6% Organic Matter		
	"LOROX" L (pts)	+	"Lasso"* (qts)	"LOROX" L (pts)	+	"Lasso"* (qts)
Coarse Sandy loam	¾ to 1½	+	2	1½ to 2⅖	+	2½
Medium Loam, Silt loam, Silt, Sandy clay, Sandy clay loam	1⅛ to 2	+	2½	2 to 3	+	3
Fine Silty clay, Silty clay loam, Clay, Clay loam	1¼ to 2¼	+	2½	2¼ to 3½	+	3

*For improved control of annual grasses and volunteer small grains, add as per labeling:

Herbicide	Recropping/Restrictions
"Lasso"	Soybeans or field corn may be replanted within 4 months; after 4 months, see "Replanting."
or "Dual"	Soybeans or field corn may be replanted within 4 months; for rotation crops see "Replanting" and follow instructions on "Dual" label. Do not graze or feed forage from treated areas to livestock.
or "Surflan"	Do not use treated vines for feed or forage. Soybeans may be replanted within 4 months; after 4 months, see "Replanting" and do not plant potatoes within 12 months.

For control of black nightshade apply with paraquat or "Roundup" as shown below:

Soil Texture	Products Per Acre: 1 to 3% Organic Matter		
	"LOROX" L (pts)	+	"Lasso" (qts)

For control of black nightshade apply with paraquat or "Roundup" as shown below:

Soil Texture	Products Per Acre 1 to 3% Organic Matter "LOROX" L (pts)	+	"Lasso" (qts)
Coarse Sandy loam	¾ to 1½	+	2
Medium Loam, Silt loam, Silt, Sandy clay, Sandy clay loam	1 to 2	+	2½ to 3
Fine Silty clay, Silty clay loam, Clay, Clay loam	1¼ to 2¼	+	3

Replanting: Soybeans or field corn may be replanted within 4 months; after 4 months, see "Replanting".

Paraquat Combinations—Select and tank mix in water one of the above treatments; then add ¼ to ½ lbs. active ingredient per acre for control of emerged weeds. Use the higher rate for weeds 4" to 6" tall. As the last ingredient, add ½ pint surfactant or "Ortho" X-77 Spreader per 100 gals of spray mixture. Maintain constant agitation. Use 20 to 60 gals of water per acre. Use the higher gallonage for dense stubble or vegetation.

"Roundup" Combinations—Select and tank mix one of the above treatments. As last ingredient, add 1½ quarts of "Roundup" per acre for control of emerged annual weeds or 2 to 4 quarts per acre for control of emerged perennial and annual weeds. Use 20 to 30 gals of water per acre.

"LOROX" L + "Bronco"—Do not add paraquat or "Roundup" to this combination.

Soil Texture	Product Per Acre ½ to 5% Organic Matter in Soil(a) "LOROX" L (pts)	+	"Bronco"(b) (qts)
Coarse Sandy loam	¾ to 2	+	3¼ to 5
Medium Loam, Silt loam,Silt, Sandy clay, Sandy clay loam	1⅛ to 2½	+	4 to 5
Fine Silty clay, Silty clay loam, Clay, Clay loam	1¼ to 3	+	4 to 5

(a) Do not use on soils with less than ½% organic matter nor on sand, loamy sand, or muck soils as crop injury may result.

(b) Use the higher rate for dense stubble, heavy crop residue, or heavy weed population.

Replanting Soybeans: If initial seeding fails to produce a stand, treated fields may be replanted to soybeans; do not rework soil; do not retreat field with a second application as injury to the crop may result. Do not replant treated areas to any crop other than soybeans within four months after treatment as injury to subsequent crops may result. See "Replanting".

SOYBEANS—CONVENTIONAL OR MINIMUM OR NO-TILLAGE

Directed Postemergence Application: Apply "LOROX" L alone or as a tank mixture with 2,4-DB, as a directed spray to cover weed foliage with minimum contact of the soybean plant. Do not spray higher than 3" on the soybean stem or crop injury may result. Do not spray over top of soybean plants. For broadcast application, use a single flood-type spray nozzle ("K" series or equivalent) per middle mounted on an oiling shoe or gauge wheel. For band treatment, use two nozzles per row mounted on an oiling shoe or gauge wheels, one of each side of row. To avoid spray drift, which may cause crop injury, do not exceed nozzle pressure of 25 psi nor use nozzle tips smaller than 8002T-Jet (or equivalent) and do not spray under windy conditions. Add 1 pint surfactant for each 25 gals spray mixture. For best results, use a preemergence treatment (such as "LOROX" L) or cultivation to control early weed growth and to increase the differential between height of soybeans and weeds.

NOTE: Do not use on soils with less than ½% organic matter. Do not apply more than 2 pints "LOROX" L per acre per season for postemergence treatments. Do not apply within 60 days of harvest. Do not feed soybean forage or hay to livestock from fields treated postemergence. Harvested soybeans may be used for food, feed or oil purposes. See "Replanting". Do not reenter treated areas for 24 hours following application unless protective clothing is worn.

"LOROX" L Alone—For soybeans at least 8" high and when weeds do not exceed 2" in height, apply ½ to 1 pint per acre. Make a second application at same rate if new flush of weeds occur.

For soybeans at least 12" high and when weeds do not exceed 4" in height, make a single application of 1 to 2 pints per acre. Alternatively, make a split application of 1 pint per acre followed by a second application at the same rate after 1 week or later.

"LOROX" L + 2,4-DB—For soybeans at least 8" high and when weeds do not exceed 4" in height, apply 1 pint "LOROX" L plus ⅕ pint 2,4-DB (13 fl oz "Butyrac" 200 or 1 pint "Butoxone") per acre. A second application may be made if needed, but do not make more than 2 applications per season.

NONCROP WEED CONTROL

For short-term control of annual weeds or non-cropland areas such as roadsides and fence rows—apply 2 to 6 pints "LOROX" L per acre in 40 to 100 gals of water. For best results, apply shortly before weed growth begins or at early seedling stage of growth. For control of established annual weeds, add surfactant at rate of 2 qts per 100 gals of spray mixture and apply as a thorough coverage spray during periods when daily temperatures exceed 70°F. and before weed growth exceeds 8" in height.

Do not reenter treated areas for 24 hours following application unless protective clothing is worn.

STORAGE AND DISPOSAL

STORAGE: Keep from freezing. Store product in original container only, away from other pesticides, fertilizer, food or feed.

PRODUCT DISPOSAL: Do not contaminate water, food or feed by storage or disposal. Wastes resulting from the use of this product may be disposed of on site or at approved waste disposal facility.

CONTAINER DISPOSAL: Triple rinse (or equivalent) the container. Then offer for recycling or reconditioning, or puncture and dispose of in a sanitary landfill, or by incineration, or, if allowed by state and local authorities, by burning. If burned, stay out of smoke.

NOTICE OF WARRANTY

Du Pont warrants that this product conforms to the chemical description on the label thereof and is reasonably fit for purposes stated on such label only when used in accordance with directions under normal use conditions. It is impossible to eliminate all risks inherently associated with the use of this product. Crop injury, ineffectiveness or other unintended consequences may result because of such factors as weather conditions, presence of other materials, or the manner of use or application, all of which are beyond the control of Du Pont. In no case shall Du Pont be liable for consequential, special or indirect damages resulting from the use or handling of this product. All such risks shall be assumed by the Buyer. DU PONT MAKES NO WARRANTIES OF MERCHANTABILITY OR FITNESS FOR A PARTICULAR PURPOSE NOR ANY OTHER EXPRESS OR IMPLIED WARRANTY EXCEPT AS STATED ABOVE.

1 Registered trademark of Union Carbide Agricultural Products Company, Inc. "Amiben" contains 2 lbs chloramben per gal; "Butyrac" 200 contains 2 lbs 2,4-DB per gal.

2 Registered trademark of Ciba-Geigy Corporation, "Dual" 8E contains 8 lbs metolachlor per gal.

3 Registered trademark of Monsanto Company, "Lasso" contains 4 lbs alachlor per gal. "Roundup" contains 4 lbs glyphosate per gal. "Bronco" contains 2.6 lbs alachlor and 1.04 lbs glyphosate per gal. "Ramrod" contains 4 lbs propachlor per gal.

4 Paraquat contains 2 lbs active per gal, paraquat is a restricted use pesticide.

5 Registered trademark of American Cyanamid Company, "Prowl" contains 4 lbs pendimethalin per gal.

6 Registered trademark of Elanco Products Company, "Surflan" 75W contains 75% W.P. oryazalin, "Surflan" AS contains 4 lbs oryazalin per gal. "Treflan" contains 4 lbs trifluralin per gal.

7 Registered trademark of Rhone-Poulenc Inc., "Butoxone" contains 1.75 lbs dimethylamine salt and 2 lbs isooctyl ester of 2,4-DB per gal.

8 Registered trademark of Westvaco Corporation.

9 Registered trademark of ICI United States, Inc.

AG-3256 8049 4/13/89

This is a specimen label. DuPont is not responsible for the accuracy of the information contained herein. As labels are subject to revision, always carefully read and follow the label on the product container.

SEE SUPPLEMENT FOR NEW LABEL INFORMATION

MANZATE® 200 DF

FUNGICIDE
DISPERSIBLE GRANULES

	BY WEIGHT
ACTIVE INGREDIENT:	75%
A coordination product of zinc ion and manganese ethylene bisdithiocarbamate in which the ingredients are:	
Manganese‡	15%
Zinc‡	1.9%
Ethylene bisdithiocarbamate ion ($C_4H_6N_2S_4$) =	58.1%
INERT INGREDIENTS:	25%
TOTAL	100%

This product contains the toxic inert paraformaldehyde

EPA Reg No. 352-449

KEEP OUT OF REACH OF CHILDREN

PRECAUTIONARY STATEMENTS

HAZARDS TO HUMANS AND DOMESTIC ANIMALS

CAUTION!

MAY IRRITATE EYES, NOSE, THROAT AND SKIN. MAY BE HARMFUL IF INHALED OR SWALLOWED.

Avoid breathing dust or spray mist. Avoid contact with skin, eyes and clothing. Keep away from fire or sparks.

During mixing, loading, and application wear a long-sleeve shirt, long pants or coveralls, socks, shoes, chemical resistant gloves, and goggles or face shield. During mixing and loading a chemical resistant apron must also be worn. See Directions For Use.

Do not apply this product in such a manner as to directly or through drift expose workers, other persons or animals. The area being treated must be vacated by unprotected persons.

FIRST AID: In case of contact, flush skin or eyes with plenty of water; for eyes, get medical attention.

For medical emergencies involving this product, call toll free 1-800-441-3637.

ENVIRONMENTAL HAZARDS

This product is toxic to fish. Drift and runoff from treated areas may be hazardous to aquatic organisms in neighboring areas. Do not apply directly to water or wetlands (swamps, bogs, marshes, and potholes) except as specified for the labeled use on cranberry. Cover or incorporate spilled treated seed. Do not contaminate water by cleaning by disposing of equipment washwaters.

DIRECTIONS FOR USE

It is a violation of federal law to use this product in a manner inconsistent with its labeling.

"MANZATE" 200 DF should be used only in accordance with recommendations on this folder or in separate published Du Pont recommendations available through local dealers.

Du Pont will not be responsible for losses or damages resulting from use of this product in any manner not specifically recommended by Du Pont. User assumes all risks associated with such nonrecommended use.

"MANZATE" 200, a dispersible granule containing mancozeb, is recommended for use as a spray for the control of many important plant diseases.

WORKER SAFETY RULES

Keep all unprotected persons, children, livestock, and pets away from treated area or where there is a danger of drift.

Do not rub eyes or mouth with hands. See First Aid.

PERSONAL PROTECTIVE EQUIPMENT—For Mixers, Loaders, Applicators and Early Reentry Workers.

HANDLE THIS PRODUCT ONLY WHEN WEARING THE FOLLOWING PROTECTIVE CLOTHING AND EQUIPMENT: a long-sleeve shirt and long pants or a coverall that covers all parts of the body except the head, hands, and feet; chemical resistant gloves; shoes, socks, and goggles or a face shield. During mixing and loading, a chemical resistant apron must also be worn.

During application from a tractor with a completely enclosed cab with positive pressure filtration, or aerially with an enclosed cockpit, a long-sleeve shirt and long pants may be worn in place of the above protective clothing. Chemical resistant gloves must be available in the cab or cockpit and worn while exiting.

IMPORTANT! Before removing gloves, wash them with soap and water. Always wash hands, face, and arms with soap and water before eating, smoking or drinking. Always wash hands and arms with soap and water before using the toilet.

After work take off all clothes and shoes. Shower using soap and water. Wear only clean clothes. Do not use contaminated clothing. Wash protective clothing and protective equipment with soap and water after each use. Personal clothing worn during use must be laundered separately from household articles. Clothing and protective equipment heavily contaminated or drenched with mancozeb must be destroyed according to state and local regulations.

HEAVILY CONTAMINATED OR DRENCHED CLOTHING CANNOT BE ADEQUATELY DECONTAMINATED.

During aerial application, human flaggers are prohibited unless in totally enclosed vehicles.

Written or oral warnings must be given to workers who are expected to be in a treated area or in an area about to be treated with this product. When oral warnings are given, warnings shall be given in a language customarily understood by workers. Oral warnings must be given if there is reason to believe that written warnings cannot be understood by workers. Warnings must include the following information: "CAUTION. Area treated with 'MANZATE' 200 on (date of application). Do not enter without appropriate protective clothing until sprays have dried or dusts have settled. In case of contact, flush skin or eyes with plenty of water; for eyes, get medical attention."

REENTRY STATEMENTS

For Agricultural Uses

After sprays have dried do not enter or allow entry into treated areas until the 24-hour reentry interval has expired unless wearing the personal protective equipment listed on this label.

For Turf Uses

Do not enter treated areas without protective clothing until sprays have dried.

AS A SPRAY (Ground or Aerial Equipment)—Apply "MANZATE" 200 DF at the rate shown; use sufficient water to provide thorough coverage, usually 20 to 100 gals per acre for ground equipment and 5 to 8 gals per acre for aircraft. Add "MANZATE" 200 DF slowly to water in the spray tank with agitation, or premix thoroughly in separate holding tank for concentrate or aircraft sprayers. Continuous agitation is required to keep the product in suspension. Add a spreader-sticker spray adjuvant if needed.

CHEMIGATION: Apply "MANZATE" 200 DF Fungicide only through sprinkler including center pivot, lateral move, end tow, side (wheel) roll, traveler, big gun, solid set or hand move irrigation systems only on crops specifically designated in the "Directions For Use" for that particular crop. Do not apply "MANZATE" 200 DF through any other type of irrigation system or on any other crops.

Crop injury, lack of effectiveness, or illegal pesticide residues in the crop can result from nonuniform distribution of treated water.

If you have questions about calibration, you should contact State Extension Service Specialists, equipment manufacturers or other experts.

Do not connect an irrigation system (including greenhouse systems) used for pesticide application to a public water system unless the pesticide label-prescribed safety devices for public water systems are in place.

A person knowledgeable of the chemigation system and responsible for its operation, or under the supervision of the responsible person, shall shut the system down and make necessary adjustments should the need arise.

Specific Instructions for Public Water Systems:

1. Public water system means a system for the provision to the public of piped water for human consumption if such system has at least 15 service connections or regularly serves an average of at least 25 individuals daily at least 60 days out of the year.
2. Chemigation systems connected to public water systems must contain a functional, reduced-pressure zone, backflow preventer (RPZ) or the functional equivalent in the water supply line upstream from the point of pesticide introduction. As an option to the RPZ, the water from the public water system should be discharged into a reservoir tank prior to pesticide introduction. There shall be a complete physical break (air gap) between the outlet end of the fill pipe and the top or overflow rim of the reservoir tank of at least twice the inside diameter of the fill pipe.
3. The pesticide injection pipeline must contain a functional, automatic, quick-closing check valve to prevent the flow of fluid back toward the injection pump.
4. The pesticide injection pipeline must contain a functional, normally closed, solenoid-operated valve located on the intake side of the injection pump and connected to the system interlock to prevent fluid from being withdrawn from the supply tank when the irrigation system is either automatically or manually shut down.
5. The system must contain functional interlocking controls to automatically shut off the pesticide injection pump when the water pump motor stops, or in cases where there is no water pump, when the water pressure decreases to the point where pesticide distribution is adversely affected.

6. Systems must use a metering pump, such as a positive displacement injection pump (e.g., diaphragm pump) effectively designed and constructed of materials that are compatible with pesticides and capable of being fitted with a system interlock.
7. Do not apply when wind speed favors drift beyond the area intended for treatment.

Specific Instructions for Sprinkler Irrigation Systems:

1. The system must contain a functional check valve, vacuum relief valve, and low pressure drain appropriately located on the irrigation pipeline to prevent water source contamination from backflow.
2. The pesticide injection pipeline must contain a functional, automatic, quick-closing check valve to prevent the flow of fluid back toward the injection pump.
3. The pesticide injection pipeline must also contain a functional, normally closed, solenoid-operated valve located on the intake side of the injection pump and connected to the system interlock to prevent fluid from being withdrawn from the supply tank when the irrigation system is either automatically or manually shut down.
4. The system must contain functional interlocking controls to automatically shut off the pesticide injection pump when the water pump motor stops.
5. The irrigation line or water pump must include a functional pressure switch which will stop the water pump motor when the water pressure decreases to the point where pesticide distribution is adversely affected.
6. Systems must use a metering pump, such as a positive displacement injection pump (e.g., diaphragm pump) effectively designed and constructed of materials that are compatible with pesticides and capable of being fitted with a system interlock.
7. Do not apply when wind speed favors drift beyond the area intended for treatment.
8. Good agitation is required in the injection tank.
9. In moving systems, apply specific dosage of "MANZATE" 200 DF as a continuous injection. In nonmoving systems inject "MANZATE" 200 DF for 15 to 30 minutes at end of cycle. Use the least amount of water possible consistent with uniform coverage.
10. Mix the amount of "MANZATE" 200 DF needed for acreage to be treated into the quantity of water determined during prior calibration. For moving systems inject into the system continuously for one complete revolution of the field. For nonmoving systems inject into system for the time established during calibration.
11. Stop injection equipment after treatment is completed and continue to operate irrigation equipment until all "MANZATE" 200 DF is flushed from system.

FIELD CROPS
APPLICATION RESTRICTIONS

CROP	DISEASE/PEST	RATE	TIMING/INTERVAL	LAST APPLICATION TO HARVEST	LIVESTOCK GRAZING/FEEDING	COMMENTS
Barley, Oats, Rye, Wheat	Helminthosporium leaf blight Septoria glume spot Septoria leaf spot Tan spot Wheat leaf rust (*Puccinia* spp.)	2 lbs/acre	Begin when disease threatens or at tillering-to-jointing stage. Repeat every 7 to 10 days, maximum of 3 applications.	26 days	Do not graze livestock in treated areas within 26 days after application.	
Corn (field, popcorn)	Helminthosporium leaf blight Rust	1½ lbs/acre	Begin when disease threatens. Repeat every 4 to 7 days.	40 days.		
Corn (sweet, hybrid seed production)	Helminthosporium leaf blight Rust	1½ lbs/acre	Begin when disease threatens. Repeat every 4 to 7 days.	7 days	Do not graze or feed forage or fodder to livestock.	
Cotton (Southwest only)	Rust	1¼ to 2 lbs/acre	Begin when rust first appears in the area. Repeat at 10 to 14 day interval.	Do not apply after bolls open	Do not graze or feed gin trash.	
Oats (see barley)						
Peanuts	Cercospora leaf spot Rust	1 to 2 lbs/acre	Begin when disease threatens. Repeat at 7 to 14 day interval.		Do not feed vine hay to livestock.	
Peanuts (Tank mix with Benlate® Fungicide)	Ascochyta web blotch Cercospora leaf spot Rust	1½ lbs plus 4 ozs "Benlate"/acre	Begin 35 to 40 days after planting or when disease threatens. Repeat at 7 to 14 day intervals (7 to 10 days for rust).	14 days	Do not graze or feed treated vines, hay or hulls to livestock.	Spray oil may be added at ½ to 1 qt/acre. Do not use "Benlate" where benzimidazole-resistant Cercospora is present.
Rye (see barley)						
Sugar beets	Cercospora leaf spot	1½ to 2 lbs/acre	Begin when disease first threatens. Repeat at 7 to 10 day intervals.	14 days	Do not feed treated sugar beet tops to livestock.	
Tobacco - Plant bed	Anthracnose Blue mold	½ lb/100 gals (2 tsp/gal), 3 to 5 gals/100 sq yds of plant bed)	Begin when plants are the size of a dime or when blue mold threatens. Repeat at 3 to 14 day intervals through transplanting.			
Tobacco - Field	Blue mold	1 to 1½ lbs/acre	Begin when blue mold is reported in the area. Repeat at 3 to 7 day intervals.			
Wheat (see barley)						

FRUIT AND NUT CROPS APPLICATION RESTRICTIONS

CROP	DISEASE/PEST	RATE	TIMING/INTERVAL	LAST APPLICATION TO HARVEST	LIVESTOCK GRAZING/FEEDING	COMMENTS
Apples, Crabapples, Quince	Bitter rot Black rot Brown rot Cedar-apple rust Fly speck Scab Sooty Blotch	1 to 2 lbs/100 gals up to 10 lbs/acre	Begin with green tip or delayed dormant spray. Repeat every 7 to 14 days throughout the season.	Crabapple and quince—15 days. Apples—21 days in AL, AR, DE, GA, IL, IN, KS, KY, MD, MO, NJ, NC, OH, PA, SC, TN, VA, WV; 30 days in all other states.	Do not graze livestock in treated orchards	Add spreader-sticker during last 3 cover sprays
Apples, Tank mix w/"Benlate" Fungicide	Bitter rot Black rot Brown rot Cedar-apple rot Fly Speck Powdery mildew Quince rust Scab Sooty blotch	1 lb plus 2 to 3 ozs "Benlate"/100 gals	Apply at ½" green tip. Repeat at 7 to 14 day intervals (or as needed) through the cover sprays.	Apples 21 days in AL, AR, DE, GA, IL, IN, KS, KY, MD, MO, NJ, NC, OH, PA, SC, TN, VA, WV; 30 days in all other states.	Do not graze livestock in treated orchards.	1 qt spray oil/100 gals may be added for varieties more susceptible to powdery mildew, or for scab if an application is missed during an infection period.
Bananas	Sigatoka *(Mycosphaerella)*	2 to 4 lbs/acre in water or water/oil emulsion or oil alone.	Begin when leaves first appear and repeat every 2 to 3 weeks or as required.			Add spreader-sticker when using water as carrier.
Caprifig	Endosepsis *(Fusarium)* Mold	4 lbs/100 gals	Prepare mamme figs by making a shallow cut through the eye and then hand dividing to avoid wasp injury. Submerge mamme figs in the continuously agitated suspension for at least 15 minutes. Drain before placement in trees.			Use fresh dipping suspension after treating 4 to 5 batches of figs.
Crabapple (see Apple)						
Cranberries	Fruit rot	3 to 6 lbs/acre	Begin at mid-bloom. Repeat at 7 to 10 day intervals.	30 days		
Grapes	Black rot Bunch rot Dead arm Downy mildew	1½ to 4 lbs/acre	Apply when shoots are ½ to 1½" long; 3 to 5" long. & 8 to 10" long. Continue at 10 to 14 day intervals.	66 days (in California do not apply after fruit set)		
Papayas	Anthracnose *(Colletotricum)* Phytophthora fruit rot Black spot *(Cercospora)*	2 to 3 lbs/acre (minimum 50 GPA)	Begin at flowering; treat central column crown, blossom area and developing fruit. Repeat at 7 to 14 day intervals.			Add spreader-sticker
Pears	Bitter rot Black rot Brown rot Cedar-apple rust Fly speck Scab Sooty blotch	1 to 2 lbs/100 gals up to 10 lbs/acre	Begin with green tip or delayed dormant spray. Repeat every 7 to 14 days throughout the season.	15 days	Do not graze livestock in treated orchards.	Add spreader-sticker during last 3 cover sprays.
	Pear psylla nymph (Pacific NW)	1½ to 2 lbs/100 gals up to 10 lbs/acre	Apply any time after bloom or postharvest when high populations of pear psylla nymphs are present. Repeat every 2 to 4 weeks.	15 days when spray oils are used post-bloom, do not apply "MANZATE" 200 DF sooner than 7 days after the last oil application.	Do not graze livestock in treated orchards.	
Pineapple	Phytophthora heart rot	3.2 lbs/10 gals Use up to 100 gals of suspension to treat clean propagation material for one acre.	Dip planning material in fungicide suspension prior to planting. Stir frequently to prevent settling.			Prepare new suspension when ⅔ of volume is used or sooner if noticeably discolored by soil from plant material.
Quince (see apples)						

VEGETABLE CROPS APPLICATION RESTRICTIONS

CROP	DISEASE/PEST	RATE	TIMING/INTERVAL	LAST APPLICATION TO HARVEST	LIVESTOCK GRAZING/FEEDING	COMMENTS
Asparagus	Rust	2 lbs/acre	Begin when rust threatens. Repeat every 10 days as needed.	Apply on on ferns after spears have been harvested		

VEGETABLE CROPS
APPLICATION RESTRICTIONS

CROP	DISEASE/PEST	RATE	TIMING/INTERVAL	LAST APPLICATION TO HARVEST	LIVESTOCK GRAZING/FEEDING	COMMENTS
Crown (planting stock)	Crown rot	1 lb per 100 gals	Dip clean, loosely packed crowns into continuously agitated fungicide suspension for 5 minutes. Drain and plant as soon as possible.			Wash dirty crowns before dip treatment. Replace suspension in clean tank when discolored by soil.
Carrots	Alternaria leaf spot Cercospora leaf spot	1½ to 2 lbs/acre	Begin when disease threatens. Repeat at 7 to 10 day intervals.	7 days	Do not use treated tops for food or feed.	
Celery, Fennel	Early & late blights	2 lbs/acre	Begin in plant beds at emergence. Repeat every 3 to 5 days in plant bed, at 7-day intervals in the field.	Celery—14 days Fennel—7		Remove excess residues by stripping, trimming and washing at harvest.
Corn (Sweet)	Helminthosporium leaf blight Rust	1½ lbs/acre	Begin when disease threatens. Repeat every 4 to 7 days.	7 days	Do not graze or feed forage or fodder to livestock.	
Cucumbers Melons Squash (summer & winter)	Alternaria leaf spot Anthracnose Cercospora leaf spot Downy mildew Gummy stem blight Pythium fruit rot Scab	2 to 3 lbs/acre (winter squash—1½ to 2 lbs/acre)	Begin at 2-leaf stage. Repeat every 5 to 7 days.	5 days		
Fennel (see celery)						
Melons (see cucumbers)						
Onions (dry bulb)	Botrytis leaf blight Downy mildew Neck rot Purple blotch	3 lbs/acre	Begin when disease is first reported in the area. Repeat every 7 days.	7 days		Do not allow spray or drift to contact exposed bulbs.
	Smut	3 lbs/acre in 75 to 125 gals/acre (29,000 linear feet of furrow)	Apply as a furrow drench to muck soils at planting to protect seedlings.			
Potatoes	Early & late blights	1 to 2 lbs/acre	Begin when plants are 4 to 6" high. Repeat every 7 to 10 days.			
	Early blight	1 to 2 lbs/acre plus Triphenyl Tin Hydroxide (TPTH) at ½ to full labeled use rate				
Potato (seed piece)	Fusarium seed piece decay Common scab (seedborne)	2½ lbs/100 gals	Dip treat, drain, and plant as soon as possible following treatment.		Do not use treated pieces for food or feed purposes.	If seed pieces are to be held before planting, spread out to dry in a cool place.
Squash (summer and winter) see cucumber						
Tomatoes	Anthracnose Early & late blights Gray leaf mold Gray leaf spot	1½ to 3 lbs/acre	Begin when seedlings emerge. Repeat every 7 days.	5 days		
	Bacterial leaf spot	1½ lbs plus 4 lbs fixed coppers in 100 gals/acre	Begin when seedlings emerge. Repeat every 3 to 4 days.			

GRASSES (TURF)
APPLICATION RESTRICTIONS

CROP	DISEASE/PEST	RATE	TIMING/INTERVAL	LAST APPLICATION TO HARVEST	LIVESTOCK GRAZING/FEEDING	COMMENTS
Turf grasses	Algae	6 ozs in 3 to 5 gals/1000 sq ft (16 lbs in 130-220 gals/acre)	Begin when algae first appears. Repeat at 7-day intervals as long as condition persists.		Do not graze treated areas or feed clippings to livestock.	

GRASSES (TURF)
APPLICATION RESTRICTIONS

CROP	DISEASE/PEST	RATE	TIMING/INTERVAL	LAST APPLICATION TO HARVEST	LIVESTOCK GRAZING/FEEDING	COMMENTS
	Leaf stem, and strip rust	3 to 4 ozs in 3 to 5 gals/1000 sq ft (8 to 11 lbs in 130 to 220 gals/acre)	Begin when disease threatens. Repeat at 7 to 10 day intervals as long as disease persists.			
	Leaf spot (*Helminthosporium* spp.) *Rhizoctonia solani* brown patch	3 to 4 ozs in 3 to 5 gals/1000 sq ft (8 to 11 lbs in 130 to 220 gals/acre)	Begin when disease threatens. **Tees and greens:** Repeat at 7 to 10 day intervals. **Fairways:** Repeat at 7 to 14 day intervals. **Home lawns:** Repeat at 14 to 21 day intervals. Continue as long as disease persists.			When conditions are unusually favorable for disease, use 6 to 8 ozs/1000 sq ft (16 to 22 lbs/acre) and reduce intervals to 3 to 5 days.

SEED TREATMENT

CROP	DISEASES	SEED TREATMENT RATE—APPLY AS A SLURRY		COMMENTS
		Ozs/Bu	Ozs/100 lb	
Barley	Bunt, Covered smut, Damping-off, Fake loose smut, Seed decay, Seedling blights	1.3 to 2.0	2.7 to 4.2	LABEL TREATED SEED "Do not use for food, feed or oil purposes. This seed treated with 'MANZATE' 200 Fungicide."
Corn	Damping off, seed rot, seedling blight	1.5 to 3.0	2.7 to 5.4	
Cotton Acid Delinted	Damping-off, Seedling blights	—	3	
Cotton Reginned	Damping-off, Seedling blights	—	6	
Flax	Seed decay, Seedling blights, Damping-off	2.0 to 4.0	3.6 to 7.1	
Oats	Damping-off, Seedling blights, Seed decay, Smuts	1.3 to 2.0	4.0 to 6.3	
Peanuts (shelled)	Damping-off, Seed rots, Seedling blights	2.0 to 4.0	8.0 to 16.0	
Rice	Achlya, other soil and seedborne fungi causing seed rot and reduced seedling vigor.	—	2.0 to 4.0 Apply before, during, or after soaking in water.	The incorporation of a dye into the treating slurry is recommended. Dye must color seed to distinguish as treated.
Rye	Bunt, Covered smut, Damping-off, Seed decay, Seedling blights	1.3 to 2.0	2.3 to 3.6	
Safflower	*Puccinia carthami* (which causes foot-and-rot disease and foliage rust disease)	—	2.0	
Sorghum	Covered kernel smut, Damping-off, Seedling blights, Seed rots	1.5 to 2.5	2.7 to 4.5	
Tomatoes	Damping-off, Seedling blights, Seed rots	—	8.0	
Wheat	Bunt, Covered smut, Damping-off, Seed decay, Seedling blights	1.3 to 2.0	2.2 to 3.3	

TREATED PLANTS MUST NOT BE USED FOR FOOD OR FEED PURPOSES

FLOWERS, FOLIAGE PLANTS, AND ORNAMENTALS

Apply in the field, nursery or greenhouses as a thorough coverage spray, using 1 to 2 lbs "MANZATE" 200 DF per 100 gals of water (1½ to 3 tsp per gal). For hard-to-wet plants, add a suitable wetting agent. Begin applications when new growth starts or when disease first threatens and continue at 7 to 10 day intervals between applications. Use higher rate and shorter intervals during periods of excessive wetness and rapid plant growth.

DuPont "MANZATE" 200 DF is recommended for use on certain flower, foliage and ornamental plants listed in the table below for control of the following diseases and pathogens:

Alternaria—leaf spots and flower blights
*Anthracnose—*Colletotrichum* or *Elsinoe* or *Gloeosporium* or *Glomerella* or *Gnomonia*
Aschochyta
*Blackspot or rose or elm
Blight—*Phytophthora*
Blister—*Taphrina*
Botryosphaeria
Botrytis blights, gray mold
Brown rot of *Prunus* spp.—*Monilinia*
Cephalosporium
**Cercospora*
**Collectotrichum*
**Coryneum*
Curvularia
Cylindrocladium
**Cylindrosporium*
**Dactylaria*
Didymellina
**Diplocarpon*
Downy mildews—*Plasmopara* or *Peronospora*
**Elsinoe*
**Entomosporium*
Fabraea
Flower blights
Fusarium
Fusicladium
*Gloesporium
*Gnomonia
Gray mold—*Botrytis*
Guignardia
**Gymnosporangium*
*Leafspots
**Leptosphaeria*
Lophodermium
Macrophoma
**Melampsora*
**Melampsoridium*
Monochaetia
Mycosphaerella
**Pestalotia*
Phomopsis—Phragmidium
**Phyllosticta*
Phytophthora
Pseudonectria
Puccinia
Ramularia
Rhizoctonia
**Rhytisma*
*Rusts—*Gymnosporangium*, *Melampsora*, *Melampsoridium*, *Phragmidium*, *Puccinia*, *Uromyces*
*Scab—*Fusicladium* or *Venturia* or *Sphaceloma*
Scirrhia
**Septoria*
*Shothole—*Coryneum*
**Sphaeropsis*
Stemphylium
Taphrina
Uromyces

* denotes leaf spots

ORNAMENTALS FOR WHICH "MANZATE" 200 DF IS RECOMMENDED

COMMON NAME	*LATIN NAME*	COMMON NAME	*LATIN NAME*	COMMON NAME	*LATIN NAME*
African violet	Saintpaulia ionantha	Ficus	Ficus benjamina	Pilea	Pilea spp.
	Aglaonema commutatum		Ficus pumila	Pine, Norfolk Island	Aracecaria
Almond	Prunus amygdalus	Fir, Douglas	Pseudotsuga menziesii	Pine White	Pinus strobus
Andromeda, Mountain	Pieris floribunda	Fir, Noble	Abies procera	Pittosporium	Pittosporium tobira
Anthurium	Anthurium spp.	Firethorn	Pyracantha coccinea	Plane Tree	Platanus occidentalis
Arborvitae	Thuja occidentalis	Fittonia	Fittonia argyroneura	Poinsettia*	Euphorbia pulcherrima
Areca Palm	Chrysalidocar lutescens	Geranium	Pelargonium hortorum	Prayer Plant	Maranta kerchoveana
Ash	Fraxinus spp.	Gladiolus	Gladiolus spp.	Primrose	Primula spp.
Ash, Mountain	Surbus spp.	Gold Dust Tree	Aucuba japonica	Privet	Ligustrum spp.
Aster, China	Callistephus chinensis	Gypsophila	Gyposphila paniculata	Protea	Leucospermum cordifolium
Azalea	Rhododendron spp.	Hawthorn	Crataegus spp.	Prune	Prunus spp.
Baby's Breath	Gypsophila paniculata	Hemp-bowstring	Sansevieria trifasciata	Pussywillow	Salix babylonica
Begonia	Begonia spp.	Holly	Ilex spp.	Pyracantha	Pyracantha waterii
Birch	Betula pendula	Holly, Chinese	Ilex cornuta	Redwood (seedlings)	Dequoia sempervirons
Boxwood	Buxus microphylla	Holly, Japanese	Ilex crenata	Rhododendron	Rhododendron spp.
	Buxus sempervirons	Hollyhock	Althaea rosea	Rose	Rosa spp.
Buckeye	Aesculus spp.		Alcea spp.	Sand Cherry	Prunus cistena
Buffalo Berry	Sherperdia argentea	Horsechestnut	Aesculus hippocastanum	Santolina	Santolina spp.
Camellia	Camellia japonica	Iris	Iris spp.	Schefflera	Schefflera arboricola
Carnation	Dianthus caryophyllus	Ivy	Hendera helix		Schefflera actinophylla
Cedar, Red	Juniperus virginiana	Juniper	Juniperus chinensis	Sequoia, Giant (seedlings)	Sequoia gigantia
Christmas Cactus	Zygocactus truncatus		Juniperus horizontalis	Snapdragons	Antirrhinum majus
Chrysanthemum	Chrysanthemum spp.		plumosa Juniperus sabina	Spindletree, Winged	Euonymus alatus
Cordyline	Dracaena spp	Laurel, Mountain	Kalmia latifolia	Spruce, Alberta (dwarf)	Picea glauca conica
Cottoneaster	Cottoneaster divaricata	Lavender Cotton	Santolina chamaecyparissus	Spruce, Engelmann	Picea engelmannii
	Cottoneaster horizontalis	Lily	Lilium spp.	Spruce, Norway	Picea abies
	perpusilla	Lily, Easter	Lilium longiflorum	Spurge, Japanese	Pachysandra terminalis
Crabapple	Malus ioensis		*Liriope muscari*	Statice	Limonium sinuatum
	Crassula argentea	Madagascar Palm	Chrysalidocar lutescens	Strawflower	Helichrysum bracteatum
Croton	Codiaeum varigatum	Magnolia	Magnolia stellata	Sumac, Skunkbush	Rhus trilobata
	Codiaeum punctatum	Mahonia	Mahonia nervosa	Sycamore	Platanus spp.
	aureaum	Maple	Acer spp.	Syngonium	Syngonium podophyllum
Cyprus, Arizona	Cypressus spp.	Maple, Red	Acer rubrum	Velvet plant	Gynura spp.
Dahlia	Dahlia spp.	Maple, Sugar	Acer saccharum	Venus Flytrap	Dionaea muscipula
Daisy, Transvall	Gerbera spp.	Nannyberry	Viburnum lentago	Verbena	Verbena spp.
Dieffenbachia	Dieffenbachia picta	Nephthytis	Nephthytis afzelii	Viburnum	Viburnum davidi
Dogwood	Cornus amomum	Oak	Quercus borealis		Viburnum opulus
	Cornus Florida		Quercus rubrum	Walnut	Juglans spp.
	Cornus racemosa		Quercus velutina	Walnut, Black	Juglans nigra
	Cornus sericea	Osmanthus	Osmanthus heterophyllus	Willow	Salix spp.
	Cornus stolonifera	Pansy	Viola spp.	Yucca	Yucca elephantipes
Dracaena	Dracaena marginata	Peony	Paeonia spp.		Yucca filamentosa
	Dracaena sanderiana	Peperomia	Peperomia spp.	Zebra plant	Aphelandra squarrosa
Dusty Miller	Centaurea cineraria	Periwinkle	Vinca spp.	Zinnia	Zinnia elegans
Elm, American	Ulmus americana	Petunia	Petunia hybrida		
Elm, Chinese	Ulmus parvifolia	Philodendron	Philodendron spp.		
Euonymus	Euonymus japonica	Photinia	Photinia fraseri		
Fern	Nephrolepus exaltata	Pick-a-back	Tolmiea menziesii		
	Polystichum adiantiforme	Pieris	Pieris spp.		
	Pieris ensiformis				

* Do not exceed 1½ lbs/100 gallons.

STORAGE AND DISPOSAL

STORAGE: Important—Never allow "MANZATE" 200 DF to become wet during storage. This may lead to certain chemical changes which will reduce the effectiveness of "MANZATE" 200 DF as a fungicide and create vapors which may be flammable. Keep container closed when not in use. Store product in original container only, away from other pesticides, fertilizer, food or feed.

PESTICIDE DISPOSAL: Do not contaminate water, food or feed by storage or disposal. Wastes resulting from the use of this product may be disposed of on site or at an approved waste disposal facility.

CONTAINER DISPOSAL Completely empty bag into application equipment. Then dispose of empty bag in a sanitary landfill or by incineration, or, if allowed by state and local authorities, by burning. If burned, stay out of smoke.

NOTICE OF WARRANTY

Du Pont warrants that this product conforms to the chemical description on the label thereof and is reasonably fit for purposes stated on such label only when used in accordance with the directions under normal use conditions. It is impossible to eliminate all risks inherently associated with the use of this product. Crop injury, ineffectiveness or other unintended consequences may result because of such factors as weather conditions, presence of other materials, or the manner of use or application, all of which are beyond the control of Du Pont. In no case shall Du Pont be liable for consequential, special or indirect damages resulting from the use or handling of this product. All such risks shall be assumed by the Buyer. DU PONT MAKES NO WARRANTIES OF MERCHANTABILITY OR FITNESS FOR A PARTICULAR PURPOSE NOR ANY OTHER EXPRESS OR IMPLIED WARRANTY EXCEPT AS STATED ABOVE.

NOTICE TO BUYER: Purchase of this material does not confer any rights under patents of countries outside of the United States.

AG-2854 8028 2/11/88

SEE SUPPLEMENT FOR NEW LABEL INFORMATION

OUST®
HERBICIDE
DISPERSIBLE GRANULES

ACTIVE INGREDIENT:	BY WEIGHT
Sulfometuron methyl [Methyl 2-[[[[(4,6-dimethyl-2-pyrimidinyl)-amino]carbonyl]amino]sulfonyl]benzoate]	75%
INERT INGREDIENTS:	25%
TOTAL	100%

EPA Reg. No. 352-401
U.S. Pat. 4,394,506

KEEP OUT OF REACH OF CHILDREN

PRECAUTIONARY STATEMENTS

HAZARDS TO HUMANS AND DOMESTIC ANIMALS

CAUTION!

MAY IRRITATE EYES, NOSE, THROAT AND SKIN.

Avoid breathing dust or spray mist. Avoid contact with skin, eyes and clothing.

If in eyes, immediately flush with plenty of water and get medical attention. If on skin, immediately flush with plenty of water and get medical attention if irritation persists.

For medical emergencies involving this product, call toll free 1-800-441-3637.

ENVIRONMENTAL HAZARDS

Do not apply directly to wetlands or any body of water. Do not contaminate water by cleaning of equipment or disposal of wastes.

IMPORTANT

DO NOT USE ON FOOD OR FEED CROPS. Injury to or loss of desirable trees or other plants may result from failure to observe the following: Do not apply "OUST" (except as recommended), or drain or flush equipment on or near desirable trees or other plants, or on areas where their roots may extend or in locations where the chemical may be washed or moved into contact with their roots.

Do not use on lawns, walks, driveways, tennis courts or similar areas. Keep from contact with fertilizers, insecticides, fungicides and seeds.

Do not treat powdery, dry soil or light, sandy soils when there is little likelihood of rainfall soon after treatment; in areas where sensitive crops are grown, do not apply "OUST" to steep banks or treat before soil has been settled by rain as off-target movement by wind or water may occur. Do not apply to impervious substrates or when ground is frozen.

Do not apply to any body of water. Do not apply where runoff water may flow onto agricultural land, as injury to crops may result. Do not apply "OUST" during periods of intense rainfall or to soils saturated with water as off-target movement may occur.

Do not apply this product through any type of irrigation system.

Do not allow spray to drift onto adjacent crops or other desirable plants or trees as injury may occur.

Following an "OUST" application, the spray tank used should not be used for other than noncrop applications. This is extremely important, as low rates of "OUST" can kill or severely injure most crops.

GENERAL INFORMATION

Du Pont "OUST" Herbicide is a dispersible granule to be mixed in water and applied as a spray for control of many annual and perennial grasses and broadleaf weeds on noncropland areas. It is noncorrosive, nonflammable, nonvolatile and does not freeze.

"OUST" may be used for general weed control on industrial noncrop sites and for selective weed control in certain types of unimproved turf grasses on noncropland.

"OUST" can also be used for selective weed control in forest site preparation and release of several types of pines.

"OUST" may be applied preemergence or postemergence to weeds; best results are obtained if application is made before or during early stages of weed growth. "OUST" should be used during seasons when rainfall occurs as moisture is required to move "OUST" into the root zone of weeds before they develop an established root system. Preemergence treatments control or suppress weeds through root uptake while postemergence works through root uptake and foliar uptake. Under limited rainfall conditions, "OUST" may not provide satisfactory control of hard-to-kill perennials.

"OUST" is absorbed by both roots and foliage of weeds, resulting in visual effects progressing from growth inhibition to reddish-purplish coloration, chlorosis, necrosis, vein discoloration and death of terminals. Initial effects are usually seen 2 to 3 weeks following application; however, the final effects are evident at about 4 to 6 weeks after application. Warm, moist conditions following treatment promote the activity of "OUST" while cold, dry conditions delay the activity of "OUST." Weeds hardened off by cold weather or drought stress may not be fully controlled.

Degree of control and duration of effect depend on rate used, weed spectrum, weed size, growing conditions at and following time of treatment, soil pH, precipitation, and soil organic matter.

DIRECTIONS FOR USE

It is a violation of federal law to use this product in a manner inconsistent with its labeling.

"OUST" Herbicide should be used only in accordance with recommendations on this label or in separate published Du Pont recommendations available through local dealers.

Du Pont will not be responsible for losses or damages resulting from the use of this product in any manner not specifically recommended by Du Pont. User assumes all risks associated with such nonrecommended use.

Du Pont "OUST" Herbicide is recommended for preemergence and postemergence control of many annual and perennial grasses and broadleaf weeds. "OUST" must be applied in ground equipment as a selective or nonselective spray treatment of noncropland areas such as airports, fencerows, highways, industrial turf, lumberyards, petroleum tank farms, pipeline rights-of-way, plant sites, railroads, roadside turf, storage areas and utility rights-of-way consistent with restrictions under "Important." Treatment of powdery dry soil, steep banks and impervious substrates should be avoided; treated soils should be left undisturbed to reduce the potential for "OUST" movement by soil erosion due to wind or water. Injury to crops may occur if treated soil is washed, blown or moved onto land used to produce crops.

"OUST" is also recommended for conifer site preparation and release (except California). Treatment may be applied by ground equipment or helicopter to these forest sites.

"OUST" may be applied preemergence or postemergence to weeds, but for best results, apply postemergence to young, actively growing weeds under favorable moisture conditions at any time of the year, except when ground is frozen.

Rate selection is based on weed species, weed size and soil texture. Use the higher rates on established plants and on fine textured soils and lower rates on smaller weeds and coarse textured soils.

SELECTIVE WEED CONTROL

For use on unimproved industrial turf, on roadsides, and on other noncropland areas where bermudagrass (common and coastal varieties), bahiagrass, crested wheatgrass and smooth brome are well established, and are desired as ground covers.

BERMUDAGRASS RELEASE

For control of many annual and perennial broadleaf weeds and grasses use the following rates and application timings (do not use surfactant):

Late Spring and Early Summer: Apply 1 to 2 oz of "OUST" per acre for control of Carolina geranium, fescue, foxtail, goldenrod, spotted spurge, and wild carrot after bermudagrass is well established, usually about 30 days after initial spring flush. If regrowth of weeds occurs and additional control is desired, repeat treatment at 1 to 2 oz during late spring and summer. After weeds become established, best results occur when applications are made 1 to 2 weeks following mowing.

Alternatively, in Arkansas, Louisiana and Mississippi, 1 to 2 ounces "OUST" per acre may be tank mixed with 3 to 4 lbs of MSMA per acre and applied to well established bermudagrass during the summer; the combination includes control of additional weeds listed on the MSMA package label. Two or more additional applications of MSMA (without "OUST") at 14-day intervals may be necessary to maintain weed control.

For short-term control of johnsongrass in the U.S., apply 2 to 3 oz per acre after bermudagrass is well established, usually about 30 days after initial spring flush. Re-treat with 2 to 3 oz if additional control is desired or if regrowth occurs.

Late Fall and Early Winter: Apply 1 to 4 oz of "OUST" per acre for control of winter annuals such as Carolina geranium, common chickweed, fescue, little barley and wild blackberry; use the lower rate on small seedling weeds and higher rates on larger weeds.

BAHIAGRASS RELEASE AND SEEDHEAD SUPPRESSION

Apply 1 oz of "OUST" per acre (without surfactant) for control of fescue, goldenrod, spotted spurge and wild carrot. Rates as low as 1/2 oz per acre reduce bahiagrass growth and suppress seedheads when applied before seedhead emergence in the spring or following mowing.

SMOOTH BROME RELEASE AND SUPPRESSION

For control of downy brome, foxtail, goldenrod, kochia, Russian thistle and prickly lettuce in well established stands of smooth brome, apply 1 oz of "OUST" per acre (without surfactant) in the spring at an early stage of plant growth. This application will also result in smooth brome seedhead reduction, as well as growth suppression and reduced plant vigor.

CRESTED WHEATGRASS RELEASE AND SUPPRESSION

For control of downy brome, kochia, Russian thistle and prickly lettuce in well established stands of crested wheatgrass on noncrop sites, apply 1 oz "OUST" per acre (without surfactant) during the rainy season or spring. This application may also result in growth suppression of the crested wheatgrass.

NOTE: Temporary discoloration or top kill of bahiagrass, bermudagrass and smooth brome may result following an application of "OUST."

Browning and degree of stand reduction may increase in areas with warm climates. Do not apply "OUST" to industrial turf that is under stress from drought, insects, disease, cold temperatures or late spring frost, as injury may result. Do not apply "OUST" to grasses within 1 year of planting as stand reduction may occur.

CONIFER SITE PREPARATION AND RELEASE (EXCEPT CALIFORNIA)

"OUST" provides both foliar and residual control of many herbaceous grass and broadleaf weeds in jack pine, loblolly pine, red pine, slash pine and Virginia pine reforestation areas.

When used as described, "OUST" controls many herbaceous plants such as: Crabgrass, dog fennel, fescue, fireweed (willowweed), goldenrod, horseweed, Kentucky bluegrass, nutsedge (yellow), panicums (broadleaf, fall, narrow), pokeweed, ragweed, white snakeroot and yellow sweetclover.

Preplant - Site preparation (trees not present)/**Postplant** - release (trees present) Make application from spring through late fall; do not apply to frozen ground.

Apply 2-8 ounces of "OUST" per acre (2-4 ounces per acre on red pine) using appropriate ground or air (helicopter only) equipment. Add surfactant at 0.25% by volume for site preparation. Do not use surfactant for pine release (trees present).

NOTE: Do not apply by air within 200 feet of any homestead, agricultural land or other desirable plantings. Extreme care must be taken to prevent drift or runoff to homesteads, desirable plantings, agricultural land or any body of water. Do not apply when weather conditions favor drift from treated areas.

Use extreme care when applying in areas adjacent to any body of water. Keep out of lakes, streams, ponds, reservoirs or any body of water. Do not apply where conifers are suffering from loss of vigor caused by insects, diseases, drought, winter damage or other stresses, as injury may result. "OUST" is not recommended for use on poorly drained or marshy sites, but it may be used where pines have been planted in beds. "OUST" applications may result in damage to Douglas fir and Ponderosa pine. Use the lower rates on coarse textured soils (i.e., loamy sands, sandy loams) and higher rates on fine textured soils (i.e., sandy clay loams and silty clay loams).

NONSELECTIVE WEED CONTROL

"OUST" is recommended for use alone and in certain tank mix combinations for general weed control on noncrop, industrial sites such as airport, fence rows, roadsides and associated rights-of-ways, lumberyards, petroleum tank farms, pipeline and utility rights-of-ways, pumping installations, railroads, storage areas, plant sites and other similar areas. Apply with ground equipment as follows:

PREEMERGENCE TO EARLY POSTEMERGENCE* TREATMENTS

Apply 3 to 5 ounces "OUST" per acre for control of:

Broadleaves

Bouncingbet	Kochia	Sowthistle (annual)
Burclover	Little mallow	Sunflower
Carolina geranium	Mustard	Sweet clover
Common chickweed	Ox-eye daisy	Tansymustard
Common dandelion	Pepperweed	Tansy ragwort
Common speedwell	Pigweed	Tumble mustard
Common yarrow	Prickly lettuce	Vetch
Crimson clover	Purple starthistle	Wild carrot
Dogfennel	Ragweed	Yellow rocket
Hoary cress (whitetop)	Russian thistle	

Grasses
(Up to 6 to 12″ in Height)

Alta fescue	Kentucky bluegrass
Annual bluegrass	Little barley
Annual ryegrass	Red brome
Bahiagrass	Red fescue
Barnyardgrass	Reed canarygrass
Downy brome	Ripgut brome
Fescue	Ryegrass
Foxtails	Smooth brome
Foxtail barley	Springletop (annual)
Indiangrass	Wheat (volunteer)
Italian ryegrass	

* Add surfactant at 0.25% by volume to improve control of emerged weeds.

EARLY POSTEMERGENCE TREATMENTS

Apply 6 to 12 ounces "OUST" per acre plus surfactant at 0.25% by volume for control of:

Broadleaves

Bedstraw	Goldenrod	Mayweed
Canada thistle	Groundsel	Muskthistle
Curly dock	Hemlock	Poison ivy
Dewberry	Honeysuckle	Spanishneedles
Fiddleneck (tarweed)	Horsetail (Equisetum)	Turkey mullein
Redstem filaree		Virginia pepperweed
Fleabane	Jerusalem artichoke	Wild blackberry
	Kudzu (use 12 oz. rate)	

Grasses

Johnsongrass**
Yellow nutsedge

** For short-term control of johnsongrass of up to 3 months, apply "OUST" at 2 to 5 ounces per acre early postemergence to the johnsongrass. Re-treat when regrowth occurs or if further control of the actively growing johnsongrass is desired.

NOTE: Use the higher level of recommended dosage ranges under the following conditions: Heavy weed growth; on soils containing more than 2½% organic matter; and on high soil moisture areas such as along road edges or railroad shoulders.

For broader spectrum weed control than offered by "OUST" alone, tank mixtures are recommended with Du Pont Escort® or Hyvar® X, or Karmex®, or Telar®, or Velpar® Herbicides; or "Roundup" [2] or 2,4-D. Refer to package labels and determine the weeds controlled that best fit the weed problem.

Apply the "OUST" plus companion herbicide at the rates and timing shown on the package labels for the weeds involved. For application method and other use specifications, use the most restrictive directions for the intended combination.

Do not tank mix "OUST" with Du Pont Hyvar® XL Herbicide.

SPRAY PREPARATION

Fill the spray tank half full of water, and with agitator running, add the proper amount of "OUST." Finish adding the required amount of water. Continuous agitation is required to keep the product in suspension. Use mechanical or bypass agitation to thoroughly mix the spray solution. After initial mixing, do not use excessive agitation when using a surfactant as foaming problems may result. To avoid foaming problems, use antifoam agents.

"OUST" spray preparations may degrade in acid solutions if not used in 48 hours; they are stable in alkaline solutions. Thoroughly reagitate before using.

"OUST" may be tank mixed with most herbicides (except "Hyvar" XL); to insure compatibility, test the desired mixture prior to use.

For nonselective postemergence applications, a nonionic surfactant of at least 80% active ingredient may be added to improve wetting and/or contact activity on weeds. To minimize drift, a drift control agent may be added at the recommended label rate; a drift control agent must be used when spraying in areas near cropland or desirable vegetation. Follow closely the mixing instructions for a drift control agent; when using cold water, be sure the additive is thoroughly dispersed.

APPLICATION TECHNIQUES

INDUSTRIAL NONCROP WEED CONTROL

Apply with ground equipment using extreme care to prevent spray drift; use coarse spray nozzles, adequate spray volume and avoid excessive application pressure, sprayer speed and application under windy conditions to avoid spray drift. Select a spray volume that will assure thorough coverage and uniform spray pattern. Do not use less than 15 gallons of water per acre. Spray volumes of 20 to 40 gallons per acre are preferred. Spray pressures of 25 to 35 psi are adequate. Before spraying "OUST," calibrate the spray equipment to determine the quantity of water necessary to uniformly cover the vegetation and soil in a measured area to be treated. The properly calibrated sprayer should be operated at the proper speed and rate of delivery during application of "OUST" alone or in tank mixtures. Avoid overlapping and shut off spray booms while starting, turning, slowing or stopping when spraying industrial turf as injury to the desired species may result. Do not apply "OUST" by aircraft.

CONIFER SITE PREPARATION AND RELEASE

"OUST" may be applied by ground equipment (as previously described) or by air (helicopter only) to recommended forest sites. Calibrate the helicopter spray system to insure uniform distribution of liquid over the area to be treated; use 5 to 15 gallons of water per acre. To reduce drift when applying by air, use precise application spray equipment (jet nozzles directed backwards [D4 or larger] or a "Microfoil" [1] boom or equivalent). When compatible with equipment, addition of a spray thickening agent is recommended to further minimize drift.

1Trademark of Union Carbide Agricultural Products Company
2Trademark of Monsanto Company

NOTICE TO BUYER: Purchase of this material does not confer any rights under patents of countries outside of the United States.

STORAGE AND DISPOSAL

STORAGE: Store product in original container only, away from other pesticides, fertilizer, food or feed.

PESTICIDE DISPOSAL: Do not contaminate water, food or feed by storage or disposal. Wastes resulting from the use of this product may be disposed of on site or at approved waste disposal facility.

CONTAINER DISPOSAL: Triple rinse (or equivalent) the container. Then offer for recycling or reconditioning, or puncture and dispose of in a sanitary landfill, or by incineration, or, if allowed by state and local authorities, by burning. If burned, stay out of smoke.

NOTICE OF WARRANTY

Du Pont warrants that this product conforms to the chemical description on the label thereof and is reasonably fit for purposes stated on such label only when used in accordance with directions under normal use conditions. It is impossible to eliminate all risks inherently associated with use of this product. Injury to adjacent crops, ineffectiveness or other unintended consequences may result because of such factors as weather conditions, presence of other materials, or the manner of use or application, all of which are beyond the control of Du Pont. In no case shall Du Pont be liable for consequential, special or indirect damages resulting from the use or handling of this product. All such risks shall be assumed by the Buyer. DU PONT MAKES NO WARRANTIES OF MERCHANTABILITY OR FITNESS FOR A PARTICULAR PURPOSE NOR ANY OTHER EXPRESS OR IMPLIED WARRANTY EXCEPT AS STATED ABOVE.

AG-2983 8048 4/11/88

This is a specimen label. DuPont is not responsible for the accuracy of the information contained herein. As labels are subject to revision, always carefully read and follow the label on the product container.

PINNACLE®

HERBICIDE
DRY FLOWABLE

ACTIVE INGREDIENT	BY WEIGHT
Methyl 3-[[[[(4-methoxy-6-methyl-1,3,5-triazin-2-yl)amino]carbonyl]amino]sulfonyl]-2-thiophenecarboxylate	25.0%
INERT INGREDIENTS	75.0%
TOTAL	100%

EPA Reg. No. 352-525 U.S. Patent 4,481,029

KEEP OUT OF REACH OF CHILDREN
CAUTION

PRECAUTIONARY STATEMENTS

HAZARDS TO HUMANS AND DOMESTIC ANIMALS

CAUTION! Causes moderate eye irritation. Harmful if absorbed through skin. Avoid contact with skin, eyes or clothing. Wash thoroughly with soap and water after handling.

STATEMENT OF PRACTICAL TREATMENT

If in eyes: Flush eyes with plenty of water. Call a physician if irritation persists.

If on skin: Wash with plenty of soap and water.

For medical emergencies involving this product, call toll free 1-800-441-3637.

ENVIRONMENTAL HAZARDS

Do not apply directly to water or wetlands (swamps, bogs, marshes and potholes). Do not contaminate water by cleaning of equipment or disposal of wastes.

DIRECTIONS FOR USE

It is a violation of federal law to use this product in a manner inconsistent with its labeling.

"PINNACLE" Herbicide should be used only in accordance with recommendations on this label or in separately published Du Pont recommendations.

Du Pont will not be responsible for losses or damages resulting from the use of this product in any manner not specifically recommended by Du Pont.

IMPORTANT

Injury to or loss of desirable trees or vegetation may result from failure to observe the following:

Do not apply or drain or flush equipment on or near desirable trees or other plants, or on areas where their roots may extend, or in locations where the chemical may be washed or moved into contact with their roots.

Do not use on lawns, walks, driveways, tennis courts or similar areas. Prevent drift of spray to desirable plants. Do not contaminate any body of water. Thoroughly clean all traces of Du Pont "PINNACLE" Herbicide from application equipment immediately after use. (Refer to sprayer cleanup section of this label and the referenced supplemental bulletin.)

Du Pont "PINNACLE" Herbicide is a dispersible granule formulation to be mixed with water and sprayed broadcast for selective postemergence weed control in soybeans.

SPRAY TANK PREPARATION

"PINNACLE" should be thoroughly mixed with water in the spray tank before adding any other material (e.g. surfactant, crop oil concentrate or nitrogen fertilizer). Use a minimum of 10 gallons of water per acre for ground application and a minimum of 5 gallons per acre for aerial application. Agitation is required for uniform mixing and application. Apply "PINNACLE" spray preparation within 24 hours of product mixing or product degradation may occur. Thoroughly reagitate before using if allowed to settle.

APPLICATION RECOMMENDATION

"PINNACLE" can be applied anytime after the first trifoliate soybean leaf has fully expanded up to 60 days before soybean harvest.

Application should be made by ground or air to young actively growing weeds.

Applications of "PINNACLE" must include a surfactant at rates of 0.125% v/v(1 pt. per 100 gallons). Use only EPA approved surfactants authorized for use on food crops. Use a good surfactant of at least 80% active ingredient. The addition of liquid nitrogen fertilizer at the rate of 1 gallon per acre is required for control of certain species.

GROUND APPLICATION

Use flat fan nozzles at 25-40 psi or hollow cone nozzles at 40-60 psi. Adjust the boom height for good weed coverage. Use a minimum of 10 gallons of water per acre.

AERIAL APPLICATION

Use orifice discs, cores and nozzle types and arrangements that will provide for optimum spray distribution and maximum coverage at 5 to 10 GPA. Do not apply during inversion conditions, when winds are gusty, or when other conditions will favor poor coverage and/or off target spray movement. Use a minimum of 5 gallons of water per acre.

NOTE: Do not apply this product through any type of irrigation system.

APPLICATION RATE

Use ¼ ounce "PINNACLE" per acre.

Weeds Controlled:	Weeds Suppressed:
Lambsquarters	Cocklebur
Pigweed *spp*	Jimson weed
Smartweed, annual	Wild Sunflower
Velvetleaf*	

TANK MIX APPLICATION WITH "CLASSIC" HERBICIDE

"PINNACLE" can be tank mixed and applied with "Classic" herbicide to improve control of certain broadleaf weeds. "Classic" rates to be used with "PINNACLE" and recropping restrictions are limited by soil pH.

(1) Soil pH 7.2 or less:
"PINNACLE" ¼ ounce per acre plus "Classic" ¼ ounce per acre

Weeds Controlled:	Weeds Suppressed:
Lambsquarters	Morningglory
Smartweed, annual	Entireleaf
Pigweed, spp	Ivyleaf
Velvetleaf*	Pitted
Cocklebur	Smallflower
Wild Sunflower	Tall
Jimsonweed	Common Ragweed
Mustard	

(2) Soil pH 7.2-7.6:
"PINNACLE" ¼ ounce per acre plus "Classic" ⅙ ounce per acre

Weeds Controlled:	Weeds Suppressed:
Lambsquarters	Common Ragweed
Smartweed, annual	Jimsonweed
Pigweed, spp	Mustard
Velvetleaf*	
Cocklebur	
Wild Sunflower	

IMPORTANT: DO NOT USE "PINNACLE" PLUS "CLASSIC" TANK MIX ON SOILS ABOVE PH 7.6

* Requires the addition of 1 gallon of liquid nitrogen fertilizer per acre to the spray solution.

Weed suppression is a visual reduction in weed competition (reduced population, size, and/or vigor) as compared to an untreated area. Degree of control can be increased by treating weeds when actively growing (not under stress).

NOTE: Do not allow spray to drift onto adjacent crops or land, as injury to plants other than soybeans, wheat or barley will likely occur. When spraying close to adjacent crops or plants, do everything possible to reduce the occurrence of off target spray movement such as:

Increase spray volume (minimum 10 GPA by air, 20 GPA by ground).
Use lower spray pressure (but not less than 20 psi).
Add an appropriate drift control agent.
Stop spraying if wind speed becomes excessive.

CROP ROTATION

Any crop may be planted 45 days after the application of "PINNACLE" applied alone.

Rotational Crop Guidelines for "PINNACLE" Plus "Classic" Tank Mixes

Crop:	Rotational Interval in Months:
Soybeans	Anytime
Cereal Grains	3
Ryegrass	3
Field Corn	9
Cotton	9
Alfalfa	9
Clover	9
Dry Beans	9
Sorghum	9
Peanut	9
Rice	9
Tobacco (transplant)	9
Tomato (transplant)	9

All other crops require a successful field bioassay**. Suggested time before initiating the bioassay is 9 months.

** Field Bioassay: The season before planting any crop not listed above, a successful field bioassay must be completed. The field bioassay will detect small quantities of "Classic" which can remain in the soil and injure rotational crops. A successful field bioassay means growing to maturity the crop(s) intended for production the following year. The test strip should cross the whole field including knolls and low areas.

SPRAYER CLEANUP

To avoid subsequent injury to crops other than soybeans immediately after spraying and prior to spraying other crops, thoroughly remove all traces of "PINNACLE" from mixing and spray equipment as follows:

1. Drain tank, then flush tank boom and hoses with clean water for a minimum of 10 minutes.
2. Fill the tank with clean water then add 1/2 gallon chlorine bleach (containing 5 1/4% sodium hypochlorite) per 100 gallons of water. Flush solution through boom and hoses, then allow to sit for 15 minutes with agitation, then drain.
3. Repeat Step 2.
4. Nozzles and screens should be removed and cleaned separately. To remove traces of chlorine bleach, rinse the tank thoroughly with clean water and flush through hoses and boom.

NOTE: To reduce the amount of water required in the above procedure, see separate Du Pont Bulletin, "Reduced Volume Cleanout Procedure for Large Sprayers."

CAUTION: Do not use chlorine bleach with ammonia. All traces of liquid fertilizer containing ammonia, ammonium nitrate or ammonium sulfate must be rinsed with water from the mixing and application equipment before adding chlorine bleach solution. Failure to do so will release a gas with a musty chlorine odor which can cause eye, nose, throat and lung irritation. Do not clean equipment in an enclosed area.

PRECAUTIONS

Do not graze or feed forage, hay or straw from treated areas to livestock.

STORAGE AND DISPOSAL

STORAGE: Store product in original container only, away from other pesticides, fertilizer, food or feed.

PRODUCT DISPOSAL: Do not contaminate water, food or feed by storage or disposal. Waste resulting from the use of this product may be disposed of on site or at an approved waste disposal facility.

CONTAINER DISPOSAL: Triple rinse (or equivalent) the container. Then offer for recycling or reconditioning, or puncture and dispose of in a sanitary landfill, or by incineration, or, if allowed by state and local authorities, by burning. If burned, stay out of smoke.

NOTICE TO BUYER: Purchase of this material does not confer any rights under patents of countries outside of the United States.

NOTICE OF WARRANTY

Du Pont warrants that this product conforms to the chemical description on the label thereof and is reasonably fit for purposes stated on such label only when used in accordance with directions under normal use conditions. It is impossible to eliminate all risks inherently associated with the use of this product. Crop injury, ineffectiveness, or other unintended consequences may result because of such factors as weather conditions, presence of other materials, or the manner of use or application, all of which are beyond the control of Du Pont. In no case shall Du Pont be liable for consequential, special or indirect damages resulting from the use or handling of this product. All such risks shall be assumed by the buyer. DU PONT MAKES NO WARRANTIES OF MERCHANTABILITY OR FITNESS FOR A PARTICULAR PURPOSE NOR ANY OTHER EXPRESSED OR IMPLIED WARRANTY EXCEPT AS STATED ABOVE.

AG - 3263 8049 4/28/89

This is a specimen label. DuPont is not responsible for the accuracy of the information contained herein. As labels are subject to revision, always carefully read and follow the label on the product container.

PREVIEW®

HERBICIDE DISPERSIBLE GRANULES

	BY WEIGHT
ACTIVE INGREDIENT:	75%
Metribuzin [4-Amino-6-(1,1-dimethylethyl)-3-(methylthio)-1,2,4-triazin-5(4H)-one]	68.5%
Ethyl 2-[[[[(4-chloro-6-methoxypyrimidin-2-yl)-amino]carbonyl]amino]sulfonyl]benzoate	6.5%
INERT INGREDIENTS:	25%
TOTAL	100%

EPA Reg. No. 352-448
U.S. Patent 4,346,220; 4,547,215; 4,394,506 & 4,645,530

KEEP OUT OF REACH OF CHILDREN

PRECAUTIONARY STATEMENTS

HAZARDS TO HUMANS AND DOMESTIC ANIMALS

CAUTION!

HARMFUL IF SWALLOWED. CAUSES MODERATE EYE INJURY.

Avoid breathing dust or spray mist. Avoid contact with eyes. Wash thoroughly with soap and water after handling. Remove and wash contaminated clothing before reuse.

STATEMENT OF PRACTICAL TREATMENT

If in eyes: Flush eyes with plenty of water. Call a physician if irritation persists.

If swallowed: Call a physician or Poison Control Center. Drink 1 or 2 glasses of water and induce vomiting by touching back of throat with finger. Do not induce vomiting or give anything by mouth to an unconscious person.

For medical emergencies involving this product, call toll free 1-800-441-3637.

ENVIRONMENTAL HAZARDS

Do not apply directly to water or wetlands. Do not contaminate water by cleaning of equipment or disposal of wastes.

IMPORTANT

Injury to or loss of desirable trees or vegetation may result from failure to observe the following: Do not apply or drain or flush equipment on or near desirable trees or other plants, or on areas where their roots may extend, or in locations where the chemical may be washed or moved into contact with their roots. Do not use on lawns, walks, driveways, tennis courts or similar areas. Prevent drift of spray to desirable plants. Do not contaminate any body of water. Keep from contact with fertilizers, insecticides, fungicides and seeds during storage.

Thoroughly clean all traces of "PREVIEW" from application equipment immediately after use and prior to spraying crops other than soybeans. Failure to remove even small amounts of "PREVIEW" from application equipment may result in injury to subsequently sprayed crops.

Prior to using "PREVIEW", consideration should be given to crop rotation plans. Crops other than soybeans can be extremely sensitive to low concentrations of "PREVIEW" remaining in the soil the next planting season. Choice of rotational crop is restricted following application of "PREVIEW". (See "ROTATIONAL CROP GUIDELINES".)

Do not apply if soil pH is greater than 6.8.

Metribuzin is a chemical which can travel (seep or leach) through soil and can contaminate groundwater which may be used as drinking water. Metribuzin has been found in groundwater as a result of agricultural use. Users are advised not to apply metribuzin where the water table (groundwater) is close to the surface and where the soils are very permeable; i.e., well drained soils such as loamy sands. Your local agricultural agencies can provide further information on the type of soil in your area and the location of groundwater.

GENERAL INFORMATION

Du Pont "PREVIEW" Herbicide is a dispersible granule formulation to be mixed with water and sprayed for selective preemergence and preplant incorporated weed control in soybeans. When applied according to the instructions on this label, it will control many broadleaf weeds and provide partial control of nutsedge.

Preplant incorporated and preemergence applications of "PREVIEW" require rainfall or sprinkler irrigation to activate the herbicide. Degree of control and duration of effect depend on: Rate used, weed spectrum, growing conditions at and following time of treatment, soil pH, texture, organic matter, moisture and precipitation.

Because most crops are highly sensitive to "PREVIEW", all direct or indirect contact (such as spray drift) to crops or to land scheduled to be planted to crops other than soybeans should be avoided.

BIOLOGICAL ACTIVITY

"PREVIEW" rapidly inhibits the growth of susceptible weeds. Following activation of preplant incorporation or preemergence treatment, susceptible weeds may germinate and emerge, but growth then ceases and leaves become yellow 3-5 days after emergence. Death of leaf tissue and growing point will follow in some species while others will remain green but stunted and noncompetitive. "PREVIEW" provides partial control of some annual grasses when used preplant or preemergence but other products may be needed to ensure adequate grass control.

DIRECTIONS FOR USE

It is a violation of federal law to use this product in a manner inconsistent with its labeling.

Do not apply this product through any type of irrigation system.

EQUIPMENT-SPRAY VOLUMES

Apply uniformly by ground equipment with a properly calibrated low pressure (20 to 40 psi) fixed-boom sprayer equipped with fan-type nozzles or other carefully calibrated nozzles and screens no finer than 50 mesh. Use 10 to 40 gals of water per acre.

Continuous agitation in the spray tank is required to keep the material in suspension. Avoid overlap and shut off spray booms while starting, turning, slowing or stopping, as injury to the crop may result.

EARLY PREPLANT OR PREPLANT INCORPORATED

(See rate chart)

"PREVIEW" can be applied at planting or up to 30 days prior to planting either preemergence or preplant incorporated. If incorporated, uniformly incorporate the treatment no deeper than the top 1-2" of soil prior to planting soybeans. Use equipment suited to proper incorporation into the top 1-2 inches of soil, e.g.: Do-all, field cultivator, roterra. "PREVIEW" can be tank mixed with "Command"[10] and applied preplant incorporated. Refer to the "Command" label for specific use instructions. If tank-mixed with a grass herbicide such as a DNA like "Treflan"[1], "Sonalan"[1], or acetanilide such as "Dual"[2], or "Lasso"[3], follow label instructions for proper incorporation of the grass herbicide into the top 1 to 2 inches of soil. Improper soil incorporation can result in erratic weed control or crop injury. Rainfall* (~1") is required to activate preplant incorporated herbicides such as "PREVIEW" before weed emergence.

"PREVIEW" + "Treflan", "Prowl"[4], "Dual" or "Sonalan" — Apply within 2 weeks of planting.
"PREVIEW" + "Treflan" (Early PPI) — Apply within 30 days of planting.
"PREVIEW" + "Lasso" — Apply within 7 days of planting.
"PREVIEW" + "Command" — Apply within 30 days of planting.

PREEMERGENCE (See rate chart)

Preemergence application: Apply before soybeans emerge. Do not spray overtop of emerged soybeans, as severe injury or death of crop will occur.

Preemergence activity of "PREVIEW" requires rainfall* to activate. Best results are obtained if treatment is made to moist soil and followed by rainfall or irrigation (~1") before weeds germinate. On dry soil, more moisture is required for activation (1 to 2") before weed emergence. If moisture is insufficient to activate the herbicide, a rotary hoeing or shallow cultivation should be made after emergence of crop while weeds are small enough to be controlled by mechanical means. "PREVIEW" can be applied preemergence in combination with "Lasso", "Dual", "Prowl", "Surflan" or following the use of a preplant grass material such as "Treflan" or "Sonalan". Consult labels for rates and use instructions.

*Rainfall required to activate ppi or pre treatments depends on soil moisture when activating moisture is received. Several small rainfalls of less than 1/4" are not as beneficial as one large rain of 1/2 inch or more for activation.

MINIMUM TILLAGE OR NO-TILL (See rate chart)

"PREVIEW" applied preemergence to soybeans but postemergence to broadleaf weeds and small annual grasses will provide some burndown of existing weeds as well as normal preemergence broadleaf weed control. When used for burndown, add 0.25% v/v (1 qt./100 gal.) surfactant such as Surfactant WK or crop oil concentrate at 0.25% v/v (1 qt./100 gal.). For added postemergence control, especially on grasses, "PREVIEW" can be applied with "Roundup"[3], paraquat, or "Bronco"[3]. For added preemergence grass control, "PREVIEW" can be mixed with "Dual", "Surflan" or "Lasso". For added postemergence broadleaf weed control, "PREVIEW" may be tank mixed with 2,4-D low volatile ester. Do not apply 2,4-D ester within 30 days of planting. Consult labels for rates and use instructions. Maintain constant agitation of spray mixture.

RATE CHART

Apply as follows:

Soil Texture	Ounces of "Preview" Per Acre	
	Percent Organic Matter in Soil*	
	½-3%	3-5%
Coarse: Loamy sand, Sandy loam	6	7
Medium:** Loam, Silty loam, Silt	7	8
Fine: Silty clay loam, Clay loam, Clay	8	9-10

* Do not apply to soil with less than ½% O.M.
Do not apply "PREVIEW" to soils with pH greater than 6.8.
** For medium soils 1½ to 3% O.M. in MO and KS, use 7-8 oz./A.

Dosage rates listed are broadcast rates. For band applications use proportionally less.

Plant seed 1 1/2" to 2" deep on a flat or raised seedbed only, or crop injury may occur.

Soybean stunting may occur if rainfall is excessive during soybean germination and emergence. Injury is more prevalent under poor drainage or compacted conditions or when soil is saturated for long periods of time. Soybeans outgrow stunting once favorable growing conditions return.

Soybean varieties such as Altona, Coker 102 and 156, Govan, Semmes, Tracy, Vansoy, Terra Vig 505 and 606, AP 71, NKS 1884, Agropro 55, Asgrow 6520, Maple Amber, Portage, and Vinton 81 are more sensitive to "PREVIEW". Injury may occur if "PREVIEW" is used on these varieties. Before use on any soybean variety, tolerance to "PREVIEW" must first be determined. Injury to soybeans may occur if "PREVIEW" is used on soils having a calcareous surface layer, or if used in conjunction with soil-applied organic phosphate pesticides such as "Dasanit"[5], "Disyston"[5], "Mocap"[6], Nemacur[5], Thimet[4], parathion or "Lorsban"[7].

NOTE: Injury may occur if atrazine was applied on the soil the year before use of "PREVIEW". Seedling disease, cold weather, deep planting (more than 2"), excessive moisture, high salt concentration or drought may weaken soybean seedlings and increase the possibility of crop injury.

WEEDS CONTROLLED

The following weeds will be controlled at the above rates both preplant incorporated and preemergence:

Carpetweed	Pigweed species	Purslane
Cocklebur*	Palmer amaranth	Ragweed, common
Hophornbeam copperleaf	Redroot	Smartweeds, annual
	Smooth	Spotted spurge
Jimsonweed	Spiny amaranth	Sunflower*
Lambsquarters	Tall waterhemp	Velvetleaf
Mustards	Prickly sida (teaweed)	

PARTIAL CONTROL

Burcucumber	Morningglory, annual*	Nutsedge species
Crabgrass species		Panicum species
Eastern black nightshade	Tall	Ragweed, giant*
	Ivyleaf	
Foxtail species	Pitted	
Johnsongrass, seedling	Entireleaf	

*Large seeded weeds, germinating deep in the soil, such as morningglory, sunflower, cocklebur and giant ragweed or weeds with subsequent flushes may require a cultivation or postemergence application of Du Pont "Classic" Herbicide for season-long control. Consult "Classic" label for application directions and use restrictions.

NOTE: Consult "Lasso", "Dual", "Prowl", "Treflan", "Sonalan", "Surflan", or "Command" labels for additional weeds controlled when "PREVIEW" is used in conjunction with these herbicides.

SPRAY TANK PREPARATION

Fill tank 1/4 full with water. Start agitation system, add "PREVIEW" and continue adding water. Add separately each additional component of any tank mix while adding water. Continue agitation throughout. If poor mixing should occur with any component, premix the component with two parts water before adding to the spray tank. A fertilizer solution may be used in the spray mixture. Small quantities should be tested for compatibility by the following procedures before full scale mixing.

1. Put 1 pint of fertilizer solution in quart jar.
2. Mix 2 teaspoons "PREVIEW" with 2 tablespoons of water; mix thoroughly and add to fertilizer solution.
3. Close jar and shake well.
4. If other herbicides are to be used in the mixture, premix 2 teaspoons of wettable powders or 1 teaspoon of liquids with 2 tablespoons of water; add to "PREVIEW"/fertilizer solution mixture.

5. Close jar and shake well.
6. Watch mixture for several seconds; check again in 30 minutes.
7. If mixture does not separate, foam, gel or become lumpy, it may be used.
8. Mixing ability may be improved by adding compatibility agents such as Kalo Laboratories "Compex"[8] or Witco Chemicals "Sponto"[9] 168D. Follow directions on container.

Provided the above procedure shows the mixture to be compatible, prepare the tank mixture as follows: Add the fertilizer solution to the spray tank first; with the agitator running, add the required amount of "PREVIEW" and thoroughly mix. For tank mixtures with other herbicides, follow directions above.

Use "PREVIEW" spray preparations the same day as mixed or product degradation may occur. Thoroughly reagitate before using if allowed to settle.

When tank mixing with other herbicides, all applicable directions, restrictions and precautions for the additional herbicides are also to be followed.

ROTATIONAL CROP GUIDELINES

Crops that can be planted after using "PREVIEW" are listed below. Choice of rotational crop is determined by time interval elapsed since last application.

Crops	Time Interval Before Replanting Months
Soybeans	anytime
Wheat, Barley, Ryegrass	4
Field Corn	10**
Alfalfa	10
Dry beans	12
Sorghum	12
Clover	12
Tobacco (transplant)	10
Tomato (transplant)	10

All other crops require a successful field bioassay***
The suggested interval before initiating a field bioassay is 10 months.

**For Applications of "PREVIEW" made after June 15, extend recrop interval 1 month.

***Field Bioassay - The season before planting any crop not listed above, a successful field bioassay must be completed. The field bioassay will detect small quantities of "PREVIEW" which can remain in the soil and injure rotational crops. A successful field bioassay means growing to maturity a test strip of the crop(s) intended for production the following year. The test strip should cross the whole field including knolls and low areas.

Combination Treatment: When recommended tank mix combinations are used, consult the product label for replanting and recropping instructions and observe the directions that are the most restrictive.

SPRAYER CLEANUP

To avoid subsequent injury to crops other than soybeans, immediately after spraying and prior to spraying other crops, thoroughly remove all traces of "PREVIEW" from mixing and spray equipment. Be careful to rinse both the spray tank, boom and clean nozzles and screens. For more information; consult supplemental bulletin on sprayer cleanout following use of Du Pont "Classic", "Canopy", "Gemini", "PREVIEW" or "Lorox Plus" Herbicides.

PRECAUTIONS

Do not apply "PREVIEW" Herbicide to soils with a pH greater than 6.8.

CAUTION: Soil pH varies greatly even within the same field. pH variations as much as 2 pH units are common when the soil has a calcareous sub soil. Determine soil pH by laboratory analysis using a 1:1 soil:water suspension.

Do not graze treated fields or harvest for forage or hay.

Use of "PREVIEW" may affect rotational crops. Follow "ROTATIONAL CROP GUIDELINES".

Do not use on sand, muck, or peat soils, nor on soils with less than 1/2% organic matter, as crop injury may occur.

Do not apply "PREVIEW" after the soybean crop has emerged.

If "Scepter"[4], "Pursuit"[4] or any product containing the active ingredients imazaquin or imazethapyr (such as "Squadron"[4], "Partner"[3]) is applied the same year as "PREVIEW", do not plant crops other than soybeans for at least 15 months from the last application. Crops listed in the "'PREVIEW' Rotational Crop Guidelines" may be planted after 15 months; all other crops require a successful field bioassay.

Do not apply to land that has been or will be treated with "Ally", "Glean" or "Finesse" herbicides in the states of Nebraska, Kansas, North Dakota and South Dakota.

1Trademark of Elanco Products Co., Indianapolis, IN
2Trademark of Ciba-Geigy Corp., Greensboro, NC
3Trademark of Monsanto Agricultural Products Co., St. Louis, MO
4Trademark of American Cyanamid Corp., Princeton, NJ
5Trademark of Mobay Chemical Corp., Kansas City, MO
6Trademark of Rhone-Poulenc, Inc., Monmouth Junction, NJ
7Trademark of Dow Chemical Co., Midland, MI
8Trademark of Kalo Agricultural Chemicals, Inc., Overland Park, KS
9Trademark of Witco Chemical Corp., Houston, TX
10Trademark of FMC Corp., Philadelphia, PA

NOTICE TO BUYER: Purchase of this material does not confer any rights under patents of countries outside of the United States.

STORAGE AND DISPOSAL

Storage: Store product in original container only, away from other pesticides, fertilizer, food or feed.

Pesticide Disposal: Do not contaminate water, food or feed by storage or disposal. Wastes resulting from the use of this product may be disposed of on site or at an approved waste disposal facility.

Container Disposal: Triple rinse (or equivalent). Then offer for recycling or reconditioning, or puncture and dispose of in a sanitary landfill, or incineration, or, if allowed by state and local authorities, by burning. If burned, stay out of smoke.

NOTICE OF WARRANTY

Du Pont warrants that this product conforms to the chemical description on the label thereof and is reasonably fit for the purposes stated on such label only when used in accordance with the directions under normal use conditions. It is impossible to eliminate all risks inherently associated with the use of this product. Crop injury, ineffectiveness or other unintended consequences may result because of such factors as weather conditions, presence of other materials, or the manner of use or application, all of which are beyond the control of Du Pont. In no case shall Du Pont be liable for consequential, special or indirect damages resulting from the use or handling of this product. All such risks are assumed by the buyer. DU PONT MAKES NO WARRANTIES OF MERCHANTABILITY OR FITNESS FOR A PARTICULAR PURPOSE NOR ANY OTHER EXPRESS OR IMPLIED WARRANTY EXCEPT AS STATED ABOVE.

AG-3229 8098 4/11/88

PYDRIN®

INSECTICIDE
2.4 EMULSIBLE CONCENTRATE

RESTRICTED USE PESTICIDE

For retail sale to and use only by Certified Applicators or persons under their direct supervision and only for those uses covered by the Certified Applicator's certification.

ACTIVE INGREDIENT	BY WEIGHT
Cyano(3-phenoxyphenyl)methyl-4-chloro-alpha-(1-methylethyl)benzeneacetate	30%
INERT INGREDIENTS:	70%
TOTAL	100%

This product contains 2.4 lbs. of PYDRIN per gallon.

EPA Reg. No. 352-485 EPA Est. No. 5905-AR-1

KEEP OUT OF REACH OF CHILDREN

WARNING

May be irritating to skin, eyes, nose and throat. Causes substantial but temporary eye injury. May be fatal if swallowed.

AVISO

PRECAUCION AL USUARIO: Si usted no lee ingles, no use este producto hasta que le etiqueta haya sido explicada ampliamente.

PRECAUTIONARY STATEMENTS

HAZARDS TO HUMANS AND DOMESTIC ANIMALS

WARNING!

May be fatal if swallowed. Do not breathe vapor or spray mist. Causes eye irritation. Do not get in eyes, on skin or on clothing. Wear goggles or face shield and rubber gloves when handling.

Remove and wash clothing before reuse.

STATEMENT OF PRACTICAL TREATMENT

If less than 2.0 ml per kg of body weight is swallowed, do not induce vomiting. If vomiting occurs spontaneously, keep head below hips to prevent aspiration of liquid into lungs. If more than 2.0 ml per kg body weight is swallowed and vomiting has not occurred, emesis should be induced with supervision by a physician or professional staff. If symptoms such as loss of gag reflex, convulsions or unconsciousness occur before emesis, gastric lavage should be considered following intubation with a cuffed endotracheal tube.

For eye contact flush eyes with plenty of water for at least 15 minutes while holding eyelids open. Get medical attention.

For skin contact wash with soap and water immediately. After drying, apply vitamin E cream or oil if available. If not available, apply vegetable oil liberally over painful areas. The oil or cream may be used repeatedly until relief is achieved. Get medical help if irritation persists.

If inhaled remove victim to fresh air and provide oxygen if breathing is difficult. Give artificial respiration if not breathing. Get medical attention.

For medical emergencies involving this product, call toll free 1-800-441-3637.

ENVIRONMENTAL HAZARDS

This pesticide is toxic to wildlife and extremely toxic to fish. Use with care when applying in areas adjacent to any body of water. Do not apply directly to water. Do not apply when weather conditions favor drift from treated areas. Do not contaminate water by cleaning of equipment or disposal of wastes. Apply this product only as specified on this label.

This product is highly toxic to bees exposed to direct treatment on blooming crops or weeds. Do not apply PYDRIN or allow it to drift to blooming crops or weeds while bees are actively visiting the treatment area. Additional information may be obtained by consulting your Cooperative Extension Service.

WILDLIFE PROTECTION IS EVERYONE'S RESPONSIBILITY

DIRECTIONS FOR USE

It is a violation of Federal Law to use this product in a manner inconsistent with the labeling.

REENTRY STATEMENTS

Do not apply this product in such a manner as to directly or through drift expose workers or other persons. The area being treated must be vacated by unprotected persons.

Do not enter treated areas without protective clothing until sprays have dried.

Because certain states may require more restrictive reentry intervals for various crops treated with this product, consult your State Department of Agriculture for further information.

Written or oral warnings must be given to workers who are expected to be in a treated area or in an area about to be treated with this product. (Indicate specific oral warnings which inform workers of areas or fields that may not be entered without specific protective clothing, period of time field must be vacated and appropriate actions to take in case of accidental exposure.) When oral warnings are given, warnings shall be given in a language customarily understood by workers. Oral warnings must be given if there is reason to believe that written warnings cannot be understood by workers. Written warnings must include the following information: "WARNING! Area treated with PYDRIN Insecticide on (date of application). Do not enter without appropriate protective clothing until sprays have dried. (Insert here action to take in case of accidental exposure. See front panel STATEMENT OF PRACTICAL TREATMENT for details.)"

GENERAL DIRECTIONS

PYDRIN Insecticide 2.4 Emulsible Concentrate contains 2.4 pounds of active ingredient per gallon. For the applications given below, mix the required amount of PYDRIN in sufficient diluent to provide uniform coverage (Refer to Use Tables). PYDRIN may be applied by ground or aerial application equipment. Do not enter treated areas before spray is dry.

Chemigation. Apply this product only through sprinkler including center pivot, lateral move, end tow, side (wheel) row, traveler, big gun, solid set, or hand move irrigation systems. Do not apply this product through any other type of irrigation system. "PYDRIN" 2.4 EC may be premixed in a supply tank with water, oil, fertilizer, or other appropriate tank mixed agricultural chemicals. Agitation may be necessary. Application should be in sufficient water and of sufficient duration to apply the recommended rate evenly to the entire treated area. No run-off can be permitted during chemigation. Crop injury, lack of effectiveness, or illegal pesticide residues in the crop can result from nonuniform distribution of treated water. Do not apply when wind speed favors drift beyond the area intended for treatment.

If you have questions about calibration, you should contact State Extension Service specialists, equipment manufacturers or other experts. A person knowledgeable of the chemigation system and responsible for its operation, or under the supervision of the responsible person, shall shut the system down and make necessary adjustments should the need arise.

Do not connect an irrigation system (including greenhouse systems) used for "PYDRIN" 2.4 EC application to a public water.

The system must contain a functional check valve, vacuum relief valve, and low pressure drain appropriately located on the irrigation pipeline to prevent water source contamination from backflow. The pesticide injection pipeline must contain a functional, automatic, quick-closing check valve to prevent the flow of fluid back toward the injection pump. The pesticide injection pipeline must also contain a functional, normally closed, solenoid-operated valve located on the intake side of the injection pump and connected to the system interlock to prevent fluid from being withdrawn from the supply tank when the irrigation system is either automatically or manually shut down. The system must contain functional interlocking controls to automatically shut-off the pesticide injection pump when the water pump motor stops. The irrigation line or water pump must include a functional pressure switch which will stop the water pump motor when the water pressure decreases to the point where pesticide distribution is adversly affected. Systems must use a metering pump, such as a positive displacement injection pump (e.g., diaphragm pump) effectively designed and constructed of materials that are compatible with pesticides and capable of being fitted with a system interlock.

Do not enter treated areas before crop had dried.

Crop Rotation. Do not plant root crops other than those in this label within 12 months after last application. Root crops in this label and all other rotation crops may be planted immediately following last application.

COTTON

CROP	INSECT	DOSAGE LB ai/ ACRE	DOSAGE FL OZ/ ACRE	NUMBER ACRES TREATED WITH GALLON	FURTHER USE INSTRUCTIONS	DAYS TO HARVEST
Cotton	Pink Bollworm	0.05-0.1	2⅔-5⅓	48-24	May be applied in water or nonvolatile vegetable oils. When using oil, apply a total spray volume of 1 or more quarts/acre. When diluted with water, apply no less than one gallon per acre by air (min. 3 gal/a in Arizona and 5 gallons per acre in California) or 4 gallons per acre by ground. Repeat as necessary throughout season to maintain control. For the control of Boll Weevil, a 3-5 day interval between sprays may be required. Apply no more than 2 lbs/a to a crop in a growing season. Do not graze livestock on treated fields or feed treated trash. *Low rate to be used only for light infestations. **Aids in the suppression.	21
	Cotton Leaf Perforator	0.1	5⅓	24		
	Cotton Bollworm	0.08*-0.20	4¼-10⅔	30-12		
	Tobacco Budworm, Lygus Bugs, Cabbage Looper, Beet Armyworm, Boll Weevil, Whitefly**, Cutworms, Saltmarsh Caterpillar, Thrips (on seedling cotton), Plant Bugs, Leafhoppers, Fleahoppers, Grasshoppers, Cotton Leafworm	0.1-0.2	5⅓-10⅔	24-12		

COTTON — TANK MIX APPLICATIONS
READ AND FOLLOW ALL PRECAUTIONS AND RESTRICTIONS ON THE PRODUCT LABELING.

TANK MIX COMBINATION	INSECT	DOSAGE	FURTHER USE INSTRUCTIONS	DAYS TO HARVEST
LANNATE L/NUDRIN 1.8 Methomyl Insecticide Solution	Cotton Bollworm and Tobacco Budworm eggs	PYDRIN (from cotton table) with ½ pt/acre LANNATE L/NUDRIN 1.8		21

TANK MIX COMBINATION	INSECT	DOSAGE	FURTHER USE INSTRUCTIONS	DAYS TO HARVEST
LANNATE L/NUDRIN 90 Methomyl Insecticide Water Soluble Powder (with adjuvant)	Cotton Bollworm and Tobacco Budworm eggs	PYDRIN (from cotton table) with 1/8 lb/acre LANNATE L/NUDRIN 90	Add adjuvant (X-77, Bio-88, Unite, Triton B-1956 or equivalent*) to spray tank, then PYDRIN, then LANNATE L/NUDRIN 90; agitate until thoroughly mixed.	

* Compatibility of alternate should be determined before use.

FIELD CROPS

CROP	INSECT	DOSAGE* LB ai/ ACRE	DOSAGE* FL OZ/ ACRE	NUMBER ACRES TREATED WITH GALLON	FURTHER USE INSTRUCTIONS	DAYS TO HARVEST
Field Corn	PYDRIN 2.4 EC may be tank-mixed with the following herbicides when insect populations require control concurrent with the need for weed control. Follow all herbicidal label directions regarding proper herbicide usage. Jar-test compatibility should be conducted before field mixtures are made and applications initiated. BLADEX 80WP[1] ATRAZINE 80 WP BICEP 4.5L[2] DUAL 8E[2] BLADEX 4L[1] ATRAZINE 4L BANVEL 4EC[3] ROUNDUP [4] GRAMOXONE SUPER[5] LASSO 4EC[4]					21
	Cutworm Grasshoppers	0.1-0.2	5 1/3-10 2/3	24-12	Apply as necessary to maintain control.	
	Western Bean Cutworm	0.05-0.1	2 2/3-5 1/3	48-24	Apply as necessary to maintain control. Application must be made before larvae enter the ear.	
	Corn Earworm	0.1-0.2	5 1/3-10 2/3	24-12	First application should be at or before silking. Subsequent applications should be made at 3-5 day intervals until silking is completed. Repeat applications may be applied if economically damaging populations exist.	
	Stalk Borer Flea Beetles	0.1-0.2	5 1/3-10 2/3	24-12	Application must be made early in migration from grassy areas to corn, before borers enter the plant. More than one application may be necessary.	
	Corn Rootworm (adult control)	0.1-0.2	5 1/3-10 2/3	24-12	Spray at first sign of silk feeding. Reinfestation may occur. Should reinfestation occur, repeat as necessary to maintain control.	
	Chinch Bugs	0.1-0.2	5 1/3-10 2/3	24-12	Over the top sprays, as used for control of ear and stalk infesting corn pests, are not adequate for chinch bug control. It is very important that the spray be directed at the base of the plant through the use of drop nozzles or some other mechanism. Repeat as necessary to control migrating chinch bugs.	
	European Corn Borer	0.15-0.2	8.0-10 2/3	16-12	First brood: spray while eggs are in the blackhead stage or before the larvae enter the whorl. Application by ground equipment is suggested. Good coverage of both upper and lower leaf surfaces is essential. This can be accomplished with drop nozzles over the row and on each side of the corn plant. Multiple applications may be required when egg laying is prolonged or where moderate to heavy populations are present. A higher rate is recommended for moderate to heavy populations. Proper coverage by ground equipment usually requires 20-30 gallons of carrier. Once larvae enter the whorl, foliar sprays will not provide adequate control. Second brood: make applications when sufficient egg masses are found. Spray when eggs are in the blackhead stage or starting to hatch. When egg laying is prolonged or a third generation is present, additional sprays may be required. A higher rate is recommended for moderate to heavy populations. Good coverage above, below, and in the ear zone is essential. This usually requires 2-3 gallons of carrier by air. If ground equipment is used, drop nozzles on each side of the plant will provide best coverage.	
	Southwestern Corn Borer	0.1-0.2	5 1/3-10 2/3	24-12	Apply as necessary to maintain control. Two or more applications may be required. Higher rates (0.15-0.2 lb. a.i.) are recommended for moderate to heavy populations.	
	Armyworm (True Armyworm)	0.1-0.2	5 1/3-10 2/3	24-12	Apply as necessary to maintain control.	
	Corn Leaf Aphid Oat Bird-Cherry Aphid	0.1-0.2	5 1/3-10 2/3	24-12	For optimum results, direct the spray at the aphid population so as to achieve maximum coverage of the exposed insects. Aphids not contacted by the spray, such as in whorls and leaf axils, may not be adequately controlled.	21
	Adult Corn Rootworm European Corn Borer	"PYDRIN" and Parathions from respective corn labels			PYDRIN 2.4 EC Insecticide may be tank-mixed with Parathion (methyl and/or ethyl) and applied on field corn (and seed corn) where supplemental control of Adult Corn Rootworm and/or European Corn Borer is desired in conjunction with insects controlled by PYDRIN Insecticide when used alone. For supplemental control of second brood European Corn	

CROP	INSECT	DOSAGE* LB ai/ ACRE	FL OZ/ ACRE	NUMBER ACRES TREATED WITH GALLON	FURTHER USE INSTRUCTIONS	DAYS TO HARVEST
*Do not exceed 1.0 lb. a.i. per acre per season.					Borer, make applications when sufficient egg masses are found. Spray when eggs are in the blackhead stage or starting to hatch. When egg laying is prolonged or a third generation is present, additional sprays may be required. A higher rate is recommended for moderate to heavy populations. Good coverage above, below, and in the ear zone is essential. This usually requires 2-3 gallons of carrier by air. If ground equipment is used, drop nozzles on each side of the plant will provide best coverage. For supplemental control of Adult Corn Rootworm, spray at first sign of silk feeding. Reinfestation may occur. Should reinfestation occur, repeat as necessary to maintain control. Refer to the PYDRIN 2.4 EC and Parathion (methyl or ethyl) labels for appropriate rates of the individual products for controlling the respective insects.	
Seed Corn	For specific insect control recommendations refer to Field Corn (above). Follow directions carefully.				Multiple applications and/or shortened intervals between sprays must be used to insure proper insect control. Do not exceed 1.0 lb a.i. per acre per season.	21
Pop Corn	For specific insect control recommendations refer to Field Corn (above). Follow directions carefully.				Multiple applications and/or shortened intervals between sprays must be used to insure proper insect control. Do not exceed 2.0 lbs a.i. per acre per season.	1
Peanuts	Corn Earworm, Red-necked Peanut Worm, Potato Leafhopper, Velvetbean Caterpillar	0.05-0.1	2⅔-5⅓	48-24	Repeat as necessary to maintain control. Do not feed or graze livestock on treated vines. Do not exceed 0.6 lb a.i. per acre per season.	21
	Beet Armyworm, Granulate Cutworm, Grasshoppers, Cutworms	0.1-0.2	5⅓-10⅔	24-12	**Aids in control.	
	Lesser Cornstalk Borer** Fall Armyworm***	0.2	10⅔	12	***Aids in control. For heavy infestations use of Lannate/Nudrin Insecticide is recommended. Please refer to the Lannate/Nudrin label for use instructions.	
Soybean	Mexican Bean Beetle, Green Cloverworm, Velvetbean Caterpillar, Potato Leafhopper, Saltmarsh Caterpillar Woolybear Caterpillar	0.05-0.1	2⅔-5⅓	48-24	Repeat as necessary to maintain control. Do not feed or graze livestock on treated plants. Do not exceed 0.8 lb a.i. per acre per season. When applying in nonvolatile vegetable oils use a total spray volume of 1 or more quarts.	21
	Cabbage Looper, Corn Earworm, Southern Green Stinkbug, BeanLeaf Beetle, Cutworms, Three-cornered Alfalfa Hopper, Beet Armyworm, Grasshoppers, Japanese Beetle (adult)	0.1-0.2	5⅓-10⅔	24-12		
Sugarcane	Sugarcane Borer	0.1-0.2	5⅓-10⅔	24-12	Repeat as necessary to maintain control. Do not exceed 0.8 lb. a.i./acre per season.	21
Sunflower	Sunflower Beetle	0.05-0.1	2⅔-5⅓	48-24	Repeat as necessary to maintain control. Do not exceed 0.8 lb. a.i./acre per season.	28
	Grasshoppers, Cutworms Heliothis (complex), Sunflower Moth, Sunflower Seed Weevil, Banded Sunflower Moth, Beet Armyworm, Sunflower Maggot, Leafhoppers, Sunflower Stem Weevil	0.1-0.2	5⅓-10⅔	24-12		

FRUITS AND NUTS

DILUTE SPRAY: Apply specified dosage per 100 gallons of water in a uniform spray applied to the point of drip with conventional ground equipment. Do not exceed maximum number of gallons per acre indicated.

NOTE: In order to apply the correct amount of PYDRIN insecticide to your orchard you must know the number of gallons of water needed to spray one acre of your trees to the point of drip. If you do not already know this gallonage, you should conduct a test to determine it. If you do not know how to conduct such a test with your equipment, you should ask assistance from your equipment dealer or State Extension specialist.

CONCENTRATE SPRAY: Apply specified dosage per acre in no less than 30 gals. of water per acre by ground. When applying PYDRIN by air, consult your Cooperative Extension Service for application guidelines.

		DOSAGE					
		CONCENTRATE SPRAY		DILUTE SPRAY	NUMBER ACRES TREATED WITH GALLON		
CROP	INSECT	LB ai/ ACRE	FL OZ/ ACRE	FL OZ/ 100 GALS		FURTHER USE INSTRUCTIONS	DAYS TO HARVEST
Apples	Oblique Banded Leafroller, Codling Moth, Variegated Leafroller, White Apple Leafhopper, Tentiform Leaf Miner, San Jose Scale (fruit infestations only), Apple Aphid, Tufted Apple Budmoth, Plum Curculio, Oriental Fruit Moth, Apple Maggot, Red-Banded Leafroller, Lesser Appleworm, Green Fruitworm, Rosy Apple Aphid, Plant Bugs (Tarnished Plant Bug, Stink Bugs), Periodical Cicada	0.1-0.3	5⅓-16	2⅔-8	24-8	Repeat as necessary to maintain control. Do not feed or graze livestock on treated orchard floors. Do not exceed 2.1 lbs. a.i. per acre per season. For dilute spray apply 200-600 gals per acre, but do not exceed 16 fl. oz. of PYDRIN 2.4 EC per acre per treatment Plant Bug, Rosy Apple Aphid Control—Time of application is critical in achieving control. Use prebloom and post bloom spray timings recommended by State Extension Services.	21
Stone Fruits Peaches Apricots Nectarines Plums Prune Plums Cherries	Peach Twig Borer, Oriental Fruit Moth, Peach Tree Borer, Plum Curculio, Lesser Peach Tree Borer, Periodical Cicada, Cherry Fruit Fly, Western Cherry Fruit Fly, American Plum Borer, Black Cherry Aphid, Plum Leafrollers Leafhoppers Green Fruitworm Plant Bugs (Tarnished Plant Bug, Stink Bugs)	0.1-0.3	5⅓-16	2⅔-8	24-8	Repeat as necessary to maintain control. Do not exceed 1.5 lbs a.i. per acre per season with no more than 1.2 lbs. a.i. per acre per season between bloom and harvest. For dilute spray apply 200-400 gals per acre, but do not exceed 16 fl. oz. of PYDRIN 2.4 EC per acre per treatment. Do not graze livestock on treated orchard floors. Peach Tree Borer, American Plum Borer, Lesser Peach Tree Borer Control—Apply as directed trunk and scaffold limb spray. Thorough coverage of trunk and scaffold limbs is required. Peach Twig Borer, Plum Curculio, Oriental Fruit Moth, Cherry Fruit Fly, Western Cherry Fruit Fly, Leafrollers, Black Cherry Aphid, Periodical Cicada Control— Apply by ground sprayer to achieve thorough coverage of all aerial portions of the tree. Plant Bug Control— Time of application is critical in achieving control. Use prebloom and post bloom spray timings recommended by State Extension Services.	14
Pear (dormant)	Pear Psylla	0.2-0.4	10⅔-21⅓	8-14	12-6	Apply during dormant to prebloom (white bud) stage only. Do not apply more than 0.8 lb a.i. per acre per season. Do not graze orchard floor. For dilute spray apply 150-250 gals per acre, but do not exceed 21⅓ fl. oz. of PYDRIN 2.4 EC per acre per treatment.	
Pear	Pear Psylla, Codling Moth, Green Fruitworm, Plum Curculio, Leafrollers, Pear Slug, Periodical Cicada	0.1-0.3	5⅓-16	2⅔-8	24-8	Repeat as necessary to maintain control. Do not exceed 1.5 lbs. a.i. per acre per season, nor more than 0.9 lb a.i. per acre between bloom and harvest. Do not feed or graze livestock on treated orchard floors. For dilute spray apply 200-600 gals per acre, but do not exceed 16 fl ozs of PYDRIN 2.4 EC per acre per treatment.	28
Almonds	Navel Orangeworm, Peach Twig Borer	0.2-0.4	10⅔-21⅓		12-6	Repeat as needed to maintain control but do not exceed 0.8 lb. a.i. per acre per season. Do not graze livestock on treated orchard floors.	21
Filberts	Filbert Worm Oblique Banded Leafroller	0.2-0.4	10⅔-21⅓	5⅓-10⅔	12-6	Make first application after emergence of filbert worm moths in early summer. Do not apply a second treatment earlier than three weeks after the first. Do not apply more than 0.8 lb a.i. per acre per season. For dilute spray, apply 200-400 gals. per acre, but do not	21

CROP	INSECT	DOSAGE: CONCENTRATE SPRAY: LB ai/ ACRE	FL OZ/ ACRE	DILUTE SPRAY: FL OZ/ 100 GALS	NUMBER ACRES TREATED WITH GALLON	FURTHER USE INSTRUCTIONS	DAYS TO HARVEST
						exceed 21⅓ fl oz of PYDRIN 2.4 EC per acre per treatment. Do not graze livestock in treated orchards.	
Pecans	Pecan Weevil, Hickory Shuckworm, Pecan Nut Casebearer, Pecan Aphids, Pecan Spittlebug	0.1-0.3	5⅓- 16	2⅔-8	24-8	Repeat as necessary to maintain control. Do not feed or graze livestock on treated orchard floors. Do not exceed 1.2 lbs. a.i. per acre per season. For dilute spray apply 200-600 gals per acre, but do not exceed 16 fl. oz. of PYDRIN 2.4 EC per acre per treatment.	21
	Pecan Stem Phylloxera Pecan Leaf Phylloxera					Correct timing of spray applications is critical in achieving optimum control of leaf and stem phylloxera. Consult local spray recommendations for correct times of application.	
Walnuts	Walnut Aphid Walnut Husk Fly Codling Moth Navel Orangeworm	0.2-0.4	10⅔- 21⅓	5⅓	12-6	Repeat as necessary to maintain control. Do not exceed 0.8 lb. a.i. per acre per season. For dilute sprays, apply 200-400 gals per acre. Do not feed or graze livestock on treated crop floors. NOTE: Use of baits in Walnut Husk fly sprays is recommended where endorsed by local Agricultural Extension Service. CONCENTRATE SPRAY: Apply specified dosage per acre in no less than 30 gals. of water per acre by ground or 5 gals. of water per acre by air.	21

VEGETABLE CROPS

CROP	INSECT	DOSAGE: LB ai/ ACRE	FL OZ/ ACRE	NUMBER ACRES TREATED WITH GALLON	FURTHER USE INSTRUCTIONS	DAYS TO HARVEST
Artichokes	Artichoke Plume Moth	0.1	5⅓	24	Repeat as necessary to maintain control, but not more often than each 14 days. Apply no more than 0.3 lb. a.i./acre between bud formation and harvest of an individual fruit. Apply in a minimum of 10 gallons per acre by air and 50-200 gallons per acre by ground (use sufficient water to obtain coverage without excessive runoff).	7
Broccoli, Cabbage Cauliflower	Imported Cabbageworm	0.05-0.1	2⅔-5⅓	48-24	Repeat as necessary to maintain control. Do not exceed 1.6 lbs. a.i. per acre per season.	3
	Cabbage Looper, Beet Armyworm, Alfalfa Looper, Flea Beetle, Grasshopper, Cutworm	0.1-0.2	5⅓-10⅔	24-12		
Carrots	Aster Leafhopper Cutworms Leafhoppers	0.1-0.2	5⅓-10⅔	24-12	Apply as needed for control, but do not exceed 2.0 lb ai/acre per season. Use ground application only.	7
	Carrot Weevil*	0.2	10⅔	12	*Begin treatment when weevils become active.	
Collards	Imported Cabbageworm, Cabbage Looper, Beet Armyworm, Alfalfa Looper, Flea Beetle, Grasshopper, Cutworm	0.1-0.2	5⅓-10⅔	24-12	Repeat as necessary to maintain control. Do not feed livestock on treated plant parts. Do not exceed 0.8 lbs. a.i. per acre per season. Use ground application equipment only.	7
Cucumber, Melons (Cantaloupe, Honeydew Melons, Watermelon, Muskmelon), Pumpkin, Squash (Summer, Winter)	Pickleworm, Rindworms, Leafhoppers, Corn Earworm, Cabbage Looper, Squash Bug, Cutworms (seedling spray), Cucumber Beetle (adult control), Grasshoppers, Plant Bugs (Lygus Bugs, Stink Bugs), Squash Vine Borer	0.1-0.2	5⅓-10⅔	24-12	Repeat as necessary to maintain control. Do not exceed 1.0 lb a.i. per acre per season. Do not apply on zucchini squash, as injury may occur.	3
Beans, Snap	Mexican Bean Beetle, Saltmarsh Caterpillar	0.05-0.1	2⅔-5⅓	48-24	Repeat as needed to maintain control but do not exceed 0.8 lb/ai acre per season. Do not allow	3

CROP	INSECT	DOSAGE LB ai/ ACRE	DOSAGE FL OZ/ ACRE	NUMBER ACRES TREATED WITH GALLON	FURTHER USE INSTRUCTIONS	DAYS TO HARVEST
					livestock to graze treated bean fields. Do not harvest treated bean vines for livestock forage, fodder, or hay.	
	Potato Leafhopper, European Cornborer, Corn Earworm, Cucumber Beetle (adult control) Flea Beetle, Green Cloverworm, Cutworm (seedling spray), Grasshopper, Pea Aphid, Leafhopper, Cabbage Looper, Western Bean Cutworm, Velvet Bean Caterpillar Beet, Armyworm, Cucumber Beetle, Corn Rootworm (adult control)	0.1-0.2	5⅓-10⅔	24-12		
Dry Beans Dry Peas	Mexican Bean Beetle, Saltmarsh Caterpillar	0.05-0.1	2⅔-5⅓	48-24	Repeat as necessary to maintain control. Do not exceed 0.8 lb a.i. per acre per season. Do not feed or graze livestock on treated vines.	21
	Corn Earworm, Potato Leafhopper, Pea Aphid, Grasshoppers, Leafhoppers, Cowpea Curculio, Cutworms, Cabbage Looper, Western Bean Cutworm, Velvetbean Caterpillar, Beet Armyworm, Cucumber Beetle, Corn Rootworm (adult control), Painted Lady Butterfly (larvae), Green Cloverworm	0.1-0.2	5⅓-10⅔	24-12		
Eggplant	Colorado Potato Beetle, Flea Beetles, Corn Earworm, Loopers, European Corn Borer	0.1-0.2	5⅓-10⅔	24-12	Begin application when insects or their damage is observed and repeat at 7 to 10 day intervals to achieve control. Do not exceed 1.4 lbs. a.i. per acre per season.	7
Green Peas	Pea Aphid, Green Cloverworm	0.05-0.1	2⅔-5⅓	48-24	Repeat as necessary to maintain control but do not exceed 0.4 lb ai/acre per season. Do not feed treated pea vines to livestock.	3
	Cabbage Looper, Celery Looper, Alfalfa Looper, Cutworms, Armyworm, Corn Earworm, Imported Cabbageworm, Alfalfa Caterpillar	0.1-0.2	5⅓-10⅔	24-12		
Peppers	European Corn Borer, Fall Armyworm, Beet Armyworm, Flea Beetles, Corn Earworm, Loopers, Colorado Potato Beetle, Pepper Weevil, Cucumber Beetle (adults), Southern Armyworm	0.1-0.2	5⅓-10⅔	24-12	Begin application when insects or their damage is observed and repeat at 7 to 10 day intervals to achieve control. Do not exceed 1.4 lbs. a.i. per acre per season.	7
Potato	Potato Psyllid, Potato Tuberworm	0.05-0.1	2⅔-5⅓	48-24	Repeat as necessary to maintain control. Do not graze livestock on treated vines. Do not exceed 1.4 lbs. a.i. per acre per season.	7
	Fleabeetles, Tarnished Plant Bug, Cabbage Looper, Potato Aphid, Potato Leafhopper, Cucumber Beetle (adult control), Grasshoppers, European Corn Borer, Beet Armyworm, Western Yellow-striped Armyworm, Buckthorn Aphid, Cutworms	0.1-0.2	5⅓-10⅔	24-12		
	Colorado Potato Beetle	0.1-0.2	5⅓-10⅔	24-12		
	Long Island, NY	0.2	10⅔	12		
Radishes	Armyworms, beetles	0.1-0.2	5⅓-10⅔	24-12	Apply as necessary to maintain control. Do not apply more than 0.4 a.i. per acre per season.	
Sweet Corn	Corn Earworm, Southwestern Corn Borer, European Corn Borer, Beet Armyworm, Flea Beetles, Sap Beetles (adults), Banded Cucumber Beetle, Tarnished Plant Bug	0.1-0.2	5⅓-10⅔	24-12	Do not exceed 2.0 lb a.i. per acre per season. Begin applications either before or at time of silking and continue as necessary to maintain control. For protection against corn borers other than at silking time, treat when necessary to maintain control. For additional information consult directions for use under field corn.	
	Stalk Borer, Cutworms, Corn Rootworms (adult control), Grasshoppers, Chinch Bugs, Armyworm	0.1-0.2	5⅓-10⅔	24-12	For additional information consult directions for use under field corn.	

CROP	INSECT	DOSAGE: LB ai/ ACRE	DOSAGE: FL OZ/ ACRE	NUMBER ACRES TREATED WITH GALLON	FURTHER USE INSTRUCTIONS	DAYS TO HARVEST
	Western Bean Cutworm	0.05-0.1	2⅔-5⅓	48-24		
	Corn Leaf Aphid, Oat Bird-Cherry Aphid	0.1-0.2	5⅓-10⅔	24-12	For optimum results, direct the spray at the aphid population so as to achieve maximum coverage of the exposed insects. Aphids not contacted by the spray, such as in whorls and leaf axils, may not be adequately controlled.	
Tomato	Tomato Hornworm	0.05-0.1	2⅔-5⅓	48-24	Repeat as necessary to maintain control. Do not feed or graze livestock on treated vines. Do not exceed 2.0 lbs. a.i. per acre per season.	1
	Tomato Pinworm, Flea Beetle, Tomato Fruitworm, Colorado Potato Beetle, Cutworms, Cabbage Looper, Southern Armyworm, Sugarbeet Armyworm, Western Yellow-Striped Armyworm, Grasshopper, Whitefly, Potato Aphid	0.1-0.2	5⅓-10⅔	24-12	(Note: Some bleaching or spotting may occur on the foliage of young plants. This does not affect yield or fruit quality.)	
	Vegetable Leafminer*	0.2	10⅔	12		

SPECIALTY USES

CROP	INSECT	DOSAGE: LB ai/ ACRE	DOSAGE: FL OZ/ ACRE	NUMBER ACRES TREATED WITH GALLON	FURTHER USE INSTRUCTIONS
Non-Crop Land (excluding public land such as forests, parks, or recreational areas.)	Grasshoppers, Saltmarsh Caterpillar	0.05-0.1	2⅔-5⅓	48-24	Spray non-crop land adjacent to tilled areas to control migrating insects (Grasshoppers, Armyworms) which are a threat to crops. Repeat as necessary to maintain control but do not exceed 2 lbs. active ingredient per acre per year. Do not feed treated crop to livestock. Do not spray ditch banks or areas adjacent to water.
	Armyworms, Army Cutworms Chinch Bugs	0.1-0.2	5⅓-10⅔	24-12	

SITE	INSECT	DOSAGE: HIGH VOLUME SPRAYS: LB ai/ 100 GALS.	HIGH VOLUME SPRAYS: FL OZ/ 100 GALS.	HIGH VOLUME SPRAYS: PYDRIN 2.4 EC/ GAL OF SPRAY	LOW VOLUME SPRAYS (MISTBLOWER) LB ai/A	FURTHER USE INSTRUCTIONS
Christmas Tree Plantings, Pine Seed Orchards, Forest Tree Nurseries	Nantucket Pine Tip Moth, Spittlebugs, Pineleaf Chermid, Redheaded Pine Sawfly, Balsam Woolly Adelgid, Pine Chafer, Red Pine Sawfly, Pine Conelet Bug, European Pine Sawfly, Cranberry Girdler (adult control), Pine Needle Midge, Balsam Twig Aphid	0.1-0.2	5⅓-10⅔	⅓-⅔ Teaspoon	0.1-0.2	Apply as needed to maintain control. Spray sufficient gallonage to obtain good coverage of entire tree. High volume sprays on Virginia Pine may cause some yellowing of needles, low volume (Mistblower) applications at the rate of 0.1 to 0.2 lb. ai/acre are recommended to avoid this problem.
	Pales Weevil, Northern Pine Weevil	0.1-0.2	5⅓-10⅔	⅓-⅔ Teaspoon	Apply as needed to maintain control. Dilute formulation in kerosene, instead of water, and spray tree stumps to control developing larvae. Use sufficient spray volume to thoroughly cover stump.	
	Coneworms, Seedbugs, Seed Chalcid	11 fl oz/100 gals water for high volume sprayers. 3.5 pts/100 gals water for low volume sprayers. 0.75 lb. ai/acre application in not less than 10 gals of water for aerial applications.			Apply first application within 1 week of female flower closure or peak pollen flight for Webbing Coneworm control. For other Coneworms and Seedbugs, apply first application within 30 days following female flower closure. Repeat application at intervals of 4 weeks but do not exceed 6.3 lbs. ai/ac/yr. For Seed Chalcid control, apply when all cones are pendant, and repeat at 1-2 week intervals for 2 or more sprays, or as needed. Apply approximately 5-10 gals of the 11 fl oz/100 gal dilution per tree with high volume sprayers. With low volume sprayers apply 100 gals of the 3.5 pts/100 gal dilution per acre. Do not graze or harvest cover crop. Avoid contact with open water.	

SITE	INSECT	DOSAGE: HIGH VOLUME SPRAYS: LB ai/ 100 GALS.	FL OZ/ 100 GALS.	PYDRIN 2.4 EC/ GAL OF SPRAY	LOW VOLUME SPRAYS (MISTBLOWER) LB ai/A	FURTHER USE INSTRUCTIONS
					Some foliar chlorosis may occur on Douglas Fir when concentrated (low volume) sprays are employed. To avoid this problem use high volume sprays.	

[1] BLADEX(R) is a registered trademark of E.I. Du Pont de Nemours & Co. (Inc.)
[2] "Bicep" and "Dual" are registered trademarks of Ciba-Geigy Corp.
[3] "Banvel" is a registered trademark of Sandoz Crop Protection Corp.
[4] "Lasso" and "Roundup" are registered trademarks of Monsanto Agricultural Co.
[5] Gramoxone Super is a registered trademark of ICI Americas Inc.

STORAGE AND DISPOSAL

Do not contaminate water, food, or feed by storage or disposal.

STORAGE: Store in a secure, dry and temperate area. Keep container closed when not in use. Store in original container. Do not store near food or feed. Do not use or store around the home. Avoid contact with water. In case of spill or leak, soak up with sand, earth or synthetic absorbent (do not use alkaline absorbents) and dispose of wastes in compliance with local, State and Federal regulations.

PESTICIDE DISPOSAL: Pesticide wastes are toxic. Improper disposal of excess pesticide, spray mixture, or rinsate is a violation of Federal Law. If these wastes cannot be disposed of by use according to label instructions contact your State Pesticide or Environmental Control Agency, or the Hazardous Waste representative at the nearest EPA Regional Office for guidance.

CONTAINER DISPOSAL: Triple rinse (or equivalent). Then offer for recycling or reconditioning if container reuse is permitted. If container reuse is prohibited, puncture metal containers and dispose of in a sanitary landfill, or by other approved State and local procedures. Dispose of plastic containers in a sanitary landfill, or by incineration if allowed by State and local authorities. Dispose of glass containers in a sanitary landfill, or by other approved State and local procedures.

In case of a significant spill, call CHEMTREC (800) 424-9300.

NOTICE OF WARRANTY

Du Pont warrants that this product conforms to the chemical description on the label thereof and is reasonably fit for the purposes stated on such label only when used in accordance with the directions under normal use conditions. It is impossible to eliminate all risks inherently associated with the use of this product. Crop injury, ineffectiveness or other unintended consequences may result because of such factors as weather conditions, presence of other materials, or the manner of use or application, all of which are beyond the control of Du Pont. In no case shall Du Pont be liable for consequential, special or indirect damages resulting from the use or handling of this product. All such risks shall be assumed by the buyer. DU PONT MAKES NO WARRANTIES OF MERCHANTABILITY OR FITNESS FOR A PARTICULAR PURPOSE NOR ANY OTHER EXPRESS OR IMPLIED WARRANTY EXCEPT AS STATED ABOVE.

AG-3815 8088 10/26/87

This is a specimen label. DuPont is not responsible for the accuracy of the information contained herein. As labels are subject to revision, always carefully read and follow the label on the product container.

SAVEY®

MITICIDE
RECOMMENDED FOR AGRICULTURAL/COMMERCIAL USE
WETTABLE POWDER

ACTIVE INGREDIENT:

Trans-5-(4-chlorophenyl)-N-cyclohexyl-4-methyl-2-oxothiazolidine-3-carboxamide	50%
INERT INGREDIENTS	50%
TOTAL	100%

EPA Reg. No. 352-531

KEEP OUT OF REACH OF CHILDREN
CAUTION

PRECAUTIONARY STATEMENTS

HAZARDS TO HUMANS AND DOMESTIC ANIMALS

CAUTION! MAY IRRITATE EYES, NOSE, THROAT AND SKIN; HARMFUL IF INHALED

Avoid breathing dust. Avoid contact with skin, eyes or clothing. Wash thoroughly with soap and water after handling and before eating or smoking. Remove contaminated clothing and wash with soap and water before reuse. Do not apply this product in such a manner as to directly or through drift expose workers, other persons or animals. The area must be vacated by all unprotected persons prior to application.

STATEMENT OF PRACTICAL TREATMENT

If in eyes: Flush with plenty of water. Get medical attention if irritation persists.

If on skin: Wash thoroughly with soap and water. Get medical attention if irritation persists.

If inhaled: Remove victim to fresh air. Apply artificial respiration if necessary.

If swallowed: Call a physician or Poison Control Center. Drink 1 or 2 glasses of water and induce vomiting by touching back of throat with finger. Do not induce vomiting or give anything by mouth to an unconscious person.

For medical emergencies involving this product, call toll free 1-800-441-3637.

ENVIRONMENTAL HAZARDS

This product is toxic to fish and aquatic invertebrates. Do not apply directly to water or wetlands (swamps, bogs, marshes, and potholes). Drift and runoff from treated areas may be hazardous to aquatic organisms in neighboring areas. Do not contaminate water when disposing of equipment washwaters.

GENERAL INFORMATION

Du Pont "Savey" Miticide is recommended for use as a foliar application in pears. It is a 50% wettable powder formulation to be mixed with water and applied as a spray. It is noncorrosive to equipment.

"Savey" is an effective miticide providing control of major economically-important mites that affect pear. The degree and duration of control is dependent on rate used, growth stage of the mite, species of mite, and climatic conditions under which the material is being applied.

"Savey" Miticide is a highly active mite ovicide and larvicide. Control is achieved from either direct contact with the spray or from contact with treated plant surfaces. "Savey" also affects adult female mites by rendering their eggs inviable.

IMPORTANT

Wear clean clothes daily. Wash thoroughly after handling and before eating or smoking. Keep animals and unprotected persons out of operational areas during treatment and while weather conditions favor drift from areas treated. Not for use or storage in or around the home.

DIRECTIONS FOR USE

It is a violation of federal law to use this product in a manner inconsistent with its labeling.

Du Pont "Savey" Miticide should be used only in accordance with recommendations on this label. Du Pont will not be responsible for losses or damages resulting from use of this product in a manner not specifically recommended by Du Pont. User assumes all risks associated with such nonrecommended use.

"Savey" is a 50% wettable powder to be diluted with water for application by ground equipment. Use only in commercial planting; do not use in home plantings.

Do not apply this product through any type of irrigation system.

PROTECTIVE CLOTHING FOR "SAVEY" MITICIDE

USE ONLY WHEN WEARING THE FOLLOWING PROTECTIVE CLOTHING AND EQUIPMENT DURING MIXING/LOADING, APPLICATION, REPAIR AND CLEANING OF MIXING, LOADING, AND APPLICATION EQUIPMENT, DISPOSAL OF PESTICIDE, AND EARLY RE-ENTRY INTO TREATED AREAS BEFORE SPRAYS HAVE DRIED: Protective suit of one or two pieces covering all parts of the body except head, hands and feet; chemical resistant gloves; chemical resistant shoes (or chemical resistant shoe covers or chemical resistant boots).

IF APPLICATION IS PERFORMED USING A CLOSED CAB THE FOLLOWING PROTECTIVE CLOTHING AND EQUIPMENT MAY BE WORN AS AN ALTERNATIVE; Long sleeve shirt and long-legged pants; shoes and socks. Chemical resistant gloves must be available in the cab or cockpit and must be worn when exiting. This clothing is inadequate to protect you during equipment repair or cleaning, reentry or pesticide disposal.

IMPORTANT! If pesticide comes in contact with skin, wash off with soap and water. Always wash hands, face, and arms with soap and water before smoking, eating, drinking, or toileting.

AFTERWORK: BEFORE REMOVING GLOVES, WASH THEM WITH SOAP AND WATER. TAKE OFF ALL WORK CLOTHES AND SHOES. SHOWER USING SOAP AND WATER. WEAR CLEAN CLOTHES. DO NOT RE-USE CONTAMINATED CLOTHING. PERSONAL CLOTHING WORN DURING WORK MUST BE LAUNDERED SEPARATELY FROM HOUSEHOLD ITEMS. STORE PROTECTIVE CLOTHING SEPARATE FROM PERSONAL CLOTHING. CLEAN OR LAUNDER PROTECTIVE CLOTHING AFTER EACH USE.

PROTECTIVE CLOTHING AND EQUIPMENT THAT BECOMES HEAVILY CONTAMINATED MUST BE DESTROYED ACCORDING TO STATE AND LOCAL REGULATIONS.

HEAVILY CONTAMINATED CLOTHING CANNOT BE ADEQUATELY DECONTAMINATED.

REENTRY STATEMENT

Do not apply this product in such a manner as to directly or through drift expose workers or other persons. The area being treated must be vacated by unprotected persons.

Do not enter treated areas without appropriate protective clothing until sprays have dried. Because certain states may require more restrictive reentry intervals for areas treated with this product, consult your State Department of Agriculture for futher information. Written or oral warnings must be given to workers who are expected to be in a treated area or in an area about to be treated with this product. When oral warnings are given, warnings shall be given in a language customarily understood by workers.

Oral warnings must be given if there is reason to believe that written warnings cannot be understood by workers. Written (or oral) warnings must include the following information: "CAUTION - Area treated with "Savey" on (date of application). Do Not Enter Treated Areas Without Appropriate Protective Clothing Until Sprays Have Dried."

MANAGING MITE RESISTANCE

Repeated use of miticides with similar modes of action may lead to buildup of resistant strains of mites. For good resistance management practices, Du Pont recommends that the ovicide/larvicide "Savey" be applied only once per year. If "Savey" has been applied, do not apply another ovicide such as "Apollo" in the same year. If "Apollo" has been applied, do not apply "Savey" in the same year. If additional miticide applications are needed, use a miticide with a different mode of action such as the contact adulticides "Vendex" or "Vydate".

IMPORTANT

PROPER TIMING - With its activity being predominately ovicidal/larvicidal, "Savey" should be applied after bloom but prior to adult mite buildup (0 - 0.5 mites/leaf). If adult mites are present (greater than 0.5 mites/leaf), apply "Savey" in combination with a registered contact adulticide such as "Vendex" or "Vydate".

PROPER MIXING

Fill tank 1/2 full with water and, with agitator running, add proper amount of "Savey". Finish adding the required amount of water. Continuous agitation is required to keep the product in suspension.

Compatibility - Before full-scale mixing of "Savey" with other products, test mix small proportionate quantities of each to insure compatibility. Tests have shown "Savey" may be incompatible with some EC formulations. Addition of a compatibility agent such as "Unite" may improve compatibility with those products.

SPRAY VOLUME

Ground Application: Use sufficient water to obtain thorough, uniform coverage; 150-800 gallons per acre for dilute application, 50-150 gallons per acre for concentrated applications.

APPLICATION RATES

Apply the lower rates on low mite egg infestation levels and the high rates on moderate to high mite egg infestation levels or to larger trees. Use one application per growing season.

The following mites should be controlled at the given rate range:

Crop	Mites	Rate oz. Form/Acre	Treatment Schedule
Pear	European red	4.0 – 6.0	See
	Two-Spotted spider	4.0 – 6.0	Proper
	McDaniel spider	4.0 – 6.0	Timing

Do not apply more than a total of 6 oz. formulated product per acre in one growing season.

RESTRICTIONS

Do not graze or feed livestock on cover crops growing in treated areas. Do not apply within 28 days of harvest.

Do not use water from treated orchards for irrigating other crops.

STORAGE AND DISPOSAL

Storage: Store product in original container only away from other pesticides, fertilizer, food or feed.

Product Disposal: Do not contaminate water, food or feed by storage or disposal. Wastes resulting from the use of this product may be disposed of on-site or at an approved waste disposal facility.

Container Disposal: Completely empty container into application equipment. Then dispose of empty container in a sanitary landfill or by incineration, or if allowed by state and local authorities, by burning. If burned, stay out of smoke.

NOTICE TO BUYER - Purchase of this material does not confer any rights under patents of countries outside of the United States.

NOTICE OF WARRANTY

Du Pont warrants that this product conforms to the chemical description on the label thereof and is reasonably fit for the purposes stated on such label only when used in accordance with the directions under normal use conditions. It is impossible to eliminate all risks inherently associated with the use of the product. Crop injury, ineffectiveness, or other unintended consequences may result because of such factors as weather conditions, presence of other materials, or the manner of use or application, all of which are beyond the control of Du Pont. In no case shall Du Pont be liable for consequential, special or indirect damages resulting from the use or handling of this product. All such risks shall be assumed by the buyer. DU PONT MAKES NO WARRANTIES OF MERCHANTABILITY OR FITNESS FOR A PARTICULAR PURPOSE NOR ANY OTHER EXPRESSED OR IMPLIED WARRANTY EXCEPT AS STATED ABOVE.

AG - 3248 8049 4/13/89

This is a specimen label. DuPont is not responsible for the accuracy of the information contained herein. As labels are subject to revision, always carefully read and follow the label on the product container.

SINBAR®

HERBICIDE
WETTABLE POWDER

ACTIVE INGREDIENT	BY WEIGHT
Terbacil[3-tert-butyl-5-chloro-6-methyluracil]	80%
INERT INGREDIENTS:	20%
TOTAL	100%

EPA Reg. No. 352-317
U.S. Pats. 3,235,357 & 3,352,862

KEEP OUT OF REACH OF CHILDREN
CAUTION

PRECAUTIONARY STATEMENTS

HAZARDS TO HUMANS AND DOMESTIC ANIMALS

CAUTION! MAY IRRITATE EYES, NOSE, THROAT, AND SKIN.

Avoid breathing dust or spray mist. Avoid contact with skin, eyes, and clothing. Wash thoroughly with soap and water after handling.

Do not apply this product in such a manner as to directly or through drift expose workers or other persons. The area being treated must be vacated by unprotected persons.

STATEMENT OF PRACTICAL TREATMENT

If in eyes: Flush with plenty of water; get medical attention if irritation persists.

If on skin: Flush with plenty of soap and water.

For medical emergencies involving this product, call toll free 1-800-441-3637.

ENVIRONMENTAL HAZARDS

Keep out of lakes, streams or ponds. Do not contaminate water by cleaning of equipment or disposal of wastes.

IMPORTANT

Injury to or loss of desirable trees or other plants may result from failure to observe the following:

Do not apply (except as recommended for crop use), or drain or flush equipment on or near desirable trees or other plants, or on areas where their roots may extend, or in locations where the chemical may be washed or moved into contact with their roots. Do not use on lawns, walks, driveways, tennis courts, or similar areas. Do not use in home planting of fruits, nuts or other crops nor in citrus, apple, peach or pecan orchards interplanted with other trees or desirable plants. Prevent drift of dry powder or spray to desirable plants. Do not contaminate any body of water. Keep from contact with fertilizers, insecticides, fungicides, and seeds. Thoroughly clean all traces of "SINBAR" from application equipment immediately after use. Flush tank, pump, hoses, and boom with several changes of water removing nozzle tips and screens (clean these parts separately).

GENERAL INFORMATION

Du Pont "SINBAR" Herbicide is a wettable powder to be mixed in water and applied as a spray for selective weed control in alfalfa, apples, peaches, blueberries, caneberries, citrus, mint, pecans, strawberries and sugarcane. It is non-volatile, non-flammable, and non-corrosive to equipment.

"SINBAR" controls susceptible weeds for an extended period of time; the degree of control and duration of effect will vary with the amount of chemical applied, soil texture, rainfall and other conditions. Soils high in clay or organic matter require higher dosages than soil low in clay or organic matter to obtain equivalent herbicide performance. Moisture is required to activate the chemical; best results occur if rainfall (or sprinkler irrigation) occurs within 2 weeks of application.

Observe all cautions and limitations on labeling of all products used in mixtures.

DIRECTIONS FOR USE

It is a violation of federal law to use this product in a manner inconsistent with its labeling.

Du Pont "SINBAR" Herbicide should be used only in accordance with recommendations on this label, or in separate published Du Pont recommendations available through local dealers.

Du Pont will not be responsible for losses or damages resulting from use of this product in any manner not specifically recommended by Du Pont. User assumes all risk associated with such non-recommended use.

Do not use in Kern County, California.

Do not enter treated areas without protective clothing until sprays have dried. Because certain states may require more restrictive reentry intervals for various crops treated with this product, consult your State Department of Agriculture for further information.

Written or oral warnings must be given to workers who are expected to be in a treated area or in an area about to be treated with this product. When oral warnings are given, warnings shall be given in a language customarily understood by workers. Oral warnings must be given if there is reason to believe that written warnings cannot be understood by workers. Warnings must include the following information: "CAUTION. Area treated with 'SINBAR' on (date of application). Do not enter without appropriate protective clothing until sprays have dried. In case of contact, flush skin or eyes with plenty of water; for eyes, get medical attention."

SELECTIVE USE IN CROPS

WEED CONTROL

Best results are obtained if application is made shortly before or shortly after weed growth begins; if dense growth is present, remove tops and spray the ground. Control of perennial grasses may be improved by cultivation prior to treatment; otherwise, avoid working the soil as long as weed control continues or else effectiveness of the treatment may be reduced. For alfalfa, the soil should not be disturbed. See "Crops" section for recommended use rates for specific crops.

"SINBAR", at rates of 1/2 to 1 1/2 lbs. per acre (alfalfa), controls annual weeds including species of chickweed, henbit, lambsquarters, mustard, tansymustard, crabgrass, and foxtail, and such weeds as downy brome (cheatgrass), marestail, peppergrass, prickly lettuce, shepherdspurse, snoweed, ryegrass, wild barley, and yellow rocket. Treatment will not control established perennial bindweeds, johnsongrass and Canada thistle.

Rates of 1 to 4 lbs. per acre control johnsongrass seedlings and annual weeds including species of chickweed, clover, crabgrass, crotolaria, seedling dandelion, fiddleneck, foxtail, henbit, knotweed, lambsquarters, mustard, nightshade, panicum, plantain, pigweed, purslane, ragweed, sandbur (sandspur), signalgrass (Brachiaria), smartweed, and wild geranium and such weeds as American burnweed (fireweed, Erechites hieracifolia), barnyardgrass (watergrass), bluegrass (Poa annua), China lettuce, crowfootgrass, dogfennel, Flora's paintbrush, Florida pusley, guineagrass, horseweed, jimsonweed, junglerice, natalgrass, and orchardgrass. In addition, treatment usually provides partial control of species of groundsel, horsenettle, quackgrass, red sorrel, and yellow nutsedge.

Rates of 2 to 4 lbs. per acre (blueberries) control weeds including species of cinquefoil, hawkweed and such weeds as annual sedge, perennial ryegrass, quackgrass, redroot (Lachnanthes caroliniana), and red sorrel.

Rates of 4 to 10 lbs. per acre are usually required for control of bermudagrass and for effective control of horsenettle, quackgrass, red sorrel, and yellow nutsedge; repeated annual applications may also be necessary. Partial control of perennial weeds such as purple nutsedge, paragrass, and torpedograss usually accompanies treatment at the maximum dosage rates. Effects on perennial weeds are slow to appear, usually progressing over a period of several months.

EQUIPMENT—SPRAY VOLUMES AND PRESSURES

Do not apply this product through any type of irrigation system.

Apply with a fixed-boom power sprayer properly calibrated to a constant speed and rate of delivery. Use sufficient water (minimum 20 gals. per acre) to provide thorough and uniform coverage of the ground. On alfalfa sugarcane and mint, preemergence broadcast applications may be made by aircraft (5 to 10 gals. spray per acre).

Continuous agitation in the spray tank is required to keep the material in suspension. Avoid overlapping, and shut off spray booms while starting, turning, slowing or stopping, or injury to the crop may result.

SPRAY PREPARATION

Mix proper amount of "SINBAR" into necessary volume of water.

USE RATES

All dosages of "SINBAR" are expressed as broadcast rates; for band treatment, use proportionately less. Where a range of dosages is given, use the lower rate on coarse textured soils (low in clay or organic matter) and the higher rate on fine textured soils (high in clay or organic matter).

SOIL LIMITATIONS

Crop injury may result from failure to observe the following: Unless otherwise directed, do not use on sand, loamy sand or gravelly soils, nor on soils low in organic matter (less than 1%).

REPLANTING

Unless otherwise directed, do not replant treated areas to any crop within 2 years after last application as injury to subsequent crops may result.

CROPS

ALFALFA

Treat only stands established for 1 year or more. Do not use on seedling alfalfa nor on alfalfa-grass mixtures or other mixed stands. Do not use with surfactants.

Do not apply on snow-covered or frozen ground as injury to the crop may result.

U.S. (EXCEPT NORTHEAST)

Make a single application of 1/2 to 1 1/2 lbs. per acre in the fall after plants become dormant or in the spring before new growth starts. For semi-dormant and non-dormant varieties, apply in fall or winter after last cutting or in the spring before new growth starts. Apply before or after emergence of weeds but before they are 2" tall or across.

Do not apply to established stands after new growth starts in the spring, as injury to the crop may result. Do not use in California south of Interstate 80.

NORTHEAST

Apply "SINBAR" at the rate of 3/4 to 1 1/2 lbs per acre (ground application only) to alfalfa that is dormant (fall through winter) or in the spring before initial new growth exceeds 2 inches in height or to stubble after cutting, following hay removal and before regrowth exceeds 2 inches in height. Two applications may be made per crop year, but should be at least 60 days apart and should not exceed a total of 2 1/4 pounds.

For winter annual weeds and early germinating summer annuals, applications of "SINBAR" during dormancy or before new spring growth exceeds 2 inches in height provides the best results. Late germinating annual grasses and broadleaf weeds are best controlled with an after cutting treatment. Where an early application is made for control of winter annual weeds and early germinating summer annuals but where late germinating grasses and broadleaves are expected to be a problem, a second application (after cutting) may be applied for improved results.

Temporary yellowing may occur when applications are made to new alfalfa growth.

APPLES, PEACHES

Use "SINBAR" alone, or apply as a tank mixture with Du Pont "Karmex" Herbicide. Make a single band or broadcast application as a directed spray; avoid contact of foliage and fruit with spray or mist. Apply either in the spring or after harvest in the fall before weeds emerge or during early seedling stage of weed growth. Where complete weed control to harvest is desired, additional weed control measures may be required during the growing season.

Where crop is grown under furrow irrigation or under raised-berm flood irrigation (trees 4" to 6" above waterline), apply only as a band treatment. Do not treat trees planted in the bottom of irrigation furrows or trees grown under flat flood or basin irrigation, and do not use on eroded areas where subsoil or tree roots are exposed, as injury to trees may result. Do not graze or feed forage from treated areas to livestock.

"SINBAR" Alone—Use only under trees established in the orchard for at least 3 years.

	Lbs. "SINBAR" Per Acre	
Soil Texture	1 to 2% Organic Matter	More Than 2% Organic Matter
Sandy loam	2	3
Loam, silt loam, silt	2 1/2	3 1/2
Clay loam, clay	3	4

"SINBAR" + "Karmex" —Use only under trees established in the orchard for at least 2 years.

Soil Texture	Lbs. Product Per Acre 1 to 2% Organic Matter "SINBAR" + "Karmex"		More Than 2% Organic Matter "SINBAR" + "Karmex"	
Sandy loam	1	+ 1	1½	+ 1½
Loam, silt loam, silt	1½	+ 1½	2	+ 2
Clay loam, clay	2	+ 2	2	+ 2

ASPARAGUS

Direct Seeded—Plant seed 1½ inches deep in coarse soils and 1 inch deep in fine soils. During planting operation, spray activated charcoal (such as Aqua NuChar[1] or Gro-Safe[2]) as a 1 inch band on soil surface directly over rows at a rate of 300 lbs. per acre (broadcast basis equivalent to 15 lbs. per acre where row spacing is 20 inches). Follow with "SINBAR" as a single spray at 1 to 2 lbs. per acre. Use the lower rate on coarser soils and the higher rate on finer soils.

Established Beds—Apply 1½ to 3 lbs. "SINBAR" per acre prior to spear emergence. Application may be made immediately after clean cutting. Use the lower rate on coarser soils and the higher rate on finer soils. Do not apply more than 3 lbs. per acre.

Apply before weeds emerge or to small weeds (½ to 2 inches tall or across).

High organic soils absorb "SINBAR" so that it is substantially inactivated as a soil residual herbicide. On these soils, weed control is provided by postemergence foliar uptake only. Apply up to 3 lbs. "SINBAR" per acre and a maximum of 2 applications per year.

Note: Do not use on areas where subsoil or roots are exposed. Do not use on plants that are diseased or lacking in vigor, as injury to the crop may result. Do not harvest within 5 days after application. Treated areas may be planted to asparagus one year after application. Otherwise, do not replant to any other crop within two years after last application.

[1] Aqua Nuchar — Registered trademark of Westvaco Corp.
[2] Gro-Safe — Registered Trademark of ICI Americas, Inc.

BLUEBERRIES

Treat only plantings established for 1 year or more. Make a single band or broadcast application to ground beneath bushes; avoid contact of foliage and fruit with spray or mist. Apply either in the spring or after harvest in the fall before weeds emerge or during early seeding stage of weed growth. Do not use on eroded areas where subsoil or roots are exposed, nor on plants that are diseased or lacking in vigor, as injury to the plants may result. Treated areas may be planted to blueberries one year after last application.

U.S.

Soil Texture	Lbs. "SINBAR" in Min. 25 Gals. Water Per Acre 1 to 3% Organic Matter	More Than 3% Organic Matter
Sand, Loamy Sand	Do not use	2
Sandy Loam, Loam, Silt Loam, Silt, Sandy Clay,	2	3
Sandy Clay Loam Silty Clay, Silty Clay Loam, Clay,	3	4
Clay Loam	4	4

MAINE, MARYLAND, NEW JERSEY

Apply as directed above, or, for broader spectrum weed control apply as a tank mix with "Karmex" DF:

Soil Texture	Lbs. Product Per Acre 1 to 3% Organic Matter "SINBAR" + "Karmex"		More Than 3% Organic Matter "SINBAR" + "Karmex"	
Sand, loamy Sand	DO NOT USE		2	+ 2
Sandy Loam	2	+ 2	2-½	+ 2
Loam, Silt Loam, Silt, Sandy Clay, Sandy Clay Loam	2-½	+ 2	3	+ 2
Silty Clay, Silty Clay Loam, Clay, Clay Loam	3	+ 2	3	+ 2

Do not replant areas treated with "SINBAR" + "Karmex" to crops other than blueberries within two years after aplication, as injury to those crops may result.

CANEBERRIES

(Blackberries, Boysenberries, Dewberries, Loganberries, Raspberries, Youngberries) Treat only plantings established for 1 year or more. Make a single band or broadcast application of 1 to 2 lbs. per acre in a minimum of 20 gals. of water to the soil beneath the canes in the fall or early spring before fruitset and before weeds emerge or during the early stage or weed growth. Do not spray foliage nor use on eroded areas where subsoil or roots are exposed, nor apply to plants that are diseased or lacking vigor as injury to the plant may result. Treated areas may be planted to mint, alfalfa, blueberries, apples, peaches, or strawberries 1 year after last treatment. Do not replant to other crops within 2 years of last application. Do not apply within 70 days before harvest.

CITRUS

Apply as a band or broadcast treatment as a directed spray; avoid contact of foliage and fruit with spray or mist. Temporary yellowing of citrus leaves may occur following treatment. Treatment may be used on coarse soils but is not to be used on sands low in organic matter (less than 1%), on poorly drained soils, nor on trees planted in irrigation furrows, as injury to trees may result. Treated areas may be planted to citrus trees one year after last application. Do not graze or feed forage from treated areas to livestock.

Trees Established for One Year—Use 2 to 4 lbs. per acre for annual weeds; apply just before or just after weeds germinate.

Trees Established for Two Years or More—For annual weeds, apply 2 to 4 lbs. per acre. For perennial weeds, use as follows:

Arizona, California—For bermudagrass, make a single application of 6 to 8 lbs. per acre during the period from winter to early summer; or make a split application of 3 to 4 lbs. per acre in the fall and again in the spring.

Florida—For bermudagrass, use 5 lbs. per acre in the spring or early summer; if needed, repeat application the same season but do not apply more than 10 lbs. per acre per year as injury to trees may result.

Texas—For johnsongrass and bermudagrass, make a single application in the spring or early summer of 4 to 6 lbs. per acre on sandy loam soils and 8 to 10 lbs. per acre on clay loam soils.

MINT (Peppermint, Spearmint)

Preemergence Application: Make a single broadcast application of 1 to 2 lbs. per acre before mint emerges.

Midwest: Apply in the spring just after the last cultivation.

Pacific Northwest:

West of Cascade Mountains—Apply in spring or fall after last cultivation.

East of Cascade Mountains—Apply any time after the first settling rain in the fall, but not when ground is frozen. If moisture is inadequate to activate the chemical, irrigation by sprinkler must follow fall or winter treatment as soon as soil thaws in the spring (or promptly after spring treatment) or poor weed control may result.

Postemergence Application (Midwest and Pacific Northwest): Apply 1 to 1½ lbs. per acre before weeds are 2" tall (or across) and grasses are 1" tall (or across). Add ½ to 1 pt. surfactant to each 25 gals. of spray; non-phytotoxic superior-type spray oil may be substituted at the rate of 1 gal. per acre. If preemergence treatment has not been applied, a second postemergence application may be made but do not exceed 2 lbs. total per acre per season. Do not apply within 60 days of harvest.

Preemergence + Postemergence: Apply as directed above. Do not apply more than 2 lbs. total per acre per season; do not apply within 60 days of harvest.

Note: For either newly planted roots or established mint, soil must be well prepared before preemergence application. Crop injury and/or poor weed control may result if application is made to ground which is cloddy or compacted, resulting in exposed or improperly covered roots. Do not apply to newly planted roots that are diseased or lacking in vigor, nor on thinly covered or exposed subsoil areas as injury to the crop may result. Use of insecticides in fields where "SINBAR" is applied may result in injury to the mint; observe use limitations on insecticide labels. Treated areas may be planted to mint one year after last application.

PECANS

Use "SINBAR" alone, or apply as a tank mixture with Du Pont "Karmex" Herbicide. Use only under trees established in the grove for at least 1 year. Make a single band or broadcast application as a directed spray, using 30 gals. (min.) of water per acre. Avoid contact of foliage and nut with spray or mist. Apply in the spring before weeds emerge or during the early seedling stage of weed growth.

Soil Texture	Lbs. Product Per Acre "SINBAR" Alone -OR-	Tank Mixture "SINBAR" + "Karmex"	
Sandy loam	2	1½	+ 1½
Loam, silt loam, silt	2½	1¾	+ 1¾
Clay loam, clay	3	2	+ 2

Do not use on eroded areas where subsoil or roots are exposed, nor on trees that are diseased or lacking in vigor or on trees planted in irrigation furrows as injury to the trees may result. Do not graze or feed forage from treated areas to livestock.

STRAWBERRIES (except California and Florida)

Use only on plants established in the field for at least 6 months (matted row system). Make a single broadcast application of 1/2 to 1/4 lbs. per acre; use the lower rate on coarse-textured soils. Apply immediately after post-harvest renovation (complete removal of old leaves is essential) but before new growth begins; or apply in the dormant season from fall to late winter. Treat before weeds are 2" tall or across.

Treatment may reduce runner production or plant stand. Varieties such as Guardian, Darrow, and Honeoye may be sensitive to "SINBAR"; determine varietal tolerance prior to adoption as field practice to prevent possible crop injury. Do not use on soils less than 2% organic matter and do not apply over top of newly planted strawberries as crop injury may result. Alfalfa, apples, blueberries, citrus, mint, peaches, strawberries, and sugar cane may be planted 1 year after last application. Do not plant any other crops within 2 years after application as injury to the crop may occur.

SUGARCANE

Because sugarcane varieties vary in their resistance to herbicides, determine tolerance to "SINBAR" prior to adoption as field practice to prevent possible crop injury. Do not use on varieties which are known to be susceptible to herbicide, such as 48-103 (Louisiana), 50-28 (Texas), 53-263 (Hawaii) or P.R. 1048 (Puerto Rico). Do not use where cane is grown on thinly covered sub-soils or rocky areas as injury to cane may result. Do not replant treated areas to any crop other than sugarcane or pineapple within 2 years after last application as injury to subsequent crops may result.

Louisiana—For best control of seedling johnsongrass; apply broadcast at 2 lbs. per acre in the fall after planting and before cane emerges; repeat application at same rate in the early spring before weeds emerge. Alternatively, for control of most seedling weeds in fall-planted or stubble cane not treated in the fall with "SINBAR", apply broadcast in the spring at 4 lbs. per acre. Use 1/3 of above broadcast rates when band-treating 1/3 of the area.

As a layby treatment immediately after last cultivation, apply 1/2 lb. per acre in a 30" band as a directed spray to row middles; do not apply over top of cane as injury to the crop may result.

Texas—Apply broadcast at 1 to 2 lbs. per acre in the fall to stubble cane or to plant cane before emergence; repeat application at same rate in early spring. Use 1/3 of above broadcast rates when band-treating 1/3 of the area.

Hawaii—Make a single preemergence broadcast application of 1 to 2 1/2 lbs. per acre on plant or ratoon cane before cane emerges. Longevity of control is enhanced if application is made during the relatively dry season (March through October).

Puerto Rico—Make a single preemergence broadcast application of 1 to 2 1/2 lbs. per acre on plant cane only.

NOTICE TO BUYER—Purchase of this material does not confer any rights under patents of countries outside of the United States.

STORAGE AND DISPOSAL

STORAGE: Store product in original container only, away from other pesticides, fertilizer, food, or feed.

PESTICIDE DISPOSAL: Do not contaminate water, food, or feed by storage or disposal. Wastes resulting from use of this product may be disposed of on site or at an approved waste disposal facility.

CONTAINER DISPOSAL: Completely empty bag into application equipment. Then dispose of empty bag in a sanitary landfill or by incineration, or, if allowed by State and local authorities, by burning. If burned, stay out of smoke.

NOTICE OF WARRANTY

Du Pont warrants that this product conforms to the chemical description on the label thereof and is reasonably fit for the purposes stated on such label only when used in accordance with the directions under normal use conditions. It is impossible to eliminate all risks inherently associated with the use of this product. Crop injury, ineffectiveness, or other unintended consequences may result because of such factors as weather conditions, presence of other materials, or the manner of use or application, all of which are beyond the control of Du Pont. In no case shall Du Pont be liable for consequential, special or indirect damages resulting from the use or handling of this product. All such risks shall be assumed by the buyer. DU PONT MAKES NO WARRANTIES OF MERCHANTABILITY OR FITNESS FOR A PARTICULAR PURPOSE NOR ANY OTHER EXPRESS OR IMPLIED WARRANTY EXCEPT AS STATED ABOVE.

AG-3240 8019 1/3/89

This is a specimen label. DuPont is not responsible for the accuracy of the information contained herein. As labels are subject to revision, always carefully read and follow the label on the product container.

TELAR®

HERBICIDE
DISPERSIBLE GRANULE

ACTIVE INGREDIENT:	BY WEIGHT
Chlorsulfuron 2-Chloro-N-[(4-methoxy-6-methyl-1,3,5-triazin-2-yl)- aminocarbonyl]benzenesulfonamide	75%
INERT INGREDIENTS:	25%
TOTAL	100%

EPA Reg. No. 352-404

U.S. Pat. 4,127,405

KEEP OUT OF REACH OF CHILDREN
CAUTION

PRECAUTIONARY STATEMENTS

HAZARDS TO HUMANS AND DOMESTIC ANIMALS

CAUTION!

MAY IRRITATE EYES, NOSE, THROAT AND SKIN.

Avoid breathing dust or spray mist. Avoid contact with skin, eyes and clothing.

In case of contact with eyes, immediately flush with plenty of water. Get medical attention if irritation persists. Wash thoroughly after handling. Remove and wash contaminated clothing before reuse.

For medical emergencies involving this product, call toll free 1-800-441-3637.

ENVIRONMENTAL HAZARDS

Do not apply directly to any body of water. Do not contaminate water by cleaning of equipment or disposal of wastes.

IMPORTANT

DO NOT USE ON FOOD OR FEED CROPS. Injury to or loss of desirable trees or other plants may result from failure to observe the following: Do not apply "TELAR" (except as recommended), or drain or flush equipment on or near desirable trees or other plants, or on areas where their roots may extend or in locations where the chemical may be washed or moved into contact with their roots.

Do not use on lawns, walks, driveways, tennis courts or similar areas. Keep from contact with fertilizers, insecticides, fungicides and seeds.

Do not treat powdery, dry soil and light, sandy soils when there is little likelihood of rainfall soon after treatment; in areas where sensitive crops are grown, do not treat before soil has been settled by rain as off-target movement may occur. Do not apply when ground is frozen.

Do not apply to any body of water. Do not apply where runoff water may flow onto agricultural land, as injury to crops may result. Do not apply "TELAR" during periods of intense rainfall or to soils saturated with water as off-target movement may occur.

Do not apply this product through any type of irrigation system.

Do not allow spray to drift onto adjacent crops or other desirable plants or trees as injury may occur.

Following a "TELAR" application, do not use sprayer for application to crops. This is extremely important, as low rates of "TELAR" can kill or severely injure most crops.

GENERAL INFORMATION

Du Pont "TELAR" Herbicide is a dispersible granule to be mixed in water and applied as a spray for control of many broadleaf weeds on non-cropland. It is noncorrosive, nonflammable, nonvolatile and does not freeze.

"TELAR" may be applied preemergence or postemergence to weeds; best results are obtained if application is made before or during early stages of weed growth. "TELAR" should be used during seasons when rainfall occurs as moisture is required to move "TELAR" into the root zone of weeds before they develop an established root system. Preemergence treatments control or suppress weeds through root uptake while postemergence works through root uptake and foliar uptake. Under limited rainfall conditions, "TELAR" may not provide satisfactory control of hard-to-kill perennials.

"TELAR" is absorbed by both roots and foliage of weeds, resulting in visual effects progressing from growth inhibition to reddish-purplish coloration, chlorosis, necrosis, vein discoloration and death of terminals. Initial effects are usually seen 2 or 3 weeks following application; however, the final effects are evident at about 4 to 6 weeks after application. Warm, moist conditions following treatment promote the activity of "TELAR" while cold, dry conditions delay the activity of "TELAR." Weeds hardened off by cold weather or drought stress may not be fully controlled.

Degree of control and duration of effect depend on a) rate used, b) weed spectrum, c) weed size, d) growing conditions at and following time of treatment, e) soil pH, f) precipitation, and g) soil organic matter.

DIRECTIONS FOR USE

It is a violation of federal law to use this product in a manner inconsistent with its labeling.

"TELAR" Herbicide should be used only in accordance with recommendations on this label or in separate published Du Pont recommendations available through local dealers.

Du Pont will not be responsible for losses or damages resulting from the use of this product in any manner not specifically recommended by Du Pont. User assumes all risks associated with such nonrecommended use.

Du Pont "TELAR" Herbicide is recommended for preemergence and postemergence control of many broadleaf weeds. "TELAR" may be applied as a selective or non-selective treatment to non-cropland areas such as airports, fencerows, highways, industrial turf, lumberyards, petroleum tank farms, pipeline rights-of-way, plant sites, railroads, roadside turf, storage areas, and utility rights-of-way.

"TELAR" may be applied preemergence or postemergence to the weeds, but for best results, apply postemergence to young, actively growing weeds (less than 2" tall or 2" across) at any time of the year, except when ground is frozen.

Treated soils should be left undisturbed to reduce the potential for "TELAR" movement by soil erosion due to wind or water. Injury to crops may occur when treated soil is blown or moved onto land used to produce crops.

Rate selection is based on weed species, weed size and soil texture. Use the higher rates on established plants and on fine textured soils and lower rates on smaller weeds and coarse textured soils.

SELECTIVE WEED CONTROL

Bahiagrass, bermudagrass, bluegrass, fescue, wheatgrass and smooth brome have demonstrated good tolerance to "TELAR." For selective weeding, "TELAR" may be applied at the rate of 1/4 1 oz per acre per year to bahiagrass, bermudagrass, bluegrass and wheatgrass and at the rate of 1/4 to 1/2 oz per acre per year to fescue and smooth brome. The higher rates and/or the addition of surfactant at 1 qt./100 gals water may result in temporary chlorosis of desirable grasses; application of "TELAR" may also suppress grass growth and inhibit grass seedhead formation. Desirable grasses should be well established at time of application or injury may result.

NONSELECTIVE WEED CONTROL

Apply "TELAR" at the rates of 1/4 to 3 oz per acre per year to control many annual and perennial broadleaf weeds in non-cropland areas. The addition of a surfactant at 1 qt./100 gals water for postemergence applications is recommended to enhance activity.

Since the degree of control and duration of effect will vary with the amount of chemical applied, soil texture, soil pH, soil organic matter, weed size, rainfall and other conditions, it is suggested that users limit their first use to small areas.

WEEDS CONTROLLED AND USE RATES

	1/4 to 1/2 ozs per acre	
Annual sowthistle	Field pennycress	Redroot pigweed
Blue mustard	Henbit	Smooth pigweed
Common chickweed	London rocket	Treacle mustard
Common speedwell	Prostrate pigweed	Wild mustard
Fiddleneck (tarweed)		
	1/2 to 1 ozs per acre	
Bouncingbet	Common speedwell*	Sweetclover*
Bur beakchervil	Dandelion*	Tumble mustard
Buttercup	Goldenrod	Turkey mullein*
Canada Thistle*†	Groundsel	Whitetop (hoary cress)†
Common lambsquarters	Musk thistle	Wild parsnip
Common sunflower	Russian knapweed†	
	1 to 3 ozs per acre	
Annual ryegrass (Lolium spp)*	Common yarrow	Prostrate knotweed*
Aster	Corn spurry	Puncturevine*
Bedstraw	Cow cockle	Red clover
Black mustard	Curly dock	Russian thistle
Burclover	Dyer's woad	Scouringrush (*Equisetum* spp)*
Canada thistle†	Flixweed	Tansymustard
Common cinquefoil	Foxtail (*Setaria* spp)*	White clover
Common mallow	Horsetail (*Equisetum* spp)*	Wild carrot
Common mullen*	Kochia	Wild garlic/wild onion
Common ragweed*	Prickly lettuce	Yellow starthistle*

* Partial control only

† Prebloom to bloom and fall rosette are recommended timings. Surfactant is required at 1 to 2 qts/100 gallons water.

HOW TO USE

SPRAY PREPARATION: Fill the spray tank half full of water and with agitator running, add the proper amount of "TELAR." Finish adding the required amount of water. Continuous agitation is required to keep the product in suspension. For postemergence applications, a nonionic surfactant of at least 80% active ingredient may be added to improve wetting and/or contact activity. To minimize drift, a drift control agent may be added at the recommended label rate. "TELAR" spray preparations may degrade in acid solutions if not used in 24 hours; they are stable in alkaline solutions. Thoroughly reagitate before using.

APPLICATION TECHNIQUE: Before spraying, calibrate the equipment to determine the quantity of water necessary to uniformly cover the vegetation and soil in a measured area to be treated. Use a fixed boom, off center nozzle or boomless straight stream nozzles properly calibrated to a constant speed of travel and rate of delivery. Avoid overlapping and shut off spray booms while starting, turning, slowing or stopping when spraying industrial turf as injury to the desired species may result. Extreme care must be taken to prevent drift.

SPRAY VOLUME: Apply "TELAR" in ground equipment. Select a spray volume that will assure thorough coverage and uniform spray pattern. Do not use less than 10 gallons of water per acre. Spray volumes of 20 to 40 gallons per acre are preferred. Spray pressure of 25 to 35 psi are adequate.

AGITATION: Use mechanical or bypass agitation to thoroughly mix the spray solution. After initial mixing, do not use excessive agitation when using a surfactant as foaming problems may result. To avoid foaming problems, use antifoam agents.

SPRAYER CLEANUP: Immediately after spraying thoroughly remove all traces of "TELAR" from mixing and spray equipment as follows:

1. Drain tank, then flush tank, boom and hoses with clean water for a minimum of 10 minutes.
2. Fill the tank with clean water, then add 1/2 gallon chlorine bleach (containing 5 1/4% sodium hypochlorite) per 100 gallons of water. Turn on sprayer long enough to flush through boom and hoses. Stop spraying but keep agitator working in the tank for 15 minutes, then drain.
3. Repeat Step 2.
4. Nozzles and screeens should be removed and cleaned separately. To remove traces of chlorine bleach, rinse the tank thoroughly with clean water and flush through hoses and boom.

NOTE: This sprayer cleanup procedure is only effective for "TELAR" and for general uses specified under "Directions for Use"; sprayer should not be used on plants such as fine turf and ornamentals.

NOTE TO BUYER: Purchase of this material does not confer any rights under patents of countries outside of the United States.

STORAGE AND DISPOSAL

STORAGE: Store product in original container only, away from other pesticides, fertilizer, food or feed.

PESTICIDE DISPOSAL: Do not contaminate water, food or feed by storage or disposal. Wastes resulting from the use of this product may be disposed of on site or at an approved waste disposal facility.

CONTAINER DISPOSAL: Triple rinse (or equivalent) the container. Then offer for recycling or reconditioning, or puncture and dispose of in a sanitary landfill, or incineration, or, if allowed by state and local authorities, by burning. If burned, stay out of smoke.

NOTICE OF WARRANTY: Du Pont warrants that this product conforms to the chemical description on the label thereof and is reasonably fit for purposes stated on such label only when used in accordance with directions under normal use conditions. It is impossible to eliminate all risks inherently associated with the use of this product. Injury to adjacent crops, ineffectiveness or other unintended consequences may result because of such factors as weather conditions, presence of other materials, or the manner of use or application, all of which are beyond the control of Du Pont. In no case shall Du Pont be liable for consequential, special or indirect damages resulting from the use or handling of this product. All such risks shall be assumed by the buyer. DU PONT MAKES NO WARRANTIES OF MERCHANTABILITY OR FITNESS FOR A PARTICULAR PURPOSE NOR ANY OTHER EXPRESS OR IMPLIED WARRANTY EXCEPT AS STATED ABOVE.

AG-3197 8078 2/10/88

This is a specimen label. DuPont is not responsible for the accuracy of the information contained herein. As labels are subject to revision, always carefully read and follow the label on the product container.

TERSAN® 1991 DF

TURF FUNGICIDE
DISPERSIBLE GRANULES

ACTIVE INGREDIENT:

Benomyl [Methyl 1-butylcarbamoyl)-2-benzimidazolecarbamete]	50%
INERT INGREDIENTS:	50%
TOTAL	100%

U.S. Pats. 3,541,213 & 3,631,176 EPA Reg. No. 352-507

KEEP OUT OF REACH OF CHILDREN

PRECAUTIONARY STATEMENTS

HAZARDS TO HUMANS AND DOMESTIC ANIMALS

CAUTION!

MAY IRRITATE EYES, NOSE, THROAT AND SKIN.

Avoid breathing dust or spray mist. Avoid contact with skin, eyes and clothing. Wash thoroughly after using.

Do not apply this product in such a manner as to directly or through drift, expose workers or other persons. The area being treated must be vacated by unprotected persons.

Protective clothing must be worn during periods of exposure. See Directions for Use.

This product may cause a temporary allergic skin reaction in a few susceptible persons. This condition should be treated as an allergic dermatitis. There is no evidence of aftereffects or permanent injury.

First Aid: In case of contact, flush skin or eyes with plenty of water; for eyes, get medical attention.

For medical emergencies involving this product, call toll free 1-800-441-3637.

ENVIRONMENTAL HAZARDS

This pesticide is toxic to fish. Do not apply directly to water. Do not apply where runoff is likely to occur. Drift and runoff from treated areas may be hazardous to fish in adjacent areas. Do not contaminate water by cleaning of equipment or disposal of wastes. Do not apply when weather conditions favor drift from areas treated.

PHYSICAL OR CHEMICAL HAZARDS

Keep away from fire or sparks.

DIRECTIONS FOR USE

It is a violation of federal law to use this product in a manner inconsistent with its labeling.

Du Pont "TERSAN" 1991 DF should be used only in accordance with recommendations on this label, or in separate published Du Pont recommendations available through local dealers.

Du Pont will not be responsible for losses or damages resulting from use of this product in any manner not specifically recommended by Du Pont. User assumes all risks associated with such nonrecommended use.

Do not enter treated areas without protective clothing until sprays have dried. Because certain states may require more restrictive reentry intervals for various crops treated with this product, consult your State Department of Agriculture for further information.

Written or oral warnings must be given to workers who are expected to be in a treated area or in an area about to be treated with this product. When oral warnings are given, warnings shall be given in a language customarily understood by workers. Oral warnings must be given if there is reason to believe that written warnings cannot be understood by workers. Warnings must include the following information: "CAUTION. Area treated with 'TERSAN' 1991 DF on (date of application). Do not enter without appropriate protective clothing until sprays have dried. In case of contact, flush skin or eyes with plenty of water; for eyes, get medical attention."

"TERSAN" 1991 DF Turf Fungicide is recommended for use as a spray for the control of dollar spot, large brown patch, Fusarium patch (pink snow mold), Fusarium blight and stripe smut on golf courses, parks, athletic fields, and commercial turf areas. It is also recommended for the control of certain diseases of roses, flowers, ornamentals, shade trees, and bulbs.

If treatment is not effective following use of "TERSAN" 1991 DF as recommended, a resistant strain of the fungus may be present. If treatment is ineffective due to the presence of a benomyl resistant strain, then neither "TERSAN" 1991 DF nor any other benzimidazole or thiophanate type fungicide will effectively control that disease; consideration should be given to prompt use of other types of suitable fungicides.

The repeated exclusive use of "TERSAN" 1991 DF may lead to buildup of resistant strains of fungi and loss of disease control. A spray program using other fungicides may delay resistant strain buildup. Consult your state extension specialist or official state recommendations for guidance on your particular crop and disease control situation.

NOTE: Do not tank mix or alternate "TERSAN" 1991 DF with benzimidazole or thiophanate products such as "Mertect"[1], "Topsin"[2], "Fungo"[3], or "3336".

WORK SAFETY RULES

Keep all unprotected persons, children, livestock and pets away from treated area or where there is danger of drift.

During mixing and loading of the concentrated product wear the following protective clothing and equipment:

Long sleeve shirt and long legged pants; chemical resistant gloves; chemical resistant apron; chemical resistant shoes, shoe coverings, or boots.

Wear the following protective clothing during application, equipment repair and equipment cleaning:

Long sleeve shirt and long legged pants; chemical resistant gloves; chemical resistant shoes, shoe coverings, or boots.

Important! Before removing gloves, wash them with soap and water. Always wash hands, face and arms with soap and water before smoking, eating, drinking, or toileting.

After work, take off all clothes and shoes. Shower using soap and water. Wear only clean clothes. Do not use contaminated clothing. Wash protective clothing and protective equipment with soap/detergent and water after each use. Personal and protective clothing worn during use must be laundered separately from household articles. Clothing or protective equipment heavily contaminated or drenched with benomyl must be destroyed according to state and local regulations. Heavily contaminated or drenched clothing cannot be adequately decontaminated.

CHEMIGATION: Apply "TERSAN" 1991 DF only through sprinkler including center pivot, lateral move, end tow, side (wheel) roll, traveler, big gun, solid set or hand move irrigation systems only on crops specifically designated in the "Directions For Use" for that particular crop. Do not apply "TERSAN" 1991 DF through any other type of irrigation system or on any other crops.

Crop injury, lack of effectiveness, or illegal pesticide residues in the crop can result from nonuniform distribution of treated water.

If you have questions about calibration, you should contact State Extension Service Specialists, equipment manufacturers or other experts.

Do not connect an irrigation system (including greenhouse systems) used for pesticide application to a public water system unless the pesticide label-prescribed safety devices for public water systems are in place.

A person knowledgeable of the chemigation systems and responsible for its operation, or under the supervision of the responsible person, shall shut the system down and make necessary adjustments should the need arise.

Specific Instructions for Public Water Systems:

1. Public water system means a system for the provision to the public of piped water for human consumption if such system has at least 15 service connections or regularly serves an average of at least 25 individuals daily at least 60 days out of the year.
2. Chemigation systems connected to public water systems must contain a functional, reduced-pressure zone, backflow preventer (RPZ) or the functional equivalent in the water supply line upstream from the point of pesticide introduction. As an option to the RPZ, the water from the public water system should be discharged into a reservoir tank prior to pesticide introduction. There shall be a complete physical break (air gap) between the outlet end of the fill pipe and the top or overflow rim of the reservoir tank of at least twice the inside diameter of the fill pipe.
3. The pesticide injection pipeline must contain a functional, automatic, quick-closing check valve to prevent the flow of the fluid back toward the injection pump.
4. The pesticide injection pipeline must contain a functional, normally closed, solenoid-operated valve located on the intake side of the injection pump and connected to the system interlock to prevent fluid from being withdrawn from the supply tank when the irrigation system is either automatically or manually shut down.
5. The system must contain functional interlocking controls to automatically shut off the pesticide injection pump when the water pump motor stops, or in cases where there is no water pump, when the water pressure decreases to the point where pesticide distribution is adversely affected.
6. Systems must use a metering pump, such as a positive displacement injection pump (e.g., diaphragm pump) effectively designed and constructed of materials that are compatible with pesticides and capable of being fitted with a system interlock.
7. Do not apply when wind speed favors drift beyond the area intended for treatment.

Specific Instructions for Sprinkler Irrigation Systems:

1. The system must contain a functional check valve, vacuum relief valve, and low pressure drain appropriately located on the irrigation pipeline to prevent water source contamination from backflow.
2. The pesticide injection pipeline must contain a functional, automatic, quick-closing check valve to prevent the flow of fluid back toward the injection pump.
3. The pesticide injection pipeline must also contain a functional, normally closed, solenoid-operated valve located on the intake side of the

injection pump and connected to the system interlock to prevent fluid from being withdrawn from the supply tank when the irrigation system is either automatically or manually shut down.

4. The system must contain functional interlocking controls to automatically shut off the pesticide injection pump when the water pump motor stops.
5. The irrigation line or water pump must include a functional pressure switch which will stop the water pump motor when the water pressure decreases to the point where pesticide distribution is adversely affected.
6. Systems must use a metering pump, such as a positive displacement injection pump (e.g., diaphragm pump) effectively designed and constructed of materials that are compatible with pesticides and capable of being fitted with a system interlock.
7. Do not apply when wind speed favors drift beyond the area intended for treatment.
8. Good agitation is required in the injection tank.
9. In moving systems, apply specified dosage of "TERSAN" 1991 DF as a continuous injection. In nonmoving systems inject "TERSAN" 1991 DF for 15 to 30 minutes at end of cycle. Use the least amount of water possible consistent with uniform coverage.
10. Mix the amount of "TERSAN" 1991 DF needed for acreage to be treated into the quantity of water determined during prior calibration. For moving systems inject into system continuously for one complete revolution of the field. For nonmoving systems inject into system for the time established during calibration.
11. Stop injection equipment after treatment is completed and continue to operate irrigation equipment until all "TERSAN" 1991 DF is flushed from system.

TURF—NOTE: The repeated and exclusive use of "TERSAN" 1991 DF has led to the development of resistant strains of dollar spot in some locations and may lead to the development of resistance in other strains of fungi. The use of "TERSAN" 1991 DF as a tank mixture with "Daconil"[4] 2787 Fungicide, "Daconil"[4] 2787 Flowable Fungicide, or Du Pont TERSAN® LSR Turf Fungicide may delay the development of resistant strains of fungi. Apply recommended amounts of "TERSAN" 1991 DF plus either "Daconil" 2787, or "TERSAN" LSR in sufficient water to obtain thorough coverage, usually 3 to 5 gals per 1,000 sq ft. When treating golf greens, always treat aprons and approaches.

"TERSAN" 1991 DF plus "Daconil" 2787: Dollar Spot and Brown Patch—Apply 1 oz "TERSAN" 1991 DF plus 3 fl oz "Daconil" 2787 Flowable (or 2 oz "Daconil" 2787—75% wettable powder) per 1,000 sq ft. Make the initial application when conditions favor disease development and repeat the treatment on 10- to 14-day intervals as needed.

"TERSAN" 1991 DF plus "Daconil" 2787: Helminthosporium leafspot and Brown Patch (and Dollar Spot, if present)—Apply 1 oz "TERSAN" 1991 DF plus 3 to 6 fl oz "Daconil" 2787 Flowable (or 2 to 4 oz "Daconil" 2787—75% wettable powder) per 1,000 sq ft. Make the initial application when conditions favor disease development and repeat the treatment on 7- to 10-day intervals as needed. Use the higher rate of "Daconil" 2787 in the combination if disease conditions are particularly severe.

"TERSAN" 1991 DF plus "Tersan" LSR: Helminthosporium leafspot and dollar spot and/or brown patch; dollar spot and brown patch; brown patch—Apply 1 oz "TERSAN" 1991 DF plus 3-4 ozs "Tersan" LSR for each 1,000 sq ft of turfgrass area to be treated. Make initial application when disease first appears and continue applications on 7-to 14-day intervals for as long as needed. Use the shorter intervals and higher use rate when conditions are unusually favorable for disease development.

"TERSAN" 1991 DF in other spray programs—Apply recommended amounts of "TERSAN" 1991 DF in sufficient water to obtain thorough coverage, usually 2 to 5 gals per 1,000 sq ft. When treating golf greens, always treat aprons and approaches.

Dollar Spot (Sclerotinia homoeocarpa)—Apply at rate of 1 oz per 1,000 sq ft at earliest appearance of disease and continue at 10- to 14-day intervals as long as needed.

Large Brown Patch (Rhizoctonia solani)—Apply at rate of 2 oz per 1,000 sq ft at earliest appearance of disease and continue at 10- to 14-day intervals as long as needed. When conditions are unusually favorable for development of disease, reduce interval to 5 to 7 days.

Fusarium Patch (Fusarium nivale)—A cool season disease—Apply at rate of 2 oz per 1,000 sq ft at earliest appearance of disease and continue at 10- to 14-day intervals as long as needed.

Fusarium Blight (Fusarium roseum and F. tricinctum)—A warm season disease—Apply at rate of 5 to 8 oz per 1000 sq ft at earliest appearance of disease and repeat 10 to 14 days later. Immediately after application, water sufficiently to thoroughly wet soil to a depth of 1 inch below any mat or thatch present.

Stripe Smut (Ustilago striformis)—Make a single application at rate of 6 oz per 1,000 sq ft in October or in early spring before grass begins growth. After application, water turf sufficiently to carry product to base of plants.

NOTE: Where Helminthosporium diseases are a problem, apply "Tersan" LSR Turf Fungicide two or three times in the early spring at rate of 3 to 4 oz per 1,000 sq ft at 7- to 10-day intervals.

Do not graze or feed clippings from treated areas to livestock.

ROSES, FLOWERS, ORNAMENTALS, SHADE TREES—Field and Greenhouse: Foliar Spray—Begin applications when disease first appears and repeat at 10- to 14-day intervals throughout the growing season; shorten interval during humid, rainy weather. Use at the following rates: **½ lb per 100 gals** (1 tablespoonful per 2 gals)—for Powdery Mildew, Botrytis Gray Mold. **1 lb per 100 gals**—for Anthracnose (for shade trees and woody ornamentals, begin at bud break and make 2 or 3 additional applications at 10- to 14-day intervals); Black Spot of roses; Cercospora, Entomosporium, Ramularia, and Septoria Leafspots; Ascochyta and Phomopsis blights; Didymellina Leafspot of iris; Cornynespora Leafspot of Ligustrum; Ovulinia Blight of azalea and rhododendron (begin as flowers open); Scab of pyracantha and flowering crab. Addition of a surfactant to the spray mixture improves distribution of the spray on hard-to-wet plants such as roses. For aerial application, use ½ to 1 lb per acre.

Drench Treatment—Botrytis, Fusarium, Rhizoctonia and Sclerotinia stem, crown and root rots on herbaceous annuals, perennials and bedding plants; Cylindrocladium and Thielaviopsis rots on woody ornamentals such as azaleas, rhododendrons, conifers, and poinsettias—Use 1 lb per 100 gals; apply as a drench or heavy spray (1 to 2 pts per sq ft) after transplanting into propagation beds or containers. Repeat at 2- to 4-week intervals during periods favorable for disease. "TERSAN" 1991 does not control *Pythium spp.* or *Phytophthora spp.*

Preplant Dip Treatment—For diseases listed under Drench Treatment, use 1 lb per 100 gals of water; immerse plants or cuttings for 10 to 15 min; remove and allow to drain.

BULBS (Easter Lily, Tulip, Gladiolus, Daffodil, Iris): Fusarium and Penicillium Rots—Use 1⅔ lbs per 100 gals of water (2 tablespoons per gal). Soak cleaned bulbs for 15 to 30 minutes in warm dip (80 to 85° F.), preferably within 48 hours after digging. Dry bulbs after treatment. If bulbs are for forcing, treat after bulbs have been heat-cured.

STORAGE AND DISPOSAL

STORAGE: Never allow "TERSAN" 1991 DF to become wet during storage. This may lead to certain chemical changes which will reduce the effectiveness of "TERSAN" 1991 DF as a fungicide. Keep container tightly closed when not in use. Store product in original container only, away from other pesticides, fertilizer, food or feed.

PESTICIDE DISPOSAL: Do not contaminate water, food, or feed by storage or disposal. Wastes resulting from the use of this product may be disposed of on site or at an approved waste disposal facility.

CONTAINER DISPOSAL: Completely empty bag into application equipment. Then dispose of empty bag in a sanitary landfill or by incineration, or, if allowed by state and local authorities, by burning. If burned, stay out of smoke.

NOTICE TO BUYER: Purchase of this material does not confer any rights under patents of countries outside of the United States.

NOTICE OF WARRANTY

Du Pont warrants that this product conforms to the chemical description on the label thereof and is reasonably fit for purposes stated on such label only when used in accordance with the directions under normal use conditions. It is impossible to eliminate all risks inherently associated with the use of this product. Ineffectiveness or other unintended consequences may result because of such factors as weather conditions, presence of other materials, or the manner of use or application, all of which are beyond the control of Du Pont. In no case shall Du Pont be liable for consequential, special or indirect damages resulting from the use or handling of this product. All such risks shall be assumed by the Buyer. DU PONT MAKES NO WARRANTIES OF MERCHANTABILITY OR FITNESS FOR A PARTICULAR PURPOSE NOR ANY OTHER EXPRESS OR IMPLIED WARRANTY EXCEPT AS STATED ABOVE.

[1] Registered trademark of Merck & Co.
[2] Registered trademark of Nippon Soda Co., Japan
[3] Registered trademark of Mallincrodt Inc.
[4] Registered trademark of Diamond Shamrock Co.

AG-3728 8058 4/28/88

This is a specimen label. DuPont is not responsible for the accuracy of the information contained herein. As labels are subject to revision, always carefully read and follow the label on the product container.

TUPERSAN®

HERBICIDE
WETTABLE POWDER

FOR CONTROL OF CERTAIN ANNUAL WEED GRASSES IN TURF

ACTIVE INGREDIENT:	
Siduron[(2-methylcyclohexyl)-3-phenylurea]	50%
INERT INGREDIENTS:	50%
TOTAL	100%

EPA REG. NO. 352-307

U.S. Patent 3,309,192

KEEP OUT OF REACH OF CHILDREN
CAUTION

PRECAUTIONARY STATEMENTS

HAZARDS TO HUMANS AND DOMESTIC ANIMALS

CAUTION!

MAY IRRITATE EYES, NOSE, THROAT, AND SKIN.

Avoid breathing dust or spray mist. Avoid contact with skin, eyes, and clothing. Do not apply this product in such a manner as to directly or through drift expose workers or other persons. The area being treated must be vacated by unprotected persons.

For medical emergencies involving this product, call toll free 1-800-441-3637.

ENVIRONMENTAL HAZARDS

This product is toxic to fish. Do not apply directly to water or wetlands. Do not apply where runoff is likely to occur. Do not apply when weather conditions favor drift from areas treated.

GENERAL INFORMATION

Du Pont "TUPERSAN" Herbicide is a wettable powder to be mixed in water and applied as a spray for selective pre-emergence control of certain annual weed grasses in turf areas such as golf course fairways, lawns, parks, roadsides, and turf grown for grass seed or sod production. It is highly effective for the control of crabgrass (smooth and hairy), and may be used on newly seeded areas as well as on established turf. Because of its high degree of selectivity, "TUPERSAN" may be applied at time of seeding without causing injury to germinating seeds of many common temperate region grasses.

"TUPERSAN" is non-corrosive to equipment, non-volatile, and non-flammable.

DIRECTIONS FOR USE

It is a violation of federal law to use this product in a manner inconsistent with its labeling.

"TUPERSAN" should be used only in accordance with recommendations on this label, or in separate published Du Pont recommendations available through local dealers.

Du Pont will not be responsible for losses or damages resulting from use of this product in any manner not specifically recommended by Du Pont. User assumes all risks associated with such non-recommended use.

Do not enter treated areas without protective clothing until sprays have dried. Because certain states may require more restrictive reentry intervals for various crops treated with this product, consult your State Department of Agriculture for further information.

Written or oral warnings must be given to workers who are expected to be in a treated area or in an area about to be treated with this product. When oral warnings are given, warnings shall be given in a language customarily understood by workers. Oral warnings must be given if there is reason to believe that written warnings cannot be understood by workers. Warnings must include the following information: "CAUTION. Area treated with 'TUPERSAN' on (date of application). Do not enter without appropriate protective clothing until sprays have dried. In case of contact, flush skin or eyes with plenty of water, for eyes, get medical attention."

"TUPERSAN" is recommended for pre-emergence control of annual weed grasses, such as crabgrass (smooth and hairy), foxtail, and barnyardgrass, in newly seeded or established plantings of the following grasses only: bluegrass, fescue, redtop, smooth brome, perennial ryegrass, orchardgrass, zoysia, and the following strains of bentgrass: Penncross, Seaside, Highland, Astoria, Nimisila, C-1, C-7 and C-19. Do not use on other bentgrass strains nor on Bermuda grass, as injury to turf may result. The effect of "TUPERSAN" on golf course greens may vary with turf management practices and environmental conditions; therefore, its use on golf greens is not recommended, as injury to turf may result. Treatment will not control annual bluegrass (Poa annua), clover, or most broadleaf weeds.

"TUPERSAN" may be applied at any stage of turf development, including new seedings. At least 1/2 inch of water must be provided by rainfall or irrigation within 3 days after treatment in order to carry the chemical into the root zone of germinating weeds.

AMOUNT TO USE—WHEN TO APPLY

Apply at the following rates, using sufficient water, at least 100 gals. per acre (2 1/2 gals. per 1000 sq. ft.), to provide thorough and uniform coverage. Use the lower rates on light, sandy soils and the higher rates on soils high in clay and/or organic matter. At least 1/2 inch of water must be provided by rainfall or irrigation within 3 days after treatment.

New Spring Plantings Apply as final operation following seeding. Use 4 to 12 lbs. "TUPERSAN" per acre (1 1/2 to 4 1/2 ozs. per 1000 sq. ft.). When no more than 6 lbs. are used, re-treat about one month later with an additional 4 to 6 lbs. per acre.

New Fall Plantings and Established Turf Apply in the spring just before expected emergence of annual weed grasses. Use 16 to 24 lbs. "TUPERSAN" per acre (6 to 9 ozs. per 1000 sq. ft.).

EQUIPMENT

Except for small areas, use a fixed-boom power sprayer properly calibrated to a constant speed and rate of delivery. Make certain spray equipment is clean (free of scale, rust, dirt, oil, and pesticide deposits). Openings in screens should be 50 mesh or coarser. Material must be kept in suspension at all times by continuous agitation; agitate by mechanical or hydraulic means in the spray tank. If by-pass or return line is used, it should terminate at bottom of tank to minimize foaming. Spray booms must be shut off while starting, turning, slowing or stopping, or injury to the turf may result. For small areas, tank-type hand sprayer may be used; shake or stir frequently.

Do not apply this product through any type of irrigation system.

NOTE: Use only on turf areas; do not apply to crops used for food or feed purposes.

NOTICE TO BUYER—Purchase of this material does not confer any rights under patents of countries outside of the United States.

STORAGE AND DISPOSAL

STORAGE: Store product in original container only, away from other pesticides, fertilizer, food, or feed.

PESTICIDE DISPOSAL: Do not contaminate water, food, or feed by storage or disposal. Wastes resulting from the use of this product may be disposed of on site or at an approved waste disposal facility.

CONTAINER DISPOSAL: Completely empty bag into application equipment. Then dispose of empty bag in a sanitary landfill or by incineration, or if allowed by State and local authorities, by burning. If burned, stay out of smoke.

NOTICE OF WARRANTY

Du Pont warrants that this product conforms to the chemical description on the label thereof and is reasonably fit for purposes stated on such label only when used in accordance with the directions under normal use conditions. It is impossible to eliminate all risks inherently associated with the use of this product. Ineffectiveness or other unintended consequences may result because of such factors as weather conditions, presence of other materials, or the manner of use or application, all of which are beyond the control of Du Pont. In no case shall Du Pont be liable for consequential, special or indirect damages resulting from the use or handling of this product. All such risks shall be assumed by the Buyer. DU PONT MAKES NO WARRANTIES OF MERCHANTABILITY OR FITNESS FOR A PARTICULAR PURPOSE NOR ANY OTHER EXPRESS OR IMPLIED WARRANTY EXCEPT AS STATED ABOVE.

AG-3165 8038 8/18/80

This is a specimen label. DuPont is not responsible for the accuracy of the information contained herein. As labels are subject to revision, always carefully read and follow the label on the product container.

VELPAR®

HERBICIDE
SOLUBLE POWDER

ACTIVE INGREDIENT:	BY WEIGHT
Hexazinone [3-cyclohexyl-6-(dimethylamino)-1-methyl-1,3,5-triazine-2,4(1H,3H)-dione]	90%
INERT INGREDIENTS	10%
TOTAL	100%

U.S. Pats. 3,902,887 & 3,983,116

EPA Reg. No. 352-378

KEEP OUT OF REACH OF CHILDREN

PRECAUTIONARY STATEMENTS

HAZARDS TO HUMANS AND DOMESTIC ANIMALS

WARNING!

MAY IRRITATE EYES, NOSE, THROAT AND SKIN.

AVISO: PRECAUCION AL USUARIO: Si usted no lee ingles, no use este producto hasta que la etiqueta haya sido explicada ampliamente.

Avoid breathing dust or spray mist. Avoid contact with skin, eyes and clothing.

Do not apply this product in such a manner as to directly or through drift expose workers or other persons. The area being treated must be vacated by unprotected persons.

In case of contact, flush skin or eyes with plenty of water; for eyes, get medical attention.

For medical emergencies involving this product, call toll free 1-800-441-3637.

ENVIRONMENTAL HAZARDS

Do not apply directly to water or wetlands. Do not apply where runoff is likely to occur. Do not contaminate water by cleaning of equipment or disposal of wastes.

IMPORTANT

Wear goggles when use or handling conditions favor eye exposure or if eye discomfort is encountered.

Injury to or loss of desirable trees or other plants may result from failure to observe the following: Do not apply or drain or flush equipment on or near desirable trees or other plants, or on areas where their roots may extend, or in locations where the chemical may be washed or moved into contact with their roots. Do not use on lawns, walks, driveways, tennis courts, or similar areas. Prevent drift of dry powder or spray to desirable plants. Keep from contact with fertilizers, insecticides, fungicides and seeds.

Thoroughly clean all traces of "VELPAR" from application equipment immediately after use. Flush tank, pump, hoses and boom with several changes of water after removing nozzle tips and screens (clean these parts separately).

GENERAL INFORMATION

Du Pont "VELPAR" Herbicide is a water-soluble powder to be mixed in water and applied as a spray for selective weed control in certain crops; Christmas trees, for selective weed and brush control in reforestation areas; and for non-selective weed and brush control on noncropland areas. It is non-volatile, non-flammable and non-corrosive to equipment.

"VELPAR" is an effective general herbicide providing both contact and residual control of many annual and biennial weeds, woody plants and, except for Johnsongrass, is effective for control of most perennial weeds. It is absorbed through the roots and foliage.

Moisture is required to activate "VELPAR" in the soil. Best results are obtained when weeds are less than 2 inches in height or diameter, soil is moist at time of application and 1/4 to 1/2 inch of rainfall occurs within 2 weeks after application. Foliar activity on weeds is most effective under conditions of high temperature (above 80°F), high humidity and good soil moisture. Foliar activity when vegetation is dormant, semi-dormant, or under stress may be reduced.

On herbaceous plants, symptoms usually appear within 2 weeks after application under warm, humid conditions, while 4 to 6 weeks may be required when weather is cool. If rainfall after application is inadequate to activate "VELPAR" in the soil, plants may recover from contact effects and continue to grow.

On woody plants, symptoms usually appear within 3 to 6 weeks after sufficient rainfall has carried the herbicide into the root zone during periods of active growth. Defoliation and subsequent refoliation may occur, but susceptible plants are killed.

The degree of control and duration of effect will vary with the amount of chemical applied, rainfall, temperature, weed species, soil texture and other conditions.

DIRECTIONS FOR USE

It is a violation of federal law to use this product in a manner inconsistent with its labeling.

Du Pont "VELPAR" Herbicide should be used only in accordance with recommendations on this label, or in separate published Du Pont recommendations available through local dealers.

Du Pont will not be responsible for losses or damages resulting from use of this product in any manner not specifically recommended by Du Pont. User assumes all risk associated with such non-recommended use.

Do not enter treated areas without protective clothing until sprays have dried. Because certain states may require more restrictive reentry intervals for various crops treated with this product, consult your State Department of Agriculture for further information.

Written or oral warnings must be given to workers who are expected to be in a treated area or in an area about to be treated with this product. When oral warnings are given, warnings shall be given in a language customarily understood by workers. Oral warnings must be given if there is reason to believe that written warnings cannot be understood by workers. Warnings must include the following information: "WARNING. Area treated with "VELPAR" on (date of application). Do not enter without appropriate protective clothing until sprays have dried. In case of contact, flush skin or eyes with plenty of water; for eyes, get medical attention."

Do not apply this product through any type of irrigation system.

APPLICATION:

"VELPAR" may be applied by ground equipment, and where permitted, aerial equipment. Use rates, minimum spray gallonage and other application information are described in the various use categories.

For best results, apply "VELPAR" just before or soon after weed emergence. Do not apply to frozen or snow covered soil. Avoid overlapping and shut off spray booms while starting, turning, slowing or stopping, as injury to the crops may result.

Always calibrate equipment prior to use.

RATE SELECTION:

Where a rate range is shown, use the higher levels of the dosage range under the following conditions; on hard-to-control species; on fine-textured soils; on soils containing more than 5% organic matter or carbon. Use the lower levels of the dosage range on coarse-textured soils; on soils low in organic matter. Refer to specific use categories for use rate ranges.

SURFACTANTS:

When emerged seedling weeds are present, the addition of a surfactant improves contact action by increased wetting and absorption by foliage. For forestry and noncrop sites, use a non-ionic surfactant of at least 80% active ingredient (1 qt/100 gals spray). Similarly, use an agricultural surfactant approved for use on growing crops as X-77 with "VELPAR" for treatment of emerged weeds in dormant alfalfa, pineapple and sugarcane. Use of "VELPAR" plus surfactant on growing crops (including trees) may cause injury to the crops.

MIXING:

Before spraying, calibrate equipment to determine the quantity of water necessary to uniformly and thoroughly cover the vegetation and soil in a measured area to be treated. Make sure the volume of water is sufficient to completely dissolve the "VELPAR" using the Water Temperature/Solubility Table as a guide. Use of concentrations greater than shown in the Water Temperature/Solubility Table may result in nozzle plugging and uneven distribution.

Water Temperature/Solubility Table

Water Temperature Degrees F.	Solubility Max. Lbs. "VELPAR" per 100 Gals.
75	23
70	20
65	18
60	16
55	14
50	13
40	12

Add the proper amount of "VELPAR" into a spray tank filled with the amount of water to be used and dissolve using sufficient agitation to prevent the powder from setting. Continue agitation for approximately 10 minutes or until the powder is dissolved. For application with a handgun sprayer, use sufficient water (50 to 200 gallons per acre) for thorough coverage. Where applicable, add surfactant and/or anti-foam agent as last ingredient in the tank.

ALFALFA:

Du Pont "VELPAR" Herbicide is recommended for control of certain weeds in established hay alfalfa except in the states of Montana, North Dakota, South Dakota and Wyoming. It is also recommended for control of certain weeds in established seed alfalfa in the state of California.

Best results are obtained when 1/2 to 1" of rainfall or sprinkler irrigation occurs within two weeks after application, when soil is moist at time of application, and when weeds have not germinated or are less than 2" tall or across.

Since the effect of "VELPAR" on alfalfa varies with soil conditions, uniformity of application and environmental conditions, it is suggested that growers limit their first use to small areas.

Treat only stands of alfalfa established for one year or for one growing season, provided the alfalfa root system is **well established** and the crop is healthy, vigorous and not under stress from weather conditions, insect or disease damage. Do not use on alfalfa-grass mixtures or on other mixed stands.

At elevations above 3,000 ft. do not use "VELPAR" on alfalfa that has been established with a cover or nurse crop until the alfalfa has gone through two summer growing seasons.

NON-DORMANT AND SEMI-DORMANT VARIETIES:
Make a single application of "VELPAR" during winter months when alfalfa plants are in the least active stage of growth; see "Use Rates" for dosage.

In states of Connecticut, Delaware, Illinois, Indiana, Iowa, Kentucky, Maine, Maryland, Massachusetts, Michigan, Minnesota, Missouri, New Hampshire, New Jersey, New York, Ohio, Pennsylvania, Rhode Island, Vermont, Virginia, West Virginia and Wisconsin, make a single application of ½ to 1½ pounds per acre in the spring before new growth exceeds 2 inches in height or to stubble after cutting following hay removal and before regrowth exceeds 2 inches in height. Severe alfalfa injury may result following an after-cutting application if the regrowth is more than 2 inches high, or significant stubble is left after alfalfa cutting, and/or the air temperature is above 90°F.

DORMANT VARIETIES:
Make a single application of "VELPAR" after it becomes dormant and before new growth begins in the spring; see "Use Rates" for dosage.

Do not exceed ¾ pounds per acre on alfalfa less than one year old.

SEED ALFALFA (California Only):
Do not use on fields with sandy loam or loamy sand soils having less than 1% organic matter.

Do not exceed ½ pound per acre on fields with sandy loam or loamy sand soils having between 1 and 2% organic matter.

Do not exceed ½ pound per acre on seed alfalfa that has been established for only one growing season.

If abnormally dry conditions exist following application, restrict first irrigation to not more than ½ acre inch of water. See "Use Rates" for other situations.

Application and Spray Volumes:
Apply "VELPAR" using a fixed-boom power sprayer or aerial equipment. Before spraying, calibrate equipment to determine the quantity of water necessary to uniformly and thoroughly cover the vegetation. The sprayer must be properly calibrated to a constant speed and rate of delivery. Avoid overlapping and shut off spray booms while starting, turning, slowing or stopping to prevent crop injury.

Add the "VELPAR" to a water-filled tank. Agitate for approximately 10 minutes, vigorously enough to prevent settling in order to dissolve the product. Apply in at least 20 gallons of water per acre by ground and 5 to 15 gallons of water per acre by air, but use at least 10 gallons of water for each 1 pound of "VELPAR".

Use Rates:
Select the appropriate dose for soil texture and organic matter content as follows:

Soil Texture	"VELPAR" (Pounds/Acre) Soils With < 1% Organic Matter	Soils With 1 to 5% Organic Matter	Soils With > 5% Organic[1] Matter
Coarse Loamy sand, sandy loam	½ to ¾	½ to ¾	1 to 1½
Medium Loam, silt loam, silt, clay loam, sandy clay loam	½ to ¾	¾ to 1½	1 to 1½
Fine Silty clay loam, sandy clay, silty clay, clay	¾ to 1½	¾ to 1½	1 to 1½

[1] On soils high in organic matter (> 5%), the effectiveness of "VELPAR" can be significantly reduced and weed control may be unsatisfactory.

IMPORTANT: Do not use on gravelly or rocky soils, exposed subsoils, hardpan, sand, poorly drained or alkali soils.

Crop injury, including mortality, may result in fields having restricted root growth due to non-uniform soil profiles such as gravel bases, hardpans, and clay lenses.

Do not add surfactant to "VELPAR" when treating non-dormant alfalfa.

For dormant alfalfa, use a surfactant approved for use on growing crops (such as X-77) at 1 qt/100 gallons spray solution.

Weeds Controlled:
"VELPAR", when applied preemergence or early postemergence to the weeds, will control the following species:

¼ to ½ pound/acre:
Tansy mustard (Colorado and Kansas only)

½ to 1 pound/acre:
Annual bluegrass
Blue mustard
Cheatgrass (downy brome)
Common chickweed
Common groundsel
Dogfennel (mayweed)
English catchfly
Fiddleneck (tarweed)
Field pennycress
Filaree
Flixweed
Jim Hill mustard
London rocket
Miners lettuce
Salsify
Seedling foxtail barley
Seedling orchardgrass
Shepherdspurse
Spurry
Tansy mustard (Except Colorado & Kansas)
Wild radish
Yellow rocket

1 to 1½ pounds/acre:
Annual ryegrass
Henbit
Quackgrass*
Seedling alfalfa*
White cockle*
Common dandelion (spring only)*
False dandelion*
Foxtail (Setari spp.)*
Mexican tea*
Ivyleaf speedwell
Perennial bluegrass* (i.e., Canada bluegrass)
Prickly lettuce*
Purslane speedwell

* Partial control. Observable reduction in weed competition as compared to an untreated area. Degree of partial control can vary with soil type, environmental factors, and size of weeds at the time of application.

Late spring and between cutting applications of ½ to 1 pound/acre will control the following weeds: (For Northeastern and Midwestern States only)

Common lambsquarters
Crabgrass
Fleabane
Foxtail (Setaria spp.)
Jimsonweed
Redroot pigweed

REPLANTING: (Following Alfalfa):
Corn may be planted 12 months after the last treatment, except in areas of low rainfall (20 inches or less) where corn may be planted 12 months after the last treatment, provided the use rate did not exceed ¾ pounds per acre. Do not plant treated areas to any other crop within two years after treatment as crop injury may result.

Do not plant seed alfalfa areas to any other crop within two years after treatment, as crop injury may result.

NOTE: Because crop injury may result, do not apply to actively growing alfalfa nor to stubble between cuttings, except as recommended for northeastern and midwestern states. Do not apply to alfalfa-grass mixtures; do not apply to alfalfa growing under stress from insect, disease, or winter injury, or showing symptoms of other stress conditions; do not use on hardpan, alkali or poorly drained soils. Extremely heavy rainfall or more than 1 inch of irrigation after application may result in crop injury or non-performance. Do not apply to snow covered or frozen ground. Do not use in low desert valleys in California or Arizona. Temporary alfalfa yellowing may occur following herbicide applications. Do not use in the states of Montana, North Dakota, South Dakota and Wyoming.

Do not graze or feed treated forage or hay to livestock within 30 days following application.

FORESTRY:

Since the effect of "VELPAR" on conifers varies with soils, uniformity of application, and environmental conditions, it is suggested that growers limit their first use to small areas.

Because injury (stunting) to conifers may result; do not use in nurseries, seed beds or ornamental plantings; do not use on trees which show poor vigor because of insect, disease or winter injury, or show symptoms of other stress conditions; do not add surfactant if applied over the top of conifers; do not use on poorly drained soils. In addition, conifer injury may result where severe winter stress, disease, or insect damage follows application, or heavy rains occur soon after application.

CHRISTMAS TREE PLANTATIONS:

"VELPAR" is recommended for control of certain weeds where the following species are grown:

Austrian pine	Noble fir	Scotch pine
Grand fir		Sitka spruce

Table 1

Soil Texture*	Pounds "VELPAR" Per Acre[3]	
	First Year Plantings[1]	Established Trees[2]
Gravelly or rocky soils, exposed subsoils, clay knobs, sand, or sandy soil with 85% or more sand	DO NOT USE	DO NOT USE
Loamy sand, sandy loam (50-85% sand)**	1	1 to 1¼
Loam, silt loam, silt and sandy clay loam	1 to 1¼	1¼ to 1¾
Silty clay loam, clay loam, sandy clay, silty clay, and clay	1¼ to 1½	1¾ to 2

Moisture is necessary for full activation of "VELPAR". Best results are obtained when weeds are less than 2 inches in height or diameter, soil is moist at time of application and ½ to 1 inch of rainfall occurs within 2 weeks after application.

* Do not use on any soil containing less than 1% organic matter; do not use on loamy sand or sandy loam with less than 2% organic matter.

** On soils high in organic matter, 5% or greater, the effectiveness of "VELPAR" can be significantly reduced and weed control may be unsatisfactory.

[1] Apply only to transplant stock that is 2 years old or more (1 year old, for loblolly pine). Allow at least two months after transplanting before applying "VELPAR", then apply only if rainfall has settled the soil around the base of the transplants.

[2] Rates recommended are for trees that have been established for 1 year or more.

[3] Use rates in excess of those specified for the soil texture may result in injury to the conifers.

USE RATES: East of the Rocky Mountains:

DO NOT USE IN:

Connecticut	New Hampshire	Pennsylvania	West Virginia
Delaware	New Jersey	Rhode Island	Texas
Maine	New York	Vermont	
Maryland	North Carolina	Virginia	

Weeds Controlled:

Annual bluegrass	Bromegrass	Common ragweed	Oxeye daisy
Barnyardgrass	Catsear (False dandelion)	Fleabane	Pennsylvania smartweed
Bentgrass	Common groundsel	Foxtail	Velvetgrass
			Wild carrot

Treatment Provides Partial Control of:

Crabgrass	Fescue	Horseweed	Ryegrass
Curly dock	Fireweed (willowweed)	Orchardgrass	
Dandelion	Goldenrod		

See **Table 1** for rate selection

Apply as a broadcast spray in the spring prior to conifer bud break. If application is made after bud break, use directional spray equipment and prevent contact of conifer foliage or injury may result. Rates expressed are for broadcast application. Select a spray volume that will ensure a thorough and uniform application, usually 25 gallons per acre (see mixing instructions under Directions For Use). For band application, use proportionately less; for example, use ½ of the broadcast rates when treating a 3 ft. band where row spacing is 6 ft.

USE RATES: West of Rocky Mountains:

In addition to trees listed above, "VELPAR" is recommended for use in Douglas Fir.

Weeds Controlled at 1 to 1½ lbs per acre:

Annual bluegrass	Bromegrass	Fleabane	Oxeye daisy
Barnyardgrass	False dandelion (catsear)	Common groundsel	Velvetgrass

Treatment at 2 lbs/acre Provides Partial Control Of:

Crabgrass	Fescue	Orchardgrass	Wild carrot
Curlydock	Goldenrod	Ryegrass	Willowweed (fireweed)
Dandelion	Heath aster	Smartweed	

"VELPAR" may be applied by air or ground equipment. Select a spray volume that will ensure a thorough and uniform application (see mixing instructions under Directions For Use). For aerial application use at least 8 gallons per acre; for ground application use enough water for thorough coverage, usually 25 gals/acre. Use mechanical or by-pass agitation to thoroughly mix the spray solution. After initial mixing, do not use excessive agitation as foaming problems may result. To avoid foaming problems, use anti-foam agents and minimal agitation. Avoid overlapping and shut off spray booms while starting, turning, slowing or stopping, as injury to conifers may result.

Rainbelt (areas of high spring rainfall): For best results, apply in the spring when weeds are actively growing, but prior to bud break on conifers. If application is made after bud break, use directional spray equipment so that spray does not contact new growth as injury may result.

Snowbelt (areas of low spring rainfall): For best results, apply in the fall before soil freezes or in the spring in anticipation of rainfall. Weed control results from spring applications will be dependent on sufficient rainfall following treatment to activate "VELPAR". Spring applications should be made after snow cover melts, but before bud break occurs.

Rate selection is based on soil texture. Do not use rates in excess of those listed for a particular soil texture as injury to conifers may result. To determine soil texture and organic matter content, have soil samples classified by a reputable laboratory.

See **Table 1** for use rates.

REFORESTATION AREAS:

"VELPAR" is recommended for use in site preparation and conifer release areas. Refer to use categories for specific recommendations for conifer species, rate, timing and species controlled by geographic area.

SITE PREPARATION:

East of Rocky Mountains:

"VELPAR" is recommended (ground or aerial) in forestry site preparation areas where these species will be grown:

Austrian pine	Ponderosa pine	Slash pine
Balsam fir	Red pine	Spruce
Black spruce	Red spruce	Virginia pine
Loblolly pine	Scotch pine	White spruce
Longleaf pine	Shortleaf pine	

Plants Controlled:

When used as directed, "VELPAR" provides both contact and residual control of:

Annual bluegrass	Common groundsel	Oxeye daisy
Asters	Common ragweed	Pennsylvania smartweed
Barnyardgrass	Elksedge	Pinegrass
Bentgrass	Fleabane	Velvetgrass
	Foxtail	Wild carrot
Ash	Hawthorne	Sweetgum
Aspens (big tooth, trembling)	Hazel	Willows
Balsam poplar	Hickory	Wild cherry
Birch	Oaks	
Elm		

Plants Partially Controlled:

Black gum	Fireweed (Willowweed)	Honeysuckle
Canada thistle	Flowering dogwood	Horseweed
Crabgrass	Fescue	Orchardgrass
Curley dock	Goldenrod	Quackgrass
Dandelion	Heath aster	Red maple
False dandelion (catsear)	Perennial ryegrass	Sourwood

Apply 2 to 6 lbs of "VELPAR" per acre. For aerial application use at least 8 gallons per acre; for ground application use enough water for thorough coverage, usually 25 gals/acre. On fine textured soils and hard-to-kill species use the higher rates of "VELPAR"; for coarser soils use a lower rate. To reduce drift, use a precise application spray boom and/or add a spray thickening agent. If a spray thickening agent is used in the tank mixture, follow all label directions.

Add the "VELPAR" to a water filled tank and mix thoroughly. A surfactant may then be added. After thorough mixing, a defoaming agent may be added. Agitate thoroughly to disperse the defoaming agent.

Combination with "Tordon" 101[1]:

Apply as a foliar treatment for woody plant control for pine site preparation. In addition to the species listed above, this treatment controls woody plants such as:

Black gum	Prunus spp.	Sourwood
Pine	Sassafras	

Apply 2 to 6 lbs of "VELPAR" plus 1 to 2 gallons of "Tordon" 101 per acre.

Use 1½ to 2 gallons of "Tordon" 101 per acre when the major components of the hardwood composition to be controlled are species other than elm, oak or sweetgum.

When oak, elm, sweetgum, blackberry and prunus predominate, use 3 to 6 lbs of "VELPAR" plus 1 gallon of "Tordon" 101.

When blackgum, dogwood, sourwood and red maple predominate, use 2 to 6 lbs of "VELPAR" plus 1½ to 2 gallons of "Tordon" 101.

"VELPAR", alone or in tank mixture with "Tordon" 101, should be applied during the period from late spring to early summer after the trees have reached full leaf, but before leaf tissue hardens; or, in Maine, Michigan, Minnesota, New Hampshire, New York, Vermont and Wisconsin, after bud break. Do not burn treated sites after initial contact (browning) effect on foliage (ten weeks or more). Burn only after brush has completely defoliated at least once from sufficient root uptake of "VELPAR", usually 60 to 90 days after adequate rainfall.

[1] Registered trademark of Dow Chemical Company.

NOTE:
1. In the South, pines planted sooner than six months after treatment with "Tordon" 101 may be injured. For all conifers, the waiting period between treatment and planting should be 8 to 9 months in the lake states and the Northeastern U.S.
2. Do not use on wet or poorly drained sites.
3. Before using "Tordon" 101, read and carefully observe the cautionary statements and all other information appearing on the product label. Note in particular that "Tordon" 101 is a **Restricted Use Pesticide,** for retail sale to and use only by certified applicators or persons under their direct supervision and only for those uses covered by the certified applicators certification.

CONIFER RELEASE:

Southern, Northern States:

Recommended for conifer release where the following species are grown:

Balsam fir	Norway spruce	Shortleaf Pine
Black spruce	Red pine	Virginia pine
Loblolly pine	Red spruce	White spruce

Plants Controlled:

Herbaceous Plants

Aster	Crabgrass	Fireweed	Quackgrass*
Bracken fern	Dog fennel	Foxtail	Ryegrass*
Common groundsel	Goldenrod	Panicums	
Common ragweed	Horseweed	Wild carrot	

Woody Plants:

Ash	Box elder	Hawthorne	Sumac*
Aspen	Brambles	Hazel	Sweetgum*
Balsam poplar	Cherry (black and pin)	Honeysuckle	Willow
Birch	Dogwood*	Oaks (red, black and white)	
Blackgum*	Elm	Red maple*	

* Considered hard-to-kill. Partial control or suppression only.

Apply 4 to 8 lbs of "VELPAR" per acre (red pine - 4 to 11 lbs, loblolly pine - 5 to 12 lbs, shortleaf pine - 8 lbs) by air or ground. For aerial application use at least 8 gallons per acre; for ground application use enough water for thorough coverage, usually 25 gals/acre. Do not use a surfactant. The preferred pine stage of growth for application is between growth spurts, if more than one occurs. Some conifer needles may be discolored, however, full recovery can be expected the following spring. Mortality of some pine trees may occur. At least 1 to 2 inches of rainfall are needed for soil activation.

Southern States:

"VELPAR" should be applied during the period from early spring (one-half leaf) until early summer after the hardwood trees have reached full leaf stage, but before leaf tissue has hardened.

Northern States:

Apply from late spring, at or near conifer bud break, until early summer. For maximum contact activity on woody brush, apply after brush is in ¾ leaf stage.

NOTE:
1. Do not apply where conifers are suffering from loss of vigor caused by insects, disease, drought, winter damage, or other stresses, as injury may result.
2. Excessive injury may result to conifers less than 4 years of age from transplanting on coarse-textured soils, and less than 3 years of age from transplanting on medium and fine-textured soils.
3. Poor weed and brush control may occur from use on poorly drained or marshy sites.
4. Poor weed and brush control may occur from applications made when the soil is saturated with water and rain is imminent within 24 hours.
5. Severe injury may result from use on Jack pine, White pine or larch.
6. Use the lower rates on coarse-textured soils (i.e., sand, loamy sand, sandy loam).
7. Use the intermediate rates on medium-textured soil (i.e., loam, silt loam, silt).
8. Use the higher rates on medium-fine and fine-textured soils (i.e., sandy clay loam, clay loam, silty clay loam, sandy clay, silty clay).
9. Use on gravelly or rocky soils and exposed subsoils may cause conifer injury or mortality.
10. On tracts of land where various soil types are present and "VELPAR" rate selection is difficult, conifer damage or less than expected vegetation suppression may occur due to rate differences required for various soil types.

SITE PREPARATION, CONIFER RELEASE:

West of Rocky Mountains:

"VELPAR" is recommended for use in forestry site preparation and conifer release where the following species are grown:

Douglas fir	Noble fir	Sitka spruce
Grand fir	Ponderosa pine	White fir
Jeffrey Pine		

Plants Controlled:

Annual bluegrass	Common groundsel	Oxeye daisey
Barnyardgrass	Fleabane	Velvet grass
Bromegrass	Foxtail	Wild carrot
Catsear (false dandelion)		
Deerbrush ceanothus	Snowbrush ceanothus (varnishleaf ceanothus)	
Greenleaf manzanita	Squawcarpet	

Plants partially controlled:

Dandelion	Heath aster	Smartweed
Fescue	Orchardgrass	Willowweed (fireweed)

Goldenrod	Ryegrass

Apply as a pre-plant or post-plant broadcast or band application. For band applications, use proportionately less; for example, use 1/2 the broadcast rate when treating a 3 ft. band where row spacing is 6 ft.

For herbaceous control, use 1 to 2 lbs per acre and 1 1/2 to 3 lbs per acre for brush control. Use the lower rates on coarse-textured soils and soils low in organic matter; use the higher rates on fine textured soils and soils high in organic matter.

For brush control, applications should be made to brush seedlings or sprouts when less than 18″ in height or diameter. "VELPAR" effects may be slow to appear and maximum results will be seen 12 to 24 months following the application.

Weed control is dependent on sufficient rainfall following treatment to activate "VELPAR". Apply "VELPAR" by air or ground equipment. Select a spray volume that will insure a thorough and uniform application, at least 8 gals. of spray per acre by air and usually 25 gals. of spray per acre by ground (see mixing instructions under Directions For Use). Use mechanical or by-pass agitation to thoroughly mix the spray solution. After initial mixing, do not use excessive agitation as foaming problems may result. To avoid foaming problems, use anti-foam agents and minimal agitation.

Rainbelt (areas of high spring rainfall): For best results, apply in the late winter or spring when weeds are actively growing, but prior to bud break on conifers. If application is made after bud break, use directional spray equipment so spray does not contact new growth, as injury may result.

Snowbelt (areas of low spring rainfall): For best results, apply in the fall before soil freezes, or in the spring in anticipation of rainfall. Weed control results from spring applications will be dependent on sufficient rainfall following treatment to activate "VELPAR". Spring applications should be made after snow cover melts, but before bud break occurs.

NOTE:
1. Use on gravelly or rocky soils, exposed subsoils, clay knobs, sand or sandy soils with 85% or more sand may cause conifer injury or mortality and poor weed and brush control may occur.
2. Conifer injury or mortality may occur on any soil containing less than 1% organic matter.
3. On soils high in organic matter (greater than 5%) effectiveness of "VELPAR" is reduced and results may be unsatisfactory.
4. Use on loamy sand or sandy loam with less than 2% organic matter may cause conifer injury except for Ponderosa and Jeffrey pine.
5. On first year plantations, "VELPAR" may only be applied (either pre-plant or post-plant) to transplant stock that is 2 years old (2-0) or more, except for 1 year old (1-0) for Ponderosa and Jeffrey pine.
6. When applying "VELPAR" after transplanting, wait until rainfall has settled the soil around the base of the transplants before making the treatment.
7. Damage may occur to Incense cedar, Sugar pine and Giant Sequoia.
8. Heavy duff or slash present at time of application may result in poor weed or brush control.

GENERAL WEED AND BRUSH CONTROL—NONCROPLAND

"VELPAR" is recommended for general weed and brush control in noncropland areas such as railroads, highways, utility and pipeline right-of-ways, petroleum tank farms, storage areas, industrial-plant sites, drainage ditch banks and other similar areas.

For broadcast treatments (ground equipment only), apply "VELPAR" as a spray just before or soon after weed emergence. Do not apply to frozen or snow-covered soil. To improve wetting and contact action, add a nonionic surfactant at 0.25% by volume.

Use a sprayer properly calibrated to a constant speed and rate of delivery. For application with a handgun sprayer, use sufficient water (50 to 200 gallons per acre) for thorough coverage.

WEED CONTROL:

SHORT-TERM CONTROL (up to 3 months):

Apply "VELPAR" at the rate of 2 to 5 pounds per acre in at least 25 gallons of water per acre for control of many annual, biennial and perennial weeds. Species controlled:

Barnyardgrass	Fiddleneck	Paragrass
Bindweed*	Filaree	Pigweed
Bouncing bet*	Fleabane	Purslane
Bromegrass	Goatsbeard vine	Quackgrass
Buffalograss*	Goldenrod	Ryegrass
Burdock	Lespedeza	Smartweed
Cocklebur	Milkweed*	Spurge
Crabgrass	Mustard	Trumpet creeper*
Crown vetch	Nutsedge*	Wild oats*
Curly dock*	Orchardgrass*	Wild parsnip
Dandelion*	Oxalis	Wild star thistle
Dogbane*		

* Considered hard-to-kill; partial control or suppression only.

SEASON-LONG CONTROL:

Apply "VELPAR" at the rate of 6 to 12 pounds per acre in at least 25 gallons of water per acre. Controls:

Bahiagrass*	Dogfennel*	Natalgrass
Bermudagrass*	Fescue*	Plantain
Bluegrass	Fingergrass	Prickly Lettuce
Broomsedge	Foxtail	Ragweed
Camphorweed	Guineagrass	Smutgrass
Canada thistle*	Heath aster	Spanish needles
Chickweed	Honeysuckle	Vaseygrass
Clovers	Lantana	Wild blackberry*
Dewberry	Marestail	Wild carrot

* Considered hard-to-kill; partial control or suppression only.

BRUSH CONTROL:

Apply "VELPAR" at the rate of 4 to 8 pounds per acre as a broadcast spray in at least 40 gallons of spray per acre between late winter and early summer. Species controlled:

Hawthorne	Oak	Sumac
Hickory	Poplar	Willow
Maple		

SELECTIVE WEED CONTROL—NONCROPLAND AREAS

BERMUDAGRASS AND BAHIAGRASS IN S.E. AND S.W. STATES:

"VELPAR" is recommended for selective weed control in established stands of Bermudagrass and Bahiagrass in noncropland areas in the following states:

Alabama	Louisiana	South Carolina
Arkansas	Mississippi	Tennessee
Florida	Oklahoma	Texas
Georgia	North Carolina	

Use only on noncropland areas such as railroads, highways, utility and pipeline right-of-ways, petroleum tank farms, storage areas, industrial plant sites, drainage ditches and similar areas.

Since the effect of "VELPAR" on bermudagrass and bahiagrass varies with soils, varieties, uniformity of application and environmental conditions, it is suggested that initial applications be limited to small areas.

Use only in stands of bermudagrass and bahiagrass established for at least one year; do not treat newly sprigged or sodded areas as severe injury to desirable grasses may result. Temporary injury (chlorosis and leaf burn) to bermudagrass and bahiagrass can be expected after application. Plants will recover from this injury within several weeks if conditions are favorable for growth, while weeds will be killed.

Make a single application per year of 3/4 to 1 1/4 pounds per acre when soil is warm and moist and weeds are actively growing. Use the lower rate on coarse-textured soils (clay loam, clay) and on soils high in organic matter (over 5%). Apply with a fixed boom power sprayer properly calibrated to a constant speed and rate of delivery, using at least 25 gallons of spray per acre. Avoid overlapping and shut off spray booms while starting, turning, slowing or stopping, as injury to desirable grasses may result.

Moisture is required to activate "VELPAR" in the soil. Best results are obtained when weeds are less than 2 inches in height or diameter and actively growing, soil is moist at time of application, and 1/2 to 1 inch of rainfall occurs within 2 weeks after application. Foliar effects on weeds are most pronounced under conditions of high temperature (above 80° F) and high humidity. Symptoms usually occur within two weeks after application under these conditions, while 4 to 6 weeks may be required when weather is cool. If rainfall after treatment is insufficient to promote root uptake, weeds may recover from foliar effects and continue to grow. Extremely heavy rainfall after treatment may result in poor weed control and/or injury to desirable grass, especially if application is made to dry soil.

NOTE: As injury to bermudagrass and bahiagrass may result, do not apply or drain or flush equipment on or near desirable trees or other plants, or on areas where their roots may extend, or in locations where the chemical may be washed or moved into contact with their roots.

PINEAPPLE:

"VELPAR" is recommended (ground application) for use in pineapple for the control or suppression of the following weeds:

Ageratum	Dallisgrass	Morningglory
Balsam apple	Guinea grass	Oxalis
Castor bean	Jungle rice	Popolo
Crabgrass	Kao haole*	Richardsonium
Crotalaria	Mauna loa*	Vaseygrass

* Suppression

Use a sprayer properly calibrated to a constant speed and rate of delivery. Mix the proper amount of "VELPAR" in water.

Intercrop Period:
Apply as broadcast spray in 100 to 400 gallons of water per acre at the rate of 1/4 to 2 pounds per acre (plus surfactant). For aerial application use at least 10 gallons water per acre.

Post mulch preplant:
Apply as a broadcast spray in 100 to 400 gallons of water per acre at the rate of 1/4 to 2 pounds per acre (plus surfactant).

Post plant but before planting material starts active growth:
Apply as a broadcast spray in 100 to 400 gallons of water per acre at the rate of 1/4 to 2 pounds per acre (plus surfactant).
Post plant after planting material starts to grow should be made when weed growth has escaped control by other herbicide applications.

Post plant crop harvest but prior to forcing first ratoon:
Apply as a broadcast spray in 100 to 400 gallons of water per acre at the rate of 1/4 to 2 pounds per acre (plus surfactant).

Directed post-emergence (pineapple and weeds) interspace application:
Apply as directed spray 3 to 10 months after planting in 50 to 200 gallons of water per acre (broadcast basis), at the rate of 1/4 to 2 pounds per acre (broadcast basis) using a stroller boom or knapsack (plus surfactant).

Directed spot treatments for perennial grasses before floral induction:
Spray perennial grasses post-emergence to wet (50 to 200 gallons per acre depending on size) with 1 to 2 pounds per 100 gallons of water as a spot treatment (plus surfactant).

Do not exceed 6 pounds "VELPAR" per acre per crop cycle.

Apply "VELPAR" at 2 to 4 pounds per acre in 100 to 400 gallons of water for field edges and roadsides.

NOTE: Use the lower rates on coarse-textured soils or in areas where rainfall exceeds 65 inches per year.
Use the higher rates on fine-textured soils or in areas where rainfall is less than 65 inches per year.
Add surfactant at .25% by volume of water.

SUGARCANE:

DuPont "VELPAR" Herbicide is recommended for selective weed control in sugarcane. Since the effect of "VELPAR" on sugarcane varies with soils, uniformity of application, and environmental conditions, it is suggested that growers limit their first use to small areas. Tolerance of new cane varieties to "VELPAR" should be determined prior to adoption as field practice.

Weeds Controlled:

- Alexandergrass
- Amaranth (slender, smooth)
- American burnweed (fireweed)
- Balsam apple
- Barnyardgrass
- Carolina geranium
- Castorbean
- Chickweed
- Crabgrass (hairy, large, smooth)
- Crotalaria (fuzzy, showy)
- Cuphea (tarweed)
- Dallisgrass
- Fingergrass (radiate, swollen)
- Flora's paintbrush
- Foxtail (bristly, yellow)
- Goosegrass
- Guineagrass
- Henbit
- Job tears
- Johnsongrass (from seed)
- Junglerice
- Lambsquarter
- Morningglory (hairy, threelobe)
- Mustard (wild)
- Oxalis
- Panicum (brownleaf, browntop, Texas, millet)
- Paspalum (ricegrass, sour)
- Pigweed (common, smooth)
- Popolo
- Purslane
- Sandbur
- Sensitive plant (Hila Hila)
- Signalgrass (broadleaf)
- Sowthistle
- Spanish needle
- Sprangletop
- Spurge (prostrate, graceful)
- Sunflower
- Vaseygrass
- Waltheria (Hialoa)

Partial Control:

- Ageratum
- Bermudagrass
- Itchgrass

Apply a single treatment per year using a fixed-boom power sprayer and a minimum of 25 gals per acre, unless otherwise directed.

Florida:
Apply 1/2 to 1 pound "VELPAR" per acre on sandy soils as a pre or directed post-emergence treatment using the lower rate on soils low in organic matter and the higher rate on soils high in organic matter. On muck soils apply 1 to 2 pounds of "VELPAR" per acre pre-emergence or 1/2 to 1 pound per acre post-emergence, using the higher rate on larger weeds. Post-emergence applications should be directed to cover the weeds and soil while minimizing crop contact.

Hawaii:
Apply "VELPAR" pre or post-emergence at the following rates for the indicated soil texture:

Soil Texture	Pounds "VELPAR" Per Acre (Plus surfactant .25% by volume)
Coarse-Textured Soils (sand, loamy sand, sandy loam)	1/2 to 1
Medium-Textured Soils (loam, silt loam, silty clay loam)	1/2 to 2
Fine-Textured Soils (clay, gray hydromorphic clay)	2 to 4

Use the higher levels of the recommended dosage ranges on soils high in organic matter. Do not apply more than twice the highest recommended rate for the indicated soil texture per crop cycle.

For pre-emergence use only, application may be made with aerial equipment using at least 10 gallons of spray per acre.

For spot treatments of emerged weeds, "VELPAR" may be applied with a knapsack sprayer in concentrations of 1/2 to 2 pounds per 100 gallons of water. Apply in sufficient volume to thoroughly wet weed foliage, but do not exceed 100 gallons of spray per treated acre. Use the lower concentrations on coarse-textured soils which are low in organic matter and the higher concentrations on fine-textured soils which are high in organic matter.

Louisiana:
Apply 1/2 to 1 pound of "VELPAR" per acre broadcast in the fall before sugarcane emerges or in the spring before active cane tillering begins. Use 1/3 of the above broadcast rates when band treating 1/3 of the area. Fall treatments of 1/2 to 3/4 pound per acre may be followed by a spring treatment of 1/2 to 3/4 pound per acre. Do not apply more than 1 1/2 pounds per year. Use the higher levels of the recommended dosage range on fine-textured soils.

Puerto Rico:
Apply 1 to 2 pounds of "VELPAR" per acre pre-emergence to the cane or 1/2 to 1 pound per acre post-emergence. Use the lower rates on coarse-textured soils and the higher rates on fine-textured soils (high in clay or organic matter).

For spot treatment of emerged weeds, "VELPAR" may be applied with a knapsack sprayer in concentrations of 1/2 to 1 pound per 100 gallons of water. Use the lower concentration on coarse-textured soils and the higher concentration on fine-textured soils. Apply a sufficient volume of spray to thoroughly wet weed foliage. Do not exceed 100 gallons of spray per treated area.

Texas:
Apply 1/2 to 1 pound of "VELPAR" per acre. On plant cane, apply before cane emerges or as a directed layby treatment. On stubble cane, apply pre-emergence (up to the 3 leaf stage) or as a directed layby treatment. A pre or early post-emergence treatment may be followed by a layby treatment, provided at least 60 days have elapsed and 3 inches of rainfall or sprinkler irrigation have occurred since the first treatment. Use the following rates for the soil texture:

Soil Texture	Pounds "VELPAR" Per Acre Pre-emergence	+	Layby
Coarse-textured soils* (sandy loam)	1/2	+	1/2
Medium-textured soils (loam, silt loam)	3/4	+	3/4
Fine-textured soils (clay loam)	1	+	1

* With at least 2% organic matter.

NOTE: Because injury to sugarcane may result: Do not use on cane which shows poor vigor because of insect, disease or winter injury, or shows symptoms of other stress conditions such as drought stress; do not add surfactant if applied over the top of cane; do not use on gravelly or rocky soils, thinly covered subsoils, nor on coarse-textured soils (sands to sandy loams) with less than 1% organic matter. Temporary chlorosis of the crop may result from application over emerged cane. Do not apply over the top of actively growing cane. Applications during active cane growth should be directed to cover the weeds and soil while minimizing crop contact. Do not use on varieties which are known to be susceptible to weed killers. Do not plant any crop other than sugarcane within 18 months of the last application of "VELPAR". Do not feed sugarcane forage to livestock.

Extremely heavy rainfall after application may result in poor weed control and/or crop injury; especially if the application is made to dry soil.

On dormant cane, addition of surfactant is permissible and will increase weed control of emerged weeds. In Hawaii, surfactant is recommended for all uses.

STORAGE AND DISPOSAL

STORAGE: Store product in original container only, away from other pesticides, fertilizer, food or feed.

PESTICIDE DISPOSAL: Do not contaminate water, food or feed by storage or disposal. Wastes resulting from the use of this product may be disposed of on site or at an approved waste disposal facility.

CONTAINER DISPOSAL: Completely empty bag into application equipment. Then dispose of empty bag in a sanitary landfill or by incineration, or, if allowed by state and local authorities, by burning. If burned, stay out of smoke.

NOTICE OF WARRANTY

Du Pont warrants that this product conforms to the chemical description on the label thereof and is reasonably fit for purposes stated on such label only when used in accordance with the directions under normal use conditions. It is impossible to eliminate all risks inherently associated with the use of this product. Crop injury, ineffectiveness or other unintended consequences may result because of such factors as weather conditions, presence of other materials, or the manner of use or application, all of which are beyond the control of Du Pont. In no case shall Du Pont be liable for consequential, special or indirect damages resulting from the use or handling of this product. All such risks shall be assumed by the Buyer. DU PONT MAKES NO WARRANTIES OF MERCHANTABILITY OR FITNESS FOR A PARTICULAR PURPOSE NOR ANY OTHER EXPRESS OR IMPLIED WARRANTY EXCEPT AS STATED ABOVE.

NOTICE TO BUYER: Purchase of this material does not confer any rights under patents of countries outside of the United States.

AG-3202 8088 10/28/87

This is a specimen label. DuPont is not responsible for the accuracy of the information contained herein. As labels are subject to revision, always carefully read and follow the label on the product container.

VELPAR® L

HERBICIDE
WATER DISPERSIBLE LIQUID

Contains 2 lbs active ingredient per gallon

ACTIVE INGREDIENT:	BY WEIGHT
Hexazinone [3-cyclohexyl-6-(dimethylamino)-1-methyl-1,3,5-triazine -2,4(1H,3H)-dione]	25%
INERT INGREDIENTS	75%
TOTAL	100%

U.S. Pat. Nos. 3,902,887 & 3,983,116

EPA Reg. No. 352-392

KEEP OUT OF REACH OF CHILDREN

PRECAUTIONARY STATEMENTS

HAZARDS TO HUMANS AND DOMESTIC ANIMALS

DANGER!

CAUSES EYE DAMAGE.

Do not get in eyes. Wear goggles or face shield and rubber gloves when handling. Avoid contact with skin and clothing. Avoid breathing spray mist. Harmful if swallowed.

In case of contact, immediately flush eyes or skin with plenty of water for at least 15 minutes. For eyes, call a physician. Get medical attention if skin irritation persists. Remove and wash contaminated clothing before re-use.

For medical emergencies involving this product, call toll free 1-800-441-3637.

ENVIRONMENTAL HAZARDS

Do not apply directly to water or wetlands. Do not apply where runoff is likely to occur. Do not contaminate water when cleaning equipment or by careless disposal of wastes.

PHYSICAL OR CHEMICAL HAZARDS

FLAMMABLE. Keep away from heat, sparks, and open flames. Keep container closed.

IMPORTANT

Injury to or loss of desirable trees or other plants may result from failure to observe the following: Do not apply except as recommended or drain or flush equipment on or near desirable trees or other plants, or on areas where their roots may extend, or in locations where the chemical may be washed or moved into contact with their roots. Do not use on lawns, walks, driveways, tennis courts, or similar areas. Prevent drift of spray to desirable plants. Keep from contact with fertilizers, insecticides, fungicides, and seeds.

Thoroughly clean all traces of "VELPAR" L from application equipment immediately after use. Flush tank, pump, hoses and boom with several changes of water after removing nozzle tips and screens (clean these parts separately).

GENERAL INFORMATION

Du Pont "VELPAR" L Herbicide is a water-dispersible liquid to be mixed in water and applied as a spray for selective weed control in certain crops, Christmas trees, for selective weed and brush control in reforestation areas, and for non-selective weed and brush control on noncropland areas. It may also be applied undiluted as a basal soil treatment for brush control on noncropland areas and in reforestation areas, or by tree injection for selective brush control. It is non-corrosive to equipment.

"VELPAR" L is an effective general herbicide providing both contact and residual control of many weeds. At rates recommended for noncropland areas, it provides control of many annual and biennial weeds, brush, woody vines, and, except for johnsongrass, is effective for control of most perennial weeds. It is absorbed through the roots and foliage. For brush control, soil application or injection is most effective.

Moisture is required to activate "VELPAR" L in the soil and best results are obtained when weeds are less than 2 inches in height or diameter, soil is moist at time of application and 1/4 to 1/2 inch of rainfall occurs within 2 weeks after application. Foliar activity on weeds is most effective under conditions of high temperature (80°F), high humidity and good soil moisture. Foliar activity when vegetation is dormant, semi-dormant, or under stress may be reduced.

On herbaceous plants, symptoms usually appear within 2 weeks after application under warm, humid conditions, while 4 to 6 weeks may be required when weather is cool. If rainfall after application is inadequate to activate "VELPAR" L in the soil, plants may recover from contact effects and continue to grow.

On woody plants, symptoms usually appear 3 to 6 weeks after sufficient rainfall has carried the herbicide into the root zone during periods of active growth. Defoliation and subsequent refoliation may occur, but susceptible plants are killed. The degree of control and duration of effect will vary with the amount of chemical applied, rainfall, temperature, weed and brush species, soil texture and other conditions.

DIRECTIONS FOR USE

It is a violation of federal law to use this product in a manner inconsistent with its labeling.

Du Pont "VELPAR" L should be used only in accordance with recommendations on this label, or in separate published Du Pont recommendations available through local dealers.

Du Pont will not be responsible for losses or damages resulting from use of this product in any manner not specifically recommended by Du Pont. User assumes all risks associated with such non-recommended use.

Do not apply this product in such a manner as to directly or through drift expose workers or other persons. The area being treated must be vacated by unprotected persons.

Do not enter treated areas without protective clothing until sprays have dried. Because certain states may require more restrictive reentry intervals for various crops treated with this product, consult your State Department of Agriculture for further information.

Written or oral warnings must be given to workers who are expected to be in a treated area or in an area about to be treated with this product. When oral warnings are given, warnings shall be given in a language customarily understood by workers. Oral warnings must be given if there is reason to believe that written warnings cannot be understood by workers. Warnings must include the following information: "DANGER. Area treated with "Velpar" L on (date of application). Do not enter without appropriate protective clothing until sprays have dried. In case of contact, flush skin or eyes with plenty of water; for eyes, get medical attention."

Do not apply this product through any type of irrigation system.

APPLICATION:

"VELPAR" L may be applied by ground equipment, and where permitted, aerial equipment. Use rates, minimum spray gallonage and other application information is described in the various use categories. **For all applications, use at least 5 gallons of water for each gallon of "VELPAR" L.**

For best results, apply "VELPAR" L just before or soon after weed emergence. Do not apply to frozen or snow covered soil. Avoid overlapping and shut off spray booms while starting, turning, slowing or stopping, as injury to the crops may result.

Always calibrate application equipment prior to use.

RATE SELECTION:
Where a rate range is shown, use the higher levels of the dosage range under the following conditions; on hard-to-control species; on fine-textured soils; on soils containing more than 5% organic matter or carbon. Use the lower levels of the dosage range on coarse-textured soils; on soils low in organic matter. Refer to specific use categories for use rate ranges.

SURFACTANTS:
When emerged seedling weeds are present, the addition of a surfactant improves contact action by increased wetting and absorption by foliage. For forestry and noncrop sites, use a non-ionic surfactant of at least 80% active ingredient at 0.25% by volume of water (1 qt/100 gals spray). Similarly, use an agricultural surfactant approved for use on growing crops such as X-77 with "VELPAR" for treatment of emerged weeds in dormant alfalfa. Use of "VELPAR" L plus surfactant on growing crops (including trees) may cause injury to the crops.

MIXING:
Add the proper amount of "VELPAR" L to the amount of water to be used and use mechanical or by-pass agitation to thoroughly mix the spray solution. A surfactant may be added to improve wetting and/or contact activity if the use directions do not prohibit the addition of a surfactant cleared for application to growing crops. After initital mixing, do not use excessive agitation as foaming problems by result. To avoid foaming problems, use anti-foam agents and minimal agitation.
Where applicable, add surfactant and/or anti-foam agent as last ingredient in the tank.

ALFALFA:
Du Pont "VELPAR" L is recommended for control of certain weeds in established alfalfa hay except in the states of Montana, North Dakota, South Dakota and Wyoming. It is also recommended for control of certain weeds in established seed alfalfa in the state of California.

Best results are obtained when 1/2 to 1" of rainfall or sprinkler irrigation occurs within two weeks after application, when soil is moist at time of application, and when weeds have not germinated or are less than 2" tall or across.

Since the effect of "VELPAR" L on alfalfa varies with soil conditions, uniformity of application, and environment conditions, it is suggested that growers limit their first use to small areas.

Treat only stands of alfalfa established for one year or for one growing season, provided the alfalfa root system is **well established** and the crop is healthy, vigorous and not under stress from weather conditions, insect or disease damage. Do not use on seedling alfalfa, alfalfa-grass mixtures or other mixed stands.

At elevations above 3,000 ft., do not use "VELPAR" L on alfalfa that has been established with a cover or nurse crop until the alfalfa has gone through two summer growing seasons.

NON-DORMANT AND SEMI-DORMANT VARIETIES:
Make a single application during winter months when alfalfa plants are in the least active stage of growth; see "Use Rates" for dosages.

In states of Connecticut, Delaware, Illinois, Indiana, Iowa, Kentucky, Maine, Maryland, Massachusetts, Michigan, Minnesota, Missouri, New Hampshire, New Jersey, New York, Ohio, Pennsylvania, Rhode Island, Vermont, Virginia, West Virginia, Wisconsin, makes a single application of 2 to 6 pints per acre in the spring before new growth exceeds 2" in height or to stubble after cutting following hay removal and before regrowth exceeds 2" in height. Severe alfalfa injury may result following an after-cutting application if the regrowth is more than 2" high, or significant stubble is left after alfalfa cutting, and/or the air temperature is above 90°F.

DORMANT VARIETIES:
Make a single application after it becomes domant and before new growth begins in the spring; see "Use Rates" for dosages.

Do not exceed 3 pints per acre on alfalfa less than one year old.

SEED ALFALFA (California only):
Do not use on fields with sandy loam or loamy sand soils having less than 1% organic matter.

Do not exceed 2 pints per acre on fields with sandy loam or loamy sand soils having between 1 and 2% organic matter.

Do not exceed 2 pints per acre on seed alfalfa that has been established for only one growing season.

If abnormally dry conditions exist following application, restrict first irrigation to not more than 1/2 acre inch of water.

See "Use Rates" for other situations.

Application and Spray Volumes:
Apply "VELPAR" L using a fixed boom power sprayer or aerial equipment. Before spraying, calibrate equipment to determine the quantity of water necessary to uniformly and thoroughly cover the vegetation. The sprayer must be properly calibrated to a constant speed and rate of delivery. Avoid overlapping and shut off spray booms when starting, turning, slowing or stopping to prevent crop injury.

Add the "VELPAR" L to a water-filled tank, mix thoroughly. Apply in at least 20 gallons of water per acre by ground and 5 to 10 gallons of water per acre by air, but use at least 5 gallons of water for every 1 gallon of "VELPAR" L. To avoid foaming problems, anti-foam agents which are cleared for use on growing crops, may be added to the spray mixture.

Use Rates:
Select the appropriate dose for soil texture and organic matter content from table below.

	"VELPAR" L (Pints/Acre)		
Soil Texture	**Soils With <1% Organic Matter**	**Soils with 1 to 5% Organic Matter**	**Soils with >5% Organic[1] Matter**
Coarse Loamy sand, sandy loam	2-3	2-3	4-6
Medium Loam, silt loam, silt, clay loam, sandy clay loam	2-3	3-6	4-6
Fine Silty clay loam, sandy clay, silty clay, clay	3-6	3-6	4-6

[1] On soils high in organic matter ($\geq$5%), the effectiveness of "VELPAR" L can be significantly reduced and weed control may be unsatisfactory.

IMPORTANT— Do not use on gravelly or rocky soils, exposed subsoils, hardpan, sand, poorly drained or alkali soils. Crop injury, including mortality, may result in fields having restricted root growth due to non-uniform soil profiles such as gravel bases, hardpans and clay lenses. Do not add surfactant to "VELPAR" L when treating non-dormant alfalfa. For domant alfalfa, use a surfactant approved for use on crops (such as X-77) at 1 qt/100 gallons spray solution.

Weed Controlled:
"VELPAR" L, when applied preemergence or early postemergence to the weeds, will control the following species:

1 to 2 Pints/Acre

Tansy-mustard (Colorado and Kansas only)

2 to 4 Pints/Acre

Annual bluegrass	Field pennycress	Seedling orchardgrass
Blue mustard	Filaree	Shepherdspurse
Cheatgrass (downy brome)	Flixweed	Spurry
	Jim Hill mustard	Tansy-mustard
Common chickweed	London rocket	Wild radish
Common groundsel	Miners lettuce	Yellow rocket
Dogfennel (mayweed)	Salsify	
English catchfly	Seedling foxtail barley	
Fiddleneck (tarweed)		

4 to 6 Pints/Acre

Annual ryegrass	Ivyleaf speedwell	Purslane speedwell
Henbit		
Common dandelion (spring only)*		Prickly lettuce*
False dandelion		Quackgrass*
Foxtail (*Setaria* spp.)*		Seedling alfalfa*
Mexican tea*		White cockle*
Perennial bluegrass (i.e., Canada bluegrass)*		

* Partial control. Observable reduction in weed competition as compared to an untreated area. Degree of partial control can vary with soil type, environmental factors, and size of weeds at the time of application.

Late Spring and Between Cutting Applications of 2 to 6 Pints/Acre Controls:

(For Northeastern and Midwestern States Only)

Common lambsquarters	Foxtail (*Setaria* spp.)
Crabgrass	Jimsonweed
Fleabane	Redroot pigweed

IMPREGNATION ON DRY BULK FERTILIZER (Except California and Arizona)

Dry bulk fertilizer may be impregnated or coated with "VELPAR" L for application to established alfalfa. All recommendations, cautions and special precautions on this label must be followed along with state regulations relating to dry bulk fertilizer blending, impregnating and labeling. Keep "VELPAR" L from heat, sparks and open flame. (Refer to Material Safety Data Sheet for "VELPAR" L.)

If fertilizer materials are excessively dusty, use a suitable additive to reduce dust prior to impregnation, as dusty fertilizer will result in poor distribution during application. The dry fertilizer must be properly impregnated and uniformly applied to the alfalfa to avoid crop injury and/or poor weed control.

To impregnate, use a system consisting of a conveyor or closed drum which is used for dry bulk fertilizer blending. Any commonly used fertilizer can be impregnated with "VELPAR" L, except potassium nitrate or sodium nitrate. Do not use on limestone.

Apply using a minimum of 250 pounds dry bulk fertilizer per acre and up to a maximum of 450 pounds per acre. To impregnate or coat the dry bulk fertilizer with "VELPAR" L, the delivery nozzles must be directed to deliver a fine spray towards the fertilizer for thorough coverage while avoiding spray contact with mixing equipment. Uniform impregnation of "VELPAR" L to dry bulk fertilizer will vary, and if the absorptivity is not adequate, the use of an absorptive powder may be required to produce a dry, free-flowing mixture. Microcel E (Johns Manville Product Corporation) is the recommended absorbent powder. When another herbicide is used with "VELPAR" L, mix and impregnate immediately.

Apply immediately after impregnation unless experience has shown that impregnated fertilizer can be stored without becoming lumpy and difficult to spread.

Select the rate of "VELPAR" L to apply per acre from the appropriate section of this label and refer to the rate table to determine the amount of which is to be impregnated on a ton of dry bulk fertilizer based on the amount of fertilizer which will be distributed in one acre.

Rate Chart for Impregnating Fertilizer with "VELPAR" L
"VELPAR" L Rate Per Acre

Fertilizer Rate/Acre	2 Pints	3 Pints	4 Pints	6 Pints
250 pounds	16 pts/ton	24 pts/ton	32 pts/ton	48 pts/ton
300 pounds	13.4 pts/ton	20 pts/ton	26.8 pts/ton	40.2 pts/ton
350 pounds	11.4 pts/ton	17.2 pts/ton	22.8 pts/ton	34.2 pts/ton
400 pounds	10 pts/ton	15 pts/ton	20 pts/ton	30 pts/ton
450 pounds	8.8 pts/ton	13.2 pts/ton	17.6 pts/ton	26.4 pts/ton

For rates other than those listed, use the following formula to calculate the amounts of "VELPAR" L to be impregnated per ton of dry fertilizer.

$$\text{Pints "VELPAR" L Per acre} \times \frac{2000}{\text{Lbs Fertilizer Per Acre}} = \text{Pints "VELPAR" L Per Ton of Fertilizer}$$

Application:

Uniform application of "VELPAR" L which has been impregnated on dry fertilizer is essential for satisfactory weed control. Accurate calibration of fertilizer application equipment is essential for uniform distribution to surface. The recommended method of application is to apply ½ the recommended rate and overlap 50 percent. This results in the best distribution pattern.

REPLANTING (Following alfalfa):

Corn may be planted 12 months after the last treatment, except in areas of low rainfall (20" or less) where corn may be planted 12 months after the last treatment, provided the use rate did not exceed 3 pints per acre. Do not plant treated areas to any other crop within two years after treatment, as crop injury may result.

Do not plant seed alfalfa areas to any crop within two years after treatment, as crop injury may result.

NOTE: Because crop injury may result, do not apply to actively growing alfalfa nor to stubble between cuttings, except as recommended for northeastern* and midwestern* states. Do not apply to alfalfa-grass mixtures; do not apply to alfalfa growing under stress from insect, disease, or winter injury, or showing symptoms of other stress conditions; do not use on hardpan, alkali or poorly drained soils. Extremely heavy rainfall or more than 1" or irrigation after application may result in crop injury or non-performance. Do not apply to snow-covered or frozen ground. Do not use in low desert valleys in California or Arizona. Temporary alfalfa yellowing may occur following herbicide applications. Do not use in the states of Montana, North Dakota, South Dakota and Wyoming.

Do not graze or feed forage or hay to livestock within 30 days following application.

FORESTRY:

Since the effect of "VELPAR" L on conifers varies with soils, uniformity of application, and environment conditions, it is suggested that growers limit their first use to small areas.

Because injury (stunting) to conifers may result; do not use in nurseries, seed beds or ornamental plantings; do not use on trees which show poor vigor because of insect, disease or winter injury, or show symptoms of other stress conditions; do not add surfactant if applied over the top of conifers; do not use on poorly drained soils. In addition, conifer injury may result where severe winter stress, disease, or insect damage follows application, or heavy rains occur soon after application.

CHRISTMAS TREE PLANTATIONS:

"VELPAR" L is recommended for control of certain weeds where the following species are grown:

Austrian pine	Noble fir	Scotch pine
Grand fir	Ponderosa pine	Sitka spruce
Loblolly pine		

USE RATES:

TABLE 1

Soil Texture*	Pints "VELPAR" L Per Acre[3] First Year Plantings[1]	Pints "VELPAR" L Per Acre[3] Established Trees[2]
Gravelly or rocky soils, exposed subsoils, clay knobs, sand, or sandy soil with 85% or more sand	DO NOT USE	DO NOT USE
Loamy sand, sandy loam (50-85% sand)**	4	4 to 5
Loam, silt loam, silt and sandy clay loam	4 to 5	5 to 7
Silty clay loam, clay loam, sandy clay, silty clay, and clay	5 to 6	7 to 8

Moisture is necessary for full activation of "VELPAR" L. Best results are obtained when weeds are less than 2 inches in height or diameter, soil is moist at time of application and ½ to 1 inch of rainfall occurs within 2 weeks after application.
* Do not use on any soil containing less than 1% organic matter; do not use on loamy sand or sandy loam with less than 2% organic matter.
** On soils high in organic matter, 5% or greater, the effectiveness of "VELPAR" L can be significantly reduced and weed control may be unsatisfactory.
[1] Apply only to transplant stock that is 2 years old or more (1 year old, for loblolly pine). Allow at least two months after transplanting before applying "VELPAR" L, then apply only if rainfall has settled the soil around the base of the transplants.
[2] Rates recommended are for trees that have been established for 1 year or more.
[3] Use rates in excess of those specified for the soil texture may result in injury to the conifers.

East of the Rocky Mountains:

DO NOT USE IN:

Connecticut	Pennsylvania
Delaware	Rhode Island
Maine	Texas
Maryland	Vermont
New Hampshire	Virginia
New Jersey	West Virginia
New York	
North Carolina	

Weeds Controlled:

Annual bluegrass	Common ragweed
Barnyardgrass	Fleabane
Bentgrass	Foxtail
Bromegrass	Oxeye daisy
Catsear (False dandelion)	Pennsylvania smartweed
Common groundsel	Velvetgrass
	Wild carrot

Treatment Provides Partial Control of:

Crabgrass	Goldenrod
Curly dock	Horseweed
Dandelion	Orchardgrass
Fescue	Ryegrass
Fireweed (willowweed)	

See **Table 1** for rate selection

Apply as a broadcast spray in the spring prior to conifer bud break. If application is made after bud break, use directional spray equipment and prevent contact of conifer foliage or injury may result. Rates expressed are broadcast. For band application, use proportionately less; for example, use ½ of the broadcast rates when treating a 3 ft. band band where row spacing is 6 ft.

West of Rocky Mountains:

In addition to trees listed above, "VELPAR" L is recommended for use in Douglas Fir.

Weeds Controlled at 4 to 6 pts/acre:

Annual bluegrass	Fleabane
Barnyardgrass	Common groundsel
Bromegrass	Oxeye daisy
False dandelion (catsear)	Velvetgrass

Treatment at 8 pts/acre Provides Partial Control Of:

Crabgrass	Orchardgrass
Curlydock	Ryegrass
Dandelion	Smartweed
Fescue	Wild carrot
Goldenrod	Willowweed (fireweed)
Heath aster	

"VELPAR" L may be applied by air or ground equipment. Select a spray volume that will ensure a thorough and uniform application, but not less than 5 gallons of water for each gallon of "VELPAR" L. Use mechanical or by-pass agitation to thoroughly mix the spray solution. After initial mixing, do not use excessive agitation as foaming problems may result. To avoid foaming problems, use anti-foam agents and minimal agitation. Avoid overlapping and shut off spray booms while starting, turning, slowing or stopping, as injury to conifers may result.

Rainbelt (areas of high spring rainfall): For best results, apply in the spring when weeds are actively growing, but prior to bud break on conifers. If application is made after bud break, use directional spray equipment so that spray does not contact new growth as injury may result.

Snowbelt (areas of low spring rainfall): For best results, apply in the fall before soil freezes or in the spring in anticipation of rainfall. Weed control results from spring applications will be dependent on sufficient rainfall following treatment to activate "VELPAR" L. Spring applications should be made after snow cover melts, but before bud break occurs.

Rate selection is based on soil texture. Do not use rates in excess of those listed for a particular soil texture as injury to conifers may result. To determine soil texture and organic matter content, have soil samples classified by a reputable laboratory.

See **Table 1** for use rates.

REFORESTATION AREAS:

"VELPAR" L is recommended for use in site preparation and conifer release areas. Refer to use categories for specific recommendations for conifer species, rate, timing, and species controlled by geographic area.

Selective Brush Control:

Injection:

Inject 1 milliliter of undiluted "VELPAR" L through the bark of undesirable trees. Injections should be made at 4" intervals around the circumference of the tree. When using tubular injection equipment, inject near the ground level, or when using the "Hypo-Hatchet" Tree Injector or similar device, inject at waist height. Treatment should be made in the summer.

Species Controlled:

Black cherry	Red maple
Oaks (including live oak)	Sweetgum

Basal Soil Treatment:

Apply "VELPAR" L undiluted with an exact-delivery handgun applicator. This equipment delivers a thin stream of a predetermined volume when triggered. Apply at the rate of 2 to 4 milliliters for each 1" of stem diameter at breast height of trees to be controlled. Direct the treatment to the soil within 3 ft. of the root collar of the trees to be controlled. When treating large stems and more than one delivery of "VELPAR" L is needed per stem, make applications on opposite sides of the stem.

Species Controlled:

Aspen	Sumac
Elm	Wild cherry
Oaks	Willow

Grid Application: Undiluted basal treatment.

"VELPAR" L is recommended for use in site preparation, pre-harvest control of hardwoods, and pine release for growth enhancement of loblolly pine, longleaf pine, shortleaf pine, and slash pine.

Plants Controlled:

Black cherry	Hawthorn	Red Maple*
Blackgum*	Hickory*	Sweetgum
Dogwood*	Oak	Sumac*
Elm		

* Considered hard-to-kill.

Apply 3 to 12 qts of undiluted "VELPAR" L per acre for pre-harvest and site preparation, or 2 to 9 qts of undiluted "VELPAR" L per acre for pine release. Make the application using a "grid" pattern and rate (ml per spot) as suggested in the table.

Do not apply "VELPAR" L spots closer than 36" to seedlings or directly upslope from seedlings. Use an exact delivery handgun applicator and calibrate for precise delivery of the undiluted product.

Selection of the rate per acre and "grid" pattern will depend on soil texture and woody plant composition. Use the higher rates on fine-textured soils and when the major component of the hardwoods is considered hard-to-kill. Use the lower rates on medium to coarse-textured soils where elm, blackcherry, oak and sweetgum are dominant.

Application Patterns and Rates For Undiluted "VELPAR" L

For Site Preparation and Pre-Harvest Control of Hardwoods

	ML/Spot	Grid (Ft)	Qts/Acre
Coarse	1.0	4′ × 4′	3
	3.0	4′ × 6′	6
Medium/Fine	3.0	3′ × 6′	8
	2.33	3′ × 3′	12

For Pine Release

	ML/Spot	Grid (Ft)	Qts/Acre
Coarse	2.0	6′ × 7′	2+
	2.5	5′ × 6′	3
Medium/Fine	2.0	4′ × 6′	4
	2.33	3′ × 6′	6
	1.5	3′ × 3′	8

\+ Use on deep sands with pines four years of age.

"VELPAR" L should be applied during the period from late winter (bud break) to early summer. Do not burn treated sites after initial foliage browing is evident. Burn only after brush has completely defoliated at least once, from sufficient root uptake of the herbicide. Inconsistent results may occur if the site is burned 3 to 6 months prior to treatment.

NOTE: Use on 1 year old or 4 year old or older seedlings is acceptable; however, injury may result from use on 2 and 3 year seedlings where root growth is extensive but hardiness is lacking.

Results may be unsatisfactory where stump spouts of less than one year's growth predominate.

SITE PREPARATION:

East of Rocky Mountains:

"VELPAR" L is recommended (ground or aerial) in forestry site preparation areas where these species will be grown:

Austrian pine	Ponderosa pine	Slash pine
Balsam fir	Red pine	Spruce
Black spruce	Red spruce	Virginia pine
Loblolly pine	Scotch pine	White spruce
Longleaf pine	Shortleaf pine	

Plants Controlled:

When used as directed, "VELPAR" L provides both contact and residual control of:

Annual bluegrass	Common groundsel	Oxeye daisy
Asters	Common ragweed	Pennsylvania smartweed
Barnyardgrass	Elksedge	Pinegrass
Bentgrass	Fleabane	Velvetgrass
	Foxtail	Wild carrot
Ash	Hawthorne	Sweetgum
Aspens (big tooth, trembling)	Hazel	Willows
	Hickory	Wild cherry
Balsam poplar	Oaks	
Birch		
Elm		

Plants Partially Controlled:

Canada thistle	Fireweed (Willowweed)	Honeysuckle
Crabgrass		Horseweed
Curly dock	Flowering dogweed	Orchardgrass
Dandelion	Fescue	Quackgrass
False dandelion (carsear)	Goldenrod	Red maple
	Health aster	Sourwood
Blackgum	Perennial ryegrass	

Apply 1 to 3 gallons of "VELPAR" L per acre. For aerial application use at least 5 gallons per acre; for ground application use enough water for thorough coverage, usually 25 gals/acre. On fine textured soils and hard-to-kill species use the higher rates of "VELPAR" L; for coarser soils use a lower rate. To reduce drift, use a precise application spray boom and/or add a spray thickening agent. If a spray thickening agent is used in the tank mixture, follow all label directions.

Add the "VELPAR" L to a water filled tank and mix thoroughly. A surfactant may then be added. After thorough mixing, a defoaming agent may be added. Agitate thoroughly to disperse the defoaming agent.

Combination with Tordon 101[1]:

Apply as a foliar treatment for woody plant control for pine site preparation. In addition to the species listed above, this treatment controls woody plants such as:

Black gum	Prunus spp.	Sourwood
Pine	Sassafras	

Apply 1 to 3 gallons of "VELPAR" L plus 1 to 2 gallons of Tordon 101 per acre.

Use 1½ to 2 gallons of Tordon 101 per acre when the major components of the hardwood composition to be controlled are species other than elm, oak, or sweetgum.

When oak, elm, sweetgum, blackberry and prunus predominate, use 1½ to 3 gallons of "VELPAR" L plus 1 gallon of Tordon 101.

When blackgum, dogwood, sourwood and red maple predominate, use 1 to 3 gallons of "VELPAR" L plus 1½ to 2 gallons of Tordon 101.

"VELPAR" L, alone or in tank mixture with Tordon 101, should be applied during the period from late spring to early summer after the trees have reached full leaf, but before leaf tissue hardens; or, in Maine, Michigan, Minnesota, New Hampshire, New York, Vermont and Wisconsin, after bud break. Do not burn treated sites after initial contact (browning) effect on foliage (ten weeks or more). Burn only after brush has completely defoliated at least once from sufficient root uptake of "VELPAR" L, usually 60 to 90 days after adequate rainfall.

[1] Registered trademark of Dow Chemcial Company.

NOTE:
1. In the South, pines planted sooner than six months after treatment with Tordon 101 may be injured. For all conifers, the waiting period between treatment and planting should be 8 to 9 months in the lake states and the Northeastern U.S.
2. Do not use on wet or poorly drained sites.
3. Before using Tordon 101, read and carefully observe the cautionary statements and all other information appearing on the product label. Note in particular that Tordon 101 is a **Restricted Use Pesticide,** for retail sale to and use only by certified applicators or persons under their direct supervision and only for those uses covered by the certified applicators certification.

CONIFER RELEASE:

Southern, Northern States:

Recommended for conifer release where the following species are grown:

Balsam fir	Norway spruce	Shortleaf Pine
Black spruce	Red pine	Virginia pine
Loblolly pine	Red spruce	White spruce

Plants Controlled:

Herbaceous Plants:

Aster	Horseweed
Bracken fern	Fireweed
Common groundsel	Foxtail
Common ragweed	Panicums
Crabgrass	Quackgrass*
Dog fennel	Ryegrass*
Goldenrod	Wild carrot

Woody Plants:

Ash	Elm
Aspen	Hawthorne
Balsam poplar	Hazel
Birch	Honeysuckle
Blackgum*	Oaks (red, black and white)
Box elder	Red maple*
Brambles	Sumac*
Cherry (black and pin)	Sweetgum*
Dogwood*	Willow

* Considered hard-to-kill. Partial control or suppression only.

Apply 2 to 4 qts of "VELPAR" L per acre (red pine–2 to 8 qts, loblolly pine–1½ to 6 qts, shortleaf pine–4 qts) by air or ground. Do not use a surfactant.

The preferred pine stage of growth for application is between growth spurts, if more than one occurs. Some conifer needles may be discolored, however, full recovery can be expected the following spring. Mortality of some pine trees may occur. At least 1 to 2 inches of rainfall are needed for soil activation.

Southern States:

"VELPAR" L should be applied during the period from early spring (one-half leaf) until early summer after the hardwood trees have reached full leaf stage, but before leaf tissue has hardened.

Northern States:

Apply from late spring, at or near conifer bud break, until early summer. For maximum contact activity on woody brush, apply after brush is in ¾ leaf stage.

NOTE:
1. Do not apply where conifers are suffering from loss of vigor caused by insects, disease, drought, winter damage, or other stresses, as injury may result.
2. Excessive injury may result to conifers less than 4 years of age from transplanting on coarse-textured soils, and less than 3 years of age from transplanting on medium and fine-textured soils.
3. Poor weed and brush control may occur from use on poorly drained or marshy sites.
4. Poor weed and brush control may occur from applications made when the soil is saturated with water and rain is imminent within 24 hours.
5. Severe injury may result from use on Jack pine, White pine, or larch.
6. Use the lower rates on coarse-textured soils (i.e., sand, loamy sand, sandy loam).
7. Use the intermediate rates on medium-textured soil (i.e., loam silt loam, silt).
8. Use the higher rates on medium-fine and fine-textured soils (i.e., sandy clay loam, clay loam, silty clay loam, sandy clay, silty clay).
9. Use on gravelly or rocky soils and exposed subsoils may cause conifer injury or mortality.
10. On tracts of land where various soil types are present and "VELPAR" L rate selection is difficult, conifer damage or less than expected vegetation suppression may occur due to rate differences required for various soil types.

SITE PREPARATION, CONIFER RELEASE:

West of Rocky Mountains:

"VELPAR" L is recommended for use in forestry site preparation and conifer release where the following species are grown:

Douglas fir	Noble fir	Sitka spruce
Grand fir	Ponderosa pine	White fir
Jeffrey pine		

Plants Controlled:

Annual bluegrass	Common groundsel	Oxeye daisey
Barnyardgrass	Fleabane	Velvet grass
Bromegrass	Foxtail	Wild carrot
Catsear (false dandelion)		

Deerbrush ceanothus	Snowbrush ceanothus (varnishleaf ceanothus)
Greenleaf manzanita	Squawcarpet

Plants partially controlled:

Dandelion	Heath aster	Smartweed
Fescue	Orchardgrass	Willowweed (fireweed)
Goldenrod	Ryegrass	

Apply as a pre-plant or post-plant broadcast or band application. For band applications, use proportionately less; for example, use ½ the broadcast rate when treating a 3 ft. band where row spacing is 6 ft.

For herbaceous control, use 2 to 4 quarts per acre and 3 to 6 quarts per acre for brush control. Use the lower rates on coarse-textured soils and soils low in organic matter; use the higher rates on fine textured soils and soils high in organic matter.

For brush control, applications should be made to brush seedlings or sprouts when less than 18" in height or diameter. "VELPAR" L effects may be slow to appear and maximum results will be seen 12 to 24 months following the application.

Rainbelt (areas of high spring rainfall): For best results, apply in the late winter or spring when weeds are actively growing, but prior to bud break on conifers. If application is made after bud break, use directional spray equipment so spray does not contact new growth, as injury may result.

Weed control is dependent on sufficient rainfall following treatment to activate "VELPAR" L. Apply "VELPAR" L by air or ground equipment. Select a spray volume that will insure a thorough and uniform application, at least 8 gals. of spray per acre by air and usually 25 gals. of spray per acre by ground, but not less than 5 gals. of water for each gallon of "VEL-

PAR" L. Use mechanical or by-pass agitation to thoroughly mix the spray solution. After initial mixing, do not use excessive agitation as foaming problems may result. To avoid foaming problems, use anti-foam agents and minimal agitation.

Snowbelt (areas of low spring rainfall): For best results, apply in the fall before soil freezes, or in the spring in anticipation of rainfall. Weed control results from spring applications will be dependent on sufficient rainfall following treatment to activate "VELPAR" L. Spring applications should be made after snow cover melts, but before bud break occurs.

NOTE:
1. Use on gravelly or rocky soils, exposed subsoils, clay knobs, sand or sandy soils with 85% or more sand may cause conifer injury or mortality and poor weed and brush control may occur.
2. Conifer injury or mortality may occur on any soil containing less than 1% organic matter.
3. On soils high in organic matter (greater than 5%) effectiveness of "VELPAR" L is reduced and results may be unsatisfactory.
4. Use on loamy sand or sandy loam with less than 2% organic matter may cause conifer injury except for Ponderosa and Jeffrey pine.
5. On first year plantations, "VELPAR" L may only be applied (either pre-plant or post-plant) to transplant stock that is 2 years old (2-0) or more, except for 1 year old (1-0) for Ponderosa and Jeffrey pine.
6. When applying "VELPAR" L after transplanting, wait until rainfall has settled the soil around the base of the transplants before making the treatment.
7. Damage may occur to Incense cedar, Sugar pine and Giant Sequoia.
8. Heavy duff or slash present at time of application may result in poor weed or brush control.
9. Poor weed and brush control may occur from use on poorly drained or marshy sites.
10. Poor weed and brush control may occur from applications made when soil is saturated with water and rain is imminent within 24 hours.

GENERAL WEED AND BRUSH CONTROL—NONCROPLAND AREAS

"VELPAR" L is recommended for general weed and brush control in noncropland areas such as railroads, highways, utility and pipeline rights-of-ways, petroleum tank farms, storage areas, industrial plant sites, drainage ditchbanks and other similar areas.

For broadcast treatments (ground equipment only), apply "VELPAR" L as a spray just before or soon after weed emergence. Do not apply to frozen or snow-covered soil. Before spraying, calibrate equipment to determine the quantity of water necessary to uniformly and thoroughly cover the vegetation and soil in a measured area to be treated. Use at least 5 gallons of water for each gallon of "VELPAR" L.

Add the proper amount of "VELPAR" L to a spray tank filled with the amount of water to be used and mix thoroughly. A surfactant at .25% by volume may then be added to improve wetting and/or contact activity. After thoroughly mixing the spray solution, a suitable defoaming agent may also be added should foaming become a problem. Agitate thoroughly to disperse the defoaming agent.

Use a sprayer properly calibrated to a constant speed and rate of delivery. For application with a handgun sprayer, use sufficient water (50 to 200 gals per acre) for thorough coverage.

WEED CONTROL

SHORT-TERM CONTROL (up to 3 months):

Apply "VELPAR" L at the rate of 1 to 2½ gallons per acre for control of many annual, biennial and perennial weeds:

Barnyardgrass	Curly dock*	Milkweed*
Bindweed*	Dandelion*	Mustard
Bouncing bet*	Dogbane*	Nutsedge*
Bromegrass	Fiddleneck	Orchardgrass*
Buffalograss*	Filaree	Oxalis
Burdock	Fleabane	Paragrass
Cocklebur	Goatsbeard vine	Pigweed
Crabgrass	Goldenrod	Purslane
Crown vetch	Lespedeza	Quackgrass
Ryegrass	Smartweed	Spurge
Trumpet creeper*	Wild oats*	Wild parsnip
		Wild star thistle

* Considered hard-to-kill; partial control or suppression only.

SEASON-LONG CONTROL:

Apply "VELPAR" L at the rate of 3 to 6 gallons per acre. Species controlled:

Bahiagrass*	Dogfennel*	Natalgrass
Bermudagrass*	Fescue*	Plantain
Bluegrass	Fingergrass	Prickly lettuce
Broomsedge	Foxtail	Ragweed
Camphorweed	Guineagrass	Smutgrass
Canada thistle*	Heath aster	Spanish nettle
Chickweed	Honeysuckle	Vaseygrass
Clovers	Lantana	Wild blackberry*
Dewberry	Marestail	Wild carrot

* Considered hard-to-kill; partial control or suppression only.

BRUSH CONTROL:

Apply "VELPAR" L as a broadcast treatment or an undiluted basal soil treatment for the control of many woody species. Apply to the soil during the period between late winter and early summer. In areas where the soil remains frozen during the winter, and spring rains are usually inadequate for soil activation, a fall or winter treatment may be applied before the soil freezes.

BROADCAST TREATMENT:

Apply 2 to 4 gallons of "VELPAR" L per acre as a coarse spray with water. Direct the spray to the soil beneath woody plants to be controlled.

BASAL SOIL TREATMENT

Apply "VELPAR" L undiluted with an exact-delivery handgun applicator. This equipment delivers a thin stream of a pre-determined volume when triggered. Apply at the rate of 2 to 4 milliliters for each 1" of stem diameter at breast height. Direct the treatment to the soil within 3 ft of the root collar of woody plants to be controlled. When treating large stems and more than one delivery of "VELPAR" L is needed per stem, make applications on opposite sides of the stem.

Plants Controlled:

Blackgum	Green Ash	Persimmon
Blackcherry	Hawthorn	Sumac
Elm	Hickory	Sweetgum
Flowering dogwood	Oaks (except live oak)	Willow

LACING AND STREAKING TREATMENT:

"VELPAR" L is recommended for control of undesirable woody plants in noncropland areas by the lacing and streaking method.

Lacing and streaking is a method which employs a straight stream application of a concentrated solution of "VELPAR" L in water to the soil. Apply to soil during the period from late winter and early summer. Moisture is necessary to activate "VELPAR" L. In areas where the soil normally remains frozen during the winter, and spring rains are usually inadequate for activation, a fall or winter treatment may be applied before the soil freezes. Do not apply to frozen or snow-covered soil.

Apply 2 to 6 gallons of "VELPAR" L per acre. Prepare by mixing 1 gallon of "VELPAR" L with 5 or more gallons of water. Use lower rates on sands and higher rates on clay and/or hard-to-kill species. Direct the spray to the soil in a serpentine pattern using a handgun so that the swath on the soil is 6 to 12 inches wide at the base of the brush to intercept the major portion of the root system and swaths are 2 to 4 feet apart.

Plants Controlled:

Alder	Hawthorne	Southern red oak
American elm	Hickory	Sumac
Ash	Huisache	Sweet bay
Black cherry	Locust	Sweet gum
Blackgum	Mesquite	Water oak
Chinaberry*	Mulberry	Wild dogwood
Chinese elm	Oak	Wild plum
Chinese tallow	Privet	Willow
Eastern red cedar*	Sassafrass*	Winged elm
Hackberry	Sourwood	Yellow poplar

* Considered hard-to-kill. Partial control or suppression only.

NOTE: For effective brush control and prevention of damage of desirable vegetation: do not apply to brush standing in water; do not use water from treated ditches for irrigation; do not use where roots of desirable trees or shrubs are immediately adjacent to the treated area.

NOTICE OF WARRANTY

Du Pont warrants that this product conforms to the chemical description on the label thereof and is reasonably fit for purposes stated on such label only when used in accordance with directions under normal use conditions. It is impossible to eliminate all risks inherently associated with the use of this product. Crop injury, ineffectiveness or other unintended consequences may result because of such factors as weather conditions, presence of other materials or the manner of use or application, all of which are beyond the control of Du Pont. In no case shall Du Pont be liable for consequential, special or indirect damages resulting from the use or handling of this product. All such risks shall be assumed by the Buyer. DU PONT MAKES NO WARRANTIES OF MERCHANTABILITY OR FITNESS FOR A PARTICULAR PURPOSE NOR ANY OTHER EXPRESS OR IMPLIED WARRANTY EXCEPT AS STATED ABOVE.

STORAGE AND DISPOSAL

STORAGE: Do not subject to temperatures below 32°F. Store product in original container only, away from other pesticides, fertilizer, food or feed.

PESTICIDE DISPOSAL: Do not contaminate water, food or feed by storage or disposal. Pesticide wastes are acutely hazardous. Improper disposal of excess pesticide, spray mixture, or rinsate is a violation of federal law. If these wastes cannot be disposed of by use according to label instructions, contact your State Pesticide or Environmental Control Agency, or the Hazardous Waste representative at the nearest EPA regional office for guidance.

CONTAINER DISPOSAL: Triple rinse (or equivalent) the container. Then offer for recycling or reconditioning, or puncture and dispose of in a sanitary landfill, by incineration, or, if allowed by state and local authorities, by burning. If burned, stay out of smoke.

NOTICE TO BUYER: Purchase of this material does not confer any rights under patents of countries outside of the United States.

AG-3203 8127/8088 10/28/87

This is a specimen label. DuPont is not responsible for the accuracy of the information contained herein. As labels are subject to revision, always carefully read and follow the label on the product container.

VELPAR® ULW®

HERBICIDE
SOLUBLE GRANULES

ACTIVE INGREDIENT:	BY WEIGHT
Hexazinone [3-cyclohexyl-6-(dimethylamino)-1-methyl,3,5-triazine-2,4(1H,3H)-dione]	75%
INERT INGREDIENTS	25%
TOTAL	100%

EPA Reg. No. 352–450

U.S. Pats. 3,902,887 & 3,983,116

KEEP OUT OF REACH OF CHILDREN

PRECAUTIONARY STATEMENTS

PELIGRO

PRECAUCION AL USUARIO: Si usted no lee ingles, no use este producto hasta que la etiqueta haya sido explicada ampliamente.

HAZARDS TO HUMANS AND DOMESTIC ANIMALS

DANGER!

CAUSES EYE DAMAGE.

Do not get in eyes. Wear goggles or face shield and rubber gloves when handling. Avoid contact with skin and clothing. Avoid breathing dust or spray mist. Harmful if swallowed.

In case of contact, immediately flush eyes or skin with plenty of water for at least 15 minutes. For eyes, call a physician. Get medical attention if skin irritation persists. Remove and wash contaminated clothing before re-use.

For medical emergencies involving this product, call toll free 1–800–441–3637.

ENVIRONMENTAL HAZARDS

Do not apply directly to water or wetlands. Do not apply where runoff is likely to occur. Do not contaminate water by cleaning of equipment or disposal of wastes.

IMPORTANT—Injury to or loss of desirable trees or other plants may result from failure to observe the following: Do not apply except as recommended or drain or flush equipment on or near desirable trees or other plants, or on areas where their roots may extend, or in locations where the chemical may be washed or moved into contact with their roots. Do not use on lawns, walks, driveways, tennis courts, or similar areas. Prevent drift of granule to desirable plants. Keep from contact with fertilizers, insecticides, fungicides and seeds. Thoroughly clean all traces of "Velpar" ULW Herbicide from application equipment immediately after use. Wash equipment thoroughly with water.

GENERAL INFORMATION

"Velpar" ULW Herbicide is a special granular product to be distributed in a broadcast pattern on the soil surface in reforestation areas for control of undesirable woody and herbaceous plants. Rainfall dissolves the chemical (hexazinone) from the granules and moves the herbicide into the root zone where it is absorbed during periods of active plant growth.

Symptoms of herbicidal activity generally appear within 3 to 4 weeks for herbaceous plants and 4 to 8 weeks for woody plants, after sufficient rainfall has carried the herbicide into the root zone. Defoliation and subsequent refoliation may occur, but susceptible plants are killed.

DIRECTIONS FOR USE

It is a violation of federal law to use this product in a manner inconsistent with its labeling.

"Velpar" ULW should be used only in accordance with recommendations on this label.

Du Pont will not be responsible for losses or damages resulting from use of this product in any manner not specifically rcommended by Du Pont. User assumes all risk associated with such non-recommended use.

HOW TO APPLY

Apply the granules using Du Pont ULW Applicator aerial equipment. Calibrate the equipment to distribute the granules uniformly over the area to be treated. Wear gloves when handling granules to prevent contact with skin. Apply before or during the period of active growth, when rainfall can be expected to activate the chemical in the soil. Do not apply to frozen soil, or to water saturated soils when rainfall is expected within 24 hours. Distribution across treatment swath must be uniform. Do not apply this product through any type of irrigation system.

FOR CONTROL OF CERTAIN WEED AND BRUSH SPECIES FOR FORESTRY SITE PREPARATION AND CONIFER RELEASE

"Velpar" ULW is recommended for forestry site preparation and conifer release in areas of greater than 20 inches of annual rainfall east of the Rocky Mountains where loblolly, longleaf, red, shortleaf, slash and Virginia pines are grown. "Velpar" ULW is also recommended for site preparation where white spruce will be planted.

When used as directed, "Velpar" ULW controls herbaceous weeds such as annual bluegrass, barnyardgrass, bentgrass, bromegrass, catsear, common groundsel, common ragweed, false-dandelion, fleabane, foxtail, oxeye daisy, Pennsylvania smartweed, velvetgrass, and wild carrot. Treatment provides partial control of curly dock, dandelion, willowweed (fireweed), fescue, goldenrod, horseweed, orchardgrass, quackgrass and ryegrass.

"Velpar" ULW, when used as directed also controls many woody plants including American elder, ashes* (white, green), aspens (big tooth, trembling), balsam poplar, birch, blackcherry, blackgum*, brambles (such as blackberry, dewberry, raspberry), cherry*, chokecherry*, dogwood* (flowering), eastern red cedar*, elm, hawthorne, hazel*, hickory*, honeysuckle, hornbeam, mulberry, multiflora rose, red maple*, oaks (such as post, southern red, turkey, water and white), Russian olive, sourwood*, sumac*, sweetgum, wild plum and willow.

* considered hard-to-kill

APPLY AT THE FOLLOWING RATES FOR SOIL TEXTURE:

Pounds Product Per Acre

Soil Texture	Site Preparation[1]	Conifer Release[2]
Sand, Loamy Sand, Sandy Loam	2½ to 4	1 to 2
Loam, Sandy Clay Loam, Silt Loam	4 to 5⅓	2 to 3
Clay Loam, Sandy Clay, Silty Clay Loam, Silt, Silty Clay, Clay	5⅓ to 8	3 to 4

NOTE: Use the lower rates on coarse-textured soils and soils low in organic matter; use the higher rates on fine-textured soils and soils high in organic matter. Use the higher rate when hard-to-kill species predominate.

"Velpar" ULW is not recommended on poorly drained soils or marshy sites. "Velpar" ULW results may be slow to appear and maximum results will be seen 12 to 24 months following application.

[1] **In Site Preparation Areas,** where burning is desired, burn only after residual stand has completely defoliated at least twice, allowing for sufficient root uptake of "Velpar" ULW usually 60 to 90 days after rainfall.

[2] **In Conifer Release Areas,** some mortality of conifers may be experienced when application is erratic or sapling vigor is lacking due to damage by insects, disease, drought, winter stress or other stress. Excessive injury may result when applications are made to loblolly pine that are less than 4 years of age since transplanting on coarse textured soils or 3 years of age since transplanting on fine textured soils. Application to gravelly, rocky or exposed subsoil may result in conifer injury. When soil texture varies abruptly, inconsistent results may occur.

TIMING: For best results, apply in spring when weeds and brush are growing, after the danger of a killing frost has passed (for target species) and when sufficient rainfall for activation of the herbicide is expected.

STORAGE AND DISPOSAL

STORAGE: Do not contaminate water, food or feed by storage or disposal.

PESTICIDE DISPOSAL: Pesticide wastes are acutely hazardous. Improper disposal of excess pesticide spray mixture or rinsate is a violation of Federal Law. If these wastes cannot be disposed of by use according to label instructions, contact your State Pesticide or Environmental Control Agency, or the Hazardous Waste representative at the nearest EPA Regional Office for guidance.

CONTAINER DISPOSAL: Completely empty bag by shaking and tapping sides and bottom to loosen clinging particles. Empty residue into application equipment, then, dispose of bags in a sanitary landfill or by incineration, or, if allowed by state and local authorities, by burning. If burned, stay out of smoke.

NOTICE OF WARRANTY

Du Pont warrants that this product conforms to the chemical description on the label thereof and is reasonably fit for purposes stated on such label only when used in accordance with directions under normal use conditions. It is impossible to eliminate all risk inherently associated with use of this product. Ineffectiveness or other unintended consequences may result because of such factors as weather conditions, presence of other materials or the manner of use or application, all of which are beyond the control of Du Pont. In no case shall Du Pont be liable for consequential, special or indirect damages resulting from the use or handling of this product. All such risks shall be assumed by the buyer. DU PONT MAKES NO WARRANTIES OF MERCHANTABILITY OR FITNESS for a particular purpose nor any other express or implied warranty except as stated above.

NOTICE TO BUYER

Purchase of this material does not confer any rights under patents of countries outside of the United States.

AG–3205 8088 1/13/88

This is a specimen label. DuPont is not responsible for the accuracy of the information contained herein. As labels are subject to revision, always carefully read and follow the label on the product container.

VENDEX® 4L

MITICIDE

ACTIVE INGREDIENTS:	BY WEIGHT
Fenbutatin-oxide[Hexakis (2-methyl-2-phenylpropyl)-distannoxane]	42.0%
INERT INGREDIENTS:	58.0%
TOTAL	100.0%

Contains 4 lbs. active ingredient per gallon

This product contains the toxic inert ingredient formaldehyde.

EPA Reg. No. 352-493

U.S. Patent No. 3,657,451

KEEP OUT OF REACH OF CHILDREN

DANGER!PELIGRO

PRECAUCION AL USUARIO: Si usted no lee ingles, no use este producto hasta que la etiqueta haya sido explicada ampliamente.

STATEMENT OF PRACTICAL TREATMENT

If In Eyes: Hold eyelids open and flush with plenty of water. Call a physician.

If On Skin: Wash with plenty of soap and water. Get medical attention if irritation persists.

If Swallowed: Drink 1 or 2 glasses of water and induce vomiting by touching back of throat with finger. Do not induce vomiting or give anything by mouth to an unconscious person.

If Inhaled: Remove victim to fresh air. If not breathing, give artificial respiration preferably mouth-to-mouth. Get medical attention.

PRECAUTIONARY STATEMENTS

HAZARDS TO HUMANS AND DOMESTIC ANIMALS

DANGER! CORROSIVE. THIS PRODUCT MAY CAUSE SKIN IRRITATION, RESPIRATORY IRRITATION, AND EYE DAMAGE. Do not get in eyes, on skin, or on clothing. Wear suitable eye protection, such as goggles or safety glasses with side shields or their equivalent, when pouring product from container. During commercial application or other prolonged exposure, use goggles and a face mask. Wear clean protective clothing that covers the body well.

Do not graze or feed animals on cover crops grown in treated area. Avoid contamination of food, feedstuffs, and domestic water supplies.

Harmful if swallowed. Avoid contamination of food. Avoid breathing spray mist. Wash thoroughly with soap and water after handling, spraying and before eating or smoking. Note Reentry Statements Section.

NOTE TO PHYSICIAN: Probable mucosal damage may contraindicate the use of gastric lavage. Measures against circulatory shock, respiratory depression, and convulsion may be needed.

For medical emergencies involving this product, call toll free 1-800-441-3637.

ENVIRONMENTAL HAZARDS

This product is toxic to fish. Do not apply directly to water or wetlands (swamps, bogs, marshes, and potholes). Drift from runoff from treated areas may be hazardous to aquatic organisms in neighboring areas. Do not contaminate water when disposing of equipment washwaters.

DIRECTIONS FOR USE

It is a violation of Federal law to use this product in a manner inconsistent with its labeling.

Do not apply this product through any type of irrigation system.

SHAKE WELL BEFORE USING.

"VENDEX" 4L Miticide is recommended for the control of a wide range of plant feeding mites, including strains that are resistant to other miticides. "VENDEX" 4L may be applied when honeybees are present and when beneficial mites are important in integrated control programs. Shake "VENDEX" 4L container to mix contents thoroughly before adding to spray tank. "VENDEX" 4L Miticide can readily be dispersed in water and can be used in conventional dilute sprayers or concentrate sprayers. Agitation is required during mixing and spraying. Thorough and complete coverage of foliage and fruit is necessary for optimum mite control.

In order to apply the correct amount of VENDEX dilute spray to your orchard, you must know the number of gallons of water needed to spray one acre of your trees to the point of drip. If you do not know this gallonage, you should conduct a test to determine it. Equipment dealers may be of assistance in conducting these tests.

Apply this product only as recommended on this label by ground application.

REENTRY STATEMENTS

Do not apply this product in such a manner as to directly or through drift expose workers or others persons. The area being treated must be vacated by unprotected persons.

Do not enter treated fields within 24 hours of application unless long pants and long-sleeved shirt (coveralls); chemical resistant gloves; shoes; and goggles or a face shield are worn.

Because certain states may require more restrictive reentry intervals for various crops treated with this product, consult your State Department of Agriculture for further information.

Written or oral warnings must be given to workers who are expected to be in a treated area or in an area about to be treated with this product. (Indicate specific oral warnings which inform workers of areas or fields that may not be entered without specific protective clothing, period of time field must be vacated and appropriate actions to take in case of accidental exposure.)

When oral warnings are given, warnings shall be given in a language customarily understood by workers. Oral warnings must be given if there is reason to believe that written warnings cannot be understood by workers.

Written warnings must include the following information: "DANGER! Area treated with "VENDEX" Miticide on (date of application). Do not enter without appropriate protective clothing until sprays have dried. (Insert here actions to take in case of accidental exposure. See STATEMENT OF PRACTICAL TREATMENT for details.)"

MANAGING MITE RESISTANCE

As with all acaricides, repeated exclusive use of VENDEX may lead to the build-up of resistant strains of mites in some crops. If more than one application of an acaricide is needed to control heavy and prolonged populations of mites, the choice of products with different modes of action should be considered. Additional application(s) of VENDEX during the growing season should be considered only after a minimum spray interval in excess of one generation of mites. Close observation and monitoring of field populations of mites should be an integral part of any resistance management program. If less than economic levels of control are observed and resistance is suspected, it is recommended that an acaricide with an alternate mode of action be used. In this eventuality, promptly consult your local State Agricultural Extension Service for advice on the selection of alternative materials.

DOSAGE AND CROP RECOMMENDATIONS

CROP	MITES CONTROLLED	FOR CONCENTRATE SPRAYS VENDEX 4L PER ACRE [1]	FOR DILUTE SPRAYS VENDEX 4L PER 100 GAL.	REMARKS
Apples	Apple Rust Carpini Spider European Red McDaniel Spider Two-Spotted Spider	1 to 3 pints	4 to 8 fl. oz. (1/4 to 1/2 pt.)	(1) Do not apply more than 600 gals. dilute spray per acre. (2) Apply when mites first appear. Repeat as necessary to maintain control. (3) Do not apply more than 4 times per season including a pre-bloom spray; apply this product no more than 3 times between petal fall and harvest. (4) Do not apply within 14 days of harvest.
Pears	European Red McDaniel Spider Pear Rust Two-Spotted Spider	1 to 3 pints	4 to 8 fl. oz. (1/4 to 1/2 pt.)	
Citrus	Citrus Red Citrus Rust Texas Citrus Yuma Spider Two-Spotted Spider	2 to 4 pints	4 to 8 fl. oz. (1/4 to 1/2 pt.)	(1) Do not apply more than 1600[2] gals. dilute spray per acre. (2) Apply when mites first appear. Repeat as necessary to maintain control. (3) Do not apply more than 2 sprays in any 12-month period. Do not apply at less than 60 day intervals. (4) Do not apply within 7 days of harvest.
Eggplant	Clover's Mite Two-Spotted Spider Mite	2 to 4 pints		(1) Apply when mites appear. Repeat as necessary to maintain control. Thorough coverage to all leaf surfaces is required for best control. (2) In Florida, apply when needed or every 1-2 weeks. (3) Do not exceed 6 applications per season with a total of no more than 6 pounds of active ingredient per acre per year. (4) Do not harvest within 3 days of last application.
Grapes	Pacific Spider European Red Two-Spotted Spider	1 to 2.5 pints	8 to 16 fl. oz. (1/2 to 1 pt.)	(1) Do not apply more than 250 gals. per acre. (2) Apply when mites first appear. Repeat as necessary to maintain control but not more often than once every 21 days. (3) Do not apply more than 2 times per season. (4) Do not apply within 28 days of harvest.
Peaches **Nectarines** **Plums** **Prunes**	European Red McDaniel Spider Pacific Spider Two-Spotted Spider Big Beaked Plum Peach Silver Plum Nursery	1 to 2 pints	4 to 8 fl. oz. (1/4 to 1/2 pt.)	(1) Do not apply more than 400 gals. dilute spray per acre. (2) Apply when mites first appear. Repeat as necessary to maintain control. (3) Do not apply more than 2 times per season. (4) Do not apply within 14 days of harvest.
Cherries (including both sweet and sour varieties)	European Red McDaniel Spider Cherry Rust	1.5 to 3 pints	4 to 8 fl. oz. (1/4 to 1/2 pt.)	(1) Do not apply more than 600 gals. dilute spray per acre. (2) Apply when mites first appear. Repeat as necessary to maintain control. (3) Do not apply more than 2 sprays while fruit is on the trees. (4) Do not apply within 14 days of harvest.
Almonds	Pacific Spider European Red Two-Spotted Spider Hickory Spider	1 to 2.5 pints	4 to 8 fl. oz. (1/4 to 1/2 pt.)	(1) Do not apply more than 500 gals. per acre. (2) Apply when mites first appear. Repeat as necessary to maintain control. (3) Do not apply more than 2 times per season. (4) Do not apply within 14 days of harvest.
Pecans	Pecan Hickory			
Walnuts	Two-Spotted Spider European Red Pacific Spider			
Strawberries	Two-Spotted Spider	1.5 to 2 pints	12 to 16 fl. oz. (3/4 to 1 pt.)	(1) Apply in 150-200 gal. per acre. (2) Apply when mites first appear. Repeat as necessary to maintain control but not more than 4 times per season. (3) Do not apply within 1 day of harvest. (4) Good coverage is essential to good mite control. Adjust spray volume and nozzle placement to adjust good coverage of top and underside of leaves.
Green house and outdoor ornamentals (including nurserystock, flowers, and plants grown for propagation purposes.)	Spruce Spider Olingonychus Two-Spotted Spider	—	8 to 16 fl. oz. (1/2 to 1 pt.)	(1) Do not add oil to the spray solution. (2) Apply when mites first appear. Repeat as necessary to maintain control. Frequent repeat applications may cause the appearance of visible spray residues on foliage. (3) Apply to the foliage only of chrysanthemums (pre-bloom) and poinsettias (prebract). (4) Under green house conditions, foliage and flowers of certain species may demonstrate sensitivity to repeat applications. If in doubt, make test applications prior to general spraying. Occasional minor sensitivity has been observed on the following species.[3]

[1] For Concentrate Sprays — Adjust the dosage to apply an amount not exceeding that used in a dilute spray.

[2] Do not exceed 4 pints VENDEX 4L per acre per application.

[3] Salix melanostachys (common willow), Asplenium bulbiferum (spleenwort garden fern), Dryopteris erythrosora (wood fern), Cercis canadensis (common redbud), Camellia japonica (red garden camellia), Pellaea rotundifolia (button fern), Davallia fejeensis (rabbit foot fern), Asparagus meyeri (Meyers asparagus fern), Asplenium nidus (birdnest fern), Adiantum cuneatum (maidenhair fern), Celosia argentea (cockscomb), Verbena hortensis (verbena), Ageratum houstonianum (floss flower), Rosa sp. (common rose — some varieties esp. yellow).

STORAGE AND DISPOSAL

STORAGE: Store in a secure, dry and temperate area. Keep container closed when not in use. Do not store near food or feed. Do not use or store around the home. Avoid contact with water. In case of spill or leak, soak up with sand, earth or synthetic absorbent and dispose of wastes in compliance with local, State and Federal regulations.

In case of a significant spill, call CHEMTREC: (800) 424-9300.

PESTICIDE DISPOSAL: Pesticide wastes are acutely hazardous. Improper disposal of excess pesticide, spray mixture, or rinsate is a violation of Federal law. If these wastes cannot be disposed according to label instructions, contact your State Pesticide or Environmental Control Agency or Hazardous Waste Representative at the nearest EPA Regional Office for guidance.

CONTAINER DISPOSAL: Triple rinse (or equivalent) and offer for recycling or reconditioning, or dispose of in a sanitary landfill or by incineration if permitted by State and local authorities.

NOTICE OF WARRANTY

Du Pont warrants that this product conforms to the chemical description on the label thereof and is reasonably fit for the purposes stated on such label only when used in accordance with the directions under normal use conditions. It is impossible to eliminate all risks inherently associated with the use of this product. Crop injury, ineffectiveness or other unintended consequences may result because of such factors as weather conditions, presence of other materials, or the manner of use or application, all of which are beyond the control of Du Pont. In no case shall Du Pont be liable for consequential, special or indirect damages resulting from the use or handing of this product. All such risks shall be assumed by the buyer. DU PONT MAKES NO WARRANTIES OF MERCHANTABILITY OR FITNESS FOR A PARTICULAR PURPOSE NOR ANY OTHER EXPRESS OR IMPLIED WARRANTY EXCEPT AS STATED ABOVE.

AG-3274 8089 7/31/89

This is a specimen label. DuPont is not responsible for the accuracy of the information contained herein. As labels are subject to revision, always carefully read and follow the label on the product container.

VENDEX® 50WP

MITICIDE

Water Soluble Bag

ACTIVE INGREDIENTS	BY WEIGHT
Fenbutatin-oxide[Hexakis(2-methyl-2-phenylpropyl)-distannoxane]. .	50%
INERT INGREDIENTS:	50%
TOTAL	100%

EPA Reg. No. 352-480

U.S. Patent No. 3,657,451

KEEP OUT OF REACH OF CHILDREN

DANGER!

PELIGRO

PRECAUCION AL USUARIO: Si usted no lee ingles, no use este producto hasta que la etiqueta haya sido explicada ampliamente.

STATEMENT OF PRACTICAL TREATMENT

If In Eyes: Hold eyelids open and flush with plenty of water. Call a physician.

If On Skin: Wash with plenty of soap and water. Get medical attention if irritation persists.

If Swallowed: Drink 1 or 2 glasses of water and induce vomiting by touching back of throat with finger. Do not induce vomiting or give anything by mouth to an unconscious person.

If Inhaled: Remove victim to fresh air. If not breathing, give artificial respiration preferably mouth-to-mouth. Get medical attention.

PRECAUTIONARY STATEMENTS

HAZARDS TO HUMANS AND DOMESTIC ANIMALS

DANGER! CORROSIVE. THIS PRODUCT MAY CAUSE SKIN IRRITATION, RESPIRATORY IRRITATION, AND EYE DAMAGE. Do not get in eyes, on skin, or on clothing. Wear suitable eye protection, such as goggles or safety glasses with side shields or their equivalent, when handling product. During commercial application or other prolonged exposure, wear goggles. Wear clean protective clothing that covers the body well. Avoid handling the inner bag as moisture will cause breakage.

Do not graze or feed animals on cover crops grown in treated areas. Avoid contamination of food, feedstuffs, and domestic water supplies.

Harmful if swallowed. Avoid contamination of food. Avoid breathing spray mist. Wash thoroughly with soap and water after handling, spraying, and before eating or smoking. Note Reentry Statements Section.

NOTE TO PHYSICIAN: Probable mucosal damage may contraindicate the use of gastric lavage. Measures against circulatory shock, respiratory depression, and convulsion may be needed.

For medical emergencies involving this product, call toll free 1-800-441-3637.

ENVIRONMENTAL HAZARDS

This product is toxic to fish. Do not apply directly to water or wetlands (swamps, bogs, marshes, and potholes). Drift from runoff from treated areas maybe hazardous to aquatic organisms in neighboring areas. Do not contaminate water when disposing of equipment washwaters.

CAREFULLY OPEN ENVELOPE AND IMMEDIATELY DROP THE INNER BAG INTO THE SPRAY TANK. DO NOT OPEN OR HANDLE THE INNER BAG. MOISTURE WILL CAUSE BREAKAGE.

DIRECTIONS FOR USE

It is a violation of Federal law to use this product in a manner inconsistent with its labeling.

Do not apply this product through any type of irrigation system.

VENDEX 50WP Miticide is recommended for the control of a wide range of plant feeding mites, including strains that are resistant to other miticides. VENDEX 50WP may be applied when honeybees and beneficial mites are present. VENDEX 50WP Miticide can readily be dispersed in water and can be used in conventional dilute sprayers or concentrate sprayers. Agitation is required during mixing and spraying. Thorough and complete coverage of foliage and fruit is necessary for optimum mite control.

In order to apply the correct amount of VENDEX dilute spray to your orchard, you must know the number of gallons of water needed to spray one acre of your trees to the point of drip. If you do not know this gallonage, you should conduct a test to determine it. Equipment dealers may be of assistance in conducting these tests.

Apply this product only as recommended on this label by ground application.

REENTRY STATEMENTS

Do not apply this product in such a manner as to directly or through drift expose workers or other persons. The area being treated must be vacated by unprotected persons.

Do not enter treated fields within 24 hours of application unless long pants and long-sleeved shirt (coveralls); chemical resistant gloves; shoes; and goggles or a face shield are worn.

Because certain states may require more restrictive reentry intervals for various crops treated with this product, consult your State Department of Agriculture for further information.

Written or oral warnings must be given to workers who are expected to be in a treated area or in an area about to be treated with this product. (Indicate specific oral warnings which inform workers of areas or fields that may not be entered without specific protective clothing, period of time field must be vacated and appropriate actions to take in case of accidental exposure.)

When oral warnings are given, warnings shall be given in a language customarily understood by workers. Oral warnings must be given if there is reason to believe that written warnings cannot be understood by workers. Written warnings must include the following information: "DANGER! Area treated with VENDEX Miticide on (date of application). Do not enter without appropriate protective clothing until sprays have dried. (Insert here actions to take in case of accidental exposure. See STATEMENT OF PRACTICAL TREATMENT for details.)"

MANAGING MITE RESISTANCE

As with all acaricides, repeated exclusive use of VENDEX may lead to the build-up of resistant strains of mites in some crops. If more than one application of an acaricide is needed to control heavy and prolonged populations of mites, the choice of products with different modes of action should be considered. Additional application(s) of VENDEX during the growing season should be considered only after a minimum spray interval in excess of one generation of mites. Close observation and monitoring of field populations of mites should be integral part of any resistance management program. If less than economic levels of control are observed and resistance is suspected, it is recommended that an acaricide with an alternate mode of action be used. In this eventuality, promptly consult your local State Agricultural Extension Service for advice on the selection of alternative materials.

DOSAGE AND CROP RECOMMENDATIONS[1]

CROP	MITES CONTROLLED	FOR CONCENTRATE SPRAY VENDEX 50[2] POUNDS PER ACRE	FOR CONCENTRATE SPRAY VENDEX 50[2] WATER SOLUBLE BAGS PER ACRE	FOR DILUTE SPRAY VENDEX 50 OZS. PER 100 GALS.	FOR DILUTE SPRAY VENDEX 50 WATER SOLUBLE BAGS PER 400 GALS.	REMARKS
CROP	MITES CONTROLLED	POUNDS PER ACRE	WATER SOLUBLE BAGS PER ACRE	OZS. PER 100 GALS.	WATER SOLUBLE BAGS PER 400 GALS.	REMARKS
Apples	Apple Rust Carpini Spider European Red McDaniel Spider Two-Spotted Spider	1 to 3	1 to 3	4 to 8	1 to 2	(1) Do not apply more than 600 gals. dilute spray per acre. (2) Apply when mites first appear. Repeat as necessary to maintain control. (3) Do not apply more than 4 times per season including a pre-bloom spray; apply this product no more than 3 times between petal fall and harvest. (4) Do not apply within 14 days of harvest.
Pears	European Red McDaniel Spider Pear Rust Two-Spotted Spider	1 to 3	1 to 3	4 to 8	1 to 2	
Citrus	Citrus Red Citrus Rust Texas Citrus Two-Spotted Spider Yuma Spider	2 to 4[3]	2 to 4	4 to 8	1 to 2	(1) Do not apply more than 1600[3] gals. dilute spray per acre. (2) Apply when mites first appear. Repeat as necessary to maintain control. (3) Do not apply more than 2 sprays in any 12-month period. Do not apply at less than 60 day intervals. (4) Do not apply within 7 days of harvest.
Eggplant	Clover Two-Spotted Spider	2 to 4	2 to 4			(1) Apply when mites appear. Repeat as necessary to maintain control. Thorough coverage to all leaf surfaces is required for best control. (2) In Florida, apply when needed or every 1 to 2 weeks. (3) Do not exceed 6 applications per season with a total of no more than 6 pounds of active ingredient per acre per year. (4) Do not harvest within 3 days of last application.
Grapes	Pacific Spider European Red Two-Spotted Spider	1 to 2.5	1 to 2.5[5]	8 to 16	2 to 4	(1) Do not apply more than 250 gals. per acre. (2) Apply when mites first appear. Repeat as necessary to maintain control but not more often than once every 21 days. (3) Do not apply more than 2 times per season. (4) Do not apply within 28 days of harvest.
Peaches **Nectarines** **Plums** **Prunes**	European Red McDaniel Spider Pacific Spider Two-Spotted Spider Big Beaked Plum Peach Silver Plum Nursery	1 to 2	1 to 2	4 to 8	1 to 2	(1) Do not apply more than 400 gals. dilute spray per acre (600 gals. for cherries). (2) Apply when mites first appear. Repeat as necessary to maintain control. (3) Do not apply more than 2 times per season. (4) Do not apply within 14 days of harvest.
Cherries (including both sweet and sour varieties)	European Red McDaniel Spider Cherry Rust	1.5 to 3	1.5[5] to 3	4 to 8	1 to 2	
Strawberries	Two-Spotted Spider	1.5 to 2	1.5[5] to 2	12 to 16	3 to 4	(1) Apply in 150-200 gals. per acre. (2) Apply when mites first appear. Repeat as necessary to maintain control but not more than 4 times per season. (3) Do not apply within 1 day of harvest. (4) Good coverage is essential to good mite control. Adjust spray volume and nozzle placement to assure coverage of top and underside of leaves.
Papayas[6]	Carmine Citrus Red Texas Citrus Broad Red and Black Flat	—	—	4 to 8	1 to 2	(1) Use only as dilute spray. Do not apply more than 400 gals. dilute spray per acre. (2) Apply when mites first appear. Repeat as necessary to maintain control. (3) Do not apply more than once per month to bearing trees. Do not apply more than 9 times per season. (4) Do not apply within 7 days of harvest.
Almonds	Pacific Spider European Red Two-Spotted Spider Hickory Spider	1 to 2.5	1 to 2.5[5]	4 to 8	1 to 2	(1) Do not apply more than 500 gals. per acre. (2) Apply when mites first appear. Repeat as necessary to maintain control. (3) Do not apply more than 2 times per season. (4) Do not apply within 14 days of harvest.
Pecans	Pecan Hickory					
Walnuts	Two-Spotted Spider European Red Pacific Spider					

CROP	MITES CONTROLLED	FOR CONCENTRATE SPRAY VENDEX 50[2]		FOR DILUTE SPRAY VENDEX 50		REMARKS
		POUNDS PER ACRE	WATER SOLUBLE BAGS PER ACRE	OZS. PER 100 GALS.	WATER SOLUBLE BAGS PER 400 GALS.	
Greenhouse & Outdoor Ornamentals (including nursery stock, flowers and plants grown for propagation purposes)	Spruce Spider Olingonychus Two-Spotted Spider	—	—	8 to 16	2 to 4	(1) Do not add oil to the spray solution. (2) Apply when mites first appear. Repeat as necessary to maintain control. Frequent repeat applications may cause the appearance of visible spray residues on foliage. (3) Apply to the foliage only of chrysanthemums (pre-bloom) and poinsettias (prebract). (4) Under green house conditions, foliage and flowers of certain species may demonstrate sensitivity to repeat applications. If in doubt, make test applications prior to general spraying. Occasional minor sensitivity has been observed on the following species.[4]
Established Landscape Ornamentals (Commercial application to established landscape ornamentals including trees, shrubs, flowering ornamentals, bedding plants, annuals, and perennials.)	Two-spotted Spider Clover	—	—	8 to 16	2 to 4	(1) Do not add oil to the spray solution. (2) Apply when mites first appear. Repeat as necessary to maintain control. Do not apply more than 4 times per year. (3) Under extreme weather conditions, foliage and flowers of certain species may demonstrate sensitivity to repeat applications. If in doubt, make test applications prior to general spraying. Occasional minor sensitivity has been observed on the following species.[4] (4) Do not enter without appropriate protective clothing until sprays have dried.

[1] This product is packaged in WATER SOLUBLE BAGS - drop required number of inner bags - unopened - into the spray tank. Do not handle inner bag; moisture will cause breakage.

[2] For Concentrate Sprays - Adjust the dosage to apply an amount not exceeding that used in a dilute spray.

[3] Do not exceed 4 lb. of VENDEX 50 WP per acre per application.

[4] Salix melanostachys (common willow), Asplenium bulbiferum (spleenwort garden fern), Dryopteris erythrosora (wood fern), Cercis canadensis (common redbud), Camellia japonica (red garden camellia), Pellaea rotundifolia (button fern), Davallia fejeensis (rabit foot fern), Asparagus meyeri (Meyers asparagus fern), Asplenium nidus (birdnest fern), Adiantum cuneatum (maidenhair fern), Celosia argentea (cockscomb), Verbena hortensis (verbena), Ageratum houstonianum (floss flower), Rosa sp. (common rose—some varieties esp. yellow).

[5] Fractions are used here for the purpose of conversion only. Do not use partial bags.

[6] Not registered for use in California.

STORAGE AND DISPOSAL

STORAGE: Store in a secure, dry and temperate area. Keep container closed when not in use. Do not store near food or feed. Do not use or store around the home. Avoid contact with water. In case of spill or leak, soak up with sand, earth or synthetic absorbent and dispose of wastes in compliance with local, State and Federal regulations. In case of a significant spill, call CHEMTREC (800) 424-9300.

PRODUCT DISPOSAL: Pesticide wastes are acutely hazardous. Improper disposal of excess pesticide, spray mixture, or rinsate is a violation of Federal law. If these wastes cannot be disposed according to label instructions, contact your State Pesticide or Environmental Control Agency or Hazardous Waste Representative at the nearest EPA Regional Office for guidance.

CONTAINER DISPOSAL: Completely empty contents of envelope into application equipment. Then dispose of empty envelope in a sanitary landfill, or, if allowed by State and local authorities, by burning. If burned, stay out of smoke.

NOTICE OF WARRANTY

Du Pont warrants that this product conforms to the chemical description on the label thereof and is reasonably fit for the purposes stated on such label only when used in accordance with the directions under normal use conditions. It is impossible to eliminate all risks inherently associated with the use of this product. Crop injury, ineffectiveness or other unintended consequences may result because of such factors as weather conditions, presence of other materials, or the manner of use or application, all of which are beyond the control of Du Pont. In no case shall Du Pont be liable for consequential, special or indirect damages resulting from the use or handling of this product. All such risks shall be assumed by the buyer. DU PONT MAKES NO WARRANTIES OF MERCHANTABILITY OR FITNESS FOR A PARTICULAR PURPOSE NOR ANY OTHER EXPRESS OR IMPLIED WARRANTY EXCEPT AS STATED ABOVE.

AG-3267 8069 5/23/89

This is a specimen label. DuPont is not responsible for the accuracy of the information contained herein. As labels are subject to revision, always carefully read and follow the label on the product container.

VYDATE® L

INSECTICIDE/NEMATICIDE

WATER SOLUBLE LIQUID

RESTRICTED USE PESTICIDE

Due to Acute Toxicity And Toxicity to Birds and Mammals.

For retail sale to and use only by Certified Applicators or persons under their direct supervision and only for those uses covered by the Certified Applicator's certification.

ACTIVE INGREDIENT:	BY WEIGHT
Oxamyl [Methyl N'N'-dimethyl-N-[(methylcarbamoyl)oxy]-1-thiooxamimidate	24%
INERT INGREDIENTS	76%
TOTAL	100%

1 GALLON CONTAINS 2 LBS ACTIVE INGREDIENT PER GALLON.

U.S. Pat. 3,658,870

EPA Reg. No. 352–372

KEEP OUT OF REACH OF CHILDREN

DANGER–POISON

PELIGRO

PRECAUCION AL USUARIO: Si usted no lee ingles, no use este producto hasta que la etiqueta haya sido explicado ampliamente.

STATEMENT OF PRACTICAL TREATMENT

If swallowed: Call a physician or Poison Control Center. Drink 1 or 2 glasses of water and induce vomiting by touching back of throat with finger. Do not induce vomiting or give anything by mouth to an unconscious person.

If inhaled: Remove from exposure and have patient lie down and keep quiet. If patient is not breathing, start artificial respiration immediately. Never give anything by mouth to an unconscious person.

In case of contact, wash skin with plenty of soap and water; for eyes, flush with water for 15 minutes and get medical attention; remove and wash contaminated clothing before re-use.

ATROPINE IS AN ANTIDOTE—SEEK MEDICAL ATTENTION AT ONCE IN ALL CASES OF SUSPECTED POISONING

If warning symptoms appear (see **WARNING SYMPTOMS**), get medical attention.

For medical emergencies involving this product, call toll free 1-800-441-3637.

See Precautionary Statements

PRECAUTIONARY STATEMENTS

HAZARDS TO HUMANS AND DOMESTIC ANIMALS

DANGER

POISON

DANGER–POISON

CONTAINS METHANOL. MAY BE FATAL OR CAUSE BLINDNESS IF SWALLOWED.

MAY BE FATAL IF ABSORBED THROUGH SKIN OR INHALED.

Do not breathe vapors or spray mist. Do not get in eyes, on skin, or on clothing. When mixing spray, loading, applying, or otherwise handling, wear protective clothing, goggles, and a mask or respirator jointly approved by the Mine Safety and Health Administration (formerly the U.S. Bureau of Mines) and by the National Institute for Occupational Safety and Health. Wear clean clothes daily. Wash thoroughly after handling and before eating or smoking. Pilot should not assist in the mixing and loading operation.

Do not apply this product in such a manner as to directly or through drift expose workers, other persons or animals. The area being treated must be vacated by unprotected persons.

WARNING SYMPTOMS—Oxamyl poisoning produces effects associated with antichlolinesterase activity which may include weakness, blurred vision, headache, nausea, abdominal cramps, discomfort in the chest, constriction of pupils, sweating, slow pulse, muscle tremors.

NOTE TO PHYSICIAN

TREATMENT—Atropine sulfate should be used for treatment. Administer repeated doses, 1.2 to 2.0 mg intravenously every 10 to 30 minutes until full atropinization is achieved. Maintain atropinization until the patient recovers. Artificial respiration or oxygen may be necessary. Allow no further exposure to any cholinesterase inhibitor until recovery is assured.

Do not use 2-PAM for exposure to "Vydate" L alone. However, for exposure to combinations of "Vydate" L and organophosphorous insecticides, 2-PAM may be used as required to supplement the atropine sulfate treatment. Do not use morphine.

For medical emergencies involving this product, call toll free 1–800–441–3637.

ENVIRONMENTAL HAZARDS

This pesticide is toxic to aquatic organisms (fish and invertebrates) and extremely toxic to birds. Cover or disc spill areas. Birds in treated areas may be killed. Do not apply directly to water or wetlands (swamps, bogs, marshes, and potholes). Drift and runoff may be hazardous to aquatic organisms in neighboring areas. Do not contaminate water when disposing of equipment washwaters.

This product is highly toxic to bees exposed to direct treatment or residues on blooming crops or weeds. Do not apply this product or allow to drift to blooming crops or weeds if bees are visiting the treatment area.

GROUND WATER ADVISORY—Residues of "Vydate" L can seep or leach through soil and can contaminate ground water which may be used for drinking. Users are advised not to apply "Vydate" L where the water table is close to the surface and where soils are very permeable, i.e., well-drained soils such as loamy sands. Local agricultural Agencies can provide information on the soil type in your area and the location of the ground water.

PHYSICAL AND CHEMICAL HAZARDS

Flammable. Keep away from heat, sparks, and open flame. Keep container closed. Use with adequate ventilation.

DIRECTIONS FOR USE

It is a violation of federal law to use this product in a manner inconsistent with its labeling.

Du Pont "Vydate" L Insecticide/Nematicide should be used only in accordance with recommendations on this label or in separate Du Pont recommendations available through local dealers.

Du Pont will not be responsible for losses or damages resulting from use of this product in any manner not specifically recommended by Du Pont. User assumes all risks associated with such non-recommended use.

"Vydate" L is a water soluble liquid to be diluted with water for application. Use only in commercial and farm plantings; do not use in home plantings. Do not plant crops other than those with registered "Vydate" L uses within 6 months after the last application.

Do not use in Suffolk and Nassau Counties, Long Island, New York.

Do not enter treated areas for 48 hours after application unless protective clothing is worn. Because certain states may require more restrictive reentry intervals for various crops treated with this product, consult your State Department of Agriculture for further information.

Written or oral warnings must be given to workers who are expected to be in a treated area or in an area about to be treated with this product. When oral warnings are given, warnings shall be given in a language customarily understood by workers. Oral warnings must be given if there is reason to believe that written warnings cannot be understood by workers. Written (or oral) warnings must include the following information: "DANGER—POISON. Area treated with 'Vydate' L on (date of application). Do not enter treated areas for 48 hours after application unless protective clothing is worn. If warning symptoms appear, refer to Statement of Practical Treatment on front panel of 'Vydate' L label and seek medical attention at once."

If mixing or loading the concentrate with a closed system, a long-sleeve shirt and long-legged pants may be substituted for the protective suit and respirator. Appropriate protective clothing and equipment must still be worn during application, equipment repair, equipment cleaning, early reentry, and disposal. WEAR THE FOLLOWING PROTECTIVE CLOTHING AND EQUIPMENT DURING HANDLING, MIXING, LOADING, APPLICATION, EQUIPMENT AND CLEANING, DURING EARLY REENTRY TO TREATED AREAS, AND DURING DISPOSAL OF THE PESTICIDE.

Wear a protective suit of one or two pieces that covers all parts of the body except the head, hands, and feet. Wear chemical-resistant gloves, chemical-resistant apron, and chemical-resistant shoes, shoe coverings or boots. Wear goggles or face shield and a pesticide respirator approved by the National Institute for Occupational Safety and Health (NIOSH) and/or the Mine Safety Health Enforcement Administration (MSHA). A chemical resistant hat or hood must be worn if the application is likely to result in overhead exposure, such as during air blast applications.

During application from a tractor with a completely enclosed cab or aerially with an enclosed cockpit, a long-sleeve shirt and long-legged pants may be worn in place of the above protective clothing. Chemical resistant gloves and an approved respirator must be worn when exiting. Note that this clothing is inadequate to protect you during equipment repair, equipment cleaning, reentry, or during pesticide disposal.

IMPORTANT! Always wash hands, face and arms with soap and water before eating, smoking, drinking or toileting.

AFTER WORK: Wash gloves with soap and water before removing. Take off all clothing and shoes. Shower using soap and water. Wear only clean clothes. Do not use contaminated clothing. Wash protective clothing and equipment with soap and water after each use, personal and protective clothing worn during use must be laundered separately from household articles. Clothing and protective equipment heavily contaminated or drenched with oxamyl must be destroyed according to state and local regulations. HEAVILY CONTAMINATED OR DRENCHED CLOTHING CANNOT BE ADEQUATELY DECONTAMINATED.

Respirators should be cleaned and cartridges replaced according to instructions included with the respirator. Replace gloves frequently.

DURING AERIAL APPLICATION, HUMAN FLAGGERS MUST BE IN A TOTALLY ENCLOSED VEHICLE.

Spray Preparation—Fill spray tank 1/4 to 1/2 full of water. Add "Vydate" L directly to tank. Mix thoroughly while adding remaining water. No further agitation is necessary. Spray mix should not be stored overnight in spray tank.

Chemigation—Refer to supplemental labeling entitled, "Vydate® L Use in Chemigation Systems" for use directions for chemigation. Do not apply this product through any irrigation system unless the supplemental labeling on chemigation is followed.

FRUITS

APPLES

SPOTTED TENTIFORM LEAFMINER—Apply 1 to 2 pts. "Vydate" L per 100 gals. water as follows:

1st Brood Leafminer—Adults and Larvae—Make one treatment at pink (before bloom—no petals opened).

2nd Brood Leafminer—Larvae—Make an application when there are an average of two or more larvae per leaf in the sap-feeding stage. For best results, the application should be made before the larvae enter into the tissue-feeding stage.

EUROPEAN RED MITE AND TWOSPOTTED SPIDER MITE—

Established Populations of Mites—Apply 1 to 2 pts. "Vydate" L per 100 gals. water as needed.

Low Populations (Preventive Control)—Apply 1/4 to 1/2 pt. "Vydate" L per 100 gals. water before mites reach damaging numbers and continue every 10 to 14 days to maintain control.

WHITE APPLE LEAFHOPPER—Apply 1 to 2 pts. "Vydate" L per 100 gals. water every 10 to 14 days as needed. Begin application when significant numbers of leafhoppers are present.

Apply "Vydate" L in sufficient water to obtain thorough coverage but not in excess of 400 gals. per acre.

Note: Do not apply at bloom or within 30 days after bloom, or fruit thinning may occur. Do not apply to apples within 14 days of harvest. Do not graze livestock in treated orchards.

CITRUS

CITRUS RUST MITE—Apply 1/4 to 1 pt. "Vydate" L per 100 gals. water. Spray to run-off using up to 400 gals. of water per acre. Do not apply more than 2 qts. of product per acre.

Begin applications as soon as significant infestations are found and continue on an as-needed basis. Under light to moderate pressure, apply on 4- to 6-week intervals or as needed; whereas under moderate to heavy pressure, apply every 2 to 3 weeks as long as the infestation continues.

DO NOT MAKE MORE THAN 6 APPLICATIONS PER YEAR

CITRUS THRIPS—Apply 1 to 2 qts of "Vydate" L per acre in sufficient water (100 to 500 gals. per acre) to give good coverage. Apply in the early spring before bloom when the new growth is about 3 to 4 inches long. Make additional applications as needed. Applications at petal fall may be critical to prevent fruit scarring. Applications are also recommended during mid-summer to protect new growth on young trees.

Note: Do not apply to citrus within 7 days of harvest. Do not graze livestock in treated orchards.

PEARS

Use not registered in California

EUROPEAN RED MITE, McDANIEL MITE, TWOSPOTTED SPIDER MITE, PEAR RUST MITE—Apply 3 to 4 qts. "Vydate" L per acre in 100 to 600 gals. of water by ground. For best results use a dilute application.

Make application when mites first appear and repeat as needed to maintain control. Use lower rate on light infestations and higher rate on heavy infestations.

Note: Do not apply at bloom or within 30 days after full bloom, or fruit thinning may occur: Do not apply within 14 days of harvest. This product has been tested on Bartlett and d'Anjou varieties of pears without russeting. Its use on other varieties should be on a small scale until the possibility of russeting has been evaluated. Do not graze livestock in treated orchards.

PINEAPPLE

Use not registered in California

RENIFORM AND ROOT KNOT NEMATODES—

Preplant or Planting Treatment—Apply 2 to 4 gals. "Vydate" L per acre broadcast within 1 week prior to planting and incorporate 4 to 6 inches into the soil. Alternatively, apply 1 to 4 gals. "Vydate" L per acre broadcast, or apply 1/2 to 4 gals. per acre directly to the soil via drip irrigation injection, within 1 week after planting.

Subsequent Treatments—Apply 1/2 to 2 gals. "Vydate" L per acre by ground equipment as a foliar spray at 2 to 4 week intervals in sufficient water to obtain thorough coverage. Begin applications when pineapple roots begin to grow following planting. Alternatively, apply 1/4 to 2 gals. "Vydate" L per acre directly to the soil via drip irrigation injection at 2, 4, or 8 week intervals. For best results subsequent applications should be made after preplant or planting application of "Vydate" L or a standard fumigant. Best results occur under optimum soil moisture conditions.

Note: Do not apply more than 12 gals. "Vydate" L per acre on the plant crop or more than 6 gals. per acre of the ratoon crop. Do not apply within 30 days of harvest.

VEGETABLES

CARROTS

Use not registered in California

ROOT KNOT (EXCEPT JAVANESE), LESION, STING, SPIRAL AND STUNT NEMATODES—

Preplant Treatment—Apply 2 to 4 gals. of "Vydate" L per acre in a minimum of 20 gals. water per acre as a broadcast treatment of the soil. Apply within one week of planting. Thoroughly incorporate to a depth of 4 to 6 inches in the soil.

In-Furrow Treatment—Apply 1 to 2 gals. of "Vydate" L per acre in a minimum of 20 gals. water per acre in the seed furrow during the planting operation.

CELERY (Florida)

SERPENTINE LEAFMINERS (EXCEPT LIRIOMYZA TRIFOLII)— Apply 2 to 4 pts. "Vydate" L per acre as a foliar spray using sufficient water (minimum 100 gals.) to obtain uniform coverage. Treat when insect first appears and repeat at 5- to 7-day intervals or as needed. Use the low rates for light infestations, the intermediate rates for heavier infestations, and the highest rates for severe infestations.

Note: Do not apply within 14 days of harvest.

CUCUMBER, CANTALOUPE, HONEYDEW MELON, WATERMELON, SQUASH, PUMPKIN

ROOT KNOT (EXCEPT JAVANESE), LESION, STING AND STUNT NEMATODES—

Preplant and Planting Treatment—Apply 1 to 2 gals. "Vydate" L per acre (broadcast); for band treatment, use proportionately less. Following application and before planting, incorporate "Vydate" L 2 to 4 inches into the soil. Use the low rate for light infestations.

Foliar Treatment—Apply 2 to 4 pts. "Vydate" L per acre in sufficient water to obtain uniform coverage of foliage. Make the first application 2 to 4 weeks after planting and repeat 2 to 3 weeks later. Use the lower rate on light infestations. The best results follow usage of "Vydate" L as a soil treatment as described above.

LEAFMINERS (Liriomyza spp.)—Apply 2 to 4 pts. "Vydate" L per acre in sufficient water to obtain uniform coverage of foliage. Make applications when insects first appear and repeat weekly or as needed. Apply the low rate for light infestations of insects and the high rate for severe infestations.

Where leafminer infestations occur annually, initiate treatment schedule 2 to 4 weeks after planting and repeat weekly or as needed.

Note: Do not apply within 1 day of harvest.

EGGPLANT

APHIDS, COLORADO POTATO BEETLE, LEAFMINERS, AND MITES—Apply 1 to 2 qts. "Vydate" L per acre by ground equipment as a foliar spray when insects first appear. Repeat at 1 to 3 week intervals as needed.

NEMATODES (EXCEPT IN CALIFORNIA)—Apply 1 gal. "Vydate" L per acre as a band treatment 2 to 3 weeks after transplanting and again 4 weeks later. Two to four weeks following these soil treatments, make 2 foliar applications of 2 qts. per acre by ground equipment at 1 to 2 week intervals.

Note: Do not apply within 1 day of harvest when using only foliar applications or within 7 days of harvest using soil with foliar applications.

PEPPER (BELL)

Use not registered in California

ROOT KNOT (EXCEPT JAVANESE), STING, RING, STUBBY ROOT, AND STUNT NEMATODES—

Transplant Water Treatment: Supplemental Foliar Treatments—Apply 2 pts. "Vydate" L per acre. Add the "Vydate" L to the transplant water, and use a minimum of 200 gals. per acre. Under very high nematode populations, begin with a transplant water treatment and supplement with foliar sprays. For the best results, make the first foliar application 14 days after transplanting and continue in accordance with the foliar treatment described below.

Note: Do not apply as a transplant water treatment during periods of slow plant growth such as when temperatures fall below 45° F, or crop injury may result.

GREEN PEACH APHID, VEGETABLE LEAFMINER, AND PEPPER WEEVIL—

Foliar Treatments—Apply 2 to 4 pts. "Vydate" L per acre in sufficient water (minimum 100 gals.) to obtain uniform coverage. Make applications when insects first appear and repeat at 1- to 2-week intervals or as needed. Apply low rates for light infestations of insects. Use highest recommended rates and shorter intervals for controlling severe infestations.

Note: Do not apply within 7 days of harvest.

POTATOES (EXCEPT IN CALIFORNIA)

ROOT KNOT (EXCEPT JAVANESE), STING, LESION AND RING NEMATODES (EXCEPT WEST OF THE ROCKIES); APHIDS, COLORADO POTATO BEETLE, FLEA BEETLES, POTATO LEAFHOPPER, AND TARNISHED PLANT BUG—

Preplant Treatment—For control of nematodes and early season control of the above insects, apply either one of the following:

In-Furrow Treatment—Apply 1 to 2 gals. of "Vydate" L per acre in a minimum of 20 gals. water per acre in the seed furrow during the planting operation.

Broadcast Treatment—Apply 2 to 4 gals. of "Vydate" L per acre in a minimum of 20 gals. of water per acre as a band or broadcast treatment of the soil within one week of planting. Thoroughly incorporate to a depth of 4 to 6 inches in the soil.

Scout fields to determine when early season control diminishes and use registered foliar insecticides including "Vydate" L to retain control.

Foliar Treatments—For control of the above insects, apply 2 to 4 pts. "Vydate" L per acre, except in the Midwest and Northeast where 1 to 4 pts. "Vydate" L per acre may be used for Colorado potato beetle. Use sufficient water (minimum 4 gals. per acre by air) to obtain uniform coverage. Make applications when insects first appear and repeat at 5- to 7-day intervals or as needed. Apply the low rates for light infestations of insects. Use the higher rates for severe infestations.

Note: Do not apply within 7 days of harvest.

SWEET POTATOES

Use not registered in California

ROOT KNOT (EXCEPT JAVANESE) AND SPIRAL NEMATODES—

Preplant Treatment—Apply 2 to 3 gals. of "Vydate" L per acre in a minimum of 20 gals. of water per acre as a broadcast treatment of the soil; for band treatment, use proportionately less. Apply within one week of planting. Thoroughly incorporate to a depth of 4 to 6 inches in the soil.

In-furrow Treatment—As an alternate to the above, apply 1 to 2 gals. of "Vydate" L per acre in a minimum of 200 gals. of water per acre in the transplant water during the planting of slips.

Note: Do not apply as a transplant water treatment during periods of slow plant growth such as when temperatures fall below 45° F, or crop injury may result.

TOMATOES

SERPENTINE LEAFMINERS (EXCEPT LIRIOMYZA TRIFOLII)— Apply 2 to 4 pts. "Vydate" L per acre in sufficient water (minimum 100 gals.) to obtain uniform coverage. Make application when insect first appears and repeat at 5- to 7-day intervals or as needed.

Apply the low rates for light infestations of insects. Use intermediate rates for heavier infestations of insects. Use the highest recommended rates for controlling severe infestation.

Note: Do not apply more than 6 gals. per acre per season. Do not apply within 1 day of harvest.

FIELD CROPS

COTTON

COTTON LEAFPERFORATOR—Apply 1 to 4 pts. "Vydate" L per acre in sufficient water to obtain thorough coverage. Make initial applications when damaging populations begin to build, and continue on a 6 to 8 day schedule for up to 4 applications.

BOLL WEEVIL, COTTON FLEAHOPPER, AND TARNISHED PLANT BUG—Apply 1/2 to 1 pt. "Vydate" L per acre in sufficient water to obtain thorough coverage. Begin applications when damaging populations appear and repeat as needed to maintain control.

Note: Do not apply within 21 days of harvest. Do not graze or feed treated cotton to livestock

PEANUTS

Use not registered in California

ROOT KNOT (EXCEPT JAVANESE), STING, RING, AND LESION NEMATODES, AND THRIPS—

Soil Treatment—Apply 2 to 6 qts. of "Vydate" L in a 7" to 12" band using a minimum of 10 gals. of water per acre. Thoroughly incorporate with a rotary tiller to a depth of 3" to 5" immediately after application. Use the higher rate and wider band for severe infestations. Peanuts should be planted within 24 hours after application. Alternatively, "Vydate" L may be applied in a 7" band immediately behind planting. Use 2 to 6 qts. in a minimum of 10 gals. water per acre.

Foliar Treatment—Foliar treatment is to be used only following soil fumigation, or following preplant or at planting soil application of "Vydate" L or other contact nematicides. Make 2 foliar applications of "Vydate" L at 2 to 4 pts. per acre in 20 to 40 gals. of water. Apply first spray 3 weeks after emergence and the second spray 6 weeks after emergence. For best results, concentrate the spray on the row using 3 cone-type nozzles positioned over and to each side of the row. Thorough spray coverage is important.

SOYBEANS

Use not registered in California

ROOT KNOT (EXCEPT JAVANESE), LESION, RING, STUNT, LANCE, AND CYST NEMATODES—

In-furrow Band Treatment—Apply 2 to 4 pts. "Vydate" L in 10 to 20 gals. of water per acre (based on 36" row spacing) at planting. Spray over open drill row at the juncture where the seed is covered with soil to assure mixing of "Vydate" L with the soil around the seed.

Incorporated Band Treatment—Apply 2 to 4 pts. "Vydate" L in 10 to 20 gals. water per acre to a 7" to 10" band in the drill area (based on 36" row spacing) as a preplant treatment (up to 1 week before planting) or as a treatment at planting. Incorporate the "Vydate" L 2" to 4" into the soil in the seed zone.

Incorporated Broadcast Treatment—Apply 1 to 2 gals. of "Vydate" L in 10 to 20 gals. of water per acre as a preplant treatment (up to 1 week before planting) or as a treatment at planting. Incorporate the "Vydate" L 2" to 4" into the soil in the seed zone.

Note: Do not cut for hay or feed treated forage to livestock. Do not apply "Vydate" L to areas with severe infestations of nematodes such as where injury to plants is manifested by severe stunting and chlorosis, as "Vydate" L will not control severe infestations.

TOBACCO

ROOT KNOT (EXCEPT JAVANESE) AND LESION NEMATODES AND FLEA BEETLES—Measure "Vydate" L accurately as over-application may cause stunting of tobacco.

"Vydate" L may be used:

• As a preplant soil incorporation treatment. • Alone as a transplant water treatment for 2 to 4 weeks of nematode and flea beetle control. • As a transplant water treatment following "Vydate" L, or other contact nematicides, preplant fumigants, or multipurpose fumigants and standard flea beetle treatments. • As a foliar treatment to extend early season nematode and flea beetle control following the use of "Vydate" L in transplant water or the soil use of "Vydate" L, or other contact nematicides, preplant fumigants, or multipurpose fumigants and standard flea beetle treatments.

Soil Treatment—"Vydate" L may be applied to the soil as a band treatment or it may be broadcast, disced, and bedded. For best results, the tobacco should be transplanted within 24 hours after soil treatment; however, if the transplant water treatment is used, tobacco may be transplanted within one week after the soil treatment.

Row Treatment—Apply 1 to 3 gals. of "Vydate" L in an 18- to 24-inch band in a minimum of 20 gals. of water per acre of tobacco (12,000 row feet). Use the lower rates for lower infestations, the higher rates and wider band for severe infestations. Thoroughly incorporate with a rotary tiller to a depth of 4 to 6 inches.

Broadcast and Bed Treatment—Apply a broadcast spray of 1 to 3 gals. per acre in a minimum of 40 gals. of water. Use lower rates for lower infestations and the higher rates for severe infestations. Incorporate thoroughly to a depth of 4 to 6 inches and bed the field in such a manner that only treated soil is used to form the beds.

Transplant Water Treatment—Apply "Vydate" L at the following recommended rates in the transplant water depending on the number of plants per acre. Use a minimum of 100 gals. of transplant water per acre.

Plants/Acre	Rate in Pts./Acre	Rate in Fl. Oz./Acre
9,000	2	32
7,900	1 3/4	28
6,800	1 1/2	24
5,600	1 1/4	20

For small quantities mix 1 tablespoonful "Vydate" L per 10 gals. of water using 1/2 pint of solution/plant.

Foliar Treatment—Make 2 foliar applications of 4 pts. "Vydate" L per acre of tobacco. Include a spreader sticker. Apply a directed spray in 20 gals. of water 2 weeks after transplanting. Apply a second directed spray in 20 to 40 gals. of water (depending on plant size) 4 to 5 weeks after transplanting.

For best results, use 3 spray nozzles positioned over and to each side of the row so that the entire plant is covered with spray.

ORNAMENTALS, NURSERY STOCK, NON-BEARING FRUIT TREES

For use only in commercial greenhouses and nurseries. Do not apply this product to interior plantscapes or potted plants in or around domestic areas such as homes, shopping malls, or commercial buildings.

"Vydate" L is recommended for use on the following ornamentals, nursery stock and non-bearing fruit trees for control of insects, mites, and nematodes:

Apple*	Boxwood	Citrus*	Dracaena	Hoya	Peach*	Philodendron	Snapdragon
Azalea	Cherry*	Croton	Gardenia	Leatherleaf Fern	Pear*	Rose	Strawberry
Bird's-Nest Fern	Chrysanthemum	Dieffenbachia Picta	Gladiolus	Maranta	Peony	Sansevieria	(Nurseries)*†

* Non-bearing trees and strawberries that will not bear fruit within 12 months after application.

†Note: Certain varieties of strawberries are sensitive to "Vydate" L exhibiting severe phytotoxic symptoms. Known sensitive varieties that should not be treated with "Vydate" L include Bounty, Canoga (NY 1362), Cavalier, Darrow, Dunlop, Earlidawn, Earliglow, Empire, Holiday, Marlate, Midland, Midway, Mic Mac, NY 1409, Ogallalla, Redglow, Redstar, Sequoia, Stark Red, Sunrise, and Viestar. However, other varieties may also be sensitive. Therefore, "Vydate" L should be used initially only on a small scale to determine crop sensitivity prior to large scale field application.

See separate Du Pont literature for additional plants.

INSECTS AND MITES including aphids, fungus fly, fungus gnats, Japanese beetle, leafhoppers, leafminers, mealybugs, royal palm bug, scales, thrips, white fly, and mites:

Foliar Treatment—Mix 2 to 4 pts. "Vydate" L per 100 gals. and apply up to 200 gals. of this mixture per acre or apply 4 to 8 pts. "Vydate" L per acre in a maximum of 600 gals. water per acre. For small areas, mix 3 to 6 tablespoonfuls "Vydate" L per 5 to 14 gals. water and apply to 1,000 sq. ft. Apply as needed to maintain control.

NEMATODES including root knot (except Javanese), sting, lesion, and burrowing nematodes:

"Vydate" L may be used as a pretransplant soil treatment, as a drench or as a root, corm, or bulb dip. It may also be used as a foliar treatment alone or as a supplement to the above "Vydate" L or other nematicide treatments. When transplanting, plants should be set into soil within 24 hours after application.

Preplant Soil Incorporation Treatment—Apply 3 to 4 gals. "Vydate" L in a minimum of 20 gals. of water per acre. Thoroughly incorporate with a rotary tiller to a depth of 4 to 8 inches immediately after application. If applied in a band, use proportionately less material.

Soil Mix Treatment—Mix $2\frac{3}{4}$ fl. ozs. of "Vydate" L in 10 gals. of water. Spray $2\frac{1}{2}$ to $3\frac{1}{3}$ gals. of the dilute mix on to 1 cubic yard of soil while tumbling in a soil mixer (use 1 to $1\frac{1}{3}$ pts. for 1 bushel of soil).

Foliar Treatment—As a foliar treatment, mix 2 to 4 pts. "Vydate" L per 100 gals. of water and apply as a dilute spray, not to exceed 8 pts. per acre. Apply on a 2- to 3-week schedule for 4 applications. Apply first spray at first full leaf or when plant is in active growth phase. Do not apply to plants under water stress or to plants not actively growing. Include a spreader sticker.

Liquid Drench—Mix 1 pt. of "Vydate" L in 100 gals. of water ($1\frac{1}{2}$ tablespoons/5 gals.) Apply drenches at a rate of 4 to 8 oz. per 6-inch pot or 2 to 4 oz. per 4-inch pot.

Root, Corm, or Bulb Dip—Mix 2 to 4 pts. of "Vydate" L in 100 gals. of water (3 to 6 tablespoons/5 gals.). Place plant material in solution for 1 to 30 minutes. Use the higher rate and longer soaking time for more severe nematode infestation.

Note: Do not treat apple, cherry, peach, or pear planting stock by root dip.

Note: Use only on commercial plantings; do not use on home plantings. Since varieties of ornamentals are numerous, continually changing, and may respond differently to "Vydate" L, test "Vydate" L on a small scale before proceeding to large-scale application. Varietal response may also vary if "Vydate" L is mixed with other products.

STORAGE AND DISPOSAL

STORAGE: Do not subject to temperatures below 32° F. Store product in original container only, away from other pesticides, fertilizer, food or feed. Not for use or storage in or around the home.

PESTICIDE DISPOSAL: Do not contaminate water, food, or feed by storage or disposal. Pesticide wastes are acutely hazardous. Improper disposal of excess pesticide, spray mixture, or rinsate is a violation of Federal Law. If these wastes cannot be disposed of by use according to label instructions, contact your State Pesticide or Environmental Control Agency, or the Hazardous Waste representative at the nearest EPA Regional Office for guidance.

CONTAINER DISPOSAL: Triple rinse (or equivalent) the container. Then offer for recycling or reconditioning, or puncture and dispose of in a sanitary landfill, or, if allowed by state and local authorities, by burning. If burned, stay out of smoke.

NOTICE OF WARRANTY

Du Pont warrants that this product conforms to the chemical description on the label thereof and is reasonably fit for the purposes stated on such label only when used in accordance with the directions under normal use conditions. It is impossible to eliminate all risks inherently associated with the use of this product. Crop injury, ineffectiveness, or other unintended consequences may result because of such factors as weather conditions, presence of other materials, or the manner of use or application, all of which are beyond the control of Du Pont. In no case shall Du Pont be liable for consequential, special or indirect damages resulting from the use or handling of this product. All such risks shall be assumed by the buyer. DU PONT MAKES NO WARRANTIES OF MERCHANTABILITY OR FITNESS FOR A PARTICULAR PURPOSE NOR ANY OTHER EXPRESS OR IMPLIED WARRANTY EXCEPT AS STATED ABOVE.

NOTICE TO BUYER—Purchase of this material does not confer any rights under patents of countries outside of the United States.

AG-3237 8128 12/16/88

This is a specimen label. DuPont is not responsible for the accuracy of the information contained herein. As labels are subject to revision, always carefully read and follow the label on the product container.

VYDATE® C

INSECTICIDE/NEMATICIDE WATER SOLUBLE LIQUID

RESTRICTED USE PESTICIDE

Due to Acute Toxicity And Toxicity to Birds and Mammals.
For retail sale to and use only by Certified Applicators or persons under their direct supervision and only for those uses covered by the Certified Applicator's certificate.

REGISTERED FOR USE IN THE FOLLOWING STATES ONLY: ARIZONA, ARKANSAS, LOUISIANA, MISSISSIPPI, MISSOURI, OKLAHOMA, TENNESSEE, TEXAS

ACTIVE INGREDIENT	BY WEIGHT
Oxamyl [Methyl N'N'-dimethyl-N-[(methylcarbamoyl) oxy]-1-thiooxamimidate]	24%
INERT INGREDIENTS:	76%
TOTAL	100%

CONTAINS 2 LBS. ACTIVE INGREDIENT PER GALLON.

EPA Reg. No. 352-372 EPA Est. No. 352-TX-1
U.S. Pat. 3,658,870

KEEP OUT OF REACH OF CHILDREN

DANGER–POISON

DANGER
PELIGRO

PRECAUCION AL USUARIO Si usted no lee ingles, no use este producto hasta que la etiqueta haya sido explicado ampliamente.

STATEMENT OF PRACTICAL TREATMENT

If swallowed: Call a physician or Poison Control Center. Drink 1 or 2 glasses of water and induce vomiting by touching back of throat with finger. Do not induce vomiting or give anything by mouth to an unconscious person.

If inhaled: Remove from exposure and have patient lie down and keep quiet. If patient is not breathing, start artificial respiration immediately. Never give anything by mouth to an unconscious person.

In case of contact, wash skin with plenty of soap and water; for eyes, flush with water for 15 minutes and get medical attention; remove and wash contaminated clothing before re-use.

ATROPINE IS AN ANTIDOTE—SEEK MEDICAL ATTENTION AT ONCE IN ALL CASES OF SUSPECTED POISONING

If warning symptoms appear (see WARNING SYMPTOMS above), get medical attention.

For medical emergencies involving this product, call toll free 1-800-441-3637.

PRECAUTIONARY STATEMENTS

HAZARDS TO HUMANS AND DOMESTIC ANIMALS

DANGER–POISON

CONTAINS METHANOL.
MAY BE FATAL OR CAUSE BLINDNESS IF SWALLOWED. MAY BE FATAL IF ABSORBED THROUGH SKIN OR INHALED.

Do not breathe vapors or spray mist. Do not get in eyes, on skin, or on clothing. When mixing spray, loading, applying, or otherwise handling, wear protective clothing, goggles, and a mask or respirator jointly approved by the Mine Safety and Health Administration (formerly the U.S. Bureau of Mines) and by the National Institute for Occupational Safety and Health. Wear clean clothes daily. Wash thoroughly after handling and before eating or smoking. Pilot should not assist in the mixing and loading operation.

Do not apply this product in such a manner as to directly or through drift expose workers, other persons or animals. The area being treated must be vacated by unprotected persons.

WARNING SYMPTOMS—Oxamyl poisoning produces effects associated with anticholinesterase activity which may include weakness, blurred vision, headache, nausea, abdominal cramps, discomfort in the chest, constriction of pupils, sweating, slow pulse, muscle tremors.

NOTE TO PHYSICIAN

TREATMENT—Atropine sulfate should be used for treatment. Administer repeated doses, 1.2 to 2.0 mg intravenously every 10 to 30 minutes until full atropinization is achieved. Maintain atropinization until the patient recovers. Artificial respiration or oxygen may be necessary. Allow no further exposure to any cholinesterase inhibitor until recovery is assured.

Do not use 2-PAM for exposure to "VYDATE" C alone. However, for exposure to combinations of "VYDATE" C and organophosphorous insecticides, 2-PAM may be used as required to supplement the atropine sulfate treatment. Do not use morphine.

For medical emergencies involving this product, call toll free 1-800-441-3637.

ENVIRONMENTAL HAZARDS

This pesticide is toxic to aquatic organisms (fish and invertebrates) and extremely toxic to birds. Cover or disc spill areas. Birds in treated areas may be killed. Do not apply directly to water or wetlands (swamps, bogs, marshes, and potholes). Drift and runoff may be hazardous to aquatic organisms in neighboring areas. Do not contaminate water when disposing of equipment washwaters.

This product is highly toxic to bees exposed to direct treatment or residues on blooming crops or weeds. Do not apply this product or allow to drift to blooming crops or weeds if bees are visiting the treatment area.

GROUND WATER ADVISORY—Residues of "VYDATE" C can seep or leach through soil and can contaminate ground water which may be used for drinking. Users are advised not to apply "VYDATE" C where the water table is close to the surface and where soils are very permeable, i.e., well-drained soils such as loamy sands. Local agricultural Agencies can provide information on the soil type in your area and the location of the ground water.

PHYSICAL OR CHEMICAL HAZARDS

Flammable. Keep away from heat, sparks, and open flame. Keep container closed. Use with adequate ventilation.

DIRECTIONS FOR USE

It is a violation of federal law to use this product in a manner inconsistent with its labeling.

Du Pont "VYDATE" C Insecticide/Nematicide should be used only in accordance with recommendations on this label or in separate Du Pont recommendations available through local dealers.

Du Pont will not be responsible for losses or damages resulting from use of this product in any manner not specifically recommended by Du Pont. User assumes all risks associated with such non-recommended use.

"VYDATE" C is a water soluble liquid to be diluted with water for application. Use only in commercial and farm plantings; do not use in home plantings. Do not plant crops other than those with registered "VYDATE" L uses within 6 months after the last application.

Do not enter treated areas for 48 hours after application unless protective clothing is worn. Because certain states may require more restrictive reentry intervals for various crops treated with this product, consult your State Department of Agriculture for further information.

Written or oral warnings must be given to workers who are expected to be in a treated area or in an area about to be treated with this product. When oral warnings are given, warnings shall be given in a language customarily understood by workers. Oral warnings must be given if there is reason to believe that written warnings cannot be understood by workers. Written (or oral) warnings must include the following information: "DANGER—POISON. Area treated with 'VYDATE' C on (date of application). Do not enter treated areas for 48 hours after application unless protective clothing is worn. If warning symptoms appear, refer to Standard of Practical Treatment on front panel of 'VYDATE' C label and seek medical attention at once."

If mixing or loading the concentrate with a closed system, a long-sleeve shirt and long-legged pants may be substituted for the protective suit and respirator. Appropriate protective clothing and equipment must still be worn during application, equipment repair, equipment cleaning, early reentry, and disposal.

WEAR THE FOLLOWING PROTECTIVE CLOTHING AND EQUIPMENT DURING HANDLING, MIXING, LOADING, APPLICATION, EQUIPMENT AND CLEANING, DURING EARLY REENTRY TO TREATED AREAS, AND DURING DISPOSAL OF THE PESTICIDE.

Wear a protective suit of one or two pieces that covers all parts of the body except the head, hands, and feet. Wear chemical-resistant gloves, chemical-resistant apron, and chemical-resistant shoes, shoe coverings or boots. Wear goggles or face shield and a pesticide respirator approved by the National Institute for Occupational Safety and Health (NIOSH) and/or the Mine Safety Health Enforcement Administration (MSHA). A chemical-resistant hat or hood must be worn if the application is likely to result in overhead exposure, such as during air blast applications.

During application from a tractor with a completely enclosed cab or aerially with an enclosed cockpit, a long sleeve shirt and long-legged pants may be worn in place of the above protective clothing. Chemical-resistant gloves and an approved respirator must be worn when exiting. Note that this clothing is inadequate to protect you during equipment repair, equipment cleaning, reentry, or during pesticide disposal.

IMPORTANT! Always wash hands, face and arms with soap and water before eating, smoking, drinking or toileting.

AFTER WORK: Wash gloves with soap and water before removing. Take off all clothing and shoes. Shower using soap and water. Wear only clean clothes. Do not use contaminated clothing. Wash protective clothing and equipment with soap and water after each use; personal and protective clothing worn during use must be laundered separately from household articles. Clothing and protective equipment heavily contaminated or drenched with oxamyl must be destroyed according to state and local regulations.

HEAVILY CONTAMINATED OR DRENCHED CLOTHING CANNOT BE ADEQUATELY DECONTAMINATED.

Respirators should be cleaned and cartridges replaced according to instructions included with the respirator. Replace gloves frequently.

DURING AERIAL APPLICATION, HUMAN FLAGGERS MUST BE IN A TOTALLY ENCLOSED VEHICLE.

Spray Preparation - Fill spray tank 1/4 to 1/2 full of water. Add "VYDATE" C directly to tank. Mix thoroughly while adding remaining water. No further agitation is necessary. Spray mix should not be stored overnight in spray tank.

Chemigation - Refer to supplemental labeling entitled, " 'VYDATE' L Use in Chemigation Systems" for use directions for chemigation. Do not apply this product through any irrigation system unless the supplemental labeling on chemigation is followed.

COTTON

COTTON LEAF PERFORATOR

Apply 1 to 4 pts. "VYDATE" C per acre in sufficient water to obtain thorough coverage. Make initial applications when damaging populations begin to build, and continue on a 6 to 8 day schedule for up to 4 applications.

BOLL WEEVIL, COTTON FLEAHOPPER, AND TARNISHED PLANT BUG

Apply 1/2 to 1 pt. "VYDATE" C per acre in sufficient water to obtain thorough coverage. Begin applications when damaging populations appear and repeat as needed to maintain control.

Note: Do not apply within 21 days of harvest. Do not graze or feed treated cotton to livestock.

STORAGE AND DISPOSAL

STORAGE: Do not subject to temperatures below 32°F. Store product in original container only, away from other pesticides, fertilizer, food or feed. Not for use or storage in or around the home.

PESTICIDE DISPOSAL: Do not contaminate water, food, or feed by storage or disposal. Pesticide wastes are acutely hazardous. Improper disposal of excess pesticide, spray mixture, or rinsate is a violation of Federal Law. If these wastes cannot be disposed of by use according to label instructions, contact your State Pesticide or Environmental Control Agency, or the Hazardous Waste representative at the nearest EPA Regional Office for guidance.

CONTAINER DISPOSAL: Triple rinse (or equivalent) the container. Then offer for recycling or reconditioning, or puncture and dispose of in a sanitary landfill, or, if allowed by state and local authorities, by burning. If burned, stay out of smoke.

NOTICE TO BUYER—Purchase of this material does not confer any rights under patents of countries outside of the United States.

NOTICE OF WARRANTY

Du Pont warrants that this product conforms to the chemical description on the label thereof and is reasonably fit for the purposes stated on such label only when used in accordance with the directions under normal use conditions. It is impossible to eliminate all risks inherently associated with the use of this product. Crop injury, ineffectiveness, or other unintended consequences may result because of such factors as weather conditions, presence of other materials, or the manner of use or application, all of which are beyond the control of Du Pont. In no case shall Du Pont be liable for consequential, special or indirect damages resulting from the use or handling of this product. All such risks shall be assumed by the buyer. DU PONT MAKES NO WARRANTIES OF MERCHANTABILITY OR FITNESS FOR A PARTICULAR PURPOSE NOR ANY OTHER EXPRESS OR IMPLIED WARRANTY EXCEPT AS STATED ABOVE.

AG-3241 8019/8079 12/16/88

This is a specimen label. DuPont is not responsible for the accuracy of the information contained herein. As labels are subject to revision, always carefully read and follow the label on the product container.

Elanco Products Company

Ag Chem Marketing
4040 Vincennes Circle, Suite 400
Indianapolis, IN 46268
(317) 276-2000

BALAN® E.C.

HERBICIDE

A selective herbicide for the preemergence control of annual grasses and broadleaf weeds

ACTIVE INGREDIENT:	
N-butyl-N-ethyl-α,α,α-trifluoro-2,6-dinitro-*p*-toluidine	19.4%
INERT INGREDIENTS:	80.6%
TOTAL	100.0%

Contain 1½ pounds active ingredient per gallon

BALAN®—(benefin, Elanco)

EPA Reg. No. 1471-55 ID 5023

Literature revised November 11, 1987
Pl 2265 AMP

PRECAUTIONARY STATEMENTS

KEEP OUT OF REACH OF CHILDREN

DANGER

HAZARDS TO HUMANS AND DOMESTIC ANIMALS

Corrosive, causes irreversible eye damage. Do not get in eyes, on skin, or on clothing. Causes skin irritation. Harmful if swallowed, inhaled or absorbed through the skin. Avoid breathing spray mist or vapors. BALAN E.C. may cause skin sensitization reactions in certain individuals. Wear protective clothing such as goggles or face shield, and coveralls, a long-sleeved shirt and impermeable gloves when handling this product. Wash thoroughly with soap and water after handling. Remove contaminated clothing and wash before reuse.

STATEMENTS OF PRACTICAL TREATMENT

If in eyes: Hold eyelids open and flush with a steady, gentle stream of water for 15 minutes. See an ophthalmologist immediately.

If swallowed: Call a physician or Poison Control Center. This product contains an aromatic hydrocarbon and can be extremely harmful if swallowed. If this happens and the patient is conscious, immediately administer activated charcoal (6–8 heaping teaspoonfuls) with water. If patient is unconscious, do not induce vomiting.

If on skin: Wash with plenty of soap and water. Get medical attention if irritation develops.

If inhaled: Remove individual to fresh air. If breathing difficulty occurs, get medical attention. If not breathing, provide cardiopulmonary resuscitation assistance (mouth-to-mouth) and call a physician.

ENVIRONMENTAL HAZARDS

This product is toxic to fish. Do not apply directly to water or wetlands. Do not contaminate water by cleaning of equipment or disposal of wastes. Drift and runoff from treated areas may be hazardous to aquatic organisms in neighboring areas.

STORAGE AND DISPOSAL

Storage: Avoid freezing. Store above 40°F. If frozen, poor weed control may result. Do not store near heat or open flame. Store in original container only. Tanks that have been used for phenoxy-type herbicides should not be used for BALAN. In case of leak or spill, use absorbent materials to contain liquids and dispose as waste.

Pesticide Disposal: Do not contaminate water, food, or feed by storage or disposal. Pesticide wastes are toxic. Improper disposal of excess pesticide, spray mixture, or rinsate is a violation of Federal Law. If these wastes cannot be disposed of by use according to label instructions, contact your State Pesticide or Environmental Control Agency, or the Hazardous Waste representative at the nearest EPA Regional Office for guidance.

Metal Container Disposal: Triple rinse (or equivalent). Then offer for recycling, or reconditioning, or puncture and dispose of in a sanitary landfill, or by other procedures approved by state and local authorities.

Plastic Container Disposal: Triple rinse (or equivalent). Then offer for recycling or reconditioning, or puncture and dispose of in a sanitary landfill, or, if allowed by state and local authorities, by burning. If burned, stay out of smoke.

Bulk/Mini-bulk Tank Cleaning: Triple rinse (or equivalent) and wash with appropriate cleaners before reusing.

GENERAL USE PRECAUTIONS

Applied according to directions and under normal growing conditions, BALAN will not harm the treated crop. Over application may result in stunting, crop injury or soil residue. Uneven application or improper soil incorporation of BALAN can result in erratic weed control or crop injury. Seedling disease, cold weather, deep planting, excessive moisture, high salt concentration or drought may weaken crop seedlings and increase the possibility of damage from BALAN. Under these conditions, delayed crop development or reduced yields may result.

In the arid, irrigated areas of the Western United States—Arizona, California, Idaho, Montana, Nevada, Oregon, Utah, Washington and Wyoming—certain crops are susceptible to injury when planted in soil previously treated with BALAN. To avoid such injury, do not plant wheat, barley, rye, other grasses or onion within 10 months following a BALAN application. Milo (grain sorghum), corn, oats, spinach, red beet, sugar beet, other root crops should not be planted within 12 months following a BALAN application.

DIRECTIONS FOR USE

BALAN is a preemergence herbicide which is soil incorporated to provide season long control of many annual grasses and broadleaf weeds throughout the growing season. Incorporation of BALAN helps assure effective weed control. BALAN control weeds as they germinate.

It is a violation of Federal Law to use this product in a manner inconsistent with its labeling.

Soil Texture Guide

The amount of BALAN applied will vary with the soil texture. A fine textured soil generally requires more BALAN than a coarse texture soil. BALAN is not recommended for use on muck or other soils high in organic matter content.

Refer to the following table to determine soil classification.

Soil Texture	Soil Classification
Coarse	Sand Loamy sand Sandy loam
Medium	Loam Silty clay loam* Silt loam Silt Sandy clay loam*
Fine	Clay Clay loam Silty clay loam* Silty clay Sandy clay Sandy clay loam*

*Silty clay loam and sandy clay loam soils are transitional soils and may be classified as either medium or fine textured soils. If silty clay loam or sandy clay loams are predominately sand or silt, they are usually classified as medium textured soils; if predominately clay, they are usually classified as fine textured soils.

Soil Preparation

Destroy existing weeds before a BALAN application. Chop and throughly mix crop residues into the soil to a depth of 4 to 6 inches by deep plowing or discing before a BALAN application. Use machinery that breaks up large clods before a BALAN application.

General Mixing Directions

Use the following instructions for mixing BALAN E.C. alone in water and tank mixed in water.

General Mixing Instructions in Water Alone: Start with a clean sprayer. Fill the spray tank 1/3 to 1/2 full with clean water. Start agitation. Add the correct quantity of BALAN, continue agitation, and finish filling the tank.

General Tank Mix Instructions: Vigorous, continuous agitation is required for all tank mixes. (Sparger pipe agitators generally provide the best agitation in spray tanks). During filling, to prevent foaming, avoid stirring or splashing air into the mixture by placing the end of the fill pipe below the surface of the water in the spray tank. Do not allow the mixture to siphon back into the water source.

Mixing order: Fill the tank 1/4 to 1/3 full with clean water. Start the agitation. Add dry flowables, flowables (F), liquids (L), aqueous suspensions (AS), or wettable powders (WP) to the water and agitate until the product(s) are completely dispersed in the water. Allow additional mixing and dispersion time when using dry flowable products. Continue agitation and fill tank to 3/4 full. Add BALAN, mix thoroughly, then add any solution (S) type formulations, agitate and finish filling. Maintain agitation from filling through application. If spraying and agitation must be stopped before the tank is empty, the materials may settle to the bottom. In this case, it is important to resuspend all of the material in the bottom of the tank before continuing the spray application. Sometimes it is more difficult to resuspend settled material than it is to suspend them originally. A sparger agitator is particularly useful for this purpose.

Read and carefully follow all label instructions for each material added to the tank. Premixing dry and flowable formulations with water (slurrying) and pouring the slurry through a 20 or 35 mesh wetting screen in the top of the tank will help assure good initial dispersion in the tank water. Line screens in the tank should be no finer than 50 mesh (100 mesh is finer than 50 mesh).

If a buildup of material is seen on the walls of the spray tank, wash the tank with soapy water between fillings. Rinse and continue the spraying operation. Clean tank, lines, and screens thoroughly after use.

Application Directions

Apply BALAN anytime after January 1 when the soil can be worked. Add the recommended amount of BALAN to clean water in the spray tank during the filling operation. (See General Mixing Instructions). Apply in 5 to 40 gallons of spray volume per acre (broadcast basis). Use any properly calibrated low pressure herbicide sprayer that will apply the spray uniformly. As the amount of spray volume decreases, the importance of accurate calibration and uniform application increases. Check the sprayer daily to insure proper calibration and uniform application. Apply BALAN to the soil surface and incorporate within the recommended time.

BALAN should not be applied to soils which are wet or are subject to prolonged periods of flooding as poor weed control may result.

Chemigation Directions

Refer to supplemental labeling entitled "BALAN E.C. Chemigation Instructions" for use directions for chemigation. Do not apply this product through any irrigation system unless the supplemental labeling on chemigation is followed.

Incorporation Directions

BALAN must be incorporated one time within 8 hours after application, except in certain areas of the Western United States (see Special Note at the bottom of next page). A second incorporation is required with most equipment (see page 11, under "Recommended equipment includes", for specific instructions). If BALAN is applied to a wet, warm soil surface or if the wind velocity is 10 mph or higher, variable weed control may result from delaying the first incorporation beyond 8 hours in the Eastern United States and 4 hours in the Western United States. Incorporation should place the BALAN into the top 2 to 3 inches of the final seedbed. Place BALAN to a depth of 3 inches for more effective control of Texas panicum. Generally, incorporation equipment will mix BALAN approximately half as deep as the equipment is run. For example, a disc running 4 inches deep will incorporate BALAN into approximately the top 2 inches.

Special Note: In the Western United States, extremely high temperatures and intense sunlight may be present at the time of application. Therefore, BALAN must be incorporated into the soil within 4 hours after application to prevent loss of activity.

Incorporation Equipment

Use machinery that mixes BALAN thoroughly with the soil. Shallow incorporation with implements set to mix BALAN less than 2 inches deep in the final seedbed may result in erratic weed control. Use of incorporation equipment not listed below may result in poor or erratic weed control and/or crop injury.

Recommended equipment includes:

Disc set to cut 4 to 6 inches deep and operated in 2 different directions at 4 to 6 mph. A tandem or double disc operated one time does not provide adequate incorporation.

Power take off driven equipment (tillers, cultivators, hoes) set to incorporate BALAN in the top 2 to 3 inches of the final seedbed. P.T.O. driven equipment should be operated one time at a speed not greater than 4 mph.

Precaution: Power take off driven equipment used alone as the incorporation implement in coarse soils may cause an uneven distribution of BALAN. This occurs when the wheel track depression is below the penetrating depth of the tines on the tiller. This may result in an excessive amount of BALAN in the tractor wheel depression area and poor weed control in the diluted areas.

Where this condition exists, a tandem disc should be used on coarse soils for the incorporation of BALAN.

Cultivation After Planting

Soil treated with BALAN may be shallowly cultivated or rotary hoed without reducing the weeds control activity of BALAN. Do not cultivate deeper than the BALAN treated layer of soil since this may bring untreated soil to the surface and poor weed control may result.

WEEDS AND GRASSES CONTROLLED BY BALAN

BALAN will not control established weeds.

Grasses Controlled

Annual blue grass	*Poa annua*
Barnyardgrass	*Echinochloa* sp.
Watergrass	
Broadleaf signalgrass	*Brachiaria platyphylla*
(See Page 17 for special instructions)	
Crabgrasses	*Digitaria* spp.
Large crabgrass	
Smooth crabgrass	
Crowfootgrass	*Dactyloctenium aegyptium*
Fall panicum	*Panicum dichotomiflorum*
Spreading panicgrass	
(See Page 17 for special instructions)	
Foxtails	*Setaria* spp.
Bottlegrass	
Bristlegrass	
Giant foxtail	
Green foxtail	
Pigeongrass	
Robust foxtail	
Yellow foxtail	
Goosegrass	*Eleusine indica*
Silver crabgrass	
Silvergrass	
Wiregrass	
Yardgrass	
Johnsongrass	*Sorghum halepense*
Seedling only	
Junglerice	*Echinochloa colonum*
Ryegrass	*Lolium multiflorum*
Annual ryegrass	
Italian ryegrass	
Sandbur	*Cenchrus incertus*
Burgrass	
Texas panicum	*Panicum texanum*
Buffalograss	
Colorado grass	

Broadleaf Weeds Controlled

Carpetweed	*Mollugo verticillata*
Chickweed	*Stellaria media*
Florida purslane	*Richardia scabra*
Florida pusley	
Mexican clover	
Pusley	
Knotweed	*Polygonum aviculare*
Lambsquarters	*Chenopodium album*
Pigsweeds	*Amaranthus* spp.
Carelessweed	
Prostrate pigweed	
Redroot	
Rough pigweed	
Spiny pigweed	
Purslane	*Portulaca oleracea*
Redmaids	*Calandrinia caulescens*

CROP RECOMMENDATIONS

These recommendations are given as the broadcast rates per acre of BALAN. For band applications, use proportionately less.

Alfalfa, Birdsfoot Trefoil, Clover (Alsike, Ladino, Red)

Apply and incorporate BALAN before seeding at a broadcast rate per acre of 3 quarts on coarse and medium soils and 4 quarts on fine soils. Do not apply after seeding.

Lettuce (Direct Seeded)

Apply and incorporate BALAN before seeding at a broadcast rate per acre of 3 quarts on coarse and medium soils and 4 quarts on fine soils. Do not apply after seeding.

Peanut—BALAN Alone

Apply and incorporate BALAN before planting at a broadcast rate per acre of 3-4 quarts on coarse and medium soils and 4 quarts on fine soils. Use the higher rate on coarse and medium soils where the weed population may be heavy.

Peanut—BALAN/Dual® Tank Mix

The BALAN/Dual tank mix controls the annual grasses and broadleaf weeds listed for BALAN alone and these additional weeds:

Foxtail millet	Southwest cupgrass
Prairie cupgrass	Witchgrass
Red rice	Yellow nutsedge

Follow recommended procedures for soil preparation, incorporation and cultivation of BALAN.

Ground application: Apply BALAN/Dual in 10 to 40 gallons of spray volume per acre.

Aerial application: Apply BALAN/Dual in a minimum of 5 gallons of spray volume per acre. BALAN/Dual may be applied and incorporated up to 14 days prior to planting.

		Broadcast Rates Per Acre	
		Dual 8E	
Soil Texture	BALAN	Less Than 3% O.M.	3% O.M. or Greater
	(quarts)	(pints)	(pints)
Coarse	3–4	1½–2	2
Medium	3–4	2–2½	2–2½
Fine	4	2–2½	2½–3

Note: Follow all cautions and precautions on the Dual label.

Peanut—BALAN/Vernam® Tank Mix

BALAN/Vernam controls the annual grasses and broadleaf weeds listed for BALAN alone and this additional weed:

Nutsedge (Nutgrass)

Follow recommended procedures for soil preparation and cultivation of BALAN. Apply BALAN/Vernam in 10 to 40 gallons of spray volume per acre. BALAN/Vernam can be applied up to 10 days prior to planting. Incorporate *immediately* to prevent loss of the Vernam. Follow suggested incorporation procedures for BALAN.

Broadcast Rates Per Acre

Soil Texture	BALAN	Vernam 7E
	(quarts)	(pints)
Coarse	3–4	2⅓
Medium	3–4	2⅓
Fine	4	3

Note: Follow all cautions and precautions on the Vernam label.

Peanut—Fall Panicum and Broadleaf Signalgrass Control

Fall panicum control—BALAN Alone: In the states of Alabama, Florida, Georgia, North Carolina, South Carolina and Virginia, apply BALAN at a broadcast rate of 4 quarts per acre on both coarse and medium soils.

Fall panicum control—BALAN/Vernam: For additional weed control, apply a broadcast rate of 4 quarts of BALAN and 2⅓ pints of Vernam on coarse soils and 4 quarts of BALAN and 3 pints of Vernam on medium soils. Follow incorporation directions for BALAN/Vernam on page 16.

Broadleaf signalgrass control: Apply BALAN at a broadcast rate of 4 quarts per acre on both coarse and medium soils.

Tobacco—(Transplant Burley and Dark Tobacco)

Apply and incorporate BALAN before transplanting at a broadcast rate per acre of 3 quarts on coarse and medium soils and 4 quarts on fine soils. Set plants so that roots extend below the BALAN treated layer of soil.

BALAN is usually compatible with Ridomil®. To assure the compatibility of BALAN with Ridomil, pour the products into a small container of water in the correct proportions. After thorough mixing, let stand for five minutes. If the combination remains mixed or can be remixed readily, the mixture is compatible.

Note: Read the Ridomil label carefully before using.

APPLICATION WITH DRY BULK FERTILIZER

Dry bulk fertilizer may be impregnated or coated with BALAN. Application of dry bulk fertilizer impregnated with BALAN has provided weed and grass control equal to the same rates of BALAN applied in liquid carriers.

Follow BALAN label recommendations regarding rates per acre, crops, incorporation directions, special instructions, cautions and special precautions.

Individual state regulations relating to dry bulk fertilizer blending, registration, labeling and application are the responsibility of the individual and/or company selling the fertilizer and chemical mixture.

Limitations

Apply a minimum of 200 pounds per acre of dry fertilizer impregnated with BALAN at the recommended rate. Most dry fertilizers can be used for BALAN impregnation. When coated ammonium nitrate and limestone are used alone, do not impregnate with BALAN. These materials will not absorb the herbicide. Blends containing mixtures of these materials can be impregnated.

Impregnation

Use any closed drum, belt, ribbon or other commonly used dry bulk fertilizer blender. Nozzles used to spray the BALAN onto the fertilizer should be placed to provide uniform spray coverage. If less than 6 pints of BALAN are mixed per ton of fertilizer, add water to the BALAN to give a total volume of at least 6 pints per ton.

Rates

Check the crop section to determine the rate per acre of BALAN. Calculate the amount of BALAN to be impregnated on a ton of dry bulk fertilizer based on the amount of fertilizer which will be applied per acre.

$$\text{Quarts of BALAN Per Acre} \times \frac{2000}{\text{Lbs. Fertilizer}} = \text{Quarts of BALAN Per Ton of Fertilizer}$$

Application

Spread the fertilizer/chemical mixture normally with a properly calibrated applicator. Be certain the material is applied uniformly to the soil surface.

Incorporation

Follow recommended incorporation procedures for BALAN.

APPLICATION WITH LIQUID FERTILIZERS

BALAN may be mixed with most liquid fertilizer materials. Combination of BALAN with solutions and suspension type fertilizers has provided weed and grass control equal to the same rates of BALAN applied in water. Follow BALAN label recommendations regarding rates per acre, crops, incorporation directions, special instructions, cautions and special precautions.

Individual state regulations relating to liquid fertilizer mixing, registration, labeling and application are the responsibility of the individual and/or company selling the fertilizer and chemical mixture.

Liquid Fertilizer Mixing Instructions

BALAN E.C. in Liquid Fertilizer: Emulsifiable concentrates, such as BALAN E.C., can be mixed with liquid fertilizers. In all cases, continuous agitation is required to prevent the BALAN from rising to the surface as an oily layer. When necessary, (see Liquid Fertilizer Compatibility Test on page 21), a compatibility agent can be used to cause the BALAN to emulsify properly (i.e., have a milky appearance rather than oily layer). The use of compatibility agents is especially important when tank mixing emulsifiable concentrates (EC) with dry flowables, wettable powders (WP), flowables (F), liquids (L), aqueous suspensions (AS), or solutions (S) in liquid fertilizer. If the emulsion is not properly formed, and the E.C. rises to the surface of the fertilizer as an oil ("oils out"), the oil may combine with the wettable powder, flowable, or suspension to form oily curds (viscous phase) which is difficult to redisperse. Any one of the compatibility agents listed below is helpful in causing emulsifiable concentrates to form non-oiling mixtures with liquid fertilizers. These compatibility agents can be used at rates as low as 1½ to 2 pints per ton of liquid fertilizer and should be mixed well with the fertilizer before adding the emulsifiable concentrate. Read the label on the compatibility agent and follow the directions.

1. Sponto 168D (Witco Chemicals Co., Chicago, IL)
2. Compat (Farm Chemicals, Inc., Aberdeen, NC)
3. Unite (Hopkins Ag Chemical, Madison, WI)
4. T-Mulz 734-2 (Thompson-Hayward Chemical Co., Kansas City, MO)
5. Rigo Compatibility Agent (Rigo Company, Buckner, KY)
6. Amoco Spray Mate™ (Amoco Oil Co., Chicago, IL)
7. Kem-Link (Universal Coop, Minneapolis, MN)

Each of the above is a phosphate ester type surfactant designed to be used with liquid fertilizers. They usually do not work well as compatibility agents in tank mixtures in plain water.

Testing for Tank Mix Compatibility in Liquid Fertilizers

Emulsifiable concentrates (EC) alone or in tank mixture with dry flowables, wettable powders (WP), liquids (L), flowables (F), aqueous suspensions (AS), or solutions (S) may not combine properly with some fluid fertilizer materials. Small quantities should always be tested before full-scale mixing. This will determine whether a compatibility agent is needed, and which agent does the best job. The seven agents listed above have been thoroughly tested. There are many others surfactants on the market which were not designed for use with liquid fertilizers. Use the following test to select the correct agent for your mixture.

1. Put 1 pint of the liquid fertilizer in a quart jar.
2. Add 1 to 4 teaspoons of the dry flowable, WP, L, F or AS formulation (depending on the recommended rate per acre) to the liquid fertilizer. Close jar and agitate until dispersed evenly in the fertilizer. If the materials do not disperse well, it may be necessary to slurry the chemicals in water before adding to the fertilizer.

DISCLAIMER OF WARRANTIES

Elanco Products Company makes no warranties concerning this product or its use, which extend beyond the description on the label. Elanco also expressly disclaims all warranties, express or implied in relation to any use of this product in combination or sequential use with any product not specifically recommended in writing by Elanco Products Company. All other warranties, expressed or implied, including but not limited to, any implied warranties of merchantability and fitness for a particular purpose, are disclaimed. All statements concerning this product apply only when used as directed. This disclaimer of warranties does not apply when prohibited by operation of law.

LIMITATION OF DAMAGES

The liability of Elanco Products Company, whether in contract, warranty, tort, negligence, strict liability or otherwise, shall not exceed the return of the amount of the purchase price of this product and under no circumstances shall Elanco be liable for special, indirect or consequential damages. If BALAN E.C. is used in combination or sequential use recommended by Elanco Products Company, the liability of Elanco shall in no manner extend to any damage, loss of injury not directly caused by the inclusion of BALAN E.C. in such combination or sequential use. This limitation of damages does not apply where prohibited by operation of law.

Dual® (metolachlor, Ciba-Geigy)
Vernam® (vernolate, Stauffer)
Ridomil® (metalaxyl, Ciba-Geigy)

This is a specimen label. Elanco is not responsible for the accuracy of the information contained herein. As labels are subject to revision, always carefully read and follow the label on the product container.

BALAN®
DRY FLOWABLE HEBICIDE

A selective herbicide for the preemergence control of annual grasses and broadleaf weeds

ACTIVE INGREDIENT:
N-butyl-*N*-ethyl-α,α,α- trifluoro-2,6-dinitro-*p*-toluidine 60%
INERT INGREDIENTS: 40%
TOTAL 100%

Contains 0.6 pounds active ingredient per pound
*BALAN®—(benefin, Elanco)
EPA Reg. No. 1471-140 ID 5529

Literature revised August 13, 1987
PI 2686 AMP

PRECAUTIONARY STATEMENTS

KEEP OUT OF REACH OF CHILDREN
CAUTION

HAZARDS TO HUMANS AND DOMESTIC ANIMALS

Harmful if swallowed, inhaled or absorbed through the skin. Avoid breathing dust or spray mist, and contact with skin, eyes, or clothing.

First Aid: In case of contact, immediately flush eyes or skin with plenty of water. If swallowed drink large quantity of water and induce vomiting by placing finger in back of throat. Never give anything by mouth to unconscious person.

If accidentally swallowed, or if irritation persists, get medical attention. If inhaled, remove victim to fresh air. Give artificial respiration, and get medical attention.

ENVIRONMENTAL HAZARDS

This product is toxic to fish. Do not contaminate any body of water by direct application, cleaning of equipment or disposal of wastes.

STORAGE AND DISPOSAL

Storage: Store in original container only. Do not store in direct sunlight. Do not store at temperatures above 120°F. If stored above 120°F, BALAN Dry Flowable could lose some of its dispersing properties, resulting in difficult application, poor weed control or crop injury. In case of leak or spill, contain material and dispose as waste.

Pesticide Disposal. Do not contaminate water, food or feed by storage or disposal. Wastes resulting from the use of this product may be disposed of on site or at an approved waste disposal facility.

Container Disposal: Triple rinse (or equivalent). Then offer for recycling or reconditioning, or puncture and dispose of in a sanitary landfill, or, if allowed by state and local authorities, by burning. If burned, stay out of smoke.

GENERAL USE PRECAUTIONS

Applied according to directions and under normal growing conditions, BALAN will not harm the treated crop. Over application may result in stunting, crop injury or soil residue. Uneven application or improper soil incorporation of BALAN can result in erratic weed control or crop injury. Seedling disease, temperature extremes, deep planting, excessive moisture, high salt concentration or drought may weaken crop seedlings and increase the possibility of damage from BALAN. Under these conditions, delayed crop development or reduced yields may result.

In the arid, irrigated areas of the Western United States—Arizona, California, Idaho, Montana, Nevada, Oregon, Utah, Washington and Wyoming—certain crops are susceptible to injury when planted in soil previously treated with BALAN. To avoid such injury, do not plant wheat, barley, rye, other grasses or onion within 10 months following a BALAN application. Milo (grain sorghum), corn, oats, spinach, red beet, sugar beet and other root crops should not be planted within 12 months following a BALAN application.

DIRECTIONS FOR USE

BALAN is a preemergence herbicide which is soil incorporated to provide season long control of many annual grasses and broadleaf weeds throughout the growing season. Incorporation of BALAN helps assure effective weed control. BALAN controls weeds as they germinate.

SOIL TEXTURE GUIDE

The amount of BALAN applied will vary with the soil texture. A fine textured soil generally requires more BALAN than a coarse textured soil. BALAN is not recommended for use on muck or other soils high in organic matter content.

Refer to the following table to determine soil classification.

Soil Texture	Soil Classification
Coarse	Sand Loamy sand Sandy loam
Medium	Loam Silty clay loam* Silt loam Silt Sandy clay loam*
Fine	Clay Clay loam Silty clay loam* Silty clay Sandy clay Sandy clay loam*

*Silty clay loam and sandy clay loam soils are transitional soils and may be classified as either medium or fine textured soils. If silty clay loam or sandy clay loams are predominately sand or silt, they are usually classified as medium textured soils; if predominately clay, they are usually classified as fine textured soils.

Soil Preparation

Destroy existing weeds before a BALAN application. Chop and thoroughly mix crop residues into the soil to a depth of four (4) to six (6) inches by deep plowing or discing before a BALAN application. Use machinery that breaks up large clods before a BALAN application.

General Sprayer Information

Vigorous, continuous agitation is required for all dry flowables. Sparger pipe agitation generally provides the best agitation. As when using any dry flowable pesticide, the sprayer in-line strainer should be no finer than a 16 mesh (16 mesh has larger openings than 50 mesh).

1. Be sure sprayer tank is clean and not contaminated with any material, as crop injury or sprayer clogging may result.
2. Be certain that agitation system is working properly and creates a **rippling or rolling action on the liquid surface.**
3. Use a pump with capacity to (a) provide sufficient agitation in tank to keep mixture in suspension, and (b) to provide to the agitator bypass at all times.
4. Centrifugal pumps are preferred since they provide propeller shear action for dispersing and mixing this product.
5. Use screens to protect the pump and to prevent nozzles from clogging. Screens placed on suction side of pump should be 16 mesh or coarser. Use 50 mesh or coarser screens between the pump and boom and when required at the nozzles.
6. Empty tank as completely as possible before refilling to prevent buildup of oil or emulsifiable concentrate residue (in the case of tank mixes). If an oil or emulsifiable concentrate film starts to build up in the tank, drain and clean with strong detergent solution.
7. It is good practice to clean sprayer tank thoroughly by flushing system with water containing a detergent, then refill with clean water.

GENERAL MIXING DIRECTIONS

Use the following instructions for mixing BALAN Dry Flowable alone in water and tank mixed in water.

Mixing Instructions In Water Alone:

1. Be sure sprayer is clean and not contaminated with any foreign material.
2. Fill tank one-fourth (1/4) full with clean water and start agitation.
3. Be certain that agitation system is working properly. Vigorous, continuous agitation throughout the spray tank is required with dry flowable formulations. Agitation should be sufficient to create a rippling or rolling action on the liquid surface. Sparger pipe agitation generally provides the best agitation in spray tanks.
4. To prevent foaming, avoid stirring or splashing air into the mixture by placing the end of the fill pipe below the surface of the water in the spray tank during the filling process.
5. **Pour product directly from container into tank.** Controlled addition of the product is important to assure adequate wetting and mixing to prevent clogging of screens and outlet ports.
6. Continue filling tank with water until full.

Tank-Mix Instructions:

1. Add BALAN Dry Flowable to the spray tank following the instructions above for BALAN alone.
2. Fill the spray tank to 90% full with water.
3. Add other formulations in this order, other dry flowables, wettable powders (WP), aqueous suspensions (A.S.), flowables (F), liquids (L), solutions (S) and emulsifiable or liquid concentrates (E.C., L.C.) in this order **allowing two to three minutes between the addition of each product,** and agitate continuously until the product(s) are completely dis-

persed in the water. Each chemical added to a spray tank needs time to either wet or mix with water and other chemicals in the spray tank. To insure proper mixing, time and agitation is required between the addition of each of the additional products.

4. Continue filling tank with water until full.

General Instructions and Precautions:

1. For either BALAN Dry Flowable alone or in tank-mix combination maintain agitation from filling through application. If spraying and agitation is stopped before the tank is empty, the materials may settle to the bottom. In this case, it is important to resuspend again all the material in the bottom of the tank before continuing the spray application. Sometimes it is more difficult to resuspend settled material than it is to suspend it originally. A sparger agitator is particularly useful for this purpose.
2. In-line screens in the tank should be no finer than 16 mesh (16 mesh has bigger openings than 50 mesh).
3. If a buildup of material is seen on the walls of the spray tank, wash the tank with soapy water between fillings. Rinse and continue the spraying operation. Clean tank, lines and screens thoroughly after use.
4. Read and carefully follow all label instructions for each material added to the tank.

APPLICATION DIRECTIONS

Apply BALAN anytime after January 1 when the soil can be worked. Apply in at least five (5) gallons of spray volume per acre (broadcast basis). Use any properly calibrated low pressure herbicide sprayer that will apply the spray uniformly. As the amount of spray volume decreases, the importance of accurate calibration and uniform application increases. Check the sprayer daily to insure proper calibration and uniform application. Apply BALAN to the soil surface and incorporate within the recommended time.

BALAN should not be applied to soils which are wet or are subject to prolonged periods of flooding as poor weed control may result.

Do not apply BALAN Dry Flowable through any type of irrigation system.

INCORPORATION DIRECTIONS

BALAN must be incorporated one time within eight (8) hours after application, except in certain areas of the Western United States (see Special Note below). A second incorporation is required with most equipment (refer to Incorporation Equipment below for specific instructions). If BALAN is applied to a wet, warm soil surface or if the wind velocity is 10 mph or higher, variable weed control may result from delaying the first incorporation beyond eight (8) hours in the Eastern United States and four (4) hours in the Western United States. Incorporation should place the BALAN into the top two (2) to three (3) inches of the final seedbed. Place BALAN to a depth of three (3) inches for more effective control of Texas panicum. Generally, incorporation equipment will mix BALAN approximately half as deep as the equipment is run. For example, a disc running four (4) inches deep will incorporate BALAN into approximately the top two (2) inches.

Special Note: In the Western United States, extremely high temperatures and intense sunlight may be present at the time of application. Therefore, BALAN must be incorporated into the soil within four (4) hours after application to prevent loss of activity.

INCORPORATION EQUIPMENT

Use machinery that mixes BALAN thoroughly with the soil, Shallow incoration with implements set to mix BALAN less than two (2) inches deep in the final seedbed may result in erratic weed control. Use of incorporation equipment not listed may result in poor or erratic weed control and/or crop injury.

Recommended equipment includes:

Disc set to cut four (4) to six (6) inches deep and operated in two (2) different directions at four (4) to six (6) mph. A tandem or double disc operated one time does not provide adequate incorporation.

Power take off driven equipment (tillers, cultivators, hoes) set to incorporate BALAN in the top two (2) to three (3) inches of the final seedbed. P.T.O. driven equipment should be operated one time at a speed not greater than four (4) mph.

Precaution: Power take off driven equipment used alone as the incorporation implement in coarse soils may cause an uneven distribution of BALAN. This occurs when the wheel track depression is below the penetrating depth of the tines on the tiller. This may result in an excessive amount of BALAN in the tractor wheel depression area and poor weed control in the diluted areas.

Where this condition exists, a tandem disc should be used on coarse soils for the incorporation of BALAN.

Cultivation After Planting

Soil treated with BALAN may be shallowly cultivated or rotary hoed without reducing the weed control activity of BALAN. Do not cultivate deeper than the BALAN treated layer of soil since this may bring untreated soil to the surface and poor weed control may result.

WEEDS AND GRASSES CONTROLLED BY BALAN

BALAN will not control established weeds.

Grasses Controlled	
Annual blue grass	*Poa annua*
Barnyardgrass	*Echinochloa crusgalli*
Watergrass	
Broadleaf signalgrass	*Brachiaria platyphylla*
Crabgrasses	*Digitaria* spp.
Large crabgrass	
Smooth crabgrass	
Crowfootgrass	*Dactyloctenium aegyptium*
Fall panicum	*Panicum*
Spreading panicgrass	*dichotomiflorum*
Foxtails	*Setaria* spp.
Bottlegrass	
Bristlegrass	
Giant foxtail	
Green foxtail	
Pigeongrass	
Robust foxtail	
Yellow foxtail	
Goosegrass	*Eleusine indica*
Silver crabgrass	
Silvergrass	
Wiregrass	
Yardgrass	
Johnsongrass	*Sorghum halepense*
Seedling only	
Junglerice	*Echinochloa colonum*
Ryegrass	*Lolium multiflorum*
Annual ryegrass	
Italian ryegrass	
Sandbur	*Cenchrus incertus*
Burgrass	
Texas panicum	*Panicum texanum*
Buffalograss	
Colorado grass	
Broadleaf Weeds Controlled	
Carpetweed	*Mollugo verticillata*
Chickweed	*Stellaria media*
Florida purslane	*Richardia scabra*
Florida pusley	
Mexican clover	
Pusley	
Knotweed	*Polygonum aviculare*
Lambsquarters	*Chenopodium album*
Pigweeds	*Amaranthus* spp.
Carelessweed	
Prostrate pigweed	
Redroot	
Rough pigweed	
Spiny pigweed	
Purslane	*Portulaca oleracea*
Redmaids	*Calandrinia caulescens*

CROP RECOMMENDATIONS

These recommendations are given as the broadcast rates per acre of BALAN. For band application, use proportionately less.

Alfalfa, Birdsfoot Trefoil, Clover (Alsike, Ladino, Red)—BALAN Dry Flowable Alone

Apply and incorporate BALAN Dry Flowable before seeding at a broadcast rate per acre of 2 pounds on coarse and medium soils and 2½ pounds on fine soils. Do not apply after seeding.

Lettuce (Direct Seeded)—BALAN Dry Flowable Alone

Apply and incorporate BALAN Dry Flowable before seeding at a broadcast rate per acre of 2 pounds on coarse and medium soils and 2½ pounds on fine soils. Do not apply after seeding.

Peanut—BALAN Dry Flowable Alone

Apply and incorporate BALAN before planting at a broadcast rate per acre of 2 to 2½ pounds on coarse and medium soils and 2½ pounds on fine soils. Use the higher rate on coarse and medium soils where the weed population may be heavy.

Peanut—BALAN Dry Flowable/Dual® Tank Mix

The BALAN/Dual tank mix controls the annual grasses and broadleaf weeds listed for BALAN alone and these additional weeds:

Foxtail millet	Southwest cupgrass
Prairie cupgrass	Witchgrass
Red rice	Yellow nutsedge

Follow recommended procedures for soil preparation, incorporation and cultivation of BALAN.

Ground application: Apply BALAN/Dual in 10 to 40 gallons of spray volume per acre.

Aerial application: Apply BALAN/Dual in a minimum of 5 gallons of spray volume per acre. BALAN/Dual may be applied and incorporated up to 14 days prior to planting.

Broadcast Rates Per Acre

Soil Texture	BALAN Dry Flowable	Dual 8E Less Than 3% O.M.	Dual 8E 3% O.M. or Greater
	(pounds)	(pints)	(pints
Coarse	2–2½	1½–2	2
Medium	2–2½	2–2½	2–2½
Fine	2½	2–2½	2½–3

Note: Follow all cautions and precautions on the Dual label.

Peanut—BALAN Dry Flowable/Vernam® Tank Mix

BALAN/Vernam controls the annual grasses and broadleaf weeds listed for BALAN alone and this additional weed:

Nutsedge (Nutgrass)

Follow recommended procedures for soil preparation and cultivation of BALAN. Apply BALAN/Vernam in 10 to 40 gallons of spray volume per acre. BALAN/Vernam can be applied up to 10 days prior to planting. Incorporate *immediately* to prevent loss of the Vernam. Follow suggested incorporation procedures for BALAN.

Broadcast Rates Per Acre

Soil Texture	BALAN Dry Flowable	Vernam 7E
	(pounds)	(pints)
Coarse	2–2½	2⅓
Medium	2–2½	2⅓
Fine	2½	3

Note: Follow all cautions and precautions on the Vernam label.

Peanut—Fall Panicum and Broadleaf Signalgrass Control

Fall panicum control—BALAN Dry Flowable Alone: In the states of Alabama, Florida, Georgia, North Carolina, South Carolina and Virginia, apply BALAN at a broadcast rate of 2½ pounds per acre on both coarse and medium soils.

Fall panicum control—BALAN Dry Flowable/Vernam: For additional weed control, apply a broadcast rate of 2½ pounds of BALAN and 2⅓ pints of Vernam on coarse soils and 2½ pounds of BALAN and 3 pints of Vernam on medium soils. Follow incorporation directions for BALAN/Vernam on page 17.

Broadleaf signalgrass control—BALAN Dry Flowable Alone: Apply BALAN at a broadcast rate of 2½ pounds per acre on both coarse and medium soils.

Tobacco—(Transplant Burley and Dark Tobacco)—BALAN Dry Flowable Alone:

Apply and incorporate BALAN before transplanting at a broadcast rate per acre of 2 pounds on coarse and medium soils and 2½ pounds on fine soils. Set plants so that roots extend below the BALAN treated layer of soil.

BALAN is usually compatible with Ridomil®. To assure the compatibility of BALAN with Ridomil, pour the products into a small container of water in the correct proportions. After thorough mixing, let stand for five minutes. If the combination remains mixed or can be remixed readily, the mixture is compatible.

Note: Read the Ridomil label carefully before using.

LIQUID FERTILIZER MIXING INSTRUCTIONS

BALAN Dry Flowable in Liquid Fertilizer

BALAN Dry Flowable can be mixed with liquid fertilizer. In all cases, vigorous and continuous agitation is required to mix BALAN Dry Flowable with the liquid fertilizer. BALAN Dry Flowable may be applied alone in liquid fertilizer without a compatibility agent. The use of a compatibility agent is especially important when tank mixing BALAN Dry Flowable with emulsifiable concentrates (EC) and liquid concentrates (LC). See the liquid fertilizer compatibility test below. If the emulsion is not properly formed, the EC/LC may rise to the surface of the fertilizer as an oil ("oils out"), the oil may combine with the BALAN Dry Flowable, wettable powder liquid flowable or aqueous suspension to form oily curds (viscous phase) which are difficult to disperse.

Any one of the compatibility agents listed below is helpful in causing emulsifiable and liquid concentrates to form non-oiling mixtures with liquid fertilizers. These compatibility agents can be used at rates as low as one and one-half (1½) to two (2) pints per ton of liquid fertilizer and should be mixed well with the fertilizer containing BALAN Dry Flowable before adding the emulsifiable concentrate. Read the label on the compatibility agent and follow the directions.

1. Sponto 168D (Witco Chemicals Co., Chicago, IL)
2. Compat (Farm Chemicals, Inc., Aberdeen, NC)
3. Unite (Hopkins Ag Chemical, Madison, WI)
4. T-Mulz 734-2 (Thompson-Hayward Chemical Co., Kansas City, MO)
5. Amoco Spray Mate™ (Amoco Oil Co., Chicago, IL)

Each of the above is a phosphate ester type surfactant designed to be used with liquid fertilizers. They usually do not work well as compatibility agents in tank mixtures in plain water.

Testing for Tank Mix Compatibility in Liquid Fertilizers:

Tank mixtures with BALAN Dry Flowable, other dry flowables, wettable powders (WP), aqueous suspensions (AS), flowables (F), liquids (L), solutions (S), or emulsifiable concentrates (EC) may not combine properly with some fluid fertilizer materials. Small quantities should always be tested before full-scale mixing. This will determine whether a compatibility agent is needed, and which agent does the best job. The five (5) agents listed above have been thoroughly tested. There are many other surfactants on the market which were not designed for use with liquid fertilizers. Use the following test to select the correct agent for your mixture.

1. Put one (1) pint of the liquid fertilizer in a quart jar.
2. Add one (1) to four (4) teaspoonful(s) of BALAN Dry Flowable with an equal amount of the WP, AS, F, L, S or EC formulation (depending on the recommended rate per acre) to the liquid fertilizer. Close jar and agitate until dispersed evenly in the fertilizer. If the materials do not disperse well, it may be necessary to slurry the chemicals in water before adding to the fertilizer.
3. Observe the jar for about 10 minutes. If the materials rise to the surface and form a thick layer (oily curds) which will not disperse when agitated, a compatibility agent is needed. If the mixture is easily dispersed to its original state with slight agitation, no agent is needed, but good agitation must be provided in the fertilizer spray tank.
4. If the need for a compatibility agent is shown in Step 3, using a clean quart jar, start at Step 1 above, add one-half (½) teaspoonful of the compatibility agent to the liquid fertilizer, mix well, then repeat Steps 2 and 3.

   An effective compatibility agent will cause the mixture to remain uniformly mixed with little or no separating or oil rising to the surface for one half (½) hour or longer. If slight separation does occur, two (2) to three (3) inversions of the jar should give a uniform remix. If oily curds form which will not disperse, more agent or another agent should be tried.

Use a clean jar for each test. The compatible mixture will have a uniform appearance and will be relatively easy to keep mixed with gentle agitation of the jar.

Application

Spread the fertilizer/chemical mixture normally with a properly calibrated applicator. Be certain the material is applied uniformly to the soil surface.

Incorporation

Follow recommended incorporation procedures for BALAN.

BALAN DRY FLOWABLE PERFORMANCE GUARANTEE A LIMITED WARRANTY OF BALAN WEED AND GRASS CONTROL

Elanco Products Company warrants that BALAN Dry Flowable will control the weeds and grasses as indicated on its current label when used according to label directions and subject to the following limitations:

1. Purchaser must notify Elanco promptly if a lack of commercially acceptable control occurs. Such notice must be given within forty-five (45) days after the first planting or application of BALAN Dry Flowable to the treated crop, whichever occurs later.

2. An Elanco representative must be satisfied that Purchaser used BALAN Dry Flowable according to the label directions.

3. The Purchaser must provide an Elanco representative with an invoice or other satisfactory record which shows the price, quantity of BALAN Dry Flowable, and date purchased.

4. Elanco's liability will be limited to a refund of the purchase price of the BALAN Dry Flowable applied to the acreage on which weed control was not satisfactory.

5. Void outside the U.S.A.

DISCLAIMER OF WARRANTIES

The warranty printed above is the only warranty applicable to this product. All other warranties, express or implied, including but not limited to the implied warranties of merchantability and fitness for a particular purpose are disclaimed. This disclaimer of warranties does not apply where prohibited by operation of law.

LIMITATION OF DAMAGES

Elanco's liability, whether in contract, warranty, tort, negligence, strict liability or otherwise shall not exceed the return of the amount of the purchase price of the BALAN Dry Flowable and under no circumstances shall Elanco be liable for special indirect or consequential damages. This limitation of damages does not apply where prohibited by operation of law.

INHERENT RISKS OF USE

Failure to carefully follow the directions for use of BALAN Dry Flowable may result in unsatisfactory weed control or crop injury. Factors such as plant disease, temperature extremes, deep planting, excessive moisture, high salt concentration, drought, or immediate intense irrigation may also result in unsatisfactory weed control or crop injury.

Dual®—(metolachlor, Ciba-Geigy)
Vernam®—(vernolate, Stauffer)
Ridomil®—(metalaxyl, Ciba-Geigy)

This is a specimen label. Elanco is not responsible for the accuracy of the information contained herein. As labels are subject to revision, always carefully read and follow the label on the product container.

PAARLAN® E.C.

HERBICIDE FOR TOBACCO

A Selective Herbicide for the Pre-emergence Control of Annual Grasses and Broadleaf Weeds.

ACTIVE INGREDIENT:

isopropalin* (2,6-dinitro-*N,N*-dipropylcumidine)	68.6%
INERT INGREDIENTS:	31.4%
TOTAL	100.0%

Contains 1½ pounds active ingredient per quart

EPA Reg. No. 1471-79 EPA Est. No. 1471-IN-2
PI 1612 AMP

PAARLAN®—the registered trademark for Elanco Products isopropalin.
ID 5030

CAUTION
KEEP OUT OF REACH OF CHILDREN

DIRECTIONS FOR USE

PAARLAN is a preemergence herbicide which is soil incorporated to provide season long control of many annual grasses and broadleaf weeds. Incorporation of PAARLAN helps assure effective weed control. PAARLAN controls weeds as they germinate.

SOIL TEXTURE GUIDE

The amount of PAARLAN to be applied will vary with the soil texture. A fine textured soil generally requires more PAARLAN than a coarse textured soil. PAARLAN is not recommended for use on muck or other soils containing more than 10% organic matter. Refer to the following table to determine soil texture grouping.

Soil Texture	Soil Classification	Soil Texture	Soil Classification
Coarse	Sand Loamy sand Sandy loam	Fine	Clay Clay loam Silty clay loam* Silty clay Sandy clay Sandy clay loam*
Medium	Loam Silty clay loam* Silt loam Silt Sandy clay loam*		

* Silty clay loam and sandy clay loam soils are transitional soils and may be classified as either medium or fine textured soils. If silty clay loam or sandy clay loams are predominately sand or silt, they are usually classified as medium textured soils; if predominately clay, they are usually classified as fine textured soils.

SOIL PREPARATION

Destroy existing weeds before a PAARLAN application. Chop and thoroughly mix crop residues into the soil to a depth of 4 to 6 inches by deep plowing or discing before a PAARLAN application. Use machinery that breaks up large clods before a PAARLAN application.

GENERAL MIXING DIRECTIONS

Use the following instructions for mixing PAARLAN E.C. alone in water and tank mixed in water.

General Mixing Instructions in Water Alone:

Start with a clean sprayer. Fill the spray tank ⅓ to ½ full with clean water. Start agitation. Add the correct quantity of PAARLAN E.C., continue agitation, and finish filling the tank.

General Tank Mix Instructions:

Vigorous, continuous agitation is required for all tank mixes. (Sparger pipe agitators generally provide the best agitation in spray tanks). During filling, to prevent foaming, avoid stirring or splashing air into the mixture by placing the end of the fill pipe below the surface of the water in the spray tank. Do not allow the mixture to siphon back into the water source.

Mixing order: Fill the tank ¼ to ⅓ full with clean water. Start the agitation. Add dry flowables, wettable powders (WP), aqueous suspensions (AS), flowables (F), and liquids (L) to the water and agitate until the product(s) are completely dispersed in the water. Allow additional mixing and dispersion time when using dry flowable products. Continue agitation and fill tank to ¾ full. Add the PAARLAN E.C., mix thoroughly, then add any solution (S) products, agitate and finish filling. Maintain agitation during filling through application. If spraying and agitation must be stopped before the tank is empty, the materials may settle to the bottom. In this case, it is important to resuspend all of the material in the bottom of the tank before continuing the spray application. Sometimes it is more difficult to resuspend settled material than it is to suspend them originally. A sparger agitator is particularly useful for this purpose.

Read and carefully follow all label instructions for each material added to the tank. Premixing dry and flowable formulations with water (slurrying) and pouring the slurry through a 20 or 35 mesh wetting screen in the top of the tank will help assure good initial dispersion in the tank water. Line screens in the tank should be no finer than 50 mesh (100 mesh is finer than 50 mesh).

If a buildup of material is seen on the walls of the spray tank, wash the tank with soapy water between fillings. Rinse and continue the spraying operation. Clean tank, lines, and screens thoroughly after use.

APPLICATION DIRECTIONS

Add the recommended amount of PAARLAN to clean water in the spray tank during the filling operation. (See General Mixing Instructions). Apply in 20 to 40 gallons of spray volume per acre (broadcast basis). Use any properly calibrated low pressure herbicide sprayer that will apply the spray uniformly. As the amount of spray volume decreases, the importance of accurate calibration and uniform application increases. Check the sprayer daily to insure proper calibration and uniform application. Apply PAARLAN to the soil surface and incorporate within the recommended time. PAARLAN should not be applied to soils which are wet or are subject to prolonged periods of flooding as poor weed control may result. Do not apply PAARLAN E.C. through any type of irrigation system.

INCORPORATION DIRECTIONS

Before planting, PAARLAN must be incorporated one time within 8 hours after application. A second incorporation is required with most equipment (see Incorporation Equipment, below, for specific instructions). If PAARLAN is applied to a wet, warm soil surface or if the wind velocity is 10 mph or higher, variable weed control may result from delaying the first incorporation beyond 8 hours. Incorporation should place the PAARLAN into the top 2 to 3 inches of the final seedbed. Generally, incorporation equipment will mix PAARLAN approximately half as deep as the equipment is run. For example, a disc running 4 inches deep will incorporate PAARLAN into approximately the top 2 inches.

Bedded Culture

For effective weed control, PAARLAN should be incorporated into the top 2 to 3 inches of the final seedbed. Knock off beds to planting height before application of PAARLAN and incorporation of bedded ground. Avoid removal of treated soil from the seedbed before or during the planting operation.

Incorporation Equipment

Use machinery that mixes PAARLAN thoroughly with the soil. Shallow incorporation with implements set to mix PAARLAN less than 2 inches deep in the final seedbed may result in erratic weed control. Use of incorporation equipment not listed below may result in poor or erratic weed control and/or crop injury.

Recommended equipment includes: Disc set to cut 4 to 6 inches deep and operated in 2 different directions at 4 to 6 mph. A tandem or double disc operated one time does not provide adequate incorporation. Power take off driven equipment (tillers, cultivators, hoes) set to incorporate PAARLAN in the top 2 to 3 inches of the final seedbed. P.T.O. driven equipment should be operated one time at a speed not greater than 4 mph.

Cultivation After Planting

Soil treated with PAARLAN may be shallowly cultivated or rotary hoed without reducing the weed control of activity of PAARLAN. Do not cultivate deeper than the PAARLAN treated layer of soil. This may bring untreated soil to the surface and poor weed control may result. In areas of bedded culture, weeds may germinate in the bottom of the furrow due to the removal of PAARLAN treated soil. These weeds should be controlled by cultivation. Shallow cultivation, rotary hoeing and hand hoeing do not reduce the weed control activity of PAARLAN.

WEEDS AND GRASSES CONTROLLED

PAARLAN will not control established weeds.

	Grasses Controlled
Barnyardgrass (Watergrass)	*Echinochloa* sp.
Crabgrass (Large crabgrass) (Small crabgrass) (Smooth crabgrass)	*Digitaria* spp.
Crowfootgrass	*Dactyloctenium aegyptium*
Foxtail (Giant foxtail)	*Setaria* spp.

(Green foxtail)
(Yellow foxtail)
(Bottlegrass)
(Robust white)
(Giant robust green)

Goosegrass	*Eleusine indica*
(Silver crabgrass)	
(Silvergrass)	
(Wiregrass)	
(Yardgrass)	
Johnsongrass	*Sorghum halepense*
(seedling only)	
Ryegrass, annual	*Lolium multiflorum*
(Italian ryegrass)	
(Ryegrass)	
Field sandbur	*Cenchrus incertus*
Fall panicum	*Panicum dichotomiflorum*
Texas panicum	*Panicum texanum*
(Buffalograss)	
(Colorado grass)	

Broadleaf Weeds Controlled

Carpetweed	*Mullugo verticillata*
Florida pusley	*Richardia scabra*
(Florida purslane)	
(Mexican clover)	
(Pusley)	
Common lambsquarters	*Chenopodium album*
Common purslane	*Portulaca oleracea*
Pigweed	
Redroot pigweed	
(Carelessweed)	
(Rough pigweed)	
Prostrate pigweed	
Spiny amaranth	
Poorjoe	*Diodia teres*

These recommendations are given as broadcast rates for PAARLAN. For band application, use proportionately less. Apply PAARLAN up to 5 weeks before planting or transplanting. No waiting period is necessary. Planting or transplanting may be done the same day as a PAARLAN application.

TRANSPLANT TOBACCO

Air-cured (Burley, Maryland, Dark)—

PAARLAN Alone: Apply and incorporate PAARLAN before transplanting at a broadcast rate per acre of 2 pints on all soil textures. Do not apply PAARLAN after transplanting.

PAARLAN Tank Mix with insecticides/nematicides: PAARLAN at 2 pints per acre can be tank mixed with Dasanit®, Diazinon®, Furadan®, Mo-Cap®, and Mo-Cap Plus for effective weed, insect and/or nematode control.

Flue-cured—

PAARLAN Alone: Apply and incorporate PAARLAN before transplanting at a broadcast rate per acre of 2 pints on coarse and medium soils at 2⅔ pints on fine soils.

PAARLAN Tank Mix with insecticides/nematicides: PAARLAN at 2 pints per acre can be tank mixed with Dasanit + Di-Syston, Mo-Cap, Mo-Cap Plus, Diazinon, Di-Syston, Dasanit, and Furadan for effective weed, insect and/or nematode control.

PAARLAN is usually compatible with Ridomil®. To assure the compatibility of PAARLAN with Ridomil, pour the products into a small container of water in the correct proportion. After thorough mixing, let stand for five minutes. If the combination remains mixed or can be remixed readily, the mixture is compatible.

Precaution

Read all product labels carefully before using. Note all directions, cautions, precautions, and special precautions. Note: Application time before transplanting varies with different products. Read all labels carefully for this information.

APPLICATION WITH DRY FERTILIZERS IMPREGNATION/RATES

PAARLAN may be impregnated on most dry fertilizers. Individual state regulations relating to dry fertilizer mixing, registration, labeling and applications are the responsibility of the individual and/or company selling the fertilizer and PAARLAN mixture. Use any closed drum, belt, ribbon or other commonly used dry bulk fertilizer blender. The nozzle or nozzles used to spray the PAARLAN on to the fertilizer should be placed to provide uniform spray coverage. If less than 6 pints of PAARLAN are mixed per ton of fertilizer, add water to the PAARLAN to give a total volume of at least 6 pints per ton. See the following table for amounts of PAARLAN and water to be premixed based on the amount of fertilizer to be applied per acre.

PAARLAN Impregnation Volume—For PAARLAN at 2 Pints/Acre

Fertilizer Rate (Pounds/Acre)	1000	1200	1400	1600	1800	2000
PAARLAN Rate (Pints/Acre)	2	2	2	2	2	2
Pints of Water to be Premixed with PAARLAN	1	1⅔	2⅓	2⅔	3⅓	4
Total Volume PAARLAN + Water (Pints)	3	3⅔	4⅓	4⅔	5⅓	6

Application

Spread the fertilizer/PAARLAN mixture with a properly calibrated applicator. Be certain the material is applied uniformly to the soil surface.

Incorporation

Follow normal PAARLAN incorporation procedures.

APPLICATION WITH LIQUID FERTILIZERS

PAARLAN may be mixed with most liquid fertilizer materials. Individual state regulations relating to liquid fertilizer mixing, registration, labeling and applications are the responsibility of the individual and/or company selling the fertilizer and PAARLAN mixture.

Liquid Fertilizer Mixing Instructions

PAARLAN E.C. in Liquid Fertilizer: Emulsifiable concentrates, such as PAARLAN E.C., can be mixed with most liquid fertilizers. In all cases, continuous agitation is required to prevent the PAARLAN from rising to the surface as an oily layer. When necessary, (see Liquid Fertilizer Compatibility Test below), a compatibility agent can be used to cause the PAARLAN E.C. to emulsify properly (i.e., have a milky appearance rather than oily layer). The use of compatibility agents is especially important when tank mixing emulsifiable concentrates (EC) with dry flowables, wettable powders (WP), aqueous suspensions (AS), flowables (F), liquids (L), or solutions (S) in liquid fertilizer.

If the emulsion is not properly formed, and the E.C. rises to the surface of the fertilizer as an oil ("oils out"), the oil may combine with the wettable powder, flowable, or suspension to form oily curds (viscous phase) which is difficult to redisperse. Any one of the compatibility agents listed below is helpful in causing emulsifiable concentrates to form non-oiling mixtures with liquid fertilizers. These compatibility agents can be used at rates as low as 1 ½ to 2 pints per ton of liquid fertilizer and should be mixed well with the fertilizer before adding the emulsifiable concentrate. Read the label on the compatibility agent and follow the directions.

1. Sponto 168D (Witco Chemicals Co., Chicago, IL)
2. Compat (Farm Chemicals, Inc., Aberdeen, NC)
3. Unite (Hopkins Ag Chemical, Madison, WI)
4. T-Mulz 734-2 (Thompson-Hayward Chemical Co.)
5. Rigo Compatibility Agent (Rigo Company, Buckner, KY)
6. Amoco Spray Mate (Amoco Oil Co., Chicago, IL)
7. Kem-Link (Universal Coop, Minneapolis, MN)

Each of the above is a phosphate ester type surfactant designed to be used with liquid fertilizers. They usually do not work well as compatibility agents in tank mixtures in plain water.

Testing for Tank Mix Compatibility in Liquid Fertilizers:

Emulsifiable concentrates alone or in tank mixture with dry flowables, wettable powders (WP), aqueous suspensions (AS), flowables (F), liquids (L), or solutions (S) may not combine properly with some fluid fertilizer materials. Small quantities should always be tested before full-scale mixing. This will determine whether a compatibility agent is needed, and which agent does the best job. The seven agents listed above have been thoroughly tested. There are many other surfactants on the market which were not designed for use with liquid fertilizers. Use the following test to select the correct agent for your mixture.

1. Put 1 pint of the liquid fertilizer in a quart jar.
2. Add 1 to 4 teaspoonfuls of the dry flowable, WP, AS, F, or L formulation (depending on the recomended rate per acre) to the liquid fertilizer. Close jar and agitate until dispersed evenly in the fertilizer. If the materials do not disperse well, it may be necessary to slurry the chemicals in water before adding to the fertilizer.
3. After dispersing the materials (Step 2), add 3 to 4 teaspoonfuls of the PAARLAN to the jar and shake well. Add solution herbicides to the mixture last and agitate. Observe the jar for about 10 minutes. If the materials rise to the surface and form a thick layer (oily curds) which will not redisperse when agitated, a compatibility agent is needed. If the mixture is easily redispersed to its original state with slight agitation, no agent is needed but good agitation must be provided in the fertilizer spray tank.

4. If the need for a compatibility agent is shown in Step 3: using a clean quart jar, start at Step 1 above, add 1/2 teaspoonful of the compatibility agent to the liquid fertilizer, mix well, then repeat Steps 2 and 3.

An effective compatibility agent will cause the mixture to remain uniformly mixed with little or no separating or oil rising to the surface for one half hour or longer. If slight separation does occur, 2 to 3 inversions of the jar should give a uniform remix. If oily curds form which will not redisperse, more agent or another agent should be tried. Use a clean jar for each test. The compatible mixture will have a uniform appearance and will be relatively easy to keep mixed with gentle agitation of the jar.

Application

Spread the PAARLAN/fertilizer with a properly calibrated applicator. Be certain the material is applied uniformly to the soil surface.

Incorporation

Follow recommended incorporation procedures for PAARLAN.

PRECAUTIONARY STATEMENTS

HAZARDS TO HUMANS AND DOMESTIC ANIMALS

CAUTION

Harmful if swallowed or absorbed through the skin. Avoid contact with skin, eyes, or clothing. In case of contact, flush with water. Do not contaminate foodstuffs or feeds.

First Aid—If swallowed: Call a physician or Poison Control Center. Drink 1 or 2 glasses of water and induce vomiting by touching back of throat with finger. Do not induce vomiting or give anything by mouth to an unconscious person. If on skin: Wash with plenty of soap and water. If in eyes: Flush with plenty of water. Get medical attention.

ENVIRONMENTAL HAZARDS

Direct contamination of any body of water with this emulsifiable concentrate may kill fish. Do not contaminate any body of water by direct application, cleaning of equipment or disposal of wastes.

STORAGE AND DISPOSAL

Storage: Do not store near heat or open flame. Store in original container only. In case of leak or spill, use absorbent materials to contain liquids and dispose as waste.

Pesticide Disposal: Do not contaminate water, food or feed by storage or disposal. Pesticide wastes are toxic. Improper disposal of excess pesticide, spray mixture, or rinsate is a violation of Federal Law. If these wastes cannot be disposed of by use according to label instructions, contact your State Pesticide or Environmental Control Agency, or the Hazardous Waste representative at the nearest EPA Regional Office for guidance.

Container Disposal: Triple rinse (or equivalent). Then offer for recycling or reconditioning or puncture and dispose of in a sanitary landfill, or by other procedures approved by state and local authorities.

General Use Precautions

Applied according to directions and under normal growing conditions, PAARLAN will not harm the treated crop. Over application, uneven application or improper soil incorporation of PAARLAN can result in erratic weed control, crop injury, delayed emergence or soil residue. Seedling disease, cold weather, deep planting, excessive moisture, high salt concentration or drought may weaken crop seedlings or transplants and increase the possibility of damage from PAARLAN. Under these conditions, delayed crop development or reduced yields may result.

PAARLAN® E.C. PERFORMANCE GUARANTEE
A LIMITED WARRANTY OF PAARLAN WEED AND GRASS CONTROL

1. Purchaser must notify Elanco promptly if a lack of commercially acceptable control occurs. Such notice must be given within forty-five (45) days after the first planting or application of PAARLAN E.C. to the treated crop, whichever occurs later.
2. An Elanco representative must be satisfied that Purchaser used PAARLAN E.C. according to the label directions.
3. The Purchaser must provide an Elanco representative with an invoice or other satisfactory record which shows the price and quantity of PAARLAN E.C. purchased.
4. Elanco's liability will be limited to a refund of the purchase price of the PAARLAN E.C. applied to the acreage on which weed control was not satisfactory.

DISCLAIMER OF WARRANTIES

The warranty printed above is the only warranty applicable to this product. All other warranties, express or implied, including but not limited to, the implied warranties of merchantability and fitness for a particular purpose are disclaimed. This disclaimer of warranties does not apply where prohibited by operation of law.

LIMITATION OF DAMAGES

Elanco's liability, whether in contract, warranty, tort, negligence, strict liability or otherwise shall not exceed the return of the amount of the purchase price of the PAARLAN E.C. and under no circumstances shall Elanco be liable for special, indirect or consequential damages. This limitation of damages does not apply where prohibited by operation of law.

INHERENT RISKS OF USE

Failure to carefully follow the directions for use of PAARLAN E.C. may result in unsatisfactory weed control or crop injury. Factors such as plant disease, deep planting, cold weather, excessive moisture, high salt concentration, improper fertilizer placement, drought, or transplant quality may also result in unsatisfactory weed control or crop injury.

PAARLAN® (isopropalin, Elanco)
Dasanit® (fensulfothion, Bayer GmbH)
Diazinon® (diazinon, Ciba-Geigy)
Di-Syston® (disulfoton, Bayer GmbH)
Furadan® (carbofuran, FMC)
Mo-Cap® (ethoprop, Mobil)
Ridomil® (metalaxyl, Ciba-Geigy)

Literature revised August 12, 1987
Pl 1612 AMP

This is a specimen label. Elanco is not responsible for the accuracy of the information contained herein. As labels are subject to revision, always carefully read and follow the label on the product container.

RUBIGAN® E.C.

FUNGICIDE

A locally systemic fungicide for the control of scab, powdery mildew and rusts of apples; scab and powdery mildew of pear; scab and powdery mildew of pecan and powdery mildew of grape.

ACTIVE INGREDIENT:	
fenarimol. α-(2-chlorophenyl)-α-(4-chlorophenyl)-5-pyrimidinemethanol	12.0%
INERT INGREDIENTS:	88.0%
TOTAL	100.0%

Contains petroleum distillate.

Contains 1 pound active ingredient per gallon.
RUBIGAN® —(fenarimol, Elanco)
EPA Registration No. 1471-146 EPA Est. 34704-MS-1
ID 5700

KEEP OUT OF REACH OF CHILDREN

WARNING

See additional precautionary statements inside booklet. See directions for use inside booklet.

AVISO

PRECAUCION AL USUARIO: Si usted no lee inglés, no usé este producto hasta que la etiqueta haya sido explicada ampliamente.

PI 3321 AMP

PRECAUTIONARY STATEMENTS

KEEP OUT OF REACH OF CHILDREN

HAZARDS TO HUMANS AND DOMESTIC ANIMALS

WARNING AVISO

PRECAUCION AL USUARIO: Si usted no lee inglés, no usé este producto hasta que la etiqueta haya sido explicada ampliamente.

Causes substantial but temporary eye injury. Causes skin irritation. Harmful if swallowed, inhaled or absorbed through the skin. Do not get in eyes, on skin or on clothing. Avoid breathing vapor or spray mist. Wear goggles, face shield or safety glasses, and protective clothing such as coveralls, a long-sleeved shirt, and impermeable gloves (i.e. butyl rubber) when handling this product. Wash thoroughly with soap and water after handling. Remove contaminated clothing and wash before use.

Do not apply this product in such a manner as to directly or through drift expose workers or other persons. The area being treated must be vacated by unprotected persons.

STATEMENT OF PRACTICAL TREATMENT

If in eyes: Flush with plenty of water. Call a physician.

If swallowed: Drink promptly large quantities of milk, egg whites, or gelatin solution, or if these are not available, drink large quantities of water. Avoid alcohol. Get medical attention.

If on skin: Immediately wash with plenty of soap and water. Get medical attention if irritation develops.

If inhaled: Remove individual to fresh air. If not breathing, give artificial respiration, preferably mouth-to-mouth. Get medical attention.

ENVIRONMENTAL HAZARDS

This product is toxic to fish. Do not apply directly to water or wetlands. Drift and runoff from treated areas may be hazardous to aquatic organisms in neighboring areas. Do not apply when weather conditions favor runoff or drift from the target area.

STORAGE AND DISPOSAL

Storage: Avoid freezing. Do not use or store near heat or open flame. Store in original container only. In case of leak or spill, use absorbent materials to contain liquids and dispose as waste.

Pesticide Disposal: Do not contaminate water, food, or feed by storage or disposal. Waste resulting from the use of this product may be disposed of on site or at an approved waste disposal facility.

Container Disposal: Triple rinse (or equivalent). Then offer for recycling or reconditioning, or, puncture and dispose of in a sanitary landfill, or incineration, or, if allowed by state and local authorities, by burning. If burned, stay out of smoke.

Directions for Use—Read All Directions Carefully Before Applying.

It is a violation of Federal Law to use this product in a manner inconsistent with its labeling.
Do not enter treated areas without protective clothing until sprays have dried.

Note to User

Because certain states may require more restrictive reentry intervals for various crops treated with this product, consult your State Department of Agriculture for further information.

Written or oral warnings must be given to workers who are expected to be in a treated area or in an area about to be treated with this product. (Indicate specific oral warnings which inform workers of areas or fields that may not be entered without specific protective clothing, period of time field must be vacated and appropriate actions to take in case of accidental exposure.) When oral warnings are given, warnings shall be given in a language customarily understood by workers. Written warnings must include the following information: WARNING. Area treated with RUBIGAN on (date of application). Do not enter without appropriate protective clothing until sprays have dried. In case of eye contact, immediately flush eyes with plenty of water and get medical attention. In case of eye contact immediately flush eyes with plenty of water and get medical attention. In case of skin contact, immediately wash with plenty of soap and water. Get medical attention if irritation occurs and persists. If inhaled and discomfort occurs, move to fresh air. Get medical attention if breathing difficulty occurs.

Diseases Controlled

Apple Scab, Powdery mildew, Rusts	Grape, Powdery mildew
Pecan Scab, Powdery mildew	Pear Scab, Powdery Mildew

Precaution: Do not graze cover crops or cut for silage or hay.

Maximum Rate Per Season

Do not apply more than 84 fl. ounces of RUBIGAN E.C. per acre per season on apple, pear and pecan; or 19 fl. ozs. per acre per season on grapes.

Chemigation

Do not apply RUBIGAN through any type of irrigation system.

MIXING DIRECTIONS

Add RUBIGAN E.C. to at least a one-half (1/2) full spray tank and continually agitate during the completion of filling and spraying operations.

When tank-mixed with other products, dry flowables, wettable powders or flowables (aqueous suspensions) should be added first to a half-full spray tank and then add the required amount of RUBIGAN E.C. followed by other emulsifiable concentrates or solutions. Continually agitate during the completion of filling and spraying operations.

APPLES: EASTERN UNITED STATES (EAST OF THE ROCKY MOUNTAINS/CONTINENTAL DIVIDE)

Use Rates and Application Schedule

General: Climatic conditions, varietal susceptibility and disease history of the orchard should be considered when choosing a use rate and spray schedule program for RUBIGAN.

Apply RUBIGAN on a **regular schedule,** an **extended regular combination schedule,** or a **post-infection schedule** or a **powdery mildrew and/or rust schedule.** Whichever schedule is used, the RUBIGAN application prior to switching to a program utilizing protectant fungicides alone should include a protectant fungicide (a nonsterol inhibiting fungicide) in tank-mix combination with RUBIGAN.

Use ground application equipment only.

A minimum of 3 oz/100 gals. dilute should be used for applications based on tree row volume calculations. Due to variability of spraying conditions, spray coverage and drying times, no less than 4 oz. of RUBIGAN in combination with a protectant fungicide should be used per acre in tree row volume sprays.

Regular Schedule For Prevention Of Apple Scab

Apply RUBIGAN alone starting at green tip. Continue applications on a 7 day schedule.

Use Rates

Apply RUBIGAN of 6 to 12 fl. ozs. per acre when used alone on a regular 7 day schedule. Under severe scab infection conditions, use the higher end of the RUBIGAN E.C. rate range when used alone. Do not exceed 12 fl. ozs. per acre.

Spray Volume Per Acre

Dilute Spray (200 or more gallons per acre): Apply RUBIGAN E.C. alone or in tank-mix combination at the rate of 3 or more fl. ozs. per 100 gallons of water in a uniform spray applied to run-off with conventional ground spray equipment.

Concentrate Spray (less than 200 gallons per acre); Apply RUBIGAN E.C. at 6 to 12 fl. ozs. per acre.

Extended Regular Schedule For Prevention Of Apple Scab

RUBIGAN may be applied in tank-mix combination with a protectant fungicide at its recommended use rate on an extended regular schedule to control apple scab. Application should start at green tip and be continued at intervals not to exceed 10 days.

Use Rates—RUBIGAN + Protectant Fungicide

Under light scab infection conditions, use RUBIGAN E.C. at 4 or more fl. ozs. per acre in tank-mix combination with protectant fungicide. When moderate to severe scab infection conditions occur and growth of new tissue is rapid, use 6 or more fl. ozs. per acre and shorter spray intervals, i.e., 7 days, when used in tank-mix combination with protectant fungicides. Do not exceed 12 fl. ozs. per acre.

Spray Volume Per Acre

Dilute Spray (200 or more gallons per acre): Apply RUBIGAN E.C. alone or in tank-mix combination at the rate of 3 or more fl. ozs. per 100 gallons of water in a uniform spray applied to run-off with conventional ground spray equipment.

Concentrate Spray (less than 200 gallons per acre): Apply RUBIGAN E.C. at 4 to 12 fl. ozs. per acre when used in combination with another protectant fungicide.

Post-Infection Schedule

RUBIGAN may be applied on a post-infection schedule. Application should be made within 96 hours (4 days) from the beginning of the **first** primary scab infection period and followed by a second application 7 days later. **Growers who do not have access to accurate on farm apple scab infection period data should not use the post-infection schedule.**

Use Rates

Apply RUBIGAN at 9 to 12 fl. ozs. per acre within 96 hours from the beginning of apple scab infection period, and follow with a second application at 6 to 9 fl. ozs. per acre 7 days later. When severe scab infection periods occur, use the higher end of the RUBIGAN rate range for each application. Do not exceed 12 fl. ozs. per acre.

Spray Volume Per Acre

Dilute Spray (200 or more gallons per acre): Apply RUBIGAN E.C. alone or in tank-mix combination at the rate of 3 or more fl. ozs. per 100 gallons of water in a uniform spray applied to run-off with conventional ground spray equipment.

Concentrate Spray (less than 200 gallons per acre): Apply RUBIGAN E.C. at 6 to 12 fl. ozs. per acre.

Powdery Mildew and/or Rust Schedule

Use Rates

For the control of powdery mildew or rusts, use RUBIGAN E.C. alone or in combination at 6 to 9 fl. ozs. per acre for light to moderate infection levels or 9 to 12 fl. ozs. per acre for severe infection levels starting at green tip. Do not allow the spray interval to exceed 14 days. Do not exceed 12 fl. ozs. per acre.

Spray Volume Per Acre

Dilute Spray (200 or more gallons per acre): Apply RUBIGAN E.C. alone or in tank-mix combination at the rate of 3 or more fl. ozs. per 100 gallons of water in a uniform spray applied to run-off with conventional ground spray equipment.

Concentrate Spray (less than 200 gallons per acre): Apply RUBIGAN E.C. at 6 to 12 fl. ozs. per acre.

Preharvest Interval

Do not apply RUBIGAN E.C. within 30 days of harvest.

APPLES: WESTERN UNITED STATES (WEST OF THE ROCKY MOUNTAINS/CONTINENTAL DIVIDE)

Use Rates

Do not apply less than 9 fl. ozs. or more than 12 fl. ozs. of RUBIGAN E.C. per acre. Do not exceed 12 fl. ozs. per acre.

Spray Volume Per Acre

Dilute Spray (200 or more gallons per acre): Apply RUBIGAN E.C. at the rate of 3 or more fl. ozs. per 100 gallons of water in a uniform spray applied to the point of run-off with conventional ground spray equipment.

Concentrate Spray (less than 200 gallons per acre), use ground application equipment only: Apply RUBIGAN E.C. at 9 to 12 fl. ozs. per acre.

Application Schedule

Apply RUBIGAN using either a **regular schedule,** an **extended regular schedule** or a **post-infection schedule** for the control of apple scab and powdery mildew. Whichever schedule is used, the RUBIGAN application prior to switching to a program utilizing protectant fungicides alone should include a protectant fungicide (a nonsterol inhibiting fungicide) in tank-mix combination with RUBIGAN.

Regular Schedule: For the control of apple scab, apply RUBIGAN starting at green tip. Continue applications on a 7 to 10 day schedule. Use the higher rates and shorter intervals when disease conditions are severe.

Extended Regular Schedule: RUBIGAN may be applied in tank-mix combination with a protectant fungicide at its recommended use rate. Application should start at green tip and continue on intervals not to exceed 10 days when using the extended regular schedule to control apple scab. Use the higher combination rates and shorter spray intervals, i.e., 7 days, when disease conditions are severe.

Post-infection Schedule: RUBIGAN may be applied on a post-infection schedule. The first application should be made within 96 hours (4 days) from the beginning of the first scab infection period. Growers who do not have access to accurate apple scab infection period data should not use the post-infection schedule.

Post-infection applications of RUBIGAN should be based on the following procedures explained in paragraphs 1 to 3 below which determine the proper timing for RUBIGAN applications. Paragraph 1 will always determine the proper timing of the first RUBIGAN application. To determine the proper timing of subsequent applications, the situations explained in paragraphs 2 and 3 must be considered.

1. The first RUBIGAN application should be made within 96 hours (4 days) from the beginning of the first primary apple scab infection period.

2. If one or more scab infection periods occur **within** 3 days after a RUBIGAN application, then RUBIGAN should be applied **within 7 days of the previous** RUBIGAN application. Do not allow the spray interval to exceed 7 days in this situation.

3. If no apple scab infection period occurs within 3 days **after** a RUBIGAN application, then apply RUBIGAN within 96 hours (4 days) after the beginning of the **next** scab infection period.

For the control of powdery mildew of apple, RUBIGAN should be applied on the following schedule as indicated by the stage of crop development.

1. Green-tip 2. Prepink/pink 3. Petal fall 4. First Cover

For the control of powdery mildew, do not exceed 14 days between RUBIGAN applications.

Preharvest Interval

Do not apply RUBIGAN E.C. within 30 days of harvest.

Pears: Eastern United States (east of the Rocky Mountains/Continental Divide)

Use Rates and Application Schedule

General: Climatic conditions, varietal susceptibility and disease history of the orchard should be considered when choosing a use rate and spray schedule program for RUBIGAN.

Apply RUBIGAN on a regular schedule, an extended regular combination schedule, or a post-infection schedule or a powdery mildew schedule. Whichever schedule is used, the RUBIGAN application prior to switching to a program utilizing protectant fungicides alone should include a protectant fungicide (a nonsterol inhibiting fungicide) in tank-mix combination with RUBIGAN.

Use ground application equipment only.

A minimum of 3 oz./100 gals. dilute should be used for applications based on tree row volume calculations. Due to variability of spraying conditions, spray coverage and drying times, no less than 4 oz. of RUBIGAN in combination with a protectant fungicide should be used per acre in tree row volume sprays.

Regular Schedule For Prevention of Pear Scab

Apply RUBIGAN alone starting at green tip. Continue applications on a 7 day schedule.

Use Rates

Apply RUBIGAN at 6 to 12 fl. ozs. per acre when used alone on a regular 7 day schedule. Under severe scab infection conditions, use the higher end of the RUBIGAN E.C. rate range when used alone. Do not exceed 12 fl. ozs. per acre.

Spray Volume Per Acre

Dilute Spray (200 or more gallons per acre): Apply RUBIGAN E.C. alone or in tank-mix combination at the rate of 3 or more fl. ozs. per 100 gallons of water in a uniform spray applied to run-off with conventional ground spray equipment.

Concentrate Spray (less than 200 gallons per acre): Apply RUBIGAN E.C. at 6 to 12 fl. ozs. per acre.

Extended Regular Schedule For Prevention Of Pear Scab

RUBIGAN may be applied in tank-mix combination with a protectant fungicide at its recommended use rate on an extended regular schedule to control pear scab. Application should start at green tip and be continued at intervals not to exceed 10 days.

Use Rates—RUBIGAN + Protectant Fungicide

Under light scab infection conditions, use RUBIGAN E.C. at 4 or more fl. ozs. per acre in tank-mix combination with protectant fungicide. When moderate to severe scab infection conditions occur and growth of new tissue is rapid, use 6 or more fl. ozs. per acre and shorter spray intervals, i.e., 7 days, when used in tank-mix combination with protectant fungicides. Do not exceed 12 fl. ozs. per acre.

Spray Volume Per Acre

Dilute Spray (200 or more gallons per acre): Apply RUBIGAN E.C. alone or in tank-mix combination at the rate of 3 or more fl. ozs. per 100 gallons of water in a uniform spray applied to run-off with conventional ground spray equipment.

Concentrate Spray (less than 200 gallons per acre): Apply RUBIGAN E.C. at 4 to 12 fl. ozs. per acre when used in combination with another protectant fungicide.

Post-Infection Schedule

RUBIGAN may be applied on a post-infection schedule. Application should be made within 96 hours (4 days) from the beginning of the first primary scab infection period and followed by a second application 7 days later. Growers who do not have access to accurate on farm pear scab infection period data should not use the post-infection schedule.

Use Rates

Apply RUBIGAN at 9 to 12 fl. ozs. per acre within 96 hours from the beginning of pear scab infection period, and follow with a second application at 6 to 9 fl. ozs. per acre 7 days later. When severe scab infection periods occur, use the higher end of the RUBIGAN rate range for each application. Do not exceed 12 fl. ozs. per acre.

Spray Volume Per Acre

Dilute Spray (200 or more gallons per acre): Apply RUBIGAN E.C. alone or in tank-mix combination at the rate of 3 or more fl. ozs. per 100 gallons of water in a uniform spray applied to run-off with conventional ground spray equipment.

Concentrate Spray (less than 200 gallons per acre): Apply RUBIGAN E.C. at 6 to 12 fl. ozs. per acre.

Powdery Mildew

Use Rates

For the control of powdery mildew, use RUBIGAN E.C. alone or in combination at 6 to 9 fl. ozs. per acre for light to moderate infection levels or 9 to 12 fl. ozs. per acre for severe infection levels starting at green tip. Do not allow the spray interval to exceed 14 days. Do not exceed 12 fl. ozs. per acre.

Spray Volume Per Acre

Dilute Spray (200 or more gallons per acre): Apply RUBIGAN E.C. alone or in tank-mix combination at the rate of 3 or more fl. ozs. per 100 gallons of water in a uniform spray applied to run-off with conventional ground spray equipment.

Concentrate Spray (less than 200 gallons per acre): Apply RUBIGAN E.C. at 6 to 12 fl. ozs. per acre.

Preharvest Interval

Do not apply RUBIGAN E.C. within 30 days of harvest.

Pears: Western United States (west of the Rocky Mountains/Continental Divide)

Use Rates

Do not apply less than 9 fl. ozs. or more than 12 fl. ozs. of RUBIGAN E.C. per acre.

Spray Volume Per Acre

Dilute Spray (200 or more gallons per acre). Apply RUBIGAN E.C. at the rate of 3 or more fl. ozs. per 100 gallons of water in a uniform spray applied to the point of run-off with conventional ground spray equipment. Do not exceed 12 fl. ozs. per acre.

Concentrate Spray (less than 200 gallons per acre), use ground application equipment only: Apply RUBIGAN E.C. at 9 to 12 fl. ozs. per acre.

Application Schedule

For the control of pear scab and powdery mildew, apply RUBIGAN E.C. starting at green tip and continue applications on a 7 to 10 day schedule. Use the higher rates and shorter intervals when disease conditions are severe. The RUBIGAN E.C. application prior to switching to a program utilizing protectant fungicides alone should include a protectant fungicide (a nonsterol inhibiting fungicide with limited curative activity) in tank-mix combination with RUBIGAN E.C.

Post-Infection Schedule: RUBIGAN may be applied on a post-infection schedule. The first application should be made within 96 hours (4 days) from the beginning of the first scab infection period. Growers who do not have access to accurate pear scab infection period data should not use the post-infection schedule.

Post-infection applications of RUBIGAN should be based on the following procedures explained in paragraphs 1 to 4 below which determine the proper timing for RUBIGAN applications. Paragraph 1 will always determine the proper timing of the first RUBIGAN application. To determine the proper timing of subsequent applications, the situations explained in paragraphs 2, 3 and 4 must be considered.

1. The first RUBIGAN application should be made within 96 hours (4 days) from the beginning of the first primary pear scab infection period.

2. If one or more scab infection periods occur within 3 days after a RUBIGAN application, then RUBIGAN should be applied within 7 days of the previous RUBIGAN application. Do not allow the spray interval to exceed 7 days in this situation.
3. If no pear scab infection period occurs within 3 days after a RUBIGAN application, then apply RUBIGAN within 96 hours (4 days) after the beginning of the next scab infection period.
4. If powdery mildew is a problem, do not allow the spray interval to exceed 14 days.

For the control of powdery mildew of pear, RUBIGAN should be applied on the following schedule as indicated by the stage of crop development.

1. Tight Cluster 2. Pink 3. Petal Fall 4. First Clover

Preharvest Interval

Do not apply RUBIGAN E.C. within 30 days of harvest.

PECANS

Use Rates

Do not apply less than 9 fluid ounces or more than 12 fluid ounces of RUBIGAN E.C. per acre.

Use the higher range of the RUBIGAN E.C. rates during conditions suitable for heavy disease infection (i.e., for scab, several hours of leaf wetting at temperatures above 50°F).

Spray Volume Per Acre

Dilute Spray (200 or more gallons per acre): Apply RUBIGAN E.C. at the rate of 3 or more fl. ozs. per 100 gallons of water in a uniform spray applied to the point of run-off with conventional ground spray equipment. Do not exceed 12 fl. ozs. per acre.

Concentrate Spray (less than 200 gallons per acre), use ground application equipment only: Apply RUBIGAN E.C. at 9 to 12 fl. ozs. per acre.

Application Schedule

For the control of scab and powdery mildew, apply RUBIGAN at 7 to 10 day schedule starting at bud break.

Applications should continue up to and including the point where nuts are up to one-half inch (1/2") long. Thereafter, RUBIGAN should be used in tank-mix combination with a protectant fungicide at its recommended use rate. The RUBIGAN application **prior** to switching to a program utilizing a protectant-only application should include a protectant fungicide (a non-sterol inhibiting fungicide with limited curative activity) as a tank-mix combination.

Do not use RUBIGAN on the Wichita variety.

Preharvest Interval

Do not apply RUBIGAN within 30 days of harvest.

GRAPES

Use Rates

RUBIGAN E.C. should be ground applied at the rate of 2-6 fl. ozs. per acre.

Application Schedule

Prebloom 18-24" Shoot	Full Bloom	Post Bloom
fl. oz./acre 2-3	fl. oz./acre 4	fl. oz./acre 6
2nd Post Bloom	Spray Interval	Minimum Spray Concentration
fl. oz./acre 6	days 14-18	fl. oz./100 gal. of water 3.0

Lower rates are recommended for early season applications when disease pressure is less intense. Applications of RUBIGAN at rates higher than recommended at prebloom may result in slight foliar injury consisting of leaf cupping and darker green leaf color.

Maximum Rates

Do not apply more than 6 fl. ozs. of RUBIGAN E.C. per acre per application or more than 19 fl. ozs. per season.

Preharvest Interval

Do not apply RUBIGAN E.C. within 30 days of harvest.

DISCLAIMER OF WARRANTIES

Elanco Products Company makes no warranties concerning this product or its use, which extend beyond the description on the label. Elanco also expressly disclaims all warranties, express or implied, in relation to any use of this product in combination or sequential use with any other product not specifically recommended in writing by Elanco Products Company. All other warranties, expressed or implied, including but not limited to, any implied warranties of merchantability and fitness for a particular purpose, are disclaimed. All statements concerning this product apply only when used as directed. This disclaimer of warranties does not apply where prohibited by operation of law.

LIMITATION OF DAMAGES

The liability of Elanco Products Company, whether in contract, warranty, tort, negligence, strict liability or otherwise, shall not exceed the return of the amount of the purchase price of this product and under no circumstances shall Elanco be liable for special, indirect or consequential damages. If RUBIGAN is used in a combination or sequential use recommended by Elanco Products Company, the liability of Elanco shall in no manner extend to any damage, loss or injury not directly caused by the inclusion of RUBIGAN in such combination or sequential use. This limitation of damages does not apply where prohibited by operation of law.

This is a specimen label. Elanco is not responsible for the accuracy of the information contained herein. As labels are subject to revision, always carefully read and follow the label on the product container.

SONALAN® E.C.

HERBICIDE

A selective herbicide for preemergence control of certain annual grasses and broadleaf weeds

ACTIVE INGREDIENT:

ethalfluralin *N*-ethyl-*N*-(2-methyl-2-propenyl)-2,6-dinitro-4-(trifluoromethyl)benzenamine	36.1%
INERT INGREDIENTS:	63.9%
TOTAL	100.0%

Contains 3 pounds active ingredient per gallon.

(SONALAN®—ethalfluralin, Elanco)

EPA Reg. No. 1471-122 ID 5036

Literature revised December 2, 1987
PI 2662 AMP

PRECAUTIONARY STATEMENTS

KEEP OUT OF REACH OF CHILDREN

WARNING

HAZARDS TO HUMANS AND DOMESTIC ANIMALS

SONALAN E.C. formulation is irritating to the eyes and skin. Avoid breathing vapors. Avoid contact with skin, eyes, or clothing. Do not contaminate foodstuffs or feed. Harmful if swallowed or absorbed through the skin. Use protective clothing such as coveralls, long sleeved shirt, and impermeable gloves when handling or using this product.

First Aid: In case of contact, immediately flush eyes or skin with plenty of water. For eyes, call a physician. Remove and wash contaminated clothing before reuse.

ENVIRONMENTAL HAZARDS

This pesticide is toxic to fish. Do not apply directly to any body of water or wetlands. Do not contaminate water by cleaning of equipment or disposal of wastes. Runoff or erosion from treated areas may be hazardous to fish in neighboring areas.

STORAGE AND DISPOSAL

Storage: Avoid freezing. Store above 40°F (5°C). If frozen, poor weed control may result. Do not use or store near heat or open flame. Store in original container only. In case of leak or spill, use absorbent materials to contain liquids and dispose as waste.

Pesticide Disposal: Pesticide, spray mixture, or rinse water that cannot be used according to label instructions must be disposed of according to applicable federal, state or local procedures. Do not contaminate water, food or feed by storage or disposal. Pesticide wastes are toxic. Improper disposal of excess pesticide, spray mixture, or rinsate is a violation of Federal Law. If these wastes cannot be disposed of by use according to label instructions, contact your State Pesticide or Environmental Control Agency, or the Hazardous Waste representative at the nearest EPA Regional Office for guidance.

Plastic Container Disposal: Triple rinse (or equivalent). Then offer for recycling or reconditioning, or puncture and dispose of in a sanitary landfill, or, if allowed by state and local authorities, by burning. If burned, stay out of smoke.

Metal Container Disposal: Triple rinse (or equivalent). Then offer for recycling or reconditioning, or puncture and dispose of in a sanitary landfill, or by other procedures approved by state and local authorities.

Bulk/Mini-bulk Tank Handling: Tanks that have been used for phenoxy-type herbicides should not be used for SONALAN. To clean, triple rinse (or equivalent) and wash with appropriate cleaners before reusing.

General Agricultural Precaution

Poor weed control and/or crop injury may result if directions are not followed. Uneven application, over application or improper incorporation can result in erratic weed control or injury to the treated or rotational crops.

Do not graze or forage crop grown in treated soil or cut for hay or silage.

Do not apply SONALAN to soils which are wet, cloddy, or subject to prolonged periods of flooding because poor weed control or crop injury may result.

Do not apply to any area not specified on the label.

Do not use SONALAN on any crop grown in the state of Montana.

Rotational Restrictions

Replanting

If replanting is required, replant only crops listed on the SONALAN label.

Sugar Beets or Red Beets

Do not plant sugar beets or red beets within thirteen (13) months following an application of three (3) pints or more per acre of SONALAN. Where less than three (3) pints of SONALAN have been applied, sugar beets or redbeets may be planted eight (8) months after an application of SONALAN, if the SONALAN treated soil is moldboard plowed to a depth of at least twelve (12) inches prior to planting.

Spinach or Oats

In California and Arizona, do not plant spinach or oats within eight (8) months following an application of three (3) pints per acre or more of SONALAN.

DIRECTIONS FOR USE

It is a violation of Federal Law to use this product in a manner inconsistent with its labeling.

SONALAN is a preemergence herbicide that controls many annual grasses and broadleaf weeds. SONALAN controls weeds as they germinate. SONALAN will not control established weeds.

MIXING AND APPLICATION DIRECTIONS

SONALAN Alone

Start with a clean spray tank. Fill sprayer one-third (1/3) to one-half (1/2) full with clean water or liquid fertilizer. Start agitation. Add correct quantity of SONALAN, continue agitation and finish filling the tank.

SONALAN Tank Mix

Vigorous, continuous agitation is required for all tank mixes. Sparger pipe agitators generally provide the best agitation in spray tanks. To prevent foaming, avoid stirring or splashing air into the mixture during filling by placing the end of the fill pipe below the surface of the water in the spray tank. Do not allow the mixture to siphon back into the water source.

Mixing order: Fill the tank one-fourth (1/4) to one-third (1/3) full with clean water or liquid fertilizer. (See page 9 for additional liquid fertilizer mixing instructions). Start the agitation. Add in the following order: dry flowables (DF), wettable powders (WP), aqueous suspensions (AS), flowables (F), or liquids (L) to the water and agitate until the product(s) are completely dispersed. Allow additional mixing and dispersion time when using dry flowable (DF) products. Continue agitation and fill tank to three-fourths (3/4) full, add the SONALAN or other emulsifiable concentrates (EC) and mix thoroughly. Then add any solution (S) formulations, agitate and finish filling. Maintain agitation during filling and through application. If spraying and agitation must be stopped before the tank is empty, the materials may settle to the bottom. In this case, it is important to resuspend all of the material in the bottom of the tank before continuing the spray application. A sparger agitator is particularly useful for this purpose. Sometimes it is more difficult to resuspend settled material than it is to suspend it originally.

Read and carefully follow all label instructions for each material added to the tank. Premixing dry and flowable formulations with water (slurrying) and pouring the slurry through a 20 or 35 mesh wetting screen in the top of the tank will help assure good initial dispersion in the tank. Line screens in the tank should be no finer than 50 mesh (100 mesh is finer than 50 mesh).

If a buildup of material on the walls of the spray tank is observed, wash the tank with soapy water between fillings. Rinse and continue the spraying operation. Clean the tank, lines, and screens thoroughly after use. Check the sprayer daily to ensure proper calibration and uniform application. Do not apply SONALAN when the wind can cause drifting of spray particles which can result in non-uniform application.

APPLICATION

Ground Application: Apply SONALAN in at least five (5) gallons of water or liquid fertilizer per acre (broadcast basis), using any properly calibrated, low pressure herbicide sprayer that will apply the spray uniformly.

Chemigation: Refer to supplemental labeling entitled "SONALAN E.C. Chemigation Instructions" for use directions for chemigation. Do not apply this product through any irrigation system unless the supplemental labeling on chemigation is followed.

Application with Liquid Fertilizer

SONALAN may be mixed with most liquid fertilizer materials. Combinations of SONALAN with solutions and suspension type fertilizers have provided weed and grass control equal to the same rates of SONALAN applied in water. Follow SONALAN label recommendations regarding rates per acre, crops, incorporation directions, special instructions, cautions and special precautions.

Individual state regulations relating to liquid fertilizer mixing, registration, labeling and applications are the responsibility of the individual and/or company selling the fertilizer and chemical mixture.

Liquid Fertilizer Mixing Instructions

Emulsifiable concentrates, such as SONALAN, can be mixed with liquid fertilizers. In all cases, continuous agitation is required to prevent the SONALAN from rising to the surface as an oily layer. When necessary, (see Liquid Fertilizer Compatibility Test), a compatibility agent can be used to ensure that the SONALAN emulsifies properly (i.e., has a milky appearance rather than an oily layer). The use of compatibility agents is especially important when tank mixing emulsifiable concentrates (EC) with dry flowables (DF), wettable powders (WP), flowables (F), liquids (L), aqueous suspensions (AS), or solutions (S) in liquid fertilizer. If the emulsion is not properly formed, and the EC rises to the surface of the fertilizer as an oil ("oils out"), the oil may combine with the wettable powder, flowable, or suspension to form oily curds (viscous phase) which are difficult to disperse.

Any one of the following compatibility agents is helpful in causing emulsifiable concentrates to form non-oiling mixtures with liquid fertilizers. These compatibility agents can be used at rates as low as one and one-half (1 1/2) to two (2) pints per ton of liquid fertilizer and should be mixed well with the fertilizer before adding the emulsifiable concentrate.

Read the label on the compatibility agent and follow the directions.

1. Sponto 168D (Witco Chemicals Co., Chicago, IL)
2. Compat (Farm Chemicals, Inc., Aberdeen, NC)
3. Unite (Hopkins Ag Chemical, Madison, WI)
4. T-Mulz 734-2 (Thompson-Hayward Chemical Co., Kansas City, MO)
5. Rigo Compatibility Agent (Rigo Company, Buckner, KY)
6. Amoco Spray Mate™ (Amoco Oil Co., Chicago, IL)
7. Kim-Link (Universal Coop, Minneapolis, MN)
8. Blendex™ (Helena Chemical Co., Cayce-West Columbia, SC)
9. Spray-Aide™ (Miller Chemical and Fertilizer Corp., Hanover, PA).

Each of the above is a phosphate ester type surfactant designed to be used with liquid fertilizers. They usually do not work well as compatibility agents in tank mixtures in water.

Testing for Tank Mix Compatibility in Liquid Fertilizers:

Emulsifiable concentrates alone or in tank mixture with dry flowables (DF), wettable powders (WP), liquid (L), flowables (F), aqueous suspensions (AS), or solutions (S), may not combine properly with some fluid fertilizer materials. Small quantities should always be tested before full-scale mixing. This will determine whether a compatibility agent is needed, and which agent does the best job. The nine (9) agents listed above have been thoroughly tested. There are many other surfactants on the market which were not designed for use with liquid fertilizers.

Use the following test to select the correct agent for your mixture:

1. Put one (1) pint of the liquid fertilizer in a quart jar.

2. Add one (1) to four (4) teaspoonful(s) of the DF, WP, L, F, or AS formulations (depending on the recommended rate per acre) to the liquid fertilizer. Close jar and agitate until dispersed evenly in the fertilizer. If the materials do not disperse well, it may be necessary to slurry the chemicals in water before adding to the fertilizer.

3. After dispersing the materials (Step 2), add three (3) to four (4) teaspoonfuls of the SONALAN to the jar and shake well. Add solution herbicides to the mixture last and agitate. Observe the jar for about 10 minutes. If the materials rise to the surface and form a thick layer (oily curds) which will not disperse when agitated, a compatibility agent is needed. If the mixture is easily dispersed to its original state with slight agitation, no agent is needed, but good agitation must be provided in the fertilizer spray tank.

4. If the need for a compatibility agent is shown in Step 3, using a clean quart jar, start at Step 1 above, add one-half (1/2) teaspoonful of the compatibility agent to the liquid fertilizer, mix well, then repeat Steps 2 and 3.

An effective compatibility agent will cause the mixture to remain uniformly mixed with little or no separating or oil rising to the surface for one-half (1/2) hour or longer. If slight separation does occur, two (2) to three (3) inversions of the jar should give a uniform remix. If oily curds form which will not disperse, more agent or another agent should be tried.

Use a clean jar for each test. The compatible mixture will have a uniform appearance and will be relatively easy to keep mixed with gentle agitation of the jar.

Application with Dry Bulk Fertilizer

Dry bulk fertilizer may be impregnated or coated with SONALAN. Application of dry bulk fertilizer impregnated with SONALAN has provided weed and grass control equal to the same rates of SONALAN applied in liquid carriers.

Follow SONALAN label recommendations regarding rates per acre, crops, special instructions, cautions and special precautions. For best results, SONALAN should be incorporated two times when applied impregnated on dry bulk fertilizers. The second incorporation must be delayed at least five (5) days after the first, completed prior to planting, and should be run in a different direction from the first. Follow other SONALAN label recommendations for soil incorporation.

Individual state regulations relating to dry bulk fertilizer blending, registration, labeling and application are the responsibility of the individual and/or company selling the fertilizer and chemical mixture.

Limitations: Apply a minimum of 200 pounds per acre of dry fertilizer impregnated with SONALAN at the recommended rate. Most dry fertilizers can be used for SONALAN impregnation. When coated ammonium nitrate and/or limestone are used alone, do not impregnate with SONALAN. These materials will not absorb the herbicide. Blends containing mixtures of these materials can be impregnated.

Impregnation: Use any closed drum, belt, ribbon or other commonly used dry bulk fertilizer blender. Nozzles used to spray the SONALAN onto the fertilizer should be placed to provide uniform spray coverage. Add water to the SONALAN to give a total volume of at least six (6) pints per ton of fertilizer.

SONALAN—PREPLANT INCORPORATED APPLICATION

Weeds Controlled

SONALAN applied preplant soil incorporated will control the following weeds:

Grasses:

Annual bluegrass	*Poa annua*
Barnyardgrass	*Echinochloa crus-galli*
Crabgrass	*Digitaria* spp.
(Large crabgrass)	
(Smooth crabgrass)	
Foxtail	*Setaria* spp.
(Giant foxtail)	
(Green foxtail)	
(Foxtail millet)	
(Pigeongrass)	
(Yellow foxtail)	
Goosegrass	*Eleusine indica*
Johnsongrass (from seed)	*Sorghum halepense*
Junglerice	*Echinochloa colonum*
Oat, wild[1]	*Avena fatua*
Panicum, fall	*Panicum dichotomiflorum*
Panicum Texas	*Panicum texanum*
(Buffalograss)	
(Colorado grass)	
Ryegrass, Italian	*Lolium miltiflorum*
(Annual ryegrass)	
Sandbur, Field	*Cenchrus incertus*
Shattercane	*Sorghum bicolor*
(Wild cane)	
Signalgrass, broadleaf	*Brachiaria platyphylla*
(Brachiaria)	
Witchgrass	*Panicum capillare*

Broadleaf Weeds:

Buckwheat, wild	*Polygonum convolvulus*
Carpetweed	*Mollugo verticillata*
Catchfly, conical	*Silene conica*
Chickweed, common	*Stellaria media*
Fiddleneck tarweed	*Amsinckia lycopsoides*
Groundcherry[2] (annual)	
(Lanceleaf groundcherry)	*Physalis lanceifolia*
(Wright's groundcherry)	*Physalis Wrightii*
Henbit	*Lamium amplexicaule*
Kochia	*Kochia scoparia*
Lambsquarters, common	*Chenopodium album*
Nightshade[2] (annual)	*Solanum* spp.
(Black nightshade)	
(Eastern black nightshade)[3]	
(Hairy nightshade)	
Pigweed	*Amaranthus* spp.
(Prostrate pigweed)	
(Redroot pigweed)	
(Smooth pigweed)	
(Tumble pigweed)	
Purslane, common	*Portulaca oleracea*
Pusley, Florida	*Richardia scabra*
Rockpurslane, redmaids	*Calandrinia caulescens*
Thistle, Russian	*Salsola kali*

[1] SONALAN will provide wild oat control except those germinating below the SONALAN treated zone.

[2] See special instructions for control in dry bean.

[3] See special instructions for partial control in soybean.

SOIL PREPARATION—GENERAL DIRECTIONS

SONALAN should be mixed uniformly into the top two (2) to three (3) inches of soil.

Factors that influence uniform application and incorporation of SONALAN are:

Crop Residues or Existing Weeds: Ground cover, such as crop residues or existing weeds, can interfere with the incorporation of SONALAN into the soil. A manageable level of such ground cover will allow the SONALAN to be uniformly mixed into the top two (2) to three (3) inches of soil. If the level of the ground cover is such that this cannot be done, the soil must be tilled prior to the application of SONALAN.

Roughness: The soil surface should be smooth enough so that you can operate a sprayer and incorporation equipment efficiently and at speeds which ensure a uniform application and incorporation of SONALAN.

General Soil Conditions: To assure uniform incorporation of SONALAN, soil moisture conditions should be such that large clods can be broken up during the incorporation process.

Soil Mixing Directions

Use incorporation equipment that thoroughly and uniformly mixes SONALAN into the top two (2) to three (3) inches of the seedbed, or erratic weed control and/or crop injury may result.

Incorporation Before Planting

Apply SONALAN up to three (3) weeks prior to planting. SONALAN should be thoroughly and uniformly mixed into the top two (2) to three (3) inches of the seedbed within two (2) days after application.

Incorporation in Bedded Culture

Application prior to bedding: Apply SONALAN and mix thoroughly into the top two (2) to three (3) inches of soil. The bedding operation provides additional mixing. Do not expose untreated soil during post-bedding operations.*

Application after bedding: Knock off beds to planting height before applying SONALAN. Apply SONALAN and mix thoroughly with equipment that will conform to the bed shape. Do not leave untreated soil exposed.*

* Avoid removal of treated soil from the seedbed before or during the planting operation. This would expose untreated soil, allowing weeds to germinate in the drill row.

Cultivation after Planting

Soil treated with SONALAN may be shallowly cultivated without reducing the weed control activity of SONALAN. Do not cultivate deeper than the treated soil since this may bring untreated soil to the surface, and poor weed control may result.

Soil Texture Guide

The amount of SONALAN applied will vary with the soil texture and organic matter. A fine textured soil will require more SONALAN per acre than a coarse soil. Choose the proper rate based on the following soil texture group and specific crop recommendations. Do not exceed recommended rates.

Soil Texture Group	Soil Classification
Coarse soils: (Light)	Sand, loamy sand, sandy loam
Medium soils:	Loam, silty clay loam*, silt loam, silt, sandy clay loam*
Fine soils: (Heavy)	Clay, clay loam, silty clay loam*, silty clay, sandy clay, sandy clay loam*

* Silty clay loam and sandy clay loam soils are transitional soils and may be classified as either medium or fine textured soils. If silty clay loam or sandy clay loam soils are predominately sand or silt, they are usually classified as medium textured soils. If they are predominately clay, they are usually classified as fine textured soils.

These recommendations are given as the broadcast rates of SONALAN per acre. Where a rate range is shown, use the lower rate for coarser textured soils or soils with lower organic matter. Use the higher rate, where a rate range is shown, on soils containing more than five percent (5%) organic matter. Where soil texture is variable within the same field, use the higher recommended SONALAN rate for the coarser textured soil. SONALAN is not recommended for use on soils containing more than 10% organic matter.

CROP RECOMMENDATIONS

Soybean—SONALAN Alone

General weed control—except groundcherry and nightshade:

SONALAN should be applied as a preplant incorporated treatment. Follow recommended soil preparation, application, and incorporation procedures for SONALAN.

For broader spectrum weed control, SONALAN may be used in combination with Amiben, Basagran, Blazer, Canopy, Classic, Command, Dual, Eptam, Gemini, Lasso, Lexone, Lorox, Lorox Plus, Preview, Sencor, Tackle or Vernam. Read the combination product label for directions, cautions, and limitations prior to use.

Broadcast Rates per Acre

Soil Texture	SONALAN
	(pints)
Coarse	1½-2
Medium	2-2½
Fine	2½-3

Eastern Black Nightshade—Partial Control:
SONALAN should be applied as a preplant incorporated treatment for partial control of nightshade in medium and fine textured soils. Follow recommended soil preparation and application procedures for SONALAN. Two incorporation passes are required for partial control of eastern black nightshade. For broader spectrum weed control, SONALAN may be used in combination with Amiben, Basagran, Blazer, Dual, Lasso, Lexone, Lorox, Sencor and Vernam. Read the combination product label for directions, cautions and limitations prior to use.

Broadcast Rates per Acre

Soil Texture	SONALAN
	(pints)
Medium	3
Fine	3½

Precaution: Soybeans should be planted no more than two (2) inches deep after early season adverse weather conditions have passed, especially when using higher rate programs. Excessively deep planting or cool, wet weather early in the growth cycle causes additional stress to the soybean plant. These factors may result in reduced stands, delayed maturity and reduced yields.

Soybean—SONALAN/Amiben® Tank Mix
SONALAN/Amiben should be applied as a preplant incorporated tank mix. The SONALAN/Amiben combination effectively controls the weeds listed for SONALAN alone for general weed control plus these additional weeds:

- Black nightshade
- Common ragweed
- Pennsylvania smartweed
- Velvetleaf
- Wild mustard

Broadcast Rates per Acre

Soil Texture	SONALAN: General Weed Control	SONALAN: Partial Control of Eastern Black Nightshade	Amiben: Amiben 2S	Amiben: Amiben DS
	(pints)	(pints)	(quarts)	(pounds)
Coarse	1¼-2	N/A	4-6**	2.4-3.6**
Medium	1¾-2½	3*	4-6	2.4-3.6
Fine	2¼-3	3½*	4-6	2.4-3.6

* Two incorporation passes are required for partial control of eastern black nightshade.
** Use the higher rate where heavy populations of smartweed, velvetleaf, ragweed, wild mustard or black nightshade are expected. Read the Amiben label for additional weeds controlled, directions, cautions and limitations prior to use.

Soybean—SONALAN/Canopy® Tank Mix (Do Not Use in California)
SONALAN/Canopy should be applied as a preplant incorporated tank mix. The SONALAN/Canopy combination effectively controls the weeds listed for SONALAN alone for general weed control plus these additional weeds:

Controlled

- Common cocklebur
- Florida beggarweed
- Hemp sesbania
- Hophornbeam copperleaf
- Jimsonweed
- Morningglory
  - (Entireleaf)
  - (Ivyleaf)
  - (Pitted)
  - (Smallflower)
  - (Tall)
- Prickly sida (Teaweed)
- Ragweed
  - (Common)
  - (Giant)
- Sicklepod
- Smartweed
- Spotted spurge
- Sunflower
- Velvetleaf (Buttonweed)

Additional weeds partially controlled by this combination include:

- Purple nutsedge
- Yellow nutsedge

Large seeded weeds, germinating deep in the soil, such as morningglory, sicklepod, cocklebur and giant ragweed or weeds with subsequent flushes may require a cultivation or an application of a postemergence herbicide for season-long control.

Follow recommended soil preparation, application and incorporation procedures for SONALAN. Plant soybeans within two (2) weeks after application.

Broadcast Rates per Acre

Soil Texture	SONALAN	Canopy Dispersible Granules: 0.5-3% O.M.	Canopy Dispersible Granules: 3-5% O.M.
	(pints)	(ounces)	(ounces)
Coarse	1¼-2	6- 8	8- 10
Medium	1¾-2½	8- 10	10- 12
Fine	2¼-3	10- 12	12- 14

Do not apply Canopy to soil with less than ½% organic matter.

Use higher rates of Canopy on soils with higher organic matter or heavy pressure from large deep germinating weed seeds.

Where Canopy is applied, plant soybean seed 1½″ to 2″ deep on a flat or raised seedbed only, or crop injury may occur.

Soybean injury may occur where Canopy is applied if excessive rainfall occurs after application but before soybeans germinate.

Precautions: Read the Canopy label carefully for cautions and precautions relating to environmental hazards, planting of rotation crops, sprayer contamination and cleanup, soil *p*H, organic matter and soil texture use restrictions, soybean variety planting restrictions, restrictions where Atrazine was used the previous year, restrictions concerning use with organic phosphate pesticides, grazing restrictions and other directions, precautions and limitations before applying the SONALAN/Canopy tank-mix.

Read the Canopy label for additional weeds controlled, directions, cautions and limitations prior to use.

Soybean—SONALAN/Command® Tank Mix (Do Not Use in California)
SONALAN/Command should be applied as a preplant incorporated tank mix treatment. The SONALAN/Command combination effectively controls the weeds listed for SONALAN alone for general weed control plus:

- Velvetleaf (Buttonweed)

Additional weeds partially controlled by this combination include:

- Annual morningglory
- Common ragweed
- Eastern black nightshade
- Jimsonweed
- Pennsylvania smartweed
- Prickly sida (Teaweed)
- Venice mallow

Control of these weeds may be erratic, ranging from poor to excellent depending upon soil temperature, time of weed germination, depth of weed seed in the soil and the amount of timing of soil moisture. Control may be improved with timely cultivation.

Application Directions: Tank mixed SONALAN/Command should be applied as a preplant incorporated treatment. Apply SONALAN/Command up to three (3) weeks prior to planting.

Do not apply aerially or through irrigation equipment.

Apply SONALAN/Command using a finished spray volume of 10 to 40 gallons of water per acre. **NOTE:** The use of an agriculturally approved drift reducing additive is required at finished spray volumes of 10-15 gallons per acre. Use nozzles suitable for broadcast boom application of herbicides. Coarse sprays are less likely to drift out of the target area than fine sprays.

Incorporation Directions: SONALAN/Command must be incorporated immediately after application. SONALAN/Command should be thoroughly and uniformly mixed into the top 2 to 3 inches of the final seed bed. Application to overly moist or wet soils will also increase the potential for off-site movement of Command vapors and may result in poor soil incorporation and unsatisfactory weed control. These directions must be followed to reduce the potential for off-site movement of Command vapors and potential injury to desirable vegetation including adjacent crops, trees and ornamentals.

Broadcast Rates per Acre

Soil Texture	SONALAN	Command 4 E.C.
	(pints)	(pints)
Coarse	1¼-2	¾
Medium	1¾-2½	1⅛
Fine	2¼-3	1½

Special Preaution: Off-site movement of spray drift or vapors of Command can cause foliar whitening or yellowing of adjacent crops, trees and ornamental plants which is usually temporary in nature but can result in permanent injury or death of the plants if exposure is excessive. Prior to making application of this product, read and strictly follow all precautions and application instructions on the Command label.

Read the Command label carefully for additional directions, cautions and limitations prior to use.

Soybean—SONALAN/Command® Lexone® or SONALAN/Command®/Sencor® Tank Mix (Do Not Use in California)
SONALAN plus Command and Lexone or Sencor should be applied as a preplant incorporated tank mix treatment. SONALAN/Command/Lexone or Sencor controls those weeds listed for SONALAN alone for general weed control plus these additional weeds:

Common ragweed	Spotted spurge
Hemp sesbania	Velvetleaf
Jimsonweed	Venice mallow
Pennsylvania smartweed	Wild mustard
Prickly sida (Teaweed)	

Additional weeds partially controlled by this combination include:

Annual morningglory	Giant ragweed
Common cocklebur	

Control of these weeds may be erratic, ranging from poor to excellent depending upon soil temperature, time of weed germination, depth of weed seed in the soil and the amount and timing of soil moisture. Control may be improved with timely cultivation.

Application Directions: Tank mixed SONALAN/Command/Lexone or Sencor should be applied as a preplant incorporated treatment. Apply SONALAN/Command/Lexone or Sencor up to two (2) weeks prior to planting.

Do not apply aerially or through irrigation equipment.

Apply SONALAN/Command/Lexone or Sencor using a finished spray volume of 10 to 40 gallons of water per acre. **NOTE:** The use of an agriculturally approved drift reducing additive is required at finished spray volumes of 10-15 gallons per acre. Use nozzles suitable for broadcast boom application of herbicides. Coarse sprays are less likely to drift out of the target area than fine sprays.

Incorporation Directions: SONALAN/Command/Lexone or Sencor must be incorporated immediately after application. SONALAN/Command/Lexone or Sencor should be thoroughly and uniformly mixed into the top 2 to 3 inches of the final seed bed. Application to overly moist or wet soils will also increase the potential for off-site movement of Command vapors and may result in poor soil incorporation and unsatisfactory weed control. These directions must be followed to reduce the potential for off-site movement of Command vapors and potential injury to desirable vegetation including adjacent crops, trees and ornamentals.

Follow other soil preparation application and incorporation procedures described on the SONALAN label.

Broadcast Rates per Acre

Soil Texture	SONALAN	Command 4 E.C.	Lexone 4L or Sencor 4F	Lexone Dry Flowable or Sencor Dry Flowable
	(pints)	(pints)	(pints)	(pounds)
Coarse	1¼-2	½	⅓-½[1]	¼-⅓[1]
Medium	1¾-2½	¾	½-¾	⅓-½
Fine	2¼-3	1⅛	¾-1	½-⅔

[1] Use the higher rate in the rate range in areas where weed populations are dense, for control of Venice mallow and wild mustard and for best control of common cocklebur, annual morningglory and giant ragweed.

Special Precaution: Off-site movement of spray drift or vapors of Command can cause foliar whitening or yellowing of adjacent crops, trees and ornamental plants which is usually temporary in nature but can result in permanent injury or death of the plants if exposure is excessive. Prior to making application of this product, read and strictly follow all precautions and application instructions on the Command label.

Read the Command, Lexone or Sencor labels carefully for additional directions, cautions and limitations prior to use.

Soybean—SONALAN/Dual® Tank Mix

SONALAN/Dual should be applied as a preplant incorporated tank mix. The SONALAN/Dual combination effectively controls the weeds listed for SONALAN alone for general weed control plus these additional weeds:

Black nightshade	Red rice
Galinsoga	Southwestern cupgrass
Prairie cupgrass	Yellow nutsedge

Additional weeds partially controlled by this combination include:

Hairy nightshade	Volunteer sorghum

Follow recommended soil preparation procedures for SONALAN. Apply the tank mix in at least ten (10) gallons of water per acre up to fourteen (14) days prior to planting. Incorporate the tank mix within two (2) days after application in the top two (2) inches of soil.

Broadcast Rates per Acre

Soil Texture	SONALAN General Weed Control	SONALAN Partial Control of Eastern Black Nightshade	Dual 8E Less than 3% O.M.	Dual 8E 3% O.M. or greater
	(pints)	(pints)	(pints)	(pints)
Coarse	1¼-2	N/A	1½-2	2
Medium	1¾-2½	3*	2-2½	2-2½
Fine	2¼-3	3½*	2-2½	2½-3

* Two incorporation passes are required for partial control of eastern black nightshade.

Read the Dual label for additional weeds controlled, directions, cautions and limitations prior to use.

Soybean—SONALAN/Lasso® Tank Mix

SONALAN/Lasso should be applied as a preplant incorporated tank mix. The SONALAN/Lasso combination effectively controls the weeds listed for SONALAN alone for general weed control plus these additional weeds:

Black nightshade[a]	Red rice[b]
Galinsoga	Yellow nutsedge[b]
Hairy nightshade[a]	

Additional weeds partially controlled by this combination include:

Common ragweed	Smartweed
Florida beggarweed	Spiny sida (teaweed)

Follow recommended soil preparation recommendations for SONALAN. Apply SONALAN/Lasso in a minimum of fifteen (15) gallons of water or liquid fertilizer per acre and incorporate thoroughly into the top two (2) inches of soil. Apply the tank mix up to seven (7) days prior to planting.

[a] or [b] Check beneath the rate tables for additional Lasso directions.

[a] Use a minimum of two and one-half (2½) quarts of Lasso to control these weeds.

[b] Use three (3) to four (4) quarts of Lasso per acre applied alone or, on medium and fine textured soils, in tank mix combination preplant incorporated only for control of these weeds.

Broadcast Rates per Acre

Soil Texture	SONALAN General Weed Control	SONALAN Partial Control of Eastern Black Nightshade	Lasso or Lasso Micro-Tech
	(pints)	(pints)	(quarts)
Coarse	1¼-2	N/A	2½
Medium	1¾-2½	3*	3-3½
Fine	2¼-3	3½*	3½-4

* Two incorporation passes are required for partial control of eastern black nightshade.

Read the Lasso label for additional weeds controlled, directions, cautions and limitations prior to use.

Soybean—SONALAN/Lexone® or SONALAN/Sencor® Tank Mix

SONALAN plus Lexone or Sencor should be applied as a preplant incorporated tank mix. Do not allow SONALAN or SONALAN/Sencor tank mix to remain in the tank overnight. The SONALAN/Lexone or SONALAN/Sencor combination effectively controls the weeds listed for SONALAN alone for general weed control plus these additional weeds:

Common ragweed	Prickly sida
Hemp sesbania	Velvetleaf
Jimsonweed	Venice mallow
Pennsylvania smartweed	Wild mustard

Control of cocklebur, morningglory, and giant ragweed may be erratic, ranging from poor to excellent depending upon soil temperature, time of weed germination, depth of weed seed in the soil and the amount and timing of soil moisture. Control may be improved with timely cultivation.

Follow recommended soil preparation, application, and incorporation directions for SONALAN.

Broadcast Rates per Acre

Soil Texture	SONALAN General Weed Control	SONALAN Partial Control of Eastern Black Nightshade	Lexone or Sencor: Lexone 4L or Sencor 4	Lexone or Sencor: Lexone Dry Flowable or Sencor Dry Flowable
	(pints)	(pints)	(pints)	(pounds)
Coarse	1¼-2	N/A	½**	⅓**
Medium	1¾-2½	3*	¾	½
Fine	2¼-3	3½*	1	⅔

* Two incorporation passes are required for partial control of eastern black nightshade.

** Do not use Lexone/Sencor on coarse soils with less than one percent (1%) organic matter.

Additional Precaution: For Lexone/Sencor tank mixes, do not plant any crop other than soybeans within four (4) months after treatment. Over application, uneven application, or improper soil incorporation may result in crop injury, soil residue, or erratic weed control. Seedling disease, cold weather, deep planting, excessive moisture, soil *p*H over seven and one-half (7.5), high salt concentration or drought may weaken crop seedlings and increase possibility of damage from the SONALAN/Lexone or SONALAN/Sencor tank mix. Under these conditions, delayed crop development or reduced yields may result. Read the Lexone and Sencor labels for directions, cautions and limitations prior to use. Do not use the foliage from soybeans treated with Lexone or Sencor for feed or forage.

Soybean—SONALAN/Preview® Tank Mix (Do Not Use in California)

SONALAN/Preview should be applied as a preplant incorporated tank mix. The SONALAN/Preview combination controls the weeds listed for SONALAN alone for general weed control plus these additional weeds:

Annual smartweeds
Common cocklebur
Common ragweed
Hophornbeam copperleaf
Jimsonweed
Mustards
Pigweed
(Palmer amaranth)
(Smooth)
(Tall waterhemp)
Prickly sida
(Teaweed)
Spotted spurge
Sunflower
Velvetleaf

Additional weeds partially controlled by this combination include:

Burcucumber
Eastern black nightshade
Giant ragweed
Morningglory
(Tall)
(Ivyleaf)
(Pitted)
(Entireleaf)
Nutsedge species

Large seeded weeds, germinating deep in the soil, such as morningglory, sicklepod, cocklebur and giant ragweed or weeds with subsequent flushes may require a cultivation or an application of a postemergence herbicide for season-long control.

Follow recommended soil preparation, application and incorporation procedures for SONALAN. Plant soybeans within two (2) weeks after application.

Broadcast Rates per Acre

Texture	SONALAN	Preview Dispersible Granules 0.5 - 3% O.M.	Preview Dispersible Granules 3 - 5% O.M.
	(pints)	(ounces)	(ounces)
Coarse	1¼-2	6	7
Medium	1¾-2½	7	8
Fine	2¼-3	8	9-10

Do not apply Preview to soil with less than ½% organic matter.

Use higher rates of Preview on soils with higher organic matter or heavy pressure from large deep germinating weed seeds.

Where Preview is applied, plant soybean seed 1½″ to 2″ deep on a flat or raised seedbed only, or crop injury may occur.

Soybean injury may occur where Preview is applied if excessive rainfall occurs after application but before soybeans germinate.

Precautions: Read the Preview label carefully for cautions and precautions relating to environmental hazards, planting of rotation crops, sprayer contamination and cleanup, soil *p*H, organic matter and soil texture use restrictions, soybean variety planting restrictions, restrictions where atrazine was used the previous year, restrictions concerning use with organophosphate pesticides, grazing restrictions and other directions, precautions and limitations before applying the SONALAN/Preview tank mix.

Read the Preview label for additional weeds controlled, directions, cautions and limitations prior to use.

Soybean—SONALAN/Lasso/Lexone or SONALAN/Lasso/Sencor Tank Mix

SONALAN plus Lasso and Lexone or Sencor should be applied as a preplant incorporated tank mix treatment. Do not allow the SONALAN/Lasso/Lexone or Sencor tank mix to remain in the tank overnight.

SONALAN/Lasso/Lexone or Sencor controls those weeds listed for SONALAN alone for general weed control plus these additional weeds:

Black nightshade[a]
Crowfootgrass
Galinsoga
Hairy nightshade[a]
Jimsonweed
Mallow, Venice
Mustard, wild
Pennsylvania smartweed
Red rice[b]
Sesbania
(Coffeebean)
(Coffeeweed)
Shepherdspurse
Spotted spurge
Spurred anoda
Velvetleaf
Yellow nutsedge[b]

Additional weeds partially controlled by this combination include:

Common ragweed
Florida beggarweed
Spiny sida (teaweed)

Follow recommended soil preparation recommendations for SONALAN. Apply the tank mixes in a minimum of fifteen (15) gallons of water or liquid fertilizer per acre. Set equipment to mix the soil no deeper than four inches. Apply the tank mix up to seven (7) days prior to planting.

[a] Use a minimum of two and one-half (2½) quarts of Lasso per acre to control these weeds.

[b] Use three (3) to four (4) quarts of Lasso per acre on medium and fine textured soils to control these weeds.

BROADCAST RATES PER ACRE

SOIL TEXTURE	SONALAN: GENERAL WEED CONTROL	SONALAN: PARTIAL CONTROL OF EASTERN BLACK NIGHTSHADE	LASSO OR LASSO MICRO-TECH	LEXONE OR SENCOR: LEXONE 4L OR SENCOR 4	LEXONE OR SENCOR: LEXONE DRY FLOWABLE OR SENCOR DRY FLOWABLE
	(pints)	(pints)	(quarts)	(pints)	(pounds)
Coarse	1¼-2	N/A	2½	½**	⅓**
Medium	1¾-2½	3*	3-3½	¾	½
Fine	2¼-3	3½*	3½-4	1	⅔

* Two incorporation passes are required for partial control of eastern black nightshade.

** Do not use Lexone or Sencor on coarse soils with less than one (1) percent organic matter.

Read the Lasso, Lexone or Sencor label for additional weeds controlled, directions, cautions and limitations prior to use.

Soybean—SONALAN/Dual/Lexone or SONALAN/Dual/Sencor Tank Mix

SONALAN plus Dual and Lexone or Sencor should be applied as a preplant incorporated tank mix treatment. Do not allow the SONALAN/Dual/Lexone or Sencor tank mix to remain in the tank overnight.

SONALAN/Dual/Lexone or Sencor controls those weeds listed for SONALAN alone for general weed control plus these additional weeds:

Black nightshade
Crowfootgrass
Galinsoga
Jimsonweed
Pennsylvania smartweed
Prairie cupgrass
Red rice
Sesbania
(Coffeebean)
(Coffeeweed)
Shepherdspurse
Southwestern
cupgrass
Spotted spurge
Spurred anoda
Velvetleaf
Venice mallow
Wild mustard
Yellow nutsedge

Additional weeds partially controlled by this combination include:

Hairy nightshade
Volunteer sorghum

Follow recommended soil preparation recommendations for SONALAN. Apply the tank mixes in a minimum of ten (10) gallons of water or liquid fertilizer per acre. Mix the tank mix into the top two (2) inches of soil. Apply the tank mixes up to fourteen (14) days prior to planting.

BROADCAST RATES PER ACRE

Use the recommended rates listed for SONALAN and Dual for the two-way combination; add the recommended rate of Lexone or Sencor for the SONALAN/Dual/Lexone or Sencor tank mix.

	SONALAN		DUAL		LEXONE OR SENCOR	
SOIL TEXTURE	GENERAL WEED CONTROL	PARTIAL CONTROL OF EASTERN BLACK NIGHTSHADE	LESS THAN 3% O.M.	3% O.M. OR GREATER	LEXONE 4L OR SENCOR 4	LEXONE DRY FLOWABLE OR SENCOR DRY FLOWABLE
	(pints)	(pints)	(pints)	(pints)	(pints)	(pounds)
Coarse	1¼-2	N/A	1½-2	2	½**	⅓**
Medium	1¾-2½	3*	2-2½	2-2½	¾	½
Fine	2¼-3	3½	2-2½	2½-3	1	⅔

* Two incorporation passes are required for partial control of eastern black nightshade.
** Do not use Sencor or Lexone on coarse soils with less than one (1) percent organic matter.
Read the Dual, Lexone and Sencor labels for additional weeds controlled, directions, cautions and limitations prior to use.

Soybean—SONALAN/Vernam® Tank Mix

SONALAN plus Vernam should be applied as a preplant incorporated tank mix treatment.

SONALAN/Vernam effectively controls the weeds listed for SONALAN alone for general weed control plus these additional weeds:

- Annual morningglory
- Purple nutsedge
- Sicklepod
- Velvetleaf
- Yellow nutsedge

Apply SONALAN/Vernam in at least ten (10) gallons per acre of water or liquid fertilizer. Follow recommended soil preparation procedures. Incorporate the tank mix immediately following application. Read the Vernam label for additional directions, cautions and limitations prior to use.

Broadcast Rates per Acre

	SONALAN		Vernam 7E
Soil Texture	General Weed Control	Partial Control of Eastern Black Nightshade	
	(pints)	(pints)	(pints)
Coarse	1¼-2	N/A	2⅓
Medium	1¾-2½	3*	3
Fine	2¼-3	3½*	3

* Two incorporation passes are required for partial control of eastern black nightshade.

Read the Vernam label for additional weeds controlled, directions, cautions and limitations prior to use.

Soybean-SONALAN Preplant Incorporated Followed by Overlay Treatments (Do Not Use in California)

SONALAN should be applied as a preplant incorporated treatment. Additional weeds tolerant to SONALAN may be controlled by using overlay preemergence applications of Amiben®, Canopy®, Dual®, Gemini™, Lasso®, Lexone®, Lorox®, Lorox® Plus, Preview® or Sencor®. Consult these product labels for additional weeds controlled, application directions and precautions before use.

Soybean-SONALAN Preplant Incorporated Followed by Postemergence Treatments (Do Not Use in California)

SONALAN should be applied as a preplant incorporated treatment. Additional weeds tolerant to SONALAN may be controlled by using postemergence applications of Basagran®, Blazer®, Classic®, or Tackle®. Consult these product labels for additional weeds controlled, application directons and precautions before use.

Dry Bean-SONALAN Alone (Does not include soybean)

SONALAN should be applied as a preplant incorporated treatment. Follow recommended soil preparation, application, and incorporation procedures for SONALAN.

General weed control—except groundcherry and nightshade: Broadcast Rates per Acre

Soil Texture	SONALAN
	(pints)
Coarse	1½-2
Medium	2-2½
Fine	2½-3

For broader spectrum control, SONALAN may be used in combination with Amiben, Dual, Eptam or Lasso. Read the combination product label for directions, cautions and limitations prior to use.

Groundcherry and Nightshade control (annual)-dry bean:

Two incorporation passes are required for groundcherry and nightshade control.

Do not use these SONALAN rates for garden bean, green bean, snap bean, or string bean.

Broadcast Rates per Acre

Soil Texture	SONALAN
	(pints)
Coarse	3-3½
Medium	3½-4
Fine	4-4½

For broader spectrum control, SONALAN may be used in combination with Amiben, Dual, Eptam or Lasso. Read the combination product label for directions, cautions and limitations prior to use.

Dry Bean—SONALAN/Amiben® Tank Mix or Overlay

SONALAN/Amiben should be applied as a preplant incorporated tank mix or SONALAN preplant incorporated followed by an Amiben overlay.

The SONALAN/Amiben combination effectively controls the weeds listed for SONALAN alone for general weed control plus these additional weeds:

- Black nightshade
- Common ragweed
- Pennsylvania smartweed
- Velvetleaf
- Wild mustard

Follow recommended soil preparation, application, and incorporation procedures for SONALAN.

Broadcast Rates per Acre

	SONALAN		Amiben**	
Soil Texture	General Weed Control	Nightshade and Groundcherry Control*	2S	DS
	(pints)	(pints)	(quarts)	(pounds)
Coarse	1¼-2	3-3½	4-6	2.4-3.6
Medium	1¾-2½	3½-4	4-6	2.4-3.6
Fine	2¼-3	4-4½	4-6	2.4-3.6

* Two incorporation passes are required for groundcherry or nightshade control.
** Use the higher rate where a heavy population of smartweed, velvetleaf, ragweed, wild mustard, or black nightshade are expected. Read the Amiben label for directions, cautions and limitations prior to use.

Read the Amiben label for additional weeds controlled, directions, cautions and limitations prior to use.

Dry Bean-SONALAN/Dual® Tank Mix or Overlay

SONALAN/Dual should be applied as a preplant incorporated tank mix or SONALAN preplant incorporated followed by a Dual overlay. The SONALAN/Dual combination effectively controls the weeds listed for SONALAN alone for general weed control plus these additional weeds:

- Black nightshade
- Galinsoga
- Prairie cupgrass
- Red rice
- Southwestern cupgrass
- Yellow nutsedge

Additional weeds partially controlled by this combination include:

- Hairy nightshade
- Volunteer sorghum

Follow recommended soil preparation procedures for SONALAN. Apply the tank mix in at least ten (10) gallons of water per acre up to fourteen (14) days prior to planting. Incorporate the tank mix within two (2) days after application in the top two (2) inches of soil. When Dual is applied as an overlay to SONALAN, apply Dual after planting (behind the planter) prior to crop and weed emergence.

Broadcast Rates per Acre

	SONALAN		Dual 8E	
Soil Texture	General Weed Control	Nightshade and Groundcherry Control*	Less than 3% O.M.	3% O.M. or greater
	(pints)	(pints)	(pints)	(pints)
Coarse	1¼-2	3-3½	1½-2	2

Soil Texture	SONALAN General Weed Control	SONALAN Nightshade and Groundcherry Control*	Dual 8E Less than 3% O.M.	Dual 8E 3% O.M. or greater
Medium	1³/₄-2¹/₂	3¹/₂-4	2-2¹/₂	2-2¹/₂
Fine	2¹/₄-3	4-4¹/₂	2-2¹/₂	2¹/₂-3

* Two incorporation passes are required for groundcherry or nightshade control.

Read the Dual label for additional weeds controlled, directions, cautions and limitations prior to use.

Dry Bean-SONALAN/Eptam®, SONALAN/EPTC Tank Mixes

SONALAN plus Eptam or SONALAN plus EPTC should be applied as a preplant incorporated tank mix treatment.

The SONALAN/Eptam or other SONALAN/EPTC tank mixes effectively control the weeds listed for SONALAN alone for general weed control plus these additional weeds:

Nightshade (black and hairy)
Nutsedge (purple and yellow)

Follow recommended soil preparation and application directions for SONALAN. Incorporate the tank mix immediately after application.

Broadcast Rates per Acre

Soil Texture	SONALAN General Weed Control	SONALAN Nightshade and Groundcherry Control*	Eptam 7E or Other EPTC 7E Formulations
	(pints)	(pints)	(pints)
Coarse	1¹/₄-2	3-3¹/₂	3¹/₂
Medium	1³/₄-2¹/₂	3¹/₂-4	3¹/₂
Fine	2¹/₄-3	4-4¹/₄	3¹/₂

* Two incorporation passes are required for groundcherry or nightshade control.

Read the Eptam or other EPTC labels for additional weeds controlled, directions, cautions and limitations prior to use.

Dry Bean-SONALAN/Lasso® Tank Mix

(For use only west of the Mississippi River, except in Kern County, California)

SONALAN/Lasso should be applied as a preplant incorporated tank mix. The SONALAN/Lasso combination effectively controls the weeds listed for SONALAN alone for general weed control plus these additional weeds:

Black nightshade[a]	Red rice[a]
Florida pusley	Witchgrass
Galinsoga	Yellow nutsedge[a]
Hairy nightshade[a]	

[a] Check beneath the rate table for additional Lasso directions.

[a] Use a minimum of two and one-half (2¹/₂) quarts of Lasso on coarse soils and three (3) quarts to medium and fine soils to control these weeds.

Additional weeds partially controlled by this combination include:

Common ragweed	Smartweed
Florida beggarweed	Spiny sida (teaweed)

SONALAN/Lasso tank mix: Follow recommended soil preparation recommendations for SONALAN. Apply SONALAN/Lasso in a minimum of fifteen (15) gallons of water per acre and incorporate thoroughly into the top two (2) inches of soil. Apply the tank mix up to seven (7) days prior to planting.

Broadcast Rates per Acre

Soil Texture	SONALAN General Weed Control	SONALAN Nightshade and Groundcherry Control*	Lasso or Lasso Micro-Tech
	(pints)	(pints)	(quarts)
Coarse	1¹/₄-2	3-3¹/₂	2¹/₂-3
Medium	1³/₄-2¹/₂	3¹/₂-4	2¹/₂-3
Fine	2¹/₄-3	4-4¹/₂	2¹/₂-3

* Two incorporation passes are required for groundcherry or nightshade control.

Read the Lasso label for additional weeds controlled, directions, cautions and limitations prior to use.

Dry Bean-SONALAN Preplant Incorporated Followed by Basagran® Postemergence Application

SONALAN should be applied preplant incorporated followed by a postemergence application of Basagran.

The SONALAN/Basagran combination effectively controls the weeds listed for SONALAN alone for general weed control plus these additional weeds:

Cocklebur	Ragweed
Jimsonweed	Tropic croton
Prickly sida	Velvetleaf

Consult the Basagran label for specific weed growth stage directions to achieve best results.

Broadcast Rates per Acre

Soil Texture	SONALAN General Weed Control	SONALAN Nightshade and Groundcherry Control*	Basagran**
	(pints)	(pints)	(pints)
Coarse	1¹/₂-2	3-3¹/₂	2
Medium	2-2¹/₂	3¹/₂-4	2
Fine	2¹/₂-3	4-4¹/₂	2

* Two incorporation passes are required for groundcherry or nightshade control.
** Apply Basagran as follows for optimum results:

Do not treat earlier than weed stage indicated on Basagran label. Add oil concentrate according to label use directions.

Read the Basagran label for additional weeds controlled, directions, cautions and limitations prior to use.

Peanut-SONALAN Alone

SONALAN should be applied as a preplant incorporated treatment. Follow recommended soil preparation, application and incorporation procedures for SONALAN.

Broadcast Rates per Acre

Soil Texture	SONALAN
	(pints)
Coarse	1¹/₂-2
Medium	2-2¹/₂
Fine	2¹/₂-3

Do not graze or forage crop grown in treated soil or cut for hay or silage.

Peanut-SONALAN/Dual Tank Mix

SONALAN/Dual should be applied as a preplant incorporated tank mix treatment. The SONALAN/Dual combination effectively controls the weeds listed for SONALAN alone for general weed control plus these additional weeds:

Galinsoga	Southwestern cupgrass
Prairie cupgrass	Yellow nutsedge
Red rice	

Follow recommended soil preparation procedures for SONALAN. Apply the tank mix in at least ten (10) gallons of water or liquid fertilizer per acre up to fourteen (14) days prior to planting. Incorporate the tank mix within two (2) days after application in the top two (2) to three (3) inches of soil.

Broadcast Rates per Acre

Soil Texture	SONALAN	+	Dual 8E Southeast	Dual 8E OK,TX,NM
	(pints)		(pints)	(pints)
Coarse	1¹/₄-2		1¹/₂-2	1¹/₄-2
Medium	1³/₄-2¹/₂		1¹/₂-2	1¹/₄-2
Fine	2¹/₄-3		1¹/₂-2	1¹/₄-2

Read the Dual label for additional directions, cautions and limitations prior to use.

Peanut-SONALAN/Lasso Tank Mix

SONALAN/Lasso should be applied as a preplant incorporated tank mix treatment. The SONALAN/Lasso combination effectively controls the weeds listed for SONALAN alone for general weed control plus these additional weeds:

Galinsoga	Yellow nutsedge
Red rice	

Follow recommended soil preparation procedures for SONALAN. Apply the tank mix in at least ten (10) gallons of water or liquid fertilizer per acre up to seven (7) days prior to planting. Incorporate the tank mix within two (2) days after application in the top two (2) to three (3) inches of soil.

Broadcast Rates per Acre

Soil Texture	SONALAN	+	Lasso 4EC
	(pints)		(quarts)
Coarse	1¹/₄-2		3
Medium	1³/₄-2¹/₂		3
Fine	2¹/₄-3		3

Read the Lasso label for additional directions, cautions and limitations prior to use.

Peanut-SONALAN/Vernam Tank Mix

SONALAN/Vernam should be applied as a preplant incorporated tank mix treatment. The SONALAN/Vernam combination effectively controls the weeds listed for SONALAN alone for general weed control plus these additional weeds:

Annual morningglory	Sicklepod
Purple nutsedge	Yellow nutsedge

Apply SONALAN/Vernam in at least ten (10) gallons per acre of water or liquid fertilizer. Follow recommended soil preparation procedures for SONALAN. SONALAN/Vernam can be applied up to ten (10) days prior to planting. Incorporate the tank mix immediately following application uniformly in the top two (2) to three (3) inches of soil.

Broadcast Rates per Acre

Soil Texture	SONALAN	+	Vernam 7E
	(pints)		(pints)
Coarse	1¼-2		2⅓
Medium	1¾-2½		3
Fine	2¼-3		3

Read the Vernam label for additional directions, cautions and limitations prior to use.

Sunflower-SONALAN Alone

SONALAN should be applied as a preplant incorporated treatment. Follow recommended soil preparation, application and incorporation procedures for SONALAN.

For broader specturm weed control, SONALAN may be used in combination with Amiben and Eptam. Read the combination product label for directions, cautions and limitations prior to use.

General weed control—except groundcherry and nightshade
Broadcast Rates per Acre

Soil Texture	SONALAN
	(pints)
Coarse	1½-2
Medium	2-2½
Fine	2½-3

Precaution: Do not graze of forage crop grown in treated or cut for hay or silage.

Sunflower-SONALAN/Amiben® Tank Mix or Overlay

SONALAN/Amiben should be applied as a preplant incorporated tank mix treatment, or SONALAN preplant incorporated followed by an Amiben overlay treatment. The SONALAN/Amiben combination effectively controls the weeds control plus these additional weeds:

Black nightshade	Velvetleaf
Common ragweed	Wild mustard
Pennsylvania smartweed	

Broadcast Rates per Acre

Soil Texture	SONALAN General Weed Control	Amiben 2S	Amiben DS
	(pints)	(quarts)	(pounds)
Coarse	1¼-2	4-6*	2.4-3.6*
Medium	1¾-2½	4-6	2.4-3.6
Fine	2¼-3	4-6	2.4-3.6

(SONALAN heads the General Weed Control column; Amiben heads the Amiben 2S and Amiben DS columns.)

* Use the higher rate where heavy populations of black nightshade, ragweed, smartweed, velvetleaf or wild mustard are expected.

Precaution: Do not graze or forage crop grown in treated soil or cut for hay or silage.

Read the Amiben label for additional weeds controlled, directions, cautions and limitations prior to use.

Sunflower-SONALAN/Eptam® or SONALAN/EPTC Tank Mix for Use Only in North Dakota, South Dakota and Minnesota.

SONALAN/Eptam or SONALAN/EPTC should be applied as a preplant incorporated tank mix treatment.

The SONALAN/Eptam tank mix effectively controls the weeds listed for SONALAN alone for general weed controls plus these additional weeds:

Annual Weeds
- Annual morningglory
- Bermudagrass (seedling)
- Black nightshade
- Corn spurry
- Fiddleneck
- Goosefoot nettleleaf
- Hairy nightshade
- Rescuegrass
- Ryegrass (Italian)
- Shepherdspurse
- Tumble pigweed
- Volunteer grains (barley, oat, wheat)

Perennial Weeds
- Bermudagrass
- Purple nutsedge
- Yellow nutsedge

Follow recommended soil preparation directions for SONALAN. Apply the tank mix in at least ten (10) gallons of water or liquid fertilizer per acre. Apply and immediatley incorporate the tank mix prior to planting in the spring.

Broadcast Rates per Acre

Soil Texture	SONALAN General Weed Control	Eptam 7E or other EPTC 7E formulations
	(pints)	(pints)
Coarse	1¼-2	2½-3½
Medium	1¾-2½	2½-3½
Fine	2¼-3	2½-3½

Precaution: Do not graze or forage crop grown in treated soil or cut for hay or silage.

Read the Eptam or other EPTC labels for additional weeds controlled, directions, cautions and limitations prior to use.

Dry Peas-SONALAN Alone

SONALAN should be applied as a preplant incorporated treatment. Follow recommended soil preparation, application and incorporation procedures for SONALAN.

General weed control—except groundcherry and nightshade:

Broadcast Rates per Acre

Soil Texture	SONALAN
	(pints)
Coarse	1½-2
Medium	2

Precaution: Do not exceed these SONALAN rates for weed control in dry peas or crop injury may occur.

Not all varieties of peas have been tested for tolerance to SONALAN. Consult your seed contractor before applying SONALAN to determine the tolerance of a particular pea variety. Application of SONALAN to non-tolerant pea varieties may result in crop injury and reduced yields. Do not graze of forage crop grown in treated soil or cut for hay or silage.

SONALAN PERFORMANCE GUARANTEE

A Limited Warranty of SONALAN Weed and Grass Control

Elanco Products Company warrants that SONALAN will provide commercially acceptable control of the weeds and grasses as indicated on its current label when used according to label directions and subject to the following limitations:

1. Purchaser must notify Elanco promptly if a lack of commercially acceptable control occurs. Notice must be given within forty-five (45) days after the first planting or after application of SONALAN to the treated crop, whichever occurs later.

2. An Elanco representative must be satisfied that Purchaser used SONALAN according to the label directions.

3. The Purchaser must provide an Elanco representative with an invoice or other satisfactory record which shows the price and quantity of SONALAN purchased.

4. Elanco's liability will be limited to a refund of the purchase price of the SONALAN applied to the acreage on which commercially acceptable control was not obtained.

5. Void outside the U.S.A.

DISCLAIMER OF WARRANTIES

The warranty printed above is the only warranty applicable to this product. All other warranties, express or implied, including but not limited to, any implied warranties or merchantability and fitness for a partial purpose, are disclaimed. This disclaimer of warranties does not apply where prohibited by operation of law.

LIMITATION OF DAMAGES

Elanco's liability, whether in contract, warranty, tort, negligence, strict liability or otherwise shall not exceed the return of the amount of the purchase price of the SONALAN and under no circumstances shall Elanco be liable for special, indirect or consequential damages. This limitation of damages does not apply where prohibited by operation of law.

INHERENT RISKS OF USE

Failure to carefully follow the directions for use of SONALAN, as well as other factors such as seedling disease, cold weather, deep planting, excessive moisture, high salt concentration, sodic soils or drought may result in unsatisfactory weed control or crop injury. Reduced yields may result under these conditions.

(Amiben®—chloramben, Union Carbide)
(Basagran® —bentazon, BASF)
(Blazer®—acifluofen, Rohm and Hass)
(Canopy® —chlorimuron ethyl plus metribuzin, DuPont)
(Classic® —chlorimuron ethyl, DuPont)

(Command® —FMC)
(Dual® —metolachlor, Ciba-Geigy)
(Eptam® —EPTC, Stauffer)
(Gemini® —chlorimuron ethyl plus linuron, DuPont)
(Lasso® —alachlor, Monsanto)
(Lasso® Micro-Tech™ —alachlor, Monsanto)
(Lexone® —metribuzin, DuPont)
(Lorox® —linuron, DuPont)
(Lorox® Plus—chlorimuron ethyl plus linuron, DuPont)
(Preview® —chlorimuron ethyl plus metribuzin, DuPont)
(Sencor® —metribuzin, Bayer)
(Tackle® —acifluorfen, Rhone-Poulenc)
(Vernam® —vernolate, Stauffer)

SUPPLEMENTAL LABELING

SONALAN® E.C. (EPA REG. NO. 1471-122)

SONALAN APPLIED THROUGH CHEMIGATION SYSTEMS FOR WEED CONTROL IN PEANUTS

Sonalan E.C. may be applied through properly equipped chemigation systems for weed control in peanuts. Refer to supplemental labeling entitled "Sonalan E.C. Chemigation Instructions" for use directions for chemigation. Do not apply Sonalan E.C. through any irrigation system unless supplemental labeling on chemigation is followed.

Apply Sonalan E.C. in sprinkler irrigation equal to 1/2 to 1 inch of water. Planting and application should occur as soon as possible after the last tillage operation. Sonalan must be applied within 2 days after planting prior to crop or weed emergence. Sonalan does not control established weeds. Soil incorporation is not required when Sonalan is applied through chemigation systems.

Broadcast Rates per Acre

Soil Texture	SONALAN E.C. (pints)
Coarse	1½ —2
Medium	2 —2½
Fine	2½ —3

Cultivation:

Soil treated with Sonalan may be shallow-cultivated without reducing weed control activity.

Precautions:

Read the Sonalan label for weeds controlled, additional directions, precautions and limitations prior to use.

Sonalan® E.C. (ethalfluralin, Elanco)
EA 8122 2/88

SUPPLEMENTAL LABELING

SONALAN® E.C. (EPA REG. NO. 1471-122)

SONALAN E.C. CHEMIGATION INSTRUCTIONS

Sonalan E.C. may be applied through properly equipped chemigation systems for weed control in certain crops. See supplemental use directions for specific crops for which chemigation with Sonalan is recommended. Read and follow all label instructions outlined below concerning chemigation before applying Sonalan E.C.

General Use Precautions

- Read the Sonalan E.C. label for directions, precautions and limitations prior to use.
- Do not apply Sonalan Dry Flowable through any chemigation system.

General Chemigation Directions

Apply this product only through continuously moving center pivot, lateral move or end tow sprinkler irrigation systems equipped for chemigation. Do not apply this product through any other type of irrigation system.

Crop injury, lack of effectiveness, or illegal pesticide residues in the crop can result from nonuniform distribution of treated water.

If you have questions about calibration, you should contact State Extension Service specialists, equipment manufacturers or other experts.

Do not connect an irrigation system (including greenhouse systems) used for chemigation application to a public water system.

A person knowledgeable of the chemigation system and responsible for its operation, or under the supervision of the responsible person, shall shut the system down and make necessary adjustments should the need arise.

Sprinkler Chemigation Directions

The following directions must be followed for all recommended sprinkler irrigation systems (center pivot, lateral move and end tow):

1. The system must contain a functional check valve, vacuum relief valve, and low pressure drain appropriately located on the irrigation pipeline to prevent water source contamination from backflow.
2. The pesticide injection pipeline must contain a functional automatic, quick-closing check valve to prevent the flow of fluid back toward the injection pump.
3. The pesticide injection pipeline must also contain a functional, normally closed, solenoid-operated valve located on the intake side of the injection pump and connected to the system interlock to prevent fluid from being withdrawn from the supply tank when the irrigation system is either automatically or manually shut down.
4. The system must contain functional interlocking controls to automatically shut off the pesticide injection pump when the water pump motor stops.
5. The irrigation line or water pump must include a functional pressure switch which will stop the water pump motor when the water pressure decreases to the point that pesticide distribution is adversely affected.
6. Systems must use a metering pump, such as a positive displacement injection pump (e.g., diaphragm pump) effectively designed and constructed of materials that are compatible with pesticides and capable of being fitted with a system interlock.
7. Do not apply when wind speed favors drift beyond the area intended for treatment. Use of low trajectory angle sprinklers or drop nozzles will decrease the likelihood of drift.
8. Sonalan E.C. should be injected continuously throughout chemigation water application period. The chemigation metering pump should be checked periodically during application to ensure proper operation.
9. The injection metering pump must be calibrated as specified by the manufacturer.
10. During chemigation, maintain agitation in supply tank at all times.
11. Sonalan E.C. may cause some staining of plastic hoses and tanks.
12. Apply Sonalan E.C. in sprinkler irrigation equal to 1/2 to 1 inch of water.

Chemigation System Calibration

The following is a sample calculation for the use of Sonalan E.C.

- Assume, in this example, 135 acres is to be covered by a chemigation treatment.
- Product required, assuming 2 pints per acre is 270 pints (33.75 gallons).
- Add 33.75 gallons of product directly to the injection supply tank.
- Adjust the injection system to deliver 33.75 gallons during the time required to apply 1 inch of water to 135 acres.
- If the irrigation system requires 20 hours to apply 1 inch of water to 135 acres, the injection rate is 1.69 gallons per hour and is calculated as follows:

  33.75 gallons ÷ 20 hours - 1.69 gallons per hour
  [1.69 gallons = 216 fluid ounces (fl. oz.)]

  Proper calibration requires the injection pump to be adjusted to deliver 3.6 fl. oz. per minute and is calculated as follows:
  216 fl. oz. per hr. ÷ 60 min. per hr. = 3.6 fl. oz. per min.

Mixing Directions

Undiluted:

When used alone, the injection of undiluted Sonalan E.C. is recommended in chemigation systems. For undiluted use, the metering pump, supply tank, and any associated equipment must be thoroughly clean and dry before Sonalan E.C. is added to the system for injection. When injecting undiluted Sonalan, maintain continuous agitation in the supply tank.

Diluted:

Sonalan E.C. may be diluted if required to achieve accurate calibration for existing equipment. Partially fill the injection supply tank with a volume of water equal to the amount of Sonalan E.C. required. Do not add water to Sonalan E.C. Start agitation. Add the required amount of Sonalan E.C. to the supply tank and continue mixing while filling the tank to the final volume required by the injection pump calibration. When injecting diluted Sonalan E.C., maintain continuous agitation in supply tank.

Precautions:

Read the Sonalan label for weeds controlled, additional directions, precautions and limitations prior to use.

Sonalan® E.C., Dry Flowable (ethalfluralin, Elanco)
EA 8119 2/88

SUPPLEMENTAL LABELING

SONALAN® E.C. (EPA REG. NO. 1471-122)
SONALON E.C. FOR CONTROL OF WOOLLY CUPGRASS

Apply and incorporate SONALAN E.C. according to label directions.

Weeds controlled:

In addition to weeds listed on the SONALAN E.C. label, SONALAN E.C. controls woolly cupgrass when used as directed.

Precautions:

Read the SONALAN label for additional directions, precautions and limitations before use.

SONALAN® E.C. (ethalfluralin, Elanco)

EA 9190 3/89

SURFLAN® A.S.

HERBICIDE

A preemergence surface-applied herbicide for the control of most annual grasses and certain broadleaf weeds.

ACTIVE INGREDIENT: oryzalin* (3,5-dinitro-N^4,N^4-dipropylsulfanilamide)	40.4%
INERT INGREDIENTS:	59.6%
TOTAL	100.0%

Contains four pounds active ingredient per gallon.

EPA Reg. No. 1471-112 ID 5500

*SURFLAN®—(oryzalin, Elanco)

Literature revised April 18, 1988
PI 2658 AMP

KEEP OUT OF REACH OF CHILDREN

CAUTION

PRECAUTIONARY STATEMENTS

HAZARDS TO HUMANS AND DOMESTIC ANIMALS

Harmful if swallowed, inhaled or absorbed through the skin. Avoid breathing spray mist and contact with skin or eyes. In case of contact, immediately flush eyes or skin with plenty of water. Get medical attention if irritation develops. SURFLAN A.S. may cause skin sensitization reactions in certain individuals. Use eye protection and protective clothing such as coveralls, a long sleeved shirt, shoes and impermeable gloves when handling this product.

ENVIRONMENTAL HAZARDS

Do not apply directly to water or wetlands. Do not contaminate water by cleaning of equipment or disposal of wastes. Cover or incorporate spills.

STORAGE AND DISPOSAL

Storage: Store in original container only. In case of leak or spill, use absorbent materials to contain liquids and dispose as waste.

Pesticide Disposal: Do not contaminate water, food or feed by storage or disposal. Pesticide wastes are toxic. Improper disposal of excess pesticide, spray mixture, or rinsate is a violation of Federal Law. If these wastes cannot be disposed of by use according to label instructions, contact your State Pesticide or Environmental Control Agency, or the Hazardous Waste representative at the nearest EPA Regional Office for Guidance.

Container Disposal: Triple rinse (or equivalent). Then offer for recycling or reconditioning, or puncture and dispose of in a sanitary landfill, or incineration, or, if allowed by state and local authorities, by burning. If burned, stay out of smoke.

Precaution

Do not graze or feed forage from treated fields or orchards to livestock. Poor weed control or crop injury may result if directions are not followed. Over application may result in crop injury or soil residue. Do not plant any root crop for 12 months following a SURFLAN application.

WEEDS AND GRASSES CONTROLLED

Weeds Controlled in Fruit, Nut Crops and Vineyards by SURFLAN

SURFLAN applied in Fruit, Nut Crops and Vineyards controls the following weeds:

Grasses	
Barley, little	*Hordeum pusillum*
Barnyardgrass (Watergrass)	*Echinochloa crus-galli*
Bluegrass, annual (Poa)	*Poa annua*
Brachiaria (Signalgrass)	*Brachiaria* sp.
Browntop panicum	*Panicum fasciculatum*
Crabgrasses (Large crabgrass) (Smooth crabgrass)	*Digitaria* spp.
Crowfootgrass	*Dactyloctenium aegyptium*
Cupgrass	*Eriochloa gracilis*
Downy brome	*Bromus tectorum*
Fall panicum (Spreading panicgrass)	*Panicum dichotomiflorum*
Foxtails (Bottlegrass) (Bristlegrass) (Pigeongrass) (Giant foxtail) (Green foxtail) (Robust foxtail) (Yellow foxtail)	*Setaria* spp.
Goosegrass (Silver crabgrass) (Wiregrass) (Yardgrass)	*Eleusine indica*
Guineagrass	*Panicum maximum*
Johnsongrass (Seedling only)	*Sorghum halepense*
Junglerice	*Echinochloa colonum*
Lovegrass, Mexican	*Eragrostis mexicana*
Lovegrass, orcutt	*Eragrostis orcuttiana*
Oat, wild	*Avena fatua*
Ryegrass, annual (Italian)	*Lolium multiflorum*
Sandbur, field	*Chenchrus incertus*
Sprangletop, red	*Leptochloa filiformis*
Texas panicum (Buffalograss) (Coloradograss)	*Panicum texanum*
Witchgrass	*Panicum capillare*
Broadleaf Weeds	
Carpetweed	*Mollugo verticillata*
Chickweed	*Stellaria media*
Cudweed	*Gnaphalium chilense*
Fiddleneck, coast	*Amsinckia intermedia*
Flordia pusley (Florida purslane) (Mexican clover) (Pusley)	*Richardia scabra*
Henbit	*Lamium amplexicaule*
Knotweed, prostrate	*Polygonum aviculare*
Lambsquarters	*Chenopodium album*
Pigweeds (Carelessweed) (Prostrate pigweed) (Redroot pigweed) (Rough pigweed) (Smooth pigweed) (Spiny pigweed) (Tumble pigweed)	*Amaranthus* spp.
Puncturevine	*Tribulus terrestris*
Purslane, common	*Portulaca oleracea*
Rockpurslane, redmaids	*Calandrinia caulescens*
Shepherdspurse	*Capsella bursa-pastoris*

SURFLAN provides partial control or suppression of :

Filaree, red and white stem	Nightshade, black
Groundsel, common	Prickly sida (teaweed)
Ladysthumb	Ragweed, common
Lettuce, prickly	Ragweed, giant (horseweed)
Mallow, common	Rocket, London
Milkweed, climbing	Smartweed, annual
Morningglory, annual	Sowthistle, annual
Mustard, black	Spurge, prostrate
Mustard, wild	Spurge, spotted
	Velvetleaf

DIRECTIONS FOR USE

SURFLAN is a preemergence, surface applied herbicide for the control of most annual grasses and certain broadleaf weeds. General rule: The more coarse the soil or the lower the organic matter, the lower the rate.

Soil Texture Guide

Refer to the following guide to determine your soil texture.

Soil Texture	Soil Classification
Coarse (light) soils	Sand Loamy sand Sandy loam
Medium soils	Loam Silt loam Silty clay loam* Silt Sandy clay loam*
Fine (heavy) soils	Clay Clay loam Silty clay loam*

Soil Texture	Soil Classification
	Silty clay Sandy clay Sandy clay loam*

*Silty clay loam and sandy clay loam soils are transitional soils and may be classified as either medium or heavy textured soils. If silty clay loam or sandy clay loam soils are predominantly sand or silt, they are usually classified as medium textured soils. If predominantly clay, they are usually classified as fine textured soils.

MIXING DIRECTIONS

SURFLAN Alone in Water

Start with a clean spray tank. Fill sprayer one-third (1/3) to one-half (1/2) full with clean water. Start agitation. Shake container well and add correct quantity of SURFLAN, continue agitation and finish filling the tank.

SURFLAN Tank Mix in Water

Vigorous, continuous agitation is required for all tank mixes. Sparger pipe agitators generally provide the best agitation in spray tanks. Avoid stirring or splashing air into the mixture during filling to prevent foaming. To do this, place the end of the fill pipe below the surface of the water in the spray tank. Do not allow the mixture to siphon back into the water source.

Mixing order: Fill the tank three-fourths (3/4) full with clean water. Start the agitation. Add dry flowables (DF) or wettable powders (WP) and agitate until the products are completely dispersed in the water. Allow additional mixing and dispersion time when using dry flowable type products. Shake container well and add SURFLAN A.S., dry flowables, or other liquids and agitate until completely dispersed in the water. Next add any solution (S) products, mix thoroughly. Add emulsifiable concentrate (EC) products last, agitate and finish filling the tank. Maintain agitation during filling and through application. If spraying and agitation must be stopped before the tank is empty, the materials may settle to the bottom. In this case, it is important to again suspend all of the material in the bottom of the tank before continuing the spray application. A sparger agitator is particularly useful for this purpose. Sometimes it is more difficult to re-suspend settled material than it is to suspend it originally.

Read and carefully follow all label instructions for each material added to the tank. Premixing dry and flowable formulations with water (slurrying) and pouring the slurry through a 20 or 35 mesh wetting screen in the top of the tank will help assure good initial dispersion in the tank water. Line screens in the tank should be no finer than 50 mesh (100 mesh is finer than 50 mesh).

If you see a buildup of material on the walls of the spray tank, wash the tank with soapy water between fillings. Rinse and continue the spraying operation. Clean the tank, lines, and screens thoroughly after use.

Soil Preparation

Destroy existing weeds before or during a SURFLAN application. The soil surface should be level enough to allow a sprayer to be operated efficiently and at speeds appropriate for uniform application.

Application

Ground Application: Apply in 20 to 40 gallons of water per acre (broadcast basis), using any properly calibrated low pressure herbicide sprayer that will apply the spray uniformly. Use herbicide tips and screens no finer than 50 mesh in nozzle and in-line strainers. As the amount of water used (spray volume) decreases, the importance of accurate calibration and uniform application increases. Check the sprayer daily to ensure proper calibration and uniform application. Agitate thoroughly before and during application. Avoid boom overlaps that will increase rates above those recommended.

Aerial Application: Use a standard aerial herbicide boom sprayer. Aerial spray equipment should be calibrated to apply the proper amount of SURFLAN A.S. alone or in tank mix combinations in two (2) to 10 gallons of spray mixture per acre. Nozzle screens and in-line strainers should be no finer than 50 mesh for tank mix combinations. SURFLAN A.S. (alone or in tank mix combinations) mixes readily with water for these concentrate aerial sprays; however, constant vigorous agitation that sweeps the contents from the bottom of the spray tank up into the main body of the liquid is required to maintain a uniform suspension until spray tank is empty. Avoid overlap of spray patterns. Do not apply when weather conditions favor drift from target area.

Chemigation: Refer to supplemental labeling entitled "SURFLAN A.S. Chemigation Instructions" for use directions for chemigation. Do not apply this product through any irrigation system unless the supplemental labeling on chemigation is followed.

FRUIT, NUT CROPS AND VINEYARDS (NONBEARING AND BEARING)—SURFLAN ALONE

SURFLAN applied as a preemergence herbicide controls many annual grasses and broadleaf weeds in several fruit, nut crops and vineyards. It is recommended on the following crops:

Almond	Grapefruit	Pear
Apple	Kiwi	Pecan
Apricot	Lemon	Pistachio
Avocado	Macadamia Nut	Plum
Cherry	Nectarine	Pomegranate
Fig	Olive	Prune
Filbert	Orange	Walnut, English
Grape	Peach	
Small fruits:		
Blackberry	Currants	Gooseberry
Blueberry	Dewberry	Loganberry
Boysenberry	Elderberry	Raspberry

See page 5 for a listing of weeds controlled by SURFLAN alone.

General Directions: Existing vegetation should be destroyed either by tillage or by a tank mix with a contact or translocated herbicide. To ensure the compatibility of SURFLAN with the contact or translocated herbicide, pour the products into small containers of water in the correct proportions. After thorough mixing, let stand for five (5) minutes and check for compatibility. If the combination remains mixed or can be remixed easily, the mixture is compatible. Read the label for all directions, cautions and precautions.

Broadcast Rates Per Acre

Soil Texture	SURFLAN A.S.	
		(quarts)
All Soil Textures	Short Term	
	(2-4 months)	2
	Long Term	
	(6-8 months)	4
	(8-12 months)	6

SURFLAN is not recommended for soils containing more than five percent (5%) organic matter.

Band Treatment: The amount of product needed for bank treatment may be calculated by the formula:

$$\frac{\text{band width in inches}}{\text{row width}} \times \frac{\text{broadcast}}{\text{rate per acre}} = \frac{\text{amount needed per}}{\text{acre of field}}$$

Mixing and Application: Fill the spray tank three-fourths (3/4) full with clean water. Add the recommended amount of SURFLAN to water. Mix thoroughly. Use 20 to 40 gallons of water per acre (broadcast basis). Apply with a properly calibrated low pressure herbicide sprayer that will uniformly apply the spray. Check the sprayer daily for the proper calibration and uniform application. Provide good mixing of the spray suspension in the tank before and during application.

Activation: A single one-half inch (1/2") to one inch (1") rainfall or sprinkler irrigation to move SURFLAN into the weed seed germination zone is required to activate the herbicide. If weeds begin to emerge a shallow cultivation of one to two inches (1" to 2") will destroy existing weeds and place SURFLAN in the zone of weed seed germination. The higher water amount is generally required on fine textured/high organic matter soil.

Precaution: Poor weed control may result if directions are not followed. Over application may result in crop injury or excessive soil residue. Apply the spray directly to the orchard or vineyard floor.

FRUIT, NUT CROPS AND VINEYARDS—TANK MIX RECOMMENDATIONS:

SURFLAN Tank Mix Reference Table

Refer to the following table to determine which herbicides may be applied to your crop. Choose the SURFLAN combination based on your orchard or vineyard weeds to be controlled. Refer to the tank mix section for specific recommendations and precautions. Always read the label before application. * = Approved Crop

	SURFLAN	Casoron	Goal	Hyvar X	Karmex	Krovar I/II	Gramoxone Super	Princep	Roundup	Sinbar	Solicam
Almond	*	*	*				*	*	*		*
Apple	*	*			*		*	*	*	*	*
Apricot	*		*				*				*
Avocado	*						*	*	*		
Cherry	*	*					*	*	*		*
Fig	*						*				
Filbert	*	*					*	*	*		*
Grape	*	*	*		*		*	*	*		

	SURFLAN	Casoron	Goal	Hyvar X	Karmex	Krovar I/II	Gramoxone Super	Princep	Roundup	Sinbar	Solicam
Grapefruit	*			*	*	*	*	*	*	*	*
Kiwi	*										
Lemon	*			*	*	*	*	*	*	*	*
Macadamia Nut	*				*			*			
Nectarine	*	*	*				*				*
Olive	*				*		*	*			
Orange	*			*	*	*	*	*	*	*	*
Peach	*	*	*		*		*	*		*	*
Pear	*	*			*		*	*	*		*
Pecan	*	*			*		*	*	*	*	
Pistachio	*								*		
Plum	*	*	*				*	*			*
Pomegranate	*										
Prune	*		*				*				*
Small fruits											
Blackberry	*	*			*		*	*		*	
Blueberry	*	*			*		*	*		*	
Boysenberry	*				*		*	*		*	
Currants	*										
Dewberry	*				*					*	
Elderberry	*										
Gooseberry	*				*						
Loganberry	*				*			*		*	
Raspberry	*	*			*		*	*		*	
Walnut, English	*	*			*		*	*	*		*

SURFLAN/Casoron® Tank Mix

The SURFLAN/Casoron tank mix controls the following weeds in addition to those controlled by SURFLAN alone:

Barley, wild	Jerusalem oak	Primrose, evening
Camphorweed	Maypops	Radish, wild
Coffeeweed	Melon, citron	Ragweed
Dandelion	Milkweed vine	Rosarypea
Deadnettle, red	Miners lettuce	Smartweed
Dogfennel	Mustard, wild	Spanishneedles
Gisekia	Natalgrass	Teaweed
Goosefoot	Peppergrass	Thistle, bull
Groundsel	Pineappleweed	Thistle, Russian
Horsetail	Plantain	Woodsorrel, yellow

SURFLAN/Casoron tank mix may be used for weed control in the following crops:

Apple	Filbert	Pecan
Almond	Grape	Plum
Blackberry	Nectarine	Raspberry
Blueberry	Peach	Walnut, English
Cherry	Pear	

General Directions: Follow recommended application procedures for SURFLAN. Apply SURFLAN/Casoron in at least 20 gallons of water per acre to ensure uniform coverage. If the application is followed by one-half inch (1/2″) to one inch (1*) of overhead irrigation for activation, use the lower Casoron rate. SURFLAN/Casoron must be activated; check the SURFLAN label for activation directions.

Broadcast Rates Per Acre

Crop	SURFLAN A.S.	Casoron W-50
	(quarts)	(pounds)
Apple, Almond, Blueberry, Cherry, Grape, Nectarine, Peach, Pear, Plum, Filbert, Pecan, English Walnut	2-6	8-12
Do not use on pecan seedlings within six (6) months of planting. May be used on bearing, non-bearing and nursery stock. Do not apply until four (4) weeks after transplanting. Do not make application within one (1) month of harvest of nuts or stone fruits.		
Blackberry and Raspberry	2-6	8
Use only on established plantings. Do not apply during new shoot emergence.		

Casoron Precautions: Do not use in seed beds, transplant or cutting beds or greenhouses. Do not apply until six (6) months after rooting of cuttings in the field, or after planting rooted cuttings. Do not apply on fruit stock within three (3) months prior to or following grafting or budding of root stocks or planting of new grafts. Do not apply more than the maximum recommended amount of Casoron W-50 per growing season. Do not use on "St. Lucie fine sand", "Arzell fine sand", or other light sand soil. Do not graze livestock in treated areas. Store in a tightly closed container in a dry place.

Caution: Read the SURFLAN and Casoron labels for directions, cautions and limitations prior to use.

SURFLAN/Goal®/Gramoxone Super Tank Mix
(For use in California only)

SURFLAN can be tank mixed with Goal 1.6E and Gramoxone Super.

SURFLAN/Goal provides preemergence control of the following weeds in addition to those controlled by SURFLAN alone:

Burclover	London rocket
Filaree, broadleaf	Mallow, little (Cheeseweed) (Malva)
Filaree, redstem	
Groundsel, common	
Lettuce, prickly	Redmaids
	Sowthistle, annual

Gramoxone Super may be tank mixed with SURFLAN/Goal to provide nonselective, postemergence control of existing emerged annual weeds and provide top kill of existing perennial weeds.

The tank mix is recommended for application to dormant crops including:

Almond	Peach
Apricot	Plum
Grape	Prune
Nectarine	

General Directions: Best preemergence activity from Goal is obtained when the tank mix is applied to a relatively weed free established berm or soil surface.

Broadcast Rates Per Acre

Crop	SURFLAN A.S.	Goal 1.6E	Gramoxone Super
	(quarts)	(quarts)	(pints)
All crops listed	2-6	5	3

Precautions:

•Dormancy determines the suitability and safety of using Goal 1.6E on tree fruit/grapes. Should bud swell occur prior to the dates listed, Goal 1.6E must not be used.

•Do not apply Goal 1.6E (either alone or in any combination recommended on this label) to tree or vine plantings before October 1 or after February 15. Do not apply Goal 1.6E (either alone or in any combination recommended on this label) to grapes in the Coachella Valley before October 1 or after December 31.

•Do not apply more than five (5) quarts (two pounds active) per broadcast acre of Goal 1.6E in one (1) season. Goal 1.6E should not be applied to grapes established less than three (3) years.

•Do not apply Goal 1.6E to grapes which are not staked or trellised.

•Do not feed or graze areas treated with Goal 1.6E to animals.

•Do not contaminate irrigation water or water used for domestic purposes with Goal 1.6E.

•Goal 1.6E or any of the combinations recommended on this label should be applied to only healthy growing trees and vines.

•Do not apply Goal 1.6E or any combination containing Goal 1.6E, that is recommended on this label, when weather conditions favor drift.

•Apply Goal 1.6E only to dormant fruit or nut trees or grape vines. Do not apply when foliage, buds or fruit/nuts are present.

•Direct Goal 1.6E spray toward the base of tree or vines. Avoid direct plant contact.

Mixing and Application: Apply SURFLAN/Goal in 20-40 gallons of water per acre. Fill the spray tank three-fourths (3/4) full. Agitate continuously. Add the recommended amount of SURFLAN to the spray tank. Add the recommended amount of Goal. Finish filling the tank with water to proper level. Do not allow the tank mix to remain in the tank without agitation.

Apply SURFLAN/Goal/Gramoxone Super in at least 50 gallons of water per acre. Use Ortho X-77® or Triton® AG-98 when tank mixing SURFLAN A.S./Goal/Gramoxone Super. Ortho X-77 and Triton AG-98 improve the compatibility of these herbicides when mixed together. Fill the spray tank one third to one half (1/3 to 1/2) full with clean water. Add either one (1) quart of X-77 or two (2) quarts of Triton AG-98 per 100 gallons of spray mixture, with the agitation running, add the recommended amount of SURFLAN and mix thoroughly. Continue filling the spray tank and add Goal, then Gramoxone Super. If a non-ionic spreader is being used, add it to the tank last.

Activation: A single one-half inch (1/2") to one inch (1") rain or sprinkler irrigation to move SURFLAN/Goal into the weed seed germination zone is required to activate the herbicide. If weeds begin to emerge, a shallow cultivation of one to two inches (1" to 2") will destroy existing weeds and place SURFLAN in the zone of weed seed germination. The higher amount of water is generally required on fine textured/high organic matter soil.

Caution: Read the SURFLAN, Goal and Gramoxone Super labels for additional limitations, directions and cautions before applying these products.

SURFLAN/Hyvar® X Tank Mix

The SURFLAN/Hyvar X tank mix controls the following annual and perennial weeds in addition to those weeds controlled by SURFLAN alone:

Annual Weeds	Perennial Weeds
Mullein, turkey	Bermudagrass (All States)
Mustard	Johnsongrass (Texas)
Natalgrass (red top)	Nutsedge (Texas and California)
Sedge	Pangolagrass (Florida)
Sprangletop	Paragrass (Florida)
	Torpedograss (Florida)

The SURFLAN/Hyvar X tank mix is recommended on the following crops:

Citrus (Orange, Grapefruit, Lemon)

General Directions: Apply the tank mix in at least 40 gallons of water per acre. Use sufficient water to provide thorough and uniform coverage of the ground. Apply Hyvar X as a spray just before or during the period of active plant growth. If dense growth is present, results will be improved if vegetation is removed before treatment. Do not apply when ground is frozen.

Activation: Moisture is necessary to activate SURFLAN/Hyvar X tank mix; best results are obtained if moisture is supplied by rainfall or overhead sprinkler irrigation within two (2) weeks after application. Check activation instructions for SURFLAN alone.

Broadcast Rates Per Acre

Crop	SURFLAN A.S.	Hyvar X
Citrus (Trees established for four (4) years or more)	(quarts)	(pounds)
Annual Weeds	2-6	2-4
Apply any time of the year, preferably shortly before weed growth begins, when adequate moisture is available.		
Perennial Weeds	2-6	4-8
Apply before weed growth begins. If dense growth is present, remove tops and spray the ground. Note: Effects on perennial weeds from Hyvar X are slow to appear, usually progressing over a period of several months. Check the Hyvar X label for more complete directions and precautionary statements prior to use for perennial weed control.		

Precaution: Avoid contact of foliage and fruit with Hyvar X spray or mist. Temporary yellowing of citrus leaves may occur following Hyvar X treatment. Do not use on citrus in Kern County, California.

Because injury to citrus trees may result: do not use Hyvar X on soils low in organic matter less than one percent (1%) nor on poorly drained soils; do not apply more than eight (8) pounds per acre per year; do not treat trees planted in irrigation furrows; do not treat diseased trees such as those with foot rot. Do not use Hyvar X in citrus orchards interplanted with other trees or desirable plants, nor in home citrus plantings or in areas where roots of other valuable plants or trees may extend, as plant injury may result. Treated areas may be planted to citrus trees one (1) year after last Hyvar X application; do not replant to other crops within two (2) years after last application, as plant injury may result.

SURFLAN/Karmex® Tank Mix

The SURFLAN/Karmex tank mix controls the following weeds in addition to those controlled by SURFLAN alone:

Ageratum	Kochia	Puncturevine
Buckwheat, wild	Kyllinga	Radish, wild
Buttonweed	Marigold	Ricegrass, red
Clover, Mexican	Milkweed, climbing	Ryegrass, annual
Dayflower	Morningglory, annual	Smartweed
Dogfennel	Mustard, black	Sowthistle, annual
Fescue rattail	Mustard, wild	Spanishneedles
Gromwell	Orchardgrass	Speedwell, corn
Groundcherry annual	Paintbrush Flora's	Spurry, corn
Hawksbeard	Pepperweed	Tansymustard
Horseweed	Pineappleweed	Tobacco, rabbit
Knawel	Pokeweed	Velvetgrass

The SURFLAN/Karmex Tank mix is recommended on the following crops:

Apple	Orange
Caneberries	Peach
Grape	Pear
Grapefruit	Pecan
Lemon	Walnut, English

Broadcast Rates Per Acre

Crop	SURFLAN A.S.	Karmex 80W
	(quarts)	(pounds)
Apple	2-6	2-4
Use Karmex only under trees established in the orchard for at least one (1) year; do not treat varieties grafted on full-dwarf root stocks.		
Caneberries	2-6	2-4
(Read the Karmex label for state and crop recommendations). Use Karmex only in fields established for at least one (1) year. Do not apply Karmex to berries interplanted with fruit trees. Do not apply to plants whose roots are exposed as injury may result.		

Grape

New York and Pennsylvania—Perennial Grasses:

Use Karmex only in established vineyards (at least four years old) for spot control of perennial grasses such as orchardgrass, quackgrass and ryegrass. Apply Karmex in the spring as a band treatment to ridged soil two to four inches (2" to 4") high under the trellis at the rate of eight (8) to 12 lbs. per acre. Band width should not exceed 30 inches. Do not apply Karmex more than once every four (4) years. Use Karmex only on heavy soils such as loams, silt loams and clay loams. Do not use Karmex in areas where grape roots are shallow or exposed because of high bedrock, poor drainage, or erosion as injury to grapes may result.

U.S. (Except New York and Pennsylvania)—Perennial Grasses:

Apply Karmex only to established vineyards (at least three years old) as a band treatment to grape rows. Do not apply Karmex to vines with trunks less than one and one-half inches (1½") in diameter as injury may result. On soils low in clay or organic matter (one to two percent) (1% to 2%), severe plant injury may result from Karmex application if heavy rainfall or more than one inch

Crop	SURFLAN A.S.	Karmex 80W
	(quarts)	(pounds)
(1") of irrigation occurs soon after treatment. This risk must be assumed by the user.		
East of Rocky Mountains:		
1 to 2% organic matter	2-6	2-3
2 to 5% organic matter	2-6	3-6
West of Rocky Mountains:	2-6	2-4
(Apply Karmex in November, December, or January)		
Orange, Grapefruit and Lemon		
Do not apply Karmex under citrus trees that have been subjected to freezing within six (6) months. Use Karmex only under trees established in the grove for at least one (1) year.		
Arizona (except Yuma area) and California (except Imperial and Coachella Valleys)	2-6	2-4

Crop	SURFLAN A.S.	Karmex 80W
Florida (oranges and grapefruit)	2-6	4-8
	(quarts)	(pounds)
Texas (oranges and grapefruit)	2-6	2-4
Peach	2-6	2-5
Use Karmex only under trees established in the orchard for at least three (3) years. Do not apply Karmex within three (3) months of harvest; in the far West, do not apply Karmex within eight (8) months of harvest.		
Pear	2-6	2-4
Use Karmex only under trees established in the orchard for at least one (1) year. Do not treat varieties grafted on full-dwarf root stocks.		
Pecan	2-6	2-4
Use only under trees established in the grove for at least three (3) years. Do not use on soils with less than one- half percent (1/2%) organic matter.		
Walnut (English) California	2-6	2-3
Use only under trees established in the orchard for at least one (1) year.		

Mixing and Application: Fill the spray tank three-fourths (3/4) full with clean water and add the recommended amount of Karmex 80W. Mix thoroughly. Then add the recommended amount of SURFLAN A.S., mix thoroughly and fill the tank. Vigorous, continuous agitation is required until spray tank is empty to keep the powders in suspension. Use 20 to 40 gallons of water per acre (broadcast basis). Apply with a properly calibrated low pressure herbicide sprayer that will apply the spray uniformly. Check the sprayer daily for proper calibration and to ensure uniform application. Provide good mixing of the spray suspension in the tank before and during application.

Activation: A single one-half inch (1/2") to one inch (1") rain or sprinkler irrigation is required to activate the herbicides. If weeds begin to emerge, a shallow cultivation of one to two inches (1" to 2") will destroy existing weeds and place SURFLAN and Karmex in the zone of weed seed germination.

Precaution: Poor weed control may result if directions are not followed. Over application may result in crop injury or excessive soil residue. Apply the spray directly to the orchard or vineyard floor.

Caution: Read the Karmex label carefully for directions, cautions, and precautions before applying this product.

SURFLAN/Krovar® I Tank Mix

The SURFLAN/Krovar I tank mix effectively controls the following weeds in addition to those weeds controlled by SURFLAN alone:

Filaree	Nightshade (annual)
Fleabane	Pineappleweed
Horseweed	Ragweed
Lettuce, wild	Shepherdspurse
Mustard, wild	Sowthistle (annual)
Natalgrass (red top)	Spanishneedles

The SURFLAN/Krovar I tank mix is recommended for use on the following crops: Citrus (Orange, Lemon, Grapefruit)

General Directions: Apply any time of the year when rainfall or overhead irrigation is available to activate the tank mix, preferably just before the weeds have germinated.

Use only where trees have been established in the orchard for at least three (3) years. Use sufficient water (over 40 gallons/acre) to obtain uniform coverage.

Broadcast Rates Per Acre

Crop	SURFLAN A.S.	Krovar I
	(quarts)	(pounds)
Citrus (Trees established in the orchard for at least three (3) years)	2-6	3-6
Use for control of annual weeds in California, Arizona, Texas, Louisiana, and Florida only. Check the Krovar I label for complete directions for use and precautionary statements.		

Activation and Cultivation: A single one-half inch (1/2") to one inch (1") rain or sprinkler irrigation to move the tank mix combination into the weed seed germination zone is required within two (2) weeks after application to activate the herbicides. If weeds begin to emerge, a shallow cultivation of one to two inches (1" to 2") will destroy existing weeds and place SURFLAN and Krovar I in the zone of weed seed germination. Citrus treated can be shallowly cultivated without loss of activity.

Precaution: Avoid contact of foliage and fruit with Krovar I spray or mist. Avoid overlapping, and shut off spray boom while starting, turning, slowing or stopping, or injury to trees may result. Continuous agitation in the spray tank is required to keep the material in suspension. Best results are obtained if application is made shortly before weed growth begins; if dense growth is present, remove tops and spray ground.

Temporary yellowing of citrus leaves may occur following Krovar I treatment. Because injury to citrus trees may result: do not use Krovar I on soils with less than one percent (1%) organic matter (one and one-half percent (1 1/2%) in Florida), poorly drained soils, gravelly soils, nor thinly covered or exposed subsoils; do not treat trees planted in irrigation furrows; do not treat diseased trees such as those with foot rot.

Do not replant Krovar I treated areas to any crop within two (2) years after last application as injury to subsequent crops may result, except that citrus trees may be planted one (1) year after the last application.

SURFLAN/Krovar® II tank Mix

The SURFLAN/Krovar II tank mix effectively controls the following annual and perennial weeds in addition to those controlled by SURFLAN alone:

Annual Weeds and Vines	Perennial Weeds
Balsam apple*	Bahiagrass
Milkweed vine*	Bermudagrass
Natalgrass (red top)	Nutsedge
Nightshade (annual)	Pangolagrass
Ragweed	Paragrass
Spanishneedles	Torpedograss

* Seedling stage in Florida only.

The SURFLAN/Krovar II tank mix is recommended for weed control in the following crops:

Citrus (Orange Lemon, Grapefruit)

General Directions: Use only where trees have been established in the orchard for at least four (4) years. Apply any time during the year when rainfall or overhead irrigation is available to activate the tank mix prior to weed germination. Moisture is necessary to activate the herbicides. Check the SURFLAN directions for activation moisture amounts. The tank mix should be activated within two (2) weeks after application. To obtain uniform coverage, use sufficient water (over 40 gallons/acre).

Broadcast Rates Per Acre

Crop	SURFLAN A.S.	Krovar II
	(quarts)	(pounds)
Citrus (Use only where trees have been established in the orchard for at least four (4) years)	2-6	2-8
Consult the Krovar II label for complete directions for use for annual, perennial and seedling vine control, cautions and limitations prior to use.		

Precautions: Avoid contact of foliage and fruit with Krovar II spray or mist. Avoid overlapping, and shut off spray boom while starting, turning, slowing or stopping, or injury to trees may result. Continuous agitation in the spray tank is required to keep the material in suspension. Best results are obtained if application is made shortly before weed growth begins; if dense growth is present, remove tops and spray the ground. Do not use Krovar II on citrus in Kern County, California.

Temporary yellowing of citrus leaves may occur following Krovar II treatment. Because injury to citrus trees may result: do not use Krovar II on soils low in organic matter (less than 1%), poorly drained soils, gravelly soils, nor thinly covered or exposed subsoils; do not apply more than 10 lbs. per acre per year (Florida, 12 lbs.); do not treat trees planted in irrigation furrows; do not treat diseased trees such as those with foot rot.

Do not replant Krovar II treated areas to any crop within two (2) years after last application as injury to subsequent crops may result, except that citrus trees may be planted one (1) year after the last application.

SURFLAN/Gramoxone® Super Tank Mix

SURFLAN may be tank mixed with Gramoxone Super. SURFLAN provides preemergence control of certain annual grass and broadleaf weeds while Gramoxone Super provides non-selective postemergence control of annual grass and broadleaf weeds and top kill and suppression of perennial weeds.

The tank mix is recommended on the following crops:

Almond	Grape	Peach
Apple	Grapefruit	Pear
Apricot	Lemon	Pecan
Avocado	Nectarine	Plum
Cherry	Olive	Prune
Fig	Orange	Walnut, English
Filbert		

Small fruits:

Blackberry	Boysenberry
Blueberry	Raspberry

General Directions: SURFLAN/Gramoxone Super may be applied any time throughout the growing seasons. Choose an application schedule based on the specific weeds to be controlled by the SURFLAN/Gramoxone Super tank mix.

Broadcast Rates Per Acre

Crop	SURFLAN A.S.	Gramoxone Super
	(quarts)	(pints)
All crops listed on page 35.	2-6	3-5*

Precaution: Do not allow Gramoxone Super spray to contact green stems, fruit or foliage as injury may result. Do not spray Gramoxone Super under windy conditions and use a shield for young trees or vines. Do not allow animals to graze on Gramoxone Super treated areas. Do not apply Gramoxone Super when nuts to be harvested are on the ground.

* Add a non-ionic spreader at eight (8) oz. per 100 gallons spray volume

Mixing and Application: Add the recommended amount of SURFLAN to the spray tank according to directions for these tank mixes. Add Gramoxone Super last prior to filling the tank. Agitate continuously during filling and through application. Apply in at least 30 gallons of water per acre. Apply when the weeds and grasses are succulent and the new growth is from one inch (1") to six inches 6") high. Retreatment or spot treatment may be necessary for late germinating weeds and grasses.

Activation: A single one-half inch (1/2") to one inch (1") rain or sprinkler irrigation to move SURFLAN into the weed seed germination zone is required to activate the herbicide. If weeds begin to emerge, a shallow cultivation of one to two inches (1" to 2") will destroy existing weeds and place SURFLAN in the zone of weed seed germination.

Caution: Read the Gramoxone Super label carefully for directions, cautions and precautions before applying this product.

SURFLAN/Princep® (simazine) Tank Mix

The SURFLAN/Princep tank mix controls the following weeds in addition to those controlled by SURFLAN alone.

- Brome, downy (cheat)
- Fescue, rattail
- Fireweed
- Hairgrass, silver
- Knawel
- Milkweed, climbing
- Morningglory, annual
- Mustard, black
- Mustard, wild
- Nightshade
- Paintbrush, Flora's
- Prickly lettuce
- Ragweed
- Ryegrass, annual
- Shieldscress
- Smartweed
- Spanishneedles
- Tansymustard
- Wiregrass

The tank mix is recommended on the following crops:

- Almond
- Apple
- Avocado
- Caneberries
- Cherry
- Filbert
- Grape
- Grapefruit
- Lemon
- Orange
- Peach
- Pear
- Pecan
- Plum
- Walnut, English

Where a rate range of application rates for Princep is given, use the lower rate on coarser textured soil and soil lower in organic matter; use the high rate on finer textured soils and soil higher in organic matter.

Broadcast Rates Per Acre

Crop	SURFLAN A.S.	Princep 80W
	(quarts)	(pounds)
Apple, Pear and Sour Cherry	2-6	2½-5
Almond and Peach California only Apply two (2) to four (4) foot band on each side of tree row. For peach in other states, see recommendations for peach, plum, and sweet cherry.	2-6	1¼-2½

Do not apply Princep to trees established in the grove less than three (3) years. Do not apply Princep to the Mission (Texas) variety of almonds. Do not apply Princep to almond trees propagated on plum rootstocks. Do not replant almond or peaches in soil treated with Princep for 12 months after treatment. Do not apply Princep on gravelly, sandy or loamy sand soil or on soil with less than one percent (1%) organic matter. Do not apply Princep to areas where water will accumulate.

Broadcast Rates Per Acre

Crop	SURFLAN A.S.	Princep 80W
	(quarts)	(pounds)
Avocado California	2-6	2½-5

Do not apply Princep to areas where water will accumulate.

Crop	SURFLAN A.S.	Princep 80W
Caneberries	2-6	2½-5

Apply two and one-half (2½) to five (5) pounds Princep per acre in the spring or apply a split application of two and one-half (2½) pounds per acre in the spring plus two and one-half (2½) pounds per acre in the fall. Apply in a minimum of 40 gallons of water per acre. On plantings less than six (6) months old, use one-half (½) the above Princep rate. To control quackgrass, apply five (5) pounds Princep per acre in the fall or split the application applying two and one-half (2½) pounds per acre in the fall plus two and one-half (2½) pounds per acre in the spring, when quackgrass is actively growing. Do not apply Princep when fruit is present.

Crop	SURFLAN A.S.	Princep 80W
Filbert Washington and Oregon	2-6	2½-5

Apply Princep at two and one-half (2½) to five (5) pounds per acre in the fall or apply a split application of two and one-half (2½) pounds per acre in the spring and two and one-half (2½) pounds per acre in the fall.

Broadcast Rates Per Acre

Crop	SURFLAN A.S.	Princep 80W
	(quarts)	(pounds)

Precaution: If trees are planted on a hillside, excessive soil erosion may result from the elimination of weeds. Do not apply on gravelly, sandy or loamy sand soil. Do not apply when nuts are on the ground during the harvest period.

Crop	SURFLAN A.S.	Princep 80W
Grape	2-6	2½-5

Apply Princep any time between harvest and early spring.

Precaution: Do not use Princep in vineyards established less than three (3) years. Do not apply Princep on gravelly, sandy or loamy sand soil.

Crop	SURFLAN A.S.	Princep 80W
Grapefruit, Lemon and Orange Arizona (lemon and orange only)	2-6	2 Spring 2 Fall
California (except Imperial, Coachella and Palo Verde Valleys)	2-6	2½-5
Florida (grapefruit and orange only)	2-6	8-12
Texas	2-6	5-6

All areas—do not use Princep in nurseries. Do not apply Princep to bedded grapefruit, lemon, or orange. To avoid possible injury, do not apply Princep to trees under stress from freeze damage for one (1) year after the freeze.

Broadcast Rates Per Acre

Crop	SURFLAN A.S.	Princep 80W
	(quarts)	(pounds)
Peach Plum and Sweet Cherry	2-6	2-5

Apply Princep in late fall to early spring prior to weed emergence.

Peach: Use Princep only in Arkansas, Louisiana, Missouri, Oklahoma, Texas and the states east of the Mississippi.

Plum and Sweet Cherry: Use Princep only in Missouri and states east of the Mississippi River except Tennessee.

Crop	SURFLAN A.S.	Princep 80W
Pecan		
Sand and loamy sand	2-6	Do Not Use
Sandy loam	2-6	2-5
Loam or clay soil low in organic matter	2-6	2½-3½
Clay soil high in organic matter	2-6	3½-5

Note: Do not apply Princep when nuts are on the ground. Do not allow animals to graze in Princep treated areas.

Do not use Princep in Arizona, California, New Mexico or west of the Pecos River in Texas as injury will occur. Do not make Princep applications to transplanted trees that have been established less than two (2) years in the grove as injury may occur.

Broadcast Rates Per Acre

Crop	SURFLAN A.S.	Princep 80W
	(quarts)	(pounds)
Walnut, English	2-6	2½-5

Do not apply Princep to gravelly, sandy or loamy sand soils. Leveling and furrowing operations after application will lessen effectiveness of weed control. Do not apply Princep when nuts are on the ground.

Mixing and Application: Fill the spray tank three-fourths (3/4) full of clean water. Follow instructions on the Princep 80W label for making a slurry of the Princep 80W in a small amount of water in a separate container, then pour the slurry into the spray tank. Mix thoroughly. Then add the recommended amount of SURFLAN A.S. Mix thoroughly and continue filling the tank. Continuous, vigorous agitation is needed to keep the powders in suspension. Use 20 to 40 gallons of water per acre (broadcast basis). Apply with a properly calibrated low pressure herbicide sprayer that will apply the spray uniformly. Check the sprayer daily for proper calibration to ensure uniform application. Provide good mixing of the spray suspension in the tank before and during application.

Activation: A single one-half inch (1/2") to one inch (1") rain or sprinkler irrigation to move the tank mix combination into the weed seed germination zone is required to activate the herbicides. If weeds begin to emerge, a shallow cultivation of one to two inches (1" to 2") will destroy existing weeds and place SURFLAN and Princep in the zone of weed seed germination.

Precaution: Poor weed control may result if directions are not followed. Over application may result in crop injury or soil residues. Apply the spray directly to the orchard or vineyard floor.

Caution: Read the Princep label carefully for directions, cautions and precautions before applying this product.

SURFLAN/Roundup® Tank Mix

SURFLAN may be tank mixed with Roundup. In this tank mix, SURFLAN provides preemergence control of certain annual grass and broadleaf weeds while the Roundup provides non-selective postemergence control of many annual and perennial weeds. This tank mix is recommended on the following crops:

Almond	Grape	Pear
Apple	Grapefruit	Pecan
Avocado	Lemon	Pistachio
Cherry (non-bearing)	Orange	Walnut, English
Filbert		

See page 5 for a listing of weeds controlled by SURFLAN alone.

General Directions: SURFLAN/Roundup may be applied any time throughout the growing season. Choose an application schedule based on the specific weeds to be controlled by the SURFLAN/Roundup tank mix. Consult the Roundup label for specific weed growth stage directions to achieve best results. Karmex and Princep may be tank mixed with SURFLAN/Roundup to provide preemergence control of those weeds listed. Use Roundup with the SURFLAN/Karmex tank mix for non-selective, postemergence control of annual weeds only. Use Roundup with the SURFLAN/Princep tank mix for non-selective, postemergence control of annual and perennial weeds.

Broadcast Rates Per Acre

All Crops	SURFLAN A.S.	Roundup
Existing annual grass and broadleaf weeds:	(quarts)	(quarts)
Less than 6" tall	2-6	1
Over 6" tall	2-6	1½
Existing perennial weeds	2-6	2-5

Tree Crops	Preharvest Interval for Roundup Application
Almond	21 days
Apple	14 days
Avocado	14 days
Cherry	1 year
Filbert	21 days
Grapefruit	14 days
Lemon	14 days
Orange	14 days
Pear	14 days
Pecan	21 days
Pistachio	21 days
Walnut	21 days

Note: Repeat treatments may be necessary to control weeds regenerating from underground parts or seed. The combined total of all Roundup treatments must not exceed 10.6 quarts per acre per year. Do not feed or forage areas treated with Roundup for eight (8) weeks after application.

Precaution: Extreme care must be exercised to avoid contact of Roundup spray, drift or mist with green foliage, green bark or bark of trees established less than two (2) years, suckers, or fruit of desirable trees, crops, plants or other desirable vegetation. Spray contact with other than matured bark on the main trunk can result in serious localized or translocated damage. If weeds have been mowed, allow regrowth to reach the recommended stage of growth prior to application of this product. Do not apply Roundup to weeds under drought stress.

Vines	Preharvest Interval for Roundup Application
Grape (Raisin, Table, Wine)	14 days

Note: Roundup applications should not be made when green shoots or canes or foliage are in the spray zone. (In the Northeast and Great Lakes regions, Roundup applications must be made prior to the end of bloom stage of grapes to avoid injury.)

If repeat Roundup treatments are necessary for weed control in vineyards, do not exceed a total of 10.6 quarts of Roundup per acre per year. Do not treat grapes with Roundup between 14 days before harvest to fall dormancy when no green vegetation, canes or shoots exist.

Precaution: Do not allow Roundup spray, drift or mist to contact green foliage, green bark, suckers or vines and renewals less than three (3) years of age. Roundup spray contact other than with mature bark of the main trunk, can result in serious localized or translocated damage.

Mixing and Application: Add the recommended amount of SURFLAN, or SURFLAN and other residual herbicides to the spray tank according to directions for these tank mixes. Add Roundup during the last step of the filling operation. Agitate continuously during filling and throughout application. Apply in 20 to 60 gallons of water per acre in a properly calibrated low pressure herbicide sprayer. Check the sprayer daily for proper calibration and uniform application.

Activation: A single one-half inch (1/2") to one inch (1") rain or sprinkler irrigation to move SURFLAN into the weed seed germination zone is required to activate the herbicide. If weeds begin to emerge, a shallow cultivation of one to two inches (1" to 2") will destroy existing weeds and place SURFLAN in the zone of weed seed germination.

Caution: Read the Roundup label carefully for directions, cautions and precautions before applying this product.

SURFLAN/Sinbar® Tank Mix

SURFLAN/Sinbar is recommended for weed control in the following crops:

Apple	Orange
Peach	Grapefruit
Blueberry	Lemon
Caneberries:	Pecan
Blackberry	
Boysenberry	
Dewberry	
Loganberry	
Raspberry	

SURFLAN/Sinbar controls weeds in addition to those controlled by SURFLAN alone. As the amount of Sinbar applied per acre increases, so does the number of additional weeds controlled. Consult the Sinbar label for application rate required for the specific additional weeds to be controlled.

General Directions: Follow recommended application procedures for SURFLAN. Apply SURFLAN/Sinbar in at least 20 gallons per acre or more to ensure a uniform application. Moisture is necessary to activate the tank mix. Consult the SURFLAN label for activation information.

Broadcast Rates per Acre

	SURFLAN A.S.	Sinbar 80W*
	(quarts)	(pounds)
Apple, Peach	2-6	2-4
Apply either in the spring or after harvest in the fall before weeds emerge. Where the crop is grown under furrow irrigation or under raised-berm flood irrigation (trees four to six inches (4" to 6") above water line), apply Sinbar only as a band treatment. Do not treat trees planted in the bottom of irrigation furrows or trees grown under flat flood or basin irrigation, and do not use on eroded areas where subsoil or tree roots are exposed, as injury to trees may result.		
Blueberry	2-6	2-4
Apply in a minimum of 25 gallons of water per acre. Treat only plantings established for one (1) year or more. Make a single band or broadcast application to ground beneath bushes: avoid contact of foliage and fruit with spray or mist. Apply either in the spring or after harvest in the fall before weeds emerge. Do not use Sinbar on eroded areas where subsoil or roots are exposed, nor on plants that are diseased or lacking in vigor, as injury to the plants may result. Sinbar treated areas may be planted to blueberries one (1) year after last application.		
Caneberries: Blackberry Boysenberry, Dewberry Loganberry, Raspberry	2-6	1-2
Treat only plantings established for one (1) year or more. Make a single band or broadcast application to the soil beneath the canes in the fall or early spring before fruit set and before weeds emerge. Do not spray foliage nor use on eroded areas where subsoil or roots are exposed, nor apply to plants that are diseased or lacking vigor as injury to the plant may result. Sinbar treated areas may be planted to caneberries one (1) year after last application. Do not apply Sinbar within 70 days before harvest.		
Citrus: Orange, Grapefruit, Lemon Trees established for one (1) year:		
	2-6	2-4
Trees established for two (2) years or more:		
Arizona and California	2-6	2-8

	SURFLAN A.S.	Sinbar 80W*
	(quarts)	(pounds)
Florida and Texas	2-6	2-10
Apply as a band or broadcast treatment as a directed spray: avoid contact of foliage and fruit with spray or mist. Temporary yellowing of citrus leaves may occur following Sinbar treatment. Sinbar may be used on coarse soils but is not to be used on sands low in organic matter (less than one percent) (1%), on poorly drained soils, or on trees planted in irrigation furrows, as injury to trees may result. Sinbar treated areas may be planted to citrus trees one (1) year after last application.		

* Sinbar application rates may vary with soil texture, weeds to be controlled, length of control desired and age of crop.

Caution: Read the Sinbar label for additional directions for use, cautions and special precautions prior to use.

SURFLAN/Solicam® Tank Mix

The SURFLAN/Solicam tank mix controls the following weeds in addition to those weeds controlled by SURFLAN alone:

Barley, wild	Filaree (red and whitestem)
Brome, downy	Fingergrass, feather
Brome, ripgut	Grama grass, sixweeks
Bursage, annual	London rocket
Cupgrass, southwestern	Pineappleweed
Dandelion, false	Thistle, Russian and sow

Weeds suppressed by this tank mix in addition to those suppressed by SURFLAN alone include:

Bermudagrass	Nutsedge, purple
Horseweed	Quackgrass
Nightshade, silverleaf	

SURFLAN/Solicam is recommended on the following crops:*

Almond	Nectarine
Apple	Peach
Apricot	Pear
Cherry	Plum
Citrus	Prune
Filbert	Walnut

* Apply Solicam on all of the above crops only in California, Oregon and Washington. In Arizona, use on citrus only. Use only on crops in the Imperial, Coachella or Palo Verde Valleys of California.

Orchard Establishment Requirements: Use SURFLAN/Solicam on crops listed above that have been established at least 18 months in the field with exception of citrus, where the trees should be established in the field for 12 months. In all cases, the soil should be settled and firm at the time of application. Do not use on nursery stock.

General Directions: Apply the tank mix to the soil surface from fall to early spring before weeds emerge, except on those crops where otherwise specified. The orchard floor must be free of weeds and must be relatively free of plant residues. Solicam does not control established weeds.

Apply as a directed spray. Avoid contact with fruit or foliage. Make only one application per year.

Activation: If no rainfall occurs within two weeks after application, the tank mix must be activated by sprinkler irrigation. Check activation instructions for SURFLAN alone.

Broadcast Rates Per Acre

Use the lower Solicam rates on coarse soils. Use the higher rates on medium to fine soils or soils having organic matter content of greater than 3%.

Crop	SURFLAN A.S.	Solicam 80W
	(quarts)	(pounds)
Almond	2-6	2½-5
Apply from October through January		
Precaution: Do not use on coarse textured soils such as sand, loamy sand, gravelly sand and fine sand.		
Apple, Apricot, Filbert, Nectarine, Peach, Pear, Plum, Prune, Walnut	2-6	2½-5
Cherry	2-6	2½-5
Precaution: Do not use on coarse textured soils such as sand, loamy sand, gravelly sand and fine sand.		
Citrus	2-6	2½-5
Apply any time during the year according to local practice.		
Precaution: Do not use where citrus is interplanted with palm trees.		

Solicam Precautions: Do not apply when nuts or fruit are on the ground at harvest. Do not graze livestock on treated areas. Do not feed or graze cover crops grown in treated orchards. Read the Solicam label for directions, cautions and limitations prior to use.

SURFLAN APPLICATION WITH LIQUID FERTILIZERS

General Instructions

SURFLAN may be mixed with most liquid fertilizer materials. The combination of SURFLAN with solutions and suspension-type fertilizers has provided weed and grass control equal to the same rates of SURFLAN applied in water. Follow SURFLAN label recommendations regarding rates per acre, crops, application directions, special instructions, cautions and special precautions. Individual state regulations relating to liquid fertilizer mixing, registration, labeling and applications are the responsibility of the individual and/or company selling the fertilizer and chemical mixture.

Testing for Tank Mix Compatibility in Liquid Fertilizers:

SURFLAN alone or in tank mixture with dry flowables (DF), wettable powders (WP), liquids (L), flowables (F), aqueous suspensions (AS), solutions (S) or emulsifiable concentrates (EC) may not combine properly with some fluid fertilizer materials. Small quantities should always be tested before full-scale mixing. This will determine whether a compatibility agent is needed and which agent does the best job. The nine (9) agents listed on page 53 have been thoroughly tested. There are many other surfactants on the market which were not designed for use with liquid fertilizers. Use the following test to select the correct agent for your mixture:

1. Put one (1) pint of liquid fertilizer in a quart jar.
2. Add one (1) to four (4) teaspoonsful of the dry flowables (DF), WP, L, F, or AS formulation (depending on the recommended rate per acre) to the liquid fertilizer. Close jar and agitate until the materials are dispersed evenly in the fertilizer. If the materials do not disperse well, it may be necessary to slurry the chemicals in water before adding to the fertilizer.
3. After dispersing the materials (Step 2), add three (3) to four (4) teaspoonsful of any emulsifiable formulation (EC) to the jar and shake well. Add solution herbicides (S) to the mixture last and agitate. Observe the jar for about 10 minutes. If the materials rise to the surface and form a thick layer (oily curds) which will not disperse again when agitated, a compatibility agent is needed. If the mixture is easily dispersed to its original state with slight agitation, no agent is needed but good agitation must be provided in the fertilizer spray tank.
4. If the need for a compatibility agent is shown in Step 3: Using a clean quart jar, start at Step 1 above, add one-half (½) teaspoon of the compatibility agent to the liquid fertilizer, mix well, then repeat Steps 2 and 3.

An effective compatibility agent will cause the mixture to remain uniformly mixed with little or no separating or oil rising to the surface for one-half (½) hour or longer. If slight separation does occur, two (2) to three (3) inversions of the jar should give a uniform remix. If oily curds form which will not disperse, more agent or another agent should be tried.

Use a clean jar for each test. The compatible mixture will have a uniform appearance and will be relatively easy to keep mixed with gentle agitation of the jar.

Liquid Fertilizer Mixing Instructions

SURFLAN in Liquid Fertilizer

Aqueous suspensions such as SURFLAN A.S. can be mixed with liquid fertilizers. The use of compatibility agents is especially important when tank mixing emulsifiable concentrates (EC) with dry flowables (DF), wettable powders (WP), flowables (F), liquids (L), aqueous suspensions (AS) or solutions (S) in liquid fertilizer. These compatibility agents can be used at rates as low as one and one-half (1½) to two (2) pints per ton of liquid fertilizer and should be mixed well with the fertilizer before adding the herbicides. Read the label on the compatibility agent and follow the directions.

1. Sponto 168D (Witco Chemicals Co., Chicago, IL)
2. Compat (Farm Chemicals, Inc., Aberdeen, NC)
   Do not use in California.
3. Unite (Hopkins Ag Chemical, Madison, WI)
4. T-Mulz 734-2 (Thompson-Hayward Chemical Co., Kansas City, KS)
   Do not use in California.
5. Rigo Compatibility Agent (Rigo Company, Buckner, KY)
6. Amoco Spray Mate™ (Amoco Oil Co., Chicago, IL)
   Do not use in California.
7. Kem-Link (Universal Coop, Minneapolis, MN)
8. Blendex™ (Helena Chemical Co., Cayce-West, Columbia, SC)
9. Spray-Aide™ (Miller Chemical and Fertilizer Corp., Hanover, PA)

All of the agents listed above are phosphate ester-type surfactants designed to be used with liquid fertilizers. They usually do not work well as compatibility agents in tank mixtures in plain water.

Application Instructions

Spread the SURFLAN/fertilizer mixture normally with a properly calibrated applicator. Be certain the material is applied uniformly to the soil surface.

SURFLAN APPLICATION WITH DRY BULK FERTILIZERS

General Instructions

Dry bulk fertilizers that do not contain nitrates may be impregnated or coated with SURFLAN A.S. Application of dry bulk fertilizers impregnated with SURFLAN A.S. has provided weed and grass control equal to the same rates of SURFLAN applied in water.

All SURFLAN label recommendations regarding rates per acre, approved crops, incorporation, special instructions, cautions and special precautions must be followed.

Apply a minimum of 200 pounds per acre of dry fertilizer impregnated with SURFLAN A.S. at the recommended rates. Most commonly used dry fertilizers can be used for SURFLAN A.S. impregnation except those containing nitrate fertilizers.

Precautions: Do not impregnate nitrate containing fertilizer blends with SURFLAN A.S. or Dry Flowable; the combination may result in explosive mixtures. Combining SURFLAN Dry Flowable with any dry bulk fertilizer may result in explosive mixtures.

Impregnation rule of thumb: Add no more than 1 quart of SURFLAN A.S. per 100 pounds of dry bulk fertilizer; i.e., if applying 300 pounds of fertilizer, up to 3 quarts may be impregnated.

Impregnation—Use any closed drum, belt, ribbon or other commonly used dry bulk fertilizer blender. Apply uniformly to the fertilizer.

Rates—Check specific crop recommendations for the rate of SURFLAN A.S. per acre. See the rate table which follows to determine the amount of SURFLAN A.S. to be impregnated into a ton of dry bulk fertilizer based on the amount of fertilizer which will be applied per acre. (See rate chart).

Application—Spread the fertilizer/chemical mixture with a properly calibrated applicator. Be certain the material is applied uniformly to the soil surface.

Rate Chart for Impregnating Fertilizer with SURFLAN A.S.
(SURFLAN Added to a Ton of Fertilizer)

Fertilizer Rate per	SURFLAN A.S. Rate per Acre				
Acre	½ quart	¾ quart	1 quart	1½ quarts	2 quarts
	(Quarts Surflan A.S. per Ton of Fertilizer)				
200 pounds	5	7½	10	15	20
250 pounds	4	6	8	12	16
300 pounds	3⅓	5	6⅔	10	13⅓
350 pounds	2¾	4¼	5¾	8⅓	11½
450 pounds	2¼	3⅓	4½	6⅔	9

For rates other than those listed above, use the following formula to calculate the amount of SURFLAN A.S. to be impregnated on a ton of dry bulk fertilizer:

$$\text{Quarts SURFLAN A.S. per Acre} \times \frac{2000}{\text{lbs. Fertilizer per Acre}} = \text{Quarts SURFLAN A.S. per Ton of Fertilizer}$$

All individual state regulations relating to dry bulk fertilizer blending, registration, labeling and application are the responsibility of the individual and/or company selling the fertilizer and chemical mixture.

SURFLAN PERFORMANCE GUARANTEE

(A Limited Warranty of Weed and Grass Control)

Elanco Products Company **warrants** that SURFLAN will control the weeds and grasses as indicated on its current label when used according to label directions and subject to the following limitations:

1. Purchaser must notify Elanco promptly if a lack of commercially acceptable control occurs. Such notice must be given within 30 days after the first planting or application of SURFLAN to the treated crop, whichever occurs later.
2. An Elanco representative must be satisfied that Purchaser used SURFLAN according to label directions.
3. The product guarantee will extend to the following situation only. Any recommended application where a single one-half inch (½″) rainfall or sprinkler irrigation occurs prior to weed seedling emergence.
4. The Purchaser must provide an Elanco representative with an invoice or other satisfactory record which shows the price and quantity of SURFLAN purchased.
5. Elanco's liability will be limtied to a refund of the purchase price of the SURFLAN applied to the acreage on which commercially acceptable weed control was not obtained.
6. Void outside the U.S.A.

DISCLAIMER OF WARRANTIES

Elanco Products Company makes no warranties concerning this product or its use, which extend beyond the description on the label. Elanco also expressly disclaims all warranties, express or implied, in relation to any use of this product in combination or sequential use with any other product not specifically recommended in writing by Elanco Products Company. All other warranties, expressed or implied, including but not limited to, any implied warranties of merchantability and fitness for a particular purpose, are disclaimed. All statements concerning this product apply only when used as directed. This disclaimer of warranties does not apply where prohibited by operation of law.

LIMITATION OF DAMAGES

The liability of Elanco Products Company whether in contract, warranty, tort, negligence, strict liability or otherwise, shall not exceed the return of the amount of the purchase price of this product and under no circumstances shall Elanco be liable for special, indirect or consequential damages. If SURFLAN A.S. is used in a combination or sequential use recommended by Elanco Products Company the liability of Elanco shall in no manner extend to any damage, loss or injury not directly caused by the inclusion of SURFLAN A.S. in such combination or sequential use. This limitation of damages does not apply where prohibited by operation of law.

INHERENT RISKS OF USE

Failure to carefully follow the directions for use of SURFLAN may result in unsatisfactory weed control or crop injury. Factors such as seedling disease, foliage disease, cold weather, deep planting, excessive moisture, high salt concentration or drought may also result in unsatisfactory weed control or crop injury.

Casoron® (dichlobenil, Phillips-Duphar)
Goal® (oxyflurofen, Rohm and Haas)
Gramoxone® Super (Paraquat, ICI)
Hyvar® X (bromacil, duPont)
Karmex® (diuron, duPont)
Krovar® I and II (bromacil/diuron, duPont)
Princep® (simazine, Ciba-Geigy)
Roundup® (glyphosate, Monsanto)
Sinbar® (terbacil, duPont)
Solicam® (norflurazon, Sandoz)

This is a specimen label. Elanco is not responsible for the accuracy of the information contained herein. As labels are subject to revision, always carefully read and follow the label on the product container.

TREFLAN® E.C.

HERBICIDE

A selective herbicide for the preemergence control of annual grasses and broadleaf weeds

ACTIVE INGREDIENT:
trifluralin*(α,α,α,-trifluoro-2,6-dinitro-*N,N*-dipropyl-*p*-toluidine) 44.5%
INERT INGREDIENTS: 55.5%
TOTAL: 100.0%

Contains 4 pounds active ingredient per gallon.

***Treflan**®—(trifluralin, Elanco)

Manufactured using the processes of U.S. Patents 4,120,905 and 4,226,789.

EPA Reg. No. 1471-35 ID 5011

Literature revised July 14, 1987
PI 2745 AMP

PRECAUTIONARY STATEMENTS

KEEP OUT OF REACH OF CHILDREN
WARNING

HAZARDS TO HUMANS AND DOMESTIC ANIMALS

Do not get in eyes. Avoid contact with skin and clothing. Harmful if swallowed or absorbed through the skin. Treflan E.C. may cause skin sensitization reactions in certain individuals. Use eye protection and protective clothing such as coveralls, a long sleeved shirt, and impermeable gloves when handling this product.

First Aid: In case of contact, immediately flush eyes or skin with plenty of water. For eyes, call a physician. Remove and wash contaminated clothing before reuse.

ENVIRONMENTAL HAZARDS

This pesticide is toxic to fish. Do not apply directly to water or wetlands (swamps, bogs, or marshes). Drift or runoff from treatment areas may be hazardous to aquatic organisms in neighboring aquatic sites. Do not contaminate water by cleaning equipment or disposal of waste.

STORAGE AND DISPOSAL

Storage: Store in original container only. Avoid freezing. Store above 40°F. If frozen, poor weed control may result. Do not store near heat or open flame. In case of leak or spill, use absorbent materials to contain liquids and dispose as waste.

Disposal: Do not contaminate water, food or feed by storage or disposal. Wastes resulting from the use of this product may be disposed of on site or at an approved waste disposal facility.

Plastic Container Disposal: Triple rinse (or equivalent). Then offer for recycling or reconditioning, or puncture and dispose of in a sanitary landfill, or incineration, or, if allowed by state and local authorities, by burning. If burned, stay out of smoke.

Metal Container Disposal: Triple rinse (or equivalent). Then offer for recycling, or reconditioning, or puncture and dispose of in a sanitary landfill, or by other procedures approved by state and local authorities.

Special Precautions

Applied according to directions and under normal growing conditions, Treflan will not harm the treated crop. Overapplication may result in crop injury or a soil residue. Uneven application or improper soil incorporation of Treflan can result in erratic weed control or crop injury. Seedling disease, cold weather, deep planting, excessive moisture, high salt concentration or drought may weaken crop seedlings and increase the possibility of damage from Treflan. Under these conditions, delayed crop development or reduced yields may result.

In Arizona, Colorado, California, Idaho, Nevada, New Mexico, Oregon, Utah, Washington and Wyoming, sugarbeets, red beets or spinach should not be planted for 12 months after a spring application or for 14 months after a fall application of Treflan. Plow the land to a depth of 12 inches prior to planting sugar beets to prevent the possibility of crop injury. Sorghum (milo), proso millet, corn, oats and annual or perennial grass crops or grass mixtures should not be planted for 14 months after a spring application or for 16 months after a fall application of Treflan to avoid crop injury. If land has not been irrigated, do not plant any of these crops for 18 months after a spring application or 20 months after a fall application of Treflan.

In those portions of Kansas, Nebraska, North Dakota, Oklahoma, South Dakota and Texas where at least 20 inches of irrigation and/or rainfall (total) was used to produce the crop, sorghum, oats and annual or perennial grass crops or grass mixtures should not be planted for 12 months after an application of Treflan.

If less than 20 inches of total water was used to produce the crop, do not plant sorghum, proso millet, oats and annual or perennial grass crops or grass mixtures for 18 months after an application of Treflan. Cool, wet weather conditions during the early stage of growth may increase the possibility of injury to sorghum.

In all other areas receiving greater than 20" rainfall per year, moldboard plow before planting sugar beets where a spring application of Treflan was made the previous season.

Vegetable crops other than those listed on this label should not be planted within 5 months following the application of Treflan.

Do not use Treflan on any crop grown in Pecos county, or Reeves county, Texas or the state of Montana.

DIRECTIONS FOR USE

Treflan is a preemergence herbicide which is incorporated into the soil to provide long-lasting control of many annual grasses and broadleaf weeds. Treflan controls weeds as they germinate. Treflan will not control established weeds.

WEEDS AND GRASSES CONTROLLED BY TREFLAN

Grasses

Annual bluegrass	*Poa annua*
Barnyardgrass (Watergrass)	*Echinochloa* sp.
Brachiaria (Signalgrass)	*Brachiaria* sp.
Bromegrass (Cheatgrass) (Downy brome)	*Bromus tectorum*
Cheat (Chess)	*Bromus secalinus*
Crabgrass (Large crabgrass) (Smooth crabgrass)	*Digitaria* spp.
Foxtail (Bottlegrass) (Bristlegrass) (Giant foxtail) (Green foxtail) (Foxtail millet) (Pigeongrass) (Robust foxtail) (Yellow foxtail)	*Setaria* spp.
Goosegrass (Silver crabgrass) (Silvergrass) (Wiregrass) (Yardgrass)	*Eleusine indica*
Johnsongrass (from seed) (Rhizome—see page 23 for special instructions for control in cotton and see page 34 for special instructions for control in soybeans)	*Sorghum halepense*
Junglerice	*Echinochloa colonum*
Panicum	
Fall panicum (Spreading panicgrass—see page 23 for special instructions in cotton and see page 31 for special instructions in soybeans)	*Panicum dichotomiflorum*
Guineagrass (See page 63 for special instructions)	*Panicum maximum*
Texas panicum (Buffalograss) (Coloradograss)	*Panicum texanum*
Itchgrass (Raoulgrass) (See page 64 for special instructions)	*Rottboellia exaltata*
Red rice (See page 32 for suppression or partial control directions)	*Oryza sativa*
Sandbur (Burgrass)	**Cenchrus incertus*
Sprangletop	*Leptochloa filiformis*
Stinkgrass (Lovegrass)	*Eragrostis cilianensis*
Wild cane (Shattercane) (See page 36 for special instructions)	*Sorghum bicolor*
Woolly cupgrass	*Eriochloa villosa*

Broadleaf Weeds

Carpetweed	*Mollugo verticillata*
Chickweed	*Stellaria media*
Field bindweed (See page 69 for special instructions)	*Convolvulus arvensis*
Florida pusley (Florida purslane) (Mexican clover) (Pusley)	*Richardia scabra*
Goosefoot	*Chenopodium hybridum*
Henbit (fall application only)	*Lamium amplexicaule*
Knotweed	*Polygonum aviculare*
Kochia (Fireweed) (Mexican fireweed)	*Kochia scoparia*
Lambsquarters	*Chenopodium album*
Pigweed (Carelessweed) (Prostrate pigweed) (Redroot) (Rough pigweed) (Spiny pigweed)	*Amaranthus* spp.
Puncturevine (Western U.S. only) (Caltrop) (Goathead)	*Tribulus terrestris*
Purslane	*Portulaca oleracea*
Russian thistle (Tumbleweed)	*Salsola kali*
Stinging nettle (Nettle)	*Urtica dioica*

SOIL PREPARATION

Crop Residues or Existing Weeds: Ground cover, such as crop residues or existing weeds, can interfere with the incorporation of Treflan into the soil. A manageable level of such ground cover will allow the Treflan to be uniformly incorporated into the top 2 to 3 inches of soil. If the level of the ground cover is such that this cannot be done, you must till the soil prior to the application of Treflan.

Roughness: The soil surface should be smooth enough so that you can operate a sprayer and incorporation equipment efficiently and at speeds which insure a uniform application and incorporation of Treflan.

General Soil Conditions: To assure uniform incorporation of Treflan, soil moisture conditions should be such that large clods can be broken up during the incorporation process.

Soil Texture Guide

The amount of Treflan you apply will vary with the soil texture and organic matter. A fine textured soil will require more Treflan per acre than a coarse soil. Choose the proper rate for each application based on the following soil texture group and specific crop recommendations. Do not exceed recommended rates.

Soil Texture	Soil Classification
Coarse Soils: (Light)	Sand, loamy sand, sandy loam
Medium Soils:	Loam, silty clay loam*, silt loam, silt, sandy clay loam*
Fine Soils:	Clay, clay loam, silty clay loam*, silty clay, sandy clay, sandy clay loam*

**Silty clay loam and sandy clay loam soils are transitional soils and may be classified as either medium or fine textured soils. If silty clay loam or sandy clay loam soils are predominately sand or silt, they are usually classified as medium textured soils. If they are predominately clay, they are usually classified as fine textured soils.*

MIXING AND APPLICATION DIRECTIONS

Treflan E.C. Alone in Water

Start with a clean spray tank. Fill sprayer 1/3 to 1/2 full with clean water. Start agitation. Add correct quantity of Treflan, continue agitation and finish filling the tank.

Treflan E.C. Tank Mix in Water

Vigorous, continuous agitation is required for all tank mixes. (Sparger pipe agitators generally provide the best agitation in spray tanks). Avoid stirring or splashing air into the mixture during filling to prevent foaming. To do this, place the end of the fill pipe below the surface of the water in the spray tank. Do not allow the mixture to siphon back into the water source.

Mixing order: Fill the tank 1/4 to 1/3 full with clean water. Start the agitation. Add dry flowables, wettable powders (WP), aqueous suspensions (AS), flowables (F), and liquids (L) to the water and agitate until the product(s) are completely dispersed in the water. Allow additional mixing and dispersion time when using dry flowable products. Continue agitation and fill tank to 3/4 full, add the Treflan E.C., mix thoroughly. Then add any solution (S) formulations, agitate and finish filling. Maintain agitation during filling and through application. If spraying and agitation must be stopped before the tank is empty, the materials may settle to the bottom. In this case it is important to resuspend all of the material in the bottom of the tank before continuing the spray application. A sparger agitator is particularly useful for this purpose. Sometimes it is more difficult to resuspend settled material than it is to suspend it originally.

Read and carefully follow all label instructions for each material added to the tank. Premixing dry and flowable formulations with water (slurrying) and pouring the slurry through a 20 or 35 mesh wetting screen in the top of the tank will help assure good initial dispersion in the tank water. Line screens in the tank should be no finer than 50 mesh (100 mesh is finer than 50 mesh).

If you see a buildup of material on the walls of the spray tank, wash the tank with soapy water between fillings. Rinse and continue the spraying operation. Clean the tank, lines, and screens thoroughly after use.

As the spray volume decreases, the importance of accurate calibration and uniform application increases. Check the sprayer daily to insure proper calibration and uniform application. Do not apply Treflan when the wind can cause drifting of spray particles which can result in non-uniform application. Treflan should not be applied to soils which are wet or are subject to prolonged periods of flooding as poor weed control may result.

Ground application: Apply Treflan in 5 to 40 gallons of water or liquid fertilizer per acre (broadcast basis), using any properly calibrated, low pressure herbicide sprayer that will apply the spray uniformly.

Aerial application: Apply Treflan in 5 to 10 gallons of water or liquid fertilizer per acre. Adjust pump pressure, nozzle arrangements, speed and height to provide a uniform application to the soil surface. Use swath markers or flagmen to assure proper application spray widths.

Chemigation: Refer to supplemental labeling entitled "Treflan E.C. Chemigation Instructions" for use directions for chemigation. Do not apply this product through any irrigation system unless the supplemental labeling on chemigation is followed.

INCORPORATION DIRECTIONS

Incorporation Equipment—General Directions

Use incorporation equipment that mixes Treflan into the top 2 to 3 inches of the final seedbed, or erratic weed control and/or crop injury may result. Incorporation equipment such as a disc will mix Treflan approximately half as deep as the equipment is set to operate. For example, a disc set to cut 4 inches deep will incorporate most of the Treflan within the top 2 inches of soil.

Incorporation Before Planting

Treflan must be incorporated one time within 24 hours after application. Then any time prior to planting, a second incorporation is necessary, this time running the equipment in a different direction from the first. You should incorporate the Treflan uniformly into the top 2 to 3 inches of the final seedbed.

Incorporation After Planting

Check specific crop for incorporation directions after planting.

Incorporation in Bedded Culture

For effective weed control, Treflan needs to be incorporated into the top 2 to 3 inches of the final seedbed.

Application prior to bedding: Apply Treflan and incorporate it one time with recommended equipment. The bedding operation serves as the second incorporation. Do not expose untreated soil during post-bedding operations.*

Application after bedding: Knock off beds to planting height before applying Treflan. Apply Treflan and incorporate it with recommended equipment that will conform to the bed shape. Do not leave untreated soil exposed.*

** Avoid removal of treated soil from the seedbed before or during the planting operation. This would expose untreated soil, allowing weeds to germinate in the drill row.*

Recommended Equipment

Any recommended incorporation tool may be used alone or in combination with any other recommended tool. Two incorporation passes are required unless specifically stated. The second incorporation should not be deeper than the first.

Disc: set to cut 4 to 6 inches deep and operate at 4 to 6 mph.

Field Cultivator: set to cut 3 to 4 inches deep and operate at 5 mph or more. A field cultivator is defined as an implement with 3 to 4 rows of sweeps, spaced at intervals of 7 inches or less and staggered so that no soil is left unturned. Chisel points should not be used.

Combination Seedbed Conditioners: set to cut 3 to 4 inches deep and operate at a speed of at least 5 mph. These implements are defined as three or more tillage devices combined and used as a single tool. For example, 2 to 3 rows of field cultivator C- or S-shaped shanks with an effective sweep spacing of 6 to 9 inches (staggered so that no soil is left unturned), followed by a spike-tooth or flextine harrow, followed by a ground-driven reel or basket.

Rolling Culitvator: set to cut 2 to 4 inches deep and operate at 6 to 8 mph. Rolling cultivators are adequate for use on coarse and medium textured soils only, except when used in sugarcane where the rolling cultivator may be used on fine textured soils.

Bed Conditioner (Do-All): set to cut 2 to 4 inches deep and operate at 4 to 6 mph. The do-all is adequate for use on coarse and medium textured soils only. When using the do-all in bedded culture, only one incorporation pass is required. However, two passes with a do-all are required in flat planted culture.

Mulch Treader (other similar disc-type implements): set to cut 3 to 4 inches deep and operate at 5 to 8 mph.

P.T.O. Driven Equipment (tillers, cultivators, hoes): adjust to incorporate Treflan into the top 2 to 3 inches of the seedbed with rotors spaced to provide a clean sweep of the soil. Only one incorporation is necessary. P.T.O driven equipment should not be operated at a speed greater than 4 mph.

Other equipment, including the flexible tine-tooth harrow (Flextine, Melroe), is also recommended but only for the special programs for which it is specified in this label.

CULTIVATION AFTER PLANTING

Soil treated with Treflan may be shallow cultivated without reducing the weed control activity of Treflan. Do not cultivate deeper than the treated soil since this may bring untreated soil to the surface, and poor weed control may result.

CROP RECOMMENDATIONS

These recommendations are given as the broadcast rates of Treflan per acre. For band applications, decrease the amount of Treflan in proportion to the amount of surface treated per acre. Apply any time after January 1 when the soil can be worked and is suitable for good incorporation. Treflan can be applied in the fall—see specific crop for recommendations. For general fall application directions where specific recommendations are not given, see page 74. Where a rate range is shown, use the lower rate for coarser soils with lower organic matter. Treflan should not be used on soils containing more than 10% organic matter.

Cotton—Treflan Alone

Treflan can be applied and incorporated before or at planting, immediately after planting, or at layby.

Cotton—Preemergence: Broadcast Rates Per Acre

Soil Texture	Treflan E.C.	
	Areas receiving less than 20" average annual rainfall*	Areas receiving greater than 20" average annual rainfall*
	(pints)	(pints)
Coarse	1	1
Medium	1 1/4-1 1/2	1 1/2
Fine	1 1/2	2

**Use $1\frac{1}{2}$ pints per acre on coarse and medium textured soils and 2 pints on fine soils with 2-5% organic matter; use 2 to $2\frac{1}{2}$ pints on all soils with 5-10% organic matter.*

Cotton—Postplant:

When incorporating Treflan after planting (postplant), be careful not to disturb the seed.

Cotton—Layby:

Apply and incorporate Treflan any time up to layby, but not less than 90 days before harvest. Direct the layby applications onto the soil between the rows and beneath emerged cotton plants. Use the same rates as for a preemergence application.

Cotton—Fall Application:

Apply and incorporate Treflan any time between October 15 and December 31. The ground may be left flat or bedded-up over winter. On bedded ground, knock beds down to desired height before planting, moving some treated soil from beds into furrows. Where soil is left flat over winter, be careful not to turn up untreated soil during spring bedding operations. Destroy established weeds during seedbed preparation. If weeds become established in furrows due to uncovering of untreated soil during bedding, destroy these weeds before planting. In the fall, do not apply Treflan to soils which are wet or subject to prolonged periods of flooding.

Broadcast Rates per Acre—Fall application only:

In Alabama, Arkansas, northern Florida, Georgia, Louisiana, Mississippi, southeastern Missouri bootheel, North Carolina, New Mexico, Oklahoma, South Carolina, Tennessee and Texas, apply and incorporate Treflan at a broadcast rate of 2 pints per acre on coarse and medium soils and $2\frac{1}{2}$ pints on fine soils.

In Arizona, California and Nevada, apply and incorporate Treflan at a broadcast rate of $1\frac{1}{2}$ pints per acre on coarse soils; 2 pints on medium soils; and $2\frac{1}{2}$ pints on fine soils.

For cotton grown in other states, apply and incorporate Treflan at a broadcast rate of 1 pint per acre on coarse soils; $1\frac{1}{2}$ pints on medium soils; 2 pints on fine soils; $1\frac{1}{2}$ pints on coarse soils with 2-5% organic matter; and 2 to $2\frac{1}{2}$ pints on soils with 5-10% organic matter.

Cotton—Special Use Directions

Cotton—Fall panicum:

Apply and incorporate Treflan at the broadcast rate of 2 pints per acre on both coarse and medium soils.

Cotton—Rhizome johnsongrass:

In all cotton-producing states except Arizona and California, you can obtain commercially acceptable control of rhizome johnsongrass with a double rate program which you apply for 2 consecutive years in accordance with the following directions.

Soil Preparation—Proper preparation of the soil before application is very important for satisfactory results. Use a chisel plow or similar implement to bring rhizomes to the top of the soil. Then follow with a disc two times before application to cut the rhizomes into small (2 to 3 inch) pieces. This should also destroy any emerged johnsongrass.

Application—Choose the one application program that best fits your cultural practices:

Spring Application—Apply Treflan any time before planting in the spring for 2 years in a row. Use a broadcast rate of 2 pints per acre on coarse soils; 3 pints on medium soils; and 4 pints on fine soils,

OR

Fall Application—Apply Treflan between October 15 and December 31 for 2 years in a row at the same rates as a spring application for the control of rhizome johnsongrass.

Incorporation—Deep incorporation is essential for good rhizome johnsongrass control. Incorporate Treflan thoroughly with a disc set to cut 4 to 6 inches deep and operate at 4 to 6 mph. Two passes are necessary, with the second pass in a different direction from the first.

Cultivation—Some johnsongrass plants will escape. Timely cultivations during the crop season to remove escaped plants are necessary to obtain commercially acceptable control. You cannot obtain commercially acceptable control with only 1 year of double rate Treflan use.

Crop Rotation—In the season following a double rate treatment, plant only rice and those crops for which Treflan can be applied as a preplant treatment or injury may result.

Cotton—Pigweed and seedling johnsongrass control:

In Alabama, Arkansas, Florida, Georgia, Louisiana, Mississippi, southeastern Missouri, North Carolina, South Carolina, Tennessee and southern Virginia, Treflan may be applied preplant at a broadcast rate of 1 to $1\frac{1}{2}$ pints per acre on coarse soils; $1\frac{1}{2}$ to 2 pints on medium soils; and 2 pints on fine soils. Exception: in the State of Louisiana, 3 pints per acre are recommended on fine soils.

Cotton—Additional weed and grass control (Texas Gulf Coast):

In the Texas Gulf Coast counties of Brazoria, Calhoun, Chambers, Fort Bend, Galveston, Harris, Jackson, Jefferson, Liberty, Matagorda, Orange, Victoria, Waller and Wharton, Treflan may be applied up to 2 weeks before planting at a broadcast rate of $1\frac{1}{2}$ pints per acre on coarse soils; 2 pints on medium soils; and 3 pints on fine soils.

Cotton Precautions

Cotton should be planted after early season adverse weather conditions have passed, especially when using higher rate programs. Cool, wet weather early in the growth cycle causes additional stress to the cotton plant. This may result in reduced stands, delayed maturity and reduced yields.

Cotton—Tank Mixes and Overlays

Follow recommended soil preparation and incorporation procedures for Treflan.

Cotton—Treflan/Caparol tank mix for cotton grown in California, Arizona, New Mexico and west Texas:

The Treflan/Caparol combination will control certain grasses and broadleaf weeds listed for Treflan alone plus the following weeds:

Annual morningglory	Prickly sida (Teaweed)
Groundcherry (Annual)	Ragweed
Malva	Smartweed
Mustard	Wild oat

The tank mix also controls shallow germinating seedlings of cocklebur and coffeeweed.

Broadcast Rates Per Acre

Soil Texture	Treflan E.C.	Caparol 80W
	(pints)	(pounds)
Coarse	1	2*
Medium	$1\frac{1}{4}$-$1\frac{1}{2}$	$2\frac{1}{2}$
Fine	2	$2\frac{1}{2}$

**Do not use on sands and loamy sands. For band applications use proportionately less.*

Mixing Directions: Carefully follow the procedures on the Caparol 80W label for making a slurry and adding it to a partially filled tank of water. After the Caparol is thoroughly mixed with the partially filled tank of water, add the Treflan E.C. and continue filling. Agitate continuously during the filling and spraying operation.

Avoid leaving the spray mixture in the tank without constant agitation. If bypass agitation is used, the bypass line should stop at the bottom of the tank to minimize foaming.

Additional Precautions: The combination of Treflan/Caparol should not be used in the cut areas of newly leveled fields, in areas of excess salt, or where flooding over the beds is likely to happen. Do not plant cotton in tractor wheel depressions. These conditions may cause crop injury. On mulch-planted cotton, water back only after cotton seedlings are well established.

Crop Rotations: Cabbage, okra, onions and peas may be planted in the fall after a spring application of Treflan plus Caparol. Winter barley, winter rye and winter wheat can be planted in the fall also, if they are plowed down and not used for food or feed. Refer to the Caparol label for directions, cautions and precautions.

Cotton—Treflan/Cotoran tank mix (except in Arizona and California):

Follow recommended soil preparation and incorporation procedures for Treflan.

The Treflan/Cotoran tank mix effectively controls all the annual grasses and broadleaf weeds listed for Treflan alone plus these additional weeds:

Buttonweed	Ragweed
Cocklebur	Ryegrass
Groundcherry, Wright	Sesbania
Jimsonweed	Sicklepod
Morningglory	Smartweed
Prickly sida (Teaweed)	Tumbleweed

Broadcast Rates Per Acre

Soil Texture	Treflan E.C.	Cotoron 80W
	(pints)	(pounds)
Coarse	1	$1\frac{1}{4}$
Medium	$1\frac{1}{2}$	2
Fine	2	$2\frac{1}{2}$

Mixing Directions: Carefully follow the procedures on the Cotoran label for making a Cotoran slurry and adding it to a partially filled tank of water. After the Cotoran is thoroughly mixed with the partially filled tank of water, add the Treflan and continue filling. Agitate continuously throughout the filling and application operations. Do not leave spray mixture in tank without constant agitation. If bypass agitation is used, the bypass line should stop at the bottom of the tank to minimize foaming. Apply in 15-40 gallons of water per acre.

Additional Precautions: Do not plant crops other than cotton on the treated land within 6 months after the application of Treflan plus Cotoran, or crop injury may result. Do not feed foliage from treated cotton plants or gin trash to livestock. Do not mix Treflan plus Cotoran with liquid fertilizer.

West Texas Only: Do not use the tank mix of Treflan plus Cotoran on sandy, loamy sand or fine sandy loam soils. Do not use on cotton planted in furrows.

Arkansas, Louisiana, and Mississippi Only: Use 1 pound Cotoran 80W in tank mix with Treflan on sandy loam soils low in organic matter.

New Mexico: Do not plant treated land with crops other than cotton until 1 year after the last application. Do not use on sandy or coarse textured soils of less than 1% organic matter.

Cotton—Treflan preplant followed by Cotoran overlay:

Apply and incorporate Treflan as recommended for the specific soil texture. Then apply Cotoran as a preemergence surface treatment at 1¼ to 2½ pounds per acre. On light silt and sandy soils low in organic matter, use the lower rate.

Additional Precaution: Refer to the Cotoran label for all cautions and precautions when using this preemergence surface treatment.

Cotton—Treflan preplant followed by Karmex overlay:

For cotton grown east of the Mississippi River plus Arkansas, southeastern Missouri, Louisiana and Eastern Texas, incorporate Treflan before planting. Then follow with a preemergence application of Karmex. This will effectively control all the weeds contolled by Treflan alone plus these additional weeds:

Annual groundcherry	Shepherdspurse
Annual morningglory	Velvetgrass
Dogfennel	Wild lettuce
Pennycress	Wild mustard
Ragweed	

Broadcast Rates Per Acre

Soil Texture	Treflan E.C.	Karmex 80W
	(pints)	(pounds)
Coarse	1	⅓
Medium	1½	⅔
Fine	2	1

Additional Precautions: Do not use Karmex on soils with less than 1% organic matter as crop injury may result. Do not allow livestock to graze on cotton treated with Karmex. Consult the Karmex label for additional instructions, cautions and precautions.

Soybean—Treflan Alone

Soybean—Preemergence:

Follow recommended soil preparation, application, and incorporation procedures for Treflan.

Broadcast Rates Per Acre

Soil Texture	Treflan E.C.*
	(pints)
Coarse	1
Medium	1½
Fine	2

**Use 1½ pints per acre on coarse and medium textured soils and 2 pints on fine soils with 2-5% organic matter; use 2-2½ pints on all soils with 5-10% organic matter.*

Soybean—Fall Application:

Apply and incorporate Treflan anytime between October 15 and December 31. Ground may be left flat or bedded-up over winter. On bedded ground, knock beds down to desired height before planting, moving some treated soil from tops into furrows. Where soil is left flat over winter, take care during spring bedding operations to prevent turning up untreated soil. Destroy established weeds during seedbed preparation. If weeds become established in furrows due to uncovering of untreated soil during listing, destroy these weeds before planting. Do not fall apply Treflan to soils which are wet, are subject to prolonged periods of flooding, or where rice was grown the previous year.

For soybeans grown in Alabama, Arkansas, northern Florida, Georgia, Louisiana, Mississippi, southeastern Missouri Bootheel, North Carolina, Oklahoma, South Carolina, Tennessee and Texas, apply and incorporate Treflan at a broadcast rate of 2 pints per acre on coarse and medium soils and 2½ pints on fine soils.

For soybeans grown in states other than those listed above, apply and incorporate Treflan at a broadcast rate of 1 pint per acre on coarse soils; 1½ pints on medium soils; 2 pints on fine soils; 1½ pints on coarse soils with 2-5% organic matter; and 2 to 2½ pints on soils with 5-10% organic matter.

Soybean—Special Use Directions

Soybean—Fall panicum:

Apply Treflan at the broadcast rate of 2 pints per acre on both coarse and medium soils.

Soybean—Pigweed and seedling johnsongrass:

In Alabama, Arkansas, Florida, Georgia, Kansas, Louisiana, Mississippi, Missouri, Nebraska, North Carolina, Oklahoma, South Carolina, Tennessee and southern Virginia, Treflan may be applied at a broadcast rate of 1 to 1½ pints per acre on coarse soils; 1½ to 2 pints on medium soils; and 2-2½ pints on fine soils. Exception: in the state of Louisiana, 3 pints per acre are recommended on fine soils.

Soybean—Additional weed and grass control (Texas Gulf Coast):

In the Texas Gulf Coast counties of Brazoria, Calhoun, Chambers, Fort Bend, Galveston, Harris, Jackson, Jefferson, Liberty, Matagorda, Orange, Victoria, Waller and Wharton, Treflan may be applied up to 2 weeks before planting at a broadcast rate of 1½ pints per acre on coarse soils; 2 pints on medium soils; and 3 pints on fine soils.

Soybean—Charcoal soils in Arkansas, Louisiana and Mississippi:

Newly cleared land often contains high organic matter (5-10%) and charcoal from burning debris. This charcoal and/or organic matter tends to bind Treflan and reduce its weed control activity. Under these conditions, higher rates of Treflan are necessary for weed control. Increased rates, however, can cause crop injury if charcoal or organic matter is not present to bind some of the Treflan. In the burn row a high level of charcoal is present; consequently, poor weed control may result even with an increased rate of Treflan.

Apply Treflan at the broadcast rate of 1½ to 2½ pints per acre on coarse soils; 2½ pints on medium soils; and 3 pints on fine textured soils. Follow recommended soil preparation, application and incorporation procedures for Treflan.

Soybean—Red rice in Arkansas, Louisiana, Mississippi and Texas only:

You can obtain suppression or partial control of red rice when you apply Treflan at the following recommended rates. Follow recommended soil preparation and incorporation procedures for Treflan.

Apply and incorporate Treflan in the spring before planting.

Broadcast Rates Per Acre

Soil Texture	Treflan E.C. Application Year 1	Treflan E.C. Application Year 2
	(pints)	(pints)
Coarse	2	1
Medium	3	1½
Fine	4	2
Coarse soils with 2 to 5% organic matter	3	1½
Soils with 5 to 10% organic matter	4	2-2½

If a combination of high organic matter (5-10%) and charcoal are present in the soil, apply Treflan the second year at the following rates for charcoal soils in Arkansas, Louisiana and Mississippi:

Soil Texture	Treflan E.C.
	(pints)
Coarse	1½-2½
Medium	2½
Fine	3

For more information on charcoal soils see page 32.

Crop Rotation: The program for red rice control in soybeans is a 2-year program. Use the rates listed for first year application and plant soybeans. The second year use the normal Treflan rates listed for your soil type and charcoal level and plant only those crops for which Treflan has been registered as a preplant treatment, or crop injury may result. Do not plant rice the second year. Rice may be planted the third year.

Soybean—Rhizome johnsongrass (Eastern United States and the state of Texas):

You can obtain commercially acceptable control of rhizome johnsongrass with a double rate program applied for 2 consecutive years in accordance with the following directions.

Soil Preparation—Proper preparation of the soil before application is very important for satisfactory results. Use a chisel plow or similar implement to bring rhizomes to the top of the soil. Then follow with a disc two times before application to cut the rhizomes into small (2 to 3 inch) pieces and to destroy any emerged johnsongrass.

Application—Choose the one application program that bests fits your cultural practices:

Spring Application—Apply Treflan anytime in the spring before planting for 2 years in a row. Use a broadcast rate of 2 pints per acre on coarse soils; 3 pints on medium soils; 4 pints on fine soils; 3 pints on coarse soils with 2-5% organic matter; and 4 pints on soils with 5-10% organic matter.

OR

Fall Application—Apply Treflan between October 15 and December 31 for 2 years in a row at the same rates as a spring application for the control of rhizome johnsongrass.

OR

Split Application—Apply Treflan at the same rate in both the spring and fall for 2 years in a row using the rates in the following table.

Broadcast Rates Per Acre

Soil Texture	Treflan E.C. Spring and Fall
	(pints)
Coarse	1
Medium	1½

Soil Texture	Treflan E.C. Spring and Fall
Fine	2
Coarse soils with 2-5% organic matter	1½
Soils with 5 to 10% organic matter	2

Incorporation—Deep incorporation is essential for good rhizome johnsongrass control. Incorporate Treflan thoroughly with a disc set to cut 4 to 6 inches deep and operate at 4 to 6 mph. Two passes are necessary, with the second pass in a different direction from the first.

Cultivation—Some johnsongrass plants will escape. Timely cultivations during the crop season to remove escaped plants are necessary to obtain commercially acceptable control.

Crop Rotation: In the season following a double rate treatment, plant only rice and those crops for which Treflan can be applied as a preplant treatment or injury may result.

Soybean—Rhizome johnsongrass - Treflan/Sencor or Treflan/Lexone tank mix:

Treflan/Sencor or Treflan/Lexone may be used for rhizone johnsongrass control and for the control of those weeds listed for Treflan alone. For the additional weeds controlled by Sencor or Lexone in tank mix, see page 38. Follow procedures for soil preparation, incorporation, and cultivation recommended in the Soybean - Rhizome johnsongrass section (see page 36).

Apply Treflan/Sencor or Lexone up to two weeks before planting for two consecutive years at the following broadcast rates per acre:

Soil Texture	Treflan E.C.	Lexone 50WP / 4L or Sencor 50WP / 4	or	Lexone (dry flowable) or Sencor (dry flowable
	(pints)	(pounds/pints)		(pounds)
Coarse*	2	½		⅓
Medium	3	¾		½
Fine	4	1		⅔

**Do not use on coarse soils with less than 1% organic matter.*

Read and follow all additional precautions listed for the Treflan/Sencor or Treflan/Lexone tank mix (seee page 37).

Soybean—Wild cane (shattercane):

Follow recommended soil preparation and application procedures for Treflan.

Wild cane (shattercane) can germinate throughout the growing season and from greater soil depth than most other weed seeds. Commercially acceptable control of wild cane can be obtained with the following increased rates of Treflan.

Apply Treflan at the broadcast rate of 1 pint per acre on a coarse soil: 2 pints on a medium soil; and 2½ pints on a fine textured soil.

Incorporation—Deep incorporation is essential for good wild cane control. Incorporate Treflan thoroughly with a disc set to cut 4 to 6 inches deep and operate at 4 to 6 mph. Two passes are necessary, with the second pass in a different direction from the first.

Cultivation—Cultivations during the crop season will also contribute to control.

Soybean Precautions:

Soybeans should be planted after early season adverse weather conditions have passed, especially when using higher rate programs. Cool, wet weather early in the growth cycle causes additional stress to the soybean plant. This may result in reduced stands, delayed maturity and reduced yields.

Soybean—Tank-Mix, Overlay and Postemergence Recommendations

Soybean—Treflan/Sencor or Treflan/Lexone tank mix:

The Treflan/Sencor or Treflan/Lexone tank mix effectively controls the annual grasses and broadleaf weeds controlled by Treflan alone plus these additional weeds:

Jimsonweed	Ragweed, common
Mallow, Venice (Flower-of-an-hour)	Sesbania, hemp
Mustard, wild	Smartweed, Pennsylvania
Prickly sida	Velvetleaf

Control of cocklebur, morningglory and giant ragweed (horseweed) may be erratic. Control may be improved with timely cultivation. Where cocklebur is a serious problem, an overlay of Sencor or Lexone may be preferred to the Treflan/Sencor or Treflan/Lexone tank mix.

Follow recommended procedures for soil preparation, incorporation, and cultivation of Treflan. Mix according to instructions on page 13. The Treflan/Sencor or Treflan/Lexone tank mix can be applied for 2 weeks before planting up to planting.

Broadcast Rates Per Acre

Soil Texture	Treflan E.C.	Lexone 50WP / 4L or Sencor 50WP / 4	or	Lexone (dry flowable) or Sencor (dry flowable)
	(pints)	(pounds/pints)		(pounds)
Coarse*	1	½		⅓
Medium	1½	¾		½
Fine	2	1		⅔

**Do not use Sencor/Lexone on coarse soils with less than 1% organic matter.*

Precautions: Do not plant any crop other than soybeans within 4 months after treatment. Overapplication, uneven application, or improper soil incorporation may result in crop injury, herbicide residue, or erratic weed control. Additional stress factors are seedling disease, cold weather, deep planting, excessive moisture, soil *p*H over 7.5, high salt concentration, or drought. Any of these may weaken crop seedlings and increase possibility of damage from the tank mix. These additional factors may also delay crop development or reduce yields when Sencor or Lexone is applied. Observe all cautions and limitations on the Sencor and Lexone labels. Do not use the foliage from soybeans treated with the Treflan/Sencor or Treflan/Lexone tank mix for feed or forage.

Additional Sencor and Lexone precautions: Do not use Lexone or Sencor on Tracy, Semmes, Altona, Vansoy or Coker 102 soybeans. These varieties are sensitive to Lexone or Sencor, and crop injury may result. Seed must be planted at least 1½ inches but not more than 2 inches below the soil surface before a Sencor or Lexone application. Do not apply Sencor or Lexone at these rates more than once per season. Do not replant areas treated with Sencor or Lexone with any crop other than soybeans within 4 months after treatment. Injury to soybeans may occur if you use Lexone or Sencor on soils having a calcarous surface or *p*H of 7.5 or higher, or if you use them in conjunction with soil-applied organic phosphate pesticides.

Soybean—Treflan/Canopy Tank-Mix (Do Not Use in California)

Apply Treflan Canopy as a preplant incorporated tank-mix treatment. Follow recommended soil preparation, application and incorporation procedures for Treflan. Plant soybeans within two (2) weeks after application.

Treflan/Canopy tank-mix controls the following weeds in addition to those controlled by Treflan alone (excluding special rate programs).

Controlled

Cocklebur, common	Prickly sida (Teaweed)
Florida beggarweed	Ragweed (Common) (Giant)
Hemp sesbania	Sicklepod
Hophornbeam copperleaf	Smartweed
Jimsonweed	Spotted spurge
Morningglory (Entireleaf) (Ivyleaf) (Pitted) (Smallflower) (Tall)	Sunflower
	Velvetleaf

Partially Controlled

Purple nutsedge
Yellow nutsedge

Large seeded weeds, germinating deep in the soil, such as morningglory, sicklepod, cocklebur and giant ragweed or weeds with subsequent flushes may require a cultivation or an application of a postemergence herbicide for season-long control.

Broadcast Rates Per Acre

Soil Texture	Treflan E.C.	Canopy Dispersible Granules 0.5-3% O.M.	Canopy Dispersible Granules 3-5% O.M.
	(pints)	(ounces)	(ounces)
Coarse	1	6 - 8	8 - 10
Medium	1½	8 - 10	10 - 12
Fine	2	10 - 12	12 - 14

Do not apply Canopy to soil with less than ½% organic matter.

Use higher rates of Canopy on soils with higher organic matter or heavy pressure from large deep germinating weed seeds.

Where Canopy is applied plant soybean seed 1½" to 2" deep on a flat or raised seedbed only, or crop injury may occur.

Soybean injury may occur where Canopy is applied if excessive rainfall occurs after application but before soybeans germinate.

Precautions: Read the Canopy label carefully for cautions and precautions relating to environmental hazards, planting of rotation crops, sprayer contamination and cleanup, soil *p*H, organic matter and soil texture use restrictions, soybean variety planting restrictions, restrictions where Atrazine or Scepter were used the previous year, restrictions concerning use with organic phosphate pesticides, grazing restrictions and other directions, precautions and limitations before applying the Treflan/Canopy tank-mix.

Soybean—Treflan/Preview Tank-Mix (Do Not Use in California)

Apply Treflan/Preview as a plant incorporated tank-mix treatment. Follow recommended soil preparation, application and incorporation procedures for Treflan. Plant soybeans within two (2) weeks after application.

Treflan/Preview tank-mix controls the following weeds in addition to those controlled by Treflan alone (excluding special rate programs).

Controlled

- Cocklebur, common
- Hophornbeam copperleaf
- Jimsonweed
- Mustards
- Pigweed
  - (Palmer amaranth)
  - (Smooth)
  - (Tall waterhemp)
- Prickly sida (Teaweed)
- Ragweed, common
- Smartweeds, annual
- Spotted spurge
- Sunflower
- Velvetleaf

Partially Controlled

- Burcucumber
- Eastern Black nightshade
- Morningglory
  - (Tall)
  - (Ivyleaf)
  - (Pitted)
  - (Entireleaf)
- Nutsedge species
- Ragweed, giant

Large seeded weeds, germinating deep in the soil, such as morningglory, sicklepod, cocklebur and giant ragweed or weeds with subsequent flushes may require a cultivation or an application of a postemergence herbicide for season-long control.

Broadcast Rates Per Acre

Soil Texture	Treflan E.C.	Preview Dispersible Granules 0.5-3% O.M.	Preview Dispersible Granules 3-5% O.M.
	(pints)	(ounces)	(ounces)
Coarse	1	6	7
Medium	1 1/2	7	8
Fine	2	8	9 - 10

Do not apply Preview to soil with less than 1/2% organic matter.

Use higher rates of Preview on soils with higher organic matter or heavy pressure from large deep germinating weed seeds.

Where Preview is applied, plant soybean seed 1 1/2" to 2" deep on a flat or raised seedbed only, or crop injury may occur.

Soybean injury may occur where Preview is applied if excessive rainfall occurs after application but before soybeans germinate.

Precautions: Read the Preview label carefully for cautions and precautions relating to environmental hazards, planting of rotation crops, sprayer contamination and cleanup, soil *p*H, organic matter and soil texture use restrictions, soybean variety planting restrictions, restrictions where Atrazine was used the previous year, restrictions concerning use with organic phosphate pesticides, grazing restrictions and other directions, precautions or limitations before applying the Treflan/Preview tank-mix.

Soybean—Treflan/Amiben tank mix:

Amiben may be applied several days prior to planting as a broadcast tank mix with Treflan. Weeds controlled by this tank mix, in addition to those controlled by Treflan alone, are smartweed, velvetleaf and ragweed. The tank mixture should be used as a spring preplant incorporated treatment.

Apply Treflan as a tank mix with Amiben.

Broadcast Rates Per Acre

Soil Texture	Treflan E.C.	Amiben 2S	Amiben DS
	(pints)	(quarts)	(pounds)
Coarse	1	4-6*	2.4-3.6*
Medium	1 1/2	4-6	2.4-3.6
Fine	2	4-6	2.4-3.6

* *Use the higher rate where you expect heavy populations of smartweed, velvetleaf, ragweed, wild mustard or black nightshade. Do not use on muck or charcoal soils. Read and observe all directions and cautions on the Amiben label.*

Soybean—Treflan/Amiben/Sencor or Lexone tank mix:

The Treflan/Amiben/Sencor or Lexone tank mix effectively controls all weeds listed for Treflan/Amiben and Treflan/Sencor or Lexone tank mixes.

Follow recommended soil preparation, application, and incorporation procedures for Treflan. The Treflan/Amiben/Sencor or Lexone tank mix may be applied from several days prior to planting up to planting in 10 to 40 gallons of water per acre. Use screens no finer than 50 mesh.

Apply the Treflan/Amiben/Sencor or Lexone tank mix at the following broadcast rate per acre.

Soil Texture	Treflan E.C.	Amiben 2S	Lexone 50WP / 4L or Sencor 50WP / 4	or Lexone (dry flowable) or Sencor (dry flowable)
	(pints)	(quarts)	(pounds/pints)	(pounds)
Coarse[a]	1	3-4[b]	1/2	1/3
Medium	1 1/2	3-4[b]	1/2-3/4[c]	1/3-1/2[c]
Fine	2	4-5	3/4[c]	1/2[c]

[a] *Do not use Sencor or Lexone on coarse soils with less than 1% organic matter.*

[b] *Use the higher rate of Amiben when velvetleaf or black nightshade is a problem.*

[c] *On Clarion/Webster soils in Minnesota and Iowa or on similar alkaline (calcareous) soils with a pH of 7.5 or above, apply Sencor or Lexone at the rates listed below:*

Soil Texture	Lexone 50WP / 4L or Sencor 50WP / 4	or Lexone (dry flowable) or Sencor (dry flowable)
	(pounds/pints)	(pounds)
Medium	1/2	1/3
Fine	1/2-3/4[d]	1/3-1/2[d]

[d] *Use the higher rate only where soil pH is less than 7.5 and where weed pressure is heavy.*

Additional precautions: The Treflan/Amiben/Sencor or Lexone tank mix will not harm the treated crop when you apply it according to directions and under normal growing conditions. However, overapplication, uneven application or improper soil incorporation of the tank mix can result in erratic weed control or crop injury. Additional stress factors are seedling disease, cold weather, deep planting, excessive moisture, soil *p*H over 7.5, high salt concentration, or drought. These additional factors may weaken crop seedlings and increase the possibility of damage from the tank mix. These additional factors may also delay crop development or reduce yields. Observe all cautions and limitations of all products used in mixtures. Do not use the foliage from soybeans treated with the Treflan/Amiben/Sencor or Lexone tank mix for feed or forage.

Soybean—Treflan/Vernam tank mix:

Treflan/Vernam tank mix effectively controls those weeds listed for Treflan alone plus these additional weeds:

- Annual morningglory
- Coffeeweed
- Purple nutsedge (nutgrass)
- Velvetleaf
- Yellow nutsedge (nutgrass)

Follow recommended soil preparation procedures for Treflan. You may apply the Treflan/Vernam tank mix up to 10 days prior to planting. Incorporate the tank mix immediately after application. Apply Treflan/Vernam at these rates:

Broadcast Rates Per Acre

Soil Texture	Treflan E.C.	Vernam 7E
	(pints)	(pints)
Coarse	1	1 3/4-2 1/3
Medium	1 1/2	2 1/3-3*
Fine	2	3-3 1/2

* *For nutsedge, wild cane and velvetleaf control, use the higher rate of 3 pints per acre on medium textured soils.*

Soybean—Treflan preplant incorporated followed by Overlay treatments (Do not use in California)

Apply Treflan as a preplant incorporated treatment. Additional weeds tolerant to Treflan may be controlled by using overlay preemergence applications of Amiben, Canopy, Dual, Gemini, Lasso, Lexone, Lorox, Lorox Plus, Preview or Sencor. Consult these product labels for additional weeds controlled, application directions and precautions before use.

Soybean—Treflan preplant incorporated followed by Post-emergence treatments (Do not use in California)

Apply Treflan as a preplant incorporated treatment. Additional weeds tolerant to Treflan may be controlled by using post-emergence applications of Basagran, Blazer, Classic or Tackle. Consult these product labels for additional weeds controlled, application directions and precautions before use.

Alfalfa—Established

In areas receiving less than 20" average annual rainfall per year, apply Treflan to established alfalfa stands at a broadcast rate of 1 1/2 pints per acre on coarse soils and 2 pints on medium and fine soils. Use incorporation equipment that will insure thorough soil mixing with minimum damage to the established alfalfa.

Asparagus—Established

Follow recommended soil preparation, application and incorporation procedures for Treflan.

Treflan can be applied to established asparagus as a single or as a split application. In the winter or early spring, apply Treflan to asparagus after ferns are removed but before spear emergence. Or, apply after harvest in the late spring or early summer before ferning begins. Treflan will suppress volunteer seedling asparagus and field bindweed if you use the following recommended rates and application schedules.

Follow recommended soil preparation, application and incorporation procedures for Treflan.

Broadcast Rates Per Acre

Soil Texture	Treflan E.C. Split Application: Before Harvest + After Harvest	Treflan E.C. Single Application: Before Harvest or After Harvest
	(pints)	(pints)
Coarse	1 + 1	2 or 2
Medium	1½ + 1½	3 or 3
Fine	2 + 2	4 or 4

* *In any single calendar year, the maximum Treflan to apply is 2 pints per acre on coarse soils; 3 pints on medium soils; and 4 pints on fine soils.*

For the Following Crop Grouping, Use the Rate Table Below.

Carrot

Castor Bean

Celery—(Direct seeded and transplant in areas receiving less than 20" average annual rainfall.)

Cole Crops—Transplant

Apply and incorporate prior to transplanting only. (Broccoli, Brussels Sprout, Cabbage and Cauliflower) See next section for direct seeded.

Okra

Pepper—Transplant

Apply and incorporate prior to transplanting only.

Southern Pea—(Before planting only.)

Apply and incorporate Treflan before planting, at planting or immediately after planting, unless otherwise indicated.

Broadcast Rates Per Acre

Soil Texture	Treflan E.C.: Areas receiving less than 20" average annual rainfall*	Treflan E.C.: Areas receiving greater than 20" average annual rainfall*
	(pints)	(pints)
Coarse	1	1
Medium	1¼-1½	1½
Fine	1½	2

* *Use 1½ pints per acre on coarse and medium textured soils and 2 pints on fine soils with 2-5% organic matter; use 2 pints on all soils with 5-10% organic matter.*

Cole Crops—Direct Seeded (Broccoli, Brussels Sprout, Cabbage and Cauliflower) See previous section for transplant.

For direct-seeded cole crops, apply and incorporate Treflan before planting at a broadcast rate of 1 pint per acre on coarse and medium soils and 1½ pints on fine soils and soils with 2-5% organic matter. Direct-seeded cole crops have exhibited marginal tolerance to recommended rates of Treflan. Stunting or reduced stands may occur.

Cucurbits—Postplant Emerged (Cantaloupe, Cucumber and Watermelon)

Western United States including Texas:

Apply Treflan as a directed spray to the soil between the rows and beneath plants which are in the 3 to 4 true leaf stage.

Broadcast Rates Per Acre

Soil Texture	Treflan E.C.: Areas receiving less than 20" average annual rainfall*	Treflan E.C.: Areas receiving greater than 20" average annual rainfall*
	(pints)	(pints)
Coarse	1	1
Medium	1¼-1½	1½
Fine	1½	2

* *Use 1½ pints on coarse and medium textured soils and 2 pints on fine soils with 2-5% organic matter; use 2 pints on all soils with 5-10% organic matter.*

Set incorporation equipment to throw treated soil around plants during incorporation.

Dry Bean—Treflan Alone

Apply and incorporate Treflan before planting using the following rates:

Broadcast Rates Per Acre

Soil Texture	Treflan E.C.: Areas receiving less than 20" average annual rainfall*	Treflan E.C.: Areas receiving greater than 20" average annual rainfall*
	(pints)	(pints)
Coarse	1	1
Medium	1¼-1½	1½
Fine	1½	2

* *Use 1½ pints per acre on coarse and medium textured soils and 2 pints on fine soils with 2-5% organic matter; use 2 pints on all soils with 5-10% organic matter.*

Dry Bean—Treflan/Eptam tank mix:

The Treflan/Eptam tank mix effectively controls all the following weeds in addition to those weeds listed for Treflan (see page 10):

Henbit (Spring applications)	Oat, wild
Nightshade, black	Ragweed, common
Nightshade, hairy	Smartweed, Pennsylvania
Nutsedge	Velvetleaf (Buttonweed)

Follow recommended soil preparation and incorporation procedures for Treflan. The Treflan/Eptam tank mix should be applied from 2 days before planting up to planting. Incorporate immediately after application.

Broadcast Rates Per Acre

Soil Texture	Treflan E.C.: Areas receiving less than 20" average annual rainfall*	Treflan E.C.: Areas receiving greater than 20" average annual rainfall*	Eptam 7E
	(pints)	(pints)	(pints)
Coarse	1	1	2½-3½**
Medium	1¼-1½	1½	2½-3½
Fine	1½	2	2½-3½

* *Use 1½ pints per acre on coarse and medium textured soils and 2 pints on fine soils with 2-5% organic matter; use 2 pints on all soils with 5-10% organic matter.*

** *Use Eptam 7E at a rate of 2½ pints per acre to control annual grasses; 3½ pints to control nutsedge and additional broadleaf weeds.*

Precaution: Read the Eptam label before using. Observe all cautions and limitations of all products used in mixtures. The combination of Treflan and Eptam should not be used on soybeans, black-eyed peas (beans), lima beans and other flatpodded beans, except Romano. Do not use the foliage from a crop treated with the Treflan/Eptam tank mix for feed or for grazing.

Dry Bean—Fall application for dry bean grown in Idaho, Oregon, and Washington:

Apply and incorporate Treflan any time between October 15 and December 31 at a broadcast rate of 1 pint per acre on coarse soils; 1¼ to 1½ pints on medium soils; and 1½ pints on fine soils. Destroy established weeds during seedbed preparation.

For the Following Crop Grouping, Use the Rate Listed Below.

Beans—(Guar and Mungbean)

Greens—Turnip greens grown for processing, Collard, Kale and Mustard greens.

Mustard—Grown for seed or processing for food in Minnesota, and North Dakota:

Apply and incorporate Treflan before planting at 1 pint per acre on coarse soils and 1½ pints on medium and fine soils.

Beans—(Lima Bean and Snap Bean):

Apply and incorporate Treflan before planting at a broadcast rate of 1 pint per acre on coarse and medium soils and 1½ pints on fine soils.

Corn (Field Corn) and Grain Sorghum (Milo)

Apply Treflan to field corn or grain sorghum (8 inches or taller) as an over-the-top or directed spray to effectively control weeds listed for Treflan.

Soil Preparation—Cultivate before a Treflan application to insure loose, friable soil, to remove established weeds, and to cover the base of plants with soil.

Application Directions—Treflan should be applied and incorporated at the recommended rates for the soil texture when the crop is well established (8 inches or taller). Treflan may be applied either as an over-the-top spray or as a directed spray. Drop nozzles should be used if foliage prevents uniform coverage of soil surface. Soil incorporation may be accomplished with only one pass of a sweep-type cultivator or a properly adjusted rolling cultivator.

The sweep-type cultivator should have 3 to 5 sweeps per row middle and be operated at 6 to 8 mph. Set the middle sweeps so as to avoid exposing untreated soil. Adjust the incorporation tools to prevent crop injury.

Broadcast Rates Per Acre

Soil Texture	Treflan E.C.
	(pints)
Coarse	¾-1*
Medium	1-1½
Fine	1½-2

Use the lower rates when you anticipate light weed pressure and the higher rates when you anticipate heavy weed pressure.

* Corn only: *Apply 1 to 1½ pints per acre in Alabama, Florida, Georgia, North Carolina, South Carolina and Virginia to control fall panicum and Texas panicum.*

Precaution: Do not apply Treflan to corn grown for seed. Do not apply Treflan to corn or sorghum as a preplant or preemergence treatment, or crop injury may occur.

Hops

Apply and incorporate Treflan while the crop is dormant. Use a broadcast rate of 1 pint per acre on coarse soils; 1¼ to 1½ pints on medium soils; and 1½ pints on fine soils and soils with 2-10% organic matter.

Mint—(Established Peppermint and Spearmint)

Apply Treflan at a rate of 1 pint per acre on coarse soils; 1¼ pints on medium soils; and 1½ pints on fine soils. Use incorporation equipment that will insure thorough soil mixing with minimum damage to the crop.

Pea—(Dry and English)—Treflan Alone

Apply and incorporate Treflan before planting at a rate of 1 pint per acre on coarse and medium soils and 1½ pints on fine soils.

Pea—Treflan/Far-Go tank mix for pea in Idaho, Oregon and Washington:

The tank mix combination of Treflan plus Far-Go will provide control of wild oat in addition to other annual grasses and broadleaf weeds controlled by Treflan.

Application Rates: Broadcast ¾ pint of Treflan per acre on coarse and medium soils; 1 pint of Treflan on fine soils. Use 1¼ quarts of Far-Go per acre for all soil textures.

Incorporation Directions: Apply and incorporate up to 3 weeks before planting. Follow recommended incorporation procedures for Treflan.

Precaution: Do not apply to lentils. Leaf crinkling and delayed maturity of peas may occur, particularly on clay points in the northwest; but this is usually more than offset by a reduction of wild oat. Do not use foliage from treated peas for feed or forage. Refer to the cautions, precautions and directions on the Far-Go label.

Pea—Fall application to dry pea and English pea in Idaho, Oregon and Washington:

Apply and incorporate Treflan any time between October 15 and December 31 at a broadcast rate of 1 pint per acre on coarse soils; 1¼ to 1½ pints on medium soils; and 1½ pints on fine soils. Destroy established weeds during seedbed preparation. Do not apply Treflan in the fall to soils which are wet or are subject to prolonged periods of flooding.

Peanut—(Spanish Peanut in Texas and Oklahoma)

Apply and incorporate Treflan before planting, at planting or immediately after planting at a broadcast rate of 1 pint per acre on coarse soils. When incorporating after planting, take care not to disturb the seed.

Peanut—Treflan/Vernam tank mix (Spanish Peanut in Texas and Oklahoma):

Treflan/Vernam tank mix effectively controls those weeds listed for Treflan alone plus these additional weeds:

Annual morningglory	Velvetleaf
Coffeeweed	Yellow nutsedge
Purple nutsedge (nutgrass)	(nutgrass)

Follow the recommended soil preparation procedures for Treflan. You may apply the Treflan/Vernam tank mix up to 10 days prior to planting. Incorporate the tank mix immediately after application. Apply Treflan/Vernam at these rates:

Broadcast Rates Per Acre

Soil Texture	Treflan E.C.	Vernam 7E
	(pints)	(pints)
Coarse	1	2⅓

Potato—(All states except Maine)

Apply and incorporate Treflan after planting, before emergence, or immediately following dragoff or after the potato plants have fully emerged.

Broadcast Rates Per Acre

Soil Texture	Treflan E.C.: Areas receiving less than 20″ average annual rainfall*	Treflan E.C.: Areas receiving greater than 20″ average annual rainfall*
	(pints)	(pints)
Coarse	1	1
Medium	1¼-1½	1½
Fine	1½	2

* *Use 1½ pints per acre on coarse and medium soils with 2-5% organic matter; use 2 pints on all soils with 5-10% organic matter.*

Set incorporation equipment so that the bed and furrow will be uniformly covered with a layer of treated soil. If the layer of treated soils is not uniform and the herbicide is concentrated over the bed, potato emergence may be retarded, and stem brittleness can occur. When applying and incorporating Treflan after potato plants have fully emerged, do not completely cover the foliage with treated soil. Likewise, do not completely cover foliage at subsequent cultivations. Be careful that incorporation machinery does not damage potato seed pieces or elongating sprouts.

Potato—Split application in Idaho, Oregon and Washington:

On all soils, apply and incorporate ¾ pint of Treflan per acre before planting and ¾ pint after planting when potato plants have fully emerged. Do not apply to soils containing 2% or more organic matter. Follow incorporation directions listed above for application to potato after planting.

Potato—Treflan/Eptam tank mix for potatoes grown in Kansas, Nebraska, North Dakota, Oklahoma, South Dakota and Texas:

The Treflan/Eptam tank mix effectively controls the following weeds in addition to those weeds controlled by Treflan.

Henbit (Spring applications)	Oat, wild
Nightshade, black	Ragweed, common
Nightshade, hairy	Smartweed, Pennsylvania
Nutsedge	Velvetleaf (Buttonweed)

Follow recommended soil preparation and application procedures for Treflan. The Treflan/Eptam tank mix may be applied after planting, but prior to crop emergence. In areas where potatoes are normally dragged off, the Treflan/Eptam tank mix should be applied and incorporated up to or immediately following drag off.

Broadcast Rates Per Acre

Soil Texture	Treflan E.C.: Areas receiving less than 20″ average annual rainfall*	Treflan E.C.: Areas receiving greater than 20″ average annual rainfall*	Eptam 7E
	(pints)	(pints)	(pints)
Coarse	1	1	1¾-7**
Medium	1-1½	1-1½	1¾-7
Fine	1-1½	1-2	1¾-7

* *Use 1½ pints per acre on coarse and medium soils with 2-5% organic matter; use 2 pints on all soils with 5-10% organic matter.*

** *Use the higher rate of Eptam 7E for nutsedge control.*

Precaution: Read the Eptam label before using. Observe cautions and limitations of products used in mixtures. Do not graze or feed forage to livestock from fields treated with Treflan/Eptam tank mix.

Potato—Treflan/Eptam application before planting in Washington, Idaho and Oregon:

Treflan/Eptam may be applied before planting at a broadcast rate of ¾ pint of Treflan per acre and 3½ pints of Eptam 7E per acre on all soil textures. Incorporate immediately.

Precaution: Do not use this tank mix both before and after planting in the same season. Read the Eptam label before using. Observe all cautions and limitations on labeling of all products used in mixtures. Do not use foliage from treated crops for feed or forage.

Safflower

Follow recommended soil preparation, application and incorporation procedures for Treflan.

Apply and incorporate Treflan in the spring before planting or in the fall between October 15 and December 31.

Broadcast Rates Per Acre

Soil Texture	Treflan E.C.: Areas receiving less than 20″ average annual rainfall*	Treflan E.C.: Areas receiving greater than 20″ average annual rainfall*
	(pints)	(pints)
Coarse	1	1
Medium	1¼-1½	1½
Fine	1½	2

* *Use 1½ pints per acre on coarse and medium textured soils and 2 pints on fine soils with 2-5% organic matter; use 2 to 2½ pints on all soils with 5-10% organic matter.*

Safflower—Fall application in Arizona, California, Idaho, Nevada, Oregon, Utah, Washington and Wyoming:

Apply and incorporate Treflan any time between October 15 and December 31. Ground may be left flat or bedded-up over winter. On bedded ground, knock beds down to desired height before planting, moving some treated soil from tops into furrows. Where soil is left flat over winter, take care during spring bedding operations to prevent turning up untreated soil. Destroy established weeds during seedbed preparation. If weeds become established in furrows due to uncovering of untreated soil during listing, destroy these weeds before planting. Apply and incorporate Treflan at a broadcast rate of $1^1/_2$ pints per acre on coarse soils; 2 pints on medium soils; and $2^1/_2$ pints on fine soils.

Do not apply Treflan in the fall to soils which are wet or are subject to prolonged periods of flooding.

Sugar Beet

Apply Treflan as a broadcast, overtop spray when plants are between 2 and 6 inches tall at a rate of 1 pint per acre on coarse soils and $1^1/_4$ to $1^1/_2$ pints on medium and fine soils. Use the higher rate for medium and fine soils in areas receiving more than 20" average annual rainfall. Set incorporation machinery to throw treated soil toward the plants in the row. Be careful that incorporation machinery does not damage the sugar beet taproot.

Precaution: Exposed beet roots should be covered with soil before a Treflan application to reduce the possibility of girdling.

Sugar Beet—Incorporation with a tine-tooth harrow in California, Colorado, Idaho, Nebraska, Oregon, Texas, Utah, Washington and Wyoming:

A properly operated tine-tooth harrow (Flextine or Melroe) can incorporate Treflan for effective weed control in sugar beet. Operate the tine-tooth harrow 2 times over the field in opposing directions at a speed of 3 to 6 mph. Set the harrow to cut 1 or 2 inches deep. Be careful that the tine-tooth harrow does not damage the sugar beet taproot. Follow recommended application procedures and broadcast rates per acre for sugar beet (see preceding paragraph).

Sugarcane—(Plant Cane)

Apply and incorporate Treflan twice a year at a broadcast rate of 2 to 4 pints per acre for all soil textures. Make the first Treflan application in the fall on firmly packed beds immediately after the seed pieces are planted. Make the second Treflan application in the spring before or shortly after the cane emerges. Loosen rain-packed beds 2 to 3 inches deep before the spring application. Take care that incorporation machinery does not damage the seed pieces or emerging shoots.

Sugarcane—Postplant in Hawaii for control of most annual grasses, including guineagrass:

Surface apply Treflan after planting (for plant cane) or after harvesting (for ratoon cane), before weeds and cane emerge. Use a broadcast rate of 6 to 8 pints per acre for all soil textures. In plant cane, the beds should be formed or rolled before application. In ratoon cane, the crop residue should be removed before application. If large amounts of crop residues are present, Treflan will not be effective. Apply Treflan just before anticipated rainfall or sprinkle irrigate immediately after application.

Sugarcane—Applications up to layby for plant cane or ratoon cane grown in Louisiana or Texas:

Apply and incorporate Treflan at a broadcast rate of 2 to 4 pints per acre for all soil textures. Do this in the spring from before or shortly after the cane emerges up to layby. Apply the Treflan after the beds have been shaved or false shaved. Loosen rain-packed beds 2 to 3 inches deep before application. Be careful that incorporation machinery does not damage seed pieces or emerging shoots. You may use a rolling cultivator or bed chopper to incorporate Treflan layby applications in sugarcane on all soil textures. Follow normal incorporation directions for the rolling cultivator. Set bed chopper to cut 3 to 4 inches deep and operate at 4 to 6 mph. Two incorporation passes are necessary.

Sugarcane—Itchgrass (Raoulgrass) control in Louisiana:

Apply and incorporate Treflan on either plant or ratoon cane at a broadcast rate of 4 pints per acre for all soil textures. Follow the directions above for sugarcane layby application in Louisiana and Texas.

Sunflower—Treflan Alone

Apply and incorporate Treflan in the spring or in the fall between October 15 and December 31. Follow recommended soil preparation, application and incorporation procedures for Treflan.

Broadcast Rates Per Acre

Soil Texture	Treflan E.C. — Areas receiving less than 20" average annual rainfall*	Treflan E.C. — Areas receiving greater than 20" average annual rainfall*
	(pints)	(pints)
Coarse	1	1
Medium	$1^1/_4$-$1^1/_2$	$1^1/_2$
Fine	$1^1/_2$	2

** Use $1^1/_2$ to 2 pints per acre on coarse and medium soils with 2-5% organic matter and 2 pints on all soils with 5-10% organic matter.*

Sunflower—Treflan/Amiben tank mix or overlay:

Follow recommended soil preparation, application and incorporation procedures for Treflan.

Amiben may be applied in a band or broadcast over sunflowers at planting in fields where Treflan has been incorporated prior to planting. Or, the Treflan/Amiben tank mix may be incorporated prior to planting. Treflan/Amiben tank mix improves mustard, smartweed, velvetleaf and ragweed control in addition to those weeds controlled by Treflan alone.

Apply Treflan/Amiben or Treflan with an Amiben overlay at the following broadcast rates per acre:

Broadcast Rates Per Acre

Soil Texture	Treflan E.C.	Amiben 2S
	(pints)	(quarts)
Coarse	1	4
Medium	$1^1/_2$	4-6*
Fine	2	4-6*

** For best control of mustard, common ragweed or black nightshade, use the 6 quart rate.*

In coarse textured soils, heavy rains on the incorporated Amiben may move it below the weed seed germination zone, and erratic weed control may result. If sufficient rain does not fall within 7 days after a preemergence application of Amiben, but there is enough soil moisture to germinate weeds and grasses, a light cultivation with a rotary hoe or similar tool will uproot these small broadleaf weeds and grasses. The shallow mixing of Amiben in the surface soil will not interfere with the action of Amiben when rains come.

Tomato

For direct-seeded tomato, apply Treflan at blocking or thinning as a directed spray to the soil between rows and beneath the plants, and incorporate. For transplant tomato, apply and incorporate Treflan before transplanting. Do not apply Treflan after transplanting.

Broadcast Rates Per Acre

Soil Texture	Treflan E.C. — Areas receiving less than 20" average annual rainfall*	Treflan E.C. — Areas receiving greater than 20" average annual rainfall*
	(pints)	(pints)
Coarse	1	1
Medium	$1^1/_4$-$1^1/_2$	$1^1/_2$
Fine	$1^1/_2$	1

** Use $1^1/_2$ pints per acre on coarse and medium textured soils and 2 pints on fine soils with 2-5% organic matter; use 2 pints on all soils with 5-10% organic matter.*

Fruit and Nut Crops and Vineyards

For areas receiving more than 20" average annual rainfall.

For new plantings of vineyards, citrus and pecan trees, apply and incorporate Treflan before planting at a broadcast rate of 1 pint per acre on coarse soils; $1^1/_2$ pints on medium soils; 2 pints on fine soils; $1^1/_2$ pints on fine soils with 2-5% organic matter; and 2 pints on soils with 5-10% organic matter.

For non-bearing, established plantings of citrus and pecan trees and bearing plantings of grapefruit, lemon, orange, pecan, tangelo and tangerine trees, apply Treflan at a broadcast rate of 2 to 4 pints per acre for all soil textures.

For areas receiving less than 20" average annual rainfall.

For new plantings of almond, apricot, citrus, nectarine, peach, pecan and walnut trees, apply and incorporate Treflan before planting at a broadcast rate of 1 pint per acre on coarse soils; $1^1/_4$ to $1^1/_2$ pints on medium soils; $1^1/_2$ pints on fine soils; $1^1/_2$ to 2 pints on soils with 2-5% organic matter; and 2 pints on soils with 5-10% organic matter.

For new plantings of vineyards, apply and incorporate Treflan before planting at a broadcast rate of 1 to $1^1/_2$ pints per acre on coarse soils; $1^1/_2$ to 3 pints on medium soils; and 3 to 4 pints on fine soils or soils with 2-10% organic matter. Do not use more than 2 pints per acre on heat-treated grape rootings.

For postplant applications on bearing or non-bearing, established plantings of vineyards and almond, apricot, grapefruit, lemon, nectarine, orange, peach, plum, prune, tangelo, tangerine and walnut trees, apply Treflan at a broadcast rate of 2 to 4 pints per acre for all soil textures. Do not apply to vineyards within 60 days of harvest.

In established plantings, apply Treflan as a directed spray to the soil and use incorporation methods not injurious to the trees or vines.

Fruit and Nut Crops and Vineyards—Rhizome johnsongrass control:

For areas receiving less than 20" average annual rainfall.

You can obtain commercially acceptable control of rhizome johnsongrass with postplant applications in bearing and non-bearing, established plantings of vineyards and almond, apricot, grapefruit, lemon, nectarine, orange, peach, pecan, tangelo, tangerine and walnut trees with a Treflan program when applied for 2 years in a row.

Soil Preparation—Work the soil thoroughly to bring the rhizomes nearer the surface.

Application—Apply Treflan at a broadcast rate of 2 quarts per acre on all soil textures each year for 2 years in a row. Do not apply to vineyards within 60 days of harvest.

Incorporation—Incorporate Treflan thoroughly with a disc set to cut 4 to 6 inches deep and operate at 4 to 6 mph. Two incorporation passes are necessary, with the second pass in a different direction from the first.

Cultivation—Some johnsongrass plants will escape. Timely cultivations are necessary to obtain commercially acceptable control. You cannot obtain commercially acceptable control with only 1 year of Treflan use.

Precautions: Do not use the 2 quart rate on new plantings, or crop injury may result. Do not interplant orchards or vineyards with other crops. If the treated vineyards and orchards are diverted to other crop uses, plant only those crops for which Treflan has been registered as a preplant treatment for the next cropping season.

Fruit and Nut Crops and Vineyards—Bindweed control in California:
Treflan can be used for the control of field bindweed in vineyards and for almond, apricot, grapefruit, lemon, nectarine, orange, peach, pecan, tangelo, tangerine and walnut trees.

Apply Treflan at a broadcast rate of 4 pints per acre on all soil textures. Treflan must be applied in the spring with a specially designed spray blade which applies a thin, concentrated layer at a soil depth of 4 to 6 inches. This layer of Treflan prevents bindweed shoots from emerging.

Land Preparation—Destroy all weeds and grasses with soil tillage before applying Treflan. This tillage is necessary to prevent trash from interfering with the operation of the spray blade.

Equipment —This operation requires a spray blade capable of running 4 to 6 inches below the surface of the soil. The spray blade should be equippped with nozzles located under the blade and directed so that the Treflan spray will be trapped under the soil which is flowing over the blade as it is pulled through the soil. Use a sufficient number of nozzles with spacing that will uniformly apply the Treflan underground in a thin, horizontal layer.

Application—Apply Treflan in 40 to 80 gallons of water per acre. Operate the spray blade at a depth of 4 to 6 inches.

Precaution: Some soils develop cracks as they dry after rainfall or irrigation. Field bindweed may emerge if the cracks extend through the Treflan layer. Prevent or eliminate cracks by shallow discing or other tillage. Avoid deep tillage which disturbs the subsurface layer. Cultivation or tillage also aids the control of germinating seeds.

Wheat (Winter) — (Idaho, Oregon and Washington)
Treflan may be applied for preplant preemergence control of cheatgrass and other annual grasses and broadleaf weeds controlled by Treflan. The growth, development and yield of winter wheat will not be adversely affected, provided the seed is placed below the zone of soil treated with Treflan.

Apply Treflan any time during a period from 3 weeks up to immediately prior to planting. Broadcast Treflan at a rate of 1½ pints per acre on coarse and medium soils and 2 pints on fine soils.

Incorporation Directions—Incorporate Treflan into the soil with a flexible tine-tooth harrow (Flextine, Melroe) set to cut 1 to 2 inches deep and operate at 3 to 6 mph. Incorporate one time within 24 hours after application, followed by a second incorporation in a different direction from the first prior to planting. Do not till the soil with a disc after the Treflan has been incorporated with a flexible tine harrow.

Seeding Directions—Use only a deep furrow or semi-deep furrow drill that will place the seed below the zone of soil into which Treflan has been incorporated.

Precaution: Wheat planted in direct contact with treated soil may suffer crop injury in the form of delayed emergence and development.

Wheat (Winter)—Fallow soil application in Washington and Oregon:
Treflan applied and shallowly incorporated into fallow soil up to four months ahead of planting will control cheatgrass and certain annual grasses and broadleaf weeds. The growth, development, or yield will not be adversely affected as long as the seed is placed below the zone of soil treated with Treflan. Use deep or semi-deep furrow drills.

Broadcast rates are 1½ pints per acre on coarse and medium soils and 2 pints on fine soils. Apply Treflan any time from May to September prior to the fall planting of winter wheat.

Incorporation—Incorporate Treflan with a flexible tine-tooth harrow (Flextine or Melroe) set to cut 1 to 2 inches deep and operated at 3 to 6 mph. For thorough incorporation, two passes of the equipment in different directions over the field are necessary. Incorporate one time within 24 hours after application followed by a second incorporation prior to seeding. Do not till the soil with a disc after Treflan has been applied with a flexible tine harrow.

Precaution: Use only deep furrow or semi-deep furrow drills. Place seed below the zone of soil into which Treflan has been incorporated. Do not plant wheat directly into the zone of soil treated with Treflan; injury to the crop or a delay in its emergence and development may occur.

Wheat (Spring), Durum and Barley—Treflan Alone
Treflan is recommended as a postplant incorporated treatment to control foxtail (pigeongrass).

Apply Treflan at a broadcast rate of 1 pint per acre on coarse and medium soils and 1½ pints on fine soils.

Plant 2 to 3 inches deep in a well-tilled seedbed. Apply Treflan after seeding but before the crop emerges. To incorporate, use flex-tine or diamond harrows operated two times in different directions, at speeds of at least 5 mph. Incorporate by operating equipment 1 to 1½ inches deep. Application and the first incorporation should be done in the same operation if possible. Both incorporations must be done within 24 hours.

Wheat (Spring), Durum and Barley—Treflan/Far-Go tank mix:
Treflan/Far-Go applied as a postplant incorporated treatment will control foxtail (pigeongrass) and wild oat.

Plant 2 to 3 inches deep in a well-tilled seedbed. Apply Treflan/Far-Go after seeding but before crop emerges. To incorporate, use flex-tine or diamond harrows two times, operated in different directions, at speeds of at least 5 mph. Incorporate by operating equipment 1 to 1½ inches deep. Application and the first incorporation should be done in the same operation if possible. If not, incorporate immediately after application.

Broadcast Rates Per Acre

Soil Texture	Treflan E.C. Barley Durum Spring Wheat	Far-Go Durum Spring Wheat	Far-Go Barley
	(pints)	(pints)	(pints)
Coarse	1	2	2½
Medium	1	2	2½
Fine	1½	2	2½

Precaution: Overapplication may result in crop injury. Read the Far-Go label carefully before using.

FALL APPLICATION

General: (Areas receiving more than 20" average annual rainfall)
See specific crop for recommendations. For all crops for which there are no specific fall application instructions and for which Treflan is recommended as a preemergence application, use the rates listed for spring applications. Do not apply Treflan in the fall for sugarbeets, potatoes and direct-seeded tomatoes.

Apply and incorporate Treflan any time between October 15 and December 31. Ground may be left flat or bedded-up over winter. On bedded ground, knock beds down to desired height before planting, moving some treated soil from beds into furrows. Where soil is left over winter, be careful not to turn up untreated soil during spring bedding operations. Destroy established weeds during seedbed preparation. If weeds become established in furrows due to uncovering of untreated soil during bedding, destroy these weeds before planting. Do not apply Treflan in the fall to soils which are wet, are subject to prolonged periods of flooding or where rice was grown the previous year.

FERTILIZER USE DIRECTIONS APPLICATION WITH LIQUID FERTILIZERS

Treflan may be mixed with most liquid fertilizer materials. The combination of Treflan with solutions and suspension-type fertilizers has provided weed and grass control equal to the same rates of Treflan applied in water. Follow Treflan label recommendations regarding rates per acre, crops, incorporation directions, special instructions, cautions and special precautions.

Individual state regulations relating to liquid fertilizer mixing, registration, labeling and applications are the responsibility of the individual and/or company selling the fertilizer and chemical mixture.

Testing for Tank Mix Compatibility in Liquid Fertilizers:
Treflan alone or in tank mixture with dry flowables, wettable powders (WP), aqueous suspensions (AS), flowables (F), liquids (L), or solutions (S) may not combine properly with some fluid fertilizer materials. Small quantities should always be tested before full-scale mixing. This will determine whether a compatibility agent is needed, and which agent does the best job. The seven agents listed on page 78 have been thoroughly tested. There are many other surfactants on the market which were not designed for use with liquid fertilizers. Use the following test to select the correct agent for your mixture.

1. Put 1 pint of the liquid fertilizer in a quart jar.

2. Add 2 to 4 teaspoonful(s) of the dry flowable, WP, AS, F or L formulation (depending on the recommended rate per acre) to the liquid fertilizer. Close jar and agitate until the materials are dispersed evenly in the fertilizer. If the materials do not disperse well, it may be necessary to slurry the chemicals in water before adding to the fertilizer.

3. After dispersing the materials (Step 2), add 3 to 4 teaspoonsful of Treflan to the jar and shake well. Add solution herbicides to the mixture last and agitate. Observe the jar for about 10 minutes. If the materials rise to the surface and form a thick layer (oily curds) which will not redisperse when agitated, a compatibility agent is needed. If the mixture is easily redispersed to its original state with slight agitation, no agent is needed but good agitation must be provided in the fertilizer spray tank.

4. If the need for a compatibility agent is shown in Step 3: Using a clean quart jar, start at Step 1 above, add ½ teaspoon of the compatibility agent to the liquid fertilizer, mix well, then repeat Steps 2 and 3.

An effective compatibility agent will cause the mixture to remain uniformly mixed with little or no separating or oil rising to the surface for one half hour or longer. If slight separation does occur, 2 to 3 inversions of the jar should give a uniform remix. If oily curds form which will not redisperse, more agent or another agent should be tried.

Use a clean jar for each test. The compatible mixture will have a uniform appearance and will be relatively easy to keep mixed with gentle agitation of the jar.

Liquid Fertilizer Mixing Instructions
Treflan E.C. in Liquid Fertilizer

General—Emulsifiable concentrates, such as Treflan E.C., can be mixed with liquid fertilizers. In all cases, continuous agitation is required to prevent the Treflan from rising to the surface as an oily layer. When necessary, (see Testing for Tank Mix Compatibility in Liquid Fertilizers, on page 71) a compatibility agent can be used to cause the Treflan to emulsify properly (i.e., have a milky appearance rather than oily layer). The use of compatibility agents is especially important when tank mixing emulsifiable concentrates (E.C.) with dry flowables, wettable powders (WP), aqueous suspensions (AS), flowables (F), liquids (L), or solutions (S) in liquid fertilizer. If the emulsion is not properly formed and the Treflan rises to the surface of the fertilizer as an oil ("oils out"), the oil may combine with the wettable powder, flowable, or suspension to form oily curds (viscous phase) which is difficult to redisperse. Any one of the compatibility agents listed below is helpful in causing liquid concentrates to form non-oiling mixtures with liquid fertilizers. These compatibility agents can be used at rates as low as 1½ to 2 pints per ton of liquid fertilizer and should be mixed well with the fertilizer before adding the liquid concentrate. Read the label on the compatibility agent and follow the directions.

1. Sponto 168D (Witco Chemicals Co., Chicago, IL)
2. Compat (Farm Chemicals, Inc., Aberdeen, NC)
3. Unite (Hopkins Ag Chemical, Madison, WI)
4. T-Mulz 734-2 (Thompson-Hayward Chemical Co., Kansas City, MO)
5. Rigo Compatibility Agent (Rigo Company, Buckner, KY)
6. Amoco Spray Mate™ (Amoco Oil Co., Chicago, IL)
7. Kem-Link (Universal Coop, Minneapolis, MN)

All of the above are phosphate ester-type surfactants designed to be used with liquid fertilizers. They usually do not work as compatibility agents in tank mixtures in plain water.

Application—Spread the fertilizer/pesticide mixture with a properly calibrated applicator. Be certain the material is applied uniformly to the soil surface.

Incorporation—Follow normal Treflan incorporation procedures.

TREFLAN APPLICATION WITH DRY BULK FERTILIZERS

General—Dry bulk fertilizers may be impregnated or coated with Treflan. Application of dry bulk fertilizers impregnated with Treflan has provided weed and grass control equal to the same rates of Treflan applied in water.

All Treflan label recommendations regarding rates per acre, approved crops, incorporation, special instructions, cautions and special precautions must be followed.

Apply a minimum of 200 pounds per acre of dry fertilizer impregnated with Treflan at the recommended rates. Any commonly used dry fertilizers can be used for Treflan impregnation except coated ammonium nitrate and straight limestone. These materials will not absorb the herbicide. Blends containing mixtures of these materials can be impregnated.

Impregnation—Use any closed drum, belt, ribbon or other commonly used dry bulk fertilizer blender. Apply Treflan uniformly to the fertilizer.

Rates

Check specific crop recommendations for the rate of Treflan per acre. See the rate table which follows to determine the amount of Treflan to be impregnated into a ton of dry bulk fertilizer based on the amount of fertilizer which will be applied per acre. (See rate chart on page 80).

Application

Spread the fertilizer/chemical mixture with a properly calibrated applicator. Be certain the material is applied uniformly to the soil surface.

Incorporation

Follow Treflan incorporation procedures.

Rate Chart for Impregnating Fertilizer with Treflan E.C.
(Treflan Added to a Ton of Fertilizer)

Fertilizer Rate Per Acre	Treflan Rate Per Acre				
	1 pint	1½ pints	2 pints	3 pints	4 pints
200 pounds	5 qts. per ton	7½ qts. per ton	10 qts. per ton	15 qts. per ton	20 qts. per ton
250 pounds	4 qts. per ton	6 qts. per ton	8 qts. per ton	12 qts. per ton	16 qts. per ton
300 pounds	3⅓ qts. per ton	5 qts. per ton	6⅔ qts. per ton	10 qts. per ton	13⅓ qts. per ton
350 pounds	2¾ qts. per ton	4¼ qts. per ton	5¾ qts. per ton	8½ qts. per ton	11½ qts. per ton
400 pounds	2½ qts. per ton	3¾ qts. per ton	5 qts. per ton	7½ qts. per ton	10 qts. per ton
450 pounds	2¼ qts. per ton	3⅓ qts. per ton	4½ qts. per ton	6⅔ qts. per ton	9 qts. per ton

For rates other than those listed above, use the following formula to calculate the amount of Treflan to be impregnated on a ton of dry bulk fertilizer:

$$\text{Pints Treflan Per Acre} \times \frac{1000}{\text{Lbs. Fertilizer Per Acre}} = \text{Quarts Treflan Per Ton of Fertilizer}$$

All individual state regulations relating to dry bulk fertilizer blending, registration, labeling and application are the responsibility of the individual and/or company selling the fertilizer and chemical mixture.

TREFLAN PERFORMANCE GUARANTEE — A LIMITED WARRANTY OF TREFLAN WEED AND GRASS CONTROL

Elanco Products Company warrants that Treflan will control the weeds and grasses as indicated on its current label when used according to label directions and subject to the following limitations:

1. Purchaser must notify Elanco promptly if a lack of commercially acceptable control occurs. Notice must be given within forty-five (45) days after the first planting or after application of Treflan to the treated crop, whichever occurs later.
2. An Elanco representative must be satisfied that Purchaser used Treflan according to the label directions.
3. The Purchaser must provide Elanco representative with an invoice or other satisfactory record which shows the price and quantity of Treflan purchased.
4. Elanco's liability will be limited to a refund of the purchase price of the Treflan applied to the acreage on which commercially acceptable control was not obtained.
5. Void outside the USA.

DISCLAIMER OF WARRANTIES

The warranty printed above is the only warranty applicable to this product. All other warranties, expressed or implied, including, but not limited to, any implied warranties of merchantability and fitness for a particular purpose, are disclaimed. This disclaimer of warranties does not apply where prohibited by operation of law.

LIMITATION OF DAMAGES

Elanco's liability, whether in contract, warranty, tort, negligence, strict liability or otherwise, shall not exceed the return of the amount of the purchase price of the Treflan and under no circumstances shall Elanco be liable for special, indirect or consequential damages. This limitation of damages does not apply where prohibited by operation of law.

INHERENT RISKS OF USE

Failure to carefully follow the directions for use of Treflan, as well as other factors such as seedling disease, cold weather, deep planting, excessive moisture, high salt concentration or drought, may result in unsatisfactory weed control or crop injury. Reduced yields may result under these conditions.

Amiben®—(chloramben, Union Carbide)
Basagran®—(BASF Wyandotte)
Blazer®—(acifluorfen, Rohm and Haas)
Canopy®—(chlorimuron ethyl plus metribuzin, E. I. duPont)
Caparol®—(prometryn, Ciba-Geigy)
Classic®—(chlorimuron ethyl, E. I. duPont)
Cotoran®—(fluometuron, Ciba-Geigy)
Dual®—(metolachlor, Ciba-Geigy)
Eptam®—(EPTC, Stauffer)
Far-Go®—(traillate, Monsanto)
Gemini™—(chlorimuron ethyl plus linuron, E. I. duPont)
Karmex®—(diuron, E. I. duPont)
Lasso®—(alachlor, Monsanto)
Lasso® Micro-Tech™—(alachlor, Monsanto)
Lexone®—(metribuzin, E. I. duPont)
Lorox®—(linuron, E. I. duPont)
Lorox® Plus—(chlorimuron ethyl plus linuron, E. I. duPont)
Preview®—(chlorimuron ethyl plus metribuzin, E. I. duPont)
Sencor®—(metribuzin, Bayer, GmbH)
Tackle®—(acifluorfen, Rhone-Poulenc)
Vernam®—(vernolate, Stauffer)

SUPPLEMENTAL LABELING

TREFLAN® EC (EPA REG. NO. 1471-35)
TREFLAN® M.T.F.™ (EPA REG. NO. 1471-116)
TREFLAN® 5 (EPA REG. NO. 1471-120)
TREFLAN® TR+10® (EPA REG. NO. 1471-143)

TREFLAN FOR FOXTAIL CONTROL IN SPRING SEEDED BARLEY USED AS A COVER CROP IN THE ACREAGE CONSERVATION RESERVE PROGRAM

TREFLAN may be applied in the spring as a preplant soil incorporated treatment for foxtail control in spring seeded barley grown on land used in acreage conservation reserve programs.

Apply and incorporate TREFLAN following recommended procedures in the product labels. TREFLAN E.C. and TREFLAN M.T.F. should be applied at the rate of one (1) pint per acre on coarse-textured soils and one and one-half (1½) pints per acre on medium- and fine-textured soils. TRE-

FLAN 5 should be applied at the rate of eight-tenths (0.8) pints per acre on coarse-textured soils and one and two-tenths (1.2) pints per acre on medium- and fine-textured soils. TREFLAN TR-10 should be applied at the rate of five (5) pounds per acre on coarse-textured soils and seven and one-half (7½) pounds per acre on medium- and fine-textured soils. Do not exceed this rate or crop injury may occur.

Planting Directions

Barley should be seeded approximately two (2) inches deep.

Precaution

Use of this practice may result in a slight stand reduction. Follow the most severe grazing restrictions imposed either by the pesticide label or by the USDA Acreage Conservation Reserve Program, whichever is longest. Consult the local ASC office or other state agency to determine the period of the USDA grazing restriction.

Read the TREFLAN labels for additional directions, precautions and limitations before use.

TREFLAN® E.C., M.T.F.™, 5, TR-10® (trifluralin, Elanco)

EA 8163 5/88

SUPPLEMENTAL LABELING

TREFLAN® E.C. (E.P.A. REG. NO. 1471-35)
TREFLAN® M.T.F.™ (E.P.A. REG. NO. 1471-116)
TREFLAN® 5 (E.P.A. REG. NO. 1471-120)

TREFLAN for PARTIAL CONTROL OR SUPPRESSION of Annual Brome Species (Cheatgrass, Downy Brome, Japanese Brome, Hairy Chess) And Jointed Goatgrass in Winter Wheat in Colorado, Kansas, Nebraska, and Wyoming

TREFLAN may be applied as a preplant incorporated treatment for PARTIAL CONTROL OR SUPPRESSION of annual Brome species (cheatgrass, downy brome, Japanese brome, hairy chess) and jointed goatgrass in winter wheat in Colorado, Kansas, Nebraska, and Wyoming.

Apply TREFLAN anytime during a period from three (3) weeks before planting up to immediately prior to planing.

Broadcast Rates Per Acre

Soil Texture	Treflan E.C./M.T.F. (pints)	Treflan 5 (pints)
Coarse	1 - 1½	0.8 - 1.2
Medium	1 - 1½	0.8 - 1.2
Fine	1½	1.2

Use the higher rate in the rate range where heavy weed populations are anticipated or where medium to high crop residues are present.

Incorporation and Planting Directions

TREFLAN should be incorporated with tillage equipment (flexible tine-tooth harrow or springtooth harrow such as Flex-Tyne® or Crust Buster®) that mixes the soil no more than 1-2 inches deep. The grain drill (disc drill or hoe drill) can serve as the incorporation implement. Do not use discs, undercutters or field cultivators for incorporation. Incorporate one time within 24 hours after application. Use a grain drill that will place the seed below the zone of soil into which TREFLAN has been incorporated.

One pass incorporation is adequate; however, where the grain drill is used as the incorporation tool, mounting a springtooth harrow in front of the drill can enhance performance. Where a tillage tool is used to incorporate prior to planting, THE WHEAT MUST BE SEEDED BELOW THE TREFLAN TREATED SOIL OR CROP INJURY MAY RESULT. The wheat seed should be placed at least 1½-2 inches deep.

Precautions

1. Crop injury in the form of delayed emergence and development may result from planting wheat in direct contact with treated soil.
2. Do not use undercutters, field cultivators, chisel plows or discs as incorporation tools. Any implement that incorporates TREFLAN deeper than the seeded wheat will contribute to crop injury.
3. Use of seeding equipment that does not place the seed below the TREFLAN treated soil layer will result in crop injury.
4. Use of TREFLAN in accordance with this label may result in some crop stand reduction but does not normally adversely affect yield.
5. Heavy rainfall prior to wheat emergence can cause soil compaction and soil crusting resulting in delayed emergence, stand reduction, stunting and yield loss.
6. Read the TREFLAN label for additional precautions and limitations before use.

TREFLAN® (trifluralin, Elanco)

EA9238 9/89

SUPPLEMENTAL LABELING

TREFLAN® E.C. (E.P.A. REG. NO. 1471-35)
TREFLAN® M.T.F.™ (E.P.A. REG. NO. 1471-116)
TREFLAN® 5™ (E.P.A. REG. NO. 1471-120)

TREFLAN for Weed Control in Forage Legumes Used as Cover Crops in the Acreage Conservation Reserve Program

TREFLAN can be used a a preplant incorporated broadcast applied treatment for preemergence control of many annual grasses and broadleaf weeds in direct seeded forage legumes.

Apply and incorporate TREFLAN following recommended procedures in the label. Apply TREFLAN at the following broadcast rates per acre:

Soil Texture	Treflan E.C./M.T.F. (Pints)	Treflan 5 (Pints)
Coarse	1	0.8
Medium	1-1½	0.8-1.2
Fine	1½	1.2

Follow the most severe grazing restrictions imposed either by the pesticide label or by the USDA Conservation Use Program, whichever is longest. Consult the local ASC committee or other State Agency to determine the period of the USDA grazing restriction.

Precaution:

Some stand reductions may occur with this use; however, excellent weed control will allow time for establishment of a quality stand.

TREFLAN® E.C., M.T.F.™, 5™ (trifluralin, Elanco)

EA8123 2/88

TREFLAN® TR-10 for Dodder Control in Established Alfalfa

SLN #CA 870029

REGISTRATION FOR SPECIAL LOCAL NEED AMENDED*
FOR DISTRIBUTION AND USE ONLY WITHIN CALIFORNIA

Trade Name: TREFLAN TR-10 **EPA Reg. No. 1471-143-AA**

Firm Name: Elanco Products Company

DIRECTIONS FOR USE

It is a violation of federal law to use this product in a manner inconsistent with its labeling.

Location: Statewide

Crop/Site/Commodity: Alfalfa

Target Pest/Problem: Dodder Weed

Dosage: Apply 20 lbs. product (2 lbs. a.i.) per acre prior to weed emergence. Apply an additional 20 lbs. product (2 lbs. a.i.) per acre approximately 60 days later.

A single rainfall or overhead sprinkler irrigation of 0.5 inches or more or flood irrigation after application is required to activate TREFLAN. If this does not occur within 10 days after application, TREFLAN should be activated using incorporation equipment that will insure thorough soil mixing with minimum damage to the established alfalfa.

Dilution Rate: See "Dosage" above

Method of Application: Ground or Aerial

Frequency/Timing of Application: Apply the first application to established alfalfa stands prior to weed emergency. Make a second application approximately 60 days later.

Field Reentry After Application: Do not enter treated area until spray residues have dried.

Preharvest Interval: N/A

Other Requirements: 1. In the season following a 20 pound per acre TREFLAN TR-10 treatment, where established alfalfa is to be rotated to another crop, plant only those crops for which TREFLAN can be applied as a preplant treatment or injury may result.

Valid until withdrawn, suspended or cancelled by the EPA, the manufacturer, the 24(c) registrant, or the Department of Food and Agriculture.

All applicable directions, restrictions, and precautions on the EPA-registered label(s) are to be followed.

This labeling must be in the possession of the user at the time of pesticide application.

*This supersedes the SLN issued July 15, 1987.

Authorization must be obtained from the county agricultural commissioner prior to this use. This does not constitute a recommendation of the Department of Food and Agriculture and will not prevent quarantine action if illegal residues are found on or in the crop.

Neither the Department nor the county agricultural commissioner, the manufacturer or forumulator makes any warranty of merchantability, fitness of purpose, or otherwise, expressed or implied, concerning the use of a pesticide in accordance with these provisions. The user and/or grower acknowledges the preceding disclaimer and accepts liability for any possible damage resulting from this use. Do not use in mixture with other pesticides unless provided for in the labeling. Trial on a small area to check out unanticipated problems is suggested.

24(c) Registrant: U.C. Cooperative Extension - Los Angeles County
1110 West Avenue J, Room 5
Lancaster, CA 93534

EPA SLN No.: CA 870029

This is a specimen label. Elanco is not responsible for the accuracy of the information contained herein. As labels are subject to revision, always carefully read and follow the label on the product container.

TREFLAN® M.T.F.™

HERBICIDE

A selective herbicide for the preemergence control of annual grasses and broadleaf weeds

ACTIVE INGREDIENT:
trifluralin* (α,α,α-trifluoro-2,6-dinitro-*N,N*-dipropyl-*p*-toluidine) . . 41.2%
INERT INGREDIENTS: 58.8%
TOTAL: 100.0%

Contains 4 pounds active ingredient per gallon.

***Treflan®, M.T.F.™**—(trifluralin, multiple temperature formulation, Elanco)

Manufactured using the processes of U.S. Patents 4,120,905 and 4,226,789.

EPA Reg. No. 1471-116
ID 5560

Literature revised July 30, 1987
PI 2696 AMP

PRECAUTIONARY STATEMENTS

KEEP OUT OF REACH OF CHILDREN

CAUTION

HAZARDS TO HUMANS AND DOMESTIC ANIMALS

Avoid contact with skin, eyes and clothing. Avoid breathing vapors. Harmful if swallowed, inhaled, or absorbed through the skin. Treflan M.T.F. may cause skin sensitization reactions in certain individuals. Use eye protection and protective clothing such as coveralls, a long sleeved shirt, and impermeable gloves when handling this product. Do not contaminate foodstuffs or feeds.

First Aid: In case of contact, immediately flush eyes or skin with plenty of water. For eyes, call a physician. Remove and wash contaminated clothing before reuse.

ENVIRONMENTAL HAZARDS

This pesticide is toxic to fish. Do not apply directly to water or wetlands (swamps, bogs, or marshes). Drift or runoff from treatment areas may be hazardous to aquatic organisms in neighboring aquatic sites. Do not contaminate water by cleaning equipment or disposal of waste.

STORAGE AND DISPOSAL

Storage: Store in original container only. May be stored in unheated facilities. Do not store near heat or open flame. In case of leak or spill, use absorbent materials to contain liquids and dispose as waste.

Disposal: Do not contaminate water, food or feed by storage or disposal. Wastes resulting from the use of this product may be disposed of on site or at an approved waste disposal facility.

Plastic Container Disposal: Triple rinse (or equivalent). Then offer for recycling or reconditioning, or puncture and dispose of in a sanitary landfill, or incineration, or, if allowed by state and local authorities, by burning. If burned, stay out of smoke.

Metal Container Disposal: Triple rinse (or equivalent). Then offer for recycling, or reconditioning, or puncture and dispose of in a sanitary landfill, or by other procedures approved by state and local authorities.

SPECIAL PRECAUTIONS

Applied according to directions and under normal growing conditions, Treflan will not harm the treated crop. Overapplication may result in crop injury or a soil residue. Uneven application or improper soil incorporation of Treflan can result in erratic weed control or crop injury. Seedling disease, cold weather, deep planting, excessive moisture, high salt concentration or drought may weaken crop seedlings and increase the possibility of damage from Treflan. Under these conditions, delayed crop development or reduced yields may result.

In Arizona, Colorado, California, Idaho, Nevada, New Mexico, Oregon, Utah, Washington and Wyoming, sugarbeets, red beets or spinach should not be planted for 12 months after a spring application or for 14 months after a fall application of Treflan. Plow the land to a depth of 12 inches prior to planting sugar beets to prevent the possibility of crop injury. Sorghum (milo), proso millet, corn, oats and annual or perennial grass crops or grass mixtures should not be planted for 14 months after a spring application or for 16 months after a fall application of Treflan to avoid crop injury. If land has not been irrigated, do not plant any of these crops for 18 months after a spring application or 20 months after a fall application of Treflan.

In those portions of Kansas, Nebraska, North Dakota, Oklahoma, South Dakota and Texas where at least 20 inches of irrigation and/or rainfall (total) was used to produce the crop, sorghum, oats and annual or perennial grass crops or grass mixtures should not be planted for 12 months after an application of Treflan.

If less than 20 inches of total water was used to produce the crop, do not plant sorghum, proso millet, oats and annual or perennial grass crops or grass mixtures for 18 months after an application of Treflan. Cool, wet weather conditions during the early stage of growth may increase the possibility of injury to sorghum.

In all other areas receiving greater than 20" rainfall per year, moldboard plow before planting sugar beets where a spring application of Treflan was made the previous season. Also note planting restrictions listed in the section on control of rhizome johnsongrass and other higher rate programs.

Vegetable crops other than those listed on this label should not be planted within 5 months following the application of Treflan.

Do not use Treflan on any crop grown in Pecos county, or Reeves county, Texas or the state of Montana.

Do not apply Treflan M.T.F. through any type of irrigation system.

DIRECTIONS FOR USE

Treflan is a preemergence herbicide which is incorporated into the soil to provide long-lasting control of many annual grasses and broadleaf weeds. Treflan control weeds as they germinate. Treflan will not control established weeds.

Weeds and Grasses Controlled by Treflan

Grasses	
Annual bluegrass	*Poa annua*
Barnyardgrass (Watergrass)	*Echinochloa* sp.
Brachiaria (Signalgrass)	*Brachiaria* sp.
Bromegrass (Cheatgrass) (Downy brome)	*Bromus tectorum*
Cheat (Chess)	*Bromus secalinus*
Crabgrass (Large crabgrass) (Smooth crabgrass)	*Digitaria* spp.
Foxtail (Bottlegrass) (Bristlegrass) (Giant foxtail) (Green foxtail) (Foxtail millet) (Pigeongrass) (Robust foxtail) (Yellow foxtail)	*Setaria* spp.
Goosegrass (Silver crabgrass) (Silvergrass) (Wiregrass) (Yardgrass)	*Eleusine indica*
Johnsongrass (from seed) (Rhizome—see page 23 for special instructions for control in cotton and see page 32 for special instructions for control in soybeans)	*Sorghum halepense*
Junglerice	*Echinochloa colonum*
Panicum	
Fall panicum (Spreading panicgrass—see page 23 for special instructions in cotton and see page 31 for special instructions in soybeans)	*Panicum dichotomiflorum*
Guineagrass (See page 64 for special instructions)	*Panicum maximum*
Texas panicum (Buffalograss) (Coloradograss)	*Panicum texanum*

Itchgrass — *Rottboellia exaltata*
(Raoulgrass)
(See page 65 for special instructions)
Red rice — *Oryza sativa*
(See page 33 for suppression or partial control directions)
Sandbur — *Cenchrus incertus*
(Burgrass)
Sprangletop — *Leptochloa filiformis*
Stinkgrass — *Eragrostis cilianensis*
(Lovegrass)
Wild cane — *Sorghum bicolor*
(Shattercane)
(See page 37 for special instructions)
Woolly cupgrass — *Eriochloa villosa*

Broadleaf Weeds

Carpetweed — *Mollugo verticillata*
Chickweed — *Stellaria media*
Field bindweed — *Convolvulus arvensis*
(See page 70 for special instructions)
Florida pusley — *Richardia scabra*
(Florida purslane)
(Mexican clover)
(Pusley)
Goosefoot — *Chenopodium hybridum*
Henbit — *Lamium amplexicaule*
(fall application only)
Knotweed — *Polygonum aviculare*
Kochia — *Kochia scoparia*
(Fireweed)
(Mexican fireweed)
Lambsquarters — *Chenopodium album*
Pigweed — *Amaranthus* spp.
(Carelessweed)
(Prostrate pigweed)
(Redroot)
(Rough pigweed)
(Spiny pigweed)
Puncturevine — *Tribulus terrestris*
(Western U.S. only)
(Caltrop)
(Goathead)
Purslane — *Portulaca oleracea*
Russian thistle — *Salsola kall*
(Tumbleweed)
Stinging nettle — *Urtica dioica*
(Nettle)

SOIL PREPARATION

Crop Residues or Existing Weeds: Ground cover, such as crop residues or existing weeds, can interfere with the incorporation of Treflan into the soil. A manageable level of such ground cover will allow the Treflan to be uniformly incorporated into the top 2 to 3 inches of soil. If the level of the ground cover is such that this cannot be done, you must till the soil prior to the application of Treflan.

Roughness: The soil surface should be smooth enough so that you can operate a sprayer and incorporation equipment efficiently and at speeds which insure a uniform application and incorporation of Treflan.

General Soil Conditions: To assure uniform incorporation of Treflan, soil moisture conditions should be such that large clods can be broken up during the incorporation process.

Soil Texture Guide

The amount of Treflan you apply will vary with the soil texture and organic matter. A fine textured soil will require more Treflan per acre than a coarse soil. Choose the proper rate for each application based on the following soil texture group and specific crop recommendations. Do not exceed recommended rates.

Soil Texture	Soil Classification
Coarse soils: (Light)	Sand, loamy sand, sandy loam
Medium Soils:	Loam, silty clay loam*, silt loam, silt, sandy clay loam*
Fine Soils:	Clay, clay loam, silty clay loam*, silty clay, sandy clay, sandy clay loam*

**Silty clay loam and sandy clay loam soils are transitional soils and may be classified as either medium or fine textured soils. If silty clay loam or sandy clay loam soils are predominately sand or silt, they are usually classified as medium textured soils. If they are predominately clay, they are usually classified as fine textured soils.*

MIXING AND APPLICATION DIRECTIONS

TREFLAN M.T.F. ALONE IN WATER

Start with a clean spray tank. Fill sprayer $\frac{1}{3}$ to $\frac{1}{2}$ full with clean water. Start agitation. Add correct quantity of Treflan, continue agitation and finish filling the tank.

TREFLAN M.T.F. TANK MIX IN WATER

Vigorous, continuous agitation is required for all tank mixes. (Sparger pipe agitators generally provide the best agitation in spray tanks). Avoid stirring or splashing air into the mixture during filling to prevent foaming. To do this, place the end of the fill pipe below the surface of the water in the spray tank. Do not allow the mixture to siphon back into the water source.

Mixing Order: Fill the tank $\frac{1}{4}$ to $\frac{1}{3}$ full with clean water. Start the agitation. Add dry flowables, wettable powders (WP), aqueous suspensions (AS), flowables (F), and liquids (L) to the water and agitate until the product(s) are completely dispersed in the water. Allow additional mixing and dispersion time when using dry flowable products. Continue agitation and fill tank to $\frac{3}{4}$ full, add the Treflan M.T.F., mix thoroughly. Then add any solution (S) formulations, agitate and finish filling. Maintain agitation during filling and through application. If spraying and agitation must be stopped before the tank is empty, the materials may settle to the bottom. In this case, it is important to resuspend all of the material in the bottom of the tank before continuing the spray application. A sparger agitator is particularly useful for this purpose. Sometimes it is more difficult to resuspend settled material than it is to suspend it originally.

Read and carefully follow all label instructions for each material added to the tank. Premixing dry and flowable formulations with water (slurrying) and pouring the slurry through a 20 or 35 mesh wetting screen in the top of the tank will help assure good initial dispersion in the tank water. Line screens in the tank should be no finer than 50 mesh (100 mesh is finer than 50 mesh).

If you see a buildup of material on the walls of the spray tank, wash the tank with soapy water between fillings. Rinse and continue the spraying operation. Clean the tank, lines, and screens thoroughly after use.

As the spray volume decreases, the importance of accurate calibration and uniform application increases. Check the sprayer daily to insure proper calibration and uniform application. Do not apply Treflan when the wind can cause drifting of spray particles which can result in non-uniform application. Treflan should not be applied to soils which are wet or are subject to prolonged periods of flooding as poor weed control may result.

Ground application: Apply Treflan in 5 to 40 gallons of water or liquid fertilizer per acre (broadcast basis), using any properly calibrated, low pressure herbicide sprayer that will apply the spray uniformly.

Aerial application: Apply Treflan in 5 to 10 gallons of water or liquid fertilizer per acre. Adjust pump pressure, nozzle arrangements, speed and height to provide a uniform application to the soil surface. Use swath markers or flagmen to assure proper application spray widths.

INCORPORATION DIRECTIONS

Incorporation Equipment—General Directions

Use incorporation equipment that mixes Treflan into the top 2 to 3 inches of the final seedbed, or erratic weed control and/or crop injury may result. Incorporation equipment such as a disc will mix Treflan approximately half as deep as the equipment is set to operate. For example, a disc set to cut 4 inches deep will incorporate most of the Treflan within the top 2 inches of soil.

Incorporation Before Planting

Treflan must be incorporated one time within 24 hours after application. Then any time prior to planting, a second incorporation is necessary, this time running the equipment in a different direction from the first. You should incorporate the Treflan uniformly into the top 2 to 3 inches of the final seedbed.

Incorporation After Planting

Check specific crop for incorporation directions after planting.

Incorporation In Bedded Culture

For effective weed control, Treflan needs to be incorporated into the top 2 to 3 inches of the final seedbed.

Application prior to bedding: Apply Treflan and incorporate it one time with recommended equipment. The bedding operation serves as the second incorporation. Do not expose untreated soil during post-bedding operations.*

Application after bedding: Knock off beds to planting height before applying Treflan. Apply Treflan and incorporate it with recommended equipment that will conform to the bed shape. Do not leave untreated soil exposed.*

** Avoid removal of treated soil from the seedbed before or during the planting operation. This would expose untreated soil, allowing weeds to germinate in the drill row.*

RECOMMENDED EQUIPMENT

Any recommended incorporation tool may be used alone or in combination with any other recommended tool. Two incorporation passes are required unless specifically stated. The second incorporation should not be deeper than the first.

Disc: set to cut 4 to 6 inches deep and operate at 4 to 6 mph.

Field Cultivator: set to cut 3 to 4 inches deep and operate 5 mph or more. A field cultivator is defined as an implement with 3 to 4 rows of sweeps, spaced at intervals of 7 inches or less and staggered so that no soil is left unturned. Chisel points should not be used.

Combination Seedbed Conditioners: set to cut 3 to 4 inches deep and operate at a speed of at least 5 mph. These implements are defined as three or more tillage devices combined and used as a single tool. For example, 2 to 3 rows of field cultivator C- or S-shaped shanks with an effective sweep spacing of 6 to 9 inches (staggered so that no soil is left unturned), followed by a spike-tooth or flextine harrow, followed by a ground-driven reel or basket.

Rolling Cultivator: set to cut 2 to 4 inches deep and operate at 6 to 8 mph. Rolling cultivators are adequate for use on coarse and medium textured soils only, except when used in sugarcane where the rolling cultivator may be used on fine textured soils.

Bed Conditioner (Do-All): set to cut 2 to 4 inches deep and operate at 4 to 6 mph. The do-all is adequate for use on coarse and medium textured soils only. When using the do-all in bedded culture, only one incorporation pass is required. However, two passes with a do-all are required in flat planted culture.

Mulch Treader (other similar disc-type implements): set to cut 3 to 4 inches deep and operate at 5 to 8 mph.

P.T.O. Driven Equipment (tillers, cultivators, hoes): adjust to incorporate Treflan into the top 2 to 3 inches of the seedbed with rotors spaced to provide a clean sweep of the soil. Only one incorporation is necessary. P.T.O. driven equipment should not be operated at a speed greater than 4 mph.

Other equipment, including the flexible tine-tooth harrow (Flextine, Melroe), is also recommended but only for the special programs for which it is specified in this label.

CULTIVATION AFTER PLANTING

Soil treated with Treflan may be shallow cultivated without reducing the weed control activity of Treflan. Do not cultivate deeper than the treated soil since this may bring untreated soil to the surface, and poor weed control may result.

CROP RECOMMENDATIONS

These recommendations are given as the broadcast rates of Treflan per acre. For band applications, decrease the amount of Treflan in proportion to the amount of surface treated per acre. Apply any time after January 1 when the soil can be worked and is suitable for good incorporation. Treflan can be applied in the fall—see specific crop for recommendations. For general fall application directions where specific recommendations are not given, see page 76. Where a rate range is shown, use the lower rate for coarser soils or soils with lower organic matter. Treflan should not be used on soils containing more than 10% organic matter.

COTTON—TREFLAN ALONE

Treflan can be applied and incorporated before or at planting, immediately after planting, or at layby.

Cotton—Preemergence: Broadcast Rates Per Acre

	Treflan M.T.F.	
Soil Texture	Areas receiving less than 20" average annual rainfall*	Areas receiving greater than 20" average annual rainfall*
	(pints)	(pints)
Coarse	1	1
Medium	1¼-1½	1½
Fine	1½	2

**Use 1½ pints per acre on coarse and medium textured soils and 2 pints on fine soils with 2-5% organic matter; use 2 to 2½ pints on all soils with 5-10% organic matter.*

Cotton—Postplant:

When incorporating Treflan after planting (postplant), be careful not to disturb the seed.

Cotton—Layby:

Apply and incorporate Treflan any time up to layby, but not less than 90 days before harvest. Direct the layby applications onto the soil between the rows and beneath emerged cotton plants. Use the same rates as for a preemergence application.

Cotton—Fall Application:

Apply and incorporate Treflan any time between October 15 and December 31. The ground may be left flat or bedded-up over winter. On bedded ground, knock beds down to desired height before planting, moving some treated soil from beds into furrows. Where soil is left flat over winter, be careful not to turn up untreated soil during spring bedding operations. Destroy established weeds during seedbed preparation. If weeds become established in furrows due to uncovering of untreated soil during bedding, destroy these weeds before planting. In the fall, do not apply Treflan to soils which are wet or subject to prolonged periods of flooding.

Broadcast Rates per Acre—Fall application only:

In Alabama, Arkansas, northern Florida, Georgia, Louisiana, Mississippi, southeastern Missouri bootheel, North Carolina, New Mexico, Oklahoma, South Carolina, Tennessee and Texas, apply and incorporate Treflan at a broadcast rate of 2 pints per acre on coarse and medium soils and 2½ pints on fine soils.

In Arizona, California and Nevada, apply and incorporate Treflan at a broadcast rate of 1½ pints per acre on coarse soils; 2 pints on medium soils; and 2½ pints on fine soils.

For cotton grown in other states, apply and incorporate Treflan at a broadcast rate of 1 pint per acre on coarse soils; 1½ pints on medium soils; 2 pints on fine soils; 1½ pints on coarse soils with 2-5% organic matter; and 2 to 2½ pints on soils with 5-10% organic matter.

COTTON—SPECIAL USE DIRECTIONS

Cotton—Fall panicum:

Apply and incorporate Treflan at the broadcast rate of 2 pints per acre on both coarse and medium soils.

Cotton—Rhizome johnsongrass:

In all cotton-producing states except Arizona and California, you can obtain commercially acceptable control of rhizome johnsongrass with a double rate program which you apply for 2 consecutive years in accordance with the following directions.

Soil Preparation—Proper preparation of the soil before application is very important for satisfactory results. Use a chisel plow or similar implement to bring rhizomes to the top of the soil. Then follow with a disc two times before application to cut the rhizomes into small (2 to 3 inch) pieces. This should also destroy any emerged johnsongrass.

Application—Choose the one application program that best fits your cultural practices:

Spring Application—Apply Treflan any time before planting in the spring for 2 years in a row. Use a broadcast rate of 2 pints per acre on coarse soils; 3 pints on medium soils; and 4 pints on fine soils,

OR

Fall Application—Apply Treflan between October 15 and December 31 for 2 years in a row at the same rates as a spring application for the control of rhizome johnsongrass.

Incorporation—Deep incorporation is essential for good rhizome johnsongrass control. Incorporate Treflan thoroughly with a disc set to cut 4 to 6 inches deep and operate at 4 to 6 mph. Two passes are necessary, with the second pass in a different direction from the first.

Cultivation—Some johnsongrass plants will escape. Timely cultivations during the crop season to remove escaped plants are necessary to obtain commercially acceptable control. You cannot obtain commercially acceptable control with only 1 year of double rate Treflan use.

Crop Rotation—In the season following a double rate treatment, plant only rice and those crops for which Treflan can be applied as a preplant treatment or injury may result.

Cotton—Pigweed and seedling johnsongrass control:

In Alabama, Arkansas, Florida, Georgia, Louisiana, Mississippi, southeastern Missouri, North Carolina, South Carolina, Tennessee and southern Virginia, Treflan may be applied preplant at a broadcast rate of 1 to 1½ pints per acre on coarse soils; 1½ to 2 pints on medium soils; and 2 pints on fine soils. Exception: in the State of Louisiana, 3 pints per acre are recommended on fine soils.

Cotton—Additional weed and grass control (Texas Gulf Coast):

In the Texas Gulf Coast counties of Brazoria, Calhoun, Chambers, Fort Bend, Galveston, Harris, Jackson, Jefferson, Liberty, Matagorda, Orange, Victoria, Waller and Wharton, Treflan may be applied up to 2 weeks before planting at a broadcast rate of 1½ pints per acre on coarse soils; 2 pints on medium soils; and 3 pints on fine soils.

Cotton Precautions

Cotton should be planted after early season adverse weather conditions have passed, especially when using higher rate programs. Cool, wet weather early in the growth cycle causes additional stress to the cotton plant. This may result in reduced stands, delayed maturity and reduced yields.

Cotton—Tank Mixes and Overlays

Follow recommended soil preparation and incorporation procedures for Treflan.

Cotton—Treflan/Caparol tank mix for cotton grown in California, Arizona, New Mexico and west Texas:

The Treflan/Caparol combination will control certain grasses and broadleaf weeds listed for Treflan alone plus the following weeds:

Annual morningglory	Prickly sida (Teaweed)
Groundcherry (Annual)	Ragweed
Malva	Smartweed
Mustard	Wild oat

The tank mix also controls shallow germinating seedlings of cocklebur and coffeeweed.

Broadcast Rates Per Acre

Soil Texture	Treflan M.T.F.	Caparol 80W
	(pints)	(pounds)
Coarse	1	2*
Medium	1¼-1½	2½
Fine	2	2½

**Do not use on sands and loamy sands. For band applications use proportionately less.*

Mixing Directions: Carefully follow the procedures on the Caparol 80W label for making a slurry and adding it to a partially filled tank of water. After the Caparol is thoroughly mixed with the partially filled tank of water, add the Treflan M.T.F. and continue filling. Agitate continuously during the filling and spraying operation.

Avoid leaving the spray mixture in the tank without constant agitation. If bypass agitation is used, the bypass line should stop at the bottom of the tank to minimize foaming.

Additional Precautions: The combination of Treflan/Caparol should not be used in the cut areas of newly leveled fields, in areas of excess salt, or where flooding over the beds is likely to happen. Do not plant cotton in tractor wheel depressions. These conditions may cause crop injury. On mulch-planted cotton, water back only after cotton seedlings are well established.

Crop Rotations: Cabbage, okra, onions and peas may be planted in the fall after a spring application of Treflan plus Caparol. Winter barley, winter rye and winter wheat can be planted in the fall also, if they are plowed down and not used for food or feed. Refer to the Caparol label for directions, cautions and precautions.

Cotton—Treflan/Cotoran tank mix (except in Arizona and California):
Follow recommended soil preparation and incorporation procedures for Treflan.

The Treflan/Cotoran tank mix effectively controls all the annual grasses and broadleaf weeds listed for Treflan alone plus these additional weeds:

Buttonweed	Ragweed
Cocklebur	Ryegrass
Groundcherry, Wright	Sesbania
Jimsonweed	Sicklepod
Morningglory	Smartweed
Prickly sida (Teaweed)	Tumbleweed

Broadcast Rates Per Acre

Soil Texture	Treflan M.T.F.	Cotoron 80W
	(pints)	(pounds)
Coarse	1	1¼
Medium	1½	2
Fine	2	2½

Mixing Directions: Carefully follow the procedures on the Cotoran label for making a Cotoran slurry and adding it to a partially filled tank of water. After the Cotoran is thoroughly mixed with the partially filled tank of water, add the Treflan and continue filling. Agitate continuously throughout the filling and application operations. Do not leave spray mixture in tank without constant agitation. If bypass agitation is used, the bypass line should stop at the bottom of the tank to minimize foaming. Apply in 15-40 gallons of water per acre.

Additional Precautions: Do not plant crops other than cotton on the treated land within 6 months after the application of Treflan plus Cotoran, or crop injury may result. Do not feed foliage from treated cotton plants or gin trash to livestock. Do not mix Treflan plus Cotoran with liquid fertilizer.

West Texas Only: Do not use the tank mix of Treflan plus Cotoran on sandy, loamy sand or fine sandy loam soils. Do not use on cotton planted in furrows.

Arkansas, Louisiana, and Mississippi Only: Use 1 pound Cotoran 80W in tank mix with Treflan on sandy loam soils low in organic matter.

New Mexico: Do not plant treated land with crops other than cotton until 1 year after the last application. Do not use on sandy or coarse textured soils of less than 1% organic matter.

Cotton—Treflan preplant followed by Cotoran overlay:
Apply and incorporate Treflan as recommended for the specific soil texture. Then apply Cotoran as a preemergence surface treatment at 1¼ to 2½ pounds per acre. On light silt and sandy soils low in organic matter, use the lower rate.

Additional Precaution: Refer to the Cotoran label for all cautions and precautions when using this preemergence surface treatment.

Cotton—Treflan preplant followed by Karmex overlay:
For cotton grown east of the Mississippi River plus Arkansas, southeastern Missouri, Louisiana and Eastern Texas, incorporate Treflan before planting. Then follow with a preemergence application of Karmex. This will effectively control all the weeds controlled by Treflan alone plus these additional weeds:

Annual groundcherry	Shepherdspurse
Annual morningglory	Velvetgrass
Dogfennel	Wild lettuce
Pennycress	Wild mustard
Ragweed	

Broadcast Rates Per Acre

Soil Texture	Treflan M.T.F.	Karmex 80W
	(pints)	(pounds)
Coarse	1	⅓
Medium	1½	⅔
Fine	2	1

Additional Precautions: Do not use Karmex on soils with less than 1% organic matter as crop injury may result. Do not allow livestock to graze on cotton treated with Karmex. Consult the Karmex label for additional instructions, cautions and precautions.

SOYBEAN—TREFLAN ALONE

Soybean—Preemergence:
Follow recommended soil preparation, application, and incorporation procedures for Treflan.

Broadcast Rates Per Acre

Soil Texture	Treflan M.T.F.*
	(pints)
Coarse	1
Medium	1½
Fine	2

**Use 1½ pints per acre on coarse and medium textured soils and 2 pints on fine soils with 2-5% organic matter; use 2-2½ pints on all soils with 5-10% organic matter.*

Soybean—Fall Application:
Apply and incorporate Treflan anytime between October 15 and December 31. Ground may be left flat or bedded-up over winter. On bedded ground, knock beds down to desired height before planting, moving some treated soil from tops into furrows. Where soil is left flat over winter, take care during spring bedding operations to prevent turning up untreated soil. Destroy established weeds during seedbed preparation. If weeds become established in furrows due to uncovering of untreated soil during listing, destroy these weeds before planting. Do not fall apply Treflan to soils which are wet, are subject to prolonged periods of flooding, or where rice was grown the previous year.

For soybeans grown in Alabama, Arkansas, northern Florida, Georgia, Louisiana, Mississippi, southeastern Missouri Bootheel, North Carolina, Oklahoma, South Carolina, Tennessee and Texas, apply and incorporate Treflan at a broadcast rate of 2 pints per acre on coarse and medium soils and 2½ pints on fine soils.

For soybeans grown in states other than those listed above, apply and incorporate Treflan at a broadcast rate of 1 pint per acre on coarse soils; 1½ pints on medium soils; 2 pints on fine soils; 1½ pints on coarse soils with 2-5% organic matter; and 2 to 2½ pints on soils with 5-10% organic matter.

SOYBEAN—SPECIAL USE DIRECTIONS

Soybean—Fall panicum:
Apply Treflan at the broadcast rate of 2 pints per acre on both coarse and medium soils.

Soybean—Pigweed and seedling johnsongrass:
In Alabama, Arkansas, Florida, Georgia, Kansas, Louisiana, Mississippi, Missouri, Nebraska, North Carolina, Oklahoma, South Carolina, Tennessee and southern Virginia, Treflan may be applied at a broadcast rate of 1 to 1½ pints per acre on coarse soils; 1½ to 2 pints on medium soils; and 2-2½ pints on fine soils. Exception: in the state of Louisiana, 3 pints per acre are recommended on fine soils.

Soybean—Additional weed and grass control (Texas Gulf Coast):
In the Texas Golf Coast counties of Brazoria, Calhoun, Chambers, Fort Bend, Galveston, Harris, Jackson, Jefferson, Liberty, Matagorda, Orange, Victoria, Waller and Wharton, Treflan may be applied up to 2 weeks before planting at a broadcast rate of 1½ pints per acre on coarse soils; 2 pints on medium soils; and 3 pints on fine soils.

Soybean—Charcoal soils in Arkansas, Louisiana and Mississippi:
Newly created land often contains high organic matter (5-10%) and charcoal from burning debris. This charcoal and/or organic matter tends to bind Treflan and reduce its weed control activity. Under these conditions, higher rates of Treflan are necessary for weed control. Increased rates, however, can cause crop injury if charcoal or organic matter is not present to bind some of the Treflan. In the burn row a high level of charcoal is present; consequently, poor weed control may result even with an increased rate of Treflan.

Apply Treflan at the broadcast rate of 1½ to 2½ pints per acre on coarse soils; 2½ pints on medium soils; and 3 pints on fine textured soils. Follow recommended soil preparation, application and incorporation procedures for Treflan.

Soybean—Red rice in Arkansas, Louisiana, Mississippi and Texas only: You can obtain suppression or partial control of red rice when you apply Treflan at the following recommended rates. Follow recommended soil preparation and incorporation procedures for Treflan.

Apply and incorporate Treflan in the spring before planting.

Broadcast Rates Per Acre

Soil Texture	Treflan M.T.F. Application Year 1	Treflan M.T.F. Application Year 2
	(pints)	(pints)
Coarse	2	1
Medium	3	1½
Fine	4	2
Coarse soils with 2 to 5% organic matter	3	1½
Soils with 5 to 10% organic matter	4	2-2½

If a combination of high organic matter (5-10%) and charcoal are present in the soil, apply Treflan the second year at the following rates for charcoal soils in Arkansas, Louisiana and Mississippi:

Broadcast Rates Per Acre

Soil Texture	Treflan M.T.F.*
	(pints)
Coarse	1½-2½
Medium	2½
Fine	3

For more information on charcoal soils see page 32.

Crop Rotation: The program for red rice control in soybeans is a 2-year program. Use the rates listed for first year application and plant soybeans. The second year use the normal Treflan rates listed for your soil type and charcoal level and plant only those crops for which Treflan has been registered as a preplant treatment, or crop injury may result. Do not plant rice the second year. Rice may be planted the third year.

Soybean—Rhizome johnsongrass (Eastern United States and the state of Texas):

You can obtain commercially acceptable control of rhizome johnsongrass with a double rate program applied for 2 consecutive years in accordance with the following directions.

Soil Preparation—Proper preparation of the soil before application is very important for satisfactory results. Use a chisel plow or similar implement to bring rhizomes to the top of the soil. Then follow with a disc two times before application to cut the rhizomes into small (2 to 3 inch) pieces and to destroy any emerged johnsongrass.

Application—Choose the one application program that best fits your cultural practices:

Spring Application—Apply Treflan anytime in the spring before planting for 2 years in a row. Use a broadcast rate of 2 pints per acre on coarse soils; 3 pints on medium soils; 4 pints on fine soils; 3 pints on coarse soils with 2-5% organic matter; and 4 pints on soils with 5-10% organic matter.

OR

Fall Application—Apply Treflan between October 15 and December 31 for 2 years in a row at the same rates as a spring application for the control of rhizome johnsongrass.

OR

Split Application—Apply Treflan at the same rate in both the spring and fall for 2 years in a row using the rates in the following table.

Broadcast Rates Per Acre

Soil Texture	Treflan M.T.F. Spring and Fall
	(pints)
Coarse	1
Medium	1½
Fine	2
Coarse soils with 2-5% organic matter	1½
Soils with 5-10% organic matter	2

Incorporation—Deep incorporation is essential for good rhizome johnsongrass control. Incorporate Treflan thoroughly with a disc set to cut 4 to 6 inches deep and operate at 4 to 6 mph. Two passes are necessary, with the second pass in a different direction from the first.

Cultivation—Some johnsongrass plants will escape. Timely cultivations during the crop season to remove escaped plants are necessary to obtain commercially acceptable control.

Crop rotation: In the season following a double rate treatment, plant only rice and those crops for which Treflan can be applied as a preplant treatment or injury may result.

Soybean—Rhizome johnsongrass - Treflan/Sencor or Treflan/Lexone tank mix:

Treflan/Sencor or Treflan/Lexone may be used for rhizome johnsongrass control and for the control of those weeds listed for Treflan alone. For the additional weeds controlled by Sencor or Lexone in tank mix, see page 38.

Follow procedures for soil preparation, incorporation, and cultivation recommended in the Soybean - Rhizome johnsongrass section (see page 34).

Apply Treflan/Sencor or Lexone up to two weeks before planting for two consecutive years at the following broadcast rates per acre:

Broadcast Rates Per Acre

Soil Texture	Treflan M.T.F.	Lexone 50 WP / 4L or Sencor 50WP / 4	or	Lexone (dry flowable) or Sencor (dry flowable)
	(pints)	(pounds/pints)		(pounds)
Coarse*	2	½		⅓
Medium	3	¾		½
Fine	4	1		⅔

**Do not use on coarse soils with less than 1% organic matter.*

Read and follow all additional precautions listed for the Treflan/Sencor or Treflan/Lexone tank mix (see page 38).

Soybean—Wild cane (shattercane):

Follow recommended soil preparation and application procedures for Treflan.

Wild cane (shattercane) can germinate throughout the growing season and from greater soil depth than most other weed seeds. Commercially acceptable control of wild cane can be obtained with the following increased rates of Treflan.

Apply Treflan at the broadcast rate of 1 pint per acre on a coarse soil; 2 pints on a medium soil; and 2½ pints on a fine textured soil.

Incorporation—Deep incorporation is essential for good wild cane control. Incorporate Treflan thoroughly with a disc set to cut 4 to 6 inches deep and operate at 4 to 6 mph. Two passes are necessary, with the second pass in a different direction from the first.

Cultivation—Cultivations during the crop season will also contribute to control.

Soybean Precautions:

Soybeans should be planted after early season adverse weather conditions have passed, especially when using higher rate programs. Cool, wet weather early in the growth cycle causes additional stress to the soybean plant. This may result in reduced stands, delayed maturity and reduced yields.

Soybean—Tank-Mix, Overlay and Postemergence Recommendations

Soybean—Treflan/Sencor or Treflan/Lexone tank mix:

The Treflan/Sencor or Treflan/Lexone tank mix effectively controls the annual grasses and broadleaf weeds controlled by Treflan alone plus these additional weeds:

Jimsonweed	Ragweed, common
Mallow, Venice (Flower-of-an-hour)	Sesbania, hemp
Mustard, wild	Smartweed, Pennsylvania
Prickly sida	Velvetleaf

Control of cocklebur, morningglory and giant ragweed (horseweed) may be erratic. Control may be improved with timely cultivation. Where cocklebur is a serious problem, an overlay of Sencor or Lexone may be preferred to the Treflan/Sencor or Treflan/Lexone tank mix.

Follow recommended procedures for soil preparation, incorporation, and cultivation of Treflan. Mix according to instructions on page 15. The Treflan/Sencor or Treflan/Lexone tank mix can be applied from 2 weeks before planting up to planting.

Broadcast Rates Per Acre

Soil Texture	Treflan M.T.F.	Lexone 50 WP / 4L or Sencor 50WP / 4	or	Lexone (dry flowable) or Sencor (dry flowable)
	(pints)	(pounds/pints)		(pounds)
Coarse*	1	½		⅓
Medium	1½	¾		½
Fine	2	1		⅔

**Do not use Sencor/Lexone on coarse soils with less than 1% organic matter.*

Precautions: Do not plant any crop other than soybeans within 4 months after treatment. Overapplication, uneven application, or improper soil incorporation may result in crop injury, herbicide residue, or erratic weed control. Additional stress factors are seedling disease, cold weather, deep planting, excessive moisture, soil *p*H over 7.5, high salt concentration, or drought. Any of these may weaken crop seedlings and increase possibility of damage from the tank mix. These additional factors may also delay crop development or reduce yields when Sencor or Lexone is applied. Observe all cautions and limitations on the Sencor and Lexone labels. Do not use the foliage from soybeans treated with the Treflan/Sencor or Treflan/Lexone tank mix for feed or forage.

Additional Sencor and Lexone precautions: Do not use Lexone or Sencor on Tracy, Semmes, Altona, Vansoy or Coker 102 soybeans. These varieties are sensitive to Lexone or Sencor, and crop injury may result. Seed must be planted at least 1½ inches but not more than 2 inches below the soil surface before a Sencor or Lexone application. Do not apply Sencor or Lexone at these rates more than once per season. Do not replant areas treated with Sencor or Lexone with any crop other than soybeans within 4 months after treatment. Injury to soybeans may occur if you use Lexone or Sencor on soils having a calcarous surface or *p*H of 7.5 or higher, or if you use them in conjunction with soil-applied organic phosphate pesticides.

Soybean—Treflan/Canopy Tank-Mix (Do Not Use in California)

Apply Treflan Canopy as a preplant incorporated tank-mix treatment. Follow recommended soil preparation, application and incorporation procedures for Treflan. Plant soybeans within two (2) weeks after application.

Treflan/Canopy tank-mix controls the following weeds in addition to those controlled by Treflan alone (excluding special rate programs).

Controlled

- Cocklebur, common
- Florida beggarweed
- Hemp sesbania
- Hophornbeam copperleaf
- Jimsonweed
- Morningglory
  - (Entireleaf)
  - (Ivyleaf)
  - (Pitted)
  - (Smallflower)
  - (Tall)
- Prickly sida
  - (Teaweed)
- Ragweed
  - (Common)
  - (Giant)
- Sicklepod
- Smartweed
- Spotted spurge
- Sunflower
- Velvetleaf

Partially Controlled

- Purple nutsedge
- Yellow nutsedge

Large seeded weeds, germinating deep in the soil, such as morningglory, sicklepod, cocklebur and giant ragweed or weeds with subsequent flushes may require a cultivation or an application of a postemergence herbicide for season-long control.

Broadcast Rates Per Acre

Soil Texture	Treflan M.T.F.	Canopy Dispersible Granules 0.5–3% O.M.	Canopy Dispersible Granules 3–5% O.M.
	(pints)	(ounces)	(ounces)
Coarse	1	6 – 8	8 – 10
Medium	1½	8 – 10	10 – 12
Fine	2	10 – 12	12 – 14

Do not apply Canopy to soil with less than ½% organic matter.

Use higher rates of Canopy on soils with higher organic matter or heavy pressure from large deep germinating weed seeds.

Where Canopy is applied plant soybean seed 1½" to 2" deep on a flat or raised seedbed only, or crop injury may occur.

Soybean injury may occur where Canopy is applied if excessive rainfall occurs after application but before soybeans germinate.

Precautions: Read the Canopy label carefully for cautions and precautions relating to environmental hazards, planting of rotation crops, sprayer contamination and cleanup, soil *p*H, organic matter and soil texture use restrictions, soybean variety planting restrictions, restrictions where Atrazine was used the previous year, restrictions concerning use with organic phosphate pesticides, grazing restrictions and other directions, precautions and limitations before applying the Treflan/Canopy tank-mix.

Soybean—Treflan/Preview Tank-Mix (Do Not Use in California)

Apply Treflan/Preview as a preplant incorporated tank-mix treatment. Follow recommended soil preparation, application and incorporation procedures for Treflan. Plant soybeans within two (2) weeks after application.

Treflan/Preview tank-mix controls the following weeds in addition to those controlled by Treflan alone (excluding special rate programs).

Controlled

- Cocklebur, common
- Hophornbeam copperleaf
- Jimsonweed
- Mustards
- Pigweed
  - (Palmer amaranth)
  - (Smooth)
  - (Tall waterhemp)
- Prickly sida
  - (Teaweed)
- Ragweed, common
- Smartweeds, annual
- Spotted spurge
- Sunflower
- Velvetleaf

Partially Controlled

- Burcucumber
- Eastern Black nightshade
- Morningglory
  - (Tall)
  - (Ivyleaf)
  - (Pitted)
  - (Entireleaf)
- Nutsedge species
- Ragweed, giant

Large seeded weeds, germinating deep in the soil, such as morningglory, sicklepod, cocklebur and giant ragweed or weeds with subsequent flushes may require a cultivation or an application of a postemergence herbicide for season-long control.

Broadcast Rates Per Acre

Soil Texture	Treflan M.T.F.	Preview Dispersible Granules 0.5 - 3% O.M.	Preview Dispersible Granules 3 - 5% O.M.
	(pints)	(ounces)	(ounces)
Coarse	1	6	7
Medium	1½	7	8
Fine	2	8	9 – 10

Do not apply Preview to soil less than ½% organic matter.

Use higher rates of Preview on soils with higher organic matter or heavy pressure from large deep germinating weed seeds.

Where Preview is applied, plant soybean seed 1½" to 2" deep on a flat or raised seedbed only, or crop injury may occur.

Soybean injury may occur where Preview is applied if excessive rainfall occurs after application but before soybeans germinate.

Precautions: Read the Preview label carefully for cautions and precautions relating to environmental hazards, planting of rotation crops, sprayer contamination and cleanup, soil *p*H, organic matter and soil texture use restrictions, soybean variety planting restrictions, restrictions where Atrazine was used the previous year, restrictions concerning use with organophosphate pesticides, grazing restrictions and other directions, precautions or limitations before applying the Treflan/Preview tank-mix.

Soybean—Treflan/Amiben tank mix:

Amiben may be applied several days prior to planting as a broadcast tank mix with Treflan. Weeds controlled by this tank mix, in addition to those controlled by Treflan alone, are smartweed, velvetleaf and ragweed. The tank mixture should be used as a spring preplant incorporated treatment.

Apply Treflan as a tank mix with Amiben.

Broadcast Rates Per Acre

Soil Texture	Treflan M.T.F.	Amiben 2S	Amiben DS
	(pints)	(quarts)	(pounds)
Coarse	1	4-6*	2.4-3.6*
Medium	1½	4-6	2.4-3.6
Fine	2	4-6	2.4-3.6

**Use the higher rate where you expect heavy populations of smartweed, velvetleaf, ragweed, wild mustard or black nightshade. Do not use on muck or charcoal soils. Read and observe all directions and cautions on the Amiben label.*

Soybean—Treflan/Amiben/Sencor or Lexone tank mix:

The Treflan/Amiben/Sencor or Lexone tank mix effectively controls all weeds listed for Treflan/Amiben and Treflan/Sencor or Lexone tank mixes.

Follow recommended soil preparation, application, and incorporation procedures for Treflan. The Treflan/Amiben/Sencor or Lexone tank mix may be applied from several days prior to planting up to planting in 10 to 40 gallons of water per acre. Use screens no finer than 50 mesh.

Apply the Treflan/Amiben/Sencor or Lexone tank mix at the following broadcast rate per acre.

Broadcast Rates Per Acre

Soil Texture	Treflan M.T.F.	Amiben 2S	Lexone 50WP/4L or Sencor 50WP/4	or	Lexone (dry flowable) or Sencor (dry flowable)
	(pints)	(quarts)	(pounds/pints)		(pounds)
Coarse[a]	1	3 – 4[b]	½		⅓
Medium	1½	3-4[b]	½-¾[c]		⅓-½[c]
Fine	2	4-5	¾[c]		½[c]

a *Do not use Sencor or Lexone on coarse soils with less than 1% organic matter.*

b *Use the higher rate of Amiben when velvetleaf or black nightshade is a problem.*

c *On Clarion/Webster soils in Minnesota and Iowa or on similar alkaline (calcareous) soils with a pH of 7.5 or above, apply Sencor or Lexone at the rates listed on page 47:*

Broadcast Rates Per Acre

Soil Texture	Lexone 50WP/4L or Sencor 50WP/4	or	Lexone (dry flowable) or Sencor (dry flowable)
	(pounds/pints)		pounds)
Medium	1/2		1/3
Fine	1/2-3/4[d]		1/3- 1/2[d]

[d] *Use the higher rate only where soil pH is less than 7.5 and where weed pressure is heavy.*

Additional precautions: The Treflan/Amiben/Sencor or Lexone tank mix will not harm the treated crop when you apply it according to directions and under normal growing conditions. However, overapplication, uneven application or improper soil incorporation of the tank mix can result in erratic weed control or crop injury. Additional stress factors are seedling disease, cold weather, deep planting, excessive moisture, soil pH over 7.5, high salt concentration, or drought. These additional factors may weaken crop seedlings and increase the possibility of damage from the tank mix. These additional factors may also delay crop development or reduce yields. Observe all cautions and limitations of all products used in mixtures. Do not use the foliage from soybeans treated with the Treflan/Amiben/Sencor or Lexone tank mix for feed or forage.

Soybean—Treflan/Vernam tank mix:

Treflan/Vernam tank mix effectively controls those weeds listed for Treflan alone plus these additional weeds:

Annual morningglory	Velvetleaf
Coffeeweed	Yellow nutsedge
Purple nutsedge (nutgrass)	(nutgrass)

Follow recommended soil preparation procedures for Treflan. You may apply the Treflan/Vernam tank mix up to 10 days prior to planting. Incorporate the tank mix immediately after application. Apply Treflan/Vernam at these rates:

Broadcast Rates Per Acre

Soil Texture	Treflan M.T.F.	Vernam 7E
	(pints)	(pints)
Coarse	1	1 3/4-2 1/3
Medium	1 1/2	2 1/3-3*
Fine	2	3-3 1/2

**For nutsedge, wild cane and velvetleaf control, use the higher rate of 3 pints per acre on medium textured soils.*

Soybean—Treflan preplant incorporated followed by Overlay treatments (Do not use in California)

Apply Treflan as a preplant incorporated treatment. Additional weeds tolerant to Treflan may be controlled by using overlay preemergence applications of Amiben, Canopy, Dual, Gemini, Lasso, Lexone, Lorox, Lorox Plus, Preview or Sencor. Consult these product labels for additional weeds controlled, application directions and precautions before use.

Soybean—Treflan preplant incorporated followed by Post-emergence treatments (Do not use in California)

Apply Treflan as a preplant incorporated treatment. Additional weeds tolerant to Treflan may be controlled by using post-emergence applications of Basagran, Blazer, Classic or Tackle. Consult these product labels for additional weeds controlled, application directions and precautions before use.

ALFALFA—ESTABLISHED

In areas receiving less than 20″ average annual rainfall per year, apply Treflan to established alfalfa stands at a broadcast rate of 1 1/2 pints per acre on coarse soils and 2 pints on medium and fine soils. Use incorporation equipment that will insure thorough soil mixing with minimum damage to the established alfalfa.

ASPARAGUS—ESTABLISHED

Follow recommended soil preparation, application and incorporation procedures for Treflan.

Treflan can be applied to established asparagus as a single or as a split application. In the winter or early spring, apply Treflan to asparagus after ferns are removed but before spear emergence. Or, apply after harvest in the late spring or early summer before ferning begins. Treflan will suppress volunteer seedling asparagus and field bindweed if you use the following recommended rates and application schedules.

Follow recommended soil preparation, application and incorporation procedures for Treflan.

Broadcast Rates Per Acre

Soil Texture	Treflan* M.T.F.					
	Split Application			Single Application		
	Before Harvest	+	After Harvest	Before Harvest	or	After Harvest
	(pints)			(pints)		
Coarse	1	+	1	2	or	2
Medium	1 1/2	+	1 1/2	3	or	3
Fine	2	+	2	4	or	4

**In any single calendar year, the maximum Treflan to apply is 2 pints per acre on coarse soils; 3 pints on medium soils; and 4 pints on fine soils.*

FOR THE FOLLOWING CROP GROUPING, USE THE RATE TABLE BELOW.

Carrot

Castor Bean

Celery—(Direct seeded and transplant in areas receiving less than 20″ average annual rainfall.)

Cole Crops—Transplant

Apply and incorporate prior to transplanting only. (Broccoli, Brussels Sprout, Cabbage and Cauliflower) See next section for direct seeded.

Okra

Pepper—Transplant

Apply and incorporate prior to transplanting only.

Southern Pea—(Before planting only.)

Apply and incorporate Treflan before planting, at planting or immediately after planting, unless otherwise indicated.

Broadcast Rates Per Acre

Soil Texture	Treflan M.T.F. Areas receiving less than 20″ average annual rainfall*	Treflan M.T.F. Areas receiving greater than 20″ average annual rainfall*
	(pints)	(pints)
Coarse	1	1
Medium	1 1/4-1 1/2	1 1/2
Fine	1 1/2	2

**Use 1 1/2 pints per acre on coarse and medium textured soils and 2 pints on fine soils with 2-5% organic matter; use 2 pints on all soils with 5-10% organic matter.*

Cole Crops—Direct Seeded (Broccoli, Brussels Sprout, Cabbage and Cauliflower) See above section for transplant.

For direct-seeded cole crops, apply and incorporate Treflan before planting at a broadcast rate of 1 pint per acre on coarse and medium soils and 1 1/2 pints on fine soils and soils with 2-5% organic matter. Direct-seeded cole crops have exhibited marginal tolerance to recommended rates of Treflan. Stunting or reduced stands may occur.

Cucurbits—Postplant Emerged (Cantaloupe, Cucumber and Watermelon)

Western United States including Texas:

Apply Treflan as a directed spray to the soil between the rows and beneath plants which are in the 3 to 4 true leaf stage.

Broadcast Rates Per Acre

Soil Texture	Treflan M.T.F. Areas receiving less than 20″ average annual rainfall*	Treflan M.T.F. Areas receiving greater than 20″ average annual rainfall*
	(pints)	(pints)
Coarse	1	1
Medium	1 1/4-1 1/2	1 1/2
Fine	1 1/2	2

**Use 1 1/2 pints on coarse and medium textured soils and 2 pints on fine soils with 2-5% organic matter; use 2 pints on all soils with 5-10% organic matter.*

Set incorporation equipment to throw treated soil around the plants during incorporation.

Dry Bean—Treflan Alone

Apply and incorporate Treflan before planting using the following rates:

Broadcast Rates Per Acre

Soil Texture	Treflan M.T.F. Areas receiving less than 20″ average annual rainfall*	Treflan M.T.F. Areas receiving greater than 20″ average annual rainfall*
	(pints)	(pints)
Coarse	1	1
Medium	1 1/4-1 1/2	1 1/2
Fine	1 1/2	2

**Use 1 1/2 pints per acre on coarse and medium textured soils and 2 pints on fine soils with 2-5% organic matter; use 2 pints on all soils with 5-10% organic matter.*

Dry Bean—Treflan/Eptam tank mix:

The Treflan/Eptam tank mix effectively controls all the following weeds in addition to those weeds listed for Treflan (see page 10):

Henbit (Spring applications)	Oat, wild
Nightshade, black	Ragweed, common
Nightshade, hairy	Smartweed, Pennsylvania
Nutsedge	Velvetleaf (Buttonweed)

Follow recommended soil preparation and incorporation procedures for Treflan. The Treflan/Eptam tank mix should be applied from 2 days before planting up to planting. Incorporate immediately after application.

Broadcast Rates Per Acre

Soil Texture	Treflan M.T.F. Areas receiving less than 20" average annual rainfall*	Treflan M.T.F. Areas receiving greater than 20" average annual rainfall*	Eptam 7E
	(pints)	(pints)	(pints)
Coarse	1	1	2½-3½**
Medium	1¼-1½	1½	2½-3½
Fine	1½	2	2½-3½

**Use 1½ pints per acre on coarse and medium textured soils and 2 pints on fine soils with 2-5% organic matter; use 2 pints on all soils with 5-10% organic matter.*

**Use Eptam 7E at a rate of 2½ pints per acre to control annual grasses; 3½ pints to control nutsedge and additional broadleaf weeds.

Precaution: Read the Eptam label before using. Observe all cautions and limitations of all products used in mixtures. The combination of Treflan and Eptam should not be used on soybeans, black-eyed peas (beans), lima beans and other flatpodded beans, except Romano. Do not use the foliage from a crop treated with the Treflan/Eptam tank mix for feed or for grazing.

Dry Bean—Fall application for dry bean grown in Idaho, Oregon, and Washington:

Apply and incorporate Treflan any time between October 15 and December 31 at a broadcast rate of 1 pint per acre on coarse soils; 1¼ to 1½ pints on medium soils; and 1½ pints on fine soils. Destroy established weeds during seedbed preparation.

FOR THE FOLLOWING CROP GROUPING, USE THE RATE LISTED BELOW.

Beans—(Guar and Mungbean)

Greens—Turnip greens grown for processing, Collard, Kale and Mustard greens.

Mustard—Grown for seed or processing for food in Minnesota, and North Dakota:

Apply and incorporate Treflan before planting at 1 pint per acre on coarse soils and 1½ pints on medium and fine soils.

Beans—(Lima Bean and Snap Bean)

Apply and incorporate Treflan before planting at a broadcast rate of 1 pint per acre on coarse and medium soils and 1½ pints on fine soils.

Corn (Field Corn) and Grain Sorghum (Milo)

Apply Treflan to field corn or grain sorghum (8 inches or taller) as an over-the-top or directed spray to effectively control weeds listed for Treflan.

Soil Preparation—Cultivate before a Treflan application to insure loose, friable soil, to remove established weeds, and to cover the base of plants with soil.

Application Directions—Treflan should be applied and incorporated at the recommended rates for the soil texture when the crop is well established (8 inches or taller). Treflan may be applied either as an over-the-top spray or as a directed spray. Drop nozzles should be used if foliage prevents uniform coverage of soil surface. Soil incorporation may be accomplished with only one pass of a sweep-type cultivator or a properly adjusted rolling cultivator.

The sweep-type cultivator should have 3 to 5 sweeps per row middle and be operated at 6 to 8 mph. Set the middle sweeps so as to avoid exposing untreated soil. Adjust the incorporation tools to prevent crop injury.

Broadcast Rates Per Acre

Soil Texture	Treflan M.T.F.
	(pints)
Coarse	¾-1*
Medium	1-1½
Fine	1½-2

Use the lower rates when you anticipate light weed pressure and the higher rates when you anticipate heavy weed pressure.

*Corn only: *Apply 1 to 1½ pints per acre in Alabama, Florida, Georgia, North Carolina, South Carolina and Virginia to control fall panicum and Texas panicum.*

Precaution: Do not apply Treflan to corn grown for seed. Do not apply Treflan to corn or sorghum as a preplant or preemergence treatment, or crop injury may occur.

Hops

Apply and incorporate Treflan while the crop is dormant. Use a broadcast rate of 1 pint per acre on coarse soils; 1¼ to 1½ pints on medium soils; and 1½ pints on fine soils and soils with 2-10% organic matter.

Mint—(Established Peppermint and Spearmint)

Apply Treflan at a rate of 1 pint per acre on coarse soils; 1¼ pints on medium soils; and 1½ pints on fine soils. Use incorporation equipment that will insure thorough soil mixing with minimum damage to the crop.

Pea—(Dry and English)—Treflan Alone

Apply and incorporate Treflan before planting at a rate of 1 pint per acre on coarse and medium soils and 1½ pints on fine soils.

Pea—Treflan/Far-Go tank mix for pea in Idaho, Oregon and Washington:

The tank mix combination of Treflan plus Far-Go will provide control of wild oat in addition to other annual grasses and broadleaf weeds controlled by Treflan.

Application Rates: Broadcast ¾ pint of Treflan per acre on coarse and medium soils; 1 pint of Treflan on fine soils. Use 1¼ quarts of Far-Go per acre for all soil textures.

Incorporation Directions: Apply and incorporate up to 3 weeks before planting. Follow recommended incorporation procedures for Treflan.

Precaution: Do not apply to lentils. Leaf crinkling and delayed maturity of peas may occur, particularly on clay points in the northwest; but this is usually more than offset by a reduction of wild oat. Do not use foliage from treated peas for feed or forage. Refer to the cautions, precautions and directions on the Far-Go label.

Pea—Fall application to dry pea and English pea in Idaho, Oregon and Washington:

Apply and incorporate Treflan any time between October 15 and December 31 at a broadcast rate of 1 pint per acre on coarse soils; 1¼ to 1½ pints on medium soils; and 1½ pints on fine soils. Destroy established weeds during seedbed preparation. Do not apply Treflan in the fall to soils which are wet or are subject to prolonged periods of flooding.

Peanut—(Spanish Peanut in Texas and Oklahoma)

Apply and incorporate Treflan before planting, at planting or immediately after planting at a broadcast rate of 1 pint per acre on coarse soils. When incorporating after planting, take care not to disturb the seed.

Peanut—Treflan/Vernam tank mix (Spanish Peanut in Texas and Oklahoma):

Treflan/Vernam tank mix effectively controls those weeds listed for Treflan alone plus these additional weeds:

Annual morningglory	Velvetleaf
Coffeeweed	Yellow nutsedge
Purple nutsedge (nutgrass)	(nutgrass)

Follow recommended soil preparation procedures for Treflan. You may apply the Treflan/Vernam tank mix up to 10 days prior to planting. Incorporate the tank mix immediately after application. Apply Treflan/Vernam at these rates:

Broadcast Rates Per Acre

Soil Texture	Treflan M.T.F.	Vernam 7E
	(pints)	(pints)
Coarse	1	2⅓

Potato—(All states except Maine)

Apply and incorporate Treflan after planting, before emergence, or immediately following dragoff or after the potato plants have fullly emerged.

Broadcast Rates Per Acre

Soil Texture	Treflan M.T.F. Areas receiving less than 20" average annual rainfall"	Treflan M.T.F. Areas receiving greater than 20" average annual rainfall"
	(pints)	(pints)
Coarse	1	1
Medium	1¼-1½	1½
Fine	1½	2

Use 1½ pints per acre on coarse and medium soils with 2-5% organic matter; use 2 pints on all soils with 5-10% organic matter.

Set incorporation equipment so that the bed and furrow will be uniformly covered with a layer of treated soil. If the layer of treated soils is not uniform and the herbicide is concentrated over the bed, potato emergence may be retarded, and stem brittleness can occur. When applying and incorporating Treflan after potato plants have fully emerged, do not completely cover the foliage with treated soil. Likewise, do not completely cover foliage at subsequent cultivations. Be careful that incorporation machinery does not damage potato seed pieces or elongating sprouts.

Potato —Split application in Idaho, Oregon and Washington:

On all soils, apply and incorporate ¾ pint of Treflan per acre before planting and ¾ pint after planting when potato plants have fully emerged. Do not apply to soils containing 2% or more organic matter. Follow incorporation directions listed above for application to potato after planting.

Potato —Treflan/Eptam tank mix for potatoes grown in Kansas, Minnesota, Nebraska, North Dakota, Oklahoma, South Dakota and Texas:

The Treflan/Eptam tank mix effectively controls the following weeds in addition to those weeds controlled by Treflan.

Henbit (Spring applications)	Oat, wild
	Ragweed, common
Nightshade, black	Smartweed, Pennsylvania

Nightshade, hairy — Velvetleaf, (Buttonweed)
Nutsedge

Follow recommended soil preparation and application procedures for Treflan. The Treflan/Eptam tank mix may be applied after planting, but prior to crop emergence. In areas where potatoes are normally dragged off, the Treflan/Eptam tank mix should be applied and incorporated up to or immediately following drag off.

Broadcast Rates Per Acre

Soil Texture	Treflan M.T.F.		
	Areas receiving less than 20" average annual rainfall*	Areas receiving greater than 20" average annual rainfall*	Eptam 7E
	(pints)	(pints)	(pints)
Coarse	1	1	1³/₄-7**
Medium	1-1¹/₂	1-1¹/₂	1³/₄-7
Fine	1-1¹/₂	1-2	1³/₄-7

**Use 1¹/₂ pints per acre on coarse and medium soils with 2-5% organic matter; use 2 pints on all soils with 5-10% organic matter.*

***Use the higher rate of Eptam 7E for nutsedge control.*

Precaution: Read the Eptam label before using. Observe cautions and limitations of products used in mixtures. Do not graze or feed forage to livestock from fields treated with Treflan/Eptam tank mix.

Potato —Treflan/Eptam application before planting in Washington, Idaho and Oregon:

Treflan/Eptam may be applied before planting at a broadcast rate of ³/₄ pint of Treflan per acre and 3¹/₂ pints of Eptam 7E per acre on all soil textures. Incorporate immediately.

Precaution: Do not use this tank mix both before and after planting in the same season. Read the Eptam label before using. Observe all cautions and limitaitons on labeling of all products used in mixtures. Do not use foliage from treated crops for feed or forage.

Safflower

Follow recommended soil preparation, application and incorporation procedures for Treflan.

Apply and incorporate Treflan in the spring before planting or in the fall between October 15 and December 31.

Broadcast Rates Per Acre

Soil Texture	Treflan M.T.F.	
	Areas receiving less than 20" average annual rainfall*	Areas receiving greater than 20" average annual rainfall*
	(pints)	(pints)
Coarse	1	1
Medium	1¹/₄-1¹/₂	1¹/₂
Fine	1¹/₂	2

**Use 1¹/₂ pints per acre on coarse and medium textured soils and 2 pints on fine soils with 2-5% organic matter; use 2 to 2¹/₂ pints on all soils with 5-10% organic matter.*

Safflower—Fall application in Arizona, California, Idaho, Nevada, Oregon, Utah, Washington and Wyoming:

Apply and incorporate Treflan any time between October 15 and December 31. Ground may be left flat or bedded-up over winter. On bedded ground, knock beds down to desired height before planting, moving some treated soil from tops into furrows. Where soil is left flat over winter, take care during spring bedding operations to prevent turning up untreated soil. Destroy established weeds during seedbed preparation. If weeds become established in furrows due to uncovering of untreated soil during listing, destroy these weeds before planting. Apply and incorporate Treflan at a broadcast rate of 1¹/₂ pints per acre on coarse soils; 2 pints on medium soils; and 2¹/₂ pints on fine soils.

Do not apply Treflan in the fall to soils which are wet or are subject to prolonged periods of flooding.

Sugar Beet

Apply Treflan as a broadcast, overtop spray when plants are between 2 and 6 inches tall at a rate of 1 pint per acre on coarse soils and 1¹/₄ to 1¹/₂ pints on medium and fine soils. Use the higher rate for medium and fine soils in areas receiving more than 20" average annual rainfall. Set incorporation machinery to throw treated soil toward the plants in the row. Be careful that incorporation machinery does not damage the sugar beet taproot.

Precaution: Exposed beet roots should be covered with soil before a Treflan application to reduce the possibility of girdling.

Sugar Beet—Incorporation with a tine-tooth harrow in California, Colorado, Idaho, Nebraska, Oregon, Texas, Utah, Washington and Wyoming:

A properly operated tine-tooth harrow (Flextine or Melroe) can incorporate Treflan for effective weed control in sugar beet. Operate the tine-tooth harrow 2 times over the field in opposing directions at a speed of 3 to 6 mph. Set the harrow to cut 1 or 2 inches deep. Be careful that the tine-tooth harrow does not damage the sugar beet taproot. Follow recommended application procedures and broadcast rates per acre for sugar beet (see preceding paragraph).

Sugarcane—(Plant Cane)

Apply and incorporate Treflan twice a year at a broadcast rate of 2 to 4 pints per acre for all soil textures. Make the first Treflan application in the fall on firmly packed beds immediately after the seed pieces are planted. Make the second Treflan application in the spring before or shortly after the cane emerges. Loosen rain-packed beds 2 to 3 inches deep before the spring application. Take care that incorporation machinery does not damage the seed pieces or emerging shoots.

Sugarcane—Postplant in Hawaii for control of most annual grasses, including guineagrass:

Surface apply Treflan after planting (for plant cane) or after harvesting (for ratoon cane), before weeds and cane emerge. Use a broadcast rate of 6 to 8 pints per acre for all soil textures. In plant cane, the beds should be formed or rolled before application. In ratoon cane, the crop residue should be removed before application. If large amounts of crop residues are present, Treflan will not be effective. Apply Treflan just before anticipated rainfall or sprinkle irrigate immediately after application.

Sugarcane—Applications up to layby for plant cane or ratoon cane grown in Louisiana or Texas:

Apply and incorporate Treflan at a broadcast rate of 2 to 4 pints per acre for all soil textures. Do this in the spring from before or shortly after the cane emerges up to layby. Apply the Treflan after the beds have been shaved or false shaved. Loosen rain-packed beds 2 to 3 inches deep before application. Be careful that incorporation machinery does not damage seed pieces or emerging shoots. You may use a rolling cultivator or bed chopper to incorporate Treflan layby applications in surgarcane on all soil textures. Follow normal incorporation directions for the rolling cultivator. Set bed chopper to cut 3 to 4 inches deep and operate at 4 to 6 mph. Two incorporation passes are necessary.

Sugarcane—Itchgrass (Raoulgrass) control in Louisiana:

Apply and incorporate Treflan on either plant or ratoon cane at a broadcast rate of 4 pints per acre for all soil textures. Follow the directions above for sugarcane layby application in Louisiana and Texas.

Sunflower—Treflan Alone

Apply and incorporate Treflan in the spring or in the fall between October 15 and December 31. Follow recommended soil preparation, application and incorporation procedures for Treflan.

Broadcast Rates Per Acre

Soil Texture	Treflan M.T.F.	
	Areas receiving less than 20" average annual rainfall*	Areas receiving greater than 20" average annual rainfall*
	(pints)	(pints)
Coarse	1	1
Medium	1¹/₄-1¹/₂	1¹/₂
Fine	1¹/₂	2

**Use 1¹/₂ to 2 pints per acre on coarse and medium soils with 2-5% organic matter and 2 pints on all soils with 5- 10% organic matter.*

Sunflower—Treflan/Amiben tank mix or overlay:

Follow recommended soil preparation, application and incorporation procedures for Treflan.

Amiben may be applied in a band or broadcast over sunflowers at planting in fields where Treflan has been incorporated prior to planting. Or, the Treflan/Amiben tank mix may be incorporated prior to planting. Treflan/Amiben tank mix improves mustard, smartweed, velvetleaf and ragweed control in addition to those weeds controlled by Treflan alone.

Apply Treflan/Amiben or Treflan with an Amiben overlay at the following broadcast rates per acre:

Broadcast Rates Per Acre

Soil Texture	Treflan M.T.F.	Amiben 2S
	(pints)	(quarts)
Coarse	1	4
Medium	1¹/₂	4-6*
Fine	2	4-6*

**For best control of mustard, common ragweed or black nightshade, use the 6 quart rate.*

In coarse textured soils, heavy rains on the incorporated Amiben may move it below the weed seed germination zone, and erratic weed control may result. If sufficient rain does not fall within 7 days after a preemergence application of Amiben, but there is enough soil moisture to germinate weeds and grasses, a light cultivation with a rotary hoe or similar tool will uproot these small broadleaf weeds and grasses. The shallow mixing of Amiben in the surface soil will not interfere with the action of Amiben when rains come.

Tomato

For direct-seeded tomato, apply Treflan at blocking or thinning as a directed spray to the soil between rows and beneath the plants, and incorporate. For transplant tomato, apply and incorporate Treflan before transplanting. Do not apply Treflan after transplanting.

Broadcast Rates Per Acre

Soil Texture	Treflan M.T.F	
	Areas receiving less than 20" average annual rainfall*	Areas receiving grater than 20" average annual rainfall*
	(pints)	(pints)
Coarse	1	1
Medium	1¼-1½	1½
Fine	1½	2

**Use 1½ pints per acre on coarse and medium textured soils and 2 pints on fine soils with 2-5% organic matter; use 2 pints on all soils with 5-10% organic matter.*

Fruit and Nut Crops and Vineyards

For areas receiving more than 20" average annual rainfall.

For new planting of vineyards, citrus and pecan trees, apply and incorporate Treflan before planting at a broadcast rate of 1 pint per acre on coarse soils; 1½ pints on medium soils; 2 pints on fine soils; 1½ pints on fine soils with 2-5% organic matter; and 2 pints on soils with 5-10% organic matter.

For non-bearing, established plantings of citrus and pecan trees and bearing plantings of grapefruit, lemon, orange, pecan, tangelo and tangerine trees, apply Treflan at a broadcast rate of 2 to 4 pints per acre for all soil textures.

For areas receiving less than 20" average annual rainfall.

For new plantings of almond, apricot, citrus, nectarine, peach, pecan and walnut trees, apply and incorporate Treflan before planting at a broadcast rate of 1 pint per acre on coarse soils; 1¼ to 1½ pints on medium soils; 1½ pints on fine soils; 1½ to 2 pints on soils with 2-5% organic matter, and 2 pints on soils with 5-10% organic matter.

For new plantings of vineyards, apply and incorporate Treflan before planting at a broadcast rate of 1 to 1½ pints per acre on coarse soils; 1½ to 3 pints on medium soils; and 3 to 4 pints on fine soils or soils with 2-10% organic matter. Do not use more than 2 pints per acre on heat-treated grape rootings.

For postplant applications on bearing or nonbearing, established plantings of vineyards and almond, apricot, grapefruit, lemon, nectarine, orange, peach, plum, prune, tangelo, tangerine and walnut trees, apply Treflan at a broadcast rate of 2 to 4 pints per acre for all soil textures. Do not apply to vineyards within 60 days of harvest.

In established plantings, apply Treflan as a directed spray to the soil and use incorporation methods not injurious to the trees or vines.

Fruit and Nut Crops and Vineyards—Rhizome johnsongrass control:

For areas receiving less than 20" average annual rainfall.

You can obtain commercially acceptable control of rhizome johnsongrass with postplant applications in bearing and non-bearing, established plantings of vineyards and almond, apricot, grapefruit, lemon, nectarine, orange, peach, pecan, tangelo, tangerine and walnut trees with a Treflan program when applied for 2 years in a row.

Soil Preparation—Work the soil thoroughly to bring the rhizomes nearer the surface.

Application—Apply Treflan at a broadcast rate of 2 quarts per acre on all soil textures each year for 2 years in a row. Do not apply to vineyards within 60 days of harvest.

Incorporation—Incorporate Treflan thoroughly with a disc set to cut 4 to 6 inches deep and operate at 4 to 6 mph. Two incorporation passes are necessary, with the second pass in a different direction from the first.

Cultivation—Some johnsongrass plants will escape. Timely cultivations are necessary to obtain commercially acceptable control. You cannot obtain commercially acceptable control with only 1 year of Treflan use.

Precautions: Do not use the 2 quart rate on new plantings, or crop injury may result. Do not interplant orchards or vineyards with other crops. If the treated vineyards and orchards are diverted to other crop uses, plant only those crops for which Treflan has been registered as a preplant treatment for the next cropping season.

Fruit and Nut Crops and Vineyards—Bindweed control in California:

Treflan can be used for the control of field bindweed in vineyards and for almond, apricot, grapefruit, lemon, nectarine, orange, peach, pecan, tangelo, tangerine and walnut trees.

Apply Treflan at a broadcast rate of 4 pints per acre on all soil textures. Treflan must be applied in the spring with a specially designed spray blade which applies a thin, concentrated layer at a soil depth of 4 to 6 inches. This layer of Treflan prevents bindweed shoots from emerging.

Land Preparation—Destroy all weeds and grasses with soil tillage before applying Treflan. This tillage is necessary to prevent trash from interfering with the operation of the spray blade.

Equipment—This operation requires a spray blade capable of running 4 to 6 inches below the surface of the soil. The spray blade should be equipped with nozzles located under the blade and directed so that the Treflan spray will be trapped under the soil which is flowing over the blade as it is pulled through the soil. Use a sufficient number of nozzles with spacing that will uniformly apply the Treflan underground in a thin, horizontal layer.

Application—Apply Treflan in 40 to 80 gallons of water per acre. Operate the spray blade at a depth of 4 to 6 inches.

Precaution: Some soils develop cracks as they dry after rainfall or irrigation. Field bindweed may emerge if the cracks extend through the Treflan layer. Prevent or eliminate cracks by shallow discing or other tillage. Avoid deep tillage which disturbs the subsurface layer. Cultivation or tillage also aids the control of germinating seeds.

Wheat (Winter)—(Idaho, Oregon and Washington)

Treflan may be applied for preplant preemergence control of cheatgrass and other annual grasses and broadleaf weeds controlled by Treflan. The growth, development and yield of winter wheat will not be adversely affected, provided the seed is placed below the zone of soil treated with Treflan.

Apply Treflan any time during a period from 3 weeks up to immediately prior to planting. Broadcast Treflan at a rate of 1½ pints per acre on coarse and medium soils and 2 pints on fine soils.

Incorporation Directions—Incorporate Treflan into the soil with a flexible tine-tooth harrow (Flextine, Melroe) set to cut 1 to 2 inches deep and operate at 3 to 6 mph. Incorporate one time within 24 hours after application, followed by a second incorporation in a different direction from the first prior to planting. Do not till the soil with a disc after the Treflan has been incorporated with a flexible tine harrow.

Seeding Directions—Use only a deep furrow or semi- deep furrow drill that will place the seed below the zone of soil into which Treflan has been incorporated.

Precaution: Wheat planted in direct contact with treated soil may suffer crop injury in the form of delayed emergence and development.

Wheat (Winter)—Fallow soil application in Washington and Oregon:

Treflan applied and shallowly incorporated into fallow soil up to four months ahead of planting will control cheatgrass and certain annual grasses and broadleaf weeds. The growth, development, or yield will not be adversely affected as long as the seed is placed below the zone of soil treated with Treflan. Use deep or semi-deep furrow drills.

Broadcast rates are 1½ pints per acre on coarse and medium soils and 2 pints on fine soils. Apply Treflan any time from May to September prior to the fall planting of winter wheat.

Incorporation—Incorporate Treflan with a flexible tine-tooth harrow (Flextine or Melroe) set to cut 1 to 2 inches deep and operated at 3 to 6 mph. For thorough incorporation, two passes of the equipment in different directions over the field are necessary. Incorporate one time within 24 hours after application followed by a second incorporation prior to seeding. Do not till the soil with a disc after Treflan has been applied with a flexible tine harrow.

Precaution: Use only deep furrow or semi-deep furrow drills. Place seed below the zone of soil into which Treflan has been incorporated. Do not plant wheat directly into the zone of soil treated with Treflan; injury to the crop or a delay in its emergence and development may occur.

Wheat (Spring), Durum and Barley—Treflan Alone

Treflan is recommended as a postplant incorporated treatment to control foxtail (pigeongrass).

Apply Treflan at a broadcast rate of 1 pint per acre on coarse and medium soils and 1½ pints on fine soils.

Plant 2 to 3 inches deep in a well-tilled seedbed. Apply Treflan after seeding but before the crop emerges. To incorporate, use flex-tine or diamond harrows operated two times in different directions, at speeds of at least 5 mph. Incorporate by operating equipment 1 to 1½ inches deep. Application and the first incorporation should be done in the same operation if possible. Both incorporations must be done within 24 hours.

Wheat (Spring), Durum, Barley (Fall Application)—Foxtail/Pigeongrass control:

Treflan may be fall applied for foxtail/pigeongrass control in spring wheat, durum and barley planted the following spring. Treflan may be applied to ground that has a manageable trash level, has been fallowed or pre-tilled. The first incorporation is required within 24 hours after application. A second incorporation is required prior to planting to destroy emerged weeds and to ensure an even distribution of Treflan treated soil.

Broadcast Rates Per Acre

Soil Texture	Treflan M.T.F.
	(pints)
Coarse	1
Medium	1
Fine	1½

Incorporation Directions—Any of the following tools are recommended for fall incorporation. The disc or field cultivator may be used for the spring incorporation pass. Care should be taken to operate the tool at a more shallow depth than the fall incorporation.

1. Chisel plow: *May be used for the first pass only.* Operate at 4-5 inches deep at 4-6 mph. A chisel plow is defined as having 3 rows of up to

18-inch sweeps on no greater than 12- inch centers. Stagger sweeps so that no soil is left unturned.
2. Tandem disc: Operate at 3-4 inches deep at 4-6 mph.
3. Field cultivator: Operate at 3-4 inches deep at 5 mph or more. A field cultivator is defined as having 3 to 4 rows of sweeps with "c" or "s" shaped shanks, spaced 7 inches or less and staggered so that no soil is left unturned.

Planting Directions—Wheat, durum, or barley should be approximately 2 inches deep.

Precaution—While use of this practice may result in a stand reduction, slight stand reductions do not normally affect yield.

Wheat (Spring), Durum and Barley—Treflan/Far-Go tank mix:

Treflan/Far-Go applied as a postplant incorporated treatment will control foxtail (pigeongrass) and wild oat.

Plant 2 to 3 inches deep in a well-tilled seedbed. Apply Treflan/Far-Go after seeding but before crop emerges. To incorporate, use flex-tine or diamond harrows two times, operated in different directions, at speeds of at least 5 mph. Incorporate by operating equipment 1 to 1½ inches deep. Application and the first incorporation should be done in the same operation if possible. If not, incorporate immediately after application.

Broadcast Rates Per Acre

Soil Texture	Treflan M.T.F. Barley Durum Spring Wheat	Far-Go Durum Spring Wheat	Far-Go Barley
	(pints)	(pints)	(pints)
Coarse	1	2	2½
Medium	1	2	2½
Fine	1½	2	2½

Precaution: Overapplication may result in crop injury. Read the Far-Go label carefully before using.

Fall Application

General: (Areas receiving more than 20" average annual rainfall)

See specific crop for recommendations. For all crops for which there are no specific fall application instructions and for which Treflan is recommended as a preemergence application, use the rates listed for spring applications. Do not apply Treflan in the fall for sugarbeets, potatoes and direct-seeded tomatoes.

In most states apply and incorporate Treflan any time between October 15 and December 31. In Minnesota, North and South Dakota, apply and incorporate Treflan anytime between September 1 and December 31. Ground may be left flat or bedded-up over winter. On bedded ground, knock beds down to desired height before planting, moving some treated soil from beds into furrows. Where soil is left over winter, be careful not to turn up untreated soil during spring bedding operations. Destroy established weeds during seedbed preparation. If weeds become established in furrows due to uncovering of untreated soil during bedding, destroy these weeds before planting. Do not apply Treflan in the fall to soils which are wet, are subject to prolonged periods of flooding, or there rice was grown the previous year.

FERTILIZER USE DIRECTIONS
APPLICATION WITH LIQUID FERTILIZERS

Treflan may be mixed with most liquid fertilizer materials. The combination of Treflan with solutions and suspension-type fertilizers has provided weed and grass control equal to the same rates of Treflan applied in water. Follow Treflan label recommendations regarding rates per acre, crops, incorporation directions, special instructions, cautions and special precautions.

Individual state regulations relating to liquid fertilizer mixing, registration, labeling and applications are the responsibility of the individual and/or company selling the fertilizer and chemical mixture.

Testing for Tank Mix Compatibility in Liquid Fertilizers:

Treflan alone or in tank mixture with dry flowables, wettable powders (WP), aqueous suspensions (AS), flowables (F), liquids (L), or solutions (S) may not combine properly with some fluid fertilizer materials. Small quantities should always be tested before full-scale mixing. This will determine whether a compatibility agent is needed, and which agent does the best job. The seven agents listed on page 80 have been thoroughly tested. There are many other surfactants on the market which were not designed for use with liquid fertilizers. Use the following test to select the correct agent for your mixture.

1. Put 1 pint of the liquid fertilizer in a quart jar.

2. Add 1 to 4 teaspoonful(s) of the dry flowable, WP, AS, F or L formulation (depending on the recommended rate per acre) to the liquid fertilizer. Close jar and agitate until the materials are dispersed evenly in the fertilizer. If the materials do not disperse well, it may be necessary to slurry the chemicals in water before adding to the fertilizer.

3. After dispersing the materials (Step 2), add 3 to 4 teaspoonsful of Treflan to the jar and shake well. Add solution herbicides to the mixture last and agitate.Observe the jar for about 10 minutes. If the materials rise to the surface and form a thick layer (oily curds) which will not redisperse when agitated, a compatibility agent is needed. If the mixture is easily re-dispersed to its original state with slight agitation, no agent is needed but good agitation must be provided in the fertilizer spray tank.

4. If the need for a compatibility agent is shown in Step 3: Using a clean quart jar, start at Step 1 above, add ½ teaspoon of the compatibility agent to the liquid fertilizer, mix well, then repeat Steps 2 and 3.

An effective compatibility agent will cause the mixture to remain uniformly mixed with little or no separating or oil rising to the surface for one half hour or longer. If slight separation does occur, 2 to 3 inversions of the jar should give a uniform remix. If oily curds form which will not redisperse, more agent or another agent should be tried.

Use a clean jar for each test. The compatible mixture will have a uniform appearance and will be relatively easy to keep mixed with gentle agitation of the jar.

Liquid Fertilizer Mixing Instructions
Treflan M.T.F. in Liquid Fertilizer

General—Emulsifiable concentrates, such as Treflan M.T.F., can be mixed with liquid fertilizers. In all cases, continuous agitation is required to prevent the Treflan from rising to the surface as an oily layer. When necessary, (See Testing for Tank Mix Compatibility in Liquid Fertilizers, on page 77) a compatibility agent can be used to cause the Treflan to emulsify properly (i.e., have a milky appearance rather than oily layer). The use of compatibilty agents is especially important when tank mixing emulsifiable concentrates (E.C.) with dry flowables, wettable powders (WP), aqueous suspensions (AS), flowables (F), liquids (L), or solutions (S) in liquid fertilizer. If the emulsion is not properly formed and the Treflan rises to the surface of the fertilizer as an oil ("oils out"), the oil may combine with the wettable powder, flowable, or suspension to form oily curds (viscous phase) which is difficult to redisperse. Any one of the compatibility agents listed below is helpful in causing liquid concentrates to form non-oiling mixtures with liquid fertilizers. These compatibility agents can be used at rates as low as 1½ to 2 pints per ton of liquid fertilizer and should be mixed well with the fertilizer before adding the liquid concentrate. Read the label on the compatibility agent and follow the directions.

1. Sponto 168D (Witco Chemicals Co., Chicago, IL)
2. Compat (Farm Chemicals, Inc., Aberdeen, NC)
3. Unite (Hopkins Ag Chemical, Madison, WI)
4. T-Mulz 734-2 (Thompson-Hayward Chemical Co., Kansas City, MO)
5. Rigo Compatibility Agent (Rigo Company, Buckner, KY)
6. Amoco Spray Mate™ (Amoco Oil Co., Chicago, IL)
7. Kem-Link (Universal Coop, Minneapolis, MN)

All of the above are phosphate ester-type surfactants designed to be used with liquid fertilizers. They usually do not work as compatibility agents in tank mixtures in plain water.

Application—Spread the fertilizer/pesticide mixture with a properly calibrated applicator. Be certain the material is applied uniformly to the soil surface.

Incorporation—Follow normal Treflan incorporation procedures.

TREFLAN APPLICATION WITH DRY BULK FERTILIZERS

General—Dry bulk fertilizers may be impregnated or coated with Treflan. Application of dry bulk fertilizers impregnated with Treflan has provided weed and grass control equal to the same rates of Treflan applied in water.

All Treflan label recommendations regarding rates per acre, approved crops, incorporation, special instructions, cautions and special precautions must be followed.

Apply a minimum of 200 pounds per acre of dry fertilizer impregnated with Treflan at the recommended rates. Any commonly used dry fertilizers can be used for Treflan impregnation except coated ammonium nitrate and straight limestone. These materials will not absorb the herbicide. Blends containing mixtures of these materials can be impregnated.

Impregnation—Use any closed drum, belt, ribbon or other commonly used dry bulk fertilizer blender. Apply Treflan uniformly to the fertilizer.

Rates

Check specific crop recommendations for the rate of Treflan per acre. See the rate table which follows to determine the amount of Treflan to be impregnated into a ton of dry bulk fertilizer based on the amount of fertilizer which will be applied per acre. (See rate chart on page 82).

Application

Spread the fertilizer/chemical mixture with a properly calibrated applicator. Be certain the material is applied uniformly to the soil surface.

Incorporation

Follow Treflan incorporation procedures.

RATE CHART FOR IMPREGNATING FERTILIZER WITH TREFLAN M.T.F.
(TREFLAN ADDED TO A TON OF FERTILIZER)

FERTILIZER RATE PER ACRE	TREFLAN RATE PER ACRE				
	1 PINT	1½ PINTS	2 PINTS	3 PINTS	4 PINTS
200 pounds	5 qts. per ton	7½ qts. per ton	10 qts. per ton	15 qts. per ton	20 qts. per ton
250 pounds	4 qts. per ton	6 qts. per ton	8 qts. per ton	12 qts. per ton	16 qts. per ton
300 pounds	3⅓ qts. per ton	5 qts. per ton	6⅔ qts. per ton	10 qts. per ton	13⅓ qts. per ton
350 pounds	2¾ qts. per ton	4¼ qts. per ton	5¾ qts. per ton	8½ qts. per ton	11½ qts. per ton
400 pounds	2½ qts. per ton	3¾ qts. per ton	5 qts. per ton	7½ qts. per ton	10 qts. per ton
450 pounds	2¼ qts. per ton	3⅓ qts. per ton	4½ qts. per ton	6⅔ qts. per ton	9 qts. per ton

For rates other than those listed above, use the following formula to calculate the amount of Treflan to be impregnated on a ton of dry bulk fertilizer:

$$\text{Pints Treflan Per Acre} \times \frac{1000}{\text{Lbs. Fertilizer Per Acre}} = \text{Quarts Treflan Per Ton of Fertilizer}$$

All individual state regulations relating to dry bulk fertilizer blending, registration, labeling and application are the responsibility of the individual and/or company selling the fertilizer and chemical mixture.

TREFLAN PERFORMANCE GUARANTEE—A LIMITED WARRANTY OF TREFLAN WEED AND GRASS CONTROL

Elanco Products Company warrants that Treflan will provide commercially acceptable control of the weeds and grasses as indicated on its current label when used according to label directions and subject to the following limitations:

1. Purchaser must notify Elanco promptly if a lack of commercially acceptable control occurs. Notice must be given within forty-five (45) days after the first planting or after application of Treflan to the treated crop, whichever occurs later.
2. An Elanco representative must be satisfied that Purchaser used Treflan according to the label directions.
3. The Purchaser must provide Elanco representative with an invoice or other satisfactory record which shows the price and quantity of Treflan purchased.
4. Elanco's liability will be limited to a refund of the purchase price of the Treflan applied to the acreage on which commercially acceptable control was not obtained.
5. Void outside the USA.

DISCLAIMER OF WARRANTIES

The warranty printed above is the only warranty applicable to this product. All other warranties, expressed or implied, including, but not limited to, any implied warranties of merchantability and fitness for a particular purpose, are disclaimed. This disclaimer of warranties does not apply where prohibited by operation of law.

LIMITATION OF DAMAGES

Elanco's liability, whether in contract, warranty, tort, negligence, strict liability or otherwise, shall not exceed the return of the amount of the purchase price of the Treflan and under no circumstances shall Elanco be liable for special, indirect or consequential damages. This disclaimer of damages does not apply where prohibited by operation of law.

INHERENT RISKS OF USE

Failure to carefully follow the directions for use of Treflan, as well as other factors such as seedling disease, cold weather, deep planting, excessive moisture, high salt concentration or drought, may result in unsatisfactory weed control or crop injury. Reduced yields may result under these conditions.

Amiben®—(chloramben, Union Carbide)
Basagran®—(BASF Wyandotte)
Blazer®—(acifluorfen, Rohm and Haas)
Canopy®—(chlorimuron ethyl plus metribuzin, E. I. duPont)
Caparol®—(prometryn, Ciba-Geigy)
Classic®—(chlorimuron ethyl, E. I. duPont)
Cotoran®—(fluometuron, Ciba-Geigy)
Dual®—(metolachlor, Ciba-Geigy)
Eptam®—(EPTC, Stauffer)
Far-Go®—(triallate, Monsanto)
Gemini™—(chlorimuron ethyl plus linuron, E. I. duPont)
Karmex®—(diuron, E. I. duPont)
Lasso®—(alachlor, Monsanto)
Lasso® Micro-Tech™—(alachlor, Monsanto)
Lexone®—(metribuzin, E. I. duPont)
Lorox®—(linuron, E. I. duPont)
Lorox® Plus—(chlorimuron ethyl plus linuron, E. I. duPont)
Preview®—(chlorimuron ethyl plus metribuzin, E. I. duPont)
Sencor®—(metribuzin, Bayer, GmbH)
Tackle®—(acifluorfen, Rhone-Poulenc)
Vernam®—(vernolate, Stauffer)

This is a specimen label. Elanco is not responsible for the accuracy of the information contained herein. As labels are subject to revision, always carefully read and follow the label on the product container.

TREFLAN 5®

HERBICIDE CONCENTRATED FORMULA

A selective herbicide for the preemergence control of annual grasses and broadleaf weeds

ACTIVE INGREDIENT:
trifluralin* (α,α,α-trifluoro-2,6-dinitro-*N,N*-dipropyl-*p*-toluidine) 50.8%
INERT INGREDIENTS: 49.2%
TOTAL 100.0%

Contains 5 pounds active ingredient per gallon.

***TREFLAN® 5**—(trifluralin, Elanco)

Manufactured using the processes of U.S. Patents 4,120,905 and 4,226,789.

EPA Reg. No. 1471-120 ID 5555

Literature revised August 11, 1987
PI 2943 AMP

PRECAUTIONARY STATEMENTS

KEEP OUT OF REACH OF CHILDREN
CAUTION

HAZARDS TO HUMANS AND DOMESTIC ANIMALS

Causes eye irritation. Do not get in eyes. Avoid contact with skin and clothing. Harmful if swallowed or absorbed through the skin. TREFLAN 5 may cause skin sensitization reactions in certain individuals. Use eye protection and protective clothing such as coveralls, a long sleeved shirt, and impermeable gloves when handling this product. Do not contaminate foodstuffs or feeds.

First Aid: In case of contact, immediately flush eyes or skin with plenty of water. For eyes, call a physician. Remove and wash contaminated clothing before reuse.

ENVIRONMENTAL HAZARDS

This pesticide is toxic to fish. Do not apply directly to water or wetlands (swamps, bogs, or marshes). Drift or runoff from treatment areas may be hazardous to aquatic organisms in neighboring aquatic sites. Do not contaminate water by cleaning equipment or disposal of waste.

STORAGE AND DISPOSAL

Storage: Avoid freezing. Store above 40°F. If frozen, poor weed control may result. Do not store near heat or open flame. Store in original container only. In case of leak or spill, use absorbent materials to contain liquids and dispose as waste.

Pesticide Disposal: Do not contaminate water, food or feed by storage or disposal. Wastes resulting from the use of this product may be disposed of on site or at an approved waste disposal facility.

Plastic Container Disposal: Triple rinse (or equivalent). Then offer for recycling or reconditioning, or puncture and dispose of in a sanitary landfill, or incineration, or, if allowed by state and local authorities, by burning. If burned, stay out of smoke.

Metal Container Disposal: Triple rinse (or equivalent). Then offer for recycling, or reconditioning, or puncture and dispose of in a sanitary landfill, or by other procedures approved by state and local authorities.

Bulk/Mini-bulk Tank Handling: Tanks that have been used for phenoxy-type herbicides should not be used for TREFLAN 5. To clean, triple rinse (or equivalent) and wash with appropriate cleaners before reusing.

Special Precautions

Applied according to directions and under normal growing conditions, TREFLAN will not harm the treated crop. Overapplication may result in crop injury or a soil residue. Uneven application or improper soil incorporation of TREFLAN can result in erratic weed control or crop injury. Seedling disease, cold weather, deep planting, excessive moisture, high salt concentration or drought may weaken crop seedlings and increase the possibility of damage from TREFLAN. Under these conditions, delayed crop development or reduced yields may result.

In Arizona, Colorado, California, Idaho, Nevada, New Mexico, Oregon, Utah, Washington and Wyoming, sugarbeets, red beets or spinach should not be planted for 12 months after a spring application or for 14 months after a fall application of TREFLAN. Plow the land to a depth of 12 inches prior to planting sugar beets to prevent the possibility of crop injury. Sorghum (milo), proso millet, corn, oats and annual or perennial grass crops or grass mixtures should not be planted for 14 months after a spring application or for 16 months after a fall application of TREFLAN to avoid crop injury. If land has not been irrigated, do not plant any of these crops for 18 months after a spring application or 20 months after a fall application of TREFLAN.

In those portions of Kansas, Nebraska, North Dakota, Oklahoma, South Dakota and Texas where at least 20 inches of irrigation and/or rainfall (total) was used to produce the crop, sorghum, oats and annual or perennial grass crops or grass mixtures should not be planted for 12 months after an application of TREFLAN.

If less than 20 inches of total water was used to produce the crop, do not plant sorghum, proso millet, oats and annual or perennial grass crops or grass mixtures for 18 months after an application of TREFLAN. Cool, wet weather conditions during the early stage of growth may increase the possibility of injury to sorghum.

In all other areas receiving greater than 20" rainfall per year, moldboard plow before planting sugar beets where a spring application of TREFLAN was made the previous season. Also note planting restrictions listed in the section on control of rhizome johnsongrass and other higher rate programs.

Vegetable crops other than those listed on this label should not be planted within 5 months following the application of TREFLAN.

Do not use TREFLAN or any crop grown in Pecos County, or Reeves County, Texas, or the state of Montana.

Do not apply TREFLAN 5 through any type of irrigation system.

DIRECTIONS FOR USE

TREFLAN 5 is a preemergence herbicide which is incorporated into the soil to provide long-lasting control of many annual grasses and broadleaf weeds. TREFLAN controls weeds as they germinate. TREFLAN will not control established weeds.

Weeds and Grasses Controlled by TREFLAN 5

Grasses

Annual bluegrass — *Poa annua*
Barnyardgrass — *Echinochloa* sp.
(Watergrass)
Brachiaria — *Brachiaria* sp.
(Signalgrass)
Bromegrass — *Bromus tectorum*
(Cheatgrass)
(Downy brome)
Cheat — *Bromus secalinus*
(Chess)
Crabgrass — *Digitaria* spp.
(Large crabgrass)
(Smooth crabgrass)
Foxtail — *Setaria* spp.
(Bottlegrass)
(Bristlegrass)
(Giant foxtail)
(Green foxtail)
(Foxtail millet)
(Pigeongrass)
(Robust foxtail)
(Yellow foxtail)
Goosegrass — *Eleusine indica*
(Silver crabgrass)
(Silvergrass)
(Wiregrass)
(Yardgrass)
Johnsongrass (from seed) *Sorghum halepense*
(Rhizome—see page 23 for special instructions for control in cotton and see page 32 for special instructions for control in soybeans)
Junglerice — *Echinochloa colonum*
Panicum
Fall panicum — *Panicum dichotomiflorum*
(Spreading panicgrass—see page 23 for special instructions in cotton and see page 31 for special instructions in soybeans)
Guineagrass — *Panicum maximum*
(See page 64 for special instructions)
Texas panicum — *Panicum texanum*
(Buffalograss)
(Coloradograss)
Itchgrass — *Rottboellia exaltata*
(Raoulgrass)
(See page 65 for special instructions)
Red rice — *Oryza sativa*
(See page 33 for suppression or partial control directions)
Sandbur — *Cenchrus incertus*
(Burgrass)
Sprangletop — *Leptochloa filiformis*
Stinkgrass — *Eragrostis cilianensis*
(Lovegrass)
Wild cane — *Sorghum bicolor*
(Shattercane)
(See page 37 for special instructions)
Woolly cupgrass — *Eriochloa villosa*

Broadleaf Weeds

Carpetweed — *Mollugo verticillata*
Chickweed — *Stellaria media*
Field bindweed — *Convolvulus arvensis*
(See page 71 for special instructions)
Florida pusley — *Richardia scabra*
(Florida purslane)
(Mexican clover)
(Pusley)
Goosefoot — *Chenopodium hybridum*
Henbit — *Lamium amplexicaule*
(fall application only)
Knotweed — *Polygonum aviculare*
Kochia — *Kochia scoparia*
(Fireweed)
(Mexican fireweed)
Lambsquarters — *Chenopodium album*
Pigweed — *Amaranthus* spp.
(Carelessweed)
(Prostrate pigweed)
(Redroot)
(Rough pigweed)
(Spiny pigweed)
Puncturevine — *Tribulus terrestris*
(Western U.S. only)
(Caltrop)
(Goathead)
Purslane — *Portulaca oleracea*
Russian thistle — *Salsola kali*
(Tumbleweed)
Stinging nettle — *Urtica dioica*
(Nettle)

Soil Preparation

Crop Residues or Existing Weeds: Ground cover, such as crop residues or existing weeds, can interfere with the incorporation of TREFLAN into the soil. A manageable level of such ground cover will allow the TREFLAN to be uniformly incorporated into the top 2 to 3 inches of soil. If the level of the ground cover is such that this cannot be done, you must till the soil prior to the application of TREFLAN.

Roughness: The soil surface should be smooth enough so that you can operate a sprayer and incorporation equipment efficiently and at speeds which insure a uniform application and incorporation of TREFLAN.

General Soil Conditions: To assure uniform incorporation of TREFLAN, soil moisture conditions should be such that large clods can be broken up during the incorporation process.

Soil Texture Guide

The amount of TREFLAN you apply will vary with the soil texture and organic matter. A fine textured soil will require more TREFLAN per acre than a coarse soil. Choose the proper rate for each application based on the following soil texture group and specific crop recommendations. Do not exceed recommended rates.

Soil Texture	Soil Classification
Coarse Soils: (Light)	Sand, loamy sand, sandy loam

Soil Texture	Soil Classification
Medium Soils:	Loam, silty clay loam*, silt loam, silt, sandy clay loam*
Fine Soils:	Clay, clay loam, silty clay loam*, silty clay, sandy clay, sandy clay loam*

**Silty clay loam and sandy clay loam soils are transitional soils and may be classified as either medium or fine textured soils. If silty clay loam or sandy clay loam soils are predominately sand or silt, they are usually classified as medium textured soils. If they are predominately clay, they are usually classified as fine textured soils.*

Mixing and Application Directions

TREFLAN 5 Alone in Water

Start with a clean spray tank. Fill sprayer 1/3 to 1/2 full with clean water. Start agitation. Add correct quantity of TREFLAN, continue agitation and finish filling the tank.

TREFLAN 5 Tank Mix in Water

Vigorous, continuous agitation is required for all tank mixes (Sparger pipe agitators generally provide the best agitation in spray tanks). Avoid stirring or splashing air into the mixture during filling to prevent foaming. To do this, place the end of the fill pipe below the surface of the water in the spray tank. Do not allow the mixture to siphon back into the water source.

Mixing order: Fill the tank 1/4 to 1/3 full with clean water. Start the agitation. Add dry flowables, wettable powders (WP), aqueous suspensions (AS), flowables (F), and liquids (L) to the water and agitate until the product(s) are completely dispersed in the water. Allow additional mixing and dispersion time when using dry flowable products. Continue agitation and fill tank to 3/4 full, add the TREFLAN 5, mix thoroughly. Then add any solution (S) formulations, agitate and finish filling. Maintain agitation during filling and through application. If spraying and agitation must be stopped before the tank is empty, the materials may settle to the bottom. In this case, it is important to resuspend all of the material in the bottom of the tank before continuing the spray application. A sparger agitator is particularly useful for this purpose. Sometimes it is more difficult to resuspend settled material than it is to suspend it originally.

Read and carefully follow all label instructions for each material added to the tank. Premixing dry and flowable formulations with water (slurrying) and pouring the slurry through a 20 or 35 mesh wetting screen in the top of the tank will help assure good initial dispersion in the tank water. Line screens in the tank should be no finer than 50 mesh (100 mesh is finer than 50 mesh).

If you see a buildup of material on the walls of the spray tank, wash the tank with soapy water between fillings. Rinse and continue the spraying operation. Clean the tank, lines, and screens thoroughly after use.

As the spray volume decreases, the importance of accurate calibration and uniform application increases. Check the sprayer daily to insure proper calibration and uniform application. Do not apply TREFLAN when the wind can cause drifting of spray particles which can result in non-uniform application. TREFLAN should not be applied to soils which are wet or are subject to prolonged periods of flooding as poor weed control may result.

Ground application: Apply TREFLAN in 5 to 40 gallons of water or liquid fertilizer per acre (broadcast basis), using any properly calibrated, low pressure herbicide sprayer that will apply the spray uniformly.

Aerial application: Apply TREFLAN in 5 to 10 gallons of water or liquid fertilizer per acre. Adjust pump pressure, nozzle arrangements, speed and height to provide a uniform application to the soil surface. Use swath markers or flagmen to assure proper application spray widths.

INCORPORATION DIRECTIONS

Incorporation Equipment—General Directions

Use incorporation equipment that mixes TREFLAN into the top 2 to 3 inches of the final seedbed, or erractic weed control and/or crop injury may result. Incorporation equipment such as a disc will mix TREFLAN approximately half as deep as the equipment is set to operate. For example, a disc set to cut 4 inches deep will incorporate most of the TREFLAN within the top 2 inches of soil.

Incorporation Before Planting

TREFLAN must be incorporated one time within 24 hours after application. Then any time prior to planting, a second incorporation is necessary, this time running the equipment in a different direction from the first. You should incorporate the TREFLAN uniformly into the top 2 to 3 inches of the final seedbed.

Incorporation After Planting

Check specific crop for incorporation directions after planting.

Incorporation in Bedded Culture

For effective weed control, TREFLAN needs to be incorporated into the top 2 to 3 inches of the final seedbed.

Application prior to bedding: Apply TREFLAN and incorporate it one time with recommended equipment. The bedding operation serves as the second incorporation. Do not expose untreated soil during post-bedding operations.*

Application after bedding: Knock off beds to planting height before applying TREFLAN. Apply TREFLAN and incorporate it with recommended equipment that will conform to the bed shape. Do not leave untreated soil exposed.*

** Avoid removal of treated soil from the seedbed before or during the planting operation. This would expose untreated soil, allowing weeds to germinate in the drill row.*

Recommended Equipment

Any recommended incorporation tool may be used alone or in combination with any other recommended tool. Two incorporation passes are required unless specifically stated. The second incorporation should not be deeper than the first.

Disc: set to cut 4 to 6 inches deep and operate at 4 to 6 mph.

Field Cultivator: set to cut 3 to 4 inches deep and operate at 5 mph or more. A field cultivator is defined as an implement with 3 to 4 rows of sweeps, spaced at intervals of 7 inches or less and staggered so that no soil is left unturned. Chisel points should not be used.

Combination Seedbed Conditioners: set to cut 3 to 4 inches deep and operate at a speed of at least 5 mph. These implements are defined as three or more tillage devices combined and used as a single tool. For example, 2 to 3 rows of field cultivator C- or S-shaped shanks with an effective sweep spacing of 6 to 9 inches (staggered so that no soil is left unturned), followed by a spike-tooth or flextine harrow, followed by a ground-driven reel or basket.

Rolling Cultivator: set to cut 2 to 4 inches deep and operate at 6 to 8 mph. Rolling cultivators are adequate for use on coarse and medium textured soils only, except when used in sugarcane where the rolling cultivator may be used on fine textured soils.

Bed Conditioner (Do-All): set to cut 2 to 4 inches deep and operate at 4 to 6 mph. The do-all is adequate for use on coarse and medium textured soils only. When using the do-all in bedded culture, only one incorporation pass is required. However, two passes with a do-all are required in flat planted culture.

Mulch Treader (other similar disc-type implements): set to cut 3 to 4 inches deep and operate at 5 to 8 mph.

P.T.O. Driven Equipment (tillers, cultivators, hoes): adjust to incorporate TREFLAN into the top 2 to 3 inches of the seedbed with rotors spaced to provide a clean sweep of the soil. Only one incorporation is necessary. P.T.O. driven equipment should not be operated at a speed greater than 4 mph.

Other equipment, including the flexible tine-tooth harrow (Flextine, Melroe), is also recommended but only for the special programs for which it is specified in this label.

CULTIVATION AFTER PLANTING

Soil treated with TREFLAN may be shallow cultivated without reducing the weed control activity of TREFLAN. Do not cultivate deeper than the treated soil since this may bring untreated soil to the surface, and poor weed control may result.

CROP RECOMMENDATIONS

These recommendations are given as the broadcast rates of TREFLAN per acre. For band applications, decrease the amount of TREFLAN in proportion to the amount of surface treated per acre. Apply any time after January 1 when the soil can be worked and is suitable for good incorporation. TREFLAN can be applied in the fall—see specific crop for recommendations. For general fall application directions where specific recommendations are not given, see page 77. Where a rate range is shown, use the lower rate for coarser soils or soils with lower organic matter. TREFLAN should not be used on soils containing more than 10% organic matter.

Cotton—TREFLAN Alone

TREFLAN can be applied and incorporated before or at planting, immediately after planting, or at layby.

Cotton—Preemergence: Broadcast Rates Per Acre

	TREFLAN 5	
Soil Texture	Areas receiving less than 20" average annual rainfall*	Areas receiving greater than 20" average annual rainfall*
	(pints)	(pints)
Coarse	0.8	0.8
Medium	1.0-1.2	1.2
Fine	1.2	1.6

** Use 1.2 pints per acre on coarse and medium textured soils and 1.6 pints on fine soils with 2-5% organic matter; use 1.6 to 2 pints on all soils with 5-10% organic matter.*

Cotton—Postplant:

When incorporating TREFLAN after planting (postplant), be careful not to disturb the seed.

Cotton—Layby:

Apply and incorporate TREFLAN any time up to layby, but not less than 90 days before harvest. Direct the layby applications onto the soil between the rows and beneath emerged cotton plants. Use the same rates as for a preemergence application.

Cotton—Fall Application:

Apply and incorporate TREFLAN any time between October 15 and December 31. The ground may be left flat or bedded-up over winter. On bedded ground, knock beds down to desired height before planting, moving some treated soil from beds into furrows. Where soil is left flat over winter, be careful not to turn up untreated soil during spring bedding operations. Destroy established weeds during seedbed preparation. If weeds become established in furrows due to uncovering of untreated soil during bedding, destroy these weeds before planting. In the fall, do not apply TREFLAN to soils which are wet or subject to prolonged periods of flooding.

Broadcast Rates per Acre—Fall application only:

In Alabama, Arkansas, northern Florida, Georgia, Louisiana, Mississippi, southeastern Missouri bootheel, North Carolina, New Mexico, Oklahoma, South Carolina, Tennessee and Texas, apply and incorporate TREFLAN at a broadcast rate of 1.6 pints per acre on coarse and medium soils and 2 pints on fine soils.

In Arizona, California and Nevada, apply and incorporate TREFLAN at a broadcast rate of 1.2 pints per acre on coarse soils; 1.6 pints on medium soils; and 2 pints on fine soils.

For cotton grown in other states, apply and incorporate TREFLAN at a broadcast rate of 0.8 pint per acre on coarse soils; 1.2 pints on medium soils; 1.6 pints on fine soils; 1.2 pints on coarse soils with 2-5% organic matter; and 1.6 to 2 pints on soils with 5-10% organic matter.

Cotton—Special Use Directions

Cotton—Fall panicum:

Apply and incorporate TREFLAN at the broadcast rate of 1.6 pints per acre on both coarse and medium soils.

Cotton—Rhizome johnsongrass:

In all cotton-producing states except Arizona and California, you can obtain commercially acceptable control of rhizome johnsongrass with a double rate program which you apply for 2 consecutive years in accordance with the following directions.

Soil Preparation—Proper preparation of the soil before application is very important for satisfactory results. Use a chisel plow or similar implement to bring rhizomes to the top of the soil. Then follow with a disc two times before application to cut the rhizomes into small (2 to 3 inch) pieces. This should also destroy any emerged johnsongrass.

Application—Choose the one application program that best fits your cultural practices:

Spring Application—Apply TREFLAN any time before planting in the spring for 2 years in a row. Use a broadcast rate of 1.6 pints per acre on coarse soils; 2.4 pints on medium soils; and 3.2 pints on fine soils,

OR

Fall Application—Apply TREFLAN between October 15 and December 31 for 2 years in a row at the same rates as a spring application for the control of rhizome johnsongrass.

Incorporation—Deep incorporation is essential for good rhizome johnsongrass control. Incorporate TREFLAN thoroughly with a disc set to cut 4 to 6 inches deep and operate at 4 to 6 mph. Two passes are necessary, with the second pass in a different direction from the first.

Cultivation—Some johnsongrass plants will escape. Timely cultivations during the crop season to remove escaped plants are necessary to obtain commercially acceptable control. You cannot obtain commercially acceptable control with only 1 year of double rate TREFLAN use.

Crop Rotation—In the season following a double rate treatment, plant only rice and those crops for which TREFLAN can be applied as a preplant treatment or injury may result.

Cotton—Pigweed and seedling johnsongrass control:

In Alabama, Arkansas, Florida, Georgia, Louisiana, Mississippi, southeastern Missouri, North Carolina, South Carolina, Tennessee and southern Virginia, TREFLAN may be applied preplant at a broadcast rate of 0.8 to 1.2 pints per acre on coarse soils; 1.2 to 1.6 pints on medium soils; and 1.6-2 pints on fine soils. Exception: in the State of Louisiana, 2.4 pints per acre are recommended on fine soils.

Cotton—Additional weed and grass control (Texas Gulf Coast):

In the Texas Gulf Coast counties of Brazoria, Calhoun, Chambers, Fort Bend, Galveston, Harris, Jackson, Jefferson, Liberty, Matagorda, Orange, Victoria, Waller and Wharton, TREFLAN may be applied up to 2 weeks before planting at a broadcast rate of 1.2 pints per acre on coarse soils; 1.6 pints on medium soils; and 2.4 pints on fine soils.

Cotton Precautions

Cotton should be planted after early season adverse weather conditions have passed, especially when using higher rate programs. Cool, wet weather early in the growth cycle causes additional stress to the cotton plant. This may result in reduced stands, delayed maturity and reduced yields.

Cotton—Tank Mixes and Overlays

Follow recommended soil preparation and incorporation procedures for TREFLAN.

Cotton—TREFLAN/Caparol tank mix for cotton grown in California, Arizona, New Mexico and west Texas:

The TREFLAN/Caparol combination will control certain grasses and broadleaf weeds listed for TREFLAN alone plus the following weeds:

Annual morningglory	Prickly sida (Teaweed)
Groundcherry (Annual)	Ragweed
Malva	Smartweed
Mustard	Wild oat

The tank mix also controls shallow germinating seedlings of cocklebur and coffeeweed.

Broadcast Rates Per Acre

Soil Texture	TREFLAN 5	Caparol 80W
	(pints)	(pounds)
Coarse	0.8	2*
Medium	1.0-1.2	2½
Fine	1.6	2½

* *Do not use on sands and loamy sands. For band applications use proportionately less.*

Mixing Directions: Carefully follow the procedures on the Caparol 80W label for making a slurry and adding it to a partially filled tank of water. After the Caparol is thoroughly mixed with the partially filled tank of water, add the TREFLAN 5 and continue filling. Agitate continuously during the filling and spraying operation.

Avoid leaving the spray mixture in the tank without constant agitation. If bypass agitation is used, the bypass line should stop at the bottom of the tank to minimize foaming.

Additional Precautions: The combination of TREFLAN/Caparol should not be used in the cut areas of newly leveled fields, in areas of excess salt, or where flooding over the beds is likely to happen. Do not plant cotton in tractor wheel depressions. These conditions may cause crop injury. On mulch-planted cotton, water back only after cotton seedlings are well established.

Crop Rotations: Cabbage, okra, onions and peas may be planted in the fall after a spring application of TREFLAN plus Caparol. Winter barley, winter rye and winter wheat can be planted in the fall also, if they are plowed down and not used for food or feed. Refer to the Caparol label for directions, cautions and precautions.

Cotton—TREFLAN/Cotoran tank mix (except in Arizona and California):

Follow recommended soil preparation and incorporation procedures for TREFLAN.

The TREFLAN/Cotoran tank mix effectively controls all the annual grasses and broadleaf weeds listed for TREFLAN alone plus these additional weeds:

Buttonweed	Ragweed
Cocklebur	Ryegrass
Groundcherry, Wright	Sesbania
Jimsonweed	Sicklepod
Morningglory	Smartweed
Prickly sida (Teaweed)	Tumbleweed

Broadcast Rates Per Acre

Soil Texture	TREFLAN 5	Cotoron 80W
	(pints)	(pounds)
Coarse	0.8	1¼
Medium	1.2	2
Fine	1.6	1½

Mixing Directions: Carefully follow the procedures on the Cotoran label for making a Cotoran slurry and adding it to a partially filled tank of water. After the Cotoran is thoroughly mixed with the partially filled tank of water, add the TREFLAN and continue filling. Agitate continuously throughout the filling and application operations. Do not leave spray mixture in tank without constant agitation. If bypass agitation is used, the bypass line should stop at the bottom of the tank to minimize foaming. Apply in 15-40 gallons of water per acre.

Additional Precautions: Do not plant crops other than cotton on the treated land within 6 months after the application of TREFLAN plus Cotoran, or crop injury may result. Do not feed foliage from treated cotton plants or gin trash to livestock. Do not mix TREFLAN plus Cotoran with liquid fertilizer.

West Texas Only: Do not use the tank mix of TREFLAN plus Cotoran on sandy, loamy sand or fine sandy loam soils. Do not use on cotton planted in furrows.

Arkansas, Louisiana, and Mississippi Only: Use 1 pound Cotoran 80W in tank mix with TREFLAN on sandy loam soils low in organic matter.

New Mexico: Do not plant treated land with crops other than cotton until 1 year after the last application. Do not use on sandy or coarse textured soils of less than 1% organic matter.

Cotton—TREFLAN preplant followed by Cotoran overlay:
Apply and incorporate TREFLAN as recommended for the specific soil texture. Then apply Cotoran as a preemergence surface treatment at 1¼ to 2½ pounds per acre. On light silt and sandy soils low in organic matter, use the lower rate.

Additional Precaution: Refer to the Cotoran label for all cautions and precautions when using this preemergence surface treatment.

Cotton—TREFLAN preplant followed by Karmex overlay:
For cotton grown east of the Mississippi River plus Arkansas, southeastern Missouri, Louisiana and Eastern Texas, incorporate TREFLAN before planting. Then follow with a preemergence application of Karmex. This will effectively control all the weeds controlled by TREFLAN alone plus these additional weeds:

Annual groundcherry	Shepherdspurse
Annual morningglory	Velvetgrass
Dogfennel	Wild lettuce
Pennycress	Wild mustard
Ragweed	

Broadcast Rates Per Acre

Soil Texture	TREFLAN 5	Karmex 80W
	(pints)	(pounds)
Coarse	0.8	⅓
Medium	1.2	⅔
Fine	1.6	1

Additional Precautions: Do not use Karmex on soils with less than 1% organic matter as crop injury may result. Do not allow livestock to graze on cotton treated with Karmex. Consult the Karmex label for additional instructions, cautions and precautions.

Soybean—TREFLAN Alone

Soybean—Preemergence:
Follow recommended soil preparation, application, and incorporation procedures for TREFLAN.

Broadcast Rates Per Acre

Soil Texture	TREFLAN 5*
	(pints)
Coarse	0.8
Medium	1.2
Fine	1.6

* *Use 1.2 pints per acre on coarse and medium textured soils and 1.6 pints on fine soils with 2-5% organic matter; use 1.6-2 pints on all soils with 5-10% organic matter.*

Soybean—Fall Application:
Apply and incorporate TREFLAN anytime between October 15 and December 31. Ground may be left flat or bedded-up over winter. On bedded ground, knock beds down to desired height before planting, moving some treated soil from tops into furrows. Where soil is left flat over winter, take care during spring bedding operations to prevent turning up untreated soil. Destroy established weeds during seedbed preparation. If weeds become established in furrows due to uncovering of untreated soil during listing, destroy these weeds before planting. Do not fall apply TREFLAN to soils which are wet, are subject to prolonged periods of flooding, or where rice was grown the previous year.

For soybeans grown in Alabama, Arkansas, northern Florida, Georgia, Louisiana, Mississippi, southeastern Missouri Bootheel, North Carolina, Oklahoma, South Carolina, Tennessee and Texas, apply and incorporate TREFLAN at a broadcast rate of 1.6 pints per acre on coarse and medium soils and 2 pints on fine soils.

For soybeans grown in states other than those listed above, apply and incorporate TREFLAN at a broadcast rate of 0.8 pint per acre on coarse soils; 1.2 pints on medium soils; 1.6 pints on fine soils; 1.2 pints on coarse soils with 2-5% organic matter; and 1.6 to 2 pints on soils with 5-10% organic matter.

Soybean—Special Use Directions

Soybean—Fall panicum:
Apply TREFLAN at the broadcast rate of 1.6 pints per acre on both coarse and medium soils.

Soybean—Pigweed and seedling johnsongrass:
In Alabama, Arkansas, Florida, Georgia, Kansas, Louisiana, Mississippi, Missouri, Nebraska, North Carolina, Oklahoma, South Carolina, Tennessee and southern Virginia, TREFLAN may be applied at a broadcast rate of 0.8 to 1.2 pints per acre on coarse soils; 1.2 to 1.6 pints on medium soils; and 1.6 pints on fine soils. Exception: in the state of Louisiana, 2.4 pints per acre are recommended on fine soils.

Soybean—Additional weed and grass control (Texas Gulf Coast):
In the Texas Gulf Coast counties of Brazoria, Calhoun, Chambers, Fort Bend, Galveston, Harris, Jackson, Jefferson, Liberty, Matagorda, Orange, Victoria, Waller and Wharton, TREFLAN may be applied up to 2 weeks before planting at a broadcast rate of 1.2 pints per acre on coarse soils; 1.6 pints on medium soils; and 2.4 pints on fine soils.

Soybean—Charcoal soils in Arkansas, Louisiana and Mississippi:
Newly cleared land often contains high organic matter (5-10%) and charcoal from burning debris. This charcoal and/or organic matter tends to bind TREFLAN and reduce its weed control activity. Under these conditions, higher rates of TREFLAN are necessary for weed control. Increased rates, however, can cause crop injury if charcoal or organic matter is not present to bind some of the TREFLAN. In the burn row a high level of charcoal is present; consequently, poor weed control may result even with an increased rate of TREFLAN.

Apply TREFLAN at the broadcast rate of 1.6 to 2 pints per acre on coarse soils; 2 pints on medium soils; and 2.4 pints on fine textured soils. Follow recommended soil preparation, application and incorporation procedures for TREFLAN.

Soybean—Red rice in Arkansas, Louisiana, Mississippi and Texas only:
You can obtain suppression or partial control of red rice when you apply TREFLAN at the following recommended rates. Follow recommended soil preparation and incorporation procedures for TREFLAN.

Apply and incorporate TREFLAN in the spring before planting.

Broadcast Rates Per Acre

Soil Texture	TREFLAN 5	
	Application Year 1	Application Year 2
	(pints)	(pints)
Coarse	1.6	0.8
Medium	2.4	1.2
Fine	3.2	1.6
Coarse soils with 2 to 5% organic matter	2.4	1.2
Soils with 5 to 10% organic matter	3.2	1.6-2

If a combination of high organic matter (5-10%) and charcoal are present in the soil, apply TREFLAN the second year at the following rates for charcoal soils in Arkansas, Louisiana and Mississippi:

Broadcast Rates Per Acre

Soil Texture	TREFLAN 5
	(pints)
Coarse	1.2-2
Medium	2
Fine	2.4

For more information on charcoal soils see page 32.

Crop Rotation: The program for red rice control in soybeans is a 2-year program. Use the rates listed for first year application and plant soybeans. The second year use the normal TREFLAN rates listed for your soil type and charcoal level and plant only those crops for which TREFLAN has been registered as a preplant treatment, or crop injury may result. Do not plant rice the second year. Rice may be planted the third year.

Soybean—Rhizome johnsongrass (Eastern United States and the state of Texas):
You can obtain commercially acceptable control of rhizome johnsongrass with a double rate program applied for 2 consecutive years in accordance with the following directions.

Soil Preparation—Proper preparation of the soil before application is very important for satisfactory results. Use a chisel plow or similar implement to bring rhizomes to the top of the soil. Then follow with a disc two times before application to cut the rhizomes into small (2 to 3 inch) pieces and to destroy any emerged johnsongrass.

Application—Choose the one application program that best fits your cultural practices:

> **Spring Application**—Apply TREFLAN anytime in the spring before planting for 2 years in a row. Use a broadcast rate of 1.6 pints per acre on coarse soils; 2.4 pints on medium soils; 3.2 pints on fine soils; 2.4 pints on coarse soils with 2-5% organic matter; and 3.2 pints on soils with 5-10% organic matter.
>
> *OR*
>
> **Fall Application**—Apply TREFLAN between October 15 and December 31 for 2 years in a row at the same rates as a spring application for the control of rhizome johnsongrass.
>
> *OR*
>
> **Split Application**—Apply TREFLAN at the same rate in both the spring and fall for 2 years in a row using the rates in the following table.

Broadcast Rates Per Acre

Soil Texture	TREFLAN 5 Spring and Fall
	(pints)
Coarse	0.8
Medium	1.2
Fine	1.6
Coarse soils with 2-5% organic matter	1.2
Soils with 5-10% organic matter	1.6

Incorporation—Deep incorporation is essential for good rhizome johnsongrass control. Incorporate TREFLAN thoroughly with a disc set to cut 4 to 6 inches deep and operate at 4 to 6 mph. Two passes are necessary, with the second pass in a different direction from the first.

Cultivation—Some johnsongrass plants will escape. Timely cultivations during the crop season to remove escaped plants are necessary to obtain commercially acceptable control.

Crop Rotation: In the season following a double rate treatment, plant only rice and those crops for which TREFLAN can be applied as a preplant treatment or injury may result.

Soybean—Rhizome johnsongrass - TREFLAN/Sencor or TREFLAN/Lexone tank mix:

TREFLAN/Sencor or TREFLAN/Lexone may be used for rhizome johnsongrass control and for the control of those weeds listed for TREFLAN alone. For the additional weeds controlled by Sencor or Lexone in tank mix, see page 38. Follow procedures for soil preparation, incorporation, and cultivation recommended in the Soybean - Rhizome johnsongrass section (see page 34).

Apply TREFLAN/Sencor or Lexone up to two weeks before planting for two consecutive years at the following broadcast rates per acre:

Broadcast Rates Per Acre

Soil Texture	TREFLAN 5	Lexone 50WP/4L or Sencor 50WP/4	or Lexone (dry flowable) or Sencor (dry flowable)
	(pints)	(pounds/pints)	(pounds)
Coarse*	1.6	1/2	1/3
Medium	2.4	3/4	1/2
Fine	3.2	1	2/3

** Do not use on coarse soils with less than 1% organic matter.*

Read and follow all additional precautions listed for the TREFLAN/Sencor or TREFLAN/Lexone tank mix (see page 39).

Soybean—Wild cane (shattercane):

Follow recommended soil preparation and application procedures for TREFLAN.

Wild cane (shattercane) can germinate throughout the growing season and from greater soil depth than most other weed seeds. Commercially acceptable control of wild cane can be obtained with the following increased rates of TREFLAN.

Apply TREFLAN at the broadcast rate of 0.8 pint per acre on a coarse soil; 1.6 pints on a medium soil; and 2 pints on a fine textured soil.

Incorporation—Deep incorporation is essential for good wild cane control. Incorporate TREFLAN thoroughly with a disc set to cut 4 to 6 inches deep and operate at 4 to 6 mph. Two passes are necessary, with the second pass in a different direction from the first.

Cultivation—Cultivations during the crop season will also contribute to control.

Soybean Precautions

Soybeans should be planted after early season adverse weather conditions have passed, especially when using higher rate programs. Cool, wet weather early in the growth cycle causes additional stress to the soybean plant. This may result in reduced stands, delayed maturity and reduced yields.

Soybean—Tank-Mix, Overlay and Postemergence Recommendations

Soybean—TREFLAN/Sencor or TREFLAN/Lexone tank mix:

The TREFLAN/Sencor or TREFLAN/Lexone tank mix effectively controls the annual grasses and broadleaf weeds controlled by TREFLAN alone plus these additional weeds:

Jimsonweed	Ragweed, common
Mallow, Venice (Flower-on-an-hour)	Sesbania, hemp
Mustard, wild	Smartweed, Pennsylvania
Prickly sida	Velvetleaf

Control of cocklebur, morningglory and giant ragweed (horseweed) may be erratic. Control may be improved with timely cultivation. Where cocklebur is a serious problem, an overlay of Sencor or Lexone may be preferred to the TREFLAN/Sencor or TREFLAN/Lexone tank mix.

Follow recommended procedures for soil preparation, incorporation, and cultivation of TREFLAN. Mix according to instructions on page 15. The TREFLAN/Sencor or TREFLAN/Lexone tank mix can be applied from 2 weeks before planting up to planting.

Broadcast Rates Per Acre

Soil Texture	TREFLAN 5	Lexone 50WP/4L or Sencor 50WP/4	or Lexone (dry flowable) or Sencor (dry flowable)
	(pints)	(pounds/pints)	(pounds)
Coarse*	0.8	1/2	1/3
Medium	1.2	3/4	1/2
Fine	1.6	1	2/3

** Do not use Sencor/Lexone on coarse soils with less than 1% organic matter.*

Precautions: Do not plant any crop other than soybeans within 4 months after treatment. Overapplication, uneven application, or improper soil incorporation may result in crop injury, herbicide residue, or erratic weed control. Additional stress factors are seedling disease, cold weather, deep planting, excessive moisture, soil *p*H over 7.5, high salt concentration, or drought. Any of these may weaken crop seedlings and increase possibility of damage from the tank mix. These additional factors may also delay crop development or reduce yields when Sencor or Lexone is applied. Observe all cautions and limitations on the Sencor and Lexone labels. Do not use the foliage from soybeans treated with the TREFLAN/Sencor or TREFLAN/Lexone tank mix for feed or forage.

Additional Sencor and Lexone precautions: Do not use Lexone or Sencor on Tracy, Semmes, Altona, Vansoy or Coker 102 soybeans. These varieties are sensitive to Lexone or Sencor, and crop injury may result. Seed must be planted at least 1 1/2 inches but not more than 2 inches below the soil surface before a Sencor or Lexone application. Do not apply Sencor or Lexone at these rates more than once per season. Do not replant areas treated with Sencor or Lexone with any crop other than soybeans within 4 months after treatment. Injury to soybeans may occur if you use Lexone or Sencor on soils having a calcarous surface or *p*H of 7.5 or higher, or if you use them in conjunction with soil-applied organic phosphate pesticides.

Soybean—TREFLAN/Canopy Tank-Mix (Do Not Use in California)

Apply TREFLAN/Canopy as a preplant incorporated tank-mix treatment. Follow recommended soil preparation, application and incorporation procedures for TREFLAN. Plant soybeans within two (2) weeks after application.

TREFLAN/Canopy tank-mix controls the following weeds in addition to those controlled by TREFLAN alone (excluding special rate programs).

Controlled

Cocklebur, common	Prickly sida (Teaweed)
Florida beggarweed	Ragweed (Common) (Giant)
Hemp sesbania	Sicklepod
Hophornbeam copperleaf	Smartweed
Jimsonweed	Spotted spurge
Morningglory (Entireleaf) (Ivyleaf) (Pitted) (Smallflower) (Tall)	Sunflower
	Velvetleaf

Partially Controlled

Purple nutsedge
Yellow nutsedge

Large seeded weeds, germinating deep in the soil, such as morningglory, sicklepod, cocklebur and giant ragweed or weeds with subsequent flushes may require a cultivation or an application of a postemergence herbicide for season-long control.

Broadcast Rates Per Acre

Soil Texture	TREFLAN 5	Canopy Dispersible Granules 0.5-3% O.M.	Canopy Dispersible Granules 3-5% O.M.
	(pints)	(ounces)	(ounces)
Coarse	0.8	6-8	8-10
Medium	1.2	8-10	10-12
Fine	1.6	10-12	12-14

Do not apply Canopy to soil with less than 1/2% organic matter.

Use higher rates of Canopy on soils with higher organic matter or heavy pressure from large deep germinating weed seeds.

Where Canopy is applied plant soybean seed 1 1/2" to 2" deep on a flat or raised seedbed only, or crop injury may occur.

Soybean injury may occur where Canopy is applied if excessive rainfall occurs after application but before soybeans germinate.

Precautions: Read the Canopy label carefully for cautions and precautions relating to environmental hazards, planting of rotation crops, sprayer contamination and cleanup, soil *p*H, organic matter and soil texture use restrictions, soybean variety planting restrictions, restrictions where Atrazine was used the previous year, restrictions concerning use with organic phosphate pesticides, grazing restrictions and other directions, precautions and limitations before applying the TREFLAN/Canopy tank-mix.

Soybean—TREFLAN/Preview Tank-Mix (Do Not Use in California)

Apply TREFLAN/Preview as a preplant incorporated tank-mix treatment. Follow recommended soil preparation, application and incorporation procedures for TREFLAN. Plant soybeans within two (2) weeks after application.

TREFLAN/Preview tank-mix controls the following weeds in addition to those controlled by TREFLAN alone (excluding special rate programs).

Controlled

Cocklebur, common	Prickly sida (Teaweed)
Hophornbeam	

copperleaf
Jimsonweed
Mustards
Pigweed
(Palmer amaranth)
(Smooth)
(Tall waterhemp)
Ragweed, common
Smartweeds, annual
Spotted spurge
Sunflower
Velvetleaf

Partially Controlled

Burcucumber
Eastern Black nightshade
Morningglory
(Tall)
(Ivyleaf)
(Pitted)
(Entireleaf)
Nutsedge species
Ragweed, giant

Large seeded weeds, germinating deep in the soil, such as morningglory, sicklepod, cocklebur and giant ragweed or weeds with subsequent flushes may require a cultivation or an application of a postemergence herbicide for season-long control.

Broadcast Rates Per Acre

Soil Texture	TREFLAN 5	Preview Dispersible Granules 0.5-3% O.M.	Preview Dispersible Granules 3-5% O.M.
	(pints)	(ounces)	(ounces)
Coarse	0.8	6	7
Medium	1.2	7	8
Fine	1.6	8	9-10

Do not apply Preview to soil with less than 1/2% organic matter.

Use higher rates of Preview on soils with higher organic matter or heavy pressure from large deep germinating weed seeds.

Where Preview is applied, plant soybean seed 1 1/2" to 2" deep on a flat or raised seedbed only, or crop injury may occur.

Soybean injury may occur where Preview is applied if excessive rainfall occurs after application but before soybeans germinate.

Precautions: Read the Preview label carefully for cautions and precautions relating to environmental hazards, planting of rotation crops, sprayer contamination and cleanup, soil *p*H, organic matter and soil texture use restrictions, soybean variety planting restrictions, restrictions where Atrazine was used the previous year, restrictions concerning use with organic phosphate pesticides, grazing restrictions and other directions, precautions or limitations before applying the TREFLAN/Preview tank-mix.

Soybean—TREFLAN/Amiben tank mix:

Amiben may be applied several days prior to planting as a broadcast tank mix with TREFLAN. Weeds controlled by this tank mix, in addition to those controlled by TREFLAN alone, are smartweed, velvetleaf and ragweed. The tank mixture should be used as a spring preplant incorporated treatment.

Apply TREFLAN as a tank mix with Amiben.

Broadcast Rates Per Acre

Soil Texture	TREFLAN 5	Amiben 2S	Amiben DS
	(pints)	(quarts)	(pounds)
Coarse	0.8	4-6*	2.4-3.6*
Medium	1.2	4-6	2.4-3.6
Fine	1.6	4-6	2.4-3.6

* *Use the higher rate where you expect heavy populations of smartweed, velvetleaf, ragweed, wild mustard or black nightshade. Do not use on muck or charcoal soils. Read and observe all directions and cautions on the Amiben label.*

Soybean—TREFLAN/Amiben/Sencor or Lexone tank mix:

The TREFLAN/Amiben/Sencor or Lexone tank mix effectively controls all weeds listed for TREFLAN/Amiben and TREFLAN/Sencor or Lexone tank mixes.

Follow recommended soil preparation, application, and incorporation procedures for TREFLAN. The TREFLAN/Amiben/Sencor or Lexone tank mix may be applied from several days prior to planting up to planting in 10 to 40 gallons of water per acre. Use screens no finer than 50 mesh.

Apply the TREFLAN/Amiben Sencor or Lexone tank mix at the following broadcast rate per acre.

Broadcast Rates Per Acre

Soil Texture	TREFLAN 5	Amiben 2S	Lexone 50WP/4L or Sencor 50WP/4	or	Lexone (dry flowable) or Sencor (dry flowable)
	(pints)	(quarts)	(pounds/pints)		(pounds)
Coarse[a]	0.8	3-4[b]	1/2		1/3
Medium	1.2	3-4[b]	1/2-3/4[c]		1/3-1/2[c]
Fine	1.6	4-5	3/4[c]		1/2[c]

[a] *Do not use Sencor or Lexone on coarse soils with less than 1% organic matter.*

[b] *Use the higher rate of Amiben when velvetleaf or black nightshade is a problem.*

[c] *On Clarion/Webster soils in Minnesota and Iowa or on similar alkaline (calcareous) soils with a pH of 7.5 or above, apply Sencor or Lexone at the rates listed on page 45:*

Broadcast Rates Per Acre

Soil Texture	Lexone 50WP/4L or Sencor 50WP/4	or	Lexone (dry flowable) or Sencor (dry flowable)
	(pounds/pints)		(pounds)
Medium	1/2		1/3
Fine	1/2-3/4[d]		1/3-1/2[d]

[d] *Use the higher rate only where soil pH is less than 7.5 and where weed pressure is heavy.*

Additional precautions: The TREFLAN/Amiben/Sencor or Lexone tank mix will not harm the treated crop when you apply it according to directions and under normal growing conditions. However, overapplication, uneven application or improper soil incorporation of the tank mix can result in erratic weed control or crop injury. Additional stress factors are seedling disease, cold weather, deep planting, excessive moisture, soil *p*H over 7.5, high salt concentration, or drought. These additional factors may weaken crop seedlings and increase the possibility of damage from the tank mix. These additional factors may also delay crop development or reduce yields. Observe all cautions and limitations of all products used in mixtures. Do not use the foliage from soybeans treated with the TREFLAN/Amiben/Sencor or Lexone tank mix for feed or forage.

Soybean—TREFLAN/Vernam tank mix:

TREFLAN/Vernam tank mix effectively controls those weeds listed for TREFLAN alone plus these additional weeds:

Annual morningglory
Coffeeweed
Purple nutsedge (nutgrass)
Velvetleaf
Yellow nutsedge (nutgrass)

Follow recommended soil preparation procedures for TREFLAN. You may apply the TREFLAN/Vernam tank mix up to 10 days prior to planting. Incorporate the tank mix immediately after application. Apply TREFLAN/Vernam at these rates:

Broadcast Rates Per Acre

Soil Texture	TREFLAN 5	Vernam 7E
	(pints)	(pints)
Coarse	0.8	1 3/4-2 1/3
Medium	1.2	2 1/3-3*
Fine	1.6	3-3 1/2

* *For nutsedge, wild cane and velvetleaf control, use the higher rate of 3 pints per acre on medium textured soils.*

Soybean—TREFLAN preplant incorporated followed by Overlay treatments (Do not use in California)

Apply TREFLAN as a preplant incorporated treatment. Additional weeds tolerant to TREFLAN may be controlled by using overlay preemergence applications of Amiben, Canopy, Dual, Gemini, Lasso, Lexone, Lorox, Lorox Plus, Preview or Sencor. Consult these product labels for additional weeds controlled, application directions and precautions before use.

Soybean—TREFLAN preplant incorporated followed by Post-emergence treatments (Do not use in California)

Apply TREFLAN as a preplant incorporated treatment. Additional weeds tolerant to TREFLAN may be controlled by using post-emergence applications of Basagran, Blazer, Classic or Tackle. Consult these product labels for additional weeds controlled, application directions and precautions before use.

Alfalfa—Established

In areas receiving less than 20" average annual rainfall per year, apply TREFLAN to established alfalfa stands at a broadcast rate of 1.2 pints per acre on coarse soils and 1.6 pints on medium and fine soils. Use incorporation equipment that will insure thorough soil mixing with minimum damage to the established alfalfa.

Asparagus—Established

Follow recommended soil preparation, application and incorporation procedures for TREFLAN.

TREFLAN can be applied to established asparagus as a single or as a split application. In the winter or early spring, apply TREFLAN to asparagus after ferns are removed but before spear emergence. Or, apply after harvest in the late spring or early summer before ferning begins. TREFLAN will suppress volunteer seedling asparagus and field bindweed if you use the following recommended rates and application schedules.

Follow recommended soil preparation, application and incorporation procedures for TREFLAN.

Broadcast Rates Per Acre

	TREFLAN 5*			
	Split Application		Single Application	
Soil Texture	Before Harvest	+ After Harvest	Before Harvest	or After Harvest
	(pints)		(pints)	
Coarse	0.8 + 0.8		1.6 or 1.6	
Medium	1.2 + 1.2		2.4 or 2.4	
Fine	1.6 + 1.6		3.2 or 3.2	

** In any single calendar year, the maximum TREFLAN 5 to apply is 1.6 pints per acre on coarse soils; 2.4 pints on medium soils; and 3.2 pints on fine soils.*

For the Following Crop Grouping, Use the Rate Table Below.

Carrot

Castor Bean

Celery—(Direct seeded and transplant in areas receiving less than 20″ average annual rainfall.)

Cole Crops—Transplant

Apply and incorporate prior to transplanting only. (Broccoli, Brussels Sprout, Cabbage and Cauliflower) See next section for direct seeded.

Okra

Pepper—Transplant

Apply and incorporate prior to transplanting only.

Southern Pea—(Before planting only.)

Apply and incorporate TREFLAN before planting, at planting or immediately after planting, unless otherwise indicated.

Broadcast Rates Per Acre

	TREFLAN 5	
Soil Texture	Areas receiving less than 20″ average annual rainfall*	Areas receiving greater than 20″ average annual rainfall*
	(pints)	(pints)
Coarse	0.8	0.8
Medium	1.0-1.2	1.2
Fine	1.2	1.6

** Use 1.2 pints per acre on coarse and medium textured soils and 1.6 pints on fine soils with 2-5% organic matter; use 1.6 pints on all soils with 5-10% organic matter.*

Cole Crops—Direct Seeded (Broccoli, Brussels Sprout, Cabbage and Cauliflower) See above section for transplant.

For direct-seeded cole crops, apply and incorporate TREFLAN before planting at a broadcast rate of 0.8 pint per acre on coarse and medium soils and 1.2 pints on fine soils and soils with 2-5% organic matter. Direct-seeded cole crops have exhibited marginal tolerance to recommended rates of TREFLAN. Stunting or reduced stands may occur.

Cucurbits—Postplant Emerged (Cantaloupe, Cucumber and Watermelon)

Western United States Including Texas: Apply TREFLAN as a directed spray to the soil between the rows and beneath plants which are in the 3 to 4 true leaf stage.

Broadcast Rates Per Acre

	TREFLAN 5	
Soil Texture	Areas receiving less than 20″ average annual rainfall*	Areas receiving greater than 20″ average annual rainfall*
	(pints)	(pints)
Coarse	0.8	0.8
Medium	1.0-1.2	1.2
Fine	1.2	1.6

** Use 1.2 pints on coarse and medium textured soils and 1.6 pints on fine soils with 2-5% organic matter; use 1.6 pints on all soils with 5-10% organic matter.*

Set incorporation equipment to throw treated soil around the plants during incorporation.

Dry Bean—TREFLAN Alone

Apply and incorporate TREFLAN before planting using the following rates:

Broadcast Rates Per Acre

	TREFLAN 5	
Soil Texture	Areas receiving less than 20″ average annual rainfall*	Areas receiving greater than 20″ average annual rainfall*
	(pints)	(pints)
Coarse	0.8	0.8
Medium	1.0-1.2	1.2
Fine	1.2	1.6

** Use 1.2 pints per acre on coarse and medium textured soils and 1.6 pints on fine soils with 2-5% organic matter; use 1.6 pints on all soils with 5-10% organic matter.*

Dry Bean—TREFLAN/Eptam tank mix:

The TREFLAN/Eptam tank mix effectively controls all the following weeds in addition to those listed for TREFLAN (see page 11):

Henbit (Spring applications)	Oat, wild
	Ragweed, common
Nightshade, black	Smartweed, Pennsylvania
Nightshade, hairy	
Nutsedge	Velvetleaf (Buttonweed)

Follow recommended soil preparation and incorporation procedures for TREFLAN. The TREFLAN/Eptam tank mix should be applied from 2 days before planting up to planting. Incorporate immediately after application.

Broadcast Rates Per Acre

	TREFLAN 5		
Soil Texture	Areas receiving less than 20″ average annual rainfall*	Areas receiving greater than 20″ average annual rainfall*	Eptam 7E
	(pints)	(pints)	(pints)
Coarse	0.8	0.8	2½-3½**
Medium	1.0-1.2	1.2	2½-3½
Fine	1.2	1.6	2½-3½

** Use 1.2 pints per acre on coarse and medium textured soils and 1.6 pints on fine soils with 2-5% organic matter; use 1.6 pints on all soils with 5-10% organic matter.*

*** Use Eptam 7E at a rate of 2½ pints per acre to control annual grasses; 3½ pints to control nutsedge and additional broadleaf weeds.*

Precaution: Read the Eptam label before using. Observe all cautions and limitations of all products used in mixtures. The combination of TREFLAN and Eptam should not be used on soybeans, black-eyed peas (beans), lima beans and other flatpodded beans, except Romano. Do not use the foliage from a crop treated with the TREFLAN/Eptam tank mix for feed or for grazing.

Dry Bean—Fall application for dry bean grown in Idaho, Oregon, and Washington:

Apply and incorporate TREFLAN any time between October 15 and December 31 at a broadcast rate of 0.8 pint per acre on coarse soils; 1 to 1.2 pints on medium soils; and 1.2 pints on fine soils. Destroy established weeds during seedbed preparation.

For the Following Crop Grouping, Use the Rate Listed Below.

Beans—(Guar and Mungbean)

Greens—Turnip greens grown for processing, Collard, Kale and Mustard greens.

Mustard—Grown for seed or processing for food in Minnesota, and North Dakota:

Apply and incorporate TREFLAN before planting at 0.8 pint per acre on coarse soils and 1.2 pints on medium and fine soils.

Beans—(Lima Bean and Snap Bean):

Apply and incorporate TREFLAN before planting at a broadcast rate of 0.8 pint per acre on coarse and medium soils and 1.2 pints on fine soils.

Corn (Field Corn) and Grain Sorghum (Milo):

Apply TREFLAN to field corn or grain sorghum (8 inches or taller) as an over-the-top or directed spray to effectively control weeds listed for TREFLAN.

Soil Preparation—Cultivate before a TREFLAN application to insure loose, friable soil, to remove established weeds, and to cover the base of plants with soil.

Application Directions—TREFLAN should be applied and incorporated at the recommended rates for the soil texture when the crop is well established (8 inches or taller). TREFLAN may be applied either as an over-the-top spray or as a directed spray. Drop nozzles should be used if foliage prevents uniform coverage of soil surface. Soil incorporation may be accomplished with only one pass of a sweep-type cultivator or a properly adjusted rolling cultivator.

The sweep-type cultivator should have 3 to 5 sweeps per row middle and be operated at 6 to 8 mph. Set the middle sweeps so as to avoid exposing untreated soil. Adjust the incorporation tools to prevent crop injury.

Broadcast Rates Per Acre

Soil Texture	TREFLAN 5
	(pints)
Coarse	0.6-0.8*
Medium	0.8-1.2
Fine	1.2-1.6

Use the lower rates when you anticipate light weed pressure and the higher rates when you anticipate heavy weed pressure.

* Corn only: *Apply 0.8 to 1.2 pints per acre in Alabama, Florida, Georgia, North Carolina, South Carolina and Virginia to control fall panicum and Texas panicum.*

Precaution: Do not apply TEFLAN to corn grown for seed. Do not apply TREFLAN to corn or sorghum as a preplant or preemergence treatment, or crop injury may occur.

Hops

Apply and incorporate TREFLAN while the crop is dormant. Use a broadcast rate of 0.8 pint per acre on coarse soils; 1 to 1.2 pints on medium soils; and 1.2 pints on fine soils and soils with 2-10% organic matter.

Mint—(Established Peppermint and Spearmint)

Apply TREFLAN at a rate of 0.8 pint per acre on coarse soils; 1.0 pints on medium soils; and 1.2 pints on fine soils. Use incorporation equipment that will insure thorough soil mixing with minimum damage to the crop.

Pea—(Dry and English)—TREFLAN Alone

Apply and incorporate TREFLAN before planting at a rate of 0.8 pint per acre on coarse and medium soils and 1.2 pints on fine soils.

Pea—TREFLAN/Far-Go tank mix for pea in Idaho, Oregon and Washington:

The tank mix combination of TREFLAN plus Far-Go will provide control of wild oat in addition to other annual grasses and broadleaf weeds controlled by TREFLAN.

Application Rates:

Broadcast 0.6 pint of TREFLAN per acre on coarse and medium soils; 0.8 pint of TREFLAN on fine soils. Use 1¼ quarts of Far-Go per acre for all soil textures.

Incorporation Directions:

Apply and incorporate up to 3 weeks before planting. Follow recommended incorporation procedures for TREFLAN.

Precaution: Do not apply to lentils. Leaf crinkling and delayed maturity of peas may occur, particularly on clay points in the northwest; but this is usually more than offset by a reduction of wild oat. Do not use foliage from treated peas for feed or forage. Refer to the cautions, precautions and directions on the Far-Go label.

Pea—Fall application to dry pea and English pea in Idaho, Oregon and Washington:

Apply and incorporate TREFLAN any time between October 15 and December 31 at a broadcast rate of 0.8 pint per acre on coarse soils; 1 to 1.2 pints on medium soils; and 1.2 pints on fine soils. Destroy established weeds during seedbed preparation. Do not apply TREFLAN in the fall to soils which are wet or are subject to prolonged periods of flooding.

Peanut—(Spanish Peanut in Texas and Oklahoma)

Apply and incorporate TREFLAN before planting, at planting or immediately after planting at a broadcast rate of 0.8 pint per acre on coarse soils. When incorporating after planting, take care not to disturb the seed.

Peanut—TREFLAN/Vernam tank mix (Spanish Peanut in Texas and Oklahoma):

TREFLAN/Vernam tank mix effectively controls those weeds listed for TREFLAN alone plus these additional weeds:

Annual morningglory	Velvetleaf
Coffeeweed	Yellow nutsedge
Purple nutsedge (nutgrass)	(nutgrass)

Follow recommended soil preparation procedures for TREFLAN. You may apply the TREFLAN/Vernam tank mix up to 10 days prior to planting. Incorporate the tank mix immediately after application. Apply TREFLAN/Vernam at these rates:

Broadcast Rates Per Acre

Soil Texture	TREFLAN 5	Vernam 7E
	(pints)	(pints)
Coarse	0.8	2⅓

Potato—(All states except Maine)

Apply and incorporate TREFLAN after planting, before emergence, or immediately following dragoff or after the potato plants have fully emerged.

Broadcast Rates Per Acre

	TREFLAN 5	
Soil Texture	Areas receiving less than 20" average annual rainfall*	Areas receiving greater than 20" average annual rainfall*
	(pints)	(pints)
Coarse	0.8	0.8
Medium	1.0-1.2	1.2
Fine	1.2	1.6

* *Use 1.2 pints per acre on coarse and medium soils with 2-5% organic matter; use 1.6 pints on all soils with 5-10% organic matter.*

Set incorporation equipment so that the bed and furrow will be uniformly covered with a layer of treated soil. If the layer of treated soils is not uniform and the herbicide is concentrated over the bed, potato emergence may be retarded, and stem brittleness can occur. When applying and incorporating TREFLAN after potato plants have fully emerged, do not completely cover the foliage with treated soil. Likewise, do not completely cover foliage at subsequent cultivations. Be careful that incorporation machinery does not damage potato seed pieces or elongating sprouts.

Potato —Split application in Idaho, Oregon and Washington:

On all soils, apply and incorporate 0.6 pint of TREFLAN per acre before planting and 0.6 pint after planting when potato plants have fully emerged. Do not apply to soils containing 2% or more organic matter. Follow incorporation directions listed above for application to potato after planting.

Potato —TREFLAN/Eptam tank mix for potatoes grown in Kansas, Minnesota, Nebraska, North Dakota, Oklahoma, South Dakota and Texas:

The TREFLAN/Eptam tank mix effectively controls the following weeds in addition to those weeds controlled by TREFLAN.

Henbit (Spring applications)	Oat, wild
Nightshade, black	Ragweed, common
Nightshade, hairy	Smartweed, Pennsylvania
Nutsedge	Velvetleaf (Buttonweed)

Follow recommended soil preparation and application procedures for TREFLAN. The TREFLAN/Eptam tank mix may be applied after planting, but prior to crop emergence. In areas where potatoes are normally dragged off, the TREFLAN/Eptam tank mix should be applied and incorporated up to or immediately following drag off.

Broadcast Rates Per Acre

	TREFLAN 5		
Soil Texture	Areas receiving less than 20" average annual rainfall*	Areas receiving greater than 20" average annual rainfall*	Eptam 7E
	(pints)	(pints)	(pints)
Coarse	0.8	0.8	1¾-7**
Medium	0.8-1.2	0.8-1.2	1¾-7
Fine	0.8-1.2	0.8-1.6	1¾-7

* *Use 1.2 pints per acre on coarse and medium soils with 2-5% organic matter; use 1.6 pints on all soils with 5-10% organic matter.*
** *Use the higher rate of Eptam 7E for nutsedge control.*

Precaution: Read the Eptam label before using. Observe cautions and limitations of products used in mixtures. Do not graze or feed forage to livestock from fields treated with TREFLAN/Eptam tank mix.

Potato —TREFLAN/Eptam application before planting in Washington, Idaho and Oregon:

TREFLAN/Eptam may be applied before planting at a broadcast rate of 0.6 pint of TREFLAN per acre and 3½ pints of Eptam 7E per acre on all soil textures. Incorporate immediately.

Precaution: Do not use this tank mix both before and after planting in the same season. Read the Eptam label before using. Observe all cautions and limitations on labeling of all products used in mixtures. Do not use foliage from treated crops for feed or forage.

Safflower

Follow recommended soil preparation, application and incorporation procedures for TREFLAN.

Apply and incorporate TREFLAN in the spring before planting or in the fall between October 15 and December 31.

Broadcast Rates Per Acre

	TREFLAN 5	
Soil Texture	Areas receiving less than 20" average annual rainfall*	Areas receiving greater than 20" average annual rainfall*
	(pints)	(pints)
Coarse	0.8	0.8
Medium	1.0-1.2	1.0-1.2
Fine	1.2	1.6

* *Use 1.2 pints per acre on coarse and medium textured soils and 1.6 pints on fine soils with 2-5% organic matter; use 1.6 to 2 pints on all soils with 5-10% organic matter.*

Safflower—Fall application in Arizona, California, Idaho, Nevada, Oregon, Utah, Washington and Wyoming:

Apply and incorporate TREFLAN any time between October 15 and December 31. Ground may be left flat or bedded-up over winter. On bedded ground, knock beds down to desired height before planting, moving some treated soil from tops into furrows. Where soil is left flat over winter, take care during spring bedding operations to prevent turning up untreated soil. Destroy established weeds during seedbed preparation. If weeds become established in furrows due to uncovering of untreated soil during listing, destroy these weeds before planting. Apply and incorporate TREFLAN at a broadcast rate of 1.2 pints per acre on coarse soils; 1.6 pints on medium soils; and 2 pints on fine soils.

Do not apply TREFLAN in the fall to soils which are wet or are subject to prolonged periods of flooding.

Sugar Beet

Apply TREFLAN as a broadcast, overtop spray when plants are between 2 and 6 inches tall at a rate of 0.8 pint per acre on coarse soils and 1 to 1.2 pints on medium and fine soils. Use the higher rate for medium and fine soils in areas receiving more than 20" average annual rainfall. Set incorporation machinery to throw treated soil toward the plants in the row. Be careful that the incorporation machinery does not damage the sugar beet taproot.

Precaution: Exposed beet roots should be covered with soil before a TREFLAN application to reduce the possibility of girdling.

Sugar Beet—Incorporation with a tine-tooth harrow in California, Colorado, Idaho, Nebraska, Oregon, Texas, Utah, Washington and Wyoming:

A properly operated tine-tooth harrow (Flextine or Melroe) can incorporate TREFLAN for effective weed control in sugar beet. Operate the tine-tooth harrow 2 times over the field in opposing directions at a speed of 3 to 6 mph. Set the harrow to cut 1 or 2 inches deep. Be careful that the tine-tooth harrow does not damage the sugar beet taproot. Follow recommended application procedures and broadcast rates per acre for sugar beet (see preceding paragraph).

Sugarcane—(Plant Cane)

Apply and incorporate TREFLAN twice a year at a broadcast rate of 1.6 to 3.2 pints per acre for all soil textures. Make the first TREFLAN application in the fall on firmly packed beds immediately after the seed pieces are planted. Make the second TREFLAN application in the spring before or shortly after the cane emerges. Loosen rain-packed beds 2 to 3 inches deep before the spring application. Take care that incorporation machinery does not damage the seed pieces or emerging shoots.

Sugarcane—Postplant in Hawaii for control of most annual grasses, including guineagrass:

Surface apply TREFLAN after planting (for plant cane) or after harvesting (for ratoon cane), before weeds and cane emerge. Use a broadcast rate of 4.8 to 6.4 pints per acre for all soil textures. In plant cane, the beds should be formed or rolled before application. In ratoon cane, the crop residue should be removed before application. If large amounts of crop residues are present, TREFLAN will not be effective. Apply TREFLAN just before anticipated rainfall or sprinkle irrigate immediately after application.

Sugarcane—Applications up to layby for plant cane or ratoon cane grown in Louisiana or Texas:

Apply and incorporate TREFLAN at a broadcast rate of 1.6 to 3.2 pints per acre for all soil textures. Do this in the spring from before or shortly after the cane emerges up to layby. Apply the TREFLAN after the beds have been shaved or false shaved. Loosen rain-packed beds 2 to 3 inches deep before application. Be careful that incorporation machinery does not damage seed pieces or emerging shoots. You may use a rolling cultivator or bed chopper to incorporate TREFLAN layby applications in sugarcane on all soil textures. Follow normal incorporation directions for the rolling cultivator. Set bed chopper to cut 3 to 4 inches deep and operate at 4 to 6 mph. Two incorporation passes are necessary.

Sugarcane—Itchgrass (Raoulgrass) control in Louisiana:

Apply and incorporate TREFLAN on either plant or ratoon cane at a broadcast rate of 3.2 pints per acre for all soil textures. Follow the directions above for sugarcane layby application in Louisiana and Texas.

Sunflower—TREFLAN Alone

Apply and incorporate TREFLAN in the spring or in the fall between October 15 and December 31. Follow recommended soil preparation, application and incorporation procedures for TREFLAN.

Broadcast Rates Per Acre

Soil Texture	TREFLAN 5: Areas receiving less than 20" average annual rainfall*	TREFLAN 5: Areas receiving greater than 20" average annual rainfall*
	(pints)	(pints)
Coarse	0.8	0.8
Medium	1.0-1.2	1.2
Fine	1.2	1.6

* *Use 1.2 to 1.6 pints per acre on coarse and medium soils with 2-5% organic matter and 1.6 pints on all soils with 5-10% organic matter.*

Sunflower—TREFLAN/Amiben tank mix or overlay:

Follow recommended soil preparation, application and incorporation procedures for TREFLAN.

Amiben may be applied in a band or broadcast over sunflowers at planting in fields where TREFLAN has been incorporated prior to planting. Or, the TREFLAN/Amiben tank mix may be incorporated prior to planting. TREFLAN/Amiben tank mix improves mustard, smartweed, velvetleaf and ragweed control in addition to those weeds controlled by TREFLAN alone.

Apply TREFLAN/Amiben or TREFLAN with an Amiben overlay at the following broadcast rates per acre:

Broadcast Rates Per Acre

Soil Texture	TREFLAN 5	Amiben 2S
	(pints)	(quarts)
Coarse	0.8	4
Medium	1.2	4-6*
Fine	1.6	4-6*

* *For best control of mustard, common ragweed or black nightshade, use the 6 quart rate.*

In coarse textured soils, heavy rains on the incorporated Amiben may move it below the weed seed germination zone, and erratic weed control may result. If sufficient rain does not fall within 7 days after a preemergence application of Amiben, but there is enough soil moisture to germinate weeds and grasses, a light cultivation with a rotary hoe or similar tool will uproot these small broadleaf weeds and grasses. The shallow mixing of Amiben in the surface soil will not interfere with the action of Amiben when rains come.

Tomato

For direct-seeded tomato, apply TREFLAN at blocking or thinning as a directed spray to the soil between rows and beneath the plants, and incorporate. For transplant tomato, apply and incorporate TREFLAN before transplanting. Do not apply TREFLAN after transplanting.

Broadcast Rates Per Acre

Soil Texture	TREFLAN 5: Areas receiving less than 20" average annual rainfall*	TREFLAN 5: Areas receiving greater than 20" average annual rainfall*
	(pints)	(pints)
Coarse	0.8	0.8
Medium	1.0-1.2	1.2
Fine	1.2	1.6

* *Use 1.2 pints per acre on coarse and medium textured soils and 1.6 pints on fine soils with 2-5% organic matter; use 1.6 pints on all soils with 5-10% organic matter.*

Fruit and Nut Crops and Vineyards

For areas receiving more than 20" average annual rainfall.

For new plantings of vineyards, citrus and pecan trees, apply and incorporate TREFLAN before planting at a broadcast rate of 0.8 pint per acre on coarse soils; 1.2 pints on medium soils; 1.6 pints on fine soils; 1.2 pints on fine soils with 2-5% organic matter; and 1.6 pints on soils with 5-10% organic matter.

For non-bearing, established plantings of citrus and pecan trees and bearing plantings of grapefruit, lemon, orange, pecan, tangelo and tangerine trees, apply TREFLAN at a broadcast rate of 1.6 to 3.2 pints per acre for all soil textures.

For areas receiving less than 20" average annual rainfall.

For new plantings of almond, apricot, citrus, nectarine, peach, pecan and walnut trees, apply and incorporate TREFLAN before planting at a broadcast rate of 0.8 pint per acre on coarse soils; 1.2 pints on medium soils; 1.6 pints on fine soils; 1.2 pints on fine soils with 2-5% organic matter; and 1.6 pints on soils with 5-10% organic matter.

For new plantings of vineyards, apply and incorporate TREFLAN before planting at a broadcast rate of 0.8 to 1.2 pints per acre on coarse soils; 1.2 to 2.4 pints on medium soils; and 2.4 to 3.2 pints on fine soils or soils with 2-10% organic matter. Do not use more than 1.6 pints per acre on heat-treated grape rootings.

For postplant applications on bearing or non-bearing, established plantings of vineyards and almond, apricot, grapefruit, lemon, nectarine, orange, peach, plum, prune, tangelo, tangerine and walnut trees, apply TREFLAN at a broadcast rate of 1.6 to 3.2 pints per acre for all soil textures. Do not apply to vineyards within 60 days of harvest.

In established plantings, apply TREFLAN as a directed spray to the soil and use incorporation methods not injurious to the trees or vines.

Fruit and Nut Crops and Vineyards—Rhizome johnsongrass control:

For areas receiving less than 20" average annual rainfall.

You can obtain commercially acceptable control of rhizome johnsongrass with postplant applications in bearing and non-bearing, established plantings of vineyards and almond, apricot, grapefruit, lemon, nectarine, orange, peach, pecan, tangelo, tangerine and walnut trees with a TREFLAN program when applied for 2 years in a row.

Soil Preparation—Work the soil thoroughly to bring the rhizomes nearer the surface.

Application—Apply TREFLAN at a broadcast rate of 1.6 quarts per acre on all soil textures each year for 2 years in a row. Do not apply to vineyards within 60 days of harvest.

Incorporation—Incorporate TREFLAN thoroughly with a disc set to cut 4 to 6 inches deep and operate at 4 to 6 mph. Two incorporation passes are necessary, with the second pass in a different direction from the first.

Cultivation—Some johnsongrass plants will escape. Timely cultivations are necessary to obtain commercially acceptable control. You cannot obtain commercially acceptable control with only 1 year of TREFLAN use.

Precautions: Do not use the 1.6 quart rate on new plantings, or crop injury may result. Do not interplant orchards or vineyards with other crops. If the treated vineyards and orchards are diverted to other crop uses, plant only those crops for which TREFLAN has been registered as a preplant treatment for the next cropping season.

Fruit and Nut Crops and Vineyards—Bindweed control in California:
TREFLAN can be used for the control of field bindweed in vineyards and for almond, apricot, grapefruit, lemon, nectarine, orange, peach, pecan, tangelo, tangerine and walnut trees.

Apply TREFLAN at a broadcast rate of 3.2 pints per acre on all soil textures. TREFLAN must be applied in the spring with a specially designed spray blade which applies a thin, concentrated layer at a soil depth of 4 to 6 inches. This layer of TREFLAN prevents bindweed shoots from emerging.

Land Preparation—Destroy all weeds and grasses with soil tillage before applying TREFLAN. This tillage is necessary to prevent trash from interfering with the operation of the spray blade.

Equipment—This operation requires a spray blade capable of running 4 to 6 inches below the surface of the soil. The spray blade should be equipped with nozzles located under the blade and directed so that the TREFLAN spray will be trapped under the soil which is flowing over the blade as it is pulled through the soil. Use a sufficient number of nozzles with spacing that will uniformly apply the TREFLAN underground in a thin, horizontal layer.

Application—Apply TREFLAN in 40 to 80 gallons of water per acre. Operate the spray blade at a depth of 4 to 6 inches.

Precaution: Some soils develop cracks as they dry after rainfall or irrigation. Field bindweed may emerge if the cracks extend through the TREFLAN layer. Prevent or eliminate cracks by shallow discing or other tillage. Avoid deep tillage which disturbs the subsurface layer. Cultivation or tillage also aids the control of germinating seeds.

Wheat (Winter)—(Idaho, Oregon and Washington)
TREFLAN may be applied for preplant preemergence control of cheatgrass and other annual grasses and broadleaf weeds controlled by TREFLAN. The growth, development and yield of winter wheat will not be adversely affected, provided the seed is placed below the zone of soil treated with TREFLAN.

Apply TREFLAN any time during a period from 3 weeks up to immediately prior to planting. Broadcast TREFLAN at a rate of 1.2 pints per acre on coarse and medium soils and 1.6 pints on fine soils.

Incorporation Directions—Incorporate TREFLAN into the soil with a flexible tine-tooth harrow (Flextine, Melroe) set to cut 1 to 2 inches deep and operate at 3 to 6 mph. Incorporate one time within 24 hours after application, followed by a second incorporation in a different direction from the first prior to planting. Do not till the soil with a disc after the TREFLAN has been incorporated with a flexible tine harrow.

Seeding Directions—Use only a deep furrow or semi-deep furrow drill that will place the seed below the zone of soil into which TREFLAN has been incorporated.

Precaution: Wheat planted in direct contact with treated soil may suffer crop injury in the form of delayed emergence and development.

Wheat (Winter)—Fallow soil application in Washington and Oregon:
TREFLAN applied and shallowly incorporated into fallow soil up to four months ahead of planting will control cheatgrass and certain annual grasses and broadleaf weeds. The growth, development, or yield will not be adversely affected as long as the seed is placed below the zone of soil treated with TREFLAN. Use deep or semi-deep furrow drills.

Broadcast rates are 1.2 pints per acre on coarse and medium soils and 1.6 pints on fine soils. Apply TREFLAN any time from May to September prior to the fall planting of winter wheat.

Incorporation—Incorporate TREFLAN with a flexible tine-tooth harrow (Flextine or Melroe) set to cut 1 to 2 inches deep and operated at 3 to 6 mph. For thorough incorporation, two passes of the equipment in different directions over the field are necessary. Incorporate one time within 24 hours after application followed by a second incorporation prior to seeding. Do not till the soil with a disc after TREFLAN has been applied with a flexible tine harrow.

Precaution: Use only deep furrow or semi-deep furrow drills. Place seed below the zone of soil into which TREFLAN has been incorporated. Do not plant wheat directly into the zone of soil treated with TREFLAN; injury to the crop or a delay in its emergence and development may occur.

Wheat (Spring), Durum and Barley—TREFLAN Alone
TREFLAN is recommended as a postplant incorporated treatment to control foxtail (pigeongrass).

Apply TREFLAN at a broadcast rate of 0.8 pint per acre on coarse and medium soils and 1.2 pints on fine soils.

Plant 2 to 3 inches deep in a well-tilled seedbed. Apply TREFLAN after seeding but before the crop emerges. To incorporate, use flex-tine or diamond harrows operated two times in different directions, at speeds of at least 5 mph. Incorporate by operating equipment 1 to 1½ inches deep. Application and the first incorporation should be done in the same operation if possible. Both incorporations must be done within 24 hours.

Wheat (Spring), Durum, Barley (Fall Application)—Foxtail/Pigeongrass control:
TREFLAN may be fall applied for foxtail/pigeongrass control in spring wheat, durum and barley planted the following spring. TREFLAN may be applied to ground that has a manageable trash level, has been fallowed or pre-tilled. The first incorporation is required within 24 hours after application. A second incorporation is required prior to planting to destroy emerged weeds and to ensure an even distribution of TREFLAN treated soil.

Broadcast Rates Per Acre

Soil Texture	TREFLAN 5
	(pints)
Coarse	0.8
Medium	0.8
Fine	1.2

Incorporation Directions—Any of the following tools are recommended for fall incorporation. The disc or field cultivator may be used for the spring incorporation pass. Care should be taken to operate the tool at a more shallow depth than the fall incorporation.

1. Chisel plow: *May be used for the first pass only.* Operate at 4-5 inches deep at 4-6 mph. A chisel plow is defined as having 3 rows of up to 18-inch sweeps on no greater than 12-inch centers. Stagger sweeps so that no soil is left unturned.
2. Tandem disc: Operate at 3-4 inches deep at 4-6 mph.
3. Field cultivator: Operate at 3-4 inches deep at 5 mph or more. A field cultivator is defined as having 3 to 4 rows of sweeps with "c" or "s" shaped shanks, spaced 7 inches or less and staggered so that no soil is left unturned.

Planting Directions—Wheat, durum, or barley should be approximately 2 inches deep.

Precaution—While use of this practice may result in a stand reduction, slight stand reductions do not normally affect yield.

Wheat (Spring), Durum and Barley—TREFLAN/Far-Go tank mix:
TREFLAN/Far-Go applied as a postplant incorporated treatment will control foxtail (pigeongrass) and wild oat.

Plant 2 to 3 inches deep in a well-tilled seedbed. Apply TREFLAN/Far-Go after seeding but before crop emerges. To incorporate, use flex-tine or diamond harrows two times, operated in different directions, at speeds of at least 5 mph. Incorporate by operating equipment 1 to 1½ inches deep. Application and the first incorporation should be done in the same operation if possible. If not, incorporate immediately after application.

Broadcast Rates Per Acre

Soil Texture	TREFLAN 5 Barley Durum Spring Wheat	Far-Go Durum Spring Wheat	Far-Go Barley
	(pints)	(pints)	(pints)
Coarse	0.8	2	2½
Medium	0.8	2	2½
Fine	1.2	2	2½

Precaution: Overapplication may result in crop injury. Read the Far-Go label carefully before using.

FALL APPLICATION

General: (Areas receiving more than 20" average annual rainfall)

See specific crop for recommendations. For all crops for which there are no specific fall application instructions and for which TREFLAN is recommended as a preemergence application, use the rates listed for spring applications. Do not apply TREFLAN in the fall for sugarbeets, potatoes and direct-seeded tomatoes.

In most states apply and incorporate TREFLAN any time between October 15 and December 31. In Minnesota, North and South Dakota, apply and incorporate TREFLAN anytime between September 1 and December 31. Ground may be left flat or bedded-up over winter. On bedded ground, knock beds down to desired height before planting, moving some treated soil from beds into furrows. Where soil is left over winter, be careful not to turn up untreated soil during spring bedding operations. Destroy established weeds during seedbed preparation. If weeds become established in furrows due to uncovering of untreated soil during bedding, destroy these weeds before planting. Do not apply TREFLAN in the fall to soils which are wet, are subject to prolonged periods of flooding, or where rice was grown the previous year.

FERTILIZER USE DIRECTIONS APPLICATION WITH LIQUID FERTILIZERS

TREFLAN 5 may be mixed with most liquid fertilizer materials. The combination of TREFLAN with solutions and suspension-type fertilizers has provided weed and grass control equal to the same rates of TREFLAN applied in water. Follow TREFLAN label recommendations regarding rates per acre, crops, incorporation directions, special instructions, cautions and special precautions.

Individual state regulations relating to liquid fertilizer mixing, registration, labeling and applications are the responsibility of the individual and/or company selling the fertilizer and chemical mixture.

Testing for Tank Mix Compatibility in Liquid Fertilizers:

TREFLAN alone or in tank mixture with dry flowables, wettable powders (WP), aqueous suspensions (AS), flowables (F), liquids (L), or solutions (S) may not combine properly with some fluid fertilizer materials. Small quantities should always be tested before fullscale mixing. This will determine whether a compatiblity agent is needed, and which agent does the best job. The seven agents listed on page 81 have been thoroughly tested. There are many other surfactants on the market which were not designed for use with liquid fertilizers. Use the following test to select the correct agent for your mixture.

1. Put 1 pint of the liquid fertilizer in a quart jar.

2. Add 1 to 4 teaspoonful(s) of the dry flowable, WP, A, F or L formulation (depending on the recommended rate per acre) to the liquid fertilizer. Close jar and agitate until the materials are dispersed evenly in the fertilizer. If the materials do not disburse well, it may be necessary to slurry the chemicals in water before adding to the fertilizer.

3. After dispersing the materials (Step 2), add 3 to 4 teaspoonsful of TREFLAN to the jar and shake well. Add solution herbicides to the mixture last and agitate. Observe the jar for about 10 minutes. If the materials rise to the surface and form a thick layer (oily curds) which will not redisperse when agitated, a compatibility agent is needed. If the tank mixture is easily redispersed to its original state with slight agitation, no agent is needed but good agitation must be provided in the fertilizer spray tank.

4. If the need for a compatibility agent is shown in Step 3: Using a clean quart jar, start at Step 1 above, add 1/2 teaspoon of the compatibility agent to the liquid fertilizer, mix well, then repeat Steps 2 and 3.

An effective compatibility agent will cause the mixture to remain uniformly mixed with little or no separating or oil rising to the surface for one half hour or longer. If slight separation does occur, 2 to 3 inversions of the jar should give a uniform remix. If oily curds form which will not redisperse, more agent or another agent should be tried.

Use a clean jar for each test. The compatible mixture will have a uniform appearance and will be relatively easy to keep mixed with gentle agitation of the jar.

Liquid Fertilizer Mixing Instructions TREFLAN 5 in Liquid Fertilizer

General—Emulsifiable concentrates, such as TREFLAN 5, can be mixed with liquid fertilizers. In all cases, continuous agitation is required to prevent the TREFLAN from rising to the surface as an oily layer. When necessary, (see Testing for Tank Mix Compatibility in Liquid Fertilizers, on page 78) a compatibility agent can be used to cause the TREFLAN to emulsify properly (i.e., have a milky appearance rather than oily layer). The use of compatibility agents is especially important when tank mixing emulsifiable concentrates (E.C.) with dry flowables, wettable powders (WP), aqueous suspensions (AS), flowables (F), liquids (L), or solutions (S) in liquid fertilizer. If the emulsion is not properly formed and the TREFLAN rises to the surface of the fertilizer as an oil ("oils out"), the oil may combine with the wettable powder, flowable, or suspensions to form oily curds (viscous phase) which is difficult to redisperse. Any one of the compatibility agents listed on page 81 is helpful in causing liquid concentrates to form non-oiling mixtures with liquid fertilizers. These compatibility agents can be used at rates as low as 1 1/2 to 2 pints per ton of liquid fertilizer and should be mixed well with the fertilizer before adding the liquid concentrate. Read the label on the compatibility agent and follow the directions.

1. Sponto 168D (Witco Chemicals Co., Chicago, IL)
2. Compat (Farm Chemicals, Inc., Aberdeen, NC)
3. Unite (Hopkins Ag Chemical, Madison, WI)
4. T-Mulz 734-2 (Thompson-Hayward Chemical Co., Kansas City, MO)
5. Rigo Compatibility Agent (Rigo Company, Buckner, KY)
6. Amoco Spray Mate™ (Amoco Oil Co., Chicago, IL)
7. Kem-Link (Universal Coop, Minneapolis, MN)

All of the above are phosphate ester-type surfactants designed to be used with liquid fertilizers. They usually do not work as compatibility agents in tank mixtures in plain water.

Application—Spread the fertilizer/pesticide mixture with a properly calibrated applicator. Be certain the material is applied uniformly to the soil surface.

Incorporation—Follow normal TREFLAN incorporation procedures.

TREFLAN 5 APPLICATION WITH DRY BULK FERTILIZERS

General—Dry bulk fertilizers may be impregnated or coated with TREFLAN. Application of dry bulk fertilizers impregnated with TREFLAN has provided weed and grass control equal to the same rates of TREFLAN applied in water.

All TREFLAN label recommendations regarding rates per acre, approved crops, incorporation, special instructions, cautions and special precautions must be followed.

Apply a minimum of 200 pounds per acre of dry fertilizer impregnated with TREFLAN at the recommended rates. Any commonly used dry fertilizers can be used for TREFLAN impregnation except coated ammonium nitrate and straight limestone. These materials will not absorb the herbicide. Blends containing mixtures of these materials can be impregnated.

Impregnation—Use any closed drum, belt, ribbon or other commonly used dry bulk fertilizer blender. Apply TREFLAN uniformly to the fertilizer.

Rates

Check specific crop recommendations for the rate of TREFLAN per acre. See the rate table which follows to determine the amount of TREFLAN to be impregnated into a ton of dry bulk fertilizer based on the amount of fertilizer which will be applied per acre. (See rate chart on page 83.)

Application

Spread the fertilizer/chemical mixture with a properly calibrated applicator. Be certain the material is applied uniformly to the soil surface.

Incorporation

Follow TREFLAN incorporated procedures.

Rate Chart for Impregnating Fertilizer with TREFLAN 5 (TREFLAN Added to a Ton of Fertilizer)

Fertilizer Rate Per Acre	TREFLAN Rate Per Acre				
	0.8 pint	1.2 pints	1.6 pints	2.4 pints	3.2 pints
200 pounds	4 qts. per ton	6 qts. per ton	8 qts. per ton	12 qts. per ton	16 qts. per ton
250 pounds	3.2 qts. per ton	4.8 qts. per ton	6.4 qts. per ton	9.6 qts. per ton	12.8 qts. per ton
300 pounds	2.7 qts. per ton	4 qts. per ton	5.6 qts. per ton	8 qts. per ton	10.8 qts. per ton
350 pounds	2.3 qts. per ton	3.6 qts. per ton	4.8 qts. per ton	6.8 qts. per ton	9.2 qts. per ton
400 pounds	2 qts. per ton	3 qts. per ton	4 qts. per ton	6 qts. per ton	8 qts. per ton
450 pounds	1.8 qts. per ton	2.7 qts. per ton	3.6 qts. per ton	5.4 qts. per ton	7.2 qts. per ton

For rates other than those listed above, use the following formula to calculate the amount of TREFLAN to be impregnated on a ton of dry bulk fertilizer:

$$\text{Pints TREFLAN Per Acre} \times \frac{1000}{\text{Lbs. Fertilizer Per Acre}} = \text{Quarts TREFLAN Per Ton of Fertilizer}$$

All individual state regulations relating to dry bulk fertilizer blending, registration, labeling and application are the responsibility of the individual and/or company selling the fertilizer and chemical mixture.

TREFLAN 5 PERFORMANCE GUARANTEE—A LIMITED WARRANTY OF TREFLAN WEED AND GRASS CONTROL

Elanco Products Company warrants that TREFLAN will provide commercially acceptable control of the weeds and grasses as indicated on its current label when used according to label directions and subject to the following limitations:

1. Purchaser must notify Elanco promptly if a lack of commercial acceptable control occurs. Notice must be given within forty-five (45) days after the first planting or after application of TREFLAN to the treated crop, whichever occurs later.
2. An Elanco representative must be satisfied that Purchaser used TREFLAN according to the label directions.
3. The Purchaser must provide Elanco representative with an invoice or other satisfactory record which shows the price and quantity of TREFLAN purchased.
4. Elanco's liability will be limited to a refund of the purchase price of the TREFLAN applied to the acreage on which commercially acceptable control was not obtained.
5. Void outside the USA.

DISCLAIMER OF WARRANTIES

The warranty printed above is the only warranty applicable to this product. All other warranties, expressed or implied, including, but not limited to any implied warranties of merchantability and fitness for a particular purpose, are disclaimed. This disclaimer of warranties does not apply where prohibited by operation of law.

LIMITATION OF DAMAGES

Elanco's liability whether in contract, warranty, tort, negligence, strict liability or otherwise, shall not exceed the return of the amount of the purchase price of the TREFLAN and under no circumstances shall Elanco be liable for special, indirect or consequential damages. This limitation of damages does not apply where prohibited by operation of law.

INHERENT RISKS OF USE

Failure to carefully follow the directions for use of TREFLAN, as well as other factors such as seedling disease, cold weather, deep planting, excessive moisture, high salt concentration or drought, may result in unsatisfactory weed control or crop injury. Reduced yields may result under these conditions.

Amiben®—(chloramben, Union Carbide)
Basagran®—(BASF Wyandotte)
Blazer®—(acifluorfen, Rohm and Haas)
Canopy®—(chlorimuron ethyl plus metribuzin, E.I. duPont)
Caparol®—(prometryn, Ciba-Geigy)
Classic®—(chlorimuron ethyl, E. I. duPont)
Cotoran®—(fluometuron, Ciba-Geigy)
Dual®—(metolachlor, Ciba-Geigy)
Eptam®—(EPTC, Stauffer)
Far-Go®—(triallate, Monsanto)
Gemini™—(chlorimuron ethyl plus linuron, E. I. duPont)
Karmex®—(diuron, E. I. duPont)
Lasso®—(alachlor, Monsanto)
Lasso® Micro-Tech™—(alachlor, Monsanto)
Lexone®—(metribuzin, E. I. duPont)
Lorox®—(linuron, E. I. duPont)
Lorox® Plus—(chlorimuron ethyl plus linuron, E. I. duPont)
Preview®—(chlorimuron ethyl plus metribuzin, E. I. duPont)
Sencor®—(metribuzin, Bayer, GmbH)
Tackle®—(acifluorfen, Rhone-Poulenc)
Vernam®—(vernolate, Stauffer)

This is a specimen label. Elanco is not responsible for the accuracy of the information contained herein. As labels are subject to revision, always carefully read and follow the label on the product container.

TREFLAN® TR-10

GRANULES
HERBICIDE

A selective herbicide for the preemergence control of annual grasses and broadleaf weeds

ACTIVE INGREDIENT:	
trifluralin* (*a,a,a*-trifluoro-2,6-dinitro-*N,N*-dipropyl-*p*-toluidine)	10.0%
INERT INGREDIENTS:	90.0%
TOTAL	100.0%

Contains 5 pounds active ingredient per 50 pound bag.

***TREFLAN®, TR-10®**—(trifluralin containing 10 percent active ingredient, Elanco)

Manufactured using the processes of U.S. Patents 4,120,905 and 4,226,789.

EPA Reg. No. 1471-143 ID 5535

Literature revised July 28, 1987
PI 3260 AMP

PRECAUTIONARY STATEMENTS

KEEP OUT OF REACH OF CHILDREN
CAUTION

HAZARDS TO HUMANS AND DOMESTIC ANIMALS

Harmful if swallowed or absorbed through the skin. Avoid breathing dust. Avoid contact with skin, eyes or clothing. In case of skin or eye contact, flush with plenty of water. If irritation persists, get medical attention.

ENVIRONMENTAL HAZARDS

This pesticide is toxic to fish. Do not apply directly to water or wetlands (swamps, bogs, or marshes). Drift or runoff from treatment areas may be hazardous to aquatic organisms in neighboring aquatic sites. Do not contaminate water by cleaning of equipment or disposal of waste.

STORAGE AND DISPOSAL

Storage: Store in original container only. In case of leak or spill, contain material and dispose as waste.

Pesticide Disposal: Do not contaminate water, food or feed by storage or disposal. Waste resulting from the use of this product may be disposed of on site or at an approved waste disposal facility.

Container Disposal: Completely empty bag by shaking and tapping sides and bottom to loosen clinging particles. Empty residue into application equipment. Then dispose of bags in a sanitary landfill or by incineration if allowed by State and local authorities. If burned, stay out of smoke.

SPECIAL PRECAUTIONS

Applied according to directions and under normal growing conditions, TREFLAN will not harm the treated crop. Overapplication may result in crop injury or a soil residue. Uneven application or improper soil incorporation of TREFLAN can result in erratic weed control or crop injury. Seedling disease, cold weather, deep planting, excessive moisture, high salt concentration or drought may weaken crop seedlings and increase the possibility of damage from TREFLAN. Under these conditions, delayed crop development or reduced yields may result.

In Arizona, Colorado, California, Idaho, Nevada, New Mexico, Oregon, Utah, Washington and Wyoming, sugar beets, red beets or spinach should not be planted for 12 months after a spring application or for 14 months after a fall application of TREFLAN. Plow the land to a depth of 12 inches prior to planting sugar beets to prevent the possibility of crop injury. Sorghum (milo), proso millet, corn, oats and annual or perennial grass crops or grass mixtures should not be planted for 14 months after a spring application or for 16 months after a fall application of TREFLAN to avoid crop injury. If land has not been irrigated, do not plant any of these crops for 18 months after a spring application or 20 months after a fall application of TREFLAN.

In those portions of Kansas, Nebraska, North Dakota, Oklahoma, South Dakota and Texas where at least 20 inches of irrigation and/or rainfall (total) was used to produce the crop, sorghum, oats and annual or perennial grass crops or grass mixtures should not be planted for 12 months after an application of TREFLAN.

If less than 20 inches of total water was used to produce the crop, do not plant sorghum, proso millet, oats and annual or perennial grass crops or grass mixtures for 18 months after an application of TREFLAN. Cool, wet weather conditions during the early stage of growth may increase the possibility of injury to sorghum.

In all other areas receiving greater than 20" rainfall per year, moldboard plow before planting sugar beets where a spring application of TREFLAN was made the previous season.

Vegetable crops other than those listed on this label should not be planted within 5 months following the application of TREFLAN.

Do not use TREFLAN on any crop grown in Pecos County or Reeves County, Texas, or the state of Montana.

DIRECTIONS FOR USE

TREFLAN is a preemergence herbicide which is incorporated into the soil to provide long-lasting control of many annual grasses and broadleaf weeds. TREFLAN controls weeds as they germinate. TREFLAN will not control established weeds.

It is a violation of Federal Law to use this product in a manner inconsistent with its labeling.

Weeds and Grasses Controlled by TREFLAN

Grasses	
Annual bluegrass	*Poa annua*
Barnyardgrass (Watergrass)	*Echinochloa* sp.
Brachiaria (Signalgrass)	*Brachiaria* sp.
Bromegrass (Cheatgrass) (Downy brome)	*Bromus tectorum*
Cheat (Chess)	*Bromus secalinus*
Crabgrass (Large crabgrass) (Smooth crabgrass)	*Digitaria* spp.
Foxtail (Bottlegrass) (Bristlegrass) (Giant foxtail) (Green foxtail) (Foxtail millet) (Pigeongrass) (Robust foxtail) (Yellow foxtail)	*Setaria* spp.
Goosegrass (Silver crabgrass) (Silvergrass) (Wiregrass) (Yardgrass)	*Eleusine indica*
Johnsongrass (from seed)	*Sorghum halepense*
Junglerice	*Echinochloa colonum*
Sandbur (Burgrass)	*Cenchrus incertus*
Sprangletop	*Leptochloa filiformis*
Stinkgrass (Lovegrass)	*Eragrostis cilianensis*
Wild cane (Shattercane)	*Sorghum bicolor*
Wild oat[1]	*Avena fatua*
Woolly cupgrass	*Erichloa villosa*
Broadleaf Weeds	
Carpetweed	*Mollugo verticillata*
Chickweed	*Stellaria media*

Florida pusley (Florida purslane) (Mexican clover) (Pusley)	*Richardia scabra*
Goosefoot	*Chenopodium hybridum*
Henbit (fall application only)	*Lamium amplexicaule*
Knotweed	*Polygonum aviculare*
Lambsquarters	*Chenopodium album*
Pigweed (Carelessweed) (Prostrate pigweed) (Redroot pigweed) (Rough pigweed) (Spiny pigweed)	*Amaranthus* spp.
Puncturevine (Western U.S. only) (Caltrop) (Goathead)	*Tribulus terrestris*
Purslane	*Portulaca oleracea*
Stinging nettle (Nettle)	*Urtica dioica*

[1] TREFLAN applied as a preplant incorporated (PPI) treatment will provide partial control of wild oats. This claim is for all PPI uses except fall applications for spring seeded cereals at foxtail (pigeongrass) control rates.

SOIL PREPARATION

Crop Residues or Existing Weeds: Apply TREFLAN to soil that has a manageable trash level, been fallowed, pre-tilled or to soil that has standing stubble. A manageable level of such ground cover would allow the TREFLAN to be uniformly incorporated into the top 2 to 3 inches of the final seedbed.

Roughness: The soil surface should be smooth enough so that application and incorporation equipment can be operated efficiently and at speeds which ensure a uniform application and incorporation of TREFLAN.

General Soil Conditions: To assure uniform incorporation of TREFLAN, soil moisture conditions should be such that large clods can be broken up during the incorporation process.

Soil Texture Guide

The amount of TREFLAN applied will vary with the soil texture and organic matter. A fine textured soil will require more TREFLAN per acre than a coarse soil. Choose the proper rate for each application based on the following soil texture group and specific crop recommendations. Do not exceed recommended rates.

Soil Texture	Soil Classification
Coarse Soils (Light)	Sand, loamy sand, sandy loam
Medium Soils	Loam, silty clay loam*, silt loam, silt, sandy clay loam*
Fine Soils (Heavy)	Clay, clay loam, silty clay loam*, silty clay, sandy clay, sandy clay loam*

**Silty clay loam and sandy clay loam soils are transitional soils and may be classified as either medium or fine textured soils. If silty clay loam or sandy clay loam soils are predominately sand or silt, they are usually classified as medium textured soils. If they are predominately clay, they are usually classified as fine textured soils.*

APPLICATION DIRECTIONS

Apply TREFLAN with a properly calibrated herbicide applicator that will apply the granules uniformly. Calibrate the applicator according to the manufacturer's directions and check frequently during application to be certain that equipment is working properly. Avoid concentration of the material in narrow bands.

Freezing will not adversely affect this product. If product is frozen at time of application, agitate or thaw to product a free-flowing granule.

INCORPORATION DIRECTIONS

Incorporation Equipment—General Directions

Use incorporation equipment that mixes TREFLAN into the top 2 to 3 inches of the final seedbed, or erratic weed control and/or crop injury may result. Incorporation equipment such as a disc will mix TREFLAN approximately half as deep as the equipment is set to operate. For example, a disc set to cut 4 inches deep will incorporate most of the TREFLAN within the top 2 inches of the final seedbed.

Incorporation Before Planting

TREFLAN must be incorporated one time within 24 hours after application. **The second incorporation must be delayed at least five (5) days after the first,** completed prior to planting, and should be run in a different direction from the first. Incorporate the TREFLAN uniformly into the top 2 to 3 inches of the final seedbed.

Incorporation After Planting

Check specific crop for incorporation directions after planting.

Incorporation in Bedded Culture

For effective weed control, TREFLAN needs to be incorporated into the top 2 to 3 inches of the final seedbed.

Application prior to bedding: Apply TREFLAN and incorporate it one time with recommended equipment. The bedding operation serves as the second incorporation. Do not expose untreated soil during post-bedding operations.*

Application after bedding: Knock off beds to planting height before applying TREFLAN. Apply TREFLAN and incorporate it with recommended equipment that will conform to the bed shape. Do not leave untreated soil exposed.*

**Avoid removal of treated soil from the seedbed before or during the planting operation. This would expose untreated soil, allowing weeds to germinate in the drill row.*

Recommended Equipment

Any recommended incorporation tool may be used alone or in combination with any other recommended tool. Two incorporation passes are required unless specifically stated. The second incorporation should not be deeper than the first.

Disc: Set to cut 4 to 6 inches deep and operate at 4 to 6 mph.

Field Cultivator: Set to cut 3 to 4 inches deep and operate at 5 mph or more. A field cultivator is defined as an implement with 3 to 4 rows of sweeps, spaced at intervals of 7 inches or less and staggered so that no soil is left unturned. Chisel points should not be used.

Chisel Plow: (For use in northern Great Plains).

The chisel plow may be used for the first incorporation pass only. Any other recommended tool may be used for the second pass for row crops. The chisel plow may be used for any tillage or incorporation pass in the Summer Fallow program. Operate 4-5 inches deep at 4-6 mph. A chisel plow is defined as having 3 rows of 14 to 18 inch sweeps on no greater than 12 inch center. Stagger sweeps so that no soil is left unturned. Chisel points should not be used.

Combination Seedbed Conditioners: Set to cut 3 to 4 inches deep and operate at a speed of at least 5 mph. These implements are defined as three or more tillage devices combined and used as a single tool. For example, 2 to 3 rows of field cultivator "c" or "s" shaped shanks with an effective sweep spacing of 6 to 9 inches (staggered so that no soil is left unturned), followed by a spike-tooth or flextine harrow, followed by a ground-driven reel or basket.

Rolling Cultivator: Set to cut 2 to 4 inches deep and operate at 6 to 8 mph. Rolling cultivators are adequate for use on coarse and medium textured soils only.

Mulch Treader (other similar disc-type implements): Set to cut 3 to 4 inches deep and operate at 5 to 8 mph.

P.T.O. Driven Equipment (tillers, cultivators, hoes): Adjust to incorporate TREFLAN into the top 2 to 3 inches of the seedbed with rotors spaced to provide a clean sweep of the soil. Only one incorporation is necessary. P.T.O. driven equipment should not be operated at a speed greater than 4 mph.

Other equipment, including the flexible tine-tooth harrow, (Flextine, Melroe), is also recommended but only for the special programs for which it is specified in this label.

CULTIVATION AFTER PLANTING

Soil treated with TREFLAN may be shallow cultivated without reducing the weed control activity of TREFLAN. Do not cultivate deeper than the treated soil since this may bring untreated soil to the surface, and poor weed control may result.

CROP RECOMMENDATIONS

These recommendations are given as the broadcast rates of TREFLAN per acre. Apply any time after January 1 when the soil can be worked and is suitable for good incorporation—see specific crop for recommendations. Where a rate range is shown, use the lower rate for coarser soils or soils with lower organic matter. TREFLAN should not be used on soils containing more than 10% organic matter or on muck soils.

Alfalfa—Established

(For Barnyardgrass, Crabgrass, Foxtail, Junglerice, Sandbur and Cupgrass Control)

TREFLAN TR-10 can be activated for barnyardgrass, crabgrass, foxtail, junglerice, sandbur and cupgrass control in established alfalfa either by rainfall/overhead sprinkler irrigation, flood irrigation or by mechanical incorporation.

Apply TREFLAN TR-10 to established alfalfa stands prior to weed emergence at a broadcast rate of 20 pounds per acre for all soil textures. TREFLAN TR-10 may be applied using properly calibrated ground or aerial application equipment. A single rainfall or overhead sprinkler irrigation of 0.5 inches or more or flood irrigation after application is required to activate TREFLAN. If this does not occur within three (3) days after application, TREFLAN should be activated using incorporation equipment that will ensure thorough soil mixing with minimum damage to the established alfalfa.

Precaution: In the season following a 20 pound per acre TREFLAN TR-10 treatment, where established alfalfa is to be rotated to another crop, plant only those crops for which TREFLAN can be applied as a preplant treatment or injury may result.

Read the TREFLAN label for additional directions, precautions and limitations prior to use.

For the Following Crop Grouping, Use the Rate Table Below.

Castor Bean

Cole Crops—Transplant

(Broccoli, Brussels Sprout, Cabbage and Cauliflower). See page 19 for direct seeded.

Apply and incorporate prior to transplanting only.

Okra

Pepper—Transplant

Apply and incorporate prior to transplanting only.

Southern Pea—(Before planting only.)

Apply and incorporate TREFLAN before planting, at planting or immediately after planting, unless otherwise indicated.

Broadcast Rates Per Acre

Soil Texture	TREFLAN TR-10: Areas receiving less than 20" average annual rainfall*	TREFLAN TR-10: Areas receiving greater than 20" average annual rainfall*
	(pounds)	(pounds)
Coarse	5	5
Medium	6¼-7½	7½
Fine	7½	10

**Use 7½ pounds per acre on coarse and medium textured soils and 10 pounds on fine soils with 2-5% organic matter; use 10 pounds on all soils with 5-10% organic matter.*

Cole Crops—Direct Seeded

(Broccoli, Brussels Sprout, Cabbage and Cauliflower). See page 18 for transplant.

For direct seeded sole crops, apply and incorporate TREFLAN before planting at a broadcast rate of 5 pounds per acre on coarse and medium soils and 7½ pounds on fine soils and soils with 2-5% organic matter. Direct-seeded cole crops have exhibited marginal tolerance to recommended rates of TREFLAN. Stunting or reduced stands may occur.

Cucurbits—Postplant Emerged

(Cantaloupe, Cucumber and Watermelon)

Western United States including Texas:

Apply TREFLAN as a broadcast granule to the soil between the rows and beneath plants which are in the 3 to 4 true leaf stage.

Broadcast Rates Per Acre

Soil Texture	TREFLAN TR-10: Areas receiving less than 20" average annual rainfall*	TREFLAN TR-10: Areas receiving greater than 20" average annual rainfall*
	(pounds)	(pounds)
Coarse	5	5
Medium	6¼-7½	7½
Fine	7½	10

**Use 7½ pounds per acre on coarse and medium textured soils and 10 pounds on fine soils with 2-5% organic matter; use 10 pounds on all soils with 5-10% organic matter.*

Set incorporation equipment to throw treated soil around the plants during incorporation.

Fruit and Nut Crops and Vineyards

For areas receiving less than 20" average annual rainfall.

For new plantings of almond, apricot, citrus, nectarine, peach, pecan and walnut trees, apply and incorporate TREFLAN before planting at a broadcast rate of 5 pounds per acre on coarse soils; 6¼ to 7½ pounds on medium soils; 7½ pounds on fine soils; 7½ to 10 pounds on soils with 2-5% organic matter; and 10 pounds on soils with 5-10% organic matter.

For new plantings of vineyards, apply and incorporate TREFLAN before planting at a broadcast rate of 5 to 7½ pounds per acre on coarse soils; 7½ to 15 pounds on medium soils; and 15 to 20 pounds on fine soils or soils with 2 to 10% organic matter. Do not use more than 10 pounds per acre on heat-treated grape rootings.

For postplant applications on bearing or nonbearing, established plantings of vineyards and almond, apricot, grapefruit, lemon, nectarine, orange, peach, pecan, plum, prune, tangelo, tangerine and walnut trees, apply TREFLAN at a broadcast rate of 10 to 20 pounds per acre for all soil textures. Do not apply to vineyards within 60 days of harvest.

In established plantings, apply TREFLAN to the soil and use incorporation methods not injurious to the trees or vines.

Cotton—TREFLAN Alone

TREFLAN can be applied and incorporated before or at planting, immediately after planting, or at layby.

Cotton—Preemergence:

Broadcast Rates Per Acre

Soil Texture	TREFLAN TR-10: Areas receiving less than 20" average annual rainfall*	TREFLAN TR-10: Areas receiving greater than 20" average annual rainfall*
	(pounds)	(pounds)
Coarse	5	5
Medium	6¼-7½	7½
Fine	7½	10

**Use 7½ pounds per acre on coarse and medium textured soils and 10 pounds on fine soils with 2-5% organic matter; use 10 pounds on all soils with 5-10% organic matter.*

Cotton—Postplant:

When incorporating TREFLAN after planting (postplant), be careful not to disturb the seed.

Cotton—layby:

Apply and incorporate TREFLAN any time up to layby, but not less than 90 days before harvest. Direct the layby applications onto the soil between the rows and beneath emerged cotton plants. Use the same rates as for a preemergence application.

Cotton—Fall Application:

Apply and incorporate TREFLAN any time between October 15 and December 31. The ground may be left flat or bedded-up over winter. On bedded ground, knock beds down to desired height before planting, moving some treated soil from beds into furrows. Where soil is left flat over winter, be careful not to turn up untreated soil during spring bedding operations. Destroy established weeds during seedbed preparation. If weeds become established in furrows due to uncovering of untreated soil during bedding, destroy these weeds before planting. In the fall, do not apply TREFLAN to soils which are wet or subject to prolonged periods of flooding.

Broadcast Rates per Acre—Fall Application Only:

In Alabama, Arkansas, northern Florida, Georgia, Louisiana, Mississippi, southeastern Missouri, Bootheel, North Carolina, New Mexico, Oklahoma, South Carolina, Tennessee and Texas, apply and incorporate TREFLAN at a broadcast rate of 10 pounds per acre on coarse and medium soils and 12½ pounds on fine soils.

In Arizona, California and Nevada, apply and incorporate TREFLAN at a broadcast rate of 7½ pounds per acre on coarse soils; 10 pounds on medium soils; and 12½ pounds on fine soils.

For cotton growth in other states, apply and incorporate TREFLAN at a broadcast rate of 5 pounds per acre on coarse soils; 7½ pounds on medium soils; 10 pounds on fine soils; 7½ pounds on coarse soils with 2-5% organic matter; and 10 to 12½ pounds on soils with 5-10% organic matter.

Dry Bean—TREFLAN Alone

Apply and incorporate TREFLAN before planting using the following rates:

Broadcast Rates Per Acre

Soil Texture	TREFLAN TR-10: Areas receiving less than 20" average annual rainfall*	TREFLAN TR-10: Areas receiving greater than 20" average annual rainfall*
	(pounds)	(pounds)
Coarse	5	5
Medium	6¼-7½	7½
Fine	7½	10

*Use 7½ pounds per acre on coarse and medium textured soils and 10 pounds on fine soils with 2-5% organic matter; use 10 pounds on all soils with 5-10% organic matter.

Dry Bean—Fall Application

Apply and incorporate TREFLAN any time between October 15 and December 31 at a broadcast rate of 5 pounds per acre on coarse soils; 6¼ to 7½ pounds on medium soils; and 7½ pounds on fine soils. Destroy established weeds during seedbed preparation.

Flax—Fall Application

Apply and incorporate TREFLAN in the fall from September 1 to December 31.

Broadcast Rates Per Acre

Soil Texture	TREFLAN TR-10
	(pounds)
Coarse	5
Medium	7½
Fine	10

Special Instructions for Flax

1. Incorporation operations or any other tillages performed in the spring prior to seeding should be relatively shallow so as to maintain a firm seedbed, and the seedbed should be packed just prior to seeding.

2. Seeding should be done with a press drill or hoe drill. Seed into a moist seedbed no more than 1½ inches deep.

3. Flax should not be seeded until the seedbed has warmed up.

Beans—(Lima Bean and Snap Bean):

Apply and incorporate TREFLAN before planting at a broadcast rate of 5 pounds per acre on coarse and medium soils and 7½ pounds on fine soils.

Mustard—Grown for seed or processing for food in Minnesota, North Dakota and South Dakota.

Apply and incorporate TREFLAN before planting at a broadcast rate of 5 pounds per acre on coarse soils and 7½ pounds on medium and fine soils.

Peanut—(Spanish Peanut in Texas and Oklahoma):

Apply and incorporate TREFLAN before planting, at planting or immediately after planting at a broadcast rate of 5 pounds per acre on coarse soils. When incorporating after planting, take care not to disturb the seed.

Rape

Apply and incorporate TREFLAN in the spring before planting or in the fall after September 1.

Broadcast Rates Per Acre

Soil Texture	TREFLAN TR-10
	(pounds)
Coarse	5
Medium	7½
Fine	10

Safflower

Follow recommended soil preparation, application and incorporation procedures for TREFLAN.

Apply and incorporate TREFLAN in the spring before planting or in the fall between October 15 and December 31.

Broadcast Rates Per Acre

	TREFLAN TR-10	
Soil Texture	Areas receiving Less than 20" average annual rainfall*	Areas receiving greater than 20" average annual rainfall*
	(pounds)	(pounds)
Coarse	5	5
Medium	6¼-7½	7½
Fine	7½	10

*Use 7½ pounds per acre on coarse and medium textured soils and 10 pounds on fine soils with 2-5% organic matter; use 10 to 12½ pounds on all soils with 5-10% organic matter.

Safflower—Fall Application—(Arizona, California, Idaho, Nevada, Oregon, Utah, Washington and Wyoming):

Apply and incorporate TREFLAN any time between October 15 and December 31. Ground may be left flat or bedded-up over winter. On bedded ground, knock beds down to desired height before planting, moving some treated soil from tops into furrows. Where soil is left flat over winter, take care during spring bedding operations to prevent turning up untreated soil. Destroy established weeds during seedbed preparation. If weeds become established in furrows due to uncovering of untreated soil during listing, destroy these weeds before planting.

Apply and incorporate TREFLAN at a broadcast rate of 7½ pounds per acre on coarse soils; 10 pounds on medium soils; and 12½ pounds on fine soils.

Do not apply TREFLAN in the fall to soils which are wet or are subject to prolonged periods of flooding.

Soybean—TREFLAN Alone

Soybean—Preemergence:

Follow recommended soil preparation, application, and incorporation procedures for TREFLAN.

Broadcast Rates Per Acre

Soil Texture	TREFLAN TR-10*
	(pounds)
Coarse	5
Medium	7½
Fine	10

*Use 7½ pounds per acre on coarse and medium textured soils and 10 pounds on fine soils with 2-5% organic matter; use 10-12½ pounds on all soils with 5-10% organic matter.

Soybean—Fall Application:

Apply and incorporate TREFLAN any time between October 15 and December 31. Ground may be left flat or bedded-up over winter. On bedded ground, knock beds down to desired height before planting, moving some treated soil from tops into furrows. Where soil is left flat over winter, take care during spring bedding operations to prevent turning up untreated soil. Destroy established weeds during seedbed preparation. If weeds become established in furrows due to uncovering of untreated soil during listing, destroy these weeds before planting. Do not apply TREFLAN in the fall to soils which are wet, are subject to prolonged periods of flooding, or where rice was grown the previous year.

For soybeans grown in Alabama, Arkansas, northern Florida, Georgia, Louisiana, Mississippi, southeastern Missouri Bootheel, North Carolina, Oklahoma, South Carolina, Tennessee and Texas, apply and incorporate TREFLAN at a broadcast rate of 10 pounds per acre on coarse and medium soils and 12½ pounds on fine soils.

For soybeans grown in states other than those listed above, apply and incorporate TREFLAN at a broadcast rate of 5 pounds per acre on coarse soils; 7½ pounds on medium soils; 10 pounds on fine soils; 7½ on coarse soils with 2-5% organic matter; and 10-12½ pounds on soils with 5-10% organic matter.

Sugar Beet

Broadcast TREFLAN overtop when plants are between 2 and 6 inches tall at a rate of 5 pounds per acre on coarse soils and 6¼ to 7½ pounds on medium and fine soils. Use the higher rate for medium and fine soils in areas receiving more than 20" average annual rainfall. Set incorporation machinery to throw treated soil toward the plants in the row. Be careful that incorporation machinery does not damage the sugar beet taproot.

Precaution: Exposed beet roots should be covered with soil before a TREFLAN application to reduce the possibility of girdling.

Sunflower—TREFLAN Alone

Apply and incorporate TREFLAN in the spring or in the fall between October 15 and December 31. Follow recommended soil preparation, application and incorporation procedures for TREFLAN.

Broadcast Rates Per Acre

	TREFLAN TR-10	
Soil Texture	Areas receiving less than 20" average annual rainfall*	Areas receiving greater than 20" average annual rainfall*
	(pounds)	(pounds)
Coarse	5	5
Medium	6¼-7½	7½
Fine	7½	10

*Use 7½ to 10 pounds per acre on coarse and medium textured soils with 2-5% organic matter and 10 pounds on all soils with 5-10% organic matter.

Tomato

For direct seeded tomato, apply TREFLAN at blocking or thinning to the soil between rows and beneath the plants, and incorporate. For transplant tomato, apply and incorporate TREFLAN before transplanting. Do not apply TREFLAN after transplanting.

Broadcast Rates Per Acre

	TREFLAN TR-10	
Soil Texture	Areas receiving less than 20" average annual rainfall*	Areas receiving greater than 20" average annual rainfall*
	(pounds)	(pounds)
Coarse	5	5
Medium	6¼-7½	7½
Fine	7½	10

*Use 7½ pounds per acre on coarse and medium textured soils and 10 pounds on fine soils with 2-5% organic matter; use 10 pounds on all soils with 5-10% organic matter.

Wheat (Spring), Durum and Barley—Summer Fallow

TREFLAN may be applied to coarse, medium and fine textured soils for control of labeled weeds in the summer fallow period and for pigeongrass (foxtail) control in wheat, durum, and barley seeded the following spring.

TREFLAN TR-10 may be applied to ground that has a manageable trash level been fallow, pre-tilled, or to soil left with a stubble cover. The first incorporation is required within 24 hours after application. The second incorporation and subsequent tillage may be done whenever necessary to destroy resistant weed growth during the remainder of the fallow year. During the fallow year, susceptible weeds may not be controlled until after the second incorporation.

Broadcast Rates Per Acre (All Soil Textures)

	TREFLAN TR-10	
Application Date	Areas Subject to less than 10 inches Annual Rainfall	All Other Areas
	(pounds)	(pounds)
April 15 - April 30	8¾	10
May 1 - May 31	8¾-7½	10-8¾
June 1 - June 30	7½-6¼	8¾- 7½
July 1 - July 31	6¼-5	7½-6¼
August 1 - August 31	5	6¼-5

Where a rate range is shown, use the higher rate per acre during the early part of an application period and the lower rate per acre during the latter part of an application period.

Incorporation Directions: Incorporate TREFLAN one (1) time within twenty-four (24) hours after application. Care should be taken to ensure that the second incorporation and subsequent tillages are more shallow than the first. Incorporate and till the TREFLAN treated ground with the following equipment:

1. Chisel plow: Operate at 4-5 inches deep at 4-6 mph. A chisel plow is defined as having 3 rows of 14 to 18 inch sweeps on no greater than 12 inch center. Stagger sweeps so that no soil is left unturned. Chisel points should not be used.

2. Tandem disc: Operate at 3-4 inches deep at 4-6 mph.

3. Field cultivator: Operate at 3-4 inches deep at 5 mph or more. A field cultivator is defined as having 3 to 4 rows of sweeps with "c" or "s" shaped shanks, spaced 7 inches or less and staggered so that no soil is left unturned.

Seeding Directions: Wheat, durum, or barley should be seeded approximately two (2) inches deep.

Precautions: While use of this practice may result in a stand reduction, slight stand reductions do not normally affect yield.

Read the TREFLAN label for additional directions, precautions and limitations prior to use.

Wheat (Spring), and Durum—Fall Application—Foxtail/Pigeongrass Control

For the control of foxtail(pigeongrass), apply TREFLAN at a broadcast rate of 5 pounds per acre on coarse and medium soils and 7½ pounds on fine soils.

Barley—Fall Application:

General weed control (except for special rate or use programs) in Minnesota, North Dakota and South Dakota.

Apply TREFLAN at a broadcast rate of 5 pounds per acre on coarse soils and 7½ pounds on medium and fine soils. TREFLAN applied at these rates also provide partial control or suppression of kochia, Russian thistle and wild oats.

Incorporation Directions: Any of the following tools are recommended for fall incorporation. The disc or field cultivator may be used for the spring incorporation pass. Care should be taken to operate the tool at a more shallow depth than the fall incorporation.

1. Chisel plow: May be used for first pass only. Operate at 4-5 inches deep at 4-6 mph. A chisel plow is defined as having 3 rows of 14 to 18 inch sweeps on no greater than 12 inch center. Stagger sweeps so that no soil is left unturned. Chisel points should not be used.

2. Tandem disc: Operate at 3-4 inches deep at 4-6 mph.

3. Field cultivator: Operate at 3-4 inches deep at 5 mph or more. A field cultivator is defined as having 3 to 4 rows of sweeps with "c" or "s" shaped shanks, spaced 7 inches or less and staggered so that no soil is left unturned.

Planting Directions: Wheat, durum, or barley should be seeded approximately two (2) inches deep.

Precautions: While use of this practice may result in a stand reduction, slight stand reductions do not normally affect yield.

Wheat (Winter)—(Idaho, Oregon and Washington):

TREFLAN may be applied for preplant preemergence control of cheatgrass and other annual grasses and broadleaf weeds controlled by TREFLAN.

Apply TREFLAN any time up to three (3) weeds prior to planting. Broadcast TREFLAN at a rate of 7½ pounds per acre on coarse and medium soils and 10 pounds on fine soils.

Incorporation Directions: Incorporate TREFLAN into the soil with a flexible tine-tooth harrow, (Flextine or Melroe), set to cut one (1) to two (2) inches deep and operated at 3 to 6 mph. Incorporate one (1) time within 24 hours after application, followed by a second incorporation at least five (5) days after the first pass in a different direction prior to planting. Do not till the soil with a disc after the TREFLAN has been incorporated with a flexible tine harrow.

Seeding Directions: Use only a deep furrow or semi-deep furrow drill that will place the seed below the zone of soil into which TREFLAN has been incorporated.

Precaution: Do not plant wheat directly into the zone of soil treated with TREFLAN as injury to the crop (delay in emergence or stand reduction) may occur. Delayed emergence or slight stand reductions do not normally affect yield.

Wheat (Winter)—Fallow Soil Application (Idaho, Oregon and Washington)

TREFLAN applied and shallowly incorporated into fallow soil up to four (4) months ahead of planting will control cheatgrass and certain annual grasses and broadleaf weeds.

Broadcast rates are 7½ pounds per acre on coarse and medium soils and 10 pounds on fine soils. Apply TREFLAN any time from May to September prior to the fall planting of winter wheat.

Incorporation Directions: Incorporate TREFLAN with a flexible tine-tooth harrow, (Flextine or Melroe), set to cut one (1) to two (2) inches deep and operate at three (3) to six (6) mph. For thorough incorporation, two (2) passes of the equipment in different directions over the field are necessary. Incorporate one (1) time within 24 hours after application followed by a second incorporation at least five (5) days after the first prior to seeding. Do not till the soil with a disc after TREFLAN has been incorporated with a flexible tine harrow.

Seeding Directions: Use only a deep furrow or semi-deep furrow drill that will place the seed below the zone of soil into which TREFLAN has been incorporated.

Precaution: Do not plant wheat directly into the zone of soil treated with TREFLAN as injury to the crop (delay in emergence or stand reduction) may occur. Delayed emergence or slight stand reductions do not normally affect yield.

Fall Application

General: (Areas receiving more than 20″ average annual rainfall). See specific crop for recommendations. For all crops for which there are no specific fall application instructions and for which TREFLAN is recommended as a preemergence application, use the rates listed for spring applications. Do not apply TREFLAN in the fall for sugar beets, potatoes and direct-seeded tomatoes.

In most states: Apply and incorporate TREFLAN any time between October 15 and December 31.

In California, Minnesota, North and South Dakota: Apply and incorporate TREFLAN any time between September 1 and December 31. Ground may be left flat or bedded-up over winter. On bedded ground, knock beds down to desired height before planting, moving some treated soil from beds into furrows. Where soil is left over winter, be careful not to turn up untreated soil during spring bedding operations. Destroy established weeds during seedbed preparation.

If weeds become established in furrows due to uncovering of untreated soil during bedding, destroy these weeds before planting. Do not apply TREFLAN in the fall to soils which are wet, are subject to prolonged periods of flooding, or where rice was grown the previous year.

TREFLAN PERFORMANCE GUARANTEE A LIMITED WARRANTY OF TREFLAN WEED AND GRASS CONTROL

Elanco Products Company warrants that TREFLAN will provide commercially acceptable control of the weeds and grasses as indicated on its current label when used according to label directions and subject to the following limitations:

1. Purchaser must notify Elanco promptly if a lack of commercially acceptable control occurs. Notice must be given within forty-five (45) days after the first planting or after application of TREFLAN to the treated crop, whichever occurs later.

2. An Elanco representative must be satisfied that Purchaser used TREFLAN according to the label directions.

3. The Purchaser must provide an Elanco representative with an invoice or other satisfactory record which shows the price and quantity of TREFLAN purchased.

4. Elanco's liability will be limited to a refund of the purchase price of the TREFLAN applied to the acreage on which commercially acceptable control was not obtained.

5. Void outside the USA.

DISCLAIMER OF WARRANTIES

The warranty printed above is the only warranty applicable to this product. All other warranties, expressed or implied, including, but not limited to, any implied warranties of merchantability and fitness for a particular purpose, are disclaimed. This disclaimer of warranties does not apply where prohibited by operation of law.

LIMITATION OF DAMAGES

Elanco's liability, whether in contract, warranty, tort, negligence, strict liability or otherwise, shall not exceed the return of the amount of the purchase price of the TREFLAN and under no circumstances shall Elanco be liable for special, indirect or consequential damages. This disclaimer of damages does not apply where prohibited by operation of law.

INHERENT RISKS OF USE

Failure to carefully follow the directions for use of TREFLAN, as well as other factors such as seedling disease, cold weather, deep planting, excessive moisture, high salt concentration or drought, may result in unsatisfactory weed control or crop injury. Reduced yields may result under these conditions.

This is a specimen label. Elanco is not responsible for the accuracy of the information contained herein. As labels are subject to revision, always carefully read and follow the label on the product container.

Fermenta ASC Corporation

5966 Heisley Road
P.O. Box 8000
Mentor, OH 44061-8000
(216) 357-4100

ANSAR® 6.6

For Postemergence Weed Control

Product Contains 6.6 Pounds MSMA Per Gallon

ACTIVE INGREDIENT:	
Monosodium Acid Methanearsonate	51.0%
INERT INGREDIENTS:	49.0%
TOTAL	100.0%
Total Arsenic, all in water-soluble form, expressed as elemental	23.6%

EPA Reg. No. 50534-16

ANSAR is a registered trademark of Fermenta ASC Corporation.

KEEP OUT OF REACH OF CHILDREN

CAUTION

See side panel for additional precautionary statements.

PRECAUTIONARY STATEMENTS

HAZARDS TO HUMANS AND DOMESTIC ANIMALS

CAUTION

Harmful if swallowed or absorbed through the skin. Avoid contact with eyes, skin or clothing. Avoid breathing vapors or spray mist.

If swallowed, induce vomiting immediately by giving two glasses of water and sticking finger down throat. **Never** give anything by mouth to an unconscious person. Seek medical attention.

In case of contact, flush eyes with plenty of water for at least 15 minutes. Seek medical attention.

Wash exposed skin gently with plenty of soap and water.

Keep children and domestic animals off treated areas until this material has been washed into the soil.

Wear protective clothing when handling or applying this product including long pants, long-sleeve shirt, and impermeable gloves and boots. Mixer-loaders should include an apron and full-face shield when handling or mixing concentrate. Flagmen should be fully protected during spray operations or mechanical flagmen used. Pilots and ground spray rig applicators should wear a mask or respirator approved by the Mining Enforcement and Safety Administration and the National Institute for Occupational Safety and Health.

ENVIRONMENTAL HAZARDS

Do not apply directly to water. **Do not** feed treated foliage to livestock or graze treated areas. **Do not** spray or allow drift onto vegetables, ornamentals or other desirable plants. **Do not** apply when weather conditions favor drift from target area. **Do not** contaminate water by cleaning of equipment or disposal of wastes.

ANSAR® 6.6

Read entire label before using this product.

GENERAL INFORMATION

ANSAR 6.6 is a postemergence herbicide for use in cotton, turf and non-crop areas. **ANSAR 6.6** does not contain a surfactant (wetting agent). A suitable non-ionic surfactant should be added to spray tank. Local conditions and recommendations vary; consult local agricultural experiment station or extension service weed specialists for recommendations in your area.

Mixing Instructions: ANSAR 6.6 herbicide must be thoroughly dissolved. Fill the spray tank with half the required amount of water and, with pump or agitator operating, add the recommended amount of **ANSAR 6.6** herbicide and continue filling the tank with balance of water. Add surfactant and mix thoroughly before using.

Do not store spray solution in tank overnight. Clean application equipment thoroughly after use by flushing with water in a safe place.

Application Methods: ANSAR 6.6 should be applied with a low-volume, properly calibrated sprayer having satisfactory pumping and bypass action. **Adjust Nozzles to Keep Spray Off Cotton Foliage and in a Manner to Allow Maximum Coverage of Weeds.** Adequate coverage is very important for effective weed control.

Apply during warm weather when weeds are in an active stage of growth.

DIRECTIONS FOR USE

It is a violation of Federal law to use this product in a manner inconsistent with its labeling.

Do not apply this product through any type of irrigation equipment.

Cotton: ANSAR 6.6 is useful as a directed postemergence spray for controlling weeds in cotton such as barnyardgrass, cocklebur, smooth and large crabgrass, dallisgrass, goosegrass, Johnsongrass and nutsedge. Apply as a **Directed** broadcast spray at rate of 40 gallons per acre of a solution containing 3 pints of **ANSAR 6.6** plus 1½ to 2 quarts of suitable surfactant, cleared for application to growing crops, in 50 gallons of water when weeds are small. For band applications, apply 1 gallon of above diluted spray per acre for each 1 inch band width to be treated of cotton grown on 40 inch row spacing. A second or repeat application, if needed, should be timed about 1 to 3 weeks after the first application. Keep spray off cotton foliage.

Apply Only When Cotton is 3 inches High to First Bloom. Do Not Apply After First Bloom.

Slight burning and reddish discoloration of cotton foliage may occasionally be seen following recommended treatment; however, cotton plants will develop normally and yield will not be affected.

Ornamental Turfgrass: ANSAR 6.6 can be used for selective control of bahiagrass, barnyardgrass, chickweed, smooth and large crabgrass, dallisgrass, nutsedge, sandbur and wood sorrell with little or no injury to well-established, actively growing turfgrasses. Mow turfgrass to height of 1 to 1½ inches before treatment. Mix 1 fluid ounce (2 tablespoons) of ANSAR 6.6 plus 2 to 3 fluid ounces of suitable surfactant in 2½ gallons of water for application to an area of 1,000 square feet. Application should be uniform and thorough to adequately wet all undesirable plants. Two or more repeat treatment at 14 day intervals may be necessary. Make applications during warm weather when temperature is between 80 and 90°F. **Do not** water turf for at least 24 hours after application. Turfgrasses may be temporarily discolored. Bermudagrass, bluegrasses and zoysiagrass have shown tolerance to properly applied **ANSAR 6.6.** Injury may result if applied to bentgrasses and fescues. **Do not** apply to St. Augustinegrass, carpetgrass, centipedegrass or to Dichondra. **Do not** reseed until 2 weeks after last application. **Do not** apply with hose-end applicators.

Non-crop: ANSAR 6.6 is effective in controlling barnyardgrass, cocklebur, smooth and large crabgrass, dallisgrass, goosegrass, Johnsongrass, nutsedge and many similar weeds on drainage ditch banks, fence rows, rights-of-way, storage yards and similar non-crop areas. Application should be made when weeds are small and conditions are favorable for good weed growth. Mix at a rate of 3 pints of **ANSAR 6.6** plus 1½ to 2 quarts of suitable surfactant in 50 gallons of water. Spray undesirable vegetation thoroughly to point of run-off. Adequate coverage and complete wetting of foliage is important for effective control. Repeat applications may be necessary if regrowth occurs. **Do not** contaminate waters, used for domestic purposes, or by animals, including wildlife and aquatic life, or water used for irrigation purposes.

STORAGE AND DISPOSAL

Do not contaminate water, food or feed by storage or disposal. Open dumping is prohibited.

PESTICIDE DISPOSAL: Wastes resulting from the use of this product may be disposed of on site or at an approved waste disposal facility.

CONTAINER DISPOSAL: Do not reuse empty container. Triple rinse (or equivalent) and offer for recycling or reconditioning, or puncture and dispose of in a sanitary landfill, or incinerate, or, if allowed by state and local authorities, by burning. If burned, stay out of smoke.

WARRANTY AND LIMITATION OF DAMAGES

Seller warrants that this material conforms to its chemical description and is reasonably fit for the purposes stated on the label when used in accordance with directions under normal conditions of use and Buyer assumes the risk of any use contrary to such directions. SELLER MAKES NO OTHER EXPRESS OR IMPLIED WARRANTY, INCLUDING ANY OTHER EXPRESS OR IMPLIED WARRANTY OF FITNESS OR OF MERCHANTABILITY. AND NO AGENT OF SELLER IS AUTHORIZED TO DO SO EXCEPT IN WRITING WITH A SPECIFIC REFERENCE TO THIS WARRANTY. In no event shall Seller's liability for any breach of warranty exceed the purchase price of the material as to which a claim is made.

BRAVO® 500

AGRICULTURAL FUNGICIDE

ACTIVE INGREDIENT:

Chlorothalonil (tetrachloroisophthalonitrile)	40.4%
INERT INGREDIENTS:	59.6%
TOTAL	100.0%

Contains 4.17 Pounds Chlorothalonil Per Gallon (500 grams per liter)

Covered under U.S. Patent No. 3,290,353, No. 3,331,735 and No. 3,948,636

EPA Reg. No. 50534-8

A Broad-Spectrum Agricultural Fungicide

Read entire label carefully and use only as directed.

KEEP OUT OF REACH OF CHILDREN

WARNING - AVISO

Precaucion al Usuario: Si usted no lee ingles, no use este producto hasta que la etiqueta haya sido explicado ampliamente.

See precautionary statements.

BRAVO 500

PRECAUTIONARY STATEMENTS

HAZARDS TO HUMANS AND DOMESTIC ANIMALS

WARNING

Causes eye irritation. May cause skin irritation. May be a potential skin sensitizer. **DO NOT** get in eyes. Wear goggles or face shield when handling this product. Avoid prolonged contact with skin. Wash exposed areas of skin with soap and warm water after handling or using. **DO NOT** take internally. Avoid breathing spray mist.

DO NOT apply this product in such a manner as to directly, or through drift, expose workers or other persons. The area being treated must be vacated by unprotected persons.

Note to User: This product may produce temporary allergic side effects characterized by redness of the eyes, mild bronchial irritation and redness or rash on exposed skin areas. Persons having allergic reaction should contact a physician.

FIRST AID

In case of contact with eyes, immediately flush eyes with plenty of water and continue for 15 minutes. Seek medical attention for eyes immediately. Wash exposed areas of skin with soap and warm water after handling or using.

Note to Physician: Persons having a temporary allergic reaction respond to treatment with antihistamines or steroid creams and/or systemic steroids.

ENVIRONMENTAL HAZARDS

This product is toxic to fish, aquatic invertebrates, and marine/estuarine organisms. Runnoff from treated areas may be hazardous to aquatic organisms in neighboring areas. **DO NOT** apply directly to water. **DO NOT** apply when weather conditions favor drift from treated areas.

Apply only to areas specified on label.

BRAVO 500

DIRECTIONS FOR USE

It is a violation of Federal law to use this product in a manner inconsistent with its labeling.

Note to User Wear long sleeve shirt, long pants, and gloves while mixing, loading and applying this product.

BRAVO 500 is an excellent fungicide when used according to label directions for control of a broad spectrum of plant diseases. **BRAVO 500** can be used effectively in dilute or concentrate sprays. Thorough, uniform coverage is essential for disease control. **DO NOT** combine **BRAVO 500** in the spray tank with pesticides, surfactants or fertilizers, unless your prior use has shown the combination physically compatible, effective and noninjurious under your conditions of use. **Do not** combine **Bravo 500** with Dipel 4L, Triton AG-98 or Triton B-1956 as phytotoxicity may result from the combination when applied to some crops on this label.

Dipel is a registered trademark of Abbott Laboratories. Triton AG-98 and Triton B-1956 are trademarks of Rohm and Haas Company.

NOTE: Slowly invert container several times to assure uniform mixture. The required amount of **BRAVO 500** should be added slowly into the spray tank during filling. With concentrate sprays, pre-mix the required amount of **BRAVO 500** in a clean container and add to the spray tank as it is being filled. Keep agitator running when filling spray tank and during spray operations. DO NOT use on greenhouse grown crops.

Dosage rates on this label indicate pints of **BRAVO 500** per acre unless otherwise stated. Under conditions favoring disease development the high rate specified and shortest application interval should be used.

Note to User: DO NOT rotate to crops other than those listed on labels within 12 months of the last treatment. After 12 months from the last application, leafy vegetables (i.e. spinach, lettuce, kale, etc.) may also be rotated.

REENTRY: Note to User: DO NOT enter treated area to perform hand labor within 24 hours of application unless protective clothing is worn.

Because certain states may require more restrictive reentry intervals for various crops treated with this product, consult your State Department of Agriculture for further information.

Written or oral warnings must be given to workers who are expected to be in a treated area or in an area about to be treated with this product. When oral warnings are given, warnings shall be given in a language customarily understood by workers. Oral warnings must be given if there is reason to believe that written warnings cannot be understood by workers. Oral and written warnings must include the following information: "**WARNING.** Area treated with **BRAVO 500** on (date of application). **DO NOT** enter without appropriate protective clothing for 24 hours. In case of accidental exposure, wash exposed area with plenty of water and get medical attention. For further information see 'Precautionary Statements' on the label."

FIELD AND ROW CROPS:

Apply **Bravo 500** in sufficient water to obtain adequate coverage of foliage. Gallonage to be used will vary with crop and amount of plant growth. Spray volume usually will range from 20 to 150 gallons per acre (200 to 1400 liters per hectare) for dilute sprays and 5 to 10 gallons per acre (50 to 100 liters per hectare) for concentrate ground sprays and aircraft applications. Both ground and aircraft methods of application are recommended unless specific directions are given for a crop. Application through sprinkler irrigation systems is not recommended unless specific directions are given for a crop. See application and calibration instruction below.

APPLICATION AND CALIBRATION TECHNIQUES FOR SPRINKLER IRRIGATION

Apply this product only through center pivot, motorized lateral move, traveling gun, solid set and portable (wheel move, side roll, and tow, or hand move) irrigation system(s). **Do not** apply this product through any other type of irrigation system.

Crop injury, lack of effectiveness, or illegal pesticide residues in the crop can result from non-uniform distribution of treated water.

If you have questions about calibration, you should contact State Extension Service specialists, equipment manufacturers or other experts.

Do not apply this product through irrigation systems connected to a public water system. 'Public water system' means a system for the provision to the public of piped water for human consumption if such system has at least 15 service connections or regularly serves an average of at least 25 individuals daily at least 60 days per year.

Controls for both irrigation water and pesticide injection systems must be functionally interlocked, so as to automatically terminate pesticide injection when the irrigation water pump motor stops. A person knowledgeable of the irrigation system and responsible for its operation shall be present so as to discontinue pesticide injection and make necessary adjustments, should the need arise.

The irrigation water pipeline must be fitted with a functional, automatic, quick-closing check valve to prevent the flow of treated irrigation water back toward the water source. The pipeline must also be fitted with a vacuum relief valve and low pressure drain, located between the irrigation water pump and the check valve, to prevent back-siphoning of treated irrigation water into the water source.

Always inject BRAVO 500 into irrigation water after it discharges from the irrigation pump and after it passes through the check valve. Never inject pesticides into the intake line on the suction side of the pump.

Pesticide injection equipment must be fitted with a functional, normally closed, solenoid-operated valve located on the intake side of the injection pump. Interlock this valve to the power system, so as to prevent fluid from being withdrawn from the chemical supply tank when the irrigation system is either automatically or manually turned off.

The pesticide injection pipeline must contain a functional, automatic, quick-closing check valve to prevent the flow of fluid back toward the injection pump.

The irrigation line or water pump must include a functional pressure switch which will stop the water pump motor when the water pressure decreases to the point where pesticide distribution is adversely affected.

Spray mixture in the chemical supply tank must be agitated at all times, otherwise settling and uneven application may occur. Do not apply when wind speed favors drift beyond the area intended for treatment.

Bravo 500 may be used through two basic types of sprinkler irrigation systems as outlined in Sections A and B below. Determine which type of system is in place, then refer to the appropriate directions provided for each type.

A. **Center Pivot, Motorized Lateral Move and Traveling Gun Irrigation Equipment**
   For injection of pesticides, these continuously moving systems must use a positive displacement injection pump, of either diaphragm or piston type, constructed of materials that are compatible with pesti-

cides and capable of being fitted with a system interlock and capable of injection at pressures approximately 2 - 3 times those encountered within the irrigation water line. Venturi applicator units cannot be used on these systems.

Fill chemical supply tank of injection equipment with water. Operate system for one complete revolution or run across the field, measuring time required, amount of water injected, and acreage covered. Thoroughly mix recommended amount of **Bravo 500** for acreage to be covered into same amount of water used during calibration and inject into system continuously for one revolution or run. Mixture in the chemical supply tank must be continuously agitated during the injection run. Shut off injection equipment after one revolution or run, but continue to operate irrigation system until **Bravo 500** has been cleared from last sprinkler head.

B. Solid Set and Portable (Wheel Move, Side Roll, End Tow, or Hand Move) Irrigation Equipment

With stationary systems, an effectively designed in-line venturi applicator unit is preferred which is constructed of materials that are compatible with pesticides; however, a positive-displacement pump can also be used.

Determine acreage covered by sprinkler. Fill tank of injection equipment with water and adjust flow to use contents over a thirty to forty-five minute period. Mix desired amount of **Bravo 500** for acreage to be covered with water so that the total mixture of **Bravo 500** plus water in the injection tank is equal to the quantity of water used during calibration and operate entire system at normal pressures recommended by the manufacturer of injection equipment used for amount of time established during calibration. No agitation should be required. **Bravo 500** can be injected at the beginning or end of the irrigation cycle or as a separate application. Stop injection equipment after treatment is completed and continue to operate irrigation system until **Bravo 500** has been cleared from last sprinkler head.

Metric Conversion Guide

2¹/₈ pints equals 1.0 liter   4¹/₄ pints equals 2.0 liters

CROP	DISEASES	RATE PER ACRE	APPLICATION DIRECTIONS
Bean (Snap)	Rust	2 to 4¹/₄ pts.	Use in sufficient water to obtain adequate coverage. Begin applications during early bloom stage or when disease first threatens and repeat at weekly intervals or as necessary to maintain control. **DO NOT** apply within 7 days of harvest. **DO NOT** graze treated areas or feed treated plant parts to livestock.
	Botrytis Blight (gray mold)	4¹/₄ pts.	
Beans (Dry) Navy, Pinto Kidney, Lima, Blackeye	Rust Anthracnose, Downy mildew, Cercospora leaf spot (blackeye only)	2 to 3 pts.	Use in sufficient water to obtain adequate coverage. Begin applications during early bloom stage and repeat at 7 to 10 day intervals. For use only on beans harvested dry with pods removed. **DO NOT** apply within 6 weeks before harvest. **DO NOT** allow livestock to graze in treated areas or feed treated plant parts to livestock.
Cabbage, Chinese Cabbage (tight-headed varieties only) Cauliflower Broccoli, Chinese Broccoli, Brussels Sprouts	Alternaria leaf spot, Downy mildew	2¹/₄ pts.	Use in sufficient water to obtain adequate coverage. Begin applications after transplants are set in field, or shortly after emergence of field-seeded crop, or when conditions favor disease development. Repeat at 7 to 10 day intervals or as necessary to maintain control. Do not apply within 7 days of harvest to Chinese cabbage or Chinese broccoli.
	Ring spot (California only)	2³/₄ pts.	For field seeded brussels sprouts, begin applications at time of early sprout development or when conditions favor disease development. Repeat at 7 to 10 day intervals or as necessary to maintain control.
Carrot	Cercospora (Early) blight. Alternaria (Late) blight	2¹/₄ to 2³/₄ pts.	Use in sufficient water to obtain adequate coverage. Start applications when disease threatens and repeat at 7 to 10 day intervals or as necessary to maintain control. **BRAVO 500** may be applied through sprinkler irrigation equipment (solid set, portable wheel move or center pivot systems only). See calibration directions preceding this section.
Celery	Cercospora (Early) blight Septoria (Late) blight Basal stalk rot (Rhizoctonia solani)	1¹/₂ to 2¹/₈ pts. semiweekly or 3 to 4¹/₄ pts. weekly	Use 1¹/₂ to 2¹/₈ pts. per acre on a 3 to 5 day spray schedule or 3 to 4¹/₄ pts. per acre on a 7 day schedule. Start applications when transplants are set in the field. Apply in sufficient water to obtain adequate coverage. **BRAVO 500** may be applied through sprinkler irrigation equipment (solid set, portable wheel move or center pivot systems only.) See calibration directions preceding this section. **DO NOT** apply within 7 days of harvest.
	Pink Rot (Suppression)	4¹/₄ pts.	
	Early blight Late blight	2¹/₄ to 2³/₄ pts./ 100 gal.	For celery seedbeds, apply 125 gallons per acre twice weekly or as needed to maintain control. Start applications shortly after crop emergence. Use the higher rate under severe disease conditions.
Corn (Sweet) Corn growth for seed	Helminthosporium leaf blights, Rust	1¹/₈ to 2³/₄ pts.	Use in sufficient water to obtain adequate coverage. Begin applications when conditions favor disease development and repeat at 4 to 7 day intervals or as required to maintain control. Under severe disease conditions, use 2¹/₄ to 2³/₄ pts. per acre. **DO NOT** apply within 14 days of harvest. **DO NOT** apply to sweet corn to be processed. **DO NOT** allow livestock to graze in treated fields. **DO NOT** ensile treated corn or use as livestock forage.
Cranberry	Fruit rots, Lophodermium leaf/twig blight	6 to 10 pts.	Apply at late bloom and repeat at 10 to 14 day intervals. Under severe disease conditions use the 10 pts/acre rate on a 10 day schedule. **DO NOT** apply **BRAVO 500** more than 3 times per season, or within 50 days before harvest. **DO NOT** apply to bogs when flooded or allow release of irrigation water from bogs for at least 3 days following application. **BRAVO 500** may be applied through sprinkler irrigation equipment. Use 300 gallons of water per acre through solid set systems only. See calibration directions preceding this section.
Cucurbits: Cucumber, Cantaloupe, Muskmelon, Honeydew melon, Watermelon, Squash, Pumpkin	Anthracnose, Downy mildew, Target spot	2¹/₄ to 2³/₄ pts.	Use in sufficient water to obtain adequate coverage. Begin applications when plants are in first true leaf stage or when conditions are favorable for disease development. Repeat applications at 7 day intervals. Under severe disease conditions, shorten spray interval.
	Cercospora leafspot, Gummy stem blight (black rot), Alternaria leaf blight, Scab, Powdery mildew (Spaerotheca only)	2³/₄ to 4¹/₄ pts.	**Bravo 500** may be applied through sprinkler irrigation equipment (solid set, portable wheel move or center pivot systems only). See calibration directions preceding this section.
	Cucumber belly rot (*Rhizoctonia solani*)	12 pts.	Use **BRAVO 500** in sufficient water to obtain runoff to soil surface. Make a single application when vines begin to form. **BRAVO 500** may be applied through sprinkler irrigation equipment as directed above.

CROP	DISEASES	RATE PER ACRE	APPLICATION DIRECTIONS
Grasses Grown for Seed	Stem rust, Leaf rust, Stripe rust, Septoria leafspot, Glume blotch, Bipolaris and Drechslera leafspots	1½ to 2 pts.	Use in sufficient water to obtain adequate covrage. Begin applications during stem elongation when conditions favor disease development. Re-apply at flag (top) leaf emergence and repeat applications at 14 day intervals. **DO NOT** apply within 14 days of harvest. **DO NOT** allow livestock to graze in treated areas or feed treated plant parts to livestock.
	Selenophoma (eyespot)	1½ to 3 pts.	
Mint	Rust, Septoria leaf spot	2 pts.	Use in sufficient water to obtain adequate coverage, normally 20 to 150 gallons per acre for dilute sprays and 5 to 10 gallons per acre for concentrate ground and aircraft applications. Begin applications when emerging plants are 4-8 inches high. Repeat applications at 7 to 10 day intervals or as necessary to maintain control. **DO NOT** apply more than 3 times per season. **DO NOT** apply within 80 days of harvest. **DO NOT** feed fresh or extracted mint hay from treated fields to livestock. Based on available residue data, use of **BRAVO 500** on mint is restricted to Indiana, Michigan and Wisconsin.
Onion (dry bulb)	Botrytis leaf blight/blast, Botrytis neck rot (suppression), Purple blotch	1½ to 3 pts.	Apply in sufficient water to obtain thorough coverage of tops. **Bravo 500** is recommended for use with disease monitoring systems which adjust fungicide rates and frequency of application according to disease hazard. Apply as follows: **Low Disease Hazard & Prior to Infection** / **Low Disease Hazard & Some Disease Present** / **High Disease Hazard** **Rate per Acre:** 1½ pt. / 2 pts. / 3 pts. **Frequency:** 10 days / 7 to 10 days / 7 days For suppression of neck rot (*Botrytis* spp.) during storage, a minimum of three weekly applications prior to lifting, using 2 to 3 pints of **Bravo 500** per acre, is recommended. **Do not** apply within 7 days of harvest.
Onion (green bunching), Leek, Shallot, Garlic, Onion grown for seed	Botrytis leaf blight/blast, Purple blotch, Downy mildew (suppression)	2 to 4¼ pts.	Use in sufficient water to obtain thorough coverage of tops. Begin applications prior to favorable infection periods, and repeat at 7 to 10 day intervals for as long as conditions favor disease. Use the high rate and a 7 day schedule of applications when heavy dew or rain persist. **Do not** apply within 7 days of harvest on garlic. **Do not** apply more than 3 times per season or within 14 days of harvest on green bunching onions, leeks or shallots. If additional disease control is needed before harvest, use another registered fungicide.
Papaya	Alternaria fruit spot, Anthracnose, Stem end rot	3 to 5¾ pts.	Apply with ground equipment only, in sufficient water to obtain adequate coverage of fruit and leaves. Begin treatment when conditions favor development of disease and continue treatments at 14 day intervals until weather conditions no longer favor disease development. **DO NOT** graze livestock in treated area or feed processing by-products to livestock.
Parsnip	Alternaria leafspot, Downy mildew, Anthracnose, Botrytis blight (gray mold), Bottom rot (Rhizoctonia)	2 to 3 pts.	Apply in sufficient water to obtain adequate coverage. Make the first application at the first sign of disease or when conditions are favorable for infection. Continue applications on a 7 to 10 day schedule. **DO NOT** apply more than 4 times per season or within 10 days of harvest. **DO NOT** feed treated plant parts to livestock.
Passion Fruit (Hawaii only)	Alternaria fruit and leaf spot (passion fruit brown spot)	2¾ pts.	Apply with ground equipment in sufficient water to obtain adequate coverage of fruit and leaves. Begin treatment when fruit spots appear (April to July) and continue treatments at 14 day intervals until weather conditions no longer favor disease development. **DO NOT** graze in treated area of feed vines or processing by-product to livestock used for food.
Peanut	Cercospora (early) leafspot Cercosporidium (late) leafspot	1½ to 2⅛ pts.	Apply in sufficient water for coverage when leaf wetness first occurs or 30 to 40 days after planting. Repeat at 10 to 14 day intervals. When conditions favor late leafspot or when rust or web blotch occur, apply 2⅛ pts. per acre at 10 day intervals for the remainder of the season. **DO NOT** apply within 14 days of harvest. **DO NOT** allow livestock to graze in treated areas. **DO NOT** feed hay or threshings from treated fields to livestock. **BRAVO 500** may be applied through sprinkler irrigation equipment. Use 2⅛ pints per acre in solid set, portable wheel move, center pivot or traveling gun sprinkler irrigation equipment. See calibration directions preceding this section.
	Rust, Web blotch	2⅛	
Potato	Late blight, Early blight, Botrytis vine rot	1 pt.	Apply as a banded treatment directed over the foliage, beginning when vines are first exposed and leaf wetness occurs. Repeat applications at 7 to 10 day intervals until vines close, then increase rate as described below.
		—then—	
		1½ to 2⅛ pt./A	In addition to the early-season application above, as vines close between rows increase water carrier volume to cover the denser canopy and begin broadcast application patterns. Continue applications at 7 to 10 day intervals. Use the highest registered rate weekly when disease conditions are severe. **BRAVO 500** may be applied through sprinkler irrigation equipment (solid set, portable wheel move, center pivot or motorized lateral move systems only). Do not exceed a 10 day interval between applications when using this technique. Follow calibration and application directions preceding this section.
Soybean Determinate (Southern) Varieties	Anthracnose, Diaporthe pod and stem blight, Frogeye leafspot (Cercospora sojina), Purple seed stain, Cercospsora leaf blight (Cercospora kikuchii), Septoria brown spot		Apply in sufficient water to obtain complete coverage, using at least five gallons of water per acre for aerial application. Use the three application program in areas having a history of moderate to severe disease intensity. **Bravo 500** may be applied through sprinkler irrigation equipment. Follow application and calibration directions preceding this section. **NOTE: DO NOT** exceed total of 3 applications per season. **DO NOT** apply within 6 weeks of harvest. **DO NOT** feed treated parts to livestock or allow grazing in treated fields.
		2.0 to 3½ pts.	**Two Application Program** —Make the first application at early pod set (R3 stage, when majority of pods are ⅛ to ⅜ inch in length) and the second at beginning of seed formation (R5) which occurs about 14 days later.

CROP	DISEASES	RATE PER ACRE	APPLICATION DIRECTIONS
		1½ to 2¾ pts.	**Three application program** - Make the first application at the beginning of flowering (R1), the second at early pod set (R3) and the third at beginning of seed formation (R5)
	Stem canker *(Diaporthe phaseolorum var. caulivora)*	1½ pts.	Apply in 10 to 20 gallons of water per acre, as a band treatment directing spray to provide coverage of entire plant. Make the first application at time of emergence of the second trifoliate leaves (V2). If conditions favor stem canker disease make a second and a third application. Make all applications at 10 to 14 day intervals.
Soybean Indeterminate (Northern) Varieties	Anthracnose, Diaporthe pod and stem blight, Frogeye leafspot *(Cercospora sojina)*, Purple seed stain Cercospora leaf blight *(Cercospora kikuchii)*, Septoria brown spot		Apply in sufficient water to obtain complete coverage, using at least five gallons of water per acre for aerial application. Use the three application program in areas having a history of moderate to severe disease intensity. **BRAVO 500** may be applied through sprinkler irrigation equipment. Follow application and calibration directions preceding this section. **NOTE: DO NOT** exceed total of 3 applications per season. **DO NOT** apply within 6 weeks of harvest. **DO NOT** feed treated parts to livestock or allow grazing in treated fields.
		2.0 to 3½ pts.	**Two application program** - Make the first application when largest pods are 1 to 1½ inches in length and make the second application 14 days later. **BRAVO 500** may be co-applied with **Benlate® 50WP** as a tank mix for disease control in indeterminate (northern) soybeans. Use 1½ pts. of **BRAVO 500** plus 8 ounces of **Benlate 50WP** per acre. Make the first application when pods near the top of plants are ½ to 1 inch in length and a second application 14 days later.
		1½ to 2¾ pts.	**Three application program** - Make the first application one week after first flowering and continue applications at 14 day intervals. ® Benlate is a registered trademake of E.I. DuPont de Nemours and Co., Inc.
Tomato	FOLIAGE (apply every 7–10 days): Early blight, Late blight, Gray leafspot, Gray leaf mold, Septoria leafspot, Target spot	2 to 3 pts.	Apply in sufficient water to obtain adequate coverage. Begin applications when dew or rain occur and disease threatens. Use the highest rate and shortest interval specified when disease conditions are severe. **BRAVO 500** may be combined in the spray tank with EPA-registered pesticide products that claim copper as the active ingredient and are labeled for control of bacterial diseases of tomatoes. Check the copper manufacturer's label for specific instructions, precautions and limitations prior to mixing with **BRAVO 500. DO NOT** use with Copper-Count* N in concentrated spray suspensions.
	FRUIT (apply every 7–14 days beginning at fruit set): Anthracnose, Alternaria fruit rot (black mold), Rhizoctonia fruit rot, Botrytis gray mold, Late blight fruit rot	3 to 4 pts.	**BRAVO 500** may be applied through sprinkler irrigation equipment (solid set or portable wheel move systems only). See calibration directions preceding this section.

** Copper-Count is a registered trademark of Mineral Research & Development Corp.*

TREE AND ORCHARD CROPS

Apply **BRAVO 500** in sufficient water and with proper calibration to obtain uniform coverage of tree canopy. Application with ground equipment is preferable to aerial application because ground applications generally give better coverage of the tree canopy. If application with ground equipment is not feasible, **BRAVO 500** may be applied with aircraft using at least 20 gallons per acre. When concentrate sprays are used or when treating non-bearing or immature trees, the lower rate of **BRAVO 500** listed may be used. **DO NOT** allow livestock to graze in treated areas. The following spray volumes are recommended as gallons of spray per acre:

Crop	Spray Volume (Gallons per Acre)
Peach, Nectarine, Apricot, Tart Cherry, Plum, Prune	20 (concentrate) to 300 (full dilute)

Crop	Spray Volume (Gallons per Acre)	
Sweet Cherry	20 (concentrate) to 400 (full dilute)	
Conifers:	**Dilute**	**Concentrate**
Forest stands	Not used	10 to 20 (aircraft)
Christmas trees	100	10 to 50 (aircraft or ground equipment)
Nursery beds	100	5 to 10 (ground equipment only)

CROP	DISEASES	BRAVO 500 RATE PER ACRE	BRAVO 500 RATE PER 100 GAL*	APPLICATION DIRECTIONS
Peach, Nectarine, Apricot, Cherry, Plum, Prune	Leaf curl, Coryneum blight (shothole)	4½ to 6 pts.	1½ to 2 pts.	For best control of both diseases apply at leaf fall in late autumn, using sufficient water and proper sprayer calibration to obtain uniform coverage. When conditions favor high disease levels use the high rate of application and apply once or twice more in mid to late winter before budswell. If the leaf fall application is not practical, application of **BRAVO 500** for control of leaf curl may be made at any time prior to budswell the following spring. Where Coryneum blight (shothole) occurs, also apply at budbreak to protect newly emerging leaves and at shuck split to prevent fruit infections.
	Brown rot blossom blight Lacy (russet) scab (plum/prune)	4½ to 8 pts.	1½ to 2 pts.	Use 6 to 8 pints per acre on trees taller than 20 ft. and 4½ to 6 pints per acre on smaller trees. Make one application at popcorn (pink, red or early white bud) and a second application at full bloom. If weather conditions favor disease development, make an additional application at petal fall. In addition to the bloom applications listed above, make one application at shuck-split. **DO NOT** apply **BRAVO 500** after shuck-split and before harvest. If additional disease control is needed before harvest, use another registered fungicide.
	Cherry leafspot; peach, nectarine, apricot scab	4½ to 6 pts.	1½ to 2 pts.	For control of cherry leafspot after harvest, make one application to foliage within 7 days after fruit is removed. In orchards with a history of high leafspot incidence, make a second application 10–14 days later.
Conifers	Swiss needlecast	4 to 8 pts.	4 to 8 pts.	Single application technique: In Christmas tree plantations or forest stands make one application in the spring when new shoot growth is ½ to 2 inches in length.

CROP	DISEASES	BRAVO 500 RATE PER ACRE	BRAVO 500 RATE PER 100 GAL*	APPLICATION DIRECTIONS
	Scleroderris canker (pines) Swiss needlecast	2 to 4 pts.	2 to 4 pts.	Make the first application in spring when new shoot growth is ½ to 2 inches in length. Make additional applications at 3 to 4 week intervals until conditions no longer favor disease development. For use in nursery beds, apply the highest rate specified on a 3 week schedule.
	Sirococcus tip blight	3 to 5 pts.	3 to 5 pts.	
	Rhizosphaera needlecast (spruces) Scirrhia brown spot (pines)	8 pts.	8 pts.	
	Cyclaneusma and Lophodermium needlecasts (pines)	4 to 8 pts.	4 to 8 pts.	Apply in early spring prior to budbreak. Repeat applications at approximately 6 to 8 week intervals, until spore release ceases in late fall. Apply monthly during periods of frequent rainfall, and where Lophodermium infections occur during dormancy (Pacific Northwest). During drought periods, applications may be suspended, then resumed upon next occurrence of needle wetness.
	Rhabdocline needlecast (Douglas-fir)	2 to 4 pts.	2 to 4 pts.	Apply at budbreak and repeat at 3 to 4 week intervals until needles are fully elongated and conditions no longer favor disease development. In plantations of mixed provenance, or when irregular budbreak occurs, apply weekly until all trees have broken bud, then every 3 to 4 weeks as specified above. In nursery beds, use the high rate on a 3 week schedule.
	Botrytis seedling blight Phoma twig blight	2 to 4 pts.	2 to 4 pts.	Begin applications in nursery beds when seedlings are 4 inches tall and when cool, moist conditions favor disease development. Make additional applications at 7 to 14 day-intervals as long as disease favorable conditions persist.

* Volumetric rates to be used only with full dilute spray volume specified on this label for tree and orchard crops.

SPECIAL LOCAL NEEDS PURSUANT TO SECTION 24(c) OF AMENDED FIFRA AND STATE REGISTRATIONS

CROP	REGISTRATION	STATES	SLN OR REG. No.	DIRECTIONS FOR USE
Forest Tree Nursery and Greenhouse Management	Control of Botrytis spp.	Oregon	OR-770060	Apply 2¾ pts. per 100 gallons of water for control of Botrytis spp. on conifer seedlings such as Douglas fir, noble fir, western hemlock, lodgepole pine and ponderosa pine. Begin application when seedlings are 4 inches tall or when disease threatens. Repeat applications at 7 to 14 day intervals as required to maintain disease control. **CAUTIONS:** Enter treated greenhouse only after spray has settled. Wear protective clothing to avoid skin contact in treated greenhouses until spray has dried.
Sugarbeet grown for seed	Control of Downy mildew	Oregon	OR-810032	Use 2½ pts. per acre in sufficient water to obtain adequate coverage, generally 25 to 50 gallons per acre. Make the first application in mid-October when disease conditions becomes favorable, and make additional applications as favorable disease conditions warrant, through April of the following year. **DO NOT** allow livestock to graze in treated areas. **DO NOT** feed treated crop residues to livestock. **DO NOT** apply **BRAVO 500** to sugarbeet plants after bolting and seedhead emergence has occurred.

STORAGE AND DISPOSAL

DO NOT contaminate water, food or feed by storage or disposal. Open dumping is prohibited.

STORAGE: Store in a cool place. Protect from excessive heat.

PESTICIDE DISPOSAL:

Pesticide wastes are toxic. Improper disposal of excess pesticide, pesticide spray or rinsate is a violation of Federal Law. If these wastes cannot be disposed of by use acording to label instructions, contact your State Pesticide or Environmental Control Agency, or the Hazardous Waste representative at the nearest EPA Regional Office for guidance.

CONTAINER DISPOSAL:

DO NOT reuse empty container. Triple rinse (or equivalent). Then offer for recycling or reconditioning, or puncture and dispose of in a sanitary landfill, or incineration, or, if allowed by state and local authorities, by burning. If burned, stay out of smoke.

WARRANTY AND LIMITATIONS OF DAMAGES

Seller warrants that this material conforms to its chemical description and is reasonably fit for the purposes stated on the label when used in accordance with directions under normal conditions of use and Buyer assumes the risk of any use contrary to such directions. SELLER MAKES NO OTHER EXPRESS OR IMPLIED WARRANTY, INCLUDING ANY OTHER EXPRESS OR IMPLIED WARRANTY OF FITNESS OR OF MERCHANTABILITY, AND NO AGENT OF SELLER IS AUTHORIZED TO DO SO EXCEPT IN WRITING WITH A SPECIFIC REFERENCE TO THIS WARRANTY. In no event shall Seller's liability for any breach of warranty exceed the purchase price of the material as to which a claim is made.

BRAVO® W-75

AGRICULTURAL FUNGICIDE

75% Wettable Powder

ACTIVE INGREDIENT:

Chlorothalonil (tetrachloroisophthalonitrile)* 75.0%

INERT INGREDIENTS: 25.0%

TOTAL 100.0%

* Covered under U.S. Patent No. 3,290,353 and No. 3,331,735

EPA Reg. No. 50534-23

KEEP OUT OF REACH OF CHILDREN

DANGER-PELIGRO

See precautionary statements.

Precaucion al Usuario: Si usted no lee ingles, no use este producto hasta que la etiqueta haya sido explicado ampliamente.

PRECAUTIONARY STATEMENTS

HAZARDS TO HUMANS AND DOMESTIC ANIMALS

DANGER

Corrosive, causes severe eye damage. May be a potential skin sensitizer. **DO NOT** get in eyes; wear goggles or eye shield when handling this product. In case of contact with eyes, flush with plenty of water immediately for 15 minutes. Seek medical attention for eyes immediately. Avoid contact with skin or clothing. Wash exposed areas of skin with soap and warm water after handling or using. **DO NOT** take internally. Avoid breathing dust or spray mist.

DO NOT apply this product in such a manner as to directly, or through drift, expose workers or other persons. The area being treated must be vacated by unprotected persons.

Note to User: This product may produce temporary allergic side effects characterized by redness of the eyes, mild bronchial irritation and redness or rash on exposed skin areas. Persons having allergic reaction should contact a physician.

FIRST AID

Note to Physician: Persons having an allergic reaction respond to treatment with antihistamines or steroid creams and/or systemic steroids.

ENVIRONMENTAL HAZARDS

This product is toxic to fish, aquatic invertebrates, and marine/estuarine organisms. Runoff from treated areas may be hazardous to aquatic organisms in neighboring areas. **DO NOT** apply directly to water. **DO NOT** apply when weather conditions favor drift from treated areas.

Apply only to areas specified on label.

BRAVO W-75

A Broad-Spectrum Agricultural Fungicide.

Read entire label carefully and use only as directed.

GENERAL INFORMATION

BRAVO W-75 is an excellent fungicide when used according to label directions for control of a broad spectrum of plant diseases. **BRAVO W-75** can be used effectively in dilute or concentrate sprays. Thorough, uniform coverage is essential for disease control.

DO NOT combine **BRAVO W-75** in the spray tank with pesticides, surfactants or fertilizers, unless your prior use has shown the combination physically compatible and noninjurious under your conditions of use. The required amount of **BRAVO W-75** should be added slowly into the spray tank during filling. With concentrate sprays, premix the required amount of **BRAVO W-75** in a clean container and add to the spray tank as it is being filled. Keep agitator running when filling spray tank and during spray operations. **DO NOT** use on greenhouse-grown crops.

DIRECTIONS FOR USE

It is a violation of Federal law to use this product in a manner inconsistent with its labeling.

Note to User: This product may produce temporary allergic side effects characterized by redness of the eyes, mild bronchial irritation and redness or rash on exposed skin areas. Persons having allergic reaction should contact a physician.

Note to User: Wear long sleeve shirt, long pants, and gloves while mixing, loading and applying this product.

Dosage rates on this label indicate pounds of **BRAVO W-75** per acre, unless otherwise stated. Applications should be made in sufficient water to obtain adequate coverage of foliage. Gallonage to be used will vary with crop and amount of plant growth. Spray volume usually will range from 20 to 150 gallons per acre for dilute sprays and 5 to 10 gallons per acre for concentrate ground sprays and aircraft applications.

Note to User: DO NOT rotate to crops other than those listed on labels within 12 months of the last treatment. After 12 months from the last application, leafy vegetables (i.e. spinach, lettuce, kale, etc.) may also be rotated.

REENTRY: Note to User: **DO NOT** enter treated area to perform hand labor within 24 hours of application unless protective clothing is worn.

Because certain states may require more restrictive reentry intervals for various crops treated with this product, consult your State Department of Agriculture for further information.

Written or oral warnings must be given to workers who are expected to be in a treated area or in an area about to be treated with this product. When oral warnings are given, warnings shall be given in a language customarily understood by workers. Oral warnings must be given if there is reason to believe that written warnings cannot be understood by workers. Oral and written warnings must include the following information: "**DANGER.** Area treated with **BRAVO W-75** on (date of application). **DO NOT** enter without appropriate protective clothing for 24 hours. In case of accidental exposure, wash exposed area with plenty of water and get medical attention. For further information see 'Precautionary Statements' on the label."

APPLICATION AND CALIBRATION TECHNIQUES FOR SPRINKLER IRRIGATION

Apply this product only through center pivot, motorized lateral move, traveling gun, solid set and portable (wheel move, side roll, end tow, or hand move) irrigation system(s). **Do not** apply this product through any other type of irrigation system.

Crop injury, lack of effectiveness, or illegal pesticide residues in the crop can result from non-uniform distribution of treated water.

If you have questions about calibration, you should contact State Extension Service specialists, equipment manufacturers or other experts.

Do not apply this product through irrigation systems connected to a public water system. 'Public water system' means a system for the provision to the public of piped water for human consumption if such system has at least 15 service connections or regularly serves an average of at least 25 individuals daily at least 60 days per year.

Controls for both irrigation water and pesticide injection systems must be functionally interlocked, so as to automatically terminate pesticide injection when the irrigation water pump motor stops. A person knowledgeable of the irrigation system and responsible for its operation shall be present so as to discontinue pesticide injection and make necessary adjustments, should the need arise.

The irrigation water pipeline must be fitted with a functional, automatic, quick-closing check valve to prevent the flow of treated irrigation water back toward the water source. The pipeline must also be fitted with a vacuum relief valve and low pressure drain, located between the irrigation water pump and the check valve, to prevent back-siphoning of treated irrigation water into the water source.

Always inject BRAVO W-75 into irrigation water after it discharges from the irrigation pump and after it passes through the check valve. Never inject pesticides into the intake line on the suction side of the pump.

Pesticide injection equipment must be fitted with a functional, normally closed solenoid-operated valve located on the intake side of the injection pump. Interlock this valve to the power system, so as to prevent fluid from being withdrawn from the chemical supply tank when the irrigation system is either automatically or manually turned off.

The pesticide injection pipeline must contain a functional, automatic, quick-closing check valve to prevent the flow of fluid back toward the injection pump.

The irrigation line or water pump must include a functional pressure switch which will stop the water pump motor when the water pressure decreases to the point where pesticide distribution is adversely affected.

Spray mixture in the chemical supply tank must be agitated at all times, otherwise settling and uneven application may occur. Do not apply when wind speed favors drift beyond the area intended for treatment.

Posting of areas to be chemigated is required when: 1) any part of a treated area is within 300 feet of sensitive areas such as residential areas, labor camps, businesses, day care centers, hospitals, in-patient clinics, nursing homes or any public areas such as schools, parks, playgrounds, or other public facilities not including public roads; or 2) when the chemigated area is open to the public.

Posting must conform to the following requirements. Treated areas shall be posted with signs at all usual points of entry and along likely routes of approach from the listed sensitive areas. When there are no usual points of entry, signs must be posted in the corners of the treated areas and in any other location affording maximum visibility to sensitive areas. The printed side of the sign should face away from the treated area towards the sensitive area. The signs shall be printed in English. Signs must be posted prior to application and must remain posted until foliage has dried and soil surface water has disappeared. Signs may remain in place indefinitely as long as they are composed of materials to prevent deterioration and maintain legibility for the duration of the posting period.

All words shall consist of letters at least 2½ inches tall, and all letters and the symbol shall be a color which sharply contrasts with their immediate background. At the top of the sign shall be the words KEEP OUT, followed by an octagonal stop sign symbol at least 8 inches in diameter containing the word STOP. Below the symbol shall be the words PESTICIDES IN IRRIGATION WATER.

Bravo W-75 may be used through two basic types of sprinkler irrigation systems as outlined in Sections A and B below. Determine which type of system is in place, then refer to the appropriate directions provided for each type.

A. Center Pivot, Motorized Lateral Move and Traveling Gun Irrigation Equipment

For injection of pesticides, these continuously moving systems must use a positive displacement injection pump, of either diaphragm or piston type, constructed of materials that are compatible with pesticides and capable of being fitted with a system interlock and capable of injection at pressures approximately 2-3 times those encountered within the irrigation water line. Venturi applicator units cannot be used on these systems.

Fill chemical supply tank of injection equipment with water. Operate system for one complete revolution or run across the field, measuring time required, amount of water injected, and acreage covered. Thoroughly mix recommended amount of **Bravo W-75** for acreage to be covered into same amount of water using during calibration and inject into system continuously for one revolution or run. Mixture in the chemical supply tank must be continuously agitated during the injection run. Shut off injection equipment after one revolution or run, but continue to operate irrigation system until **Bravo W-75** has been cleared from last sprinkler head.

B. Solid Set and Portable (Wheel Move, Side Roll, End Tow, or Hand Move) Irrigation Equipment

With stationary systems, an effectively designed in-line venturi applicator unit is preferred which is constructed of materials that are compatible with pesticides; however, a positive-displacement pump can also be used.

Determine acreage covered by sprinkler. Fill tank of injection equipment with water and adjust flow to use contents over a thirty to forty-five minute period. Mix desired amount of **Bravo W-75** for acreage to be covered with water so that the total mixture of **Bravo W-75** plus water in the injection tank is equal to the quantity of water used during calibration and operate entire system at normal pressures recommended by the manufacturer of injection equipment used for amount of time established during calibration. Agitation is recommended. **Bravo W-75** can be injected at the beginning or end of the irrigation cycle or as a separate application. Stop injection equipment after treatment is completed and continue to operate irrigation system until **Bravo W-75** has been cleared from last sprinkler head.

Bean (Snap): Rust-Use 3 lbs. per acre in sufficient water to obtain adequate coverage. Begin applications during early bloom stage or when disease first threatens and repeat at weekly intervals or as necessary to maintain control. **DO NOT** apply within 7 days of harvest. **DO NOT** graze treated areas or feed treated plant parts to livestock.

Cabbage, Cauliflower, Broccoli, Brussels Sprouts: Alternaria leaf spot (Southeastern States only) and downy mildew-Use 1½ lbs. per acre in sufficient water to obtain adequate coverage. Begin applications after transplants are set in field, or shortly after emergence of field-seeded crop or when conditions favor disease development. Repeat at 7 to 10 day intervals or as necessary to maintain control. Ring spot in field-seeded brussels sprouts (California only)-Use 2 lbs. per acre. Begin applications at time of early sprout development or when conditions favor disease development. Repeat at 7 to 10 day intervals or as necessary to maintain control.

Carrot: Early blight and late blight-Use 1½ to 2 lbs. per acre in sufficient water to obtain adequate coverage. Start applications when disease threatens and repeat at 7 to 10 day intervals or as necessary to maintain control.

Celery: Early blight, late blight, basal stalk rot *(Rhizoctonia solani)* -Use 1 to 1½ lbs. per acre on a 3 to 5 day spray schedule, or use 2 to 3 lbs. per acre on a 7 day schedule. Suppression of pink rot-Use 3 lbs. per acre on a 7 day schedule. Start applications when transplants are set in the field. Apply **BRAVO W-75** in sufficient water to obtain adequate coverage. **DO NOT** apply within 7 days of harvest.

For control of early blight and late blight in celery seedbeds, use 1½ to 2 lbs. **BRAVO W-75** per 100 gallons of water and apply 125 gallons per acre twice weekly or as needed to maintain control. Start applications shortly after crop emergence. Use the higher rate under severe disease conditions.

Corn (Sweet): Helminthosporium leaf blight-Use 1½ to 2 lbs. per acre in sufficient water to obtain adequate coverage. Begin applications when conditions favor disease development and repeat at 4 to 7 day intervals or as required to maintain control. **DO NOT** apply within 14 days of harvest. For use on fresh market crop only. **DO NOT** apply to sweet corn to be processed. **DO NOT** allow livestock to graze in treated fields. **DO NOT** ensile treated corn or use as livestock forage.

Cucumber: Target spot (Florida only), anthracnose and downy mildew-Use 1½ to 2 lbs per acre. Powdery mildew (except Southwestern States), gummy stem blight, leaf blight and scab-Use 2 to 3 lbs per acre. For suppression of fruit rot *(Rhizoctonia solani)*-Use 3 lbs per acre. Apply **BRAVO W-75** in sufficient water to obtain adequate coverage. Begin applications when plants are in first true leaf stage or when conditions are favorable for disease development. Repeat applications at 7 day intervals. Under severe disease conditions, shorten spray interval.

Cantaloupe, Muskmelon, Honeydew: Anthracnose and downy mildew-Use 1½ to 2 lbs. per acre. Powdery mildew (except in Southwestern States), Cercospora leaf spot, gummy stem blight, leaf blight and scab-Use 2 to 3 lbs. per acre. Apply **BRAVO W-75** in sufficient water to obtain adequate coverage. Begin applications when plants are in first true leaf stage or when conditions are favorable for disease development. Repeat applications at 7 day intervals. Under severe disease conditions, shorten spray interval.

Watermelon, Squash, Pumpkin: Anthracnose and downy mildew-Use 1½ to 2 lbs. per acre. Cercospora leaf spot, gummy stem blight, leaf blight and scab-Use 2 to 3 lbs. per acre. Apply **BRAVO W-75** in sufficient water to obtain adequate coverage. Begin applications when plants are in first true leaf stage or when conditions are favorable for disease development. Repeat applications at 7 day intervals. Under severe disease conditions shorten spray interval.

Onion (Green and dry bulb): Botrytis leaf blight/blast, Downy mildew (Suppression) and Purple blotch-Use 1½ to 3 lbs. per acre. Apply before disease appears and repeat at 7 to 10 day intervals. Under severe disease conditions, use the higher application rate and shorten the interval between applications. **DO NOT** apply on dry bulb onions within 7 days of harvest or on green onions within 14 days of harvest. **DO NOT** apply more than 3 times per season on green onions. If more than 14 days remain before harvest of green onions after 3 applications of **BRAVO W-75** have been made, it is recommended that another labeled fungicide be used to complete the application season. **DO NOT** apply to sweet Spanish onions.

Peanuts: Early leaf spot *(Cercospora)*, late leaf spot *(Cercosporidium)* -Use 1 to 1½ lbs. per acre. Rust-Use 1½ lbs. per acre. Apply **BRAVO W-75** in sufficient water to obtain adequate coverage. Start applications when disease first appears and repeat at 10 to 14 day intervals or as necessary to maintain control. Under severe disease conditions, use the 1½ lbs. per acre rate. **DO NOT** apply within 14 days of harvest. **DO NOT** allow livestock to graze treated areas. **DO NOT** feed hay or threshings from treated fields to livestock.

Potato: Early blight, late blight, Botrytis vine rot *(Botrytis spp.)* -Use 1 to 1½ lbs. per acre in sufficient water to obtain adequate coverage. Begin applications when plants are 6 inches high or when disease threatens and continue at 7 to 10 day intervals or as needed to maintain disease control. Under severe disease conditions, use the 1½ lbs. per acre rate and shorten spray interval.

Tomato: Early blight, late blight, gray leaf spot, gray leaf mold-Use 1½ to 2 lbs. per acre. Anthracnose-Use 2 to 3 lbs. per acre. Rhizoctonia fruit rot and Botrytis gray mold-Use 3 lbs. per acre. Apply **BRAVO W-75** in sufficient water to obtain adequate coverage. Begin applications when disease threatens and repeat at 7 to 10 day intervals. Under severe disease conditions, shorten spray interval. Alternaria fruit rot or black-mold (California only)-Make three applications at the rate of 3 lbs. per acre on a 10 to 14 day schedule beginning 5 to 6 weeks before anticipated harvest.

Bluegrass, Perennial Ryegrass Grown For Seed: Stem rust, leaf rust, and stripe rust-Use 1 to 1½ lbs. per acre in sufficient water to obtain adequate coverage. Begin applications at the time of seed head elongation (early to mid-April) or when conditions favor disease development. Repeat at 7 to 10 day intervals or as necessary to maintain control. Use the 1 lb. per acre rate under a light incidence of rust. Under severe disease conditions, use the 1.5 lb. per acre rate at a 7 day spray interval. **DO NOT** apply within 14 days of harvest. **DO NOT** allow livestock to graze in treated areas. **DO NOT** feed hay or threshings from treated fields to livestock.

DIRECTIONS FOR USE ON CONIFER TREES

Apply **BRAVO W-75** in sufficient water and with proper calibration to obtain uniform coverage of tree canopy. Application through either ground equipment or aircraft is recommended except where otherwise indicated. When concentrate sprays are used the lower rate per acre of **BRAVO W-75** listed may be used.

The following spray volumes are recommended as gallons of spray per acre:

SPRAY VOLUME (Gallons per Acre)

Crop	Dilute	Concentrate
CONIFERS:		
Forest stands	Not used	10 to 20 (aircraft)
Christmas trees	100	10 to 50 (aircraft or ground equipment)
Nursery beds	100	5 to 10 (ground equipment only)

		BRAVO W-75 RATE PER		
CROP	DISEASES	ACRE	100 GAL*	APPLICATION DIRECTIONS
Conifers	Swiss needlecast	2¾ to 5½ lbs.	2¾ to 5½ lbs.	Single application technique: In Christmas tree plantations or forest stands make one application in the spring when new shoot growth is ½ to 2 inches in length.
	Scleroderris canker (pines) Swiss needlecast and Rhabdocline needlecast (Douglas-fir)	1½ to 2¾ lbs.	1½ to 2¾ lbs.	Make the first application in spring when new shoot growth is ½ to 2 inches in length. Make additional applications at 3 to 4 weeks intervals until conditions no longer favor disease development. For use in nursery beds, apply the highest rate specified on a 3 week schedule.
	Sirococcus tip blight	2 to 3½ lbs.	2 to 3½ lbs.	
	Rhizosphaera needlecast	5½ lbs.	5½ lbs.	

CROP	DISEASES	BRAVO W-75 RATE PER ACRE	BRAVO W-75 RATE PER 100 GAL*	APPLICATION DIRECTIONS
	(spruces) Scirrhia brown spot (pines)			
	Lophodermium needlecast	1½ to 2¾ lbs.	1½ to 2¾ lbs.	North Central and Northeastern States: Begin applications in mid-July to early August before infection occurs. Make additional applications at 3 to 4 weeks intervals until conditions no longer favor disease development. For use in nursery beds, apply 2¾ lbs. per acre on a 3 week schedule.
	Botrytis seedling blight Phoma twig blight	1½ to 2¾ lbs.	1½ to 2¾ lbs.	Begin applications in nursery beds when seedlings are 4 inches tall and when cool, moist conditions favor disease development. Make additional applications at 7 to 14 day intervals as long as disease favorable conditions persist.

** Volumetric rates to be used only with full dilute spray volume specified on this label for tree crops.*

STORAGE AND DISPOSAL

DO NOT contaminate water, food or feed by storage or disposal. Open dumping is prohibited.

STORAGE: Store in a dry place.

PESTICIDE DISPOSAL: Pesticide wastes are toxic. Improper disposal of excess pesticide, pesticide spray or rinsate is a violation of Federal Law. If these wastes cannot be disposed of by use according to label instructions, contact your State Pesticide or Environmental Control Agency, or the Hazardous Waste representative at the nearest EPA Regional Office for guidance.

CONTAINER DISPOSAL: Completely empty bag into application equipment. Then dispose of empty bag in a sanitary landfill or by incineration, or, if allowed by State and local authorities, by burning. If burned, stay out of smoke.

WARRANTY AND LIMITATION OF DAMAGES

Seller warrants that this material conforms to its chemical description and is reasonably fit for the purposes stated on the label when used in accordance with directions under normal conditions of use and Buyer assumes the risk of any use contrary to such directions. SELLER MAKES NO OTHER EXPRESS OR IMPLIED WARRANTY, INCLUDING ANY OTHER EXPRESS OR IMPLIED WARRANTY OF FITNESS OR OF MERCHANTABILITY, AND NO AGENT OF SELLER IS AUTHORIZED TO DO SO EXCEPT IN WRITING WITH A SPECIFIC REFERENCE TO THIS WARRANTY. In no event shall Seller's liability for any breach of warranty exceed the purchase price of the material as to which a claim is made.

BRAVO® 720

AGRICULTURAL FUNGICIDE

ACTIVE INGREDIENT:
Chlorothalonil (tetrachloroisophthalonitrile)* 54.0%
INERT INGREDIENTS: 46.0%

TOTAL 100.0%

Contains 6.0 Pounds Chlorothalonil Per Gallon (720 grams per liter)

EPA Reg. No. 50534-188

* Covered under U.S. Patent No. 3,290,353, No. 3,948,636 and No. 3,331,735

KEEP OUT OF REACH OF CHILDREN

WARNING—AVISO

See precautionary statements.

Precaucion al Usuario: Si usted no lee ingles, no use este producto hasta que la etiqueta haya sido explicado ampliamente.

A BROAD SPECTRUM

Agricultural Fungicide

Read entire label carefully and use only as directed.

PRECAUTIONARY STATEMENTS

HAZARDS TO HUMANS AND DOMESTIC ANIMALS

WARNING

Cause eye irritation. May cause skin irritation. May be a potential skin sensitizer.

DO NOT get in eyes. Wear goggles or face shield when handling this product.

Avoid prolonged contact with skin. Wash exposed areas of skin with soap and warm water after handling or using.

Avoid breathing spray mist.

DO NOT take internally.

Note to User: This product may produce temporary allergic side effects characterized by redness of the eyes, mild bronchial irritation and redness or rash on exposed skin areas. Persons having allergic reaction should contact a physician.

DO NOT apply this product in such a manner as to directly, or through drift, expose workers or other persons. The area being treated must be vacated by unprotected persons.

FIRST AID:

In case of contact with eyes, immediately flush eyes with plenty of water and continue for 15 minutes. Seek medical attention for eyes immediately.

Wash exposed areas of skin with soap and warm water after handling or using.

Note to physician: Persons having a temporary allergic reaction respond to treatment with antihistamines or steroid creams and/or systemic steroids.

ENVIRONMENTAL HAZARDS

This product is toxic to fish, aquatic invertebrates, and marine/estuarine organisms. Runoff from treated areas may be hazardous to aquatic organisms in neighboring areas. **DO NOT** apply directly to water. **DO NOT** apply when weather conditions favor drift from treated areas.

Apply only to areas specified on label.

DIRECTIONS FOR USE

It is a violation of Federal law to use this product in a manner inconsistent with its labeling.

Note to User: Wear long sleeve shirt, long pants, and gloves while mixing, loading and applying this product.

BRAVO 720 is an excellent fungicide when used according to label directions for control of a broad spectrum of plant diseases.

BRAVO 720 can be used effectively in dilute or concentrate sprays. Thorough, uniform coverage is essential for disease control.

DO NOT use on greenhouse-grown crops.

DO NOT combine **BRAVO 720** in the spray tank with pesticides, surfactants or fertilizers, unless your prior use has shown the combination physically compatible, effective and noninjurious under your conditions of use.

DO NOT combine **BRAVO 720** with Dipel 4L, Triton AG-98, or Triton B-1956 as phytotoxicity may result from the combination when applied to some crops on this label.

Dipel is a registered trademark of Abbott Laboratories. Triton and Triton B-1956 are trademarks of Rohm and Haas Company.

NOTE: Slowly invert container several times to assure uniform mixture.

Dosage rates on this label indicate pints of **BRAVO 720** per acre, unless otherwise stated. Under conditions favoring disease development the high rate specified and shortest application interval should be used.

The required amount of **BRAVO 720** should be added slowly into the spray tank during filling. With concentrate sprays, pre-mix the required amount of **BRAVO 720** in a clean container and add to the spray tank as it is being filled. Keep agitator running when filling spray tank and during spray operations.

Note to User: DO NOT rotate to crops other than those listed on labels within 12 months of the last treatment. After 12 months from the last application, leafy vegetables (i.e. spinach, lettuce, kale, etc.) may also be rotated.

REENTRY:

Note to User: DO NOT enter treated area to perform hand labor within 24 hours of application unless protective clothing is worn.

Because certain states may require more restrictive reentry intervals for various crops treated with this product, consult your State Department of Agriculture for further information.

Written or oral warnings must be given to workers who are expected to be in a treated area or in an area about to be treated with this product. When oral warnings are given, warnings shall be given in a language customarily understood by workers. Oral warnings must be given if there is reason to believe that written warnings cannot be understood by workers. Oral and written warnings must include the following information: **WARNING.** Area treated with **BRAVO 720** on (date of application). **DO NOT** enter without appropriate protective clothing for 24 hours. In case of accidental exposure, wash exposed area with plenty of water and get medical attention. For further information see 'Precautionary Statements' on the label."

FIELD AND ROW CROPS:

Apply **BRAVO 720** in sufficient water to obtain adequate coverage of foliage. Gallonage to be used will vary with crop and amount of plant growth. Spray volume usually will range from 20 to 150 gallons per acre (200 to 1400 liters per hectare) for dilute sprays and 5 to 10 gallons per acre (50 to 100 liters per hectare) for concentrate ground sprays and aircraft applications. Both ground and aircraft methods of application are recommended unless specific directions are given for a crop. Application through sprinkler irrigation systems is not recommended unless specific directions are given for a crop. See application and calibration instruction below.

Application and Calibration Techniques for Sprinkler Irrigation

Apply this product only through center pivot, motorized lateral move, traveling gun, solid set and portable (wheel move, side roll, end tow, or hand move) irrigation systems(s). **Do not** apply this product through any other type of irrigation system.

Crop injury, lack of effectiveness, or illegal pesticide residues in the crop can result from non-uniform distribution of treated water.

If you have questions about calibration, you should contact State Extention Service specialists, equipment manufacturers or other experts.

Do not apply this product through irrigation systems connected to a public water system. 'Public water system' means a system for the provision to the public of piped water for human consumption if such system has at least 15 service connections or regularly serves an average of at least 25 individuals daily at least 60 days per year.

Controls for both irrigation water and pesticide injection systems must be functionally interlocked, so as to automatically terminate pesticide injection when the irrigation water pump motor stops. A person knowledgeable of the irrigation system and responsible for its operation shall be present so as to discontinue pesticide injection and make necessary adjustments, should the need arise.

The irrigation water pipeline must be fitted with a functional, automatic, quick-closing check valve to prevent the flow of treated irrigation water back toward the water source. The pipeline must also be fitted with a vacuum relief valve and low pressure drain, located between the irrigation water pump and the check valve, to prevent back-siphoning of treated irrigation water into the water source.

Always inject BRAVO 720 into irrigation water after it discharges from the irrigation pump and after it passes through the check valve. Never inject pesticides into the intake line on the suction side of the pump.

Pesticide injection equipment must be fitted with a functional, normally closed, solenoid-operated valve located on the intake side of the injection pump. Interlock this valve to the power system, so as to prevent fluid from being withdrawn from the chemical supply tank when the irrigation system is either automatically or manually turned off.

The pesticide injection pipeline must contain a functional, automatic, quick-closing check valve to prevent the flow of fluid back toward the injection pump.

The irrigation line or water pump must include a functional pressure switch which will stop the water pump motor when the water pressure decreases to the point where pesticide distribution is adversely affected.

Spray mixture in the chemical supply tank must be agitated at all times, otherwise settling and uneven application may occur. Do not apply when wind speed factors drift beyond the area intended for treatment.

BRAVO 720 may be used through two basic types of sprinkler irrigation systems as outlined in Sections A and B below. Determine which type of system is in place, then refer to the appropriate directions provided for each type.

A. Center Pivot, Motorized Lateral Move and Traveling Gun Irrigation Equipment

For injection of pesticides, these continuously moving systems must use a positive displacement injection pump, of either diaphragm or piston type, constructed of materials that are compatible with pesticides and capable of being fitted with a system interlock and capable of injection at pressures approximately 2–3 times those encountered within the irrigation water line. Venturi applicator units cannot be used on these systems.

Fill chemical supply tank of injection equipment with water. Operate system for one complete revolution or run across the field, measuring time required, amount of water injected, and acreage covered. Thoroughly mix recommended amount of **BRAVO 720** for acreage to be covered into same amount of water used during calibration and inject into system continuously for one revolution or run. Mixture in the chemical supply tank must be continuously agitated during the injection run. Shut off injection equipment after one revolution or run, but continue to operate irrigation system until **BRAVO 720** has been cleared from last sprinkler head.

B. Solid Set and Portable (Wheel Move, Side Roll, End Tow, or Hand Move) Irrigation Equipment

With stationary systems, an effectively designed in-line venturi applicator unit is preferred which is constructed of materials that are compatible with pesticides; however, a positive-displacement pump can also be used.

Determine acreage covered by sprinkler. Fill tank of injection equipment with water and adjust flow to use contents over a thirty to forty-five minute period. Mix desired amount of **BRAVO 720** for acreage to be covered with water so that the total mixture of **BRAVO 720** plus water in the injection tank is equal to the quantity of water used during calibration and operate entire system at normal pressures recommended by the manufacturer of injection equipment used for amount of time established during calibration. No agitation should be required. **BRAVO 720** can be injected at the beginning or end of the irrigation cycle or as a separate application. Stop injection equipment after treatment is completed and continue to operate irrigation system until **BRAVO 720** has been cleared from last sprinkler head.

CROP	DISEASES	RATE PER ACRE	APPLICATION DIRECTIONS
Bean (Snap)	Rust	1³/₈ to 3 pts	Use in sufficient water to obtain adequate coverage. Begin applications during early bloom stage or when disease first threatens and repeat at weekly intervals or as necessary to maintain control. **DO NOT** apply within 7 days of harvest. **DO NOT** graze treated areas or feed treated plant parts to livestock.
	Botrytis blight (gray mold)	3 pts	
Beans (Dry) Navy, Pinto, Kidney, Lima, Blackeye	Rust, Anthracnose, Downy mildew, Cercospora leaf spot (blackeye only)	1³/₈ to 2 pts	Use in sufficient water to obtain adequate coverage. Begin applications during early bloom stage and repeat at 7 to 10 day intervals. For use only on beans to be harvested dry with pods removed. **DO NOT** apply within 6 weeks before harvest. **DO NOT** allow livestock to graze in treated areas or feed treated plant parts to livestock.
Cabbage, Chinese Cabbage (tight-headed varieties only), Cauliflower, Broccoli, Chinese Broccoli, Brussels Sprouts	Alternaria leaf spot, Downy mildew	1½ pts	Use in sufficient water to obtain adequate coverage. Begin applications after transplants are set in field, or shortly after emergence of field-seeded crop, or when conditions favor disease development. Repeat at 7 to 10 day intervals or as necessary to maintain control. Do not apply within 7 days of harvest to Chinese cabbage or Chinese broccoli.
	Ring spot (California only)	2 pts	For field-seeded brussels sprouts, begin applications at time of early sprout development or when conditions favor disease development. Repeat at 7 to 10 day intervals or as necessary to maintain control.
Carrot	Cercospora (early) blight, Alternaria (late) blight	1½ to 2 pts	Use in sufficient water to obtain adequate coverage. Start applications when disease threatens and repeat at 7 to 10 day intervals or as necessary to maintain control. **BRAVO 720** may be applied through sprinkler irrigation equipment (solid set, portable wheel move, motorized lateral move or center pivot systems only). See calibration directions preceding this section.

CROP	DISEASES	RATE PER ACRE	APPLICATION DIRECTIONS
Celery	Cercospora (early) blight, Septoria (late) blight, Basal Stalk rot *(Rhizoctonia solani)*	1 to 1½ pts or 2 to 3 pts.	Use 1 to 1½ pts per acre on a 3 to 5 day spray schedule or 2 to 3 pts per acre on a 7 day schedule. Start applications when transplants are set in the field. Apply in sufficient water to obtain adequate coverage. **BRAVO 720** may be applied through sprinkler irrigation equipment (solid set, portable wheel move, motorized lateral move or center pivot systems only). See calibration directions preceding this section. **DO NOT** apply within 7 days of harvest.
	Pink rot (Suppression-7 Day schedule)	3 pts	
	Early blight, Late blight	1½ to 2 pts per 100 gal.	For celery seedbeds, apply 125 gallons per acre twice weekly or as needed to maintain control. Start applications shortly after crop emergence. Use the higher rate under severe disease conditions.
Corn (sweet), Corn grown for seed	Helminthosporium leaf blights, Rust	¾ to 2 pts	Use in sufficient water to obtain adequate coverage. Begin applications when conditions favor disease development and repeat at 4 to 7 day intervals or as required to maintain control. Under severe disease conditions, use 1½ to 2 pts per acre. **DO NOT** apply within 14 days of harvest. **DO NOT** apply to sweet corn to be processed. **DO NOT** allow livestock to graze in treated fields. **DO NOT** ensile treated corn or use as livestock forage.
Cranberry	Fruit rots, Lophodermium leaf-twig blight	4 to 7 pts	Apply at early bloom and repeat at 10 to 14 day intervals. Under severe disease conditions use the 7 pts acre rate on a 10 day schedule. **DO NOT** apply **BRAVO 720** more than 3 times per season, or within 50 days before harvest. **DO NOT** apply to bogs when flooded or allow release of irrigation water from bogs for at least 3 days following application. **BRAVO 720** may be applied through sprinkler irrigation equipment. Use 300 gallons of water per acre through solid set systems only. See calibration directions preceding this section.
Cucurbits: Cucumber, Cantaloupe, Muskmelon, Honeydew melon, Watermelon, Squash, Pumpkin	Anthracnose, Downy mildew, Target spot	1½ to 2 pts	Use in sufficient water to obtain adequate coverage. Begin applications when plants are in first true leaf stage or when conditions are favorable for disease development. Repeat applications at 7 day intervals. Under severe disease conditions, shorten spray interval. **BRAVO 720** may be applied through sprinkler irrigation equipment (solid set, portable wheel move or center pivot systems only). See calibration directions preceding this section.
	Cercospora leafspot, Gummy stem blight (black rot), Alternaria leaf blight, Scab, Powdery mildew (Spaerotheca only)	2 to 3 pts	
	Cucumber belly rot *(Rhizoctonia solani)*	8¼ pts	Use **BRAVO 720** in sufficient water to obtain runoff to soil surface. Make a single application when vines begin to form. **BRAVO 720** may be applied through sprinkler irrigation equipment as directed above.
Grasses Grown for Seed	Stem rust, Leaf rust, Stripe rust, Septoria leafspot, Glume blotch, Bipolaris and Drechslera leafspots	1 to 1½ pts	Use in sufficient water to obtain adequate coverage. Begin applications during stem elongation when conditions favor disease development. Re-apply at flag (top) leaf emergence and repeat applications at 14-day intervals. **DO NOT** apply within 14 days of harvest. **DO NOT** allow livestock to graze in treated areas or feed treated plant parts to livestock.
	Selenophoma (eyespot)	1 to 2 pts	
Mint	Rust, Septoria leaf spot	1⅜ pts	Use in sufficient water to obtain adequate coverage, normally 20 to 150 gallons per acre for dilute sprays and 5 to 10 gallons per acre for concentrate ground and aircraft applications. Begin applications when emerging plants are 4-8 inches high. Repeat applications at 7 to 10 day intervals or as necessary to maintain control. **DO NOT** apply more than 3 times per season. **DO NOT** apply within 80 days of harvest. **DO NOT** feed fresh or extracted mint hay from treated fields to livestock. Based on available residue data, use of **BRAVO 720** on mint is restricted to Indiana, Michigan and Wisconsin.
Onion (dry bulb)	Botrytis leaf blight/blast, Botrytis neck rot (suppression), Purple blotch	1 to 2 pts	Apply in sufficient water to obtain thorough coverage of tops. **BRAVO 720** is recommended for use with disease monitoring systems which adjust fungicide rates and frequency of application according to disease hazard. Apply as follows: **Low Disease Hazard & Prior to Infection** / **Low Disease Hazard & Some Disease Present** / **High Disease Hazard** **Rate per Acre:** 1 pt. / 1⅜ pts. / 2 pts. **Frequency:** 10 days / 7 to 10 days / 7 days For suppression of neck rot (*Botrytis* spp.) during storage, a minimum of three weekly applications prior to lifting, using 1⅜ to 2 pints of **BRAVO 720** per acre, is recommended. **Do not** apply within 7 days of harvest.
Onion (green bunching), Leek, Shallot, Garlic, Onion grown for seed	Botrytis leaf blight/blast, Purple blotch, Downy mildew (suppression)	1½ to 3 pts	Use in sufficient water to obtain thorough coverage of tops. Begin applications prior to favorable infection periods, and repeat at 7 to 10 day intervals for as long as conditions favor disease. Use the high rate and a 7 day schedule of applications when heavy dew or rain persist. **Do not** apply within 7 days of harvest on garlic. **Do not** apply more than 3 times per season or within 14 days of harvest on green bunching onions, leeks or shallots. If additional disease control is needed before harvest, use another registered fungicide.

CROP	DISEASES	RATE PER ACRE	APPLICATION DIRECTIONS
Papaya	Alternaria fruit spot, Anthracnose Stem end rot	2 to 4 pts	Apply with ground equipment only, in sufficient water to obtain adequate coverage of fruit and leaves. Begin treatment when conditions favor development of disease and continue treatments at 14 day intervals until weather conditions no longer favor disease development. **DO NOT** graze livestock in treated area or feed processing by-products to livestock.
Parsnip	Alternaria leaf spot, Downy mildew, Anthracnose, Botrytis blight (gray mold), Bottom rot (Rhizoctonia)	1½ to 2 pts	Apply in sufficient water to obtain adequate coverage. Make the first application at the first sign of disease or when conditions are favorable for infection. Continue applications on a 7 to 10 day schedule. **DO NOT** apply more than 4 times per season or within 10 days of harvest. **DO NOT** feed treated plant parts to livestock.
Passion Fruit (Hawaii only)	Alternaria fruit and leaf spot, (passion fruit brown spot)	2 pts	Apply with ground equipment in sufficient water to obtain adequate coverage of fruit and leaves. Begin treatment when fruit spots appear (April to July) and continue treatments at 14 day intervals until weather conditions no longer favor disease development. **DO NOT** graze in treated area or feed vines or processing by-products to livestock used for food.
Peanut	Early leaf spot (Cercospora), Late leaf spot (Cercosporidium)	1 to 1½ pts	Apply in sufficient water for coverage when leaf wetness first occurs or 30 to 40 days after planting: Repeat at 10 to 14 day intervals. When conditions favor late leaf spot or when rust or web blotch occur, apply 1½ pints per acre at 10 day intervals for the remainder of the season. **DO NOT** apply within 14 days of harvest. **DO NOT** allow livestock to graze in treated areas. **DO NOT** feed hay or threshings from treated fields to livestock. **BRAVO 720** may be applied through sprinkler system irrigation equipment. Use 1½ pints per acre in solid set, portable wheel move, center pivot, motorized lateral move or traveling gun sprinkler irrigation equipment. See calibration directions preceding this section.
	Rust, Web blotch	1½ pts	
Potato	Late blight, Early blight, Botrytis vine rot	¾ pt —then— 1 to 1½ pt	Apply as a banded treatment directed over the foliage, beginning when vines are first exposed and leaf wetness occurs. Repeat applications at 7 to 10 day intervals until vines close, then increase rate as described below. In addition to the early-season applications above, as vines close between rows increase water carrier volume to cover the denser canopy and begin broadcast application patterns. Continue applications at 7 to 10 day intervals. Use the highest registered rate weekly when disease conditions are severe. **BRAVO 720** may be applied through sprinkler irrigation equipment (solid set, portable wheel move, center pivot or motorized lateral move systems only). **Do not** exceed a 10 day interval between applications when using this technique. Follow calibration and application directions preceding this section.
Soybean Determinate (Southern) Varieties	Anthracnose, Diaporthe pod and stem blight, Frogeye leaf spot *(Cercospora sojina),* Purple seed stain, Cercospora leaf blight, *(Cercospora kikuchii),* Septoria brown spot		Apply in sufficient water to obtain complete coverage, using at least five gallons of water per acre for aerial application. Use the three application program in areas having a history of moderate to severe disease intensity. **BRAVO 720** may be applied through sprinkler irrigation equipment. Follow application and calibration directions preceding this section. NOTE: **DO NOT** exceed total of 3 applications per season. **DO NOT** apply within 6 weeks of harvest. **DO NOT** feed treated parts to livestock or allow grazing in treated fields.
		1½ to 2½ pts	**Two application program**—Make the first application at early pod set (R3 stage, when majority of pods are ⅛ to ⅜ inch in length) and the second at beginning of seed formation (R5) which occurs about 14 days later.
		1 to 2 pts	**Three application program**—Make the first application at the beginning of flowering (R1), the second at early pod set (R3) and the third at beginning of seed formation (R5).
	Stem canker *(Diaporthe phaseolorum var. caulivora)*	1 pt	Apply in 10 to 20 gallons of water per acre, as a band treatment directing spray to provide coverage of entire plant. Make the first application at time of emergence of the second trifoliate leaves (V2). If conditions favor stem canker disease make a second and a third application. Make all applications at 10 to 14 day intervals.
Soybean Indeterminate (Northern) Varieties	Anthracnose, Diaporthe pod and stem blight, Frogeye leaf spot *(Cercospora sojina)* Purple seed stain, Cercospora leaf blight *(Cercospora kikuchii),* Septoria brown spot		Apply in sufficient water to obtain complete coverage, using at least five gallons of water per acre for aerial application. Use the three application program in areas having a history of moderate to severe disease intensity. **BRAVO 720** may be applied through sprinkler irrigation equipment. Follow application and calibration directions preceding this section. NOTE: **DO NOT** exceed total of 3 applications per season. **DO NOT** apply within 6 weeks of harvest. **DO NOT** feed soybean hay or threshings from treated fields to livestock.
		1½ to 2½ pts	**Two application program**—Make the first application when the largest pods are 1 to 1½ inches in length and make the second application 14 days later. **BRAVO 720** may be coapplied with **Benlate® 50 WP** as a tank mix for disease control in indeterminate (northern) soybeans. Use 1 pt of **BRAVO 720** plus 8 ounces of **Benlate 50WP** per acre. Make the first application when pods near the top of plants are ½ to 1 inch in length and a second application 14 days later.
		1 to 2 pts	**Three application program**—Make the first application one week after first flowing and continue applications at 14 day intervals.

® Benlate is a registered trademark of E.I. DuPont de Nemours and Co., Inc.

Tomato	FOLIAGE (apply every 7-10 days): Early blight, Late blight, Gray leaf spot, Gray leaf mold, Septoria leaf spot, Target spot	1⅜ to 2 pts	Apply in sufficient water to obtain adequate coverage. Begin applications when dew or rain occur and disease threatens. Use the highest rate and shortest interval specified when disease conditions are severe. **BRAVO 720** may be combined in the spray tank with EPA-registered pesticide products that claim copper as the active ingredient and are labeled for control of bacterial disease of tomatoes. Check the copper manufacturer's label for specific instructions, precautions and limitations prior to mixing with **BRAVO 720. DO NOT** use with **Copper-Count®N** in concentrated spray suspensions.

CROP	DISEASES	RATE PER ACRE	APPLICATION DIRECTIONS
	Fruit (apply every 7-14 days beginning at fruit set): Anthracnose, Alternaria fruit rot (black mold), Botrytis gray mold, Late blight fruit rot, Rhizoctonia fruit rot	2 to 3 pts	**BRAVO 720** may be applied through sprinkler irrigation equipment (solid set or portable wheel move systems only). See calibration directions preceding this section.

** Copper-Count is a registered trademark of Mineral Research & Development Corporation*

TREE AND ORCHARD CROPS

Apply **BRAVO 720** in sufficient water and with proper calibration to obtain uniform coverage of tree canopy. Application with ground equipment is preferable to aerial application because ground applications generally give better coverage of the tree canopy. If application with ground equipment is not feasible, **BRAVO 720** may be applied with aircraft using at least 20 gallons per acre. When concentrate sprays are used or when treating non-bearing or immature trees, the lower rate of **BRAVO 720** listed may be used. **DO NOT** allow livestock to graze in treated areas. The following spray volumes are recommended as gallons of spray per acre:

CROP	SPRAY VOLUME (Gallons per Acre)	
Peach, Nectarine, Apricot, Tart Cherry, Plum, Prune	20 (concentrate) to 300 (full dilute)	
Sweet Cherry	20 (concentrate) to 400 (full dilute)	
Conifers	**Dilute**	**Concentrate**
Forest stands	Not used	10 to 20 (aircraft)
Christmas trees	100	10 to 50 (aircraft or ground equipment)
Nursery beds	100	5 to 10 (ground equipment only)

CROP	DISEASES	BRAVO 720 RATE PER ACRE	BRAVO 720 RATE PER 100 GAL*	APPLICATION DIRECTIONS
Peach, Nectarine, Apricot, Cherry, Plum, Prune	Leaf curl, Coryneum blight (shothole)	$3\frac{1}{8}$ to $4\frac{1}{8}$ pts	1 to $1\frac{3}{8}$ pts	For best control of both diseases apply at leaf fall in late autumn, using sufficient water and proper sprayer calibration to obtain uniform coverage. When conditions favor high disease levels use the high rate of application and apply once or twice more in mid to late winter before budswell. If the leaf fall application is not practical, application of **BRAVO 720** for control of leaf curl may be made at any time prior to budswell the following spring. Where Coryneum blight (shothole) occurs, also apply at budbreak to protect newly emerging leaves and at shuck split to prevent fruit infections.
	Brown rot blossom blight, Lacy (russet) scab (plum/prune)	$3\frac{1}{8}$ to $5\frac{1}{2}$ pts	1 to $1\frac{3}{8}$ pts	Use $4\frac{1}{8}$ to $5\frac{1}{2}$ pints per acre on trees taller than 20 ft. and $3\frac{1}{8}$ to $4\frac{1}{8}$ pints per acre on smaller trees. Make one application at popcorn (pink, red or early white bud) and a second application at full bloom. If weather conditions favor disease development, make an additional application at petal fall.
	Cherry leaf spot; Peach, Nectarine, Apricot scab	$3\frac{1}{8}$ to $4\frac{1}{8}$ pts	1 to $1\frac{3}{8}$ pts	In addition to the bloom applications listed above, make one application at shuck split. **DO NOT** apply **BRAVO 720** after shuck split and before harvest. If additional disease control is needed before harvest, use another registered fungicide. For control of cherry leaf spot after harvest, make one application to foliage within 7 days after fruit is removed. In orchards with a history of high leaf spot incidence, make a second application 10-14 days later.
Conifers	Swiss needlecast	$2\frac{3}{4}$ to $5\frac{1}{2}$ pts	$2\frac{3}{4}$ to $5\frac{1}{2}$ pts	**Single application technique:** In Christmas tree plantations or forest stands make one application in the spring when new shoot growth is $\frac{1}{2}$ to 2 inches in length.
	Scleroderris canker (pines) Swiss needlecast	$1\frac{1}{2}$ to $2\frac{3}{4}$ pts	$1\frac{1}{2}$ to $2\frac{3}{4}$ pts	Make the first application in spring when new shoot growth is $\frac{1}{2}$ to 2 inches in length. Make additional applications at 3 to 4 week intervals until conditions no longer favor disease development. For use in nursery beds, apply the highest rate specified on a 3 week schedule.
	Sirococcus tip blight	2 to $3\frac{1}{2}$ pts	2 to $3\frac{1}{2}$ pts	
	Rhizosphaera needlecast (spruces) Scirrhia brown spot (pines)	$5\frac{1}{2}$ pts	$5\frac{1}{2}$ pts	
	Cyclaneusma and Lophodermium needlecasts (pines)	$2\frac{3}{4}$ to $5\frac{1}{2}$ pts	$2\frac{3}{4}$ to $5\frac{1}{2}$ pts	Apply in early spring prior to budbreak. Repeat applications at approximately 6 to 8 week intervals, until spore release ceases in late fall. Apply monthly during periods of frequent rainfall, and where Lophodermium infections occur during dormancy (Pacific Northwest). During drought periods, applications may be suspended, then resumed upon next occurrence of needle wetness.
	Rhabdocline needlecast (Douglas-fir)	$1\frac{1}{2}$ to $2\frac{3}{4}$ pts	$1\frac{1}{2}$ to $2\frac{3}{4}$ pts	Apply at budbreak and repeat at 3 to 4 week intervals until needles are fully elongated and conditions no longer favor disease development. In plantations of mixed provenance, or when irregular budbreak occurs, apply weekly until all trees have broken bud, then every 3 to 4 weeks as specified above. In nursery beds, use the high rate on a 3 week schedule.
	Botrytis seedling blight Phoma twig blight	$1\frac{1}{2}$ to $2\frac{3}{4}$ pts	$1\frac{1}{2}$ to $2\frac{3}{4}$ pts	Begin applications in nursery beds when seedlings are 4 inches tall and when cool, moist conditions favor disease development. Make additional applications at 7 to 14 day intervals as long as disease favorable conditions persist.

*Volumetric rates to be used only with full dilute spray volume specified on this label for tree and orchard crops.

STORAGE AND DISPOSAL

DO NOT contaminate water, food or feed by storage or disposal. Open dumping is prohibited.

STORAGE: Store in a cool place. Protect from excessive heat.

PESTICIDE DISPOSAL: Pesticide wastes are toxic. Improper disposal of excess pesticide, pesticide spray or rinsate is a violation of Federal Law. If these wastes cannot be disposed of by use according to label instructions, contact your State Pesticide or Environmental Control Agency, or the Hazardous Waste representative at the nearest EPA Regional Office for guidance.

CONTAINER DISPOSAL: DO NOT reuse empty container. Triple rinse (or equivalent). Then offer for recycling or reconditioning, or puncture and dispose of in a sanitary landfill, or incineration, or, if allowed by state and local authorities, by burning. If burned, stay out of smoke.

WARRANTY AND LIMITATION OF DAMAGES

Seller warrants that this material conforms to its chemical description and is reasonably fit for the purposes stated on the label when used in accordance with directions under normal conditions of use and Buyer assumes the risk of any use contrary to such directions. SELLER MAKES NO OTHER EXPRESS OR IMPLIED WARRANTY, INCLUDING ANY OTHER EXPRESS OR IMPLIED WARRANTY OF FITNESS OR OF MERCHANTABILITY, AND NO AGENT OF SELLER IS AUTHORIZED TO DO SO EXCEPT IN WRITING WITH A SPECIFIC REFERENCE TO THIS WARRANTY. In no event shall Seller's liability for any breach of warranty exceed the purchase price of the material as to which a claim is made.

BRAVO® 90DG

AGRICULTURAL FUNGICIDE

(90% Water Dispersible Granules)

ACTIVE INGREDIENT:	
Chlorothalonil (tetrachloroisophthalonitrile)	90.0%
INERT INGREDIENT:	10.0%
TOTAL	100.0%

*Covered under U.S. Patent No. 3,290,353 and No. 3,331,735

EPA Reg. No. 50534-157

KEEP OUT OF REACH OF CHILDREN

DANGER—PELIGRO

Precaucion al Usuario: Si usted no lee ingles, no use este producto hasta que la etiqueta haya sido explicado ampliamente.

See precautionary statements.

A BROAD SPECTRUM

AGRICULTURAL FUNGICIDE

Read entire label carefully and use only as directed.

PRECAUTIONARY STATEMENTS

HAZARDS TO HUMANS AND DOMESTIC ANIMALS

DANGER

Corrosive, causes severe eye damage. May cause skin irritation. May be a potential skin sensitizer.

DO NOT get in eyes. Wear goggles or face shield when handling this product.

Avoid prolonged contact with skin. Wash exposed areas of skin with soap and warm water after handling or using.

Avoid breathing spray mist.

DO NOT take internally.

Note to User: This product may produce temporary allergic side effects characterized by redness of the eyes, mild bronchial irritation and redness or rash on exposed skin areas. Persons having allergic reaction should contact a physician.

DO NOT apply this product in such a manner as to directly, or through drift, expose workers or other persons. The area being treated must be vacated by unprotected persons.

FIRST AID

In case of contact with eyes, immediately flush eyes with plenty of water and continue for 15 minutes. Seek medical attention for eyes immediately.

Wash exposed areas of skin with soap and warm water after handling or using.

Note to physician: Persons having an allergic reaction respond to treatment with antihistamines or steroid creams and/or systemic steroids.

ENVIRONMENTAL HAZARDS

This product is toxic to fish, aquatic invertebrates, and marine/estuarine organisms. Runoff from treated areas may be hazardous to aquatic organisms in neighboring areas. **DO NOT** apply directly to water. **DO NOT** apply when weather conditions favor drift from treated areas. Apply only to areas specified on label.

DIRECTIONS FOR USE

It is a violation Federal law to use this product in a manner inconsistent with its labeling.

Note to User: Wear long sleeve shirt, long pants, and gloves while mixing, loading and applying this product.

BRAVO 90DG is an excellent fungicide when used according to label directions for control of a broad spectrum of plant diseases.

BRAVO 90DG can be used effectively in dilute or concentrate sprays. Thorough, uniform coverage is essential for disease control.

Do not use on greenhouse-grown crops.

Do not combine **BRAVO 90DG** in the spray tank with pesticides, surfactants or fertilizers, unless your prior use has shown the combination physically compatible, effective and noninjurious under your conditions of use. Do not combine **BRAVO 90DG** with Dipel 4L, Triton AG-98, or Triton B-1956 as phytotoxicity may result from the combination when applied to some crops on this label.

Dipel is a registered trademark of Abbott Laboratories. Triton and Triton B-1956 are trademarks of Rohm and Haas Company.

Dosage rates on this label indicate pounds of **BRAVO 90DG** per acre, unless otherwise stated. Under conditions favoring disease development the high rate specified and shortest application interval should be used.

The required amount of **BRAVO 90DG** should be added slowly into the spray tank during filling. With concentrate sprays, pre-mix the required amount of **BRAVO 90DG** in a clean container and add to the spray tank as it is being filled. Keep agitator running when filling spray tank and during spray operations.

Note to User: DO NOT rotate to crops other than those listed on labels within 12 months of the last treatment. After 12 months from the last application, leafy vegetables (i.e., spinach, lettuce, kale, etc.) may also be rotated.

Reentry: Note to User: **DO NOT** enter treated area to perform hand labor within 24 hours of application unless protective clothing is worn.

Because certain states may require more restrictive reentry intervals for various crops treated with this product, consult your State Department of Agriculture for further information.

Written or oral warnings must be given to workers who are expected to be in a treated area or in an area about to be treated with this product. When oral warnings are given, warnings shall be given in a language customarily understood by workers. Oral warnings must be given if there is reason to believe that written warnings cannot be understood by workers. Oral and written warnings must include the following information. "**DANGER,** Area treated with **BRAVO 90DG** on (date of application). **DO NOT** enter without appropriate protective clothing for 24 hours. In case of accidental exposure, wash exposed area with plenty of water and get medical attention. For further information see 'Precautionary Statements' on the label."

APPLICATION AND CALIBRATION TECHNIQUES FOR SPRINKLER IRRIGATION

Apply this product only through center pivot, motorized lateral move, traveling gun, solid set or portable (wheel move, side roll, end tow, or hand move) irrigation system(s). **Do not** apply this product through any other type of irrigation system.

Crop injury, lack of effectiveness, or illegal pesticide residues in the crop can result from non-uniform distribution of treated water.

If you have questions about calibration, you should contact State Extension Service specialists, equipment manufacturers or other experts.

Do not apply this product through irrigation systems connected to a public water system. 'Public water system' means a system for the provision to the public of piped water for human consumption if such system has at least 15 service connections or regularly serves an average of at least 25 individuals daily at least 60 days per year.

Controls for both irrigation water and pesticide injection systems must be functionally interlocked, so as to automatically terminate pesticide injection when the irrigation water pump motor stops. A person knowledgeable of the irrigation system and responsible for its operation shall be present so as to discontinue pesticide injection and make necessary adjustments, should the need arise.

The irrigation water pipeline must be fitted with a functional, automatic, quick-closing check valve to prevent the flow of treated irrigation water back toward the water source. The pipeline must also be fitted with a vacuum relief valve and low pressure drain, located between the irrigation water pump and the check valve, to prevent back-siphoning of treated irrigation water into the water source.

Always inject BRAVO 90DG into irrigation water after it discharges from the irrigation pump and after it passes through the check valve. Never inject pesticides into the intake line on the suction side of the pump.

Pesticide injection equipment must be fitted with a functional, normally closed, solenoid-operated valve located on the intake side of the injection pump. Interlock this valve to the power system, so as to prevent fluid from being withdrawn from the chemical supply tank when the irrigation system is either automatically or manually turned off.

The pesticide injection pipeline must contain a functional, automatic, quick-closing check valve to prevent the flow of fluid back toward the injection pump. The irrigation line or water pump must include a functional pressure switch which will stop the water pump motor when the water pressure decreases to the point where pesticide distribution is adversely affected.

Spray mixture in the chemical supply tank must be agitated at all times, otherwise settling and uneven application may occur. Do not apply when wind speed favors drift beyond the area intended for treatment.

Posting of areas to be chemigated is required when 1) any part of a treated area is within 300 feet of sensitive areas such as residential areas, labor camps, businesses, day care centers, hospitals, in-patient clinics, nursing homes or any public areas such as schools, parks, playgrounds, or other public facilities not including public roads, or 2) when the chemigated area is open to the public.

Posting must conform to the following requirements. Treated areas shall be posted with signs at all usual points of entry and along likely routes of approach from the listed sensitive areas. When there are no usual points of entry, signs must be posted in the corners of the treated areas and in any other location affording maximum visibility to sensitive areas. The printed side of the sign should face away from the treated area towards the sensitive area. The signs shall be printed in English. Signs must be posted prior to application and must remain posted until foliage has dried and soil surface water has disappeared. Signs may remain in place indefinitely as long as they are composed of materials to prevent deterioration and maintain legibility for the duration of the posting period.

All words shall consist of letters at least 2 1/2 inches tall, and all letters and the symbol shall be a color which sharply contrasts with their immediate background. At the top of the sign shall be the words KEEP OUT, followed by an octagonal stop sign symbol at least 8 inches in diameter containing the word STOP. Below the symbol shall be the words PESTICIDES IN IRRIGATION WATER.

BRAVO 90DG may be used through two basic types of sprinkler irrigation systems as outlined in Sections A and B below. Determine which type of system is in place, then refer to the appropriate directions provided for each type.

A. Center Pivot, Motorized Lateral Move and Traveling Gun Irrigation Equipment

For injection of pesticides, these continuously moving systems must use a positive displacement injection pump, of either diaphragm or piston type, constructed of materials that are compatible with pesticides and capable of being fitted with a system interlock and capable of injection at pressures approximately 2-3 times those encountered within the irrigation water line. Venturi applicator units cannot be used on these systems.

Fill chemical supply tank of injection equipment with water. Operate system for one complete revolution or run across the field, measuring time required, amount of water injected, and acreage covered. Thoroughly mix recommended amount of **BRAVO 90DG** for acreage to be covered into same amount of water used during calibration and inject into system continuously for one revolution or run. Mixture in the chemical supply tank must be continuously agitated during the injection run. Shut off injection equipment after one revolution or run, but continue to operate irrigation system until **BRAVO 90DG** has been cleared from last sprinkler head.

B. Solid Set and Portable (Wheel Move, Side Roll, End Tow, or Hand Move) Irrigation Equipment

With stationary systems, an effectively designed in-line venturi applicator unit is preferred which is constructed of materials that are compatible with pesticides; however, a positive-displacement pump can also be used.

Determine acreage covered by sprinkler. Fill tank of injection equipment with water and adjust flow to use contents over a thirty to forty-five minute period. Mix desired amount of **BRAVO 90DG** for acreage to be covered with water so that the total mixture of **BRAVO 90DG** plus water in the injection tank is equal to the quantity of water used during calibration and operate entire system at normal pressures recommended by the manufacturer of injection equipment used for amount of time established during calibration. Agitation is recommended. **BRAVO 90DG** can be injected at the beginning or end of the irrigation cycle or as a separate application. Stop injection equipment after treatment is completed and continue to operate irrigation system until **BRAVO 90DG** has been cleared from last sprinkler head.

FIELD AND ROW CROPS

Apply **BRAVO 90DG** in sufficient water to obtain adequate coverage of foliage. Gallonage to be used will vary with crop and amount of plant growth. Spray volume usually will range from 20 to 150 gallons per acre (200 to 1400 liters per hectare) for dilute sprays and 5 to 10 gallons per acre (50 to 100 liters per hectare) for concentrate ground sprays and aircraft applications. Both ground and aircraft methods of application are recommended unless specific directions are given for a crop. Application through sprinkler irrigation systems is not recommended unless specific directions are given for a crop. See application and calibration instruction below.

CROP	DISEASES	RATE PER ACRE	APPLICATION DIRECTIONS
Bean (Snap)	Rust	1 1/8 to 2 1/2 lb	Use in sufficient water to obtain adequate coverage. Begin applications during early bloom stage or when disease first threatens and repeat at weekly intervals or as necessary to maintain control. **DO NOT** apply within 7 days of harvest. **DO NOT** graze treated areas or feed treated plant parts to livestock.
	Botrytis blight (gray mold)	2 1/2 lb	
Beans (Dry)	Rust, Anthracnose, Downy mildew, Cercospora leaf spot (blackeye only)	1 1/8 to 1 3/4 lb	Use in sufficient water to obtain adequate coverage. Begin applications during early bloom stage and repeat at 7 to 10 day intervals. For use only on beans harvested dry with pods removed. **DO NOT** apply within 6 weeks before harvest. **DO NOT** allow livestock to graze in treated areas or feed treated plant parts to livestock.
Cabbage, Chinese Cabbage (tight-headed varieties only), Cauliflower, Broccoli, Chinese Broccoli, Brussels Sprouts	Alternaria leaf spot, Downy mildew	1 1/4 lb	Use in sufficient water to obtain adequate coverage. Begin applications after transplants are set in field, or shortly after emergence of field-seeded crop, or when conditions favor disease development. Repeat at 7 to 10 day intervals or as necessary to maintain control. Do not apply within 7 days of harvest to Chinese cabbage or Chinese broccoli.
	Ring spot (California only)	1 1/2 lb	For field-seeded brussel sprouts, begin applications at time of early sprout development or when conditions favor disease development. Repeat at 7 to 10 day intervals or as necessary to maintain control.
Carrot	Cercospora (Early) blight, Alternaria (Late) blight	1 1/4 to 1 1/2 lb	Use in sufficient water to obtain adequate coverage. Start applications when disease threatens and repeat at 7 to 10 day intervals or as necessary to maintain control. **BRAVO 90DG** may be applied through sprinkler irrigation equipment. See calibration directions preceding this section.
Celery	Cercospora (Early) blight, Septoria (Late) blight	7/8 to 1 1/4 lb.	Use 7/8 to 1 1/4 lb per acre on a 3 to 5 day spray schedule or 1 3/4 to 2 1/2 lb per acre on a 7 day schedule. Start applications when transplants are set in the field. Apply in sufficient water to obtain adequate coverage. **DO NOT** apply within 7 days of harvest. **BRAVO 90DG** may be applied through sprinkler irrigation equipment. See Calibration directions preceding this section.
	Basal stalk rot (*Rhizoctonia solani*)	1 3/4 to 2 1/2 lb	
	Early blight Late blight	1 1/4 to 1 1/2 lb per 100 gal.	For celery seedbeds, apply 125 gallons per acre twice weekly or as needed to maintain control. Start applications shortly after crop emergence. Use the higher rate under severe disease conditions.
Corn (sweet), Corn grown for seed	Helminthosporium leaf blights, Rust	5/8 to 1 1/2 lb	Use in sufficient water to obtain adequate coverage. Begin applications when conditions favor disease development and repeat at 4 to 7 day intervals or as required to maintain control. Under severe disease conditions, use 1 1/4 to 1 1/2 lb per acre. **DO NOT** apply within 14 days of harvest. **DO NOT** apply to sweet corn to be processed. **DO NOT** allow livestock to graze in treated fields. **DO NOT** ensile treated corn or use as livestock forage.
Cranberry	Fruit rots, Lophodermium leaf/twig blight	3 1/2 to 5 3/4 lbs	Apply at early bloom and repeat at 10 to 14 day intervals. Under severe disease conditions use the 5 3/4 lb/acre rate on a 10 day schedule. **DO NOT** apply **BRAVO 90DG** more than 3 times per season, or within 50 days before harvest. **DO NOT** apply to bogs when flooded or allow release of irrigation water from bogs for at least 3 days following application.

CROP	DISEASES	RATE PER ACRE	APPLICATION DIRECTIONS
			BRAVO 90DG may be applied through sprinkler irrigation equipment. Use 300 gallons of water per acre through solid set systems only. See calibration directions preceding this section.
Cucurbits: Cucumber, Cantaloupe, Muskmelon, Honeydew melon, Watermelon, Squash, Pumpkin	Anthracnose, Downy mildew, Target spot	1¼ to 1½ lbs	Use in sufficient water to obtain adequate coverage. Begin applications when plants are in first true leaf stage or when conditions are favorable for disease development. Repeat applications at 7 day intervals. Under severe disease conditions, shorten spray interval. **BRAVO 90DG** may be applied through sprinkler irrigation equipment (solid set, portable wheel move or center pivot systems only). See calibration directions preceding this section.
	Cercospora leafspot, Gummy stem blight (black rot), Alternaria leaf blight, Scab, Powdery mildew (Sphaerotheca only)	1½ to 2½ lbs	
	Cucumber belly rot *(Rhizocatonia solani)*	7 lbs	Use **BRAVO 90DG** in sufficient water to obtain runoff to soil surface. Make a single application when vines begin to form. **BRAVO 90DG** may be applied through sprinkler irrigation equipment as directed above.
Grasses Grown for Seed	Stem rust, Leaf rust, Stripe rust Septoria leafspot, Glume blotch, Bipolaris and Drechslera leafspots	⅞ to 1⅛ lbs	Use in sufficient water to obtain adequate coverage. Begin applications during stem elongation when conditions favor disease development. Re-apply at flag (top) leaf emergence and repeat applications at 14 day intervals. **DO NOT** apply within 14 days of harvest. **DO NOT** allow livestock to graze in treated areas or feed treated plant parts to livestock.
	Selenophoma (eyespot)	⅞ to 1¾ lbs	
Mint	Rust, Septoria leaf spot	1⅛ lb	Use in sufficient water to obtain adequate coverage. Begin applications when emerging plants are 4-8 inches high. Repeat applications at 7 to 10 day intervals or as necessary to maintain control. **DO NOT** apply more than 3 times per season. **DO NOT** apply within 80 days of harvest. **DO NOT** feed fresh or extracted mint hay from treated fields to livestock. Based on available residue data, use of **BRAVO 90DG** on mint is restricted to Indiana, Michigan and Wisconsin.
Onion (dry bulb)	Botrytis leaf blight/blast, Botrytis neck rot (suppression), Purple blotch	⅞ to 1¾ lbs	Apply in sufficient water to obtain thorough coverage of tops. **BRAVO 90DG** is recommended for use with disease monitoring systems which adjust fungicide rates and frequency of application according to disease hazard. Apply as follows: (columns: **Low Disease Hazard & Prior to Infection** / **Low Disease Hazard & Some Disease Present** / **High Disease Hazard**) **Rate per Acre:** ⅞ lb / 1⅛ lbs / 1¾ lbs **Frequency:** 10 days / 7 to 10 days / 7 days For suppression of neck rot (*Botrytis* spp.) during storage, a minimum of three weekly applications prior to lifting, using 1⅛ to 1¾ lbs. of **BRAVO 90DG** per acre, is recommended. **DO NOT** apply within 7 days of harvest.
Onion (green bunching), Leek, Shallot, Garlic, Onion grown for seed	Botrytis leaf blight/blast, Purple blotch, Downy mildew (suppression)	1⅛ to 2¼ lbs	Use in sufficient water to obtain thorough coverage of tops. Begin applications prior to favorable infection periods, and repeat at 7 to 10 day intervals for as long as conditions favor disease. Use the high rate and a 7 day schedule of applications when heavy dew or rain persist. **DO NOT** apply within 7 days of harvest on garlic. **DO NOT** apply more than 3 times per season or within 14 days of harvest on green bunching onions, leeks or shallots. If additional disease control is needed before harvest, use another registered fungicide.
Papaya	Alternaria fruit spot, Anthracnose, Stem end rot	1¾ to 3¼ lb	Apply with ground equipment only, in sufficient water to obtain adequate coverage of fruit and leaves. Begin treatment when conditions favor development of disease and continue treatments at 14 day intervals until weather conditions no longer favor disease development. **DO NOT** graze livestock in treated area or feed processing by-products to livestock.
Parsnip	Alternaria leaf spot, Downy mildew, Anthracnose, Botrytis blight (gray mold), Bottom rot *(Rhizoctonia)*	1⅛ to 1¾ lb	Apply in sufficient water to obtain adequate coverage. Make the first application at the first sign of disease or when conditions are favorable for infection. Continue applications on a 7 to 10 day schedule. **DO NOT** apply more than 4 times per season or within 10 days of harvest. **DO NOT** feed treated plant parts to livestock.
Passion Fruit (Hawaii only)	Alternaria fruit and leaf spot, (passion fruit brown spot)	1½ lb	Apply with ground equipment in sufficient water to obtain adequate coverage of fruit and leaves. Begin treatment when fruit spots appear (April to July) and continue treatments at 14 day intervals until weather conditions no longer favor disease development. **DO NOT** graze in treated area or feed vines or processing by-products to livestock used for food.
Peanut	*Cercospora* (early) leaf spot *Cercosporidium* (late) leaf spot	1 to 1¼ lb	Apply in sufficient water for adequate coverage when leaf wetness first occurs and before disease symptoms appear, generally 30 to 40 days after planting; repeat at 10 to 14 day intervals. When late leaf spot prevails or when rust or web blotch occur, apply 1¼ lb per acre at 10 day intervals for the remainder of the season. **DO NOT** apply within 14 days of harvest. **DO NOT** allow livestock to graze in treated areas. **DO NOT** feed hay or threshings from treated fields to livestock. **BRAVO 90DG** may be applied through sprinkler irrigation equipment. See calibration directions preceding this section.
	Rust, Web blotch	1¼ lb	
Potato	Late blight, Early blight, Botrytis vine rot	⅝ lb —then— ⅞ to 1¼ lb/A	Apply as a banded treatment directed over the foliage, beginning when vines are first exposed and leaf wetness occurs. Repeat applications at 7 to 10 day intervals until vines close, then increase rate as described below. In addition to the early-season applications above, as vines close between rows increase water carrier volume to cover the denser canopy and begin broadcast application patterns. Continue applications at 7 to 10 day intervals. Use the highest registered rate weekly when disease conditions are severe. **BRAVO 90DG** may be applied through sprinkler irrigation equipment (solid set,

CROP	DISEASES	RATE PER ACRE	APPLICATION DIRECTIONS
			portable wheel move, center pivot or motorized lateral move systems only). **DO NOT** exceed a 10 day interval between applications when using this technique. Follow calibration and application directions preceding this section.
Soybean Determinate (Southern) Varieties	Anthracnose, Diaporthe pod and stem blight, Frogeye leaf spot *(Cercospora sojina)*, Purple seed stain, Cercospora leaf blight *(Cercospora kikuchii)*, Septoria brown spot		Apply in sufficient water to obtain complete coverage, using at least five gallons of water per acre for aerial application. Use the three application program in areas having a history of moderate to severe disease intensity. **BRAVO 90DG** may be applied through sprinkler irrigation equipment. Follow application and calibration directions preceding this section. **DO NOT** exceed total of 3 applications per season. **DO NOT** apply within 6 weeks of harvest. **DO NOT** feed treated plant parts to livestock or allow grazing in treated fields.
		1 1/8 to 2 lb	**Two application program—**Make the first application at early pod set (R3) stage, when majority of pods are 1/8 to 3/8 inch in length) and the second at beginning of seed formation (R5) which occurs about 14 days later.
		7/8 to 1 1/2 lb	**Three application program—**Make the first application at the beginning of flowering (R1), the second at early pod set (R3) and the third at beginning of seed formation (R5).
	Stem canker *(Diaporthe phaseolorum* var. *caulivora)*	7/8	Apply in 10 to 20 gallons of water per acre, as a band treatment directing spray to provide coverage of entire plant. Make one application at emergence of second trifoliate leaves (V2). If conditions favor continued infection, make additional applications at 10 to 14 day intervals.
Soybean Indeterminate (Northern) Varieties	Anthracnose, Diaporthe pod and stem blight, Frogeye leaf spot *(Cercospora sojina)*, Purple seed stain, Cercospora leaf blight *(Cercospora kikuchii)*, Septoria brown spot		Apply in sufficient water to obtain complete coverage, using at least five gallons of water per acre for aerial application. Use the three application program in areas having a history of moderate to severe disease intensity. **BRAVO 90DG** may be applied through sprinkler irrigation equipment. Follow application and calibration directions preceding this section. **DO NOT** apply more than 3 times per season or within 6 weeks of harvest. **DO NOT** feed treated plant parts to livestock or allow grazing in treated fields.
		1 1/8 to 2 lb	**Two application program—**Make the first application when largest pods are 1 to 1 1/2 inches in length and make the second application 14 days later. **BRAVO 90DG** may be co-applied with **Benlate* 50WP** as a tank mix for disease control in indeterminate (northern) soybeans. Use 7/8 of **BRAVO 90DG** plus 8 ounces of **Benlate 50WP** per acre. Make the first application when pods near the top of plants are 1/2 to 1 inch in length and a second application 14 days later.
		7/8 to 1 1/2 lb	**Three application program—**Make the first application one week after first flowering and continue applications at 14 day intervals.
Tomato	FOLIAGE (apply every 7-10 days): Early blight, Late blight, Gray leaf spot, Gray leaf mold, Septoria leaf spot, Target spot	1 1/8 to 1 3/4 lb	Apply in sufficient water to obtain adequate coverage. Begin applications when dew or rain occur and disease threatens. Use the highest rate and shortest interval specified when disease conditions are severe. **BRAVO 90DG** may be combined in the spray tank with EPA-registered pesticide products that claim copper as the active ingredient and are labeled for control of bacterial diseases of tomatoes. Check the copper manufacturer's label for specific instructions, precautions and limitations prior to mixing with **BRAVO 90DG. DO NOT** use with **Copper-Count*N** in concentrated spray suspensions. **BRAVO 90DG** may be applied through sprinkler irrigation equipment (solid set or portable wheel move systems only). See calibration directions preceding this section. *Copper-Count is a registered trademark of Mineral Research & Development Corp.
	FRUIT (apply every 7-14 days beginning at fruit set): Anthracnose, Alternaria fruit rot (black mold), Rhizoctonia fruit rot, Late blight fruit rot, Botrytis gray mold	1 3/4 to 2 1/4 lb	

*Benlate is a registered trademark of E.I. DuPont de Nemours and Co., Inc.

TREE AND ORCHARD CROPS

Apply **BRAVO 90DG** in sufficient water and with proper calibration to obtain uniform coverage of tree canopy. Application with ground equipment is preferable to aerial application because ground applications generally give better coverage of the tree canopy. If application with ground equipment is not feasible, **BRAVO 90DG** may be applied, with aircraft using at least 20 gallons of spray per acre. When concentrate sprays are used or when treating non-bearing or immature trees, the lower rate of **BRAVO 90DG** listed may be used. **DO NOT** allow livestock to graze in treated areas. The following spray volumes are recommended as gallons of spray per acre:

CROP	SPRAY VOLUME (Gallons per Acre)	
Peach, Nectarine, Apricot, Tart Cherry, Plum, Prune	20 (concentrate) to 300 (full dilute)	
Sweet Cherry	20 (concentrate) to 400 (full dilute)	
Conifers	**Dilute**	**Concentrate**
Forest stands	Not used	10 to 20 (aircraft)
Christmas trees	100	10 to 50 (aircraft or ground equipment)
Nursery beds	100	5 to 10 (ground equipment only)

CROP	DISEASES	BRAVO 90DG RATE PER ACRE	BRAVO 90DG RATE PER 100 GAL*	APPLICATION DIRECTIONS
Peach, Nectarine, Apricot, Cherry, Plum, Prune	Leaf curl, Coryneum blight (shothole)	2½ to 3½ lb	⅞ to 1⅛ lb	For best control of both diseases apply at leaf fall in late autumn, using sufficient water and proper sprayer calibration to obtain uniform coverage. When conditions favor high disease levels use the high rate of application and apply once or twice more in mid to late winter before budswell. If the leaf fall application is not practical, application of **BRAVO 90DG** for control of leaf curl may be made at any time prior to budswell the following spring. Where Coryneum blight (shothole) occurs, also apply at budbreak to protect newly emerging leaves and at shuck-split to prevent fruit infections.
	Brown rot blossom blight, Lacy (russet) scab (plum/prune)	2½ to 4½ lb	⅞ to 1⅛ lb	Use 3½ to 4½ lb per acre on trees taller than 20 feet and 2½ to 3½ per acre on smaller trees. Make one application at popcorn (pink, red or early white bud) and a second application at full bloom. If weather conditions favor disease development, make an additional application at petal fall.
	Cherry leaf spot; Peach, Nectarine, Apricot scab	2½ to 4½ lb	⅞ to 1⅛ lb	In addition to the bloom applications listed above, make one application at shuck-split. **DO NOT** apply **BRAVO 90DG** after shuck-split and before harvest. If additional disease control is needed before harvest, use another registered fungicide. For control of cherry leaf spot after harvest, make one application to foliage within 7 days after fruit is removed. In orchards with a history of high leaf spot incidence, make a second application 10-14 days later.
Conifers	Swiss needlecast	2¼ to 4½ lb	2¼ to 4½ lb	Single application technique: In Christmas tree plantations or forest stands make one application in the spring when new shoot growth is ½ to 2 inches in length.
	Scleroderris canker (pines), Swiss needlecast	1⅛ to 2¼ lb	1⅛ to 2¼ lb	Make the first application in spring when new shoot growth is ½ to 2 inches in length. Make additional applications at 3 to 4 week intervals until conditions no longer favor disease development. For use in nursery beds, apply the highest rate specified on a three week schedule.
	Sirococcus tip blight	1¾ to 3 lb	1¾ to 3 lb	
	Rhizosphaera needlecast (spruces), Scirrhia brown spot(pines)	4½ lb	4½ lb	
	Rhabdocline needlecast (Douglas-fir)	1⅛ to 2¼ lb	1⅛ to 2¼ lb	Apply at budbreak and repeat at 3 to 4 week intervals until needles are fully elongated and conditions no longer favor disease development. In plantations of mixed provenance, or when irregular budbreak occurs, apply weekly until all trees have broken bud, then every 3 to 4 weeks as specified above. In nursery beds, use the high rate on a 3 week schedule.
	Cyclaneusma and Lophodermium needlecasts (pines)	2¼ to 4½ lb	2¼ to 4½ lb	Apply in early spring prior to budbreak. Repeat applications at approximately 6 to 8 week intervals, until spore release ceases in late fall. Apply monthly during periods of frequent rainfall, and where Lophodermium infections occur during dormancy (Pacific Northwest). During drought periods, applications may be suspended, then resumed upon next occurrence of needle wetness.
	Botrytis seedling blight, Phoma twig blight	1⅛ to 2¼ lb	1⅛ to 2¼ lb	Begin applications in nursery beds when seedlings are 4 inches tall and when cool, moist conditions favor disease development. Make additional applications at 7 to 14 day intervals as long as disease favorable conditions persist.

*Volumetric rates to be used only with full dilute spray volume specified on this label for tree and orchard crops.

STORAGE AND DISPOSAL

DO NOT contaminate water, food or feed by storage or disposal. Open dumping is prohibited.

STORAGE: Store in a dry place.

PESTICIDE DISPOSAL:

Pesticide wastes are toxic. Improper disposal of excess pesticide, pesticide spray or rinsate is a violation of Federal Law. If these wastes cannot be disposed of by use according to label instructions, contact your State Pesticide or Environmental Control Agency, or the Hazardous Waste representative at the nearest EPA Regional Office for guidance.

CONTAINER DISPOSAL:

Completely empty bag into application equipment. Then dispose of empty bag in a sanitary landfill or by incineration, or if allowed by State and local authorities, by burning. If burned, stay out of smoke.

WARRANTY AND LIMITATION OF DAMAGES

Seller warrants that this material conforms to its chemical description and is reasonably fit for the purposes stated on the label when used in accordance with directions under normal conditions of use and Buyer assumes the risk of any use contrary to such directions. SELLER MAKES NO OTHER EXPRESS OR IMPLIED WARRANTY, INCLUDING ANY OTHER EXPRESS OR IMPLIED WARRANTY OF FITNESS OR OF MERCHANTABILITY, AND NO AGENT OF SELLER IS AUTHORIZED TO DO SO EXCEPT IN WRITING WITH A SPECIFIC REFERENCE TO THIS WARRANTY. In no event shall Seller's liability for any breach of warranty exceed the purchase price of the material as to which a claim is made.

BUENO® 6

HERBICIDE

MSMA Plus Surfactant For Postemergence Weed Control

Contains 6.0 Pounds MSMA Per Gallon

ACTIVE INGREDIENT:

Monosodium Acid Methanearsonate	48.3%
INERT INGREDIENTS:	51.7%
TOTAL	100.0%
Total Arsenic, all in water-soluble form, expressed as elemental	22.38%

EPA Reg. No. 50534-6

KEEP OUT OF REACH OF CHILDREN

CAUTION

See precautionary statements.

Bueno 6

PRECAUTIONARY STATEMENTS

HAZARDS TO HUMANS AND DOMESTIC ANIMALS

CAUTION

Harmful if swallowed or absorbed through the skin. Avoid contact with eyes, skin or clothing. Avoid breathing vapors or spray mist.

If swallowed, induce vomiting immediately by giving two glasses of water and sticking finger down throat. **Never** give anything by mouth to an unconscious person. Seek medical attention.

In case of contact, flush eyes with plenty of water for at least 15 minutes. Seek medical attention.

Wash exposed skin gently with plenty of soap and water.

Keep children and domestic animals off treated areas until this material has been washed into the soil.

Wear protective clothing when handling or applying this product including long pants, long-sleeve shirt, and impermeable gloves and boots. Mixer-loaders should include an apron and full-face shield when handling or mixing concentrate. Flagmen should be fully protected during spray operations or mechanical flagmen used. Pilots and ground spray rig applicators should wear a mask or respirator approved by the Mining Enforcement and Safety Administration and the National Institute for Occupational Safety and Health.

ENVIRONMENTAL HAZARDS

Do not apply directly to water. **Do not** feed treated foliage to livestock or graze treated areas. **Do not** spray or allow drift onto vegetables, ornamentals or other desirable plants. **Do not** apply when weather conditions favor drift from target area. **Do not** contaminate water by cleaning of equipment or disposal of wastes.

BUENO 6

Read entire label before using this product.

GENERAL INFORMATION

BUENO 6 is a postemergence herbicide for use in cotton, turf, fruit, nuts and non-crop areas. **BUENO 6** contains a surfactant (wetting agent) in proper ratio for best results. It is not necessary to add additional surfactant to the spray tank. Local conditions and recommendations vary; consult local agricultural experiment station or extension service weed specialists for recommendations in your area.

A partial list of weeds, controlled with **BUENO 6** includes:

Barnyardgrass	Crabgrass	Nutsedge
Brachiaria spp.	Dallisgrass	Pigweed
Chickweed	Goosegrass	Puncturevine
Cocklebur	Johnsongrass	Sandbur

Mixing Instructions: Fill the spray tank with half the required amount of water and, with pump or agitator operating, add the recommended amount of **BUENO 6** herbicide and continue filling tank with the balance of water and mix thoroughly.

Do not store spray solution in tank overnight. Clean application equipment thoroughly after use by flushing with water in a safe place.

Application Methods: BUENO 6 should be applied with a low-pressure calibrated sprayer having satisfactory pumping and bypass action. **Adjust Nozzles to Keep Spray Off Cotton Foliage and in a Manner to Allow Maximum Coverage of Weeds.** Adequate coverage is very important for effective weed control.

Apply during warm weather when weeds are in an active stage of growth.

DIRECTIONS FOR USE

It is a violation of Federal law to use this product in a manner inconsistent with its labeling.

Do not apply this product through any type of irrigation equipment.

Cotton: BUENO 6 is useful as a directed postemergence spray for control of the weeds listed above. Mix **BUENO 6** at the rate of 2⅔ pints in 40 gallons of water for application as a **Directed** broadcast spray to one acre when weeds are small. For band applications, apply 1 gallon of above diluted spray per acre for each 1 band width to be treated of cotton grown on 40 inch row spacing. A second or repeat application, if needed, should be timed about 1 to 3 weeks after the first application. Keep spray off cotton foliage.

Apply Only When Cotton is 3 Inches High to First Bloom. Do Not Apply After First Bloom.

Slight burning and reddish discoloration of cotton foliage may occasionally be seen following recommended treatment; however, cotton plants will develop normally and yield will not be affected.

Emergency Preplant Application: When planting of cotton to prepared seedbeds has been delayed and weeds have emerged, one application of **BUENO 6** at the above rate can be made prior to seeding.

NOTE: Applications to cotton in Florida should be confined to band treatments.

Ornamental Turfgrass: BUENO 6 can be used for selective control of Bahiagrass, barnyardgrass, chickweed, smooth and large crabgrass, dallisgrass, nutsedge, sandbur and wood sorrel with little or no injury to well-established, actively growing turfgrasses. Mow turfgrass to height of 1 to 1½ inches before treatment. Mix 1 fluid ounce (2 tablespoons) of **BUENO 6** in 2½ gallons of water for application to an area of 1,000 square feet. **Established bermudagrass and zoysiagrass:** Mix 1 to 2 fluid ounces (2 to 4 tablespoons) of **BUENO 6** in 2½ gallons of water for application to an area of 1,000 square feet. Application should be uniform and thorough to adequately wet all undesirable plants. Two or more repeat treatments at 14 day intervals may be necessary. Make applications during warm weather when temperature is between 80 and 90°F. **Do not** water turf for at least 24 hours after application. Turfgrasses may be temporarily discolored. Bermudagrass, bluegrasses and zoysiagrass have shown tolerance to properly applied **BUENO 6.** Injury may result if applied to bentgrasses and fescues. **Do not** apply to St. Augustinegrass, carpetgrass, centipedegrass or to Dichondra. **Do not** reseed until 2 weeks after last application. **Do not** apply with hose-end applicators.

Non-Bearing Fruit and Nuts: BUENO 6 is effective as a directed post-emergence spray control of the above listed weeds in non-bearing almond, walnut, grapefruit, lemon, lime, orange and tangerine orchards and plantings of non-bearing apples, cherries, peaches, pears, plums and prunes. Mix at a rate of 2⅔% pints **BUENO 6** in 100 gallons of water and apply to an area of one acre. For spot treatment of weeds, mix 2⅔ pints of **BUENO 6** in 50 gallons of water and apply to point of run-off. Application should be made when weeds are small, during warm weather and when conditions are favorable for good weed growth. If regrowth occurs, repeat applications should be made, but not more than three applications per year. **Do not** allow spray solution to contact foliage, stems or bark of trees or vines. **Do not** use around trees or vines from which crops will be harvested within one year. **DO NOT** graze treated areas. In Florida, use only as a spot treatment.

Non-Crop: BUENO 6 is effective in controlling the above listed weeds and many similar weeds on drainage ditchbanks, fence rows, rights-of-way, storage yards and similar non-crop areas. Application should be made when weeds are small and conditions are favorable for good weed growth. Mix at a rate of 2½ to 2⅔ pints of **BUENO 6** in 40 gallons of water. Spray undesirable vegetation thoroughly to point of run-off. Adequate coverage and complete wetting of foliage is important for effective control. Repeat applications may be necessary if regrowth occurs. **Do not** contaminate waters used for domestic purposes, or by animals including wildlife and aquatic life, or water used for irrigation purposes.

SPECIAL LOCAL NEEDS PURSUANT TO SECTION 24(c) OF AMENDED FIFRA DIRECTIONS FOR USE

CROP	REGISTRATIONS	STATES	SLN NO.	DIRECTIONS FOR USE
Bluegrass, Fescue, Ryegrass	Control of wild oats and certain other broadleaf and grassy weeds in bluegrass, fescue and ryegrass grown for seed only.	Idaho Oregon Washington	ID-770019 OR-770025 WA-770025	Apply 3 to 4 quarts **BUENO 6** per acre in sufficient water for good coverage. Application can be made anytime after weeds emerge and before grass has reached boot stage. **Do not** USE MORE THAN ONE APPLICATION PER YEAR. **Do not** APPLY AFTER BOOT STAGE. **Do not** FEED TREATED FOLIAGE OR SEED PRODUCED FROM TREATED AREAS.
Citrus	Citrus - Bearing	California	CA-790189	**BUENO 6** should be applied as a directed application in citrus orchards, such as orange, grapefruit, tangerine, lemon and lime orchards. It should be applied at the rate of 2⅔ to 5⅓ pints per acre. Mix **BUENO 6** at the rate of 2⅔ pints in 75 gallons of water. Apply as a directed spray in interspaces and around base of trees. Spray unwanted vegetation to just short of run-off. If regrowth occurs, reapply as required; however, **Do not** exceed 3 applications per year. **Do not** allow spray solution to contact fruit, leaves, stem or bark. Use a shield, if necessary, for nursery plantings or young trees. SPECIAL PRECAUTIONS - **Do not** allow spray or spray drift to contact adjacent crops or injury will occur. Apply only on still days when weather conditions do not favor drift from areas being treated. Applications should be made in accordance with state and Federal regulations.
Cotton	Application on Cotton	South Carolina	SC-770015	POSTEMERGENT APPLICATION ON COTTON: A single ground or aircraft application may be made as an over-the-top broadcast spray to actively growing cotton as a salvage operation when cotton is 3 to 6 inches high or up to early first square stage, whichever occurs first. Mix **BUENO 6** at a rate of 1 to 1⅓ pints in 40 gallons of water for ground application, or in 5 to 10 gallons of water for aircraft application and apply using a properly calibrated sprayer to one acre. APPLY AS AN OVER-THE-TOP BROADCAST SPRAY ONLY WHEN COTTON IS 3 TO 6 INCHES HIGH OR UP TO EARLY FIRST SQUARE STAGE, WHICHEVER OCCURS FIRST. **Do not** APPLY AS AN OVER-THE-TOP BROADCAST SPRAY AFTER COTTON IS MORE THAN 6 INCHES HIGH OR AFTER EARLY FIRST SQUARE STAGE. **Do not** make more than two applications total of

CROP	REGISTRATIONS	STATES	SLN NO.	DIRECTIONS FOR USE
				either DSMA or MSMA (or a combination) per season. See SPECIAL PRECAUTIONS below. Slight burning and reddish discoloration of cotton foliage may occasionally be seen following recommended treatment; however, cotton plants will develop normally. SPECIAL PRECAUTIONS - **Do not** feed treated forage to livestock or graze treated areas. **Do not** allow spray or spray drift to contact adjacent crops or injury and residues will result. Apply only still days when weather conditions **do not** favor drift from areas being treated. Follow other precautions listed on this label. Aircraft applications of **BUENO 6** should only be made by applicators experienced in the use of herbicides, and applications should be made in accordance with state and Federal regulations.
Cotton	Application of Cotton	Georgia Louisiana Missouri New Mexico North Carolina Tennessee Texas	GA-770008 LA-880008 MO-770005 NM-780004 NC-770018 TN-830017 TX-770026	**BUENO 6** is useful for the control of weeds listed on the product label and many similar weeds. Application can be made as follows. 1. Preplant or postplant up to cracking of soil before cotton emergence using ground or aircraft equipment. 2. Postemergent, over-the-top, when cotton is 3 to 6 inches high or up to early first square stage, whichever occurs first, using ground or aircraft equipment. 3. Postemergence as a directed spray with ground equipment when cotton is 3 inches high to first bloom. PREPLANT OR POSTPLANT UP TO CRACKING APPLICATION ON COTTON: A single ground or aircraft application of BUENO 6 can be made to prepared cotton seedbeds when planting has been delayed and weeds have emerged, or as a postplant treatment, BUT NO LATER THAN INITIAL CRACKING of soil in field before emergence of cotton. Planting of cotton may immediately follow the pre-plant application. Mix at the rate of $2\frac{2}{3}$ pints of **BUENO 6** in 40 gallons of water for broadcast ground application, or in 5 to 10 gallons of water for aircraft application and apply using a properly calibrated sprayer to one acre. See SPECIAL PRECAUTIONS below. POSTEMERGENCE, OVER-THE-TOP, APPLICATION ON COTTON: Application may be made using ground or aircraft equipment as an over-the-top broadcast spray to actively growing cotton as a salvage operation when cotton is 3 to 6 inches high or up to early first square stage, whichever occurs first. Mix **BUENO 6** at rate of 1 to $1\frac{1}{3}$ pints in 40 gallons of water for ground applications, or in 5 to 10 gallons of water for aircraft application and apply using a properly calibrated sprayer to one acre. A second application, if needed, should be made 1 to 3 weeks after the first treatment. APPLY AS AN OVER-THE-TOP BROADCAST SPRAY ONLY WHEN COTTON IS 3 TO 6 INCHES OR UP TO EARLY FIRST SQUARE STAGE, WHICHEVER OCCURS FIRST. **DO NOT APPLY AS AN OVER-THE-TOP BROADCAST SPRAY AFTER COTTON IS MORE THAN 6 INCHES HIGH OR AFTER EARLY FIRST SQUARE STAGE WHICHEVER OCCURS FIRST.** See SPECIAL PRECAUTIONS below. POSTEMERGENT, DIRECTED, APPLICATION ON COTTON: Mix **BUENO 6** at the rate of $2\frac{2}{3}$ pints in 40 gallons of water for application as a DIRECTED broadcast spray to one acre when weeds are small. For band applications, apply 1 gallon of above diluted spray per acre for each 1-inch band width to be treated of cotton grown on 40-inch row spacing. Keep spray off cotton foliage. APPLY ONLY WHEN COTTON IS 3 INCHES HIGH TO FIRST BLOOM. **DO NOT** APPLY AFTER FIRST BLOOM. Slight burning and reddish discoloration of cotton foliage may occasionally be seen following recommended treatement; however, cotton plants will develop normally. SPECIAL PRECAUTIONS—**DO NOT** allow spray or spray drift to contact adjacent crops or injury will result. Apply only on still days when weather conditions **DO NOT** favor drift from areas being treated. Follow other precautions listed on this label. Aircraft applications of **BUENO 6** should only be made by applicators experienced in the use of herbicides, and applications should be made in accordance with state and Federal regulations.
Cotton	Application of Cotton	California	CA-770095	**BUENO 6** is useful for the control of weeds listed on the product label and many similar weeds. Application can be made as: 1. Preplant or postplant up to cracking of soil before cotton emergence using ground or aircraft equipment. 2. Postemergent, over-the-top, when cotton is 3 to 6 inches high or up to early first stage, whichever occurs first, using ground or aircraft equipment. 3. Postemergence as a directed spray with ground equipment when cotton is 3 inches high to first bloom. PREPLANT OR POSTPLANT UP TO CRACKING APPLICATION

CROP	REGISTRATIONS	STATES	SLN NO.	DIRECTIONS FOR USE
				ON COTTON: A single ground or aircraft application of **BUENO 6** can be made to prepared cotton seedbeds when planting has been delayed and weeds have emerged, or as a postplant treatment. BUT NO LATER THAN INITIAL CRACKING of soil in field before emergence of cotton. Planting of cotton may immediately follow the pre-plant application. Mix at the rate of $2\frac{2}{3}$ pints of **BUENO 6** in 40 gallons of water for broadcast application, or in 5 to 10 gallons of water for aircraft application, and apply using a properly calibrated sprayer to one acre. See SPECIAL PRECAUTIONS below. POSTEMERGENT, OVER-THE-TOP, APPLICATION ON COTTON: One or two over-the-top applications may be applied by ground or aircraft equipment to actively growing cotton as a salvage operation when crop is 3 to 6 inches high or to early first square, whichever occurs first. **ONE APPLICATION:** Use $2\frac{2}{3}$ pints **BUENO 6** in 40 gallons of water per acre for ground application, or in 5 to 10 gallons of water for aircraft application. **DO NOT REPEAT THIS TREATMENT. TWO APPLICATIONS:** Use $1\frac{1}{3}$ pints **BUENO 6** in 40 gallons of water per acre for ground application, or in 5 to 10 gallons of water for aircraft application. Apply second application, if needed, using the $1\frac{1}{3}$ pint rate, 1 to 3 weeks following the first treatment. Use a properly calibrated sprayer. DO NOT APPLY OVER-THE-TOP SPRAYS AFTER COTTON IS MORE THAN 6 INCHES HIGH OR AFTER EARLY FIRST SQUARE STAGE, WHICHEVER OCCURS FIRST. **POSTEMERGENT, DIRECTED, APPLICATION ON COTTON:** Mix **BUENO 6** at the rate of $2\frac{2}{3}$ pints in 40 gallons of water for application as a DIRECTED broadcast spray to one acre when weeds are small. For band applications, apply 1 gallon of above diluted spray per acre for each 1-inch band width to be treated of cotton grown on 40-inch row spacing. Keep spray off cotton foliage. APPLY ONLY WHEN COTTON IS 3 INCHES HIGH TO FIRST BLOOM. **DO NOT** APPLY AFTER FIRST BLOOM. Slight burning and reddish discoloration of cotton foliage may occasionally be seen following recommended treatment; however, cotton will develop normally. SPECIAL PRECAUTIONS—**DO NOT** allow spray or spray drift to contact adjacent crops or injury will result. Apply only on still days when weather conditions **DO NOT** favor drift from areas being treated. Follow other precautions listed on this label. Aircraft applications of **BUENO 6** should only be made by applicators experienced in use of herbicides, and applications should be made in accordance with state and Federal regulations.

STORAGE AND DISPOSAL

Do not contaminate water, food or feed by storage or disposal. Open dumping is prohibited.

PESTICIDE DISPOSAL: Wastes resulting from the use of this product may be disposed of on site or at an approved waste disposal facility.

CONTAINER DISPOSAL: Do not reuse empty container. Triple rinse (or equivalent) and offer for recycling or reconditioning, or puncture and dispose of in a sanitary landfill, or incineration, or, if allowed by state and local authorities, by burning. If burned, stay out of smoke.

WARRANTY AND LIMITATION OF DAMAGES

Seller warrants that this material conforms to its chemical description and is reasonably fit for the purpose stated on the label when used in accordance with directions under normal conditions of use and Buyer assumes the risk of any use contrary to such directions. SELLER MAKES NO OTHER EXPRESS OR IMPLIED WARRANTY, INCLUDING ANY OTHER EXPRESS OR IMPLIED WARRANTY OF FITNESS OR OF MERCHANTABILITY, AND NO AGENT OF SELLER IS AUTHORIZED TO DO SO EXCEPT IN WRITING WITH A SPECIFIC REFERENCE TO THIS WARRANTY. In no event shall Seller's liability for any breach of warranty exceed the purchase price of the material as to which a claim is made.

DACAMINE® 4D

WEED KILLER

ACTIVE INGREDIENTS:

N-Oleyl-1,3-Propylenediamine Salt of 2,4-Dichlorophenoxyacetic Acid*	31.7%
2,4-Dichlorophenoxyacetic Acid*	21.6%
INERT INGREDIENTS:	46.7%
TOTAL	100.0%

*2,4-Dichlorophenoxyacetic Acid Equivalent: 39.6%, 3.6 lb. gal.

*Isomer specific by AOAC Method

EPA Reg. No. 50534-13

KEEP OUT OF REACH OF CHILDREN
CAUTION

See precautionary statements.

PRECAUTIONARY STATEMENTS

HAZARDS TO HUMANS AND DOMESTIC ANIMALS

CAUTION

Harmful if swallowed or absorbed through the skin. If swallowed, **Do not** induce vomiting. If vomiting occurs spontaneously, keep airway clear. Never give anything by mouth to an unconscious person. Seek medical attention. **Causes eye and skin irritation. Do not** get in eyes. In case of contact, flush eyes with plenty of water for at least 15 minutes and seek medical attention. **Do not** allow contact with skin or clothing. Wash exposed skin gently with plenty of soap and water. Avoid breathing spray mist.

ENVIRONMENTAL HAZARDS

Do not apply directly to water. **Do not** contaminate irrigation ditches or water used for domestic purposes. Use care to avoid spray contact or drift to 2,4-D susceptible plants such as cotton, tomatoes, flowers, grapes, fruit trees and ornamentals. **Do not** permit spray mist containing **DACAMINE 4D** to drift onto them, since even very small quantities of the spray, which may not be visible, can cause severe injury during both growing and dormant periods. **Do not** spray when the wind is blowing towards susceptible crops or ornamental plants. Use coarse sprays to minimize drift. With ground equipment, spray drift can be lessened by keeping the spray boom as low as possible; by applying 20 gallons or more of spray per acre; by using no more than 20 pounds spraying pressure with flat fan or flooding flat fan nozzle tips; by spraying when wind velocity is low; and by stopping all spraying when wind exceeds 8 miles per hour. **Do not** apply with hollow cone-type insecticide or other nozzles that produce a fine droplet spray. With aircraft application, drift can be lessened by applying not less than 5 gallons of spray per acre; by using no more than 20 pounds spray pressure at the nozzles; by using nozzles which produce a coarse spray pattern; and by spraying only when the wind velocity is less than 5 miles per hour. Flush sprayer out on suitable non-crop area after use. **Do not** use the same spray equipment for applying other materials to 2,4-D susceptible crops as injury may result.

DACAMINE 4D

Read entire label before using this product.

GENERAL INFORMATION

DACAMINE 4D is an oil-soluble diamine form of 2,4-D which is formulated to be used only with water. It combines the advantages of both the ester and amine forms of 2,4-D. **DACAMINE 4D** contains both the non-volatile features of amines and the effectiveness of esters. It is designed to control weeds in croplands, pear and apple orchards and lawn areas. The following is a partial list of weeds controlled by **DACAMINE 4D:**

Artemisia (common mugwort)	**Leafy spurge**	
Canada Thistle	**Milkweed**	
Field Bindweed (creeping jenny)	**Russian Knapweed**	
Horse Nettle	**Yarrow**	
Annual chickweed	Henbit	Ragweed (common, giant western)
Beggartick	Ironweed	
Bitterweed	Jimsonweed	Shepherdspurse
Blackeyed Susan	Knotweed	Slender aster
Broomweed	Kochia	Snow-on-the-mountain
Buckthorn plantain	Lambsquarters	Sour dock
Bull thistle	Mallows	Spike rush
Burdock	Marestail	Spurge
Carpetweed	Marshelder	Star thistle
Chicory	Mexicanweed	Sumac
Cocklebur	Morningglory	Sunflower
Coffeebean	Mustards	Tarweed
Croton	Nettle, stinging	Velvetleaf
Curly dock	Pennycress	Wild carrot
Curly indigo	Pennywort	Wild cucumber
Dandelion	Peppergrass	Wild lettuce
Dayflower	Pigweed	Wild parsnip
Fanweed	Plaintain	Wild radish
Fiddleneck	Poison Ivy	Willow
Flower-of-an-hour	Poorjoe	Wintercress
Gooseweed	Puncturevine	Yankeeweed
Hemp		

Note: Local conditions, crop varieties and application regulations vary and may affect use of this herbicide. Consult local agricultural experiment station or extension service weed specialists and state regulatory agencies for recommendations in your area.

Aerial application may be of use for control of weeds on certain crops where there would be no danger of drift to susceptible crops. Applications should only be made by applicators experienced in the use of 2,4-D formulations. Regulations governing aerial application of herbicides are in effect in many states.

Consult local regulatory agencies concerning requirements before making applications.

Treating Small Areas: One tablespoonful of **DACAMINE 4D** in 1½ gallons of water is almost equal to 1 quart in 100 gallons.

Note: When stored at temperatures below freezing, it may be necessary to warm contents to 45°F and mix thoroughly before using.

DIRECTIONS FOR USE

It is a violation of Federal law to use this product in a manner inconsistent with its labeling.

Do not apply this product through any type of irrigation equipment.

Preparation of the Spray: Fill the spray tank with half the required amount of water, then add the recommended amount of **DACAMINE 4D** and continue filling spray tank with balance of water. Keep agitator running when filling spray tank and during spray operations. Use enough spray volume for uniform coverage by ground or air application. For ground application, when weeds are 1 to 3 inches tall, use 15 to 25 gallons of water per acre. When weeds are taller than 3 inches, use 25 to 50 gallons of water per acre. For aerial application, when weeds are 1 to 3 inches tall, use 5 to 10 gallons of water per acre. When weeds are taller than 3 inches, use 10 to 15 gallons of water per acre. In any case, use the same amount of **DACAMINE 4D** recommended per acre. For crop uses, **Do not** mix with oil, surfactant or other adjuvants unless specifically recommended on this label as this may reduce selectivity to crops resulting in crop damage.

Time of Application: Best results are obtained when **DACAMINE 4D** is used on young weeds that are actively growing. Applications of lower listed rates to susceptible annual weeds usually will be satisfactory, but for perennial weeds and other conditions, such as in very dry areas where kill is difficult, use the higher recommended rates. When used as a selective spray on crops, the stage of growth of the crop must be considered. Some woody plants and weeds are hard to kill and repeat applications may be necessary.

Small Grains (Wheat, Barley, Rye): Apply **DACAMINE 4D** in sufficient water for uniform coverage on small grains when fully tillered or stooled (4 to 8 inches tall), but before the "boot" stage. Crop injury may result if applied earlier than "tiller" or from early "boot" through the "milk" stage. **Do not** use on grains undersown with legumes, such as alfalfa or clovers, except where some legume injury can be tolerated. **Do not** graze or feed forage from treated fields within 2 weeks after treatment. See chart for recommended rates. For late season use to control large succulent weeds, apply 1 to 2 pints per acre in sufficient water for uniform coverage on small grain from dough stage to harvest-use only when weeds threaten to interfere with harvest operations. **Do not** use treated straw for livestock feed.

Rice: Apply **DACAMINE 4D** when the rice is in the late tillering stage, but before it has started to joint. This should be when the rice seedlings are about 6 to 10 weeks old. Damage to rice may occur if **DACAMINE 4D** is applied after rice is 12 weeks old. Treatment after flooding is usually safer than treatment before flooding. See chart for recommended rates.

Corn: Postemergence—Apply **DACAMINE 4D** from emergence to tasseling. When spraying corn above 10 inches in height, use nozzle extensions ("corn drops"), directing the spray at base of the corn plant to keep the spray off the leaves as much as possible. **Do not** apply from tassel emergence to dough stage. Crop injury is more likely to occur if corn is growing rapidly under high temperature and high soil moisture conditions. Under such conditions, use the lowest recommended rate. Delay cultivation for 8 to 10 days after application to reduce stalk breakage resulting from temporary brittleness caused by 2,4-D. Hybrids vary in tolerance to 2,4-D. Consult local agricultural experiment station or extension service weed specialists regarding the use of 2,4-D on your specific hybrid. See chart for recommended rates.

Sorghum: Apply **DACAMINE 4D** when sorghum plants are 6 to 12 inches tall and secondary roots are well established. Use drop nozzles when crop is over 10 inches tall. **Do not** apply from flowering to dough stage. See chart for recommended rates.

Sugarcane: Apply as a postemergence application in spring after cane emerges and through layby. See chart for recommended rates.

Amount of Dacamine 4D per Acre

Crop (See Detailed Directions Above)	For Average Conditions	For Dry Conditions as in Western States*
Wheat, Barley, Rye		
Annual Weeds	½ to 1 pint	1 to 2 pints
Perennial Weeds	1 pint	1¼ to 2 pints
Rice	1 to 2 pints	
Corn**		
Postemergence	½ pint	½ to ¾ pint
Sorghum**		
Postemergence	½ pint	½ to ¾ pint
Sugarcane	1 to 2 quarts	

* Arizona, California, Idaho, Montana, Nevada, Oregon, Utah, Washington and Wyoming

** If only rows or bands are treated, leaving middles unsprayed, reduce dosage rate per crop acre proportionate to the ground area actually sprayed.

Forage Crops (Pastures and Rangelands): to control many broadleaved weeds in pastures, meadows and rangelands, use 1 quart per acre of **DACAMINE 4D** in sufficient water to provide for uniform application. Treat pastures when weeds are growing actively. **Do not** apply when grass is in "boot" to "milk" stage if grass seed production is desired. **Do not** apply to newly-seeded pastures until grass is well established or after heading begins. Most legumes are usually injured or killed at the rates recommended. For control of wild garlic and wild onion in pastures, apply 2 to 3 quarts per acre in very early spring and repeat annually until weeds are eradicated. **Do not** graze dairy animals on treated areas within 7 days after application.

Lawns, Golf Courses and Similar Turf: Apply 1½ to 2 pints of **DACAMINE 4D** per acre in 100 gallons of water or sufficient water to provide adequate uniform coverage. For treating small areas, use 1½ tablespoonsful of **DACAMINE 4D** in 2 gallons of water per 1,000 sq. ft. **Do not** apply during periods of drought. **Do not** apply **DACAMINE 4D** to newly-seeded turf until grasses are well established and after making at least two mowings. Bare areas can be reseeded in 3 to 6 weeks. Bentgrass varieties vary in tolerance to **DACAMINE 4D** under different conditions of growth. Use rates no higher than 2 pints per acre on bentgrass and apply only where experience has shown application to be safe. Injury may result if applied to St. Augustinegrass, centipedegrass, carpetgrass, clovers and Dichondra.

Pear and Apple Orchards: Apply at rate of 1 to 2 quarts **DACAMINE 4D** per acre in sufficient water to give good coverage and thorough wetting of weeds. Use the higher rate for control of perennial weeds. Repeat applications may be necessary to control some woody plants and perennial weeds. **Do not** apply to bare ground areas as injury may result. When bare areas do occur, spot treat by directly spraying only at undesirable plants. **Do not** apply immediately before irrigation. Withhold irrigation for three (3) days following application. Best results are obtained when applied within two (2) days following irrigation and when weeds are growing actively. **Do not** allow spray drift to contact foliage, fruit, stems or trunks of fruit trees. Use low pressure and coarse nozzles, and spray on still days to minimize drift.

Woody Plant Control: To control 2,4-D susceptible plants such as alder, buckbrush, elderberry, sumac and willow in waste areas or non-crop land, such as powerlines and highway rights-of-way, vacant lots and fence rows, use 4 to 6 quarts in 100 gallons of water in spring and early summer. For treating small areas use 1½ to 2 ounces of **DACAMINE 4D** in one gallon of water. Wet throughly all parts of the plants, including foliage and stems to point of runoff. **Do not** treat during periods of severe drought or in early fall when leaves have lost their healthy green color. Hard-to-kill species may need retreatment the following season.

General Weed Control: On vacant lots, roadsides, along fence rows and around buildings, use 1 to 2 quarts of **DACAMINE 4D** in 100 gallons of water per acre. Thoroughly wet all foliage to run-off. Application should be made to weeds in an early active stage of growth.

SPECIAL LOCAL NEEDS PURSUANT TO SECTION 24(c) OF AMENDED FIFRA

CROP	REGISTRATIONS	STATES	SLN NO.	DIRECTIONS FOR USE
Rice, Stubble Crop	Control of Certain broadleaf weeds*	Louisiana	LA-830022	For control of certain broadleaf weeds such as red stem, dayflower, ducksalad, northern jointvetch (curly indigo), and hemp sesbania (coffeebean, coffeeweed, tall indigo), and suppression of alligator weed. Apply **DACAMINE 4D**, 1 to 2 pints per acre to vigorously growing weeds within two weeks following harvest of the first crop. To insure adequate spray coverage, apply when weeds have fully emerged through crop residue left from first harvest. **Do not** apply **DACAMINE 4D** after rice has reached the panicle-differentiation stage of growth, or crop injury may result. Weeds that are fully grown at time of first harvest or under drought stress may not be controlled by this application.

STORAGE AND DISPOSAL

Do not contaminate water, food or feed by storage or disposal. Open dumping is prohibited.

PESTICIDE DISPOSAL:

Pesticide wastes are toxic. Improper disposal of excess pesticide, spray mixture or rinsate is a violation of Federal Law. If these wastes cannot be disposed of by use according to label instructions, contact your State Pesticide or Environmental Control Agency or the Hazardous Wastes Representative at the nearest EPA Regional Office for guidance.

CONTAINER DISPOSAL:

Do not reuse empty container. Triple rinse (or equivalent) and offer for recycling or reconditioning or puncture and dispose of in a sanitary landfill, or by other procedures approved by state and local authorities.

WARRANTY AND LIMITATION OF DAMAGES

Seller warrants that this material conforms to its chemical description and is reasonably fit for the purposes stated on the label when used in accordance with directions under normal conditions of use and Buyer assumes the risk of any use contrary to such directions. SELLER MAKES NO OTHER EXPRESS OR IMPLIED WARRANTY, INCLUDING ANY OTHER EXPRESS OR IMPLIED WARRANTY OF FITNESS OR OF MERCHANTABILITY, AND NO AGENT OF SELLER IS AUTHORIZED TO DO SO EXCEPT IN WRITING WITH A SPECIFIC REFERENCE TO THIS WARRANTY. In no event shall Seller's liability for any breach of warranty exceed the purchase price of the material as to which a claim is made.

DACONATE® 6

MSMA Plus Surfactant for Postemergence Weed Control

Product Contains 6.0 Pounds MSMA Per Gallon

ACTIVE INGREDIENT:	
Monosodium Acid Methanearsonate	48.3%
INERT INGREDIENTS:	51.7%
TOTAL	100.0%
Total Arsenic, all in water-soluble form, expressed as elemental	22.3%

EPA Reg. No. 50534-5

KEEP OUT OF REACH OF CHILDREN

CAUTION

See precautionary statements.

PRECAUTIONARY STATEMENTS

HAZARDS TO HUMANS AND DOMESTIC ANIMALS

CAUTION

Harmful if swallowed or absorbed through the skin. Avoid contact with eyes, skin or clothing. Avoid breathing vapors or spray mist.

If swallowed, induce vomiting immediately by giving two glasses of water and sticking finger down throat. **Never** give anything by mouth to an unconscious person. Seek medical attention.

In case of contact, flush eyes with plenty of water for at least 15 minutes. Seek medical attention.

Wash exposed skin gently with plenty of soap and water.

Keep children and domestic animals off treated areas until this material has been washed into the soil.

Wear protective clothing when handling or applying this product including long pants, long-sleeve shirt, and impermeable gloves and boots. Mixer-loaders should include an apron and full-face shield when handling or mixing concentrate. Flagmen should be fully protected during spray operations or mechanical flagmen used. Pilots and ground spray rig applicators should wear a mask or respirator approved by the Mining Enforcement and Safety Administration and the National Institute for Occupational Safety and Health.

ENVIRONMENTAL HAZARDS

Do not apply directly to water. **Do not** feed treated foliage to livestock or graze treated areas. **Do not** spray or allow drift onto vegetables, ornamentals or other desirable plants. **Do not** apply when weather conditions favor drift from target area. **Do not** contaminate water by cleaning of equipment or disposal of wastes.

DACONATE 6

Read entire label before using this product.

GENERAL INFORMATION

DACONATE 6 is a postemergence herbicide for use in cotton, turf and non-crop areas. **DACONATE 6** contains a surfactant (wetting agent) in proper ratio for best results. It is not necessary to add additional surfactant to the spray tank. Local conditions and recommendations vary; consult local agricultural experiment station or extension service weed specialists for recommendations in your area.

Mixing Instructions: DACONATE 6 herbicide must be thoroughly dissolved. Fill the spray tank with half the required amount of water and, with pump or agitator operating, add the recommended amount of **DACONATE 6** herbicide and continue filling the tank with balance of water.

Do not store spray solution in tank overnight. Clean application equipment thoroughly after use by flushing with water in a safe place.

Application Methods: DACONATE 6 should be applied with a low-volume, properly calibrated sprayer having satisfactory pumping and bypass action. **Adjust Nozzles to Keep Spray Off Cotton Foliage and In a Manner to Allow Maximum Coverage of Weeds.** Adequate coverage is very important for effective weed control.

Apply during warm weather when weeds are in an active stage of growth.

DIRECTIONS FOR USE

It is a violation of Federal law to use this product in a manner inconsistent with its labeling.

Do not apply this product through any type of irrigation system.

Cotton: DACONATE 6 is useful as a directed postemergence spray for controlling weeds in cotton such as barnyardgrass, cocklebur, smooth and large crabgrass, dallisgrass, goosegrass, johnsongrass and nutsedge. Apply as a **Directed** broadcast spray at rate of 40 gallons per acre of a solution containing 2⅔ pints of **DACONATE 6** in 40 gallons of water when weeds are small. For band applications, apply 1 gallon of above diluted spray per acre for each 1 inch band width to be treated of cotton grown on 40 inch row spacing. A second or repeat application, if needed, should be timed about 1 to 3 weeks after the first application. Keep spray off cotton foliage.

Apply Only When Cotton is 3 Inches High to First Bloom. Do Not Apply After First Bloom.

Slight burning and reddish discoloration of cotton foliage may occasionally be seen following recommended treatment; however, cotton plants will develop normally and yield will not be affected.

Ornamental Turfgrass: DACONATE 6 can be used for selective control of Bahiagrass, barnyardgrass, chickweed, smooth and large crabgrass, dallisgrass, nutsedge, sandbur and wood sorrel with little or no injury to well-established, actively growing turfgrasses. Mow turfgrass to height of 1 to 1½ inches before treatment. Mix 1 fluid ounce (2 tablespoons) of **DACONATE 6** in 2½ gallons of water for application to an area of 1,000 square feet. **Established bermudagrass and zoysiagrass:** Mix 1 to 2 fluid ounces (2 to 4 tablespoons) of **DACONATE 6** in 2½ gallons of water for application to an area of 1,000 square feet. Application should be uniform and thorough to adequately wet all undesirable plants. Two or more repeat treatments of 14 day intervals may be necessary. Make applications during warm weather when temperature is between 80 and 90°F. **Do not** water turf for at least 24 hours after application. Turfgrass may be temporarily discolored. Bermudagrass, bluegrass and zoysiagrass have shown tolerance to properly applied **DACONATE 6.** Injury may result if applied to bentgrass and fescues. **Do not** apply to St. Augustinegrass, carpetgrass, centipedegrass or to Dichondra. **Do not** reseed until 2 weeks after last application. **Do not** apply with hose-end applicators.

Non-Crop: DACONATE 6 is effective in controlling barnyardgrass, cocklebur, smooth and large crabgrass, dallisgrass, goosegrass, johnsongrass, nutsedge and many similar weeds on drainage ditchbanks, fence rows, rights-of-way, storage yards and similar non-crop areas. Application should be made when weeds are small and conditions are favorable for good weed growth. Mix at a rate of 2½ to 2⅔ pints of **DACONATE 6** in 40 gallons of water. Spray undesirable vegetation thoroughly to point of run-off. Adequate coverage and complete wetting of foliage is important for effective control. Repeat applications may be necessary if regrowth occurs. **Do not** contaminate water used for domestic purposes, or by animals including wildlife and aquatic life, or water used for irrigation purposes.

STORAGE AND DISPOSAL

Do not contaminate water, food or feed by storage or disposal. Open dumping is prohibited.

PESTICIDE DISPOSAL: Wastes resulting from the use of this product may be disposed of on site or at an approved waste disposal facility.

CONTAINER DISPOSAL: Do not reuse empty container. Triple rinse (or equivalent) and offer for recycling or reconditioning, or puncture and dispose of in a sanitary landfill, or incinerate, or, if allowed by state and local authorities, by burning. If burned, stay out of smoke.

WARRANTY AND LIMITATION OF DAMAGES

Seller warrants that this material conforms to its chemical description and is reasonably fit for the purposes stated on the label when used in accordance with directions under normal conditions of use and Buyer assumes the risk of any use contrary to such directions. SELLER MAKES NO OTHER EXPRESS OR IMPLIED WARRANTY, INCLUDING ANY OTHER EXPRESS OR IMPLIED WARRANTY OF FITNESS OR OF MERCHANTABILITY, AND NO AGENT OF SELLER IS AUTHORIZED TO DO SO EXCEPT IN WRITING WITH A SPECIFIC REFERENCE TO THIS WARRANTY. In no event shall Seller's liability for any breach of warranty exceed the purchase price of the material as to which a claim is made.

DACONIL 2787®

FLOWABLE FUNGICIDE

Turf Care®

Turf and Ornamental Fungicide

Contains 4.17 Pounds Chlorathalonil Per Gallon (500 grams per liter)

ACTIVE INGREDIENTS:
Chlorothalonil (tetrachloroisophthalonitrile)* 40.4%
INERT INGREDIENTS: 59.6%
TOTAL 100.00%

*Covered under U.S. Patent No. 3,290,353, No. 3,331,735 and No. 3,948,636

EPA Reg. No. 50534–9

KEEP OUT OF REACH OF CHILDREN

WARNING

See precautionary statements.

DACONIL 2787 FLOWABLE FUNGICIDE

PRECAUTIONARY STATEMENTS

HAZARDS TO HUMANS AND DOMESTIC ANIMALS

WARNING

Causes eye irritation. May be a potential skin sensitizier. **DO NOT** get in eyes. Wear goggles or eye shield when handling this product. In case of contact with eyes, flush with plenty of water immediately for 15 minutes. Seek medical attention for eyes immediately. Avoid contact with skin or clothing. Wash exposed areas of skin with soap and warm water after handling or using. **DO NOT** take internally. Avoid breathing spray mist.

Note to User: This product may produce temporary allergic side effects characterized by redness of the eyes, mild bronchial irritation and redness or rash on exposed skin areas. Persons having allergic reaction should contact a physician.

FIRST AID

Note to Physician: Persons having allergic reaction respond to treatment with antihistamines or steroid creams and/or systemic steroids.

ENVIRONMENTAL HAZARDS

This product is toxic to fish, aquatic invertebrates, and marine/estuarine organisms. Runoff from treated areas may be hazardous to aquatic organisms in neighboring areas. **DO NOT** apply directly to water. **DO NOT** apply when weather conditions favor drift from treated areas.

Apply only to areas specified on label.

DACONIL 2787 FLOWABLE FUNGICIDE

A Broad–Spectrum Fungicide for Use on Golf Course Tees, Greens and Fairways, Ornamental Turfgrass and Ornamentals.

Read entire label and use only as directed.

GENERAL INFORMATION

DACONIL 2787 Flowable Fungicide is formulated for use on golf course tees, greens and fairways, ornamental turfgrass and ornamental herbs, shrubs and trees. It is highly effective for the control of a broad spectrum of turf and ornamental plant diseases when it is used according to the directions on this product label. Thorough, uniform coverage of plant surfaces is essential for good disease control. **Note:** Slowly invert container several times to assure uniform mixture. Add the required amount of **DACONIL 2787** to the spray tank while filling with water. Keep agitator running when filling spray tank and during spray operations. **DO NOT** combine **DACONIL 2787** in the spray tank with pesticides, surfactants or fertilizers unless your prior use has shown the combination physically compatible, effective and noninjurious under your conditions of use.

DIRECTIONS FOR USE

It is a violation of Federal law to use this product in manner inconsistent with its labeling.

Note to User: This product may produce temporary allergic side effects characterized by redness of the eyes, mild bronchial irritation and redness or rash on exposed skin areas. Persons having allergic reaction should contact a physician.

Note to User: Wear long sleeve shirt, long pants, and gloves while mixing, loading and applying this product.

Turf: DO NOT mow or water after treatment until spray deposit on turfgrass is thoroughly dry; **DACONIL 2787** should always be used in conjunction with good turf management practices.

Golf Course Fairways: Apply **DACONIL 2787 Flowable Fungicide** in 30 to 40 gallons of water per acre. Begin applications when conditions favor disease development and repeat applications as long as these conditions persist. Under severe disease conditions use the highest rate and shortest interval corresponding with the application schedule selected from the table below.

Disease	Interval of Application	Application Rate Per Acre
Sclerotinia dollar spot	7-10 days	4 to 8 pts.
	14-21 days	8 to 14 pts.
Helminthosporium leaf spot	7-10 days	8 pts.
	14-21 days	8 to 14 pts.
Rhizoctonia brown patch	7-14 days	8 to 14 pts.
Anthracnose	7-14 days	6 to 12 qts.

Golf Course Tees, Greens and Ornamental Turfgrass: Apply **DACONIL 2787** in an adequate amount of water to provide complete coverage. This amount may vary from 2 to 10 gallons (approximately 8 to 38 liters) per 1,000 square feet. See below for suggested rates and timing. Under severe disease conditions, use the curative rates and spray on a 7 day schedule.

Disease	Interval of Application	RATE—Liquid Ounces* Per 1,000 Sq. Ft.	
		Preventive**	Curative***
Anthracnose	7-14 Days	5-9	—
Copper spot	7-10 Days	6-9	9-11
Curvularia leaf spot	7-10 Days	3-6	6-11
Dollar spot	7-14 Days	3-6	6-11
Gray leaf spot	7-10 Days	3-6	6-11
Helminthosporium leaf spot and melting out	7-10 Days	3-6	6-11
Large brown patch	7-10 Days	3-6	6-11
Red thread	7-10 Days	3-9	9-11
Stem rust of bluegrass	7-14 Days	6-9	9-11
DICHONDRA			
Alternaria leaf spot (California only)	7-14 Days	6-9	9-11

* One liquid ounce is approximately equal to 29 milliliters.

** Recommended rates for preventing disease establishment, use lower rates when disease conditions are light to moderate, higher indicated rates when conditions are severe.

*** Rates for use on a 7 day schedule when disease is present. Higher indicated rate should be applied under severe conditions.

Turfgrasses: Gray Snow Mold caused by *Tuphula* spp.-Apply in sufficient water to obtain adequate coverage (2 to 10 gallons per 1,000 square feet). Apply 8 to 16 fluid ounces of **DACONIL 2787 Flowable Fungicide** per 1,000 square feet of turf area. Application must be made before snow cover in autumn. Use the higher rate if turf layer remains frozen prior to snow cover. If snow cover is intermittent or lacking during the winter, re-apply

DACONIL 2787 at 8 fluid ounces per 1,000 square feet at monthly intervals until gray snow mold conditions no longer prevail. In areas where pink snow mold (Gerlachia or Fusarium patch) is likely to occur, apply **DACONIL 2787** at 8 fluid ounces per 1,000 square feet in combination with either Tersan* 1991 50WP at 2 ounces per 1,000 square feet or Chipco** 26019 50WP at 4 ounces per 1,000 square feet of turf area.

Fusarium (Gerlachia) Patch: For control of Fusarium patch only in areas where snow cover is intermittent or lacking during the winter, apply 8 to 14 fluid ounces of **DACONIL 2787 Flowable Fungicide** per 1,000 sq. ft. of turf area. Begin applications in late autumn and reapply at 21 to 28 day intervals until conditions favorable for Fusarium patch no longer prevail.

*Tersan is a registered trademark of E.I. du Pont de Nemours & Company, Inc.

**Chipco is a registered trademark of Rhone-Poulenc, Inc.

Ornamentals and Conifers: Apply **DACONIL 2787 Flowable Fungicide** at a rate of 2 pints per 100 gallons of water unless other directions are given in the tables below. Begin applications as directed for each species and disease condition cited and repeat on a 7 to 14 day schedule until conditions are no longer favorable for disease development. During periods when conditions favor severe disease incidence, generally cloudy or wet weather, use the higher rate specified and the shortest indicated interval between applications.

DACONIL 2787 may be used in greenhouses. Applicators and attending personnel should wear protective clothing including long sleeves, gloves, goggles or eye shield plus a face-fitting respirator specifically designed to remove organic pesticide vapors and particulates. **DO NOT** use mistblowers or high pressure spray equipment when making applications of **DACONIL 2787** in greenhouses. Use of **DACONIL 2787** is recommended for disease control on the ornamentals and conifers listed in the tables below. Use of **DACONIL 2787** is not recommended for species that are not listed.

ORNAMENTALS:

Species	Diseases Controlled	Suggested First Application
BROADLEAF SHRUBS AND TREES:		
Ash (Fraxinus)	Cercospora, Cercosporidium, Cylindrosporium leaf spots	Spring bud break
Azalea,* Rhododendron*	Phytophthora die-back; Ovulinia flower blight	New leaf emergence; Early bloom
Buckeye, Horsechestnut	Leaf blotch, Anthracnose	Spring bud break
Cherry-Laurel	Cercospora leaf spot	Petal fall
Crabapple	Scab, Cedar-apple rust, Sphaeropsis leaf spot	Spring bud break
Dogwood	Septoria leaf spot	Early bloom
Euonymus	Anthracnose	Spring bud break
Firethorn	Scab	Spring bud break
Flowering Almond, Quince, Sand Cherry	Monilinia blossom/branch blight	Early bloom
Hawthorn	Rust, Fabraea leaf spot	Pre-bloom
Holly	Rhizoctonia web blight	Warm, moist conditions
Mountain Laurel	Cercospora leaf spot	Spring bud break
Oak (red group only)	Taphrina blister, Actinopelte leaf spot, Anthracnose	Dormant budswell
Oregon-Grape (Mahonia)	Rust	Spring bud break
Photinia	Fabraea (Entomosporium) leaf spot	Spring bud break
Pieris (Andromeda)	Phytophthora die-back	New leaf emergence
Poplar	Marssonina leaf spot	Spring bud break
Privet	Cercospora leaf spot	Prolonged wet conditions
Sycamore, Planetree	Anthracnose	Spring bud break
Viburnum	Powdery mildew	Mid-summer
BULBS AND FLOWERING PLANTS:		
Carnation	Alternaria leaf spot/branch rot; Botrytis flower blight	Transplant of cuttings; Cool, moist conditions
Chrysanthemum Daisy	Mycosphaerella ray blight, Septoria leaf spot	Transplant of cuttings;
	Botrytis flower blight (gray mold)	Pre-bloom
Geranium	Botrytis blight, rust	Cool, moist conditions
Gladiolus	Curvularia leaf/flower spot, Botrytis leaf/flower spot	Early propagation
Hollyhock	Rust	Early seedling stage
Hydrangea* (foliage only)	Cercospora and Septoria leaf spots, Rust	Early propagation
Iris	Botrytis blossom blight, Didymellina leaf spot	Cool, moist conditions
Lily	Botrytis gray mold	Pre-bloom
Petunia*	Phytophthora blight (foliar phase), Botrytis blight	Pre-bloom
Rose (use 1½ pt. per 100 gallons)	Black spot, Botrytis blight	Spring bud break
Statice	Anthracnose, Cercospora, Alternaria, Botrytis leaf blights	Spring bud break
Zinnia	Powdery mildew	First sign of disease
FOLIAGE PLANTS:		
Dracaena	Fusarium leaf spot	Pre-transplant
Pachysandra (Use 4 pt. per 100 gallons)	Volutella leaf blight	Spring bud break
Leatherleaf fern	Ascochyta blight, Cersospora leaf spot, Cylindrocladium leaf spot, Rhizoctonia blight	Spring bud break
Parlor palm (Chamaedorea)	Bipolaris (Helminthosporium) leaf spot	Cool, moist conditions
Prayer plant (Maranta)	Helminthosporium leaf spot	Early propagation
Oyster plant (Rhoeo)	Tan leaf spot	Early propagation
Syngonium	Cephalosporium leaf spot	Warm, moist conditions
Philodendron	Phytophthora blight, Dactylaria leaf spot	Moist conditions

* Discoloration of blooms has been noted on certain varieties when applications are made during flowering.

CONIFERS:

Diseases Controlled	DACONIL Flowable Rate/Acre	Application Directions
Scleroderris Canker (pines), Swiss Needlecast and Rhabdocline Needlecast (Douglas-fir)	2 to 4 pts.	Make first application in spring when new shoot growth is ½ to 2 inches in length. Make additional applications at 3 to 4 week intervals until conditions no longer favor disease development. For use in nursery beds, apply the highest rate specified on a 3 week schedule.
Sirococcus Tip Blight	3 to 5 pts.	
Rhizosphaera Needlecast (spruces), Scirrhia brown spot (pines)	8 pts.	
Lophodermium Needlecast	2 to 4 pts.	North Central and Northeastern States: Begin applications in mid-July to early August before infection occurs. Make additional applications at 3 to 4 week intervals until conditions no longer favor disease development. For use in nursery beds, apply 4 pts. per acre on a 3 week schedule.
Botrytis seedling blight, Phoma twig blight	2 to 4 pts.	Begin applications in nursery beds when seedlings are 4 inches tall and when cool, moist conditions favor disease development. Make additional applications at 7 to 14 day intervals as long as disease favorable conditions persist.

APPLICATION AND CALIBRATION TECHNIQUES FOR SPRINKLER IRRIGATION ON SOD FARMS, ORNAMENTAL HERBS, SHRUBS AND TREES:

Apply this product only through center pivot, motorized lateral move, traveling gun, solid set and portable (wheel move, side roll, end tow, or hand move) irrigation system(s). **DO NOT** apply this product through any other type of irrigation system. **DO NOT** use **DACONIL 2787 Flowable Fungicide** through sprinkler irrigation equipment on golf courses.

Crop injury, lack of effectiveness, or illegal pesticide residues in the crop can result from non-uniform distribution of treated water.

If you have questions about calibration, you should contact State Extension Service specialists, equipment manufacturers or other experts.

DO NOT apply this product through irrigation systems connected to a public water system. 'Public water system' means a system for the provision to the public of piped water for human consumption if such system has at least 15 service connections or regularly serves an average of at least 25 individuals daily at least 60 days per year.

Controls for both irrigation water and pesticide injection systems must be functionally interlocked, so as to automatically terminate pesticide injection when the irrigation water pump motor stops. A person knowledgeable of the irrigation system and responsible for its operation shall be present so as to discontinue pesticide injection and make necessary adjustments, should the need arise.

The irrigation water pipeline must be fitted with a functional, automatic, quick-closing check valve to prevent the flow of treated irrigation water back toward the water source. The pipeline must also be fitted with a vacuum relief valve and low pressure drain, located between the irrigation water pump and the check valve, to prevent back-siphoning of treated irrigation water into the water source.

Always inject DACONIL 2787 FLOWABLE FUNGICIDE into irrigation water after it discharges from the irrigation pump and after it passes through the check valve. Never inject pesticides into the intake line on the suction side of the pump.

Pesticide injection equipment must be fitted with a functional, normally closed solenoid-operated valve located on the intake side of the injection pump. Interlock this valve to the power system, so as to prevent fluid from being withdrawn from the chemical supply tank when the irrigating system is either automatically or manually turned off.

The pesticide injection pipeline must contain a functional, automatic, quick-closing check valve to prevent the flow of fluid back toward the injection pump.

The irrigation line or water pump must include a functional pressure switch which will stop the water pump motor when the water pressure decreases to the point where pesticide distribution is adversely affected.

Spray mixture in the chemical supply tank must be agitated at all times, otherwise settling and uneven application may occur. Do not apply when wind speed favors drift beyond the area intended for treatment.

DACONIL 2787 Flowable Fungicide may be used through two basic types of sprinkler irrigation systems as outlined in Sections A and B below. Determine which type of system is in place, then refer to the appropriate directions provided for each type.

A. Center Pivot, Motorized Lateral Move and Traveling Gun Irrigation Equipment

For injection of pesticides, these continuously moving systems must use a positive displacement injection pump, of either diaphragm or piston type, constructed of materials that are compatible with pesticides and capable of being fitted with a system interlock and capable of injection at pressures approximately 2-3 times those encountered within the irrigation water line. Venturi applicator units cannot be used on these systems.

Fill chemical supply tank of injection equipment with water. Operate system for one complete revolution or run across the field, measuring time required, amount of water injected, and acreage covered. Thoroughly mix recommended amount of **DACONIL 2787 Flowable Fungicide** for acreage to be covered into same amount of water used during calibration and inject into system continuously for one revolution or run. Mixture in the chemical supply tank must be continuously agitated during the injection run. Shut off injection equipment after one revolution or run, but continue to operate irrigation system until **DACONIL 2787 Flowable Fungicide** has been cleared from last sprinkler head.

B. Solid Set and Portable (Wheel Move, Side Roll, End Tow, or Hand Move) Irrigation Equipment

With stationary systems, an effectively designed in-line venturi applicator unit is preferred which is constructed of materials that are compatible with pesticides; however, a positive-displacement pump can also be used.

Determine acreage covered by sprinkler. Fill tank of injection equipment with water and adjust flow to use contents over a thirty to forty-five minute period. Mix desired amount of **DACONIL 2787 Flowable Fungicide** for acreage to be covered with water so that the total mixture of **DACONIL 2787 Flowable Fungicide** plus water in the injection tank is equal to the quantity of water used during calibration and operate entire system at normal pressures recommended by the manufacturer of injection equipment used for amount of time established during calibration. No agitation should be required. **DACONIL 2787 Flowable Fungicide** can be injected at the beginning or end of the irrigation cycle or as a separate application. Stop injection equipment after treatment is completed and continue to operate irrigation system until **DACONIL 2787 Flowable Fungicide** has been cleared from last sprinkler head.

STORAGE AND DISPOSAL

DO NOT contaminate water, food or feed by storage or disposal. Open dumping is prohibited.

STORAGE: Store in a cool place. Protect from excessive heat.

PESTICIDE DISPOSAL: Pesticide wastes are toxic. Improper disposal of excess pesticide, pesticide spray or rinsate is a violation of Federal Law. If these wastes cannot be disposed of by use according to label instructions, contact your State Pesticide or Environmental Control Agency, or the Hazardous Waste representative at the nearest EPA Regional Office for guidance.

CONTAINER DISPOSAL: Triple rinse (or equivalent). Then offer for recycling or reconditioning, or puncture and dispose of in a sanitary landfill, or incineration, or, if allowed by state and local authorities, by burning. If burned, stay out of smoke.

WARRANTY AND LIMITATION OF DAMAGES

Seller warrants that this material conforms to its chemical description and is reasonably fit for the purposes stated on the label when used in accordance with directions under normal conditions of use and Buyer assumes the risk of any use contrary to such directions. SELLER MAKES NO OTHER EXPRESS OR IMPLIED WARRANTY, INCLUDING ANY OTHER EXPRESS OR IMPLIED WARRANTY OF FITNESS OR OF MERCHANTABILITY, AND NO AGENT OF SELLER IS AUTHORIZED TO DO SO EXCEPT IN WRITING WITH A SPECIFIC REFERENCE TO THIS WARRANTY. In no event shall Seller's liability for any breach of warranty exceed the purchase price of the material as to which a claim is made.

DACONIL 2787® FUNGICIDE

TURF CARE® (75% Wettable Powder)

ACTIVE INGREDIENT:	
Chlorothalonil (tetrachloroisophthalonitrile)*	75.0%
INERT INGREDIENTS:	25.0%
TOTAL	100.0%

*Covered under U.S. Patent No. 3,290,353 and No. 3,331,735

EPA Reg. No. 50535-4

KEEP OUT OF REACH OF CHILDREN

DANGER

See precautionary statements.

DACONIL 2787 FUNGICIDE

PRECAUTIONARY STATEMENTS

HAZARDS TO HUMANS AND DOMESTIC ANIMALS

DANGER

Corrosive, causes severe eye damage. May be a potential skin sensitizer. **DO NOT** get in eyes. Wear goggles or eye shield when handling this product. In case of contact with eyes, flush with plenty of water immediately for 15 minutes. Seek medical attention for eyes immediately. Avoid contact with skin or clothing. Wash exposed areas of skin with soap and warm water after handling or using. **DO NOT** take internally. Avoid breathing dust or spray mist.

Note to User: This product may produce temporary allergic side effects characterized by redness of the eyes, mild bronchial irritation and redness or rash on exposed skin areas. Persons having allergic reaction should contact a physician.

FIRST AID:

Note to physician: Persons having an allergic reaction respond to treatment with antihistamines or steroid creams and/or systemic steroids.

ENVIRONMENTAL HAZARDS

This product is toxic to fish, aquatic invertebrates, and marine/estuarine organisms. Runoff from treated areas may be hazardous to aquatic organisms in neighboring areas. **DO NOT** apply directly to water. **DO NOT** apply when weather conditions favor drift from treated areas. Apply only to areas specified on label.

DACONIL 2787 FUNGICIDE

A Broad-Spectrum Fungicide for Use on Golf Course Tees, Greens and Fairways; Ornamental Turfgrass and Ornamentals.

Read entire label and use only as directed.

GENERAL INFORMATION

DACONIL 2787 Fungicide is formulated for use on golf course tees, greens and fairways, ornamental turfgrass and ornamental herbs, shrubs and trees. It is highly effective for the control of a broad spectrum of turf and ornamental plant diseases when it is used according to the directions on this product label. Thorough, uniform coverage of plant surfaces is essential for good disease control. Add the required amount of **DACONIL 2787** to the spray tank while filling with water. Keep agitator running when filling spray tank and during spray operations. **DO NOT** combine **DACONIL 2787** in the spray tank with pesticides, surfactants or fertilizers unless your prior use has shown the combination physically compatible, effective and noninjurious under your conditions of use.

DIRECTIONS FOR USE

It is a violation of Federal law to use this product in a manner inconsistent with its labeling.

Note to User: This product may produce temporary allergic side effects characterized by redness of the eyes, mild bronchial irritation and redness or rash on exposed skin areas. Persons having allergic reaction should contact a physician.

Note to User: Wear long sleeve shirt, long pants, and gloves while mixing, loading and applying this product.

Turf: DO NOT mow or water after treatment until spray deposit on turfgrass is thoroughly dry; **DACONIL 2787** should always be used in conjunction with good turf management practices.

Golf Course Fairways: Apply **DACONIL 2787 Fungicide** in 30 to 40 gallons of water per acre. Begin applications when conditions favor disease development and repeat applications as long as these conditions persist. Under severe disease conditions use the highest rate and shortest interval corresponding with the application schedule selected from the table below.

Disease	Interval of Application	Application Rate Per Acre
Sclerotinia dollar spot	7-10 Days	2 3/4 to 5 1/2 lbs.
	14-21 Days	5 1/2 to 10 lbs.
Helminthosporium leaf spot	7-10 Days	5 1/2 lbs.
	14-21 Days	5 1/2 to 10 lbs.
Rhizoctonia brown patch	7-14 Days	5 1/2 to 10 lbs.
Anthracnose	7-14 Days	8 to 16 lbs.

Golf Course Tees, Greens and Ornamental Turfgrass: Apply **DACONIL 2787 Fungicide** in an adequate amount of water to provide complete coverage. This amount may vary from 2 to 10 gallons per 1,000 square feet. See below for suggested rates and timing. Under severe disease conditions, use the curative rates and spray on a 7 day schedule.

Disease	Interval of Application	Rate-Ounces Per 1,000 Sq. Ft. Preventive*	Curative**
Anthracnose	7-14 Days	3-6	-
Copper spot	7-10 Days	4-6	6-8
Curvularia leaf spot	7-10 Days	2-4	4-8
Dollar spot	7-14 Days	2-4	4-8
Gray leaf spot	7-10 Days	2-4	4-8
Helminthosporium leaf spot and melting out	7-10 Days	2-4	4-8
Large brown patch	7-10 Days	2-4	4-8
Red thread	7-10 Days	2-6	6-8
Stem rust of bluegrass	7-14 Days	4-6	6-8
DICHONDRA			
Alternaria leaf spot (California only)	7-14 Days	4-6	6-8

* Recommended rates for preventing disease establishment, use lower rates when disease conditions are light to moderate, higher indicated rates when conditions are severe.

** Rates for use on a 7 day schedule when disease is present. Higher indicated rate should be applied under severe conditions.

Turfgrasses: Gray Snow Mold caused by *Typhula* spp.-Apply in sufficient water to obtain adequate coverage (2 to 10 gallons per 1,000 square feet). Apply 5 1/2 to 11 ounces of **DACONIL 2787 Fungicide** per 1,000 square feet of turf area. Application must be made before snow cover in autumn. Use the higher rate if turf layer remains frozen prior to snow cover. If snow cover is intermittent or lacking during the winter, re-apply **DACONIL 2787** at 5 1/2 ounces per 1,000 square feet at monthly intervals until gray snow mold conditions no longer prevail. In areas where pink snow mold (Gerlachia or Fusarium patch) is likely to occur, apply **DACONIL 2787** at 5 1/2 ounces per 1,000 square feet in combination with either Tersan* 1991 50WP at 2 ounces per 1,000 square feet or Chipco** 26019 50WP at 4 ounces per 1,000 square feet of turf area.

Fusarium (Gerlachia) Patch: For control of Fusarium patch only in areas where snow cover is intermittent or lacking during the winter, apply 5 1/2 to 10 ounces of **DACONIL 2787 Fungicide** per 1,000 sq. ft. of turf area.

Begin applications in late autumn and re-apply at 21 to 28 day intervals until conditions favorable for Fusarium patch no longer prevail.

* Tersan is a registered trademark of E.I. du Pont de Nemours & Company, Inc.

** Chipco is a registered trademark of Rhone-Poulenc Inc.

Ornamentals and Conifers: Apply **DACONIL 2787 Fungicide** at a rate of 1 1/2 pounds per 100 gallons of water unless other directions are given in the tables below. Begin applications as directed for each species and disease condition cited and repeat on a 7 to 14 day schedule until conditions are no longer favorable for disease development. During periods when conditions favor severe disease incidence, generally cloudy or wet weather, use the higher rate specified and the shortest indicated interval between applications.

DACONIL 2787 may be used in greenhouses. Applicators and attending personnel should wear protective clothing including long sleeves, gloves, goggles or eye shield plus a face-fitting respirator specifically designed to remove organic pesticide vapors and particles. **DO NOT** use mistblowers or high pressure spray equipment when making applications of **DACONIL 2787** in greenhouses. Use of **DACONIL 2787** is recommended for disease control on the ornamentals and conifers listed in the tables below. Use of **DACONIL 2787** is not recommended for species that are not listed.

ORNAMENTALS

Species	Diseases Controlled	Suggested First Application
BROADLEAF SHRUBS AND TREES		
Ash (Fraxinus)	Cercospora Cercosporidium, Cylindrosporium leaf spots	Spring bud break
Azalea,* Rhododendron*	Phytophthora die-back; Ovulinia flower blight	New leaf emergence Early bloom
Buckeye, Horsechestnut	Leaf blotch, Anthracnose	Spring bud break
Cherry-Laurel	Cercospora leaf spot	Petal fall
Crabapple	Scab, Cedar-apple rust, Sphaeropsis leaf spot	Spring bud break
Dogwood	Septoria leaf spot	Early bloom
Euonymus	Anthracnose	Spring bud break
Firethorn	Scab	Spring bud break
Flowering Almond, Quince, Sand Cherry	Monilinia blossom/branch blight	Early bloom
Hawthorn	Rust, Fabraea leaf spot	Pre-bloom
Holly	Rhizoctonia web blight	Warm, moist conditions
Mountain Laurel	Cercospora leaf spot	Spring bud break
Oak (red group only)	Taphrina blister, Actinopelte leaf spot, Anthracnose	Dormant budswell
Oregon-Grape (Mahonia)	Rust	Spring bud break
Photinia	Fabraea (Entomosporium) leaf spot	Spring bud break
Pieris (Andromeda)	Phytophthora die-back	New leaf emergence
Poplar	Marssonina leaf spot	Spring bud break
Privet	Cercospora leaf spot	Prolonged wet conditions
Sycamore, Planetree	Anthracnose	Spring bud break
Viburnum	Powdery mildew	Mid-summer
BULBS AND FLOWERING PLANTS:		
Carnation	Alternaria leaf spot/branch rot; Botrytis flower blight	Transplant of cuttings: Cool. moist conditions
Chrysanthemum/Daisy	Mycosphaerella ray blight, Septoria leaf spot;	Transplant of cuttings
	Botrytis flower blight (gray mold)	Pre-bloom
Geranium	Botrytis blight, rust	Cool, moist conditions
Gladiolus	Curvularia leaf/flower spot, Botrytis leaf flower spot	Early propagation
Hollyhock	Rust	Early seedling stage
Hydrangea*(foliage only)	Cercospora and Septoria leaf spots, Rust	Early propagation

Species	Diseases Controlled	Suggested First Application
Iris	Botrytis blossom blight, Didymellina leaf spot	Cool, moist conditions
Lily	Botrytis gray mold	Pre-bloom
Petunia*	Phytophthora blight (foliar phase). Botrytis blight	Pre-bloom
Rose (Use 1 lb. per 100 gallons)	Black spot, Botrytis blight	Spring bud break
Statice	Anthracnose, Cercospora, Alternaria, Botrytis leaf blights	Spring bud break
Zinnia	Powdery mildew	First sign of disease
FOLIAGE PLANTS:		
Dracaena	Fusarium leaf spot	Pre-transplant
Pachysandra (Use 3 lb. per 100 gallons)	Volutella leaf blight	Spring bud break
Leatherleaf fern	Ascochyta blight, Cercospora leaf spot, Cylindrocladium leaf spot, Rhizoctonia blight	Spring bud break
Parlor palm (Chamaedorea)	Bipolaris (Helminthosporium) leaf spot	Cool, moist conditions
Prayer plant (Maranta)	Helminthosporium leaf spot	Early propagation
Oyster plant (Rhoeo)	Tan leaf spot	Early propagation
Syngonium	Cephalosporium leaf spot	Warm, moist conditions
Philodendron	Phytophthora blight, Dactylaria leaf spot	Moist conditions

* Discoloration of blooms has been noted on certain varieties when applications are made during flowering.

CONIFERS

Diseases Controlled	DACONIL 2787 Rate/Acre	Application Direction
Scleroderris Canker (pines), Swiss Needlecast and Rhabdocline Needlecast (Douglas-fir)	1½ to 2 ¾ lbs.	Make first application in spring when new shoot growth is ½ to 2 inches in length. Make additional applications at 3 to 4 week intervals until conditions no longer favor disease development. For use in nursery beds, apply the highest rate specified on a 3 week schedule.
Sirococcus Tip Blight	2 to 3½ lbs.	
Rhizosphaera Needlecast (spruces), Scirrhia brown spot (pines)	5½ lbs.	
Lophodermium Needlecast	1½ to 2¾ lbs.	North Central and Northeastern States: Begin applications in mid-July to early August before infection occurs. Make additional applications at 3 to 4 week intervals until conditions no longer favor disease development. For use in nursery beds, apply 2¾ lbs. per acre on a 3 week schedule.
Botrytis seedling blight, Phoma twig blight	1½ to 2¾ lbs.	Begin application in nursery beds when seedlings are 4 inches tall and when cool, moist conditions favor disease development. Make additional applications at 7 to 14 day intervals as long as disease favorable conditions persist.

APPLICATION AND CALIBRATION TECHNIQUES FOR SPRINKLER IRRIGATION ON SOD FARMS, ORNAMENTAL HERBS, SHRUBS AND TREES:

Apply this product only through center pivot, motorized lateral move, traveling gun, solid set and portable (wheel move, side roll, end tow, or hand move) irrigation system(s). **DO NOT** apply this product through any other type of irrigation system. **DO NOT** use **DACONIL 2787 Fungicide** through sprinkler irrigation equipment on golf courses.

Crop injury, lack of effectiveness, or illegal pesticide residues in the crop can result from non-uniform distribution of treated water.

If you have questions about calibration, you should contact State Extension Service specialists, equipment manufacturers or other experts.

DO NOT apply this product through irrigation systems connected to a public water system. 'Public water system' means a system for the provision to the public of piped water for human consumption if such system has at least 15 service connections or regularly serves an average of at least 25 individuals daily at least 60 days per year.

Controls for both irrigation water and pesticide injection systems must be functionally interlocked, so as to automatically terminate pesticide injection when the irrigation water pump motor stops. A person knowledgeable of the irrigation system and responsible for its operation shall be present so as to discontinue pesticide injection and make necessary adjustments, should the need arise.

The irrigation water pipeline must be fitted with a functional, automatic, quick-closing check valve to prevent the flow of treated irrigation water back toward the water source. The pipeline must also be fitted with a vacuum relief valve and low pressure drain, located between the irrigation water pump and the check valve, to prevent back-siphoning of treated irrigation water into the water source.

Always inject DACONIL 2787 FUNGICIDE into irrigation water after it discharges from the irrigation pump and after it passes through the check valve. Never inject pesticides into the intake line on the suction side of the pump.

Pesticide injection equipment must be fitted with a functional, normally closed solenoid-operated valve located on the intake side of the injection pump. Interlock this valve to the power system, so as to prevent fluid from being withdrawn from the chemical supply tank when the irrigation system is either automatically or manually turned off.

The pesticide injection pipeline must contain a functional, automatic, quick-closing check valve to prevent the flow of fluid back toward the injection pump.

The irrigation line or water pump must include a functional pressure switch which will stop the water pump motor when the water pressure decreases to the point where pesticide distribution is adversely affected.

Spray mixture in the chemical supply tank must be agitated at all times, otherwise settling and uneven application may occur. Do not apply when wind speed favors drift beyond the area intended for treatment.

Posting of areas to be chemigated is required when: 1) any part of a treated area is within 300 feet of sensitive areas such as residential areas, labor camps, businesses, day care centers, hospitals, in-patient clinics, nursing homes or any public areas such as schools, parks, playgrounds, or other public facilities not including public roads; or 2) when the chemigated area is open to the public, such as retail greenhouses.

Posting must conform to the following requirements. Treated areas shall be posted with signs at all usual points of entry and along likely routes of approach from the listed sensitive areas. When there are no usual points of entry, signs must be posted in the corners of the treated areas and in any other location affording maximum visibility to sensitive areas. The printed side of the sign should face away from the treated area towards the sensitive area. The signs shall be printed in English. Signs must be posted prior to application and must remain posted until foliage has dried and soil surface water has disappeared. Signs may remain in place indefinitely as long as they are composed of materials to prevent deterioration and maintain legibility for the duration of the posting period.

All words shall consist of letters at least 2½ inches tall, and all letters and the symbol shall be a color which sharply contrasts with their immediate background. At the top of the sign shall be the words KEEP OUT, followed by an octagonal stop sign symbol at least 8 inches in diameter containing the word STOP. Below the symbol shall be the words PESTICIDES IN IRRIGATION WATER.

DACONIL 2787 Fungicide may be used through two basic types of sprinkler irrigation systems as outlined in Sections A and B below. Determine which type of system is in place, then refer to the appropriate directions provided for each type.

A. Center Pivot, Motorized Lateral Move and Traveling Gun Irrigation Equipment

For injection of pesticides, these continuously moving systems must use a positive displacement injection pump, of either diaphragm or piston type, constructed of materials that are compatible with pesticides and capable of being fitted with a system interlock and capable of injection at pressures approximately 2-3 times those encountered within the irrigation water line. Venturi applicator units cannot be used on these systems.

Fill chemical supply tank of injection equipment with water. Operate system for one complete revolution or run across the field, measuring time required, amount of water injected, and acreage covered. Thoroughly mix recommended amount of **DACONIL 2787 Fungicide** for acreage to be covered into same amount of water used during calibration and inject into system continuously for one revolution or run. Mixture in the chemical supply tank must be continuously agitated during the injection run. Shut off injection equipment after one revolution or run, but continue to operate irrigation system until **DACONIL 2787 Fungicide** has been cleared from last sprinkler head.

B. Solid Set and Portable (Wheel Move, Side Roll, End Tow, or Hand Move) Irrigation Equipment

With stationary systems, an effectively designed in-line venturi applicator unit is preferred which is constructed of materials that are compatible with pesticides; however, a positive-displacement pump can also be used.

Determine acreage covered by sprinkler. Fill tank of injection equipment with water and adjust flow to use contents over a thirty to forty-five minute period. Mix desired amount of **DACONIL 2787 Fungicide** for acreage to be covered with water so that the total mixture of **DACONIL 2787 Fungicide** plus water in the injection tank is equal to the quantity of water used during calibration and operate entire system at normal pressures recommended by the manufacturer of injection equipment used for amount of time established during calibration. Agitation is recommended. **DACONIL 2787 Fungicide** can be injected at the beginning or end of the irrigation cycle or as a separate application. Stop injection equipment after treatment is completed and continue to operate irrigation system until **DACONIL 2787 Fungicide** has been cleared from last sprinkler head.

STORAGE AND DISPOSAL

DO NOT contaminate water, food or feed by storage or disposal. Open dumping is prohibited.

STORAGE: Store in a dry place.

PESTICIDE DISPOSAL: Pesticide wastes are toxic. Improper disposal of excess pesticide, pesticide spray or rinsate is a violation of Federal Law. If these wastes cannot be disposed of by use according to label instructions, contact your State Pesticide or Environmental Control Agency, or the Hazardous Waste representative at the nearest EPA Regional Office for guidance.

CONTAINER DISPOSAL: Completely empty bag into application equipment. Then dispose of empty bag in a sanitary landfill or by incineration, or, if allowed by State and local authorities, by burning. If burned, stay out of smoke.

WARRANTY AND LIMITATION OF DAMAGES

Seller warrants that this material conforms to its chemical description and is reasonably fit for the purposes stated on the label when used in accordance with directions under normal conditions of use and Buyer assumes the risk of any use contrary to such directions. SELLER MAKES NO OTHER EXPRESS OR IMPLIED WARRANTY, INCLUDING ANY OTHER EXPRESS OR IMPLIED WARRANTY OF FITNESS OR OF MERCHANTABILITY, AND NO AGENT OF SELLER IS AUTHORIZED TO DO SO EXCEPT IN WRITING WITH A SPECIFIC REFERENCE TO THIS WARRANTY. In no event shall Seller's liability for any breach of warranty exceed the purchase price of the material as to which a claim is made.

DACONIL 2787® WDG

TURF CARE®
TURF AND ORNAMENTAL FUNGICIDE

(90% Water Dispersible Granule)

ACTIVE INGREDIENT:

Chlorothalonil (tetrachloroisophthalonitrile)*	90.0%
INERT INGREDIENTS	10.0%
TOTAL	100.0%

EPA Est. No. 37429-GA-1
EPA Reg. No. 50534-195

* Covered under U.S. Patent Nos. 3,290,353 and 3,331,735

KEEP OUT OF REACH OF CHILDREN
DANGER

See side panel for additional precautionary statements.

Daconil 2787® WDG

A Broad Spectrum Fungicide for Use on Golf Course Tees, Greens and Fairways, Ornamental Turfgrass, Ornamental Herbs, Shrubs and Trees.

Read entire label carefully and use only as directed.

PRECAUTIONARY STATEMENTS

HAZARDS TO HUMANS AND DOMESTIC ANIMALS

DANGER

Corrosive, causes severe eye damage. Do not get in eyes. Wear goggles or face shield when handling this product.

Causes skin irritation. May be a potential skin sensitizer. Avoid prolonged contact with skin. Wear long sleeve shirt, long pants, and gloves while mixing, loading and applying this product. Wash exposed areas of skin with soap and warm water after handling or using. Launder clothing worn during mixing, loading and spraying, before re-wearing.

Do not take internally.

Avoid breathing spray mist.

Note to User: This product may produce temporary allergic side effects characterized by redness of the eyes, mild bronchial irritation and redness or rash on exposed skin areas. Persons having allergic reaction should contact a physician.

First Aid:

In case of contact with eyes, flush with plenty of water immediately for 15 minutes. Seek medical attention for eyes immediately. Call physician.

Note to Physician: Persons having an allergic reaction respond to treatment with antihistamines or steroid creams and/or systemic steroids.

ENVIRONMENTAL HAZARDS

This product is toxic to fish, aquatic invertebrates, and marine/estuarine organisms. Runoff from treated areas may be hazardous to aquatic organisms in neighboring areas. **Do not** apply directly to water. **Do not** apply when weather conditions favor drift from areas treated. **Do not** contaminate water when disposing of equipment washwaters.

Apply only to areas specified on label.

GENERAL INFORMATION

Daconil 2787 WDG is formulated for use on golf course tees, greens and fairways, ornamental turfgrass and ornamental herbs, shrubs and trees. It is highly effective for the control of a broad spectrum of turf and ornamental plant diseases when it is used according to the directions on this product label. Thorough, uniform coverage of plant surfaces is essential for good disease control.

DIRECTIONS FOR USE

It is a violation of Federal law to use this product in a manner inconsistent with its labeling.

Add the required amount of **Daconil 2787 WDG** to the spray tank while filling with water. Keep agitator running when filling spray tank and during spray operations.

Do not combine **Daconil 2787 WDG** in the spray tank with pesticides, surfactants or fertilizers unless your prior use has shown the combination physically compatible, effective and non-injurious under your conditions of use. **Do not** combine **Daconil 2787 WDG** with Dipel® 4L, Triton® B-1956® or Triton AG-98, as phytotoxicity may result from the combination when applied to some crops on this label.

Dipel is a registered trademark of Abbott Laboratories; Triton and B-1956 are registered trademarks of Rohm and Haas Company.

TURF: Do not mow or water after treatment until spray deposit on turfgrass is thoroughly dry; **Daconil 2787 WDG** should always be used in conjunction with good turf management practices.

Golf Course Fairways: Apply **Daconil 2787 WDG** in 30 to 40 gallons of water per acre. Begin applications when conditions favor disease development and repeat applications as long as these conditions persist. Under severe disease conditions use the highest rate and shortest interval corresponding with the application schedule selected from the table below.

Disease	Application Interval	Application Rate Per Acre
Sclerotinia dollar spot	7-10 Days 14-21 Days	2¼ to 4½ lbs 4½ to 8 lbs
Helminthosporium leafspot	7-10 Days 14-21 Days	4½ lbs 4½ to 8 lbs
Rhizoctonia brown patch	7-14 Days	4½ to 8 lbs
Anthracnose	7-14 Days	7 to 14 lbs

Golf Course Tees and Greens and Ornamental Turfgrass: Apply **Daconil 2787 WDG** in an adequate amount of water to provide complete coverage. This amount may vary from 2 to 10 gallons per 1,000 square feet. See below for suggested rates and timing. Under severe disease conditions, use the curative rates and spray on a 7 day schedule.

Disease	Application Interval	RATE Ounces Per 1,000 sq. ft. Preventive*	Curative**
Anthracnose	7-14 Days	$2\frac{1}{2}$ to 5	–
Copper spot	7-10 Days	$3\frac{1}{2}$ to 5	5 to $6\frac{1}{2}$
Curvularia leafspot	7-10 Days	$1\frac{3}{4}$ to $3\frac{1}{2}$	$3\frac{1}{2}$ to $6\frac{1}{2}$
Dollar spot	7-14 Days	$1\frac{3}{4}$ to $3\frac{1}{2}$	$3\frac{1}{2}$ to $6\frac{1}{2}$
Gray leafspot	7-10 Days	$1\frac{3}{4}$ to $3\frac{1}{2}$	$3\frac{1}{2}$ to $6\frac{1}{2}$
Helminthosporium leafspot and melting out	7-10 Days	$1\frac{3}{4}$ to $3\frac{1}{2}$	$3\frac{1}{2}$ to $6\frac{1}{2}$
Large brown patch	7-10 Days	$1\frac{3}{4}$ to $3\frac{1}{2}$	$3\frac{1}{2}$ to $6\frac{1}{2}$
Red thread	7-10 Days	$1\frac{3}{4}$ to 5	5 to $6\frac{1}{2}$
Stem rust of bluegrass	7-14 Days	$3\frac{1}{2}$ to 5	5 to $6\frac{1}{2}$
DICHONDRA			
Alternaria leafspot (California only)	7-14 Days	$3\frac{1}{2}$ to 5	5 to $6\frac{1}{2}$

* Recommended rates for preventing disease establishment; use lower rate when disease conditions are light to moderate, higher indicated rates when conditions are severe.

** Rates for use on a 7 day schedule when disease is present. Higher indicated rate should be applied under severe conditions.

Turfgrasses: Gray Snow Mold caused by *Typhula* spp.—Apply in sufficient water to obtain adequate coverage (2 to 10 gallons per 1,000 sq. ft.). Apply $4\frac{1}{2}$ to 9 ounces of **Daconil 2787 WDG** per 1,000 sq. ft. of turf area. Application must be made before snow cover in autumn. Use the higher rate if turf layer remains frozen prior to snow cover. If snow cover is intermittent or lacking during the winter, re-apply **Daconil 2787 WDG** at $4\frac{1}{2}$ ounces per 1,000 sq. ft. at monthly intervals until gray snow mold conditions no longer prevail. In areas where pink snow mold (Gerlachia or Fusarium patch) is likely to occur, apply **Daconil 2787 WDG** at $4\frac{1}{2}$ ounces per 1,000 sq. ft. in combination with either Tersan* 1991 50WP at 2 ounces per 1,000 sq. ft. or Chipco** 26019 50WP at 4 ounces per 1,000 sq. ft. of turf area.

*Tersan is a registered trademark of E.I. Dupont de Nemours & Company, Inc.
**Chipco is a registered trademark of Rhone-Poulenc Inc.

Fusarium (Gerlachia) **Patch:** For control of Fusarium patch only in areas where snow cover is intermittent or lacking during the winter, apply $4\frac{1}{2}$ to 8 ounces of **Daconil 2787 WDG** per 1,000 sq. ft. of turf area. Begin applications in late autumn and re-apply at 21 to 28 day intervals until conditions favorable for Fusarium patch no longer prevail.

ORNAMENTALS AND CONIFERS: Apply **Daconil 2787 WDG** at rate of $1\frac{1}{4}$ pounds per 100 gallons of water unless other directions are given in the tables below. Begin applications as directed for each species and disease condition cited and repeat on a 7 to 14 day schedule until conditions are no longer favorable for disease development. During periods when conditions favor severe disease incidence, generally cloudy or wet weather, use the higher rate specified and the shortest indicated interval between applications.

Daconil 2787 WDG may be used in greenhouses. Applicators and attending personnel should wear protective clothing including long sleeves, gloves, goggles or face shield plus a face-fitting respirator specifically designed to remove organic pesticide vapors and particulates. **Do not** use mistblowers or high pressure spray equipment when making applications of **Daconil 2787 WDG** in greenhouses.

Use of **Daconil 2787 WDG** is recommended for disease control on the ornamentals and conifers listed in the tables below. Use of **Daconil 2787 WDG** is not recommended for species that are not listed.

ORNAMENTALS:

Species	Diseases Controlled	Suggested First Application
BROADLEAF SHRUBS AND TREES:		
Ash (Fraxinus)	Cercospora, Cercosporidium, Cylindrosporium leafspots	Spring bud break
Azalea,* Rhododendron*	Phytophthora die-back, Ovulinia flower blight	New leaf emergence; Early bloom
Buckeye, Horsechestnut	Leaf blotch, Anthracnose	Spring bud break
Cherry–Laurel	Cercospora leafspot	Petal fall
Crabapple	Scab, Cedar-apple rust, Sphaeropsis leafspot	Spring bud break
Dogwood	Septoria leafspot	Early bloom
Euonymus	Anthracnose	Spring bud break
Firethorn	Scab	Spring bud break
Flowering Almond, Quince, Sand Cherry	Monilinia blossom/branch blight	Early bloom
Hawthorn	Rust, Fabraea leafspot	Pre-bloom
Holly	Rhizoctonia web blight	Warm, moist conditions
Mountain Laurel	Cercospora leafspot	Spring bud break
Oak (red group only)	Taphrina blister, Actinopelte leafspot, Anthracnose	Dormant budswell
Oregon–Grape (Mahonia)	Rust	Spring bud break
Photinia	Fabraea (Entomosporium) leafspot	Spring bud break
Pieris (Andromeda)	Phytophthora die-back	New leaf emergence
Poplar	Marssonina leafspot	Spring bud break
Privet	Cercospora leafspot	Prolonged wet conditions
Sycamore, Planetree	Anthracnose	Spring bud break
Viburnum	Powdery mildew	Mid-summer
BULBS AND FLOWERING PLANTS:		
Carnation	Alternaria leafspot/branch rot	Transplant of cuttings
	Botrytis flower blight	Cool, moist conditions
Chrysanthemum/Daisy	Mycosphaerella ray blight, Septoria leafspot	Transplant of cuttings
	Botrytis flower blight (gray mold)	Pre-bloom
Geranium	Botrytis blight, rust	Cool, moist conditions
Gladiolus	Curvularia leaf/flower spot, Botrytis leaf/flower spot	Early propagation
Hollyhock	Rust	Early seedling stage
Hydrangea* (foliage only)	Cercospora and Septoria leafspots, Rust	Early propagation
Iris	Botrytis blossom blight, Didymellina leafspot, Ink spot	Cool, moist conditions
Lily, Crocus, Daffodil, Narcissus, Tulip	Botrytis blight (gray mold, fire, measles) Stagonospora leaf scorch	Pre-bloom
Petunia*	Phytophthora blight (foliar phase), Botrytis blight	Pre-bloom
Rose (Use $\frac{7}{8}$ lb. per 100 gallons)	Black spot, Botrytis blight	Spring bud break
Statice	Anthracnose, Cercospora, Alternaria, Botrytis leaf blights	Spring bud break
Zinnia	Powdery mildew	First sign of disease

*Discoloration of blooms has been noted on certain varieties when applications are made during flowering.

FOLIAGE PLANTS:		
Dracaena	Fusarium leafspot	Pre-transplant
Pachysandra (Use $2\frac{1}{4}$ lb per 100 gallons)	Volutella leaf blight	Spring bud break
Leatherleaf fern	Ascochyta blight, Cercospora leafspot, Cylindrocladium leafspot, Rhizoctonia blight	Spring bud break
Parlor palm (Chamaedorea)	Bipolaris (Helminthosporium) leafspot	Cool, moist conditions
Prayer plant (Maranta)	Helminthosporium leafspot	Early propagation
Oyster plant (Rhoeo)	Tan leafspot	Early propagation
Syngonium	Cephalosporium leafspot	Warm, moist conditions
Philodendron	Phytophthora blight Dactylaria leafspot	Moist conditions

Conifers:

Diseases Controlled	Daconil 2787 WDG Rate/Acre	Application Directions
Rhabdocline needlecast (Douglas-fir)	$1\frac{1}{8}$ to $2\frac{1}{4}$ lbs	Apply at budbreak and repeat at 3 to 4 week intervals until needles are fully elongated and conditions no longer favor disease development. In plantations of mixed provenance, or when irregular budbreak occurs, apply weekly until all trees have broken bud, then every 3 to 4 weeks as specified

Diseases Controlled	Daconil 2787 WDG Rate/Acre	Application Directions
		above. In nursery beds, use the high rate on a 3 week schedule.
Scleroderris Canker (pines), Swiss Needlecast (Douglas-fir)	1 1/8 to 2 1/4 lbs.	Make first application in spring when new shoot growth is 1/2 to 2 inches in length. Make additional applications at 3 to 4 week intervals until conditions no longer favor disease development. In nursery beds, apply the highest rate specified on a 3 week schedule.
Sirococcus Tip Blight	1 3/4 to 3 lbs.	
Rhizosphaera Needlecast (spruces), Scirrhia brown spot (pines)	4 1/2 lbs.	
Cyclaneusma and Lophodermium needlecasts (pines)	2 1/4 to 4 1/2 lbs.	Apply in early spring prior to budbreak. Repeat applications at approximately 6 to 8 week intervals until spore release ceases in late fall. Apply monthly during periods of frequent rainfall, and where Lophodermium infections occur during dormancy (Pacific Northwest). During drought periods, applications may be suspended, then resumed upon next occurrence of needle wetness.
Botrytis seedling blight, Phoma twig blight	1 1/8 to 2 1/4 lbs	Begin applications in nursery beds when seedlings are 4 inches tall and when cool, moist conditions favor disease development. Make additional applications at 7 to 14 day intervals as long as disease favorable conditions persist.

APPLICATION AND CALIBRATION TECHNIQUES FOR SPRINKLER IRRIGATION ON ORNAMENTAL HERBS, SHRUBS AND TREES:

Apply **Daconil 2787 WDG** only through center pivot, motorized lateral move, solid set or portable (side roll, end tow or hand move) irrigation system(s). **Do not** apply this product through any other type of irrigation system. **Do not** use **Daconil 2787 WDG** through sprinkler irrigation equipment on golf courses.

Crop injury or lack of effectiveness can result from non-uniform distribution of treated water.

If you have questions about calibration, you should contact State Extension Service specialists, equipment manufacturers or other experts.

Do not apply this product through irrigation systems connected to a public water system. 'Public water system' means a system for the provision to the public of piped water for human consumption if such system has at least 15 service connections or regularly serves an average of at least 25 individuals daily at least 60 days per year.

Controls for both irrigation water and pesticide injection systems must be functionally interlocked, so as to automatically terminate pesticide injection when the irrigation water pump motor stops. A person knowledgeable of the irrigation system and responsible for its operation shall be present so as to discontinue pesticide injection and make necessary adjustments, should the need arise.

The irrigation water pipeline must be fitted with a functional, automatic, quick-closing check valve to prevent the flow of irrigation water back toward the water source. The pipeline must also be fitted with a vacuum relief valve and low pressure drain, located between the irrigation water pump and the check valve, to prevent back-siphoning of irrigation water into the water source.

Always inject Daconil 2787 WDG into irrigation water after it discharges from the irrigation pump and after it passes through the check valve. Never inject pesticides into the intake line on the suction side of the pump.

Pesticide injection equipment must be fitted with a functional, normally closed, solenoid-operated valve located on the intake side of the injection pump. Interlock this valve to the power system, so as to prevent fluid from being withdrawn from the chemical supply tank when the irrigation system is either automatically or manually turned off.

The pesticide injection pipeline must contain a functional, automatic, quick-closing check valve to prevent the flow of fluid back toward the injection pump. The irrigation line or water pump must include a functional pressure switch which will stop the water pump motor when the water pressure decreases to the point where pesticide distribution is adversely affected.

Spray mixture in the chemical supply tank must be agitated at all times, otherwise settling and uneven application may occur. Do not apply when wind speed favors drift beyond the area intended for treatment.

Posting of areas to be chemigated is required when: 1) any part of a treated area is within 300 feet of sensitive areas such as residential areas, labor camps, businesses, day care centers, hospitals, in-patient clinics, nursing homes or any public areas such as schools, parks, playgrounds, or other public facilities not including public roads; or 2) when the chemigated area is open to the public.

Posting must conform to the following requirements. Treated areas shall be posted with signs at all usual points of entry and along likely routes of approach from the listed sensitive areas. When there are no usual points of entry, signs must be posted in the corners of the treated areas and in any other location affording maximum visibility to sensitive areas. The printed side of the sign should face away from the treated area towards the sensitive area. The signs shall be printed in English. Signs must be posted prior to application and must remain posted until foliage has dried and soil surface water has disappeared. Signs may remain in place indefinitely as long as they are composed of materials to prevent deterioration and maintain legibility for the duration of the posting period.

All words shall consist of letters at least 2 1/2 inches tall, and all letters and the symbol shall be a color which sharply contrasts with their immediate background. At the top of the sign shall be the words KEEP OUT, followed by an octagonal stop sign symbol at least 8 inches in diameter containing the word STOP. Below the symbol shall be the words PESTICIDES IN IRRIGATION WATER

Daconil 2787 WDG may be used through two basic types of sprinkler irrigation systems as outlined in Sections A and B below. Determine which type of system is in place, then refer to the appropriate directions provided for each type.

A. Center Pivot, Motorized Lateral Move and Traveling Gun Irrigation Equipment

For injection of pesticides, these continuously moving systems must use a positive displacement injection pump, of either diaphragm or piston type, constructed of materials that are compatible with pesticides and capable of being fitted with a system interlock and capable of injection at pressures approximately 2-3 times those encountered within the irrigation water line. Venturi applicator units cannot be used on these systems.

Fill chemical supply tank of injection equipment with water. Operate system for one complete revolution or run across the field, measuring time required, amount of water injected, and acreage covered. Thoroughly mix recommended amount of **Daconil 2787 WDG** for acreage to be covered into same amount of water used during calibration and inject into system continuously for one revolution or run. Mixture in the chemical supply tank must be continuously agitated during the injection run. Shut off injection equipment after one revolution or run, but continue to operate irrigation system until **Daconil 2787 WDG** has been cleared from last sprinkler head.

B. Solid Set and Portable (Wheel Move, Side Roll, End Tow, or Hand Move) Irrigation Equipment

With stationary systems, an effectively designed in-line venturi applicator unit is preferred which is constructed of materials that are compatible with pesticides; however, a positive-displacement pump can also be used.

Determine acreage covered by sprinkler. Fill tank of injection equipment with water and adjust flow to use contents over a thirty to forty-five minute period. Mix desired amount of **Daconil 2787 WDG** for acreage to be covered with water so that the total mixture of **Daconil 2787 WDG** plus water in the injection tank is equal to the quantity of water used during calibration and operate entire system at normal pressures recommended by the manufacturer of injection equipment used for amount of time established during calibration. Agitation is recommended. **Daconil 2787 WDG** can be injected at the beginning or end of the irrigation cycle or as a separate application. Stop injection equipment after treatment is completed and continue to operate irrigation system until **Daconil 2787 WDG** has been cleared from last sprinkler head.

STORAGE AND DISPOSAL

Do not contaminate water, food or feed by storage or disposal. Open dumping is prohibited.

Storage: Store in a dry place.

Pesticide Disposal: Pesticide wastes are toxic. Improper disposal of excess pesticide, pesticide spray or rinsate is a violation of Federal Law. If these wastes cannot be disposed of by use according to label instructions, contact your State Pesticide or Environmental Control Agency, or the Hazardous Waste representative at the nearest EPA Regional Office for guidance.

Container Disposal: Triple rinse (or equivalent). Then offer for recycling or reconditioning, or puncture and dispose of in a sanitary landfill, or incineration, or, if allowed by state and local authorities, by burning. If burned, stay out of smoke.

WARRANTY AND LIMITATION OF DAMAGES

Seller warrants that this material conforms to its chemical description and is reasonably fit for the purposes stated on the label when used in accordance with directions under normal conditions of use and Buyer assumes the risk of any use contrary to such directions. **Seller makes no other express or implied warranty, including any other express or implied warranty of Fitness or of Merchantability, and no agent of Seller is authorized to do so except in writing with a specific reference to this warranty.** In no event shall Seller's liability for any breach of warranty exceed the purchase price of the material as to which a claim is made.

DACTHAL® W-75

HERBICIDE

(75% Wettable Powder)

ACTIVE INGREDIENT:	
Dimethyl tetrachloroterephthalate	75.0%
INERT INGREDIENTS:	25.0%
TOTAL	100.0%

EPA Reg. No. 50534-1

KEEP OUT OF REACH OF CHILDREN

CAUTION

See precautionary statements.

PRECAUTIONARY STATEMENTS

HAZARDS TO HUMANS AND DOMESTIC ANIMALS

CAUTION

Causes moderate eye irritation. Avoid contact with eyes or clothing. Wash thoroughly with soap and water after handling.

STATEMENT OF PRACTICAL TREATMENT

If in Eyes: Flush with plenty of water. Call a physician if irritation persists.

If Swallowed: Do not induce vomiting. If vomiting occurs spontaneously, keep airway clear. Never give anything by mouth to an unconscious person. Seek medical attention.

If on Skin: Wash thoroughly with soap and water.

ENVIRONMENTAL HAZARDS

Do not apply directly to water or wetlands (swamps, bogs, marshes, and potholes). **Do not** contaminate water when disposing of equipment wastewater.

DACTHAL W-75 HERBICIDE

A selective preemergence herbicide for control of annual grasses and certain broadleaved weeds in vegetable crops and ornamental turf. **Read entire label carefully and use only as directed.**

GENERAL INFORMATION

Dacthal W-75 is a 75 percent wettable powder formulation for use in preemergence application for control of crabgrass and other annual grasses and certain broadleaved weeds on mineral soils in vegetables, strawberries, agronomic crops, ornamentals and turf.

DIRECTIONS FOR USE

It is a violation of Federal law to use this product in a manner inconsistent with its labeling.

DO NOT APPLY IN SUFFOLK COUNTY, LONG ISLAND, NEW YORK.

Do Not apply when wind speed favors drift beyond the area intended for treatment.

Protective Clothing Required: Wear a long-sleeved shirt, long-legged pants, chemical resistant gloves, shoes and socks when handling (mixing, loading or applying) pesticides or when cleaning and repairing equipment used during mixing, loading or application of pesticides.

In case of contact with skin, wash thoroughly with soap and water. Always wash hands, face and arms with soap and water after using pesticides before smoking, eating, drinking or toileting. Shower after work and do not reuse clothing worn during handling (mixing, loading or applying) pesticides or when cleaning and repairing equipment used during mixing, loading or application of pesticides until clothing has been thoroughly laundered. Clothing worn during work must be laundered separately from household articles before reuse.

Reentry: For agricultural crops and commercial vegetable production, **Do not** enter treated areas for 24 hours after application unless wearing long sleeved shirt and long pants.

Preparation and Application of the Spray: Dacthal W-75 should be thoroughly mixed in the spray tank with water to provide for broadcast application of at least 20 gallons of water per acre, depending on equipment and local practices. In order to insure uniform suspension in the spray tank, **Dacthal 75-W** should be pre-mixed in a clean container at the rate of about 12 pounds to 3 gallons of water. This mixture should be stirred until a slurry is formed and then added to the spray tank during or after filling. Adequate agitation should be maintained during filling and spraying to keep material in uniform suspension. Use nozzles suitable for applying wettable powders so that the spray can be uniformly distributed over the properly prepared soil surface. Any screen or strainer in spray equipment should be no finer than 50 mesh to prevent clogging. It is essential that the sprayer be properly calibrated before making applications. Standard low pressure (30–50 psi) boom-type sprayers are recommended.

RECOMMENDATIONS-Any Application of Dacthal Should Be Made Prior to Weed Seed Germination

CROPS	WHEN TO APPLY	RATE	REMARKS
Broccoli, brussels sprouts, cauliflower, cabbage, Chinese broccoli, Chinese cabbage (tight-headed varieties only)	At seeding or transplanting	*See Rates of Application and Weeds Controlled	Apply uniformly to the soil as a spray. **DACTHAL** can be sprayed directly over transplants without injury. If weeds have emerged, soil should be clean cultivated or weeded prior to application. Can be preplant incorporated.
Field beans, snap beans, mungbeans, Southern peas (blackeyed)	At seeding	*See Rates of Application and Weeds Controlled	Apply uniformly to the soil as a spray at time of planting. **DO NOT** graze on treated areas or feed plant refuse, vines, or threshings to livestock. Can be preplant incorporated.
Seeded melons (cantaloupe, honeydew, watermelons), cucumbers, squash (summer and winter)	When plants have 4-5 true leaves (See Remarks)	*See Rates of Application and Weeds Controlled	Apply only when plants have 4-5 true leaves, are well established, and growing conditions are favorable for good plant growth. (If applied earlier than recommended and/or growing conditions are unfavorable, crop injury may result.) If weeds have emerged, crop should be cultivated and weeded prior to the **DACTHAL** application. Incorporation not recommended.
Collards, kale, mustard greens, turnips (greens and roots)	At seeding	*See Rates of Application and Weeds Controlled	Apply uniformly to the soil as a spray at time of seeding. Can be preplant incorporated.
Garlic	At seeding or transplanting and/or at layby	*See Rates of Application and Weeds Controlled	Apply uniformly to the soil as a spray. **DACTHAL** can be sprayed directly over transplants without injury. If weeds emerge prior to layby (when garlic plants have 3 to 5 true leaves) the crop should be cultivated or weeded before application. Preplant incorporation is not recommended. **DO NOT** graze on treated areas or feed plant refuse to livestock.
Horseradish	At planting	*See Rates of Application and Weeds Controlled	Apply uniformly to the soil as a spray at time of seeding. Preplant incorporation not recommended.

RECOMMENDATIONS-Any Application of Dacthal Should Be Made Prior to Weed Seed Germination

CROPS	WHEN TO APPLY	RATE	REMARKS
Onions	At seeding or transplanting and/or at layby	*See preemergence rate and weeds controlled. In sandy loam soils, however, a maximum preemergence rate of 10 pounds W-75 per acre is recommended.	Apply uniformly to the soil as a spray. **DACTHAL** can be sprayed directly over transplants without injury. A layby application can be made on onions either alone or in addition to a **DACTHAL** preemergence treatment up to 14 weeks after planting at rates up to 14 lb./A on any soil type. If weeds emerge prior to layby, the onions should be cultivated or weeded prior to application. Preplant incorporation is not recommended
Potatoes (whole or cut pieces)	At planting, drag-off or layby	*See Rates of Application and Weeds Controlled	Apply uniformly to the soil as a spray at planting or following planting. If the tops of the beds are to be dragged off, **DACTHAL** should be applied after drag-off. Layby applications can be made up to 9 weeks after planting. If weeds have emerged before application, the crop should be weeded or cultivated prior to application. The initial sprinkler irrigation following **DACTHAL** application should not exceed 1 inch. In the Pacific Northwest under dry conditions, **DACTHAL** should not be applied until just prior to the initial sprinkler irrigation. Under furrow irrigation in the Pacific Northwest, shallow incorporation (no greater than 1/2 inch deep) while the sprouts are still well below the soil surface, may be desirable. Preplant incorporation not recommended. **DACTHAL** not recommended for use on potatoes in California.
Radish	At seeding or up to three-leaf stage	6 to 14 lbs.W-75 per acre	Apply uniformly to the soil as a spray in 20 to 30 gallons of water per acre. If weeds have emerged, soil should be clean cultivated or weeded prior to application to permit uniform coverage of soil. Preplant incorporation not recommended. **DO NOT** graze on treated areas or feed plant refuse to livestock. **DO NOT** harvest within 25 days of application. Based upon available residue data, the use of **DACTHAL** for weed control in radishes is limited to the State of California.
Sweet potatoes and yams	At transplanting and layby	*See Rates of Application and Weeds Controlled	Apply uniformly to the soil as a spray at transplanting. **DACTHAL** can be sprayed directly over transplants without injury. Layby application can be made up to 6 weeks after transplanting. If weeds are present, the crop should be weeded or cultivated prior to **DACTHAL** application. Incorporation of layby applications not recommended.
Tomatoes, eggplant and peppers	4-6 weeks after transplanting or on direct seeded plants at 4-6 inches in height. (See Remarks)	*See Rates of Application and Weeds Controlled	Application uniformly to the soil should be confined to a period of 4-6 weeks after transplanting. Plants should be well established and growing conditions favorable for good plant growth. **DACTHAL** can be sprayed directly over the transplants without injury. If weeds have emerged, the crop should be cultivated or weeded prior to making the **DACTHAL** application. **DO NOT** apply to seeded plants until plants are 4-6 inches in height, well established and growing conditions are favorable for good plant growth.
Strawberries: New Plantings	At transplanting	12 pounds W-75 per acre	Apply uniformly to the soil as a spray. For new beds apply 12 pounds at transplanting. Can be preplant incorporated.
Strawberries: Established Plantings	Fall and Early Spring	8-12 pounds W-75 per acre	Application of 8-12 pounds per acre may be applied to control late summer or early fall germinating weeds. Application to established plantings should be made in fall and early spring. **DO NOT** apply after first bloom through harvest. These applications may be made directly over the plants without injury.
Cotton	At seeding and layby	*See Rates of Application and Weeds Controlled	Apply uniformly to the soil at the time of planting. Higher rates of 10-14 pounds are recommended for areas under irrigation. Layby application at 6-10 pounds can be made, but not after peak bloom or within 45 days of harvest. If weeds are present, the cotton should be weeded or cultivated prior to making layby application. Can be preplant incorporated.

Note: Rates can be reduced by banding. See chart on banding.

Layby used in this table refers to the time just prior to the stage when crop plant size would prevent further cultivation.

Aircraft Application: It is recommended that **DACTHAL W-75** be applied in a minimum of 10 gallons of spray per acre. Aircraft should be equipped with pumps with sufficient capacity to deliver at least 10 gallons per acre. Placement of nozzles should provide a uniform spray pattern. **DACTHAL W-75** should be pre-mixed in a nurse tank or mixing tank equipped with mechanical or by-pass agitation. Mixture can then be pumped into the aircraft. Screens on mixing tank should be no finer than 50 mesh. Adequate agitation should be maintained in the mixing tank as well as the aircraft throughout the spraying operation. Regulations governing aerial application of herbicides are in effect in many states. Consult local regulatory agencies concerning requirements before making applications. Avoid spray drift to adjacent fields: particular care should be exercised when adjacent fields contain edible crops or those that will be fed or grazed by livestock. Spray only on still days.

A minimum of one-third to one-half inch of water, either as rain or irrigation, is necessary to activate **DACTHAL W-75** and thereby obtain preemergence control of weeds specified. Normally, water must be applied within 3 to 5 days of application if sufficient rainfall does not occur.

However, when surface soil is dry but subsurface soil contains enough moisture to germinate weeds, water should be applied immediately. If furrow irrigation is used, the top of the bed must be thoroughly wet. If rainfall during this period does not occur and irrigation cannot be made, shallow incorporation, when recommended, will aid in weed control.

Soil Preparation: Prior to making application, the soil should be cultivated to provide a weed-free and uniform surface since **DACTHAL** herbicide is effective only when applied before weed seed germination occurs. Unless incorporated, **DACTHAL** should be applied within two to three days following cultivation, to prevent subsurface germinating weeds from becoming established. When cultivating between rows, use shields on cultivator to prevent untreated soil from covering treated bands. Cultural practices, such as forming crop beds, should be completed prior to **DACTHAL W-75** application in order to avoid exposing untreated soil in the bottom of the furrow and concentrating the sprayed soil within the bed.

Soil Incorporation: (**DO NOT** incorporate deeper than 2 inches.) Shallow soil incorporation, when recommended (see specific crop recommendation), often gives more consistent results. Depending on crop and time of application, several incorporation methods may be used.

Preplant Incorporation: This type of incorporation often provides better weed control on irrigated land, and also on non-irrigated land when dry weather is anticipated following planting. When recommended, **DACTHAL** can be sprayed on the soil surface and then incorporated into the soil using a disc harrow or related tillage equipment. To insure complete incorporation, double disc in two directions. For incorporating banded treatments, hooded power driven rotary tillers or ground tillers are essential.

Delayed Preemergent Incorporation: For crops where preplant incorporation can be used (see recommendation). Incorporation at this time should be used when adequate rainfall or irrigation is not available within 3 to 5 days after planting. Lightly incorporate (1/2 to 1 inch deep) with drag harrow, rotary hoe or other suitable tillage equipment.

Layby Incorporation: (Just prior to stage when plant growth would prevent further cultivation.) Cultivate soil to remove established weeds. Apply **DACTHAL** over top of foliage and lightly incorporate with suitable equipment.

Rates of Application and Weeds Controlled: DACTHAL W-75 is recommended at rates of 6-14 lbs./A depending on soil type and weeds to be controlled. The following table outlines general recommendation for the use of **DACTHAL W-75** on various soil types. The maximum rate for each soil type should be used when moderately susceptible weeds are present. Specific recommendations for each crop should be consulted. **DACTHAL W-75** should not be applied to certain crops under conditions of cool, wet weather when plants are not actively growing. Refer to specific label recommendations for seeded melons, tomatoes, peppers, onions and eggplants for further details.

Soil Type	Organic Matter %	Broadcast Rate of DACTHAL W-75/A
Light-sand or sandy loam	0 - 3	6- 8 pounds
Medium-silt loam	0 - 3	8- 10 pounds
Heavy-clay loam	0 - 3	10- 14 pounds
Heavy-silt loam	3 - 5	12- 14 pounds

The following weeds are grouped according to their susceptibility to **DACTHAL W-75.**

Susceptible (controlled at recommended rates): Common lambsquarters, smooth crabgrass, large crabgrass, lovegrass, carpetweed, witchgrass, purslane, Florida pusley, yellow foxtail, green foxtail, common chickweed.

Moderately Susceptible (maximum rates required for effective control): Redroot pigweed, sandbur, dodder, barnyardgrass, goosegrass, browntop panicum, groundcherry, annual bluegrass, johnsongrass (from seed), nodding spurge, spotted spurge and prostrate spurge.

Tolerant: Velvetleaf, common ragweed, johnsongrass established from rhizomes, mustards, jimsonweed, galinsoga, smartweed, nutsedge.

Replanting: Replanting crops other than those included on this label in **DACTHAL** treated soil within 8 months of application may result in crop injury. If replanting is required because of an early crop failure, the planting of onions, seeded cucurbits, potatoes, tomatoes, eggplants or peppers at this time may result in crop injury. However, all crops on this label may be planted following harvest of a **DACTHAL** treated crop.

Ornamental Turf: Application at the rate of 14 lbs. of **DACTHAL W-75** in 40 to 100 gallons of water per acre (1 lb./3 to 7 gals./3000 sq. ft.) gives excellent control of crabgrass and other weeds as listed on this label. One application applied in early spring before weed seed germination usually provides seasonal control. Crabgrass seed germination usually coincides with the time when Forsythia blossoms start to fall. In areas which commonly experience stands of late-germinating crabgrass and other annual weedy grasses, a second application at half the regular rate can be safely made two months after the first application. In order to control annual bluegrass *(Poa annua)* and other erratic or late-germinating grasses and weeds in northern states, a late summer or early fall application before weed seed germination of 20 pounds of **DACTHAL W-75** in 40 to 100 gallons of water per acre (1/2 lb./1 to 2 gals./1000 sq. ft.) should supplement the early spring application. Earlier applications are required in southern United States. Consult agricultural experiment station or extension service weed specialist for date *Poa annua* seed germinates in your area and make application at least two weeks prior to this date.

Early spring application of **DACTHAL** may be made to new turfgrass seedlings after the grasses have exhibited a uniform greening of the newly sprouted grass, preferably when 1 to 2 inches in height. Such **DACTHAL** applications give control of crabgrass without injury to newly emerged grass. Fall sowing of permanent grasses can follow early spring application of **DACTHAL.**

Where some spring seeding is necessary, there should be a delay of approximately 60 days after the application of most preemergence herbicides, including **DACTHAL.** This delay increases the survival of desirable grasses.

DACTHAL is not recommended for use on putting-greens. **DACTHAL** is not recommended for use on bentgrasses mowed at putting-green height. For control of creeping speedwell *(Veronica filiformis)*, apply 16 pounds of **DACTHAL W-75** in 40 to 100 gallons of water per acre as a postemergence spray when creeping speedwell is growing vigorously. Spring or early fall applications are usually suitable, provided temperatures are between 65° and 90°F, and there is ample soil moisture. Thorough coverage is essential. Delayed control of creeping speedwell is to be expected and will be manifested first by a gradual loss of color and vigor.

Nursery Stock: DACTHAL W-75 can be applied to a wide range of nursery stock at the rate of 14 to 16 pounds in 50 to 100 gallons of water per acre of area treated (1 pound in 3-7 gallons of water/3000 square feet).

Applications should be made to soil recently cultivated to a uniform texture. These can be made immediately following lining-out of stock. With established plantings, application should be made following proper cultivation to remove existing weeds. Where possible, this should be done early in the spring. Late summer application may prove beneficial for control of fall germinating weeds, if made following cultivation. Weed control up to three months or more may be expected following proper application.

FOR USE ON TURF:	TO CONTROL	WHEN TO APPLY	RATE OF DACTHAL W-75	REMARKS
Established turf	Crabgrass	Early Spring, before crabgrass and other Spring annual weed seeds germinate	14 lbs./Acre or 1 lb./3 to 7 gals./3000 sq. ft.	Apply when Forsythia blossoms. A second application at half the recommended rate may be made 2 months after the first application if necessary.
	Annual bluegrass (Poa annua)	Late summer or early fall in Northern states before weed seed germination. Consult agricultural experiment station or extension weed specialist for date **Poa annua** germinates in your area and make application at least two weeks prior to this date.	20 lbs./Acre or 1/2 lb./1 to 2 gals./1000 sq. ft.	Should be used following an early spring application for crabgrass control.
	Creeping Speedwell (Veronica filiformis)	As a postemergence spray in spring or early fall when creeping speedwell is growing vigorously, there is ample soil moisture and temperature is between 70° and 90°F.	16 lbs./Acre or 1/3 lb./1 to 2 gals./1000 sq. ft.	Apply spray to obtain thorough coverage of foliage. Delayed control of creeping speedwell, approximately 30 days, is to be expected and will be noticed first by a gradual loss of color and vigor.
	Spotted Spurge or Prostrate Spurge (Euphorbia maculata, E. supina)	Mid-April or before seed germination of annual weeds. Repeat application in 6 to 8 weeks.	14 to 16 lbs./Acre or 5 to 6 ounces/1 to 2 gals./1000 sq. ft.	Apply first application before germination of annual weed seeds, such as crabgrass. Repeat application at the same rate 6 to 8 weeks later. In addition to control of spotted spurge or prostrate spurge, crabgrass control is also obtained.
Newly seeded turf	Crabgrass	After new turfgrass seedlings have exhibited a greening	14 lbs./Acre or 1 lb./3 to 7 gals./3000 sq. ft.	Allows crabgrass control without injury to the new turf.

FOR USE ON TURF:	TO CONTROL	WHEN TO APPLY	RATE OF DACTHAL W-75	REMARKS
		of the newly-sprouted grass (about 1 to 2 inches high).		

Caution: After **DACTHAL** treatment, wait approximately 60 days before any new seeding is undertaken.

DO NOT use on Dichondra

See recommendations on bentgrass before using on this species.

DACTHAL W-75 is specifically recommended for weed control with the following:

Abelia
Ageratum
Alyssum
Andromeda, Jap.
Arborvitae
Ash
Aster
Azalea
Babysbreath
Barberry
Bellflower
Birch
Bleedingheart
Bloodleaf
Boxwood
Bugloss
Camellia
Candytufts
Chestnut
Chrysanthemum
Cinquefoil
Coleus
Columbine
Coneflower, purple
Coreopsis
Coralbells
Cosmos
Cotoneaster
Cottonwood
Crabapple
Cuphea
Cypress
Dahlia
Delphinium
Deutzia
Dogwood
Elaeagnus
Elm
Euonymus
Evening primrose
Feverfew
Fir
Forget-me-not
Forsythia
Four-o'Clock
Foxglove
Gaillardia
Geranium
Gladiolus
Golddust
Gum
Hawthorn
Heath, pink
Holly
Honeysuckle
Hydrangea
Iris
Ivy, Baltic
Ivy, Boston
Ivy, English
Ivy, Wilson
Juniper
Lantana
Larkspur, candle
Lavendercotton
Lilac
Lily
Locust
Lupine
Magnolia
Maple
Marguerite, golden
Marigold
Mockorange
Morningglory
Moss, rose
Mother-of-Thyme
Mountain Laurel
Mourning Bride
Nasturtium
Oak
Orpine
Pachistima
Pachysandra
Peony
Petunia
Pine
Pittosporum
Podocarpus
Poker plant
Poplar
Privet
Privet, variegated
Redbud
Rhododendron
Rose
Russian olive
Sage, scarlet
Snapdragon
Spiderwort, Va.
Spirea
Spruce
Stonecrop
Strawflower
Sundrops
Sunflower
Sweet pea
Sycamore
Tree peony
Tulip tree
Viburnum
Violet, African
Walnut
Weigela
Willow
Wormwood
Yarrow, fernleaf
Yew
Zinnia

Do not use on the following plants:

Bugleweed
Button pink
Carnation
Geum
Germander
Mesembryanthemum
Pansy
Phlox
Sweet William
Telanthera

Amount of DACTHAL W-75 To Use in Treating Small Areas

Area to be treated*	Recommended rate of DACTHAL W-75 lb./A	Amount of DACTHAL W-75 to Apply	Struck-off (level) Tablespoons
1000 sq. ft. (From 1 to 2 gallons of water should be used to spray 1000 sq. ft.)	14	5 oz.	22
	12(heavy soil)	4½ oz.	19
	10	3½ oz.	16
	8(medium soil)	3 oz.	13
	6(sandy soil)	2 oz.	10
250 sq. ft. (From ¼ to ½ gallon of water should be used to spray 250 sq. ft.)	14	1¼ oz.	5½
	12(heavy soil)	1 oz.	4½
	10	¾-1 oz.	4
	8(medium soil)	¾ oz.	3
	6(sandy soil)	½ oz.	2½

Note: In treating small areas it is important to use care in measuring out required quantity for area to be treated, properly mixing it in water, and keeping it adequately agitated during application.

* Other size areas to be treated can be calculated from the above table.

For example:

To treat 5000 sq. ft. at the 10 lbs./acre rate, use 80 tbsp. (5 × 16) in 5 to 10 gallons of water (5 × 1 to 2)

To treat 600 sq. ft. at the 12 lbs./acre rate, use 11.4 tbsp. (0.6 × 19) in 0.6 to 1.2 gallons of water (0.6 × 1 to 2).

Rates of application can be reduced by banding.

See conversion chart.

Conversion Chart for DACTHAL W-75

For a 12-inch band of spray, use this amount of **DACTHAL W-75** per acre.

	If overall rate is:			
Row Width	6 lbs. per acre	8 lbs. per acre	10 lbs. per acre	12 lbs. per acre
24 inches	3 pounds	4 pounds	5 pounds	6 pounds
32 inches	2½ pounds	3 pounds	3¾ pounds	4½ pounds
36 inches	2 pounds	2⅔ pounds	3⅓ pounds	4 pounds
40 inches	1¾ pounds	2½ pounds	3 pounds	3⅔ pounds

Special Precautions: DO NOT feed treated foliage to livestock or graze treated areas. Applied according to directions and under conditions favorable to good plant growth, **DACTHAL W-75** will not harm crops for which its use is recommended. However, conditions such as high salt concentration, seedling disease, cold weather, deep planting, excessive moisture or drought may injure or weaken crops normally tolerant to **DACTHAL W-75**, thereby increasing the possibility of herbicide damage. Under any of these conditions, one or more of the following may result: delayed crop development, reduced yields, or reduced quality.

SPECIAL LOCAL NEEDS PURSUANT TO SECTION 24(c) OF AMENDED FIFRA

CROP	REGISTRATIONS	STATES	SLN OR REG. NO.	DIRECTIONS FOR USE
Radish	Application on radish grown for seed	Washington	WA-820004	Apply uniformly to the soil as a spray. Incorporate **DACTHAL** thoroughly into the soil to a depth of 3 to 4 inches using a rotary tiller or double disk. **DO NOT** allow seed screenings to be used for feed. **DO NOT** graze on treated areas or feed plant refuse to livestock.
Upland Cress	Control of crabgrass and other annual grasses & certain broadleaved weeds in upland cress	North Carolina Tennessee	NC-810030 TN-870009	Apply uniformly to the soil as a spray at time of seeding. Can be preplant incorporated.

APPLICATION AND CALIBRATION TECHNIQUES FOR SPRINKLER IRRIGATION

Apply **Dacthal W-75** only through center pivot, motorized lateral move, solid set or portable (side roll, end tow or hand move) irrigation system(s). **Do not** apply this product through any other type of irrigation system.

Crop injury or lack of effectiveness can result from non-uniform distribution of treated water.

If you have questions about calibration, you should contact State Extension Service specialists, equipment manufacturers or other experts.

Do not apply this product through irrigation systems connected to a public water system. 'Public water system' means a system for the provision to the public of piped water for human consumption if such system has at least 15 service connections or regularly serves an average of at least 25 individuals daily at least 60 days per year.

A person knowledgeable of the chemigation system and responsible for its operation, or under the supervision of the responsible person, shall shut the system down and make necessary adjustments should the need arise.

The system must contain a functional check valve, vacuum relief valve, and low pressure drain appropriately located on the irrigation pipeline to prevent water source contamination from backflow.

The pesticide injection pipeline must contain a functional, automatic quick-closing check valve to prevent the flow of fluid back toward the injection pump.

The pesticide injection pipeline must also contain a functional, normally closed, solenoid-operated valve located on the intake side of the injection pump and connected to the system interlock to prevent fluid from being withdrawn from the supply tank when the irrigation system is either automatically or manually shut down.

Always inject **Dacthal W-75** into irrigation water after it discharges from the irrigation pump and after it passes through the check valve. Never inject pesticides into the intake line on the suction side of the pump.

The system must contain functional interlocking controls to automatically shut off the pesticide injection pump when the water pump motor stops.

The irrigation line or water pump must include a functional pressure switch which will stop the water pump motor when the water pressure decreases to the point where pesticide distribution is adversely affected.

Systems must use a metering pump, such as a positive displacement injection pump (e.g., diaphragm pump) effectively designed and constructed of materials that are compatible with pesticides and capable of being fitted with a system interlock.

Do not apply when wind speed favors drift beyond the area intended for treatment.

Spray mixture in the chemical supply tank must be agitated at all times, otherwise settling and uneven application may occur.

Dacthal W-75 may be used through two basic types of sprinkler irrigation systems as outlined in Sections A and B below. Determine which type of system is in place, then refer to the appropriate directions provided for each type. The tank containing **Dacthal W-75** must be equipped with effective mechanical agitation to keep the wettable powder in suspension. Premixing **Dacthal W-75** as instructed under ground application will insure that a uniform suspension is obtained. The injection equipment which contains the **Dacthal W-75** suspension should be connected to the discharge side of the irrigation pump or other pressurized equipment attached to the irrigation line.

A. Center Pivot and Motorized Lateral Move Irrigation Equipment

For injection of pesticides, these continuously moving systems must use a positive displacement injection pump, of either diaphragm or piston type, constructed of materials that are compatible with pesticides and capable of injection at pressures approximately 2-3 times those encountered within the irrigation water line. Venturi applicator units cannot be used on these systems.

Fill chemical supply tank of injection equipment with water. Operate system for one complete revolution or run across the field, measuring time required, amount of water injected, and acreage covered. Thoroughly mix recommended amount of **Dacthal W-75** for acreage to be covered into same amount of water used during calibration and inject into system continuously for one revolution or run. Mixture in the chemical supply tank must be continuously agitated during the injection run. Shut off injection equipment after one revolution or run, but continue to operate irrigation system until **Dacthal W-75** has been cleared from last sprinkler head.

B. Solid Set and Portable (Side Roll, End Tow, or Hand Move) Irrigation Equipment

Determine acreage covered by sprinkler. Fill tank of injection equipment with water and adjust flow to use contents over a thirty to forty-five minute period. Mix desired amount of **Dacthal W-75** for acreage to be covered into quantity of water used during calibration and operate entire system at normal pressures recommended by the manufacturer of injection equipment used for amount of time established during calibration. **Dacthal W-75** can be injected at the beginning or end of the irrigation cycle or as a separate application. Stop injection equipment after treatment is completed and continue to operate irrigation system until **Dacthal W-75** has been cleared from last sprinkler head.

STORAGE AND DISPOSAL

DO NOT contaminate water, food or feed by storage or disposal. Open dumping is prohibited.

STORAGE: Store in a dry place.

PESTICIDE DISPOSAL:
Wastes resulting from the use of this product may be disposed of at an approved waste disposal facility. **Do not** contaminate water when disposing of equipment washwaters.

CONTAINER DISPOSAL:
Completely empty bag into application equipment and dispose of empty bag in a sanitary landfill or by incineration, or, if allowed by state and local authorities, by burning. If burned, stay out of smoke.

WARRANTY AND LIMITATION OF DAMAGES

Seller warrants that this material conforms to its chemical description and is reasonably fit for the purposes stated on the label when used in accordance with directions under normal conditions of use and Buyer assumes the risk of any use contrary to such directions. SELLER MAKES NO OTHER EXPRESS OR IMPLIED WARRANTY, INCLUDING ANY OTHER EXPRESS OR IMPLIED WARRANTY OF FITNESS OR OF MERCHANTIBILITY, AND NO AGENT OF SELLER IS AUTHORIZED TO DO SO EXCEPT IN WRITING WITH A SPECIFIC REFERENCE TO THIS WARRANTY. In no event shall Seller's liability for any breach of warranty exceed the purchase price of the material as to which a claim is made.

DACTHAL® W-75 TURF

TURF CARE®

Turf and Ornamental Herbicide (75% Wettable Powder)

ACTIVE INGREDIENT:

Dimethyl tetrachloroterephthalate	75.0%
INERT INGREDIENTS:	25.0%
TOTAL	100.0%

EPA Reg. No. 50534-20

KEEP OUT OF REACH OF CHILDREN
CAUTION

See precautionary statements.

DACTHAL W-75 Turf

PRECAUTIONARY STATEMENTS

HAZARDS TO HUMANS AND DOMESTIC ANIMALS
CAUTION

Causes moderate eye irritation. Avoid contact with eyes or clothing. Wash thoroughly wih soap and water after handling.

STATEMENT OF PRACTICAL TREATMENT

If in Eyes: Flush with plenty of water. Call a physician if irritation persists.

If Swallowed: Do not induce vomiting. If vomiting occurs spontaneously, keep airway clear. Never give anything by mouth to an unconscious person. Seek medical attention.

If on Skin: Wash thoroughly with soap and water.

ENVIRONMENTAL HAZARDS

Do not apply directly to water or wetlands (swamps, bogs, marshes, and potholes). **Do not** contaminate water when disposing of equipment wastewater.

DACTHAL W-75 Turf

A Selective Preemergence Herbicide for Nurseries and Professional Ornamental Turf Management

Read entire label carefully and use only as directed.

GENERAL INFORMATION

DACTHAL W-75 Turf is a 75 percent wettable powder formulation for use in preemergence application for control of crabgrass and other annual grasses and certain broadleaved weeds on mineral soils in ornamental turf and nursery stock.

The following weeds are grouped according to their susceptibility to **DACTHAL W-75 Turf.**

Susceptible (controlled at recommended rates): Common lambsquarters, smooth crabgrass, large crabgrass, lovegrass, carpetweed, witchgrass, purslane, Florida pusley, yellow foxtail, green foxtail, common chickweed.

Moderately Susceptible (Maximum rates required for effective control): Redroot pigweed, sandbur, dodder, barnyardgrass, goosegrass, browntop panicum, groundcherry, annual bluegrass, johnsongrass (from seed), nodding spurge, spotted spurge and prostrate spurge.

Tolerant: Velvetleaf, common ragweed, johnsongrass established from rhizomes, mustards, jimsonweed, galinsoga, smartweed, nutsedge.

DACTHAL W-75 Turf may be used as a postemergence for the control of creeping speedwell.

DIRECTIONS FOR USE

It is a violation of Federal law to use this product in a manner inconsistent with its labeling.

DO NOT APPLY IN SUFFOLK COUNTY, LONG ISLAND, NEW YORK.

Do Not apply when wind speed favors drift beyond the area intended for treatment.

Protective Clothing Required: Wear a long-sleeved shirt, long-legged pants, chemical resistant gloves, shoes and socks when handling (mixing, loading or applying) pesticides or when cleaning and repairing equipment used during mixing, loading or application of pesticides.

In case of contact with skin, wash thoroughly with soap and water. Always wash hands, face and arms with soap and water after using pesticides before smoking, eating, drinking or toileting. Shower after work and do not reuse clothing worn during handling (mixing, loading or applying) pesticides or when cleaning and repairing equipment used during mixing, loading or application of pesticides until clothing has been thoroughly laundered. Clothing worn during work must be laundered separately from household articles before reuse.

Preparation and Application of the Spray: DACTHAL W-75 Turf should be thoroughly mixed in the spray tank with water to provide for broadcast application of at least 20 gallons of water per acre, depending on equipment and local practices. In order to insure uniform suspension in the spray tank, **DACTHAL W-75 Turf** should be premixed in a clean container at the rate of about 12 pounds to 3 gallons of water. This mixture should be stirred until a slurry is formed and then added to the spray tank during or after filling. Adequate agitation should be maintained during filling and spraying to keep material in uniform suspension. Use nozzles suitable for applying wettable powder so that the spray can be uniformly distributed over turf or properly prepared soil surface. Any screen or strainer in spray equipment should be no finer than 50 mesh to prevent clogging. It is essential that the sprayer be properly calibrated before making applications. Standard low pressure (30-50 psi) boom type sprayers are recommended.

A minimum of one-fourth inch of water, either as rain or irrigation, is necessary to activate **DACTHAL W-75 Turf,** and thereby obtain preemergence control of weeds specified. Normally, water must be applied within 3 to 5 days of application if sufficient rainfall does not occur. However, when surface soil is dry but subsurface soil contains enough moisture to germinate weeds, water must be applied immediately.

Ornamental Turf: Application at the rate of 14 lbs. of **DACTHAL W-75 Turf** in 40 to 100 gallons of water per acre (1 lb./3 to 7 gals./3000 sq. ft.) gives excellent control of crabgrass and other weeds as listed on this label. One application applied in early spring before weed seed germination usually provides seasonal control. Crabgrass seed germination usually coincides with the time when Forsythia blossoms start to fall. In areas which commonly experience stands of late-germinating crabgrass and other annual weedy grasses, a second application at half the regular rate can be safely made two months after the first application. In order to control annual bluegrass *(Poa annua)* and other erratic or late-germinating grasses and weeds in Northern states, a late summer or early fall application before weed seed germination of 20 pounds of **DACTHAL W-75 Turf** in 40 to 100 gallons of water per acre (½ lb./1 to 2 gals./1000 sq. ft.) should supplement the early spring application. Earlier applications are required in Southern States. Consult agricultural experiment station or extension service weed specialist for date *Poa annua* seed germinates in your area and make application at least two weeks prior to date.

Early spring application of **DACTHAL W-75 Turf** may be made to new turfgrass seedlings after the grasses have exhibited a uniform greening of the newly sprouted grass, preferably when 1 to 2 inches in height. Such **DACTHAL W-75 Turf** application, gives control of crabgrass without injury to newly emerged grass. Fall sowing of permanent grasses can follow early spring applications of **DACTHAL W-75 Turf.**

Where some spring seeding is necessary, there should be a delay of approximately 60 days after the application of most preemergence herbicides, including **DACTHAL W-75 Turf.** This delay increases the survival of desirable grasses.

DACTHAL W-75 Turf is not recommended for use on putting greens.

DACTHAL W-75 Turf is not recommended for use on bentgrasses mowed at putting green height. **DO NOT** feed treated foliage to livestock or graze treated areas.

For control of creeping seedwell *(Veronica filliformis),* apply 16 pounds of **DACTHAL W-75 Turf** in 40 to 100 gallons of water per acre as a post-emergence spray when creeping speedwell is growing vigorously. Spring or early fall applications are usually suitable, provided temperatures are between 65° and 90°F. and there is ample soil moisture. Thorough coverage is essential. Delayed control of creeping speedwell is to be expected. Foliage turns darker green following application and remains so for about 30 days when leaves curl and the plants die.

FOR USE ON ORNAMENTAL TURF:	TO CONTROL	WHEN TO APPLY	RATE OF DACTHAL W-75 TURF	REMARKS
Established turf	Crabgrass	Early spring, before crabgrass and other spring annual weed seeds germinate	14 lbs./Acre or 1 lb./3 to 7 gals./3000 sq. ft.	Apply when Forsythia blossoms. A second application at half the recommended rate may be made 2 months after the first application, if necessary.
	Annual bluegrass (Poa annua)	Late summer or early fall in northern states before weed seed germination. Consult agricultural experiment station or extension service weed specialist for date *Poa annua* germinates in your area and make application at least 2 weeks prior to this date.	20 lbs./Acre or ½ lb./1 to 2 gals./1000 sq. ft.	Should be used following an early spring application for crabgrass control.
	Creeping speedwell (Veronica filiformis)	As a postemergence spray in spring or early fall when creeping speedwell is growing vigorously, there is ample soil moisture and temperature is between 65° and 90°F.	16 lbs./Acre or ⅓ lb./1 to 2 gals./1000 sq. ft.	Apply spray to obtain thorough coverage of foliage. Delayed control of creeping speedwell, approximately 30 days, is to be expected. Foliage turns darker green following application and remains so for about 30 days when leaves curl and the plants die.
	Spotted Spurge or Prostrate Spurge (Euphorbia maculata, E. supina)	Mid-April or before seed germination of annual weeds. Repeat application in 6 to 8 weeks.	14 to 16 lbs./Acre or 5 to 6 ozs. in 1 to 2 gals./1000 sq. ft.	Apply first application before germination of annual weed seeds, such as crabgrass. Repeat application at the same rate 6 to 8 weeks later. In addition to control of spotted spurge or prostrate spurge, crabgrass control is also obtained.
Newly-seeded turf	Crabgrass	After new turfgrass seedlings have exhibited a greening of the newly sprouted grass (about 1 to 2 inches high).	14 lbs./Acre or 1 lb./3 to 7 gals./3000 sq. ft.	Allow crabgrass control without injury to the new turf.

Caution: After **DACTHAL W-75** treatment, wait approximately 60 days before any new seeding is undertaken. **DO NOT** use on Dichondra. See recommendations on bentgrass before using on this species.

Nursery Stock: DACTHAL W-75 Turf can be applied to a wide range of nursery stock at the rate of 14 to 16 pounds in 50 to 100 gallons of water per acre of area treated (1 pound in 3-7 gallons of water/3000 square feet).

Applications should be made to soil recently cultivated to a uniform texture. These can be made immediately following lining-out of stock. With established plantings, application should be made following proper cultivation to remove existing weeds. Where possible, this should be done early in the spring. Late summer application may prove beneficial for control of fall germinating weeds, if made, following cultivation. Weed control up to three months or more may be expected following proper application.

DACTHAL W-75 Turf is specifically recommended for weed control with the following:

Abelia	Cinquefoil	Feverfew	Ivy, English	Nasturtium	Spiderwort, Va.
Ageratum	Coleus	Fir	Ivy, Wilson	Oak	Spirea
Alyssum	Columbine	Forget-me-not	Juniper	Orpine	Spruce
Andromeda, Jap.	Coneflower, purple	Forsythia	Lantana	Pachistima	Stonecrop
Arborvitae	Coreopsis	Four-O-Clock	Larkspur, candle	Pachysandra	Strawflower
Ash	Coralbells	Foxglove	Lavendercotton	Peony	Sundrops
Aster	Cosmos	Gaillardia	Lilac	Petunia	Sunflower
Azalea	Cotoneaster	Geranium	Lily	Pine	Sweet pea
Babysbreath	Cottonwood	Gladiolus	Locust	Pittosporum	Sycamore
Barberry	Crabapple	Golddust	Lupine	Podocarpus	Tree peony
Bellflower	Cuphea	Goldentuff	Magnolia	Poker plant	Tulip tree
Birch	Cypress	Gum	Maple	Poplar	Viburnum
Bleedingheart	Dahlia	Hawthorn	Marguerite, golden	Privet	Violet, African
Bloodleaf	Delphinium	Heath, pink	Marigold	Privet, variegated	Walnut
Boxweed	Deutzia	Holly	Mockorange	Redbud	Weigela
Bugloss	Dogwood	Honeysuckle	Morningglory	Rhododendron	Willow
Camellia	Elaeagnus	Hydrangea	Moss, rose	Rose	Wormwood
Candytuft	Elm	Iris	Mother-of-Thyme	Russian olive	Yarrow, fernleaf
Chestnut	Euonymus	Ivy, Baltic	Mountain Laurel	Sage, scarlet	Yew
Chrysanthemum	Evening primrose	Ivy, Boston	Mourning Bride	Snapdragon	Zinnia

DO NOT use on the following plants:

Bugleweed	Geum	Pansy	Telanthera
Button pink	Germander	Phlox	
Carnation	Mesembryanthemum	Sweet William	

Amount of DACTHAL W-75 Turf to Use in Treating Small Areas

Area to be treated*	Recommended rate of DACTHAL W-75 Turf lb./A	Amount of DACTHAL W-75 Turf to Apply	Struck-off (level) Tablespoons
1000 sq. ft. (From 1 to 2 gallons of water should be used to spray 1000 sq. ft.)	14 (heavy soil)	5 oz.	22
	12 (heavy soil)	4½ oz.	19
	10 (medium soil)	3½ oz.	16
	8 (medium soil)	3 oz.	13
	6 (sandy soil)	2 oz.	10
250 sq. ft. (From ¼ to ½ gallon of water should be used to spray 250 sq. ft.)	14 (heavy soil)	1¼ oz.	5½
	12 (heavy soil)	1 oz.	4½
	10 (medium soil)	¾-1 oz.	4
	8 (medium soil)	¾ oz.	3
	6 (sandy soil)	½ oz.	2½

Note: In treating small areas it is important to use care in measuring out required quantity for area to be treated, properly mixing it in water, and keeping it adequately agitated during application.

* Other size areas to be treated can be calculated from the above table.

For example:

To treat 5000 sq. ft. at the 10 lbs./acre rate, use 5 cups or 80 tbsp. (5 × 16) in 5 to 10 gallons of water (5 × 1 to 2).

To treat 600 sq. ft. at the 12 lbs./acre rate, use 11.4 tbsp. (0.6 × 19) in 0.6 to 1.2 gallons of water (0.6 × 1 to 2).

APPLICATION AND CALIBRATION TECHNIQUES FOR SPRINKLER IRRIGATION

Apply **DACTHAL W-75 Turf** only through center pivot, motorized lateral move, solid set or portable (side roll, end tow, or hand move) irrigation system(s). **Do not** apply this product through any other type of irrigation system.

Crop injury or lack of effectiveness can result from non-uniform distribution of treated water.

If you have questions about calibration, you should contact State Extension Service specialists, equipment manufacturers or other experts.

Do not apply this product through irrigation systems connected to a public water system. "Public water system" means a system for the provision to the public of piped water for human consumption if such system has at least 15 service connections or regularly serves an average of at least 25 individuals daily at least 60 days per year.

A person knowledgeable of the chemigation system and responsible for its operation, or under the supervision of the responsible person, shall shut the system down and make necessary adjustments should the need arise.

The system must contain a functional check valve, vacuum relief valve, and low pressure drain appropriately located on the irrigation pipeline to prevent water source contamination from backflow.

The pesticide injection pipeline must contain a functional, automatic, quick-closing check valve to prevent the flow of fluid back toward the injection pump.

The pesticide injection pipeline must also contain a functional, normally closed, solenoid-operated valve located on the intake side of the injection pump and connected to the system interlock to prevent fluid from being withdrawn from the supply tank when the irrigation system is either automatically or manually shut down.

Always inject **DACTHAL W-75 Turf** into irrigation water after it discharges from the irrigation pump and after it passes through the check valve. Never inject pesticides into the intake line on the suction side of the pump.

The system must contain functional interlocking controls to automatically shut off the pesticide injection pump when the water pump motor stops. The irrigation line or water pump must include a functional pressure switch which will stop the water pump motor when the water pressure decreases to the point where pesticide distribution is adversely affected.

Systems must use a metering pump, such as a positive displacement injection pump (e.g., diaphragm pump) effectively designed and constructed of materials that are compatible with pesticides and capable of being fitted with a system interlock.

Do not apply when wind speed favors drift beyond the area intended for treatment.

Spray mixture in the chemical supply tank must be agitated at all times, otherwise settling and uneven application may occur.

DACTHAL W-75 Turf may be used through two basic types of sprinkler irrigation systems as outlined in Sections A and B below. Determine which type of system is in place, then refer to the appropriate directions provided for each type. The tank containing **DACTHAL W-75 Turf** must be equipped with effective mechanical agitation to keep the wettable powder in suspension. Premixing **DACTHAL W-75 Turf** as instructed under ground application will insure that a uniform suspension is obtained. The injection equipment which contains the **DACTHAL W-75 Turf** suspension should be connected to the discharge side of the irrigation pump or other pressurized equipment attached to the irrigation line.

A. Center Pivot and Motorized Lateral Move Irrigation Equipment

For injection of pesticides, these continuously moving systems must use a positive displacement injection pump, of either diaphragm or piston type, constructed of materials that are compatible with pesticides and capable of injection at pressures approximately 2-3 times those encountered within the irrigation water line. Venturi applicator units cannot be used on these systems.

Fill chemical supply tank of injection equipment with water. Operate system for one complete revolution or run across the field, measuring time required, amount of water injected, and acreage covered. Thoroughly mix recommended amount of **DACTHAL W-75 Turf** for acreage to be covered into same amount of water used during calibration and inject into system continuously for one revolution or run. Mixture in the chemical supply tank must be continuously agitated during the injection run. Shut off injection equipment after one revolution or run, but continue to operate irrigation system until **DACTHAL W-75 Turf** has been cleared from last sprinkler head.

B. Solid Set and Portable (Side Roll, End Tow, or Hand Move) Irrigation Equipment

Determine acreage covered by sprinkler. Fill tank of injection equipment with water and adjust flow to use contents over a thirty to forty-five minute period. Mix desired amount of **DACTHAL W-75 Turf** for acreage to be covered into quantity of water used during calibration and operate entire system at normal pressures recommended by the manufacturer of injection equipment used for amount of time established during calibration. **DACTHAL W-75 Turf** can be injected at the beginning or end of the irrigation cycle or as a separate application. Stop injection equipment after treatment is completed and continue to operate irrigation system until **DACTHAL W-75 Turf** has been cleared from last sprinkler head.

STORAGE AND DISPOSAL

DO NOT contaminate water, food or feed by storage or disposal. Open dumping is prohibited.

STORAGE: Store in a dry place.

PRESTICIDE DISPOSAL:

Wastes resulting from the use of this product may be disposed of on site or at an approved waste disposal facility. **Do not** contaminate water when disposing of equipment washwaters.

CONTAINER DISPOSAL:

Completely empty bag into application equipment and dispose of empty bag in a sanitary landfill or by incineration, or, if allowed by state and local authorities, by burning. If burned, stay out of smoke.

WARRANTY AND LIMITATION OF DAMAGES

Seller warrants that this material conforms to its chemical description and is reasonably fit for the purposes stated on the label when used in accordance with directions under normal conditions of use and Buyer assumes the risk of any use contrary to such directions. SELLER MAKES NO OTHER EXPRESS OR IMPLIED WARRANTY, INCLUDING ANY OTHER EXPRESS OR IMPLIED WARRANTY OF FITNESS OR OF MERCHANTABILITY, AND NO AGENT OF SELLER IS AUTHORIZED TO DO SO EXCEPT IN WRITING WITH A SPECIFIC REFERENCE TO THIS WARRANTY. In no event shall Seller's liability for any breach of warranty exceed the purchase price of the material as to which a claim is made.

DSMA LIQUID

DSMA Plus Surfactant For Postemergence Weed Control

Product Contains 3.6 Pounds Disodium Methanearsonate Hexahydrate Equivalent Per Gallon

ACTIVE INGREDIENT:	
Disodium Methanearsonate	21.8%
INERT INGREDIENTS:	78.2%
TOTAL	100.0%
Total Arsenic, all in water-soluble form, expressed as elemental	8.9%

EPA Reg. No. 50534-27

KEEP OUT OF REACH OF CHILDREN

CAUTION

See precautionary statements.

PRECAUTIONARY STATEMENTS

HAZARDS TO HUMANS AND DOMESTIC ANIMALS

CAUTION

Harmful if swallowed or absorbed through the skin. Avoid contact with eyes, skin or clothing. Avoid breathing vapors or spray mist.

If swallowed, induce vomiting immediately by giving two glasses of water and sticking finger down throat. **Never** give anything by mouth to an unconscious person. Seek medical attention.

In case of contact, flush eyes with plenty of water for at least 15 minutes. Seek medical attention.

Wash exposed skin gently with plenty of soap and water.

Keep children and domestic animals off treated areas until this material has been washed into the soil.

Wear protective clothing when handling or applying this product including long pants, long-sleeve shirt, and impermeable gloves and boots. Mixer-loaders should include an apron and full-face shield when handling or mixing concentrate. Flagmen should be fully protected during spray operations or mechanical flagmen used. Pilots and ground spray rig applicators should wear a mask or respirator approved by the Mining Enforcement and Safety Administration and the National Institute for Occupational Safety and Health.

ENVIRONMENTAL HAZARDS

Do not apply directly to water. **Do not** feed treated foliage to livestock or graze treated areas. **Do not** spray or allow drift onto vegetables, ornamentals or other desirable plants. **Do not** apply when weather conditions favor drift from target area. **Do not** contaminate water by cleaning of equipment or disposal of wastes.

DSMA LIQUID

Read entire label before using this product.

GENERAL INFORMATION

DSMA Liquid is a postemergence herbicide for use in cotton, turf and non-crop areas. **DSMA Liquid** contains a surfactant (wetting agent) in proper ratio for best results. It is not necessary to add additional surfactant to the spray tank. Local conditions and recommendations vary; consult local agricultural experiment station or extension service weed specialists for recommendations in your area.

Mixing Instructions: DSMA Liquid herbicide must be thoroughly dissolved. Fill the spray tank with half the required amount of water and, with pump or agitator operating, add the recommended amount of **DSMA Liquid** herbicide and continue filling the tank with balance of water.

Do not store spray solution in tank overnight. Clean application equipment thoroughly after use by flushing with water in a safe place.

Application Methods: DSMA Liquid should be applied with a low-volume, properly calibrated sprayer having satisfactory pumping and bypass action. **Adjust Nozzles to Keep Spray Off Cotton Foliage and in a Manner to Allow Maximum Coverage of Weeds.** Adequate coverage is very important for effective weed control.

Apply during warm weather when weeds are in an active stage of growth.

DIRECTIONS FOR USE

It is a violation of Federal law to use this product in a manner inconsistent with its labeling.

Do not apply this product through any type of irrigation equipment.

Cotton: DSMA Liquid is useful as directed postemergence spray for controlling weeds in cotton such as barnyardgrass, cocklebur, smooth and large crabgrass, dallisgrass, goosegrass, johnsongrass and nutsedge. Apply as a **Directed** broadcast spray at rate of 40 gallons per acre of a solution containing 1 gallon of **DSMA Liquid** in 40 gallons of water when weeds are small. For band applications, apply 1 gallon of above diluted spray per acre for each 1 inch band width to be treated of cotton grown on 40 inch row spacing. A second or repeat application, if needed, should be timed about 1 to 3 weeks after the first application. Keep spray off cotton foliage.

Apply Only When Cotton is 3 Inches High to First Bloom. Do not Apply After First Bloom. Slight burning and reddish discoloration of cotton foliage may occasionally be seen following recommended treatment; however, cotton plants will develop normally and yield will not be affected.

Ornamental Turfgrass: DSMA Liquid can be used for selective control of bahiagrass, barnyardgrass, chickweed, smooth and large crabgrass, dallisgrass, nutsedge, sanbur and wood sorrell with little or no injury to well-established, actively growing turfgrasses. Mow turfgrass to height of 1 to 1½ inches before treatment. Mix 3 fluid ounces (6 tablespoons) of **DSMA Liquid** in 2½ gallons of water for application to an area of 1,000 square feet. Application should be uniform and thorough to adequately wet all undesirable plants. Two or more repeat treatments at 14 day intervals may be necessary. Make applications during warm weather when temperature is between 80 and 90°F. **Do not** water turf for at least 24 hours after application. Turfgrasses may be temporarily discolored. Bermudagrass, bluegrasses and zoysiagrass have shown tolerance to properly applied **DSMA Liquid.** Injury may result if applied to bentgrasses and fescues. **Do not** apply to St. Augustinegrass, carpetgrass, centipedegrass or to Dichondra.

Do not reseed until 2 weeks after last application. **Do not** apply with hose-end applicators.

Non-crop: DSMA Liquid is effective in controlling barnyardgrass, cocklebur, smooth and large crabgrass, dallisgrass, goosegrass, johnsongrass, and nutsedge and many similar weeds on drainage ditchbanks, fence rows, rights-of-way, storage yards and similar non-crop areas. Application should be made when weeds are small and conditions are favorable for good weed growth. Mix at a rate of 1 gallon of **DSMA Liquid** in 40 gallons of water. Spray undesirable vegetation thoroughly to point of run-off. Adequate coverage and complete wetting of foliage is important for effective control. Repeat applications may be necessary if regrowth occurs. **Do not** contaminate waters used for domestic purposes, or by animals including wildlife and aquatic life, or water used for irrigation purposes.

SPECIAL LOCAL NEEDS PURSUANT TO SECTION 24(c) OF AMENDED FIFRA AND STATE REGISTRATIONS

CROP	REGISTRATIONS	STATES	SLN NO.	DIRECTIONS FOR USE
Cotton	Application on Cotton	Arkansas Louisiana Tennessee	AR-880003 LA-880010 TN-830016	**DSMA Liquid** is useful for the control of weeds listed on the product label and many similar weeds. Application can be made as follows: 1. Preplant or postplant up to cracking of soil before cotton emergence using ground or aircraft equipment. 2. Postemergence, over-the-top, when cotton is 3 to 6 inches high or up to early first square stage, whichever occurs first, using ground or aircraft equipment. 3. Postemergence as a directed spray with ground equipment when cotton is 3 inches high to first bloom. PREPLANT OR POSTPLANT UP TO CRACKING APPLICATION ON COTTON: A single ground or aircraft application of **DSMA Liquid** can be made to prepared cotton seedbeds when planting has been delayed and weeds have emerged, or as a postplant treatment, BUT NO LATER THAN INITIAL CRACKING of soil in field before emergence of cotton. Planting of cotton may immediately follow the preplant application. Mix at the rate of 1 gallon of **DSMA Liquid** in 40 gallons of water for broadcast ground application or in 5 to 10 gallons of water for aircraft application and apply using a properly calibrated sprayer to one acre. See SPECIAL PRECAUTIONS below. POSTEMERGENT, OVER-THE-TOP, APPLICATION ON COTTON: Application may be made using ground or aircraft equipment as an over-the-top broadcast spray to actively growing cotton as a salvage operation when cotton is 3 to 6 inches high or up to early first square stage, whichever occurs first. Mix **DSMA Liquid** at rate of 3 to 4 pints in 40 gallons of water for ground application, or in 5 to 10 gallons of water for aircraft application and apply using properly calibrated sprayer to one acre. A second or repeat application, if needed, should be made 1 to 3 weeks after the first treatment. APPLY AS AN OVER-THE-TOP BROADCAST SPRAY ONLY WHEN COTTON IS 3 TO 6 INCHES HIGH OR UP TO EARLY FIRST SQUARE STAGE, WHICHEVER OCCURS FIRST. **DO NOT** APPLY AS AN OVER-THE-TOP BROADCAST SPRAY AFTER COTTON IS MORE THAN 6 INCHES HIGH OR AFTER EARLY FIRST SQUARE STAGE. See SPECIAL PRECAUTIONS below. POSTEMERGENT, DIRECTED, APPLICATION ON COTTON: Apply as DIRECTED broadcast spray at a rate of 40 gallons per acre of a solution containing 1 gallon of **DSMA Liquid** in 40 gallons of water when weeds are small. For band applications, apply 1 gallon of above diluted spray per acre for each 1-inch band width to be treated of cotton grown on 40-inch row spacing. Keep spray off cotton foliage. APPLY ONLY WHEN COTTON IS 3 INCHES HIGH TO FIRST BLOOM. **DO NOT** APPLY AFTER FIRST BLOOM. Slight burning and reddish discoloration of cotton foliage may occasionally be seen following recommended treatment; however, cotton plants will develop normally. SPECIAL PRECAUTIONS - **DO NOT** allow spray or spray drift to contact adjacent crops or injury will result. Apply only on still days when weather conditions **DO NOT** favor drift from areas being treated. Follow other precautions listed on this label. Aircraft applications of **DSMA Liquid** should only be made by applicators experienced in use of herbicides, and applications should be made in accordance with state and Federal regulations.
Cotton	Application on Cotton	Georgia Missouri New Mexico North Carolina Texas	GA-770007 MO-770006 NM-780005 NC-770017 TX-770025	**DSMA Liquid** is useful for the control of weeds listed on the product label and and many similar weeds. Application can be made as follows: 1. Preplant or postplant up to cracking of soil before cotton emergence using ground or aircraft equipment. 2. Postemergent, over-the-top, when cotton is 3 to 6 inches high or up to early first square stage, whichever occurs first, using ground or aircraft equipment. 3. Postemergence as a directed spray with ground equipment when cotton is 3 inches high to first bloom. PREPLANT OR POSTPLANT UP TO CRACKING APPLICATION ON COTTON: A single ground or aircraft application of **DSMA Liquid** can be made to prepared cotton seedbeds when planting has been delayed and weeds have emerged, or as a postplant treatment. BUT NO LATER THAN INITIAL CRACKING of soil in field before emergence of cotton. Planting of cotton may immediately follow the preplant application. Mix at the rate of 1 gallon of **DSMA Liquid** in 40 gallons of water for broadcast ground application, or in 5 to 10 gallons of water for aircraft application and apply using a properly calibrated sprayer to one acre. See SPECIAL PRECAUTIONS below. POSTEMERGENT, OVER-THE-TOP, APPLICATION ON COTTON: Application may be made using ground or aircraft equipment as an over-the-top broadcast spray to actively growing cotton as a salvage operation when cotton is 3 to 6 inches high or up to early first square stage, whichever occurs first. Mix **DSMA Liquid** at rate of 2 to 3 quarts in 40 gallons of water for ground application and apply using a properly calibrated sprayer to one acre. A second or repeat application, if needed, should be made 1 to 3 weeks after the first treatment. APPLY AS AN OVER-THE-TOP BROADCAST SPRAY ONLY WHEN COTTON IS 3 TO 6 INCHES HIGH OR UP TO EARLY FIRST SQUARE STAGE, WHICHEVER OCCURS FIRST. **DO NOT** APPLY AS AN OVER-THE-TOP BROADCAST SPRAY AFTER COTTON IS MORE THAN 6 INCHES HIGH OR AFTER EARLY FIRST SQUARE STAGE. See SPECIAL PRECAUTIONS below. POSTEMERGENT, DIRECTED, APPLICATION ON COTTON: Apply as DIRECTED broadcast spray at a rate of 40 gallons per acre of a solution containing 1 gallon of **DSMA Liquid** in 40 gallons of water when weeds are small. For band applications, apply 1 gallon of above diluted spray per acre for each 1-inch band width to

CROP	REGISTRATIONS	STATES	SLN NO.	DIRECTIONS FOR USE
				be treated of cotton grown on 40-inch row spacing. Keep spray off cotton foliage. APPLY ONLY WHEN COTTON IS 3 INCHES HIGH TO FIRST BLOOM. **DO NOT** APPLY AFTER FIRST BLOOM. Slight burning and reddish discoloration of cotton foliage may occasionally be seen following recommended treatment; however, cotton plants will develop normally. SPECIAL PRECAUTIONS—**DO NOT** allow spray or spray drift to contact adjacent crops or injury will result. Apply only on still days when weather conditions **DO NOT** favor drift from areas being treated. Follow other precautions listed on this label. Aircraft applications of **DSMA Liquid** should only be made by applicators experienced in use of herbicides, and applications should be made in accordance with state and Federal regulations.
Cotton	Application on Cotton	Alabama	AL-770010	**DSMA Liquid** is useful for the control of weeds listed on the product label and many similar weeds. Application can be made as: 1. Preplant or postplant up to cracking of soil before cotton emergence using ground or aircraft equipment. 2. Postemergent, over-the-top, when cotton is 3 to 6 inches high or up to early first square stage, whichever occurs first, using ground or aircraft equipment. 3. Postemergence as a directed spray with ground equipment when cotton is 3 inches high to first bloom. PREPLANT OR POSTPLANT UP TO CRACKING APPLICATION ON COTTON: A single ground or aircraft application of **DSMA Liquid** can be made to prepared cotton seedbeds when planting has been delayed and weeds have emerged, or as a postplant treatment. BUT NO LATER THAN INITIAL CRACKING of soil in field before emergence of cotton. Planting of cotton may immediately follow the preplant application. Mix at the rate of 1 gallon of **DSMA Liquid** in 40 gallons of water for broadcast ground application, or in 5 to 10 gallons of water for aircraft application and apply using a properly calibrated sprayer to one acre. See SPECIAL PRECAUTIONS below. POSTEMERGENT, OVER-THE-TOP, APPLICATION ON COTTON: A single ground or aircraft application may be made as an over-the-top broadcast spray to actively growing cotton as a SALVAGE operation when cotton is 3 to 6 inches high or up to early first square stage, whichever occurs first. Mix **DSMA Liquid** at rate of 3 to 4 pints in 40 gallons of water for ground application, or in 5 to 10 gallons of water for aircraft application and apply using a properly calibrated sprayer to one acre. APPLY AS AN OVER-THE-TOP BROADCAST SPRAY ONLY WHEN COTTON IS 3 TO 6 INCHES HIGH OR UP TO EARLY FIRST SQUARE STAGE, WHICHEVER OCCURS FIRST. **DO NOT** APPLY AS AN OVER-THE-TOP BROADCAST SPRAY AFTER COTTON IS MORE THAN 6 INCHES HIGH OR AFTER EARLY FIRST SQUARE STAGE. See SPECIAL PRECAUTIONS below. POSTEMERGENT, DIRECTED, APPLICATION ON COTTON: Apply as DIRECTED broadcast spray at a rate of 40 gallons per acre of a solution containing 1 gallon of **DSMA Liquid** in 40 gallons of water when weeds are small. For band applications, apply 1 gallon of above diluted spray per acre for each 1-inch band width to be treated of cotton grown on 40-inch row spacing. Keep spray off cotton foliage. APPLY ONLY WHEN COTTON IS 3 INCHES HIGH TO FIRST BLOOM. **DO NOT** APPLY AFTER FIRST BLOOM. Slight burning and reddish discoloration of cotton foliage may occasionally be seen following recommended treatment; however, cotton plants will develop normally. SPECIAL PRECAUTIONS - **DO NOT** allow spray or spray drift to contact adjacent crops or injury will result. Apply only on still days when weather conditions **DO NOT** favor drift from areas being treated. Follow other precautions listed on this label. Aircraft applications of **DSMA Liquid** should only be made by applicators experienced in use of herbicides, and applications should be made in accordance with state and Federal regulations.
Cotton	Application on Cotton	Arizona	AZ-770024	**DSMA Liquid** is useful for the control of weeds listed on the product label and many similar weeds. Application can be made as: 1. Postemergent, over-the-top, when cotton is 3 to 6 inches high or up to early first square stage, whichever occurs first, using ground or aircraft equipment. 2. Postemergence as a directed spray with ground equipment when cotton is 3 inches high to first bloom. POSTEMERGENT, OVER-THE-TOP, APPLICATION ON COTTON: A single ground or aircraft application may be made as an over-the-top broadcast spray to actively growing cotton as a salvage operation when cotton is 3 to 6 inches high or up to early first square stage, whichever occurs first. Mix **DSMA Liquid** at rate of 2 to 3 quarts in 40 gallons of water for ground application, or in 5 to 10 gallons of water for aircraft application and apply using a properly calibrated sprayer to one acre. APPLY AS AN OVER-THE-TOP BROADCAST SPRAY ONLY WHEN COTTON IS 3 TO 6 INCHES HIGH OR UP TO EARLY FIRST SQUARE STAGE, WHICHEVER OCCURS FIRST. **DO NOT** APPLY AS AN OVER-THE-TOP BROADCAST SPRAY AFTER COTTON IS MORE THAN 6 INCHES HIGH OR AFTER EARLY FIRST SQUARE STAGE. See SPECIAL PRECAUTIONS below. POSTEMERGENT, DIRECTED, APPLICATION ON COTTON: Apply as DIRECTED broadcast spray at a rate of 40 gallons per acre of a solution containing 1 gallon of **DSMA Liquid** in 40 gallons of water when weeds are small. For band applications, apply 1 gallon of above diluted spray per acre for each 1-inch band width to be treated of cotton grown on 40-inch row spacing. Keep spray off cotton foliage. APPLY ONLY WHEN COTTON IS 3 INCHES HIGH TO FIRST BLOOM. **DO NOT** APPLY AFTER FIRST BLOOM. Slight burning and reddish discoloration of cotton foliage may occasionally be seen following recommended treatment; however, cotton plants will develop normally. SPECIAL PRECAUTIONS—**DO NOT** allow spray or spray drift to contact adjacent crops or injury will result. Apply only on still days when weather conditions **DO NOT** favor drift from areas being treated. Follow other precautions listed on this label. Aircraft applications of **DSMA Liquid** should only be made by applicators experienced in use of herbicides, and applications should be made in accordance with state and Federal regulations.

CROP	REGISTRATIONS	STATES	SLN NO.	DIRECTIONS FOR USE
Cotton	Application on Cotton	California	CA-770096	**DSMA Liquid** is useful for the control of weeds listed on the product label and many similar weeds. Application can be made as: 1. Preplant or postplant up to cracking of soil before cotton emergence using ground or aircraft equipment. 2. Postemergent, over-the-top, when cotton is 3 to 6 inches high or up to early first square stage, whichever occurs first, using ground or aircraft equipment. 3. Postemergence as a directed spray with ground equipment when cotton is 3 inches high to first bloom. PREPLANT OR POSTPLANT UP TO CRACKING APPLICATION ON COTTON: A single ground or aircraft application of **DSMA Liquid** can be made to prepared cotton seedbeds when planting has been delayed and weeds have emerged, or as a postplant treatment. BUT NO LATER THAN INITIAL CRACKING of soil in field before emergence of cotton. Planting of cotton may immediately follow the preplant application. Mix at the rate of 1 gallon of **DSMA Liquid** in 40 gallons of water for broadcast ground application, or in 5 to 10 gallons of water for aircraft application and apply using a properly calibrated sprayer to one acre. See SPECIAL PRECAUTIONS below. POSTEMERGENT, OVER-THE-TOP, APPLICATION ON COTTON: One or two over-the-top applications may be applied by ground or aircraft equipment to actively growing cotton as a salvage operation when crop is 3 to 6 inches high or to early first square stage, whichever occurs first. ONE APPLICATION: Use 1 gallon **DSMA Liquid** in 40 gallons of water per acre for ground application, or in 5 to 10 gallons of water for aircraft application. **DO NOT** REPEAT THIS TREATMENT. TWO APPLICATIONS: Use 2 to 3 quarts **DSMA Liquid** in 40 gallons of water per acre for ground application, or in 5 to 10 gallons of water for aircraft application. Apply second applications, if needed, using the 2 to 3 quart rate, 1 to 3 weeks following the first treatment. Use a properly calibrated sprayer. **DO NOT APPLY OVER-THE-TOP SPRAYS AFTER COTTON IS MORE THAN 6 INCHES HIGH OR AFTER EARLY FIRST SQUARE STAGE, WHICHEVER OCCURS FIRST.** POSTEMERGENT, DIRECTED, APPLICATION ON COTTON: Apply as DIRECTED broadcast spray at a rate of 40 gallons per acre of a solution containing 1 gallon of **DSMA Liquid** in 40 gallons of water when weeds are small. For band applications, apply 1 gallon of above diluted spray per acre for each 1-inch band width to be treated of cotton grown on 40-inch row spacing. Keep spray off cotton foliage. APPLY ONLY WHEN COTTON IS 3 INCHES HIGH TO FIRST BLOOM. Slight burning and reddish discoloration of cotton foliage may occasionally be seen following recommended treatment; however, cotton plants will develop normally. SPECIAL PRECAUTIONS - **DO NOT** allow spray or spray drift to contact adjacent crops or injury will result. Apply only on still days when weather conditions **DO NOT** favor drift from areas being treated. Follow other precautions listed on this label. Aircraft applications of **DSMA Liquid** should only be made by applicators experienced in use of herbicides, and applications should be made in accordance with state and Federal regulations.
Cotton	Application on Cotton	Oklahoma	OK-780002	**DSMA Liquid** is useful for the control of weeds listed on the product label and many similar weeds. POSTEMERGENT, OVER-THE-TOP, APPLICATION ON COTTON. Application may be made using ground or aircraft equipment as an over-the-top broadcast spray to actively growing cotton as a salvage operation when cotton is 3 to 6 inches high or up to early first square stage, whichever occurs first. Mix **DSMA Liquid** at rate of 2 to 3 quarts in 40 gallons of water for ground application, or in 5 to 10 gallons of water for aircraft application and apply using a properly calibrated sprayer to one acre. A second or repeat application, if needed, should be made 1 to 3 weeks after first treatment. APPLY AS AN OVER-THE-TOP BROADCAST SPRAY ONLY WHEN COTTON IS 3 TO 6 INCHES HIGH OR UP TO EARLY FIRST SQUARE STAGE, WHICHEVER OCCURS FIRST. **DO NOT** APPLY AS AN OVER-THE-TOP BROADCAST SPRAY AFTER COTTON IS MORE THAN 6 INCHES HIGH OR AFTER EARLY FIRST SQUARE STAGE. See SPECIAL PRECAUTIONS below. Slight burning and reddish discoloration of cotton foliage may occasionally be seen following recommended treatment; however, cotton plants will develop normally. SPECIAL PRECAUTIONS - **DO NOT** allow spray or spray drift to contact adjacent crops or injury will result. Apply only on still days when weather conditions **DO NOT** favor drift from areas being treated. **DO NOT** feed treated forage or gin trash to ruminant animals or graze treated areas. Follow other precautions listed on this label. Aircraft applications of **DSMA Liquid** should only be made by applicators experienced in use of herbicides, and applications should be made in in accordance with state and Federal regulations.
Cotton	Application on Cotton	South Carolina	SC-770014	**DSMA Liquid** is useful for the control of weeds listed on the product label. POSTEMERGENT, APPLICATION ON COTTON. A single ground or aircraft application may be made as an over-the-top broadcast spray to actively growing cotton as a salvage operation when cotton is 3 to 6 inches high or up to early first square stage, whichever occurs first. Mix **DSMA Liquid** at rate of 3 to 4 pints in 40 gallons of water for ground application, or in 5 to 10 gallons of water for aircraft application and apply using a properly calibrated sprayer to one acre. APPLY AS AN OVER-THE-TOP BROADCAST SPRAY ONLY WHEN COTTON IS 3 TO 6 INCHES HIGH OR UP TO EARLY FIRST SQUARE STAGE, WHICHEVER OCCURS FIRST. **DO NOT** APPLY AS AN OVER-THE-TOP BROADCAST SPRAY AFTER COTTON IS MORE THAN 6 INCHES HIGH OR AFTER EARLY FIRST SQUARE STAGE. **DO NOT** make more than two applications total of either DSMA or MSMA or a combination) per season. See SPECIAL PRECAUTIONS below. Slight burning and reddish discoloration of cotton foliage may occasionally be seen following recommended treatment; however, cotton plants will develop normally. SPECIAL PRECAUTIONS - **DO NOT** feed treated forage to livestock or graze treated areas. **DO NOT** allow spray or spray drift to contact adjacent crops or injury and residues will result. Apply only on still days when weather conditions **DO NOT** favor drift from

CROP	REGISTRATIONS	STATES	SLN NO.	DIRECTIONS FOR USE
				areas being treated. Follow other precautions listed on this label. Aircraft applications of **DSMA Liquid** should only be made by applicators experienced in use of herbicides, and applications should be made in accordance with state and Federal regulations.

STORAGE AND DISPOSAL

Do not contaminate water, food or feed by storage or disposal. Open dumping is prohibited.

PESTICIDE DISPOSAL:

Wastes resulting from the use of this product may be disposed of on site or at an approved waste disposal facility.

CONTAINER DISPOSAL:

Do not reuse empty container. Triple rinse (or equivalent) and offer for recycling or reconditioning or puncture and dispose of in a sanitary landfill, or incinerate, or if allowed by state and local authorities, by burning. If burned, stay out of smoke.

WARRANTY AND LIMITATION OF DAMAGES

Seller warrants that this material conforms to its chemical description and is reasonably fit for the purposes stated on the label when used in accordance with directions under normal conditions of use and Buyer assumes the risk of any use contrary to such directions. SELLER MAKES NO OTHER EXPRESS OR IMPLIED WARRANTY, INCLUDING ANY OTHER EXPRESS OR IMPLIED WARRANTY OF FITNESS OR OF MERCHANTABILITY, AND NO AGENT OF SELLER IS AUTHORIZED TO DO SO EXCEPT IN WRITING WITH A SPECIFIC REFERENCE TO THIS WARRANTY. In no event shall Seller's liability for any breach of warranty exceed the purchase price of the material as to which a claim is made.

FRIGATE®

AGRICULTURAL ADJUVANT

Containing: 6.6 lb./gal. Fatty Amine Ethoxylate
Constituents effective as a spray adjuvant 84.8 percent

KEEP OUT OF REACH OF CHILDREN
DANGER

See side panel for additional precautionary statements.

Read entire label carefully and use as directed.

PRECAUTIONARY STATEMENTS

HAZARDS TO HUMANS AND DOMESTIC ANIMALS

DANGER

Causes irreversible eye damage. Causes skin irritation.

Flammable.

Do not get in eyes. Avoid contact with skin and clothing. Wear goggles or face shield when handling this product.

Wash exposed areas of skin thoroughly after use.

FIRST AID:

Eyes: In case of contact with eyes, immediately flush eyes with plenty of water for at least 15 minutes. Call a physician.

Skin: In case of contact with skin, flush with water.

Ingestion: If swallowed, have conscious person drink several glasses of water, then induce vomiting by tickling back of throat with finger. Keep airway clear. Never give anything by mouth to an unconscious person. Seek medical attention.

ENVIRONMENTAL HAZARDS

Harmful to fish. **DO NOT** contaminate ponds, waterways or ditches with chemical or used container. Store in original container, tightly closed, in a cool, safe place with adequate ventilation. Keep away from heat, sparks and flame. Wash out container thoroughly, puncture and dispose of in a sanitary landfill or incineration, or, if allowed by state and local authorities, by burning. If burned, stay out of smoke.

DECONTAMINATION OF SPRAYING EQUIPMENT

Follow manufacturer's label recommendations for the ROUNDUP* (glyphosate) product for cleanup of spraying equipment following use.

PRINCIPAL USES

FRIGATE is recommended for use with ROUNDUP (glyphosate) when the use of a nonionic surfactant is recommended on the ROUNDUP label.

FRIGATE is recommended for use with ROUNDUP to improve wetting of foliage, enhance uptake of glyphosate and permits ROUNDUP to provide more consistent control of quackgrass and johnsongrass. Use of **FRIGATE** with ROUNDUP will also enhance ROUNDUP's control of annual weeds in the following:

Fallow and Reduced Tillage Systems
Minimum Tillage Systems
Noncrop Uses

DIRECTIONS FOR USE

Follow all precautionary statements, directions for use and general product information appearing on both the ROUNDUP and **FRIGATE** labels when using **FRIGATE** in combination with ROUNDUP.

A tank-mix of **FRIGATE** with ROUNDUP improves ROUNDUP'S control of quackgrass (*Agropyron repens*) and johnsongrass (*Sorghum halepense*). The recommended use of **FRIGATE** in combination with ROUNDUP improves initial burndown of quackgrass and reduces the amount of regrowth the following season. **FRIGATE** tank-mixed with ROUNDUP will also permit ROUNDUP to provide superior control of other weeds where the addition of a nonionic surfactant to ROUNDUP is recommended on the ROUNDUP label. Refer to the ROUNDUP label to identify the optimum stage of growth necessary to obtain weed control when utilizing the addition of 0.5% **FRIGATE** by total spray volume.

RATE OF USE

Use **FRIGATE** at 0.5% (volume/volume) in the final spray mixture with ROUNDUP at the recommended rate.

FRIGATE should be used at a concentration of 4 pts./100 gal. in the final spray solution. For conventional application with ROUNDUP in spray volume of 20 gal. per acre, the rate is 0.8 pt. **FRIGATE** per acre in 20 gallons of water. For the 10-gal. spray volume, use 0.4 pt. **FRIGATE** per acre. When using ROUNDUP on hard-to-wet weed species, use 1% **FRIGATE** (1 gal./100 gal.)

MIXING

Prepare spray in accordance with ROUNDUP label directions for tank-mixing with surfactants. Fill the mixing or spray tank with most of the required amount of water. Under agitation, add the required amount of ROUNDUP near the end of the spray tank filling process. Maintain good agitation while adding the recommended volume of **FRIGATE** to the spray tank. After adding the **FRIGATE,** continue to agitate the spray mixture thoroughly for at least five (5) minutes **before spraying.** Agitation for ten (10) minutes is recommended if the tank water temperature is less than 50°F.

PLEASE NOTE: When spray tank agitation is minimal and the tank water temperature is less than 50°., premix one part FRIGATE with two parts warm water (greater than 75°F) and thoroughly mix before addition to the spray tank.

WARRANTY AND LIMITATION OF DAMAGES

Seller warrrants that this material conforms to its chemcial description and is reasonably fit for the purposes stated on the label when used in accordance with directions under normal conditions of use and Buyer assumes the risk of any use contrary to such directions. SELLER MAKES NO OTHER EXPRESS OR IMPLIED WARRANTY, INCLUDING ANY OTHER EXPRESS OR IMPLIED WARRANTY OF FITNESS OR OF MERCHANTABILITY, AND NO AGENT OF SELLER IS AUTHORIZED TO DO SO. In no event shall Seller's liability for any breach of warranty exceed the purchase price of the material as to which a claim is made.

2 PLUS 2 (MCPP + 2,4-D AMINE)

TURF CARE® HERBICIDE

ACTIVE INGREDIENTS:

Dimethylamine Salt of 2,4-Dichlorophenoxyacetic acid*	24.5%
Dimethylamine Salt of 2-(2-Methyl-4-Chlorophenoxy) Propionic Acid**	24.2%
INERT INGREDIENTS:	51.3%
TOTAL	100.0%

* 2,4-Dichlorophenoxyacetic Acid Equivalent: 20.4%, 1.9 lbs./gal.
** 2-(2-Methyl-4-Chlorophenoxy) Propionic Acid: 20.0%, 1.8 lbs./gal.
*/** Isomer specific by AOAC Method

EPA Reg. No. 50534-11

KEEP OUT OF REACH OF CHILDREN
CAUTION

See precautionary statements.

PRECAUTIONARY STATEMENTS

HAZARDS TO HUMANS AND DOMESTIC ANIMALS

CAUTION

Harmful if swallowed or absorbed through the skin. **Causes eye and skin irritation.** Avoid breathing spray mist. Avoid contact with eyes, skin or clothing. Wash thoroughly after handling. Keep container closed. If swallowed, have conscious person drink several glasses of water, then induce vomiting by tickling back of throat with finger. Keep airway clear. Never give anything by mouth to an unconscious person. Seek medical attention. In case of contact, immediately flush eyes with plenty of water for at least 15 minutes holding eyelids apart to ensure flushing of the entire surface. Seek medical attention. Wash exposed skin gently with plenty of soap and water. Seek medical attention.

ENVIRONMENTAL HAZARDS

Do not apply directly to water. **Do not** contaminate water by cleaning of equipment or disposal of wastes. **Do not** contaminate irrigation ditches or water used for domestic purposes. Use care to avoid spray contact or drift to 2,4-D susceptible plants such as cotton, tomatoes, flowers, grapes, fruit trees and ornamentals. **Do not** permit spray mist containing **2 PLUS 2 (MCPP + 2,4-D AMINE)** to drift onto them, since even very small quantities of the spray, which may not be visible, can cause severe injury during both growing and dormant periods. **Do not** spray when the wind is blowing towards susceptible crops or ornamental plants. Use coarse sprays to minimize drift. Spray drift can be lessened by keeping the spray boom as low as possible; by applying 20 gallons or more of spray per acre; by using no more than 20 pounds spray pressure with flat or flooding flat fan nozzle tips; by spraying when wind velocity is low; and by stopping all spraying when wind exceeds 8 miles per hour. **Do not** apply with hollow cone-type insecticide or other nozzles that produce a fine droplet spray. Flush sprayer on suitable non-crop area after use. **Do not** use the same spray equipment for applying other materials to 2,4-D susceptible crops as injury may result.

2 PLUS 2 (MCPP + 2,4-D AMINE)

Read entire label before using this product

GENERAL INFORMATION

2 PLUS 2 (MCPP + 2,4-D AMINE) is a postemergence herbicide especially formulated for use on turfgrass for selective control of many broadleaf weeds commonly growing in lawns without injury to desirable grasses. This chemical may be used on all types of turf including Merion, Kentucky bluegrass, bermudagrass, fescues and rye grasses. (This product has been used successfully on bentgrass; however, it should not be used unless prior tests have shown it to be nondamaging under the particular conditions of use. To eradicate broadleaf weeds, it is better to use another selective herbicide for fine lawn grasses).

NOTE: MCPP acts more slowly than 2,4-D so that some weeds will be controlled before others. Allow 3 to 4 weeks for complete weed control action to occur. The following is a partial list of weeds controlled by **2 PLUS 2 (MCPP + 2,4-D AMINE)**:

Buckhorn	Lambsquarters	Red clover
Clover	Mallow	*Red sorrel
Common chickweed	Morningglory	*Speedwell
Curly dock	Mouseear chickweed	Stitchwort
*Dandelion	Pigweed	White clover
*English daisy	Plantain	*Wild garlic
Ground ivy	Purslane	*Wild onion
Knotweed	Ragweed	Yarrow

* Use high rate on chart for weed marked with an asterisk (*) in the above list.

NOTE: Local conditions and application regulations vary and may affect use of this herbicide. Consult local agricultural experiment station or extension service weed specialists and state regulatory agencies for recommendations in your area.

NOTE: When stored at temperatures below freezing, it may be necessary to warm contents to 45°F and mix thoroughly before using.

DIRECTIONS FOR USE

It is a violation of Federal law to use this product in a manner inconsistent with its labeling.

Do not apply this product through any type of irrigation equipment.

Preparation of the Spray: This product must be mixed with water prior to use. For best results use 2 to 4 gallons of water per 1,000 sq. ft. for smaller areas, 20 to 40 gallons of water per acre for larger areas. Apply by means of a spray to give uniform coverage. Repeat applications may be necessary especially with heavy infestation of the less susceptible weeds indicated with an asterisk in above list.

Either before or after application, it is wise to properly fertilize the lawn so as to encourage the lawn to heal as rapidly as possible, filling in the bare areas where the weeds have died away. Once the weeds have been removed, the bare areas should be reseeded. Bare areas may be reseeded 3 to 4 weeks after application.

Time of Application: Apply in spring or fall when weeds are actively growing. Knotweed is controlled more easily in spring or early summer when weeds are young. **Do not** apply when turf is suffering from drought conditions or during very hot weather or damage to grass may result. **Do not** mow for several days before or after application. Withold sprinkling for 24 hours after application. **Do not** apply during periods of heavy rainfall. **Do not** use on newly seeded areas. **Do not** use if you want clover in your turf. In some areas bent, carpet, buffalo and St. Augustine grasses may be injured; therefore, **Do not** use on turf where these grasses predominate.

Amount to Apply	Per 1,000 Sq. Ft.	Per Acre
Low rate	2 fluid ounces	2½ quarts
High rate	2½ fluid ounces	3 quarts

STORAGE AND DISPOSAL

DO NOT contaminate water, food or feed by storage or disposal. Open dumping is prohibited.

PESTICIDE DISPOSAL:

Pesticide wastes are toxic. Improper disposal of excess pesticide, spray mixture or rinsate is a violation of Federal Law. If these wastes cannot be disposed of by use according to label instructions, contact your State Pesticide or Environmental Control Agency or the Hazardous Waste Representative at the nearest EPA Regional Office for guidance.

CONTAINER DISPOSAL:

DO NOT reuse empty container. Triple rinse (or equivalent) and offer for recycling or reconditioning or puncture and dispose of in a sanitary landfill, or by other procedures approved by state and local authorities.

WARRANTY AND LIMITATION OF DAMAGES

Seller warrants that this material conforms to its chemical description and is reasonably fit for the purposes stated on the label when used in accordance with directions under normal conditions of use and Buyer assumes the risk of any use contrary to such directions. SELLER MAKES NO OTHER EXPRESS OR IMPLIED WARRANTY, INCLUDING ANY OTHER EXPRESS OR IMPLIED WARRANTY OF FITNESS OR OF MERCHANTABILITY, AND NO AGENT OF SELLER IS AUTHORIZED TO DO SO EXCEPT IN WRITING WITH A SPECIFIC REFERENCE TO THIS WARRANTY. In no event shall Seller's liability for any breach of warranty exceed the purchase price of the material as to which a claim is made.

FMC Corporation

Agricultural Chemical Group
2000 Market Street
Philadelphia, PA 19103
(215) 299-6000

Ammo® 2.5 EC

INSECTICIDE

RESTRICTED USE PESTICIDE

For retail sale to, and use only by, Certified Applicators or persons under their direct supervision, and only for those uses covered by the Certified Applicator's certification.

ACTIVE INGREDIENTS:
*Cypermethrin** 30.6%
INERT INGREDIENTS***: 69.4%
TOTAL 100.0%

*(±) α-Cyano (3-phenoxyphenyl)methyl (±) *cis/trans* 3-(2,2-dichloroethenyl)-2,2 dimethylcyclopropane carboxylate

***Cis/trans* ratio: Max. 55% (±) cis and min. 45% (±) trans

*** Contains xylene range aromatic solvents.

Contains 2.5 pounds cypermethrin per gallon.

EPA Reg. No. 279-3027 EPA Est. No. 279-Code 537
U.S. Patent No. 4,024,163

KEEP OUT OF REACH OF CHILDREN

CAUTION

See other panels for additional precautionary information.

FIRST AID:

If swallowed: Do not cause vomiting. Call a physician.

If on skin: Wash thoroughly with soap and water.

If in eyes: Flush thoroughly with water. If signs of irritation persist, contact physician.

If inhaled: Exposure to vapor or dust may cause tingling or numbness of exposed areas, especially the face, but this is temporary and requires no treatment. Lung irritation from breathing vapors is possible, and medical attention should be obtained if it is severe.

NOTE TO PHYSICIAN:

Like the natural pyrethrins, the synthetic derivative is expected to have relatively minor toxicity in humans; in fact, any significant acute toxic effects are more likely from a carrier by hydrocarbon solvent. Consequently, induction of vomiting may increase the likelihood of the most important toxic potential, chemical pneumonia, and so should either be avoided or done only under medical supervision. Ingestion of a large amount calls for gastric lavage, with care (Trendelenburg position, suction available, cuffed endotracheal tube if patient is unconscious) to avoid intrapulmonary aspiration. A saline cathartic (sodium or magnesium sulfate), 15–30 gm. dissolved in water should be given, as should 15–20 gm. activated charcoal as a slurry in water. Digestible fats, oils or alcohol may increase absorption and so should be avoided. Skin contact (vapor or powder) may be followed by transient tingling or numbness, usually of the face, but this subsides without treatment.

For Emergency Assistance Call 716-735-3765

PRECAUTIONARY STATEMENTS

HAZARDS TO HUMANS (AND DOMESTIC ANIMALS)

CAUTION

Harmful if absorbed through skin, inhaled or swallowed. Causes eye irritation. Avoid breathing vapor or spray mist. Avoid contact with skin, may cause sensitization reaction in some individuals. Avoid contact with eyes or clothing. Wash thoroughly with soap and water after handling. Remove contaminated clothing and wash before reuse.

Product should be opened and poured in a well ventilated area. Applicators must wear protective clothing. Mixers and loaders must wear a full face shield, impermeable gloves, a rubber apron, boots and protective clothing. When mixing/loading and spraying, wear midforearm to elbow length chemical-resistant gloves, boots or overshoes, one piece overalls which have long sleeves and long pants, and face shield or goggles. Mixer/loaders should in addition wear a chemical-resistant apron.

Any article of clothing worn while applying this product must be cleaned before reusing. Clothing which has been drenched or heavily contaminated should be disposed of in accordance with State or local regulations.

Instead of clothing and equipment specified above, the applicator can use an enclosed tractor cab with filtered air supply.

ENVIRONMENTAL HAZARDS

This product is highly toxic to bees exposed to direct treatment or blooming crops or weeds. Do not apply this product or allow it to drift to blooming crops or weeds while bees are actively visiting the treatment area.

This product is extremely toxic to fish. Use with care when applying in areas adjacent to any body of water. Do not apply directly to water. Do not apply when weather conditions favor drift from treated areas. Do not contaminate water by cleaning of equipment, or disposal of wastes. Do not contaminate water when disposing of equipment washwaters.

PHYSICAL/CHEMICAL HAZARDS

Do not use or store near heat or open flame.

DIRECTIONS FOR USE

It is a violation of Federal law to use this product in a manner inconsistent with its labeling.

CHEMIGATION USE DIRECTIONS

Apply this product only through sprinkler including center pivot, lateral move, end tow, side (wheel) roll, traveler, big gun, solid set, or hand move irrigation systems. Do not apply this product through any other type of irrigation system. Do not connect an irrigation system (including greenhouse systems) used for pesticide application to a public water system.

Crop injury, lack of effectiveness, or illegal residues in the crop can result from nonuniform distribution of treated water. If you have questions about calibration, you should contact State Extension Service specialists, equipment manufacturers, or other experts. A person knowledgeable of the chemigation system and responsible for its operation, or under the supervision of the responsible person, shall shut the system down and make necessary adjustments should the need arise.

The system must contain a functional check valve, vacuum relief valve, and low pressure drain approximately located on the irrigation pipeline to prevent water source contamination from backflow.

The pesticide injection pipeline must contain a functional, automatic, quick-closing check valve to prevent the flow of fluid back toward the injection pump.

The pesticide injection pipeline must also contain a functional, normally closed, solenoid-operated valve located on the intake side of the injection pump and connected to the system interlock to prevent fluid from being withdrawn from the supply tank when the irrigation system is either automatically or manually shut down.

The system must contain functional interlocking controls to automatically shut off the pesticide injection pump when the water pump motor stops.

The irrigation line or water pump must include a functional pressure switch which will stop the water pump motor when the water pressure decreases to the point where pesticide distribution is adversely affected.

Systems must use a metering pump, such as a positive displacement injection pump (e.g., diaphragm pump) effectively designed and constructed of materials that are compatible with pesticides and capable of being fitted with a system interlock.

Do not apply when wind speed favors drift beyond the area intended for treatment.

Ammo® 2.5 EC insecticide should be applied continuously for the duration of the water application. Ammo 2.5 EC should be diluted in sufficient volume to insure accurate application over the area to be treated. When using chemigation, a minimum of 0.5 inches per acre of irrigation water is recommended. Agitation is not required when a suitable diluent is used.

STORAGE AND DISPOSAL

PESTICIDE STORAGE

Do not store below 10°F, (–12°C).

Keep out of reach of children and animals. Store in original containers only. Store in a cool, dry place and avoid excessive heat. Carefully open containers. After partial use, replace lids and close tightly. Do not put concentrate or dilute material into food or drink containers. Do not contaminate other pesticides, fertilizers, water, food, or feed by storage or disposal.

In case of spill, avoid contact, isolate area and keep out animals and unprotected persons. Confine spills. Call FMC collect: (716) 735-3765.

To confine spill: If liquid, dike surrounding area or absorb with sand, cat litter or commercial clay. If dry material, cover to prevent dispersal. Place damaged package in a holding container. Identify contents.

PESTICIDE DISPOSAL

Pesticide wastes are toxic. Improper disposal of excess pesticide, spray mixture, or rinsate is a violation of Federal Law. If these wastes cannot be disposed of by use according to label instructions, contact your State Pesticide or Environmental Control Agency, or the Hazardous Waste representative at the nearest EPA Regional Office for guidance.

CONTAINER DISPOSAL

Metal Containers: Triple rinse (or equivalent). Then offer for recycling or reconditioning, or puncture and dispose of in a sanitary landfill, or by other procedures approved by state and local authorities. Do not cut or weld metal containers.

Plastic Containers: Triple rinse (or equivalent). Then offer for recycling or reconditioning, or puncture and dispose of in a sanitary landfill, or incineration, or, if allowed by State and local authorities, by burning. If burned, stay out of smoke.

GENERAL INSTRUCTIONS

RE-ENTRY STATEMENTS

Do not treat areas while unprotected humans or domestic animals are present in the treatment area.

Do not allow entry into treated areas without protective clothing until sprays have dried.

State re-entry regulations will apply if they are more restrictive than those stated on this label. Consult with State regulatory officials.

ROTATIONAL CROPS

Rotational crops may be planted within 30 days after last application.

Use low rate under light to moderate infestation. Higher rates should be used under heavy insect pressure. The rate of application is variable according to insect pressure, timing of spray and field scouting.

Crop	Pest	Dosage Lb Ai/A	Dosage Fl Oz/A	Remarks
Cotton	Cutworms Tobacco Thrips Soybean (banded) Thrips	0.025 to 0.1	1.3 to 5	Ammo® 2.5 EC may be applied in water, or refined vegetable oil. When water is used, apply a minimum of one gallon of finished spray per acre by air or five gallons of finished spray with ground equipment. When applying in water by air, one quart of emulsified oil may be substituted for one quart of water in the finished spray. When using oil, use a minimum of one quart per acre in the finished spray.
	Beet Armyworm Fall Armyworm Yellow Striped Armyworm Cotton Bollworm Pink Bollworm Tobacco Budworm Tarnished Plant Bug Lygus Bugs Saltmarsh Caterpillar Cotton Fleahopper Cotton Leafperforator Cabbage Looper Boll Weevil Other Plant Bugs White Flies*	0.04 to 0.1	2 to 5	Ammo 2.5 EC may be injected into overhead sprinkler irrigation water provided 1) an antibackflow check valve is present between the injection port and the water source, 2) a check valve is present in the line to prevent irrigation water from entering the chemical supply tank and 3) the irrigation injection system has interlocking on-off switches. For boll weevil control, apply Ammo 2.5 EC at a 3 to 4 day interval until pest numbers are reduced to acceptable levels. A maximum of 0.6 lb. active ingredient may be applied per acre per season. Ammo 2.5 EC may be applied within 14 days of harvest. Do not graze or feed cotton for forage.
Lettuce (Head)	Beet Armyworm Cabbage Looper Alfalfa Looper	0.025-0.1	1.28 to 5	Apply in water as necessary for insect control using a minimum of 15 gallons of finished spray per acre with ground equipment and 5 gallons per acre by air. Lower rates of Ammo 2.5 EC should be used with light to moderate insect pressure. Higher rates should be used to control heavy to extremely heavy insect populations. Do not apply more than 0.6 lb active ingredient per acre per season. Do not apply within 5 days of harvest. Do not plant rotational crops within 30 days of last application.
Pecans	Hickory Shuckworm Yellow Pecan Aphid Black Pecan Aphid Pecan Nut Casebearer Pecan Weevil	0.06 to 0.10	3 to 5	Applications at the lower rate should be made when pest populations are low. Rates should be increased as the pest pressure increases. Apply by ground equipment to the point of drip. Use 100 gallons of dilute spray per acre for smaller trees. For larger trees which require higher gallonage to achieve adequate coverage, apply in 200 to 300 gallons of water. In order to calculate the correct number of gallons of water needed to spray one acre of your trees to the point of drip, you may need to conduct a test. If you do not know how to conduct such a test with your equipment, you should request assistance from your equipment dealer. Up to 0.6 lb active ingredient per acre per crop season may be applied prior to shuck split. Do not apply within 21 days of harvest. Do not graze livestock in treated orchards or cut treated cover crops for feed.

*Aids in control

TANK-MIXTURE

Ammo® 2.5 EC insecticide may be applied in tank mixtures with other products approved for use on cotton. Observe all restrictions and precautions which appear on the labels of these products.

For improved control of aphids and mites, Ammo 2.5 EC may be tank-mixed with Curacron® 6E. The tank-mixture is most effective when pest populations are first observed. Subsequent applications may be needed to maintain control.

DEALERS SHOULD SELL IN ORIGINAL PACKAGES ONLY.

TERMS OF SALE OR USE: On purchase of this product buyer and user agree to the following conditions:

WARRANTY: FMC makes no warranty, expressed or implied, concerning the use of this product other than indicated on the label. Except as so warranted, the product is sold as is. Buyer and user assume all risk of use and/or handling and/or storage of this material when such use and/or handling and/or storage is contrary to label instructions.

DIRECTIONS AND RECOMMENDATIONS: Follow directions carefully. Timing and method of application, weather and crop conditions, mixture with other chemicals not specifically recommended and other influencing factors in the use of this product are beyond the control of the seller and are assumed by the buyer at his own risk.

USE OF PRODUCT: FMC's recommendations for the use of this product are based upon tests believed to be reliable. The use of this product being beyond the control of the manufacturer, no guarantee, expressed or implied, is made as to the effects of such or the results to be obtained if not used in accordance with directions or established safe practice.

DAMAGES: Buyer's or user's exclusive remedy for damages for breach of warranty or negligence shall be limited to direct damages not exceeding the purchase price paid and shall not include incidental or consequential damages.

Curacron®—Ciba-Geigy Corporation
Ammo® and FMC—FMC trademarks (537-6/15/89-A) 7/89

This is a specimen label. FMC is not responsible for the accuracy of the information contained herein. As labels are subject to revision, always carefully read and follow the label on the product container.

Ammo® WSB

INSECTICIDE

RESTRICTED USE PESTICIDE

For retail sale to, and use only by, Certified Applicators or persons under their direct supervision, and only for those uses covered by the Certified Applicator's certification.

EPA Reg. No. 279–3084 EPA Est.

Active Ingredients	
*Cypermethrin **	38.7%
Inert Ingredients	61.3%
TOTAL	100.0%

*(±) α-Cyano (3-phenoxyphenyl)methyl (±) *cis/trans* 3-(2,2-dichloroethenyl)-2,2 dimethylcyclopropane carboxylate

***Cis/trans* ratio: Max. 55% (±) cis and min. 45% (±) trans

U.S. Patent No. 4,024,163
Code 1194

KEEP OUT OF REACH OF CHILDREN

WARNING

See other panels for additional precautionary information.

FIRST AID

If inhaled: Remove to fresh air. Apply artificial respiration if necessary.

If swallowed: Drink 1 or 2 glasses of water and induce vomiting by touching the back of the throat with a finger or blunt object. Do not induce vomiting or give anything by mouth to an unconscious person.

If in eyes: Flush eyes with plenty of water. Obtain medical attention if irriation occurs and persists.

If on skin: Wash with plenty of soap and water.

Note to Physician: Ammo WSB is a pyrethroid insecticide containing the active ingredient cypermethrin. Digestible fats, oils or alcohol may increase absorption and so should be avoided. Skin contact (vapor or powder) may be followed by transient tingling or numbness, usually of the face, but this subsides without treatment.

For Emergency Assistance, Call 716–735–3765.

PRECAUTIONARY STATEMENTS

HAZARDS TO HUMANS (AND DOMESTIC ANIMALS)

Warning

Harmful if swallowed or absorbed through the skin. Causes eye irritation. Do not get in eyes, on skin or on clothing. Avoid breathing vapor or spray mist. May cause allergic skin reactions. Wash thoroughly with soap and water after handling. Remove contaminated clothing and wash before reuse.

ENVIRONMENTAL HAZARDS

This product is highly toxic to bees exposed to direct treatment or blooming crops or weeds. Do not apply this product or allow it to drift to blooming crops or weeds while bees are actively visiting the treatment area.

This product is highly toxic to fish. Use with care when applying in area adjacent to any body of water. Do not apply directly to water. Do not apply when weather conditions favor drift from treated areas. Do not contaminate water when disposing of equipment washwaters.

DIRECTIONS FOR USE

It is a violation of Federal law to use this product in a manner inconsistent with its labeling.

Do not remove package from container except for immediate use.

STORAGE AND DISPOSAL

Pesticide Storage

Keep out of reach of children and animals. Store in original containers only. Store in a cool dry place and avoid excess heat. Do not store at temperatures below 32°F (0°C). Rough handling may cause breakage, especially at low temperatures. Allow to warm above 50°(10°C) for use. Do not allow inner bags to become wet during storage. Do not handle inner bag with wet hands or wet gloves. Do not put concentrate or dilute material into food or drink containers. Do not contaminate other pesticides, fertilizers, water, food, or feed by storage or disposal.

In case of spill, avoid contact, isolate area and keep out animals and unprotected persons. Confine spills. Call FMC collect: (716) 735–3765.

To confine spill: If liquid, dike surrounding area or absorb with sand, cat litter, or commercial clay. If dry material, cover to prevent dispersal. Place damaged package in a holding container. Identify contents.

Pesticide Disposal

Pesticide wastes are toxic. Improper disposal of excess pesticide, spray mixture, or rinsate is a violation of Federal law. If these wastes cannot be disposed of by use according to label instructions, contact your State Pesticide or Environmental Control Agency, or the Hazardous Waste representative of the nearest EPA Regional Office for guidance.

Container Disposal

When all water soluble bags are used, the outer package should be clean and may be disposed of in a sanitary landfill, by incineration or, if allowed by state and local authorities, by burning. If burned, stay out of smoke. If the outer container contacts formulated product in any way, it must be triple rinsed with clean water. Add rinsate to the spray tank and dispose of the outer package as described above.

CHEMIGATION USE DIRECTIONS

Apply this product only through sprinkler, including center pivot, lateral move, end tow, side (wheel) roll, traveler, big gun, solid set, or hand move irrigation systems. Do not apply this product through any other type of irrigation system. Do not connect an irrigation system (including greenhouse systems) used for pesticide application to a public water system.

Crop injury, lack of effectiveness, or illegal pesticide residues in the crop can result from nonuniform distribution of treated water. If you have questions about calibration, you should contact State Extension Service specialists, equipment manufacturers, or other experts. A person knowledgeable of the chemigation system and responsible for its operation, or under the supervision of the responsible person, shall shut the system down and make necessary adjustments should the need arise.

The system must contain a functional check valve, vacuum relief valve, and low pressure drain approximately located on the irrigation pipeline to prevent water source contamination from backflow.

The pesticide injection pipeline must contain a functional, automatic, quick-closing check valve to prevent the flow of fluid back toward the injection pump.

The pesticide injection pipeline must also contain a functional, normally closed, solenoid-operated valve located on the intake side of the injection pump and connected to the system interlock to prevent fluid from being withdrawn from the supply tank when the irrigation system is either automatically or manually shut down.

The system must contain functional interlocking controls to automatically shut off the pesticide injection pump when the water pump motor stops.

The irrigation line or water pump motor must include a functional pressure switch which will stop the water pump motor when the water pressure decreases to the point where pesticide distribution is adversely affected.

Systems must use a metering pump, such as a positive displacement injection pump (e.g., diaphragm pump) effectively designed and constructed of materials that are compatible with pesticides and capable of being fitted with a system interlock.

Do not apply when wind speed favors drift beyond the area intended for treatment.

For Ammo WSB mixing instructions, see general instructions. Ammo WSB should be diluted in sufficient volume of water to ensure accurate application over the area to be treated. Add the recommended number of water soluble bag(s) to the supply tank. Maintain sufficient agitation during both mixing and application to ensure that the bags dissolve and that there is uniformity of the supply tank suspension. Hydraulic or mechanical agitation is recommended. The solution of Ammo WSB should be applied continuously for the duration of the application. When using chemigation, a minimum of 0.5 inch per ace of irrigation water is recommended.

GENERAL INSTRUCTIONS

Re-Entry Statements

Do not treat areas while unprotected humans or domestic animals are present in the treatment area.

Do not allow entry into treated areas without protective clothing until sprays have dried.

State re-entry regulations will apply if they are more restrictive than those stated on this label. Consult with state regulatory officials.

The product in this outer package is a wettable powder formulation of the insecticide cypermethrin packaged in a translucent water soluble bag. Do not allow the inner bag to become wet before adding to spray or nurse tank. Do not handle the inner bag with wet hands or wet gloves. Rough handling may cause breakage. Do not store at temperatures below 32°F (0°C). Allow to warm above 50°F (10°C) before use. Cooler water temperatures increase the time needed for the inner bag to dissolve completely.

Ammo WSB insecticide is intended for dilution with water for spray application. Do not use strainers finer than 50 mesh size. Determine the number of water soluble bag(s) to make up necessary spray suspension.

Use a minimum spray volume of 5 gallons of water per acre. Fill the spray or nurse tank 1/2 full with water. Open the outer wrapper of this product and immediately dump required contents into spray tank. Allow the water soluble bag(s) to dissolve completely. Start hydraulic or mechanical agitation. Air agitation is not recommended. Mix thoroughly to fully disperse and suspend the wettable powder. Fill the spray tank with the required amount of water. Maintain agitation during storage in nurse or supply tank and during application. Apply Ammo WSB when insects appear or feeding is noticed. Repeat the application as necessary to maintain control. Ammo WSB may be applied by both ground and aerial equipment. Use sufficient water to obtain full coverage.

Mix as needed; do not store diluted material.

Rotational Crops

Rotational crops may be planted within 30 days after last application.

Use low rate under light to moderate infestation. Higher rates should be used under heavy insect pressure. The rate of application is variable according to insect pressure, timing of spray and field scouting.

Calculate the number of bags needed for the recommended rate and number of acres to be treated by using the following formula:

$$\frac{\text{Recommended Rate for Pest (Active/Acre)} \times \text{\# Acres to be Treated with Tankload}}{0.05\ \text{(amt. ai in each bag)}} = \text{\# of Ammo Water Soluble Bags to use for acres to be treated}$$

Do not break bags. When the equation equals less than a whole number of bags, use the next higher whole number of bags. Do not exceed 2 bags per acre per application.

CROP	PEST	DOSAGE* LB/AI/A	BAG/A	REMARKS
Cotton	Cutworms Tobacco Thrips Soybean (banded) Thrips Beet Armyworm Fall Armyworm Yellow Striped Armyworm Cotton Bollworm Pink Bollworm Tobacco Budworm Tarnished Plant Bug Lygus Bug Saltmarsh Caterpillar Cotton Fleahopper Cotton Leaf-perforator Cabbage Looper Boll Weevil Other Plant Bugs White Flies**	0.05 to 0.1	1 to 2	Ammo WSB may be diluted in water only. Apply a minimum of 5 gallons of finished spray per acre by air or ground application. For boll weevil control, apply Ammo WSB at 3 to 4 day intervals until pest numbers are reduced to acceptable levels. A maximum of 0.6 lb active ingredient may be applied per acre per season. Ammo WSB may be applied within 14 days of harvest. Do not graze or feed cotton for forage.
Lettuce (Head)	Beet Armyworm Cabbage Looper Alfalfa Looper	0.05–0.1	1 to 2	Apply in water as necessary for insect control using a minimum of 15 gallons of finished spray per acre with ground equipment and 5 gallons per acre by air. Lower rates of Ammo WSB should be used with light to moderate insect pressure. Higher rates should be used to control heavy to extremely heavy insect populations. Do not apply more than 0.6 lb ai/A per season. Do not apply within 5 days of harvest.
Pecans	Hickory Shuckworm Yellow Pecan Aphid Black Pecan Aphid Pecan Nut Casebearer Pecan Weevil	0.05 to 0.1	1 to 2	Applications at the lower rate should be made when pest populations are low. Rates should be increased as the pest pressure increases. Apply by ground equipment to the point of drip. Use 100 gallons of dilute spray per acre for smaller trees. For larger trees which require higher gallonage to achieve adequate coverage, apply in 200 to 300 gallons of water. In order to calculate the correct number of gallons of water needed to spray one acre of your trees to the point of drip, you may need to conduct a test. If you do not know how to conduct such a test with your equipment, you should request assistance from your equipment dealer. Up to 0.6 lb ai/A per crop season may be applied prior to shuck split. Do not apply within 21 days of harvest. Do not graze livestock in treated orchards or cut treated cover crops for feed.

*1 bag = .05 lb ai

**Aids in control

Dealers Should Sell in Original Packages Only.

Terms of Sale or Use: On purchase of this product buyer and user agree to the following conditions:

Warranty: FMC makes no warranty, expressed or implied, concerning the use of this product other than indicated on the label. Except as so warranted, the product is sold as is. Buyer and user assume all risk of use and/or handling and/or storage of this material when such use and/or handling and/or storage is contrary to label instructions.

Directions and Recommendations: Follow directions carefully. Timing and method of application, weather and crop conditions, mixture with other chemicals not specifically recommended and other influencing factors in the use of this producut are beyond the control of the seller and are assumed by the buyer at his own risk.

Use of Product: FMC's recommendations for the use of this product are based upon tests believed to be reliable. The use of this product being beyond the control of the manufacturer, no guarantee, expressed or implied, is made as to the effects of such or the results to be obtained if not used in accordance with directions or established safe practice.

Damages: Buyer's or user's exclusive remedy for damages for breach of warranty or negligence shall be limited to direct damages not exceeding the purchase price paid and shall not include incidental or consequential damages.

FMC—FMC Trademark (1194–6/15/89–A)

This is a specimen label. FMC is not responsible for the accuracy of the information contained herein. As labels are subject to revision, always carefully read and follow the label on the product container.

Capture® 2 EC

INSECTICIDE/MITICIDE

RESTRICTED USE PESTICIDE

Toxic to fish and aquatic organisms.

For retail sale to and use only by certified applicators, or persons under their direct supervision and only for the uses covered by the certified applicator's certification.

ACTIVE INGREDIENT:	BY WT.
Bifenthrin, (2 methyl[1,1'-biphenyl]-3-yl) methyl 3-(2-chloro-3,3,3-trifluoro-1-propenyl)-2,2-dimethyl-cyclopropanecarboxylate*	25.1%
INERT INGREDIENTS:**	74.9%
TOTAL	100.00%

EPA Reg. No. 279-3069 EPA Est. 279-Code 777

*Cis isomers 97% minimum, trans isomers 3% maximum.
**Contains xylene range aromatic solvents.

This product contains 2 pounds active ingredient per gallon.
U.S. Patent No. 4,238,505

KEEP OUT OF REACH OF CHILDREN
WARNING

This label must be in the possession of the user at the time of application.

STATEMENT OF PRACTICAL TREATMENT

If Swallowed: Get medical attention. Call a Poison Control Center or physician promptly for advice. Describe Precautionary Statements and Note to Physician on the label. Do not induce vomiting unless advised by a physician or qualified medical advisor. Do not give anything by mouth to an unconscious person.

If Inhaled: Remove victim to fresh air. If not breathing, give artificial respiration, preferably mouth-to-mouth. Get medical attention.

If on Skin: Wash with plenty of soap and water. Get medical attention.

If in Eyes: Flush with plenty of water. Call a physician if irritation persists.

Note to Physician:
Pesticide Hotline (800) 858-7378. This product is a pyrethroid. If large amounts have been ingested, the stomach and intestines should be evacuated. Treatment is symptomatic and supportive. Digestible fats, oils, or alcohol may increase absorption and so should be avoided.

For Emergency Assistance call 716-735-3765

See other panels for additional precautionary information.

PRECAUTIONARY STATEMENTS

HAZARDS TO HUMANS (AND DOMESTIC ANIMALS)

Warning

May be fatal if swallowed. Harmful if inhaled, or absorbed through skin. Causes moderate eye irritation. Avoid breathing vapor or spray mist. Avoid contact with skin, eyes or clothing. Wash thoroughly with soap and water after handling. Remove contaminated clothing and wash contaminated clothing before reuse.

Applicators must wear long sleeve shirt and trousers. Mixers and loaders must wear long sleeve shirt, trousers, chemical resistant gloves, and goggles or face shield.

ENVIRONMENTAL HAZARDS

This pesticide is extremely toxic to fish and aquatic invertebrates. Drift and runoff from treated areas may be hazardous to aquatic organisms in neighboring areas. Do not apply directly to water or wetlands (swamps, bogs, marshes and potholes). Do not contaminate water when disposing of equipment washwaters.

This product is highly toxic to bees exposed to direct treatment on blooming crops or weeds. Do not apply this product or allow it to drift while bees are actively visiting the treatment area.

PHYSICIAL/CHEMICAL HAZARDS

Do no use or store near heat or open flame.

DIRECTIONS FOR USE

It is a violation of Federal law to use this product in a manner inconsistent with its labeling.

REENTRY STATEMENTS

Do not apply this product in such a manner as to directly or through drift expose workers or other persons. The area being treated must be vacated by unprotected persons. Do not enter treated areas without protective clothing until sprays have dried. Because certain states may require more restrictive reentry intervals for crops treated with this product, consult your State Department of Agriculture for further information.

CHEMIGATION USE DIRECTIONS

Apply this product only through sprinkler including center pivot, lateral move, end tow, side (wheel) roll, traveler, big gun, solid set, or hand move irrigation systems. Do not apply this product through any other type of irrigation system. Do not connect an irrigation system (including greenhouse systems) used for pesticide application to a public water system.

Crop injury, lack of effectiveness, or illegal residues in the crop can result from nonuniform distribution of treated water. If you have questions about calibration, you should contact State Extension Service specialists, equipment manufacturers, or other experts. A person knowledgeable of the chemigation system and responsible for its operation, or under the supervision of the responsible person, shall shut the system down and make necessary adjustments should the need arise.

The system must contain a functional check valve, vacuum relief valve, and low pressure drain appropriately located on the irrigation pipeline to prevent water source contamination from backflow.

The pesticide injection pipeline must contain a functional, automatic, quick-closing check valve to prevent the flow of fluid back toward the injection pump.

The pesticide injection pipeline must also contain a functional, normally closed, solenoid-operated valve located on the intake side of the injection pump and connected to the system interlock to prevent fluid from being withdrawn from the supply tank when the irrigation system is either automatically or manually shut down.

The system must contain functional interlocking controls to automatically shut off the pesticide injection pump when the water pump motor stops.

The irrigation line or water pump must include a functional pressure switch which will stop the water pump motor when the water pressure decreases to the point where pesticide distribution is adversely affected.

Systems must use a metering pump, such as a positive displacement injection pump (e.g., diaphragm pump) effectively designed and constructed of materials that are compatible with pesticides and capable of being fitted with a system interlock.

Do not apply when wind speed favors drift beyond the area intended for treatment.

Capture® 2 EC insecticide/miticide should be applied continuously for the duration of the water application. Capture 2 EC should be diluted in sufficient volume to insure accurate application over the area to be treated. When using chemigation, a minimum of 0.5 inch per acre of irrigation water is recommended. Agitation is not required when a suitable diluent is used.

STORAGE AND DISPOSAL

PESTICIDE STORAGE

Do not freeze. Do not store below 40° F. If crystals are observed, warm material to above 60° F by placing container in warm location. Shake or roll container periodically to redissolve solids.

Keep out of reach of children and animals. Store in original containers only. Store in a cool, dry place and avoid excess heat. Carefully open containers. After partial use, replace lids and close tightly. Do not put concentrate or dilute material into food or drink containers. Do not contaminate other pesticides, fertilizers, water, food, or feed by storage or disposal.

In case of spill, avoid contact, isolate area and keep out animals and unprotected persons. Confine spills. Call FMC collect: (716) 735-3765.

To confine spill: If liquid, dike surrounding area or absorb with sand, cat litter or commercial clay. If dry material, cover to prevent dispersal. Place damaged package in a holding container. Identify contents.

PESTICIDE DISPOSAL

Pesticide wastes are toxic. Improper disposal of excess pesticide, spray mixture, or rinsate is a violation of Federal law. If these wastes cannot be disposed of by use according to label instructions, contact your State Pesticide or Environmental Control Agency, or the Hazardous Waste representative at the nearest EPA Regional Office for guidance.

CONTAINER DISPOSAL

Metal or Plastic Container: Triple rinse (or equivalent) then offer for recycling or reconditioning, or puncture and dispose of in a sanitary landfill, or by other procedures approved by State and local authorities. Do not cut or weld metal containers.

U-Turn® Container. Do not rinse container. Do not empty remaining formulated product. Do not break seals. Return intact to point of purchase.

ENDANGERED SPECIES RESTRICTIONS

This product may pose a hazard to endangered aquatic species. Do not apply this product within 100 yards of aquatic habitat in the following counties:

Alabama

Colbert, Greene, Jackson, Lamar, Lauderdale, Limestone, Madison, Marshall, Morgan, Pickens and Sumter.

Arizona

Graham, Maricopa, Mohave, Pima, Pinal and Santa Cruz

Arkansas

Benton, Clay, Clark, Cross, Lawrence, Lee, Poinsette, Polk, Randolph, Sharp and St. Francis

California

Butte, Colusa, Glenn, Imperial, Kern, Merced, Modoc, Riverside, Sacramento, Solano, Sutter, Tehema and Yolo

Florida

Broward, Dade, Glades and Palm Beach

Kentucky

Ballard, Butler, Edmundson, Green, Hart, Jackson, Laurel, Livingston, Marshall, McCracken, McCreary, Pulaski, Rockcastle, Taylor, Warren and Wayne

Mississippi

Claiborne, Copiah, Hinds, Itawamba, Lowndes, Monroe and Noxubee

Missouri

Barry, Benton, Camden, Christian, Dallas, Greene, Hickory, Jasper, Lawrence, Miller, Newton, Osage, Polk, St. Clair, Stone and Webster.

Nevada

Clark

New Mexico

Chaves, Debacca and Eddy

North Carolina

Edgecombe, Nash and Pitt

Ohio

Pickaway

Oklahoma

Delaware, McCurtain and Pushmataha

Oregon

Lake

Tennessee

Bedford, Blount, Claiborne, Decatur, Franklin, Hancock, Hardin, Hickman, Knox, Lawrence, Lincoln, Loudon, Marshall, Maury, Meigs, Monroe, Rhea, Roane, Scott, Sequatchie, Smith, Sullivan and Wayne

Texas

Bastrop, Burleson, Comal, Harris, Hays, Jeff Davis, Pecos and Reeves

Utah

Utah and Washington

Virginia

Lee, Russell, Scott, Smyth, Tazewell, Washington and Wise

SPECIAL EQUIPMENT

Capture 2 EC should be mixed and loaded using systems that comply with accepted standards that identify a closed mixing-loading system. That system should be designed and operated in such a manner to minimize human exposure to the formulated product or the mixed spray solution of Capture. Follow State directions for closed systems. For more details, contact your State Extension Service.

Application Instructions

Rate of application is variable according to pest pressure, timing of sprays, and field scouting. Use lower rates under light to moderate infestations; higher rates under heavy insect pressure and for mite control. Arid climates generally require higher rates.

Crop	Pest	Dosage: Pound Active Per Acre	Dosage: Ounces 2 EC Per Acre	Remarks
Cotton	European Corn Borer Soybean (Banded) Thrips Tobacco Thrips	0.02-0.1	1.3-6.4	Capture 2 EC may be applied in water or refined vegetable oil (soybean/cottonseed).
	Boll Weevil Bollworm Cabbage Looper Cotton Aphid Cotton Fleahopper Cotton Leafperforator Cutworms Fall Armyworm Lygus Spp. Plant Bugs Saltmarsh Caterpillar Southern Garden Leafhopper Stink Bugs Tobacco Budworm Whitefly Yellow Striped Armyworm	0.04-0.1	2.6-6.4	Application in Water: Apply in a minimum of 5 gallons per acre with ground equipment or 1 gallon per acre by aircraft. When applying by air, 1 quart of emulsified oil may be substituted for one quart of water in the finished spray. ULV Application: Apply the recommended rate of Capture 2 EC in refined vegetable oil in a minimum of 1 quart of finished spray per acre with aircraft calibrated to give adequate coverage. To Control Boll Weevil: Apply Capture 2 EC at an interval of 3 to 4 days until pest numbers are reduced to acceptable levels.
	Carmine Spider Mite Twospotted Spider Mite Beet Armyworm Pink Bollworm	0.06-0.1	3.8-6.4	To Control Mites and Aphids: Apply when pests first appear. Repeat as necessary to maintain control. Higher rates will be required once a damaging threshold is established.

Do not apply more than 0.5 pound active per acre per season.

Do not apply within 14 days of harvest.

Do not graze livestock in treated areas or cut treated crops for feed.

Rotational Crops

Rotational crops may be planted no sooner than 30 days after last application. Straw may not be used for food or feed.

TANK-MIXTURE

Capture 2 EC insecticide/miticide may be applied in tank-mixtures with other products approved for use on cotton. Observe all restrictions and precautions which appear on the labels of these products.

DEALERS SHOULD SELL IN ORIGINAL PACKAGES ONLY.

TERMS OF SALE OR USE: On purchase of this product buyer and user agree to the following conditions:

WARRANTY: FMC makes no warranty, expressed or implied, concerning the use of this product other than indicated on the label. Except as so warranted, the product is sold as is. Buyer and user assume all risk of use and/or handling and/or storage of this material when such use and/or handling and/or storage is contrary to label instructions.

DIRECTIONS AND RECOMMENDATIONS: Follow directions carefully. Timing and method of application, weather and crop conditions, mixture with other chemicals not specifically recommended and other influencing factors in the use of this product are beyond the control of the seller and are assumed by the buyer at his own risk.

USE OF PRODUCT: FMC's recommendations for the use of this product are based upon tests believed to be reliable. The use of this product being beyond the control of the manufacturer, no guarantee, expressed or implied, is made as to the effects of such or the results to be obtained if not used in accordance with directions or established safe practice.

DAMAGES: Buyer's or user's exclusive remedy for damages for breach of warranty or negligence shall be limited to direct damages not exceeding the purchase price paid and shall not include incidental or consequential damages.

(777-8/5/88-A)

This is a specimen label. FMC is not responsible for the accuracy of the information contained herein. As labels are subject to revision, always carefully read and follow the label on the product container.

Carbamate WDG

FUNGICIDE

For Agricultural or Commercial Use Only

ACTIVE INGREDIENT:	BY WT.
Ferbam (Ferric dimethyl dithiocarbamate)	76.0%
INERT INGREDIENTS:	24.0%
TOTAL	100.0%

EPA Reg. No. 45728-7-279 EPA Est., Code 93

CAUTION

KEEP OUT OF REACH OF CHILDREN

For Emergency Assistance Call 716-735-3765.

See other panels for additional precautionary information.

STATEMENT OF PRACTICAL TREATMENT

It in eyes: Flush affected eyes with water for 15 minutes. Get medical attention.

If on skin: Remove any contaminated clothing. Wash exposed skin with plenty of soap and water. If person feels unwell, get medical attention immediately.

If inhaled: Move exposed person to fresh air and allow to rest.

If swallowed: If conscious, wash out mouth and give patient plenty of water. Never give anything by mouth to an unconscious person. Get medical attention immediately, taking this labeled container with you. If medical aid is not readily available, induce vomiting and repeat until vomit fluid is clear.

PRECAUTIONARY STATEMENTS

HAZARDS TO HUMANS AND DOMESTIC ANIMALS

CAUTION

May cause irritation of eyes, nose, throat and skin! Avoid breathing dust or spray mist. Avoid contact with eyes, skin or clothing.

ENVIRONMENTAL HAZARDS

This product is toxic to fish. Keep out of lakes, streams, ponds, tidal marshes and estuaries. Do not apply where runoff is likely to occur. Do not apply when weather conditions favor drift from areas treated. Do not contaminate water by cleaning of equipment, or disposal of wastes. Apply this product only as specified on this label. Do not contaminate water when disposing of equipment wash waters.

Do not allow this material to drift onto neighboring crops or non-crop areas or use in a manner or at a time other than in accordance with label directions because animal, plant or crop injury, illegal residues or other undesirable results may occur.

PHYSICAL/CHEMICAL HAZARDS

Keep packages or dust away from sparks, flames or hot surfaces.

DIRECTIONS FOR USE

It is a violation of Federal Law to use this product in a manner inconsistent with its labeling.

General Use Information

Carbamate WDG (76% water-dispersible granules) is a broad spectrum fungicide for the control of certain diseases in fruit trees, smallfruits and berries, ornamentals, conifers, and tobacco.

FAILURE TO FOLLOW DIRECTIONS AND PRECAUTIONS ON THIS LABEL MAY RESULT IN CROP INJURY, POOR DISEASE CONTROL, OR ILLEGAL RESIDUES.

REENTRY AND WORKER PROTECTION STATEMENTS

Do not apply this product in such a manner as to directly or through drift expose workers or other persons. The area being treated must be vacated by unprotected persons.

Do not enter treated areas without protective clothing until sprays have dried.

Protective clothing means, at least, a hat or other suitable head covering, a long sleeved shirt and long legged trousers or a coverall type garment (all of closely woven fabric covering the body, including the arms and legs), shoes and socks.

To avoid inhalation when using this product, the use of an efficient mask or respirator is recommended. Eye protection is also recommended.

Wash hands and exposed skin thoroughly after handling and before eating, drinking, or smoking. Remove any contaminated clothing and launder before reuse.

Because certain states may require more restrictive reentry intervals for various crops treated with this product, consult your State Department of Agriculture for further information.

Written or oral warnings must be given to workers who are expected to be in a treated area or in an area about to be treated with this product. Inform workers of areas or fields that may not be entered without specific protective clothing, period of time field must be vacated and appropriate actions to take in case of accidental exposure. An example of such information is given under written warnings. When oral warnings are given, warnings shall be given in a language customarily understood by workers. Oral warnings must be given if there is reason to believe that written warnings cannot be understood by workers. Written warnings must include the following information:

CAUTION

Area treated with Carbamate WDG on (insert date). Do not enter without appropriate protective clothing until sprays have dried (insert State Department of Agriculture's re-entry interval, if more restrictive).

In case of accidental exposure to pesticide spray or dust, wash the skin thoroughly with soap and water. Remove contaminated clothing and wash before reuse. If in eyes, flush with plenty of water. If inhaled, go to an area where the pesticide has not been applied. Get medical attention if needed.

STORAGE AND DISPOSAL

PESTICIDE STORAGE

Keep away from fire and sparks.

Keep out of reach of children and animals. Store in original containers only. Store in a cool, dry place and avoid excess heat. Carefully open containers. After partial use, fold and roll back bags, clamp and close tightly. Do not put concentrate or dilute material into food or drink containers. Do not contaminate other pesticides, fertilizers, water, food or feed by storage or disposal.

In case of spill, avoid contact, isolate area and keep out animals and unprotected persons. Confine spills. Call FMC collect: (716) 735-3765.

To confine spill: If liquid, dike surrounding area or absorb with sand, cat litter or commercial clay. If dry material, cover to prevent dispersal. Place damaged package in a holding container. Identify contents.

PESTICIDE DISPOSAL

Pesticide wastes are toxic. Improper disposal of excess pesticide, spray mixture, or rinsate is a violation of Federal Law. If these wastes cannot be disposed of by use according to label instructions, contact your State Pesticide or Environmental Control Agency, or the Hazardous Waste representative at the nearest EPA Regional Office for guidance.

CONTAINER DISPOSAL

Completely empty bag into application equipment. Then dispose of empty bag in a sanitary landfill or by incineration, or, if allowed by State and local authorities, by burning. If burned, stay out of smoke.

Mixing Instructions: wear rubber gloves, long sleeves, and eye protection when mixing, and loading this product into spray equipment. Be sure equipment is clean before use; if it is contaminated with any material, crop injury or sprayer clogging may result. Calibrate spray equipment before use. Fill the spray tank one-fourth to one-third full with clean water. Start agitation (proper agitation creates a rippling or rolling action on the liquid surface). Slowly add Carbamate WDG at the recommended rate. Add water to fill spray tank. Mix thoroughly and apply at once. Maintain sufficient agitation during mixing and spraying operations to ensure a uniform spray mixture.

IMPORTANT: When spraying, thorough coverage is necessary to provide good disease control. Make new dilutions for each spraying. Consult State Agricultural Experiment Station recommendations or State Extension Service Specialist for spray schedules for timing and frequency of applications in your area.

When tank mixing other products with Carbamate WDG, follow the proper sequence of adding products to the spray tank. Wettable powders or water-dispersible granules should be added to the water in the tank first, followed by flowable products, with emulsifiable concentrates added last. Provide sufficient agitation during mixing and application to maintain a uniform emulsion.

Carbamate WDG is compatible with most commonly used insecticides and fungicides except Bordeaux mixture, lime sulfur, fixed copper and hydrated lime. When preparing tank mixes, user should consult spray compatibility charts or State Extension Service Specialist prior to actual use.

To assure the compatibility of Carbamate WDG with other products, pour the products into a small container of water in the correct proportions. After thorough mixing, let stand for five minutes. If the combination remains mixed or can be remixed readily, the mixture is compatible.

BEFORE TANK MIXING CARBAMATE WDG WITH OTHER REGISTERED PRODUCTS FOR ANY USE ON THIS LABEL, READ THE LABEL OF THE TANK MIX PARTNER TO BE CERTAIN IT IS LABELED FOR USE ON THE PARTICULAR CROP AND THAT USE PATTERNS ARE COMPATIBLE WITH THOSE OF CARBAMATE WDG. OBSERVE ALL CAUTIONS AND LIMITATIONS ON LABELING OF ALL PRODUCTS USED IN MIXTURES.

APPLICATION EQUIPMENT

Aerial Application Procedures

Avoid application under conditions where uniform coverage cannot be obtained or where excessive spray drift may occur. Do not apply directly to humans or animals. Flagmen and loaders should wash thoroughly before easting, drinking, or smoking, and at the end of each day's operation.

Application through Irrigation Systems

Carbamate WDG alone or in combination with other fungicides which are registered for application through irrigation systems, may be applied through irrigation systems. Apply this product only through center pivot, solid set, moving wheel, micro-sprinkler or drip irrigation systems. Do not apply this product through any other type of irrigation system. Crop injury, lack of effectiveness, or illegal pesticide residues in the crop can result from non-uniform distribution of treated water. If you have questions about calibration, you should contact State Extension Service specialists, equipment manufacturers or other experts. Do not connect an irrigation system (including greenhouse systems) used for pesticide application to a public water system unless the pesticide label-prescribed safety devices for public water systems are in place. A person knowledgeable of the chemigation system and responsible for its operation, or under the supervision of the responsible person, shall shut the system down and make necessary adjustments should the need arise.

Operating Instructions (Sprinkler and Drip Irrigation Systems)

1. The system must contain a functional check valve, vacuum relief valve, and low pressure drain appropriately located on the irrigation pipeline to prevent water-source contamination from backflow.
2. The pesticide injection pipeline must also contain a functional, automatic, quick-closing check valve to prevent the flow of fluid back toward the injection pump.
3. The pesticide injection pipeline must also contain a functional, normally closed, solenoid-operated valve located on the intake side of the injection pump and connected to the system interlock to prevent fluid from being withdrawn from the supply tank when the irrigation system is either automatically or manually shut down.
4. The system must contain functional interlocking controls to automatically shut off the pesticide injection pump when the water pump motor stops.
5. The irrigation line or water pump must include a functional pressure switch which will stop the water pump motor when the water pressure decreases to the point where pesticide distribution is adversely affected.
6. Systems must use a metering pump, such as a positive displacement injection pump (e.g., diaphragm pump) effectively designed and constructed of materials that are compatible with pesticides and capable of being fitted with a system interlock.
7. Do not apply when wind speed favors drift beyond the area intended. (Applies to sprinkler systems.)

Application instructions

Carbamate WDG must be applied on the schedule specified in the specific crop use recommendations, not according to the irrigation schedule.

With the exception of cranberries. Carbamate WDG has not bee sufficiently tested when applied through irrigation systems to assure consistent product performance for all labeled uses. The following calibration and application techniques are provided for user reference, but do not constitute a warranty of fitness for application through sprinkler or drip irrigation equipment. Users should check with state and local regulatory agencies for potential use restrictions before applying any agricultural chemical through sprinkler or drip irrigation equipment.

- Center Pivot Irrigation Equipment (Use only with drive systems which provide uniform water distribution).
- Determine the size of the area to be treated.
- Determine the time required to apply 1/4-inch of water over the area to be treated when the system and injection equipment are operated at normal pressures recommended by the equipment manufacturer. Run the system at 80-95% of the manufacturer's rated capacity.
- Using water, determine the injection pump output when operated at normal line pressure.
- Determine the amount of Carbamate WDG required to treat the area covered by the irrigation system.
- Add the required amount of Carbamate WDG and sufficient water to meet the injection time requirements to the solution tank.
- Maintain constant solution tank agitation during the injection period.
- Continue to operate the system until the Carbamate WDG solution has cleared the sprinkler head.

Solid Set and Moving Wheel Irrigation Equipment

- Determine the acreage covered by the sprinklers.
- Fill injector solution tank with water and adjust flow rate to use the contents over a 10 to 30-minute interval.
- Determine the amount of Carbamate WDG required to treat the area covered by the irrigation system.
- Add the required amount of Carbamate WDG into the same quantity of water used to calibrate the injection period.
- Operate the system at the same pressure and time interval established during the calibration.
- Inject Carbamate WDG at the end of the irrigation cycle or as a separate application to maximize the effectiveness of the fungicide.
- Stop injection equipment after treatment is completed. Continue to operate the system until the Carbamate WDG solution has cleated the last sprinkler head.

Micro-sprinkler or Drip Irrigation Systems

General Instructions

- Each run of the irrigation system must be calibrated separately to determine the time it takes water to move through the system and to make sure all emitters in the system are putting out the same amount of water.
- Only pressure injection or venturi equipment are recommended.
- Figure the area to be treated in each irrigation run.
- Measure the output of each of the emitters or drip tubes closest to and farthest from the injector site.

• For calibration, substitute a concentrated detergent (such as Wisk) or a soluble fertilizer for the Carbamate WDG in the injector tank. The detergent will bubble as it leaves the emitters. The time period over which bubbles occur should be checked for both the closest and furthest emitters. If these times are not within two minutes of each other, adjust the dilution ratio and/or the injection rate.

- If a soluble fertilizer is used, measure the time intervals with a salt bridge. If a drip system is being calibrated, substitute soluble fertilizer for the Carbamate WDG in the injector and measure the time intervals with a salt bridge.

Step-by-Step Instructions

- Before starting to calibrate, operate the system until all the emitters are putting out at equal flow rates or until the system is operating at full pressure.
- Make up an indicator solution of detergent or fertilizer, using the same ratio to be used when mixinf Carbamate WDG.
- Set the injector to apply the indicator solution at the injection rate to be used in the actual Carbamate WDG.
- Attach a 5-inch length of flexible tubing over the emitter closest to the injection point, another length over the emitter farthest away. Both emitters should be monitored to determine the time intervals that the indicator solutions are observed.
- Begin injecting the indicator solution. Direct the flow from the tubes at the emitters into a small container. Begin timing when the indicator solution is first detected, stop timing when the indicator solutions are no longer detected.
- If the period of detection of the indicator solution between the wto emitters are within two minutes of each other, comparable coverage will be obtained. If they are not, make adjustments by increasing the dilution ratio, using more water per part of Carbamate WDG, or adjust the injector to a slower flow rate.
- Once the system is calibrated, dilute the needed amount of Carbamate WDG with water using a minimum of 10 parts water to 1 part Carbamate WDG.
- Do not begin to inject Carbamate WDG into the system until all emitters are producing equal flow rates or until the system is at full pressure.
- Inject the Carbamate WDG into the system at the end of the irrigation in 1/2-1 inch of irrigation water.

RECOMMENDATIONS

Do not apply within the time interval between last application and harvest which is indicated in days by the number in ( ) following each crop.

Under low disease conditions, minimum label reates per application can be used while maximum label rates and shortened spray schedules are recommended for severe or threatening disease conditions.

THE FOLLOWING DILUTION RATES ARE BASED ON 100 GALLONS OF SPRAY.

APPLES (7): For early season control of Scab, Cedar Rust, Blotch—Use 1 to 1 1/2 pounds of this material, or 1/2 pound plus 3 pounds of wettable sulfur per 100 gallons of water in prebloom, calyx and early cover sprays. For Rust control, make an additional bloom application using 1/2 pound per 100 gallons of water. For late season control of Scab, Black Rot, Bitter Rot, Sooty Blotch, Fly Speck, and Brooks' Spot—Use 3/4 pound per 100 gallons of water. Where Bitter Rot and Brooks' Spot are severe problems increase dosage to 1 1/2 to 2 pounds per 100 gallons of water. On Golden Delicious, Jonathan, Baldwin and other varieties of Apples which are especially apt to Russet, some injury may result from the use of this type of material if used in the early season or bloom applications. The best time for use is in the cover sprays.

When used in combination with sulfur, do not apply within 10 days to 2 weeks of an oil spray nor just prior to or during periods of excessively high temperatures.

Cherries (0): Brown Rot, Leaf Spot—Use 1 1/2 pounds per 100 gallons of water in petal fall and cover sprays, or alternately, use 1 pound of Carbamate WDG plus 3 pounds of wettable sulfur per 100 gallons of water in petal fall and cover spray.

To aid in control of leaf spot, apply 1½ pounds of Carbamate WDG per 100 gallons of water immediately after harvest but prior to leaf drop.

CITRUS (O) (Grapefruit, Lemons, Limes, Oranges, Tangerines, Tangelos): Scab—Use 1½ pounds per 100 gallons of water. Do not apply more than 1000 gallons of spray per acre. Apply during prebloom period and when ⅔ of petals have fallen. May be applied in late summer or early fall if a heavy flush of growth appears.

Citrus (Rough Lemon Nusery Stock only): Scab—Use 1½ to 2.6 pounds per 100 gallons. Apply at 7 to 10 day intervals and after heavy rains during growing periods.

CRANBERRIES: Fruit Rots, Lophodermium Twig Blight—Use 2 pounds per 100 gallons of water. Begin applications early in blossoming period and repeat at two week intervals for a total of 5 applications. Do not apply within 28 days after mid-bloom. Carbamate WDG may be applied through sprinkler irrigation equipment. Use 300 gallons of water per acre through solid set systems only. See calibration directions preceding this section. Fairy Ring—Use 9 pounds per 100 gallons of water and apply 1 gallon of this mixture to a one foot square area. Treat an area 3 feet beyond the advancing line of dead vines and 2 feet within this line. Apply in the fall immediately after harvest.

Restriction: Do not use water from treated cranberry bogs for irrigating other crops.

GRAPES (7): Black Rot—Use 2 pounds per 100 gallons of water, applying just before bloom, just after bloom and as disease conditions warrant. Do not make more than 3 applications per season.

PEACHES (21): Leaf Curl—Use 1½ pounds per 100 gallons of water. Apply during the dormant period in the fall after leaves drop or in the spring before buds begin to swell. If Leaf Curl has been severe, make two applications, one in fall and one in spring during the dormant period.

PEARS (7): Scab, Sooty Blotch—Use 1½ pounds per 100 gallons of water in pink, calyx, first and second cover sprays, and 1 pound in summer sprays.

TOBACCO (Plant Beds): Blue Mold—Use 2 to 3 pounds per 100 gallons of water. Apply at the rate of 3 gallons per 100 square yards when plants are small, increasing to 6 gallons when plants are ready for transplanting. Begin applications when plants are the size of a dime, or when Blue Mold is reported in the area, and repeat twice weekly until plants are transplanted.

Conifers

Fusiform Rust (in pine seedling plant beds): Use 1 to 2 pounds per 100 gallons of water and apply at 3 to 7-day intervals as needed, beginning when plants emerge.

Southern Cone Rust (Cronartium stroblinum) in Slash and Long Leaf Pines: Use 2 pounds plus 4 to 8 ounces of suitable spreader-sticker. Begin at time of flowering when strobili are emerging from bud scales. Repeat at 5-day intervals until conelet stage. Direct spray to flower-bearing portions of tree crowns.

Botrytis (in nursery conifer seedling crops): Use 1½ to 2 pounds per 100 gallons of water. Use a hydraulic sprayer and add a suitable spreader-sticker to the tank mix. Spray for complete coverage; wetting the foliage thoroughly. Begin applications when cool, moist conditions favor disease development. Repeat at 7 to 10 day intervals to maintain control as long as disease favorable conditons persist.

Spruce Cone Rust in Spruce: Use 2 pounds per 100 gallons of water at pollination. Spray to run off. To ensure all trees are treated at pollination, applications may be required 1 week before and after pollination.

Melampsora Needle Rust in Douglas Fir Christmas trees. Use 2 pounds per 100 gallons of water. Make a single application in the spring when new shoot growth is ½ to 2 inches in length.

Puccineastrum Needle Rust in Shasta Fir Christmas trees: Use 2 pounds per 100 gallons of water. Make a single application in the spring when new shoot growth is ½ to 2 inches in length.

Lophodermium Needle Cast in Conifers: Use 2 pounds per 100 gallons of water. Begin application in spring or early summer before infection occurs. Repeat after heavy rains and at 10 to 14 day intervals as long as needed.

Ornamentals

Directions, unless otherwise specified, include: Use 1 to 1½ pounds per 100 gallons of water, applying at 7 to 10 day intervals during growing season to control the diseases on the ornamentals listed below. The black residue from use of this product may be undesirable on some ornamentals.

Asters, Begonias, Columbine, Cyclamens, Delphiniums, Geraniums, Golden Glow, Hollyhocks, Hydrangeas, Iris, Ivy, Laurel, Lilacs, Morning Glory, Oriental Poppy, Palm (Areca), Phlox, Primrose, Snapdragons, Sweet Peas, Verbene, Zinnias: Leaf Spot.

Ornamental Cuttings: Damping-off, Rots—Use 1 to 1½ pounds per 100 gallons. Drench soil and dip cuttings before setting. Do not use on Hydrangea cuttings.

Azaleas: Leaf Spot, Petal Blight—Apply at 3 to 4 day intervals during bloom to plants, flowers, and litter around plants.

Cacti and Succulents: Drechslera Stem Rot—Use 2 to 3 pounds per 100 gallons.

Camellia: Petal Blight—Apply as a monthly soil drench and to soil and litter around plants at 3 to 4 day intervals during bloom.

Carnations, Chrysanthemums: Septoria Leaf Spot.

Carnations, Delphiniums, Geraniums, Hollyhocks, Marigolds, Morning Glory, Phox, Primrose, Roses, Snapdragons: Rust.

Carnations, Scheffleras: Alternaria blight.

Cyclamens: Botrytis Rot.

Gardenias: Botrytis Rot, Stemphylium and Septoria Leaf Spots—Apply at 3 to 4 day intervals.

Gladiolus: Botrytis Blight, Leaf Spot—Use 2 pounds per 100 gallons.

Juniper: Cedar Rust.

Lilies: Botrytis (fire)—Use 1½ pounds per 100 gallons of water. Begin applications when disease first appears and repeat at 10 to 14 day intervals as long as disease conditons exist. Planting densities may affect the level of control.

Orchids: Rust—Use 2 pounds per 100 gallons.

Peonies, Poinsettia, Tulips: Botrytis Blight.

Roses: Black Spot, Blossom Blight—For Black Spot, begin when first leaves unfold.

Roses, Snapdragons: Anthracnose.

DEALERS SHOULD SELL IN ORIGINAL PACKAGES ONLY.

TERMS OF SALE OR USE: On purchase of this product buyer and user agree to the following conditions:

WARRANTY: FMC makes no warranty, expressed or implied, concerning the use of this product other than indicated on the label. Except as so warranted, the product is sold as is. Buyer and user assume all risk of use and/or handling and/or storage of this material when such use and/or handling and/or storage is contrary to label instructions.

DIRECTIONS AND RECOMMENDATIONS: Follow directions carefully. Timing and method of application, weather and crop conditions, mixture with other chemicals not specifically recommended and other influencing factors in the use of this product are beyond the control of the seller and are assumed by buyer at his own risk.

USE OF PRODUCT: FMC's recommendations for the use of this product are based upon tests believed to be reliable. The use of this product being beyond the control of the manufacturer, no guarantee, expressed or implied, is made as to the effects of such or the results to be obtained if not used in accordance with directions or established safe practice.

DAMAGES: Buyer's or user's exclusive remedy for damages for breach of warranty or negligence shall be limited to direct damages not exceeding the purchase price paid and shall not include incidental or consequential damages.

FMC—FMC Trademarks

(93-6/8/88-A)

This is a specimen label. FMC is not responsible for the accuracy of the information contained herein. As labels are subject to revision, always carefully read and follow the label on the product container.

Command® 4 EC

HERBICIDE

FOR AGRICULTURAL OR COMMERCIAL USE ONLY
NOT FOR SALE OR USE IN CALIFORNIA

ACTIVE INGREDIENT:	BY WT.
Clomazone 2-(2-Chlorophenyl) methyl-4, 4-dimethyl-3-isoxazolidinone	47.1%
INERT INGREDIENTS:*	52.9%
TOTAL	100.0%

* Contains xylene range aromatic solvents

Contains 4 pounds of active ingredient per gallon

EPA Reg. No. 279-3053 — EPA Est. 279-
U.S. Patent No. 4,405,357 — Code 618

KEEP OUT OF REACH OF CHILDREN

WARNING

FIRST AID

If in eyes: Flush immediately with plenty of water. Get medical attention as soon as possible.

If swallowed: Drink promptly large quantities of milk or water. Avoid alcohol. Do not induce vomiting. Call a physician.

If inhaled: Remove to fresh air. If not breathing, give artificial respiration, preferably mouth-to-mouth. Get medical attention as soon as possible.

If on skin: Remove contaminated clothing. Wash skin with plenty of soap and water.

For Emergency Assistance Call Collect: (716) 735-3765

PRECAUTIONARY STATEMENTS

HAZARDS TO HUMANS (AND DOMESTIC ANIMALS)

WARNING

Causes eye injury (corneal opacity) that is temporary. Do not get in eyes or on clothing. Harmful if swallowed, inhaled, or absorbed through skin. Avoid breathing vapor or spray mist. Wear goggles or face shield when measuring, mixing or loading. Avoid contact with skin. Wash thoroughly with soap and water after handling. Remove contaminated clothing and wash before reuse.

ENVIRONMENTAL HAZARDS

Do not apply directly to water or wetlands. Do not apply when weather conditions favor drift from the area treated. Do not apply where runoff is likely to occur. Do not contaminate water by cleaning of equipment or disposal of wastes. Apply this product only as specified on this label.

See other panels for additional precautionary information.

SPECIAL PRECAUTION

Off-site movement of spray drift or vapors of COMMAND® 4 EC herbicide can cause foliar whitening or yellowing of some plants. Prior to making applications, read and strictly follow all precautions and application instructions on this label.

PHYSICAL/CHEMICAL HAZARDS

Do not use or store near heat or open flame.

DIRECTIONS FOR USE

It is a violation of federal law to use this product in a manner inconsistent with its labeling.

Do not apply this product through any type of irrigation system.

STORAGE AND DISPOSAL

PESTICIDE STORAGE

Do not freeze. Do not store below 40°F. If solid crystals are observed, warm material to above 60°F by placing container in warm location. Shake or roll container periodically to redissolve solids.

Keep out of reach of children and animals. Store in original containers only. Store in a dry place. Carefully open containers. After partial use, replace lids and close tightly. Do not put concentrate or dilute material into food or drink containers. Do not contaminate other pesticides, fertilizers, water, food or feed by storage or disposal.

In case of spill, avoid contact, isolate area and keep out animals and unprotected persons. Confine spills. Call FMC Collect (716) 735-3765.

To confine spill: Dike surrounding area or absorb with sand, cat litter or commercial clay. Place damaged package in a holding container. Identify contents.

PESTICIDE DISPOSAL

Pesticide wastes are toxic. Improper disposal of excess pesticide, spray mixture or rinsate is a violation of Federal law. If these wastes cannot be disposed of by use according to label instructions, contact your State Pesticide or Environmental Control Agency or the Hazardous Waste representative at the nearest EPA Regional Office for guidance.

CONTAINER DISPOSAL

Triple rinse (or equivalent). Then offer for recycling or reconditioning, or puncture and dispose of in a sanitary landfill or by incineration, or, if allowed by state and local authorities, by burning. If burned, stay out of smoke.

Returnable/Refillable Sealed Containers
Do not rinse container. Do not empty remaining formulated product. Do not break seals. Return intact to point of purchase.

IMPORTANT

FAILURE TO OBSERVE THE APPLICATION PRECAUTIONS SECTION OF THIS LABEL MAY RESULT IN INJURY TO DESIRABLE VEGETATION

- Desirable plants including some species of trees, shrubs, flowers, agronomic crops, and fruits and vegetables are sensitive to COMMAND® 4 EC herbicide.
- Foliar contact with spray drift or vapors may cause whitening of sensitive plants. Symptoms are generally temporary in nature but may persist on some plants.

SPRAYER CLEANUP

Do not drain or flush equipment on or near desirable trees or other plants, or in areas where their roots may extend or in locations where the chemical may be washed or move into contact with their roots. Do not contaminate any body of water including irrigation water that may be used on other crops. Carefully follow sprayer clean-up instructions noted below to prevent spray tank residues from damaging other crops.

Sprayer equipment should be thoroughly rinsed to remove residues of herbicide that might injure other subsequently sprayed crops. The steps below should be followed for the thorough rinsing of sprayer equipment following applications of COMMAND® 4 EC herbicide or tank mixes of COMMAND 4 EC with other labeled products.

1) Drain any remaining spray solution from tank, pump, hoses and boom and discard in an approved manner (See Note that follows).
2) Clean tank and fittings by:
   * Thoroughly hosing down the inside walls of the spray tank with a quantity of water equal to 1/8 of the total tank capacity and operating the pump to circulate this solution through the sprayer system for 15 minutes.
   * Washing down the outside surfaces of equipment.
   * Removing nozzle tip and screen from end nozzle in each boom section and allowing several gallons of rinsate solution to flush completely through boom (collect rinsate while flushing).
3) Thoroughly drain remaining rinsate solution from tank, pump and hoses. Combine with boom flushings and dispose of all rinsates from this first rinsing in an approved manner (see Note that follows).
   * When switching from water dilutions to applications utilizing crop oil or liquid fertilizer as a carrier, a small volume of crop oil or liquid fertilizer should be flushed through the tank, pump, hoses, and boom prior to the next use. Dispose of crop oil or liquid fertilizer rinsate in an approved manner (see Note for local, state and Federal guidelines).
4) Remove the remaining nozzle tips, and screens and the line filter and wash in a pail of warm soapy water, thoroughly rinse and replace.
5) Hose down the inside walls of the spray tank a second time and circulate this solution using the same procedure as noted in #2 above.
6) If the next use of the sprayer will be for applying a preemergent or preplant incorporated pesticide on any crop for which COMMAND 4 EC is registered, rinsate from this second rinsing may be utilized by diluting with the water for the next pesticide load:

HOWEVER

If the next use of the sprayer will be a postemergence applied pesticide on any crop, drain rinsate solution from this second rinsing. Retain rinsate solution for use only with a soil incorporated pesticide to be applied on a crop for which COMMAND 4 EC is registered. Refill tank (after draining second rinsate solution) in accordance with postemergence product label directions.

NOTE: Dispose of excess spray mixture and/or *rinsate from first tank rinsing* by application to cropland as described on this label. If excess spray mixture and/or *rinsate from first rinsing* cannot be disposed of according to label instructions, dispose of in compliance with local, state and federal guidelines. Contact your state pesticide or Environmental Control Agency or the Hazardous Waste representative at the nearest EPA regional office for guidance.

GENERAL MIXING INSTRUCTIONS

Care must be taken when mixing COMMAND 4 EC herbicide. Avoid mixing in areas adjacent to desirable plants. (See Section headed **"IMPORTANT"**)

COMMAND 4 EC Alone: Mix COMMAND 4 EC with water or liquid fertilizer in the following manner: Fill the spray tank one-half to three-fourths full with water or liquid fertilizer, add the proper amount of COMMAND 4 EC, then add the rest of the water or liquid fertilizer. Provide sufficient agitation during mixing and application to maintain a uniform spray mixture.

Tank Mixtures: Fill spray tank one-fourth to one-third full with water; with agitator operating add the recommended amount of ingredients using the following order: dry formulations (e.g., wettable powders, dry flowables) first, liquid suspensions (e.g., flowables) next and finally liquid (e.g., EC's). Mix thoroughly and fill tank one-half full continuing agitation. Add COMMAND 4 EC herbicide to tank while maintaining agitation. Complete filling the sprayer tank with water. Where use of a surfactant is recommended, add as the last ingredient to the spray tank. Maintain agitation during filling, mixing and application. When using drift reducing agents, follow specific product label instructions for order of addition to spray tank.

Fertilizer Spray Mixtures: Applications of COMMAND 4 EC alone, or with recommended tank mixtures, in conjunction with fertilizer solutions may be used unless use directions specifically state otherwise. Small quantities should be tested for compatibility by the following procedure before mixing tankful quantities.

1) Put 1 pint of fertilizer solution in a quart jar.
2) Add the appropriate amount of herbicide based on the tables below. If more than one product is to be used, add each separately using the following sequence: dry formulations (e.g., wettable powders, dry flowables) first, liquid suspensions (e.g., flowables) next and finally liquids (e.g., EC's).

Herbicide	Rate/Acre	Amount to be Added Per Pint of Fluid Fertilizer*
Wettable	1/2 pound	3/4 level teaspoon
Powders	1 pound	1 1/2 teaspoons
or	2 pounds	3 level teaspoons
Dry Flowables	3 pounds	4 1/2 level teaspoons
(Dry Granules)		

Herbicide	Rate/Acre	Amount to be Added Per Pint of Fluid Fertilizer*
Emulsifiable	1 pint	½ level teaspoon
Concentrates	1 quart	1 level teaspoon
or	2 quarts	2 level teaspoons
Flowables	3 quarts	3 level teaspoons

* Based on a spray volume of 25 gpa. For lower or higher spray volumes, adjust fluid fertilizer quantity accordingly.

3) Close jar and shake well.
4) Watch mixture for several seconds, again in 5 minutes and again after 30 minutes. If herbicide/fertilizer combination remains mixed or can be remixed readily (i.e., does not permanently separate, foam, gel or become lumpy), the mixture is compatible and can be mixed in full volumes and sprayed.

If the mixture is compatible, prepare spray by adding fertilizer solution to the tank first, then follow directions noted below:

1) **COMMAND 4 EC Alone:** Dilute with 2 parts of water and add slowly to spray tank of fertilizer with agitator running. Maintain agitation during application.
2) **COMMAND 4 EC plus Tank Mixtures:** Dilute individual products with two (2) parts of water, then add to the spray tank of fertilizer, while maintaining agitation, using the following order—slurry of dry formulations (wettable powders, dry flowables) first, diluted liquid formulations (EC's, flowables) second. Continue agitation during application.

APPLICATION PRECAUTIONS

Geographic Restrictions: COMMAND 4 EC must be applied only as a soil incorporated treatment in the following states, unless there is an approved SLN (24C) for counties within these states:

Connecticut	Maryland	Ohio
Delaware	Massachusetts	Pennsylvania
Illinois	Michigan	Rhode Island
Indiana	Minnesota	Vermont
Iowa	New Hampshire	Virginia
Kentucky	New Jersey	West Virginia
Maine	New York	Wisconsin

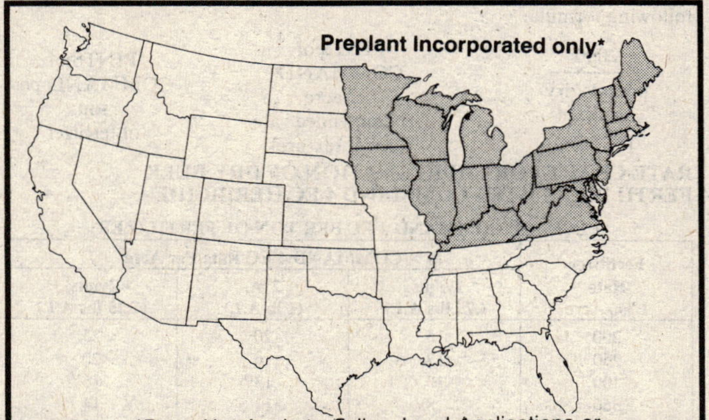

*Does Not Apply to Fallow Land Applications or Counties for which an Approved SLN (24C) is on File

For Soil Incorporated Applications:
See Geographic Restrictions above.
Do not apply COMMAND® 4 EC within 1,000 feet of the areas listed below:
Towns and Subdivisions
Commercial Vegetable Production*
Commercial Fruit Production
Commercial Nurseries
Commercial Greenhouses
* except sweet corn

Plants such as the following may show symptoms of foliar whitening or yellowing if contacted by COMMAND 4 EC herbicide. It is recommended that, prior to application, adjacent properties be checked and that spraying within 100 feet of such plants be avoided.

Trees (Deciduous)
- Apple (inc. fruit & ornamental types)
- Ash (Green, White, Mountain)
- Basswood
- Boxelder
- Catalpa
- Cherry (inc. fruit & ornamental types)
- Cottonwood
- Elm
- Ginkgo
- Hackberry
- Mulberry
- Peach
- Pear (inc. fruit & ornamental types)
- Pecan
- Poplar
- Russian olive
- Tree-of-Heaven
- Tulip tree
- Walnut trees
- Willow species

Trees (Evergreen)
- Fir species
- Spruce species

Shrubs & Vines
- Azalea
- Burningbush (Winged Euonymus)
- Grape
- Honeysuckle
- Roses
- Yew

Agronomic Crops
- Alfalfa
- Oats

Vegetable and Flower Plants

Others
- Ferns
- Herbs
- Strawberry
- Raspberry
- Blackberry

Apply COMMAND 4 EC herbicide only to surfaces that will be soil incorporated. Do not apply COMMAND 4 EC herbicide to non-field areas including fence rows, waterways, ditches, and roadsides.

For Preemergent Applications: See Geographic Restrictions above.

Except as restricted above (see Geographic Restrictions), COMMAND may be used as a preemergent soil applied treatment without incorporation, however, additional precautions, beyond those required for a soil preplant incorporated treatment, must be taken as follows:

Do not apply COMMAND 4 EC within 1,500 feet of the following areas. Towns and Subdivisions. Commercial Fruit or Vegetable* Production, Commercial Greenhouses or Nurseries.

Prior to application, adjacent properties must be checked, and spraying within 1,500 feet of desirable plants must be avoided.
* except sweet corn.

Spray Drift Precautions:
Non-target spray drift of COMMAND 4 EC herbicide should be avoided to prevent whitening of desirable vegetation. Drift is influenced by many factors which include wind speed, spray pressure, particle size, nozzle type, and boom height.

- Do not apply when weather conditions favor drift. If wind speeds exceed 10 miles per hour, a drift reducing additive must be used.
- A minimum spray volume of 10 gallons per acre is recommended with appropriate nozzle types and sizes that produce coarser sprays.
- The use of agriculturally approved drift reducing additives is recommended for application volumes of 15 to 40 gallons per acre when spraying in the proximity of desirable plants (see list above).
- The use of an agriculturally approved drift reducing additive is **required** at finished spray volumes of 10 to 15 gallons per acre.
- Use minimum nozzle pressure and boom height while maintaining uniform spray pattern.
- Do not exceed 40 psi spray pressure.
- Selection and proper use of spray equipment is critical in minimizing spray drift. The following table suggests pressures, flow rates, and nozzle sizes for drift reduction using various nozzle types.

Suggested Nozzle Types, Minimum Size and Recommended Pressure Ranges for Minimizing Drift

Nozzle	Pressure Range (PSI)	Minimum Flow Rate Within Pressure Range (GPM)	Minimum Nozzle Size
Flat-Fan	15-30	0.3	#4*
LP-flat-fan	10-25	0.3	#3
Even flat-fan**	15-30	0.3	#4
Flood	10-25	0.3	#2.5*
Whirl-chamber	5-20	0.3	#5
Raindrop	15-40	0.15	#2
Wide angle full cone	15-40	0.3	#5

* Refers to tip number such as 8004 or LF2.5.
** Recommended for banded application.

Rotational Cropping Precautions: Temporary whitening or yellowing of leaves may occur on approved rotational crops where undesirable soil residues of COMMAND 4 EC herbicide exist. The following conditions may result in undesirable soil residues, and greater potential for injury to rotational crops:

1) Exceeding label recommended rates.
2) Overapplication resulting through use of worn nozzles, excessive overlapping of spray swaths, failing to shut off spray booms when turning (end row areas), or slowing or stopping sprayer.

Refer to Rotational Guidelines and replanting instructions of specific crops for additional crop planting information.

SPECIAL PRECAUTION

Off-site movement of spray drift or vapors of COMMAND® 4 EC herbicide can cause foliar whitening or yellowing of some plants. Prior to making applications, read and strictly follow all precautions and application instructions on this label.

GENERAL APPLICATION INSTRUCTIONS

Do not apply aerially or through irrigation equipment.

Ground Applications

Broadcast or Banded Applications: Apply COMMAND 4 EC alone or in tank mix combinations by ground equipment using a finished spray volume of 10 to 40 gallons of water per acre. **NOTE: The use of an agriculturally approved drift reducing additive is required at finished spray volumes of 10 to 15 gallons per acre.** Use nozzles suitable for broadcast boom or banded application of herbicides. Coarse sprays are less likely to drift out of the target area than fine sprays. See "APPLICATION PRECAUTIONS" section for specific recommendations to reduce spray drift. For COMMAND 4 EC tank mixtures with wettable powder or dry flowable formulations, nozzle screens and strainers should be no finer than 50-mesh.

Banded Applications—Calculate the rates and volumes required by using the following formulas:

$$\frac{\text{Band width in inches}}{\text{Row width in inches}} \times \text{Broadcast Rate Per Acre} = \text{Band rate per acre}$$

$$\frac{\text{Band width in inches}}{\text{Row width in inches}} \times \text{Broadcast Volume Per Acre} = \text{Band volume per acre}$$

Follow broadcast minimum gallonage requirements adjusted to band width. Do not exceed 40 psi spray pressure.

Refer to Spray Drift Precautions for suggested pressures, flow rates, and nozzle sizes for drift reduction using various nozzle types. For banded applications an even flat-fan nozzle is recommended.

Soil Incorporated Application: Apply COMMAND 4 EC to the soil surface and uniformly incorporate. Select equipment suited for proper incorporation. Shallow incorporation normally results in better weed control than a deep incorporation.

In states where soil incorporation is required (see Geographical Restrictions) application and immediate incorporation to a maximum depth of 3 inches is required unless the soil is dry. On dry soils, incorporation to a maximum depth of 3 inches **must** be completed within 8 hours of COMMAND 4 EC herbicide application.

Soil must be in good tilth to allow for thorough mixing of the soil. Application to wet soils will increase the potential for off-site movement of COMMAND 4 EC herbicide vapors and may result in poor soil incorporation and unsatisfactory weed control.

INCORPORATION EQUIPMENT

ONE PASS INCORPORATION

Suggested Implements

For one pass incorporation use C or S shank field cultivators equipped with a leveling device such as a spike or coil tine harrow or rolling basket attachment. Field cultivator must have 3 or 4 rows of sweeps spaced at intervals of 7 inches or less and staggered so that no soil is left unturned. Equipment design, including any drag attachments, must be adequate to avoid soil ridging which may result in streaked or reduced weed control. The implement should be operated at 6 to 8 mph and set to cut 3 to 4 inches deep to incorporate herbicide 1 to 2 inches.

Combination equipment (three or more tillage devices combined and used as a simple tool) should be operated at a depth of 3 to 4 inches and a minimum speed of 6 mph.

PTO-drive seed bed preparation equipment should be set to cut 2 inches deep and operated at 4 mph or less.

TWO PASS INCORPORATION

Suggested Implements

When using two pass incorporation, the second pass should be at an angle to the first pass.

Disc set to cut 3 to 4 inches deep at 4 to 6 mph. Blades should be less than 20 inches in diameter and spaced 9 inches or less apart.

Rolling cultivator set to cut 2 to 3 inches deep at 6 to 8 mph (use on coarse or medium textured soils only).

Field cultivator with 3 to 4 rows of sweeps spaced at 7-inch intervals or less and set to cut 3 inches deep at 5 mph or greater.

Cultivation After Planting

Soil treated with COMMAND 4 EC may be shallow-cultivated, rotary-hoed or hand-hoed without reducing the weed control activity of COMMAND 4 EC. Do not cultivate deeper than the COMMAND 4 EC treated layer of soil since this may bring untreated soil to the surface and poor weed control may result.

COMMERCIAL IMPREGNATION AND APPLICATION OF COMMAND 4 EC ON DRY BULK FERTILIZERS

COMMAND herbicide may be impregnated on dry bulk fertilizers. When applied as directed, COMMAND/dry bulk fertilizer mixtures provide weed control equal to that provided by the same rates of COMMAND applied in water.

The COMMAND/fertilizer mixtures must be soil incorporated in all states listed under Geograohic Restrictions of the Application Precautions section of this label.

Impregnation: Apply using a minimum of 200 pounds of dry bulk fertilizer per acre and up to a maximum of 450 pounds per acre with the recommended amount of COMMAND 4 EC herbicide per acre. Use a closed rotary-drum mixer or a similar type of closed blender equipped with suitable spray equipment. The spray nozzle(s) should be positioned to provide a uniform, fine spray pattern over the tumbling fertilizer for thorough coverage. The physical properties of fertilizers vary, particularly in liquid absorptive capacity. When absorptivity is sufficient, simple spray impregnation of the fertilizer with COMMAND provides a satisfactory, dry mixture. If the absorptive capacity is inadequate, use of a highly absorptive powder is required to provide a dry, flowable mixture. Microcel E (Johns-Manvill Products Corporation) is a recommended absorbent powder. Generally less than 2% by weight of Microcel E is required. DO NOT impregnate COMMAND 4EC onto straight coated ammonium nitrate or straight limestone because these materials will not absorb the herbicide. Dry fertilizer blends containing mixtures of ammonium nitrate or limestone may be impregnated with COMMAND.

The amount of COMMAND actually required in the preparation of individual fertilizer mixtures should be determined carefully for each production operation. This is necessary to ensure that the amount of pesticide actually contained in the mixture applied to the soil represents the correct rate of use. Bulk fertilizer impregnated with COMMAND 4 EC herbicide should be applied immediately, not stored. Care should be taken to ensure spreading procedures in the field. Do not place fertilizer treated with COMMAND in areas which cannot be incorporated, if incorporation is required within the Geographic Restrictions of this label

All state regulations, labeling, etc., of the mixtures are the responsibility of the seller.

For those rates not listed in the following table, calculate the amount of COMMAND to be impregnated on a ton of dry bulk fertilizer using the following formula:

$$\frac{2000}{\text{pounds dry fertilizer per acre}} \times \text{PINTS of COMMAND per acre (recommended rate for soil texture)} = \text{PINTS of COMMAND per ton of fertilzer}$$

RATE CHART FOR IMPREGNATION OF DRY BULK FERTILIZERS WITH COMMAND 4 EC HERBICIDE

PINTS OF COMMAND 4 EC PER TON OF FERTILIZER

Fertilizer Rate Lbs./Acre	COMMAND 4 EC Rate Per Acre		
	1½ pts. (.75 lbs A.I.)	2 pt. (1 lb A.I.)	2½pts. (1.25 lbs A.I.)
200	15	20	25
250	12	16	20
300	10	13⅓	16⅔
350	8½	11½	14
400	7½	10	12½
450	6⅔	9	11

If fertilizer materials are excessively dusty, use diesel oil or other suitable additive to reduce dust prior to impregnation as dusty fertilizer will result in poor distribution during application. Crop injury and/or poor weed control may occur where the impregnated fertilizer is not uniformly applied.

SPECIAL PRECAUTIONS FOR DRY BULK FERTILIZER IMPREGNATION AND APPLICATION WITH COMMAND 4 EC

- All equipment used to apply COMMAND should be thoroughly cleaned immediately following use to ensure no contamination results which could cause injury to non-labeled crops or desirable vegetation. Refer to "Sprayer Cleanup" section for additional details and disposal of rinsates.
- Equipment used to physically transport COMMAND treated fertilizer, including boots, augers, conveyers, bins, etc. should be covered to prevent loss of fine particles and subjected to cleanup procedures previously described.
- Mixing of COMMAND herbicide and dry fertilizer should be conducted in a blender which will ensure that airborne particles and potential for volatilization are avoided.
- Bulk containers should be tightly covered while the product is being transported and applied to reduce chances of COMMAND volatilization or product loss.
- It is recommended that Dry Bulk Fertilizer impregnation operations using COMMAND not be conducted within ¼ mile of residential areas

or areas where COMMAND symptomology on desirable vegetation would not be tolerated.

COMMAND GEORGRAPHICAL AREAS
For Rates, Weed Control, and Crop Rotation

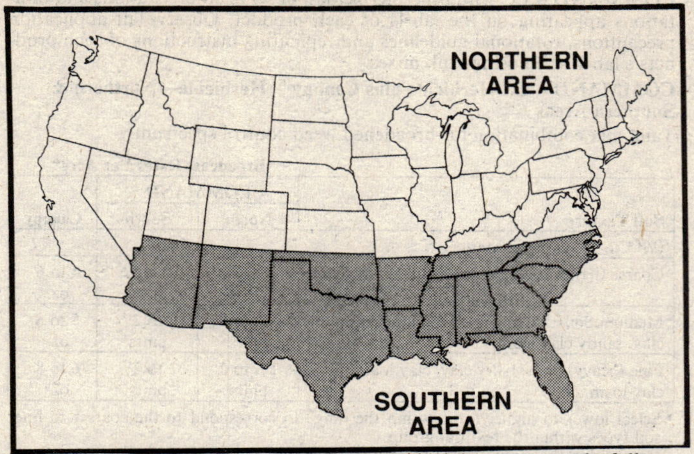

* Southern area includes Missouri bootheel, which incorporates the following counties. Butler, Dunklin, Mississippi, New Madrid, Pemiscot, Scott, and Stoddard.

SOYBEANS

GENERAL INFORMATION

COMMAND 4 EC herbicide may be utilized as a preemergent soil applied or soil incorporated treatment for the control of annual grass and broadleaf weeds in soybeans.

COMMAND 4 EC must be soil incorporated in the following states unless an approved 24(C) is on file for counties within that state (check with your local dealer): Connecticut, Delaware, Illinois, Indiana, Iowa, Kentucky, Maine, Maryland, Massachusetts, Michigan, Minnesota, New Hampshire, New Jersey, New York, Ohio, Pennsylvania, Rhode Island, Vermont, Virginia, West Virginia, and Wisconsin.

COMMAND 4 EC herbicide may be tank mixed with other herbicides to broaden the weed control spectrum compared to the products applied alone. COMMAND 4 EC may be tank mixed with Canopy®, Dual®, Lasso®, Lexone®, Lorox®, Lorox Plus®, Preview®, Prowl®, Scepter®, Sencor®, Sonalan®, Squadron®, Treflan®, Tri-Scept®, or other registered herbicides. Observe all precautions, instructions, and rotational cropping guidelines of each product's label when tank mixing. Water or liquid fertilizer may be used as a carrier for COMMAND 4 EC when applied alone, or when tank mixed with the herbicides listed above, unless use directions specifically state otherwise.

REPLANTING INSTRUCTIONS

If initial seeding of soybeans fails to produce a stand, soybeans may be replanted in fields treated with COMMAND 4 EC alone (or with recommended tank mixtures). Do not retreat field with a second application of COMMAND 4 EC. When tank mixing with a labeled product refer to the soybean replant instructions for that product. Do not replant treated fields with any crop at intervals which are inconsistent with the Rotational Guidelines on this label.

Northern Area—See Map

COMMAND® 4EC HERBICIDE
APPLIED ALONE
RATES AND WEEDS CONTROLLED[1]

1.0 PT.[2] (0.50 LB. A.I.)	1.5 PT. (0.75 LB. A.I.)	2.0 PT. (1.0 LB. A.I.)
GRASSES	GRASSES	GRASSES
Suppression of some annual grasses	Barnyardgrass Crabgrass —Large —Smooth Field Sandbur Foxtail —Giant —Green —Robust —Yellow Goosegrass Panicum —Common —Fall —Texas Seedling Johnsongrass*	All grasses controlled at 1.5 pints *plus:* Bermudagrass* Black Seeded Plantain Broadleaf Signalgrass Cupgrass —Southwestern —Woolly* Seedling Johnsongrass Shattercane* Wild Proso Millet*
BROADLEAVES	BROADLEAVES	BROADLEAVES
Velvetleaf Spurred Anoda	Velvetleaf Spurred Anoda *plus:* Common Ragweed* Galinsoga Jimsonweed* Lambsquarter Prickly Sida Purslane Smartweed* Ladysthumb* Pennsylvania* Venice Mallow	All broadleaves controlled at 1.5 pints *plus:* Cocklebur* Common Ragweed Jimsonweed Kochia Smartweed —Ladysthumb —Pennsylvania Tropic Croton

* Partially controlled.

[1] Higher label rates (minimum of 2 pints per acre) are recommended for fields that have previously been in conservation programs, or where heavy weed pressure is anticpated. Select lower to higher rates based on lighter to heavier soil types.

[2] For heavy weed pressure or heavy soils use 1.25 pints per acre to control Velvetleaf or Spurred Anoda.

NOTE: When using COMMAND 4 EC more than 30 days prior to planting, the highest label rate should be used.

ROTATIONAL CROP GUIDELINES

Rotate to crops as listed below, otherwise crop injury may occur.

1.0 PT. (0.50 LB. A.I.)	1.5 PT (0.75 LB. A.I.)	2.0 PT. (1.0 LB. A.I.)
ANYTIME	ANYTIME	ANYTIME
Soybeans Pumpkins Peas	Soybeans Pumpkins	Soybeans Pumpkins
9 MONTHS	9 MONTHS	9 MONTHS
Corn —Field —Pop —Seed —Sweet Cotton Cucurbits Dry Beans Peanuts Peppers Potatoes Rice Snap Beans Sorghum Sugar Beets Sweet Potatoes Tobacco Tomatoes (Transplanted)	Same as 1.0 pint, *plus:* Peas	Same as 1.5 pints, *except* Seed Corn and Sweet Corn require a 12 month rotation.
12 MONTHS	12 MONTHS	12 MONTHS
Cabbage Tomatoes (all) Wheat	Same as 1.0 pint.	Same as 1.5 pints, *plus:* Seed Corn Sweet Corn
16 MONTHS	16 MONTHS	16 MONTHS
All crops	All crops	All crops

Cover crops, however, may be planted anytime but stand reductions may occur in some areas. Do not graze or harvest for food or feed cover crops planted less than 9 months after COMMAND 4 EC treatment. Do not allow livestock to graze on treated soybean vines or feed treated vines or vine trash to livestock.

Southern Area—See Map

COMMAND® 4 EC HERBICIDE
APPLIED ALONE
RATES AND WEEDS CONTROLLED[1]

1.0 PT.[2] (0.50 LB. A.I.)	1.5 PT. (0.75 LB. A.I.)	2.0 PT. (1.0 LB. A.I.)	2.5 PT. (1.25 LB. A.I.)
GRASSES	GRASSES	GRASSES	GRASSES
Suppression of some annual grasses	Barnyardgrass Broadleaf Signalgrass Crabgrass —Large —Smooth Foxtail —Giant —Green Goosegrass Panicum —Common —Fall —Texas	All grasses controlled at 1.5 pints *plus:* Cupgrass —Southwestern —Woolly Field Sandbur Bermudagrass* Red Rice* Itchgrass[3]	All grasses controlled at 2.0 pints *plus:* Foxtail —Purple —Robust —Yellow Shattercane* Wild Proso Millet*

1.0 PT.[2] (0.50 LB. A.I.)	1.5 PT. (0.75 LB. A.I.)	2.0 PT. (1.0 LB. A.I.)	2.5 PT. (1.25 LB. A.I.)
	Seedling Johnsongrass Field Sandbur*		
BROADLEAVES Velvetleaf Spurred Anoda	BROADLEAVES Velvetleaf Spurred Anoda *plus:* Common Ragweed* Jimsonweed* Lambsquarter Pennsylvania Smartweed* Prickly Sida Purslane Redweed Venice Mallow	BROADLEAVES All broadleaves controlled at 1.5 pints *plus:* Cocklebur Common Ragweed Dayflower Florida Beggarweed Florida Pusley Jimsonweed Kochia* Redvine* Tropic Croton Wild Poinsettia[4]	BROADLEAVES All broadleaves controlled at 2.0 pints *plus:* Balloonvine* Black Nightshade* Curly Dock Joint Vetch* Jungle Rice* Kochia Morninglory —Pitted Pennsylvania Smartweed Pigweed —Redroot* —Smooth* Prostrate Spurge

* Partially Controlled

[1] Higher label rates (a minimum of 2 pints per acre) are recommended for fields that have previously been in conservation programs, or where heavy weed pressure is anticipated. Select lower to higher rates based on lighter to heavier soil types. For fields that have previously been in conservation programs, a minimum application rate of 2½ pints per acre for cocklebur or common ragweed control is recommended.

[2] For heavy weed pressure or heavy soils use 1.25 pints per acre to control Velvetleaf or Spurred Anoda.

[3] For control of Itchgrass, do not incorporate COMMAND 4 EC. For high weed pressure and heavy soils use 2.5 pints.

[4] Use in AR/LA/MS/TX/OK only—for high pressure and heavy soils use 2.5 pints.

NOTE: When using COMMAND 4 EC more than 30 days prior to planting, the highest label rate should be used.

ROTATIONAL CROP GUIDELINES

Rotate to crops as listed below, otherwise crop injury may occur.

1.0 PT. (0.50 LB. A.I.)	1.5 to 2.0 PT. (0.75 to 1.0 LB. A.I.)	2.5 PT. (1.25 LB. A.I.)
ANYTIME Soybeans Pumpkins Peas	ANYTIME Soybeans Pumpkins	ANYTIME Soybeans
9 MONTHS Corn —Field —Pop —Seed —Sweet Cotton Cucurbits Dry Beans Peanuts Potatoes Pumpkins Rice Snap Beans Sorghum Sugar Beets Sweet Potatoes Tobacco Tomatoes (Transplanted)	9 MONTHS Same as 1.0 pint, *plus:* Peas	9 MONTHS Same as 1.5 to 2.0 pints, *plus:* Pumpkins
12 MONTHS All crops	12 MONTHS All crops	12 MONTHS All crops

Cover crops, however, may be planted anytime but stand reductions may occur in some areas. Do not graze or harvest for food or feed cover crops planted less than 9 months after COMMAND 4 EC treatment. Do not allow livestock to graze on treated soybean vines or feed treated vines or vine trash to livestock.

COMMAND® 4 EC HERBICIDE APPLIED IN COMBINATIONS FOR SOYBEANS

COMMAND 4 EC herbicide may be tank mixed with other soybean herbicides to broaden the weed control spectrum compared to the products applied alone. In the sections that follow, FMC recommends a combination of products and application rates to broaden weed control spectrum. FMC makes no claims as to weed control for any combination of rates or products other than as recommended in this label. Additional recommendations for rate combinations with other herbicides not included in the following sections may become available subsequent to the date of publication for this label. Please check with your dealer for more recent FMC technical bulletins and label supplements. When applying mixtures of COMMAND 4 EC with other herbicides, observe all precautions and limitations appearing on the labels of each product. Observe all application precautions, rotational guidelines and replanting instructions of each product's label when using tank mixes.

COMMAND 4 EC Herbicide plus Canopy® Herbicide—Northern & Southern Areas

Tank mix combination for broadened weed control spectrum.

Broadcast Rates Per Acre*

Soil Texture	COMMAND North	COMMAND South	Canopy
(½** to 5% Organic Matter)			
Coarse (light) Soils: (loamy sand, sandy loam)	1½ to 2 pints	2 to 2½ pints	4 to 6 oz.
Medium Soils: (loam, silt, silt loam, sandy clay, sandy clay loam)	1½ to 2 pints	2 to 2½ pints	5 to 8 oz.
Fine (heavy) Soils: (silty clay, clay loam, silty clay loam, clay)	1½ to 2 pints	2 to 2½ pints	6 to 8 oz.

* Select lower to higher rates within the range to correspond to the coarser to finer soil types within the textural group. Rate within the range should correspond to anticipated weed pressure.

** Do not apply to soils less than ½% Organic Matter.

WEEDS CONTROLLED:

Grass Weeds

All species listed under the section COMMAND Applied Alone Rates and Weeds Controlled.

Broadleaf Weeds

All species listed under the section COMMAND Applied Alone Rates and Weeds Controlled plus the following:

Cocklebur**
Copperleaf, hophornbeam*
Hemp sesbania
Morningglory* (Annual, Ivyleaf, Entireleaf, Pitted, Tall)
Pigweed (Redroot, Smooth)
Sicklepod*

* Partially Controlled

** Large seeded weeds, germinating deep in the soil, such as morningglory, sicklepod, cocklebur or weeds with subsequent flushes may require a cultivation or an application of a postemergence herbicide for season long control.

When applying mixtures of COMMAND 4 EC plus Canopy, observe all cautions and limitations appearing on the labels of each product. Special attention must be paid to sections regarding planting depth, phosphate pesticide interaction, previous pesticides, and sensitive soybean varieties on the Canopy label. Observe Application Precautions, Rotational Guidelines and Replanting Instructions of each product's label when using tank mixtures.

COMMAND 4 EC Herbicide plus Dual® 8 E Herbicide—Northern & Southern Areas

Tank-mix combination for broadened weed control spectrum.

Broadcast Rates Per Acre

Soil Texture	COMMAND North	COMMAND South	Dual 8 E
Coarse (light) Soils: (loamy sand, sandy loam)	1½ to 2 pints	2 to 2½ pints	1½ to 2 pints
Medium Soils: (loam, silt, silt loam, sandy clay, sandy clay loam)	1½ to 2 pints	2 to 2½ pints	2 to 2½ pints
Fine (heavy) Soils: (silty clay, clay loam, silty clay loam, clay)	1½ to 2 pints	2 to 2½ pints	2 to 2½ pints

WEEDS CONTROLLED:

Grass Weeds

All species listed under the section COMMAND Applied Alone Rates and Weeds Controlled plus the following:

Nutsedge
Red Rice*
Witchgrass

Broadleaf Weeds

All species listed under the section COMMAND Applied Alone Rates and Weeds Controlled plus the following:

Black Nightshade
Carpetweed
Pigweed

* Use higher rates in areas with heavy pressure.

When applying mixtures of COMMAND 4EC plus Dual, observe all cautions and limitations appearing on the labels of each product. Observe all application precautions, rotational guidelines and replanting instructions of each product's label when using tank mixtures.

COMMAND 4 EC Herbicide plus Lasso® or Lasso® Micro-Tech—Northern & Southern Areas.

A tank-mix combination of COMMAND 4 EC herbicide and Lasso or Lasso Micro-Tech herbicide will broaden the weed control spectrum compared to either product used alone.

	Broadcast Rates Per Acre		
	COMMAND		Lasso*/Lasso
Soil Texture	North	South	Micro-Tech
Coarse (light) Soils: (loamy sand, sandy loam)	1½ to 2 pints	2 to 2½ pints	2 to 2½ quarts
Medium Soils: (loam, silt, silt loam, sandy clay, sandy clay loam)	1½ to 2 pints	2 to 2½ pints	2½ to 3 quarts
Fine (heavy) Soils: (silty clay, clay loam, silty clay loam, clay)	1½ to 2 pints	2 to 2½ pints	2½ to 3 quarts

* Select lower to higher rates within the ranges noted based on lighter to heavier soil types (e.g., sand, loamy sand, sandy loam) within a textural group.

WEEDS CONTROLLED:

Grass Weeds	Broadleaf Weeds
All species listed under the section COMMAND Applied Alone Rates and Weeds Controlled.	All species listed under the section COMMAND Applied Alone Rates and Weeds Controlled plus the following:
	Black Nightshade
	Carpetweed
	Pigweed

When applying mixtures of COMMAND 4 EC herbicide plus Lasso or Lasso Micro-Tech, observe all cautions and limitations appearing on the labels of each product. Observe all application precautions, rotational guidelines and replanting instructions of each product's label when using tank mixtures.

COMMAND 4 EC Herbicide plus Lorox®

A preemergence tank mix combination of COMMAND 4 EC herbicide and Lorox herbicide will broaden the week control sprectrum compared to either product alone.

The tank mix combination may *not* be incorporated. Where COMMAND 4 EC must be incorporated (See Geographical Restrictions), apply Lorox as an overlay treatment.

	Broadcast Rates Per Acre*		
	COMMAND		Lorox
Soil Texture**	North	South	DF†
Coarse (light) Soils: (loamy sand**, sandy loam)	1½ to 2 pints	2 to 2⅙ pints	⅔ to 1 lb.
Medium Soils: (loam, silt, silt loam, sandy clay, sandy clay loam)	1½ to 2 pints	2 to 2⅙ pints	1 to 1½ lbs.
Fine (heavy) Soils: (silty clay, clay loam, silty clay loam, clay)	1½ to 2 pints	2 to 2½ pints	1¼ to 2½ lbs.

* Select lower to higher rates within the ranges noted based on lighter to heavier soil types (e.g., loam, silt loam, sandy clay, sandy clay loam) within a textural group.

† Plant seed at least 1¾ inches deep or Lorox injury to soybeans may result. Refer to Lorox label for other planting and application precautions, one pint of Lorox L equals one pound of Lorox DF, use equivalent rates.

** Do not use on sand, loamy sand, gravelly soils, or exposed subsoils or soils with less than 1% organic matter. Use only on loamy sand with over 1% organic matter.

WEEDS CONTROLLED:

Grass Weeds	Broadleaf Weeds
All species listed under COMMAND Applied Alone Rates and Weeds Controlled section	All species listed under COMMAND Applied Alone Rates and Weeds Controlled section plus the following:
	Carpetweed
	Chickweed
	Pigweed (Redroot, Smooth)
	Wild Mustard
	Wild Radish

When applying mixtures of COMMAND plus Lorox observe all cautions and limitations appearing on the labels of each product. Observe application precautions, rotational crop/grazing and feeding restrictions and replanting instructions sections of this label when using tank mixtures.

COMMAND 4 EC Herbicide plus Lorox Plus®

A preemergence tank mix combination of COMMAND 4 EC herbicide and Lorox Plus herbicide will broaden the weed control spectrum compared to either product alone.

The tank mix combination may *not* be incorporated. Where COMMAND 4 EC must be incorporated (See Geographical Restrictions), apply Lorox Plus as an overlay treatment.

	Broadcast Rates Per Acre*		
	COMMAND		Lorox
Soil Texture**	North	South	Plus†
Coarse (light) Soils: (loamy sand**, sandy loam)	1½ to 2 pints	2 to 2½ pints	9 to 12 oz.
Medium Soils: (loam, silt, silt loam, sandy clay, sandy clay loam)	1½ to 2 pints	2 to 2½ pints	10 to 14 oz.
Fine (heavy) Soils: (silty clay, clay loam, silty clay loam, clay)	1½ to 2 pints	2 to 2½ pints	11 to 14 oz.

* Select lower to higher rates within the ranges noted based on lighter to heavier soil types (e.g., loam, silt loam, sandy clay, sandy clay loam) within a textural group.

† Plant seed at least 1¾ inches deep or Lorox injury to soybeans may result. Refer to Lorox Plus label for other planting and application precautions.

** Do not use on sand, loamy sand, gravelly soils, or exposed subsoils or soils with less than 1% organic matter. Use only on loamy sand with over 1% organic matter.

WEEDS CONTROLLED:

Grass Weeds	Broadleaf Weeds
All species listed under COMMAND Applied Alone Rates and Weeds Controlled section	All species listed under COMMAND Applied Alone Rates and Weeds Controlled section plus the following:
	Black Nightshade*
	Carpetweed
	Cocklebur**
	Giant Ragweed*
	Morningglory* (Annual, Ivyleaf, Entireleaf, Pitted, Tall)
	Mustards
	Pigweed (Redroot, Smooth)

* Partially Controlled

** Large seeded weeds, germinating deep in the soil, such as morningglory, sicklepod, cocklebur or weeds with subsequent flushes may require a cultivation or an application of a postemergence herbicide for season long control.

When applying mixtures of COMMAND plus Lorox observe all cautions and limitations appearing on the labels of each product. Observe application precautions, rotational crop/grazing and feeding restrictions and replanting instructions sections of this label when using tank mixtures.

COMMAND 4 EC Herbicide plus Metribuzin (Sencor® or Lexone®)—Northern Area Only

A tank mix combination of COMMAND 4 EC herbicide and metribuzin will broaden the weed control spectrum compared to either product used alone.

	Broadcast Rates Per Acre*	
Soil Texture**	COMMAND	Sencor DF Lexone DF†
Coarse (light) Soils: (loamy sand**, sandy loam)	1½ to 2 pints	¼ to ⅓ lb.
Medium Soils: (loam, silt, silt loam, sandy clay, sandy clay loam)	1½ to 2 pints	¼ to ½ lb.
Fine (heavy) Soils: (silty clay, clay loam, silty clay loam, clay)	1½ to 2 pints	⅓ to ½ lb.

* Select lower to higher rates within the ranges noted based on lighter to heavier soil types (e.g., loam, silt loam, sandy clay, sandy clay loam) within a textural group.

† For pints Sencor 4 or Lexone 4L, multiply pounds DF by 1.5.

** Do not use COMMAND 4 EC in combination with metribuzin on soils with less than 0.5% organic matter, on sand, or on loamy sand with less than 2% organic matter, or on soils having a calcareous surface area or a pH of 7.5 or higher as metribuzin injury to soybeans may occur.

NOTE: Metribuzin injury to soybeans may occur under excess moisture conditions, cool weather, or compacted soil areas.

WEEDS CONTROLLED:

Grass Weeds	Broadleaf Weeds
All species listed under the section COMMAND Applied Alone Rates and Weeds Controlled.	All species listed under the section COMMAND Applied Alone Rates and Weeds Controlled, plus the following:
	Carpetweed
	Pigweed* (Redroot, Smooth)
	Wild mustard

* Use the higher rate of metribuzin for a given soil type.

When applying mixtures of COMMAND® 4 EC plus metribuzin, observe all cautions and limitations appearing on the labels of each product. Observe all application precautions, rotational guidelines and replanting instructions of each product's label when using tank mixtures.

COMMAND 4 EC Herbicide plus Metribuzin (Sencor® or Lexone®)—Southern Area Only

A tank mix combination of COMMAND 4 EC herbicide and metribuzin will broaden the weed control spectrum compared to either product used alone.

Soil Texture**	Broadcast Rates Per Acre* COMMAND	Sencor DF† Lexone DF
Coarse (light) Soils: (loamy sand**, sandy loam)	2 to 2½ pints	¼ to ⅓ lb.
Medium Soils: (loam, silt, silt loam, sandy clay, sandy clay loam)	2 to 2½ pints	¼ to ½ lb.
Fine (heavy) Soils: (silty clay, clay loam, silty clay loam, clay)	2 to 2½ pints	⅓ to ⅔ lb.

* Select lower to higher rates within the ranges noted based on lighter to heavier soil types (e.g., loam, silt loam, sandy clay, sandy clay loam) within a textural group.

† For pints Sencor 4 or Lexone 4L, multiply pounds DF by 1.5.

** Do not use COMMAND 4 EC in combination with metribuzin on soils with less than 0.5% organic matter, on sand, or on loamy sand with less than 2% organic matter, or on soils having a calcareous surface area or a pH of 7.5 or higher as metribuzin injury to soybeans may occur. Use ⅔ to 1 pound metribuzin on silty clay and heavy clay soils of the Mississippi Delta.

NOTE: Metribuzin injury to soybeans may occur under excess moisture conditions, cool weather, or compacted soil areas.

WEEDS CONTROLLED:

Grass Weeds
All species listed under the section COMMAND Applied Alone Rates and Weeds Controlled.

Broadleaf Weeds
All species listed under the section COMMAND Applied Alone Rates and Weeds Controlled, plus the following:

Carpetweed
Hemp sesbania (Coffeeweed)
Pigweed (Redroot, Smooth)
Sicklepod†*
Wild Mustard

† Reduced Competition
* Use the higher rate of metribuzin for a given soil type.

When applying mixtures of COMMAND 4 EC plus metribuzin, observe all cautions and limitations appearing on the labels of each product. Observe all application precautions, rotational guidelines and replanting instructions of each product's label when using tank mixtures.

COMMAND 4 EC Herbicide plus Preview®—Northern Area Only

A tank mix combination of COMMAND 4 EC herbicide and Preview herbicide will broaden the weed control spectrum compared to either product used alone.

Soil Texture	Broadcast Rates Per Acre* COMMAND	Preview
Coarse (light) Soils: (loamy sand, sandy loam)	1½ to 2 pints	5 to 6 oz.
Medium Soils: (loam, silt, silt loam, sandy clay, sandy clay loam)	1½ to 2 pints	6 to 7 oz.
Fine (heavy) Soils: (silty clay, clay loam, silty clay loam, clay)	1½ to 2 pints	6 to 8 oz.

* Select lower to higher rates within the ranges noted based on lighter to heavier soil types (e.g., loam, silt loam, sandy clay, sandy clay loam) within a textural group. Do not use COMMAND 4 EC in combination with Preview on soils with less than 0.5% organic matter, or on soils with pH greater than 6.8 or on soils having a calcareous surface area as Preview injury to rotational crops may occur.

WEEDS CONTROLLED:

Grass Weeds
All species listed under the section COMMAND Applied Alone Rates and Weeds Controlled.

Broadleaf Weeds
All species listed under the section COMMAND Applied Alone Rates and Weeds Controlled plus the following:

Burcucumber*
Carpetweed
Cocklebur*
Copperleaf, hophornbeam
Giant Ragweed*
Morningglory* (Annual, Tall, Ivyleaf, Pitted, Entireleaf)
Mustards
Pigweed
Spurge, spotted
Sunflower*

***NOTE:** Control of morningglory, sunflower, cocklebur and giant ragweed may be erratic ranging from poor to excellent depending upon soil temperature, time of weed seed germination, depth of weed seed in the soil, and the amount and timing of soil moisture. Control may be improved with timely cultivation.

When applying mixtures of COMMAND 4 EC plus Preview, observe all cautions and limitations appearing on the labels of each product. Observe all application precautions, rotational guidelines and replanting instructions of each product's label when using tank mixtures.

COMMAND 4 EC Herbicide plus Prowl® 4 EC Herbicide—Northern & Southern Areas

A tank-mix combination of COMMAND 4 EC and Prowl will broaden the weed control spectrum of either product used alone.

Soil Texture	Broadcast Rates Per Acre* COMMAND North	COMMAND South	Prowl 4 EC
Coarse (light) Soils: (loamy sand, sandy loam)	1½ to 2 pints	2 to 2½ pints	1 to 1½ pints
Medium Soils: (loam, silt, silt loam, sandy clay, sandy clay loam)	1½ to 2 pints	2 to 2½ pints	2 pints
Fine (heavy) Soils: (silty clay, clay loam, silty clay loam, clay)	1½ to 2 pints	2 to 2½ pints	2 to 2½ pints

* Select lower to higher rates within the ranges noted based on lighter to heavier soil types (e.g., loam, silt loam, sandy clay, sandy clay loam) within a textural group.

WEEDS CONTROLLED:

Grass Weeds
All species listed under the section COMMAND Applied Alone Rates and Weeds Controlled.

Broadleaf Weeds
All species listed under the section COMMAND Applied Alone Rates and Weeds Controlled plus the following:

Carpetweed
Pigweed (Redroot, Smooth)

When applying mixtures of COMMAND 4 EC plus Prowl, observe all cautions and limitations appearing on the labels of each product. Observe all application precautions, rotational guidelines and replanting instructions of each product's label when using tank mixtures.

COMMAND 4 EC Herbicide plus Scepter® Herbicide—Northern Area Only

See Scepter label for appropriate geographic restrictions.

A tank mix combination of COMMAND 4 EC and Scepter will broaden the weed control spectrum of either product used alone.

NOTE: The Scepter label *does not allow rotation to Corn, Wheat, Barley, or Oats* prior to 18 months after last Scepter application *when Scepter is applied preemergent or preplant incorporated within the Northern Use Area as defined on the Scepter label.* See Scepter label for appropriate geographic restrictions.

Soil Texture	Broadcast Rates Per Acre* COMMAND	Scepter
All Soil Types	1½ to 2 pints	⅓ to ½ pint

* Use higher rates when heavier weed pressure is anticipated.

WEEDS CONTROLLED:

Grass Weeds
All species listed under the section COMMAND Applied Alone Rates and Weeds Controlled.

Broadleaf Weeds
All species listed under the section COMMAND Applied Alone Rates and Weeds Controlled plus the following:

Black Nightshade
Cocklebur
Common Sunflower
Giant Ragweed*
Pigweed spp.*

* Partially controlled

In the event of a crop loss due to weather conditions, soybeans can be replanted. DO NOT work the soil deeper than two inches, or make additional COMMAND® 4 EC or COMMAND-Scepter® tank mix applications.

NOTE: Adequate soil moisture is important for optimum tank mix applications.

When applying mixtures of COMMAND 4 EC plus Scepter, observe all cautions and limitations appearing on the labels of each product. Observe all application precautions, rotational guidelines and replanting instructions of each product's label when using tank mixtures.

COMMAND 4 EC Herbicide plus Scepter Herbicide—Southern Area Only

See Scepter label for appropriate geographic restrictions.

A tank mix combination of COMMAND 4 EC and Scepter will broaden the weed control spectrum of either product used alone.

	Broadcast Rates Per Acre*	
Soil Texture	COMMAND	Scepter*
All Soil Types	2 to 2½ pints	½ to ⅔ pint

* Use higher rates when heavier weed pressure is anticipated.

WEEDS CONTROLLED:

Grass Weeds

All species listed under the section COMMAND Applied Alone Rates and Weeds Controlled.

Broadleaf Weeds

All species listed under the section COMMAND Applied Alone Rates and Weeds Controlled plus the following:

Black Nightshade
Cocklebur
Common Sunflower
Pigweed spp.
Wild Poinsettia

In the event of a crop loss due to weather conditions, soybeans can be replanted. DO NOT work the soil deeper than two inches, or make additional COMMAND 4 EC or COMMAND-Scepter tank mix applications.

NOTE: Adequate soil moisture is important for optimum tank mix applications.

When applying mixtures of COMMAND® EC plus Scepter®, observe all cautions and limitations appearing on the labels of each product. Observe all application precautions, rotational guidelines and replanting instructions of each product's label when using tank mixtures.

COMMAND 4 EC Herbicide plus Sonalan® EC Herbicide—Northern Area Only

A tank-mix combination of COMMAND 4 EC and Sonalan will broaden the weed control spectrum of either product used alone.

	Broadcast Rates Pre Acre*	
Soil Texture	COMMAND	Sonalan EC
Coarse (light) Soils: (loamy sand, sandy loam)	1½ to 2 pints	1¼ to 2 pints
Medium Soils: (loam, silt, silt loam, sandy clay, sandy clay loam)	1½ to 2 pints	1¾ to 2½ pints
Fine (heavy) Soils: (silty clay, clay loam, silty clay loam, clay)	1½ to 2 pints	2¼ to 3 pints

* Select lower to higher rates within the range to correspond to the coarser to finer soil types within the textural group.
Rates within the range should correspond to anticipated weed pressure.

WEEDS CONTROLLED:

Grass Weeds

All species listed under the section COMMAND Applied Alone Rates and Weeds Controlled.

Broadleaf Weeds

All species listed under the section COMMAND Applied Alone Rates and Weeds Controlled plus the following:

Carpetweed
Pigweed (Redroot, Smooth)

When applying mixtures of COMMAND 4 EC plus Sonalan, observe all cautions and limitations appearing on the labels of each product. Observe all application precautions, rotational guidelines and replanting instructions of each product's label when using tank mixtures.

COMMAND 4 EC Herbicide plus Squadron—Northern Area Only

See Squadron label for appropriate geographic restrictions.

A tank-mix combination of COMMAND 4 EC and Squadron, will broaden the weed control spectrum of either product used alone.

	Broadcast Rates Per Acre	
Soil Texture	COMMAND	Squadron*
All Soil Types	1½ to 2 pints	1½ to 2¼ pints

* Do not exceed 1½ pints in Northern Area (as defined on the Squadron label). Use higher rate of Squadron in areas of heavy cocklebur infestation.

WEEDS CONTROLLED:

Grass Weeds

All species listed under the section COMMAND Applied Alone Rates and Weeds Controlled.

Broadleaf Weeds

All species listed under the section COMMAND Applied Alone Rates and Weeds Controlled, plus the following:

Black Nightshade
Cocklebur
Common Sunflower
Giant Ragweed*
Pigweed spp.

* Partially controlled

When applying mixtures of COMMAND 4 EC plus Squadron observe all cautions and limitations appearing on the labels of each product. Observe all application precautions, rotational guidelines and replanting instructions of each product's label when using tank mixtures.

COMMAND 4 EC Herbicide plus Squadron—Southern Area Only

See Squadron label for appropriate geographic restrictions.

A tank-mix combination of COMMAND 4 EC and Squadron, will broaden the weed control spectrum of either product used alone.

	Broadcast Rates Per Acre	
Soil Texture	COMMAND	Squadron*
All Soil Types	2 to 2½ pints	2¼ to 3 pints

* Use higher rate of Squadron in areas of heavy pigweed and cocklebur infestation.

WEEDS CONTROLLED:

Grass Weeds

All species listed under the section COMMAND Applied Alone Rates and Weeds Controlled.

Broadleaf Weeds

All species listed under the section COMMAND Applied Alone Rates and Weeds Controlled, plus the following:

Black Nightshade
Cocklebur
Common Sunflower
Pigweed spp.
Wild Poinsettia

When applying mixtures of COMMAND 4 EC plus Squadron observe all cautions and limitations appearing on the labels of each product. Observe all application precautions, rotational guidelines and replanting instructions of each product's label when using tank mixtures.

COMMAND 4 EC Herbicide plus Treflan® EC (Trifluralin)—Northern & Southern Areas

A tank-mix combination of COMMAND 4 EC herbicide and Treflan EC herbicide will broaden the weed control spectrum compared to either product used alone.

	Broadcast Rates Per Acre*		
	COMMAND		Treflan EC
Soil Texture	North	South	
Coarse (light) Soils: (loamy sand, sandy loam)	1½ to 2 pints	2 to 2½ pints	1 pint
Medium Soils: (loam, silt, silt loam, sandy clay, sandy clay loam)	1½ to 2 pints	2 to 2½ pints	1 to 1½ pints
Fine (heavy) Soils: (silty clay, clay loam, silty clay loam, clay)	1½ to 2 pints	2 to 2½ pints	1 to 1½ pints

* Select lower to higher rates within the ranges noted based on lighter to heavier soil types (e.g., loam, silt loam, sandy clay, sandy clay loam) within a textural group.

WEEDS CONTROLLED:

Grass Weeds

All species listed under the section COMMAND Applied Alone Rates and Weeds Controlled.

Broadleaf Weeds

All species listed under the section COMMAND Applied Alone Rates and Weeds Controlled, plus the following:

Carpetweed
Chickweed
Field Bindweed
Pigweed (Redroot, Smooth)

When applying mixtures of COMMAND 4 EC plus Treflan, observe all cautions and limitations appearing on the labels of each product. Observe all application precautions, rotational guidelines and replanting instructions of each product's label when using tank mixtures.

COMMAND 4 EC Herbicide plus Tri-Scept Herbicide—Northern Area Only

See Tri-Scept label for appropriate geographic restrictions.

A tank-mix combination of COMMAND 4 EC and Tri-Scept will broaden the weed control spectrum of either product used alone.

	Broadcast Rates Per Acre	
Soil Texture	COMMAND	Tri-Scept*
All Soil Types	1½ to 2 pints	1⅙ to 1¾ pints

* Do not exceed 1⅙ pints in Northern Area (as defined on the Tri-Scept label). Use higher rate of Tri-Scept in areas of heavy cocklebur infestation.

WEEDS CONTROLLED:

Grass Weeds	Broadleaf Weeds
All species listed under the section COMMAND Applied Alone Rates and Weeds Controlled.	All species listed under the section COMMAND Applied Alone Rates and Weeds Controlled, plus the following: Black Nightshade, Cocklebur, Common Sunflower, Giant Ragweed*, Pigweed spp.*

* Partially controlled

When applying mixtures of COMMAND 4 EC plus Tri-Scept observe all cautions and limitations appearing on the labels of each product. Observe all application precautions, rotational guidelines and replanting instructions of each product's label when using tank mixtures.

COMMAND 4 EC Herbicide plus Tri-Scept—
Southern Area Only

See Tri-Scept label for appropriate geographic restrictions.

A tank mix combination of COMMAND 4 EC and Tri-Scept, will broaden the weed control spectrum of either product used alone.

	Broadcast Rates Per Acre	
Soil Texture	**COMMAND**	**Tri-Scept***
All Soil Types	2 to 2½ pints	1¾ to 2⅓ pints

* Use higher rate of Tri-Scept in areas of heavy pigweed and cocklebur infestation.

WEEDS CONTROLLED:

Grass Weeds	Broadleaf Weeds
All species listed under the section COMMAND Applied Alone Rates and Weeds Controlled.	All species listed under the section COMMAND Applied Alone Rates and Weeds Controlled, plus the following: Black Nightshade, Cocklebur, Common Sunflower, Pigweed spp., Wild Poinsettia

When applying mixtures of COMMAND 4 EC plus Tri-Scept observe all cautions and limitations appearing on the labels of each product. Observe all application precautions, rotational guidelines and replanting instructions of each product's label when using tank mixtures.

COMMAND 4 EC HERBICIDE APPLICATION FOLLOWED BY PREEMERGENCE TREATMENTS

Preemergence herbicides approved for use on soybeans may be applied following treatments of COMMAND 4 EC alone or in tank mix combinations for control of additional weed species listed on the preemergence product labels. Read and follow the precautionary statements, directions for use, rates of application and all other information that appear on the product labels. **DO NOT** apply compounds containing metribuzin after a soil incorporated COMMAND 4 EC plus Lexone® or Sencor® tank mix treatment, unless split applications are used where the pounds of active ingredient do not exceed the total allowed.

DO NOT apply Canopy, Preview, Pursuit Plus, Scepter, Squadron, or Tri-Scept herbicides after COMMAND 4 EC plus any of those products in a tank mix treament, as these other herbicides share a common mode of action and may cumulatively cause injury to soybeans or rotational crops.

COMMAND 4 EC HERBICIDE APPLICATION FOLLOWED BY POSTEMERGENCE TREATMENTS

Postemergence herbicides approved for use on soybeans may be applied following treatments of COMMAND 4 EC alone or in tank mix combinations for control of emerged weeds as listed on the postemergence product labels. Read and follow the precautionary statements, directions for use, rates of application and all other information appearing on the product labels.

DO NOT apply Classic, Pursuit, or Scepter as a postemergence treatment after COMMAND 4 EC plus Canopy, Preview, Pursuit Plus, Scepter, Squadron, or Tri-Scept in a tank mix treatment, unless the other product labels specifically recommend such a treatment, as those other herbicides share a common mode of action and may cumulatively cause injury to soybeans or rotational crops.

COMMAND 4 EC HERBICIDE FOLLOWED BY ACIFLUORFEN (BLAZER® 2L OR TACKLE®) HERBICIDE

Following an application of COMMAND 4 EC or tank mixture with COMMAND 4 EC, a postemergence application of acifluorfen will control the following emerged broadleaf weeds:

Broadleaf Weeds

Black nightshade
Cocklebur
Hemp sesbania
Pigweed (redroot)
Purple Moonflower
Pitted Morningglory
Wild mustard

Apply acifluorfen at a rate of 1 pint per acre with .25% crop oil concentrate to actively growing weeds at no more than the 4-leaf growth stage (do not count cotyledonary leaves, but only the fully developed true leaves). This timing generally correlates to soybean growth stages of the first to third trifoliate leaves.

NOTE: Application of COMMAND 4 EC herbicide generally retards the development rate of weeds which may extend the period where acifluorfen can be effectively applied.

Read and follow all precautions, restrictions, and warnings on all product labels.

COMMAND 4 EC HERBICIDE FOLLOWED BY SCEPTER® HERBICIDE

Following an application of COMMAND 4 EC or tank-mixtures with COMMAND 4 EC, a postemergence application of Scepter herbicide will control the following emerged broadleaf weeds:

Broadleaf Weeds

Cocklebur, common
Pigweed
Wild Poinsettia

Apply Scepter after crop emergence but before weeds exceed a height of 12 inches (6 inches for wild poinsettia). Apply at a broadcast rate of ⅓ pints per acre. Apply when weeds are actively growing. DO NOT apply Scepter postemergence when soybeans and weeds have been subjected to stress conditions such as temperature and moisture extremes. The total amount of Scepter should not exceed one half pint per acre per season in northern areas, or ⅔ pint per acre per season in southern areas as defined on the Scepter label. Squadron and Tri-Scept both include the same active ingredient as Scepter.

For postemergence applications, the addition of a nonionic surfactant or crop oil concentrate is required. The nonionic surfactant should contain at least 80% active ingredient and should be applied at a rate of 2 pints per 100 gallons of spray mixture. Apply the crop oil concentrate (COC) at the rate stated on the COC label.

Read and follow all precautionary statements and restrictions, directions for use and all other information appearing on the Scepter label.

COMMAND® 4 EC HERBICIDE APPLICATION— MINIMUM TILLAGE OR NO-TILLAGE SOYBEANS

COMMAND 4 EC herbicide may be used alone, or in combination with Sencor®, Lexone®, Lorox®, or Lorox Plus® herbicides, plus paraquat or Roundup® herbicide for preemergence (to crop and weeds) and postemergence (to weeds, but prior to crop emergence) control of broadleaf and grass weeds where soybeans will be planted into a cover crop or in previous crop residues (e.g., corn, small grain stubble).

WEEDS CONTROLLED:

When used as directed, the paraquat portion of the tank mix controls most emerged annual weeds and suppresses many perennial weeds. Roundup combinations will control emerged annual and perennial weeds when applied as directed on the Roundup label. COMMAND 4 EC or COMMAND 4 EC plus Sencor, Lexone or Lorox provides preemergence control of the weeds listed in the "COMMAND 4 EC Applied Alone" and "COMMAND 4 EC Applied in Combinations" portions of this label.

APPLICATION AND MIXING INSTRUCTIONS

Paraquat Combinations: Tank mix COMMAND 4 EC alone or with Sencor/Lexone, Lorox or Lorox Plus (observe mixing instructions in the General Mixing instructions portion of this label, as well as individual product label use directions). Add 1 to 2 pints paraquat (based on products containing 2 pounds active ingredient per gallon) per acre. Use lower rate for annual weeds less than 4 inches tall and the higher rate for weeds 4 to 6 inches tall. As the last ingredient, add a suitable surfactant (refer to paraquat label.)

Roundup Combinations: Tank mix COMMAND 4 EC alone or with Sencor/Lexone, Lorox or Lorox Plus (observe mixing instructions in the General Mixing Instructions portion of this label, as well as individual product label use directions). Add 1½ quarts of Roundup per acre for control of emerged annual weeds or 2 to 4 quarts per acre for control of emerged perennial and annual weeds.

Apply in 20 to 60 gallons of water per acre. Use the higher gallonage for dense stubble or vegetation.

PUMPKINS

GENERAL INFORMATION

COMMAND 4 EC herbicide may be utilized as a preemergent soil applied treatment for the control of annual grass and broadleaf weeds in pumpkins.

COMMAND 4 EC must be soil incorporated in the following states unless an approved 24(C) is on file for counties within that state (check with your local dealer): Connecticut, Delaware, Illinois, Indiana, Iowa, Kentucky, Maine, Maryland, Massachusetts, Michigan, Minnesota, New Hampshire, New Jersey, New York, Ohio, Pennsylvania, Rhode Island, Vermont, Virginia, West Virginia, and Wisconsin.

NOTE: When utilizing COMMAND 4 EC as a soil incorporated treatment, incorporate COMMAND to a depth of 1 inch or less and place the seed below the chemical barrier when planting.

COMMAND 4 EC herbicide may be tank mixed with other pumpkin herbicides to broaden the weed control spectrum compared to the products applied alone. When applying mixtures of COMMAND 4 EC herbicide plus other pumpkin herbicides, observe all cautions and limitations appearing on the labels of each product. Observe all application precautions, rotational guidelines and replanting instructions of each product's label when using tank mixtures. Water or liquid fertilizer may be used as a carrier for COMMAND 4 EC when applied alone, or when tank mixed with the herbicides listed above unless use directions specifically state otherwise.

SPECIAL PRECAUTION

Off-site movement of spray drift or vapors of COMMAND 4 EC herbicide can cause foliar whitening or yellowing of some plants. Prior to making applications, read and strictly follow all precautions and application instructions on this label.

REPLANTING INSTRUCTIONS

If initial seeding of pumpkins fails to produce a stand, pumpkins may be replanted in fields treated with COMMAND 4 EC alone. Do not retreat field with a second application of COMMAND 4 EC. When tank mixing with a labeled product refer to the pumpkin replant instructions for that product. Do not replant treated fields with any crop at intervals which are inconsistent with the Rotational Guidelines on this label.

COMMAND 4 EC APPLIED ALONE
RATES AND WEEDS CONTROLLED

Do not exceed a rate of 2 pints per acre. Weeds controlled (by rate) are as listed under Soybean for Northern and Southern areas.

ROTATIONAL CROP GUIDELINES

Rotate to crops as listed under Soybean Rotational Crop Guidelines, otherwise crop injury may occur.
Cover crops, however, may be planted anytime but stand reductions may occur in some areas. Do not graze or harvest for food or feed cover crops planted less than 9 months after COMMAND 4 EC treatment. Do not allow livestock to graze on treated pumpkin vines or feed treated vines or vine trash to livestock.

SUCCULENT PEAS

GENERAL INFORMATION

COMMAND 4 EC herbicide may be utilized as preemergent soil applied treatment for the control of annual grass and broadleaf weeds in peas.

COMMAND 4 EC must be soil incorporated in the following states unless an approved 24(C) is on file for counties within that state (check with your local dealer): Connecticut, Delaware, Illinois, Indiana, Iowa, Kentucky, Maine, Maryland, Massachusetts, Michigan, Minnesota, New Hampshire, New Jersey, New York, Ohio, Pennsylvania, Rhode Island, Vermont, Virginia, West Virginia, and Wisconsin.

NOTE: When utilizing COMMAND 4 EC as a soil incorporated treatment, incorporate COMMAND to a depth of 1 inch or less and place the seed below the chemical barrier when planting.

COMMAND 4 EC herbicide may be tank mixed with other herbicides registered for use in peas to broaden the weed control spectrum compared to the products applied alone. When applying mixtures of COMMAND 4 EC plus other succulent pea herbicides, observe all cautions and limitations appearing on the labels of each product. Observe all application precautions, rotational guidelines and replanting instructions of each product's label when using tank mixes. Water or liquid fertilizer may be used as a carrier for COMMAND 4 EC when applied alone, or when tank mixed with other succulent pea herbicides unless use directions specifically state otherwise.

SPECIAL PRECAUTION

Off-site movement of spray drift or vapors of COMMAND® 4 EC herbicide can cause foliar whitening or yellowing of some plants. Prior to making applications, read and strictly follow all precautions and application instructions on this label.

REPLANTING INSTRUCTIONS

If initial seeding of peas fails to produce a stand, peas may be replanted in fields treated with COMMAND 4 EC alone (or with recommended tank mixtures). Do not retreat field with a second application of COMMAND 4 EC. When tank mixing with a labeled product refer to the replant instructions for that product. Do not replant treated fields with any crop at intervals which are inconsistent with the Rotational Guidelines on this label.

COMMAND 4 EC HERBICIDE APPLIED ALONE

FOR PEAS ONLY: Apply only at 1 pint (0.5 Lb. A.I.) Per Acre.

RATE AND WEEDS CONTROLLED

1 PT. (0.5 LB. A.I.) (for all soil types)

GRASSES	BROADLEAVES
Barnyardgrass*	Galinsoga spp*
Crabgrass (Large, Smooth)	Lambsquarters*
Field Sandbur*	Prickly Side*
Foxtail (Green, Giant, Robust)	Spurred Anoda
Goosegrass	Velvetleaf
Panicum (Fall, Common)	Venice Mallow*
Seedling Johnsongrass*	

* Partially Controlled

Rotate to crops as listed under Soybean Rotational Crop Guidelines, otherwise crop injury may occur.

Cover crops, however, may be planted at anytime but stand reductions may occur in some areas. Do not graze or harvest for food or feed cover crops planted less than 9 months after COMMAND 4 EC treatment. Do not allow livestock to graze on treated pumpkin vines or feed treated vines or vine trash to livestock.

FALLOW LAND

GENERAL INFORMATION

COMMAND® 4 EC is a herbicide which may be utilized as preemergence surface applied treatment to control many annual grass and broadleaf weeds on fallow land. COMMAND 4 EC may be tank mixed with other herbicides registered for fallow cropland use (e.g., atrazine) to broaden the weed control spectrum compared to the products applied alone.

SPECIAL PRECAUTION

Off-site movement of spray drift or vapors of COMMAND® 4 EC herbicide can cause foliar whitening or yellowing of some plants. Prior to making applications, read and strictly follow all precautions and application instructions on this label.

Additional Precaution For Fallow Land Applications Only: Do not apply within 500 feet of Emerged Winter Wheat or within 1,000 feet of the areas listed below:

Towns and Subdivisions
Commercial Vegetable Production*
Commercial Fruit Production
Commercial Nurseries
Commercial Greenhouses
* except sweet corn

Supplemental Spray Drift Precautions For Fallow Land Applications: Non-target spray drift of COMMAND 4 EC herbicide should be avoided to prevent whitening of desirable vegetation. Specific recommendations and requirements for minimizing spray drift are included the Application Precautions section of this label. For Fallow Land applications only:

* The use of agriculturally approved drift reducing additive is *required* at finished spray volumes of 5 to 15 gallons per acre and is *recommended* for application volumes of 15 to 40 gallons per acre when spraying in the proximity of desirable plants (see list under Application Precautions).

ROTATIONS CROP/GRAZING AND FEEDING RESTRICTIONS:

Do not plant wheat sooner than nine (9) months after a late summer or fall application. Do not allow livestock to graze on treated fields.

APPLICATION RATES:

Use COMMAND 4 EC herbicide alone at a rate of 1½ to 2 pints (24 to 32 ounces) per acre in a surface applied broadcast application for preemergent control of weeds. Where Glean is registered and was applied emergent control of weeds. Where Glean is registered and was applied rate of 1½ pints (24 ounces) per acre of COMMAND 4 EC is recommended. However, the additional weed control benefit from Glean may not last throughout the following spring (see Glean label). Use the higher rate of COMMAND 4 EC when heavier weed infestations are anticipated or for longer residual control.

In areas where the winter wheat-fallow-winter wheat cropping system is practiced, make application after wheat harvest. For optimum control, make applications after August 15 but before October 31. COMMAND 4 EC can also be used in areas where spring wheat is planted following an 18 month fallow period.

Treatments should be made only to fields where good weed control has been maintained in the wheat crop through wheat harvest. Treatment recommendations are only designated to control weeds that have recently emerged and those that germinate throughout the fall, winter and spring period. Do not apply in situations where dense weed growth exists.

WEEDS CONTROLLED:

Grass Weeds	Broadleaf Weeds
Cheat grass	Common lambsquarters
Downy bromegrass	Common purslane
Jointed goatgrass	Kochia
Volunteer barley	Prickly lettuce
Volunteer rye	Tansy mustard
Volunteer wheat	Tumble mustard
	Wild buckwheat

COMMAND 4 EC APPLIED IN COMBINATIONS

Tank mix combinations of COMMAND 4 EC herbicide with atrazine will broaden the weed control spectrum and length of control compared to the herbicides used alone.

COMMAND 4 EC PLUS ATRAZINE

COMMAND 4 EC at 1 to 1½ pints (16 to 24 ounces) per acre plus 1 to 2 pints of Atriazine 4L or equivalent rates of Atriazine 90 DF and 80 W. Weeds controlled in addition to those noted under COMMAND 4 EC alone will depend on atrazine rate used (see atrazine label). Use the higher rate of COMMAND 4 EC when heavier weed infestations are anticipated or for longer residual control.

There is significant risk of atrazine carryover if not used according to local Experiment Station and label recommendations. Consult your local Experiment Station for recommended atrazine rates for use in fallow in your area.

Refer to the atrazine label for planting, cropping and other restrictions. Follow all precautions, statements, and restrictions on this and the atrazine label.

Mixing Instructions:

Add atrazine with agitator operating. Mix thoroughly and fill tank ½ full continuing agitation. Add COMMAND 4 EC herbicide to tank while maintaining agitation. Complete filing the sprayer tank with water. Maintain agitation during mixing and application.

Application Instructions:

Follow the General Application Instructions as for COMMAND 4 EC alone.

DEALERS SHOULD SELL IN ORIGINAL PACKAGES ONLY.

TERMS OF SALE OR USE: On purchase of this product buyer and user agree to the following conditions:

WARRANTY: FMC makes no warranty, expressed or implied, concerning the use of this product other than indicated on the label. Except as so warranted, the product is sold as is. Buyer and user assume all risk of use and/or handling and/or storage of this material when such use and/or handling and/or storage is contrary to label instructions.

DIRECTIONS AND RECOMMENDATIONS: Follow directions carefully. Timing and method of application, weather and crop conditions, mixture with other chemicals not specifically recommended and other influencing factors in the use of this product are beyond the control of the seller and are assumed by the buyer at his own risk.

USE OF PRODUCT: FMC's recommendations for the use of this product are based upon tests believed to be reliable. The use of this product being beyond the control of the manufacturer, no guarantee, expressed or implied, is made as to the effects of such or the results to be obtained if not used in accordance with directions or established safe practice.

DAMAGES: Buyer's or user's exclusive remedy for damages for breach of warranty or negligence shall be limited to direct damages not exceeding the purchase price paid and shall not include incidental or consequential damages.

COMMAND® and FMC—Trademarks of FMC Corporation.
CANOPY®, GLEAN®, LEXONE®, LOROX®, LOROX PLUS® and PREVIEW®—Trademarks of E.I. duPont de Nemours and Co., Inc.
OPTION—Trademark of Hoechst-Roussel Agri-Vet Company
SONALAN®, TREFLAN®—Trademarks of Elanco Products Company
LASSO® and ROUNDUP®—Trademark of Monsanto Company
SENCOR®—Trademark of Parent Company of Farbenfabriken Bayer GmbH, Leverkusen
PROWL®, SCEPTER®, SQUADRON® and TRI-SCEPT®—Trademarks of American Cyanamid Company
DUAL®—Trademark of Ciba-Geigy Corporation
BLAZER®—Trademark of BASF AG
TACKLE®—Trademark of Rhone-Poulenc, Inc.

(618-1/11/88-A)

8/89 (618-6/6/89-A)

This is a specimen label. FMC is not responsible for the accuracy of the information contained herein. As labels are subject to revision, always carefully read and follow the label on the product container.

Commence® EC

HERBICIDE

A selective herbicide for the control of annual grasses and broadleaf weeds in soybeans. Do not use in California.

For Agricultural or Commercial Use Only

ACTIVE INGREDIENTS:	BY WT.
(α,α,α-trifluoro-2, 6-dinitro-N,N-dipropyl-p-toluidine)	33.2%
2-(2-Chlorophenyl)methyl -4,4-dimethyl-3- isoxazolidinone	24.9%
INERT INGREDIENTS:**	41.9%
TOTAL	100.0%

Contains a total of 5.25 pounds active ingredient per gallon

* U.S. Patent No. 4,405,357

** Contains xylene range aromatic solvents

EPA Reg. No. 279-3074 EPA Est. 279-Code 1097

Commence herbicide is serviced and promoted by **FMC Corporation** and **Elanco Products Company.**

KEEP OUT OF REACH OF CHILDREN

WARNING

FIRST AID

If in eyes: Hold eyelids open and flush immediately with large amounts of water. Call a physician immediately.

If swallowed: Do not induce vomiting. Call a physician or poison control center. If available, administer activated charcoal (6-8 heaping teaspoonfuls) with a large quantity of water. Do not give anything by mouth to an unconscious person. Immediately transport to a medical care facility and see a physician.

If on skin: Wash with plenty of soap and water. Get medical attention if irritation develops.

If inhaled: Remove individual to fresh air. If not breathing, provide cardiopulmonary resuscitation assistance and get medical attention.

Notes to physician: This product contains an aromatic hydrocarbon and can be extremely harmful if swallowed. Aspiration of this product may produce a severe pneumonitis. Stomach lavage with a cuffed endotracheal tube in place and immediate administration of activated charcoal, 6-8 heaping teaspoonfuls with water, should be considered. Treatment is otherwise symptomatic and supportive.

For Emergency Assistance Call Collect: (716) 735-3765.

See other panels for additional precautionary information.

PRECAUTIONARY STATEMENTS

HAZARDS TO HUMANS (AND DOMESTIC ANIMALS)

WARNING

Causes substantial but temporary eye injury. Do not get in eyes. Wear goggles, face shield, or safety glasses when handling. Harmful if swallowed or inhaled. Avoid breathing spray mist and contact with skin or clothing. Commence EC may cause skin sensitization reactions in certain individuals. Use protective clothing such as coveralls, a long-sleeved shirt and impermeable gloves when handling this product. Wash thoroughly with soap and water after handling. Remove contaminated clothing and wash before reuse.

ENVIRONMENTAL HAZARDS

This pesticide is toxic to fish. Do not apply directly to water or wetlands (swamps, bogs or marshes). Drift or runoff from treatment areas may be hazardous to aquatic organisms in neighboring aquatic sites. Do not contaminate water by cleaning of equipment or disposal of wastes.

SPECIAL PRECAUTION

Off-site movement of spray drift or vapors of Commence EC herbicide can cause foliar whitening or yellowing of some plants. Prior to making applications, read and strictly follow all precautions and application instructions on this label.

DIRECTIONS FOR USE

It is a violation of Federal law to use this product in a manner inconsistent with its labeling.

Do not apply this product through any type of irrigation system.

STORAGE AND DISPOSAL

PESTICIDE STORAGE

Do not freeze. Do not store below 40°F. If solid crystals are observed, warm material to about 60°F by placing container in warm location. Shake or roll container periodically to redissolve solids.

Keep out of reach of children and animals. Store in original containers only. Store in a dry place. Carefully open containers. After partial use, replace lids and close tightly. Do not put concentrate or dilute material into food or drink containers. Do not contaminate other pesticides, fertilizers, water, food or feed by storage or disposal.

In case of spill, avoid contact, isolate area and keep out animals and unprotected persons. Confine spills. Call FMC collect (716) 735-3765.

To confine spill: Dike surrounding area or absorb with sand, cat litter or commercial clay. Place damaged package in a holding container. Identify contents.

PESTICIDE DISPOSAL

Pesticide wastes are toxic. Improper disposal of excess pesticide, spray mixture or rinsate is a violation of Federal law. If these wastes cannot be disposed of by use according to label instructions, contact your State Pesticide or Environmental Control Agency or the Hazardous Waste representative at the nearest EPA Regional Office for guidance.

CONTAINER DISPOSAL

Plastic Containers: Triple rinse (or equivalent). Then offer for recycling or reconditioning, or puncture and dispose of in a sanitary landfill, or by incineration, or, if allowed by state and local authorities, by burning. If burned, stay out of smoke.

Metal Containers: Triple rinse (or equivalent). Then offer for recycling or reconditioning, or puncture and dispose of in a sanitary landfill, or by other procedures approved by state and local authorities. Do not cut or weld metal containers.

Returnable/Refillable Sealed Containers: Do not rinse container. Do not empty remaining formulated product. Do not break seals. Return intact to point of purchase.

GENERAL INFORMATION

Commence EC selective herbicide must be utilized as a soil incorporated treatment for the control of annual grass and broadleaf weeds in soybeans.

Commence EC herbicide may be tank mixed with or followed by overlay or postemergence treatments of other soybean herbicides to broaden weed control spectrum compared to the products applied alone. Commence EC may be tank mixed with Lexone®, Preview®, Canopy®, Scepter® and Sencor® herbicides and applied preplant incorporated. Water or liquid fertilizer may be used as a carrier for Commence EC when applied alone, or when tank mixed with the herbicides listed above unless use directions specifically state otherwise.

IMPORTANT

FAILURE TO OBSERVE THE APPLICATION PRECAUTIONS SECTION OF THIS LABEL MAY RESULT IN INJURY TO DESIRABLE VEGETATION.

- Desirable plants including some species of trees, shrubs, flowers, agronomic crops, and fruits and vegetables are sensitive to Commence EC herbicide.
- Foliar contact with spray drift or vapors may cause whitening or yellowing of sensitive plants. Symptoms are generally temporary in nature but may persist on some plants.

SPRAYER CLEANUP

Do not drain or flush equipment on or near desirable trees or other plants, or in areas where their roots may extend or in locations where the chemical may be washed or moved into contact with their roots. Do not contaminate any body of water including irrigation water that may be used on other crops. Carefully follow sprayer clean-up instructions noted below to prevent spray tank residues from damaging other crops.

Sprayer equipment should be thoroughly cleaned to remove all traces of herbicide that might injure other subsequently sprayed crops. The steps below are suggested for the thorough cleaning of spray equipment following applications of Commence EC herbicide or tank mixes of Commence EC with other labeled products.

1) Drain any remaining spray solution from tank and discard in an approved manner (See Note below).
2) Thoroughly wash down the inside as well as outside surfaces of equipment while filling the spray tank half full of water. Recycle water solution through the equipment for five minutes and dispose of in an approved manner (see Note below).
3) Fill tank with water while adding 1 quart of bleach and 1 pint of detergent for every 25 gallons of water. Operate the pump to circulate the solution through the sprayer system for 15 to 20 minutes and discharge a small amount of the solution through the boom and nozzles. Let the solution stand for several hours, preferably overnight.
4) Start spray system up, recirculate for 15 minutes, then flush the solution out of spray tank through the boom.
   - When switching from water dilutions to applications utilizing crop oil or liquid fertilizer as a carrier, a small volume of crop oil or liquid fertilizer should be flushed through the tank, pump, hoses, and boom prior to the next use. Dispose of crop oil or liquid fertilizer rinsate in an approved manner (see Note for local, state and Federal guidelines).
5) Remove the nozzles, screens, and line filter and wash in a pail of warm, soapy water.
6) Flush the system with two tankfuls of water.

NOTE: Dispose of excess spray mixture and/or rinsates by application and incorporation to cropland as described on this label. If excess spray mixture and/or rinsates cannot be disposed of according to label instructions, contact your State Pesticide or Environmental Control Agency or the Hazardous Waste representative at the nearest EPA Regional Office for guidance.

GENERAL MIXING CONDITIONS

Commence EC Alone: Start with a clean spray tank. Fill sprayer 1/3 to 1/2 full with clean water or liquid fertilizer. Start agitation. Add correct quantity of Commence EC, continue agitation and finish filling the tank.

Tank Mixtures: Vigorous, continuous agitation is required for all tank mixes. Sparger pipe agitators generally provide the best agitation in spray tanks. To prevent foaming, avoid stirring or splashing air into the mixture during filling by placing the end of the fill pipe below the surface of the water in the spray tank. Do not allow the mixture to siphon back into the water source.

Mixing Order: Fill the tank 1/4 to 1/3 full with clean water or liquid fertilizer. (See next page for additional liquid fertilizer mixing instructions.) Start the agitation. Add in the following order: dry flowables (DF), wettable powders (WP), aqueous suspensions (AS), flowables (F), and liquids (L) to the water and agitate until the product(s) are completely dispersed. Allow additional mixing and dispersion time when using dry flowable products. Continue agitation and fill tank to 3/4 full, add the Commence EC, mix thoroughly. Maintain agitation during filling and through application. If spraying and agitation must be stopped before the tank is empty, the materials may settle to the bottom. In this case, it is important to resuspend all of the material in the bottom of the tank before continuing the spray application. A sparger agitator is particularly useful for this purpose. Sometimes it is more difficult to resuspend settled material than it is to suspend originally.

Read and carefully follow all label instructions for each material added to the tank. Premixing dry and flowable formulations with water (slurrying) and pouring the slurry through a 20 to 35 mesh wetting screen in the top of the tank will help assure good initial dispersion in the tank water. Line screens in the tank should be no finer than 50 mesh (100 mesh is finer than 50 mesh).

If a buildup of material on the walls of the spray tank is observed, wash the tank with soapy water between fillings. Rinse and continue the spraying operation. Clean the tank, lines, and screens thoroughly after use.

As the spray volume per acre decreases, the importance of accurate calibration and uniform application increases. Check the sprayer daily to ensure proper calibration and uniform application. Do not apply Commence EC when the wind can cause drifting of spray particles which can result in non-uniform application. When using drift reducing agents, follow specific product label instructions for order of addition to spray tank.

Liquid Fertilizer Mixing Directions: Emulsifiable concentrates, such as Commence EC, can be mixed with liquid fertilizers. In all cases, continuous agitation is required to prevent the Commence EC from rising to the surface as an oily layer. When necessary (see Liquid Fertilizer Compatibility Test below), a compatibility agent can be used to ensure that the Commence EC emulsifies properly (i.e., has a milky appearance rather than an oily layer). The use of compatibility agents is especially important when tank mixing emulsifiable concentrates (EC) with dry flowables (DF), wettable powders (WP), flowables (F), liquids (L), aqueous suspensions (AS), or solutions (S), in liquid fertilizer. If the emulsion is not properly formed, and the EC rises to the surface of the fertilizer as an oil ("oils out"), the oil may combine with the wettable powder, flowable, or suspension to form oily curds (viscous phase) which are difficult to disperse.

Any one of the compatibility agents listed below is helpful in causing emulsifiable concentrates to form non-oiling mixtures with liquid fertilizers. These compatibility agents can be used at rates as low as one and one-half (1½) to two (2) pints per ton of liquid fertilizer and should be mixed well with the fertilizer before adding the emulsifiable concentrate.

Read the label on the compatibility agent and follow the directions.

1. Sponto 168D (Witco Chemicals, Co., Chicago, IL)
2. Compat™ (Farm Chemicals Inc., Aberdeen, NC)
3. Unite (Hopkins Ag Chemical, Madison, WI)
4. T-Mulz 734-2 (Thompson-Hayward Chemical Co., Kansas City, MO)
5. Rigo Compatibility Agent (Rigo Company, Buckner, KY)
6. Amoco Spray Mate™ (Amoco Oil Co., Chicago, IL)
7. Kem-Link™ (Universal Coop., Minneapolis, MN)

Each of the above is a phosphate ester type surfactant designed to be used with liquid fertilizers. They usually do not work well as compatibility agents in tank mixtures in water.

Testing for Tank Mix Compatibility in Liquid Fertilizers: Emulsifiable concentrates alone or in tank mixture with dry flowables (DF), wettable powders (WP), liquids (L), flowables (F), aqueous suspensions (AS), or solutions (S), may not combine properly with some fluid fertilizer materials. Small quantities should always be tested before full-scale mixing. This will determine whether a compatibility agent is needed, and which agent does the best job. The compatibility agents listed above have been thoroughly tested. There are many other surfactants on the market which were not designed for use with liquid fertilizers.

Use the following test to select the correct agent for your mixture:

1. Put one (1) pint of the liquid fertilizer in a quart jar.

2. Add one (1) to four (4) teaspoonful(s) of the DF, WP, L, F, or AS, formulation (depending on the recommended rate per acre) to the liquid fertilizer. Close jar and agitate until dispersed evenly in the fertilizer.
   If the materials do not disperse well, it may be necessary to slurry the chemicals in water before adding to the fertilizer.
3. After dispersing the materials (Step 2), add three (3) to four (4) teaspoonfuls of the Commence® EC to the jar and shake well. Add solution herbicides to the mixture last and agitate. Observe the jar for about 10 minutes. If the materials rise to the surface and form a thick layer (oily curds), which will not disperse when agitated, a compatibility agent is needed. If the mixture is easily dispersed to its original state with slight agitation, no agent is needed, but good agitation must be provided in the fertilizer spray tank.
4. If the need for a compatibility agent is shown in Step 3, using a clean quart jar, start at Step 1 above, add one-half (1/2) teaspoonful of the compatibility agent to the liquid fertilizer, mix well, then repeat Steps 2 and 3.

An effective compatibility agent will cause the mixture to remain uniformly mixed with little or no separating or oil rising to the surface for one-half (1/2) hour or longer. If slight separation does occur, two (2) or three (3) inversions of the jar should give a uniform remix. If oil curds form which will not disperse, more agent or another agent should be tried.

Use a clean jar for each test. The compatible mixture will have a uniform appearance and will be relatively easy to keep mixed with gentle agitation of the jar.

APPLICATION PRECAUTIONS:

Do not apply Commence® EC within 1,000 feet of the areas listed below:
Towns and Subdivisions
Commercial Vegetable Production*
Commercial Fruit Production
Commercial Nurseries
Commercial Greenhouses
*except sweet corn

Plants such as the following may show symptoms of foliar whitening or yellowing if contacted by Commence EC. Symptoms are usually temporary in nature but may result in permanent injury if the exposure is excessive. It is recommended that, prior to application, adjacent properties be checked and that spraying within 100 feet of such plants be avoided.

Trees (Deciduous)
- Apple (inc. fruit & ornamental types)
- Ash (Green, White, Mountain)
- Basswood
- Boxelder
- Catalpa
- Cherry (inc. fruit & ornamental types)
- Cottonwood
- Elm
- Ginkgo
- Hackberry
- Mulberry
- Peach
- Pear (inc. fruit & ornamental types)
- Pecan
- Poplar
- Russian olive
- Tree-of-Heaven
- Tulip tree
- Walnut trees
- Willow species

Trees (Evergreen)
- Fir species
- Spruce species

Shrubs & Vines
- Azalea
- Burningbush (winged Euonymus)
- Grape
- Honeysuckle
- Roses
- Yews

Agronomic Crops
- Alfalfa
- Oats

Vegetables and Flower plants

Others
- Ferns
- Herbs
- Strawberry
- Raspberry
- Blackberry

Apply Commence EC only to surfaces that will be incorporated. Do not apply Commence EC to non-field areas including fence rows, waterways, ditches, and road sides.

Spray Drift Precautions:

Non-target drift of Commence EC herbicide should be avoided to prevent whitening of desirable vegetation. Drift is influenced by many factors which include wind speed, spray pressure, particle size, nozzle type, and boom height.

- Do not apply when weather conditions favor drift. If wind speeds exceed 10 miles per hour, a drift reducing additive must be used.
- A minimum spray volume of 10 gallons per acre is recommended with appropriate nozzle types and sizes that produce coarser sprays.
- The use of agriculturally approved drift reducing additives is recommended for application volumes of 15-40 gallons per acre when spraying in the proximity of desirable plants (see list above).
- The use of an agriculturally approved drift reducing additive is **required** at finished spray volumes of 10-15 gallons per acre.
- Use minimum spray pressure and boom height maintaining uniform spray pattern.
- Do not exceed 40 psi spray pressure.
- Selection and proper use of spray equipment is critical in minimizing spray drift. The following table suggests pressures, flow rates, and nozzle sizes for drift reduction using various nozzle types.

Suggested Nozzle Types, Minimum Size and Recommended Pressure Ranges for Minimizing Drift.

Nozzle	Pressure Range (PSI)	Minimum Flow Rate Within Pressure Range (GPM)	Minimum Nozzle Size
Flat-fan	15-30	0.3	#4*
LP-flat-fan	10-25	0.3	#3
Even flat-fan**	15-30	0.3	#4
Flood	10-25	0.3	#2.5*
Whirl-chamber	5-20	0.3	#5
Raindrop	15-40	0.15	#2
Wide angle full cone	15-40	0.3	#5

*Refers to tip number such as 8004 or LF2.5.
**Recommended for banded application.

ROTATIONAL CROPPING PRECAUTIONS: Temporary whitening or yellowing of leaves may occur on approved rotational crops where undesirable soil residues of Commence 4 EC herbicide exist. The following conditions may result in undesirable soil residues, and greater potential for injury to rotational crops:
1) Exceeding label recommended rates.
2) Overapplication resulting through use of worn nozzles, excessive overlapping of spray swaths, failing to shut off spray booms when turning (end row areas), or slowing or stopping sprayer.

Refer to Rotational Guidelines and replanting instructions of specific crops for additional crop planting information.

GENERAL USE PRECAUTIONS

Application and Environmental Conditions:

Applied according to directions and under normal growing conditions Commence EC will not harm the treated crop. Overapplication may result in crop injury or a soil residue. Uneven application or improper soil incorporation of Commence EC can result in erratic weed control or crop injury. Seedling disease, cold weather, deep planting, excessive moisture, high salt concentration or drought may weaken crop seedlings and increase the possibility of damage from Commence EC. Under these conditions, delayed crop development or reduced yields may result. Application to soils with pH of 6.0 or lower may result in undesirable soil residues and greater potential for injury to rotational crops.

REPLANTING INSTRUCTIONS

If initial seeding of soybeans fails to produce a stand, soybeans may be replanted in fields treated with Commence EC alone (or with recommended tank mixtures). Do not retreat field with a second application of Commence EC. When tank mixing with a labeled product refer to the soybean replant instructions for that product. Do not replant treated fields with any crop at intervals which are inconsistent with the Rotational Guidelines on this label.

Rotational Crop/Grazing and Feeding Restrictions:

The following rotational crops may be planted nine (9) months after the application of Commence EC. Do not rotate to any crops other than those listed below as crop injury may occur.

Corn (Field, Sweet, Pop, Seed)	Potatoes
Cotton	Rice
Cucurbits	Snap Beans
Dry Beans	Sweet Potatoes
Peanuts	Tobacco
Peas	Tomatoes (Transplanted)
Peppers	

In those areas where at least twenty (20) inches of irrigation and/or rainfall (total) was used to produce the soybean crop, sorghum should not be planted for twelve (12) months after an application of Commence EC. If less than (20) inches of total water was used to produce the soybean crop, do not plant sorghum for eighteen (18) months after an application of Commence EC. Cool, wet weather conditions during the early stage of growth may increase the possibility of injury to sorghum.

In areas receiving greater than twenty (20) inches of rainfall per year, moldboard plow at least twelve (12) inches deep before planting sugar beets as a rotational crop. Do not rotate to sugar beets for thirteen (13) months after an application of Commence EC if less than twenty (20) inches of water was used to produce the soybean crop.

NOTE: Do not rotate to wheat, oats, barley, rye, or alfalfa in the fall of the year of application or in the spring of the following year as crop injury may occur. Cover crops may be planted anytime but stand reduction may occur. Do not graze or harvest these cover crops for food or feed. Do not allow livestock to graze on treated soybean vines or feed treated vines or vine trash to livestock.

GENERAL APPLICATION INSTRUCTIONS

This Product Must Be Applied As A Preplant Incorporated Treatment.

Do not apply aerially or through irrigation equipment.

Ground Applications

Broadcast Application: Apply Commence EC alone or in tank mix combinations by ground equipment using a finished spray volume of 10 to 40 gallons of water per acre. **NOTE: The use of an agriculturally approved drift reducing additive is required at finished spray volumes of 10 to 15 gallons per acre.** Use nozzles suitable for broadcast boom application of herbicides. Coarse sprays are less likely to drift out of the target area than fine sprays. See "APPLICATION PRECAUTIONS" Section for specific recommendations to reduce spray drift.

INCORPORATION DIRECTIONS

General Directions: Use incorporation equipment that thoroughly mixes Commence EC into the top 2 to 3 inches of the final seedbed, or erratic weed control and/or crop injury may result. Incorporation equipment such as a disc will mix Commence EC approximately half as deep as the equipment is set to operate. For example, a disc set to cut 4 inches deep will incorporate most of the Commence EC within the top 2 inches of soil.

Incorporation: Application and immediate incorporation to a depth of 2 to 3 inches is required unless the soil surface is dry. On dry soils, incorporation to a depth of 2 to 3 inches **MUST** be completed within 8 hours of Commence EC herbicide application. Soil must be in good tilth to allow for thorough mixing of the soil. Application to overly moist or wet soils will increase the potential for off-site movement of Commence EC herbicide vapors and may result in poor soil incorporation and unsatisfactory weed control.

A second incorporation is necessary, unless specifically stated, this time running the equipment in a different direction from the first. Incorporate the Commence EC uniformly into the top 2 to 3 inches of the final seedbed. Commence EC may be applied up to 3 weeks prior to planting.

Recommended Equipment

Any recommended incorporation tool may be used alone or in combination with any other recommended tool.

Disc: Set to cut 4 to 6 inches deep and operate at 4 to 6 mph.

Field Cultivator: Set to cut 3 to 4 inches deep and operate at 5 mph or more. A field cultivator is defined as an implement with 3 to 4 rows of sweeps, spaced at intervals of 7 inches or less and staggered so that no soil is left unturned. Chisel points should not be used.

Combination Seedbed Conditioners: Set to cut 3 to 4 inches deep and operate at a speed of at least 5 mph. These implements are defined as three or more tillage devices combined and used as a single tool. For example, C- or S-shaped shanks with an effective sweep spacing of 6 to 9 inches (staggered so that no soil is left unturned), followed by a spike-tooth or flextine harrow, followed by a ground-driven reel or basket. Only one incorporation is necessary.

Rolling Cultivator: Set to cut 2 to 4 inches deep and operate at 6 to 8 mph. Rolling cultivators are adequate for use on coarse and medium textured soils only.

Bed Conditioner (Do-All): Set to cut 2 to 4 inches deep and operate at 4 to 6 mph. The Do-All is adequate for use on coarse and medium textured soils only.

Mulch Treader (other similar disc-type implements): Set to cut 3 to 4 inches deep and operate at 5 to 8 mph.

P.T.O. Driven Equipment (tillers, cultivators, hoes): Adjust to incorporate Commence EC into the top 2 to 3 inches of the seedbed with rotors spaced to provide a clean sweep of the soil. Only one incorporation is necessary. P.T.O. driven equipment should not be operated greater than 4 mph.

CULTIVATION AFTER PLANTING

Soil treated with Commence EC herbicide may be shallow cultivated without reducing the weed control activity of Commence EC. Do not cultivate deeper than the treated soil since this may bring untreated soil to the surface and poor weed control may result.

WEEDS CONTROLLED BY COMMENCE EC:

Grass Weeds	
Annual bluegrass	*Poa annua*
Barnyardgrass (Watergrass)	*Echinochloa* spp.
Brachiaria (Signalgrass)	*Brachiaria* spp.
Bromegrass (Cheatgrass) (Downy brome)	*Bromus tectorum*
Cheat (Chess)	*Bromus secalinus*
Crabgrass (Large crabgrass) (Smooth crabgrass)	*Digitaria* spp.
Foxtail (Bottlegrass) (Bristlegrass) (Giant foxtail) (Green foxtail) (Foxtail millet) (Pigeon grass) (Robust foxtail) (Yellow foxtail)	*Setaria* spp.
Goosegrass (Silver crabgrass) (Silvergrass) (Wiregrass) (Yardgrass)	*Eleusine indica*
Johnsongrass (from seed)	*Sorghum halepense*
Junglerice	*Echinochloa colonum*
Panicum, fall	*Panicum dichotomiflorum*
Panicum, Texas (Buffalograss) (Coloradograss)	*Panicum texanum*
Sandbur (Burgrass)	*Cenchrus incertus*
Shattercane*	*Sorghum bicolor*
Sprangletop (Lovegrass)	*Leptochola filifornis*
Stinkgrass (Lovegrass)	*Eragrostis cilianensis*
Woolly cupgrass	*Eriochloa villosa*
Broadleaf Weeds	
Carpetweed	*Mollugo verticillata*
Chickweed	*Stellaria media*
Florida pusley (Florida purslane) (Mexican clover) (Pusley)	*Richardia scabra*
Goosefoot	*Chenopodium hybridum*
Knotweed	*Polygonum aviculare*
Kochia (Fireweed) (Mexican fireweed)	*Kochia scoparia*
Lambsquarters	*Chenopodium album*
Pigweed (Carelessweed) (Prostrate pigweed) (Redroot) (Rough pigweed) (Spiny pigweed)	*Amaranthus* spp.
Potatoweed (Smallflower	*Galinsoga spp.*
Purslane	*Portulaca oleracea*
Russian Thistle (Tumbleweed)	*Salssola kali*
Stinging nettle (Nettle)	*Urtica dioica*
Velvetleaf (Buttonweed)	*Abutilon theophrasti*
Venice mallow	*Hibiscus trionium*

Commence EC will provide partial control or suppression of the following weeds:

Jimsonweed	Ragweed, common
Morninglory, annual	Redweed
Prickley sida (Teaweed)	Smartweed, Pennsylvania
	Spurred Anoda

Control of these weeds may be erratic, ranging from poor to excellent depending upon soil temperature, time of weed germination, depth of weed seed in the soil and the amount and timing of soil moisture. Control may be improved with timely cultivation.

*Two pass incorporation required.

SOIL PREPARATION

Crop residues or Existing Weeds: Ground cover, such as crop residues or existing weeds, can interfere with the incorporation of Commence EC into the soil. A manageable level of such ground cover will allow the Commence EC to be uniformly incorporated into the top 2 to 3 inches of soil. If the level of the ground cover is such that this cannot be done, till the soil prior to the application of Commence EC.

Roughness: The soil surface should be smooth enough to operate the sprayer and incorporation equipment efficiently and at speeds which insure a uniform application and incorporation of Commence EC.

General Soil Conditions: To assure uniform incorporation of Commence EC, soil moisture conditions should be such that large clods can be broken up during the incorporation. Application to overly moist or wet soils will increase the potential for off-site movement of Commence EC herbicide vapors and may result in poor soil incorporation and unsatisfactory weed control.

SOIL TEXTURE GUIDE

The amount of Commence EC you apply will vary with the soil texture and organic matter. A fine textured soil will require more Commence EC per acre than a coarse soil. Choose the proper rate for each application based on the following soil texture group. Do not exceed recommended rates.

Soil Texture	Soil Classification
Coarse (light) Soils: Medium Soils:	Sand, loamy sand, sandy loam Loam, silty clay loam*, silt loam, silt, sandy clay loam*
Fine Soils:	Clay, clay loam, silty clay loam*, silty clay, sandy clay, sandy clay loam*

* Silty clay loam and sandy clay loam soils are transitional soils and may be classified as either medium or fine textured soils. If silty clay loam or sandy clay loam soils are predominantly sand or silt, they are usually classified as medium textured soils. If they are predominantly clay, they are usually classified as fine textured soils.

SPECIAL PRECAUTION

Off-site movement of spray drift or vapors of Commence® EC herbicide can cause foliar whitening or yellowing of some plants. Prior to making applications, read and strictly follow all precautions and application instructions on this label.

SOYBEAN APPLICATION RATES

Commence EC—Alone

Broadcast Rates Per Acre

Soil Texture	Commence EC (Pints)
Coarse	1¾ to 2*
Medium	2 to 2¼*
Fine	2⅔

*Where rate range exists, select lower to higher rates within the ranges noted for lighter to heavier soil types within a textural group.

Commence EC—Tank Mix with Canopy® Herbicide.

The Commence EC/Canopy tank mix controls the annual grasses and broadleaf weeds controlled by Commence EC alone plus these additional weeds:

Cocklebur*	Smartweeds, Annual
Hophornbeard Copperleaf	Spotted Spurge
Jimsonweed	Sunflower*
Mustards	Spurred Anoda
Prickly Sida (Teaweed)	Ragweed, Common
Morningglory, Annual	Ragweed, Giant*
Sicklepod*	

Partial Control:

Eastern Black Nightshade	Nutsedge Species

Soil Texture	Broadcast Rates Per Acre Commence EC (pints)	Canopy (oz)
Coarse	1⅓ pts	6
Medium	2 pts	6 to 8*
Fine	2⅔ pts	7 to 10*

***NOTE:** Use the higher rates when heavier weed pressure is anticipated. Large seeded weeds, germinating deep in the soil, such as cocklebur, morningglory, sicklepod, giant ragweed, and common sunflower or weeds with subsequent flushes may require a cultivation or an application of postemergence herbicide.

Additional Precautions:

Read the Canopy label carefully for cautions and precautions relating to environmental hazards, planting of rotation crops, sprayer contamination and cleanup, soil pH, organic matter and soil texture use restrictions, soybean variety planting restrictions, restrictions where Atrazine® or Sceptor were used the previous year, restrictions concerning use with organic phosphate pesticides, grazing restrictions and other directions, precautions and limitations.

Commence EC—Tank Mix with Lexone® or Sencor® Herbicides

The Commence EC/Lexone or Sencor tank mix controls the annual grasses and broadleaf weeds controlled by Commence EC alone plus these additional weeds:

Jimsonweed	Smartweed, Pennsylvania
Mustard	Spotted Spurge
Prickly Sida (Teaweed)	Spurred Anoda
Ragweed, common	
Sesbania, hemp	

Commence EC/Lexone or Sencor tank mix also provides partial control or suppression of common cocklebur, annual morningglory and giant ragweed. Control of these weeds may be erratic ranging from poor to excellent depending upon soil temperature, time of weed seed germination, depth of weed seed in the soil and the amount and timing of soil moisture. Control may be improved with timely cultivation.

Broadcast Rates Per Acre

Soil Texture	Commence EC (pts)	Lexone 4L or Sencor 4F (pts)	Lexone DF or Sencor DF (lbs)
Coarse	1⅓ pts	⅓ to ½	¼ to ⅓
Medium	2 pts	½ to ¾	⅓ to ½
Fine	2⅔ pts	¾	½

NOTE: Use the higher rate in the rate range for Lexone or Sencor where weed populations are dense or for the control of Wild mustard. Also, for best control of common cocklebur, annual morningglory and giant ragweed, use the higher rate in the rate range for Sencor and Lexone on coarse soils.

Additional Precautions: Do not use Commence EC in combination with Lexone or Sencor on soils with less than 0.5% organic matter, on sand, or on loamy sand with less than 2% organic matter, or on soils having a calcareous surface area or a pH of 7.5 or higher as Lexone or Sencor injury to soybeans may occur.

Commence EC—Tank Mix with Preview® herbicide.

The Commence EC/Preview tank mix controls the annual grasses and broadleaf weeds controlled by Commence EC alone plus these additional weeds:

Cocklebur*	Prickly Sida (teaweed)
Common Ragweed	Smartweed, Annual
Hophornbeam Copperleaf	Spurred Anoda
Jimsonweed	Spotted Spurge
Mustards	Sunflower*

Partial Control:

Eastern Black Nightshade	Nutsedge Species
Ragweed, Giant*	

Soil Texture	Commence EC (pts.)	Preview (pts.)
Coarse	1⅓ pts.	6
Medium	2 pts.	6 to 7
Fine	2⅔ pts.	8*

***NOTE:** Use the higher rates when heavier weed pressure is anticipated. Large seeded weeds, germinating deep in the soil, such as cocklebur and common sunflower or weeds with subsequent flushes may require a cultivation or an application of postemergence herbicide.

Additional Precautions:

Read the Preview label carefully for cautions and precautions relating to environmental hazards, planting of rotation crops, sprayer contamination and cleanup, soil pH, organic matter and soil texture use restrictions, soybean variety planting restrictions, restrictions where Atrazine® or Scepter were used the previous year, restrictions concerning use with organic phosphate pesticides, grazing restrictions and other directions, precautions and limitations.

Commence EC—Tank Mix with Scepter® Herbicide.

The Commence EC/Scepter tank mix controls the annual grasses and broadleaf weeds controlled by Commence EC alone plus these additional weeds:

Eastern Black Nightshade	Jimsonweed
Common Cocklebur	Palmer amaranth
Common Ragweed	Pennsylvania smartweed
Common Sunflower	Prickly Sida (teaweed)

Partial Control:

Mustards
Giant Ragweed
Wild Poinsettia

Soil Texture	Commence EC (pts./fl. oz.)	Scepter (pts.)
Coarse	1⅓ pts.	⅓
Medium	2 pts.	⅓

Soil Texture	Commence EC (pts./fl. oz.)	Scepter (pts.)
Fine	2⅔ pts.	⅓

Additional Precautions:

1. Do not use the Commence/Septer tank-mix in the "Northern Use Area" as defined by the Scepter label.
2. The use of Scepter is limited to those states listed on the Scepter label.
3. Certain severe restrictions apply to corn, wheat and other rotational crops following an application of Scepter. Be sure to refer to the Scepter label for complete details of these and other restrictions.

Read the Scepter label carefully for cautions and precautions relating to environmental hazards, planting of rotation crops, sequential program uses of Scepter, harvest restrictions following postemergence treatments of Scepter, use of Scepter in conjunction with Classic®, Canopy®, or Gemini®, grazing restrictions and other directions, precautions and limitations before applying Scepter. The use of Scepter is limited to those states listed on the Scepter Label.

Commence EC—Overlay Treatments

Preemergence herbicides approved for use on soybeans may be applied following preplant incorporated treatments of Commence EC alone or in tank mix combinations with Lexone, Preview, Scepter, Sencor, or Canopy for control of additional weed species listed on the preemergence product labels. Read and follow the precautionary statements, directions for use, rates of application and all other information that appears on the product labels. For overlay treatments with Scepter, see Additional Precautions Section under COMMENCE EC tank mix with Scepter Herbicide.

Commence EC—Postemergence Treatments

Postemergence herbicides approved for use on soybeans may be applied following preplant incorporated treatments of Commence EC alone or in tank mix combinations with Lexone, Preview, Scepter, Sencor, or Canopy for control of emerged weeds as listed on the postemergence product labels. Read and follow the precautionary statements, directions for use, rates of application and all other information appearing on the product labels.

COMMENCE EC HERBICIDE FOLLOWED BY ACIFLUORFEN (BLAZER® OR TACKLE®) HERBICIDE

Following a soil incorporated application of Commence EC or tank mixture with Commence EC, a postemergence application of acifluorfen will control the following emerged broadleaf weeds:

Broadleaf Weeds
Common Cocklebur
Hemp sesbania
Eastern Black nightshade
Wild mustard
Purple Moonflower
Pitted Morningglory

Apply acifluorfen at a rate of 1 pint per acre with .25% crop oil concentrate to actively growing weeds at no more than the 4-leaf growth stage (do not count cotyledonary leaves, but only the fully developed true leaves). This timing generally correlates to soybean growth stages of the first to third trifoliate leaves.

NOTE: Application of Commence® EC herbicide generally retards the development rate of weeds which may extend the period where acifluorfen can be effectively applied.

Read and follow all precautions, restrictions, and warnings on all product labels.

COMMENCE EC HERBICIDE FOLLOWED BY SCEPTER® HERBICIDE

Following a soil incorporated application of Commence EC or tank-mixtures with Commence EC, a postemergence application of Scepter herbicide will control the following emerged broadleaf weeds:

Broadleaf Weeds
Cocklebur
Pigweed
(Palmer)
(Smooth)
(Tall Waterhemp)

Apply Scepter after crop emergence but before weeds exceed a height of 12 inches. Apply at a broadcast rate of ⅓ pints per acre. Apply when weeds are actively growing. DO NOT apply Scepter postemergence when soybeans and weeds have been subjected to stress conditions such as temperature and moisture extremes. The total amount of Scepter should not exceed one half pint per acre per season.

For postemergence applications, the addition of a nonionic surfactant or crop oil concentrate is required. The nonionic surfactant approved for use on growing crops should contain at least 80% active ingredient and should be applied at a rate of 2 pints per 100 gallons of spray mixture. Apply the crop oil concentrate (COC) at the rate stated on the COC label.

Additional Precautions:

Read the Scepter label carefully for cautions and precautions relating to environmental hazards, planting of rotation crops, sequential program uses of Scepter, harvest restrictions following postemergence treatments of Scepter, use of Scepter in conjunction with Classic®, Canopy®, Preview®, or Gemini®, grazing restrictions and other directions, precautions and limitations before applying Scepter. The use of Scepter is limited to those states listed on the Scepter label. If a Commence/Scepter preplant incorporated tank mix was used as the initial treatment, this sequential postemergence application can only be used in Alabama, Arkansas, Florida, Georgia, Louisiana, Mississippi, North Carolina, Oklahoma, South Carolina, Tennessee and Texas.

COMMERCIAL IMPREGNATION AND APPLICATION OF COMMENCE EC ON DRY BULK FERTILIZERS

Commence herbicide may be impregnated on dry bulk fertilizers. When applied as directed, Commence/dry bulk fertilizer mixtures provide weed control equal to that provided by the same rates of Commence EC applied in water.

The Commence/fertilizer mixtures must be soil incorporated. For best results, Commence should be incorporated two times when applied impregnated on dry bulk fertilizers. The second incorporation must be delayed at least (5) days after the first, completed prior to planting, and should be run in a different direction from the first. Follow other Commence label recommendations for soil incorporation.

Impregnation: Apply using a minimum of 200 pounds of dry bulk fertilizer per acre and up to a maximum of 450 pounds per acre with the recommended amount of Commence EC herbicide per acre. Use a closed rotary-drum mixer or a similar type of closed blender equipped with suitable spray equipment. The spray nozzle(s) should be positioned to provide a uniform, fine spray pattern over the tumbling fertilizer for thorough coverage. The physical properties of fertilizers vary, paticularly in liquid absorptive capacity. When absorptivity is sufficient, simple spray impregnation of the fertilizer with Commence provides a satisfactory, dry mixture. If the absorptive capacity is inadequate, use of a highly absorptive powder is required to provide a dry, flowable mixture. Microcel E (Johns-Manville Products Corporation) is a recommended absorbent powder. Generally less than 2% by weight of Microcel E is required. **DO NOT** impregnate Commence EC onto straight coated ammonium nitrate or straight limestone because these materials will not absorb the herbicide. Dry fertilizer blends containing mixtures of ammonium nitrate or limestone may be impregnated with Commence.

The amount of Commence actually required in the preparation of individual fertilizer mixtures should be determined carefully for each production operation. This is necessary to ensure that the amount of pesticide actually contained in the mixture applied to the soil represents the correct rate of use. Bulk fertilizer impregnated with Commence EC herbicide should be applied immediately, not stored. Care should be taken to ensure spreading procedures in the field do not place fertilizer treated with Commence in areas which cannot be incorporated. All state regulations, labeling, etc., of the mixtures are the responsibility of the seller.

For those rates listed in the following table, calculate the amount of Commence EC to be impregnated on a ton of dry bulk fertilizer using the following formula:

$$\text{2000 pounds dry fertilizer per acre} \times \text{PINTS of Commence per acre (recommended rate for soil texture)} = \text{PINTS of Commence per ton of fertilizer}$$

RATE CHART FOR IMPREGNATION OF DRY BULK FERTILIZERS WITH COMMENCE EC HERBICIDE

PINTS OF Commence EC PER TON OF FERTILIZER

Fertilizer	Commence EC Rate Per Acre			
Rate Lbs./Acre	1¾ pts. (1.15 lbs A.I.)	2 pt. (1.31 lbs A.I.)	2¼ pts. (1.48 lbs A.I.)	2⅔ pts. (1.75 lbs A.I.)
200	17½	20	22½	26⅔
250	14	16	18	21⅓
300	11⅔	13⅓	15	17⅘
350	10	11⅖	12⅘	15¼
400	8¾	10	11¼	13⅓
450	7¾	8⅘	10	11⅘

If fertilizer materials are excessively dusty, diesel oil or other suitable additive to reduce dust prior to impregnation as dusty fertilizer will result in poor distribution during application. Crop injury and/or poor weed control may occur where the impregnated fertilizer is not uniformly applied.

SPECIAL PRECAUTIONS

- All equipment used to apply Commence should be thoroughly cleaned immediately following use to ensure no contamination results which could cause injury to non labeled crops or desirable vegetation. Refer to "Sprayer Cleanup" section for additional details and disposal of rinsates.
- Equipment used to physically transport Commence treated fertilizer, including boots, augers, conveyers, bins, etc. should be covered to prevent loss of fine particles and subjected to cleanup procedures previously described.

- Mixing of Commence herbicide and dry fertilizer should be conducted in a blender which will ensure airborne particles and potential for volatilization are avoided.
- Bulk containers should be tightly covered while the product is being transported and applied to reduce chances of Commence volatilization of product loss.
- It is recommended that Dry Bulk Fertilizer impregnation operations using Commence not be conducted within 1/4 mile of residential areas or areas where Commence symptomology on desirable vegetation would not be tolerated.

DEALERS SHOULD SELL IN ORIGINAL PACKAGES ONLY.

TERMS OF SALE OR USE: On purchase of this product buyer and user agree to the following conditions:

WARRANTY: FMC makes no warranty, expressed or implied, concerning the use of this product other than indicated on the label. Except as so warranted, the product is sold as is. Buyer and user assume all risk of use and/or handling and/or storage of this material when such use and/or handling and/or storage is contrary to label instructions.

DIRECTIONS AND RECOMMENDATIONS: Follow directions carefully. Timing and method of application, weather and crop conditions, mixture with other chemicals not specifically recommended and other influencing factors in the use of this product are beyond the control of the seller and are assumed by the buyer at his own risk.

USE OF PRODUCT: FMC's recommendations for the use of this product are based upon tests believed to be reliable. The use of this product being beyond the control of the manufacturer, no guarantee, expressed or implied, is made as to the effects of such or the results to be obtained if not used in accordance with directions or established safe practice.

DAMAGES: Buyer's or user's exclusive remedy for damages for breach of warranty or negligence shall be limited to direct damages not exceeding the purchase price paid and shall not include incidental or consequential damages.

COMMENCE® and FMC—Trademarks of FMC Corporation
CANOPY®, CLASSIC®, GEMINI®, LEXONE® and PREVIEW®—Trademarks of E.I. duPont de Nemours and Co., Inc.
SENCOR®—Trademark of Parent Company of Farbenfabriken Bayer GmbH, Leverkusen
SCEPTER®—Trademark of American Cyanamid Company
BLAZER®—Trademark of BASF AG
TACKLE®—Trademark of Rhone-Poulenc, Inc.

(1097-4/25/89-A)
10/88

This is a specimen label. FMC is not responsible for the accuracy of the information contained herein. As labels are subject to revision, always carefully read and follow the label on the product container.

Dusting Sulfur

FUNGICIDE-INSECTICIDE

ACTIVE INGREDIENT:	BY WT.
Sulfur	98.00%
INERT INGREDIENTS:	2.00%
TOTAL	100.00%

The material in this bag is guaranteed to pass 95% through a 325 mesh screen.

EPA Reg. No. 279-1721 ZB EPA Est., 279-
Code 30749

KEEP OUT OF REACH OF CHILDREN
CAUTION

PRECAUTIONARY STATEMENTS

HAZARDS TO HUMANS (AND DOMESTIC ANIMALS)

Caution

Harmful if swallowed, inhaled, or absorbed through skin. Avoid breathing dust or spray mist. Avoid contact with eyes, skin, and clothing. Wash thoroughly with soap and water after handling. Remove contaminated clothing and wash before reuse.

STATEMENT OF PRACTICAL TREATMENT

If swallowed: Drink one or two glasses of water and induce vomiting by touching back of throat with finger. Do not induce vomiting or give anything by mouth to an unconscious person.

If in eyes: Flush with plenty of water. Get medical attention if irritation persists.

If on skin: Wash with plenty of soap and water. Get medical attention.

See other panels for additional precautionary information.

For Emergency Assistance Call: (716) 735-3765

ENVIRONMENTAL HAZARDS

Do not apply directly to water. Do not apply where runoff is likely to occur. Do not apply when weather conditions favor drift from areas treated. Do not contaminate water by cleaning of equipment or disposal of wastes. Apply this product only as specified on this label.

PHYSICAL OR CHEMICAL HAZARDS

Keep away from heat, sparks, or flame. Do not smoke while applying this product.

DIRECTIONS FOR USE

It is a violation of Federal law to use this product in a manner inconsistent with its labeling.

Do not apply this product in such a manner as to directly or through drift expose workers or other persons. The area being treated must be vacated by unprotected persons.

Do not enter treated areas for 24 hours unless appropriate protective clothing is worn.

Protective clothing means, at least, a hat or other suitable head covering, a long sleeved shirt and long legged trousers or a coverall type garment (all of closely woven fabric covering the body, including the arms and legs), shoes and boots.

Because certain states may require more restrictive reentry intervals for various crops treated with this product, consult your State Department of Agriculture for further information.

Written or oral warnings must be given to workers who are expected to be in a treated area or in an area about to be treated with this product. Farm workers of areas or fields that may not be entered without specific protective clothing, period of time field must be vacated and appropriate actions to take in case of accidental exposure. An example of such information is given under written warnings. When oral warnings are given, warnings shall be given in a language customarily understood by workers. Oral warnings must be given if there is reason to believe that written warnings cannot be understood by workers. Written warnings must include the following information:

CAUTION

Area treated with Dusting Sulfur on (insert date). Do not enter treated area for 24 hours unless appropriate protective clothing is worn (insert State Department of Agriculture's re-entry interval, if more restrictive).

In case of accidental exposure to pesticide spray or dust, wash the skin thoroughly with soap and water. Remove contaminated clothing and wash before reuse. If in eyes, flush with plenty of water. If inhaled, go to an area where the pesticide has not been applied. Get medical attention if needed.

STORAGE AND DISPOSAL

PESTICIDE STORAGE

Keep away from fire and sparks.

Keep out of reach of children and animals. Store in original containers only. Store in a cool, dry place and avoid excess heat. Carefully open containers. After partial use, fold and roll back bags, clamp and close tightly. Do not put concentrate or dilute material into food or drink containers. Do not contaminate other pesticides, fertilizers, water, food or feed by storage or disposal.

In case of spill, avoid contact, isolate area and keep out animals and unprotected persons. Confine spills. Call FMC collect: (716) 735-3765.

To confine spill: If liquid, dike surrounding area or absorb with sand, cat litter or commercial clay. If dry material, cover to prevent dispersal. Place damaged package in a holding container. Identify contents.

PESTICIDE DISPOSAL

Wastes resulting from the use of this product may be disposed of on site or at an approved waste disposal facility.

CONTAINER DISPOSAL

Completely empty bag into application equipment. Then dispose of empty bag in a sanitary landfill or by incineration, or, if allowed by State and local authorities, by burning. If burned, stay out of smoke.

Application of dusting and spraying materials should not be made just prior to or during periods of excessively high temperatures. Do not make applications within two weeks of an oil spray.

On Citrus do not make an oil application within 60 days prior to or after a sulfur application.

APPLES, PEARS: Scab, **ALMONDS, CHERRIES, PEACHES, NECTARINES, PLUMS, PRUNES:** Brown Rot, Cherry Leaf Spot (Cherries only), Peach Silver Mite and Rust, **APPLES, NECTARINES, PEACHES:** Powdery Mildew—Use 40 to 50 pounds per acre in prebloom, bloom and post-bloom periods. Apply when disease threatens during rain periods and repeat at 7 to 10 day intervals. On Pears do not apply after the fruit begins to turn down. Sulfur may injure certain varieties of apples, check tolerance prior to using.

GRAPES: To control Powdery Mildew—Use 8 to 15 pounds of Dusting Sulfur per acre. Begin applications when shoots are 6 to 8 inches long and repeat at 7 to 10 day intervals. Concord and other Labrusca type grapes may be injured by sulfur.

BLACKBERRIES, BOYSENBERRIES, GOOSEBERRIES, LOGANBERRIES, RASPBERRIES, STRAWBERRIES. To control Powdery Mildew—Use 10 to 20 pounds of Dusting Sulfur per acre. Apply when disease first appears and repeat at 7 to 10 day intervals.

CITRUS: Citrus Thrips, Citrus Flat Mite, Yuma Spider Mite, Sixspotted Mite and Citrus Rust Mite—Use 100 pounds per acre. Apply when pests first appear and repeat as necessary.

VEGETABLE, FIELD AND ORNAMENTAL CROPS: Apply recommended rates of Dusting Sulfur when pests or disease first appear and repeat as necessary. For disease control keep new growth protected.

BEANS: To control Powdery Mildew, Rust, Thrips, Spider Mites—Use Dusting Sulfur at 15 to 30 pounds per acre. For Rust control start applications as soon as the first true leaves form and continue at weekly intervals until blossom time. May cause injury to some varieties. Check tolerance prior to application.

CARROTS: To control Petrobia Mite—Use Dusting Sulfur at 30 to 35 pounds per acre.

COTTON: To control Srawberry Spider Mite—Use Dusting Sulfur at 35 to 40 pounds per acre when mites first appear and repeat as necessary.

ONIONS: To control Petrobia Mite—Use Dusting Sulfur at 26 to 30 pounds per acre.

ORNAMENTALS (flowers, shrubs, trees): To control Rust and Powdery Mildew—Use Dusting Sulfur at a rate that will completely cover foliage and repeat at 5 to 10 day intervals.

PEAS, TURNIPS, RUTABAGAS: To control Powdery Mildew—Use Dusting Sulfur at 15 to 20 pounds per acre. Repeat at 7 to 14 day intervals.

PISTACHIOS: To control Citrus Flat Mite—Use 50 pounds of Dusting Sulfur per acre. Apply when mites first appear and repeat as necessary. May be applied by ground or air.

TOMATOES: To control Tomato Russet Mite—Use Dusting Sulfur at 30 to 40 pounds per acre.

DEALERS SHOULD SELL IN ORIGINAL PACKAGES ONLY.

TERMS OF SALE OR USE: On purchase of this product buyer and user agree to the following conditions:

WARRANTY: FMC makes no warranty, expressed or implied, concerning the use of this product other than indicated on the label. Except as so warranted, the product is sold as is. Buyer and user assume all risk of use and/or handling and/or storage of this material when such use and/or handling and/or storage is contrary to label instructions.

DIRECTIONS AND RECOMMENDATIONS: Follow directions carefully. Timing and method of application, weather and crop conditions, mixture with other chemicals not specifically recommended and other influencing factors in the use of this product are beyond the control of the seller and are assumed by the buyer at his own risk.

USE OF PRODUCT: FMC's recommendations for the use of this product are based upon tests believed to be reliable. The use of this product being beyond the control of the manufacturer, no guarantee, expressed or implied, is made as to the effects of such or the results to be obtained if not used in accordance with directions or established safe practice.

DAMAGES: Buyer's or user's exclusive remedy for damages for breach of warranty or negligence shall be limited to direct damages not exceeding the purchase price paid and shall not include incidental or consequential damages.

FMC—FMC Trademark
(30749-1/87)

This is a specimen label. FMC is not responsible for the accuracy of the information contained herein. As labels are subject to revision, always carefully read and follow the label on the product container.

Elgetol®

FUNGICIDE AND APPLE THINNER

For Agricultural or Commercial Use Only

ACTIVE INGREDIENT:	BY WT.
Sodium dinitro-o-cresylate	19.0%
INERT INGREDIENTS:	81.0%
TOTAL	100.0%

This product contains the toxic inert ingredient Sodium Dichromate.

Contains the equivalent of 1.625 lbs. DNOC per gallon.

Contains Nitrated Phenols

EPA Reg. No. 279-1853 EPA Est., 279–
Code 542

KEEP OUT OF REACH OF CHILDREN

DANGER–POISON

PELIGRO

PRECAUCION AL USUARIO: Si usted no lee ingles, no use este producto hasta que la etiqueta le haya sido explicada ampliamente.

STATEMENT OF PRACTICAL TREATMENT

If swallowed, drink 1 or 2 glasses of water and induce vomiting by touching back of throat with finger or blunt object. This may be followed by drinking a teaspoon of baking soda dissolved in a glass of warm water and induce vomiting. Do not induce vomiting or give anything by mouth to an unconscious person. Get medical attention.

If in eye, flush with plenty of water for at least 15 minutes. Get medical attention.

If on skin, remove contaminated clothing. Wash with rubbing alcohol followed with soap and water.

Note to Physician: Sodium dinitro-o-cresylate is a metabolic stimulant. Treat symptomatically.

For Emergency Assistance Call: 716-735-3765

See Other Panels for Additional Precautionary Information.

PRECAUTIONARY STATEMENTS

HAZARDS TO HUMANS (AND DOMESTIC ANIMALS)

DANGER

Poisonous if swallowed or absorbed through skin. Do not breathe spray mist. Do not get in eyes, on skin or on clothing. Keep away from children and domestic animals.

ENVIRONMENTAL HAZARDS

This product is toxic to fish and wildlife. Birds feeding in treated areas may be killed. Keep out of lakes, streams or ponds. Do not contaminate water by cleaning of equipment or disposal of wastes.

Do not apply when weather conditions favor drift from treated areas.

PHYSICAL OR CHEMICAL HAZARDS

The dry solid material of Elgetol is flammable. Thoroughly clean equipment after use.

DIRECTIONS FOR USE

It is a violation of Federal law to use this product in a manner inconsistent with its labeling.

Do not apply this product through any type of irrigation system.

Do not apply this product is such a manner as to directly or through drift expose workers or other persons. The area being treated must be vacated by unprotected persons.

Do not enter treated areas without protective clothing until sprays have dried.

Protective clothing means, at least, a hat or other suitable head covering, a long sleeved shirt and long legged trousers or a coverall type garment (all of closely woven fabric covering the body, including the arms and legs), shoes and socks.

Because certain states may require more restrictive reentry intervals for various crops treated with this product, consult your State Department of Agriculture for further information.

Written or oral warnings must be given to workers who are expected to be in a treated area or in an area about to be treated with this product. Inform workers of areas or fields that may not be entered without specific protective clothing, period of time field must be vacated and appropriate actions to take in case of accidental exposure. An example of such information is given under written warnings. When oral warnings are given, warnings shall be given in a language customarily understood by workers. Oral warnings must be given if there is reason to believe that written warnings cannot be understood by workers. Written warnings must include the following information:

DANGER

Area treated with Elgetol® fungicide on (insert date). Do not enter without appropriate protective clothing until sprays have dried (insert State Department of Agriculture's re-entry interval, if more restrictive).

In case of accidental exposure to pesticide spray or dust, wash the skin thoroughly with soap and water. Remove contaminated clothing and wash before reuse. If in eyes, flush with plenty of water. If inhaled, go to an area where the pesticide has not been applied. Get medical attention if needed.

STORAGE AND DISPOSAL

PESTICIDE STORAGE

Do not store in or around the home.

Keep out of reach of children and animals. Store in original containers only. Store in a cool, dry place and avoid excess heat. Carefully open containers. After partial use, replace lids and close tightly. Do not put concentrate or dilute material into food or drink containers. Do not contaminate other pesticides, fertilizers, water, food, or feed by storage or disposal.

In case of spill, avoid contact, isolate area and keep out animals and unprotected persons. Confine spills. Call FMC collect: (716)735-3765.

To confine spill: If liquid, dike surround area or absorb with sand, cat litter or commercial clay. If dry material, cover to prevent dispersal. Place damaged package in a holding container. Identify contents.

PESTICIDE DISPOSAL

Pesticide wastes are acutely hazardous. Improper disposal of excess pesticide, spray mixture, or rinsate is a violation of Federal law. If these wastes cannot be disposed of by use according to label instructions, contact your State Pesticide or Environmental Control Agency, or the Hazardous Waste representative at the nearest EPA Regional Office for guidance.

CONTAINER DISPOSAL

Triple rinse (or equivalent). Then offer for recycling or reconditioning, or puncture and dispose of in a sanitary landfill, or by other procedures approved by state and local authorities. Do not cut or weld metal containers.

Not For Use or Storage in or Around the Home.

Elgetol is a supersaturated solution. Dissolve the entire contents of this container in water. If smaller portions are needed, prepare a stock solution by dissolving 1 gallon of Elgetol in 7 gallons of water. Each gallon of solution will contain the equivalent of 1 pint of Elgetol.

Elgetol contains no oils. Therefore, oil any leather packings and washers to prevent drying.

Elgetol is compatible with and may be used with dormant oils for a dual purpose spray.

Read the entire label before using this product. If there is doubt regarding the development of your trees or the clarity of the instructions or recommendations, consult your State Experimental Station or your supplier before you spray.

Clean spray equipment before using Elgetol. Use spray equipment which will assure complete coverage. Complete coverage is essential for uniform control. Rain occurring after the application is made does not reduce the effectiveness of the treatment if the spray has an opportunity to dry.

APPLE (Dormant Application Only): **Bud Moth, Oyster Shell Scale, Scurfy Scale**—Use 2 quarts per 100 gallons of water. **Green Apple Aphid, Grain Aphid, Rosy Apple Aphid**—Use 1½ to 2 quarts per 100 gallons of water. Apply only as a dilute spray and use no more than 600 gallons per acre. **Apple Scab**—Use 2 quarts per 100 gallons of water and apply at rate of 500 to 600 gallons per acre to destroy overwintering spores which are found on dead leaves on orchard floor. Thorough coverage of all leaves is essential.

MUSHROOM HOUSES (For Control of Diseases Between Crop): **Brown Spot, Bubble, Green Mildew, Lipstick Mold**—Use 4 quarts per 100 gallons of water. Make thorough applications to internal walls, ceilings, floor and to both sides of bedboards. Spray only when houses are empty. DO NOT SPRAY on filled or producing beds. This treatment will not corrode steam pipes.

BLOSSOM THINNING

Complete wetting of all flowers is essential for best results. Apply the recommended amount of Elgetol in sufficient water to treat one acre. Do not use concentrated spray solution—much better results will be obtained with a full volume spray. **Apples:** Delicious and Winesap—Use ⅔ pint in 100 gallons of water. Golden Delicious, Jonathan and Yellow Newton—Use ⅔ to 1⅓ pints per 100 gallons of water. Do not apply more than 900 gallons of spray per acre. Apply when 70 to 80% of blossoms are open. Do not apply after 90% of blossoms are open. **Note**—The above blossom thinning directions for use are intended primarily for the Pacific Northwest area. For use in other geographical areas. Consult your Agricultural Experiment Station for recommended rates.

DEALERS SHOULD SELL IN ORIGINAL PACKAGES ONLY

TERMS OF SALE OR USE On purchase of this product buyer and user agree to the following conditions:

WARRANTY FMC makes no warranty, expressed or implied, concerning the use of this product other than indicated on the label. Except as so warranted, the product is sold as is. Buyer and user assume all risk of use and/or handling and/or storage of this material when such use and/or handling and/or storage is contrary to label instructions.

DIRECTIONS AND RECOMMENDATIONS Follow directions carefully. Timing and method of application, weather and crop conditions, mixture with other chemicals not specifically recommended, and other influencing factors in the use of this product are beyond the control of the seller and are assumed by buyer at his own risk.

USE OF PRODUCT FMC's recommendations for the use of this product are based upon tests believed to be reliable. The use of this product being beyond the control of the manufacturer, no guarantee, expressed or implied, is made as to the effects of such or the results to be obtained if not used in accordance with directions or established safe practice.

DAMAGES Buyer's or user's exclusive remedy for damages for breach of warranty or negligence shall be limited to direct damages not exceeding the purchase price paid and shall not include incidental or consequential damages.

Elgetol and **FMC**—FMC Trademarks

(542-5/87)

This is a specimen label. FMC is not responsible for the accuracy of the information contained herein. As labels are subject to revision, always carefully read and follow the label on the product container.

Ethion 4 Miscible

INSECTICIDE-MITICIDE

For Agricultural or Commercial Use Only

ACTIVE INGREDIENTS:	BY WT.
Ethion (0,0,0',0'-Tetraethyl S,S'-methylene bisphosphorodithioate)	46.5%
Petroleum Distillate:	32.0%
INERT INGREDIENTS:	21.5%
TOTAL	100.0%

Contains 4 pounds Ethion per gallon.

EPA Reg. No. 279-1254 EPA Est., 279-Code 1321

KEEP OUT OF REACH OF CHILDREN

WARNING

AVISO

PRECAUCION AL USUARIO: Si usted no lee ingles, no use este producto hasta que la etiqueta le haya sido explicado ampliamente.

STATEMENT OF PRACTICAL TREATMENT

If swallowed: Call a physician or Poison Control Center. Drink 1 or 2 glasses of water and induce vomiting by touching back of throat with finger. Do not induce vomiting or give anything by mouth to an unconscious person.

If on skin: Wash with plenty of soap and water. Get medical attention.

If inhaled: Remove victim to fresh air. If not breathing, give artificial respiration, preferably mouth-to-mouth. Get medical attention.

If in eyes: Flush with plenty of water. Get medical attention if irritation persists.

Antidote

NOTE TO PHYSICIAN:

Active ingredient is a cholinesterase inhibitor. Patient should be kept under observation for 48 hours. Atropine is antidotal. Give 4 mg. Atropine sulfate intramuscularly or intravenously as soon as cyanosis is overcome. Repeat every 15 minutes until patient is fully atropinized. Pralidoxine (2-PAM) is also antidotal and should be administered in conjunction with Atropine. Never give morphine.

For Emergency Assistance Call 716-735-3765.

See Side Panel for Additional Precautionary Information.

PRECAUTIONARY STATEMENTS

HAZARDS TO HUMANS (AND DOMESTIC ANIMALS)

WARNING

May be fatal if swallowed, inhaled or absorbed through skin. Do not breathe vapor or spray mist. Do not get in eyes, on skin or on clothing. Wear protective clothing and rubber gloves. Wash thoroughly with soap and water after handling. Remove contaminated clothing and wash before reuse.

ENVIRONMENTAL HAZARDS

This product is toxic to fish, wildlife and shrimp. Do not apply directly to water or wetlands. Drift and runoff from treated areas may be hazardous to aquatic organisms in neighboring areas. Do not contaminate water by cleaning of equipment or disposal of wastes and containers.

This product is toxic to bees and should not be applied when bees are actively visiting the area.

Do not allow this material to drift onto neighboring crops or noncrop areas or use in a manner or at a time other than in accordance with label directions because animal, plant or crop injury, illegal residues or other undesirable results may occur.

PHYSICAL/CHEMICAL HAZARDS

Do not use or store near heat or open flame.

DIRECTIONS FOR USE

It is a violation of Federal Law to use this product in a manner inconsistent with its labeling.

Do not apply this product through any type of irrigation system.

Do not apply this product in such a manner as to directly or through drift expose workers or other persons. The area being treated must be vacated by unprotected persons.

Do not enter treated areas for 24 hours unless appropriate protective clothing is worn.

Protective clothing means, at least, a hat or other suitable head covering, a long sleeved shirt and long legged trousers or a coverall type garment (all of closely woven fabric covering the body, including the arms and legs), shoes and socks.

Because certain states may require more restrictive reentry intervals for various crops treated with this product, consult your State Department of Agriculture for further information.

Written or oral warnings must be given to workers who are expected to be in a treated area or in an area about to be treated with this product. Inform workers of areas or fields that may not be entered without specific protective clothing, period of time field must be vacated and appropriate actions to take in case of accidental exposure. An example of such information is given under written warnings. When oral warnings are given, warnings shall be given in a language customarily understood by workers. Oral warnings must be given if there is reason to believe that written warnings cannot be understood by workers. Written warnings must include the following information:

WARNING

Area treated with Ethion 4 Miscibleon (insert date). Do not enter without appropriate protective clothing for 24 hours (insert State Department of Agriculture's re-entry interval, if more restrictive).

In case of accidental exposure to pesticide spray or dust, wash the skin thoroughly with soap and water. Remove contaminated clothing and wash before reuse. If in eyes, flush with plenty of water. If inhaled, go to an area where the pesticide has not been applied. Get medical attention if needed.

STORAGE AND DISPOSAL

PESTICIDE STORAGE

Do not use or store near heat, open flame or hot surfaces.

Keep out of reach of children and animals. Store in original containers only. Store in a cool, dry place and avoid excess heat. Carefully open containers. After partial use, replace lids and close tightly. Do not put concentrate or dilute material into food or drink containers. Do not contaminate other pesticides, fertilizers, water, food, or feed by storage or disposal.

In case of spill, avoid contact, isolate area and keep out animals and unprotected persons. Confine spills. Call FMC collect: (716) 735-3765.

To confine spill: If liquid, dike surrounding area or absorb with sand, cat litter or commercial clay. If dry material, cover to prevent dispersal. Place damaged package in a holding container. Identify contents.

PESTICIDE DISPOSAL

Pesticide wastes are toxic. Improper disposal of excess pesticide, spray mixture, or rinsate is a violation of Federal Law. If these wastes cannot be disposed of by use according to label instructions, contact your State Pesticide or Environmental Control Agency, or the Hazardous Waste representative at the nearest EPA Regional Office for guidance.

CONTAINER DISPOSAL

Triple rinse (or equivalent). Then offer for recycling or reconditioning, or puncture and dispose of in a sanitary landfill, or by incineration, or, if allowed by state and local authorities, by burning. If burned, stay out of smoke.

Do not use in undiluted form.

Do not use if material does not emulsify.

Apply the listed amount per acre when insects first appear and repeat as required to maintain effective control. Use in sufficient water for thorough coverage, unless otherwise noted for listed crops. Coverage of upper and lower leaf surfaces is essential for good control. Observe use limitations. If insect control is required beyond the following use patterns, supplement the control program with other suitable pesticides. When applying this material by aircraft, mix the recommended amount with sufficient water to provide a minimum of 1 gallon finished spray per acre. Where more than 2 quarts are recommended, mix with sufficient water to provide a minimum of finished spray equal to twice the amount of ETHION MISCIBLE used. Observe days interval between last application and harvest indicated by number in ( ) following the crop.

CHERRIES: Black Cherry Aphid, San Jose Scale — 0.5 to .75 pt. plus 2 gals. dormant oil per 100 gals. water. Apply during dormant or delayed dormant period ("Superior" type oil of 60 or 70 second viscosity preferred for delayed dormant use).

GRAPEFRUIT (0), LEMONS (21), LIMES (21), ORANGES (0), TANGELOS (0), TANGERINES (0): To control Citrus Red Mite, Citrus Rust Mite, Sixspotted Mite, Texas Citrus Mite, Black Scale, Brown Soft Scale, California Red Scale, Chaff Scale, Florida Red Scale, Glover Scale, Purple Scale, Snow Scale, Yellow Scale, Citrus Whitefly. Mites—Use 0.5 to 0.75 pt. in 100 gals. water. Mites, Scales, and Citrus Whitefly—Use 0.75 pt. per 100 gals. water combined with 0.5 to 0.7% summer oil **(Florida and Texas areas)** or 0.5 to 0.75 pt. per 100 gals. water combined with 1 to 2 gals. summer oil **(West Coast)**. Where Snow Scale is a problem, spray trunk, limbs and twigs thoroughly. For Whitefly, apply shortly after Whiteflies have laid their eggs. Do not apply more than 2000 gals. spray per acre. Do not apply more than once on Lemons and Limes; on other citrus crops listed, do not apply within 90 days of previous application. On Tangerines, do not apply more than twice per crop. Do not apply oil within 3 weeks of a sulfur application. Do not use oil on Navel Oranges.

HOME GROUNDS (For Use By Trained Pest Control Operators) For outside use only. Clover (Brown Almond) Mite infesting houses—1 pt. per 2500 square feet area. Mix well with sufficient water to cover 2500 sq. ft. Spray areas of mite infestation from the bottom of first floor windows down to the ground and 15 to 20 ft. away from building. Keep children and animals off treated areas until spray has completely dried.

MELONS (0): Melon Leafminer, Serpentine Leafminer, Twospotted Spider Mite—1 pt. per acre. Melon Leafhopper—1.5 to 2 pts. per acre. Petrobia Mite **(California only)**, Cucumber Beetle—1 to 2 pts. in a minimum of 10 gals. water per acre.

SUMMER SQUASH (3): Serpentine Leafminer, Twospotted Spider Mite—1 pt. per acre.

Turf: (Commercial Use)—Chinch Bugs, Sod Webworm—Use 0.33 to 0.5 pt. per 1000 sq. ft. of turf applied in 20 gals. of water or 1.67 to 2.5 pts. in 100 gals. of water to treat 5000 sq. ft. To treat one acre, use 1.75 to 2.5 gals. applied in 800 gals. of water. It is important that the spray penetrate through the mat of turf to reach the insects. It helps to have the turf well irrigated prior to treatment. **Caution:** Keep children and animals off treated areas until this material has been thoroughly watered into soil and the grass has completely dried.

DEALERS SHOULD SELL IN ORIGINAL PACKAGES ONLY.

TERMS OF SALE OR USE: On purchase of this product buyer and user agree to the following conditions:

WARRANTY: FMC makes no warranty, expressed or implied, concerning the use of this product other than indicated on the label. Except as so warranted, the product is sold as is. Buyer and user assume all risk of use and/or handling and/or storage of this material when such use and/or handling and/or storage is contrary to label instructions.

DIRECTIONS AND RECOMMENDATIONS: Follow directions carefully. Timing and method of application, weather and crop conditions, mixture with other chemicals not specifically recommended and other influencing factors in the use of this product are beyond the control of the seller and are assumed by the buyer at his own risk.

USE OF PRODUCT: FMC's recommendations for the use of this product are based upon tests believed to be reliable. The use of this product being beyond the control of the manufacturer, no guarantee, expressed or implied, is made as to the effects of such or the results to be obtained if not used in accordance with directions or established safe practice.

DAMAGES: Buyer's or user's exclusive remedy for damages for breach of warranty or negligence shall be limited to direct damages not exceeding the purchase price paid and shall not include incidental or consequential damages.

FMC-FMC Trademark

(1321-8/14/86-A)

This is a specimen label. FMC is not responsible for the accuracy of the information contained herein. As labels are subject to revision, always carefully read and follow the label on the product container.

Funginex® 1.6 EC

FUNGICIDE

For Agricultural and Commercial Use Only

ACTIVE INGREDIENT:

Triforine (N,N'-[1,4-piperazinediyl-bis-(2,2,2-trichloroethylidene)]-bis-[formamide]) 18.2%

INERT INGREDIENTS: 81.8%

TOTAL 100.0%

EPA Reg. No. 21137-4-279 EPA Est. 279-

Code 3913

This product contains 1.6 lb. of Triforine per gallon.

KEEP OUT OF REACH OF CHILDREN

DANGER

PELIGRO

See Other Panels for Additional Precautionary Information.

PRECAUCION AL USUARIO: Si usted no lee ingles, no use este producto hasta que la etiqueta haya sido explicada ampliamente.

PRECAUTIONARY STATEMENTS

HAZARDS TO HUMANS AND DOMESTIC ANIMALS

DANGER

Corrosive, causes eye damage. Do not get in eyes. Wear goggles or face shield. Harmful if swallowed. Do not inhale spray mist. Avoid contact with skin. After handling, wash thoroughly with soap and water.

STATEMENT OF PRACTICAL TREATMENT

If in eyes, flush with plenty of water for 15 minutes and get medical attention.

If swallowed, drink 1 or 2 glasses of water and induce vomiting by touching back of throat with finger. Do not induce vomiting or give anything by mouth to an unconscious person. Get medical attention.

If inhaled, remove patient from contaminated area and get medical attention.

If on skin, remove contaminated clothing and wash skin with soap and water.

For Emergency Assistance Call 716-735-3765.

ENVIRONMENTAL HAZARDS

Do not apply directly to water. Do not contaminate water by cleaning of equipment or disposal of wastes. Apply this product only as specified on this label.

DIRECTIONS FOR USE

It is a violation of Federal law to use this product in a manner inconsistent with its labeling. For use as a fungicide for the control of certain important plant diseases.

Do not apply this product in such a manner as to directly or through drift expose workers or other persons. The area being treated must be vacated by unprotected persons.

Do not enter treated areas without protective clothing until sprays have dried.

Protective clothing means, at least, a hat or other suitable head covering, a long sleeved shirt and long legged trousers or a coverall type garment (all of closely woven fabric covering the body, including the arms and legs), shoes and socks.

Because certain states may require more restrictive re-entry intervals for various crops treated with this product, consult your State Department of Agriculture for further information.

Written or oral warnings must be given to workers who are expected to be in a treated area or in an area about to be treated with this product. Inform workers of areas or fields that may not be entered without specific protective clothing, period of time field must be vacated and appropriate actions to take in case of accidental exposure. An example of such information is given under written warnings. When oral warnings are given, warnings shall be given in a language customarily understood by workers. Oral warnings must be given if there is reason to believe that written warnings cannot be understood by workers. Written warnings must include the following information:

DANGER

Area treated with Funginex® 1.6 EC on (insert date). Do not enter without appropriate protective clothing until sprays have dried (insert State Department of Agriculture's re-entry interval, if more restrictive).

In case of accidental exposure to pesticide spray or dust, wash the skin thoroughly with soap and water. Remove contaminated clothing and wash before reuse. If in eyes, flush with plenty of water. If inhaled, go to an area where the pesticide has not been applied. Get medical attention if needed.

CHEMIGATION USE DIRECTIONS

Do not apply this product through any irrigation system unless the directions for use for chemigation or sprinkler irrigation are followed.

Apply this product only through the following type of system: sprinkler including center pivot, lateral move, end tow, side (wheel) roll, traveler, big gun, solid set or hand move. Do not apply this product through any other type of irrigation system.

Crop injury, lack of effectiveness or illegal pesticide residues in the crop can result from nonuniform distribution of treated water.

If you have questions about calibration, you should contact State Extension Service specialists equipment manufacturers or other experts.

Do not connect an irrigation system (including greenhouse systems) used for pesticide application to a public water system unless the pesticide label-prescribed safety devices for public water systems are in place.

A person knowledgeable of the chemigation system and responsible for its operation or under the supervision of the responsible person, shall shut the system down and make necessary adjustments should the need arise.

When mixing with other pesticides or fluid fertilizers agitation is recommended for mixing.

Posting of areas to be chemigated is required when 1) any part of a treated area is within 300 feet of sensitive areas such as residential areas, labor camps, businesses, day care centers, hospitals, in-patient clinics, nursing homes or any public areas such as schools, parks, playgrounds or other public facilities not including public roads, or 2) when the chemigated area is open to the public such as golf courses or retail greenhouses.

Posting must conform to the following requirements. Treated areas shall be posted with signs at all usual points of entry and along likely routes of approach from the listed sensitive areas. When there are no usual points of entry, signs must be posted in the corners of the treated areas and in any other location affording maximum visibility to sensitive areas. The printed side of the sign should face away from the treated area towards the sensitive area. The signs shall be printed in English. Signs must be posted prior to application and must remain posted until foliage has dried and soil surface water has disappeared. Signs may remain in place indefinitely as long as they are composed of materials to prevent deterioration and maintain legibility for the duration of the posting period.

All words shall consist of letters at least 2½ inches tall, and all letters and the symbol shall be a color which sharply contrasts with their immediate background. At the top of the sign shall be the words KEEP OUT, followed by an octagonal stop sign symbol at least 8 inches in diameter containing the word STOP. Below the symbol shall be the words PESTICIDES IN IRRIGATION WATER.

Public water system means a system for the provision to the public of piped water for human consumption if such system has at least 15 service connections or regularly serves an average of at least 25 individuals daily at least 60 days out of the year.

Chemigation systems connected to public water systems must contain a functional, reduced-pressure zone, backflow preventer (RPZ) or the functional equivalent in the water supply line upstream from the point of pesticide introduction. As an option to the RPZ, the water from the public water system should be discharged into a reservoir tank prior to pesticide introduction. There shall be a complete physical break (air gap) between the outlet end of the fill pipe and the top or overflow rim of the reservior tank of at least twice the inside diameter of the fill pipe.

SPRINKLER IRRITATION

The system must contain a functional check valve, vacuum relief valve and low pressure drain appropriately located on the irrigation pipeline to prevent water source contamination from backflow.

The pesticide injection pipeline must contain a functional, automatic, quick-closing check valve to prevent the flow of fluid back toward the injection pump.

The pesticide injection pipeline must also contain a functional, normally closed, solenoid-operated valve located on the intake side of the injection pump and connected to the system interlock to prevent fluid from being withdrawn from the supply tank when the irrigation system is either automatically or manually shut down.

The system must contain functional interlocking controls to automatically shut off the pesticide injection pump when the water pump motor stops.

The irrigation line or water pump must include a functional pressure switch which will stop the water pump motor when the water pressure decreases to the point where pesticide distribution is adversely affected. Systems must use a metering pump, such as a positive displacement injection pump (e.g., diaphragm pump) effectively designed and constructed of materials that are compatible with pesticides and capable of being fitted with a system interlock.

Do not apply when wind speed favors drift beyond the area intended for treatment.

STORAGE AND DISPOSAL

PESTICIDE STORAGE

Do not store below 32°F, (0°C).

Do not use or store near heat, open flame or hot surfaces.

Keep out of reach of children and animals. Store in original containers only. Store in a cool, dry place and avoid excess heat. Carefully open containers. After partial use, replace lids and close tightly. Do not put concentrate or dilute material into food or drink containers. Do not contaminate other pesticides, fertilizers, water, food, or feed by storage or disposal.

In case of spill, avoid contact, isolate area and keep out animals and unprotected persons. Confine spill by diking surrounding area or absorbing with sand, cat litter or commercial clay. Place damaged package in a holding container. Identify products.

PESTICIDE DISPOSAL

Wastes resulting from the use of this product may be disposed of on site or at an approved waste disposal facility. Do not contaminate water, food or feed by storage or disposal.

CONTAINER DISPOSAL

Triple rinse (or equivalent). Then offer for recycling or reconditioning, or puncture and dispose of in a sanitary landfill, or incineration, or, if allowed by state and local authorities, by burning. If burned, stay out of smoke.

USE PRECAUTIONS

Do not graze animals in treated orchards.

As timing of fungicide applications for disease control varies due to climatic and other conditions, consult Agricultural Experiment Station or State Extension Service specialist.

Do not use this material if it cannot be applied according to the use pattern on this label.

Do not mix Funginex with wetting agents, spreader-stickers or other adjuvants.

Do not let spray mixture stand in tank overnight.

ALMONDS: Brown Rot Blossom Blight (California Only) Apply a mixed solution of 12 fl. oz. of Funginex per 100 gallons of water; spray to run off. Or, for low volume application, apply a mixed solution of 36 to 48 oz. of Funginex in 50 to 200 gallons of water per acre. Make the first application at pink bud and the second at 50% to 100% bloom. Do not exceed two applications. Do not apply after petal fall.

APPLES: Scab (Venturia inaequalis), Powdery Mildew (Podosphaera leucotricha), and Rust (Gymnosporangium spp.) For full coverage spray only, mix 10 fl. oz. of Funginex per 100 gallons and apply to run-off. For low volume sprayers, apply 36 to 40 fl. oz. of undiluted Funginex per acre per application in sufficient water (50-200 gallons of water per acre). For aerial application, apply 36 to 40 fl. oz. of Funginex in a minimum of 20 gallons of water per acre. Complete coverage is essential to insure adequate control. Make first application at 1/2 inch green tip and repeat every 7 days for a preventive control program. Do not apply after petal fall. Do not exceed a total of 5 applications. Consult Agricultural Experiment Station or State Extension Service specialist for use of Funginex in apple scab monitoring control program.

APRICOTS, CHERRIES, NECTARINES, PEACHES, PLUMS and PRUNES: Brown Rot Blossom Blight (Monilinia spp.) For full coverage spray only, mix 12 to 16 fl. oz. of Funginex per 100 gallons and apply to run-off. For low volume sprayers apply 36 to 48 fl. oz. of undiluted Funginex per acre per application in sufficient water (50 to 200 gallons of water per acre). For aerial application, apply 36 to 48 fl. oz. of Funginex in a minimum of 20 gallons of water per acre. Complete coverage is essential to insure adequate control. Make first application at early bloom (peaches, nectarines: pink bud; apricots: red bud; cherries, plums, prunes: white bud or popcorn). Repeat after 50% bloom. If necessary, depending upon the length of the bloom period and conditions favoring brown rot blossom blight development, make a third application at early petal fall. Alternately, if warm, wet weather prevails, apply the two or three applications at 2 to 4 day intervals beginning at early bloom, since blossom period will be shortened. Do not exceed three sprays of Funginex during the blossom period. The higher rate of Funginex is only necessary under conditions of severe disease pressure:

California Only: For full coverage spray only, mix 12 fl. oz. of Funginex per 100 gallons and apply to run-off. For low volume sprayers, apply 36 to 48 fl. oz. of undiluted Funginex per acre per application in sufficient water (50 to 200 gallons of water per acre). For aerial application, apply 36 to 48 fl. oz. of Funginex in a minimum of 20 gallons of water per acre. Complete coverage is essential to insure adequate control. Make first application on peaches and nectarines at pink bud to 5% bloom, on apricots at red bud, on cherries, plums and prunes at popcorn or white bud, followed by a second application at 50-100% bloom. Do not exceed two sprays of Funginex during the blossom period.

APRICOTS, NECTARINES, PEACHES: Brown Rot, Fruit Rot (Monilinia spp.) For full coverage spray only, mix 12 to 16 fl. oz. of Funginex per 100 gallons and apply to run-off. For low volume sprayers, apply 36 to 48 fl. oz. of undiluted Funginex per acre per application in sufficient water (50 to 200 gallons of water per acre). For aerial application, apply 36 to 48 fl. oz. of Funginex in a minimum of 20 gallons of water per acre. Complete coverage is essential to insure adequate control. Make the first application 2 or 3 weeks before harvest and repeat in 5 to 10 days. Make a third application just prior to harvest. Do not exceed three sprays of Funginex during the pre-harvest period. The higher rate of Funginex is only necessary under conditions of severe disease pressure.

California Only: For full coverage spray only, mix 12 fl. oz. of Funginex per 100 gallons and apply to run-off. For low volume sprayers, apply 36 to 48 fl. oz. of undiluted Funginex per acre per application in sufficient water (50 to 200 gallons of water per acre). For aerial application, apply 36 to 48 fl. oz. of Funginex in a minimum of 20 gallons of water per acre. Complete coverage is essential to insure adequate control. Make first application 2 or 3 weeks before harvest and repeat in 5 to 10 days. Do not exceed two sprays of Funginex during the pre-harvest period.

ASPARAGUS: Asparagus Rust (California and Arizona only) Apply 10-20 fl oz. of Funginex per acre in 20 to 50 gallons of water for ground application or in 5 gal. of water for aerial application. Apply at 7-14 day intervals. Adjust application rate and intervals depending on the severity of rust infection and climatic conditions favorable for rust sporulation. For application through sprinkler irrigation systems, apply in 0.12 acre inch of water through sprinkler systems during the last few minutes of irrigation in 150 to 200 gallons of water per acre. Apply to aspargus ferns only. Do not make more than seven applications. Do not harvest spears within 24 weeks of the last fern application.

HIGHBUSH BLUEBERRIES: Mummyberry Disease (Monilinia vacciniicorymbosi), Pacific and Mid-Western States Apply 24 fl. oz. of Funginex per acre in 20 to 50 gallons of water for ground application or in 5 gallons of water for aerial application. Make the first application at leaf bud break and repeat in 7 to 10 days. Make the third application at pink bud stage and repeat in 7 to 10 days at early bloom. For the last application, apply 16 fl. oz. of Funginex per acre in 20 to 50 gallons of water for ground application or in 5 gallons of water for aerial application. Make the last application between full bloom and early petal fall. Do not make more than five applications from leaf bud break to early petal fall.

Eastern Seaboard States (For primary infection only) Apply 24 fl. oz. of Funginex per acre in 20 to 50 gallons of water for ground application or in 5 gallons of water for aerial application. Make the first application at leaf bud break and repeat in 7 to 10 days. Make the last application at pink bud stage. Do not make more than three applications from leaf bud break to pink bud stage. Application of Funginex during or beyond early bloom may result in fruit russetting.

DEALERS SHOULD SELL IN ORIGINAL PACKAGE ONLY.

TERMS OF SALE OR USE: On purchase of this product buyer and user agree to the following conditions:

WARRANTY: FMC makes no warranty, expressed or implied, concerning the use of this product other than indicated on the label. Except as so warranted, the product is sold as is. Buyer and user assume all risk of use and/or handling and/or storage of this material when such use and/or handling and/or storage is contrary to label instructions.

DIRECTIONS AND RECOMMENDATIONS: Follow directions carefully. Timing and method of application, weather and crop conditions, mixture with other chemicals not specifically recommended and other influencing factors in the use of this product are beyond the control of the seller and are assumed by buyer at his own risk.

USE OF PRODUCT: FMC's recommendations for the use of this product are based upon tests believed to be reliable. The use of this product being beyond the control of the manufacturer, no guarantee, expressed or implied, is made as to the effects of such or the results to be obtained if not used in accordance with directions or established safe practice.

DAMAGES: Buyer's or user's exclusive remedy for damages for breach of warranty or negligence shall be limited to direct damages not exceeding the purchase price paid and shall not include incidental or consequential damages.

FMC is a registered trademark of FMC Corporation. (3913-3/88)

This is a specimen label. FMC is not responsible for the accuracy of the information contained herein. As labels are subject to revision, always carefully read and follow the label on the product container.

Furadan® 3 G

INSECTICIDE

RESTRICTED USE PESTICIDE

Due To Avian Toxicity

For retail sale to and use only by certified applicators or persons under their direct supervision, and only for those uses covered by the certified applicator's certification.

For Agricultural or Commercial Use Only

ACTIVE INGREDIENTS:

* Carbofuran	3.00%
INERT INGREDIENTS:	97.00%
TOTAL	100.00%

* 2,3-Dihydro-2,2-dimethyl-7-benzofuranyl methylcarbamate

EPA Reg. No. 279-2792 EPA Est., No.
Code 2836
U.S. Patent No. 3,474,171

KEEP OUT OF REACH OF CHILDREN

WARNING

STATEMENT OF PRACTICAL TREATMENT

If swallowed: drink 1 or 2 glasses of water and induce vomiting by touching back of throat with finger. Do not induce vomiting or give anything by mouth to an unconscious person. Get medical attention.

If in eyes: flush with plenty of water for at least 15 minutes. Get medical attention.

If on skin: wash immediately with plenty of soap and water. Remove contaminated clothing and wash before reuse.

ANTIDOTE

Note to Physician: Carbofuran is reversible cholinesterase inhibitor. Do not use oximes such as 2-PAM. Start by giving 2 mg. atropine intramuscularly. According to clinical response, continue until signs of atropinization occur (dry mouth or dilated pupils). If in eye, instill one drop of homatropine.

For Emergency Assistance Call 716-735-3765.

PRECAUTIONARY STATEMENTS

HAZARDS TO HUMANS (AND DOMESTIC ANIMALS)

WARNING

Poisonous if swallowed. May be fatal or harmful as a result of skin or eye contact or by breathing dust. Causes cholinesterase inhibition. Warning symptoms of poisoning include weakness, headache, sweating, nausea, vomiting, diarrhea, tightness in chest, blurred vision, pinpoint eye pupils, abnormal flow of saliva, abdominal cramps, and unconsciousness. Atropine sulfate is antidotal.

Wear long-sleeved clothing and protective gloves when handling. Wash hands and face before eating or smoking. Bathe at the end of the work day. Change clothing daily and wash before reuse.

See side panel for additional precautionary measures.

ENVIRONMENTAL HAZARDS

This pesticide is toxic to fish, birds and other wildlife. Birds feeding on treated area may be killed. Birds killed by carbofuran pose a hazard to hawks and other birds-of-prey; bury or otherwise dispose of dead birds to prevent poisoning of other wildlife. Cover or incorporate granules in spill areas. Runoff from treated areas may be hazardous to fish in neighboring areas. Fish may be killed at recommended rates. Do not contaminate wells, wetlands or any body of water by cleaning of equipment or disposal of waste.

Carbofuran is a chemical which can travel (seep or leach) through soil and can contaminate ground water which may be used as drinking water. Carbofuran has been found in ground water as a result of agricultural use. Users are advised not to apply carbofuran where the water table (ground water) is close to the surface and where the soils are very permeable, i.e., well-drained soils such as loamy sands. Your local agricultural agencies can provide further information on the type of soil in your area and the location of ground water.

NOTE TO USER

Do not allow this material to drift onto neighboring crops or non-crop areas or use in a manner or at a time other than in accordance with label directions because animal, plant or crop injury, illegal residues or other undesirable results may occur.

DIRECTIONS FOR USE

It is a violation of Federal law to use this product in a manner inconsistent with its labeling.

Do not re-enter treated fields less than 24 hours following application.

Do not use this product on Long Island, N.Y.

STORAGE AND DISPOSAL

PESTICIDE STORAGE

Do not store in or around the home.

Keep out of reach of children and animals. Store in original containers only. Store in a cool, dry place and avoid excess heat. Carefully open containers. After partial use, fold and roll back bags, clamp and close tightly. Do not put concentrate or dilute material into food or drink containers. Do not contaminate other pesticides, fertilizers, water, food or feed by storage or disposal.

In case of spill, avoid contact, isolate area and keep out animals and unprotected persons. Confine spills. Call FMC collect: (716) 735-3765.

To confine spill: If liquid, dike surrounding area or absorb with sand, cat litter or commercial clay. If dry material, cover to prevent dispersal. Place damaged package in a holding container. Identify contents.

PESTICIDE DISPOSAL

Pesticide wastes are acutely hazardous. Improper disposal of excess pesticide, spray mixture, or rinsate is a violation of Federal Law. If these wastes cannot be disposed of by use according to label instructions, contact your State Pesticide or Environmental Control Agency, or the Hazardous Waste representative at the nearest EPA Regional Office for guidance.

CONTAINER DISPOSAL

Completely empty bag into application equipment. Then dispose of empty bag in a sanitary landfill or by incineration, or, if allowed by State and local authorities, by burning. If burned, stay out of smoke.

Do not plant with any crop for which carbofuran treatment is not registered for at least 10 months.

Do not rotate with any crop on soil treated at greater than 3.0 pounds active ingredient per acre for at least 10 months.

RICE: (Arkansas, Louisiana, Mississippi, Missouri and Texas only) Rice Water Weevil and Mosquitoes—Apply 15 to 20 pounds of Furadan 3G per acre one day before or within 21 days after permanent flooding. Do not make more than one application per season. Apply by air or ground equipment. The subsequent use of propanil herbicide may result in crop injury. For effective larval control of the dark rice field mosquito, application must be made 2 to 4 days after the field has been flooded.

DEALERS SHOULD SELL IN ORIGINAL PACKAGES ONLY.

TERMS OF SALE OR USE: On purchase of this product buyer and user agree to the following conditions:

WARRANTY: FMC makes no warranty, expressed or implied, concerning the use of this product other than indicated on the label. Except as so warranted, the product is sold as is. Buyer and user assume all risk of use and/or handling and/or storage of this material when such use and/or handling and/or storage is contrary to label instructions.

DIRECTIONS AND RECOMMENDATIONS: Follow directions carefully. Timing and method of application, weather and crop conditions, mixture with other chemicals not specifically recommended, and other influencing factors in the use of this product are beyond the control of the seller and are assumed by the buyer at his own risk.

USE OF PRODUCT: FMC's recommendations for the use of this product are based upon tests believed to be reliable. The use of this product being beyond the control of the manufacturer, no guarantee, expressed or implied, is made to the effects of such or the results to be obtained if not used in accordance with directions or established safe practice.

DAMAGES: Buyer's or user's exclusive remedy for damages for breach of warranty or negligence shall be limited to direct damages not exceeding the purchase price paid and shall not include incidental or consequential damages.

Furadan and FMC —FMC Trademarks (2836–5/86)

This is a specimen label. FMC is not responsible for the accuracy of the information contained herein. As labels are subject to revision, always carefully read and follow the label on the product container.

Furadan® 4 F

INSECTICIDE-NEMATICIDE

RESTRICTED USE PESTICIDE

Due to acute oral and inhalation toxicity. For retail sale to and application only by certified applicators or personnel under their direct supervision.

ACTIVE INGREDIENT:	
* Carbofuran	44.0%
INERT INGREDIENTS:	56.0%
TOTAL	100.0%

* 2-3-Dihydro-2, 2-dimethyl-7-benzofuranyl methylcarbamate.

This product contains 4 lbs. of carbofuran per gallon.

EPA Reg. No. 279-2876 ZA EPA Est., 279-Code 279

U.S. Patent No. 3,474,171

KEEP OUT OF REACH OF CHILDREN

DANGER–POISON

PELIGRO

PRECAUCION AL USUARIO: Si usted no lee ingles, no use este producto hasta que la etiqueta le haya sido explicada ampliamente.

STATEMENT OF PRACTICAL TREATMENT

If swallowed: Drink 1 or 2 glasses of water and induce vomiting by touching back of throat with finger. Do not induce vomiting or give anything by mouth to an unconscious person. Get medical attention.

If inhaled: Remove to fresh air. Call a physician immediately.

If in eyes: Flush with plenty of water for at least 15 minutes. Get medical attention.

If on skin: Wash skin immediately with soap and water. Remove contaminated clothing and wash before reuse.

ANTIDOTE

NOTE TO PHYSICIAN: Carbofuran is a reversible cholinesterase inhibitor. Do not use oximes such as 2-PAM. Start by giving 2 mg. atropine intramuscularly. According to clinical response, continue until signs of atropinization occur (dry mouth or dilated pupils). If in eye, instill one drop of homatropine.

For Emergency Assistance Call 716–735–3765.

See other panels for additional precautionary information.

PRECAUTIONARY STATEMENTS

HAZARDS TO HUMANS (& DOMESTIC ANIMALS)

DANGER

Poisonous if swallowed or inhaled. May be fatal or harmful as a result of skin or eye contact or by breathing spray mist. Causes cholinesterase inhibition. Warning symptoms of poisoning include weakness, headache, sweating, nausea, vomiting, diarrhea, tightness in chest, blurred vision, pinpoint eye pupils, abnormal flow of saliva, abdominal cramps, and unconsciousness. Atropine sulfate is antidotal.

Wear a pesticide respirator jointly approved by the Mining Enforcement and Safety Administration (formerly the U.S. Bureau of Mines) and by the National Institute for Occupational Safety and Health under the provisions of 30 CFR Part II. Wear goggles or face shield for eye protection when measuring or mixing. Wear long sleeved clothing and protective gloves when handling. Wash hands and face before eating or smoking.

Bathe at the end of the work day. Change clothing daily and wash before reuse.

Keep all unprotected persons out of the operating area or vicinity where there may be danger of drift.

ENVIRONMENTAL HAZARDS

This product is toxic to fish, birds and other wildlife. Birds feeding on treated areas may be killed. For waterfowl protection, do not apply immediately before or during irrigation, or on fields in proximity of waterfowl nesting areas, or on fields where waterfowl are known to repeatedly feed. Drift and runoff from treated areas may be hazardous to fish in neighboring areas. Do not apply directly to water.

Notice: It is a federal offense to use any pesticide in a manner that results in the death of a member of an endangered species.

The use of Furadan 4 F may pose a hazard to the following Federally designated endangered/threatened species known to be found in certain areas within the named locations.

Attwater's Greater Prairie Chicken—Texas counties including: Aransas, Austin, Brazoria, Colorado, Galveston, Goliad, Harris, Refugio and Victoria

Aleutian Canada Goose—California counties including: Colusa, Merced, Stanislaus and Sutter

Kern Primrose Sphinx Moth—Walker Basin of Kern County, California

This product may not be used in areas where adverse impact on the Federally designated endangered/threatened species noted above is likely. Prior to making applications, the user of this product must determine that no such species are located in or immediately adjacent to the area to be treated. If the user is in doubt whether or not the above named endangered species may be affected, he should contact either the regional U.S. Fish and Wildlife Service office (Endangered Species Specialist) or personnel of the State Fish and Game office.

This product is highly toxic to bees exposed to direct treatment or residues on crops. Do not apply this product or allow it to drift to blooming crops or weeds if bees are visiting the treatment area. Protective information may be obtained from your Cooperative Agricultural Extension Service.

Carbofuran is a chemical which can travel (seep or leach) through soil and can contaminate groundwater which may be used as drinking water. Carbofuran has been found in groundwater as a result of agricultural use. Users are advised not to apply carbofuran where the water table (groundwater) is close to the surface and where the soils are very permeable, i.e., well-drained soils such as loamy sands. Your local agricultural agencies can provide further information on the type of soil in your area and the location of groundwater.

DIRECTIONS FOR USE

It is a violation of Federal law to use this product in a manner inconsistent with its labeling.

Do not apply this product through any type of irrigation system.

Do not use this product on Long Island, N.Y.

Do not apply this product in such a manner as to directly or through drift expose workers or other persons. The area being treated must be vacated by unprotected persons.

SHAKE WELL BEFORE USING

If prolonged intimate contact with corn and or sorghum foliage will result, do not reenter treated fields within 14 days of application without wearing proper protective clothing. For all other situations, do not reenter fields less than 24 hours following application, unless appropriate protective clothing is worn.

Protective clothing means, at least, a hat or other suitable head covering, a long sleeved shirt and long legged trousers or a coverall type garment (all of closely woven fabric covering the body, including the arms and legs), shoes and socks.

Because certain states may require more restrictive reentry intervals for various crops treated with this product, consult your State Department of Agriculture for further information.

Written or oral warnings must be given to workers who are expected to be in a treated area or in an area about to be treated with this product. Inform workers of areas or fields that may not be entered without specific protective clothing, period of time field must be vacated and appropriate actions to take in case of accidental exposure. An example of such information is given under written warnings. When oral warnings are given, warnings shall be given in a language customarily understood by workers. Oral warnings must be given if there is reason to believe that written warnings cannot be understood by workers. Written warnings must include the following information:

DANGER

Areas treated with Furadan® 4 F insecticide–nematicide on (insert date). Do not enter without appropriate protective clothing for 24 hours (insert State Department of Agriculture's reentry interval, if more restrictive).

In case of accidental exposure to pesticide spray, wash the skin thoroughly with the soap and water. Remove contaminated clothing and wash before reuse. If in eyes, flush with plenty of water. If inhaled, go to an area where the pesticide has not been applied. Get medical attention if needed.

STORAGE AND DISPOSAL

PESTICIDE STORAGE

Not for use or storage in or around the house.

Do not store below 35°F, (2°C).

Keep out of reach of children and animals. Store in original containers only. Store in cool, dry place and avoid excess heat. Carefully open containers. After partial use, replace lids and close tightly. Do not put concentrate or dilute material into food or drink containers. Do not contaminate other pesticides, fertilizers, water, food or feed by storage or disposal.

In case of spill, avoid contact, isolate area and keep out animals and unprotected persons. Confine spills. Call FMC collect: (716) 735–3765.

To confine spill: If liquid, dike surrounding area or absorb with sand, cat litter or commercial clay. If dry material, cover to prevent dispersal. Place damaged package in a holding container. Identify contents.

PESTICIDE DISPOSAL

Pesticide wastes are acutely hazardous. Improper disposal of excess pesticide, spray mixture, or rinsate is a violation of Federal law. If these wastes cannot be disposed of by use according to label instructions, contact your State Pesticide or Environmental Control Agency, or the Hazardous Waste representative at the nearest EPA Regional Office for guidance.

CONTAINER DISPOSAL

Metal Containers: Triple rinse (or equivalent). Then offer for recycling or reconditioning, or puncture and dispose of in a sanitary landfill, or by other procedures approved by state and local authorities. Do not cut or weld metal containers.

Plastic Containers: Triple rinse (or equivalent). Then offer for recycling or reconditioning, or puncture and dispose of in a sanitary landfill, or incineration, or, if allowed by state and local authorities, by burning. If burned, stay out of smoke.

GENERAL INSTRUCTIONS

Do not plant with any crop for which carbofuran treatment is not registered for at least 10 months.

Do not rotate with any crop on soil treated at greater than 3.0 pounds active ingredient per acre for at least 10 months.

ALFALFA: Alfalfa Weevil Larvae, Egyptian Alfalfa Weevil, Pea Aphid, and in N.Y. State for Snout Beetle control—Apply the amount of Furadan 4 F, indicated in the chart, when feeding is noticed or when insects appear. Alfalfa Weevil Adult—Apply 1 to 2 pints per acre when insects appear. Alfalfa Blotch Leafminer (Northeastern States only) and Potato Leafhopper. Apply 1 to 2 pints per acre when insects appear. Lygus Bugs—Apply 2 pints per acre prior to bloom. Grasshoppers—Apply 1/4 to 1/2 pint per acre when grasshopper feeding is noticed. For control of Blue Alfalfa Aphid (nymphs and wingless adults)—Apply Furadan 4 F at 1/2 to 1 pint per acre when insect feeding is noticed or when insects appear. Do not apply more than twice per season. Do not apply more than once per cutting. Do not use more than 1 pint per acre in the second application. Apply only to fields planted to pure stands of alfalfa. Do not move bees to alfalfa fields within 7 days of application. Observe the indicated number of days after application before cutting or grazing.

Pints of Furadan 4 F Per Acre	Do not cut or graze within
1/2	7 days
1	14 days
2	28 days

Minimum gallonage requirements. Ten gallons of finished spray per acre with ground equipment, two gallons per acre with aircraft, except Blue Alfalfa Aphid use 30 gallons of finished spray per acre with ground equipment, 5 gallons per acre with aircraft.

For waterfowl protection do not apply on fields in proximity of waterfowl nesting areas and/or fields where waterfowl are known to repeatedly feed.

COTTON: Thrips—Use Furadan 4 F at 2.5 fluid ounces per 1,000 linear feet of row (1 quart per acre with 40 inch row spacing). Apply in the seed furrow at planting. Furadan 4 F may be mixed with water or liquid fertilizer. When Furadan 4 F is used with liquid fertilizers, premix one part of Furadan 4 F with two parts of water. Check physicial compatibility before mixing large quantities. Add this premix to the tank of fertilizer along with the rinsings from the premixing container. Maintain agitation in the tank after mixing and during application. Do not mix until ready to use. Do not feed cotton forage.

FIELD CORN, POPCORN, SWEET CORN—AT PLANTING: Corn Rootworms, Flea Beetles, and to aid in the control of first generation European Corn Borer—Use 2.5 fluid ounces of Furadan 4 F per 1,000 linear feet of row (1 quart per acre with 40 inch row spacing). Apply at planting, as a 7 inch band over the row or inject on each side of the row by mixing with water or liquid fertilizers.

Southern Corn Rootworm, Flea Beetles, Seedcorn Maggot, Wireworms, and to aid in the control of first generation European Corn Borer and Armyworm for approximately 4 to 6 weeks after planting—Use 2.5 fluid ounces of Furadan® 4 F insecticide-nematicide per 1,000 linear feet of row (1 quart per acre with 40 inch row spacing). Apply at planting directly into the seed furrrow.

Furadan 4 F may be mixed with water or liquid fertilizers. If Furadan 4 F is used with liquid fertilizers, premix one part of Furadan 4 F with two parts of water. Check physical compatibility before mixing large quantities. Add this premix to the tank of fertilizer along with the rinsings from the premixing container. Maintain agitation in the tank after mixing and during application. Do not mix until ready to use. Do not feed forage within 30 days of last application.

FIELD CORN, POPCORN—POST PLANT: Corn rootworms (northern, southern and western)—use 2.5 fluid ounces of Furadan 4 F per 1,000 linear feet of row (1 quart per acre with 40 inch row spacing). Apply as a post emergent spray by banding over the row, or by side dressing or basal spraying both sides of the row after corn emerges. Control will generally be improved if the treatment is cultivated into the soil. Do not feed forage within 30 days of last application.

FIELD CORN—FOLIAR APPLICATION

European Corn Borer—Use 1½ to 2 pints of Furadan 4 F per acre as a foliar spray when corn borer eggs begin to hatch. Use the higher rate for heavier pest infestations. For treatment with aerial equipment, apply as a broadcast spray using a minimum of 1 gallon of finished spray per acre. For treatment with ground equipment, direct the spray into the corn whorl for first brood and into the ear zone for second brood, using a minimum of 10 gallons of finished spray per acre. Repeat if necessary. Observe all precautions listed below.

Southwestern Corn Borer—Use Furadan 4 F at 1 to 2 pints per acre as a broadcast foliar spray when eggs begin to hatch. Rate used will depend on desired residual activity. If infestation continues, retreat in 7 days after a 1 pint application and within 14 days after a 2 pint application. Apply in sufficient water for thorough coverage using a minimum of 1 gallon of finished spray per acre with air equipment or 10 gallons of finished spray per acre with ground equipment. Observe all precautions listed below.

Banks Grass Mites (suppression)—Furadan 4 F when applied at 2 pints per acre for the control of European or Southwestern Corn Borers will also suppress Banks Grass Mites.

Grasshoppers—Use Furadan 4 F as a foliar spray at a ¼ to ½ pint per acre when insects appear or feeding is noticed.

—Use the ¼ pint rate under light to moderate population levels (0 –14 grasshoppers per sq. yd.).

—Use the ½ pint rate under more severe population levels (15 or more grasshoppers per sq. yd.).

Apply in sufficient water for thorough coverage using a minimum of 2 gallons of finished spray per acre with air equipment or 10 gallons of finished spray per acre with ground equipment. Observe all precautions listed below.

Do not make more than two applications per season at the 1½ to 2 pint use rate. Do not make more than four applications per season at the 1 pint use rate. Do not forage, cut or harvest within 30 days of last application. Do not apply on seed corn less than 14 days prior to detasseling or rogueing. If prolonged intimate contact will result, do not reenter treated field within 14 days of application without wearing proper protective clothing.

SWEET CORN—FOLIAR APPLICATION (Machine Harvested Only): European Corn Borer—For control of second generation borers apply 1 pint Furadan 4 F per acre. Make first application just prior to first silking and repeat at weekly intervals. Do not make more than four (4) applications per season. Do not apply within seven (7) days of harvest. If prolonged, intimate contact will result, do not reenter treated field within 14 days of application without wearing proper protective clothing. Do not graze or harvest stalks within 21 days of last application.

Do not make a foliar application if more than 12 ozs. of Furadan 10 G per 1,000 linear feet of row (10 lbs. per acre with 40" row spacing) or 8 ozs. of Furadan 15 G per 1,000 linear feet of row (6.7 lbs. per acre with 40" row spacing) or 2.5 fluid ounces of Furadan 4 F per 1,000 linear feet of row (1 quart per acre with 40 inch row spacing) were used in an at-planting application.

Minimum gallonage requirements: 10 gallons of finished spray per acre with ground equipment, 2 gallons per acre with aircraft.

GRAPES (For use in California): Nematodes (Root Knot and Dagger) and Grape Phylloxera—Apply Furadan® 4 F insecticide-nematicide at 2½ gallons per acre as a broadcast treatment only, to the soil surface between the vine rows and immediately incorporate by mechanical means. Do not apply within 200 days of harvest. Remove dense weed growth prior to treatment. Do not use on soils of pH 8.0 or greater. Do not make more than one application per crop year.

Wear protective clothing including rubber gloves and impervious foot covering at time of application.

ORNAMENTALS—CONTAINER GROWN

Root Weevil Larvae—Prepare a stock solution by thoroughly mixing 1 to 2 fluid ounces of Furadan 4 F per 100 gallons of water. Apply as a soil drench in sufficient volume to saturate the entire soil profile within each container. The following guidelines give approximate amounts to use on various sized containers. Make a single application when larvae are present (usually from July to mid-October in outdoor growing areas). Later application may be less effective due to lower temperatures and/or the presence of more mature larvae.

Container Size	Amount of Stock Solution Per Container
6 inches diameter	1 pint
8 inches diameter	1 to 2 pints
10 inches diameter	2 to 4 pints
12 inches diameter	3 to 6 pints
16 inches diameter	4 to 8 pints

Early marginal necrosis or leaf drop may occur on Hydrangea or Birch. Not all species or varieties of ornamentals have been tested. Before treating large numbers of plants of a particular variety, treat a few plants and observe prior to full scale application.

Do not handle treated plants, soil or containers that are wet with spray.

Application of Furadan 4 F through overhead sprinkler equipment is prohibited.

PINE SEEDLINGS: Pales Weevil and Pitch-Eating Weevil in pine plantations—Apply a 1% (W/W) active Furadan clay slurry (see following for preparation) to the roots of pine seedlings prior to transplanting. Treat seedlings by dipping roots or use any other suitable means which allows thorough coating. Keep roots moist until transplanted. Prepare the slurry as follows: Add 1.6 ounces (2½ tablespoons) of Furadan 4 F to ½ gallon of water. Mix thoroughly. Add 2 pounds of pulverized kaolin clay (pH 4.5) to this suspension. Mix thoroughly. This is sufficient to treat the roots of 150 to 200 seedlings. Adequate ventilation is required for indoor treatment. Wear protective unlined rubber gloves at time of application. Users should avoid contamination of the inside of the gloves. If contamination of the inside of the gloves is suspected, the gloves should be immediately discarded.

POTATOES: Potato Tuberworm (Virginia only); Colorado Potato Beetle, European Corn Borer, Potato Flea Beetle, Potato Leafhopper—Use 1 to 2 pints of Furadan 4 F in sufficient water to treat one acre. Apply when insects first appear and repeat as necessary to maintain control. Do not make more than 8 foliar applications per season. Do not apply more than 6 pints to the foliage if either Furadan 10 G or Furadan 15 G was used at planting. Do not apply more than 2 pints per application. Do not apply within 14 days of harvest. Minimum gallonage requirements: 10 gallons of finished spray per acre with ground equipment and 3 gallons per acre with aircraft.

Do not use this product on Long Island, New York.

For waterfowl protection do not apply:

—immediately before or during furrow irrigation

—on fields in proximity of waterfowl nesting areas

—on fields where waterfowl are known to repeatedly feed

SMALL GRAINS (Wheat, Oats, Barley): Grasshoppers—Use Furadan 4 F at ¼ to ½ pint per acre when insects appear. Cereal Leaf Beetle—Use ½ pint per acre when insects appear. Apply before heads emerge from boot. Do not make more than two applications per season. Do not feed treated forage to livestock.

Minimum gallonage requirements: Ten gallons of finished spray per acre with ground equipment, two gallons per acre with aircraft.

For waterfowl protection do not apply on fields in proximity of waterfowl nesting areas and/or on fields where waterfowl are known to repeatedly feed.

SOYBEANS—AT PLANTING APPLICATION: Nematodes (Root-Knot, Cyst, Stunt, Ring, Sting, Spiral, Lesion, Lance, Dagger, Stubby Root)—Use 3.75 to 5 fluid ounces per 1,000 linear feet of row (3 to 4 pints per acre with 40-inch row spacing). Apply at planting as a 12-inch surface band. Soybeans may be grazed or cut for forage 30 days or later following an at planting application.

SOYBEANS—FOLIAR APPLICATION: Mexican Bean Beetle—Use Furadan 4 F at ½ to 1 pint per acre when insects appear. Grasshoppers—Use Furadan 4 F at ¼ to ½ pint per acre when insects appear. Do not use Furadan 4 F as a foliar application if Furadan 10 G, Furadan 15 G, Furadan 4 F was applied to soybeans at planting time. Do not make more than 2 foliar applications per season. Do not apply within 21 days of harvest. Do not graze or feed foliar-treated forage to livestock or cut for silage or hay. Minimum gallonage requirements: 20 gallons of finished spray per acre with ground equipment, 1½ gallons per acre with aircraft.

STRAWBERRY: (For use in Washington & Oregon) Root Weevils—Use Furadan 4 F at 2 to 4 pints per acre (2.6 to 5.1 fluid ounces per 1,000 linear feet of row for 42 inch row spacing). Apply as a 10 to 12 inch band over the row after last harvest but before October 1st. Do not make more than one application per season. Do not apply if berries are present.

SUGARCANE: Sugarcane Borer—Apply 1 to 1½ pints Furadan 4 F per acre using ground or aerial equipment. Check sugarcane fields weekly, beginning in early June and continuing through August. Make first application only after visible joints form and 5% or more of the plants are infested with young larvae feeding in or under the leaf sheath and which have not bored into the stalks. Repeat whenever field checks indicate the infestation exceeds 5 percent. Do not apply within 17 days of harvest. Do not use in Hawaii.

SUNFLOWERS (Confectionary and Oil)—At Planting

Sunflower Stem Weevil, Sunflower Beetle, Grasshoppers—Use 2.5 to 5.0 fluid ounces of Furadan 4 F per 1,000 linear feet of row (1.4 to 2.8 quarts per acre with 30 inch row spacing). Apply directly into the seed furrow—OR—Apply in a 7 inch band. Furadan 4 F may be mixed with water or liquid fertilizer. When Furadan 4 F is used with liquid fertilizers, premix one part of Furadan 4 F with two parts of water. Check physical compatibility before mixing large quantities. Add this premix to the tank of fertilizer along the rinsings from the premixing container. Maintain agitation in the tank after mixing and during application. Do not mix until ready to use.

Use the higher rate when heavier insect infestations are anticipated.

SUNFLOWERS (Confectionary and Oil)—Foliar Application

Sunflower Moth, Banded Sunflower Moth, Stem Weevils, Seed Weevils—Use Furadan 4 F at 1 pint per acre. Grasshoppers—Use Furadan 4 F at 1/4 to 1 pint per acre. Sunflower Beetle—Use Furadan 4 F at 1/4 to 1/2 pint per acre. Apply as a foliar spray when insects appear. When a rate range is indicated, use higher rate for heavier insect infestations. Repeat applications as necessary but not more than four times per season. Apply a minimum of two gallons of finished spray per acre with aircraft and 10 gallons by ground equipment. If prolonged, intimate contact will result, do not reenter treated field within 14 days of application without wearing proper protective clothing. Do not harvest crop within 28 days of last application.

TOBACCO—BURLEY: Flea Beetles—Use 1 gallon of Furadan 4 F per acre. Apply as a broadcast spray over the soil surface prior to transplanting and incorporate into the top 3 inches of soil with a suitable device.

TOBACCO—FLUE-CURED: Flea Beetles, Wireworms, and to aid in the control of Budworms and Root Knot Nematodes—Use 1 1/2 gallons of Furadan 4 F per acre— OR—Use 1 gallon of Furadan 4 F per acre for control of Flea Beetles only. Apply as a broadcast spray over the soil surface prior to forming beds. Incorporate into the top 3 inches of soil. Form beds and plant as usual. This product may induce flecking of the bottom or lower leaves.

DEALERS SHOULD SELL IN ORIGINAL PACKAGES ONLY.

TERMS OF SALE OR USE: On purchase of this product buyer and user agree to the following conditions:

WARRANTY: FMC makes no warranty, expressed or implied, concerning the use of this product other than indicated on the label. Except as so warranted, the product is sold as is. Buyer and user assume all risk of use and/or handling and/or storage of this material when such use and/or handling and/or storage is contrary to label instructions.

DIRECTIONS AND RECOMMENDATIONS: Follow directions carefully. Timing and method of application, weather and crop conditions, mixture with other chemicals not specifically recommended, and other influencing factors in the use of this product are beyond the control of the seller and are assumed by buyer at his own risk.

USE OF PRODUCT: FMC's recommendations for the use of this product are based upon tests believed to be reliable. The use of this product being beyond the control of the manufacturer, no guarantee, expresssed or implied, is made as to the effects of such or the results to be obtained if not used in accordance with directions or established safe practice.

DAMAGES: Buyer's or user's exclusive remedy for damages for breach of warranty or negligence shall be limited to direct damages not exceeding the purchase price paid and shall not include incidental or consequential damages.

Furadan and FMC—FMC trademarks (279-4/7/86-A)

This is a specimen label. FMC is not responsible for the accuracy of the information contained herein. As labels are subject to revision, always carefully read and follow the label on the product container.

Furadan® 5 G

INSECTICIDE

RESTRICTED USE PESTICIDE

Due To Avian Toxicity

For retail sale to and use by certified applicators or persons under their direct supervision, and only for those uses covered by the certified applicator's certification.

For Agricultural or Commercial Use Only

ACTIVE INGREDIENTS:	
* Carbofuran	5.0%
INERT INGREDIENTS:	95.0%
TOTAL	100.0%

* 2,3-Dihydro-2, 2-dimethyl-7-benzofuranyl methylcarbamate

EPA Reg No. 279-2874 AA EPA Est., Code 3214

U.S. Patent No. 3,474,171

KEEP OUT OF REACH OF CHILDREN

WARNING

STATEMENT OF PRACTICAL TREATMENT

If swallowed: drink 1 or 2 glasses of water and induce vomiting by touching back of throat with finger. Do not induce vomiting or give anything by mouth to an unconscious person. Get medical attention.

If in eyes: flush with plenty of water for at least 15 minutes. Get medical attention.

If on skin: wash immediately with plenty of soap and water. Remove contaminated clothing and wash before reuse.

ANTIDOTE

NOTE TO PHYSICIAN:

Carbofuran is a reversible cholinesterase inhibitor. Do not use oximes such as 2-PAM.

Start by giving 2 mg. atropine intramuscularly. According to clinical response, continue until signs of atropinization occur (dry mouth or dilated pupils).

If in eye, instill one drop of homatropine.

For Emergency Assistance Call 716–735–3765.

See Side Panel For Additional Precautionary Information.

PRECAUTIONARY STATEMENTS

HAZARDS TO HUMANS (AND DOMESTIC ANIMALS)

WARNING

Poisonous if swallowed. May be fatal or harmful as a result of skin or eye contact or by breathing dust. Causes cholinesterase inhibition. Warning symptoms of poisoning include weakness, headache, sweating, nausea, vomiting, diarrhea, tightness in chest, blurred vision, pinpoint eye pupils, abnormal flow of saliva, abdominal cramps, and unconsciousness. Atropine sulfate is antidotal.

Wear long-sleeved clothing and protective gloves when handling. Wash hands and face before eating or smoking. Bathe at the end of the work day. Change clothing daily and wash before reuse.

ENVIRONMENTAL HAZARDS

This pesticide is toxic to fish, birds and other wildlife. Birds feeding on treated areas may be killed. Birds killed by carbofuran pose a hazard to hawks and other birds-of-prey; bury or otherwise dispose of dead birds to prevent poisoning of other wildlife. Cover or incorporate granules in spill areas. Runoff from treated areas may be hazardous to fish in neighboring areas. Fish may be killed at recommended rates. Do not contaminate wells, wetlands or any body of water by cleaning of equipment or disposal of waste.

Carbofuran is a chemical which can travel (seep or leach) through soil and can contaminate ground water which may be used as drinking water. Carbofuran has been found in ground water as a result of agricultural use. Users are advised not to apply carbofuran where the water table (ground water) is close to the surface and where the soils are very permeable, i.e., well-drained soils such as loamy sands. Your local agricultural agencies can provide further information on the type of soil in your area and the location of ground water.

DIRECTIONS FOR USE

It is a violation of Federal law to use this product in a manner inconsistent with its labeling.

Do not re-enter treated fields less than 24 hours following application.

Do not use this product on Long Island, N.Y.

STORAGE AND DISPOSAL

PESTICIDE STORAGE

Do not store in or around the home.

Keep out of reach of children and animals. Store in original containers only. Store in a cool, dry place and avoid excess heat. Carefully open containers. After partial use, fold and roll back bags, clamp and close tightly. Do not put concentrate or dilute material into food or drink containers. Do not contaminate other pesticides, fertilizers, water, food or feed by storage or disposal.

In case of spill, avoid contact, isolate area and keep out animals and unprotected persons. Confine spills. Call FMC collect: (716) 735–3765.

To confine spill: If liquid, dike surrounding area or absorb with sand, cat litter or commercial clay. If dry material, cover to prevent dispersal. Place damaged package in a holding container. Identify contents.

PESTICIDE DISPOSAL

Pesticide wastes are acutely hazardous. Improper disposal of excess pesticides, spray mixture, or rinsate is a violation of Federal Law. If these wastes cannot be disposed of by use according to label instructions, contact your State Pesticide or Environmental Control Agency, or the Hazardous Waste representative at the nearest EPA Regional Office for guidance.

CONTAINER DISPOSAL

Completely empty bag into application equipment. Then dispose of empty bag in a sanitary landfill or by incineration, or, if allowed by State and local authorities, by burning. If burned, stay out of smoke.

Do not plant with any crop for which carbofuran treatment is not registered for at least 10 months.

Do not rotate with any crop on soil treated at greater than 3.0 pounds active ingredient per acre for at least 10 months.

RICE: To control Rice Water Weevil and Dark Rice Field Mosquito in Arkansas, Louisiana, Mississippi and Texas apply 10 pounds of Furadan® 5 G insecticide per acre one day before or within 21 days after flooding. Do not make more than one application per season. The subsequent use of propanil herbicide may result in crop injury.

For effective control of the dark rice field mosquito, application must be made within 2 to 4 days after flooding.

To control Rice Water Weevil in California apply 10 pounds of Furadan 5 G per acre prior to flooding and planting. Do not make more than one application per season. The subsequent use of propanil herbicide may result in crop injury.

SUGARCANE: Sugarcane borer—Apply 10 to 15 pounds of Furadan 5 G per acre using ground or aerial equipment. Check sugarcane fields weekly, beginning in early June and continuing through August. Make first application only after visible joints form and 5% or more of the plants are infested with young larvae feeding in or under the leaf sheath and which have not bored into the stalks. Repeat whenever field checks indicate the infestation exceeds 5 percent. Do not apply within 17 days of harvest. Do not use in Hawaii.

DEALERS SHOULD SELL IN ORIGINAL PACKAGES ONLY.

TERMS OF SALES OR USE: On purchase of this product buyer and user agree to the following conditions:

WARRANTY: FMC makes no warranty, expressed or implied, concerning the use of this product other than indicated on the label. Except as so warranted, the product is sold as is. Buyer and user assume all risk of use and/or handling and/or storage of this material when such use and/or handling and/or storage is contrary to label instructions.

DIRECTIONS AND RECOMMENDATIONS: Follow directions carefully. Timing and method of application, weather and crop conditions, mixture with other chemicals not specifically recommended and other influencing factors in the use of this product are beyond the control of the seller and are assumed by buyer at his own risk.

USE OF PRODUCT: FMC's recommendations for the use of this product are based upon tests believed to be reliable. The use of this product being beyond the control of the manufacturer, no guarantee, expressed or implied is made as to the effects of such or the results to be obtained if not used in accordance with directions or established safe practice.

DAMAGES: Buyer's or user's exclusive remedy for damages for breach of warranty or negligence shall be limited to direct damages not exceeding the purchase price paid and shall not include incidental or consequential damages.

Furadan and FMC—FMC Trademarks 3214–5/86

This is a specimen label. FMC is not responsible for the accuracy of the information contained herein. As labels are subject to revision, always carefully read and follow the label on the product container.

Furadan® 15G

INSECTICIDE-NEMATICIDE

RESTRICTED USE PESTICIDE

Due to Avian Toxicity

For retail sale to and use only by certified applicators or persons under their direct supervision, and only for those uses covered by the certified applicator's certification.

For Agricultural or Commercial Use Only

ACTIVE INGREDIENT:	
*Carbofuran	15.0%
INERT INGREDIENTS:	85.0%
TOTAL	100.0%

*2,3-Dihydro-2,2-dimethyl-7-benzofuranyl methylcarbamate

EPA Reg. No. 279-3023 EPA Est., 279-
Code 3669 U.S. Patent No. 3,474,171

WARNING

KEEP OUT OF REACH OF CHILDREN

PRECAUTIONARY STATEMENTS

HAZARDS TO HUMANS (AND DOMESTIC ANIMALS)

WARNING

Poisonous if swallowed. May be fatal or harmful as a result of skin or eye contact or by breathing dust. Causes cholinesterase inhibition. Warning symptoms of poisoning include weakness, headache, sweating, nausea, vomiting, diarrhea, tightness in chest, blurred vision, pinpoint eye pupils, abnormal flow of saliva, abdominal cramps, and unconsciousness. Atropine sulfate is antidotal.

Wear long-sleeved clothing and protective gloves when handling. Wash hands and face before eating or smoking. Bathe at the end of the work day. Change clothing daily and wash before reuse.

STATEMENT OF PRACTICAL TREATMENT

If swallowed: drink 1 to 2 glasses of water and induce vomiting by touching back of throat with finger. Do not induce vomiting or give anything by mouth to an unconscious person. Get medical attention.

If in eyes: flush with plenty of water for at least 15 minutes. Get medical attention.

If on skin: wash immediately with soap and water. Remove contaminated clothing and wash before reuse.

Antidote

Note to Physician: Carbofuran is a reversible cholinesterase inhibitor. Do not use oximes such as 2-PAM. Start by giving 2 mg. atropine intramuscularly. According to clinical response, continue until signs of atropinization occur (dry mouth or dilated pupils). If in eye, instill one drop of homatropine.

For Emergency Assistance Call 716-735-3765.

See other panels for additional precautionary information.

ENVIRONMENTAL HAZARDS

This pesticide is toxic to fish, birds and other wildlife. Birds feeding on treated areas may be killed. Birds killed by carbofuran pose a hazard to hawks and other birds-of-prey; bury or otherwise dispose of dead birds to prevent poisoning of other wildlife. Cover or incorporate granules in spill areas. Runoff from treated areas may be hazardous to fish in neighboring areas. Do not apply directly to water. Do not contaminate wells, wetlands or any body of water by cleaning of equipment or disposal of waste.

Notice: It is a Federal offense to use any pesticide in a manner that results in the death of a member of an endangered species.

The use of FURADAN 15 G may pose a hazard to the following Federally designated endangered/threatened species known to be found in certain areas within the named locations.

Attwater's Greater Prairie Chicken—Texas counties including: Aransas, Austin, Brazoria, Colorado, Galveston, Goliad, Harris, Refugio and Victoria

Aleutian Canada Goose—California counties including Colusa, Merced, Stanislaus and Sutter

Kern Primrose Sphinx Moth—Walker Basin of Kern County, California.

This product may not be used in areas where adverse impact on the Federally designated endangered threatened species, noted above, is likely. Prior to making applications, the user of this product must determine that no such species are located in or immediately adjacent to the area to be treated. If the user is in doubt whether or not the above named endangered species may be affected, he should contact either the regional U.S. Fish and Wildlife Service office (Endangered Species Specialist) or personnel of the State Fish and Game office.

Carbofuran is a chemical which can travel (seep or leach) through soil and can contaminate ground water which may be used as drinking water. Carbofuran has been found in ground water as a result of agricultural use. Users are advised not to apply carbofuran where the water table (ground water) is close to the surface and where the soils are very permeable, i.e., well-drained soils such as loamy sands. Your local agricultural agencies can provide further information on the type of soil in your area and the location of ground water.

DIRECTIONS FOR USE

It is a violation of Federal Law to use this product in a manner inconsistent with its labeling.

Do not use this product on Long Island, N.Y.

STORAGE AND DISPOSAL

PESTICIDE STORAGE

Do not store in or around the home.

Keep out of reach of children and animals. Store in original containers only. Store in a cool, dry place and avoid excess heat. Carefully open containers. After partial use, fold and roll back bags, clamp and close tightly. Do not put concentrate or dilute material into food or drink containers. Do not contaminate other pesticides, fertilizers, water, food or feed by storage or disposal.

In case of spill, avoid contact, isolate area and keep out animals and unprotected persons. Confine spills. Call FMC collect: (716) 735-3765.

To confine spill: If liquid, dike surrounding area or absorb with sand, cat litter or commercial clay. If dry material, cover to prevent dispersal. Place damaged package in a holding container. Identify contents.

PESTICIDE DISPOSAL

Pesticide wastes are acutely hazardous. Improper disposal of excess pesticide, spray mixture, or rinsate is a violation of Federal Law. If these wastes cannot be disposed of by use according to label instructions, contact your State Pesticide or Environmental Control Agency, or the Hazardous Waste representative at the nearest EPA Regional Office for guidance.

CONTAINER DISPOSAL

Completely empty bag into application equipment. Then dispose of empty bag in a sanitary landfill or by incineration, or, if allowed by State and local authorities, by burning. If burned, stay out of smoke.

GENERAL USE INSTRUCTIONS

If prolonged intimate contact with corn and/or sorghum foliage will result, do not re-enter treated fields within 14 days of application without wearing proper protective clothing. For all other situations, do not re-enter fields less than 24 hours following application.

Cover or incorporate granules in spill areas.

Do not plant with any crop for which carbofuran treatment is not registered for at least 10 months.

Do not rotate with any crop on soil treated at greater than 3.0 pounds active ingredient per acre for at least 10 months.

Specific Use Instructions

FIELD CORN, POPCORN, SWEET CORN—Conventional, Minimum & No Tillage

Corn Rootworms (Northern, Southern & Western), Flea Beetles, Wireworms, aid in the control of White Grubs and first generation European Corn Borer, reduce losses due to Stalk Rot by reducing the incidence of insect wounds which permit entry by the Stalk Rot Fungi; Armyworm control for approximately 4 to 6 weeks after planting and suppression of Cutworms—Use 8 ounces of FURADAN® 15G insecticide-nematicide per 1,000 linear feet of row (6.7 lbs. per acre with 40 inch row spacing). Early season suppression of Common Stalk Borer—Use 16 to 24 ounces per 1,000 linear feet of row (13.3 to 20 lbs. per acre with 40 inch row spacing). Apply in a 7 inch band ahead of the planter press wheel and incorporate into the top 1 inch of soil by use of special covering devices or by dragging a short length of chain.—OR—Direct the granules into the planter shoe with the seed.—OR—Place the applicator tube directly behind the planter shoe so that the granules drop into the furrow and mix with the covering soil.

Lesion Nematodes (Northeast and Midwest Corn States including Colorado and Wyoming)—Use 8 to 16 ounces per 1,000 linear feet of row (6.7 to 13.3 lbs. per acre with 40 inch row spacing). Apply "at planting," as above (in a covered band or in the seed furrow).

European Corn Borer (first generation)—Use 16 to 24 ounces per 1,000 linear feet of row (13.3 to 20 lbs. per acre with 40 inch row spacing) "at planting" as above (in a covered band or in the seed furrow).

Seed Corn Maggot—Use 8 ounces per 1,000 linear feet of row (6.7 lbs. per acre with 40 inch row spacing). Southwestern Corn Borer (second to third generation)—Use 12 to 24 ounces per 1,000 linear feet of row (10 to 20 lbs. per acre with 40 inch row spacing). Southern Corn Billbug (Southeastern states), and to aid in the control of Leafhoppers and thereby reduce losses due to Maize Chlorotic Dwarf Virus and Corn Stunt disease—Use 16 to 24 ounces per 1,000 linear feet of row (13.3 to 20 lbs. per acre with 40 inch row spacing). Apply into the seed furrow as noted above.

Nematodes (Sting, Stunt, Stubby Root, Root-Knot, Dagger, Lesion, Lance and Spiral)—Use 12 to 16 ounces per 1,000 linear feet of row (10 to 13.3 lbs. per acre with 40 inch row spacing). Apply at planting time in a 7 to 15 inch band and incorporate into the top 3 inches of soil.

FIELD CORN, POPCORN—Post Plant

Northern and Western Corn Rootworm—Use 8 ounces of FURADAN 15G per 1,000 linear feet of row (6.7 lbs. per acre with 40 inch row spacing) by banding over the row or by side dressing both sides of the row and cultivating into the soil.

FIELD CORN—Foliar Application

Southwestern Corn Borers and European Corn Borers—Use 6.7 lbs. per acre. Apply with aircraft by broadcasting over the corn plants or with ground equipment by directing the granules into the corn whorls. Apply when eggs begin to hatch. Do not make a foliar application if FURADAN 15G was applied at more than 8 ounces per 1,000 linear feet of row (6.7 lbs. per acre with 40 inch row spacing) at planting. Do not make more than two foliar applications per season.

Corn Insecticide Applicator Setting

This product is a dense, free flowing insecticide requiring accurate equipment calibration to apply the recommended amount per acre. Always calibrate your planter at the speed which you plan to travel during the planting operation. The following are approximate settings for guidance in calibration:

Applicator Gauge Setting

Application Rate	6.7 lbs./A	10 lbs./A	13.3 lbs./A
Planter Speed (MPH)	**4 6 8**	**4 6 8**	**4 6 8**
Equipment			
Allis Chalmers	3 3 3	5 5 5	7 7 7
Gandy	14 18 21	18 22 25	22 25 29
IHC Cyclo Planter			
Gate	1 1 2	1 2 2	2 2 3
Dial	7 9 2	9 3 8	2 8 5
John Deere			
Range	1 2 2	1 2 2	2 2 2
Notch	18 1 2	20 2 4	1 4 8
John Deere			
Max. Emerge (even number notches—0-10-20-30-40)	10 14 19	14 21 25	19 25 30
Max. Emerge (odd number notches—5-15-25-35)	9 12 14	12 15 18	14 18 21
Noble			
Round Opening	8 12 16	12 17 22	16 22 28
Rectangular Opening	5 7 9	7 10 13	9 13 17

Corn Application Rates

When the granular application equipment is set to deliver FURADAN 15G at a rate of 8 ounces per 1,000 linear feet, the total amount applied per acre will vary depending on row spacing. Use the table below to determine your FURADAN needs for each acre.

Row Spacing (Inches)	40"	38"	36"	30"	28"
FURADAN 15G (lbs./acre)	6.7	7.0	7.3	8.7	9.3

COTTON

Thrips—Use FURADAN 15G insecticide-nematicide at 8 ounces per 1,000 linear feet of row (6.7 lbs. per acre with 40 inch row spacing). Apply in the seed furrow at planting. Do not feed cotton forage.

CRANBERRIES

For control of Black Vine Weevils—Use FURADAN 15G at 13.3 lbs. per acre. For control of adult weevils and larvae, spread granules uniformly over soil surface using a rotary spreader. Make one application about June 15 to 25 and repeat one month later. Incorporate granules by following each application with sprinkler irrigation. For use only on cranberry bogs that are sprinkler irrigated and are not flooded for purposes of irrigation or harvest. Do not apply within 60 days of harvest. For use only in the State of Washington.

CUCURBITS (Cucumber, Melons, Squash, Pumpkins)

Nematodes (Root-Knot, Sting, Lance)—For use in Southeastern States—Use FURADAN 15G at the rate of 1½ lbs. per 1,000 linear feet of row (13.3 lbs. per acre based on 60 inch row spacing). Apply in a 12 to 15 inch band and incorporate into the top 3 inches of soil.

GRAPES (For Use in California)

For control of Nematodes (Root-Knot, Dagger and Lesion) and Grape Phylloxera—Broadcast FURADAN 15G at 66.7 lbs. per acre over the soil surface between the vine rows and immediately incorporate. Do not apply within 200 days of harvest. Remove dense weed growth prior to treatment. Do not use on soils of pH 8.0 or greater. Retreat as necessary but do not make more than one application per crop year.

PEANUTS

Do not feed treated peanut forage to dairy animals or animals being finished for slaughter.

In-furrow at planting application method:

Thrips, Potato Leafhopper and as an aid in the control of Southern Corn Rootworm in Southeastern States—Use in-furrow at planting, 3.3 to 6.7 lbs. per acre (36 inch row spacing); 3.5 to 6.7 ounces per 1,000 linear feet of row.

Row banding application method:

Nematodes (Root-Knot, Sting, Stunt, Lesion, Ring) in Southeastern States and in the Southwestern States of Texas and Oklahoma; and in the Southeastern States to aid in the control of Potato Leafhopper and Southern Corn Rootworm and for control of Thrips—Apply FURADAN® 15G insecticide-nematicide in a band over the row prior to planting and incorporate 3 inches into the soil. Use the rate indicated below for various band widths. Additional control measures may be needed late in the season for Potato Leafhopper.

Band Width	Rate/Acre (36 inch rows)	Rate/1000 linear feet
12 inches	13.3 to 26.7 lbs.	0.9 to 1.8 lbs.
18 inches	20 to 33.3 lbs.	1.33 to 2.3 lbs.

Banded Pegging—Time Application Method

For control of Southern Corn Rootworm in the Southeastern States—Use FURADAN 15G at 7.3 to 14.7 ounces per 1,000 linear feet (6.7 to 13.3 lbs. per acre with 36 inch row spacing). Apply on a 14 to 18 inch band over the row when pegs begin to form or up to 30 days after pegging begins (approximately 60 to 90 days after planting). Incorporate granules immediately if vine growth permits. Do not apply FURADAN 15G at pegging *if FURADAN 15G was used in a band at planting.*

PEPPERS

Green Peach Aphid—(Delmarva Peninsula and South Jersey only). European Corn Borer—Make two applications of FURADAN® 15G by side-dressing on either one or both sides of the row. Make a first application two to four weeks after transplanting using 13.3 lbs. per acre (equivalent to 1/2 lb. per 500 linear feet with 38 inch rows). Make the second application four to six weeks later using 20 lbs. per acre (0.7 lb. per 500 linear feet). Do not make the second application within 21 days of harvest.

POTATOES

Colorado Potato Beetle, European Corn Borer, Potato Flea Beetle, Potato Leafhopper, Tuber Flea Beetle; Green Peach Aphid (Western and Northeastern states including the Delmarva Peninsula); Potato Tuberworm (Virginia only); and to aid in the control of Wireworm—Use of 1 1/2 lbs. of FURADAN 15G per 1,000 linear feet of row (20 lbs. per acre, 38-inch row spacing). Apply directly into the bottom of the furrow during the planting operation.

SORGHUM

Chinch Bug, Corn Leaf Aphid, Flea Beetle, Greenbug, Lesion Nematode, Seedcorn Maggot, Southern Corn Rootworm, Wireworm, Yellow Sugarcane Aphid and aids in the control of White Grubs and Cutworms—Use only on grain sorghum or sorghum grown for forage. Do not graze or harvest sorghum grown for forage within 75 days of application.

For Corn Leaf Aphid, Greenbug, Yellow Sugarcane Aphid—Use 6 to 8 ounces of FURADAN 15G per 1000 linear feet of row (5 to 6.7 lbs. per acre with 40 inch row spacing) applied directly into the seed furrow—OR—Use 8 ounces of FURADAN 15G per 1000 linear feet of row (6.7 lbs. per acre with 40 inch row spacing, applied in a 7 inch band over the row and incorporate the granules into the top 1 inch of soil by the use of a press wheel, dragging a short length of chain or suitable device.

For Chinch Bug, Flea Beetle, Southern Corn Rootworm, Lesion Nematode, Wireworm, Seedcorn Maggot (in furrow application only), and to aid in the control of White Grubs and Cutworms—Use 8 ounces of FURADAN 15G per 1000 linear feet of row (6.7 lbs. per acre with 40 inch row spacing). Apply at planting as noted above.

Equivalent pounds of FURADAN 15G per acre for various row spacings are as follows:

	Amount of FURADAN 15G per 1,000 Linear Feet of Row	Pounds of FURADAN 15G per Acre for Various Row Spacings				
		40"	36"	34"	30"	20"
In Furrow	6 Ounces	5	5.5	5.8	6.7	10
	8 Ounces	6.7	7.3	7.7	8.7	13.3
7" Band	8 Ounces	6.7	7.3	7.7	8.7	13.3

When a rate range is indicated, use the higher rate when heavier pest infestations are anticipated.

SOYBEANS

Nematodes (Root-Knot, Cyst, Stunt, Ring, Sting, Spiral Lesion, Lance, Dagger, Stubby-Root)—Use FURADAN 15G at 12 to 16 ounces per 1,000 linear feet (equivalent to an application of 10 to 13.3 lbs. per acre on a 40-inch row spacing) at planting in a 7-inch band and mix with the covering soil.

—OR—

Apply FURADAN 15G at 12 ounces per 1,000 linear feet (equivalent to an application of 10 lbs. per acre on a 40-inch row spacing). Apply at planting by directing the granules into the planter shoe with the seed or by placing the applicator tube directly behind the planter shoe so that the granules drop into the furrow and mix with the covering soil.

Mexican Bean Beetle (larvae)—Use FURADAN 15G at 12 ounces per 1,000 linear feet (equivalent to 10 lbs. per acre on a 40-inch row spacing) at planting in a 7-inch band and mix with the covering soil.

—OR—

Apply FURADAN 15G at 8 tO 12 ounces per 1,000 linear feet (equivalent to an application of 6.7 to 10 lbs. per acre on a 40-inch row spacing). Apply at planting by directing the granules into the planter shoe with the seed or by placing the applicator tube into the furrow and mix with the covering soil.

SUGAR BEET

Sugar Beet Root Maggot—Use 9 ounces of FURADAN 15G per 1,000 linear feet of row (13.3 lbs. per acre using 22 inch row spacing). Apply granules in a 6 to 7 inch band and incorporate into the top 1 inch of soil using a suitable device or use 4.5 ounces of FURADAN 15G per 1,000 linear feet of row (6.7 lbs. per acre using 22 inch row spacing) by applying granules in a "modified in-furrow" application. (Drop granules from a tube in a narrow band of 1 to 2 inches in front of the planter disks and allow some granules to fall into seed furrow with seed.)

Sugar Beet Application Rates

When the granular application equipment is set to deliver FURADAN 15G at rates of 4.5, or 9 ounces per 1,000 linear feet, the total amount applied per acre will vary depending on row spacing. Use the table below to determine your FURADAN needs for each acre.

Ounces of FURADAN 15G per 1,000 Linear Feet of Row	Pounds of FURADAN 15G per Acre for Various Row Spacings			
	20"	22"	24"	30"
4.5	7.5	6.5	6	5
9.0	15	13.5	12	10

SUGARCANE

Apply FURADAN 15G at the rate of 1.7 lbs. to 3.3 lbs. per 1,000 linear feet of row (equivalent to 13.3 to 26.7 lbs. per acre based on 60 inch row spacings). Do not use in Hawaii.

Plant cane: Nematodes (Root-Knot and Stunt), Wireworm—Apply the above rate as a 15 inch band directly over the planted cane and cover with soil.

Stubblecane: Wireworm (Florida only), Nematodes (Root-Knot and Stunt)—Apply the above rate as a 15 inch band over the stubble row. Use any time after harvest until regrowth reaches 18 inches. Cover with a 1 to 2 inch layer of soil. Covering with more than 2 inches of soil may reduce stand.

SUNFLOWERS (Confectionary and Oil)

Sunflower Stem Weevils, Grasshoppers—Use 8 to 16 ounces of FURADAN® 15G insecticide-nematicide per 1,000 linear feet of row (8.7 to 17.4 lbs. per acre with 30 inch row spacing). Apply at planting by directing the granules into the planter shoe with the seed or by placing the applicator tube directly behind the planter shoe so that the granules drop into the furrow and mix with the covering soil—OR—apply at planting in a 7 inch band and mix with the covering soil. Sunflower Beetle—Use 8 to 16 ounces of FURADAN 15G per 1,000 linear feet of row (8.7 to 17.4 lbs. per acre with 30 inch row spacing) and apply in furrow as noted above—OR—use 16 ounces of FURADAN 15G per 1,000 linear feet of row (17.4 lbs. per acre with 30 inch row spacing) and apply at planting in a 7 inch band as noted above. When a rate range is indicated, use the higher rate when heavier insect infestation is anticipated.

TOBACCO—ALL TYPES EXCEPT FLUE-CURED

For control of Flea Beetles, Hornworms and to aid in the control of Budworms—Use FURADAN® 15G insecticide-nematicide at 40 lbs. per acre. Broadcast granules over the soil surface prior to transplanting and incorporate into the top 3 inches with a suitable device.

For full season control of Flea Beetles only—Use FURADAN 15G at 20 to 26.7 lbs. per acre—OR—for full season control of Flea Beetles and approximately 60 days control of Hornworms—Use FURADAN 15G at 26.7 lbs. per acre. Apply broadcast as noted above.

TOBACCO—FLUE-CURED

For control of Stunt Nematodes, Flea Beetles, Hornworms, Wireworms and to aid in the control of Root-Knot Nematodes and Budworms—Use FURADAN® 15G at 40 lbs. per acre. Broadcast granules over the soil surface prior to transplanting and incorporate into the top 3 inches with a suitable device. Form beds and plant as usual. This product may induce flecking of the bottom or lower leaves.

For full season control of Flea Beetles and approximately 60 days control of Hornworms—Use FURADAN 15G at 26.7 lbs. per acre. Apply broadcast as noted above. Note leaf flecking caution listed above.

COTTONWOOD

Cottonwood Twig Borer, Cottonwood Leaf Beetle, Clearwing Borer—**In Commercial Plantations—**Use 6.7 to 13.3 lbs. per acre (12 foot row spacing) or 1.8 to 3.5 lbs. per 1,000 linear feet of row. Apply during June or July to the root zone of the cutting by the use of a granular subsoil applicator as a sidedress 10 to 12 inches from the trees in a continuous band on both sides of the trees. **In Nursery Plantings—**use 26.7 lbs. per acre (40 inch row spacing) or 2 lbs. per 1,000 linear feet of row. Apply during May or June to the root zone of the cutting, side-dressing 10 to 12 inches from the trees in a continuous band on both sides of the trees.

PINE SEEDLINGS IN PINE PLANTATIONS

Pales Weevil, Pitch-Eating Weevil, Nantucket Pine Tip Moth and Pitch Pine Tip Moth—Use FURADAN 15G at 7 grams (3/4 teaspoon) per seedling at transplanting. Distribute the granules in a 6 inch radius on the soil around each seedling. Cover granules with soil. Do not use at planting if seedlings were treated with FURADAN 4F slurry root dip.

PINE SEED ORCHARDS

Seedbugs, Coneworms, Coneborers—Apply FURADAN 15G at 2.7 to 5.3 oz. per inch of tree diameter. Cone Beetles—Use 5.3 ounces per inch of tree diameter. Broadcast within the drip area of the tree and incorporate with a suitable device. Make one (1) application in winter or early spring. Do not graze or feed orchard plants or plant parts.

SIBERIAN ELM

Elm Leaf Beetle control by commercial applicators, only and in the States of Colorado, Nebraska, Kansas, New Mexico, Arizona, Utah and Wyoming only—Apply FURADAN® 15G before insects appear in the spring, at the rate of 1/3 ounce per inch of trunk circumference. Measure the circumference (girth) of the tree 3 to 4 feet above the ground. Dig one hole for each inch of girth, 3 to 9 inches deep under the drip line of the tree. Thoroughly soak areas under the tree. Place granules required in each hole. Replace soil plug immediately and compress soil. Keep treated areas soaked with water for 14 days.

If Girth of tree is	Locate Holes away from trunk	Amount of FURADAN 15G
1 to 10 inches	1 ft.	1/3 to 3 1/3 oz.
14 to 20 inches	3 ft.	4.7 to 6.7 oz.
24 to 80 inches	6 ft. and 12 ft. (alternating)	8 to 13.3 oz.
44 to 80 inches	10 ft. and 12 ft. (alternating)	14.7 to 26.7 oz.
80 inches and larger	12 ft. and 20 ft. (alternating)	26.7 oz.

DEALER SHOULD SELL IN ORIGINAL PACKAGES ONLY.

TERMS OF SALE OR USE: On purchase of this product buyer and user agree to the following conditions:

WARRANTY: FMC makes no warranty, expressed or implied, concerning the use of this product other than indicated on the label. Except as so warranted, the product is sold as is. Buyer and user assume all risk of use and/or handling and/or storage of this material when such use and/or handling and/or storage is contrary to label instructions.

DIRECTIONS AND RECOMMENDATIONS: Follow directions carefully. Timing and method of application, weather and crop conditions, mixture with other chemicals not specifically recommended and other influencing factors in the use of this product are beyond the control of the seller and are assumed by buyer at his own risk.

USE OF PRODUCT: FMC's recommendations for the use of this product are based upon tests believed to be reliable. The use of this product being beyond the control of the manufacturer, no guarantee, expressed or implied, is made as to the effects of such or the results to be obtained if not used in accordance with directions or established safe practice.

DAMAGES: Buyer's or User's exclusive remedy for damages for breach of warranty or negligence shall be limited to direct damages not exceeding the purchase price paid and shall not include incidental or consequential damages.

FURADAN and FMC—FMC trademarks (3669-7/88)

This is a specimen label. FMC is not responsible for the accuracy of the information contained herein. As labels are subject to revision, always carefully read and follow the label on the product container.

Golden Leaf Tobacco Spray

INSECTICIDE

For Agricultural or Commercial Use Only.

ACTIVE INGREDIENTS:	BY WT.
*Endosulfan (Hexachlorohexahydromethano-2,4,3-benzodioxathiepin oxide)	33.7%
INERT INGREDIENTS:	66.3%
TOTAL	100.0%

*Thiodan

Thiodan 3 E.C. contains 3 pounds of Endosulfan per gallon.
Contains Xylene Range Aromatic Solvents.

EPA Reg. No. 279–2924 EPA Est., 279–
Code 3079

KEEP OUT OF REACH OF CHILDREN

DANGER–POISON

PELIGRO

See Other Panels for Additional Precautionary Information.

PRECAUCION AL USUARIO: Si usted no lee ingles, no use este producto hasta que la etiqueta le haya sido explicado ampliamente.

STATEMENT OF PRACTICAL TREATMENT

If swallowed: Call a physician or Poison Control Center immediately. If possible, vomiting should be induced under medical supervision. Drink one or two glasses of water and induce vomiting by giving one (1) ounce of syrup of ipecac, if available, or by touching the back of the throat with a finger. Do not induce vomiting or give anything by mouth to a person who is unconscious or convulsing.

If inhaled: Remove victim to fresh air. Apply artificial respiration if indicated.

If on skin: Remove contaminated clothing and wash skin with soap and water.

If in eyes: Flush eyes with plenty of water. Call a physician immediately.

Note to Physicians: Endosulfan is a central nervous system stimulant absorbable by mouth, inhalation or through contact with skin. It may cause convulsions. There is no specific antidote. Diazepam I.V. is the drug of choice. Barbituric acid derivatives such as Phenobarbital may be used additionally. A neuromuscular blocking agent may be used if convulsions persist. This type of drug may be used *only* if complete control of respiration can be maintained. Epinephrine derivatives are absolutely contraindicated.

For Emergency Assistance Call 716–735–3765.

PRECAUTIONARY STATEMENTS

HAZARDS TO HUMANS (& DOMESTIC ANIMALS)

Danger

Fatal if swallowed. May be fatal if inhaled or absorbed through skin. Do not breathe vapors or spray mist. Causes eye damage. Do not get in eyes, on skin, or on clothing. Applicators must wear a long sleeve shirt, trousers, unlined waterproof gloves made of natural rubber or neoprene, full foot covering (such as shoes or boots), head covering (such as a hat), and a pesticide mask or respirator jointly approved by the Mine Safety and Health Administration (MSHA) and the National Institute for Occupational Safety and Health (NIOSH) under the provisions of 30 CFR Part 11 for protection against endosulfan. Mixers and/or loaders must wear a chemical resistant apron and goggles or face shield in addition to all of the protective equipment specified for applicators. Do not apply or allow to drift to areas occupied by unprotected humans or beneficial animals. Wash thoroughly with soap and water after handling and before smoking or eating. Remove contaminated clothing and wash with soap and hot water before reuse.

ENVIRONMENTAL HAZARDS

This product is toxic to fish, birds, and other wildlife. Birds feeding on treated areas may be killed. Do not apply directly to water or wetlands. Due to the risk of runoff and drift, do not apply within a distance of 300 feet of lakes, ponds, streams, and estuaries. Shrimp and crab may be killed at application rates recommended on this label. Do not apply where fish, shrimp, crab, and other aquatic life are important resources. This product must not be used in areas where impact on threatened endangered species is likely. Contact your State Fish and Game Agency before applying this product. Apply this product only as specified on this label.

This pesticide is toxic to bees exposed to direct application. Applications should be timed to coincide with periods of minimum bee activity, usually between late evening and early morning.

PHYSICAL OR CHEMICAL HAZARDS

Do not use, pour, spill or store near heat or open flame.

DIRECTIONS FOR USE

It is a violation of Federal law to use this product in a manner inconsistent with its labeling.

Do not apply this product through any type of irrigation system.

Do not apply this product in such a manner as to directly or through drift expose workers or other persons. The area being treated must be vacated by unprotected persons.

Do not enter treated areas without protective clothing for 24 hours.

Protective clothing means, at least, a hat or other suitable head covering, a long sleeved shirt and long legged trousers or a coverall type garment (all of closely woven fabric covering the body, including the arms and legs), shoes and socks.

Because certain states may require more restrictive reentry intervals for various crops treated with this product, such as the 48 hour reentry interval established by California, consult your State Department of Agriculture for further information.

Written or oral warnings must be given to workers who are expected to be in a treated area or in an area about to be treated with this product. Inform workers of areas or fields that may not be entered without specific protective clothing, period of time field must be vacated and appropriate actions to take in case of accidental exposure. An example of such information is given under written warnings. When oral warnings are given, warnings shall be given in a language customarily understood by workers. Oral warnings must be given if there is reason to believe that written warnings cannot be understood by workers. Written warnings must include the following information:

DANGER

Area treated with endosulfan on (insert date). Do not enter without appropriate protective clothing for 24 hours (insert State Department of Agriculture's reentry interval, if more restrictive).

In case of accidental exposure to pesticide spray, wash the skin thoroughly with soap and water. Remove contaminated clothing and wash before reuse. If in eyes, flush with plenty of water. If inhaled, go to an area where the pesticide has not been applied. Get medical attention if needed.

STORAGE AND DISPOSAL

PESTICIDE STORAGE

Do not store in or around the home.

Do not store below 20°F, (−7°C).

Do not use or store near heat, open flame or hot surfaces.

Keep out of reach of children and animals. Store in original containers only. Store in a cool, dry place and avoid excess heat. Carefully open containers. After partial use, replace lids and close tightly. Do not put concentrate or dilute material into food or drink containers. Do not contaminate other pesticides, fertilizers, water, food, or feed by storage or disposal.

In case of spill, avoid contact, isolate area and keep out animals and unprotected persons. Confine spills. Call FMC collect: (716) 735-3765.

To confine spill: If liquid, dike surrounding area or absorb with sand, cat litter or commercial clay. If dry material, cover to prevent dispersal. Place damaged package in a holding container. Identify contents.

PESTICIDE DISPOSAL

Pesticide wastes are acutely hazardous. Improper disposal of excess pesticide, spray mixture, or rinsate is a violation of Federal law. If these wastes cannot be disposed of by use according to label instructions, contact your State Pesticide or Environmental Control Agency, or the Hazardous Waste representative at the nearest EPA Regional Office for guidance.

CONTAINER DISPOSAL

Triple rinse (or equivalent). Then offer for recycling or reconditioning, or puncture and dispose of in a sanitary landfill, or by other procedures approved by state and local authorities. Do not cut or weld metal containers.

GENERAL INSTRUCTIONS

Not For Use Or Storage In Or Around The Home.

Do not use in undiluted form.

Apply the listed amount per acre when insects first appear and repeat as required, unless otherwise noted, to maintain effective control. Use in sufficient water for thorough coverage of listed crops, unless otherwise noted. Coverage of upper and lower leaf surfaces is essential for good control. For ground application, apply recommended amount of pesticide in a minimum of 10 gallons of water per acre on vegetable or row crops. Observe use limitations. If insect control is required beyond the following use patterns, supplement the control program with other suitable pesticides.

When applying this material by aircraft, mix the recommended amount with sufficient water to provide a minimum of 1 gallon of finished spray per acre on vegetable and field crops, unless otherwise noted, and a minimum of 20 gallons of finished spray per acre on fruit and nut trees and on vines. Where more than 2 quarts of this material are recommended, mix with sufficient water to provide a minimum of finished spray equal to twice the amount of Thiodan® 3 E.C. insecticide used.

Do not plant root crops other than carrots, potatoes, sweet potatoes, and sugar beets as follow-up crops. Observe days interval between last application and harvest indicated by number in ( ) following the crop.

Tobacco (5): Budworm, Cabbage Looper, Flea Beetle, Green Peach Aphid, Hornworm—**Seed Bed:** Use ⅔ quart per 100 gallons of water and apply about 6 gallons of finished spray per 100 square yards. **Plant Bed:** Green June Bug Larvae—Use ⅔ pint per 100 gallons of water drenching at a rate of 1 gallon per square yard. **Field:** Uses ⅔ to 1⅓ quarts per acre to control above insects. Stink Bugs—Use ⅔ quart per 100 gallons of water. Do not exceed 3 applications per year.

DEALERS SHOULD SELL IN ORIGINAL PACKAGES ONLY.

TERMS OF SALE OR USE: On purchase of this product buyer and user agree to the following conditions:

WARRANTY: FMC makes no warranty, expressed or implied, concerning the use of this product other than indicated on the label. Except as so warranted, the product is sold as is. Buyer and user assume all risk of use and/or handling and/or storage of this material when such use and/or handling and/or storage is contrary to label instructions.

DIRECTIONS AND RECOMMENDATIONS: Follow directions carefully. Timing and method of application, weather and crop conditions, mixture with other chemicals not specifically recommended and other influencing factors in the use of this product are beyond the control of the seller and are assumed by the buyer at his own risk.

USE OF PRODUCT: FMC's recommendations for the use of this product are based upon tests believed to be reliable. The use of this product being beyond the control of the manufacturer, no guarantee, expressed or implied, is made as to the effects of such or the results to be obtained if not used in accordance with directions or established safe practice.

DAMAGES: Buyer's or user's exclusive remedy for damages for breach of warranty or negligence shall be limited to direct damages not exceeding the purchase price paid and shall not include incidental or consequential damages.

Thiodan® is a registered trademark of Hoechst AG.
FMC—FMC Trademark
(3079–11/2/87–A) (6/88)

This is a specimen label. FMC is not responsible for the accuracy of the information contained herein. As labels are subject to revision, always carefully read and follow the label on the product container.

Methyl Parathion 2 Thiodan® 3 EC

INSECTICIDE

RESTRICTED USE PESTICIDE

Due to very high acute toxicity to Humans and Birds

For retail sale to and use only by certified applicator or persons under their direct supervision and only for those uses covered by the certified applicator's certification. Direct supervision for this product is defined as the certified applicator being physically present during application, mixing, loading, repair and cleaning of application equipment. Commercial certified applicators must also ensure that all persons involved in these activities are informed of the precautionary statements.

ACTIVE INGREDIENTS:	
Methyl Parathion	19.4%
Related Compounds	1.0%
**Endosulfan	31.0%
INERT INGREDIENTS:	48.6%
TOTAL	100.0%

*O,O—dimethyl O-4-nitrophenyl phosphorothioate

**Hexachlorohexahydromethano-2,4,3-benzodioxathiepin oxide

***Contains xylene range aromatic solvents

Contains 2 pounds methyl parathion and 3 pounds endosulfan per gallon.

EPA Reg. No. 279-2149 EPA Est., 279-

Code 31297

STOP! READ THE LABEL

CAN KILL YOU

KEEP OUT OF REACH OF CHILDREN

DANGER–POISON

PELIGRO

PRECAUCION AL USUARIO: Si usted no lee ingles, no use este producto hasta que la etiqueta le haya sido explicado ampliamente.

DANGER—POISON—PRECAUTIONS

POISONOUS IF SWALLOWED

This product can kill you if swallowed even in small amounts: spray mist or dust may be fatal if swallowed.

POISONOUS BY SKIN CONTACT

Poisonous if touched by hands or spilled or splashed on skin, in eyes or on clothing (liquid goes through clothes.)

POISONOUS IF BREATHED

Breathing vapors, spray mist or dust may be fatal.

USE ONLY WHEN WEARING THE FOLLOWING PROTECTIVE CLOTHING AND EQUIPMENT DURING MIXING/LOADING, APPLICATION, REPAIR, AND CLEANING OF APPLICATION EQUIPMENT, DISPOSAL OF PESTICIDE AND EARLY REENTRY INTO TREATED AREAS: protective suit of one or two pieces covering all parts of the body except the head, hands, and feet; chemical resistant gloves; chemical resistant shoes (or chemical resistant shoe coverings or chemical resistant boots); goggles or face shield; hood or wide brimmed hat; NIOSH approved respirator.

POISONOUS TO FISH AND WILDLIFE

POISON SIGNS (Symptoms)

Methyl Parathion is a very dangerous poison. It rapidly enters the body on contact with all skin surfaces and eyes. Clothing wet with this material must be removed immediately. Exposed person must receive prompt medical treatment or they may die.

Some of the signs and symptoms of poisoning are: Headache, nausea, vomiting, cramps, weakness, blurred vision, pin-point pupils, tightness in chest, labored breathing, nervousness, sweating, watering of eyes, drooling or frothing of mouth and nose, muscle spasms and coma.

WORK SAFETY RULES

The National Pesticide Telecommunication Network is available for recommendations regarding poisoning management, emergency treatment, and other information regarding the toxicity of Methyl Parathion. The toll free number for the National Pesticide Telecommunication Network is 1-800-858-7378.

If handled indoors provide mechanical exhaust ventilation. Keep all unprotected persons, children, livestock, and pets away from treated area or where there is danger of drift. Do not rub eyes or mouth with hands. If you feel sick in any way, STOP work and get help right away. See First Aid (Practical Treatment) section.

HANDLE THE CONCENTRATE ONLY WHEN WEARING THE FOLLOWING PROTECTIVE CLOTHING AND EQUIPMENT:

Wear a protective suit of one or two pieces that covers all parts of the body except the head, hands, and feet. Wear chemical resistant gloves, chemical resistant apron, and chemical resistant shoes, shoe coverings, or boots. Wear goggles or a face shield. Wear a pesticide respirator approved by the National Institute for Occupational Safety and Health under the provisions of 30 CFR part II.

IF MIXING/LOADING IS PERFORMED USING A CLOSED SYSTEM, THE FOLLOWING PROTECTIVE CLOTHING AND EQUIPMENT MAY BE WORN AS AN ALTERNATIVE: long-sleeve shirt; long-legged pants; chemical resistant gloves; chemical resistant apron; shoes and socks. Goggles or face shield must be worn when the system is under pressure. All other protective clothing and equipment required for use with open systems must be available nearby.

IF APPLICATION IS PERFORMED USING AN ENCLOSED CAB OR COCKPIT, THE FOLLOWING PROTECTIVE CLOTHING AND EQUIPMENT MAY BE WORN AS AN ALTERNATIVE: long-sleeve shirt and long-legged pants; shoes and socks. Chemical resistant gloves must be available in the cab or cockpit and must be worn during entry to and exit from the application vehicle. All other protective clothing and equipment required for use during application must be available in the cab and must be worn when exiting the cab into treated area. When used for this purpose, contaminated clothing may not be brought back into the cab unless in an enclosure such as a plastic bag. REMEMBER—THIS CLOTHING IS INADEQUATE TO PROTECT YOU DURING REPAIR AND CLEANING OF APPLICATION EQUIPMENT AND EARLY REENTRY TO TREATED AREAS! REFER TO PROTECTIVE CLOTHING AND EQUIPMENT REQUIREMENTS ABOVE.

IMPORTANT! If pesticide comes in contact with skin, wash off with soap and water and contact a physician immediately. Always wash hands, face, and arms with soap and water before smoking, eating, drinking, or toileting.

AFTER WORK: before removing gloves, wash them with soap and water. Take off all work clothes and shoes. Shower using soap and water. Wear only clean clothes when leaving job—do not wear contaminated clothing. Personal clothing worn during work must be stored and laundered separately from protective clothing and household articles. Store protective clothing separately from personal clothing. Clean or launder protective clothing after each use. Respirators must be cleaned and filters replaced according to instructions included with the respirators. Protective clothing and protective equipment heavily contaminated or drenched with Methyl Parathion must be destroyed according to state and local regulations. HEAVILY CONTAMINATED OR DRENCHED CLOTHING CANNOT BE ADEQUATELY DECONTAMINATED. DURING AERIAL APPLICATION, HUMAN FLAGGERS ARE PROHIBITED.

FIRST AID TREATMENT

Call a doctor (physician), clinic or hospital immediately. Explain that the victim has been exposed to Methyl Parathion and describe his condition.

If breathing has stopped, start artificial respiration immediately and maintain until doctor sees victim.

If swallowed, drink 1 or 2 glasses of water and induce vomiting by touching back of throat with finger. Do not induce vomiting or give anything by mouth to an unconscious person. Get medical attention.

In case of contact, immediately flush eyes or skin with plenty of water for at least 15 minutes while removing contaminated clothing and shoes. See doctor immediately.

NOTE TO PHYSICIAN

*Since any symptoms from absorption of this combination product are likely to be those of Methyl Parathion intoxication, treatment should be directed towards the relief of cholinesterase inhibition.

Antidote—administer atropine sulfate in large doses, TWO to FOUR mg. intravenously or intramuscularly as soon as cyanosis is overcome. Repeat at 5 to 10 minute intervals until signs of atropinization appear. 2-PAM chloride is also antidotal and may be administered in conjunction with atropine. DO NOT GIVE MORPHINE OR TRANQUILIZERS. Parathion is a strong cholinesterase inhibitor affecting the central and peripheral nervous systems and producing cardiac and respiratory depression. At first sign of pulmonary edema, the patient should be given supplemental oxygen and treated symptomatically. Continued absorption of the poison may occur and fatal relapses have been reported after initial improvement; VERY CLOSE SUPERVISION OF THE PATIENT IS INDICATED FOR AT LEAST 48 HOURS.

FOR EMERGENCY ASSISTANCE CALL 716-735-3765

POST TREATED AREA

Consult your State Agricultural Extension Service or Experiment Station regarding posting treated areas.

ENVIRONMENTAL HAZARDS

This pesticide is highly toxic to aquatic invertebrates and wildlife. Birds in treated areas may be killed. Do not apply directly to water or wetlands (swamps, marshes, bogs, and potholes). Due to the risk of runoff and drift, do not apply within a distance of 300 feet of lakes, ponds, streams and estuaries. Runoff and drift from target areas may be hazardous to aquatic organisms in adjacent aquatic sites. Do not contaminate water by cleaning of equipment or disposal of wastes.

This product is highly toxic to bees exposed to direct treatment or residues on blooming crops or weeds. Do not apply this product or allow it to drift to blooming crops or weeds if bees are visiting the treatment area.

PHYSICAL OR CHEMICAL HAZARDS

Do not use or store near heat or open flame.

DIRECTIONS FOR USE

It is a violation of Federal law to use this product in a manner inconsistent with its labeling.

Do not apply this product through any type of irrigation system.

Do not apply this product in such a manner as to directly or through drift expose workers or other persons. The area being treated must be vacated by unprotected persons.

Reentry into treated fields is prohibited for 48 hours after application, unless all protective clothing and equipment required for early reentry as described elsewhere on the label are worn.

Because certain states may require more restrictive reentry intervals for various crops treated with this product, consult your State Department of Agriculture for further information.

Written or oral warnings must be given to workers who are expected to be in a treated area or in an area about to be treated with this product. Inform workers of areas or fields that may not be entered without specific protective clothing, period of time field must be vacated and appropriate actions to take in case of accidental exposure. An example of such information is given under written warnings. When oral warnings are given, warnings shall be given in a language customarily understood by workers. Oral warnings must be given if there is reason to believe that written warnings cannot be understood by workers. Written warnings must include the following information:

DANGER

Area treated with Methyl Parathion 2 THIODAN® on (insert date).
Do not enter without appropriate protective clothing for 48 hours (insert State Department of Agriculture's re-entry interval, if more restrictive).

In case of accidental exposure to pesticide spray or dust, wash the skin thoroughly with soap and water. Remove contaminated clothing and wash before reuse. If in eyes, flush with plenty of water. If inhaled, go to an area where the pesticide has not been applied. Get medical attention if needed.

STORAGE AND DISPOSAL

PESTICIDE STORAGE

Do not store below 20°F, (−7°C).

Do not use or store near heat, open flame or hot surfaces.

Keep out of reach of children and animals. Store in original containers only. Store in a cool, dry place and avoid excess heat. Carefully open containers. After partial use, replace lids and close tightly. Do not put concentrate or dilute material into food or drink containers. Do not contaminate other pesticides, fertilizers, water, food, or feed by storage or disposal.

In case of spill, avoid contact, isolate area and keep out animals and unprotected persons. Confine spills. Call FMC collect: (716) 735-3765.

To confine spill: If liquid, dike surrounding area or absorb with sand, cat litter or commercial clay. If dry material, cover to prevent dispersal. Place damaged package in a holding container. Identify contents.

PESTICIDE DISPOSAL

Pesticide wastes are acutely hazardous. Improper disposal of excess pesticide, spray mixture, or rinsate is a violation of Federal law. If these wastes cannot be disposed of by use according to label instructions, contact your State Pesticide or Environmental Control Agency, or the Hazardous Waste representative at the nearest EPA Regional Office for guidance.

CONTAINER DISPOSAL

Triple rinse (or equivalent). Then offer for recycling or reconditioning, or puncture and dispose of in a sanitary landfill, or by other procedures approved by state and local authorities. Do not cut or weld metal containers.

GENERAL INSTRUCTIONS

Not for use or storage in or around the home.

Do not use in undiluted form.

Apply when insects first appear and repeat as necessary. Apply the recommended rates in sufficient water for thorough coverage. (By ground, apply a minimum of 10 gallons; by air, apply a minimum of 1 gallon of finished spray per acre). Observe days interval between last application and harvest indicated by number in parenthesis ( ) following the crop.

Workers entering treated fields within 24 hours after application should wear protective clothing.

BEANS, SUCCULENT AND DRY (except Lima Beans) (21): Bean Leaf Skeletonizer, Black Bean Aphid, Cucumber Beetles, Green Stink Bug and Mexican Bean Beetle—Use 2/3 to 1 1/3 quarts per acre. Aphids, Armyworms, Flea Beetles and Leafhoppers—Use 1 1/3 quarts per acre. Do not exceed 3 applications per season. Do not feed treated threshings to livestock or allow livestock to graze in treated fields. Do not apply if the crop or weeds in the treatment area are in bloom.

BROCCOLI (7), BRUSSELS SPROUTS (14), CABBAGE (21), CAULIFLOWER (14): Aphids, Armyworm, Cabbage Looper, Cross-Striped Cabbageworm, Diamond-back Moth Larvae, Flea Beetles, Harlequin Bug, Imported Cabbageworm, Leafhoppers, Stink Bugs—Use 1 quart per acre.

CARROTS (15): Green Peach Aphid, Armyworm, Flea Beetle, Leafhoppers—Use 1 1/3 quarts per acre. Do not exceed 2 applications per season. Do not use treated tops for food or feed.

CELERY (whole plant) (15): Aphids (including green peach aphid), Cabbage Looper, Armyworms, Flea Beetles, Leafhoppers—Use 1 1/3 quarts per acre. Do not exceed 3 applications per season.

COTTON: Aphids, Armyworms, Boll Weevil, Bollworm, Cabbage Looper, Cotton Leafworm, Cotton Leaf Perforator, Flea Hoppers, Lygus Bugs, Stink Bugs, and Thrips—Use 2 quart per acre. Do not exceed 2 applications per year or a maximum of 4 quarts per acre per year. Do not apply after bolls open. Do not graze meat or dairy animals in treated fields. Workers entering fields within 24 hours after application should wear protective clothing.

LETTUCE (21): Cabbage Looper, Diamond-back Moth Larvae, Green Peach Aphid, Imported Cabbageworm—Use 1 1/3 quarts per acre. On Head Lettuce, do not exceed 3 applications after thinning. Remove wrapper leaves at harvest. On Leaf Lettuce, do not exceed 2 applications per season. All Uses: Do not feed livestock. Do not exceed 4 quarts per acre per year.

PEPPERS (15): Armyworm, Flea Beetles, Green Peach Aphid, Flea Beetles, Hornworms, Leafhopper, Pepper Maggot—Use 1 1/3 quarts per acre. Do not exceed 3 applications per year or a maximum of 4 quarts per acre per year.

POTATOES (5): Aphids, Colorado Potato Beetles, European Corn Borer, Plant Bugs (including Shield shaped Bugs), Potato Flea Beetle, Potato Leafhopper, Potato Tuberworm, Southern Armyworm Three-lined Potato Beetle, Potato Psyllid, Mites—Use 1 to 1 1/3 quarts per acre. All uses: Do not exceed 4 applications per year or a maximum of 4 quarts per acre per year.

SPINACH (21): Armyworms, Crown Mite, Flea Beetles, Green Peach Aphid, Leafhoppers—Use 1 quart per acre. Do not exceed one application per season.

TOMATOES (15): Blister Beetle, Colorado Potato Beetle, Flea Beetles, Green Peach Aphid, Tomato Hornworm, Whitefly—Use 3/4 to 1 1/3 quarts per acre. Cabbage Looper, Stink Bugs—Use 1 to 1 1/3 quarts per acre. Tomato Russet Mite, Tomato Fruitworm, Armyworms—Use 1 1/3 quarts per acre. All uses: Do not exceed 5 applications per year or a maximum of 4 quarts per acre per year.

DEALERS SHOULD SELL IN ORIGINAL PACKAGES ONLY.

TERMS OF SALE OR USE: On purchase of this product buyer and user agree to the following conditions:

WARRANTY: FMC makes no warranty, expressed or implied, concerning the use of this product other than indicated on the label. Except as so warranted, the product is sold as is. Buyer and user assume all risk of use and/or handling and/or storage of this material when such use and/or handling and/or storage is contrary to label instructions.

DIRECTIONS AND RECOMMENDATIONS. Follow directions carefully. Timing and method of application, weather and crop conditions, mixture with other chemicals not specifically recommended and other influencing factors in the use of this product are beyond the control of the seller and are assumed by the buyer at his own risk.

USE OF PRODUCT: FMC's recommendations for the use of this product are based upon tests believed to be reliable. The use of this product being beyond the control of the manufacturer, no guarantee, expressed or implied, is made as to the effects of such or the results to be obtained if not used in accordance with directions or established safe practice.

DAMAGES: Buyer's or user's exclusive remedy for damages for breach of warranty or negligence shall be limited to direct damages not exceeding the purchase price paid and shall not include incidental or consequential damages.

**Thiodan is a registered trademark of Hoechst AG.
FMC—FMC Trademark
(31297-7/12/89-A)

This is a specimen label. FMC is not responsible for the accuracy of the information contained herein. As labels are subject to revision, always carefully read and follow the label on the product container.

Option® 1 EC

HERBICIDE

For selective postemergence annual and perennial grass control in soybeans.

ACTIVE INGREDIENT:
fenoxaprop-ethyl:(±)-ethyl 2-[4-[(6-chloro 2-benzoxazolyl)oxy]phenoxy]propanoate 12.50%*
INERT INGREDIENTS: 87.50%
TOTAL 100.00%

*Equivalent to 1.00 pound of active ingredient per gallon

EPA Reg. No. 8340-23-279 EPA Est. 34704
Code 1152

KEEP OUT OF REACH OF CHILDREN

WARNING

STATEMENT OF PRACTICAL TREATMENT

In case of eye contact, immediately flush with plenty of water for 15 minutes. Get medical attention. If swallowed, do not induce vomiting. Option 1 EC Herbicide contains petroleum distillates. Call a physician. Vomiting should be supervised by a physician because of the possible pulmonary damage via aspiration of the solvent.

If possible bring this container and labeling to the attending physician.

For emergency assistance call (201) 231-2000.

PRECAUTIONARY STATEMENTS

HAZARDS TO HUMANS (AND DOMESTIC ANIMALS)

WARNING

May cause substantial but temporary eye injury. Do not get in eyes. Wear goggles or face shield and impermeable rubber gloves (such as neoprene or PVC) while mixing. Harmful if swallowed, absorbed through skin or inhaled. Do not take internally. Avoid inhalation of vapor or spray mist. Remove contaminated clothing and wash before reuse. Wash thoroughly with soap and water after handling.

ENVIRONMENTAL HAZARDS

This pesticide is toxic to fish. Do not apply directly to water. Do not apply when weather conditions favor runoff or drift. Do not contaminate arable land and/or water when disposing of equipment washwaters.

STORAGE AND DISPOSAL

Do not contaminate water, food, or feed by storage or disposal.

Do not store over 100°F or below 10°F. Do not use or store near heat or open flame.

PESTICIDE DISPOSAL: Pesticide wastes are toxic. Improper disposal of excess pesticide, spray mixture, or rinsate is a violation of Federal law. If these wastes cannot be disposed of by use according to label instructions, contact your State Pesticide or Environmental Control Agency, or the hazardous waste representative at the nearest EPA Regional Office for guidance.

CONTAINER DISPOSAL: Empty containers should be triple rinsed into the spray tank during the spray operation. Then offer for recycling or reconditioning, or puncture and dispose of in a sanitary landfill, or incineration, or, if allowed by state and authorities, by burning. If burned, stay out of smoke.

GEOGRAPHICAL USE

For use only in the following states: Pennsylvania, Ohio, Michigan, Indiana, Illinois, Iowa, Wisconsin, Nebraska, Kansas, New York, Colorado, Missouri (*excluding* Ripley, Butler, Stoddard, Scott, Mississippi, New Madrid, Pemiscot and Dunklin Counties), Minnesota (*excluding* Kittson, Roseau, Lake of the Woods, Clearwater, Marshall, Pennington, Red Lake, Polk, Mahnomen, Norman, Clay, Becker, Otter Tail, Wilkin, Traverse, Grant, Douglas, Pope, Stevens, Big Stone and Swift Counties) and South Dakota (*excluding* Brookings, Kingsbury, Beadle, Hand, Hyde, Hughes, and all other counties north of Highway 14 and east of the Missouri River).

GENERAL INFORMATION

OPTION 1 EC Herbicide is an emulsifiable concentrate for use in the selective postemergence control of annual and perennial grassy weeds in soybeans. Thorough spray coverage of emerged grasses is important. Visible effects begin as a general chlorosis (yellowing) followed by death of the weed. Visible injury of the grasses is evident approximately 4-10 days after application (dependent upon environmental conditions); but complete kill of the target grass will take 12-21 days.

Since many grass crops are sensitive to Option 1 EC Herbicide, including sorghum and corn, avoid all direct or indirect contact to neighboring fields.

Option 1 EC Herbicide does not control broadleaf weeds or sedges.

DIRECTIONS FOR USE

It is a violation of Federal law to use this product in a manner inconsistent with its labeling. Do not apply until you have read the entire label.

APPLICATION INFORMATION:

Ground Application: OPTION 1 EC Herbicide should be applied in a minimum of ten (10) gallons of water per broadcast acre. Flat fan or hollow cone nozzles are recommended. Use a minimum pressure of forty (40) pounds per square inch. Under dense weed/crop canopies, high spray pressure is very important for obtaining thorough coverage; therefore, use higher spray pressure under these conditions.

Air Application: OPTION 1 EC Herbicide should be applied in a minimum of five (5) gallons of water per broadcast acre. Uniform spray coverage is essential when using aircraft, and is achieved by the use of spray droplet size (150-300 micron range). DO NOT USE raindrop nozzles. Aerial applications with this product should be made at a height which provides the most effective swath width for the aircraft.

DO NOT APPLY by aircraft when wind speeds exceed eight (8) mph. Avoid all direct or indirect contact to neighboring fields.

TIMING OF APPLICATION:

OPTION® 1 EC Herbicide will control grasses at most growth stages; but for best production oriented grass control, applications should be made to young, vigorously growing grassy weeds. Follow the recommendations for time of application from the rate/grass recommendation chart below. Earlier applications of OPTION 1 EC Herbicide (before all grasses have emerged) could result in late flushes of grass reinfesting the treated area.

Rate and Grass Recommendation Chart

Grass Species	Optimum Stage of Growth*	Crop Oil Concentrate	Application Rate**	
			Pts./A	Fl. Oz./A
Annual Grasses				
Giant Foxtail (*Setaria faberii*)	3-6″	always add 1 qt. with ground equipment or 1 pt. with aerial application	0.8	13
Green Foxtail (*Setaria viridis*)	3-6″		″	″
Volunteer Corn (*Zea mays*)	10-24″		″	″
Wild Proso Millet (*Panicum miliaceum*)	5-10″	″	″	″
Johnsongrass, Seedling (*Sorghum halepense*)	4-10″	″	″	″
Wild Cane/Shattercane (*Sorghum bicolor*)	6-12″	″	″	″
Barnyardgrass (*Echinochloa crusgalli*)	3-6″	OPTIONAL	1.2	19
Broadleaf Signalgrass (*Brachiaria platyphylla*)	″	″	″	″
Fall Panicum (*Panicum dichotomiflorum*)	″	″	″	″
Bristle Foxtail (*Setaria verticillata*)	″	″	″	″

Grass Species	Optimum Stage of Growth*	Crop Oil Concentrate	Application Rate** Pts./A	Fl. Oz./A
Purple Foxtail *(Setaria viridis robusta purpurea)*	"	"	"	"
Robust Foxtail *(Setaria viridis robusta alba)*	"	"	"	"
Goosegrass *(Eleusine indica)*	"	"	"	"
Jungle Rice *(Echinochloa colonum)*	"	"	"	"
Southwestern Cupgrass *(Eriochloa gracilis)*	"	"	"	"
Sprangletop *(Leptochloa filiformis)*	"	"	"	"
Texas panicum *(Panicum texanum)*	"	"	"	"
Wild oats *(Avena fatua)*	"	"	"	"
Witchgrass *(Panicum capillare)*	"	"	"	"
Wooly Cupgrass *(Eriochlea villosa)*	"	"	"	"
Wirestem muhly *(Muhlenbergia frondosa)*	3-6"	Always add COC	1.2	19
Yellow Foxtail *(Setaria lutescens)*	"	"	"	"
Large Crabgrass *(Digitaria sanguinalis)*	1-2"	Always add COC	1.2	19
Smooth Crabgrass *(Digitaria ischaemum)*	"	"	"	"
Itchgrass *(Rottboellia exaltata)*	3-6"	OPTIONAL	1.2	19
Perennial Grasses				
Johnsongrass from rhizomes *(Sorghum halepense)*	10-20"	DO NOT ADD OIL	1.2	19
Second application if needed (regrowth) (A timely cultivation may override necessity of a second application.)	10-20"		0.8	13

* See control of larger annual grasses.

** When controlling mixed populations of grassy weeds, always use the rate that will control the least susceptible species.

CONTROL OF LARGER ANNUAL GRASSES

Always use crop oil concentrate when controlling annual grasses that have grown beyond the recommended stages of growth. When grasses are large, the control may take somewhat longer. Yield reductions due to weed competition may result from these late applications.

ADDITIVES

The addition of non-phytotoxic oil concentrate to the spray solution at one (1) quart per acre for ground applications and one (1) pint per acre for aerial applications may improve the herbicidal activity of OPTION 1 EC Herbicide. A non-phytotoxic oil should be added when the grassy weeds are under drought stress or when a less than optimum application timing is necessary (see Timing of Application Section). Add a non-phytotoxic oil concentrate or a once-refined vegetable oil or soybean oil concentrate containing 15-20% approved emulsifiers cleared for application to growing crops. Crop oil concentrates vary in their viscosity; therefore, it is important to maintain constant agitation while the spray mixture is in the spray tank.

The increased speed of foliage burn resulting from the addition of crop oil concentrate may reduce the translocation of OPTION 1 EC Herbicide to the johnsongrass roots and rhizomes. Therefore, the addition of non-phytotoxic oil is *not* recommended for the control of rhizome johnsongrass.

TANK MIX RECOMMENDATIONS

OPTION 1 EC Herbicide may be tank mixed with Basagran® Herbicide, Blazer® or Tackle® 2AS in a postemergence program for broad spectrum weed control in soybeans. Tank mix applications are to be used only when both the annual grass and broadleaf weeds are in the proper stage of growth as specified on each respective herbicide label. When tank mixing, always follow the use directions in accordance with the respective label. No label dosage rates should be exceeded.

Water Volume and Spray Pressure

Ground Equipment: For the tank mix, use a minimum of 20 gallons per acre of total spray solution and a minimum pressure of 40 PSI. Use standard high pressure hollow cone or flat-fan nozzles. Do not use flood nozzles.

Aerial Equipment: For tank mixes, use a minimum of 5 gallons per acre of total spray solution and a minimum pressure of 40 PSI.

Mixing: Fill the spray tank half full with water while the agitator is running. Add the recommended amount of Option 1 EC Herbicide followed by the tank mix component. Then add the remaining amount of water.

BASAGRAN

OPTION 1 EC Herbicide should be applied at a rate of 1.2 to 1.6 pints per acre and Basagran at a rate of 1.5 to 2.0 pints per acre. The choice of rates of each product and activities is dependent on the weed size and weed spectrum present. Refer to the Basagran label to identify the proper rate and additives for control of the species and size of the broadleaf weeds present.

The Option 1 EC Herbicide rates for tank mixing with Basagran are 1.2 pints per acre when the annual grassy weeds are 1 to 4 inches tall, and 1.6 pints per acre when the annual grassy weeds are 4 to 6 inches tall or less than 2 tillers. DO NOT use this tank mix if the annual grassy weeds have developed more than 2 tillers or are larger than 6 inches tall. For the control of shattercane 6 to 12 inches tall and broadleaf weeds that are on the Basagran label, tank mix OPTION 1 EC Herbicide at a rate of 1.2 pints per acre with Basagran at 1.5 to 2.0 pints per acre. DO NOT use this tank mix to control rhizome johnsongrass. Sequential applications of OPTION 1 EC Herbicide and Basagran may be necessary if the stages of the grass and broadleaf weeds are not within tank mix label recommendations at the same time.

When tank mixing Option 1 EC Herbicide and Basagran, always add crop oil concentrate and other labeled additives at the appropriate labeled rate.

Special Note: When tank mixing Basagran and Option 1 EC Herbicide, a minimum rate of Option at 1.2 pints per acre is recommended.

BLAZER OR TACKLE 2AS

Option 1 EC Herbicide should be tank mixed at a rate of 1.6 pints per acre. Blazer or Tackle 2AS should be tank mixed at a rate of 1.5 to 2.0 pints per acre. In no instances should crop oil concentrate or a surfactant be used with this tank mix. This tank mix should not be used for the control of rhizome Johnsongrass.

The stage of growth of both the annual grassy weeds and the broadleaf weeds should conform to the directions on each product label. The tank mix of Option 1 EC Herbicide plus Blazer or Tackle 2AS should not be applied after the annual grasses have begun tillering. Whenever the grass and broadleaf weeds are not in the proper stage of growth according to this tank mix label, a sequential application should be utilized. When Option 1 EC Herbicide is applied first, a waiting period of 3 days is necessary before applying Blazer or Tackle 2AS. When Blazer or Tackle 2AS is applied first, a waiting period of 7 days is necessary before applying Option 1 EC Herbicide.

Special Note; The mixture of Option 1 EC Herbicide plus Blazer or Tackle 2AS may only suppress velvetleaf.

SPECIAL NOTES

1. Annual ryegrass (*Lolium* sp.), quackgrass (*Agropyron repens*) and Bermudagrass (*Cynodon dactylon*) are not controlled by Option 1 EC Herbicide.
2. Rainfall within one hour of an application may cause a reduction in grass control.
3. If a new flush of grass occurs, either a timely cultivation or a second application of Option 1 EC Herbicide may be necessary.
4. DO NOT cultivate within four days before or after an Option 1 EC Herbicide application.
5. Always clean sprayer thoroughly before and after any pesticide application.
6. Apply Option 1 EC Herbicide as a spot treatment in a 1% volume/volume solution with water (e.g., 32 fluid ounces per 25 gallons of water). (See instructions for use in the Ground Application Section.)
7. Option 1 EC Herbicide is not phytotoxic to soybeans at any growth stage. For best results, it should be applied according to the development of the annual grassy weeds as noted, but before the bloom stage of the soybean crop.
8. DO NOT graze or feed treated forage, hay or straw.
9. Application of Option 1 EC Herbicide to grasses under stress (e.g., drought), may result in reduced control.
10. DO NOT plant any rotational crop in an Option 1 EC Herbicide treated field for 30 days after application (120 days for small grains).
11. DO NOT apply Option 1 EC Herbicide less than 90 days before harvesting soybeans.
12. DO NOT make more than two applications of Option 1 EC Herbicide per growing season and do not apply more than two pints (0.25 pounds of active ingredient) per acre per growing season.
13. DO NOT apply this product through any irrigation system.
14. Read and follow restrictions and limitations on the Basagran Herbicide and/or Blazer/Tackle labels as applicable. The most restrictive labeling applies in tank mixes.

ACREAGE CONSERVATION RESERVE (SET-ASIDE)
Option 1 EC Herbicide may be used to control annual grassy weeds in acreage conservation reserve (set-aside) acres. This acreage is often seeded to the following cover crops: clover, alfalfa, tall fescue, bromegrass, and ryegrass. Special note: Timothy and orchardgrass are sensitive to Option 1 EC Herbicide. The cover crops listed above have excellent tolerance to Option 1 EC Herbicide at 1.2 to 1.6 pints per acre. Select the proper rate from the Rate and Grass Recommendation Chart found in this label.

SPECIAL NOTES:

1. DO NOT harvest or graze cover crops treated with Option 1 EC Herbicide.
2. DO NOT apply to cover crops such as oats, sorghum, sudangrass, and timothy as injury may occur.

DEALERS SHOULD SELL IN ORIGINAL PACKAGES ONLY.

TERMS OF SALE OR USE: On purchase of this product buyer and user agree to the following conditions:

WARRANTY: Neither FMC nor Hoechst-Roussel makes any warranty, expressed or implied, concerning the use of this product other than indicated on the label. Except as so warranted, the product is sold as is. Buyer and user assume all risk of use and/or handling and/or storage is contrary to label instructions.

DIRECTIONS AND RECOMMENDATIONS: Follow directions carefully. Timing and method of application, weather and crop conditions, mixture with other chemicals not specifically recommended and other influencing factors in the use of this product are beyond the control of the seller and are assumed by the buyer at his own risk.

USE OF PRODUCT: The recommendations for the use of this product are based upon tests believed to be reliable. The use of this product being beyond the control of the manufacturer and seller, no guarantee, expressed or implied, is made as to the effects of such or the results to be obtained if not used in accordance with directions or established safe practice.

DAMAGES: Buyer's or user's exclusive remedy for damages for breach of warranty or negligence shall be limited to direct damages not exceeding the purchase price paid and shall not include incidents or consequential damages.

The name and logo Hoechst are registered trademarks of Hoechst AG.
The name and logo Roussel are registered trademarks of Roussel Uclaf S.A.
Option® is a registered trademark of Hoechst-Roussel Agri-Vet Company.
Basagran® is a registered trademark of BASF AG.
Blazer® is a registered trademark of BASF Corporation.
Tackle® is a registered trademark of Rhone-Poulenc Incorporated.
FMC® is a registered trademark of FMC corporation.
1/89

This is a specimen label. FMC is not responsible for the accuracy of the information contained herein. As labels are subject to revision, always carefully read and follow the label on the product container.

Pounce® 1.5 G
INSECTICIDE

RESTRICTED USE PESTICIDE
For retail sale to and use only by certified applicators or persons under their direct supervision, and only for those uses covered by the certified applicator's certification.

ACTIVE INGREDIENT:	BY WT.
*Permethrin**	1.50%
INERT INGREDIENTS:	98.50%
TOTAL	100.00%

*(3-Phenoxyphenyl) methyl (±) *cis-trans* 3-(2,2-dichloroethenyl) -2,2-dimethylcyclopropanecarboxylate
***cis/trans* ratio: Max. 55% (±) *cis* and min. 45% (±) *trans*

EPA Reg. No. 279-3059 EPA Est. 279-
Code 692
U.S. Patent No. 4,024,163

KEEP OUT OF REACH OF CHILDREN
CAUTION
STATEMENT OF PRACTICAL TREATMENT

If swallowed, drink 1 or 2 glasses of water and induce vomiting by touching back of throat with finger. Do not induce vomiting or give anything by mouth to an unconscious person. Get medical attention.

If in eyes, flush with plenty of water. Get medical attention if irritation persists.

If on skin, wash with soap and water.

For Emergency Assistance Call (716) 735-3765

PRECAUTIONARY STATEMENTS

HAZARDS TO HUMAN (AND DOMESTIC ANIMALS)

CAUTION

Harmful if swallowed or absorbed through skin. Causes moderate eye irritation. Avoid contact with eyes, skin or clothing. Wash thoroughly after handling. Remove and wash clothing before reuse.

ENVIRONMENTAL HAZARDS

This product is extremely toxic to fish. Use with care when applying in areas adjacent to any body of water. Do not apply directly to water. Keep out of lakes, ponds or streams. Do not contaminate water when disposing of equipment washwaters.

DIRECTIONS FOR USE

It is a violation of Federal law to use this product in a manner inconsistent with its labeling.

STORAGE AND DISPOSAL

PESTICIDE STORAGE

Do not store in or around the home.

Keep out of reach of children and animals. Store in original containers only. Store in a cool, dry place and avoid excess heat. Carefully open containers. After partial use, fold and roll back bags, clamp and close tightly. Do not put concentrate or dilute material into food or drink containers. Do not contaminate other pesticides, fertilizers, water, food or feed by storage or disposal.

In case of spill, avoid contact, isolate area and keep out animals and unprotected persons. Confine spills. Call FMC collect: (716) 735-3765.

To confine spill: If liquid, dike surrounding area or absorb with sand, cat litter or commercial clay. If dry material, cover to prevent dispersal. Place damaged package in a holding container. Identify contents.

PESTICIDE DISPOSAL

Wastes resulting from the use of this product may be disposed of on site or at an approved waste disposal facility.

CONTAINER DISPOSAL

Completely empty bag into application equipment. Then dispose of empty bag in a sanitary landfill or by incineration, or, if allowed by State and local authorities, by burning. If burned, stay out of smoke.

GENERAL INSTRUCTIONS

ALMONDS: To control ants—Use Pounce 1.5 G at a rate of 26 pounds (0.4 pound active) per acre. Apply as a broadcast treatment by ground equipment or aircraft when ants appear. Overhead moisture following application will enhance activity.

Do not apply more than 0.8 pound active per acre during hull split. Do not apply more than 2 pounds active ingredient per acre per season. Do not apply within seven days of harvest. Do not graze livestock in treated orchards.

FIELD CORN, POPCORN, FIELD CORN GROWN FOR SEED

Preemergent Use: To control Armyworms and Cutworms—Use Pounce 1.5 G at a rate of 6.7 to 13.3 pounds (0.1 to 0.2 pound active) per acre in the time period from five days prior to planting up to emergence of the crop. Apply either as a broadcast using ground or aerial equipment or as a 7-inch band. For the band application use Pounce 1.5 G at a rate of 8 to 16 ounces per 1,000 linear feet of row.

Foliar Use: To control Armyworms, Cutworms, European Corn Borer, Southwestern Corn Borer, Western Bean Cutworm, and Stalk Borers—Use Pounce 1.5 G at a rate of 6.7 to 13.3 pounds (0.1 to 0.2 pound active) per acre. When treating for Armyworms, Cutworms, European Corn Borer, Southwestern Corn Borer and Western Bean Cutworm apply when eggs begin to hatch. For treatment of Stalk Borers, Pounce 1.5 G must be applied when or shortly before the Stalk Borer larvae are moving into the corn from the surrounding weeds and grasses. Mowing or burn-down herbicide is suggested to initiate movement of Stalk Borer larvae. Apply by aircraft broadcasting over corn plants or with ground equipment by directing the granule into the whorl.

Do not apply more than 0.4 pound active per acre after brown silk stage. Do not exceed a total of 1.0 pound active per acre per season.

PISTACHIOS: To control ants—Use Pounce 1.5 G at a rate of 26 pounds (0.4 pound active) per acre. Apply as a broadcast treatment by ground equipment or aircraft when ants appear. Overhead moisture following application will enhance activity.

Do not apply more than 0.8 pound active per acre per crop season. Do not apply 10% hull split. Nuts may be harvested on the day of application. Do not graze livestock in treated orchards.

SWEET CORN

Preemergent Use: To control Armyworms and Cutworms—Use Pounce 1.5 G at a rate of 6.7 to 13.3 pounds (0.1 to 0.2 pound active) per acre in the time period from five days prior to planting up to emergence of the crop. Apply either as a broadcast using ground or aerial equipment or as a 7-inch band. For the band application use Pounce 1.5 G at a rate of 8 to 16 ounces per 1,000 linear feet of row for 40-inch row spacing.

Foliar Use: To control Armyworms, Cutworms, European Corn Borer, Western Bean Cutworm, Southwestern Corn Borer and Stalk Borer—Use Pounce 1.5 G at a rate of 6.7 to 13.3 pounds (0.1 to 0.2 pound active) per acre. When treating for Armyworms, Western Bean Cutworm, Southwestern Corn Borer, and European Corn Borer apply when eggs begin to hatch. For treatment of Stalk Borer, Pounce 1.5 G must be applied when or shortly before Stalk Borer larvae are moving into sweet corn from the surrounding weeds and grasses. Mowing or burn-down herbicide is suggested to initiate movement of Stalk Borer larvae. Apply by aircraft by broadcasting over sweet corn plants or with ground equipment by directing the granule into the whorl.

Apply every five days or as needed. Do not apply more than 1.2 pounds active ingredient per acre per crop season. Do not apply within 1 day of harvest of ears or forage or livestock grazing.

CORN INSECTICIDE APPLICATOR SETTINGS

Pounce 1.5 G is a dense, free flowing insecticide requiring accurate equipment calibration to apply the recommended amount per acre. Calibrate your planter to the speed at which you plan to travel during planting operation. The following are approximate settings for guidance in calibration:

Application Rate Pounds/Acre	6.7			10.0			13.3		
Ounces/1000 Linear Feet	8			12			16		
Planter Speed (mph)	4	6	8	4	6	8	4	6	8
Equipment:				Gauge Setting					
Gandy Model 901	14	17	20	17	21	25	20	25	28
John Deere MaxEmerge									
(Even Notch)	8	12	16	12	17	22	15	22	27
(Odd Notch)	7	10	12	10	13	16	12	16	19
1-H Case	1.3	1.5	1.7	1.5	1.7	2.1	1.7	2.2	2.8

CORN APPLICATION RATES

When the granular application equipment is set to deliver Pounce 1.5 G at a rate of 12 ounces per 1,000 linear feet (0.15 pound active per acre) the total amount applied per acre will vary depending on row spacings.

Use table below to determine the Pounce 1.5 G needs for each acre:

Row Spacings (inches)	40	38	36	32	30	28
Pounce 1.5 G (pounds per acre)	9.8	10.3	10.9	12.2	13.0	14.0

DEALERS SHOULD SELL IN ORIGINAL PACKAGES ONLY.

TERMS OF SALE OR USE: On purchase of this product buyer and user agree to the following conditions:

WARRANTY: FMC makes no warranty, expressed or implied, concerning the use of this product other than indicated on the label. Except as so warranted, the product is sold as is. Buyer and user assume all risk of use and/or handling and/or storage of this material when such use and/or handling and/or storage is contrary to label instructions.

DIRECTIONS AND RECOMMENDATIONS: Follow directions carefully. Timing and method of application, weather and crop conditions, mixture with other chemicals not specifically recommended and other influencing factors in the use of this product are beyond the control of the seller and are assumed by the buyer at his own risk.

USE OF PRODUCT: FMC's recommendations for the use of this product are based upon tests believed to be reliable. The use of this product being beyond the control of the manufacturer, no guarantee, expressed or implied, is made as to the effects of such or the results to be obtained if not used in accordance with directions or established safe practice.

DAMAGES: Buyer's or user's exclusive remedy for damages for breach of warranty or negligence shall be limited to direct damages not exceeding the purchase price paid and shall not include incidental or consequential damages.

Pounce® and FMC—FMC trademarks (692-9/28/88-A)

This is a specimen label. FMC is not responsible for the accuracy of the information contained herein. As labels are subject to revision, always carefully read and follow the label on the product container.

Pounce® 3.2 EC

INSECTICIDE

RESTRICTED USE PESTICIDE

For retail sale to and use only by certified applicators or persons under their direct supervision, and only for those uses covered by the certified applicator's certification.

ACTIVE INGREDIENTS:
*Permethrin** . . . 38.4%
INERT INGREDIENTS:*** . . . 61.6%
TOTAL . . . 100.0%

*(3-Phenoxyphenyl) methyl (±) *cis-trans* 3 -(2,2-dichloroethenyl)-2,2-dimethylcyclopropanecarboxylate
***cis/trans* ratio: Max. 55% (±) *cis* and min. 45% (±) *trans*
***Contains xylene range aromatic solvents.

Contains 3.2 pounds permethrin per gallon.

EPA Reg. No. 279-3014 EPA Est., 279-Code 3510

U.S. Patent No. 4,024,163

KEEP OUT OF REACH OF CHILDREN

CAUTION

STATEMENT OF PRACTICAL TREATMENT

If in eyes: Flush eyes with water for 15 minutes. Call a physician.

If swallowed: Do not induce vomiting. Call a physician.

If on skin: Remove contaminated clothing. Wash with plenty of soap and water. Get medical attention if irritation persists.

Note to Physician: Vomiting should be supervised by a physician or the professional staff because of the possible pulmonary damages by aspiration of the solvent.

For Emergency Assistance Call 716-735-3765.

PRECAUTIONARY STATEMENTS

HAZARDS TO HUMANS (& DOMESTIC ANIMALS)

CAUTION

Harmful if swallowed or absorbed through skin. Causes moderate eye irritation. Avoid contact with eyes, skin or clothing. Wash thoroughly after handling. Remove and wash clothing before reuse.

See other panels for additional precautionary information.

ENVIRONMENTAL HAZARDS

This product is highly toxic to bees exposed to direct treatment or residues on crops or weeds. Do not apply this product or allow it to drift to crops or weeds on which bees are actively foraging. Additional information may be obtained from your Cooperative Extension Service.

This product is extremely toxic to fish. Use with care when applying in areas adjacent to any body of water. Do not apply directly to water. Do not apply when weather conditions favor drift from treated areas. Do not contaminate water when disposing of equipment washwaters.

PHYSICAL/CHEMICAL HAZARDS

Do not use or store near heat or open flame.

DIRECTIONS FOR USE

It is a violation of federal law to use this product in a manner inconsistent with its labeling.

Do not apply this product in such a manner as to directly or through drift expose workers or other persons. The area being treated must be vacated by unprotected persons.

Do not enter treated areas without protective clothing until sprays have dried.

Protective clothing means, at least, a hat or other suitable head covering, a long sleeved shirt and long legged trousers or a coverall type garment (all of closely woven fabric covering the body, including the arms and legs), shoes and socks.

Because certain states may require more restrictive reentry intervals for various crops treated with this product, consult your State Department of Agriculture for further information.

Written or oral warnings must be given to workers who are expected to be in a treated area or in an area about to be treated with this product. Inform workers of areas or fields that may not be entered without specific protective clothing, period of time field must be vacated and appropriate actions to take in case of accidental exposure. An example of such information is given under written warnings. When oral warnings are given, warnings should be given in a language customarily understood by workers. Oral warnings must be given if there is reason to believe that written warnings cannot be understood by workers. Written warnings must include the following information:

CAUTION

Area treated with Pounce® 3.2 EC on (insert date). Do not enter without appropriate protective clothing until sprays have dried (insert State Department of Agriculture's re-entry interval, if more restrictive).

In case of accidental exposure to pesticide spray or dust, wash the skin thoroughly with soap and water. Remove contaminated clothing and wash before reuse. If in eyes, flush with plenty of water. If inhaled, go to an area where the pesticide has not been applied. Get medical attention if needed.

CHEMIGATION USE DIRECTIONS

Apply this product only through sprinkler including center pivot, lateral move, end tow, side (wheel) roll, traveler, big gun, solid set, or hand move irrigation systems. Do not apply this product through any other type of irrigation system. Do not connect an irrigation system (including greenhouse systems) used for pesticide application to a public water system.

Crop injury, lack of effectiveness, or illegal residues in the crop can result from nonuniform distribution of treated water. If you have questions about calibration, you should contact State Extension Service specialists, equipment manufacturers, or other experts. A person knowledgeable of the chemigation system and responsible for its operation, or under the supervision of the responsible person, shall shut the system down and make necessary adjustments should the need arise.

The system must contain a functional check valve, vacuum relief valve, and low pressure drain appropriately located on the irrigation pipeline to prevent water source contamination from backflow.

The pesticide injection pipeline must contain a functional, automatic, quick-closing check valve to prevent the flow of fluid back toward the injection pump.

The pesticide injection pipeline must also contain a functional, normally closed, solenoid-operated valve located on the intake side of the injection pump and connected to the system interlock to prevent fluid from being withdrawn from the supply tank when the irrigation system is either automatically or manually shut down.

The system must contain functional interlocking controls to automatically shut off the pesticide injection pump when the water pump motor stops.

The irrigation line or water pump must include a functional pressure switch which will stop the water pump motor when the water pressure decreases to the point where pesticide distribution is adversely affected.

Systems must use a metering pump, such as a positive displacement injection pump (e.g., diaphragm pump) effectively designed and constructed of materials that are compatible with pesticides and capable of being fitted with a system interlock.

Do not apply when wind speed favors drift beyond the area intended for treatment.

Pounce® 3.2 EC insecticide should be applied continuously for the duration of the water application. POUNCE 3.2 EC should be diluted in sufficient volume to insure accurate application over the area to be treated. When using chemigation, a minimum of 0.5 inch per acre of irrigation water is recommended. Agitation is not required when a suitable diluent is used.

STORAGE AND DISPOSAL

PESTICIDE STORAGE

Do not store below 10°F, (−12°C).

Do not use or store near heat, open flame or hot surfaces.

Keep out of reach of children and animals. Store in original containers only. Store in a cool, dry place and avoid excess heat. Carefully open containers. After partial use, replace lids and close tightly. Do not put concentrate or dilute material into food or drink containers. Do not contaminate other pesticides, fertilizers, water, food, or feed by storage or disposal.

In case of spill, avoid contact, isolate area and keep out animals and unprotected persons. Confine spills. Call FMC collect: (716) 735-3765.

To confine spill: If liquid, dike surrounding area or absorb with sand, cat litter or commercial clay. If dry material, cover to prevent dispersal. Place damaged package in a holding container. Identify contents.

PESTICIDE DISPOSAL

Pesticide wastes are toxic. Improper disposal of excess pesticide, spray mixture, or rinsate is a violation of Federal law. If these wastes cannot be disposed of by use according to label instructions, contact your State Pesticide or Environmental Control Agency, or the Hazardous Waste representative at the nearest EPA Regional Office for guidance.

CONTAINER DISPOSAL

Metal Containers: Triple rinse (or equivalent) then offer for recycling or reconditioning, or puncture and dispose of in a sanitary landfill, or by other procedures approved by State and local authorities. Do not cut or weld metal containers.

Plastic Containers: Triple rinse (or equivalent). Then offer for recycling or reconditioning, or dispose of in a sanitary landfill, or incineration, or, if allowed by State and local authorities, by burning. If burned, stay out of smoke.

ENDANGERED SPECIES RESTRICTIONS

Aerial Application: Do not use within 100 yards of aquatic habitats.

Ground Application: Do not use within 20 yards of aquatic habitats.

GENERAL INSTRUCTIONS

POUNCE® 3.2 EC is a 3.2 pounds per gallon formulation of the insecticide permethrin. Apply POUNCE 3.2 EC when insects appear or feeding is noticed. The higher rate should be used as pest populations increase. Repeat the application as necessary to maintain control. POUNCE 3.2 EC may be applied by both ground and aerial equipment. Use sufficient water to obtain full coverage. With the exception of crops listed below, rotational crops should not be planted within 60 days of last application.

This label must be in the possession of the user at the time of application.

ALFALFA: To control Alfalfa Caterpillar, Armyworms, Cutworms, Loopers, Webworms, Blue Alfalfa Aphid, Green Peach Aphid, Pea Aphid, Green Cloverworm, Spotted Alfalfa Aphid, and Velvetbean Caterpillar—Use POUNCE 3.2 EC at a rate of 2 to 8 ounces (0.05 to 0.2 pound active) per acre. To control Alfalfa Weevil, Egyptian Alfalfa Weevil, Cucumber Beetle, Meadow Spittlebug, Plant Bugs (including Lygus spp.), Potato Leafhopper, and Stink Bugs—Use POUNCE 3.2 EC at a rate of 4 to 8 ounces (0.1 to 0.2 pound active) per acre.

Use higher recommended dosage for increased pest pressure or for increased residual pest control. Apply in a minimum of 1 gallon of finished spray per acre by aircraft or 10 gallons of finished spray per acre with ground equipment. Do not apply more than 0.2 pound active per cutting. When rates of 0.1 pound active per acre or less are used, application may be made on day of harvest. When rates greater than 0.1 pound active per acre are used, do not apply within 14 days of harvest.

ALMONDS: To control Navel Orangeworm, Peach Twig Borer—Use POUNCE 3.2 EC at a rate of 8 to 16 ounces (0.2 to 0.4 pounds active) per acre. Apply when insects appear. Apply in a minimum of 15 gallons of finished spray per acre by aircraft or 20 gallons of finished spray per acre with ground equipment.

To control ants—Apply POUNCE 3.2 EC at a rate of 16 ounces (0.4 pound active) per acre by ground equipment in a minimum of 15 gallons of finished spray per acre. Application should follow mowing of weed growth to insure maximum coverage of soil surface. Overhead moisture following application will enhance activity.

Do not apply more than 0.8 pound active per acre during hull split. Do not apply more than 2 pounds active per acre per season. Do not apply within seven days of harvest. Do not graze livestock in treated orchards.

APPLES: To control Plum Curculio, Redbanded Leafroller, Rosy Apple Aphid, Spotted Tentiform Leafminer, Tarnished Plant Bug, White Apple Leafhopper, Oblique Banded Leafroller and Green Fruitworm—Use POUNCE 3.2 EC at a rate of 4 to 8 ounces (0.1 to 0.2 pounds active) per acre with ground equipment only. Apply in 25 to 400 gallons of finished spray per acre when insects appear and repeat as required to maintain control. Do not make more than three applications per season. Do not apply after petal fall.

ARTICHOKE: To control Artichoke Plume Moth and Leafminers—Use POUNCE 3.2 EC at a rate of 4 to 12 ounces (0.1 to 0.3 pound active) per acre. Apply with ground equipment using sufficient water to obtain uniform coverage. Apply as needed. Do not apply more than 1.5 pounds active per acre per season. Buds may be harvested on the day of application.

ASPARAGUS: To control Cutworms and Asparagus Beetle—Use POUNCE 3.2 EC at the rate of 2 to 4 ounces (0.05 to 0.1 lbs. active) per acre. Apply by ground equipment using sufficient water to obtain uniform coverage. Do not apply more than 0.4 pound active ingredient per acre per season. Do not apply within 1 day of harvest.

For post harvest application, use POUNCE 3.2 EC at a rate of 4 ounces (0.1 lbs. active) per acre. Apply to the fern stage of the asparagus plant after spear harvest when larval and adult stage of the Asparagus Beetle, Tarnished Plant Bug and Lygus Bugs and the adult stage of the Japanese* beetle are present.

*Not for control of this insect in California.

AVOCADO: To control Avocado Caterpillar, Avocado Lace Bug, Avocado Leafhopper, Avocado Leafroller, Orange Tortrix, Avocado Looper, Omnivorous Looper, Avocado Tree Girdler, Avocado Whitefly, Brown Soft Scale, Spanworm, Twig Borers, Mirids, Scale Crawlers, and Thrips—Use POUNCE 3.2 EC at a rate of 8 ounces (0.2 pound active) per acre. Apply with ground equipment in a minimum of 20 gallons of water per acre. Apply when insects first appear and repeat at 7 to 10 day intervals as needed to provide control. Do not apply more than 1.2 pounds active per acre per season. Do not apply within 7 days of harvest.

BROCCOLI, CHINESE BROCCOLI (gia ian, white flowering broccoli), BRUSSELS SPROUTS, CAULIFLOWER: To control Beet Armyworm, Cabbage Looper, Imported Cabbageworm and Diamondback Moth—Use POUNCE 3.2 EC at a rate of 2 to 4 ounces (0.05 to 0.1 pound active) per acre. Apply in a minimum of 3 gallons of finished spray per acre by aircraft and 20 gallons per acre by ground equipment. Do not make more than 8 applications per season. Do not apply within one day of harvest.

BULB ONIONS AND GARLIC: To control Onion Thrips and Armyworms—Use POUNCE 3.2 EC at a rate of 6 to 12 ounces (o.15 to 0.3

pound active) per acre. To control Cutworms, Leafminers, Onion Maggots (adults), and Stink Bugs—Use POUNCE 3.2 EC at a rate of 4 to 12 ounces (0.1 to 0.3 pound active) per acre.

Apply in a minimum of 20 gallons per acre with ground equipment or in a minimum of 5 gallons per acre by aircraft. Begin applications when pests appear and repeat as necessary to maintain control. Use the higher label rates as Onion Thrips population increases and avoid rescue situations.

Do not apply more than 2.4 pounds active per acre per season. Do not apply within 1 day of harvest. Do not graze livestock in treated areas or cut treated crops for feed.

CABBAGE, CHINESE CABBAGE (tight-heading varieties only): To control Cabbage Looper, Diamondback Moth, Imported Cabbageworm, and Southern White Butterfly—Use POUNCE 3.2 EC at a rate of 2 to 8 ounces (0.05 to 0.2 pound active) per acre. To control Beet Armyworm, Cutworms, and Flea Beetle—Use POUNCE 3.2 EC at a rate of 4 to 8 ounces (0.1 to 0.2 pound active) per acre. Apply in a minimum of 1 gallon of finished spray per acre by aircraft or 20 gallons per acre with ground equipment. Do not apply more than 1 pound active ingredient per acre per season. Do not apply within 1 day of harvest.

CANTALOUPES: To control Aphids, Leafminers, Squash Bug, and Whiteflies—Use POUNCE 3.2 EC at a rate of 8 ounces (0.2 pound active) per acre. To control Cucumber Beetle, Cutworms, Melon Worm, Pickleworm, and Squash Vine Borer—Use POUNCE 3.2 EC at a rate of 4 to 8 ounces (0.1 to 0.2 pounds active) per acre. Apply a minimum of 4 gallons of finished spray per acre by air or 20 gallons of finished spray per acre with ground equipment. Do not apply more than 1.6 pounds active per acre per season. Do not apply within one day of harvest.

CELERY, FLORENCE FENNEL (sweet anise, sweet fennel, finochio): For general use directions refer to the **Leafy Vegetable** crop grouping. In California do not apply more than 2.0 pounds active per acre per season.

CHERRIES: To control Lesser Peach Tree Borer, Rose Chafer, Green Fruitworm, Red-banded Leafroller, Plum Curculio and Tarnished Plant Bug—Use POUNCE 3.2 EC insecticide at a rate of 1 to 2 ounces (0.025 to 0.05 pounds active) per 100 gallons of water. Do not exceed 400 gallons per acre. Apply when insects appear. East of the Rockies, do not exceed 6 applications per season, with no more than 4 applications after petal fall. West of the Rockies, do not apply more than 4 applications per season, with no more than 3 applications after petal fall. Do not apply within 3 days of harvest. Do not graze livestock or feed crop forage from treated orchards.

CHRYSANTHEMUMS (Greenhouse): For control of Liromyza Leafminer flies, use POUNCE 3.2 EC at a rate of 20 fluid ounces (0.5 pound active) per 100 gallons (1 teaspoon per gallon). Avoid spraying the blooms. POUNCE 3.2 EC may be applied on a weekly schedule.

Caution: POUNCE 3.2 EC has demonstrated excellent plant safety; however, not all cultivars have been tested. Before treating large numbers of plants of a particular cultivar, treat a few plants and observe prior to full scale applications.

COLLARDS AND TURNIPS: To control Beet Armyworm, Corn Earworm, Southern White Butterfly, Leafminer, Leaf-hoppers, European Corn Borer, Fall Armyworm, Cutworms, Green Cloverworm, Southern Armyworm, Tobacco Budworm, Vegetable Leafminer, Cabbage Looper, Imported Cabbageworm, and Diamondback Moth—Use POUNCE 3.2 EC at a rate of 2 to 4 ounces (0.05 to 0.1 pound active) per acre. Do not apply within 1 day of harvest. Do not graze treated areas or feed crop refuse to livestock. Apply with ground equipment only. For use on Collards in AZ, GA, NC, OK, SC, and TX and on Turnips in FL, GA, IN, SC, TX, and WA. Do not apply more than 0.8 pound active ingredient per acre per season on Collards and Turnips (TX & IN). Do not apply more than 0.4 pound active ingredient per acre per season on Turnips (SC, GA, FL & WA).

CONIFERS (Container and Field Grown): For control of Nantucket Pine Tip Moth use POUNCE 3.2 EC at a rate of 4 to 8 fluid ounces (0.1 to 0.2 pound active) per acre. POUNCE 3.2 EC may be diluted in a non-volatile vegetable oil or water in a minimum of 1 gallon of finished spray per acre using equipment calibrated to give adequate coverage. Begin application when the adults appear and repeat at 5 to 7 day intervals or as needed throughout the season.

COTTON: Preemergent Use: To control Cutworms—Use POUNCE 3.2 EC at a rate of 4 to 8 ounces (0.1 to 0.2 pound active) per acre. Apply prior to planting or preemergent to crop. Use a minimum of 1 gallon of water per acre by aircraft or 5 gallons of water per acre by ground equipment. POUNCE 3.2 EC has been found to be effective and compatible in tank mixtures with herbicides and liquid fertilizers.

Foliar Use: To control Beet Armyworm (California and Arizona only), Bollworm, Cabbage Looper, Cotton Fleahopper, Cotton Leaf Perforator, Cutworms, Lygus Bugs, Pink Bollworm, Tarnished Plant Bug and Tobacco Budworm—Use POUNCE 3.2 EC at a rate of 4 to 8 ounces (0.1 to 0.2 pound active) per acre. To controll Boll Weevil, alternate POUNCE 3.2 EC at 4 ounces (0.1 pound active) per acre with POUNCE 3.2 EC at 2 ounces (0.05 pound active) per acre every 3 to 4 days until control is achieved. Apply the above rates in a minimum of 1 gallon of water per acre by aircraft or 5 gallons of water per acre with ground equipment. POUNCE 3.2 EC may also be applied using either refined non-volatile vegetable oil for control of the above named insects. When applying in water by aircraft, 1 quart of oil may be substituted for 1 quart of water per gallon of finished spray. POUNCE 3.2 EC should be diluted with oil and applied in a minimum of 1 quart total volume/acre using equipment calibrated to give adequate coverage. To control Bollworm and Tobacco Budworm use 2 to 8 ounces (0.05 to 0.2 pound active) per acre. The 2 ounce rate per acre should be used only when insect infestation levels are light and cotton is closely scouted before and after application.

Do not apply more than 3.0 pounds active ingredient per acre per season. Do not apply within 14 days of harvest. Do not graze or feed cotton forage.

COTTON TANK-MIX: POUNCE 3.2 EC insecticide plus Tank Mixtures (all cotton growing states). Mix POUNCE 3.2 EC at 4.0 to 8.0 ounces (0.1 to 0.2 pound active) per acre with other insecticides recommended on this label for control of cotton insect pests. POUNCE 3.2 EC has been found to be effective and compatible in tank mixtures with the following products:

Methyl Parathion	Guthion®
EPN plus Methyl Parathion	Azinphos-Methyl
Methomyl	Malathion

When using POUNCE 3.2 EC in tank mixtures, observe all restrictions and precautions which appear on the labels of these products. Apply tank mixtures immediately after mixing.

EGGPLANTS: To control Colorado Potato Beetle—Use POUNCE 3.2 EC at the rate of 8 ounces (0.2 lbs. active) per acre. To control Vegetable Leafminer—Use POUNCE 3.2 EC at the rate of 4 to 8 ounces (0.1 to 0.2 lbs. active) per acre. Apply using sufficient water to obtain uniform coverage. Apply as needed. Do not exceed 16 applications. Do not apply within 3 days of harvest.

FIELD CORN, POPCORN, FIELD CORN GROWN FOR SEED: Preemergent Use: To control Armyworms, Cutworms, and Stalk Borers—Use POUNCE 3.2 EC at a rate of 4 to 8 ounces (0.1 to 0.2 pounds active) per acre in the time period from five days prior to planting up to emergence of the crop. Apply either as a banded or broadcast spray in a minimum of 10 gallons finished spray per acre with ground equipment or 1 gallon per acre by air. Use a higher rate of POUNCE 3.2 EC when incorporating into the soil. POUNCE 3.2 EC has been found compatible and may be tank mixed with the following herbicides:

Aatrex®	Lasso®	Prowl®
Banvel®	Lorox®	Ramrod®
Bladex®	Paraquat CL	Roundup®
Dual®	Princep®	Sutan®
Eradicane®		

When using POUNCE 3.2 EC in tank mixtures, observe all restrictions and precautions which appear on the labels of these products. Provide constant agitation during mixing and application to keep the mixture in suspension. Label dosage rates should not be exceeded.

Foliar Use: For control of Armyworm, Corn Earworm, Corn Rootworm Beetles, Cutworms, European Corn Borer, Flea Beetle, Southwestern Corn Borer and Stalk Borers—Use POUNCE 3.2 EC at a rate of 4 to 8 ounces (0.1 to 0.2 pounds active) per acre prior to the brown silk stage. For control of Western Bean Cutworm—Use POUNCE 3.2 EC at a rate of 2 to 4 ounces (0.05 to 01 pound active) per acre prior to the brown silk stage.

When treating for stalk borer, POUNCE 3.2 EC must be applied when or shortly before the stalk borer larvae are moving into the corn from surrounding weeds and grasses. Mowing or burndown herbicide are suggested to initiate movement. For control of Corn Earworm apply just before silking and continue as necessary to maintain control.

Apply a minimum of 1 gallon of finished spray per acre by air or 10 gallons per acre with ground equipment.

FILBERTS: To control Filbertworm and Oblique Banded Leafroller—Use POUNCE 3.2 EC at a rate of 8 to 16 ounces (0.2 to 0.4 pounds active) per acre. For full coverage application apply 0.05 to 0.1 pounds active per 100 gallons (based on 400 gallons finished spray per acre), spray to runoff; OR, for low volume application apply 0.2 to 0.4 pounds active per acre (50 to 200 gallons finished spray per acre). For aerial application apply 0.2 to 0.4 pounds active in a minimum of 15 gallons of finished spray per acre. Apply when insects appear. Do not apply more than 1.6 pounds active per acre per season. Do not apply within 14 days of harvest. Do not graze livestock in treated orchards or feed cover crops from treated orchards to livestock.

HEAD LETTUCE: For general use directions refer to the **Leafy Vegetable** crop grouping.

HORSERADISH: To control Imported Crucifer Weevil *(Baris lepidii)*.

As a spring preplant dip—Use a 0.1% active solution (2 pints, 1 ounce of POUNCE 3.2 EC per 100 gallons). Soak sets for 30 minutes and air-dry before planting.

For foliar application, use POUNCE 3.2 EC at a rate of 8 ounces (0.2 pound active) per acre. Apply by ground equipment in a minimum spray volume of 20 gallons per acre. Make up to 3 foliar applications as needed to control weevil adults during oviposition.

Do not apply within 22 days of harvest.

LEAFY VEGETABLES (except Brassica): Amaranth; arrugula; celery; celtuce; chervil; corn salad; chrysanthemum (edible-leaved and garland); cress (garden and upland); dandelion; dock; endive; fennel; lettuce (leaf and head); orach; parsley; purslane (garden and winter); rhubarb; spinach; Swiss chard: To control Beet Armyworm, Corn Earworn, Cutworms, European Corn Borer, Fall Armyworm, Green Cloverworm, Southern Armyworm, Tobacco Budworm and Vegetable Leafminer—Use POUNCE 3.2 EC at a rate of 4 to 8 ounces (0.1 to 0.2 pound active) per acre. To control Alfalfa Looper, Cabbage Looper and Leafhoppers—Use POUNCE 3.2 EC at a rate of 2 to 8 ounces (0.05 to 0.2 pound active) per acre. Apply every 3 to 5 days or as needed by air or ground. Use sufficient water to obtain full coverage of foliage. Do not apply more than 2.0 pounds active ingredient per acre per season. Can be applied up to 1 day of harvest. Do not graze treated areas or feed crop refuse to livestock.

ORNAMENTAL NURSERY STOCK (Field Grown): To control Bagworm, Beet Armyworm, Cabbage Looper, Citrus Thrips, *Heliothis spp.*, Lace Bug, Leafhoppers, Leafminers and Whiteflies—Use 4 to 8 ounces per 100 gallons of water. POUNCE 3.2 EC may be used to control specified pests on non-edible ornamentals and non-bearing plants of fruiting species.

CAUTION: Marginal leaf burn may occur on Salvia, Dieffenbachia and Pteris Fern. Application to blooming plants may cause browning of petals. POUNCE 3.2 EC has demonstrated excellent plant safety; however, not all species and varieties have been tested. Before treating large numbers of plants of a particular variety, treat a few plants and observe prior to full scale application.

PAPAYA (Florida Only): To control Papaya Fruit Fly, Papaya Webworm, Papaya Whitefly, Brown Soft Scale, Scale Crawlers, Aphids, and Mealybug—Use POUNCE 3.2 EC at a rate of 8 ounces (0.2 pound active) per acre. Apply with ground equipment in a minimum of 20 gallons of water per acre. Apply when insects first appear and repeat at 7 to 10 day intervals as needed to provide control. Do not apply more than 1.2 pounds active per acre per season. Do not apply within 7 days of harvest.

PEACHES: To control Lesser Peach Tree Borer, Peach Twig Borer, Rose Chafer, Plum Curculio, Tarnished Plant Bug, Green Fruitworm, Oriental Fruit Moth—Use POUNCE 3.2 EC at a rate of 2 to 4 ounces (0.05 to 0.1 pound active) per 100 gallons of water. Do not exceed 400 gallons per acre. Apply when insects appear. Make applications with ground equipment, using sufficient water to obtain uniform coverage. Do not apply within 7 days of harvest. Do not make more than 8 applications per season. Do not graze livestock in treated orchards.

PEARS (Dormant through Delayed Dormant): To control Pear Psylla—Apply POUNCE 3.2 EC at a rate of 8 to 16 ounces (0.2 to 0.4 pound active) per acre. POUNCE 3.2 EC may be combined with 2 to 8 gallons of spray oil per acre. Apply during the dormant through delayed dormant growth periods only. Apply in 3 to 20 gallons of finished spray per acre by aircraft and 25 to 400 gallons per acre by ground equipment. Do not make more than two applications per season.

PEARS (Summer): To control Pear Psylla, Green Fruitworm, Codling Moths—Use POUNCE 3.2 EC insecticide at a rate of 8 to 16 ounces (0.2 to 0.4 pound active) per acre. Use in sufficient water to achieve runoff and apply by ground equipment. Do not make more than 3 applications per season. Do not apply within 14 days of harvest.

PEPPERS (Bell): To control Vegetable Leafminer, Cabbage Looper, Flea Beetle, Corn Earworm, Pepper Weevil—Use POUNCE 3.2 EC at a rate of 4 to 8 ounces (0.1 to 0.2 lbs. active) per acre. To control European Corn Borer—Use POUNCE 3.2 EC at the rate of 8 ounces (0.2 lbs. active) per acre. Apply using sufficient water to obtain uniform coverage. Apply as needed. Do not exceed 8 sprays per season. Do not apply within 3 days of harvest.

PINE SEED ORCHARDS: To control Coneworms, Seed Bugs—Use POUNCE 3.2 EC at the following rates:

For high volume sprayers: Use 8 ounces in 100 gallons of water (0.025% dilution by weight). Apply 5 to 10 gallons of finished spray per tree.

For low volume sprayers: Use 42 ounces in 100 gallons of water (0.125% dilution by weight). Apply 100 gallons per acre.

For aerial application: Use 30 ounces in not less than 10 gallons of water per acre.

To control Webbing Coneworm—make first application within 1 week of female flower closure or peak pollen flight.

To control other coneworms and seed bugs—make first application within 30 days following female flower closure.

Repeat applications at 4 week intervals, but do not apply more than 6 applications. Do not graze or harvest cover crop. Avoid contact with open water.

PISTACHIOS: To control Peach Twig Borer, Navel Orangeworm, Plant Bugs, Stink Bugs, and Leaf-footed Bugs—Use POUNCE 3.2 EC at a rate of 8 to 16 ounces (0.2 to 0.4 pound active) per acre. Use sufficient water to obtain full coverage of foliage.

To control ants—Apply POUNCE 3.2 EC at a rate of 16 ounces (0.4 pound active) per acre by ground equipment in a minimum of 15 gallons of finished spray per acre. Application should follow mowing of weed growth to insure maximum coverage of the soil surface. Overhead moisture following application will enhance activity.

Do not apply more than 0.8 pound active per acre per crop season. Do not apply after 10% hull split. Nuts may be harvested on the day of application. Do not graze livestock in treated orchards.

POTATOES: To control Beet Armyworm, Cabbage Looper, Colorado Potato Beetle, Cutworms, European Corn Borer, Potato Flea Beetle, Potato Leafhopper, Potato Psyllid, Potato Tuberworm, Aster Leafhopper and Tarnished Plant Bug. Use POUNCE 3.2 EC at a rate of 4 to 8 ounces (0.1 to 0.2 pound active) per acre. Apply the above rates in a minimum of 3 gallons of water per acre by aircraft and 20 gallons of water per acre with ground equipment. POUNCE 3.2 EC may also be applied using refined non-volatile vegetable oil for control of the above-named insects. POUNCE 3.2 EC should be diluted with oil and applied in a minimum of one quart total volume per acre using equipment calibrated to give adequate coverage. Do not make more than 12 applications per season. Do not apply within 7 days of harvest. Do not graze or feed potato forage.

PUMPKINS: To control Aphids, Leafminers, Squash Bug, and Whiteflies—Use POUNCE 3.2 EC at a rate of 8 ounces (0.2 pound active) per acre. To control Cucumber Beetle, Cutworms, Melon Worm, Pickleworm and Squash Vine Borer—Use POUNCE 3.2 EC at a rate of 4 to 8 ounces (0.1 to 0.2 pound active) per acre. Apply a minimum of 4 gallons of finished spray per acre by air or 20 gallons of finished spray per acre with ground equipment. Do not apply more than 0.8 pound active per acre per season. Do not apply within 1 day of harvest.

RANGE GRASS: To control Range Caterpillar—Use POUNCE 3.2 EC at a rate of 0.4 ounces (0.01 lbs. active) per acre. Apply using sufficient water to obtain uniform coverage. Do not apply more than once per year. Cattle may be present during application. Do not harvest or feed hay to livestock.

ROSES (Field Grown): For control of *Heliothis sp.*, use 4 to 8 fluid ounces (0.1 to 0.2 pound active) per acre. POUNCE 3.2 EC may be applied in 5 gallons of finished spray per acre with ground equipment and 1 gallon per acre by air.

ROSES (Greenhouse): For control of Beet Armyworm, Cabbage Looper and Omniverous Leafroller, use POUNCE 3.2 EC insecticide at a rate of 8 fluid ounces (0.2 pound active) per 100 gallons of water.

Caution: POUNCEis not phytotoxic to the following varieties of greenhouse roses: Ballena, Bettina, Cara Mia, Coquette, Excitement, Forever Yours, G. Wave, Jack Frost, Jr. Bridesmaid, Matador, Paul's Pink, Samantha, Seventeen, Sonia, Town Crier, Tropicana and Visa. Other varieties may vary in their sensitivity to POUNCE 3.2 EC, and a small number of plants should be treated under local conditions to determine plant safety prior to commercial use.

SOYBEANS: To control Cabbage Looper, Green Cloverworm, Mexican Bean Beetle, Soybean Looper, Cutworms, Velvetbean Caterpillar, Bean Leaf Beetle and Potato Leafhopper—Use POUNCE 3.2 EC at a rate of 2 to 4 ounces (0.05 to 0.1 pound active) per acre. To control Beet Armyworm and Corn Earworm—Use POUNCE 3.2 EC at a rate of 4 to 8 ounces (0.1 to 0.2 pound active) per acre. Apply a minimum of 1 gallon of finished spray per acre by air or 5 gallons with ground equipment.

POUNCE 3.2 EC may also be applied using refined non-volatile vegetable oil for control of the above named insects. POUNCE 3.2 EC should be diluted with oil and applied in a minimum of 1 quart total volume/acre using equipment calibrated to give adequate coverage. When applying in water by aircraft, 1 quart of oil may be substituted for 1 quart of water per gallon of finished spray.

Do not apply more than 0.4 pound active ingredient per acre per season. Do not apply within 60 days of harvest. Do not graze or feed soybean forage or hay.

SPINACH: Refer to **Leafy Vegetable** crop grouping for general use directions.

SWEET CORN: To control Corn Earworm, Corn Rootworm Beetle*, Cutworms, European Corn Borer, Fall Armyworm, Southern Armyworm, Flea Beetle, Leafhoppers, and Stalk Borers—Use POUNCE 3.2 EC at a rate of 4 to 8 ounces (0.1 to 0.2 pound active) per acre. Apply every 3 to 5 days or as needed. Do not apply more than 1.2 pounds active per acre per season. Do not apply within 1 day of harvest of ears or forage or livestock grazing.

*Pest does not occur on this crop in California.

SWEET CORN (Florida only) fresh market use only: To control Corn Earworm, European Corn Borer, Fall Armyworm, Southern Armyworm—Use 4 to 10 ounces (0.1 to 0.25 pound active) per acre. Apply every 5 days or as needed.

Do not apply more than 3.25 pounds active ingredient per acre per crop season. Do not graze livestock in treated areas or harvest for forage.

TOMATOES (Only For Application To Tomatoes Grown In Florida For Final Marketing As Fresh Tomatoes)
To control Beet Armyworm, Cabbage Looper, Colorado Potato Beetle, Granulate Cutworm, Hornworms, Vegetable Leafminers, Southern Armyworm, Tomato Fruitworm and Tomato Pinworm—Use POUNCE 3.2 EC at a rate of 2 to 8 ounces (0.05 to 0.2 pound active) per acre. Apply by aerial or ground equipment in sufficient water to obtain uniform coverage. Do not apply more than 1.2 pound active ingredient per acre per season. Can be applied up to the day of harvest. Do not apply to cherry tomatoes or other varieties which produce fruit less than one inch in diameter.

WALNUTS: To control Navel Orangeworm, Codling Moth and Walnut Husk Fly—Use POUNCE 3.2 EC at a rate of 8 to 16 ounces (0.2 to 0.4 pounds active) per acre. For full coverage application apply 0.05 to 0.1 pounds active per 100 gallons (based on 400 gallons finished spray per acre), spray to run-off; OR, for low volume application apply 0.2 to 0.4 pounds active per acre (50 to 200 gallons finished spray per acre). For aerial application apply 0.2 to 0.4 pounds active in a minimum of 20 gallons of finished spray per acre. Do not apply more than 1.6 pounds active ingredient per acre per season. Apply when insects appear. Do not apply within 1 day of harvest. Do not graze livestock in treated orchards or feed cover crops from treated orchards to livestock.

PREMISES SPRAY
For agricultural use only.
Spray directly to walls and ceiling as residual surface treatment only.
Do not treat manure or litter. Avoid contamination of feed and water.
Do not apply directly to livestock or poultry.

For Application in	Target Insects	Method of Applic.	Dilute	Applic. Rate
Dairies, Barns, Feedlots, Stables, Poultry Houses, Swine and Livestock Houses	House Flies, Stable Flies and other Manure Breeding Flies. Also aids in the reduction of Cockroaches, Mosquitoes and Spiders.	Sprayer	4 oz. to 12.5 gallons water	1 gal. per 750 sq. ft. of surface

Re-treat as necessary, but not more often than once every 2 weeks. The use of any residual fly spray should be supplemental with proper manure management and general sanitation to reduce or eliminate fly breeding site.

Treatment of Preconstruction Lumber and Logs

GENERAL INFORMATION
POUNCE 3.2 EC insecticide should be diluted with water. To prepare the spray, dilute POUNCE 3.2 EC as shown in the following spray dilution chart:

Spray Dilution Chart

Gallons of Spray Mixture Desired	Gallons of POUNCE 3.2 EC Insecticide To Use		
	0.5% Solution	0.75% Solution	1.0% Solution
40	½	¾	1
80	1	1½	2
200	2½	3¾	5
400	5	7½	10
800	10	15	20

Directions for Application
To protect unseasoned lumber and logs from wood destroying insects, such as Termites, Carpenter Ants and Beetles (Ambrosia, Powder-post, Old house borers and others), totally treat wood with a 0.5% to 1.0% solution of Pounce® 3.2 EC. This solution can be applied by various methods including spraying, brushing, dipping, and pressure treatment. Frequent monitoring of dip and pressure systems are necessary to insure that the desired level of POUNCE 3.2 EC is maintained. Wood can be handled after treatment when dry.

1. For dip treatments, the wood should be totally submersed in the solution until thoroughly wet and then allowed to dry in a suitable location. Dipping solutions to which POUNCE 3.2 EC has been added should be agitated before use if left unused for long periods of time. Sediment, debris and other deposits should be periodically cleaned from the tank.

2. For pressure treatments, the wood should be placed in the treatment chamber, the POUNCE 3.2 EC solution added and the system pressurized up to 250 psi for up to one hour depending on the density and type of wood treated. After the pressure is released and the system drained, the wood should be placed in a suitable location for drying.

3. For spray treatments, the wood should be sprayed thoroughly including back and ends.

4. For brush treatments, all parts of wood surfaces should be thoroughly treated.

DEALERS SHOULD SELL IN ORIGINAL PACKAGES ONLY.

TERMS OF SALE OR USE: On purchase of this product buyer and user agree to the following conditions:

WARRANTY: FMC makes no warranty, expressed or implied, concerning the use of this product other than indicated on the label. Except as so warranted, the product is sold as is. Buyer and user assume all risk of use and/or handling and/or storage of this material when such use and/or handling and/or storage is contrary to label instructions.

DIRECTIONS AND RECOMMENDATIONS: Follow directions carefully. Timing and method of application, weather and crop conditions, mixture with other chemicals not specifically recommended and other influencing factors in the use of this product are beyond the control of the seller and are assumed by the buyer at his own risk.

USE OF PRODUCT: FMC's recommendations for the use of this product are based upon tests believed to be reliable. The use of this product being beyond the control of the manufacturer, no guarantee, expressed or implied, is made as to the effects of such or the results to be obtained if not used in accordance with directions or established safe practice.

DAMAGES: Buyer's or user's exclusive remedy for damages for breach of warranty or negligence shall be limited to direct damages not exceeding the purchase price paid and shall not include incidental or consequential damages.

AAtrex, Dual and Princep—Ciba Geigy trademarks
Banvel—Velsicol Chemical trademark
Bladex—Shell Oil trademark
Eradicane and Sutan—Stauffer trademarks
Guthion—Bayer AG trademark
Lasso, Ramrod and Roundup—Monsanto trademarks
Lorox—E.I. DuPont de Nemours trademark
Prowl—American Cyanamid trademark
POUNCEand FMC—FMC trademark (3510-3/2/89-A 7/89)

This is a specimen label. FMC is not responsible for the accuracy of the information contained herein. As labels are subject to revision, always carefully read and follow the label on the product container.

Pounce® 25 WP

INSECTICIDE

RESTRICTED USE PESTICIDE

For retail sale to and use only by certified applicators or persons under their direct supervision, and only for those uses covered by the certified applicator's certification

ACTIVE INGREDIENTS:	BY WT.
*Permethrin**	25.0%
INERT INGREDIENTS:	75.0%
TOTAL	100.0%

*(3-Phenoxyphenyl)methyl(±)cis-trans 3-(2,2-dichloroethenyl)-2,2-dimethylcyclopropanecarboxylate
**cis/trans ratio: Max. 55% (±) cis and min. 45% (±) trans

EPA Reg. No. 279-3051 EPA Est. 279-Code 96

U.S. Patent No. 4,024,163

KEEP OUT OF REACH OF CHILDREN

WARNING

AVISO

PRECAUCION AL USUARIO: Si usted no lee ingles, no use este producto hasta que la etiqueta le haya sido explicado ampliamente.

STATEMENT OF PRACTICAL TREATMENT

If in eyes: Flush eyes with water for 15 minutes. Get medical attention if irritation persists.

If swallowed: Drink 1 or 2 glasses of water and induce vomiting by touching back of throat with finger. Do not induce vomiting or give anything by mouth to an unconscious person.

For Emergency Assistance Call 716-735-3765

PRECAUTIONARY STATEMENTS

HAZARDS TO HUMANS (AND DOMESTIC ANIMALS)

WARNING

Do not get in eyes, on skin or on clothing. Wash thoroughly after handling.

ENVIRONMENTAL HAZARDS

This pesticide is extremely toxic to fish. Use with care when applying in areas adjacent to any body of water. Do not apply directly to water. Do not apply when weather conditions favor drift from treated areas. Do not contaminate water when disposing of equipment washwaters. This product is highly toxic to bees exposed to direct treatment or residues on crops or weeds. Do not apply this product or allow it to drift to crops or weeds on which bees are actively foraging. Additional information may be obtained from your Cooperative Extension Service. Apply this product only as specified on this label.

See other panels for additional precautionary statements.

DIRECTIONS FOR USE

It is a violation of Federal law to use this product in a manner inconsistent with its labeling.

Do not apply this product in such a manner as to directly or through drift expose workers or other persons. The area being treated must be vacated by unprotected persons.

Do not enter treated areas without protective clothing until sprays have dried.

Protective clothing means, at least, a hat or other suitable head covering, a long sleeved shirt and long legged trousers or a coverall type garment (all of closely woven fabric covering the body, including the arms and legs), shoes and socks.

Because certain states may require more restrictive reentry intervals for various crops being treated with this product, consult your State Department of Agriculture for further information.

Written or oral warnings must be given to workers who are expected to be in a treated area or in an area about to be treated with this product. Inform workers of areas or fields that may not be entered without specific protective clothing, period of time field must be vacated and appropriate actions to take in case of accidental exposure. An example of such information is given under written warnings. When oral warnings are given, warnings shall be given in a language customarily understood by workers. Oral warnings must be given if there is reason to believe that written warnings cannot be understood by workers. Written warnings must include the following information:

WARNING

Area treated with Pounce 25 WP insecticide on (insert date). Do not enter without appropriate protective clothing until sprays have dried (insert State Department of Agriculture's re-entry interval, if more restrictive).

In case of accidental exposure to pesticide spray or dust, wash the skin thoroughly with soap and water. Remove contaminated clothing and wash before reuse. If in eyes, flush with plenty of water. If inhaled, go to an area where the pesticide has not been applied. Get medical attention if needed.

CHEMIGATION USE DIRECTIONS

Apply this product only through sprinkler including center pivot, lateral move, end tow, side (wheel) roll, traveler, big gun, solid set, or hand move irrigation systems. Do not apply this product through any other type of irrigation system. Do not connect an irrigation system (including greenhouse systems) used for pesticide applications to a public water system.

Crop injury, lack of effectiveness, or illegal residues in the crop can result from nonuniform distribution of treated water. If you have questions about calibration, you should contact State Extension Service specialists, equipment manufacturers, or other experts. A person knowledgeable of the chemigation system and responsible for its operation, or under the supervision of the responsible person, shall shut the system down and make necessary adjustments should the need arise.

The system must contain a functional check valve, vacuum relief valve, and low pressure drain appropriately located on the irrigation pipeline to prevent water source contamination from backflow.

The pesticide injection pipeline must contain a functional, automatic, quick-closing check valve to prevent the flow of fluid back toward the injection pump.

The pesticide injection pipeline must also contain a functional, normally closed, solenoid-operated valve located on the intake side of the injection pump and connected to the system interlock to prevent fluid from being withdrawn from the supply tank when the irrigation system is either automatically or manually shut down.

The system must contain functional interlocking controls to automatically shut off the pesticide injection pump when the water pump motor stops.

The irrigation line or water pump must include a functional pressure switch which will stop the water pump motor when the water pressure decreases to the point where pesticide distribution is adversely affected.

Systems must use a metering pump, such as a positive displacement injection pump (e.g., diaphragm pump) effectively designed and constructed of materials that are compatible with pesticides and capable of being fitted with a system interlock.

Do not apply when wind speed favors drift beyond the area intended for treatment.

Pounce® 25 WP mixes readily with water to form a suspension. Pounce 25 WP should be diluted in sufficient volume of water to insure accurate application over the area to be treated. Mix the required amount of Pounce 25 WP with a small quantity of water and add this premix to the supply tank with the required amount of water. Maintain sufficient agitation during both mixing and application to ensure uniformity of the supply tank. Hydraulic or mechanical agitation is recommended. Pounce 25 WP should be applied continuously for the duration of the water application. When using chemigation, a minimum of 0.5 inch per acre of irrigation water is recommended.

STORAGE AND DISPOSAL

PESTICIDE STORAGE

Keep out of reach of children and animals. Store in original containers only. Store in a cool, dry place and avoid excess heat. Carefully open containers. After partial use, fold and roll back bags, clamp and close tightly. Do not put concentrate or dilute material into food or drink containers. Do not contaminate other pesticides, fertilizers, water, food or feed by storage or disposal.

In case of spill, avoid contact, isolate area and keep out animals and unprotected persons. Confine spills. Call FMC collect: (716) 735-3765.

To confine spill: If liquid, dike surrounding area or absorb with sand, cat litter or commecial clay. If dry material, cover to prevent dispersal. Place damaged package in a holding container. Identify contents.

PESTICIDE DISPOSAL

Wastes resulting from the use of this product may be disposed of on site or at an approved waste disposal facility.

CONTAINER DISPOSAL

Paper Bag: Completely empty bag into application equipment. Then dispose of empty bag in a sanitary landfill or by incineration, or, if allowed by State and local authorities, by burning. If burned, stay out of smoke.

Plastic Container: Triple rinse (or equivalent). Then offer for recycling or reconditioning, or dispose of in a sanitary landfill, or incineration, or, if allowed by State and local authorities, by burning. If burned, stay out of smoke.

ENDANGERED SPECIES RESTRICTIONS

Aerial Application: Do not use within 100 yards of aquatic habitats.

Ground Application: Do not use within 20 yards of aquatic habitats.

GENERAL INSTRUCTIONS

Pounce 25 WP insecticide is a 25% wettable powder formulation of insecticide permethrin. Apply Pounce 25 WP when insects appear or feeding is noticed. The higher rate should be used as pest populations increase. Repeat the application as necessary to maintain control. Pounce 25 WP may be applied by both ground and aerial equipment. Use sufficient water to obtain full coverage. With the exception of crops listed below, rotational crops should not be planted within 60 days of last application.

Mix as needed; do not store diluted material.

ALFALFA: To control Alfalfa Caterpillar, Armyworms, Cutworms, Loopers, Webworms, Blue Alfalfa Aphid, Green Peach Aphid, Pea Aphid, Green Cloverworm, Spotted Alfalfa Aphid, and Velvetbean Caterpillar—Use Pounce 25 WP at a rate of 3.2 to 12.8 ounces (0.05 to 0.2 pound active) per acre. To control Alfalfa Weevil, Egyptian Alfalfa Weevil, Cucumber Beetle, Meadow Spittlebug, Plant Bugs (including Lygus spp.), Potato Leafhopper, and Stink Bugs—Use Pounce 25 WP at a rate of 6.4 to 12.8 ounces (0.1 to 0.2 pound active) per acre.

Use higher recommended dosage for increased pest pressure or for increased residual pest control. Apply in a minimum of 1 gallon of finished spray per acre by aircraft or 10 gallons of finished spray per acre with ground equipment. Do not apply more than 0.2 pound active per cutting. When rates of 0.1 pound active per acre or less are used, application may be made on day of harvest. When rates greater than 0.1 pound active per acre are used, do not apply within 14 days of harvest.

ALMONDS: To control Navel Orangeworm, Peach Twig Borer—Use Pounce 25 WP at a rate of 12.8 to 25.6 ounces (0.2 to 0.4 pounds active) per acre. Do not make more than 2 applications during hull split. Do not apply more than 2.0 pounds active ingredient per acre per season. Apply in a minimum of 15 gallons of finished spray per acre by aircraft or 20 gallons of finished spray per acre with ground equipment. Do not apply within 7 days of harvest. Do not graze livestock in treated orchards.

APPLES: To control Plum Curculio, Redbanded Leafroller, Rosy Apple Aphid, Spotted Tentiform Leafminer, Tarnished Plant Bug, White Apple Leafhopper, Oblique Banded Leafroller and Green Fruitworm—Use Pounce 25 WP at a rate of 6.4 to 12.8 ounces (0.1 to 0.2 pounds active) per acre with ground equipment only. Apply in 25 to 400 gallons of finished spray per acre. Do not apply more than 0.6 pounds active ingredient per acre per season. Do not apply after petal fall.

ARTICHOKE: To control Artichoke Plume Moth and Leafminers—Use POUNCE 25 WP at a rate of 6.4 to 19.2 ounces (0.1 to 0.3 pound active) per acre. Apply with ground equipment using sufficient water to obtain uniform coverage. Apply as needed. Do not apply more than 1.5 pounds active per acre per season. Buds may be harvested on the day of application.

ASPARAGUS: To control Cutworms and Asparagus Beetle—Use Pounce 25 WP at the rate of 3.2 to 6.4 ounces (0.05 to 0.1 pound active) per acre. Apply by ground equipment as needed. Do not apply more than 0.4 pounds active ingredient per acre per season. Do not apply within 1 day of harvest.

For post-harvest application—Use Pounce 25 WP at a rate of 6.4 ounces (0.1 pound active) per acre. Apply to the fern stage of the asparagus plant after spear harvest when larval and adult stage of the Asparagus Beetle, Tarnished Plant Bug, Lygus Bugs and Japanese Beetle* are present.

*Not for control of this insect in California

AVOCADO: To control Avocado Caterpillar, Avocado Lace Bug, Avocado Leafhopper, Avocado Leafroller, Orange Tortrix, Avocado Looper, Omnivorous Looper, Avocado Tree Girdler, Avocado Whitefly, Brown Soft Scale, Spanworm, Twig Borers, Mirids, Scale Crawlers, and Thrips—Use POUNCE 25 WP at a rate of 12.8 ounces (0.2 pound active) per acre. Apply with ground equipment in a minimum of 20 gallons of water per acre. Apply when insects first appear and repeat at 7 to 10 day intervals as needed to provide control. Do not apply more than 1.2 pounds active per acre per season. Do not apply within 7 days of harvest.

BROCCOLI, CHINESE BROCCOLI (gai ian, white flowering broccoli), BRUSSELS SPROUTS, CAULIFLOWER: To control Beet Armyworm, Cabbage Looper, Cutworms, Imported Cabbageworm and Diamondback Moth—Use Pounce 25 WP at a rate of 3.2 to 6.4 ounces (0.05 to 0.1 pound active) per acre. Apply in a minimum of 3 gallons of finished spray per acre by aircraft or 20 gallons per acre by ground equipment. Do not apply more than 0.8 pound active ingredient per acre per season. Do not apply within 1 day of harvest.

BULB ONIONS AND GARLIC: To control Onion Thrips and Armyworms—Use POUNCE 25 WP at a rate of 9.6 to 19.2 ounces (0.15 to 0.3 pound active) per acre. To control Cutworms, Leafminers, Onion Maggots (adults), and Stink Bugs—Use Pounce 25 WP at a rate of 6.4 to 19.2 ounces (0.1 to 0.3 pound active) per acre.

Apply in a minimum of 20 gallons per acre with ground equipment or in a minimum of 5 gallons per acre by aircraft. Begin applications when pests appear and repeat as necessary to maintain control. Use the higher label rates as Onion Thrips population increases and avoid rescue situations.

Do not apply more than 2.4 pounds active per acre per season. Do not apply within 1 day of harvest. Do not graze livestock in treated areas or cut treated crops for feed.

CABBAGE, CHINESE CABBAGE (tight-heading varieties only): To control Beet Armyworm, Cutworms, Cabbage Looper, Diamondback Moth and Imported Cabbageworm—Use Pounce 25 WP at a rate of 3.2 to 6.4 ounces (0.05 to 0.1 pound active) per acre. Apply in a minimum of 1 gallon of finished spray per acre by aircraft or 20 gallons per acre with ground equipment. Do not apply more than 1.0 pound of active ingredient per acre per season. Do not apply within 1 day of harvest.

CANTALOUPES: To control Aphids, Leafminers, Squash Bug, and Whiteflies—Use Pounce 25 WP at a rate of 12.8 ounces (0.2 pound active) per acre. To control Cucumber Beetle, Cutworms, Melon Worm, Pickleworm, and Squash Vine Borer—Use Pounce 25 WP at a rate of 6.4 to 12.8 ounces (0.1 to 0.2 pound active) per acre. Apply a minimum of 4 gallons of finished spray per acre by air or 20 gallons of finished spray per acre with ground equipment. Do not apply more than 1.6 pounds active per acre per season. Do not apply within one day of harvest.

CELERY, FLORENCE FENNEL (sweet anise, sweet fennel, finochio): For general use directions refer to the **Leafy Vegetable** crop grouping.

CHERRIES: To control Lesser Peach Tree Borer, Rose Chafer, Green Fruitworm, Red-banded Leafroller, Plum Curculio and Tarnished Plant Bug—Use Pounce 25 WP at a rate of 1.6 to 3.2 ounces (0.025 to 0.05 pounds active) per 100 gallons of water. Do not exceed 400 gallons per acre. Apply when insects appear. East of the Rockies, do not exceed 6 applications per season, with no more than 4 applications after petal fall. West of the Rockies, do not apply more than 4 applications per season, with no more than 3 applications after petal fall. Do not apply within 3 days of harvest. Do not graze livestock or feed crop forage from treated orchards.

CHRYSANTHEMUMS (Greenhouse*): For control of Liromyza Leafminer flies—Use Pounce 25 WP at a rate of 32 ounces (0.5 pound active) per 100 gallons. Avoid spraying the blooms. Pounce may be applied on a weekly schedule. CAUTION: Cultivars may vary in sensitivity and a small number of plants should be treated to determine plant safety prior to commercial use.

*Not for use in California

COLLARDS AND TURNIPS: To control Beet Armyworm, Corn Earworm, Southern White Butterfly, Leafminer, Leaf-hoppers, European Corn Borer, Fall Armyworm, Cutworms, Green Cloverworm, Southern Armyworm, Tobacco Budworm, Vegetable Leafminer, Cabbage Looper, Imported Cabbageworm, and Diamondback Moth—Use POUNCE 25 WP at a rate of 3.2 to 6.4 ounces (0.05 to 0.1 pound active) per acre. Do not apply within 1 day of harvest. Do not graze treated areas or feed crop refuse to livestock. Apply with ground equipment only. For use on Collards in AZ, GA, NC, OK, SC, and TX and on Turnips in FL, GA, IN, SC, TX, and WA. Do not apply more than 0.8 pound active ingredient per acre per season on Collards and Turnips (TX & IN). Do not apply more than 0.4 pound active ingredient per acre per season on Turnips (SC, GA, FL & WA).

CONIFERS (Container and Field Grown): For control of Nantucket Pine Tip Moth—Use Pounce 25 WP at a rate of 6.4 to 12.8 ounce (0.1 to 0.2 pound active) per acre. Begin application when the adults appear and repeat at 5 to 7 day intervals or as needed throughout the season.

COTTON: Preemergent Use: To control Cutworms—Use Pounce 25® WP insecticide at a rate of 6.4 to 12.8 ounces (0.1 to 0.2 pound active) per acre. Apply prior to planting or preemergent to crop.

Foliar Use: To control Beet Armyworm (California and Arizona only), Bollworm, Cabbage Looper, Cotton Fleahopper, Cotton Leaf Perforator, Cutworms, Lygus Bugs, Pink Bollworm, Tarnished Plant Bug and Tobacco Budworm—Use Pounce 25 WP at a rate of 6.4 to 12.8 ounces (0.1 to 0.2 pound active) per acre. To control Boll Weevil, alternate Pounce 25 WP at 6.4 ounces (0.1 pound active) per acre with Pounce 25 WP at 3.2 ounces (0.05 pound active) per acre every 3 to 4 days until control is achieved.

Use a minimum of 1 gallon of water per acre by aircraft or 5 gallons of water per acre by ground equipment. Do not apply more than 3.0 pounds active ingredient per season. Do not apply within 14 days of harvest. Do not graze or feed cotton forage.

EGGPLANTS: To control Colorado Potato Beetle—Use Pounce 25 WP at the rate of 12.8 ounces (0.2 pound active) per acre. To control Vegetable Leafminer—Use Pounce 25 WP at the rate of 6.4 to 12.8 ounces (0.1 to 0.2 pound) active per acre. Do not apply more than 3.2 pounds active ingredient per acre per season. Do not apply within 3 days of harvest.

FIELD CORN, POPCORN, FIELD CORN GROWN FOR SEED: Preemergent Use: To control Armyworm and Cutworms—Use Pounce 25 WP at a rate of 6.4 to 12.8 ounces (0.1 to 0.2 pound active) per acre in the time period from five days prior to planting up to emergence of the crop. Apply as a broadcast spray with ground equipment in a minimum of 20 gallons of finished spray per acre.

Foliar Use: For control of Cutworms, European Corn Borer, Flea Beetle, Southwestern Corn Borer, Stalk Borer, Corn Rootworm Beetle, and Armyworm—Use Pounce 25 WP at a rate of 6.4 to 12.8 ounces (0.1 to 0.2 pound active) per acre prior to ear formation. Ear formation occurs when pollination is completed as evidenced by initiation of brown silk in field to be treated. Apply a minimum of 1 gallon of finished spray per acre by air or 10 gallons per acre with ground equipment. When treating for stalk borer, Pounce 25 WP must be applied when or shortly before the stalk borer larvae are moving into the corn from surrounding weeds and grasses. Mowing or burn-down herbicide is suggested to initiate movement.

FILBERTS: To control Filbertworm and Oblique Banded Leafroller—Use Pounce 25 WP at a rate of 12.8 to 25.6 ounces (0.2 to 0.4 pound active) per acre. For full coverage application apply 0.05 to 0.1 pound active per 100 gallons (based on 400 gallons finished spray per acre), spray to runoff; or, for low volume application apply 0.2 to 0.4 pound active per acre (50 to 200 gallons finished spray per acre). For aerial application apply 0.2 to 0.4 pound active in a minimum of 15 gallons of finished spray per acre. Apply when insects appear. Do not apply more than 1.6 pounds active ingredient per acre per season. Do not apply within 14 days of harvest. Do not graze livestock in treated orchards or feed cover crops from treated orchards to livestock.

HEAD LETTUCE: For general use directions refer to the **Leafy Vegetable** crop grouping.

HORSERADISH: To control Imported Crucifer Weevil (Baris lepidii). For foliar application—Use Pounce 25 WP at a rate of 12.8 ounces (0.2 pound active) per acre. Apply by ground equipment in a minimum spray volume of 20 gallons per acre. Make up to 3 foliar applications as needed to control weevil adults during oviposition.

Do not apply within 22 days of harvest.

LEAFY VEGETABLES (except Brassica): Amaranth; arrugula; celery; celtuce, chervil; corn salad; chrysanthemum (edible-leaved and garland); cress, (garden and upland); dandelion; dock; endive; fennel; lettuce (leaf and head); orach; parsley; purslane (garden and winter); rhubarb; spinach; Swiss chard: To control Beet Armyworm, Corn Earworm, Cutworms, European Corn Borer, Fall Armyworm, Green Cloverworm, Southern Armyworm, Tobacco Budworm and Vegetable Leafminer—Use Pounce 25 WP at a rate of 6.4 to 12.8 ounces (0.1 to 0.2 pound active) per acre. To control Alfalfa Looper, Cabbage Looper and Leafhoppers—Use Pounce 25 WP at a rate of 3.2 to 12.8 ounces (0.05 to 0.2 pounds active) per acre. Apply every 3 to 5 days or as needed by air or ground. Use sufficient water to obtain full coverage of foliage. Do not apply more than 2.0 pounds active ingredient per acre per season. Can be applied up to 1 day of harvest. Do not graze treated areas or feed crop refuse to livestock.

ORNAMENTAL NURSERY STOCK (Field Grown): To control Beet Armyworm, Cabbage Looper, Citrus Thrips, Heliothis spp., Leafhoppers, Leafminers and Whiteflies—Use 6.4 to 12.8 ounces (0.1 to 0.2 pound active) per 100 gallons of water. Pounce 25 WP may be used to control specified pests on non-edible ornamentals and non-bearing plants of fruiting species.

CAUTION: Pounce 25 WP has demonstrated excellent plant safety, however, not all species and varieties have been tested. Before treating large numbers of plants of a particular variety, treat a few plants and observe prior to full-scale application. Marginal leaf burn may occur on Salvia, Dieffenbachia and Pteris Fern. Application to blooming plants may cause browning of petals.

PAPAYA (Florida only): To control Papaya Fruit Fly, Papaya Webworm, Papaya Whitefly, Brown Soft Scale, Scale Crawlers, Aphids, and Mealybug—Use Pounce 25 WP at a rate of 12.8 ounces (0.2 pound active) per acre. Apply with ground equipment in a minimum of 20 gallons of water per acre. Apply when insects first appear and repeat at 7 to 10 day intervals as needed to provide control. Do not apply more than 1.2 pounds active per acre per season. Do not apply within 7 days of harvest.

PEACHES: To control Lesser Peach Tree Borer, Peach Twig Borer, Rose Chafer, Plum Curculio, Tarnished Plant Bug, Green Fruitworm, Oriental Fruit Moth—Use POUNCE 25 WP at a rate of 3.2 to 6.4 ounces (0.05 to 0.1 pound active) per 100 gallons of water. Do not exceed 400 gallons per acre. Apply when insects appear. Make applications with ground equipment, using sufficient water to obtain uniform coverage. Do not apply within 7 days of harvest. Do not make more than 8 applications per season. Do not graze livestock in treated orchards.

PEARS (Dormant through Delayed Dormant): To control Pear Psylla—Use Pounce 25 WP at a rate of 12.8 to 25.6 ounces (0.2 to 0.4 pound active) per acre. Apply during the dormant through delayed dormant growth periods only. Apply in 3 to 20 gallons of finished spray per acre by aircraft and 25 to 400 gallons per acre by ground equipment. Do not apply more than 0.8 pound active ingredient per season. Do not graze orchard floor.

PEARS (Summer): To control Pear Psylla, Green Fruitworm, Codling Moths—Use Pounce 25 W at a rate of 12.8 to 25.6 ounces (0.2 to 0.4 pound active) per acre. Use in sufficient water to achieve runoff and apply by ground equipment. Do not apply more than 1.2 pounds active ingredient per acre per season. Do not apply within 14 days of harvest.

PEPPERS (Bell): To control Vegetable Leafminer, Cabbage Looper, Cutworms, Flea Beetle, Corn Earworm, Pepper Weevil—Use Pounce 25 WP at a rate of 6.4 to 12.8 ounces (0.1 to 0.2 pound active) per acre. To control European Corn Borer—Use Pounce 25 WP at a rate of 12.8 ounces (0.2 pound active) per acre. Do not apply more than 1.6 pounds of active ingredient per acre per season. Do not apply within 3 days of harvest.

PINE SEED ORCHARDS: To control Coneworms, Seed Bugs—Use Pounce 25 WP at the following rates:

For high volume sprayers: Use 13 ounces (0.8 pound) in 100 gallons of water (0.025% dilution by weight). Apply 5 to 10 gallons of finished spray per tree.

For low volume sprayers: Use 66 ounces (4.1 pounds) in 100 gallons of water (0.125% dilution by weight). Apply 100 gallons per acre.

For aerial application: Use 48 ounces (3 pounds) in not less than 10 gallons of water per acre.

To control Webbing Coneworm—make first application within 1 week of female flower closure or peak pollen flight.

To control other coneworms and seed bugs—make first application within 30 days following female flower closure.

Repeat applications at 4 week intervals, but do not apply more than 6 application. Do not graze or harvest cover crop. Avoid contact with open water.

PISTACHIOS: To control Peach Twig Borer and Navel Orangeworm—Use Pounce 25 WP at a rate of 12.8 to 25.6 ounces (0.2 to 0.4 pound active) per acre. Do not apply more than 0.8 pound active ingredient per acre per season. Do not apply after 10% hull split. Nuts may be harvested on the day of application. Do not graze livestock in treated orchards.

POTATOES: To control Beet Armyworm, Cabbage Looper, Colorado Potato Beetle, Cutworms, European Corn Borer, Potato Flea Beetle, Potato Leafhopper, Potato Psyllid, Potato Tuberworm, Aster Leafhopper and Tarnished Plant Bug. Use Pounce 25 WP at a rate of 6.4 to 12.8 ounces (0.1 to 0.2 pound active) per acre. Apply in a minimum of 3 gallons of finished spray per acre by air or 20 gallons of finished spray per acre using ground equipment. Do not apply more than 2.4 pounds active ingredient per acre per season. Do not apply within 7 days of harvest. Do not graze or feed potato forage.

PUMPKINS: To control Aphids, Leafminers, Squash Bug, and Whiteflies—Use Pounce 25 WP at a rate of 12.8 ounces (0.2 pound active) per acre. To control Cucumber Beetle, Cutworms, Melon Worm, Pickleworm and Squash Vine Borer—Use Pounce 25 WP at a rate of 6.4 to 12.8 ounces (0.1 to 0.2 pound active) per acre. Apply a minimum of 4 gallons of finished spray per acre by air or 20 gallons of finished spray per acre with ground equipment. Do not apply more than 0.8 pound active ingredient per acre per season. Do not apply within one day of harvest.

RANGE GRASS: To control Range Caterpillar—Use Pounce 25 WP at a rate of 0.64 ounces (0.01 pound active) per acre. Do not apply more than once per year. Cattle may be present during application. Do not harvest or feed hay to livestock.

ROSES (Field Grown): For control of Heliothis spp., use Pounce 25 WP at a rate of 6.4 to 12.8 fluid ounces (0.1 to 0.2 pound active) per acre. Apply in 5 gallons of finished spray per acre with ground equipment and 1 gallon per acre by air.

ROSES (Greenhouse): For control of Beet Armyworm, Cabbage Looper and Omnivorous Leafroller—Use Pounce 25 WP at a rate of 12.8 ounces (0.2 pound active) per 100 gallons of water.

CAUTION: Varieties may vary in their sensitivity to Pounce 25 WP, and a small number of plants should be treated under local conditions to determine plant safety prior to commercial use.

SOYBEANS: To control Cabbage Looper, Green Cloverworm, Mexican Bean Beetle, Soybean Looper, Velvetbean Caterpillar, Bean Leaf Beetle and Potato Leafhopper—Use Pounce® 25 WP at 3.2 to 6.4 ounces (0.05 to 0.1 pound active) per acre. To control Beet Armyworm and Corn Earworm—Use Pounce 25 WP at a rate of 6.4 to 12.8 oz. (0.1 to 0.2 pound active) per acre. Apply a minimum of 1 gallon of finished spray per acre by air or 5 gallons with ground equipment. Do not apply more than 0.4 pound active ingredient per acre per season. Do not apply within 60 days of harvest. Do not graze or feed soybean forage or hay.

SPINACH: For general use directions refer to the **Leafy Vegetable** crop grouping.

SWEET CORN: To control Aster Leafhopper, Corn Earworm, European Corn Borer, Fall Armyworm, Flea Beetle, Corn Rootworm Beetle*, and Cutworms—Use Pounce 25 WP at a rate of 6.4 to 12.8 ounces (0.1 to 0.2 pound active) per acre. Apply every 3 to 5 days or as needed. Do not apply more than 1.2 pounds active ingredient per acre per crop season. Do not apply within 1 day of harvest of ears or forage or livestock grazing.

*Not for control of this insect in California

SWEET CORN (Florida Only) fresh market use only: To control Aster Leafhopper, Cutworms, Corn Earworm, European Corn Borer, Fall Armyworm, Southern Armyworm—Use 6.4 to 16 ounces (0.1 to 0.25 pound active) per acre. Apply every 3 to 5 days or as needed.

Do not apply more than 3.25 pounds active ingredient per acre per crop season. Do not graze livestock in treated areas or harvest for forage.

TOMATOES (Only for Application to Tomatoes Grown in Florida for Final Marketing as Fresh Tomatoes): To control Beet Armyworm, Cabbage Looper, Colorado Potato Beetle, Granulate Cutworm, Hornworms, Vegetable Leafminer, Southern Armyworm, Tomato Fruitworm and Tomato Pinworm—Use Pounce 25 WP at a rate of 3.2 to 12.8 ounces (0.05 to 0.2 pound active) per acre. Do not apply more than 1.2 pound active ingredient per acre per season. Can be applied up to the day of harvest. Do not apply to cherry tomatoes or other varieties which produce fruit less than one inch in diameter.

WALNUTS: To control Navel Orangeworm, Codling Moth and Walnut Husk Fly—Use Pounce 25 WP at a rate of 12.8 to 25.6 ounces (0.2 to 0.4 pound active) per acre. For full coverage application apply 0.05 to 0.1 pound active per 100 gallons (based on 400 gallons finished spray per acre), spray to runoff; or, for low volume application apply 0.2 to 0.4 pound active per acre (50 to 200 gallons finished spray per acre). For aerial application apply 0.2 to 0.4 pound active in a minimum of 20 gallons of finished spray per acre. Apply when insects appear. Do not apply more than 1.6 pounds active per acre per season. Do not apply within 1 day of harvest. Do not graze livestock in treated orchards or feed cover crops from treated orchards to livestock.

PREMISES SPRAY

For agricultural use only.

Spray directly to walls and ceiling as residual surface treatment only.
Do not treat manure or litter. Avoid contamination of feed and water.
Do not apply directly to livestock or poultry.

Shake well before measuring.

For Application in	Target Insects	Method of Applic.	Dilute	Applic. Rate
Dairies, Barns, Feedlots, Stables, Poultry Houses, Swine and Livestock Houses	House Flies, Stable Flies and other Manure Breeding Flies. Also aids in the reduction of Cockroaches, Mosquitoes and Spiders.	Sprayer	6 oz.* to 11 gal. water or 8 level tablespoons to 3 gal. water	1 gal. per 750 sq. ft. of surface

* 1 oz. of this powder equals 5 level tablespoonfuls. Shake canister before measuring. Make up only as required.

Re-treat as necessary, but not more often than once every 2 weeks. The use of any residual fly spray should be supplemental with proper manure management and general sanitation to reduce or eliminate fly breeding site.

Rate Conversion Chart

Pounds Active per Acre	Formulation Ounces per Acre	Formulation Pounds per Acre
0.05	3.2	0.2
0.10	6.4	0.4
0.125	8.0	0.5
0.15	9.6	0.6
0.175	11.2	0.7
0.20	12.8	0.8

DEALERS SHOULD SELL IN ORIGINAL PACKAGES ONLY.

TERMS OF SALE OR USE: On purchase of this product buyer and user agree to the following conditions:

WARRANTY: FMC makes no warranty, expressed or implied, concerning the use of this product other than indicated on the label. Except as so warranted, the product is sold as is. Buyer and user assume all risk of use and/or handling and/or storage of this material when such use and/or handling and/or storage is contrary to label instructions.

DIRECTIONS AND RECOMMENDATIONS: Follow directions carefully. Timing and method of application, weather and crop conditions, mixture with other chemicals not specifically recommended and other influencing factors in the use of this product are beyond the control of the seller and are assumed by the buyer at his own risk.

USE OF PRODUCT: FMC's recommendations for the use of this product are based upon tests believed to be reliable. The use of this product being beyond the control of the manufacturer, no guarantee, expressed or implied, is made as to the effects of such or the results to be obtained if not used in accordance with directions or established safe practice.

DAMAGES: Buyer's or user's exclusive remedy for damages for breach of warranty or negligence shall be limited to direct damages not exceeding the purchase price paid and shall not include incidental or consequential damages.

Pounce and FMC—FMC trademarks (96-3/2/89-A) 8/89

This is a specimen label. FMC is not responsible for the accuracy of the information contained herein. As labels are subject to revision, always carefully read and follow the label on the product container.

Talstar® 10 WP

INSECTICIDE/MITICIDE

ACTIVE INGREDIENT:	BY WT.
Bifenthrin, (2 methyl[1,1'-biphenyl]-3-yl) methyl 3-(2-chloro-3,3,3-trifluoro-1-propenyl)-2,2-dimethylcyclopropanecarboxylate*	10.0%
INERT INGREDIENTS:	90.0%
TOTAL	100.0%

* Cis isomers 97% minimum, trans isomers 3% maximum.

U.S. Patent No. 4,238,505

EPA Reg. No. 279-3057 EPA Est. 279-Code 806

KEEP OUT OF REACH OF CHILDREN

WARNING

See other panels for additional precautionary information.

STATEMENT OF PRACTICAL TREATMENT

If swallowed: Get medical attention. Call a Poison Control Center or physician promptly for advice. Describe Precautionary Statements and Notes to Physician statements on the label. Do not induce vomiting unless advised by a physician or qualified medical advisor. Do not give anything by mouth to an unconscious person.

If inhaled: Remove victim to fresh air. If discomfort persists, obtain medical attention.

If on skin: Wash with plenty of soap and water. Get medical attention.

If in eyes: Flush with plenty of water. Call a physician if irritation persists.

Note to Physician: Pesticide Hotline (800) 858-7378. This product is a pyrethroid. If large amounts have been ingested, the stomach and intestine should be evacuated. Treatment is symptomatic and supportive. Digestible fats, oils, or alcohol may increase absorption and so should be avoided.

For Emergency Assistance Call 716-735-3765.

This label must be in the possession of the user at the time of application.

PRECAUTIONARY STATEMENTS

HAZARDS TO HUMANS (& DOMESTIC ANIMALS)

WARNING

May be fatal if swallowed. Harmful if inhaled or absorbed through skin. Causes moderate eye irritation. Avoid breathing vapor or spray mist. Avoid contact with skin, eyes or clothing. Remove contaminated clothing and wash before reuse. Wash thoroughly with soap and water after handling and before eating, drinking or using tobacco.

ENVIRONMENTAL HAZARDS

This pesticide is toxic to fish and aquatic invertebrates. Do not apply directly to water. Drift and runoff from treated areas may be hazardous to fish in adjacent sites. Do not contaminate water by cleaning of equipment or disposal of equipment washwaters.

DIRECTIONS FOR USE

It is a violation of Federal law to use this product in a manner inconsistent with its labeling.

Do not apply this product through any type of irrigation system.

STORAGE AND DISPOSAL

PESTICIDE STORAGE

Keep out of reach of children and animals. Store in original containers only. Store in a cool, dry place and avoid excess heat. Carefully open containers. After partial use replace lids and close tightly. Do not put concentrate or dilute material into food or drink containers. Do not contaminate other pesticides, fertilizers, water, food or feed by storage or disposal.

In case of spill, avoid contact, isolate area and keep out animals and unprotected persons. Confine spills. Call FMC collect: (716) 735-3765.

To confine spill: If liquid, dike surrounding area or absorb with sand, cat litter or commercial clay. If dry material, cover to prevent dispersal. Place damaged package in a holding container. Identify contents.

PESTICIDE DISPOSAL

Pesticide wastes are toxic. Do not contaminate water, food or feed by storage or disposal. Improper disposal of excess pesticide, spray mixture, or rinsate is a violation of Federal Law. Dispose of excess or waste pesticide by use according to label directions, or contact your State Pesticide or Environmental Control Agency, or the Hazardous Waste representative at the nearest EPA Regional Office for guidance.

CONTAINER DISPOSAL

Plastic Container: Triple rinse (or equivalent). Then offer for recycling or reconditioning, or disposed of in a sanitary landfill, or incineration, or, if allowed by State and local authorities, by burning. If burned, stay out of smoke.

TALSTAR® 10 WP Insecticide/Miticide

Greenhouse Grown
Ornamental Trees, Shrubs, Plant, and Flowers, Non-bearing Fruit and Nut Trees

DIRECTIONS FOR USE

Pest	lb ai/ 10 gal	Rate oz/ 10 gal*	scoops/ 10 gal	Comments
Aphids Armyworms Bagworm Brown Soft Scale Broad Mite California Red Scale (crawlers) Corn Earworm Cucumber Beetle Cutworms Elm Leaf Beetle Fall Webworm Flea Beetles Fungus Gnat Heliothis spp. Lacebugs Leafhoppers Leaf-feeding Caterpillars Loopers Mealybugs Oriental Fruit Moth Pine Needle Scale (crawlers) Pine Tip Moths Plant Bugs including Lygus spp.) San Jose Scale (crawlers) Whiteflies	0.004 to 0.02	0.64 to 3.2	1¾ to 9	Apply the specified rate as a full coverage foliar spray. Repeat as necessary to achieve control, using higher rates as pest pressure and foliage area increases. Do not handle the plants until the spray has dried. To control Bagworm: Apply when larvae begin to hatch. Spray larvae directly. Applications made when larvae are young will be most effective. To control scale crawlers and twig borers: Treat trunks, stems and twigs in addition to plant foliage. Certain cultivars may be sensitive to the final spray solution. A small number of plants should be treated prior to application of the entire planting. Use of an alternate class of chemistry in a treatment program is recommended to prevent or delay pest resistance.

TALSTAR® 10 WP Insecticide/Miticide

Greenhouse Grown
Ornamental Trees, Shrubs, Plant, and Flowers,
Non-bearing Fruit and Nut Trees

DIRECTIONS FOR USE

Pest	Rate lb ai/ 10 gal	Rate oz/ 10 gal*	Rate scoops/ 10 gal	Comments
Citrus Thrips Beet Armyworm European Red Mite Leafrollers Spider Mites Thrips Twig Borers	0.006 to 0.02	0.96 to 3.2	2 1/4 to 9	
Japanese Beetle Leafminers Pecan Leaf Scorch Mite	0.01 to 0.02	1.6 to 3.2	4 1/2 to 9	

* 1 Ounce of TALSTAR 10 WP by weight = 6 Tablespoons
1 pound of TALSTAR 10 WP = 5.6 cups
1 scoop (30 cc) = .36 oz TALSTAR 10 WP
1 scoop = 2 Tablespoons

Rate lb ai	TALSTAR® 10 WP/Gallons of Finished Spray					
	1	5	10	30	50	100
0.004	1t	5 1/2t	3 3/4T	10 3/4T	1C	2 1/4C
0.006	1 3/4t	2 3/4t	5 1/2T	1C	1 3/4C	3 1/2C
0.010	2 3/4t	4 1/2T	9T	1 3/4C	2 2/4C	5 2/3C
0.015	4t	6 3/4t	13 1/2T	2 1/2C	4 1/4C	8 1/2C
0.020	5 1/2t	9T	1C	3 1/2C	5 2/3C	11 1/4C

t = teaspoon
T = Tablespoon
C = Cups

DEALERS SHOULD SELL IN ORIGINAL PACKAGES ONLY.

TERMS OF SALE OR USE: On purchase of this product buyer and user agree to the following conditions:

WARRANTY: FMC makes no warranty, expressed or implied, concerning the use of this product other than indicated on the label. Except as so warranted, the product is sold as is. Buyer and user assume all risk of use and/or handling and/or storage of this material when such use and/or handling and/or storage is contrary to label instructions.

DIRECTIONS AND RECOMMENDATIONS: Follow directions carefully. Timing and method of application, weather and crop conditions, mixture with other chemicals not specifically recommended, and other influencing factors in the use of this product are beyond the control of the seller and are assumed by the buyer at his own risk.

USE OF PRODUCT: FMC's recommendations for the use of this product are based upon tests believed to be reliable. The use of this product being beyond the control of the manufacturer, no guarantee, expressed or implied, is made to the effects of such or the results to be obtained if not used in accordance with directions or established safe practice.

DAMAGES: Buyer's or user's exclusive remedy for damages for breach of warranty or negligence shall be limited to direct damages not exceeding the purchase price paid and shall not include incidental or consequential damages.

Talstar and **FMC**—FMC Trademarks
(806-10/14/88-A)

SUPPLEMENTAL LABELING

Plant Safety

Talstar® 10 WP insecticide/miticide has been tested on many species of ornamental plants. The species listed below have shown no phytotoxic response to Talstar® 10 WP; however not all cultivars have been tested. It is possible that certain cultivars may be sensitive and it is advised that a small number of plants be treated prior to application to the entire planting. Talstar® 10 WP may leave visible residues when used at high rates.

Talstar 10 WP—Plant Safety

Scientific Name	Common Name
Abies balsamea 'nana'	fir, balsam
Acalypha hispida	chenille plant
Acer griseum	maple, osakazuki
Acer palmatum	maple
Ageratum houstianum 'blue blazer'	floss flower
Anthurium	anthurium
Antirrhinum majus 'dwarf mixed'	snapdragon
Antirrhinum majus 'super tetra mix'	snapdragon
Asparagus sprengeri	asparagus sprengeri
Azalea kurume	azalea, coral
Azalea sp	azalea
Begonia semperflorens 'white'	begonia
Berberis thubergii 'atropurpureum nana'	barbarry, japanese
Berberis thubergii 'rosy glow'	barbarry, japanese
Brassaia actinphylla	schefflera
Brassaia microphylla	schefflera
Brassica oleracea	ornamental kale
Buxus japonica	boxwood, japanese
Caladium sp	caladium
Caladium sp	mother-in-law plant
Calendula officinalis	calendula
Callistephus chinensis 'crego giant mixed'	aster
Cedrus atlantica 'aurea'	cedar, atlas
Cedrus atlantica 'glauca pendula'	cedar, atlas
Cedrus atlantica 'glauca'	cedar, atlas
Cedrus deodar 'albospica'	cedar, deodar
Celosia argenta 'dwarf mixed'	celosia
Centaurea cinerarea	dusty miller
Chamaecyparis obtusa 'nana gracilis'	cypress, hinoki
Chamaecyparis obtusa 'nana lutea'	cypress, hinoki
Chamaecyparis pisifera 'filifera aurea nana'	cypress, sawarra
Chamaecyparis thyoides 'andeleyensis'	cypress, white
Chamaerops humilis	Mediterranean fan palm
Chrysanthemum morifolium 'capri'	chrysanthemum
Chrysanthemum morifolium 'charm'	chrysanthemum
Chrysanthemum morifolium 'circus'	chrysanthemum
Chrysanthemum morifolium 'echo'	chrysanthemum
Chrysanthemum morifolium 'engarde'	chrysanthemum
Chrysanthemum morifolium 'jade'	chrysanthemum
Chrysanthemum morifolium 'spirit'	chrysanthemum
Chrysanthemum morifolium 'golden times'	chrysanthemum
Chrysanthemum morifolium 'hurricane'	chrysanthemum
Chrysanthemum morifolium 'paragon'	chrysanthemum
Chrysanthemum morifolium 'pert'	chrysanthemum
Chrysanthemum morifolium 'polaris'	chrysanthemum
Chrysanthemum morifolium 'radiant'	chrysanthemum
Chrysanthemum morifolium 'songster'	chrysanthemum
Chrysanthemum morifolium 'tan'	chrysanthemum
Codiaeum variegatum 'pictum'	croton
Coleus hybridus	coleus
Cordyline terminalis 'baby doll'	cordyline
Cornus alba 'elegantissima'	dogwood, red barked
Corylusavellana sp 'contorta'	corylusavellana
Continus coggygria 'royal purple'	smoke tree
Cotoneaster apiculata	cotoneaster
Cotoneaster microphylla 'cochleata'	cotoneaster
Crossandra infundibuliformis	crossandra
Cyclamen sp	cyclamen
Cynodon dactylon	bermuda grass
Dahlia sp	dahlia
Dianthus caryophyllus 'enfant de nice mixed'	carnation
Dianthus caryophyllus 'georgia ann'	carnation
Dianthus caryophyllus 'light pink sam's pride'	carnation
Dianthus caryophyllus 'mandy'	carnation
Dianthus caryophyllus 'princess ann'	carnation
Dianthus chinesis	pinks
Dianthus plumarius	pinks
Dieffenbachia sp	dumb cane
Dorotheanthus bellidiformis	mesembryanthemum
Dracaena marginata	dracaena
Epipremnum aureum	pothos ivy

Scientific Name	Common Name
Eribotrya sp	loquat
Euonymus japonica 'Aureo variegata'	Japanese spindle tree
Euonymus japonica 'silver king'	Japanese spindle tree
Euphorbia pulcherrima 'c-1 red'	poinsettia
Euphorbia pulcherrima 'dark red annette hegg'	poinsettia
Euphorbia pulcherrima 'v-14 glory'	poinsettia
Ficus benjamina	fig
Ficus lyrata	fiddleleaf fig
Freesia sp	freesia
Fuchsia sp	fuchsia
Gardenia jasminoides 'August beauty'	gardenia
Geranium sp	geranium
Gerbera	gerber daisy
Hedera sp	hedera
Hibiscus sp	hibiscus
Hibiscus syriacus 'variegated'	hibiscus
Ilex cornuta 'burfordii'	holly, Chinese
Ilex sp	holly
Impatiens sp	impatiens
Juniperus commis 'effusa'	juniper, common
Juniperus horizontalis 'bar harbor'	juniper, creeping
Juniperus horizontalis 'blue chip'	juniper, creeping
Juniperus horizontalis 'wiltoni'	juniper, creeping
Juniperus sabina 'buffalo'	juniper, savin
Juniperus scopulorum 'table top blue'	cedar, California red
Juniperus squamata 'blue star'	juniper
Juniperus x-media sea green	juniper
Kalanchoe sp	kalanchoe
Kalmia latifolia	mountain laurel
Lantana sp 'radians'	shrub verbena
Leucophyllum frutescens	Texas sage
Lilium sp	lily
Liquidaber styraciflua	sweet gum
Liqustrum recurrifolia	privet
Liqustrum sinesis	privet
Liqustrum vicary	privet, vicary
Lithodora sp	lithodora
Lobelia 'blue spy'	lobelia
Lobelia 'crystal palace'	lobelia
Lobularia maritima 'white'	alyssum
Maranta leuconeura	prayer plant
Monstera deliciosa	monstera
Nandina domestica	nandina
Nephrolepsis exaltata	boston fern
Olea europaea	olive
Pelegonium sp	pelegonium
Petunia 'harmony boy'	petunia
Petunia 'lacy sails'	petunia
Petunia 'white cascade'	petunia
Phlox paniculata 'perennial'	phlox
Photinia faseri 'red topped'	photina
Picea abies 'little gem'	spruce, norway
Picea abies 'nidiformis'	spruce, norway
Picea abies 'pendula'	spruce, norway
Picea glauca 'conica'	spruce, white
Picea omorika 'nana'	spruce
Picea pungens 'globosa'	spurce, colorado
Picea pungens 'hoopsi'	spruce, colorado
Picea pungens 'special'	spruce, colorado
Pinus beldrichi 'leucodermis'	pine
Pinus eldarica	pine, mondell
Pinus mugo	pine, mugo
Pinus strobus	pine, white
Pinus sylverstris	pine, scotch
Pittosporum sp	pittosporum
Pittosporum tobira	pittosporum, japanese
Polystichum adiantiforme	leatherleaf fern
Populus deltoides	cottonwood
Potentilla fruticosa 'goldfinger'	shrubby cinquefoil
Potentilla fruticosa 'sunset'	shrubby cinquefoil
Primula malacoides 'pacifica mix'	primrose
Primula sp	primrose
Pseudotsugo menziesii 'fletccheri'	fir, douglas
Pteris biaurita	fern, pteris
Pyracantha coccinea 'gnome'	firethorn
Pyracantha coccinea 'lalander'	firethorn
Pyracantha coccinea 'lalandi'	firethorn
Quercus falcata	oak, southern red
Quercus nigra	oak, water
Quercus virginiana	oak, live
Raphiolepis indica 'Jack Evans'	india hawthorn
Rosa sp 'excitement'	rose
Rosa sp 'golden wave'	rose
Rosa sp 'matador'	rose
Rosa sp 'paul's pink'	rose
Rosa sp 'royalty'	rose
Rosa sp 'samantha'	rose
Rosa sp 'seventeen'	rose
Rosa sp 'sonia'	rose
Rosa sp 'sterling silver'	rose
Rosa sp 'white mystery'	rose
Saintpaulia sp	african violets
Salvia 'St. John's fire'	salvia
Salvia 'victoria'	salvia
Schulumbergera bridgesii	christmas cactus
Sciadoptys verticillata	pine, parasol
Senecio cruentus	cinararia
Sinningia speciosa	gloxinia
Spiraea sp	bridal veil
Stenotaphrum secundatum	st. augustine grass
Tagetes patula 'bonanza flame'	marigold
Tagetes patula 'first lady orange'	marigold
Tagetes patula 'orange boy'	marigold
Taxodium distichum	cypress, bald
Taxus cuspidata	yew, japanese
Thuja occidentalis 'hetz midget'	arbor vitea
Thuja occidentalis 'rheingold'	arbor vitea
Thuja occidentalis 'smaragd'	arbor vitea
Tolmiea menziesii	piggyback plant
Tradescantia sp	wandering jew
Tsuga canadensis 'gentsch white'	hemlock, canadian
Tsuga canadensis 'pendula'	hemlock, canadian
Tsuga diversifolia	hemlock
Tsuga heterophylla	hemlock, western
Tsuga mertensiana	hemlock, mountain
Verbena hortensis 'ideal bedding mix'	verbena
Verbena hortensis 'mixed'	verbena
Verbena hybrida 'blaze'	verbena
Vibernum burkwoodi	vibernum
Vibernum opulis 'sterile'	vibernum
Vinca sp 'little blanche'	vinca
Vinca sp 'periwinkle'	vinca
Viola tricolor 'mixed'	pansy
Zantedeschia elliottiana	lily, calla

This is a specimen label. FMC is not responsible for the accuracy of the information contained herein. As labels are subject to revision, always carefully read and follow the label on the product container.

Thiodan® 2 C.O. EC

INSECTICIDE

For Agricultural or Commercial Use Only

EPA Reg. No. 279–2659 EPA Est. 279–

ACTIVE INGREDIENT:	BY WT.
Endosulfan; Hexachlorohexahydromethano–2,4,3–benzodioxathiepin 3–oxide	23.1%
*Inert Ingredients:	76.9%
TOTAL	100.0%

Thiodan 2 C.O. EC contains 2 pounds Endosulfan (Thiodan) per gallon.

*Contains Xylene Range Aromatic Solvents.
Code 31994

KEEP OUT OF REACH OF CHILDREN

WARNING

AVISO

PRECAUCION AL USUARIO: Si usted no lee ingless, no use este producto hasta que la etiqueta le haya sido explicada ampliamente.

STATEMENT OF PRACTICAL TREATMENT

If swallowed: Call a physician or Poison Control Center immediately. If possible, vomiting should be induced under medical supervision. Drink one or two glasses of water and induce vomiting by touching the back of the throat with a finger. Do not induce vomiting or give anything by mouth to a person who is unconscious or convulsing.

If inhaled—Remove victim to fresh air. Apply respiration if indicated.

If on skin—Remove contaminated clothing and wash skin with soap and water.

If in eyes—Flush eyes with plenty of water. Call a physician immediately.

Note to Physician: Endosulfan is a central nervous system stimulant absorbable by mouth, inhalation or through contact with skin. It may cause convulsions. There is no specific antidote. Diazepam by injection is the drug of choice. Barbituric acid derivatives such as Phenobarbital may be used additionally. Epinephrine derivatives are absolutely contraindicated. Solvent may present aspiration hazard. Gastric lavage may be indicated if product was taken internally.

For Emergency Assistance Call 716—735–3765.

See other panels for additional precautionary information.

PRECAUTIONARY STATEMENTS

HAZARDS TO HUMANS (AND DOMESTIC ANIMALS)

Warning

May be fatal if swallowed, inhaled or absorbed through skin. Causes moderate eye irritation. Avoid contact with skin, eyes, or clothing. Do not breathe vapor. Wear waterproof rubber gloves and a pesticide mask or respirator jointly approved by the Mine Safety and Health Administration (MSHA) and the National Institute for Occupational Safety and Health (NIOSH) for pesticide application under the provision of 30 CFR Part 11 for protection against endosulfan. Do not apply or allow to drift to areas occupied by unprotected humans or beneficial animals. Wash thoroughly with soap and water after handling and before smoking or eating. Remove contaminated clothing and wash before reuse.

ENVIRONMENTAL HAZARDS

This product is toxic to fish, birds and other wildlife. Birds feeding on treated areas may be killed. Do not apply directly to water or wetlands. Due to risk of runoff and drift, do not apply within a distance of 300 feet of lakes, ponds, streams, and estuaries. Shrimp and crab may be killed at application rates recommended on this label. Do not apply where fish, shrimp, crab and other aquatic life are important resources. This product must not be used in areas where impact on threatened endangered species is likely. Contact your State Fish and Game Agency before applying this product. Apply this product only as specified on this label.

This pesticide is toxic to bees exposed to direct application. Applications should be timed to coincide with periods of minimum bee activity, usually between late evening and early morning.

PHYSICAL OR CHEMICAL HAZARDS

Do not use or store near heat or open flame.

DIRECTIONS FOR USE

It is a violation of federal law to use this product in a manner inconsistent with its labeling.

Do not apply this product through any type of irrigation system.

Do not apply this product in such a manner as to directly or through drift expose workers or other persons. The area being treated must be vacated by unprotected persons.

Do not enter treated areas without protective clothing for 24 hours.

Protective clothing means, at least, a hat or other suitable head covering, a long sleeved shirt and long legged trousers or a coverall type garment (all of closely woven fabric covering the body, including the arms and legs), shoes and socks.

Because certain states may require more restrictive reentry intervals for various crops treated with this product, such as the 48 hour reentry interval established by California, consult your State Department of Agriculture for further information.

Written or oral warnings must be given to workers who are expected to be in a treated area or in an area about to be treated with this product. Inform workers of areas or fields that may not be entered without specific protective clothing, period of time field must be vacated and appropriate actions to take in case of accidental exposure. An example of such information is given under written warnings. When oral warnings are given, warnings shall be given in a language customarily understood by workers. Oral warnings must be given if there is reason to believe that written warnings cannot be understood by workers. Written warnings must include the following information:

WARNING

Area treated with endosulfan on (insert date). Do not enter without appropriate protective clothing for 24 hours (insert State Department of Agriculture's reentry interval, if more restrictive).

In case of accidental exposure to pesticide spray, wash the skin thoroughly with soap and water. Remove contaminated clothing and wash before reuse. If in eyes, flush with plenty of water. If inhaled, go to an area where the pesticide has not been applied. Get medical attention if needed.

STORAGE AND DISPOSAL

Pesticide Storage

Do not store below 0°F, (–18°C).

Do not use or store near heat, open flame or hot surfaces.

Keep out of reach of children and animals. Store in original containers only. Store in a cool, dry place and avoid excess heat. Carefully open containers. After partial use, replace lids and close tightly. Do not put concentrate or dilute material into food or drink containers. Do not contaminate other pesticides, fertilizers, water, food, or feed by storage or disposal.

In case of spill, avoid contact, isolate area and keep out animals and unprotected persons. Confine spills. Call FMC collect: (716) 735–3765.

To confine spill: If liquid, dike surrounding area or absorb with sand, cat litter or commercial clay. If dry material, cover to prevent dispersal. Place damaged package in a holding container. Identify contents.

Pesticide Disposal

Pesticide wastes are acutely hazardous. Improper disposal of excess pesticide, spray mixture, or rinsate is a violation of Federal Law. If these wastes cannot be disposed of by use according to label instructions, contact your State Pesticide or Environmental Control Agency, or the Hazardous Waste representative at the nearest EPA Regional Office for guidance.

Container Disposal

Triple rinse (or equivalent) then offer for recycling or reconditioning, or puncture and dispose of in a sanitary landfill, or by other procedures approved by state and local authorities. Do not cut or weld metal containers.

Not For Use Or Storage In Or Around The Home

Apply the listed amount per acre when insects first appear and repeat as required, unless otherwise noted, to maintain effective control. Use in sufficient water for thorough coverage, unless otherwise noted, of listed crops. Coverage of upper and lower leaf surfaces is essential for good control. For ground application, apply recommended amount of pesticide in a minimum of 10 gallons of water per acre on vegetable or row crops. Observe use limitations. If insect control is required beyond the following use patterns, supplement the control program with other suitable pesticides.

When applying this material by aircraft, mix the recommended amount with sufficient water to provide a minimum of 1 gallon finished spray per acre on vegetable and field crops, unless otherwise noted, and a minimum of 20 gallons of finished spray per acre on fruit and nut trees and on vines. Where more than 2 quarts are recommended, mix with sufficient water to provide a minimum of finished spray equal to twice the amount of Thiodan® 2 C.O. EC insecticide used.

Do not plant root crops other than carrots, potatoes, sweet potatoes, and sugar beets as follow-up crops. Observe days interval between last application and harvest indicated by number in ( ) following the crop name.

Almonds: Peach Twig Borer—Use 4 to 5 quarts per acre for control of early season infestation. Apply during popcorn, pink or petal fall in a minimum of 200 gallons dilute or 40 gallons per acre concentrate. (For areas less than one acre use 1 quart per 100 gallons). Do not exceed one application per season. Do not graze livestock on orchard crops or grasses in treated areas.

Apples: Apple Aphid, Rosy Apple Aphid—Use 1 quart per 100 gallons of water or 4 to 5 quarts per acre. Do not make more than 3 applications during the fruiting period nor apply within 21 days of harvest; or do not apply within 30 days of harvest with no limit on the number of applications. Apple Rust Mite—Use 1½ to 2 pints per 100 gallons of water or 6 to 8 quarts per acre. Woolly Apple Aphid—Use 1 quart per 100 gallons (8 quarts per acre). If 5 quarts or less are applied per acre may be applied within 30 days of harvest with no limit on number of applications; or if no more than 3 applications are made during the fruiting period do not apply within 21 days of harvest. Do not apply more than twice during the fruiting period. Do not feed pomace to livestock. Do not apply within 30 days of harvest. Do not feed cull fruits to animals nor allow livestock to graze treated orchards.

Apricots, Nectarines: Catfacing Insects, Green Peach Aphid, Peach Twig Borer—Use 1 quart per 100 gallons of water or 4 to 5 quarts per acre. Peach Silver Mite—Use 1 pint per 100 gallons of water or 4 to 5 quarts per acre. Do not make more than 2 applications during the fruiting period. Do not apply within 30 days of harvest. Peachtree Borer—Use 1$^1/_2$ quarts per 100 gallons of water (1 quart per 100 gallons of water in Western Areas). Spray all bark areas from ground level to scaffold limbs. Do not apply within 21 days of harvest. Do not feed cull fruits to animals nor allow livestock to graze treated orchards.

Artichokes (7): Artichoke Aphid, Green Peach Aphid—Use 1$^1/_2$ quarts per acre. Artichoke Plume Moth—Use 2 quarts per acre.

Barley, Oats, Rye, Wheat: Army Cutworms—Use 1 quart per acre. Do not apply after heads begin to form. Do not feed treated forage to dairy animals or animals being finished for slaughter. For airplane application, use 2 quarts in 2 gallons diesel fuel oil per acre.

Beans, Dry and Succulent (3): Bean Leafskeletonizer, Black Bean Aphid, Cucumber Beetles, Green Stink Bug, Mexican Bean Beetle—Use 1 to 2 quarts per acre. Do not make more than 3 applications per season. Do not feed treated threshings to livestock or allow to graze in treated fields. Do not use on lima beans that are to be removed from the field for processing.

Blueberries: Blueberry Bud Mite—Use 3 quarts per acre in 300 gallons of water. Apply immediately after harvest, and repeat 6 to 8 weeks later. Do not apply after buds are well formed.

Broccoli (7), Brussels Sprouts (14), Cabbage (7), Cauliflower (14): Cabbage Aphid, Cabbage Looper, Cross-Striped Cabbageworm, Diamondback Moth Larvae, Flea Beetle, Harlequin Bug, Imported Cabbageworm, Stink Bug—Use 3 to 4 pints per acre.

Carrots (7): Green Peach Aphid—Use 1 to 2 quarts per acre. Do not make more than 2 applications. Do not use tops for food or feed.

Celery, Whole Plant; (7): Green Peach Aphid—Use 1 quart per acre. **(4):** Green Peach Aphid, Cabbage Looper—Use 1 to 2 quarts per acre. Do not make more than 3 applications.

Cherries (21): Eyespsotted Bud Moth **(Pacific Northwest Only)**—Use 1 quart per 100 gallons of water. Apply during "popcorn" stage of growth. Fruittree Leafroller **(Pacific Northwest Only)**—Use 1 quart per 100 gallons of water. Apply during pre-pink stage of growth. Black Cherry Aphid, Plum Nursery Mite—Use 1 quart per 100 gallons of water (4 to 5 quarts per acre). Mineola Moth **(Michigan Only)**—Use 2 quarts per 100 gallons. Apply in delayed dormant period. Do not feed cull fruits to animals nor allow livestock to graze treated orchards.

Do not make more than 2 applications after shuck split.

Cherry, Plum, and Peach Nursery Stock: Peachtree Borer—Use 1 gallon per 40 gallons of water. Mix thoroughly. Immerse the bundles of trees so that the roots and crowns are covered well above the grafting bud scar. If trees are not planted immediately, allow to dry before returning them to storage. Wear rubber gloves during the dipping operation.

Citrus—Non-Bearing Trees and Nursery Stock: Citrus Aphid—Use 1 pint per 100 gallons of water or 5 quarts per acre. Do not apply to bearing trees or trees that will bear fruit within 12 months. NOTE: Shrimp and crab may be killed at application rates recommended on this label. Do not apply where these are important resources.

Collards (21): Cabbage Looper, Diamondback Moth Larvae, Harlequin Bug, Imported Cabbageworm—Use 3 pints per acre. Do not make more than 1 application.

Corn—Seed Crop Only: Corn Earworm—Use 2 to 3 quarts per acre. Direct spray to area of cornsilks. Corn Leaf Aphid—Use 2 quarts per acre. Do not apply more than 5 times. Do not feed treated forage or ensilage to livestock or allow livestock to graze in treated fields. Do not use seed for food, feed, or oil purposes.

Corn—Sweet (0): For Fresh Vegetable Use Only. Corn Leaf Aphid—Use 2 to 3 quarts per acre. Corn Earworm—Use 3 quarts per acre. For Corn Earworm, apply at 2 to 3 day intervals beginning when silks first appear and continuing until they begin to dry. Do not apply more than 5 times. Do not apply to sweet corn to be processed. Do not feed treated forage or ensilage to livestock or allow livestock to graze in treated fields.

Cotton: Boll Weevil—Use 1 to 2$^1/_2$ quarts per acre. Bollworm, Cabbage Looper, Lygus Bugs, Stink Bugs, Cotton Leafperforator—Use 2 quarts per acre. Thrips—Use 3 quarts per acre. Do not apply after bolls open. Do not graze meat or dairy animals in treated fields. Do not feed treated plant parts to meat or dairy animals.

Eggplants (1): Blister Beetle, Colorado Potato Beetle, Flea Beetle, Green Peach Aphid, Green Stink Bug, Whitefly—Use 1 quart per acre. Do not exceed 0.5 lbs. active ingredient endosulfan per acre.

Grapes (7): Grape Leafhopper—Use 1 quart per 100 gallons of water or 2 to 3 quarts per acre. Rose Chafer—Use 1 quart per 100 gallons of water. Do not apply to Concord variety as severe injury is likely to occur. Ground application equipment preferred.

Kale (21): Cabbage Flea Beetle, Imported Cabbageworm, Harlequin Bug—Use 1$^1/_2$ quarts per acre. Apply only once during growing season.

Lettuce (14): Cabbage Looper, Diamondback Moth Larvae, Green Peach Aphid, Imported Cabbageworm—Use 2 quarts per acre. On Head Lettuce do not make more than 3 applications after thinning

Remove outside leaves at harvest. Do not feed crop refuse to livestock. On Leaf Lettuce do not make more than 2 applications per season.

Macadamia Nuts (0): Southern Green Stink Bug—Use 2 quarts per 100 gallons. Do not graze livestock on orchard crops or grasses in treated areas.

Melons (0), Cucumbers (0), Squash (Summer-Winter) (0): Aphids, Cucumber Beetles, Striped Cabbage Flea Beetle, Striped Flea Beetle, Melonworm, Pickleworm, Squash Bug, Squash Beetle, Rindworm (watermelon), Squash Vine Borer—Use 1 to 2 quarts per acre. For Squash Vine Borer control, apply weekly to flower buds, stems and vines beginning when moths first appear. **Melons only:** Cabbage Looper, Omnivorous Leafroller—Use 2 quarts per acre.

Mustard Greens (21), Turnip Greens (21): Aphids, Cabbage Looper, Imported Cabbageworm—Use 3 pints per acre. Do not make more than 1 application. Do not apply to turnips grown for roots.

Peaches: Black Cherry Aphid, Black Peach Aphid, Catfacing Insects, Green Peach Aphid, Peach Twig Borer, Rusty Plum Aphid—Use 1 quart per 100 gallons of water or 4 to 5 quarts per acre. Peach Silver Mite—Use 1 pint per 100 gallons of water or 4 to 5 quarts per acre. Do not make more than 2 applications during the fruiting period. Do not apply within 30 days of harvest. Peachtree Borer—Use 3 pints per 100 gallons of water (2 pints per 100 gallons of water in Western Areas). Spray all bark areas from ground level to scaffold limbs. Do not apply within 21 days of harvest. Do not feed cull fruits to animals nor allow livestock to graze treated orchards.

Pears (7): Apple Rust Mite, Consperse Stink Bug—

Ground treatment: Use 1 quart per 100 gallons of water; 200 to 400 gallons per acre applied to orchard floor and around trees prior to bloom. Do not graze livestock in treated orchards. Pear Rust Mite—Use 1 quart per 100 gallons or 4 to 5 quarts per acre. **Foliar treatment:** Use 1 quart per 100 gallons of water or 4 to 5 quarts per acre. Pearleaf Blister Mite—Use 1 to 2 pints per 100 gallons of water applied as a post-harvest or dormant treatment. Do not make more than 2 applications during the fruiting period. **(21):** Pear Psylla—Use 5 quarts in sufficient water to treat 1 acre. (Minimums of 10 gallons per acre for aerial application, 40 gallons per acre for semi-concentrate and 300 gallons per acre for dilute). Apply when adults are first observed or nymphs are small and repeat to maintain control. Do not apply more than five times during the fruiting period. If no more than two applications are made during the fruiting period, may be applied within 7 days of harvest. To control second generation Pear Psylla apply 3 pints per 100 gallons of water, not to exceed 330 gallons of spray per acre. Apply when nymphs are still small and repeat if necessary in 7 to 10 days. Do not apply more than 2 applications during the fruiting season. Note: Aerial application does not result in satisfactory control and should only be employed if impossible to apply by ground. Do not feed cull fruits to animals nor allow livestock to graze treated orchards.

Peas, Garden (Grown for Seed Only): Pea Aphid, Pea Weevil—2 to 3 pints per acre. Do not exceed 2 applications during the fruiting period. Use only on peas to be harvested by combine. Do not feed treated vines to livestock or allow livestock to graze in treated fields.

Peas, Garden (Succulent) Northwest Only (5): Pea Aphid, Pea Weevil—Use 2 quarts per acre. Do not apply more than twice. Use only on peas to be harvested by combine. Do not feed treated vines to livestock or allow livestock to graze in treated fields.

Peas, Southern Field (Succulent type, including Black-eyed Peas, Crowder Peas, and Southern Peas) (3): Cowpea Curculio—Use 1 to 2 quarts per acre. Make 3 applications at 5 day intervals starting when pods are $^1/_2$ inch long. Do not feed treated threshings to livestock or allow livestock to graze in treated fields.

Pecans: Black Pecan Aphid—Use 3 pints per 100 gallons. Do not apply after shuck split. Do not graze meat or dairy animals in treated groves. Do not graze livestock on orchard crops or grasses in treated areas.

Peppers: Green Peach Aphid, Flea Beetles, Hornworms, Pepper Maggot—Use 1 to 2 quarts per acre. May be applied within 1 day of harvest at the 1 quart per acre rate. At a rate exceeding 1 quart, do not apply within 4 days of harvest.

Pineapples (7): For fresh market only—Pineapple Fruit Mite—Use 1 gallon per acre. Apply at intervals of 7 to 10 days if necessary, particularly during the 40-day period of blooming. Do not make more than 4 applications. Do not feed treated forage or pineapple by-products to livestock.

Plums (7), Prunes (7): Fruittree Leafroller **(Pacific Northwest Only)**—Use 1 quart per 100 gallons. Apply during pre-pink stage of growth. Hop Aphid, Leaf Curl Plum Aphid, Peach Twig Borer, Plum Nursery Mite, Thistle Aphid, Thistle Plum—Use 1 quart per 100 gallons or 4 to 5 quarts per acre. For aphids, apply when eggs hatch during prebloom or petal fall. Summer applications should be made before leaves curl. Lesser Peachtree Borer—Use 3 pints per 100 gallons or 4 to 5 quarts per acre. Apply spray to thoroughly wet trunks and main branches. Peachtree Borer—Use 3 pints per 100 gallons. (2 to 3 pints in Western areas). Spray thoroughly to cover all bark areas from ground to scaffold limbs. Do not use more than 5 quarts per acre. Do not apply within 21 days of harvest. (See under **Cherries** for nursery stock dip treatment for Peachtree Borer). Do not feed cull fruits to animals nor allow livestock to graze treated orchards.

Potatoes (0): Aphids, Armyworms, Colorado Potato Beetles, Green Stink Bug, Leaffooted Bug, Plant Bugs, Potato Flea Beetle, Potato Leafhopper, Potato Tuberworm, Whitefly—Use 1 to 2 quarts per acre. European Corn Borer, Potato Psyllid—Use 3 to 4 pints per acre. False Chinch Bug—Use 2 quarts per acre. Threelined Potato Beetle—Use 1 quart per acre.

Pumpkins (1): Aphids—Use 1 quart per acre. Cucumber Beetles, Squash Vine Borer—Use 1-2 quarts per acre. For Squash Vine Borer, apply weekly to flower buds, stems, and vines beginning when moths first appear.

Safflower: Green Peach Aphid—Use 2 quarts per acre. Do not apply after flower heads open.

Spinach (21): Green Peach Aphid, Crown Mite—Use 1½ quarts per acre. Apply only once during growing season.

Strawberries (4): Cyclamen Mite—Use 1 quart per 100 gallons. Apply 800 gallons per acre on West Coast; 400 gallons per acre in other areas. Do not reapply within 35 days when fruit is present. Meadow Spittlebug, Strawberry Aphid—Use 1 quart per 100 gallons or 2 quarts per acre. Apply a minimum of 150 gallons of water per acre. Tarnished Plant Bug—Use 1 quart per 100 gallons or 2 quarts per acre. Apply a minimum of 150 gallons of water per acre. Do not reapply within 15 days or more than twice within a 35-day period when fruit is present. Garden Symphylan—**(Northwest Only—aids in reducing damage)**—Use 2 quarts per 100 gallons plant dip. Mix thoroughly. Dip entire plant. When emersing bundles of plants, make certain any trapped air is forced out to assure thorough wetting of the entire plant. Drain and allow plants to dry before setting them out in the field. Wear rubber gloves during dipping operation.

Sugar Beets (0): Green Peach Aphid—Use 1 to 2 quarts per acre. Meadow Spittlebug—Use 1 quart per acre. Sugar-beet Webworm—Use 3 to 4 pints per acre. Do not feed sugar-beet tops to livestock.

Sunflower (0): Sunflower Moth—Use 2 quarts per acre. Do not make more than 3 applications. Make first application at onset of bloom and make two more applications at 4 to 7 day intervals. Do not feed treated forage to livestock.

Sweet Potato (South-Central States Only): Banded Cucumber Beetle Larvae (aids in the control of)—Use 2 to 4 quarts per acre broadcast, or ⅔ to 1⅓ quarts per acre on a 16 inch band over the row (48 inch row spacing). Work into the soil to a depth of approximately 3 inches. Treatment should be made just prior to plant set. Do not feed treated potatoes to livestock.

Tobacco (5): Budworm, Cabbage Looper, Flea Beetle, Green Peach Aphid, Hornworm—**Seed Bed:** Use 1 quart per 100 gallons. Apply about 6 gallons per 100 square yards. **Field:** Use 1 to 2 quarts per acre. Apply in sufficient water for uniform coverage. Stink Bugs—Use 2 to 3 quarts per 100 gallons. **Plant Bed:** Green June Bug Larvae—Use 1 pint per 100 gallons, drenching at rate of 1 gallon per square yard.

Tomatoes (Field and Greenhouse) (1): Blister Beetles, Colorado Potato Beetle, Flea Beetles, Green Peach Aphid—Use 1 quart per acre. Cabbage Looper—Use 3 pints per acre. Stink Bugs—Use 3 to 4 pints per acre. Tomato Fruitworm, Tomato Russet Mite, Yellowstriped Armyworm—Use 2 quarts per acre. Tomato Hornworm, Webworm **(Calif.)**—Use 1 to 2 quarts per acre. Whitefly—Use 1 quart per 100 gallons of water using 100 to 200 gallons finished spray per acre.

Walnuts: To control Walnut Aphid—Use 3–4 quarts per acre. (Apply in a minimum of 150 gallons of water per acre.) Do not apply after husksplit. Do not graze livestock on orchard crops or grasses in treated areas.

Commercially Grown Ornamentals

Ornamental Plants

Leatherleaf Fern (Leather Holly Fern): Leatherleaf Fern Borer—Use 1 quart per 100 gallons. Begin treatment when first larval feeding is observed in the mid-vein area at the base of the leaflets. Repeat at intervals of 2 to 3 weeks as necessary to maintain control.

Ornamentals (Greenhouse and Outdoors): Aphids, Cyclamen Mite, Rose Chafer, Whitefly—Use 1 quart per 100 gallons of water. On chrysanthemums, best results will be obtained if applied before plants flower. Do not use on White Birch Trees. Do not apply to "Bonnafon Deluxe," "Fred Shoesmith" and "White Knight" chrysanthemums as injury may result.

Ornamental Trees and Shrubs

Dogwood, Lilac: Dogwood Borer, Lilac Borer—Use 1 to 2 quarts per 100 gallons. Apply in early June and repeat in 10-14 days. Drench all bark areas down to ground level.

Douglas Fir (Grown for Ornamentals) Pacific Northwest Only: Cooley Spruce Gall Aphid, Douglas Fir Needle Midge—Use 1 quart per 100 gallons or 2 teaspoonfuls per gallon. To control Gall Aphid—Apply when white cottony tufts appear. To control Needle Midge—Apply in late April or early May just before buds open.

Elm: Aphids—Use 1 quart per 100 gallons. **Elm Leaf Beetle (New England Area Only)**—Use 2 quarts per 100 gallons. Apply when about 50% of eggs have hatched. Repeat as needed.

Pines (Austrian, Jack, Red, Scotch, White): Zimmerman Pine Moth—Use 3 pints per 100 gallons. Applications should be made in mid-April and, if necessary, in late fall. Spray with enough pressure to thoroughly wet bark and main stem, especially where branches join main stem.

Spruce (New England Area Only): Spruce Gall Aphid—Use 1 quart per 100 gallons. Apply in late April or early May when aphids are present but before galls are formed.

Taxus: Taxus Bud Mite—Use 1 quart per 100 gallons. Make 3 to 5 applications as required beginning in mid-May. A thorough treatment of foliage, twigs and bark is essential. Black Vine Weevil—Use 2 quart per 100 gallons. Spray thoroughly and drench surface of soil under plants. Apply when weevils first appear, repeat in 14 days.

Dealers Should Sell in Original Packages Only.

Terms of Sale or Use: On purchase of this product buyer and user agree to the following conditions:

Warranty: FMC makes no warranty, expressed or implied, concerning the use of this product other than indicated on the label. Except as so warranted, the product is sold as is. Buyer and user assume all risk of use and/or handling and/or storage of this material when such use and/or handling and/or storage is contrary to label instructions.

Directions and Recommendations: Follow directions carefully. Timing and method of application, weather and crop conditions, mixture with other chemicals not specifically recommended and other influencing factors in the use of this product are beyond the control of the seller and are assumed by buyer at his own risk.

Use of Product: FMC's recommendations for the use of this product are based upon tests believed to be reliable. The use of this product being beyond the control of the manufacturer, no guarantee, expressed or implied, is made as to the effects of such or the results to be obtained if not used in accordance with directions or established safe practice.

Damages: Buyer's or user's exclusive remedy for damages for breach of warranty or negligence shall be limited to direct damages not exceeding the purchase price paid and shall not include incidental or consequential damages.

Thiodan is a registered trademark of Hoechst AG

FMC-FMC Trademark (31994–2/12/88–A)

This is a specimen label. FMC is not responsible for the accuracy of the information contained herein. As labels are subject to revision, always carefully read and follow the label on the product container.

Thiodan® 3 E.C.

INSECTICIDE

For Agricultural or Commercial Use Only.

ACTIVE INGREDIENTS:	BY WT.
* Endosulfan (Hexachlorohexahydromethano-2,4,3-benzodioxathiepin oxide)	33.7%
INERT INGREDIENTS:	66.3%
TOTAL	100.0%

* Thiodan

Thiodan 3 E.C. contains 3 pounds of Endosulfan per gallon.

Contains Xylene Range Aromatic Solvents.

EPA Reg. No. 279-2924 EPA Est., 279-

Code 3212

KEEP OUT OF REACH OF CHILDREN

DANGER–POISON

PELIGRO

PRECAUCION AL USUARIO: Si usted no lee ingles, no use este producto hasta que la etiqueta le haya sido explicada ampliamente.

STATEMENT OF PRACTICAL TREATMENT

If swallowed: Call a physician or Poison Control Center immediately. If possible, vomiting should be induced under medical supervision. Drink one or two glasses of water and induce vomiting by giving one (1) ounce of syrup of ipecac, if available, or by touching the back of the throat with a finger. Do not induce vomiting or give anything by mouth to a person who is unconscious or convulsing.

If inhaled: Remove victim to fresh air. Apply artificial respiration if indicated.

If on skin: Remove contaminated clothing and wash skin with soap and water.

If in eyes: Flush eyes with plenty of water. Call a physician immediately.

Note to Physician: Endosulfan is a central nervous system stimulant absorbable by mouth, inhalation or through contact with skin. It may cause convulsions. There is no specific antidote. Diazepam I.V. is the drug of choice. Barbituric acid derivatives such as Phenobarbital may be used additionally. A neuromuscular blocking agent may be used if convulsions persist. This type of drug may be used *only* if complete control of respiration can be maintained. Epinephrine derivatives are absolutely contraindicated.

For Emergency Assistance Call 716-735-3765

See other panels for additional precautionary information.

PRECAUTIONARY STATEMENTS

HAZARDS TO HUMANS (& DOMESTIC ANIMALS)

DANGER

Fatal if swallowed. May be fatal if inhaled or absorbed through skin. Do not breathe vapors or spray mist. Causes eye damage. Do not get in eyes, on skin, or on clothing.

Applicators must wear a long sleeve shirt, trousers, unlined waterproof gloves made of natural rubber or neoprene, full foot covering (such as shoes or boots), head covering (such as a hat), and a pesticide mask or respirator jointly approved by the Mine Safety and Health Administration (MSHA) and the National Institute for Occupational Safety and Health (NIOSH) under the provisions of 30 CFR Part 11 for protection against endosulfan.

Mixers and/or loaders must wear a chemical resistant apron and goggles or face shield in addition to all the protective equipment specified for applicators. Do not apply or allow to drift to areas occupied by unprotected humans or beneficial animals. Wash thoroughly with soap and water after handling and before smoking or eating. Remove contaminated clothing and wash with soap and hot water before reuse.

ENVIRONMENTAL HAZARDS

This product is toxic to fish, birds, and other wildlife. Birds feeding on treated areas may be killed. Do not apply directly to water or wetlands. Due to the risk of runoff and drift, do not apply within a distance of 300 feet of lakes, ponds, streams, and estuaries. Shrimp and crab may be killed at application rates recommended on this label. Do not apply where fish, shrimp, crab and other aquatic life are important resources. This product must not be used in areas where impact on threatened endangered species is likely. Contact your State Fish and Game Agency before applying this product. Apply this product only as specified on this label.

This pesticide is toxic to bees exposed to direct application. Applications should be timed to coincide with periods of minimum bee activity, usually between late evening and early morning.

PHYSICAL OR CHEMICAL HAZARDS

Do not use, pour, spill or store near heat or open flame.

DIRECTIONS FOR USE

It is a violation of Federal law to use this product in a manner inconsistent with its labeling.

Do not apply this product through any type of irrigation system.

Do not apply this product in such a manner as to directly or through drift expose workers or other persons. The area being treated must be vacated by unprotected persons.

Do not enter treated areas without protective clothing for 24 hours.

Protective clothing means, at least, a hat or other suitable head covering, a long sleeved shirt and long legged trousers or a coverall type garment (all of closely woven fabric covering the body, including the arms and legs), shoes and socks.

Because certain states may require more restrictive reentry intervals for various crops treated with this product, such as the 48 hour reentry interval established by California, consult your State Department of Agriculture for further information.

Written or oral warnings must be given to workers who are expected to be in a treated area or in an area about to be treated with this product. Inform workers of areas or fields that may not be entered without specific protective clothing, period of time field must be vacated and appropriate actions to take in case of accidental exposure. An example of such information is given under written warnings. When oral warnings are given, warnings shall be given in a language customarily understood by workers. Oral warnings must be given if there is reason to believe that written warnings cannot be understood by workers. Written warnings must include the following information:

DANGER

Area treated with endosulfan on (insert date). Do not enter without appropriate protective clothing for 24 hours (insert State Department of Agriculture's reentry interval, if more restrictive).

In case of accidental exposure to pesticide spray, wash the skin thoroughly with soap and water. Remove contaminated clothing and wash before reuse. If in eyes, flush with plenty of water. If inhaled, go to an area where the pesticide has not been applied. Get medical attention if needed.

STORAGE AND DISPOSAL

PESTICIDE STORAGE

Do not store in or around the home.

Do not store below 20°F, (−7°C).

Do not use or store near heat, open flame or hot surfaces.

Keep out of reach of children and animals. Store in original containers only. Store in a cool, dry place and avoid excess heat. Carefully open containers. After partial use, replace lids and close tightly. Do not put concentrate or dilute material into food or drink containers. Do not contaminate other pesticides, fertilizers, water, food, or feed by storage or disposal.

In case of spill, avoid contact, isolate area and keep out animals and unprotected persons. Confine spills. Call FMC collect: (716) 735-3765.

To confine spill: If liquid, dike surrounding area or absorb with sand, cat litter or commercial clay. If dry material, cover to prevent dispersal. Place damaged package in a holding container. Identify contents.

PESTICIDE DISPOSAL

Pesticide wastes are acutely hazardous. Improper disposal of excess pesticide, spray mixture, or rinsate is a violation of Federal Law. If these wastes cannot be disposed of by use according to label instructions, contact your State Pesticide or Environmental Control Agency, or the Hazardous Waste representative at the nearest EPA Regional Office for guidance.

CONTAINER DISPOSAL

Triple rinse (or equivalent), then offer for recycling or reconditioning, or puncture and dispose of in a sanitary landfill, or by other procedures approved by state and local authorities. Do not cut or weld metal containers.

GENERAL INSTRUCTIONS

Not For Use Or Storage In Or Around The Home.

Do not use in undiluted form.

Apply the listed amount per acre when insects first appear and repeat as required, unless otherwise noted, to maintain effective control. Use in sufficient water for thorough coverage of listed crops, unless otherwise noted. Coverage of upper and lower leaf surfaces is essential for good control. For ground application, apply recommended amount of pesticide in a minimum of 10 gallons of water per acre on vegetable or row crops. Observe use limitations. If insect control is required beyond the following use patterns, supplement the control program with other suitable pesticides.

When applying this material by aircraft, mix the recommended amount with sufficient water to provide a minimum of 1 gallon of finished spray per acre on vegetable and field crops, unless otherwise noted, and a minimum of 20 gallons of finished spray per acre on fruit and nut trees and on vines. Where more than 2 quarts of this material are recommended, mix with sufficient water to provide a mimimum of finished spray equal to twice the amount of Thiodan® 3 E.C. insecticide used.

Do not plant root crops other than carrots, potatoes, sweet potatoes, and sugar beets as follow-up crops. Observe days interval between last application and harvest indicated by number in ( ) following the crop.

ALFALFA (Grown for Forage) (21): Meadow Spittlebug—Apply ⅔ pint per acre one week after eggs begin to hatch or when first small masses of froth are produced in the crowns of the plant which is usually about the time the alfalfa is 4 to 6 inches tall. Apply in a minimum of 10 gallons of water per acre by ground or 2 gallons of water per acre by air. Do not apply more than once per cutting nor exceed a maximum of 3 applications per year. Do not graze or cut forage within 21 days of treatment.

APPLES (21): Apple Aphid, Rosy Apple Aphid, Woolly Apple Aphid, Apple Rust Mite, Green Fruitworm, Tarnished Plant Bug, first generation White Apple Leafhopper—Use ⅔ per 100 gallons of water or a maximum of 3⅓ quarts per acre. Applications made at pink and/or petal fall provide best control of Tarnished Plant Bug and Green Fruitworm. For best control of first generation White Apple Leafhopper, apply when nymphs first appear on leaves. If necessary, prior to petal fall, use 1 quart per 100 gallons to control Apple Aphids and Rosy Apple Aphids. Do not exceed 3 applications per year or a miaximum of 4 quarts total per acre per year. Do not feed pomace from treated apples to livestock. Do not feed cull fruits to animals or allow livestock to graze in treated orchards.

APPLES (30): Apple Aphid, Rosy Apple Aphid, Woolly Apple Aphid, Apple Rust Mite, Green Fruitworm. Tarnished Plant Bug, first generation White Apple Leafhopper—Use ⅔ quart per 100 gallons or a maximum of 4 quarts per acre. Do not exceed 2 applications per year or a maximum of 4 quarts per acre per year. Do not feed pomace from treated apples to livestock. Do not feed cull fruits to animals or allow livestock to graze treated orchards.

APRICOTS (21), NECTARINES (21), PEACHES (21): Peach Tree Borer, Lesser Peach Tree Borer—Use 1 quart per 100 gallons, **West Coast—**Use ⅔ to 1 quart per 100 gallons. Do not exceed 2 applications per year. **Southeastern States—**Use 2 to 3⅓ quarts per 100 gallon post-harvest, not to exceed 4 quarts per acre. Best control is obtained with a single application during the first week of September. Spray all bark areas from ground level to lower scaffold limbs. **All uses:** Do not feed cull fruits to animals or allow livestock to graze in treated orchards.

APRICOTS (30), NECTARINES (30), PEACHES (30): Catfacing insects (Stinkbug type), Green Peach Aphid, Black Cherry Aphid, Black Peach

Aphid, Rusty Plum Aphid, Peach Twig Borer, Green Fruitworm, Peach Silver Mite—Use 2/3 quart per 100 gallons or a maximum of 2 2/3 to 3 1/3 quarts per acre. **West Coast**—Use up to a maximum of 4 quarts per acre. **All Uses:** Do not exceed 2 applications per year or a maximum of 4 quarts per acre per year. Do not feed cull fruits to animals or allow livestock to graze treated orchards.

ARTICHOKES (7): Artichoke Aphid, Green Peach Aphid—Use 1 to 1 1/3 quarts per acre. Artichoke Plume Moth—Use 1 1/3 quarts per acre. Apply in a minimum of 20 gallons of water per acre by ground and in a minimum of 10 gallons of water per acre by air. Do not exceed 2 applications per year.

BARLEY, OATS, RYE, WHEAT: Army Cutworm—Use 2/3 quart per acre. For aerial application, apply in 2 gallons of diesel fuel oil per acre. Cereal Leaf Beetle (Illinois, Indiana, Michigan, and Ohio only)—Use 1/3 to 2/3 quart per acre. Apply when small larvae are readily found in the field. For aerial application use a minimum of 1 to 2 gallons of water per acre. Do not apply after heads begin to form. Do not exceed 2 applications per season. Do not feed treated forage to livestock.

BEANS (Succulent and Dry, Except Lima Beans) (3): Black Bean Aphid, Bean Leafskeletonizer, Cucumber Beetles, Green Stink Bug, Mexican Bean Beetle—Use 2/3 to 1 1/3 quarts per acre. Western Bean Cutworm—Use 1 1/3 quarts per acre. Do not exceed 3 applications per year. Do not feed threshings or allow livestock to graze in treated fields.

BLUEBERRIES: Blueberry Bud Mite—Use 2 quarts per acre in 300 gallons of water. Apply immediately after harvest and repeat 6 to 8 weeks later. Do not apply after buds are well formed. Do not exceed 2 applications per year.

BROCCOLI (7), BRUSSELS SPROUTS (14), CABBAGE (7), CAULIFLOWER (14): Cabbage Aphid, Cabbage Looper, Cross-Striped Cabbage Worm, Diamondback Moth larvae, Flea Beetle, Harlequin Bug, Imported Cabbage Worm, Stink Bugs—Use 1 to 1 1/3 quarts per acre. Armyworms, Cutworms—Use 1 1/3 quarts per acre. Do not exceed 2 applications per year.

CARROTS (7): Green Peach Aphid—Use 2/3 to 1 1/3 quarts per acre. Do not use tops for food or feed. Do not exceed 1 application per year.

CELERY (4): Green Peach Aphid, Cabbage Looper—Use 2/3 to 1 1/3 quarts per acre. Do not exceed one application per season.—OR **Celery (7):** Green Peach Aphid—Use 2/3 quart per acre. Do not exceed 2 applications per season.

CHERRIES (21): Black Cherry Aphid, Plum Rust (Nursery) Mite—Use 2/3 quart per 100 gallons; 2 2/3 to 3 1/3 quarts per acre. **Pacific Northwest Only—**Eyespotted Bud Moth, Fruit Tree Leafroller—Use 2/3 quarts per 100 gallons; 2 2/3 to 3 1/3 quarts per acre. For Bud Moth control apply at "popcorn" stage and for leafroller control apply during pre-pink stage of growth. **Michigan Only—**For the control of Mineola Moth use 1 1/3 quarts per 100 gallons in the delayed dormant period.**All uses:** Do not exceed 2 applications per year or a maximum of 4 quarts per acre per year. Do not feed cull fruits to animals or allow livestock to graze in treated orchards.

CHERRY, PEACH, PLUM NURSERY STOCK DIP: Peachtree Borer—Thoroughly mix 2 2/3 quarts per 40 gallons of water. Immerse the bundles of trees so that the roots and crowns are covered well above the grafting bud scar. Plant immediately, or dry before returning to storage. Wear rubber gloves during the dipping operation.

CITRUS, NURSERY AND NON-BEARING STOCK: Citrus Aphid—Use 2/3 quart per 100 gallons; a maximum of 3 1/3 quarts per acre. Do not apply to bearing trees or trees that will bear fruit within twelve months. Do not exceed 2 applications per year or a maximum of 4 quarts per acre per year.

COLLARDS (21): Cabbage Looper, Diamondback Moth larvae, Harlequin Bug, Imported Cabbageworm—Use 1 quart per acre. Apply only once per season.

COTTON: Thrips—Use 2 quarts per acre. Do not apply after bolls open. Do not graze meat or dairy animals in treated fields. Do not exceed 4 quarts per year.

COTTON (Eastern Use Only): Boll Weevil—Use 2/3 to 1 2/3 quarts per acre. Bollworm, Cabbage Looper, Cotton Leafperforator, Lygus Bugs, Stink Bugs—Use 1 1/3 quarts per acre. Do not apply after bolls open. Do not graze dairy or meat animals in treated fields. Do not exceed 4 quarts per year.

COTTON (Western Use Only): Boll Weevil—Use 2/3 to 2 quarts per acre. Bollworm, Cabbage Looper, Cotton Leafperforator, Lygus Bugs, Stink Bugs—Use 1 1/3 to 2 quarts per acre. Do not apply after bolls open. Do not graze dairy or meat animals in treated fields. Do not exceed 4 quarts per year.

CUCUMBERS (2), MELONS (2), SUMMER AND WINTER SQUASH (2): Aphids, Cucumber Beetles, Striped Flea Beetle, Melonworm, Pickleworm, Rindworm (on Watermelons), Squash Beetle, Squash Bug, Squash Vine Borer—Use 2/3 to 1 1/3 quarts per acre. For Vine Borer control, apply to flower buds, stems, and vines beginning when moths first appear. For Cabbage Looper and Omniverous Leafroller on Melons use 1 1/3 quarts per acre. Do not exceed 3 applications per year.

EGGPLANT (1): Colorado Potato Beetle, Flea Beetle, Green Peach Aphid, Whitefly, Blister Beetle, Green Stink Bug—Use 2/3 quarts per acre. Do not exceed two applications per year.

GRAPES (7): Grape Leafhopper, Rose Chafer, Grape Phylloxera (leaf form)—Use 2/3 quart per 100 gallons; 1 1/3 to 2 quarts per acre per year.Do not exceed 3 applications per year or a maximum of 4 quarts per acre per year. Do not apply to Concord variety as severe injury is likely to occur. Ground application is preferred.

KALE (21): Cabbage Flea Beetle, Harlequin Bug, Imported Cabbageworm—Use 1 quart per acre. Apply only once per season.

LETTUCE (14): Cabbage Looper, Green Peach Aphid, Diamondback Moth larvae, Imported Cabbageworm—Use 1 to 1 1/3 quarts per acre. On Head Lettuce, do not exceed 3 applications after thinning. Remove wrapper leaves at harvest. On Leaf Lettuce, do not exceed 2 applications per season. **All Uses:** Do not feed crop refuse to livestock. Do not exceed 4 quarts per acre per year.

MUSTARD GREENS (21): Aphids, Cabbage Looper, Imported Cabbageworm—Use a maximum of 1 quart per acre. Do not exceed one application per season.

PEARS (7): Green Fruitworm, Tarnished Plant Bug—Use 2/3 quart per 100 gallons; 2 2/3 to 3 1/3 quarts per acre. Apply at white bud or petal fall. Pear Leaf Blister Mite—Apply 1/3 to 2/3 quarts per 100 gallons of water as a post harvest or dormant treatment. Pear Psylla—Use 3 1/3 quarts per acre in a minimum of 10 gallons of water by aerial application, in 40 gallons of water for semi-concentrate, or in 300 gallons of water for dilute application. Apply when adults are first observed or nymphs are small and repeat to maintain control. Consperse Stink Bug, Pear Rust Mite (foliar treatment)—Use 2/3 quart per 100 gallons; 2 2/3 to 3 1/3 quarts per acre. Stink Bugs must be wet by spray to obtain control. Consperse Stink Bug (soil treatment)—Use 2/3 quart per 100 gallons; 200 to 400 gallons per acre applied to orchard floor around trees prior to bloom. **All Uses:** Do not exceed 2 applications per year or a maximum of 4 quarts per acre per year. Do not feed cull fruits to animals or allow livestock to graze treated orchards.

Note: Aerial application does not result in satisfactory control and should only be employed if impossible to apply by ground.

PEAS, SOUTHERN (Black-eyed Peas, Crowder Peas, Southern Peas) (3): Cowpea Curculio—Use 2/3 to 1 1/3 quarts per acre. Make 3 applications at 5-day intervals starting when pods are 1/2-inch long. Do not feed treated threshings to livestock or allow livestock to graze in treated fields. Do not exceed 3 applications per year or a maximum of 4 quarts per acre per year.

PEAS, SEED CROP ONLY (1): Pea Aphid, Pea Weevil—Use 2/3 to 1 quart per acre. Do not exceed 2 applications per year. Use only on Peas to be harvested by combine. Do not feed treated vines or threshings to livestock or allow livestock to graze in treated fields.

PECANS: Black Pecan Aphid, Pecan Nut Casebearer, Spittlebug—Use 1 quart per 100 gallons. For Casebearer, apply when eggs of first generation appear on tips of the young nuts. Another application may be required after the second generation of eggs is deposited. For Spittlebug, apply when first leaves are half grown and repeat as required. Pecan Leaf Phylloxera—Use 2/3 to 1 quart per 100 gallons when nymphs appear and before they are enclosed in plant tissue. For high populations, use the higher rate. Do not apply after shuck split. Do not exceed 2 applications per year or a maximum of 4 quarts per acre per year. Do not graze livestock on orchard crops or grasses in treated areas.

PEPPERS (1): Green Peach Aphid, Flea Beetle, Hornworms, Pepper Maggot—Use 2/3 quart per acre. **Peppers (4):** Green Peach Aphid, Flea Beetles, Hornworms, Pepper Maggot—Use 1 1/3 quarts per acre. Do not exceed 2 applications per season.

PINEAPPLE (for fresh market only) (7): Pineapple Fruit Mite—Use 2 to 2 2/3 quarts per acre. Apply at intervals of 7 to 10 days if necessary, particularly during the 40 day period of blooming. Do not exceed 2 applications per year or a maximum of 4 quarts per acre per year. Do not feed forage or pineapple by-products to livestock.

PLUMS (7), PRUNES (7): Fruittree Leafroller **(Pacific Northwest Only)—**Use 2/3 quart per 100 gallons; 2 2/3 to 3 1/3 quarts per acre. Apply during pre-pink stage of growth. Hop Aphid, Leafcurl Plum Aphid, Plum Rust (Nursery) Mite Thistle Aphid—Use 2/3 quart per 100 gallons; 2 2/3 to 3 1/3 quarts per acre. For aphids, apply when eggs hatch during pre-bloom or petal fall. Summer applications should be made before leaves curl. Lesser Peachtree Borer—Use 1 quart per 100 gallons; 2 2/3 to 3 1/3 quarts per acre. Apply spray to thoroughly wet the trunk and main branches. Peachtree Borer—Use 1 quart per 100 gallons per acre except **West Coast** use 2/3 to 1 quart per 100 gallons. Spray thoroughly to cover all bark areas from ground to scaffold limbs. Peach Twig Borer—Use 2/3 quart per 100 gallons; 2 2/3 to 3 1/3 quarts per acre. **All Uses:** Do not exceed 2 applications per year or a maximum of 4 quarts per acre per year. Do not graze livestock on orchard crops or grasses in treated areas.

POTATOES (1): Threelined Potato Beetle—Use 2/3 quart per acre. Aphids, Armyworms, Colorado Potato Beetle, Green Stink Bug, Leaffooted Bug, Potato Flea Beetles, Potato Leafhopper, Plant Bugs, Potato Tuberworm, Whitefly—Use 2/3 to 1 1/3 quarts per acre. European Corn Borer, Potato Psyllid—Use 1 to 1 1/3 quarts per acre. False Chinch Bug—Use 1 1/3 quarts per acre. Threelined Potato Beetle—Use 2/3 quarts per acre.

Do not exceed 6 applications per year or a maximum of 4 quarts per acre per year.

PUMPKINS (1): Aphids, Cucumber Beetle, Squash Vine Borer—Use 2/3 to 1 1/3 quart per acre. For Vine Borer, apply to flower buds, stems and vines beginning when moths first appear. Do not exceed 6 applications per year or a maximum of 4 quarts per acre per year.

SAFFLOWER: Green Peach Aphid—Use 1 1/3 quarts per acre. Do not apply after flower heads open. Do not exceed 2 applications per year.

SPINACH (21): Crown Mite, Green Peach Aphid—Use 1 quart per acre. Do not exceed one application per season.

STRAWBERRIES (4): Tarnished Plant Bug, Meadow Spittlebug, Strawberry Aphid—Use 1 1/3 quarts per acre. Do not reapply within 15 days or more than twice during a 35 day period when fruit is present. Cyclamen Mite—Use 2 2/3 quarts per acre in 400 gallons of water. Thoroughly wet the foliage, stem and crown of the plant. For multiple applications, do not apply at intervals less than 35 days when fruit is present. Do not exceed 3 applications per year or a maximum of 4 quarts per acre per year.

STRAWBERRIES—Northwest Use Only: Garden Symphylan (aids in reducing damage)—Use 1 1/3 quarts per 100 gallons (plant dip). Mix thoroughly. Dip entire plant. When immersing bundles of plants, make certain any trapped air is forced out to assure thorough wetting of entire plant. Drain and allow plants to dry before setting them out in the field. Wear rubber gloves during the dipping operation.

SUGAR BEETS (30): Meadow Spittlebug—Use 2/3 quart per acre. Green Peach Aphid—Use 2/3 to 1 1/3 quarts per acre. Sugar Beet Webworm—Use 1 to 1 1/3 quarts per acre. Make one or two applications as needed for control. **Western Use Only** Meadow Spittlebug—Use 2/3 to 1 1/3 quarts per acre. **All Uses:** Do not feed treated tops to livestock. Do not exceed 4 quarts per acre per year.

SUNFLOWERS (1): Sunflower Moth—Use 1 1/3 quarts per acre. Do not exceed 3 applications. Make the first application at onset of bloom. Allow 7 days between repeat applications. Do not feed treated forage to livestock.

SWEET CORN (Fresh Vegetable Use Only) (1): Corn Leaf Aphid—Use 1 1/3 quarts per acre. Corn Earworm—Use 2 quarts per acre beginning when silks first appear and continuing until they begin to dry. Allow 5 days between Thiodan* applications. Do not exceed 3 applications per year or a maximum of 4 quarts per acre per year. Do not apply to Sweet Corn to be processed. Do not feed treated forage or ensilage to livestock or allow livestock to graze in treated fields.

SWEET POTATOES (1): Sweet Potato Flea Beetle—Use 2/3 quart per acre. Begin applications shortly after transplanting or as soon as Flea Beetles appear. Do not exceed 3 applications per year.

TOBACCO (5): Budworm, Cabbage Looper, Flea Beetle, Green Peach Aphid, Hornworm—**Seed Bed:** Use 2/3 quart per 100 gallons of water and apply about 6 gallons of finished spray per 100 square yards. **Plant Bed:** Green June Bug larvae—Use 2/3 pint per 100 gallons of water drenching at a rate of 1 gallon per square yard. **Field:** Uses 2/3 to 1 1/3 quarts per acre to control above insects. Stink Bugs—Use 2/3 quart per 100 gallons of water. Do not exceed 6 applications per year or a maximum of 4 quarts per acre per year.

TOMATOES (Field and Greenhouse) (2): Blister Beetle, Colorado Potato Beetle, Flea Beetle, Green Peach Aphid, Hornworm; **California Only.** Webworm—Use 2/3 to 1 1/3 quarts per acre. Cabbage Looper, Stink Bugs—Use 2/3 to 1 1/3 quarts per acrea. Tomato Fruitworm, Tomato Russet Mite, Yellow-striped Armyworm—Use 1 1/3 quarts per acre. Whitefly—Use 2/3 quart per 100 gallons in 100 to 200 gallons of water. Do not exceed 6 applications per acre per year or a maximum of 4 quarts per acre per year.

WALNUTS: Walnut Aphid—Use 2 to 3 1/3 quarts per acre. Do not apply after husk split. Do not exceed 2 applications or 4 quarts per acre per year. Do not graze livestock on orchard crops of grasses in treated areas.

COMMERCIALLY GROWN ORNAMENTALS

ORNAMENTAL PLANTS

LEATHERLEAF FERN: Leatherleaf Fern Borer—Use 2/3 quart per 100 gallons. Begin treatment when first larval feeding is observed in the midvein area at the base of the leaflets. Repeat at intervals of 2 to 3 weeks as necessary. Do not exceed 3 applications per year.

ORNAMENTALS (Greenhouse, and Out-of-Doors): Aphids, Cyclamen Mite, Whitefly, Rose Chafer—Use 2/3 quart per 100 gallons. On chrysanthemums, best results will be obtained if applied before plants flower. Do not exceed 4 quarts per year. Do not apply to "Bonnafon Deluxe", "Fred Shoemsmith", and "White Knight" chrysanthemums as injury may result. Do not use on Birch trees. In greenhouse applications, wear a mask or respirator approved by MSA and OSHA for protection against endosulfan.

ORNAMENTAL TREES AND SHRUBS

DOGWOOD, LILAC: Dogwood Borer, Lilac Borer—Use 2/3 to 1 1/3 quarts per 100 gallons. Apply in early June and repeat in 10 to 14 days. Drench all bark areas down to the ground level. Do not exceed 4 quarts per acre per year.

DOUGLAS FIR (Grown for Ornamentals Nursery Stock or Christmas trees.) —Pacific Northwest Only: Cooley Spruce Gall Adelgid, Douglas Fir Needle Midge—Use 2/3 quart per 100 gallons (1 1/3 teaspoons per gallon). To control Gall Adelgid, apply when white, cottony tufts appear. To control Needle Midge, apply in late April or early May just before buds open. Do not exceed 4 quarts per acre per year.

PINES (Austrian, Jack, Red, Scotch, White): Zimmerman Pine Moth—Use 1 quart per 100 gallons of water. Apply in mid-April and again if necessary in late fall. Thoroughly wet bark and main stem, especially where branches join main stem. Do not exceed 4 quarts per acre per year.

SHADE TREES (except Birch), SHRUBS: Aphids—2/3 quart per 100 gallons of water. Do not exceed 4 quarts per acre per year.

SPRUCE: (New England area only): Spruce Gall Aphid—Use 2/3 quart per 100 gallons of water. Apply in late April or early May when aphids are present but before galls are formed. Do not exceed 4 quarts per acre per year.

TAXUS: Taxus Bud Mite—Use 2/3 quart per 100 gallons. Make 3 to 5 applications beginning in mid-May. Thoroughly spray foliage, twigs, and bark. Black Vine Weevil—Use 1 1/3 quarts per 100 gallons of water. Spray thoroughly and drench soil under plants. Apply when beetles first appear. Repeat in 14 days. **All Uses:** Do not exceed 4 quarts per acre per year.

DEALERS SHOULD SELL IN ORIGINAL PACKAGES ONLY.

TERMS OF SALE OR USE: On purchase of this product buyer and user agree to the following conditions:

WARRANTY: FMC makes no warranty, expressed or implied, concerning the use of this product other than indicated on the label. Except as so warranted, the product is sold as is. Buyer and user assume all risk of use and/or handling and/or storage of this material when such use and/or handling and/or storage is contrary to label instructions.

DIRECTIONS AND RECOMMENDATIONS: Follow directions carefully. Timing and method of application, weather and crop conditions, mixture with other chemicals not specifically recommended and other influencing factors in the use of this product are beyond the control of the seller and are assumed by the buyer at his own risk.

USE OF PRODUCT: FMC's recommendations for the use of this product are based upon tests believed to be reliable. The use of this product being beyond the control of the manufacturer, no guarantee, expressed or implied, is made as to the effects of such or the results to be obtained if not used in accordance with directions or established safe practice.

DAMAGES: Buyer's or user's exclusive remedy for damages for breach of warranty or negligence shall be limited to direct damages not exceeding the purchase price paid and shall not include incidental or consequential damages.

(3212-4/27/88–A)
(4/88)

Thiodan® is a registered trademark of Hoechst AG.
FMC-FMC Trademark

This is a specimen label. FMC is not responsible for the accuracy of the information contained herein. As labels are subject to revision, always carefully read and follow the label on the product container.

Thiodan® 50 WP

INSECTICIDE

For Agricultural or Commercial Use Only

ACTIVE INGREDIENT:	BY WT.
*Endosulfan: (Hexachlorohexahydromethano-2, 4,3-benzodioxathiepin 3-oxide)	50.0%
INERT INGREDIENTS:	50.0%
TOTAL	100.0%

*Thiodan

EPA Reg. No. 279-1380AA EPA Est., 279-
Code 190

KEEP OUT OF REACH OF CHILDREN

DANGER–POISON

PELIGRO

PRECAUCION AL USUARIO: Si usted no lee ingles, no use este producto hasta que la etiqueta le haya sido explicada ampliamente.

STATEMENT OF PRACTICAL TREATMENT

If swallowed: Call a physician or Poison Control Center immediately. If possible, vomiting should be induced under medical supervision. Drink one or two glasses of water and induce vomiting by giving one (1) ounce of syrup of ipecac, if available, or by touching the back of the throat with a finger. Do not induce vomiting or give anything by mouth to a person who is unconscious or convulsing.

If inhaled: Remove victim to fresh air. Apply artificial respiration if indicated.

If on skin: Remove contaminated clothing and wash skin with soap and water.

If in eyes: Flush eyes with plenty of water. Call a physician immediately.

Note to Physician: Endosulfan is a central nervous system stimulant absorbable by mouth, inhalation or through contact with skin. It may cause convulsions. There is no specific antidote. Diazepam I.V. is the drug of choice. Barbituric acid derivatives such as Phenobarbital may be used additionally. A neuromuscular blocking agent may be used if convulsions persist. This type of drug may be used *only* if complete control of respiration can be maintained. Epinephrine derivatives are absolutely contraindicated.

For Emergency Assistance Call 716-735-3765.

See other panels for additional precautionary information.

PRECAUTIONARY STATEMENTS

HAZARDS TO HUMANS (AND DOMESTIC ANIMALS)

DANGER

Fatal if swallowed or inhaled. Harmful if absorbed through skin. Causes moderate eye irritation. Avoid contact with skin, eyes, or clothing. Do not breathe vapor. Applicators must wear a long sleeve shirt, trousers, unlined waterproof gloves made of natural rubber or neoprene, full foot covering (such as shoes or boots), head covering (such as a hat), and a pesticide mask or respirator jointly approved by the Mine Safety and Health Administration (MSHA) and the National Institute for Occupational Safety and Health (NIOSH) under the provisions of 30 CFR Part 11 for protection against endosulfan. Mixers and/or loaders must wear a chemical resistant apron and goggles or face shield in addition to all of the protective equipment specified for applicators. Do not apply or allow to drift to areas occupied by unprotected humans or beneficial animals. Wash thoroughly with soap and water after handling and before smoking or eating. Remove contaiminated clothing and wash before reuse.

ENVIRONMENTAL HAZARDS

This product is toxic to fish, birds, and other wildlife. Birds feeding on treated areas may be killed. Do not apply directly to water or wetlands. Due to the risk of runoff and drift, do not apply within a distance of 300 feet of lakes, ponds, streams, and estuaries. Shrimp and crab may be killed at application rates recommended on this label. Do not apply where fish, shrimp, crab, and other aquatic life are important resources. This product must not be used in areas where impact on threatened endangered species is likely. Contact your State Fish and Game Agency before applying this product. Apply this product only as specified on this label.

This pesticide is toxic to bees exposed to direct application. Applications should be timed to coincide with periods of minimum bee activity, usually between late evening and early morning.

DIRECTIONS FOR USE

It is a violation of Federal law to use this product in a manner inconsistent with its labeling.

Do not apply this product through any type of irrigation system.

Do not apply this product in such a manner as to directly or through drift expose workers or other persons. The area being treated must be vacated by unprotected persons. When mixing, loading, and applying this product wear unlined waterproof gloves and a MSHA/OSHA approved pesticide mask or respirator.

RE-ENTRY

Do not enter treated areas without appropriate protective clothing for 24 hours.

Protective clothing means, at least, a hat or other suitable head covering, a long sleeved shirt and long legged trousers or a coverall type garment (all of closely woven fabric covering the body, including the arms and legs), shoes and socks.

Because certain states may require more restrictive reentry intervals for various crops treated with this product such as the 48 hour reentry interval established by California, consult your State Department of Agriculture for further information.

Written or oral warnings must be given to workers who are expected to be in a treated area or in an area about to be treated with this product. Inform workers of areas or fields that may not be entered without specific protective clothing, period of time field must be vacated and appropriate actions to take in case of accidental exposure. An example of such information is given under written warnings. When oral warnings are given, warnings shall be given in a language customarily understood by workers. Oral warnings must be given if there is reason to believe that written warnings cannot be understood by workers. Written warnings must include the following information:

DANGER

Area treated with endosulfan on (insert date). Do not enter without appropriate protective clothing for 24 hours (insert State Department of Agriculture's reentry interval, if more restrictive).

In case of accidental exposure to pesticide spray or dust, wash the skin thoroughly with soap and water. Remove contaminated clothing and wash before reuse. If in eyes, flush with plenty of water. If inhaled, go to an area where the pesticide has not been applied. Get medical attention if needed.

STORAGE AND DISPOSAL

PESTICIDE STORAGE

Do not store in or around the home.

Keep out of reach of children and animals. Store in original containers only. Store in a cool, dry place and avoid excess heat. Carefully open containers. After partial use, fold and roll back bags, clamp and close tightly. Do not put concentrate or dilute material into food or drink containers. Do not contaminate other pesticides, fertilizers, water, food or feed by storage or disposal.

In case of spill, avoid contact, isolate area and keep out animals and unprotected persons. Confine spills. Call FMC collect: (716) 735-3765.

To confine spill: If liquid, dike surrounding area or absorb with sand, cat litter or commercial clay. If dry material, cover to prevent dispersal. Place damaged package in a holding container. Identify contents.

PESTICIDE DISPOSAL

Pesticide wastes are acutely hazardous. Improper disposal of excess pesticide, spray mixture, or rinsate is a violation of Federal Law. If these wastes cannot be disposed of by use according to label instructions, contact your State Pesticide or Environmental Control Agency, or the Hazardous Waste representative at the nearest EPA Regional Office for guidance.

CONTAINER DISPOSAL

Completely empty bag into application equipment. Then dispose of empty bag in a sanitary landfill or by incineration, or, if allowed by State and local authorities, by burning. If burned, stay out of smoke.

GENERAL INSTRUCTIONS

Not for use or storage in or around the home.

Apply the listed amount per acre when insects first appear and repeat as required, unless otherwise noted, to maintain effective control. Use in sufficient water for thorough coverage. Coverage of upper and lower leaf surfaces is essential for good control. For ground application, apply recommended amount of pesticide in a minimum of 10 gallons of water per acre on vegetable or row crops. Observe use limitations. If insect control is required beyond the following use patterns, supplement the control program with other suitable pesticides.

When applying this material by aircraft, mix the recommended amount with sufficient water to provide a minimum of 1 gallon of finished spray per acre on vegetable and field crops, unless otherwise noted, and a minimum of 20 gallons of finished spray per acre on fruit and nut trees and on vines. When applying 4 pounds or more of Thiodan® 50 WP, mix in a minimum of 10 gallons of water.

Do not plant root crops other than carrots, potatoes, sweet potatoes, and sugar beets as follow-up crops. Observe days interval between last application and harvest indicated by number in [ ] following the crop name.

ALFALFA (Grown for Forage) (2): Meadow Spittlebug—apply 1/2 pound per acre one week after eggs begin to hatch or when first small masses of froth are produced in the crowns of the plant which is usually about the time the Alfalfa is 4 to 6 inches tall. Apply in a minimum of 10 gallons of water per acre by ground or 2 gallons of water per acre by air. Do not apply more than once per cutting nor exceed a maximum of 3 applications per year. Do not graze or cut forage within 21 days of treatment.

APPLES (21): Apple Aphid, Rosy Apple Aphid, Woolly Apple Aphid, Apple Rust Mite, Green Fruitworm, Tarnished Plant Bug, first generation White Apple Leafhopper—Use 1 pound per 100 gallons of water or a maximum of 5 pounds per acre. Applications made at pink and/or petal fall provide best control of Tarnished Plant Bug and Green Fruitworms. For best control of first generation White Apple Leafhopper, apply when nymphs first appear on leaves. If necessary, prior to petal fall, use 1 1/2 pounds per 100 gallons to control Apple Aphids and Rosy Apple Aphids. Do not exceed 3 applications per year or a maximum of 6 pounds total per acre per year. Do not feed cull fruits to animals nor allow livestock to graze in treated orchards.

APPLES (30): Apple Aphid, Rosy Apple Aphid, Woolly Apple Aphid, Apple Rust Mite, Green Fruitworm, Tarnished Plant Bug, first generation White Apple Leafhopper—Use 1 pound per 100 gallons or a maximum of 6 pounds per acre. Do not exceed 2 applications per year or a maximum of 6 pounds per acre per year. Do not feed pomace from treated apples to livestock. Do not feed cull fruits to animals nor allow livestock to graze treated orchards.

APRICOTS (21), NECTARINES (21), PEACHES (21): Peach Tree Borer, Lesser Peach Tree Borer—Use 1 1/2 pounds per 100 gallons. **West Coast**—Use 1 to 1 1/2 pounds per 100 gallons. Do not exceed 2 applications per year. **Southeastern States**—Use 3 to 5 pounds per 100 gallons post-harvest, not to exceed 6 pounds per acre. Best control is obtained with a single application during the first week of September. Spray all bark areas from ground level to lower scaffold limbs. **All Uses:** Do not feed cull fruits to animals or allow livestock to graze in treated orchards.

APRICOTS (30), NECTARINES (30), PEACHES (30): Catfacing insects (Stinkbug type), Green Peach Aphid, Black Cherry Aphid, Black Peach Aphid, Rusty Plum Aphid, Peach Twig Borer, Green Fruitworm, Peach Silver Mite—Use 1 pound per 100 gallons or a maximum of 4 to 5 pounds per acre. **West Coast**—Do not exceed 2 applications per year or a maximum of 6 pounds per acre per year. Do not feed cull fruits to animals or allow livestock to graze treated orchards.

BARLEY, OATS, RYE, WHEAT: Army Cutworm—Use 1 pound per acre. For aerial application, apply in 2 gallons of diesel fuel oil per acre. Cereal Leaf Beetle (Illinois, Indiana, Michigan, and Ohio only)—Use 1/2 to 1 pound per acre. Apply when small larvae are readily found in the field. For aerial application use a minimum of 1 to 2 gallons of water per acre. Do not apply after heads begin to form. Do not exceed 2 applications per year. Do not feed treated forage to livestock.

BEANS (Succulent and Dry), SOUTHERN FIELD PEAS (3): Black Bean Aphid, Bean Leafskeletonizer, Cucumber Beetle, Green Stink Bug, Mexican Bean Beetle, Cowpea Curculio—Use 1 to 2 pounds per acre. For control of Cowpea Curculio make 3 applications at 5-day intervals, starting when the pods are 1/2-inch long. Do not exceed 3 applications per year. Do not feed treated threshings to livestock or allow livestock to graze in treated fields. Do not use on Lima Beans that are to be removed from the field for processing.

BLUEBERRIES: Blueberry Bud Mite—Use 3 pounds per acre in 300 gallons of water. Apply immediately after harvest and repeat 6 to 8 weeks later. Do not apply after buds are well formed. Do not exceed 2 applications per year.

BROCCOLI (7), BRUSSELS SPROUTS (14), CABBAGE (7), CAULIFLOWER (14): Cabbage Aphid, Cabbage Looper, Cross-Striped Cabbage Worm, Diamond-back Moth larvae, Flea Beetle, Harlequin Bug, Imported Cabbage Worm, Stink Bugs—Use 1 1/2 to 2 pounds per acre. Armyworm, Cutworm—Use 2 pounds per acre. Do not exceed 2 applications per year.

CARROTS (7): Green Peach Aphid—Use 1 to 2 pounds per acre. Do not use tops for food or feed. Do not exceed 1 application per season.

CELERY (4): Green Peach Aphid, Cabbage Looper—Use 1 to 2 pounds per acre. Do not exceed one application per season. **-OR- Celery (7):** Green Peach Aphid—Use 1 pound per acre. Do not exceed 2 applications per season.

CHERRIES (21): Peach Tree Borer, Lesser Peach Tree Borer—Use 1 1/2 pounds per 100 gallons. **West Coast**—Use 1 to 1 1/2 pounds per 100 gallons. Thoroughly wet all bark from ground to scaffold limbs. Black Cherry Aphid, Plum Nursery Mite, Green Fruitworm—Use 1 pound per 100 gallons or a maximum of 4 to 5 pounds per acre. **Pacific Northwest Only**—Eyespotted Bud Moth, Fruit Tree Leafroller—Use 1 pound per 100 gallons at "popcorn" stage for Bud Moth control and during pre-pink stage of growth for leafroller control. **Michigan only**—For the control of Mineola Moth use 2 pounds per 100 gallons in the delayed dormant period. **All Uses:** Do not exceed 2 applications per year or a maximum of 6 pounds per acre per year. Do not feed cull fruits to animals or allow livestock to graze in treated orchards.

CHERRY, PEACH, PLUM NURSERY STOCK DIP: Peachtree Borer—Use 4 pounds per 40 gallons of water. Mix thoroughly. Wear rubber gloves during dipping operation. Immerse trees so that roots and crowns are covered well above grafting bud scar. Plant immediately or dry before returning stock to storage.

CITRUS, NURSERY AND NON-BEARING STOCK: Citrus Aphid—Use 1/2 pound per 100 gallons; a maximum of 5 pounds per acre. Do not apply to bearing trees or trees that will bear fruit within twelve months. Do not exceed 2 applications or a maximum of 6 pounds per acre per year.

COLLARDS (21): Cabbage Looper, Diamondback Moth larvae, Harlequin Bug, Imported Cabbageworm—Use 1 1/2 pounds per acre. Do not exceed one application per season.

COTTON: Thrips—Use 3 pounds per acre. Do not apply after bolls open. Do not graze meat or dairy animals in treated fields. Do not exceed 6 pounds per year.

COTTON (Eastern Use Only): Boll Weevil—Use 1 to 2 1/2 pounds per acre. Bollworm, Cabbage Looper, Cotton Leafperforator, Lygus Bugs, Stink Bugs—Use 2 pounds per acre. Do not apply after bolls open. Do not graze dairy or meat animals in treated fields. Do not exceed 6 pounds per year.

COTTON (Western Use Only): Boll Weevil—Use 1 to 3 pounds per acre. Bollworm, Cabbage Looper, Cotton Leafperforator, Lygus Bugs, Stink Bugs—Use 2 to 3 pounds per acre. Do not apply after bolls open. Do not graze dairy or meat animals in treated fields. Do not exceed 6 pounds per year.

CUCUMBERS (2), MELONS (2), SUMMER AND WINTER SQUASH (2): Aphids, Cucumber Beetles, Striped Flea Beetle, Melonworm, Pickleworm, Rindworm (on Watermelons), Squash Beetle, Squash Bug, Squash Vine Borer—Use 1 to 2 pounds per acre. For Vine Borer control, apply to flower buds, stems and vines beginning when moths first appear. For Cabbage Looper and Omniverous Leafroller on melons use 2 pounds per acre. Do not exceed 3 applications per year.

EGGPLANT (1): Colorado Potato Beetle, Flea Beetle, Green Peach Aphid, Whitefly, Blister Beetle, Green Stink Bug—Use 1 pound per acre. Do not exceed two applications per year.

FILBERTS (1): Filbert Aphid, Filbert Leafroller—Use 1 pound per 100 gallons or 3 to 4 pounds per acre. Do not exceed 2 applications per year or a maximum of 6 pounds per acre per year. Do not graze livestock on orchard crops or grasses in treated areas.

GRAPES (7): Grape Leafhopper, Rose Chafer, Grape Phylloxera (leaf form)—Use 1 pound per 100 gallons or 2 to 3 pounds per acre. Do not exceed 3 applications per year or a maximum of 6 pounds per acre per year. Do not use on Baco No. 1, Baco Noir, Concord, Seibel 7053 (Chancellor), Seibel 8357 (Colobel), Seibel 13053 (Cascade) varieties of grapes as severe injury is likely to occur.

KALE (21): Cabbage Flea Beetle, Imported Cabbageworm, Harlequin Bug—Use 1 1/2 pounds per acre. Do not exceed one application per season.

LETTUCE (14): Cabbage Looper, Green Peach Aphid, Diamondback Moth larvae, Imported Cabbageworm—Use 1 1/2 to 2 pounds per acre. On Head Lettuce, do not exceed 3 applications after thinning. Remove wrapper leaves at harvest. On Leaf Lettuce, do not exceed 2 applications per season. **All Uses**—Do not feed crop refuse to livestock. Do not exceed 6 pounds per acre per year.

MUSTARD GREENS (21): Aphids, Cabbage Looper, Imported Cabbageworm—Use 1 1/2 pounds per acre. Do not exceed one application per season.

PEARS (7): Green Fruitworm, Tarnished Plant Bug—Use 1 pound per 100 gallons; 4 to 5 pounds per acre. Apply at white bud or petal fall. Pear Leaf Blister Mite—Apply 1/2 to 1 pound per 100 gallons as a post harvest or dormant treatment. Pear Psylla—Use 5 pounds per acre in a minimum of 10 gallons of water by aerial application, in 40 gallons of water for semi-concentrate, or in 300 gallons of water for dilute application. Apply when adults are first observed or nymphs are small and repeat to maintain control. Consperse Stink Bug, Pear Rust Mite (foliar treatment)—Use 1 pound per 100 gallons; 4 to 5 pounds per acre. Stink Bugs must be wet by spray to obtain control. Consperse Stink Bug (Soil Treatment)—Use 1 pound per 100 gallons; 200 to 400 gallons per acre applied to orchard floor and around trees prior to bloom. **All Uses:** Do not exceed 2 applications per year or a maximum of 6 pounds per acre per year. Do not feed cull fruits to animals or allow livestock to graze treated orchards.

NOTE: Aerial application does not result in satisfactory control and should only be employed if impossible to apply by ground.

PEAS, SEED CROP ONLY (1): Pea Aphid, Pea Weevil—Use 1 to 1 1/2 pounds per acre. Do not exceed 2 applications per year. Use only on Peas to be harvested by combine. Do not feed treated vines or threshings to livestock or allow livestock to graze in treated fields.

PECANS: Black Pecan Aphid, Pecan Nut Casebearer, Spittlebug—Use 1 1/2 pounds per 100 gallons. For Casebearer, apply when eggs of first generation appear on tips of the young nuts. Another application may be required after the second generation of eggs is deposited. For Spittlebug, apply when first leaves are half grown and repeat as required. Pecan Leaf Phylloxera—Use 1 to 1 1/2 pounds per 100 gallons when nymphs appear and before they are enclosed in plant tissue. For high populations, use the higher rate. Do not apply after shuck split. Do not exceed 2 applications per year or a maximum of 6 pounds per acre per year. Do not graze livestock on orchard crops or grasses in treated areas.

PEPPERS (1): Green Peach Aphid, Flea Beetles, Hornworms, Pepper Maggot—Use 1 pound per acre. **Peppers (4)** Green Peach Aphid, Flea Beetles, Hornworms, Pepper Maggot—Use 2 pounds per acre. Do not exceed 2 applications per season.

PINEAPPLE (for fresh market only) (7): Pineapple Fruit Mite—Use 3 to 4 pounds per acre. Apply at intervals of 7 to 10 days if necessary, particularly during the 40 day period of blooming. Do not exceed 2 applications per year or a maximum of 6 pounds per acre per year. Do not feed treated forage or pineapple by-products to livestock.

PLUMS (7), PRUNES (7): Fruittree Leafroller **(Pacific Northwest Only)**—Use 1 pound per 100 gallons; 4 to 5 pounds per acre. Apply during pre-pink stage of growth. Hop Aphid, Leafcurl Plum Aphid, Plum Rust (Nursery) Mite, Thistle Aphid—Use 1 pound per 100 gallons; 4 to 5 pounds per acre. For Aphids, apply when eggs hatch during pre-bloom or petal fall. Summer applications should be made before leaves curl. Lesser Peachtree Borer—Use 1 1/2 pounds per 100 gallons; 4 to 5 pounds per acre. Apply spray to thoroughly wet trunk and main branches. Peachtree Borer—Use 1 1/2 pounds per 100 gallons; 4 to 5 pounds per acre except West Coast use 1 to 1 1/2 pounds per 100 gallons. Spray thoroughly to cover all bark areas from ground to scaffold limbs. Peach Twig Borer—Use 4 to 5 pounds per acre. **All Uses:** Do not exceed 2 applications per year or a maximum of 6 pounds per acre per year. Do not graze orchard crops or grasses in treated areas.

POTATOES (1): Threelined Potato Beetle—Use 1 pound per acre. Aphids, Colorado Potato Beetle, Green Stink Bug, Leaffooted Bug, Potato Flea Beetles, Potato Leafhopper, Potato Tuberworm, Southern Armyworm, Whitefly—Use 1 to 2 pounds per acre. European Corn Borer, Potato Psyllid—Use 1 1/2 to 2 pounds per acre. False Chinch Bug—Use 2 pounds per acre. Do not exceed 6 applications per year or a maximum of 6 pounds per acre per year.

PUMPKINS (1): Aphids, Cucumber Beetle, Squash Vine Borer—Use 1 to 2 pounds per acre. For Vine Borer apply to flower buds, stems and vines beginning when moths first appear. Do not exceed 6 applications per year or a maximum of 6 pounds per acre per year.

SAFFLOWER: Green Peach Aphid—Use 2 pounds per acre. Do not apply after flower heads open. Do not exceed 2 applications per year.

SPINACH (21): Green Peach Aphid, Crown Mite —Use 1½ pounds per acre. Do not exceed one application per season.

STRAWBERRIES (4): Tarnished Plant Bug, Meadow Spittlebug, Strawberry Aphid — Use 2 pounds per acre. Do not reapply within 15 days or more than twice during a 35 day period when fruit is present. Cyclamen Mite —Use 4 pounds per acre in 400 gallons of water. Thoroughly wet the foliage, stem and crown of the plant. For multiple applications, do not apply at intervals less than 35 days during the fruiting period. Do not exceed 3 applications per year or a maximum of 6 pounds per acre per year.

STRAWBERRIES (4):—Northwest Use Only — Garden Symphylan (aids in reducing damage)—Use 2 pounds per 100 gallons (plant dip). Mix thoroughly. Dip entire plant. When immersing bundles of plants, make certain any trapped air is forced out to assure thorough wetting of entire plant. Drain and allow plants to dry before setting them out in the field. Wear rubber gloves during dipping operation.

SUGAR BEETS (30): Meadow Spittlebug—Use 1 pound per acre. Green Peach Aphid—Use 1 to 2 pounds per acre. Sugar Beet Webworm—Use 1½ to 2 pounds per acre. Make 1 or 2 applications per year as needed for control. **Western Use Only**—Meadow Spittlebug—Use 1 to 2 pounds per acre. **All Uses:** Do not feed treated tops to livestock. Do not exceed 6 pounds per acre per year.

SUNFLOWERS (1): Sunflower Moth—Use 2 pounds per acre. Do not exceed three applications. Make the first application per year. Make first application at the onset of bloom. Allow 7 days between repeat applications. Do not feed treated forage to livestock.

SWEET CORN (Fresh Vegetable Use Only) (1): Corn Leaf Aphid—Use 2 pounds per acre. Corn Earworm—Use 3 pounds per acre beginning when silks first appear and continuing until they begin to dry. Allow 5 days between Thiodan applications. Do not exceed 3 applications per year or a maximum of 6 pounds per acre per year. Do not apply to Sweet Corn to be processed. Do not feed treated forage or ensilage to livestock or allow livestock to graze in treated fields.

SWEET POTATOES (1): Sweet Potatoe Flea Beetle—Use 1 pound per acre. Begin applications shortly after transplanting or as soon as Flea Beetles appear. Do not exceed 3 applications per year.

TOBACCO (5): Budworm, Cabbage Looper, Flea Beetle, Green Peach Aphid, Hornworm—SEED BED—Use 1 pound per 100 gallons of water. Apply about 6 gallons of finished spray per 100 square yards. PLANT BED—Green June Bug larvae—Use ½ pound per 100 gallons of water drenching at a rate of 1 gallon per square yard. FIELD—Use 1 to 2 pounds per acre to control above insects. Stink Bugs—Use 2 to 3 pounds per acre. Do not exceed 6 applications per year or a maximum of 6 pounds per acre per year.

TOMATOES (Field and Greenhouse) (2): Blister Beetle, Colorado Potato Beetle, Flea Beetle, Green Peach Aphid, Hornworm, **California only:** Webworm—Use 1 to 2 pounds per acre. Cabbage Looper, Stink Bugs—Use 1½ to 2 pounds per acre. Tomato Fruitworm, Yellowstriped Armyworm—Use 2 pounds per acre. Whitefly—Use 1 pound per 100 gallons; 100 to 200 gallons of water per acre. Do not exceed 6 applications per year or a maximum of 6 pounds per acre per year.

WALNUTS: Walnut Aphid—Use 3 to 4 pounds per acre. Do not apply after husk split. Do not exceed 2 applications or 6 pounds per acre per year. Do not graze livestock on orchard crops or grasses in treated areas.

COMMERCIALLY GROWN ORNAMENTALS

ORNAMENTAL PLANTS

LEATHERLEAF FERN: Leatherleaf Fern Borer—Use 1 pound per 100 gallons of water. Begin treatment when first larvae feeding is observed in the mid-vein area at the base of the leaflets. Repeat at intervals of 2 to 3 weeks as necessary. Do not exceed 3 applications per year.

ORNAMENTALS (Greenhouse and Out of Doors): Aphids, Cyclamen Mite, Whitefly, Rose Chafer—Use 1 pound per 100 gallons of water. On Chrysanthemums, best results will be obtained if applied before plants flower. Do not exceed a maximum of 6 pounds per acre per year. Do not apply to Birch trees or to "Bonnafon Deluxe", "Fred Shoesmith", and "White Knight" chrysanthemums as injury may result. In greenhouse applications, wear a mask or respirator approved by MSA and OSHA for protections against endosulfan.

ORNAMENTAL TREES AND SHRUBS

DOGWOOD, Lilac: Dogwood Borer, Lilac Borer—Use 1 to 2 pounds per 100 gallons of water. Apply in early June and repeat in 10 to 14 days. Drench all bark areas down to the ground level. Do not exceed 6 pounds per acre per year.

DOUGLAS FIR (Grown for Ornamentals, Nursery Stock or Christmas trees) Pacific Northwest Only: Cooley Spruce Gall Adelgid, Douglas Fir Needle Midge, Aphids—Use 1 pound per 100 gallons of water (1 tablespoon per gallon). For Gall Adelgid control apply when white cotton tufts appear. For Needle Midge control apply in late April or early May just before buds open. Do not exceed 6 pounds per acre per year.

PINES: (Austrian, Jack, Red, Scotch, White): Zimmerman Pine Moth—Use 1½ pounds per 100 gallons of water. Apply in mid-April and again, if necessary, in late fall. Thoroughly wet bark and main stem, especially where branches join main stem. Do not exceed 6 pounds per acre per year.

SHADE TREES (except Birch), SHRUBS: Aphids—Use 1 pound per 100 gallons of water. Do not exceed 6 pounds per acre per year.

SPRUCE: (New England area only) Spruce Gall Aphid—Use 1 pound per 100 gallons of water. Apply in late April or early May when aphids are present but before galls are formed. Do not exceed 6 pounds per acre per year.

TAXUS: Taxus Bud Mite—Use 1 pound per 100 gallons of water. Apply 3 to 5 applications beginning in mid-May. Thoroughly spray foliage, twigs and bark. Black Vine Weevil—Use 2 pounds per 100 gallons of water. Spray thoroughly and drench soil under plants. Apply when weevils first appear. Repeat in 14 days. **All Uses:** Do not exceed 6 pounds per acre per year.

DEALERS SHOULD SELL IN ORIGINAL PACKAGES ONLY.

TERMS OF SALE OR USE: On purchase of this product buyer and user agree to the following conditions:

WARRANTY: FMC makes no warranty, expressed or implied, concerning the use of this product other than indicated on the label. Except as so warranted, the product is sold as is. Buyer and user assume all risk of use and/or handling and/or storage of this material when such use and/or handling and/or storage is contrary to label instructions.

DIRECTIONS AND RECOMMENDATIONS: Follow directions carefully. Timing and method of application, weather and crop conditions, mixture with other chemicals not specifically recommended and other influencing factors in the use of this product are beyond the control of the seller and are assumed by buyer at his own risk.

USE OF PRODUCT: FMC's recommendations for the use of this product are based upon tests believed to be reliable. The use of this product being beyond the control of the manufacturer, no guarantee, expressed or implied, is made as to the effects of such or the results to be obtained if not used in accordance with directions or established safe practice.

DAMAGES: Buyer's or user's exclusive remedy for damages for breach of warranty or negligence shall be limited to direct damages not exceeding the purchase price paid and shall not include incidental or consequential damages.

Thiodan® is a registered trademark of Hoechst Ag.

(190-4/27/88-A) 4/88

FMC is a registered trademark of FMC Corporation.

This is a specimen label. FMC is not responsible for the accuracy of the information contained herein. As labels are subject to revision, always carefully read and follow the label on the product container.

Wettable Sulfur

FUNGICIDE-INSECTICIDE

For Agricultural or Commercial Use Only

ACTIVE INGREDIENT:	BY WT.
Sulfur	92.0%
INERT INGREDIENTS:	8.0%
TOTAL	100.0%

The material in this bag is guaranteed to pass 95% through a 325 mesh screen.

EPA Reg. No. 279-387AA EPA Est., 279-Code 30042

CAUTION

KEEP OUT OF REACH OF CHILDREN

PRECAUTIONARY STATEMENTS

HAZARDS TO HUMANS (AND DOMESTIC ANIMALS)

Caution

Harmful if swallowed, inhaled or absorbed through skin. Avoid breathing dust or spray mist. Avoid contact with skin, eyes or clothing. Wash thoroughly with soap and water after handling. Remove contaminated clothing and wash before reuse.

STATEMENT OF PRACTICAL TREATMENT

If swallowed: Drink one or two glasses of water and induce vomiting by touching back of throat with finger. Do not induce vomiting or give anything to an unconscious person.

If in eyes: Flush with plenty of water. Get medical attention if irritation persists.

If on skin: Wash with plenty of soap and water. Get medical attention.

See other panels for additional precautionary information.

For Emergency Assistance Call (716) 735-3765

ENVIRONMENTAL HAZARDS

Do not apply directly to water. Do not apply where runoff is likely to occur. Do not apply when weather conditions favor drift from areas treated. Do not contaminate water by cleaning of equipment or disposal of wastes. Apply this product only as specified on this label.

PHYSICAL OR CHEMICAL HAZARDS

Keep away from heat, sparks or flame. Do not smoke while applying this product.

DIRECTIONS FOR USE

It is a violation of Federal Law to use this product in a manner inconsistent with its labeling.

Do not apply this product through any type of irrigation system.

Do not apply this product in such a manner as to directly or through drift expose workers or other persons. The area being treated must be vacated by unprotected persons.

Do not enter treated areas for 24 hours unless appropriate protective clothing is worn.

Protective clothing means, at least, a hat or other suitable head covering, a long sleeved shirt and long legged trousers or a coverall type garment (all of closely woven fabric covering the body, including the arms and legs), shoes and socks.

Because certain states may require more restrictive reentry intervals for various crops treated with this product, consult your State Department of Agriculture for further information.

Written or oral warnings must be given to workers who are expected to be in a treated area or in an area about to be treated with this product. Inform workers of areas or fields that may not be entered without specific protective clothing, period of time field must be vacated and appropriate actions to take in case of accidental exposure. An example of such information is given under written warnings. When oral warnings are given, warnings shall be given in a language customarily understood by workers. Oral warnings must be given if there is reason to believe that written warnings cannot be understood by workers. Written warnings must include the following information:

CAUTION

Area treated with Wettable Sulfur on (insert date). Do not enter treated areas for 24 hours unless appropriate protective clothing is worn (insert State Department of Agriculture's re-entry interval, if more restrictive).

In case of accidental exposure to pesticide spray or dust, wash the skin thoroughly with soap and water. Remove contaminated clothing and wash before reuse. If in eyes, flush with plenty of water. If inhaled, go to an area where the pesticide has not been applied. Get medical attention if needed.

STORAGE AND DISPOSAL

PESTICIDE STORAGE

Keep away from fire and sparks.

Keep out of reach of children and animals. Store in original containers only. Store in a cool, dry place and avoid excess heat. Carefully open containers. After partial use, fold and roll back bags, clamp and close tightly. Do not put concentrate or dilute material into food or drink containers. Do not contaminate other pesticides, fertilizers, water, food or feed by storage or disposal.

In case of spill, avoid contact, isolate area and keep out animals and unprotected persons. Confine spills. Call FMC collect: (716) 735-3765.

To confine spill: If liquid, dike surrounding area or absorb with sand, cat litter or commercial clay. If dry material, cover to prevent dispersal. Place damaged package in a holding container. Identify contents.

PESTICIDE DISPOSAL

Wastes resulting from the use of this product may be disposed of on site or at an approved waste disposal facility.

CONTAINER DISPOSAL

Completely empty bag into application equipment. Then dispose of empty bag in a sanitary landfill or by incineration, or, if allowed by State and local authorities, by burning. If burned, stay out of smoke.

Applications of material containing sulfur should not be made prior to or during periods of excessively high summer temperatures or within 2 weeks of an oil spray.

To mix Wettable Sulfur, fill the spray tank with water until the agitator blades are covered. With the agitator running, sift in the required amount of Wettable Sulfur. Continue filling and keep agitator running during spraying.

ALMONDS: Brown Rot, Scab—Use 6 pounds per 100 gallons of water in pre-bloom, bloom and post-bloom periods. Apply when disease threatens during rain periods and repeat at 7 to 10 day intervals.

APPLES AND PEARS: For Scab or Powdery Mildew control, use as a spray in pre-blossom. Calyx and Early cover sprays. Use 6 to 8 pounds of Wettable Sulfur or 4 to 6 pounds of Wettable Sulfur plus 2 to 3 pounds Polysulphide Compound or 2 gallons of Liquid Lime Sulfur Solution per 100 gallons of dilute spray. Sulfur may injure certain varieties of Apples. Check specific varietal tolerance to sulfur prior to using. **Pears only:** Do not apply after fruit begins to turn down. **Fall Clean-up Spray:** Blister Mites, European Red Mites and Twospotted Spider Mites—Use 6 pounds of Wettable Sulfur plus 6 pounds of Polysulphide Compound or 5 gallons of Liquid Lime Sulfur Solution to each 100 gallons of dilute spray.

BLACKBERRIES, BOYSENBERRIES, GOOSEBERRIES, LOGANBERRIES, RASPBERRIES, STRAWBERRIES: To control Powdery Mildew—Use 2 pounds Wettable Sulfur per 100 gallons of water. Apply when disease first appears and repeat at 7 to 10 day intervals or as necessary to maintain control.

CITRUS: Citrus Flat Mite, Yuma Spider Mite, Sixspotted Mite and Citrus Rust Mite—Use 5 to 12 pounds per 100 gallons of water (75 to 125 pounds per acre). Apply when mites appear and repeat as necessary. Do not make an oil application within 60 days prior to or after a sulfur application.

GRAPES: Powdery Mildew—Use 2 to 5 pounds per 100 gallons of dilute spray. Use as a spray beginning when shoots are 6 to 8 inches long and repeat at 7 to 10 day intervals. Ground application preferred. Concord and other Labrusa type grapes may be injured by sulfur.

PISTACHIOS: Citrus Flat Mite—Use 50 pounds per acre. Apply when mites first appear and repeat as necessary. May be applied by ground or air.

PLUMS, PRUNES, PEACHES, NECTARINES, CHERRIES: Powdery Mildew, Silver Mite, Cherry Leaf Spot, Prune Rust and Brown Rot—Use 5 to 10 pounds per 100 gallons of water. (By ground apply a minimum of 200 gallons of water per acre.) Apply in the pre-bloom, bloom and post-bloom periods. Apply when disease threatens during rain periods and repeat at 7 to 10 day intervals.

TOMATOES: Tomato Russet Mite—Use 12 to 25 pounds per acre. Apply with ground equipment at the rate of 60 to 250 gallons of water per acre. Use only application equipment which has paddle type agitation. Continuous and complete agitation is necessary.

DEALERS SHOULD SELL IN ORIGINAL PACKAGES ONLY.

TERMS OF SALE OR USE: On purchase of this product buyer and user agree to the following conditions:

WARRANTY: FMC makes no warranty, expressed or implied, concerning the use of this product other than indicated on the label. Except as so warranted, the product is sold as is. Buyer and user assume all risk of use and/or handling and/or storage of this material when such use and/or handling and/or storage is contrary to label instructions.

DIRECTIONS AND RECOMMENDATIONS: Follow directions carefully. Timing and method of application, weather and crop conditions, mixture with other chemicals not specifically recommended and other influencing factors in the use of this product are beyond the control of the seller and are assumed by the buyer at his own risk.

USE OF PRODUCT: FMC's recommendations for the use of this product are based upon tests believed to be reliable. The use of this product being beyond the control of the manufacturer, no guarantee, expressed or implied, is made as to the effects of such or the results to be obtained if not used in accordance with directions or established safe practice.

DAMAGES: Buyer's or user's exclusive remedy for damages for breach of warranty or negligence shall be limited to direct damages not exceeding the purchase price paid and shall not include incidental or consequential damages.

FMC—FMC Trademark

(30042-4/87)

This is a specimen label. FMC is not responsible for the accuracy of the information contained herein. As labels are subject to revision, always carefully read and follow the label on the product container.

Ziram 76 WDG

FUNGICIDE

For Agricultural or Commercial Use Only

ACTIVE INGREDIENT:	BY WT.
Ziram (zinc dimethyl dithiocarbamate)	76.0%
INERT INGREDIENTS:	24.0%
TOTAL	100%
Zinc expressed as metalic	16.25%

EPA Reg. No. 45728-12-279 EPA Est.
Code 628

KEEP OUT OF REACH OF CHILDREN

CAUTION

PRECAUCION AL USUARIO: Si usted no lee ingles, no use este producto hasta que la etiqueta le haya sido explicado ampliamente.

PRECAUTIONARY STATEMENTS

HAZARDS TO HUMANS (AND DOMESTIC ANIMALS)

Caution

May be harmful if swallowed or inhaled. May cause irritation of eyes, nose, throat and skin. Avoid contact with eyes, skin or clothing. Avoid breathing dust or spray mist. In case of contact, flush with plenty of water, for eyes, get medical attention.

For Emergency Assistance Call 716–735-3765.

See Side Panel For Additional Precautionary Statements.

ENVIRONMENTAL HAZARDS

This product is toxic to fish and wildlife. Do not apply directly to water. Do not apply where runoff is likely to occur. Do not apply when weather conditions favor drift from areas treated. Do not contaminate water when disposing of equipment washwaters. Apply this product only as specified on this label.

PHYSICAL/CHEMICAL HAZARDS

Keep away from fire and sparks.

DIRECTIONS FOR USE

It is a violation of Federal Law to use this product in a manner inconsistent with its labeling.

REENTRY AND WORKER PROTECTION STATEMENT

Do not apply this product in such a manner as to directly or through drift expose workers or other persons. The area being treated must be vacated by unprotected persons.

Do not enter treated areas without protective clothing until sprays have dried.

Protective clothing means, at least, a hat or other suitable head covering, a long sleeved shirt and long legged trousers or a coverall type garment (all of closely woven fabric covering the body, including the arms and legs), shoes and socks. Wash hands and exposed skin thoroughly after handling and before eating, drinking or smoking. Remove any contaminated clothing and launder before re-use.

Because certain states may require more restrictive re-entry intervals for various crops treated with this product, consult your State Department of Agriculture for further information.

Written or oral warnings must be given to workers who are expected to be in a treated area or in an area about to be treated with this product. Inform workers of areas or fields that may not be entered without specific protective clothing, period of time field must be vacated and appropriate actions to take in case of accidental exposure. An example of such information is given under written warnings. When oral warnings are given, warnings shall be given in a language customarily understood by workers. Oral warnings must be given if there is reason to believe that written warnings cannot be understood by workers. Written warnings must include the following information:

CAUTION

Area treated with Ziram 76 WDG on (insert date). Do not enter without appropriate protective clothing until sprays have dried (insert State Department of Agriculture's reentry interval, if more restrictive).

In case of accidental exposure to pesticide spray or dust, wash the skin thoroughly with soap and water. Remove contaminated clothing and wash before reuse. If in eyes, flush with plenty of water. If inhaled, go to an area where the pesticide has not been applied. Get medical attention if needed.

STORAGE AND DISPOSAL

PESTICIDE STORAGE

Keep away from fire and sparks.

Keep out of reach of children and animals. Store in original containers only. Store in a cool, dry place and avoid excess heat. Carefully open containers. After partial use, fold and roll back bags, clamp and close tightly. Do not put concentrate or dilute material into food or drink containers.

Do not contaminate other pesticides, fertilizers, water, food or feed by storage or disposal.

In case of spill, avoid contact, isolate area and keep out animals and unprotected persons. Confine spills. Call FMC collect: (716) 735-3765.

To confine spill: If liquid, dike surrounding area or absorb with sand, cat litter or commercial clay. If dry material, cover to prevent dispersal. Place damaged package in a holding container. Identify contents.

PESTICIDE DISPOSAL

Wastes resulting from the use of this product may be disposed of on site or at an approved waste disposal facility.

CONTAINER DISPOSAL

Completely empty bag into application equipment. Then dispose of empty bag in a sanitary landfill or by incineration, or, if allowed by State and local authorities, by burning. If burned, stay out of smoke.

GENERAL USE INFORMATION

ZIRAM 76 WDG is broad spectrum fungicide for the control of certain diseases in fruit and nut crops, ornamentals, and pine seedings.

FAILURE TO FOLLOW DIRECTIONS AND PRECAUTIONS ON THIS LABEL MAY RESULT IN CROP INJURY, POOR DISEASE CONTROL, OR ILLEGAL RESIDUES.

Application Equipment and Rates

Use the recommended rates for concentrate, dilute and aerial applications. Ground application is preferred, especially when full canopy of foliage is developed.

Be sure spray equipment is clean before use; if it is contaminated with any material, crop injury or sprayer clogging may result. Calibrate spray equipment before use.

For deciduous fruits and almonds, apply the recommended rate in a minimum of 50 gallons per acre by ground. By air, use 20 gallons per acre, or sufficient, water to provide complete coverage.

Under low disease conditions, minimum label rates per application can be used while maximum label rates and shortened spray schedules are recommended for severe or threatening disease conditons.

Do not apply this product through any type of irrigation system.

MIXING INSTRUCTIONS

Wear rubber gloves, long sleeves, and eye protection when mixing and loading this product into spray equipment.

To mix, fill the spray tank about 2/3 full of water, and with agitator running, slowly sift in the required amount of ZIRAM 76 WDG. Finish filling the tank with water and let mixing continue in the tank until thoroughly mixed before applying. Continue agitation and apply immediately.

IMPORTANT: When spraying, thorough coverage is necessary to provide good disease control. Make new dilutions for each spraying. Consult State Agricultural Experiment Stations recommendations or State Extension Service Specialist for spray schedules for timing and frequency of applications in your area.

RECOMMENDATIONS

ALMONDS: Brown Rot, Shot-Hole (Coryneum Blight), Leaf Blight—Use 8 pounds per acre in pre-bloom, blossom and petal fall periods. Scab—Use 8 pounds per acre at petal fall and 5 weeks after petal fall. Do not use later than 5 weeks after petal fall.

APPLES: Scab—Use 6 to 12 pounds per acre in pre-blossom, blossom and early cover stages. Bull's Eye Rot (Oregon and Washington)—Use 6 to 12 pounds per acre beginning 12 to 14 days after calyx stage and repeat as necessary in summer sprays. Perennial Canker (Oregon)—Use 6 to 12 pounds per acre in pre-blossom and calyx sprays.

APRICOTS: Brown Rot, Blossom and Twig Blight, Shot-Hole (Coryneum Blight)—Use 8 to 10 pounds per acre in pre-bloom, blossom and petal fall through early cover sprays.

CHERRIES: Brown Rot, Blossom Blight—Use 4 to 5 pounds per acre in popcorn, full blossom, petal fall and shuck stages and approximately 2 weeks after shuck fall. Do not apply within 7 days of harvest.

CONIFERS: Fusiform Rust in pine seedlings—Use 2 pounds per 100 gallons of water at 3 to 6 day intervals from plant emergence to mid-June.

Melampsora Needle Rust in Douglas Fir Christmas Trees—Use 2 pounds per 100 gallons of water. Make a single application in the spring when new shoot growth is 1/2 to 2 inches in length.

Puccineastrum Needle Rust in Shasta Fir Christmas Trees—Use 2 pounds per 100 gallons of water. Make a single application in the spring when new shoot growth is 1/2 to 2 inches in length.

NECTARINES, PEACHES: Peach Blight (Coryneum) and Leaf Curl—Use 8 to 12 pounds per acre in the fall before heavy rains. For Leaf Curl only, use 8 to 12 pounds per acre as a dormant application prior to bud swell. Brown Rot and Peach Blight (Coryneum)—Use 8 to 10 pounds per acre in pre-bloom, blossom and petal fall sprays. May be used in early cover and before harvest. Remove excess residue by washing or wiping.

ORNAMENTALS: Directions, unless otherwise specified, include: Use the specified rate per 100 gallons of water applying at 7 to 10 day intervals during growing season to control the disease on the ornamentals listed below.

Asters: Rust—Use 1 pound in 12 gallons. Begin application in early season and continue at 7 day intervals until blossoms form.

Azaleas, Camellias: Petal Blight—Use 1½ to 2 pounds. Begin application with first bloom. Repeat at 3 to 5 day intervals through bloom. Apply as a drench to mulch and litter beneath and around bushes.

Carnations, Chrysanthemums, Marigolds, Morning Glory, Snapdragons: Rust—1⅔ to 2 pounds.

Carnations, Chrysanthemums, Dahlias, Gladiolus, Snapdragons: Leaf Spot—Use 1⅔ to 2 pounds, apply at weekly intervals.

Cyclamens, Gardenias: Botrytis, Leaf Spot—Use 1½ to 2 pounds.

Dahlias, Delphinium, Golden Glow, Hollyhocks, Phlox: Powder Mildew—Use 1⅔ pounds apply at weekly intervals.

Lilies, Peonies, Poinsettias, Tulips: Botrytis—Use 1½ to 2 pounds.

Roses: Black Spot, Powdery Mildew—Use 1⅔ pounds. Begin application when leaves unfold. Repeat at 7 to 10 day intervals throughout season.

ORNAMENTAL FRUIT VARIETIES: Use 2 to 3 pounds per 100 gallons of water.

NOTE: DO NOT USE FRUIT FOR HUMAN CONSUMPTION.

Apricots, Sweet Cherries, Wild Cherries, Choke Cherries, Nectarines, Peaches, Flowering Plums (Non-edible only), European Laurels, Cherry Laurels, Poplars, Maples, Oak, Birch, Alder, Elm, American Hornbeam, California Buckeys: Shoot and Twig Blight, Shot Hole Fungus, Leaf Curl, Cherry Witches Broom, Plum Pockets, Leaf Blister, Calkin Hypertrophy of Alder—Apply as a thorough coverage spray in the Dormant period prior to bud swell.

Apples, Flowering Crab Apples, Mountain Ash, Oregon Crab, Pear, Quince, and Service Berry: Anthracnose—Apply as a thorough coverage spray in the fall and dormant periods.

Apples, Peach, Pear, Poplar and Willow: Europeran Canker—Apply as a thorough coverage spray int he fall and dormant periods.

PEARS: Scab—Use 6 to 12 pounds per acre in pre-blossom, blossom and early cover sprays. Storage Rots (Oregon)—Use 6 to 12 pounds per acre in pre-harvest spray.

PECANS: (Except California) Scab—Use 8 to 10 pounds per acre applying four times at 3 to 4 week intervals starting when tips of small nuts become brown. Under severe disease conditions, a fifth application in August may be required. Do not apply after August.

DEALERS SHOULD SELL IN ORIGINAL PACKAGES ONLY

TERMS OF SALE OR USE: On purchase of this product buyer and user agree to the following conditions:

WARRANTY: FMC makes no warranty, expressed or implied, concerning the use of this product other than indicated on the label. Except as so warranted, the product is sold as is. Buyer and user assume all risk of use and/or handling and/or storage of this material when such use and/or handling and/or storage is contrary to label instructions.

DIRECTIONS AND RECOMMENDATIONS: Follow directions carefully. Timing and method of application, weather and crop conditions, mixture with other chemicals not specifically recommended and other influencing factors in the use of this product are beyond the control of the seller and are assumed by buyer at his own risk.

USE OF PRODUCT: FMC's recommendations for the use of this product are based upon tests believed to be reliable. The use of this product being beyond the control of the manufacturer, no guarantee, expressed or implied, is made as to the effects of such or the results to be obtained if not used in accordance with directions or established safe practice.

DAMAGES: Buyer's or user's exclusive remedy for damages for breach of warranty or negligence shall be limited to direct damages not exceeding the purchase price paid and shall not include incidental or consequential damages.

FMC—FMC Trademark
(628-4/3/89-A

This a specimen label. FMC is not responsible for the accuracy of the information contained herein. As labels are subject to revision, always carefully read and follow the label on the product container.

Hoechst-Roussel Agri-Vet Company

Route 202-206 North
P.O. Box 2500
Somerville, NJ 08876-1258
(201) 231-2000

ACCLAIM® 1EC

HERBICIDE

For Selective Postemergence Annual and Perennial Grass Control in Turfgrass Including Sod Farms, Commercial and Residential Turf, and Ornamentals.

Not for use in California and Hawaii

ACTIVE INGREDIENT:
fenoxaprop-ethyl:(±)-ethyl
2-[4-[(6-chloro-2-benzoxazolyl) oxy]
phenoxy]propanoate 12.50%*
INERT INGREDIENTS: 87.50%
TOTAL 100.00%

*Equivalent to 1.00 pound of active ingredient per gallon

EPA REG. NO. 8340-18-54382

KEEP OUT OF REACH OF CHILDREN

WARNING

SEE ADDITIONAL PRECAUTIONARY STATEMENTS AT BEGINNING OF LABEL

PRECAUTIONARY STATEMENTS

WARNING

HAZARDS TO HUMANS AND DOMESTIC ANIMALS

May cause substantial but temporary eye injury. Do not get in eyes. Wear goggles or face shield and impermeable rubber gloves (such as neoprene or PVC) while mixing. Harmful if swallowed, absorbed through skin or inhaled. Do not take internally. Avoid inhalation of vapor or spray mist. Remove contaminated clothing and wash before reuse. Wash thoroughly with soap and water after handling.

ENVIRONMENTAL HAZARDS

This pesticide is toxic to fish. Do not apply directly to water.
Do not apply when weather conditions favor runoff or drift.
Do not contaminate arable land and/or water when disposing of equipment washwaters.

STORAGE AND DISPOSAL

Do not contaminate water, food or feed by storage or disposal.

Do not store over 100°F or below 10°F. Do not use or store near heat or open flame.

Pesticide Disposal: Pesticide wastes are toxic. Improper disposal of excess pesticide, spray mixture, or rinsate is a violation of Federal Law. If these wastes cannot be disposed of by use according to label instructions, contact your state Pesticide or Environmental Control Agency, or the Hazardous Waste representative at the nearest EPA Regional Office for guidance.

Container Disposal: Empty containers should be triple rinsed into the spray tank during the spray operation. Then offer for recycling or reconditioning, or puncture and dispose of in a sanitary landfill, or incinerator, or, if allowed by state and local authorities, by burning. If burned, stay out of smoke.

STATEMENT OF PRACTICAL TREATMENT

In case of eye contact, immediately flush with plenty of water for 15 minutes. Get medical attention.

If swallowed, do not induce vomiting. Acclaim 1EC Herbicide contains petroleum distillates. Call a physician. Vomiting should be supervised by a physician because of the possible pulmonary damage via aspiration of the solvent.

If possible, bring this container and labeling to the attending physician. For emergency assistance call (201) 231-2000.

IMPORTANT NOTICE: DISCLAIMER

Read "IMPORTANT NOTICE: DISCLAIMER" before buying or using. If terms are not acceptable, return at once unopened. HOECHST-ROUSSEL AGRI-VET COMPANY warrants only that the product conforms to the chemical dscription on the label and is reasonably fit for the purpose stated on the label when used in accordance with the directions under normal conditions of use. This warranty does not extend to the use of this product contrary to label instructions or under abnormal conditions, or under conditions not reasonably foreseeable to HOECHST-ROUSSEL AGRI-VET COMPANY and user assumes the risk of any such use. HOECHST-ROUSSEL AGRI-VET COMPANY MAKES NO OTHER WARRANTY, EXPRESSED OR IMPLIED, INCLUDING ANY IMPLIED WARRANTY OF FITNESS FOR A PARTICULAR PURPOSE OR OF MERCHANTABILITY. In no case shall HOECHST-ROUSSEL AGRI-VET COMPANY be liable for consequential, special, indirect or incidental damages resulting from the use or handling of this product. The foregoing conditions of sale and warranty can be varied only by an agreement in writing signed by a duly authorized representative of HOECHST-ROUSSEL AGRI-VET COMPANY.

GENERAL INFORMATION

Acclaim 1EC Herbicide is an emulsifiable concentrate (EC) formulation containing one pound of active ingredient per gallon. Acclaim 1EC Herbicide is used for postemergence control of annual and perennial grassy weeds in established turfgrass and ornamental species. Acclaim 1EC Herbicide does not control broadleaf weeds or sedges.

Acclaim 1EC Herbicide is absorbed through the foliage. Thorough coverage is important. Visual injury on the grassy weed is evident approximately 4-10 days after application (dependent upon environmental conditions) but control of the grassy weeds will take 12-21 days. Visible effects begin as a general chlorosis (yellowing) and reddening of the leaf surface followed by death of the weed.

DIRECTIONS FOR USE

It is a violation of Federal law to use this product in a manner inconsistent with its labeling.

DIRECTIONS FOR USE IN TURFGRASS SPECIES

Acclaim 1EC Herbicide can be applied as a broadcast application to all cultivars of the following turfgrass species grown throughout the United States:

Kentucky Bluegrass	*(Poa pratensis)*
Perennial Ryegrass	*(Lolium perenne)*
Fine Fescue	*(Festuca rubra)*
Tall Fescue	*(Festuca arundinaceae)*
Annual Bluegrass	*(Poa annua)*
Zoysiagrass	*(Zoysia spp.)*

Weed Species Controlled

Acclaim 1EC Herbicide will control the following annual grassy weeds:

Smooth Crabgrass	*(Digitaria ischaemum)*
Hairy Crabgrass	*(Digitaria sanguinalis)*
Goosegrass, Silver Crabgrass	*(Eleusine indica)*
Barnyardgrass	*(Echinochloa crusgalli)*
Foxtail Species	*(Setaria spp.)*
Panicum Species	*(Panicum spp.)*
Johnsongrass (seedlings)	*(Sorghum halepense)*
Sprangletop	*(Leptochloa spp.)*

Weed Species Suppressed

Acclaim 1EC Herbicide will temporarily suppress the growth of the following perennial grassy weeds:

Common Bermudagrass	*(Cynodon dactylon)*
Johnsongrass (rhizome)	*(Sorghum halepense)*

Acclaim 1EC Herbicide will suppress the growth of Common Bermudagrass and rhizome Johnsongrass. Apply Acclaim 1EC Herbicide at 23 ounces/acre (0.53 ounces/1000 sq. ft.) as soon as Common Bermudagrass or rhizome Johnsongrass begins growth. Repeat applications of Acclaim 1EC Herbicide every 28-35 days to continue growth suppression of these species. Do not apply more than 3 applications or a total of 135 ounces per acre (3.09 ounces/1000 sq. ft.) per growing season.

Application Information

Acclaim 1EC Herbicide only controls grasses which are emerged at the time of spraying. Young actively growing grassy weeds are more easily controlled than larger grassy weeds. Refer to the Acclaim 1EC Herbicide application rate and timing recommendation chart for application rates and timing of application. Avoid applications to turfgrass under stress due to lack of moisture, chemical injury or temperature extremes.

Pressurized Hydraulic Sprayer Application

Apply Acclaim 1EC Herbicide by means of a pressurized hydraulic sprayer using 30-60 PSI and 30-60 gallons of water per acre (0.7 to 1.4 gallons of finished spray per 1,000 sq. ft.). *Thorough spray coverage is essential for optimum results.* Flat fan nozzles are recommended.

High Volume Handgun Spray Application

If a low pressure (< 30 psi), high volume (> 60 gpa) handgun type sprayer is utilized, the addition of a surfactant at 32 ounces per 100 gallons is recommended. *Thorough coverage is essential for optimum results.*

Hand-Held Sprayer Application for Spot Treatments

For hand-held pump-type sprayers, add 0.34-0.53 fluid ounces of Acclaim 1EC Herbicide per gallon of water. Apply the lower rate when grassy weeds have less than two tillers. Spray target weeds thoroughly to the point of spray deposit runoff. *Optimum results occur when spray coverage is thorough.* The addition of a surfactant will improve coverage. A dye or colorant may be added to avoid spray pattern overlap.

Mowing Information

Acclaim 1EC Herbicide is a systemic herbicide. Spray Acclaim 1EC Herbicide when maximum leaf area is present for optimum activity. Do not mow treated areas for at least 24 hours to allow time for penetration and translocation into the grassy weeds. The effectiveness of Acclaim 1EC Herbicide is reduced when grass clippings interfere with thorough spray coverage of the target weeds.

Repeat Application Information

Repeat applications of Acclaim 1EC Herbicide at 10-45 fluid ounces (0.23-1.02 ounces/1000 sq. ft.) per acre (refer to application and timing chart) may be required under the following conditions:

(a) Continued grassy weed seed germination after the initial application

(b) When dense populations of weeds prevent thorough spray coverage of the entire leaf surface of target weeds

(c) For suppression of Common Bermudagrass and rhizome Johnsongrass

Do not apply more than 3 applications or a total of 135 ounces per acre (3.09 ounces/1000 sq. ft.) per growing season. Avoid applying repeat applications of Acclaim 1EC Herbicide sooner than 14 days after the previous application. Repeat Acclaim 1EC Herbicide applications should only be made to grassy weeds with vigorous growth. If turfgrass stunting occurred with the previous application, avoid applying until turfgrass recovery is complete (unless temporary stunting can be tolerated).

DIRECTIONS FOR USE OF ACCLAIM® 1EC HERBICIDE IN TANK MIXES

A. Preemergence Residual Herbicides: A tank mix application of Acclaim 1EC Herbicide and residual herbicides may be made to provide residual annual grass control. Apply after grassy weed germination. Always follow the use directions in accordance with the respective labels. No label dosage rates should be exceeded.

Apply Acclaim 1EC Herbicide at rates of *10-32 ounces/acre or .023-.073 ounces/1000 sq. ft.* when tank mixing with the following products:

	Product/Acre	Ounces/1000 sq. ft.
benefin DF	2-3 pounds	1.35-1.85 ounces
bensulide 4E	1.88-3.1 gallons	5.6-9.4 ounces
Dacthal[1] W75	6-20 pounds	2.2-7.0 ounces
Pre-M[2] 60 DG	2.5-5.0 pounds	0.9-1.8 ounces
Tupersan[3] 50W	16-24 pounds	6-9 ounces

Comments:

Refer to the stage of growth chart to determine the proper Acclaim 1EC Herbicide use rate. Also see Note 2.

B. Postemergence Broadleaf Herbicides: A tank mix application of Acclaim 1EC Herbicide and broadleaf herbicides may be made to provide broadleaf weed and annual grassy weed control. Apply after grassy weed germination. Always follow the use directions in accordance with the respective labels. No label dosage rates should be exceeded.

Apply Acclaim 1EC Herbicide at rates of *10-45 ounces/acre or 0.23-1.02 ounces/1000 sq. ft.* when tank mixing with the following products. Apply the 10 ounce rate only when tank mixing with preemergence products.

	Product/Acre	Ounces/1000 sq. ft.
dicamba 4E	0.25- 0.50 pint	0.09-0.18 ounces
Turflon[4] Amine 3EC	1- 4 pints	0.37-1.50 ounces
Turflon[4] Ester 4EC	1-2 pints	0.37-0.73 ounces
Breakthru[5]	0.125-0.50 lb. of total active ingredient per acre	0.25-1.00 ounces

Comments

Refer to the stage of growth chart to determine the proper Acclaim 1EC Herbicide use rate. Also see Note 2.

Always apply the following tank mixes of Acclaim 1EC Herbicide with pendimethalin at 2.5 pounds/acre or 0.9 ounces/1000 sq. ft. *Apply before grassy weed tillering* at a rate of *32 ounces/acre or 0.73 ounces/1000 sq. ft.* when mixing with the following products. Apply by means of a pressurized hydraulic sprayer using 30-60 psi and 30-60 gpa. Thorough spray coverage is important. Flat fan nozzles are recommended.

	Product/Acre	Ounces/1000 sq. ft.
Lesco Three Way	3.3- 4.0 pints	1.2-1.5 ounces
Super Trimec[6]	2-3 pints	0.75-1.10 ounces
Trimec[6] Turf Ester	2-3 pints	0.75-1.10 ounces
Trimec[6] 992	2-3 pints	0.75-1.10 ounces
Turflon[4] D	1-2 quarts	0.73-1.50 ounces
Turflon[4] II Amine	1-2 quarts	0.73-1.50 ounces
2,4-D amine 4	1.5-2.0 pints	0.60-0.73 ounces

Comments:

See Notes 2 and 4 when applying Acclaim 1EC Herbicide with these products or similar products.

C. Insecticides: Acclaim 1EC Herbicide may be tank mixed with the following insecticides only by or under the direct supervision of certified commercial applicators responsible for pest control programs as specified on the specific insecticide label. Apply after grassy weed germination. Always follow the use directions in accordance with the respective labels. No label dosage rates should be exceeded.

Apply Acclaim 1EC Herbicide at rates of *15-45 ounces/acre or 0.34-1.02 ounces/1000 sq. ft.* when tank mixing with the following products:

	Product/Acre	Ounces/1000 sq. ft.
diazinon 4EC		2-6 ounces
Dursban[4] 4E	1- 4 quarts	0.75-3.00 ounces
Dursban[4] 50W	2-8 pounds	0.75-3.00 ounces
Oftanol[7] 2	2- 4 quarts	1.5-3.0 ounces
Triumph[8]		0.75-1.5 ounces
Turcam[9]		1-2 ounces

Comments:

Refer to the stage of growth chart to determine the proper Acclaim 1EC Herbicide use rate. Also see Note 2.

D. Fungicides: A tank mix application of Acclaim 1EC Herbicide and fungicides may be made to provide disease and annual grass control. Apply after grassy weed germination. Always follow the use directions in accordance with the respective labels. No label dosage rates should be exceeded.

Apply Acclaim 1 EC Herbicide at rates of *15-45 ounces/acre or 0.34-1.02 ounces/1000 sq. ft.* when tank mixing with the following products:

	Product/Acre	Ounces/1000 sq. ft.
Chipco[10] 26019		1-2 ounces
Daconil[1] 2787 F	4-24 pints	3-11 ounces
Daconil[1] 2787 W75	2.75-16 pounds	2-8 ounces
mancozeb 80 WP		4-8 ounces
Tersan[3] 1991		1-2 ounces

Comments:

Refer to the stage of growth chart to determine the proper Acclaim 1EC Herbicide use rate.

E. Liquid Fertilizers: Acclaim 1EC Herbicide may be tank mixed with liquid nitrogen fertilizers (up to one pound/1000 sq. ft.) and iron chelates.

Notes

1. Stunting and/or chlorosis may occur on succulent, rapidly growing Kentucky Bluegrass. Symptoms are temporary and will subside.
2. Adequate soil moisture will enhance the performance of Acclaim 1EC Herbicide. Avoid application of Acclaim 1EC Herbicide under **DROUGHT STRESS** conditions. Thorough irrigation one to two days before or within one day after application will be helpful under these conditions. Rainfall or irrigation within one hour of application may cause a reduction in grass control. Allow spray deposit to dry before irrigation.
3. **DO NOT APPLY** Acclaim 1EC Herbicide to spring-seeded Kentucky Bluegrass established less than 4 growing months. **DO NOT APPLY** Acclaim 1EC Herbicide to seedling fine leaf fescue, tall fescue, or perennial ryegrass less than 4 weeks old.
4. Broadleaf herbicides such as 2,4-D or MCP may reduce the effectiveness of Acclaim 1EC Herbicide. Tank mixes with 2,4-D or MCP containing herbicides are most effective on untillered crabgrass. **Avoid tank mixing Acclaim 1EC Herbicide with 2,4-D or MCP containing herbicides after crabgrass has tillered.**
5. Acclaim 1EC Herbicide is a selective grassy weed herbicide and has little or no activity on broadleaf plants and sedges.
6. Acclaim 1EC Herbicide may be applied to newly plugged zoysiagrass. Fescues and ryegrass may be overseeded immediately following Acclaim 1EC Herbicide applications. All other turf species should be overseeded after a 21-day waiting period.
7. **DO NOT APPLY** this product through any type of irrigation system.
8. **DO NOT APPLY** Acclaim 1EC Herbicide to sod within 4 weeks before cutting for transplanting or within 4 weeks after transplanting.

ACCLAIM 1EC HERBICIDE APPLICATION RATE AND TIMING RECOMMENDATION CHART

Annual Grassy Weed Growth Stage	Fluid Oz/ Acre	Fluid Oz/ 1,000 sq. ft.
Untillered	10*	0.23
Untillered	15	0.34
1 tiller to 2 tiller	23	0.53
3 tiller to 4 tiller	32	0.73
4 tiller to 5 tiller	45**	1.02

*Apply this rate only when tank mixing with preemergence herbicides.

* *Avoid use of this rate on Kentucky Bluegrass, unless temporary injury can be tolerated.

DIRECTIONS FOR USE IN ORNAMENTALS

Acclaim 1EC Herbicide can be applied for control of grassy weeds to a wide range of ornamental species grown throughout the United States. Plant tolerance may vary according to environmental conditions. Determine tolerance prior to extensive use.

Acclaim 1EC Herbicide may be applied as a broadcast spray in the following plantings:

DECIDUOUS AND EVERGREEN TREES AND SHRUBS

Common Name	Scientific Name
Azalea	*Rhododendron*
Barberry	*Berberis*
Boxwood, Korean	*Buxus*
Firethorn	*Pyracantha*
Hawthorn	*Crataegus*
Holly, American	*Ilex*
Holly, Japanese	*Ilex*
Hydrangea	*Hydrangea*
Juniper, Blue Rug	*Juniperus*
Juniper, Dwarf Shore	*Juniperus*
Juniper, Parsonii	*Juniperus*
Juniper, San Jose	*Juniperus*
Juniper, Wiltoni	*Juniperus*
Magnolia	*Magnolia*
Maple, Red	*Acer*
Oleander	*Nerium*
Pine, White	*Pinus*
Privet	*Ligustrum*
Rhododendron	*Rhododendron*
Rose	*Rosa*
Viburnum	*Virburnum*
Weigela	*Weigela*
Willow	*Salix*
Yew	*Taxus*

HERBACEOUS & FLOWERING PLANTS

Common Name	Scientific Name
African Daisy	*Osteospermum*
Ajuga	*Ajuga*
Astilbe	*Astilbe*
Baby Blue Eyes	*Nemophila*
Baby's Breath	*Gypsophila*
Begonia	*Begonia*
Bellflower	*Campanula*
Black-Eyed Susan	*Rudbeckia*
Bleeding Heart	*Dicentra*
Blue Bells	*Mertensia*
Blue Flax	*Lobelia*
California Poppy	*Eschscholzia*
Candytuft	*Iberis*
Catchfly	*Silene*
Chrysanthemum	*Chrysanthemum*
Coleus	*Coleus*
Columbine	*Aquilegia*
Coneflower, Purple	*Echinacea*
Coreopis	*Coreopis*
Corn Poppy	*Papaver*
Cosmos	*Cosmos*
Creeping Zinnia	*Sanvitalia*
Daisy, Shasta	*Chrysanthemum*
Dames Rocket	*Hesperis*
Dwarf Columbine	*Aquilegia*
Dwarf Cornflower	*Centaurea*
English Wallflower	*Erysimum*
Forget-Me-Not	*Myosotis*
Forget-Me-Not	*Anchusa*
Gaillardia	*Gaillardia*
Garland Chrysanthemum	*Chrysanthemum*
Gazania	*Gazania*
Geranium	*Pelargonium*
Gilia	*Gilia*
Iris	*Iris*
Leopards-Bane	*Doronicum*
Liatrus	*Liatrus*
Maiden Pinks	*Dianthus*
Missouri Primrose	*Oenothera*
Mountain Phlox	*Phloxhera*
Petunia	*Petunia*
Pimpernel	*Anagallis*
Prairie Coneflower	*Ratibida*
Red Yarrow	*Achillea*
Rocket Larkspur	*Delphinium*
Rocky Mt. Penstemon	*Penstemon*
Roman Chamomile	*Matricaria*
Scarlet Flax	*Linum*
Siberian Wallflower	*Erysimum*
Small Burnet	*Sanguisorba*
Snapdragon	*Antirrhinum*
Snow-in-Summer	*Cerastium*
Soapwort	*Saponaria*
Spurred Snapdragon	*Antirrhinum*
Starflower	*Trientalis*
Statice, Annual	*Statice*
Sweet Alyssum	*Lobularia*
Sweet William	*Dianthus*
Tall Primrose	*Oenothera*
Tall Godetia	*Godetia*
Tidy Tips	*Layia*
White Yarrow	*Achillea*
Wild Thyme	*Thymus*
Zinnia	*Zinnia*

Broadcast use of Acclaim® 1EC Herbicide on the following ornamental plants will result in phytotoxicity:

Common Name	Scientific Name
Juniper, Barharbor	*Juniperus*
Philodendron	*Philodendron*
Salvia	*Salvia*
Podocarpus	*Podocarpus*
Pittosporium	*Pittosporium*

WEED SPECIES CONTROLLED

Acclaim 1EC Herbicide will control the following annual grassy weeds:

Smooth Crabgrass	*(Digitaria ischaemum)*
Hairy Crabgrass	*(Digitaria sanguinalis)*
Goosegrass, Silver Crabgrass	*(Eleusine indica)*
Barnyardgrass	*(Echinochloa crusgalli)*
Foxtail Species	*(Setaria* spp.)
Panicum Species	*(Panicum* spp.)
Johnsongrass (seedlings)	*(Sorghum halepense)*
Sprangletop	*(Leptochloa* spp.)

WEED SPECIES SUPPRESSED

Acclaim 1EC Herbicide will temporarily suppress the growth of the following perennial grassy weeds:

Common Bermudagrass	*(Cynodon dactylon)*
Johnsongrass (rhizome)	*(Sorghum halepense)*

Acclaim 1EC Herbicide will suppress the growth of Common Bermudagrass and rhizome Johnsongrass. Apply Acclaim 1EC Herbicide at 23 ounces/acre (0.53 ounces/1000 sq. ft.) as soon as Common Bermudagrass or rhizome Johnsongrass begins growth. Repeat applications of Acclaim every 28-35 days to continue growth suppression of these species.

APPLICATION INFORMATION

Acclaim 1EC Herbicide only controls grasses which are emerged at the time of spraying. Young, actively growing grassy weeds are more easily controlled than larger grassy weeds. Refer to the Acclaim 1EC Herbicide application rate and timing recommendation chart for application rates and timing of application. Avoid applications to ornamentals under stress due to lack of moisture, chemical injury or temperature extremes.

Pressurized Hydraulic Sprayer Application

Apply Acclaim 1EC Herbicide by means of a pressurized hydraulic sprayer using 30-60 PSI and 30-60 gallons of water per acre (0.7 to 1.4 gallons of finished spray per 1,000 sq. ft.). **Thorough spray coverage is essential for optimum results.** Flat fan nozzles are recommended.

High Volume Handgun Spray Application

If a low pressure (< 30 psi), high volume (> 60 gpa) handgun type sprayer is utilized, the addition of a surfactant at 32 ounces per 100 gallons is recommended. **Thorough coverage is essential for optimum results.**

Hand-Held Sprayer Application for Spot Treatments

For hand-held pump-type sprayers, add 0.34-0.53 fluid ounces of Acclaim 1EC Herbicide per gallon of water. Apply the lower rate when grassy weeds have less than two tillers. Spray target weeds thoroughly to the point of spray deposit runoff. **Optimum results occur when spray coverage is thorough.** The addition of a surfactant will improve coverage. A dye or colorant may be added to avoid spray pattern overlap.

Repeat Application Information

Repeat applications of Acclaim 1EC Herbicide at 10-45 fluid ounces (0.23-1.02 ounces/1000 sq. ft.) per acre (refer to application and timing chart) may be required under the following conditions:

(a) Continued grassy weed seed germination after the initial application

(b) When dense populations of weeds prevent thorough spray coverage of the entire leaf surface of target weeds

(c) For supression of Common Bermudagrass and rhizome Johnsongrass

Do not apply more than 3 applications or a total of 135 onces per acre (3.09 ounces/1000 sq. ft.) per growing season. Avoid applying repeat applications of Acclaim 1EC Herbicide sooner than 14 days after the previous application. Repeat Acclaim 1EC Herbicide applications should only be made to grassy weeds with vigorous growth. If stunting or injury occurred with the previous application, avoid applying until turfgrass recovery is complete (unless temporary stunting can be tolerated).

NOTES

1. Stunting and/or chlorosis may occur on succulent, rapidly growing ornamentals. Symptoms are temporary and will subside.
2. Adequate soil moisture will enhance the performance of Acclaim 1EC Herbicide. Avoid application of Acclaim 1EC Herbicide under **DROUGHT STRESS** conditions. Thorough irrigation one to two days before or within one day after application will be helpful under these conditions. Rainfall or irrigation within one hour of application may cause a reduction in grass control. Allow spray deposit to dry before irrigation.
3. **DO NOT APPLY** this product through any type of irrigation system.

ACCLAIM 1EC HERBICIDE APPLICATION RATE AND TIMING RECOMMENDATION CHART

Annual Grassy Weed Growth Stage	Fluid Oz/ Acre	Fluid Oz/ 1,000 sq. ft.
Untillered	15	0.34
1 tiller to 2 tiller	23	0.53
3 tiller to 4 tiller	32	0.73
4 tiller to 5 tiller	45	1.02

Trademark Information

[1] Registered trademark of SDS Biotech Corporation
[2] Registered trademark of Lesco, Inc.
[3] Registered trademark of E.I. duPont de Nemours & Company
[4] Registered trademark of The Dow Chemical Company
[5] Registered trademark of The Andersons
[6] Registered trademark of PBI/Gordon Corporation
[7] Registered trademark of the Parent Company of Farbenfabriken Bayer GmbH, Leverkusen
[8] Registered trademark of Ciba-Geigy Corporation
[9] Registered trademark of Nor-Am Chemical Company
[10] Registered trademark of Rhone-Poulenc Incorporated

TH 8603001 (2/89)

"This is a specimen label. Hoechst-Roussel Agri-Vet Company is not responsible for the accuracy of the information contained herein, as labels are subject to revision. Always read and follow the use instructions on the container label".

HOELON® 3EC
HERBICIDE

RESTRICTED USE PESTICIDE

For retail sale to and use only by certified applicators or persons under their direct supervision and only for those uses covered by the certified applicator's certification.

For Control of Wild Oats and Other Annual Grasses in Wheat (Including Durum), Barley, Lentils, Dry Peas (Field Peas) Including Austrian Winter Peas, Flax, and Acreage Conservation Reserve (Set-Aside)

ACTIVE INGREDIENT*
Diclofop-methyl: methyl 2-[4-(2,4-dichlorophenoxy).
phenoxy] propanoate 35.49%
Inert Ingredients 64.51%
TOTAL 100.00%

*Equivalent to 3.0 pounds of active ingredient per gallon.

8/8/89

EPA Reg. No. 8340-20-54382

The name and logo HOECHST are registered trademarks of Hoechst AG.
The name and logo ROUSSEL are registered trademarks of Roussel Uclaf S.A.

KEEP OUT OF REACH OF CHILDREN
DANGER

See Additional Precautionary Statements at End of Label.

GENERAL INFORMATION

Hoelon 3EC Herbicide is a highly effective herbicide for the control of a broad spectrum of annual grassy weeds in wheat (including durum), barley, lentils, dry field peas (including Austrian Winter Peas), flax, and acreage conservation reserve. Hoelon 3EC Herbicide is a three pound active ingredient per gallon emulsifiable concentrate. Depending on the crop and the grass species being controlled, Hoelon 3EC Herbicide may be applied preplant incorporated (to wheat only), preemergence, or postemergence. Read and carefully follow all label directions specified in this labeling.

For the use of Hoelon 3EC Herbicide in California, see Supplemental Labeling.

Application Equipment Notes

Ground

Whether applying Hoelon 3EC Herbicide preplant incorporated, preemergence, or postemergence, thorough uniform coverage of the soil or target weeds is critical to achieving satisfactory results. The use of flat fan nozzles, spaced 10 or 20 inches apart across the boom, is recommended for optimum coverage with ground equipment. Ground applications should be in at least 10 gallons of water carrier per acre, using a minimum of 40 PSI.

Aerial

Aerial applications must be applied in at least 5 gallons of water carrier per acre. Use spray nozzle tips and sufficient pressure to provide a uniform pattern and a median droplet size of 200-350 microns. DO NOT wide swath the application; Hoelon 3EC Herbicide is a contact material when used postemergence, and highly immobile in the soil when applied preplant incorporated or preemergence. DO NOT make aerial applications when wind is above 5 mph or within 100 feet of a lake, stream, drainage basin, tidal marsh or estuary.

Only clean water, free of suspended matter or other contaminants, should be used when applying Hoelon 3EC Herbicide.

General Notes

1. Hoelon 3EC Herbicide does not control broadleaf weeds or perennial grassy weeds.
2. DO NOT allow livestock to graze treated fields. DO NOT harvest forage, hay or straw from treated fields prior to grain harvest.
3. Hoelon 3EC Herbicide can be tank mixed with broadleaf herbicides listed on this label for broad spectrum weed control.
4. DO NOT apply phenoxy, phenoxy-type, or Banvel® Herbicide within 5 days of a Hoelon 3EC Herbicide application, or reduced grassy weed control will occur.
5. DO NOT apply Hoelon 3EC Herbicide through any type of irrigation system.

When controlling mixed infestations of grassy weeds, always use the rate that will control the least susceptible species, but DO NOT exceed the maximum rate for each registered crop use.

DIRECTIONS FOR USE

IT IS A VIOLATION OF FEDERAL LAW TO USE THIS PRODUCT IN A MANNER INCONSISTENT WITH ITS LABELING. THIS LABELING MUST BE IN THE POSSESSION OF THE USER AT THE TIME OF THE APPLICATION.

DIRECTIONS FOR THE PREPLANT INCORPORATED USE OF HOELON 3EC HERBICIDE ON WINTER WHEAT:

This product is registered for the preplant incorporated use to control Downy Brome, Ripgut Brome, Japanese Brome, Windgrass and Bulbous Bluegrass in the states of:

Colorado	Nebraska
Idaho	Oregon
Kansas	South Dakota
Montana	Utah
	Washington

(The general information section must also be read for complete use instructions.)

Application Rate Information

The use rate of Hoelon 3EC Herbicide to control Downy Brome *(Bromus tectorum),* Ripgut Brome *(Bromus rigidus),* Japanese Brome *(Bromus japonicus),* Windgrass *(Apera spica - venti),* and Bulbous Bluegrass *(Poa bulbosa)* will vary, depending on soil texture, soil organic matter content, soil moisture, and density of grass populations.

HOELON 3EC HERBICIDE APPLICATION RATES:

	Broadcast Rate in Pints/A	
Soil Texture	% Organic Matter ½-2%	% Organic Matter >2%
Coarse (Sandy Loam, Loamy Sand)	2.0 - 2⅔	2⅔
Medium (Loam, Silt Loam, Silt)	2.0 - 2⅔	2⅔ SUPPRESSION ONLY

Use the highest recommended rate, according to the chart above, when heavy populations of annual grass (more than 50 plants per square foot) are anticipated (based on field history) and/or if soils are dry at the time of application. On soils with a medium texture and greater than 2% organic matter, the 2⅔ pint rate may only provide suppression of the grassy weeds listed on the preplant incorporated section of this label.

Incorporation Directions

Hoelon 3EC Herbicide is highly immobile on the soil and must be mechanically incorporated into the annual grass germination zone to help insure satisfactory control. Incorporate Hoelon 3EC Herbicide into the soil to a maximum depth of 2 inches (do not set incorporation equipment to cut more than 4 inches deep). Incorporate one time within 48 hours after application, followed by a second incorporation, in a different direction from the first, prior to seeding.

Incorporation Equipment

If the seedbed is free of trash and clods at the time of application, the use of a spiketooth, springtooth or flextine harrow, operated at 3-6 mph, is recommended. If the seedbed is trashy at the time of application, use an implement which is capable of incorporating the trash in the top two inches of the soil. A skew treader operated at 4-6 mph is recommended. Chisel plows — may be used for first pass incorporation only. Operate at 3-4 inch depth at 4-6 mph. A chisel plow is defined as having 3 rows of up to 18 inch sweeps on no greater than 12 inch centers. Sweeps should be staggered so that no soil is left unturned. Field cultivators should be set to cut 3-4 inches deep and operated at a speed of at least 5 mph. A field cultivator used to incorporate Hoelon 3EC Herbicide should have 3-4 rows of sweeps with "C" or "S" shaped shanks, spaced 7 inches or less and staggered so that no soil is left unturned. Conventional, deep furrow, or semi-deep furrow drills may be used for seeding.

Preplant Incorporated Special Notes

1. DO NOT apply Hoelon 3EC Herbicide preplant incorporated to barley, as barley damage will occur.
2. DO NOT tank mix Hoelon 3EC Herbicide with any other herbicides for preplant incorporation unless specifically recommended in this label.
3. Hoelon 3EC Herbicide may be tank mixed with liquid fertilizers. Follow label recommendations concerning rates and incorporation. Before mixing Hoelon 3EC Herbicide with fertilizer, perform a compatibility test in a quart jar. If the materials do not readily mix, do not tank mix Hoelon 3EC Herbicide with the fertilizer.

DIRECTIONS FOR THE PREEMERGENCE USE OF HOELON 3EC HERBICIDE ON WINTER WHEAT

This product is registered for preemergence use on winter wheat to control annual ryegrass in the states of:

Alabama	Louisiana	Oregon	Washington
Arkansas	Maryland	South Carolina	West Virginia
Delaware	Mississippi	Tennessee	
Florida	North Carolina	Texas	
Georgia	Oklahoma	Virginia	

(The general information section must also be read for complete use instructions.)

HOELON 3EC HERBICIDE APPLICATION RATES:

	Broadcast Rate in Pints/A	
Soil Texture	**% Organic Matter ½-2%**	**% Organic Matter >2%**
Coarse (Sandy Loam, Loamy Sand)	2.0	2⅔
Fine (Loam, Silt Loam, Silt)	2⅔	2⅔

Use the 2⅔ pint/acre rate of Hoelon® 3EC Herbicide in fields that have a history of heavy annual ryegrass pressure. If rainfall does not occur within seven (7) days after application, reduced control may occur.

Preemergence Special Instructions

1. DO NOT apply Hoelon 3EC Herbicide preemergence to barley as barley damage will occur.
2. Hoelon 3EC Herbicide may be mixed with liquid fertilizers when making preemergence applications. Follow label recommendations concerning Hoelon 3EC Herbicide rates. Before tank mixing Hoelon 3EC Herbicide with fertilizer, perform a compatibility test in a quart jar. If the materials do not readily mix, do not tank mix Hoelon 3EC Herbicide with the fertilizer.
3. Hoelon 3EC Herbicide applied preemergence surface will not provide adequate brome control or control wild oats.

DIRECTIONS FOR THE POSTEMERGENCE USE OF HOELON 3EC HERBICIDE ON WINTER WHEAT, SPRING WHEAT, AND DURUM:

This product is registered for postemergence use on wheat and durum in the following states:

Alabama	Florida	Maryland	Nevada	Oregon	Utah
Arizona	Georgia	Minnesota	New Mexico	South Carolina	Virginia
Arkansas	Idaho	Mississippi	North Carolina	South Dakota	Washington
Colorado	Kansas	Montana	North Dakota	Tennessee	West Virginia
Delaware	Louisiana	Nebraska	Oklahoma	Texas	Wyoming

(The general information section must also be read for complete use instructions.)

Apply Hoelon 3EC herbicide as a broadcast postemergence spray at the rates listed below.

AMOUNT OF HOELON 3EC HERBICIDE PER ACRE (PINTS)[1] RELATIVE TO GROWTH STAGE OF ANNUAL GRASSY WEEDS

SUSCEPTIBLE ANNUAL GRASSY WEEDS	1-3 LEAF	3-4 LEAF	4-5 LEAF	5 LEAF-2 TILLERS
Annual Ryegrass (Italian)[2] *(Lolium multiflorum)*	1⅓	1⅓	2.0	2⅔
Broadleaf Signalgrass (suppression) *(Brachiaria platyphylla)*	2⅔			
Crabgrass[3] *(Digitaria sanguinalis)*	2⅔			
Persian Darnel *(Lolium persicum)*	2⅔			
Fall Panicum *(Panicum dichotomiflorum)*				
Barnyardgrass (watergrass) *(Echinochloa crusgalli)*				
Giant Foxtail *(Setaria faberii)*				DO
Green Foxtail (pigeongrass) *(Setaria viridis)*				NOT APPLY
Yellow Foxtail (pigeongrass)[3] *(Setaria lutescens)*				Reduced Grassy Weed Control
Wild Oat *(Avena fatua)*	2.0 to 2⅔	2⅔		Will Occur
Itchgrass (Raoulgrass) *(Rottboellia exaltata)*				
Volunteer Corn *(Zea mays)*				
Witchgrass (suppression) *(Panicum capillare)*				
Smallseed Canarygrass (suppression)[4] *(Phalaris minor)*				

SUSCEPTIBLE ANNUAL GRASSY WEEDS	AMOUNT OF HOELON 3EC HERBICIDE PER ACRE (PINTS)[1] RELATIVE TO GROWTH STAGE OF ANNUAL GRASSY WEEDS			
	1-3 LEAF	3-4 LEAF	4-5 LEAF	5 LEAF-2 TILLERS
Hooded Canarygrass[4] *(Phalaris paradoxa)*				

[1] When controlling mixed populations of grassy weeds, always use the rate that will control the least susceptible species.
[2] Annual ryegrass may also be controlled preemergence in winter wheat. See appropriate recommendation section in this labeling.
[3] For best control of crabgrass and yellow foxtail, application should be made before the second leaf fully emerges.
[4] See specific control recommendations for Smallseed and Hooded Canarygrass in this section of the labeling.

Postemergence Special Instructions

When using Hoelon 3EC Herbicide postemergence in wheat to control annual grassy weed(s), and the weeds do not include Annual Ryegrass, Broadleaf Signalgrass, Crabgrass, Persian Darnel or Yellow Foxtail, apply the 2.0 pint per acre rate when the annual grassy weeds are in the 1-3 leaf stage of growth, and growing conditions are optimal. In the northcentral region of the United States (east of the Rocky Mountains) in spring wheat and durum under normal growing conditions, the 1-3 leaf stage of wild oat plants generally occurs 16-26 days after planting. In the northwest region of the United States in spring wheat and durum (west of the Rocky Mountains) under normal growing conditions the 1-3 leaf stage of wild oat plants generally occurs 18-28 days after planting. In either region, under normal growing conditions, wild oat plants will add an additional leaf every four days. In either region, begin checking your fields 10-12 days after planting to assure correct application timing. Higher use rates of Hoelon 3EC Herbicide should be used when the majority of annual grassy weeds are not clearly in the specified stage of growth, or if growth is retarded due to adverse growing conditions.

Crop Oil Concentrate

In spring wheat (including durum) and winter wheat at the lower labeled use rates (1¹/₃ to 2²/₃ pints per acre) it may be helpful to use 1 pint to 1 quart of crop oil concentrate per acre approved for use on growing crops. Use a crop oil concentrate containing a blend of 80 percent (minimum) petroleum or vegetable base oil and the remaining composed of a tolerance-exempt surfactant. Some slight wheat yellowing may be noted when crop oil concentrates are added to Hoelon 3EC Herbicide.

For more detailed information on determining the low use rates in combination with crop oil concentrates, refer to the Hoechst-Roussel Agri-Vet Company Technical Bulletin No. 110, "Low Rate Technology." Contact your local dealer or Hoechst-Roussel Agri-Vet Company representative for this Technical Bulletin.

Broadleaf Tank Mixes - For Postemergence Applications in Wheat

All broadleaf herbicides should be used in accordance with the label limitations and precautions for each product. No label dosage rates should be exceeded. For best results, tank mixes should be used when growing conditions (air temperature and soil moisture) are optimum, and grasses are in the 1-3 leaf stage of growth with light to moderate infestations.

Hoelon® 3EC Herbicide Plus Buctril® Herbicide

Hoelon 3EC Herbicide may be tank mixed with Buctril Herbicide for broadleaf weed control in spring wheat (including durum) and winter wheat. Hoelon 3EC Herbicide should be used at a rate of 2.0 to 2²/₃ pints per acre (depending on size of annual grasses) as specified on this label. Buctril should be mixed at a rate of 1.0 to 1¹/₂ pints per acre, according to the directions specified on the respective label.

Hoelon 3EC Herbicide Plus Harmony Herbicide

A tank of Hoelon 3EC Herbicide + Harmony can be applied for wild oat and broadleaf weed control in spring wheat, durum, and winter wheat. The use rate of Hoelon should be 2²/₃ pints per acre in wheat and durum. The maximum use rate of Harmony when tank mixed with Hoelon in spring cereals is ¹/₂ oz. product per acre.

Wild oat control may be reduced under conditions of soil moisture stress or extremely heavy wild oat populations.

Reduced control of foxtail is likely when tank mixing with Harmony. When Foxtail is the major grassy weed in the field, DO NOT tank mix Hoelon 3EC Herbicide + Harmony. Use sequential treatments.

In wheat and durum, crop oil concentrate can be added to a Hoelon 3EC Herbicide + Harmony tank mix.

Hoelon 3EC Herbicide Plus Buctril Herbicide Plus MCPA (ester) 4EC

Low rates of MCPA (ester) 4EC may be added to a tank mix of Hoelon 3EC Herbicide plus Buctril Herbicide. Hoelon 3EC Herbicide should be used at a rate of 2²/₃ pints per acre. When using Buctril Herbicide, 1.0 pint per acre should be tank mixed with MCPA (ester) 4EC at 1¹/₂ ounces per acre. MCPA (ester) 4EC rates in excess of 1¹/₂ ounces per acre may reduce the grassy weed control provided by Hoelon 3EC Herbicide. No other formulations of MCPA should be substituted for the 4.0 pound active ingredient ester formulation.

Hoelon 3EC Herbicide Plus Buctril Plus Glean - Winter Wheat Only

A three-way tank mix of Hoelon 3EC Herbicide + Buctril + Glean can be applied for wild oat control and enhanced broadleaf weed control in winter wheat. The use rate of Hoelon 3EC Herbicide ranges from 2.0 pints per acre. At least 2²/₃ pints per acre are required when wild oats are larger than the three-leaf stage, infestations are heavy, or soil moisture is lacking. Buctril should be used at 1 to 1¹/₂ pints per acre, and the maximum use rate of Glean is ¹/₆ oz. per acre.

Fertilizer Tank Mixes Postemergence in Wheat

Hoelon 3EC Herbicide applied postemergence in wheat may only be tank mixed with liquid nitrogen fertilizer (28-32%) if the timing for use is compatible. The concentration of liquid nitrogen fertilizer to water should not exceed 50% of the total carrier in the spray tank. When mixing with liquid nitrogen fertilizers as a partial carrier, crop burn may occur. This condition is intensified under higher temperatures (greater than 75°F daytime) and/or low soil moisture. This mixture should not be used when these environmental conditions are present. Thorough spray coverage is essential when using Hoelon 3EC Herbicide and flat fan type nozzles are recommended for ground applications. If flood jet nozzles are used, they must be spaced no greater than 40 inches apart, using a maximum TK-15, with a minimum of 40 PSI and ground speeds at application not to exceed 10 MPH. DO NOT use raindrop type nozzles.

DO NOT use less than 2.0 pints per acre Hoelon 3EC Herbicide when tank mixing with liquid nitrogen.

In all of the above-labeled tank mix options, ground applications should be made with a minimum of 10 gallons of water per acre, and air applications with a minimum of 5 gallons of water per acre. If other broadleaf herbicides are applied within 5 days, reduced grassy weed control may occur.

DIRECTIONS FOR THE POSTEMERGENCE CONTROL OF SMALLSEED AND HOODED CANARYGRASS IN WHEAT IN ARIZONA ONLY

Apply Hoelon 3EC Herbicide as a broadcast postemergence spray at the rates listed below:

	AMOUNT OF HOELON 3EC HERBICIDE PER ACRE (PINTS) RELATIVE TO GROWTH STAGE OF ANNUAL GRASSY WEEDS	
SUSCEPTIBLE ANNUAL GRASSY WEEDS	STAGE OF GROWTH	RATE-PINTS/ACRE
Hooded Canarygrass[1] *(Phalaris paradoxa)*	1-3 leaves	2.0 to 2²/₃
Smallseed Canarygrass (Suppression)[2] *(Phalaris minor)*	1-2 leaves	2²/₃

[1] DO NOT add crop oil concentrate.
[2] Under dry conditions add 1 pint to 1 quart per acre of crop oil concentrate.

DIRECTIONS FOR THE POSTEMERGENCE USE OF HOELON 3EC HERBICIDE ON BARLEY

This product is registered for postemergence use on barley in the following states:

Arizona	Kansas	Nebraska	North Dakota	South Dakota	Washington
Colorado	Maryland	Nevada	Oklahoma	Texas	West Virginia
Delaware	Minnesota	New Mexico	Oregon	Utah	Wyoming

Idaho Montana North Carolina South Carolina Virginia

(The general information section must also be read for complete use instructions.)
Apply Hoelon 3EC Herbicide as a broadcast postemergence spray at the rates listed below.

SUSCEPTIBLE ANNUAL GRASSY WEEDS	AMOUNT OF HOELON 3EC HERBICIDE PER ACRE (PINTS)[1] RELATIVE TO GROWTH STAGE OF ANNUAL GRASSY WEEDS			
	1-3 LEAF	**4 LEAF**	**5 LEAF**	**5 LEAF - 2 TILLERS**
Annual Ryegrass (Italian)[2] *(Lolium multiflorum)*	1⅓	1⅓	2.0 pints	2⅔
Broadleaf Signalgrass (suppression) *(Brachiaria platyphylla)*	2⅔			
Crabgrass[3] *(Digitaria sanguinalis)*	2⅔			
Persian Darnel *(Lolium persicum)*	2⅔			
Fall Panicum *(Panicum dichotomiflorum)*				
Barnyardgrass (watergrass) *(Echinochloa crusgalli)*				
Giant Foxtail *(Setaria faberii)*				DO NOT APPLY
Green Foxtail (pigeongrass) *(Setaria viridis)*				REDUCED GRASSY WEED CONTROL WILL OCCUR
Yellow Foxtail (pigeongrass)[3] *(Setaria lutescens)*	2.0 to 2⅔			
Wild Oat *(Avena fatua)*				
Itchgrass (Raoulgrass) *(Rottboellia exaltata)*				
Volunteer Corn *(Zea mays)*				
Witchgrass (suppression) *(Panicum capillare)*				
Hooded Canarygrass *(Phalaris paradoxa)*				

[1] When controlling mixed populations of grassy weeds, always use the rate that will control the least susceptible species.
[2] Annual ryegrass may also be controlled preemergence in winter wheat. See appropriate recommendation section in this labeling.
[3] For best control of crabgrass and yellow foxtail, application should be made before the second leaf fully emerges.

HOELON® 3EC HERBICIDE RECOMMENDATIONS FOR SPRING BARLEY IN THE NORTH CENTRAL, SOUTHWEST AND WESTERN REGIONS OF THE UNITED STATES

When using Hoelon 3EC Herbicide to control annual grassy weeds and the weeds do not include annual ryegrass, broadleaf signalgrass, crabgrass or Persian darnel, apply the 2.0 pint per acre rate when the annual grassy weeds are in the 1-3 leaf stage of growth and growing conditions are optimal. The higher rate of 2⅔ pints per acre should be used (see chart) when the majority of grassy weeds is not clearly in the specified stage of growth or if growth is retarded due to a lack of moisture. Thorough spray coverage is essential.

Weather Considerations

In spring seeded barley, if one or a combination of the following environmental conditions are present or prolonged preceding a Hoelon 3EC Herbicide application, crop injury may occur.

(1) COLD TEMPERATURES—If daily minimum temperatures reach 40°F or less for three (3) consecutive days before application.

(2) MOISTURE CONDITIONS
a) Water-logged or poorly drained fields.
b) Moisture content is at field capacity.

If the above adverse environmental conditions are present, barley is already under severe environmental stress; do not compound the stress with any chemical applications. With respect to the aforegoing statement: buyer specifically assumes all risks should cold and/or wet weather conditions occur prior to or during the Hoelon 3EC Herbicide application.

For more detailed information on the use of Hoelon 3EC Herbicide in spring planted barley, refer to the Hoechst-Roussel Agri-Vet Company Technical Release - Hoelon 3EC Herbicide Recommendations for Barley, available through your local chemical supplier or your Hoechst-Roussel Agri-Vet Company representative.

Broadleaf Tank Mixes for Postemergence Applications in Spring Barley

All broadleaf herbicides should be used in accordance with the label limitations and precautions for each product. No label dosage rates should be exceeded. For best results, tank mixes should be used when growing conditions (air temperature and soil moisture) are optimum and grasses are in the 1-3 leaf stage of growth with light to moderate infestations (less than 25 wild oat plants per square foot).

Hoelon 3EC Herbicide Plus Buctril® Herbicide

Hoelon 3EC Herbicide may be tank mixed with Buctril Herbicide for broadleaf weed control in spring barley. Hoelon 3EC Herbicide should be used at a rate of 2.0 to 2⅔ pints per acre (depending on the size of annual grasses, environmental conditions, and density of grass populations) as specified on this label. Buctril should be mixed at a rate of 1 to 1½ pints per acre, according to the directions specified on the respective label.

Hoelon 3EC Herbicide Plus Harmony® Tank Mixes

A tank mix of Hoelon 3EC Herbicide + Harmony can be applied for wild oat and broadleaf weed control in barley. In barley, 2⅔ pints of Hoelon 3EC Herbicide per acre is recommended. The maximum use rate of Harmony when tank mixed with Hoelon 3EC Herbicide in spring cereals is ½ oz. product per acre.

Wild oat control may be reduced under conditions of soil moisture stress or extremely heavy wild oat populations.

Reduced control of foxtail is likely when tank mixing with Harmony. When foxtail is the major grassy weed in the field, DO NOT tank mix Hoelon + Harmony. Use sequential treatments.

Hoelon 3EC Herbicide Plus Buctril® Herbicide Plus MCPA (ester) 4EC

Low rates of MCPA (ester) 4EC may be added to a tank mix of Hoelon 3EC Herbicide plus Buctril 4EC Herbicide. Hoelon 3EC Herbicide should be used at a rate of 2⅔ pints per acre. When using Buctril Herbicide, 1.0 pint per acre should be tank mixed with MCPA (ester) 4EC at 1½ oz. per acre. MCPA (ester) 4EC rates in excess of 1½ oz. per acre may reduce the grassy weed control provided by Hoelon 3EC Herbicide. No other formulations of MCPA should be substituted for the 4.0 pound active ingredient ester formulation.

Special Notes for Spring Barley

1) DO NOT tank mix Hoelon 3EC Herbicide with Glean®.
2) DO NOT tank mix Hoelon 3EC Herbicide with crop oil concentrate.
3) DO NOT tank mix Hoelon 3EC Herbicide with liquid fertilizers.
4) Hoelon 3EC Herbicide is registered for use on all varieties of spring seeded barley.
5) Hoelon 3EC Herbicide should be applied prior to barley tiller initiation.
6) Do not apply preemergence to barley.

HOELON 3EC HERBICIDE RECOMMENDATIONS FOR WINTER BARLEY IN THE MIDATLANTIC AND SOUTHEAST REGIONS OF THE UNITED STATES

When using Hoelon 3EC Herbicide to control annual ryegrass in fall seeded barley, use the 1⅓ pints per acre rate when the ryegrass is in the 1 to 4 leaf stage of growth. The low rate of 1⅓ pints per acre should only be used under optimum growing conditions. If the crop and annual ryegrass is under drought stress, increase the rate of Hoelon 3EC Herbicide from 2.0-2⅔ pints per acre, depending on the severity of infestation and size of annual ryegrass. For mixed grass infestations, refer to the rate and timing chart for grasses controlled in barley.

Broadleaf Tank Mixes for Postemergence Applications in Winter Barley

See broadleaf tank mixes approved for use on spring barley.

Fertilizer Tank Mixes Postemergence in Winter Barley

Hoelon 3EC Herbicide may be tank mixed with liquid nitrogen fertilizer (e.g. 28% or 32%) if the timing for use is compatible. The concentration of liquid nitrogen fertilizer to water should not exceed 50% of the total carrier in the spray tank. When mixing with liquid nitrogen fertilizers as a partial carrier, crop burn may occur. This condition is intensified under high temperatures (greater than 75°F daytime) and/or low soil moisture. This mixture should not be used when these environmental conditions are present. Thorough spray coverage is essential when using Hoelon 3EC Herbicide and flat fan type nozzles are recommended for ground applications. If flood jet nozzles are used, they must be spaced no greater than 40 inches apart, using a maximum TK-15 with a minimum of 40 PSI and ground speeds at application not to exceed 10 mph. DO NOT use raindrop type nozzles. Use a minimum of 2.0 pints per acre of Hoelon 3EC Herbicide when tank mixing with liquid nitrogen fertilizer.

Special Notes for Winter Barley

1) DO NOT tank mix Hoelon 3EC Herbicide with Glean®.
2) DO NOT tank mix Hoelon 3EC Herbicide with crop oil concentrate.
3) Hoelon 3EC Herbicide is registered for use on the following winter barley varieties - Milton, Boone, Henry and Sussex.
4) Apply Hoelon 3EC Herbicide after tiller initiation, but prior to jointing of the winter barley.
5) DO NOT apply preemergence to winter barley.

DIRECTIONS FOR THE POSTEMERGENCE USE OF HOELON 3EC HERBICIDE ON DRY FIELD PEAS (INCLUDING AUSTRIAN WINTER PEAS) AND LENTILS.

This product is registered for postemergence use on dry field peas in the following states: Idaho, Oregon, and Washington.

This product is registered for postemergence use on lentils in the following states: Idaho and Washington.

(The general information section must also be read for complete use instructions.)
Apply Hoelon® 3EC Herbicide as a broadcast postemergence spray at the rates listed below.

SUSCEPTIBLE ANNUAL GRASSY WEEDS	AMOUNT OF HOELON 3EC HERBICIDE PER ACRE (PINTS)[1] RELATIVE TO GROWTH STAGE OF ANNUAL GRASSY WEEDS			
	1-2 LEAF	3-4 LEAF	4-5 LEAF	5 LEAF - 2 TILLERS
Annual Ryegrass (Italian)[2] *(Lolium multiflorum)*	1 1/3	1 1/3	2.0	2 2/3
Barnyardgrass (Watergrass) *(Echinochloa crusgalli)*	2.0	2 2/3	DO NOT APPLY	
Green Foxtail *(Setaria viridis)*	2.0	2 2/3	REDUCED GRASSY WEED CONTROL WILL OCCUR	
Wild Oat *(Avena fatua)*	2.0	2 2/3		

[1] When controlling mixed populations of annual grassy weeds, always use the rate which will control the least susceptible species.
[2] Annual ryegrass may also be controlled preemergence in winter wheat. See appropriate recommendation section in this labeling.

When using Hoelon 3EC Herbicide to control annual grassy weeds, and the weeds do not include annual ryegrass, Hoelon 3EC Herbicide should be applied at a rate of 2.0 to 2 2/3 pints per acre. Apply the 2.0 pint per acre rate only when the annual grassy weeds are in the 1-2 leaf stage of growth and growing conditions are normal. The higher rate should be used when the development of the annual grassy weeds are not clearly in the specified stages of growth, or if growth is retarded due to adverse growing conditions.

Special Notes For Dry Field Peas and Lentils

1) DO NOT apply more than one application of Hoelon 3EC Herbicide in a growing season.
2) DO NOT allow livestock to graze treated fields. DO NOT harvest *pea vine* forage or hay from treated fields for livestock feed.
3) DO NOT tank mix with any broadleaf herbicides for use on dry field peas and lentils.
4) DO NOT tank mix with liquid fertilizers.
5) DO NOT apply by air on dry field peas and lentils.
6) Hoelon 3EC Herbicide may be tank mixed with crop oil concentrate at 1-2 pints per acre for use on dry field peas.
7) DO NOT tank mix Hoelon 3EC Herbicide with crop oil concentrate for use on lentils.
8) Apply prior to the bloom stage of dry field peas and lentils.

DIRECTIONS FOR THE POSTEMERGENCE USE OF HOELON 3EC HERBICIDE ON FLAX

This product is registered for postemergence use on flax in the following states: Minnesota, Montana, North Dakota and South Dakota.

(The general information section must also be read for complete use instructions.)

Apply Hoelon 3EC Herbicide as a broadcast postemergence spray at the rates listed below:

Susceptible Annual Grassy Weeds	Amount of Hoelon 3EC Herbicide Per Acre (Pints)[1] Relative To Growth Stage of Annual Grassy Weeds	
	1-3 Leaf	3-4 Leaf
Wild Oat *(Avena fatua)*	2.0	2 2/3
Green Foxtail *(Setaria viridis)*	2.0	2 2/3
Yellow Foxtail[2] *(Setaria lutescens)*	2.0	2 2/3
Barnyardgrass (Watergrass) *(Echinochloa crusgalli)*	2.0	2 2/3
Persian Darnel *(Lolium persicum)*	2 2/3	DO NOT APPLY

[1] When controlling mixed populations of annual grassy weeds, always use the rate which will control the least susceptible species.
[2] For best control of yellow foxtail, application should be made before the second leaf fully emerges.

When using Hoelon 3EC Herbicide in flax to control annual grassy weeds (and the weeds do not include Persian darnel) apply the 2.0 pint per acre rate when the annual grassy weeds are in the 1-3 leaf stage and growing conditions are optimum. The 2 2/3 pint per acre rate should be used if the grassy weeds are not clearly in the specified stage of growth, or if growth is retarded due to adverse growing conditions.

Special Notes for Flax

1) DO NOT apply more than one application in a growing season.
2) DO NOT tank mix Hoelon 3EC Herbicide with liquid nitrogen fertilizer.
3) DO NOT tank mix Hoelon 3EC Herbicide with crop oil concentrate.
4) DO NOT allow livestock to graze treated fields. DO NOT feed flax straw treated with Hoelon 3EC Herbicide.
5) DO NOT apply Hoelon 3EC Herbicide within 70 days of flax harvest.

DIRECTIONS FOR THE USE OF HOELON 3EC HERBICIDE ON ACREAGE CONSERVATION RESERVE

This product is registered for the postemergence use on acreage conservation reserve in the following states:

Delaware	Kansas	Michigan	New Jersey	S. Dakota
Illinois	Kentucky	Minnesota	N. Dakota	Virginia
Indiana	Louisiana	Missouri	Ohio	Wisconsin
Iowa	Maryland	Nebraska	Pennsylvania	

(The general information section must also be read for complete use instructions.)

Apply Hoelon 3EC Herbicide as a broadcast postemergence spray at the rates listed below.

SUSCEPTIBLE ANNUAL GRASSY WEEDS	AMOUNT OF HOELON 3EC HERBICIDE PER ACRE (PINTS)[1] RELATIVE TO GROWTH STAGE OF ANNUAL GRASSY WEEDS			
	1-3 LEAF	3-4 LEAF	4-5 LEAF	5 LEAF-2 TILLERS
Annual Ryegrass (Italian)[2] *(Lolium multiflorum)*	1⅓	1⅓	2.0	2⅔
Broadleaf Signalgrass (suppression) *(Brachiaria platyphylla)*	2⅔			
Crabgrass[3] *(Digitaria sanguinalis)*	2⅔			
Witchgrass (suppression) *(Panicum capillare)*	2⅔			
Fall Panicum *(Panicum dichotomiflorum)*				
Barnyardgrass (watergrass) *(Echinochloa crusgalli)*				DO NOT APPLY
Giant Foxtail *(Setaria faberii)*	2.0 to 2⅔	2⅔		
Green Foxtail (pigeongrass) *(Setaria viridis)*				
Yellow Foxtail (pigeongrass)[3] *(Setaria lutescens)*				REDUCED GRASSY WEED CONTROL WILL OCCUR
Wild Oat *(Avena fatua)*				
Itchgrass (Raoulgrass) *(Rottboellia exaltata)*	2.0 (1-24")	2⅔ (24-36")		
Volunteer Corn *(Zea mays)*	2.0 to 2⅔	2⅔		

[1] When controlling mixed populations of grassy weeds, always use the rate that will control the least susceptible species.
[2] Annual ryegrass may also be controlled preemergence in winter wheat. See appropriate recommendation section in this labeling.
[3] For best control of crabgrass and yellow foxtail, application should be made before the second leaf fully emerges.

Apply Hoelon® 3EC Herbicide at the higher labeled rates when the majority of annual grassy weeds is not clearly in the specified stages of growth, or if growth is retarded due to adverse growing conditions.

Crop Oil Concentrate

Under adverse growing conditions (drought), the addition of 1 to 2 pints of crop oil concentrate may result in improved weed control. Use a crop oil concentrate containing a blend of 80% (minimum) petroleum or vegetable base oil, with the remaining composed of a tolerance-exempt surfactant.

DIRECTIONS FOR THE USE OF HOELON 3EC HERBICIDE ON ACREAGE CONSERVATION RESERVE (SET-ASIDE)

(The general information section must also be read for complete use instructions.)

Hoelon 3EC Herbicide may be used to control the annual grassy weeds in acreage conservation reserve or set-aside programs. This acreage is often seeded to wheat, clover, or alfalfa as a cover crop. These cover crops have excellent tolerance to Hoelon 3EC Herbicide. Use 2.0 pints to 2⅔ pints per acre when the annual grassy weeds are in the 1 to 4 leaf stage as noted in the use directions section of this labeling. In most situations, 2.0 pints of Hoelon 3EC Herbicide plus 2 pints of crop oil concentrate will offer commercially acceptable control of annual grasses in acreage conservation reserve programs. Apply early to young, actively growing grassy weeds. DO NOT harvest or graze cover crops treated with Hoelon 3EC Herbicide. DO NOT apply to cover crops such as oats, sorghum, and sudangrass, as injury may occur.

KEEP OUT OF REACH OF CHILDREN
DANGER

PRECAUTIONARY STATEMENTS

HAZARDS TO HUMANS AND DOMESTIC ANIMALS

Causes eye and skin damage. Do not get in eyes or on skin or clothing. Wear goggles and respirator when handling. Avoid contact or inhalation of spray mist. Wash thoroughly after using and change clothing. Do not take internally.

The following protective apparel must be worn during all loading and mixing operations and when applied by ground equipment: impermeable rubber gloves (such as neoprene or PVC), waterproof boots, impermeable pants and shirts, goggles and a cartridge type respirator. Aerial pilots are not to be involved in the mixing and loading unless the aforementioned equipment is worn. Pilots must wear a cartridge type respirator during application unless the aircraft is equipped with an air filter system.

The Environmental Protection Agency has concluded that diclofop-methyl, the active ingredient in this product, produced tumors in laboratory mice. This effect was not observed in other species tested. The user must read and follow all precautionary statements on this label. Give particular attention to protective clothing requirements.

ENVIRONMENTAL HAZARDS

This pesticide is toxic to fish. Use with care when applying in areas adjacent to any body of water. Keep out of lakes, streams, ponds, drainage basins, tidal marshes and estuaries. Do not apply when weather conditions favor runoff or drift. Avoid direct application or drift of spray material to water surfaces. Do not apply within 100 feet of aquatic habitat. Do not contaminate arable land and/or water when disposing of equipment washwaters.

STATEMENT OF PRACTICAL TREATMENT

In case of skin or eye contact, immediately flush with plenty of water for at least 15 minutes.

If swallowed, DO NOT INDUCE VOMITING, but seek medical attention immediately

Note to Physician: Aspiration hazard may contraindicate the use of gastric lavage.

For emergency assistance call (201) 231-2000.

STORAGE AND DISPOSAL

Do not contaminate water, food, or feed by storage or disposal.

Do not use or store near heat or open flame.

PESTICIDE DISPOSAL: Pesticide wastes are toxic. Improper disposal of excess pesticide, spray mixture, or rinsate is a violation of Federal Law. If these wastes cannot be disposed of by use according to label instructions, contact your State Pesticide or Environmental Control Agency, or the Hazardous Waste representative at the nearest EPA Regional Office for guidance.

CONTAINER DISPOSAL: Triple rinse (or equivalent). Then offer for recycling or reconditioning, or puncture and dispose of in a sanitary landfill, or by other procedures approved by state and local authorities.

Do not store below 20°F.

IMPORTANT NOTICE: DISCLAIMER:

Read "IMPORTANT NOTICE: DISCLAIMER" before buying or using. If terms are not acceptable, return at once unopened. HOECHST CELANESE CORPORATION warrants only that the product conforms to the chemical description on the label and is reasonably fit for the purpose stated on the label when used in accordance with the directions under normal conditions of use. This warranty does not extend to the use of this product contrary to label instructions or under abnormal conditions, or under conditions not reasonably foreseeable to HOECHST CELANESE CORPORATION AND USER ASSUMES THE RISK OF ANY SUCH USE. HOECHST CELANESE CORPORATION MAKES NO OTHER WARRANTY, EXPRESS OR IMPLIED, INCLUDING ANY IMPLIED WARRANTY OF FITNESS FOR A PARTICULAR PURPOSE OR OF MERCHANTABILITY. In no case shall HOECHST CELANESE CORPORATION be liable for consequential, special, indirect or incidental damages resulting from the use or handling of this product. The aforegoing conditions of sale and warranty can be varied only by an agreement in writing signed by a duly authorized representative of HOECHST CELANESE CORPORATION.

Hoelon® 3EC Herbicide is a registered trademark of Hoechst AG.
Buctril® Herbicide is a registered trademark of Rhone-Poulenc Inc., Agrochemical Division.
Glean® Herbicide is a registered trademark of E. I. DuPont deNemours and Company, Inc.

TH8501007 (10/88)

"This is a specimen label. Hoechst-Roussel Agri-Vet Company is not responsible for the accuracy of the information contained herein, as labels are subject to revision. Always read and follow the use instructions on the container label"

HORIZON®

1EC HERBICIDE

For Selective Postemergence Annual and Perennial Grass Control in Rights-of-Way

ACTIVE INGREDIENT:	
fenoxaprop-ethyl:(±)-ethyl 2-[4-[(6-chloro-2-benzoxazolyl) oxy]phenoxy]propanoate	12.50%*
INERT INGREDIENTS:	87.50%
TOTAL	100.00%

*Equivalent to 1 pound of active ingredient per gallon

EPA REG. NO. 8340-18-54382

KEEP OUT OF REACH OF CHILDREN

WARNING

SEE ADDITIONAL PRECAUTIONARY STATEMENTS AT THE BEGINNING OF LABEL

NOT APPROVED FOR USE IN THE STATES OF CALIFORNIA AND HAWAII.

PRECAUTIONARY STATEMENTS

WARNING

HAZARDS TO HUMANS AND DOMESTIC ANIMALS

May cause substantial but temporary eye injury. Do not get in eyes. Wear goggles or face shield and impermeable rubber gloves (such as neoprene or PVC) while mixing. Harmful if swallowed, absorbed through skin or inhaled. Do not take internally. Avoid inhalation of vapor or spray mist. Remove contaminated clothing and wash before reuse. Wash thoroughly with soap and water after handling.

ENVIRONMENTAL HAZARDS

This pesticide is toxic to fish. Do not apply directly to water. Do not apply when weather conditions favor runoff or drift. Do not contaminate arable land and/or water when disposing of equipment washwaters.

STORAGE AND DISPOSAL

Do not contaminate water, food or feed by storage or disposal.

Do not store over 100°F or below 10°F.

Do not use or store near heat or open flame.

Pesticide Disposal: Pesticide wastes are toxic. Improper disposal of excess pesticide, spray mixture, or rinsate is a violation of Federal Law. If these wastes cannot be disposed of by use according to label instructions, contact your State Pesticide or Environmental Control Agency, or the Hazardous Waste representative at the nearest EPA Regional Office for guidance.

Container Disposal: Empty containers should be triple rinsed into the spray tank during the spray operation. Then offer for recycling or reconditioning, or puncture and dispose of in a sanitary landfill, or incineration, or, if allowed by state and local authorities, by burning. If burned, stay out of smoke.

STATEMENT OF PRACTICAL TREATMENT

In case of eye contact, immediately flush with plenty of water for 15 minutes. Get medical attention.

If swallowed, do not induce vomiting. Horizon® 1EC Herbicide contains petroleum distillates. Call a physician. Vomiting should be supervised by a physician because of the possible pulmonary damage via aspiration of the solvent.

If possible bring this container and labeling to the attending physician. For emergency assistance call (201) 231-2000.

IMPORTANT NOTICE: DISCLAIMER

Read "IMPORTANT NOTICE: DISCLAIMER" before buying or using. If terms are not acceptable, return at once unopened. HOECHST-ROUSSEL AGRI-VET COMPANY warrants only that the product conforms to the chemical description on the label and is reasonably fit for the purpose stated on the label when used in accordance with the directions under normal conditions of use. This warranty does not extend to the use of this product contrary to label instructions or under abnormal conditions, or under conditions not reasonably foreseeable to HOECHST-ROUSSEL AGRI-VET COMPANY, and user assumes the risk of any such use. HOECHST-ROUSSEL AGRI-VET COMPANY MAKES NO OTHER WARRANTY, EXPRESSED OR IMPLIED, INCLUDING ANY IMPLIED WARRANTY OF FITNESS FOR A PARTICULAR PURPOSE OR OF MERCHANTABILITY. In no case shall HOECHST-ROUSSEL AGRI-VET COMPANY be liable for consequential, special, indirect or incidental damages resulting from the use or handling of this product. The foregoing conditions of sale and warranty can be varied only by an agreement in writing signed by a duly authorized representative of HOECHST-ROUSSEL AGRI-VET COMPANY.

GENERAL INFORMATION

Horizon 1EC Herbicide is an emulsifiable concentrate formulation containing one pound of active ingredient per gallon. Horizon 1EC Herbicide is used for postemergence control of annual and perennial grassy weeds in several established grass species. Horizon 1EC Herbicide does not control broadleaf weeds or sedges.

Since Horizon 1EC Herbicide is absorbed primarily through the foliage, thorough coverage is necessary. Visual injury of the grassy weed is evident approximately 4-10 days after application (dependent upon environmental conditions) but control of the grassy weeds will take 12-21 days. Visible effects begin as a general chlorosis (yellowing) and reddening of the leaf surface followed by death of the weed.

DIRECTIONS FOR USE

It is a violation of Federal law to use this product in a manner inconsistent with its labeling. This label must be in the possession of the user at the time of application.

Horizon 1EC Herbicide can be applied to all cultivars of the following cool season grass species grown throughout the United States:

Smooth Brome (*Bromus inermis*)
Perennial Ryegrass (*Lolium perenne*)
Fine Fescue (*Festuca rubra* spp.)
Tall Fescue (*Festuca arundinaceae*)
Common Bermudagrass (*Cynodon dactylon*)
Wheatgrass (*Agropyron* spp.)

Horizon 1EC Herbicide can be applied as a broadcast spray for control of grassy weeds in the following wildflower species:

Common Name	Scientific Name
African Daisy	*Dimorphotheca*
Annual Indian Blanket	*Gaillardia*
Baby Blue Eyes	*Nemophila*
Baby's Breath	*Gypsophila*
Birds Eyes	*Gilia*
Black-Eyed Susan	*Rudbeckia*
Blue Bells	*Mertensia*
Blue Flax	*Linum*
California Poppy	*Eschscholzia*
Catchfly	*Silene*
Corn Poppy	*Papaver*
Cosmos	*Cosmos*
Creeping Zinnia	*Sanvitalia*
Dames Rocket	*Hesperis*
Dwarf Columbine	*Aquilegia*
Dwarf Cornflower	*Centaurea*
Dwarf Lance Leafed Coreopsis	*Coreopsis*
English Wallflower	*Cheiranthus*
Forget-Me-Not	*Anchusa*
Garland Chrysanthemums	*Chrysanthemum*
Gilia	*Gilia*
Glove Gilia	*Gilia*
Johnny Jump Up	*Viola*
Lemon Mint	*Monarda*
Maiden Pinks	*Dianthus*
Missouri Primrose	*Oenohera*
Mountain Phlox	*Phlox*
Pimpernel	*Anagallis*
Prairie Coneflower	*Ratibida*
Purple Coneflower	*Echinacea*
Red Yarrow	*Achillea*
Rocket Larkspur	*Delphinium*
Rocky Mountain Penstemon	*Penstemon*
Roman Chamomile	*Matricaria*
Scarlet Flax	*Linum*
Siberian Wallflower	*Erysimum*
Small Burnet	*Sanguisorba*
Soapwort	*Saponaria*
Spurred Snapdragon	*Linaria*
Starflower	*Trientalis*
Sweet Allyssum	*Lobularia*
Sweet William	*Dianthus*
Tall Evening Primrose	*Oenothera*
Tall Godetia	*Godetia*
Tall Plains Coreopsis	*Coreopsis*
Tidy Tips	*Layia*

White Yarrow	*Achillea*
Wild Thyme	*Thymus*

Horizon 1EC Herbicide will control the following grassy weeds:

- Smooth Crabgrass (*Digitaria ischaemum*)
- Hairy Crabgrass (*Digitaria sanguinalis*)
- Goosegrass/Silver Crabgrass (*Eleusine indica*)
- Barnyardgrass (*Echinochloa crus-galli*)
- Foxtail Species (*Setaria* spp.)
- Panicum Species (*Panicum* spp.)
- Johnsongrass (*Sorghum halepense*)
- Wild Oats (*Avena fatua*)
- Roughstalk Bluegrass (*Poa trivialis*)

Application Information

Horizon® 1EC Herbicide only controls grasses which are emerged at the time of spraying. Young actively growing grassy weeds are more easily controlled than the larger grassy weeds. Refer to the Horizon 1EC Herbicide application rate and timing recommendation chart for application rates and timing of application. Avoid applications to grass under drought stress.

Spray Application For Spot Treatments

Add 3 pints of Horizon 1EC Herbicide per 100 gallons of water and spray target weed thoroughly to the point of spray deposit runoff. Optimum results occur when spray pressure is sufficient to provide retention of spray deposit on target weed.

Highway Rights-of-Way

Apply with ground equipment using 30-100 gallons of water per acre and 30-60 PSI of water to uniformly cover the vegetation in the area to be treated. Use a fixed boom, off-center nozzles or boomless straight stream nozzles properly calibrated to a constant speed of travel and rate of delivery. Allow mowed areas to regrow for 7-14 days before applying Horizon 1EC Herbicide.

Repeat Application Information

Repeat applications of Horizon 1EC Herbicide at 12-45 fluid ounces per acre (refer to application and timing chart) may be required under the following conditions:

(a) Continued grassy weed seed germination after the initial application

(b) When dense populations of weeds prevent thorough spray coverage of the entire leaf surface of the target weeds

Do not apply more than three applications or a total of 135 ounces per acre (3.09 ounces/1000 sq.ft.) per growing season.

Avoid applying a second application of Horizon 1EC Herbicide sooner than 14 days after the first application. A second Horizon 1EC Herbicide application should only be made to grassy weeds with vigorous regrowth.

Surfactants and Crop Oils

The addition of a non-ionic surfactant or paraffin-based crop oil at a rate of 32 ounces per 100 gallons of spray solution is recommended for spray volumes between 30 and 100 gallons per acre. Thorough spray coverage is extremely important for optimum results.

NOTES:

1. Adequate soil moisture will enhance the performance of Horizon 1EC Herbicide. Reduced control may occur with Horizon 1EC Herbicide applied under *DROUGHT STRESS* conditions.

2. Rainfall within one hour of application may cause a reduction in grass control.

3. Broadleaf herbicides containing 2,4-D will reduce the effectiveness of Horizon 1EC Herbicide. Tank mixes with Telar, Escort, or Garlon are recommended. Broadleaf herbicides containing 2,4-D may be applied 5 days before or after a Horizon 1EC Herbicide application. *DO NOT TANK MIX* with any herbicides when applying Horizon 1EC Herbicide to wildflower species.

4. Horizon 1EC Herbicide is a selective grassy weed herbicide, and has little or no activity on broadleaf plants or sedges.

5. Fescues and Ryegrass may be overseeded immediately following Horizon 1EC Herbicide applications. All other grass species should be overseeded after a 21-day waiting period.

6. Applications to Bermudagrass may result in temporary injury. Bermudagrass should be well established at the time of application or severe injury may result.

HORIZON USE RATES PER ACRE

Grassy Weed Growth Stage	Fluid Ounces Per Acre	Pints Per Acre
ANNUAL GRASSES		
1-2 Tillers	19-24	1.2-1.5
2-4 Tillers	24-32	1.5-2.0
4-5 Tillers	32-45	2.0-2.8
Grassy Weed Growth Stage	Fluid Ounces Per 100 Gallons	Pints Per 100 Gallons
SEEDLING JOHNSONGRASS		
12-24"	19-24	1.2-1.5
RHIZOME JOHNSONGRASS		
18-48 Inches	32-45	2.0-2.8
Repeat Applications	12-45	0.8-2.8

TH 8707002-(2/89)

SUPPLEMENTAL LABELING

HORIZON 1EC HERBICIDE

EPA SLN NO. OR-870016

FOR DISTRIBUTION AND USE ONLY WITHIN THE STATE OF OREGON UNDER A SPECIAL LOCAL NEED AUTHORIZATION UNDER FIFRA SECTION 24(c)

FOR SELECTIVE POSTEMERGENCE ANNUAL AND PERENNIAL GRASS CONTROL IN TURFGRASS SEED PRODUCTION

GENERAL INFORMATION

Horizon 1EC Herbicide is an emulsifiable concentrate formulation containing one pound of active ingredient per gallon. Horizon 1EC Herbicide is used for postemergence control of annual and perennial grassy weeds in several established perennial grass crops. Horizon 1EC Herbicide does not control broadleaf weeds or sedges.

Since Horizon 1EC Herbicide is absorbed primarily through the foliage, thorough coverage is necessary. Visual injury of the grassy weed is evident approximately 4–10 days after application (dependent upon environmental conditions), but control of the grassy weeds will take 12–21 days. Visible effects begin as a general chlorosis (yellowing) and reddening of the leaf surface followed by death of the weed.

DIRECTIONS FOR USE

It is a violation of Federal law to use this product in a manner inconsistent with its labeling. This label must be in the possession of the user at the time of application.

Horizon 1EC Herbicide can be applied to all cultivars of the following cool season grass species grown for seed that have been established for at least one (1) month:

- Perennial Ryegrass *(Lolium perenne)*
- Fine Fescue *(Festuca rubra* spp.)
- Tall Fescue *(Festuca arundinaceae)*

Horizon 1EC Herbicide will control the following grassy weeds:

- Smooth Crabgrass *(Digitaria ischaemum)*
- Hairy Crabgrass *(Digitaria sanguinalis)*
- Goosegrass/Silver Crabgrass *(Eleusine indica)*
- Barnyardgrass *(Echinochloa crus-galli)*
- Foxtail Species *(Setaria* spp.*)*
- Panicum Species *(Panicum* spp.*)*
- Johnsongrass *(Sorghum halepense)*
- Wild Oats *(Avena fatua)*
- Roughstalk Bluegrass *(Poa trivialis)*

APPLICATION INFORMATION

Horizon 1EC Herbicide only controls grasses which are emerged at the time of spraying. Young actively growing grassy weeds are more easily controlled than the larger grassy weeds. Refer to the Horizon 1EC Herbicide application rate and timing recommendation chart for application rates and timing of application. Avoid applications to these grass crops under drought stress.

Hand-Held Pump Sprayer Application For Spot Treatments

For hand-held pump-type sprayers, add 0.5 fluid ounce of Horizon 1EC Herbicide per gallon of water and spray target weed thoroughly to the point of spray deposit runoff. **Optimum results occur when *spray pressure* is sufficient to provide retention of spray deposit on target weed.**

TURFGRASS SEED PRODUCTION

A. Ground Application: Horizon 1EC Herbicide should be applied in a minimum of ten (10) gallons of water per broadcast acre. Flat-fan or hollow-cone nozzles are recommended. Use a minimum pressure of forty (40) PSI. Under dense weed canopies, high spray pressure is very important for obtaining thorough coverage; therefore, use higher spray pressure under these conditions.

B. Air Application: Horizon 1EC Herbicide should be applied in a minimum of 10 gallons of water per broadcast acre. Uniform spray coverage is essential when using aircraft and is achieved by the use of spray droplet size (200-350 micron range). DO NOT USE raindrop nozzles. Aerial applications with this product should be made at a height which provides the most effective swath width for the aircraft.

DO NOT APPLY by aircraft when wind speeds exceed 8 mph. Avoid all direct or indirect contact to neighboring fields.

NOTES:

1. DO NOT APPLY Horizon 1EC Herbicide to Kentucky Bluegrass seed production fields.

2. Avoid application of Horizon 1EC Herbicide under DROUGHT STRESS conditions. Irrigation several days prior to application may be helpful under these conditions.

3. Rainfall or irrigation within one hour of application may cause a reduction in grass control. Allow spray deposit to dry before irrigation.
4. Broadleaf herbicides such as dicamba and phenoxy-type herbicides reduce the effectiveness of Horizon 1EC Herbicide. Broadleaf herbicides may be applied 5 days before or after a Horizon 1EC Herbicide application.
5. Horizon 1EC Herbicide is a selective grassy weed herbicide and has little or no activity on broadleaf plants or sedges.
6. Because Horizon 1EC Herbicide controls only a limited number of the weedy grasses that are commonly found in turfgrass grown for seed in Oregon, it is important that the user apply other registered, broad spectrum herbicides for the control of seedling grasses during the fall. Species *not* controlled by Horizon® 1EC Herbicide include:

Annual Bluegrass *(Poa annua)*
California brome *(Bromus carinatus)*
Rattail Fescue *(Vulpia myuros)*

7. DO NOT APPPLY this product through any type of irrigation system.
8. DO NOT graze or feed fresh hay, hay, or seed screenings to livestock.
9. DO NOT APPLY fertilizer within 2 weeks prior to Horizon 1EC Herbicide application.
10. Under cool and wet conditions, wheel tracking may be observed following Horizon 1EC Herbicide application.

HORIZON 1EC HERBICIDE APPLICATION RATE AND TIMING RECOMMENDATION CHART

Weed Species	Stage of Growth of Grassy Weeds	fl. oz/Acre	Pints/Acre
Smooth Crabgrass Hairy Crabgrass Goosegrass/ Silver Crabgrass Barnyardgrass Foxtail Species Panicum Species Johnsongrass	2-leaf to 1-tiller	19	1.2
	1-tiller to 2-tiller	24	1.5
Roughstalk Bluegrass	2 inches to 7 inches	24–32	1.5–2.0
In perennial ryegrass, apply after grasses break winter dormancy and ryegrass is 3–6" tall. Rate is dependent upon environmental conditions and size, density, and maturity of weeds. Application should be made between February 20 and April 1.			
Wild Oats	2-leaf to 1-tiller	19	1.2
Maximum rate for wild oat control in seedling annual and perennial ryegrass. Only for Gulf, Common, Barspectra, and Promenade cultivars of annual ryegrass.			
Wild Oats	1-tiller to 3-tiller	24	1.5
Rate for larger wild oats in tall and fine fescues.			

TH 8707001–(1/89)

"This is a specimen label. Hoechst-Roussel Agri-Vet Company is not responsible for the accuracy of the information contained herein, as labels are subject to revision. Always read and follow the use instructions on the container label".

RELY®
HERBICIDE

For Selective Postemergence Broadleaf Weed Control in Rice Only in the State of Arkansas under the Authority of Section 18 of FIFRA

ACTIVE INGREDIENT:
Triclopyr (3,5,6-trichloro-2-pyridinyloxyacetic acid), as the triethylamine salt 44.4%
INERT INGREDIENTS: 55.6%
TOTAL. 100.0%
ACID EQUIVALENT: Triclopyr - 31.8% - 3 lb/gal

Section 18 Exemption (Arkansas) 89-AR-04
This specific exemption expires August 1, 1989.
The name and logo HOECHST are registered trademarks of Hoechst AG.
The name and logo ROUSSEL are registered trademarks of Roussel Uclaf S.A.

KEEP OUT OF REACH OF CHILDREN
DANGER
PELIGRO

PRECAUCION AL USARIO:
Si usted no lee ingles, no use este producto hasta que la etiqueta le haya sido explicada ampliamente.

PRECAUTIONARY STATEMENTS

DANGER

HAZARDS TO HUMANS AND DOMESTIC ANIMALS
Causes eye damage and skin irritation. Do not get in eyes, on skin or on clothing. Wear goggles or face shield and rubber gloves when handling. Harmful if swallowed. Avoid contamination of food. Wash thoroughly after handling.

ENVIRONMENTAL HAZARDS
Do not apply directly to a body of water outside the treated rice field. Do not contaminate water by cleaning of equipment or disposal of wastes.

STORAGE AND DISPOSAL
DO NOT contaminate water, food or feed by storage or disposal.

Store above 28°F or agitate before use. DO NOT use or store near heat or open flame.

Pesticide Disposal: Pesticide wastes are acutely hazardous. Improper disposal of excess pesticide, spray mixture, or rinsate is a violation of Federal Law. If these wastes cannot be disposed of by use according to label instructions, contact your State Pesticide or Environmental Control Agency, or the hazardous waste representative at the nearest EPA regional office for guidance.

Container Disposal: Triple rinse (or equivalent). Then offer for recycling or reconditioning, or puncture and dispose of in a sanitary landfill, or by other procedures approved by state and local authorities. Plastic containers may also be disposed of by incineration, or, if allowed by state and local authorities, by burning. If burned, stay out of smoke.

Prohibitions: Open dumping is prohibited.

General: Consult Federal, state or local disposal authorities for approved alternative procedures.

NOTICE: Read the entire label. Use only according to label directions.

STATEMENT OF PRACTICAL TREATMENT
In case of eye contact, immediately flush eyes with plenty of water for at least 15 minutes and get medical attention promptly. Flush skin with water and get medical attention if signs of irritation develop. Remove and wash contaminated clothing before reuse. If swallowed, promptly drink a large quantity of milk, egg whites, gelation solution or, if these are not available, drink large quantities of water. AVOID ALCOHOL. Call a physician.

In case of an emergency endangering life or property involving this product, call (201) 231-2000.

NOTE TO PHYSICIAN: Probable mucosal damage may contraindicate use of gastric lavage.

IMPORTANT NOTICE: DISCLAIMER
Read "IMPORTANT NOTICE: DISCLAIMER" before buying or using. If terms are not acceptable, return at once unopened. HOECHST-ROUSSEL AGRI-VET COMPANY warrants only that the product conforms to the chemical description on the label and is reasonably fit for the purpose stated on the label when used in accordance with the directions under normal conditions of use. This warranty does not extend to the use of this product contrary to label instructions or under abnormal conditions, or under conditions not reasonably foreseeable to HOECHST-ROUSSEL AGRI-VET COMPANY and user assumes the risk of any such use. HOECHST-ROUSSEL AGRI-VET COMPANY MAKES NO OTHER WARRANTY, EXPRESSED OR IMPLIED, INCLUDING ANY IMPLIED WARRANTY OF FITNESS FOR A PARTICULAR PURPOSE OR OF MERCHANTABILITY. In no case shall HOECHST-ROUSSEL AGRI-VET COMPANY be liable for consequential, special, indirect or incidental damages resulting from the use or handling of this product. The foregoing conditions of sale and warranty can be varied only by an agreement in writing signed by a duly authorized representative of HOECHST-ROUSSEL AGRI-VET COMPANY.

GENERAL INFORMATION
Rely® Herbicide is a postemergence systemic herbicide for the control of Northern jointvetch (*Aeschynomene virginica*), Morningglory species (*Ipomoea* spp.), and Hemp sesbania (*Sesbania exaltata*) in rice only. Rely Herbicide controls certain broadleaf weeds through foliar uptake; therefore, thorough coverage of target weeds is important. DO NOT allow spray to drift onto adjacent broadleaf crops as crop injury may occur.

DIRECTIONS FOR USE
It is a violation of Federal Law to use this product in a manner inconsistent with its labeling. DO NOT apply until you have read the entire label. For use only in the state of Arkansas under the Authority of Section 18 of FIFRA as amended. DO NOT apply to any crop other than rice. Read and follow all label directions.

RICE
Rice is tolerant to postemergence applications of Rely Herbicide from the 4-leaf stage to the 1/2" internode elongation stage of rice development. Postemergence applications of the higher rates of Rely Herbicide may result in temporary rice injury that appears as leaf chlorosis or stunting. The rice will normally recover from these symptoms in 2-4 weeks.

Rice Varieties

Rely Herbicide has shown selectivity when tested on the following rice varieties when applied from the 4-leaf stage to the 1/2" internode elongation stage: Lemont, Newbonnet, Tebonnet, Mars, and Skybonnet. DO NOT apply to Bond, Starbonnet, and Lebonnet, as the selectivity to Rely Herbicide has not been determined at this time.

APPLICATION INFORMATION

Rely Herbicide should be applied at 2/3 to 1 pint (0.25-0.375 lb. a.e./A) to actively growing broadleaf weeds when the rice growth stage is in the 4-leaf to 1/2" internode elongation stage of development. DO NOT apply to rice in the booting, or subsequent stages. Apply 1 pint when broadleaf weeds are large or as a postflood application.

AERIAL APPLICATION

Rely Herbicide should be applied in a minimum of 10 gallons of water per acre at a height which is the most effective swath width for the aircraft. Fixed-wing aircraft or helicopters should have a well designed spray system that produces a uniform spray pattern and minimizes spray drift.

GROUND APPLICATION

Rely Herbicide should be applied by ground application in a minimum of ten (10) gallons of water per broadcast acre. Flat fan nozzles are recommended. A spray pressure of twenty (20) PSI is recommended. Utilize a well designed spray system that produces a uniform spray pattern and one that will minimize spray drift.

AVOID SPRAY DRIFT

Applications should be made only when there is little or no hazard from spray drift. Very small quantities of spray, which may not be visible, may seriously injure susceptible plants. DO NOT spray when wind is blowing toward susceptible crops or ornamental plants that are near enough to be injured. It is suggested that a continuous smoke column at the spray site be used to detect air movement, lapse conditions, or temperature inversions. If the smoke layers or indicates a potential of hazardous spray drift, do not spray.

Drift can be lessened by:

1. Applying a coarse spray
2. Selecting the minimum spray pressure which provides adequate plant coverage
3. Using a spray boom no longer than 3/4 the rotor or wing length
4. Spraying only when wind velocities are low
5. Adding an approved drift control agent to the spray tank for aerial applications to rice near susceptible crops. Nalco-Trol or Sta-put may be added to the spray tank, following manufacturers' recommendations.

TANK MIX RECOMMENDATIONS

Rely Herbicide may be tank mixed with Stam® M-4 Herbicide, Propanex® 4, or Cedar Propanil® when grass control is desired. Use 2/3 to 1 pint (0.25-0.375 lb. a.e./A) of Rely Herbicide in combination with 3 to 4 quarts (3-4 lbs./A) of Stam M-4 Herbicide, Propanex 4, or Cedar Propanil per acre. This tank mixture may be applied after a satisfactory stand of rice has been established (minimum 4-leaf stage of rice). When tank mixing, always follow the use directions and precautions in accordance with each respective label. No label dosage rates should be exceeded. DO NOT add a surfactant with this tank mix.

ADDITIVES

The addition of a non-ionic surfactant such as X-77, AG-98, or similar EPA agriculturally approved wetting agent at 1/4%-1/2% by volume is recommended for increased broadleaf weed control when Rely Herbicide is used alone.

WATER MANAGEMENT

For preflood applications, the rice should be in the 4-leaf stage or larger. A shallow flood may be applied 24 hours following the Rely Herbicide application. If the weeds are drought stressed, flush the field before applying Rely Herbicide.

For postflood applications, treatments should be made only when the weeds are well emerged above the surface of the water. Weeds submerged at the time of the application will not be controlled. Weeds larger than 20-24" in size may not be adequately controlled. DO NOT raise water levels for at least 24 hours after the application.

SPECIAL NOTES

1. Tank mix Rely® Herbicide only with the pesticides recommended on this label.
2. Applications of Rely Herbicide after the 1/2" internode elongation stage of rice development may cause injury to the rice.
3. Applications to ditches used to transport irrigation water are prohibited.
4. DO NOT apply where runoff or irrigation water may flow onto agricultural land other than rice fields as injury to crops may occur.
5. DO NOT graze lactating dairy animals or harvest hay from treated areas for one year following treatment. Withdraw livestock from forage treated with Rely Herbicide at least 3 days before slaughter during the year of treatment.
6. DO NOT rotate treated land to crops other than rice for 6 months following treatment.
7. Apply no more than 1 pint (0.375 lb.) per acre per growing season.
8. DO NOT apply Rely Herbicide directly to, or otherwise permit it, to come into direct contact with grapes, tobacco, vegetable crops, cotton, soybeans, flowers, shrubs, desirable trees, or other desirable broadleaf plants.
9. DO NOT apply later than 45 days before harvesting rice.
10. DO NOT apply this product through any type of irrigation system.
11. DO NOT apply within 1/4 mile of cotton or soybean fields.

Stam® M-4 Herbicide is a registered trademark of Rohm and Haas Company.

Propanex® 4 is a registered trademark of Crystal Chemical Company.

Cedar Propanil® is a registered trademark of Cedar Chemical Company.

TH8909003 (5/89)

"This is a specimen label. Hoechst-Roussel Agri-Vet Company is not responsible for the accuracy of the information contained herein, as labels are subject to revision. Always read and follow the use instructions on the container label".

SCOUT X-TRA™
INSECTICIDE

RESTRICTED USE PESTICIDE

TOXIC TO FISH AND AQUATIC ORGANISMS

For retail sale to and use only by certified applicators or persons under their direct supervision and only for those uses covered by the certified applicator's certification.

NOT FOR SALE OR DISTRIBUTION AFTER DECEMBER 31, 1989
NOT FOR USE AFTER DECEMBER 31, 1990

ACTIVE INGREDIENT: (1R, 3S) 3 [(1' RS) (1', 2', 2', 2', -tetrabromoethyl)]-2, 2-dimethylcyclopropanecarboxylic acid (S)-alpha-cyano-3-phenoxybenzyl ester	10.30%*
INERT INGREDIENTS**:	89.70%
TOTAL	100.00%

* Equivalent to 0.9 pound of active ingredient per gallon
** Contains petroleum distillates.

EPA REG. NO. 34147-3-54382

KEEP OUT OF REACH OF CHILDREN
DANGER
PELIGRO

PRECAUCION AL USUARIO: Si usted no lee ingles, no use este producto hasta que le etiqueta haya sido explicado ampliamente.

See statement of practical treatment and additional precautionary statements at beginning of label.

The name and logo HOECHST are registered trademarks of Hoechst AG.
The name and logo ROUSSEL are registered trademarks of Roussel Uclaf S.A.

PRECAUTIONARY STATEMENTS

DANGER

HAZARDS TO HUMANS AND DOMESTIC ANIMALS

Corrosive, causes irreversible eye damage. Wear goggles while handling. Can cause irritation to skin such as burning and tingling. Do not get in eyes or on skin or on clothing. Wear protective gloves while handling. May be fatal if swallowed. Harmful if inhaled or absorbed through skin. Do not breathe vapor. If material gets on skin, the affected area should be thoroughly washed with plenty of soap and water. Remove and wash clothing before reuse.

RE-ENTRY STATEMENTS

Do not apply this product in such a manner as to directly or through drift expose workers or other persons. The area being treated must be vacated by unprotected persons. Do not enter treated areas without protective clothing until sprays have dried. Because certain states may require more restrictive re-entry intervals for various crops treated with this product, consult your State Department of Agriculture for further information.

Written or oral warnings must be given to workers who are expected to be in a treated area or in an area about to be treated with this product. Oral warnings should include those statements found under the label heading "Precautionary Statements" and "Re-entry Statements" included on this label. When oral warnings are given, warnings shall be given in a language customarily understood by workers. Oral warnings must be given if there is reason to believe that written warnings cannot be understood by workers. Written warnings must include the following information: DANGER, area treated with SCOUT X-TRA™ Insecticide on (date of application). Do not enter without appropriate protective clothing until sprays have dried. **If eyes are contaminated,** wash with flowing water for at least 15 minutes. **If on skin,** remove contaminated clothing and wash skin immediately with soap and warm water. **If swallowed,** do not induce vomiting. Call a physician. (See Precautionary Statements on label.)

ENVIRONMENTAL HAZARDS

This pesticide is extremely toxic to fish and other aquatic organisms, including crab and shrimp. Use with care when applying in areas adjacent to any body of water. Do not apply directly to water or wetlands. Do not apply when weather conditions favor drift from treated area. Drift or runoff from treated fields may kill fish and other aquatic organisms. Do not contaminate water when disposing of equipment washwaters.

This pesticide is highly toxic to bees exposed to direct treatment. Do not apply it or allow it to drift to crops or weeds on which bees are actively foraging.

Notify state and/or Federal authorities and HOECHST-ROUSSEL AGRI-VET COMPANY immediately if you observe any adverse environmental effects due to the use of SCOUT X-TRA™ Insecticide.

NOTICE: It is a Federal offense to use any pesticide in a manner that results in the death of a member of an endangered species.

STATEMENT OF PRACTICAL TREATMENT

If Swallowed: Do not induce vomiting. Call a physician. Gastric lavage should be supervised by trained personnel because of the possible pulmonary damage via aspiration of the solvent. Contains petroleum distillates. If possible, bring the container and labeling to attending physician.

If On Skin: Wash with plenty of soap and water. Get medical attention if irritation persists.

If In Eyes: Flush eyes with plenty of water. Get medical attention immediately.

If Inhaled: Remove victim to fresh air. Apply artificial respiration if necessary.

For emergency assistance call (201) 231-2000.

IMPORTANT NOTICE: DISCLAIMER

Read "IMPORTANT NOTICE: DISCLAIMER" before buying or using. If terms are not acceptable, return at once unopened. HOECHST-ROUSSEL AGRI-VET COMPANY warrants only that the product conforms to the chemical description on the label and is reasonably fit for the purpose stated on the label when used in accordance with the directions under normal conditions of use. This warranty does not extend to the use of this product contrary to label instructions or under abnormal conditions, or under conditions not reasonably foreseeable to HOECHST-ROUSSEL AGRI-VET COMPANY and user assumes the risk of any such use. HOECHST-ROUSSEL AGRI-VET COMPANY MAKES NO OTHER WARRANTY, EXPRESSED OR IMPLIED, INCUDING ANY IMPLIED WARRANTY OF FITNESS FOR A PARTICULAR PURPOSE OR OF MERCHANTABILITY. In no case shall HOECHST-ROUSSEL AGRI-VET COMPANY be liable for consequential, special, indirect or incidental damages resulting from the use or handling of this product. The foregoing conditions of sale and warranty can be varied only by an agreement in writing signed by a duly authorized representative of HOECHST-ROUSSEL AGRI-VET COMPANY.

STORAGE AND DISPOSAL

Do not contaminate water, food or feed by storage or disposal.

Do not store over 100°F or below 10°F. Do not use or store near heat or open flame.

Pesticide Disposal: Pesticide wastes are toxic. Improper disposal of excess pesticide, spray mixture, or rinsate is a violation of Federal law. If these wastes cannot be disposed of by use according to label instructions, contact your State Pesticide or Environmental Control Agency, or the hazardous waste representative at the nearest EPA Regional Office for guidance.

Container Disposal: Empty containers should be triple rinsed into the spray tank during the spray operation. Then offer for recycling or reconditioning, or puncture and dispose of in a sanitary landfill, or incineration, or, if allowed by state and local authorities, by burning. If burned, stay out of smoke.

DIRECTIONS FOR USE

It is a violation of Federal Law to use this product in a manner inconsistent with this label.

Dilute SCOUT X-TRA™ Insecticide in water or at least once-refined crop oil according to the directions shown, and use as a foliar spray on cotton and soybeans at the rates listed. Uniform coverage is essential. Begin application when economic infestations occur based on local information, and repeat as needed. Do not enter treated field before spray is dried.

Crop	Insect	Lbs. ai/A	Fluid Ounces Per Acre*	Dilution Requirements
Cotton	Lygus Bug, Cotton Leaf-perforator, Cutworm, Leafhoppers, Cotton Fleahoppers, Plant Bugs, Thrips, Cotton Leafworm	.016-.02	2.29-2.90	**Minimum gallonage requirements:** For aerial applications use a minimum of one (1) gallon of water per acre or one (1) quart of at least once-refined crop oil per acre. **Optimum gallonage requirements:** For aerial applications under hot, dry environmental conditions or when the crop canopy is dense, gallonage requirements may need to be increased to 3-5 gallons of water **plus** 1 quart of crop oil per acre. For ground applications, use a minimum of four (4) gallons of water per acre for thorough coverage of the foliage.
	Cotton Bollworm, Tobacco Budworm, Pink Bollworm, Boll Weevil, Cabbage Looper, Soybean Looper, Fall Armyworm, Beet Armyworm, European Corn Borer, Brown Stinkbug, Southern Green Stinkbug, Mites*, Aphids*, Whitefly*	.018-.024	2.56-3.33	

* Suppression only.

Crop	Insect	Lbs. ai/A	Fluid Ounces Per Acre*	Dilution Requirements
Soybeans	Mexican Bean Beetle, Green Cloverworm, Velvet Bean Caterpillar, Leaf Hopper, Corn Earworm	.012-.016	1.73- to 2.33	**Minimum gallonage requirements:** For aerial applications use a minimum of one (1) gallon of water per acre or one (1) quart of at least once-refined crop oil per acre. **Optimum gallonage requirements:** For aerial applications under hot, dry environmental conditions or when the crop canopy is dense, gallonage requirements may need to be increased to 3-5 gallons of water **plus** 1 quart of crop oil per acre. For ground applications use a minimum of five (5) gallons of water per acre for thorough coverage of the foliage.
	Fall Armyworm, Beet Armyworm, Stinkbug, Soybean Looper, Cabbage Looper, Cutworm, Grasshopper	.016-.024	2.33 to 3.33	

* Make the first application as required before insect populations increase to economic significance. Additional applications may be made if insect populations rebuild.

Conversion Table for SCOUT X-TRA™ Insecticide

Pounds ai Per Acre	Fluid Ounces Per Acre	Pints Per Acre	Acres Treated Per Gallon
.012	1.71	0.107	75.0
.016	2.29	0.143	56.0
.018	2.56	0.160	50.0
.02	2.85	0.178	45.0
.024	3.37	0.211	38.0

Tank Mix Recommendations

SCOUT X-TRA™ Insecticide may be tank mixed with the following insecticides in accordance with label limitations and precautions (refer to the label of the companion insecticides). Label dosage rates should not be exceeded.

Lannate®	Guthion®	Lorsban™
	malathion	methyl parathion

Note:

1. *Do NOT* graze or harvest treated soybean forage, straw or hay for livestock feed.

2. *Do NOT* apply more than 27 fluid ounces per acre (0.19 pound active ingredient per acre) on cotton and 17 fluid ounces per acre (0.12 pound active ingredient per acre) on soybeans in one growing season.

3. The combination of Guthion® and synthetic pyrethroids may increase the toxicity to mammals. Accordingly, all safety precautions noted on the Guthion label should be closely followed.

4. *Do NOT* apply within 28 days of harvest for cotton and 21 days of harvest for soybeans.

5. *Do NOT* apply this product through any type of irrigation system.

Guthion® is a registered trademark of the Parent Company of Farbenfabriken Bayer GmbH, Leverkusen.
Lannate® is a registered trademark of the E.I. duPont de Nemours & Company.
Lorsban™ is a trademark of the Dow Chemical Company.

T18802014 (3/89)

"This is a specimen label. Hoechst-Roussel Agri-Vet Company is not responsible for the accuracy of the information contained herein, as labels are subject to revision. Always read and follow the use instructions on the container label".

TILLER™

HERBICIDE

For the Selective Postemergence Control of Green and Yellow Foxtail Species in Hard Red Spring Wheat and Only in the State of North Dakota Under the Authority of Section 18 of FIFRA

ACTIVE INGREDIENTS:	
fenoxaprop-ethyl: (±)-ethyl 2-[4-[(6-chloro-2-benzoxazolyl)oxy] phenoxy]propanoate	8.85%*
2,4-D, isooctylester: 2-ethylhexyl-2,4-dichlorophenoxyacetate	10.27%**
MCPA, isooctylester/isooctyl-2-methyl-4-chlorophenoxyacetate	32.85%***
INERT INGREDIENTS:	48.03%
TOTAL	100.00%

*Equivalent to 0.75 pound of pure fenoxaprop-ethyl per gallon
**Equivalent to 0.58 pound of pure 2,4-dichlorophenoxyacetic acid per gallon
***Equivalent to 1.75 pounds of pure 2-methyl-4-chlorophenoxyacetic acid per gallon

The name and logo HOECHST are registered trademarks of Hoechst AG.
The name and logo ROUSSEL are registered trademarks of Roussel Uclaf S.A.

KEEP OUT OF REACH OF CHILDREN

DANGER

Section 18 Exemption (North Dakota)
89-ND-01

PRECAUTIONARY STATEMENTS

DANGER

HAZARDS TO HUMANS AND DOMESTIC ANIMALS

Corrosive. Causes irreversible eye damage. Do not get in eyes. Avoid contact with skin. Wear goggles or face shield and impermeable gloves (such as neoprene or PVC) while mixing. Harmful if swallowed, absorbed through skin or inhaled. Do not take internally. Avoid inhalation of vapor or spray mist. Remove contaminated clothing and wash before reuse. May cause allergic skin reactions. In case of skin contact, wash skin with soap and water. Get medical attention if irritation persists.

ENVIRONMENTAL HAZARDS

This pesticide is toxic to fish and aquatic invertebrates. Drift or runoff may adversely affect nontarget plants. Do not apply directly to water or wetlands (swamps, bogs, marshes, and potholes). Do not contaminate water when disposing of equipment washwaters.

This product can reach groundwater from mixing and loading. To minimize groundwater contamination from spills during mixing, loading and cleaning of equipment, take the following steps:

Mixing and Loading: When mixing, loading or applying this product, wear chemical resistant gloves. Wash nondisposable gloves thoroughly with soap and water before removing. The mixing and loading of spray mixtures into the spray equipment must be carried out on an impervious pad (i.e., concrete slab, plastic sheeting) large enough to catch any spilled material. If spills occur, contain the spill by using an absorbent material (e.g., sand, earth or synthetic absorbent). Dispose of the contaminated absorbent material by placing in a plastic bag and following disposal instructions on this label.

Cleaning of Equipment: When cleaning equipment, do not pour the washwater on the ground; spray or drain over a large area away from wells and other water sources.

STATEMENT OF PRACTICAL TREATMENT

In case of eye contact, immediately flush with plenty of water for 15 minutes. Get medical attention. If on skin, wash with plenty of soap and water. If swallowed, do not induce vomiting. Tiller Herbicide contains petroleum distillates. Call a physician. Vomiting should be supervised by a physician because of the possible pulmonary damage via aspiration of the solvent.

If possible, bring this container and labeling to the attending physician. For emergency assistance, call (201) 231-2000.

STORAGE AND DISPOSAL

Do not contaminate water, food, or feed by storage or disposal. Keep container tightly closed when not in use. Do not store near other pesticides and seeds.

Do not store over 100°F or below 10°F. Do not use or store near heat or open flame.

Pesticide Disposal: Pesticide wastes are toxic. Improper disposal of excess pesticide, spray mixture, or rinsate is a violation of Federal Law. If these wastes cannot be disposed of by use according to label instructions, contact your State Pesticide or Environmental Control Agency, or the hazardous waste representative at the nearest EPA Regional Office for guidance.

Container Disposal: Empty containers should be triple rinsed into the spray tank during the spray operation. Then offer for recycling or reconditioning, or puncture and dispose of in a sanitary landfill, or incineration, or, if allowed by state and local authorities by burning. If burned, stay out of smoke.

Tiller Herbicide contains the herbicides 2,4-D and MCPA as two of the active ingredients. Therefore, do not enter treated areas without protective clothing until sprays have dried. Written or oral warnings must be given to workers who are expected to be in a treated area or in an area about to be treated with this product. The precautionary statements should be read to workers as well as the instruction not to enter until sprays have dried. When oral warnings are given, warnings shall be given in a language customarily understood by workers. Written warnings must include the following information: Caution. Areas treated with Tiller on **GIVE DATE.** Do not enter without appropriate protective clothing until sprays have dried.

IMPORTANT NOTICE: DISCLAIMER

Read "Important Notice: Disclaimer" before buying or using. If terms are not acceptable, return at once unopened. HOECHST-ROUSSEL AGRI-VET COMPANY warrants only that the product conforms to the chemical description on the label and is reasonably fit for the purpose stated on the label when used in accordance with the directions under normal conditions of use. This warranty does not extend to the use of this product contrary to label instructions or under abnormal conditions, or under conditions not reasonably foreseeable to HOECHST-ROUSSEL AGRI-VET COMPANY, and user assumes the risk of any such use. HOECHST-ROUSSEL AGRI-VET COMPANY MAKES NO OTHER WARRANTY, EXPRESSED OR IMPLIED, INCLUDING ANY IMPLIED WARRANTY OF FITNESS FOR A PARTICULAR PURPOSE OR OF MERCHANTABILITY. In no case shall HOECHST-ROUSSEL AGRI-VET COMPANY be liable for consequential, special, indirect or incidental damages resulting from the use or handling of this product. The foregoing conditions of sale and warranty can be varied only by an agreement in writing signed by a duly authorized representative of HOECHST-ROUSSEL AGRI-VET COMPANY.

GENERAL INFORMATION

Tiller Herbicide is a broad-spectrum postemergence herbicide for the control of green and yellow foxtail species in hard red spring wheat. Tiller Herbicide should not be applied to barley or durum wheat. Applications to wheat that is not tillered may result in malformed spikes.

DIRECTIONS FOR USE

It is a violation of Federal Law to use this product in a manner inconsistent with its labeling. Do not apply until you have read the entire label. **For use only in the state of North Dakota under the authority of Section 18 of FIFRA as amended.**

APPLICATION INFORMATION

Ground Application: Do not apply when weather conditions favor drift. Spray drift can be lessened by keeping the spray boom as low as possible, by applying 10 gallons or more of spray per acre, by using no more than 40 pounds spraying pressure with flat fan nozzle tips, by spraying when wind velocity is low, and by stopping all spraying when wind exceeds 6 to 7 miles per hour. Do not apply with hollow cone type insecticide or other nozzles that produce a fine-droplet spray. Under dense weed/crop canopies, high spray volume is very important for obtaining thorough coverage; therefore, use higher spray volume under these conditions.

Tiller Herbicide contains the herbicides 2,4-D and MCPA as two of the active ingredients. Therefore, spray equipment used in applying this product should be thoroughly cleaned before using for any other purpose. Use repeated flushings with soap and warm water or suitable chemical cleaner.

It is best to use a separate sprayer for application of insecticides and fungicides.

TIMING OF APPLICATION

The time of application is determined by the growth stage of the wheat. Apply after the wheat begins to tiller (1-2 tiller: 4-leaf stage on the main clone) but prior to the beginning of wheat jointing. DO NOT spray wheat in the boot to dough stage.

Tiller Herbicide will control foxtail species in the 3-leaf to 2-tiller stage of growth. Applications should be made to young, vigorously growing weeds.

Tiller Herbicide, when applied as recommended, controls the following annual weeds:

Rate Recommendation Chart

Grass Species		Pints Per Acre	Fl. Oz. Per Acre
Green Foxtail	*Setaria viridis)*	0.95	15.0
Mixed Population of Green Foxtail Yellow Foxtail	*(Setaria lutescens)*	1.2	19.0

SPECIAL NOTES

1. For use in hard red spring wheat only. Do not apply to durum wheat, barley, rye, or tame oats.
2. Rainfall within one (1) hour of an application may cause a reduction in weed control.
3. Application of Tiller™ Herbicide to weeds under stress (e.g., drought) may result in reduced weed control.
4. Do Not graze or feed treated forage, hay or straw.
5. Do Not mix Tiller Herbicide with other pesticides, fertilizers, or adjuvants. The addition of broadleaf herbicides may reduce the effectiveness of Tiller Herbicide for annual grass control. Broadleaf herbicides may be applied five (5) days before or after a Tiller Herbicide application.
6. Do Not apply Tiller Herbicide within 60 days of harvest.
7. Do Not apply more than one (1) application in a growing season; do Not apply more than 1.2 pints per acre per growing season.
8. Do Not apply this product through any irrigation system.
9. Do Not apply this product aerially.
10. Tiller Herbicide contains the herbicides 2,4-D and MCPA as two of the active ingredients. Therefore, this product **may** kill or seriously injure many desirable forms of vegetation. Do not apply directly to flowers, fruits, vegetables, grapes, ornamentals, cotton, or other desirable plants. Do not use when there is hazard from drifting mists. (Coarse sprays are less likely to drift.) Vapors from this product may injure susceptible plants in the immediate vicinity.
11. Tiller Herbicide contains the herbicides 2,4-D and MCPA as two of the active ingredients. Excessive amounts of this product in the soil may temporarily inhibit seed germination and plant growth.

TH8908006 (3/89)

"This is a specimen label. Hoechst-Roussel Agri-Vet Company is not responsible for the accuracy of the information contained herein, as labels are subject to revision. Always read and follow the use instructions on the container label".

WHIP® 1EC

HERBICIDE

For Selective Postemergence Annual and Perennial Grass Control in Rice, Soybeans and Acreage Conservation Reserve (Set-Aside)

ACTIVE INGREDIENT:
fenoxaprop-ethyl:(±)-ethyl
2-[4-[(6-chloro-2-benzoxazolyl)
oxy]phenoxy]propanoate 12.50%*
INERT INGREDIENTS: 87.50%
TOTAL 100.00%

*Equivalent to 1.00 pound of active ingredient per gallon

EPA REG. NO. 8340-23-54382

The name and logo HOECHST are registered trademarks of Hoechst AG.
The name and logo ROUSSEL are registered trademarks of Roussel Uclaf S.A.

KEEP OUT OF REACH OF CHILDREN

WARNING

SEE ADDITIONAL PRECAUTIONARY STATEMENTS AT THE BEGINNING OF LABEL

PRECAUTIONARY STATEMENTS

WARNING

HAZARDS TO HUMANS AND DOMESTIC ANIMALS

May cause substantial but temporary eye injury. Do not get in eyes. Wear goggles or face shield and impermeable rubber gloves (such as neoprene or PVC) while mixing. Harmful if swallowed, absorbed through skin or inhaled. Do not take internally. Avoid inhalation of vapor or spray mist. Remove contaminated clothing and wash before reuse. Wash thoroughly with soap and water after handling.

ENVIRONMENTAL HAZARDS

This pesticide is toxic to fish. Do not apply directly to a body of water outside of the treated rice field. Do not apply when weather conditions favor runoff or drift. Do not contaminate arable land and/or water when disposing of equipment washwaters.

ENDANGERED SPECIES RESTRICTIONS

The use of Whip 1EC Herbicide on rice is restricted to protect the endangered fat pocketbook pearly mussel *(Potamilus capax)* and its habitat. Use is prohibited in the following areas of Arkansas:

Mississippi County—Within the basin that drains directly into the Right Hand Chute of Little River, south of Big Lake National Wildlife Refuge.

Poinsett County—Between Crowley's Ridge and the levee east of the Right Hand Chute of Little River and the St. Francis Floodway. Use is also prohibited west of Route 140 and north of Route 63 at the siphon near Marked Tree. The prohibited area does *not* include the area bounded by Arkansas Highway 373 on the west, Highway 63 on the east, and Highway 14 on the south.

Cross, St. Francis and Lee Counties—Between Crowley's Ridge and the levee east of the Right Hand Chute of Little River and the St. Francis Floodway as far south as the confluence of L'Anguille River (Lee County).

STORAGE AND DISPOSAL

Do not contaminate water, food or feed by storage or disposal.

Do not store over 100°F or below 10°F. Do not use or store near heat or open flame.

Pesticide Disposal: Pesticide wastes are toxic. Improper disposal of excess pesticide, spray mixture, or rinsate is a violation of Federal Law. If these wastes cannot be disposed of by use according to label instructions, contact your State Pesticide or Environmental Control Agency, or the hazardous waste representative at the nearest EPA regional office for guidance.

Container Disposal: Empty containers should be triple rinsed into the spray tank during the spray operation. Then offer for recycling or reconditioning, or puncture and dispose of in a sanitary landfill, or incineration, or, if allowed by state and local authorities, by burning. If burned, stay out of smoke.

STATEMENT OF PRACTICAL TREATMENT

In case of eye contact, immediately flush with plenty of water for 15 minutes. Get medical attention.

If swallowed, do not induce vomiting. Whip 1EC Herbicide contains petroleum distillates. Call a physician. Vomiting should be supervised by a physician because of the possible pulmonary damage via aspiration of the solvent.

If possible bring this container and labeling to the attending physician. For emergency assistance call (201) 231-2000.

IMPORTANT NOTICE: DISCLAIMER

Read "IMPORTANT NOTICE: DISCLAIMER" before buying or using. If terms are not acceptable, return at once unopened. HOECHST-ROUSSEL AGRI-VET COMPANY warrants only that the product conforms to the chemical description on the label and is reasonably fit for the purpose stated on the label when used in accordance with the directions under normal conditions of use. This warranty does not extend to the use of this product contrary to label instructions or under abnormal conditions, or under conditions not reasonably foreseeable to HOECHST-ROUSSEL AGRI-VET COMPANY, and user assumes the risk of any such use. HOECHST-ROUSSEL AGRI-VET COMPANY MAKES NO OTHER WARRANTY, EXPRESSED OR IMPLIED, INCLUDING ANY IMPLIED WARRANTY OF FITNESS FOR A PARTICULAR PURPOSE OR OF MERCHANTABILITY. In no case shall HOECHST-ROUSSEL AGRI-VET COMPANY be liable for consequential, special, indirect or incidental damages resulting from the use or handling of this product. The foregoing conditions of sale and warranty can be varied only by an agreement in writing signed by a duly authorized representative of HOECHST-ROUSSEL AGRI-VET COMPANY.

GENERAL INFORMATION

Whip 1EC Herbicide is an emulsifiable concentrate for use in the selective postemergence control of annual and perennial grassy weeds in soybeans and rice. Thorough spray coverage of emerged grasses is important. Visible effects begin as a general chlorosis (yellowing) followed by death of the weed. Visible injury of the grasses is evident approximately 4-10 days after application (dependent upon environmental conditions); but complete kill of the target grass will take 12-21 days.

Since many grass crops are sensitive to Whip 1EC Herbicide, including sorghum and corn, avoid all direct or indirect contact to neighboring fields.

Whip 1EC Herbicide does not control broadleaf weeds or sedges.

DIRECTIONS FOR USE

It is a violation of Federal law to use this product in a manner inconsistent with its labeling. Do not apply until you have read the entire label.

SOYBEANS

APPLICATION INFORMATION

A. Ground Application: Whip 1EC Herbicide should be applied in a minimum of ten (10) gallons of water per broadcast acre. Flat fan or hollow cone nozzles are recommended. Use a minimum pressure of forty (40) pounds per square inch. Under dense weed/crop canopies, high spray pressure is very important for obtaining thorough coverage; therefore, use higher spray pressure under these conditions.

B. Air Application: Whip 1EC Herbicide should be applied in a minimum of five (5) gallons of water per broadcast acre. Uniform spray coverage is essential when using aircraft and is achieved by the use of spray droplet size (150-300 micron range). DO NOT USE raindrop nozzles. Aerial applications with this product should be made at a height which provides the most effective swath width for the aircraft.

DO NOT APPLY by aircraft when wind speeds exceed eight (8) mph. Avoid all direct or indirect contact to neighboring fields.

Timing of Application

Whip 1EC Herbicide will control grasses at most growth stages; but for best production oriented grass control, applications should be made to young, vigorously growing grassy weeds. Follow the recommendations for time of application from the adjacent rate and grass recommendation chart. Earlier applications of Whip 1EC Herbicide (before all grasses have emerged) could result in late flushes of grass reinfesting the treated area.

Control of Larger Annual Grasses

In soybeans always use crop oil concentrate when controlling annual grasses that have grown beyond the recommended stages of growth. When grasses are large, the control may take somewhat longer. Yield reductions due to weed competition may result from these late applications.

Additives

The addition of nonphytotoxic oil concentrate to the spray solution at one (1) quart per acre for ground applications and one (1) pint per acre for aerial applications may improve the herbicidal activity of Whip 1EC Herbicide. A nonphytotoxic oil should be added when the grassy weeds are under drought stress or when a less than optimum application timing is necessary (see Timing of Application section). Add a nonphytotoxic oil concentrate or a once-refined vegetable oil or soybean oil concentrate containing 15-20% approved emulsifiers cleared for application to growing crops. Crop oil concentrates vary in their viscosity; therefore, it is important to maintain constant agitation while the spray mixture is in the spray tank.

The increased speed of foliage burn resulting from the addition of crop oil concentrate may reduce the translocation of Whip 1EC Herbicide to the Johnsongrass roots and rhizomes. Therefore, the addition of nonphytotoxic oil is *not* recommended for the control of rhizome Johnsongrass.

Tank Mix Recommendations for Soybeans

Whip® 1EC Herbicide may be tank mixed with Basagran Herbicide, Blazer or Tackle 2AS in a postemergence program for broad spectrum weed control in soybeans. Tank mix applications are to be used only when both the annual grass and broadleaf weeds are in the proper stage of growth as specified on each respective herbicide label. When tank mixing, always follow the use directions in accordance with the respective label. No label dosage rates should be exceeded.

RATE AND GRASS RECOMMENDATION CHART

Grass Species	Optimum Stage of Growth*	Crop Oil Concentrate	Application Rate** pts./A	fl. oz./A
ANNUAL GRASSES				
Giant Foxtail (*Setaria faberii*)	3-6"	Always add 1 qt. with ground 1 pt. with aerial application	0.8	13
Green Foxtail (*Setaria viridis*)	3-6"			
Volunteer Corn (*Zea mays*)	10-24"			
Wild Proso Millet (*Panicum miliaceum*)	5-10"			
Johnsongrass, Seedling (*Sorghum halepense*)	4-10"			
Wild Cane/Shattercane (*Sorghum bicolor*)	6-12"			
Barnyardgrass (*Echinochloa crus-galli*) Broadleaf Signalgrass (*Brachiaria platyphylla*) Fall Panicum (*Panicum dichotomiflorum*) Bristle Foxtail (*Setaria verticillata*) Purple Foxtail (*Setaria viridis robusta purpurea*) Robust Foxtail (*Setaria viridis robusta alba*) Goosegrass (*Eleusine indica*) Jungle Rice (*Echinochloa colonum*) Southwestern Cupgrass (*Eriochloa gracilis*) Sprangletop (*Leptochloa filiformis*) Texas Panicum (*Panicum texanum*) Wild Oats (*Avena fatua*) Witchgrass (*Panicum capillare*) Wooly Cupgrass (*Eriochloa villosa*)	3-6"	OPTIONAL	1.2	19
Wirestem muhly (*Muhlenbergia frondosa*) Yellow Foxtail (*Setaria lutescens*)	3-6"	Always add COC	1.2	19
Large Crabgrass (*Digitaria sanguinalis*) Smooth Crabgrass (*Digitaria ischaemum*)	1-2"	Always add COC	1.2	19
Itchgrass (*Rottboellia exaltata*)	3-6"	OPTIONAL	1.2	19
PERENNIAL GRASSES				
Johnsongrass from rhizomes (*Sorghum halepense*)	10-20"		1.2	19
Second application if needed (regrowth) (A timely cultivation may override necessity of a second application.)	10-20"	DO NOT ADD OIL	0.8	13

*See Control of Larger Annual Grasses section.

**When controlling mixed populations of grassy weeds, always use the rate that will control the least susceptible species.

Water Volume and Spray Pressure

Ground Equipment: For the tank mix, use a minimum of 20 gallons per acre of total spray solution and a minimum pressure of 40 PSI. Use standard high pressure hollow cone or flat fan nozzles. Do not use flood nozzles.

Aerial Equipment: For tank mixes, use a minimum of 5 gallons per acre of total spray solution and a minimum pressure of 40 PSI.

Mixing: Fill the spray tank half full with water while the agitator is running. Add the recommended amount of Whip 1EC Herbicide followed by the tank mix component. Then add the remaining amount of water.

BASAGRAN®

Whip 1EC Herbicide should be applied at a rate of 1.2 to 1.6 pints per acre and Basagran at a rate of 1.5 to 2.0 pints per acre. The choice of rates of each product and additives is dependent on the weed size and weed spectrum present. Refer to the Basagran label to identify the proper rate and additives for control of the species and size of the broadleaf weeds present.

The Whip 1EC Herbicide rates for tank mixing with Basagran are 1.2 pints per acre when the annual grassy weeds are 1 to 4 inches tall, and 1.6 pints per acre when the annual grassy weeds are 4 to 6 inches tall or less than 2 tillers. DO NOT use this tank mix if the annual grassy weeds have developed more than 2 tillers or are larger than 6 inches tall. For the control of shattercane 6 to 12 inches tall and broadleaf weeds that are on the Basagran label, tank mix Whip 1EC Herbicide at a rate of 1.2 pints per acre with Basagran at 1.5 to 2.0 pints per acre. DO NOT use this tank mix to control rhizome Johnsongrass. Sequential applications of Whip 1EC Herbicide and Basagran may be necessary if the stages of the grass and broadleaf weeds are not within tank mix label recommendations at the same time.

For Use On Soybeans Only: When tank mixing Whip 1EC Herbicide and Basagran, always add crop oil concentrate or other labeled additives at the appropriate labeled rate.

Special Note: When tank mixing Basagran and Whip 1EC Herbicide, a minimum rate of Whip at 1.2 pints per acre is recommended.

BLAZER® or TACKLE 2AS

Whip 1EC Herbicide should be tank mixed at a rate of 1.6 pints per acre. Blazer or Tackle 2AS should be tank mixed at a rate of 1.5 to 2.0 pints per acre. In no instances should crop oil concentrate or a surfactant be used with this tank mix. This tank mix should not be used for the control of rhizome Johnsongrass.

The stage of growth of both the annual grassy weeds and the broadleaf weeds should conform to the directions on each product label. The tank mix of Whip 1EC Herbicide plus Blazer or Tackle 2AS should not be applied after the annual grasses have begun tillering. Whenever the grass and broadleaf weeds are not both in the proper stage of growth according to this tank mix label, a sequential application should be utilized. When Whip 1EC Herbicide is applied first, a waiting period of 3 days is necessary before applying Blazer or Tackle 2AS. When Blazer or Tackle 2AS is applied first, a waiting period of 7 days is necessary before applying Whip 1EC Herbicide.

Special Note: The mixture of Whip 1EC Herbicide plus Blazer or Tackle 2AS may only suppress velvetleaf.

SPECIAL NOTES FOR SOYBEANS

1. Annual ryegrass (*Lolium* sp.), quackgrass (*Agropyron repens*) and Bermudagrass (*Cynodon dactylon*) are not controlled by Whip 1EC Herbicide.
2. Rainfall within one hour of an application may cause a reduction in grass control.
3. If a new flush of grass occurs, either a timely cultivation or a second application of Whip 1EC Herbicide may be necessary.
4. DO NOT cultivate within four days before or after a Whip 1EC Herbicide application.
5. ALWAYS clean sprayer thoroughly before and after any pesticide application.
6. Apply Whip 1EC Herbicide as a spot treatment in a 1% volume/volume solution with water (e.g., 32 fluid ounces per 25 gallons of water). (See instructions for use in the Ground Application section.)
7. Whip 1EC Herbicide is not phytotoxic to soybeans at any growth stage but, for best results, it should be applied according to the development of the annual grassy weeds as noted, but before the bloom stage of the soybean crop.
8. DO NOT graze or feed treated forage, hay or straw.
9. Application of Whip 1EC Herbicide to grasses under stress (e.g., drought), may result in reduced control.
10. DO NOT plant any rotational crop in a Whip 1EC Herbicide treated field for 30 days after application (120 days for small grains.)
11. DO NOT apply Whip 1EC Herbicide less than 90 days before harvesting soybeans.
12. DO NOT make more than two applications of Whip 1EC Herbicide per growing season, and do not apply more than two pints (0.25 pound of active ingredient) per acre per growing season.
13. DO NOT apply this product through any irrigation system.
14. Read and follow restrictions and limitations on the Basagran Herbicide and/or Blazer/Tackle labels as applicable. The most restrictive labeling applies in tank mixes.

RICE

Rice is tolerant to postemergence applications of Whip® 1EC Herbicide from the 4-leaf to the late tillering stage of rice development. Always plant high quality seed in order to obtain a uniform germination and a good rice stand. Postemergence applications may result in temporary rice injury that appears as leaf chlorosis and stunting. The rice will normally recover from these symptoms in two to four weeks.

Use instructions, particularly water management, must be followed to minimize rice injury. Read and follow all label directions carefully.

APPLICATION INFORMATION

Rice fields should be level and free of large clods to obtain uniform germination of rice and grassy weeds and to insure uniform flood levels. If necessary, fields may be flushed prior to treatment. If fields are flushed prior to treatment, flush in sufficient time so that the rice and grass are actively growing at time of treatment. Allow sufficient time for water to drain from the paddy before the Whip 1EC Herbicide application.

Rice injury may occur if applications of Whip 1EC Herbicide are made within three (3) days following periods of inclement (rainy, foggy, cloudy) weather. Do not apply Whip 1EC Herbicide within 14 days following the activation of fertilizer.

A. Air Application: Apply aerially using a minimum of ten (10) gallons of water per broadcast acre. It is recommended to increase the gallonage to obtain thorough coverage when a dense weed canopy is present. Uniform spray coverage is essential when using aircraft, and is achieved by the use of a spray droplet size ranging from 150 to 300 microns. A hydraulic boom-nozzle system that will apply 10 gallons of water per acre with a minimum pressure of 20 pounds per square inch is recommended. Best results are obtained with D-8 nozzles. DO NOT USE raindrop nozzles. Aerial applications with this product should be made at a height which provides the most effective swath width for the aircraft, but no lower than 8 feet from the rice crop.

DO NOT APPLY by aircraft when wind speeds exceed eight (8) mph. Avoid all direct or indirect contact to neighboring fields.

SPECIAL NOTES:

1. Thoroughly clean mixing vat and airplane by rinsing with clean water before Whip 1EC Herbicide is added.
2. It is important to calibrate the spray equipment with Whip 1EC Herbicide in the spray solution. The spray swath width and total volume per acre may vary when compared to other rice herbicides.
3. A flow meter is recommended to obtain proper water volume (gpa).

B. Ground Application: Apply in a minimum of ten (10) gallons of water per broadcast acre. Flat fan or hollow cone nozzles are recommended. Use a minimum pressure of 40 PSI. Under dense weed/crop canopies, high spray pressure and increased gallonage is very important in obtaining thorough coverage.

While broadleaf crops, such as cotton and soybeans, are very tolerant to Whip 1EC Herbicide, grassy crops, such as corn and sorghum, are extremely sensitive, and drift onto these grassy crops must be avoided.

To insure thorough coverage and to avoid drift, DO NOT APPLY when the wind speed exceeds eight (8) mph.

Timing of Application

For Arkansas, Louisiana, Mississippi, Missouri, and Texas

When recommended water management practices are followed (see Water Management section), optimal conditions for controlling grass usually occur when the rice is in the 4-leaf to late tillering stage of development (but prior to panicle initiation). However, applications should be made following the Rate and Grass Recommendation Chart for Rice.

RATE AND GRASS RECOMMENDATION CHART FOR RICE

Grass Species	Amount of Whip per acre (pints). Relative to Stage of Annual Grass Weeds		
	1-5 leaf or 1-5 inches	5 leaf-2 tiller or 5-10 inches	> 2 tiller or > 10 inches
Sprangletop (*Leptochloa* spp.), Barnyardgrass, watergrass (*Echinochloa crus-galli*), Broadleaf Signalgrass (*Brachiaria platyphylla*), Goosegrass (*Eleusine indica*), Jungle Rice (*Echinochloa colonum*), Crabgrass (*Digitaria* spp.), Johnsongrass (10-15") (*Sorghum halepense*), Giant Foxtail (*Setaria faberii*), Fall Panicum (*Panicum dichotomiflorum*)	1.2 pts./A	1.6 pts./A	SUPPRESSION
Red Rice* (*Oryza sativa*)	1.2-1.6 pts./A	Not Recommended	

*For suppression of red rice, apply Whip 1EC Herbicide at 1.2-1.6 pints/A when the red rice is in the 4-leaf stage of growth.

Early Season Applications in Texas

For early season applications when the daily minimum temperatures are below 60°F for 3 consecutive days, tank mix Whip 1EC Herbicide at 0.9 pint per acre with Basagran Herbicide at 1.5 pints per acre. For this early season tank mix application, the annual grassy weeds should be actively growing and should be no larger than the 4-leaf stage of growth.

WATER MANAGEMENT—IMPORTANT INSTRUCTIONS

The following paddy flood program must be used:

PREFLOOD:

1. Rice fields must be level. If desirable, fields may be flushed prior to treatment. To expose existing grasses, allow sufficient time for water to drain from the field before the Whip 1EC Herbicide application.
2. When to apply the paddy flood:
a. When the rice is less than eight (8) inches in height, do not flood fields for at least seven (7) days after the Whip 1EC Herbicide application.
b. When the rice is greater than eight (8) inches in height, the fields can be flooded in 4-5 days following the application.
3. The water depth (flush or flood) should not exceed 25% of the rice height for 21 days after the Whip 1EC Herbicide application. A deep flood can be applied anytime after 21 days following treatment.

POST-FLOOD:

1. The rice should have at least one tiller before making a postflood application with Whip 1EC Herbicide.
2. Water levels at the time of application should cover no more than 25% of the rice and annual grass foliage.

3. Reflood to a normal depth two to three days after the application.

4. Whip 1EC Herbicide may only provide suppression of annual grassy weeds having developed beyond the 2-tiller stage of growth.

Control of Other Weeds

Tank Mix Recommendations:

Whip 1EC Herbicide may be tank mixed with Basagran Herbicide in a postemergence program for broad spectrum weed control in rice. When tank mixing, always follow the use directions in accordance with the respective label. Refer to the Basagran Herbicide label to identify the proper rate for control of the weed species and size of the weeds present. No label dosage rate should be exceeded.

The Whip 1EC Herbicide rates for tank mixes with Basagran are 1.2 pints per acre, when the annual grassy weeds are 1 to 3 leaf (1-4 inches) and 1.6 pints per acre when the annual grassy weeds are 4 leaf to 1 tiller (4-8 inches). Do not tank mix Whip 1EC Herbicide and Basagran when the annual grassy weeds have developed more than 1 tiller or if the weeds are under drought stress. Sequential applications may be necessary if the growth stage of the grasses and broadleaf weeds are not within the tank mix label recommendations at the same time.

In preflood applications, a new flush of broadleaf and grassy weeds may occur under certain environmental conditions before the field receives permanent flood 7 days later; therefore, appropriate control measures may be required.

Mixing Instructions

Fill the spray tank half full with water while the agitator is running. Add the recommended amount of Whip 1EC Herbicide followed by Basagran. Then add the remaining amount of water.

Sequential Treatments

Applications of other herbicides may be made sequentially to Whip 1EC Herbicide. A six (6) day interval is recommended between applications of Whip 1EC Herbicide and other pesticide applications.

SPECIAL NOTES FOR RICE

1. DO NOT add a crop oil concentrate to the spray solution when treating rice as rice injury may occur.

2. Rainfall within one hour of an application may reduce the grass control.

3. DO NOT make more than two applications of Whip 1EC Herbicide per growing season and do not apply more than 2.4 pints per acre per growing season.

4. DO NOT apply Whip 1EC Herbicide after the late tillering stage of the rice crop development (but prior to panicle initiation).

5. ALWAYS clean spray system thoroughly before and after any pesticide application. Avoid using any water that is contaminated with other pesticides.

6. DO NOT graze or feed rice straw to livestock.

7. DO NOT plant any rotational crop in a Whip 1EC Herbicide treated field for 30 days after application (120 days for small grains).

8. DO NOT apply Whip 1EC Herbicide in areas where catfish and crayfish are commercially cultivated.

9. DO NOT use rice irrigation water to irrigate crops not registered for use with Whip® 1EC Herbicide within 14 days of the last application of this product.

10. DO NOT apply Whip 1EC Herbicide less than 65 days before harvesting rice.

11. DO NOT apply Whip 1EC Herbicide within 14 days following the activation of fertilizer.

12. Applications of Whip 1EC Herbicide to grasses under drought stress may result in reduced control.

13. DO NOT tank mix Whip 1EC Herbicide with Blazer, Propanil, Ordram, phenoxy herbicides, or liquid fertilizers.

14. DO NOT apply Whip 1EC Herbicide within seven (7) days following a Furadan application.

15. DO NOT use on the rice varieties Mars and Leah, as damage to these varieties may occur.

16. Whip 1EC Herbicide has shown selectivity when tested on the following rice varities: Newbonnet, Lemont, Skybonnet, Tebonnet, Bond, Gulfmont, Rexmont, Labelle, Starbonnet, L-201, Newrex and CB-801.

ACREAGE CONSERVATION RESERVE (SET-ASIDE)

Whip 1EC Herbicide may be used to control annual grassy weeds in acreage conservation reserve (set-aside) acres. This acreage is often seeded to the following cover crops: clover, alfalfa, tall fescue, bromegrass, and ryegrass. Special note: Timothy and orchardgrass are sensitive to Whip 1EC Herbicide. The cover crops listed above have excellent tolerance to Whip 1EC Herbicide at 1.2 to 1.6 pints per acre. Select the proper rate from the Rate and Grass Recommendation Chart found in the Soybean Section of this label.

SPECIAL NOTES:

1. DO NOT harvest or graze cover crops treated with Whip 1EC Herbicide.

2. DO NOT apply to cover crops such as oats, sorghum, sudangrass, and Timothy as injury may occur.

Basagran® is a registered trademark of BASF AG.
Blazer® is a registered trademark of BASF Corporation.
Tackle® is a registered trademark of Rhone-Poulenc Incorporated.

SUPPLEMENTAL LABELING

WHIP® 1EC HERBICIDE

EPA Reg. No. 8340-23-54382

FOR SELECTIVE POSTEMERGENCE ANNUAL AND PERENNIAL GRASS CONTROL IN COTTON AND PEANUTS

IMPORTANT

Before using this product, read and carefully observe all applicable directions, restrictions, and precautionary statements on the EPA-approved Whip 1EC Herbicide container and use directions label.

GENERAL INFORMATION

Whip 1EC Herbicide is an emulsifiable concentrate for use in the selective postemergence control of annual and perennial grassy weeds in cotton and peanuts. Thorough spray coverage of emerged grasses is important. Visible effects begin as a general chlorosis (yellowing) followed by death of the weed. Visible injury of the grasses is evident approximately 4-10 days after application (dependent upon environmental conditions); but complete kill of the target grass will take 12-21 days.

Since many grass crops are sensitive to Whip 1EC Herbicide, including sorghum and corn, avoid all direct or indirect contact with neighboring fields.

Whip 1EC Herbicide does not control broadleaf weeds or sedges.

DIRECTIONS FOR USE

It is a violation of Federal law to use this product in a manner inconsistent with its labeling. Do not apply until you have read the entire label. All applicable directions, restrictions, and precautions on the EPA label are to be followed. This labeling must be in the possession of the user at the time of pesticide application.

APPLICATION INFORMATION

For Ground Application Only

Broadcast: Whip 1EC Herbicide should be applied in a minimum of ten (10) gallons of water per broadcast acre. Flat fan or hollow cone nozzles are recommended. Use a minimum pressure of forty (40) pounds per square inch. Under dense weed/crop canopies, high spray pressure is very important for obtaining thorough coverage; therefore, use higher spray pressure under these conditions.

Spot Treatment: Whip 1EC Herbicide may be applied for the control of grasses through knapsack sprayers or high-volume equipment utilizing handguns or other suitable nozzle arrangements. Apply to actively growing grasses. Apply to the foliage of grasses on a spray-to-wet basis. DO NOT spray to the point of runoff. Spray coverage should be uniform and complete.

Whip 1EC Herbicide should be applied at 0.8 to 1.6 pints/A depending on the specific weed species and its stage of development. In a mixed grassy weed population, always apply the higher recommended rate. Follow the Rate and Grass Recommendation Chart on the Whip 1EC Herbicide Label.

ADDITIVES

The addition of a nonphytotoxic oil concentrate to the spray solution at 1 quart/A will improve the herbicidal activity of Whip 1EC Herbicide. The nonphytotoxic oil concentrate (once-refined vegetable oil or soybean oil concentrate) should contain 15-20% approved emulsifiers.

The increased speed of foliage burn on rhizome Johnsongrass from the addition of a crop oil concentrate may reduce the translocation of Whip 1EC Herbicide to the rhizomes. Therefore, the addition of a nonphytotoxic oil is not recommended for the control of rhizome Johnsongrass.

SPECIAL NOTES FOR COTTON AND PEANUTS

1. Whip 1EC Herbicide is not phytotoxic to cotton or peanuts at any stage of growth. For best results, it should be applied according to the development of the annual grassy weeds as noted, but before the bloom stage of cotton and prior to peg initiation of peanuts.
2. Do not apply Whip 1EC Herbicide less than 60 days before harvesting peanuts and 40 days before harvesting cotton.
3. In peanuts do not make more than two applications of Whip 1EC Herbicide per growing season, and do not apply more than 2 pints (0.25 lb. ai) per acre per growing season.
4. In cotton do not make more than three applications of Whip 1EC Herbicide per growing season, and do not apply more than 2.8 pints (0.35 lb. ai) per acre per growing season.

The name and logo HOECHST are registered trademarks of Hoechst AG.
The name and logo ROUSSEL are registered trademarks of Roussel Uclaf S.A.

TH8904025 (5/89)

"This is a specimen label. Hoechst-Roussel Agri-Vet Company is not responsible for the accuracy of the information contained herein, as labels are subject to revision. Always read and follow the use instructions on the container label".

ICI Americas Inc.

Agricultural Products
Wilmington, DE 19897
(302) 886-1000

ACTELLIC® 5E

INSECTICIDE

For Control Of Insects On Stored Corn And Grain Sorghum

ACTIVE INGREDIENT:
Pirimiphos-methyl
O-[2-(diethylamino)-6-methyl-4-pyrimidinyl]
O,O-dimethyl phosphorothioate 57%
INERT INGREDIENTS: 43%
TOTAL 100%

ACTELLIC Insecticide contains 5 pounds active ingredient per gallon.

EPA Reg. No. 10182-79

KEEP OUT OF REACH OF CHILDREN
WARNING

PRECAUTIONARY STATEMENTS

HAZARDS TO HUMANS AND DOMESTIC ANIMALS

WARNING

Causes eye irritation: Do not get in eyes. Wear goggles or face shield when handling.

Harmful if swallowed, inhaled or absorbed through the skin: Avoid breathing vapors or spray mist. Keep container closed. Wear rubber gloves when handling concentrate.

Causes skin irritation: Do not get on skin or clothing. Wash thoroughly with soap and water after handling. Remove contaminated clothing and wash before reuse.

STATEMENT OF PRACTICAL TREATMENT

If In Eyes: Immediately flush with plenty of water for at least 15 minutes. Call a physician if irritation persists.

If swallowed: Do not induce vomiting. Call a physician immediately.

If on skin: Wash with plenty of soap and water. Get medical attention if irritation persists.

If inhaled: Remove to fresh air. If not breathing, give artifical respiration, preferably mouth-to-mouth. If breathing is difficult, give oxygen. Call a physician.

Note to physician: Inducing vomiting as first aid for this substance may result in increased risk of chemical pneumonia or pulmonary edema caused by aspiration of the hydrocarbon solvent. Vomiting should be induced only under professional supervision.

Atropine is antidotal for cholinesterase inhibition. Give 2 mg. of atropine intramuscularly or subcutaneously and repeat if symptoms of poisoning reappear. Pralidoxime (2 PAM) may be given if available. Give 1 gram intramuscularly or intravenously and repeat as necessary.

FOR 24-HOUR EMERGENCY MEDICAL ASSISTANCE CALL 1-800-F-A-S-T-M-E-D (327-8633).
FOR CHEMICAL EMERGENCY: Spill, leak, fire, exposure, or accident call CHEMTREC 1-800-424-9300.

ENVIRONMENTAL HAZARDS

Keep out of lakes, streams and ponds. Do not contaminate water by cleaning of equipment or disposal of waste.

PHYSICAL AND CHEMICAL HAZARDS

Do not use or store near heat or open flame. (Combustible Liquid—Class III A: DOL Reg. 29 CFR 1910.106)

DIRECTIONS FOR USE

It is a violation of Federal law to use this product in a manner inconsistent with its labeling.

Application Directions

Apply spray solution uniformly with suitable applicating equipment as grain enters storage or export containers. Select nozzle size and operating pressure to adjust the application rate with the flow of grain. Prepare desired dilution of ACTELLIC in spray tank and mix well. Make sure spray solution stays mixed, especially in large holding tanks. Apply the spray solution as a coarse, uniform, low pressure spray in order to minimize drift. Shield nozzle if necessary.

Apply the specified amount of ACTELLIC 5E in 5 gallons of water or 3-5 gallons of at least twice refined soybean oil to each 30 tons (60,000 pounds) of grain according to the following rates:

	Fluid Ounces	ppm Concentration
Shelled Corn (fieldcorn, popcorn)	9.2-12.3	6-8
Grain Sorghum	9.2-12.3	6-8

INSECTS CONTROLLED WITH ACTELLIC 5E

Coleoptera

Cigarette beetle	Sawtoothed beetle
Confused flour beetle	Granary weevil
Corn sap beetle	Maize weevil
Flat grain beetle	Merchant grain beetle
Hairy fungus beetle	Rice weevil
Red flour beetle	Lesser grain borer*

Lepidoptera

Angoumois grain moth
Indian meal moth
Almond moth

*Aids in suppression at 12.3 ounce rate.

REMARKS AND PRECAUTIONS

Grain treated with ACTELLIC at rates indicated on this label may be used immediately for any food or feed purpose.

- Do not make more than one application per crop.
- Calibrate spray equipment and periodically check in order to ensure proper spray rates.
- Do not store diluted ACTELLIC in spray tank more than 48 hours before use.
- Avoid severely dusty application sites.

Grain	No. of Bushels Approximately Equal to 30 Tons of Grain
Corn (fieldcorn, popcorn)	1,071
Grain sorghum	1,071

STORAGE AND DISPOSAL

Prohibitions: Do not contaminate water, food or feed by storage or disposal. Open dumping is prohibited. Do not reuse empty container.

Storage: Store in a cool place and protect from freezing. If accidentally frozen, reconstitute by gently warming to room temperature and thoroughly mixing contents. Keep container closed when not in use.

Pesticide Disposal: Pesticide wastes are toxic. Improper disposal of excess pesticide, spray mixture, or rinsate is a violation of federal law. If these wastes cannot be disposed of by use according to label instructions, contact your State Pesticide or Environmental Control Agency, or the Hazardous Waste representative at the nearest EPA Regional Office for guidance.

Container Disposal: Metal Containers: Triple rinse (or equivalent). Then offer for recycling or reconditioning or puncture and dispose of in a sanitary landfill, or by other procedures approved by state and local authorities.

Plastic Containers: Triple rinse (or equivalent). Then offer for recycling or reconditioning, or puncture and dispose of in a sanitary landfill, by incineration, or if allowed by state and local authorities, by burning. If burned, stay out of smoke.

NOTICE TO BUYER AND USER:

Seller warrants that this product conforms to the chemcial description on the label and is reasonably fit for the purposes stated on the label when used in accordance with directions under normal conditions of use. This warranty does not extend to the use of this product contrary to label instructions, or under abnormal use conditions or under conditions not reasonably foreseeable to Seller, and Buyer and User assume the risk of any such use. SELLER DISCLAIMS ALL OTHER WARRANTIES, EXPRESSED OR IMPLIED, INCLUDING ANY WARRANTY OF FITNESS OR MERCHANTABILITY. SELLER SHALL NOT BE LIABLE FOR CONSEQUENTIAL, SPECIAL OR INDIRECT DAMAGES RESULTING FROM THE USE OR HANDLING OF THIS PRODUCT AND SELLER'S SOLE LIABILITY AND BUYER'S AND USER'S EXCLUSIVE REMEDY SHALL BE LIMITED TO THE REFUND OF THE PURCHASE PRICE.

A-1R

The information and specimen labels contained in this booklet are correct as of the publication date but are subject to change without notice. For current information, contact ICI Americas Inc.

AMBUSH® 25W

INSECTICIDE
WETTABLE POWDER

RESTRICTED USE PESTICIDE

For retail sale to and use only by Certified Applicators, or persons under their direct supervision, and only for those uses covered by the Certified Applicator's certification.

ACTIVE INGREDIENT:	
Permethrin	
(3-Phenoxyphenyl) methyl (±)-cis, trans-3-(2,2-dichloroethenyl)-2,2-dimethylcyclopropanecarboxylate*	25.0%
INERT INGREDIENTS:	75.0%
TOTAL	100.0%

*Cis/trans ratio: Min. 35% (±) cis and max. 65% (±) trans.

EPA Reg. No. 10182-35 U.S. Patent No. 4,024,163

KEEP OUT OF REACH OF CHILDREN
WARNING
AVISO

PRECAUCION AL USUARIO:

Si usted no lee ingles, no use este producto hasta que la etiqueta le haya sido explicado ampliamente.

STATEMENT OF PRACTICAL TREATMENT

IF IN EYES: Flush with water for 15 minutes. Get medical attention if irritation persists. IF SWALLOWED: Drink 1 or 2 glasses of water and induce vomiting by touching back of throat with finger. Do not induce vomiting or give anything by mouth to an unconscious person.

FOR 24-HOUR EMERGENCY MEDICAL ASSISTANCE CALL 1-800-F-A-S-T-M-E-D (327-8633).
FOR CHEMICAL EMERGENCY: Spill, leak, fire, exposure, or accident call CHEMTREC 1-800-424-9300

PRECAUTIONARY STATEMENTS

HAZARDS TO HUMANS AND DOMESTIC ANIMALS

WARNING

MAY CAUSE EYE IRRITATION. HARMFUL IF SWALLOWED.

Do not get in eyes, on skin or on clothing. Wash thoroughly after handling.

ENVIRONMENTAL HAZARDS

This pesticide is extremely toxic to fish. Use with care when applying in areas adjacent to any body of water. Do not apply directly to water. Do not apply when weather conditions favor drift from treated areas. Do not contaminate water when disposing of equipment washwaters. This product is highly toxic to bees exposed to direct treatment or residues on crops or weeds. Do not apply AMBUSH 25W or allow it to drift to crops or weeds on which bees are actively foraging. Additional information may be obtained from your Cooperative Extension Service. Apply this product only as specified on this label.

DIRECTIONS FOR USE

It is a violation of Federal law to use this product in a manner inconsistent with its labeling.

GENERAL INSTRUCTIONS:

Apply AMBUSH 25W as shown in "Directions for Application." Use sufficient water to obtain full coverage of foliage. Timing and frequency of applications should be based upon insect populations reaching economic thresholds. The higher rates may be required under conditions of heavy infestation. Do not apply more than 12 lbs per acre per season (3 lbs ai/acre). Crops in this label may be planted immediately after last treatment. Do not plant other crops within 60 days after last application. Any other crop may be planted if it is not harvested or grazed.

REENTRY STATEMENTS:

Do not apply this product in such a manner as to directly or through drift expose workers or other persons. The area being treated must be vacated by unprotected persons.

Do not enter treated areas without protective clothing until sprays have dried.

Protective clothing means, at least, a hat or other suitable head covering, a long sleeved shirt and long-legged trousers or a coverall type garment (all of closely woven fabric covering the body, including the arms and legs), shoes and socks.

Because certain states may require more restrictive reentry intervals for various crops treated with this product, consult your State Department of Agriculture for further information.

Written or oral warnings regarding use of protective clothing and accidental exposure must be given to workers who are expected to be in treated areas or in areas about to be treated. Inform workers of areas or fields that may not be entered without specific protective clothing, period of time field must be vacated and appropriate actions to take in case of accidental exposure. An example of such information is given under written warnings. When oral warnings are given, warnings shall be given in a language customarily understood by workers. Oral warnings must be given if there is reason to believe that written warnings cannot be understood by workers. Written warnings must include the following information.

"**Warning:** Area treated with AMBUSH 25W Insecticide on (insert date). Do not enter without appropriate protective clothing until sprays have dried (if more restrictive, insert State Department of Agriculture reentry interval).

In case of accidental exposure to pesticide spray, wash the skin thoroughly with soap and water. Remove contaminated clothing and wash before reuse. If in eyes, flush with plenty of water. Get medical attention if needed."

ENDANGERED SPECIES RESTRICTIONS

For Aerial Application: Do not use within 100 yards of aquatic habitats.

For Ground Application: Do not use within 20 yards of aquatic habitats.

CHEMIGATION:

Sprinkler Irrigation Application

Apply AMBUSH 25W insecticide at rates and timing described elsewhere in this label. Apply with center pivot or continuous-move equipment distributing 1/2 acre-inch or less during treatment. If stationary (solid sets, handlines or wheel lines other than continuous-move) are used, AMBUSH 25W insecticide should be injected into no more than the last 20-30 minutes of the set. Do not apply when winds are greater than 10-15 mph to avoid drift or wind skips. Thorough coverage of foliage is required for good control. Good agitation should be maintained during the entire application period.

Use Precautions—Sprinkler Irrigation Application

A. Apply this product only through (sprinkler including center pivot, lateral move, end tow, side (wheel) roll, traveler, big gun, solid set, or hand move) irrigation system(s). Do not apply this product through any other type of irrigation system.

B. Crop injury, lack of effectiveness, or illegal pesticide residues in the crop can result from nonuniform distribution of treated water.

C. If you have questions about calibration, you should contact State Extension Service specialists, equipment manufacturers or other experts.

D. Do not connect an irrigation system (including greenhouse systems) used for pesticide application to a public water system unless the pesticide label-prescribed safety devices for public water systems are in place.

E. A person knowledgeable of the chemigation system and responsible for its operation, or under the supervision of the responsible person, shall shut the system down and make necessary adjustments should the need arise.

F. The system must contain a functional check valve, vacuum relief valve, and low pressure drain appropriately located on the irrigation pipeline to prevent water source contamination from backflow.

G. The pesticide injection pipeline must contain a functional, automatic, quick-closing check valve to prevent the flow of fluid back toward the injection pump.

H. The pesticide injection pipeline must also contain a functional, normally closed, solenoid-operated valve located on the intake side of the injection pump and connected to the system interlock to prevent fluid from being withdrawn from the supply tank when the irrigation system is either automatically or manually shut down.

I. The system must contain functional interlocking controls to automatically shut off the pesticide injection pump when the water pump motor stops.

J. The irrigation line or water pump must include a functional pressure switch which will stop the water pump motor when the water pressure decreases to the point where pesticide distribution is adversely affected.

K. Systems must use a metering pump, such as a positive displacement injection pump (e.g., diaphragm pump) effectively designed and constructed of materials that are compatible with pesticides and capable of being fitted with a system interlock.

L. Do not apply when wind speed favors drift beyond the area intended for treatment.

M. Do not apply through chemigation systems connected to public water systems.

Directions for Application

Crop	Target Pests	Rate lbs ai/A (oz/A)	Remarks
Alfalfa	Alfalfa Caterpillar Armyworm, Cutworm, Looper, Webworm, Potato Leafhopper, Blue Alfalfa Aphid, Green Peach Aphid, Pea Aphid, Green Cloverworm, Spotted Alfalfa Aphid, Velvetbean Caterpillar	0.05–0.2 (3.2–12.8)	Use higher recommended dosage for increased pest pressure or for increased residual pest control. Apply in a minimum of 1 gallon of finished spray per acre by aircraft or 10 gallons of finished spray per acre with ground equipment. Do not apply more than 0.2 pound active per cutting. When rates of 0.1 pound active per acre or less are used, application may be made on day of harvest. When rates greater than 0.1 pound active per acre are used, do not apply within 14 days of harvest. Avoid application when honey bees are actively foraging by applying during the early morning or during the evening hours.
	Alfalfa Weevil, Egyptian Alfalfa Weevil, Cucumber Beetle, Meadow Spittlebug, Plant Bugs including Lygus spp., Stink Bugs	0.1–0.2 (6.4–12.8)	
Almonds	Navel Orangeworm, Peach Twig Borer	0.2-0.4 (12.8-25.6)	Apply by ground or air[c] equipment using sufficient water to obtain uniform coverage. Can be applied up to 7 days before harvest. Do not apply more than 2.0 lbs. a.i. per acre per season. Do not graze livestock in treated orchards.
Apples	Apple Aphid, Redbanded Leafroller, Oblique-banded Leafroller, Plum Curculio, White Apple Leafhopper, Spotted Tentiform Leafminer, Tarnished Plant Bug	0.1-0.4 (6.4-25.6)	Apply by ground equipment only, using 25-400 gallons of spray per acre. Spray to wet all foliage. Do not apply after petal fall.
Artichokes	Artichoke Plume Moth	0.1-0.3 (6.4-19.2)	Apply by ground equipment using sufficient water to obtain uniform coverage. Apply as needed. Do not apply more than 1.5 lbs. a.i. per acre per season. Buds may be harvested on the day of application.
Asparagus	Cutworm Complex, Asparagus Beetle	0.05-0.1 (3.2-6.4)	Apply by ground equipment using sufficient water to obtain uniform coverage. Apply as needed. Do not apply more than 0.4 lbs a.i. per acre per season. Do not apply within 3 days of harvest.
Asparagus Ferns	Japanese Beetle, Asparagus Beetle, Lygus Bugs	0.1 (6.4)	Apply to the fern stage of the asparagus plant after spear harvest when larval and adult stage of the asparagus beetle, lygus bugs and the adult stage of the Japanese beetle are present.
Avocados[c]	Thrips, Leafhoppers, Mirid bugs, larvae	0.2 (12.8)[1] oz./100gal. 12.8	Apply by ground equipment. Spray to wet all foliage. Apply when insects appear and repeat at 7-10 day intervals as needed to provide control. Do not make more than 6 applications per season. Do not apply within 7 days of harvest. Do not graze livestock in treated areas. Do not feed cover crops from treated areas to livestock.
Cantaloupes	Squash bug, Pickleworm, Leafminers, Cutworms	0.1-0.2 (6.4-12.8)	Apply by ground equipment using sufficient water to obtain full coverage of foliage. Make applications when insects appear and repeat weekly as needed. Do not apply more than 1.6 lbs. a.i. per acre per season. Do not apply within 1 day of harvest.
Celery (Refer to Leafy Vegetables for application directions)			
Cherries	Lesser Peach Tree Borer, Rose Chafer, Green Fruitworm, Redbanded Leafroller, Plum Curculio, Tarnished Plant Bug	Rate lbs ai/ 100 gal 0.025-0.05 (1.6-3.2)	Do not exceed 400 gallons per acre. Apply when insects appear. In the East, do not exceed 6 applications per season, with no more than 4 applications after petal fall. In the West, do not apply more than 4 applications per season, with no more than 3 applications after petal fall. Do not apply within 3 days of harvest. Do not graze livestock in treated orchards. Do not feed forage to livestock.
Christmas Trees[c] (container and field grown)	Nantucket Pine Tip Moth	0.1-0.2 (6.4-12.8)	Apply first application when adults appear and repeat at 5-7 day intervals or as needed throughout the season. Use sufficient water to obtain full coverage of foliage.
Chrysanthemums (Florida only)	Vegetable Leafminer	0.1-0.2 (6.4-12.8	Apply by ground.
COLE CROPS Broccoli	Cabbage Looper, Imported Cabbageworm, Diamondback Moth, Cabbage Aphid[a]	0.05-0.1 (3.2-6.4)	Apply every 5 days or as needed by air or ground. Use sufficient water to obtain full coverage of foliage. Do not apply more than 0.8 lbs. a.i. per acre per season. Do not apply within 1 day of harvest.

Crop	Target Pests	Rate lbs ai/A (oz/A)	Remarks
Brussels Sprouts	Cabbage Looper, Imported Cabbageworm, Diamondback Moth, Cabbage Aphid[a]	0.05-0.1 (3.2-6.4)	Apply as needed by air or ground. Use sufficient water to obtain full coverage of foliage. Do not apply more than 0.8 lbs. a.i. per acre per season. Do not apply within 1 day of harvest.
Cabbage	Cabbage Looper, Imported Cabbageworm, Diamondback Moth, Cabbage Aphid[a]	0.05-0.2 (3.2-12.8)	Apply every 5-10 days or as needed by air or ground. Use sufficient water to obtain full coverage of foliage. Do not apply more than 1.0 lb. a.i. per acre per season. Do not apply within 1 day of harvest.
Cauliflower	Cabbage Looper, Imported Cabbageworm, Diamondback Moth, Cabbage Aphid[a]	0.05-0.1 (3.2-6.4)	Apply every 5 days or as needed by air or ground. Use sufficient water to obtain full coverage of foliage. Do not apply more than 0.8 lbs. a.i. per acre per season. Do not apply within 1 day of harvest.
Chinese Broccoli[2] (gai lon, white flowering)	Cabbage Looper, Imported Cabbageworm, Diamondback Moth, Cabbage Aphid[a]	0.05–0.1 (3.2–6.4)	Apply every 5 days or as needed by air or ground. Use sufficient water to obtain full coverage of foliage. Do not apply more than 0.8 lbs a.i. per acre per season. Do not apply within 1 day of harvest.
Chinese Cabbage (Tight-heading varieties only)	Cabbage Looper, Imported Cabbageworm, Diamondback Moth, Cabbage Aphid[a]	0.05-0.2 (3.2-12.8)	Apply every 5-10 days or as needed by air or ground. Use sufficient water to obtain full coverage of foliage. Do not apply more than 1.0 lb. a.i. per acre per season. Do not apply within 1 day of harvest.
Collards (NC, SC, GA, AZ, OK, TX only)	Cabbage Looper Imported Cabbageworm Diamondback Moth Cabbage Aphid[a]	0.05-0.1 (3.2-6.4)	Apply by ground only. Spray using sufficient water to obtain uniform coverage. Apply as needed. Do not exceed 8 applications. Do not apply within 1 day of harvest. Do not graze treated areas or feed crop refuse to livestock.
Conifers[c] (container and field grown)	Nantucket Pine Tip Moth	0.1-0.2 (6.4-12.8)	Apply first application when adults appear and repeat at 5-7 day intervals or as needed throughout the season. Use sufficient water to obtain full coverage of foliage.
Cotton	Bollworm, Tobacco Budworm, Pink Bollworm, Lygus Bugs, Cabbage Looper, Cotton Leafperforator, Boll Weevil, Cotton Fleahopper, Saltmarsh Caterpillar, Whitefly,[b] Cotton Aphid,[b] Cutworms	0.1-0.2 (6.4-12.8)	Apply as required by scouting reports, usually at intervals of 5 to 7 days. Apply by air or ground. Use sufficient water to obtain full coverage of foliage. For control of boll weevils spray on a 3 to 5 day schedule or as needed. Do not apply within 14 days of harvest. The higher rates of AMBUSH 25W (0.15 to 0.2 lbs. a.i. per acre) may be required under conditions of heavy lepidopterous worm infestation. AMBUSH 25W aids in the reduction of lepidopterous egg hatch by contact action. Do not graze livestock in treated areas. Do not apply more than 3.0 lbs. a.i. per acre per season.
Eggplant	Colorado Potato Beetle	0.2 (12.8)	Apply by ground or air equipment using sufficient water to obtain uniform coverage. Apply as needed. Do not apply more than 3.0 lbs a.i. per acre per season. Do not apply within 3 days of harvest.
	Vegetable Leafminer, Cabbage Looper, Flea Beetle	0.1-0.2 (6.4-12.8)	
Field Corn/ Popcorn/Field Corn Grown for Seed (Preemergent Use)	Armyworm, Cutworms	0.1-0.2 (6.4-12.8)	Apply in the time period from 5 days prior to planting up to emergence of the crop. Apply as a broadcast spray in a minimum of 20 gallons of finished spray/acre with ground equipment.*
Field Corn/ Popcorn/Field Corn Grown for Seed (Foliar Use)	Cutworms, European Corn Borer, Armyworm, Southwestern Corn Borer, Flea Beetle, Corn Earworm,*** Corn Rootworm Bettle,*** Tobacco Budworm***	0.1-0.2 (6.4-12.8)	Apply prior *to the brown silk stage.* Apply a minimum of 1 gallon of finished spray/acre by air and 10 gallons/acre with ground equipment.*
	Stalk Borer		When treating for stalk borer, AMBUSH 25W must be applied when or shortly before the stalk borer larvae are moving into the corn from surrounding weeds and grasses. Mowing or burndown herbicide are suggested to initiate movement.*
Filberts	Filbertworm, Oblique Banded Leafroller	0.2-0.4 (12.8-25.6)	Apply when insects appear. Do not apply more than 1.6 lbs. a.i. per acre per season. Apply in a minimum of 15 gallons of finished spray per acre with ground equipment. Do not apply within 14 days of harvest. Do not graze livestock in treated orchards.

Crop	Target Pests	Rate lbs ai/A (oz/A)	Remarks
Leafy Vegetables** (except Brassica)	Vegetable Leafminer, Cabbage Looper, Cutworm Complex, Fall Armyworm, Beet Armyworm, Corn Earworm, Green Cloverworm, Leafhoppers, European Corn Borer	0.1-0.2 (6.4-12.8)	Apply every 3-5 days or as needed by air or ground. Use sufficient water to obtain full coverage of foliage. Do not apply more than 2.0 lbs. a.i. per acre per season. Can be applied up to 1 day of harvest. Do not graze treated areas or feed crop refuse to livestock.
Lettuce (Refer to Leafy Vegetables for application directions)			
Mushrooms (Mushroom houses & adjacent premise areas)	Mushroom Flies (Sciarid and Phorid Adults)	Apply 6 oz.[d] to 11 gal. water or 8 level tablespoons to 3 gal. water. As a guide, use 1 gal. spray per 750 sq. ft. (400-1,400 sq. ft.)	Spray directly to walls and ceilings as residual surface treatment only. Spray to point of runoff. Use AMBUSH 25W prior to filling house, during cooldown, during spawning, up to pinning and between breaks. Do not use when mushrooms are present. Treat as needed when flies appear. Do not make more than 20 applications prior to pinning of first break; apply no more than two applications between each break. Do not apply more than 30 applications total per crop of 5 breaks. Do not apply within 3 days of harvest.
Onions (Dry Bulb) and Garlic[c]	Onion Thrips Armyworms	0.15-0.3 (9.6-19.2)	Apply in a minimum of 20 gals of water per acre with ground equipment as a foliar spray. Begin applications when pests appear and repeat as necessary to maintain control. Use the higher label rates as Onion Thrips population increases and avoid rescue situations. Do not apply more than 2.4 pounds active per acre per season. Do not apply within 1 day of harvest. Do not graze livestock in treated areas or cut treated crops for feed.
	Cutworms Leafminers Onion Maggots (adults) Stink Bugs	0.1-0.3 (6.4-19.2)	
Ornamentals Nursery Stock[c] (field-grown)	Beet Armyworm Cabbage Looper Citrus Thrip Heliothis Spp. Leafhoppers Leafminers Whiteflies Cutworms	(lbs ai/100 gal) 0.1-0.2 (6.4-12.8)	Marginal leaf burn may occur on Salvia, Dieffenbachia and Pteris Fern. Application to blooming plants may cause browning of petals. Not all species

			and varieties have been tested. Before treating large numbers of plants of a particular variety, treat a few plants and observe prior to full scale application. Apply to nonbearing plants of fruiting species.
Papaya (Florida only)	Fruit fly	0.1 (12.8)[1] oz./100 gal. / 6.4	Apply by ground equipment. Spray to wet all foliage. Apply when insects appear and repeat at 7-10 day intervals as needed to provide control. Do not make more than 6 applications per season. Do not apply within 7 days of harvest. Do not graze livestock in treated areas. Do not feed cover crops from treated areas to livestock.
Peaches	Peach Twig Borer, Oriental Fruit Moth, Green Fruitworm, Lesser Peachtree Borer, Plum Curculio, Tarnished Plantbug, Rose Chafer	0.1-0.4 (6.4-25.6)	Apply by ground equipment using 25-400 gallons of spray per acre. Spray to wet all foliage. Can be applied up to 7 days before harvest. Do not apply more than 3.0 lbs. a.i. per acre per season. Do not graze livestock in treated orchards.
Pears (Dormant)	Pear Psylla	0.2-0.4 (12.8-25.6)	Apply at a rate of 0.8 to 1.6 lbs. (0.2 to 0.4 lbs. active) per acre. May be combined with 2 to 8 gallons of spray oil per acre. Apply during the dormant through delayed dormant growth periods only. Apply in 3 to 20 gallons of finished spray per acre by aircraft and 25 to 400 gallons per acre by ground equipment. Do not apply more than 0.8 lbs. a.i. per acre per season. Do not graze orchard floor.
Pears (Pre-Bloom/ Seasonal	Pear Psylla, Codling Moth, Green Fruitworm	0.2-0.4 (12.8-25.6)	Apply two pre-bloom sprays followed by three seasonal sprays. Pre-bloom sprays can be applied from dormant through bud burst stages. Seasonal sprays can be applied from petal fall up to 14 days of harvest. Do not apply more than 2.0 lbs. a.i. per acre per season.
Peppers (Bell)	Vegetable Leafminer, Cabbage Looper, Flea Beetle, Pepper Weevil	0.1-0.2 (6.4-12.8)	Apply by ground or air equipment using sufficient water to obtain uniform coverage. Apply as needed. Do not apply more than 1.6 lbs. a.i.
	European Corn Borer	0.2 (12.8)	

Crop	Target Pests	Rate lbs ai/A (oz/A)	Remarks
			per acre per season. Do not apply within 3 days of harvest.
Pistachios	Peach Twig Borer, Navel Orangeworm, Chinch Bug	0.2-0.4 (12.8-25.6)	Apply at a rate of 0.8 to 1.6 lbs. (0.2- 0.4 lb active) per acre. Use sufficient water to obtain full coverage of foliage. Do not apply more than 0.8 lbs. a.i. per acre per season. Do not apply after 10 percent hull split. Nuts may be harvested on the day of application. Do not graze livestock in treated orchards.
Potatoes	Colorado Potato Beetle, Potato Leafhopper, Aster Leafhopper, Potato Flea Beetle, Potato Aphid, Potato Tuberworm, Potato Psyllid, Cabbage Looper	0.05-0.2 (3.2-12.8)	Apply as needed by air or ground. Use sufficient water to obtain full coverage of foliage. Do not apply more than 2.4 lbs. a.i. per acre per season. Can be applied up to 7 days prior to harvest.
Pumpkins	Squash Bug, Melonworm	0.2 (12.8)	Apply by ground or air equipment using sufficient water to obtain uniform coverage. Apply as needed. Do not exceed 4 applications. Do not apply within 1 day of harvest.
Roses[c] (field-grown)	Heliothis Spp.	0.1-0.2 (6.4-12.8)	Apply in 5 gallons of finished spray per acre with ground equipment and 1 gallon per acre by air.
Roses[c] (greenhouse)	Beet Armyworm Cabbage Looper Omnivorous Leafroller	8 oz/100 gals water	AMBUSH is not phytotoxic to the following varieties of greenhouse roses: Ballena, Bettina, Cara Mia, Coquette, Excitement, Forever Yours, G. Wave, Jack Frost, Jr. Bridesmaid, Matador, Paul's Pink, Samantha, Seventeen, Town Crier, Tropicana and Visa. Other varieties may vary in their sensitivity to AMBUSH, and a small number of plants should be treated under local conditions to determine plant safety prior to commercial use.
Soybeans	Green Cloverworm, Mexican Bean Beetle, Bean Leaf Beetle, Cabbage Looper, Potato Leafhopper, Velvetbean Caterpillar, Saltmarsh Caterpillar	0.05-0.1 (3.2-6.4)	Apply by air or ground. Do not apply more than 0.4 lbs. a.i. per acre per season. Can be applied up to 60 days before harvest. Do not graze treated areas or harvest for forage or hay.
	Corn Earworm, Soybean Looper[c] Beet Armyworm	0.1-0.2 (6.4-12.8)	
Spinach (Refer to Leafy Vegetables for application directions)			
Sweet Corn (Florida only-corn grown for the fresh market)	Corn Earworm, European Corn Borer, Fall Armyworm, Southern Armyworm, Corn Rootworm Adults, Western Bean Cutworm, Aster Leafhopper	0.1-0.25 (6.4-16.0)	Apply every 5 days or as needed by air or ground. Do not apply more than 3.25 lbs. a.i. per acre per crop season. Do not graze livestock in treated areas or harvest for forage. AMBUSH 25W may be mixed with LANNATE® (methomyl) in accordance with label limitations and precautions. Label dosage rates should not be exceeded.
Tomatoes (only for application to tomatoes grown in Florida and Puerto Rico for final marketing as fresh tomatoes.)	Tomato Fruitworm, Cabbage Looper, Tomato Pinworm, Vegetable Leafminer, Beet Armyworm, Southern Armyworm, Colorado Potato Beetle, Granulate Cutworm, Hornworm	0.05-0.2 (3.2-12.8)	Apply by air or ground in sufficient water to obtain uniform coverage. Do not more than 1.2 lbs. a.i./acre per crop season. Can be applied up to the day of harvest. Do not apply as a band. Do not use on tomato varieties in which the mature tomatoes will be less than 1 inch in diameter (such as cherry tomatoes).
Turnips (SC, GA, FL, WA only) (TX & IN only)	Cabbage Looper Imported Cabbageworm Diamondback Moth Cabbage Aphid[a]	0.05-0.1 (3.2-6.4)	Apply by ground equipment only. Spray using sufficient water to obtain uniform coverage. Apply as needed. Do not exceed 4 applications. Do not apply within 1 day of harvest. Do not graze treated areas or feed crop refuse to livestock. Do not exceed 8 applications.
Walnuts	Navel Orangeworm, Codling Moth, Walnut Husk Fly	0.2-0.4 (12.8-25.6)	Apply when insects appear. Do not apply more than 1.6 lbs. a.i. per acre per season. Apply in a minimum of 20 gallons of finished spray per acre by aircraft or 100 gallons of finished spray per acre with ground equipment. Do not apply within 1 day of harvest. Do not graze livestock in treated orchards.

* AMBUSH 25W may be applied in tank mixtures with the following herbicides in accordance with label limitations and precautions. Label dosage rates should not be exceeded.
LASSO®/Atrazine Pre-mix; LASSO®4E & ME; DUAL®8E; Atrazine 4L & 80W; BLADEX®80W & 4L; PRINCEP®; PRINCEP®CALIBER®90; 2,4-D: ROUNDUP®, BRONCO®, BANVEL®, EXTRAZINE®
NOTE: Do not tank-mix AMBUSH 25W® Insecticide with BICEP®, GRAMOXONE® SUPER.

** Amaranth (leafy amaranth, Chinese spinach, tampala); arrugula (Roquette); celery; celtuce, chervil; corn salad; chrysanthemum, edible-leaved; chrysanthemum, garland; cress, garden; cress, upland (yellow rocket, winter cress); dandelion; dock (sorrel); endive (escarole); fennel, Florence; lettuce (head and leaf); orach; parsley; purslane, garden; purslane, winter, rhubarb; spinach; spinach, fine; spinach, New Zealand; Swiss chard.

*** Not for control of this insect in California.

[a] Aids in suppression.

[b] Only suppression can be expected when utilized on a 5-7 day schedule.

[c] Currently not registered for use in California.

[d] 1 oz. of this powder equals 5 level teaspoons.

[e] Maximum use rate in California is 0.1 lbs. a.i./Acre.

[1] Do not use more than 12.8 oz. AMBUSH 25W/A. The final spray should be no more concentrated than 12.8 oz./100 gallons for avacados and 6.4 oz./100 gallons for papayas. Since tree size and spacing will affect the amount of spray needed for dilute sprays, the applicator must know the number of gallons of water/A needed to

spray the trees to the point of drip. If you do not know this gallonage, you should conduct a test to determine it. If you do not know how to conduct such a test with your equipment, you should ask for assistance from your equipment dealer. For dilute sprays (i.e., sprayed to the point of runoff) 12.8 oz. AMBUSH 25W should be dilute with the gallonage determined by this test.

STORAGE AND DISPOSAL

Prohibitions: Do not contaminate water food or feed by storage or disposal. Do not reuse empty container.

Storage: Mix as needed. Do not store diluted material. Keep container closed when not in use. Do not store near food or feed. In case of spill on floor or paved surfaces, sweep up and remove to chemical waste area.

Pesticide Disposal: Wastes resulting from the use of this product may be disposed of on site or at an approved waste disposal facility.

Container Disposal: Completely empty bag into application equipment. Then dispose of empty bag in a sanitary landfill or by incineration, or, if allowed by state and local authorities, by burning. If burned, stay out of smoke.

RATE CONVERSION CHART

Pounds Active/Acre	Formulation Ounces/Acre	Formulation Lbs./Acre
0.05	3.2	0.2
0.10	6.4	0.4
0.125	8.0	0.5
0.15	9.6	0.6
0.175	11.2	0.7
0.20	12.8	0.8

NOTICE TO BUYER AND USER:

Seller warrants that this product conforms to the chemical description on the label and is reasonably fit for the purposes stated on the label when used in accordance with directions under normal conditions of use. This warranty does not extend to the use of this product contrary to label instructions, or under abnormal use conditions, or under conditions not reasonably foreseeable to Seller, and Buyer and User assume the risk of any such use. SELLER DISCLAIMS ALL OTHER WARRANTIES EXPRESSED OR IMPLIED, INCLUDING ANY WARRANTY OF FITNESS OR MERCHANTABILITY. SELLER SHALL NOT BE LIABLE FOR CONSEQUENTIAL, SPECIAL OR INDIRECT DAMAGES RESULTING FROM THE USE OR HANDLING OF THIS PRODUCT AND SELLER'S SOLE LIABILITY AND BUYER'S AND USER'S EXCLUSIVE REMEDY SHALL BE LIMITED TO THE REFUND OF THE PURCHASE PRICE.

AMBUSH® 25W Insecticide and GRAMOXONE® SUPER Herbicide are registered trademarks of ICI Americas Inc.
BICEP®, CALIBER®, DUAL®, and PRINCEP are registered trademarks of Ciba-Geigy Corp.
ATRAZINE®, BLADEX®, EXTRAZINE® and LANNATE® are registered trademarks of E.I. du Pont de Nemours & Co.
LASSO, ROUNDUP and BRONCO are registered trademarks of Monsanto Company.
BANVEL is a registered trademark of Sandoz Crop Protection Corp.
DE 12-88 011689MEW01 01-0804-367 A-1R

The information and specimen labels contained in this booklet are correct as of the publication date but are subject to change without notice. For current information, contact ICI Americas Inc.

AMBUSH® 25W

INSECTICIDE
Water Soluble Pack

RESTRICTED USE PESTICIDE

For retail sale to and use only by Certified Applicators, or persons under their direct supervision, and only for those uses covered by the Certified Applicator's certification.
Complete Directions For Use and Use Precautions

ACTIVE INGREDIENT:
Permethrin
(3-Phenoxyphenyl) methyl (±)-cis, trans-3-(2,2-dichloroethenyl)-2,2-dimethylcyclopropanecarboxylate* 25.0%
INERT INGREDIENTS 75.0%
TOTAL 100.0%
*Cis/trans ratio: Min. 35%(±) cis and max. 65%(±) trans.
EPA Reg. No. 10182-110

KEEP OUT OF REACH OF CHILDREN
AVISO
WARNING

PRECAUCION AL USUARIO:
Si usted no lee ingles, no use este producto hasta que la etiqueta le haya sido explicado ampliamente.

STATEMENT OF PRACTICAL TREATMENT

IF IN EYES: Flush with water for 15 minutes. Get medical attention if irritation persists.

IF SWALLOWED: Drink 1 or 2 glasses of water and induce vomiting by touching back of throat with finger. Do not induce vomiting or give anything by mouth to an unconscious person.

FOR 24-HOUR EMERGENCY MEDICAL ASSISTANCE, CALL 1-800-F-A-S-T-M-E-D (327-8633).

FOR CHEMICAL EMERGENCY: Spill, leak, fire, exposure, or accident call CHEMTREC 1-800-424-9300.

PRECAUTIONARY STATEMENTS

HAZARDS TO HUMANS AND DOMESTIC ANIMALS

WARNING
MAY CAUSE EYE IRRITATION. HARMFUL IF SWALLOWED.

Do not get in eyes, on skin or on clothing. Wash thoroughly after handling.

ENVIRONMENTAL HAZARDS

This pesticide is extremely toxic to fish. Use with care when applying in areas adjacent to any body of water. Do not apply directly to water. Do not apply when weather conditions favor drift from treated areas. Do not contaminate water when disposing of equipment wash waters. This product is highly toxic to bees exposed to direct treatment or residues on crops or weeds. Do not apply AMBUSH 25W or allow it to drift to crops or weeds on which bees are actively foraging. Additional information may be obtained from your Cooperative Extension Service. Apply this product only as specified on this label.

MIXING INSTRUCTIONS

AMBUSH 25W insecticide is intended for dilution with water for spray application. Prepare no more spray mixture than is needed for the immediate operation. Thoroughly clean spray equipment before using this product. Agitate thoroughly before and during application. Good bypass agitation is sufficient to keep AMBUSH 25W insecticide thoroughly mixed. Do not let spray mixture stand overnight in the spray tank. Rinse spray tank at the end of the day.

Fill spray tank 1/4 to 1/2 full of water. With the agitator running, drop the required number of unopened soluble packs of AMBUSH 25W insecticide into the tank. Continue agitation while adding the remainder of the water.

DIRECTIONS FOR USE

It is a violation of Federal law to use this product in a manner inconsistent with its labeling.

Do not apply this product through any type of irrigation system.

GENERAL INSTRUCTIONS

Apply AMBUSH 25W as shown in "Directions for Application". Use sufficient water to obtain full coverage of foliage. Timing and frequency of applications should be based upon insect populations reaching economic thresholds. The higher rates may be required under conditions of heavy infestation. Do not apply more than 12 lbs per acre per season (3 lbs ai/acre). Crops in this label may be planted immediately after last treatment. Do not plant other crops within 60 days after last application. Any other crop may be planted if it is not harvested or grazed.

REENTRY STATEMENTS

Do not apply this product in such a manner as to directly or through drift expose workers or other persons. The area being treated must be vacated by unprotected persons.

Do not enter treated areas without protective clothing until sprays have dried.

Protective clothing means, at least, a hat or other suitable head covering, a long sleeved shirt and long-legged trousers or a coverall type garment (all of closely woven fabric covering the body, including the arms and legs), shoes and socks.

Because certain states may require more restrictive reentry intervals for various crops treated with this product, consult your State Department of Agriculture for further information.

Written or oral warnings regarding use of protective clothing and accidental exposure must be given to workers who are expected to be in treated areas or in areas about to be treated. Inform workers of areas or fields that may not be entered without specific protective clothing, period of time field must be vacated and appropriate actions to take in case of accidental exposure. An example of such information is given under written warnings. When oral warnings are given, warnings shall be given in a language customarily understood by workers. Oral warnings must be given if there is reason to believe that written warnings cannot be understood by workers. Written warnings must include the following information:

"**WARNING:** Area treated with AMBUSH 25W Insecticide on (insert date). Do not enter without appropriate protective clothing until sprays have dried (if more restrictive, insert State Department of Agriculture reentry interval). In case of accidental exposure to pesticide spray, wash the skin thoroughly with soap and water. Remove contaminated clothing and wash before reuse. If in eyes, flush with plenty of water. Get medical attention if needed."

ENDANGERED SPECIES RESTRICTIONS

For Aerial Application: Do not use within 100 yards of aquatic habitats.

For Ground Application: Do not use within 20 yards of aquatic habitats.

DIRECTIONS FOR APPLICATION

Crop	Target Pests	Rate lbs ai/A (oz/A)	Remarks
Alfalfa	Alfalfa Caterpillar, Armyworm, Cutworm, Looper, Webworm, Potato Leafhopper, Blue Alfalfa Aphid, Green Peach Aphid, Pea Aphid, Green Cloverworm, Spotted Alfalfa Aphid, Velvetbean Caterpillar,	0.05-0.2 (3.2-12.8)	Use higher recommended dosage for increased pest pressure or for increased residual pest control. Apply in a minimum of 1 gallon of finished spray per acre by aircraft or 10 gallons of finished spray per acre with ground equipment. Do not apply more than 0.2 pound active per cutting. When rate of 0.1 pound active per acre or less are used, application may be made on day of harvest. When rates greater than 0.1 pound active per acre are used, do not apply within 14 days of harvest. Avoid application when honey bees are actively foraging by applying during the early morning or during the evening hours.
	Alfalfa Weevil, Egyption Alfalfa Weevil, Cucumber Beetle, Meadow Spittlebug, Plant bugs including Lygus spp., Stink bugs	0.1-0.2 (6.4-12.8)	
Almonds	Navel Orangeworm, Peach Twig Borer	0.2-0.4 (12.8-25.6)	Apply by ground equipment using sufficient water to obtain uniform coverage. Can be applied up to 7 days before harvest. Do not apply more than 2.0 lbs. a.i. per acre per season. Do not graze livestock in treated orchards.
Apples	Apple Aphid, Redbanded Leafroller, Oblique-banded Leafroller, Plum Curculio, White Apple Leafhopper, Spotted Tentiform Leafminer, Tarnished Plant Bug	0.1-0.4 (6.4-25.6)	Apply by ground equipment only, using 25-400 gallons of spray per acre. Spray to wet all foliage. Do not apply after petal fall.
Artichokes	Artichoke Plume Moth	0.1-0.3 (6.4-19.2)	Apply by ground equipment using sufficient water to obtain uniform coverage. Apply as needed. Do not apply more than 1.5 lbs. a.i. per acre per season. Buds may be harvested on the day of application.
Asparagus	Cutworm Complex, Asparagus Beetle	0.05-0.1 (3.2-6.4)	Apply by ground equipment using sufficient water to obtain uniform coverage. Apply as needed. Do not apply more than 0.4 lbs. a.i. per acre per season. Do not apply within 3 days of harvest.
Asparagus Fern	Japanese Beetle, Asparagus Beetle, Lygus Bugs	0.1 (6.4)	Apply to the fern stage of the asparagus plant after spear harvest when larval and adult stage of the asparagus beetle, lygus bugs and the adult stage of the Japanese beetle are present.
Cantaloupes (Refer to Cucurbit Vegetables for application directions)			
Celery (Refer to Leafy Vegetables for application directions)			
Cherries[c]	Lesser Peach Tree Borer, Rose Chafer, Green Fruitworm, Redbanded Leafroller, Plum Curculio, Tarnished Plant Bug	lbs ai/100 gal 0.025-0.05 (1.6-3.2)	Do not exceed 400 gallons per acre. Apply when insects appear. In the East, do not exceed 6 applications per season, with no more than 4 applications after petal fall. In the West, do not apply more than 4 applications per season, with no more than 3 applications after petal fall. Do not apply within 3 days of harvest. Do not graze livestock in treated orchards. Do not feed forage to livestock.
Christmas Trees[c] (container and field-grown)	Nantucket Pine Tip Moth	0.1-0.2 (6.4-12.8)	Apply first application when adults appear and repeat at 5-7 day intervals or as needed throughout the season. Use sufficient water to obtain full coverage of foliage.
Chrysanthemums (Florida only)	Vegetable Leafminer	0.1-0.2 (6.4-12.8)	Apply by ground.
COLE CROPS Broccoli	Cabbage Looper, Imported Cabbageworm Diamondback Moth, Cabbage Aphid[a]	0.05-0.1 (3.2-6.4)	Apply every 5 days or as needed by air or ground. Use sufficient water to obtain full coverage of foliage. Do not apply more than 0.8 lbs. a.i. per acre per season. Do not apply within 1 day of harvest.
Brussels Sprouts	Cabbage Looper, Imported Cabbageworm, Diamondback Moth, Cabbage Aphid[a]	0.05-0.1 (3.2-6.4)	Apply as needed by air or ground. Use sufficient water to obtain full coverage of foliage. Do not apply more than 0.8 lbs. a.i. per acre per season. Do not apply within 1 day of harvest.

Crop	Target Pests	Rate lbs ai/A (oz/A)	Remarks
Cabbage	Cabbage Looper, Imported Cabbageworm, Diamondback Moth Cabbage Aphid[a]	0.05-0.2 (3.2-12.8)	Apply every 5-10 days or as needed by air or ground. Use sufficient water to obtain full coverage of foliage. Do not apply more than 1.0 lb. a.i. per acre per season. Do not apply within 1 day of harvest.
Cauliflower	Cabbage Looper, Imported Cabbageworm, Diamondback Moth, Cabbage Aphid[a]	0.05-0.1 (3.2-6.4)	Apply every 5 days or as needed by air or ground. Use sufficient water to obtain full coverage of foliage. Do not apply more than 0.8 lbs. a.i. per acre per season. Do not apply within 1 day of harvest.
Chinese Broccoli[c] (gai lon, white flowering	Cabbage Looper, Imported Cabbageworm Diamondback Moth, Cabbage Aphid[a]	0.05-0.1 (3.2-6.4)	Apply every 5 days or as needed by air or ground. Use sufficient water to obtain full coverage of foliage. Do not apply more than 0.8 lbs. a.i. per acre per season. Do not apply within 1 day of harvest.
Chinese Cabbage (Tightheading varieties only)	Cabbage Looper, Imported Cabbageworm Diamondback Moth, Cabbage Aphid[a]	0.05-0.2 (3.2-12.8)	Apply every 5-10 days or as needed by air or ground. Use sufficient water to obtain full coverage of foliage. Do not apply more than 1.0 lb. a.i. per acre per season. Do not apply within 1 day of harvest.
Conifers[c] (container and field-grown)	Nantucket Pine Tip Moth	0.1-0.2 (6.4-12.8)	Apply first application when adults appear and repeat at 5-7 day intervals or as needed throughout the season. Use sufficient water to obtain full coverage.
Cotton	Bollworm, Tobacco Budworm, Pink Bollworm, Lygus Bugs, Cabbage Looper, Cotton Leafperforator, Boll Weevil, Cotton Fleahopper, Saltmarsh Caterpillar, Whitefly,[b] Cotton Aphid,[b] Cutworms	0.1-0.2 (6.4-12.8)	Apply as required by scouting reports, usually at intervals of 5 to 7 days. Apply by air or ground. Use sufficient water to obtain full coverage of foliage. For control of boll weevils spray on a 3 to 5 day schedule or as needed. Do not apply within 14 days of harvest. The higher rates of AMBUSH 25W (0.15 to 0.2 lbs. a.i. per acre) may be required under conditions of heavy lepidopterous worm infestation. AMBUSH 25W aids in the reduction of lepidopterous egg hatch by contact action. Do not graze livestock in treated areas. Do not apply more than 3.0 lbs. a.i. per acre per season.
Cucurbit Vegetables+	Leafminers and Squash Bugs	0.2 (12.8)	Apply a minimum of 4 gallons of finished spray per acre by air or 20 gallons of finished spray per acre with ground equipment. Do not apply more than 1.6 pounds active ingredient per acre per season. AMBUSH may be applied up to harvest.
	Cabbage Looper, Cucumber Beetle, (adults), Cutworms, Melon Worm, Pickleworm, Plant Bugs, (including Lygus and Stink Bugs), Rindworms and Squash Vine Borer	0.1-0.2 (6.4-12.8)	
Eggplant	Colorado Potato Beetle	0.2 (12.8)	Apply by ground or air equipment using sufficient water to obtain uniform coverage. Apply as needed. Do not apply more than 3.0 lbs a.i. per acre per season. Do not apply within 3 days of harvest.
	Vegetable Leafminer, Cabbage Looper, Flea Beetle	0.1-0.2 (6.4-12.8)	
Field Corn/ Popcorn/Field Corn Grown for Seed (Preemergent Use)	Armyworm, Cutworms	0.1-0.2 (6.4-12.8)	Apply in the time period from 5 days prior to planting up to emergence of the crop. Apply as a broadcast spray in a minimum of 20 gallons of finished spray/acre with ground equipment*.

Crop	Target Pests	Rate lbs ai/A (oz/A)	Remarks
Field Corn/ Popcorn Field Corn Grown for Seed (Foliar Use)	Cutworms, European Corn Borer, Armyworm, Southwestern Corn Borer, Flea Beetle, Corn Earworm,*** Corn Rootworm Beetle,*** Tobacco Budworm,***	0.1-0.2 (6.4-12.8)	Apply prior to the brown silk stage. Apply a minimum of 1 gallon of finished spray/acre by air and 10 gallons/acre with ground equipment.*
	Stalk Borer		When treating for stalk borer, AMBUSH 25W must be applied when or shortly before the stalk borer larvae are moving into the corn from surrounding weeds and grasses. Mowing or burndown herbicide are suggested to initiate movement.*
Filberts	Filbertworm, Oblique Banded Leafroller	0.2-0.4 (12.8-25.6)	Apply when insects appear. Do not apply more than 1.6 lbs. a.i. per acre per season. Apply in a minimum of 15 gallons of finished spray per acre with ground equipment. Do not apply within 14 days of harvest. Do not graze livestock in treated orchards.
Leafy Vegetables** (except Brassica)	Vegetable Leafminer, Cabbage Looper, Cutworm Complex, Fall Armyworm, Beet Armyworm, Corn Earworm, Green Cloverworm, Leafhoppers, European Corn Borer	0.1-0.2 (6.4-12.8)	Apply every 3-5 days or as needed by air or ground. Use sufficient water to obtain full coverage of foliage. Do not apply more than 2.0 lbs. a.i. per acre per season. Can be applied up to 1 day of harvest. Do not graze treated areas or feed crop refuse to livestock.
Lettuce (Refer to Leafy Vegetables for application directions)			
Mushrooms (Mushroom houses & adjacent premise areas)	Mushroom Flies (Sciarid and Phorid Adults)	Apply 6 oz.[d] to 11 gal. water or 8 level tablespoons to 3 gal. water. As a guide, use 1 gal. spray per 750 sq. ft. (400-1,400 sq. ft.)	Spray directly to walls and ceilings as residual surface treatment only. Spray to point of runoff. Use AMBUSH 25W prior to filling house, during cooldown, during spawning, up to pinning and between breaks. Do not use when mushrooms are present. Treat as needed when flies appear. Do not make more than 20 applications prior to pinning of first break; apply no more than two applications between each break. Do not apply more than 30 applications total per crop of 5 breaks. Do not apply within 3 days of harvest.
Onions (Dry Bulb) and Garlic[c]	Onion Thrips Armyworms	0.15-0.3 (9.6-19.2)	Apply in a minimum of 20 gals of water per acre with ground equipment as a foliar spray. Begin applications when pests appear and repeat as necessary to maintain control. Use the higher label rates as Onion Thrips population increases and avoid rescue situations. Do not apply more than 2.4 pounds active per acre per season. Do not apply within 1 day of harvest. Do not graze livestock in treated areas or cut treated crops for feed.
	Cutworms Leafminers Onion Maggots (adults) Stink Bugs	0.1-0.3 (6.4-19.2)	
Ornamentals Nursery Stock[c] (field-grown)	Beet Armyworm Cabbage Looper Citrus Thrip Heliothis Spp. Leafhoppers Leafminers Whiteflies Cutworms	0.1-0.2 (lbs ai/100 gal) (6.4-12.8)	Marginal leaf burn may occur on Salvia, Dieffenbachia and Pteris Fern. Application to blooming plants may cause browning of petals. Not all species and varieties have been tested. Before treating large numbers of plants of a particular variety, treat a few plants and observe prior to full scale application. Apply to nonbearing plants of fruiting species.
Peaches	Peach Twig Borer, Oriental Fruit Moth, Green Fruitworm, Lesser Peachtree Borer, Plum Curculio, Tarnished Plantbug, Rose Chafer	0.1-0.4 (6.4-25.6)	Apply by ground equipment using 25-400 gallons of spray per acre. Spray to wet all foliage. Can be applied up to 7 days before harvest. Do not apply more than 3.0 lbs. a.i. per acre per season. Do not graze livestock in treated orchards.

Crop	Target Pests	Rate lbs ai/A (oz/A)	Remarks
Pears (Dormant)	Pear Psylla	0.2-0.4 (12.8-25.6)	Apply at a rate of 0.8 to 1.6 lbs. (0.2 to 0.4 lbs. active) per acre. Apply during the dormant through delayed dormant growth periods only. Apply in 3 to 20 gallons of finished spray per acre by aircraft and 25 to 400 gallons per acre by ground equipment. Do not apply more than 0.8 lbs. a.i. per acre per season. Do not graze orchard floor.
Pears (Pre-Bloom/ Seasonal)	Pear Psylla, Codling Moth, Green Fruitworm,	0.2-0.4 (12.8-25.6)	Apply two pre-bloom sprays followed by three seasonal sprays. Pre-bloom sprays can be applied from dormant through bud burst stages. Seasonal sprays can be applied from petal fall up to 14 days of harvest. Do not apply more than 2.0 lbs. a.i. per acre per season.
Peppers (Bell)	Vegetable Leafminer, Cabbage Looper, Flea Beetle, Pepper Weevil	0.1-0.2 (6.4-12.8)	Apply by ground or air equipment using sufficient water to obtain uniform coverage. Apply as needed. Do not apply more than 1.6 lbs. a.i. per acre per season. Do not apply within 3 days of harvest.
	European Corn Borer	0.2 (12.8)	
Pistachios	Peach Twig Borer, Navel Orangeworm, Chinch Bug	0.2-0.4 (12.8-25.6)	Apply at a rate of 0.8 to 1.6 lbs. (0.2-0.4 lb active) per acre. Use sufficient water to obtain full coverage of foliage. Do not apply more than 0.8 lbs. a.i. per acre per season. Do not apply after 10 percent hull split. Nuts may be harvested on the day of application. Do not graze livestock in treated orchards.
Potatoes	Colorado Potato Beetle, Potato Leafhopper, Aster Leafhopper, Potato Flea Beetle, Potato Aphid, Potato Tuberworm, Potato Psyllid, Cabbage Looper	0.05-0.2 (3.2-12.8)	Apply as needed by air or ground. Use sufficient water to obtain full coverage of foliage. Do not apply more than 2.4 lbs. a.i. per acre per season. Can be applied up to 7 days prior to harvest.
Pumpkins	(Refer to Cucurbit Vegetables for application directions)		
Roses[c] (field-grown)	Heliothis Spp.	0.1-0.2 (6.4-12.8)	Apply in 5 gallons of finished spray per acre with ground equipment and 1 gallon per acre by air.
Roses[c] (greenhouse)	Beet Armyworm Cabbage Looper Omnivorous Leafroller	8 oz/100 gals water	AMBUSH is not phytotoxic to the following varieties of greenhouse roses: Ballena, Bettina, Cara Mia, Coquette, Excitement, Forever Yours, G. Wave, Jack Frost, Jr. Bridesmaid, Matador, Paul's Pink, Samantha, Seventeen, Town Crier, Tropicana and Visa. Other varieties may vary in their sensitivity to AMBUSH, and a small number of plants should be treated under local conditions to determine plant safety prior to commercial use.
Soybeans	Green Cloverworm, Mexican Bean Beetle, Bean Leaf Beetle, Cabbage Looper, Potato Leafhopper, Velvetbean Caterpillar, Saltmarsh Caterpillar	0.05-0.1 (3.2-6.4)	Apply by air or ground. Do not apply more than 0.4 lbs. a.i. per acre per season. Can be applied up to 60 days before harvest. Do not graze treated areas or harvest for forage or hay.
	Corn Earworm, Beet Armyworm Soybean Looper,[c]	0.1-0.2 (6.4-12.8)	
Spinach (Refer to Leafy Vegetables for application directions)			
Sweet Corn/ Sweet Corn Grown For Seed	Corn Earworm, European Corn Borer, Fall Armyworm, Corn Rootworm Adults, Aster Leafhopper	0.1-0.2 (6.4-12.8)	Apply every 5 days or as needed by air or ground. Do not apply more than 1.2 lbs. a.i. per acre per season. Do not apply within 1 day of harvest of ears or forage or livestock grazing. AMBUSH 25W may be mixed with LANNATE® (methomyl) in accordance with label limitations and precautions. Label dosage rates should not be exceeded.
Sweet Corn (Florida only-corn grown for the fresh market)	Corn Earworm, European Corn Borer, Fall Armyworm, Southern Armyworm, Corn Rootworm Adults, Western Bean Cutworm, Aster Leafhopper	0.1-0.25 (6.4-16.0)	Apply every 5 days or as needed by air or ground. Do not apply more than 3.25 lbs. a.i, per acre per crop season. Do not graze livestock in treated areas or harvest for forage. AMBUSH 25W may be mixed with LANNATE® (methomyl) in

Crop	Target Pests	Rate lbs ai/A (oz/A)	Remarks
			accordance with label limitations and precautions. Label dosage rates should not be exceeded.
Tomatoes (Only for application to tomatoes grown in Florida and Puerto Rico for final marketing as fresh tomatoes.)	Tomato Fruitworm, Cabbage Looper, Tomato Pinworm, Vegetable Leafminer, Beet Armyworm, Southern Armyworm, Colorado Potato Beetle, Granulate Cutworm, Hornworm	0.05-0.2 (3.2-12.8)	Apply by air or ground in sufficient water to obtain uniform coverage. Do not apply more than 1.2 lbs. a.i./acre per crop season. Can be applied up to the day of harvest. Do not apply as a band. Do not use on tomato varieties in which the mature tomatoes will be less than 1 inch in diameter (such as cherry tomatoes).
Walnuts	Navel Orangeworm, Codling Moth, Walnut Husk Fly	0.2-0.4 (12.8-25.6)	Apply when insects appear. Do not apply more than 1.6 lbs. a.i. per acre per season. Apply in a minimum of 20 gallons of finished spray per acre by aircraft or 100 gallons of finished spray per acre with ground equipment. Do not apply within 1 day of harvest. Do not graze livestock in treated orchards.

* AMBUSH 25W may be applied in tank mixtures with the following herbicides in accordance label limitations and precautions. Label dosage rates should not be exceeded. LASSO® / Atrazine Pre-mix; LASSO® 4E & ME; DUAL® 8E; Atrazine 4L & 80W; BLADEX® 80W & 4L; PRINCEP®; PRINCEP® CALIBER® 90; 2,4-D; ROUNDUP®, BRONCO®, BANVEL®, EXTRAZINE® NOTE: Do not tank-mix AMBUSH® 25W Insecticide with BICEP®, GRAMOXONE® SUPER.

** Amaranth (leafy amaranth, Chinese spinach, tampala); arrugula (Roquette); celery; celtuce, chervil; corn salad; chrysanthemum, edible-leaved; chrysanthemum, garland; cress, garden; cress, upland (yellow rocket, winter cress); dandelion; dock (sorrell); endive (escarole); fennel, Florence; lettuce (head and leaf); orach; parsley; purslane, garden; purslane, winter; rhubarb; spinach; spinach, fine; spinach, New Zealand; Swiss chard.

*** Not for control of this insect in California.

+ Balsam pear (Bitter Melon); Chinese Waxgourd (Chinese preserving melon); citron melon; cucumber; gherkin; gourds, edible; melons, including hybrids such as cantaloupe, casaba, crenshaw, honeydew melons, honey balls, mango melon, muskmelon, Persian melon; pumpkin.

[a] Aids in suppression.

[b] Only suppression can be expected when utilized on a 5-7 day schedule.

[c] Currently not registered for use in California.

[d] 1oz. of this powder equals 5 level teaspoons.

[e] Maximum use rate in California is 0.1 lbs. a.i./Acre.

STORAGE AND DISPOSAL

PROHIBITIONS: Do not contaminate water, food or feed by storage or disposal. Do not reuse empty container.

STORAGE: Mix as needed. Do not store diluted material. Do not store near food or feed. In case of spill on floor or paved surfaces, sweep up and remove to chemical waste area.

Keep water soluble pack in original foil bag and store in a cool dry place. Do not store below 32°F. Handle carefully when stored at temperatures less than 50°F to avoid breakage of pack.

PESTICIDE DISPOSAL: Wastes resulting from the use of this product may be disposed of on site or at an approved waste disposal facilility.

CONTAINER DISPOSAL: When all packs are used, dispose of foil bags in a sanitary landfill, by incineration, or, if allowed by state and local authorities, by burning. If burned, stay out of smoke.

Read label to determine rate of AMBUSH 25W insecticide to be applied. Consult the table below to determine number of acres a 1 lb. or 2 lb. water soluble pack will treat at the desired rate. Base on number of acres to be sprayed per tankload, drop in the required number of unopened packs.

AMBUSH 25W insecticide treatment rate (lb. a.i. per acre)								
	.05	.10	.15	.20	.25	.30	.35	.40
# acres/ 1 lb.	5	2½	1 ⅔	1¼	1	⅚	5/7	⅝
# acres/ 2 lb.	10	5	2 ⅓	2½	2	1 ⅔	1 3/7	1¼

AMBUSH® 25W Insecticide and GRAMOXONE® SUPER Herbicide are registered trademarks of ICI Americas Inc.
BICEP®, CALIBER®, DUAL®, and are registered trademarks of Ciba-Geigy Corp.
ATRAZINE®, BLADEX®, EXTRAZINE® and LANNATE® are registered trademarks of E.I. du Pont de Nemours & Co.
LASSO, ROUNDUP and BRONCO are registered trademarks of Monsanto Company. BANVEL is a registered trademark of Sandoz Crop Protection Corp.

NOTICE TO BUYER AND USER

Seller warrants that this product conforms to the chemical description on the label and is reasonably fit for the purposes stated on the label when used in accordance with directions under normal conditions of use. This warranty does not extend to the use of this product contrary to label instructions, or under abnormal use conditions, or under conditions not reasonably foreseeable to Seller, and Buyer and User assume the risk of any such use. SELLER DISCLAIMS ALL OTHER WARRANTIES EXPRESSED OR IMPLIED, INCLUDING ANY WARRANTY OF FITNESS OR MERCHANTABILITY. SELLER SHALL NOT BE LIABLE FOR CONSEQUENTIAL, SPECIAL OR INDIRECT DAMAGES RESULTING FROM THE USE OR HANDLING OF THIS PRODUCT AND SELLER'S SOLE LIABILITY AND BUYER'S AND USER'S EXCLUSIVE REMEDY SHALL BE LIMITED TO THE REFUND OF THE PURCHASE PRICE.

A-2R
01-0804-172

The information and specimen labels in this booklet are correct as of the publication date but are subject to change without notice. For current information, contact ICI Americas Inc.

AMBUSH®
INSECTICIDE

RESTRICTED USE PESTICIDE

For retail sale to and use only by Certified Applicators, or persons under their direct supervision, and only for those uses covered by the Certified Applicator's certification.

ACTIVE INGREDIENT:
Permethrin
(3-Phenoxyphenyl) methyl (±)-cis, trans-3-
(2,2-dichloroethenyl)-2,2- dimethylcyclopropanecarboxylate* . . . 25.6%
INERT INGREDIENTS: 74.4%
TOTAL 100.0%

*Cis/trans ratio: Min. 35% (±) cis and max. 65% (±) trans. AMBUSH contains 2 pounds active ingredient per gallon and is an emulsifiable concentrate.

EPA Reg. No. 10182-18 U.S. Patent No. 4,024,163

KEEP OUT OF REACH OF CHILDREN
WARNING
AVISO

PRECAUCION AL USUARIO:

Si usted no lee ingles, no use este producto hasta que la etiqueta le haya sido explicado ampliamente.

PRECAUTIONARY STATEMENTS

HAZARDS TO HUMANS AND DOMESTIC ANIMALS
WARNING

HARMFUL IF SWALLOWED OR ABSORBED THROUGH SKIN. MAY CAUSE EYE IRRITATION.

Avoid contact with eyes, skin or clothing. Avoid breathing vapor or spray mist. Wash thoroughly after handling.

STATEMENT OF PRACTICAL TREATMENT

IF SWALLOWED: Do not induce vomiting. Call a physician. Vomiting should be supervised by a physician or the professional staff because of the possible pulmonary damages via aspiration of the solvent. IF IN EYES:

Flush with plenty of water. Get medical attention if irritation persists. IF ON SKIN: Wash with plenty of soap and water.

FOR 24-HOUR EMERGENCY MEDICAL ASSISTANCE, CALL 1-800-F-A-S-T-M-E-D (327-8633).
FOR CHEMICAL EMERGENCY: Spill, leak, fire, exposure, or accident call CHEMTREC 1-800-424-9300.

ENVIRONMENTAL HAZARDS

This pesticide is extremely toxic to fish. Use with care when applying in areas adjacent to any body of water. Do not apply directly to water. Do not apply when weather conditions favor drift from treated areas. Do not contaminate water when disposing of equipment washwaters. This product is highly toxic to bees exposed to direct treatment or residues on crops or weeds. Do not apply AMBUSH, or allow it to drift to crops or weeds on which bees are actively foraging. Additional information may be obtained from your Cooperative Extension Service. Apply this product only as specified on this label.

PHYSICAL OR CHEMICAL HAZARDS

Do not use or store near heat or open flame.

DIRECTIONS FOR USE

It is a violation of Federal law to use this product in a manner inconsistent with its labeling.

General Instructions:

Apply AMBUSH as shown in "Directions for Application." Use sufficient water to obtain full coverage of foliage. Timing and frequency of applications should be based upon insect populations reaching economic thresholds. The higher rates may be required under conditions of heavy infestation. Do not apply more than 12 pints per acre per season (3 lbs ai/acre). Crops in this label may be planted immediately after last treatment. Do not plant other crops within 60 days after last application. Any other crop may be planted if it is not harvested or grazed.

Reentry Statements:

Do not apply this product in such a manner as to directly or through drift expose workers or other persons. The area being treated must be vacated by unprotected persons.

Do not enter treated areas without protective clothing until sprays have dried.

Protective clothing means, at least, a hat or other suitable head covering, a long sleeved shirt and long-legged trousers or a coverall type garment (all of closely woven fabric covering the body, including the arms and legs), shoes and socks.

Because certain states may require more restrictive reentry intervals for various crops treated with this product, consult your State Department of Agriculture for further information.

Written or oral warnings regarding use of protective clothing and accidental exposure must be given to workers who are expected to be in treated areas or in areas about to be treated. Inform workers of areas or fields that may not be entered without specific protective clothing, period of time field must be vacated and appropriate actions to take in case of accidental exposure. An example of such information is given under written warnings. When oral warnings are given, warnings shall be given in a language customarily understood by workers. Oral warnings must be given if there is reason to believe that written warnings cannot be understood by workers. Written warnings must include the following information.

"Warning: Area treated with AMBUSH Insecticide on (insert date). Do not enter without appropriate protective clothing until sprays have dried (if more restrictive, insert State Department of Agriculture reentry interval).

In case of accidental exposure to pesticide spray, wash the skin thoroughly with soap and water. Remove contaminated clothing and wash before re-use. If in eyes, flush with plenty of water. Get medical attention if needed."

ENDANGERED SPECIES RESTRICTIONS

For Aerial Application: Do not use within 100 yards of aquatic habitats.

For Ground Application: Do not use within 20 yards of aquatic habitats.

CHEMIGATION:

Sprinkler Irrigation Application

Apply AMBUSH insecticide at rates and timing described elsewhere in this label. Apply with center pivot or continuous-move equipment distributing 1/2 acre-inch or less during treatment. If stationary (solid sets, hand-lines or wheel lines other than continuous-move) are used, AMBUSH insecticide should be injected into no more than the last 20-30 minutes of the set. Do not apply when winds are greater than 10-15 mph to avoid drift or wind skips. Thorough coverage of foliage is required for good control. Good agitation should be maintained during the entire application period.

Use Precautions—Sprinkler Irrigation Application

A. Apply this product only through (sprinkler including center pivot, lateral move, end tow, side [wheel] roll, traveler, big gun, solid set, or hand move) irrigation system(s). Do not apply this product through any other type of irrigation system.

B. Crop injury, lack of effectiveness, or illegal pesticide residues in the crop can result from nonuniform distribution of treated water.

C. If you have questions about calibration, you should contact State Extension Service specialists, equipment manufacturers or other experts.

D. Do not connect an irrigation system (including greenhouse systems) used for pesticide application to a public water system unless the pesticide label-prescribed safety devices for public water systems are in place.

E. A person knowledgeable of the chemigation system and responsible for its operation, or under the supervision of the responsible person, shall shut the system down and make necessary adjustments should the need arise.

F. The system must contain a functional check valve, vacuum relief valve, and low pressure drain appropriately located on the irrigation pipeline to prevent water source contamination from backflow.

G. The pesticide injection pipeline must contain a functional, automatic, quick-closing check valve to prevent the flow of fluid back toward the injection pump.

H. The pesticide injection pipeline must also contain a functional, normally closed, solenoid-operated valve located on the intake side of the injection pump and connected to the system interlock to prevent fluid from being withdrawn from the supply tank when the irrigation system is either automatically or manually shut down.

I. The system must contain functional interlocking controls to automatically shut off the pesticide injection pump when the water pump motor stops.

J. The irrigation line or water pump must include a functional pressure switch which will stop the water pump motor when the water pressure decreases to the point where pesticide distribution is adversely affected.

K. Systems must use a metering pump, such as a positive displacement injection pump (e.g., diaphragm pump) effectively designed and constructed of materials that are compatible with pesticides and capable of being fitted with a system interlock.

L. Do not apply when wind speed favors drift beyond the area intended for treatment.

M. Do not apply through chemigation systems connected to public water systems.

Directions for Application

Crop	Target Pests	Rate lbs ai/A (fl oz/A)	Remarks
Alfalfa	Alfalfa Caterpillar Armyworm, Cutworm, Looper, Webworm, Potato Leafhopper, Blue Alfalfa Aphid, Green Peach Aphid, Pea Aphid, Green Cloverworm, Spotted Alfalfa Aphid, Velvetbean Caterpillar	0.05-0.2 (3.2-12.8)	Use higher recommended dosage for increased pest pressure or for increased residual pest control. Apply in a minimum of 1 gallon of finished spray per acre by aircraft or 10 gallons of finished spray per acre with ground equipment. Do not apply more than 0.2 pound active per cutting. When rates of 0.1 pound active per acre or less are used, application may be made on day of harvest. When rates greater than 0.1 pound active per acre are used, do not apply within 14 days of harvest. Avoid application when honey bees are actively foraging by applying during the early morning or during the evening hours.
	Alfalfa Weevil, Egyptian Alfalfa Weevil, Cucumber Beetle, Meadow Spittlebug, Plant Bugs including Lygus spp., Stink Bugs	0.1-0.2 (6.4-12.8)	
Almonds	Naval Orangeworm, Peach Twig Borer	0.2-0.4 (12.8-25.6)	Apply by ground equipment using sufficient water to obtain uniform coverage. Can be applied up to 7 days before harvest. Do not apply more than 2.0 lbs. a.i. per acre per season. Do not graze livestock in treated orchards.

Crop	Target Pests	Rate lbs ai/A (fl oz/A)	Remarks
Apples	Apple Aphid, Redbanded Leafroller, Oblique-banded Leafroller, Plum Curculio, White Apple Leafhopper, Spotted Tentiform Leafminer, Tarnished Plant Bug	0.1-0.4 (6.4-25.6)	Apply by ground equipment only, using 25-400 gallons of spray per acre. Spray to wet all foliage. Do not apply after petal fall.
Artichokes	Artichoke plume moth	0.1-0.3 (6.4-19.2)	Apply by ground equipment using sufficient water to obtain uniform coverage. Apply as needed. Do not apply more than 1.5 lbs. a.i. per acre per season. Buds may be harvested on the day of application.
Asparagus	Cutworm Complex, Asparagus Beetle	0.05-0.1 (3.2-6.4)	Apply by ground equipment using sufficient water to obtain uniform coverage. Apply as needed. Do not apply more than 0.4 lbs. a.i. per acre per season. Do not apply within 3 days of harvest.
Asparagus Ferns	Japanese Beetle, Asparagus Beetle, Lygus Bugs	0.1 (6.4)	Apply to the fern stage of the asparagus plant after spear harvest when larval and adult stage of the asparagus beetle, lygus bugs and the adult stage of the Japanese beetle are present.
Avocados[c]	Thrips, Leafhoppers, Mirid bugs, larvae	0.2 (12.8)[1] <u>fl.oz./100 gal.</u> 12.8	Apply by ground equipment. Spray to wet all foliage. Apply when insects appear and repeat at 7–10 day intervals as needed to provide control. Do not make more than 6 applications per season. Do not apply within 7 days of harvest. Do not graze livestock in treated areas. Do not feed cover crops from treated areas to livestock.
Cantaloupes	Squash bug, Pickleworm, Leafminers, Cutworms	0.1-0.2 (6.4-12.8)	Apply by ground equipment using sufficient water to obtain full coverage of foliage. Make applications when insects appear and repeat weekly as needed. Do not apply more than 1.6 lbs. a.i. per acre per season. Do not apply within 1 day of harvest
Celery (Refer to Leafy Vegetables for application directions)			
Cherries	Lesser Peach Tree Borer, Rose Chafer, Green Fruitworm, Redbanded Leafroller, Plum Curculio, Tarnished Plant Bug	Rate lbs ai/ 100 gal. 0.025-0.05 (1.6-3.2)	Do not exceed 400 gallons per acre. Apply when insects appear. In the East, do not exceed 6 applications per season, with no more than 4 applications after petal fall. In the West, do not apply more than 4 applications per season, with no more than 3 applications after petal fall. Do not apply within 3 days of harvest. Do not graze livestock in treated orchards. Do not feed forage to livestock.
Christmas Trees[c] (container and field grown)	Nantucket Pine Tip Moth	0.1-0.2 (6.4-12.8)	Apply first application when adults appear and repeat at 5-7 day intervals or as needed throughout the season. Use sufficient water to obtain full coverage of foliage.
Chrysanthemums (Florida only)	Vegetable Leafminer	0.1-0.2 (6.4-12.8)	Apply by ground.
COLE CROPS Broccoli	Cabbage Looper, Imported Cabbageworm, Diamondback Moth, Cabbage Aphid[a]	0.05-0.1 (3.2-6.4)	Apply every 5 days or as needed by air or ground. Use sufficient water to obtain full coverage of foliage. Do not apply more than 0.8 lbs. a.i. per acre per season. Do not apply within 1 day of harvest.
Brussels Sprouts	Cabbage Looper, Imported Cabbageworm, Diamondback Moth, Cabbage Aphid[a]	0.05-0.1 (3.2-6.4)	Apply as needed by air or ground. Use sufficient water to obtain full coverage of foliage. Do not apply more than 0.8 lbs. a.i. per acre per season. Do not apply within 1 day of harvest.
Cabbage	Cabbage Looper, Imported Cabbageworm, Diamondback Moth, Cabbage Aphid[a]	0.05-0.2 (3.2-12.8)	Apply every 5-10 days or as needed by air or ground. Use sufficient water to obtain full coverage of foliage. Do not apply more than 1.0 lb. a.i. per acre per season. Do not apply within 1 day of harvest.
Cauliflower	Cabbage Looper, Imported Cabbageworm, Diamondback Moth, Cabbage Aphid[a]	0.05-0.1 (3.2-6.4)	Apply every 5 days or as needed by air or ground. Use sufficient water to obtain full coverage of foliage. Do not apply more than 0.8 lbs. a.i. per acre per season. Do not apply within 1 day of harvest.
Chinese broccoli[c] (gai lon, white flowering)	Cabbage Looper, Imported Cabbageworm, Diamondback Moth Cabbage Aphid[a]	0.05-0.1 (3.2-6.4)	Apply every 5 days or as needed by air or ground. Use sufficient water to obtain full coverage of foliage. Do not apply more than 0.8 lbs a.i. per acre per season. Do not apply within 1 day of harvest.

Crop	Target Pests	Rate lbs ai/A (fl oz/A)	Remarks
Chinese Cabbage (Tight-heading varieties only)	Cabbage Looper, Imported Cabbageworm, Diamondback Moth, Cabbage Aphid[a]	0.05-0.2 (3.2-12.8)	Apply every 5-10 days or as needed by air or ground. Use sufficient water to obtain full coverage of foliage. Do not apply more than 1.0 lb. a.i. per acre per season. Do not apply within 1 day of harvest.
Collards (NC, SC, GA AZ, OK, TX only)	Cabbage Looper Imported Cabbageworm Diamondback Moth Cabbage Aphid[a]	0.05-0.1 (3.2-6.4)	Apply by ground only. Spray using sufficient water to obtain uniform coverage. Apply as needed. Do not exceed 8 applications. Do not apply within 1 day of harvest. Do not graze treated areas or feed crop refuse to livestock.
Conifers (container and field grown)	Nantucket Pine Tip Moth	0.1-0.2 (6.4-12.8)	Apply first application when the adults appear and repeat at 5-7 day intervals or as needed throughout the season. AMBUSH may be diluted in a nonvolatile vegetable oil or in water in a minimum of one gallon of finished spray per acre using equipment calibrated to give adequate coverage.
Cotton	Bollworm, Tobacco Budworm, Pink Bollworm, Cabbage Looper, Cotton Leafperforator, Boll Weevil, Cotton Fleahopper, Saltmarsh Caterpillar, Whitefly,[b] Cotton Aphid,[b] Cutworms	0.1-0.2 (6.4-12.8)	Apply as required by scouting reports, usually at intervals of 5 to 7 days. Apply by air or ground. Use sufficient water to obtain full coverage of foliage. AMBUSH may also be applied by air diluted with a vegetable oil such as cottonseed oil or soybean oil. Agricultural mineral oils (crop oils) are not recommended. Apply using a minimum of 1 quart of finished spray per acre. For control of boll weevils, spray on a 3 to 5 day schedule or as needed. Do not apply within 14 days of harvest. The higher rates of AMBUSH (0.15 to 0.2 lbs. a.i. per acre) may be required under conditions of heavy lepidopterous worm infestation. AMBUSH aids in the reduction of lepidopterous egg hatch by contact action. Do not graze livestock in treated areas. Do not apply more than 3.0 lbs. a.i. per acre per season.
Eggplant	Colorado Potato Beetle Vegetable Leafminer, Cabbage Looper, Flea Beetle	0.2 (12.8) 0.1-0.2 (6.4-12.8)	Apply by ground or air equipment using sufficient water to obtain uniform coverage. Apply as needed. Do not apply more than 3.0 lbs a.i. per acre per season. Do not apply within 3 days of harvest.
Field Corn/ Popcorn/Field Corn Grown for Seed (Preemergent Use)	Armyworm, Cutworms	0.1-0.2 (6.4-12.8)	Apply in the time period from 5 days prior to planting up to emergence of the crop. Apply as a broadcast spray in a minimum of 20 gallons of finished spray/acre with ground equipment.**
Field Corn/ Popcorn/ Field Corn Grown for Seed (Foliar Use)	Cutworms, European Corn Borer, Armyworm, Southwestern Corn Borer, Flea Beetle, Corn Earworm,*** Corn Rootworm Beetle,*** Tobacco Budworm*** Stalk Borer	0.1-0.2 (6.4-12.8)	Apply *prior to the brown silk stage.* Apply a minimum of 1 gallon of finished spray/acre by air and 10 gallons/acre with ground equipment.** When treating for stalk borer, AMBUSH must be applied when or shortly before the stalk borer larvae are moving into the corn from surrounding weeds and grasses. Mowing or burndown herbicide are suggested to initiate movement.**
Filberts	Filbertworm, Oblique Banded Leafroller	0.2-0.4 (12.8-25.6)	Apply when insects appear. Do not apply more than 1.6 lbs. a.i. per acre per season. Apply in a minimum of 15 gallons of finished spray per acre by aircraft or 20 gallons of finished spray per acre with ground equipment. Do not apply within 14 days of harvest. Do not graze livestock in treated orchards.
Horseradish	Imported Crucifer Weevil (*Baris lepidii*) control	0.1% ai solution[d] in a spring preplant dip and 0.2 lb ai/A (12.8) as a foliar spray.	Soak sets for 30 min. and air-dry before planting; make up to 3 foliar applications as needed to control weevil adults during oviposition. Apply by ground equipment in a minimum spray volume of 20 gallons per acre. Do not apply within 22 days of harvest.
Leafy Vegetables* (except *Brassica*)	Vegetable Leafminer, Cabbage Looper, Cutworm Complex, Fall Armyworm, Beet Armyworm, Corn Earworm, Green Cloverworm, Leafhoppers, European Corn Borer	0.1-0.2 (6.4-12.8)	Apply every 3-5 days or as needed by air or ground. Use sufficient water to obtain full coverage of foliage. Do not apply more than 2.0 lbs. a.i. per acre per season. Can be applied up to 1 day of harvest. Do not graze treated areas or feed crop refuse to livestock.
Lettuce (Refer to Leafy Vegetables for application directions)			

Crop	Target Pests	Rate lbs ai/A (fl oz/A)	Remarks
Mushrooms (Mushroom houses & adjacent premise areas)	Mushroom Flies (Sciarid and Phorid Adults)	Apply as a fogging or aerosol treatment at the rate of 3.2 to 4.0 ounces (0.05- 0.0625 lb ai) per 30 ounces of water or suitable diluent. Use 1 quart of solution per standard double house (35,000 cu. ft.: 8,000 sq. ft.).	Preparation of the building prior to fogging: (1) Close all doors, windows and ventilators, (2) Lock or barricade all entrances, turn off pilot lights, post warning signs, and take precautions to prevent persons and animals from entering the area. Use prior to filling house, during cool-down, during spawning, up to pinning and between breaks. Do not use when mushrooms are present. Treat once daily or as needed when flies appear. Do not make more than 20 applications prior to pinning of first break; apply no more than two applications between each break. Do not apply more than 30 applications total per crop of 5 breaks. Do not apply within 3 days of harvest. Length of exposure time should be limited to 1 hour; then ventilate the house. Use fans to ventilate in houses that do not have forced air circulation. Wear full-faced gas mask with canister type recommended for general insecticide protection.
Onions (Dry Bulb) and Garlic[c]	Onion Thrips Armyworms	0.15-0.3 (9.6-19.2)	Apply in a minimum of 20 gallons of water per acre with ground equipment as foliar spray. Begin applications when pests appear and repeat as necessary to maintain control. Use the higher label rates as onion thrips population increases and avoid rescue situations. Do not apply more than 2.4 lbs active per acre per season. Do not apply within 1 day of harvest. Do not graze livestock in treated areas or cut treated crops for feed.
	Cutworms Leafminers Onion Maggots (adult) Stink bugs	0.1-0.3 (6.4-19.2)	
Ornamentals Nursery Stock[c] (field-grown)	Beet Armyworm Cabbage Looper Citrus Thrip Heliothis Spp. Leafhoppers Leafminers Whiteflies Cutworms	Rate lbs ai/100 gal. 0.1-0.2 (6.4-12.8)	Marginal leaf burn may occur on Salvia, Dieffenbachia and Pteris Fern. Application to blooming plants may cause browning of petals. Not all species and varieties have been tested. Before treating large numbers of plants of a particular variety, treat a few plants and observe prior to full scale application. Apply to nonbearing plants of fruiting species.
Papaya[c] (Florida only)	Fruit Fly	0.1 (12.8)[1] oz./100 gal. 6.4	Apply by ground equipment. Spray to wet all foliage. Apply when insects appear and repeat at 7-10 day intervals as needed to provide control. Do not make more than 6 applications per season. Do not apply within 7 days of harvest. Do not graze livestock in treated areas. Do not feed cover crops from treated areas to livestock.
Peaches	Peach Twig Borer, Oriental Fruit Moth, Green Fruitworm, Lesser Peachtree Borer, Plum Curculio, Tarnished Plantbug, Rose Chafer	0.1-0.4 (6.4-25.6)	Apply by ground equipment using 25-400 gallons of spray per acre. Spray to wet all foliage. Can be applied up to 7 days before harvest. Do not apply more than 3.0 lbs. a.i. per acre per season. Do not graze livestock in treated orchards.
Pears (Dormant)	Pear Psylla	0.2-0.4 (12.8-25.6)	Apply at a rate of 0.8 to 1.6 pints (0.2 to 0.4 lbs. active) per acre. May be combined with 2 to 8 gallons of spray oil per acre. Apply during the dormant through delayed dormant growth periods only. Apply in 3 to 20 gallons of finished spray per acre by aircraft and 25 to 400 gallons per acre by ground equipment. Do not apply more than 0.8 lbs. a.i. per acre per season. Do not graze orchard floor.
Pears (Pre-Bloom/ Seasonal)	Pear Psylla, Codling Moth, Green Fruitworm	0.2-0.4 (12.8-25.6)	Apply two pre-bloom sprays followed by three seasonal sprays. Pre-bloom sprays can be applied from dormant through bud burst stages. Seasonal sprays can be applied from petal fall up to 14 days of harvest. Do not apply more than 2.0 lbs. a.i. per acre per season.
Peppers (Bell)	Vegetable Leafminer, Cabbage Looper, Flea Beetle, Pepper Weevil	0.1-0.2 (6.4-12.8)	Apply by ground or air equipment using sufficient water to obtain uniform coverage. Apply as needed. Do not apply more than 1.6 lbs. a.i.
	European Corn Borer	0.2 (12.8)	

Crop	Target Pests	Rate lbs ai/A (fl oz/A)	Remarks
			per acre per season. Do not apply within 3 days of harvest.
Pine Seed Orchard	Coneworms, Seed Bugs	13 fl oz/ 100 gals water (0.025% dilution by wt) for high volume sprayers; 68 fl oz/ 100 gals water (0.125% dilution by wt) for low volume sprayers; 48 fl oz./ A/application in not less than 10 gals of water for aerial applications	Apply first application within 1 week of female flower closure or peak pollen flight for webbing coneworm control. For other coneworms and seed bugs, apply first application within 30 days following female flower closure. Repeat applications at intervals of 4 weeks, but do not apply more than 6 applications. Apply approximately 5-10 gals of the 0.025% dilution per tree with high volume sprayers. With low volume sprayers, apply 100 gallons of the 0.125% dilution per acre. Do not graze or harvest cover crop. Avoid contact with open water.
Pistachios	Peach Twig Borer, Navel Orangeworm, Chinch Bug	0.2-0.4 (12.8-25.6)	Apply at a rate of 0.8 to 1.6 pints (0.2-0.4 lb active) per acre. Use sufficient water to obtain full coverage of foliage. Do not apply more than 0.8 lbs. a.i. per acre per season. Do not apply after 10 percent hull split. Nuts may be harvested on the day of application. Do not graze livestock in treated orchards.
Potatoes	Colorado Potato Beetle, Potato Leafhopper, Aster Leafhopper, Potato Flea Beetle, Potato Aphid, Potato Tuberworm, Potato Psyllid, Cabbage Looper	0.05-0.2 (3.2-12.8)	Apply as needed by air or ground. Use sufficient water to obtain full coverage of foliage. Do not apply more than 2.4 lbs. a.i. per acre per season. Can be applied up to 7 days prior to harvest.
Pumpkins	Squash bug, Melonworm	0.2 (12.8)	Apply by ground or air equipment using sufficient water to obtain uniform coverage. Apply as needed. Do not exceed 4 applications. Do not apply within 1 day of harvest.
Range Grass	Range Caterpillar	0.01 (0.64)	Apply by ground or air using sufficient volume to obtain uniform coverage. Do not apply more than once per year. Cattle may be present during application. Do not harvest or feed hay to livestock.
Roses[c] (field-grown)	Heliothis Spp.	0.1-0.2 (6.4-12.8)	Apply in 5 gallons of finished spray per acre with ground equipment and 1 gallon per acre by air.
Roses[c] (greenhouse)	Beet Armyworm Cabbage Looper Omnivorous Leafroller	8 fl oz/100 gals water	AMBUSH is not phytotoxic to the following varieties of greenhouse roses: Ballena, Bettina, Cara Mia, Coquette, Excitement, Forever Yours, G. Wave, Jack Frost, Jr. Bridesmaid, Matador, Paul's Pink, Samantha, Seventeen, Town Crier, Tropicana and Visa. Other varieties may vary in their sensitivity to AMBUSH, and a small number of plants should be treated under local conditions to determine plant safety prior to commercial use.
Soybeans	Green Cloverworm, Mexican Bean Beetle, Bean Leaf Beetle, Cabbage Looper, Potato Leafhopper, Velvetbean Caterpillar, Saltmarsh Caterpillar	0.05-0.1 (3.2-6.4)	Apply by air or ground. Do not apply more than 0.4 lbs. a.i. per acre per season. Can be applied up to 60 days before harvest. Do not graze treated areas or harvest for forage or hay.
	Corn Earworm, Soybean Looper[c], Beet Armyworm	0.1-0.2 (6.4-12.8)	
Spinach (Refer to Leafy Vegetables for application directions)			
Sweet corn	Corn Earworm, European Corn Borer, Fall Armyworm, Corn Rootworm Adults, Western Bean Cutworm, Aster Leafhopper	0.1-0.2 (6.4-12.8)	Apply every 5 days or as needed by air or ground. Do not apply more than 1.2 lbs. a.i. per acre per season. Do not apply within 1 day of harvest of ears or forage or livestock grazing. AMBUSH may be mixed with LANNATE® (methomyl) in accordance with label limitations and precautions. Label dosage rates should not be exceeded.
Sweet Corn (Florida only-corn grown for the fresh market)	Corn Earworm, European Corn Borer, Fall Armyworm, Southern Armyworm, Corn Rootworm Adults, Western Bean Cutworm, Aster Leafhopper	0.1-0.25 (6.4-16.0)	Apply every 5 days or as needed by air or ground. Do not apply more than 3.25 lbs. a.i. per acre per crop season. Do not graze livestock in treated areas or harvest for forage. AMBUSH may be mixed with LANNATE® (methomyl) in accordance with label limitations and precautions. Label dosage rates should not be exceeded.

Crop	Target Pests	Rate lbs ai/A (fl oz/A)	Remarks
Tomatoes (Only for application to tomatoes grown in Florida and Puerto Rico for final marketing as fresh tomatoes.)	Tomato Fruitworm, Cabbage Looper, Tomato Pinworm, Vegetable Leafminer, Beet Armyworm, Southern Armyworm, Colorado Potato Beetle, Granulate Cutworm, Hornworm	0.05-0.2 (3.2-12.8)	Apply by air or ground in sufficient water to obtain uniform coverage. Do not apply more than 1.2 lbs. a.i./acre per crop season. Can be applied up to the day of harvest. Do not apply as a band. Do not use on tomato varieties in which the mature tomatoes will be less than 1 inch in diameter (such as cherry tomatoes).
Turnips (SC, GA, FL, WA only)	Cabbage Looper Imported Cabbageworm Diamond Back Moth Cabbage Aphid[a]	0.05-0.1 (3.2-6.4)	Apply by ground equipment only. Spray using sufficient water to obtain uniform coverage. Apply as needed. Do not exceed 4 applications. Do not apply within 1 day of harvest. Do not graze treated areas or feed crop refuse to livestock.
(TX & IN only)			Do not exceed 8 applications.
Walnuts	Naval Orangeworm, Codling Moth, Walnut Husk Fly	0.2-0.4 (12.8-25.6)	Apply when insects appear. Do not apply more than 1.6 lbs. a.i. per acre per season. Apply in a minimum of 20 gallons of finished spray per acre by aircraft or 100 gallons of finished spray per acre with ground equipment. Do not apply within 1 day of harvest. Do not graze livestock in treated orchards.

* Amaranth (leafy amaranth, Chinese spinach, tampala); arrugula (Roquette); celery, celtuce; chervil; corn salad; chrysanthemum, edible-leaved; chrysanthemun, garland; cress, garden; cress, upland (yellow rocket, winter cress); dandelion; dock (sorrel); endive (escarole); fennel, Florence; lettuce (head and leaf); orach; parsley; purslane, garden; purslane, winter; rhubarb; spinach; spinach, fine; spinach, New Zealand; Swiss chard.

** AMBUSH may be applied in tank-mixtures with the following herbicides in accordance with label limitations and precautions. Label dosage rates should not be exceeded. GRAMOXONE® SUPER; BICEP®; LASSO®/Atrazine Pre-mix; LASSO® 4E & ME; DUAL® 8E; Atrazine 4L & 80W; BLADEX® 80W & 4L; PRINCEP®; PRINCEP® CALIBER 90®; 2,4-D; ROUNDUP®; BRONCO®; BANVEL®; EXTRAZINE®

*** Not for control of this insect in California.

[a] Aids in suppression.

[b] Only suppression can be expected when utilized on a 5-7 day schedule.

[c] Currently not registered for use in California.

[d] 0.1% ai solution = 3⅓ pt formulation per 100 gallons of water.

[e] Maximum use rate in California is 0.1 lbs. a.i./Acre.

[f] Do not use more than 12.8 oz. AMBUSH/A. The final spray should be no more concentrated than 12.8 oz./100 gallons for avocados and 6.4 oz./100 gallons for papayas. Since tree size and spacing will affect the amount of spray needed for dilute sprays, the applicator must know the number of gallons of water/A needed to spray the trees to the point of drip. If you do not know this gallonage, you should conduct a test to determine it. If you do not know how to conduct such a test with your equipment, you should ask for assistance from your equipment dealer. For dilute sprays (i.e., sprayed to the point of runoff) 12.8 oz. AMBUSH should be dilute with the gallonage determined by this test.

STORAGE AND DISPOSAL

Prohibitions: Do not contaminate water, food or feed by storage or disposal. Open dumping is prohibited. Do not reuse empty container.

Storage: Keep container closed when not in use. Do not store near food or feed. Protect from freezing. In case of spill or leak on floor or paved surfaces, soak up with sand, earth or synthetic absorbent. Remove to chemical waste area.

Pesticide Disposal: Pesticide wastes are toxic. Improper disposal of excess pesticides, spray mixture or rinsate is a violation of federal law. If these wastes cannot be disposed of by use according to label instructions, contact your State Pesticide or Environmental Control Agency, or the Hazardous Waste representative at the nearest EPA Regional Office for guidance.

Container Disposal: Metal Containers: Triple rinse (or equivalent). Then offer for recycling or reconditioning, or puncture and dispose of in a sanitary landfill, or by other procedures approved by state and local authorities.

Plastic Containers: Triple rinse (or equivalent). Then offer for recycling or reconditioning, or puncture and dispose of in a sanitary landfill, or by incineration, or, if allowed by state and local authorities by burning. If burned, stay out of smoke.

Glass Containers: Triple rinse (or equivalent). Then dispose of in a sanitary landfill or by other approved state and local procedures.

RATE CONVERSION CHART

Pounds Active/Acre	Fluid Ounces/Acres	Pints/Acre	Acres Treated/ Gallon
0.05	3.2	0.2	40.0
0.10	6.4	0.4	20.0
0.125	8.0	0.5	16.0
0.15	9.6	0.6	13.3
0.175	11.2	0.7	11.4
0.20	12.8	0.8	10.0

NOTICE TO BUYER AND USER:

Seller warrants that this product conforms to the chemical description on the label and is reasonably fit for the purposes stated on the label when used in accordance with directions under normal conditions of use. This warranty does not extend to the use of this product contrary to label instructions, or under abnormal use conditions, or under conditions not reasonably foreseeable to Seller, and Buyer and User assume the risk of any such use. SELLER DISCLAIMS ALL OTHER WARRANTIES EXPRESSED OR IMPLIED, INCLUDING ANY WARRANTY OF FITNESS OR MERCHANTABILITY. SELLER SHALL NOT BE LIABLE FOR CONSEQUENTIAL, SPECIAL OR INDIRECT DAMAGES RESULTING FROM THE USE OR HANDLING OF THIS PRODUCT AND SELLER'S SOLE LIABILITY AND BUYER'S AND USER'S EXCLUSIVE REMEDY SHALL BE LIMITED TO THE REFUND OF THE PURCHASE PRICE.

AMBUSH® Insecticide and GRAMOXONE® SUPER Herbicide are registered trademarks of ICI Americas Inc.

LANNATE® is a registered trademark of E.I. du Pont de Nemours & Co.

LASSO; ROUNDUP; and BRONCO are registered trademarks of Monsanto Company.

BICEP, DUAL; PRINCEP; and CALIBER are registered trademarks of Ciba-Geigy Corp.

BLADEX; ATRAZINE; and EXTRAZINE are registered trademarks of E.I. du Pont de Nemours & Co., Inc.

BANVEL is a registered trademark of Sandoz Corp Protection Corp.

A-IR
01-0804-290

The information and specimen labels contained in this booklet are correct as of the publication date but are subject to change without notice. For current information, contact ICI Americas Inc.

ARROSOLO® 3-3E

An Emulsifiable Liquid Herbicide For Use On Rice Grown in the Southern Region Only.

ACTIVE INGREDIENTS:	
S-ethyl hexahydro-1*H*-azepine-1-carbothioate	33.1%
N-(3,4-dichlorophenyl) propanamide	33.1%
INERT INGREDIENTS:	33.8%
TOTAL	100.0%

Contains 6 Pounds Active Ingredients Per Gallon

EPA Reg. No. 10182-260

KEEP OUT OF REACH OF CHILDREN

WARNING

STATEMENT OF PRACTICAL TREATMENT

Call a Poison Center or Physician Immediately

If a known exposure occurs or is suspected, immediately start the procedures given below and contact a Poison Center, physician or the nearest hospital.

NOTE: Be sure to advise the person contacted that the product is a weak cholinesterase inhibitor in experimental animals. Describe the type and extent of exposure, the victim's symptoms, and follow the advice given.

Note to medical personnel: If cholinesterase inhibition is suspected, atropine by injection is antidotal. It has not been determined if pralidoxime chloride (2-PAM, protopam) is antidotal for the product.

If swallowed: Immediately give several glasses of water but *do not* induce vomiting. If vomiting occurs, give fluids again. Have a physician determine if condition of patient will permit induction of vomiting or evacuation of stomach. Do not give anything by mouth to an unconscious or convulsing person.

If in eyes: Immediately flush eyes with large amounts of water for at least 15 minutes. Hold eyelids apart during flushing to ensure rinsing of the entire surface of the eye and lids with water. Get medical attention immediately.

If on skin: Flush affected areas with plenty of water for several minutes. Remove contaminated clothing and shoes. Get medical attention if irritation occurs.

If inhaled: Remove to fresh air. Seek medical attention if respiratory irritation occurs or breathing becomes difficult.

FOR 24-HOUR EMERGENCY MEDICAL ASSISTANCE CALL 1-800-F-A-S-T-M-E-D (327-8633).
FOR CHEMICAL EMERGENCY: Spill, leak, fire, exposure, or accident Call CHEMTREC 1-800-424-9300.

PRECAUTIONARY STATEMENTS

HAZARDS TO HUMANS AND DOMESTIC ANIMALS

WARNING

Causes substantial but temporary eye irritation. Wear goggles or face shield. Harmful if swallowed. Avoid contact with skin, eyes and clothing.

Wash with soap and water after use. Avoid breathing spray mist. Remove contaminated clothing and wash before reuse.

ENVIRONMENTAL HAZARDS

This pesticide is extremely toxic to fish. Drift and runoff from treated areas may be hazardous to aquatic organisms in neighboring areas. Do not apply directly to water except as specified on this label. Do not contaminate water when disposing of equipment washwaters. Do not contaminate water used for domestic use.

Water drained from treated rice fields must not be used to irrigate other crops or released within 1/2 mile upstream of a potable water intake in flowing water (i.e., river, stream, etc.) or within 1/2 mile of a potable water intake in a standing body of water such as a lake, pond or reservoir.

PHYSICAL/CHEMICAL HAZARDS

Do not use or store near heat or open flame.

GENERAL USE PRECAUTIONS

Read all label directions before using. Do not overdose.

ARROSOLO 3-3E should be used only for recommended purposes and at recommended rates.

ARROSOLO 3-3E may cause crop injury under extreme soil or climatic conditions or if directions are not followed.

Except as specified on this label, do not apply other herbicides, oils, surfactants or liquid fertilizers in conjunction with ARROSOLO 3-3E as crop injury may occur.

Avoid spray drift to all other crops and non-target areas.

Do not apply carbamate or organophosphorus insecticides in combination with ARROSOLO 3-3E.

Do not apply carbamate or organophosphorus insecticides within 7 days before or after ARROSOLO 3-3E application. Severe injury or kill of rice plants may result from tank mix combination or separate sprays of ARROSOLO 3-3E and certain insecticides.

Do not apply ARROSOLO 3-3E when rice is exhibiting stress from very cold temperatures, drought conditions, injury from previously applied pesticides, diseases or insects.

ARROSOLO 3-3E should only be applied to fields which have been drained of flood water to expose susceptible weeds.

To avoid injury do not apply when the air temperature is above 100°F.

Weed control may be impaired if ARROSOLO 3-3E is applied when the air temperature is below 70°F.

For best results apply ARROSOLO 3-3E to actively growing weeds in the 1-2 leaf stage.

Do not apply if rainfall is anticipated within 6 hours after application.

Apply only when wind velocity is low, 0-5 mph.

Do not apply more than 9 pounds actual ORDRAM per acre per season when more than one method of application is used.

GENERAL APPLICATION DIRECTIONS

ARROSOLO 3-3E is a selective broad spectrum postemergence rice herbicide for control of annual grasses and broadleaf weeds.

All equipment used in the application of ARROSOLO 3-3E should be carefully calibrated before use and checked frequently during application to be sure the application is uniform. Avoid overlaps that will increase ARROSOLO 3-3E dosage above recommended limits, because crop injury may occur. Apply ARROSOLO 3-3E in 20 to 50 gallons of water per acre by ground equipment and in 10 to 15 gallons of water per acre by airplane.

Shake well before using.

DIRECTIONS FOR USE AND RECOMMENDATIONS

It is a violation of Federal law to use this product in a manner inconsistent with its labeling.

Do not apply this product through any type of irrigation system.

POST-EMERGENCE, PRE-FLOOD: For use on dry-seeded and water-seeded rice.

For control of Barnyardgrass, Sprangletop, Dayflower and Broadleaf Signalgrass. Apply ARROSOLO 3-3E at 3 to 4 quarts per acre depending on weed size as shown in the table below. For best results, apply ARROSOLO 3-3E when weeds are in the 1-2 leaf stage. Flood the rice within 5 days after application to prevent weed reinfestation.

Weeds Controlled by ARROSOLO 3-3E and Rate Chart

Weeds		Maximum Leaf Stage 3 Quarts	Maximum Leaf Stage 4 Quarts
Barnyardgrass	(*Echinochloa* spp.)	1-2	3-4
Sprangletop (maximum 1 in. tall)	(*Leptochloa* spp.)	1-2	** 1-2
Dayflower, common	(*Commelina communis*)	—	1-2
Signalgrass, broadleaf	(*Brachiaria platyphylla*)	1-2	3-4
Crabgrass, large	(*Digitaria sanguinalis*)	—	1-early 4
Croton, woolly	(*Croton capitatus*)	—	*
Foxtail	(*Setaria* spp.)	—	1-early 4
Goosegrass	(*Eleusine indica*)	—	1-early 4
Junglerice	(*Echinochloa colonum*)	—	1-early 4
Gulf Cockspur	(*Echinochloa crus-pavonis*)	—	1-early 4
Paragrass	(*Panicum purpurascens*)	—	1-early 4
Pigweed, redroot	(*Amaranthus retroflexus*)	—	*
Redweed	(*Melochia corchorifolia*)	—	*
Dock, curly	(*Rumex crispus*)	—	*
Horned beakrush	(*Rhynchospora corniculata*)	—	*
Hemp Sesbania or Coffee Bean	(*Sesbania exaltata*)	—	*
Hoorahgrass or Fringerush	(*Fimbristylis* spp.)	—	1-early 4
Texasweed	(*Caperonia palustris*)	—	*
Texas Millet	(*Panicum Texanum*)	—	1-early 4
Spikerush	(*Eleocharis* spp.)	—	1-early 4

* For control of seedling broadleaf weeds.
** Suppression only in the states of Arkansas, Mississippi, Louisiana and Missouri.

For Bulk and Mini-Bulk Containers

STORAGE AND DISPOSAL

Storage: Do not use or store near heat or open flame. Do not store near seeds and fertilizer. Keep container closed when not in use. Open dumping is prohibited.

Pesticide Disposal: Wastes resulting from the use of this product may be disposed of on site or at an approved waste disposal facility.

Container Disposal: Reseal container and offer for reconditioning, or triple rinse (or equivalent) and offer for recycling or reconditioning, or clean in accordance with manufacturer's instructions.

FOR MINI-BULK CONTAINERS

Container Precautions: Before refilling, inspect thoroughly for damage, such as cracks, punctures, bulges, dents, abrasions or worn threads on closure devices. REFILL ONLY WITH ARROSOLO 3-3E. The contents of this container cannot be completely removed by cleaning. Refilling with materials other than ARROSOLO 3-3E will result in contamination and may weaken container. After filling and before transporting, check for leaks. Do not refill or transport damaged or leaking container. CONTAINER IS NOT SAFE FOR FOOD, FEED OR DRINKING WATER!

NOTICE TO BUYER AND USER

Seller warrants that this product conforms to the chemical description on the label and is reasonably fit for the purposes stated on the label when used in accordance with directions under normal conditions of use. This warranty does not extend to the use of this product contrary to label instructions, or under abnormal use conditions, or under conditions not reasonably foreseeable to Seller, and Buyer and User assume the risk of any such use. SELLER DISCLAIMS ALL OTHER WARRANTIES EXPRESSED OR IMPLIED INCLUDING ANY WARRANTY OF FITNESS OR MERCHANTABILITY. SELLER SHALL NOT BE LIABLE FOR CONSEQUENTIAL, SPECIAL OR INDIRECT DAMAGES RESULTING FROM THE USE OR HANDLING OF THIS PRODUCT AND SELLER'S SOLE LIABILITY AND BUYER'S AND USER'S EXCLUSIVE REMEDY SHALL BE LIMITED TO THE REFUND OF THE PURCHASE PRICE.

This product is sold only for uses stated on its label. No express or implied license is granted to use or sell this product under any patent in any country except as specified. Country: United States of America. Patent No.: 3,573,031.

A-4R

The information and specimen labels in this booklet are correct as of the publication date but are subject to change without notice. For current information, contact ICI Americas Inc.

BETASAN® 12.5-G

Granular Selective Herbicide for Pre-Emergence Control of Crabgrass and Annual Bluegrass in Turf, Ornamentals, and Ground Covers.

Only for Sale to, Use, and Storage by Servicepersons.

ACTIVE INGREDIENT:	
Bensulide: *S*-(*0,0*-Diisopropyl phosphorodithioate) ester of *N*-(2-mercaptoethyl) benzenesulfonamide	12.5%
INERT INGREDIENTS:	87.5%
TOTAL	100.0%

Contains 12.5 pounds of active ingredient per 100 pounds.

EPA Reg. No. 10182–175

KEEP OUT OF REACH OF CHILDREN
WARNING

PRECAUTIONARY STATEMENTS

HAZARDS TO HUMANS AND DOMESTIC ANIMALS

WARNING

- Harmful if swallowed.
- Avoid breathing dust.
- Causes substantial but temporary eye injury. Do not get in eyes, on skin, or on clothing. Wear goggles or safety glasses when handling.
- Wash thoroughly with soap and water after handling and before eating or smoking.
- Wash all contaminated clothing with soap and hot water before reuse.

ENVIRONMENTAL HAZARDS

- This product is toxic to fish. Do not apply directly to any body of water. Do not apply where runoff is likely to occur. Do not contaminate water by cleaning of equipment or disposal of wastes.
- Do not apply when weather conditions favor drift from areas treated.
- Apply this product only as specified on this label.

STATEMENT OF PRACTICAL TREATMENT

If a known exposure occurs or is suspected, immediately start the procedures given below and simultaneously contact a Poison Center, a physician, or the nearest hospital.

NOTE: Be sure to advise the person contacted that the active ingredient in this compound is a cholinesterase inhibitor. Describe the type of exposure, the victim's symptoms, and follow the advice given.

NOTE TO MEDICAL PERSONNEL: Should cholinesterase inhibition occur, atropine by injection is antidotal. 2-PAM (Protopam Chloride) is also antidotal when administered early and in conjunction with atropine.

If swallowed: Give several glasses of water and induce vomiting by placing a finger on the back of the victim's tongue. Give fluids until vomitus is clear. Do not induce vomiting or give anything by mouth to an unconscious or convulsing person.

For skin contact: Flush all affected areas with plenty of water for several minutes. Remove and clean contaminated clothing and shoes. Seek medical attention if irritation occurs.

For eye contact: Immediately flush eyes with large amounts of running water for at least 15 minutes. Hold eyelids apart to ensure rinsing of the entire surface of the eye and lids with water. Get medical attention immediately.

If inhaled: Remove to fresh air. Seek medical attention if respiratory irritation occurs or if breathing becomes difficult.

FOR 24-HOUR EMERGENCY MEDICAL ASSISTANCE CALL 1-800-F-A-S-T-M-E-D (327-8633).
FOR CHEMICAL EMERGENCY: Spill, leak, fire, exposure, or accident call CHEMTREC 1-800-424-9300.

GENERAL USE PRECAUTIONS

- READ ALL DIRECTIONS BEFORE USING.
- BETASAN 12.5-G should be used only for recommended uses and at recommended rates.
- Sweep up spills on walks or patios and hose area.
- Do not allow children or pets on treated lawns until BETASAN 12.5-G has been watered into the soil and grass is completely dry.
- BETASAN 12.5-G is recommended for use on mineral soils only.
- Do not apply peat moss to lawn, ground covers or established ornamentals before treating with BETASAN 12.5-G.

PRODUCT INFORMATION

BETASAN 12.5-G is recommended for use on home lawns, parks and golf courses, on ornamental plants and on ground covers for control of crabgrass, annual bluegrass and other annual weeds listed.

Application of BETASAN 12.5-G must precede emergence of weeds from the soil. Crabgrass will emerge from the soil from four to six weeks before it is visible above lawn sod. If weeds are visible above the soil, it is too late to apply BETASAN 12.5-G.

Remove leaves, dead or tall grass, and other debris before applying BETASAN 12.5-G to turf areas. Any weeds present around ornamentals must be removed or worked into the soil before applying BETASAN 12.5-G.

WEEDS CONTROLLED

Smooth Crabgrass *(Digitaria ischaemum)*
Hairy Crabgrass *(Digitaria sanguinalis)*
Redroot Pigweed *(Amaranthus retroflexus)*
Barnyardgrass (Watergrass) *(Echinochloa* spp.)
Lambsquarters *(Chenopodium album)*
Goosegrass *(Eleusine indica)*
Annual Bluegrass *(Poa annua)*
Shepherdspurse *(Capsella Bursa-pastoris)*
Deadnettle (Henbit) *(Lamium* spp.)

DIRECTIONS FOR USE

It is a violation of Federal law to use this product in a manner inconsistent with its labeling. Avoid contamination of food, feed, or water supplies.

TIMING OF APPLICATION

The most effective program for control of crabgrass, goosegrass, annual bluegrass and other annual weeds is one specifically designed for local conditions. Two or more applications per year may be necessary, using the recommended rates given below for a particular weed problem. For additional information regarding timing and rate of application, consult weed specialists of the State Agricultural Extension Service or the State Agricultural Experiment Station.

Grass Lawns: BETASAN 12.5-G should be applied only to well established grass lawns. BETASAN 12.5-G can be used on the following established turf grasses:

Bahia	Centipede	Poa trivialis
Bentgrass spp.	Fescue spp.	Saint Augustine
Perennial Bluegrass spp.	Pensacola Bahia	Red Top
Bermudagrass spp.	Perennial Ryegrass	Zoysia

Dichondra Lawns: BETASAN 12.5-G can be applied to dichondra lawns at the time of seeding or any time thereafter.

Ornamentals and Ground Covers: BETASAN 12.5-G should be applied only to the following well established ornamentals and ground covers:

Abelia	Daphne	Primrose
Ajuga	Freesia	Privet
Alyssum	Gazania	Pyracantha
Aster	Gladiolus	Ranunculus
Azalea	Holly	Sandankwa
Azara	Hypericum	Star Jasmine
Bachelor's-Button	Iceplant	Sedum
Boxwood	Ivy	Stock
Calendula	Juniper	Sweet Pea
Campanula	Marigold	Tobira
Candy-tuft	Monterey Cypress	Tulip
Coral Bell	Monterey Pine	Wall Flower
Daffodil	Myrtle	Wild Strawberry
Dahlia	Narcissus	Xylosma
Daisy	Pachysandra	Zinnia
	Pansy	
	Periwinkle	

For control of crabgrass, apply BETASAN 12.5-G at the recommended rate (see RATES OF APPLICATION) any time from fall (September or October) through early spring, for spring and summer control. With a late winter or any early spring application, the application must be made prior to crabgrass emergence (see PRODUCT INFORMATION).

For control of annual bluegrass and other annual weeds, apply BETASAN 12.5-G at the recommended rate (see RATES OF APPLICATION) before the weeds emerge from the soil.

Where germination of crabgrass, annual bluegrass and other weeds extends into the late summer and early fall, a second application of BETASAN 12.5-G at the recommended rate should be made four to five months after the initial application.

BETASAN 12.5-G should be applied with a lawn spreader or any other applicator that will give uniform coverage. A more uniform application can be made by spreading half the required amount of BETASAN 12.5-G over the area to be treated, then applying the remainder at right angles or crosswise. Avoid misses by overlapping slightly. Sprinkle the area with water for 10 to 15 minutes after application to wash the BETASAN 12.5-G down into the soil where the weed seeds are located.

RATES OF APPLICATION

Apply a **maximum** of 200 pounds of BETASAN 12.5-G per acre (4½ pounds per 1,000 square feet) per year in two or more applications depending on the weed species and degree of infestations. If weed specialists are unavailable to provide specific local recommendations, follow these directions:

Crabgrass Only: Apply 80 pounds of BETASAN 12.5-G uniformly on one acre (40 pounds per one-half acre; 1.9 pounds per 1,000 sq. ft.).

Annual Bluegrass, Goosegrass and Other Annual Weeds: Apply 100 pounds of BETASAN 12.5-G uniformly on one acre (50 pounds per one-half acre; 2.4 pounds per 1,000 sq. ft.).

BETASAN 12.5-G RATE TABLE AND SPREADER SETTINGS

Make of Spreader	Desired Rate Per Acre	Desired Rate Per 1,000 Sq. Ft.	Spreader Settings Once Over	Spreader Settings Twice Over	Spread Width
Cyclone	80 lbs.	1.9 lbs.	2.6	2.3	6½ ft.
	100 lbs.	2.4 lbs.	2.8	2.5	6½ ft.
Scott	100 lbs.	2.4 lbs.	3.5	2.5	
Central					
= 1615A	100 lbs.	2.4 lbs.	4	3	
= 1665	100 lbs.	2.4 lbs.	5¼	4½	

RESEEDING

Any material that provides effective control of annual grasses may affect new lawn grass seedlings which germinate after treatment. Proper fertilization and watering will encourage the existing turf grass to fill in the thin or bare spots formerly occupied by crabgrass.

The following procedures are suggested in order to minimize any stunting of turf grasses which may occur when reseeding into BETASAN 12.5-G treated areas:

1. Do not reseed within 4 months after application.
2. Renovate the area to be reseeded by raking thoroughly to scarify the soil surface. Where practical, mix peat in with the soil to provide an optimum seed bed.
3. When turf diseases or physiological disorders necessitate reseeding sooner than 4 months after BETASAN 12.5-G application, apply powdered activated charcoal prior to reseeding to deactivate the BETASAN 12.5-G.

Apply the activated charcoal at a rate of 7 pounds in 14 gallons of water per 1,000 square feet (300 pounds in 600 gallons of water per acre). Following application, irrigate the turf with sufficient water to wash the charcoal into the soil.

Reseed no sooner than 7 days after applying charcoal.

STORAGE AND DISPOSAL

• Do not contaminate water, food, or feed by storage or disposal. **Storage:** Store in a cool dry place. **Pesticide Disposal:** Wastes resulting from the use of this product may be disposed of on site or at an approved waste disposal facility. **Container Disposal:** Completely empty bag into application equipment. Then dispose of empty bag in a sanitary landfill or by incineration, or, if allowed by State and local authorities, by burning. If burned, stay out of smoke. **Spill or Leak:** Small spills can be handled routinely. Use adequate ventilation and wear an air-supplied respirator to prevent inhalation. Wear suitable protective clothing to prevent skin and eye contact. Use the following procedures: 1. Sweep up spilled material being careful not to create dust. 2. Place sweepings in an appropriate chemical waste container. 3. Generously cover the contaminated areas with a common household detergent. Using a stiff brush and small amounts of water, work the detergent into the spill material forming a slurry. Do not splatter on one's self or bystanders and completely avoid skin or eye contact with this material. Brush the slurry into cracks and crevices and allow to stand for 2–3 minutes. 4. Spread a suitable absorbent such as clay, sawdust or kitty litter on the slurried liquid. Shovel the absorbed material into the container. 5. Repeat if necessary. 6. Flush the area with water while observing proper environmental considerations. 7. Seal container and dispose of in approved pesticide landfill. Large spills should be handled according to a predetermined plan. For assistance in developing a plan, contact ICI Americas Inc., Agricultural Products, Wilmington, Delaware 19897.

NOTICE TO BUYER AND USER

Seller warrants that this product conforms to the chemical description on the label and is reasonably fit for the purposes stated on the label when used in accordance with directions under normal conditions of use. This warranty does not extend to the use of this product contrary to label instructions, or under abnormal use conditions, or under conditions not reasonably foreseeable to Seller, and Buyer and User assume the risk of any such use. SELLER DISCLAIMS ALL OTHER WARRANTIES, EXPRESSED OR IMPLIED, INCLUDING ANY WARRANTY OF FITNESS OR MERCHANTABILITY. SELLER SHALL NOT BE LIABLE FOR CONSEQUENTIAL, SPECIAL OR INDIRECT DAMAGES RESULTING FROM THE USE OR HANDLING OF THIS PRODUCT AND SELLER'S SOLE LIABILITY AND BUYER'S AND USER'S EXCLUSIVE REMEDY SHALL BE LIMITED TO THE REFUND OF THE PURCHASE PRICE.

A-3R

05–0630–010

The information and specimen labels in this booklet are correct as of the publication date but are subject to change without notice. For current information, contact ICI Americas Inc.

BETASAN® 2.9-E

EMULSIFIABLE LIQUID

Selective Herbicide for Pre-Emergence Control of Crabgrass and Annual Bluegrass in Turf, Ornamentals, and Ground Covers.

ACTIVE INGREDIENT:
Bensulide:
S-(*O,O*-Diisopropyl phosphorodithioate)
ester of *N*-(2-mercaptoethyl) benzenesulfonamide 34.8%
INERT INGREDIENTS: 65.2%
TOTAL 100.0%

Contains 2.9 Pounds Of Active Ingredient Per Gallon

EPA Reg. No. 10182-170

KEEP OUT OF REACH OF CHILDREN CAUTION

PRECAUTIONARY STATEMENTS

HAZARDS TO HUMANS AND DOMESTIC ANIMALS

CAUTION

- Harmful if swallowed or absorbed through the skin
- Avoid breathing spray mist
- Avoid contact with skin, eyes, and clothing
- Wash all contaminated clothing with soap and hot water before reuse.
- Avoid contamination of food, feed, or water supplies.

ENVIRONMENTAL HAZARDS

- This product is toxic to fish. Keep out of lakes, streams, and ponds. Do not apply where runoff is likely to occur. Do not contaminate water by cleaning of equipment or disposal of wastes.
- Do not apply when weather conditions favor drift from area treated
- Apply this product only as specified on this label.

PHYSICAL HAZARDS

- Combustible. Do not use, store, pour or spill near heat or open flame. Keep container closed when not in use.

STATEMENT OF PRACTICAL TREATMENT

If a known exposure occurs or is suspected, immediately start the procedures given below and simultaneously contact a Poison Center, a physician, or the nearest hospital.

NOTE: Be sure to advise the person contacted that the active ingredient in this compound is a cholinesterase inhibitor. Describe the type of exposure, the victim's symptoms, and follow the advice given.

NOTE TO MEDICAL PERSONNEL: Should cholinesterase inhibition occur atropine by injection is antidotal. 2-PAM (Protopam Chloride) is also antidotal when administered early and in conjunction with atropine.

If swallowed: Immediately give several glasses of water but *do not* induce vomiting. If vomiting does occur, give fluids again. Have a physician determine if condition of patient will permit induction of vomiting or evacuation of stomach. Do not give anything by mouth to an unconscious or convulsing person.

For skin contact: Flush all affected areas with plenty of water for several minutes. Remove and clean contaminated clothing and shoes. Seek medical attention if irritation occurs.

For eye contact: Flush eyes with large amounts of running water for at least 15 minutes. Hold eyelids apart to ensure rinsing of the entire surface of the eye and lids with water. Get medical attention immediately.

If inhaled: Remove to fresh air. Seek medical attention if respiratory irritation occurs or if breathing becomes difficult.

FOR 24-HOUR EMERGENCY MEDICAL ASSISTANCE CALL 1-800-F-A-S-T-M-E-D (327-8633).
FOR CHEMICAL EMERGENCY: Spill, leak, fire, exposure, or accident call CHEMTREC 1-800-424-9300.

For other product information, contact your local ICI Americas Inc. Distributor or see Material Safety Data Sheet.

GENERAL USE PRECAUTIONS

- READ ALL DIRECTIONS BEFORE USING.
- MIX WELL BEFORE DILUTING WITH WATER.
- BETASAN 2.9-E should be used only for recommended uses and at recommended rates.
- Do not allow children or pets on treated lawns until BETASAN 2.9-E has been watered into the soil and grass is completely dry.
- BETASAN 2.9-E is recommended for use on mineral soils only.
- Do not apply peat moss to lawn, ground covers or established ornamentals before treating with BETASAN 2.9-E.

PRODUCT INFORMATION

BETASAN 2.9-E is recommended for use on home lawns, parks and golf courses, on ornamental plants and on ground covers for control of crabgrass, annual bluegrass and other annual weeds listed.

Application of BETASAN 2.9-E must precede emergence of weeds from the soil. Crabgrass will emerge from the soil from four to six weeks before it is visible above lawn sod. If weeds are visible above the soil, it is too late to apply BETASAN 2.9-E. Remove leaves, dead or tall grass and other debris, or thoroughly work them into the soil before applying BETASAN 2.9-E.

WEEDS CONTROLLED

Annual Bluegrass *(Poa annua)*	Lambsquarters *(Chenopodium album)*
Barnyardgrass (Watergrass) *(Echinochloa* spp.)	Redroot Pigweed *(Amaranthus retroflexus)*
Deadnettle (Henbit)*(Lamium* spp.)	Shepherdspurse *(Capsella bursa-pastoris)*
Foxtails *(Setaria* spp.)	Smooth Crabgrass *(Digitaria ischaemum)*
Goosegrass *(Eleusine indica)*	
Hairy Crabgrass *(Digitaria sanguinalis)*	

DIRECTIONS FOR USE

It is a violation of Federal Law to use this product in a manner inconsistent with its labeling.

Do not apply this product through any type of irrigation system.

TIMING OF APPLICATION

Grass Lawns: BETASAN 2.9-E should be applied only to well established grass lawns. BETASAN 2.9-E can be used on the following established turf grasses:

Bahia	Fescue spp.	Poa trivialis
Bentgrass spp.	Pensacola Bahia	Saint Augustine
Bermudagrass spp.	Perennial Bluegrass spp.	Red Top
Centipede	Perennial Ryegrass	Zoysia

Dichondra Lawns: BETASAN 2.9-E can be applied to dichondra lawns at the time of seeding or any time thereafter.

Ornamentals and Ground Covers: BETASAN 2.9-E should be applied only to the following well established ornamentals and ground covers:

Abelia	Freesia	Primrose
Ajuga	Gazania	Privet
Alyssum	Gladiolus	Pyracantha
Aster	Holly	Ranunculus
Azalea	Hypericum	Sandankwa
Azara	Ice Plant	Sedum
Bachelor's-Button	Ivy	Star Jasmine
Boxwood	Juniper	Stock
Calendula	Marigold	Sweet Pea
Campanula	Monterey Cypress	Tobira
Candy-Tuft	Monterey Pine	Tulip
Coral Bell	Myrtle	Wall Flower
Daffodil	Narcissus	Wild Strawberry
Dahlia	Pachysandra	Xylosma
Daisy	Pansy	Zinnia
Daphne	Periwinkle	

For control of crabgrass, apply BETASAN 2.9-E at the recommended rate (see RATES OF APPLICATION) any time from fall (September or October) through early spring for spring and summer control. With a late winter or an early spring application, the application must be made prior to crabgrass emergence (see PRODUCT INFORMATION).

For control of annual bluegrass and other annual weeds, apply BETASAN 2.9-E at the recommended rate (see RATES OF APPLICATION) before weeds emerge from the soil.

Consult weed specialists of the State Agricultural Experiment Station or the State Agricultural Extension Service for additional information as the timing of applications needed will vary with local conditions.

BETASAN 2.9-E should be applied with spray equipment that will provide uniform application such as a hose end sprayer or 1 to 3 gallon pump sprayer.

RATES OF APPLICATION

Crabgrass Only: Apply 7²/₃ to 10 fluid ounces of BETASAN 2.9-E in 1 to 3 gallons of water uniformly on 1,000 square feet.

Annual Bluegrass and Other Annual Weeds: Apply 12³/₄ fluid ounces of BETASAN 2.9-E in 1 to 3 gallons of water uniformly on 1,000 square feet.

Sprinkle with water for 10 to 15 minutes after application to wash the BETASAN 2.9-E off the grass and down into the soil where the weed seeds are located.

Where germination of crabgrass, annual bluegrass and other annual weeds extends into the late summer and early fall, a second application of BETASAN 2.9-E at recommended rates should be made four to five months after the initial application.

RESEEDING

Any material that provides effective control of annual grasses may affect new lawn grass seedlings which germinate after treatment. Proper fertilization and watering will encourage the existing turf grass to fill in the thin or bare spots formerly occupied by crabgrass.

The following procedures are suggested in order to minimize any stunting of turf grasses which may occur when reseeding into BETASAN-treated areas:

1. Do not reseed within four months after application.
2. Renovate the area to be reseeded by raking thoroughly and mix sterilized (weed-free) organic mulch (50-50 by volume) into the top two inches of soil.
3. When turf diseases or physiological disorders necessitate reseeding sooner than 4 months after BETASAN 2.9-E application, apply powdered activated charcoal prior to reseeding to deactivate the BETASAN 2.9-E. Apply the activated charcoal at a rate of 7 pounds in 14 gallons of water per 1,000 square feet (300 pounds in 600 gallons of water per acre). Following application, irrigate the turf with sufficient water to wash the charcoal into the soil. Reseed no sooner than 7 days after applying the charcoal.

STORAGE AND DISPOSAL

- Do not contaminate water, food, or feed by storage or disposal.

Storage: Store only in original container in an area that can be locked and is out of reach of children. Protect from temperatures below 32°F. Product crystallizes at lower temperatures. Warm or store at higher temperatures and mix to redissolve crystals and assure uniformity before use. Do not store near sources of heat or open flame. **Pesticide Disposal:** Securely wrap original container in several layers of newspaper and discard in trash. **Container Disposal:** Do not reuse empty bottle. Rinse thoroughly before discarding in trash. **Spill or Leak:** Soak up pooled liquid with a suitable absorbent such as clay, sawdust or kitty litter. Sweep up absorbed material and wrap in several layers of newspaper and discard in trash. Generously cover contaminated area with a slurry of common household laundry detergent and water. Using a stiff brush, work the slurry into cracks and crevices. Allow to stand for 2-3 minutes then flush with water. Repeat if necessary.

NOTICE TO BUYER AND USER

Seller warrants that this product conforms to the chemical description on the label and is reasonably fit for the purposes stated on the label when used in accordance with directions under normal conditions of use. This warranty does not extend to the use of this product contrary to label instructions, or under abnormal use conditions, or under conditions not reasonably foreseeable to Seller, and Buyer and User assume the risk of any such use. SELLER DISCLAIMS ALL OTHER WARRANTIES, EXPRESSED OR IMPLIED, INCLUDING ANY WARRANTY OF FITNESS OR MERCHANTABILITY. SELLER SHALL NOT BE LIABLE FOR CONSEQUENTIAL, SPECIAL OR INDIRECT DAMAGES RESULTING FROM THE USE OR HANDLING OF THIS PRODUCT AND SELLER'S SOLE LIABILITY AND BUYER'S AND USER'S EXCLUSIVE REMEDY SHALL BE LIMITED TO THE REFUND OF THE PURCHASE PRICE.

A-4R

05-0630-006

The information and specimen labels in this booklet are correct as of the publication date but are subject to change without notice. For current information, contact ICI Americas Inc.

BETASAN® 3.6-G

Granular Selective Herbicide for Pre-emergence Control of Crabgrass and Annual Bluegrass in Turf, Ornamentals, and Ground Covers.

ACTIVE INGREDIENT:

Bensulide: *S-(0,0*-Diisopropyl phosphorodithioate) ester of *N*-(2-mercaptoethyl) benzenesulfonamide	3.6%
INERT INGREDIENTS:	96.4%
TOTAL	100.0%

Contains 3.6 pounds of active ingredient per 100 pounds.

EPA Reg. No. 10182–183

KEEP OUT OF REACH OF CHILDREN
CAUTION

PRECAUTIONARY STATEMENTS

HAZARDS TO HUMANS AND DOMESTIC ANIMALS

CAUTION

- Harmful if swallowed.
- Avoid breathing dust.
- Causes moderate eye irritation. Do not get in eyes, on skin, or on clothing.

- Wash thoroughly after handling and before eating or smoking.
- Wash all contaminated clothing with soap and hot water before reuse.
- Avoid contamination of food, feed, or water supplies.

ENVIRONMENTAL HAZARDS

- This product is toxic to fish. Do not apply directly to any body of water. Do not apply where runoff is likely to occur. Do not contaminate water by cleaning of equipment or disposal of wastes.
- Do not apply when weather conditions favor drift from areas treated.

STATEMENT OF PRACTICAL TREATMENT

If a known exposure occurs or is suspected, immediately start the procedures given below and simultaneously contact a Poison Center, a physician, or the nearest hospital.

NOTE: Be sure to advise the person contacted that the active ingredient in this compound is a cholinesterase inhibitor. Describe the type of exposure, the victim's symptoms, and follow the advice given.

NOTE TO MEDICAL PERSONNEL: Should cholinesterase inhibition occur, atropine by injection is antidotal. 2-PAM (Protopam Chloride) is also antidotal when administered early and in conjunction with atropine.

If swallowed, give several glasses of water and induce vomiting by placing a finger on the back of the victim's tongue. Give fluids until vomitus is clear. Do not induce vomiting or give anything by mouth to an unconscious or convulsing person.

For skin contact, flush all affected areas with plenty of water for several minutes. Remove and clean contaminated clothing and shoes. Seek medical attention if irritation occurs.

For eye contact, immediately flush eyes with large amounts of running water for at least 15 minutes. Hold eyelids apart to ensure rinsing of the entire surface of the eye and lids with water. Get medical attention immediately.

If inhaled, remove to fresh air. Seek medical attention if respiratory irritation occurs or if breathing becomes difficult.

FOR 24-HOUR EMERGENCY MEDICAL ASSISTANCE CALL 1-800-F-A-S-T-M-E-D (327-8633).
FOR CHEMICAL EMERGENCY: Spill, leak, fire, exposure, or accident call CHEMTREC 1-800-424-9300.

GENERAL USE PRECAUTIONS

- READ ALL DIRECTIONS BEFORE USING.
- BETASAN 3.6-G should be used only for recommended uses and at recommended rates.
- Do not allow children or pets on treated lawns until BETASAN 3.6-G has been watered into the soil and grass is completely dry.
- BETASAN 3.6-G is recommended for use on mineral soils only.
- Do not apply peat moss to lawn, ground covers or established ornamentals before treating with BETASAN 3.6-G.

PRODUCT INFORMATION

BETASAN 3.6-G is recommended for use on home lawns, parks and golf courses, on ornamental plants and on ground covers for control of crabgrass, annual bluegrass and other annual weeds listed.

Application of BETASAN 3.6-G must precede emergence of weeds from the soil. Crabgrass will emerge from the soil from four to six weeks before it is visible above lawn sod. If weeds are visible above the soil, it is too late to apply BETASAN 3.6-G.

Remove leaves, dead or tall grass, and other debris before applying BETASAN 3.6-G to turf areas. Any weeds present around ornamentals must be removed or worked into the soil before applying BETASAN.

WEEDS CONTROLLED

Smooth Crabgrass *(Digitaria ischaemum)*
Hairy Crabgrass *(Digitaria sanguinalis)*
Redroot Pigweed *(Amaranthus retroflexus)*
Barnyardgrass (Watergrass) *(Echinochloa* spp.)
Lambsquarters *(Chenopodium album)*
Goosegrass *(Eleusine indica)*
Annual Bluegrass *(Poa annua)*
Shepherdspurse *(Capsella bursa-pastoris)*
Deadnettle (Henbit) *(Lamium* spp.)

DIRECTIONS FOR USE

It is a violation of Federal law to use this product in a manner inconsistent with its labeling.

TIMING OF APPLICATION

Grass Lawns: BETASAN 3.6-G should be applied only to well established grass lawns. BETASAN 3.6-G can be used on the following established turf grasses:

Bahia	Centipede	Poa trivialis
Bentgrass spp.	Fescue spp.	Saint Augustine
Bermudagrass spp.	Pensacola Bahia	Red Top
Perennial Bluegrass spp.	Perennial Ryegrass	Zoysia

Dichondra Lawns: BETASAN 3.6-G can be applied to dichondra lawns at the time of seeding or any time thereafter.

Ornamentals and Ground Covers: BETASAN 3.6-G should be applied only to the following well established ornamentals and ground covers:

Abelia	Freesia	Primrose
Ajuga	Gazania	Pyracantha
Alyssum	Gladiolus	Ranunculus
Aster	Hypericum	Sandankwa
Azalea	Ice Plant	Sedum
Azara	Ivy	Star Jasmine
Bachelor's-Button	Juniper	Stock
Boxwood	Marigold	Sweet Pea
Calendula	Monterey Cyprus	Tobira
Campanula	Monterey Pine	Tulip
Candy-Tuft	Myrtle	Wall Flower
Coral Bell	Narcissus	Wild Strawberry
Daffodil	Pachysandra	Zinnia
Dahlia	Pansy	
Daisy	Periwinkle	

For control of crabgrass, apply BETASAN 3.6-G at the recommended rate (see RATES OF APPLICATION) anytime from fall (September or October) through early spring, for spring and summer control. With a late winter or an early spring application, the application must be made prior to crabgrass emergence (see PRODUCT INFORMATION).

For control of annual bluegrass and other annual weeds, apply BETASAN 3.6-G at the recommended rate (see RATES OF APPLICATION) before the weeds emerge from the soil.

Consult weed specialists of the State Agricultural Experiment Station or the State Agricultural Extension Service for additional information, as the timing and rate of application needed will vary with local conditions.

Where germination of crabgrass, annual bluegrass, and other weeds extends into the late summer and early fall, a second application of BETASAN 3.6-G at the recommended rates should be made four to five months after the initial application.

BETASAN 3.6-G should be applied with a lawn spreader or any other applicator that will give uniform coverage. A more uniform application can be made by spreading half the required amount of BETASAN 3.6-G over the area to be treated, then applying the remainder at right angles or crosswise. Avoid misses by overlapping slightly.

RATES OF APPLICATION

Crabgrass Only: Apply 12 to 16 pounds of BETASAN 3.6-G uniformly on 2,500 square feet.

Annual Bluegrass and Other Annual Weeds: Apply 20 pounds of BETASAN 3.6-G uniformly on 2,500 square feet.

Sprinkle the area with water for 10 to 15 minutes after application to wash the BETASAN 3.6-G down into the soil where the weed seeds are located.

RESEEDING

Any material that provides effective control of annual grasses may affect new lawn grass seedlings which germinate after treatment. Proper fertilization and watering will encourage the existing turf grass to fill in the thin or bare spots formerly occupied by crabgrass.

The following procedures are suggested in order to minimize any stunting of turf grasses which may occur when reseeding into BETASAN-treated areas.

1. Do not reseed within 4 months after application.
2. Renovate the area to be reseeded by raking thoroughly to scarify the soil surface. Where practical, mix peat in with the soil to provide an optimum seed bed.
3. When turf diseases or physiological disorders necessitate reseeding sooner than 4 months after BETASAN 3.6-G application, apply powdered activated charcoal prior to reseeding to deactivate the BETASAN 3.6-G.

Apply the activated charcoal at a rate of 7 pounds in 14 gallons of water per 1,000 square feet (300 pounds in 600 gallons of water per acre). Following application, irrigate the turf with sufficient water to wash the charcoal into the soil.
Reseed no sooner than 7 days after applying the charcoal.

STORAGE AND DISPOSAL

Do not contaminate water, food or feed by storage or disposal. **Storage:** Store only in original container in a cool dry place that can be locked and is out of reach of children. **Pesticide Disposal:** Securely wrap original container in several layers of newspaper and discard in trash. **Container Disposal:** Do not reuse empty bag. Discard bag in trash. **Spill or Leak:** Small spills can be handled routinely. Use adequate ventilation and suitable protective clothing and eye protection to prevent skin and eye contact. Sweep or shovel up spilled material and wrap in several layers of newspaper and discard in trash. Wash spill area with household detergent and flush with water to remove any residue.

NOTICE TO BUYER AND USER

Seller warrants that this product conforms to the chemical description on the label and is reasonably fit for the purposes stated on the label when used in accordance with directions under normal conditions of use. This warranty does not extend to the use of this product contrary to label in-

structions, or under abnormal use conditions, or under conditions not reasonably foreseeable to Seller, and Buyer and User assume the risk of any such use. SELLER DISCLAIMS ALL OTHER WARRANTIES, EXPRESSED OR IMPLIED, INCLUDING ANY WARRANTY OF FITNESS OR MERCHANTABILITY. SELLER SHALL NOT BE LIABLE FOR CONSEQUENTIAL, SPECIAL OR INDIRECT DAMAGES RESULTING FROM THE USE OR HANDLING OF THIS PRODUCT AND SELLER'S SOLE LIABILITY AND BUYER'S AND USER'S EXCLUSIVE REMEDY SHALL BE LIMITED TO THE REFUND OF THE PURCHASE PRICE.

A-3R

05–0630–008

The information and specimen labels in this booklet are correct as of the publication date but are subject to change without notice. For current information, contact ICI Americas Inc.

BETASAN® 4-E LF

EMULSIFIABLE LIQUID

Selective Herbicide for Pre-Emergence Control of Crabgrass and Annual Bluegrass in Turf, Ornamentals, and Ground Covers.

ACTIVE INGREDIENT:
Bensulide:
S-(*O,O*-Diisopropyl phosphorodithioate)

ester of *N*-(2-mercaptoethyl) benzenesulfonamide	46.0%
INERT INGREDIENTS:	54.0%
TOTAL	100.0%

Contains 4 pounds of active ingredient per gallon.

EPA Reg. No. 10182-163

KEEP OUT OF REACH OF CHILDREN
CAUTION

PRECAUTIONARY STATEMENTS

HAZARDS TO HUMANS AND DOMESTIC ANIMALS

CAUTION

- Harmful if swallowed or absorbed through the skin.
- Avoid breathing spray mist.
- Causes moderate eye irritation. Do not get in eyes, on skin, or on clothing.
- Wash thoroughly after handling and before eating or smoking.
- Wash all contaminated clothing with soap and hot water before reuse.
- Avoid contamination of food, feed, or water supplies.

ENVIRONMENTAL HAZARDS

This product is toxic to fish. Do not apply directly to any body of water. Do not contaminate water by cleaning of equipment or disposal of wastes. Do not apply when weather conditions favor drift from area treated.

PHYSICAL HAZARDS

Combustible. Do not use, store, pour or spill near heat or open flame. Keep container closed when not in use.

STATEMENT OF PRACTICAL TREATMENT

If a known exposure occurs or is suspected, immediately start the procedures given below and simultaneously contact a Poison Center, a physician, or the nearest hospital.

NOTE: Be sure to advise the person contacted that the active ingredient in this compound is a cholinesterase inhibitor. Describe the type of exposure, the victim's symptoms, and follow the advice given.

NOTE TO MEDICAL PERSONNEL: Should cholinesterase inhibition occur, atropine by injection is antidotal. 2-PAM (Protopam Chloride) is also antidotal when administered early and in conjunction with atropine.

If swallowed: Immediately give several glasses of water but *do not* induce vomiting. If vomiting does occur, give fluids again. Have a physician determine if condition of patient will permit induction of vomiting or evacuation of stomach. Do not give anything by mouth to an unconscious or convulsing person.

For skin contact: Flush all affected areas with plenty of water for several minutes. Remove and clean contaminated clothing and shoes. Seek medical attention if irritation occurs.

For eye contact: Flush eyes with large amounts of running water for at least 15 minutes. Hold eyelids apart to ensure rinsing of the entire surface of the eye and lids with water. Get medical attention immediately.

If inhaled: Remove to fresh air. Seek medical attention if respiratory irritation occurs or if breathing becomes difficult.

FOR 24-HOUR EMERGENCY MEDICAL ASSISTANCE CALL 1-800-F-A-S-T-M-E-D (327-8633).
FOR CHEMICAL EMERGENCY: Spill, leak, fire, exposure, or accident call CHEMTREC 1-800-424-9300.

For other product information, contact your local ICI Americas Inc. Distributor or see Material Safety Data Sheet.

GENERAL USE PRECAUTIONS

READ ALL DIRECTIONS BEFORE USING.
MIX WELL BEFORE DILUTING WITH WATER.
BETASAN 4-E LF should be used only for recommended uses and at recommended rates.
Do not allow children or pets on treated lawns until grass is completely dry.
BETASAN 4-E LF is recommended for use on mineral soils only.
Do not apply peat moss to lawn, ground covers, or established ornamentals before treating with BETASAN 4-E LF.

PRODUCT INFORMATION

BETASAN 4-E LF is recommended for use on home lawns, parks and golf courses, on ornamental plants and on ground covers for control of crabgrass, annual bluegrass and other annual weeds listed.

Application of BETASAN 4-E LF must precede emergence of weeds from the soil. Crabgrass will emerge from the soil from 4 to 6 weeks before it is visible above lawn sod. If weeds are visible above the soil, it is too late to apply BETASAN 4-E LF.

Remove leaves, dead tall grass, and other debris before applying BETASAN 4-E LF to turf areas. Any weeds present around ornamentals must be removed or worked into the soil before applying BETASAN 4-E LF.

Weeds Controlled

Annual Bluegrass *(Poa annua)*	Hairy Crabgrass *(Digitaria sanguinalis)*
Barnyardgrass (Watergrass) *(Echinochloa* spp.)	Lambsquarters *(Chenopodium album)*
Deadnettle (Henbit)*(Lamium* spp.)	Redroot Pigweed *(Amaranthus retroflexus)*
Fall Panicum *(Panicum dichotomiflorum)*	Shepherdspurse *(Capsella bursa-pastoris)*
Foxtails *(Setaria* spp.)	Smooth Crabgrass *(Digitaria ischaemum)*
Goosegrass *(Eleusine indica)*	

DIRECTIONS FOR USE

It is a violation of Federal Law to use this product in a manner inconsistent with its labeling.

Do not apply this product through any type of irrigation system.

TIMING OF APPLICATION

The most effective program for control of crabgrass, goosegrass, annual bluegrass and other annual weeds is one specifically designed for local conditions. Two or more applications per year may be necessary, using the recommended rates given below for a particular weed problem. For additional information regarding timing and rate of application, consult weed specialists of the State Agricultural Extension Service or the State Agricultural Experiment Station.

Grass Lawns: BETASAN 4-E LF should be applied only to well established grass lawns. It can be used on the following established turf grasses:

Bahia	Perennial Bluegrass spp.
Bentgrass spp.	Perennial Ryegrass
Bermudagrass spp.	Poa trivialis
Centipede	Saint Augustine
Fescue spp.	Red Top
Pensacola Bahia	Zoysia

Dichondra Lawns: BETASAN 4-E LF can be applied to dichondra lawns at the time of seeding or any time thereafter.

Ornamentals and Ground Covers: BETASAN 4-E LF should be applied only to the following well established ornamentals and ground covers:

Abelia	Freesia	Primrose
Ajuga	Gazania	Privet
Alyssum	Gladiolus	Pyracantha
Aster	Holly	Ranunculus
Azalea	Hypericum	Sandankwa
Azara	Ice Plant	Sedum
Bachelor's-Button	Ivy	Star Jasmine
Boxwood	Juniper	Stock
Calendula	Marigold	Sweet Pea
Campanula	Monterey Cypress	Tobira
Candy-Tuft	Monterey Pine	Tulip
Coral Bell	Myrtle	Wall Flower
Daffodil	Narcissus	Wild Strawberry
Dahlia	Pachysandra	Xylosma
Daisy	Pansy	Zinnia
Daphne	Periwinkle	

For control of crabgrass, apply BETASAN 4-E LF at the recommended rate (see RATES OF APPLICATION) any time from fall (September or October) through early spring for spring and summer control. With a late winter or an early spring application, the application must be made prior to crabgrass emergence (see PRODUCT INFORMATION).

For control of annual bluegrass and other annual weeds, apply BETASAN 4-E LF at the recommended rate (see RATES OF APPLICATION) before weeds emerge from the soil.

Where germination of crabgrass, annual bluegrass and other annual weeds extends into the late summer and early fall, a second application of BETASAN 4-E LF at the recommended rates should be made 4 to 5 months after the initial application.

BETASAN 4-E LF should be applied with spray equipment that will provide uniform application such as a knapsack sprayer, hose proportioner, or pressure sprayer.

RATES OF APPLICATION

Apply a *maximum* of 6¼ gallons (50 pints) of BETASAN 4-E LF per acre (18 fluid ounces per 1,000 square feet) per year in two or more applications depending on the weed species and degree of infestation.

If weed specialists are unavailable to provide specific local recommendations, follow these directions:

Crabgrass Only: Apply 1⅞ gallons (15 pints) to 2.5 gallons (20 pints) of BETASAN 4-E LF in 80 to 100 gallons of water uniformly on one acre (5.6 to 7.3 fluid ounces per 1,000 square feet).

Rate of Application for the *John Doe* Liquid Applicator: Use 22 to 29 fluid ounces in 2½ gallons of water and apply to 4,000 square feet of lawn. Water thoroughly after application.

Annual Bluegrass and Other Annual Weeds: Apply 3⅛ gallons of BETASAN 4-E LF in 80 to 100 gallons of water uniformly on one acre (9.4 fluid ounces per 1,000 square feet).

When a weed control program involves two to three applications of BETASAN for crabgrass, goosegrass and bluegrass, the rate of application for goosegrass and annual bluegrass may be reduced to 1⅞ gallons (15 pints) to 2.5 gallons (20 pints) per acre (5.6 to 7.3 fluid ounces per 1,000 square feet), using the higher rate for heavier infestations.

RESEEDING

Any material that provides effective control of annual grasses may affect new lawn grass seedlings which germinate after treatment. Proper fertilization and watering will encourage the existing turf grass to fill in the thin or bare spots formerly occupied by crabgrass.

The following procedures are suggested in order to minimize any stunting of turf grasses which may occur when reseeding into BETASAN-treated areas:

1. Do not reseed within 4 months after application.
2. Renovate the area to be reseeded by raking thoroughly to scarify the soil surface. Where practical, mix peat in with the soil to provide an optimum seed bed.
3. When turf diseases or physiological disorders necessitate reseeding sooner than 4 months after BETASAN 4-E LF application, apply powdered activated charcoal prior to reseeding to deactivate the BETASAN 4-E LF.

Apply the activated charcoal at a rate of 7 pounds in 14 gallons of water per 1,000 square feet (300 pounds in 600 gallons of water per acre).

Following application, irrigate the turf with sufficient water to wash the charcoal into the soil.

Reseed no sooner than 7 days after applying the charcoal.

STORAGE AND DISPOSAL

- **Do not contaminate water, food, or feed by storage or disposal.**

STORAGE: Protect from temperatures below 42°F. At temperatures below 42°F product may crystallize. Dissolve crystals with quiet storage at warm temperatures (75°F). Redissolution rate: 40°-48°F (40-60 days); or 60°F (1-3 weeks); 70°-80°F (1-3 days). Roll container and agitate contents to provide a uniform product prior to use. **Pesticide Disposal:** Wastes resulting from the use of this product may be disposed of on site or at an approved waste disposal facility. **Container Disposal:** Triple rinse (or equivalent). Then offer for recycling or reconditioning, or puncture and dispose of in a sanitary landfill, or by other procedures approved by State and local authorities. **Spill or Leak:** Small spills can be handled routinely. Use adequate ventilation and wear an air-supplied respirator to prevent inhalation. Wear suitable protective clothing to prevent skin and eye contact. Use the following procedures: **1.** Spread a suitable absorbent such as clay, sawdust or kitty litter on the spilled material. **2.** Place sweepings in an appropriate chemical waste container. **3.** Generously cover the contaminated areas with a common household detergent. Using a stiff brush and small amounts of water, work the detergent into the spill material forming a slurry. Do not splatter on one's self or bystanders and completely avoid skin or eye contact with this material. Brush the slurry into cracks and crevices and allow to stand for 2-3 minutes. **4.** Spread absorbents on the slurried liquid and shovel the absorbed material into the container. **5.** Repeat if necessary. **6.** Flush the area with water while observing proper environmental considerations. **7.** Seal container and dispose of in approved pesticide landfill.

Large spills should be handled according to a predetermined plan. For assistance in developing a plan, contact Agricultural Products, ICI Americas Inc., Wilmington, DE 19897.

NOTICE TO BUYER AND USER

Seller warrants that this product conforms to the chemical description on the label and is reasonably fit for the purposes stated on the label when used in accordance with directions under normal conditions of use. This warranty does not extend to the use of this product contrary to label instructions, or under abnormal use conditions, or under conditions not reasonably foreseeable to Seller, and Buyer and User assume the risk of any such use. SELLER DISCLAIMS ALL OTHER WARRANTIES, EXPRESSED OR IMPLIED, INCLUDING ANY WARRANTY OF FITNESS OR MERCHANTABILITY. SELLER SHALL NOT BE LIABLE FOR CONSEQUENTIAL, SPECIAL OR INDIRECT DAMAGES RESULTING FROM THE USE OR HANDLING OF THIS PRODUCT AND SELLER'S SOLE LIABILITY AND BUYER'S AND USER'S EXCLUSIVE REMEDY SHALL BE LIMITED TO THE REFUND OF THE PURCHASE PRICE.

This product is sold only for uses stated on its label. No express or implied license is granted to use or sell this product under any patent in any country except as specified: Country: United States of America

A-7R

05-0630-007

The information and specimen labels in this booklet are correct as of the publication date but are subject to change without notice. For current information, contact ICI Americas Inc.

BETASAN® 7-G

Granular Selective Herbicide for Pre-Emergence Control of Crabgrass and Annual Bluegrass in Turf, Ornamentals, and Ground Covers.
Only for Sale to, Use, and Storage by Servicepersons.

ACTIVE INGREDIENT:	
Bensulide: *S*-(*0,0*-Diisopropyl phosphorodithioate) ester of *N*-(2-mercaptoethyl) benzenesulfonamide	7.0%
INERT INGREDIENTS:	93.0%
TOTAL	100.0%

Contains 7 pounds of active ingredient per 100 pounds.

EPA Reg. No. 10182-240

KEEP OUT OF REACH OF CHILDREN WARNING

PRECAUTIONARY STATEMENTS

HAZARDS TO HUMANS AND DOMESTIC ANIMALS

WARNING

- Harmful if swallowed.
- Avoid breathing dust.
- Causes substantial but temporary eye injury. Do not get in eyes, on skin, or on clothing. Wear goggles or safety glasses when handling.
- Wash thoroughly with soap and water after handling and before eating or smoking.
- Wash all contaminated clothing with soap and hot water before reuse.

ENVIRONMENTAL HAZARDS

- This product is toxic to fish. Do not apply directly to any body of water. Do not apply where runoff is likely to occur. Do not contaminate water by cleaning of equipment or disposal of wastes.
- Do not apply when weather conditions favor drift from areas treated.
- Apply this product only as specified on this label.

STATEMENT OF PRACTICAL TREATMENT

If a known exposure occurs or is suspected, immediately start the procedures given below and simultaneously contact a Poison Center, a physician, or the nearest hospital.

NOTE: Be sure to advise the person contacted that the active ingredient in this compound is a cholinesterase inhibitor. Describe the type of exposure, the victim's symptoms, and follow the advice given.

NOTE TO MEDICAL PERSONNEL: Should cholinesterase inhibition occur, atropine by injection is anitdotal. 2-PAM (Protopam Chloride) is also antidotal when administered early and in conjunction with atropine.

If Swallowed: Give several glasses of water and induce vomiting by placing a finger on the back of the victim's tongue. Give fluids until vomitus is clear. Do not induce vomiting or give anything by mouth to an unconscious or convulsing person.

For Skin Contact: Flush all affected areas with plenty of water for several minutes. Remove and clean contaminated clothing and shoes. Seek medical attention if irritation occurs.

For Eye Contact: Immediately flush eyes with large amounts of running water for at least 15 minutes. Hold eyelids apart to ensure rinsing of the entire surface of the eye and lids with water. Get medical attention immediately.

If Inhaled: Remove to fresh air. Seek medical attention if respiratory irritation occurs or if breathing becomes difficult.

FOR 24-HOUR EMERGENCY MEDICAL ASSISTANCE CALL 1-800-F-A-S-T-M-E-D (327-8633).
FOR CHEMICAL EMERGENCY: Spill, leak, fire, exposure, or accident call CHEMTREC 1-800-424-9300.

GENERAL USE PRECAUTIONS

- READ ALL DIRECTIONS BEFORE USING.
- BETASAN 7-G should be used only for recommended uses and at recommended rates.
- Do not allow children or pets on treated lawns until BETASAN 7-G has been watered into the soil and grass is completely dry.
- BETASAN 7-G is recommended for use on mineral soils only.
- Do not apply peat moss to lawn, ground covers or established ornamentals before treating with BETASAN 7-G.

PRODUCT INFORMATION

BETASAN 7-G is recommended for use on home lawns, parks and golf courses, on ornamental plants and on ground covers for control of crabgrass, annual bluegrass and other annual weeds listed.

Application of BETASAN 7-G must precede emergence of weeds from the soil. Crabgrass will emerge from the soil from four to six weeks before it is visible above lawn sod. If weeds are visible above the soil, it is too late to apply BETASAN 7-G.

Remove leaves, dead or tall grass, and other debris before applying BETASAN 7-G to turf areas. Any weeds present around ornamentals must be removed or worked into the soil before applying BETASAN 7-G.

WEEDS CONTROLLED

Smooth Crabgrass *(Digitaria ischaemum)*
Hairy Crabgrass *(Digitaria sanguinalis)*
Redroot Pigweed *(Amaranthus retroflexus)*
Barnyardgrass (Watergrass) *(Echinochloa* spp.)
Lambsquarters *(Chenopodium album)*
Goosegrass *(Eleusine indica)*
Annual Bluegrass *(Poa annua)*
Shepherdspurse *(Capsella Bursa-pastoris)*
Deadnettle (Henbit) *(Lamium* spp.)

DIRECTIONS FOR USE

It is a violation of Federal law to use this product in a manner inconsistent with its labeling. Avoid contamination of food, feed, or water supplies.

TIMING OF APPLICATION

The most effective program for control of crabgrass, goosegrass, annual bluegrass and other annual weeds is one specifically designed for local conditions. Two or more applications per year may be necessary, using the recommended rates given below for a particular weed problem. For additional information regarding timing and rate of application, consult weed specialists of the State Agricultural Extension Service or the State Agricultural Experiment Station.

Grass Lawns: BETASAN 7-G should be applied only to well established grass lawns. BETASAN 7-G can be used on the following established turf grasses:

Bahia	Centipede	Poa trivialis
Bentgrass spp.	Fescue spp.	Saint Augustine
Bermudagrass spp.	Pensacola Bahia	Red Top
Perennial Bluegrass spp.	Perennial Ryegrass	Zoysia

Dichondra Lawns: BETASAN 7-G can be applied to dichondra lawns at the time of seeding or any time thereafter.

Ornamentals and Ground Covers: BETASAN 7-G should be applied only to the following well established ornamentals and ground covers:

Abelia	Daphne	Primrose
Ajuga	Freesia	Privet
Alyssum	Gazania	Pyracantha
Aster	Gladiolus	Ranunculus
Azalea	Holly	Sandankwa
Azara	Hypericum	Sedum
Bachelor's-Button	Ice Plant	Star Jasmine
Boxwood	Ivy	Stock
Calendula	Juniper	Sweet Pea
Campanula	Marigold	Tobira
Candy-Tuft	Monterey Cypress	Tulip
Coral Bell	Monterey Pine	Wall Flower
Daffodil	Myrtle	Wild Strawberry
Dahlia	Narcissus	Xylosma
Daisy	Pachysandra	Zinnia
	Pansy	
	Periwinkle	

For control of crabgrass, apply BETASAN 7-G at the recommended rate (see RATES OF APPLICATION) any time from fall (September or October) through early spring, for spring and summer control. With a late winter or an early spring application, the application must be made prior to crabgrass emergence (see PRODUCT INFORMATION).

For control of annual bluegrass and other annual weeds, apply BETASAN 7-G at the recommended rate (see RATES OF APPLICATION) before the weeds emerge from the soil.

Where germination of crabgrass, annual bluegrass and other weeds extends into the late summer and early fall, a second application of BETASAN 7-G at the recommended rate should be made four to five months after the initial application.

BETASAN 7-G should be applied with a lawn spreader or any other applicator that will give uniform coverage. A more uniform application can be made by spreading half of the required amount of BETASAN 7-G over the area to be treated, then applying the remainder at right angles or crosswise. Avoid misses by overlapping slightly.

RATES OF APPLICATION

Apply a **maximum** of 357 pounds of BETASAN 7-G per acre (8.2 pounds per 1,000 square feet) per year in two or more applications depending on the weed species and degree of infestation. If weed specialists are unavailable to provide specific local recommendations, follow these directions:

Crabgrass Only: Apply 107 to 143 pounds of BETASAN 7-G uniformly on one acre (2.4–3.2 pounds per 1,000 square feet).

Annual Bluegrass, Goosegrass and Other Annual Weeds: Apply 180 pounds of BETASAN 7-G uniformly on one acre (4.1 pounds per 1,000 square feet).

Sprinkle the area with water for 10 to 15 minutes after application to wash the BETASAN 7-G down into the soil where the weed seeds are located.

When a weed control program involves two to three applications of BETASAN 7-G for crabgrass, goosegrass and annual bluegrass the rate of application for goosegrass and annual bluegrass may be reduced to 107 pounds to 143 pounds per acre, using the higher rate for heavier infestations.

BETASAN 7-G RATE TABLE AND SPREADER SETTINGS

Make of Spreader	Desired Rate Per Acre	Desired Rate Per 1,000 Sq. Ft.	Spreader Settings Once Over	Twice Over	Spread Width
Cyclone	107 lbs.	2.4 lbs.	3.3	3.0	6½ ft.
	180 lbs.	4.1 lbs.	3.8	3.5	6½ ft.
Scott	107 lbs.	2.4 lbs.	2.8	2.2	20 in.
	180 lbs.	4.1 lbs.	3.8	3.1	20 in.

RESEEDING

Any material that provides effective control of annual grasses may affect new lawn grass seedlings which germinate after treatment. Proper fertilization and watering will encourage the existing turf grass to fill in the thin or bare spots formerly occupied by crabgrass.

The following procedures are suggested in order to minimize any stunting of turf grasses which may occur when reseeding into BETASAN 7-G treated areas:

1. Do not reseed within 4 months after application.
2. Renovate the area to be reseeded by raking thoroughly to scarify the soil surface. Where practical, mix peat in with the soil to provide an optimum seed bed.
3. When turf diseases or physiological disorders necessitate reseeding sooner than 4 months after BETASAN 7-G application, apply powdered activated charcoal prior to reseeding to deactivate the BETASAN 7-G.

Apply the activated charcoal at a rate of 7 pounds per 1,000 square feet in 14 gallons of water (300 pounds per acre in 600 gallons of water).

Following application, irrigate the turf with sufficient water to wash the charcoal into the soil.

Reseed no sooner than 7 days after applying the charcoal.

STORAGE AND DISPOSAL

- Do not contaminate water, food, or feed by storage or disposal. **Storage:** Store in a cool dry place. **Pesticide Disposal:** Wastes resulting from the use of this product may be disposed of on site or at an approved waste disposal facility. **Container Disposal:** Completely empty bag into application equipment. Then dispose of empty bag in a sanitary landfill or by incineration, or, if allowed by State and local authorities, by burning, If burned, stay out of smoke. **Spill or Leak:** Small spills can be handled routinely. Use adequate ventilation and wear an air-supplied respirator to prevent inhalation. Wear suitable protective clothing to prevent skin and eye contact. Use the following procedures: 1. Sweep up spilled material being careful not to create dust. 2. Place sweepings in an appropriate chemical waste container. 3. Generously cover the contaminated areas with a common household detergent. Using a stiff brush and a small amount of water, work the detergent into the spill material forming a slurry. Do not splatter on one's self or bystanders and completely avoid skin or eye contact with this material. Brush the slurry into cracks and crevices and allow to stand for 2–3 minutes. 4. Spread a suitable absorbent such as clay, sawdust or kitty litter on the slurried liquid. Shovel the absorbed material into the container. 5. Repeat if necessary. 6. Flush the area with water while observing proper environmental considerations. 7. Seal container and dispose of in approved pesticide landfill. Large spills should be handled according to a predetermined plan. For assistance in developing a plan, contact ICI Americas Inc., Agricultural Products, Wilmington, Delaware 19897

NOTICE TO BUYER AND USER

Seller warrants that this product conforms to the chemical description on the label and is reasonably fit for the purposes stated on the label when used in accordance with directions under normal conditions of use. This warranty does not extend to the use of this product contrary to label instructions, or under abnormal use conditions, or under conditions not reasonably foreseeable to Seller, and Buyer and User assume the risk of any such use. SELLER DISCLAIMS ALL OTHER WARRANTIES, EXPRESSED OR IMPLIED, INCLUDING ANY WARRANTY OF FITNESS OR MERCHANTABILITY. SELLER SHALL NOT BE LIABLE FOR CONSEQUENTIAL, SPECIAL OR INDIRECT DAMAGES RESULTING FROM THE USE OR HANDLING OF THIS PRODUCT AND SELLER'S SOLE LIABILITY AND BUYER'S AND USER'S EXCLUSIVE REMEDY SHALL BE LIMITED TO THE REFUND OF THE PURCHASE PRICE.

A-4R

05–0630–009

The information and specimen labels in this booklet are correct as of the publication date but are subject to change without notice. For current information, contact ICI Americas Inc.

CYCLONE®

HERBICIDE

RESTRICTED USE PESTICIDE

For retail sale to and use only by Certified Applicators or persons under their direct supervision and only for those uses covered by the Certified Applicator's certification.

ACTIVE INGREDIENT:	BY WT.
Paraquat dichloride (1,1'-dimethyl- 4,4'-bipyridinium dichloride)	29.1%
INERT INGREDIENTS:	70.9%
TOTAL	100.0%

Contains 2 lbs. Paraquat cation per gallon.
Contains Stench (Odor) and Emetic

EPA Reg. No. 10182-111

KEEP OUT OF REACH OF CHILDREN

DANGER–POISON

ONE SWALLOW CAN KILL! Symptoms Are Prolonged and Painful. Onset of Symptoms May Be Delayed For Up to 3 Days After Swallowing.

PELIGRO

PRECAUCION AL USUARIO: Si usted no lee ingles, no use este producto hasta que la etiqueta le haya sido explicada ampliamente.

PRECAUTIONARY STATEMENTS

HAZARDS TO HUMANS AND DOMESTIC ANIMALS

DANGER POISON

Can Kill If Swallowed. May Be Harmful or Fatal If Absorbed Through Skin or Inhaled. Harmful to Skin and Eyes.

WORKER SAFETY RULES

DO NOT get on skin, eyes or clothing. DO NOT inhale spray mist. WASH SPLASHES from skin and eyes immediately. REMOVE and wash contaminated clothing. WASH before eating, smoking and drinking. WEAR full face shield, rubber gloves and apron and waterproof footwear when pouring concentrate or when exposure to concentrate is possible. WEAR waterproof footwear when spraying outside of an enclosed cab or aircraft or when contacting vegetation wet with spray.

Do Not Use or Store In or Around the Home.

Do Not Remove Contents except for Immediate Use.

Never Put into Food, Drink or Other Containers.

STATEMENT OF PRACTICAL TREATMENT

If Swallowed: Immediately induce vomiting by inserting finger in throat. Administer fluids and induce further vomiting. Get to a hospital or physician fast. PROMPT TREATMENT IS ESSENTIAL and must be initiated before signs and symptoms of injury appear.

Note to Physicians: Call ICI Americas Medical Emergency Information Center 800/327-8633 (Dial 1-800-F-A-S-T-M-E-D) at any hour to obtain toxicology and medical management consultation, a supply of bentonite clay, and paraquat analysis. Symptoms following ingestion may be delayed up to 3 days. To be effective, treatment for paraquat poisoning must begin **Immediately.** Treatment consists of binding paraquat in the gut with suspensions of bentonite clay or charcoal and removal of paraquat from the blood by prolonged charcoal hemoperfusion or continuous hemodialysis.

Skin Contact: Immediately wash with soap and water. Remove and wash all contaminated clothing with soap and hot water before reuse. Prolonged contact will cause severe irritation. Contact with irritated skin, or a cut, or repeated contact with intact skin may result in poisoning.

Eye Contact: Immediately wash eyes with water for at least 15 minutes and get medical attention. Contact may cause eye injury.

Inhalation: Immediately get away from spray mist. Exposure may cause irritation, nose bleeds and may lead to poisoning. Stop and check spray procedure. Follow Worker Safety Rules.

ENVIRONMENTAL SAFETY CAUTIONS

Wildlife: This product is toxic to wildlife. Birds and other wildlife in treated areas may be killed. Do not apply directly to water or wetlands (swamps, bogs, marshes and potholes). Do not contaminate water by cleaning of equipment or disposal of waste.

Drift: CYCLONE is a contact herbicide that kills all green plant tissue. Do not apply under conditions involving possible drift to food, forage, or other plantings that might be damaged or the crops thereof rendered unfit for sale, use or consumption. Do not apply when weather conditions favor drift from areas treated. For aerial applicators use, do not apply during periods of thermal inversion to avoid drift.

Chemical Hazard: This product reacts with aluminum to produce hydrogen gas which may form a highly combustible gas mixture. Such gas could flash or explode, causing serious personal injury, if ignited by open flame, spark, welder's torch, lighted cigarette, or other ignition source. Do not mix or store in containers, spray tanks, nurse tanks, or such systems made of aluminum or having aluminum fittings. After dilution with water to spray concentrations, spray systems made of, or containing aluminum parts, may be used but should not be used for extended (overnight) storage.

Equipment/Containers: Flush all spray equipment with water after use each day. CYCLONE is corrosive to aluminum. Aluminum spray equipment and aluminum aircraft structures that are exposed to spray solution or spray drift should be flushed thoroughly with water immediately after use.

DIRECTIONS FOR USE

It is a violation of Federal law and may endanger the applicator or third parties to use this product without complying with all label directions and warnings.

READ ENTIRE LABEL. USE STRICTLY IN ACCORDANCE WITH DANGER STATEMENTS AND DIRECTIONS, AND WITH APPLICABLE STATE AND FEDERAL REGULATIONS.

NOTE: Do Not Use Around Home Gardens, Schools, Recreational Parks, or Playgrounds.

Do not apply this product through any type of irrigation system.

Do not combine CYCLONE herbicide with liquid fertilizer solutions except as specified below.

DO NOT combine with suspension type liquid fertilizers as the activity of CYCLONE herbicide will be reduced.

Use Spreader (non-ionic)

Do not use dirty or muddy water for CYCLONE herbicide at dilution as CYCLONE herbicide will be inactivated.

SMALL GRAINS

Barley, Wheat (Preplant or Preemergence Treatment) — Apply 1 to 4 pts. CYCLONE herbicide per sprayed acre as a band treatment over the crop row, or as a broadcast treatment prior to, during, or after planting—but before emergence of the crop. **Aerial Application—**5 to 10 gals. of water per acre. **Ground Application—**Use 20 to 60 gals. diluted spray per acre. Add Spreader (non-ionic) at 8 to 32 fl. oz. per 100 gals. of diluted spray. Seedbeds should be formed as far ahead of planting and treatment as possible to permit maximum weed and grass emergence. Seeding should be done with a minimum amount of soil disturbance. Weeds and grasses emerging after application will not be controlled. Crop plants emerged at time of application may be killed. Do not graze treated areas or harvest for forage or hay.

SORGHUM (Grain)

Preplant or Preemergence Treatment—Apply 2 to 4 pts. CYCLONE herbicide per sprayed acre as a band treatment over the crop row, or as a broadcast treatment prior to, during, or after planting—but before emergence of the crop. **Aerial Application—**5 to 10 gals. of water per acre. **Ground Application—**Use 20 to 60 gals. diluted spray per acre. Add Spreader (non-ionic) at 8 to 32 fl. oz. per 100 gals. of diluted spray. Seedbeds should be formed as far ahead of planting and treatment as possible to permit maximum weed and grass emergence. Seeding should be done with a minimum amount of soil disturbance. Weeds and grasses emerging after application will not be controlled. Crop plants emerged at time of application will be killed.

WHEAT: CHEMICAL FALLOW

For Desiccation of Most Emerged Annual Broadleaf Weeds and Grasses and Suppression of Many Emerged Perennial Weeds During the Fallow Period(s) of the Wheat-Fallow, Wheat-Annual Crop Fallow, Wheat-Annual Crop and Continuous Wheat Cropping Systems—

Apply 1½ to 2 pts. CYCLONE per acre as a broadcast spray.

Aerial Application—Apply 3 to 10 gals. spray mix per acre.

Ground Application—Use 10 to 60 gals. spray mix per acre. As density of stubble, crop residue or weeds increases, spray gallonage should be increased within the recommended range for complete coverage. Add Spreader (non-ionic) at 1 to 4 pt. per 100 gals. of diluted spray. For best results by ground, use flat-fan nozzles spaced at 20 inch intervals with boom height adjusted to obtain a 30% spray pattern overlap. Use only clean water or nitrogen solutions as a carrier. Weeds and grasses emerging after application and regrowth from perennial weeds will not be controlled. Weeds taller than 6 inches at application may not be controlled. The addition of 2,4-D and BANVEL® may aid in the suppression of perennial broadleaf weeds and annual broadleaf weeds that have made excessive growth before application. Refer to 2,4-D or BANVEL labels for directions, limitations and cautions. For extended weed control during the fallow period, tank mix with a registered residual herbicide or herbicide combination. Refer to residual herbicide label(s) for directions, limitations, cautions and for a list of weeds controlled by preemergence activity of the residual herbicide(s). Add wettable powder herbicide(s) to spray tank first, mixing thoroughly in carrier. Then add liquid herbicide(s), CYCLONE and Spreader (non-ionic). Constant agitation recommended at all times.

CONSERVATION RESERVE AND FEDERAL SET-ASIDE PROGRAMS

For control of existing cover crops and/or emerged broadleaf weeds and grasses in compliance with the Federal Conservation Reserve Program or Federal set-aside programs.

Apply 1½ to 2 pints of CYCLONE herbicide per acre. Always add a nonionic surfactant at one pint per 100 gallons of diluted spray. Apply in a minimum of 10 gallons per acre by ground or 3 gallons per acre by air to obtain thorough coverage. For best results by ground, use flat-fan nozzles spaced at 20 inch intervals with boom height adjusted to obtain a 30% spray pattern overlap. For improved emerged weed control or extended weed control, CYCLONE herbicide may be tank-mixed with other registered herbicides. Refer to the label(s) of each herbicide for directions, limitations, cautions and for a list of weeds controlled.

REENTRY STATEMENTS

DO NOT enter treated areas without protective clothing until sprays have dried.

AVOID working in spray mist. If there is risk of exposure wear goggles and approved face mask capable of filtering spray droplets.

KEEP all unprotected persons out of operating areas or vicinity where there may be danger of drift.

Certain states may require more restrictive reentry intervals; consult your State Department of Agriculture for further information.

Written or oral warnings regarding use of protective clothing and accidental exposure must be given to workers who are expected to be in treated areas or in areas about to be treated.

Important: The effect of swallowing CYCLONE is far more severe than that which usually results from skin contact or inhalation of spray mist. This product should not cause injury if used according to label instructions. Follow the label **as if your life depends on it.**

STORAGE AND DISPOSAL

Prohibitions: Do not contaminate water, food or feed by storage, disposal or cleaning of equipment. Open dumping is prohibited.

Storage: Store at temperatures above 32°F. Do not contaminate feed, foodstuffs or drinking water. Do not store next to feed or food, or transport in or on vehicles containing foodstuffs or feeds. For help with any spill, leak, fire or exposure involving this material, call CHEMTREC 800/424-9300.

Pesticide Disposal: This product is acutely hazardous. Improper disposal of excess pesticide, spray mixture, or rinsate is a violation of Federal law. If these wastes cannot be disposed of by use according to label instructions, contact your State Pesticide or Environmental Control Agency, or the Hazardous Waste representative at the nearest EPA Regional Office for guidance.

Container Disposal: Triple rinse (or equivalent). Do not reuse container. Offer for recycling or reconditioning, or puncture and dispose of in a sanitary landfill, or dispose of by other procedures allowed by State and local authorities.

NOTICE TO BUYER AND USER

Seller warrants that this product conforms to the chemical description on the label and is reasonably fit for the purposes stated on the label when used in accordance with directions under normal conditions of use. This warranty does not extend to the use of this product contrary to label instructions, or under abnormal use conditions, or under conditions not reasonably foreseeable to Seller, and Buyer and User assume the risk of any such use. SELLER DISCLAIMS ALL OTHER WARRANTIES, EXPRESSED OR IMPLIED, INCLUDING ANY WARRANTY OF FITNESS OR MERCHANTABILITY, SELLER SHALL NOT BE LIABLE FOR CONSEQUENTIAL, SPECIAL OR INDIRECT DAMAGES RESULTING FROM THE USE OR HANDLING OF THIS PRODUCT AND SELLER'S SOLE LIABILITY AND BUYER'S AND USER'S EXCLUSIVE REMEDY SHALL BE LIMITED TO THE REFUND OF THE PURCHASE PRICE.

BANVEL®—Reg. TM of Sandez Protection Corp.

04-0799-192R1 C/H A-2R

The information and specimen labels in this booklet are correct as of the publication date but are subject to change without notice. For current information, contact ICI Americas Inc.

CYMBUSH® 3E

INSECTICIDE

RESTRICTED USE PESTICIDE

For retail sale to and use only by Certified Applicators, or persons under direct supervision, and only for those uses covered by the Certified Applicator's Certification.

ACTIVE INGREDIENT:

Cypermethrin
(±) α-cyano-(3-phenoxyphenyl) methyl (±)-cis, trans-3-(2,2-dichloroethenyl)-2, 2-dimethylcyclopropanecarboxylate* 35.6%
INERT INGREDIENTS: 64.4%

TOTAL 100.0%

*Cis/trans ratio: 45/55 ± 10

CYMBUSH 3E contains 3 pounds active ingredient per gallon.

EPA Reg. No. 10182-65 U.S. Patent No. 4,024,163

KEEP OUT OF REACH OF CHILDREN

PELIGRO

DANGER

PRECAUCION AL USUARIO:
Si usted no lee ingles, no use este producto hasta que la etiqueta le haya sido explicado ampliamente.

PRECAUTIONARY STATEMENTS

HAZARDS TO HUMANS AND DOMESTIC ANIMALS

DANGER

CORROSIVE. CAUSES EYE BURNS AND SKIN IRRITATION. MAY BE FATAL IF SWALLOWED. HARMFUL IF INHALED OR ABSORBED THROUGH SKIN. MAY CAUSE ALLERGIC SKIN REACTIONS. Do not get in eyes, on skin or on clothing. Avoid breathing vapor or spray mist. Wash thoroughly after handling.

Wear protective clothing when spraying. Wear protective clothing, impermeable gloves, full face shield, rubber apron and boots when handling or mixing concentrate. When mixing/loading and spraying, wear mid-forearm to elbow length chemical-resistant gloves, boots or overshoes, one piece overalls which have long sleeves or long pants, and face shield or goggles. Mixers/loaders should in addition wear a chemical-resistant apron.

Any article of clothing worn while applying this product must be cleaned before reusing. Clothing which has been drenched or heavily contaminated should be disposed of in accordance with state or local regulations.

Instead of clothing and equipment specified above, the applicator can use an enclosed tractor cab with filtered air supply.

STATEMENT OF PRACTICAL TREATMENT

IF SWALLOWED: Do not induce vomiting. Call a physician immediately.

IF IN EYES: Immediately flush eyes with plenty of water for at least 15 minutes. Get medical attention.

IF ON SKIN: Immediately remove contaminated clothing and wash with plenty of soap and water. Get medical attention if irritation persists.

Note to Physician: Inducing vomiting as first aid for this substance may result in increased risk of chemical pneumonia or pulmonary edema caused by aspiration of the hydrocarbon solvent. Vomiting should be induced only under professional supervision.

FOR 24-HOUR EMERGENCY MEDICAL ASSISTANCE CALL 1-800-F-A-S-T-M-E-D (327-8633).
FOR CHEMICAL EMERGENCY: Spill, leak, fire, exposure, or accident call CHEMTREC 1-800-424-9300.

ENVIRONMENTAL HAZARDS

This pesticide is extremely toxic to fish. Use with care when applying in areas adjacent to any body of water. Do not apply directly to water. Do not apply when weather conditions favor drift from treated areas. Do not contaminate water when disposing of equipment washwaters. This product is highly toxic to bees exposed to direct treatment or residues on blooming crops or weeds. Do not apply CYMBUSH or allow it to drift to blooming crops or weeds if bees are visiting treatment area. Additional information may be obtained from your Cooperative Extension Service. Apply this product only as specified on this label.

PHYSICAL OR CHEMICAL HAZARDS

Do not use or store near heat or open flame.

DIRECTIONS FOR USE

It is a violation of Federal law to use this product in a manner inconsistent with its labeling.

Apply CYMBUSH 3E as shown in the following chart, using sufficient spray to obtain full coverage of foliage. Apply every 5 to 7 days or as needed. Timing and frequency of applications should be based upon insect populations reaching locally determined economic thresholds. The use of less than recommended rates should only occur under conditions of light to moderate insect infestations.

REENTRY STATEMENTS

Do not treat areas while unprotected humans or domestic animals are present in the treatment area.

Do not allow entry into treated areas without protective clothing until sprays have dried.

State reentry regulations will apply if they are more restrictive than those stated on this label. Consult with State regulatory officials.

CHEMIGATION:

Sprinkler Irrigation Application

Apply CYMBUSH 3E at rates and timing described elsewhere in this label. Apply with center pivot or continuous-move equipment distributing 1/2 acre-inch or less during treatment. If stationary (solid sets, handlines or wheel lines other than continuous move) are used, CYMBUSH 3E should be injected into no more than the last 20-30 minutes of the set. Do not apply when winds are greater than 10-15 mph to avoid drift or wind skips. Thorough coverage of foliage is required for good control. Good agitation should be maintained during the entire application period.

Use Precautions—Sprinkler Irrigation Application

A. Apply this product only through (sprinkler including center pivot, lateral move, end tow, side (wheel) roll, traveler, big gun, solid set, or hand move) irrigation system(s). Do not apply this product through any other type of irrigation system.

B. Crop injury, lack of effectiveness, or illegal pesticide residues in the crop can result from nonuniform distribution of treated water.

C. If you have questions about calibration, you should contact State Extension Service specialists, equipment manufacturers or other experts.

D. Do not connect an irrigation system (including greenhouse systems) used for pesticide application to a public water system unless the pesticide label-prescribed safety devices for public water systems are in place.

E. A person knowledgeable of the chemigation system and responsible for its operation, or under the supervision of the responsible person, shall shut the system down and make necessary adjustments should the need arise.

F. The system must contain a functional check valve, vacuum relief valve, and low pressure drain appropriately located on the irrigation pipeline to prevent water source contamination from backflow.

G. The pesticide injection pipeline must contain a functional, automatic, quick-closing check valve to prevent the flow of fluid back toward the injection pump.

H. The pesticide injection pipeline must also contain a functional, normally closed, solenoid-operated valve located on the intake side of the injection pump and connected to the system interlock to prevent fluid from being withdrawn from the supply tank when the irrigation system is either automatically or manually shut down.

I. The system must contain functional interlocking controls to automatically shut off the pesticide injection pump when the water pump motor stops.

J. The irrigation line or water pump must include a functional pressure switch which will stop the water pump motor when the water pressure decreases to the point where pesticide distribution is adversely affected.

K. Systems must use a metering pump, such as a positive displacement injection pump (e.g., diaphragm pump) effectively designed and constructed of materials that are compatible with pesticides and capable of being fitted with a system interlock.

L. Do not apply when wind speed favors drift beyond the area intended for treatment.

M. Do not apply through chemigation systems connected to public water systems.

Posting Requirements

Posting of areas to be chemigated is required when 1) any part of a treated area is within 300 feet of sensitive areas such as residential areas, labor camps, businesses, day care centers, hospitals, in-patient clinics, nursing homes or any public areas such as schools, parks, playgrounds, or other public facilities not including public roads, or 2) when the chemigated area is open to the public such as golf courses or retail greenhouses.

Posting must conform to the following requirements. Treated areas shall be posted with signs at all usual points of entry and along likely routes of approach from the listed sensitive areas. When there are no usual points of entry, signs must be posted in the corners of the treated areas and in any other location affording maximum visibility to sensitive areas. The printed side of the sign should face away from the treated area towards the sensitive area. The signs shall be printed in English. Signs must be posted prior to application and must remain posted until foliage has dried and soil surface water has disappeared. Signs may remain in place indefinitely as long as they are composed of materials to prevent deterioration and maintain legibility for the duration of the posting period.

All words shall consist of letters at least 2 1/2 inches tall, and all letters and the symbol shall be a color which sharply contrasts with their immediate background. At the top of the sign shall be the words KEEP OUT, followed by an octagonal stop sign symbol at least 8 inches in diameter containing the word STOP. Below the symbol shall be the words PESTICIDES IN IRRIGATION WATER.

Spray Recommendations

Crop	Target Pests	Rate (lbs ai/A)	Remarks
Cotton	Cotton Bollworm Tobacco Budworm Boll Weevil Pink Bollworm Lygus Bug Cabbage Looper Cotton Leafperforator Cutworm spp. Soybean Thrips Tobacco Thrips Beet Armyworm Flea Beetle Cotton Fleahopper Saltmarsh Caterpillar Whitefly* Cotton Aphid*	0.06-0.10 (2.56-4.27 fl oz/A)	Apply as required by scouting reports, usually at intervals of 5 to 7 days. For boll weevil control, spray on a 3 to 4 day schedule. Apply by air or ground. Use sufficient water to obtain full coverage of foliage. Do not apply more than 1.0 lb. ai per acre per season. In areas of the country where insects have a history of being highly susceptible to insecticides, available data indicate rates as low as 0.04 lb ai/acre will control light infestations of bollworm, budworm, pink bollworm, cabbage looper, and lygus. NOTE: 1). The use of rates lower than 0.06 lb ai/A should be used only in conjunction with intense field monitoring. 2). Do not use less than 0.06 lb ai/A on initial (first) application because of the variation in size and age of insects present. CYMBUSH 3E may also be applied by air diluted with at least a once refined vegetable oil such as cottonseed oil or soybean oil. Apply using 1-2 quarts of finished spray per acre. CYMBUSH 3E may be applied in tank mixture with the following insecticides in accordance with label limitations and precautions. Label dosage rate should not be exceeded. Chlordimeform, CURACRON®**, Malathion, Methyl Parathion, LANNATE®, LARVIN®**, GUTHION®, LORSBAN® Do not apply within 14 days of harvest. Do not graze livestock in treated areas.
Lettuce (Head)	Cabbage Looper Alfalfa Looper Black Cutworm Granulate Cutworm	0.025-0.10 (1.07-4.27)	Apply in water as necessary for insect control using a minimum of 15 gallons of finished spray per acre with ground equipment or 5 gallons per acre by air. Lower rates of CYMBUSH should be used under light to moderate insect pressure. Higher rates should be used to control heavy insect populations.
	Tobacco Budworm Corn Earworm Beet Armyworm	0.04-0.10 (1.71-4.27)	

			Do not apply more than 1.5 lb ai/A per season. Do not apply within 5 days of harvest. Do not plant rotational crops within 30 days of the last application.
Pecans	Hickory Shuckworm Yellow Pecan Aphid Black Pecan Aphid Pecan Nut Casebearer Pecan Weevil	0.06-0.10 (2.56-4.27 fl oz/A)	Apply by ground equipment to the point of drip. The CYMBUSH 3E rates are based on a standard of 100 gallons of dilute spray per acre for smaller trees. For larger trees which could require 200-300 gallons of spray per acre, apply at the rate of 0.06-0.10 lb ai/200-300 gallons of water. In order to calculate the correct gallons of water needed to spray one acre of your trees to the point of drip, you may need to conduct a test. If you do not know how to conduct such a test with your equipment, you should request assistance from your equipment dealer. Can be applied only prior to shuck split and a PHI of at least 21 days must be observed. Do not apply more than 0.8 lb ai per acre per crop season. Do not graze livestock on treated ground or cut treated cover crops for feed. CYMBUSH may be applied in tank mixtures with the following pesticides in accordance with label limitations and precautions.** Label dosage rates should not be exceeded. BENLATE®**, CYPREX®**, DI-SYSTON®**, Duter**, EPN**, Ethion**, LORSBAN®**, Super Tin**, SYSTOX®**, TOPSIN®**, ZOLONE®**

* Aids in Suppression

** Currently, not registered for use in California for tankmix use with CYMBUSH 3E.

Rate Conversion Chart

Pounds Active/Acre	Fluid Ounces/Acre	Pints/Acre	Acres Treated/Gallon
0.06	2.56	0.16	50.0
0.08	3.41	0.21	37.5
0.10	4.27	0.27	30.0

STORAGE AND DISPOSAL

Prohibitions: Do not contaminate water, food or feed by storage or disposal. Open dumping is prohibited. Do not reuse empty container.

Storage: Keep container closed when not in use. Do not store near food or feed. Protect from freezing. In case of spill or leak on floor or paved surfaces, soak up with sand, earth or synthetic absorbent. Remove to chemical waste area.

Pesticide Disposal: Pesticide wastes are acutely hazardous. Improper disposal of excess pesticide, spray mixture or rinsate is a violation of Federal law. If these wastes cannot be disposed of by use according to label instructions, contact your State Pesticide or Environmental Control Agency, or the hazardous waste representative at the nearest EPA Regional Office for guidance.

Container Disposal: Metal Containers: Triple rinse (or equivalent). Then offer for recycling or reconditioning, or puncture and dispose of in a sanitary landfill, or by other procedures approved by state and local authorities.

Plastic Containers: Triple rinse (or equivalent). Then offer for recycling or reconditioning, or puncture and dispose of in a sanitary landfill, or by incineration, or, if allowed by state and local authorities, by burning. If burned, stay out of smoke.

Glass Containers: Triple rinse (or equivalent). Then dispose of in a sanitary landfill or by other approved state and local procedures.

NOTICE TO BUYER AND USER:

Seller warrants that this product conforms to the chemical description on the label and is reasonably fit for the purposes stated on the label when used in accordance with directions under normal conditions of use. This warranty does not extend to the use of this product contrary to label instructions, or under abnormal use conditions, or under conditions not reasonably foreseeable to Seller, and Buyer and User assume the risk of any such use. SELLER DISCLAIMS ALL OTHER WARRANTIES, EXPRESSED OR IMPLIED, INCLUDING ANY WARRANTY OF FITNESS OR MERCHANTABILITY. SELLER SHALL NOT BE LIABLE FOR CONSEQUENTIAL, SPECIAL OR INDIRECT DAMAGES RESULTING FROM THE USE OR HANDLING OF THIS PRODUCT AND SELLER'S SOLE LIABILITY AND BUYER'S AND USER'S EXCLUSIVE REMEDY SHALL BE LIMITED TO THE REFUND OF THE PURCHASE PRICE.

BENLATE® and LANNATE®—Reg. TMs of E.I. du Pont de Nemours & Co. Inc.
CYMBUSH®—Reg. TM of ICI Americas Inc.
CYPREX®—Reg. TM of American Cyanamid Co.
CURACRON®—Reg. TM of Ciba-Geigy Corporation.
DI-SYSTON® and SYSTOX®—Reg. TMs of Mobay Corporation
GUTHION®—Reg. TM of Parent Company of Farbenfabriken Bayer Gmb. H. Leverkusen.
LARVIN®—Reg. TM of Rhone-Poulenc, Inc.
LORSBAN®—Reg. TM of the Dow Chemical Company
TOPSIN®—Reg. TM of Pennwalt Corporation
ZOLONE®—Reg. TM of Rhone-Poulenc, Inc.

A-1R

The information and specimen labels contained in this booklet are correct as of the publication date but are subject to change without notice. For current information, contact ICI Americas Inc.

DEMON® EC

INSECTICIDE

Crack and Crevice and Spot Spray for Cockroach Control. Only for Sale to, Use and Storage by Professional Pest Control Operators.

ACTIVE INGREDIENT:	
Cypermethrin (±)α-cyano-(3-phenoxyphenyl) methyl (±)-*cis*, *trans*-3-(2,2-dichloroethenyl)-2, 2- dimethylcyclopropanecarboxylate*	25.3%
INERT INGREDIENTS:**	74.7%
TOTAL	100.0%

* *Cis/trans* ratio: 45/55 ± 10

** Contains xylene range aromatic solvent.

DEMON EC contains 2 pounds active ingredient per gallon.

EPA Reg. No. 10182-105
U.S. Patent No. 4,024,163

KEEP OUT OF REACH OF CHILDREN
WARNING

PRECAUTIONARY STATEMENTS

HAZARDS TO HUMANS AND DOMESTIC ANIMALS

WARNING

MAY BE FATAL IF SWALLOWED. CAUSES EYE AND SKIN IRRITATION. HARMFUL IF ABSORBED THROUGH SKIN. MAY CAUSE ALLERGIC SKIN REACTIONS. Do not get in eyes, on skin or on clothing. Wear protective clothing when spraying. Wear protective clothing, impermeable gloves and full face shield or goggles when handling or mixing concentrate. Avoid breathing vapors or spray mist. Wash thoroughly with soap and water after handling. Remove contaminated clothing and wash before reuse.

STATEMENT OF PRACTICAL TREATMENT

IF SWALLOWED: Do not induce vomiting. Call a physician or Poison Control Center immediately.

IF IN EYES: Flush eyes with plenty of water. Call a physician if irritation persists.

IF ON SKIN: Wash with plenty of soap and water. Get medical attention if irritation persists.

IF INHALED: Remove victim to fresh air. Apply artificial respiration if necessary.

NOTE TO PHYSICIAN

Inducing vomiting as first aid for this substance may result in increased risk of chemical pneumonia or pulmonary edema caused by aspiration of the hydrocarbon solvent. Vomiting should be induced only under professional supervision.

FOR 24-HOUR EMERGENCY MEDICAL ASSISTANCE CALL 1-800-F-A-S-T-M-E-D (327-8633).
FOR CHEMICAL EMERGENCY: Spill, leak, fire, exposure, or accident call CHEMTREC 1-800-424-9300.

ENVIRONMENTAL HAZARDS

This product is extremely toxic to fish. Do not contaminate water by disposing of equipment washwaters. Do not apply directly to any body of water. Apply this product only as specified on this label. Care should be used when spraying to avoid fish and reptile pets.

PHYSICAL OR CHEMICAL HAZARDS

Do not use or store near heat or open flame.

DIRECTIONS FOR USE

It is a violation of Federal law to use this product in a manner inconsistent with its labelling.

For residual pest control in and on buildings and structures and their immediate surroundings and on modes of transport. Permitted areas of use include industrial buildings, houses, apartment buildings, laboratories, buses, greenhouses and the nonfood/feed areas of stores, warehouses, vessels, railcars, trucks, trailers, aircraft, schools, nursing homes, hospitals, restaurants, hotels, and food manufacturing, processing and servicing establishments.

DEMON EC insecticide is intended for dilution with water for spray application.

Fill sprayer with the desired volume of water and add DEMON EC insecticide. Close and shake before use in order to insure proper mixing. Shake or reagitate sprayer before use if spraying is interrupted. Make up only as required; repeat treatments when needed.

Pests		Concentration of Active Ingredient	Dilution Rate
Ants Centipedes Cockroaches (Maintenance) Crickets Firebrats	Flies* Millipedes Pillbugs Silverfish Sowbugs	.1%	½ fl. oz. per 1 gal. water
Bees Cockroaches (Clean-out) Spiders Wasps		.2%	1 fl. oz. per 1 gal. water

* Not for indoor use in California for fly control.

Cockroaches, Spiders, Crickets, Silverfish and Firebrats: Apply as a coarse, low pressure spray to areas where these pests hide, such as baseboards, corners, storage areas, closets, around water pipes, doors and windows, attics and eaves, behind and under refrigerators, cabinets, sinks, furnaces, and stoves, the underside of shelves, drawers and similar areas. Pay particular attention to cracks and crevices; also see OUTDOOR USE.

Ants: Apply to any trails, around doors and windows and other places where ants may be found. Refer to barrier treatment directions to prevent infestation; also see OUTDOOR USE.

Wasps and Bees: Application to nests should be made late in evening when insects are at rest. Thoroughly spray nest and entrance and surrounding areas where insects alight; also see OUTDOOR USE.

Sowbugs, Pillbugs, Millipedes and Centipedes: Apply around doors and windows and other places where these pests may be found or where they may enter premises. Spray baseboards, storage areas and other locations. Refer to barrier treatment directions to prevent infestation; also see OUTDOOR USE.

Flies*: Apply directly to walls, ceilings, window screens, and other fly resting areas as a residual surface treatment. May be used inside residential buildings as well as in and around carports, garages, and storage sheds. Also see OUTDOOR USE.

Food Handling Establishments: Places other than private residences in which food is held, processed, prepared or served:

Nonfood Areas: DEMON EC insecticide may be used as a general, spot or crack and crevice treatment in nonfood areas. All areas where insects hide or through which insects may enter should be treated. Examples of nonfood areas include garbage rooms, lavatories, floor drains (to sewers), entries and vestibules, offices, locker rooms, machine rooms, boiler rooms, garages, mop closets and storage (after canning or bottling).

Food Areas: DEMON EC is not labeled for use in food areas. Do not use in any area where food or feed is stored, prepared or consumed. Do not use in food areas of food processing plants, restaurants or other areas where food is commercially prepared or processed. Do not use in serving areas while food is exposed or facility is in operation. Serving areas are where prepared food are served such as dining rooms but excluding areas where foods may be prepared or held. In the home all food processing surfaces and utensils should be covered during treatment or thoroughly washed before use. Exposed food should be covered or removed. Examples of food areas are areas for receiving, storage, packing (canning, bottling, wrapping, boxing), preparing, edible waste storage and enclosed processing systems (mills, dairies, edible oils, syrups). Serving areas (when food is exposed and facility is in operation) also would be considered a food area.

Outdoor Use: For control of ants, bees, centipedes, cockroaches, crickets, millipedes, sowbugs, pillbugs, spiders, flies and wasps. Apply as a residual spray to surfaces of buildings, porches, screen, window frames, eaves, patios, lawns, refuse dumps, garages in either hand or power sprayer and in other areas where these pests are found.

Barrier Treatment (except cockroaches): To help prevent infestation of buildings, apply to a band of soil and vegetation 6 to 10 feet wide around and adjacent to the building. Also, treat the building foundation to a height of 2 to 3 feet where pests are active and may find entrance. Apply as a coarse spray to thoroughly and uniformly wet the band area, using 1 gallon of spray mix per 400 square feet.

Note: Do not use water base sprays of DEMON EC in conduits, motor housings, junction boxes, switch boxes, or other electrical equipment because of possible shock hazard. Let surfaces dry before allowing children and pets to contact surfaces. Protect aquariums from spray mist. Thoroughly wash out sprayer and screen with water and detergent before using DEMON EC insecticide. Do not use this product with oil.

Do not apply this product to edible crops.

STORAGE AND DISPOSAL

Prohibitions: Do not contaminate water, food or feed by storage or disposal. Open dumping is prohibited. Do not reuse empty container.

Storage: Keep container closed when not is use. Do not store near food or feed. Protect from freezing. In case of spill or leak on floor or paved surfaces, soak up with sand, earth or synthetic absorbent. Remove to chemical waste area.

Pesticide Disposal: Pesticide wastes are acutely hazardous. Improper disposal of excess pesticide, spray mixture or rinsate is a violation of Federal law. If these wastes cannot be disposed of by use according to label instructions, contact your State Pesticide or Environmental Control Agency, or the Hazardous Waste representative of the nearest EPA Regional Office for guidance.

Container Disposal: *Metal Containers:* Triple rinse (or equivalent). Then offer for recycling or reconditioning, or puncture and dispose of in a sanitary landfill, or by other procedures approved by state and local authorities. *Plastic Containers:* Triple rinse (or equivalent). Then offer for recycling or reconditioning, or puncture and dispose of in a sanitary landfill, or incineration, or, if allowed by state and local authorities, by burning. If burned, stay out of smoke. *Glass Containers:* Triple rinse (or equivalent). Then dispose of in a sanitary landfill or by other approved state and local procedures.

NOTICE TO BUYER AND USER

Seller warrants that this product conforms to the chemical description on the label and is reasonably fit for the purposes stated on the label when used in accordance with directions under normal conditions of use. This warranty does not extend to the use of this product contrary to label instructions, or under abnormal use conditions, or under conditions not reasonably foreseeable to Seller, and Buyer and User assume the risk of any such use. SELLER DISCLAIMS ALL OTHER WARRANTIES, EXPRESSED OR IMPLIED, INCLUDING ANY WARRANTY OF FITNESS OR MERCHANTABILITY. SELLER SHALL NOT BE LIABLE FOR CONSEQUENTIAL, SPECIAL OR INDIRECT DAMAGES RESULTING FROM THE USE OR HANDLING OF THIS PRODUCT AND SELLER'S SOLE LIABILITY AND BUYER'S AND USER'S EXCLUSIVE REMEDY SHALL BE LIMITED TO THE REFUND OF THE PURCHASE PRICE.

05-0864-057 A-1R

This is a specimen label. ICI Americas Inc. is not responsible for the accuracy of the information contained herein. As labels are subject to revision, always carefully read and follow the label on the product container.

DEMON® TC

INSECTICIDE

Only for Sale to, Use and Storage by Professional Pest Control Operators. Soil Treatment to Establish a Barrier Against Termites.

ACTIVE INGREDIENT:

Cypermethrin (±) α-cyano-(3-phenoxyphenyl) methyl (±)-*cis, trans*-3(2,2-dichloroethenyl-2,2,-dimethyl-cyclopropanecarboxylate*	25.3%
INERT INGREDIENTS:	74.7%
TOTAL	100.0%

**Cis/Trans* ratio: Min. 45% (±) *cis* and max. 55% (±) *trans.*

DEMON TC contains 2 pounds active ingredient per gallon.

EPA Reg. No. 10182-107
U.S. Patent No. 4,024,163

KEEP OUT OF REACH OF CHILDREN WARNING

PRECAUTIONARY STATEMENTS

HAZARDS TO HUMANS AND DOMESTIC ANIMALS

WARNING

- May Be Fatal If Swallowed.
- Causes Eye and Skin Irritation.
- Harmful If Absorbed Through Skin.
- May Cause Allergic Skin Reaction.

Do not get in eyes, on skin or on clothing. Wear protective clothing when spraying. Wear protective clothing, impermeable gloves and full face shield or goggles when handling or mixing concentrate. Avoid breathing vapors or spray mist. Wash thoroughly with soap and water after handling. Remove contaminated clothing and wash before reuse.

STATEMENT OF PRACTICAL TREATMENT

IF IN EYES: Flush with plenty of water. Get medical attention if irritation persists. IF ON SKIN: Wash with plenty of soap and water. IF SWALLOWED: Do not induce vomiting. Call a physician or Poison Control Center. Vomiting should be supervised by a physician or the professional staff because of the possible pulmonary damages from aspiration of the solvent. IF INHALED: Remove victim to fresh air. Apply artificial respiration if necessary.

FOR 24-HOUR EMERGENCY MEDICAL ASSISTANCE CALL 1-800-F-A-S-T-M-E-D (327-8633).
FOR CHEMICAL EMERGENCY: Spill, leak, fire, exposure, or accident call CHEMTREC 1-800-424-9300.

ENVIRONMENTAL HAZARDS

This pesticide is extremely toxic to fish. Use with care when applying in areas adjacent to any body of water. Do not apply directly to water. Do not apply when weather conditions favor drift from treated areas. Do not contaminate water by cleaning of equipment or disposal of wastes. This product is highly toxic to bees exposed to direct treatment or residues on crops or weeds. Do not apply DEMON TC, or allow it to drift to crops or weeds on which bees are actively foraging. Additional information may be obtained from your Cooperative Extension Service. Apply this product only as specified on this label.

PHYSICAL OR CHEMICAL HAZARDS

Do not use or store near heat or open flame.

GENERAL INFORMATION ON THE USE OF THIS PRODUCT

Chemicals for soil treatment are used to establish a barrier against termite attack. The chemical emulsion must be adequately dispersed in the soil to provide a barrier between the wood in the structure and the termite colonies in the soil.

For the effective use of this product, it is necessary that the service technician be familiar with current control practices including trenching, rodding, subslab injection, and low-pressure spray applications. These techniques must be correctly employed to prevent or control infestations by subterranean termite species of *Reticulitermes, Zootermopsis, Heterotermes* and *Coptotermes.* Choice of appropriate procedures should include consideration of such variable factors as the design of the structure, water table, soil type, soil compaction, grade conditions, location and type of domestic water supplies and drainage systems. The biology and behavior of the termite species involved are important factors to be known as well as suspected location of the colony and severity of the infestation within the structure to be protected.

Effective termite control also includes elimination of termite access to moisture by recommending repair of faulty construction grade and/or plumbing. Remove all wood and cellulose containing debris in contact with soil from crawl spaces, porches, and around foundations.

For advice concerning current control practices with relation to the specific local conditions, consult resources in structural pest control and the State regulatory agency.

DIRECTIONS FOR USE—SUBTERRANEAN TERMITE CONTROL

It is a violation of Federal law to use this product in a manner inconsistent with its labeling. This labeling must be in the possession of the user at the time of application. Avoid contamination of public and private water supplies by following these precautions:

- Use anti-backflow equipment or procedures.
- Do not treat soil beneath structures that contain wells or cisterns.
- Care must be taken to avoid runoff. Do not treat soil that is water-saturated or frozen.

Consult State and local specifications for recommended distance of treatment areas from wells. Refer to Federal Housing Administration Specifications for guidance on preconstruction treatments.

After Treatment: Securely plug all holes drilled in construction elements of commonly occupied areas of structures, including unfinished basements, enclosed porches, garages, and workshops.

PRECONSTRUCTION SUBTERRANEAN TERMITE CONTROL

Effective preconstruction subterranean termite control requires the establishment of an unbroken vertical and/or horizontal chemical barrier between wood in the structure and the termite colonies in the soil.

To meet F.H.A. termite-proofing requirements, follow the latest edition of the Housing and Urban Development (H.U.D.) Minimum Property Standards.

Use a 0.25%–0.50% emulsion* for subterranean termites. For a 0.25% emulsion mix 1 gallon of DEMON TC in 99 gallons of water. For a 0.50% emulsion mix 2 gallons of DEMON TC in 98 gallons of water.

* Maximum use rate approved in CA is 0.25% emulsion.

After grading is completed and prior to pouring of the slab, slab-supported or constructed porches, and other critical areas, make the following treatments:

Horizontal Barriers: Horizontal barriers may be established in areas intended for covering such as floors, porches, and other critical areas; application shall be made by a low-pressure spray (less than 50 p.s.i.).

Apply the emulsion at the rate of 1 gallon per 10 square feet to fill dirt. If fill is washed gravel or other coarse material, apply the emulsion at the rate of 1½ gallons per 10 square feet.

If concrete slabs cannot be poured over soil the same day it has been treated a waterproof cover, such as polyethylene sheeting, should be placed over the soil. This is not necessary if foundation walls have been installed around the treated soil.

Vertical Barriers: Vertical barriers may be established around the base of foundations, plumbing, back-filled soil against foundation walls, and other critical areas; applications may be made by rodding and/or trenching. Apply the emulsion at the rate of 4 gallons per 10 linear feet per foot of depth. For example, a footing 3 feet deep would require 12 gallons of emulsion per 10 linear feet.

Outside and inside perimeter applications must be made by rodding and/or trenching. When rodding from grade or from the bottom of a shallow trench, rod holes should be spaced in a manner that will allow for application of a continuous chemical barrier. Rod holes should not extend beneath the top of the footings.

A trench need not be wider than 6 inches. Rod from the base of a shallow trench to the top of the footings. Low-pressure spray (less than 50 p.s.i.) may be used to treat soil which will be replaced in the trench. Mix the emulsion with the soil as it is being replaced in the trench.

Soil should be treated around sewer lines, plumbing, or around any other utility extending from the soil through a slab.

Hollow Masonry Units of the Foundation: In preconstruction situations in which horizontal barrier application is not made to soil prior to pouring the footing treatment may be made through masonry voids to establish a continuous chemical barrier at the top of the footing. Apply at the rate of 2 gallons per 10 linear feet. Apply the emulsion so it will reach to the footing.

Crawl Spaces: For crawl spaces apply at the rate of 4 gallons of emulsion per 10 linear feet per foot of depth from grade to the top of the footing. Application may be made by rodding and/or trenching. Treatment should include both sides of foundation and around all piers and pipes.

- Rod holes should be spaced (about 1 foot) to provide a continuous chemical barrier.
- Trench need not be wider than 6 inches nor below the foundation. The emulsion should be mixed with the soil as it is being replaced in the trench.

POSTCONSTRUCTION TREATMENTS

Use a 0.25%–0.50% emulsion* for subterranean termites. For a 0.25% emulsion mix 1 gallon of DEMON TC in 99 gallons of water. For a 0.50% emulsion mix 2 gallons of DEMON TC in 98 gallons of water.

* Maximum use rate approved in CA is 0.25% emulsion.

Postconstruction applications may be made by subslab injection, rodding, and/or trenching using low pressure spray (25 p.s.i.).

Do not apply emulsion until location of heat or air conditioning ducts, vents, and water and sewer (or plumbing) lines are known and identified. Caution must be taken to avoid contamination of these structural elements and airways.

Slab-On-Ground: Apply the emulsion at the rate of 4 gallons per 10 linear feet. Application shall be made by sub-slab injection, trenching, or horizontal rodding.

Treat the soil from grade to the top of the footing along the outside and, where necessary, along the inside of the foundation perimeter. Treatment may also be required along one side of a partition wall (especially where the wall is connected to the floor by fixtures inserted in the slab) and along cracks, expansion joints, and other critical areas.

Drill holes should be spaced about 10 to 24 inches* apart to provide a continuous chemical barrier. (For best results, application should be made with a lateral dispersion nozzle.)

* Minimum approved in CA is 12 inches.

Where necessary, drill through the foundation walls from the outside and inject the chemical just beneath the slab or along the inside of the foundation.

Along the outside of the foundation walls where shallow foundations exist (1 foot or less), dig a narrow trench approximately 6 inches wide and not below the top of the footing. Apply the emulsion at the rate of 2 gallons per 10 linear feet. As the soil is being replaced into the trench, apply another 2 gallons per 10 linear feet to the backfill. When making soil applications to the foundations extending deeper than 1 foot, follow instructions under: BASEMENTS—OUTSIDE PERIMETER. (See exception for monolithic slabs immediately following.) Note: For monolithic slab construction a vertical barrier may be established along the outside of foundation walls from grade to the bottom of the monolithic poured concrete foundation. Where the foundation extends deeper than one foot, rod holes should not extend beneath the bottom of the monolithic poured concrete foundation.

Hollow Masonry Units of Foundation Walls: Treatment may be made through masonry voids to establish a continuous chemical barrier at the top of the footing. Apply at the rate of 2 gallons per 10 linear feet of footing. Where this treatment is necessary, access holes must be drilled *below* the sill plate and should be through a lower mortar joint as close as possible to the footing.

Basements: Apply the emulsion at the rate of 4 gallons per 10 linear feet per foot of depth from the grade to the top of the footing. For example, a footing 3 feet deep would require 12 gallons of emulsion per 10 linear feet.

Application shall be made by subslab injection, trenching, and/or rodding.

Inside: Treatment may be required along inside of foundation walls and along one side of interior partition walls (or bearing walls) especially where the wall is connected by fixtures inserted in the floor. Application may also be necessary around sewer pipes, floor drains, conduits, or any cracks in the basement floor.

Drill holes should be spaced about 10 to 24 inches* apart to provide a continuous chemical barrier. Note: Sandy soils will tend to give less lateral dispersion than clay soils. Spacing should be determined by soil type.

* Minimum approved in CA is 12 inches.

Outside Perimeter: Applications must be made by rodding and/or trenching. When rodding from grade or from the bottom of a shallow trench, rod holes should be spaced in a manner that will allow for application of a continuous chemical barrier. Rod holes should not extend beneath the top of the footings.

A trench need not be wider than 6 inches. Rod from the base of a shallow trench to the top of the footings. Low-pressure spray (25 p.s.i.) may be used to treat soil which will be replaced in the trench. Mix the emulsion with the soil as it is being replaced in the trench.

Hollow Masonry Units of the Foundation and/or Basement Wall (below grade): Treat so as to make a continuous chemical barrier in the voids. Apply the emulsion at the rate of 2 gallons per 10 linear feet. Apply the emulsion so it will reach to the footing.

Crawl Spaces: Apply the emulsion at the rate of 4 gallons per 10 linear feet per foot of depth from the grade to the top of the footing. For example, a footing 3 feet deep would require 12 gallons of emulsion per 10 linear feet. For best results application should be made by rodding and/or trenching.

Treat both sides of foundation and around all piers and pipes. When rodding from grade or from the bottom of a shallow trench, rod holes should be spaced in a manner that will allow for application of a continuous chemical barrier. Rod holes should not extend beneath the top of the footings.

A trench need not be wider than 6 inches. Rod from the base of a shallow trench to the top of the footings. Low-pressure spray (25 p.s.i.) may be used to treat soil which will be replaced in the trench. Mix the emulsion with the soil as it is being replaced in the trench.

To prevent subterranean termites from constructing tubes from soil to crawl space wood members above, an overall soil treatment of this product may be applied.* Remove all cellulose debris before application of 1 gallon per 10 square feet overall. Use fans to exhaust crawl space air when working in a confined space. Wear appropriate protective clothing, gloves, and unvented goggles.

* Not approved for use in California.

If adequate ventilation is not available in the crawl space wear a respirator approved by the Mine Safety and Health Administration during treatment. It is recommended that inadequately ventilated crawl spaces be brought into compliance with FHA Minimum Property Standards specifying 1 square foot of ventilator opening per 150 square feet of crawl space area.

Bath Traps: Where there is exposed soil beneath and around plumbing/waste pipe entrances through a concrete slab, this soil may be treated with 0.5% dilution of this product.

An access door for inspection and treatment should be cut and installed if not already present. After inspection and removal of any wood (form boards) or cellular debris, treat the soil by rodding and/or flooding with a 0.25% emulsion of this product.

Posts, Poles, and Other Constructions: (Not approved for use in California)

Application may be made to create a chemical barrier in the soil around wooden construction such as signs and landscape ornamentation.

Use 1 gallon of emulsion per foot of depth for poles and posts less than six inches in diameter. For larger poles, use 1 1/2 gallons of emulsion per foot of depth. For larger constructions, use 4 gallons per linear feet per foot of depth.

For treatments made during installation, the emulsion may be applied to the soil as it is replaced around the pole or post. Previously installed poles and posts may be treated by subsurface injection or treated by gravity-flow through holes made from the bottom of a trench around the pole or post. Treat on all sides to create a continuous chemical barrier. Apply to a depth of six inches below the bottom of the wood.

Excavation Technique: If treatment must be made in difficult situations such as near wells, cisterns, along fieldstone or rubble walls, along faulty foundation walls, and around pipes and utility lines which lead downward from the structure to a well or pond, application may be made in the following manner:

1) Trench and remove soil to be treated onto heavy plastic sheeting or similar material.

2) Treat the soil at the rate of 4 gallons of emulsion per 10 linear feet per foot of depth of the trench. Mix the emulsion thoroughly into the soil taking care to prevent liquid from running off the liner.

3) After the treated soil has absorbed the liquid emulsion, replace the soil in the trench.

Prior to using this technique near wells or cisterns, consult State, local, or Federal agencies for information regarding approved treatment practices in your area.

Treatment of Wood in Place for Control of Termites, Carpenter Ants, Carpenter Bees and Wood Infesting Beetles

(Not approved for use in California.)

In addition to subsurface applications, this product may be used for treating infested wood in place. It can be applied to wood by crack and crevice tool, coarse fan spray or injection. Overall broadcast spray applications must be limited to attics, crawl spaces, unfinished basements and similar generally unoccupied areas. In occupied indoor areas, treat wood trim and exposed beams by brush or coarse spray directed only onto the wood to be treated.

IMPORTANT: Do not apply emulsion until location of heat pipes, ducts, water and sewer lines and electrical conduits are known and identified. Caution must be taken to avoid puncturing and injection into these structural elements. Do not apply into electrical fixtures, switches, or sockets.

Remove pets, birds, and cover aquariums before spraying. Do not permit humans or pets to contact treated surfaces until the spray has dried.

During any applications to overhead interior areas of structures, cover surfaces below with plastic sheeting or similar material.

Do not use in food areas of food handling establishments, restaurants or other areas where food is commercially prepared or processed. Do not use in serving areas while food is exposed or facility is in operation. Serving areas are areas where prepared foods are served such as dining rooms but excluding areas where food may be prepared or held. In the home all food processing surfaces and utensils should be covered during treatment or thoroughly washed before reuse. Cover exposed food.

Non-food areas are areas such as garbage rooms, lavatories, floor drains (to sewers), entries and vestibules, offices, locker rooms, machine rooms, boiler rooms, garages, mop closets and storage (after canning or bottling).

Not for use in USDA meat and poultry plants.

NOTE:

For above ground treatments, use a 0.1% concentration. To prepare a 0.1% emulsion, add 0.5 fluid ounces of concentrate to 1 gallon of water. To prepare 50 gallons of emulsion, add 0.2 gallons (25.6 oz) of concentrate to 49.8 gallons of water. To prepare 100 gallons of emulsion, add 0.4 gallons (51.2 oz) of concentrate in 99.6 gallons of water. Use this spray at the rate of 1 gallon of diluted spray per 1000 square feet of surface area.

Control of Wood Investing Beetles: (Not approved for use in California.)

To control wood infesting insects such as powderpost beetle (Lyctidae), false powderpost beetles (Bostrichidae), deathwatch beetles (Anobiidae), old house borers (Cerambycidae) and ambrosia beetles (Scolytidae) in homes and other structures, apply as an emulsion containing .1% to .25% DEMON TC. For treatment of small areas, apply by brushing the emulsion evenly on wood surfaces. For large or overhead areas, apply as a coarse spray to the point of runoff. When spraying overhead interior areas of homes, apartment buildings, etc. spray to the point of runoff, cover all surfaces below the area being sprayed with plastic sheeting or other material which could be disposed of by placing in trash if contamination from dripping occurs. Sprayed surfaces should be avoided until spray has totally dried. Do not use in structures occupied by animals to be used for food purposes or which produce products for human consumption.

Termites Above Ground:* For control of termites, subterranean aerial colonies, Formosan aerial colonies or drywood termites in localized areas of infested wood in structures, apply a 0.1% emulsion to voids and galleries in damaged wood and in spaces between wooden members of a structure and between wood and foundations where wood is vulnerable. Application may be made to inaccessible areas by drilling, and then injecting the emulsion with a crack and crevice injector into the damaged wood or void spaces. Application to attics, crawl spaces, unfinished basements, or man-made voids may be made with a coarse fan spray of 0.1% emulsion to control workers and winged reproductive forms of termites in mud shelter tubes. This type of application is not intended to be a substitute for soil treatment for extensive infestation of drywood termites or other wood-infesting insects.

For termites active inside trees, utility poles and/or fence posts, drill to find the interior infested cavity and inject 0.1% emulsion using treatment tool with a splashback guard.

Termite carton nests in trees or building voids may be injected with 0.25%–0.50% emulsion using a pointed injection tool. Multiple injection points to varying depths may be necessary. It is desirable to physically remove carton nest material from building voids when such nests are found.

Carpenter Ants:* For control of carpenter ants in houses and other structures, apply as a 0.25% emulsion for protection up to 5 weeks, a 0.5% emulsion for protection up to 11 weeks, and a 1.0% emulsion for protection up to 1 year, with retreatment semiannually as needed, around doors and windows and other places where carpenter ants enter the premises and where they crawl. Spray into cracks and crevices or through openings or small drilled holes into voids where these ants or their nests are present. Use no more that a sufficient amount of coarse spray to cover the area to the point of runoff. Do not exceed 1 gallon of dilute emulsion per 1000 square feet of treated surface.

For carpenter ants active inside trees, utility poles and/or fence posts, drill to find the interior infested cavity and inject 0.25% emulsion for protection up to 1 week using a treatment tool with a splashback guard. Reapply under heavy reinfestation pressure.

Firewood Protection From Carpenter Ants:* Prior to laying in firewood, soil beneath the cord(s) may be treated with a 0.25%–0.50% emulsion at 1 gallon per 10 square feet to prevent carpenter ant infestation.

Carpenter Bees:* Use a 0.1% emulsion for control of carpenter bees. Liquid may be sprayed directly into gallery entrance holes. Following treatment, the entrance holes may be left open 24 hours to be certain that returning adult bees are killed. When there is no activity, the hole may be closed with wood putty.

* Not approved for use in California.

RETREATMENT

Retreatment for subterranean termites should be made when there is evidence of reinfestation subsequent to the initial treatment, or there has been a disruption of the chemical barrier in the soil due to construction, excavations, landscaping, etc. Retreatment should be made as a spot application to these areas.

Retreatments may be made to vulnerable areas in accordance with the application techniques described above. This application should be made as a spot treatment to these areas. Routine or annual retreatment of the entire premises should be avoided.

OUTDOOR USE

For control of ants, bees, centipedes, cockroaches, crickets, firebrats, flies, millipedes, pillbugs, silverfish, sowbugs, spiders, wasps. Apply a 0.1% solution as a residual spray to surfaces of building, porches, screens, window frames, eaves, patios, lawns, refuse dumps, garages with either hand or power sprayer and in other areas where these pests are found.

BARRIER TREATMENTS

To help prevent infestation of buildings, apply a 0.1% solution to a band of soil and vegetation 6 to 10 feet wide around and adjacent to the building. Also, treat the building foundation to a height of 2 to 3 feet when pests are active and may find entrance. Apply as a coarse spray to thoroughly and uniformly wet the band areas, using 1 gallon of spray mix per 400 square feet.

Do not apply this product to edible crops.

RECOMMENDATION FOR CONTROL OF CLUSTER FLY

Apply DEMON TC to the outside of the structure. Mix 1 oz. DEMON TC to 10 ozs. deodorized lightweight mineral oil; apply to 1,000 square feet of wall area. Use a fogging apparatus which delivers the material in a strong air carrier, producing a small particle size. The apparatus should be held within three feet of the surface being treated. Apply only when air movement is less than 2 miles per hour.

The surface treated should be dry at the time of application. Attics and unoccupied lofts should be treated at the same time and at the same rate.

Generally the north side of structures need not be treated since flies seldom enter from areas not sunlit.

STORAGE AND DISPOSAL

Prohibitions: Do not contaminate water, food or feed by storage or disposal. Open dumping is prohibited. Do not reuse empty container.

Storage: Keep container closed when not in use. Do not store near food or feed. Protect from freezing. In case of spill or leak on floor or paved surfaces, soak up with sand, earth or synthetic absorbent. Remove to chemical waste area.

Pesticide Disposal: Pesticide wastes are toxic. Improper disposal of excess pesticide, spray mixture or rinsate is a violation of Federal law. If these wastes cannot be disposed of by use according to label instructions, contact your State Pesticide or Environmental Control Agency, or the Hazardous Waste representative at the nearest EPA Regional Office for guidance.

Container Disposal: METAL CONTAINERS: Triple rinse (or equivalent). Then offer for recycling or reconditioning, or puncture and dispose of in a sanitary landfill, or by other procedures approved by state and local authorities. PLASTIC CONTAINERS: Triple rinse (or equivalent). Then offer for recycling or reconditioning, or puncture and dispose of in a sanitary landfill, or incineration, or, if allowed by state and local authorities, by burning. If burned, stay out of smoke. GLASS CONTAINERS: Triple rinse (or equivalent). Then dispose of in a sanitary landfill or by other approved state and local procedures.

NOTICE TO BUYER AND USER

Seller warrants that product conforms to the chemical description on the label and is reasonably fit for the purposes stated on the label when used in accordance with directions under normal conditions of use. This warranty does not extend to the use of this product contrary to label instructions, or under abnormal use conditions, or under conditions not reasonably foreseeable to Seller, and Buyer and User assume the risk of any such use. SELLER DISCLAIMS ALL OTHER WARRANTIES, EXPRESSED OR IMPLIED, INCLUDING ANY WARRANTY OF FITNESS OR MERCHANTABILITY. SELLER SHALL NOT BE LIABLE FOR CONSEQUENTIAL, SPECIAL OR INDIRECT DAMAGES RESULTING FROM THE USE OR HANDLING OF THIS PRODUCT AND SELLER'S SOLE LIABILITY AND BUYER'S AND USER'S EXCLUSIVE REMEDY SHALL BE LIMITED TO THE REFUND OF THE PURCHASE PRICE.

05-0864-028 R2 A-1R

The information and specimen labels in this booklet are correct as of the publication date but are subject to change without notice. For current information, contact ICI Americas Inc.

DEMON® WP

INSECTICIDE

Crack and Crevice and Spot Spray for Cockroach Control. Only for Sale to, Use and Storage by Pest Control Operators.

ACTIVE INGREDIENT:

Cypermethrin (±)α-cyano-(3-phenoxyphenyl)methyl (±)-*cis, trans*-3-(2,2-dichloroethenyl)-2, 2- dimethylcyclopropanecarboxylate*	40.0%
INERT INGREDIENTS	60.0%
TOTAL	100.0%

* *Cis/trans* ratio: 45/55 ± 10

EPA Reg. No. 10182-71

KEEP OUT OF REACH OF CHILDREN

WARNING

PRECAUTIONARY STATEMENTS

HAZARDS TO HUMANS AND DOMESTIC ANIMALS

WARNING

HARMFUL IF SWALLOWED OR ABSORBED THROUGH SKIN. CAUSES EYE IRRITATION—DO NOT GET IN EYES, OR SKIN OR ON CLOTHING.

Avoid contact with eyes, skin or clothing. Avoid breathing vapor or spray mist. Wash thoroughly with soap and water after handling. May cause allergic skin reactions.

STATEMENT OF PRACTICAL TREATMENT

IF IN EYES: Flush with plenty of water. Get medical attention if irritation persists. IF ON SKIN: Wash with plenty of soap and water. IF SWALLOWED: Drink 1 or 2 glasses of water and induce vomiting by touching back of throat with finger. IF INHALED: Remove victim to fresh air. Apply artificial respiration if necessary.

FOR 24-HOUR EMERGENCY MEDICAL ASSISTANCE CALL 1-800-F-A-S-T-M-E-D (327-8633).
FOR CHEMICAL EMERGENCY: Spill, leak, fire, exposure, or accident call CHEMTREC 1-800-424-9300.

ENVIRONMENTAL HAZARDS

This product is extremely toxic to fish. Do not contaminate water by disposing of equipment washwaters. Do not apply directly to any body of water. Apply this product only as specified on this label. Care should be used when spraying to avoid fish and reptile pets.

DIRECTIONS FOR USE

It is a violation of Federal law to use this product in a manner inconsistent with its labeling.

For residual pest control in and on buildings and structures and their immediate surroundings and on modes of transport. Permitted areas of use include industrial buildings, houses, apartment buildings, laboratories, buses, greenhouses and the nonfood/feed areas of stores, warehouses, vessels, railcars, trucks, trailers, aircraft, schools, nursing homes, hospitals, restaurants, hotels and food manufacturing, processing and servicing establishments.

Authorized by the USDA for use in inedible product areas in Federally inspected meat, poultry, and egg processing plants.

DEMON WP insecticide is intended for dilution with water for spray application. Do not use strainers finer than 50 mesh size. Fill sprayer with the desired volume of water and add DEMON WP insecticide. Close and shake before use in order to insure proper suspension of the wettable powder. Shake or reagitate sprayer before use if spraying is interrupted. Make up only as required, repeat treatments when needed.

Pests		Concentration of Active Ingredient	Dilution Rate
Ants Centipedes Cockroaches (Maintenance) Crickets Firebrats	Flies* Millipedes Pillbugs Silverfish Sowbugs	.1%	1 scoop** per 1 gal. water
Bees Cockroaches (Clean-out) Spiders Wasps	Fire Ants (Mounds)† Carpenter Ants† Sod Webworms†	.2%	2 scoops** per 1 gal. water

* Not for indoor use in California for fly control.
** 1 scoop = 0.33 ounces of DEMON WP. Shake container well before measuring.
† Not registered in California for use to control this insect.

Cockroaches, Spiders, Crickets, Silverfish and Firebrats:
Apply as a coarse, low pressure spray to areas where these pests hide, such as baseboards, corners, storage areas, closets, around water pipes, doors and windows, attics and eaves, behind and under refrigerators, cabinets, sinks, furnaces, and stoves, the underside of shelves, drawers and similar areas. Pay particular attention to cracks and crevices; also see OUTDOOR USE.

Ants: Apply to any trails, around doors and windows and other places where ants may be found. Apply barrier treatments to prevent infestation as described below; also see OUTDOOR USE.

Wasps and Bees: Application to nests should be made late in evening when insects are at rest. Thoroughly spray nest and entrance and surrounding areas where insects alight, also see OUTDOOR USE.

Sowbugs, Pillbugs, Millipedes and Centipedes: Apply around doors and windows and other places where these pests may be found or where they may enter premises. Spray baseboards, storage areas and other locations. Apply barrier treatments to prevent infestation as described below; also see OUTDOOR USE.

Flies*: Apply directly to walls, ceilings, window screens, and other fly resting areas as a residual surface treatment. May be used inside residential buildings as well as in and around carports, garages, and storage sheds. Also see OUTDOOR USE.

* Not for indoor use in California for fly control.

Fire Ants (Mounds)†: For individual fire ant mounds, apply DEMON WP as a drench. Gently sprinkle (watering can) a total of one gallon over the surface of mounds up to 12 inches in diameter and two gallons for larger mounds. Thoroughly wet mound and surrounding area. Treat new mounds as they appear. Pressurized sprays may disturb the ants and cause migration. If possible, applications should be made on warm days after recent rainfall. All colonies in the vicinity should be treated, including those that have not yet constructed a mound.

Carpenter Ants†: Apply as a coarse wet spray to the point of runoff around doors, door frames, windows and other areas where these ants enter premises or hide; also see OUTDOOR USE.

Sod Webworms†: Spray when pests first appear; retreat when needed. Delay watering or mowing the treated area for 12 to 24 hours after treatment. Use a minimum of five gallons spray mix per 1000 square feet when spraying.

† Not registered for use in California.

Food Handling Establishments: Places other than private residences in which food is held, processed, prepared or served:

Nonfood Areas: DEMON WP insecticide may be used as a general, spot or crack and crevice treatment in nonfood areas. All areas where insects hide or through which insects may enter should be treated. Examples of nonfood areas include garbage rooms, lavatories, floor drains (to sewers), entries and vestibules, offices, locker rooms, machine rooms, boiler rooms, garages, mop closets and storage (after canning or bottling).

Food Areas: DEMON WP is not labeled for use in food areas. Do not use in food areas of food processing plants, restaurants or other areas where food is commercially prepared or processed. Do not use in serving areas while food is exposed or facility is in operation. Serving areas are where prepared food are served such as dining rooms but excluding areas where foods may be prepared or held. In the home all food processing surfaces and utensils should be covered during treatment or thoroughly washed before use. Exposed food should be covered or removed. Examples of food areas are areas for receiving, storage, packing (canning, bottling, wrapping, boxing) preparing, edible waste storage and enclosed processing systems (mills, dairies, edible oils, syrups). Serving areas (when food is exposed and facility is in operation) also would be considered a food area.

Outdoor Use: For control of ants, bees, centipedes, cockroaches, crickets, millipedes, sowbugs, pillbugs, spiders, flies and wasps. Apply as a residual spray to surfaces of buildings, porches, screens, window frames, eaves, patios, lawns, refuse dumps, garages in either hand or power sprayer and in other areas where these pests are found.

Barrier Treatments (except cockroaches): To help prevent infestation of buildings, apply to a band of soil and vegetation 6 to 10 feet wide around and adjacent to the building. Also, treat the building foundation to a height of 2 to 3 feet where pests are active and may find entrance. Apply as a coarse spray to thoroughly and uniformly wet the band area, using 1 gallon of spray mix per 400 square feet.

Note: Do not use water base sprays of DEMON in conduits, motor housings, junction boxes, switch boxes, or other electrical equipment because of possible shock hazard. Let surfaces dry before allowing children and pets to contact surfaces. Protect aquariums from spray mist. Thoroughly wash out sprayer and screen with water and detergent before using DEMON WP insecticide. Do not use this product with oil.

Do not apply this product to edible crops.

STORAGE AND DISPOSAL

Prohibitions: Do not contaminate water, food or feed by storage or disposal. Do not reuse empty container.

Storage: Mix as needed. Do not store diluted material. Keep container closed when not in use. Do not store near food or feed. In case of spill on floor or paved surfaces, sweep up and remove to chemical waste area.

Pesticide Disposal: Wastes resulting from the use of this product may be disposed of on site or at an approved waste disposal facility.

Container Disposal: Completely empty jar into application equipment. Then puncture and dispose of empty jar in a sanitary landfill, by incineration or, if allowed by state and local authorities, by burning. If burned, stay out of smoke.

NOTICE TO BUYER AND USER

Seller warrants that this product conforms to the chemical description on the label and is reasonably fit for the purposes stated on the label when used in accordance with directions under normal conditions of use. This warranty does not extend to the use of this product contrary to label instructions, or under abnormal use conditions, or under conditions not reasonably foreseeable to Seller, and Buyer and User assume the risk of any such use. SELLER DISCLAIMS ALL OTHER WARRANTIES EXPRESSED OR IMPLIED INCLUDING ANY WARRANTY OF FITNESS OR MERCHANTABILITY. SELLER SHALL NOT BE LIABLE FOR CONSEQUENTIAL, SPECIAL OR INDIRECT DAMAGES RESULTING FROM THE USE OR HANDLING OF THIS PRODUCT AND SELLER'S SOLE LIABILITY AND BUYER'S AND USER'S EXCLUSIVE REMEDY SHALL BE LIMITED TO THE REFUND OF THE PURCHASE PRICE.

05-0864-059 R2 A-2R

The information and specimen labels in this booklet are correct as of the publication date but are subject to change without notice. For current information, contact ICI Americas Inc.

DEMON® WP

INSECTICIDE IN WATER-SOLUBLE PACKETS

Crack and Crevice and Spot Spray for Cockroach Control. Only for Sale to, Use and Storage by Pest Control Operators.

ACTIVE INGREDIENT:
CYPERMETHRIN
(±)α-cyano-(3-phenoxyphenyl)methyl
(±)-*cis, trans*-3-(2,2-dichloroethenyl)-2,
2-dimethylcyclopropanecarboxylate* 35.6%
INERT INGREDIENTS 64.4%

TOTAL 100.0%

**Cis/trans* ratio: 45/55 ± 10

EPA Reg. No. 10182-100

KEEP OUT OF REACH OF CHILDREN
WARNING

PRECAUTIONARY STATEMENTS

HAZARDS TO HUMANS AND DOMESTIC ANIMALS

WARNING

HARMFUL IF SWALLOWED OR ABSORBED THROUGH SKIN. CAUSES EYE IRRITATION—DO NOT GET IN EYES, ON SKIN OR ON CLOTHING.

Avoid contact with eyes, skin or clothing. Avoid breathing vapors or spray mist. Wash thoroughly with soap and water after handling. May cause allergic skin reactions.

STATEMENT OF PRACTICAL TREATMENT

IF IN EYES: Flush with plenty of water. Get medical attention if irritation persists. IF ON SKIN: Wash with plenty of soap and water. IF SWALLOWED: Drink 1 or 2 glasses of water and induce vomiting by touching back of throat with finger. IF INHALED: Remove victim to fresh air. Apply artificial respiration if necessary.

FOR 24-HOUR EMERGENCY MEDICAL ASSISTANCE CALL 1-800-F-A-S-T-M-E-D (327-8633).
FOR CHEMICAL EMERGENCY: Spill, leak, fire, exposure, or accident call CHEMTREC 1-800-424-9300.

ENVIRONMENTAL HAZARDS

This product is extremely toxic to fish. Do not contaminate water by disposing of equipment washwaters. Do not apply directly to any body of water. Apply this product only as specified on this label. Care should be used when spraying to avoid fish and reptile pets.

DIRECTIONS FOR USE

It is a violation of Federal law to use this product in a manner inconsistent with its labeling.

For residual pest control in and on buildings and structures and their immediate surroundings and on modes of transport. Permitted areas of use include industrial buildings, houses, apartment buildings, laboratories, buses, greenhouses, and the nonfood/feed areas of stores, warehouses, vessels, railcars, trucks, trailers, aircraft, schools, nursing homes, hospitals, restaurants, hotels and food manufacturing, processing and servicing establishments.

Do not allow packets to become wet before adding to the spray tank. Wear dry rubber gloves when handling packets. Minimize handling of packets since excessive handling may cause rupture.

DEMON WP insecticide is intended for dilution with water for spray application. Do not use strainers finer than 50 mesh size. Fill sprayer to 1/2 of the desired volume. Add water-soluble packet(s) of DEMON WP insecticide per the table below. Add remaining water needed and wait for 2 minutes for packet(s) to dissolve. Close and shake for 30 seconds before use. Shake again before use if spraying is interrupted. Make up only as required, repeat treatments when needed.

Pests		Concentration of Active Ingredient	Dilution Rate
Ants Centipedes Cockroaches (Maintenance) Crickets Firebrats	Flies* Millipedes Pillbugs Silverfish Sowbugs	.1%	1 packet** per 1 gal. water
Bees Cockroaches (Clean-out) Spiders	Fire Ant (Mounds)† Carpenter Ants†	.2%	2 packets** per 1 gal. water
Wasps	Sod Webworms†	.2%	2 packets per 1 gal. water

*Not for indoor use in California for fly control.
**1 packet = 0.33 ounces of DEMON WP.
†Not registered in California for use to control this insect.

Cockroaches, Spiders, Crickets, Silverfish and Firebrats: Apply as a coarse, low pressure spray to areas where these pests hide, such as baseboards, corners, storage areas, closets, around water pipes, doors and windows, attics and eaves, behind and under refrigerators, cabinets, sinks, furnaces, and stoves, the underside of shelves, drawers and similar areas. Pay particular attention to cracks and crevices; also see OUTDOOR USE.

Ants: Apply to any trails, around doors and windows and other places where ants may be found. Apply barrier treatments to prevent infestation as described below; also see OUTDOOR USE.

Wasps and Bees: Application to nests should be made late in evening when insects are at rest. Thoroughly spray nest and entrance and surrounding areas where insects alight, also see OUTDOOR USE.

Sowbugs, Pillbugs, Millipedes and Centipedes: Apply around doors and windows and other places where these pests may be found or where they may enter premises. Spray baseboards, storage areas and other locations. Apply barrier treatments to prevent infestation as described below; also see OUTDOOR USE.

Flies*: Apply directly to walls, ceilings, window screens, and other fly resting areas as a residual surface treatment. May be used inside residential buildings as well as in and around carports, garages, and storage sheds. Also see OUTDOOR USE.

Fire Ants (Mounds): For individual fire ant mounds, apply DEMON WP as a drench. Gently sprinkle (watering can) a total of one gallon over the surface of mounds up to 12 inches in diameter and two gallons for larger mounds. Thoroughly wet mound and surrounding area. Treat new mounds as they appear. Pressurized sprays may disturb the ants and cause migration. If possible, applications should be made on warm days after recent rainfall. All colonies in the vicinity should be treated, including those that have not yet constructed a mound.

Carpenter Ants: Apply as a coarse wet spray to the point of runoff around doors, door frames, windows and other areas where these ants enter premises or hide; also see OUTDOOR USE.

Sod Webworms: Spray when pests first appear; retreat when needed. Delay watering or mowing the treated area for 12 to 24 hours after treatment. Use a minimum of five gallons spray mix per 1000 square feet when spraying.

Food Handling Establishments: Places other than private residences in which food is held, processed, prepared or served:

Nonfood Areas: DEMON WP insecticide may be used as a general spot or crack and crevice treatment in nonfood areas. All areas where insects hide or through which insects may enter should be treated. Examples of nonfood areas include garbage rooms, lavatories, floor drains (to sewers), entries and vestibules, offices, locker rooms, machine rooms, boiler rooms, garages, mop closets and storage (after canning or bottling).

Food Areas: DEMON WP is not labeled for use in food areas. Do not use in food areas of food processing plants, restaurants or other areas where food is commercially prepared or processed. Do not use in serving areas while food is exposed or facility is in operation. Serving areas are where prepared food are served such as dining rooms but excluding areas where foods may be prepared or held. In the home all food processing surfaces and utensils should be covered during treatment or thoroughly washed before use. Exposed food should be covered or removed. Examples of food areas are areas for receiving, storage, packing (canning, bottling, wrapping, boxing) preparing, edible waste storage and enclosed processing systems (mills, dairies, edible oils, syrups). Serving areas (when food is exposed and facility is in operation) also would be considered a food area.

Outdoor Use: For control of ants, bees, centipedes, cockroaches, crickets, millipedes, sowbugs, pillbugs, spiders, flies and wasps. Apply as a residual spray to surfaces of buildings, porches, screens, window frames, eaves, patios, lawns, refuse dumps, garages in either hand or power sprayer and in other areas where these pests are found.

Barrier Treatments (except cockroaches): To help prevent infestation of buildings, apply to a band of soil and vegetation 6 to 10 feet wide around and adjacent to the building. Also, treat the building foundation to a height of 2 to 3 feet where pests are active and may find entrance. Apply as a coarse spray to thoroughly and uniformly wet the band area, using 1 gallon of spray mix per 400 square feet.

Note: Do not use water base sprays of DEMON in conduits, motor housings, junction boxes, switch boxes, or other electrical equipment because of possible shock hazard. Let surfaces dry before allowing children and pets to contact surfaces. Protect aquariums from spray mist. Thoroughly wash out sprayer and screen with water and detergent before using DEMON WP insecticide. Do not use this product with oil.

Do not apply this product to edible crops.

STORAGE AND DISPOSAL

Prohibitions: Do not contaminate water, food or feed by storage or disposal. Do not reuse empty container.

Storage: Mix as needed. Do not store diluted material. Do not store near food or feed. In case of spill on floor or paved surfaces, sweep up and remove to chemical waste area. Keep water soluble packets in original foil package and store in a cool dry place. Do not store below 32°F. Handle carefully when stored at temperatures less than 50°F to avoid breakage of packets.

Pesticide Disposal: Wastes resulting from the use of this product may be disposed of on site or at an approved waste disposal facility.

Container Disposal: When all packets are used dispose of foil package in a sanitary landfill, by incineration or, if allowed by state and local authorities, by burning. If burned stay out of smoke.

NOTICE TO BUYER AND USER

Seller warrants that this product conforms to the chemical description on the label and is reasonably fit for the purposes stated on the label when used in accordance with directions under normal conditions of use. This warranty does not extend to the use of this product contrary to label instructions, or under abnormal use conditions, or under conditions not reasonably foreseeable to Seller, and Buyer and User assume the risk of any such use. SELLER DISCLAIMS ALL OTHER WARRANTIES, EXPRESSED OR IMPLIED, INCLUDING ANY WARRANTY OF FITNESS OR MERCHANTABILITY. SELLER SHALL NOT BE LIABLE FOR CONSEQUENTIAL, SPECIAL OR INDIRECT DAMAGES RESULTING FROM THE USE OR HANDLING OF THIS PRODUCT AND SELLER'S SOLE LIABILITY AND BUYER'S AND USER'S EXCLUSIVE REMEDY SHALL BE LIMITED TO THE REFUND OF THE PURCHASE PRICE.

05-0864-060 R2 A-2R

The information and specimen labels in this booklet are correct as of the publication date but are subject to change without notice. For current information, contact ICI Americas Inc.

DEMON® WP

INSECTICIDE IN WATER-SOLUBLE 4 OZ. PACKETS

Crack and Crevice and Spot Spray for Cockroach Control. Only for Sale to, Use and Storage by Professional Pest Control Operators.

KEEP OUT OF REACH OF CHILDREN
WARNING

See Below for Additional Precautionary Statements and Statement of Practical Treatment.

EPA Reg. No. 10182-100

EPA Est. No. 33967-NJ-1

Net Contents Four/4 oz (113.4 gm) packets

ACTIVE INGREDIENT:
Cypermethrin
(±)α-cyano-(3-phenoxyphenyl)methyl
(±)-*cis, trans*-3-(2,2-dichloroethenyl)-2,
2-dimethylcyclopropanecarboxylate* 35.6%
INERT INGREDIENTS 64.4%

TOTAL 100.0%

**Cis/trans* ratio: 45/55 ± 10

PRECAUTIONARY STATEMENTS

HAZARDS TO HUMANS AND DOMESTIC ANIMALS

WARNING

HARMFUL IF SWALLOWED OR ABSORBED THROUGH SKIN. CAUSES EYE IRRITATION—DO NOT GET IN EYES, ON SKIN OR ON CLOTHING.

Avoid contact with eyes, skin or clothing. Avoid breathing vapor or spray mist. Wash thoroughly with soap and water after handling. May cause allergic skin reactions.

STATEMENT OF PRACTICAL TREATMENT

IF IN EYES: Flush with plenty of water. Get medical attention if irritation persists.

IF ON SKIN: Wash with plenty of soap and water.

IF SWALLOWED: Drink 1 or 2 glasses of water and induce vomiting by touching back of throat with finger.

IF INHALED: Remove victim to fresh air. Apply artificial respiration if necessary.

For 24-hour emergency assistance, call ICI Americas Medical Emergency Information Center 1-800-327-8633.

ENVIRONMENTAL HAZARDS

This product is extremely toxic to fish. Do not contaminate water by cleaning of equipment or disposal of wastes. Do not apply directly to any body of water. Apply this product only as specified on this label. Care should be used when spraying to avoid fish and reptile pets.

In case of a significant spill, call CHEMTREC 800/424-9300.

DIRECTIONS FOR USE

It is a violation of Federal law to use this product in a manner inconsistent with its labeling.

For residual pest control in and on buildings and structures and their immediate surroundings and on modes of transport. Permitted areas of use include industrial buildings, houses, apartment buildings, laboratories, buses, greenhouses, and the nonfood/feed areas of stores, warehouses, vessels, railcars, trucks, trailers, aircraft, schools, nursing homes, hospitals, restaurants, hotels and food manufacturing, processing and servicing establishments.

Authorized by the USDA for use in inedible product areas in Federally inspected meat, poultry, and egg processing plants.

Do not allow packets to become wet before adding to the spray tank. Wear dry rubber gloves when handling packets. Minimize handling of packets since excessive handling may cause rupture.

DEMON WP insecticide is intended for dilution with water for spray application. Do not use strainers finer than 50 mesh size. Fill sprayer to 1/2 of the desired volume. Add water-soluble packet(s) of DEMON WP insecticide per the table below. Add remaining water needed and wait for 2 minutes for packet(s) to dissolve. Close and shake for 30 seconds before use. Shake again before use if spraying is interrupted. Make up only as required, repeat treatments when needed.

Pests		Concentration of Active Ingredient	Dilution Rate
Ants Flies* Centipedes Crockroaches (Maintenance) Crickets	Firebrats Millipedes Pillbugs Silverfish Sowbugs	.1%	1 packet** per 12 gal. water
Bees Cockroaches (Clean-out) Spiders Wasps	Fire Ant (Mounds)† Carpenter Ants† Sod Webworms†	.2%	2 packets** per 12 gal. water

*Not for indoor use in California for fly control.
**1 packet = 4 ounces of DEMON WP.
†Not registered in California for use to control this insect.

Cockroaches, Spiders, Crickets, Silverfish and Firebrats: Apply as a coarse, low pressure spray to areas where these pests hide, such as baseboards, corners, storage areas, closets, around water pipes, doors and windows, attics and eaves, behind and under refrigerators, cabinets, sinks, furnaces, and stoves, the underside of shelves, drawers and similar areas. Pay particular attention to cracks and crevices; also see OUTDOOR USE.

Ants: Apply to any trails, around doors and windows and other places where ants may be found. Apply barrier treatments to prevent infestation as described below; also see OUTDOOR USE.

Wasps and Bees: Application to nests should be made late in evening when insects are at rest. Thoroughly spray nest and entrance and surrounding areas where insects alight, also see OUTDOOR USE.

Sowbugs, Pillbugs, Millipedes and Centipedes: Apply around doors and windows and other places where these pests may be found or where they may enter premises. Spray baseboards, storage areas and other locations. Apply barrier treatments to prevent infestation as described below; also see OUTDOOR USE.

Flies*: Apply directly to walls, ceilings, window screens, and other fly resting areas as a residual surface treatment. May be used inside residential buildings as well as in and around carports, garages, and storage sheds. Also see OUTDOOR USE.

* Not for indoor use in California for fly control.

Fire Ants (Mounds)†: For individual fire ant mounds, apply DEMON WP as a drench. Gently sprinkle (watering can) a total of one gallon over the surface of mounds up to 12 inches in diameter and two gallons for larger mounds. Thoroughly wet mound and surrounding area. Treat new mounds as they appear. Pressurized sprays may disturb the ants and cause migration. If possible, applications should be made on warm days after recent rainfall. All colonies in the vicinity should be treated, including those that have not yet constructed a mound.

Carpenter Ants†: Apply as a coarse wet spray to the point of runoff around doors, door frames, windows and other areas where these ants enter premises or hide; also see OUTDOOR USE.

Sod Webworms†: Spray when pests first appear; retreat when needed. Delay watering or mowing the treated area for 12 to 24 hours after treatment. Use a minimum of five gallons spray mix per 1000 square feet when spraying.

† Not registered for use in California.

Food Handling Establishments: Places other than private residences in which food is held, processed, prepared or served:

Nonfood Areas: DEMON WP insecticide may be used as a general spot or crack and crevice treatment in nonfood areas. All areas where insects hide or through which insects may enter should be treated. Examples of nonfood areas include garbage rooms, lavatories, floor drains (to sewers), entries and vestibules, offices, locker rooms, machine rooms, boiler rooms, garages, mop closets and storage (after canning or bottling).

Food Areas: DEMON WP is not labeled for use in food areas. Do not use in edible product areas of food processing plants, restaurants or other areas where food is commercially prepared or processed. Do not use in serving areas while food is exposed. Examples of food areas are areas for receiving, storage, packing (canning, bottling, wrapping, boxing) preparing, edible waste storage and enclosed processing systems (mills, dairies, edible oils, syrups). Serving areas (when food is exposed and facility is in operation) also would be considered a food area.

Outdoor Use: For control of ants, bees, centipedes, cockroaches, crickets, millipedes, sowbugs, pillbugs, spiders, flies and wasps. Apply as a residual spray to surfaces of buildings, porches, screens, window frames, eaves, patios, lawns, refuse dumps, garages in either hand or power sprayer and in other areas where these pests are found.

Barrier Treatments (except cockroaches): To help prevent infestation of buildings, apply to a band of soil and vegetation 6 to 10 feet wide around and adjacent to the building. Also, treat the building foundation to a height of 2 to 3 feet where pests are active and may find entrance. Apply as a coarse spray to thoroughly and uniformly wet the band area, using 1 gallon of spray mix per 400 square feet.

Note: Do not use water base sprays of DEMON in conduits, motor housings, junction boxes, switch boxes, or other electrical equipment because of possible shock hazard. Let surfaces dry before allowing children and pets to contact surfaces. Protect aquariums from spray mist. Thoroughly wash out sprayer and screen with water and detergent before using DEMON WP insecticide. Do not use this product with oil.

Do not apply this product to edible crops.

STORAGE AND DISPOSAL

Prohibitions: Do not contaminate water, food or feed by storage or disposal. Do not reuse empty container.

Storage: Mix as needed. Do not store diluted material. Do not store near food or feed. In case of spill or leak on floor or paved surfaces, sweep up and remove to chemical waste area. Keep water soluble packets in original foil package and store in a cool dry place. Do not store below 32°F. Handle carefully when stored at temperatures less than 50°F to avoid breakage of packets.

Pesticide Disposal: Wastes resulting from the use of this product may be disposed of on site or at an approved waste disposal facility.

Container Disposal: When all packets are used dispose of foil package in a sanitary landfill, by incineration, or, if allowed by state and local authorities, by burning. If burned, stay out of smoke.

NOTICE TO BUYER AND USER

Seller warrants that this product conforms to the chemical description on the label and is reasonably fit for the purposes stated on the label when used in accordance with directions under normal conditions of use. This warranty does not extend to the use of this product contrary to label instructions, or under abnormal use conditions, or under conditions not reasonably foreseeable to Seller, and Buyer and User assume the risk of any such use. SELLER DISCLAIMS ALL OTHER WARRANTIES, EXPRESSED OR IMPLIED, INCLUDING ANY WARRANTY OF FITNESS OR MERCHANTABILITY. SELLER SHALL NOT BE LIABLE FOR CONSEQUENTIAL, SPECIAL OR INDIRECT DAMAGES RESULTING FROM THE USE OR HANDLING OF THIS PRODUCT AND SELLER'S SOLE LIABILITY AND BUYER'S AND USER'S EXCLUSIVE REMEDY SHALL BE LIMITED TO THE REFUND OF THE PURCHASE PRICE.

A-1R

05-0864-061 R1

The information and specimen labels in this booklet are correct as of the publication date but are subject to change without notice. For current information, contact ICI Americas Inc.

DEVRINOL® 10-G

SELECTIVE HERBICIDE-GRANULES

For Use On Certain Citrus, Nuts, Pome Fruits, Small Fruits, Stone Fruits, Strawberries and Grapes

ACTIVE INGREDIENT:	
N,N-diethyl-2-(1-naphthalenyloxy)-propionamide	10.0%
INERT INGREDIENTS:	90.0%
TOTAL	100.0%

Contains 10 lbs. of active ingredient per 100 pounds.

EPA Reg. No. 10182-253

KEEP OUT OF REACH OF CHILDREN
CAUTION

STATEMENT OF PRACTICAL TREATMENT

First Aid: If an adverse reaction occurs after exposure, promptly start the recommended procedures below. Seek medical attention if further treatment is required.

In case of contact: Flush eyes and skin with plenty of water for several minutes. Get medical attention if eye or skin irritation occurs.

If swallowed: Immediately give several glasses of water and induce vomiting by gagging the victim with a finger placed on the back of the victim's tongue. Give fluids until vomitus is clear. If victim is unconscious or convulsing, do not induce vomiting or give anything by mouth.

If inhaled: Remove to fresh air. Seek medical attention if respiratory irritation occurs or if breathing becomes difficult.

FOR 24-HOUR EMERGENCY ASSISTANCE CALL 1-800-F-A-S-T-M-E-D (327-8633).
FOR CHEMICAL EMERGENCY: Spill, leak, fire, exposure, or accident call CHEMTREC 1-800-424-9300.

PRECAUTIONARY STATEMENTS

HAZARDS TO HUMANS AND DOMESTIC ANIMALS

CAUTION

MAY BE HARMFUL IF SWALLOWED. Avoid contact with eyes, skin and clothing. Avoid inhalation of dust.

ENVIRONMENTAL HAZARDS

Do not apply to any body of water. Do not contaminate water by cleaning of equipment or disposal of wastes. Do not apply when weather conditions favor drift from areas treated. Do not contaminate irrigation water or water used for domestic purposes.

GENERAL PRECAUTIONS

- Read all label directions before using.
- Use only for recommended purposes and at recommended rates. Do not overdose.
- DEVRINOL 10-G is recommended for use on mineral soils only. Not recommended for use on soils with over 10% organic matter.
- Treating ground heavily covered with leaves or trash may adversely affect weed control.
- Do not apply to frozen ground.
- To avoid injury to crops not specified on this label, do not plant until 12 months after the last DEVRINOL 10-G application. A deep disc plow or moldboard plow operation must be carried out before planting.

GENERAL DIRECTIONS FOR USE

It is a violation of Federal law to use this product in a manner inconsistent with its labeling.

DEVRINOL 10-G is a selective surface-applied herbicide for use in certain types of newly planted or established crops. DEVRINOL 10-G does not control established weeds. All weed growth and crop stubble in the fields must be thoroughly worked into the soil prior to planting. Application equipment should be carefully calibrated before use and checked frequently during application to be sure equipment is free from clogging and is delivering a uniform pattern. Apply the recommended rate of DEVRINOL 10-G by ground or air equipment.

NOTE: All rates are given on a broadcast basis, per acre. Reduce rates proportionately for band or strip treatment.

Weeds Controlled by DEVRINOL 10-G

NOTE: DEVRINOL 10-G does not control established weeds.

A surface application of 40 lbs. DEVRINOL 10-G will control the following weeds:

Annual Grasses:	Annual Broadleaf Weeds:
Barnyardgrass, Junglerice (*Echinochloa* spp.)	Carpetweed (*Mollugo verticillata*)
Wild Barley (*Hordeum leporinum*)	Common Chickweed (*Stellaria media*)
Annual Bluegrass (*Poa annua*)	Coast Fiddleneck (*Amsinckia*)
Bromes (*Bromus*)	

Cheatgrass *(B. secalinus)*
Downy Brome *(B. tectorum)*
Ripgut Brome *(B. rigidus)*
Soft Chess *(B. mollis)*
Large (Hairy) Crabgrass *(Digitaria sanguinalis)*
Smooth Crabgrass *(Digitaria ischaemum)*
Southwestern Crabgrass *(Eriochloa gracilis)*
Foxtails *(Setaria)*
Bristly Foxtail *(S. verticilla)*
Giant Foxtail *(S. faberi)*
Green Foxtail *(S. viridis)*
Yellow Foxtail *(S. lutescens)*
Goosegrass *(Eleusine indica)*
Seedling Johnsongrass *(Sorghum halepense)*
Wild Oats *(Avena fatua)*
Panicums *(Panicum)*
Guineagrass *(P. maximum)*
Fall Panicum *(P. dichotomiflorum)*
Texas Panicum *(P. texanum)*
Witchgrass *(P. capillare)*
Italian Ryegrass *(Lolium miltiflorum)*
Sandbur (*Cenchrus* spp.)
Red Sprangletop *(Leptochloa filiformis)*
Stinkgrass *(Eragrostis cilianensis)*
Filaree (*Erodium* spp.)
Groundsel (*Senecio* spp.)
Prostrate Knotweed *(Polygonum aviculare)*
Lambsquarters (*Chenopodium* album)
Prickly Lettuce *(Lactuca serriola)*
Little Mallow* *Malva parviflora)*
Redroot Pigweed *(Amaranthus retroflexus)*
Pineappleweed *(Matricaria matricarioides)*
Common Purslane *(Portulaca oleracea)*
Horse Purslane *(Trianthema postullacastrum)*
Common Ragweed** *(Ambrosia artemisiieolia)*
Annual Sowthistle *(Sonchus oleraceus)*
* From Seed
** Suppression

NOTE: See Recommendations Section for weeds controlled in Cranberries.

DIRECTIONS FOR USE

For control of all weeds listed, apply 40 pounds of DEVRINOL 10-G per acre to a weed free soil surface. Apply in the fall through early spring prior to weed emergence. For best results, the treatment must reach the zone of weed seed germination. If rainfall does not occur the treatment must be shallowly incorporated or irrigated-in (sprinkler or flood) following application with sufficient water to wet the soil to a depth of 2 to 4 inches. In the Western and Pacific Northwest Regions treatments made in fall through early spring must be incorporated or irrigated-in if no rainfall occurs within two weeks of application. For treatments made other than in fall through early spring in these Regions, and for all treatments applied in the Northern, Southwestern and Southeastern Regions regardless of season, the treatment must be incorporated or irrigated-in within 24 hours if no rainfall occurs.

Avoid contact with fruit or foliage. Do not apply when fruit or nuts are on the ground during the harvest period. Do not graze treated areas.

Recommendations

Application may be made immediately after planting, except where treatment is restricted to established crops only. For these uses, apply to stands that have been established in the field for at least one year.

Trees and Vines: Apples, Apricots, Cherries, Filberts, Grapefruit, Grapes, Lemons, Nectarines, Oranges, Peaches, Pears, Pecans, Plums, Prunes, Tangerines and Walnuts. Apply 40 pounds DEVRINOL 10-G per acre to a weed free soil surface.

Strawberries: Apply 40 pounds DEVRINOL 10-G per acre to a weed free soil surface. Delay application until the desired number of daughter plants have become established. Do not apply from bloom through harvest.

Bush Fruit: Blueberries and Caneberries (Blackberries, Raspberries, Boysenberries, Loganberries). Apply 40 pounds DEVRINOL 10-G per acre to a weed free soil surface.

Cranberries: CAUTION: Do not apply in the springtime to bogs that do not have sprinkler systems for frost control.

Massachusetts (Established Bogs Only): For control of nutsedge (Cyperus dentatus) and Rice Cutgrass (Leersia oryzoides), apply 60-90 pounds DEVRINOL 10-G per acre to a weed free soil surface before spring growth begins or in the fall after harvest.

New Jersey (Established Bogs Only): For control of nutsedge (Cyperus dentatus), Rice Cutgrass (Leersia oryzoides) and Purplestem Beggarticks (Bidens cannata), apply 60-90 pounds DEVRINOL 10-G per acre to a weed free soil surface before spring growth begins or in the fall after harvest.

Oregon and Washington (Established Bogs Only): For control of Clover (Trifolium fimbriatum), Aster (Aster douglasii), Silverweed Cinquefoil* (Silverleaf) (Potentilla anserina), Red Sorrel (Sheep Sorrel) (Rumex acetosella) and Birdsfoot Trefoil (Lotus corniculatus):

In peat beds: Apply 150 pounds DEVRINOL 10-G per acre to a weed free soil surface before spring growth begins in Mid-February or before the end of March. Do not apply after buds start to swell.

In sand beds: Apply 40-90 pounds DEVRINOL 10-G per acre to a weed free soil surface before spring growth begins in Mid-February or before the end of March. Use the lower rate on one-year old beds only. Do not apply after buds start to swell. For further information, consult your local Ag Extension Service.

* Suppression only.

Wisconsin: For control of Purplestem Beggarticks (Bidens connata) and Rice Cutgrass (Leersia oryzoides) in:

New Plantings: Apply 30 pounds DEVRINOL 10-G per acre to a weed free soil surface in the spring after setting vines.

Established Plantings: Apply 60 pounds DEVRINOL 10-G per acre to a weed free soil surface before spring growth begins or in the fall after harvest. Do not apply when bogs are under winter flood.

Mint: Established mint (Pacific Northwest only). Apply 40 pounds DEVRINOL 10-G per acre to a weed free soil surface.

STORAGE AND DISPOSAL

Prohibitions: Do not contaminate water, food or feed by storage or disposal.

Storage: Keep container closed when not in use. Store in a cool dry place. Do not store near seeds, fertilizer or foodstuffs.

Pesticide Disposal: Wastes resulting from the use of this product may be disposed of on site or at an approved waste disposal facility.

Container Disposal: Completely empty bag into application equipment. Then dispose of empty bag in a sanitary landfill or by incineration, or, if allowed by State and local authorities, by burning. If burned, stay out of smoke.

NOTICE TO BUYER AND USER:

Seller warrants that this product conforms to the chemical description on the label and is reasonably fit for the purposes stated on the label when used in accordance with directions under normal conditions of use. This warranty does not extend to the use of this product contrary to label instructions, or under abnormal use conditions, or under conditions not reasonably foreseeable to Seller, and Buyer and User assume the risk of any such use. SELLER DISCLAIMS ALL OTHER WARRANTIES EXPRESSED OR IMPLIED INCLUDING ANY WARRANTY OF FITNESS OR MERCHANTABILITY. SELLER SHALL NOT BE LIABLE FOR CONSEQUENTIAL, SPECIAL OR INDIRECT DAMAGES RESULTING FROM THE USE OR HANDLING OF THIS PRODUCT AND SELLER'S SOLE LIABILITY AND BUYER'S AND USER'S EXCLUSIVE REMEDY SHALL BE LIMITED TO THE REFUND OF THE PURCHASE PRICE.

This product is sold only for uses stated on its label. No express or implied license is granted to use or sell this product under any patent in any country except as specified: Country: United States Patent No.: 3,480,671; 3,718,455.

A-3R

The information and specimen labels in this booklet are correct as of the publication date but are subject to change without notice. For current information, contact ICI Americas Inc.

DEVRINOL® 2-E

SELECTIVE HERBICIDE — EMULSIFIABLE LIQUID

For use on Tomatoes, Peppers, Strawberries and Tobacco

ACTIVE INGREDIENT:	
N,N-diethyl-2-(1-naphthalenyloxy)-propionamide	21.8%
INERT INGREDIENTS:	78.2%
TOTAL	100.0%

Contains 2 pounds active ingredient per gallon

EPA Reg. No. 10182-219

KEEP OUT OF REACH OF CHILDREN

DANGER

- Corrosive • Causes eye damage and skin irritation.
- Do not get in eyes, on skin, or on clothing.
- Wear goggles or face shield and rubber gloves when handling.
- Harmful if swallowed or absorbed through the skin.
- Avoid contamination of food or feed.
- Do not use, store, spill, or pour near heat or open flame.

FIRST AID

CALL A POISON CENTER OR A PHYSICIAN IMMEDIATELY

If a known exposure occurs or is suspected, immediately start the procedures given below. Simultaneously contact a Poison Center, a physician or the nearest hospital. NOTE: Be sure to advise the person contacted that the product contains a Hydrocarbon Solvent. Describe the type of exposure, the victim's symptoms, and follow the advice given.

If swallowed, immediately give several glasses of water but *do not* induce vomiting. If vomiting does occur, give fluids again. Have a physician determine if condition of patient will permit induction of vomiting or evacuation of stomach. Do not give anything by mouth to an unconscious or convulsing person.

In case of contact, immediately flush eyes and skin with large amounts of running water for at least 15 minutes while removing contaminated clothing and shoes. For eyes, hold eyelids apart during the flushing to ensure rinsing of the entire surface of the eye and lids with water. Get medical attention immediately. Wash clothing before reuse.

If inhaled, remove to fresh air. If victim is breathing, oxygen may be given from a demand-type or continous-flow inhaler, preferably with a physician's advice. If not breathing, clear victim's airway and start mouth-to-mouth artificial respiration. Get medical attention immediately.

FOR 24-HOUR EMERGENCY MEDICAL ASSISTANCE CALL 1-800-F-A-S-T-M-E-D (327-8633).
FOR CHEMICAL EMERGENCY: Spill, leak, fire, exposure, or accident call CHEMTREC 1-800-424-9300.

PRECAUTIONARY STATEMENTS

HAZARDS TO HUMANS & DOMESTIC ANIMALS

DANGER

Corrosive Causes Eye Damage and Skin Irritation

- Do not get in eyes, on skin, or on clothing.
- Wear goggles or face shield and rubber gloves when handling.
- Harmful if swallowed or absorbed through the skin.
- Avoid contamination of food or feed.
- Do not use, store, spill, or pour near heat or open flame.

ENVIRONMENTAL HAZARDS

- Do not apply directly to water or wetlands.
- Do not apply when weather conditions favor drift from areas treated.
- Do not contaminate water by cleaning of equipment or disposal of wastes. Apply this product as specified on this label.
- Do not contaminate water used for domestic purposes.
- Avoid contamination of food and feed.

GENERAL PRECAUTIONS

- Read all label directions before using. Use only for recommended purposes at recommended rates. Do not overdose.
- Do not contaminate irrigation water or water used for domestic purposes.
- Do not store near seeds or fertilizer.
- Do not apply more than once per season.
- Do not use with PVC plastic fittings, nozzles, couplers and/or tubing.
- DEVRINOL is recommended for use on mineral soils only. Not recommended for use on soils with over 10% organic matter.
- Spraying on ground heavily covered with leaves or trash may adversely affect weed control.

DIRECTIONS FOR USE

It is a violation of Federal law to use this product in a manner inconsistent with its labeling.

Do not apply this product through any type of irrigation system.

DEVRINOL 2-E is a selective, surface-applied herbicide for use in strawberries. It is also recommended for weed control in certain vegetables and tobacco, applied and incorporated prior to planting.

DEVRINOL 2-E controls weeds before they emerge. It does not control established weeds. All weed growth and crop stubble in fields intended for vegetables or tobacco must be thoroughly worked into the soil prior to planting.

Spray equipment should be carefully calibrated before use and checked frequently during application to be sure nozzles are free from clogging and are delivering a uniform spray pattern. Apply the recommended rate of DEVRINOL 2-E in 20 to 100 gallons of water per acre.

STRAWBERRIES

Weeds Controlled

NOTE: DEVRINOL does not control established weeds.

Annual Grasses:	Annual Broadleaf Weeds:
Annual Bluegrass *(Poa annua)*	Chickweed *(Stellaria media)*
Barnyardgrass-Watergrass *(Enchinochloa* spp.)	Common Fiddleneck *(Amsinckia intermedia)*
	Common Knotweed *(Polygonum aviculare)*
Bristly Foxtail *(Setaria verticilla)*	Common Purslane *(Portulaca oleracea)*
Large (Hairy) Crabgrass *(Digitaria sanguinalis)*	Common Sowthistle *(Sonchus oleraceus)*
	Filaree *(Erodium* spp.)
Ripgut Brome *(Bromus rigidus)*	Groundsel *(Senecio* spp.)
Southwestern Cupgrass *(Eriochloa gracilis)*	Little Mallow* *(Malva parviflora)*
Wild Barley *(Hordeum leporinum)*	Lambsquarters *(Chenopodium album*
Wild Oats *(Avena fatua)*	Pineappleweed *(Matricaria matricarioides)*
	Prickley Lettuce *(Lactuca scariola)*
	Redroot Pigweed *(Amaranthus retroflexus)*

*From seed.

Recommendations

NOTE: All rates given on a broadcast basis. Reduce rates proportionately for band or strip treatment.

Strawberries (Post Transplant and Established):
For control of all weeds listed, apply 2 gallons DEVRINOL 2-E per broadcast acre to the soil surface. Apply in the fall through early spring, prior to weed emergence. Do not apply DEVRINOL 2-E to frozen ground. Do not apply DEVRINOL from bloom through harvest. For best results, the treatment must reach the zone of weed seed germination. Treatments made other than in November through February must be irrigated-in within 24 hours (sprinkler or flood) with sufficient water to wet the soil to a depth of at least 2 to 4 inches. Irrigation or shallow incorporation is recommended for treatments made from November through February if no rainfall occurs within one week of application.

VEGETABLES & TOBACCO

Weeds Controlled

NOTE: DEVRINOL 2-E does not control established weeds.

Annual Grasses:	Annual Broadleaf Weeds:
Barnyardgrass *(Echinochloa* spp.)	Carpetweed *(Mollugo verticillata)*
Crabgrass *(Digitaria* spp.)	Common Purslane *(Portulaca oleracea)*
Fall Panicum *(Panicum dichotomiflorum)*	Lambsquarters *(Chenopodium album)*
Goosegrass *(Eleusine indica)*	
Seedling Johnsongrass *(Sorghum halepense)*	Redroot Pigweed *(Amaranthus retroflexus)*

DIRECTIONS FOR USE

On flat-planted or bedded crops apply, before planting, to well worked soil that is dry enough to permit thorough incorporation to a depth of 1 to 2 inches. Incorporate the same day as applied, using equipment which will result in uniform incorporation of the herbicide to the desired depth.

After harvest or prior to planting of succeeding crops, a deep moldboard or disc plowing operation must be carried out.

To avoid injury, do not seed to alfalfa, small grains, sorghum, corn, lettuce of sugarbeets for 12 months after DEVRINOL 2-E application.

Recommendations

NOTE: All rates are given on a broadcast basis. Reduce rates proportionately for band or strip treatment.

Tomatoes (Direct Seeded and Transplanted):
Apply 1/2 to 1 gallon DEVRINOL 2-E per broadcast acre. Use the lower rate on light soil (coarse textured-sandy) and the higher rate on heavy soil (fine textured-clay).

Peppers (Direct Seeded and Transplanted):
Apply 1/2 to 1 gallon DEVRINOL 2-E per broadcast acre. Use the lower rate on light soil (coarse textured-sandy) and the higher rate on heavy soil (fine textured-clay).

Tobacco (Transplanted):
Before transplanting, apply 1/2 to 1 gallon DEVRINOL 2-E per broadcast acre. Use the lower rate on light soils (coarse textured-sandy) and the higher rate on heavy soil (fine textured-clay).

DEVRINOL 2-E/TILLAM® 6-E
Tank Mix Combination

A tank mix combination of DEVRINOL 2-E and TILLAM 6-E can be applied and mixed (incorporated) into the soil before planting for weed control in tomatoes and tobacco.

IMPORTANT:
Observe all cautions and limitations on labeling of all products used in combination.

Make only one application per season.

In each case where a range of rates is given, the lower rates should be used on light (coarse) textured soils and soils low in organic matter. The higher rates should be used on heavy (fine) textured soils.

After harvest or prior to planting of succeeding crops, a deep moldboard or disc plowing operation must be carried out.

To avoid injury do not seed to alfalfa, small grains, sorghum, corn, lettuce or sugarbeets for 12 months after application of the DEVRINOL 2-E/ TILLAM 6-E tank mix.

Weeds Controlled

The tank mix is recommended for control of the following weeds:

Annual Grasses:	Annual Broadleaf Weeds:
Barnyardgrass-Watergrass (*Echinochloa* spp.)	Blackeyed Susan; Sweet Suzy *(Rudbeckia serotina)*
Bermudagrass *(Cynodon dactylon)*	Common Purslane *(Portulaca oleracea)*
Crabgrass (*Digitaria* spp.)	Deadnettle; Henbit *(Lamium amplexicaule)*
Foxtail-yellow, giant, green (*Setaria* spp.)	Florida Pusley *(Richardia scabra)*
	Hairy Nighshade *(Solanum villosum)*
Goosegrass *(Eleusine indica)*	Lambsquarters *(Chenopodium album)*
Millet (*Setaria* spp.)	Nettleleaf Goosefoot *(Chenopodium murale)*
Signalgrass *(Brachiaria piligera)*	Prostrate Pigweed *(Amaranthus graecizans)*
Wild Oats *(Avena fatua)*	Redroot Pigweed *(Amaranthus retroflexus)*
	Shepherdspurse *(Capsella bursapastoris)*

PERENNIAL WEEDS:

Yellow Nutsedge *(Cyperus escultentus)*
Purple Nutsedge *(Cyperus rotundus)*

Existing stands of nutsedge must be turned under and chopped up thoroughly before treatment.

Directions For Use

The recommended tank mix controls weeds by interfering with normal seed germination and seedling development. It does not control established weeds. All weed growth and crop stubble should be thoroughly worked into the soil before treatment.

Tank Mixing Directions:

Be sure the sprayer tank, boom and nozzle have been thoroughly flushed with clean water if another pesticide has been used previously.

Sufficient jet or mechanical agitation should be provided during the filling operation and during application. Fill the spray tank at least one-fourth full of clean water and while pump and agitator are running, add the recommended amounts of the herbicides to the tank. Fill the tank with water to the desired level. Do not allow the mixture to stand in the spray tank without agitation.

Allow the pump to run several minutes before application to insure proper suspension and mixing. The tank-mix should not be left in the spray tank for prolonged periods; batches should be mixed and applied the same day.

Application Directions:

The soil should be well-worked and dry enough to permit good soil mixing (incorporation). Pour the recommended amounts of DEVRINOL 2-E and TILLAM 6-E into the spray tank during the filling operations. Apply in 25 to 50 gallons of water per acre.

For soil incorporated application, use a properly calibrated, low-pressure, boom-type sprayer having good agitation.

In irrigated areas, do not apply before pre-irrigation. For all band applications, reduce dosage proportionately depending upon row spacing and band to be treated. Do not use on rocky ground because thorough incorporation is not possible.

Incorporation Directions:

The herbicides must be mixed (incorporated) into the soil IMMEDIATELY after application to prevent loss by evaporation. Whenever possible, application and incorporation should be done in the same operation.

Incorporation Before Planting:

Use power-driven cultivation equipment set to cut to a depth of two to three inches in coarse soils and three to four inches in heavy soils OR a tandem disc set to cut to a depth of six inches, followed by a spike-toothed harrow or some other leveling device which extends beyond ends of the disc. For thorough mixing, disc in two different directions (cross disc).

Incorporation At Planting:

Use power-driven cultivation equipment set to cut to a depth of two to three inches OR, on coarse soils only, ground-driven tillers (rolling cultivators, rotary hoe, row wheels, etc.) set to cut to a depth of 1½ inches and operated at 6 to 8 mph. When using ground-driven tillers, precultivate if soil is crusted.

Planting:

Crops should be planted or seeded immediately after application and incorporation. Do not use a drag behind the planter as it may concentrate the herbicides over the seed row and cause crop injury.

Cultural Practices Following Application:

Shallow cultivation may be necessary to control those susceptible weeds that escape control as well as those weeds not susceptible to either herbicide. Do not cultivate deeper than the depth of herbicide incorporation.

When cultivating fields where the tank mix has been banded in the row, use shields to prevent the movement of untreated soil into the treated row.

Recommendations

Tomatoes (Direct Seeded (California Only) and Transplant):
Apply ½ to 1 gallon DEVRINOL 2-E plus 2⅔ to 4 quarts TILLAM 6-E per broadcast acre. Use the lower rates on light soils (coarse textured-sandy) and the higher rate on heavy soils (fine textured-clay). Do not use hotcaps on transplanted tomatoes.

Tobacco (Transplanted):
Before transplanting, apply ½ gallon DEVRINOL 2-E plus 2⅔ quarts TILLAM 6-E per broadcast acre.

DEVRINOL 2-E/TILLAM 6-E/DYFONATE 4-EC
Tank Mix Combination

A tank mix combination of DEVRINOL 2-E, DYFONATE 4-EC and TILLAM 6-E can be applied and mixed into the soil before planting tomatoes or tobacco, for control of weeds and soil insects. Refer to DEVRINOL/TILLAM Tank Mix Combination for list of weeds controlled. Refer to the DYFONATE 4-EC label for list of insects controlled.

NOTE: DYFONATE 4-EC is a Restricted Use Pesticide. Observe all cautions and limitations on labeling of all products used in combinations. Make only one application per season. Do not contaminate water by cleaning of equipment or disposal of wastes.

Directions for Use

The recommended tank mix controls weeds by interfering with normal seed germination and seedling development. It does not control established weeds. All weed growth and crop stubble should be thoroughly worked into the soil before treatment.

For GENERAL USE PRECAUTIONS and DIRECTIONS FOR USE refer to DEVRINOL 2-E/TILLAM 6-E Tank Mix Recommendations for methods of tank mixing, application and incorporation.

Recommendations

Tomatoes (Direct Seeded (California Only) and Transplanted): Apply ½ to 1 gallon DEVRINOL 2-E plus 2⅔ to 4 qts. TILLAM 6-E plus 2 qts. DYFONATE 4-EC for control of symphylans or 4 qts. for control of symphylans and wireworms in 25 to 50 gallons of water per broadcast acre. Use the lower DEVRINOL 2-E and TILLAM 6-E rates on light soils (coarse textured-sandy) and the higher rates on heavy soils (fine textured-clay). Do not use hot caps on transplanted tomatoes.

Tobacco (Transplanted): Before transplanting apply ½ gallon DEVRINOL 2-E plus 2⅔ qts. TILLAM 6-E plus 2 qts. DYFONATE 4-E for control of wireworms in 25 to 50 gallons of water per broadcast acre.

STORAGE AND DISPOSAL

Prohibitions: Do not contaminate water, food, or feed by storage or disposal. Open dumping is prohibited. Do not reuse container.

Storage: Store in a cool dry place. Do not store below 20°. Keep container closed when not in use.

Pesticide Disposal: Pesticide wastes are acutely hazardous. Improper disposal of excess pesticide, spray mixture or rinsate is a violation of Federal Law. If these wastes cannot be disposed of by use according to label instructions, contact your State Pesticide or Environmental Control Agency, or the Hazardous Waste representative at the nearest EPA Regional Office for guidance.

Container Disposal: Triple rinse (or equivalent). Then offer for recycling or reconditioning, or puncture and dispose of in a sanitary landfill, or by other procedures approved by State and local authorities.

NOTICE TO BUYER AND USER

Seller warrants that this product conforms to the chemical description on the label and is reasonably fit for the purposes stated on the label when used in accordance with directions under normal conditions of use. This warranty does not extend to the use of this product contrary to label instructions, or under abnormal use conditions, or under conditions not reasonably foreseeable to Seller, and Buyer and User assume the risk of any such use. SELLER DISCLAIMS ALL OTHER WARRANTIES EXPRESSED OR IMPLIED INCLUDING ANY WARRANTY OF FITNESS OR MERCHANTABILITY. SELLER SHALL NOT BE LIABLE FOR CONSEQUENTIAL, SPECIAL OR INDIRECT DAMAGES RESULTING FROM THE USE OR HANDLING OF THIS PRODUCT AND SELLER'S SOLE LIABILITY AND BUYER'S AND USER'S EXCLUSIVE REMEDY SHALL BE LIMITED TO THE REFUND OF THE PURCHASE PRICE.

This product is sold only for uses stated on its label. No express or implied license is granted to use or sell this product under any patent in any country except as specified:
Country: United States of America Patent No.: 3,480,671; 3,718,455.

A-12R

The information and specimen labels in this booklet are correct as of the publication date but are subject to change without notice. For current information, contact ICI Americas Inc.

DEVRINOL® 2-G

ORNAMENTAL SELECTIVE HERBICIDE

A Pre-emergence Herbicide for Weed Control in Home Flower Gardens and Landscape Plantings. Use in Deciduous and Evergreen Trees and Shrubs, Ground Covers, Perennial Flowers, Fruit and Nut Crops, Container Grown Plants, and Turfgrass

ACTIVE INGREDIENT:
N,N-diethyl-2-(1-naphthalenyloxy)-propionamide 2.0%
INERT INGREDIENTS: 98.0%
TOTAL 100.0%

Contains 2 Pounds Active Ingredient per 100 Pounds
EPA Reg. No. 10182-243

KEEP OUT OF REACH OF CHILDREN
CAUTION

PRECAUTIONARY STATEMENTS

HAZARDS TO HUMANS AND DOMESTIC ANIMALS

CAUTION

MAY BE HARMFUL IF SWALLOWED. Do not breathe dust. Avoid contact with skin, eyes and clothing. Wash thoroughly with soap and water after use.
- Avoid Contamination of Food and Feed.

ENVIRONMENTAL HAZARDS
- Do not apply directly to lakes, streams, and/or ponds.
- Do not contaminate irrigation water or water used for domestic purposes.

GENERAL PRECAUTIONS
- Read all instructions before using.
- Use only for recommended purposes and at recommended rates. Do not overdose.

STATEMENT OF PRACTICAL TREATMENT

If an adverse reaction occurs after exposure, promptly initiate the recommended procedures below. If further treatment is required, seek medical attention.

In case of contact: Flush eyes and skin with plenty of water for several minutes. Get medical attention if eye or skin irritation occurs.

If swallowed: Immediately give several glasses of water and induce vomiting by placing a finger on the back of the victim's tongue. Give fluids until vomitus is clear. Do not induce vomiting or give anything by mouth to an unconscious or convulsing person.

If inhaled: Remove to fresh air. Seek medical attention if respiratory irritation occurs or if breathing becomes difficult.

FOR 24-HOUR EMERGENCY MEDICAL ASSISTANCE CALL 1-800-F-A-S-T-M-E-D (327-8633).
FOR CHEMICAL EMERGENCY: Spill, leak, fire, exposure, or accident call CHEMTREC 1-800-424-9300.

For other product information, contact your local ICI Americas Distributor or see Material Safety Data Sheet.

WHAT THIS PRODUCT DOES

DEVRINOL 2-G Ornamental controls certain annual grassy and broadleaf weeds as they germinate. It does not control established weeds.

The weeds controlled are listed below:
Weeds Controlled by Devrinol 2-G

Annual Grasses:

Barnyardgrass, Junglerice (*Echinochloa* spp.)
Wild Barley (*Hordeum leporinum*)
Annual Bluegrass (*Poa annua*)
Bromes (*Bromus*)
 Cheatgrass (*B. secalinus*)
 (*Digitaria ischaemum*)
Southwestern Cupgrass (*Eriochloa gracilis*)
Foxtails (*Setaria*)
 Bristly Foxtail (*S. verticilla*) Giant Foxtail (*S. faberi*)
 Green Foxtail (*S. viridis*)
 Yellow Foxtail (*S. lutescens*)
Goosegrass (*Eleusine indica*)
Seedling Johnsongrass (*Sorghum halepense*)
 Downy Brome (*B. tectorum*)
 Ripgut Brome (*B. rigidus*)
 Soft Chess (*B. mollis*)
Large (Hairy) Crabgrass *Digitaria sanguinalis*)
Smooth Crabgrass
Wild Oats (*Avena fatua*)
Panicums (*Panicum*)
 Guineagrass (*P. Paximum*)
 Fall Panicum (*P. dichotomiflorum*)
 Texas Panicum (*P. texanum*)
 Witchgrass (*P. capillare*)
Italian Ryegrass (*Lolium multiflorum*)
Sandbur (*Cenchrus* spp.)
Red Sprangletop (*Leptochloa filiformis*)
Stinkgrass (*Eragrostis cilianensis*)

Annual Broadleaf Weeds:

Carpetweed (*Mollugo verticillata*)
Common Chickweed (*Stellaria media*)
Coast Fiddleneck (*Amsinckia intermedia*)
Filaree (*Erodium* spp.)
Groundsel (*Senecio* spp.)
Prostrate Knotweed (*Polygonum aviculare*)
Lambsquarters (*Chenopodium album*)
Prickly Lettuce (*Lactuca serriola*)
Little Mallow*(*Malva parviflora*)
Redroot Pigweed (*Amaranthus retroflexus*)
Pineappleweed (*Matricaria matricarioides*)
Common Purslane (*Portulaca oleracea*)
Horse Purslane (*Trianthema portullacastrum*)
Common Ragweed** (*Ambrosia artemisiifolia*)
Annual Sowthistle (*Sonchus oleraceus*)

** From seed*
*** Suppression*

WHERE TO USE THIS PRODUCT

DEVRINOL 2-G Ornamental can be applied over small ornamental plants or around larger plants in rose gardens, container plantings, ornamental landscapes and ground covers including dichondra.

Use DEVRINOL 2-G Ornamental for weed control in the following plants:

Deciduous and Evergreen Trees

Acer	Eucalyptus	Picea
Ash	Ficus	Pin Oak
Birch	Fir	Pinus
Black Walnut	Gleditsia	Pittosporum
Cedar	Hemlock	Podocarpus
Citrus	Japanese Larch	Populus
Crataegus	Laburnum	Prunus
Cypress	Malus	Pears
Dogwood	Olive (Olea)	Pistachios
Douglas fir	Palm (Washingtonia robusta)	

Deciduous and Evergreen Shrubs

Abelia	Forsythia	Oleander
Asparagus (Sprengeri)	Gardenia	Osmanthus
	Heather (Calluna)	Photinia
Aucuba	Hibiscus	Pittosporum
Azalea	Honeysuckle (Lonicera)	Podocarpus
Bottlebrush		Pyracantha
Bougainvillea	Hypericum	Raphiolepis
Buxus	Ilex	Rhododendron
Camellia	Juniperus	Rosa
Cotoneaster	Leucothe	Star Jasmine
Crepe Myrtle	Ligustrum	Taxus
Dogwood	Nandina	Viburnum
Euonymus	Natal Plum	Xylosema

Ground Covers

Ajuga	Gazania	Liriope
Carpobrotus	Hedera	Osteosperma

	Ground Covers	
Delosperma	Hypericum	Pachysandra
Dichondra	Iceplant	Sedum
Erysimum	Ivy	Vinca
	Lantana	

	Container Stock	
Abelia	Eucalyptus	Pachysandra
Agapanthus	Euonymus	Palm (Washing-
Ageratum	Fir	tonia robusta)
Asparagus	Forsythia	Photinia
Azalea	Gardenia	Pinus Spp.
Bottlebrush	Geranium	Pittosporum
Bougainvillea	Heather	Podocarpus
Buxus	Hedera	Pyracantha
Camellia	Hibiscus	Raphiolepis
Cedar	Ilex	Rhododendron
Cotoneaster	Juniperus	Rosa
Crepe Myrtle	Leucothe	Sedum
Cypress	Ligustrum	Star Jasmine
Dogwood	Liriope	Thuja
Douglas Fir	Nandina	Vinca
		Xylosema

	Flowers	
African Daisy	Dahlia	Narcissus
Aster	Daisy	Petunia
Bird of Paradise	Geranium	Plantain Lily
Chrysanthemum	Gladiolus	(Hasta)
		Zinnia

	Fruit & Nuts	
Almond	Lemon	Pistachio
Apple	Nectarine	Plum
Apricot	Orange	Prune
Cherry	Peach	Tangerine
Filbert	Pear	Walnut
Grape	Pecan	

Strawberry:

(Application should be delayed until the desired number of daughter plants have become established. Do not apply from bloom through harvest.)

WHEN TO USE THIS PRODUCT

DEVRINOL 2-G Ornamental should be used before weeds germinate. It will not control established weeds.

Application can be made anytime of the year to weed-free soil. Fall and winter applications will provide excellent weed control the following spring and summer.

In containers, apply after setting and watering the plants.

HOW TO USE THIS PRODUCT

It is a violation of Federal law to use this product in a manner inconsistent with its labeling.

Apply this product as specified on this label.

DEVRINOL 2-G Ornamental should be spread evenly on the soil surface using a shaker can or granular herbicide applicator.

Do not apply when weather conditions favor drift from areas treated.

When weed infestation is heavy or when broadleaves are the primary weeds, use one pound on 145 square feet (6.9 lbs./1,000 sq. ft.). For annual grassy weeds or when weed infestation is light, use one pound on 218 square feet (4.6 lbs./1,000 sq. ft.).

DEVRINOL 2-G Ornamental should be incorporated evenly into the soil using a hand cultivator to a depth of 1-2 inches or irrigation or natural moisture (rainfall or snow) 1-2 inches within 2-3 days) for optimum results.

For weed control in dichondra, apply at the time of seeding or to establish dichondra including dichondra sod. Irrigate with a minimum of one inch of water.

RECOMMENDATIONS: TURF GRASS

(For Use on Warm Season Grasses Only)

DEVRINOL 2-G is recommended for annual bluegrass, crabgrass and goosegrass control in the following established turf grasses: bahia, bermudagrass, centipede, St. Augustine and fescue.

TIMING OF APPLICATION

For control of crabgrass and goosegrass, apply DEVRINOL 2-G at the recommended rate. Application should be made 2-3 weeks prior to weed emergence. DEVRINOL 2-G should be applied with a gravity flow drop spreader or a cyclone type spreader. A more uniform application can be made by spreading half the required amount over the area to be treated, then applying the remainder at right angles.

DEVRINOL 2-G requires a minimum of one half (1/2) inch of rainfall or irrigation 2-3 days after application for adequate control. If irrigation methods are not available, the timing of the application of DEVRINOL 2-G should be moved back several weeks to allow for some rainfall between application and weed germination.

RATES OF APPLICATION

For Crabgrass Control: Apply 100 to 150 lbs. of DEVRINOL 2-G uniformly per acre (2.52-3.38* lbs. per 1,000 sq. ft.). Use the higher rate when applying before January 1 (not applicable north of South Carolina or when weed infestation is expected to be heavy).

For Goosegrass Control: Apply 100 to 150 lbs. of DEVRINOL 2-G per acre (2.25-3.38 lbs per 1,000 sq. ft.). Goosegrass usually starts to germinate 6 to 7 weeks after the start of crabgrass germination.

Split Application For Crabgrass and Goosegrass Control: When both of these weeds are to be controlled, apply 100 lbs. of DEVRINOL 2-G spread uniformly per acre (2.25 lbs. per 1,000 sq. ft.) for crabgrass and another 100 lbs. of DEVRINOL 2-G per acre for goosegrass. Apply 8-10 weeks apart. Timing should be adjusted to the germination of weeds in your area. Split applications give longer and better weed control than one application.

For Annual Bluegrass (Poa Annua) Control: Annual bluegrass is a winter annual in most of the warm season turf regions; therefore, application of DEVRINOL 2-G should be made in the fall before the annual bluegrass germinates. Apply 100-150 lbs. of DEVRINOL 2-G per acre (2.25-3.38 lbs. per 1,000 sq. ft.).

Reseeding: Do not reseed or overseed within six months after application.

* Where a rate range is given, use the higher rate when weed infestation is expected to be heavy.

STORAGE AND DISPOSAL

Do not contaminate water, food or feed by storage or disposal. **Storage:** Keep container closed when not in use. Store in a cool, dry place. Do not store near seeds, fertilizer or food stuffs. **Pesticide Disposal:** Wastes resulting from the use of this product may be disposed of on site or at an approved waste disposal facility. **Container Disposal:** Completely empty bag into application equipment. Then dispose of empty bag in a sanitary landfill or by incineration, or, if allowed by State and local authorities, by burning. If burned, stay out of smoke.

NOTICE TO BUYER AND USER

Seller warrants that this product conforms to the chemical description on the label and is reasonably fit for the purposes stated on the label when used in accordance with directions under normal conditions of use. This warranty does not extend to the use of this product contrary to label instructions, or under abnormal use conditions, or under conditions not reasonably foreseeable to Seller, and Buyer and User assume the risk of any such use. SELLER DISCLAIMS ALL OTHER WARRANTIES, EXPRESSED OR IMPLIED, INCLUDING ANY WARRANTY OF FITNESS OR MERCHANTABILITY. SELLER SHALL NOT BE LIABLE FOR CONSEQUENTIAL, SPECIAL OR INDIRECT DAMAGES RESULTING FROM THE USE OR HANDLING OF THIS PRODUCT AND SELLER'S SOLE LIABILITY AND BUYER'S AND USER'S EXCLUSIVE REMEDY SHALL BE LIMITED TO THE REFUND OF THE PURCHASE PRICE.

This product is sold only for uses stated on its label. No express or implied license is granted to use or sell this product under any patent in any country except as specified: Country: United States of America. Patent No.: 3,480,671; 3,718,455

03-0622-015 A-5R

This is a specimen label. ICI Americas Inc. is not responsible for the accuracy of the information contained herein. As labels are subject to revision, always carefully read and follow the label on the product container.

DEVRINOL® 50-DF

SELECTIVE HERBICIDE — DRY FLOWABLE

For use on certain Citrus, Nuts, Pome fruits, Small fruits, Stone fruits, Vegetables and Tobacco

ACTIVE INGREDIENT:	
N,N-Diethyl-2-(1-naphthalenyloxy)-propionamide	50%
INERT INGREDIENTS:	50%
TOTAL	100%

Contains 50 lbs. of active ingredient per 100 lbs.

EPA Reg. No. 10182-258

KEEP OUT OF REACH OF CHILDREN

DANGER

STATEMENT OF PRACTICAL TREATMENT

FIRST AID

Immediately start the procedures given below and contact a Poison Center, a physician, or the nearest hospital. Report the type and extent of exposure, describe the victim's symptoms and follow the advice given.

If swallowed: Immediately give several glasses of water and induce vomiting by gagging the victim with a finger placed on the back of the victim's tongue. Give fluids until vomitus is clear. If victim is unconscious or convulsing, do not induce vomiting or give anything by mouth.

For skin contact: Flush all affected areas with plenty of water for several minutes. Remove and clean contaminated clothing and shoes before reuse. Seek medical attention if skin irritation occurs.

For eye contact: Immediately flush eyes with large amounts of running water for 15 minutes. Hold eyelids apart to ensure rinsing of the entire surface of the eye and lids with water. Get medical attention immediately.

If inhaled: Remove to fresh air. Seek medical attention if respiratory irritation occurs or if breathing becomes difficult.

FOR 24-HOUR EMERGENCY MEDICAL ASSISTANCE CALL 1-800-F-A-S-T-M-E-D (327-8633).
FOR CHEMICAL EMERGENCY: Spill, leak, fire, exposure, or accident call CHEMTREC 1-800-424-9300.

For other product information, contact your local ICI Americas Distributor or see Material Safety Data Sheet.

PRECAUTIONARY STATEMENTS

HAZARDS TO HUMANS AND DOMESTIC ANIMALS

DANGER

CORROSIVE, CAUSES IRREVERSIBLE EYE DAMAGE

Harmful if swallowed. Wear goggles or face shield when handling. Avoid contact with skin, eyes and clothing. Wash thoroughly with soap and water after handling. Remove contaminated clothing and wash before reuse.

ENVIRONMENTAL HAZARDS

- Do not apply directly to water or wetlands.
- Do not contaminate water by cleaning of equipment or disposal of wastes. Apply this product as specified on this label.
- Do not contaminate water used for domestic purposes.
- Avoid Contamination of Food and Feed.

GENERAL PRECAUTIONS

- Read all instructions before using.
- Use only for recommended purposes and at recommended rates. Do not overdose.
- Except where specifically noted, DEVRINOL 50-DF is not recommended for use on soil containing more than 10% organic matter.
- Spraying on ground heavily covered with leaves or trash may adversely affect weed control.
- Do not apply to frozen ground.
- To minimize injury to succeeding crops not listed on this label, a deep disc plow or moldboard plow operation (minimum 10 inches) must be carried out prior to planting. To avoid injury to crops not specified on the label, do not plant until 12 months after the last application of DEVRINOL 50-DF.
- When applied at rates no greater than 2 lbs. a.i./A, rotational crops which are not registered on this label may be planted no sooner than 12 months after application when harvest and use of mature crop(s) is only for food or feed purposes. Crops are not to be grazed or harvested for food or feed purposes earlier than maturity.

GENERAL DIRECTIONS FOR USE

It is a violation of Federal law to use this product in a manner inconsistent with its labeling.

DEVRINOL 50-DF is a selective surface-applied or incorporated herbicide for use in certain types of newly planted or established crops. DEVRINOL 50-DF does not control established weeds. All weed growth and crop stubble in the fields must be thoroughly worked into the soil prior to planting. Spray equipment should be carefully calibrated before use and checked frequently during application to be sure nozzles are operating properly and are delivering a uniform spray pattern. Apply the recommended rate of DEVRINOL 50-DF in 20 to 100 gallons of water per acre in a low pressure boom-type sprayer.

Be sure the spray tank, boom and nozzle have been thoroughly flushed with clean water before use. Sufficient jet or mechanical agitation should be provided during the filling operation and during application.

When mixing DEVRINOL 50-DF alone, or in combination with other recommended wettable powders, make a slurry and add to a nearly filled spray tank with adequate agitation. This will enhance initial mixing. Recommended contact herbicides and spreaders should always be added to a nearly filled spray tank. Never add them to the slurry.

Whenever a rate range is recommended for DEVRINOL 50-DF, use the lower rate on light soil (coarse textured-sandy) and the higher rate on heavy soil (fine textured-clay).

APPLICATIONS THROUGH IRRIGATION SYSTEMS

DEVRINOL 50-DF may be applied at the recommended rates through irrigation water for pre-emergence control of listed weeds in grapes and tree crops (Western Region Only). Do not allow treatment to contact foliage or crop. Remove existing weed growth before application. Meter the DEVRINOL 50-DF into the irrigation water during the entire period. Apply with sufficient water to wet the soil to a depth of 2-4 inches. Good agitation should be maintained during the entire application period. Avoid run-off.

USE PRECAUTIONS:

A. Apply this product only through sprinkler including center pivot, lateral move, end tow, side (wheel) roll, traveler, solid set, or hand move, flood (basin); furrow; border irrigation systems. Do not apply this product through any other type of irrigation system.
B. Crop injury, lack of effectiveness, or illegal pesticide residues in the crop can result from non-uniform distribution of treated water.
C. If you have questions about calibration, you should contact State Extension Service specialists, equipment manufacturers or other experts.
D. Do not connect an irrigation system (including greenhouse systems) used for pesticide application to a public water system unless the pesticide label-prescribed safety devices for public water systems are in place.
E. A person knowledgeable of the chemigation system and responsible for its operation, or under the supervision of the responsible person, shall shut the system down and make necessary adjustments should the need arise.

FOR SPRINKLER IRRIGATION

1. The system must contain a functional check valve, vacuum relief valve, and low pressure drain appropriately located on the irrigation pipeline to prevent water source contamination from backflow.
2. The pesticide injection pipeline must contain a functional, automatic, quick-closing check valve to prevent the flow of fluid back toward the injection pump.
3. The pesticide injection pipeline must also contain a functional, normally closed, solenoid-operated valve located on the intake side of the injection pump and connected to the system interlock to prevent fluid from being withdrawn from the supply tank when the irrigation system is either automatically or manually shut down.
4. The system must contain functional interlocking controls to automatically shut off the pesticide injection pump when the water pump motor stops.
5. The irrigation line or water pump must include a functional pressure switch which will stop the water pump motor when the water pressure decreases to the point where pesticide distribution is adversely affected.
6. Systems must use a metering pump, such as a positive displacement injection pump (e.g., diaphragm pump) effectively designed and constructed of materials that are compatible with pesticides and capable of being fitted with a system interlock.
7. Do not apply when wind speed favors drift beyond the area intended for treatment.

FOR FLOOD (BASIN), FURROW AND BORDER IRRIGATION

1. Systems using a gravity flow pesticide dispensing system must meter the pesticide into the water at the head of the field and downstream of a hydraulic discontinuity such as a drop structure of weir box to decrease potential for water source contamination from backflow if water flow stops.
2. Systems utilizing a pressurized water and pesticide injection system must meet the following requirements:
   a. The system must contain a functional check valve, vacuum relief valve, and low pressure drain appropriately located on the irrigation pipeline to prevent water source contamination from backflow.
   b. The pesticide injection pipeline must contain a functional, automatic, quick-closing check valve to prevent the flow of fluid back toward the injection pump.

c. The pesticide injection pipeline must also contain a functional, normally closed, solenoid-operated valve located on the intake side of the injection pump and connected to the system interlock to prevent fluid from being withdrawn from the supply tank when the irrigation system is either automatically or manually shut down.
d. The system must contain functional interlocking controls to automatically shut off the pesticide injection pump when the water pump motor stops.
e. The irrigation line or water pump must include a functional pressure switch which will stop the water pump motor when the water pressure decreases to the point where pesticide distribution is adversely affected.
f. Systems must use a metering pump, such as a positive displacement injection pump (e.g., diaphragm pump) effectively designed and constructed of materials that are compatible with pesticides and capable of being fitted with a system interlock.

NOTE: All rates are given on a broadcast basis, per acre. Reduce rates proportionately for band or strip treatment.

Posting Requirements

Posting of areas to be chemigated is required when 1) any part of a treated area is within 300 feet of sensitive areas such as residential areas, labor camps, businesses, day care centers, hospitals, in-patient clinics, nursing homes or any public areas such as schools, parks, playgrounds, or any other public facilities not including public roads, or 2) when the chemigated area is open to the public such as golf courses or retail greenhouses.

Posting must conform to the following requirements. Treated areas shall be posted with signs at all usual points of entry and along likely routes of approach from listed sensitive areas. When there are no usual points of entry, signs must be posted in the corners of the treated areas and in any other location affording maximum visibility to sensitive areas. The printed side of the sign should face away from the treated area towards the sensitive area. The signs shall be printed in English. Signs must be posted prior to application and must remain posted until foliage has dried and soil surface water has disappeared. Signs may remain in place indefinately as long as they are composed of materials to prevent deterioration and maintain legibility for the duration of the posting period.

All words shall consist of letters at least 2½ inches tall, and all letters and the symbol shall be a color which sharply contrasts with the immediate background. At the top of the sign shall be the words KEEP OUT, followed by an octagonal stop sign symbol at least 8 inches in diameter containing the word STOP. Below the symbol shall be the words PESTICIDES IN IRRIGATION WATER.

WEEDS CONTROLLED BY 8 LBS. DEVRINOL 50-DF PER ACRE SURFACE APPLICATION

NOTE: DEVRINOL 50-DF does not control established weeds.

ANNUAL GRASSES

Barnyardgrass, Junglerice	*(Echinochloa* spp.)
Wild Barley	*(Hordeum leporinum)*
Annual Bluegrass	*(Poa annua)*
Bromes	*(Bromus)*
Cheatgrass	*(B. secalinus)*
Downey Brome	*(B. tectorum)*
Ripgut Brome	*(B. rigidus)*
Soft Chess	*(B. mollis)*
Large (Hairy) Crabgrass	*(Digitaria sanguinallis)*
Smooth Crabgrass	*(Digitaria ischaemum)*
Southwestern Cupgrass	*(Eriochloa gracillis)*
Foxtails	*(Setaria)*
Bristly Foxtail	*(S. verticillia)*
Giant Foxtail	*(S. faberii)*
Green Foxtail	*(S. viridis)*
Yellow Foxtail	*(S. lutescens)*
Goosegrass	*(Eleusine indica)*
Seeding Johnsongrass	*(Sorghum halepense)*
Wild Oats	*(Avena fetua)*
Panicums	*(Panicum)*
Guineagrass	*(P. paximum)*
Fall Pancium	*(P. dichotomiflorum)*
Texas Pancium	*(P. texanum)*
Witchgrass	*(P. capillare)*
Italian Ryegrass	*(Lolium multiflorum)*
Sandbur	*(Cenchrus* spp.)
Red Sprangletop	*(Leptochloa fillformis)*
Stinkgrass	*(Eragrostis cilianensis)*

ANNUAL BROADLEAF WEEDS

Carpetweed	*(Mollugo verticillata)*
Common Chickweed	*(Stellaria media)*
Coast Fiddleneck	*(Amsinckia intermedia)*
Filaree	*(Erodium,* spp.)
Groundsel	*(Senecio* spp.)
Prostrate Knotweed	*(Polygonum aviculare)*
Lambsquarters	*(Chenopodium album)*
Prickly Lettuce	*(Lactuca serriola)*
Little Mallow*	*(Malva parviflora)*
Redroot Pigweed	*(Amaranthus retroflexus)*
Pineappleweed	*(Matricaria matricardiodes)*
Common Purslane	*(Portulaca oleracea)*
Horse Purslane	*(Trianthema portullacastrum)*
Common Ragweed**	*(Ambrosia artemislifolia)*
Annual Sowthistle	*(Sonchus oleraceus)*

*From seed
**Suppression

DIRECTIONS FOR USE

NOTE: All rates are given on a broadcast basis. Reduce rates proportionately for band or strip treatment.

For control of all weeds listed, apply 8 pounds DEVRINOL 50-DF per broadcast acre to the soil surface. Apply in the Fall through early Spring prior to weed emergence. Do not apply DEVRINOL 50-DF to frozen ground. For best results, the treatment must reach the zone of weed seed germination. If rainfall does not occur the treatment must be shallowly incorporated or irrigated–in following application (sprinkler or flood) with sufficient water to wet the soil to a depth of 2 to 4 inches.

ALLOWABLE TIME FOR TREATMENT TO REMAIN ON SOIL SURFACE (BY REGION):

WEST: Three weeks (November through February) or 24 hours (remainder of year).

PACIFIC NORTHWEST: Three weeks (Fall through Early Spring) or 24 hours (remainder of year).

OTHER REGIONS: 24 hours.

Use as a directed spray and avoid contact with fruit or foliage. Do not apply when fruit or nuts are on the ground during the harvest period. Do not graze treated areas.

Regional Use Map

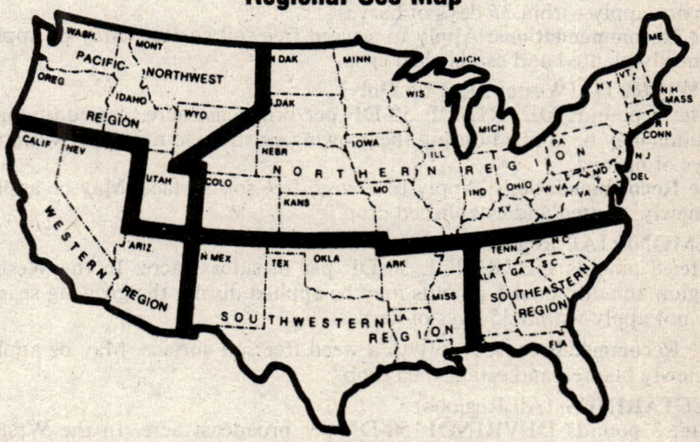

FRUIT AND NUT CROPS SURFACE APPLICATION RECOMMENDATIONS

DEVRINOL 50-DF is recommended for use on the following crops. Refer to regional map under "DIRECTIONS FOR USE."

ALMONDS: (All Regions)
Rate: 8 pounds DEVRINOL 50-DF per broadcast acre. In the Western region an additional 8 pounds may be applied during the growing season. Do not apply within 35 days of harvest.
Use Recommendations: Apply to a weed free soil surface. May be applied to newly planted and established crop.

APPLES: (All Regions)
Rate: 8 pounds DEVRINOL 50-DF per broadcast acre. In the Western Region an additional 8 pounds may be applied during the growing season. Do not apply within 35 days of harvest.
Use Recommendations: Apply to a weed free soil surface. May be applied to newly planted and established crop.

APRICOTS: (All Regions)
Rate: 8 pounds DEVRINOL 50-DF per broadcast acre. In the Western Region an additional 8 pounds may be applied during the growing season. Do not apply within 35 days of harvest.
Use Recommendations: Apply to a weed free soil surface. May be applied to newly planted and established crop.

AVOCADOS: (Western Region Only)
Rate: 8 pounds DEVRINOL 50-DF per broadcast acre. An additional 8 pounds may be applied during the growing season. Do not apply within 35 days of harvest.
Use Recommendations: Apply to a weed free soil surface. May be applied to established crop only.

BLUEBERRIES: (All Regions)

Rate: 8 pounds DEVRINOL 50-DF per broadcast acre.
Use Recommendations: Apply to a weed free soil surface. May be applied to newly planted and established crop.

CANEBERRIES (BLACKBERRIES, RASPBERRIES, BOYSENBERRIES AND LOGANBERRIES): (All Regions)
Rate: 8 pounds DEVRINOL 50-DF per broadcast acre.
Use Recommendations: Apply to a weed free soil surface. May be applied to newly planted and established crop.

CHERRIES: (All Regions)
Rate: 8 pounds DEVRINOL 50-DF per broadcast acre. In the Western Region an additional 8 pounds may be applied during the growing season. Do not apply within 35 days of harvest.
Use Recommendations: Apply to a weed free soil surface. May be applied to newly planted and established crop.

FIGS: (All Regions)
Rate: 8 pounds DEVRINOL 50-DF per broadcast acre. In the Western Region an additional 8 pounds may be applied during the growing season. Do not apply within 35 days of harvest.
Use Recommendations: Apply to a weed free soil surface. May be applied to newly planted and established crop.

FILBERTS: (All Regions)
Rate: 8 pounds DEVRINOL 50-DF per broadcast acre. In the Western Region an additional 8 pounds may be applied during the growing season. Do not apply within 35 days of harvest.
Use Recommendations: Apply to a weed free soil surface. May be applied to newly planted and established crop.

GRAPEFRUIT: (All Regions)
Rate: 8 pounds DEVRINOL 50-DF per broadcast acre. In the Western Region an additional 8 pounds may be applied during the growing season. Do not apply within 35 days of harvest.
Use Recommendations: Apply to a weed free soil surface. May be applied to newly planted and established crop.

GRAPES: (All Regions)
Rate: 8 pounds DEVRINOL 50-DF per broadcast acre. In the Western Region an additional 8 pounds may be applied during the growing season. Do not apply within 35 days of harvest.
Use Recommendations: Apply to a weed free soil surface. May be applied to newly planted and established crop.

KIWIFRUIT: (Western Region Only)
Rate: 8 pounds DEVRINOL 50-DF per broadcast acre. An additional 8 pounds may be applied during the growing season. Do not apply within 35 days of harvest.
Use Recommendations: Apply to a weed free soil surface. May be applied to newly planted and established crop.

LEMONS: (All Regions)
Rate: 8 pounds DEVRINOL 50-DF per broadcast acre. In the Western Region an additional 8 pounds may be applied during the growing season. Do not apply within 35 days of harvest.
Use Recommendations: Apply to a weed free soil surface. May be applied to established crop.

NECTARINES: (All Regions)
Rate: 8 pounds DEVRINOL 50-DF per broadcast acre. In the Western Region an additional 8 pounds may be applied during the growing season. Do not apply within 35 days of harvest.
Use Recommendations: Apply to a weed free soil surface. May be applied to newly planted and established crop.

OLIVES: (Western Region Only)
Rate: 8 pounds DEVRINOL 50-DF per broadcast acre. An additional 8 pounds may be applied during the growing season. Do not apply within 35 days of harvest.
Use Recommendations: Apply to a weed free soil surface. May be applied to newly planted and established crop.

ORANGES: (All Regions)
Rate: 8 pounds DEVRINOL 50-DF per broadcast acre. In the Western Region an additional 8 pounds may be applied during the growing season. Do not apply within 35 days of harvest.
Use Recommendations: Apply to a weed free soil surface. May be applied to newly planted and established crop.

PEACHES: (All Regions)
Rate: 8 pounds DEVRINOL 50-DF per broadcast acre. In the Western Region an additional 8 pounds may be applied during the growing season. Do not apply within 35 days of harvest.
Use Recommendations: Apply to a weed free soil surface. May be applied to newly planted and established crop.

PEARS: (All Regions)
Rate: 8 pounds DEVRINOL 50-DF per broadcast acre. In the Western Region an additional 8 pounds may be applied during the growing season. Do not apply within 35 days of harvest.
Use Recommendations: Apply to a weed free soil surface. May be applied to newly planted and established crop.

PECANS: (All Regions)
Rate: 8 pounds DEVRINOL 50-DF per broadcast acre. In the Western Region an additional 8 pounds may be applied during the growing season. Do not apply within 35 days of harvest.
Use Recommendations: Apply to a weed free soil surface. May be applied to newly planted and established crop.

PERSIMMONS: (Western Region Only)
Rate: 8 pounds DEVRINOL 50-DF per broadcast acre. An additional 8 pounds may be applied during the growing season. Do not apply within 35 days of harvest.
Use Recommendations: Apply to a weed free soil surface. May be applied to established crop only.

PISTACHIOS: (All Regions)
Rate: 8 pounds DEVRINOL 50-DF per broadcast acre.
Use Recommendations: Apply to a weed free soil surface. May be applied to newly planted and established crop.

PLUMS: (All Regions)
Rate: 8 pounds DEVRINOL 50-DF per broadcast acre. In the Western Region an additional 8 pounds may be applied during the growing season. Do not apply within 35 days of harvest.
Use Recommendations: Apply to a weed free soil surface. May be applied to newly planted and established crop.

PRUNES: (All Regions)
Rate: 8 pounds DEVRINOL 50-DF per broadcast acre. In the Western Region an additional 8 pounds may be applied during the growing season. Do not apply within 35 days of harvest.
Use Recommendations: Apply to a weed free soil surface. May be applied to newly planted and established crop.

STRAWBERRIES: (All Regions)
Rate: 8 pounds DEVRINOL 50-DF per broadcast acre.
Use Recommendations: Apply to a weed free soil surface. May be applied to newly transplanted and established crop. Delay application until desired number of daughter plants have become established. Do not apply from bloom through harvest.

TANGERINES: (All regions)
Rate: 8 pounds DEVRINOL 50-DF per broadcast acre. In the Western Region an additional 8 pounds may be applied during the growing season. Do not apply within 35 days of harvest.
Use Recommendations: Apply to a weed free soil surface. May be applied to newly planted and established crop.

WALNUTS: (All Regions)
Rate: 8 pounds DEVRINOL 50-DF per broadcast acre. In the Western Region an additional 8 pounds may be applied during the growing season. Do not apply within 35 days of harvest.
Use Recommendations: Apply to a weed free soil surface. May be applied to newly planted and established crop.

OTHER CROPS SURFACE APPLICATION RECOMMENDATIONS

ARTICHOKES: (Western Region Only)
Rate: 8 pounds DEVRINOL 50-DF per broadcast acre.
Use Recommendations: Apply to a weed free soil surface. May be applied to established crop only.

ASPARAGUS: (All Regions)
Rate: 8 pounds DEVRINOL 50-DF per broadcast acre.
Use Recommendations: Apply to a weed free soil surface. May be applied to established crop only.

Mint: (Pacific Northwest Only)
Rate: 8 pounds DEVRINOL 50-DF per broadcast acre.
Use Recommendations: Apply to a weed free soil surface. May be applied to newly planted and established crop.

Where recommended for use on established crops, apply DEVRINOL 50-DF only to stands that have been established for at least one growing season. Where recommended for use on newly planted crops, application may be made immediately after planting young trees, vines, and bushes.

DEVRINOL 50-DF may be applied at the recommended rates through irrigation water for pre-emergence control of listed weeds in grapes and tree crops (Western Region Only). Do not allow treatment to contact foliage or crop. Remove existing weed growth before application. Apply at the start of irrigation set and with sufficient water to wet the soil to a depth of 2-4 inches. Avoid run-off. These irrigation systems should be equipped with automatic shutoff devices and valves to prevent back flow into the water source.

WEEDS CONTROLLED BY 2 TO 4 LBS. DEVRINOL 50-DF PREPLANT INCORPORATED APPLICATION

NOTE: DEVRINOL 50-DF does not control established weeds.

ANNUAL GRASSES	
Barnyardgrass, Junglerice	(*Echinochloa* spp.)
Crabgrass	(*Digitaria* spp.)
Foxtails	(*Setaria*)
Giant Foxtail	(*S. faberi*)
Green Foxtail	(*S. viridis*)
Yellow Foxtail	(*S. lutescens*)

Goosegrass	*(Eleusine indica)*
Seedling Johnsongrass	*(Sorghum halepense)*
Panicums	*(Panicum)*
Fall Panicum	*(P. dichotomiflorum)*
Witchgrass	*(P. capillare)*
Stinkgrass	*(Eragrostis cilianensis)*

ANNUAL BROADLEAF WEEDS

Carpetweed	*(Mollugo verticillata)*
Purple Cudweed	*(Gnaphalium purpureum)*
Galinsoga*	*(Galinsoga* spp.)
Lambsquarters	*(Chenopodium album)*
Redroot Pigweed	*(Amaranthus retroflexus)*
Common Purslane	*(Portulaca oleracea)*
Florida Pusley	*(Richardis scabra)*
Common Ragweed**	*(Ambrosia artemisiifolia)*

* Controlled at 4 lb. DEVRINOL 50-DF

** Suppression

DIRECTIONS FOR USE OF DEVRINOL 50-DF PRE-PLANT INCORPORATED

On flat-planted or bedded crops apply, before planting, to well-worked soil that is dry enough to permit thorough incorporation to a depth of 1 to 2 inches. Incorporate the same day as applied, using equipment which will result in uniform incorporation of the herbicide to the desired depth.

After harvest or prior to planting of succeeding crops, a deep moldboard or disc plowing operation must be carried out.

When applied at rates no greater than 2 lbs. a.i,/A, rotational crops which are not registered on this label may be planted no sooner than 12 months after application when harvest and use of the mature crop(s) only is for food or feed purposes. Crops are not to be grazed or harvested for food or feed purposes earlier than maturity.

MELONS, TOBACCO AND VEGETABLES PRE-PLANT INCORPORATED APPLICATION

RECOMMENDATIONS

NOTE: All rates are given on a broadcast basis. Reduce rates proportionately for band or strip treatment.

COLE CROPS—Broccoli, Brussels Sprouts, Cauliflower: (California Only)
Rate: 2-4 pounds DEVRINOL 50-DF per broadcast acre.
Use Recommendations: Apply to a weed free soil surface. May be applied to direct seeded and transplanted crop, pre-plant incorporated or as a surface spray after direct seeding or transplanting. Shallow incorporate no deeper than seeding depth or sprinkler irrigate within 24 hours using sufficient water to wet the soil to a depth of 2-4 inches. Use the lower rate on light soil (course textured-sandy) and the higher rate on heavy soil (fine textured-clay).

COLE CROPS—Broccoli, Brussels Sprouts, Cauliflower: (All regions except Pacific Northwest)
Rate: 2-4 pounds DEVRINOL 50-DF per broadcast acre.
Use Recommendations: Apply to a weed free soil surface. May be applied to direct seeded and transplanted crop, pre-plant incorporated or as a surface spray. Shallow incorporate to a depth of 1-2 inches or sprinkler irrigate within 24 hours using sufficient water to wet the soil to a depth of 2-4 inches. Use the lower rate on light soil (coarse textured-sandy) and the higher rate on heavy soil (fine textured-clay).

EGGPLANTS: (All Regions)
Rate: Before transplanting, apply 2-4 pounds DEVRINOL 50-DF per broadcast acre.
Use Recommendations: Apply to a weed free soil surface. May be applied to transplanted crop only. Use the lower rate on light soil (coarse textured-sandy) and the higher rate on heavy soil (fine textured-clay).

MELONS: (except cantaloupes) and Squash: (All Regions)
Rate: 2-4 pounds DEVRINOL 50-DF per broadcast acre.
Use Recommendations: Apply to a weed free soil surface. May be applied to direct seeded crop. Use the lower rate on light soil (coarse textured-sandy) and the higher rate on heavy soil (fine textured-clay).

PEPPERS: (All Regions)
Rate: 2-4 pounds DEVRINOL 50-DF per broadcast acre.
Use Recommendations: Apply to a weed free soil surface. May be applied to direct seeded and transplanted crop. Use the lower rate on light soil (coarse textured-sandy) and the higher rate on heavy soil (fine textured-clay).

TOMATOES: (All Regions)
Rate: 2-4 pounds DEVRINOL 50-DF per broadcast acre.
Use Recommendations: Apply to a weed free soil surface. May be applied to direct seeded and transplanted crop. Use the lower rate on light soil (coarse textured-sandy) and the higher rate on heavy soil (fine textured-clay).

TOBACCO: (All Regions)
Rate: Before transplanting, apply 2-4 pounds DEVRINOL 50-DF per broadcast acre.
Use Recommendations: Apply to a weed free soil surface. Use the lower rate on light soil (coarse textured-sandy) and the higher rate on heavy soil (fine textured-clay). In Southeast only: Apply up to 3 weeks before transplanting. No waiting period is necessary.

POSTTRANSPLANT SURFACE AND LAYBY SURFACE APPLICATIONS
FOR USE IN TRANSPLANTED TOBACCO (Virginia and Southeastern Region Only)

Weeds Controlled

See list of weeds controlled under DEVRINOL 50-DF preplant incorporated application.

Directions for Use

At Transplanting: Apply the recommended amount of DEVRINOL 50-DF to the soil surface immediately after transplanting. May be applied over the top of transplants.

At Layby: Apply the recommended amount of DEVRINOL 50-DF in a band immediately after the layby cultivation. Apply as a directed spray to the row middles.

Use 20 to 50 gallons of spray per acre. For best results the treatment must reach the zone of weed seed germination. If rainfall does not occur within 24 hours after application, the treatment must be shallowly incorporated or irrigated-in with sufficient overhead irrigation to wet the soil to a depth of 2 to 4 inches.

RECOMMENDATIONS

Rate: Apply 2 to 4 pounds DEVRINOL 50-DF per broadcast acre.

Use Recommendations: Apply to a weed free soil surface. Reduce rates proportionately for band application. Treatments may follow a preplant incorporate application of DEVRINOL 50-DF or DEVRINOL 50-DF/TILLAM® 6-E. Do not apply more than 4 pounds DEVRINOL 50-DF in any one season.

Rotational Crop Restrictions

When DEVRINOL 50-DF is used on tobacco posttransplant or layby at 2 to 4 pounds, surface applied, small grains may be seeded in rotation in the Fall to prevent soil erosion. These grains may be stunted but not otherwise adversely affected.

No tolerance is established for residues of DEVRINOL 50-DF in small grains. Rotational cover crops must be destroyed by plowing under or otherwise prevented from reaching food or feed production.

If rotating out of tobacco the following Spring, to avoid injury to crops not specified on this label do not plant until 12 months after the last DEVRINOL 50-DF application.

When applied at rates no greater than 2 lbs. a.i./A, rotational crops which are not registered on this label may be planted no sooner than 12 months after application when harvest and use of the mature crop(s) only is for food or feed purposes. Crops are not to be grazed or harvested for food or feed purposes earlier than maturity.

TANK MIXTURES
FRUIT AND NUT CROPS

IMPORTANT

Observe all cautions and limitations on labeling of all products used in combination.

Make only one application per season.

Do not contaminate water by cleaning of equipment or disposal of wastes. Apply these combinations only as specified on this label.

For a tank mix application, apply the recommended dosage (see directions for use) in 20 to 100 gallons of water per acre.

In any tank mix including GRAMOXONE* SUPER, add an approved non-ionic spreader to the tank water, such as ORTHO** X-77 spreader at the rate of 8 oz. per 100 gallons.

* Registered trademark of ICI Americas Inc.

** Registered trademark of Chevron Chemical Company

DEVRINOL 50-DF/GRAMOXONE SUPER COMBINATION

DEVRINOL 50-DF will not control emerged weeds. GRAMOXONE SUPER may be used to control existing weeds when applied prior to, or in combination with DEVRINOL 50-DF.
Refer to the section on DEVRINOL 50-DF surface application for a list of weeds controlled.

DIRECTIONS FOR USE

Follow use directions for DEVRINOL 50-DF surface application. Read the GRAMOXONE SUPER label for timing with respect to stage of weed growth and for application methods recommended to minimize crop injury.

For control of all weeds listed, apply 8 pounds DEVRINOL 50-DF and 3 to 5 pints GRAMOXONE SUPER per broadcast acre.

RECOMMENDATIONS

Consult the GRAMOXONE SUPER label for recommended uses in orchards and vineyards. Then read the crop recommendations for DEVRINOL 50-DF surface application to insure that the combination may be used for a particular crop in your state. Recommended *only* for the uses specified on the GRAMOXONE SUPER and DEVRINOL 50-DF labels. Do not apply this combination to caneberries.

DEVRINOL 50-DF/SIMAZINE 80-WP COMBINATION
(California and Pacific Northwest Only)

DEVRINOL 50-DF may be used in combination with SIMAZINE 80-WP for control of additional weeds not controlled by DEVRINOL 50-DF alone. This tank mix does not control emerged weeds.

Additional weeds controlled by DEVRINOL 50-DF/SIMAZINE 80-W tank mix, surface applied, at recommended rates:

California burclover	*(Medicago polymorpha)*
Henbit	*(Lamium amplexicaule)*
London rocket	*(Sisymbrium irio)*
Marestail (horseweed)	*(Conyza canadensis)*
Miners lettuce	*(Montia perfoliata)*
Panicle willoweed	*(Epiloblum paniculatum)*
Redmaids (rockpurslane)	*(Calandrinia caulescens)*

DIRECTIONS FOR USE

For control of all weeds listed, apply 8 pounds DEVRINOL 50-DF and 0.6 to 1.25 pounds SIMAZINE 80-WP per broadcast acre. Limit treatment of almonds, grapes and peaches to a 4 to 8 foot band centered on the tree row, reducing the dosage proportionately. Apply in the fall through early spring, prior to weed emergence. For best results, the treatment must reach the zone of weed seed germination. Treatments made other than in November through February must be shallowly incorporated or irrigated-in the same day as applied (sprinkler or flood) with sufficient water to wet the soil to a depth of 2 to 4 inches. Irrigation or shallow incorporation is recommended for treatments made from November through February if no rainfall occurs within three weeks of application.

Read directions on the SIMAZINE label concerning nozzle and screen sizes.

Use as a directed spray and avoid contact with fruit or foliage. Do not apply when fruit or nuts are on the ground during the harvest period. Do not graze treated areas.

RECOMMENDATIONS

For use on almonds, apples, avocados, grapefruit, grapes, lemons, olives, oranges, peaches, pears, and walnuts in California and apples, filberts, grapes and walnuts in the Pacific Northwest.

NOTE: To avoid possible crop injury follow these precautions.

- Apples and Pears. Make application only to orchards where trees have been established one year or more.
- Almonds and Peaches (California Only). Do not treat trees established in the orchard less than 3 years. Do not treat the Mission variety of almonds. Do not apply to almond trees propagated on plum rootstocks. Do not replant almonds or peaches in treated soil. Do not apply on gravelly, sandy, or loamy sand soil, or on soil with less than 1% organic matter. Do not treat areas where water will accumulate (flood or furrow irrigation water should not be applied to any treated areas or allow to stand on treated areas).
- Filberts (Oregon and Washington Only). Do not apply on gravelly, sand or loamy sand soil.
- Grapes. Do not use in vineyards established less than three years.
- Grapefruit, Lemons, and Oranges (California Only). Do not use in the Imperial, Coachella, or Palo Verde Valleys. Do not use in nurseries. Do not apply to bedded grapefruit, lemons, or oranges. Do not apply to trees under stress from freeze damage for one year after the freeze.
- Walnuts. Do not apply to sandy soil. Leveling and furrowing operations after application will lessen effectiveness of weed control.

DEVRINOL 50-DF/SIMAZINE 80-WP/GRAMOXONE SUPER COMBINATION
(California and Pacific Northwest Only)

DEVRINOL 50-DF/SIMAZINE 80-WP will not control emerged weeds. GRAMOXONE SUPER may be used to control existing weeds when applied prior to, or in combination with DEVRINOL/SIMAZINE.

For weeds controlled refer to the weed lists for DEVRINOL 50-DF and DEVRINOL/SIMAZINE tank mix surface applied.

For control of all weeds listed, apply 8 pounds DEVRINOL 50-DF, 0.6 to 1.25 pounds SIMAZINE 80-WP, and 3 to 5 pints GRAMOXONE SUPER per broadcast acre.

Follow the use directions and recommendations for the DEVRINOL 50-DF/SIMAZINE 80-WP combination. Consult the GRAMOXONE SUPER label for timing with respect to stage of weed growth and for application methods recommended to minimize crop injury.

RECOMMENDATIONS

For crop recommendations, read the section on DEVRINOL/SIMAZINE.

DEVRINOL 50-DF/KARMEX* 80-WP COMBINATION

DEVRINOL 50-DF may be used in combination with KARMEX 80-WP for improved control of annual broadleaf weeds. This tank mix does not control emerged weeds.

* Registered trademark of E.I. DuPont De Nemours & Company

DIRECTIONS FOR USE

NOTE: All rates are given on a broadcast basis. Reduce rates proportionately for band or strip treatment.

For control of all weeds listed, apply 8 pounds DEVRINOL 50-DF and 1.25 pounds KARMEX 80-WP per broadcast acre to the soil surface. Apply in the Fall through early Spring prior to weed emergence. Do not apply to frozen ground. For best results, the treatment must reach the zone of weed seed germination. If rainfall does not occur, the treatment must be shallowly incorporated or irrigated-in following application (sprinkler or flood) with sufficient water to wet the soil to a depth of 2 to 4 inches. In the Western and Pacific Northwest Regions treatment made in November through February must be incorporated-in or irrigated-in if no rainfall occurs within three weeks of application.

For treatments made other than in November through February in these regions, and for all treatments applied in the Northern, Southwestern, and Southeastern Regions regardless of season, the treatment must be incorporated-in or irrigated-in within 24 hours if no rainfall occurs.

Use as a directed spray and avoid contact with fruit or foliage. Do not apply when fruit is on the ground during the harvest period. Do not graze treated areas.

DEVRINOL 50-DF/GOAL* 1.6-E COMBINATION
(California Only)

DEVRINOL 50-DF may be used in combination with GOAL 1.6-E for control of additional weeds not controlled by DEVRINOL 50-DF alone.

Additional weeds controlled preemergence by DEVRINOL 50-DF/GOAL 1.6-E tank mix, applied preemergence to a relatively weed-free established berm or soil surface, at recommended rates:

Burclover	*(Medicago hipide)*
Filaree, (Redstem,	*(Erodium cicutarium)*
Broadleaf)	*(Erodium botrys)*
Henbit	*(Lamium amplexicaule)*
London Rocket	*(Sisymbrium irio)*
Redmaids	*(Calandrinia caulescens)*
Shepherdspurse	*(Capsella bursapastoris)*

* Registered trademark of Rohm and Haas Company

DIRECTIONS FOR USE—DORMANT APPLICATION,
(California Only)

For control of all weeds listed, apply 8 pounds of DEVRINOL 50-DF and 5.0 quarts GOAL 1.6-E (2.0 lb active) per broadcast acre. Spray should be directed to the soil and the base of dormant trees or vines. Use of a low pressure sprayer equipped with a break-away boom and flat fan or off-center (OC) nozzle is recommended. For best results, the treatment must reach the zone of weed germination. Irrigation sufficient to wet the soil to a minimum depth of 2 to 4 inches is recommended for treatments made from November through February 15 if no rainfall occurs within three weeks of application.

NOTE: Applications made in October must be irrigated in (sprinkler or flood) the same day as applied with sufficient water to wet the soil to a minimum depth of 2 to 4 inches.

Cultural practices which result in redistribution or disturbance of the soil surface after treatment will decrease the herbicidal effectiveness of GOAL 1.6-E. For best results from the DEVRINOL 50-DF/GOAL 1.6E tank mix combination, application should be made to established berms or soil surfaces which are left undisturbed during the time period for which weed control is desired.

Read the GOAL 1.6-E label for specific use directions and precautions.

RECOMMENDATIONS

For use on almond, apricot, cherry, fig, grape, nectarine, peach, pear, pistachio, plum, prune and walnut in California only.

NOTE: To avoid possible crop injury follow these precautions:

- Apply only to dormant fruit or nut trees or grape vines. Do not apply when foliage, buds or fruit/nuts are present.
- Grapes: Do not apply to grapes established less than 3 years. Do not apply to grapes which are not staked or trellised.

RECOMMENDATIONS

For use on the following crops: Apples and peaches (All Regions). Citrus (California and Arizona only). Grapes, pears and walnuts (California only).

NOTE: To avoid possible crop injury follow these precautions:

- **Apples and Pears:** Use only under trees established in the orchard for at least 1 year; do not treat varieties grafted on full-dwarf root stocks.
- **Citrus:** Use only under trees established in the grove for at least 1 year.
- **Grapes:** Apply only to established vineyards (at least 3 years old) as a band treatment to grape rows. Do not apply to vines with trunks less than 1½" in diameter as injury may result. On soils low in clay or organic matter (1 to 2%), severe plant injury may result if heavy rainfall or more than one inch of irrigation occurs soon after treatment. This risk must be assumed by the user.
- **Peaches:** Use only under trees established in the orchard for at least 2 years.
- **Walnuts:** Use only under trees established in the orchard for at least 1 year.

DEVRINOL 50-DF/GOAL 1.6-E/GRAMOXONE SUPER COMBINATION
(California Only)

DEVRINOL 50-DF/GOAL 1.6-E/GRAMOXONE SUPER is recommended for residual preemergence control of susceptible broadleaf and grassy weeds. GRAMOXONE SUPER is included in the tank mix to control existing emerged weeds.

For weeds controlled refer to the weed lists for DEVRINOL 50-DF and DEVRINOL 50-DF/GOAL 1.6-E tank mix surface applied.

DIRECTIONS FOR USE

For control of all weeds listed, apply 8 pounds DEVRINOL 50-DF, 5 quarts GOAL 1.6-E (2.0 lb. active) and 3 to 5 pints GRAMOXONE SUPER per broadcast acre.

Follow the use directions and recommendations for the DEVRINOL 50-DF/GOAL 1.6-E combination. Consult the GRAMOXONE SUPER label for timing with respect to stage of weed growth and for application methods recommended to minimize crop injury.

RECOMMENDATIONS

For use on almond, apricot, cherry, fig, grape, nectarine, peach, pear, plum, prune and walnut in California only.

DEVRINOL 50-DF/SINBAR* 80-WP COMBINATION

DEVRINOL 50-WP may be used in combination with SINBAR 80-WP for improved control of annual broadleaf weeds. This tank mix does not control emerged weeds.

*Registered trademark of E. I. DuPont De Nemours & Company.

DIRECTIONS FOR USE

NOTE: All rates are given on a broadcast basis. Reduce rates proportionately for band or strip treatment.

For control of all weeds listed, apply 8 pounds DEVRINOL 50-DF AND 1.25 pounds SINBAR 80-WP per broadcast acre to the soil surface. Apply in the Fall through early Spring prior to weed emergence. Do not apply to frozen ground. For best results, the treatment must reach the zone of weed seed germination. If rainfall does not occur the treatment must be shallowly incorporated or irrigated-in following application (sprinkler or flood) with sufficient water to wet the soil to a depth of 2 to 4 inches. In the Western and Pacific Northwest Regions treatment made in November through February must be incorporated-in or irrigated-in if no rainfall occurs within three weeks of application. For treatments made other than in November through February in these regions, and for all treatments applied in the Northern, Southwestern, and Southeastern Regions regardless of season, the treatment must be incorporated-in or irrigated-in within 24 hours if no rainfall occurs.

Use as a directed spray and avoid contact with fruit or foliage. Do not apply when fruit is on the ground during the harvest period. Do not graze treated areas.

RECOMMENDATIONS

For use on apples, peaches, and citrus (restrict use on citrus to California only).

NOTE: To avoid possible crop injury follow these precautions:

- **Apples and Peaches:** Where crop is grown under furrow irrigation or under raised-berm flood irrigation (trees 4" to 6" above waterline), apply only as a band treatment. Do not treat trees planted in the bottom of irrigation furrows or trees grown under flat flood or basin irrigation, and do not use on eroded areas where subsoil or tree roots are exposed, as injury to trees may result. Use only on trees established in the orchard for at least two years.
- **Citrus:** Temporary yellowing of citrus leaves may occur following treatment. Treatment may be used on coarse soils but is not to be used on sands low in organic matter (less than 1%), on poorly drained soils, nor on trees planted in irrigation furrows, as injury to trees may result. Treated areas may be planted to citrus trees one year after the last application. Use only on trees established in the grove for at least one year.

TANK MIXTURES
VEGETABLES AND TOBACCO

IMPORTANT

Observe all cautions and limitations on labeling of all products used in combination.

Make only one application per season.

Do not contaminate water by cleaning of equipment or disposal of wastes.

In each case where a range of rates is given, the lower rates should be used on light (coarse) textured soils and soils low in organic matter. The higher rates should be used on heavy (fine) textured soils.

After harvest or prior to planting of succeeding crops, a deep moldboard or disc plowing operation must be carried out. To avoid injury to crops not specified on the label, do not plant until 12 months after the last DEVRINOL 50-DF application. When applied at rates no greater than 2 lbs. a.i./A, rotational crops which are not registered on this label may be planted no sooner than 12 months after application when harvest and use of the mature crop(s) only is for food or feed purposes. Crops are not to be grazed or harvested for food or feed purposes earlier than maturity.

DIRECTIONS FOR USE

The recommended tank mix controls weeds by interfering with normal seed germination and seedling development. It does not control established weeds. All weed growth and crop stubble should be thoroughly worked into the soil before treatment.

TANK MIX DIRECTIONS

Follow directions given in the sections DEVRINOL 50-DF GENERAL DIRECTIONS FOR USE AND TANK MIXTURES—IMPORTANT.

At labeled spray volumes of 20-50 gallons per acre on transplanted tobacco, DEVRINOL 50-DF is physically compatible with DYFONATE®, DASANIT*, DIAZINON**, NEMACUR*, RIDOMIL** and TILLAM®.

* Registered trademarks of Mobay Chemical Corporation
** Registered trademarks of Ciba-Geigy Corporation.

DEVRINOL 50-DF/TILLAM 6-E TANK MIX COMBINATION APPLIED PRE-PLANT INCORPORATED

A tank mix combination of DEVRINOL 50-DF and TILLAM 6-E can be applied and mixed (incorporated) into the soil before planting for weed control in tomatoes and tobacco.

ADDITIONAL WEEDS CONTROLLED BY DEVRINOL 50-DF/TILLAM 6-E TANK MIX AT RECOMMENDED RATES.

ANNUAL GRASSES

Bermudagrass	*(Cynodon dactylon)*
Narrowleaf	*(Brachiaria piligera)*
Signalgrass	*(Avena fatua)*
Wild Oats	*(Setaria italica)*
Millet, Foxtail	

ANNUAL BROADLEAF WEEDS

Henbit (Deadnettle)	*(Lamium amplexicaule)*
Nettleleaf Goosefoot	*(Chenopodium murale)*
Hairy Nightshade*	*(Solanum villosum)*
Blackeyedsusan	*(Sweet Suzy)*
Shepherdspurse	*(Rudbeckia serotina)*
	(Capsella bursa-pastoris)

* Suppression (partial control)

PERENNIAL WEEDS

Purple Nutsedge	*(Cyperus rotundus)*
Yellow Nutsedge	*(Cyperus esculentus)*

Existing stands of nutsedge must be turned under and chopped up thoroughly before treatment.

APPLICATION DIRECTIONS

The soil should be well-worked and dry enough to permit good soil mixing (incorporation). Pour the recommended amounts of DEVRINOL 50-DF and TILLAM 6-E into a spray tank during the filling operations. Apply in 25 to 50 gallons of water per acre.

For soil incorporated applications, use a properly calibrated low-pressure, boom-type sprayer having good agitation.

In irrigated areas, do not apply before pre-irrigation. For all band applications, reduce dosage proportionately depending upon row spacing and band to be treated. Do not band application on rocky ground because thorough incorporation is not possible.

INCORPORATION DIRECTIONS
The herbicides must be mixed (incorporated) into the soil IMMEDIATELY after application to prevent loss by evaporation. Whenever possible, application and incorporation should be done in the same operation.

INCORPORATION BEFORE PLANTING
Use power-driven cultivation equipment set to cut to a depth of 2 to 3 inches in coarse soils and 3 to 4 inches in heavy soils OR a tandem disc set to cut to a depth of 6 inches, followed by a spike-tooth harrow or some other leveling device which extends beyond ends of the disc. For thorough mixing, disc in two different directions (cross disc).

INCORPORATION AT PLANTING
Use power-driven cultivation equipment set to cut to a depth of 2 to 3 inches.

PLANTING
Crops should be planted or seeded immediately after application and incorporation. Do not use a drag behind behind the planter as it may concentrate the herbicides over the seed row and cause crop injury

CULTURAL PRACTICES FOLLOWING APPLICATION
Shallow cultivation may be necessary to control those susceptible weeds that escape control as well as those weeds not susceptible to either herbicide. Do not cultivate deeper than the depth of herbicide incorporation.

When cultivating fields where the tank mix has been banded in the row, use shields to prevent the movement of untreated soil into the treated row.

RECOMMENDATIONS
TOMATOES (Direct Seeded [California Only] and Transplanted): Apply 2 to 4 pounds DEVRINOL 50-DF plus 2⅔ to 4 quarts TILLAM 6-E per broadcast acre. Use the lower rates on light soils (coarse textured-sandy) and the higher rate on heavy soils (fine textured-clay). Do not use hotcaps on transplanted tomatoes.

TOBACCO (Transplanted): Before transplanting, apply 2 pounds DEVRINOL 50-DF plus 2⅔ quarts TILLAM 6-E. This combination may be tank mixed with FURADAN* 4-F or MOCAP** 6-E at recommended rates. Read the FURADAN 4-F and MOCAP 6-E labels for recommended uses, directions for use, precautions and limitations. When filling the spray tank follow these orders of mixing: DEVRINOL 50-DF slurry, FURADAN 4-F, TILLAM 6-E, OR, DEVRINOL 50-DF slurry, TILLAM 6-E, MOCAP 6-E.

* Registered trademark of FMC Corporation
** Registered trademark of Rhone-Poulenc Chemical Company.

DEVRINOL 50-DF/TILLAM 6-E/DYFONATE [4-E, 4-EC or 4-ED] TANK MIX COMBINATION
A tank mix combination of DEVRINOL 50-DF, DYFONATE 4-E, 4-EC or 4-ED and TILLAM 6-E can be applied and mixed into the soil before transplanting tomatoes and tobacco, for control of weeds and soil insects. Refer to DEVRINOL 50-DF/TILLAM 6-E Tank Mix Combination for list of weeds controlled.

NOTE: DYFONATE 4-E, 4-EC AND 4-ED are Restricted Use Pesticides. Observe all cautions and limitations on labeling of all products used in combinations. Make only one application per season. Do not contaminate water by cleaning equipment or disposal of wastes.

DIRECTIONS FOR USE
The recommended tank mix controls weeds by interfering with normal seed germination and seedling development. It does not control established weeds. All weed growth and crop stubble should be thoroughly worked into the soil before treatment.

For GENERAL USE PRECAUTIONS and DIRECTIONS FOR USE refer to DEVRINOL 50-DF/TILLAM 6-E Tank Mix Recommendations for methods of tank mixing, application and incorporation. When using DEVRINOL 50-DF, make a slurry and add to a nearly filled spray tank with adequate agitation.

RECOMMENDATIONS
TOMATOES (Direct Seeded [California Only] and Transplanted): Apply 2 to 4 pounds DEVRINOL 50-DF plus 2⅔ to 4 qts. TILLAM 6-E plus 2 qts. DYFONATE 4-E, 4-EC or 4-ED for control of symphylans or 4 qts. for control of symphylans and wireworms in 25 to 50 gallons of water per broadcast acre. Use the lower DEVRINOL 50-DF and TILLAM 6-E rates on light soils (coarse textured-sandy) and the higher rates on heavy soils (fine textured-clay). Do not use hotcaps on transplanted tomatoes.

TOBACCO (Transplanted): Before transplanting apply 2 pounds DEVRINOL 50-DF plus 2⅔ qts. TILLAM 6-E plus 2 qts DYFONATE 4-E, 4-EC OR 4-ED for control of wireworms in 25 to 50 gallons of water per broadcast acre.

STORAGE AND DISPOSAL
Do not contaminate water, food or feed by storage or disposal.

STORAGE: Keep container closed when not in use. Store in a cool dry place. Do not store near seeds, fertilizer or food stuffs.

PESTICIDE DISPOSAL: Pesticide wastes are acutely hazardous. Improper disposal of excess pesticide, spray mixture, or rinsate is a violation of Federal law. If these wastes cannot be disposed of by use according to label instructions, contact your State Pesticide or Environmental Control Agency, or the Hazardous Waste representative at the nearest EPA Regional Office for guidance.

CONTAINER DISPOSAL: PLASTIC BAGS—Do not reuse. Completely empty bag into application equipment. Then dispose of empty bag in a sanitary landfill or by incineration, or, if allowed by State and local authorities, by burning. If burned, stay out of smoke.

PLASTIC CONTAINERS: Triple rinse (or equivalent). Then offer for recycling or reconditioning, or puncture and dispose of in a sanitary landfill, or incineration, or, if allowed by state and local authorities, by burning. If burned, stay out of smoke.

NOTICE TO BUYER AND USER
Seller warrants that this product conforms to the chemical description on the label and is reasonably fit for the purposes stated on the label when used in accordance with directions under normal conditions of use. This warranty does not extend to the use of this product contrary to label instruction, or under abnormal use conditions, or under conditions not reasonably foreseeable to Seller, and Buyer and User assume the risk of any such use. SELLER DISCLAIMS ALL OTHER WARRANTIES EXPRESSED OR IMPLIED INCLUDING ANY WARRANTY OF FITNESS OR MERCHANTABILITY. SELLER SHALL NOT BE LIABLE FOR CONSEQUENTIAL, SPECIAL OR INDIRECT DAMAGES RESULTING FROM THE USE OR HANDLING OF THIS PRODUCT AND SELLER'S SOLE LIABILITY AND BUYER'S AND USER'S EXCLUSIVE REMEDY SHALL BE LIMITED TO THE REFUND OF THE PURCHASE PRICE.

A-2R

The information and specimen labels in this booklet are correct as of the publication date but are subject to change without notice. For current information, contact ICI Americas Inc.

DEVRINOL® 50-WP

SELECTIVE HERBICIDE WETTABLE POWDER

For use on certain Citrus, Nuts, Pome fruits, Small fruits, Stone fruits, Vegetables and Tobacco

ACTIVE INGREDIENT:	
N,N-Diethyl-2-(1-naphthalenyloxy)-propionamide	50.0%
INERT INGREDIENTS:	50.0%
TOTAL	100.0%

Contains 50 lbs. of active ingredient per 100 lbs.
EPA Reg. No. 10182-205

KEEP OUT OF REACH OF CHILDREN
CAUTION

STATEMENT OF PRACTICAL TREATMENT
First Aid Immediately start the procedures given below. If further treatment is required, contact a Poison Center, a physician, or the nearest hospital.

If swallowed: Immediately give several glasses of water and induce vomiting by gagging the victim with a finger placed on the back of the victim's tongue. Give fluids until vomitus is clear. If victim is unconscious or convulsing, do not induce vomiting or give anything by mouth.

For skin contact: Flush all affected areas with plenty of water for several minutes. For eye contact, hold eyelids apart and flush with plenty of water for 15 minutes. Remove and clean contaminated clothing and shoes before reuse. Seek medical attention if skin or eye irritation persists.

If inhaled: Remove to fresh air. Seek medical attention if respiratory irritation occurs or if breathing becomes difficult.

For other product information, contact your local ICI Americas Inc. distributor or see Product Safety Information Sheet.

FOR 24-HOUR EMERGENCY MEDICAL ASSISTANCE CALL 1-800-F-A-S-T-M-E-D (327-8633).
FOR CHEMICAL EMERGENCY: Spill, leak, fire, exposure, or accident call CHEMTREC 1-800-424-9300.

PRECAUTIONARY STATEMENTS

HAZARDS TO HUMANS AND DOMESTIC ANIMALS
CAUTION
MAY BE HARMFUL IF SWALLOWED
Do not breathe spray mist. Avoid contact with skin, eyes and clothing. Wash thoroughly with soap and water after use.

ENVIRONMENTAL HAZARDS
- Do not apply directly to water or wetlands.
- Do not contaminate water by cleaning of equipment or disposal of wastes. Apply this product as specified on this label.
- Do not contaminate water used for domestic purposes.

- Avoid Contamination of Food and Feed.

General Precautions

- Read all instructions before using.
- Use only for recommended purposes and at recommended rates. Do not overdose.
- Except where specifically noted, DEVRINOL 50-WP is not recommended for use on soil containing more than 10% organic matter.
- Spraying on ground heavily covered with leaves or trash may adversely affect weed control.
- Do not apply to frozen ground.
- To minimize injury to succeeding crops not listed on this label, a deep disc plow or moldboard plow operation (minimum 10 inches) must be carried out prior to planting. To avoid injury to crops not specified on the label, do not plant until 12 months after the last application of DEVRINOL 50-WP.
- When applied at rates no greater than 4 lbs. DEVRINOL 50-WP per acre, rotational crops which are not registered on this label may be planted no sooner than 12 months after application when harvest and use of mature crop(s) is only for food or feed purposes. Crops are not to be grazed or harvested for food or feed purposes earlier than maturity.

GENERAL DIRECTIONS FOR USE

It is a violation of Federal law to use this product in a manner inconsistent with its labeling.

DEVRINOL 50-WP is a selective surface-applied or incorporated herbicide for use in certain types of newly planted or established crops. DEVRINOL 50-WP does not control established weeds. All weed growth and crop stubble in the fields must be thoroughly worked into the soil prior to planting. Spray equipment should be carefully calibrated before use and checked frequently during application to be sure nozzles are operating properly and are delivering a uniform spray pattern. Apply the recommended rate of DEVRINOL 50-WP in 20 to 100 gallons of water per acre in a low pressure boom-type sprayer.

Be sure the spray tank, boom and nozzle have been thoroughly flushed with clean water before use. Sufficient jet or mechanical agitation should be provided during the filling operation and during application.

When mixing DEVRINOL 50-WP alone, or in combination with other recommended wettable powders, make a slurry and add to a nearly filled spray tank with adequate agitation. This will enhance initial mixing. Recommended contact herbicides and spreaders should always be added to a nearly filled spray tank. Never add them to the slurry.

APPLICATIONS THROUGH IRRIGATION SYSTEMS

DEVRINOL 50-WP may be applied at the recommended rates through irrigation water for pre-emergence control of listed weeds in grapes and tree crops (Western Region Only). Do not allow treatment to contact foliage or crop. Remove existing weed growth before application. Meter the DEVRINOL 50-WP into the irrigation water during the entire period. Apply with sufficient water to wet the soil to a depth of 2–4 inches. Good agitation should be maintained during the entire application period. Avoid run-off.

Use Precautions:

A. Apply this product only through sprinkler including center pivot, lateral move, end tow, side (wheel) roll, traveler, solid set, or hand move, flood (basin); furrow; border irrigation systems. Do not apply this product through any other type of irrigation system.

B. Crop injury, lack of effectiveness, or illegal pesticide residues in the crop can result from non-uniform distribution of treated water.

C. If you have questions about calibration, you should contact State Extension Service specialists, equipment manufacturers or other experts.

D. Do not connect an irrigation system (including greenhouse systems) used for pesticide application to a public water system unless the pesticide label-prescribed safety devices for public water systems are in place.

E. A person knowledgeable of the chemigation system and responsible for its operation, or under the supervision of the responsible person, shall shut the system down and make necessary adjustments should the need arise.

For Sprinkler Irrigation

1. The system must contain a functional check value, vacuum relief valve, and low pressure drain appropriately located on the irrigation pipeline to prevent water source contamination from backflow.
2. The pesticide injection pipeline must contain a functional, automatic, quick-closing check valve to prevent the flow of fluid back toward the injection pipe.
3. The pesticide injection pipeline must also contain a functional, normally closed, solenoid-operated valve located on the intake side of the injection pump and connected to the system interlock to prevent fluid from being withdrawn from the supply tank when the irrigation system is either automatically or manually shut down.
4. The system must contain functional interlocking controls to automatically shut off the pesticide injection pump when the water pump motor stops.
5. The irrigation line or water pump must include a functional pressure switch which will stop the water pump motor when the water pressure decreases to the point where pesticide distribution is adversely affected.
6. Systems must use a metering pump, such as a positive displacement injection pump (e.g., diaphragm pump) effectively designed and constructed of materials that are compatible with pesticides and capable of being fitted with a system interlock.
7. Do not apply when wind speed favors drift beyond the area intended for treatment.

For Flood (Basin), Furrow and Border Irrigation

1. Systems using a gravity flow pesticide dispensing system must meter the pesticide into the water at the head of the field and downstream of a hydraulic discontinuity such as a drop structure of weir box to decrease potential for water source contamination from backflow if water flow stops.
2. Systems utilizing a pressurized water and pesticide injection system must meet the following requirements:

a. The system must contain a functional check valve, vacuum relief value, and low pressure drain appropriately located on the irrigation pipeline to prevent water source contamination from backflow.

b. The pesticide injection pipeline must contain a functional, automatic, quick-closing check valve to prevent the flow of fluid back toward the injection pump.

c. The pesticide injection pipeline must also contain a functional, normally closed, solenoid-operated valve located on the intake side of the injection pump and connected to the system interlock to prevent fluid from being withdrawn from the supply tank when the irrigation system is either automatically or manually shut down.

d. The system must contain functional interlocking controls to automatically shut off the pesticide injection pump when the water pump moter stops.

e. The irrigation line or water pump must include a functional pressure switch which will stop the water pump motor when the water pressure decreases to the point where pesticide distribution is adversely affected.

f. Systems must use a metering pump, such as a positive displacement injection pump (e.g., diaphragm pump) effectively designed and constructed of materials that are compatible with pesticides and capable of being fitted with a system interlock.

NOTE: All rates are given on a broadcast basis, per acre. Reduce rates proportionately for band or strip treatment.

WEEDS CONTROLLED BY 4 TO 8 LBS. DEVRINOL 50-WP PER ACRE (CALIFORNIA ONLY)

Surface Application

NOTE: DEVRINOL 50-WP does not control established weeds.

Barnyardgrass, Junglerice	(*Echinochloa* spp.)
Annual Bluegrass	(*Poa annua*)
Large (Hairy) Crabgrass	(*Digitaria sanguinalis*)
Red Sprangletop	(*Leptochloa filiformis*)

DIRECTIONS FOR USE

For control of all weeds listed, apply 4 to 8 pounds DEVRINOL 50-WP per broadcast acre to the soil surface. Apply in the fall through early spring prior to emergence. Do not apply DEVRINOL 50-WP to frozen ground. For best results, the treatment must reach the zone of weed seed germination. If rainfall does not occur the treatment must be shallowly incorporated or irrigated-in following application (sprinkler or flood) with sufficient water to wet the soil to a depth of 2-4 inches.

RECOMMENDATIONS

DEVRINOL 50-WP is recommended for use on the following crops: Almonds, Apples, Apricots, Cherries, Figs, Filberts, Grapefruit, Grapes, Kiwifruit, Lemons, Nectarines, Olives, Oranges, Peaches, Pears, Pecans, Persimmons, Pistachios, Plums, Pomegranates, Prunes, Tangerines and Walnuts.

WEEDS CONTROLLED BY 8 LBS. DEVRINOL 50-WP PER ACRE

Surface Application

NOTE: DEVRINOL 50-WP does not control established weeds.

Annual Grasses	
Barnyardgrass, Junglerice	(*Echinochloa* spp.)
Foxtail Barley	(*Hordeum jubatum*)
Wild Barley	(*Hordeum leporinum*)
Annual Bluegrass	(*Poa Annua*)
Bromes	(*Bromus*)
Cheatgrass	(*B. secalinus*)
Downy Brome	(*B. tectorum*)
Ripgut Brome	(*B. rigidus*)
Soft Chess	(*B. mollis*)
Canarygrass	(*Phalaris canariensis*)
Large (Hairy) Crabgrass	(*Digitaria sanguinalis*)

Smooth Crabgrass	(*Digitaria ischaemum*)
Southwestern Cupgrass	(*Erichloa gracilis*)
Foxtails	(*Setaria*)
Bristly Foxtail	(*S. verticilla*)
Giant Foxtail	(*S. faberi*)
Green Foxtail	(*S. viridis*)
Yellow Foxtail	(*S. lutescens*)
Goosegrass	(*Eleusine indica*)
Seedling Johnsongrass	(*Sorghum halepense*)
Wild Oats	(*Avena fatua*)
Panicums	(*Panicum*)
Guineagrass	(*P. paximum*)
Fall Panicum	(*P. dichotomiflorum*)
Texas Panicum	(*P. texanum*)
Witchgrass	(*P. capillare*)
Italian Ryegrass	(*Lolium multiflorum*)
Sandbur	(*Cenchrus* spp.)
Mexican Sprangletop	(*Leptochloa uninervia*)
Red Sprangletop	(*Leptochloa filiformis*)
Stinkgrass	(*Eragrostis cilianensis*)

Annual Broadleaf Weeds

Carpetweed	(*Mollugo verticillata*)
Common Chickweed	(*Stellaria media*)
Coast Fiddleneck	(*Amsinckia intermedia*)
Cudweed, Purple	(*Gnaphalium purpureum*)
Redstem Filaree	(*Erodium circulatorium*)
Common Groundsel	(*Senecio vulgaris*)
Prostrate Knotweed	(*Polygonum aviculare*)
Lambsquarters	(*Chenopodium album*)
Prickly Lettuce	(*Lactuca serriola*)
Little Mallow*	(*Malva parviflora*)
Redroot Pigweed	(*Amaranthus retroflexus*)
Pineappleweed	(*Matricaria matricarioides*)
Common Purslane	(*Portulaca oleracea*)
Horse Purslane	(*Trianthema portullacastrum*)
Common Ragweed**	(*Ambrosia artemisiifolia*)
Annual Sowthistle	(*Sonchus oleraceus*)

* From seed
** Suppression

DIRECTIONS FOR USE

NOTE: All rates are given on a broadcast basis. Reduce rates proportionately for band or strip treatment.

For control of all weeds listed, apply 8 pounds DEVRINOL 50-WP per broadcast acre to the soil surface. Apply in the Fall through early Spring prior to weed emergence. Do not apply DEVRINOL 50-WP to frozen ground. For best results, the treatment must reach the zone of weed seed germination. If rainfall does not occur the treatment must be shallowly incorporated or irrigated—in following application (sprinkler or flood) with sufficient water to wet the soil to a depth of 2 to 4 inches.

Allowable Time for Treatment To Remain On Soil Surface (by Region):

West: Three weeks (November through February) or 24 hours (remainder of year).

Pacific Northwest: Three weeks (Fall through Early Spring) or 24 hours (remainder of year).

Other Regions: 24 hours.

Use as a directed spray and avoid contact with fruit or foliage. Do not apply when fruit or nuts are on the ground during the harvest period. Do not graze treated areas.

Regional Use Map

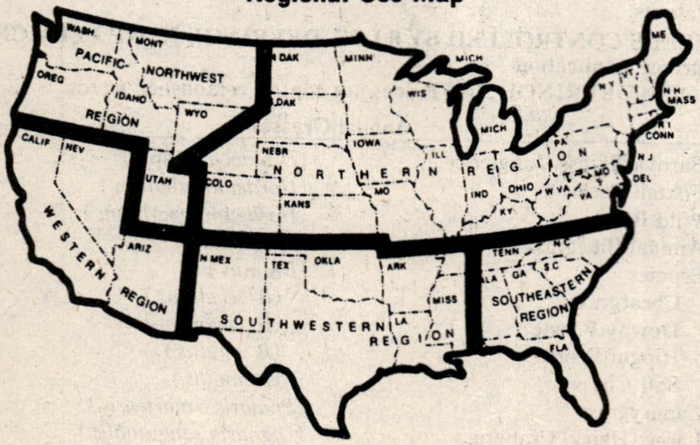

FRUIT AND NUT CROPS SURFACE APPLICATION

Recommendations

DEVRINOL 50-WP is recommended for use on the following crops. Refer to regional map under "DIRECTIONS FOR USE."

ALMONDS: (All Regions)

Rate: 8 pounds DEVRINOL 50-WP per broadcast acre. In the Western region an additional 8 pounds may be applied during the growing season. Do not apply within 35 days of harvest.

Use Recommendations: Apply to a weed free soil surface. May be applied to newly planted and established crop.

APPLES: (All Regions)

Rate: 8 pounds DEVRINOL 50-WP per broadcast acre. In the Western Region an additional 8 pounds may be applied during the growing season. Do not apply within 35 days of harvest.

Use Recommendations: Apply to a weed free soil surface. May be applied to newly planted and established crop.

APRICOTS: (All Regions)

Rate: 8 pounds DEVRINOL 50-WP per broadcast acre. In the Western Region an additional 8 pounds may be applied during the growing season. Do not apply within 35 days of harvest.

Use Recommendations: Apply to a weed free soil surface. May be applied to newly planted and established Crop.

AVOCADOS: (Western Region Only)

Rate: 8 pounds DEVRINOL 50-WP per broadcast acre. An additional 8 pounds may be applied during the growing season. Do not apply within 35 days of harvest.

Use Recommendations: Apply to a weed free soil surface. May be applied to established crop only.

BLUEBERRIES: (All Regions)

Rate: 8 pounds DEVRINOL 50-WP per broadcast acre.

Use Recommendations: Apply to a weed free soil surface. May be applied to newly planted and established crop.

CANEBERRIES (blackberries, raspberries, boysenberries and loganberries): (All Regions)

Rate: 8 pounds DEVRINOL 50-WP per broadcast acre.

Use Recommendations: Apply to a weed free soil surface. May be applied to newly planted and established crop.

CHERRIES: (All Regions)

Rate: 8 pounds DEVRINOL 50-WP per broadcast acre. In the Western Region an additional 8 pounds may be applied during the growing season. Do not apply within 35 days of harvest.

Use Recommendations: Apply to a weed free soil surface. May be applied to newly planted and established crop.

FIGS: (All Regions)

Rate: 8 pounds DEVRINOL 50-WP per broadcast acre. In the Western Region an additional 8 pounds may be applied during the growing season. Do not apply within 35 days of harvest.

Use Recommendations: Apply to a weed free soil surface. May be applied to newly planted and established crop.

FILBERTS: (All Regions)

Rate: 8 pounds DEVRINOL 50-WP per broadcast acre. In the Western Region an additional 8 pounds may be applied during the growing season. Do not apply within 35 days of harvest.

Use Recommendations: Apply to a weed free soil surface. May be applied to newly planted and established Crop.

GRAPEFRUIT: (All Regions)

Rate: 8 pounds DEVRINOL 50-WP per broadcast acre. In the Western Region an additional 8 pounds may be applied during the growing season. Do not apply within 35 days of harvest.

Use Recommendations: Apply to a weed free soil surface. May be applied to newly planted and established crop.

GRAPES: (All Regions)

Rate: 8 pounds DEVRINOL 50-WP per broadcast acre. In the Western Region an additional 8 pounds may be applied during the growing season. Do not apply within 35 days of harvest.

Use Recommendations: Apply to a weed free soil surface. May be applied to newly planted and established crop.

KIWIFRUIT: (Western Region Only)

Rate: 8 pounds DEVRINOL 50-WP per broadcast acre. An additional 8 pounds may be applied during the growing season. Do not apply within 35 days of harvest.

Use Recommendations: Apply to a weed free soil surface. May be applied to newly planted and established crop.

LEMONS: (All Regions)

Rate: 8 pounds DEVRINOL 50-WP broadcast acre. In the Western Region an additional 8 pounds may be applied during the growing season. Do not apply within 35 days of harvest.

Use Recommendations: Apply to a weed free soil surface. May be applied to newly planted and established crop.

NECTARINES: (All Regions)
Rate: 8 pounds DEVRINOL 50-WP per broadcast acre. In the Western Region an additional 8 pounds may be applied during the growing season. Do not apply within 35 days of harvest.
Use Recommendations: Apply to a weed free soil surface. May be applied to newly planted and established crop.

OLIVES: (Western Region Only)
Rate: 8 pounds DEVRINOL 50-WP per broadcast acre. An additional 8 pounds may be applied during the growing season. Do not apply within 35 days of harvest.
Use Recommendations: Apply to a weed free soil surface. May be applied to newly planted and established crop.

ORANGES: (All Regions)
Rate: 8 pounds DEVRINOL 50-WP per broadcast acre. In the Western Region an additional 8 pounds may be applied during the growing season. Do not apply within 35 days of harvest.
Use Recommendations: Apply to a weed free soil surface. May be applied to newly planted and established crop.

PEACHES: (All Regions)
Rate: 8 pounds DEVRINOL 50-WP per broadcast acre. In the Western Region an additional 8 pounds may be applied during the growing season. Do not apply within 35 days of harvest.
Use Recommendations: Apply to a weed free soil surface. May be applied to newly planted and established crop.

PEARS: (All Regions)
Rate: 8 pounds DEVRINOL 50-WP per broadcast acre. In the Western Region an additional 8 pounds may be applied during the growing season. Do not apply within 35 days of harvest.
Use Recommendations: Apply to a weed free soil surface. May be applied to newly planted and established crop.

PECANS: (All Regions)
Rate: 8 pounds DEVRINOL 50-WP per broadcast acre. In the Western Region an additional 8 pounds may be applied during the growing season. Do not apply within 35 days of harvest.
Use Recommendations: Apply to a weed free soil surface. May be applied to newly planted and established crop.

PERSIMMONS: (Western Region Only)
Rate: 8 pounds DEVRINOL 50-WP per broadcast acre. An additional 8 pounds may be applied during the growing season. Do not apply within 35 days of harvest.
Use Recommendations: Apply to a weed free soil surface. May be applied to established crop only.

PISTACHIOS: (All Regions)
Rate: 8 pounds DEVRINOL 50-WP per broadcast acre.
Use Recommendations: Apply to a weed free soil surface. May be applied to newly planted and established crop.

PLUMS: (All Regions)
Rate: 8 pounds DEVRINOL 50-WP per broadcast acre. In the Western Region an additional 8 pounds may be applied during the growing season. Do not apply within 35 days of harvest.
Use Recommendations: Apply to a weed free soil surface. May be applied to newly planted and established crop.

POMEGRANATES: (Western Region Only)
Rate: 8 pounds DEVRINOL 50-WP per broadcast acre. Do not apply within 6 months of harvest.
Use Recommendations: Apply to a weed free soil surface. May be applied to established crop only.

PRUNES: (All Regions)
Rate: 8 pounds DEVRINOL 50-WP per broadcast acre. In the Western Region an additional 8 pounds may be applied during the growing season. Do not apply within 35 days of harvest.
Use Recommendations: Apply to a weed free soil surface. May be applied to newly planted and established crop.

RHUBARB: (Pacific Northwest Only)
Rate: 8 pounds DEVRINOL 50-WP per broadcast acre.
Use Recommendations: Apply to a weed free soil surface in Winter when rhubarb is dormant.

STRAWBERRIES: (All Regions)
Rate: 8 pounds DEVRINOL 50-WP per broadcast acre.
Use Recommendations: Apply to a weed free soil surface. May be applied to newly transplanted and established crop. Delay application until desired number of daughter plants have become established. Do not apply from bloom through harvest.

TANGERINES: (All Regions)
Rate: 8 pounds DEVRINOL 50-WP per broadcast acre. In the Western Region an additional 8 pounds may be applied during the growing season. Do not apply within 35 days of harvest.
Use Recommendations: Apply to a weed free soil surface. May be applied to newly planted and established crop.

WALNUTS: (All Regions)
Rate: 8 pounds DEVRINOL 50-WP per broadcast acre. In the Western Region an additional 8 pounds may be applied during the growing season. Do not apply within 35 days of harvest.
Use Recommendations: Apply to a weed free soil surface. May be applied to newly planted and established crop.

OTHER CROPS SURFACE APPLICATION

Recommendations

ARTICHOKES: (Western Region Only)
Rate: 8 pounds DEVRINOL 50-WP per broadcast acre.
Use Recommendations: Apply to a weed free soil surface. May be applied to established crop only.

ASPARAGUS: (All Regions)
Rate: 8 pounds DEVRINOL 50-WP per broadcast acre.
Use Recommendations: Apply to a weed free soil surface. May be applied to established crop only.

MINT: (Pacific Northwest Only)
Rate: 8 pounds DEVRINOL 50-WP per broadcast acre.
Use Recommendations: Apply to a weed free soil surface. May be applied to a newly planted and established crop.

Where recommended for use on established crops, apply DEVRINOL 50-WP only to stands that have been established for at least one growing season. Where recommended for use on newly planted crops, application may be made immediately after planting young trees, vines, and bushes.

DEVRINOL 50-WP may be applied at the recommended rates through irrigation water for pre-emergence control of listed weeds in grapes and tree crops (Western Region Only). Do not allow treatment to contact foliage or crop. Remove existing weed growth before application. Apply at the start of irrigation set and with sufficient water to wet the soil to a depth of 2-4 inches. Avoid run-off. These irrigation systems should be equipped with automatic shutoff devices and valves to prevent back flow into the water source.

WEEDS CONTROLLED BY 2 TO 4 LBS. DEVRINOL 50-WP PREPLANT

Incorporated Application

NOTE: DEVRINOL 50-WP does not control established weeds.

Annual Grasses	
Barnyardgrass, Junglerice	(*Echinochloa* spp.)
Crabgrass	(*Digitaria* spp.)
Foxtails	(*Setaria*)
Giant Foxtail	(*S. faberi*)
Green Foxtail	(*S. viridis*)
Yellow Foxtail	(*S. lutescens*)
Goosegrass	(*Eleusine indica*)
Seedling Johnsongrass	(*Sorghum halepense*)
Panicums	(*Panicum*)
Fall Panicum	(*P. dichotomiflorum*)
Witchgrass	(*P. capillare*)
Stinkgrass	(*Eragrostis cilianensis*)
Annual Broadleaf Weeds	
Carpetweed	(*Mollugo verticillata*)
Purple Cudweed	(*Gnaphalium purpureum*)
Galinsoga*	(*Galinsoga* spp.)
Lambsquarters	(*Chenopodium album*)
Redroot Pigweed	(*Amaranthus retroflexus*)
Smooth Pigweed	(*Amaranthus hybridus*)
Common Purslane	(*Portulaca oleracea*)
Florida Pusley	(*Richardia scabra*)
Common Ragweed**	(*Ambrosia artemisiifolia*)

* Controlled at 4 lb. DEVRINOL 50-WP
** Suppression

DIRECTIONS FOR USE OF DEVRINOL 50-WP PRE-PLANT INCORPORATED

On flat-planted or bedded crops apply, before planting, to well-worked soil that is dry enough to permit thorough incorporation to a depth of 1 to 2 inches. Incorporate the same day as applied, using equipment which will result in uniform incorporation of the herbicide to the desired depth.

Whenever a rate range is recommended for DEVRINOL 50-WP, use the lower rate on light soil (coarse textured-sandy) and the higher rate on heavy soil (fine textured-clay).

After harvest or prior to planting of succeeding crops, a deep moldboard or disc plowing operation must be carried out.

When applied at rates no greater than 2 lbs. a.i./A, rotational crops which are not registered on this label may be planted no sooner than 12 months after application when harvest and use of the mature crop(s) only is for food or feed purposes. Crops are not to be grazed or harvested for food or feed purposes earlier than maturity.

MELONS, TOBACCO AND VEGETABLES PRE-PLANT INCORPORATED APPLICATION

Recommendations

NOTE: All rates are given on a broadcast basis. Reduce rates proportionately for band or strip treatment.

COLE CROPS—Broccoli, Brussels Sprouts, Cauliflower: (California Only)

Rate: 1-2 pounds DEVRINOL 50-WP per broadcast acre.

Use Recommendations: For suppression of Annual Bluegrass, Barnyardgrass, Common Groundsel, Common Purslane and Redroot Pigweed. Apply to a weed free soil surface.

May be applied to direct seeded and transplanted crop, preplant incorporated or as a surface spray after planting. Shallow incorporate to a depth of 1-2 inches or sprinkler irrigate within 24 hours using sufficient water to wet the soil to a depth of 2-4 inches. Use the lower rate on light soil (coarse textured-sandy) and the higher rate on heavy soil (fine textured-clay).

COLE CROPS—Broccoli, Brussels Sprouts, Cabbage, Cauliflower: (Northern Region Only)

Rate: 2 pounds DEVRINOL 50-WP per broadcast acre.

Use Recommendations: Apply to a weed free surface. May be applied to direct seeded and transplanted crop, pre-plant incorporated or as a surface spray after planting. Shallow incorporate no deeper than seeding depth or sprinkler irrigate within 24 hours using sufficient water to wet the soil to a depth of 2-4 inches.

COLE CROPS—Broccoli, Brussels Sprouts, Cauliflower: (Pacific Northwest Only)

Rate: 2-4 pounds DEVRINOL 50-WP per broadcast acre.

Use Recommendations: Apply to a weed free surface. May be applied to direct seeded and transplanted crop, pre-plant incorporated or as a surface spray after planting. Shallow incorporate no deeper than seeding depth or sprinkler irrigate within 24 hours using sufficient water to wet the soil to a depth of 2-4 inches. Use the lower rate on light soil (coarse textured-sandy) and the higher rate on heavy soil (fine textured-clay).

COLE CROPS—Broccoli, Brussels Sprouts, Cabbage, Cauliflower: (Southeast Region Only)

Rate: 4 pounds DEVRINOL 50-WP per broadcast acre.

Use Recommendations: Apply to a weed free surface. May be applied to direct seeded (except in Florida) and transplanted crop, as a surface spray after planting. Sprinkler irrigate within 24 hours using sufficient water to wet the soil to a depth of 2-4 inches.

COLE CROPS—Broccoli, Brussels Sprouts, Cabbage, Cauliflower: (Southwest Region)

Rate: 2-4 pounds DEVRINOL 50-WP per broadcast acre.

Use Recommendations: Apply to a weed free soil surface. May be applied to direct seeded and transplanted crop, preplant incorporated or as a surface spray after planting. Incorporate the surface spray to a shallow depth of one inch, avoiding the seed placement when mixing the soil. Use the lower rate on light soil (coarse textured-sandy) and the higher rate on heavy soil (fine textured-clay).

EGGPLANTS: (All Regions)

Rate: Before transplanting, apply 2-4 pounds DEVRINOL 50-WP per broadcast acre.

Use Recommendations: Apply to a weed free soil surface. May be applied to transplanted crop only. Use the lower rate on light soil (coarse textured-sandy) and the higher rate on heavy soil (fine textured-clay).

MELONS (Except cantaloupes): (Texas, South of Highway I-10 Only)

Rate: 2-4 pounds DEVRINOL 50-WP per broadcast acre.

Use Recommendations: Apply to a weed free soil surface. May be applied to direct seeded crop as a preemergence surface treatment only. Incorporate the surface spray to a shallow depth of one inch, avoiding the seed placement when mixing the soil. Use the lower rate on light soil (coarse textured-sandy) and the higher rate on heavy soil (fine textured-clay).

PEPPERS: (All Regions)

Rate: 2-4 pounds DEVRINOL 50-WP per broadcast acre.

Use Recommendations: Apply to a weed free soil surface. May be applied to direct seeded and transplanted crop. Use the lower rate on light soil (coarse textured-sandy) and the higher rate on heavy soil (fine textured-clay).

TOMATOES: (All Regions)

Rate: 2-4 pounds DEVRINOL 50-WP per broadcast acre.

Use Recommendations: Apply to a weed free soil surface. May be applied to direct seeded and transplanted crop. Use the lower rate on light soil (coarse textured-sandy) and the higher rate on heavy soil (fine textured-clay).

TOBACCO: (All Regions—Except North and South Carolina)

Rate: Before transplanting, apply 2-4 pounds DEVRINOL 50-WP per broadcast acre.

Use Recommendations: Apply to a weed free soil surface. Use the lower rate on light soil (coarse textured-sandy) and the higher rate on heavy soil (fine textured-clay). In Southeast only (except North and South Carolina): Apply up to 3 weeks before transplanting. No waiting period is necessary.

POSTTRANSPLANT SURFACE AND LAYBY SURFACE APPLICATIONS FOR USE IN TRANSPLANTED TOBACCO (KENTUCKY, MARYLAND, VIRGINIA AND SOUTHEASTERN REGION ONLY)

Weeds Controlled

See list of weeds controlled under DEVRINOL 50-WP preplant incorporated application.

DIRECTIONS FOR USE

At Transplanting: Apply the recommended amount of DEVRINOL 50-WP to the soil surface immediately after transplanting. May be applied over the top of transplants.

At Layby: Apply the recommended amount of DEVRINOL 50-WP in a band immediately after the layby cultivation. Apply as a directed spray to the row middles.

Use 20 to 50 gallons of spray per acre. For best results the treatment must reach the zone of weed seed germination. If rainfall does not occur within 5 days after application, the treatment must be shallowly incorporated or irrigated-in with sufficient overhead irrigation to wet the soil to a depth of 2 to 4 inches.

Recommendations

Rate: Apply 2 to 4 pounds DEVRINOL 50-WP per broadcast acre.

Use Recommendations: Apply to a weed free soil surface. Reduce rates proportionately for band application. Treatments may follow a preplant incorporated application of DEVRINOL 50-WP or DEVRINOL 50-WP/TILLAM® 6-E. Do not apply more than 4 pounds DEVRINOL 50-WP in any one season.

Rotational Crop Restrictions

When DEVRINOL 50-WP is used on tobbaco posttransplant or laybay at 2 to 4 pounds, surface applied, small grains may be seeded in rotation in the Fall to prevent soil erosion. These grains may be stunted but not otherwise adversely affected.

No tolerance is established for residues of DEVRINOL 50-WP in small grains. Rotational cover crops must be destroyed by plowing under or otherwise prevented from reaching food or feed production.

If rotating out of tobacco the following Spring, to avoid injury to crops not specified on this label do not plant until 12 months after the last DEVRINOL 50-WP application.

When applied at rates no greater than 4 lbs. DEVRINOL 50-WP per acre, rotational crops which are not registered on this label may be planted no sooner than 12 months after application when harvest and use of the mature crop(s) only is for food or feed purposes. Crops are not to be grazed or harvested for food or feed purposes earlier than maturity.

TANK MIXTURES FRUIT AND NUT CROPS

Important

Observe all cautions and limitations on labeling of all products used in combination.

Make only one application per season.

Do not contaminate water by cleaning of equipment or disposal of wastes. Apply these combinations only as specified on this label.

For a tank mix application, apply the recommended dosage (see directions for use) in 20 to 100 gallons of water per acre.

In any tank mix including GRAMOXONE SUPER*, add an approved nonionic spreader to the tank water, such as ORTHO** X-77 spreader, at the rate of 8 oz. per 100 gallons.

*GRAMOXONE SUPER is a registered trademark of ICI Americas, Inc.

**ORTHO is a registered trademark of Chevron Chemical Company.

DEVRINOL 50-WP/GRAMOXONE SUPER COMBINATION

DEVRINOL 50-WP will not control emerged weeds. GRAMOXONE SUPER may be used to control existing weeds when applied prior to, or in combination with DEVRINOL 50-WP.

Refer to the section on DEVRINOL 50-WP surface application for a list of weeds controlled.

DIRECTIONS FOR USE

Follow use directions for DEVRINOL 50-WP surface application. Read the GRAMOXONE SUPER label for timing with respect to stage of weed growth and for application methods recommended to minimize crop injury.

For control of all weeds listed, apply 8 pounds DEVRINOL 50-WP and 1 to 2 quarts GRAMOXONE SUPER per broadcast acre.

Recommendations

Consult the GRAMOXONE SUPER label for recommended uses in orchards and vineyards. Then read the crop recommendations for DEVRINOL 50-WP surface application to insure that the combination may be used for a particular crop in your state. Recommended *only* for the uses specified on the GRAMOXONE SUPER and DEVRINOL 50-WP labels. Do not apply this combination to caneberries.

DEVRINOL 50-WP/GOAL* 1.6-E COMBINATION (California Only)

DEVRINOL 50-WP may be used in combination with GOAL 1.6-E for control of additional weeds not controlled by DEVRINOL 50-WP alone.

Additional weeds controlled preemergence by DEVRINOL 50-WP/GOAL 1.6-E tank mix, applied preemergence to a relatively weed-free established berm or soil surface, at recommended rates:

Burclover	(*Medicago hipide*)
Filaree, (Redstem,	(*Erodium cicutarium*)
Broadleaf)	(*Erodium botrys*)
Henbit	(*Lamium amplexicaule*)
London Rocket	(*Sisymbrium irio*)
Redmaids	(*Calandrinia caulescens*)
Sheperdspurse	(*Capsella bursa-pastoris*)

*GOAL is a registered trademark of Rohm and Haas Company

DIRECTIONS FOR USE—DORMANT APPLICATION, (California Only)

For control of all weeds listed, apply 8 pounds of DEVRINOL 50-WP and 5.0 quarts GOAL 1.6-E (2.0 lb active) per broadcast acre. Spray should be directed to the soil and the base of dormant trees or vines. Use of a low pressure sprayer equipped with a breakaway boom and flat fan or offcenter (OC) nozzle is recommended. For best results, the treatment must reach the zone of weed germination. Irrigation sufficient to wet the soil to a minimum depth of 2 to 4 inches is recommended for treatments made from November through February 15 if no rainfall occurs within three weeks of application.

NOTE: Applications made in October must be irrigated in (sprinkler or flood) the same day as applied with sufficient water to wet the soil to a minimum depth of 2 to 4 inches.

Cultural practices which result in redistribution or disturbance of the soil surface after treatment will decrease the herbicidal effectiveness of GOAL 1.6-E. For best results from the DEVRINOL 50-WP/GOAL 1.6-E tank mix combination, application should be made to established berms or soil surfaces which are left undisturbed during the time period for which weed control is desired.

Read the GOAL 1.6-E label for specific use directions and precautions.

Recommendations

For use on almond, apricot, cherry, fig, grape, nectarine, peach, pear, pistachio, plum, prune and walnut in California only.

NOTE: To avoid possible crop injury follow these precautions:

- Apply only to dormant fruit or nut trees or grape vines. Do not apply when foliage, buds or fruit/nuts are present.
- Grapes: Do not apply to grapes established less than 3 years. Do not apply to grapes which are not staked or trellised.

DEVRINOL 50-WP/GOAL 1.6-E/GRAMOXONE SUPER COMBINATION (California Only)

DEVRINOL 50-WP/GOAL 1.6-E/GRAMOXONE SUPER is recommended for residual preemergence control of susceptible broadleaf and grassy weeds. GRAMOXONE SUPER is included in the tank mix to control existing emerged weeds.

For weeds controlled refer to the weed lists for DEVRINOL 50-WP and DEVRINOL 50-WP/GOAL 1.6-E tank mix surface applied.

DIRECTIONS FOR USE

For control of all weeds listed, apply 8 pounds DEVRINOL 50-WP, 5 quarts GOAL 1.6-E (2.0 lb active) and 1 quart GRAMOXONE SUPER per broadcast acre.

Follow the use directions and recommendations for the DEVRINOL 50-WP/GOAL 1.6-E combination. Consult the GRAMOXONE SUPER label for timing with respect to stage of weed growth and for application methods recommended to minimize crop injury.

Recommendations

For use on almond, apricot, cherry, fig, grape, nectarine, peach, pear, plum, prune and walnut in California only.

DEVRINOL 50-WP/KARMEX* 80-WP COMBINATION

DEVRINOL 50-WP may be used in combination with KARMEX 80-WP for improved control of annual broadleaf weeds. This tank mix does not control emerged weeds.

*KARMEX is a registered trademark of E.I. DuPont De Nemours & Company.

DIRECTIONS FOR USE

NOTE: All rates are given on a broadcast basis. Reduce rates proportionately for band or strip treatment.

For control of all weeds listed, apply 8 pounds DEVRINOL 50-WP and 1.25 pounds KARMEX 80-WP per broadcast acre to the soil surface. Apply in the Fall through early Spring prior to weed emergence. Do not apply to frozen ground. For best results, the treatment must reach the zone of weed seed germination. If rainfall does not occur, the treatment must be shallowly incorporated or irrigated-in following application (sprinkler or flood) with sufficient water to wet the soil to a depth of 2 to 4 inches. In the Western and Pacific Northwest Regions treatment made in November through February must be incorporated-in or irrigated-in if no rainfall occurs within three weeks of application.

For treatments made other than in November through February in these regions, and for all treatments applied in the Northern, Southwestern, and Southeastern Regions regardless of season, the treatment must be incorporated-in or irrigated-in within 24 hours if no rainfall occurs.

Use as a directed spray and avoid contact with fruit or foliage. Do not apply when fruit is on the ground during the harvest period. Do not graze treated areas.

Recommendations

For use on the following crops: Apples and peaches (All Regions). Citrus (California and Arizona only). Grapes, pears and walnuts (California only).

NOTE: To avoid possible crop injury follow these precautions:

- **Apples and Pears:** Use only under trees established in the orchard for at least 1 year; do not treat varieties grafted on full-dwarf root stocks.
- **Citrus:** Use only under trees established in the grove for at least 1 year.
- **Grapes:** Apply only to established vineyards (at least 3 years old) as a band treatment to grape rows. Do not apply to vines with trunks less than 1½" in diameter as injury may result. On soils low in clay or organic matter (1 to 2%), severe plant injury may result if heavy rainfall or more than one inch of irrigation occurs soon after treatment. This risk must be assumed by the user.
- **Peaches:** Use only under trees established in the orchard for at least 2 years.
- **Walnuts:** Use only under trees established in the orchard for at least 1 year.

DEVRINOL 50-WP/ROUNDUP* COMBINATION

ROUNDUP may be used to control existing weeds when applied prior to or in combination with DEVRINOL 50-WP.

*ROUNDUP is a registered trademark of Monsanto Company.

DIRECTIONS FOR USE

Follow use directions for DEVRINOL 50-WP surface application. Read the ROUNDUP label for timing with respect to stage of weed growth and for application methods. For control of all weeds listed, apply 8 pounds of DEVRINOL 50-WP in 20 to 100 gallons of water per broadcast acre. For application rates of ROUNDUP, refer to the ROUNDUP label.

Recommendations

Consult the ROUNDUP label for recommended uses in orchards and vineyards. Then read the crop recommendations for DEVRINOL 50-WP surface application to insure that the combination may be used for a particular crop in your state. Recommended only for the uses specified on the ROUNDUP and DEVRINOL 50-WP labels.

DEVRINOL 50-WP/SIMAZINE 80-WP COMBINATION (California and Pacific Northwest Only)

DEVRINOL 50-WP may be used in combination with SIMAZINE 80-WP for control of additional weeds not controlled by DEVRINOL 50-WP alone. This tank mix does not control emerged weeds.

Additional weeds controlled by DEVRINOL 50-WP/SIMAZINE 80-WP tank mix, surface applied, at recommended rates:

California Burclover	(*Medicago polymorpha*)
Henbit	(*Lamium amplexicaule*)
London Rocket	(*Sisymbrium irio*)
Marestail (Horseweed)	(*Conyza canadensis*)
Miners Lettuce	(*Montia perfoliata*)
Panicle Willoweed	(*Epilobium paniculatum*)
Redmaids (Rockpurslane)	(*Calandrinia caulescens*)

DIRECTIONS FOR USE

For control of all weeds listed, apply 8 pounds DEVRINOL 50-WP and 0.6 to 1.25 pounds SIMAZINE 80-WP per broadcast acre. Limit treatment of almonds, grapes and peaches to a 4 to 8 foot band centered on the tree row, reducing the dosage proportionately. Apply in the Fall through early Spring, prior to weed emergence. For best results, the treatment must reach the zone of weed seed germination. Treatments made other than in November through February must be shallowly incorporated or irrigated in the same day as applied (sprinkler or flood) with sufficient water to wet the soil to a depth of 2 to 4 inches. Irrigation or shallow incorporation is recommended for treatments made from November through February if no rainfall occurs within three weeks of application.

Read directions on the SIMAZINE 80-WP label concerning nozzle and screen sizes.

Use as a directed spray and avoid contact with fruit or foliage.

Do not apply when fruit or nuts are on the ground during the harvest period. Do not graze treated areas.

Recommendations

For use on almonds, apples, avocados, grapefruit, grapes, lemons, olives, oranges, peaches, pears, and walnuts in California and apples, filberts, grapes and walnuts in the Pacific Northwest.

NOTE: To avoid possible crop injury follow these precautions.

• **Apples and Pears.** Make application only to orchards where trees have been established one year or more.

• **Almonds and Peaches** (California Only). Do not treat trees established in the orchard less than 3 years. Do not treat the Mission variety of almonds. Do not apply to almond trees propagated on plum rootstocks. Do not replant almonds or peaches in treated soil. Do not apply on gravelly, sandy, or loamy sand soil, or on soil with less than 1% organic matter. Do not treat areas where water will accumulate (flood or furrow irrigation water should not be applied to any treated areas or allow to stand on treated areas).

• **Filberts** (Oregon and Washington Only). Do not apply on gravelly, sandy or loamy sand soil.

• **Grapes.** Do not use in vineyards established less than three years.

• **Grapefruit, Lemons, and Oranges** (California Only). Do not use in the Imperial, Coachella, or Palo Verde Valleys. Do not use in nurseries. Do not apply to bedded grapefruit, lemons, or oranges. Do not apply to trees under stress from freeze damage for one year after the freeze.

• **Walnuts.** Do not apply to sandy soil. Leveling and furrowing operations after application will lessen effectiveness of weed control.

DEVRINOL 50-WP/SIMAZINE 80-WP/GRAMOXONE SUPER COMBINATION (California and Pacific Northwest Only)

DEVRINOL 50-WP/SIMAZINE 80-WP will not control emerged weeds. GRAMOXONE SUPER may be used to control existing weeds when applied prior to, or in combination with DEVRINOL 50-WP/SIMAZINE 80-WP.

For weeds controlled refer to the weed lists for DEVRINOL 50-WP and DEVRINOL 50-WP/SIMAZINE 80-WP tank mix surface applied.

For control of all weeds listed, apply 8 pounds DEVRINOL 50-WP, 0.6 to 1.25 pounds SIMAZINE 80-WP, and 1 to 2 quarts GRAMOXONE SUPER per broadcast acre.

Follow the use directions and recommendations for the DEVRINOL 50-WP/SIMAZINE 80-WP combination. Consult the GRAMOXONE SUPER label for timing with respect to stage of weed growth and for application methods recommended to minimize crop injury.

Recommendations

For crop recommendations, read the section on DEVRINOL 50-WP/SIMAZINE 80-WP.

DEVRINOL 50-WP/SINBAR* 80-WP COMBINATION

DEVRINOL 50-WP may be used in combination with SINBAR 80-WP for improved control of annual broadleaf weeds. This tank mix does not control emerged weeds.

*SINBAR is a registered trademark of E. I. DuPont De Nemours & Company.

DIRECTIONS FOR USE

NOTE: All rates are given on a broadcast basis. Reduce rates proportionately for band or strip treatment.

For control of all weeds listed, apply 8 pounds DEVRINOL 50-WP and 1.25 pounds SINBAR 80-WP per broadcast acre to the soil surface. Apply in the Fall through early Spring prior to weed emergence. Do not apply to frozen ground. For best results, the treatment must reach the zone of weed seed germination. If rainfall does not occur the treatment must be shallowly incorporated or irrigated-in following application (sprinkler or flood) with sufficient water to wet the soil to a depth of 2 to 4 inches. In the Western and Pacific Northwest Regions treatment made in November through February must be incorporated-in or irrigated-in if no rainfall occurs within three weeks of application. For treatments made other than in November through February in these regions, and for all treatments applied in the Northern, Southwestern, and Southeastern Regions regardless of season, the treatment must be incorporated-in or irrigated-in within 24 hours if no rainfall occurs. Use as a directed spray and avoid contact with fruit or foliage. Do not apply when fruit is on the ground during the harvest period. Do not graze treated areas.

Recommendations

For use on apples, peaches and citrus (restrict use on citrus to California only) and mint (restrict use on mint to Pacific Northwest only).

NOTE: To avoid possible crop injury follow these precautions:

• **Apples and Peaches:** Where crop is grown under furrow irrigation or under raised-berm flood irrigation (trees 4″ to 6″ above waterline), apply only as a band treatment. Do not treat trees planted in the bottom of irrigation furrows or trees grown under flat flood or basin irrigation, and do not use on eroded areas where subsoil or tree roots are exposed, as injury to trees may result. Use only on trees established in the orchard for at least two years.

• **Citrus:** Temporary yellowing of citrus leaves may occur following treatment. Treatment may be used on coarse soils but is not to be used on soils low in organic matter (less than 1%), on poorly drained soils, nor on trees planted in irrigation furrows, as injury to trees may result. Treated areas may be planted to citrus trees one year after last application. Use only on trees established in the grove for at least one year.

• **Mint:** Apply in the Fall through early spring as a post-harvest treatment after last cultivation.

TANK MIXTURES VEGETABLES AND TOBACCO

Important

Observe all cautions and limitations on labeling of all products used in combination.

Make only one application per season.

Do not contaminate water by cleaning of equipment or disposal of wastes.

In each case where a range of rates is given, the lower rates should be used on light (coarse) textured soils and soils low in organic matter. The higher rates should be used on heavy (fine) textured soils.

After harvest or prior to planting of succeeding crops, a deep moldboard or disc plowing operation must be carried out. To avoid injury to crops not specified on the label, do not plant until 12 months after the last DEVRINOL 50-WP application.

When applied at rates no greater than 4 lbs. DEVRINOL 50-WP per acre, rotational crops which are not registered on this label may be planted no sooner than 12 months after application when harvest and use of the mature crop(s) only is for food or feed purposes. Crops are not to be grazed or harvested for food or feed purposes earlier than maturity.

DIRECTIONS FOR USE

The recommended tank mix controls weeds by interfering with normal seed germination and seedling development. It does not control established weeds. All weed growth and crop stubble should be thoroughly worked into the soil before treatment.

TANK MIX DIRECTIONS

Follow directions given in the sections DEVRINOL 50-WP GENERAL DIRECTIONS FOR USE AND TANK MIXTURES—IMPORTANT.

At labeled spray volumes of 20-50 gallons per acre on transplanted tobacco, DEVRINOL 50-WP is physically compatible with DYFONATE®, DASANIT*, DIAZINON**, NEMACUR*, RIDOMIL** and TILLAM.

*DASANIT and NEMACUR are registered trademarks of Mobay Chemical Corporation

**DIAZINON and RIDOMIL are registered trademarks of Ciba-Geigy Corporation

DEVRINOL 50-WP/TILLAM 6-E TANK MIX COMBINATION APPLIED PRE-PLANT INCORPORATED

A tank mix combination of DEVRINOL 50-WP and TILLAM 6-E can be applied and mixed (incorporated) into the soil before planting for weed control in tomatoes and tobacco.

Additional Weeds Controlled by DEVRINOL 50-WP/TILLAM 6-E Tank Mix at Recommended Rates

Annual Grasses	
Bermudagrass	(*Cynodon dactylon*)
Narrowleaf Signalgrass	(*Brachiaria piligera*)
Wild Oats	(*Avena fatua*)
Millet, Foxtail	(*Setaria italica*)
Annual Broadleaf Weeds	
Henbit (Deadnettle)	(*Lamium amplexicaule*)
Nettleleaf Goosefoot	(*Chenopodium murale*)
Hairy Nightshade*	(*Solanum villosum*)
Blackeyedsusan	(*Rudbeckia serotina*)
Shepherdspurse	(*Capsella bursa-pastoris*)
*Suppression (partial control)	
Perennial Weeds	
Purple Nutsedge	(*Cyperus rotundus*)
Yellow Nutsedge	(*Cyperus esculentus*)

Existing stands of nutsedge must be turned under and chopped up thoroughly before treatment.

Application Directions

The soil should be well-worked and dry enough to permit good soil mixing (incorporation). Pour the recommended amounts of DEVRINOL 50-WP and TILLAM 6-E into a spray tank during the filling operations. Apply in 25 to 50 gallons of water per acre.

For soil incorporated applications, use a properly calibrated, low-pressure, boom-type sprayer having good agitation.

In irrigated areas, do not apply before pre-irrigation. For all band applications, reduce dosage proportionately depending upon row spacing and band to be treated. Do not band application on rocky ground because thorough incorporation is not possible.

Incorporation Directions

The herbicides must be mixed (incorporated) into the soil IMMEDIATELY after application to prevent loss by evaporation. Whenever possible, application and incorporation should be done in the same operation.

Incorporation before Planting

Use power-driven cultivation equipment set to cut to a depth of 2 to 3 inches in coarse soils and 3 to 4 inches in heavy soils OR a tandem disc set to cut a depth of 6 inches, followed by a spiketooth harrow or some other leveling device which extends beyond ends of the disc. For thorough mixing, disc in two different directions (cross disc).

Incorporation at Planting
Use power-driven cultivation equipment set to cut to a depth of 2 to 3 inches.

Planting
Crops should be planted or seeded immediately after application and incorporation. Do not use a drag behind the planter as it may concentrate the herbicides over the seed row and cause crop injury.

Cultural Practices following Application
Shallow cultivation may be necessary to control those susceptible weeds that escape control as well as those weeds not susceptible to either herbicide. Do not cultivate deeper than the depth of herbicide incorporation.

When cultivating fields where the tank mix has been banded in the row, use shields to prevent the movement of untreated soil into the treated row.

Recommendations
Tomatoes (Direct Seeded [California Only] and Transplanted): Apply 2 to 4 pounds DEVRINOL 50-WP plus 2⅔ to 4 quarts TILLAM 6-E per broadcast acre. Use the lower rates on light soils (coarse textured—sandy) and the higher rate on heavy soils (fine textured—clay). Do not use hotcaps on transplanted tomatoes.

Tobacco (Transplanted): Before transplanting, apply 2 pounds DEVRINOL 50-WP plus 2⅔ quarts TILLAM 6-E. This combination may be tank mixed with FURADAN* 4-F or MOCAP** 6-E at recommended rates. Read the FURADAN 4-F and MOCAP 6-E labels for recommended uses, directions for use, precautions and limitations. When filling the spray tank follow these orders of mixing: DEVRINOL 50-WP slurry, FURADAN 4-F, TILLAM 6-E OR, DEVRINOL 50-WP slurry, TILLAM 6-E, MOCAP 6-E.

*FURADAN is a registered trademark of FMC Corporation
**MOCAP is a registered trademark of Rhone-Poulenc Chemical Company

DEVRINOL 50-WP/TILLAM 6-E/DYFONATE [4-E, 4-EC or 4-ED] TANK MIX COMBINATION

A tank mix combination of DEVRINOL 50-WP, DYFONATE 4-E, 4-EC or 4-ED and TILLAM 6-E can be applied and mixed into the soil before transplanting tomatoes and tobacco, for control of weeds and soil insects. Refer to DEVRINOL 50-WP/TILLAM 6-E Tank Mix Combination for list of weeds controlled.

NOTE: DYFONATE 4-E, 4-EC and 4-ED are Restricted Use Pesticides. Observe all cautions and limitations on labeling of all products used in combinations. Make only one application per season. Do not contaminate water by cleaning equipment or disposal of wastes.

DIRECTIONS FOR USE
The recommended tank mix controls weeds by interfering with normal seed germination and seedling development. It does not control established weeds. All weed growth and crop stubble should be thoroughly worked into the soil before treatment.

For GENERAL USE PRECAUTIONS and DIRECTIONS FOR USE refer to DEVRINOL 50-WP/TILLAM 6-E Tank Mix Recommendations for methods of tank mixing, application and incorporation. When using DEVRINOL 50-WP, make a slurry and add to a nearly filled spray tank with adequate agitation.

Recommendations
Tomatoes (Direct Seeded) [California Only] and Transplanted): Apply 2 to 4 pounds DEVRINOL 50-WP plus 2⅔ to 4 qts TILLAM 6-E plus 2 qts DYFONATE 4-E, 4-EC or 4-ED for control of symphylans or 4 qts for control of symphylans and wireworms in 25 to 50 gallons of water per broadcast acre. Use the lower DEVRINOL 50-WP and TILLAM 6-E rates on light soils (coarse textured-sandy) and the higher rates on heavy soils (fine textured-clay). Do not use hotcaps on transplated tomatoes.

Tobacco (Transplanted): Before transplanting apply 2 pounds DEVRINOL 50-WP plus 2⅔ qts TILLAM 6-E plus 2 qts DYFONATE 4-E, 4-EC or 4-ED for control of wireworms in 25 to 50 gallons of water per broadcast acre.

STORAGE AND DISPOSAL
Do not contaminate water, food or feed by storage or disposal.

Storage: Keep container closed when not in use. Store in a cool dry place. Do not store near seeds, fertilizer or food stuffs.

Pesticide Disposal: Wastes resulting from the use of this product may be disposed of on site or at an approved waste disposal facility.

Container Disposal: Do not reuse. Completely empty bag into application equipment. Then dispose of empty bag in a sanitary landfill or by incineration, or, if allowed by State and local authorities, by burning. If burned, stay out of smoke.

NOTICE TO BUYER AND USER:
Seller warrants that this product conforms to the chemical description on the label and is reasonably fit for the purposes stated on the label when used in accordance with directions under normal conditions of use. This warranty does not extend to the use of this product contrary to label instructions, or under abnormal use conditions, or under conditions not reasonably foreseeable to Seller, and Buyer and User assume the risk of any such use. SELLER DISCLAIMS ALL OTHER WARRANTIES EXPRESSED OR IMPLIED INCLUDING ANY WARRANTY OF FITNESS OR MERCHANTABILITY. SELLER SHALL NOT BE LIABLE FOR CONSEQUENTIAL, SPECIAL OR INDIRECT DAMAGES RESULTING FROM THE USE OR HANDLING OF THIS PRODUCT AND SELLER'S SOLE LIABILITY AND BUYER'S AND USER'S EXCLUSIVE REMEDY SHALL BE LIMITED TO THE REFUND OF THE PURCHASE PRICE.

This product is sold only for uses stated on its label. No express or implied license is granted to use or sell this product under any patent in any country except as specified: Country: United States of America Patent No: 3,480,671; 3,718,455

A-18MR

The information and specimen labels in this booklet are correct as of the publication date but are subject to change without notice. For current information, contact ICI Americas Inc.

DEVRINOL® 50-WP

ORNAMENTAL SELECTIVE HERBICIDE

A Pre-Emergence Herbicide for Weed Control in Container—Grown Ornamentals, Field-Grown Nursery Stock, Liner Stock, Ground Covers and Turfgrass

DEVRINOL 50-WP Ornamental Controls Many Annual Grasses and Broadleaf Weeds

ACTIVE INGREDIENT:	
N,N-Diethyl-2-(1-naphthalenyloxy)-propionamide	50.0%
INERT INGREDIENTS:	50.0%
TOTAL	100%

Contains 50 Pounds Active Ingredient Per 100 Pounds

EPA Reg. No. 10182-231

KEEP OUT OF REACH OF CHILDREN

CAUTION

PRECAUTIONARY STATEMENTS

HAZARDS TO HUMANS AND DOMESTIC ANIMALS

CAUTION

MAY BE HARMFUL IF SWALLOWED. Do not breathe dust. Avoid contact with skin, eyes and clothing. Wash thoroughly with soap and water after use.

STATEMENT OF PRACTICAL TREATMENT
Immediately start the procedures given below. If further treatment is required, contact a Poison Center, a physician, or the nearest hospital.

If swallowed: immediately give several glasses of water and induce vomiting by gagging the victim with a finger placed on the back of the victim's tongue. Give fluids until vomitus is clear. Do not induce vomiting or give anything by mouth to an unconscious or convulsing person.

For skin contact: Flush all affected areas with plenty of water for several minutes. For eye contact, hold eyelids apart and flush with plenty of water for 15 minutes. Remove and clean contaminated clothing and shoes before reuse. Seek medical attention if skin or eye irritation occurs.

If inhaled: Remove to fresh air. Seek medical attention if respiratory irritation occurs or if breathing becomes difficult.

FOR 24-HOUR EMERGENCY MEDICAL ASSISTANCE CALL 1-800-F-A-S-T-M-E-D (327-8633).
FOR CHEMICAL EMERGENCY: Spill, leak, fire, exposure, or accident call CHEMTREC 1-800-424-9300.

ENVIRONMENTAL HAZARDS
- Do not apply directly to water or wetlands.
- Do not contaminate water when disposing of equipment washwaters. Apply this product as specified on this label.
- Do not apply when weather conditions favor drift from areas treated.
- Do not contaminate water used for domestic purposes.
- Avoid contamination of food and feed.

GENERAL PRECAUTIONS
- Read all instructions before using.
- Use only for recommended purposes and at recommended rates. Do not overdose.

GENERAL DIRECTIONS FOR USE
It is a violation of Federal law to use this product in a manner inconsistent with its labeling.

1. DEVRINOL 50-WP ORNAMENTAL does not control established weeds. It must be applied to freshly weeded soil before weeds germinate or during Fall and Winter months.

2. DEVRINOL 50-WP ORNAMENTAL can be applied to *newly planted* container stock after soil has settled from first watering, field grown nursery stock, dichondra and to established plants.
3. DEVRINOL 50-WP ORNAMENTAL needs mechanical incorporation (such as power tiller), adequate irrigation or natural moisture (rainfall or snow) (1-2 inches within 2-3 days) for optimum results.
4. DEVRINOL 50-WP ORNAMENTAL is recommended on several different container potting soil mixtures. These mixtures include those containing 30 percent soil and/or sand and mixtures containing no soil. (Example: perlite, peat, bark, sawdust and sand.)
5. Do not apply this product through any type of irrigation system.

WEEDS CONTROLLED BY DEVRINOL 50-WP ORNAMENTAL

Annual Grasses	
Barnyardgrass, Junglerice	(*Echinochloa* spp.)
Wild Barley	(*Hordeum leporinum*)
Annual Bluegrass	(*Poa annua*)
Bromes	(*Bromus*)
Cheatgrass	(*B. secalinus*)
Downy Brome	(*B. tectorum*)
Ripgut Brome	(*B. rigidus*)
Soft Chess	(*B. mollis*)
Large (Hairy) Crabgrass	(*Digitaria sanguinalis*)
Smooth Crabgrass	(*Digitaria ischaemum*)
Southwestern Cupgrass	(*Eriochloa gracilis*)
Foxtails	(*Setaria*)
Bristly Foxtail	(*S. verticilla*)
Giant Foxtail	(*S. faberi*)
Green Foxtail	(*S. viridis*)
Yellow Foxtail	(*S. lutescens*)
Goosegrass	(*Eleusine indica*)
Seedling Johnsongrass	(*Sorghum halepense*)
Wild Oats	(*Avena fatua*)
Panicums	(*Panicum*)
Guineagrass	(*P. paximum*)
Fall Panicum	(*P. dichotomiflorum*)
Texas Panicum	(*P. texanum*)
Witchgrass	(*P. capillare*)
Italian Ryegrass	(*Lolium multiflorum*)
Sandbur	(*Cenchrus* spp.)
Red Sprangletop	(*Leptochloa filiformis*)
Stinkgrass	(*Eragrostis cilianensis*)
Annual Broadleaf Weeds	
Carpetweed	(*Mollugo verticillata*)
Common Chickweed	(*Stellaria media*)
Coast Fiddleneck	(*Amsinckia intermedia*)
Filaree	(*Erodium* spp.)
Groundsel	(*Senecio* spp.)
Prostrate Knotweed	(*Polygonum aviculare*)
Lambsquarters	(*Chenopodium album*)
Prickly Lettuce	(*Lactuca serriola*)
Little Mallow*	(*Malva parviflora*)
Redroot Pigweed	(*Amaranthus retroflexus*)
Pineappleweed	(*Matricaria matricarioides*)
Common Purslane	(*Portulaca oleracea*)
Horse Purslane	(*Trianthema portullacastrum*)
Common Ragweed**	(*Ambrosia artemisiifolia*)
Annual Sowthistle	(*Sonchus oleraceus*)

* From seed
** Suppression

RECOMMENDATIONS

NURSERY STOCK—FIELD GROWN

Consult list of plant species on which DEVRINOL 50-WP ORNAMENTAL can be used.

Apply DEVRINOL 50-WP ORNAMENTAL as a broadcast treatment over young nursery stock or as a directed application to larger stock either as a band or broadcast application.

Application can be made anytime of the year to weed free soil; however, since DEVRINOL 50-WP ORNAMENTAL requires 1-2 inches of rainfall or irrigation soon after application for adequate weed control, application should be timed to take advantage of rainfall if irrigation is not available. Fall and Winter applications will provide excellent weed control the following Spring and Summer. A thorough incorporation of DEVRINOL 50-WP ORNAMENTAL into the soil to a depth of 1-2 inches will provide excellent weed control with less rainfall required.

One application of DEVRINOL 50-WP ORNAMENTAL should provide adequate weed control for a growing season in temperate areas.

CONTAINER GROWN ORNAMENTALS

Consult list of plant species on which DEVRINOL 50-WP ORNAMENTAL can be used.

Apply DEVRINOL 50-WP ORNAMENTAL as a broadcast application over the containers.

Normal irrigation or rainfall (if a Fall or Winter application is made in cold areas) should provide sufficient water for satisfactory weed control.

HOME GROUNDS, HIGHWAY AND INDUSTRIAL ORNAMENTALS PLANTINGS

Use an overall (broadcast) treatment over young ornamentals or as a directed application where larger plants are growing. Irrigate soon after application or apply just before expected rain. Fall through Spring applications are preferred if irrigation is not available or if application is to be made in lower Summer rainfall areas.

GROUND COVERS

Apply DEVRINOL 50-WP ORNAMENTAL as a broadcast treatment over the ground covers. Irrigate immediately with a minimum of 1 inch of water or apply when rainfall is expected.

DICHONDRA

Apply DEVRINOL 50-WP ORNAMENTAL at the time of seeding, or to established dichondra including dichondra sod. Irrigate with a minimum of 1 inch of water.

RECOMMENDATIONS FOR USE ON THESE ORNAMENTALS AND NURSERY STOCK AND CONTAINER STOCK

Field Grown

Deciduous and Evergreen Trees		
Acer	Eucalyptus	Picea
Ash	Ficus	Pin Oak
Birch	Fir	Pinus
Black Walnut	Gleditsia	Pittosporum
Cedar	Hemlock	Podocarpus
Citrus	Japanese Larch	Populus
Crataegus	Laburnum	Prunus
Cypress	Malus	Pears
Dogwood	Olive (Olea)	Pistachios
Douglas Fir	Palm (Washingtonia robusta)	
Deciduous and Evergreen Shrubs		
Abelia	Forsythia	Osmanthus
Asparagus (Sprengeri)	Gardenia	Photinia
	Heather (Calluna)	Pittosporum
Aucuba	Hibiscus	Podocarpus
Azalea	Honeysuckle (Lonicera)	Pyracantha
Bottlebrush	Hypericum	Raphiolepis
Bougainvillea	Ilex	Rhododendron
Buxus	Juniperus	Rosa
Camellia	Leucothe	Star Jasmine
Cotoneaster	Ligustrum	Taxus
Crepe Myrtle	Nandina	Viburnum
Dogwood	Natal Plum	Xylosema
Euonymus	Oleander	
Fruit & Nuts		
Almond	Lemon	Pistachio
Apple	Nectarine	Plum
Apricot	Orange	Prune
Cherry	Peach	Tangerine
Filbert	Pear	Walnut
Grape	Pecan	

Strawberry (Application should be delayed until the desired number of daughter plants have become established. Do not apply from bloom through harvest.)

Ground Covers		
Ajuga	Gazania	Liriope
Carpobrotus	Hedera	Osteosperma
Delosperma	Hypericum	Pachysandra
Dichondra	Iceplant	Sedum
Erysimum	Ivy	Vinca
	Lantana	
Container Stock		
Abelia	Eucalyptus	Pachysandra
Agapanthus	Euonymus	Palm (Washing-

Container Stock

Ageratum	Fir	tonia robusta)
Asparagus	Forsythia	Photinia
Azalea	Gardenia	Pinus Spp.
Bottlebrush	Geranium	Pittosporum
Bougainvillea	Heather	Podocarpus
Buxus	Hedera	Pyracantha
Camellia	Hibiscus	Rhododendron
Cedar	Ilex	Raphiolepis
Cotoneaster	Juniperus	Rosa
Crepe Myrtle	Leucothe	Sedum
Cypress	Ligustrum	Star Jasmine
Dogwood	Liriope	Thuja
Douglas Fir	Nandina	Vinca
		Xylosema

Flowers

African Daisy	Dahlia	Narcissus
Aster	Daisy	Petunia
Bird of Paradise	Geranium	Plantain Lily (Hosta)
Chrysanthemum	Gladiolus	Zinnia

APPLICATION RATES

Annual Weeds

For annual grasses and broadleaf weeds apply 8 to 12 lbs. of DEVRINOL 50-WP ORNAMENTAL per broadcast acre (3 to 4.4 oz. per 2½ gallons water per 1,000 sq. ft.). Use the higher rate when weed populations are heavy or when broadleaves are the primary weeds.

RECOMMENDATIONS: TURF GRASS

(For Use on Warm Season Grasses Only)

DEVRINOL 50-WP ORNAMENTAL is recommended for annual bluegrass, crabgrass and goosegrass control in the following established turf grasses: bahia, bermudagrass, centipede, St. Augustine and fescue.

TIMING OF APPLICATION

For control of crabgrass and goosegrass, apply DEVRINOL 50-WP ORNAMENTAL at the recommended rate. Application should be made 2-3 weeks prior to weed emergence. DEVRINOL 50-WP ORNAMENTAL should be mixed with enough water to give adequate coverage of the area to be treated (but no less than 80 gallons per acre). Use a lawn type spray gun.

DEVRINOL 50-WP ORNAMENTAL requires a minimum of one half (½) inch of rainfall or irrigation 2-3 days after application for adequate control. If irrigation methods are not available, the timing of the application of DEVRINOL 50-WP ORNAMENTAL should be moved back several weeks to allow for some rainfall between application and weed germination.

RATES OF APPLICATION

For Crabgrass Control: Apply 4 to 6 lbs. of DEVRINOL 50-WP ORNAMENTAL per acre in 80-120 gallons of water (1.5-2.25 ozs. per 2½ gallons of water per 1,000 sq. ft.). Use the higher rate when applying before January 1 (not applicable north of South Carolina) or when weed infestation is expected to be heavy.

For Goosegrass Control: Apply 4 to 6 lbs. of DEVRINOL 50-WP ORNAMENTAL per acre in 80-120 gallons of water (1.5-2.25 ozs. per 2½ gallons of water per 1,000 sq. ft.). Goosegrass usually starts to germinate 6 to 7 weeks after the start of crabgrass germination.

Split Application for Crabgrass and Goosegrass Control: When both of these weeds are to be controlled, apply 4 lbs. of DEVRINOL 50-WP ORNAMENTAL spread uniformly per acre (1.5 ozs. per 1,000 sq. ft.) for crabgrass and another 4 lbs. of DEVRINOL 50-WP ORNAMENTAL per acre for goosegrass. Apply 8-10 weeks apart. Timing should be adjusted to the germination of weeds in your area. Split applications give longer and better weed control than one application.

For Annual Bluegrass (Poa Annua) Control: Annual bluegrass is a Winter annual in most of the warm season turf regions; therefore, application of DEVRINOL 50-WP ORNAMENTAL should be made in the Fall before the annual bluegrass germinates. Apply 4-6 lbs. of DEVRINOL 50-WP ORNAMENTAL per acre in 80-120 gallons of water (1.5-2.25 ozs. per 2½ gallons of water per 1,000 sq. ft.).

TANK MIXTURES

Observe all cautions and limitations on labeling of all products used in combination.

Make only one application per season.

Do not contaminate water by cleaning of equipment or disposal of wastes. Apply these combinations only as specified on this label.

For a tank mixture application, apply the recommended dosage (See Directions for Use) in 80 gallons of water per acre.

GENERAL TANK MIX DIRECTIONS

Proper care must be used in preparing combination tank mixtures to avoid incompatibility problems. Fill the spray tank at least ⅔ full of water. Start agitation and continue moderate agitation throughout mixing and application. Add, mix and disperse wettable powders first and flowable products next. Make sure each product is thoroughly wetted and dispersed prior to addition of the next product. Complete filling the tank with water, mix thoroughly, and maintain moderate agitation to provide a uniform mix during application. The tank mix combination should not be left in the spray tank for prolonged periods; batches should be mixed and applied the same day. All return lines to the tank must discharge below liquid level.

DEVRINOL 50-WP ORNAMENTAL/BETASAN 4-E COMBINATION

DEVRINOL 50-WP ORNAMENTAL may be used as a tank mix combination with BETASAN 4-E for crabgrass control in the following established turf grasses; bahia, bermudagrass, centipede, St. Augustine and fescue.

DIRECTIONS FOR USE

Apply 4 lbs. of DEVRINOL 50-WP ORNAMENTAL and 5 quarts of BETASAN 4-E per acre using a minimum of 80 gallons of water. Application should be made at least 2-3 weeks before the earliest germination date for crabgrass in your area.

The DEVRINOL 50-WP ORNAMENTAL/BETASAN 4-E combination requires a minimum of one half (½) inch of rainfall or irrigation 2-3 days after application for adequate control. If irrigation is not available, the application should be delayed several weeks to allow for some rainfall between application and weed germination.

Tank Mixing Directions

Add the recommended amount of DEVRINOL 50-WP ORNAMENTAL to the spray tank while filling with water. After the DEVRINOL 50-WP ORNAMENTAL is thoroughly wetted out and dispersed in all of the water, then add the recommended amount of BETASAN 4-E and mix to provide a uniform dispersion. Apply in at least 80 gallons of water per acre.

Reseeding: Do not reseed or overseed within six months after application.

STORAGE AND DISPOSAL

Do not contaminate water, food or feed by storage or disposal. **Storage:** Keep container closed when not in use. Store in a cool, dry place. Do not store near seeds, fertilizer or food stuffs. **Pesticide Disposal:** Wastes resulting from the use of this product may be disposed of on site or at an approved waste disposal facility. **Container Disposal:** Completely empty bag into application equipment. Then dispose of empty bag in a sanitary landfill or by incineration, or, if allowed by State and local authorities, by burning. If burned, stay out of smoke.

NOTICE TO BUYER AND USER

Seller warrants that this product conforms to the chemical description on the label and is reasonably fit for the purposes stated on the label when used in accordance with directions under normal conditions of use. This warranty does not extend to the use of this product contrary to label instructions, or under abnormal use conditions, or under conditions not reasonably foreseeable to Seller, and Buyer and User assume the risk of any such use. SELLER DISCLAIMS ALL OTHER WARRANTIES, EXPRESSED OR IMPLIED, INCLUDING ANY WARRANTY OF FITNESS OR MERCHANTABILITY. SELLER SHALL NOT BE LIABLE FOR CONSEQUENTIAL, SPECIAL OR INDIRECT DAMAGES RESULTING FROM THE USE OR HANDLING OF THIS PRODUCT AND SELLER'S SOLE LIABILITY AND BUYER'S AND USER'S EXCLUSIVE REMEDY SHALL BE LIMITED TO THE REFUND OF THE PURCHASE PRICE.

This product is sold only for uses stated on its label. No express or implied license is granted to use or sell this product under any patent in any country except as specified: Country: United States of America. Patent No.:3,480,671;3,718,455

03-0622-014 R1 A-7R

This is a specimen label. ICI Americas Inc. is not responsible for the accuracy of the information contained herein. As labels are subject to revision, always carefully read and follow the label on the product container.

DEVRINOL® 5-G

ORNAMENTAL SELECTIVE HERBICIDE

A Pre-Emergence Herbicide for Weed Control in Container—Grown Ornamentals, Field-Grown Nursery Stock, Liner Stock, Ground Covers and Turfgrass.

DEVRINOL 5-G Ornamental Controls Many Annual Grasses and Broadleaf Weeds

ACTIVE INGREDIENT:	
N,N-Diethyl-2-(1-naphthalenyloxy)-propionamide	5.0%
INERT INGREDIENTS:	95.0%
TOTAL	100.0%

Contains 5 pounds active ingredient per 100 pounds.

EPA Reg. No. 10182-242

KEEP OUT OF REACH OF CHILDREN
CAUTION

PRECAUTIONARY STATEMENTS

HAZARDS TO HUMANS AND DOMESTIC ANIMALS

CAUTION

MAY BE HARMFUL IF SWALLOWED. Do not breathe dust. Avoid contact with skin, eyes and clothing. Wash thoroughly with soap and water after use.

ENVIRONMENTAL HAZARDS

- Do not apply directly to water or wetlands.
- Do not contaminate water by cleaning of equipment, or disposal of wastes. Apply this product as specified on this label.
- Do not apply when weather conditions favor drift from areas treated.
- Do not contaminate water used for domestic purposes.
- Avoid contamination of food and feed.

GENERAL PRECAUTIONS

- Read all instructions before using.
- Use only for recommended purposes and at recommended rates. Do not overdose.

STATEMENT OF PRACTICAL TREATMENT

If an adverse reaction occurs after exposure, promptly initiate the recommended procedures below. If further treatment is required, seek medical attention.

In case of contact: Flush eyes and skin with plenty of water for several minutes. Get medical attention if eye or skin irritation occurs.

If swallowed: Immediately give several glasses of water and induce vomiting by placing a finger on the back of the victim's tongue. Give fluids until vomitus is clear. Do not induce vomiting or give anything by mouth to an unconscious or convulsing person.

If inhaled: Remove to fresh air. Seek medical attention if respiratory irritation occurs or if breathing becomes difficult.

FOR 24-HOUR EMERGENCY MEDICAL ASSISTANCE CALL 1-800-F-A-S-T-M-E-D (327-8633).

FOR CHEMICAL EMERGENCY: Spill, leak, fire, exposure, or accident call CHEMTREC 1-800-424-9300.

GENERAL DIRECTIONS FOR USE

It is a violation of Federal law to use this product in a manner inconsistent with its labeling.

1. DEVRINOL 5-G ORNAMENTAL does not control established weeds. It must be applied to freshly weeded soil before weeds germinate or during fall and winter months.
2. DEVRINOL 5-G ORNAMENTAL can be applied to *newly planted* container stock after soil has settled from first watering, field grown nursery stock, dichondra and to established plants.
3. DEVRINOL 5-G ORNAMENTAL needs mechanical incorporation (such as power tiller), adequate irrigation or natural moisture (rainfall or snow) (1-2 inches within 2-3 days) for optimum results.
4. DEVRINOL 5-G ORNAMENTAL is recommended on several different container potting soil mixtures. These mixtures include those containing 30 percent soil and/or sand and mixtures containing no soil. (Example: perlite, peat, bark, sawdust and sand.)

Weeds Controlled by Devrinol 5-G Ornamental

Annual Grasses:	Annual Broadleaf Weeds:
Barnyardgrass, Junglerice (*Echinochloa* spp.)	Carpetweed (*Mollugo verticillata*)
Wild Barley (*Hordeum leporinum*)	Common Chickweed (*Stellaria media*)
Annual Bluegrass (*Poa annua*)	Coast Fiddleneck (*Amsinckia intermedia*)
Bromes (*Bromus*)	Filaree (*Erodium* spp.)
Cheatgrass (*B. secalinus*)	Groundsel (*Senecio* spp.)
Downy Brome (*B. tectorum*)	Prostrate Knotweed (*Polygonum aviculare*)
Ripgut Brome (*B. rigidus*)	Lambsquarters (*Chenopodium album*)
Soft Chess (*B. mollis*)	Prickly Lettuce (*Lactuca serriola*)
Large (Hairy) Crabgrass (*Digitaria sanguinalis*)	Little Mallow*(*Malva parviflora*)
Smooth Crabgrass (*Digitaria ischaemum*)	Redroot Pigweed (*Amaranthus retroflexus*)
Southwestern Cupgrass (*Eriochloa gracilis*)	Pineappleweed (*Matricaria matricarioides*)
Foxtails (*Setaria*)	Common Purslane (*Portulaca oleracea*)
Bristly Foxtails (*S. verticilla*)	Horse Purslane (*Trianthema portullacastrum*)
Giant Foxtails (*S. faberi*)	Common Ragweed** (*Ambrosia artemisiifolia*)
Green Foxtails (*S. viridis*)	Annual Sowthistle (*Sonchus oleraceus*)
Yellow Foxtails (*S. lutescens*)	
Goosegrass (*Eleusine indica*)	
Seedling Johnsongrass (*Sorghum halepense*)	
Wild Oats (*Avena fatua*)	
Panicums (*Panicum*)	
Guineagrass (*P. paximum*)	
Fall Panicum (*P. dichotomiflorum*)	
Texas Panicum (*P. texanum*)	
Witchgrass (*P. capillare*)	
Italian Ryegrass (*Lolium multiflorum*)	
Sandbur (*Cenchrus* spp.)	
Red Sprangletop (*Leptochloa filiformis*)	
Stinkgrass (*Eragrostis cilianensis*)	

* From seed
** Suppression

RECOMMENDATIONS

Nursery Stock—Field Grown: Consult list of plant species on which DEVRINOL 5-G ORNAMENTAL can be used.

Apply DEVRINOL 5-G ORNAMENTAL as a broadcast treatment over young nursery stock or as a directed application to larger stock either as a band or broadcast treatment.

Application can be made anytime of the year to weed free soil; however, since DEVRINOL 5-G ORNAMENTAL requires 1-2 inches of rainfall or irrigation soon after application for adequate weed control, application should be timed to take advantage of rainfall if irrigation is not available. Fall and winter applications will provide excellent weed control the following spring and summer.

A thorough incorporation of DEVRINOL 5-G ORNAMENTAL into the soil to a depth of 1-2 inches will provide excellent weed control with less rainfall required.

One application of DEVRINOL 5-G ORNAMENTAL should provide adequate weed control for a growing season in temperate areas.

Container Grown Ornamentals: Consult list of plant species on which DEVRINOL 5-G ORNAMENTAL can be used.

Apply DEVRINOL 5-G ORNAMENTAL as a broadcast application over the containers.

Normal irrigation or rainfall (if a fall or winter application is made in cold areas) should provide sufficient water for satisfactory weed control.

Home Grounds, Highway and Industrial Ornamentals Plantings: Use an overall (broadcast) treatment over young ornamentals or as a directed application where larger plants are growing. Irrigate soon after application or apply just before expected rain. Fall through spring applications are preferred if irrigation is not available or if application is to be made in lower summer rainfall areas.

Ground Covers: Apply DEVRINOL 5-G ORNAMENTAL as a broadcast treatment over the ground covers. Irrigate immediately with a minimum of 1 inch of water or apply when rainfall is expected.

Dichondra: Apply DEVRINOL 5-G ORNAMENTAL at the time of seeding, or to established dichondra including dichondra sod. Irrigate with a minimum of 1 inch of water.

APPLICATION RATES

Annual Weeds: For annual grasses and broadleaf weeds apply 80 to 120 lbs. of DEVRINOL 5-G ORNAMENTAL per broadcast acre (1.8 to 2.8 lbs. per 1,000 sq. ft.).

Use the higher rate when weed populations are heavy or when broadleaves are the primary weeds.

Use DEVRINOL 5-G Ornamental on These Ornamentals and Nursery Stock and Container Stock

Field Grown

Deciduous And Evergreen Trees

Acer	Eucalyptus	Picea
Ash	Ficus	Pin Oak
Birch	Fir	Pinus
Black Walnut	Gleditsia	Pittosporum
Cedar	Hemlock	Podocarpus
Citrus	Japanese Larch	Populus
Crataegus	Laburnum	Prunus
Cypress	Malus	Pears
Dogwood	Olive (Olea)	Pistachios
Douglas Fir	Palm (Washingtonia robusta)	

Deciduous And Evergreen Shrubs

Abelia	Gardenia	Osmanthus
Asparagus (Sprengeri)	Heather (Calluna)	Photinia
Aucuba	Hibiscus	Pittosporum
Azalea	Honeysuckle (Lonicera)	Podocarpus
Bottlebrush	Hypericum	Pyracantha
Bougainvillea	Ilex	Raphiolepis
Buxus	Juniperus	Rhododendron
Camellia	Leucothe	Rosa
Cotoneaster	Ligustrum	Star Jasmine
Crepe Myrtle	Nandina	Taxus
Dogwood	Natal Plum	Viburnum
Euonymus	Oleander	Xylosema
Forsythia		

Fruits & Nuts

Almond	Lemon	Pistachio
Apple	Nectarine	Plum
Apricot	Orange	Prune
Cherry	Peach	Tangerine
Filbert	Pear	Walnut
Grape	Pecan	

Strawberry (Application should be delayed until the desired number of daughter plants have become established. Do not apply from bloom through harvest.)

Ground Covers

Ajuga	Hedera	Osteosperma
Carpobrotus	Hypericum	Pachysandra
Delosperma	Iceplant	Sedum
Dichondra	Ivy	Vinca
Erysimum	Lantana	
Gazania	Liriope	

Container Stock

Abelia	Eucalyptus	Pachysandra
Agapanthus	Euonymus	Palm (Washingtonia robusta)
Ageratum	Fir	Photinia
Asparagus	Forsythia	Pinus spp.
Azalea	Gardenia	Pittosporum
Bottlebrush	Geranium	Podocarpus
Bougainvillea	Heather	Pyracantha
Buxus	Hedera	Raphiolepis
Camellia	Hibiscus	Rhododendron
Cedar	Ilex	Rosa
Cotoneaster	Juniperus	Sedum
Crepe Myrtle	Leucothe	Star Jasmine
Cypress	Ligustrum	Thuja
Dogwood	Liriope	Vinca
Douglas Fir	Nandina	Xylosema

Flowers

African Daisy	Dahlia	Narcissus
Aster	Daisy	Petunia
Bird of Paradise	Geranium	Plantain Lily (Hosta)
Chrysanthemum	Gladiolus	Zinnia

RECOMMENDATIONS: TURF GRASS (FOR USE ON WARM SEASON GRASSES ONLY)

DEVRINOL 5-G is recommended for annual bluegrass, crabgrass and goosegrass control in the following established turf grasses: bahia, bermudagrass, centipede, St. Augustine and fescue.

TIMING OF APPLICATION

For control of crabgrass and goosegrass, apply DEVRINOL 5-G at the recommended rate. Application should be made 2-3 weeks prior to weed emergence. DEVRINOL 5-G should be applied with a gravity flow drop spreader or a cyclone type spreader. A more uniform application can be made by spreading half the required amount over the area to be treated, then applying the remainder at right angles.

DEVRINOL 5-G requires a minimum of one half (1/2) inch of rainfall or irrigation 2-3 days after application for adequate control. If irrigation methods are not available, the timing of the application of DEVRINOL 5-G should be moved back several weeks to allow for some rainfall between application and weed germination.

RATES OF APPLICATION

For Crabgrass Control: Apply 40 to 60 lbs. of DEVRINOL 5-G uniformly per (0.9-1.35* lbs. per 1,000 sq. ft.). Use the higher rate when applying before January 1 (not applicable north of South Carolina or when weed infestation is expected to be heavy).

For Goosegrass Control: Apply 40 to 60 lbs. of DEVRINOL 5-G per acre (0.9-1.35 lbs. per 1,000 sq. ft.). Goosegrass usually starts to germinate 6 to 7 weeks after the start of crabgrass germination.

Split Application for Crabgrass and Goosegrass Control: When both of these weeds are to be controlled, apply 40 lbs. of DEVRINOL 5-G spread uniformly per acre (0.9 lbs. per 1,000 sq. ft.) for crabgrass and another 40 lbs. of DEVRINOL 5-G per acre for goosegrass. Apply 8-10 weeks apart. Timing should be adjusted to the germination of weeds in your area. Split applications give longer and better weed control than one application.

For Annual Bluegrass (Poa Annua) Control: Annual bluegrass is a winter annual in most of the warm season turf regions; therefore, application of DEVRINOL 5-G should be made in the fall before the annual bluegrass germinates. Apply 40-60 lbs. of DEVRINOL 5-G per acre (0.9-1.35 lbs. per 1,000 sq. ft.).

* Where a rate range is given, use the higher rate when weed infestation is expected to be heavy.)

Reseeding: Do not reseed or overseed within six months after application.

BETASAN 7-G/DEVRINOL 5-G Ornamental Sequential Applications

BETASAN-DEVRINOL sequential applications are recommended for crabgrass control in the following turf grasses: bahia, bermudagrass, centipede, St. Augustine and fescue. Apply 72 lbs. of BETASAN 7-G per acre (1,6 lbs/1,000 sq. ft.) followed by 40 lbs. of DEVRINOL 5-G ORNAMENTAL per acre (0.9 lbs./1,000 sq. ft.) Applications should be made 2-3 weeks before the normal germination date for crabgrass in your area. The BETASAN 7-G/DEVRINOL 5-G ORNAMENTAL combination requires a minimum of one half (1/2) inch of rainfall or irrigation 2-3 days after application for adequate control. If irrigation is not available, the application should be delayed several weeks to allow for some rainfall between application and weed germination.

Reseeding: Do not reseed or overseed within six months after application.

STORAGE AND DISPOSAL

Do not contaminate water, food or feed by storage or disposal. **Storage:** Keep container closed when not in use. Store in a cool, dry place. Do not store near seeds, fertilizer or food stuffs. **Pesticide Disposal:** Wastes resulting from the use of this product may be disposed of on site or at an approved waste disposal facility. **Container Disposal:** Completely empty bag into application equipment. Then dispose of empty bag in a sanitary landfill or by incineration, or, if allowed by State and local authorities, by burning. If burned, stay out of smoke.

NOTICE TO BUYER AND USER

Seller warrants that this product conforms to the chemical description on the label and is reasonably fit for the purposes stated on the label when used in accordance with directions under normal conditions of use. This warranty does not extend to the use of this product contrary to label instructions, or under abnormal use conditions, or under conditions not reasonably foreseeable to Seller, and Buyer and User assume the risk of any such use. SELLER DISCLAIMS ALL OTHER WARRANTIES, EXPRESSED OR IMPLIED, INCLUDING ANY WARRANTY OF FITNESS OR MERCHANTABILITY. SELLER SHALL NOT BE LIABLE FOR CONSEQUENTIAL, SPECIAL OR INDIRECT DAMAGES RESULTING FROM THE USE OR HANDLING OF THIS PRODUCT AND SELLER'S SOLE LIABILITY AND BUYER'S AND USER'S EXCLUSIVE REMEDY SHALL BE LIMITED TO THE REFUND OF THE PURCHASE PRICE.

03-0622-016 A-6R

The information and specimen labels in this booklet are correct as of the publication date but are subject to change without notice. For current information, contact ICI Americas Inc.

DYFONATE® 4-EC

EMULSIFIABLE LIQUID—INSECTICIDE

RESTRICTED USE PESTICIDE

Due to ACUTE DERMAL TOXICITY

For retail sale to and application only by Certified Applicators or personnel under their direct supervision and only for those covered by the certified applicator's certification.

Complete Directions For Use,
Use Precautions and Recommendations

ACTIVE INGREDIENTS:	
O-Ethyl S-phenyl ethylphosphonodithioate*	47.3%
Xylene Range Aromatic Solvent	39.1%
INERT INGREDIENTS:	13.6%
TOTAL	100.0%

(Contains 4 Pounds DYFONATE per gallon)

*Dyfonate

EPA Reg. No. 10182-212

KEEP OUT OF REACH OF CHILDREN

DANGER–POISON

PELIGRO

PRECAUCION AL USUARIO: SI USTED NO LEE INGLES, NO USE ESTE PRODUCTO HASTA QUE LA ETIQUETA LE HAYA SIDO EXPLICADA AMPLIAMENTE.

Danger—Read All Label Directions Before Using

STATEMENT OF PRACTICAL TREATMENT

Call a Physician Immediately: If a known exposure occurs, or if poisoning is suspected, do not wait for symptoms to develop. Immediately initiate the recommended procedures below. Simultaneously contact a physician, or the nearest hospital, or the nearest Poison Control Center. Inform the person contacted of the type and extent of exposure, describe the victim's symptoms, and follow the advice given.

FOR 24-HOUR EMERGENCY MEDICAL ASSISTANCE CALL 1-800-F-A-S-T-M-E-D (327-8633).

FOR CHEMICAL EMERGENCY: Spill, leak, fire, exposure, or accident call CHEMTREC 1-800-424-9300.

Note: Be sure to advise the physician that the compound is a cholinesterase inhibitor, and follow the physician's advice.

Poison Symptoms—DYFONATE 4-EC is a highly toxic cholinesterase inhibitor. Symptoms include headache, blurred vision, weakness, nausea, discomfort in the chest, abdominal cramps, vomiting, diarrhea, salivation, sweating and pinpoint pupils.

Always Call a Physician in Case of Suspected Poisoning

Atropine by Injection is Antidotal

First Aid—If patient has stopped breathing, *immediately* remove from contaminated area, then start artificial respiration, preferably by mechanical means. **If swallowed** immediately dilute the swallowed material by giving large quantities of water and induce vomiting by gagging the victim with a finger or blunt object placed on the back of the victim's tongue. Give fluids until vomitus is clear. Never give anything by mouth to an unconscious person. Call a physician or the nearest Poison Control Center. **If inhaled**, remove to fresh air. Get medical attention if respiratory irritation occurs. **For eye contact**, flush with large amounts of running water for at least 15 minutes. Hold eyelids apart to insure rinsing of the entire surface of the eyes. Get medical attention. **For skin contact**, remove contaminated clothing and immediately wash skin thoroughly for 15 minutes with soap and water. Avoid abrading the skin.

Information for Physician—Initial measures include removal of secretions, maintenance of a patent airway and, if necessary, artificial respiration. *When cyanosis is relieved*, atropine may be administered in large therapeutic doses, repeated as necessary to the point of tolerance. If symptoms warrant further treatment, protopam chloride (pralidoxime chloride, 2-PAM chloride) has shown utility as *adjunctive* therapy. Never use morphine. Continued absorption of the poison may occur, resulting in a fatal relapse after initial improvement; *very close supervision of the patient is indicated for at least 48 to 72 hours.*

IN CASE OF SPILL

ISOLATE the spill. Hold this package, other cargo and vehicles involved. IMMEDIATELY TELEPHONE (24-hour service) for emergency assistance: 800-424-9300.

PRECAUTIONARY STATEMENTS

HAZARDS TO HUMANS AND DOMESTIC ANIMALS

DANGER–POISON

Danger: Poisonous if Swallowed, Inhaled or Absorbed Through the Skin. Rapidly Absorbed Through the Skin. Do Not Get in the Eyes, on Skin, or on Clothing. Do Not Breathe Spray Mist.

Use a respirator approved by the National Institute for Occupational Safety and Health (30 CFR Part 11—Subpart M) for protection against pesticide inhalation.

Wear rubber gloves, shoes, protective clothing and goggles when handling or applying.

Wash hands, arms, and face thoroughly with soap and water after handling and before eating or smoking. Routinely shower or bathe after work and wash all clothing with soap and hot water before re-using. In case of contact, immediately remove contaminated clothing and wash thoroughly with soap and water.

Do not transport or store DYFONATE 4-EC with any food or feed intended for human or animal consumption.

Not for use or storage in or around the home.

Do not use, pour, spill or store near heat or open flame.

Do not apply or allow to drift to areas occupied by unprotected humans or beneficial animals or onto adjoining food, fiber or pasture crops. Do not contaminate food or feed.

REENTRY STATEMENT

- Do not enter treated areas for 24 hours after application unless appropriate protective clothing is worn.
- Because certain states may require more restrictive reentry intervals for various crops treated with this product, consult your State Department of Agriculture for further information.
- Do not apply this product in such a manner as to directly or through drift expose workers or other persons. The area being treated must be vacated by unprotected persons.
- Do not apply or allow to drift to areas occupied by beneficial animals or onto adjoining food, fiber or pasture crops. Do not contaminate food or feed.
- Written or oral warnings must be given to workers who are expected to be in a treated area or in an area about to be treated with this product. Oral warnings must be given if there is reason to believe that written warnings cannot be understood by workers.

When oral warnings are given, warnings shall be given in a language customarily understood by workers and must indicate the specific period of time that treated areas may not be entered without protective clothing. Oral warnings must also include appropriate first aid instructions in case, of accidental exposure. These instructions are given in the STATEMENT OF PRACTICAL TREATMENT at the beginning of this label.

Written warnings must include the following information, "WARNING. Area treated with DYFONATE® Agricultural Insecticide on (date of application). Do not enter treated areas for 24 hours after application unless appropriate protective clothing is worn (or other reentry interval if your state has a more restrictive interval for this product). If accidental exposure occurs, follow the instructions below." (Written warnings must include the STATEMENT OF PRACTICAL TREATMENT given at the beginning of this label.)

ENVIRONMENTAL HAZARDS

This pesticide is toxic to fish and wildlife. Birds feeding in treated areas may be killed. Remove or cover if spills occur. Runoff from treated areas may be hazardous to fish in neighboring areas. Do not apply directly to water or wetlands. Do not contaminate water by cleaning of equipment or disposal of waste. For guidance contact your regional office of the EPA. Apply this product only as specified on the label.

PHYSICAL OR CHEMICAL HAZARDS

Do not use or store near heat or open flame.

GENERAL PRECAUTIONS

Read all directions before using.

Do not apply to any crop, or in any manner not specified on this label.

Do not rotate with carrots.

INSECTS CONTROLLED

Black Cutworm	*(Agrotis ipsilon)**
Cabbage Maggot	*(Hylemya brassicae)*
Corn Leaf Aphid	*(Rhopalosiphum maidis)*
Garden Symphylan	*(Scutigerella immaculata)*

Onion Maggot	*(Hylemya antiqua)*
Greenbug	*(Schizaphis graminum)*
Seed Corn Maggot	*(Hylema platura)*
* (Suppression only)	
Flea Beetles:	
Elongate Flea Beetle	*(Systena elongata)*
Palestriped Flea Beetle	*(Systena blanda)*
Threespotted Flea Beetle	*(Disonycha triangularis)*
Rootworms:	
Northern Corn Rootworm	*(Diabrotica longicornis)*
Southern Corn Rootworm	*(Diabrotica undecimpunctata howardi)*
Western Corn Rootworm	*(Diabrotica virgifera)*
Wireworms:	
Corn Wireworm	*(Melanotus communis)*
Pacific Coast Wireworm	*(Limonius canus)*
Southern Potato Wireworm	*(Conoderus falli)*
Sugarbeet Wireworm	*(Limonius californicus)*
Tobacco Wireworm	*(Conoderus vespertinus)*
Western Field Wireworm	*(Limonius infuscatus)*
Wheat Wireworm	*(Agriotes mancus)*
Ctenicera lobata	
Dalopius pallidus	

DIRECTIONS FOR USE

It is a violation of Federal law to use this product in a manner inconsistent with its labeling.

Do not apply this product through any type of irrigation system.

Application should be made only with tractor mounted spray equipment or larger, custom applicator type vehicles. Application should be made by trained personnel only, using approved protective equipment. Pour the recommended amount of DYFONATE 4-EC into nearly filled spray tank. Add balance of water to fill tank. Keep agitator running during filling and spraying operations. Apply in 20 to 50 gallons of water per broadcast acre using a properly calibrated, low-pressure sprayer having good agitation.

DYFONATE 4-EC may be combined with solution, slurry, and suspension fertilizers. In addition, certain herbicides may be combined with DYFONATE 4-EC, both with and without fertilizer. Read the recommendations for specific uses. Before combining in the spray tank, physical compatibility with these fluid fertilizers must be determined. See Appendix 1 for special directions regarding these combinations. Even though found to be compatible, agitation must be maintained during filling and application operations to insure uniform distribution of material.

Preplant broadcast applications should be mixed (incorporated) into the soil to a depth of 2 to 3 inches.

RECOMMENDATIONS

BEANS (GREEN AND DRY):
For control of garden symphylans and wireworms, apply 2 quarts (for symphylans) or 1 gallon (for symphylans and wireworms) DYFONATE 4-EC per acre, broadcast prior to planting and incorporate into the soil using suitable tillage equipment. (For garden symphylans only, reduce the rate proportionately for band treatments.)

For control of seed corn maggot: Northwestern U.S.: Apply 2 quarts to 1 gallon DYFONATE 4-EC per acre, broadcast on soil surface, prior to planting and incorporate to a depth of 2 to 4 inches using suitable tillage equipment. Use higher rate where population pressure is severe. Northeastern U.S.: Apply 1 gallon DYFONATE 4-EC per acre, broadcast on soil surface, prior to planting and incorporate to a depth of 2 to 4 inches using suitable tillage equipment.

All treatments may be applied as a tank mix with EPTAM® 7-E Selective Herbicide in fluid fertilizer and/or water. Follow the appropriate regional recommendations in the crop section of the EPTAM 7-E label for broadcast, preplant incorporated applications. Use only fluid fertilizers compatible with EPTAM 7-E and DYFONATE 4-EC (see Appendix 1). Observe all use precautions and limitations on the EPTAM 7-E label.

BEANS (LIMA):
For control of garden symphylans and wireworms, apply 2 quarts (for symphylans) or 1 gallon (for symphylans and wireworms) DYFONATE 4-EC per acre, broadcast, prior to planting and incorporate into the soil using suitable tillage equipment. The higher rate is recommended on lima beans grown in California only. (For garden symphylans only, reduce the rate proportionately for band treatment.)

COLE CROPS (BROCCOLI, BRUSSELS SPROUTS, CABBAGE, CAULIFLOWER):

Garden Symphylans: Prior to or at time of seeding or transplanting apply 2 quarts DYFONATE 4-EC per acre, broadcast, and incorporate into the soil using suitable tillage equipment. (Reduce the rate proportionately for band treatments.)

Cabbage Maggot (Northeastern U.S. only)—For transplant cole crops: use 1 to 2 quarts DYFONATE 4-EC in 200 to 400 gallons of water per acre. Apply with a sprayer equipped with drop nozzles to direct a drenching spray to the base of the plants. Apply immediately after transplanting.

CORN (FIELD, SWEET AND POPCORN):
NOTE: May be used in corn grown for grain or silage.

A. Band application for control of northern, southern and western corn rootworms—Apply at planting, 3/4 to 1 quart DYFONATE 4-EC per acre in a 6 to 8 inch band over corn rows 38 to 40 inches apart.

Incorporate into the top 1/2 to 1 inch of soil by making application ahead of the press wheels or by use of special covering devices or by dragging a short length of chain behind the press wheels.

Do not place DYFONATE 4-EC in direct contact with seed.

B. Broadcast application for control of symphylans, northern, southern and western corn rootworms, and wireworms, and for suppression of black cutworm—Apply 2 quarts (for symphylans), 3 quarts (for symphylans and corn rootworms), or 4 quarts (for symphylans, corn rootworms, wireworms and black cutworm) DYFONATE 4-EC per acre prior to planting and incorporate into the soil using suitable tillage equipment. This treatment may be applied as a tank mix with ERADICANE® 6.7-E or SUTAN® + 6.7-E, selective herbicides or their recommended tank mixes with ATRAZINE 80W or 4L BLADEX* 80W or 4L. A compatible fluid fertilizer (see Appendix 1) or water may be used as carrier. Follow all directions, limitations, and use precautions on the ERADICANE 6.7-E, SUTAN + 6.7-E, ATRAZINE, and BLADEX labels. Do not tank mix DYFONATE with ERADICANE 6.7-E on sweet corn or popcorn.

C. Split boot application for control of northern and western corn rootworms (recommended in the Northcentral and Midwestern corn-growing areas only)—Apply, at planting, 1 quart DYFONATE 4-EC in 5 to 20 gallons of compatible fluid fertilizer, (see Appendix 1), per acre (40 inch row spacing) in subsurface split bands on each side of the corn row, 1 1/4 to 1 1/2 inches to each side of the seed and 1 to 2 inches deep. *Do not place DYFONATE 4-EC in direct contact with the seed.*

*BLADEX is a registered trademark of E.I. duPont de Nemours & Co., Inc.

MINT (PEPPERMINT, SPEARMENT): For control of garden symphylans apply 2 quarts DYFONATE 4-EC per acre, broadcast, and incorporate into the soil using suitable tillage equipment prior to planting or on established plantings prior to Spring growth. (Do not apply terbacil for at least 3 weeks before or after the DYFONATE 4-EC application.)

BULB ONIONS: For control of onion maggot, apply as a furrow treatment at time of planting. Based on a 20 inch row spacing, apply at a rate of 1 quart DYFONATE 4-EC per acre (1 quart per 26,135 feet of row; 3 1/4 fluid ounces per 2,613 of row; 1 1/4 fluid ounces per 1,000 feet of row). Use on organic soils only (soils containing 10% or more organic matter).

PEPPERS: For control of garden symphylans and wireworms, apply 2 quarts (for symphylans) or 4 quarts (for symphylans and wireworms) DYFONATE 4-EC per acre, broadcast, prior to planting and incorporate into the soil using suitable tillage equipment. (For garden symphylans only, reduce the rate proportionately for band treatments.)

POTATOES (IRISH):
A. Broadcast preplant application for control of garden symphylans and wireworms:

Apply 2 quarts (for symphylans) or 4 quarts (for symphylans and wireworms) DYFONATE 4-EC per acre prior to planting and incorporate into the soil using suitable tillage equipment. This treatment may also be applied in certain areas as a tank mix with EPTAM* 7-E selective herbicide in a compatible fluid fertilizer and/or water (see Appendix 1). Follow recommendations in the crop section of the EPTAM 7-E label for broadcast, preplant incorporated applications. For use in the Northern, Western, and Pacific Northwest Regions only. Observe all use precautions and limitations on the EPTAM 7-E label.

B. Banded preplant soil injection with fluid fertilizer for control of wireworms (Northwestern U.S. only):

Apply through fluid fertilizer shank immediately ahead of the planter. Set the shank to run 8 to 10 inches deep. Apply 2 quarts DYFONATE 4-EC per crop acre, based on 36" row spacing, in the recommended quantity of compatible liquid fertilizer (see Appendix 1).

NOTE: For heavy wireworm infestations, use the broadcast application only.

SORGHUM (MILO): For control of corn leaf aphid and greenbug, apply 3/4 to 1 quart DYFONATE 4-EC in 5 to 30 gallons of water per acre over the top using aerial or ground application equipment. Do not apply within 14 days of harvest nor feed or graze to livestock within 14 days of applications. Limit use to 2 applications per crop.

STRAWBERRIES: For control of symphylans apply 2 quarts DYFONATE 4-EC per acre, broadcast, and incorporate into the soil using suitable tillage equipment prior to planting. (Reduce the rate proportionately for band treatments.)

SUGAR BEETS: For control of garden symphylans and wireworms, apply 2 quarts (for symphylans) or 1 gallon (for symphylans and wireworms) DYFONATE 4-EC per acre, broadcast, prior to planting and incorporate into the soil using suitable tillage equipment. (For garden symphylans only, reduce the rate proportionately for band treatments.)

SUGARCANE: At planting—For wireworm control, apply 4 quarts DYFONATE 4-EC per acre in a 14 inch band on cane rows 60 inches apart. This is equivalent to 11 to 14½ fluid ounces of DYFONATE 4-EC per 1,000 linear feet of row. Apply over the top of sugarcane seed pieces immediately ahead of covering discs.

TABLE BEETS: For control of cabbage maggot and garden symphylans, apply 2 quarts DYFONATE 4-EC per acre, broadcast, prior to planting. Incorporate into the soil using suitable tillage equipment. (For garden symphylans only, reduce the rate proportionately for band treatments.)

TOBACCO: For control of wireworms, apply 1 to 2 quarts DYFONATE 4-EC per acre, broadcast, prior to bedding and transplanting and incorporate into the soil using suitable tillage equipment.

Flue Cured Tobaccos only: DYFONATE 4-EC in combination with TILLAM® 6-E controls wireworms and certain weeds (see TILLAM 6-E label) in tobacco. Add to spray tank and apply 1 quart DYFONATE 4-EC and 2⅔ quarts TILLAM 6-E in 20 to 50 gallons of water per acre (see Appendix 1). Use a properly calibrated, low-pressure, boom-type sprayer having good agitation. IMPORTANT: Refer to both DYFONATE 4-EC and TILLAM 6-E labels for pertinent directions for use, and use precautions and restrictions. The soil should be well worked and dry enough to permit good soil incorporation. The tank mix must be incorporated (mixed) into the soil immediately after application to prevent loss of the herbicide by evaporation. Whenever possible, application and incorporation should be done in the same operation. Use powerdriven cultivation equipment set to cut to a depth of 3" in light soil and 3" to 4" in heavy soil, or a tandem disc set to cut to a depth of 4" to 6". Cross disc at right angles for thorough mixing. The bedding operation should follow the incorporation as soon as practical. Planting may be done immediately after bedding or at a later date; however, to insure long term weed control, the crop should be planted as soon as possible after application and bedding.

TOMATOES: For control of garden symphylans, apply 2 quarts of DYFONATE 4-EC per acre, broadcast, prior to planting and incorporate into the soil using suitable tillage equipment. (Reduce the rate proportionately for band treatments.)

Appendix 1

Procedure for Testing the Tank-Mix Compatibility of DYFONATE 4-EC with Fluid Fertilizers and DYFONATE 4-EC with Herbicides and Water or Fluid Fertilizers.

This procedure is suggested for determining whether DYFONATE 4-EC and a specific fluid fertilizer and/or the specifically approved herbicide(s) may be combined for spray-tank application.

Materials Required:

1. a. DYFONATE 4-EC.
   b. DYFONATE 4-EC/herbicide combinations (EPTAM 7-E, ERADICANE 6.7-E, SUTAN + 6.7-E, TILLAM 6-E, ATRAZINE 80W or 4L, or BLADEX 80W or 4L.
2. Water or fluid fertilizer to be used.
3. Adjuvant for tank mix—Compex,* Sponto* 16B-D, Unite* or equivalent. The adjuvant which performs best depends on the specific fertilizer being considered.
4. Two one-quart, wide-mouth glass jars with tops.
5. Measuring spoons (a graduated cylinder or pipette will provide more accurate measurement).
6. Measuring cup, 8 oz. (237 mL)

*Compex is a trademark of Kalo Laboratories, Inc; Sponto 168.D is a trademark of Witco Chemical Corporation; Unite is a trademark of Hopkins Agricultural Chemical Company.

Procedure:

1. Add one pint (2 cups or approximately 473 mL) of the test water or fluid fertilizer to each of the quart jars.
2. Add adjuvant to one of the jars and mix (see rate tables).
3. Carefully add the pesticides to each jar (see rate tables) as follows:
   a. If only DYFONATE 4-EC is to be used, add it directly to each jar.
   b. If a tank mix with a herbicide emulsifiable concentrate is to be used, add the herbicide first and mix, and then add the DYFONATE 4-EC.
   c. If a tank mix with a herbicide emulsifiable concentrate and a wettable powder or flowable formulation is to be used, add and mix the pesticides in the following order:
      1) add and mix the wettable powder or flowable formulation first (premixing these in water prior to addition to fluid fertilizer often provides a more compatible mix with fluid fertilizer),
      2) add and mix the herbicide emulsifiable concentrate second,
      3) and finally, add the DYFONATE 4-EC.
4. Securely close each jar using the top and mix the contents by inverting the jar ten times.
5. Inspect the mixtures immediately and after standing for thirty minutes without mixing.
6. Dispose of jar mixes according to disposal statements contained within this label.

A mixture may be incompatible if either of the following conditions is observed:

a. an oil layer or large oil globules are observed, or
b. sludge or clumps are present.

If either mixture remains uniform for thirty minutes, the combination may be used. Should either mixture separate after thirty minutes, but remix readily with ten jar inversions, the combination can be used if good agitation is maintained in the spray tank. If nondispersible oil, sludge or clumps persist, the combination must not be used.

If the mixture with the adjuvant stays mixed, or remixes readily to form a uniform dispersion, but the one without adjuvant does not, be sure to use the adjuvant in the spray tank. Add the adjuvant first to the water or fluid fertilizer at the recommended rate. Minimize foaming by using moderate agitation.

Rate Table for DYFONATE® 4-EC/EPTAM® 7-E and ADJUVANT with the FLUID FERTILIZER**

Gallons of Fluid Fertilizer To Be Applied Per Acre	Milliliters or Teaspoons of DYFONATE 4-EC* To Be Added to 1 Pint of Fertilizer		Milliliters or Teaspoons of EPTAM 7-E* To Be Added to 1 Pint of Fertilizer	
	mL	Tsp.	mL	Tsp.
10	12	2½	6⅔	1⅓
15	8	1⅔	5	1
20	6	1¼	3⅓	⅔
25	5	1	2½	½
40	3	⅔	1⅔	⅓

* Based on a field rate of one pound active ingredient per acre in the fertilizer volumes indicated. Adjust volumes proportionately to correspond with intended field rates in terms of pounds active ingredients per acre (e.g. for a field rate of 3 pounds of actual EPTAM 7-E in 20 gallons of fertilizer per acre, add 10 mL or 2 tsp. of EPTAM 7-E to each jar for compatibility testing purposes.)

** A typical adjuvant rate of use is 3 pints per 100 gallons of fluid fertilizer. Two (2) mL or one-half (½) tsp. of adjuvant should be added to 1 pint of fertilizer in order to equal a rate of 3 pints of adjuvant per 100 gallons of fluid fertilizer.

Rate Table for DYFONATE 4-EC and ADJUVANT with the Fluid Fertilizer**

Gallons of Fluid Fertilizer To Be Applied Per Acre	Milliliters or Teaspoons of DYFONATE 4-EC* To Be Added to 1 Pint of Fertilizer	
	mL	Tsp.
10	12	2½
15	8	1⅔
20	6	1¼
25	5	1
40	3	⅔

* Based on a fluid rate of one pound active ingredient per acre in the fertilizer volumes indicated. Adjust volume proportionately to correspond with intended rate in terms of pounds active ingredient per acre (e.g. for a field rate of 2 pounds active ingredient in 10 gallons of fertilizer per acre, add 24 mL or 5 tsp. DYFONATE 4-EC to each jar for compatibility testing purposes.

** A typical adjuvant rate of use is 3 pints per 100 gallons of fluid fertilizer. Two (2) mL or one-half (½) tsp. of adjuvant should be added to 1 pint of fertilizer in order to equal a rate of 3 pints of adjuvant per 100 gallons of fluid fertilizer.

Rate Table for DYFONATE 4-EC/ERADICANE 6.7-E and DYFONATE 4-EC/SUTAN + 6.7-E and ADJUVANT with FLUID FERTILIZER**

Gallons of Fluid Fertilizer To Be Applied Per Acre	Milliliters or Teaspoons of DYFONATE 4-EC* To Be Added to 1 Pint of Fertilizer		Milliliters or Teaspoons of ERADICANE 6.7-E* or SUTAN + 6.7-E To Be Added to 1 Pint of Fertilizer	
	mL	Tsp.	mL	Tsp.
10	12	2½	7	1½
15	8	1⅔	5	1
20	6	1¼	3½	¾
25	5	1	2½	½
40	3	⅔	1¾	⅓

* Based on a field rate of one pound active ingredient per acre in the fertilizer volumes indicated. Adjust volumes proportionately to correspond with intended field rates in terms of pounds active ingredients per acre (e.g. for a field rate of 4 pounds of actual SUTAN + 6.7-E in 20 gallons of fertilizer per acre, add 14 mL or 3 tsp. of SUTAN + 6.7-E to each jar for compatibility testing purposes.)

** A typical adjuvant rate of use is 3 pints per 100 gallons of fluid fertilizer. Two (2) mL or one-half (½) tsp. of adjuvant should be added to 1 pint of fertilizer in order to equal a rate of 3 pints of adjuvant per 100 gallons of fluid fertilizer.

Rate Table for DYFONATE 4-EC/ERADICANE 6.7-E/ATRAZINE or BLADEX and DYFONATE 4-EC/SUTAN + 6.7-E/ ATRAZINE or BLADEX and ADJUVANT with water or FLUID FERTILIZER**

Gallons of Water or Fluid Fertilizer to Be Applied/Acre	10		15		20		25		40	
Measure	mL	Tsp.	mL	Tsp.	mL	Tsp.	mL	Tsp.	mL	Tsp.
Milliliters or Teaspoons of DYFONATE 4-EC* To Be Added to 1 Pint of Water or Fertilizer	12	2½	8	1⅔	6	1¼	5	1	3	⅔
Milliliters or Teaspoons of ERADICANE 6.7-E* or SUTAN + 6.7-E* To Be Added to 1 Pint of Water or Fertilizer	7	1½	5	1	3½	¾	2½	½	1¾	⅓
Milliliters or Teaspoons of ATRAZINE 80W* or BLADEX 80W* To Be Added to 1 Pint of Water or Fertilizer	22	4½	15	3	11	2¼	8⅔	1¾	5	1
Milliliters or Teaspoons of ATRAZINE 4L* or BLADEX 4L* To Be Added to 1 Pint of Water or Fertilizer	12	2½	8	1⅔	6	1¼	5	1	3	⅔

* Based on a field rate of one pound active ingredient per acre in the water or fertilizer volumes indicated. Adjust volumes proportionately to correspond with intended field rates in terms of pounds active ingredients per acre (e.g. for a field rate of 4 pounds of actual SUTAN + 6.7-E in 40 gallons of fertilizer per acre, add 7 mL or 1½ tsp. of SUTAN + 6.7-E to each jar for compatibility testing purposes, i.e., 4 times the 10-gallon rate for 1 pound equals 28 mL or 6 tsp.; then dividing by the dilution factor four [40 ÷ 10 = 4] gives 7 mL or 1½ tsp. per pint, which is equivalent to 4 pounds actual SUTAN + 6.7-E/40 gallons).

** A typical adjuvant rate of use is 3 pints per 100 gallons of fluid fertilizer. Two (2) mL or one-half (½) tsp. of adjuvant should be added to 1 pint of fertilizer in order to equal a rate of 3 pints of adjuvant per 100 gallons of fluid fertilizer.

Rate Table for DYFONATE 4-EC/TILLAM 6-E and ADJUVANT**

Gallons of Fluid Fertilizer To Be Applied Per Acre	Milliliters or Teaspoons of DYFONATE 4-EC* To Be Added to 1 Pint of Fertilizer		Milliliters or Teaspoons of TILLAM 6-E* To Be Added to 1 Pint of Fertilizer	
	mL	Tsp.	mL	Tsp.
10	12	2½	8	1⅔
15	8	1⅔	5¼	1
20	6	1¼	4	¾
25	5	1	3¼	⅔
40	3	⅔	2	⅓

* Based on a field rate of one pound active ingredient per acre in the fertilizer volumes indicated. Adjust volumes proportionately to correspond with intended field rates in terms of pounds active ingredients per acre (e.g. for a field rate of 4 pounds of actual TILLAM 6-E in 20 gallons of fertilizer per acre, add 10 mL or 2 tsp. of TILLAM 6-E to each jar for compatibility testing purposes.)

** A typical adjuvant rate of use is 3 pints per 100 gallons of fluid fertilizer. Two (2) mL or one-half (½) tsp. of adjuvant should be added to 1 pint of fertilizer in order to equal a rate of 3 pints of adjuvant per 100 gallons of fluid fertilizer.

Tank Mix Directions

1. Fill the spray tank at least ⅔ full of the appropriate carrier fluid. Be sure tank is clean and not contaminated with 2,4-D type herbicides, oil residue or other materials.
2. Start agitation and continue moderate agitation throughout mixing and application procedures.
3. Add the proper amount of adjuvant and mix (if adjuvant is to be used).
4. Next add the ATRAZINE 80W or 4L or BLADEX 80W or 4L (if it is to be combined in the tank) and mix to disperse it thoroughly.
5. Then add the herbicide emulsifiable concentrate (EPTAM 7-E, ERADICANE 6.7-E, SUTAN + 6.7-E or TILLAM 6-E) and mix thoroughly.
6. Lastly, add the DYFONATE 4-EC.
7. Complete filling the remaining portion of the tank with carrier fluid; mix thoroughly.
8. The tank-mix combination should not be left in the spray tank for prolonged periods; batches should be mixed and applied the same day.

NOTE: Constant moderate agitation is required for successful tank-mix preparation and application. Avoid excessive agitation which can cause tank-mix incompatibility. All return lines to the tank must discharge below liquid level.

STORAGE AND DISPOSAL

STORAGE: Minimum storage temperature: MINUS 40°F. Any steel container may be corroded by its contents. Dents or damage may promote corrosion. Use damaged and older containers FIRST. Inspect rusted cans for leaks.

PROHIBITIONS: Do not contaminate water, food or feed by storage or disposal. Open dumping is prohibited.

PESTICIDE DISPOSAL: Pesticide wastes are acutely hazardous. Improper disposal of excess pesticide, spray mixture, or rinsate is a violation of Federal law. If these wastes cannot be disposed of by use according to label instructions, contact your State Pesticide or Environmental Control Agency, or the Hazardous Waste representative at the nearest EPA Regional Office for guidance.

CONTAINER DISPOSAL: Triple rinse (or equivalent). Then offer for recycling or reconditioning, or puncture and dipose of in a sanitary landfull, or by other procedures approved by State and local authorities.

NOTICE TO BUYER AND USER:

Seller warrants that this product conforms to the chemical description on the label and is reasonably fit for the purposes stated on the label when used in accordance with directions under normal conditions of use. This warranty does not extend to the use of this product contrary to label instructions, or under abnormal use conditions, or under conditions not reasonably foreseeable to Seller, and Buyer and User assume the risk of any such use. SELLER DISCLAIMS ALL OTHER WARRANTIES EXPRESSED OR IMPLIED INCLUDING ANY WARRANTY OF FITNESS OR MERCHANTABILITY. SELLER SHALL NOT BE LIABLE FOR CONSEQUENTIAL, SPECIAL OR INDIRECT DAMAGES RESULTING FROM THE USE OR HANDLING OF THIS PRODUCT AND SELLER'S SOLE LIABILITY AND BUYER'S AND USER'S EXCLUSIVE REMEDY SHALL BE LIMITED TO THE REFUND OF THE PURCHASE PRICE.

This product is sold only for uses stated on its label. No express or implied license is granted to use or sell this product under any patent in any country except as specified: Country: United States of America.

A-13MR WT 4-88

The information and specimen labels in this booklet are correct as of the publication date but are subject to change without notice. For current information, contact ICI Americas Inc.

DYFONATE® II 10-G

GRANULAR INSECTICIDE

ACTIVE INGREDIENT:

O-Ethyl S-phenyl ethylphosphonodithioate*	10.0%
INERT INGREDIENTS	90.0%
TOTAL	100.0%

*DYFONATE (Contains 10 Lbs Active Ingredient Per 100 Pounds)

EPA Reg. No. 10182-180

KEEP OUT OF REACH OF CHILDREN

WARNING!

AVISO

PRECAUTION AL USUARIO: Si usted no lee ingles, no use este producto hasta que la etiqueta la haya sido explicada ampliamente.

STATEMENT OF PRACTICAL TREATMENT

FOR PESTICIDE EMERGENCY

CALL A PHYSICIAN IMMEDIATELY: If a known exposure occurs, or if poisoning is suspected, do not wait for symptoms to develop. Immediately start the procedures given below and simultaneously contact a Poison Center, a physician, or the nearest hospital.

NOTE: Be sure to advise the person contacted that the compound is a cholinesterase inhibitor. Describe the type of exposure, the victim's symptoms, and follow the advice given.

POISON SYMPTOMS: DYFONATE is a highly toxic cholinesterase inhibitor. Symptoms of poisoning include headache, blurred vision, weakness, nausea, discomfort in the chest, abdominal cramps, vomiting, diarrhea, salivation, sweating and pinpoint pupils.

Always Call a Physician in Case Of Suspected Poisoning.
Atropine by Injection is Antidotal.

If swallowed, immediately give several glasses of water and induce vomiting by placing a finger on the back of the victim's tongue. Give fluids until vomitus is clear. Do not induce vomiting or give anything by mouth to an unconscious or convulsing person.

For skin contact, flush with large amounts of running water for 15 minutes. Remove contaminated clothing and shoes. Get medical attention. Wash clothing before reuse.

For eye contact, immediately flush eyes with large amounts of running water for at least 15 minutes. Hold eyelids apart to insure rinsing of the entire surface of the eye and lids with water. Get medical attention immediately.

If inhaled, remove to fresh air. If not breathing, clear victim's airway and start mouth-to-mouth artificial respiration.

If breathing is difficult, give oxygen, preferably with a physician's advice. Get medical attention immediately.

FOR 24-HOUR EMERGENCY MEDICAL ASSISTANCE, CALL 1-800-F-A-S-T-M-E-D (327-8633).
FOR CHEMICAL EMERGENCY: Spill, leak, fire, exposure, or accident call CHEMTREC 1-800-424-9300.

INFORMATION FOR MEDICAL PERSONNEL: Initial measures include removal of secretions, maintenance of a patent airway, and if necessary, artificial respiration. When cyanosis is relieved, atropine by injection is antidotal and may be administered in large therapeutic doses, repeated as necessary to the point of tolerance. If symptoms warrant further treatment, 2-PAM (Protopam Chloride) has shown utility as adjunctive therapy. Never use morphine. Continued absorption of the poison may occur, resulting in fatal relapse after initial improvement; very close supervision of the patient is indicated for at least 48 to 72 hours.

In case of spill isolate the spill.

HOLD this package, other cargo and vehicles involved. IMMEDIATELY TELEPHONE (24-hour service) for emergency assistance: 800-424-9300.

Read all directions before using.

Do not transport or store DYFONATE 10-G with food or feed intended for human or animal consumption.

PRECAUTIONARY STATEMENTS

HAZARDS TO HUMANS AND DOMESTIC ANIMALS

WARNING

MAY BE FATAL IF SWALLOWED, INHALED OR ABSORBED THROUGH THE SKIN. RAPIDLY ABSORBED THROUGH THE SKIN. DO NOT GET IN THE EYES, ON SKIN, OR ON CLOTHING. AVOID BREATHING OF DUST.

Wear rubber gloves, shoes, protective clothing and goggles when handling or applying.

Wash hands, arms, and face thoroughly with soap and water after handling and before eating or smoking. Routinely shower or bathe after work and wash all clothing with soap and hot water before reusing. In case of contact, immediately remove contaminated clothing and wash thoroughly with soap and water.

Do not transport or store DYFONATE 10-G with any food or feed intended for human or animal consumption.

Not for use or storage in or around the home.

ENVIRONMENTAL HAZARDS

This pesticide is toxic to fish and wildlife. Birds feeding in treated areas may be killed. Remove or cover granules if spills occur. Runoff from treated areas may be hazardous to fish in neighboring areas. Do not apply directly to water or wetlands. Do not contaminate water by cleaning of equipment or disposal of waste. For guidance contact your regional office of the EPA. Apply this product only as specified on the label.

General Precautions

- Read all directions before using.
- Do not apply on any crop, or in any manner not specified on this label. Do not rotate with carrots.

DIRECTIONS FOR USE

It is a violation of Federal law to use this product in a manner inconsistent with its labeling.

REENTRY STATEMENT

- Do not enter treated areas for 24 hours after application unless appropriate protective clothing is worn.
- Because certain states may require more restrictive reentry intervals for various crops treated with this product, consult your State Department of Agriculture for further information.
- Do not apply this product in such a manner as to directly or through drift expose workers or other persons. The area being treated must be vacated by unprotected persons.
- Do not apply or allow to drift to areas occupied by beneficial animals or onto adjoining food, fiber or pasture crops. Do not contaminate food or feed.
- Written or oral warnings must be given to workers who are expected to be in a treated area or in an area about to be treated with this product. Oral warnings must be given if there is reason to believe that written warnings cannot be understood by workers.

When oral warnings are given, warnings shall be given in a language customarily understood by workers and must indicate the Specific Period of time that treated areas may not be entered without protective clothing. Oral warnings must also include appropriate first aid instructions in case of accidental exposure. These instructions are given in the STATEMENT OF PRACTICAL TREATMENT at the beginning of this label.

Written warnings must include the following information, "WARNING. Area treated with DYFONATE Agricultural Insecticide on (date of application). Do not enter treated areas for 24 hours after application unless appropriate protective clothing is worn (or other reentry interval if your state has a more restrictive interval for this product). If accidental exposure occurs, follow the instructions below." (Written warnings must include the STATEMENT OF PRACTICAL TREATMENT given at the beginning of this label.)

Gauge settings and calibration instructions

NOTE: These settings should be used as starting points only. Continually check the amount of DYFONATE® II 10-G used against a known length of row or acreage and make further adjustments accordingly. Also, check calibration occasionally to make sure equipment wear, changing conditions, etc., have not caused a change in flow rate.

GAUGE SETTINGS
DYFONATE® II 10-G 12 oz/1000 ft. of row and 22 oz/1000 ft. of row

Equipment	Speed in mph 4		6		8	
Application Rate: Oz. per 1000 ft. of row	**12**	**22**	**12**	**22**	**12**	**22**
Gandy bolt on bottom	26	34	31	41	35	48
Gandy snap-on bottom	26	34	31	41	35	47
Noble (round opening)	22	31	29	38	33	45
Noble (rectangular opening)	15	24	21	32	26	41
John Deere 1986 Maxi-Emerge II (adjustable knob-0-79)	33	50	45	58	32	68
John Deere 7000 Maxi-Emerge (Even) (0-10-20-30-40)	24	33	31	38	35	45
John Deere 7000 Maxi-Emerge (Odd) (5-15-25-35-45)	18	25	23	31	26	37
John Deere (metal box with 3 range settings and dial) Range/Notch	2/13	3/6	2/32	3/28	3/9	*
International Harvester (one gauge per hopper) Gate/Dial	2/7	4/5	3/8	6/4	4/8	8/3
International Harvester (two gauges per hopper) Gate/Dial	3/0	4/8	4/1	6/7	5/1	8/6
Buffalo (with gauge)	21	31	27	40	33	*
Buffalo (fluted-roll feed—revolutions on hexagonal adjusting nut)	3⅓	5⅚	Uniform Flow at All Speeds			
Allis Chalmers (old)	5	7	Uniform Flow at All Speeds			
Allis Chalmers (new, with Noble metering unit)	16	25	22	23	27	41
Microband - Canaan Industries	16**	27**	Uniform Flow at All Speeds			

* Unable to deliver quantity of DYFONATE® II 10-G required at speed indicated.
** Number of teeth - Silver Rotor.

EQUIVALENT LBS. DYFONATE® II 10-G PER ACRE

Ounces of DYFONATE® II 10-G per 1000 ft. of row	Row Spacing in Inches 60	40	38	36	34	32	30	22	18
6.8	3.6	5.6	5.8	6.2	6.4	6.8	7.4	10.0	12.0
10.2	5.6	8.2	8.6	9.2	9.8	10.4	11.0	15.0	18.4
12	6.6	10.0	10.4	10.8	11.6	12.2	13.0	19.8	21.8
22	12.0	18.0	19.0	20.0	21.2	22.6	24.0	32.8	40.0
24	13.0	20.0	20.6	21.8	23.0	24.4	26.2	35.6	—
55.1	30.0	—	—	—	—	—	—	—	—
73.5	40.0	—	—	—	—	—	—	—	—

RECOMMENDATIONS

CROPS	APPLICATION METHOD	PESTS CONTROLLED	RATES OF DYFONATE® II 10-G	REMARKS	APPLICATION DIRECTIONS
ASPARAGUS	Up to 30 days prior to harvest	Symphylans	Broadcast: 30 to 100 pounds per acre.	Recommended in California only. Reduce the rate proportionately for band treatments.	Broadcast: Apply DYFONATE® II 10-G and incorporate into the soil to a depth of 2 to 3 inches using suitable tillage equipment. Do not apply within 30 days of harvest.
BEANS, GREEN & DRY	Prior to planting	Symphylans	Broadcast: 20 pounds per acre.	For symphlan control, reduce the rate proportionately for band treatments.	Broadcast: Apply DYFONATE® II 10-G prior to planting and incorporate into the soil to a depth of 2 to 3 inches using suitable tillage equipment.
		Wireworms	Broadcast: 40 pounds per acre.		
		Seed Corn Maggot	Broadcast: 20 to 40 pounds per acre.	Recommended for control of seed corn maggot in the Northwest U.S. only. Use the higher rate where population pressure is severe.	
BEANS, LIMA	Prior to planting	Sympylans	Broadcast: 20 pounds per acre.	Reduce the rate proportionately for band treatments.	Broadcast: Apply DYFONATE® II 10-G prior to planting and incorporate into the soil to a depth of 2 to 3 inches using suitable tillage equipment.
COLE CROPS (BROCCOLI, BRUSSELS SPROUTS, CABBAGE AND CAULIFOWER	Prior to planting or at planting	Symphlans	Broadcast: 20 pounds per acre.	Reduce the rate proportionately for band treatments.	Broadcast: Apply DYFONATE® II 10-G prior to or at time of seeding or transplanting and incorporate into the soil to a depth of 2 to 3 inches using suitable tillage equipment.
CORN-FIELD (Dent, seed, grain, silage) SWEET, POPCORN	Prior to planting or at planting	Northern, Southern, and Western Corn Rootworm	Band: 12 oz. per 1000 linear feet. or	May be used on conventional, minimum or conservation and no-tillage systems.	Band: Apply DYFONATE® II 10-G in a 7 inch wide band using a row bander directly behind the seed furrow opener and in front of the press wheels, or in a 7 inch wide band directly behind the press wheels using drag chains or spring tines to insure incorporation. Under windy conditions the use of windshields is recommended to prevent drift.
		Suppression of Cutworms Suppression of White Grubs Seedcorn Maggot Seedcorn Beetle Wireworms	Broadcast: 40 pounds per acre.	DYFONATE® II 10-G is recommended for control of seedcorn beetle in the Central and Midwestern corn growing areas only.	Some granules can fall into the open furrow without causing harm to the seed. However, to avoid stand reduction, do not make "in-furrow" applications of DYFONATE® II 10-G.
		Lesser Cornstalk Borer	Band: 24 oz. per 1000 linear feet	Band applications of DYFONATE® II 10-G will provide suppression of wireworms. Use the broadcast application for control.	Broadcast: Apply DYFONATE® II 10-G prior to planting and incorporate into the soil to a depth of 2 to 3 inches using suitable tillage equipment.
		Symphlans	Broadcast: 20 pounds per acre		
		Maize Billbug	Broadcast: 40 pounds per acre.		
	At Cultivation	Northern, Southern, and Western Corn Rootworm	Band: 12 oz. per 1000 linear feet.	Apply DYFONATE ® II 10-G when activity of rootworm larvae is first observed.	Band: Apply DYFONATE® II 10-G in a 7 inch wide band over the corn rows. Cover the treated bands with 2 to 3 inches of soil by making the application ahead of disc hillers or cultivation equipment. Do not apply within 45 days of harvest nor feed or graze livestock within 30 days of treatment.
	Over the Whorl	European Corn, Borer, Southwestern Corn Borer Reduction of Fall Armyworm Populations	Band: 12 oz. per 1000 liear feet. or Broadcast: 10 pounds per acre.	Consult your State Agricultural Experiment Station or extension service specialist for local treatment threshold and application timing information. Scouting for insects is strongly recommended.	Band or Broadcast: Apply DYFONATE® II 10-G over the corn rows so granules fall into the whorl. May be applied broadcast, using air or ground equipment, or, in a band directed over corn rows. Do not apply within 30 days of harvest nor feed or graze livestock within 30 days of treatment.
SWEET CORN (Grown on Muck Soils)	At Planting	Wireworms	Band: 22 oz. per 1000 linear feet.	Recommended in Florida only.	Band: Apply DYFONATE® II 10-G in a 2 inch wide band over the corn rows.
ONIONS, BULB	At Planting	Onion Maggot	In-Furrow; 6 oz. per 1000 linear feet.	Use only in soils containing 10% or more organic matter.	In-Furrow: Apply DYFONATE® II 10-G into the furrow at time of planting.

CROPS	APPLICATION METHOD	PESTS CONTROLLED	RATES OF DYFONATE® II 10-G	REMARKS	APPLICATION DIRECTIONS
PEANUTS (Southeatern U.S.)	At pegging or up to 30 days After Pegging Begins	Southern Corn Rootworm, Granulate Cutworm Lesser Cornstalk Borer Suppression of White Mold (*Sclerotium rolfsii*)	Band: 22 oz. per 1000 linear feet or Broadcast: 40 pounds per acre.	For control of southern corn rootworm and granulate cutworm apply DYFONATE® II 10-G at pegging. For control of lesser cornstalk borer, apply DYFONATE® II 10-G at first signs of borer infestation after pegging begins. For heavy white mold infestations a suitable fungicide is necessary.	Band: Apply DYFONATE® II 10-G in an 18 inch wide band. Broadcast: Apply DYFONATE® II 10-G by air or ground equipment.
PEANUTS: (Southwestern U.S.)	At Pegging or up to 30 Days After Pegging Begins	Southern Corn Rootworm	Band:22 oz. per 1000 linear feet.		Band: Apply DYFONATE® II 10-G in an 18 inch wide band.
		Lesser Cornstalk Borer	Band: 9.8 to 13.7 oz. per 1000 linear feet.		Band: Apply DYFONATE® II 10-G in a 14 inch wide band.
		Burrowing Bug	Band: 11 to 22 oz. per 1000 linear feet. or Broadcast: 20 pounds per acre.	Recommended for control of burrowing bug in Texas only.	Band: Apply DYFONATE® II 10-G in an 18 inch wide band. Broadcast: Apply DYFONATE® II 10-G and lightly incorporate into the soil with 1 or 2 inches of water.
POTATOES (IRISH)	Prior to Planting or at Planting.	Wireworms	Subsurface Bands: 20 oz. per 1000 linear feet. or Broadcast: 40 pound per acre.	The use of subsurface bands is recommended in the Pacific Northwest only. This application will control light to moderate infestations of wireworms. Use the 40 pound per acre broadcast application for heavy infestations of wireworms.	Subsurface Bands: Apply DYFONATE® II 10-G in subsurface bands 3 to 4 inches on both sides of and 2 inches below the seed pieces. Broadcast: Apply DYFONATE® II 10-G prior to or at time of planting and incorporate into the soil to a depth of 2 to 3 inches using suitable tillage equipment.
		Symphylans	Broadcast: 20 pounds per acre.		
RADISH	Prior to planting or at planting	Cabbage Maggot Symphylans	Broadcast: 20 pounds per acre.	For symphylan control, reduce the rate proportionately for band treatments.	Broadcast: Apply DYFONATE® II 10-G prior to or at time of planting and incorporate into the soil to a depth of 2 to 3 inches using suitable tillage equipment.
SORGHUM (Milo)	At Planting	Northern, Southern and Wester Corn Rootworm Lesser Cornstalk Borer	Band: 12 oz. per 1000 linear feet.		Band: Apply DYFONATE® II 10-G in a 7 inch wide band using a row bander directly behind the seed furrow opener and in front of the press wheels, or in a 7 inch wide band directly behind the press wheels using drag chains or spring tines to insure corporation. Under windy conditions, the use of windshields is recommended to prevent drift. Some granules can fall into the open furrow without causing harm to the seed, However, to avoid stand reduction, do not make "in furrow" applications of DYFONATE® II 10-G.
	Over the Top	Corn Leaf Aphid Greenbug	Broadcast: 7.5 pounds per acre.	Recommended in Texas only. Limit post-emergence use to 2 applications per crop.	Broadcast: Apply DYFONATE® II 10-G over the top using aerial equipment. Use only on irrigated sorghum. Do not harvest nor feed or graze livestock within 14 days of application.
SUGAR BEETS	Prior to planting or at planting	Sugar Beet Root Maggot	Band: 6.8 to 10.5 oz per 1000 linear feet. Subsurface Band: 10.5 oz per 1000 linear feet.	Recommended in Colorado and Northwest sugar beet growing areas only. To avoid stand reduction, do not place DYFONATE® II 10-G in direct contact with the seed.	Band: Apply DYFONATE® II 10-G in a 7 inch wide band over the sugar beet rows. Incorporate lightly into the soil using suitable tillage equipment. Subsurface Band: Apply DYFONATE® II 10-G 2 inches from the seed furrow and 2 inches below the soil surface on the irrigated side of the sugar beet rows. (Recommended in the Big Horn Basin of Wyoming only) Broadcast: Apply DYFONATE® II 10-G prior to planting and incorporate into the soil to a depth of 2 to 3 inches using suitable tillage equipment.
		Symphylans	Broadcast: 20 pounds per acre.	For symphylan control, reduce the rate proportionately for band treatments.	
		Wireworms	Broadcast: 40 pounds per acre.		

CROPS	APPLICATION METHOD	PESTS CONTROLLED	RATES OF DYFONATE® II 10-G	REMARKS	APPLICATION DIRECTIONS
SUGARCANE	At Planting	Wireworms	Band: 55.1 to 73.5 oz. per 1000 linear feet.	A use rate of less than 73.5 oz. per 1000 linear feet is only recommended in Florida.	Band: Apply DYFONATE® II 10-G in a 14 inch band over the tops of the sugarcane seed pieces immediately ahead of covering discs.
	Post-Emergence	White Grub	Band: 73.4 to 110.2 oz. per 1000 feet.	Recommended in the Southwestern U.S. only. Application should be timed according to the flight of the adult beetles. Apply at any time from the peak of the adult flight up until after two weeks after flight peak.	Band: Apply DYFONATE® II 10-G over the top of the cane row and lightly incorporate as close to plants as possible. For best results, irrigate after application.
TABLE BEETS	Prior to Planting	Cabbage Maggot Symphylans	Broadcast: 20 pounds per acre.	For symphylan control, reduce the rate proportionately for band treatments.	Broadcast: Apply DYFONATE® II 10-G prior to planting and incorporate into the soil to a depth of 2 to 3 inches using suitable tillage equipment.
TOBACCO	Prior to bedding	Wireworms	Broadcast: 10 to 20 pounds per acre.		Broadcast: Apply DYFONATE® II 10-G prior to bedding and incorporate into the soil to a depth of 2 to 3 inches using suitable tillage equipment.
TOMATOES	Prior to planting	Symphylans	Broadcast: 20 pounds per acre.	Reduce the rate proportionately for band treatments.	Broadcast: Apply DYFONATE® II 10-G prior to planting and incorporate into the soil to a depth of 2 to 3 inches using suitable tilalge equipment.

NOTE: For all crops where APPLICATION METHOD is "Prior to planting," results are best when applciation is made as close as possible to planting.

STORAGE AND DISPOSAL

PROHIBITIONS: Do not contaminate water, food, or feed by storage or disposal. Open dumping is prohibited.

STORAGE: Do not transport or store with any food or feed intended for human or animal consumption. Not for use or storage in or around the home.

PESTICIDE DISPOSAL: Wastes resulting from the use of this product may be disposed of on site or at an approved waste disposal facility.

CONTAINER DISPOSAL: Completely empty bag into application equipment. Then dispose of empty bag in a sanitary landfill or by incineration, or, if allowed by State and local authorities, by burning. If burned, stay out of smoke.

NOTICE—READ CAREFULLY

NOTICE TO BUYER AND USER:

Seller warrants that this product conforms to the chemical description on the label and is reasonably fit for the purposes stated on the label when used in accordance with directions under normal conditions of use. This warranty does not extend to the use of this product contrary to label instructions, or under abnormal use conditions, or under conditions not reasonably foreseeable to Seller, and Buyer and User assume the risk of any such use. SELLER DISCLAIMS ALL OTHER WARRANTIES EXPRESSED OR IMPLIED INCLUDING ANY WARRANTY OF FITNESS OR MERCHANTABILITY. SELLER SHALL NOT BE LIABLE FOR CONSEQUENTIAL, SPECIAL OR INDIRECT DAMAGES RESULTING FROM THE USE OR HANDLING OF THIS PRODUCT AND SELLER'S SOLE LIABILITY AND BUYER'S AND USER'S EXCLUSIVE REMEDY SHALL BE LIMITED TO THE REFUND OF THE PURCHASE PRICE.

This product is sold only for uses stated on its label. No express or implied license is granted to use or sell this product under any patent in any country except as specified: Country: United States of America.

A-11MR

The information and specimen labels in this booklet are correct as of the publication date but are subject to change without notice. For current information, contact ICI Americas Inc.

DYFONATE® II 20-G

GRANULAR INSECTICIDE

Soil insecticide for protection against corn rootworm and other major pests

RESTRICTED USE PESTICIDE

Due to ACUTE ORAL TOXICITY

For retail sale to and application only by Certified Applicators or personnel under their direct supervision and only for those uses covered by the Certified Applicator's certification.

ACTIVE INGREDIENT:	
0-Ethyl S-phenyl ethylphosphonodithioate*	20.0%
INERT INGREDIENTS:	80.0%
TOTAL	100.0%

(contains 20 pounds active ingredient per 100 pounds)

*DYFONATE

EPA Reg. No. 10182-135

KEEP OUT OF REACH OF CHILDREN

DANGER–POISON

PELIGRO

PRECAUCION AL USUARIO: Si usted no lee ingles, no use este producto hasta que la etiqueta le haya sido explicado ampliamente.

Hazardous to Livestock

STATEMENT OF PRACTICAL TREATMENT

FOR PESTICIDE EMERGENCY

CALL A PHYSICIAN IMMEDIATELY: If a known exposure occurs, or if poisoning is suspected, do not wait for symptoms to develop. Immediately start the procedures given below and simultaneously contact a Poison Control Center, a physician, or the nearest hospital.

Note: Be sure to advise the person contacted that the compound is a cholinesterase inhibitor. Describe the type of exposure, the victim's symptoms, and follow the advice given.

POISON SYMPTOMS: DYFONATE II 20-G is a highly toxic cholinesterase inhibitor. Symptoms of poisoning include headache, blurred vision, weakness, nausea, discomfort in the chest, abdominal cramps, vomiting, diarrhea, salivation, sweating and pinpoint pupils.

Always call a Physician in Case of Suspected Poisoning.

Atropine by Injection Is Antidotal.

If swallowed, immediately give several glasses of water and induce vomiting by placing a finger on the back of the victim's tongue. Give fluids until vomitus is clear. Do not induce vomiting or give anything by mouth to an unconscious or convulsing person.

For skin contact, flush with large amounts of running water for 15 minutes. Remove contaminated clothing and shoes. Get medical attention. Wash clothing before reuse.

For eye contact, immediately flush eyes with large amounts of running water for at least 15 minutes. Hold eyelids apart to insure rinsing of the entire surface of the eye and lids with water. Get medical attention immediately!

If inhaled, remove to fresh air. If not breathing, clear victim's airway and start mouth-to-mouth artificial respiration.

If breathing is difficult, give oxygen, preferably with a physician's advice. Get medical attention immediately.

FOR 24-HOUR EMERGENCY MEDICAL ASSISTANCE CALL 1-800-F-A-S-T-M-E-D (327-8633).

FOR CHEMICAL EMERGENCY: Spill, leak, fire, exposure, or accident call CHEMTREC 1-800-424-9300.

INFORMATION FOR MEDICAL PERSONNEL: Initial measures include removal of secretions, maintenance of patient airway, and if necessary, artificial respiration. When cyanosis is relieved, atropine by injection is antidotal and may be administered in large therapeutic doses, repeated as necessary to the point of tolerance. If symptoms warrant further treatment, 2-PAM (Protopam Chloride) has shown utility as adjunctive therapy. Never use morphine. Continued absorption of the poison may occur, resulting in fatal relapse after initial improvement; very close supervision of the patient is indicated for at least 48 to 72 hours.

Do not transport or store Dyfonate II 20-G with food or feed intended for human or animal consumption.

SEE BACK PANEL FOR ADDITIONAL PRECAUTIONARY STATEMENTS

PRECAUTIONARY STATEMENTS

HAZARDS TO HUMANS AND DOMESTIC ANIMALS

DANGER–POISON

Hazardous to livestock—do not contaminate feed used for livestock.

Poisonous if swallowed, inhaled or absorbed through the skin. Rapidly absorbed through the skin. Do not get in the eyes, on skin, or on clothing. Avoid breathing of dust.

Wear a respirator approved by the National Institute for Occupational Safety and Health, (30 CFR Part 11—Subpart M) for protection against pesticide inhalation.

Wear rubber gloves, shoes, protective clothing and goggles when handling or applying.

Wash hands, arms, and face thoroughly with soap and water after handling and before eating or smoking. Routinely shower or bathe after work and wash all clothing with soap and hot water before reusing. In case of contact, immediately remove contaminated clothing and wash thoroughly with soap and water.

- Do not transport or store Dyfonate II 20-G with any food or feed intended for human or animal consumption.
- Not for use or storage in or around the home.

ENVIRONMENTAL HAZARDS

This pesticide is toxic to fish and wildlife. Birds feeding in treated areas may be killed. Remove or cover granules if spills occur. Runoff from treated areas may be hazardous to fish in neighboring areas. Do not apply directly to water or wetlands. Do not contaminate water by cleaning of equipment or disposal of waste. For guidance contact your regional office of the EPA. Apply this product only as specified on the label.

HOLD this package, other cargo and vehicles involved. IMMEDIATELY TELEPHONE (24-hour Service) for emergency assistance (800) 424-9300.

GENERAL PRECAUTIONS

- Read all directions before using.
- Do not apply on any crop, or in any manner not specified on this label.
- Do not rotate with carrots.

DIRECTIONS FOR USE AND RECOMMENDATIONS

It is a violation of Federal law to use this product in a manner inconsistent with its labeling.

REENTRY STATEMENT

- Do not enter treated areas for 24 hours after application unless appropriate protective clothing is worn.
- Because certain states may require more restrictive reentry intervals for various crops treated with this product, consult your State Department of Agriculture for further information.
- Do not apply this product in such a manner as to directly or through drift expose workers or other persons. The area being treated must be vacated by unprotected persons.
- Do not apply or allow to drift to areas occupied by beneficial animals or onto adjoining food, fiber or pasture crops. Do not contaminate food or feed.
- Written or oral warnings must be given to workers who are expected to be in a treated area or in an area about to be treated with this product. Oral warnings must be given if there is reason to believe that written warnings cannot be understood by workers.

When oral warnings are given, warnings shall be given in a language customarily understood by workers and must indicate the specific period of time that treated areas may not be entered without protective clothing. Oral warnings must also include appropriate first aid instructions in case of accidental exposure. These instructions are given in the STATEMENT OF PRACTICAL TREATMENT at the beginning of this label.

Written warnings must include the following information, "WARNING Area treated with DYFONATE® II 20-G Agricultural Insecticide on (date of application). Do not enter treated areas for 24 hours after application unless appropriate protective clothing is worn (or other reentry interval if your state has a more restrictive interval for this product). If accidental exposure occurs, follow the instructions below." (Written warnings must include the STATEMENT OF PRACTICAL TREATMENT given at the beginning of this label.)

NOTE: Preplant broadcast applications should be mixed (incorporated) into the soil to a depth of 2 to 3 inches.

GAUGE SETTINGS AND CALIBRATION INSTRUCTIONS

NOTE: These settings should be used as starting points only. Continually check the amount of DYFONATE II 20-G used against a known length of row or acreage and make further adjustments accordingly. Also, check calibration occasionally to make sure equipment wear, changing conditions, etc., have not caused a change in flow rate.

GAUGE SETTINGS
DYFONATE II 20-G 6 oz/1000 ft of row

	Speed in mph							
Equipment	**3**	**4**	**5**	**6**	**7**	**8**	**9**	**10**
Gandy bolt on bottom	13	15.3	17.3	19.5	21	22.5	24	25
Gandy snap-on bottom	13.5	15.5	17.5	19.5	21	22.5	24	25
Noble (round opening)	6	8	10	12	14	16	17	19
Noble (rectangular opening)	4	5	7	8	10	11	12	13
John Deere 1986 Maxi-Emerge II (adjustable knob-0-79)	12	14.3	16.5	19	22	25	28	31
John Deere 7000 Maxi-Emerge (Even) (0-10-20-30-40)	8	10	13	15	17	20	21	23
John Deere 7000 Maxi-Emerge (Odd) (5-15-25-35-45)	8	9	11	12	14	15	16	17
John Deere (metal box with 3 range settings and dial) Range/Notch	1/22	1/24	1/27	1/29	2/2	2/4	2/7	2/9
International Harvester (one gauge per hopper) Gate/Dial	1/5	1/6.5	1/7.8	1/9	2/0.5	2/2.5	2/5.0	2/7.5
International Harvester (two gauges per hopper) Gate/Dial	1/6.3	1/7.4	1/8.3	1/9.5	2/1.5	2/3.3	2/6.0	2/9.0

Equipment	Speed in mph							
	3	4	5	6	7	8	9	10
Buffalo (with gauge)		7	9	11	13	14	16	18
Buffalo (fluted-roll feed)	All speeds—4 complete revolutions of hexagonal nut							
Allis Chalmers	Gauge setting 5 will provide 6 oz. per 1000 feet of row at all speeds							

EQUIVALENT LBS. DYFONATE II 20-G PER ACRE									
Ounces of DYFONATE II 20-G per 1000 ft. of row	Row Spacing in inches								
	60	40	38	36	34	32	30	22	18
3.4	1.8	2.8	2.9	3.1	3.2	3.4	3.7	5.0	6.0
5.1	2.8	4.1	4.3	4.6	4.9	5.2	5.5	7.5	9.2
6	**3.3**	**5.0**	**5.2**	**5.4**	**5.8**	**6.1**	**6.5**	**9.9**	**10.9**
11	6.0	9.0	9.5	10.0	10.6	11.3	12.0	16.4	20.0
12	6.5	10.0	10.3	10.9	11.5	12.2	13.1	17.8	—
36.7	20.0	—	—	—	—	—	—	—	—
55.1	30.0	—	—	—	—	—	—	—	—

RECOMMENDATIONS

CROPS	APPLICATION METHOD	PESTS CONTROLLED	RATES of DYFONATE II 20-G	REMARKS	APPLICATION DIRECTIONS
COLE CROPS (BROCCOLI, BRUSSELS SPROUTS, CABBAGE AND CAULIFLOWER)	Immediately prior to planting or at planting	Cabbage Maggot	Broadcast: 20 pounds per acre.	Not recommended in the Northwestern U.S.	Broadcast: Apply DYFONATE II 20-G prior to or at time of seeding or transplanting and incorporate into the soil to a depth of 2 to 3 inches using suitable tillage equipment.
CORN—FIELD (Dent, Seed, Grain, Silage) SWEET, POPCORN	Immediately prior to planting or at planting	Northern, Southern Western Corn Rootworm	Band 6 oz. per 1000 linear feet. or Broadcast 20 pounds per acre.	May be used on conventional, minimum or conservation and no-till tillage systems.	
		Black Cutworm Suppression of White Grubs Seedcorn Maggot Seedcorn Beetle Wireworms	Band: 6 oz. per 1000 linear feet. or Broadcast: 20 pounds per acre.	With heavy infestations of cutworms, a sufficient number of larvae may survive the DYFONATE treatment to cause damage to the corn. If this occurs, an insecticide rescue treatment with another insecticide registered for this use can be made to control these surviving larvae. DYFONATE II 20-G is recommended for control of seedcorn beetle in the central and midwestern corn growing areas only.	Band: Apply DYFONATE II 20-G in a 7 inch wide band using a row bander directly behind the seed furrow opener and in front of the press wheels, or in a 7 inch wide band directly behind the press wheels using drag chains or spring tines to insure incorporation. Under windy conditions the use of windshields is recommended to prevent band drift.
		Lesser Cornstalk Borer	Band: 12 oz. per 1000 linear feet.		Some granules can fall into the open furrow without causing harm to the seed. However, to avoid stand reduction, do not make "in-furrow" applications of DYFONATE II 20-G.
		Symphylans	Broadcast: 10 pounds per acre.	Band applications of DYFONATE II 20-G will provide suppression of wireworms. Use the broadcast application for control.	
		Maize Billbug	Broadcast: 20 pounds per acre.	DYFONATE II 20-G is recommended for control of maize billbug in the Southeastern U.S. only.	Broadcast: Apply DYFONATE II 20-G prior to planting and incorporate into the soil to a depth of 2 to 3 inches using suitable tillage equipment.
	At Cultivation	Northern, Southern, and Western Corn Rootworm	Band: 6 oz. per 1000 linear feet.	Apply DYFONATE II 20-G when activity of rootworm larvae is first observed.	Band: Apply DYFONATE II 20-G in a 7 inch wide band over the corn rows. Cover the treated bands with 2 to 3 inches of soil by making the application ahead of disc hillers or cultivation equipment. Do not apply within 45 days of harvest nor feed or graze livestock within 30 days of treatment.

CROPS	APPLICATION METHOD	PESTS CONTROLLED	RATES of DYFONATE II 20-G	REMARKS	APPLICATION DIRECTIONS
	Over the Whorl	European Corn Borer, Southwestern Corn Borer Reduction of Fall Armyworm Populations	Band: 6 oz. per 1000 linear feet. or Broadcast: 5 pounds per acre.	Consult your State Agricultural Experiment Station or extension service specialist for local treatment threshold and application timing information. Scouting for insect damage is strongly recommended.	Band or Broadcast: Apply DYFONATE II 20-G over the corn rows so granules fall into the whorl. May be applied broadcast, using air or ground equipment, or, in a band directed over corn rows. Do not apply within 30 days of harvest nor feed or graze livestock within 30 days of treatment.
SWEET CORN (Grown on Muck Soils)	At Planting	Wireworms	Band: 12 oz. per 1000 linear feet.	Recommended in Florida only	Band: Apply DYFONATE II 20-G in a 2 inch wide band over the corn rows.
PEANUTS	At Pegging	Southern Corn Rootworm, Granulate Cutworm Lesser Cornstalk Borer	Band: 12 oz. per 1000 linear feet.	Recommended for the Southeastern U.S. Only. For control of southern corn rootworm and granulate cutworm apply DYFONATE II 20-G at pegging. For control of lesser cornstalk borer, apply DYFONATE II 20-G at first signs of borer infestation after pegging begins.	Band: Apply DYFONATE II20-G in an 18 inch wide band. May be used up to 30 days after pegging begins.
SORGHUM (Milo)	At Planting	Northern, Southern and Western Corn Rootworm Lesser Cornstalk Borer	Band: 6 oz. per 1000 linear feet.		Band: See Bank Application Directions for CORN-FIELD (Dent, Seed, Grain, Silage), SWEET, POPCORN
SUGAR BEETS	At Planting	Sugar Beet Root Maggot	Band: 3.4 to 5.1 oz. per 1000 linear feet.	Recommended in Colorado and the northwest sugar beet growing areas only **To avoid stand reduction, do not place DYFONATE II 20-G in direct contact with the seed.**	Band: Apply DYFONATE II 20-G in a 7 inch wide band over the sugar beet rows. Incorporate lightly into the soil using suitable tillage equipment.
SUGARCANE	Post Emergence	White Grub	Band: 36.7 to 55.1 oz. per 1000 linear feet.	Recommended in the Southwestern U.S. only. Consult your State Agricultural Experiment Station or extension service specialist for local treatment threshold and application timing information.	Band: Apply DYFONATE II 20-G over the top of the cane row and lightly incorporate as close to the plants as possible. For best results, irrigate the application.
	At Planting	Wireworms	Band: 27.2 to 36.7 oz. per 1000 linear feet.	Use only 36.7 oz. rate outside of Florida.	Apply in 14 inch band on cane rows 60 inches apart. Apply over the tops of sugar cane seed pieces immediately ahead of covering discs.

STORAGE AND DISPOSAL

PROHIBITIONS: Do not contaminate water, food, or feed by storage or disposal. Open dumping is prohibited.

STORAGE: Do not transport or store with any food or feed intended for human or animal consumption. Not for use or storage in or around the home.

PESTICIDE DISPOSAL: Pesticide wastes are acutely hazardous. Improper disposal of excess pesticide, or rinsate is a violation of Federal law. If these wastes cannot be disposed of by use according to label instructions, contact your State Pesticide or Environmental Control Agency, or the Hazardous Waste representative at the nearest EPA Regional office for guidance.

CONTAINER DISPOSAL: Completely empty bag into application equipment. Then dispose of empty bag in a sanitary landfill or by incineration, or, if allowed by State and local authorities, by burning. If burned, stay out of smoke.

NOTICE TO BUYER AND USER:

Seller warrants that this product conforms to the chemical description on the label and is reasonably fit for the purposes stated on the label when used in accordance with directions under normal conditions of use. This warranty does not extend to the use of this product contrary to label instructions, or under abnormal use conditions, or under conditions not reasonably foreseeable to Seller, and Buyer and User assume the risk of any such use. SELLER DISCLAIMS ALL OTHER WARRANTIES EXPRESSED OR IMPLIED INCLUDING ANY WARRANTY OF FITNESS OR MERCHANTABILITY. SELLER SHALL NOT BE LIABLE FOR CONSEQUENTIAL, SPECIAL OR INDIRECT DAMAGES RESULTING FROM THE USE OR HANDLING OF THIS PRODUCT AND SELLER'S SOLE LIABILITY AND BUYER'S AND USER'S EXCLUSIVE REMEDY SHALL BE LIMITED TO THE REFUND OF THE PURCHASE PRICE.

This product is sold only for uses stated on its label. No express or implied license is granted to use or sell this product under any patent in any country except as specified: Country: United States of America.

A-15MR

The information and specimen labels in this booklet are correct as of the publication date but are subject to change without notice. For current information, contact ICI Americas Inc.

EPTAM® 10-G

GRANULES— A SELECTIVE HERBICIDE

IMPORTANT

RECALIBRATE YOUR EQUIPMENT FOR THIS PRODUCT BEFORE USING

ACTIVE INGREDIENT:	
S-ethyl dipropylthiocarbamate	10.0%
INERT INGREDIENTS:	90.0%
TOTAL	100.0%

Contains 10 Pounds Active Ingredient Per 100 Pounds

EPA Reg. No. 10182-160

KEEP OUT OF REACH OF CHILDREN

CAUTION

Harmful if Swallowed

STATEMENT OF PRACTICAL TREATMENT

First Aid: Immediately start the procedures given below. If further treatment is required, contact a Poison Center, a physician or the nearest hospital.

Ingestion: If swallowed, immediately give several glasses of water and induce vomiting by placing a finger on the back of the victim's tongue. Give fluids until vomitus is clear. If victim is unconscious or convulsing, do not induce vomiting or give anything by mouth.

Eye Contact: For eye contact, immediately flush eyes with large amounts of running water for at least 15 minutes. Hold eyelids apart during flushing to ensure rinsing of the entire surface of the eye and lids with water. Get medical attention immediately.

Skin Contact: For skin contact, flush affected areas with plenty of water for several minutes.

Inhalation: If inhaled, remove to fresh air. Seek medical attention if respiratory irritation occurs or if breathing becomes difficult.

FOR 24-HOUR EMERGENCY MEDICAL ASSISTANCE CALL 1-800-F-A-S-T-M-E-D (327-8633).
FOR CHEMICAL EMERGENCY: Spill, leak, fire, exposure, or accident call CHEMTREC 1-800-424-9300.

PRECAUTIONARY STATEMENTS

HAZARDS TO HUMANS AND DOMESTIC ANIMALS

CAUTION

Harmful if Swallowed

Avoid contact with skin, eyes and clothing. Avoid breathing dust. Wash thoroughly after use. Avoid contamination of feed and food.

ENVIRONMENTAL HAZARDS

Do not apply to water or wetlands. Do not contaminate water, food or feed by storage or disposal.

Use Precautions

- READ ALL LABEL DIRECTIONS BEFORE USING.
- EPTAM 10-G should be used only for recommended purposes and at recommended rates (DO NOT OVERDOSE). Recalibrate your application equipment for this product before using.
- Applied according to directions and under normal growing conditions, EPTAM 10-G will not harm the treated crop. However, during germination and early growth, extended periods of unusually cold and wet or hot and dry weather, insect, nematode, or plant disease attack, carryover pesticide residues, the use of certain soil applied systemic insecticides, highly saline or alkaline soil conditions, improperly placed fertilizers or soil insecticides may create abnormal conditions that weaken crop seedlings. Used under these abnormal conditions, EPTAM 10-G could result in crop injury.
- Use equipment which has proven to incorporate thoroughly to the recommended depth.
- Do not contaminate water to be used on susceptible crops or for domestic purposes.
- EPTAM 10-G is recommended for use on mineral soils only.
- When applied according to directions and under conditions for normal crop growth, no harmful residues should remain beyond harvest.
- Do not apply prior to pre-irrigation.
- Keep container closed when not in use. Do not store near seeds or fertilizers.

DIRECTIONS FOR USE OF EPTAM 10-G

It is a violation of Federal law to use this product in a manner inconsistent with its labeling.

EPTAM 10-G is a selective herbicide which is mixed (incorporated) into the soil for control of weeds listed on this label. EPTAM 10-G controls weeds as their seeds germinate by interfering with normal germination and seedling development. It does not control established weeds.

Application Directions

Apply only to well-worked soil that is dry enough to permit thorough mixing with incorporation equipment. Uniformly apply the recommended rate using equipment designed for application of granular herbicides. Equipment should be carefully calibrated before use and checked frequently during application to be sure it is working properly and delivering a uniform distribution pattern. Avoid overlaps that will increase EPTAM 10-G dosage above recommended rates as plant injury may occur.

Incorporation Directions

EPTAM 10-G must be incorporated into the soil immediately to prevent loss of the herbicide. Whenever possible, application and incorporation should be done in the same operation.

Soil Mixing (Incorporation) Before Planting: Use power-driven cultivation equipment set to cut to a depth of 2 to 3 inches *or* tandem disc set to cut to a depth of 4 to 6 inches, operated at 4 to 6 mph, followed by a spike-tooth harrow or some other leveling device which extends beyond the ends of the disc. For more thorough mixing (for perennial grasses and in heavier soils) disc in two different directions (cross disc).

Incorporation at or After Planting: Use power-driven cultivation equipment set to cut to a depth of 2 to 3 inches or, on light soils only, ground-driven tillers (rolling cultivators, rotary hoe, row wheels, etc.) set to cut to a depth of 1½ inches and operated at 6 to 8 mph. When using ground driven tillers, precultivate if soil is crusted. When incorporating after planting, care must be taken not to disturb the crop seed or seeding.

Planting Directions

For pre-plant applications, seeding should be done as soon as possible after treatment to obtain a maximum period of weed control.

Cultural Practices Following Application

EPTAM 10-G is not persistent in the soil and susceptible weeds germinating later during the growing season may not be controlled. Shallow cultivation may be necessary to control these susceptible weeds that escape control by EPTAM 10-G as well as those weeds not susceptible to EPTAM 10-G. Do not cultivate deeper than the depth of the herbicide incorporation.

Weeds Controlled

Eptam 10-G will not control established weeds.

Annual Grasses:	Annual Broadleaf Weeds:
Annual Bluegrass (*Poa annua*)	Black Nightshade (*Solanum nigrum)*
Annual Ryegrass (Italian ryegrass) (*Lolium multiflorum)*	Common Chickweed (*Stellaria media)*
Barnyardgrass (Watergrass, Junglerice) (*Echinochloa* spp.)	Common Lambsquarters (*Chenopodium album)*
	Corn Spurry (*Spergula arvensis)*
Crabgrass (*Digitaria* spp.)	Deadnettle (Henbit) (*Lamium amplexicaule)*
Field Sandbur (*Cenchrus pauciflorus)*	Florida Pusley (*Richardia scabra)*
Giant Foxtail (*Setaria faberii)*	Hairy Nightshade (*Solanum sarrachoides)*
Goosegrass (*Eleusine indica)*	Nettleleaf Goosefoot (*Chenopodium murale)*
Green Foxtail (*Setaria viridis)*	Common Purslane (*Portulaca oleracea)*
Johnsongrass (Seedlings) (*Sorghum halepense)*	
Shattercane (Wildcane, Black ambercane, Chicken-corn, Scattercane) (*Sorghum bicolor)*	Prostrate Pigweed (*Amaranthus blitoides)*
Volunteer Grains (Barley, Oats, Wheat)	Redroot Pigweed (Common pigweed) (*Amaranthus retroflexus)*
Wild Oats (*Avena fatua)*	Tumble Pigweed (*Amaranthus albus)*
Yellow Foxtail (*Setaria lutescens)*	

The annual broadleaf weeds listed above will be controlled only if treatment is made when conditions are favorable for weed germination and growth.

Perennial Weeds:

Bermudagrass (*Cynodon dactylon)*

Purple Nutsedge (*Cyperus rotundus)*

Quackgrass (Couchgrass) *Agropyron repens)*

Yellow Nutsedge (*Cyperus esculentus)*

Perennial weeds must be turned under and chopped up thoroughly prior to treatment. The underground rhizomes of quackgrass and the rhizomes and stolons of bermudagrass must be cut up thoroughly so that four or less nodes remain on a strand. For the control of quackgrass and bermudagrass the disc must be set to cut 6 inches deep and the higher rates of EPTAM 10-G must be used. Consult recommendations for crops on which the higher rates may be used.

Recommendations

These recommendations are given as broadcast (overall) rates of EPTAM 10-G per acre. For band treatment, use proportionately less material per acre depending on the width of band to be treated and the crop row spacing. Do not apply EPTAM 10-G on rocky ground because thorough incorporation is not possible.

Important: The following recommendations are national or regional. National recommendations are for all regions. Regional recommendations are limited to the regions specified. Refer to the regional map below to determine your EPTAM 10-G region.

Regional Use Map

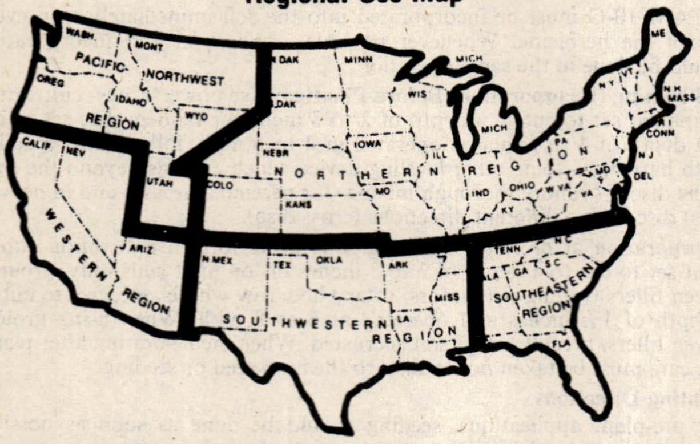

ALFALFA, BIRDSFOOT TREFOIL, CLOVER, LESPEDEZA

Temporary crop stunting and sealing of first leaves will occur if conditions for germination and crop growth are not good. Do not use EPTAM 10-G if a grass or grain nurse crop is to be planted with the legume.

National:
For control of annual grasses, broadleaf weeds and nutsedge apply and incorporate 30 pounds of EPTAM 10-G per acre just before planting.

Northern, Pacific Northwest and Western Regions Only:
Apply 40 pounds of EPTAM 10-G per acre for quackgrass suppression, and for control of nutsedge, annual broadleaf weeds and annual grasses in medium and fine textured soils.

Do not exceed 30 pounds EPTAM 10-G per acre in coarse textured soils.

BEANS, GREEN OR DRY

Do not use EPTAM 10-G on cowpeas (blackeye peas, blackeye beans), soybeans, adzuki beans, lima beans, or other flat-podded beans except romano. Under abnormal weather conditions, stunting may occur on Gratiot, Michilite, Sanilac and Seaway varieties. Do not exceed the 30 pound per acre rate on small white beans, or on green beans grown on coarse textured soils.

Apply and incorporate EPTAM 10-G just before planting, immediately after planting or at the time of last cultivation. If applied before planting, rotary hoe during or shortly after emergence of the beans to break any crust which occurs. If applied at the time of last cultivation apply to the soil before bean pods start to form, and do not feed or pasture vines to livestock within 45 days after application.

National:
Use 30 pounds EPTAM 10-G per acre for nutsedge, annual grass and broadleaf weed control.

Northern and Pacific Northwest Regions Only:
Use 40 pounds of EPTAM 10-G per acre for quackgrass suppression, and for control of seedlings johnsongrass, nutsedge, annual broadleaf weeds and annual grasses in medium and fine textured soils.

Fall Application (Dry Beans: Minnesota, North Dakota Only):
Apply and incorporate in the late Fall before the ground freezes. Use 40 pounds EPTAM 10-G per acre on coarse textured soils and 45 pounds per acre on medium and fine textured soils.

CITRUS, NURSERY STOCK AND YOUNG FIELD PLANTINGS (NON-BEARING ORANGE AND GRAPEFRUIT GROVES)

National:
After lining out, apply 30 to 60 pounds of EPTAM 10-G per acre to the soil and incorporate with cultivation equipment such as a tree hoe or rotary hoe.

Flax

Fall Application (Minnesota, North Dakota): Apply and incorporate in the late Fall before the ground freezes. Use 40 pounds EPTAM 10-G per acre on coarse textured soils and 45 pounds per acre on medium and fine textured soils.

POTATOES, IRISH

- **Preplant**

Southeast, Southwest and Western Regions Only:
Apply and incorporate 30 pounds EPTAM 10-G per acre.

Northern and Pacific Northwest Regions Only:
For annual grasses and broadleaf weeds use 30 to 60 pounds EPTAM 10-G per acre. For quackgrass suppression (or control) and control of nutsedge, annual broadleaf weeds and annual grasses in fine textured soils, use 40 to 60 pounds per acre.

Or

- **Drag-Off (Come Up, Weeding Time)**

The field must first be "dragged-off" followed by EPTAM 10-G application and incorporation. Use spike-tooth harrows or cultivation equipment for incorporation.

National:
For annual grasses and broadleaf weeds use 30 pounds of EPTAM 10-G per acre.

Northern and Pacific Northwest Regions Only:
For control of nutsedge, annual grasses and broadleaf weeds in medium and fine textured soils use 40 to 60 pounds EPTAM 10-G per acre.

- **Layby**

Apply after a clean cultivation. Apply when vines are dry and use cloth drag to knock granules off plant. Do not apply within 45 days of harvest.

Southeast, Southwest and Western Regions Only:
Apply and incorporate 30 pounds EPTAM 10-G per acre.

Northern and Pacific Northwest Regions Only:
Apply to the soil and incorporate 30 to 40 pounds EPTAM 10-G per acre after the potato plants have emerged.

Fall Application (Minnesota, North Dakota):
Apply and incorporate in the late Fall before the ground freezes. Use 40 pounds EPTAM 10-G per acre on coarse textured soils and 45 pounds per acre on medium and fine textured soils.

SAFFLOWER

National:
Apply and incorporate 30 pounds EPTAM 10-G per acre just before planting.

SUGAR BEETS

National (Post-Thinning):
Apply and incorporate 30 pounds of EPTAM 10-G per acre after thinning and clean cultivation and incorporate to a depth of 2 to 3 inches.

Fall Application (Minnesota, North Dakota):
Apply and incorporate in the late Fall before the ground freezes. Use 40 pounds EPTAM 10-G per acre on coarse textured soils and 45 pounds per acre on medium and fine textured soils.

TABLE BEETS

Pacific Northwest Region Only:
Apply and incorporate 20 pounds of EPTAM 10-G per acre just before planting. (Under normal use table beets are susceptible to injury and when EPTAM 10-G is used the seeding rate should be increased 10 percent).

SUNFLOWERS

Fall Application (Minnesota, North Dakota):
Apply and incorporate in the late Fall before the ground freezes. Use 40 pounds EPTAM 10-G per acre on coarse textured soils and 45 pounds per acre on medium and fine textured soils.

STORAGE AND DISPOSAL

Prohibitions: Do not contaminate water, food, or feed by storage or disposal. Open dumping is prohibited.

Storage: Keep container closed when not in use. Store in a cool, dry place. Do not store near seeds, fertilizers or foodstuffs.

Pesticide Disposal: Wastes resulting from the use of this product may be disposed of on site or at an approved waste disposal facility.

Container Disposal: Completely empty bag into application equipment. Then dispose of empty bag in a sanitary landfill or by incineration, or, if allowed by State and local authorities, by burning. If burned, stay out of smoke.

NOTICE TO BUYER AND USER:

Seller warrants that this product conforms to the chemical description on the label and is reasonably fit for the purposes stated on the label when used in accordance with directions under normal conditions of use. This warranty does not extend to the use of this product contrary to label instructions, or under abnormal use conditions, or under conditions not reasonably foreseeable to Seller, and Buyer and User assume the risk of any such use. SELLER DISCLAIMS ALL OTHER WARRANTIES EXPRESSED OR IMPLIED INCLUDING ANY WARRANTY OF FITNESS OR MERCHANTABILITY. SELLER SHALL NOT BE LIABLE FOR CONSEQUENTIAL, SPECIAL OR INDIRECT DAMAGES RESULTING FROM THE USE OR HANDLING OF THIS PRODUCT AND SELLER'S SOLE LIABILITY AND BUYER'S AND USER'S EXCLUSIVE REMEDY SHALL BE LIMITED TO THE REFUND OF THE PURCHASE PRICE.

This product is sold only for uses stated on its label. No express or implied license is granted to use or sell this product under any patent in any country except as specified: Country: United States of America.

A-3MR

The information and specimen labels in this booklet are correct as of the publication date but are subject to change without notice. For current information, contact ICI Americas Inc.

EPTAM® 2.3-G

GRANULES
A SELECTIVE HERBICIDE

ACTIVE INGREDIENT:

S-ethyl dipropylthiocarbamate	2.3%
INERT INGREDIENTS:	97.7%
TOTAL	100.0%

Contains 2.3 pounds active ingredient per 100 pounds.

EPA Reg. No. 10182-172

KEEP OUT OF REACH OF CHILDREN
CAUTION

PRECAUTIONARY STATEMENTS

HAZARDS TO HUMANS AND DOMESTIC ANIMALS
CAUTION

Harmful if Swallowed: Avoid contact with eyes, skin and clothing. Wash thoroughly after use. Avoid breathing dust. Do not contaminate food or feed.

ENVIRONMENTAL HAZARDS

Do not apply directly to water or wetlands. Do not contaminate water by cleaning of equipment or disposal of wastes.

STATEMENT OF PRACTICAL TREATMENT

First Aid

If a known exposure occurs or is suspected, immediately initiate the recommended procedures below. If further treatment is required contact a Poison Center, a physician, or the nearest hospital.

If swallowed: Immediately give several glasses of water and induce vomiting by gagging the victim with finger placed on the back of the victim's tongue. Give fluids until vomitus is clear. If victim is unconscious or convulsing, do not induce vomiting or give anything by mouth.

For skin contact: Flush all affected areas with plenty of water for several minutes. For eye contact, hold eyelids apart and flush with plenty of water for 15 minutes. Remove and clean contaminated clothing and shoes before reuse. Seek medical attention if skin or eye irritation occurs.

If inhaled: Remove to fresh air. Seek medical attention if respiratory irritation occurs or if breathing becomes difficult.

FOR 24-HOUR EMERGENCY MEDICAL ASSISTANCE CALL 1-800-F-A-S-T-M-E-D (327-8633).
FOR CHEMICAL EMERGENCY: Spill, leak, fire, exposure, or accident call CHEMTREC 1-800-424-9300.

OBSERVE THE FOLLOWING USE PRECAUTIONS

READ ALL LABEL DIRECTIONS BEFORE USING.

EPTAM 2.3-G should be used only for recommended purposes and at recommended rates (do not over-dose).

EPTAM 2.3-G is recommended for use on mineral soils only.

For all uses, EPTAM 2.3-G must be thoroughly mixed into the soil.

EPTAM 2.3-G may cause injury to ornamentals under certain soil and climatic conditions or if directions are not followed.

Keep container closed when not in use. Do not store near seeds or fertilizers.

EPTAM 2.3-G CAN BE USED ON THESE ORNAMENTALS

Herbaceous Plants:

Alyssum	Begonia	Marigold
Ageratum	Chrysanthemum	Nasturtium
Amaranthus	Dahlia	Pansy
Asters	Daylilies	Petunia
Balsam	Dianthus	Zinnia

Ground Covers:

Ajuga	Ice Plant	Periwinkle (vinca minor)
Gazania	Ivy	Sedum
Hypericum	Pachysandra	Strawberry (ornamental)

Evergreen and Deciduous Trees and Shrubs:

Berberis	Hemlock	Oak
Boxwood	Holly (American and Japanese)	Pieris
Camellia		Pine
Chamaecyparis	Juniper	Podocarpus
Citrus (non-bearing)	Leucothoe	Rhododendron
	Lilac	Rose
Dogwood	Linden	Spruce
Euonymus	Magnolia	Viburnum
Fir	Maple	Yew (Texas)

NOTE: All flowering bulbs, alvia, phlox, snapdragon and ornamental pepper are susceptible to injury from an application of EPTAM 2.3-G.

WEEDS CONTROLLED

EPTAM 2.3-G controls over 20 species of weeds and will keep a treated area virtually weed-free for an average growing season. (In semi-tropical areas, more than one application may be needed.)

EPTAM 2.3-G will not control established weeds.

Annual Grasses

Annual Bluegrass *(Poa annua)*
Annual Ryegrass (Italian Ryegrass) *(Lolium multiflorum)*
Barnyardgrass (Watergrass, Junglerice) *(Echinochloa* spp.)
Crabgrass *(Digitaria* spp.)
Giant Foxtail *(Setaria laberii)*
Goosegrass *(Eleusine indica)*
Green Foxtail *(Setaria viridis)*
Johnsongrass Seedlings *(Sorghum halepense)*
Lovegrass (Stinkgrass) *(Eragrostis cilianesis)*
Sandbur, Field *(Cenchrus pauciflorus)*
Volunteer Grains (Barley, Oats, Wheat)
Wild Cane (Black Amber Cane, Chicken-Corn, Scatter Cane, Shattercane) *(Sorghum bicolor)*
Wild Oats *(Avena fatua)*
Yellow Foxtail *(Setaria glauca)*

Annual Broadleaf Weeds

Black Nightshade *(Solanum nigrum)*
Common Chickweed *(Stellaria media)*
Corn Spurry *(Spergula arvensis)*
Deadnettle (Henbit) *(Lamium amplexicaule)*
Florida Purslane *(Richardia scabra)*
Hairy Nightshade *(Solanum villosum)*
Lambsquarters, Common *(Chenopodium album)*
Nettleleaf Goosefoot *(Chenopodium murale)*
Purslane, Common *(Portuluca oleracea)*
Prostrate Pigweed *(Amaranthus graecizans)*
Ragweed, Common (Michigan only) *(Ambrosia artenisifolia)*
Redroot Pigweed (Common Pigweed) *(Amaranthus retroflexus)*
Tumble Pigweed *(Amaranthus albus)*

The annual broadleaf weeds listed above will be controlled only if treatment is made when conditions are favorable forr weed germination and growth.

Perennial Weeds

Bermudagrass *(Cynodon dactylon)*
Mugwort (Chrysanthemumweed) *(Artemisia vulgaris)*
Purple Nutsedge *(Cyperus rotundus)*
Quackgrass (Couchgrass) *(Agropyron repens)*
Yellow Nutsedge *(Cyperus esculentus)*

DIRECTIONS FOR USE

It is a violation of Federal Law to use this product in a manner inconsistent with its labeling.

Soil Preparation: The soil to be treated should be loose and free of clods. All weed growth should be removed or thoroughly worked into the soil before application.

Application: The recommended rate of EPTAM 2.3-G should be applied as uniformly as possible. Apply to well worked soil that is dry enough to permit thorough mixing with incorporation equipment. Use one of the following appropriate means of application:

Granular Herbicide Applicators
(for nursery field application)
Equipment should be carefully calibrated before use and checked frequently during application.

Home Lawn Spreaders
(for applying EPTAM 2.3-G between rows of nursery stock and small areas planted with ornamentals)

Where practical, apply half the required amount of EPTAM 2.3-G over the area to be treated; then apply the remainder at right angles or crosswise.

Soil Mixing Incorporation: Immediately after application, thoroughly mix EPTAM 2.3-G into the soil to a depth of 2 to 3 inches. Mix to a depth of 6 inches for nutgrass, quackgrass, Bermudagrass and chrysanthemumweed (mugwort) control. Thorough soil mixing is necessary for good weed control. Use the following equipment or other equipment which has proven satisfactory under local conditions.

Commercial Nursery
Pre-plant broadcast (over-all) application.
Pre-plant band application.
Post-plant application: Nursery cultivator or rototiller.

Home Garden
Pre-plant application—Rototiller.
Post-plant application—For annual weed use hand rake or hoe or water in immediately after application to a depth of 2 to 3 inches. Where incorporation by hand raking is done, light watering after raking is recommended.
For perennial weeds, incorporate to a depth of 6 inches with a rototiller.

WHEN TO USE EPTAM 2.3-G

Herbaceous Plants and Ground Covers: Apply 2 weeks after transplanting or after growth starts in the spring.

Trees and Shrubs: Apply 2 weeks before transplanting balled and canned stock (only) and anytime after transplanting. Around established plants apply after growth starts in the spring.

RECOMMENDATIONS

For Annual Weed Control—Use EPTAM 2.3-G at the rate of 5 pounds per 1,000 square feet.

For Quackgrass, Nutsedge and Bermudagrass Control Around Established Trees and Shrubs: Existing stands of these perennial grasses must be turned under and chopped up thoroughly before treatment. Use EPTAM 2.3-G at the rate of 6 pounds per 1,000 square feet. Mix thoroughly into the top 6 inches of soil.

For Mugwort (Chrysanthemumweed) Control in Areas to be Planted with the following Plants: Juniper, Japanese Holly, Ivy, Pachysandra, Petunias—Use EPTAM 2.3-G at the rate of 15 pounds per 1,000 square feet. Mix thoroughly into the top 6 inches of soil. Apply 4 weeks before desired planting date.

Specialty Use—Golf Course Sand Traps: For control of annual grasses and broadleaf weeds use 1 pound EPTAM 2.3-G on 200 square feet of sand. Spread evenly with a spreader calibrated to deliver the proper amount of EPTAM 2.3-G granules. Rake into the sand and water lightly. For control of nutgrass use 1 pound EPTAM 2.3-G on 100 square feet of sand. Mix into the sand to a depth of 2 to 3 inches and water lightly for 2 to 3 minutes. Repeat as needed to maintain weed control.

Home Garden Use—Irish Potatoes and Green Beans: (Do not use EPTAM 2.3-G on blackeye peas or beans, soybeans, lima beans or other flat podded beans except Romano). Apply and mix into the soil 1 pound of EPTAM 2.3-G per 320 square feet just before planting. Thoroughly mix to a depth of 2 to 3 inches using a rototiller or hoe and rake. Plant at a normal seeding depth.

STORAGE AND DISPOSAL

Prohibitions: Do not contaminate water, food or feed by storage or disposal. Open dumping is prohibited. Do not reuse empty container. **Pesticide Disposal:** Wastes resulting from the use of this product may be disposed of on site or at an approved waste disposal facility. **Container Disposal:** Completely empty bag into application equipment. Then dispose of empty bag in a sanitary landfill or by incineration, or, if allowed by State and local authorities, by burning. If burned, stay out of smoke.

NOTICE TO BUYER AND USER

Seller warrants that this product conforms to the chemical description on the label and is reasonably fit for the purposes stated on the label when used in accordance with directions under normal conditions of use. This warranty does not extend to the use of this product contrary to label instructions, or under abnormal use conditions, or under conditions not reasonably foreseeable to Seller, and Buyer and User assume the risk of any such use. SELLER DISCLAIMS ALL OTHER WARRANTIES, EXPRESSED OR IMPLIED, INCLUDING ANY WARRANTY OF FITNESS OR MERCHANTABILITY. SELLER SHALL NOT BE LIABLE FOR CONSEQUENTIAL, SPECIAL OR INDIRECT DAMAGES RESULTING FROM THE USE OR HANDLING OF THIS PRODUCT AND SELLER'S SOLE LIABILITY AND BUYER'S AND USER'S EXCLUSIVE REMEDY SHALL BE LIMITED TO THE REFUND OF THE PURCHASE PRICE.

02-0612-008 A-2R

This is a specimen label. ICI Americas Inc. is not responsible for the accuracy of the information contained herein. As labels are subject to revision, always carefully read and follow the label on the product container.

EPTAM® 5-G

SELECTIVE HERBICIDE

A Selective Herbicide For Weed Control In Flowers And Ornamentals. For Control Of Annual Grassy Weeds And Many Annual Broadleaf Weeds.

ACTIVE INGREDIENT:	
S-ethyl dipropylthiocarbamate	5.0%
INERT INGREDIENTS:	95.0%
TOTAL	100.0%

Contains 5 pounds active ingredient per 100 pounds.

EPA Reg. No. 10182-155

KEEP OUT OF REACH OF CHILDREN
CAUTION

PRECAUTIONARY STATEMENTS

HAZARDS TO HUMANS AND DOMESTIC ANIMALS

CAUTION

Harmful if Swallowed. Avoid contact with skin, eyes or clothing. Avoid breathing dust. Wash thoroughly after use.

ENVIRONMENTAL HAZARDS

Do not apply directly to water. This product is toxic to fish. Do not contaminate water, food or feed by storage or disposal.

OBSERVE THE FOLLOWING USE PRECAUTIONS

- READ ALL LABEL DIRECTIONS BEFORE USING.
- EPTAM 5-G ORNAMENTAL should be used only for recommended purposes and at recommended rates (DO NOT OVERDOSE).
- EPTAM 5-G ORNAMENTAL is recommended for use on mineral soils only.
- For all uses, EPTAM 5-G ORNAMENTAL must be thoroughly mixed into the soil.
- EPTAM 5-G ORNAMENTAL may cause injury to ornamentals under certain soil and climatic conditions or if directions are not followed.
- Keep container closed when not in use. Do not store near seeds or fertilizers.

STATEMENT OF PRACTICAL TREATMENT

First Aid

Immediately initiate the procedures given below. If further treatment is required, contact a Poison Center, a physician or the nearest hospital.

If Swallowed: Immediately give several glasses of water and induce vomiting by placing a finger on the back of the victim's tongue. Give fluids until vomitus is clear. If victim is unconscious or convulsing, do not induce vomiting or give anything by mouth.

For Eye Contact: Immediately flush eyes with large amounts of running water for at least 15 minutes. Hold eyelids apart during flushing to ensure rinsing the entire surface of the eye and lids with water. Get medical attention immediately.

For Skin Contact: Flush affected areas with plenty of water for several minutes.

Inhalation: If inhaled, remove to fresh air. Seek medical attention if respiratory irritation occurs or if breathing becomes difficult.

FOR 24-HOUR EMERGENCY MEDICAL ASSISTANCE CALL 1-800-F-A-S-T-M-E-D (327-8633).
FOR CHEMICAL EMERGENCY: Spill, leak, fire, exposure, or accident call CHEMTREC 1-800-424-9300.

EPTAM 5-G ORNAMENTAL CAN BE USED ON THESE ORNAMENTALS

Herbaceous Plants:

Alyssum	Begonia	Marigold
Ageratum	Chrysanthemum	Nasturtium
Amaranthus	Dahlia	Pansy
Asters	Daylilies	Petunia
Balsam	Dianthus	Zinnia

Ground Covers:

Ajuga	Ice Plant	Periwinkle (vinca minor)
Gazania	Ivy	Sedum
Hypericum	Pachysandra	Strawberry (ornamental)

Evergreen and Deciduous Trees and Shrubs:

Berberis	Holly (American and Japanese)	Oak
Boxwood	Juniper	Pieris
Camellia	Leucothoe	Pine
Chamaecyparis	Lilac	Podocarpus
Citrus (non-bearing)	Linden	Rhododendron
Dogwood	Magnolia	Rose
Euonymus	Maple	Spruce
Fir		Viburnum
Hemlock		Yew (Taxus)

NOTE: All flowering bulbs, salvia, phlox, snapdragon and ornamental pepper are susceptible to injury from an application of EPTAM 5-G ORNAMENTAL.

WEEDS CONTROLLED BY EPTAM 5-G ORNAMENTAL

EPTAM 5-G ORNAMENTAL controls over 20 species of weeds and will keep a treated area virtually weed-free for an average growing season. (In semi-tropical areas, more than one application may be needed.)

EPTAM 5-G ORNAMENTAL will not control established weeds.

Annual Grasses
Annual Bluegrass *(Poa annua)*
Annual Ryegrass (Italian Ryegrass) *(Lolium multiflorum)*
Barnyardgrass (Watergrass) (Junglerice) *(Echinochloa* spp.)
Crabgrass *(Digitaria* spp.)
Giant Foxtail *(Setaria faberii)*
Goosegrass *(Eleusine indica)*
Green Foxtail *(Setaria viridis)*
Johnsongrass Seedlings *(Sorghum halepense)*

Annual Grasses
Lovegrass (Stinkgrass) *(Eragrostis cilianensis)*
Sandbur, Field *(Cenchrus pauciflorus)*
Volunteer Grains (Barley, Oats, Wheat)
Wild Cane (Black, Amber Cane, Chicken-Corn, Scatter Cane, Shattercane) *(Sorghum bicolor)*
Wild Oats *(Avena fatua)*
Yellow Foxtail *(Setaria glauca)*

Annual Broadleaf Weeds
Black Nightshade *(Solanum nigrum)*
Common Chickweed *(Stellaria media)*
Corn Spurry *(Spergula arvensis)*
Deadnettle (Henbit) *(Lamium amplexicaule)*
Florida Purslane *(Richardia scabra)*
Hairy Nightshade *(Solanum villosum)*
Lambsquarters, Common *(Chenopodium album)*
Nettleleaf Goosefoot *(Chenopodium murale)*
Purslane, Common *(Portulaca oleracea)*
Prostrate Pigweed *(Amaranthus graecizans)*
Ragweed, Common (Michigan only) *(Ambrosia artemisiifolia)*
Redroot Pigweed (Common Pigweed) *(Amaranthus retroflexus)*
Tumble Pigweed *(Amaranthus albus)*

The annual broadleaf weeds listed above will be controlled only if treatment is made when conditions are favorable for weed germination and growth.

Perennial Weeds
Bermudagrass *(Cynodon dactylon)*
Mugwort (Chrysanthemumweed) *(Artemisia vulgaris)*
Purple Nutsedge *(Cyperus rotundus)*
Quackgrass (Couchgrass) *(Agropyron repens)*
Yellow Nutsedge *(Cyperus esculentus)*

DIRECTIONS FOR USE

It is a violation of Federal law to use this product in a manner inconsistent with its labeling.

Soil Preparation: The soil to be treated should be loose and free of clods. All weed growth should be removed or thoroughly worked into the soil before application.

Application: The recommended rate of EPTAM 5-G ORNAMENTAL should be applied as uniformly as possible. Apply to well worked soil that is dry enough to permit thorough mixing with incorporation equipment. Use one of the following appropriate means of application.

GRANULAR HERBICIDE APPLICATORS (for nursery field application)
Equipment should be carefully calibrated before use and checked frequently during application.

HOME LAWN SPREADERS (for applying EPTAM 5-G ORNAMENTAL between rows of nursery stock and small areas planted with ornamentals).

Where practical, apply half the required amount of EPTAM 5-G ORNAMENTAL over the area to be treated; then apply the remainder at right angles or crosswise.

Soil Mixing Incorporation: Immediately after application, thoroughly mix EPTAM 5-G ORNAMENTAL into the soil to a depth of 2 to 3 inches. Mix to a depth of 6 inches for nutgrass, quackgrass, Bermudagrass and chrysanthemumweed (mugwort) control. Thorough soil mixing is necessary for good weed control. Use the following equipment or other equipment which has proven satisfactory under local conditions.

COMMERCIAL NURSERY
Pre-plant broadcast (over-all) application,
Pre-plant band application,
Post-plant application: Nursery cultivator or rototiller.

HOME GARDEN
Pre-plant application—Rototiller.
Post-plant application—For annual weed use hand rake or hoe or water in immediately after application to a depth of 2 to 3 inches. Where incorporation by hand raking is done, light watering after raking is recommended.

For perennial weeds, incorporate to a depth of 6 inches with a rototiller.

WHEN TO USE EPTAM 5-G ORNAMENTAL

Herbaceous Plants and Ground Covers: Apply 2 weeks after transplanting or after growth starts in the Spring.

Trees and Shrubs: Apply 2 weeks before transplanting balled and canned stock (only) and anytime after transplanting. Around established plants apply after growth starts in the Spring.

RECOMMENDATIONS

For Annual Weed Control—Use EPTAM 5-G ORNAMENTAL at the rate of 100 pounds per acre (2.3 pounds per 1,000 square feet).

For Quackgrass, Nutsedge and Bermudagrass Control Around Established Trees and Shrubs—Existing stands of these perennial grasses must be turned under and chopped up thoroughly before treatment. Use EPTAM 5-G ORNAMENTAL at the rate of 120 pounds per acre (2.75 pounds per 1,000 square feet). Mix thoroughly into the top 6 inches of soil.

For Mugwort (Chrysanthemumweed) Control in Areas to be Planted with the Following Plants: Juniper, Japanese Holly, Ivy, Pachysandra, Petunias—Use 300 pounds of EPTAM 5-G ORNAMENTAL per acre (7 pounds per 1,000 square feet). Mix thoroughly into the top 6 inches of soil. Apply 4 weeks before desired planting date.

Specialty Use—Golf Course Sand Traps—For control of annual grasses and broadleaf weeds use 2½ pounds EPTAM 5-G ORNAMENTAL on 1,000 square feet of sand. Spread evenly with a spreader calibrated to deliver the proper amount of EPTAM 5-G ORNAMENTAL granules. Rake into the sand and water lightly. For control of nutgrass use 5 pounds EPTAM 5-G ORNAMENTAL on 1,000 square feet of sand. Mix into the sand to a depth of 2 to 3 inches and water lightly for 2 to 3 minutes. Repeat as needed to maintain weed control.

STORAGE AND DISPOSAL

Prohibitions: Do not contaminate water, food or feed by storage or disposal. Open dumping is prohibited. **Storage:** Keep container closed when not in use. Store in a cool, dry place. Do not store near seeds, fertilizers or foodstuffs. **Pesticide Disposal:** Wastes resulting from the use of this product may be disposed of on site or at an approved waste disposal facility. **Container Disposal:** Completely empty bag into application equipment. Then dispose of empty bag in a sanitary landfill or by incineration, or, if allowed by State and local authorities, by burning. If burned, stay out of smoke.

NOTICE TO BUYER AND USER

Seller warrants that this product conforms to the chemical description on the label and is reasonably fit for the purposes stated on the label when used in accordance with directions under normal conditions of use. This warranty does not extend to the use of this product contrary to label instructions, or under abnormal use conditions, or under conditions not reasonably foreseeable to Seller, and Buyer and User assume the risk of any such use. SELLER DISCLAIMS ALL OTHER WARRANTIES, EXPRESSED OR IMPLIED, INCLUDING ANY WARRANTY OF FITNESS OR MERCHANTABILITY. SELLER SHALL NOT BE LIABLE FOR CONSEQUENTIAL, SPECIAL OR INDIRECT DAMAGES RESULTING FROM THE USE OR HANDLING OF THIS PRODUCT AND SELLER'S SOLE LIABILITY AND BUYER'S AND USER'S EXCLUSIVE REMEDY SHALL BE LIMITED TO THE REFUND OF THE PURCHASE PRICE.

02-0612-009 A-2R

The information and specimen labels in this booklet are correct as of the publication date but are subject to change without notice. For current information, contact ICI Americas Inc.

EPTAM® 7-E

SELECTIVE HERBICIDE EMULSIFIABLE LIQUID

Complete Directions For Use, Use Precautions, Regional Crop Recommendations

ACTIVE INGREDIENT:	
S-ethyl dipropylthiocarbamate	87.8%
INERT INGREDIENTS:	12.2%
TOTAL	100.0%

Contains 7 pounds of active ingredient per gallon

EPA Reg. No. 10182-220

KEEP OUT OF REACH OF CHILDREN

CAUTION

HARMFUL IF SWALLOWED

STATEMENT OF PRACTICAL TREATMENT

First Aid

Immediately start the procedures given below. If further treatment is required, contact a Poison Center, a physician, or the nearest hospital.

If swallowed: Immediately give several glasses of water but *do not* induce vomiting. If vomiting does occur, give fluids again. Have a physician determine if condition of patient will permit induction of vomiting or evacuation of stomach. Do not give anything by mouth to an unconscious or convulsing person.

For eye contact: Immediately flush eyes with large amounts of running water for at least 15 minutes. Hold eyelids apart to ensure rinsing of the entire surface of the eye and lids with water. Get medical attention immediately.

For skin contact: Flush all affected areas with plenty of water for several minutes. Remove and clean contaminated clothing and shoes. Seek medical attention if irritation occurs.

If inhaled: Remove to fresh air. Seek medical attention if respiratory irritation occurs or if breathing becomes difficult.

FOR 24-HOUR EMERGENCY MEDICAL ASSISTANCE CALL 1-800-F-A-S-T-M-E-D (327-8633).

FOR CHEMICAL EMERGENCY: Spill, leak, fire, exposure, or accident call CHEMTREC 1-800-424-9300.

PRECAUTIONARY STATEMENTS

HAZARDS TO HUMANS AND DOMESTIC ANIMALS

CAUTION

Harmful If Swallowed: Avoid contact with skin, eyes and clothing. Avoid inhalation of spray mist. Wear rubber gloves and fresh clothing. Wear goggles to avoid contact with eyes. Wash with soap and water immediately after use or contact. Do not contaminate food or feed.

Do not apply this product in such a manner as to directly or through drift expose workers or other persons. When applied through sprinkler irrigation method of application, the area being treated must be vacated by unprotected persons.

ENVIRONMENTAL HAZARDS

Do not contaminate water to be used on susceptible crops and ornamentals, or to be used for domestic purposes. This product is toxic to shrimp. Keep out of tidal marshes and estuaries. Do not apply where runoff is likely to occur. Do not apply when weather conditions favor drift from areas treated.

Do not contaminate water by cleaning of equipment or disposal of wastes.

GENERAL USE PRECAUTIONS

Read all label directions before using.

EPTAM 7-E should be used only for recommended purposes and recommended rates. DO NOT OVERDOSE.

EPTAM 7-E is recommended for use on mineral soils only (soils containing less than 10% organic matter).

Keep container tightly closed when not in use. Do not store near seeds or fertilizers. Store out of reach of children, pets and domestic animals.

Rinse spray equipment and empty container.

Apply this product only as specified on this label.

Special Precautions for Crop Uses

For incorporated applications, use equipment which has been proven to incorporate thoroughly to the recommended depth.

In irrigated areas, do not apply EPTAM 7-E prior to pre-irrigation.

Tank-mix this product with fungicides, insecticides or herbicides only as recommended.

When properly applied and weather conditions exist for normal plant growth through the season, EPTAM 7-E will not harm the treated crop nor should harmful soil residues remain beyond harvest. However, during germination and early growth, extended periods of unusually cold and wet or hot and dry weather, insect, nematode, or plant disease attack, carry-over soil residues of certain persistent herbicides, the use of certain soil applied systemic insecticides, highly saline or alkaline soil conditions, improperly placed fertilizers or soil insecticides may create abnormal conditions that weaken crop seedlings. Also some of these abnormal conditions may weaken established crops: alfalfa, almonds, etc. EPTAM 7-E used under these abnormal conditions could result in crop injury.

Weeds Controlled

EPTAM 7-E will not control established weeds.

Annual Grasses:

Annual Bluegrass	(*Poa annua*)
Annual Ryegrass (Italian Ryegrass)	(*Lolium multiflorum*)
Barnyardgrass (Watergrass Junglerice)	(*Echinochloa* spp.)
Bermudagrass Seedlings	(*Cynodon dactylon*)
Crabgrass	(*Digitaria* spp.)
Giant Foxtail	(*Setaria faberii*)
Goosegrass	(*Eleusine indica*)
Green Foxtail	(*Setaria viridis*)
Johnsongrass Seedlings	(*Sorghum halepense*)
Lovegrass (Stinkgrass)	(*Eragrostis cilianensis*)
Panicum, Fall	(*Panicum dichotomiflorum*)
Panicum*, Texas	(*Panicum texanum*)
Rescuegrass	(*Bromus willdenowii*)
Sandbur, Field	(*Cenchrus pauciflorus*)
Shattercane (Wild Cane)	(*Sorghum bicolor*)
Signalgrass	(*Brachiaria* spp.)
Volunteer grains (Barley, Oats, Wheat)	
Wild Oats	(*Avena fatua*)
Witchgrass*	(*Panicum capillare*)
Yellow Foxtail	(*Setaria lutescens*)

* May not be controlled at less than $3^1/_2$ pints of EPTAM 7-E.

Annual Broadleaf Weeds:

Tall Morningglory	(*Ipomoea purpurea*)
Black Nightshade	(*Solanum nigrum*)
Carpetweed	(*Mollugo verticillata*)
Chickweed, Common	(*Stellaria media*)
Corn Spurry	(*Spergula arvensis*)
Deadnettle (Henbit)	(*Lamium amplexicaule*)
Fiddleneck	(*Amsinckia* spp.)
Florida Pusley	(*Richardia scabra*)
Hairy Nightshade	(*Solanum sarrachoides*)
Lambsquarters*, Common	(*Chenopodium album*)
Nettleleaf, Goosefoot	(*Chenopodium murale*)
Purslane, Common	(*Portulaca oleracea*)
Prostrate Pigweed	(*Amaranthus blitoides*)
Prickly Sida*	(*Sida spinosa*)
Redroot Pigweed* (Common Pigweed)	(*Amaranthus retroflexus*)
Shepherdspurse	(*Capsella bursa-pastoris*)
Sicklepod*	(*Cassia obtusifolia*)
Tumble Pigweed	(*Amaranthus albus*)

The annual broadleaf weeds listed above will be controlled only if treatment is made when conditions are favorable for weed germination and growth. Broadleaf weeds may only be suppressed at less than $3^1/_2$ pints EPTAM 7-E per acre in heavier soils or under very cold soil conditions.

* May not be controlled at less than $3^1/_2$ pints of EPTAM 7-E.

Perennial Weeds:

Bermudagrass	(*Cynodon dactylon*)
Purple Nutsedge*	(*Cyperus rotundus*)
Quackgrass (Couchgrass)	(*Agropyron repens*)
Yellow Nutsedge*	(*Cyperus esculentus*)

* May not be controlled at less than $3^1/_2$ pints of EPTAM 7-E per acre.

Perennial weeds must be turned under and chopped up thoroughly prior to treatment. The underground rhizomes of quackgrass and the rhizomes and stolons of bermudagrass must be cut up thoroughly so that four or less nodes remain on a strand. For the suppression or control of quackgrass and bermudagrass the disc must be set to cut 6 inches deep. Use $4^1/_2$ to 7 pints EPTAM 7-E for quackgrass and $3^1/_2$ to 7 pints for bermudagrass. The EPTAM 7-E should be incorporated by discing or applied in the irrigation water after the rhizomes and stolons have been cut up.

Consult recommendations for crops on which these higher rates may be used. Nutsedge may not be controlled by water-run applications in heavier soils.

Tank Mix Combinations

For broader spectrum weed control and increased control of certain broadleaf weeds EPTAM 7-E may be tank mixed with the following herbicides. On dry beans, EPTAM 7-E may be tank mixed with Treflan*, or with a combination of Treflan and Amiben**. On Irish potatoes, EPTAM 7-E may be tank mixed with metribuzin (Lexone† or Sencor‡). Consult supplemental literature for exact rates and application directions.

* Teflan—Registered Trademark of Elanco Products Incorporated.

** Amiben chloramben herbicide—Registered Trademark of Rhone Poulenc Inc.

† Lexone—Registered Trademark of DuPont Company.

‡ Sencor—Registered Trademark of Mobay Chemical Corp.

DIRECTIONS FOR USE

CROP SECTION

It is a violation of Federal law to use this product in a manner inconsistent with its labeling.

EPTAM 7-E is a selective herbicide which for most uses must be mixed (incorporated), or injected subsurface into the soil, or applied in the irrigation water for control of weeds listed on this label. EPTAM 7-E controls weeds by interfering with normal germination and seedling development. It does not control established weeds.

Application Directions:

Pour the recommended amount of EPTAM 7-E into the spray tank during the filling operations. Apply in 10 to 50 gallons of water per acre using a properly calibrated, low-pressure sprayer having good agitation. The soil should be well worked and dry enough to permit good soil mixing (incorporation).

EPTAM 7-E may be combined with solution, slurry or suspension fertilizers. However, physical compatibility with these fluid fertilizers must be determined before combining in the spray tank. See Appendix 1 for special direction regarding these combinations. Even though found to be compatible, constant agitation is necessary to keep the EPTAM 7-E uniformly mixed with the fluid fertilizer.

For all band applications, reduce dosage proportionately depending upon row spacing and band width to be treated.

Impregnation on Dry Fertilizer:

EPTAM 7-E may be impregnated on dry fertilizer for use on registered crops. However, uniform distribution of the EPTAM 7-E on fertilizer particles and uniform application are necessary to assure good results. See special instructions elsewhere on this label giving directions for impregnation and use.

Incorporation Directions:

EPTAM 7-E must be incorporated into the soil immediately to prevent loss of the herbicide. Whenever possible, application and incorporation should be done in the same operation.

Soil Mixing (Incorporation) Before Planting: The following equipment commonly is used for soil mixing (incorporation) before planting:

- **Power Driven Cultivation Equipment** (recommended on all soil types) set to cut to a depth of 2 to 3 inches.
- **Tandem Discs** (recommended on all soil types) set to cut to a depth of 4 to 6 inches, operated at 4 to 6 mph followed by a spiketooth harrow or some other leveling device which extends beyond the ends of the discs. For more thorough mixing (for perennial grasses and in heavier soils) disc in two different directions (cross disc).
- **Field Cultivators** (recommended for spring application on coarse textured soils, and for fall application on all soils. Use only on soils in good tilth). Use 3 to 4 rows of sweeps, spaced at 7 inch or less intervals and staggered so that no soil is left unturned, followed by a spike-tooth harrow pulled behind the cultivator. Do not use chisel plows to incorporate. Set the cultivator to cut 4 inches deep, operated at 5 mph or more. Run the equipment over the field twice, the second run at an angle to the first.
- **Rotary Ground Driven or Spring-Tooth Cultivators** (recommended on coarse textured soils in good tilth only). Set to penetrate to a depth of 4 to 6 inches and operated at 5 to 8 mph in two different directions.

Incorporation at or After Planting: Use power driven cultivation equipment set to cut to a depth of 2 to 3 inches or on coarse textured soils only, ground driven tillers (rolling cultivators, rotary hoe, row wheels, etc.) set to cut to a depth of 1½ inches and operated at 6 to 8 mph. When incorporating after planting, care must be taken not to disturb the crop seed or seedling.

SUBSURFACE APPLICATION—AT PLANTING OR POST EMERGENCE

Apply EPTAM 7-E in 10 or more gallons of water per acre.

Special equipment designed for subsurface application MUST be used. Injector and sweep units must be rigidly mounted on the planter or cultivation unit. When using sweeps at planting they must be mounted ahead of the planters.

Soil Injection—Injector shanks must be spaced 2⅓ to 3 inches apart and mounted in staggered positions to avoid trash buildup. Set shanks to inject EPTAM 7-E 2 to 3 inches below the soil surface. The width of the band in which weed control is desired will determine the number and spacing of injector shanks required per row. (Example: Four injector shanks spaced 3 inches apart give a 12-inch band.) A broadcast application can be made by increasing the number of shanks. The two shanks adjacent to the drill row must be 1¼ to 1½ inches on either side of it, EXCEPT IN COTTON WHERE THE DISTANCE MUST BE 4 INCHES ON EITHER SIDE OF THE DRILL ROW, AND SUGAR BEETS WHERE THE DISTANCE MUST BE 2¾ INCHES ON EITHER SIDE OF THE DRILL ROW.

Covered Sweeps—Set the sweeps to run below the soil surface deep enough to cover the EPTAM 7-E with 2 to 3 inches of soil. Calibrate by measuring the spray band width at the back of the sweep, not the sweep width. For broadcast applications stagger sweeps on double tool bar so they overlap sufficiently to allow spray bands to meet.

Note: When applying with either injectors or sweeps, EPTAM 7-E must be applied deep enough to allow 2 to 3 inches of soil to remain over the treatment after the planting operations.

Planting Directions

For pre-plant applications, seeding should be done as soon as possible after treatment to obtain a maximum period of weed control.

IRRIGATION APPLICATION—POST-PLANTING AND ESTABLISHED CROPS

Meter EPTAM 7-E into the irrigation water by using a metering device that will introduce a constant flow into the water. For flood, furrow, or sprinkler irrigation meter the EPTAM 7-E into the water during the entire period OR, for sprinkler irrigation, the EPTAM 7-E may be metered into sufficient water to penetrate to a depth of 3 to 4 inches. Time this EPTAM 7-E application to insure that proper penetration of the herbicide corresponds with the end of the irrigation period. Flush the lines and then turn the water off promptly. Consult "RECOMMENDATIONS" on this label for proper timing of application for each crop for which irrigation application is recommended. A flow rate chart for EPTAM 7-E metering into water run applications will be found in Appendix 3 of this booklet.

USE PRECAUTIONS FOR SPRINKLER IRRIGATION SYSTEMS

Apply this product only through sprinkler, including center pivot, flood (basin) or furrow irrigation systems. Do not apply this product through any other type of irrigation system.

Crop injury, lack of effectiveness, or illegal pesticide residues in the crop can result from nonuniform distribution of treated water.

If you have questions about calibration, you should contact State Extension Service specialists, equipment manufacturers or other experts.

Do not connect an irrigation system (including greenhouse systems) used for pesticide application to a public water system unless the pesticide label-prescribed safety devices for public water systems are in place.

A person knowledgeable of the chemigation system and responsible for its operation or under the supervision of the responsible person, shall shut the system down and make necessary adjustments should the need arise.

Application of more than label recommended quantities of irrigation water per acre may result in decreased product performance by removing the chemical from the zone of effectiveness.

The system must contain a functional check valve, vacuum relief valve, and low pressure drain appropriately located on the irrigation pipeline to prevent water source contamination from backflow.

The pesticide injection must contain a functional, automatic, quick-closing check valve to prevent the flow of fluid back toward the injection pump.

The pesticide injection pipeline must also contain a functional, normally closed, solenoid-operated valve located on the intake side of the injection pump and connected to the system interlock to prevent fluid from being withdrawn from the supply tank when the irrigation system is either automatically or manually shut down.

The system must contain functional interlocking controls to automatically shut off the pesticide injection pump when the water pump motor stops.

The irrigation line or water pump must include a functional pressure switch which will stop the water pump motor when the water pressure decreases to the point where pesticide distribution is adversely affected.

Systems must use a metering pump, such as a positive displacement injection pump (e.g., diaphragm pump) effectively designed and constructed of materials that are compatible with pesticides and capable of being fitted with a system interlock.

Do not apply when wind speed favors drift beyond the area intended for treatment, when system connections or fittings leak, when nozzles do not provide uniform distribution or when lines containing the product must be dismantled and drained.

USE PRECAUTIONS FOR FLOOD OR FURROW IRRIGATION

Tailwater (runoff water) from flood or furrow irrigation should be recirculated or used only on other crops which are registered for this type of application.

Systems using a gravity flow pesticide dispensing system must meter the pesticide into the water at the head of the field and downstream of a hydraulic discontinuity such as a drop structure of weir box to decrease potential for water source contamination from backflow if water flow stops.

Systems utilizing a pressurized water and pesticide injection system must meet the following requirements:

The system must contain a functional check valve, vacuum relief valve, and low pressure drain appropriately located on the irrigation pipeline to prevent water source contamination from backflow.

The pesticide injection pipeline must contain a functional, automatic, quick-closing check valve to prevent the flow of fluid back toward the injection pump.

The pesticide injection pipeline must also contain a functional, normally closed, solenoid-operated valve located on the intake side of the injection pump and connected to the system interlock to prevent fluid from being withdrawn from the supply tank when the irrigation system is either automatically or manually shut down.

The system must contain functional interlocking controls to automatically shut off the pesticide injection pump when the water pump motor stops.

The irrigation line or water pump must include a functional pressure switch which will stop the water pump motor when the water pressure decreases to the point where pesticide distribution is adversely affected.

Systems must use a metering pump, such as a positive displacement injection pump (e.g., diaphragm pump) effectively designed and constructed of materials that are compatible witn pesticides and capable of being fitted with a system interlock.

Cultural Practices Following Application

Should weeds develop, a shallow cultivation or rotary hoeing will generally result in better weed control. When cultivating for any reason, it should be shallow, i.e., no more than ½ the depth of the herbicide was incorporated or injected. Pre-emergence or post-emergence herbicides may be necessary to control weeds resistant to EPTAM 7-E.

Regional Use Map

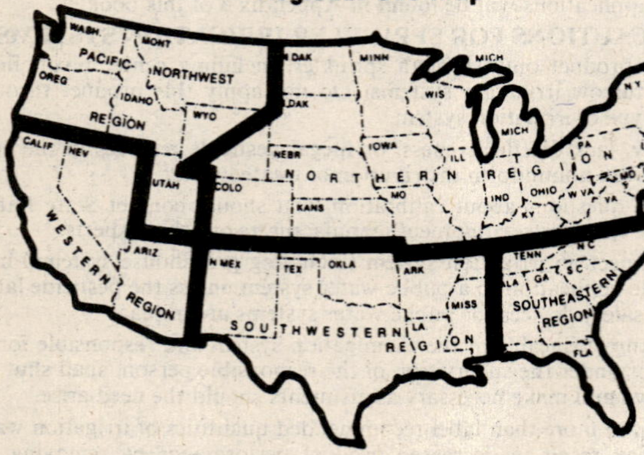

Crop Recommendations

All crop recommendations are given on a regional basis. There are five regions, as delineated on the U.S. map printed above. USE THE RECOMMENDATIONS IN YOUR REGION ONLY.

RATE CONVERSION TABLE

Dosage rates in this booklet are expressed as pints EPTAM 7-E per acre. The following table shows pints EPTAM 7-E per acre in the left column and the equivalent amount of active ingredient per acre in the center column.

Pints EPTAM 7-E Per Acre	Lb. Active Ingredient Per Acre	Acres Treated By One Gallon EPTAM 7-E
1¼	1	7
1¾	1½	4⅔
2¼	2	3½
3½	3	2⅓
4½	4	1¾
5¼	4½	1½
5¾	5	1⅖
7	6	1⅙
8½	7½	1
17	15	½

RECOMMENDATIONS

Northern Region

These recommendations are given as the broadcast (overall) rates of EPTAM 7-E per acre. For band treatment, use proportionately less material per acre depending upon the width of the band to be treated and the crop row spacing. Do not use band application on rocky ground because thorough incorporation is not possible.

Alfalfa*, Birdsfoot Trefoil, Clovers, Lespedeza: Do not use EPTAM 7-E if a grass or grain nurse crop is to be planted with the legume. Do not use on white dutch clover. Apply and incorporate 3½ to 4½ pints EPTAM 7-E per acre just before planting. Temporary crop stunting and sealing of the first leaves will occur if conditions for germination and growth are not optimum (e.g. lack of moisture), and will be relieved by irrigation or adequate rainfall.

Or Alfalfa* (For Control of Annual Grasses Growing from Seed Only): Apply and incorporate 2¼ pints EPTAM 7-E per acre just before planting. Temporary crop stunting and sealing of the first leaves will occur if conditions for germination and growth are not optimum.

And/Or Alfalfa (Established Stands): Meter 2¼ to 3½ pints EPTAM 7-E per acre into the irrigation water applied to established stands prior to weed emergence. Use the lower rate on very coarse textured soils. Do not apply within 14 days of harvesting or grazing alfalfa.

* Alfalfa is sensitive to soil residues of Atrazine. Do not use EPTAM 7-E on alfalfa if Atrazine was applied within the previous 12 months.

Ladino Clover (Established Stands): Meter 2¼ to 3½ pints EPTAM 7-E per acre into the irrigation water applied to established stands prior to weed emergence. Use the lower rate on very coarse textured soils. Do not apply within 45 days of harvesting or grazing.

Beans, Green or Dry: Do not use EPTAM 7-E on Adzuki beans, cowpeas (blackeye peas, blackeye beans), soybeans, lima beans, Mung beans, Garbanzo beans or other flat-podded beans except Romano. Under abnormal weather conditions, stunting may occur on Gratiot, Michilite, Sanilac, Seafarer, and Seaway varieties. Do not exceed 3½ pints EPTAM 7-E per acre on small white beans or green beans grown on coarse textured soils. Do not exceed 9¾ pints EPTAM 7-E per acre per crop.

Fall Application: (dry beans Minnesota, North Dakota only): Apply and incorporate in the late fall before the ground freezes. Use 4½ pints EPTAM 7-E per acre on coarse textured soils and 5¼ pints EPTAM 7-E per acre on medium and fine textured soils.

Application at Planting: Apply and incorporate just before planting or meter into the irrigation water before or immediately after planting, 3½ to 4½ pints EPTAM 7-E per acre. Rotary hoe lightly during or shortly after emergence of the beans to break any crust which occurs.

Or

Lay-by Application: At time of last cultivation for the season apply and incorporate 3½ to 4½ pints EPTAM 7-E per acre. Apply as a directed spray to the soil at the base of the plants before bean pods start to form. Do not feed or pasture vines to livestock within 45 days after application.

For control of smartweed and ragweed and for increased control of labeled broadleaf weeds, EPTAM 7-E may be followed by a pre-emergence application of Amiben* 2E, 2-EC. Apply according to label directions and observe all use precautions. Slight phytotoxicity may be noted following sequential application.

* Chloramben herbicide—Registered trademark of Rhone Poulenc Inc.

Castor Beans: Apply and incorporate 2¼ pints EPTAM 7-E per acre immediately after planting. Use a rotary hoe for incorporation. Early cultivation after EPTAM 7-E application enhances weed control.

Flax:

Fall Application: (Minnesota, North Dakota): Apply and incorporate in the late fall before the ground freezes. Use 4½ pints EPTAM 7-E per acre on coarse textured soils and 5¼ pints EPTAM 7-E per acre on medium and fine textured soils.

Potatoes, Irish: Do not exceed 7 pints EPTAM 7-E per acre per crop. The Superior variety potato is sensitive to EPTAM 7-E and under stress conditions, early season stunting may occur.

Fall Application (Minnesota, North Dakota): Apply and incorporate in the late fall before the ground freezes. Use 5¼ pints EPTAM 7-E per acre on coarse textured soils and 7 pints EPTAM 7-E per acre on medium and fine textured soils.

Pre-Plant Incorporation: Apply and incorporate 3½ to 7 pints EPTAM 7-E per acre. For quackgrass and nutgrass control use the higher rate.

Or

Drag-Off (Come Up, Weeding Time) Incorporation: Apply and incorporate 3½ to 7 pints EPTAM 7-E per acre. For nutgrass control use the higher rate. The field first must be "dragged-off," followed by EPTAM 7-E application and incorporation. Use spike-tooth harrows or cultivation equipment for incorporation.

And/Or Lay-By Incorporation: Apply and incorporate 3½ to 4½ pints EPTAM 7-E per acre to clean cultivated soil after potato plants have emerged from the soil. Apply as directed spray to the soil. Do not apply within 45 days of harvest.

Or

Irrigation: Meter up to 3½ pints EPTAM 7-E per acre into the irrigation water after clean cultivation. Do not apply within 45 days of harvest.

Safflower: Apply and incorporate 3½ pints EPTAM 7-E per acre just before planting.

Sugar Beets:

Fall Application (Minnesota, North Dakota): Apply and incorporate in the late fall before the ground freezes. Use 4½ pints EPTAM 7-E per acre on coarse textured soils and 5¼ pints EPTAM 7-E per acre on medium and fine textured soils.

Pre-Plant (Iowa, Eastern Nebraska, North Dakota, South Dakota, Minnesota, Michigan): Apply and incorporate 2¼ pints EPTAM 7-E per acre on coarse textured soils, or 3½ pints per acre on medium and fine textured soils just before planting. Injury will occur if conditions for germination and growth are not optimum.

Or

Post Emergence
(After the first true leaves have formed)

Irrigation Water: Meter 2¼ to 3½ pints EPTAM 7-E per acre into the first irrigation applied after the last cultivation for the season.

Incorporation: Apply 3½ pints EPTAM 7-E per acre after thinning and clean cultivation and incorporate to a depth of 2 to 3 inches. Treatment may be used following a fall application of EPTAM 7-E at recommended rates.

Or

Subsurface Injection: Apply 3½ pints EPTAM 7-E per broadcast acre, or in band treatment (using 2 shanks per row 5½ inches apart centered on the drill row with rows 22 inches apart) use 1¾ pints EPTAM 7-E per acre. Prior to application, a clean cultivation must be made for all existing weed growth to be destroyed.

Sunflower (Minnesota, North Dakota, South Dakota): Apply and incorporate 2½ to 3½ pints EPTAM 7-E per acre just before planting. Use lower rates on lighter soil.

Fall Application (Minnesota, North Dakota): Apply and incorporate in the late fall before the ground freezes. Use 4½ pints EPTAM 7-E per acre on coarse textured soils and 5¼ pints EPTAM 7-E per acre on medium and fine textured soils.

RECOMMENDATIONS
Southeastern Region

These recommendations are given as the broadcast (overall) rate of EPTAM 7-E per acre. For band treatment, use proportionately less material per acre depending on the width of band to be treated and the crop row spacing. Do not use band application on rocky ground because thorough incorporation is not possible.

Alfalfa*, Birdsfoot Trefoil, Clovers, Lespedeza: Do not use EPTAM 7-E if a grass or grain nurse crop is to be planted with the legume. Do not use on white dutch clover. Apply and incorporate 3½ pints EPTAM 7-E per acre just before planting. (For fall seeded alfalfa in South Carolina only, apply and incorporate 1¾ pints EPTAM 7-E per acre just before planting.) Temporary crop stunting and sealing of the first leaves will occur if conditions for germination and growth are not optimum (e.g. lack of moisture), and will be relieved by irrigation or adequate rainfall.

* Alfalfa is sensitive to soil residues of Atrazine. Do not use EPTAM 7-E on alfalfa if Atrazine was applied within the previous 12 months.

Beans, Green or Dry: Do not use EPTAM 7-E on Adzuki beans, cowpeas (blackeye peas, blackeye beans), soybeans, lima beans, Mung beans, Garbanzo beans or other flat-podded beans, except Romano. Under abnormal weather conditions stunting may occur on Gratiot, Michilite, Sanilac, Seafarer and Seaway varieties. Do not exceed 7 pints EPTAM 7-E per acre per crop.

At Planting

Pre-Plant (Flat-Planted): Use 3½ pints EPTAM 7-E per acre incorporated just before planting on dry, snap and pole beans. Rotary hoe lightly during or shortly after emergence of the beans to break any crust which occurs.

Or

Subsurface Application: Apply 2¼ pints EPTAM 7-E per acre preplant or at planting. See DIRECTIONS FOR USE.

Or

Bed Treatments:

Method A—Apply 3½ pints EPTAM 7-E per acre broadcast and disc in 6 inches deep prior to forming beds and planting.

Method B—Apply 1¾ pints EPTAM 7-E per acre broadcast (do not disc in) immediately ahead of bedding discs. Plant 7 days after treatment.

Method C—Apply as a band treatment (do not disc in) immediately ahead of bedding discs, or as a band treatment to partially formed beds or bed tops immediately in front of the re-bedding operation. Use a band rate equivalent to 2¼ pints per acre broadcast. Care should be taken not to fold in treatment.

Example: To apply EPTAM 7-E as an 18 inch band on 36-inch rows, use 1¼ pints per crop acre. Plant 7 days after application.

Note: With Methods B and C, if bed shapers (levelers) are used, the bedding up and shaping should be done so that 3 to 4 inches of soil remain over the EPTAM 7-E.

Or Lay-By

Incorporation: At the time of last cultivation apply and incorporate 3½ pints EPTAM 7-E per acre. Apply as a directed spray to the soil at the base of the plants before bean pods start to form. Do not feed or pasture vines to livestock until 45 days after application.

Citrus Nursery Stock and Young Field Plantings (Non-bearing Orange and Grapefruit Groves): After lining out, apply 3½ to 7 pints EPTAM 7-E per acre as a directed spray to the soil. Incorporate with cultivation equipment, i.e., tree hoes and rotary hoes.

Citrus (Oranges, Tangerines, Grapefruit): AFTER CLEAN CULTIVATION OR PRIOR TO WEED EMERGENCE in bearing citrus, apply 3½ pints EPTAM 7-E per acre by flood or furrow irrigation. Meter EPTAM 7-E into the water during the entire irrigation period. Do not apply within 15 days of harvest.

Cotton: Non-Irrigated Areas Only—**APPLICATION AFTER STAND IS ESTABLISHED:** Apply 2¼ pints EPTAM 7-E per broadcast acre. Use specially designed injector units or sweeps for application. If incorporated application is to be made, use power driven rotary tillers set to a depth of 2 to 3 inches. Apply after cotton has 2 to 4 leaves. Do not apply after first bolls open. DO NOT APPLY CLOSER THAN 4 INCHES EITHER SIDE OF THE COTTON DRILL.

Note: Tandem discs may be used for incorporation in the skips of skip row cotton.

Cotton is susceptible to injury from EPTAM 7-E. Follow directions for use carefully to avoid crop injury.

Pine Seedling Nurseries (Loblolly, Slash, Longleaf, Shortleaf): Apply and incorporate 7 pints EPTAM 7-E per acre 14 days prior to seeding.

Potatoes, Irish: Do not exceed 3½ pints EPTAM 7-E per acre per crop.

Caution: In Florida on winter and early spring potatoes apply only after potatoes have emerged and true leaves have formed.

Before or at Planting

Pre-Plant: Apply and incorporate 3½ pints EPTAM 7-E per broadcast acre just before planting. For incorporated applications to beds, apply as a band application and incorporate with ground or power driven tillers.

Example: In 18-inch bands on 36-inch rows, use 1¾ pints per crop acre. See DIRECTIONS FOR USE: INCORPORATION.

Or

Before Planting and Before Bed Formation: Band application: Apply as a band, equivalent to 3½ pints per acre broadcast basis. Cover with 3 to 4 inches of soil with bedding discs, middle busters or other suitable bed making equipment. Care should be taken not to fold in the band treatment.

Or

After Planting but Before Bed Formation: Apply 1¾ pints EPTAM 7-E per broadcast acre over planted crop and bed up immediately with bedding discs set to cover 3 to 4 inches of soil.

Or

After Planting and After Bed Formation: Apply EPTAM 7-E as a band at a rate equivalent to 3½ pints per acre, broadcast basis. Re-bed immediately after application with bedding discs set to cover with 3 to 4 inches of soil. Care should be taken not to fold in the band treatment.

Or

After Planting and After Bed Formation: Apply 1¾ pints EPTAM 7-E per broadcast acre. Re-bed immediately after application with bedding discs set to cover with 3 to 4 inches of soil.

Or

Drag-Off (Come Up, Weeding Time) Apply EPTAM 7-E as a band treatment after drag-off, at a rate equivalent to 3½ pints per acre (broadcast basis) and cover with bedding discs set to cover with 3 to 4 inches of soil. Care should be taken not to fold in the band treatment.

Or

Post-Emergence—Lay-By

Incorporation: Apply and incorporate EPTAM 7-E at a rate equivalent to 3½ pints per acre (broadcast basis) after potato plants have emerged from the soil. Apply as a directed spray to the soil in bands on both sides of the row.

Immediately cover the EPTAM 7-E with 3 to 4 inches of soil by re-bedding with bedding discs. Care should be taken not to fold in the band treatment. **Example:** Apply 2¼ pints EPTAM 7-E per crop acre as a directed spray to the soil in 12-inch bands on both sides of 36-inch rows. Do not apply within 45 days of harvest.

Or

Irrigation: Meter 3½ pints EPTAM 7-E per acre into the irrigation water after clean cultivation. Do not apply within 45 days of harvest.

RECOMMENDATIONS
Southwestern Region

These recommendations are given as the broadcast (overall) rate of EPTAM 7-E per acre. For band treatment, use proportionately less material per acre depending on the width of band to be treated and the crop row spacing. Do not use band application on rocky ground because thorough incorporation is not possible.

Alfalfa*, Birdsfoot Trefoil, Clovers, Lespedeza: Do not use EPTAM 7-E if a grass or grain nurse crop is to be planted with the legume. Do not use on white dutch clover. Apply and incorporate 3½ pints EPTAM 7-E per acre just before planting. Temporary crop stunting and sealing of the first leaves will occur if conditions for germination and growth are not optimum (e.g. lack of moisture), and will be relieved by irrigation or adequate rainfall.

And/Or Alfalfa (Established Stands): Meter 2¼ to 3½ pints EPTAM 7-E per acre into the irrigation water applied to established stands prior to weed emergence. Use the lower rate on very coarse textured soils. Do not apply within 14 days of harvesting or grazing alfalfa.

* Alfalfa is sensitive to soil residues of Atrazine. Do not use EPTAM 7-E on alfalfa if Atrazine was applied within the previous 12 months.

Ladino Clover (Established Stands): Meter 2¼ to 3½ pints EPTAM 7-E per acre into the irrigation water applied to established stands prior to weed emergence. Use the lower rate on very coarse textured soils. Do not apply within 45 days of harvesting or grazing.

Beans, Green or Dry: Do not use EPTAM 7-E on Adzuki beans, cowpeas (blackeye peas, blackeye beans), soybeans, lima beans, Mung beans, Garbanzo beans or other flat-podded beans except Romano. Under abnormal weather conditions, stunting may occur on Gratiot, Michilite, Sanilac, Seafarer, and Seaway varieties. Do not exceed 7 pints EPTAM 7-E per acre per crop.

Before or at Planting

Pre-Plant (Flat-Planted): Apply and incorporate 3½ pints EPTAM 7-E per acre just before planting. Rotary hoe lightly during or shortly after emergence of the beans to break any crust which occurs.

Or

Subsurface Application: Apply 3½ pints EPTAM 7-E per acre pre-plant, or at planting. See DIRECTIONS FOR USE.

Or Lay-By

At the time of the last cultivation, apply and incorporate 3½ pints EPTAM 7-E per acre, apply as a directed spray to the soil at the base of the plants before bean pods start to form. Do not feed or pasture vines to livestock until 45 days after application.

Citrus Nursery Stock and Young Field Plantings (Non-bearing Orange and Grapefruit Groves). After lining out, apply 3½ to 7 pints EPTAM 7-E per acre as a directed spray to the soil. Incorporate with cultivation equipment, i.e., tree hoes and rotary hoes.

Citrus (Oranges, Tangerines, Grapefruit): AFTER CLEAN CULTIVATION OR PRIOR TO WEED EMERGENCE in bearing citrus, apply 3½ pints EPTAM 7-E per acre by flood or furrow irrigation. Meter EPTAM 7-E into the water during the entire irrigation period. Do not apply within 15 days of harvest.

Cotton: Non-Irrigated Areas Only—**APPLICATION AFTER STAND IS ESTABLISHED:** Apply 2¼ pints EPTAM 7-E per broadcast acre. Use specially designed injector units or sweeps for application. If incorporated application is to be made, use power driven rotary tillers set to a depth of 2 to 3 inches. Apply after cotton has 2 to 4 leaves. Do not apply after first bolls open. DO NOT APPLY CLOSER THAN 4 INCHES EITHER SIDE OF THE COTTON DRILL.

Note: Tandem discs may be used for incorporation in the skips of skip row cotton.

Cotton is susceptible to injury from EPTAM 7-E. Follow directions for use carefully to avoid crop injury.

Pine Seedling Nurseries (Loblolly, Slash, Longleaf, Shortleaf): Apply and incorporate 7 pints EPTAM 7-E per acre 14 days prior to seeding.

Potatoes, Irish: Do not exceed 7 pints EPTAM 7-E per acre per crop.

Pre-Plant: Apply and incorporate 3½ to 7 pints EPTAM 7-E per acre just before planting.

Or

Drag-Off (Come Up, Weeding Time) Incorporation: Apply and incorporate 3½ to 7 pints EPTAM 7-E per acre. For nutgrass control, use the higher rate. The field first must be "dragged-off," followed by EPTAM 7-E application and incorporation. Use spike-tooth harrows or cultivation equipment for incorporation.

And/Or Lay-By

Incorporation: Apply and incorporate 3½ to 7 pints EPTAM 7-E per acre after potato plants have emerged from the soil. Apply as a directed spray to the soil. Do not apply within 45 days before harvest.

Or

Irrigation: Meter up to 3½ pints EPTAM 7-E per acre into the irrigation water after clean cultivation. Do not apply within 45 days of harvest.

Sugar Beets—Post Thinning:

Irrigation Water: Meter 2¼ to 3½ pints EPTAM 7-E per acre into the first irrigation applied after the last cultivation for the season.

Or

Incorporation: Apply and incorporate 2¼ pints EPTAM 7-E per acre after thinning and clean cultivation and incorporate to a depth of 2 to 3 inches.

Sweet Potatoes:

Pre-Plant: Apply and incorporate 2¼ pints EPTAM 7-E per acre on coarse textured soils or 3½ pints per acre on medium and fine textured soils just before planting. Incorporate to maximum depth of 3 inches.

Or

Pre-Plant—Bed-over: Apply 1¾ pints EPTAM 7-E per acre on coarse textured soils or 2¼ pints per acre on medium and fine textured soils just before planting. Treat a band width equal to ⅓ of the total distance between rows.

Soil from areas adjacent to the band that is not treated is then placed on top of the treated band with bed shaping equipment forming the bed. Band depth in finished and planted bed should be 2 to 4 inches below the bed surface. Bed over immediately after application.

Or

Pre-Plant—Bed-up: Apply 1¾ pints EPTAM 7-E per acre on coarse textured soils or 2¼ pints per acre on medium and fine textured soils just before planting. After preshaped beds have been dragged down, EPTAM 7-E is applied broadcast. Soil is then shaped into beds with bed shaping equipment so that the undisturbed EPTAM 7-E layer in the finished bed is 2 to 4 inches below the bed surface. Bed-up immediately after application.

Or

Post-Plant: Apply 8½ pints EPTAM 7-E per acre immediately after planting or within 2 days after planting slips or vine cuttings. Apply to a dry soil surface and do not mix into the soil. If sweet potatoes are irrigated, EPTAM 7-E should be applied prior to irrigation. Apply as a solid overall spray.

RECOMMENDATIONS
Pacific Northwest Region

DIRECTIONS FOR USE

Incorporation Directions:

EPTAM 7-E must be incorporated into the soil to prevent loss of the herbicide. Whenever possible, application and incorporation should be done in the same operation.

Soil Mixing (Incorporation) Directions: For semi-arid areas of Eastern Washington, Eastern Oregon and Idaho only: When application and incorporation are done in separate operations, EPTAM 7-E must be incorporated the same day as application. Application must be made on a dry soil surface (at least ½ inch deep) free from dew and incidental moisture.

Delay incorporation of dry bulk fertilizers for semi-arid areas of Eastern Washington, Eastern Oregon and Idaho only: The application and incorporation of dry bulk fertilizer impregnated with EPTAM 7-E must be carried out on the same day. Application must be made on a dry soil surface (at least ½ inch deep) free from dew and incidental moisture.

Sprinkler incorporation of EPTAM 7-E in the semi-arid areas of Eastern Washington, Eastern Oregon and Idaho only: Surface apply EPTAM 7-E after planting. The soil surface should be dry (at least ½ inch deep) and free from dew and incidental moisture. Incorporate using ½ to ¾ inch of water within 36 hours following application. The application and incorporation must be done within five days after the last tillage operation, since poor results will occur if weeds have germinated.

CROP RECOMMENDATIONS

These recommendations are given as the broadcast (overall) rate of EPTAM 7-E per acre. For band treatment, use proportionately less material per acre depending on the width of band to be treated and the crop row spacing. Do not use band application on rocky ground because thorough incorporation is not possible.

Alfalfa*, Birdsfoot Trefoil, Clovers, Lespedeza: Do not use EPTAM 7-E if grass or grain nurse crop is to be planted with the legume. Do not use on white dutch clover. Apply and incorporate 2¼ to 4½ pints EPTAM 7-E per acre just before planting. (Use the lower rate on very coarse textured soils.) Temporary crop stunting and sealing of the first leaves will occur if conditions for germination and growth are not optimum (e.g. lack of moisture), and will be relieved by irrigation or adequate rainfall.

Or Alfalfa*: Meter 2¼ to 3½ pints EPTAM 7-E per acre into the irrigation water that is applied immediately after planting. (Use the lower rate on very coarse textured soils.)

And/Or Alfalfa (Established Stands): Meter 2¼ to 3½ pints EPTAM 7-E per acre into the irrigation water applied to established stands prior to weed emergence. Use the lower rate on very coarse textured soils. Do not apply within 14 days of harvesting or grazing alfalfa.

* Alfalfa is sensitive to soil residues of Atrazine. Do not use EPTAM 7-E on alfalfa if Atrazine was applied within the previous 12 months.

Ladino Clover (Established Stands): Meter 2¼ to 3½ pints EPTAM 7-E per acre into the irrigation water applied to established stands prior to weed emergence. Use the lower rate on very coarse textured soils. Do not apply within 45 days of harvesting or grazing.

Beans, Green or Dry: Do not use EPTAM 7-E on Adzuki beans, cowpeas (blackeye peas, blackeye beans), soybeans, lima beans, Mung beans, Garbanzo beans or other flat-podded beans except Romano. Under abnormal weather conditions, stunting may occur on Gratiot, Michilite, Sanilac, Seafarer and Seaway varieties. Do not exceed 9 pints EPTAM 7-E per acre per crop. Apply and incorporate 3½ to 4½ pints EPTAM 7-E per acre just before planting. Rotary hoe lightly during or shortly after emergence of the beans to break any crust which occurs.

Or

Subsurface Application: Apply $3^1/_2$ pints EPTAM 7-E per acre pre-plant, just before planting or at planting. See DIRECTIONS FOR USE.

Or Lay-By

Incorporation: At time of last cultivation for the season, apply and incorporate $3^1/_2$ to $4^1/_2$ pints EPTAM 7-E per acre for grass and broadleaf control. Apply as a directed spray to the soil at the base of the plants before bean pods start to form. Do not feed or pasture vines to livestock until 45 days after application.

Or Lay-By

Subsurface: Prior to application, a clean cultivation must be made for all existing weed growth to be destroyed. Apply $3^1/_2$ pints EPTAM 7-E per broadcast acre or in a band treatment (using 2 shanks per row $5^1/_2$ inches apart, centered on the drill row with rows 38 inches apart) use $1^3/_4$ pints per acre. See DIRECTIONS FOR USE.

Peas, Green Processing (Western Washington Only): Apply and incorporate $2^1/_4$ pints EPTAM 7-E per acre just before planting. Early stunting of crop may occur.

Potatoes, Irish: Do not exceed 14 pints EPTAM 7-E per acre per crop.

Pre-Plant: Apply and incorporate just before planting $3^1/_2$ pints EPTAM 7-E per acre; $4^1/_2$ pints per acre for quackgrass control.

Lay-By: Apply and incorporate $3^1/_2$ pints EPTAM 7-E per acre after potato plants have emerged from the soil. Apply as a directed spray to the soil. Do not apply within 45 days of harvest.

Irrigation: Meter $3^1/_2$ pints EPTAM 7-E per acre into the irrigation water after lean cultivation. Do not apply within 45 days of harvest.

Safflower: Apply and incorporate $3^1/_2$ pints EPTAM 7-E per acre per crop.

Sugar Beets: Post Emergence (after the first true leaves have formed). Do not exceed $3^1/_2$ pints EPTAM 7-E per acre per crop except for irrigation applications where 2 applications of $3^1/_2$ pints may be made.

Incorporation: Apply $3^1/_2$ pints EPTAM 7-E per acre after thinning and clean cultivation and incorporate to a depth of 2 to 3 inches.

Or

Irrigation Water: Meter $2^1/_4$ to $3^1/_2$ pints EPTAM 7-E per acre into the irrigation water after clean cultivation. Do not exceed 7 total pints EPTAM 7-E per acre per crop. Do not apply within 49 days of harvest.

Or

Subsurface Injection: Apply $3^1/_2$ pints EPTAM 7-E per broadcast acre, or in band treatment (using 2 shanks per row $5^1/_2$ inches apart, centered on the drill row with rows 22 inches apart) use $1^3/_4$ pints EPTAM 7-E per acre. Prior to application, a clean cultivation must be made for all existing weed growth to be destroyed.

Table Beets: Apply and incorporate $2^1/_4$ pints EPTAM 7-E per acre just before planting. (Under normal use table beets are susceptible to injury from EPTAM 7-E and when EPTAM 7-E is used the seeding rate should be increased 10 per cent.)

Walnuts: After clean cultivation or prior to weed emergence on well established trees, meter $3^1/_2$ pints EPTAM 7-E per acre into the irrigation water during the entire irrigation period.

RECOMMENDATIONS

Western Region

These recommendations are given as the broadcast (overall) rate of EPTAM 7-E per acre. For band treatment, use proportionately less material per acre depending on the width of band to be treated and the crop row spacing. Do not use band application on rocky ground because thorough incorporation is not possible.

Alfalfa*, Birdsfoot Trefoil, Clovers, Lespedeza: Do not use EPTAM 7-E if a grass or grain nurse crop is to be planted with the legume. Do not use on white dutch clover. Apply and incorporate $2^1/_4$ to $4^1/_2$ pints EPTAM 7-E per acre just before planting. (Use the lower rate on very coarse textured soils.) Temporary crop stunting and sealing of the first leaves will occur if conditions for germination and growth are not optimum (e.g. lack of moisture), and will be relieved by irrigation or adequate rainfall.

Or Alfalfa*: Meter $2^1/_4$ to $3^1/_2$ pints EPTAM 7-E per acre into the irrigation water that is applied immediately after planting. (Use the lower rate on very coarse textured soils.)

Do not use EPTAM 7-E pre-emergence on rill irrigated (corrugated) alfalfa.

And/Or Alfalfa (Established Stands): Meter $2^1/_4$ to $3^1/_2$ pints EPTAM 7-E per acre into the irrigation water applied to established stands prior to weed emergence. Use the lower rate on very coarse textured soils. Limit use to one application per cutting. Up to 14 pints EPTAM 7-E per acre per year may be used if applied into the irrigation water. Do not apply within 14 days of harvesting or grazing alfalfa.

* Alfalfa is sensitive to soil residues of Atrazine. Do not use EPTAM 7-E on alfalfa if Atrazine was applied within the previous 12 months.

Ladino Clover (Established Stands): Meter $2^1/_4$ to $3^1/_2$ pints EPTAM 7-E per acre into the irrigation water applied to established stands prior to weed emergence. Use the lower rate on very coarse textured soils. Do not apply within 45 days of harvesting or grazing.

Almonds: After making the last cultivation for the season, meter $2^1/_2$ to $3^1/_2$ pints EPTAM 7-E per acre into the irrigation water. Do not make more than 2 applications per season or exceed 7 pints per acre. Do not apply within 16 days of harvest.

Beans, Green or Dry: Do not use EPTAM 7-E on Adzuki beans, cowpeas (blackeye peas, blackeye beans), soybeans, lima beans, Mung beans, Garbanzo beans or other flat-podded beans except Romano. Under abnormal weather conditions stunting may occur on Gratiot, Michilite, Sanilac, Seafarer and Seaway varieties. Do not exceed 8 pints EPTAM 7-E per acre per crop.

Pre-Plant

Apply and incorporate $3^1/_2$ pints EPTAM 7-E per acre just before planting. Rotary hoe lightly during or shortly after emergence of the beans to break any crust which occurs.

Or

Subsurface Application: Apply $3^1/_2$ pints EPTAM 7-E per acre pre-plant, just before planting or at planting. See DIRECTIONS FOR USE.

Or Lay-By

Incorporation: At time of last cultivation for the season, apply and incorporate $3^1/_2$ to $4^1/_2$ pints EPTAM 7-E per acre for grass and broadleaf control. Apply as a directed spray to the soil at the base of the plants before bean pods start to form. Do not feed or pasture vines to livestock until 45 days after application.

Or Lay-By

Subsurface: Prior to application a clean cultivation must be made for all existing weed growth to be destroyed. Apply $3^1/_2$ pints EPTAM 7-E per broadcast acre or in a band treatment (using 2 shanks per row $5^1/_2$ inches apart, centered on the drill row with rows 38 inches apart) use $1^3/_4$ pints per acre. See DIRECTIONS FOR USE.

Citrus Nursery Stock and Young Field Plantings (Non-bearing Orange, Grapefruit and Lemon Groves): After lining out, apply $3^1/_2$ to 7 pints EPTAM 7-E per acre as a directed spray to the soil. Incorporate with cultivation equipment, i.e., tree hoes and rotary hoes.

Citrus (Oranges, Tangerines, Grapefruit, Lemons): After clean cultivation or prior to weed emergence in bearing citrus, apply $3^1/_2$ pints EPTAM 7-E per acre by flood or furrow irrigation. Do not exceed $10^1/_2$ pints total EPTAM 7-E per acre per year when multiple applications are made. Do not apply within 14 days of harvest.

Potato, Irish: Do not exceed 14 pints EPTAM 7-E per acre per crop.

Pre-Plant: Apply and incorporate just before planting $3^1/_2$ pints EPTAM 7-E per acre.

Drag-Off (Come Up Weeding Time) Incorporation: Apply and incorporate $3^1/_2$ pints EPTAM 7-E per acre. The field first must be "dragged off," followed by EPTAM 7-E application and incorporation. Use spike-tooth harrows or cultivation equipment for incorporation.

And/Or Lay-By

Incorporation: Apply and incorporate $3^1/_2$ to $4^1/_2$ pints EPTAM 7-E per acre after potato plants have emerged from the soil. (Use lower rate on coarse textured soils.) Apply as a direct spray to the soil. Do not apply within 30 days of harvest.

Or

Irrigation: Meter $3^1/_2$ pints EPTAM 7-E per acre into the irrigation water after clean cultivation. Do not apply within 30 days of harvest.

Safflower: Apply and incorporate $3^1/_2$ pints EPTAM 7-E per acre just before planting.

Sugar Beets: Post Emergence (after first true leaves have formed).

Incorporation: Apply $3^1/_2$ pints EPTAM 7-E per acre after thinning and clean cultivation and incorporate to a depth of 2 to 3 inches.

Or

Irrigation Water: Meter $2^1/_4$ to $3^1/_2$ pints EPTAM 7-E per acre into the first irrigation applied after the last cultivation for the season. Two applications of $2^1/_4$ pints each should be made when beets are to be carried in the ground longer than the normal growing season.

Or

Subsurface Injection: Apply $3^1/_2$ pints EPTAM 7-E per broadcast acre, or in band treatment (using 2 shanks per row $5^1/_2$ inches apart, centered on the drill row) use $1^3/_4$ pints EPTAM 7-E per acre. Prior to application, a clean cultivation must be made for all existing weed growth to be destroyed. See DIRECTIONS FOR USE.

Tomatoes: Lay-By Application

(Northern California Counties only, i.e., Butte, Colusa, Contra Costa, Glenn, Merced (North of Highway 152), Sacramento, San Joaquin, Solano, Stanislaus, Sutter, Yolo and Yuba.) For use on tomatoes at least 3-4 inches tall; on clay and clay loam soils only. DO NOT USE ON SANDY SOILS.

Apply EPTAM 7-E as a spray to the soil surface at a rate of $3^1/_2$ pints per treated acre. Incorporate immediately. For band applications reduce rates proportionately. DO NOT APPLY WITHIN 2 INCHES OF THE CROP ROW. Do not use where grain will be planted within 90 days. Do not irrigate for at least 5 days after application. Do not apply within 21 days of harvest.

Walnuts: After clean cultivation or prior to weed emergence on well established trees, meter 3½ pints EPTAM 7-E per acre into the irrigation water during the entire irrigation period.

APPENDIX I

EPTAM 7-E with Fluid Fertilizers

The following procedure is suggested for determining whether EPTAM 7-E may be combined with a specific fluid fertilizer for spray tank application.

Material Required:

1. EPTAM 7-E.
2. Fluid fertilizer to be used.
3. Adjuvant for fertilizer tank mix: Compex*, Sponto 168-D*, Unite*, or equivalent. The adjuvant which provides the best emulsification depends on the specific fertilizer under consideration.

* Compex, Colloidal Products Incorp., Petaluma, Calif. Sponto 168-D, Retzloff Chemical Company, Houston, Texas, Unite, Hopkins Agricultural Chemical Co., Madison, Wisconsin.

4. Two one-quart, wide-mouth glass jars with lid or stopper.
5. Measuring spoons (a 25 mL pipette or graduated cylinder provides more accurate measurement).
6. Measuring cup, 8 oz. (237 mL).

Procedure:

1. Pour a pint (about 473 mL) of the fluid fertilizer into each of the quart jars.
2. Add adjuvant to one of the jars and mix (see next rate table).
3. Add the EPTAM 7-E to both jars (see next rate table).
4. Close both jars with lid or stopper and mix the contents by turning the jars upside down ten times.
5. Inspect the surface and body of the mixtures—
(A) Immediately after completing the jar inversions,
(B) After allowing the jars to stand quietly for 30 minutes,
(C) And then again after turning the jars upside down ten times.

If a uniform mix cannot be made, the mixture should not be used. If either mixture remains uniform for 30 minutes, the combination may be used. Should either mixture separate after 30 minutes, but readily remix uniformly with ten jar inversions, the mixture can be used if adequate agitation is maintained in the tank. If the mixture with adjuvant is satisfactory, but the one without adjuvant is not, be sure to use the adjuvant in the spray tank. Add the adjuvant first at a rate of 3 pints per 100 gallons of fluid fertilizer; foaming can be minimized by using moderate agitation.

If nondispersible oil, sludge or clumps of solids form in the mixtures, the combination should not be used.

Note: 1 For some combinations, premixing wettable powders in a little water in a pail before adding them to the spray tank will improve the compatability of the final mixtures with EPTAM 7-E. This technique can be tested in the smallscale jar test by premixing the wettable powder in one-eighth (⅛) cup of water prior to addition to the pint of fluid fertilizer.

Rate Table for EPTAM 7-E and ADJUVANT with the Fluid Fertilizer**

Gallons of fluid fertilizer to be applied per acre	mL or Tsp. of EPTAM 7-E* to be added to 1 pint of fertilizer	
	7-E	
	mL	Tsp.
10	7	1⅓
15	4	¾
20	3	⅔
25	3	⅔
30	2	½
40	2	½

* Based on field rate of 1 pound active ingredient per acre in the fertilizer volumes indicated. Increase volume proportionately to correspond with intended field rate in terms of pounds active ingredient per acre (e.g., for field rate of 4 pounds actual EPTAM 7-E in 40 gallons fertilizer per acre, add 8 ml or 2 tsp. EPTAM 7-E to each jar for compatibility testing purposes).

** Two (2) milliliters or one-half (½) teaspoon of adjuvant to be added to 1 pint of fluid fertilizer in order to equal the rate of 3 pints of adjuvant per 100 gallons of fluid fertilizer.

Rate Chart for the Impregnation of Dry Bulk Fertilizers with EPTAM 7-E

Fertilizer Rate Per Acre		EPTAM 7-E Rate per acre	
	3½ pts. per acre	4½ pts. per acre	7 pts. per acre
200 lbs.	17½ qts./ton	22¼ qts./ton	35 qts./ton
250 lbs.	14 qts./ton	18 qts./ton	28 qts./ton
300 lbs.	11⅔ qts./ton	15 qts./ton	23⅓ qts./ton
350 lbs.	10 qts./ton	12⅞ qts./ton	20 qts./ton
400 lbs.	8¾ qts./ton	11¼ qts./ton	17⅕ qts./ton
450 lbs.	7¾ qts./ton	10 qts./ton	15⅕ qts./ton
500 lbs.	7 qts./ton	9 qts./ton	14 qts./ton
550 lbs.	6⅓ qts./ton	8⅕ qts./ton	12⅔ qts./ton
600 lbs.	5⅞ qts./ton	7½ qts./ton	11¾ qts./ton
650 lbs.	5⅖ qts./ton	7 qts./ton	10⅘ qts./ton
700 lbs.	5 qts./ton	6⅖ qts./ton	10 qts./ton

APPENDIX II

EPTAM 7-E Impregnation on Dry Bulk Fertilizers

CAUTION: EPTAM 7-E alone or in combination with other herbicides must not be impregnated on ammonium nitrate, sodium nitrate or potassium nitrate. Such mixtures may cause explosion and fire.

All individual state regulations relating to bulk dry fertilizer blending, registration, labeling and application are the responsibility of the individual and/or company selling the fertilizer and EPTAM 7-E mixture.

EPTAM 7-E may be impregnated on many dry bulk fertilizers and applied and incorporated in the soil before planting for the control of grass and broadleaf weeds.

All EPTAM 7-E supplementary literature instructions and label recommendations regarding rates per acre, soil incorporation, application, cautions, general use precautions and other directions must be followed.

Test results have shown that EPTAM 7-E on bulk dry fertilizers gives weed control equal to EPTAM 7-E applied as a spray in water or liquid fertilizer. However, uniform impregnation of EPTAM 7-E on dry fertilizer particles and uniform application in the field are necessary to assure good results.

A minimum of 200 pounds and a maximum of 700 pounds of approved ingredients impregnated with EPTAM 7-E at the recommended rate must be applied per acre.

For impregnating EPTAM 7-E on dry fertilizers, use a closed rotary-drum mixer or similar type of closed blender equipped with suitable spray equipment. The spray nozzle (or nozzles) should be positioned inside of the mixer to provide *uniform* spray coverage of the tumbling fertilizer.

The EPTAM 7-E should be sprayed uniformly onto the fertilizer using a fine spray pattern.

The physical properties of fertilizers vary, particularly in liquid absorptive capacity. When absorptivity is sufficient, simple spray impregnation of the fertilizer with EPTAM 7-E provides a satisfactory, dry mixture.

If the absorptive capacity is inadequate, use of a highly absorptive powder is required to provide a dry, free-flowing mixture.

Microcel E (Johns-Manville Products Corp.) is the recommended absorbent powder. It should be added separately and uniformly to the prepared EPTAM 7-E fertilizer mixture, in a quantity that is sufficient to provide a suitably free-flowing mixture. Generally less than 2% by weight of Microcel E is required.

The amount of EPTAM 7-E actually required in the manufacture of individual fertilizer mixtures should be determined carefully for each production operation. This is necessary to ensure that the amount of EPTAM 7-E actually contained in the mixture applied to the soil represents the correct rate of use.

Bulk fertilizer impregnated with EPTAM 7-E *should be applied immediately,* NOT STORED. All bulk containers must be tightly covered while the product is being transported and applied to reduce chances of EPTAM 7-E loss via volatilization.

EPTAM 7-E Physical Data

Specific Gravity (20/20°C): 0.954 (typical)
Pounds/Gallon (20°C): 7.94 (typical)
Flashpoint: 190°F (Tag. Closed Cup)
Viscosity: sprayable down to minus 20°F.

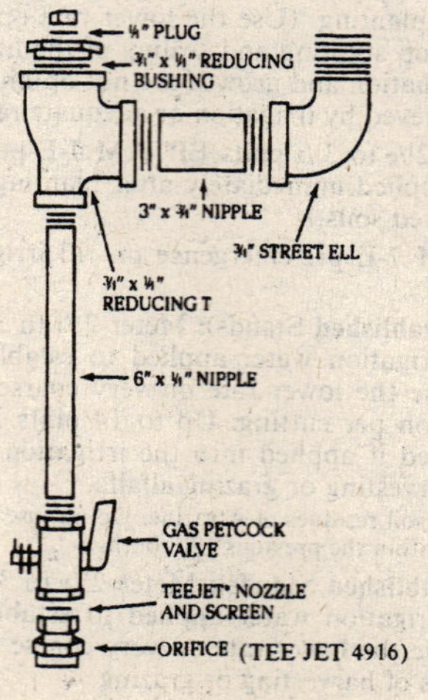

Approved Dry Fertilizer Ingredients

	N	P	K
Ammonium sulfate	21	0	0
Diammonium phosphate	18	46	0
Potassium chloride	0	0	60
Potassium sulfate	0	0	52
Super-phosphate (single)	0	20	0
Triple super-phosphate	0	46	0
Urea	45	0	0
Ammonium phosphate-sulfate	16	20	0
11-48-0	11	48	0

Note: K-Mag has been shown to be compatible with EPTAM 7-E and is approved for use.

APPENDIX III

Flow Rates for EPTAM 7-E Using Various Tee Jet* Orifices (4916)**

Tee Jet Orifice	Ounces Per Minute	cc Per Minute	Gallons Per Hour	Pounds Per Hour
.012	0.215	6.37	0.101	0.707
.014	0.286	8.45	0.134	0.938
.015	0.324	9.59	0.152	1.064
.016	0.375	11.10	0.176	1.232
.018	0.523	15.46	0.245	1.715
.020	0.610	18.04	0.286	2.002
.022	0.796	23.53	0.373	2.611
.024	0.896	26.50	0.420	2.940
.025	0.996	29.46	0.467	3.269
.026	1.111	32.87	0.521	3.647
.027	1.269	37.54	0.595	4.165
.029	1.284	37.98	0.602	4.214
.030	1.502	44.42	0.704	4.928
.032	1.641	48.52	0.769	5.383
.034	1.871	55.33	0.877	6.139
.035	2.091	61.83	0.980	6.860
.037	2.223	65.74	1.042	7.294
.039	2.539	75.08	1.190	8.330
.040	2.603	76.97	1.220	8.540
.041	2.807	83.03	1.316	9.212
.043	2.882	85.24	1.351	9.457
.045	3.334	98.61	1.563	10.941
.046	3.441	101.77	1.613	11.291
.047	3.678	108.77	1.724	12.068
.048	3.951	116.84	1.852	12.965
.051	4.102	121.32	1.923	13.461
.052	4.437	131.42	2.083	14.581
.054	4.849	143.41	2.273	15.911
.055	5.079	150.22	2.381	16.667
.057	5.333	157.73	2.500	17.500
.059	5.926	175.27	2.788	19.446
.063	6.272	185.49	2.940	20.580
.067	7.110	210.28	3.333	23.331
.070	8.205	242.65	3.846	26.922

* Registered trademark of Spraying Systems Co.

** Figures were taken at 70°F and are approximate. Be sure occasionally to measure flow in the field to make certain you have the correct orifice and because rates vary with temperature. (Flow on an .037 orifice increases from 2.2 ounces at 70°F. to 2.4 ounces at 92°F). Use a 300 mesh screen on orifice sizes below .014 and a 200 mesh screen on all others.

STORAGE AND DISPOSAL

Prohibitions: Do not contaminate water, food or feed by storage or disposal. Open dumping is prohibited.

Storage: Keep container tightly closed when not in use. Do not store near seeds, fertilizers or foodstuffs. Can be stored at temperatures as low as minus 50°F.

Pesticide Disposal: Wastes resulting from the use of this product may be disposed of on site or at an approved waste disposal facility.

Container Disposal: Triple rinse (or equivalent). Then offer for recycling or reconditioning, or puncture and dispose of in a sanitary landfill, or by other procedures approved by State and local authorities.

NOTICE TO BUYER AND USER:

Seller warrants that this product conforms to the chemical description on the label and is reasonably fit for the purposes stated on the label when used in accordance with directions under normal conditions of use. This warranty does not extend to the use of this product contrary to label instructions, or under abnormal use conditions, or under conditions not reasonably foreseeable to Seller, and Buyer and User assume the risk of any such use. SELLER DISCLAIMS ALL OTHER WARRANTIES EXPRESSED OR IMPLIED INCLUDING ANY WARRANTY OF FITNESS OR MERCHANTABILITY, SELLER SHALL NOT BE LIABLE FOR CONSEQUENTIAL, SPECIAL OR INDIRECT DAMAGES RESULTING FROM THE USE OR HANDLING OF THIS PRODUCT AND SELLER'S SOLE LIABILITY AND BUYER'S AND USER'S EXCLUSIVE REMEDY SHALL BE LIMITED TO THE REFUND OF THE PURCHASE PRICE.

A-10MR

This product is sold only for uses stated on its label. No express or implied license is granted to use or sell this product under any patent in any country except as specified: Country: United States of America.

The information and specimen labels in this booklet are correct as of the publication date but are subject to change without notice. For current information, contact ICI Americas Inc.

SUPPLEMENTAL LABELING

EPTAM 7-E SELECTIVE HERBICIDE

EPTAM 7-E/TREFLAN E.C.* TANK MIX

For Weed Control in Beans (Green and Dry), Sugar Beets and Sunflowers

A tank mix combination of EPTAM 7-E plus TREFLAN E.C. will give a broader spectrum of weed control than either product used separately.

EPA Reg. No. 10182-220

Caution: Do not graze or feed forage from treated fields to livestock. Read both the EPTAM 7-E and TREFLAN E.C. labels carefully before using. Observe all cautions and limitations on labeling of both products.

Do not contaminate water by cleaning of equipment or disposal of wastes.

*TREFLAN is the registered trademark of Elanco Products Incorporated

Weeds Controlled by the Combination

Annual Grasses	
Crabgrass	Brachiaria
Barnyardgrass (Watergrass)	Sandbur
Foxtails (including giant foxtail)	Junglerice
Johnsongrass (from seed)	Annual Bluegrass
Goosegrass	Sprangletop
Shattercane (Wild Cane)	Cheat
Stinkgrass	Wild Oat
Bromegrass	Texas Panicum**
Annual Broadleaves	
Annual Morningglory	Knotweed
Pigweeds (Spiny, Redroot)	Stinging Nettle
Lambsquarters	Puncturevine
Carpetweed	Nettleleaf Goosefoot
Russian Thistle	Chickweed
Kochia	Corn Spurry
Purslane	Henbit
Florida Purslane	Nightshade
Perennial Weed	
Nutsedge**	

** May not be controlled at less than 3½ pints per acre of EPTAM 7-E.

DIRECTIONS FOR USE

It is a violation of Federal law to use this product in a manner inconsistent with its labeling.

Mixing: Add the recommended rates of both EPTAM 7-E and TREFLAN E.C. to the spraytank during filling and mix thoroughly. Apply in 10 to 40 gallons of water per acre.

Spray Equipment: Use any properly calibrated low pressure boom type herbicide sprayer which will apply the material uniformly. Check calibration frequently during application and observe the nozzles to be sure they are delivering a uniform spray pattern.

Soil Incorporation: The EPTAM 7-E and TREFLAN E.C. combination must be incorporated (mixed) thoroughly into the top 2 to 3 inches of soil immediately after spraying. Spraying and incorporation should be accomplished in the same operation if possible. This can be done by mounting the tank and boom right on the incorporation rig. Thorough incorporation can be achieved with any of the following equipment.

Preplant: Power-driven rotary cultivators—set to cut 2 to 3 inches deep. Double disc (or double disc with spike-tooth harrow in tandem)—set to cut 3 to 6 inches deep and operate in two directions (cross disc) at 4 to 6 mph.

In Established Row Crops: Power-driven rotary cultivators—set to cut 2 to 3 inches deep. Rolling cultivators—set to cut 2 to 4 inches and operated 2 times at 6 to 8 mph. Use on coarse and medium textured soils only. Adjust equipment to throw treated soil toward the crop. Avoid mechanically damaging the crop. Shallow incorporation with implements set to cut less than 2 inches may result in erratic weed control.

Recommendations

Beans, Greens and Dry: The combination of EPTAM 7-E and TREFLAN E.C. should not be used on Adzuki beans, cowpeas (black-eyed peas, black-eyed beans) soybeans, lima beans, Mung beans, Garbanzo beans or other flat-podded beans, except Romano.

Make application before planting, using the rates listed in the following table.

Application Rates Per Broadcast Acre			
		TREFLAN E.C.	
EPTAM 7-E	Soil Type	Organic Matter Content	Rate
2½ to 3½ pints†	Coarse (sand)	0-2%	1 Pint
	Coarse (sand)	2-5%	1-1½ Pints
	Medium (loam)	0-5%	1½ Pints
	Fine (clay)	0-5%	1½ Pints
	All soil types	5.1-10%	1½ Pints

†Use the higher rate for nutsedge control

Planting Instructions

Dry Beans—Plant within 48 hours after incorporation. In the lighter soils under sprinkler irrigation, when it is necessary to irrigate beans after planting and before emergence, sufficient water should be applied to wet the soil well below the depth of planted seed.

Green Beans—Plant soon after incorporation to provide the maximum period of weed control.

Sunflowers—(Minnesota, North Dakota, South Dakota)—Apply and incorporate just before planting, using the rates listed below.

Application Rates Per Broadcast Acre				
EPTAM 7-E			TREFLAN E.C. Rate	
	Soil Type	Organic Matter Content	Minnesota Eastern Dakotas	Western Dakotas
2¼ pints	Coarse (sand)	0-2%	1 Pint	1 Pint
	Coarse (sand)	2-5%	1½-2 Pints	1½-2 Pints
	Medium (loam)	0-5%	1½ Pints	1¼-1½ Pints
	Fine (clay)	0-5%	2 Pints	1½ Pints
	All soil types	5.1-10%	2 Pints	2 Pints

SUGAR BEETS (California)—Apply as a broadcast spray over the top when plants are 2 to 6 inches tall, using the rates listed below.

Application Rates Per Broadcast Acre		
	TREFLAN E.C.	
EPTAM 7-E	Soil Type	Rate
3½ Pints	Coarse (sand)	1 Pint
	Medium (loam)	1¼-1½ Pints
	Fine (clay)	1½ Pints

Exposed beet roots should be covered with soil before application to reduce possibility of girdling. Set incorporation machinery to throw treated soil toward the plants in the row. Care should be taken that incorporation machinery does not damage the sugar beet taproot.

General Use Precautions

Before using EPTAM 7-E plus TREFLAN E.C. as a tank mix, read and observe the cautionary statements, and all other information appearing in both product labels.

SUPPLEMENTAL LABELING

EPTAM 7-E SELECTIVE HERBICIDE

EPTAM 7-E/AMIBEN* TANK MIX

FOR WEED CONTROL IN SUNFLOWERS in the States of Minnesota, North Dakota and South Dakota

A tank mix combination of EPTAM 7-E and AMIBEN will give improved control of many of the broadleaf weeds listed on the EPTAM 7-E label. The combination will also control wild mustard and kochia.

Caution: Do not graze treated fields of sunflowers or use these plants as livestock feeds.

Do not contaminate water by cleaning of equipment or disposal of wastes.

*AMIBEN is a registered trademark of Rhone Poulenc Inc.

DIRECTIONS FOR USE

It is a violation of Federal law to use this product in a manner inconsistent with its labeling.

Mixing: Pour the recommended amount of EPTAM 7-E into the spray tank during the filling operation. Then add the recommended amount of AMIBEN. Apply in 10 to 50 gallons of water per acre.

Spray Equipment: Use any properly calibrated low pressure boom type herbicide sprayer which will apply the material uniformly. Check calibration frequently during application and observe the nozzles to be sure they are delivering a uniform spray pattern.

Soil Incorporation: The EPTAM 7-E and AMIBEN combination must be incorporated (mixed) thoroughly into the top 2 to 3 inches of soil immediately after spraying. Spraying and incorporation should be accomplished in the same operation if possible. This can be done by mounting the tank and boom right on the incorporation rig.

Thorough incorporation can be achieved with any of the following equipment.

- Power Driven Cultivation Equipment (recommended on all soil types) set to cut to a depth of 2 to 3 inches.
- Tandem Discs (recommended on all soil types) set to cut to a depth of 4 to 6 inches, operated at 4 to 6 mph followed by a spike-tooth harrow or some other leveling device which extends beyond the ends of the discs. For more thorough mixing in heavier soils, disc in two different directions (cross disc).
- Field Cultivators (recommended on coarse textured soils. Use only on soils in good tilth). Use 3 to 4 rows of sweeps, spaced at 7 inch or less intervals and staggered so that no soil is left unturned, followed by a spiked-tooth harrow pulled behind the cultivator. Do not use chisels or points. Set the cultivator to cut 4 inches deep, operated at 5 mph or more. Run the equipment over the field twice, the second run at an angle to the first.
- Rotary Ground Driven or Spring-Tooth Cultivators (recommended on coarse textured soils in good tilth only). Set to penetrate to a depth of 4 to 6 inches and operated at 5 to 8 mph in two different directions.

Recommendations

Apply and incorporate just before planting, using the rates given below.

Broadcast Rates Per Acre		
Soil Texture	EPTAM 7-E	AMIBEN
Coarse (sandy)	2¼ Pints	4 Pints
Medium (loams)*	2¼- 3½ Pints	4-8 Pints
Fine (clays)	3½ Pints	8 Pints
All soil types with organic matter 3%	3⅓ Pints	8 Pints
Peat or Muck	NOT RECOMMENDED	

*Use the higher rates with greater amounts of clay and organic matter

SUPPLEMENTAL LABELING

EPTAM 7-E SELECTIVE HERBICIDE

EPTAM 7-E AND METRIBUZIN TANK MIX FOR CONTROL OF WEEDS IN IRISH POTATOES

A tank-mix combination of EPTAM 7-E and METRIBUZIN (Sencor** 4, Sencor 50W, Sencor Sprayule 75%, Lexone ‡4, Lexone 50W or Lexone Dry Flowable 75%) can be applied to Irish potatoes in the Pacific Northwestern and Northern regions of the U.S. to give a broader spectrum of weed control than either product used separately.

Caution:

It is a violation of Federal law to use this product in a manner inconsistent with its labeling.

Before using EPTAM 7-E and METRIBUZIN (Sencor, Lexone) as a tank mix, read both the EPTAM 7-E and METRIBUZIN (Sencor, Lexone) labels carefully. Observe all cautions and limitations on labeling of both products.

Do not contaminate water by cleaning of equipment or disposal of wastes.

Weeds Controlled

Annual Grasses

Annual Bluegrass
Annual Ryegrass
  (Italian Ryegrass)
Barnyardgrass
  (Watergrass, Junglerice)
Bermudagrass seedlings
Crabgrass
Foxtail
  (Giant, Green, Yellow)
Lovegrass
  (Stinkgrass)
Panicum, Fall
Panicum, Texas
Rescuegrass
Sandbur, Field
Signalgrass
Volunteer Grains
  (Barley, Oat, Wheat)

Goosegrass	Shattercane
Johnsongrass seedlings	Wild Oats
	Witchgrass

Annual Broadleaf Weeds

Annual Morningglory	Nettleleaf Goosefoot
Carpetweed	Nightshade (Hairy)
Chickweed, Common	Pennsylvania Smartweed
Cocklebur	Prickly Sida
Corn Spurry	Prostrate Pigweed
Deadnettle	Purslane, Common
(Henbit)	Ragweed, Common
Fiddleneck	Redroot Pigweed
Florida Pusley	Shepherdspurse
Jimsonweed	Sicklepod
Lambsquarters, Common	Tumble Pigweed
	Wild Mustard

** SENCOR is the TM of METRIBUZIN, manufactured by Mobay Chemical Corp.
‡ LEXONE is the TM of METRIBUZIN, manufactured by DuPont Company.

Perennial Grasses

Bermudagrass
Mugwort (Chrysanthemum weed)

DIRECTION FOR USE

Mixing: Add the recommended rates of EPTAM 7-E and MERTIBUZIN (Sencor, Lexone) to the spray tank during the filling operation. Add EPTAM 7-E last. Mix thoroughly. Apply in 20 to 30 gallons of water per acre.

Spray Equipment: Use any properly calibrated, low pressure, boom type herbicide sprayer which will apply the material uniformly. Check calibration frequently during application and observe to be sure that the nozzles are delivering a uniform spray pattern.

Soil Incorporation: The EPTAM 7-E and METRIBUZIN (Sencor, Lexone) combination must be incorporated (mixed) thoroughly into the top 2 to 3 inches of soil immediately after spraying. Spraying and incorporation should be accomplished in the same operation if possible. See the EPTAM 7-E label for complete incorporation directions.

Sprinkler Application: Apply through irrigation sprinkler system after planting as a pre-emergence application or as an early post-emergence application until potatoes are 4 to 6 inches tall. Use the appropriate rate of EPTAM 7-E and METRIBUZIN (Sencor, Lexone) as indicated on the table below.

Pre-mix the desired rate of METRIBUZIN (Sencor, Lexone) in the holding tank in 4 to 5 parts water to 1 part chemical. Add EPTAM 7-E last. Meter the chemical-water mixture into the sprinkler system at a rate proportionate to the acreage to be covered. For center pivot systems, apply 1/2 to 3/4 inch of water per acre. For solid set, wheel lines or hand lines, moisten the soil surface lightly first, then apply the herbicide mixture in 1/2 to 1 inch of water per acre. For best results, the soil should be wetted to a depth of 5 to 7 inches.

All irrigation systems used for this purpose must check valves and approved antisyphon devices to prevent back-flow into water source or supply tank and must be equipped with a power interlock to insure simultaneous shut-off of water pumps and injector.

USE PRECAUTIONS:

Apply this product only through center pivot, sprinkler irrigation systems. Do not apply this product through any other type of irrigation system.

Crop injury, lack of effectiveness, or illegal pesticide residues in the crop can result from nonuniform distribution of treated water.

If you have questions about calibration, you should contact State Extension Service specialists, equipment manufacturers or other experts.

Do not connect an irrigation system (including greenhouse systems) used for pesticide application to a public water system unless the pesticide label-prescribed safety devices for public water systems are in place.

A person knowledgeable of the chemigation system and responsible for its operation or under the supervision of the responsible person, shall shut the system down and make necessary adjustments should the need arise.

The system must contain a functional check valve, vacuum relief valve, and low pressure drain appropriately located on the irrigation pipeline to prevent water source contamination from backflow.

The pesticide injection pipeline must contain a functional, automatic, quick-closing check valve to prevent the flow of fluid back toward the injection pump.

The pesticide injection pipeline must also contain a functional, normally closed, solenoid-operated valve located on the intake side of the injection pump and connected to the system interlock to prevent fluid from being withdrawn from the supply tank when the irrigation system is either automatically or manually shut down.

The system must contain functional interlocking controls to automatically shut off the pesticide injection pump when the water pump motor stops.

The irrigation line or water pump must include a functional pressure switch which will stop the water pump motor when the water pressure decreases to the point where pesticide distribution is adversely affected.

Systems must use a metering pump, such as a positive displacement injection pump (e.g., diaphragm pump) effectively designed and constructed of materials that are compatible with pesticides and capable of being fitted with a system interlock.

Do not apply when wind speed favors drift beyond the area intended for treatment.

Application Rates:

Use the appropriate rates of EPTAM 7-E and METRIBUZIN (Sencor, Lexone) as indicated in the following table:

Application Rates Per Broadcast Acre

	Pints EPTAM 7-E	Pints Sencor 4 to Lexone 4 or Pounds Sencor 50 or Lexone 50	Pounds Sencor Sprayule 75% or Lexone Dry Flowable 75%
COARSE SOILS Sand Sandy loam Loamy sand	3 1/2	1/2	1/3
FINE SOILS Loam Silt loam Sandy clay Clay loam	4 1/2	1/2-1	1/3-2/3

Use EPTAM 7-E plus METRIBUZIN (Sencor, Lexone) post-emergence only on white-skinned potato varieties that are not early maturing.

General Use Precautions

Before using EPTAM 7-E plus METRIBUZIN (Sencor, Lexone) as a tank mix, read and observe the cautionary statements, and all other information appearing in both product labels.

SUPPLEMENTAL LABELING

EPTAM 7-E SELECTIVE HERBICIDE

EPTAM 7-E/DUAL 8E* TANK MIX FOR WEED CONTROL IN DRY BEANS IN THE NORTHERN AND PACIFIC NORTHWEST REGIONS.

A tank mix combination of EPTAM 7-E plus DUAL 8E will give better weed control than either product used separately.

* DUAL 8E is the registered trademark of Ciba-Geigy Corp.

CAUTION: Read both the EPTAM 7-E and DUAL 8E labels carefully before using this tank mix. Observe all cautions and limitations on each label.

DIRECTIONS FOR USE

It is a violation of Federal law to use these products in a manner inconsistent with their labeling.

Apply the EPTAM 7-E/DUAL 8E tank mix as a pre-plant soil incorporated treatment.

Mixing: Add the recommended rates of both EPTAM 7-E and DUAL 8E to the spray tank during filling and mix thoroughly. Apply in 10 to 40 gallons of water per acre.

Spray Equipment: Use any properly calibrated low pressure boom type herbicide sprayer which will apply the material uniformly. Check calibration frequently during application and observe the nozzles to be sure they are delivering a uniform spray pattern.

Soil Incorporation: Immediately after spraying, the EPTAM 7-E and DUAL 8E combination must be incorporated thoroughly into the top 2 to 3 inches of soil. Spraying and incorporation should be accomplished in the same operation if possible by mounting the tank and boom directly on the incorporation equipment. Incorporate the herbicide into the soil using any of the following equipment:

- **Power Driven Cultivation Equipment** (recommended on all soil types) set to cut to a depth of 2 to 3 inches.
- **Tandem Discs** (recommended on all soil types) set to cut to a depth of 4 to 6 inches, operated at 4 to 6 mph followed by a spike-tooth harrow or some other leveling device which extends beyond the ends of the discs. For more thorough mixing (for perennial grasses and in heavier soils) disc in two different directions (cross disc).
- **Field Cultivators** (recommended for spring application on coarse textured soils, and for fall application on all soils. Use only on soils in good tilth). Use 3 to 4 rows of sweeps, spaced at 7 inch or less intervals and staggered so that no soil is left unturned, followed by a spike-tooth harrow pulled behind the cultivator. Do not use chisel plows to incorporate. Set the cultivator to cut 4 inches deep, operated at 5 mph or more. Run the equipment over the field twice, the second run at an angle to the first.

Application Rates: Use the appropriate rates of EPTAM 7-E and DUAL 8E as indicated in the following table:

Application Rates Per Broadcast Acre

Soil Type	EPTAM 7-E	DUAL 8E	
		Less than 3% Organic Matter	3% or Greater Organic Matter
Coarse (sand)	3½-4½ pints	1¼ pints	1½ pints
Medium (loam)	3½-4½ pints	1½ pints	2 pints
Fine (clay)	3½-4½ pints	2 pints	2-2½ pints

Planting: Seeding should be done as soon as possible after treatment to obtain a maximum period of weed control.

General Use Precautions

Before using EPTAM 7-E plus DUAL 8E as a tank mix, read and observe the cautionary statements, and all other information appearing in both product labels.

SUPPLEMENTAL LABELING

EPTAM 7-E SELECTIVE HERBICIDE

EPTAM 7-E/SONALAN E.C.* TANK MIX FOR WEED CONTROL IN DRY BEANS IN THE PACIFIC NORTHWEST REGION.

A tank mix combination of EPTAM 7-E plus SONALAN E.C. will give a broader spectrum of weed control than either product used separately.

* SONALAN is the registered trademark of Elanco Products, Inc.

Caution: Read both the EPTAM 7-E and SONALAN E.C. labels carefully before using this tank mix. Observe all cautions and limitations on each label.

In addition to the weeds listed on the label for EPTAM 7-E alone, the following annual broadleaf weeds can be controlled with an EPTAM 7-E/SONALAN E.C. tank mix:

Kochia (*Kochia scoparia*)

Russian Thistle (*Salsoa kali*)

Wild Buckwheat (*Polygonum convoluvlus*)

DIRECTIONS FOR USE

It is a violation of Federal law to use these products in a manner inconsistent with their labeling.

Apply the EPTAM 7-E/SONALAN E.C. tank mix as a pre-plant soil incorporated treatment.

Mixing: Add the recommended rates of both EPTAM 7-E and SONALAN E.C. to the spray tank during filling and mix thoroughly. Apply in 10 to 40 gallons of water per acre.

Spray Equipment: Use any properly calibrated low pressure boom type herbicide sprayer which will apply the material uniformly. Check calibration frequently during application and observe the nozzles to be sure they are delivering a uniform spray pattern.

Soil Incorporation: Immediately after spraying, the EPTAM 7-E and SONALAN E.C. combination must be incorporated thoroughly in the top 2 to 3 inches of soil. Spraying and incorporation should be accomplished in the same operation if possible by mounting the tank and boom directly on the incorporation equipment. Incorporate the herbicides into the soil using any of the following equipment:

- **Power Driven Cultivation Equipment** (recommended on all soil types) set to cut to a depth of 2 to 3 inches.
- **Tandem Discs** (recommended on all soil types) set to cut to a depth of 4 to 6 inches, operated at 4 to 6 mph followed by a spike-tooth harrow or some other leveling device which extends beyond the ends of the discs. For more thorough mixing (for perennial grasses and in heavier soils) disc in two different directions (cross disc).
- **Field Cultivators** (recommended for spring application on coarse textured soils, and for fall application on all soils. Use only on soils in good tilth). Use 3 to 4 rows of sweeps, spaced at 7 inch or less intervals and staggered so that no soil is left unturned, followed by a spike-tooth harrow pulled behind the cultivator. Do not use chisel plows to incorporate. Set the cultivator to cut 4 inches deep, operated at 5 mph or more. Run the equipment over the field twice, the second run at an angle to the first.

Application Rates: Use the appropriate rates of EPTAM 7-E and SONALAN E.C. as indicated in the following table:

Application Rates Per Broadcast Acre

Soil Type	EPTAM 7-E	SONALAN E.C.
Coarse (sand)	3½-4½ pints	1½-2 pints
Medium (loam)	3½-4½ pints	2-2½ pints
Fine (clay)	3½-4½ pints	2½-3 pints

Planting: Seeding should be done as soon as possible after treatment to obtain a maximum period of weed control.

General Use Precautions

Before using EPTAM 7-E plus SONALAN E.C. as a tank mix, read and observe the cautionary statements, and all other information appearing in both product labels.

SUPPLEMENTAL LABELING

EPTAM 7-E SELECTIVE HERBICIDE

EPTAM 7-E/LASSO 4-E* TANK MIX FOR WEED CONTROL IN DRY BEANS IN THE NORTHERN REGION

A tank mix combination of EPTAM 7-E plus LASSO 4-E will give a broader spectrum of weed control than either product used separately.

* LASSO is the registered trademark of Monsanto Agricultural Products Co.

Caution: Read both the EPTAM 7-E and LASSO 4-E labels carefully before using this mix. Observe all cautions and limitations on each label.

In addition to the weeds listed on the label for EPTAM 7-E alone, the following annual broadleaf weeds can be controlled with an EPTAM 7-E/LASSO 4-E tank mix.

Common Ragweed (*Ambrosia artemisiifolia*)

Pennsylvania Smartweed (*Polygonum pensylvanicum*)

DIRECTIONS FOR USE

It is a violation of Federal Law to use these products in a manner inconsistent with their labeling.

Apply the EPTAM 7-E/LASSO 4-E tank mix as a pre-plant soil incorporated treatment.

Mixing: Add the recommended rates of both EPTAM 7-E and LASSO 4-E to the spray tank during filling and mix thoroughly. Apply in 10 to 40 gallons of water per acre.

Spray Equipment: Use any properly calibrated low pressure boom type sprayer which will apply the material uniformly. Check calibration frequently during operation and observe nozzles to be sure they are delivering a uniform spray pattern.

Soil Incorporation: Immediately after spraying, the EPTAM 7-E and LASSO 4-E combination must be incorporated thoroughly into the top 2 to 3 inches of soil. Spraying and incorporation should be accomplsihed in the same operation if possible by mounting the tank and boom directly on the incorporation equipment. Incorporate the herbicides into the soil using any of the following equipment:

- **Power Driven Cultivation Equipment** (recommended on all soil types) set to cut to a depth of 2 to 3 inches.
- **Tandem Discs** (recommended on all soil types) set to cut to a depth of 4 to 6 inches, operated at 4 to 6 mph followed by a spike-tooth harrow or some other leveling device which extends beyond the ends of the discs. For more thorough mixing (for perennial grasses and in heavier soils) disc in two different directions (cross disc).
- **Field Cultivators** (recommended for spring application on coarse textured soils, and for fall application on all soils. Use only on soils in good tilth). Use 3 to 4 rows of sweeps, spaced at 7 inch or less intervals and staggered so that no soil is left unturned, followed by a spike-tooth harrow pulled behind the cultivator. Do not use chisel plows to incorporate. Set the cultivator to cut 4 inches deep, operated at 5 mph or more. Run the equipment over the field twice, the second run at an angle to the first.

Application Rates: Use 2 to 3 pints of EPTAM 7-E plus 4 to 6 pints of LASSO 4-E. Use only the 4 pint rate of LASSO 4-E in Michigan. Use the higher rates of herbicides for heavy weed infestations and hard-to-control weeds.

Planting: Seeding should be done as soon as possible after treatment to obtain a maximum period of weed control.

General Use Precautions

Before using EPTAM 7-E plus LASSO 4-E as a tank mix, read and observe the cautionary statements, and all other information appearing in both product labels.

SUPPLEMENTAL LABELING

EPTAM 7-E SELECTIVE HERBICIDE

EPTAM 7-E/TREFLAN*/AMIBEN** ON DRY FIELD BEANS

** Amiben is the registered trademark of Rhone Poulenc Inc.

* Treflan is the registered trademark of Elanco Products Co.

EPTAM herbicide may be applied as a preplant soil incorproated tank mixture with Amiben and Treflan herbicides, or a tank mix of EPTAM and Treflan may be applied preplant soil incorporated and followed by a pre-emergence application of Amiben for weed control in dry beans in the Northern region of the U.S. When Amiben is applied pre-emergence, effectiveness depends on adequate rainfall after treatment. When applied according to directions, this three-way combination will give increased control of broadleaf weeds listed on the EPTAM label as well as control of additional broadleaf weeds and labeled grasses.

Weeds Controlled

Annual Grasses:

Annual Bluegrass	Johnsongrass seedlings
Barnyardgrass (Watergrass, Junglerice)	Lovegrass (Stinkgrass)
	Sandbur, Field
Bromegrass	Signalgrass, (Brachiaria)
Crabgrass	Sprangletop
Cheat	Texan Panicum†
Foxtails (Giant, Green, Yellow)	Shattercane
Goosegrass	Wild Oats

Broadleaf Weeds:

Annual Morningglory	Nettleleaf Goosefoot
Bindweed, Field	Nightshade (Black, Hairy)
Carpetweed	Pigweed
Chickweed	Puncturevine
Corn Spurry	Purslane
Florida Pusley	Ragweed
Henbit	Russian Thistle
Knotweed	Shepherdspurse
Kochia	Smartweed
Lambsquarters	Stinging Nettle

Perennial Weeds:

Nutsedge† (Purple, Yellow)

† Suppression (partial control)

DIRECTIONS FOR USE

It is a violation of Federal law to use this product in a manner inconsistent with its labeling.

Dry Field Beans

Make treatments according to the following table:

Chemical	Timing	Amounts of Formulation Per Acre
EPTAM 7-E	Preplant Soil Incorporated	2.5 pints
+		
Treflan* 4-E	Preplant Soil Incorporated	1 pint
+		
Amiben**	Preplant Soil Incorporated or Pre-emergence (surface)	6-8 pints

EPTAM/Amiben**/Treflan* or the EPTAM/Treflan tank mixes should be incorporated immediately after application with a field cultivator, tandem disc, mulch treaders, rolling cultivator or a power driven tiller. Thoroughly mix into the top 2-3 inches by incorporating twice in different directions. Refer to complete product labels for further information on incorporation procedures.

Apply in ten or more gallons of water per acre. Prepare the spray mixture by first filling the spray tank 1/2 to 2/3 full of water. With agitator running slowly add Amiben, followed by EPTAM and Treflan. Do not mix the concentrate before adding to water. Fill tank with remaining required water.

If sufficient rain does not occur within five (5) days after pre-emergence application of Amiben, a light cultivation or incorporation with a rotary hoe or similar tool is recommended.

General Use Precautions

Before using the tank mix of EPTAM/Amiben/Treflan, read and observe the cautionary statements and all other information on the EPA registered EPTAM 7-E, Treflan 4-E and Amiben labels. This labeling must be in the possession of the user at time of application.

SUPPLEMENTAL LABELING

EPTAM 7-E/RO-NEET 7-E TANK MIX FOR PRE-PLANT USE IN SUGARBEETS

DIRECTIONS FOR USE

It is a violation of Federal law to use this product in a manner inconsistent with its labeling.

FOR MICHIGAN, MINNESOTA AND THE RED RIVER VALLEY AREA OF NORTH DAKOTA

The EPTAM 7-E and RO-NEET 6-E combination is a selective tank mix which must be incorporated into the soil. This combination controls weeds by interfering with normal germination and seedling development. This tank mixture does not control established weeds. The EPTAM/RO-NEET combination will give equal to or greater control of the weeds listed on this label than either product used separately. This tankmixture can be applied only once per growing season. The combination may be used in the fall or in the spring.

Application Directions

During filling, pour the recommended rates of both EPTAM 7-E and RO-NEET 6-E into a properly calibrated, low-pressure boom sprayer having good agitation, and mix thoroughly. Apply the material uniformly in 10 to 50 gallons of water per acre. Check calibration frequently during application and observe the nozzles to ensure a uniform spray pattern. Do not use single nozzle boom-jet type sprayers. The soil should be well-worked prior to application and dry enough to permit thorough mixing with incorporation equipment.

Spring Pre-Plant Incorporated Application Rates*
(Pints/Acre)

Soil Texture	Organic Matter %	EPTAM 7-E	RO-NEET 6-E
Loam or	3-4.5	1.25	3.3
Coarser**	4.5>	1.25	4.0
Loam to	3-4.5	1.25	3.3
Clay Loam	4.5>	1.75	3.3
Clay Loam	3-4.5	1.75	3.3
or Finer	4.5>	2.25	3.3

*Make only one application per growing season. Do not apply the EPTAM 7-E/RO-NEET 6-E combination in the spring if either this combination or RO-NEET 7-E alone was applied in the fall.

**EPTAM 7-E should not be used on Loam or Coarser soils with less than 3% organic matter. RO-NEET 6-E at 4 to 5 1/3 pints/Acre should be used on these soils.

Fall Pre-Plant Incorporated Application Rates*
(Pints/Acre)

Soil Texture	Organic Matter %	EPTAM 7-E	RO-NEET 6-E
Loam or	3-4.5	1.25	4.0
Coarser	4.5>	1.25	4.0
Loam to	3-4.5	1.25	4.0
Clay Loam	4.5>	1.75	4.0
Clay Loam	3-4.5	2.25	2.67
or Finer	4.5>	2.25	4.0

*Make only one application per growing season. Do not apply the EPTAM 7-E and RO-NEET 6-E combination in the spring if either this combination or RO-NEET 7-E alone was applied in the fall.

Soil Incorporation

The EPTAM 7-E and RO-NEET 6-E tankmixture must be immediately incorporated (mixed) into the top 2 to 3 inches of soil after spraying to prevent loss of the herbicides. Whenever possible, application and incorporation should be done in the same operation. Refer to the EPTAM 7-E and RO-NEET 6-E labels for complete incorporation procedures and the use of equipment which has been proven to incorporate thoroughly to the recommended depth.

General Use Precautions

The EPTAM 7-E and RO-NEET 6-E combination may cause crop injury on very light sandy soil and when used under adverse environmental conditions that weaken crop seedlings. Refer also to EPTAM 7-E and RO-NEET 6-E labels for recommendations regarding cultural practices after application.

A tankmixture of EPTAM 7-E and RO-NEET 6-E will give equal to or greater control of the following listed weeds than either product used separately. This combination does not control established weeds.

Green foxtail	*(Setaria viridis)*
Yellow foxtail	*(Setaria glauca)*
Wild Oats	*(Avena fatua)*
Yellow nutsedge	*(Cyperus esculentus)*
Purple nutsedge	*(Cyperus rotundus)*
Lambsquarters, common*	*(Chenopodium album)*
Pigweed, redroot*	*(Amaranthus retroflexus)*

*Partial control only (suppression)

Read both the EPTAM 7-E and RO-NEET 6-E labels carefully before using this tank mix. Observe all cautions and limitations on each label. The most restrictive labeling applies in tank mixes.

SUPPLEMENTAL LABELING

EPTAM 7-E SELECTIVE HERBICIDE

EPTAM 7-E/PROWL* 4-E TANK MIX

For Weed Control in Dry Beans

A tank mix combination of EPTAM 7-E plus PROWL* 4-E will give a broader spectrum of weed control than either product used separately.

* PROWL is the registered trademark of American Cyanamid Co.

Caution: Read both the EPTAM 7-E and PROWL 4-E labels carefully before using this tank mix. Observe all cautions and limitations on each label. Do not use EPTAM 7-E plus PROWL 4-E on adzuki beans, cowpeas, (blackeye peas, blackeye beans), soybeans, lima beans, mung beans or other flat-podded beans except Romano. Under abnormal weather conditions stunting may occur on Gratiot, Michilite, Sanilac, Seafarer, and Seaway varieties. Do not exceed 3½ pints EPTAM 7-E per acre on small white beans grown on coarse textured soils.

In addition to the weeds listed on the label for EPTAM 7-E alone, the following annual broadleaf weeds can be controlled with an EPTAM 7-E/PROWL 4-E tank mix:

Annual Spurge (*Euphorbia* spp.)
Kochia (*Kochia scoparia*)

DIRECTIONS FOR USE

It is a violation of Federal law to use these products in a manner inconsistent with their labeling.

Apply the EPTAM 7-E/PROWL 4-E tank mix as a pre-plant soil incorporated treatment.

Mixing: Add the recommended rates of both EPTAM 7-E and PROWL 4-E to the spray tank during filling and mix thoroughly. Apply in 10 to 40 gallons of water per acre.

Spray Equipment: Use any properly calibrated low pressure boom type herbicide sprayer which will apply the material uniformly. Check calibration frequently during application and observe the nozzles to be sure they are delivering a uniform spray pattern.

Soil Incorporation: Immediately after spraying, the EPTAM 7-E and PROWL 4-E combination must be incorporated thoroughly into the top 2 to 3 inches of soil. Spraying and incorporation should be accomplished in the same operation if possible by mounting the tank and boom directly on the incorporation equipment.

For semi-arid areas of Eastern Washington, Eastern Oregon and Idaho only: When application and incorporation are done in separate operations, EPTAM 7-E and PROWL 4-E must be incorporated the same day as applied. Application must be made on a dry soil surface (at least ½ inch deep) free from dew and incidental moisture.

Incorporate the herbicides into the soil using any of the following equipment:

- POWER DRIVEN CULTIVATION EQUIPMENT (recommended on all soil types) set to cut to a depth of 2 to 3 inches.
- TANDEM DISCS (recommended on all soil types) set to cut to a depth of 4 to 6 inches, operated at 4 to 6 mph followed by a spike-tooth harrow or some other leveling device which extends beyond the ends of the discs. For more thorough mixing (for perennial grasses and in heavier soils) disc in two different directions (cross disc).
- FIELD CULTIVATORS (recommended for Spring application on coarse textured soils, and for Fall application on all soils. Use only on soils in good tilth). Use 3 to 4 rows of sweeps, spaced at 7-inch or less intervals and staggered so that no soil is left unturned, followed by a spike-tooth harrow pulled behind the cultivator. Do not use chisel plows to incorporate. Set the cultivator to cut 4 inches deep, operated at 5 mph or more. Run the equipment over the field twice, the second run at an angle to the first.

Application Rates: Use the appropriate rates of EPTAM 7-E and PROWL 4-E as indicated in the following table:

Application Rates Per Broadcast Acre*
Western, Southwestern and Southeastern Regions

Soil Type	EPTAM 7-E	PROWL 4-E
Coarse (sandy)	2½-4½ pints	1-1½ pints
Medium (loam)	3-4½ pints	1½-2 pints
Fine (clay)	3-4½ pints	1½-3 pints

Northern Region

Soil Type	Organic Matter	EPTAM 7-E	PROWL 4-E
Coarse (sandy)	3% or less	2½-4 pints	1-1½ pints
	more than 3%	2½-4 pints	1½ pints
Medium (loam)	3% or less	3-4½ pints	1½-2 pints
	more than 3%	3-4½ pints	1½-2 pints
Fine (clay)	3% or less	3-4½ pint	1½-2 pints
	more than 3%	3-4½ pints	2-2½ pints
Pacific Northwest Region			
Coarse (sandy)	3% or less	3½-4½ pints	1-2 pints
	more than 3%	3½-4½ pints	2 pints
Medium (loam)	3% or less	3½-4½ pints	1½-2½ pints
	more than 3%	3½-4½ pints	2½-3 pints
Fine (clay)	3% or less	3½-4½ pints	2-3 pints
	more than 3%	3½-4½ pints	3 pints

* Use the higher recommended rate of EPTAM 7-E where hairy nightshade, black nightshade or nutsedge are present.

Planting: Seeding should be done as soon as possible after treatment to obtain a maximum period of weed control.

General Use Precautions

Before using EPTAM 7-E plus PROWL 4-E as a tank mix, read and observe the cautionary statements, and all other information appearing in both product labels.

NOTICE BUYER AND USER:

Seller warrants that this product conforms to the chemical description on the label and is reasonably fit for the purposes stated on the label when used in accordance with directions under normal conditions of use. This warranty does not extend to the use of this product contrary to label instructions, or under abnormal use conditions, or under conditions not reasonably foreseeable to Seller, and Buyer and User assume the risk of any such use. SELLER DISCLAIMS ALL OTHER WARRANTIES EXPRESSED OR IMPLIED INCLUDING ANY WARRANTY OF FITNESS OR MERCHANTABILITY. SELLER SHALL NOT BE LIABLE FOR CONSEQUENTIAL, SPECIAL OR INDIRECT DAMAGES RESULTING FROM THE USE OR HANDLING OF THIS PRODUCT AND SELLER'S SOLE LIABILITY AND BUYER'S AND USER'S EXCLUSIVE REMEDY SHALL BE LIMITED TO THE REFUND OF THE PURCHASE PRICE.

This product is sold only for uses stated on its label. No express or implied license is granted to use or sell this product under any patent in any country except as specified: Country: United States of America.

A-10MR

The information and specimen labels in this booklet are correct as of the publication date but are subject to change without notice. For current information, contact ICI Americas Inc.

ERADICANE® 6.7-E

SELECTIVE HERBICIDE

for control of weeds in corn (field, sweet, silage and popcorn) including Foxtails, Seedling Johnsongrass, Nutsedge, Bermudagrass, and Suppression of Woolly Cupgrass and Wild Proso Millet

ACTIVE INGREDIENT:	
S-ethyl dipropylthiocarbamate	82.6%
INERT INGREDIENTS:	17.4%
TOTAL	100.0%

Contains 6.7 lbs. Active Ingredient Per Gallon

EPA Reg. No. 10182-223

KEEP OUT OF REACH OF CHILDREN

CAUTION

Harmful if Swallowed

STATEMENT OF PRACTICAL TREATMENT

First Aid

Immediately start the procedures below. If further treatment is required contact a Poison Center, a physician, or the nearest hospital.

If swallowed: Immediately give several glasses of water but *do not* induce vomiting. If vomiting does occur, give fluids again. Have a physician determine if condition of patient will permit induction of vomiting or evacuation of stomach. Do not give anything by mouth to an unconscious or convulsing person.

For eye contact: Immediately flush eyes with large amounts of running water for at least 15 minutes. Hold eyelids apart to ensure rinsing of the entire surface of the eye and lids with water. Get medical attention immediately.

For skin contact: Flush all affected areas with plenty of water for several minutes. Remove and clean contaminated clothing and shoes. Seek medical attention if irritation occurs.

If inhaled: Remove to fresh air. Seek medical attention if respiratory irritation occurs or if breathing becomes difficult.

FOR 24-HOUR EMERGENCY MEDICAL ASSISTANCE CALL 1-800-F-A-S-T-M-E-D (327-8633).
FOR CHEMICAL EMERGENCY: Spill, leak, fire, exposure, or accident call CHEMTREC 1-800-424-9300.

PRECAUTIONARY STATEMENTS

HAZARDS TO HUMANS AND DOMESTIC ANIMALS
CAUTION

Harmful if Swallowed—Avoid contact with skin, eyes and clothing—Avoid inhalation of spray mist—Wear rubber gloves and fresh clothing—Wear goggles to avoid contact with eyes—Wash with soap and water immediately after use or contact—Do not contaminate food or feed.

Do not apply this product in such a manner as to directly or through drift, expose workers or other persons. When applied through the sprinkler irrigation method of application, the area being treated must be vacated by unprotected persons.

ENVIRONMENTAL HAZARDS—Keep out of any body of water—Do not contaminate water by cleaning of equipment or disposal of wastes.

PHYSICAL AND CHEMICAL HAZARDS—Keep away from heat and flame.

Can be stored at temperatures as low as minus 50°F.

GENERAL USE PRECAUTIONS

READ ALL LABEL DIRECTIONS BEFORE USING.

ERADICANE 6.7-E should be used for recommended purposes and at recommended rates. (DO NOT OVERDOSE.)

ERADICANE 6.7-E is recommended for use on mineral soils only (soils containing less than 10% organic matter).

Do not use on milo or sorghum. Do not use on corn seed stock (Breeders, Foundation or Increase). ERADICANE 6.7-E *may* be used on Production seed corn.

Do not contaminate water to be used on susceptible crops and ornamentals, or for domestic purposes. Do not contaminate feed or food. Keep container tightly closed when not in use. Do not store near seeds or fertilizers.

When properly applied and weather conditions exist for normal plant growth through the season, ERADICANE 6.7-E will not harm the treated crop nor should harmful soil residues remain beyond harvest. However, during germination and early growth, extended periods of unusually cold and wet or hot and dry weather, insect, nematode, or plant disease attack, carry-over soil residues of certain persistent herbicides, the use of certain soil applied systemic insecticides may create abnormal conditions that weaken crop seedlings. ERADICANE 6.7-E used under these abnormal conditions could result in crop injury.

CONTINUOUS USE PRECAUTION. The use of ERADICANE 6.7-E in the same fields, consecutively year after year (corn on corn) can in some areas of North America lead to a decrease in the level of performance. Therefore, as a precaution, it is recommended that ERADICANE 6.7-E treated fields be rotated with another crop or herbicide classification on an annual basis. This precaution does not pertain to the Western Region (refer to regional map).

Weeds Controlled By ERADICANE 6.7-E

ERADICANE 6.7-E will not control established weeds.

Annual Grasses:	
Barnyardgrass (Watergrass, Junglerice)	(*Echinochloa* spp.)
Bermudagrass—seedlings	(*Cynodon dactylon*)
Bluegrass, annual	(*Poa annua*)
Crabgrass	(*Digitaria* spp.)
Cupgrass, woolly*	(*Eriochloa villosa*)
Foxtail, giant	(*Setaria faberi*)
Foxtail, green	(*Setaria viridis*)
Foxtail, yellow	(*Setaria lutescens*)
Goosegrass	(*Eleusine indica*)
Johnsongrass—seedlings	(*Sorghum halepense*)
Millet, wild proso*‡	(*Panicum miliaceum*)
Oats, wild	(*Avena fatua*)
Panicum, fall	(*Panicum dichotomiflorum*)
Panicum, Texas (Coloradograss)*†	(*Panicum texanum*)
Rescuegrass	(*Bromus willdenowii*)
Rice, red	(*Oryza sativa*)
Ryegrass, Italian (Annual)	(*Lolium multiflorum*)
Sandbur, field*	(*Cenchrus pauciflorus*)
Signalgrass	(*Brachiaria* spp.)
Stinkgrass (Lovegrass)	(*Eragrostis cilianensis*)

Other volunteer grains, (i.e., barley, oats, wheat)

* Cultivation is suggested in addition to ERADICANE 6.7-E treatment in fields with moderate to heavy infestations of seedling Johnsongrass, field sandbur, Texas panicum, woolly cupgrass or wild proso millet.

† Southeastern U.S. Only.

‡ Central Region only—Partial control (suppression)

Annual Broadleaf Weeds:	
Carpetweed	(*Mollugo verticillata*)
Chickweed, common	(*Stellaria media*)
Goosefoot, nettleleaf	(*Chenopodium murale*)
Fiddleneck	(*Amsinckia* spp.)
Henbit (Deadnettle)	(*Lamium amplexicaule*)
Lambsquarters, common	(*Chenopodium album*)
Morningglory, tall	(*Ipomoea purpurea*)
Nightshade, black	(*Solanum nigrum*)
Nightshade, hairy	(*Solanum sarachoides*)
Pigweed, redroot (Common Pigweed)	(*Amaranthus retroflexus*)
Pigweed, prostrate	(*Amaranthus blitoides*)
Pigweed, tumble	(*Amaranthus albus*)
Puncturevine**	(*Tribulus terrestris*)
Purslane, common	(*Portulaca oleracea*)
Pusley, Florida	(*Richardia scabra*)
Sicklepod	(*Cassia obtusifolia*)
Sida, prickly (teaweed)	(*Sida spinosa*)
Shepherdspurse	(*Capsella bursa-pastoris*)
Spurry, corn	(*Spergula arvensis*)

** Western Region Only

The annual broadleaf weeds listed above will be controlled only if treatment is made when conditions are favorable for weed seed germination and growth. Under prolonged conditions of very cold soil, control of the above broadleaf weeds may not be adequate.

Perennial Grasses:	
Bermudagrass	(*Cynodon dactylon*)
Nutsedge, purple	(*Cyperus rotundus*)
Nutsedge, yellow	(*Cyperus esculentus*)
Quackgrass*	(*Agropyron repens*)

* Partial Control (Suppression) Only

Perennial weeds must be turned under and chopped up thoroughly prior to treatment. The underground rhizomes of quackgrass and the rhizomes and stolons of bermudagrass must be cut up thoroughly into small pieces so that 4 or less nodes remain on a strand.

Consult RECOMMENDATIONS for ERADICANE 6.7-E rates necessary to control the above listed weeds.

DIRECTIONS FOR USE

It is a violation of Federal law to use this product in a manner inconsistent with its labeling.

ERADICANE 6.7-E is a selective herbicide which must be mixed (incorporated) into the soil for control of weeds listed on this label. ERADICANE 6.7-E controls weeds by interfering with normal germination and seedling development. ERADICANE 6.7-E does not control established weeds.

APPLICATION DIRECTIONS

Pour the recommended amount of ERADICANE 6.7-E into the spray tank during the filling operation. Apply in 10 to 50 gallons of water per acre using a property calibrated, low-pressure sprayer having good agitation. The soil should be dry enough to permit good soil mixing (incorporation). See Appendix I for soil mixing directions.

For band applications, calculate the amount to be applied per acre as follows:

$$\frac{\text{Band Width in Inches}}{\text{Row Width in Inches}} \times \text{Rate/A for Broadcast Treatment} = \text{Amount Needed for Band Treatment}$$

COMBINATION WITH FLUID FERTILIZERS

ERADICANE 6.7-E may be combined with fluid (solution, slurry or suspension) fertilizers. However, physical compatibility with these fertilizers must be verified before combining in the spray tank. See Appendix II for special directions regarding these combinations. Even though ERADICANE 6.7-E may be found compatible with fluid fertilizers, constant agitation is necessary to maintain uniformity of this mixture during application.

IMPREGNATION ON DRY BULK FERTILIZERS

ERADICANE 6.7-E may be impregnated on dry fertilizer and applied as the fertilizer is spread. See Appendix III for information and directions regarding impregnation and use of these combinations.

CENTER PIVOT SPRINKLER APPLICATION

Apply this product only through center pivot, sprinkler irrigation systems. Do not apply this product through any other type of irrigation system.

Crop injury, lack of effectiveness, or illegal pesticide residues in the crop can result from nonuniform distribution of treated water.

If you have questions about calibration, you should contact State Extension Service specialists, equipment manufacturers or other experts.

Do not connect an irrigation system (including greenhouse systems) used for pesticide application to a public water system unless the pesticide label-prescribed safety devices for public water systems are in place.

A person knowledgeable of the chemigation system and responsible for its operation or under the supervision of the responsible person, shall shut the system down and make necessary adjustments should the need arise.

Apply and soil incorporate ERADICANE 6.7-E, ERADICANE 6.7-E/ATRAZINE, or ERADICANE 6.7-E/BLADEX* (see TANK MIX RECOMMENDATIONS) with irrigation water before planting (see PLANTING), or immediately after planting before either corn or weeds emerge. It is important that application and incorporation be done immediately after the last tillage operation since poor results will occur if the weeds have germinated before application of the herbicide. Meter the ERADICANE 6.7-E or tank mix combinations with ATRAZINE or BLADEX into the irrigation water during the entire period. Apply in ½ to ¾ inches of water.

* BLADEX is a registered trademark of E.I. du Pont de Nemours & Co., Inc.

USE PRECAUTIONS:

The system must contain a functional check valve, vacuum relief valve, and low pressure drain appropriately located on the irrigation pipeline to prevent water source contamination from backflow.

The pesticide injection pipeline must contain a functional, automatic, quick-closing check valve to prevent the flow of fluid back toward the injection pump.

The pesticide injection pipeline must also contain a functional, normally closed, solenoid-operated valve located on the intake side of the injection pump and connected to the system interlock to prevent fluid from being withdrawn from the supply tank when the irrigation system is either automatically or manually shut down.

The system must contain functional interlocking controls to automatically shut off the pesticide injection pump when the water pump motor stops.

The irrigation line or water pump must include a functional pressure switch which will stop the water pump motor when the water pressure decreases to the point where pesticide distribution is adversely affected.

Systems must use a metering pump, such as a positive displacement injection pump (e.g., diaphragm pump) effectively designed and constructed of materials that are compatible with pesticides and capable of being fitted with a system interlock.

Application of more than recommended quantities of irrigation water per acre may result in decreased product performance by removing the chemical from the zone of effectiveness.

Do not apply when wind speed favors drift beyond the area intended for treatment, when system connections or fittings leak, when nozzles do not provide uniform distribution or when lines containing the product must be dismantled and drained.

Greater accuracy in calibration and distribution will be achieved by injecting a larger volume of a more dilute slurry per hour.

Example:

110 Acres to be covered by 4¾ pints (4 lbs. active) ERADICANE 6.7-E per acre.

110 A. × 4 lbs. = 440 lbs. active ERADICANE 6.7-E.

$\frac{440}{6.7} = 65.6$ gals. ERADICANE 6.7-E

Mixing tank size 750 gallons. Sprinkler takes 30 hours to apply ½ inch of water.

Injection rate $= \frac{750}{30} = 25$ gal./hr. or 0.4 gal./min.

Calibrate injection pump to deliver the herbicide mix into the irrigation system at the rate of *25* gal. per hr. or *0.4* gal. per minute.

Planting

Plant seed to a maximum depth of 2 inches within 2 weeks after treatment. To avoid removing ERADICANE 6.7-E from the row (with loss of weed control in the row), do not move or shape soil after treatment such as by planting in deep furrows.

Cultural Practices Following Application

Should weeds develop, a shallow cultivation or rotary hoeing will generally result in better weed control. When cultivating for any reason, it should be shallow, i.e. no more than ½ the depth the herbicide was incorporated. A sequential application of 2,4-D, BANVEL* 4E or a pre-emergence or post-emergence herbicide may be necessary to control weeds resistant to ERADICANE 6.7-E.

* BANVEL is a registered trademark of Sandoz Crop Protection Corp.

RECOMMENDATIONS

Consult the U.S. regional map below and use the recommendations in your region only.

Important: Limit use on sweet corn to 4¾ pints ERADICANE 6.7-E per acre with the exception of Minnesota and Wisconsin where up to full rate (7⅓ pints) can be used.

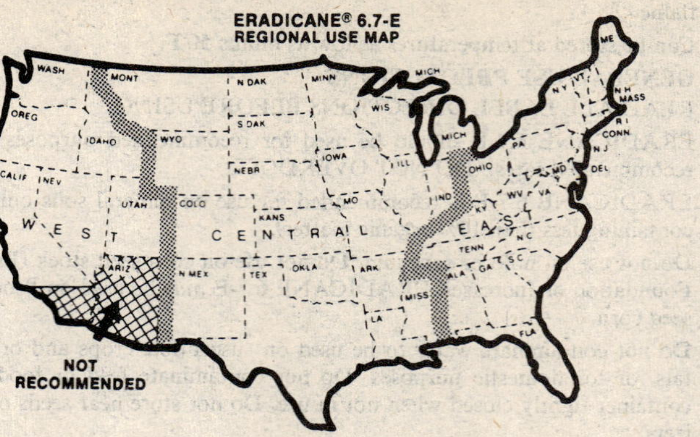

Eastern U.S.

(Note: In the Southeastern U.S. when ERADICANE 6.7-E is used for weed control in silage corn, do not seed small seeded grains after corn harvest until September.)

Central U.S.

(Note: Western slope of Colorado: Where furrow irrigation will be used when corn is flat planted, do not form furrows until corn is in the three-leaf stage.)

Western U.S.

(Note: ERADICANE 6.7-E is not recommended on corn grown in Arizona or in the 10 southernmost California counties. In California, do not use ERADICANE 6.7-E on corn planted after May 1.)

Note: Soils lacking sufficient moisture for seed germination must be preirrigated prior to ERADICANE 6.7-E application.

Note: Do not use on "Golden Jubilee" sweet corn west of the Cascade Mountains in Oregon or Washington.

Preplant Incorporation:

ERADICANE 6.7-E Broadcast Rate Per Acre in Pints

WEED	EASTERN U.S.	CENTRAL U.S.	WESTERN U.S.	RESTRICTIONS
Annual Grasses & Annual Broadleaf Weeds	4¾ to 7⅓	4¾ to 7⅓	3¾ to 6*	Use the lower rate for light infestations and in sand and light (coarse textured) soils.
Bermudagrass	4¾ to 7⅓	4¾ to 7⅓	3¾ to 7⅓*	Use the higher rate for heavy infestations and in heavier soils.
Nutsedge	4¾ to 7⅓	4¾ to 7⅓	3¾ to 7⅓	
Quackgrass	7⅓	7⅓	7⅓	For suppression. See tank mix section.
Sandbur, Field	4¾ to 7⅓	4¾ to 7⅓	4¾ to 7⅓	
Wild Oat	3¾	3¾	3¾ to 6*	
Woolly Cupgrass & Wild Proso Millet (suppression)	4¾ to 7⅓	4¾ to 7⅓	4¾ to 7⅓	Suppression. Use higher rate for heavy infestations and heavier soils.

Use the lower rates on sand and light (coarse textured) soils. Use the higher rates on heavy (fine textured) soils.

* Maximum rate in California is 4¾ pints.

Center Pivot Sprinkler Application

Apply 4¾ to 7⅓ pints of ERADICANE 6.7-E per acre for control of annual grasses and annual broadleaf weeds. ATRAZINE may be tank mixed with ERADICANE 6.7-E for sprinkler application (see TANK MIX RECOMMENDATIONS).

Subsurface Injection

(Southeastern U.S. with coarse textured soils only) Apply 4¾ pints of ERADICANE 6.7-E for control of light infestations of annual grasses, annual broadleaf weeds and nutsedge. Adjust the dosage proportionately depending on the row spacing and width to be treated. Recommended on lighter (coarse textured) soils only.

TANK MIX COMBINATIONS

For control of additional weeds and for increased control of labelled broadleaf weeds, ATRAZINE, BLADEX, PRINCEP, or a combination of ATRAZINE and BLADEX may be tank mixed with the recommended rates of ERADICANE 6.7-E and applied preplant incorporated. ERADI-

CANE 6.7-E/ATRAZINE may be applied through a center pivot irrigation system.

As an alternative, ERADICANE 6.7-E can be applied preplant incorporated at recommended rates followed by ATRAZINE applied pre-emergence to soil surface. Consult the pre-emergence application directions on the ATRAZINE label. A rotary hoeing or shallow cultivation is recommended if rainfall or sprinkler irrigation has not occurred within 10 days of the surface application. For enhanced or additional broadleaf weed control, an overlay treatment of 2,4-D may be made.

With the combination of ERADICANE 6.7-E/BLADEX/ATRAZINE, the potential hazard of ATRAZINE residues affecting certain crops the following year is reduced because a lower rate of ATRAZINE can be used.

Additional Weeds Controlled By ERADICANE 6.7-E/ATRAZINE Tank Mix and ERADICANE 6.7-E/BLADEX Tank Mix

Annual Grasses:	
Witchgrass	(*Panicum capillare*)
Annual Broadleaf Weeds:	
Cocklebur, common	(*Xanthium strumarium*)
Jimsonweed	(*Datura stramonium*)
Kochia	(*Kochia scoparia*)
Mallow, Venice	(*Hibiscus trionum*)
Mustard	(*Brassica* spp.)
Ragweed	(*Ambrosia* spp.)
Smartweed	(*Polygonum* spp.)
Thistle, Russian*	(*Salsola kali*)
Velvetleaf	(*Abutilon theophrasti*)

* Controlled by ERADICANE 6.7-E/Atrazine tank mix only.

ERADICANE 6.7-E/Atrazine tank mix will also improve control of broadleaf weeds listed under "Weeds Controlled by ERADICANE 6.7-E" if unfavorable conditions exist.

Additional Weeds Controlled By ERADICANE 6.7-E/PRINCEP Tank Mix

Annual Broadleaf Weeds:	
Cocklebur, common	(*Xanthium strumarium*)
Jimsonweed	(*Datura stramonium*)
Ragweed	(*Ambrosia* spp.)
Smartweed	(*Polygonum* spp.)
Thistle, Russian	(*Salsola Kali*)
Velvetleaf	(*Abutilon theophrasti*)

ERADICANE 6.7-E/PRINCEP tank mix will control these additional weeds and improve the control of broadleaf weeds listed under "Weeds Controlled by ERADICANE 6.7-E," if unfavorable conditions exist.

* PRINCEP is a registered trademark of Ciba Geigy Corporation.

Application Directions

Apply ERADICANE 6.7-E/ATRAZINE, ERADICANE 6.7-E/BLADEX, ERADICANE 6.7-E/PRINCEP or ERADICANE 6.7-E/BLADEX/ATRAZINE in 20-50 gallons of water per acre. It is recommended that the compatibility of tank mix combinations be tested on a small scale, for example, a jar test before actual tank mixing. See Appendix II which describes the procedures for such tests.

RECOMMENDATIONS

Broadcast Rates Per Acre:

Atrazine: Use the recommended rates of ERADICANE 6.7-E (in the Eastern and Central U.S. use 4$^3/_4$ to 7$^1/_3$ pints) plus 1$^1/_4$ to 2 pounds of ATRAZINE 80W or 2 to 3 pints of ATRAZINE 4L per acre for both tank mix (preplant incorporation or center pivot sprinkler application) and sequential application. Use the higher rate of ATRAZINE in soils with over 5% organic matter.

Bladex: Use the recommended rates of ERADICANE 6.7-E (in the Eastern and Central U.S. use 4$^3/_4$ to 7$^1/_3$ pints) plus 2 to 2$^1/_2$ pounds of BLADEX 80W or 3 to 4 pints BLADEX 4L per acre for preplant incorporation. Use the lower rate of BLADEX on sand or loam soils. Use the higher rate of BLADEX on fine textured (clay) soils or soils with 3% organic matter and above.

Princep: Use 1$^1/_4$ to 3$^3/_4$ pounds of PRINCEP 80W, 1$^1/_{10}$ to 3$^1/_3$ pounds of PRINCEP CALIBER* 90 or 2 to 6 pints of PRINCEP 4L per acre with recommended rates of ERADICANE 6.7-E preplant incorporated. Use the higher rate of PRINCEP on fine-textured soils and where heavy broadleaf weed infestations are expected.

* CALIBER is a registered trademark of Ciba Geigy Corporation.

Bladex/Atrazine: Use the recommended rate of ERADICANE 6.7-E (in the Eastern and Central U.S. use 4$^3/_4$ to 7$^1/_3$ pints) plus 1$^1/_4$ to 2$^1/_2$ pounds of BLADEX 80W or 2 to 4 pints of ATRAZINE 4L per acre. BLADEX and ATRAZINE should be used in a two for one proportion.

2,4-D or Banvel 4E: Use the amount of 2,4-D or BANVEL 4E recommended on the manufacturer's label.

Note: BLADEX is not recommended for use on sandy soils consisting of more than 70% sand and containing less than 1% organic matter. Consult the BLADEX label for more specific recommended rates on sandy soils.

Follow all the use precautions and warnings that appear on the ATRAZINE, BLADEX, PRINCEP, BANVEL and 2,4-D labels and supplemental literature.

Make only one application per crop. After a treatment including ATRAZINE, do not plant to any crop except corn until the following year or injury may occur.

Do not plant to alfalfa where more than 1.2 pounds actual ATRAZINE was used the previous year.

PRECAUTIONS TO FOLLOW WHEN USING ERADICANE 6.7-E/PRINCEP:

Do not rotate to any crop except corn until the following year or injury may occur.

After harvest, plow and thoroughly till the soil in fall or spring to minimize possible injury to spring-seeded rotational crops, regardless of rate used.

Do not apply ERADICANE 6.7-E/PRINCEP 80W, ERADICANE 6.7-E/PRINCEP CALIBER 90 or ERADICANE 6.7-E/PRINCEP 4L in the High Plains and Intermountain areas of the West (including central and western KS, western NE, western OK, and the Panhandle of TX) where rainfall is sparse and erratic or where irrigation is required.

Injury may occur to soybeans planted in northcentral and northwest IA, southcentral and southwest MN, northeast NE, southeast SD, and other areas the year following application on soils having a calcareous surface layer.

Do not plant sugar beets, tobacco, vegetables (including dry beans), spring-seeded small grains, or small-seeded legumes and grasses the year after ERADICANE 6.7-E/PRINCEP 80W, ERADICANE 6.7-E/PRINCEP CALIBER 90 or ERADICANE 6.7-E/PRINCEP 4L application or injury may occur.

Quackgrass: For improved quackgrass suppression ERADICNE 6.7-E can be preceded by a preplow application of ATRAZINE. Broadcast 2$^1/_2$ lbs. ATRAZINE 80W or 2 quarts ATRAZINE 4L per acre in the fall or spring and plow 1-3 weeks later. In the spring, follow with the recommended rate of ERADICANE 6.7-E.

Velvetleaf: For control of velvetleaf use 4$^3/_4$ to 7$^1/_3$ pints of ERADICANE 6.7-E plus 2 pounds of ATRAZINE 80W or 3 pints of ATRAZINE 4L per broadcast acre.

Tank Mixing Directions

1. Fill a thoroughly rinsed and uncontaminated spray tank at least $^2/_3$ full of clean water.
2. Start agitation and continue moderate agitation throughout mixing.
3. Mix and disperse wettable powders first, flowable products next and ERADICANE 6.7-E last.
4. Premix the BLADEX, ATRAZINE, or PRINCEP products individually in proper order (step 3 above) making sure the products are thoroughly wetted and dispersed in the tank prior to adding ERADICANE 6.7-E.
5. Add the ERADICANE 6.7-E to the tank $^2/_3$ full of water already mixed with BLADEX and ATRAZINE or PRINCEP.
6. Complete filling the remaining portion of the tank with water; mix thoroughly. Maintain moderate agitation to provide a uniform mix during application.
7. The tank-mix combination should not be left in the spray tank for prolonged periods; batches should be mixed and applied the same day.

Do not use excessive agitation as it may cause the BLADEX, ATRAZINE, or PRINCEP to settle out and set up in the spray tank. All return lines to the tank must discharge below liquid level.

(3) RATE CHART FOR IMPREGNATION OF DRY BULK FERTILIZERS (1)(2) WITH ERADICANE 6.7-E OR ERADICANE 6.7-E PLUS ATRAZINE 4L OR 80W

FERTILIZER RATE PER ACRE	ERADICANE 6.7- E RATE PER ACRE			ATRAZINE 4L RATE PER ACRE		ATRAZINE 80W RATE PER ACRE	
	3¾ PTS.	4¾ PTS.	7⅓ PTS.	2 PTS.	3 PTS.	1¼ LBS.	2 LBS.
200 lbs.	18¾ qts/ton	23¾ qts/ton	36⅔ qts/ton	10 qts/ton	15 qts/ton	12½ lbs/ton	20 lbs/ton
250 lbs.	15 qts/ton	19 qts/ton	29⅓ qts/ton	8 qts/ton	12 qts/ton	10 lbs/ton	16 lbs/ton
300 lbs.	12½ qts/ton	15⅘ qts/ton	24⅖ qts/ton	6⅔ qts/ton	10 qts/ton	8⅓ lbs/ton	13⅓ lbs/ton
350 lbs.	10¾ qts/ton	13⅗ qts/ton	21 qts/ton	5¾ qts/ton	8⅗ qts/ton	7⅕ lbs/ton	11⅖ lbs/ton
400 lbs.	9⅖ qts/ton	11⅞ qts/ton	18⅓ qts/ton	5 qts/ton	7½ qts/ton	6¼ lbs/ton	10 lbs/ton
450 lbs.	8⅓ qts/ton	10½ qts/ton	16⅓ qts/ton	4½ qts/ton	6⅔ qts/ton	5⅔ lbs/ton	8⅞ lbs/ton
500 lbs.	7½ qts/ton	9½ qts/ton	14⅔ qts/ton	4 qts/ton	6 qts/ton	5 lbs/ton	8 lbs/ton
550 lbs.	6⅘ qts/ton	8⅔ qts/ton	13⅓ qts/ton	3⅔ qts/ton	5½ qts/ton	4⅔ lbs/ton	7⅓ lbs/ton
600 lbs.	6¼ qts/ton	8 qts/ton	12⅕ qts/ton	3⅓ qts/ton	5 qts/ton	4⅕ lbs/ton	6⅔ lbs/ton
650 lbs.	5¾ qts/ton	7⅓ qts/ton	11¼ qts/ton	3⅕ qts/ton	4⅗ qts/ton	3⅞ lbs/ton	6⅕ lbs/ton
700 lbs.	5⅓ qts/ton	6⅘ qts/ton	10½ qts/ton	2⅞ qts/ton	4¼ qts/ton	3⅔ lbs/ton	5¾ lbs/ton

(4) RATE CHART FOR IMPREGNATION OF DRY BULK FERTILIZERS (1)(2) WITH ERADICANE 6.7-E OR ERADICANE 6.7-E PLUS BLADEX 4L OR 80W

FERTILIZER RATE PER ACRE	ERADICANE 6.7- E RATE PER ACRE			BLADEX 4L RATE PER ACRE		BLADEX 80W RATE PER ACRE	
	3¾ PTS.	4¾ PTS.	7⅓ PTS.	3 PTS.	4 PTS.	2 LBS.	2½ LBS.
200 lbs.	18¾ qts/ton	23¾ qts/ton	36⅔ qts/ton	15 qts/ton	20 qts/ton	20 lbs/ton	25 lbs/ton
250 lbs.	15 qts/ton	19 qts/ton	29⅓ qts/ton	12 qts/ton	16 qts/ton	16 lbs/ton	20 lbs/ton
300 lbs.	12½ qts/ton	15⅘ qts/ton	24⅖ qts/ton	10 qts/ton	13⅓ qts/ton	13⅓ lbs/ton	16⅔ lbs/ton
350 lbs.	10¾ qts/ton	13⅗ qts/ton	21 qts/ton	8⅗ qts/ton	11⅖ qts/ton	11⅖ lbs/ton	14⅓ lbs/ton
400 lbs.	9⅖ qts/ton	11⅞ qts/ton	18⅓ qts/ton	7½ qts/ton	10 qts/ton	10 lbs/ton	12½ lbs/ton
450 lbs.	8⅓ qts/ton	10½ qts/ton	16⅓ qts/ton	6⅔ qts/ton	8⅞ qts/ton	8⅞ lbs/ton	11⅕ lbs/ton
500 lbs.	7½ qts/ton	9½ qts/ton	14⅔ qts/ton	6 qts/ton	8 qts/ton	8 lbs/ton	10 lbs/ton
550 lbs.	6⅘ qts/ton	8⅔ qts/ton	13⅓ qts/ton	5½ qts/ton	7⅓ qts/ton	7⅓ lbs/ton	9⅕ lbs/ton
600 lbs.	6¼ qts/ton	8 qts/ton	12⅕ qts/ton	5 qts/ton	6⅔ qts/ton	6⅔ lbs/ton	8⅓ lbs/ton
650 lbs.	5¾ qts/ton	7⅓ qts/ton	11¼ qts/ton	4⅗ qts/ton	6⅕ qts/ton	6⅕ lbs/ton	7¾ lbs/ton
700 lbs.	5⅓ qts/ton	6⅘ qts/ton	10½ qts/ton	4¼ qts/ton	5¾ qts/ton	5¾ lbs/ton	7⅕ lbs/ton

(5) RATE CHART FOR THE IMPREGNATION OF DRY BULK FERTILIZERS WITH ERADICANE 6.7-E OR ERADICANE 6.7-E PLUS ATRAZINE (80W OR 4L) PLUS BLADEX (80W OR 4L)

FERTILIZER RATE PER ACRE	ERADICANE 6.7-E RATE PER ACRE			ATRAZINE 80W RATE PER ACRE		ATRAZINE 4L RATE PER ACRE	
	3¾ PTS. PER ACRE	4¾ PTS. PER ACRE	7⅓ PTS. PER ACRE	¾ LB PER ACRE	1¼ LBS PER ACRE	1 PT PER ACRE	2 PTS PER ACRE
200 lbs.	18¾ qts/ton	23¾ qts/ton	36⅔ qts/ton	7½ lbs/ton	12½ lbs/ton	5 qts/ton	10 qts/ton
250 lbs.	15 qts/ton	19 qts/ton	29⅓ qts/ton	6 lbs/ton	10 lbs/ton	4 qts/ton	8 qts/ton
300 lbs.	12½ qts/ton	15⅘ qts/ton	24⅖ qts/ton	5 lbs/ton	8⅓ lbs/ton	3⅓ qts/ton	6⅗ qts/ton
350 lbs.	10¾ qts/ton	13⅗ qts/ton	21 qts/ton	4⅓ lbs/ton	7⅛ lbs/ton	2⅞ qts/ton	5⅘ qts/ton
400 lbs.	9⅖ qts/ton	11⅞ qts/ton	18⅓ qts/ton	3¾ lbs/ton	6¼ lbs/ton	2½ qts/ton	5 qts/ton
450 lbs.	8⅓ qts/ton	10½ qts/ton	16⅓ qts/ton	3⅓ lbs/ton	5½ lbs/ton	2¼ qts/ton	4⅖ qts/ton
500 lbs.	7½ qts/ton	9½ qts/ton	14⅔ qts/ton	3 lbs/ton	5 lbs/ton	2 qts/ton	4 qts/ton
550 lbs.	6⅘ qts/ton	8⅔ qts/ton	13⅓ qts/ton	2¾ lbs/ton	4⅔ lbs/ton	1⅞ qts/ton	3⅔ qts/ton
600 lbs.	6¼ qts/ton	8 qts/ton	12⅕ qts/ton	2½ lbs/ton	4⅕ lbs/ton	1⅔ qts/ton	3⅓ qts/ton
650 lbs.	5¾ qts/ton	7⅓ qts/ton	11¼ qts/ton	2⅓ lbs/ton	3⅞ lbs/ton	1⅗ qts/ton	3⅛ qts/ton
700 lbs.	5⅓ qts/ton	6⅘ qts/ton	10½ qts/ton	2⅕ lbs/ton	3⅔ lbs/ton	1½ qts/ton	2⅞ qts/ton

(6) Rate Chart for the Impregnation of Dry Bulk Fertilizers with ERADICANE 6.7-E or ERADICANE 6.7-E Plus ATRAZINE (80W or 4L) Plus Bladex (80W or 4L)

Fertilizer Rate Per Acre	BLADEX 80W Rate Per Acre		BLADEX 4L Rate Per Acre	
	1¼ lbs per acre	2½ lbs per acre	1 qt per acre	2 qts per acre
200 lbs.	12½ lbs/ton	25 lbs/ton	10 qts/ton	20 qts/ton
250 lbs.	10 lbs/ton	20 lbs/ton	8 qts/ton	16 qts/ton
300 lbs.	8⅓ lbs/ton	16⅔ lbs/ton	6⅗ qts/ton	13⅓ qts/ton
350 lbs.	7⅛ lbs/ton	14¼ lbs/ton	5⅘ qts/ton	11⅖ qts/ton
400 lbs.	6¼ lbs/ton	12½ lbs/ton	5 qts/ton	10 qts/ton
450 lbs.	5½ lbs/ton	11⅛ lbs/ton	4⅖ qts/ton	8⅞ qts/ton
500 lbs.	5 lbs/ton	10 lbs/ton	4 qts/ton	8 qts/ton
550 lbs.	4⅔ lbs/ton	9⅛ lbs/ton	3⅔ qts/ton	7⅓ qts/ton
600 lbs.	4⅕ lbs/ton	8⅓ lbs/ton	3⅓ qts/ton	6⅔ qts/ton
650 lbs.	3⅞ lbs/ton	7¾ lbs/ton	3⅛ qts/ton	6⅕ qts/ton
700 lbs.	3⅔ lbs/ton	7⅕ lbs/ton	2⅞ qts/ton	5¾ qts/ton

APPENDIX I

Soil Mixing (Incorporation) Directions

ERADICANE 6.7-E must be incorporated into the soil to prevent loss of the herbicide. Whenever possible, application and incorporation should be done in the same operation. If there must be a delay (less than four hours) in incorporation of the herbicide when application and incorporation are done in separate operations, ERADICANE 6.7-E must be applied to a dry soil surface (at least ½ inch deep) and tree from dew and incidental moisture.

Pacific Northwest Region: (semi-arid areas of Eastern Washington, Eastern Oregon and Idaho only): When application and incorporation are done in separate operations, ERADICANE 6.7-E must be incorporated on the same day as applied. Application must be made on a dry soil surface (at least ½ inch deep) free from dew and incidental moisture.

Delayed Incorporation of Dry Bulk Fertilizers: The application and incorporation of ERADICANE 6.7-E must be carried out on the same day and delayed only if soil moisture conditions permit. See Appendix III for directions on the delay in soil incorporation.

Important: Thorough mixing of ERADICANE 6.7-E in the soil is necessary. Improper discing or poor tilth (large clods) may result in unsatisfactory control.

Incorporation Directions:

The following equipment commonly is used for soil mixing (incorporation) before planting.

Power-Driven Cultivation Equipment: (Recommended on all soil types previously primary tilled: *Ground speed should be adjusted with all PTO driven cultivation tools to insure thorough soil mixing.*

a. Horizontal action: Set to cut 4 to 6 inches deep. Example: Lely Roterra, Niemeyer or like equipment.
b. Vertical action: Set to cut 3 to 4 inches deep. Example: Roto-tiller, Side-winder or like equipment.

Tandem Discs (Recommended on all soil types: Set to cut to a depth of 4 to 6 inches, operated at 4 to 6 mph followed by a spike-tooth harrow or some other leveling device which extends beyond the ends of the discs. For more thorough mixing (for control or suppression of deep-germinating grasses—seedling johnsongrass, quackgrass and nutsedge), disc in two different directions. When discing the second time, the disc should be reset to cut no deeper than 4 to 6 inches.

Field Cultivators (Recommended on lighter soils in good tilth): Use 3 to 4 rows of sweeps spaced at 7 inch or less intervals and staggered so that no soil is left unturned, followed by a spike-tooth harrow pulled behind the cultivator. Do not use chisel plows or points. Set the cultivator to cut 4-6" deep, 5 mph or more. A second pass with a field cultivator at an angle to the initial pass will improve soil mixing. The field cultivator should not be used alone as an incorporation tool where control or suppression of deep-germinating weeds such as Nutgrass, Quackgrass, and Johnsongrass seedlings is desired. For control of these weeds, use a tandem disc for the initial incorporation pass.

Sprinkler Incorporation Pacific Northwest Region (Semi-arid areas of Eastern Washington, Eastern Oregon and Idaho only): Surface apply ERADICANE 6.7-E immediately after planting. The soil surface should be dry (at least ½ inch deep) and free from dew and incidental moisture. Incorporate using ½ to ¾ inch of water within 36 hours following application. The application and incorporation must be done within 5 days after the last tillage operation since poor results will occur if weeds have germinated.

APPENDIX II

Procedure for Testing the Compatibility of ERADICANE 6.7-E with Fluid Fertilizers

The following procedure is suggested for determining whether ERADICANE 6.7-E may be combined with a specific fluid fertilizer for spray tank application.

Materials Required:

1. ERADICANE 6.7-E
2. Fluid fertilizer to be used.
3. Adjuvant for fertilizer tank mix: Compex*, Sponto 168-D*, Unite*, or equivalent. The adjuvant which provides the best emulsification depends on the specific fertilizer under consideration.
4. Two one-quart, wide-mouth glass jars with lid or stopper.
5. Measuring spoons (a 25 mL pipette or graduated cylinder provides more accurate measurement).
6. Measuring cup, 8 oz. (237 mL).

*Compex, Kalo Laboratories Inc., Kansas City, Missouri: Sponto 168-D. Witco Chemical Company, Houston, Texas; Unite, Hopkins Agricultural Chemical Co., Madison, Wisconsin.

Procedure

1. Pour a pint (about 473 mL) of the fluid fertilizer into each of the quart jars.
2. Add adjuvant to one of the jars and mix (see rate table.)
3. Add the ERADICANE 6.7-E to both jars (see rate table.)
4. Close both jars with lid or stopper and mix the contents by turning the jars upside down ten times.
5. Inspect the surface and body of the mixtures—
   (A) Immediately after completing the jar inversions.
   (B) After allowing the jars to stand quietly for 30 minutes.
   (C) And then again after turning the jars upside down 10 times.

If a uniform mix cannot be made, the mixture should not be used. If either mixture remains uniform for 30 minutes, the combination may be used. Should either mixture separate after 30 minutes, but readily remix uniformly with ten jar inversions, the mixture can be used if adequate agitation is maintained in the tank. If the mixture with adjuvant is satisfactory, but the one without adjuvant is not, be sure to use the adjuvant in the spray tank. Add the adjuvant first at a rate of 3 pints per 100 gallons of fluid fertilizer, foaming can be minimized by using moderate agitation.

If nondispersible oil, sludge or clumps of solids form in the mixtures, the combination should not be used.

Note 1: For some combinations, premixing wettable powders in a little water in a pail before adding them to the spray tank will improve the compatibility of the final mixtures with ERADICANE 6.7-E. This technique can be tested in the small scale jar test by premixing the wettable powder in one-eighth (⅛) cup of water prior to addition to the pint of fluid fertilizer.

Rate Table for ERADICANE 6.7-E and Adjuvant with the Fluid Fertilizer**

Gallons of fluid fertilizer to be applied per acre	mL or tsp. of ERADICANE 6.7-E* to be added to 1 pint of fertilizer	
	mL	tsp.
10	7	1⅓
15	5	1
20	4	¾
25	3	⅔
30	2	½
40	2	½

* Based on field rate of 1 pound active ingredient per acre in the fertilizer volumes indicated. Increase volume proportionately to correspond with intended field rate in terms of pounds active ingredient per acre (e.g., for field rate of 4 pounds actual active ingredient in 40 gallons fertilizer per acre, add 8 mL or 1⅗ tsp. ERADICANE 6.7-E to each jar for compatibility testing purposes).

** Two (2) milliliters or two-fifths (⅖) teaspoon of adjuvant to be added to 1 pint of fluid fertilizer in order to equal the rate of 3 pints of adjuvant per 100 gallons of fluid fertilizer.

APPENDIX III

Impregnation on Dry Bulk Fertilizers

ERADICANE 6.7-E, ERADICANE 6.7-E/ATRAZINE (4L or 80W) or ERADICANE 6.7-E/BLADEX (4L or 80W) OR ERADICANE 6.7-E/ATRAZINE/BLADEX.

CAUTION: ERADICANE 6.7-E alone or in combination with other herbicides must not be impregnated on ammonium nitrate, potassium nitrate or sodium nitrate fertilizers. Such mixtures may cause explosion and fire.

All individual state regulations relating to bulk dry fertilizer blending, registration, labeling and application are the responsibility of the individual and/or company selling the ERADICANE 6.7-E or ERADICANE 6.7-E/ATRAZINE (4L or 80W) and/or BLADEX (4L or 80W) and fertilizer mixtures, to include 3-way tank mix.

ERADICANE 6.7-E alone and in combination with ATRAZINE and/or BLADEX may be impregnated on many dry bulk fertilizers (1) (2) and applied and incorporated in the soil before planting for the control of grass and broadleaf weeds in corn.

Test results have shown that ERADICANE 6.7-E, ERADICANE 6.7-E/ATRAZINE (4L or 80W) and/or BLADEX (4L or 80W) on many dry bulk fertilizers give weed control equal to ERADICANE 6.7-E or a tank mix of ERADICANE 6.7-E/ATRAZINE (4L or 80W) and/or BLADEX (4L or 80W) applied as a spray in water or liquid fertilizer. However, uniform impregnation of the pesticides on the dry fertilizer particles and uniform application in the field are necessary to assure good results.

A minimum of 200 pounds and a maximum of 700 pounds of approved ingredients (1) (2) impregnated with the appropriate amount of ERADICANE 6.7-E alone or in combination with ATRAZINE (4L or 80W) and/or BLADEX (4L or 80W) must be applied per acre.

For impregnating the pesticides on dry fertilizers, use a closed rotary-drum mixer or a similar type of closed blender equipped with suitable spray equipment. The spray nozzle should be positioned inside of the mixer to provide uniform spray coverage of the tumbling fertilizer. The ERADICANE 6.7-E should be sprayed uniformly onto the fertilizer using a fine spray pattern. For the ERADICANE 6.7-E/ATRAZINE or ERADICANE 6.7-E/BLADEX combination product, the ATRAZINE (4L or 80W) and/or BLADEX (4L or 80W) can be added and mixed uniformly with the fertilizer prior to the ERADICANE 6.7-E addition, or alternately, the ERADICANE 6.7-E/ATRAZINE (4L or 80W) and/or BLADEX (4L or 80W) can be mixed in the proposed use ratio to prepare a uniform slurry for joint spray impregnation.

The physical properties of fertilizers vary, particularly in liquid absorptive capacity. When absorptivity is sufficient, simple spray impregnation of the fertilizer with ERADICANE 6.7-E, ERADICANE 6.7-E/ATRAZINE (4L or 80W), and/or BLADEX (4L or 80W) provides a satisfactory, dry mixture.

If the absorptive capacity is not adequate, use of a drying agent is required to provide dry, freeflowing mixtures.

For mixtures to be used in spinning-disc applicators, Microcel E calcium silicate powder (Manville, Filtration & Minerals) is recommended for use as a drying agent.

For mixtures to be used in pneumatic applicators, Microcel E or Agsorb 16/30 RVM-MS granularclay (Oil-Dri Corporation) or Celatom MP-79 (Eagle-Picher Industries, Inc.) is recommended for use as the drying agent.

The drying agent should be added separately and uniformly to the prepared pesticide-fertilizer mixture, in a quantity that is sufficient to provide a suitable free-flowing mixture. Generally, less than 2% Microcel E or 5% Agsorb 16/30 RVM-MS and Celatom MP-79 by weight is required.

The amount of ERADICANE 6.7-E, ERADICANE 6.7-E/ATRAZINE (4L or 80W) and/or BLADEX (4L or 80W) actually required in the manufacture of individual fertilizer mixtures should be determined carefully for each production operation. This is necessary to ensure that the amount of pesticides actually contained in the mixture applied to the soil represents the correct rate of use.

Bulk fertilizer impregnated with ERADICANE 6.7-E, ERADICANE 6.7-E/ATRAZINE (4L or 80W) and/or BLADEX (4L or 80W) should be *applied immediately*, not stored. It is recommended that all bulk containers be tightly covered while the product is being transported and applied to reduce chances of ERADICANE 6.7-E loss via volatilization.

When application and incorporation are done in separate operations, ERADICANE 6.7-E impregnated on the dry fertilizer must be incorporated on the same day as applied. Application must be made on a dry soil surface (at least 1/2 inch deep) free from dew and incidental moisture.

ERADICANE 6.7-E Physical Data

Specific Gravity 68/68°F: 0.972 (typical)
Pounds/Gallon 68°F: 8.09 (typical)
Flashpoint: 186°F (Tagliabue closed cup)
Viscosity: sprayable down to minus 20°F

(1) Approved Dry Fertilizer Ingredients for use with ERADICANE 6.7-E

	N	P	K
Ammonium sulfate	21	0	0
Diammonium phosphate	18	46	0
Potassium chloride	0	0	60
Potassium sulfate	0	0	52
Single Super-phosphate	0	20	0
Treble Super-phosphate	0	46	0
Urea*	45	0	0
Ammonium phosphate-sulfate	16	20	0

*Some ureas may be phytotoxic when high rates are applied to corn. Use only urea rates known to be safe for corn application.

Note: K-Mag and 11-48-0 are also approved for use.

(2) Approved Dry Fertilizer Ingredients for use with ERADICANE 6.7-E/ATRAZINE (4L or 80W), ERADICANE 6.7-E/BLADEX (4L or 80W) or ERADICANE PLUS ATRAZINE and BLADEX

	N	P	K
Ammonium Sulfate	21	0	0
Diammonium Phosphate	18	46	0
Potassium Chloride	0	0	60
Potassium Sulfate	0	0	52
Urea*	45	0	0
Ammonium phosphate-sulfate	16	20	0

* Some ureas may be phytotoxic when high rates are applied to corn. Use only urea rates known to be safe for corn application.

Note: K-Mag and 11-48-0 are also approved for use.

Note: Do not combine ERADICANE 6.7-E plus ATRAZINE (4L or 80W) or BLADEX (4L or 80W) with single super-phosphate (0-20-0) or treble super-phosphate (0-46-0).

STORAGE AND DISPOSAL

Prohibitions: Do not contaminate water, food or feed by storage or disposal. Open dumping is prohibited. Keep away from heat and flame. **Storage:** Keep container tightly closed when not in use. Do not store near seeds, fertilizers, or foodstuffs. Can be stored at temperatures as low as minus 50°F. **Pesticide Disposal:** Waste resulting from the use of this product may be disposed of on site or at an approved waste disposal facility. **Container Disposal: Metal**—Triple rinse (or equivalent), then offer for recycling or reconditioning, or puncture and dispose of in a sanitary landfill, or by other procedures approved by state and local authorities. **Plastic**—Triple rinse (or equivalent), then offer for recycling or reconditioning, or puncture and dispose of in a sanitary landfill, or by incineration, or, if allowed by state and local authorities, by burning. If burned stay out of smoke.

NOTICE TO BUYER AND USER:

Seller warrants that this product conforms to the chemical description on the label and is reasonably fit for the purposes stated on the label when used in accordance with directions under normal conditions of use. This warranty does not extend to the use of this product contrary to label instructions, or under abnormal use conditions, or under conditions not reasonably foreseeable to Seller, and Buyer and User assume the risk of any such use. SELLER DISCLAIMS ALL OTHER WARRANTIES EXPRESSED OR IMPLIED INCLUDING ANY WARRANTY OF FITNESS OR MERCHANTABILITY. SELLER SHALL NOT BE LIABLE FOR CONSEQUENTIAL, SPECIAL OR INDIRECT DAMAGES RESULTING FROM THE USE OR HANDLING OF THIS PRODUCT AND SELLER'S SOLE LIABILITY AND BUYER'S AND USER'S EXCLUSIVE REMEDY SHALL BE LIMITED TO THE REFUND OF THE PURCHASE PRICE.

This product is sold only for uses stated on its label. No express or implied license is granted to use or sell this product under any patent in any country except as specified: Country: United States of America. Patent No. 4,021,224; 4,137,070 and other patents pending.

A-14R

The information and specimen labels in this booklet are correct as of the publication date but are subject to change without notice. For current information, contact ICI Americas Inc.

ERADICANE® EXTRA

SELECTIVE HERBICIDE

pre-plant incorporated liquid herbicide for tough weed control in corn

Complete Directions For Use, Use Precautions and Recommendations

ACTIVE INGREDIENT:
S-ethyl dipropylthiocarbamate 72.2%
INERT INGREDIENTS: 27.8%
TOTAL 100.0%

Contains 6.0 lbs. Active Ingredient Per Gallon
EPA Reg. No. 10182-244

KEEP OUT OF REACH OF CHILDREN
CAUTION

Harmful If Swallowed

STATEMENT OF PRACTICAL TREATMENT

First Aid

Immediately start the procedures below. If further treatment is required contact a Poison Center, a physician, or the nearest hospital.

If swallowed: Immediately give several glasses of water but *do not* induce vomiting. If vomiting does occur, give fluids again. Have a physician determine if condition of patient will permit induction of vomiting or evacuation of stomach. Do not give anything by mouth to an unconscious or convulsing person.

For eye contact: Immediately flush eyes with large amounts of running water for at least 15 minutes. Hold eyelids apart to ensure rinsing of the entire surface of the eye and lids with water. Get medical attention immediately.

For skin contact: Flush all affected areas with plenty of water for several minutes. Remove and clean contaminated clothing and shoes. Seek medical attention if irritation occurs.

If inhaled: Remove to fresh air. Seek medical attention if respiratory irritation occurs or if breathing becomes difficult.

FOR 24-HOUR EMERGENCY MEDICAL ASSISTANCE CALL 1-800-F-A-S-T-M-E-D (327-8633).
FOR CHEMICAL EMERGENCY: Spill, leak, fire, exposure, or accident call CHEMTREC 1-800-424-9300.

PRECAUTIONARY STATEMENTS

HAZARDS TO HUMANS AND DOMESTIC ANIMALS

CAUTION

HARMFUL IF SWALLOWED—Avoid contact with skin, eyes and clothing—Avoid inhalation of spray mist—Wear rubber gloves and fresh clothing—Wear goggles to avoid contact with eyes—Wash with soap and water immediately after use or contact—Do not contaminate food or feed.

ENVIRONMENTAL HAZARDS—Keep out of any body of water—Do not contaminate water by cleaning of equipment or disposal of wastes.

PHYSICAL AND CHEMICAL HAZARDS—Keep away from heat and flame. Do not apply this product in such a manner as to directly or through drift, expose workers or other persons. When applied through the sprinkler irrigation method of application, the area being treated must be vacated by unprotected persons.

Can be stored at temperatures as low as minus 50°F.

GENERAL USE PRECAUTIONS

READ ALL LABEL DIRECTIONS BEFORE USING.

ERADICANE EXTRA should be used for recommended purposes and at recommended rates. (DO NOT OVERDOSE.)

ERADICANE EXTRA is recommended for use on mineral soils only (soils containing less than 10% organic matter).

Do not use on milo or sorghum. Do not use on corn seed stock (Breeders, Foundation or Increase). ERADICANE EXTRA may be used on Production seed corn.

Do not contaminate water to be used on susceptible crops and ornamentals, or for domestic purposes. Do not contaminate feed or food. Keep container tightly closed when not in use. Do not store near seeds or fertilizers.

When properly applied and weather conditions exist for normal plant growth through the season, ERADICANE EXTRA will not harm the treated crop nor should harmful soil residues remain beyond harvest. However, during germination and early growth, extended periods of unusually cold and wet or hot and dry weather, insect, nematode, or plant disease attack, carry-over soil residues of certain persistent herbicides, the use of certain soil applied systemic insecticides may create abnormal conditions that weaken crop seedlings. ERADICANE EXTRA used under these abnormal conditions could result in crop injury.

Weeds Controlled by ERADICANE EXTRA

ERADICANE EXTRA will not control established weeds.

Annual Grasses:	
Barnyardgrass (Watergrass, junglerice)	(*Echinochloa* spp.)
Bermudagrass seedlings	(*Cynodon dactylon*)
Bluegrass, annual	(*Poa annua*)
Crabgrass	(*Digitaria* spp.)
Cupgrass, woolly	(*Erichloa villosa*)
Foxtail, giant	(*Setaria faberi*)
Foxtail, green	(*Setaria viridis*)
Foxtail, yellow	(*Setaria lutescens*)
Goosegrass	(*Eleusine indica*)
Johnsongrass—seedlings	(*Sorghum halepense*)
Millet, wild proso*‡	(*Panicum miliaceum*)
Oats, wild	(*Avena fatua*)
Panicum, fall	(*Panicum dichotomiflorum*)
Panicum, Texas (Coloradograss)*†	(*Panicum texanum*)
Rescuegrass	(*Bromus willdenowii*)
Rice, red	(*Oryza sativa*)
Ryegrass, annual (Italian ryegrass)	(*Lolium multiflorum*)
Sandbur, field*	(*Cenchrus pauciflorus*)
Shattercane *¶ (Wild cane)	(*Sorghum bicolor*)
Signalgrass	(*Brachiaria* spp.)
Stinkgrass (Lovegrass)	(*Eragrostis cilianensis*)

Other volunteer grains, i.e., barley, oats, wheat.

¶Partial control (suppression) see wild cane section.

*Cultivation is suggested in addition to ERADICANE EXTRA treatment in fields with moderate to heavy infestations of wild cane, sandbur, Texas panicum, or wild proso millet.

†Southeastern U.S. Only.

‡Central Region Only—Partial Control (suppression)

Annual Broadleaf Weeds:	
Carpetweed	(*Mollugo verticillata*)
Chickweed, common	(*Stellaria media*)
Fiddleneck	(*Amsinckia* spp.)
Goosefoot, nettleleaf	(*Chenopodium murale*)
Henbit (Deadnettle)	(*Lamium amplexicaule*)
Lambsquarters, common	(*Chenopodium album*)
Morningglory, tall	(*Ipomoea purpurea*)
Nightshade, black	(*Solanum nigrum*)
Nightshade, hairy	(*Solanum sarachoides*)
Pigweed, prostrate	(*Amaranthus biltoides*)
Pigweed, redroot (*Common*)	(*Amaranthus retroflexus*)
Pigweed, tumble	(*Amaranthus albus*)
Puncturevine**	(*Tribulus terrestris*)
Purslane, common	(*Portulaca oleracea*)
Pusley, Florida	(*Richardia scabra*)
Sicklepod	(*Cassia obtusifolia*)
Sida, prickly (Teaweed)	(*Sida spinosa*)
Shepherdspurse	(*Capsella bursa-pastoris*)
Spurry, corn	(*Spergula arvensis*)
Velvetleaf	(*Abutilon theophrasti*)

**Western Region Only.

The annual broadleaf weeds listed above will be controlled only if treatment is made when conditions are favorable for weed seed germination and growth. Under prolonged conditions of very cold soil, control of the above broadleaf weeds may not be adequate.

Perennial Grassses:	
Bermudagrass	(*Cynodon dactylon*)
Johnsongrass, rhizome	(*Sorghum halepense*)
Nutsedge, purple	(*Cyperus rotundus*)
Nutsedge, yellow	(*Cyperus esculentus*)
Quackgrass	(*Agropyron repens*)

Perennial weeds must be turned under and chopped up thoroughly prior to treatment. The underground rhizomes of quackgrass and johnsongrass and the rhizomes and stolons of bermudagrass must be cut up thoroughly into small pieces so that 4 or less nodes remain on a strand. Consult RECOMMENDATIONS for ERADICANE EXTRA rates necessary to control the above listed weeds.

DIRECTIONS FOR USE

It is a violation of Federal law to use this product in a manner inconsistent with its labeling.

ERADICANE EXTRA is a selective herbicide which must be mixed (incorporated) into the soil for control of weeds listed on this label. ERADICANE EXTRA controls weeds by interfering with normal germination and seedling development. ERADICANE EXTRA does not control established weeds.

Application Directions

Pour the recommended amount of ERADICANE EXTRA into the spray tank during the filling operation. Apply in 10 to 50 gallons of water per acre using a properly calibrated, low-pressure sprayer having good agitation. The soil should be well-worked and dry enough to permit good soil mixing (incorporation). See Appendix I for soil mixing directions.

For band applications, calculate the amount to be applied per acre as follows:

$$\frac{\text{Band Width in inches}}{\text{Row Width in inches}} \times \text{Rate/A for Broadcast Treatment} = \text{Amount Needed for Band Treatment}$$

Combination with Fluid Fertilizers

ERADICANE EXTRA may be combined with fluid (solution, slurry or suspension) fertilizers. However, physical compatibility with these fertilizers must be verified before combining in the spray tank. See Appendix II for special directions regarding these combinations. Even though ERADICANE EXTRA may be found compatible with fluid fertilizers, constant agitation is necessary to maintain uniformity of this mixture during application.

Impregnation on Dry Bulk Fertilizers

ERADICANE EXTRA may be impregnated on dry fertilizer and applied as the fertilizer is spread. See Appendix III for information and directions regarding impregnation and use of these combinations.

CENTER PIVOT SPRINKLER APPLICATION

Apply this product only through center pivot, sprinkler irrigation systems. Do not apply this product through any other type of irrigation system.

Crop injury, lack of effectiveness, or illegal pesticide residues in the crop can result from nonuniform distribution of treated water.

If you have questions about calibration, you should contact State Extension Service specialists, equipment manufacturers or other experts.

Do not connect an irrigation system (including greenhouse systems) used for pesticide application to a public water system unless the pesticide label-prescribed safety devices for public water systems are in place.

A person knowledgeable of the chemigation system and responsible for its operation or under the supervision of the responsible person, shall shut the system down and make necessary adjustments should the need arise.

Apply and soil incorporate ERADICANE EXTRA, ERADICANE EXTRA/ATRAZINE, (see TANK MIX RECOMMENDATIONS) with irrigation water before planting (see PLANTING), or immediately after planting before either corn or weeds emerge. It is important that application and incorporation be done immediately after the last tillage operation since poor results will occur if the weeds have germinated before application of the herbicide. Meter the ERADICANE EXTRA or tank mix combination with ATRAZINE into the irrigation water during the entire period. Apply in 1/2 to 3/4 inches of water.

USE PRECAUTIONS:

The system must contain a functional check valve, vacuum relief valve, and low pressure drain appropriately located on the irrigation pipeline to prevent water source contamination from backflow.

The pesticide injection pipeline must contain a functional, automatic, quick-closing check valve to prevent the flow of fluid back toward the injection pump.

The pesticide injection pipeline must also contain a functional, normally closed, solenoid-operated valve located on the intake side of the injection pump and connected to the system interlock to prevent fluid from being withdrawn from the supply tank when the irrigation system is either automatically or manually shut down.

The system must contain functional interlocking controls to automatically shut off the pesticide injection pump when the water pump motor stops.

The irrigation line or water pump must include a functional pressure switch which will stop the water pump motor when the water pressure decreases to the point where pesticide distribution is adversely affected.

Systems must use a metering pump, such as a positive displacement injection pump (e.g., diaphragm pump) effectively designed and constructed of materials that are compatible with pesticides and capable of being fitted with a system interlock.

Application of more than label recommended quantities of irrigation water per acre may result in decreased product performance by removing the chemical from the zone of effectiveness.

Do not apply when wind speed favors drift beyond the area intended for treatment, when system connections or fittings leak, when nozzles do not provide uniform distribution or when lines containing the product must be dismantled and drained.

Greater accuracy in calibration and distribution will be achieved by injecting a larger volume of a more dilute slurry per hour.

* BLADEX is a registered trademark of E.I. du Pont de Nemours & Co., Inc.

EXAMPLE:

110 Acres to be covered by 5⅓ pints (4 lbs active) ERADICANE EXTRA

110 A. × 4 lbs. = *440* lbs. active ERADICANE EXTRA

$\frac{440}{6.0}$ = *73.3* gals. ERADICANE EXTRA

Mixing tank size 750 gallons. Sprinkler takes 30 hours to apply ½ inch of water.

Injection rate = $\frac{750}{30}$ = *25* gal./hr. or *0.4* gal./min.

Calibrate injection pump to deliver the herbicide mix into the irrigation system at the rate of *25* gal. per hr. or *0.4* gal. per minute

PLANTING

Plant seed to a maximum depth of 2 inches within 2 weeks after treatment. To avoid removing ERADICANE EXTRA from the row (with loss of weed control in the row), do not move or shape soil after treatment such as by planting in deep furrows.

Cultural Practices Following Application

Should weeds develop, a shallow cultivation or rotary hoeing will generally result in better weed control. When cultivating for any reason, it should be shallow, i.e. no more than ½ the depth the herbicide was incorporated. A sequential application of 2,4-D, BANVEL* 4E or pre-emergence or post-emergence herbicides may be necessary to control weeds resistant to ERADICANE EXTRA.

*BANVEL is a registered trademark of Sandoz Crop Protection Corp.

RECOMMENDATIONS

Consult the U.S. regional map below and use the recommendations in your region only.

Important: Limit use on sweet corn to 5⅓ pints ERADICANE EXTRA per acre with the exception of Minnesota and Wisconsin where up to the full rate of 8 pints per acre can be used.

Eastern U.S.

(Note: In the Southeastern U.S., when ERADICANE EXTRA is used for weed control in silage corn, do not seed small seeded grains after corn harvest until September.)

Central U.S.

(Note: Western slope of Colorado: Where furrow irrigation will be used when corn is flat planted, do not form furrows until corn is in the three-leaf stage.)

Western U.S.

(Note: ERADICANE EXTRA is not recommended on corn grown in Arizona or in the 10 Southernmost California counties. In California, do not use ERADICANE EXTRA on corn planted after May 1.)

NOTE: Soils lacking sufficient moisture for seed germination must be preirrigated prior to ERADICANE EXTRA application.

NOTE: Do not use on "Golden Jubilee" sweet corn west of the Cascade Mountains in Oregon or Washington.

ERADICANE® EXTRA Regional Use Map

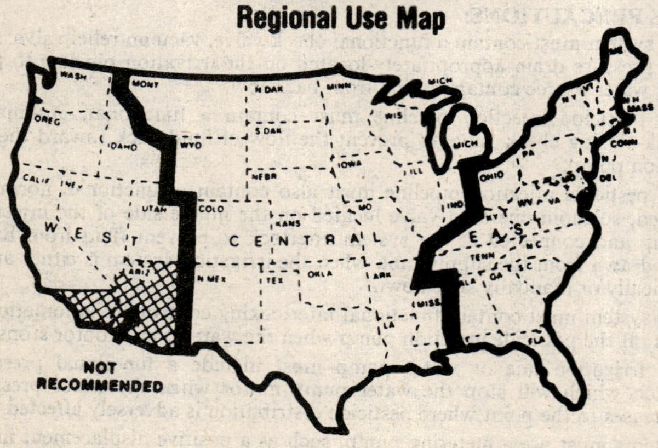

Preplant Incorporation:

Eradicane Extra Broadcast Rate Per Acre In Pints

Weed	Eastern U.S.	Central U.S.	Western U.S.	Restrictions
Annual Grasses & Annual Broadleaf Weeds	4 to 5⅓	4 to 8	4 to 5⅓	Use the lower rate for light infestations and in sand and light (coarse textured soils).
Bermudagrass	5⅓ to 8	5⅓ to 8	4 to 5⅓	Use the higher rate for heavy infestations.
Nutsedge	5⅓ to 8	5⅓ to 8	4 to 8	
Quackgrass	8	6⅔ to 8	5⅓	For control of moderate infestations and suppression of heavy infestations.
Rhizome Johnsongrass	8	8	8	Application should follow a program of plowing and discing for good control. See special instructions.
Sandbur, Field	4	5⅓ to 8	5⅓ to 8*	
Shattercane (suppression)	6⅔ to 8	6⅔ to 8	—	Partial control (suppression) only in central U.S. See special instructions.
Wild Oat	4	4	4 to 5⅓*	
Wild Proso Millet (suppression)	—	5⅓ to 8	—	Control may be improved with a combination treatment of ERADICANE EXTRA/ ATRAZINE or ERADICANE EXTRA/ BLADEX.
Velvetleaf	8	8	5⅓ to 8	See tank mix section.

*Use the lower rates on sand and light (coarse textured) soils. Use the higher rates on heavy (fine textured) soils.

Quackgrass—For improved quackgrass control ERADICANE EXTRA can be preceded by a preplow application of ATRAZINE. Broadcast 2½ lbs. ATRAZINE 80W or 2 quarts ATRAZINE 4L per acre in the fall or spring and plow 1-3 weeks later. In the spring, follow with the recommended rate of ERADICANE EXTRA.

Rhizone Johnsongrass—For control, apply and incorporate 8 pints ERADICANE EXTRA per acre. Using a program of plowing and discing, followed by ERADICANE EXTRA will give good control of seedlings and rhizomes.

In some areas discing 2 to 3 times spaced at 4 to 5 week intervals during fall and winter can be done. If this cannot be done, start with fall plowing followed by several discings in the spring before planting. The fall plowing and discing cuts rhizomes into small pieces and brings them to the surface, where both weather and ERADICANE EXTRA will kill or weaken them.

Velvetleaf—For control of velvetleaf use 5⅓ to 8 pints of ERADICANE EXTRA plus 2 pounds of ATRAZINE 80W or 3 pints of ATRAZINE 4L per broadcast acre.

Shattercane—For partial control (suppression)—Apply 6⅔ to 8 pints ERADICANE EXTRA per acre. Use the higher rate on heavy infestations or on soils with higher organic content. For enhanced control of shattercane, especially with heavy infestations, ERADICANE EXTRA should be used in combination with ATRAZINE, BLADEX or PRINCEP* (see TANK MIX COMBINATIONS).

*PRINCEP is a registered trademark of Ciba Geigy Corporation.

The following practices are *recommended* for suppression of Shattercane:

1. Fall or spring moldboard plow.
2. Before application, the soil should be well worked with a disc and dry enough to permit good soil mixing of the herbicide (incorporation). All trash should be worked into the soil.
3. Delay application until soil warms to a constant 55°F or greater temperature. Plant immediately.
4. Incorporate *immediately* and thoroughly. If application of ERADICANE EXTRA is made impregnated on dry bulk fertilizer and a delay of incorporation is expected, make certain the delay does not exceed 4 hours. If delay is expected when impregnated, make certain application is made to a dry soil surface (at least ½ inch deep) that is free from dew or incidental moisture.

5. Incorporation directions:
a. Power tillers, or
b. tandem disc set to cut 5-6" deep followed by a second pass at an angle to the initial set to cut 4" deep, or
c. tandem disc set to cut 5-6" deep followed by a field cultivator at an angle to the initial pass set to cut 4" deep.
6. Cultivate immediately if wild cane emerges. Cultivation should be shallow, no more than 1/2 the depth of incorporation of the herbicide.

Center Pivot Sprinkler Application

Apply 4 to 8 pints of ERADICANE EXTRA per acre for control of annual grasses and annual broadleaf weeds. ATRAZINE or BLADEX may be tank mixed with ERADICANE EXTRA for sprinkler application (see TANK MIX RECOMMENDATIONS).

Subsurface Injection

(Southeastern U.S. with coarse textured soils only)

Apply 4 pints of ERADICANE EXTRA for control of light infestations of annual grasses (except shattercane), annual broadleaf weeds, and nutsedge. Adjust the dosage proportionately depending on the row spacing and width to be treated.

TANK MIX COMBINATIONS

For control of additional weeds and for increased control of labelled broadleaf weeds, ATRAZINE, BLADEX, PRINCEP, or a combination of ATRAZINE and BLADEX may be tank mixed with the recommended rates of ERADICANE EXTRA and applied preplant incorpoated. ERADICANE EXTRA/ATRAZINE or ERADICANE EXTRA/BLADEX may be applied through a center pivot irrigation system.

As an alternative, ERADICANE EXTRA can be applied preplant incorporated at recommended rates followed by ATRAZINE applied pre-emergence to soil surface. Consult the pre-emergence application directions on the ATRAZINE label. A rotary hoeing or shallow cultivation is recommended if rainfall or sprinkler irrigation has not occurred within 10 days of the surface application. For enhanced or additional broadleaf weed control, an overlay treatment of 2,4-D, or BANVEL 4E may be made.

With the combination of ERADICANE EXTRA/BLADEX/ATRAZINE, the potential hazard of ATRAZINE residues affecting certain crops the following year is reduced because a lower rate of ATRAZINE can be used.

Additional Weeds Controlled by ERADICANE EXTRA/ATRAZINE and ERADICANE EXTRA/BLADEX Tank Mix

Annual Grasses:	
Witchgrass	(*Panicum capillare*)
Annual Broadleaf Weeds:	
Cocklebur, common	(*Xanthium pensylvanicum*)
Jimsonweed	(*Datura stramonium*)
Kochia	(*Kochia scoparia*)
Mallow, Venice	(*Hibiscus trionum*)
Mustard	(*Brassica* spp.)
Ragweed	(*Ambrosia* spp.)
Smartweed	(*Polygonum* spp.)
Thistle, Russian*	(*Salsola kali*)

*Controlled by ERADICANE EXTRA/ATRAZINE TANK MIX only.

ERADICANE EXTRA/ATRAZINE and ERADICANE EXTRA/BLADEX TANK MIX will control these additional weeds and improve the control of broadleaf weeds listed under "Weeds Controlled by ERADICANE EXTRA" alone, if unfavorable conditions exist.

Additional Weeds Controlled by ERADICANE EXTRA/PRINCEP Tank Mix

Annual Broadleaf Weeds:	
Cocklebur, common	(*Xanthium pensylvanicum*)
Jimsonweed	(*Datura stramonium*)
Ragweed	(*Ambrosia* spp.)
Smartweed	(*Polygonum* spp.)

ERADICANE EXTRA/PRINCEP TANK MIX will control these additional weeds and improve the control of broadleaf weeds listed under "Weeds Controlled by ERADICANE EXTRA" alone, if unfavorable conditions exist.

Application Directions

Apply ERADICANE EXTRA/ATRAZINE, ERADICANE EXTRA/BLADEX, ERADICANE EXTRA/PRINCEP or ERADICANE EXTRA/BLADEX/ATRAZINE in 20-50 gallons of water per acre. It is recommended that the compatibility of tank mix combinations be tested on a small scale, for example, a jar test before actual tank mixing. See Appendix II which describes the procedure for such tests.

Recommendations

Broadcast Rates Per Acre:

Atrazine: Use the recommended rates of ERADICANE EXTRA plus 1 1/4 to 2 pounds of ATRAZINE 80W or 2 to 3 pints of ATRAZINE 4L per acre for both tank mix (preplant incorporation or center pivot sprinkler application) and sequential application. Use the higher rate of ATRAZINE in soils with over 5% organic matter.

Bladex: Use the recommended rates of ERADICANE EXTRA plus 2 to 2 1/2 pounds of BLADEX 80W or 3 to 4 pints BLADEX 4L per acre for preplant incorporation or application through center pivot sprinkler system. Use the lower rate of BLADEX on sand or loam soils. Use the higher rate of BLADEX on fine textured (clay) soils or soils with 3% organic matter and above.

Princep: Use 1 1/4 to 3 3/4 pounds of PRINCEP 80W, 1 1/10 to 3 1/3 pounds of PRINCEP CALIBER* 90 or 2 to 6 pints of PRINCEP 4L per acre with recommended rates or ERADICANE EXTRA preplant incorporated. Use the higher rate of PRINCEP on fine-textured soils and where heavy broadleaf weed infestations are expected.

Bladex/Atrazine: Use the recommended rate of ERADICANE EXTRA plus 1 1/4 to 2 1/2 pounds of BLADEX 80W or 2 to 4 pints of BLADEX 4L and 3/4 to 1 1/4 pounds of ATRAZINE 80W or 1 to 2 pints of ATRAZINE 4L per acre. BLADEX and ATRAZINE should be used in a two to one proportion.

*CALIBER is a registered trademark of Ciba Geigy Corporation.

2,4-D or Banvel 4E: Use the amount of 2,4-D or BANVEL 4E recommended on the manufacturer's label.

NOTE: BLADEX is not recommended for use on sandy soils consisting of more than 70% sand and containing less than 1% organic matter. Consult the BLADEX label for more specific recommended rates on sandy soils.

Follow all the use precautions and warnings that appear on the ATRAZINE, BLADEX and PRINCEP labels and supplement literature.

Make only one application per crop. After a treatment including ATRAZINE, do not plant to any crop except corn until the following year or injury may occur.

Do not plant to alfalfa where more than 1.2 pounds actual ATRAZINE was used the previous year.

PRECAUTIONS TO FOLLOW WHEN USING ERADICANE EXTRA/PRINCEP:

Do not rotate to any crop except corn until the following year or injury may occur.

After harvest, plow and thoroughly till the soil in fall or spring to minimize possible injury to spring-seeded rotational crops, regardless of rate used.

Do not apply ERADICANE EXTRA/PRINCEP 80W, ERADICANE EXTRA/PRINCEP CALIBER 90 or ERADICANE EXTRA/PRINCEP 4L in the High Plains and Intermountain areas of the WEST (including central and western KS, western NE, western OK, and the Panhandle of TX) where rainfall is sparse and erratic or where irrigation is required.

Injury may occur to soybeans planted in northcentral and northwest IA, south-central and southwest MN, northeast NE, southeast SD, and other areas the year following application on soils having a calcareous surface layer.

Do not plant sugar beets, tobacco, vegetables (including dry beans), spring-seeded small grains, or small-seeded legumes and grasses the year after ERADICANE EXTRA/PRINCEP 80W, ERADICANE EXTRA/PRINCEP CALIBER 90 or ERADICANE EXTRA/PRINCEP 4L application or injury may occur.

Tank Mixing Directions

1. Fill a thoroughly rinsed and uncontaminated spray tank 2/3 full of clean water.
2. Start agitation and continue moderate agitation throughout mixing.
3. Mix and disperse wettable powders first, flowable products next and ERADICANE EXTRA last.
4. Premix the BLADEX, ATRAZINE or PRINCEP products individually in proper order (step 3 above) making sure the products are thoroughly wetted and dispersed in the tank prior to adding ERADICANE EXTRA.
5. Add the ERADICANE EXTRA to the tank 2/3 full of water already mixed with BLADEX, ATRAZINE or PRINCEP.
6. Complete filling the remaining portion of the tank with water; mix thoroughly, maintain moderate agitation to provide a uniform mix during application.
7. The tank-mix combination should not be left in the spray tank for prolonged periods; batches should be mixed and applied the same day.

Do not use excessive agitation as it may cause the BLADEX, ATRAZINE or PRINCEP to settle out and set up in the spray tank. All return lines to the tank must discharge below liquid level.

APPENDIX I

Soil Mixing (Incorporation) Directions

ERADICANE EXTRA must be incorporated into the soil to prevent loss of the herbicide. Whenever possible, application and incorporation should be done in the same operation. If there must be a delay (less than four hours) in incorporation of the herbicide when application and incorporation are done in separate operations, ERADICANE EXTRA must be applied to a dry soil surface (at least 1/2 inch deep) and free from dew and incidental moisture. (Refer to special directions for Shattercane suppression).

Pacific Northwest Region (Semi-arid areas of Eastern Washington, Eastern Oregon and Idaho only): When application and incorporation are done in separate operations, ERADICANE EXTRA must be incorporated on the same day as applied. Application must be made on a dry soil surface (at least 1/2 inch deep) free from dew and incidental moisture.

DELAYED INCORPORATION OF DRY BULK FERTILIZER: The application and incorporation of ERADICANE EXTRA must be carried out on the same day and delayed only if soil moisture conditions permit. See Appendix III for directions on the delay in soil incorporation. (Refer to special instructions for Shattercane suppression.)

IMPORTANT: Thorough mixing of ERADICANE EXTRA in the soil is necessary. The soil must be in good tilth and cross discing is recommended. Improper discing or poor tilth (large clods) may result in unsatisfactory control.

Incorporation Directions:

The following equipment commonly is used for soil mixing (incorporation) before planting.

Power-Driven Cultivation Equipment: (Recommended on all soil types previously primary tilled): *Ground speed should be adjusted with all PTO driven cultivation tools to insure thorough soil mixing.*

a. Horizontal action: set to cut 4 to 6 inches deep. Example: Lely Roterras Niemeyer or like equipment.

b. Vertical action: set to cut 3 to 4 inches deep. Example: Roto-tiller, Sidewinder or like equipment.

Tandem Discs (Recommended on all soil types): Set to cut to a depth of 4 to 6 inches, operated at 4 to 6 mph followed by a spike-tooth harrow or some other leveling device which extends beyond the ends of the discs. For more thorough mixing (for control or suppression of deep-germinating grasses—shattercane, seedling johnsongrass, quackgrass and nutsedge), disc in two different directions. When discing the second time, the disc should be reset to cut no deeper than 4 to 6 inches.

Field Cultivators (Recommended on lighter soils in good tilth): Use 3 to 4 rows of sweeps spaced at 7 inch or less intervals and staggered so that no soil is left unturned, followed by a spike-tooth harrow pulled behind the cultivator. *Do not use chisel plows.* Set the cultivator to cut 4 to 6 inches deep, operated at 5 mph or more. The field cultivator should not be used alone as an incorporation tool where control or suppression of deep-germinating weeds such as nutgrass, quackgrass, johnsongrass seedlings or shattercane is desired. For control of these weeds, use a disc for at least one of the passes over the field.

Sprinkler Incorporation Pacific Northwest Region (semi-arid areas of Eastern Washington, Eastern Oregon and Idaho only): Surface apply ERADICANE EXTRA immediately after planting. The soil surface should be dry (at least 1/2 inch deep) and free from dew and incidental moisture. Incorporate using 1/2 to 3/4 inch of water within 36 hours following application. The application and incorporation must be done within 5 days after the last tillage operation since poor results will occur if weeds have germinated.

APPENDIX II

Procedure for Testing the Compatibility of ERADICANE EXTRA with Fluid Fertilizers

The following procedure is suggested for determining whether ERADICANE EXTRA may be combined with a specific fluid fertilizer for spray tank application.

Materials Required:

1. ERADICANE EXTRA
2. Fluid fertilizer to be used.
3. Adjuvant for fertilizer tank mix: Compex*, Sponto 168-D*, Unite*, or equivalent. The adjuvant which provides the best emulsification depends on the specific fertilizer under consideration.
4. Two one-quart, wide-mouth glass jars with lid or stopper.
5. Measuring spoons (a 25 mL pipette or graduated cylinder provides more accurate measurement).
6. Measuring cup, 8 oz. (257 mL).

*Compex, Kalo Laboratories Inc., Kansas City, Missouri; Sponto 168-D, Witco Chemical Company, Houston, Texas; Unite, Hopkins Agricultural Chemical Co., Madison, Wisconsin.

Procedure:

1. Pour a pint (about 473 mL) of the fluid fertilizer into each of the quart jars.
2. Add adjuvant to one of the jars and mix (see rate table on reverse side).
3. Add the ERADICANE EXTRA to both jars (see rate table).
4. Close both jars with lid or stopper and mix the contents by turning the jars upside down ten times.
5. Inspect the surface and body of the mixtures—
(A) Immediately after completing the jar inversions.
(B) After allowing the jars to stand quietly for 30 minutes.
(C) And then again after turning the jars upside down 10 times.

If a uniform mix cannot be made, the mixture should not be used. If either mixture remains uniform for 30 minutes, the combination may be used. Should either mixture separate after 30 minutes, but readily remix uniformly with ten jar inversions, the mixture can be used if adequate agitation is maintained in the tank. If the mixture with adjuvant is satisfactory, but the one without adjuvant is not, be sure to use the adjuvant in the spray tank. Add the adjuvant first at a rate of 3 pints per 100 gallons of fluid fertilizer; foaming can be minimized by using moderate agitation. If nondispersible oil, sludge or clumps of solids form in the mixtures, the combination should not be used.

NOTE 1: For some combinations, premixing wettable powders in a little water in a pail before adding them to the spray tank will improve the compatibility of the final mixtures with ERADICANE EXTRA. This technique can be tested in the small-scale jar test by premixing the wettable powder in one-eighth (1/8) cup of water prior to addition to the pint of fluid fertilizer.

Rate Table for ERADICANE EXTRA And Adjuvantwith the Fluid Fertilizer**

Gallons of fluid fertilizer to be applied per acre	mL or tsp. of ERADICANE EXTRA* to be added to 1 pint of fertilizer	
	mL	tsp.
10	8	1 3/5
15	5	1
20	4	4/5
25	3	2/3
30	2.5	1/2
40	2	2/5

*Based on field rate of 1 pound active ingredient per acre in the fertilizer volumes indicated. Increase volume proportionately to correspond with intended field rate in terms of pounds active ingredient per acre (e.g., for field rate of 4 pounds actual active ingredient in 40 gallons fertilizer per acre, add 8 mL or 1 3/5 tsp. ERADICANE EXTRA to each jar for compatibility testing purposes).

**Two (2) milliliters or two-fifths (2/5) teaspoon of adjuvant to be added to 1 pint of fluid fertilizer in order to equal the rate of 3 pints of adjuvant per 100 gallons of fluid fertilizer.

APPENDIX III

Impregnation on Dry Bulk Fertilizers

ERADICANE EXTRA, ERADICANE EXTRA/ATRAZINE (4L or 80W) or ERADICANE EXTRA/BLADEX (4L or 80W) or ERADICANE EXTRA/ATRAZINE/BLADEX

CAUTION: ERADICANE EXTRA alone or in combination with other herbicides must not be impregnated on ammonium nitrate, potassium nitrate or sodium nitrate fertilizers. Such mixtures may cause explosion and fire.

All individual state regulations relating to bulk dry fertilizer blending, registration, labeling and application are the responsibility of the individual and/or company selling the ERADICANE EXTRA or ERADICANE EXTRA/ATRAZINE (4L or 80W) and/or BLADEX (4L or 80W) and fertilizer mixtures.

ERADICANE EXTRA alone and in combination with ATRAZINE and/or BLADEX may be impregnated on many dry bulk fertilizers (1) (2) and applied and incorporated in the soil before planting for the control of grass and broadleaf weeds in corn.

Test results have shown that ERADICANE EXTRA, ERADICANE EXTRA/ATRAZINE (4L or 80W) and/or BLADEX (4L or 80W) on many dry bulk fertilizers give weed control equal to ERADICANE EXTRA or a tank mix of ERADICANE EXTRA/ATRAZINE (4L or 80W) and/or BLADEX (4L or 80W) applied as spray in water or liquid fertilizer. However, uniform impregnation of the pesticides on the dry fertilizer particles and uniform application in the field are necessary to assure good results.

A minimum of 200 pounds and a maximum of 700 pounds of approved ingredients (1) (2) impregnated with the appropriate amount of ERADICANE EXTRA alone or in combination with ATRAZINE (4L or 80W) and/or BLADEX (4L or 80W) must be applied per acre.

For impregnating the pesticides on dry fertilizers, use a closed rotary-drum mixer or a similar type of closed blender equipped with suitable spray equipment. The spray nozzle should be positioned inside of the mixer to provide uniform spray coverage of the tumbling fertilizer. The ERADICANE EXTRA should be sprayed uniformly onto the fertilizer using a fine spray pattern. For the ERADICANE EXTRA/ATRAZINE or ERADICANE EXTRA/BLADEX combination product, the ATRAZINE (4L or 80W) and/or BLADEX (4L or 80W) can be added and mixed uniformly with the fertilizer prior to the ERADICANE EXTRA addition, or alternately, the ERADICANE EXTRA/ATRAZINE (4L or 80W) and/or BLADEX (4L or 80W) can be mixed in the proposed use ratio to prepare a uniform slurry for joint spray impregnation.

The physicial properties of fertizers vary, particularly in liquid absorptive capacity. When absorptivity is sufficient, simple spray impregnation of the fertilizer with ERADICANE EXTRA, ERADICANE EXTRA/ATRAZINE (4L or 80W), and/or BLADEX (4L or 80W) provides a satisfactory, dry mixture.

If the absorptive capacity is not adequate, use of a drying agent is required to provide dry, freeflowing mixtures.

For mixtures to be used in spinning-disc applicators, Microcel E calcium silicate powder (Manville, Filtration & Minerals) is recommended for use as a drying agent.

For mixtures to be used in pneumatic applicators, Microcel E or Agsorb 16/30 RVM-MS granularclay (Oil-Dri Corporation) or Celatom MP-79 (Eagle-Picher Industries, Inc.) is recommended for use as the drying agent.

The drying agent should be added separately and uniformly to the prepared pesticide-fertilizer mixture, in a quantity that is sufficient to provide a suitable free-flowing mixture. Generally, less than 2% Microcel E or 5% Agsorb 16/30 RVM-MS and 5% Celatom MP-79 by weight is required.

The amount of ERADICANE EXTRA, ERADICANE EXTRA/ATRAZINE (4L or 80W) and/or BLADEX (4L or 80W) actually required in the manufacture of individual fertilizer mixtures should be determined carefully for each production operation. This is necessary to ensure that the amount of pesticides actually contained in the mixture applied to the soil represents the correct rate of use.

Bulk Fertilizer impregnated with ERADICANE EXTRA, ERADICANE EXTRA/ATRAZINE (4L or 80W) and/or BLADEX (4L or 80W) should be *applied immediately,* not stored. It is recommended that all bulk containers be tightly covered while the product is being transported and applied to reduce chances of ERADICANE EXTRA loss via volatilization.

When application and incorporation are done in separate operations, ERADICANE EXTRA impregnated on the dry fertilizer must be incorporated on the same day as applied. Application must be made on a dry soil surface (at least ½ inch deep) free from dew and incidental moisture. (Refer to special instructions for suppression of Shattercane.)

ERADICANE EXTRA Physical Data
Specific Gravity 68/68°F: 0.998 (typical)
Pounds/Gallon 68°F: 8.31 (typical)
Flashpoint: 194°F (setaflash closed cup)
Viscosity: sprayable down to minus 20°F

(1) Approved Dry Fertilizer Ingredients for use with ERADICANE EXTRA

	N	P	K
Ammonium sulfate	21	0	0
Diammonium phosphate	18	46	0
Potassium chloride	0	0	60
Potassium sulfate	0	0	52
Single Super-phosphate	0	20	0
Treble Super-phosphate	0	46	0
Urea*	45	0	0
Ammonium phosphate-sulfate	16	20	0

*Some ureas may be phytotoxic when high rates are applied to corn. Use only urea rates known to be safe for corn application.

NOTE: K-Mag and 11-48-0 are also approved for use.

(2) Approved Dry Fertilizer Ingredients for use with ERADICANE EXTRA/ATRAZINE (4L or 80W) or ERADICANE EXTRA/BALDEX (4L or 80W) OR ERADICANE EXTRA PLUS ATRAZINE AND BLADEX

	N	P	K
Ammonium Sulfate	21	0	0
Diammonium Phosphate	18	46	0
Potassium Chloride	0	0	60
Potassium Sulfate	0	0	52
Urea*	45	0	0
Ammonium phosphate-sulfate	16	20	0

*Some ureas may be phytotoxic when high rates are applied to corn. Use only urea rates known to be safe for corn application.

NOTE: K-Mag and 11-48-0 are also approved for use.

NOTE: Do not combine ERADICANE EXTRA plus ATRAZINE. (4L or 80W) or BLADEX (4L or 80W) with single super-phosphate (0-20-0) or treble super-phosphate (0-46-0).

(3) RATE CHART FOR IMPREGNATION OF DRY BULK FERTILIZERS (1)(2) WITH ERADICANE EXTRA OR ERADICANE EXTRA/ATRAZINE 4L OR 80W

FERTILIZER RATE PER ACRE	ERADICANE EXTRA RATE PER ACRE			ATRAZINE 4L RATE PER ACRE		ATRAZINE 80W RATE PER ACRE	
	4 pts.	5⅓ pts.	8 pts.	2 pts.	3 pts.	1¼ lbs.	2 lbs.
200 lbs.	20 qts/ton	26⅔ qts/ton	40 qts/ton	10 qts/ton	15 qts/ton	12½ lbs/ton	20 lbs/ton
250 lbs.	16 qts/ton	21⅓ qts/ton	32 qts/ton	8 qts/ton	12 qts/ton	10 lbs/ton	16 lbs/ton
300 lbs.	13⅓ qts/ton	17¾ qts/ton	26⅔ qts/ton	6⅔ qts/ton	10 qts/ton	8⅓ lbs/ton	13⅓ lbs/ton
350 lbs.	11½ qts/ton	15¼ qts/ton	23 qts/ton	5¾ qts/ton	8⅗ qts/ton	7⅕ lbs/ton	11⅖ lbs/ton
400 lbs.	10 qts/ton	13⅓ qts/ton	20 qts/ton	5 qts/ton	7½ qts/ton	6¼ lbs/ton	10 lbs/ton
450 lbs.	9 qts/ton	11⅘ qts/ton	17⅘ qts/ton	4½ qts/ton	6⅔ qts/ton	5⅔ lbs/ton	8⅞ lbs/ton
500 lbs.	8 qts/ton	10⅔ qts/ton	16 qts/ton	4 qts/ton	6 qts/ton	5 lbs/ton	8 lbs/ton
550 lbs.	7¼ qts/ton	9¾ qts/ton	14½ qts/ton	3⅔ qts/ton	5½ qts/ton	4⅔ lbs/ton	7¼ lbs/ton
600 lbs.	6⅔ qts/ton	9 qts/ton	13⅓ qts/ton	3⅓ qts/ton	5 qts/ton	4⅕ lbs/ton	6⅔ lbs/ton
650 lbs.	6¼ qts/ton	8¼ qts/ton	12⅓ qts/ton	3⅕ qts/ton	4⅗ qts/ton	3⅞ lbs/ton	6⅕ lbs/ton
700 lbs.	5¾ qts/ton	7⅔ qts/ton	11½ qts/ton	2⅞ qts/ton	4¼ qts/ton	3⅔ lbs/ton	5¾ lbs/ton

(4) RATE CHART FOR IMPREGNATION OF DRY BULK FERTILIZERS (1)(2) WITH ERADICANE EXTRA OR ERADICANE EXTRA/BLADEX 4L OR 80W

FERTILIZER RATE PER ACRE	ERADICANE EXTRA RATE PER ACRE			BLADEX 4L RATE PER ACRE		BLADEX 80W RATE PER ACRE	
	4pts.	5⅓ pts.	8 pts.	3 pts.	4 pts.	2 lbs.	2½ lbs.
200 lbs.	20 qts/ton	26⅔ qts/ton	40 qts/ton	15 qts/ton	20 qts/ton	20 lbs/ton	25 lbs/ton
250 lbs.	16 qts/ton	21⅓ qts/ton	32 qts/ton	12 qts/ton	16 qts/ton	16 lbs/ton	20 lbs/ton
300 lbs.	13⅓ qts/ton	17¾ qts/ton	26⅔ qts/ton	10 qts/ton	13½ qts/ton	13⅓ lbs/ton	16⅔ lbs/ton
350 lbs.	11½ qts/ton	15¼ qts/ton	23 qts/ton	8⅗ qts/ton	11½ qts/ton	11⅖ lbs/ton	14⅓ lbs/ton
400 lbs.	10 qts/ton	13⅓ qts/ton	20 qts/ton	7½ qts/ton	10 qts/ton	10 lbs/ton	12½ lbs/ton
450 lbs.	9 qts/ton	11⅘ qts/ton	17⅘ qts/ton	6⅔ qts/ton	9 qts/ton	8⅞ lbs/ton	11⅕ lbs/ton
500 lbs.	8 qts/ton	10⅔ qts/ton	16 qts/ton	6 qts/ton	8 qts/ton	8 lbs/ton	10 lbs/ton
550 lbs.	7¼ qts/ton	9¾ qts/ton	14½ qts/ton	5½ qts/ton	7¼ qts/ton	7¼ lbs/ton	9⅕ lbs/ton
600 lbs.	6⅔ qts/ton	9 qts/ton	13⅓ qts/ton	5 qts/ton	6⅔ qts/ton	6⅔ lbs/ton	8⅓ lbs/ton
650 lbs.	6¼ qts/ton	8¼ qts/ton	12⅓ qts/ton	4⅗ qts/ton	6¼ qts/ton	6⅕ lbs/ton	7⅔ lbs/ton
700 lbs.	5¾ qts/ton	7⅔ qts/ton	11½ qts/ton	4¼ qts/ton	5¾ qts/ton	5¾ lbs/ton	7⅕ lbs/ton

(5) RATE CHART FOR THE IMPREGNATION OF DRY BULK FERTILIZERS WITH ERADICANE EXTRA OR ERADICANE EXTRA PLUS ATRAZINE (80W OR 4L) PLUS BLADEX (80W OR 4L)

FERTILIZER RATE PER ACRE	ERADICANE EXTRA RATE PER ACRE			ATRAZINE 80W RATE PER ACRE		ATRAZINE 4L RATE PER ACRE	
	4 pts. per acre	5 1/3 pts. per acre	8 pts. per acre	3/4 lb per acre	1 1/4 lbs per acre	1 pt. per acre	2 pts. per acre
200 lbs.	20 qts/ton	26 2/3 qts/ton	40 qts/ton	7 1/2 lbs/ton	12 1/2 lbs/ton	5 qts/ton	10 qts/ton
250 lbs.	16 qts/ton	21 1/3 qts/ton	32 qts/ton	6 lbs/ton	10 lbs/ton	4 qts/ton	8 qts/ton
300 lbs.	13 1/3 qts/ton	17 3/4 qts/ton	26 2/3 qts/ton	5 lbs/ton	8 1/3 lbs/ton	3 1/3 qts/ton	6 3/5 qts/ton
350 lbs.	11 1/2 qts/ton	15 1/4 qts/ton	23 qts/ton	4 1/3 lbs/ton	7 1/8 lbs/ton	2 7/8 qts/ton	5 4/5 qts/ton
400 lbs.	10 qts/ton	13 1/3 qts/ton	20 qts/ton	3 3/4 lbs/ton	6 1/4 lbs/ton	2 1/2 qts/ton	5 qts/ton
450 lbs.	9 qts/ton	11 4/5 qts/ton	17 4/5 qts/ton	3 1/3 lbs/ton	5 1/2 lbs/ton	2 1/4 qts/ton	4 2/5 qts/ton
500 lbs.	8 qts/ton	10 2/3 qts/ton	16 qts/ton	3 lbs/ton	5 lbs/ton	2 qts/ton	4 qts/ton
550 lbs.	7 1/4 qts/ton	9 3/4 qts/ton	14 1/2 qts/ton	2 3/4 lbs/ton	4 2/3 lbs/ton	1 7/8 qts/ton	3 2/3 qts/ton
600 lbs.	6 2/3 qts/ton	9 qts/ton	13 1/3 qts/ton	2 1/2 lbs/ton	4 1/5 lbs/ton	1 2/3 qts/ton	3 1/3 qts/ton
650 lbs.	6 1/4 qts/ton	8 1/4 qts/ton	12 1/3 qts/ton	2 1/3 lbs/ton	3 7/8 lbs/ton	1 3/5 qts/ton	3 1/8 qts/ton
700 lbs.	5 3/4 qts/ton	7 2/3 qts/ton	11 1/2 qts/ton	2 1/5 lbs/ton	3 2/3 lbs/ton	1 1/2 qts/ton	2 7/8 qts/ton

(6) Rate Chart for the Impregnation of Dry Bulk Fertilizers with ERADICANE EXTRA or ERADICANE EXTRA PLUS ATRAZINE (80W or 4L) PLUS BLADEX (80W or 4L)

Fertilizer Rate Per Acre	Bladex 80W Rate Per Acre		Bladex 4L Rate Per Acre	
	1 1/4 lbs per acre	2 1/2 lbs per acre	1 qt per acre	2 qts per acre
200 lbs.	12 1/2 lbs/ton	25 lbs/ton	10 qts/ton	20 qts/ton
250 lbs.	10 lbs/ton	20 lbs/ton	8 qts/ton	16 qts/ton
300 lbs.	8 1/3 lbs/ton	16 2/3 lbs/ton	6 3/5 qts/ton	13 1/3 qts/ton
350 lbs.	7 1/8 lbs/ton	14 1/4 lbs/ton	5 4/5 qts/ton	11 2/5 qts/ton
400 lbs.	6 1/4 lbs/ton	12 1/2 lbs/ton	5 qts/ton	10 qts/ton
450 lbs.	5 1/2 lbs/ton	11 1/8 lbs/ton	4 2/5 qts/ton	8 7/8 qts/ton
500 lbs.	5 lbs/ton	10 lbs/ton	4 qts/ton	8 qts/ton
550 lbs.	4 2/3 lbs/ton	9 1/8 lbs/ton	3 2/3 qts/ton	7 1/3 qts/ton
600 lbs.	4 1/5 lbs/ton	8 1/3 lbs/ton	3 1/3 qts/ton	6 2/3 qts/ton
650 lbs.	3 7/8 lbs/ton	7 3/4 lbs/ton	3 1/8 qts/ton	6 1/5 qts/ton
700 lbs.	3 2/3 lbs/ton	7 1/5 lbs/ton	2 7/8 qts/ton	5 3/4 qts/ton

STORAGE AND DISPOSAL

Prohibitions: Do not contaminate water, food, or feed by storage or disposal. Open dumping is prohibited. Keep away from heat and flame.

Storage: Keep container tightly closed when not in use. Do not store near seeds, fertilizers, or foodstuffs. Can be stored at temperatures as low as minus 50°F.

Pesticide Disposal: Waste resulting from the use of this product may be disposed of on site or at an approved waste disposal facility.

Container Disposal: Metal—Triple rinse (or equivalent), then offer for recycling or reconditioning, or puncture and dispose of in a sanitary landfill, or by other procedures approved by state and local authorities.

Plastic—Triple rinse (or equivalent), then offer for recycling or reconditioning, or puncture and dispose of in a sanitary landfill, or by incineration, or if allowed by state and local authorities, by burning. If burned, stay out of smoke.

NOTICE TO BUYER AND USER:

Seller warrants that this product conforms to the chemical description on the label and is reasonably fit for the purposes stated on the label when used in accordance with directions under normal conditions of use. This warranty does not extend to the use of this product contrary to label instructions, or under abnormal use conditions, or under conditions not reasonably foreseeable to Seller, and Buyer and User assume the risk of any such use. SELLER DISCLAIMS ALL OTHER WARRANTIES EXPRESSED OR IMPLIED INCLUDING ANY WARRANTY OF FITNESS OR MERCHANTABILITY. SELLER SHALL NOT BE LIABLE FOR CONSEQUENTIAL, SPECIAL OR INDIRECT DAMAGES RESULTING FROM THE USE OR HANDLING OF THIS PRODUCT AND SELLER'S SOLE LIABILITY AND BUYER'S AND USER'S EXCLUSIVE REMEDY SHALL BE LIMITED TO THE REFUND OF THE PURCHASE PRICE.

This product is sold only for uses stated on its label. No express or implied license is granted to use or sell this product under any patent in any country except as specified: Country: United States of America. Patent No. 4,021,224; 4,137,070 and other patents pending.

A-9R

The information and specimen labels in this booklet are correct as of the publication date but are subject to change without notice. For current information, contact ICI Americas Inc.

FORCE® 1.5G

INSECTICIDE

RESTRICTED USE PESTICIDE

Toxic to fish and aquatic organisms.

For retail sale to and use only by Certified Applicators, or persons under their direct supervision, and only for those uses covered by the Certified Applicator's certification.

For Use to Control Insects On Field corn, Popcorn and Seedcorn.

ACTIVE INGREDIENT

Tefluthrin

(2,3,5,6-tetrafluoro-4-methylphenyl)-methyl-(1α,3α,)-(Z)-(±)-3-(2-chloro-3,3,3-trifluoro-1-propenyl)-2,2-dimethylcyclopropanecarboxylate 1.5%

INERT INGREDIENTS. 98.5%

TOTAL. 100.0%

Contain 1.5 lbs. active ingredient per 100 lbs. of granular insecticide.

EPA Reg. No.: 10182-130

KEEP OUT OF REACH OF CHILDREN

WARNING

AVISO Precaucion Al Usario:

Si usted no lee ingles, no use este producto hasta que la etiqueta le haya sido explicada ampliamente.

See back panel for additional precautionary statements and statement of practical treatment.

PRECAUTIONARY STATEMENTS

HAZARDS TO HUMANS AND DOMESTIC ANIMALS

WARNING

CAUSES EYE INJURY. Do not get in eyes or on clothing. HARMFUL IF SWALLOWED, INHALED, OR ABSORBED THROUGH SKIN. Avoid contact with skin. Avoid breathing dust. Wear goggles or face shield when handling. Wash thoroughly with soap and water after handling. Remove contaminated clothing and wash before reuse.

STATEMENT OF PRACTICAL TREATMENT

If In Eyes: Flush eyes with plenty of water. Get medical attention if irritation persists.

If Swallowed: Call a physician or Poison Control Center. Drink 1 or 2 glasses of water and induce vomiting by touching back of throat with finger. Do not induce vomiting or give anything to an unconscious person.

If On Skin: Wash with plenty of soap and water. Get medical attention if irritation persists.

FOR 24-HOUR EMERGENCY MEDICAL ASSISTANCE CALL 1-800-F-A-S-T-M-E-D (327-8633).

FOR CHEMICAL EMERGENCY: Spill, leak, fire, exposure, or accident call CHEMTREC 1-800-424-9300.

ENVIRONMENTAL HAZARDS

This pesticide is very highly toxic to freshwater and estuarine fish and invertebrates. Drift and runoff may be hazardous to aquatic organisms in neighboring areas. Do not apply directly to water or wetlands (swamps, bogs, marshes and potholes). Do not contaminate water when disposing of equipment washwater.

DIRECTIONS FOR USE

It is a violation of Federal law to use this product in a manner inconsistent with its labeling.

Do not rotate to crops other than corn.

At Planting Band Treatment

Apply FORCE 1.5G at planting time. Place the granules in a 7-inch wide band over the row behind the planter shoe, and in front of/or behind the press wheel. Incorporate into the top one inch of soil using suitable equipment. For best wireworm control, apply the granules as an in-furrow treatment.

At Planting In-Furrow Treatment

Apply FORCE 1.5G at planting time as an in-furrow treatment for control of corn rootworms, wireworms, and seedcorn maggots. In-furrow applications may be made by 1) directing the granules into the planter shoe with the seed, of 2) placing the applicator tube directly behind the planter shoe so that the granules drop into the seed furrow.

FORCE 1.5G RATE RECOMMENDATIONS

Time of Application	Controlled	(oz/1000 ft of row)	Application
At Planting	Western Corn Rootworm Northern Corn Rootworm Southern Corn Rootworm Black Cutworm Seedcorn Maggot Seedcorn Beetle Lesser Cornstalk Borer White Grubs (Suppression) Wireworms (Suppression)	8-10 Banded or In-Furrow	Banded - Place granules in a 7-inch band directly behind planter shoe in front of/or behind the press wheel. In-Furrow - Place granules directly in the seed furrow behind the planter shoe.

ENDANGERED SPECIES RESTRICTIONS

This product may pose a hazard to endangered aquatic species. For ground application, do not apply this product within 20 yards of water (ponds, streams or lakes).

Calibration Table

Amount of FORCE 1.5G/1000 ft of row	Pounds of FORCE 1.5G Required Per Acre for Various Row Spacings					
	Linear Ft/A					
	13,068	13,756	14,520	15,374	16,335	17,424
	Row Spacing					
	40"	38"	36"	34"	32"	30"
8 oz	6.5	6.9	7.3	7.7	8.2	8.7
10 oz	8.2	8.6	9.1	9.6	10.2	10.9

STORAGE AND DISPOSAL

Do not contaminate water, food or feed by storage or disposal.

STORAGE: Do not store or use in or around the home. Store in original containers only. In case of spill on floor or paved surfaces, sweep up and remove to chemical waste area.

PESTICIDE DISPOSAL: Waste resulting from the use of this product may be disposed of on site or at an approved waste disposal facility.

CONTAINER DISPOSAL: Completely empty bag into application equipment. When bag is empty dispose of in a sanitary landfill, by incineration or, if allowed by State and local authorities, by burning. If burned, stay out of smoke.

NOTICE TO BUYER AND USE

Seller warrants that this product conforms to the chemical description on the label and is reasonably fit for the purpose stated on the label when used in accordance with directions under normal conditions of use. This warranty does not extend to the use of this product contrary to label instructions, or under abnormal use conditions, or under conditions not reasonably foreseeable to Seller, and Buyer and User assume the risk of any such use. SELLER DISCLAIMS ALL OTHER WARRANTIES, EXPRESSED OR IMPLIED, INCLUDING ANY WARRANTY OF FITNESS OR MERCHANTABILITY. SELLER SHALL NOT BE LIABLE FOR CONSEQUENTIAL, SPECIAL OR INDIRECT DAMAGES RESULTING FROM THE USE OR HANDLING OF THIS PRODUCT AND SELLER'S SOLE LIABILITY AND BUYER'S AND USER'S EXCLUSIVE REMEDY SHALL BE LIMITED TO THE REFUND OF THE PURCHASE PRICE.

A-4R

The information and specimen labels in this booklet are correct as of the publication date but are subject to change without notice. For current information, contact ICI Americas Inc.

FUSILADE® 2000

HERBICIDE

Postemergence Herbicide for Control of Perennial and Annual Grass Weeds

ACTIVE INGREDIENT:
Fluazifop-p-butyl
Butyl (R)-2-[4-[[5-trifluoromethyl)-2-pyridinyl]oxy]
phenoxypropanoate* 13%
INERT INGREDIENTS 87%

TOTAL 100%

EPA Reg. No. 10182-104

*FUSILADE 2000 contains 1 pound (+) isomer (fluazifop-p-butyl) per gallon

KEEP OUT OF REACH OF CHILDREN

CAUTION

STATEMENT OF PRACTICAL TREATMENT

IF IN EYES: Immediately flush eyes with plenty of water. Call physician if irritation persists.
IF ON SKIN: Wash with plenty of soap and water. Get medical attention if irritation persists.

FOR 24-HOUR EMERGENCY MEDICAL ASSISTANCE CALL 1-800-F-A-S-T-M-E-D (327-8633).
FOR CHEMICAL EMERGENCY: Spill, leak, fire, exposure, or accident call CHEMTREC 1-800-424-9300.

PRECAUTIONARY STATEMENTS

HAZARDS TO HUMANS AND DOMESTIC ANIMALS

CAUTION

HARMFUL IF ABSORBED THROUGH SKIN OR INHALED. CAUSES EYE IRRITATION. MAY CAUSE ALLERGIC SKIN REACTIONS. A·oid contact with skin, eyes or clothing. Avoid breathing vapor or spray mist. Wear protective clothing and PVC gloves when spraying. Wear protective clothing, PVC gloves, apron, and full face shield when handling or mixing concentrate. Remove contaminated clothing and wash before reuse. Wash thoroughly with soap and water after handling.

ENVIRONMENTAL HAZARDS

This product is toxic to fish. Do not apply directly to water or wetlands. Do not contaminate water by cleaning of equipment or disposal of wastes. Do not apply when weather conditions favor drift from target area.

PHYSICAL OR CHEMICAL HAZARDS

Do not use or store near heat or open flame.

DIRECTIONS FOR USE

It is a violation of Federal law to use this product in a manner inconsistent with its labeling.

This labeling must be in the possession of the user at the time of application.

REENTRY STATEMENTS

Do not apply this product in such a manner as to directly or through drift expose workers or other persons. The area being treated must be vacated by unprotected persons.

Do not treat areas while unprotected humans or domestic animals are present in the treatment areas. Do not allow entry into treated areas without protective clothing until sprays have dried. Because certain states may require more restrictive reentry intervals for various crops treated with this product, consult your State Department of Agriculture for further information.

Written or oral warnings must be given to workers who are expected to be in a treated area or in an area about to be treated with this product. When oral warnings are given, warnings shall be given in a language customarily understood by workers. Oral warnings must be given if there is reason to believe that written warnings cannot be understood by workers. Warnings must include the following information:

CAUTION: Area treated with FUSILADE® 2000 on (date of application). Do not enter without appropriate protective clothing until sprays have dried. In case of accidental exposure to pesticide spray, wash the skin thoroughly with soap and water. Remove contaminated clothing and wash before reuse. If in eyes, flush with plenty of water. If irritation persists, get medical attention.

GENERAL INFORMATION

Read all label directions before using.

FUSILADE® 2000 herbicide is a selective postemergence herbicide for control of annual and perennial grass weeds in apricots, asparagus (restricted to Maryland, New Jersey, Oregon, and Washington), carrots, cherries, coffee (restricted to Hawaii), cotton, endive (restricted to Florida), garlic, nectarines, onions (dry bulb) peaches, pecans, plums, prunes, rhubarb (restricted to Maryland and New Jersey), soybeans, spinach, sweet potatoes, Tobasco peppers (restricted to Louisianna) and yams. FUSILADE 2000 can also be used in fallow land, in noncrop areas, in nonbearing crops, and in ornamentals. FUSILADE 2000 will provide effective control of grass weeds in conventional tillage, minimum tillage, and no-till plantings. FUSILADE 2000 does not control broadleaf weeds or sedges (nutgrass).

FUSILADE 2000 is a systemic herbicide which moves from the treated foliage into the shoots, roots, rhizomes, stolons, and growing points (meristematic regions) of treated grass weeds.

FUSILADE 2000 is rainfast in one hour.

Control Symptoms: Thorough coverage of all weed plant foliage is important for good activity. Optimum weed control is achieved when young actively growing weeds are treated that are not under stress from moisture, temperature, low soil fertility, mechanical, or chemical injury.

Growth of treated grass weeds stops soon after application. Symptoms include loss of vigor, yellowing and/or reddening, and eventual death to the treated grass plant. Symptoms are generally observed within one week after treatment, depending on grass weed species and environmental conditions.

APPLICATION DIRECTIONS

Timing—Best control of susceptible grasses is obtained when FUSILADE 2000 is applied to actively growing grasses before they exceed the recommended growth stages shown on this label. Refer to the grass weed table for specific recommendations on weed growth stages.

Spray Additives—Only crop oil concentrate and nonionic surfactants cleared for use on growing crops under 40 CFR 180.1001 may be used in spray mixture.

ALWAYS ADD ONE OF THE FOLLOWING:

- Crop Oil Concentrate—Add a non-phytotoxic crop oil concentrate or a once-refined vegetable oil concentrate containing 15-20% approved emulsifier, at 0.5-1% v/v (1-2 pints per 25 gallons) of the finished spray volume. For aerial applications, add 1 pint of crop oil concentrate per acre.
- Nonionic Surfactant—Add nonionic surfactant containing at least 75% surface-active agent, at 0.25-0.5% v/v (1/2-1 pint per 25 gallons) of the finished spray volume for ground sprays. For aerial application, add 1 pint of surfactant per acre.

In addition to crop oil concentrate or nonionic surfactant, liquid nitrogen fertilizer (28% or similar) can be added to the spray mixture in soybeans only. This 28% liquid nitrogen fertilizer is water soluble and should be used at a rate of one gallon per acre. Liquid nitrogen fertilizers should not be used as a substitute for crop oil concentrate or nonionic surfactant in the spray mixture.

GROUND APPLICATION—Use sufficient spray volume and pressure to ensure complete coverage of the target grasses. Apply in 5-40 gallons per acre of spray mixture with spray pressures of 40-60 psi at the nozzle tip. When grass foliage is dense, use 60 psi and a minimum of 20 gallons per acre to ensure coverage of weed foliage. Do not exceed the maximum application rates for FUSILADE 2000.

DO NOT USE FLOOD TYPE OR OTHER SPRAY NOZZLE TIPS WHICH DELIVER COARSE, LARGE DROPLET SPRAYS.

DO NOT APPLY FUSILADE 2000 WITH RECIRCULATING SPRAYERS, ROPEWICK APPLICATORS, CONTROLLED DROPLET APPLICATORS (CDA) OR ANY SIMILAR DEVICES.

DO NOT APPLY THIS PRODUCT THROUGH ANY TYPE OF IRRIGATION SYSTEM.

BAND APPLICATIONS—Thorough weed coverage is important for control. Best coverage is obtained with a minimum of two nozzles, one directed to each side of the planted row. Application with a single nozzle directed over the top of the row is not recommended. Cultivation of untreated areas may be needed following band applications.

When making band applications and cultivating in the same operation, position nozzles in advance of the cultivation device. This will reduce dust in the spray area. Dust can intercept the spray, reducing weed coverage resulting in less than adequate weed control.

Calculate the amount of herbicide and water volume needed for band treatment by the following formulas:

$$\frac{\text{Band width in inches}}{\text{row width in inches}} \times \text{broadcast rate per acre} = \text{Band herbicide rate per acre}$$

$$\frac{\text{Band width in inches}}{\text{row width in inches}} \times \text{broadcast volume per acre} = \text{Band water volume per acre}$$

Band applications to perennial grasses are not recommended as reinfestation of the treated band from the untreated middle may result.

AERIAL APPLICATION—Use sufficient spray volume to ensure complete coverage of target grasses. A minimum of 5 gallons per acre should be applied. When grass foliage is dense, use a minimum of 10 gallons per acre to ensure coverage of weed foliage. Add one pint/A of crop oil concentrate or nonionic surfactant in the spray mixture.

CULTIVATION—Cultivation of treated grasses is not recommended within 7 days prior to or within 7 days after application of FUSILADE 2000 as weeds may be put under stress reducing weed control. Timely cultivation 2-3 weeks after applying FUSILADE 2000 may assist weed control.

SPOT TREATMENTS

Mix FUSILADE 2000 and crop oil concentrate or nonionic surfactant with water according to the amounts shown below. Spray to obtain thorough coverage, but do not spray to runoff. Re-treat if necessary.

Spot Spray Recommendations

To Make This Spray Volume	Add These Amounts: FUSILADE 2000	Crop/Vegetable Oil Concentrate*	or	Nonionic Surfactant
1 gallon	1½ fl. oz. (3 tbsp.)	1½ fl. oz.	or	½ fl. oz.
10 gallons	13 fl. oz.	13 fl. oz.		3 fl. oz.
25 gallons	1 quart	1 quart		½ pint
50 gallons	2 quarts	2 quarts		1 pint

*Do not use a crop oil concentrate on ornamentals. Use only nonionic surfactant on ornamentals.

SPECIAL INSTRUCTIONS FOR USE IN COFFEE

Apply FUSILADE 2000 as a postemergence ground application. Apply with sufficient spray volume to ensure complete coverage of the target grasses. Applications can be made as a broadcast, strip band, or spot spray application.

Apply 32-64 oz. per acre of FUSILADE 2000 herbicide along with 1% crop oil concentrate (2 pints per 25 gallons) or 0.25% (½ pint per 25 gallons) of a nonionic surfactant to annual and perennial grasses at the recommended growth stages in the following Tables.

SPECIAL INSTRUCTIONS FOR USE IN APRICOTS, CHERRIES, NECTARINES, PEACHES, PECANS, PLUMS, AND PRUNES

Apply FUSILADE 2000 as a directed postemergence application in the interspaces and around the base of the trees avoiding contact of spray with foliage of trees. Apply with sufficient spray volume and pressure to ensure complete coverage of target grasses. Applications can be made as a broadcast, strip band or spot spray application at rates and growth stages listed in the weed tables.

REFER TO THE GRASS WEED TABLE FOR SPECIFIC RECOMMENDATIONS ON WEED GROWTH STAGES.

GENERAL USE PRECAUTIONS

- Apply to actively growing grasses. Do not apply to grasses which are stressed due to moisture, temperature, low soil fertility, mechanical or chemical injury.
- Apply at the recommended rate to grasses at the recommended growth stages as outlined. Do not apply to grasses which have tillered, formed seed heads, or exceeded recommended growth stages.
- Apply when the first grass weed species in a mixed grass weed population reaches the recommended growth stage for treatment. Use the highest recommended rate for grasses in that group.
- Where irrigation is used as part of normal cropping practice, best results are usually obtained when FUSILADE 2000 is applied within 7 days after irrigation.
- Best perennial grass control can be obtained if rhizomes or stolons are cut up by preplant tillage practices (discing, plowing, etc.) to stimulate maximum emergence of grass shoots.
- Do not plant rotational grass crops such as corn, sorghum, and cereals within 60 days of last application of FUSILADE 2000.
- Avoid drift to all other crops and nontarget areas. Grass crops are highly susceptible to FUSILADE 2000 herbicide.
- Do not tankmix FUSILADE 2000 herbicide with other pesticides, liquid fertilizers or any other additives except as specified on this label or other supplemental labeling. Use of liquid nitrogen fertilizer (28% or similar) as recommended on this label is for use in soybeans only.
- Sequential applications of other herbicides except as specified on this label or on supplemental labeling within five days before or after FUSILADE 2000 herbicide application may result in crop injury and/or reduced grass control.
- Thoroughly clean spray tank with water and a commercial tank cleaner before and after each use.
- Do not apply FUSILADE 2000 if rainfall is expected within 1 hour.

APRICOTS, CHERRIES, NECTARINES, PEACHES, PLUMS AND PRUNES

(Not for use in California)

- Do not apply a total of more than 144 fl oz of FUSILADE 2000 Herbicide per acre per season.
- Do not harvest within 14 days of application.

ASPARAGUS

(New Jersey, Maryland, Oregon and Washington only)

- Do not apply a total of more than 96 oz. of FUSILADE 2000 herbicide per acre per season to bearing asparagus.
- Do not harvest asparagus within one day of application.

CARROTS (including seed carrots, DRY BULB ONIONS AND GARLIC

- Do not apply a total of more than 96 oz. of FUSILADE 2000 herbicide per acre per season to carrots, dry bulb onions or garlic.
- Do not harvest carrots, dry bulb onions or garlic within 45 days after application.

COFFEE (Hawaii only)

- Do not apply more than a total of 96 fl oz of FUSILADE 2000 Herbicide per acre per season to coffee.
- Do not harvest coffee within 1 day of application.

COTTON

- Do not apply a total of more than 96 oz. of FUSILADE 2000 herbicide per acre per season to cotton.
- Do not apply to cotton after boll set.
- Do not harvest cotton within 90 days of FUSILADE 2000 application.
- Do not graze or harvest for forage or hay.

ENDIVE (Florida only).

- Do not apply a total of more than 96 oz. of FUSILADE 2000 Herbicide per acre per season to endive.
- Do not harvest endive within 28 days of application.

PECANS (Not for use in California)

- Do not apply a total of more than 144 fl oz of FUSILADE 2000 Herbicide per acre per season.
- Do not harvest within 30 days of application.

RHUBARB (Maryland and New Jersey only)

- Do not apply a total of more than 72 oz. of FUSILADE 2000 herbicide per acre per season to rhubarb.
- Do not harvest rhubarb within 14 days of the spring application.

SOYBEANS

- Do not apply a total of more than 64 oz. of FUSILADE 2000 herbicide per acre per season to soybeans.
- Make the last FUSILADE 2000 application to soybeans before bloom.
- Do not graze or harvest for forage or hay.

SPINACH

- Do not apply a total of more than 96 oz. of FUSILADE 2000 herbicide per acre per season.
- Do not harvest within 30 days of application.

SWEET POTATOES AND YAMS

- Do not apply a total of more than 96 oz. of FUSILADE 2000 herbicide per acre per season to sweet potatoes and yams.
- Do not harvest within 55 days of application.

TABASCO PEPPERS (Louisiana Only)

- Do not apply a total of more than 96 oz. of FUSILADE 2000 herbicide per acre per season to Tabasco peppers.
- Do not harvest Tabasco peppers within 45 days of application.

DIRECTIONS FOR USE

ANNUAL AND PERENNIAL GRASS WEED CONTROL FOR COTTON, SOYBEANS, AND OTHER REGISTERED CROPS

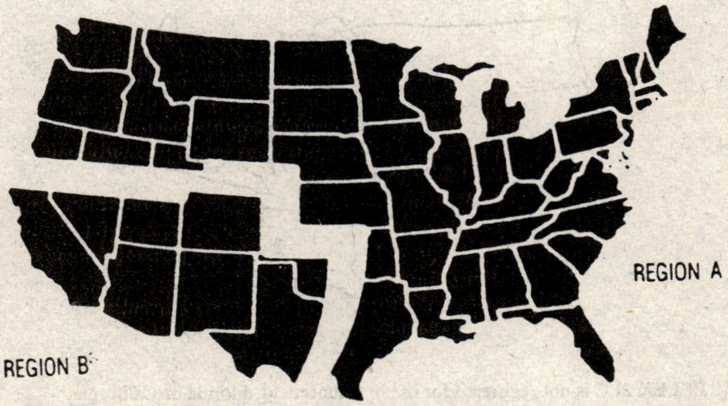

Region A—Includes the following states or portions of states where FUSILADE 2000 can be applied at recommended rates.

Alabama, Arkansas, California (Northern; see regional map), Connecticut, Delaware, Florida, Georgia, Idaho, Illinois, Indiana, Iowa, Kansas, Kentucky, Louisiana, Maine, Maryland, Massachusetts, Michigan, Minnesota, Mississippi, Missouri, Montana, Nebraska, Nevada (Northern; see regional map), New Hampshire, New Jersey, New York, North Carolina, North Dakota, Ohio, Oklahoma (east of Interstate 35), Oregon, Pennsylvania, Rhode Island, South Carolina, South Dakota, Tennessee, Texas (east of Interstate 35), Utah (Northern; see regional map), Vermont, Virginia, Washington, West Virginia, Wisconsin, and Wyoming.

Region B – Includes the following states or portions of states where FUSILADE 2000 can be applied at recommended rates.

Arizona, California (Southern; see regional map), Colorado, Hawaii, Oklahoma (west of Interstate 35), Nevada (Southern; see regional map), New Mexico, Utah (Southern; see regional map), and Texas (west of Interstate 35).

Cotton, Soybeans, and Other Registered Crops
Fusilade 2000 Herbicide Annual Grass Control Use Rate Table

Region A

Annual Grass Species[1]	Ht. (In.)	No. of Leaves Not to Exceed	Rate(s) fl. oz/A	Special[3] Rate fl. oz/A
Barnyardgrass	2-3	3	24	20
Broadleaf signalgrass	2-4	5	24	20
Crabgrass				
Large crabgrass	1-2	4	24	20
Smooth crabgrass	1-2	4	24	20
Southern Crabgrass	1-2	4	24	20
Tropical Crabgrass	1-2	4	24	20
Downy Brome[2]	2-6	4	16	
Fall Panicum	2-6	6	24	20
Field Sandbur	2-4	4	24	20
Foxtails				
Giant foxtail	2-6	4	24	20
Green foxtail	2-4	4	24	20
Yellow foxtail	2-4	4	24	20
Goosegrass	2-4	6	16	
Italian Ryegrass	2-4	4	24	20
Itchgrass	4-24	6	16	
Johnsongrass, seedling	2-8	4	12	
Junglerice	2-3	3	24	20
Red rice	½-1	2	32	20[4]
Shattercane	6-12	8	12	
Sorghum almum	6-12	8	16	
Southern sandbur	2-6	6	24	20
Texas Panicum	2-8	8	24	20
Volunteer Cereals				
V. Barley	2-6	6	16	
V. Corn	12-24	10	12	
V. Milo	6-12	4	12	
V. Oats	2-6	6	16	
V. Rye	2-6	6	16	
V. Wheat	2-6	6	16	
Wild Proso Millet	4-8	6	12	
Witchgrass	2-4	6	24	20
Wild Oats	2-6	6	16	
Wooly cupgrass	2-4	6	24	20

[1] Retreatment at the recomended rate may be needed to control later germinating grasses or if regrowth occurs.

[2] Not registered in California for use on this species.

[3] USE DIRECTIONS FOR SPECIAL RATE

Rates of FUSILADE 2000 herbicide/acre can be reduced to 20 oz. for species indicated under the following conditions:

- Application under favorable soil moisture and humidity conditions, normally within a few days after rainfall or irrigation. Avoid extreme air temperatures.
- Application at earliest growth stages indicated on rate tables.
- Application in highly competitive crop stands such as narrow row or drilled soybeans, or where cultivation is planned.
- Application to light or moderate weed densities.
- Application with 1% v/v crop oil concentrate only.
- Application alone, avoiding tankmixes with other pesticides.

[4] Most infestations require two applications at 20 oz. of FUSILADE 2000 herbicide/acre to achieve satisfactory control.

Cotton, Soybeans, and Other Registered Crops (except Coffee)
Fusilade 2000 Herbicide Annual Grass Control Use Rate Table

Region B

Annual Grass Species[1]	Ht. (In.)	No. of Leaves Not to Exceed	Rate(s) fl. oz/A
Barnyardgrass[2]	1-2	3	24
Canarygrass, Littleseed	2-4	4	24
Crabgrass, Large[3]	1-2	3	24
Johnsongrass, seedling	2-4	3	16
Junglerice[3]	2-3	3	24
Volunteer Cereals			
V. Barley	2-4	3	24
V. Corn[3]	12-18	6	24
V. Milo	2-4	4	24
V. Oats	2-4	3	24
V. Wheat	2-4	3	24
Wild Oats	2-4	4	24

[1] Retreatment at the recommended rate may be needed to control late germinating grasses or if regrowth occurs.
[2] Use 48 oz. in California and Arizona.
[3] Not labeled for control of this species in Arizona and California.

Cotton, Soybeans, and Other Registered Crops
Fusilade 2000 Herbicide Perennial Grass Control Use Rate Table

Region A

Perennial Grass Species	Application	Ht. (In.)	Rate fl. oz/A
Bermudagrass[1]	1st	4-8 (runner length)	24 oz.
	2nd	4-8	16 oz.
Quackgrass[2]	1st	6-10	24 oz.
	2nd	up to 10 in.	16 oz.
Rhizome johnsongrass[3]	1st	8-18	24 oz.
	2nd	6-12	16 oz.
Wirestem muhly[4]	1st/2nd	4-12	24 oz.

[1] Make second application to Bermudagrass if regrowth occurs (usually about 4 weeks after first application). Bermudagrass control may be improved by directing the spray beneath the crop canopy. To improve coverage, make applications in a minimum of 15 gallons per acre.
[2] Make second application 2-3 weeks after the first, but before the quackgrass exceeds 10 inches in height. Always use 1% v/v crop oil concentrate for quackgrass control. In no-till soybeans, a preplant application of GRAMOXONE® SUPER herbicide is recommended. Spot treatment is not recommended.
[3] Make first application before the boot stage. In eastern Oklahoma, the Brazos Bottoms, the Blacklands, Coastal Bend and Rio Grande areas of eastern Texas, make the first application at 8-12 inches. If new shoots emerge or regrowth occurs, make a second application at 4-6 inches.
[4] Make a second application if regrowth occurs.

Cotton, Soybeans, and Other Registered Crops (except Coffee)
Fusilade 2000 Herbicide Perennial Grass Control Use Rate Table

Region B

For best results, apply FUSILADE 2000 3 days before to 7 days after irrigation.

Grass Species	Application	Ht. (In.)	Rate fl. oz/A
Bermudagrass[1]	1st	4-8 (runner length)	32-48
	2nd	4-8	32-48
Rhizome johnsongrass[2]	1st	12-18	32-48[3]
	2nd	12-18	32-48[3]

[1] Make second application to Bermudagrass if regrowth occurs. (Usually about 4 weeks after first application.) Bermudagrass control may be improved by directing the spray beneath the crop canopy. To improve coverage, make applications in a minimum of 20 gallons per acre. In soybeans, do not apply more than 64 oz. in one season.
[2] Make first application before the boot stage. If new shoots emerge or regrowth occurs, make a second application. (Usually about 4 weeks after the first application.) In soybeans, do not apply more than 64 oz. in one season.
[3] *For California only,* make first application to Johnsongrass at 32 fl. oz/A; and second application at 24-32 fl oz/A.

FUSILADE 2000 HERBICIDE ANNUAL GRASS CONTROL GROWTH STAGES FOR COFFEE
Region B (Hawaii Only)

Annual Grass Species

Make applications when grasses are 2-8 inches tall, but before tillering and/or heading.

Barnyardgrass	Prairie cupgrass
Broadleaf signalgrass	Red Rice
Crabgrass, large	Shattercane/Wildcane/Volunteer milo
Crabgrass, smooth	Sorghum almum
Crabgrass, southern	Southern sandbur
Crabgrass, tropical	Southwestern cupgrass
Fall panicum	Texas panicum
Field sandbur	Volunteer barley
Foxtail, giant	Volunteer corn
Foxtail, green	Volunteer oats
Foxtail, yellow	Volunteer rye
Goosegrass	Volunteer wheat
Guineagrass, seedling	Wild oats
Itchgrass	Wild proso millet
Italian ryegrass	Witchgrass
Johnsongrass, seedling	Wooly cupgrass
Junglerice	

FUSILADE 2000 HERBICIDE PERENNIAL GRASS CONTROL GROWTH STAGES FOR COFFEE
Region B (Hawaii only)

Perennial Grass Species	Growth Stage
Bermudagrass	4-8″ runners
Guineagrass[1]	6-12″ tall before seedhead initiation
Kikuyagrass	4-8″ runners
Quackgrass	6-10″
Rhizome johnsongrass	8-18″ tall and before boot stage
Torpedograss[2]	3-6″ tall (1st application) 2-3″ tall (subsequent application)
Wirestem muhly	4-12″ tall and before seedhead initiation

[1] For best results, use 64 fl. ozs. per acre per application.
[2] Use 64 fl. oz. per acre per application. Three applications may be needed.

TANKMIX AND SEQUENTIAL APPLICATIONS FOR SOYBEANS

See Table 1 and 2 for Tankmix and Sequential Application Rate Table.

Fill the spray tank with half the amount of required water and add the recommended amounts of FUSILADE 2000 herbicide, an appropriate broadleaf herbicide, and proper rate of approved adjuvant while the agitator is running, and then add the remaining quantity of water.

Note: Tankmix applications sometimes have resulted in reduced grass weed control and possible increases in crop injury as compared to either product used alone. If grass regrowth occurs following an application of the tankmix or an additional flush of grasses emerge, make a second application of FUSILADE 2000 herbicide to actively growing grass weeds, as per the label recommendations. When perennial grasses are the predominant grass to be controlled, a tankmix is not recommended. Follow the directions for sequential applications of FUSILADE 2000 and the appropriate broadleaf herbicide.

USE OF FUSILADE® 2000 WITH REFLEX® 2LC* – SOYBEANS ONLY

REFLEX 2LC REGIONAL USE MAP

Refer to the REFLEX 2LC herbicide Label Use Map for specific state/county recommendations. Refer to the Application Rate Tables for recommendations by weed species and growth stage.

A maximum of 1½ pints of REFLEX 2LC herbicide may be applied per acre per season to soybeans in Region 1. A maximum of 1¼ pints of REFLEX 2LC herbicide may be applied per acre in alternate years in Region 2. A maximum of 1 pint of REFLEX 2LC herbicide may be applied per acre in alternate years in Region 3. A maximum of 1½ pints of REFLEX 2LC herbicide may be applied per acre in alternate years in Region 4. Do not apply more than one application per growing season.

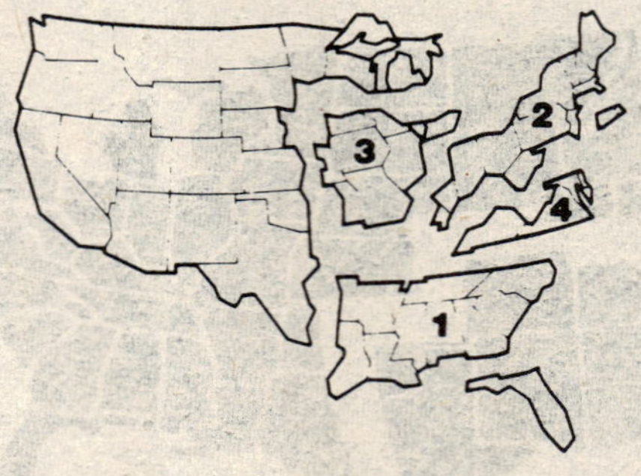

*REFLEX 2LC is not registered for use in Connecticut, Florida or California.

FUSILADE 2000 and REFLEX 2LC may be used together in a postemergence program for broad-spectrum weed control in soybeans.

FUSILADE 2000 and REFLEX 2LC may be applied sequentially or in a tankmix. The growth stage of weeds at the time of application will determine which method of application will provide the most satisfactory results.

Both FUSILADE 2000 herbicide and REFLEX 2LC herbicide should be applied to actively growing weeds. Do not apply either herbicide if weeds appear stressed due to unfavorable temperatures, drought and/or low soil fertility. Do not apply REFLEX 2LC or FUSILADE 2000 tankmix if soybeans show injury from prior herbicide applications.

METHOD I: Sequential Application – FUSILADE 2000 Followed by REFLEX 2LC

(Annual and/or perennial grass weeds at proper growth stage for treatment, prior to broadleaf weed treatment.)

Apply FUSILADE 2000 with an approved adjuvant to annual and/or perennial grass weeds at the recommended rate and growth stage listed in the APPLICATION DIRECTIONS SECTION.

FUSILADE 2000 is a systemic herbicide; therefore, when treating annual grasses, allow at least 3 days, and when treating perennial grasses, allow at least 5 days, to elapse prior to a REFLEX 2LC application. This will enable FUSILADE 2000 to be adequately translocated to the meristematic regions of the plant.

After the appropriate time interval has elapsed, apply REFLEX 2LC with an approved adjuvant at the recommended rate to susceptible broadleaf weeds as outlined on the REFLEX 2LC label.

METHOD 2: Sequential Application – REFLEX 2LC Followed by FUSILADE 2000

(Broadleaf weeds at proper growth stage for treatments, prior to annual and/or perennial grass weed treatment.)

Apply REFLEX 2LC with an approved adjuvant to susceptible broadleaf weeds at the recommended rate and growth stage.

A sequential application of FUSILADE 2000 may be applied following a REFLEX 2LC application when grasses resume active growth with the development of a new leaf. Follow the recommended rates and growth stages listed in the APPLICATION DIRECTIONS SECTION for the grass weeds being treated.

METHOD 3: Tank Mix Applicatons – FUSILADE 2000 and REFLEX 2LC

(Annual grasses and broadleaf weeds are at the proper stage of growth for treatment, as per respective labels.)

A tankmix of FUSILADE 2000 and REFLEX 2LC may be applied by ground application only at the recommended rates and growth stages to susceptible annual grass and broadleaf weed species in a manner consistent with respective labels.

Use a nonionic surfactant or crop oil concentrate in the tankmix. Follow water volume and spray pressure recommendations, as per the REFLEX 2LC label.

SEE SOYBEAN TANK MIX USE PRECAUTIONS SECTION FOR ADDITIONAL INFORMATION.

USE OF FUSILADE 2000 WITH BASAGRAN – SOYBEANS ONLY

FUSILADE 2000 and BASAGRAN may be used together in a postemergence program for broad-spectrum weed control in soybeans.

FUSILADE 2000 and BASAGRAN may be applied sequentially or in tankmix. The growth stage of weeds at the time of application will determine which method of application will provide the most satisfactory results.

Both FUSILADE 2000 and BASAGRAN should be applied to actively growing weeds. Do not apply BASAGRAN or FUSILADE 2000/BASAGRAN tankmix if soybeans show injury from prior herbicide applications. Read and observe all applicable label directions and limitations for both FUSILADE 2000 and BASAGRAN before using.

METHOD 1: Sequential Application – FUSILADE 2000 Followed by BASAGRAN herbicide

(Annual and/or perennial grass weeds at proper growth stage for treatment, prior to broadleaf weed treatment.)

Apply FUSILADE 2000 and an approved adjuvant to annual and/or perennial grass weeds at the recommended rate and growth stage listed in the APPLICATION DIRECTIONS SECTION of this label. Following the FUSILADE 2000 application, allow at least 24 hours to elapse prior to a BASAGRAN herbicide application. This will enable FUSILADE 2000 to be adequately translocated to the meristematic regions of the plant. Apply BASAGRAN to susceptible broadleaf weeds according to the BASAGRAN label.

METHOD 2: Sequential Application — BASAGRAN Followed by FUSILADE 2000

(Broadleaf weeds at proper growth stage for treatment, prior to annual and/or perennial grass weed treatment.)

Apply BASAGRAN to susceptible broadleaf weeds at the recommended rate and growth stage according to the BASAGRAN label. A sequential application of FUSILADE 2000 may be applied 24 hours after a BASAGRAN application following the recommended rates and growth stages in the APPLICATION DIRECTIONS SECTION of this label for the grass weeds being treated.

METHOD 3: Tankmix Applications – FUSILADE 2000 and BASAGRAN

(Annual grasses and broadleaf weeds are at the proper stage of growth for treatment, as per respective labels.)

A tankmix of FUSILADE 2000 and BASAGRAN may be applied at the recommended growth stages to susceptible annual grass and broadleaf weed species in a manner consistent with respective labels.

Use only a crop oil concentrate in the tankmix. Follow water volume and spray pressure recommendations, as per the BASAGRAN label. This tankmix can be applied by ground or aerial application.

SEE SOYBEAN TANKMIX USE PRECAUTIONS SECTION FOR ADDITIONAL INFORMATION.

USE OF FUSILADE 2000 WITH REFLEX 2LC + BASAGRAN* - SOYBEANS ONLY

FUSILADE 2000, REFLEX 2LC and BASAGRAN may be used together in a tankmix as a postemergence program for broad spectrum weed control in soybeans. The tankmix should be applied at the recommended rates and growth stages to susceptible annual grass and broadleaf weed species in a manner consistent with respective labels.

The FUSILADE 2000/REFLEX 2LC/BASAGRAN tankmix should be applied to actively growing weeds. Do not apply this tankmix if weeds are stressed due to unfavorable temperatures, drought and/or low soil fertility. Do not apply the FUSILADE 2000/REFLEX 2LC/BASAGRAN tankmix if soybeans show injury from prior herbicide applications.

Use a nonionic surfactant or crop oil concentrate. Apply in a minimum of 15 gallons per acre of spray mixture with spray pressures of 40-60 psi at the nozzle tip.

USE OF FUSILADE 2000 WITH BLAZER SOYBEANS ONLY

FUSILADE 2000 and BLAZER may be used together in a postemergence program for broad spectrum weed control in soybeans.

FUSILADE 2000 and BLAZER may be applied sequentially or in a tankmix. The growth stage of weeds at the time of application will determine which method of application will provide the most satisfactory results.

Both FUSILADE 2000 and BLAZER should be applied to actively growing weeds. Do not apply either herbicide if weeds appear stressed due to unfavorable temperatures, drought and/or low soil fertility. Do not apply BLAZER or TACKLE and FUSILADE 2000 as a tankmix if soybeans show injury from prior herbicide applications.

METHOD 1: Sequential Application - FUSILADE 2000 Followed by BLAZER herbicide

(Annual and/or perennial grass weeds at proper growth stage for treatment, prior to broadleaf weed treatment.)

Apply FUSILADE 2000 and an approved adjuvant to annual and/or perennial grass weeds at the recommended rate and growth stage listed in the APPLICATION DIRECTIONS SECTION.

FUSILADE 2000 is a systemic herbicide, therefore when treating annual grasses allow at least 3 days, and when treating perennial grasses allow at least 5 days to elapse prior to a BLAZER herbicide application. This will enable FUSILADE 2000 to be adequately translocated to the meristematic regions of the plant.

After the appropriate time interval has elapsed, apply BLAZER or TACKLE with an approved nonionic surfactant to susceptible broadleaf weeds according to the BLAZER label.

METHOD 2: Sequential Application - BLAZER Followed by FUSILADE 2000

(Broadleaf weeds at proper growth stage for treatment, prior to annual and/or perennial grass weed treatment.)

Apply BLAZER with an approved nonionic surfactant to susceptible broadleaf weeds at the recommended rate and growth stage.

A sequential application of FUSILADE 2000 may be applied following a BLAZER application when grasses resume active growth with the development of a new leaf. Follow the recommended rates and growth stages in the APPLICATION DIRECTIONS SECTION for the grass weeds being treated.

METHOD 3: Tankmix Applications - FUSILADE 2000 and BLAZER

(Annual grasses and broadleaf weeds are at the proper stage of growth for treatment, as per respective labels.)

A tankmix of FUSILADE 2000 and BLAZER may be applied at the recommended rates and growth stages to susceptible annual grass and broadleaf weed species in a manner consistent with respective labels.

Follow water volume and spray pressure recommendations, as per the BLAZER label. For BLAZER, use only an approved nonionic surfactant in the tankmix. This tankmix can be applied by ground or aerial application.

SEE SOYBEAN TANKMIX USE PRECAUTIONS SECTION FOR ADDITIONAL INFORMATION.

TABLE I. SOYBEAN HERBICIDE TANKMIX APPLICATION RATE TABLE[1]

REGION A[2]

TANKMIX COMBINATION	FUSILADE 2000[3] fl.oz./A	REFLEX 2LC[4] pts./A	BASAGRAN pts./A	BLAZER pts./A	NONIONIC SURFACTANT CROP OIL CONCENTRATE
FUSILADE 2000 + REFLEX 2LC	24	1-1½			0.5 - 1% COC 0.25-0.5% NIS
FUSILADE 2000 + BASAGRAN	24		1½-2		0.5 - 1% COC
FUSILADE 2000 + REFLEX 2LC + BASAGRAN	24	1-1½	1½-2		0.5 - 1% COC 0.25-0.5% NIS
FUSILADE 2000 + BLAZER	24			1½-3	0.25% NIS DO NOT USE COC WITH THIS MIXTURE

[1] Follow the recommended rate and growth stages for the grass and broadleaf weeds on the respective labels.
[2] Region A - All states except the areas of Arizona, California, Colorado, New Mexico, West Oklahoma, West Texas, and Utah as outlined in the "Directions for Use" section.
[3] Apply a minimum of 24 fl ozs per acre of FUSILADE 2000 in a tankmix.
[4] Refer to REFLEX 2LC label for maximum use rate for specified geographic regions.

TABLE 2. SOYBEAN HERBICIDE TANKMIX APPLICATION RATE TABLE[1]

REGION B[2]

TANKMIX COMBINATION	FUSILADE 2000 fl.oz./A	BASAGRAN pts./A	BLAZER pts./A	NONIONIC SURFACTANT CROP OIL CONCENTRATE
FUSILADE 2000 + BASAGRAN	24-48	1½-2		0.5 - 1% COC
FUSILADE 2000 + BLAZER	24-48		1½-3	0.25-0.5% NIS DO NOT USE COC WITH THIS MIXTURE.

[1] Follow the recommended rate and growth stages for the grass and broadleaf weeds on the respective labels.
[2] Region B - For the areas of Arizona, California, Colorado, New Mexico, West Oklahoma, West Texas, and Utah as outlined in the "Directions for Use" section.
[3] Apply a minimum of 24 fl ozs per acre of FUSILADE 2000 in a tankmix

SOYBEAN TANKMIX USE PRECAUTIONS

- Always read and follow the restrictions and limitations for all products whether used alone or in a tankmix. The most restrictive labeling of any product used applies in tank mixtures.
- Do not apply less than 24 fl. oz./A FUSILADE 2000 in a tankmix with broadleaf herbicides in soybeans.
- Do not apply tankmixes if rain is threatening. REFLEX herbicide requires a 4-hour rain-free period; BLAZER herbicide requires a 6 hour rain-free period; and BASAGRAN herbicide requires an 8-hour rain-free period for best results.
- Do not apply a total of more than 2 quarts of BASAGRAN herbicide per acre in one season to soybeans.
- A maximum of 1½ pints of REFLEX 2LC herbicide may be applied per acre per growing season to soybeans in Region 1. A maximum of 1¼ pints of REFLEX 2LC herbicide may be applied per acre in alternate years in Region 2. A maximum of 1 pint of REFLEX 2LC herbicide may be applied per acre in alternate years in Region 3. A maximum of 1½ pints of REFLEX 2LC herbicide may be applied per acre in alternate years in Region 4. (Do not make more than one application per growing season).
- Applications of REFLEX 2LC should not be made beyond 3 weeks after the soybeans have emerged.
- REFLEX 2LC can only be applied by ground application.
- Do not apply more than 4 pints (1.0 lb. active) per acre of BLAZER herbicide per growing season for soybeans.
- Do not apply BLAZER herbicide within 50 days of harvest for soybeans.
- In the case of crop failure, only soybeans may be immediately replanted following use of BASAGRAN, BLAZER, and REFLEX .
- Do not make more than one application of the FUSILADE 2000/REFLEX 2LC; FUSILADE 2000/BASAGRAN; FUSILADE 2000/BLAZER; or FUSILADE 2000/REFLEX 2LC/BASAGRAN tankmix in a single season.
- Do not make sequential or tankmix applications of any of these herbicides if the weeds appear stressed due to unfavorable temperatures, drought and/or low soil fertility.

TANKMIXES FOR COTTON

USE OF FUSILADE 2000 WITH AMBUSH®, CYMBUSH® 3E, BIDRIN 8, OR CYGON 400 INSECTICIDES - COTTON ONLY

FUSILADE 2000 and AMBUSH, CYMBUSH 3E, BIDRIN 8, or CYGON 400 may be used in a tankmix when required according to the weed and insect problems present. Apply FUSILADE 2000 and an approved adjuvant to actively growing annual and perennial grasses at the recommended rate and growth stage listed in the APPLICATION DIRECTIONS SECTION of the FUSILADE 2000 label. Apply AMBUSH, CYMBUSH 3E, BIDRIN 8, or CYGON 400 as per label recommendations for insect control based on scouting reports. Subsequent applications of AMBUSH, CYMBUSH 3E, BIDRIN 8 or CYGON 400 may be required at intervals of 5 to 7 days for insect control. Regrowth or subsequent emergence of grasses may be treated according to rate and growth stage recommendations on the FUSILADE 2000 label.

Use of crop oil concentrate or a nonionic surfactant as indicated in the APPLICATION DIRECTIONS SECTION of this label is required in the tankmix.

Load the spray tank with half the amount of required water and add the recommended amounts of FUSILADE 2000 herbicide and AMBUSH, CYMBUSH 3E, BIDRIN 8, or CYGON 400 insecticide while the agitator is running. Add crop oil concentrate or nonionic surfactant and the remaining quantity of water and mix thoroughly.

FUSILADE HERBICIDE AND COTTON INSECTICIDE TANKMIX RATE TABLE[1]

REGION A[2]

TANKMIX COMBINATION	FUSILADE fl.oz./A	AMBUSH pts./A	CYMBUSH 3E pts./A	CYGON 400 pts./A	BIDRIN 8 pts./A	NONIONIC SURFACTANT CROP OIL CONCENTRATE
FUSILADE 2000 + AMBUSH	12-24	0.4-0.8				0.5 - 1% COC 0.25 - 0.5% NIS
FUSILADE 2000 + CYMBUSH 3E	12-24		0.16-0.27			0.5 - 1% COC 0.25-0.5% NIS
FUSILADE 2000 + CYGON 400	12-24			0.25 - 0.5		0.5 - 1% COC 0.25-0.5% NIS
FUSILADE 2000 + BIDRIN 8	12-24				0.1 - 0.5	0.5 - 1% COC 0.25-0.5% NIS

[1] Follow the recommended rate and growth stages for the grass and target insect pests on the respective labels.
[2] Region A - All states except the areas of Arizona, California, Colorado, Hawaii, New Mexico, West Oklahoma, West Texas, and Utah as outlined in the "Directions for Use" section.

FUSILADE HERBICIDE AND COTTON INSECTICIDE TANKMIX RATE TABLE[1]

REGION B[2]

TANKMIX COMBINATION	FUSILADE fl.oz./A	AMBUSH pts./A	CYMBUSH 3E pts./A	CYGON 400 pts./A	BIDRIN 8 pts./A	NONIONIC SURFACTANT CROP OIL CONCENTRATE
FUSILADE 2000 + AMBUSH	16-48	0.4-0.8				0.5 - 1% COC 0.25% NIS
FUSILADE 2000 + CYMBUSH 3E	16-48		0.16-0.27			0.5 - 1% COC 0.25% NIS
FUSILADE 2000 + CYGON 400	16-48			0.25 - 0.5		0.5 - 1% COC 0.25% NS
FUSILADE 2000 + BIDRIN 8	16-48				0.1 - 0.5	0.5-1% COC 0.25% NIS

[1] Follow the recommended rate and growth stages for the grass and target insect pests on the respective labels.
[2] Region B - For the areas of Arizona, California, Colorado, Hawaii, New Mexico, West Oklahoma, West Texas, and Utah as outlined in the "Directions for Use" section

COTTON TANKMIX USE PRECAUTIONS

- AMBUSH, CYMBUSH 3E, AND BIDRIN 8 ARE RESTRICTED USE PESTICIDES FOR RETAIL SALE TO AND USE ONLY BY CERTIFIED APPLICATORS, OR PERSONS UNDER THEIR DIRECT SUPERVISION, AND ONLY FOR THOSE USES COVERED BY THE CERTIFIED APPLICATOR'S CERTIFICATION.
- Always read and follow the restrictions and limitations for all products whether used alone or in a tankmix. The most restrictive labeling of any product used applies in tank mixtures.
- For additional precautions, restrictions, limitations, and sprayer clean-up information, refer to the appropriate sections in the respective labels.
- Do not apply a total of more than 12 pints (3 lbs. ai/acre) of AMBUSH per acre per season.
- Do not apply a total of more than 1.0 lb. ai/A of CYMBUSH 3E per acre per season.
- Do not apply a total of more than (3 applications) of Bidrin 8 per acre per season.
- Do not apply more than 2 applications of Cygon 400 at the higher rate (0.5 pint/A) per season.
- Bidrin 8 may not be applied within 30 days of harvest.
- Repeat application of Cygon 400 should not be made at intervals closer than 14 days.
- Automatic flagging devices should be used when possible whenever Bidrin 8 is used.
- Do not enter area treated with Bidrin 8 for 48 hours unless appropriate clothing is worn.
- Oral or written warnings must be given workers who are expected to enter a treated area or areas about to be treated with Bidrin 8. Oral warnings must be in language customarily understood by workers.
- When applying Cygon 400, if automatic flagging is not possible, then the human flagger must wear protective clothing and respirator as specified on the Cygon 400 label.

NONFOOD USES - FALLOW LAND, NONCROP AREAS, NONBEARING CROPS, AND ORNAMENTALS.

FUSILADE 2000 herbicide can be used to control annual and perennial grass weeds in fallow land; in nonbearing groves, orchards or vineyards, during site preparation prior to transplanting; on tree farms*, and in many newly transplanted and established nongrassy ornamentals, trees, shrubs and ground covers and in noncrop areas. Noncrop areas include airports, electric transformer stations and sub stations, pipeline pumping stations, around buildings, storage yards, fence lines, parkways, roadsides, rights-of-way, and other noncropland areas. See the following Tables for specific recommended uses.

GENERAL USE PRECAUTIONS FOR NONFOOD USES
NOT FOR USE OR STORAGE IN OR AROUND THE HOME.

- Apply FUSILADE 2000 herbicide to actively growing grasses before they exceed the recommended growth stages in the following Tables.
- Spray to obtain thorough coverage of grass foliage, but not to runoff. Repeat treatments may be necessary to provide full season control.
- Use standard pesticide hollow cone or flat fan nozzles.
- Apply FUSILADE 2000 herbicide when the first grass species in a mixed grass weed population reaches the recommended growth stage for treatment.
- Do not apply FUSILADE 2000 herbicide to grasses which are stressed due to moisture, temperature, low soil fertility, mechanical or chemical injury.
- Do not apply to grasses which have tillered, formed seed heads, or exceeded recommended growth stages.
- When treating nonbearing trees and vines, avoid contact of spray with foliage by using directed sprays.
- Do not apply to foliage of ornamental trees, shrubs, or ground covers unless they are listed in Table I of the ornamental section of this label.
- Do not apply to trees, vines, ornamentals, or other listed crops that will be harvested for food within one year after application.
- Do not apply to asparagus from which spears will be harvested within one year of treatment.
- Do not graze or harvest Florigraze perennial peanuts for one year after treatment.
- Do not plant rotational grass crops such as corn, sorghum and cereals within 60 days after the last application of FUSILADE 2000.

* Use of FUSILADE 2000 on tree farms is not registered for use in California.

NONFOOD USES - FALLOW LAND, NONCROP AREAS, AND NONBEARING CROPS

Apply 32-48 oz. per acre of FUSILADE 2000 herbicide along with 1% crop oil concentrate (2 pints per 25 gallons) or 0.25% (1/2 pint per 25 gallons) of a nonionic surfactant to annual and perennial grasses at the recommended growth stages in the following Tables.

FUSILADE 2000 HERBICIDE ANNUAL GRASS CONTROL GROWTH STAGES FOR FALLOW LAND AND NONBEARING CROPS

Annual Grass Species

Make applications when grasses are 2-8 inches tall, but before tillering and/or heading.

Barnyardgrass	Prairie cupgrass
Broadleaf signalgrass	Red Rice
Crabgrass, large	Shattercane/Wildcane/Volunteer milo
Crabgrass, smooth	Sorghum almum
Crabgrass, southern	Southern sandbur
Crabgrass, tropical	Southwestern cupgrass
Fall panicum	Texas panicum
Field sandbur	Volunteer barley
Foxtail, giant	Volunteer corn
Foxtail, green	Volunteer oats
Foxtail, yellow	Volunteer rye
Goosegrass	Volunteer wheat
Guineagrass, seedling	Wild oats
Itchgrass	Wild proso millet
Italian ryegrass	Witchgrass
Johnsongrass, seedling	Woolly cupgrass
Junglerice	

FUSILADE 2000 HERBICIDE PERENNIAL GRASS CONTROL GROWTH STAGES FOR FALLOW LAND AND NONBEARING CROPS

Perennial Grass Species	Growth Stage
Bermudagrass	4-8" runners
Guineagrass	6-12" tall before seedhead initiation
Kikuyagrass*	4-8" runners
Quackgrass	6-10"
Rhizome johnsongrass	8-18" tall and before boot stage

Torpedograss**	3-6" tall (1st application) 2-3" tall (Subsequent application)
Wirestem muhly	4-12" tall and before seedhead initiation

* Not registered in California for use on this species.
** Use 48 fl. oz. per acre per application. Three applications may be needed.

FUSILADE 2000 HERBICIDE IS RECOMMENDED ON THE FOLLOWING NONBEARING CROPS

Acerola (West Indian Cherry)*	Lemons
Almonds	Limes
Apples	Macadamia
Apricots	Mango*
Asparagus	Nectarines
Avocados	Olives
Berries	Oranges
Cherries	Peaches
Coffee*	Pears
Conifers*	Pecans
Christmas tree plantings	Pistachios
Nursery beds	Plums
Seedling establishment	Pomegranates
Dates	Prunes
Figs	Tangelos
Filberts	Tangerines
Florigraze perennial peanuts* (Nursery stock only)	Walnuts
Ginseng*	
Grapes	
Grapefruit	
Guava*	
Jojoba	
Kiwi*	

* Not applicable in California.

NONFOOD USE - ORNAMENTALS

Apply 32-48 oz. per acre of FUSILADE 2000 herbicide along with 0.25% (1/2 pint per 25 gallons) of a nonionic surfactant. DO NOT USE A CROP OIL CONCENTRATE WITH FUSILADE 2000 ON ORNAMENTALS. Refer to the weed tables in the Fallow Land, Noncrop Areas, and Nonbearing Crops sections for recommended growth stages of grasses at application.

FUSILADE may be applied over-the-top or as a directed spray. Please refer to Table 1 for a list of ornamental species for which over-the-top application is appropriate. For species listed in Table II-IV, directed spray applications are necessary to minimize phytotoxicity. However, there is the possibility of damage in the range listed for each table when applied at label rates.

NOTICE TO BUYER AND USER: It is impossible to test every species and variety or cultivar of ornamental or nursery plants under all conditions. Plant tolerance of pesticides vary as conditions vary. Plant tolerance of FUSILADE 2000 at label rates has been found to be acceptable within the ranges specified for the indicated genera and species. Neither the manufacturer nor the seller has determined whether or not FUSILADE 2000 can safely be used on plants not specified on this label. The professional user should determine if FUSILADE 2000 can be used safely prior to use.

TABLE I

OVER-THE-TOP APPLICATIONS may be applied to the following ornamentals: Use Only Nonionic Surfactant on Ornamentals.

Scientific Name	Common Name/Variety
Abelia grandiflora	Glossy abelia
Abies balsamea *	Balsam fir
Abies concolor	Concolor fir
Abies procera	Noble fir
Acacia redolens	Acacia ongerops
Acacia salacina	Willow-leafed acacia
Acacia salinga	Willow acacia
Acacia schafnerii	Jim wheat acacia
Acacia stenophylla	Shoe-string acacia
Acer ginnala *	Flame amur maple
Acer palmatum	Maple/japanese
Acer platanoides	Norway maple
Acer saccharinum *	Silver maple
Acer saccharum	Sugar maple
Achillea filipendulina	Coronation gold, fernleaf yarrow
Achillea millefolium	Common yarrow
Acrecastrum romanzoffianum	Queen palm
Agapanthus africanus	Peter pan lily of the Nile
Agave americana	Century plant
Ageratum sp.	
Aglanoema commutatum	Silver queen evergreen, Fransher evergreen, Treubii ribbon evergreen
Aglanoema crispum	Painted evergreen
Alcea rosea	Hollyhock
Allium schoenoprasum	Chives
Aloe barbadensis	Barbados aloe
Aloe vera	
Aloe zanzibarica	
Alyssum sp.	
Antirrhinum majus *	Snapdragon Yellow floral carpet
Araucaria heterophylla	Norfolk Island pine
Arctostaphylos uva-ursi	Red bearberry
Arctotheca calendula	Cape weed
Arenaria verna	Sandwort
Asparagus densiflorus	Sprengeri, Myres asparagus, Sprenger asparagus
Astilbe x arendsii	False spirea
Atriplex sp.	Saltbrush
Aucuba japonica	Aucuba
Aucuba japonica variegata	
Begonia semperflorens aultoreum *	Scarletta begonia
Berberis mentorensis	Mentor barberry
Berberis thunbergii *	Crimson pygmy, Redleaf Japanese barberry
Betula pendula *	Eastern white birch
Bougainvillea sp.	Raspberry ice
Buxus microphylla koreana	Korean boxwood
Buxus microphylla var. japonica	Japanese boxwood
Buxus sempervirens	Welleri, Common boxwood
Caesalpinia cacalaco	
Caesalpinia gilliesii	Bird of paradise
Calendula officinalis *	Orange coronet
Calendula sp.	Marigold
Callistemon lanceolatus	Bottle-brush
Calluna vulgaris	Scotch heather
Camellia japonica	Camellia
Camellia sasanqua	Sasanqua camellia
Campanula carpatica	Bellflower
Carissa grandiflora	Natal plum
Carissa tuttlei	
Carnegiea gijantea	Saguaro cactus
Carpobrotus chilensis *	Trailing hottentot-fig
Cassia artemisioides	Feathery cassia
Cassia didymobotrya	African cassia
Cassia sturdii	
Celtis occidentalis *	Hackberry
Centaurea cineraria	Dusty miller
Cerastium tomentosum	Snow in summer
Ceratonia siliqua	Carob tree
Cercis canadensis	Red bud
Chaenomeles speciosa *	Flowering quince
Chamaecyparis obtusa	Cripps hinoki false cypress
Chamaecyparis lawsoniana	Allum lawson cypress
Chamaecyparis pisifera	Plumosa
Chamaerops excelsa	Windmill palm
Chamaerops humilis	Mediterranean fan palm
Chrysalidocarpus lutescens (small)	Golden fruited palm
Chrysanthemum x superbum	Shasta daisy
Cissus rhombifloia	Ellen Danica, grape ivy
Cistus hybridus	Rock rose
Citrus aurantium	Sour orange
Citrus paradisi	Grapefruit
Clerodendrum speciosum	Pagoda flower
Cleyera spp.	Cleyera
Clivia miniata	Kaffir lily

*Not applicable in California.
*Not applicable in California

Scientific Name	Common Name/Variety
Cocos plumosa	Queen palm
Codiaeum variegatum	Croton
Coffea arabica	Coffee
Coleus x hybridus *	Coleus
	Jade wizard
Convolvulus oneorum	Bush morningglory
Coprosma baueri	Mirror plant
Coprosma repens	Varigated mirror plant
Cordyline terminalis	Ti
Coreopsis verticillata	Threadleaf coreopsis
Cornus florida	Flowering dogwood
Cornus mas	Cornelian cherry dogwood
Cornus sericea	Red twig dogwood, Flaviramea
Cotoneaster apiculata	
Cotoneaster dammeri	Coral beauty, Royal beauty
Cotoneaster divaricatus	Spreading cotoneaster
Cotoneaster microphyllus	
Cotoneaster repens	Cotoneaster
Cotoneaster salicifolius franch	Willowleaf cotoneaster
Crossandra nilotica	Crossandra
Cupressus semprevirens	Italian cypress
Cycus revoluta	Sago palm
Deutzia gracilis	Slender deutzia
Dianthus barbatus	Sweet William
Dieffenbachia amoena	Giant dumbcane
Diervilla lonicera	Bush honeysuckle
Dodonea viscosa purplurea	Purple hopseed bush
Echinocatus sp.	Hedgehog cactus
Elaeagnus angustrifolia	Russian olive
Encelia farinosa	Bittle brush
Erythrina fusca	Fastadiata, Swamp immortella
Escallonia fradessii	
Escallonia rubra	
Eucalyptus microtheca	Gum-barked coolibah
Eucalyptus polyanthemos	Red box gum
Eucalyptus rostrata	Red gum
Eucalyptus rudis	Desert gum
Eucalyptus spathulata	Narrow-leaf gimlet
Eugenia myrtifolia	Brush cherry
Euonymus alata	Siebold
Euonymus fortunei	
Euonymus japonica	Silver king
Euonymus kiautschovicus	Spreading euonymus
Euphorbia lactea	Candelabra plant
Euryops pectinatus	
Fatshedera lizei	Fatshedera
Ferocactus sp.	Barrel cactus
Ficus benjamina	Exotica weeping fig
Ficus elastica decora	Rubber tree
Ficus microcarpa nitida	Indian laurel
Ficus nitida	Indian laurel
Ficus repens	Creeping fig
Forsythia intermedia	
Forsythia sp.	
Forsythia suspensa	Weeping forsythia
Forsythia x intermedia	
Fouquieria splendens	Ocotillo
Fragaria chiloensis	Ornamental strawberry
Fraxinus americana *	White ash
Fraxinus pennsylvanica *	Green ash
Fraxinus velutina	Arizona ash
Gardenia jasminoides	Gardenia, dwarf
Gardenia taitensis	Tahitian gardenia
Gazania splendens	Gazania gold rush
Gazania uniflora leucoleana	
Geijera parviflora	Australia willow
Gelsemium sempervirens	Carolina jessamine
Geranium pratense	Cranesbill
Gladiolus x hortulanus	Debbie, Jennie, Mahogany, Stargazer
Gleditsia triancanthos var. inermis	Honey locust/shade master
Graptophyllum pictum	Caricature plant
Grevillea robusta	Silk oak
Gevillea rosmarinifolia	Spider flower
Hedera canariensis	Algerian ivy
Hedera helix	English ivy
Hedera helix hahnii	Hahn's ivy
Hemerocallis hybrids	Daylily
Hemigraphis sp.	Waffle plant/Metallic plant
Hesperaloe parviflora	
Hibiscus rosa-sinensis	Chinese hibiscus
Hibiscus syriacus	Althea
Hosta lancifolia	Variegated hosta
Hydrangea paniculata	Panicle hydrangea
Hydrangea quercifolia	Oakleaf hydrangea
Iberis sempervirens	Candytuff
Ilex cornuta	Dwarf buford holly
Ilex crenata	Japanese holly
Ilex glabra	Compact inkberry
Ilex opaca	American holly
Ilex vomitoria	Dwarf yaupon/Tall yaupon
Ilex x meserveae	Meserve holly
Ilex x attenuata	Fosteri holly
Iris sp.	
Isotoma sp.	Blue star creeper
Ixora coccinea	Ixora
Jacaranda acutifolia	Jacaranda
Jacobinia ghiesbreghtiana	
Juniperus horizontalis *	Admiral
Juniperus scopulorum	Cologreen
Juniperus virginiana	Red ceder
Justicia spicigera	Firecracker, Mexican indigo
Kochia scoparia f. trychophylla	Compact burningbrush
Koelreuteria paniculata *	Golden-rain tree, Varnish tree
Lagerstroemia indica	Crepe myrtle
Lantana camera	Bush lantana
Lantana sellowiana	Purple lantana (trailing)
Leptospermum laevigatum	
Leucophyllum frutescens	Texas sage
Liatris spicata	Gay feather
Ligustrum amurense	Amur River
Ligustrum lucidum	Wax ligustrum
Ligustrum ovalifolium	Privet/california
Ligustrum texanum	Texas privet
Ligustrum x vicari	Vicari
Liquidambar styraciflua	American sweetgum
Liriope spicata	Liriope
Liriope muscari	Green liriope/Variegated liriope
Livistona chinenis	Chinese fan palm
Lonicera x morrowii	Morrow honeysuckle
Lonicera xylosteum	Emerald mound
Lysiloma thornberii	Desert fern tree
Lysimachia nummularia	Moneywort/Wandering Sally
Magnolia sp.	Oregon Grapeholly
Magnolia grandiflora	Southern magnolia
Magnolia stellata	Star magnolia
Mahonia aquifolium	Mahonia
Mahonia wagoneri *	King's Ransom
Malus floribunda	Showy crabapple
Mesembryanthemum drosanhemum hispidum	Rose ice plant
Mesembryanthemum lampranthus prductus	Purple trailing ice plant
Mesembryanthemum lampranthus spectabilis	Red spike ice plant
Metrosideros excelsus	New Zealand christmas tree
Musa maurelli	Ethiopia banana
Myoponum parvifolium	Prostrate myoponum
Myrica cerifera	Wax myrtle
Nephrolepsis exaltata *	Sword fern
Nerium oleander	Pink, variegated, petite oleander
Odocanthus sp.	
Olea europaea	Olive tree
Olneya tesota	Ironwood
Opuntia Cholla	Cholla cactus
Osmanthus fragrans	Osmanthus, tea olive
Osteospermum fruticosum alba	White african daisy

*Not applicable in California
*Not applicable in California

Scientific Name	Common Name/Variety
Pachysandra terminalis	Japanese pachysandra
Parkensonia aculeata	Green palo verde
Passiflora pfordtiii	Passion vine
Pelargonium domesticum	Geranium
Pelargonium peltatum	Ivy geranium
Pelargonium x hortorum *	Smash Hit Red geranium
Pennisetum setaceum	Red fountain grass
Peperomia scandens	Radiator plant
Petunia sp.	
Philadelphus coronarius	Aurea
Philodendron oxycardium	"Micans" velvetleaf philodendron
Philodendron selloum	
Phoenix canariensis	Canary Island date palm
Phoenix roebelenii	Pygmy date palm
Photinia x fraseri	Photinia
Phyllostachys aurea	Golden bamboo
Physocarpus fruticosa	Gold drop, Jackmanni, Abbotswood
Physocarpus opulifolius	Dwarf Ninebark, Nanus
Picea abies	Norway spruce
Picea glauca	Dwarf Alberta, Black Hills, Densata
Picea omarika	Serbian spruce
Picea pungens	Blue spruce
Pieris japonica	Lily-of-the-valley bush
Pilea nummulariifolia	Creeping Charlie
Pinus canariensis	Canary Island pine
Pinus clausa *	Sand pine
Pinus echinata *	Shortleaf pine
Pinus elliottii *	Slash pine
Pinus glabra *	Spruce pine
Pinus mugo	Dwarf Swiss mountain pine
Pinus nigra	Austrian pine/Black pine
Pinus palustris *	Longleaf pine
Pinus ponderosa	Western/Ponderosa pine
Pinus pungens *	Table-Mountain pine
Pinus resinosa	Red pine
Pinus rigida *	Pitch pine
Pinus serotina *	Pond pine
Pinus strobiformus	Mexican border pine
Pinus strobus	Eastern white pine
Pinus sylvestris	Scotch pine
Pinus taeda *	Loblolly pine
Pinus virginiana *	Virginia pine
Pittosporum tobira	Wheelers dwarf, variegated
Pittosporum phillyraeoides	Desert willow
Platanus sp. *	Sycamore
Plumbago capensis	Cane plumbago
Podocarpus gracilior	African fern pine
Podocarpus macrophylla	Yew pine
Polygonum capitatum	Pink clover
Polyscias fruticosa	Parsley panax
Polystichum capense	
Portulaca grandiflora *	Sunglo portulaca
Potentilla fructosa	Gold drop, Primrose beauty
Potentilla verna *	-
Prosopis chilensis	Chilean mesquite
*Protea compacta**	Protea
Protea cynaroides *	Giant/king protea
*Protea eximia**	Protea
*Protea neriifolia**	Oleander-leaved protea
*Protea repens**	Protea
Prumus trialoba	Flowering almond
Prunus caroliniana ompacta	Carolina cherry
Pseuderanthemum atropurpureum	Purple false eranthemum
Pseudotsuga mensiessi	Douglas fir
Pyracantha koidzumii	Lodense pyracantha
Pyracantha angustifolia	Variegated firethorn
Pyracantha coccinea	Scarlet firethorn, Lalandei
Pyracantha graberi	
Pyracantha koidzumii x coccinea	Mojave firethorn
Pyrus calleryana	Bradford pear
Quercus palustris *	Pine oak
Quercus virginiana	Oak, live
Raphiolepis indica	Pink lady
Raphiolepis umbellata	Yedda/Indian hawthorn
Rhamnus frangula	Tallhedge buckthorn
Rhododendron "Gable Hydrid"	Elizabeth Gable, Purple splendor, Rose greely
Rhododendron catawbiense	Less dark purple, Roseum elegans, Chionoides rhododendron, White catawba, Purple elegans, Roseum superbum, English roseum
Rhododendron formosa	
Rhododendron obtusum	Amoenum, Coral bells
Rhododendron sp.	Blaauw's pink, Rutherfordiana Constance, White cascade, Prize, Kluis sensation, Sweetheart, Kaempo, Royalty, Hahm red, Red wing, Gumpo white, Salmon spray, Road runner, Tradition, Gloria, Gumpo pink, Greeting, Pink pearl, Pericat, Tabor, Mother's Day Snow, Masasoit, Red ruffle, Delaware Valley white, Purple gem, H. H. Hume, Stewartstonian, Hino red, Rosebud, Fashion, Herbert, Gibraltar, Girard's rose, Boule de neige, President Lincoln
Rhododendron yedoense	Korean azalea/Poukhanense
Rhuellia californica	
Rhus lancea	African sumar standard
Rhus aromatica	Fragrant sumac
Rosa banksiae	Banksia
Rosa hybrida	Hybrid tea rose
Rosa sp.	Rose
Rosmarinus officinalis prostratus	Rosemary dwarf
Rumohra adiantiformis *	Leatherleaf fern
Salix alba	White willow
Salix babylonia *	Weeping willow
Salix matsudana	Tortuosa corkscrew willow
Salix purpurea *	Purple willow Basket willow
Salvia greggii	
Sansevieria sp.	Moon Glow
Sansevieria trifasciata	Hahaii/Mother-in-Law 's tongue
Santolina chanaecy parissus	Lavender cotton
Schefflera arboricola	Manila Ripple
Schinus molle	California pepper
Schinus terebinthifolius	Brazilean pepper
Sedum brevifolium	Green stone crop
Sedum quatemalense	Brown beam
Sedum spectabile	Sedum

*Not applicable in California
*Not applicable in California

Scientific Name	Common Name/Variety
Sedum x rubrotinctum	
Sempervivum tectorum	Hen and chickens
Simmiondsia chinensii	Jojoba
Sophora japonica *	Pagoda tree
Sorbus americana *	American mountain ash
Spiraea japonica *	Coccinea
Spiraea nipponica	Snowmound spirea
Spiraea thunbergii *	Thunberg
Spiraea x billiardii	Billiard spirea
Spiraea x bumalda	Froebelii, Crispa spirea, Gold flame, Anthony Waterer
Statice sinuata	Annual statice
Stenocarpus sinuatus	Firewheel tree
Strelitzia nicolai	Giant bird of paradise
Strelitzia reginae	Bird of paradise
Syringa meyeri	Palibin
Syringa patula	Korean lilac
Syringa villosa	James McFarlane lilac
Syzgium paniculatum	Australian bush cherry
Tagetes sp.	Marigold
Taxus cuspidata	Japanese yew
Taxus x media	Thayeri yew, Hicks yew, Dense yew
Tecoma stans angustate	Yellow bells
Tecomaria capensis	Cape honeysuckle
Ternstroemia gymnanthera	Cleyera
Thevetia peruviana	Yellow oleander tree
Thuga occidentalis	White cedar, Emerald green, Woodwardii, Techny, Pyramidalis, Globosa, Techny american arborvitae
Thuja orientalis	Berkman's
Tilia cordata *	Little-leaf linden
Trachelospermum asiaticum	Asiatic jasmine
Trachelospermum jasminoides	Star jasmine
Trifolium fragiferum	O'Conners legume
Tsuga canadensis	Eastern hemlock
Viburnum dentatum	Arrowwood viburnum
Viburnum japonicum	Viburnum/japanese snowball
Virburnum lantana *	Wayfaring tree Twistwood
Virburnum lentago *	Nannyberry
Viburnum opulus	Nanum
Viburnum odoratissium	Viburnum
Viburnum plicatum	Doublefile viburnum/tomentosum
Viburnum suspensum	Sandankwa viburnum
Viburnum trilobum	Compact cranberrybush viburnum
Viburnum x juddi	Judd
Viburnum x rhytidophylloides	Viburnum/willowwood
Vicia sp.	Crown Vetch
Vinca major	Periwinkle
Vinca minor	Myrtle, Dwarf periwinkle
Washingtonia robusta	Mexican fan palm
Weigela florida	Newport red, Pink weigelia
Xylosma senticosa	
Yucca filamentosa	Yucca
Yucca gloriosa	Spanish dagger
Yucca pendula	Weeping dagger
Zinnia sp.	

* Not applicable in California.

TABLE II

DIRECTED APPLICATIONS

- When plant growth habit allows, applications should be made as a directed spray to the ornamental plants listed below to minimize phytotoxicity.
- Limited testing of the ornamental plants listed below has shown phytotoxicity of up to 20% when FUSILADE 2000 is applied over-the-top at label rates. (Phytotoxicity can occur whenever spray comes in contact with the foliage, even during directed sprays.)

Scientific Name	Common Name/Variety
Abies fraseri	Fraser Fir
Acer rubrum	Maple/red
Ajuga variegata	Bugle weed
Ajuga reptans	Varigated ajuga
Aquilegia hybrida	Columbine
Arbutus unedo	Strawberry tree
Banksia prinotes *	Protea
Banksia victoria *	Protea
Banksia speciosa *	Protea
Bigonia tweediana	Cats claw, Yellow trumpet
Callistemom viminalis	Weeping bottle-brush
Ceonothus griseus	
Cupressocyparis leylandi	Leyland cypress
Dracaena fragans	Massangeana, dracaena
Dracaena marginata	Tricolor dracaena
Eucalyptus citriodora	Lemon-scented gum
Feijoa sellowiana	Pineapple guava
Gamolepsis chrysanthemoides	
Gazania ringens	
Gladiolus x hortulanus	Gallery
Grewia caffra	Lavender star plant
Herniaria glabra	Green carpet
Juniperus chinensis	Pfitzeriana, Nana, Spearmint, Old Gold, Prostrata, Robusta, Maney
Juniperus conferta	Blue Pacific
Juniperus horizontalis	Wiltonii, Hughes, Webberi, Variagata, Blue Ring, Youngtown Compacta
Juniperus japonica	San Jose
Juniperus sabina	Tamariseifolia, Arcadia, Scandia, Broadmoor
Juniperus scopulorum	Welchii, Pathfinder
Juniperus virginiana	Sky rocket, Grey owl
Lantana montevidensis x	White lantana
Leucothoe axillaris	Fetterbush
Ligustrum japonicum	Japanese privet
Lonicera japonica	Japanese honeysuckle
Luecospermum cordifolium *	Pincushion protea
Nandina domestica	Heavenly bamboo
Nerium oleander standard	Oleander
Ophiopogon japonicum	Mondo grass
Opuntia sp.	Prickly pear cactus
Pennistum sectaceum	Green fountain grass
Philodendron sp.	
Plumeria acuminata	Temple tree
Podocarpus macrophyllus	Maki
Potentilla verna	Spring cinquefoil
Rhododendron indicum	Indica
Rhododendron obtusum	Kurume, Eureka, Sunglow *
Rhododendron sp.	P.J.M.
Rhoeo spathacea	Oyster plant
Ruellia ciliosa	Ruellia
Salix caroliniana	Willow
Spirea x vanhouteii	Vanhoutte spirea
Syringa chinensis	Lilac
Viburnum macrocephalum	Chinese snowball

*Not applicable in California

*Not applicable in California

TABLE III
DIRECTED APPLICATIONS

- When plant growth habit allows, applications should be made as a directed spray to the ornamental plants listed below to minimize phytotoxicity.
- Limited testing of the ornamental plants listed below has shown phytotoxicity of up to 50% when FUSILADE 2000 is applied over-the-top at label rates. (Phytotoxicity can occur whenever spray comes in contact with the foliage, even during directed sprays).

Scientific Name	Common Name/Variety
Acacia farnesiana	Sweet acacia
Acacia latifolia	
Aptenia cordifolia	Hearts and flowers
Asystasia gangetica	Coromandel
Baccharis sarothroides	Desert broom
Brachychiton populneum	Bottle tree
Cassia condolioma	
Cercidium praecox	Sonoran palo verde
Citharexylum spinosum	Fiddlewood
Cordyline stricta	Cordyline
Cupaniopsis anacardioides	Carrot wood
Crassula argentea	Jade plant
Cyperus alternifolius	Umbrella plant
Dicentra spectabilis	Bleeding heart
Dracaena deremensis	Janet Craig/Warneckii
Epipremnum aureum	Marble Queen pothos
Eucalyptus nicholii	
Exococaria cochichinensis	Chinese crenate croton
Hibiseus lepenk	
Juniperus chinensis	Armstrongii
Juniperus scopulorum	Spiny Greek, Excelsa Strieta
Juniperus virginiana	Burkii
Justicia betonia	White shrimp plant
Justicia brandegeana	Shrimp plant
Justicia carnea	Kings crown
Mesembryanthemum delosperma alba	White ice plant (trailing)
Odontonema strictum	Red justicia
Oenothera berlandier	Mexican evening primrose
Pachystachys lutea	Yellow shrimp plant
Pedilanthus tithymaloides	Slipper flower
Peperomia obtusifolia	Baby rubber plant
Plectranthus australis	Swedish ivy
Polygonum capitatum	Pinkhead knotweed
Prunum avium *	Mazzard cherry
Rhododendron indicum	Formosa
Rhododendron obtusum	Hersey red, Hino pink, Hinodegeri
Rhododendron poukhanensis	Karen
Thunbergia laurifloia	Laurel-leaved thunbergia
Vaccinum ashei	Tifblue Blueberry

* Not applicable in California.

TABLE IV
DIRECTED APPLICATIONS

- When plant growth habit allows, applications should be made as a directed spray to the ornamental plants listed below to minimize phytotoxicity.
- Limited testing of the ornamental plants listed below has shown phytotoxicity greater than 50% when FUSILADE 2000 is applied over-the-top at label rates. (Phytotoxicity can occur whenever spray comes in contact with the foliage, even during directed sprays.)

Scientific Name	Common Name/Variety
Acalypha godsefeiana hertophylla	
Alsophia australis	Australia tree fern
Betula nigra	Birch/River
Catharanthus roseus	Madagascar periwinkle
Chamaecyparis pisifera	Boulevard falsecypress
Chlorophytum comosum	Spider plant
Cortederia selloana	Pampas grass
Euonymus alata	Compacta
Juniperus chinensis	Sea Green
Juniperus horizontalis	Blue Chip, Price of Wales, Bar Harbor
Juniperus scopulorum	Blue Haven
Kalanchoe fedtschenkoi	Lavender-scallops
Kalanchoe tubiflora	Chandelier plant
Physocarpus fruticosa	Katherine Dykes
Rhododendron obtusum	Hinocrimson
Setcreasea purpurea	Purple heart
Zebrina pendula	Wandering jew

*Not applicable in California

Conversion Table
FUSILADE 2000 Rate to be Applied

Lb. a.i./ Acre	Pints/ Acre	Fl. oz./ Acre	Acres/ Gallon
0.094	3/4	12	10.7
0.125	1	16	8.0
0.156	1 1/4	20	6.4
0.188	1 1/2	24	5.3
0.250	2	32	4.0
0.375	3	48	2.7

FUSILADE 2000 contains 1 pound of active ingredient per gallon.

For more information, call ICI Americas at 1-800-759-2500.

EPA registered uses as of August 1989.

APPENDIX

COMMON NAME	SCIENTIFIC NAME
Barnyardgrass	*Echinochloa crusgalli*
Bermudagrass	*Cynodon dactylon*
Broadleaf signalgrass	*Brachiaria platyphylla*
Canarygrass, littleseed	*Phalaris minor*
Crabgrass, Large	*Digitaria sanguinalis*
Crabgrass, Smooth	*Digitaria ischaemum*
Crabgrass, Southern	*Digitaria ciliaris*
Crabgrass, Tropical	*Digitaria bicornis*
Downy brome	*Bromus tectorum*
Fall Panicum	*Panicum dichotomiflorum*
Field Sandbur	*Cenchrus incertus*
Foxtail, Giant	*Setaria faberi*
Foxtail, Green	*Setaria viridis*
Foxtail, Yellow	*Setaria lutescens*
Goosegrass	*Eleusine indica*
Guineagrass, seedling	*Panicum maximum*
Italian Ryegrass	*Lolium multiflorum*
Itchgrass	*Rottboellia exaltata*
Johnsongrass, Rhizome	*Sorghum halapense*
Johnsongrass, Seedling	*Sorghum halapense*
Junglerice	*Echinochloa colonum*
Prairie cupgrass	*Eriochloa contracta*
Quackgrass	*Agropyron repens*
Red Rice	*Oryza sativa*
Shattercane	*Sorghum bicolor*
Sorghum almum	*Sorghum almum*
Southern Sandbur	*Cenchrus echinatus*
Southwestern cupgrass	*Eriochloa gracilis*
Texas Panicum	*Panicum texanum*
Volunteer Cereals	
V. Barley	*Hordeum vulgare*
V. Corn	*Zea mays*
V. Milo	*Sorghum bicolor*
V. Oats	*Avena sativa*
V. Rye	*Secale cereale*
V. Wheat	*Triticum aestivum*
Wild Proso Millet	*Panicum miliaceum*
Wild Oats	*Avena fatua*
Wirestem Muhly	*Muhlenbergia frondosa*
Witchgrass	*Panicum capillare*
Wooly cupgrass	*Eriochloa villosa*

STORAGE AND DISPOSAL

PROHIBITIONS: Do not contaminate water, food or feed by storage or disposal. Open dumping is prohibited. Do not reuse empty container.

STORAGE: Store in original container only. Keep container closed when not in use. Do not store near food or feed. In case of spill or leak on floor or paved surfaces, soak up with sand, earth or synthetic absorbent. Remove to chemical waste area.

PESTICIDE DISPOSAL: Pesticide wastes are toxic. Improper disposal of excess pesticide, spray mixture or rinsate is a violation of Federal law. If these wastes cannot be disposed of by use according to label instructions, contact your State Pesticide or Environmental Control Agency or the Hazardous Waste Representative at the nearest EPA Regional Office for guidance.

CONTAINER DISPOSAL:

METAL CONTAINERS: Triple-rinse (or equivalent). Then offer for recycling or reconditioning, or puncture and dispose of in a sanitary landfill or by other procedures approved by state and local authorities.

PLASTIC CONTAINERS: Triple-rinse (or equivalent). Then offer for recycling or reconditioning, or puncture and dispose of in a sanitary landfill or by other procedures approved by state and local authorities, by burning. If burned stay out of smoke.

GLASS CONTAINERS: Triple-rinse (or equivalent). Then offer for recycling or reconditioning, or puncture and dispose of in a sanitary landfill or by other state and local procedures.

FUSILADE 2000, GRAMOXONE SUPER, and REFLEX 2LC herbicides and AMBUSH and CYMBUSH insecticides are trademarks of ICI Americas Inc.
BASAGRAN and BLAZER herbicides are trademarks of BASF Corporation.
BIDRIN 8 insecticide is a trademark of E. I. duPont de Nemours & Company.
CYGON 400 insecticide is a trademark of American Cyanamid Company.
TACKLE herbicide is a trademark of Rhone-Poulenc.

AMBUSH, CYMBUSH, and GRAMOXONE SUPER are restricted use pesticides.

NOTICE TO BUYER AND USER

Seller warrants that this product conforms to the chemical description on the label and is reasonably fit for the purposes stated on the label when used in accordance with directions under normal conditions of use. This warranty does not extend to the use of this product contrary to label instructions, or under abnormal use conditions, or under conditions not reasonably foreseeable to Seller, and Buyer and User assume the risk of any such use. SELLER DISCLAIMS ALL OTHER WARRANTIES EXPRESSED OR IMPLIED INCLUDING ANY WARRANTY OF FITNESS OR MERCHANTABILITY. SELLER SHALL NOT BE LIABLE FOR CONSEQUENTIAL, SPECIAL OR INDIRECT DAMAGES RESULTING FROM THE USE OR HANDLING OF THIS PRODUCT AND SELLER'S SOLE LIABILITY AND BUYER'S AND USER'S EXCLUSIVE REMEDY SHALL BE LIMITED TO THE REFUND OF THE PURCHASE PRICE.

03-0761-420R2 A-3R

The information and specimen labels in this booklet are correct as of the publication date but are subject to change without notice. For current information, contact ICI Americas Inc.

GRAMOXONE® EXTRA

HERBICIDE

RESTRICTED USE PESTICIDE

For retail sale to and use only by Certified Applicators, or persons under their direct supervision, and only for those uses covered by the Certified Applicator's certification.

A Weed, Grass and Harvest Aid Herbicide Speciment Label
Complete Directions For Use—Read Worker Safety Rules
Not registered for use or sale in California.

ACTIVE INGREDIENT:	By Wt.
Paraquat dichloride (1,1'-dimethyl-4,4'- bypridinium dichloride) .	37.05%
INERT INGREDIENTS:	62.95%
TOTAL	100.0%

Contains 1.5 lbs. paraquat cation per gallon.
Contains stench (odor) and emetic.

EPA Reg. No. 10182-280

KEEP OUT OF REACH OF CHILDREN

DANGER–POISON

ONE SWALLOW CAN KILL! SYMPTOMS ARE PROLONGED AND PAINFUL. ONSET OF SYMPTOMS MAY BE DELAYED FOR UP TO 3 DAYS AFTER SWALLOWING.

PELIGRO

PRECAUTION AL USUARIO:

Si usted no lee ingles, no use este producto hasta que la etiqueta le haya sido explicada ampliamente.

- DO NOT USE OR STORE IN OR AROUND HOME.
- DO NOT REMOVE CONTENTS EXCEPT FOR IMMEDIATE USE.
- NEVER PUT INTO FOOD, DRINK OR OTHER CONTAINERS.

STATEMENT OF PRACTICAL TREATMENT

IF SWALLOWED

IMMEDIATELY induce vomiting by inserting finger in throat. Administer fluids and induce further vomiting. **Get** to a hospital or physician fast. **PROMPT TREATMENT IS ESSENTIAL** and must be initiated before signs and symptons of injury appear.

SKIN CONTACT

IMMEDIATELY wash with soap and water. Remove and wash all contaminated clothing with soap and hot water before reuse. Prolonged contact will cause severe irritation. Contact with irritated skin, or a cut, or repeated contact with intact skin may result in poisoning.

EYE CONTACT

IMMEDIATELY wash eyes with water for at least 15 minutes and get medical attention. Contact may cause eye injury.

INHALATION

IMMEDIATELY get away from spray mist. Exposure may cause irritation, nose bleeds and may lead to poisoning. Stop and check spray procedure. The odor of this product is from the stenching agent which has been added, not from paraquat. Follow Worker Safety Rules.

FOR 24-HOUR EMERGENCY MEDICAL ASSISTANCE CALL 1-800-F-A-S-T-M-E-D (327-8633).

FOR CHEMICAL EMERGENCY: Spill, leak, fire, exposure, or accident call **CHEMTREC** 1-800-424-9300.

Note to Physicians: Call at any hour to obtain toxicology and medical management consultation, a supply of bentonite clay, and paraquat analysis. Symptoms following ingestion may be delayed up to 3 days. To be effective, treatment for paraquat poisoning must begin **IMMEDIATELY.** Treatment consists of binding paraquat in the gut with suspensions of bentonite clay or charcoal and removal of paraquat from the blood by prolonged charcoal hemoperfusion or continuous hemodialysis.

PRECAUTIONARY STATEMENTS

HAZARDS TO HUMANS AND DOMESTIC ANIMALS

DANGER–POISON

Can Kill if Swallowed. May be Harmful or Fatal if Absorbed Through Skin or Inhaled. Harmful to Skin and Eyes.

WORKER SAFETY RULES

DO NOT get on skin, eyes or clothing. **DO NOT** inhale spray mist. **WASH SPLASHES** from skin and eyes immediately. **REMOVE** and wash contaminated clothing. **WASH** before eating, smoking and drinking. **WEAR** full face shield, rubber gloves and apron and waterproof footwear when pouring concentrate or when exposure to concentrate is possible. **WEAR** waterproof protective clothing and footwear when spraying outside of an enclosed cab or aircraft or when contacting vegetation wet with spray.

ENVIRONMENTAL SAFETY CAUTIONS

WILDLIFE: This product is toxic to wildlife. Birds and other wildlife in treated areas may be killed. Do not apply directly to water or wetlands (swamps, bogs, marshes, and potholes). Do not contaminate water by cleaning of equipment or disposal of waste.

DRIFT: GRAMOXONE EXTRA is a contact herbicide that kills all green plant tissue. Do not apply under conditions involving possible drift to food, forage, or other plantings that might be damaged or the crops thereof rendered unfit for sale, use, or consumption. Do not apply when weather conditions favor drift from areas treated. Do not apply during periods of thermal inversion to avoid drift.

CHEMICAL HAZARD: This product reacts with aluminum to produce hydrogen gas which may form a highly combustible gas mixture. Such gas could flash or explode, causing serious personal injury, if ignited by open flame, spark, welder's torch, lighted cigarette, or other ignition source. Do not mix or store in containers, spray tanks, nurse tanks, or such systems made of aluminum or having aluminum fittings. After dilution with water to spray concentrations, spray systems made of, or containing aluminum parts, may be used but should not be used for extended (overnight) storage.

EQUIPMENT/CONTAINER: Flush all spray equipment with water after use each day. GRAMOXONE EXTRA is corrosive to aluminum. Aluminum spray equipment and aluminum aircraft structures that are exposed to spray solution or spray drift should be flushed thoroughly with water immediately after use.

DIRECTIONS FOR USE

It is a violation of Federal law and may endanger the applicator or third parties to use this product without complying with all label directions and warnings.

READ ENTIRE LABEL. USE STRICTLY IN ACCORDANCE WITH DANGER STATEMENTS AND DIRECTIONS, AND WITH APPLICABLE STATE AND FEDERAL REGULATIONS. READ DANGER AND FIRST AID TREATMENT AND WORKER SAFETY RULES.

NOTE: DO NOT USE AROUND HOME GARDENS, SCHOOLS, RECREATIONAL PARKS, OR PLAYGROUNDS.

DO NOT APPLY THIS PRODUCT THROUGH ANY TYPE OF IRRIGATION SYSTEM.

REENTRY STATEMENTS

DO NOT enter treated areas without protective clothing until sprays have dried.

AVOID working in spray mist. If there is risk of exposure, wear goggles and approved face mask capable of filtering spray droplets.

KEEP ALL UNPROTECTED PERSONS OUT OF OPERATING AREAS OR VICINITY WHERE THERE MAY BE DANGER OF DRIFT.

Certain states may require more restrictive reentry intervals; consult your State Department of Agriculture for further information. Written or oral warnings regarding use of protective clothing and accidental exposure must be given to workers who are expected to be in treated areas or in areas about to be treated.

IMPORTANT: The effect of swallowing GRAMOXONE EXTRA is far more severe than that which usually results from skin contact or inhalation of spray mist. This product should not cause injury if used according to label instructions. Follow all these rules **as if your life depends on it.**

GENERAL INFORMATION

GRAMOXONE® EXTRA herbicide is a contact herbicide used to control or suppress a broad spectrum of emerged weeds. GRAMOXONE EXTRA controls most small annual weeds—both broadleaves and grasses, and suppresses perennial weeds by destroying green foliage. GRAMOXONE EXTRA can also be used as a crop desiccant at harvest.

GRAMOXONE EXTRA is formulated as a liquid which contains 2½ lbs. of active ingredient per gallon. The formulation is green in color and contains a non-toxic odor, and an emetic (an agent which will induce vomiting if the product is swallowed). The green color and odor are included in the formulation to help prevent accidental ingestion of GRAMOXONE EXTRA.

GRAMOXONE EXTRA is rapidly absorbed by green plant tissue and interacts with the photosynthetic process to produce superoxides which destroy the plant cells. GRAMOXONE EXTRA requires actively growing green plant tissue to function. Thorough coverage of all green foliage is essential for effective weed control. GRAMOXONE EXTRA is not as effective on drought-stressed weeds, weeds with little green foliage (i.e., mowed or cut weeds), or mature woody bark of trees and vines.

Clay and organic matter rapidly tie up GRAMOXONE EXTRA. As a result, GRAMOXONE EXTRA has no residual soil activity to affect later-planted crops or later germinating weeds.

RAINFASTNESS

Because GRAMOXONE EXTRA is rapidly absorbed by the weed foliage, rain occurring 30 minutes or more after application will have no effect on the activity of GRAMOXONE EXTRA.

APPLICATION

Since GRAMOXONE EXTRA is a contact-type herbicide, it is essential to obtain complete coverage of target weeds to get good control. Improper application technique and/or application to large, stressed, or mown weeds will usually result in unacceptable weed control. Good coverage is also essential for proper performance in harvest-aid applications. See details below for specific application instructions.

USE OF A NONIONIC SURFACTANT

ALWAYS ADD a nonionic surfactant (approved for use on growing crops) containing at least 50% surface active ingredient. The concentration of surface active ingredient in a particular surfactant will determine how much should be added to the tank. Refer to table below for specific recommendations:

Surfactant % Nonionic Active Ingredient	Rate of Surfactant** Per 100 Gallons of Water
75% or greater	1 pint*
50-74%	2 pints
Less than 50%	DO NOT USE

* 1 pint = 0.125% V/V.

** For applications where higher rates of surfactant are desired or required, additional surfactant may be used.

FAILURE TO USE A NONIONIC SURFACTANT AT RECOMMENDED RATES WILL RESULT IN REDUCED PERFORMANCE OF GRAMOXONE EXTRA.

NOZZLE SELECTION

The use of flat-fan nozzles will result in the most effective application of GRAMOXONE EXTRA. Flood nozzles are generally not as good as flat fans since they produce large uneven droplets. The use of flood nozzles may result in reduced weed control due to inadequate coverage.

Recommended Nozzles, Pressures and Setup

	Nozzle Type	
	Flat fan	Flood
Maximum Size	8	15
Spray Pressure (at nozzle)	30-50 psi	30-50 psi
Maximum Nozzle Spacing	30″	40″
Direction of Spray Pattern	Down	Down
Maximum Speed	10 mph	10 mph
Spray Overlap (at each edge)	30%	50%

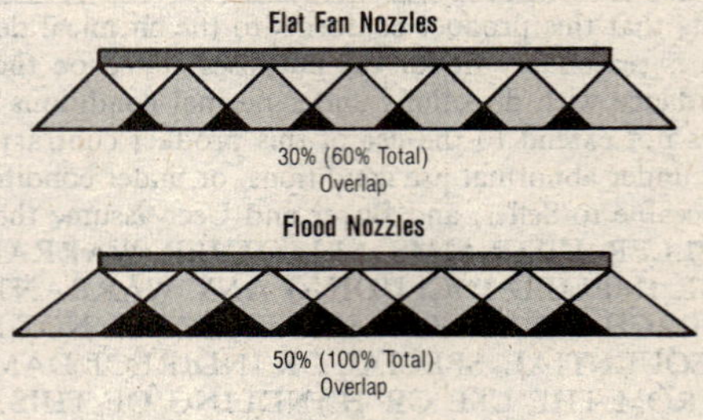

Using nozzles, pressures, or setups different from the above chart will result in reduced control.

SPRAY CARRIER

Always use clean water (free of mud or clay), clear liquid nitrogen, or complete clear liquid fertilizers as the carrier when spraying GRAMOXONE EXTRA. Muddy water, or suspension-type fertilizers containing clay, can inactivate GRAMOXONE EXTRA. Never use suspension-type fertilizers containing clay as the spray carrier. If using a complete clear liquid fertilizer containing high phosphate levels as the spray carrier, always use the higher rate of GRAMOXONE EXTRA and surfactant.

NOTE: When using liquid fertilizers such as 28% N, as a carrier, it is important that nonionic surfactant still be used with GRAMOXONE EXTRA. Liquid fertilizer carriers cannot substitute for surfactant.

RATES OF GRAMOXONE EXTRA

Follow recommended rates listed with each use of GRAMOXONE EXTRA. Use the higher label rates when weeds are dense or large. Also, use higher label rates for harvest aid when crop vegetation is dense.

SPRAY VOLUME

Follow recommended minimum spray volumes listed with each use of GRAMOXONE EXTRA. These are **minimum** volumes only, and spray volumes should be increased as necessary to obtain complete coverage of the target weed or plant without runoff from the foliage.

DRIFT MINIMIZATION

DO NOT spray if conditions of thermal inversion exist, or if wind direction and speed may cause spray drift onto adjacent areas. With both aerial and ground applications, the use of a drift control or spray deposition additive (in accordance with label directions) may reduce the potential for drift to occur. **Drift minimization is the responsibility of the applicator.** Consult with local and state agricultural authorities for information regarding avoiding or minimizing spray drift.

APPLICATION TIMING

GRAMOXONE EXTRA should be applied to emerged weeds when they are small. Weeds 1″ to 6″ in height are the easiest to control. Larger weeds may be more difficult to control. When weeds have been grazed or mowed, thus removing much of the green foliage, allow weeds to regrow to a height of 2 to 4 inches before spraying if possible. Similarly, when forage or grain crops have been harvested prior to spraying, weeds present in the field will also have been cut. To allow for adequate green foliage to remain on weeds in this situation, raise cutter bars as high as possible from the ground to cut stubble and weeds at a greater height.

BURNDOWN OF GRASS COVER CROPS OR VOLUNTEER CEREALS

When using GRAMOXONE EXTRA for control of grass cover crops or volunteer cereals, best results are obtained when GRAMOXONE EXTRA is applied **prior to tillering** or **after boot stage.** This is especially important with a wheat cover crop or volunteer wheat. Treatments made between tillering and boot stage will generally not provide complete control.

ENVIRONMENTAL CONDITIONS

GRAMOXONE EXTRA is active over a wide range of environmental conditions. Cool weather (below 55°) will slow the activity of GRAMOXONE EXTRA, as will cloudy, overcast weather, but will not affect performance.

In dry areas, dust stirred up by high winds or equipment tires can coat weed leaves and reduce GRAMOXONE EXTRA activity. Avoid applying GRAMOXONE EXTRA in extremely dusty conditions.

SPOT SPRAYING

When only small areas are to be sprayed with labeled applications, it is advantageous to mix small quantities of GRAMOXONE EXTRA. To aid in mixing small quantities, the following table should be consulted. When spraying with wand-type applicators as in spot spraying, always follow precautionary statements regarding use, protective clothing and footwear.

If the broadcast rate per acre for GRAMOXONE EXTRA is:	Add the following amount of GRAMOXONE EXTRA to 1 gallon of water
1½ pts	⅓ fl oz
2 pints	⅜ fl oz.
2½ pts	½ fl oz
3 pts	⅔ fl oz

Always add ⅓ to ½ fl. oz. of a nonionic surfactant for each gallon of spray. When spot spraying in this manner, spray to thoroughly wet the foliage, but not to the point of runoff.

TANK MIXING FOR IMPROVED BURNDOWN OF DIFFICULT WEEDS, AND RESIDUAL WEED CONTROL

PHOTOSYNTHETIC INHIBITOR HERBICIDES

Difficult weeds can often be controlled by tankmixing GRAMOXONE EXTRA with other herbicides. The addition of herbicides which are also photosynthetic inhibitors (psi) will slow the activity of GRAMOXONE EXTRA, allowing GRAMOXONE EXTRA to thoroughly distribute itself within the treated leaf. The resulting level of control is usually greater than if GRAMOXONE EXTRA was applied alone.

GRAMOXONE EXTRA may be applied in tank mixture with the following psi herbicides:

ATRAZINE	CONQUEST*	LOROX PLUS
AATREX	EXTRAZINE*	LARIAT
BICEP	GEMINI*	PREVIEW*
BLADEX	GRIFFEX*	PRINCEP
CANOPY*	LEXONE/SENCOR	LINEX/LOROX*

Refer to respective product label(s) for rates of application, directions for use, limitations, cautions and for a list of weeds controlled.

*Not for use in California

Improved Weed Control With PSI's

Control of difficult weeds listed below can be improved by the addition of a Psi herbicide.

Marestail	Tansymustard
Pennsylvania Smartweed	Prickly lettuce
Velvetleaf	Lambsquarters
Malva (Cheeseweed)	Morningglory
Fall Panicum	Cheatgrass
Giant Ragweed	Knotweed
Kochia	Volunteer wheat
Cocklebur	Perennial Weeds (suppression only)

IMPROVED CONTROL OF PERENNIAL AND ANNUAL BROADLEAF WEEDS

When perennial broadleaf weeds such as Canada thistle, bindweed, dandelion, etc. or difficult to control annual broadleaf weeds such as giant ragweed or morningglory are present, tank mixes with 2,4-D ester (cv), 2, 4-DB or BANVEL, where labeled, will help improve control. Tankmixing the amine formulation of 2,4-D with GRAMOXONE EXTRA, may result in reduced grass control.

ORDER OF TANKMIXING

In general, GRAMOXONE EXTRA tankmixes with other products should be mixed as follows:

1. Fill spray tank ½ full with clean water or other approved carriers such as clear liquid fertilizer.
2. Begin tank agitation and continue throughout mixing and spraying.
3. Add dry formulations (WP, DF, etc.) to tank.
4. Add liquid formulations (SC, EC, L, etc.) to tank.
5. Add GRAMOXONE EXTRA to tank.
6. Add nonionic surfactant to tank.
7. Fill remainder of spray tank.

Always refer to labels of other pesticide products for mixing directions and precautions which may differ from those outlined here.

Since many of the herbicides listed on this label are available in several types of formulations, it is advisable to perform a jar test to check physical compatibility.

LIMITATIONS, PRECAUTIONS AND WARNINGS

- DO NOT use around home gardens, schools, recreational parks, or playgrounds.
- In preplant and preemergence (to the crop) uses, do not apply to soils lacking clay minerals, i.e., peat, muck, pure sand, artificial planting media.
- Seedbeds and plantbeds should be formed as far ahead of planting and treatment as possible to permit maximum weed and grass emergence prior to treatment.
- Seeding or transplanting should be done with a minimum amount of soil disturbance.
- Weeds and grasses emerging after application of GRAMOXONE EXTRA will not be controlled or suppressed.
- Unless otherwise indicated, crop plants emerged at time of application may be severely injured or killed if contacted by sprays of GRAMOXONE EXTRA.

SPECIFIC USE RECOMMENDATIONS

The following tables indicate use patterns, rates, minimum spray volumes, pre-harvest intervals and other precautions, restrictions and comments specific to each crop. Read and follow directions carefully.

CROP	USE PATTERN	GRAMOXONE EXTRA RATE PER ACRE	MINIMUM TOTAL SPRAY PER ACRE	GRAZING OR PREHARVEST INTERVAL (DAYS)	PRECAUTIONS, RESTRICTIONS AND COMMENTS
ALFALFA New Seedlings (California only)	Broadcast	See Table 1 Page 28	Ground: 25 gallons Air: 5 gallons	70	• Apply during late winter or early spring. • Do not cut or harvest within 70 days after application. • Do not apply more than once during the first growing season. • Caution: Seedling alfalfa stands will be reduced and replanting may be necessary. • Not recommended for seedling alfalfa grown for seed. • Alfalfa foliage present at time of application will be burned.
Preplant or Preemergence (No-till or conventional planting)	Broadcast or Banded Over Row	2–3 pints	Ground: 10 gallons Air: 3 gallons	—	• Apply prior to emergence of the crop. • Crop plants emerged at time of application will be killed. • Seeding should be done with a minimum amount of soil disturbance.

CROP	USE PATTERN	GRAMOXONE EXTRA RATE PER ACRE	MINIMUM TOTAL SPRAY PER ACRE	GRAZING OR PREHARVEST INTERVAL (DAYS)	PRECAUTIONS, RESTRICTIONS AND COMMENTS
Dormant season on established plantings (WA, OR, CO, ID, MT, UT, NV, WY, and Del Norte, Siskiyou, Modoc, Shasta, Lassen, Plumas, Sierra and Nevada Counties in CA)	Broadcast	1½–2 pints	Ground: 20 gallons Air: 5 gallons	42	• For desiccation of weeds including ryegrass, bluegrass, cheatgrass, dogfennel, chickweed, and tansymustard. Use higher rate for cheatgrass. • Apply to established stands after the last fall cutting when crop is dormant but before spring growth starts. • Do not apply if regrowth after last fall cutting is more than 2". • Alfalfa foliage at time of application will be burned. • Do not apply more than once per season. • Do not cut or harvest within 42 days of application.
Dormant season on established plantings (East of the Rocky Mountains and North of the Southern Borders of CO, KS, MO, KY, and VA)	Broadcast	1½–2 pints	Ground: 20 gallons Air: 5 gallons	60	• For control of weeds, including bluegrass, chickweed, henbit, and downy brome and suppression of perennial weeds. • Do not apply if fall *regrowth* following last fall cutting is greater than 6". • Apply to well-established stands (at least one year old) after the crop is dormant, but before spring *regrowth* reaches one inch. • Alfalfa foliage present at the time of application will be burned which may reduce the yield of the first cutting. • Do not cut or harvest within 60 days of application. • Do not apply more than once per season.
Dormant season tankmix with Velpar L (East of the Rocky Mountains and North of the Southern borders of CO, KS, MO, KY, and VA.)	Broadcast	1–1½ pints	Ground: 20 gallons Air: 10 gallons	42	• For control of weeds such as chickweed, downy brome and tansymustard. • Use the 1 pt. rate of GRAMOXONE EXTRA when weeds and grasses are less than 4" tall. • Mix with 1 to 2 qts. of Velpar L per acre. • Use the lower rate of Velpar L on loamy sands or sandy loams. Refer to Velpar L label for directions, limitations, cautions and for a list of weeds controlled. • Apply once to established alfalfa stands during the dormant season. • Do not apply to **regrowth** that is more than 2 inches. • Do not apply following cuttings during growing season. • Do not apply to alfalfa during the first season after seeding. • Temporary chlorosis may occur on alfalfa regrowth. • Stress which may be caused in part by low fertility, disease, insects, winterkill, over cutting, drought or frost may increase the chances of crop injury. • **DO NOT USE** on gravelly or rocky soils, exposed subsoils, hardpan, sand or poorly drained alkaline soils as crop injury, including mortality, may result. • Do not cut or harvest within 42 days of application.
Dormant season Arizona and California	Broadcast	13–24 fl. oz.	Ground: 20 gallons Air: 5 gallons	60	• Apply during late fall or winter months after the last fall cutting and before first spring cutting. • Do not apply if **regrowth** after grazing or cutting is more than 2 inches. • Do not harvest within 60 days of application or excessive residues may result. • Alfalfa foliage present at the time of application will be discolored and temporarily stunted. • Total hay yield of first cutting may be reduced in alfalfa fields with severe weed infestation. This reduction will usually be directly proportionate to the loss of weed weight. • Do not apply more than once per season. • Do not apply following cuttings during the growing season.
California					• For desiccation of weeds including bluegrass, ryegrass, shepherdspurse, chickweed, tansymustard, foxtail, sowthistle and groundsel. • Use high rate if ryegrass, shepherdspurse, sowthistle or groundsel is present.
Arizona					• For desiccation of weeds, including London rocket, sowthistle, rescue brome and wild oats.

CROP	USE PATTERN	GRAMOXONE EXTRA RATE PER ACRE	MINIMUM TOTAL SPRAY PER ACRE	GRAZING OR PREHARVEST INTERVAL (DAYS)	PRECAUTIONS, RESTRICTIONS AND COMMENTS
Dormant season tankmix with Velpar L (ID, OR, UT, and WA, NV, and the northern California Counties of Del Norte, Siskiyou, Modoc, Shasta, Lassen, Plumas, Sierra, and Nevada)	Broadcast	13 –24 fl. oz.	Ground: 20 gallons Air: 10 gallons	42	• For control chickweed, downy brome and tansymustard. • Use the 13 fl. oz. rate of GRAMOXONE EXTRA when weeds and grasses are less than 4" tall. • Mix with 1 to 2 qts. of Velpar L per acre. • Use the lower rate of Velpar L on loamy sands or sandy loams. Refer to Velpar L label for directions, limitations, cautions and for a list of weeds controlled. • Apply once to established alfalfa stands during the dormant season. • Do not apply to **regrowth** that is more than 2 inches. • Do not apply following cuttings during growing season. • Do not apply to alfalfa during the first season after seeding. • Temporary chlorosis may occur on alfalfa regrowth. • Do cut or harvest within 42 days of application. • Stress which may be caused in part by low fertility, disease, insects, winterkill, over cutting, drought or frost may increase the chances of crop injury. • **DO NOT USE** on gravelly or rocky soils, exposed subsoils, hardpan, sand or poorly drained alkaline soils as crop injury, including mortality, may result.
Between-cuttings treatment in established plantings. (All states east of the Rocky Mountains) (Includes first-year alfalfa)	Broadcast	1½ pints	Ground: 20 gallons	30	• Weeds much beyond the seedling stage and the stubble of weeds cut off during harvest will be less affected by this treatment. • Apply immediately after alfalfa has been removed for hay or silage. • Do not treat more than 5 days after cutting. • **CAUTION:** First-year alfalfa stands and yields may be reduced if alfalfa is allowed to regrow more than two (2) inches. • Alfalfa foliage present at time of application will be burned. • In arid areas where moisture is limited, weed control may be reduced. • Do not cut or harvest within 30 days of application. • Make 1 to 3 applications, as needed, during the growing season. These sprays may be applied in addition to a dormant application. • For first year alfalfa, do not apply more than twice during the first growing season.
ASPARAGUS Preplant or Preemergence	Broadcast or Banded Over–Row	2–3 pints	Ground: 25 gallons Air: 5 gallons	—	• Apply prior to emergence of the crop. • Crop plants emerged at time of application will be killed.
Preemergence to established planting at least 2 years old	Broadcast or banded Over Row	2–3 pints	Ground: 25 gallons	6	• Apply prior to emergence of crop or after last harvest. • Crop plants emerged at time of planting will be killed. • May be tankmixed with Princep herbicide for residual control. Refer to Princep label for specific directions, limitations, cautions and for a list of weeds controlled.
CASSAVAS, TANIERS YAMS (Puerto Rico only) (Shielded Post directed spray) (Puerto Rico only)	Directed Spray	1½ pints	Ground: 50 gallons	90	• Apply when weeds are succulent and growth is 1–6 inches. • On cassavas and taniers, do not make more than 3 applications per crop season. • On yams do not make more than 2 applications per crop season. • Do not allow spray to contact cassavas, tanier or yam plants as injury may result. • Do not spray under windy conditions. • Do not graze treated areas or feed treated forage to livestock.

CROP	USE PATTERN	GRAMOXONE EXTRA RATE PER ACRE	MINIMUM TOTAL SPRAY PER ACRE	GRAZING OR PREHARVEST INTERVAL (DAYS)	PRECAUTIONS, RESTRICTIONS AND COMMENTS
CLOVER Dormant season on establised plantings (WA, OR, CO, ID, MT, UT, NV, WY and the counties of Del Norte, Siskiyou, Modoc, Shasta, Lassen, Plumas, Sierra, and Nevada in CA)	Broadcast	1½–2¼ pints	Ground: 20 gallons Air: 5 gallons	60	• For desiccation of ryegrass, bluegrass, cheatgrass, dogfennel, chickweed, and tansymustard. Use higher rate if cheatgrass is present. • Apply after the last fall cutting when the crop is dormant but before spring growth starts. • Do not apply if regrowth following the last fall cutting is more than 2". • Clover foliage present at the time of application will be burned. • Do not pasture animals in treated fields before first cutting. • Do not graze, cut or harvest within 60 days of application. • Do not apply more than once per season.
CONSERVATION RESERVE AND FEDERAL SET-ASIDE PROGRAMS (For use in compliance with the Federal Conservation Reserve Program or Federal set-aside programs.)	Broadcast	2½–5 pints (Currently not registered for use in California at rate higher than 2½ pts.)	Ground: 10 gallons Air: 3 gallons	–	• For improved emerged weed control or extended weed control, GRAMOXONE EXTRA may be tankmixed with other herbicides registered for this use. Refer to tankmix herbicide labels for specific directions, limitations, cautions and for a list of weeds controlled.
CORN Preplant or Preemergence (Used alone)	Broadcast or Banded Over Row	Weeds 1–3": 1½–2 pts Weeds 3–6": 2–2½ pts Weeds 6": 2½–5 pts	Ground: 20 gallons Air: 5 gallons		• Includes field, fresh, sweet, forage, fodder and popcorn. • Seedbeds should be formed as far ahead of planting and treatment as possible to permit maximum weed and grass emergence. • Seeding should be done with a minimum amount of soil disturbance. • Weeds and grasses emerging after application will not be controlled. Crop plants emerged at time of application will be killed.
Preplant or Preemergence (Tank-mixture treatments for no-till/reduced till corn	Broadcast or Banded	Weeds 1–3": 1½–2 pts Weeds 3–6": 2–2½ pts Weeds 6": 2½–5 pts	Ground: 20 gallons		• Apply as a broadcast spray before, during or after planting, but before crop emergence. • For improved burndown or residual control, GRAMOXONE EXTRA may be tankmixed with the following herbicides: 2,4-D Ester*, AAtrex/Atrazine, Banvel, Bicep, Prowl, Bladex, Dual, Extrazine*, Griffex*, Lasso, Linex*, Lorox, Modown, Princep, Simazine *Not for use in California • GRAMOXONE EXTRA may also be tankmixed with AMBUSH Insecticide. • Refer to respective product label(s) for rates of application, directions for use, limitations, cautions, and for a list of weeds or insects controlled.
Field Corn **Popcorn** **Sweet Corn***	Postemergence Directed Spray (Used Alone or Tankmix with Atrazine	1½ pints	Ground: 20 gallons		• Apply when corn is at least 10 inches tall with nozzles arranged to spray no higher than the lower 3 inches of corn stalks. • Corn plants shorter than 10 inches may be injured and not recover (corn height measured from soil surface to top of whorl).
* Currently not registered for use in California on sweet corn	Tankmix with Atrazine	1½ pints	Ground: 20 gallons	–	• For improved control of emerged weeds (especially broadleaf weeds). Add ½ to 1 lb ai/acre atrazine in a tankmix. Refer to the atrazine label for directions, limitations, cautions and for a list of weeds controlled. • Do not mix with liquid fertilizer. • Constant agitation of tankmix is necessary at all times. • Refer to other precautions, etc., in "used alone" section immediately above.
Field Corn (grain, fodder, forage) **and Popcorn only**	Postemergence directed spray USDA Witchweed Eradication Program	2½ pints	Ground: 10 gallons	–	• Initiate sprays in early July and repeat in early August if regrowth of grassy weeds occurs. • Follow application instructions in postemergence directed spray section above.

CROP	USE PATTERN	GRAMOXONE EXTRA RATE PER ACRE	MINIMUM TOTAL SPRAY PER ACRE	GRAZING OR PREHARVEST INTERVAL (DAYS)	PRECAUTIONS, RESTRICTIONS AND COMMENTS
COTTON Preplant or Preemergence (Used alone)	Broadcast	2½–5 pints	Ground: 20 gallons Air: 5 gallons	–	• Apply prior to, during or after planting, but before crop emergence. • For fallow bed treatment, beds should be preformed to permit maximum weed and grass emergence prior to treatment. • Seeding should be done with a minimum of soil disturbance.
Preplant (California only) (Used alone)	Broadcast	11–21 fl oz	Ground: 20 gallons Air: 5 gallons	–	• For control of volunteer barley in preformed seedbeds.
Preplant or Fallow Bed (Tankmix with Goal)	Broadcast	1½–5 pints	Ground or Air: 10 gallons	–	• Refer to Goal label for specific use directions and restrictions, and weeds controlled.
Preplant (Tankmix with Bladex)	Broadcast	2½–5 pints	Ground: 20 gallons Air: 5 gallons	–	• Refer to Bladex label for specific use directions and restrictions and weeds controlled.
Preplant/ Preemergence (Other Tankmixes)	Broadcast	2½–5 pints	Ground: 20 gallons Air: 5 gallons		• Apply as a broadcast spray before, during or after planting, but before crop emergence.
Other Tankmixes					• For improved burndown or residual control, GRAMOXONE EXTRA may be tankmixed with the following herbicides: – Cotoran (4L only) – Zorial* – Karmex/Direx* – Dual* – Prowl* * Not registered for use in California. • When tankmixing with Cotoran 4L, add Dawn®* dishwashing liquid to the tank at 8 oz/100 gallons water *prior* to adding the Cotoran and GRAMOXONE EXTRA. • When tankmixing with any of the herbicides listed above, refer to that product's label for specific directions and restrictions and for a list of weeds controlled.
Cotton Harvest Aid	Broadcast	See Rates Below for Each Use	See Below	See Below	**Harvest Aid Use Precautions** • Do not pasture lactating dairy animals. • Do not pasture livestock in treated fields within 15 days after treatment. • Remove livestock from treated area 30 days before slaughter. • Do not feed gin trash to livestock. • Do not apply to cotton within 3 days before harvest. • Repeat application if necessary. Do not make more than 2 applications or exceed a total of 1½ pts./Acre. • When combined with phosphate defoliants, observe livestock cautions listed on the phosphate label. • When combined with chlorate defoliants, do not pasture livestock on treated areas or feed treated foliage or gin trash. • Do not apply within 7 days when used with phosphate or chlorate defoliant.
Harvest aid for boll opening and defoliation (Tankmix with phosphate and chlorate defoilants)	Broadcast	11 fl. oz. plus 1 pt. phosphate or 1 gal chlorate	Ground: 10 gallons Air: 3 gallons	7	• Development of immature bolls will be inhibited. • Apply when 80% or more of the bolls are open and the remaining bolls to be harvested are mature. • Follow Harvest and Use Precautions listed above.
Other Tank Mixes for boll opening and defoliation of Southern cotton	Broadcast	4–6 oz.	Ground: 10 gallons Air: 3 gallons		• To aid in defoliation and opening of mature bolls, GRAMOXONE EXTRA may be tankmixed with the following products: DEF HARVADE FOLEX PREP DROPP • Apply when 60% or more of the bolls are open and the remaining bolls to be harvested are mature. • Development of immature bolls will be inhibited. • Refer to tankmix product label for rates, directions, limitations and cautions. • Follow Harvest Aid Use Precautions on page 15.

CROP	USE PATTERN	GRAMOXONE EXTRA RATE PER ACRE	MINIMUM TOTAL SPRAY PER ACRE	GRAZING OR PREHARVEST INTERVAL (DAYS)	PRECAUTIONS, RESTRICTIONS AND COMMENTS
Post Defoliation—To aid in opening of mature bolls and to desiccate green weeds. (Except AZ & CA)	Broadcast	1½–2½ pints	Ground: 20 gallons Air: 5 gallons	3	• Use higher rate if weed infestation is heavy or dense. • Apply when 75% or more of the bolls are open and remaining bolls to be harvested are mature. • Development of immature bolls will be inhibited. • Follow Harvest Aid Use Precautions on page 15.
Harvest aid for boll opening and defoliation of Western cotton (AZ & CA)	Broadcast Split Application	First application 7–11 fl. oz. plus 1–3 pts. phosphate or 1–2 gal chlorate	Ground: 10 gallons Air: 3 gallons	7	• Use higher rate of GRAMOXONE EXTRA on rank cotton. • Apply when 60% or more of the bolls are open and the remaining bolls to be harvested are mature. • Development of immature bolls will be inhibited. • Do not use more than 4.0 lbs. of actual sodium chlorate or 2.28 lbs. of actual phosphate type defoliants per acre. • Follow Harvest Aid Use Precautions listed above. • Do not use more than 6 fl. oz. of GRAMOXONE EXTRA at first application as excessive desiccation may occur.
(Tankmix with phosphate or chlorate defoliants)		Second application 11–21 fl. oz. alone **or** tankmixed with 1½–2 gal. chlorate, **or**, 1½–2 pts. phosphate		3 (Alone) or 7 (TankMix)	• In desert cotton areas or on rank vigorous cotton, use the 21 fl. oz. rate of GRAMOXONE EXTRA. • Apply when 75% or more of the bolls are open and remaining bolls to be harvested are mature. • Development of immature bolls will be inhibited. • Follow Harvest Aid Use Precautions on page 15.
Harvest aid desiccation of stripper cotton or spindle harvested cotton late season	Broadcast	1½–2½ pints	Ground: 10 gallons Air: 3 gallons	3	• May be applied as a split application. Do not make more than 2 applications or exceed a total of 2½ pts/Acre. • Apply when 85% of the bolls are open and the remaining bolls to be harvested are mature. Development of immature bolls will be inhibited. • Follow Harvest Aid Use Precautions on page 15.
Desiccation of Regrowth (except CA)	Broadcast	11–21 fl oz	Ground: 10 gallons Air: 3 gallons	3	• Use to desiccate regrowth occuring after defoliation or desiccation. • Use higher rate if regrowth is excessive. • Follow Harvest Aid Use Precautions on page 15.
EASTER LILIES (Field grown)	Preemergence	2½–5 pints	Ground: 20 gallons	–	• Do not apply more than twice per season.
GRASSES (For Seed) (For Use in Seedbed Preparation)	Preplant, At Planting	2½ pints	Ground: 20 gallons	–	• Prepare the seedbeds and allow weeds to germinate. • Apply GRAMOXONE EXTRA when weeds are at the 3 to 5 leaf stage. • Repeat applications as necessary prior to or on the day of seeding. • Do not graze treated areas or use the seed or straw from treated areas for animal feed or bedding.
GUAR (Preharvest desiccation)	Preharvest	2½ pints	Ground: 20 gallons	4	• Apply after the pods are fully mature. • Do not graze treated areas or use the treated forage for animal feed.
MINT (Peppermint, Spearmint)	Dormant Season	2–4 pints	Ground: 25 gallons	–	• For suppression of weeds such as Italian ryegrass, prickly lettuce, groundsel, chickweed, downy brome and bluegrass. • Apply when crop is dormant before spring growth begins and when weeds are less than 6 inches tall. • Do not apply more than 2 pts per acre per dormant season. • May be tankmixed with Sinbar (terbacil) weed killer for improved contact activity and residual control of Italian ryegrass, prickly lettuce and groundsel. Apply this tank mixture no more than once per season. Refer to the Sinbar label for rates, directions, and cautions and for a list of weeds controlled.
NONCROP USES	Broadcast or Spot Treatment	2½–5 pints	Ground: 50 gallons		• For use in noncrop areas such as public airports, electric transformer stations, pipeline pumping stations, around commercial buildings, storage yards and other installations, fence lines or similar noncrop areas. • Avoid contact with the foliage of ornamentals or desired plants. • Repeat as necessary.
ONIONS (seeded) and garlic	Preplant/Preemergence	2½–5 pints	Ground: 20 gallons	60 200 (CA only)	• Use the higher rate for heavy weed infestations or wild oat control. Apply only one application per season at the 3 pt per acre dosage. • Allow maximum weed and grass emergence prior to treatment but apply prior to crop emergence. • Apply a maximum of 3 pt. per acre per season.

CROP	USE PATTERN	GRAMOXONE EXTRA RATE PER ACRE	MINIMUM TOTAL SPRAY PER ACRE	GRAZING OR PREHARVEST INTERVAL (DAYS)	PRECAUTIONS, RESTRICTIONS AND COMMENTS
PASTURE RESEEDING For suppression of existing sod and undesirable emerged broadleaf weeds and grasses prior to, or at time of, planting of grasses or forage legumes or winter annuals	Broadcast or Banded Over Row	1½–2½ pts	Ground: 20 gallons Air: 3 gallons	See specific geographic recommendation	**West of Cascade and Sierra Nevada Mountains:** • Apply in October through December after first fall rains and after weeds have emerged and sod has started new growth. • For best seeding results, apply on moderately to heavily grazed areas. • Do not use in areas with heavy sod and weed growth. **East of Rocky Mountains:** • Use the 24 fl. oz. rate on vigorous or coarse sod species such as bromegrass. • See specific directions for use in the South and Southeast for suppression of bermudagrass or bahiagrass sods. • Apply prior to, or at time of seeding grasses or forage legumes such as alfalfa, clover and birdsfoot trefoil. • Apply only to grazed or mowed pastures not more than 3 inches in height at time of treatment.
Bermudagrass or Bahiagrass Sods—Late Summer or Early Fall Application for Sod Suppression when Seeding Winter Annuals:		1½ pints			**South and Southeast:** • Apply in late summer or early fall to sod not exceeding 3 inches in height.
Bermudagrass Pastures—Dormant Season Suppression of Emerged Weeds		1½ pints			**South and Southeast:** • Apply during March. • For control of little barley, apply before the mid-boot stage. • Do not mow for hay until 40 days after treatment.
For Control of Endophyte-Fungus-Infected Fescue Forage Legume/Grass Mixture and Other Grass Pastures	Broadcast (split application)	1½–2½ pts followed by 1½–2½ pts	Ground: 20 gallons		• Use split applications of 10 to 21 days apart if necessary. • Do not exceed 3 pints per acre total in preparation for reseeding. • For spring plantings, the initial application of 12-24 fl. oz. may be made the previous fall. • Apply when fescue is actively growing and no more than 4 inches high. • To reduce the infestation of endophyte-infested grass, do not allow fescue to go to seed starting with the preceding year's crop.
PEANUTS* *Not registered for use in California	Broadcast	11 oz. Second application 11 oz.	Ground: 20 gallons		• To control or suppress small (1-6") emerged annual grass and broadleaf weeds in peanuts at ground crack. A second application of 6½ fl. oz./acre may be made up to 28 days after ground crack. • Make no more than two applications per crop and do not apply a total of more than 13 fl. oz. of product per crop. • Crop foliage sprayed will be injured in the form of bronzing and crinkling but the crop will recover and develop normally.
PEANUTS (Tankmix with Basagran)	Broadcast	11 oz. Second application 11 oz.	Ground: 20 gallons		• For improved control of weeds such as cocklebur, bristly starbur, smartweed and prickly sida, tank mix GRAMOXONE EXTRA with Basagran at 1 pint/acre. • This tank mix should be applied at the ground crack stage of peanuts. A second application may be made up to 28 days after ground crack. • Make no more than two applications per crop and do not apply a total of more than 13 fl. oz. of product per crop. • Crop foliage sprayed will be injured in the form of bronzing and crinkling but the crop will recover and develop normally. • Refer to the Basagran label for specific use directions, limitations, cautions and for a list of weeds controlled. • Do not apply this tankmix if peanuts show injury (leaf phytotoxicity and/or plant stunting) produced by any other herbicide treatment as injury may be enhanced and/or prolonged. • Do not apply this tankmix during prolonged periods of drought or unreasonably cold weather as unsatisfactory weed control may result.
PEANUTS (Tankmix with Butyrac or Butoxone 200)	Broadcast	11 oz. Second application 11 oz.	Ground: 20 gallons		• For improved control of weeds such as cocklebur, sicklepod and morningglory tankmix GRAMOXONE EXTRA with 8-16 oz.(0.125-0.25 lbs) per acre of Butyrac or Butoxone 200.

CROP	USE PATTERN	GRAMOXONE EXTRA RATE PER ACRE	MINIMUM TOTAL SPRAY PER ACRE	GRAZING OR PREHARVEST INTERVAL (DAYS)	PRECAUTIONS, RESTRICTIONS AND COMMENTS
					• Make no more than two applications per crop and do not apply a total of more than 13 fl. oz. of product per crop. • Crop foliage sprayed will be injured in the form of bronzing and crinkling but the crop will recover and develop normally. • Refer to the complete Butyrac or Butoxone 200 label for specific use directions, limitations, cautions and for a list of weeds controlled.
PIGEON PEAS (Puerto Rico only)	Directed Spray	2½ pts	Ground: 20 gallons	60	• Avoid contact with pigeon pea foliage. • Cannery waste can be fed to livestock. • Do not make more than 1 application per season. • Do not graze treated areas or feed treated forage to livestock.
PINEAPPLE	Directed Spray	2½–5 pints	Ground: 50 gallons	20	• Retreatment may be necessary on more mature weeds.
POTATOES Preplant or Preemergence	Broadcast	1½–2½ pints	Ground: 30 gallons Air: 5 gallons	–	• Do not apply later than ground cracking, before potatoes have emerged.
Preplant (California only) (Used alone)	Broadcast	11–21 fl oz	Ground: 20 gallons Air: 5 gallons		• For control of volunteer barley in preformed seedbeds.
Preharvest vine killing and weed desiccation (Except Northern Alabama; Counties of Marshal, DeKalb, Jackson and Cherokee)	Broadcast	1½–2½ pints	Ground: 50 gal	3	• Begin application when leaves begin to turn yellow. • Immature or drought-stressed potato foliage are tolerant to GRAMOXONE EXTRA. Desiccation will not be complete under these conditions. • Use 1½ pts./Acre rate where quick vine kill is desired. • Use two applications of 12 fl. oz./Acre when vine growth is dense. Split applications must be applied a minimum of five days apart. • Do not pasture livestock in treated potato fields. • Do not use on muck or peat soils to avoid injury to subsequent crops. • **DO NOT** use to desiccate potato vines when potatoes are to be stored or used for seed as this use may result in tuber decomposition or failure of seed pieces to germinate and grow normally. • Do not exceed 3 pts./Acre per season.
RESIN SOAKING (Pines)	See Pages 28-29				
SMALL FRUITS Blackberries Boysenberries Raspberries Blueberries	Postemergence Directed Spray	2½–5 pints	Ground: 50 gallons	–	• Apply before emergence of new canes or shoots as injury to those canes or shoots can occur. • Apply as a coarse spray to avoid crop injury from fine spray mist.
Strawberries	Postemergence Directed Spray	2½ pints	Ground: 20 gallons	21	• Apply by directing spray between the rows and using shields to prevent spray contact with crop plants. • Do not allow spray to contact strawberry plants as injury or excessive residues may result. • Do not apply more than 3 times per season. • Do not graze livestock in treated areas.
SMALL GRAINS (Barley, wheat) Preplant or Preemergence	Broadcast or Banded Over-Row	Weeds 1–3″: 1½–2 pts Weeds 3–6″: 2–2½ pts Weeds 6″: 2½–5 pts	Ground: 20 gallons Air: 5 gallons	–	• Do not graze livestock in treated areas or harvest for forage or hay.
CHEMICAL FALLOW (Use during fallow periods in wheat-fallow, wheat-annual crop, and continuous wheat cropping systems)	Broadcast	1½–2½ pints	Ground: 10 gallons Air: 5 gallons	–	• Use higher spray volumes for better coverage as density of stubble, crop residue or weeds increase. • The addition of Banvel or 2,4-D may aid in the suppression of emerged perennial broadleaf weeds and large annual broadleaf weeds. • For extended weed control during the fallow period, tankmixes with registered residual herbicide combinations are permissible. • Refer to 2,4-D, Banvel or residual herbicide label(s) for directions, limitations, cautions and for a listing of weeds controlled. • Weeds taller than 6 inches may not be controlled.
SORGHUM (Grain)	Preplant/ Preemergence Broadcast or Band Application	Weeds 1–3″: 1½–2 pts Weeds 3–6″: 2–2½ pts Weeds 6″: 2½–5 pts	Ground: 20 gallons Air: 5 gallons	–	• Seedbeds should be formed as far ahead of planting as possible to allow maximum weed and grass emergence. • Seeding should be done with a minimum amount of soil disturbance.

CROP	USE PATTERN	GRAMOXONE EXTRA RATE PER ACRE	MINIMUM TOTAL SPRAY PER ACRE	GRAZING OR PREHARVEST INTERVAL (DAYS)	PRECAUTIONS, RESTRICTIONS AND COMMENTS
(Tankmixes with Atrazine, Milogard & 2,4-D)	Preplant or Preemergence				• GRAMOXONE EXTRA may be tankmixed with Atrazine, Igran 80W or Milogard for improved preemergence or residual weed control. The addition of 2,4-D (ester) may aid in the suppression of perennial and annual broadleaf weeds emerged at the time of application. Refer to the specific tank mix herbicide label(s) for rates, directions, limitations, and cautions and a list of weeds controlled.
	Postemergence Directed Applications	1½–2½ pints	Ground: 20 gallons		• Apply when sorghum is at least 12 inches tall when naturally standing. Do not exceed 30 psi nozzle pressure or spray under conditions which may cause excessive drift. • Use precision directed-spray application equipment adjusted so that no more than the lower 3 inches of the sorghum stalk is contacted by the application spray. • Some crop injury will occur. The degree of injury is related to the precision of application and spraying conditions.
SOYBEANS	Preplant/ Preemergence	Weeds 1–3″: 1½–2 pts Weeds 3–6″: 2–2½ pts Weeds 6″: 2½–5 pts	Ground: 20 gallons Air: 5 gallons		• Apply as a broadcast spray before, during or after planting, but before crop emergence. • For improved burndown or residual control, GRAMOXONE EXTRA may be tankmixed with the following herbicides: 2,4-DB, Canopy*, Dual, Gemini*, Goal, Modown, Lasso, Lexone, Linex*, Lorox, Lorox Plus, Preview*, Prowl, Scepter*, Sencor, Surflan, Turbo* *Not for use in California. The rate of GRAMOXONE EXTRA to be used in these tank mixtures is dependent on weed height and growing conditions. Use the highest recommended rate of GRAMOXONE EXTRA under dry conditions or where the weed canopy is dense. Refer to the specific tankmix herbicide label(s) for rates, directions, limitations, and cautions and for a list of weeds controlled. • The lower rate may be used when weeds are less than 4″ tall and a selective post-emergence spray or cultivation will be made within 3 weeks after planting. • Seeding should be done with a minimum amount of soil disturbance.
Tank mix w/2,4-D ester	Preplant/ Preemergence	Weeds 1–3″: 1½–2 pts Weeds 3–6″: 2–2½ pts Weeds 6″: 2½–5 pts	Ground: 20 gallons Air: 5 gallons		• Apply at least 30 days prior to planting. Do not use the amine formulation of 2,4-D as GRAMOXONE EXTRA activity may be reduced. • May be tank mixed with residual herbicides listed above. • Refer to the 2,4-D ester label for a list of rates, directions, limitations and cautions and for a list of weeds controlled.
Directed spray	Postemergence	⅓–⅔ pts	20 gallons		• For control of seedling johnsongrass, crabgrass, goosegrass, brachiaria, Texas millet and pigweed less than 2 inches tall, use the lower rate of GRAMOXONE EXTRA. • For control of the above grasses up to 4 inches tall or pigweed up to 3 inches tall, use the higher rate of GRAMOXONE EXTRA. • Do not treat if soybeans are less than 8 inches tall. • Use precision directed-spray application equipment adjusted so that no more than the lower 3 inches of the soybean plant is contacted by the application spray. • Do not exceed 30 psi nozzle pressure or spray under conditions which may cause excessive drift. • Some crop injury will occur. The degree of injury is related to the precision of application and spraying conditions. • Do not graze treated areas or feed treated forage to livestock. • If needed make a second and final application 7–14 days later.
	Harvest Aid	11 - 22 ozs	Ground: 20 gallons Air: 3 gallons		• Indeterminate varieties: Apply when at least 65% of the seed pods have reached a mature brown color or when seed moisture is 30% or less.

CROP	USE PATTERN	GRAMOXONE EXTRA RATE PER ACRE	MINIMUM TOTAL SPRAY PER ACRE	GRAZING OR PREHARVEST INTERVAL (DAYS)	PRECAUTIONS, RESTRICTIONS AND COMMENTS
					• Determinate varieties: Apply when plants are mature, i.e., beans are fully developed, 1/2 of leaves have dropped, and remaining leaves are yellowing. • Immature soybeans will be injured. • Mature cocklebur, especially drought-stressed plants, are tolerant to GRAMOXONE EXTRA and desiccation will not be complete. • Do not pasture livestock in treated areas within 15 days after treatment. Remove livestock from treated areas at least 30 days prior to slaughter.
SUGARCANE	Postemergence Directed Spray				**General Comments:** • Apply as a shielded or directed spray to avoid contact with cane foliage to prevent leaf burn and yield reduction. • Make a second and final application, if necessary, when new weed growth is 2–6 inches high. • Do not graze treated areas or feed treated forage to livestock.
Florida		2½ pints	Ground: 50 gallons		• For optimum results, apply in early spring (March-April) when weeds are small. • Do not apply after June 1 as cane growth may be stunted and yields reduced.
Hawaii		2½ pints	Ground: 20 gallons		• Do not apply after cane rows have closed in.
Louisiana		1¼–4 pints	Ground: 20 gallons	30	• For tiller control, apply when tillers are less than 18 inches high. • Use the higher rate for heavier weed infestations or tiller growth.
Florida & Texas	Harvest Aid	11–21 fl. oz.	Air: 4 gallons		• Use higher rate under cool, cloudy weather conditions. • Apply 3–14 days before burning and harvest.
SUNFLOWER AND SAFFLOWER Preplant or Preemergence	Broadcast or Banded Over-row	2½–5 pints	Ground: 20 gallons Air: 5 gallons	–	• Apply before, during or after planting but before crop emergence.
Preplant (California only) (Safflower only)	Broadcast	11–21 fl. oz.	Ground: 20 gal Air: 5 gal		• For control of volunteer barley in preformed seedbeds.
Preharvest Dessiccation (Oil seed sunflower varieties only)	Broadcast	1½–2½ pts	Ground: 20 gallons Air: 5 gallons	7	• Apply when sunflower seeds reach physiological maturity (when seed moisture is 35% or lower). For many varieties, this corresponds to the time when the back of the heads are yellow and the bracts are turning brown. • Do not graze treated areas or feed treated forage to livestock. • Use the higher rate when crop stands or weed infestations are heavy.
TREES AND VINES Orchards, Vineyards, Windbreak, Shade & Ornamental Trees Acerola Almonds Apples Apricots Avocados Bananas Cherries Citrus Coffee Figs Filberts Grapes Kiwi Fruit Macadamia Nuts Nectarines Olives Papayas Peaches Pears Pecans Pistachios Plums Prunes Walnuts Other shade and	Directed Spray	2½–5 pints	Ground: 30 gallons	Kiwi Fruit: 14 Pistachios: 7 All Others: —	• Do not allow spray to contact green stems (except suckers), fruit or foliage. • Use a shield or wrap plant when spraying around young trees or vines. • Do not graze treated areas. • Do not feed cover crops grown in treated areas to livestock. • Do not apply when figs, nuts or olives to be harvested are on the ground. • For grapes—treat when sucker growth is no more than 8″ long. • For kiwi fruit—Do not treat more than 3 times per year • For pistachios - Do not exceed two applications after shells split. • For mature woody weeds, perennial weeds, late germinating weeds and green suckers, retreatment or spot treatment may be necessary.

CROP	USE PATTERN	GRAMOXONE EXTRA RATE PER ACRE	MINIMUM TOTAL SPRAY PER ACRE	GRAZING OR PREHARVEST INTERVAL (DAYS)	PRECAUTIONS, RESTRICTIONS AND COMMENTS
ornamental trees such as arborvitae, ash, elm, fir, oak, pine, etc.					
Tankmixes	Directed Spray	2½–5 pts.	Ground: 30 gallons	Refer to other TM labels	• GRAMOXONE EXTRA may be tankmixed with registered residual herbicides listed below for combined emerged and residual weed control. Always refer to other herbicide label(s) for respective precautions, limitations, restrictions, dates and directions for use and weeds controlled. • GRAMOXONE EXTRA may be tank mixed with the following herbicides: Goal, Surflan Devrinol, Solican Princep, Karmex
Guava	Directed Spray	5 pints	Ground: 50 gallons	–	• Do not allow spray to contact green stems, fruit or foliage. • Do not graze treated areas. • Do not feed cover crops grown in treated areas to livestock. • For mature woody weeds, late-germinating weeds and grasses, and perennials, retreatment or spot spraying may be necessary.
Hops (ID, OR, & WA only)	Directed Spray and/or Suckering and Stripping	2½ pints	Ground: 20 gallons	14	• Do not apply to hop vines less than 6 ft. tall. • Retreatment or spot treatment may be necessary. • Do not apply more than 3 times per season. • Do not allow spray to contact green stems, foliage, flowers, or cones as injury may result. • Do not allow animals to graze in treated hopyards. • Hop vine refuse and silage may be fed to livestock. • For suckering and stripping, spray only the basal 2 ft. of the vines. Repeat as necessary. • Experience with varieties other than CASCADE, YAKIMA CLUSTER, and BULLION is limited. If using GRAMOXONE EXTRA on other varieties than these, test the use pattern on a small number of vines of each variety to determine sensitivity to injury. Do not use on unlisted varieties if unacceptable crop injury occurs.
Passion Fruit	Directed Spray	5 pints	Ground: 50 gallons		• Use a shield or wrap vine if bark is still green at application time. • If application is to be made during harvest season, pick all fruit off the ground prior to application. • Do not allow animals to graze on treated areas. • Retreatment or spot treatment may be necessary.
TYFON (New Hampshire only)	Preplant Preemergence	2½–5 pints	Ground: 20 gallons		• Seeding should be done with a minimum of soil disturbance. • Weeds and grasses emerging after treatment will not be controlled. • Crop plants emerged at time of application will be injured.
VEGETABLES (Seeded or Transplanted) Beans* (Lima, Snap) Broccoli* Cabbage* Cantaloupe* Carrots* Califlower* Chinese Cabbage* Collards* Cucumber* Eggplant* Lettuce Musk Melons Peas* Peppers Pumpkin* Squash* Sugar beets Sweet Corn* Turnips* Tomatoes Watermelon *Ground application only.	Preplant Preemergence	2½–5 pints	Ground: 20 gallons Air: 5 gallons		• Seedbeds or plantbeds should be formed as far ahead of treatment as possible to permit maximum weed emergence. • Applications can be made as a banded or broadcast treatment before, during or after planting but prior to the crop emergence. • Use the higher rate for heavier weed infestations. • Seeding or transplanting should be done with a minimum amount of soil disturbance. • Crop plants emerged at time of application will be killed.

CROP	USE PATTERN	GRAMOXONE EXTRA RATE PER ACRE	MINIMUM TOTAL SPRAY PER ACRE	GRAZING OR PREHARVEST INTERVAL (DAYS)	PRECAUTIONS, RESTRICTIONS AND COMMENTS
Tomatoes Peppers (Middle Atlantic, Southeast, South Central Regions, and Southwest to the Western Border of Texas)	Directed Spray	2½ pints	Ground: 20 gallons		• For control or suppression of emerged weeds between rows after crop establishment. • Use precision directed spray application equipment adjusted to prevent spray contact with crop plants. Do not exceed 30 psi nozzle pressure or spray under conditions which may cause excessive drift. • Apply when weeds are succulent and weed growth is less than 6 inches. • Do not apply more than 3 applications per season. • Do not allow animals to graze in treated areas.
SWEET CORN	Directed Spray	12.8 fl. oz.	Ground: 20 gallons		• Apply when corn is at least 10 inches tall. Arrange nozzles to spray no higher than the lower 3 inches of the corn plant. • Corn plants shorter than 10 inches may be injured and not recover (corn height measured from soil surface to top of whorl).
VEGETABLES (California only) Lettuce Melon Sugar beets Tomatoes	Broadcast	11–21 fl. oz.	Ground: 20 gallons Air: 5 gallons		• For control of volunteer barley in preformed seedbeds.
VEGETABLES Rhubarb	Dormant	2½–5 pints	Ground: 50 gallons		• Apply during dormant season before buds in crown begin to grow. • Do not make more than 2 applications per season.

ALFALFA

NEW SEEDLINGS—Suppression and Control of Broadleaf Weeds and Grasses in New Alfalfa Seedlings Grown for Hay (California Only).

For Control of:	Rate/Acre	
	For Suppression	For Control
Spikeweed (4 in. tall or less)	6½ fl. oz.	1 pint
Volunteer Small Grain (8 in. tall or less)	6½–13 fl. oz.	1½ pints
Fiddleneck (6 in. tall or less)	6½–13 fl. oz.	1½ pints
Shepherdspurse	13–24 fl. oz.	—
Annual Bluegrass	—	13–24 fl. oz.
Chickweed	—	13–24 fl. oz.
Red Maids (6 in. tall or less)	—	13–24 fl. oz.

Do not use the 6½ fl. oz. rate unless the alfalfa has at least 3 trifoliate leaves; the 13 fl. oz. rate unless the alfalfa has 6 trifoliate leaves; or rates over 12.8 fl. oz. unless there are 9 trifoliate leaves.

RESIN SOAKING

PINES (Loblolly, Shortleaf, Longleaf, Slash, Virginia, Pond, Pitch, and Spruce Pines).

Tree Selection—Select trees to be treated from stands on sites not subject to periods of extreme drought stress, as the desiccating effect of GRAMOXONE EXTRA Herbicide to pines is accentuated during such periods, causing a reduction in the amount of oleoresin deposited in the xylem. Select trees to be treated from vigorous, non-stagnated stands, either natural or planted. In stagnated stands or commercial timber stands, plan treating with GRAMOXONE EXTRA Herbicide not sooner than three years after a commercial thinning.

Application Directions—Apply GRAMOXONE EXTRA Herbicide diluted in water (0.2% to 5% cation, wt./wt. basis, depending on the type of treatment method) to a suitable wound in the tree trunk to bring the treatment into contact with the xylem (sapwood). One of the following methods may be used:

Bark Streaks or Cuts: This type of wound is made using a standard or rotary bark hack or a chainsaw chipping tool employed in naval stores work to remove a single one-inch wide streak of bark about 1 to 2 feet from ground level.

The total length should not exceed ⅓ of the tree circumference. Multiple streaks or cuts can result in serious girdling of the trunk and premature death of the tree. A coarse spray (about 2 to 6 ml.) of GRAMOXONE EXTRA Herbicide solution (1% to 5% cation, wt./wt. basis) should be applied to runoff to the exposed xylem, using a plastic acid bottle or a low-pressure sprayer. Wear rubber gloves when using the acid bottle. The amount of spray required per cut depends on tree circumference and the length of cut or streak (⅓ of circumference). For a 9-inch diameter tree, 3 ml. of spray will cover the 1-inch wide streak. Using 3 ml. of a 2% or 4% GRAMOXONE EXTRA Herbicide solution will result in application of 60 or 120 mg. GRAMOXONE EXTRA Herbicide per streak.

Tree Injection: Pressure injections of 1% to 4% GRAMOXONE EXTRA Herbicide solution (cation, wt./wt. basis) should be applied beneath the bark and just beneath the cambium layer. Injections should be made 1 to 2 inches apart around ⅓ of the tree circumference or no closer than 3 inches apart all around the tree.

From 0.2 ml. to 0.4 ml. of GRAMOXONE EXTRA Herbicide solution is applied per injection. Using 2% to 4% GRAMOXONE EXTRA Herbicide solution, this results in 36 to 144 mg. GRAMOXONE EXTRA Herbicide applied per 9-inch diameter tree, for holes spaced 1 inch apart around ⅓ of the tree or 3 inches apart all around the tree.

Bore Holes: ⅜ to ⅝ inch diameter holes 2 to 6 inches deep, depending on tree diameter are bored sloping slightly downward and tangentially inward rather than directly towards the center of the tree. Solutions of 0.3% to 5% GRAMOXONE EXTRA Herbicide can be used. Lower concentrations (0.3%) are usually applied at higher volumes of 15 to 35 ml. per tree, resulting in application of 45 to approximately 100 mg. GRAMOXONE EXTRA Herbicide per tree. Using 2% to 4% solutions and applying 5 ml. per single hole per tree results in application of 100 or 200 mg. GRAMOXONE EXTRA Herbicide per tree. Bore hole applications in the range of 45 to 100 mg. GRAMOXONE EXTRA Herbicide per tree have resulted in effective oleoresin induction.

Time of Treatment: Resin soaking can occur from treatments made any time of the year; however, cool season treatments under non-drought conditions usually result in less severe pine beetle infestations and longer tree life.

Interval Between Treatment and Tree Harvest: The interval between application of GRAMOXONE EXTRA Herbicide and tree harvest should be a minimum of 6 months and preferably from 12 to 24 months. Intervals of over 6 months may not be possible under conditions of drought or serious pine beetle attacks, which may make early harvest necessary. The GRAMOXONE EXTRA Herbicide treatment may encourage beetle attack, or may cause premature death of the tree. Desiccation of the xylem tissue, rather than the desired resin soaking, may occur, and is more likely at higher dosage rates.

Effect on Stem Growth:

GRAMOXONE EXTRA Herbicide treatment can result in reduced stem growth during the interval between treatment and tree harvest.

Dilution Table for GRAMOXONE EXTRA Herbicide
(1.5 lbs. cation per gallon)

Concentration of Cation Desired (Wt./Wt. Basis)	To 1½ Gal. of GRAMOXONE EXTRA Herbicide add the following No. Gals. of Water
0.2%	118.8
0.5%	46.8
1.0%	22.9
2.0%	10.9

Concentration of Cation Desired (Wt./Wt. Basis)	To 1½ Gal. of GRAMOXONE EXTRA Herbicide add the following No. Gals. of Water
3.0%	6.9
4.0%	4.9
5.0%	3.7

STORAGE AND DISPOSAL

PROHIBITIONS

Do not contaminate water, food, or feed by storage, disposal or cleaning of equipment. Open dumping is prohibited.

STORAGE

Store at temperatures above 32°F. Do not contaminate feed, foodstuffs or drinking water. Do not store next to feed or food, or transport in or on vehicles containing foodstuffs or feeds.

For help with any spill, leak, fire or exposure involving this material, call CHEMTREC 800/424-9300.

PESTICIDE DISPOSAL

This product is acutely hazardous. Improper disposal of excess pesticide, spray mixture, or rinsate is a violation of Federal Law. If these wastes cannot be disposed of by use according to label instructions, contact your State Pesticide or Enviromental Control Agency, or the Hazardous Waste representative at the nearest EPA Regional Office for guidance.

CONTAINER DISPOSAL

Triple Rinse (or equivalent). Do not reuse container. Offer for recycling or reconditioning, or puncture and dispose of in a sanitary landfill, or dispose of by other procedures allowed by state and local authorities.

Conversion Table
GRAMOXONE EXTRA Herbicide to be Applied

Pints/Ac	fl oz/Ac	lb ai/Ac	Ac/gal
—	7	.08	18.3
—	11	.13	11.6
—	21	.25	6.1
1½	24	.28	5.3
2	32	.38	4.0
2½	40	.47	3.2
3	48	.56	2.7
4	64	.75	2.0
5	80	.94	1.6

NOTICE TO BUYER AND USER

Seller warrants that this product conforms to the chemical description on the label and is reasonably fit for the purposes stated on the label when used in accordance with directions under normal conditions of use. This warranty does not extend to the use of this product contrary to label instructions, or under abnormal use conditions, or under conditions not reasonably foreseeable to Seller, and Buyer and User assume the risk of any such use. SELLER DISCLAIMS ALL OTHER WARRANTIES EXPRESSED OR IMPLIED INCLUDING ANY WARRANTY OF FITNESS OR MERCHANTABILITY. SELLER SHALL NOT BE LIABLE FOR CONSEQUENTIAL, SPECIAL OR INDIRECT DAMAGES RESULTING FROM THE USE OR HANDLING OF THIS PRODUCT AND SELLER'S SOLE LIABILITY AND BUYER'S AND USER'S EXCLUSIVE REMEDY SHALL BE LIMITED TO THE REFUND OF THE PURCHASE PRICE.

AATREX®—Reg. TM of CIBA-GEIGY Corp.
BANVEL®—Reg. TM of Sandoz Crop Protection Corp.
BICEP®—Reg. TM of CIBA-GEIGY Corp.
BLADEX®—Reg. TM of E.I. du Pont de Nemours & Co., Inc.
*CANOPY®—Reg. TM of E.I. du Pont de Nemours & Co., Inc.
*CLASSIC®—Reg. TM of E.I. du Pont de Nemours & Co., Inc.
CONQUEST®—Reg. TM of E.I. du Pont de Nemours & Co., Inc.
COTORAN®—Reg. TM of CIBA-GEIGY Corp.
*DAWN®—Reg. TM of Proctor & Gamble Co.
DEF®—Reg. TM of Bayer Ag.
DEVRINOL®—Reg. TM of ICI Americas Inc.
DROPP®—Reg. TM of NOR-AM Chem. Co.
DUAL®—Reg. TM of CIBA-GEIGY Corp.
*EXTRAZINE®—Reg. TM of E.I. du Pont de Nemours & Co., Inc.
FOLEX®—Reg. TM of Rhone Poulenc Inc.
GEMINI®—Reg. TM of E.I. du Pont de Nemours & Co., Inc.
GOAL®—Reg. TM of Rohm and Haas Co.
*GRIFFEX®—Reg. TM of Griffith Corp.
HARVADE®—Reg. TM of Uniroyal Chemical Co.
IGRAN®—Reg. TM of CIBA-GEIGY Corp.
KARMEX®—Reg. TM of E.I. du Pont de Nemours & Co., Inc.
LARIAT®—Reg. TM of Monsanto Company
LASSO®—Reg. TM of Monsanto Company
LEXONE®—Reg. TM of E.I. du Pont de Nemours & Co., Inc.
*LINEX®—Reg. TM of Griffith Corp.
LOROX®—Reg. TM of E.I. du Pont de Nemours & Co., Inc.
LOROX PLUS®—Reg. TM of E.I. du Pont de Nemours & Co., Inc.
MODOWN®—Reg. TM of Rhone Poulenc Inc.
PREP®—Reg. TM of Rhone Poulenc Inc.
PRINCEP®—Reg. TM of CIBA-GEIGY Corp.
*PREVIEW®—Reg. TM of E.I. du Pont de Nemours & Co., Inc.
PROWL®—Reg. TM of American Cyanamid Company
SENCOR®—Reg. TM of the Parent Company of Farbenfabriken Bayer GmbH, Leverkusen.
*SCEPTER®—Reg. TM of American Cyanamid Company.
SINBAR®—Reg. TM of E.I. du Pont de Nemours & Co., Inc.
SOLICAM®—Reg. TM of Sandoz Crop Protection Corp.
SURFLAN®—Reg. TM of Elanco Products Co.
TURBO®—Reg. TM of Mobay Corp.
VELPAR®—Reg. TM of E.I. du Pont de Nemours & Co., Inc.
ZORIAL®—Reg. TM of Sandoz Crop Protection Corp.

Please consult each appropriate label for specific use recommendations.
*Not for use in California

The information and specimen labels in this booklet are correct as of the publicatiion date but are subject to change without notice. For current information, contact ICI Americas Inc.

GRAMOXONE® SUPER

HERBICIDE

A WEED, GRASS AND HARVEST AID HERBICIDE

RESTRICTED USE PESTICIDE

For retail sale to and use only by Certified Applicators, or persons under their direct supervision, and only for those uses covered by the Certified Applicator's certification.

Complete Directions For Use
Read Worker Safety Rules

ACTIVE INGREDIENT:	By Wt.
Paraquat dichloride (1,1′-dimethyl-4,4′- bipyridinium dichloride)	23.2%
INERT INGREDIENTS:	76.8%
TOTAL	100.0%

Contains 1.5 lbs. paraquat cation per gallon.
Contains stench (odor) and emetic.
EPA Reg. No. 10182-103

KEEP OUT OF REACH OF CHILDREN

DANGER–POISON

ONE SWALLOW CAN KILL! SYMPTOMS ARE PROLONGED AND PAINFUL. ONSET OF SYMPTOMS MAY BE DELAYED FOR UP TO 3 DAYS AFTER SWALLOWING.

PELIGRO—PRECAUCION AL USUARIO: Si usted no lee ingles, no use este producto hasta que la etiqueta le haya sido explicado ampliamente.

- Do not use or store in or around home.
- Do not remove contents except for immediate use.
- Never put into food, drink or other containers.
- Read Danger Statements and Worker Safety Rules on side panels.
- Refer to side panel for Statement of Practical Treatment.

STATEMENT OF PRACTICAL TREATMENT

IF SWALLOWED

IMMEDIATELY induce vomiting by inserting finger in throat. Administer fluids and induce further vomiting. Get to a hospital or physician fast. PROMPT TREATMENT IS ESSENTIAL and must be initiated before signs and symptoms of injury appear.

SKIN CONTACT

IMMEDIATELY wash with soap and water. Remove and wash all contaminated clothing with soap and hot water before reuse. Prolonged contact will cause severe irritation. Contact with irritated skin, or a cut, or repeated contact with intact skin may result in poisoning.

EYE CONTACT

IMMEDIATELY wash eyes with water for at least 15 minutes and get medical attention. Contact may cause eye injury.

INHALATION

IMMEDIATELY get away from spray mist. Exposure may cause irritation, nose bleeds and may lead to poisoning. Stop and check spray procedure. The odor of this product is from the stenching agent which has been added, not from paraquat. Follow Worker Safety Rules.

Note to Physicians: Call ICI Americas Medical Emergency Information Center, 800/327-8633 (Dial 1-800-F-A-S-T-M-E-D) at any hour to obtain toxicology and medical management consultation, a supply of bentonite clay, and paraquat analysis. Symptoms following ingestion may be delayed up to 3 days. To be effective, treatment for paraquat poisoning must begin **IMMEDIATELY.** Treatment consists of binding paraquat in the gut with suspensions of bentonite clay or charcoal and removal of paraquat from the blood by prolonged charcoal hemoperfusion or continuous hemodialysis.

PRECAUTIONARY STATEMENTS

HAZARD TO HUMANS AND DOMESTIC ANIMALS

DANGER–POISON

Can Kill if Swallowed. May be Harmful or Fatal if Absorbed Through Skin or Inhaled. Harmful to Skin and Eyes.

WORKER SAFETY RULES

DO NOT get on skin, eyes or clothing.
DO NOT inhale spray mist.
WASH SPLASHES from skin and eyes immediately.
REMOVE and wash contaminated clothing.
WASH before eating, smoking and drinking.
WEAR full face shield, rubber gloves and apron and waterproof footwear when pouring concentrate or when exposure to concentrate is possible.
WEAR protective clothing and waterproof footwear when spraying outside of an enclosed cab or aircraft or when contacting vegetation wet with spray.

ENVIRONMENTAL SAFETY CAUTIONS

WILDLIFE

This product is toxic to wildlife. Birds and other wildlife in treated areas may be killed. Do not apply directly to water or wetlands (swamps, bogs, marshes, and potholes). Do not contaminate water by cleaning of equipment or disposal of wastes.

DRIFT

GRAMOXONE SUPER Herbicide is a contact herbicide that kills all green plant tissue. Do not apply under conditions involving possible drift to food, forage, or other plantings that might be damaged or the crops thereof rendered unfit for sale, use, or consumption. Do not apply when weather conditions favor drift from areas treated. For aerial applicators use, do not apply during periods of thermal inversion to avoid drift.

CHEMICAL HAZARD

This product reacts with aluminum to produce hydrogen gas which may form a highly combustible gas mixture. Such gas could flash or explode, causing serious personal injury, if ignited by open flame, spark, welder's torch, lighted cigarette, or other ignition source. Do not mix or store in containers, spray tanks, nurse tanks, or such systems made of aluminum or having aluminum fittings. After dilution with water to spray concentrations, spray systems made of, or containing aluminum parts, may be used but should not be used for extended (overnight) storage.

EQUIPMENT/CONTAINERS

Flush all spray equipment with water after use each day. GRAMOXONE SUPER is corrosive to aluminum. Aluminum spray equipment and aluminum aircraft structures that are exposed to spray solution or spray drift should be flushed thoroughly with water immediately after use.

DIRECTIONS FOR USE

It is a violation of Federal law and may endanger the applicator or third parties to use this product without complying with all label directions and warnings.

READ ENTIRE LABEL. USE STRICTLY IN ACCORDANCE WITH DANGER STATEMENTS AND DIRECTIONS, AND WITH APPLICABLE STATE AND FEDERAL REGULATIONS. Read Danger and First Aid Treatment and Worker Safety Rules listed above.

NOTE: Do Not Use Around Home Gardens, Schools, Recreational Parks, or Playgrounds. DO NOT apply this product through any type of irrigation system.

REENTRY STATEMENTS

DO NOT enter treated areas without protective clothing until sprays have dried.
AVOID working in spray mist. If there is risk of exposure, wear goggles and approved face mask capable of filtering spray droplets.
KEEP all unprotected persons out of operating areas or vicinity where there may be danger of drift.

Certain states may require more restrictive reentry intervals; consult your State Department of Agriculture for further information. Written or oral warnings regarding use of protective clothing and accidental exposure must be given to workers who are expected to be in treated areas or in areas about to be treated.

IMPORTANT: The effect of swallowing GRAMOXONE SUPER Herbicide is far more severe than that which usually results from skin contact or inhalation of spray mist. This product should not cause injury if used according to label instructions. Follow all these rules **as if your life depends on it.**

GENERAL INFORMATION

GRAMOXONE® SUPER herbicide is a contact herbicide used to control or suppress a broad spectrum of emerged weeds. GRAMOXONE SUPER controls most small annual weeds—both broadleaves and grasses, and suppresses perennial weeds by destroying green foliage. GRAMOXONE SUPER can also be used as a crop desiccant at harvest.

GRAMOXONE SUPER is formulated as a liquid which contains $1^1/_2$ lbs. of active ingredient per gallon. The formulation is green in color and contains a non-toxic odor, and an emetic (an agent which will induce vomiting if the product is swallowed). The green color and odor are included in the formulation to help prevent accidental ingestion of GRAMOXONE SUPER.

GRAMOXONE SUPER is rapidly absorbed by green plant tissue and interacts with the photosynthetic process to produce superoxides which destroy the plant cells. GRAMOXONE SUPER requires actively growing green plant tissue to function. Thorough coverage of all green foliage is essential for effective weed control. GRAMOXONE SUPER is not as effective on drought-stressed weeds, weeds with little green foliage (i.e., mowed or cut weeds), or mature woody bark of trees and vines.

Clay and organic matter rapidly tie up GRAMOXONE SUPER. As a result, GRAMOXONE SUPER has no residual soil activity to affect later-planted crops or later germinating weeds.

RAINFASTNESS

Because GRAMOXONE SUPER is rapidly absorbed by the weed foliage, rain occurring 30 minutes or more after application will have no effect on the activity of GRAMOXONE SUPER.

APPLICATION

Since GRAMOXONE SUPER is a contact-type herbicide, it is essential to obtain complete coverage of target weeds to get good control. Improper application technique and/or application to large, stressed, or mown weeds will usually result in unacceptable weed control. Good coverage is also essential for proper performance in harvest-aid applications. See below for specific application instructions.

USE OF A NONIONIC SURFACTANT

ALWAYS ADD a nonionic surfactant (approved for use on growing crops) containing at least 50% surface active ingredient. The concentration of surface active ingredient in a particular surfactant will determine how much should be added to the tank. Refer to table below for specific recommendations:

Surfactant % Nonionic Active Ingredient	Rate of Surfactant** Per 100 Gallons of Water
75% or greater	1 pint*
50-74%	2 pints
Less than 50%	DO NOT USE

* 1 pint = 0.125% V/V.

** For applications where higher rates of surfactant are desired or required, additional surfactant may be used.

FAILURE TO USE A NONIONIC SURFACTANT AT RECOMMENDED RATES WILL RESULT IN REDUCED PERFORMANCE OF GRAMOXONE SUPER.

NOZZLE SELECTION

The use of flat fan nozzles will result in the most effective application of GRAMOXONE SUPER. Flood nozzles are generally not as good as flat fans since they produce large uneven droplets. The use of flood nozzles may result in reduced weed control due to inadequate coverage.

Recommended Nozzles, Pressures and Setup

	Nozzle Type	
	Flat fan	Flood
Maximum Size	8	15
Spray Pressure (at nozzle)	30-50 P.S.I.	30-50 P.S.I.
Maximum Nozzle Spacing	30″	40″
Direction of Spray Pattern	Down	Down
Maximum Speed	10 mph	10 mph
Spray Overlap (at each edge)	30%	50%

Flat Fan Nozzles

30% (60% Total) Overlap

Flood Nozzles

50% (100% Total) Overlap

Using nozzles, pressures, or setups different from the above chart will result in reduced control.

SPRAY CARRIER

Always use clean water (free of mud or clay), clear liquid nitrogen, or complete clear liquid fertilizers as the carrier when spraying GRAMOXONE SUPER. Muddy water, or suspension-type fertilizers containing clay, can inactivate GRAMOXONE SUPER. Never use suspension-type fertilizers containing clay as the spray carrier. If using a complete clear liquid fertilizer containing high phosphate levels as the spray carrier, always use the higher rate of GRAMOXONE SUPER and surfactant.
NOTE: When using liquid fertilizers such as 28% N, as a carrier, it is important that nonionic surfactant still be used with GRAMOXONE SUPER. Liquid fertilizer carriers cannot substitute for surfactant.

RATES OF GRAMOXONE SUPER

Follow recommended rates listed with each use of GRAMOXONE SUPER. Use the higher label rates when weeds are dense or large. Also, use higher label rates for harvest aid when crop vegetation is dense.

SPRAY VOLUME

Follow recommended minimum spray volumes listed with each use of GRAMOXONE SUPER. These are *minimum* volumes only, and spray volumes should be increased as necessary to obtain complete coverage of the target weed or plant without runoff from the foliage.

DRIFT MINIMIZATION

DO NOT spray if conditions of thermal inversion exist, or if wind direction and speed may cause spray drift onto adjacent areas. With both aerial and ground applications, the use of a drift control or spray deposition additive (in accordance with label directions) may reduce the potential for drift to occur. **Drift minimization is the responsibility of the applicator.** Consult with local and state agricultural authorities for information regarding avoiding or minimizing spray drift.

APPLICATION TIMING

GRAMOXONE SUPER should be applied to emerged weeds when they are small. Weeds 1" to 6" in height are the easiest to control. Larger weeds may be more difficult to control. When weeds have been grazed or mowed, thus removing much of the green foliage, allow weeds to regrow to a height of 2 to 4 inches before spraying if possible. Similarly, when forage or grain crops have been harvested prior to spraying, weeds present in the field will also have been cut. To allow for adequate green foliage to remain on weeds in this situation, raise cutter bars as high as possible from the ground to cut stubble and weeds at a greater height.

BURNDOWN OF GRASS COVER CROPS OR VOLUNTEER CEREALS

When using GRAMOXONE SUPER for control of grass cover crops or volunteer cereals, best results are obtained when GRAMOXONE SUPER is applied **prior to tillering** or **after boot stage.** This is especially important with a wheat cover crop or volunteer wheat. Treatments made between tillering and boot stage will generally not provide complete control.

ENVIRONMENTAL CONDITIONS

GRAMOXONE SUPER is active over a wide range of environmental conditions. Cool weather (below 55°) will slow the activity of GRAMOXONE SUPER, as will cloudy, overcast weather, but will not affect performance.

In dry areas, dust stirred up by high winds or equipment tires can coat weed leaves and reduce GRAMOXONE SUPER activity. Avoid applying GRAMOXONE SUPER in extremely dusty conditions.

SPOT SPRAYING

When only small areas are to be sprayed with labeled applications, it is advantageous to mix small quantities of GRAMOXONE SUPER. To aid in mixing small quantities, the following table should be consulted. When spraying with wand-type applicators as in spot spraying, always follow precautionary statements regarding use, protective clothing and footwear.

If the broadcast rate per acre for GRAMOXONE SUPER is:	Add the following amount of GRAMOXONE SUPER to 1 gal of water
1 1/2 pts	1/2 fl oz
2 1/2 pts	3/4 fl oz
3 pts	1 fl oz
5 pts	1 2/3 fl oz.

Always add 1/3 to 1/2 fl. oz. of a nonionic surfactant for each gallon of spray. When spot spraying in this manner, spray to thoroughly wet the foliage, but not to the point of runoff.

TANK MIXING FOR IMPROVED BURNDOWN OF DIFFICULT WEEDS, AND RESIDUAL WEED CONTROL

PHOTOSYNTHETIC INHIBITOR HERBICIDES

Difficult weeds can often be controlled by tankmixing GRAMOXONE SUPER with other herbicides. The addition of herbicides which are also photosynthetic inhibitors (PSI) will slow the activity of GRAMOXONE SUPER, allowing GRAMOXONE SUPER to thoroughly distribute itself within the treated leaf. The resulting level of control is usually greater than if GRAMOXONE SUPER was applied alone.

GRAMOXONE SUPER may be applied in tank mixture with the following PSI herbicides:

ATRAZINE	CONQUEST*	LOROX PLUS
AATREX	EXTRAZINE*	LARIAT
BICEP	GEMINI*	PREVIEW*
BLADEX*	GRIFFEX*	PRINCEP
CANOPY*	LENOX/SENCOR	LINEX/LOROX*

Refer to respective product label(s) for rates of application, directions for use, limitations, cautions and for a list of weeds controlled.
* Not for use in California.

Improved Weed Control With PSI's

Control of difficult weeds listed below can be improved by the addition of a PSI herbicide.

Marestail	Prickly lettuce
Pennsylvania Smartweed	Lambsquarters
Velvetleaf	Morningglory
Malva (Cheeseweed)	Cheatgrass
Fall Panicum	Knotweed
Giant Ragweed	Volunteer wheat
Kochia	Perennial Weeds (suppression only)
Cocklebur	
Tansy Mustard	

IMPROVED CONTROL OF PERENNIAL AND ANNUAL BROADLEAF WEEDS

When perennial broadleaf weeds such as canada thistle, bindweed, dandelion, etc. or difficult to control annual broadleaf weeds such as giant ragweed or morningglory are present, tank mixes with 2,4-D ester (cv), 2, 4-DB or BANVEL, where labeled, will help improve control. Tankmixing the amine formulation of 2,4-D with GRAMOXONE SUPER, may result in reduced grass control.

ORDER OF TANKMIXING

In general, GRAMOXONE SUPER tankmixes with other products should be mixed as follows:

1. Fill spray tank 1/2 full with clean water or other approved carriers such as clear liquid fertilizer.
2. Begin tank agitation and continue throughout mixing and spraying.
3. Add dry formulations (WP, DF, etc.) to tank.
4. Add liquid formulations (SC, EC, L, etc.) to tank.
5. Add GRAMOXONE SUPER to tank.
6. Add nonionic surfactant to tank.
7. Fill remainder of spray tank.

Always refer to labels of other pesticide products for mixing directions and precautions which may differ from those outlined here.

Since many of the herbicides listed on this label are available in several types of formulations, it is advisable to perform a jar test to check physical compatibility.

LIMITATIONS, PRECAUTIONS AND WARNINGS

- DO NOT use around home gardens, schools, recreational parks, or playgrounds.
- In preplant and preemergence (to the crop) uses, do not apply to soils lacking clay minerals, i.e., peat, muck, pure sand, artificial planting media.
- Seedbeds and plantbeds should be formed as far ahead of planting and treatment as possible to permit maximum weed and grass emergence prior to treatment.
- Seeding or transplanting should be done with a minimum amount of soil disturbance.
- Weeds and grasses emerging after application of GRAMOXONE SUPER will not be controlled or suppressed.
- Unless otherwise indicated, crop plants emerged at time of application may be severely injured or killed if contacted by sprays of GRAMOXONE SUPER.

CROP	USE PATTERN	GRAMOXONE SUPER RATE PER ACRE	MINIMUM TOTAL SPRAY PER ACRE	GRAZING OR PREHARVEST INTERVAL (DAYS)	PRECAUTIONS, RESTRICTIONS AND COMMENTS
ALFALFA New Seedlings (California only)	Broadcast	See Table below.	Ground: 25 gal Air: 5 gal	70	• Apply during late winter or early spring. • Caution: Seedling alfalfa stands will be reduced and replanting may be necessary. • Not recommended for seedling alfalfa grown for seed. • Alfalfa foliage present at time of application will be burned. • Do not cut or harvest within 70 days after application. • Do not apply more than once during the first growing season.
Preplant or Preemergence (No-till or conventional planting)	Broadcast or Banded Over Row	2½–5 pints (use a minimum of 3 pts. in California.)	Ground: 10 gal Air: 3 gal	—	• Apply prior to emergence of the crop. • Crop plants emerged at time of application will be killed. • Seeding should be done with a minimum amount of soil disturbance.
Dormant season on established plantings (WA, OR, CO, ID, MT, UT, NV, WY, and Del Norte, Siskiyou, Modoc, Shasta, Lassen, Plumas, Sierra and Nevada Counties in CA)	Broadcast	2½–4 pints	Ground: 20 gal Air: 5 gal	42	• For desiccation of weeds including ryegrass, bluegrass, cheatgrass, dogfennel, chickweed, and tansymustard. Use higher rate for cheatgrass. • Apply to established stands after the last fall cutting when crop is dormant but before spring growth starts. • Do not apply if regrowth after last fall cutting is more than 2". • Alfalfa foliage at time of application will be burned. • Do not apply more than once per season. • Do not cut or harvest within 42 days of application.
Dormant season on established plantings (East of the Rocky Mountains and North of the Southern Borders of CO, KS, MO, KY, and VA)	Broadcast	2½–4 pints	Ground: 20 gal Air: 5 gal	60	• For control of weeds, including bluegrass, chickweed, henbit, and downy brome and suppression of perennial weeds. • Apply to well-established stands (at least one year old) after the crop is dormant, but before spring *regrowth* reaches one inch. • Alfalfa foliage present at the time of application will be burned which may reduce the yield of the first cutting. • Do not cut or harvest within 60 days of application. • Do not apply more than once per season. • Do not apply if fall *regrowth* following last fall cutting is greater than 6".
Dormant season tankmix with Velpar L (East of the Rocky Mountains and North of the Southern borders of CO, KS, MO, KY, and VA.)	Broadcast	1½–2½ pints	Ground: 20 gal Air: 10 gal	42	• For control of weeds such as chickweed, downy brome and tanseymustard. • Use the 1½ pt. rate of GRAMOXONE SUPER when weeds and grasses are less than 4" tall. • Mix with 1 to 2 qts. of Velpar L per acre. • Use the lower rate of Velpar L on loamy sands or sandy loams. Refer to Velpar L label for directions, limitations, cautions and for a list of weeds controlled. • Apply once to established alfalfa stands during the dormant season. • Do not apply to *regrowth* that is more than 2 inches. • Do not apply following cuttings during growing season. • Do not apply to alfalfa during the first season after seeding. • Temporary chlorosis may occur on alfalfa regrowth. • Stress which may be caused in part by low fertility, disease, insects, winterkill, over cutting, drought or frost may increase the chances of crop injury. • *DO NOT USE* on gravelly or rocky soils, exposed subsoils, hardpan, sand or poorly drained alkaline soils as crop injury, including mortality, may result. • Do not graze or harvest within 42 days of application
Dormant season Arizona and California	Broadcast	1½–2½ pints	Ground: 20 gal Air: 5 gal	60	• Apply during late fall or winter months after the last fall cutting and before first spring cutting. • Alfalfa foliage present at the time of application will be discolored and temporarily stunted. • Total hay yield of first cutting may be reduced in alfalfa fields with severe weed infestation. This reduction will usually be directly proportionate to the loss of weed weight.

CROP	USE PATTERN	GRAMOXONE SUPER RATE PER ACRE	MINIMUM TOTAL SPRAY PER ACRE	GRAZING OR PREHARVEST INTERVAL (DAYS)	PRECAUTIONS, RESTRICTIONS AND COMMENTS
--- California --- --- Arizona ---					• Do not apply more than once per season. • Do not apply following cuttings during the growing season. • Do not apply if *regrowth* after grazing or cutting is more than 2 inches. • Do not harvest within 60 days of application or excessive residues may result. • For desiccation of weeds including bluegrass, ryegrass, shepherdspurse, chickweed, tanseymustard, foxtail, sowthistle and groundsel. • Use high rate if ryegrass, shepherdspurse, sowthistle or groundsel is present. • For desiccation of weeds, including London rocket, sowthistle, rescue brome and wild oats.
Dormant season tankmix with Velpar L (ID, OR, UT, and WA, NV, and the northern California Counties of Del Norte, Siskiyou, Modoc, Shasta, Lassen, Plumas, Sierra, and Nevada*) *Currently not registered for use in California.	Broadcast	1½–2½ pints	Ground: 20 gal Air: 10 gal	42	• For control of chickweed, downy brome and tanseymustard. • Use the 1½ pt. rate of GRAMOXONE SUPER when weeds and grasses are less than 4" tall. • Mix with 1 to 2 qts. of Velpar L per acre. • Use the lower rate of Velpar L on loamy sands or sandy loams. Refer to Velpar L label for directions, limitations, cautions and for a list of weeds controlled. • Apply once to established alfalfa stands during the dormant season. • Do not apply to *regrowth* that is more than 2 inches. • Do not apply following cuttings during growing season. • Do not apply to alfalfa during the first season after seeding. • Temporary chlorosis may occur on alfalfa regrowth. • Do cut or harvest within 42 days of application. • Stress which may be caused in part by low fertility, disease, insects, winterkill, over cutting, drought or frost may increase the chances of crop injury. • *DO NOT USE* on gravelly or rocky soils, exposed subsoils, hardpan, sand or poorly drained alkaline soils as crop injury, including mortality, may result.
Between-cuttings treatment in established plantings. (All states east of the Rocky Mountains) (Includes first year alfalfa)	Broadcast	1½ pints	Ground: 20 gal	30	• Weeds much beyond the seedling stage and the stubble of weeds cut off during harvest will be less affected by this treatment. • Apply immediately after alfalfa has been removed for hay or silage. • Do not treat more than 5 days after cutting. • CAUTION: First year alfalfa stands and yields may be reduced if alfalfa is allowed to regrow more than two (2) inches. • Alfalfa foliage present at time of application will be burned. • In arid areas where moisture is limited, weed control may be reduced. • Do not cut or harvest within 30 days of application. • Make 1 to 3 applications, as needed, during the growing season. These sprays may be applied in addition to a dormant application. • For first year alfalfa, do not apply more than twice during the first growing season.
ASPARAGUS Preplant or Preemergence	Broadcast or Banded Over Row	2½–5 pints (Use a minimum of 3 pts. in California.)	Ground: 25 gal Air:* 5 gal *Currently not registered for use in California.	—	• Apply prior to emergence of the crop. • Crop plants emerged at time of application will be killed.
Preemergence to established plantings at least 2 years old	Broadcast or Banded Over Row	2½–5 pints (Use a minimum of 3 pts. in California.)	Ground: 25 gal	6	• Apply prior to emergence of crop or after last harvest. • Crop plants emerged at time of planting will be killed. • May be tank mixed with PRINCEP herbicide for residual control. Refer to PRINCEP label for specific directions, limitations, cautions and for a list of weeds controlled.

CROP	USE PATTERN	GRAMOXONE SUPER RATE PER ACRE	MINIMUM TOTAL SPRAY PER ACRE	GRAZING OR PREHARVEST INTERVAL (DAYS)	PRECAUTIONS, RESTRICTIONS AND COMMENTS
CASSAVAS, TANIERS YAMS (Puerto Rico) only Shielded Post directed spray	Directed Spray	2½ pints	Ground: 50 gal	90	• Apply when weeds are succulent and growth is 1–6 inches. • On cassavas and taniers, do not make more than 3 applications per crop season. • On yams do not make more than 2 applications per crop season. • Do not allow spray to contact cassavas, tanier or yam plants as injury may result. • Do not spray under windy conditions. • Do not graze treated areas or feed treated forage to livestock.
CLOVER Dormant season on established plantings (WA, OR, CO, ID, MT, UT, NV, WY and the counties of Del Norte, Siskiyou, Modoc, Shasta, Lassen, Plumas, Sierra, and Nevada in CA)	Broadcast	2½–4 pints	Ground: 20 gal Air: 5 gal	60	• For desiccation of ryegrass, bluegrass, cheatgrass, dogfennel, chickweed, and tanseymustard. Use higher rate if cheatgrass is present. • Apply after the last fall cutting when the crop is dormant but before spring growth starts. • Do not apply if *regrowth* following the last fall cutting is more than 2″. • Clover foliage present at the time of application will be burned. • Do not pasture animals in treated fields before first cutting. • Do not graze, cut or harvest within 60 days of application. • Do not apply more than once per season.
CONSERVATION RESERVE AND FEDERAL SET-ASIDE PROGRAMS (For use in compliance with the Federal Conservation Reserve Program or Federal set-aside programs.)	Broadcast	2½–5 pints (Currently not registered for use in California at rate higher than 2½ pts.)	Ground: 10 gal Air: 3 gal	–	• For improved emerged weed control or extended weed control, GRAMOXONE SUPER may be tankmixed with other herbicides registered for this use. Refer to tankmix herbicide labels for specific directions, limitations, cautions and for a list of weeds controlled.
CORN Preplant or Preemergence (Used Alone)	Broadcast or Banded Over Row	Weeds 1–3″: 1½–2 pts Weeds 3–6″: 2–2½ pts Weeds 6″: 2½–5 pts	Ground: 20 gal Air: 5 gal		• Includes field, fresh, sweet*, forage, fodder and popcorn. • Seedbeds should be formed as far ahead of planting and treatment as possible to permit maximum weed and grass emergence. • Seeding should be done with a minimum amount of soil disturbance. • Weeds and grasses emerging after application will not be controlled. Crop plants emerged at time of application will be killed. *Currently not registered for use in California on sweet corn.
Preplant or Preemergence Tank Mixture Treatments for No-till/Reduced Till Corn	Broadcast or Banded	Weeds 1–3″: 1½–2 pts Weeds 3–6″: 2–2½ pts Weeds 6″: 2½–5 pts	Ground: 20 gal		• Apply as a broadcast spray before, during or after planting, but before crop emergence. • For improved burndown or residual control, GRAMOXONE SUPER may be tank-mixed with the following herbicides: 2,4-D Ester*, AAtrex/Atrazine, Banvel, Bicep, Prowl*, Bladex, Dual, Extrazine*, Griffex*, Lasso, Linex*, Lorox, Modown, PRINCEP, Simazine • GRAMOXONE SUPER maay also be tankmixed with AMBUSH Insecticide. • Refer to respective product label(s) for rates of application, directions for use, limitations, cautions, and for a list of weeds or insects controlled. *Not for use in California.
FIELD CORN POPCORN SWEET CORN*	Postemergence Directed Spray Used Alone	1½ pints	Ground: 20 gal		• Apply when corn is at least 10 inches tall with nozzles arranged to spray no higher than the lower 3 inches of corn stalks. • Corn plants shorter than 10 inches may be injured and not recover (corn height measured from soil surface to top of whorl).
	Postemergence Directed Spray Tankmix with Atrazine	1½ pints	Ground: 20 gal	–	• For improved control of emerged weeds (especially broadleaf weeds). Add ½ to 1 lb ai/acre atrazine in a tankmix. Refer to the atrazine label for directions, limitations, cautions and for a list of weeds controlled.

CROP	USE PATTERN	GRAMOXONE SUPER RATE PER ACRE	MINIMUM TOTAL SPRAY PER ACRE	GRAZING OR PREHARVEST INTERVAL (DAYS)	PRECAUTIONS, RESTRICTIONS AND COMMENTS
*Currently not registered for use in California on sweet corn.					• Do not mix with liquid fertilizer. • Constant agitation of tankmix is necessary at all times. • Refer to other precautions, etc., in "used alone" section immediately above.
FIELD CORN ONLY (grain, fodder, forage, popcorn)	Postemergence directed spray USDA witchweed eradication program	2½ pints	Ground: 10 gal	–	• Initiate sprays in early July and repeat in early August if regrowth of grassy weeds occurs. • Follow application instructions in postemergence directed spray section above.
COTTON Preplant or Preemergence (Used alone)	Broadcast	2½–5 pints (Use a minimum of 3 pts. in California.)	Ground: 20 gal Air: 5 gal	–	• Apply prior to, during or after planting, but before crop emergence. • For fallow bed treatment, beds should be preformed to permit maximum weed and grass emergence prior to treatment. • Seeding should be done with a minimum of soil disturbance.
Preplant (California only) (Used alone)	Broadcast	11–21 fl oz	Ground: 20 gal Air: 5 gal	–	• For control of volunteer barley in preformed seedbeds.
Preplant or Fallow Bed (Tank-mix with Goal)	Broadcast	1½–5 pints	Ground or Air: 10 gal	–	• Refer to Goal label for specific use directions and restrictions, and weeds controlled.
Preplant (Tankmix with Bladex)	Broadcast	2½–5 pints	Ground: 20 gal Air: 5 gal	–	• Refer to Bladex label for specific use directions and restrictions and weeds controlled.
Preplant/ Preemergence	Broadcast	2½–5 pints	Ground: 20 gal Air: 5 gal		• Apply as a broadcast spray before, during or after planting, but before crop emergence.
Other Tankmixes					• For improved burndown or residual control, GRAMOXONE SUPER may be tankmixed with the following herbicides: – Cotoran (4L only) – Zorial* – Karmex/Direx* – Dual* – Prowl* • When tankmixing with Cotoran 4L, add Dawn® dishwashing liquid to the tank at 8 oz/100 gallons water *prior* to adding the Cotoran and GRAMOXONE SUPER. • When tankmixing with any of the herbicides listed above, refer to that product's label for specific directions and restrictions and for a list of weeds controlled. *Tankmix currently not registered for use in California. Dawn® Registered Trademark of Proctor & Gamble
	Harvest Aid				*Harvest Aid Use Precautions* • Remove livestock from treated area 30 days before slaughter. • Repeat application if necessary. Do not make more than 2 applications or exceed a total of 2½ pts./A. • When combined with phosphate defoliants, observe livestock cautions listed on the phosphate label. • When combined with chlorate defoliants, do not pasture livestock on treated areas or feed treated foliage or gin trash. • Do not pasture lactating dairy animals. • Do not pasture livestock in treated fields within 15 days after treatment. • Do not feed gin trash to livestock. • Do not apply to cotton within 3 days before harvest. • Do not apply within 7 days when used with phosphate or chlorate defoliant.
Harvest aid for boll opening and defoliation (Tankmix with phosphate and chlorate defoilants)	Broadcast	11 fl. oz. plus 1 pt. phosphate or 1 gal chlorate	Ground: 10 gal Air: 3 gal	7	• Development of immature bolls will be inhibited. • Apply when 80% or more of the bolls are open and the remaining bolls to be harvested are mature. • Follow Harvest and Use Precautions listed above.

CROP	USE PATTERN	GRAMOXONE SUPER RATE PER ACRE	MINIMUM TOTAL SPRAY PER ACRE	GRAZING OR PREHARVEST INTERVAL (DAYS)	PRECAUTIONS, RESTRICTIONS AND COMMENTS
Other Tankmixes for boll opening and defoliation of southern cotton	Broadcast	4–6 oz.	Ground: 10 gal. Air: 3 gal.		• To aid in defoliation and opening of mature bolls, GRAMOXONE SUPER may be tankmixed with the following products: DEF HARVADE FOLEX PREP DROPP • Apply when 60% or more of the bolls are open and the remaining bolls to be harvested are mature. • Development of immature bolls will be inhibited. • Refer to tankmix product label for rates, directions, limitations and cautions. • Follow Harvest Aid Use Precautions listed above.
Post Defoliation— To aid in opening of mature bolls and to desiccate green weeds. (Except AZ & CA)	Broadcast	1½–2½ pints	Ground: 20 gal Air: 5 gal	3	• Use higher rate if weed infestation is heavy or dense. • Apply when 75% or more of the bolls are open and remaining bolls to be harvested are mature. • Development of immature bolls will be inhibited. • Follow Harvest Aid Use Precautions listed above.
Harvest aid for boll opening and defoliation of western cotton (AZ & CA)	Broadcast Split Application	First application 7–11 fl. oz. plus 1–3 pts. phosphate or 1–2 gal chlorate	Ground: 10 gal Air: 3 gal	7	• Use higher rate of GRAMOXONE SUPER on rank cotton. • Apply when 60% or more of the bolls are open and the remaining bolls to be harvested are mature. • Development of immature bolls will be inhibited. • Do not use more than 4.0 lbs. of actual sodium chlorate or 2.28 lbs. of actual phosphate type defoliants per acre. • Follow Harvest Aid Use Precautions listed above. • Do not use more than 11 fl. oz. of GRAMOXONE SUPER at first application as excessive desiccation may occur.
(Tankmix with phosphate or chlorate defoliants)		Second application 11–21 fl. oz. alone *or* tank-mixed with 1½–2 gal. chlorate, *or,* 1½–2 pts. phosphate		3 (Alone) or 7 (Tankmix)	• In desert cotton areas or on rank vigorous cotton, use the 21 fl. oz. rate of GRAMOXONE SUPER. • Apply when 75% or more of the bolls are open and remaining bolls to be harvested are mature. • Development of immature bolls will be inhibited. • Follow Harvest Aid Use Precautions listed above.
Harvest aid desiccation of stripper cotton or spindle harvested cotton late season	Broadcast	1½–2½ pints	Ground: 10 gal Air: 3 gal	3	• May be applied as a split application. Do not make more than 2 applications or exceed a total of 2½ pts/Acre. • Apply when 85% of the bolls are open and the remaining bolls to be harvested are mature. Development of immature bolls will be inhibited. • Follow Harvest Aid Use Precautions listed above.
Desiccation of Regrowth (Except CA)	Broadcast	11–21 fl oz	Ground: 10 gal Air: 3 gal	3	• Use to desiccate regrowth occuring after defoliation or desiccation. • Use higher rate if regrowth is excessive. • Follow Harvest Aid Use Precautions listed above.
EASTER LILIES (Field grown)	Preemergence	2½–5 pints (Use a minimum of 3 pts. in California.)	Ground: 20 gal	–	• Do not apply more than twice per season.
GRASSES (For Seed) (For Use in Seedbed Preparation	Preplant, At Planting	2½ pints	Ground: 20 gal	–	• Prepare the seedbeds and allow weeds to germinate. • Apply GRAMOXONE SUPER when weeds are at the 3 to 5 leaf stage. • Repeat applications as necessary prior to or on the day of seeding. • Do not graze treated areas or use the seed or straw from treated areas for animal feed or bedding.
GUAR (Preharvest desiccation)	Preharvest	2½ pints	Ground: 20 gal	4	• Apply after the pods are fully mature. • Do not graze treated ares or use the treated forage for animal feed.
MINT (Peppermint, Spearmint)	Dormant Season	2–4 pints	Ground: 25 gal	–	• For suppression of weeds such as Italian ryegrass, prickly lettuce, groundsel, chickweed, downy brome and bluegrass. • Apply when crop is dormant before spring growth begins and when weeds are less than 6 inches tall. • Do not apply more than 4 pts per acre per dormant season. • May be tankmixed with Sinbar (terbacil) weed killer for improved contact activity and residual control of Italian ryegrass, prickly lettuce and groundsel. Apply this tank mixture no more than once per season. Refer to the Sinbar label for rates, directions, and cautions and for a list of weeds controlled.

CROP	USE PATTERN	GRAMOXONE SUPER RATE PER ACRE	MINIMUM TOTAL SPRAY PER ACRE	GRAZING OR PREHARVEST INTERVAL (DAYS)	PRECAUTIONS, RESTRICTIONS AND COMMENTS
NONCROP USES	Broadcast or Spot Treatment	2½–5 pints (Use a minimum of 3 pts. in California.)	Ground: 50 gal	–	• For use in noncrop areas such as public airports, electric transformer stations, pipeline pumping stations, around commercial buildings, storage yards and other installations, fence lines or similar noncrop areas. • Avoid contact with the foliage of ornamentals or desired plants. • Repeat as necessary.
ONIONS (seeded) and garlic	Preplant/ Preemergence	2½–5 pints (Use a minimum of 3 pts. in California.)	Ground: 20 gal	60 200 (CA only)	• Use the higher rate for heavy weed infestations or wild oat control. Apply only one application per season at the 5 pt per acre dosage. • Allow maximum weed and grass emergence prior to treatment but apply prior to crop emergence. • Apply a maximum of 5 pt per acre per season.
PASTURE RESEEDING For suppression of existing sod and undesirable emerged broadleaf weeds and grasses prior to, or at time of, planting of grasses or forage legumes or winter annuals	Broadcast or Banded Over-Row	1½–2½ pts	Ground: 20 gal Air: 3 gal	See Specific geographic recommendation	*West of Cascade and Sierra Nevada Mountains:* • Apply in October through December after first fall rains and after weeds have emerged and sod has started new growth. • For best seeding results, apply on moderately to heavily grazed areas. • Do not use in areas with heavy sod and weed growth. *East of Rocky Mountains:* • Use the 2½-pint rate on vigorous or course sod species such as bromegrass. • See specific directions for use in the south and southeast for suppression of bermudagrass or bahiagrass sods. • Apply prior to, or at time of seeding grasses or forage legumes such as alfalfa, clover and birdsfoot trefoil. • Apply only to grazed or mowed pastures not more than 3 inches in height at time of treatment.
Bermudagrass or Bahiagrass Sods - Late Summer or Early Fall Application for Sod Suppression when Seeding Winter Annuals:		1½ pts			*South and Southeast* • Apply in late summer or early fall to sod not exceeding 3 inches in height.
Bermudagrass Pastures - Dormant Season Suppression of Emerged Weeds		1½ pts			*South and Southeast* • Apply during March • For control of little barley, apply before the mid-boot stage. • Do not mow for hay until 40 days after treatment.
For Control of Endophyte-Fungus-Infected Fescue Forage Legume/Grass Mixture and Other Grass Pastures	Broadcast (Split application)	1½–2½ pts followed by 1½–2½ pts	Ground: 20 gal		• Use split applications—10 to 21 days apart if necessary. • Do not exceed 5 pints per acre total in preparation for reseeding. • For spring plantings, the initial application of 1½ to 2½ pts may be made the previous fall. • Apply when fescue is actively growing and no more than 4 inches high. • To reduce the infestation of endophyte-infested grass, do not allow fescue to go to seed starting with the preceding year's crop.
PEANUTS* *Not registered for use in California	Broadcast	11 oz. Second application 11 oz.	Ground: 20 gal		• To control or suppress small (1-6″) emerged annual grass and broadleaf weeds in peanuts at ground crack. A second application of 11 fl. oz./acre may be made up to 28 days after ground crack. • Make no more than two applications per crop and do not apply a total of more than 22 fl. oz. of product per crop. • Crop foliage sprayed will be injured in the form of bronzing and crinkling but the crop will recover and develop normally.
PEANUTS (Tankmix with Basagran)	Broadcast	11 oz. Second application 11 oz.	Ground: 20 gal		• For improved control of weeds such as cocklebur, bristly starbur, smartweed and pricklysida, tankmix GRAMOXONE SUPER with Basagran at 1 pint/acre.

CROP	USE PATTERN	GRAMOXONE SUPER RATE PER ACRE	MINIMUM TOTAL SPRAY PER ACRE	GRAZING OR PREHARVEST INTERVAL (DAYS)	PRECAUTIONS, RESTRICTIONS AND COMMENTS
					• This tankmix should be applied at the ground crack stage of peanuts. A second application may be made up to 28 days after ground crack. • Make no more than two applications per crop and do not apply a total of more than 22 fl. oz. of product per crop. • Crop foliage sprayed will be injured in the form of bronzing and crinkling but the crop will recover and develop normally. • Refer to the Basagran label for specific use directions, limitations, cautions and for a list of weeds controlled. • Do not apply this tankmix if peanuts show injury (leaf phytotoxicity and/or plant stunting) produced by any other herbicide treatment as injury may be enhanced and/or prolonged. • Do not apply this tankmix during prolonged periods of drought or unreasonably cold weather as unsatisfactory weed control may result.
PEANUTS (Tank mix with Butyrac or Butoxone 200)	Broadcast	11 oz. Second application 11 oz.	Ground: 20 gal		• For improved control of weeds such as cocklebur, sicklepod and morningglory tank mix GRAMOXONE SUPER with 8-16 oz.(0.125-0.25 lbs.) per acre of Butyrac or Butoxone 200. • Make no more than two applications per crop and do not apply a total of more than 22 fl. oz. of product per crop. • Crop foliage sprayed will be injured in the form of bronzing and crinkling but the crop will recover and develop normally. • Refer to the complete Butyrac or Butoxone 200 label for specific use directions, limitations, cautions and for a list of weeds controlled.
PIGEON PEAS (Puerto Rico only)	Directed Spray	2½ pts	Ground: 20 gal	60	• Avoid contact with pigeon pea foliage. • Cannery waste can be fed to livestock. • Do not make more than 1 application per season. • Do not graze treated areas or feed treated forage to livestock.
PINEAPPLE	Directed Spray	2½–5 pints	Ground: 50 gal	20	• Retreatment may be necessary on more mature weeds.
POTATOES Preplant or Preemergence	Broadcast	1½–2½ pints	Ground: 30 gal Air: 5 gal	–	• Do not apply later than ground cracking, before potatoes have emerged.
Preplant (California only) (Used alone)	Broadcast	11–21 fl oz	Ground: 20 gal Air: 5 gal		• For control of volunteer barley in preformed seedbeds.
Preharvest vine killing and weed desiccation (Except Northern Alabama; Counties of Marshal, DeKalb, Jackson and Cherokee)	Broadcast	1½–2½ pints	Ground: 50 gal	3	• Begin application when leaves begin to turn yellow. • Immature or drought-stressed potato foliage are tolerant to GRAMOXONE SUPER. Desiccation will not be complete under these conditions. • Use 2½ pts./Acre rate where quick vine kill is desired. • Use two applications of 1½ pts./Acre when vine growth is dense. Split applications must be applied a minimum of five days apart. • Do not pasture livestock in treated potato fields. • Do not use on muck or peat soils to avoid injury to subsequent crops. • DO NOT use to desiccate potato vines when potatoes are to be stored or used for seed as this use may result in tuber decomposition or failure of seed pieces to germinate and grow normally. • Do not exceed 5 pts./Acre per season.
RESIN SOAKING (Pines)	See information on page 15.				
SMALL FRUITS Blackberries Boysenberries Raspberries Blueberries	Postemergence Directed Spray	2½–5 pints	Ground: 50 gal	–	• Apply before emergence of new canes or shoots as injury to those canes or shoots can occur. • Apply as a course spray to avoid crop injury from fine spray mist.
Strawberries	Postemergence Directed Spray	2½ pints	Ground: 20 gal	21	• Apply by directing spray between the rows and using shields to prevent spray contact with crop plants. • Do not allow spray to contact strawberry plants as injury or excessive residues may result. • Do not apply more than 3 times per season. • Do not graze livestock in treated areas.

CROP	USE PATTERN	GRAMOXONE SUPER RATE PER ACRE	MINIMUM TOTAL SPRAY PER ACRE	GRAZING OR PREHARVEST INTERVAL (DAYS)	PRECAUTIONS, RESTRICTIONS AND COMMENTS
SMALL GRAINS (Barley, wheat) Preplant or Preemergence	Broadcast or Banded Over Row	Weeds 1–3": 1½–2 pts Weeds 3–6": 2–2½ pts Weeds 6": 2½–5 pts	Ground: 20 gal Air: 5 gal	–	• Do not graze livestock in treated areas or harvest for forage or hay.
CHEMICAL FALLOW (Use during fallow periods in wheat-fallow, wheat-annual crop, and continuous wheat cropping systems)	Broadcast	1½–2½ pints	Ground: 10 gal Air: 5 gal	–	• Use higher spray volumes for better coverage as density of stubble, crop residue or weeds increase. • The addition of Banvel or 2,4-D may aid in the suppression of emerged perennial broadleaf weeds and large annual broadleaf weeds. • For extended weed control during the fallow period, tankmixes with registered residual herbicide combinations are permissible. • Refer to 2,4-D, Banvel or residual herbicide label(s) for directions, limitations, cautions and for a listing of weeds controlled. • Weeds taller than 6 inches may not be controlled.
SORGHUM (Grain)	Preplant/ Preemergence Broadcast or Band Application	Weeds 1–3": 1½–2 pts Weeds 3–6": 2–2½ pts Weeds 6": 2½–5 pts	Ground: 20 gal Air: 5 gal	–	• Seedbeds should be formed as far ahead of planting as possible to allow maximum weed and grass emergence. • Seeding should be done with a minimum amount of soil disturbance.
(Tankmixes with Atrazine, Milogard & 2, 4-D)	Preplant or Preemergence				• GRAMOXONE SUPER may be tankmixed with Atrazine, or Milogard for improved preemergence or residual weed control. The addition of 2,4-D (ester) may aid in the suppression of perennial and annual broadleaf weeds emerged at the time of application. Refer to the specific tankmix herbicide label(s) for rates, directions, limitations, and cautions and a list of weeds controlled.
	Postemergence Directed Applications	1½–2½ pints	Ground: 20 gal		• Apply when sorghum is at least 12 inches tall when naturally standing. Do not exceed 30 psi nozzle pressure or spray under conditions which may cause excessive drift. • Use precision directed-spray application equipment adjusted so that no more than the lower 3 inches of the sorghum stalk is contacted by the application spray. • Some crop injury will occur. The degree of injury is related to the precision of application and spraying conditions.
SOYBEANS	Preplant/ Preemergence	Weeds 1–3": 1½–2 pts Weeds 3–6": 2–2½ pts Weeds 6": 2½–5 pts	Ground: 20 gal Air: 5 gal		• Apply as a broadcast spray before, during or after planting, but before crop emergence. • For improved burndown or residual control, GRAMOXONE SUPER may be tankmixed with the following herbicides: 2,4-DB, Canopy*, Dual, Gemini*, Goal, Modown, Lasso, Lexone, Linex*, Lorox, Lorox Plus, Turbo*, Preview*, Prowl, Scepter*, Sencor, Surflan *Not for use in California. The rate of GRAMOXONE SUPER to be used in these tank mixtures is dependent on weed height and growing conditions. Use the highest recommended rate of GRAMOXONE SUPER under dry conditions or where the weed canopy is dense. Refer to the specific tankmix herbicide label(s) for rates, directions, limitations, and cautions and for a list of weeds controlled. • The lower rate may be used when weeds are less than 4" tall and a selective post-emergence spray or cultivation will be made within 3 weeks after planting. • Seeding should be done with a minimum amount of soil disturbance.
Tankmix w/2,4-D ester *Not registered for tankmix use with this product in California.	Preplant	Weeds 1–3": 1½–2 pts Weeds 3–6": 2–2½ pts Weeds 6": 2½–5 pts	Ground: 20 gal Air: 5 gal		• Apply at least 30 days prior to planting. Do not use the amine formulation as GRAMOXONE SUPER activity may be reduced. • May be tankmixed with residual herbicides listed above. • Refer to the 2,4-D ester label for a list of rates, directions, limitations and cautions and for a list of weeds controlled.
Directed spray	Postemergence	⅓–⅔ pts	20 gal		• For control of seedling johnsongrass, crabgrass,

CROP	USE PATTERN	GRAMOXONE SUPER RATE PER ACRE	MINIMUM TOTAL SPRAY PER ACRE	GRAZING OR PREHARVEST INTERVAL (DAYS)	PRECAUTIONS, RESTRICTIONS AND COMMENTS
					goosegrass, brachiaria, Texas millet and pigweed less 2 inches tall, use the lower rate of GRAMOXONE SUPER. • For control of the above grasses up to 4 inches tall or pigweed up to 3 inches tall, use the higher rate of GRAMOXONE SUPER. • Do not treat if soybeans are less than 8 inches tall. • Use precision directed-spray application equipment adjusted so that no more than the lower 3 inches of the soybean plant is contacted by the application spray. • Do not exceed 30 psi nozzle pressure or spray under conditions which may cause excessive drift. • Some crop injury will occur. The degree of injury is related to the precision of application and spraying conditions. • Do not graze treated areas or feed treated forage to livestock. • If needed make a second and final application 7–14 days later.
	Harvest Aid	11 - 22 ozs	Ground: 20 gal Air: 3 gal	–	• Indeterminate varieties: Apply when at least 65% of the seed pods have reached a mature brown color or when seed moisture is 30% or less. • Determinate varieties: Apply when plants are mature, i.e., beans are fully developed, ½ of leaves have dropped, and remaining leaves are yellowing. • Immature soybeans will be injured. • Mature cocklebur, especially drought-stressed plants, are tolerant to GRAMOXONE SUPER and desiccation will not be complete. Always use the higher rate for cocklebur. • Do not pasture livestock in treated areas within 15 days after treatment. Remove livestock from treated areas at least 30 days prior to slaughter.
SUGARCANE	Postemergence Directed Spray				*General Comments:* • Apply as a shielded or directed spray to avoid contact with cane foliage to prevent leaf burn and yield reduction. • Make a second and final application, if necessary, when new weed growth is 2–6 inches high. • Do not graze treated areas or feed treated forage to livestock.
- - - Florida - - -		2½ pints	Ground: 50 gal		• For optimum results, apply in early spring (March-April) when weeds are small. • Do not apply after June 1 as cane growth may be stunted and yields reduced.
- - - Hawaii - - -		2½ pints	Ground: 20 gal		• Do not apply after cane rows have closed in.
- - - Louisiana - - -		1¼–4 pints	Ground: 20 gal	30	• For tiller control, apply when tillers are less than 18 inches high. • Use the higher rate for heavier weed infestations or tiller growth.
- - -Florida & Texas - - -	Harvest Aid	11–21 fl oz	Air: 4 gal		• Use higher rate under cool, cloudy weather conditions. • Apply 3–14 days before burning and harvest.
SUNFLOWER AND SAFFLOWER Preplant or Preemergence	Broadcast or Banded Over Row	2½–5 pints (Use a minimum of 3 pts. in California.)	Ground: 20 gal Air: 5 gal	–	• For control of volunteer barley in performed seed beds.
Preplant (California only) (Safflower only)	Broadcast	11–21 fl oz	Ground: 20 gal Air: 5 gal		• For control of volunteer barley in preformed seedbeds. • Apply before, during or after planting but before crop emergence.
Preharvest Dessiccation (All seed sunflower varieties only)	Broadcast	1½–2½ pts	Ground: 20 gal Air: 5 gal	7	• Apply when sunflower seeds reach physiological maturity (when seed moisture is 35% or lower). For many varieties, this corresponds to the time when the back of the heads are yellow and the bracts are turning brown. • Do not graze treated areas or feed treated forage to livestock • Use the higher rate when crop stands or weed infestations sre heavy.

CROP	USE PATTERN	GRAMOXONE SUPER RATE PER ACRE	MINIMUM TOTAL SPRAY PER ACRE	GRAZING OR PREHARVEST INTERVAL (DAYS)	PRECAUTIONS, RESTRICTIONS AND COMMENTS
TREES AND VINES Orchards, Vineyards, Windbreak, Shade & Ornamental Trees Acerola Almonds Apples Apricots Avocados Bananas Cherries Citrus Coffee Figs Filberts Grapes Kiwi Fruit Macadamia Nuts Nectarines Olives Papayas Peaches Pears Pecans Pistachios Plums Prunes Walnuts Other shade and ornamental trees such as arborvitae, ash, elm, fir, oak, pine, etc.	Directed Spray	2½–5 pints (Use a minimum of 3 pts in California.)	Ground: 30 gal	Kiwi Fruit: 14 Pistachios: 7 All Others: —	• For grapes - treat when sucker growth is no more than 8" long. • For kiwi fruit - Do not treat more than 3 times per year • Do not allow spray to contact green stems (except suckers), fruit or foliage. • Use a shield or wrap plant when spraying around young trees or vines. • Do not graze treated areas. • Do not feed cover crops grown in treated areas to livestock. • Do not apply when figs, nuts or olives to be harvested are on the ground. • For pistachios - Do not exceed two applications after shells split. • For mature woody weeds, perennial weeds, late germinating weeds and green suckers, retreatment or spot treatment may be necessary.
Tank Mixes	Directed Spray	2½–5 pts. (Use a minimum of 3 pts. in California.)	Ground: 30 gal.	Refer to other TM labels	• GRAMOXONE SUPER may be tankmixed with registered residual herbicides listed below for combined emerged and residual weed control. Always refer to other herbicide label(s) for respective precautions, limitations, restriction, dates and directions for use and weeds controlled. • GRAMOXONE SUPER may be tank mixed with the following herbicides: Goal* Surflan* Devrinol Solican* Princep Karmex* *Tankmix currently not registered for use in California.
Guava	Directed Spray	5 pints	Ground: 50 gal	–	• For mature woody weeds, late-germinating weeds and grasses, and perennials, retreatment or spot spraying may be necessary. • Do not allow spray to contact green stems, fruit or foliage. • Do not graze treated areas. • Do not feed cover crops grown in treated areas to livestock.
Hops (ID, OR, & WA only)	Directed Spray and/or Suckering and Stripping	2½ pints	Ground: 20 gal	14	• For suckering and stripping, spray only the basal 2 ft. of the vines. Repeat as necessary. • Do not apply to hop vines less than 6 ft. tall. • Retreatment or spot treatment may be necessary. • Do not apply more than 3 times per season. • Do not allow spray to contact green stems, foliage, flowers, or cones as injury may result. • Do not allow animals to graze in treated hopyards. • Hop vine refuse and silage may be fed to livestock. • Experience with varieties other than CASCADE, YAKIMA CLUSTER, and BULLION is limited. If using GRAMOXONE SUPER on other varieties than these, test the use pattern on a small number of vines of each variety to determine sensitivity to injury. Do not use on unlisted varieties if unacceptable crop injury occurs.
Passion Fruit	Directed Spray	5 pints	Ground: 50 gal		• Use a shield or wrap vine if bark is still green at application time. • If application is to be made during harvest season, pick all fruit off the ground prior to application. • Do not allow animals to graze on treated areas. • Retreatment or spot treatment may be necessary.

CROP	USE PATTERN	GRAMOXONE SUPER RATE PER ACRE	MINIMUM TOTAL SPRAY PER ACRE	GRAZING OR PREHARVEST INTERVAL (DAYS)	PRECAUTIONS, RESTRICTIONS AND COMMENTS
TYFON (New Hampshire only)	Preplant Preemergence	2½–5 pints	Ground: 20		• Seeding should be done with a minimum of soil disturbance. • Weeds and grasses emerging after treatment will not be controlled. • Crop plants emerged at time of application will be injured.
VEGETABLES (Seeded or Transplanted) Beans (Lima, Snap) Broccoli Cabbage Cantaloupe Carrots Cauliflower Chinese Cabbage Collards Cucumber Eggplant Lentils* Lettuce Musk Melons Peas Peppers Pumpkin Squash Sugar beets Sweet Corn Turnips Tomatoes Watermelon *Currently not registered in California for use on lentils.	Preplant Preemergence	2½–5 pints (Use a minimum of 3 pts. in California.)	Ground: 20 gal	–	• Seedbeds or plantbeds should be formed as far ahead of treatment as possible to permit maximum weed emergence. • Applications can be made as a banded or broadcast treatment before, during or after planting but prior to the crop emergence. • Use the higher rate for heavier weed infestations. • Seeding or transplanting should be done with a minimum amount of soil disturbance. • Crop plants emerged at time of application will be killed.
VEGETABLES Tomatoes Peppers (Middle Atlantic, Southeast, South Central Regions, and Southwest to the Western Border of Texas)	Directed Spray	2½ pints	Ground: 20 gal		• For control or suppression of emerged weeds between rows after crop establishment. • Use precision directed spray application equipment adjusted to prevent spray contact with crop plants. Do not exceed 30 psi nozzle pressure or spray under conditions which may cause excessive drift. • Apply when weeds are succulent and weed growth is less than 6 inches. • Do not apply more than 3 applications per season. • Do not allow animals to graze in treated areas.
VEGETABLES (California only) Lettuce Melon Sugar beets Tomatoes	Broadcast	11–21 fl oz	Ground: 20 gal Air: 5 gal		• For control of volunteer barley in preformed seedbeds.
VEGETABLES Rhubarb	Dormant	2½–5 pints (Use a minimum of 3 pts. in California.)	Ground: 50 gal		• Apply during dormant season before buds in crown begin to grow. • Do not make more than 2 applications per season.

ALFALFA

NEW SEEDLINGS—Suppression and Control of Broadleaf Weeds and Grasses in New Alfalfa Seedlings Grown for Hay (California Only).

For Control of:	Rate/Acre	
	For Suppression	For Control
Spikeweed (4 in. tall or less)	11 fl. oz	1½-2½ pts
Volunteer Small Grain (8 in. tall or less)	11 to 21 fl. oz	2½ pts
Fiddleneck (6 in. tall or less)	11 to 21 fl. oz	2½ pts
Shepherdspurse	1½-2½ pts	—
Annual Bluegrass	—	1½-2½ pts
Chickweed	—	1½-2½ pts
Red Maids	—	1½-2½ pts

(6 in. tall or less)		

Do not use the 11 fl. oz rate unless the alfalfa has at least 3 trifoliate leaves; the 1½ pt rate unless the alfalfa has 6 trifoliate leaves; or rates over 1½ pts unless there are 9 trifoliate leaves.

RESIN SOAKING

PINES (Loblolly, Shortleaf, Longleaf, Slash, Virginia, Pond, Pitch, and Spruce Pines).

Tree Selection—Select trees to be treated from stands on sites not subject to periods of extreme drought stress as the desiccating effect of GRAMOXONE SUPER Herbicide to pines is accentuated during such periods, causing a reduction in the amount of oleoresin deposited in the xylem. Select trees to be treated from vigorous, non-stagnated stands, either natural or planted. In stagnated stands or commercial timber stands, plan treating with GRAMOXONE SUPER Herbicide not sooner than three years after a commercial thinning.

Application Directions—Apply GRAMOXONE SUPER Herbicide diluted in water (0.2% to 5% cation, wt./wt. basis, depending on the type of treatment method) to a suitable wound in the tree trunk to bring the treatment into contact with the xylem (sapwood). One of the following methods may be used:

Bark Streaks or Cuts: This type of wound is made using a standard or rotary bark hack or a chainsaw chipping tool employed in naval stores work to remove a single one-inch wide streak of bark about 1 to 2 feet from ground level. The total length should not exceed 1/3 of the tree circumference. Multiple streaks or cuts can result in serious girdling of the trunk and premature death of the tree. A coarse spray (about 2 to 6 ml.) of GRAMOXONE SUPER Herbicide solution (1% to 5% cation, wt./wt. basis) should be applied to runoff to the exposed xylem, using a plastic acid bottle or a low-pressure sprayer. Wear rubber gloves when using the acid bottle. The amount of spray required per cut depends on tree circumference and the length of cut or streak (1/3 of circumference). For a 9-inch diameter tree, 3 ml. of spray will cover the 1-inch wide streak. Using 3 ml. of a 2% or 4% GRAMOXONE SUPER Herbicide solution will result in application of 60 to 120 mg GRAMOXONE SUPER Herbicide per streak.

Tree Injection: Pressure injections of 1% to 4% GRAMOXONE SUPER Herbicide solution (cation, wt./wt. basis) should be applied beneath the bark and just beneath the cambium layer. Injections should be made 1 to 2 inches apart around 1/3 of the tree circumference or no closer than 3 inches apart all around the tree. From 0.2 ml. to 0.4 ml. of GRAMOXONE SUPER Herbicide solution is applied per injection. Using 2% to 4% GRAMOXONE SUPER Herbicide solution, this results in 36 to 144 mg GRAMOXONE SUPER Herbicide applied per 9-inch diameter tree, for holes spaced 1 inch apart around 1/3 of the tree or 3 inches apart all around the tree.

Bore Holes: 3/8 to 5/8 inch diameter holes 2 to 6 inches deep, depending on tree diameter, are bored sloping slightly downward and tangentially inward rather than directly towards the center of the tree. Solutions of 0.3% to 5% GRAMOXONE SUPER Herbicide can be used. Lower concentrations (0.3%) are usually applied at higher volumes of 15 to 35 ml per tree, resulting in application of 45 to approximately 100 mg GRAMOXONE SUPER Herbicide per tree. Using 2% to 4% solutions and applying 5 ml per single hole per tree results in application of 100 or 200 mg. GRAMOXONE SUPER Herbicide per tree. Bore hold applications in the range of 45 to 100 mg GRAMOXONE SUPER Herbicide per tree have resulted in effective oleoresin induction.

Time of Treatment: Resin soaking can occur from treatments made anytime of the year; however, cool season treatments under non-drought conditions usually result in less severe pine beetle infestations and longer tree life.

Interval Between Treatment and Tree Harvest: The interval between application of GRAMOXONE SUPER Herbicide and tree harvest should be a minimum of 6 months and preferably from 12 to 24 months. Intervals of over 6 months may not be possible under conditions of drought or serious pine beetle attacks, which may make early harvest necessary. The GRAMOXONE SUPER Herbicide treatment may encourage beetle attack, or may cause premature death of the tree. Desiccation of the xylem tissue, rather than the desired resin soaking, may occur, and is more likely at higher dosage rates.

Effect on Stem Growth:
GRAMOXONE SUPER Herbicide treatment can result in reduced stem growth during the interval between treatment and tree harvest.

Dilution Table for GRAMOXONE SUPER Herbicide
(1.5 lbs. cation per gallon)

Concentration of Cation Desired (Wt./Wt. Basis)	To 1½ Gal. of GRAMOXONE SUPER Herbicide, add the following No. Gals. of Water
0.2%	118.8
0.5%	46.8
1.0%	22.9
2.0%	10.9
3.0%	6.9
4.0%	4.9
5.0%	3.7

STORAGE AND DISPOSAL

Prohibitions: Do not contaminate water, food or feed by storage, disposal or cleaning of equipment. Open dumping is prohibited.

Storage: Store at temperatures above 32°F. Do not contaminate feed, foodstuffs or drinking water. Do not store next to feed or food, or transport in or on vehicles containing foodstuffs or feeds. For help with any spill, leak, fire or exposure involving this material, call CHEMTREC 800/424-9300.

Pesticide Disposal: This product is acutely hazardous. Improper disposal of excess pesticide, spray mixture, or rinsate is a violation of Federal Law. If these wastes cannot be disposed of by use according to label instructions, contact your State Pesticide or Enviromental Control Agency, or the Hazardous Waste representative at the nearest EPA Regional Office for guidance.

Container Disposal: Triple rinse (or equivalent). Do not reuse container. Offer for recycling or reconditioning, or puncture and dispose of in a sanitary landfill, or dispose of by other procedures allowed by state and local authorities.

Conversion Table
GRAMOXONE SUPER Herbicide to be Applied

Pints/Ac	fl oz/Ac	lb ai/Ac	Ac/gal
—	7	.08	18.3
—	11	.13	11.6
—	21	.25	6.1
1½	24	.28	5.3
2	32	.38	4.0
2½	40	.47	3.2
3	48	.56	2.7
4	64	.75	2.0
5	80	.94	1.6

NOTICE TO BUYER AND USER

Seller warrants that this product conforms to the chemical description on the label and is reasonably fit for the purposes stated on the label when used in accordance with directions under normal conditions of use. This warranty does not extend to the use of this product contrary to label instructions, or under abnormal use conditions, or under conditions not reasonably foreseeable to Seller, and Buyer and User assume the risk of any such use. SELLER DISCLAIMS ALL OTHER WARRANTIES EXPRESSED OR IMPLIED INCLUDING ANY WARRANTY OF FITNESS OR MERCHANTABILITY. SELLER SHALL NOT BE LIABLE FOR CONSEQUENTIAL, SPECIAL OR INDIRECT DAMAGES RESULTING FROM THE USE OR HANDLING OF THIS PRODUCT AND SELLER'S SOLE LIABILITY AND BUYER'S AND USER'S EXCLUSIVE REMEDY SHALL BE LIMITED TO THE REFUND OF THE PURCHASE PRICE.

AATREX®—Reg. TM of CIBA-GEIGY Corp.;
BANVEL®—Reg. TM of Velsicol Chemical Corp.;
BICEP®—Reg. TM of CIBA-GEIGY Corp.;
BLADEX®—Reg. TM of E.I. du Pont de Nemours & Co., Inc.;
CANOPY®*—Reg. TM of E.I. du Pont de Nemours & Co., Inc.;
CLASSIC®*—Reg. TM of E.I. du Pont de Nemours & Co., Inc.;
CONQUEST®—Reg. TM of E.I. du Pont de Nemours & Co., Inc.;
COTORAN®—Reg. TM of CIBA-GEIGY Corp.;
DEF®—Reg. TM of Bayer Ag.;
DEVRINOL® —Reg. TM of ICI Amercias Inc.;
DROPP®—Reg. TM of NOR-AM Chem. Co.;
DUAL®—Reg. TM of CIBA-GEIGY Corp.;
EXTRAZINE®*—Reg. TM of E.I. du Pont de Nemours & Co., Inc.;
FOLEX®—Reg. TM of Rhone Poulenc Inc.;
GEMINI®*—Reg. TM of E.I. du Pont de Nemours & Co., Inc.;
GOAL®—Reg. TM of Rohm and Haas Co.;
GRIFFEX®*—Reg. TM of Griffith Corp.;
HARVADE®—Reg. TM of Uniroyal Chemical Co.;
IGRAN®—Reg. TM of CIBA-GEIGY Corp.;
KARMEX®—Reg. TM of E.I. du Pont de Nemours & Co., Inc.;
LARIAT®—Reg. TM of Monsanto Company;
LASSO®—Reg. TM of Monsanto Company;
LEXONE®—Reg. TM of E.I. du Pont de Nemours & Co., Inc.;
LINEX®*—Reg. TM of Griffith Corp.;
LOROX®—Reg. TM of E.I. du Pont de Nemours & Co., Inc.;
LOROX PLUS®—Reg. TM of E.I. du Pont de Nemours & Co., Inc.;
MODOWN®—Reg. TM of Rhone Poulenc Inc.;
PREP®—Reg. TM of Rhone Poulenc Inc.;
PRINCEP®—Reg. TM of CIBA-GEIGY Corp.;
PREVIEW®—Reg. TM of E.I. du Pont de Nemours & Co., Inc.;
PROWL®—Reg. TM of American Cyanamid Company;
SCEPTER®*—Reg. TM of American Cyanamid Company;
SENCOR®—Reg. TM of the Parent Company of Farbenfabriken Bayer GmbH, Leverkusen;
SINBAR®—Reg. TM of E.I. du Pont de Nemours & Co., Inc.;
SOLICAM®—Reg. TM of Sandoz Crop Protection Corp.;
SURFLAN®—Reg. TM Of Elanco Products Company;
TURBO™*—Reg. TM of Mobay Corporation;
VELPAR®—Reg. TM of E.I. du Pont de Nemours & Co., Inc.;
ZORIAL®—Reg. TM of Sandoz Crop Protection Corp.;
Please consult each appropriate label for specific use recommendations.
*Not for use in California.

A-2R

The information and specimen labels contained in this booklet are correct as of the publication date but are subject to change without notice. For current information, contact ICI Americas Inc.

IMIDAN® 50-WP

AGRICULTURAL INSECTICIDE— WETTABLE POWDER

FOR AGRICULTURAL USE ONLY

Complete Directions For Use, Use Precautions and Recommendations

ACTIVE INGREDIENT:
Phosmet:
N-(Mercaptomethyl) phthalimide
S-(0,0-dimethyl phosphorodithioate) 50.0%
INERT INGREDIENTS: 50.0%
TOTAL 100.0%

EPA Reg. No. 10182-173

FOR AGRICULTURAL USE ONLY

KEEP OUT OF REACH OF CHILDREN
WARNING
AVISO

PRECAUCION AL USUARIO: Si usted no lee ingles, no use este producto hasta que la etiqueta le haya sido explicada ampliamente.

STATEMENT OF PRACTICAL TREATMENT

Call a Poison Center or a Physician immediately.

If a known exposure occurs or is suspected, immediately start the procedures below. If further treatment is required contact a Poison Center, a Physician, or the nearest hospital.

Note: Advise the person contacted that the active ingredient in this product is a cholinesterase inhibitor. Describe the type and extent of exposure, the victim's symptoms and follow the advice given.

Note to Medical Personnel: Atropine by injection is antidotal for cholinesterase inhibition. Pralidoxime Chloride (2-PAM, Protopam) is also antidotal when administered early and in conjunction with atropine.

If swallowed: Immediately give several glasses of water and induce vomiting by placing finger on the back of the victim's tongue. Give fluids until vomitus is clear. If victim is unconscious or convulsing, do not induce vomiting or give anything by mouth.

If inhaled: Remove to fresh air. If breathing is difficult, oxygen may be given with a physician's advice. If not breathing, clear victim's airway, start mouth-to-mouth artificial respiration and get medical attention.

If on skin: Flush affected areas with plenty of water for several minutes. Remove and clean contaminated clothing and shoes. Get medical attention if skin irritation occurs.

If in eyes: Hold eyelids apart and flush with large amounts of running water for 15 minutes. Get medical attention if eye irritation occurs.

FOR 24-HOUR EMERGENCY MEDICAL ASSISTANCE CALL 1-800-F-A-S-T-M-E-D (327-8633).
FOR CHEMICAL EMERGENCY: Spill, leak, fire, exposure, or accident call CHEMTREC 1-800-424-9300.

PRECAUTIONARY STATEMENTS

HAZARDS TO HUMANS AND DOMESTIC ANIMALS

WARNING

- MAY BE HARMFUL OR FATAL IF SWALLOWED, INHALED OR ABSORBED THROUGH THE SKIN.
- Do not get in eyes or on skin, or on clothing.
- Do not breathe dust or spray mist.
- Wash hands thoroughly after handling and before eating or smoking.
- Wash all contaminated clothing with soap and hot water before reuse.

ENVIRONMENTAL HAZARDS

- This pesticide is extremely toxic to fish. Do not apply directly to water or wetlands (swamps, bogs, marshes, and potholes). Do not contaminate water by cleaning of equipment or disposal of wastes. Drift or runoff from treated areas may be hazardous to aquatic organisms in neighboring areas.
- This product is highly toxic to bees exposed to direct treatment on blooming crops or weeds. Do not apply this product or allow it to drift to blooming crops while bees are actively visiting the treatment areas.

USE PRECAUTIONS

- Read all precautions and directions before using. Apply this product only as specified on this label.
- IMIDAN 50-WP is compatible with most commonly used insecticides and fungicides, but is incompatible with alkaline materials such as spray lime, lime sulfur, and bordeaux mixtures. These materials will reduce the insecticidal activity of IMIDAN 50-WP.
- Insecticidal activity may also be reduced when the water in the spray solution has a pH of 7 or higher. The pH of the spray solution may be corrected by the addition of a suitable buffering or acidifying agent for optimum insecticidal activity.

DIRECTIONS FOR USE

It is a violation of Federal law to use this product in a manner inconsistent with its labeling.

Do not apply this product through any type of irrigation system.

REENTRY STATEMENT

- Do not allow worker reentry into treated fields within 24 hours of application unless appropriate protective clothing is worn. Protective clothing means, at least, a hat or other suitable head covering, a long-sleeved shirt and long-legged trousers or a coverall type garment (all closely woven fabric covering the body, including the arms and legs) shoes and socks.
- Written or oral warnings must be given to workers who are expected to be in a treated area or in an area about to be treated with this product. Oral warnings must be given if there is reason to believe that written warnings cannot be understood by workers.

When oral warnings are given, warnings shall be given in a language customarily understood by workers and must indicate the specific period of time that treated areas may not be entered without protective clothing. Oral warnings must also include appropriate first aid instructions in case of accidental exposure. These instructions are given in the STATEMENT OF PRACTICAL TREATMENT at the beginning of this label.

Written warnings must include the following information, "WARNING. Area treated with IMIDAN® Agricultural Insecticide on (date of application). Do not enter treated areas without appropriate protective clothing within 24 hours of application (or other reentry interval if your State has a more restrictive interval for this product). If accidental exposure occurs, follow the instructions below." (Written warnings must include the STATEMENT OF PRACTICAL TREATMENT given at the beginning of this label.)

PROTECTIVE CLOTHING. When mixing/loading wear mid forearm to elbow length chemical resistant gloves, long-sleeve shirt, and long pants. During the application wear long-sleeve shirt and long pants. Applicator must also wear a wide-brimmed chemical resistant hat while spraying up toward fruit and nut crops. Any article of clothing worn while applying product must be cleaned before reusing. Clothing should be laundered separately from house-hold articles. Clothing which has been drenched or has otherwise absorbed concentrated pesticide must be disposed of in a sanitary landfill, incinerated, or burned if allowed by State and local authorities.

Pour recommended amount of this material on surface of water in nearly filled spray tank. Add balance of water to fill tank. Keep agitator running during filling and spraying operations. Do not allow mixture to stand.

Applications may be made by aircraft or ground power equipment including concentrate and semiconcentrate equipment. The recommended dosages, except as specified for alfalfa, corn, cotton, peas, and pine seedlings are calculated for high volume conventional hydraulic sprayers. When concentrate and semiconcentrate sprays are used, apply the recommended rate of IMIDAN 50-WP per acre in proportionately lower volumes of water according to equipment manufacturer's directions and State spray schedules. Do not exceed the maximum rate of IMIDAN 50-WP per acre or the time limitations specified for the individual crops. [In California, use a minimum of 4 pounds IMIDAN 50-WP per acre on mature Pome and Stone Fruit Trees (3½ pounds on cherries)].

RECOMMENDATIONS:
FRUIT AND NUT CROPS

Note: Directions for applications to **dormant** fruit and nut crops are located at the end of this section.

Almonds: For control of Peach Twig Borer, apply 1 pound IMIDAN 50-WP per 100 gallons of water (up to a maximum of 6 pounds IMIDAN 50-WP per acre). Apply as a full cover spray when infestation starts. Limit use on bearing almonds to one foliar application per season. Do not apply within 30 days of harvest.

Apples: For control of Apple Maggot, Codling Moth, Elm Spanworm, Fruittree Leafroller, Green Apple Aphid, Green Fruitworm, Gypsy Moth, Japanese Beetle, Obliquebanded Leafroller, Orange Tortrix*, Plum Curculio, Redbanded Leafroller, Redhumped Caterpillar, Rosy Apple Aphid and Tarnished Plant Bug, apply 1 to 1½ pounds per 100 gallons of water (or a maximum of 8 pounds per acre per application) as a full cover spray. For heavy insect infestations use the higher dosage rate. Repeat as necessary in accordance with insect infestations and local or State spray programs. Usually suppresses European Red and Two Spotted Mites when used in a seasonal program. Do not apply within 7 days of harvest. Recommended for control of Elm Spanworm, Green Fruitworm, Gypsy Moth, Fruittree Leafroller, and Obliquebanded Leafroller in the Northeast only.

*California Only. Consult your Local County Agent or Extension Service Representative regarding proper timing of application.

Tank Mix (Northeast Only): For control of the insects listed above for APPLES plus Tentiform Leafminer, Tufted Apple Budmoth and White Apple Leafhopper, a tank mix of IMIDAN 50-WP and methomyl (LANNATE* or NUDRIN*) may be used as follows: Mix ½ to 1 pound of IMIDAN 50-WP plus ½ to 1 pint of methomyl 1.8-L or 2 to 4 oz. of methomyl 90%-SP per 100 gallons of water. Apply as a full cover spray (350-400 gallons per acre). Repeat as necessary in accordance with insect infestation and local or State spray programs. Do not use on Early MacIntosh or Wealthy varieties. Do not apply within 8 days of harvest. Do not graze/feed livestock under treated trees for 10 days after application.

*LANNATE and NUDRIN are registered trademarks of E.I. duPont de Nemours and Company, Inc.

Apricots: For control of Apple Maggot, Oriental Fruit Moth, Peach Twig Borer, Plum Curculio, and Redbanded Leafroller, apply 1 to 1½ pounds per 100 gallons water (or maximum of 6 pounds per acre per application) as a full cover spray. For heavy insect infestations use the higher dosage rate. Repeat as necessary in accordance with insect infestations and local or State spray programs. Usually suppresses European Red and Two Spotted Mites when used in a seasonal program. Do not apply within 14 days of harvest.

Blueberries (Northeastern U.S. only): For control of Blueberry Maggot, Cranberry Fruitworm, and Plum Curculio, apply 2 pounds IMIDAN 50-WP in 150 to 200 gallons of water per acre as a foliar spray. A second application may be made when indicated by insect infestations and local or State spray programs. May be applied up to 3 days prior to harvest.

Cherries, Sour (Tart): For control of Cherry Fruit Fly, Fruittree Leafroller, Japanese Beetle, Peach Twig Borer, and Plum Curculio, apply 1 pound per 100 gallons water (or maximum of 3½ pounds per acre per application) as a full cover spray. Repeat as necessary in accordance with insect infestations and local or State spray programs. Usually suppresses European Red and Two Spotted Mites when used in a seasonal program. Do not apply within 7 days of harvest. Recommended for Cherry Fruit Fly, Fruittree Leafroller and Plum Curculio in the Northeast only.

Citrus (Lemons, Oranges—California, Arizona, and Texas only): For control of California Red Scale and Brown Soft Scale, apply 1 pound IMIDAN 50-WP per 100 gallons of water plus a minimum of one quart (per 100 gallons) of a suitable spray oil according to oil manufacturer's specifications. Apply as a full cover spray using ground application equipment (up to a maximum of 30 pounds IMIDAN 50-WP per acre). Repeat as necessary in accordance with scale infestations and local or State spray programs, up to three applications per season. Do not reapply within 30-days of the previous application. Do not apply within 7 days of harvest.

Grapes (West): For control of Grape Leaffolder, Omnivorous Leafroller and Western Grapeleaf Skeletonizer, apply 2 pounds IMIDAN 50-WP in 200 gallons of water per acre. Spray both sides of each row, also tops of vines to assure adequate coverage of fruit and foliage. Applications may be made at any time between egg hatch and pupation for Leaffolder, Leafroller and Western Grapeleaf Skeletonizer. Do not apply within 7 days of harvest.

Grapes (East): For control of Grape Berrymoth, Grape Leafhopper, and Japanese Beetle, apply 2 to 3 pounds IMIDAN 50-WP in 200 gallons of water per acre. For Grape Berrymoth, apply prebloom, postbloom, first and late cover sprays as needed. For Grape Leafhopper apply when most nymphs hatch (generally coincides with Berrymoth). Use the higher rate for control of Japanese Beetle. Spray both sides of each row, also tops of vines to assure adequate coverage of fruit and foliage. Make applications as necessary to maintain control in accordance with insect infestations and local or State spray programs. Do not apply within 14 days of harvest.

Nectarines: For control of Apple Maggot, Oriental Fruit Moth, Peach Twig Borer, Plum Curculio, and Redbanded Leafroller, apply 1 to 1½ pounds per 100 gallons water (or maximum of 6 pounds per acre per application) as a full cover spray. For heavy insect infestations use the higher dosage rate. Repeat as necessary in accordance with insect infestations and local or State spray programs. Usually suppresses European Red and Two Spotted Mites when used in a seasonal program. Do not apply within 14 days of harvest. Do not use OMITE* in combination with IMIDAN 50-WP on late maturing nectarine varieties as fruit injury may result.

*OMITE is a registered tademark of Uniroyal Chemical.

Peaches: For control of Japanese Beetle, Oriental Fruit Moth, Peach Twig Borer, Plum Curculio and Tarnished Plant Bug, apply 1 to 1½ pounds per 100 gallons water (or a maximum of 6 pounds per acre per application) as a full cover spray. For heavy insect infestations use the higher dosage rate. Repeat as necessary in accordance with insect infestations and local or State spray programs. Usually suppresses European Red and Two Spotted Mites when used in a seasonal program. Do not apply within 14 days of harvest.

Pears: For control of Codling Moth, Elm Spanworm, Fruittree Leafroller, Green Apple Aphid, Gypsy Moth, Pear Psylla, Plum Curculio, Redbanded Leafroller, Rosy Apple Aphid and Tarnished Plant Bug, apply 1 to 1½ pounds per 100 gallons of water (or maximum of 10 pounds per acre per application) as a full cover spray. For heavy insect infestations use the higher dosage rate. Repeat as necessary in accordance with insect infestations and local or State programs. Usually suppresses European Red and Two Spotted Mites when used in a seasonal program. Do not apply within 7 days of harvest. Recommended for control of Elm Spanworm, Gypsy Moth and Fruittree Leafroller in Northeast only.

Pecans: For control of Pecan Nut Casebearer*, Pecan Weevil and Hickory Shuckworm, apply 1½ pounds IMIDAN 50-WP per 100 gallons of water as a full cover spray when infestation starts. Repeat as necessary in accordance with insect infestations and local or State programs. For Pecan Weevil, repeat applications should be made at 7 day intervals. Do not apply within 14 days of harvest. Do not graze livestock on cover crops grown in treated pecan groves.

*Recommended for Pecan Nut Casebearer in Texas only.

Plums, Prunes: For control of Apple Maggot, Codling Moth*, Oriental Fruit Moth, Peach Twig Borer, Plum Curculio, Redbanded Leafroller, and Redhumped Caterpillar, apply 1 to 1½ pounds per 100 gallons water (or maximum of 6 pounds per acre per application) as a full cover spray. For heavy insect infestations use the higher dosage rate. Repeat as necessary in accordance with insect infestations and local or State spray programs. Usually suppresses European Red and Two Spotted Mites when used in a seasonal program. Do not apply within 7 days of harvest.

*For control of Codling Moth in California only, apply 1 pound per 100 gallons water (or a maximum of 6 pounds per acre per application) by ground equipment.

Fruit and Nut Crops—Dormant Applications (Pacific Coast States Only)

Almonds, Apples, Apricots, Cherries, Nectarines, Peaches, Pears, Plums, Prunes . . . . San Jose Scale

Almonds, Apricots, Cherries, Nectarines, Peaches, Plums, Prunes . . . . Peach Twig Borer

Use 1 pound IMIDAN 50-WP per 100 gallons of water plus suitable spray oil according to oil manufacturer's specifications. Make thorough coverage using conventional ground sprayers.

FIELD, FORAGE, AND VEGETABLE CROPS

Alfalfa (Arizona, California, Nevada Only): For control of Common Alfalfa Weevil Larvae and Egyptian Alfalfa Weevil Larvae, apply 1½ pounds of IMIDAN 50-WP in 10 to 50 gallons of water per acre (20 to 50 gallons of water for dense stands) by ground equipment, or in 5 to 10 gallons of water per acre by aircraft. Consult your local Farm Advisor regarding the proper timing of application. Larvae should be sprayed when they are actively feeding. Do not apply more than once per cutting. Do not graze or cut for hay within 14 days of application. Do not use a spreader, sticker, or surfactant with IMIDAN 50-WP on alfalfa in Arizona, California, or Nevada. Do not apply to alfalfa in the bloom period.

For Pea Aphids, apply the recommended rate of IMIDAN 50-WP in combination with ⅓ to ⅔ pint DIMETHOATE 2.67-E or ¼ to ½ pint DIMETHOATE 4-E. Do not apply more than once per cutting. Do not graze or cut for hay within 14 days of application. Do not apply to alfalfa in the bloom period. Wear the protective clothing specified on the DIMETHOATE label when preparing and using this combination.

Alfalfa, (Except Arizona, California, Nevada): For control of Common and Egyptian Alfalfa Weevil Larvae and Adults, Potato Leafhopper*, Meadow Spittlebug, and Alfalfa Blotch Leafminer, apply 2 pounds IMIDAN 50-WP in 10 to 50 gallons of water per acre (20 to 50 gallons of water for dense stands) by ground equipment or 2 pounds IMIDAN 50-WP in 2 to 10 gallons of water per acre by aircraft. Consult your Local County Agent or Extension Service Representative regarding the proper timing of application. Larvae should be sprayed when they are actively feeding. Application for adult weevils should be timed to periods of visible activity. For Alfalfa Blotch Leafminer and Meadow Spittlebug, application should be made when first signs of infestation are visible. Do not apply more than once per cutting. Do not graze or cut for hay within 7 days of application. Do not apply to alfalfa in the bloom period.

*Recommended for control of Potato Leafhopper in Northeast and North Central States Only.

For Pea Aphids, apply the recommended rate of IMIDAN 50-WP in combination with ⅓ to ⅔ pint DIMETHOATE 2.67-E or ¼ to ½ pint DIMETHOATE 4-E. Do not apply more than once per cutting. Do not graze or cut for hay within 10 days of application. Do not apply to alfalfa in the bloom period. Wear the protective clothing specified on the DIMETHOATE label when preparing and using this combination.

Corn (Corn Belt only): For control of Corn Rootworm adult populations, apply ½ to 1 pound IMIDAN 50-WP in 2 to 5 gallons of water per acre by aircraft, or in 20 to 50 gallons of water per acre by ground equipment. Apply when beetles are present in sufficient numbers to warrant treatment for protection of the silk from adult feeding. Consult your Local County Agent or Extension Service Representative regarding proper timing of application. Do not apply within 14 days of harvest.

Cotton (Except San Joaquin Valley, CA): For control of the overwintering generation of boll weevils, apply ½ to 1 pound of IMIDAN 50-WP per acre in 5 to 20 gallons of water by ground equipment or 1 pound IMIDAN 50-WP per acre in 3 to 5 gallons of water by aircraft. Make two applications. The first application should be made at the ⅓ grown square stage and the second application 5 to 7 days later. Use the higher rate under heavy infestations. For control of first, second, or third generation Boll Weevils, apply 1½ to 2 pounds IMIDAN 50-WP in 3 to 5 gallons of water per acre by aircraft or in 5 to 20 gallons of water per acre by ground equipment. Make applications at intervals of 3 to 7 days depending upon weevil population and weevil migration in fields. Use the higher rate for heavy infestations. Check infestations regularly. Do not exceed 20 pounds IMIDAN 50-WP per acre per season. Do not apply within 21 days of harvest. Do not graze or feed forage to livestock. Use on cotton prohibited in Copiah and Claiborne Counties, MS; Lauderdale and Madison Counties, AL; and Lawrence County, TN. Do not apply within one mile of any coastal or estuarine waters. Do not apply within 100 feet of aquatic habitats.

Peas, Fresh and Dry (Pacific Northwest Only): For control of Pea Weevil and Pea Leaf Weevil, apply 1½ to 2 pounds IMIDAN 50-WP in 5 to 10 gallons of water per acre by aircraft, or in 20 to 50 gallons of water per acre by ground equipment. Apply between emergence and early pod formation when adult populations are present but before eggs are laid. Consult your Local County Agent or Extension Service Representative regarding proper timing of application. Do not apply within 7 days of harvest. Do not graze or feed forage to livestock within 7 days of application. Do not cut treated fresh pea forage for hay within 10 days of application.

Potatoes: For control of Colorado Potato Beetle, Potato Flea Beetle, and Potato Leafhopper, apply 2 pounds per acre (1½ to 2 lb. per acre for Colorado Potato Beetle and Potato Flea Beetle in the Northeast) in sufficient water to provide good coverage. Repeat applications as necessary throughout the growing season. Do not apply within 7 days of harvest. Use only on potatoes to be harvested by machine. Not recommended for use on potatoes in California.

Field Margins: For control of grasshoppers in margins of cultivated fields and forage crops, apply 3–4 pounds of IMIDAN 50-WP in 10 to 50 gallons of water per acre (20-50 gallons of water for dense stands) or in 5-10 gallons of water per acre by aircraft. Do not graze livestock in treated area. Do not harvest for food or feed.

DECIDUOUS SHADE AND ORNAMENTAL TREES AND WOODY EVERGREENS

IMIDAN 50-WP is recommended for use by commercial applicators on Deciduous Shade and Ornamental Trees and Woody Evergreens in parks, residential and recreational areas, along thoroughfares and other localized areas where infestations of Gypsy Moth, Spring Cankerworm, Elm Spanworm, Birch Leaf Miner, Eastern Tent Caterpillar, Elm Leaf Beetle Larvae, Japanese Beetle, or Redhumped Caterpillar occur. When such insects or their damage occur, apply at a rate of 1 to 1½ pounds IMIDAN 50-WP per 100 gallons of water, thoroughly wetting all parts of the affected plants to the point of runoff. For heavy insect infestations use the higher dosage rate. Repeat application as necessary to maintain insect control.

- Recommended on **Deciduous Shade and Ornamental Trees** (such as Ash, Beech, Oak, Dogwood, Willow, Hickory, Locust, Liquidambar, Hawthorne, Birch, Elm, Maple) to control: Gypsy Moth, Spring Cankerworm, Elm Spanworm, Eastern Tent Caterpillar, Elm Leaf Beetle Larvae, Japanese Beetle, and Redhumped Caterpillar.
- Recommended on **Woody Evergreens** (such as Arborvitae, Cedar, Fir, Hemlock, Juniper, Pine, Spruce, Yew) to control: Gypsy Moth.
- Recommended on **Birch Trees** to control: Birch Leaf Miner.

Choose a cool, calm period, preferably in early morning or evening. Do not apply if rain is expected or before leaf surfaces are dry.

Pine Trees (Pacific Northwest Only): For control of European, Pine Shoot Moth **(Rhyacionia buoliana)**, apply 2 pounds of IMIDAN 50-WP per acre or for individual trees at a rate of 2 pounds per 100 gallons of water. Spray trees for thorough coverage of foliage. Repeat at 2 week intervals throughout flight period of the moths.

Pine Seedlings (White, Slash and Loblolly): For control of Pales Weevil **(Hylobius pales)** and Pitch-Eating Weevil **(Pachylobius picivorus)**, use IMIDAN 50-WP as a 4% top dip, dipping down to and including the root collar only. Dip in bundles loose enough to allow solution to penetrate the bundles. Avoid coverage of roots. Swish tops in solution for 10-15 seconds to assure adequate coverage of all top growth. Drain and allow seedlings to dry before planting. The addition of an extender such as Nu-Film 17* or Plantgard** may help to retain the IMIDAN 50-WP on the foliage under high rainfall conditions.

Some slight needle burn and first year growth reduction may occur following treatment of loblolly pine.

Wear rubber gloves during treating and planting.

Dilution Table for a 4% Top Dip

To Make	5 gallons	30 gallons	50 gallons	100 gallons
Use this amount of IMIDAN 50-WP.	3½ lbs.	20 lbs.	33½ lbs.	67 lbs.

Agitate frequently to keep IMIDAN 50-WP in suspension. Five gallons of solution should be enough to treat 10,000 seedlings. Do not keep IMIDAN solution overnight. Make up a fresh batch each day that seedlings will be dipped.

*Nu-Film is a registered trademark of Miller Chemical & Fertilizer Corp.
**Plantgard is a registered trademark of Polymetrics International Inc.

STORAGE AND DISPOSAL

- Do not contaminate water, food or feed by storage or disposal.

STORAGE: Store in a cool dry place.

PESTICIDE DISPOSAL: Wastes resulting from the use of this product may be disposed of on site or at an approved waste disposal facility.

CONTAINER DISPOSAL: Completely empty bag into application equipment. Then dispose of empty bag in a sanitary landfill or by incineration, or, if allowed by State and local authorities, by burning. If burned, stay out of smoke.

SPILL OR LEAK: A small spill can be handled routinely. Use adequate ventilation and wear an air-supplied respirator to prevent inhalation. Wear suitable protective clothing and eye protection to prevent skin and eye contact. Use the following procedures:
1. Sweep up spilled material being careful not to create dust.
2. Place sweepings in an open drum.
3. Generously cover the contaminated areas with a common household detergent. Using a stiff brush and small amounts of water, work the detergent into the spill material forming a slurry. Do not splatter on one's self or bystanders, and completely avoid skin or eye contact with this material. Brush the slurry into cracks and crevices and allow to stand for 2 to 3 minutes.
4. Spread a suitable absorbent such as clay, sawdust or kitty litter on the slurried liquid. Shovel absorbed material into an open drum.
5. Repeat if necessary.
6. Flush the area with water while observing proper environmental considerations.
7. Seal drum and dispose of contaminated material in an approved pesticide landfill.

Large spills must be handled according to a predetermined plan. For assistance in developing a plan, contact ICI Americas Inc. Agricultural Products Wilmington, DE 19897

NOTICE TO BUYER AND USER:

Seller warrants that this product conforms to the chemical description on the label and is reasonably fit for the purposes stated on the label when used in accordance with directions under normal conditions of use. This warranty does not extend to the use of this product contrary to label instructions, or under abnormal use conditions, or under conditions not reasonably foreseeable to Seller, and Buyer and User assume the risk of any such use. SELLER DISCLAIMS ALL OTHER WARRANTIES EXPRESSED OR IMPLIED INCLUDING ANY WARRANTY OF FITNESS OR MERCHANTABILITY. SELLER SHALL NOT BE LIABLE FOR CONSEQUENTIAL, SPECIAL OR INDIRECT DAMAGES RESULTING FROM THE USE OR HANDLING OF THIS PRODUCT AND SELLER'S SOLE LIABILITY AND BUYER'S AND USER'S EXCLUSIVE REMEDY SHALL BE LIMITED TO THE REFUND OF THE PURCHASE PRICE.

This product is sold only for uses stated on its label. No express or implied license is granted to use or sell this product under any patent in any country except as specified. Country: United States of America.

A-12R

The information and specimen labels in this booklet are correct as of the publication date but are subject to change without notice. For current information, contact ICI Americas Inc.

KARATE®

INSECTICIDE

RESTRICTED USE PESTICIDE

Toxic to fish and aquatic organisms.
For retail sale to and use only by Certified Applicators, or persons under their direct supervision, and only for those uses covered by the Certified Applicator's certification.

ACTIVE INGREDIENT: Lambdacyhalothrin
[1α(S*),3α(Z)]-(±)-cyano
(3-phenoxyphenyl)methyl 3-(2-chloro-3,
3,3-trifluoro-1-propenyl)-2,2-
dimethylcyclopropanecarboxylate. 13.1%
INERT INGREDIENTS 86.9%

TOTAL 100.0%

KARATE contains 1 pound of active ingredient per gallon and is an emulsifiable concentrate.

EPA Reg No. 10182-96

KEEP OUT OF REACH OF CHILDREN

DANGER PELIGRO

PRECAUCION AL USUARIO: Si usted no puede leer o entender ingles, no use este producto hasta que la etiqueta le haya sido explicado ampliamente.

STATEMENT OF PRACTICAL TREATMENT

IF ON SKIN: Wash with plenty of soap and water. Get medical attention if irritation persists.

IF SWALLOWED: Do not induce vomiting. Call a physician immediately.

IF INHALED: Remove victim to fresh air. If not breathing give artificial respiration, preferably mouth-to-mouth. Get medical attention.

IF IN EYES: Hold eyelids open and flush with steady, gentle stream of water. Remove any contact lenses and repeat flushing procedure for several minutes. Get medical attention if irritation persists.

Note to Physician: Inducing vomiting as first aid for this substance may result in increased risk of chemical pneumonia or pulmonary edema caused by aspiration of the hydrocarbon solvent. Vomiting should be induced only under professional supervision.

FOR 24-HOUR EMERGENCY MEDICAL ASSISTANCE CALL 1-800-F-A-S-T-M-E-D (327-8633).
FOR CHEMICAL EMERGENCY: Spill, leak, fire, exposure, or accident call CHEMTREC 1-800-424-9300.

PRECAUTIONARY STATEMENTS

HAZARDS TO HUMANS AND DOMESTIC ANIMALS

DANGER

Corrosive, Causes Skin Damage.

May Be Fatal If Swallowed or Inhaled.

Causes Eye Injury.

Harmful If Absorbed Through Skin.

May Cause Allergic Skin Reactions.

DO NOT get on skin, in eyes or on clothing. DO NOT breathe spray mist. WEAR a mask or pesticide respirator approved by NIOSH under the provisions of 30 CFR 11. WEAR goggles or face shield, protective clothing, and rubber gloves when mixing. WASH thoroughly with soap and water after handling and before eating, drinking or using tobacco. REMOVE contaminated clothing and wash before reuse.

WEAR THE FOLLOWING PROTECTIVE CLOTHING AND EQUIPMENT WHEN HANDLING THE CONCENTRATE AND DURING APPLICATION, EQUIPMENT REPAIR, CLEANING, DISPOSAL OF THE SPRAY SOLUTION, AND DURING EARLY RE-ENTRY TO TREATED AREAS.

Wear a protective suit of one or two pieces that covers all parts of the body except the head, hands, and feet. Wear chemical resistant gloves and chemical resistant boots, shoes, or shoe coverings. Wear goggles or face-shield. A helmet with visor is acceptable during application from nonenclosed cockpits. Wear a pesticide respirator approved by the National Institute for Occupational Safety and Health under the provisions of 30 CFR 11.

During application only from a tractor with a completely enclosed cab or aerially with an enclosed cockpit, long sleeve shirt and long-legged pants may be worn in place of the above protective equipment. Chemical resistant gloves must be made available in the cab or cockpit and must be worn while exiting. This clothing is inadequate to protect you during equipment repairs, equipment cleaning, spray solution disposal or early re-entry.

IMPORTANT! Before removing gloves, wash hands with soap and water. Always wash hands before smoking, eating, drinking, or toileting.

Human flaggers must either be in totally enclosed vehicles **or** wear a protective suit, chemical-resistant gloves; a chemical-resistant hat with a wide brim or a hood; chemical-resistant shoes, shoe covers or boots; a respiratory protective device, and a face shield.

ENVIRONMENTAL HAZARDS

This pesticide is extremely toxic to fish and wildlife. Do not contaminate water when disposing of equipment washwaters. Do not apply directly to water or wetlands (swamps, bogs, marshes and potholes). Do not apply when weather conditions favor drift from treated areas. Drift and runoff from treated areas may be hazardous to aquatic organisms in neighboring areas. This product is highly toxic to bees exposed to direct treatment or residues on blooming crops or weeds. Do not apply this product or allow it to drift to blooming crops or weeds if bees are visiting the treatment area.

PHYSICAL OR CHEMICAL HAZARDS

Do not use or store near heat or open flame.

DIRECTIONS FOR USE

It is a violation of Federal law to use this product in a manner inconsistent with its labeling.

This labeling must be in the possession of the user at the time of application.

Do not apply this product through any type of irrigation system.

RE-ENTRY STATEMENTS

Do not apply this product in such a manner as to directly or through drift expose workers or other persons. The area being treated must be vacated by unprotected persons. Do not enter treated areas without protective clothing until sprays have dried. Because certain states may require more restrictive re-entry intervals for various crops treated with this product, consult your State Department of Agriculture for further information.

Apply KARATE as shown in the following chart:

COTTON SPRAY RECOMMENDATIONS

Target Pests	Rate (LB. a.i./A)	Rate (Fl. oz./A)	Remarks
Thrips Spp. Tobacco Thrips Soybean Thrips	0.01-0.2	1.28-2.56	Apply as required by scouting, usually at intervals of 5-7 days. Timing and frequency of applications should be based upon insect populations reaching locally determined economic thresholds.
Lygus Bugs Pink Bollworm Cabbage Looper Cotton Leafperforator Cutworm Spp. Saltmarsh Caterpillar Cotton Fleahopper	0.02-.03	2.56-3.84	Apply with ground or air equipment using sufficient water to obtain full coverage of foliage. Under light bollworm/budworm infestation levels, 0.02 lb. a.i./A may be applied in conjunction with intense field monitoring. For boll weevil control spray on a 3-5 day schedule.
Cotton Bollworm Tobacco Budworm Boll Weevil Fall Armyworm Beet Armyworm** European Corn Borer Brown Stink Bug Green Stink Bug Southern Green Stink Bug Twospotted Spider Mite* Cotton Aphid* Bandedwing Whitefly* Sweetpotato Whitefly*	0.025-0.04	3.20-5.12	Do not apply within 21 days of harvest. Do not apply more than 1.6 pints (0.2 lb. a.i.) per acre per season. Do not graze livestock in treated areas.

*Suppression only
* *For control of 1st and 2nd instar only.

ENDANGERED SPECIES RESTRICTIONS

This product may pose a hazard to endangered aquatic species: **Do not apply this product within 100 yards of aquatic habitats in the following counties:**

ALABAMA: Colbert, Greene, Jackson, Lamar, Lauderdale, Limestone, Madison, Marshall, Morgans, Pickens, and Sumter

ARIZONA: Graham, Maricopa, Pima, Pinal, and Santa Cruz

ARKANSAS: Benton, Clay, Clark, Cross, Lawrence, Lee, Poinsette, Polk, Randolph, Sharp, and St. Francis

FLORIDA: Broward, Dade, Glades, and Palm Beach

KENTUCKY: Ballard, Butler, Edmonson, Green, Hart, Jackson, Laurel, Livingston, Marshall, McCracken, McCreary, Pulaski, Rockcastle, Taylor, Warren, and Wayne

MISSISSIPPI: Claiborne, Copiah, Hinds, Itawamba, Lowndes, Monroe, and Noxubee

NEVADA: Clark

NORTH CAROLINA: Edgecombe, Nash, and Pitt

OKLAHOMA: Delaware, McCurtain, and Pushmataha

TENNESSEE: Bedford, Blount, Claiborne, Decatur, Franklin, Hancock, Hardin, Hickman, Knox, Lawrence, Lincoln, Loudon, Marshall, Maury, Meigs, Monroe, Rhea, Roane, Scott, Sequatchie, Smith, Sullivan, and Wayne

TEXAS: Bastrop, Burleson, Comal, Harris, Hays, Jeff Davis, Pecos and Reeves

VIRGINIA: Lee, Russell, Scott, Smyth, Tazewell, Washington, and Wise

This product may pose a hazard to the endangered **Attwater's Greater Prairie Chicken,** and is prohibited from use in the following counties:

TEXAS: Aransas, Austin, Colorado, Fort Bend, Goliad, Refugio, and Victoria

STORAGE AND DISPOSAL

Do not contaminate water, food or feed by storage and disposal.

STORAGE

Store in original containers only. Keep container closed when not in use. Do not store near food or feed. In case of spill or leak on floor or paved surfaces, soak up with sand, earth, or synthetic absorbent. Remove to chemical waste area.

PESTICIDE DISPOSAL

Pesticide wastes are acutely hazardous. Improper disposal of excess pesticide, spray mixture, or rinsate is a violation of Federal law. If these wastes cannot be disposed of by use according to label instructions, contact your

State Pesticide or Environmental Control Agency, or the hazardous-waste representative of the nearest EPA Regional Office for guidance.

CONTAINER DISPOSAL

Metal Containers: Triple rinse (or equivalent). Then offer for recycling or reconditioning, or puncture and dispose of in a sanitary landfill, or by other approved State and local procedures. **Plastic Containers:** Triple rinse (or equivalent). Then offer for recycling or reconditioning, or puncture and dispose of in a sanitary landfill, or by incineration, or, if allowed by State and local authorities, by burning. If burned, stay out of smoke. **Glass Containers:** Triple rinse (or equivalent). Then dispose of in a sanitary landfill, or by other approved State and local procedures.

NOTICE TO BUYER AND USER: Seller warrants that this product conforms to the chemical description on the label and is reasonably fit for the purpose stated on the label when used in accordance with directions under normal conditions of use. This warranty does not extend to the use of this product contrary to label instructions, or under abnormal use conditions, or under conditions not reasonably foreseeable to Seller, and Buyer and User assume the risk of any such use. SELLER DISCLAIMS ALL OTHER WARRANTIES, EXPRESSED OR IMPLIED, INCLUDING ANY WARRANTY OF FITNESS OR MERCHANTABILITY. SELLER SHALL NOT BE LIABLE FOR THE CONSEQUENTIAL, SPECIAL OR INDIRECT DAMAGES RESULTING FROM THE USE OR HANDLING OF THIS PRODUCT AND SELLER'S SOLE LIABILITY AND BUYER'S AND USER'S EXCLUSIVE REMEDY SHALL BE LIMITED TO THE REFUND OF THE PURCHASE PRICE.

Rate Conversion Chart

Lb. a.i./ Acre	Fl. oz./ Acre	Pints/ Acre	Treated/ Gallon
0.01	1.28	0.08	100
0.02	2.56	0.16	50
0.025	3.20	0.20	40
0.03	3.84	0.24	33
0.04	5.12	0.32	25

A-1R

The information and specimen labels in this booklet are correct as of the publication date but are subject to change without notice. For current information, contact ICI Americas Inc.

ORDRAM® 10-G

GRANULES — SELECTIVE HERBICIDE FOR USE ON RICE GROWN IN CALIFORNIA ONLY

ACTIVE INGREDIENT:

S-ethyl hexahydro-1 *H*-azepine-1-carbothioate	10.0%
INERT INGREDIENTS:	90.0%
TOTAL	100.0%

Contains 10 Pounds of Active Ingredient per 100 Pounds

EPA Reg. No. 10182-174

KEEP OUT OF REACH OF CHILDREN

CAUTION

Harmful If Swallowed

STATEMENT OF PRACTICAL TREATMENT

If a known exposure occurs or is suspected, immediately start the procedures given below. If further treatment is required, contact a Poison Center, a physician, or the nearest hospital.

If Swallowed: Immediately give several glasses of water and induce vomiting by gagging the victim with a finger placed on the back of the victim's tongue. Give fluids until vomitus is clear. Do not induce vomiting or give anything by mouth to an unconscious or convulsing person. Call a physician immediately.

For Eye Contact: Hold eyelids apart and flush eyes with large amounts of running water for at least 15 minutes. Get medical attention.

For Skin Contact: Wash all affected areas with plenty of soap and water. Remove and clean contaminated clothing and shoes before reuse. Seek medical attention if irritation occurs.

If Inhaled: Remove to fresh air. Seek medical attention if respiratory irritation occurs or if breathing becomes difficult.

FOR 24-HOUR EMERGENCY MEDICAL ASSISTANCE CALL 1-800-F-A-S-T-M-E-D (327-8633).

FOR CHEMICAL EMERGENCY: Spill, leak, fire, exposure, or accident call CHEMTREC 1-800-424-9300.

PRECAUTIONARY STATEMENTS

HAZARDS TO HUMANS AND DOMESTIC ANIMALS

CAUTION

Harmful If Swallowed. Avoid contact with skin, eyes and clothing. Avoid inhalation of dust. Wear rubber gloves and fresh clothing. Wear goggles to avoid contact with eyes. Wash with soap and water immediately after use or contact. Do not contaminate food and feed.

ENVIRONMENTAL HAZARDS

This product is extremely toxic to fish. Do not contaminate water used for domestic purposes. Do not contaminate water by cleaning of equipment or disposal of wastes. Do not apply directly to water except as stated on label.

USE PRECAUTIONS

READ ALL LABEL INSTRUCTIONS BEFORE USING. ORDRAM 10-G should be used only for recommended purposes and at recommended rates. ORDRAM 10-G may cause crop injury under extreme soil or climatic conditions or if directions are not followed. Do not apply any other herbicide in conjunction with ORDRAM 10-G as injury may occur. Do not mix with fertilizer.

DIRECTIONS FOR USE OF ORDRAM 10-G

It is a violation of Federal law to use this product in a manner inconsistent with its labeling.

Directions and Recommendations for Pre-Plant, Pre-Flood Use On Rice (Water-Seeded Rice Only):

For use on Rice to control Barnyardgrass (watergrass) (*Echinochloa* spp.)

Soil Preparation: Use only on soils in good tilth. Prepare the soil for seeding according to good agricultural practices. All weed growth must be thoroughly worked into the soil before treatment.

Application: When ORDRAM 10-G is applied before flooding and planting, application may be made by either air or ground equipment. Application equipment should be carefully calibrated before use and checked frequently during use to be sure application is uniform. Avoid overlaps that will increase ORDRAM 10-G dosage above recommended limits because crop injury may occur. Aerial application should be made only when the wind velocity is low, 0-10 mph.

Soil Incorporation: All applications of ORDRAM 10-G must be incorporated into the soil to a depth of 2 to 3 inches within 6 hours by mechanical incorporation. Incorporation should be done by at least two discings or heavy spike-tooth harrowings. Do not set disc to cut deeper than 4 inches.

Rate: Apply 30 pounds of ORDRAM-10 G per acre. FLOOD THE FIELD FOR SEEDING AND HOLD THE WATER FOUR TO SIX DAYS FOLLOWING SEEDING.

Directions and Recommendations for Post-Flood, Post-Emergence Use On Rice (Water-Seeded or Drilled Rice):

For use on Rice to control Barnyardgrass (watergrass) (*Echinochloa* spp.). Rice should be in the seeding stage at the time of application.

For Air Application Only: Barnyardgrass must be a minimum of 2" high but not higher than 5", and it must be at least 2/3 submerged at time of application. This water level (not less than 2") must be maintained until the Barnyardgrass dies.

Equipment should be carefully calibrated before use and checked frequently during use to be sure application is uniform. Avoid overlaps that will increase ORDRAM 10-G dosage above recommended limits because crop injury may occur. Aerial application should be made only when the wind velocity is low, 0-10 mph. To avoid uneven distribution of ORDRAM 10-G, the water flow through the field should be at a minimum at the time of application and for at least 2 hours after application.

HOLD FLOOD WATER A MINIMUM OF 4 DAYS AFTER APPLICATION.

Rate: Apply 30 pounds of ORDRAM 10-G per acre.

STORAGE AND DISPOSAL

Storage: Do not contaminate water, food, feed, fertilizer or seeds by storage or disposal. Open dumping is prohibited. Keep container closed when not in use.

Pesticide Disposal: Wastes resulting from the use of this product may be disposed of on site or at an approved waste disposal facility.

Container Disposal: Completely empty bag into application equipment. Then dispose of empty bag in a sanitary landfill or by incineration, or, if allowed by State and local authorities, by burning. If burned, stay out of smoke.

NOTICE TO BUYER AND USER:

Seller warrants that this product conforms to the chemical description on the label and is reasonably fit for the purposes stated on the label when used in accordance with directions under normal conditions of use. This warranty does not extend to the use of this product contrary to label instructions, or under abnormal use conditions, or under conditions not reasonably foreseeable to Seller, and Buyer and User assume the risk of any such use. SELLER DISCLAIMS ALL OTHER WARRANTIES EXPRESSED OR IMPLIED INCLUDING ANY WARRANTY OF FIT-

NESS OR MERCHANTABILITY. SELLER SHALL NOT BE LIABLE FOR CONSEQUENTIAL, SPECIAL OR INDIRECT DAMAGES RESULTING FROM THE USE OR HANDLING OF THIS PRODUCT AND SELLER'S SOLE LIABILITY AND BUYER'S AND USER'S EXCLUSIVE REMEDY SHALL BE LIMITED TO THE REFUND OF THE PURCHASE PRICE.

This product is sold only for uses stated on its label. No express or implied license is granted to use or sell this product under any patent in any country except as specified: Country: United States of America, Patent No: 3,573,031

A-5R

The information and specimen labels in this booklet are correct as of the publication date but are subject to change without notice. For current information, contact ICI Americas Inc.

ORDRAM® 15-G

SOUTHERN REGION
GRANULES — SELECTIVE HERBICIDE FOR USE ON RICE ONLY

FOR SEMI-BULK CONTAINERS

ACTIVE INGREDIENT:	
S-ethyl hexahydro-1*H*-azepine-1-carbothioate	15.0%
INERT INGREDIENTS:	85.0%
TOTAL	100.0%

Contains 15 Pounds of Active Ingredient per 100 Pounds

EPA Reg No. 10182-274

KEEP OUT OF REACH OF CHILDREN
CAUTION

STATEMENT OF PRACTICAL TREATMENT

If a known exposure occurs or is suspected, **immediately** start the procedures given below. If further treatment is required, contact a Poison Center, a physician, or the nearest hospital.

If swallowed: Immediately give several glasses of water and induce vomiting by gagging the victim with a finger placed on the back of the victim's tongue. Give fluids until vomitus is clear. Do not induce vomiting or give anything by mouth to an unconscious or convulsing person. Call a physician immediately.

For eye contact: Hold eyelids apart and flush with large amounts of running water for at least 15 minutes. Get medical attention.

For skin contact: Wash all affected areas with plenty of soap and water. Remove and clean contaminated clothing and shoes before reuse. Seek medical attention if irritation occurs.

If inhaled: Remove to fresh air. Seek medical attention if respiratory irritation occurs or if breathing becomes difficult.

FOR 24-HOUR EMERGENCY MEDICAL ASSISTANCE CALL 1-800-F-A-S-T-M-E-D (327-8633).
FOR CHEMICAL EMERGENCY: Spill, leak, fire, exposure, or accident call CHEMTREC 1-800-424-9300.

PRECAUTIONARY STATEMENTS

HAZARDS TO HUMANS AND DOMESTIC ANIMALS

CAUTION

HARMFUL IF SWALLOWED. Avoid contact with skin, eyes, and clothing. Avoid inhalation of dust. Wear rubber gloves and fresh clothing. Wear goggles to avoid contact with eyes. Wash with soap and water immediately after use or contact. Do not contaminate food and feed.

ENVIRONMENTAL HAZARDS: This product is extremely toxic to fish. Do not contaminate waters used for domestic purposes. Do not contaminate water when disposing of equipment washwaters. Do not apply directly to water except as stated on label.

READ ALL LABEL DIRECTIONS BEFORE USING

GENERAL USE PRECAUTIONS

READ ALL LABEL INSTRUCTIONS BEFORE USING. ORDRAM 15-G should be used only for recommended purposes and at recommended rates. ORDRAM 15-G may cause crop injury under extreme soil or climatic conditions or if directions are not followed. Do not mix any other herbicide in the same application with ORDRAM 15-G as injury may occur.

Do not apply more than 4 pounds actual ORDRAM per acre preplant or more than 5 pounds actual ORDRAM per application per acre post-flood, post-emergence. Do not apply more than 9 pounds of actual ORDRAM per acre per season.

Allow 30 days between pre-plant and post-flood treatments. Allow 30 days between the last application and harvest. Do not feed rice straw if application is made within 40 days of harvest.

Equipment used in the application of ORDRAM 15-G should be carefully calibrated before use and checked frequently during application to be sure it is working properly and delivering a uniform distribution pattern. Avoid overlaps that will increase ORDRAM 15-G dosage above recommended limits because crop injury may occur.

ORDRAM 15-G may be applied by aircraft or ground equipment. Aerial application should be made only when the wind velocity is low, 0 - 10 mph.

DIRECTIONS FOR USE OF ORDRAM 15-G

It is a violation of Federal law to use this product in a manner inconsistent with its labeling.

SOUTHERN REGION (Arkansas, Louisiana, Mississippi, Missouri [Boot-Heel Region Only] and Texas)

DIRECTIONS AND RECOMMENDATIONS FOR PRE-PLANT, PRE-FLOOD USE ON RICE (Water-Seeded Rice Only):

For control of Barnyardgrass (*Echinochloa* spp.), for suppression of Annual Sedges (*Cyperus* spp.) Broadleaf Signalgrass (*Brachiaria platyphylla*), Crabgrass (*Digitaria* spp.) and Sprangletop (*Leptochloa* spp.), and for reduced competition from Red Rice (*Oryza sativa*).

SOIL PREPARATION: Use only on soils in good tilth. Prepare the soil for seeding according to good agricultural practices. All weed growth must be thoroughly worked into the soil before treatment.

APPLICATION (Air or Ground Application): When ORDRAM 15-G is applied before flooding and planting, application may be made by either air or ground equipment.

SOIL INCORPORATION: Applications of ORDRAM 15-G must be immediately incorporated thoroughly and uniformly to a depth of 1–2 inches. Incorporation may be delayed up to 6 hours when applied to a dry soil surface. Incorporation should be done by at least one discing (pulverizing disc) and harrowing. Do not set disc to cut deeper than 4 inches.

RATE: To control Barnyardgrass and suppress Crabgrass, apply 20–25 pounds of ORDRAM 15-G per acre. To suppress Annual Sedges, Broadleaf Signalgrass and Sprangletop, and to reduce competition from Red Rice, apply 25 pounds of ORDRAM 15-G per acre. Reduced competition from Red Rice is improved with continuous floodwater management. An increased seeding rate may be desirable under water-seeded cultures.

FLOOD THE FIELD FOR SEEDING AND HOLD THE WATER FOUR TO SIX DAYS FOLLOWING SEEDING.

DIRECTIONS AND RECOMMENDATIONS FOR POST-FLOOD, PRE-PLANT USE ON RICE (Water-Seeded Rice at Seeding):

For control of Barnyardgrass (*Echinochloa* spp.) and for reduced competition from Red Rice (*Oryza sativa*).

SOIL PREPARATION: Use only on soils in good tilth. Prepare the soil for seeding according to good agricultural practices. All weed growth must be thoroughly worked into the soil before treatment. ORDRAM 15-G should be applied to murky (silty) water after working of soil.

APPLICATION: Apply ORDRAM 15-G into the flood water by air application. Follow with seeding 2–3 days after the ORDRAM 15-G application. Allow the flood water to remain stable from application until germination and rooting have occurred as in normal water-seeding procedures, usually 4 – 6 days following seeding.

RATE: To control Barnyardgrass apply 20 – 25 pounds of ORDRAM 15-G per acre, and to reduce competition from Red Rice, apply 25 pounds of ORDRAM 15-G per acre. Reduced competition from Red Rice is improved with continuous flood water management.

DIRECTIONS AND RECOMMENDATIONS FOR POST-FLOOD, POST-EMERGENCE USE ON RICE (Water-Seeded Or Dry-Seeded Rice):

For control of Barnyardgrass (*Echinochloa* spp.) and Dayflower (*Commelina communis*), and for suppression of Annual Sedges (*Cyperus* spp.), Spikerush (*Eleocharis* spp.), Brooks Paspalum (*Paspalum accuminatum*) and Sprangletop (*Leptochloa* spp.).

APPLICATION: Apply ORDRAM 15-G into the flood water by air application only.

STANDARD TREATMENT: Rice must be established at the time of application. Barnyardgrass, Dayflower, Annual Sedges, Spikerush and Sprangletop must not be more than 5 inches high. Barnyardgrass, Annual Sedges, Spikerush and Sprangletop must be at least 2/3 submerged. Dayflower must be completely submerged. Brooks Paspalum runners should be completely submerged and not longer than 12 inches. Floods should be maintained at these depths until the Barnyardgrass and/or Dayflower are controlled or until the Annual Sedges, Spikerush, Brooks Paspalum and Sprangletop are suppressed.

RATE: Apply 20 pounds of ORDRAM 15-G per acre.

FOR SUPPRESSION OF LARGE GRASSES (Suppression-stopping normal development): Apply 20 – 25 pounds of ORDRAM 15-G per acre for Barnyardgrass 5 to 12 inches in height. Apply 33 pounds of ORDRAM 15-G per acre for unusually heavy infestations of Barnyardgrass under 12 inches in height and for Barnyardgrass 12 to 24 inches in height. Apply 33 pounds of ORDRAM 15-G per acre for Sprangletop 5 to 12 inches in height. Flood water must be maintained at a 4 to 6 inch depth. Avoid overlaps because crop injury may occur.

STORAGE AND DISPOSAL

STORAGE: Do not contaminate water, food, feed, fertilizer or seeds by storage or disposal. Open dumping is prohibited. Keep container closed when not in use.

PESTICIDE DISPOSAL: Wastes resulting from the use of this product may be disposed of on site or at an approved waste disposal facility.

CONTAINER DISPOSAL: Completely empty bag into application equipment. Then dispose of empty bag in a sanitary landfill or by incineration, or, if allowed by State and local authorities, by burning. If burned, stay out of smoke.

NOTICE TO BUYER AND USER:

Seller warrants that this product conforms to the chemical description on the label and is reasonably fit for the purposes stated on the label when used in accordance with directions under normal conditions of use. This warranty does not extend to the use of this product contrary to label instructions, or under abnormal use conditions, or under conditions not reasonably forseeable to Seller, and Buyer and User assume the risk of any such use. SELLER DISCLAIMS ALL OTHER WARRANTIES EXPRESSED OR IMPLIED INCLUDING ANY WARRANTY OF FITNESS OR MERCHANTABILITY. SELLER SHALL NOT BE LIABLE FOR CONSEQUENTIAL, SPECIAL OR INDIRECT DAMAGES RESULTING FROM THE USE OR HANDLING OF THIS PRODUCT AND SELLER'S SOLE LIABILITY AND BUYER'S AND USER'S EXCLUSIVE REMEDY SHALL BE LIMITED TO THE REFUND OF THE PURCHASE PRICE.

This product is sold only for uses stated on its label. No express or implied license is granted to use or sell this product under any patent in any country except as specified: Country: United States of America. Patent No: 3,573,031.

A-1R

The information and specimen labels in this booklet are correct as of the publication date but are subject to change without notice. For current information, contact ICI Americas Inc.

ORDRAM® 8-E

SELECTIVE HERBICIDE
AN EMULSIFIABLE LIQUID HERBICIDE

For Use On Rice Grown in California Only

ACTIVE INGREDIENT:

S-ethyl hexahydro-1 *H*-azepine-1-carbothioate	90.9%
INERT INGREDIENTS	9.1%
TOTAL	100.0%

Contains 8 Pounds Active Ingredient Per Gallon

EPA Reg. No. 10182-204

KEEP OUT OF REACH OF CHILDREN
WARNING

May Be Fatal If Swallowed

STATEMENT OF PRACTICAL TREATMENT

CALL A POISON CENTER OR PHYSICIAN IMMEDIATELY

If a known exposure occurs or is suspected, immediately start the procedures given below and contact a Poison Center, physician or the nearest hospital.

NOTE: Be sure to advise the person contacted that the product is a weak cholinesterase inhibitor in experimental animals. Describe the type and extent of exposure, the victim's symptoms, and follow the advice given.

Note to medical personnel: If cholinesterase inhibition is suspected, atropine by injection is antidotal. It has not been determined if pralidoxime chloride (2-PAM, protopam) is antidotal for the product.

If swallowed: Immediately give several glasses of water but *do not* induce vomiting. If vomiting occurs, give fluids again. Have a physician determine if condition of patient will permit induction of vomiting or evacuation of stomach. Do not give anything by mouth to an unconscious or convulsing person.

If in eyes: Immediately flush eyes with large amounts of water for at least 15 minutes. Hold eyelids apart during flushing to ensure rinsing of the entire surface of the eye and lids with water. Get medical attention immediately.

If on skin: Flush affected areas with plenty of water for several minutes. Remove contaminated clothing and shoes. Get medical attention if irritation occurs.

If inhaled: Remove to fresh air. Seek medical attention if respiratory irritation occurs or breathing becomes difficult.

FOR 24-HOUR EMERGENCY MEDICAL ASSISTANCE CALL 1-800-F-A-S-T-M-E-D (327-8633).

FOR CHEMICAL EMERGENCY: Spill, leak, fire, exposure, or accident call CHEMTREC 1-800-424-9300.

PRECAUTIONARY STATEMENTS

HAZARDS TO HUMANS AND DOMESTIC ANIMALS

WARNING

MAY BE FATAL IF SWALLOWED. Avoid contact with skin, eyes and clothing. Wash with soap and water after use. Avoid breathing spray mist. Avoid contamination of feed and food. Repeated skin contact may cause cumulative effects.

ENVIRONMENTAL HAZARDS

This product is extremely toxic to fish. Do not apply directly to water except as indicated in the directions for use. Do not contaminate water by cleaning of equipment or disposal of wastes. Do not contaminate water used for domestic purposes.

GENERAL USE PRECAUTIONS

Read all label instructions before using.

ORDRAM 8-E should be used only for recommended purposes and at recommended rates. ORDRAM 8-E may cause crop injury under extreme soil or climatic conditions or if directions are not followed.

Do not apply any other herbicide in conjunction with ORDRAM 8-E as injury may occur.

ORDRAM 8-E may be combined with fluid (solution, slurry or suspension) fertilizers. However, physical compatibility with these fluid fertilizers must be determined before combining in the spray tank. See special literature giving directions for these combinations. Even though found to be compatible, constant agitation is necessary to keep the ORDRAM 8-E uniformly mixed with the fluid fertilizer.

DIRECTIONS FOR USE AND RECOMMENDATIONS

It is a violation of Federal law to use this product in a manner inconsistent with its labeling.

Do not apply this product through any type of irrigation system.

For use on rice to control watergrass (barnyardgrass); *Echinochloa* spp.

Pre-plant, Pre-Flood: (water-seeded rice only).

Soil Preparation: Prepare the soil for seeding according to good agricultural practices. All weed growth must be thoroughly worked into the soil before treatment.

For Ground Application Only: Use standard low-pressure (20-50 psi) boom-spray equipment. DO NOT USE SINGLE NOZZLE BOOM-JET TYPE SPRAYERS. Spray equipment should be carefully calibrated before use and checked frequently during application to be sure nozzles are free from clogging and are delivering a uniform spray pattern. Avoid overlaps that will increase ORDRAM 8-E dosage above recommended limits because crop injury may occur. Apply the recommended rate of ORDRAM 8-E in 20 to 50 gallons of water per acre.

Rate: Apply 3 pints ORDRAM 8-E per acre.

Soil Incorporation: Apply to dry soil in good tilth. ORDRAM 8-E must be incorporated immediately (within minutes) to prevent loss of the herbicide. Whenever possible, application and incorporation should be done in the same operation. Incorporation should be done by at least one discing or spike-tooth harrowing. Flood the field for seeding and hold the flood for 4 to 6 days following seeding.

Post-Flood, Post-Emergence: (water-seeded or drilled rice). Rice should be in the seedling stage at the time of application.

For Air Application Only: Watergrass must be a minimum of 2 inches high but not higher than 5 inches, and it must be at least ⅔ submerged at time of application. This water level (not less than 2 inches) must be maintained until the watergrass dies.

Equipment should be carefully calibrated before use and checked frequently during application to be sure application is uniform. Avoid overlaps that will increase ORDRAM 8-E dosage above recommended limits because crop injury may occur. Apply the recommended rate of ORDRAM 8-E in 5 to 10 gallons of water per acre. Aerial application should be made only when the wind velocity is low, 0 to 5 mph.

To avoid uneven distribution of ORDRAM 8-E, the water flow through the field should be at a minimum at the time of application and for at least two hours after application.

Hold flood water a minimum of 4 days after application.

Rate: Apply 3 pints ORDRAM 8-E per acre.

STORAGE AND DISPOSAL

Storage: Do not store near seeds and fertilizer. Keep container closed when not in use. Open dumping is prohibited. Do not reuse empty container.

Pesticide Disposal: Wastes resulting from the use of this product may be disposed of on site or at an approved waste disposal facility.

Container Disposal: Triple rinse (or equivalent). Then offer for recycling or reconditioning, or puncture and dispose of in a sanitary landfill, or by other procedures approved by State and local authorities.

NOTICE TO BUYER AND USER:

Seller warrants that this product conforms to the chemical description on the label and is reasonably fit for the purposes stated on the label when used in accordance with directions under normal conditions of use. This warranty does not extend to the use of this product contrary to label instructions, or under abnormal use conditions, or under conditions not reasonably foreseeable to Seller, and Buyer and User assume the risk of any such use. SELLER DISCLAIMS ALL OTHER WARRANTIES EXPRESSED OR IMPLIED INCLUDING ANY WARRANTY OF FITNESS OR MERCHANTABILITY. SELLER SHALL NOT BE LIABLE FOR CONSEQUENTIAL, SPECIAL OR INDIRECT DAMAGES RESULTING FROM THE USE OR HANDLING OF THIS PRODUCT AND SELLER'S SOLE LIABILITY AND BUYER'S AND USER'S EXCLUSIVE REMEDY SHALL BE LIMITED TO THE REFUND OF THE PURCHASE PRICE.

A-2R

The information and specimen labels in this booklet are correct as of the publication date but are subject to change without notice. For current information, contact ICI Americas Inc.

ORDRAM® 8-E

SELECTIVE HERBICIDE

Liquid Herbicide for Rice Grown in The Southern Region Only

ACTIVE INGREDIENT:	
S-ethyl hexahydro- 1 *H*-azepine-1 -carbothioate	90.9%
INERT INGREDIENTS:	9.1%
TOTAL	100.0%

Contains 8 Pounds of Active Ingredients Per Gallon

EPA Reg. No. 10182-204

KEEP OUT OF REACH OF CHILDREN

WARNING

May Be Fatal If Swallowed

STATEMENT OF PRACTICAL TREATMENT

CALL A POISON CENTER OR PHYSICIAN IMMEDIATELY

If a known exposure occurs or is suspected, immediately start the procedures given below and contact a Poison center, physician or the nearest hospital.

NOTE: Be sure to advice the person contacted that the product is a weak cholinesterase inhibitor in experimental animals. Describe the type and extent of exposure, the victim's symptoms, and follow the advice given.

Note to medical personnel: If cholinesterase inhibition is suspected, atropine by injection is antidotal. It has not been determined if pralidoxime chloride (2-PAM, protopam) is antidotal for the product.

If swallowed: Immediately give several glasses of water but *do not* induce vomiting. If vomiting occurs, give fluids again. Have a physician determine if condition of patient will permit induction of vomiting or evacuation of stomach. Do not give anything by mouth to an unconscious or convulsing person.

If in eyes: Immediately flush eyes with large amounts of water for at least 15 minutes. Hold eyelids apart during flushing to ensure rinsing of the entire surface of the eye and lids with water. Get medical attention immediately.

On the skin: Flush affected areas with plenty of water for several minutes. Remove contaminated clothing and shoes. Get medical attention if irritation occurs.

If inhaled: Remove to fresh air. Seek medical attention if respiratory irritation occurs or breathing becomes difficult.

FOR 24-HOUR EMERGENCY MEDICAL ASSISTANCE CALL 1-800-F-A-S-T-M-E-D (327-8633).
FOR CHEMICAL EMERGENCY: Spill, leak, fire, exposure, or accident call CHEMTREC 1-800-424-9300.

PRECAUTIONARY STATEMENTS

HAZARDS TO HUMANS AND DOMESTIC ANIMALS

WARNING

May Be Fatal If Swallowed: Avoid contact with skin, eyes and clothing. Wash with soap and water after use. Avoid breathing spray mist. Avoid contamination of feed and food. Repeated skin contact may cause cumulative effects.

ENVIRONMENTAL HAZARDS

This product is extremely toxic to fish. Do not apply directly to water except as indicated in the directions for use. Do not contaminate water by cleaning of equipment or disposal of wastes. Do not contaminate water used for domestic purposes.

GENERAL USE PRECAUTIONS

Read all label directions before using. Do not overdose.

ORDRAM 8-E should be used only for recommended purposes and at recommended rates. ORDRAM 8-E may cause crop injury under extreme soil or climatic conditions or if directions are not followed.

Except as specified on this label, do not apply other herbicides in conjunction with ORDRAM 8-E as injury may occur.

ORDRAM 8-E may be combined with fluid (solution, slurry, or suspension) fertilizers. However, physical compatibility with these fluid fertilizers must be determined before combining in the spray tank. See special literature giving directions for these combinations. Even though found to be compatible, constant agitation is necessary to keep the ORDRAM 8-E uniformty mixed with the fluid fertilizer.

Allow 30 days between pre-plant and post-flood treatments.

Allow 60 days between the last application and harvest.

GENERAL APPLICATION DIRECTIONS

All equipment used in the application of ORDRAM 8-E should be carefully calibrated before use and checked frequently during application to be sure the application is uniform. Avoid overlaps that will increase ORDRAM 8-E dosage above recommended limits because crop injury may occur. Apply ORDRAM 8-E in 10 to 50 gallons of water by ground equipment and 5 to 10 gallons of water by airplane except as specified below.

To avoid uneven distribution of ORDRAM 8-E in flooded fields, the water flow through the field should be at a minimum at the time of application and for at least 2 hours after application.

Do not apply more than 6 pints ORDRAM 8-E per acre per season where more than one method of application is used. Allow 30 days between applications.

For Bulk and Mini-Bulk Containers

This Label Contains Directions for Use on Rice Grown in The Southern Region Only (AR, LA, MS, MO and TX).

DIRECTIONS FOR USE AND RECOMMENDATIONS

It is a violation of Federal law to use this product in a manner inconsistent with its labeling.

Pre-Plant, Pre-Flood: For use on water-seeded rice ONLY

For control of Barnyardgrass (Watergrass) *Echinochloa* spp. ONLY.

Soil Preparation: Prepare the soil for seeding according to good agricultural practices. All weed growth must be thoroughly worked into the soil before treatment.

For Ground Application Only: Use standard low pressure (20-50 psi) boom-spray equipment. DO NOT USE SINGLE NOZZLE BOOM-JET TYPE SPRAYERS.

Application and Incorporation: Apply to dry soil in good tilth. ORDRAM 8-E must be incorporated immediately (within minutes) to prevent loss of the herbicide. Whenever possible, application and incorporation should be done in the same operation. Incorporation should be done by at least one discing or spiketooth harrowing. Flood the field for seeding and hold the flood for 4 to 6 days following seeding.

Recommended Rate: Apply 3 pints of ORDRAM 8-E per acre.

PPI Ground Applications: For use on dry-seeded rice only.

For partial control of early season grasses (barnyardgrass, sprangletop, broadleaf signalgrass and crabgrass) and nutsedge (*Cyperus* spp.)

On a thoroughly clean and prepared seed bed, apply 3-4 pints of ORDRAM 8-E. Application must be made to a dry soil surface. Use 3 pints on light soil, 3-4 pints on medium soil, and 4 pints on heavy soil. Use ground application equipment only. Incorporate immediately to a depth of 1-2″. Whenever possible, application and incorporation should be done in the same operation. When applied in fluid fertilizer, incorporation can be delayed up to 2 hours. Early flushing or flooding may improve the control received by this treatment method.

Caution: Other control methods may be needed for escaped weeds or for late season infestations.

Post-Emergence, Pre-Flood: For use on dry-seeded and water-seeded rice.

For control of Barnyardgrass, Sprangletop, Dayflower and Broadleaf Signalgrass. Apply ORDRAM 8-E as a tank mix combination with a registered propanil brand in 10-15 gallons of water per acre by airplane or in 20-50 gallons of water per acre by ground equipment using the rates below. Flood the rice within 5 days after application to prevent weed reinfestation. Follow all directions and use precautions on the propanil label. Avoid drift or accidental application to all crops, shrubs or ornamentals not in the rice field being treated.

Weeds	Maximum-Leaf Stage	5 P 8-E (Pints)	Propanil (4-E) (Pints)
Barnyardgrass (*Echinochola* spp.)	3	3	6
Sprangletop (emerged less than 1 in) (*Leptochloa* spp.)	—	3	6
Dayflower, common (*Commelina communis*)	2	2-3	6
Signalgrass broadleaf (*Brachiaria platyphylla*)	3	3	6

Post-Emergence at Flooding: For use on dry-seeded rice ONLY

For control of Barnyardgrass (Watergrass) and Dayflower under 5 inches tall.

USE PRECAUTIONS

A. Apply this product only through flood (basin) irrigation system. Do not apply this product through any other type of irrigation system.

B. Crop injury, lack of effectiveness, or illegal pesticide residues in the crop can result from nonuniform distribution of treated water.

C. If you have questions about calibration you should contact State Extension Service specialists equipment manufacturers or other experts.

D. Do not connect an irrigation system (including greenhouse systems) used for pesticide application to a public water system unless the pesticide label-prescribed safety devices for public water systems are in place.

E. A person knowledgeable of the chemigation system and responsible for its operation, or under the supervision of the responsible person, shall shut the system down and make necessary adjustments should the need arise.

FOR FLOOD IRRIGATION

1. Systems using a gravity flow pesticide dispensing system must meter the pesticide into the water at the head of the field and downstream of a hydraulic discontinuity such as a drop structure or weir box to decrease potential for water source contamination from backflow if water flow stops.
2. Systems utilizing a pressurized water and pesticide injection system must meet the following requirements:
   a. The system must contain a functional check valve, vacuum relief valve, and low pressure drain appropriately located on the irrigation pipeline to prevent water source contamination from backflow.
   b. The pesticide injection pipeline must contain a functional, automatic, quick-closing check valve to prevent the flow of fluid back toward the injection pump.
   c. The pesticide injection pipeline must also contain a functional, normally closed, solenoid-operated valve located on the intake side of the injection pump and connected to the system interlock to prevent fluid from being withdrawn from the supply tank when the irrigation system is either automatically or manually shut down.
   d. The system must contain functional interlocking controls to automatically shut off the pesticide injection pump when the water pump motor stops.
   e. The irrigation line or water pump must include a functional pressure switch which will stop the water pump motor when the water pressure decreases to the point where pesticide distribution is adversely affected.
   f. Systems must use a metering pump, such as a positive displacement injection pump (e.g. diaphragm pump) effectively designed and constructed of materials that are compatible with pesticides and capable of being fitted with a system interlock.

When establishing the permanent flood, meter the recommended rate of ORDRAM 8-E per acre into the irrigation water at the header ditch or canal where water enters the treated field. Meter the ORDRAM 8-E during the entire irrigation period using a constant head metering device.

Recommended Rate: 3 pints ORDRAM 8-E per acre.

The rice must be established at the time of application. Dayflower must be completely submerged and Barnyardgrass 2/3 submerged by the flooding operation. Flood should be maintained at the prescribed depth until Barnyardgrass and Dayflower are controlled.

Post-Emergence, Post-Flood: For use on dry-seeded and water-seeded rice.

For control of Barnyardgrass (Watergrass) ONLY.

Application: Apply the recommended rate directly to the flooded field. Apply by airplane ONLY. Apply only when wind velocity is low, 0 to 5 mph.

Recommended Rate: 3 pints ORDRAM 8-E per acre.

The Barnyardgrass must be a minimum of 2 inches high but not higher than 5 inches, and it must be at least 2/3 submerged at time of applications. This water level (not less than 2 inches) must be maintained until the Barnyardgrass dies. Applications should be made before rice is 2 inches above surface of water.

STORAGE AND DISPOSAL

Prohibitions: Do not contaminate water, food or feed by storage or disposal: Open dumping is prohibited.

Storage: Do not store near seeds and fertilizer. Keep container closed when not in use.

Pesticide Disposal: Wastes resulting from the use of this product may be disposed of on site or at an approved waste disposal facility.

Container Disposal: Reseal container and offer for reconditioning, or triple rinse (or equivalent) and offer for recycling or reconditioning or clean in accordance with manufacturers instructions.

FOR MINI-BULK CONTAINERS

Container Precautions: Before refilling, inspect thoroughly for damage, such as cracks, punctures, bulges, dents, abrasions or worn threads on closure devices.

REFILL ONLY WITH ORDRAM 8-E. The contents of this container cannot be completely removed by cleaning. Refilling with materials other than ORDRAM 8-E will result in contamination and may weaken container.

After filling and before transporting, check for leaks.

Do not refill or transport damaged or leaking container.

CONTAINER IS NOT SAFE FOR FOOD, FEED OR DRINKING WATER!

NOTICE TO BUYER AND USER:

Seller warrants that this product conforms to the chemical description on the label and is reasonably fit for the purposes stated on the label when used in accordance with directions under normal conditions of use. This warranty does not extend to the use of this product contrary to label instructions, or under abnormal use conditions, or under conditions not reasonably foreseeable to Seller, and Buyer and User assume the risk of any such use. SELLER DISCLAIMS ALL OTHER WARRANTIES EXPRESSED OR IMPLIED INCLUDING ANY WARRANTY OF FITNESS OR MERCHANTABILITY. SELLER SHALL NOT BE LIABLE FOR CONSEQUENTIAL, SPECIAL OR INDIRECT DAMAGES RESULTING FROM THE USE OR HANDLING OF THIS PRODUCT AND SELLER'S SOLE LIABILITY AND BUYER'S AND USER'S EXCLUSIVE REMEDY SHALL BE LIMITED TO THE REFUND OF THE PURCHASE PRICE.

This product is sold only for uses stated on its label. No express or implied license is granted to use or sell this product under any patent in any country except as specified: Country: *United States of America.* Patent No.: *3,573,031*

A-7R

The information and specimen labels in this booklet are correct as of the publication date but are subject to change without notice. For current information, contact ICI Americas Inc.

ORDRAM® 8-E

Supplemental Directions for Use

FOR DISTRIBUTION AND USE IN THE SOUTHWEST ONLY

EPA Reg. No. 10182-204

DIRECTIONS FOR USE

It is a violation of Federal law to use this product in a manner inconsistent with its labeling.

Pre-Plant, Incorporated Ground Application For Use on Dry-Seeded Rice Only

For partial control of early season grasses (watergrass [barnyardgrass], sprangletop, broadleaf signalgrass and crabgrass) and nutsedge (**Cyperus** spp.)

On a thoroughly clean and prepared seed bed, apply 3-4 pints of ORDRAM 8-E. Application must be made to a dry soil surface. Use 3 pints of light soil, 3-4 pints on medium soil, and 4 pints on heavy soil. Use ground application equipment only. Incorporate immediately to a depth of 1-2". Whenever possible, application and incorporation should be done in the same operation. When applied in fluid fertilizer, incorporation can be delayed up to 2 hours. Early flushing or flooding may improve the control received by this treatment method.

Caution: Other control methods may be needed for escaped weeds or for late season infestations.

All applicable directions, restrictions, and precautions on the EPA-registered label are to be followed.

This labeling must be in the possession of the user at the time of pesticide application.

NOTICE TO BUYER AND USER:

Seller warrants that this product conforms to the chemical description on the label and is reasonably fit for the purposes stated on the label when used in accordance with directions under normal conditions of use. This warranty does not extend to the use of this product contrary to label in-

structions, or under abnormal use conditions, or under conditions not reasonably foreseeable to Seller, and Buyer and User assume the risk of any such use. SELLER DISCLAIMS ALL OTHER WARRANTIES EXPRESSED OR IMPLIED INCLUDING ANY WARRANTY OF FITNESS OR MERCHANTABILITY. SELLER SHALL NOT BE LIABLE FOR CONSEQUENTIAL, SPECIAL OR INDIRECT DAMAGES RESULTING FROM THE USE OR HANDLING OF THIS PRODUCT AND SELLER'S SOLE LIABILITY AND BUYER'S AND USER'S EXCLUSIVE REMEDY SHALL BE LIMITED TO THE REFUND OF THE PURCHASE PRICE.

A-2

The information and specimen labels in this booklet are correct as of the publication date but are subject to change without notice. For current information, contact ICI Americas Inc.

Ordram® 8-E

Supplemental Directions for Use

FOR DISTRIBUTION AND USE IN THE SOUTHWEST ONLY

EPA Reg. No. 10182–204

DIRECTIONS FOR USE

It is a violation of Federal law to use this product in a manner inconsistent with its labeling.

Pre-Plant, Incorporated Ground Application For Use on Dry-Seeded Rice Only

For partial control of early season grasses (watergrass [barnyardgrass], sprangletop, broadleaf signalgrass and crabgrass) and nutsedge (**Cyperus** spp.)

On a thoroughly clean and prepared seed bed, apply 3-4 pints of ORDRAM 8-E. Application must be made to a dry soil surface. Use 3 pints on light soil, 3-4 pints on medium soil, and 4 pints on heavy soil. Use ground application equipment only. Incorporate immediately to a depth of 1–2". Whenever possible, application and incorporation should be done in the same operation. When applied in fluid fertilizer, incorporation can be delayed up to 2 hours. Early flushing or flooding may improve the control received by this treatment method.

Caution: Other control methods may be needed for escaped weeds or for late season infestations.

All applicable directions, restrictions, and precautions on the EPA-registered label are to be followed.

This labeling must be in the possession of the user at the time of pesticide application.

NOTICE TO BUYER AND USER:

Seller warrants that this product conforms to the chemical description on the label and is reasonably fit for the purposes stated on the label when used in accordance with directions under normal conditions of use. This warranty does not extend to the use of this product contrary to label instructions, or under abnormal use conditions, or under conditions not reasonably foreseeable to Seller, and Buyer and User assume the risk of any such use. SELLER DISCLAIMS ALL OTHER WARRANTIES EXPRESSED OR IMPLIED INCLUDING ANY WARRANTY OF FITNESS OR MERCHANTABILITY. SELLER SHALL NOT BE LIABLE FOR CONSEQUENTIAL, SPECIAL OR INDIRECT DAMAGES RESULTING FROM THE USE OR HANDLING OF THIS PRODUCT AND SELLER'S SOLE LIABILITY AND BUYER'S AND USER'S EXCLUSIVE REMEDY SHALL BE LIMITED TO THE REFUND OF THE PURCHASE PRICE.

A-2

The information and specimen labels in this booklet are correct as of the publication date but are subject to change without notice. For current information, contact ICI Americas Inc.

PREFAR® 4-E

Emulsifiable Liquid Selective Herbicide for Control of Annual Grasses and Broadleaf Weeds

ACTIVE INGREDIENT:	
Bensulide: S-(O,O)-diisopropyl phosphorodithioate) ester of N-(2-mercaptoethyl) benzenesulfonamide	46.0%
INERT INGREDIENTS:	54.0%
TOTAL	100.0%

Contains 4 pounds of active ingredient per gallon.

EPA Reg. No. 10182-184

KEEP OUT OF REACH OF CHILDREN

CAUTION

STATEMENT OF PRACTICAL TREATMENT

If a known exposure occurs or is suspected, immediately start the procedures given below and simultaneously contact a Poison Center, a physician, or the nearest hospital.

Note: Be sure to advise the person contacted that the active ingredient in this compound is a cholinesterase inhibitor. Describe the type of exposure, the victim's symptoms, and follow the advice given.

Note to Medical Personnel: Should cholinesterase inhibition occur, atropine by injection is antidotal. 2-PAM (Protopam Chloride) is also antidotal when administered early and in conjunction with atropine.

If Swallowed: Immediately give several glasses of water but *do not* induce vomiting. If vomiting does occur, give fluids again. Have a physician determine if condition of patient will permit induction of vomiting or evacuation of stomach. Do not give anything by mouth to an unconscious or convulsing person.

For Skin Contact: Flush all affected areas with plenty of water for several minutes. Remove and clean contaminated clothing and shoes. Seek medical attention if irritation occurs.

For Eye Contact: Flush eyes with large amounts of running water for at least 15 minutes. Hold eyelids apart to ensure rinsing of the entire surface of the eye and lids with water. Get medical attention immediately.

If Inhaled: Remove to fresh air. Seek medical attention if respiratory irritation occurs or if breathing becomes difficult.

FOR 24-HOUR EMERGENCY MEDICAL ASSISTANCE CALL 1-800-F-A-S-T-M-E-D (327-8633).
FOR CHEMICAL EMERGENCY: Spill, leak, fire, exposure, or accident call CHEMTREC 1-800-424-9300.

READ ALL LABEL DIRECTIONS BEFORE USING

PRECAUTIONARY STATEMENTS

HAZARDS TO HUMANS AND DOMESTIC ANIMALS

CAUTION

Harmful if swallowed or absorbed through the skin. Avoid breathing spray mist. Causes moderate eye irritation. Do not get in eyes, on skin, or on clothing. Wash thoroughly after handling and before eating or smoking. Wash all contaminated clothing with soap and hot water before reuse.

ENVIRONMENTAL HAZARDS

This product is toxic to fish. Do not apply directly to any body of water. Do not contaminate water by cleaning of equipment or disposal of wastes. Do not apply when weather conditions favor drift from area treated.

Physical Hazards: Combustible. Do not use, store, pour, or spill near heat or open flame. Keep container closed when not in use.

GENERAL USE PRECAUTIONS

READ ALL DIRECTIONS BEFORE USING. MIX WELL BEFORE DILUTING WITH WATER. PREFAR 4-E should be used only for recommended purposes and at recommended rates. (DO NOT OVERDOSE) Do not apply in combination with fluid fertilizers. Use on mineral soils only. Do not contaminate irrigation water.

All crops specified on the label may be planted in rotation but do not apply PREFAR 4-E more often than once every 12 months. To avoid injury to crops not specified on the label do not plant until 18 months after the last PREFAR 4-E application. Soybeans may be planted 12 months after PREFAR 4-E application.

Weeds Controlled

Annual Grasses	Annual Broadleaf Weeds
Crabgrass *(Digitaria spp.)* Foxtail *(Setaria spp.)* Jungle Rice *(Echinochloa colonum)* Watergrass *(Echinochloa crusgalli)* Goosegrass *(Eleusine indica)* Fall Panicum *(Panicum dichotomiflorum)* Sprangletop *(Leptochloa spp.)*	(Arizona, New Mexico, Texas and ten southernmost California counties only) Lambsquarters *(Chenopodium album)* Redroot Pigweed *(Amaranthus retroflexus)* Purslane *(Portulaca oleracea)*

DIRECTIONS FOR USE

It is a violation of Federal law to use this product in a manner inconsistent with its labeling. Do not apply this product through any type of irrigation system.

PREFAR 4-E is a selective herbicide which is applied pre-plant and mixed (incorporated) into the soil. On certain crops which are irrigated up, PREFAR 4-E may be applied to the soil surface after planting. PREFAR 4-E controls weeds as their seeds germinate by interfering with normal germination and seedling development.

Soil Preparation: Prepare the soil according to good agricultural practice. PREFAR 4-E does not control emerged weeds and all weed growth and crop stubble must be thoroughly worked into the soil prior to application.

Application and Soil Incorporation: Spray equipment should be carefully calibrated before use and checked frequently during application to be sure nozzles are free from clogging and are delivering a uniform spray pattern. Apply uniformly the recommended rate of PREFAR 4-E in 10 to 50 gallons of water per acre.

Pre-plant Application: Apply to well-worked soil that is dry enough to permit thorough incorporation. Incorporate to a depth of 1 to 2 inches before planting. Use power driven rotary cultivators or other equipment which will result in satisfactory incorporation.

Pre-emergence Application: (For use only on crops that will be irrigated up). After planting apply PREFAR 4-E as a surface application. Follow immediately with irrigation (sprinkler or furrow). For sprinkler irrigation apply 1 inch of water, wetting the soil at least 2 to 4 inches deep. For furrow irrigation thoroughly saturate the entire bed top (completely black out bed top).

RECOMMENDATIONS

All rates are given on a broadcast basis. Reduce rate proportionately for band treatment.

Cotton

Pre-plant: Apply 5 to 6 quarts PREFAR 4-E per broadcast acre. Apply to the flat soil surface before furrowing (listing) for the pre-planting irrigation and then continue with normal practices.

or

Apply after furrowing (listing) but before the pre-planting irrigation. Incorporate with a sectional, rolling cultivator.

or

Apply after the pre-planting irrigation but before the final seed bed preparation.

Pre-plant: (Arizona, New Mexico & Texas only) For control of watergrass, jungle rice, foxtail and crabgrass only, apply 2 quarts PREFAR 4-E per broadcast acre. Use only on light, sandy soils.

Apply to the flat soil surface and thoroughly disc the PREFAR 4-E into the soil before furrowing (listing) for the pre-planting irrigation and then continue with normal practices.

or

Apply after furrowing (listing). Incorporate with a sectioned, rolling cultivator. Follow with pre-plant irrigation.

Pre-emergence: (Use only on cotton that will be irrigated up). Apply 5 to 6 quarts PREFAR 4-E per broadcast acre as a surface application.

Cucurbits (cucumbers, summer squash, winter squash, cantaloupes, crenshaw melons, muskmelons, persian melons, pumpkins, watermelons).

Apply, pre-plant, 5 to 6 quarts PREFAR 4-E per broadcast acre to flat-planted or bedded crops. Follow directions for APPLICATION AND SOIL INCORPORATION given on this label.

Cole Crops (broccoli, brussels sprouts, cabbage, cauliflower).

Apply, pre-plant or pre-emergence, 5 to 6 quarts PREFAR 4-E per broadcast acre. Follow directions for APPLICATION AND SOIL INCORPORATION given on this label. Use pre-emergence only on cole crops to be irrigated up.

(**NOTE:** Limit use on cabbage to Southern and California/Arizona growing regions.)

Fruiting Vegetables (tomatoes [direct seeded and transplants], seeded bell peppers, chili peppers).

Apply, pre-plant or pre-emergence, 5 to 6 quarts (4 to 5 quarts for tomatoes) PREFAR 4-E per broadcast acre. Follow directions for APPLICATION AND SOIL INCORPORATION given on this label. Use pre-emergence only on crops to be irrigated up.

(**NOTE:** Limit use on bell peppers to the Southern and California/Arizona growing regions; on chili peppers to the Southwestern U.S.)

Lettuce (head or leaf)

Apply, pre-plant or pre-emergence, 5 to 6 quarts PREFAR 4-E per broadcast acre. Follow directions for APPLICATION AND SOIL INCORPORATION given on this label. Use pre-emergence only on lettuce to be irrigated up.

Root Crop Vegetables

(**NOTE:** Limit use on carrots to Texas; on bulb onions to Idaho, New Mexico, Oregon, and Texas. Use on onions in Idaho and Oregon is restricted to fall application followed by spring planting).

Carrots (Texas Only): Apply pre-plant or pre-emergence, 5 to 6 quarts PREFAR 4-E per broadcast acre. Follow directions for APPLICATION AND SOIL INCORPORATION given on this label. Use pre-emergence only on carrots to be irrigated up. Do not feed treated carrots to livestock.

Bulb Onions (New Mexico and Texas): Apply pre-plant or pre-emergence, 5 to 6 quarts PREFAR 4-E per broadcast acre. Follow directions for APPLICATION AND SOIL INCORPORATION given on this label. Use pre-emergence on onions to be irrigated up.

Bulb Onions (Idaha and Oregon): For control of watergrass, redroot pigweed and common lambsquarters, prepare a good, flat seed bed by standard cultural practices. Apply in the fall 3 quarts of PREFAR 4-E per acre in 10 to 50 gallons of water in a 10 to 12 inch band and bed-up, throwing soil from the furrows over the sprayed band of PREFAR 4-E. Allow to remain undistributed until spring. Prior to planting in spring, drag-off the bed tops with a spike-tooth harrow being careful not to drag off soil below the level of PREFAR 4-E application. Plant in the center of the beds.

Grass Seed Crops (bentgrass, bluegrass)

For *Poa annua* control in bentgrass and bluegrass seed fields: Apply 2 to 2½ gallons PREFAR 4-E per acre to the soil surface after straw removal but before weed seed germination. If field has been burned, ashes should be completely dispersed before application of PREFAR 4-E.

Do not feed straw to livestock. Do not graze livestock in treated fields.

(**NOTE:** Limit use to Oregon, Washington and Idaho.)

PREFAR 4-E/ALANAP* Tank Mix Directions

*ALANAP is a registered trademark of Uniroyal Chemical

A tank mix combination of PREFAR 4-E and ALANAP can be mixed (incorporated) into the soil before planting for weed control in cantaloupes, cucumbers, muskmelons and watermelons. This combination will control a broader spectrum of weeds than will either material used alone.

IMPORTANT—Follow all general directions for use and use precautions that appear on the ALANAP label and elsewhere on this label.

Weeds controlled by PREFAR 4-E/ALANAP combinations:

Annual Grasses:	Annual Broadleaf Weeds:
Watergrass (Barnyardgrass)	Bindweed
Crabgrass	Carpetweed
Fall Panicum	Chickweed
Foxtails	Cocklebur
Goosegrass	Galinsoga
Sprangletop	Groundcherry
	Lambsquarters
	Mustards
	Pigweed
	Purslane
	Ragweed
	Shepherdspurse
	Velvetleaf

DIRECTIONS FOR USE

Soil Preparation: Prepare the soil according to good agricultural practice. All weed growth and crop stubble must be thoroughly worked into the soil.

Tank Mixing Directions: Combine PREFAR 4-E and ALANAP in the spray tank in the recommended dosage proportions using 10 to 40 gallons of water per acre of area to be treated.

Spray equipment should be carefully calibrated before use and checked frequently during application to be sure nozzles are free from clogging and are delivering a uniform spray pattern.

Pre-plant Application and Soil Incorporation: Apply 4 to 6 quarts per acre of PREFAR 4-E plus 4 to 8 quarts of ALANAP in sufficient water (10 to 40 gallons) to provide thorough coverage. Incorporate lightly (½ to 1 inch) into the soil using suitable equipment just prior to planting. Use lower rates on light (sandy) soils and higher rates on heavier (clay) soils. Reduce the rate proportionately for band applications. Incorporation is not necessary if the application is followed by immediate irrigation.

STORAGE AND DISPOSAL

•Do not contaminate water, food, or feed by storage or disposal.

Storage: Protect from temperatures below 42°F. At temperatures below 42°F, product may crystalize. Dissolve crystals with quiet storage at warm temperatures (75°F). Redissolution rate: 40°-48°F (40 to 60 days); or 60°F (1 to 3 weeks); 70°-80°F (1-3 days). Roll container and agitate contents to provide a uniform product prior to use.

Pesticide Disposal: Wastes resulting from the use of this product may be disposed of on site or at an approved waste disposal facility.

Container Disposal: Triple rinse (or equivalent). Then offer for recycling or reconditioning, or puncture and dispose of in a sanitary landfill, or by other procedures approved by State and local authorities.

Spill or Leak: Small spills can be handled routinely. Use adequate ventilation and wear an air-supplied respirator to prevent inhalation. Wear suitable protective clothing to prevent skin and eye contact. Use the following procedures:

1. Spread a suitable absorbent such as clay, sawdust or kitty litter on the spilled material.
2. Place sweepings in an appropriate chemical waste container.
3. Generously cover the contaminated areas with a common household detergent. Using a stiff brush and small amounts of water, work the detergent into the spill material forming a slurry. Do not splatter on one's self or bystanders and completely avoid skin or eye contact with this material. Brush the slurry into cracks and crevices and allow to stand for 2-3 minutes.
4. Spread absorbents on the slurried liquid and shovel the absorbed material into the container.
5. Repeat if necessary.
6. Flush the area with water while observing proper environmental considerations.
7. Seal container and dispose of in approved pesticide landfill.

NOTICE TO BUYER AND USER:

Seller warrants that this product conforms to the chemical description on the label and is reasonably fit for the purposes stated on the label when used in accordance with directions under normal conditions of use. This warranty does not extend to the use of this product contrary to label instructions, or under abnormal use conditions, or under conditions not reasonably foreseeable to Seller, and Buyer and User assume the risk of any such use. SELLER DISCLAIMS ALL OTHER WARRANTIES EXPRESSED OR IMPLIED INCLUDING ANY WARRANTY OF FITNESS OR MERCHANTABILITY. SELLER SHALL NOT BE LIABLE FOR CONSEQUENTIAL, SPECIAL OR INDIRECT DAMAGES RESULTING FROM THE USE OR HANDLING OF THIS PRODUCT AND SELLER'S SOLE LIABILITY AND BUYER'S AND USER'S EXCLUSIVE REMEDY SHALL BE LIMITED TO THE REFUND OF THE PURCHASE PRICE.

A-4R

The information and specimen labels in this booklet are correct as of the publication date but are subject to change without notice. For current information, contact ICI Americas Inc.

REFLEX® 2LC

HERBICIDE

For Postemergence Control of Broadleaf Weeds in Soybeans

ACTIVE INGREDIENT:
Sodium salt of fomesafen
5-[2-chloro-4-(trifluoromethyl)phenoxy]-*N*-(methylsulfonyl)-2-nitrobenzamide 22.8%*
INERT INGREDIENTS: 77.2%
TOTAL 100.0%

*Equivalent to 21.7% fomesafen or 2 pounds fomesafen active ingredient per gallon.

EPA Reg. No. 10182–83

KEEP OUT OF REACH OF CHILDREN

WARNING

STATEMENT OF PRACTICAL TREATMENT

IF IN EYES Immediately flush eyes with plenty of water for at least 15 minutes. Get medical attention if irritation persists.

IF SWALLOWED Drink 1 or 2 glasses of water and incude vomiting by touching back of throat with finger. Do not induce vomiting or give anything by mouth to an unconcious person. Call a physician or Poison Control Center.

IF ON SKIN Immediately remove contaminated clothing and wash thoroughly with plenty of soap and water. Get medical attention if irritation persists.

FOR 24-HOUR EMERGENCY MEDICAL ASSISTANCE CALL 1-800-F-A-S-T-M-E-D (327-8633).

FOR CHEMICAL EMERGENCY: Spill, leak, fire, exposure, or accident call CHEMTREC 1-800-424-9300.

PRECAUTIONARY STATEMENTS

HAZARDS TO HUMANS AND DOMESTIC ANIMALS

WARNING

This product contains fomesafen which has been determined to cause tumors in laboratory animals (mice). Risks can be reduced by closely following use directions and precautions and by wearing the protective clothing specified elsewhere on this label.

CAUSES SUBSTANTIAL BUT TEMPORARY EYE INJURY. HARMFUL IF SWALLOWED, INHALED OR ABSORBED THROUGH SKIN. Do not get in eyes, on skin or on clothing. Wear protective clothing, goggles or full face shield and rubber gloves when handling or mixing concentrate. Wear protective clothing and rubber gloves when spraying. Avoid breathing vapors or spray mist. Wash thoroughly with soap and water after handling and before eating or smoking. Remove contaminated clothing and wash before reuse.

ENVIRONMENTAL HAZARDS

Do not apply directly to water. Do not contaminate water by cleaning of equipment or disposal of waste. Do not apply when weather conditions favor drift from target area.

DIRECTIONS FOR USE

It is a violation of Federal law to use this product in a manner inconsistent with its labeling.

GENERAL INFORMATION

Read all label directions before using.

REFLEX 2LC is a selective postemergence herbicide for control of broadleaf weeds in soybeans. Soybean plants are tolerant to REFLEX 2LC when applied at recommended rates. Some bronzing, crinkling or spotting of soybean leaves may occur, but soybeans soon outgrow these effects and develop normally with no yield reduction.

Thorough coverage of all weed plant foliage is important for good activity. Optimum weed control is achieved when young actively growing weeds are treated that are not under stress from moisture, temperature, low soil fertility, mechanical or chemical injury.

REFLEX 2LC is effective through contact action, therefore weeds must be thoroughly covered with spray.

Secondary flushes of certain weeds may be controlled by soil residual activity if rainfall occurs within one week after application. The extent of soil activity is dependent upon soil type, ground cover at time of application, amount of rainfall and rate of REFLEX 2LC used.

REFLEX 2LC is rainfast in 4 hours.

APPLICATION DIRECTIONS

Timing—Best control of susceptible broadleaf weeds is obtained when REFLEX 2LC is applied early to actively growing weeds. This usually occurs 14 to 21 days after planting. Refer to the weed tables for specific recommendation on weed growth stages, rates, and regions.

Spray Additives—Only crop oil concentrate and nonionic surfactants cleared for use on growing crops under 40 CFR 180.1001 may be used in spray mixture.

Always Add One of the Following:

- **Nonionic Surfactant**—Add nonionic surfactant containing at least 75% surface active agent at 0.25 to 0.5% (1/2 to 1 pint per 25 gallons) of the finished spray volume to improve contact activity.
- **Crop Oil Concentrate**—Add a nonphytotoxic crop oil concentrate, containing 15%-20% approved emulsifier, at 0.5 to 1% (1–2 pints per 25 gallons) of the finished spray volume. Crop oil concentrate can improve weed control but may slightly reduce crop tolerance.

In addition to crop oil concentrate or nonionic surfactant, liquid nitrogen fertilizer (28% or similar) can be added to the spray mixture. This 28% liquid nitrogen fertilizer is water soluble and should be used at a rate of one gallon per acre. Liquid nitrogen fertilizers should not be used as a substitute for crop oil concentrate or nonionic surfactant in the spray mixture.

In addition to crop oil concentrate or nonionic surfactant, diammonium phosphate (aqueous ammonium polyphosphate) commonly sold as a solution (10-34-0) can be added to the spray mixture. This completely water soluble material should be used at a rate of 2 pts. per acre. Diammonium phosphate should not be used as a substitute for crop oil concentrate or nonionic surfactant in the spray mixture.

MIXING—Fill the clean spray tank with 1/2 the required amount of water. Begin agitation and add the recommended amount of REFLEX 2LC herbicide and appropriate amount of approved adjuvant. Add the remaining quantity of water and allow sprayer to agitate and recycle 5-10 minutes before application.

Ground Application—Use sufficient spray volume and pressure to ensure complete coverage of the target weeds. A minimim of 10 gallons per acre of spray mixture should be used with spray pressures of 40 to 60 psi at the nozzle tip. When weed foliage is dense, use 60 psi and a minimum of 20 gallons per acre to ensure coverage of weed foliage.

Use only hollow cone or flat fan nozzles. The sprayer must be calibrated to provide the proper volume and rate per acre. In addition, the boom and nozzle height must be adjusted to provide complete coverage of all weeds. DO NOT USE FLOOD TYPE OR OTHER SPRAY NOZZLES, WHICH DELIVER COARSE, LARGE DROPLET SPRAYS.

DO NOT APPLY THIS PRODUCT THROUGH ANY TYPE OF IRRIGATION SYSTEM.

Band Applications—Thorough weed coverage is important for control. Best coverage is obtained with a minimum of two nozzles, one directed to each side of the planted row. Application with a single nozzle directed over the top of the row is not recommended. Cultivation of untreated areas may be needed following band applications. When making band applications and cultivating in the same operation, position nozzles in advance of the cultivation device. This will reduce dust in the spray area. Dust can intercept spray, reducing weed coverage, resulting in less than adequate weed control.

Calculate the amount of herbicide and water volume needed for band treatment by the following formulas:

$$\frac{\text{Band width in inches}}{\text{row width in inches}} \times \text{broadcast rate per acre} = \text{Band herbicide rate per acre}$$

$$\frac{\text{Band width in inches}}{\text{row width in inches}} \times \text{broadcast volume per acre} = \text{Band water volume per acre}$$

Cultivation—Cultivation within 7 days prior to application is not recommended. Cultivation may put weeds under stress, reducing weed control. Timely cultivation 2-3 weeks after applying REFLEX 2LC may often assist in weed control.

GENERAL USE PRECAUTIONS

- REFLEX 2LC can be applied only in the states or parts of states included in Regions 1, 2, 3 and 4. DO NOT APPLY REFLEX 2LC TO ANY FIELD IN REGIONS 2, 3 or 4 MORE THAN ONCE EVERY TWO YEARS.
- Applications of REFLEX 2LC should not be made beyond 3 weeks after the soybeans have emerged.
- A maximum of 1½ pints of REFLEX 2LC herbicide may be applied per acre per season to soybeans in Region 1. A maximum of 1¼ pints of REFLEX 2LC herbicide may be applied per acre in alternate years in Region 2. A maximum of 1 pint of REFLEX 2LC herbicide may be applied per acre in alternate years in Region 3. A maximum of 1½ pints of REFLEX 2LC herbicide may be applied per acre in alternate years in Region 4.
- Thoroughly clean spray tank with water and a commercial tank cleaner before and after each use.
- REFLEX 2LC herbicide should not be mixed with other pesticides, fertilizers or any other additive except as specified on this label or on other approved ICI Americas supplemental labeling.
- REFLEX 2LC herbicide requires a 4-hour rain free period for best results. Do not apply if rain is threatening.

  Apply to actively growing weeds. Do not apply to weeds which are under drought stress including cold temperature such as when maximum day temperature is below 70°F, or when soil temperature is below 60°F, as reduced weed control may result.
- Avoid applying REFLEX 2LC to soybeans which have been under stress from drought, hail damage, flooding or herbicide injury as increased crop injury may result.
- Avoid overlapping spray swaths, as injury may occur to rotational crops.
- Avoid drift to all other crops and nontarget areas. Crops other than soybeans may be severely injured by drift.
- Do not graze treated areas or harvest for forage or hay.

ROTATIONAL CROP RESTRICTIONS

The following rotational crops may be planted after applying REFLEX 2LC at recommended rates in soybeans.

Crop To Be Planted	Minimum Rotation Interval (Months After Last REFLEX Application)
Small grains such as wheat, barley, rye.	4
Corn, cotton, peanuts, rice.	10
To avoid crop injury do not plant sunflowers, sugar beets, sorghum or any other crop within	18

Do not graze rotated small grain crops or harvest forage or straw for livestock. In the event of a crop loss due to weather conditions soybeans can be replanted.

REFLEX 2LC Regional Use Map

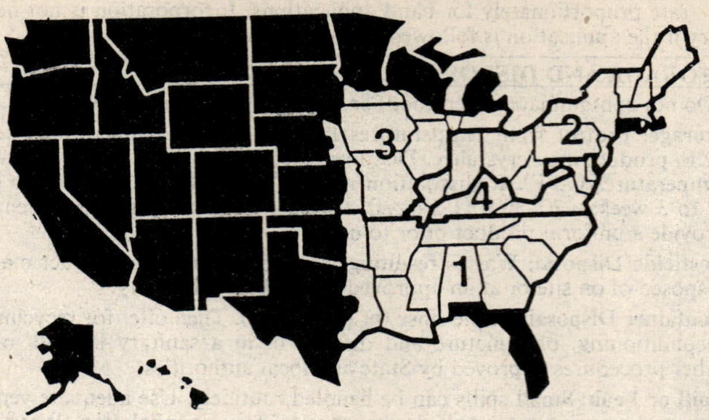

REFLEX 2LC in Soybeans
Use Rates and Weeds Controlled
Region 1

Refer to Map for Definition of Specified Geographic Regions

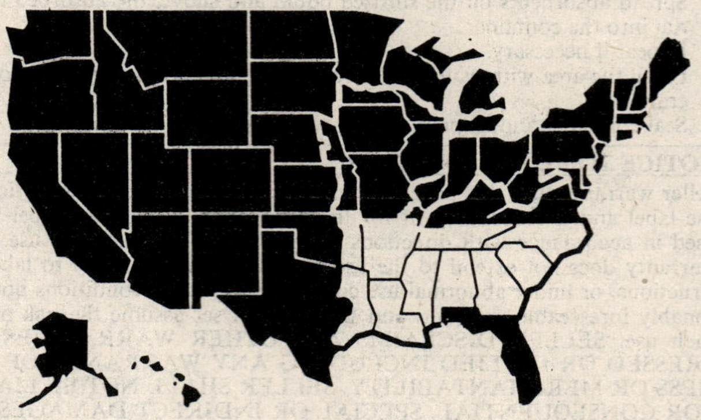

Region 1—Includes the following states or portion of states where REFLEX 2LC may be applied:
Alabama, Arkansas, Georgia, Louisiana, Mississippi, North Carolina, Oklahoma (east of U.S. Highway 75 and east of Indian Nation Parkway), South Carolina, Tennessee, and Texas (counties east of U.S. Highway 75 and Interstate 45, additional counties of Austin, Brazoria, Chambers, Colorado, Fort Bend, Harris, Matagorda, Waller, and Wharton).

APPLICATION RATES FOR WEED GROWTH STAGES

REGION 1	REFLEX 2LC RATE (Pts./Acre)[a]	
	Maximum Growth Stage Controlled At	
Weeds Controlled[b]	1 Pt./Acre # of Leaves	1½ Pt./Acre # of Leaves
Amaranth, Palmer	4	6
Amaranth, Spiny	—	2
Anoda, Spurred	—	2
Carpetweed	6" Diamter Size	Unlimited Size
Citron (Wild Watermelon)	—	4
Cocklebur, Common[c]	—	4
Copperleaf, Hophornbeam	—	4
Copperleaf, Virginia	—	4
Crotalaria, Showy	4	6
Croton, Tropic	2	4
Cucumber, Volunteer	4	6
Eclipta	—	4
Jimsonweed	4	8
Ladysthumb	2	4
Lambsquarters, Common	2	2
Mexicanweed	Suppression Only 2	Suppression Only 2
Morningglory	Suppression Only	

REGION 1	REFLEX 2LC RATE (Pts./Acre)[a]	
	Maximum Growth Stage Controlled At	
Weeds Controlled[b]	1 Pt./Acre # of Leaves	1½ Pt./Acre # of Leaves
Cypressvine	4	6
Entireleaf var.	2	3
integriuscula	Suppression Only	
Ivyleaf var.	2	3
hederacea	Suppression Only	
Purple Moonflower	2	4
Scarlet	—	4
Smallflower	—	4
Smallwhite (pitted)	2	4
Tall	2	4
(Common)	Suppression Only	
Willowleaf (Palmleaf)	—	4
Mustard, Wild	4	8
Nightshade, Black	4	4
Nutsedge, Yellow	—	Suppression Only
Pigweed, Redroot	4	6
Pigweed, Smooth	4	6
Poinsettia, Wild	—	3
Purslane, Common	6" Diameter	8" Diameter
Pusley, Florida	—	2
Ragweed, Common	4	6
Ragweed, Giant	—	4
Redweed	—	3 Suppression Only
Sesbania, Hemp	6	12
Sicklepod	—	Suppression Only Cotyledon
Sida, Prickly	—	Suppression Only Cotyledon
Smartweed, Pennsylvania	4	4
Smellmelon	—	2
Spurge, Prostrate	—	Suppression Only 1" Diameter
Spurge, Spotted	—	2 Suppression Only
Starbur, Bristly	2	4
Velvetleaf	—	2 Suppression Only
Venice Mallow	4	6
Waterhemp, Tall	—	4
Witchweed	Multi-leaf Up to 7"	Multi-leaf Up to 10"
Yellow Rocket	4	6

[a]It is necessary to use 0.25-0.5% nonionic surfactant or 0.5 to 1% crop oil concentrate.
[b]Scientific names for weeds are listed in the Appendix.
[c]Do not apply REFLEX 2LC to cotyledon stage.

REFLEX 2LC in Soybeans Use Rates and Weeds Controlled

Region 2

Refer to Map for Definition of Specified Geographic Regions

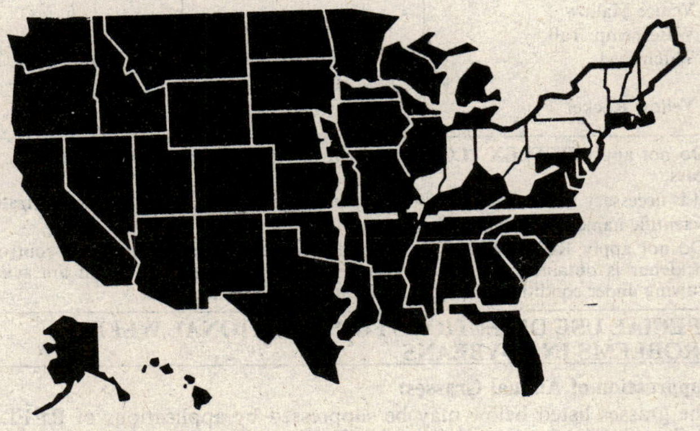

Region 2—Includes the following states or portion of states where REFLEX 2LC may be applied:
Indiana, Illinois (counties south of Interstate 70), Maine, Massachusetts, Missouri (counties of Bollinger, Butler, Cape Girardeau, Dunklin, Madison, Mississippi, New Madrid, Pemiscot, Perry, Ripley, Scott, Stoddard and Wayne), New Hampshire, New Jersey, New York, Ohio, Pennsylvania, Rhode Island, Vermont and West Virginia.

Application Rates for Weed Growth Stages

Region 2[a]	REFLEX 2LC Rate (Pts./Acre)[b]
Weeds Controlled[c]	Maximum Growth Stage at 1¼ Pt./Acre (# of Leaves)
Amaranth, Palmer	4
Carpetweed	6" Diameter
Cocklebur, Common[d]	2
Croton, Tropic	2
Cucumber, Volunteer	4
Jimsonweed	6
Ladysthumb	2
Lambsquarters, Common	2
Morningglory	
Cypressvine	4
Entireleaf var. *integriuscula*	2
Ivyleaf var. *hederacea*	2
Purple Moonflower	4
Smallwhite (Pitted)	4
Tall (Common)	2
Mustard, Wild	6
Nightshade, Black	4
Pigweed, Redroot	6
Pigweed, Smooth	6
Purslane, Common	6" diameter
Ragweed, Common	4
Sesbania, Hemp	6
Smartweed, Pennsylvania	4
Venice Mallow	6
Yellow Rocket	6

[a]Do not apply REFLEX 2LC to any field in Region 2 more than once every two years.
[b]It is necessary to use 0.25-0.5% nonionic surfactant or 0.5 to 1% crop oil concentrate.
[c]Scientific names for weeds are listed in the Appendix.
[d]Do not apply REFLEX 2LC to cotyledon stage. The most consistent control of cocklebur is obtained when applications are made to seedlings which are actively growing under conditions of high soil moisture and high humidity.

REFLEX 2LC in Soybeans Use Rates and Weeds Controlled

Region 3

Refer to Map for Definition of Specified Geographic Regions

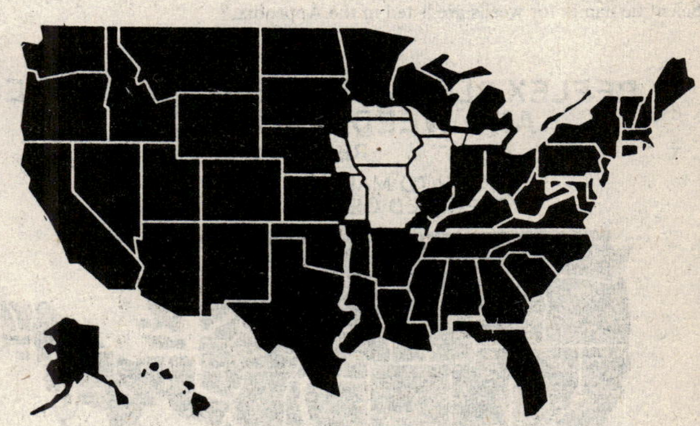

Region 3—Includes the following states or portion of states where REFLEX 2LC may be applied:
Illinois (all counties north of Interstate 70 and not listed in Region 2), Iowa, Kansas (all counties east of U.S. Highway 75), Michigan (all counties south of U.S. Highway 46), Minnesota (all counties south of Highway 212), Missouri (all counties except for those listed in Region 2), Nebraska (all counties east of U.S. Highway 81), and Wisconsin (all counties south of Interstate 94).

Application Rates for Weed Growth Stages

Region 3[a]	REFLEX 2LC Rate (Pts./Acre)[b]
Weeds Controlled[c]	Maximum Growth Stage at 1 Pt./Acre (# of Leaves)
Carpetweed	6" Diameter
Cucumber, Volunteer	4
Jimsonweed	4
Ladysthumb	2
Lambsquarters, Common	2
	Suppression Only
Morningglory	
Entireleaf var. *integriscula*	2
	Suppression Only
Ivyleaf var. *hederacea*	2
	Suppression Only
Purple Moonflower	2
Smallwhite (Pitted)	2
Tall (Common)	2
	Suppression Only
Mustard, Wild	4
Nightshade, Black	4
Pigweed, Redroot	4
Pigweed, Smooth	4
Purslane, Common	6" diameter
Ragweed, Common	4
Smartweed, Pennsylvania	4
Venice Mallow	4
Yellow Rocket	4

[a]Do not apply REFLEX 2LC to any field in Region 3 more than once every two years.
[b]It is necessary to use 0.25-0.5% nonionic surfactant or 0.5 to 1% crop oil concentrate.
[c]Scientific names for weeds are listed in the Appendix.

Region 3[a]	REFLEX 2LC Rate (Pts./Acre)[b]
Weeds Controlled[c]	Maximum Growth Stage at ¾ Pt./Acre (# of Leaves)
Carpetweed	4" diameter
Jimsonweed	2
Mustard, Wild	4
Nightshade, Black	2
Pigweed, Redroot	4
Pigweed, Smooth	4
Purslane, Common	4" diameter
Ragweed, Common	2
Venice Mallow	2
Yellow Rocket	2

[a]Do not apply REFLEX 2LC to any field in Region 3 more than once every two years.
[b]It is necessary to use 0.25-0.5% nonionic surfactant or 0.5 to 1% crop oil concentrate.
[c]Scientific names for weeds are listed in the Appendix.

REFLEX 2LC IN SOYBEANS USE RATES AND WEEDS CONTROLLED

REGION 4

REFER TO MAP FOR DEFINITION OF SPECIFIED GEOGRAPHIC REGIONS

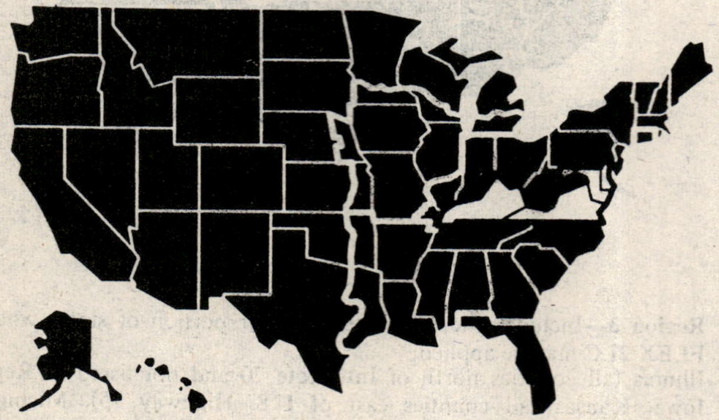

Region 4—Includes the following states where REFLEX 2LC may be applied:
Delaware, Kentucky, Maryland, and Virginia.

Application Rates for Weed Growth Stages

REGION 4[a]	REFLEX 2LC RATE (Pts./Acre)[b]	
	Maximum Growth Stage Controlled At	
Weeds Controlled[c]	1 Pt./Acre # of Leaves	1½ Pt./Acre # of Leaves
Amaranth, Palmer	4	6
Amaranth, Spiny	—	2
Anoda, Spurred	—	2
Carpetweed	6" Diamter Size	Unlimited Size
Citron (Wild Watermelon)	—	4
Cocklebur, Common[d]	—	4
Copperleaf, Hophornbeam	—	4
Copperleaf, Virginia	—	4
Crotalaria, Showy	4	6
Croton, Tropic	2	4
Cucumber, Volunteer	4	6
Eclipta	—	4
Jimsonweed	4	8
Ladysthumb	2	4
Lambsquarters, Common	2	2
	Suppression Only	Suppression Only
Mexicanweed	2	2
	Suppression Only	
Morningglory		
Cypressvine	4	6
Entireleaf var.	2	3
integriuscula	Suppression Only	
Ivyleaf var.	2	3
hederacea	Suppression Only	
Purple Moonflower	2	4
Scarlet	—	4
Smallflower	—	4
Smallwhite (pitted)	2	4
Tall (Common)	2	4
	Suppression Only	
Willowleaf (Palmleaf)	—	4
Mustard, Wild	4	8
Nightshade, Black	4	4
Nutsedge, Yellow	—	Suppression Only
Pigweed, Redroot	4	6
Pigweed, Smooth	4	6
Poinsettia, Wild	—	3
Purslane, Common	6" Diameter	8" Diameter
Pusley, Florida	—	2
Ragweed, Common	4	6
Ragweed, Giant	—	4
Redweed	—	3
		Suppression Only
Sesbania, Hemp	6	12
Sicklepod	—	Suppression Only
		Cotyledon
Sida, Prickly	—	Suppression Only
		Cotyledon
Smartweed, Pennsylvania	4	4
Smellmelon	—	2
Spurge, Prostrate	—	Suppression Only
		1" Diameter
Spurge, Spotted	—	2
		Suppression Only
Starbur, Bristly	2	4
Velvetleaf	—	2
		Suppression Only
Venice Mallow	4	6
Waterhemp, Tall	—	4
Witchweed	Multi-leaf Up to 7"	Multi-leaf Up to 10"
Yellow Rocket	4	6

[a]Do not apply REFLEX 2LC to any field in Region 4 more than once every two years.
[b]It is necessary to use 0.25-0.5% nonionic surfactant or 0.5-1% crop oil concentrate.
[c]Scientific names for weeds are listed in the Appendix (page 5).
[d]Do not apply REFLEX 2LC to cotyledon stage. The most consistent control of cocklebur is obtained when applications are made to seedlings which are actively growing under conditions of high soil moisture and high humidity.

SPECIAL USE DIRECTIONS FOR ADDITIONAL WEED PROBLEMS IN SOYBEANS

Suppression of Annual Grasses:
The grasses listed below may be suppressed by applications of REFLEX 2LC at 1-1½ pints/acre. Consult Use Rate Table for maximum rate in each region. For full-season broad-spectrum annual grass control, FUSILADE 2000 should be used alone or in tankmix with REFLEX 2LC. Consult tankmix section.

Barnyardgrass	Goosegrass
Broadleaf Signalgrass	Johnsongrass, Seedling
Crabgrass	Panicum, Fall
Foxtail	Panicum, Texas
Giant	
Green	
Yellow	

Suppression of Perennial Weeds:
Use of REFLEX 2LC at rates of 1 to 1½ pints/acre will aid in suppressing the above-ground portions of the weeds listed below until crop canopy can assist in suppression. Perennial weeds continue to regrow from underground rootstocks even if above-ground foliage is temporarily controlled or retarded. Even though REFLEX 2LC and crop competition can suppress perennial weeds for a growing season, the rootstocks will continue to live and reestablishment will occur in subsequent years.

Milkweed, Climbing	Bindweed, Field
Milkweed, Honeyvine	Bindweed, Hedge

TANKMIX AND SEQUENTIAL APPLICATIONS FOR SOYBEANS

SEE TABLES FOR TANKMIX APPLICATION RATES.

Fill the spray tank with half the amount of required water and add the recommended amounts of REFLEX 2LC, the selected tankmix herbicide, and proper rate of approved adjuvant while the agitator is running and then add the remaining quantity of water. Allow the spray mixture to agitate and recycle 5-10 minutes before application.

NOTE: Tankmix applications can result in increases in crop injury as compared to either product used alone.

It is recommended that tankmixes should be applied in a minimum of 20 gallons per acre with spray pressures of 40 to 60 psi at the nozzle tip. When foliage is dense, use 60 to 80 psi and 20 to 50 gallons per acre to ensure coverage of weed foliage.

USE OF REFLEX 2LC WITH FUSILADE 2000-SOYBEANS ONLY

REFLEX 2LC and FUSILADE 2000 may be used together in a postemergence program for broad-spectrum weed control in soybeans.

REFLEX 2LC and FUSILADE 2000 may be applied sequentially or in a tankmix. The growth stage of weeds at the time of application will determine which application will provide the most satisfactory results.

Both REFLEX 2LC and FUSILADE 2000 should be applied to actively growing weeds. Do not apply either herbicide if weeds appear stressed due to unfavorable temperatures, drought and/or low soil fertility. Do not apply REFLEX 2LC or FUSILADE 2000 if soybeans show injury from prior herbicide applications.

Method 1: Sequential Application
FUSILADE 2000 Followed by REFLEX 2LC

(Annual and/or perennial grass weeds at proper growth stage for treatment, prior to broadleaf weed treatment.)

Apply FUSILADE 2000 with an approved adjuvant to annual and/or perennial grass weeds at the recommended rate and growth stage listed in the Application Directions Section.

FUSILADE 2000 is a systemic herbicide; therefore, when treating annual grasses, allow at least 3 days, and when treating perennial grasses, allow at least 5 days, to elapse prior to a REFLEX 2LC application. This will enable FUSILADE 2000 to be adequately translocated to the meristematic regions of the plant.

After the appropriate time interval has elapsed, apply REFLEX 2LC with an approved adjuvant at the recommended rate and growth stage listed in the Application Directions Section.

Method 2: Sequential Application
REFLEX 2LC Followed by
FUSILADE 2000

(Broadleaf weeds at proper growth stage for treatments, prior to annual and/or perennial grass weed treatment.)

Apply REFLEX 2LC with an approved adjuvant susceptible broadleaf weeds at the recommended rate and growth stage.

A sequential application for FUSILADE 2000 may be applied following a REFLEX 2LC application when grasses resume active growth with the development of the new leaf. Follow the recommended rates and growth stages listed in the Application Directions Section.

Method 3: Tankmix Applications
and REFLEX 2LC FUSILADE 2000

(Annual grasses only and broadleaf weeds are at the proper stage of growth for treatment, as per the respective labels.)

A tankmix of REFLEX 2LC and FUSILADE 2000 may be applied at the recommended rates and growth stages to susceptible annual grass and broadleaf weed species in a manner consistent with respective labels.

Use a nonionic surfactant or crop oil concentrate in the tankmix. Follow water volume and spray pressure recommendations, as per the REFLEX 2LC label.

Tankmix applications have sometimes resulted in reduced grass control compared to FUSILADE 2000 used alone. If grass regrowth occurs following an application of the tankmix or an additional flush of grasses emerges, make a second application of FUSILADE 2000 to actively growing weeds. Refer to the Application Directions Section of the FUSILADE 2000 label. When perennial grasses are the predominant grass to be controlled, a tankmix is not recommended. Follow the directions for sequential applications of REFLEX and FUSILADE 2000 listed above.

SEE SOYBEAN TANKMIX USE PRECAUTIONS SECTION FOR ADDITIONAL INFORMATION.

USE OF REFLEX 2LC WITH BASAGRAN 4E - SOYBEANS ONLY

REFLEX 2LC and Basagran are selective postemergence herbicides which may be applied in a tankmix for improved postemergence control of troublesome annual broadleaf weeds in soybeans such as prickly sida, and velvetleaf. These weeds will be controlled in addition to weeds listed in the REFLEX 2LC Rate Tables.

The tankmix should be applied to actively growing weeds. Refer to the REFLEX 2LC and Basagran labels for defined environmental conditions that promote active growth.

SEE SOYBEAN TANKMIX USE PRECAUTIONS SECTION FOR ADDITIONAL INFORMATION.

USE OF REFLEX 2LC WITH FUSILADE 2000 + BASAGRAN 4E - SOYBEANS ONLY

REFLEX 2LC, FUSILADE 2000 and Basagran may be used together in a tankmix as a postemergence program for broad spectrum weed control in soybeans. The tankmix should be applied at the recommended rates and growth stages to susceptible annual grass and broadleaf weed species in a manner consistent with respective labels.

The REFLEX 2LC/FUSILADE 2000/Basagran tankmix should be applied to actively growing weeds. Do not apply this tankmix if weeds are stressed due to unfavorable temperatures, drought and/or low soil fertility. Do not apply the REFLEX 2LC/FUSILADE 2000/Basagran tankmix if soybeans show injury from prior herbicide application.

Use a nonionic surfactant or crop oil concentrate. Apply in a minimum of 15 gallons per acre of spray mixture with spray pressures of 40-60 PSI at the nozzle tip.

USE OF REFLEX 2LC WITH BUTRYAC 200 - SOYBEANS ONLY

REFLEX 2LC and Butyrac 200 may be applied in a tankmix for improved postemergence control of annual morningglory species (Entireleaf, Ivyleaf and Tall), giant ragweed and cocklebur in soybeans.

The tankmix should be applied to actively growing weeds before they exceed the growth stages described on the REFLEX 2LC label. The most consistent control is obtained when applications are made to actively growing seedlings under conditions of high soil moisture and high humidity. The most restrictive labeling of either product used applies in tank mixtures.

SEE SOYBEAN TANKMIX USE PRECAUTIONS SECTION FOR ADDITIONAL INFORMATION.

USE OF REFLEX 2LC WITH CLASSIC HERBICIDE - SOYBEANS ONLY

REFLEX 2LC herbicide and Classic herbicide may be applied in a tankmix (for improved postemergence control of cocklebur in Regions 2 and 3, and yellow nutsedge in soybeans).

REFLEX 2LC and Classic are selective postemergence herbicides which control annual broadleaf weeds. The tankmix should be applied to actively growing weeds. Refer to the REFLEX 2LC and Classic labels for defined environmental conditions that promote active growth. Do not apply either herbicide if weeds appear stressed due to unfavorable temperatures, drought, water saturated soil, and/or low soil fertility. Do not apply the REFLEX 2LC and Classic tankmix if soybeans show injury from disease, insect damage, nutrient deficiency, or prior herbicide applications.

The REFLEX 2LC and Classic tankmix may be applied by ground application only to susceptible annual broadleaf weeds at the recommended rates and growth stages in a manner consistent with respective labels. The most restrictive labeling of either product used applies in tank mixtures.

Use only a nonionic surfactant in the tankmix at the rate of 0.25% (½ pint per 25 gallons) of the finished spray volume to improve wetting and/or contact activity.

USE OF REFLEX 2LC WITH SCEPTER HERBICIDE - SOYBEANS ONLY

REFLEX 2LC herbicide and Scepter herbicide may be applied in a tankmix (for improved postemergence control of cocklebur in Regions 2 and 3).

REFLEX 2LC and Scepter are selective postemergence herbicides which control annual broadleaf weeds. The tankmix should be applied to actively growing weeds. Refer to the REFLEX 2LC and Scepter labels for defined environmental conditions that promote active growth. Do not apply either herbicide if weeds appear stressed due to unfavorable temperatures, drought, or low soil fertility. Do not apply the REFLEX 2LC and Scepter tankmix if soybeans show injury from disease, insect damage, nutrient deficiency, or prior herbicide applications.

The REFLEX 2LC and Scepter tankmix may be applied by ground application only to susceptible annual broadleaf weeds at the recommended rates and growth stages in a manner consistent with respective labels. The most restrictive labeling of either product used applies in tank mixtures.

SEE SOYBEAN TANKMIX USE PRECAUTIONS SECTION FOR ADDITIONAL INFORMATION.

TABLE 1 SOYBEAN HERBICIDE TANKMIX APPLICATION RATE TABLE[1]

TANKMIX COMBINATION	REFLEX 2LC PTS/A	FUSILADE 2000 FL. OZ./A	BASAGRAN PTS/A	BUTYRAC 200 OZ/A	CLASSIC OZ./A	SCEPTER PTS/A	NONIONIC SURFACTANT CROP OIL CONCENTRATE
REFLEX 2LC + FUSILADE 2000	¾-1½	24					0.5-1% COC 0.25-0.5% NIS
REFLEX 2LC + BASAGRAN 4E	¾-1½		1-2				COC 1 Qt/A
REFLEX 2LC + FUSILADE 2000 + BASAGRAN	¾-1½	24	1-2				0.5-1% COC .25-0.5% NIS
REFLEX 2LC + BUTYRAC 200	1-1½			2-3			0.5-1% COC 0.25% NIS
REFLEX 2LC + CLASSIC	¾-1½				½		0.25% NIS
REFLEX 2LC + SCEPTER	¾-1½					⅓	0.5-1% COC 0.25-1% NIS

[1] Follow the recommended rate and growth stages for the grass and broadleaf weeds on the respective labels.

SOYBEAN TANKMIX USE PRECAUTIONS

- Always read and follow the restrictions and limitations for all products whether used alone or in a tankmix. The most restrictive labeling of any product used applies in tank mixtures.
- Do not apply tankmixes if rain is threatening. FUSILADE requires a 1-hour rain-free period, Basagran requires an 8-hour rain-free period; and Classic requires a 4-hour rain-free period.
- Do not apply a total of more than 64 ozs. of FUSILADE 2000 per acre in one season to soybeans.
- Make the last FUSILADE 2000 application to soybeans before bloom.
- Do not apply a total of more than 2 quarts of Basagran herbicide per acre in one season to soybeans.
- Do not apply a total of more than 1½ oz. of Classic herbicide per acre in one season to soybeans.
- Do not apply Classic before crop emergence or later than 60 days before soybean maturity.
- Do not apply a total of more than 1⅓ pints of Scepter herbicide per acre in one season to soybeans.
- There must be an interval of at least 90 days between the last Scepter application and soybean harvest.
- Do not apply Butyrac 200 within 60 days of soybean harvest.
- In the case of crop failure, only soybeans may be immediately replanted following use of Basagran, REFLEX, Scepter and Classic.
- Do not make more than one application of the REFLEX/Basagran, REFLEX/FUSILADE, REFLEX/Butyrac, REFLEX/Scepter, and REFLEX/Classic tankmix in a single season.
- Butyrac 200 may form a precipitate when diluted in very hard water. Refer to Butyrac 200 directions for testing compatibility at proposed dilution rates.
- Do not make tank mix applications of any of these herbicides if the weeds appear stressed due to unfavorable temperatures, drought, and/or low soil fertility.

STORAGE AND DISPOSAL

Prohibitions: Do not contaminate water, food or feed by storage or disposal. Open dumping is prohibited. Do not reuse empty container.

Storage: Store above 32°F in original containers only. If product freezes, return to room temperature and agitate to reconstitute. Keep container closed when not in use. Do not store near food or feed. In case of spill or leak on floor or paved surfaces, soak up with sand, earth or synthetic absorbent. Remove to chemical waste area.

Pesticide Disposal: Pesticide wastes are toxic. Improper disposal of excess pesticide, spray mixture or rinsate is a violation of Federal law. If these wastes cannot be disposed of by use according to label instructions, contact your State Pesticide or Environmental Control Agency or the Hazardous Waste Representative at the nearest EPA Regional Office for guidance.

Container Disposal: Metal Containers: Triple rinse (or equivalent). Then offer for recycling or reconditioning or puncture and dispose of in a sanitary landfill or by other procedures approved by state and local authorities.

Plastic Containers: Triple rinse (or equivalent). Then offer for recycling or reconditioning or puncture and dispose of in a sanitary landfill, by incineration or, if allowed by state and local authorities, by burning. If burned, stay out of smoke.

Glass Containers: Triple rinse (or equivalent). Then offer for recycling or reconditioning or puncture and dispose of in a sanitary landfill or by other approved state and local procedures.

Butyrac 200® - Reg. TM of Rhone Poulenc Ag Company
Basagran® - Reg. TM of BASF Corporation
Classic® - Reg. TM of E.I. du Pont de Nemours & Co., Inc.
Scepter® - Reg. TM of American Cyanamid Company
FUSILADE® 2000 and RELFEX® are trademarks of an ICI Group Company.

APPENDIX

Scientific names are listed for those weeds referred to in the REFLEX 2LC label. Specific weed control recommendations are to be found in tables for Regions 1, 2, 3, and 4.

Common Name	Scientific Name
Amaranth, Palmer	*Amaranthus palmeri*
Amaranth, Spiny	*Amaranthus spinosus*
Anoda, Spurred	*Anoda cristata*
Barnyardgrass	*Echinochloa crusvgalli*
Broadleaf Signalgrass	*Brachiaria platphylla*
Carpetweed	*Mollugo verticillata*
Citron (Wild Watermelon)	*Citrullus vulgaris*
Cocklebur, Common	*Xanthium Pennsylvanicum*
Copperleaf, Hophornbeam	*Acalypha ostryaefolia*
Copperleaf, Virginia	*Acalypha virginica*
Crabgrass	*Digitaria spp.*
Crotalaria, Showy	*Crotalaria spectabilis*
Croton, Tropic	*Croton glandulosus*
Cucumber, Volunteer	*Cucumis sativas*
Eclipta	*Eclipta prostrata*
Foxtail, Giant	*Setaria faberi*
Foxtail, Green	*Setaria viridis*
Foxtail, Yellow	*Setaria lutescens*
Goosegrass	*Eleusine indica*
Jimsonweed	*Datura stramonium*
Johnsongrass, Seedling	*Sorghum halepense*
Ladysthumb	*Polygonum persicaria*
Lambsquarters, Common	*Chenopodium album*
Mexicanweed	*Caperonia castanaefolia*
Morningglory, Cypressvine	*Ipomoea quamoclit*
Entireleaf var. integriuscula	*Ipomoea hederacea*
Ivyleaf var. hederacea	*Ipomoea hederacea*
Purple Moonflower	*Ipomoea turbinata*
Scarlet	*Ipomoea coccinea*
Smallflower	*Jacquemontia tamnifolia*
Smallwhite (Pitted)	*Ipomoea iacunosa*
Tall (Common)	*Ipomoea purpurea*
Willowleaf (Palmleaf)	*Ipomoea wrightii*
Mustard, Wild	*Brassica kaber*
Nightshade, Black	*Solanum nigrum*
Nutsedge, Yellow	*Cyperus esculentus*
Panicum, Fall	*Panicum dichotomiflorum*
Panicum, Texas	*Panicum texanum*

Common Name	Scientific Name
Pigweed, Redroot	*Amaranthus retroflexus*
Pigweed, Smooth	*Amaranthus hybridus*
Poinsettia, Wild	*Euphorbia heterophylla*
Purslane, Common	*Portulaca oleracea*
Pusley, Florida	*Richardia scabra*
Ragweed Common	*Ambrosia artemisiifolia*
Ragweed, Giant	*Ambrosia trifida*
Redweed	*Melochia corchorifolia*
Sesbania, Hemp	*Sesbania exaltata*
Sicklepod	*Cassia obtusifolia*
Sida, Prickly	*Sida spinosa*
Smartweed, Pennsylvania	*Polygonum Pennsylvanicum*
Smellmelon	*Cucumis melo*
Spurge, Prostrate	*Euphorbia supina*
Spurge, Spotted	*Euphorbia maculata*
Starbur, Bristly	*Acanthospermum hispidum*
Velvetleaf	*Abutilon theophrasti*
Venice Mallow	*Hibiscus trionum*
Waterhemp, Tall	*Amaranthus tuberculatos*
Witchweed	*Striga asiatica*
Yellow Rocket	*Barbarea vulgaris*

EPA registered uses as of August 1989.

NOTICE TO BUYER AND USER:

Seller warrants that this product conforms to the chemical description on the label and is reasonably fit for the purposes stated on the label when used in accordance with directions under normal conditions of use. This warranty does not extend to the use of this product contrary to label instructions, or under abnormal use conditions, or under conditions not reasonably foreseeable to Seller, and Buyer and User assume the risk of any such use. SELLER DISCLAIMS ALL OTHER WARRANTIES EXPRESSED OR IMPLIED INCLUDING ANY WARRANTY OF FITNESS OR MERCHANTABILITY. SELLER SHALL NOT BE LIABLE FOR CONSEQUENTIAL, SPECIAL OR INDIRECT DAMAGES RESULTING FROM THE USE OR HANDLING OF THIS PRODUCT AND SELLER'S SOLE LIABILITY AND BUYER'S AND USER'S EXCLUSIVE REMEDY SHALL BE LIMITED TO THE REFUND OF THE PURCHASE PRICE.

03-0573-043R4 A-2MR

The information and specimen labels in this booklet are correct as of the publication date but are subject to change without notice. For current information, contact ICI Americas Inc.

RO-NEET® 6-E

Emulsifiable Liquid • Selective Herbicide for Sugar Beets, Table Beets & Spinach

ACTIVE INGREDIENT:	
S-Ethyl cyclohexylethylthiocarbamate	73.9%
INERT INGREDIENTS:	26.1%
TOTAL	100.0%

Contains six pounds active ingredient per gallon.

EPA Reg. No. 10182-178

Complete Directions For Use,
Use Precautions and Recommendations

KEEP OUT OF REACH OF CHILDREN

CAUTION

Harmful if Swallowed

STATEMENT OF PRACTICAL TREATMENT

First Aid: If a known exposure occurs or is suspected, immediately start the recommended procedures below. Simultaneously contact a Poison Control Center, a physician or the nearest hospital. Describe the situation and follow the advice given.

If swallowed: Immediately give large quantities of water but **DO NOT** induce vomiting. **This product contains hydrocarbon solvent.** If vomiting does occur, give fluids again. Have a physician determine if condition of patient will permit evacuation of stomach. **DO NOT** give anything by mouth to an unconscious person. Call a physician immediately.

In case of contact: Hold eyelids apart and flush eyes and skin with plenty of water for at least 15 minutes while removing contaminated clothing and shoes. Get medical attention. Wash clothing before reuse.

If inhaled: Remove to fresh air. Seek medical attention if respiratory irritation occurs.

FOR 24-HOUR EMERGENCY MEDICAL ASSISTANCE CALL 1-800-F-A-S-T-M-E-D (327-8633).
FOR CHEMICAL EMERGENCY: Spill, leak, fire, exposure, or accident call CHEMTREC 1-800-424-9300.

PRECAUTIONARY STATEMENTS

HAZARDS TO HUMANS AND DOMESTIC ANIMALS

CAUTION

Harmful if Swallowed

- Avoid contact with skin, eyes and clothing.
- Wear rubber gloves, and protective clothing.
- Wash hands, arms and face thoroughly with soap and water after handling and before eating or smoking.
- Routinely shower or bathe after work and wash all clothing with soap and hot water before re-using.
- In case of contact, immediately remove contaminated clothing.
- Avoid contamination of feed or food.

ENVIRONMENTAL HAZARDS

Do not apply directly to water. Do not contaminate water by cleaning of equipment or disposal of wastes.

Read Entire Label Before Using

Protect from temperatures below 20°F.

Product crystallizes at lower temperatures.

Warm or store at higher temperatures, and mix to redissolve crystals and assure uniformity before use.

Weeds Controlled

Annual Grasses:

Annual Bluegrass	(*Poa annua*)
Annual Ryegrass	(*Lolium multiflorum*)
Barley, Volunteer	(*Hordeum* spp.)
Barnyardgrass (Watergrass)	(*Echinochloa* spp.)
Crabgrass	(*Digitaria* spp.)
Foxtails	(*Setaria* spp.)
Wild oats	(*Avena fatua*)

Annual Broadleaf Weeds:

Black Nightshade	(*Solanum nigrum*)
Hairy Nightshade	(*Solanum villosum*)
Henbit, Deadnettle	(*Lamium* spp.)
Lambsquarters	(*Chenopodium album*)
Nettleleaf Goosefoot*	(*Chenopodium murale*)
Common Purslane	(*Portulaca oleracea*)
Redroot Pigweed	(*Amaranthus retroflexus*)
Shepherdspurse	(*Capsella bursa-pastoris*)
Small Stinging Nettle, Burning Nettle	(*Urtica urens*)

*Arizona and California excluded

The above broadleaf weeds will be controlled only if the application is made when conditions are favorable for germination. RO-NEET 6-E WILL NOT CONTROL ESTABLISHED WEEDS.

Perennial Weeds:

Yellow Nutsedge	(*Cyperus esculentus*)
Purple Nutsedge	(*Cyperus rotundus*)

Existing stands of nutsedge must be turned under and chopped thoroughly before treatment. Prior to planting, RO-NEET 6-E must be thoroughly mixed into the soil, or applied sub-surface to the soil with injectors for nutsedge control.

General Use Precautions

Not for use or storage in or around the home.

RO-NEET 6-E should be used only for recommended purposes and at recommended rates (DO NOT OVERDOSE). Applied according to directions and under normal growing conditions RO-NEET 6-E will not harm the treated crop.

During germination and early growth, extended periods of unusually cold and wet or hot and dry weather, insect, nematode, or plant disease attack, carryover pesticide residues, the use of certain soil applied systemic insecticides, highly saline or alkaline soil conditions, improperly placed fertilizers or soil insecticides may create abnormal conditions that weaken crop seedlings. RO-NEET 6-E used under these abnormal conditions could result in crop injury.

RO-NEET 6-E may cause crop injury on very light sandy soil.

RO-NEET 6-E may be combined with fluid (solution, slurry, or suspension) fertilizers. However, physical compatibility with these fluid fertilizers must be determined before combining in the spray tank. See special literature giving directions for these combinations. Even though found to be compatible, constant agitation is necessary to keep the RO-NEET 6-E evenly mixed with the fluid fertilizer.

Tank mix this product with fungicides, insecticides or herbicides only as recommended.

Do not contaminate irrigation water or water used for domestic purposes.

DIRECTIONS FOR USE

It is a violation of Federal law to use this product in a manner inconsistent with its labeling.

Do not apply this product through any type of irrigation system.

RO-NEET 6-E is a selective herbicide which is mixed (incorporated) into the soil or applied below the surface of the soil for control of weeds listed on this label. RO-NEET 6-E controls weeds by interfering with normal seed germination and seedling development. It does not control established weeds. All weed growth and crop stubble should be thoroughly worked into the soil before treatment.

Soil Preparation: Prepare the soil for seeding according to good agricultural practice. All weed growth should be thoroughly worked into the soil before treatment. DO NOT APPLY BEFORE PRE-IRRIGATION.

Application: Apply RO-NEET 6-E to well-worked soil which is dry enough to permit thorough mixing with incorporation equipment, or proper injection. Use standard low-pressure (20-50 psi) boom spray equipment. DO NOT USE SINGLE NOZZLE BOOM-JET TYPE SPRAYERS. Application equipment should be carefully calibrated before use and checked frequently during application to be sure nozzles are free from clogging and delivering a uniform pattern. Apply UNIFORMLY the recommended rate of RO-NEET 6-E in 20 to 50 gallons of water per acre. Avoid overlaps that will increase RO-NEET 6-E dosage above recommended limits because plant injury will occur.

RO-NEET 6-E with Fluid Fertilizer: RO-NEET 6-E may be combined with fluid (solution, slurry, or suspension) fertilizers. However, physical compatibility with these fluid fertilizers must be determined before combining in the spray tank. See Appendix I giving directions for these combinations. Even though found to be compatible, constant agitation is necessary to keep the RO-NEET 6-E evenly mixed with the fluid fertilizer.

Impregnation on Dry Bulk Fertilizers: RO-NEET 6-E may be impregnated on many dry bulk fertilizers for use on sugar beets. RO-NEET 6-E impregnated on dry bulk fertilizers may be applied and incorporated into the soil either in the fall (in states where fall application for RONEET 6-E is recommended) before the ground freezes or before planting. However, uniform distribution of RO-NEET 6-E on fertilizer particles and uniform application are necessary to assure good results. See Appendix II for special instructions regarding directions for impregnation and use.

INCORPORATION DIRECTIONS

Immediately (within minutes) after application thoroughly mix RO-NEET 6-E into the soil to a depth of 2 to 3 inches. (California only: Incorporate to a depth of 3 inches.) Thorough soil mixing is necessary for good weed control.

RO-NEET 6-E must be incorporated into the soil immediately to prevent loss of the herbicide. Whenever possible, application and incorporation should be done in the same operation. Use equipment which has been proven to incorporate thoroughly to the recommended depth.

Soil Mixing (Incorporation) Before Planting:

The following equipment commonly is used for soil mixing (incorporation) before planting:

For Broadcast (Overall) Application:

- **Power Driven Cultivation Equipment** (recommended on all soil types) set to cut to a depth of 2 to 3 inches.
- **Tandem Discs** (recommended on all soil types) set to cut to a depth of 4 to 6 inches, operated at 4 to 6 mph followed by a spike-tooth harrow or some other leveling device which extends beyond the ends of the discs. For more thorough mixing (for nutsedge, heavier soils) disc in two different directions (cross disc).
- **Field Cultivators** (recommended on lighter soils in good tilth only). Not recommended in California or Arizona. Use 3 to 4 rows of sweeps, spaced at 7 inch or less intervals and staggered so that no soil is left unturned, followed by a spike-tooth harrow pulled behind the cultivator. Do not use chisels or points. Set the cultivator to cut 4 inches deep, operated at 5 mph or more. Run the equipment over the field twice, the second run at an angle to the first.
- **Rotary Ground Driven or Spring-Tooth Cultivators** (recommended on lighter soils in good tilth only). Not recommended in California or Arizona. Set to penetrate to a depth of 4 to 6 inches and operated at 5 to 8 mph in two different directions.

For Band (Row) Application:

Uniformly mix to a depth of 2 to 3 inches.
- Hooded power-driven rotary tillers.
- Hooded ground-driven rotary tillers.

Rolling:

Under most conditions, the treated area should be firmly rolled between the incorporation unit and the planters.

Planting

For maximum weed control benefits, crops should be planted or seeded immediately after application. Do not use a drag behind the planter as it may concentrate RO-NEET 6-E over the seed row and cause crop injury.

CULTURAL PRACTICES FOLLOWING APPLICATION

RO-NEET 6-E is not persistent in the soil and susceptible weeds germinating later during the growing season may not be controlled. Shallow cultivation or approved or post emergence herbicides may be necessary to control these susceptible weeds that escape control by RO-NEET 6-E as well as those weeds not susceptible to RONEET 6-E. Do not cultivate deeper than the depth of the herbicide incorporation.

When cultivating fields where RO-NEET 6-E has been banded in the row, use shields to prevent the movement of untreated soil into the treated row.

Dosage Recommendations
Solid Broadcast Treatment

Crop	When to Apply	Gallons RO-NEET 6-E Per Acre	Remarks
Sugar Beets*	Pre-plant (soil incorporation) or at planting (soil injection)	1/2 To 2/3	Use on mineral soils ONLY. Use lower rate on sandy soils. Use higher dosage rate on heavier soil. Injury may result in highly saline or alkaline soils.
	Fall application	2/3	Recommended only in the states of Idaho, Minnesota, Montana, North Dakota, Oregon, Wyoming and Washington. Apply and incorporate in late fall before the ground freezes.
Table Beets	Pre-plant (soil incorporation)	1/2 to 2/3	Use on mineral soils ONLY. Use lower rate on sandy soils. Use higher dosage rate on heavier soils. When applying RO-NEET 6-E in combination with fluid fertilizer, do not apply over 150 lbs. actual nitrogen per acre.
Spinach*	Pre-plant (soil incorporation)	1/2	Use on sandy mineral soils only.

* Make only one (1) application per growing season. If RO-NEET 6-E is applied in the fall, do not reapply RO-NEET 6-E the following spring.

* Recommended on spinach in Arkansas, Connecticut, Delaware, Maine, Maryland, Massachusetts, Mississippi, New Hampshire, New Jersey, New York, Ohio, Oklahoma, Pennsylvania, Texas, Vermont, Virginia and Western Tennessee only.

BAND TREATMENT:

The rates given above are for solid coverage basis. For band treatment, reduce the amount of RO-NEET 6-E proportionately depending on the row spacing and band width to be treated. DO NOT USE BAND APPLICATION ON ROCKY GROUND BECAUSE THOROUGH INCORPORATION IS NOT POSSIBLE.

Soil injection application on sugar beets—Special equipment designed for soil injection must be used. The injector units must be rigidly mounted immediately ahead or immediately behind the planter unit.

Soil Injection—injector shanks must be spaced 2 1/2 to 3 inches apart and mounted in staggered positions to avoid trash building up on them. Injector shanks must be set to inject RO-NEET 6-E 1 1/2 to 2 inches below the soil surface. The width of the band in which weed control is desired will determine the number of injector shanks required per row (that is, four injector shanks for a 10 to 12-inch band, six injector shanks for a 15 to 18-inch band). The two injector shanks adjacent to the drill row must be 1 1/4 to 1 1/2 inches on either side of it.

Appendix I

RO-NEET 6-E with Fluid Fertilizers

The following procedure is suggested for determining whether RO-NEET 6-E may be combined with a specific fluid fetilizer for spray tank application.

Material Required:

1. RO-NEET 6-E.
2. Fluid fertilizer to be used.
3. Adjuvant for fertilizer tank mix: Compex*, Sponto 168-D*, Unite*, or equivalent. The adjuvant which provides the best emulsification depends on the specific fertilizer under consideration.
4. Two one-quart, wide-mouth glass jars with lid or stopper.
5. Measuring spoons (a 25 mL pipette or graduated cylinder provides more accurate measurement.
6. Measuring cup, 8 oz. (237 mL).

* Compex, Kalo Laboratories Inc., Kansas City, Missouri; Sponto 168-D, Witco Chemical Company, Houston, Texas; Unite, Hopkins Agricultural Chemical Company, Madison, Wisconsin.

Procedure:

1. Pour a pint (about 473 mL) of the fluid fertilizer into each of the quart jars.
2. Add adjuvant to one of the jars and mix (see rate table).
3. Add the RO-NEET 6-E to both jars (see rate table).
4. Close both jars with lid or stopper and mix the contents by turning the jars upside down ten times.
5. Inspect the surface and body of the mixtures—
   (A) Immediately after completing the jar inversions.

(B) After allowing the jars to stand quietly for 30 minutes.
(C) And then again after turning the jars upside down 10 times.

If the uniform mix cannot be made, the mixture should not be used. If either mixture remains uniform for 30 minutes, the combination may be used. Should either mixture separate after 30 minutes, but readily remix uniformly with the jar inversions, the mixture can be used if adequate agitation is maintained in the tank. If the mixture with adjuvant is satisfactory, but the one without adjuvant is not, be sure to use the adjuvant in the spray tank. Add the adjuvant first at a rate of 3 pints per 100 gallons of fluid fertilizer; foaming can be minimized by using moderate agitation.

If nondispersible oil, sludge or clumps of solids form in the mixtures, the combinations should not be used.

Rate Table for RO-NEET 6-E and Adjuvant with the Fluid Fertilizer**

Gallons of fluid fertilizer to be applied per acre	mL or Tsp. of RO-NEET 6-E* to be added to 1 pint of fluid fertilizer	
	6-E	
	mL	Tsp.
10	8	1⅔
15	5	1
20	4	¾
25	3	⅔
30	2½	½
40	2	⅖

* Based on field rate of 1 pound active ingredient per acre in the fertilizer volumes indicated. Increase volume proportionately to correspond with intended field rate in terms of pounds active ingredient per acre (e.g., for field rate of 4 pounds actual RO-NEET 6-E in 40 gallons fertilizer per acre, add 8 mL or 2 tsp. RO-NEET 6-E to each jar for compatibility testing purposes).

** Two (2) milliliters or one-half (½) teaspoon of adjuvant to be added to 1 pint of fluid fertilizer in order to equal the rate of 3 pints of adjuvant per 100 gallons of fluid fertilizer.

Appendix II

RO-NEET 6-E Impregnation on Dry Bulk Fertilizers

CAUTION: Nitrate fertilizers represent a potential explosive hazard, particularly in contact with organic substances such as RO-NEET 6-E. Do not impregnate RO-NEET 6-E on ammonium nitrate. Do not use fertilizers containing ammonium, potassium or sodium nitrate. Such mixtures may cause explosion.

All individual state regulations relating to dry bulk fertilizer blending, registration, labeling and application are the responsibility of the individual and/or company selling the RO-NEET fertilizer mixtures.

RO-NEET 6-E may be impregnated on many dry bulk fertilizers (1) and applied and incorporated into the soil before planting for the control of labeled grasses and broadleaf weeds in sugar beets.

All RO-NEET 6-E label and supplementary literature instructions and precautions regarding rates per acre, soil type and soil incorporation, application and other directions must be followed.

Test results have shown RO-NEET 6-E on many dry bulk fertilizers gives weed control equal to RO-NEET 6-E applied as a spray in water or liquid fertilizer. However, uniform impregnation of the pesticide on the dry fertilizer particles and uniform application in the field are necessary to assure good results.

A minimum of 200 pounds and a maximum of 700 pounds of approved fertilizer ingredients (1) impregnated with the appropriate amount of RO-NEET 6-E must be applied per acre.

For impregnating the pesticide on dry fertilizers, use a closed rotary-drum type mixer equipped with suitable spraying equipment. The spray nozzles should be positioned inside of the mixer to provide uniform spray coverage of the tumbling fertilizer. The RO-NEET 6-E should be sprayed uniformly onto the fertilizer using a fine spray pattern.

The physical properties of fertilizers vary, particularly in liquid absorptive capacity. When absorptivity is sufficient, simple spray impregnation of the fertilizer with RO-NEET 6-E provides a satisfactory dry mixture.

If the absorptivity is not adequate, use of a highly absorptive powder is required to provide a dry, free-flowing mixture. Microcel E (Johns-Manville Products Corporation) is the recommended absorbent powder. It should be added separately and uniformly to the prepared pesticide-fertilizer mixture, in a quantity that is sufficient to provide a suitably free-flowing mixture. Generally, less than 2% by weight of Microcel E is required.

The amount of RO-NEET 6-E actually required in the manufacture of individual fertilizer mixtures should be determined carefully for each production operation. This is necessary to ensure that the amounts of RO-NEET actually contained in the mixture applied to the soil represent the correct rates of use.

Bulk fertilizers impregnated with RO-NEET 6-E should be **applied immediately.** NOT STORED.

All bulk containers should be tightly covered while the products are being transported and applied to reduce chances of RO-NEET 6-E loss via volatilization.

(1) Approved Dry Fertilizer Ingredients for use with RO-NEET 6-E

	N	P	K
Ammonium Sulfate	21	0	0
Ammonium Phosphate-sulfate	16	20	0
Diammonium Phosphate	18	46	0
Monoammonium Phosphate	11	56	0
Potassium Chloride	0	0	60
Potassium Sulfate	0	0	52
Single Superphosphate	0	20	0
Treble Superphosphate	0	46	0
Urea	45	0	0

NOTE: SUL-PO-MAG, K-MAG and 11-48-0 have been shown to be compatible with RO-NEET 6-E and are approved for use.

RO-NEET 6-E Physical Data

Specific Gravity 20/20°C:	0.976
Pounds/Gallon, 20/20°C:	8.12
Flashpoint:	168°F (Tagliabue Closed Cup)
Viscosity:	Sprayable down to 25°F
Freeze Point:	20°F (seeded)

(2) Rate Chart for the Impregnation of Dry Bulk Fertilizers with RO-NEET 6-E:

Fertilizer Rate Per Acre	RO-NEET 6-E Rate Per Acre	
	2 qts. per acre	2⅔ qts. per acre
200 lbs.	20 quarts/ton	26⅔ quarts/ton
250 lbs.	16 quarts/ton	21⅓ quarts/ton
300 lbs.	13⅓ quarts/ton	17¾ quarts/ton
350 lbs.	11⅖ quarts/ton	15¼ quarts/ton
400 lbs.	10 quarts/ton	13⅓ quarts/ton
450 lbs.	8⅞ quarts/ton	11⅞ quarts/ton
500 lbs.	8 quarts/ton	10⅔ quarts/ton
550 lbs.	7⅓ quarts/ton	9¾ quarts/ton
600 lbs.	6⅔ quarts/ton	8⅞ quarts/ton
650 lbs.	6⅕ quarts/ton	8¼ quarts/ton
700 lbs.	5¾ quarts/ton	7⅝ quarts/ton

STORAGE AND DISPOSAL

Prohibitions: Do not contaminate water, food, or feed by storage or disposal. Open dumping is prohibited.

Storage: Keep containers closed when not in use. Do not store near seeds, fertilizers or foodstuffs. Protect from temperatures below 20°F but do not store near sources of heat or open flame.

Pesticide Disposal: Waste resulting from the use of this product may be disposed of on site or at an approved waste disposal facility.

Container Disposal: Triple rinse (or equivalent), then offer for recycling or reconditioning, or puncture and dispose of in a sanitary landfill, or by other procedures approved by State and local authorities.

Notice to Buyer and User:

Seller warrants that this product conforms to the chemical description on the label and is reasonably fit for the purposes stated on the label when used in accordance with directions under normal conditions of use. This warranty does not extend to the use of this product contrary to label instructions, or under abnormal use conditions, or under conditions not reasonably foreseeable to Seller, and Buyer and User assume the risk of any such use. SELLER DISCLAIMS ALL OTHER WARRANTIES EXPRESSED OR IMPLIED INCLUDING ANY WARRANTY OF FITNESS OR MERCHANTABILITY. SELLER SHALL NOT BE LIABLE FOR CONSEQUENTIAL, SPECIAL OR INDIRECT DAMAGES RESULTING FROM THE USE OR HANDLING OF THIS PRODUCT AND SELLER'S SOLE LIABILITY AND BUYER'S AND USER'S EXCLUSIVE REMEDY SHALL BE LIMITED TO THE REFUND OF THE PURCHASE PRICE.

A-5R

The information and specimen labels in this booklet are correct as of the publication date but are subject to change without notice. For current information, contact ICI Americas Inc.

SULFUR ALFA®

WETTABLE SULFUR

An Agricultural Insecticide and Fungicide

ACTIVE INGREDIENT:

SULFUR	97.0%
INERT INGREDIENTS:	3.0%
TOTAL	100.0%

® ALFA is a registered trademark of ICI Americas Inc.

EPA Reg. No. 10182-144

KEEP OUT OF REACH OF CHILDREN CAUTION

STATEMENT OF PRACTICAL TREATMENT

If an adverse reaction occurs after exposure: Promptly start the recommended procedures below. Seek medical attention if further treatment is required.

If a large amount of this material is swallowed: Give several glasses of water. Induce vomiting by gagging the victim with a finger placed on the back of the victim's tongue. Give fluids until vomitus is clear. Do not induce vomiting or give anything by mouth to an unconscious or convulsing person.

If inhaled: Remove to fresh air. Seek medical attention if respiratory irritation occurs or if breathing becomes difficult.

In case of contact: Flush all affected areas with plenty of water for several minutes. Remove and clean contaminated clothing and shoes.

For eye contact: Hold eyelids apart and flush with plenty of water for 15 minutes. Get medical attention if irritation occurs.

FOR 24-HOUR EMERGENCY MEDICAL ASSISTANCE CALL 1-800-F-A-S-T-M-E-D (327-8633).
FOR CHEMICAL EMERGENCY: Spill, leak, fire, exposure, or accident call CHEMTREC 1-800-424-9300.

For other product information, contact your local ICI Americas Distributor or see Material Safety Data Sheet.

PRECAUTIONARY STATEMENTS

HAZARDS TO HUMANS AND DOMESTIC ANIMALS
CAUTION

- Do not take internally.
- Avoid contact with skin, eyes, and clothing.
- Wash thoroughly after handling.

ENVIRONMENTAL HAZARDS

- Do not apply when weather conditions favor drift from treated areas.
- Do not apply directly to water or wetlands. Do not contaminate water by cleaning of equipment or disposal of wastes.

PHYSICAL AND CHEMICAL HAZARDS

BURNING SULFUR EMITS HIGHLY TOXIC FUMES. SULFUR DUST SUSPENDED IN AIR IGNITES EASILY.

Keep away from heat, sparks and flame. Avoid creating dust in handling. Keep container closed. Use with adequate ventilation. Keep from contact with oxidizing materials.

In Case of Fire: Use water spray, fog, CO_2 or sand. Wear NIOSH approved self-contained breathing apparatus.

PRODUCT INFORMATION

Sulfur will cause severe fruit and leaf injury to sulfur-sensitive crops. Do not use on apricots, Anjou and cornice pears, cranberries, cucurbitis (cucumbers, squash and some varieties of melons), fiberts, spinach, tung trees, walnuts and other sensitive plants.

Do not apply when shade temperatures exceed or are likely to exceed 90°F.

Sulfur may burn foliage and fruit during periods of high temperatures and under certain climatic conditions.

Do not use sulfur within 30 days of an oil spray (60 days on citrus).

If the crop to be treated will be used for canning or freezing purposes, consult food processor before applying sulfur.

Consult State Agricultural Extension Service or State Agricultural Experiment Stations for additional information, as the timing, and number of applications needed will vary with local conditions.

DIRECTIONS FOR USE

It is a violation of Federal law to use this product in a manner inconsistent with its labeling.

Do not apply this product through any type of irrigation system.

Measure the specified amount of this product then wash through filter screen in nearly-filled spray tank or pre-mix in a bucket before pouring into spray tank. Add balance of water to fill tank. Keep agitator running during filling and spraying operations. Do not allow mixture to stand. Do not combine with emulsifiable liquids unless previous use of the mixture has proven physically compatible and safe to plants.

Emulsifiable concentrates should always be thoroughly mixed with at least half of total water *before* adding this product.

REENTRY STATEMENT

- Do not apply this product in such a manner as to directly or through drift expose workers or other persons. The area being treated must be vacated by unprotected persons.
- Do not enter treated areas for 24 hours unless appropriate protective clothing is worn.
- Because certain states may reqire more restrictive reentry intervals for various crops treated with this product, consult your State Department of Agriculture for further information.
- Written or oral warnings must be given to workers who are expected to be in a treated area or in an area about to be treated with this product. Oral warnings must be given if there is reason to believe that written warnings cannot be understood by workers.

When oral warnings are given, warnings shall be given in a language customarily understood by workers and must indicate the specific period of time that treated areas may not be entered without protective clothing. Oral warnings must also include appropriate first aid instructions in case of accidental exposure. These instructions are given in the STATEMENT OF PRACTICAL TREATMENT at the beginning of this label.

Written warnings must include the following information. "CAUTION. Area treated with ALFA® Wettable Sulfur on (date of application). Do not enter treated areas for 24 hours unless aprpropriate protective clothing is worn (or other reentry interval if your State has a more restrictive interval for this product). If accidental exposure occurs, follow the instructions below." (Written warnings must include the STATEMENT OF PRACTICAL TREATMENT given at the beginning of this label.)

RECOMMENDATIONS

Except as specified, apply at the first sign of infection or infestation and repeat at 7 to 10 day intervals. Repeat as necessary to maintain control, following local spray practices.

APPLES: Scab, powder mildew—apply 4 to 5 pounds per 100 gallons from pink or pre-blossom spray through second cover spray. Add 1 gallon lime-sulfur in pink and calyx sprays. MacIntosh, Golden Delicious, Jonathan and certain varieties may be injured by sulfur. Use on sulfur-tolerant varieties only.

CHERRIES: Apply 4 to 6 pounds per 100 gallons. Blossom blight, brown rot—apply in the popcorn, full bloom and petal-fall stages. Leaf spot and powdery mildew—apply in petal-fall, shuck-fall and 2 weeks after shuck-fall.

CITRUS: Use 4 pounds per 100 gallons. Mite control (six-spotted, citrus, rust, citrus flat and Yuma)—apply at first sign of infestation and repeat at weekly intervals as needed to maintain control. Citrus thrips—make 3 applications (mid-March, petal-fall and 2 to 4 weeks later). Avoid applications during or preceding high temperattues. Do not apply within 2 months of an oil spray.

COTTON: Atlantic Mite—apply 4 to 6 pounds per acre in 5 to 10 gallons water by air, or 20 to 40 gallons of water by ground power equipment. Begin applications when first true leaves form or when disease threatens.

GRAPES: Rust, powdery mildew—use 2 pounds per 100 gallons. Make applications starting when shoots are 6 inches long; 12 to 15 inches long; 14 days later and 10 to 14 day intervals thereafter to harvest. Care must be taken to wet all surfaces on which the fungus is to be controlled. CAUTION: Concord and Labrusca type grapes may be injured by sulfur.

PEARS: Scab, powdery mildew—apply 4 pounds plus 1 gallon lime-sulfur per 100 gallons in the pink or pre-blossom spray. On Bartlett pears, apply 4 to 6 pounds at petal-fall through second cover sprays.

PEACH, NECTARINE: Brown rot, coryneum blight, rust, peach silver mite, powdery mildew—apply 5 to 6 pounds per 100 gallons from popcorn stage up to harvest. For brown rot control in summer sprays, apply up to 10 pounds per 100 gallons starting about 4 to 5 weeks before harvest and continuing at weekly intervals.

PLUMS, PRUNES: Brown rot, leaf spot, prune rust—use 5 to 6 pounds per 100 gallons. For brown rot of fruit, apply when fruit begins to ripen. For leaf spot and rust, apply in pink stage, bloom, petal-fall and shuck-fall. In cover period, apply as necessary.

STRAWBERRIES: Powdery mildew—use 1 to 5 pounds per 100 gallons. Apply at first sign of infection and repeat as necessary to maintain control. Discontinue applications during hot weather.

TOMATOES: Tomato russet mite—apply 3 to 6 pounds per acre depending on size of plants. Use full coverage spray and make one repeat application if needed.

ORNAMENTALS:

ROSES, PERENNIALS, SHRUBS: Black spot, powdery mildew, rust—apply 3 tablespoons per gallon (6 pounds per 100 gallons) and spray for full coverage, wetting both sides of leaves. Begin applications when new growth appears in the spring and repeat every 7 days, if necessary.

NOTE: Some burning may occur when used during warm weather.

STORAGE AND DISPOSAL

- Do not contaminate water, food, or feed by storage or disposal.

STORAGE: Store in a cool dry place.

PESTICIDE DISPOSAL: Wastes resulting from the use of this product may be disposed of on site or at an approved waste disposal facility.

CONTAINER DISPOSAL: Completely empty bag into application equipment. Then dispose of empty bag in a sanitary landfill or by incineration, or, if allowed by State and local authorities, by burning. If burned, stay out of smoke.

SPILL OR LEAK: Small spills can be handled routinely. Use adequate ventilation and wear a respirator or a dust mask to prevent inhalation. Wear suitable protective clothing and eye protection to prevent skin and eye contact. Use the following procedures:

Sweep or shovel up spilled material using a natural fiber broom and/or aluminum shovel to reduce sparking. Be careful not to create dust as an explosion could result. Place sweepings, in an appropriate chemical waste container. Seal container and dispose of in an approved facility. Flush spill area with detergent and water to remove any residue.

NOTICE TO BUYER AND USER

Seller warrants that the product conforms to the chemical description on the label and is reasonably fit for the purposes stated on the label when used in accordance with directions under normal conditions of use. This warranty does not extend to the use of the product contrary to label instructions, or under abnormal use conditions, or under conditions not reasonably foreseeable to Seller, and Buyer and User assume the risk of any such use. SELLER DISCLAIMS ALL OTHER WARRANTIES EXPRESSED OR IMPLIED INCLUDING ANY WARRANTY OF FITNESS OR MERCHANTABILITY SELLER SHALL NOT BE LIABLE FOR CONSEQUENTIAL SPECIAL OR INDIRECT DAMAGES RESULTING FROM THE USE OR HANDLING OF THIS PRODUCT AND SELLER'S SOLE LIABILITY AND BUYER'S AND USER'S EXCLUSIVE REMEDY SHALL BE LIMITED TO THE REFUND OF THE PURCHASE PRICE.

WT 4-88 A-4R

The information and specimen labels in this booklet are correct as of the publication date but are subject to change without notice. For current information, contact ICI Americas Inc.

SULFUR SIGNAL™

DUSTING SULFUR

An Agricultural Insecticide and Fungicide

ACTIVE INGREDIENT:	
SULFUR	98.0%
INERT INGREDIENTS:	2.0%
TOTAL	100.0%

EPA Reg. No. 10182-143

KEEP OUT OF REACH OF CHILDREN

CAUTION

STATEMENT OF PRACTICAL TREATMENT

If an adverse reaction occurs after exposure: Promptly start the recommended procedures below. **For 24-hour emergency medical assistance,** call (800) F-A-S-T-M-E-D (327-8633). **For chemical emergency:** Spill, leak, fire, exposure, or accident call CHEMTREC 1-800-424-9300.

If a large amount of this material is swallowed: Give several glasses of water. Induce vomiting by gagging the victim with a finger placed on the back of the victim's tongue. Give fluids until vomitus is clear. Do not induce vomiting or give anything by mouth to an unconscious or convulsing person.

If inhaled: Remove to fresh air. Seek medical attention if respiratory irritation occurs or if breathing becomes difficult.

In case of contact: Flush all affected areas with plenty of water for several minutes. Remove and clean contaminated clothing and shoes.

For eye contact: Hold eyelids apart and flush with plenty of water for 15 minutes. Get medical attention if irritation occurs.

PRECAUTIONARY STATEMENTS

HAZARDS TO HUMANS AND DOMESTIC ANIMALS

CAUTION

- Do not take internally.
- Avoid contact with skin, eyes, and clothing.
- Wash thoroughly after handling.

ENVIRONMENTAL HAZARDS

- Do not apply when weather conditions favor draft from treated areas.
- Do not apply directly to water or wetlands. Do not contaminate water by cleaning of equipment or disposal of wastes.

PHYSICAL AND CHEMICAL HAZARDS

BURNING SULFUR EMITS HIGHLY TOXIC FUMES. SULFUR DUST SUSPENDED IN AIR IGNITES EASILY.

Keep away from heat, sparks and flame. Avoid creating dust in handling. Keep container closed. Use with adequate ventilation Keep from contact with oxidizing materials.

In Case of Fire Use water spray, fog CO2 or sand. Wear NIOSH approved self-contained breathing apparatus.

PRODUCT INFORMATION

Sulfur will cause severe fruit and leaf injury to sulfur-sensitive crops. Do not use on apricots, Anjou and cornice pears, cranberries, cucurbits (cucumbers, squash and some varieties of melons), filberts, spinach, tung trees, walnuts and other sensitive plants.

Except on cotton, do not apply when shade temperatures exceed or are likely to exceed 90°F.

Sulfur may burn foilage and fruit during periods of high temperatures and under certain climatic conditions.

Do not use sulfur within 30 days of an oil spray (60 days on citrus).

If the crop to be treated will be used for canning or freezing purposes, consult food processor before applying sulfur.

Consult State Agricultural Extension Service or State Agricultural Experiment Stations for additional information, as the timing and number of applications needed will vary with local conditions.

DIRECTIONS FOR USE

It is a violation of Federal law to use this product in a manner inconsistent with its labeling.

Apply the specified amounts of this product uniformly to the foilage when air is calm, using ground power equipment or aircraft. Except as otherwise shown, apply at the first sign of infection and repeat at 7-10 day intervals as needed to maintain control.

Use the higher dosage during periods of frequent and heavy rainfall. Spray often enough to keep susceptible growth covered This may require weekly or bi-weekly applications during periods of rapid growth and frequent rainfall.

REENTRY STATEMENT

- Do not apply this product in such a manner as to directly or through drift expose workers or other persons. The area being treated must be vacated by unprotected persons.
- Do not enter treated areas for 24 hours unless appropriate protective clothing is worn.
- Because certain states may require more restrictive reentry intervals for various crops treated with this product, consult your State Department of Agriculture for further information.
- Written or oral warnings must be given to workers who are expected to be in a treated area or in an area about to be treated with this product. Oral warnings must be given if there is reason to believe that written warnings cannot be understood by workers.

When oral warnings are given, warnings shall be given in a language customarily understood by workers and must indicate the specific period of time that treated areas may not be entered without protective clothing. Oral warnings must also include appropriate first aid instructions in case of accidental exposure. These instructions are given in the STATEMENT OF PRACTICAL TREATMENT at the beginning of this label.

Written warnings must include the following information. "CAUTION Area treated with SIGNAL™ Dusting Sulfur on (date of application). Do not enter treated areas for 24 hours unless protective clothing is worn (or other reentry interval if your State has a more restrictive interval for this product). If accidental exposure occurs, follow the instructions below." (Written warnings must include the STATEMENT OF PRACTICAL TREATMENT given at the beginning of this label.)

RECOMMENDATIONS

FRUIT CROPS

APPLES: Scab, powdery mildew—Apply 30-35 lbs. per acre in the green tip to pink bud stage. MacIntosh, Golden Delicious, Jonathan and certain other varieties may be injured by sulfur Use on sulfur tolerant varieties only.

CHERRIES: Blossom blight, brown rot, powdery mildew—Apply 50 lbs. per acre in the popcorn, full bloom and petal-fall stages. Leaf spot, powdery mildew—Apply 50 lbs. per acre at petal-fall, shuck-fall and 2 weeks after shuck-fall.

CITRUS: Six-spotted and citrus rust mites—Apply 80 lbs. per acre. Citrus and Yuma spider mites—Apply 100 lbs. per acre Citrus thrips—Apply 100 lbs. per acre in three applications: mid-March, petal-fall, and 2-4 weeks later. Avoid applications during or preceeding high temperatures. Do not apply within 2 months of an oil spray.

GRAPES: Rust, powdery midlew—Apply 5-10 lbs. per acre in successive applications when shoots are 6" long, shoots 12-15" long, 14 days afterwards and 10-14 days thereafter to harvest. Caution: Concord and Labrusca type may be injured by sulfur.

PLUMS, PRUNES: Brown rot, leaf spot, prune rust—Apply 35-50 lbs. per acre. For brown rot of fruit, dust when fruit begins to ripen. For leaf spot and rust, begin at pink stage, repeat at bloom, petal-fall, shuck-fall and in cover period apply as necessary.

STRAWBERRIES: Powdery mildew—Apply 10-20 lbs. per acre.

VEGETABLES & FIELD CROPS

ALFALFA: Spider mites—Apply 15-25 lbs. per acre Caution: Sulfur may cause petal burn to open flowers.

BEANS: Powdery mildew, rust, spider mites—Apply 15-30 lbs. per acre. For rust control, start applications as soon as the first true leaves form and continue at weekly intervals until blossom time. Caution: May cause injury to some varieties of beans.

CARROTS: Petrobia mite—Apply 30-35 lbs. per acre.

CLOVER: Powdery mildew, spider mites—Apply 15-25 lbs. per acre.

COTTON: Atlantic mite, cotton rust—Apply 25-30 lbs. per acre.

GRASS SEED CROPS: Timothy mite—Apply 30 lbs. per acre.

MELONS: (Honeydew and sulfur-resistant melons only)—Powdery mildew, spider mites—Apply 20-25 lbs. per acre.

ONIONS: Petrobia mite—Apply 20-30 lbs. per acre during warm weather.

PEAS: Powdery mildew—Apply 15-20 lbs. per acre.

SUGAR BEETS: Spider mite—Apply 30-35 lbs. per acre.

TOMATOES: Tomato russet mite—Apply 15-40 lbs. per acre.

TURNIPS and RUTABAGAS: Powdery mildew—Apply 15-20 lbs. per acre.

ORNAMENTALS

Make full coverage application on both sides of leaves. Keep new growth protected.

CHRYSANTHEMUM: Powdery mildew, rust. Do not apply when flowering.

HOLLYHOCK: Rust

ROSES, PERENNIALS, SHRUBS: Black spot, powdery mildew, rust. Caution: Some burning may occur when used during warm weather.

STORAGE AND DISPOSAL

- Do not contaminate water, food, or feed by storage or disposal.

STORAGE: Store in a cool dry place.

PESTICIDE DISPOSAL: Wastes resulting from the use of this product may be disposed of on site or at an approved waste disposal facility.

CONTAINER DISPOSAL: Completely empty bag into application equipment. Then dispose of empty bag in a sanitary landfill or by incineration, or, if allowed by State and local authorities, by burning. If burned, stay out of smoke.

SPILL OR LEAK: Small spills can be handled routinely. Use adequate ventilation and wear a respirator or a dust mask to prevent inhalation. Wear suitable protective clothing and eye protection to prevent skin and eye contact. Use the following procedures.

Sweep or shovel up spilled material using a natural fiber broom and/or aluminum shovel to reduce sparking. Be careful not to create dust as an explosion could result. Place sweepings in an appropriate chemical waste container. Seal container and dispose of in an approved facility. Flush spill area with detergent and water to remove any residue.

A-2R

NOTICE TO BUYER AND USER

Seller warrants that this product conforms to the chemical description on the label and is reasonably fit for the purposes stated on the label when used in accordance with directions under normal conditions of use. This warranty does not extend to the use of this product contrary to label instructions or under abnormal use conditions or under conditions not reasonably foreseeable to Seller and Buyer and User assumes the risk of any such use. SELLER DISCLAIMS ALL OTHER WARRANTIES EXPRESSED OR IMPLIED INCLUDING ANY WARRANTY OF FITNESS OR MERCHANTABILITY. SELLER SHALL NOT BE LIABLE FOR CONSEQUENTIAL SPECIAL OR INDIRECT DAMAGES RESULTING FROM THE USE OR HANDLING OF THIS PRODUCT AND SELLER'S SOLE LIABILITY AND BUYER'S AND USER'S EXCLUSIVE REMEDY SHALL BE LIMITED TO THE REFUND OF THE PURCHASE PRICE.

The information and specimen labels in this booklet are correct as of the publication date but are subject to change without notice. For current information, contact ICI Americas Inc.

SULFUR SPECIAL ELECTRIC

REFINED SUPER-ADHESIVE DUSTING SULFUR

An Agricultural Insecticide and Fungicide

ACTIVE INGREDIENT:

SULFUR	98.0%
INERT INGREDIENTS:	2.0%
TOTAL	100.0%

GUARANTEE OF FINENESS: ALL MATERIAL CONTAINED IN THIS BAG IS GUARANTEED TO PASS 95% THROUGH A 325 MESH SCREEN.

EPA Reg. No. 10182-142

™ SPECIAL ELECTRIC is a registered trademark of ICI Americas Inc.

KEEP OUT OF REACH OF CHILDREN

CAUTION

STATEMENT OF PRACTICAL TREATMENT

If an adverse reaction occurs after exposure: Promptly start the recommended procedures below. Seek medical attention if further treatment is required.

If a large amount of this material is swallowed: Give several glasses of water. Induce vomiting by gagging the victim with a finger placed on the back of the victim's tongue. Give fluids until vomitus is clear. Do not induce vomiting or give anything by mouth to an unconscious or convulsing person.

If inhaled: Remove to fresh air. Seek medical attention if respiratory irritation occurs or if breathing becomes difficult.

In case of contact: Flush all affected areas with plenty of water for several minutes. Remove and clean contaminated clothing and shoes.

For eye contact: Hold eyelids apart and flush with plenty of water for 15 minutes. Get medical attention if irritation occurs.

FOR 24-HOUR EMERGENCY MEDICAL ASSISTANCE CALL 1-800-F-A-S-T-M-E-D (327-8633).
FOR CHEMICAL EMERGENCY: Spill, leak, fire, exposure, or accident call CHEMTREC 1-800-424-9300.

For other product information, contact your local ICI Americas Distributor or see Material Safety Data Sheet.

PRECAUTIONARY STATEMENTS

HAZARDS TO HUMANS AND DOMESTIC ANIMALS

CAUTION

- Do not take internally.
- Avoid contact with skin, eyes, and clothing.
- Wash thoroughly after handling.

ENVIRONMENTAL HAZARDS

- Do not apply when weather conditions favor draft from treated areas.
- Do not apply directly to water or wetlands. Do not contaminate water by cleaning of equipment or disposal of wastes.

PHYSICAL AND CHEMICAL HAZARDS

BURNING SULFUR EMITS HIGHLY TOXIC FUMES. SULFUR DUST SUSPENDED IN AIR IGNITES EASILY.

Keep away from heat, sparks and flame. Avoid creating dust in handling. Keep container closed. Use with adequate ventilation. Keep from contact with oxidizing materials.

In Case of Fire: Use water spray, fog, CO2 or sand. Wear NIOSH approved self-contained breathing apparatus.

PRODUCT INFORMATION

Sulfur will cause severe fruit and leaf injury to sulfur-sensitive crops. Do not use on apricots, Anjou and cornice pears, cranberries, cucurberts (cucumbers, squash and some varieties of melons), filberts, spinach, tung trees, walnuts and other sensitive plants.

Except on cotton, do not apply when shade temperatures exceed or are likely to exceed 90°F.

Sulfur may burn foilage and fruit during periods of high temperatures and under certain climatic conditions.

Do not use sulfur within 30 days of an oil spray (60 days on citrus).

If the crop to be treated will be used for canning or freezing purposes, consult food processor before applying sulfur.

Consult State Agricultural Experiment Stations or the State Agricultural Extension Service for additional information, as the timing and number of aplications needed will vary with local conditions.

DIRECTIONS FOR USE

It is a violation of Federal law to use this product in a manner inconsistent with its labeling.

Apply the specified amounts of this product uniformly to the foilage when air is calm, using ground power equipment or aircraft. Except as otherwise shown, apply at the first sign of infection and repeat at 7-10 day intervals as needed to maintain control.

REENTRY STATEMENT

- Do not apply this product in such a manner as to directly or through drift expose workers or other persons. The area being treated must be vacated by unprotected persons.
- Do not enter treated areas for 24 hours unless appropriate protective clothing is worn.
- Because certain states may require more restrictive reentry intervals for various crops treated with this product, consult your State Department of agriculture for further information.
- Written or oral warnings must be given to workers who are expected to be in a treated area or in an area about to be treated with this product. Oral warnings must be given if there is reason to believe that written warnings cannot be understood by workers.

When oral warnings are given, warnings shall be given in a language customarily understood by workers and must indicate the specific period of time that treated areas may not be entered without protective clothing. Oral warnings must also include appropriate first aid instruction in case of accidental exposure. These instructions are given in the STATEMENT OF PRACTICAL TREATMENT at the beginning of this label.

Written warnings must include the following information, "CAUTION. Area treated with SPECIAL ELECTRIC™ Dusting Sulfur on (date of application). Do not enter treated areas for 24 hours unless appropriate protective clothing is worn (or after reentry interval if your state has a more restrictive interval for this product). If accidental exposure occurs, follow the instructions below." (Written warnings must include the STATEMENT OF PRACTICAL TREATMENT given at the beginning of this label.)

RECOMMENDATIONS

FRUIT CROPS

APPLES: Scab, powdery mildew—apply 30-35 lbs. per acre in the green tip to pink bud stage. MacIntosh, Golden Delicious, Jonathan and certain other varieties may be injured by sulfur. Use on sulfur-tolerant varieties only.

CHERRIES: Blossom blight, brown rot, powdery mildew—apply 50 lbs. per acre in the popcorn, full bloom and petal-fall stages. Leaf spot, powdery mildew—apply 50 lbs. per acre at petal-fall, shuck-fall and 2 weeks after shuck-fall.

CITRUS: Six-spotted and citrus rust mites—apply 80 lbs per acre. Citrus and Yuma spider mites—apply 100 lbs. per acre. Citrus thrips—apply 100 lbs. per acre in three applications: mid-March, petal-fall, and 2-4 weeks later. Avoid applications during or preceding high temperatures. Do not apply whithin 2 months of an oil spray.

GRAPES: Rust, powdery mildew—apply 10-20 lbs per acre in successive applications: when shoots are 6" long, shoots 12-15" long, 14 days afterwards, and 10-14 days thereafter to harvest. Caution: Concord and labrusca type grapes may be injured by sulphur.

PLUMS, PRUNES: Brown rot, leaf spot, prune rust—apply 35-50 lbs. per acre. For brown rot of fruit, dust when fruit begins to ripen. For leaf spot and rust begin at pink stage, repeat at bloom, petal-fall, shuck-fall and in cover period apply as necessary.

STRAWBERRIES: Powdery mildew—apply 10-20 lbs. per acre.

VEGETABLE & FIELD CROPS

ALFALFA: Spider mites—apply 15-25 lbs. per acre. Caution: sulfur may cause petal burn to open flowers.

BEANS: Powdery mildew, rust, spider mites—apply 15-30 lbs. per acre. For rust control start applications as soon as the first true leaves form and continue at weekly intervals until blossom time. Caution: may cause injury to some varieties of beans.

CARROTS: Petrobia mite—apply 30-35 lbs. per acre.

CLOVER: Powdery mildew, spider mites—apply 15-25 lbs. per acre.

COTTON: Atlantic mite, cotton rust—apply 25-30 lbs. per acre.

GRASS SEED CROPS: Timothy mite—apply 30 lbs. per acre.

MELONS: (Honeydew and sulfur-resistant melons only)—Powdery mildew, spider mites—apply 20-25 lbs. per acre.

ONIONS: Petrobia mite—apply 20-30 lbs. per acre during warm weather.

PEAS: Powdery mildew—apply 15-20 lbs. per acre.

SUGAR BEETS: Spider mite—apply 30-35 lbs. per acre.

TOMATOES: Tomato russet mite—apply 15-40 lbs. per acre.

TURNIPS and RUTABAGAS: Powdery mildew—apply 15-20 lbs. per acre.

ORNAMENTALS:

Make full coverage application on both sides of leaves. Keep new growth protected.

HOLLYHOCK: Rust

ROSES, PERENNIALS, SHRUBS: Black spot, powdery mildew, rust. Caution: some burning may occur when used during warm weather.

STORAGE AND DISPOSAL

- Do not contaminate water, food, or feed by storage or disposal.

STORAGE: Store in a cool dry place.

PESTICIDE DISPOSAL: Wastes resulting from the use of this product may be disposed of on site or at an approved waste disposal facility.

CONTAINER DISPOSAL: Completely empty bag into application equipment. Then dispose of empty bag in a sanitary landfill or by incineration or, if allowed by State and local authorities, by burning. If burned, stay out of smoke.

SPILL OR LEAK: Small spills can be handled routinely. Use adequate ventilation and wear a respirator or a dust mask to prevent irritation. Wear suitable protective clothing and eye protection to prevent skin and eye contact. Use the following procedures.

Sweep or shovel up spilled material using a natural fiber broom and/or aluminum shovel to reduce sparking. Be careful not to create dust as an explosion could result. Place sweepings in an appropriate chemical waste container. Seal container and dispose of in an approved facility. Flush spill area with detergent and water to remove any residue.

NOTICE TO BUYER AND USER:

Seller warrants that this product conforms to the chemical description on the label and is reasonably fit for the purposes stated on the label when used in accordance with directions under normal conditions of use. This warranty does not extend to the use of this product contrary to label instructions or under abnormal use conditions, or under conditions not reasonably foreseeable to Seller and Buyer and User assumes the risk of any such use. SELLER DISCLAIMS ALL OTHER WARRANTIES EXPRESSED OR IMPLIED INCLUDING ANY WARRANTY OF FITNESS OR MERCHANTABILITY. SELLER SHALL NOT BE LIABLE FOR CONSEQUENTIAL, SPECIAL OR INDIRECT DAMAGES RESULTING FROM THE USE OR HANDLING OF THIS PRODUCT AND SELLER'S SOLE LIABILITY AND BUYER'S AND USER'S EXCLUSIVE REMEDY SHALL BE LIMITED TO THE REFUND OF THE PURCHASE PRICE.

A-5R

The information and specimen labels in this booklet are correct as of the publication date but are subject to change without notice. For current information, contact ICI Americas Inc.

SURPASS® 6.7-E

SELECTIVE HERBICIDE EMULSIFIABLE LIQUID

For Corn (Field, Sweet and Silage) Complete Directions For Use, Use Precautions and Recommendations

ACTIVE INGREDIENT:	
S-propyl dipropylthiocarbamate	83.4%
INERT INGREDIENTS:	16.6%
TOTAL	100.0%

Contains 6.7 Pounds of Active Ingredient Per Gallon

EPA Reg. No. 10182-235

KEEP OUT OF REACH OF CHILDREN

CAUTION

Harmful if Swallowed

STATEMENT OF PRACTICAL TREATMENT

If a known exposure occurs or is suspected, please start the procedures given below. If further treatment is required, contact a Poison Center, a physician, or the nearest hospital.

If swallowed: Immediately give several glasses of water but *do not* induce vomiting. If vomiting does occur, give fluids again. Have a physician determine if condition of patient will permit induction of vomiting or evacuation of stomach. Do not give anything by mouth to an unconscious or convulsing person.

For eye contact: Immediately flush eyes with large amounts of running water for at least 15 minutes. Hold eyelids apart to ensure rinsing of the entire surface of the eye and lids with water. Get medical attention immediately.

For skin contact: Flush all affected areas with plenty of water for several minutes. Remove and clean contaminated clothing and shoes. Seek medical attention if irritation occurs.

If inhaled: Remove to fresh air. Seek medical attention if respiratory irritation occurs or if breathing becomes difficult.

FOR 24-HOUR EMERGENCY MEDICAL ASSISTANCE CALL 1-800-F-A-S-T-M-E-D (327-8633).
FOR CHEMICAL EMERGENCY: Spill, leak, fire, exposure, or accident call CHEMTREC 1-800-424-9300.

CAN BE STORED AT TEMPERATURES AS LOW AS MINUS 50°F.

PRECAUTIONARY STATEMENTS

HAZARDS TO HUMANS AND DOMESTIC ANIMALS

CAUTION

Harmful if Swallowed. Avoid contact with skin, eyes and clothing. Avoid inhalation of spray mist. Wear rubber gloves, goggles and protective clothing. Wash with soap and water immediately after use or contact. Do not contaminate food and feed. Keep away from heat or flame.

Do not apply this product in such a manner as to directly or through drift, expose workers or other persons. When using the sprinkler method of application, the area being treated must be vacated by unprotected persons.

Environmental Hazards—Do not apply directly to water. Do not contaminate water by cleaning of equipment or disposal of wastes.

Weeds Controlled by SURPASS 6.7-E in Corn

SURPASS 6.7-E will not control established weeds.

Annual Grasses:	
Barnyardgrass;	
Watergrass	(*Echinochloa crus-galli*)
Crabgrass	(*Digitaria* spp.)
Fall Panicum*	(*Panicum dichotomiflorum*)
Field Sanbur*	(*Cenchrus pauciflorus*)
German Millet	(*Setaria italica*)
Giant Foxtail	(*Setaria faberi*)
Goosegrass	(*Eleusine indica*)
Green Foxtail	(*Setaria viridis*)
Seedling Johnsongrass*	(*Sorghum halepense*)
Shattercane;	
Wild Cane*†	
Black Amber Cane;	
Chicken Corn	(*Sorghum bicolor*)
Wild Oats	(*Avena fatua*)
Wild Proso Millet*	(*Panicum miliaceum*)
Witchgrass	(*Panicum capillare*)
Yellow Foxtail	(*Setaria lutescens*)

* Cultivation is suggested in addition to the SURPASS 6.7-E treatment in fields with moderate to heavy infestations of wild cane, sandbur, johnsongrass seedlings, fall panicum or wild proso millet.

† Partial control (suppression) only.

Annual Broadleaf Weeds

The broadleaf weeds listed below will be controlled only if treatment is made when conditions are favorable for weed seed germination and growth.

Tall Morningglory	(*Ipomoea purpurea*)
Carpetweed	(*Mollugo verticillata*)
Common Purslane	(*Portulaca oleracea*)
Florida Pusley	(*Richardia scabra*)
Lambsquarters	(*Chenopodium album*)
Redroot Pigweed	(*Amaranthus retroflexus*)
Sicklepod	(*Cassia obtusifolia*)
Smooth Pigweed	(*Amaranthus hybridus*)
Velvetleaf (Buttonweed)	(*Abutilon theophrasti*)
Perennial Weeds:	
Purple Nutsedge	(*Cyperus rotundus*)
Yellow Nutsedge	(*Cyperus esculentus*)
Rhizome Johnsongrass	(*Sorghum halepense*)
Quackgrass*	(*Agropyron repens*)

*Partial control (suppression) only.

Existing stands of perennial weeds must be turned under and chopped up thoroughly before treatment. Cultivation is suggested in addition to SURPASS 6.7-E treatment in fields with moderate to heavy infestations.

GENERAL PRECAUTIONS

Read all label directions before using.

Applied according to directions and under normal growing conditions, SURPASS 6.7-E will not harm the treated crop. During the germination and early stages of growth, extended periods of unusally cold and wet or hot and dry weather, insect or plant disease attack, carry-over pesticide residues, the use of certain soil-applied systemic insecticides, improperly placed fertilizers or soil insecticides may create abnormal conditions that weaken crop seedlings. SURPASS 6.7-E used under these abnormal conditions could result in crop injury.

When applied according to directions and when conditions exist for normal plant growth through the season, no harmful residues of SURPASS 6.7-E should remain beyond harvest. In the Southeastern U.S., when SURPASS 6.7-E is used for weed control in silage corn, do not seed small seeded grains after corn harvest until September.

Do not use on milo or sorghum. Do not use on corn seed stock (Breeders, Foundation, or Increase).

SURPASS 6.7-E is recommended for use on mineral soils only (soils containing less than 10% organic matter).

Do not contaminate irrigation water (used for crops other than corn) or water used for domestic purposes.

DIRECTIONS FOR USE OF SURPASS 6.7-E

IT IS A VIOLATION OF FEDERAL LAW TO USE THIS PRODUCT IN A MANNER INCONSISTENT WITH ITS LABEL.

SURPASS 6.7-E is a selective herbicide which is mixed (incorporated) or injected into the soil for control of weeds listed on this label. SURPASS 6.7-E controls annual weeds as their seeds germinate by interfering with normal germination and seedling development. It does not control established weeds. All weed growth and crop stubble should be thoroughly worked into the soil before treatment.

Impregnation of Dry Bulk Fertilizers

Dry bulk fertilizer may be impregnated with SURPASS 6.7-E (except nitrate fertilizers) and applied as the fertilizer is spread.

See Appendix III for information and directions regarding impregnation and use of these combinations.

Combination with Fluid Fertilizers

SURPASS 6.7-E may be combined with fluid (solution, slurry, or suspension) fertilizers. *However, physical compatibility with these fluid fertilizers must be determined before combination in the spray tank.* See Appendix IV for directions regarding these combinations. Even though found to be compatible, constant agitation is necessary to keep the SURPASS 6.7-E evenly mixed with the fluid fertilizer during application.

APPLICATION AND SOIL INCORPORATION DIRECTIONS

Application Directions

Pour the recommended amount of SURPASS 6.7-E into the spray tank during the filling operations. Apply in 10 to 50 gallons of water per acre using a properly calibrated, low-pressure (20-40 psi) boom type sprayer having good agitation (use the higher gallonage in arid areas). The soil should be well-worked and dry enough to permit good soil mixing (incorporation).

Incorporation Directions

SURPASS 6.7-E must be incorporated into the soil immediately to prevent loss of the herbicide. Whenever possible, application and incorporation should be done in the same operation.

See Appendix I for soil mixing (incorporation) before planting directions.

Important: Thorough mixing for SURPASS 6.7-E in the soil is necessary. The soil must be in good tilth. Improper discing or poor tilth (large clods) may result in unsatisfactory control.

INCORPORATION DIRECTIONS (PACIFIC NORTHWEST ONLY)

Mechanical Incorporation: SURPASS 6.7-E must be incorporated into the soil within 24 hours after application. The application should be made just prior to planting. The soil surface should be dry (at least 1/2 inch deep) and free from dew and incidental moisture.

Important: Thorough mixing of SURPASS 6.7-E in the soil is necessary. The soil must be in good tilth. Improper discing or poor tilth (large clods) may result in unsatisfactory control.

Sprinkler Incorporation: Surface apply SURPASS 6.7-E immediately after planting. The soil surface should be dry (at least 1/2 inch deep) and free from dew and incidental moisture. Incorporate using 1/2 to 3/4 inch of water within 36 hours following application. The application and incorporation must be done within 5 days after the last tillage operation since poor results will occur if weeds have germinated.

Soil Injection

Coulters and injector shanks must be spaced 3 1/2 inches apart. The two shanks adjacent to the drill row must be 1 3/4 to 2 inches on either side of it. Mount injectors in staggered positions to avoid trash buildup. Set injectors to inject SURPASS 6.7-E 3 inches below the soil surface. The width of the band in which weed control is desired will determine the number of injectors required per row. (Example: Four injectors spaced 3 1/2 inches apart give a 14 inch band). A broadcast application can be made by increasing the number of shanks. Seal injector openings by using a roller or chain drag behind the injectors. Apply the recommended rate of SURPASS 6.7-E in 20 to 50 gallons of water per acre.

For band applications, calculate the amount to be applied per acre as follows:

$$\frac{\text{Band width in inches}}{\text{distance between rows in inches}} \times \text{Rate/A for broadcast treatment} = \text{Amount needed for band treatment}$$

CENTER PIVOT SPRINKLER APPLICATION

Apply SURPASS 6.7-E with irrigation water before planting (see planting), or immediately after planting before corn or weeds emerge. It is important that application and incorporation be done immediately after the last tillage operation since poor results will occur if the weeds have germinated before application of the herbicide. Meter the SURPASS 6.7-E into the irrigation water during the entire period.

Apply in $^1/_2$ to 1 inch of water or sufficient water to penetrate the soil to a depth of 5 to 7 inches. If diluted, SURPASS 6.7-E should be mixed at a volume ratio of no more than 1 part SURPASS 6.7-E to 4 parts of water. Good agitation should be maintained during the entire application period.

Use Precautions for Chemigation:

A. Apply this product only through center pivot irrigation systems. Do not apply this product through any other type of irrigation system.
B. Crop injury, lack of effectiveness, or illegal pesticide residues in the crop can result from nonuniform distribution of treated water.
C. If you have questions about calibration, you should contact State Extension Service specialists, equipment manufacturers or other experts.
D. Do not connect an irrigation system (including greenhouse systems) used for pesticide application to a public water system unless the pesticide label-prescribed safety devices for public water systems are in place.
E. A person knowledgeable of the chemigation system and responsible for its operation, or under the supervision of the responsible person, shall shut the system down and make necessary adjustments should the need arise.

Use Precautions For Center Pivot Sprinkler:

1. The system must contain a functional check valve, vacuum relief valve, and low pressure drain appropriately located on the irrigation pipeline to prevent water source contamination from backflow.
2. The pesticide injection pipeline must contain a functional, automatic, quick-closing check valve to prevent the flow of fluid back toward the injection pump.
3. The pesticide injection pipeline must also contain a functional, normally closed, solenoid-operated valve located on the intake side of the injection pump and connected to the system interlock to prevent fluid from being withdrawn from the supply tank when the irrigation system is either automatically or manually shut down.
4. Inject the product with a positive displacement pump into the main line ahead of a right angle turn to insure adequate mixing.
5. The system must contain functional interlocking controls to automatically shut off the pesticide injection pump when the water pump motor stops.
6. The irrigation line or water pump must include a functional pressure switch which will stop the water pump motor when the water pressure decreases to the point where pesticide distribution is adversely affected.
7. Systems must use a metering pump, such as a positive displacement injection pump (e.g., diaphragm pump) effectively designed and constructed of materials that are compatible with pesticides and capable of being fitted with a system interlock.
8. Do not apply when wind speed favors drift beyond the area intended for treatment, when system connections or fittings leak, when nozzles do not provide uniform distribution or when lines containing the product must be dismantled and drained.

Planting

Plant corn seed to a maximum depth of 2 inches. Ideally, planting should be done within 2 weeks after preplant application. If planting must be delayed, do not delay more than 4 weeks after preplant application. To avoid removing SURPASS 6.7-E from the row (with loss of weed control in the row), do not move or shape soil after treatment, such as by planting in deep furrows. If it is necessary to work the soil following application and incorporation, but before planting, do not work the soil deeper than the depth SURPASS 6.7-E was incorporated.

Cultural Practices Following Application

Should weeds develop, a shallow cultivation or rotary hoeing will generally result in better weed control. When cultivating for any reason, it should be shallow, i.e., no more than $^1/_2$ the depth the herbicide was incorporated.

Under circumstances when a poor stand of corn is obtained after SURPASS 6.7-E treatment (caused by flooding, disease, etc.) corn can be planted again without working the field. If it is too late to plant corn again, soybeans or peanuts may be planted providing no other herbicide was used with the SURPASS 6.7-E. An interval of 4 weeks should elapse between when SURPASS 6.7-E was applied and the soybeans or peanuts are planted. Disc cultivate the soil thoroughly prior to planting the soybeans.

Dosage Recommendations

All rates are expressed as pints of formulation per broadcast acre. For all uses cultivation is recommended in addition to the SURPASS 6.7-E treatment in fields with moderate to heavy infestations of shattercane, nutgrass, quackgrass, sandbur, johnsongrass seedlings, fall panicum or wild proso millet. Note: SURPASS 6.7-E must be applied prior to May 1st on corn grown in Arizona or in the ten southern California counties.

For control of most annual grasses and broadleaf weeds: Use $3^3/_4$ to $4^3/_4$ pints of SURPASS 6.7-E. Use the higher rate on medium (silt loam, clay loam, etc.) to fine (clay) soils and for control of sandbur, johnsongrass seedlings, fall panicum and witchgrass.

For partial control (suppression) of shattercane and for control of nutsedge (nutgrass) and wild proso millet: Use $4^3/_4$ to $7^1/_3$ pints of SURPASS 6.7-E, use the higher rates on medium (silt loam, clay loam, etc.) soils, fine (clay) soils and for heavy infestations.

For partial control (suppression) of quackgrass: Use $7^1/_3$ pints of SURPASS 6.7-E.

Johnsongrass: Using a program of plowing and discing followed by SURPASS 6.7-E will give good control of seedlings and greatly reduce competition from plants coming from rhizomes.

In some areas discing 2 to 3 times spaced at 4 to 6 week intervals during fall and winter can be done. If this cannot be done, start with fall plowing followed by several discings in the spring before planting. The discing cuts rhizomes into small pieces and brings them to the surface, where the weather will kill or weaken them.

Next, apply and disc-incorporate SURPASS 6.7-E just before planting at the rate of $7^1/_3$ pints per acre. Disc a second time at right angles to the first, cultivate at least once, preferably twice.

SURPASS 6.7-E PLUS ATRAZINE (80W, 4L or AATREX*NINE-O)
Tank Mix Combination

*AATREX NINE-O, ATRAZINE 80W and ATRAZINE 4L are registered trademarks of Ciba-Geigy Corporation

A tank mix combination of SURPASS 6.7-E and ATRAZINE can be applied and mixed into the soil before planting for weed control in corn. This combination will control a broader spectrum of weeds than will either material used alone. In addition, potential hazard of ATRAZINE residues affecting certain succeeding crops is reduced because a lower rate of ATRAZINE can be used when it is combined with SURPASS 6.7-E than when it is used alone.

Follow all use precautions and warnings that appear on the ATRAZINE label, supplementary ATRAZINE literature, and elsewhere on this label.

Make only one application per crop. After treatment with the SURPASS 6.7-E/ATRAZINE combination, do not plant any crop except corn until the following year or injury may occur.

When applied as directed, the SURPASS 6.7-E/ATRAZINE tank mixture will control these weeds:

Annual Grasses:	
Barnyardgrass; Watergrass	(*Echinochloa crus-galli*)
Crabgrass	(*Digitaria* spp.)
Fall Panicum*	(*Panicum dichotomiflorum*)
Field Sandbur*	(*Cenchrus pauciflorus*)
German Millet	(*Setaria italica*)
Giant Foxtail	(*Setaria faberi*)
Goosegrass	(*Eleusine indica*)
Green Foxtail	(*Setaria viridis*)
Seedling Johnsongrass*	(*Sorghum halepense*)
Shattercane*†	(*Sorghum bicolor*)
Wild Oats	(*Avena fatua*)
Wild Proso Millet*	(*Panicum miliaceum*)
Witchgrass	(*Panicum capillare*)
Yellow Foxtail	(*Setaria glauca*)

* Cultivation is suggested in addition to the SURPASS 6.7-E/ATRAZINE treatment in fields with moderate to heavy infestations of wild cane, sandbur, johnsongrass seedlings, fall panicum or wild proso millet.

† Partial control (suppression) only.

Annual Broadleaf Weeds:	
Annual morningglory	(*Ipomoea purpurea*)
Buffalobur	(*Solanum rostratum*)
Carpetweed	(*Mollugo verticillata*)
Common Cocklebur	(*Xanthium pensylvanicum*)
Common Lambsquarters	(*Chenopodium album*)
Common Purslane	(*Portulaca oleracea*)
Florida Pusley	(*Richardia scabra*)
Horsenettle	(*Solanum carolinense*)
Jimsonweed	(*Datura stramonium*)
Ladysthumb	(*Polygonum persicaria*)
Pennsylvania Smartweed	(*Polygonum pensylvanicum*)
Prostrate Pigweed	(*Amaranthus graecizans*)
Ragweed	(*Ambrosia* spp.)
Redroot Pigweed	(*Amaranthus retroflexus*)
Sicklepod	(*Cassia obtusifolia*)
Smooth Pigweed	(*Amaranthus hybridus*)
Velvetleaf	(*Abutilon theophrasti*)
Wild Mustard	(*Brassica kaber*)

Perennial Grasses:	
Purple Nutsedge, Nutgrass	(*Cyperus rotundus*)

Yellow Nutsedge, Nutgrass	(*Cyperus esculentus*)
Rhizome Johnsongrass	(*Sorghum halepense*)
Quackgrass**	(*Agropyron repens*)

** For partial control (suppression) of quackgrass: use 7 1/3 pints of SURPASS 6.7-E and 1 1/4 to 2* pounds ATRAZINE 80W, 1 to 1 1/2 quarts ATRAZINE 4L, or 1 to 1 3/4* pounds AATREX NINE-O per acre.

Existing stands of perennial weeds must be turned under and chopped up thoroughly before treatment. Cultivation is suggested in addition to SURPASS 6.7-E/ATRAZINE treatment in fields with moderate or heavy infestations.

*Use the higher rate of ATRAZINE on soils with over 5% organic matter.

DIRECTIONS FOR USE

The SURPASS 6.7-E/ATRAZINE (80W or 4L) combination can be mixed with water or combined with fluid (solution, slurry, or suspension) fertilizers and applied as a spray, or be impregnated on dry bulk fertilizers. The SURPASS 6.7-E/AATREX NINE-O combination can be mixed with water and applied as a spray.

Dry bulk fertilizer may be impregnated with SURPASS 6.7-E/ATRAZINE (80W or 4L) (except nitrate fertilizers) and applied as the fertilizer is spread. See Appendix III for information and directions regarding impregnation and use of these combinations.

SURPASS 6.7-E/ATRAZINE (80W or 4L) combination can be combined with fluid fertilizers for applications made before planting. However, the physical compatibility with these fluid fertilizers must be determined before combining in the spray tank, and even though found to be compatible, constant agitation is necessary to keep the SURPASS 6.7-E/ATRAZINE evenly mixed with the fluid fertilizer during application. Refer to Appendix II regarding the procedure for testing the compatiblilty of SURPASS 6.7-E/ATRAZINE combination with fluid fertilizers.

See Appendix II for Tank Mix Directions.

APPLICATION AND SOIL INCORPORATION DIRECTIONS

Spray Application and Mechanical Incorporation

Apply the recommended dosage in 10 to 50 gallons of water per acre (use the higher gallonage in arid areas), using a properly calibrated, low-pressure (20-40 psi) boom-type sprayer having good, but not excessive agitation. Check frequently during application to be sure nozzles are free from clogging and are delivering a uniform spray pattern. Read directions on the ATRAZINE label concerning nozzle and screen sizes.

The herbicides must be incorporated into the soil immediately after application to prevent loss of the SURPASS. Whenever possible, application and incorporation should be done in the same operation. Apply the herbicides to well-worked soil, dry enough to permit good soil mixing (incorporation).

See Appendix I for soil mixing (incorporation) before planting directions.

Important: Thorough mixing of SURPASS 6.7-E in the soil is necessary. The soil must be in good tilth. Improper discing or poor tilth (large clods) may result in unsatisfactory control.

INCORPORATION DIRECTIONS (PACIFIC NORTHWEST ONLY)

Mechanical Incorporation: The tank mix must be incorporated into the soil within 24 hours after application. The application should be made just prior to planting. The soil surface should be dry (at least 1/2 inch deep) and free from dew and incidental moisture.

Important: Thorough mixing of SURPASS 6.7-E in the soil is necessary. The soil must be in good tilth. Improper discing or poor tilth (large clods) may result in unsatisfactory control.

Sprinkler Incorporation: Surface apply SURPASS 6.7-E/ATRAZINE immediately after planting. The soil surface should be dry (at least 1/2 inch deep) and free from dew and incidental moisture. Incorporate using 1/2 to 3/4 inch of water within 36 hours following application. The application and incorporation must be done within 5 days after the last tillage operation since poor results will occur if weeds have germinated.

Planting

Do not plant seed deeper than 2 inches. For SURPASS 6.7-E/ATRAZINE application made before planting, seeding can be done immediately, or it may be delayed up to 4 weeks. To avoid removing SURPASS 6.7-E/ATRAZINE from the row (with loss of weed control in the row), do not move or shape soil after treatment, such as by planting in deep furrows. If it is necessary to work the soil following application and incorporation, but before planting, do not work the soil deeper than the depth the herbicides were incorporated.

Do not use on milo or sorghum.

Do not use on corn seed stock (Breeders, Foundation, or Increase).

Dosage Recommendations

All rates are expressed as pints of SURPASS 6.7-E combined with quarts of ATRAZINE 4L or pounds of ATRAZINE 80W or AATREX NINE-O per broadcast acre.

Note: The SURPASS 6.7-E/ATRAZINE tank mix combination is not recommended on corn grown in Arizona or in the ten southern California counties.

For all uses cultivation is recommended in addition to the SURPASS 6.7-E/ATRAZINE treatment in fields with moderate to heavy infestations of shattercane, nutgrass, quackgrass, sandbur, johnsongrass seedlings, fall panicum or wild proso millet.

For control of annual grasses and broadleaf weeds: Use 3 3/4 to 4 3/4 pints SURPASS 6.7-E and 1 to 2* pounds of ATRAZINE 80W, 3/4 to 1 1/2* quarts ATRAZINE 4L, or 8/10 to 1 3/4 pounds AATREX NINE-O per acre. Use the higher rate of SURPASS 6.7-E on medium (silt loam, clay loam, etc.) to fine (clay) soils and for control of sandbur, johnsongrass seedlings, fall panicum and witchgrass.

For partial control (suppression) of shattercane and for control of nutsedge (nutgrass) and wild proso millet (in addition to the broadleaf weeds and annual grasses listed): Use 4 3/4 to 7 1/3 pints of SURPASS 6.7-E and 1 to 2* pounds ATRAZINE 80W or 3/4 to 1 1/2* quarts ATRAZINE 4L or 8/10 to 1 3/4* pounds AATREX NINE-O per acre. Use the higher rates of SURPASS 6.7-E on medium (silt loam, clay loam, etc.) soils, fine (clay) soils, and for heavy infestations.

For partial control (suppression) of quackgrass: Use 7 1/3 pints of SURPASS 6.7-E and 1 1/4 to 2* pounds ATRAZINE 80W or 1 to 1 1/2 quarts ATRAZINE 4L or 1 to 1 3/4* pounds AATREX NINE-O per acre.

Johnsongrass: Using a program of plowing and discing followed by SURPASS 6.7-E and ATRAZINE will give good control of seedlings and greatly reduce competition from plants coming from rhizomes.

In some areas discing 2 to 3 times spaced at 4 to 6 week intervals during fall and winter is recommended. If this cannot be done, start with fall plowing followed by several discings in the spring before planting. The discing cuts rhizomes into small pieces and brings them to the surface, where both the weather and SURPASS 6.7-E will kill or weaken them.

Next, apply and disc-incorporate SURPASS 6.7-E just before planting at the rate of 7 1/3 pints plus 1 to 2* pounds of ATRAZINE 80W or 3/4 to 1 1/2* quarts of ATRZINE 4L, or 8/10 to 1 3/4* pounds AATREX NINE-O per acre. Disc a second time at right angles to the first. Cultivate at least once, preferably twice.

*Use the higher rate of ATRAZINE on soils with over 5% organic matter.

Cultural Practices Following Application

Should weeds develop, a shallow cultivation or rotary hoeing will generally result in better weed control. When cultivating for any reason, it should be shallow, i.e., no more than 1/2 the depth the herbicides were incorporated.

For all uses cultivation is recommended in addition to the SURPASS 6.7-E/ATRAZINE treatment in fields with moderate to heavy infestations of shattercane, nutgrass, quackgrass, sandbur, johnsongrass seedlings, fall panicum or wild proso millet.

SURPASS 6.7-E/BLADEX* (80W or 4L) Combination

*BLADEX is a registered trademark of E.I. duPont de Nemours & Co., Inc.

SURPASS 6.7-E may be used in combination with BLADEX (80W or 4L) for weed control in corn. These combinations will control a broader spectrum of weeds than will either of the products used alone.

These combinations may be applied as a tank mix, preplant and soil-incorporated; or the BLADEX (80W or 4L) may be applied pre-emergence following a preplant incorporated application of SURPASS 6.7-E.

Important: Observe all cautions and limitations on the BLADEX (80W or 4L) and SURPASS 6.7-E labels.

Make only one application per crop.

Do not use on milo or sorghum. Do not use on corn seed stock (Breeders, Foundation, or Increase).

Do not contaminate water by cleaning of equipment or disposal of wastes.

Weeds Controlled by SURPASS 6.7-E/BLADEX

Annual Grasses:

Annual Bluegrass	(*Poa annua*)
Annual Fescues	(*Festuca* spp.)
Annual (Italian) Ryegrass	(*Lolium multiflorum*)
Barnyardgrass, Watergrass	(*Echinochloa crus-galli*)
Bullgrass	(*Paspalum boscianum*)
Crabgrass	(*Digitaria* spp.)
Fall Panicum*	(*Panicum dichotomiflorum*)
Field Sandbur*	(*Cenchrus pauciflorus*)
German Millet	(*Setaria italica*)
Giant Foxtail	(*Setaria faberi*)
Goosegrass	(*Eleusine indica*)
Green Foxtail	(*Setaria viridis*)
Indian Lovegrass	(*Eragrostis pilosa*)
Junglerice	(*Echinochloa colonum*)
Seedling Johnsongrass*	(*Sorghum halepense*)
Stinkgrass	(*Eragrostis cilianensis*)

Shattercane,	
Wild Cane*†,	
Black Amber Cane	
Chicken Corn	
Scatter Cane	(*Sorghum bicolor*)
Wild Oats	(*Avena fatua*)
Wild Proso Millet*	(*Panicum miliaceum*)
Witchgrass	(*Panicum capillare*)
Yellow Foxtail	(*Setaria glauca*)

* Cultivation is suggested in addition to the SURPASS 6.7- E/BLADEX treatment in fields with moderate to heavy infestations of fall panicum, sandbur, johnsongrass seedlings, wild cane or wild proso millet.

†Partial control (suppression) only.

Annual Broadleaf Weeds:

Annual Buttercup	(*Ranunculus* spp.)
Annual Groundcherry	(*Physalis lanceifolia*)
Tall Morningglory	(*Impomoea purpurea*)
Annual Sedge	(*Carex* spp.)
Black Mustard	(*Brassica nigra*)
Buffalobur	(*Solanum rostratum*)
Carpetweed	(*Mollugo verticillata*)
Common Chickweed	(*Stellaria media*)
Common Cocklebur	(*Xanthium pensylvanicum*)
Common Groundsel	(*Senecio vulgaris*)
Common Lambsquarters	(*Chenopodium album*)
Common Mallow	(*Malva neglecta*)
Common Purslane	(*Portulaca oleracea*)
Corn Spurry	(*Spergula arvensis*)
Floras Paintbrush	(*Emilia sagittata*)
Florida Pusley	(*Richardia scabra*)
Hedge Mustard	(*Sisymbrium officinale*)
Jimsonweed	(*Datura stramonium*)
Kochia	(*Kochia scoparia*)
Ladysthumb	(*Polygonum persicaria*)
Mayweed	(*Anthemis cotula*)
Pennsylvania Smartweed	(*Polygonum pensylvanicum*)
Pigweed	(*Amaranthus* spp.)
Pineappleweed	(*Matricaria matricarioides*)
Plantain	(*Plantago* spp.)
Poorjoe	(*Diodia teres*)
Prostrate Knotweed	(*Polygonum aviculare*)
Protrate Spurge	(*Euphorbia supina*)
Ragweed	(*Ambrosia* spp.)
Seedling Curly Duck	(*Rumex crispus*)
Shepherdspurse	(*Capsella bursa-pastoris*)
Sicklepod	(*Cassia obtusifolia*)
Smallflower Galinsoga	(*Galinsoga parviflora*)
Prickly Sida	(*Sida spinosa*)
Sunflower	(*Helianthus annuus*)
Tarweed	(*Hemizonia congesta*)
Velvetleaf	(*Abutilon theophrasti*)
Wild Buckwheat	(*Polygonum convolvulus*)
Wild Mustard	(*Brassica kaber*)
Wild radish	(*Raphanus raphanistrum*)
Wild turnip	(*Brassica campestris*)

Perennial Weeds:

Rhizome Johnsongrass	(*Sorghum halepense*)
Purple Nutsedge	(*Cyperus rotundus*)
Yellow Nutsedge	(*Cyperus esculentus*)
Quackgrass*	(*Agropyron repens*)

*Partial control (suppression) only

Existing stands of perennial weeds must be turned under and chopped up thoroughly before treatment. Cultivation is suggested in addition to SURPASS 6.7-E/BLADEX treatment in fields with moderate to heavy infestations.

DIRECTIONS FOR USE

The SURPASS 6.7-E/BLADEX (80W or 4L) combination can be mixed with water or combined with fluid (solution, slurry, or suspension) fertilizers and applied as a spray.

Tank Mix Directions: See Appendix II

Application and Incorporation Directions:

For a tank mix application, apply the recommended dosage (see RECOMMENDATIONS) in 10 to 50 gallons of water per acre, (use the higher gallonage in arid areas) using a properly calibrated, low-pressure type nozzle (flood, fan or swirl chamber) on boom-type sprayer having good agitation. Check frequently during application to be sure the nozzles are free from clogging and are delivering a uniform spray pattern. Read directions on the BLADEX (80W or 4L) label concerning nozzle and screen sizes.

The tank mix must be incorporated into the soil immediately after application to prevent loss of the SURPASS 6.7-E. Whenever possible, application and incorporation should be done in the same operation. Apply the herbicides to worked soil dry enough to permit good mixing.

See Appendix I for soil mixing incorporation before planting directions.

Important: Thorough mixing of SURPASS 6.7-E in the soil is necessary. The soil must be in good tilth. Improper discing or poor tilth (large clods) may result in unsatisfactory control.

Incorporation Directions (Pacific Northwest Only)

Mechanical Incorporation: The tank mix must be incorporated into the soil within 24 hours after application. The application should be made just prior to planting. The soil surface should be dry (at least 1/2 inch deep) and free from dew and incidental moisture.

Important: Thorough mixing of SURPASS 6.7-E in the soil is necessary. The soil must be in good tilth. Improper discing or poor tilth (large clods) may result in unsatisfactory control.

Sprinkler Incorporation: Surface apply SURPASS 6.7-E/BLADEX immediately after planting. The soil surface should be dry (at least 1/2 inch deep) and free from dew and incidental moisture. Incorporate using 1/2 to 3/4 inch of water within 36 hours following application. The application and incorporation must be done within 5 days after the last tillage operation since poor results will occur if weeds have germinated.

SURPASS 6.7-E/BLADEX (80W or 4L) Fluid Fertilizer Tank Mix

The SURPASS 6.7-E/BLADEX (80W or 4L) combination can be combined with fluid fertilizers for applications made before planting. However, the physical compatibility with these fluid fertilizers must be determined before combining in the spray tank, and even though found to be compatible, constant agitation is necessary to keep SURPASS 6.7-E/BLADEX evenly mixed with the fluid fertilizer during application.

Refer to Appendix IV regarding the procedure for testing the compatibility of SURPASS 6.7-E/BLADEX combination with fluid fertilizers.

For sequential application of SURPASS 6.7-E preplant incorporated followed by BLADEX (80W or 4L) pre-emergence, consult the appropriate application directions on the respective product labels. Follow the dosage recommendations given below.

Dosage Recommendations

For all uses, cultivation is recommended in addition to the SURPASS 6.7-E/BLADEX treatment in fields with moderate to heavy infestations of shattercane, nutgrass, quackgrass, sandbur, johnsongrass seedlings, fall panicum or wild proso millet.

Use the lower rate of BLADEX on coarse to medium textured soils. Use the higher rate of BLADEX on fine (fine textured-clay) soils. Consult the BLADEX label for more specific recommended rates on various soil types.

SURPASS 6.7-E/BLADEX can be applied either as a tank mix combination and incorporated into the soil or apply SURPASS 6.7-E preplant incorporated followed by BLADEX pre-emergence. If BLADEX is surface-applied on top of a preplant application of SURPASS 6.7-E, a rotary hoeing or shallow cultivation is recommended if rainfall or sprinkler irrigation has not occurred within about 10 days after the application of BLADEX.

Note: SURPASS 6.7-E must be applied prior to May 1st on corn grown in Arizona or the ten southern California counties.

Broadcast Rate Per Acre (Conventional Sprays)

For control of annual grasses and broadleaf weeds: Use 3 3/4 to 4 3/4 pints of SURPASS 6.7-E and 2 to 2 1/2 pounds of BLADEX 80W or 3 to 4 pints BLADEX 4L per acre. Use the higher rates of SURPASS 6.7-E on medium (medium fine textured-silt loam, clay loam, etc.) to fine (fine textured-clay) soils and for control of sandbur, johnsongrass seedlings, fall panicum and witchgrass.

For partial control (suppression) of shattercane and for control of nutsedge (nutgrass) and wild proso millet (in addition to the broadleaf weeds and annual grasses listed.) Use 4 3/4 to 7 1/3 pints of SURPASS 6.7-E and 2 to 2 1/2 pounds BLADEX 80W or 3 to 4 pints BLADEX 4L per acre. Use the higher rates of SURPASS 6.7-E on medium (silt loam, clay loam, etc.) soils, fine (clay) soils, and for heavy infestations.

For partial control (suppression) of quackgrass: Use 7 1/3 pints SURPASS 6.7-E and 2 to 2 1/2 pounds of BLADEX 80W or 3 to 4 pints BLADEX 4L per acre.

Johnsongrass: Using a program of plowing and discing followed by SURPASS 6.7-E and BLADEX will give good control of seedlings and greatly reduce competition from plants coming from rhizomes.

In some areas discing 2 to 3 times spaced at 4 to 6 week intervals during fall and winter is recommended. If this cannot be done, start with fall plowing followed by several discings in the spring before planting. The discing cuts rhizomes into small pieces and brings them to the surface where both the weather and SURPASS 6.7-E will kill or weaken them.

Next, apply and disc-incorporate SURPASS 6.7-E just before planting at the rate of 7 1/3 pints plus 2 to 2 1/2 pounds of BLADEX 80W or 3 to 4 pints of BLADEX 4L per acre. Disc a second time at right angles to the first. Cultivate at least once, preferably twice.

Planting

Do not plant seed deeper than 2 inches. For a SURPASS 6.7-E/BLADEX application made before planting, seeding can be done immediately or it may be delayed up to 4 weeks. To avoid removing SURPASS 6.7-E/BLADEX from the row (with loss of weed control in the row), do not move or shape soil after treatment, such as by planting in deep furrows. If it is necessary to work the soil following application and incorporation, but before planting, do not work the soil deeper than the depth the herbicides were incorporated.

Cultural Practices Following Application

Should weeds develop, a shallow cultivation or rotary hoeing will generally result in better weed control. When cultivating for any reason, it should be shallow, i.e., no more than 1/2 the depth the herbicide was incorporated.

APPENDIX I

Soil Mixing (Incorporation) Before Planting Directions

The following equipment commonly is used for soil mixing (incorporation) before planting.

Power-Driven Cultivation Equipment (Recommended on all soil types previously primary tilled): Ground speed should be adjusted with all PTO driven cultivation tools to insure thorough soil mixing.
a. Horizontal action: set to cut 4 to 6 inches deep.
b. Vertical action: set to cut 3 to 4 inches deep.

Tandem Discs (Recommended on all soil types): Set to cut to a depth of 4 to 6 inches, operated at 4 to 6 mph followed by a spike-tooth harrow or some other leveling device which extends beyond the ends of the discs. For more thorough mixing (for control or suppression of deep-germinating grasses—shattercane, wild oats, seedling johnsongrass, nutsedge and wild proso millet), disc in two different directions. When discing the second time, the disc should be reset to cut no deeper than 4 to 6 inches.

Field Cultivator (Recommended on lighter soils in good tilth): Use 3 to 4 rows of sweeps spaced at 7 inch or less intervals and staggered so that no soil is left unturned, followed by a spike-tooth harrow pulled behind the cultivator. Do not use chisel plows for incorporation. Set the cultivator to cut 4 to 6 inches deep, operated at 5 mph or more. The field cultivator should not be used alone as an incorporation tool where control or suppression of deep-germinating weeds such as nutgrass, wild proso millet, johnsongrass seedlings, shattercane or wild oats is desired. For control of these weeds, use a disc alone for at least one of the passes over the field.

APPENDIX II

Tank Mix Directions

1) Fill the spray tank at least 2/3 full of appropriate carrier. Be sure tank is clean and not contaminated with 2,4-D, oil residues or other materials.
2) Start agitation and continue moderate agitation throughout mixing.
3) Premix the BLADEX or ATRAZINE making sure it is thoroughly wetted and dispersed in the tank prior to adding SURPASS 6.7-E.
4) Add the SURPASS 6.7-E to the tank 2/3 full of carrier already mixed with BLADEX or ATRAZINE.
5) Complete filling the remaining portion of the tank with carrier; mix thoroughly, maintain moderate agitation to provide a uniform mix during application.
6) The tank mix combination should not be left in the spray tank for prolonged periods; batches should be mixed and applied the same day.

Do not use excessive agitation as it may cause the BLADEX or ATRAZINE to settle out and set up in the spray tank. All return lines to the tank must discharge below liquid level.

APPENDIX III

SURPASS 6.7-E or SURPASS 6.7-E/ATRAZINE (80W or 4L) Impregnation on Dry Bulk Fertilizers

CAUTION: SURPASS 6.7-E alone or in combination with other herbicides must not be impregnated on ammonium nitrate, potassium nitrate or sodium nitrate fertilizers. Such mixtures may cause explosion and fire.

All individual state regulations relating to dry bulk fertilizer blending, registration, labeling and application are the responsibility of the individual and/or company selling the SURPASS 6.7-E or SURPASS 6.7-E/ATRAZINE fertilizer mixtures.

SURPASS 6.7-E alone and in combination with ATRAZINE may be impregnated on many dry, bulk fertilizers (see tables 1 & 2) and applied and incorporated into the soil either in the Fall before the ground freezes or before planting for the control of certain grassy weeds by SURPASS 6.7-E and many grassy and broadleaf weeds by the SURPASS 6.7-E/ATRAZINE combination in corn.

All SURPASS 6.7-E and ATRAZINE (80W or 4L) label and supplementary literature instructions and precautions regarding rates per acre, soil-type and soil incorporation, application and other directions must be followed.

Test results have shown that SURPASS 6.7-E and SURPASS 6.7-E/ATRAZINE (80W or 4L) on many dry bulk fertilizers give weed control equal to SURPASS 6.7-E or tank mixtures of SURPASS 6.7-E/ATRAZINE (80W or 4L) applied as a spray in water or liquid fertilizer. However, uniform impregnation of the pesticides on the dry fertilizer particles and uniform application in the field are necessary to assure good results.

A minimum of 200 pounds and a maximum of 700 pounds of approved fertilizer ingredients (see tables 1 & 2) impregnated with the appropriate amount of SURPASS 6.7-E alone or in combination with ATRAZINE (80W or 4L) must be applied per acre.

For impregnating the pesticides on dry fertilizers, use a closed rotary drum type mixer equipped with suitable spraying equipment. The spray nozzles should be positioned inside of the mixer to provide uniform spray coverage of the tumbling fertilizer. The SURPASS 6.7-E should be sprayed uniformly onto the fertilizer using a fine spray pattern. For the SURPASS 6.7-E/ATRAZINE combination product, the ATRAZINE (80W or 4L) can be added and mixed uniformly with the fertilizer prior to the SURPASS 6.7-E addition, or alternatively, the SURPASS 6.7-E/ATRAZINE (80W or 4L) can be mixed in the proper use ratio to prepare a uniform slurry for joint spray impregnation.

The physical properties of fertilizers vary, particularly in liquid absorptive capacity. When absorptivity is sufficient, simple spray impregnation of the fertilizer with SURPASS 6.7-E provides a satisfactory dry mixture.

If the absorptive capacity is not adequate, use of a drying agent is required to provide dry, free-flowing mixtures.

For mixtures to be used in spinning-disc applicators, Microcel E calcium silicate powder (Manville, Filtration & Minerals) is recommended for use as a drying agent.

For mixtures to be used in pneumatic applicators, Microcel E or Agsorb 16/30 RVM-MS granular-clay (Oil-Dri Corporation) is recommended for use as the drying agent.

The drying agent should be added separately and uniformly to the prepared pesticide-fertilizer mixture, in a quantity that is sufficient to provide a suitably free-flowing mixture. Generally, less than 2% Microcel E or 5% Agsorb 16/30 RMV-MS by weight is required.

The amount of SURPASS 6.7-E or SURPASS 6.7-E/ATRAZINE (80W or 4L) actually required in the manufacture of individual fertilizer mixtures should be determined carefully for each production operation (see table 3). This is necessary to ensure that the amounts of SURPASS 6.7-E and ATRAZINE actually contained in the mixture applied to the soil represent the correct rates of use.

Bulk fertilizers impregnated with SURPASS 6.7-E or SURPASS 6.7-E/ATRAZINE (80W or 4L) should be applied immediately, NOT STORED. All bulk containers should be tightly covered while the products are being transported and applied to reduce chances of SURPASS 6.7-E loss via volatilization.

Do not combine SURPASS 6.7-E plus ATRAZINE (80W or 4L) with single superphosphate (0-20-0) or treble superphosphate (0-46-0).

SURPASS 6.7—E Physical Data

Specific Gravity 20/20°C:	0.966 (typical)
Pounds/Gallon, 20/20°C:	8.04 (typical)
Flashpoint:	> 212°F (Setaflash closed cup)
Viscosity:	Sprayable down to minus 20°F.

Table 1 Approved Dry Fertilizer Ingredients for use with SURPASS 6.7-E

	N	P	K
Ammonium Sulfate	21	0	0
Ammonium Phosphate-Sulfate	16	20	0
Diammonium Phosphate	18	46	0
Monoammonium Phosphate	11	56	0
Potassium Chloride	0	0	60
Potassium Sulfate	0	0	52
Single Superphosphate	0	20	0
Treble Superphosphate	0	46	0
Urea*	45	0	0

* Some ureas may be phytotoxic when high rates are applied to corn. Use only urea rates known to be safe for corn application.

NOTE: Sul-Po-Mag, K-Mag and 11-48-0 have been shown to be compatible with SURPASS 6.7-E and are approved for use.

Table 2 Approved Dry Fertilizer Ingredients for use with SURPASS 6.7-E/ATRAZINE (80W or 4L)

	N	P	K
Ammonium Sulfate	21	0	0
Ammonium Phosphate-Sulfate	16	20	0
Diammonium Phosphate	18	46	0
Monoammonium Phosphate	11	56	0
Potassium Chloride	0	0	60
Potassium Sulfate	0	0	52
Urea*	45	0	0

* Some ureas may be phytotoxic when high rates are applied to corn. Use only urea rates known to be safe for corn application.

NOTE: Sul-Po-Mag, K-Mag and 11-48-0 have been shown to be compatible with SURPASS 6.7-E/ATRAZINE 80W and are approved for use.

If the absorptivity is not adequate, use of a highly absorptive powder is required to provide a dry, free-flowing mixture.

Microcel E (Johns-Mansville Products Corporation) is the recommended absorbent powder. It should be added separately and uniformly to the prepared pesticide-fertilizer mixture, in a quantity that is sufficient to privide a suitably free-flowing mixture. Generally, less than 2% by weight of Microcel E is required.

APPENDIX IV

Procedure for Testing the Compatibility of SURPASS 6.7-E or SURPASS 6.7-E/ATRAZINE (80W or 4L) or SURPASS 6.7-E/BLADEX (80W or 4L) with Fluid Fertilizers.

The following procedure is suggested for determining whether the above herbicide combinations may be combined with a specific fluid fertilizer for spray tank application.

Materials Required:

1. A. SURPASS 6.7-E
   B. or SURPASS 6.7-E/ATRAZINE (80W or 4L) combination
   C. or SURPASS 6.7-E/BLADEX (80W or 4L) combination.
2. Fluid fertilizer to be used.
3. Tank mix adjuvants for fluid fertilizer combinations; Compex*, Sponto 168-D*, Uni-Mix*, Unite* or equivalent. The adjuvant which provides the best compatibility depends on the specific fertilizer involved.
   *Compex, Kalo Laboratories Inc., Kansas City, Missouri; Sponto 168-D, Witco Chemical Company, Houston, Texas; Uni-Mix, Thompson-Hayward Chemical Co., Kansas City, Kansas; Unite, Hopkins Agricultural Chemical Company, Madison, Wisconsin.
4. Two one-quart, wide-mouth glass jars with lids.

TABLE 3 RATE CHART FOR THE IMPREGNATION OF DRY BULK FERTILIZERS WITH SURPASS 6.7-E OR SURPASS 6.7-E PLUS ATRAZINE (80W or 4L)

FERTILIZER RATE PER ACRE	SURPASS 6.7-E RATE/ACRE			ATRAZINE 80WP RATE/ACRE		ATRAZINE 4L RATE/ACRE	
	3¾ PTS. PER ACRE	4¾ PTS. PER ACRE	7⅓ PTS. PER ACRE	1¼ LBS. PER ACRE	2 LBS. PER ACRE	1 QT. PER ACRE	1½ QTS. PER ACRE
200 lbs.	18¾ qts./ton	23¾ qts./ton	36⅔ qts./ton	12½ lbs./ton	20 lbs./ton	10 qts./ton	15 qts./ton
250 lbs.	15 qts./ton	19 qts./ton	29⅓ qts./ton	10 lbs./ton	16 lbs./ton	8 qts./ton	12 qts./ton
300 lbs.	12½ qts./ton	15⅘ qts./ton	24⅖ qts./ton	8⅓ lbs./ton	13⅓ lbs./ton	6⅗ qts./ton	10 qts./ton
350 lbs.	10¾ qts./ton	13⅗ qts./ton	21 qts./ton	7⅛ lbs./ton	11⅖ lbs./ton	5⅘ qts./ton	8⅗ qts./ton
400 lbs.	9⅖ qts./ton	11⅞ qts./ton	18⅓ qts./ton	6¼ lbs./ton	10 lbs./ton	5 qts./ton	7½ qts./ton
450 lbs.	8⅓ qts./ton	10½ qts./ton	16⅓ qts./ton	5½ lbs./ton	8⅞ lbs./ton	4⅖ qts./ton	6⅗ qts./ton
500 lbs.	7½ qts./ton	9½ qts./ton	14⅔ qts./ton	5 lbs./ton	8 lbs./ton	4 qts./ton	6 qts./ton
550 lbs.	6⅘ qts./ton	8⅔ qts./ton	13⅓ qts./ton	4⅔ lbs./ton	7⅓ lbs./ton	3⅔ qts./ton	5½ qts./ton
600 lbs.	6¼ qts./ton	8 qts./ton	12⅕ qts./ton	4⅕ lbs./ton	6⅔ lbs./ton	3⅓ qts./ton	5 qts./ton
650 lbs.	5¾ qts./ton	7⅓ qts./ton	11¼ qts./ton	3⅞ lbs./ton	6⅕ lbs./ton	3⅛ qts./ton	4⅗ qts./ton
700 lbs.	5⅓ qts./ton	6⅘ qts./ton	10½ qts./ton	3⅔ lbs./ton	5¾ lbs./ton	2⅞ qts./ton	4¼ qts./ton

5. Measuring spoons; (a graduated pipette or cylinder will provide more accurate measurement); One tablespoon equals 3 teaspoons, 1 teaspoon equals approximately 5 milliliters.
6. Measuring cup, 8 oz. (237 mL).

Procedure: (See Note 1):

1. Pour a pint (473 mL) of the fluid fertilizer into each of the quart jars.
2. Add adjuvant to one of the jars and mix (see next rate table).
3. Add the SURPASS 6.7-E or the desired combination of herbicides to both jars (see next four rate tables). The order of addition should be wettable powders first with mixing, flowables next with mixing, and the SURPASS 6.7-E last.
4. Close both jars with lids and mix the contents by turning the jars upside down ten times.
5. Inspect the surface and body of the mixtures—
   (A) Immediately after completing the jar inversions.
   (B) After allowing the jars to stand quietly for 30 minutes.
   (C) And then again after turning the jars upside down 10 times.

If a uniform mix cannot be made, the mixture should not be used. If either mixture remains uniform for 30 minutes, the combination may be used. Should either mixture separate after 30 minutes, but readily remix uniformly with ten jar inversions, the mixture can be used if adequate agitation is maintained in the tank. If the mixture with adjuvant is satisfactory, but the one without adjuvant is not, be sure to use the adjuvant in the spray tank. Add the adjuvant first at a rate of 3 pints per 100 gallons of fluid fertilizer; foaming can be minimized by using moderate agitation.

If nondispersible oil, sludge or clumps of solids form in the mixtures, the combination should not be used.

NOTE 1: For some combinations, premixing wettable powders in a little water in a pail before adding them to the spray tank will improve the compatibility of the final mixtures with SURPASS 6.7-E. This technique can be tested in the small scale jar test by premixing the wettable powder in one-eighth (⅛) cup of water prior to addition to the pint of fluid fertilizer.

Rate Table for SURPASS 6.7-E and Adjuvant with the Fluid Fertilizer**

Gallons of Fluid Fertilizer to be applied per acre	mL or tsp. of SURPASS 6.7-E* to be added to 1 pint of Fertilizer	
	mL	tsp.
10	7	1⅓
15	5	1
20	4	¾
25	3	⅔
30	2	½
40	2	½

* Based on field rate of 1 pound active ingredient per acre in the fertilizer volumes indicated. Increase voume proportionately to correspond with intended field rate in terms of pounds active ingredient per acre (e.g., for field rate of 4 pounds actual SURPASS 6.7-E in 40 gallons fertilizer per acre, add 8 mL. or 2 tsp. SURPASS 6.7-E to each jar for compatibility purposes).

** Two (2) milliliters or one-half (½) teaspoon of adjuvant should be added to 1 pint of fluid fertilizer in order to equal the rate of 3 pints of adjuvant per 100 gallons of fluid fertilizer.

Rate Table* for SURPASS 6.7-E/ATRAZINE (80W or 4L) and Adjuvant with Fluid Fertilizer**

Gallons of Fluid Fertilizer to be applied per acre	10	15	20	25	30	40
tsp. of SURPASS 6.7-E	5¾	3¾	2¾	2⅓	2	1½
tsp. of ATRAZINE 80W	7	4⅔	3½	2¾	2⅓	1¾
tsp. of ATRAZINE 4L	3⅔	2½	1¾	1½	1¼	1

* Based on field rates of 4 pounds actual SURPASS and 1.6 pounds actual ATRAZINE (80W) or 1.5 pounds actual ATRAZINE (4L) per acre in the fertilizer volumes indicated.

** Two (2) milliliters or one-half (½) teaspoon of adjuvant should be added to 1 pint of fluid fertilizer in order to equal the rate of 3 pints of adjuvant per 100 gallons of fluid fertilizer.

Rate Table* for SURPASS 6.7-E/BLADEX (80W or 4L) and Adjuvant with Fluid Fertilizer**

Gallons of Fluid Fertilizer to be applied per acre	10	15	20	25	30	40
tsp. of SURPASS 6.7-E	5¾	3¾	2¾	2⅓	2	1½
tsp. of BLADEX 80W	7¾	5¼	4	3	2⅔	2
tsp. of BLADEX 4L	4¾	3¼	2 ½	2	1⅔	1¼

* Based on field rates of 4 pounds actual SURPASS and 2 pounds actual BLADEX (80W or 4L) per acre in the fertilizer volumes indicated.

** Two (2) milliliters or one-half (½) teaspoon of adjuvant should be added to 1 pint of fluid fertilizer in order to equal the rate of 3 pints of adjuvant per 100 gallons of fluid fertilizer.

STORAGE AND DISPOSAL

Prohibitions: Do not contaminate water, food or feed by storage and disposal.

Pesticide Disposal: Rinse spray equipment. Wastes resulting from the use of this product may be disposed of on site or at an approved waste disposal facility.

Container Disposal: Triple rinse (or equivalent), then offer for recycling, reconditioning, or puncture and dispose of in a sanitary landfill, or by other approved State or local procedures.

NOTICE TO BUYER AND USER:

Seller warrants that this product conforms to the chemical description on the label and is reasonably fit for the purposes stated on the label when used in accordance with directions under normal conditions of use. This warranty does not extend to the use of this product contrary to label in-

structions, or under abnormal use conditions, or under conditions not reasonably foreseeable to Seller, and Buyer and User assume the risk of any such use. SELLER DISCLAIMS ALL OTHER WARRANTIES EXPRESSED OR IMPLIED INCLUDING ANY WARRANTY OF FITNESS OR MERCHANTABILITY. SELLER SHALL NOT BE LIABLE FOR CONSEQUENTIAL, SPECIAL OR INDIRECT DAMAGES RESULTING FROM THE USE OR HANDLING OF THIS PRODUCT AND SELLER'S SOLE LIABILITY AND BUYER'S AND USER'S EXCLUSIVE REMEDY SHALL BE LIMITED TO THE REFUND OF THE PURCHASE PRICE.

This product is sold only for uses stated on the label. No express or implied license is granted to use or sell this product under any patent in any country except as specified: Country: United States of America. Patent No. 4,021,224; 4,137,070.

A-8R

The information and specimen labels in this booklet are correct as of the publication date but are subject to change without notice. For current information, contact ICI Americas Inc.

SUTAN® + 10-G

GRANULES

A Selective Herbicide For Corn (Field, Sweet and Silage)

ACTIVE INGREDIENT:

S-Ethyl diisobutylthiocarbamate 10.0%

INERT INGREDIENTS: 90.0%

TOTAL 100.0%

(Contains 10 Pounds of Active Ingredient per 100 Pounds)

EPA Reg. No. 10182-182

KEEP OUT OF REACH OF CHILDREN

CAUTION

STATEMENT OF PRACTICAL TREATMENT

If a known exposure occurs or is suspected, immediately start the procedures given below. If further treatment is required, contact a Poison Center, a physician, or the nearest hospital.

If swallowed: Immediately give several glasses of water and induce vomiting by gagging the victim with a finger placed on the back of the victim's tongue. Give fluids until vomitus is clear. If victim is unconscious or convulsing, do not induce vomiting or give anything by mouth. Call a physician or nearest Poison Center immediately.

For skin contact: Flush all affected areas with plenty of water for several minutes. Remove and clean contaminated clothing and shoes before reuse.

For eye contact: Hold eyelids apart and flush with plenty of water for 15 minutes. Seek medical attention if irritation occurs.

If inhaled: Remove to fresh air. Seek medical attention if respiratory irritation occurs or if breathing becomes difficult.

FOR 24-HOUR EMERGENCY MEDICAL ASSISTANCE CALL 1-800-F-A-S-T-M-E-D (327-8633).

FOR CHEMICAL EMERGENCY: Spill, leak, fire, exposure, or accident call CHEMTREC 1-800-424-9300.

PRECAUTIONARY STATEMENTS

HAZARDS TO HUMANS AND DOMESTIC ANIMALS

CAUTION

Harmful if Swallowed

Avoid contact with skin, eyes and clothing. Avoid inhalation of dust. Wash thoroughly with soap and water after use.

ENVIRONMENTAL HAZARDS

Do not apply directly to water or wetlands. Do not contaminate water by cleaning of equipment or disposal of wastes. Cover or incorporate spills.

READ ALL LABEL DIRECTIONS BEFORE USING

General Use Precautions

Read all label directions before using. Apply this product as specified on this label.

Applied according to directions and under normal growing conditions, SUTAN + 10-G will not harm the treated crop. During germination and early stages of growth, extended periods of unusually cold and wet or hot and dry weather, insect or plant disease attack, the use of certain soil-applied systemic insecticides, improperly placed fertilizers or soil insecticides may create abnormal conditions that weaken crop seedlings. SUTAN + 10-G used under these abnormal conditions could result in crop injury.

When applied according to directions and when conditions exist for normal plant growth through the season, no harmful residues of SUTAN + 10-G should remain beyond harvest. In the **Southeastern U.S.,** when SUTAN + 10-G is used for weed control in silage corn, do not seed small seeded grains after corn harvest until September.

Do not use on milo or sorghum. Do not use on corn seed stock. (Breeders, Foundation, or Increase). SUTAN + 10-G may be used on Production seed corn.

This product is recommended for use on mineral soils only.

Do not apply in combination with fertilizers, insecticides or fungicides.

Do not contaminate food or feed.

Do not store near seeds, fertilizers, or foodstuffs.

Do not contaminate irrigation water or water used for domestic purposes.

Weeds Controlled

SUTAN + 10-G will not control established weeds.

Annual Grasses:	
Barnyardgrass; (Watergrass)	(*Echinochloa crus-galli*)
Crabgrass	(*Digitaria* spp.)
Giant Foxtail	(*Setaria faberii*)
Green Foxtail	(*Setaria viridis*)
Yellow Foxtail	(*Setaria lutescens*)
Goosegrass	(*Eleusine indica*)
Seedling Johnsongrass	(*Sorghum halepense*)
Perennial Weeds:	
Purple Nutsedge	(*Cyperus rotundus*)
Yellow Nutsedge	(*Cyperus esculentus*)

Existing stands of nutgrass must be turned under and chopped up thoroughly before treatment. Cultivation is suggested in addition to SUTAN + 10-G treatment in fields with moderate or heavy nutgrass infestations.

DIRECTIONS FOR USE

It is a violation of Federal law to use this product in a manner inconsistent with its labeling.

SUTAN + 10-G is a selective herbicide which is mixed (incorporated) into the soil for control of weeds listed on this label. SUTAN + 10-G controls annual grasses as their seeds germinate by interfering with normal germination and seedling development. It does not control established weeds. All weed growth and crop stubble should be thoroughly worked into the soil before treatment.

Application Directions

Apply only to well-worked soil that is dry enough to permit thorough mixing with incorporation equipment. Uniformly apply the recommended rate of SUTAN + 10-G using equipment designed for application of granular herbicides. Equipment should be carefully calibrated before use and checked frequently during application to be sure it is working properly and delivering a uniform distribution pattern. Avoid overlaps that will increase SUTAN + 10-G dosage above recommended rates as plant injury may occur.

In irrigated areas, do not apply SUTAN + 10-G before pre-irrigation.

Incorporation Directions

SUTAN + 10-G must be incorporated into the soil immediately after application to prevent loss of the herbicide. Whenever possible, application and incorporation should be done in the same operation.

Soil Mixing (Incorporation) Before Planting:

Use power-driven cultivation equipment set to cut to a depth of 2 to 3 inches; OR tandem discs set to cut to a depth of 4 to 6 inches operated at 4 to 6 mph, followed by a spike-tooth harrow or some other leveling device which extends beyond the ends of the disc. For thorough mixing, disc in two different directions (cross disc).

Planting

Plant seed to a maximum depth of 2 inches within 2 weeks after treatment. To avoid removing SUTAN + 10-G from the row (with loss of weed control in the row) do not move or shape soil after treatment, such as by planting in deep furrows.

Cultural Practices Following Application

SUTAN + 10-G is not persistent in the soil and susceptible weeds germinating late in the season may not be controlled. Shallow cultivation may be necessary to control these late-germinating weeds. Do not cultivate deeper than the depth of herbicide incorporation. Pre-emergence or post-emergence herbicides may be necessary to control weeds resistant to SUTAN + 10-G.

Recommendations

Do not use on any type corn in Arizona or in the ten southernmost California counties (except in Kern County where field and silage corn may be treated).

Broadcast Rates Per Acre: Apply and incorporate 30 lbs. SUTAN + 10-G per acre on light soils (coarse textured-sandy); apply 40 lbs. SUTAN + 10-G per acre on medium (medium textured-sandy loam) to heavy (fine textured-clay) soils. In the north central and midwest corn growing regions, 40 lbs. of SUTAN + 10-G per acre also is recommended on all soils with moderate to heavy infestations of nutsedge. Use 30 lbs. SUTAN + 10-G per acre on blow sands in these regions. Cultivation is suggested in addition to the SUTAN + 10-G treatment in fields with moderate or heavy infestations of nutsedge.

STORAGE AND DISPOSAL

Prohibitions: Do not contaminate water, food, or feed by storage or disposal. Open dumping is prohibited.

Storage: Keep container closed when not in use. Do not store near seeds, fertilizers, or foodstuffs.

Pesticide Disposal: Wastes resulting from the use of this product may be disposed of on site or at an approved waste disposal facility.

Container Disposal: Completely empty bag into application equipment. Then dispose of empty bag in a sanitary landfill or by incineration, or, if allowed by State and local authorities, by burning. If burned, stay out of smoke.

NOTICE TO BUYER AND USER:

Seller warrants that this product conforms to the chemical description on the label and is reasonably fit for the purposes stated on the label when used in accordance with directions under normal conditions of use. This warranty does not extend to the use of this product contrary to label instructions, or under abnormal use conditions, or under conditions not reasonably foreseeable to Seller, and Buyer and User assume the risk of any such use. SELLER DISCLAIMS ALL OTHER WARRANTIES EXPRESSED OR IMPLIED INCLUDING ANY WARRANTY OF FITNESS OR MERCHANTABILITY. SELLER SHALL NOT BE LIABLE FOR CONSEQUENTIAL, SPECIAL OR INDIRECT DAMAGES RESULTING FROM THE USE OR HANDLING OF THE PRODUCT AND SELLER'S SOLE LIABILITY AND BUYER'S AND USER'S EXCLUSIVE REMEDY SHALL BE LIMITED TO THE REFUND OF THE PURCHASE PRICE.

This product is sold only for uses stated on its label. No express or implied license is granted to use or sell this product under any patent in any country except as specified: Country: United States of America. Patent No. 4,021,224; 4,137,070.

A-2R

The information and specimen labels in this booklet are correct as of the publication date but are subject to change without notice. For current information, contact ICI Americas Inc.

SUTAN® + 6.7-E

SELECTIVE HERBICIDE EMULSIFIABLE LIQUID FOR CORN

(Field, Sweet, Silage and Popcorn)

Selective Herbicide

corn herbicide for control of annual grasses

ACTIVE INGREDIENT:	
S—Ethyl diisobutylthiocarbamate	85.1%
INERT INGREDIENTS:	14.9%
TOTAL	100.0%

Contains 6.7 Pounds of Active Ingredient Per Gallon

EPA Reg. No. 10182-222

Complete Directions For Use

Use Precautions and Recommendations

KEEP OUT OF REACH OF CHILDREN

CAUTION

HARMFUL IF SWALLOWED

STATEMENT OF PRACTICAL TREATMENT

If a known exposure occurs or is suspected, immediately initiate the recommended procedures below. If further treatment is required, contact a Poison Center, a physician, or the nearest hospital.

If swallowed: Immediately give several glasses of water but *do not* induce vomiting. If vomiting does occur, give fluids again. Have a physician determine if condition of patient will permit evacuation of stomach. Do not give anything by mouth to an unconscious or convulsing person.

For eye contact: Immediately flush eyes with large amounts of running water for at least 15 minutes. Hold eyelids apart during flushing to ensure rinsing of the entire surface of the eye and lids with water. Get medical attention immediately

For skin contact: Immediately flush all affected areas with plenty of water for at least 15 minutes. Remove contaminated clothing and shoes. Seek medical attention if skin irritation occurs. Wash clothing before reuse.

If Inhaled: Remove to fresh air. Seek medical attention if respiratory irritation occurs or if breathing becomes difficult.

FOR 24-HOUR EMERGENCY MEDICAL ASSISTANCE CALL 1-800-F-A-S-T-M-E-D (327-8633).

FOR CHEMICAL EMERGENCY: Spill, leak, fire, exposure, or accident call CHEMTREC 1-800-424-9300.

PRECAUTIONARY STATEMENTS

HAZARDS TO HUMANS AND DOMESTIC ANIMALS

CAUTION

HARMFUL IF SWALLOWED. Avoid contact with skin, eyes and clothing. Avoid breathing spray mist. Wear goggles, rubber gloves and protective clothing. Wash skin with soap and water immediately after contact. Flush eyes with water.

Do not apply this product in such a manner as to directly or through drift expose workers or other persons. When applied through sprinkler irrigation method of application, the area being treated must be vacated by unprotected persons.

ENVIRONMENTAL HAZARDS

Do not apply directly to water or wetlands. Do not contaminate water by cleaning of equipment or disposal of wastes. Cover or incorporate spills.

CAN BE STORED AT TEMPERATURES AS LOW AS MINUS 50°F

READ ALL LABEL DIRECTIONS BEFORE USING

Weeds Controlled by SUTAN + 6.7-E

SUTAN + 6.7-E will not control established weeds.

Annual Grasses:	
Barnyardgrass, Watergrass	*(Echinochloa crus-galli)*
Crabgrass	*(Digitaria spp.)*
Fall Panicum	*(Panicum dichotomiflorum)*
Field Sandbur*	*Cenchrus pauciflorus)*
Giant Foxtail	*(Setaria faberi)*
Green Foxtail	*(Setaria viridis)*
Goosegrass	*(Eleusine indica)*
Seedling Bermudagrass	*(Cynodon dactylon)*
Seedling Johnsongrass*	*(Sorghum halepense)*
Texas Panicum*†, Buffalograss, Coloradograss	*(Panicum texanum)*
Volunteer Sorghum*	*(Sorghum spp.)*
Shattercane‡, Wild Cane, Black Amber Cane, Chicken Corn, Scatter Cane	*(Sorghum bicolor)*
See Wild Cane Rate and Cultural Recommendations	
Yellow Foxtail	*(Setaria lutescens)*

‡ Partial control (suppression) only.

* Cultivation is suggested in addition to the SUTAN + 6.7-E treatment in fields with moderate to heavy infestations of shattercane, sandbur, johnsongrass seedlings, Texas panicum or volunteer sorghum.

† Southeast Region only.

Perennial Weeds:	
Purple Nutsedge	*(Cyperus rotundus)*
Rhizome Bermudagrass†	*(Cynodon dactylon)*
Rhizome Johnsongrass‡	*(Sorghum halepense)*
Yellow Nutsedge	*(Cyperus esculentus)*

‡ Partial control (suppression) only.

† Southeast Region only.

Existing stands of nutsedge must be turned under and chopped up thoroughly before treatment. Cultivation is suggested in addition to SUTAN + 6.7-E treatment in fields with moderate to heavy nutsedge infestations.

NOTE: See special instructions under "RECOMMENDATIONS" for Johnsongrass and Rhizome Bermudagrass.

General Use Precautions

Read all label directions before using.

Applied according to directions and under normal growing conditions, SUTAN + 6.7-E will not harm the treated crop. During germination and early stages of growth, extended periods of unusually cold and wet or hot and dry weather, insect or plant disease attack, carryover pesticide residues, the use of certain soil-applied systemic insecticides, improperly placed fertilizers or soil insecticides may create abnormal conditions that weaken crop seedlings. SUTAN + 6.7-E used under these abnormal conditions could result in crop injury.

When applied according to directions and when conditions exist for normal plant growth through the season, no harmful residues of SUTAN +

6.7-E should remain beyond harvest. In the Southeastern U.S., when SUTAN + 6.7-E is used for weed control in silage corn, do not seed small seeded grains after corn harvest until September.

Do not use on milo or sorghum. Do not use on corn seed stock (Breeders, Foundation, or Increase). SUTAN + 6.7-E may be used on Production seed corn.

SUTAN + 6.7-E is recommended for use on mineral soils only (soils containing less than 10% organic matter).

Do not contaminate irrigation water (used for crops other than corn) or water used for domestic purposes.

Do not contaminate food or feed.

Keep container closed when not in use. Do not store near seeds, fertilizers, or foodstuffs.

Do not use in Arizona (except at elevations of 2500 feet or higher, where field and silage corn may be treated).

Do not use in the ten southernmost California counties (except Kern County where field and silage corn may be treated).

In irrigated areas, do not apply SUTAN + 6.7-E before pre-irrigation.

DIRECTIONS FOR USE OF SUTAN + 6.7-E ON CORN (Field, Sweet, Silage and Popcorn, including Production Seed Corn)

It is a violation of Federal law to use this product in a manner inconsistent with its labeling.

SUTAN + 6.7-E is a selective herbicide which is mixed (incorporated) or injected into the soil for control of weeds listed on this label. SUTAN + 6.7-E controls annual grasses as their seeds germinate by interfering with normal germination and seedling development. It does not control established weeds.

Combination with Fluid Fertilizers

SUTAN + 6.7-E may be combined with fluid (solution, slurry, or suspension) fertilizers. *However, physical compatibility with these fluid fertilizers must be determined before combining in the spray tank.* See Appendix II for directions regarding these combinations. Even though found to be compatible, constant agitation is necessary to keep the SUTAN + 6.7-E evenly mixed with the fluid fertilizer during application.

Impregnation on Dry Bulk Fertilizers

(See Appendix IV for use on No-Till Corn).

Dry Bulk fertilizers (except nitrate fertilizers) may be impregnated or coated with SUTAN + 6.7-E. However, uniform distribution of SUTAN + 6.7-E on fertilizer particles and uniform application are necessary to assure good results. See Appendix IV for information and directions regarding impregnation and use for these combinations.

APPLICATION AND SOIL INCORPORATION DIRECTIONS

Spray Application and Mechanical Incorporation

Pour the recommended amount of SUTAN + 6.7-E into the spray tank during the filling operations. Apply in 10 to 50 gallons of water per acre using a properly calibrated, low-pressure type nozzle (flood, fan or swirl chamber) on boom type sprayer having good agitation (use the higher gallonage in arid areas). The soil should be dry enough to permit good soil mixing (incorporation).

Incorporation Directions

See Appendix I for soil mixing (incorporation) before planting directions.

SUTAN + 6.7-E must be incorporated into the soil within 4 hours (within 1 hour in Western Region) to prevent loss of the herbicide. Whenever possible, application and incorporation should be done in the same operation.

If there must be a delay (less than 4 hours, except less than 1 hour in Western Region) in incoporation of the herbicide when application and incorporation are done in separate operations, SUTAN + 6.7-E must be applied to a dry soil surface (at least 1/2 inch deep), free from dew and incidental moisture.

Loss of weed control will result from any delay in incorporation if SUTAN + 67-E is applied to a moist soil surface.

IMPORTANT: Thorough mixing of SUTAN + 6.7-E in the soil is necessary. The soil must be in good tilth. Improper discing or poor tilth (large clods) may result in unsatisfactory control.

Incorporation of Impregnated Dry Bulk Fertilizer

The application and incorporation of SUTAN + 6.7-E must be carried out on the same day and delayed only if soil moisture conditions permit. See Appendix IV for directions on the delay in soil incorporation.

CENTER PIVOT SPRINKLER APPLICATION

Apply SUTAN + 6.7-E with irrigation water before planting (see PLANTING), or immediately after planting before either corn or weeds emerge. It is important that application be done immediately after the last tillage operation since poor results will occur if the weeds have germinated before application of the herbicide. Meter the SUTAN + 6.7-E into the irrigation water during the entire period.

Apply in 1/2 to 1 inch of water or sufficient water to penetrate the soil to a depth of 5 to 7 inches. If diluted, SUTAN + 6.7-E should be mixed at a volume ratio of no more than 1 part SUTAN + 6.7-E to 4 parts of water or fluid fertilizer. Good agitation should be maintained during the entire application period.

USE PRECAUTIONS

A. Apply this product only through sprinkler including center pivot irrigation system. Do not apply this product through any other type of irrigation system.

B. Crop injury, lack of effectiveness, or illegal pesticide residues in the crop can result from nonuniform distribution of treated water.

C. If you have questions about calibration, you should contact State Extension Service specialists, equipment manufacturers or other experts.

D. Do not connect an irrigation system (inlcuding greenhouse systems) used for pesticide application to a public water system unless the pesticide label-prescribed safety devices for public water systems are in place.

E. A person knowledgeable of the chemigation system and responsible for its operation, or under the supervision of the responsible person, shall shut the system down and make necessary adjustments should the need arise.

USE DIRECTIONS

1. The system must contain a functional check valve, vacuum relief valve, and low pressure drain appropriately located on the irrigation pipeline to prevent water source contamination from backflow.
2. The pesticide injection pipeline must contain a functional, automatic, quick-closing check valve to prevent the flow of fluid back toward the injection pump.
3. The pesticide injection pipeline must also contain a functional, normally closed, solenoid-operated valve located on the intake side of the injection pump and connected to the system interlock to prevent fluid from being withdrawn from the supply tank when the irrigation system is either automatically or manually shut down.
4. The system must contain functional interlocking controls to automatically shut off the pesticide injection pump when the water pump motor stops.
5. The irrigation line or water pump must include a functional pressure switch which will stop the water pump motor when the water pressure decreases to the point where pesticide distribution is adversely affected.
6. Systems must use a metering pump, such as a positive displacement injection pump (e.g., diaphragm pump) effectively designed and constructed of materials that are compatible with pesticides and capable of being fitted with a system interlock.
7. Do not apply when wind speed favors drift beyond the area intended for treatment.

Subsurface Application Before or At Planting

(Limit use to the Southeastern U.S. only).

Special equipment designed for subsurface application *must* be used. Injector units must be rigidly mounted on, *immediately* ahead of, or *immediately* behind the planter unit.

Soil Injection: Coulters and injector shanks must be spaced 3 1/2 inches apart. The two shanks adjacent to the drill row must be 1 3/4 to 2 inches on either side of it. Mount injectors in staggered positions to avoid trash buildup. Set injectors to inject SUTAN + 6.7-E 3 inches below the soil surface. The width of the band in which weed control is desired will determine the number of injectors required per row. (Example: Four injectors spaced 3 1/2 inches apart give a 14 inch band). A broadcast application can be made by increasing the number of shanks. Seal injector openings by using a roller or chain drag behind the injectors. Appply the recommended rate of SUTAN + 6.7-E in 20 to 50 gallons of water per acre.

For band applications, calculate the amount to be applied per acre as follows:

$$\frac{\text{Band width In Inches}}{\text{Distance Between Rows In Inches}} \times \text{Rate/A for Broadcast Treatment} = \text{Amount Needed For band Treatment}$$

Planting

Plant corn seed to a maximum depth of 2 inches. Seeding can be done immediately or it may be delayed up to 2 weeks in the Southeast, Southwest, and Western regions or up to 4 weeks in the Northeast, North Central, Midwest, and Pacific Northwest regions.* To avoid removing SUTAN + 6.7-E from the row (with loss of weed control in the row), do not move or shape soil after treatment, such as by planting in deep furrows. If it is necessary to work the soil following application and incorporation, but before planting, do not work the soil deeper than the depth SUTAN + 6.7-E was incorporated.

* For Shattercane—See Special Recommendations.

Cultural Practices Following Application

Should weeds develop, a shallow cultivation or rotary hoeing will generally result in better weed control. When cultivating for any reason, it should be shallow, i.e., no more than 1/2 the depth the herbicide was incorporated or injected. A sequential application of 2,4-D or BANVEL*4E may be necessary to control weeds resistant to SUTAN + 6.7-E.

Under circumstances when a poor stand of corn is obtained after SUTAN + 6.7-E treatment (caused by flooding, disease, etc.) corn can be planted again without working the field.

* BANVEL is a registered trademark of Sandoz Crop Protection Corp.

Recommendations
For all uses, cultivation is recommended in addition to SUTAN + 6.7-E treatment in fields with moderate to heavy infestations of shattercane, nutsedge, sandbur, johnsongrass seedlings, rhizome bermudagrass, Texas panicum or volunteer sorghum.

Broadcast Rates Per Acre (Conventional Sprays and Center Pivot Sprinklers).

For control of most annual grasses: Use 4³/₄ pints of SUTAN + 6.7-E per acre.

For control of Nutsedge and Texas Panicum (in addition to the other annual grasses listed): Use 4³/₄ to 7¹/₃ pints of SUTAN + 6.7-E per acre. Use the higher rates on medium (medium fine textured-silt loam, clay loam, etc.) soils, fine (fine textured-clay) soils and for heavy infestations.

For partial control (suppression) of Shattercane (and for control of the annual grasses listed): When applied and incorporated *immediately* before planting, SUTAN + 6.7-E at the rate of 7¹/₃ pints per acre on a broadcast basis will reduce competition from shattercane. On continuous corn acreage with high seed population, rotation to another crop is recommended.

Suppression of shattercane may be improved if all the following practices are followed:
1. Fall or Spring Moldboard Plow.
2. Before application, the soil should be well worked with a disc and dry enough to permit good soil mixing of the herbicide (incorporation). All trash should be worked into the soil.
3. Delay application until soil warms to a constant 55 degrees or greater temperature. Plant immediately.
4. Incorporate thoroughly. Use recommended incorporation equipment and procedures found on this label.
5. Cultivate immediately if wild cane emerges. Cultivation should be shallower than the depth used when incorporating the herbicide.

For control of Johnsongrass and Bermudagrass (In addition to the annual grasses listed): Apply and disc-incorporate SUTAN + 6.7-E just before planting at the rate of 4³/₄ to 7¹/₃ pints per acre. Use the higher rates of SUTAN + 6.7-E on medium (medium fine textured-silt loam, clay loam, etc.) soils, fine (fine textured-clay) soils, for rhizome johnsongrass suppression, rhizome bermudagrass, and for heavy infestations. Disc a second time at right angles to the first. Cultivate at least once, preferably twice.

Johnsongrass: Using a program of plowing and discing followed by SUTAN + 6.7-E will give good control of seedlings and greatly reduce competition from plants coming from rhizomes.

In some areas discing 2 to 3 times spaced at 4 to 6 weeks intervals during Fall and Winter can be done. If this cannot be done, start with Fall plowing followed by several discings in the Spring before planting. The discing cuts rhizomes into small pieces and brings them to the surface, where both the weather and SUTAN + 67-E will kill or weaken them.

Rhizome Bermudagrass: The rhizomes and stolons of bermudagrass must be cut up thoroughly so that four or less nodes remain on a strand. For control of bermudagrass, the disc must be set to cut 6 inches deep.

Subsurface Injection Application: (Limit use to the Southeastern U.S. only) Use 3³/₄ pints SUTAN + 6.7-E per broadcast acre, reducing the dosage proportionately depending upon the row spacing and band width to be treated.

Fall Application: For use only in Minnesota, Northern Iowa (North of Interstate 80), Northeast Nebraska (North of Interstate 80 and East of State Highway 14), South Dakota (East of Missouri River), Wisconsin and Northern Illinois (North of Interstate 80). Apply and incorporate 7¹/₃ pints SUTAN + 6.7-E per acre in the late Fall before the ground freezes.

See Appendix IV for directions on use of dry bulk fertilizer applied in the Fall with SUTAN + 6.7-E.

NOTE: Do not use Fall application for suppression of shattercane.

Arizona: For control of most annual grasses in field and silage corn grown at elevations of 2500 feet or higher, use 3³/₄ to 4³/₄ pints of SUTAN + 6.7-E per acre. Use the higher rate on medium (medium fine textured silt loam, clay loam, etc.) to heavy (fine textured clay) soils. For nutsedge control, use 4³/₄ pints of SUTAN + 6.7-E per acre.

SUTAN + 6.7-E/ATRAZINE (80W, 4L or AATREX* NINE-0*) Tank Mix Directions

* AATREX trademark of Ciba-Geigy Corporation for atrazine
* NINE-0 trademark of Ciba-Geigy Corporation

A tank mix combination of SUTAN + 6.7-E and ATRAZINE can be applied and mixed into the soil before planting for weed control in corn. This combination will control a broader spectrum of weeds than will either material used alone. In addition, potential hazard of ATRAZINE residues affecting certain succeeding crops is reduced because a lower rate of ATRAZINE can be used when it is combined with SUTAN + 6.7-E than when it is used alone.

- FOLLOW ALL USE PRECAUTIONS AND WARNINGS THAT APPEAR ON THE ATRAZINE LABEL, SUPPLEMENTARY ATRAZINE LITERATURE AND ELSEWHERE ON THIS LABEL.
- MAKE ONLY ONE APPLICATION PER CROP. AFTER TREATMENT WITH SUTAN + 6.7-E/ATRAZINE COMBINATION, DO NOT PLANT ANY CROP EXCEPT CORN UNTIL THE FOLLOWING YEAR OR INJURY MAY OCCUR.
- DO NOT USE ON MILO OR SORGHUM.
- DO NOT USE ON CORN SEED STOCK (BREEDERS, FOUNDATION, OR INCREASE). SUTAN + 6.7-E/ATRAZINE MAY BE USED ON PRODUCTION SEED CORN.

Weeds Controlled by SUTAN + 6.7-E/ATRAZINE

Annual Grasses:	
Barnyardgrass, Watergrass	*(Echinochloa crus-galli)*
Broadleaf Signalgrass‡*	*(Brachiaria platyphylla)*
Crabgrass	*(Digitaria spp.)*
Fall Panicum	*(Panicum dichotomiflorum)*
Field Sandbur*	*(Cenchrus pauciflorus)*
Giant Foxtail	*(Setaria faberi)*
Green Foxtail	*(Setaria viridis)*
Goosegrass	*(Eleusine indica)*
Seedling Bermudagrass	*(Cynodon dactylon)*
Seedling Johnsongrass*	*(Sorghum halepense)*
Texas Panicum*†, Buffalograss, Coloradograss	*(Panicum texanum)*
Volunteer Sorghum*	*(Sorghum spp.)*
Shattercane*†, Wild Cane, Black Amber Cane, Chicken Corn, Scatter Cane	*(Sorghum bicolor)*
See Wild Cane Rate and Cultural Recommendations	
Wild Oats	*(Avena fatua)*
Witchgrass	*(Panicum capillare)*
Yellow Foxtail	*(Setaria glauca)*

‡ Partial control (suppression) only
* Cultivation is suggested in addition to the SUTAN + 6.7.E/ATRAZINE treatment in fields with moderate to heavy infestations of wild cane, sandbur, johnsongrass, Texas panicum, volunteer sorghum, or broadleaf signalgrass.

Annual Broadleaf Weeds:	
Annual Morningglory	*(Ipomoea purpurea)*
Carpetweed	*(Mollugo verticillata)*
Common Cocklebur	*(Xanthium pensylvanicum)*
Common Lambsquarters	*(Chenopodium album)*
Common Purslane	*(Portulaca oleracea)*
Florida Pusley	*(Richardia scabra)*
Jimsonweed	*(Datura stramonium)*
Mustards	*(Brassica spp.)*
Nightshade	*(Solanum spp.)*
Prostrate Pigweed	*(Amaranthus biltoides)*
Ragweed	*(Ambrosia spp.)*
Redroot Pigweed	*(Amaranthus retroflexus)*
Sicklepod† Coffeeweed	*(Cassia obtusifolia)*
Smartweed	*(Polygonum spp.)*
Smooth Pigweed	*(Amaranthus hybridus)*
Velvetleaf	*(Abutilon theophrasti)*

† Southeast Region only.

Perennial Weeds:	
Hemp Dogbane‡	*(Apocynum cannabinum)*
Purple Nutsedge	*(Cyperus rotundus)*
Rhizome Bermudagrass†	*(Cynodon dactylon)*
Rhizome Johnsongrass‡	*(Sorghum halepense)*
Yellow Nutsedge	*(Cyperus esculentus)*

‡ Partial control (suppression) only.
† Southeast Region only.

Existing strands of nutsedge must be turned under and chopped up thoroughly before treatment. Cultivation is suggested in addition to SUTAN + 6.7-E/ATRAZINE treatment in fields with moderate to heavy nutgrass infestations.

DIRECTIONS FOR USE

The SUTAN + 6.7-E/ATRAZINE (80W or 4L) combination can be mixed with water or combined with fluid (solution, slurry, or suspension) fertilizers and applied as a spray, or can be impregnated on dry bulk fertilizers. The SUTAN + 6.7E/AATREX NINE-0 combination can be mixed with water and applied as a spray, or can be impregnated on dry bulk fertilizers.

SUTAN +6.7-E/ATRAZINE (80W, 4L or AATREX NINE-0) Tank Mix.

See Appendix III For Tank Mix Directions.

SUTAN + 6.7.E/ATRAZINE (80W or 4L) Fluid Fertilizer Tank Mix.

The SUTAN + 6.7-E/ATRAZINE (80W or 4L) combination can be combined with fluid fertilizers for applications made before planting. *However, the physical compatibility with these fluid fertilizers must be determined before combining in the spray tank, and even though found to be compatible, constant agitation is necessary to keep the SUTAN + 6.7-E/ATRAZINE evenly mixed with the fluid fertilizer during application.*

Refere to Appendix II regarding the procedure for testing the compatibility of SUTAN + 6.7-E/ATRAZINE combination with fluid fertilizers.

Impregnation on Dry Bulk Fertilizers

(See Appendix IV for use on No-Till Corn.)

Dry bulk fertilizers (except single superphosphate, treble superphosphate and nitrate fertilizers) may be impregnated or coated with SUTAN + 6.7.E/ATRAZINE (80W, 4L or AATREX NINE-0). However, uniform distribution of SUTAN + 6.7.E/ATRAZINE (80W, 4L or AATREX NINE-0) on fertilizer particles and uniform application are necessary to assure good results. See Appendix IV for information and directions for impregnation and use for these combinations.

APPLICATION AND SOIL INCORPORATION DIRECTIONS

Spray Application and Mechanical Incorporation

Apply the recommended dosage in 10 to 50 gallons of water per acre (use the higher gallonage in arid areas) using a properly calibrated, low-pressure type nozzle (flood, fan or swirl chamber) on boom-type sprayer having good, but not excessive agitation. Check frequently during application to be sure nozzles are free from clogging and are delivering a uniform spray pattern. *Read directions on ATRAZINE label concerning nozzle and screen sizes.*

See Appendix I for Soil Mixing (incorporation) before Planting Directions.

The herbicides must be incorporatd into the soil within 4 hours (within 1 hour in Western Region) after application to prevent loss of the SUTAN + 6.7-E. Whenever possible, application and incorporation should be done in the same operation. Apply the herbicides to soil dry enough to permit good soil mixing.

If there must be a delay (less than 4 hours, except less than 1 hour in Western Region) in incorporation of the herbicides when application and incorporation are done in separate operations, SUTAN + 6.7-E must be applied to a dry soil surface (at least 1/2 inch deep), free from dew and incidental moisture.

Loss of weed control will result from any delay in incorporation if SUTAN + 6.7-E is applied to a moist soil surface.

IMPORTANT: Thorough mixing of SUTAN + 6.7-E in the soil is necessary. The soil must be in good tilth. Improper discing or poor tilth (large clods) may result in unsatisfactory control.

Incorporation of Impregnated Dry Bulk Fertilizer

The application and incorporation of SUTAN + 6.7-E must be carried out on the same day and delayed only if soil moisture conditions permit. See Appendix IV for directions on the delay in soil incorporation.

Center Pivot Sprinkler Application

Apply SUTAN + 6.7-E/ATRAZINE (80W or 4L) with irrigation water before planting (see PLANTING), or immediately after planting before either corn or weeds emerge. It is important that application be done immediately after the last tillage operation since poor results will occur if the weeds have germinated before application of the herbicide. Meter the premixed SUTAN + 6.7-E/ATRAZINE into the irrigation water during the entire period. Apply in 1/2 to 1 inch of water or sufficient water to penetrate the soil to a depth of 5 to 7 inches. Good agitation of the premixed SUTAN + 6.7-E/ATRAZINE should be maintained during the entire application period. See Appendix III for tank mixiing directions for SUTAN + 6.7-E with ATRAZINE.

Use Precautions:

A. Apply the product only through center pivot irrigation systems. Do not apply this product through any other type of irrigation system.
B. Crop injury, lack of effectiveness, or illegal pesticide residues in the crop can result from nonuniform distribution of treated water.
C. If you have questions about calibration, you should contact State Extension Service specialists, equipment manufacturers or other experts.
D. Do not connect an irrigation system (including greenhouse systems) used for pesticide application to a public water system unless the pesticide label-prescribed safety devices for public water systems are in place.
E. A person knowledgeable of the chemigation system and responsible for its operation, or under the supervision of the responsible person, shall shut the system down and make necessary adjustments should the need arise.

Use Directions

1. The system must contain a functional check valve, vacuum relief valve, and low pressure drain appropriately located on the irrigation pipeline to prevent water source contamination from backflow.
2. The pesticide injection pipeline must contain a functional, automatic, quick-closing check valve to prevent the flow of fluid back toward the injection pump.
3. The pesticide injection pipeline must also contain a functional, normally closed, solenoid-operated valve located on the intake side of the injection pump and connected to the system interlock to prevent fluid from being withdrawn from the supply tank when the irrigation system is either automatically or manually shut down.
4. The system must contain functional interlocking controls to automatically shut off the pesticide injection pump when the water pump motor stops.
5. The irrigation line or water pump must include a functional pressure switch which will stop the water pump motor when the water pressure decreases to the point where pesticide distribution is adversely affected.
6. Systems must use a metering pump, such as a positive displacement injection pump (e.g., diaphragm pump) effectively designed and constructed of materials that are compatible with pesticides and capable of being fitted with a system interlock.
7. Do not apply when wind speed favors drift beyond the area intended for treatment.

Planting

Do not plant seed deeper than 2 inches. For SUTAN + 6.7-E/ATRAZINE application made before planting, seeding can be done immediately, or it may be delayed up to 2 weeks in the Southeast, Southwest, and Western regions or up to 4 weeks in the Northeast, North Central, Midwest, and Pacific Northwest regions. To avoid removing SUTAN + 6.7-E/ATRAZINE from the row (with loss of weed control in the row), do not move or shape soil after treatment, such as by planting in deep furrows. If it is necessary to work the soil following application and incorporation, but before planting, do not work the soil deeper than the depth the herbicides were incorporated.

RECOMMENDATIONS

NOTE: The SUTAN + 6.7-E/ATRAZINE tank mix combination is not recommended on corn grown in Arizona or in the ten southernmost California counties.

For all uses, cultivation is recommended in addition to the SUTAN + 6.7-E/ATRAZINE treatment in fields with moderate to heavy infestations of shattercane, nutsedge, sandbur, johnsongrass seedlings, rhizome bermudagrass, Texas panicum or volunteer sorghum.

Broadcast Rates Per Acre [Conventional Sprays—SUTAN + 6.7-E/ATRAZINE (80W, 4L, AATREX NINE-0), Center Pivot Sprinklers—SUTAN + 6.7-E/ATRAZINE (80W, 4L)].

For control of annual grasses and broadleaf weeds: Use 3 3/4 to 4 3/4 pints SUTAN + 6.7-E and 1 to 2* pounds of ATRAZINE 80W, 3/4 to 1 1/2* quarts ATRAZINE 4L, or 4/5 to 1 3/4* pounds of AATREX NINE-0 per acre. Use the higher rate of SUTAN + 6.7-E on medium (medium fine textured-silt loam, clay loam, etc.) to fine (fine textured-clay) soils and for sandbur and johnsongrass seedling control.

For control of Nutsedge and Texas Panicum (In addition to the broadleaf weeds and other annual grasses listed): Use 4 3/4 to 7 1/3 pints of SUTAN + 6.7-E and 1 to 2* pounds ATRAZINE 80W, 3/4 to 1 1/2* quarts ATRAZINE 4L, or 4/5 to 1 3/4* pounds AATREX NINE-0 per acre. Use the higher rates of SUTAN + 6.7-E on medium (medium fine textured-silt loam, clay loam, etc.) soils, fine (fine textured-clay) soils and for heavy infestations.

For partial control (suppression) of Shattercane (and for control of the broadleaf weeds and annual grasses listed): When applied and incorporated *immediately* before planting, SUTAN + 6.7-E at the rate of 7 1/3 pints and 1 to 2* pounds ATRAZINE 80W, 3/4 to 1 1/2 quarts ATRAZINE 4L, or 4/5 to 1 3/4* pounds AATREX NINE-0 per acre will reduce competition from shattercane. On continuous corn acreage with high seed population, rotation to another crop is recommended.

NOTE: See preceding SUTAN + 6.7-E section for shattercane supplemental cultural and tillage practices.

For control of Johnsongrass and Bermudagrass (in addition to the broadleaf weeds and annual grasses listed): Apply and discincorporate SUTAN + 6.7-E just before planting at the rate of 4 3/4 to 7 1/3 pints plus 1 to 2* pounds of ATRAZINE 80W, or 3/4 to 1 1/2* quarts of ATRAZINE 4L, or 4/5 to 1 3/4* pounds AATREX NINE-0 per acre. Use the higher rate of SUTAN + 6.7-E on medium (medium fine textured-silt loam, clay loam, etc.) soils, fine (fine textured-clay) soils, for rhizome johnsongrass suppression and for heavy infestations. Disc a second time at right angles to the first. Cultivate at least once, preferably twice.

Johnsongrass: Using a program of plowing and discing followed by SUTAN + 6.7-E and ATRAZINE will give good control of seedlings and greatly reduce competition from plants coming from rhizomes.

In some areas discing 2 to 3 times spaced at 4 to 6 week intervals during Fall and Winter can be done. If this cannot be done, start with Fall plowing followed by several discings in the Spring before planting. The discing cuts rhizomes into small pieces and brings them to the surface, where both the weather and SUTAN + 6.7-E will kill or weaken them.

Rhizome Bermudagrass: The rhizomes and stolons of bermudagrass must be cut up thoroughly so that four or less nodes remain on a strand. For control of bermudagrass, the disc must be set to cut 6 inches deep.

For suppression of Broadleaf Signalgrass: For early and midseason suppression, use 7¹/₃ pints of SUTAN + 6.7-E and 1¹/₄ to 2* pounds ATRAZINE 80W, 1 to 1¹/₂* quarts ATRAZINE 4L, or 1 to 1³/₄* pounds AATREX NINE-0 per acre. Cultivation or application of a post-emergence herbicide recommended for broadleaf signalgrass will aid in extending control to late season.

For suppression of Hemp Dogbane: For early and mid-season suppression, use 7¹/₃ pints of SUTAN + 6.7-E and 1¹/₄ to 2* pounds ATRAZINE 80W, 1 to 1¹/₂* quarts ATRAZINE 4L, or 1 to 1³/₄* pounds AATREX NINE-0 per acre. Cultivation or application of a post-emergence herbicide recommended for hemp dogbane may aid in extending control beyond mid-season.

Fall Application: For use only in Minnesota, Northern Iowa (North of Interstate 80), Northeast Nebraska (North of Interstate 80 and East of State Highway 14), South Dakota (East of Missouri River), Wisconsin and Northern Illinois (North of Interstate 80). Apply and incorporate 7¹/₃ pints SUTAN +6.7-E and 2 pounds ATRAZINE 80W or 1¹/₂ quarts ATRAZINE 4L per acre in the late Fall before the ground freezes.

See Appendix IV for directions on use of dry bulk fertilizer applied in the Fall with this combination.

NOTE: Do not use Fall application for suppression of wild cane.

* Use the higher rate of ATRAZINE on soils with over 5% organic matter.

Cultural Practices Following Application

Should weeds develop, a shallow cultivation or rotary hoeing will generally result in better weed control. When cultivating for any reason, it should be shallow; i.e., no more than ¹/₂ the depth the herbicide was incorporated.

SUTAN + 6.7-E/BLADEX* [80W or 4L] Combination

* BLADEX is the registered trademark of E.I. duPont de Nemours & Co., Inc.

General Information

SUTAN + 6.7-E may be used in combination with BLADEX (80W or 4L) for weed control in corn. These combinations will control a broader spectrum of weeds than will either of the products used alone.

These combinations may be applied as a tank mix, preplant and soil-incorporated; or the BLADEX (80W or 4L) may be applied pre-emergence following a pre-plant incorporated application of SUTAN + 6.7-E.

IMPORTANT: Observe all cautions and limitations on the BLADEX (80W or 4L) and SUTAN + 6.7-E labels.

Make only one application per crop.

Do not use on milo or sorghum. Do not use on corn seed stock (Breeders, Foundation, or Increase). SUTAN + 6.7-E/BLADEX may be used on Production seed corn.

Weeds Controlled by SUTAN + 6.7-E/BLADEX

Annual Grasses:

Annual Bluegrass	*(Poa annua)*
Annual Fescues	*(Festuca spp.)*
Annual (Italian) Ryegrass	*(Lolium multiflorum)*
Barnyardgrass, Watergrass	*(Echinochloa crus-galli)*
Bullgrass	*(Paspalum boscianum)*
Crabgrass	*(Digitaria spp.)*
Fall Panicum	*(Panicum dichotomiflorum)*
Field Sandbur*	*(Cenchrus pauciflorus)*
Giant Foxtail	*(Setaria faberi)*
Goosegrass	*(Eleusine indica)*
Green Foxtail	*(Setaria viridis)*
Indian Lovegrass	*(Eragrostis pilosa)*
Junglerice	*(Echinochloa colonum)*
Seedling Bermudagrass	*(Cynodon dactylon)*
Seedling Johnsongrass*	*(Sorghum halepense)*
Stinkgrass	*(Eragrostis cilianensis)*
Texas Panicum*†, Buffalograss, Coloradograss	*(Panicum texanum)*
Volunteer Sorghum*	*(Sorghum spp.)*
Shattercane*‡, Wild Cane, Black Amber Cane, Chicken Corn, Scatter Cane	*(Sorghum bicolor)*
See Shattercane Rate and Cultural Recommendations.	
Wild Oats	*(Avena fatua)*
Witchgrass	*(Panicum capillare)*
Yellow Foxtail	*(Setaria lutescens)*

‡ Partial control (suppression) only.

* Cultivation is suggestd in addition to the SUTAN + 6.7-E/BLADEX treatment in fields with moderate to heavy infestations of wild cane, sandbur, Texas panicum, johnsongrass seedlings, or volunteer sorghum.

† Southeast Region only.

Annual Broadleaf Weeds:

Annual Buttercup	*(Ranunculus spp.)*
Annual Groundcherry	*(Physalis lanceifolia)*
Tall Morningglory	*(Ipomoea purpurea)*
Annual Sedge	*(Carex spp.)*
Black Mustard	*(Brassica nigra)*
Buffalobur	*(Solanum rostratum)*
Carpetweed	*(Mollugo verticillata)*
Common Chickweed	*(Stellaria media)*
Common Cocklebur	*(Xanthium pensylvanicum)*
Common Groundsel	*(Senecio vulgaris)*
Common Lambsquarters	*(Chenopodium album)*
Common Mallow	*(Malva neglecta)*
Common Purslane	*(Portulaca oleracea)*
Corn Spurry	*(Spergula arvensis)*
Floras Paintbrush	*(Emilia sagittata)*
Florida Pusley	*(Richadia scabra)*
Hedge Mustard	*(Sisymbrium officinale)*
Jimsonweed	*(Datura stramonium)*
Kochia	*(Kochia scoparia)*
Mayweed	*(Anthemis cotula)*
Pennsylvania Smartweed	*(Polygonum pensylvanicum)*
Pigweed	*(Amaranthus spp.)*
Pineappleweed	*(Matricaria matricarioides)*
Plantain	*(Plantago spp.)*
Poorjoe	*(Diodia teres)*
Prostrate Knotweed	*(Polygonum) aviculare)*
Prostrate Spurge	*(Euphorbia supina)*
Ragweed	*(Ambrosia spp.)*
Seedling Curly Dock	*(Rumex crispus)*
Shepherdspurse	*(Capsella bursa-pastoris)*
Sicklepod†, Coffeeweed	*(Cassia obtusifolia)*
Smallflower Galinsoga	*(Galinsoga parviflora)*
Prickly Sida	*(Sida spinosa)*
Tarweed	*(Hemizonia congesta)*
Velvetleaf	*(Abutilon theophrasti)*
Wild Buckwheat	*(Polygonum convolvulus)*
Wild Mustard	*(Brassica kaber)*
Wild Radish	*(Raphanus raphanistrum)*
Wild Turnip	*(Brassica campestris)*

† Southeast Region only.

Perennial Weeds:

Hemp Dogbane‡	*(Apocynum cannabinum)*

Perennial Weeds:

Rhizome Bermudagrass†	*(Cynodon dactylon)*
Rhizome Johnsongrass‡	*(Sorghum halepense)*
Purple Nutsedge	*(Cyperus rotundus)*
Yellow Nutsedge	*(Cyperus esculentus)*

‡ Partial Control (suppression only)
† Southeast Region only

Existing stands of nutsedge must be turned under and chopped up thoroughly before treatment. Cultivation is suggested in addition to SUTAN + 6.7-E/BLADEX treatment in fields with moderate to heavy nutsedge infestations.

DIRECTIONS FOR USE

The SUTAN + 6.7-E/BLADEX (80W or 4L) combination can be mixed with water or combined with fluid (solution, slurry, or suspension) fertilizers and applied as a spray, or it can be impregnated on dry bulk fertilizers.

Tank Mix Directions: See Appendix III

Application and Incorporation Directions:

For a tank mix application, apply the recommended dosage (see RECOMMENDATIONS) in 10 to 50 gallons of water per acre, (use the higher gallonage in arid areas) using a properly calibrated, low-pressure type nozzle (flood, fan or swirl chamber) or boom-type sprayer having good agitation. Check frequently during application to be sure the nozzles are free from clogging and are delivering a uniform spray pattern. Read directions on the BLADEX (80W or 4L) label concerning nozzle and screen sizes.

See Appendix I for Soil Mixing (incorporation) before Planting Directions.

The tank mix must be incorporated into the soil within 4 hours (within 1 hour in Western Region) after application to prevent loss of the SUTAN + 6.7-E. Whenever possible, application and incorporation should be done in the same operation. Apply the herbicides to soil dry enough to permit good soil mixing.

If there must be a delay (less than 4 hours, except less than 1 hour in Western Region) in incorporation of the herbicides when application and incorporation are done in separate operations, SUTAN + 6.7-E must be applied to a dry soil surface (at least 1/2 inch deep), free from dew and incidential moisture.

Loss of weed control will result from any delay in incorporation if SUTAN + 6.7-E is applied to a moist soil surface.

IMPORTANT: Thorough mixing of SUTAN + 6.7-E in the soil is necessary. The soil must be in good tilth. Improper discing or poor tilth (large clods) may result in unsatisfactory control.

Incorporation of Impregnated Dry Bulk Fertilizer

The application and incorporation of SUTAN + 6.7-E must be carried out on the same day and delayed only if soil moisture conditions permit. See Appendix IV for directions on the delay in soil incorporation.

SUTAN + 6.7-E/BLADEX (80W or 4L) Fluid Fertilizer Tank Mix

The SUTAN + 6.7-E/BLADEX (80W or 4L) combination can be combined with fluid fertilizers for applications made before planting. However, the physical compatibility with these fluid fertilizers must be determined before combining in the spray tank, and even though found to be compatible, constant agitation is necessary to keep the SUTAN + 6.7-E/BLADEX evenly mixed with the fluid fertilizer during application.

Refer to Appendix II regarding the procedure for testing the compatibility of SUTAN + 6.7-E/BLADEX combination with fluid fertilizers.

Impregnation on Dry Bulk Fertilizers

(See Appendix IV for use no No-Till Corn.)

Dry bulk fertilizers (except single superphosphate, treble superphosphate and nitrate fertilizers) may be impregnated or coated with SUTAN + 6.7-E/BLADEX (80W or 4L). However, uniform distribution of SUTAN + 6.7-E/BLADEX (80W or 4L) on fertilizer particles and uniform application are necessary to assure good results. See Appendix IV for information and directions for impregnation and use for these combinations.

SUTAN + 6.7-E/BLADEX (80W or 4L) Sequential Application

For sequential application of SUTAN + 6.7-E preplant incorporated followed by BLADEX (80W or 4L) pre-emergence, consult the appropriate application directions on the respective product labels. Follow the dosage recommendations given below.

RECOMMENDATIONS

NOTE: The SUTAN + 6.7-E/BLADEX tank mix combination is not recommendated on corn grown in Arizona or in the ten southernmost California counties.

For all uses, cultivation is recommended in addition to the SUTAN + 6.7-E/BLADEX treatment in fields with moderate to heavy infestations of wild cane, nutgrass, sandbur, johnsongrass seedlings, rhizome bermudagrass, Texas panicum, or volunteer sorghum.

Broadcast Rates Per Acre (Conventional Sprays and Center Pivot Sprinklers)

For control of annual grasses and broadleaf weeds: Use 3 3/4 to 4 3/4 pints of SUTAN + 6.7-E and 2 to 2 1/2 pounds BLADEX 80W or 3 to 4 pints BLADEX 4L per acre. Use the higher rates of SUTAN + 6.7-E on medium (medium fine textured-silt loam, clay loam, etc.) to fine (fine textured-clay) soils and for sandbur and johnsongrass seedling control.

For control of Nutsedge and Texas Panicum (In addition to the broadleaf weeds and annual grasses listed): Use 4 3/4 to 7 1/3 pints of SUTAN + 6.7-E and 2 to 2 1/2 pounds BLADEX 80W or 3 to 4 pints BLADEX 4L per acre.

For partial control (suppression) of Shattercane (and for control of the broadleaf weeds and annual grasses listed): When applied and incorporated immediately before planting, SUTAN + 6.7-E at the rate of 7 1/3 pints and 2 to 2 1/2 pounds BLADEX 80W or 3 to 4 pints BLADEX 4L per acre will reduce competition from shattercane. On continuous corn acreage with high cane seed population, rotation to another crop is recommended.

NOTE: See preceding SUTAN + 6.7-E section for wild cane supplemental cultural and tillage practices.

For control of Johnsongrass and Bermudagrass (in addition to the broadleaf weeds and annual grasses listed): Apply and disc-incorporate SUTAN + 6.7-E just before planting at the rate of 4 3/4 to 7 1/3 pints plus 2 to 2 1/2 pounds of BLADEX 80W or 3 to 4 pints BLADEX 4L per acre. Use the higher rate of SUTAN + 6.7-E on medium (medium fine textured-silt loam, clay loam, etc.) soils, fine (fine textured-clay) soils, for rhizome johnsongrass suppression and for heavy infestations. Disc a second time at right angles to the first. Cultivate at least once, preferably twice.

Johnsongrass: Using a program of plowing and discing followed by SUTAN + 6.7-E will give good control of seedings and greatly reduce competition from plants coming from rhizomes.

In some areas discing 2 to 3 times spaced at 4 to 6 week intervals during Fall and Winter can be done. If this cannot be done, start with Fall plowing followed by several discings in the Spring before planting. The discing cut rhizomes into small pieces and brings them to the surface, where both the weather and SUTAN + 6.7-E will kill or weaken them.

Rhizome Bermudagrass: The rhizomes and stolons of bermudagrass must be cut up thoroughly so that four or less nodes remain on a strand. For control of bermudagrass, the disc must be set to cut 6 inches deep.

For suppression of Hemp Dogbane: For early and mid-season suppression use 7 1/3 pints of SUTAN + 6.7-E and 2 to 2 1/2 pounds BLADEX 80W or 3 to 4 pints BLADEX 4L per acre. Cultivation or application of a post-emergence herbicide recommended for hemp dogbane may aid in extending control beyond mid-season.

NOTE: BLADEX in combination with SUTAN + 6.7-E is not recommended for use on coarse or medium textured soils containing less than 1% organic matter. Use the lower rate of BLADEX on coarse to medium textured soils containing 1% organic matter and above. Use the higher rate of BLADEX on fine (fine textured-clay) soils with 3% organic matter and above.

SUTAN + 6.7-E/BLADEX can be applied either as a tank mix combination and incorporated into the soil or apply SUTAN + 6.7-E preplant incorporated followed by BLADEX pre-emergence. If BLADEX is surface-applied on top of a preplant application of SUTAN + 6.7-E, a rotary hoeing or shallow cultivation is recommended if rainfall or sprinkler irrigation has not occurred within about 10 days after the application of BLADEX.

Fall Application: For use only in Minnesota, Northern Iowa (North of Interstate 80), Northeast Nebraska (North of Interstate 80 and East of State Highway 14), South Dakota (East of Missouri River), Wisconsin, and Northern Illinois (North of Interstate 80). Apply and incorporate 7 1/3 pints SUTAN + 6.7-E and 2 1/2 pounds BLADEX 80W or 4 pints BLADEX 4L per acre in the late Fall before the ground freezes.

See Appendix IV for directions on use of dry bulk fertilizer applied in the Fall with this combination.

NOTE: Do not use Fall application for suppression of shattercane.

Planting

Do not plant seed deeper than 2 inches. For SUTAN + 6.7-E/BLADEX application made before planting, seeding can be done immediately, or it may be delayed up to 2 weeks in the Southeast, Southwest, and Western regions or up to 4 weeks in the Northwest, North Central, Midwest, and Pacific Northwest regions. To avoid removing SUTAN + 6.7-E/BLADEX from the row (with loss of weed control in the row), do not move or shape soil after treatment, such as by planting in deep furrows. If it is necessary to work the soil following application and incorporation, but before planting, do not work the soil deeper than the depth the herbicides were incorporated.

Cultural Practices Following Application

Should weeds develop, a shallow cultivation or rotary hoeing will generally result in better weed control. When cultivating for any reason, it should be shallow, i.e., no more than 1/2 the depth the herbicide was incorporated.

SUTAN + 6.7-E/ATRAZINE (80W or 4L)/BLADEX (80W or 4L) 3-Way Tank Mix Directions

A tank mix combination of SUTAN + 6.7-E, ATRAZINE (80W or 4L) and BLADEX (80W or 4L) can be applied and mixed into the soil before planting for weed control in corn. This combination will control a broader spectrum of weeds than will any of these materials used alone. In addition, potential hazard of ATRAZINE residues affecting certain succeeding crops is reduced because a lower rate of ATRAZINE can be used when it is combined with SUTAN + 6.7-E/BLADEX than when it is used alone.

IMPORTANT: Observe all cautions and limitations on the ATRAZINE (80W or 4L), BLADEX (80W OR 4L), and SUTAN + 6.7-E labels.

Make only one application per crop. After treatment with SUTAN + 6.7-E/ATRAZINE/BLADEX combination, do not plant any crop except corn until the following year or injury may occur.

Do not use on milo or sorghum. Do not use on corn seed stock (Breeders, Foundation, or Increase). SUTAN + 6.7-E/ATRAZINE/BLADEX may be used on Production seed corn.

Weeds Controlled by SUTAN + 6.7-E/ATRAZINE/BLADEX Tank Mix

Annual Grasses:	
Annual Bluegrass	*(Poa annua)*
Annual Fescues	*(Festuca spp.)*
Annual (Italian) Ryegrass	*(Lolium multiflorum)*
Barnyardgrass, Watergrass	*(Echinochloa crus-galli)*
Bullgrass	*(Paspalum boscianum)*
Crabgrass	*(Digitaria spp.)*
Fall Panicum	*(Panicum dichotomiflorum)*
Field Sandbur*	*(Cenchrus pauciflorus)*
Giant Foxtail	*(Setaria faberi)*
Goosegrass	*(Eleusine indica)*
Green Foxtail	*(Setaria viridis)*
Indian Lovegrass	*(Eragrostis pilosa)*
Junglerice	*(Echinochloa colonum)*
Seedling Bermudagrass	*(Cynodon dactylon)*
Seedling Johnsongrass*	*(Sorghum halepense)*
Stinkgrass	*(Eragrostis cilianensis)*
Texas Panicum*†, Buffalograss, Coloradograss	*(Panicum texanum)*
Volunteer Sorghum*	*(Sorghum spp.)*
Shattercane*‡, Wild cane, Black Amber Cane, Chicken Corn, Scatter Cane	*(Sorghum bicolor)*
See Wild Cane Rate and Cultural Recommendations.	
Wild Oats	*(Avena fatua)*
Witchgrass	*(Panicum capillare)*
Yellow Foxtail	*(Setaria lutescens)*

‡ Partial control (suppression) only.
* Cultivation is suggested in addition to the SUTAN + 6.7-E/ATRAZINE/BLADEX treatment in fields with moderate to heavy infestations of shattercane, sandbur, Texas Panicum, johnsongrass seedlings or volunteer sorghum.
† Southeast Region Only.

Annual Broadleaf Weeds:	
Annual Buttercup	*(Ranunculus spp.)*
Annual Groundcherry	*(Physalis lanceifolia)*
Annual Morningglory	*(Ipomoea purpurea)*
Annual Sedge	*(Carex spp.)*
Buffalobur	*(Solanum rostratum)*
Carpetweed	*(Mollugo verticillata)*
Common Chickweed	*(Stellaria media)*
Common Cocklebur	*(Xanthium pensylvanicum)*
Common Groundsel	*(Senecio vulgaris)*
Common Lambsquarters	*(Chenopodium album)*
Common Mallow	*(Malva neglecta)*
Common Purslane	*(Portulaca oleracea)*
Corn Spurry	*(Spergula arvensis)*
Floras Paintbrush	*(Emilia sagittata)*
Florida Pusley	*(Richardia scabra)*
Hedge Mustard	*(Sisymbrium officinale)*
Jimsonweed	*(Datura stramonium)*
Kochia	*(Kochia scoparia)*
Mayweed	*(Anthemis cotula)*
Mustards	*(Brassica spp.)*
Nightshade	*(Solanum spp.)*
Pigweed	*(Amaranthus spp.)*
Pineappleweed	*(Matricaria matricarioides)*
Plantain	*(Plantago spp.)*
Poorjoe	*(Diodia teres)*
Prostrate Knotweed	*(Polygonum aviculare)*
Prostrate Spurge	*(Euphorbia supina)*
Ragweed	*(Ambrosia spp.)*
Seedling Curly Dock	*(Rumex crispus)*
Shepherdspurse	*(Capsella bursa-pastoris)*
Smallflower Galinsoga	*(Galinsoga parviflora)*
Smartweed	*(Polygonum spp.)*
Prickly Sida	*(Sida spinosa)*
Tarweed	*(Hemizonia congesta)*
Velvetleaf	*(Abutilon theophrasti)*
Wild Buckwheat	*(Polygonum convolvulus)*
Wild Radish	*(Raphanus raphanistrum)*
Wild Turnip	*(Brassica campestris)*

Perennial Weeds:	
Rhizome Bermudagrass†	*(Cynodon dactylon)*
Purple Nutsedge	*(Cyperus rotundus)*
Yellow Nutsedge	*(Cyperus esculentus)*
Rhizome Johnsongrass‡	*(Sorghum halepense)*

Existing stands of nutsedge must be turned under and chopped up thoroughly before treatment. Cultivation is suggested in addition to SUTAN + 6.7-E/ATRAZINE/BLADEX treatment in fields with moderate to heavy nutsedge infestations.

‡ Partial Control (suppression) Only.
† Southeast Region Only.

DIRECTIONS FOR USE

The SUTAN + 6.7-E/ATRAZINE (80W or 4L)/BLADEX (80W or 4L) combination can be mixed with water or combined with fluid (solution, slurry or suspension) fertilizers and applied as a spray, or can be impregnated on dry bulk fertilizers.

Tank Mix Directions: See Appendix III.

Application and Incorporation Directions:

For a tank mix application, apply the recommended dosage (see RECOMMENDATIONS) in 10 to 50 gallons of water per acre (use the higher gallonage in arid areas) using a properly calibrated, low-pressure type nozzle (flood, fan or swirl chamber), on boom-type sprayer having good, but not excessive agitation. Check frequently during application to be sure the nozzles are free from clogging and are delivering a uniform pattern. Read directions on the ATRAZINE (80W or 4L) and BLADEX (80W or 4L) labels concerning nozzle and screen sizes.

See Appendix I for Soil Mixing (incorporation) before Planting Directions.

The tank mix must be incorporated into the soil within 4 hours (within 1 hour in Western Region) after application to prevent loss of the SUTAN + 6.7-E. Whenever possible, application and incorporation should be done in the same operation. Apply the herbicides to soil dry enough to permit good soil mixing.

Loss of weed control will result from any delay in incorporation if SUTAN + 6.7-E is applied to a moist soil surface.

If there must be a delay (less than 4 hours, except less than 1 hour in Western Region) in incorporation of the herbicides when application and incorporation are done in separate operations, SUTAN + 6.7-E must be applied to a dry soil surface (at least 1/2 inch deep), and free from dew and incidental moisture.

IMPORTANT: Thorough mixing of SUTAN + 6.7-E in the soil is necessary. The soil must be in good tilth. Improper discing or poor tilth (large clods) may result in unsatisfactory control.

Incorporation of Impregnated Dry Bulk Fertilizer

The application and incorporation of SUTAN + 6.7-E must be carried out on the same day and delayed only if soil moisture conditions permit. See Appendix IV for directions on the delay in soil incorporation.

SUTAN + 6.7-E/ATRAZINE (80W or 4L)/BLADEX (80W or 4L) Fluid Fertilizer Tank Mix

The SUTAN + 6.7-E/ATRAZINE (80W or 4L)/BLADEX (80W or 4L) 3-way combination can be combined with fluid fertilizers for applications made before planting. However, the physical compatibility with these fertilizers must be determined before combining in the spray tank, and even though found to be compatible, constant agitation is necessary to keep the SUTAN + 6.7-E/ATRAZINE/BLADEX evenly mixed with the fluid fertilizer during application.

Refer to Appendix II regarding the procedure for testing the compatability of SUTAN + 6.7-E/ATRAZINE/BLADEX combination with fluid fertilizers.

Impregnation on Dry Bulk Fertilizers

(See Appendix IV for use no No-Till Corn.)

Dry bulk fertilizers (except single superphosphate, treble superphosphate and nitrate fertilizers) may be impregnated or coated with SUTAN + 6.7-E/ATRAZINE (80W or 4L)/BLADEX (80W or 4L). However, uniform distribution of SUTAN + 6.7-E/ATRAZINE (80W or 4L)/BLADEX (80W or 4L) on fertilizer particles and uniform application are necessary to assure good results. See Appendix IV for information and directions for impregnation and use of these combinations.

Planting

Do not plant seed deeper than 2 inches. For SUTAN + 6.7- E/ATRAZINE/BLADEX application made before planting, seeding can be done immediately, or it may be delayed up to 2 weeks in the Southeast, Southwest, and Western regions or up to 4 weeks in the Northeast, North Central, Midwest, and Pacific Northwest regions. To avoid removing SUTAN + 6.7-E/ATRAZINE/BLADEX from the row (with loss of weed control in the row), do not move or shape soil after treatment, such as by planting in deep furrows. If it is necessary to work the soil following application and incorporation, but before planting, do not work the soil deeper than the depth the herbicides were incorporated.

RECOMMENDATIONS

NOTE: The SUTAN + 6.7-E/ATRAZINE/BLADEX tank mix combination is not recommended on corn grown in Arizona or in the ten southern-most California counties.

For all uses, cultivation is recommended in addition to the SUTAN + 6.7-E/ATRAZINE/BLADEX treatment in fields with moderate to heavy infestations of wild cane, nutgrass, sandbur, johnsongrass seedlings, rhizome bermudagrass, Texas panicum, or volunteer sorghum.

Broadcast Rates Per Acre (Conventional Sprays)

For control of annual grasses and broadleaf weeds: Use 3 3/4 to 4 3/4 pints of SUTAN + 6.7-E plus 3/4 to 1 1/4 pounds of ATRAZINE 80W or 1 to 2 pints ATRAZINE 4L plus 1 1/4 to 2 1/2 pounds BLADEX 80W or 2 to 4 pints of BLADEX 4L per acre. ATRAZINE and BLADEX should be used in a one to two proportion. Use the higher rate of SUTAN + 6.7-E on medium (medium fine textured-silt loam, clay loam, etc.) to fine (fine textured clay) soils and for sandbur and johnsongrass seedling control.

For control of Nutsedge and Texas Panicum (In addition to the broadleaf weeds and annual grasses listed): Use 4 3/4 to 7 1/3 pints of SUTAN + 6.7-E plus 3/4 to 1 1/4 pounds ATRAZINE 80W or 1 to 2 pints ATRAZINE 4L plus 1 1/4 to 2 1/2 pounds BLADEX 80W or 2 to 4 pints of BLADEX 4L per acre. ATRAZINE and BLADEX should be used in a one to two proportion. Use the higher rates of SUTAN + 6.7-E on medium (medium fine textured-silt loam, clay loam, etc.) soils, fine (fine textured-clay) soils and for heavy infestations.

For partial control (suppression) of Shattercane (and for control of the broadleaf weeds and annual grasses listed): When applied and incoporated immediately before planting, SUTAN + 6.7-E at the rate of 7 1/3 pints plus 3/4 to 1 1/4 pounds ATRAZINE 80W or 1 to 2 pints ATRAZINE 4-L plus 1 1/4 to 2 1/2 pounds BLADEX 80W or 1 to 2 quarts BLADEX 4-L per acre will reduce competition from shattercane. ATRAZINE and BLADEX should be used in a one to two proportion. On continuous corn acreage with high seed population, rotation to another crop is recommended.

NOTE: See preceding SUTAN + 6.7-E section for shattercane supplemental cultural and tillage practices.

For control of Johnsongrass and Bermudagrass (in addition to the broadleaf weeds and annual grasses listed): Apply and disc-incorporate SUTAN + 6.7-E just before planting at the rate of 4 3/4 to 7 1/3 pints plus 3/4 to 1 1/4 pounds of ATRAZINE 80W or 1 to 2 pints ATRAZINE 4L plus 1 1/4 to 2 1/2 pounds of BLADEX 80W or 2 to 4 pints BLADEX 4L per acre. ATRAZINE and BLADEX should be used in a one to two proportion. Use the higher rates of SUTAN + 6.7-E on medium (medium fine textured-silt loam, clay loam, etc) soils, fine (fine textured-clay) soils, for rhizome johnsongrass suppression, and for heavy infestations. Disc a second time at right angles to the first. Cultivate at least once, preferably twice.

Rhizome Bermudagrass: The rhizomes and stolons of bermudagrass must be cut up thoroughly so that four or less nodes remain on a strand. For control of bermudagrass, the disc must be set to cut 6 inches deep.

Johnsongrass: Using a program of plowing and discing followed by SUTAN + 6.7-E/ATRAZINE/BLADEX will give good control of seedlings and greatly reduce competition from plants coming from rhizomes. In some areas discing 2 to 3 times spaced at 4 to 6 week intervals during Fall and Winter can be done. If this cannot be done, start with Fall plowing following by several discings in the Spring before planting. The discing cuts rhizomes into small pieces and brings them to the surface, where both the weather and SUTAN + 6.7-E will kill or weaken them.

NOTE: BLADEX in combination with SUTAN + 6.7-E and ATRAZINE is not recommended for use on coarse or medium textured soils containing less than 1% organic matter.

Fall Application: For use only in Minnesota, Northern Iowa (North of Interstate 80), Northeast Nebraska (North of Interstate 80 and East of State Highway 14), South Dakota (East of Missouri River), Wisconsin, and Northern Illinois (North of Interstate 80). Apply and incorporate 7 1/3 pints SUTAN + 6.7-E plus 1 1/4 pounds ATRAZINE 80W or 2 pints ATRAZINE 4L plus 2 1/2 pounds BLADEX 80W or 4 pints BLADEX 4L per acre in the late Fall before the ground freezes.

See Appendix IV for directions on use of dry bulk fertilizer applied in the Fall with this combination.

NOTE: Do not use Fall application for suppression of wild cane.

Cultural Practices Following Application

Should weeds develop, a shallow cultivation or rotary hoeing will generally result in better weed control. When cultivating for any reason, it should be shallow; i.e., no more than 1/2 the depth the herbicide was incorporated.

SUTAN + 6.7-E/PRINCEP* [4L, 80W or CALIBER*90] Combination

*PRINCEP is the registered trademark of Ciba-Geigy Corporation for simazine.
*CALIBER is the registered trademark of Ciba-Geigy Corporation.

General Information

SUTAN + 6.7-E may be used in a preplant incorporated tank mixture with PRINCEP (4L, 80W or CALIBER 90). These combinations will control a broader spectrum of weeds than will either of the products used alone.

IMPORTANT: Observe all cautions and limitations on the SUTAN + 6.7-E and PRINCEP (4L, 80W or CALIBER 90) labels.

Make only one application per crop.
Do not use on milo or sorghum.
Do not use on corn seed stock (Breeders, Foundation, or Increase). SUTAN + 6.7-E/PRINCEP may be used on Production seed corn.

Weeds Controlled by SUTAN + 6.7-E/PRINCEP

Annual Grasses:	
Barnyardgrass, Watergrass	*(Echinochloa crus-galli)*
Crabgrass	*(Digitaria spp.)*
Fall Panicum	*(Panicum dichotomiflorum)*
Field Sandbur*	*(Cenchrus pauciflorus)*
Giant Foxtail	*(Setaria faberi)*
Green Foxtail	*(Setaria viridis)*
Goosegrass	*(Eleusine indica)*
Seedling Bermudagrass	*(Cynodon dactylon)*
Seedling Johnsongrass*	*(Sorghum halepense)*
Texas Panicum*†, Buffalograss, Coloradograss	*(Panicum texanum)*
Volunteer Sorghum*	*(Sorghum spp.)*
Shattercane, Wild Cane*‡, Black Amber Cane, Chicken Corn, Scatter Cane	*(Sorghum bicolor)*
See Shattercane Rate and Cultural Recommendations.	
Witchgrass	*(Panicum capillare)*
Yellow Foxtail	*(Setaria lutescens)*

† Southeast Region only.
‡ Partial control (suppression) only.
* Cultivation is suggested in addition to the SUTAN + 6.7-E/PRINCEP treatment in fields with moderate to heavy infestations of johnsongrass, sandbur, Texas panicum, volunteer sorghum and shattercane.

Annual Broadleaf Weeds:	
Common Lambsquarters	*(Chenopodium album)*
Common Ragweed	*(Ambrosia artemisifolia)*
Florida Pusley	*(Richardia scabra)*
Redroot Pigweed	*(Amaranthus retroflexus)*
Velvetleaf	*(Abutilon theophrasti)*

Perennial Weeds:

Purple Nutsedge	*(Cyperus rotundus)*
Rhizome Bermudagrass†	*(Cynodon dactylon)*
Rhizome Johnsongrass‡	*(Sorghum halepense)*
Yellow Nutsedge	*(Cyperus esculentus)*

† Southeast Region only.
‡ Partial control (suppression) only.

Existing stands of nutsedge must be turned under and chopped up thoroughly before treatment. Cultivation is suggested in addition to SUTAN + 6.7-E/PRINCEP treatment in fields with moderate to heavy nutsedge infestations.

DIRECTIONS FOR USE

Tank Mix Directions: See Appendix III

Application and Incorporation Directions:

For a tank mix application, apply the recommended dosage (see RECOMMENDATIONS) in 20 to 50 gallons of water per acre (use the higher gallonage in arid areas), using a properly calibrated, low-pressure type nozzle (flood, fan or swirl chamber) on boom-type sprayer having good agitation. Check frequently during application to be sure the nozzles are free from clogging and are delivering a uniform spray pattern. Read directions on the PRINCEP (4L, 80W or CALIBER 90) label concerning nozzle and screen sizes.

See Appendix I for Soil Mixing (incorporation) before Planting Directions.

The tank mix must be incorporated into the soil within 4 hours (within 1 hour in Western Region) after application to prevent loss of the SUTAN + 6.7-E. Whenever possible, application and incorporation should be done in the same operation. Apply the herbicides to soil dry enough to permit good soil mixing.

If there must be delay (less than 4 hours except less than 1 hour in Western Region) in incorporation of the herbicides when application and incorporation are done in separate operations, SUTAN + 6.7-E must be applied to a soil surface that is dry (at least 1/2 inch deep), free from dew and incidental moisture.

Loss of weed control will result from any delay in incorporation if SUTAN + 6.7-E is applied to a moist soil surface.

IMPORTANT: Thorough mixing of SUTAN + 6.7-E in the soil is necessary. The soil must be in good tilth. Improper discing or poor tilth (large clods) may result in unsatisfactory control.

RECOMMENDATIONS

NOTE: The SUTAN + 6.7-E/PRINCEP tank mix combination is not recommended on corn grown in Arizona or in the ten southernmost California counties.

For all uses, cultivation is recommended in addition to the SUTAN + 6.7-E/PRINCEP treatment in fields with moderate to heavy infestations of johnsongrass seedlings, nutsedge, rhizome bermudagrass, sandbur, Texas panicum, volunteer sorghum, or shattercane.

Broadcast Rates Per Acre (Conventional Sprays)

*Use the higher rate of PRINCEP on fine-textured soils and where heavy broadleaf weed infestations are expected.

For control of annual grasses and broadleaf weeds: Use 3¾ to 4¾ pints of SUTAN + 6.7-E and 2 to 6* pints of PRINCEP 4L, 1¼ to 3¾* pounds PRINCEP 80W, or 1¹/₁₀ to 3⅓* pounds of PRINCEP CALIBER 90 per acre. Use the higher rates of SUTAN + 6.7-E on medium (medium fine textured silt loam, clay loam, etc,) to fine (fine textured-clay) soils and for sandbur and johnsongrass seedling control.

For control of Nutsedge, Texas Panicum and velvetleaf (in addition to the broadleaf weeds and annual grasses listed): Use 4¾ to 7⅓ pints of SUTAN + 6.7-E and 2 to 6* pints of PRINCEP 4L, 1¼ to 3¾* pounds PRINCEP 80W, or 1¹/₁₀ to 3⅓* pounds of PRINCEP CALIBER 90 per acre.

For partial control (suppression) of Shattercane (and for control of the broadleaf weeds and the annual grasses listed): When applied and incorporated immediately before planting SUTAN + 6.7-E at the rate of 7⅓ pints and 2 to 6* pints PRINCEP 4L, 1¼ to 3¾* pounds PRINCEP 80W, or 1¹/₁₀ to 3⅓* pounds of PRINCEP CALIBER 90 per acre will reduce competition from shattercane.

On continuous corn acreage with high seed population, rotation to another crop is recommended.

NOTE: See preceding SUTAN + 6.7-E section for shattercane supplemental cultural and tillage practices.

For control of Johnsongrass and Bermudagrass (In addition to the broadleaf weeds and annual grasses listed): Apply and disc-incorporate SUTAN + 6.7-E just before planting at the rate of 4¾ to 7⅓ pints plus 2 to 6 * pints PRINCEP 4L, 1¼ to 3¾* pounds PRINCEP 80W, or 1¹/₁₀ to 3⅓* pounds of PRINCEP CALIBER 90 per acre. Use the higher rate of SUTAN + 6.7-E on medium (medium fine textured-silt loam, clay loam, etc.) soils, fine (fine textured-clay) soils, for rhizome johnsongrass suppression and for heavy infestations. Disc a second time at right angles to the first. Cutivate at least once, preferably twice.

Johnsongrass: Using a program of plowing and discing followed by SUTAN + 6.7-E will give good control of seedlings and greatly reduce competition from plants coming from rhizomes.

In some areas discing 2 to 3 times spaced at 4 to 6 week intervals during fall and winter can be done. If this cannot be done, start with fall plowing followed by several discings in the spring before planting. The discing cuts rhizomes into small pieces and brings them to the surface, where both the weather and SUTAN + 6.7-E will kill or weaken them.

Rhizome Bermudagrass: The rhizomes and stolons of bermudagrass must be cut up thoroughly so that four or less nodes remain on a strand. For control of bermudagrass, the disc must be set to cut 6 inches deep.

Planting

Do not plant seed deeper than 2 inches. For a SUTAN + 6.7- E/PRINCEP application made before planting, seeding can be done immediately or it may be delayed up to 2 weeks in the Southeast, Southwest, and Western regions or up to 4 weeks in the Northeast, North Central, Midwest and Pacific Northwest regions. To avoid removing SUTAN + 6.7-E/PRINCEP from the row (with loss of weed control in the row), do not remove or shape soil after treatment, such as by planting in deep furrows. If it is necessary to work the soil following application and incorporation, but before planting, do not work the soil deeper than the depth the herbicides were incorporated.

Cultural Practices Following Application

Should weeds develop, a shallow cultivation or rotary hoeing will generally result in better weed control. When cultivating for any reason, it should be shallow, i.e., no more than 1/2 the depth the herbicide was incorporated.

APPENDIX I

Soil Mixing (Incorporation) Before Planting Directions

The following equipment commonly is used for soil mixing (incorporation) before planting.

Power-Driven Cultivation Equipment (Recommended on all soil types previously primary tilled). Ground speed should be adjusted with all PTO driven cultivation tools to insure thorough soil mixing.

a. Horizontal action: set to cut 4 to 6 inches deep. Example—Lely Roterra, Niemeyer or like equipment.

b. Vertical action: set to cut 3 to 4 inches deep. Example—Roto-tiller, side-winder or like equipment.

Tandem Discs (Recommended on all soil types):

Set to cut to a depth of 4 to 6 inches, operated at 4 to 6 mph followed by a spike-tooth harrow or some other leveling device which extends beyond the ends of the discs. For more thorough mixing (for control of deep-germinating grasses—shattercane, seedling johnsongrass and nutsedge), disc in two different directions. When discing the second time, the disc should be reset to cut no deeper than 4 to 6 inches.

Field Cultivator (Recommended on lighter soils in good tilth):

Use 3 to 4 rows of sweeps spaced at 7 inch or less intervals and staggered so that no soil is left unturned, followed by a spike-tooth harrow pulled behind the cultivator. Do not use chisel plows to incorporate. Set the cultivator to cut 4 to 6 inches deep, operated at 5 mph or more. The field cultivator should not be used alone as an incorporation tool where control of deep germinating weeds such as nutsedge, johnsongrass seedlings or shattercane is desired. For control of these weeds, use a disc alone for at least one of the passes over the field.

Till 'N Bed Conditioner (Recommended for bedded fields of all soil types in good tilth): Set tillers (choppers) to a depth of 4 to 6 inches. Operate at a speed of 5 to 7 mph. To insure moisture convervation and proper mixing, set the disc hippers to run only in the upper loose soil of the bed. Do not use in fields with excessive trash and/or moisture. For johnsongrass, wild cane and nutgrass control: In winter or prior to the SUTAN + application, run a row disc to chop up rhizomes and to destroy vegetating nutgrass tubers and early germinating seeds. Till 'N Hill equipment is most effective on well-peaked beds. On low beds, two passes are necessary for effective incorporation.

APPENDIX II

Procedure For Testing the Compatibility of SUTAN + 6.7-E or SUTAN + 6.7-E/ATRAZINE (80W OR 4L) or SUTAN + 6.7-E/BLADEX (80W or 4L) or SUTAN + 6.7-E/ATRAZINE (80W or 4L/BLADEX (80W or 4L) with Fluid Fertilizers.

The following procedure is suggested for determining whether the above herbicide combinations may be combined with specific fluid fertilizer for spray tank application.

Materials Required:

1. A. SUTAN + 6.7-E
   B. or SUTAN + 6.7-E/ATRAZINE (80W or 4L) combination
   C. or SUTAN + 6.7-E/BLADEX (80W or 4L) combination.
   D. or SUTAN + 6.7-E/ATRAZINE (80W or 4L)/BLADEX (80W or 4L) combination.
2. Fluid fertilizer to be used.

3. Tank mix adjuvants for fluid fertilizer combinations: Compex*, Sponto 168-D*, Unite* or equivalent. The adjuvant which provides the best compatibility depends on the specific fertilizer involved.

* Compex, Kalo Laboratories Inc., Kansas City, Missouri; Sponto 168-D, Witco Chemical Company, Houston, Texas; Unite, Hopkins Agricultural Chemical Company, Madison, Wisconsin.

4. Two one-quart, wide-mouth glass jars with lids.
5. Measuring spoons; a graduated pipette or cylinder will provide more accurate measurement; 1 tablespoon equals 3 teaspoons, 1 teaspoon equals approximately 5 milliliters.
6. Measuring cup, 8 oz. (237 mL).

Procedure: (See Note 1):

1. Pour a pint (473 mL) of the fluid fertilizer into each of the quart jars.
2. Add adjuvant to one of the jars and mix (see next rate table).
3. Add the SUTAN + 6.7-E or the desired combination of herbicides to both jars (see next four rate tables). The order of addition should be; wettable powders first with mixing, flowables next with mixing, and the SUTAN + 6.7-E last.
4. Close both jars with lids and mix the contents by turning the jars upside down ten times.
5. Inspect the surface and body of the mixtures—
   (A) Immediately after completing the jar inversions.
   (B) After allowing the jars to stand quietly for 30 minutes.
   (C) And then again after turning the jars upside down 10 times.

If a uniform mix cannot be made, the mixture should not be used. If either mixture remains uniform for 30 minutes, the combination may be used. Should either mixture separate after 30 minutes, but readily remix uniformly with ten jar inversions, the mixture can be used if adequate agitation is maintained in the tank. If the mixture with adjuvant is satisfactory, but the one without adjuvant is not, be sure to use the adjuvant in the spray tank. Add the adjuvant first at a rate of 3 pints per 100 gallons of fluid fertilizer; foaming can be minimized by using moderate agitation.

If nondispersible oil, sludge or clumps of solids form in the mixtures, the combination should not be used.

NOTE 1: For some combinations, premixing wettable powders in a little water in a pail before adding them to the spray tank will improve the compatibility of the final mixtures with SUTAN + 6.7-E. This technique can be tested in the small-scale jar test by premixing the wettable powder in one-eighth (1/8) cup of water prior to addition to the pint of fluid fertilizer.

Rate Table for SUTAN + 6.7-E and Adjuvant with the Fluid Fertilizer**

Gallons of Fluid Fertilizer to be applied per acre	mL or tsp. of SUTAN + 6.7-E* to be added to 1 pint of Fertilizer	
	mL	tsp.
10	7	1 1/3
15	5	1
20	4	3/4
25	3	2/3
30	2	1/2
40	2	1/2

* Based on field rate of 1 pound active ingredient per acre in the fertilizer volumes indicated. Increase volume proportionately to correspond with intended field rate in terms of pounds active ingredient per acre (e.g. for field rate of 4 pounds actual SUTAN in 40 gallons fertilizer per acre, add 8 mL or 2 tsp. SUTAN + 6.7-E to each jar for compatibility testing purposes).

** Two (2) milliliters or one-half (1/2) teaspoon of adjuvant should be added to 1 pint of fluid fertilizer in order to equal the rate of 3 pints of adjuvant per 100 gallons of fluid fertilizer.

Rate Table* for SUTAN + 6.7-E/ATRAZINE (80W or 4l) and Adjuvant with Fluid Fertilizer**

Gallons of Fluid Fertilizer to be applied per acre	10	15	20	25	30	40
tsp. of SUTAN + 6.7-E	5 3/4	3 3/4	2 3/4	2 1/3	2	1 1/2
tsp. of ATRAZINE 80W	7	4 2/3	3 1/2	2 3/4	2 1/3	1 3/4
tsp. of ATRAZINE 4L	3 2/3	2 1/2	1 3/4	1 1/2	1 1/4	1

* Based on field rates of 4 pounds actual SUTAN and 1.6 pounds actual ATRAZINE (80W) or 1.5 pounds actual ATRAZINE (4L) per acre in the fertilizer volumes indicated.

** Two (2) milliliters or one-half (1/2) teaspoon of adjuvant should be added to 1 pint of fluid fertilizer in order to equal the rate of 3 pints of adjuvant per 100 gallons of fluid fertilizer.

Rate Table* for SUTAN + 6.7-E/BLADEX (80W or 4L) and Adjuvant with Fluid Fertilizer**

Gallons of Fluid Fertilizer to be applied per acre	10	15	20	25	30	40
tsp. of SUTAN + 6.7-E	5 3/4	3 3/4	2 3/4	2 1/3	2	1 1/2
tsp. of BLADEX 80W	7 3/4	5 1/4	4	3	2 1/3	2
tsp. of BLADEX 4L	4 3/4	3 1/4	2 1/2	2	1 2/3	1 1/4

* Based on field rates of 4 pounds actual SUTAN and 2 pounds actual BLADEX (80W or 4L) per acre in the fertilizer volumes indicated.

** Two (2) milliliters or one-half (1/2) teaspoon of adjuvant should be added to 1 pint of fluid fertilizer in order to equal the rate of 3 pints of adjuvant per 100 gallons of fluid fertilizer.

Rate Table* for SUTAN + 6.7-E/ATRAZINE (80W or 4L)/BLADEX (80W or 4L) and Adjuvant with Fluid Fertilizer**

Gallons of Fluid Fertilizer to be applied per acre	10	15	20	25	30	40
tsp. of SUTAN + 6.7-E	5 3/4	3 3/4	2 3/4	2 1/3	2	1 1/2
tsp. of ATRAZINE 80W	2 1/4	1 1/2	1	3/4	3/4	1/2
tsp. of ATRAZINE 4L	1 1/4	3/4	2/3	1/2	1/3	1/3
tsp. of BLADEX 80W	4	2 2/3	2	1 1/2	1 1/4	1
tsp. of BLADEX 4L	2 1/2	1 2/3	1 1/4	1	3/4	2/3

* Based on field rates of 4 pounds actual SUTAN, 0.5 pounds actual ATRAZINE and 1 pound actual BLADEX per acre in the fertilizer volumes indicated.

** Two (2) milliliters or one-half (1/2) teaspoon of adjuvant should be added to 1 pint of fluid fertilizer in order to equal the rate of 3 pints of adjuvant per 100 gallons of fluid fertilizer.

APPENDIX III

Tank Mix Directions

Proper care must be used in preparing combination tank mixtures to avoid incompatibility problems. Small-scale tests should be made to establish that the mixtures are compatible (refer to Appendix II for general test procedures).

1) Fill the spray tank at least 2/3 full of the appropriate carrier, except when dry flowable products are used. These products should be thoroughly wet out, mixed and predispersed first in 1/4 to 1/2 of a tank with carrier liquid. Then the tank should be filled 2/3 full with additional carrier. Be sure the tank is clean and not contaminated with 2,4-D, oil residue or other materials.
2) Continue moderate agitation throughout mixing and application.
3) Add, mix and disperse dry flowables first, wettable powders second, liquid flowables third and SUTAN + 6.7-E last. For some combinations, premixing dry flowables and wettable powders in a little water in a pail before adding them to the spray tank will improve the compatibility of the final mixtures with SUTAN + 6.7-E.
4) Add the ATRAZINE, BLADEX and PRINCEP products individually to the spray tank in proper orders (steps 1 and 3). Make sure each product is thoroughly wetted and dispersed prior to addition of the next product. In particular, dry flowables should be thoroughly predispersed prior to addition of other pesticides.
5) Add the SUTAN + 6.7-E to the tank at least 2/3 full of carrier and mix; if ATRAZINE, BLADEX or PRINCEP are being used in the tank mix, they should be thoroughly mixed in the carrier prior to adding SUTAN + 6.7-E.
6) Complete filling the tank with the carrier; mix thoroughly, and maintain moderate agitation to provide a uniform mix during application.
7) The tank-mix combination should not be left in the spray tank for prolonged periods; batches should be mixed and applied the same day.

Do not use excessive agitation because it may cause the ATRAZINE, BLADEX or PRINCEP to settle out and set up in the spray tank. All return lines to the tank must discharge below liquid level.

APPENDIX IV

SUTAN + 6.7-E, SUTAN + 6.7-E/ATRAZINE (80W, 4L or AATREX NINE-O), SUTAN + 6.7-E/BLADEX (80W or 4L) OR SUTAN + 6.7-E/ATRAZINE (80W or 4L)/BLADEX (80W or 4L) Impregnation on Dry Bulk Fertilizers

Caution: SUTAN + 6.7-E alone or in combination with other herbicides must not be impregnated on ammonium nitrate, potassium nitrate or sodium nitrate fertilizers. Such mixtures may cause explosion and fire.

All individual state regulations relating to dry bulk fertilizer blending, registration, labeling and application are the responsibility of the individual and/or company selling the SUTAN + 6.7-E, SUTAN + 6.7-E/ATRAZINE, SUTAN + 6.7-E/BLADEX or SUTAN + 6.7-E/ATRAZINE/BLADEX fertilizer mixtures.

SUTAN + 6.7-E alone and in combination with ATRAZINE and/or BLADEX may be impregnated on many dry bulk fertilizers (1)(2) and applied and incorporated into the soil either in the Fall before the ground freezes or before planting for the control of certain grassy weeds by SUTAN + 6.7-E and many grassy and broadleaf weeds by the SUTAN + 6.7-E/ATRAZINE, SUTAN + 6.7-E/BLADEX or SUTAN + 6.7-E/ATRAZINE/BLADEX combinations in corn.

All SUTAN + 6.7-E, ATRAZINE (80W, 4L or AATREX NINE-O) and BLADEX (80W OR 4L) label and supplementary literature instructions and precautions regarding rates per acre, soil type and soil incorporation, application and other directions must be followed.

Test results have shown that SUTAN + 6.7-E, SUTAN + 6.7-E/ATRAZINE (80W, 4L or AATREX NINE-O), SUTAN + 6.7-E/BLADEX (80W or 4L) and SUTAN + 6.7-E/ATRAZINE (80W or 4L)/BLADEX (80W or 4L) on many dry bulk fertilizers give weed control equal to SU-

TAN + 6.7-E or tank mixtures of SUTAN + 6.7-E/ATRAZINE (80W, 4L or AATREX NINE-O), SUTAN + 6.7-E/BLADEX (80W or 4L) and SUTAN + 6.7-E/ATRAZINE (80W or 4L)/BLADEX (80W or 4L) applied as a spray in water or liquid fertilizer. However, uniform impregnation of the pesticides on the dry fertilizer particles and uniform application in the field are necessary to assure good results.

A minimum of 200 pounds and a maximum of 700 pounds of approved fertilizer ingredients (1)(2) impregnated with the appropriate amount of SUTAN +6.7-E alone or in combination with ATRAZINE (80W, 4L or AATREX NINE-O) and/or BLADEX (80W or 4L) must be applied per acre.

For impregnating the pesticides on dry fertilizers, use a closed rotary-drum type mixer equipped with suitable spraying equipment. The spray nozzles should be positioned inside the mixer to provide uniform spray coverage of the tumbling fertilizer. The SUTAN + 6.7-E should be sprayed uniformly onto the fertilizer using a fine spray pattern. For the SUTAN + 6.7-E/ATRAZINE (80W or 4L), SUTAN + 6.7-E/BLADEX or SUTAN + 6.7-E/ATRAZINE (80W or 4L)/BLADEX combination product the ATRAZINE (80W or 4L) and/or BLADEX (80W or 4L) can be added and mixed uniformly with the fertilizer prior to the SUTAN + 6.7-E addition, or alternatively, the SUTAN + 6.7-E/ATRAZINE (80W or 4L), SUTAN + 6.7-E/BLADEX (80W or 4L) or SUTAN + 6.7-E/ATRAZINE (80W or 4L)/BLADEX (80W or 4L) can be mixed in the proper use ratio to prepare a uniform slurry for joint spray impregnation. For the SUTAN + 6.7-E/AATREX NINE-O combination, the AATREX NINE-O should be mixed and thoroughly dispersed in the smallest amount of water feasible, and then uniformly sprayed onto the dry fertilizer. Following this, the SUTAN 6.7-E can be sprayed onto the dry fertilizer.

The physical properties of fertilizers vary, particularly in liquid absorptive capacity. When absorptivity is sufficient, simple spray impregnation of the fertilizer with SUTAN + 6.7-E provides a satisfactory dry mixture.

If the absorptivity is not adequate, use of a drying agent is required to provide a dry, free-flowing mixture.

For mixtures to be used in spinning-disc applicators, Microcel E calcium silicate powder (Manville, Filtration & Minerals) is recommended for use as a drying agent.

For mixtures to be used in pneumatic applicators, Microcel E or Agsorb 16/30 RVM-MS granular-clay (Oil-Dri Corporation) is recommended for use as the drying agent.

The drying agent should be added separately and uniformly to the prepared pesticide-fertilizer mixture, in a quantity that is sufficient to provide a suitably free-flowing mixture. Generally, less than 2% Microcel E or 5% Agsorb 16/30 RVM-MS by weight is required.

The amount of SUTAN + 6.7-E, SUTAN + 6.7-E/ATRAZINE (80W, 4L or AATREX NINE-O), SUTAN + 6.7-E/BLADEX (80W or 4L) or SUTAN + 6.7-E/ATRAZINE (80W or 4L)/BLADEX (80W or 4L) actually required in the manufacture of individual fertilizer mixtures should be determined carefully for each production operation (3)(4)(5). This is necessary to ensure that the amounts of SUTAN + 6.7-E, ATRAZINE and BLADEX actually contained in the mixture applied to the soil represent the correct rates of use.

Bulk fertilizers impregnated with SUTAN + 6.7-E, SUTAN + 6.7-E/ATRAZINE (80W, 4L or AATREX NINE-O), SUTAN + 6.7-E/BLADEX (80W or 4L) or SUTAN + 6.7-E/ATRAZINE (80W or 4L)/BLADEX (80W or 4L) should be **applied immediately,** NOT STORED. All bulk containers should be tightly covered while the products are being transported and applied to reduce chances of SUTAN + 6.7-E loss via volatilization.

Incorporation of Impregnated Fertilizers

When application and incroporation are done in separate operations, SUTAN + 6.7-E impregnated on dry fertilizer must be incorporated on the same day as applied. Application must be made on a dry soil surface (at least 1/2 inch deep) free from dew and incidental moisture.

Do not combine SUTAN + 6.7-E plus ATRAZINE (80W, 4L or AATREX NINE-O) and/or BLADEX (80W or 4L) with single superphosphate (0-20-0) or treble superphosphate (0-46-0).

DIRECTIONS FOR USE of SUTAN + 6.7-E IMPREGNATED ON DRY BULK FERTILIZER FOR APPLICATION IN NO-TILL CORN (DELAWARE, MARYLAND, NEW JERSEY, PENNSYLVANIA AND VIRGINIA ONLY)

For control of labeled weeds, SUTAN + 6.7-E may be impregnated on dry bulk fertilizer and applied preplant surface (PPS) or pre-emergence surface (PES) to untilled soil over a standing cover crop (vetch, red clover, rye, oats) or over previous corn ground having at least 90% ground cover of corn stubble remaining from previous years' crops just before or just after corn has been planted. (Refer to the list of weeds controlled by SUTAN 6.7-E.) At the time of application, the cover crop and other existing vegetation must be dry to permit impregnated fertilizer particles to sift down to the soil surface. Immediately after application and corn planting, existing vegetation must be sprayed with a knockdown type herbicide such as Paraquat that will cause the plants to quickly brown and collapse, forming a residue mulch barrier over the ground surface. Weed control will be unsatisfactory if the cover crop is too tall and interferes with uniform fertilizer coverage, or if the cover crop density is inadequate to form a good mulch cover over the soil surface.

Rates: Use 4¾ to 7⅓ pints of SUTAN + 6.7-E per acre. Use the higher rates on medium (medium fine textured—silt loam, clay loam, etc.) soils, fine (fine textured clay) soils and for heavy weed infestations. Apply a minimum of 200 pounds and a maximum of 700 pounds of approved fertilizer ingredients per acre.

Note: Atrazine and/or Bladex for broadleaf weed control may be included when the fertilizer is impregnated with SUTAN + 6.7-E or combined later in the cover crop knockdown spray.

Caution—Refer to the labels of all products used in combination for additional cautions and directions for use.

SUTAN + 6.7-E Physical Data

Specific Gravity 20/20°C:	0.946 (typical)
Pounds/Gallon 20/20°C:	7.87 (typical)
Flashpoint:	197°F (Tagliabue closed cup)
Viscosity:	Sprayable down to minus 20°F.

(1) Approved Dry Fertilizer Ingredients for use with SUTAN + 6.7-E

	N	P	K
Ammonium Sulfate	21	0	0
Ammonium Phosphate-Sulfate	16	20	0
Diammonium Phosphate	18	46	0
Monoammonium Phosphate	11	56	0
Potassium Chloride	0	0	60
Potassium Sulfate	0	0	52
Single Superphosphate	0	20	0
Treble Superphosphate	0	46	0
Urea*	45	0	0

* Some ureas may be phytotoxic when high rates are applied to corn. Use only urea rates known to be safe for corn application.

Note: Sul-po-mag, K-mag and 11-48-0 have been shown to be compatible with SUTAN + 6.7-E and are approved for use.

(2) Approved Dry Fertilizer Ingredients For Use With SUTAN + 6.7-E/ATRAZINE (80W, 4L or AATREX NINE-O), SUTAN + 6.7-E/BLADEX (80W or 4L) or SUTAN + 6.7-E/ATRAZINE (80W or 4L)/BLADEX (80W or 4L)

	N	P	K
Ammonium Sulfate	21	0	0
Ammonium Phosphate-Sulfate	16	20	0
Diammonium Phosphate	18	46	0
Monoammonium Phosphate	11	56	0
Potassium Chloride	0	0	60
Potassium Sulfate	0	0	52
Urea*	45	0	0

* Some ureas may be phytotoxic when high rates are applied to corn. Use only urea rates known to be safe for corn application.

Note: Sul-po-mag, K-mag and 11-48-0 have shown to be compatible with SUTAN + 6.7-E/ATRAZINE (80W, 4L or AATREX NINE-O), SUTAN + 6.7-E/BLADEX (80W or 4L) or SUTAN + 6.7-E-ATRAZINE (80W or 4L)/BLADEX (80W or 4L) are approved for use.

(3) RATE CHART FOR IMPREGNATION OF DRY BULK FERTILIZERS WITH SUTAN + 6.7-E OR SUTAN + 6.7-E PLUS ATRAZINE (80W, 4L or AATREX NINE-O)

FERTILIZER RATE PER ACRE	SUTAN + 6.7-E RATE/ACRE			ATRAZINE 80W RATE/ACRE			ATRAZINE 4L RATE/ACRE			AATREX NINE-O RATE/ACRE		
	3¾ PTS. PER ACRE	4¾ PTS. PER ACRE	7⅓ PTS. PER ACRE	1 LB. PER ACRE	1¼ LBS. PER ACRE	2 LBS. PER ACRE	¾ QT. PER ACRE	1 QT. PER ACRE	1½ QTS. PER ACRE	⅘ LB. PER ACRE	1 LB. PER ACRE	1¾ LBS. PER ACRE
200 lbs.	18¾ qts./ton	23¾ qts./ton	36⅔ qts./ton	10 lbs./ton	12½ lbs./ton	20 lbs./ton	7½ qts./ton	10 qts./ton	15 qts./ton	8 lbs./ton	10 lbs./ton	17½ lbs./ton
250 lbs.	15 qts./ton	19 qts./ton	29⅓ qts./ton	8 lbs./ton	10 lbs./ton	16 lbs./ton	6 qts./ton	8 qts./ton	12 qts./ton	6⅖ lbs./ton	8 lbs./ton	14 lbs./ton
300 lbs.	12½ qts./ton	15⅘ qts./ton	24⅖ qts./ton	6⅔ lbs./ton	8⅓ lbs./ton	13⅓ lbs./ton	5 qts./ton	6⅗ qts./ton	10 qts./ton	5⅓ lbs./ton	6⅔ lbs./ton	11⅔ lbs./ton
350 lbs.	10¾ qts./ton	13⅗ qts./ton	21 qts./ton	5¾ lbs./ton	7⅛ lbs./ton	11⅖ lbs./ton	4¼ qts./ton	5⅘ qts./ton	8⅗ qts./ton	4⅖ lbs./ton	5¾ lbs./ton	10 lbs./ton
400 lbs.	9⅖ qts./ton	11⅞ qts./ton	18⅓ qts./ton	5 lbs./ton	6¼ lbs./ton	10 lbs./ton	3¾qts./ton	5 qts./ton	7½ qts./ton	4 lbs./ton	5 lbs./ton	8¾ lbs./ton

FERTILIZER RATE PER ACRE	SUTAN + 6.7-E RATE/ACRE			ATRAZINE 80W RATE/ACRE			ATRAZINE 4L RATE/ACRE			AATREX NINE-O RATE/ACRE		
	3¾ PTS. PER ACRE	4¾ PTS. PER ACRE	7⅓ PTS. PER ACRE	1 LB. PER ACRE	1¼ LBS. PER ACRE	2 LBS. PER ACRE	¾ QT. PER ACRE	1 QT. PER ACRE	1½ QTS. PER ACRE	⅘ LB. PER ACRE	1 LB. PER ACRE	1¾ LBS. PER ACRE
450 lbs.	8⅓ qts./ton	10½ qts./ton	16⅓ qts./ton	4⅖ lbs./ton	5½ lbs./ton	8⅞ lbs./ton	3⅓ qts./ton	4⅖ qts./ton	6⅗ qts./ton	3⅗ lbs./ton	4⅖ lbs./ton	7¾ lbs./ton
500 lbs.	7½ qts./ton	9½ qts./ton	14⅔ qts./ton	4 lbs./ton	5 lbs./ton	8 lbs./ton	3 qts./ton	4 qts./ton	6 qts./ton	3⅕ qts./ton	4 lbs./ton	7 lbs./ton
550 lbs.	6⅘ qts./ton	8⅔ qts./ton	13⅓ qts./ton	3⅔ lbs./ton	4⅔ lbs./ton	7⅓ lbs./ton	2¾ qts./ton	3⅔ qts./ton	5½ qts./ton	2⅞ lbs./ton	3⅔ lbs./ton	6⅜ lbs./ton
600 lbs.	6¼ qts./ton	8 qts./ton	12⅕ qts./ton	3⅓ lbs./ton	4⅕ lbs./ton	6⅔ lbs./ton	2½ qts./ton	3⅓ qts./ton	5 qts./ton	2⅔ lbs./ton	3⅓ lbs./ton	5⅘ lbs./ton
650 lbs.	5¾ qts./ton	7⅓ qts./ton	11¼ qts./ton	3⅛ lbs./ton	3⅞ lbs./ton	6⅕ lbs./ton	2⅓ qts./ton	3⅛ qts./ton	4⅗ qts./ton	2½ lbs./ton	3⅛ lbs./ton	5⅜ lbs./ton
700 lbs.	5⅓ qts./ton	6⅘ qts./ton	10½ qts./ton	2⅞lbs./ton	3⅔ lbs./ton	5¾ lbs./ton	2⅛ qts./ton	2⅞ qts./ton	4¼ qts./ton	2¼ lbs./ton	2⅞ lbs./ton	5 lbs./ton

(4) RATE CHART FOR IMPREGNATION OF DRY BULK FERTILIZERS WITH SUTAN + 6.7-E OR SUTAN + 6.7-E PLUS BLADEX (80W or 4L)

FERTILIZER RATE PER ACRE	SUTAN + 6.7-E RATE/ACRE			BLADEX 80W RATE/ACRE		BLADEX 4L RATE/ACRE	
	3¾ PTS. PER ACRE	4¾ PTS. PER ACRE	7⅓ PTS. PER ACRE	2 LBS. PER ACRE	2½ LBS. PER ACRE	1½ QTS. PER ACRE	2 QTS. PER ACRE
200 lbs.	18¾ qts./ton	23¾ qts./ton	36⅔ qts./ton	20 lbs./ton	25 lbs./ton	15 qts./ton	20 qts./ton
250 lbs.	15 qts./ton	19 qts./ton	29⅓ qts./ton	16 lbs./ton	20 lbs./ton	12 qts./ton	16 qts./ton
300 lbs.	12½ qts./ton	15⅘ qts./ton	24⅖ qts./ton	13⅓ lbs./ton	16⅔ lbs./ton	10 qts./ton	13⅓ qts./ton
350 lbs.	10¾ qts./ton	13⅗ qts./ton	21 qts./ton	11⅖ lbs./ton	14¼ lbs./ton	8⅗ qts./ton	11⅖ qts./ton
400 lbs.	9⅖ qts./ton	11⅞ qts./ton	18⅓ qts./ton	10 lbs./ton	12½ lbs./ton	7½ qts./ton	10 qts./ton
450 lbs.	8⅓ qts./ton	10½ qts./ton	16⅓ qts./ton	8⅞ lbs./ton	11⅛ lbs./ton	6⅗ qts./ton	8⅞ qts./ton
500 lbs.	7½ qts./ton	9½ qts./ton	14⅔ qts./ton	8 lbs./ton	10 lbs./ton	6 qts./ton	8 qts./ton
550 lbs.	6⅘ qts./ton	8⅔ qts./ton	13⅓ qts./ton	7⅓ lbs./ton	9⅛ lbs./ton	5½ qts./ton	7⅓ qts./ton
600 lbs.	6¼ qts./ton	8 qts./ton	12⅕ qts./ton	6⅔ lbs./ton	8⅓ lbs./ton	5 qts./ton	6⅔ qts./ton
650 lbs.	5¾ qts./ton	7⅓ qts./ton	11¼ qts./ton	6⅕ lbs./ton	7¾ lbs./ton	4⅗ qts./ton	6⅕ qts./ton
700 lbs.	5⅓ qts./ton	6⅘ qts./ton	10½ qts./ton	5¾ lbs./ton	7⅕ lbs./ton	4¼ qts./ton	5¾ qts./ton

(5) RATE CHART FOR IMPREGNATION OF DRY BULK FERTILIZERS WITH SUTAN + 6.7-E OR SUTAN + 6.7-E PLUS ATRAZINE (80W or 4L) PLUS BLADEX (80W or 4L)

FERTILIZER RATE PER ACRE	SUTAN + 6.7-E RATE/ACRE			ATRAZINE 80W RATE/ACRE		ATRAZINE 4L RATE/ACRE		BLADEX 80W RATE/ACRE		BLADEX 4L RATE/ACRE	
	3¾ PTS. PER ACRE	4¾ PTS. PER ACRE	7⅓ PTS. PER ACRE	¾ LBS. PER ACRE	1¼ LBS. PER ACRE	1 PT. PER ACRE	2 PTS. PER ACRE	1¼ LBS. PER ACRE	2½ LBS. PER ACRE	1 QT. PER ACRE	2 QTS. PER ACRE
200 lbs.	18¾ qts./ton	23¾ qts./ton	36⅔ qts./ton	7½ lbs./ton	12½ lbs./ton	5 qts./ton	10 qts./ton	12½ qts./ton	25 lbs./ton	10 qts./ton	20 qts./ton
250 lbs.	15 qts./ton	19 qts./ton	29⅓ qts./ton	6 lbs./ton	10 lbs./ton	4 qts./ton	8 qts./ton	10 lbs./ton	20 lbs./ton	8 qts./ton	16 qts./ton
300 lbs.	12½ qts./ton	15⅘ qts./ton	24⅖ qts./ton	5 lbs./ton	8⅓ lbs./ton	3⅓ qts./ton	6⅗ qts./ton	8⅓ lbs./ton	16⅔ lbs./ton	6⅗ qts./ton	13⅓ qts./ton
350 lbs.	10¾ qts./ton	13⅗ qts./ton	21 qts./ton	4⅓ lbs./ton	7⅛ lbs./ton	2⅞ qts./ton	5⅘ qts./ton	7⅛ lbs./ton	14¼ lbs./ton	5⅘ qts./ton	11⅖ qts./ton
400 lbs.	9⅖ qts./ton	11⅞ qts./ton	18⅓ qts./ton	3¾ lbs./ton	6¼ lbs./ton	2½ qts./ton	5 qts./ton	6¼ lbs./ton	12½ lbs./ton	5 qts./ton	10 qts./ton
450 lbs.	8⅓ qts./ton	10½ qts./ton	16⅓ qts./ton	3⅓ lbs./ton	5½ lbs./ton	2¼ qts./ton	4⅖ qts./ton	5½lbs./ton	11⅛ lbs./ton	4⅖ qts./ton	8⅞ qts./ton
500 lbs.	7½ qts./ton	9½ qts./ton	14⅔ qts./ton	3 lbs./ton	5 lbs./ton	2 qts./ton	4 qts./ton	5 lbs./ton	10 lbs./ton	4 qts./ton	8 qts./ton
550 lbs.	6⅘ qts./ton	8⅔ qts./ton	13⅓ qts./ton	2¾ lbs./ton	4⅔ lbs./ton	1⅞ qts./ton	3⅔ qts./ton	4⅔ lbs./ton	9⅛ lbs./ton	3⅔ qts./ton	7⅓ qts./ton
600 lbs.	6¼ qts./ton	8 qts./ton	12⅕ qts./ton	2½ lbs./ton	4⅕ lbs./ton	1⅔ qts./ton	3⅓ qts./ton	4⅕ lbs./ton	8⅓ lbs./ton	3⅓ qts./ton	6⅔ qts./ton
650 lbs.	5¾ qts./ton	7⅓ qts./ton	11¼ qts./ton	2⅓ lbs./ton	3⅞ lbs./ton	1⅗ qts./ton	3⅛ qts./ton	3⅞ lbs./ton	7¾ lbs./ton	3⅛ qts./ton	6⅕ qts./ton
700 lbs.	5⅓ qts./ton	6⅘ qts./ton	10½ qts./ton	2⅕ lbs./ton	3⅔ lbs./ton	1½ qts./ton	2⅞ qts./ton	3⅔ lbs./ton	7⅕ lbs./ton	2⅞ qts./ton	5¾ qts./ton

STORAGE AND DISPOSAL

Prohibitions: Do not contaminate water, food, or feed by storage or disposal. Open dumping is prohibited.

Storage: Keep container tightly closed when not in use. Do not store near seeds, fertilizers or foodstuffs. Store out of reach of children, pets and domestic animals. Can be stored at temperatures as low as minus 50°F.

Pesticide Disposal: Rinse spray equipment. Waste resulting from the use of this product may be disposed of on site or at an approved waste disposal facility.

Container Disposal: Metal—Triple rinse (or equivalent). Then offer for recycling or reconditioning, or puncture and dispose of in a sanitary landfill, or by other procedures approved by State and local authorities.

Plastic—Triple rinse (or equivalent). Then offer for recycling or reconditioning, or puncture and dispose of in a sanitary landfill, or by incineration, or, if allowed by State and local authorities, by burning. If burned, stay out of smoke.

NOTICE TO BUYER AND USER:

Seller warrants that this product conforms to the chemical description on the label and is reasonably fit for the purposes stated on the label when used in accordance with directions under normal conditions of use. This warranty does not extend to the use of this product contrary to label instructions, or under abnormal use conditions, or under conditions not reasonably foreseeable to Seller, and Buyer and User assume the risk of any such use. SELLER DISCLAIMS ALL OTHER WARRANTIES EXPRESSED OR IMPLIED INCLUDING ANY WARRANTY OF FITNESS OR MERCHANTABILITY. SELLER SHALL NOT BE LIABLE FOR CONSEQUENTIAL, SPECIAL OR INDIRECT DAMAGES RESULTING FROM THE USE OR HANDLING OF THIS PRODUCT AND SELLER'S SOLE LIABILITY AND BUYER'S AND USER'S EXCLUSIVE REMEDY SHALL BE LIMITED TO THE REFUND OF THE PURCHASE PRICE.

This product is sold only for uses stated on its label. No express or implied license is granted to use or sell this product under any patent in any country except as specified: Country: United States of America. Patent No. 4,021,224; 4,137,070.

A-14R

The information and specimen in this booklet are correct as of the publication date but are subject to change without notice. For current information, contact ICI Americas, Inc.

SUTAZINE® +

SELECTIVE HERBICIDE

Preplant incorporated herbicide for annual grass and broadleaf weed control in corn (Field, Sweet, Silage and Popcorn)

ACTIVE INGREDIENTS:

S-ethyl diisobutylthiocarbamate	56.80%
Atrazine: 2-chloro-4(ethylamino)-6-(isopropylamino)-s- triazine	13.90%
Related Triazines	0.30%
INERT INGREDIENTS:	29.00%
TOTAL	100.00%

6-ME (Contains 6.0 pounds active ingredients per gallon.)

EPA Reg. No. 10182-248

Complete Directions For Use,
Use Precautions and Recommendations

KEEP OUT OF REACH OF CHILDREN

DANGER

Corrosive, Causes Irreversible Eye Damage

Harmful if Swallowed

STATEMENT OF PRACTICAL TREATMENT

First Aid

If a known exposure occurs or is suspected, immediately start the procedures given below. If further treatment is required, CONTACT A POISON CENTER, A PHYSICIAN, OR THE NEAREST HOSPITAL.

If swallowed: Immediately give several glasses of water and induce vomiting by gagging the victim with a finger placed on the back of the victim's tongue. Give fluids until vomitus is clear. If victim is unconscious or convulsing, do not induce vomiting or give anything by mouth.

Eye contact: Immediately flush the eyes with plenty of running water for at least 15 minutes. Hold the eyelids apart during the flushing to ensure rinsing of the entire surface of the eye and lids with water. Call a physician.

Skin contact: Flush affected areas with plenty of water for several minutes. Remove and clean any contaminated clothing and shoes. Seek medical attention if skin irritation occurs.

Inhalation: If inhaled, remove to fresh air. Seek medical attention if respiratory irritation occurs or breathing becomes difficult.

FOR 24-HOUR EMERGENCY MEDICAL ASSISTANCE CALL 1-800-F-A-S-T-M-E-D (327-8633).
FOR CHEMICAL EMERGENCY: Spill, leak, fire, exposure, or accident call CHEMTREC 1-800-424-9300.

PRECAUTIONARY STATEMENTS

HAZARDS TO HUMANS AND DOMESTIC ANIMALS

DANGER

CORROSIVE, CAUSES IRREVERSIBLE EYE DAMAGE. HARMFUL IF SWALLOWED. Avoid contact with skin, eyes and clothing. Avoid breathing spray mist. Wear goggles, rubber gloves and protective clothing when handling. Wash skin with soap and water immediately after contact. Flush eyes with water.

Do not apply this product in such a manner as to directly or through drift expose workers or other persons. When applied through sprinkler irrigation method of application, the area being treated must be vacated by unprotected persons.

ENVIRONMENTAL HAZARDS

Atrazine leaches readily and accepted label rates have been found to result in contamination of water supplies by way of groundwater. Therefore, users are advised to avoid use of atrazine in well-drained soils, particularly in areas having high groundwater tables.

This pesticide is toxic to aquatic invertebrates. Do not apply directly to water or wetlands. Runoff and drift from treated areas may be hazardous to aquatic organisms in neighboring areas. Do not contaminate water by cleaning of equipment or disposal of wastes.

Weeds Controlled by SUTAZINE +

Barnyardgrass, Watergrass	*(Echinochloa crus-galli)*
Broadleaf Signalgrass†*‡	*(Brachiaria platyphylla)*
Crabgrass	*(Digitaria* spp.*)*
Fall Panicum	*(Panicum dichotomiflorum)*
Field Sandbur*	*(Cenchrus pauciflorus)*
Giant Foxtail	*(Setaria faberi)*
Green Foxtail	*(Setaria viridis)*
Goosegrass	*(Eleusine indica)*
Seedling Bermudagrass	*(Cynodon dactylon)*
Seedling Johnsongrass*	*(Sorghum halepense)*
Texas Panicum*‡, Buffalograss, Coloradograss	*(Panicum texanum)*
Volunteer Sorghum*	*(Sorghum* spp.*)*
Shattercane*†, Wildcane, Black Amber Cane, Chicken Corn, Scatter Cane	*(Sorghum bicolor)*
See Wild Cane Rate and Cultural Recommendations	
Wild Oats	*(Avena fatua)*
Witchgrass	*(Panicum capillare)*
Yellow Foxtail	*(Setaria lutescens)*

† Partial control (suppression) only.
* Cultivation is suggested in addition to the SUTAZINE + treatment in fields with moderate to heavy infestations of wild cane, sandbur, johnsongrass, Texas panicum, volunteer sorghum, or broadleaf signalgrass.
‡ Southeast Region only.

Annual Broadleaf Weeds:

Annual Morningglory	*(Ipomoea purpurea)*
Carpetweed	*(Mollugo verticillata)*
Common Cocklebur	*(Xanthium pensylvanicum)*
Common Lambsquarters	*(Chenopodium ablum)*
Common Purslane	*(Portulaca oleracea)*
Florida Pusley	*(Richardia scabra)*
Jimsonweed	*(Datura stramonium)*
Mustards	*(Brassica* spp.*)*
Nightshade	*(Solanum* spp.*)*
Prostrate Pigweed	*(Amaranthus biltoides)*
Ragweed	*(Ambrosia* spp.*)*
Redroot Pigweed	*(Amaranthus retroflexus)*
Sicklepod‡, Coffeeweed	*(Cassia obtusifolia)*
Smartweed	*(Polygonum* spp.*)*
Smooth Pigweed	*(Amaranthus hybridus)*
Velvetleaf	*(Abutilon theophrasti)*
‡Southeast Region Only	

Perennial Weeds:

Hemp Dogbane†	*(Apocynum cannabinum)*
Purple Nutsedge	*(Cyperus rotundus)*
Rhizome Bermudagrass‡	*(Cynodon dactylon)*
Rhizome Johnsongrass†	*(Sorghum halepense)*
Yellow Nutsedge	*(Cyperus esculentus)*

† Partial control (suppression) only.
‡ Southeast Region only.

Existing stands of nutsedge must be turned under and chopped up thoroughly before treatment. Cultivation is suggested in addition to SUTAZINE+ treatment in fields with moderate to heavy infestations.

General Use Precautions

Read all label directions before using.

This product contains atrazine and thus, may not control weeds that are known, or suspected to be triazine resistant.

Applied according to directions and under normal growing conditions, SUTAZINE+ will not harm the treated crop. During germination and early stages of growth, extended periods of unusually cold and wet or hot and dry weather, insect or plant disease attack, carryover pesticide residues, the use of certain soil-applied systemic insecticides, improperly placed fertilizers or soil insecticides may create abnormal conditions that weaken crop seedlings. SUTAZINE+ used under these abnormal conditions could result in crop injury.

Make only one application per crop. After treatment with SUTAZINE+, do not plant any crop except corn until the following year or injury may occur. Do not apply after June 10.

Following corn harvest, thoroughly invert soil (deep disk or plow) in the fall or early spring to minimize injury potential to rotational springseeded crops. Corn or soybeans may be planted the spring season following treatment. Injury may occur to soybeans planted in North Central and Northwest Iowa, South Central and Southwest Minnesota, Northern Nebraska and Southeast South Dakota on soils having a calcareous surface layer. Small grains may be planted 15 months following treatment. All other atrazine-sensitive crops may be planted 18 months after application.

Do not use on milo or sorghum. Do not use on corn seed stock (Breeders, Foundation, or Increase). SUTAZINE+ may be used on Production seed corn.

SUTAZINE+ is recommended for use on mineral soils only (soils containing less than 10% organic matter).

Do not contaminate irrigation water (used for crops other than corn) or water used for domestic purposes.

Do not contaminate food or feed.

Keep container closed when not in use. Do not store near seeds, fertilizers, or foodstuffs.

Do not use in Arizona or in the ten Southernmost California counties.

In irrigated areas, do not apply SUTAZINE+ before pre-irrigation.

DIRECTIONS FOR USE OF SUTAZINE+

It is a violation of Federal law to use this product in a manner inconsistent with its labeling.

SUTAZINE+ is a selective herbicide which is mixed (incorporated) into the soil for control of weeds listed on this label. SUTAZINE+ controls annual grasses as their seeds germinate by interfering with normal germination and seedling development. It does not control established weeds.

SUTAZINE+ can be mixed with water or combined with fluid (solution, slurry, or suspension) fertilizers and applied as a spray, or can be impregnated on dry bulk fertilizers.

SUTAZINE+ Fluid Fertilizer Tank Mix

SUTAZINE+ can be combined with fluid fertilizers for applications made before planting. However, *the physical compatability with these fluid fertilizers must be determined before combining in the spray tank, and even though found to be compatible, constant agitation is necessary to keep the SUTAZINE+ evenly mixed with the fluid fertilizer during application.*

Refer to Appendix II regarding the procedure for testing the compatability of SUTAZINE+ with fluid fertilizers.

Impregnation on Dry Bulk Fertilizers

Dry bulk fertilizers (except single superphosphate, treble superphosphate and nitrate fertilizers) may be impregnated or coated with SUTAZINE+. However, uniform distribution of SUTAZINE+ on fertilizer particles and uniform application are necessary to assure good results. See Appendix III for information and directions for impregnation and use for these combinations.

APPLICATION AND SOIL INCORPORATION DIRECTIONS

Spray Application and Mechanical Incorporation

Apply the recommended dosage in 10 to 50 gallons of water per acre (use the higher volume in arid areas) using a properly calibrated, low-pressure type nozzle (flood, fan or swirl chamber) on boom-type sprayer having good, but not excessive agitation. Check frequently during application to be sure nozzles are free from clogging and are delivering a uniform spray pattern.

The herbicides must be incorporated into the soil within 4 hours (within 1 hour in Western Region) to prevent loss of the SUTAZINE+. Whenever possible, application and incorporation should be done in the same operation. Apply the herbicides to soil dry enough to permit good soil mixing.

See Appendix I for Soil Mixing (incorporation) before Planting Directions.

Loss of weed control will result from any delay in incorporation if SUTAZINE+ is applied to a moist soil surface.

If there must be a delay (less than 4 hours, except less than 1 hour in Western Region) in incorporation of the herbicides when application and incorporation are done in separate operations, SUTAZINE+ must be applied to a dry soil surface (at least 1/2 inch deep) and free from dew and incidental moisture.

Important: Thorough mixing of SUTAZINE+ in the soil is necessary. The soil must be in good tilth. Improper discing or poor tilth (large clods) may result in unsatisfactory control.

Incorporation of Impregnated Dry Bulk Fertilizer

The application and incorporation of SUTAZINE+ must be carried out on the same day and delayed only if soil moisture conditions permit. See Appendix III for directions on the delay in soil incorporation.

CENTER PIVOT SPRINKLER APPLICATION

Apply SUZATINE + with irrigation water before planting (see PLANTING), or immediately after planting before either corn or weeds emerge. It is important that application be done immediately after the last tillage operation since poor results will occur if the weeds have germinated before application of the herbicide. Meter the pre-mixed SUTAZINE + into the irrigation water during the entire period. Apply in 1/2 to 1 inch of water or sufficient water to penetrate the soil to a depth of 5 to 7 inches. Good agitation should be maintained during the entire application period.

USE PRECAUTIONS:

A. Apply this product only through sprinkler including center pivot irrigation system. Do not apply this product through any other type of irrigation system.
B. Crop injury, lack of effectiveness, or illegal pesticide residues in the crop can result from nonuniform distribution of treated water.
C. If you have questions about calibration, you should contact State Extension Service specialists, equipment manufacturers or other experts.
D. Do not connect an irrigation system (including greenhouse systems) used for pesticide application to a public water system unless the pesticide label-prescribed safety devices for public water systems are in place.
E. A person knowledgeable of the chemigation system and responsible for its operation, or under the supervision of the responsible person, shall shut the system down and make necessary adjustments should the need arise.

POSTING REQUIREMENTS

Posting of areas to be chemigated is required when 1) any part of a treated area is within 300 feet of sensitive areas such as residential areas, labor camps, businesses, day care centers, hospitals, in-patient clinics, nursing homes or any public areas such as schools, parks, playgrounds, or other public facilities not including public roads, or 2) when the chemigated area is open to the public such as golf courses or retail greenhouses.

Posting must conform to the following requirements. Treated areas shall be posted with signs at all usual points of entry and along likely routes of approach from the listed sensitive areas. When there are no usual points of entry, signs must be posted in the corners of the treated areas and in any other location affording maximum visibility to sensitive areas. The printed side of the sign should face away from the treated area towards the sensitive area. The signs shall be printed in English. Signs must be posted prior to application and must remain posted until foliage has dried and soil surface water has disappeared. Signs may remain in place indefinitely as long as they are composed of materials to prevent deterioration and maintain legibility for the duration of the posting period.

All words must consist of letters at least 2 1/2 inches tall, and all letters and the symbol shall be a color which sharply contrasts with their immediate background. At the top of the sign shall be the words KEEP OUT, followed by an octagonal stop sign symbol at least 8 inches in diameter containing the word STOP. Below the symbol shall be the words PESTICIDES IN IRRIGATION WATER.

CENTER PIVOT SPRINKLER REQUIREMENTS

1. The system must contain a functional check valve, vacuum relief valve, and low pressure drain appropriately located on the irrigation pipeline to prevent water source contamination from backflow.
2. The pesticide injection pipeline must contain a functional, automatic, quick-closing check valve to prevent the flow of fluid back toward the injection pump.
3. The pesticide injection pipeline must also contain a functional, normally closed, solenoid-operated valve located on the intake side of the injection pump and connected to the system interlock to prevent fluid from being withdrawn from the supply tank when the irrigation system is either automatically or manually shut down.
4. The system must contain functional interlocking controls to automatically shut off the pesticide injection pump when the water pump motor stops.
5. The irrigation line or water pump must include a functional pressure switch which will stop the water pump motor when the water pressure decreases to the point where pesticide distribution is adversely affected.
6. Systems must use a metering pump, such as a positive displacement injection pump (e.g. diaphragm pump) effectively designed and constructed of materials that are compatible with pesticides and capable of being fitted with a system interlock.
7. Do not apply when wind speed favors drift beyond the area intended for treatment.

Planting

Do not plant seed deeper than 2 inches. For a SUTAZINE+ application made before planting, seeding can be done immediately, or it may be delayed up to 2 weeks*. To avoid removing SUTAZINE+ from the row (with loss of weed control in the row), do not move or shape soil after treatment, such as by planting in deep furrows. If it is necessary to work the soil following application and incorporation but before planting, do not work the soil deeper than the depth the herbicides were incorporated.

*For Shattercane—See Special Recommendations.

RECOMMENDATIONS FOR USE ON CORN (FIELD, SWEET, SILAGE AND POPCORN)

Note: SUTAZINE+ is not recommended on corn grown in Arizona or in the ten Southernmost California counties.

For all uses, cultivation is recommended in addition to the SUTAZINE+ treatment in fields with moderate to heavy infestations of wild cane, nutgrass, sandbur, johnsongrass seedlings, rhizome bermudagrass, Texas panicum or volunteer sorghum.

Broadcast Rates Per Acre Conventional sprays and center pivot sprinklers.

For control of annual grasses and broadleaf weeds: Use 5.25 to 7 pints of SUTAZINE+ per acre. Use the higher rate of SUTAZINE+ on medium (medium fine textured-silt loam, clay loam, etc.) to fine (fine textured-clay) soils and for sandbur and johnsongrass seedling control.

For control of nutsedge and Texas panicum (in addition to the broadleaf weeds and other annual grasses listed): Use 7 to 10.5 pints of SUTAZINE+ per acre. Use the higher rates of SUTAZINE+ on medium (medium fine textured-silt loam, clay loam, etc.) soils, fine (fine textured-clay) soils and for heavy infestations.

For partial control (suppression) of Shattercane (and for control of the broadleaf weeds and annual grasses listed): When applied and incorporated *immediately* before planting, SUTAZINE+ at the rate of 10.5 pints per acre on a broadcast basis will reduce competition from shattercane. On continuous corn acreage with high cane seed population, rotation to another crop is recommended.

Suppression of shattercane may be improved if all the following practices are followed:

1. Fall or Spring Moldboard Plow.
2. Before application, the soil should be well worked with a disc and dry enough to permit good soil mixing of the herbicide (incorporation). All trash should be worked into the soil.
3. Delay application until soil warms to a constant 55 degrees or greater temperature. Plant immediately.

4. Incorporate thoroughly. Use recommended incorporation equipment and procedures found on this label.
5. Cultivate immediately if shattercane emerges. Cultivation should be shallower than the depth used when incorporating the herbicide.

For control of Johnsongrass and Bermudagrass (in addition to the broadleaf weeds and annual grasses listed): Apply and disc-incorporate SUTAZINE+ just before planting at the rate of 7 to 10.5 pints per acre. Use the higher rate of SUTAZINE+ on medium (medium fine textured silt loam, clay loam, etc.) soils, fine (fine textured-clay) soils, for rhizome johnsongrass suppression and for heavy infestations. Disc a second time at right angles to the first. Cultivate at least once, preferably twice.

Johnsongrass: Using a program of plowing and discing followed by SUTAZINE+ will give good control of seedlings and greatly reduce competition from plants coming from rhizomes.

In some areas discing 2 to 3 times spaced at 4 to 6 week intervals during fall and winter can be done. If this cannot be done, start with fall plowing followed by several discings in the spring before planting. The discing cuts rhizomes into small pieces and brings them to the surface, where both the weather and SUTAZINE+ will kill or weaken them.

Rhizome Bermudagrass: The rhizomes and stolons of bermudagrass must be cut up thoroughly so that four or less nodes remain on a strand. For control of bermudagrass, the disc must be set to cut 6 inches deep.

For Suppression of Broadleaf Signalgrass: For early and mid-season suppression, use 10.5 pints of SUTAZINE+ per acre. Cultivation or application of a post-emergence herbicide recommended for broadleaf signalgrass will aid in extending control to late season.

For Suppression of Hemp Dogbane: For early and mid-season suppression, use 10.5 pints of SUTAZINE+ per acre. Cultivation or application of a post-emergence herbicide recommended for hemp dogbane may aid in extending control beyond mid-season.

Fall Application: For use only in Minnesota, Northern Iowa (North of Interstate 80), Northeast Nebraska (North of Interstate 80 and East of State Highway 14), South Dakota (East of Missouri River), Wisconsin and Northern Illinois (North of Interstate 80). Apply and incorporate 10.5 pints SUTAZINE+ per acre in the late fall before the ground freezes.

NOTE: Do not use Fall application for suppression of wild cane.

Cultural Practices Following Application

Should weeds develop, a shallow cultivation or rotary hoeing will generally result in better weed control. When cultivating for any reason, it should be shallow; i.e., no more than ½ the depth the herbicide was incorporated.

APPENDIX I

Soil Mixing (Incorporation) Before Planting Directions

The following equipment commonly is used for soil mixing (incorporation) before planting.

Power-Driven Cultivation Equipment (Recommended on all soil types previously primary tilled); All PTO driven cultivation tools must be adjusted so as to have a tillage action speed (originating from internal gears and/or tractor gear) synchronized to the tractor's forward speed so that thorough soil mixing is obtained.

a. Horizontal action: set to cut 4 to 6 inches deep.
b. Vertical action: set to cut 3 to 4 inches deep.

Tandem discs (Recommended on all soil types): Set to cut to a depth of 4 to 6 inches, operated at 4 to 6 mph followed by a spike-tooth harrow or some other leveling device which extends beyond the ends of the discs. For more thorough mixing (for control of deep-germinating grasses—wild cane, seedling johnsongrass and nutsedge), disc in two different directions. When discing the second time, the disc should be reset to cut no deeper than 4 to 6 inches.

Field cultivator (Recommended on lighter soils in good tilth): Use 3 to 4 rows of sweeps spaced at 7 inch intervals and staggered so that no soil is left unturned, followed by a spike-tooth harrow pulled behind the cultivator. Do not use chisel plows to incorporate. Set the cultivator to cut 4 to 6 inches deep, operated at 5 mph or more. The field cultivator should not be used alone as an incorporation tool where control or deep-germinating weeds such as nutgrass, johnsongrass seedlings or wild cane is desired. For control of these weeds, use a disc alone for at least one of the passes over the field.

APPENDIX II

Procedure for Testing the Compatability of SUTAZINE+ with Fluid Fertilizers.

The following procedure is suggested for determining whether SUTAZINE+ may be combined with a specific fluid fertilizer for spray tank application.

Materials Required:

1. SUTAZINE+
2. Fluid fertilizer to be used.
3. Tank mix adjuvants for fluid fertilizer combinations; Compex*, Sponto 168-D*, Unite* or equivalent. The adjuvant which provides the best compatability depends on the specific fertilizer involved.
4. Two one-quart, wide-mouth glass jars with lids.
5. Measuring spoons; a graduated pipette or cylinder will provide more accurate measurement; 1 tablespoon equals 3 teaspoons, 1 teaspoon equals approximately 5 milliliters.
6. Measuring cup, 8 oz. (237 mL).

* Compex, Kalo Laboratories Inc., Kansas City, Missouri; Sponto 168-D, Witco Chemical Company, Houston, Texas; Unite, Hopkins Agricultural Chemical Company, Madison, Wisconsin.

Procedure:

1. Pour a pint (473 mL) of the fluid fertilizer into each of the quart jars.
2. Add adjuvant to one of the jars and mix (see next rate table).
3. Add the SUTAZINE+ to both jars.
4. Close both jars with lids and mix the contents by turning the jars upside down ten times.
5. Inspect the surface and body of the mixtures:
   (A) Immediately after completing the jar inversions.
   (B) After allowing the jars to stand quietly for 30 minutes.
   (C) And then again after turning the jars upside down 10 times.

If a uniform mix cannot be made, the mixture should not be used. If either mixture remains uniform for 30 minutes, the combination may be used. Should either mixture separate after 30 minutes, but readily remix uniformly with ten jar inversions, the mixture can be used if adequate agitation is maintained in the tank. If the mixture with adjuvant is satisfactory, but the one without adjuvant is not, be sure to use the adjuvant in the spray tank. Add the adjuvant at a rate of 3 pints per 100 gallons of fluid fertilizer; foaming can be minimized by using moderate agitation.

If nondispersible oil, sludge or clumps of solids form in the mixtures, the combination should not be used.

Rate Table* for SUTAZINE+ and Adjuvant with the Fluid Fertilizer**

Gallons of Fluid Fertilizer to be applied per acre	mL of SUTAZINE+ to be added to 1 pint of Fertilizer	
	mL	tsp.
10	10½	2
15	7	1⅓
20	5¼	1
25	4¼	¾
30	3½	⅔
40	2½	½

* Based on field rate of 1 pound SUTAN® and ¼ pound atrazine per acre in the fertilizer volumes indicated. Increase volume proportionately to correspond with intended field rate in terms of pounds active ingredients per acre (e.g., for field rate of 4 pounds actual SUTAN® and 1 pound atrazine in 40 gallons fertilizer per acre, add 10 mL or 2 tsp. SUTAZINE+ to each jar for compatability testing purposes).

** Two (2) milliliters or one-half (½) teaspoon of adjuvant should be added to 1 pint of fluid fertilizer in order to equal the rate of 3 pints of adjuvant per 100 gallons of fluid fertilizer.

APPENDIX III

SUTAZINE+ Impregnation on Dry Bulk Fertilizers

CAUTION: SUTAZINE+ alone or in combination with other herbicides must not be impregnated on ammonium nitrate, potassium nitrate or sodium nitrate fertilizers. Such mixtures may cause explosion and fire.

All individual state regulations relating to dry bulk fertilizer blending, registration, labeling and application are the responsibility of the individual and/or company selling the SUTAZINE+/fertilizer mixtures.

SUTAZINE+ may be impregnated on many dry bulk fertilizers (1) and applied and incorporated into the soil either in the Fall before the ground freezes or before planting for the control or many grasses and broadleaf weeds.

All SUTAZINE+ label and supplementary literature instructions and precautions regarding rates per acre, soil type and soil incorporation, application and other directions must be followed.

Test results have shown that SUTAZINE+ on many dry bulk fertilizers gives weed control equal to SUTAN+ 6.7-E or tank mixtures of SUTAN+ 6.7-E/ATRAZINE (80W or 4L) applied as a spray in water or liquid fertilizer, or impregnated on dry bulk fertilizer. However, uniform impregnation of the SUTAZINE+ on the dry fertilizer particles and uniform application in the field are necessary to assure good results.

A minimum of 200 pounds and a maximum of 700 pounds of approved fertilizer ingredients (1) impregnated with the appropriate amount of SUTAZINE+ must be applied per acre.

For impregnating the SUTAZINE+ on dry fertilizers, use a closed rotary-drum type mixer equipped with suitable spraying equipment. The spray nozzles should be positioned inside of the mixer to provide uniform spray coverage of the tumbling fertilizer. The SUTAZINE+ should be sprayed uniformly onto the fertilizer using a fine spray pattern. SUTAZINE+ is a relatively viscous material, particularly at low temperatures (20°F). At warmer temperatures (45-80°F) the viscosity is significantly lower (see viscosity data later in this appendix).

The physical properties of fertilizers vary, particularly in liquid absorptive capacity. When absorptivity is sufficient, simple spray impregnation of the fertilizer with SUTAZINE+ provides a satisfactory dry mixture.

If the absorptivity is not adequate, use of a highly absorptive powder is required to provide a dry, free-flowing mixture.

For mixtures to be used in spinning-disc applicators, Microcel E calcium silicate powder (Manville, Filtration & Minerals) is recommended for use as a drying agent.

For mixtures to be used in pneumatic applicators, Microcel E or Agsorb 16/30 RVM-MS granular-clay (Oil-Dri Corporation) is recommended for use as the drying agent.

The drying agent should be added separately and uniformly to the prepared pesticide-fertilizer mixture, in a quantity that is sufficient to provide a suitably free-flowing mixture. Generally, less than 2% Microcel E or 5% Agsorb 16/30 RVM-MS by weight is required.

The amount of SUTAZINE+ actually required in the manufacture of individual fertilizer mixtures should be determined carefully for each production operation (2). This is necessary to ensure that the amount of SUTAZINE+ actually contained in the mixture applied to the soil represents the correct rates of use.

Bulk fertilizers impregnated with SUTAZINE+ should be *applied immediately*, NOT STORED. All bulk containers should be tightly covered while the products are being transported and applied to reduce chances of SUTAN loss via volatilization.

Do not combine SUTAZINE+ with single superphosphate (0-20-0) or treble superphosphate (0-46-0).

Incorporation of Impregnated Fertilizers

When application and incorporation are done in separate operations, SUTAZINE+ impregnated on dry fertilizer must be incorporated on the same day as applied. Application must be made on a dry soil surface (at least ½ inch deep) free from dew and incidental moisture.

SUTAZINE+ Physical Data

Specific Gravity 20/20°C:	1.019(typical)
Pounds/Gallon (20°C):	8.480(typical)
Flashpoint:	> 216°F (Setaflash closed cup)
Viscosity: (typical)	1120 centipoise at 20°F
	660 centipoise at 32°F
	430 centipoise at 45°F
	275 centipoise at 60°F
	175 centipoise at 80°F
	145 centipoise at 95°F

(1) Approved Dry Fertilizer Ingredients for use with SUTAZINE+

	N	P	K
Ammonium Sulfate	21	0	0
Ammonium Phosphate-Sulfate	16	20	0
Diammonium Phosphate	18	46	0
Monoammonium Phosphate	11	56	0
Potassium Chloride	0	0	60
Potassium Sulfate	0	0	52
Urea*	45	0	0

* Some ureas may be phytotoxic when high rates are applied to corn. Use only urea rates known to be safe for corn application.

Note: Sul-Po-Mag, K-Mag and 11-48-0 have been shown to be compatible with SUTAZINE+ and are approved for use.

(2) Rate Chart for the Impregnation of Dry Bulk Fertilizers with SUTAZINE+

Fertilizer Rate/Acre	SUTAZINE Rate/Acre		
	5¼ pts. per acre	7 pts. per acre	10½ pts. per acre
200 lbs.	26¼ qts./ton	35 qts./ton	52½ qts./ton
250 lbs.	21 qts./ton	28 qts./ton	42 qts./ton
300 lbs.	17½ qts./ton	23⅓ qts./ton	35 qts./ton
350 lbs.	14½ qts./ton	20 qts./ton	30 qts./ton
400 lbs.	13¼ qts./ton	17½ qts./ton	26¼ qts./ton
450 lbs.	11⅔ qts./ton	15½ qts./ton	23⅓ qts./ton
500 lbs.	10½ qts./ton	14 qts./ton	21 qts./ton
550 lbs.	9½ qts./ton	12¾ qts./ton	19 qts./ton
600 lbs.	8¾ qts./ton	11⅔ qts./ton	17½ qts./ton
650 lbs.	8 qts./ton	10¾ qts./ton	16¼ qts./ton
700 lbs.	7½ qts./ton	10 qts./ton	15 qts./ton

STORAGE AND DISPOSAL

Prohibitions: Do not contaminate water, food or feed by storage or disposal. Open dumping is prohibited. Do not reuse empty container.

Storage: Keep container tightly closed when not in use. Do not store near seeds, fertilizers or foodstuffs. Can be stored at temperatures as low as minus 30°F.

Pesticide Disposal: Rinse spray equipment. Pesticide wastes are acutely hazardous. Improper disposal of excess pesticide, spray mixture, or rinsate is a violation of Federal Law. If these wastes cannot be disposed of by use according to label instuctions, contact your State Pesticide or Environmental Control Agency, or the Hazardous Waste representative at the nearest EPA Regional Office for guidance.

Container Disposal: Metal—Triple rinse (or equivalent). Then offer for recycling or reconditioning, or puncture and dispose of in a sanitary landfill, or by other procedures approved by State and local authorities.

Plastic—Triple rinse (or equivalent). Then offer for recycling or reconditioning, or puncture and dispose of in a sanitary landfill, or by incineration, or, if allowed by State and local authorities, by burning. If burned, stay out of smoke.

NOTICE TO BUYER AND USER:

Seller warrants that this product conforms to the chemical description on the label and is reasonably fit for the purposes stated on the label when used in accordance with directions under normal conditions of use. This warranty does not extend to the use of this product contrary to label instructions, or under abnormal use conditions, or under conditions not reasonably foreseeable to Seller, and Buyer and User assume the risk of any such use. SELLER DISCLAIMS ALL OTHER WARRANTIES EXPRESSED OR IMPLIED INCLUDING ANY WARRANTY OF FITNESS OR MERCHANTABILITY. SELLER SHALL NOT BE LIABLE FOR CONSEQUENTIAL, SPECIAL OR INDIRECT DAMAGES RESULTING FROM THE USE OR HANDLING OF THIS PRODUCT AND SELLER'S SOLE LIABILITY AND BUYER'S AND USER'S EXCLUSIVE REMEDY SHALL BE LIMITED TO THE REFUND OF THE PURCHASE PRICE.

This product is sold only for uses stated on its label. No express or implied license is granted to use or sell this product under any patent in any country except as specified: Country: United States of America. Patent No. 4,021,224;, 4,137,070.

A-5R

The information and specimen labels in this booklet are correct as of the publication date but are subject to change without notice. For current information, contact ICI Americas Inc.

SUTAZINE® + 18-6.G

SELECTIVE HERBICIDE GRANULES

for selective weed control in field, sweet, and silage corn

ACTIVE INGREDIENTS:	
S-ethyl diisobutylthiocarbamate	18.0%
Atrazine: 2-chlor-4-(ethylamino)-6-(isopropylamino)-S-triazine	5.7%
Related triazines	0.3%
INERT INGREDIENTS:	76.0%
TOTAL	100.0%

Contains 24 pounds of active ingredients per 100 pounds.

EPA Reg. No. 10182-201

KEEP OUT OF REACH OF CHILDREN

CAUTION

Harmful If Swallowed

STATEMENT OF PRACTICAL TREATMENT

If a known exposure occurs or is suspected, immediately start the procedures given below. If further treatment is required, contact a Poison Center, a physician, or the nearest hospital.

If swallowed: Immediately give several glasses of water and induce vomiting by gagging the victim with a finger placed on the back of the victim's tongue. Give fluids until vomitus is clear. If victim is unconscious or convulsing, do not induce vomiting or give anything by mouth.

For skin contact: Flush all affected areas with plenty of water for several minutes.

For eye contact: Hold eyelids apart and flush with plenty of water for 15 minutes.

Remove and clean contaminated clothing and shoes before reuse. Seek medical attention if irritation occurs.

If inhaled: Remove to fresh air. Seek medical attention if respiratory irritation occurs or if breathing becomes difficult.

FOR 24-HOUR EMERGENCY MEDICAL ASSISTANCE CALL 1-800-F-A-S-T-M-E-D (327-8633).

FOR CHEMICAL EMERGENCY: Spill, leak, fire, exposure, or accident call CHEMTREC 1-800-424-9300.

READ ALL LABEL DIRECTIONS BEFORE USING

PRECAUTIONARY STATEMENTS

HAZARDS TO HUMANS AND DOMESTIC ANIMALS

CAUTION

Harmful if Swallowed

Avoid contact with skin, eyes and clothing. Avoid inhalation of dust. Wash thoroughly with soap and water after use.

ENVIRONMENTAL HAZARDS

Do not apply directly to water or wetlands. Do not contaminate water by cleaning of equipment or disposal of wastes. Cover or incorporate spills.

GENERAL USE PRECAUTIONS

Read all label directions before using. Apply this product as specified on this label.

Make only one application per crop.

After treatment with SUTAZINE + 18-6.G, do not plant any crop except corn the following year or injury may occur.

Do not use on milo or sorghum. Do not use on corn seed stock (Breeders, Foundation, or Increase). SUTAZINE + 18-6.G may be used on Production seed corn.

This product is recommended for use on mineral soils only.

Applied according to directions and under normal growing conditions, SUTAZINE + 18-6.G will not harm the treated crop. During germination and early stages of growth, extended periods of unusually cold and wet or hot and dry weather, insect or plant disease attack, the use of certain soil-applied systemic insecticides, improperly placed fertilizers or soil insecticides, may create abnormal conditions that weaken crop seedlings. SUTAZINE + 18-6.G used under these abnormal conditions could result in crop injury.

Care should be taken to avoid use where adjacent desirable trees, shrubs, or plants may be injured as a result of the product leaching through the soil.

Do not contaminate food or feed.

Do not store near seeds, fertilizers, or foodstuffs.

Do not contaminate irrigation water or water used for domestic purposes.

Weeds Controlled

This product will not control established weeds.

Annual Grasses	Annual Broadleaf Weeds
Barnyardgrass *(Echinochloa crusgalli)*	Annual Morningglory *(Ipomoea purpurea)*
Crabgrass *(Digitaria* spp.*)*	Carpetweed *(Mollugo verticillata)*
Fall Panicum *(Panicum dichotomiflorum)*	Cocklebur Common *(Xanthium pensylvanicum)*
Giant Foxtail *(Setaria faberii)*	Jimsonweed *(Datura stramonium)*
Goosegrass *(Eleusine indicia)*	Lambsquarters *(Chenopodium album)*
Green Foxtail *(Setaria viridis)*	Mustards *(Brassica* spp.*)*
Seedling Bermudagrass *(Cynodon dactylon)*	Nightshade *(Solanum* spp.*)*
Seedling Johnsongrass *(Sorghum halepense)*	Common Purslane *(Portulaca oleracea)*
*Wild Cane *(Sorghum bicolor)*	Ragweed *(Ambrosia* spp.*)*
Wild Oats *(Avena fatua)*	Redroot Pigweed *(Amaranthus retroflexus)*
Witchgrass *(Panicum capillare)*	Prostrate Pigweed *(Amaranthus graecizans)*
Yellow Foxtail *(Setaria lutescens)*	Smooth Pigweed *(Amaranthus hybridus)*
*Texas Panicum; Buffalograss—Coloradograss *(Panicum Texanum)*	Smartweed *(Polygonum* spp.*)* Velvetleaf *(Abutilon theophrasti)* Florida Pusley *(Richardia scabra)*

* Cultivation is suggested in addition to the SUTAZINE + 18.6.G treatment in fields with heavy infestation of wild cane or Texas panicum.

Perennial Grasses
Purple Nutsedge *(Cyperus rotundus)*
Yellow Nutsedge *(Cyperus esculentus)*

Existing stands of nutsedge must be turned under and chopped up thoroughly before treatment. Cultivation is suggested in addition to the SUTAZINE + 18-6.G treatment in fields with moderate to heavy nutsedge infestations.

DIRECTIONS FOR USE

It is a violation of Federal law to use this product in a manner inconsistent with its labeling.

SUTAZINE + 18-6.G is a selective herbicide which is mixed (incorporated) into the soil for control of weeds listed on this label by interfering with normal germination and seedling development. It does not control established weeds. All weed growth and crop stubble should be thoroughly worked into the soil before treatment.

Application Directions

Apply only to well-worked soil that is dry enough to permit thorough mixing with incorporation equipment. Uniformly apply the recommended rate of SUTAZINE + 18-6.G using equipment designed for application of granular herbicides. Equipment should be carefully calibrated before use and checked frequently during application to be sure it is working properly and delivering a uniform distribution pattern. Avoid overlaps that will increase dosage above recommended rates as plant injury may occur.

In irrigated areas, do not apply before pre-irrigation.

Incorporation Directions

SUTAZINE + 18-6.G must be incorporated into the soil after application to prevent loss of the Butylate. Whenever possible, application and incorporation should be done in the same operation. When application and incorporation are done in separate operations, SUTAZINE + 18-6.G must be incorporated on the same day as applied. Application must be made on a dry soil surface (at least 1/2 inch deep) free from dew and incidental moisture.

Soil Mixing (Incorporation) Before Planting

Incorporate using power-driven cultivation equipment set to cut to a depth of 2 to 3 inches, *or* a tandem disc set to cut to a depth of 4 to 6 inches, operated at 4 to 6 mph, followed by a spike-tooth harrow or some other leveling device which extends beyond the ends of the disc. For thorough mixing, disc in two different directions (cross disc.) Prior to the second discing adjust the disc to prevent cutting deeper than 4 to 6 inches.

Planting

Do not plant seed deeper than 2 inches. For SUTAZINE + 18-6.G application made before planting, seeding can be done immediately, or it may be delayed up to 2 weeks. To avoid removing SUTAZINE + 18-6.G from the row (with loss of weed control in the row), do not move or shape soil after treatment, such as by planting in deep furrows. If it is necessary to work the soil following application and incorporation, but before planting, do not work the soil deeper than the depth the herbicides were incorporated.

Recommendations

Do not use on any type corn in Arizona or in the ten southernmost California counties (except Kern County where field and silage corn may be treated).

NOTE: All doses expressed as broadcast rates per acre. For band applications, reduce dosage proportionately depending upon the row spacing and band width to be treated.

Before Planting (Preplant) Application In The North Central, Northeastern And Midwestern Corn Growing Areas

Apply and incorporate 16 2/3 to 22 2/3 pounds of SUTAZINE + 18-6.G per broadcast acre. Use the higher rate on extremely heavy soils (fine textured-clay) with over 5% organic matter. On fields with moderate to heavy infestations of shattercane or nutsedge use the higher rate regardless of soil texture, except on blow sands and the light soils (coarse textured-sandy) of the Eastern Coastal States. Cultivation is suggested in addition to the SUTAZINE + 18-6.G treatment in fields with moderate to heavy infestations of shattercane or nutsedge.

In All Other Corn Growing Areas

Apply and incorporate 16 2/3 pounds of SUTAZINE + 18-6.G per broadcast acre. Cultivation is suggested in addition to the SUTAZINE + 18-6.G treatment in fields with moderate to heavy infestations of wild cane, nutsedge or Texas panicum.

After Planting Application NOTE: Recommended in the following states *only*—North and South Dakota, Nebraska, Kansas, Minnesota and Iowa. If *local* conditions in the above-named states make preplant SUTAZINE + 18-6.G application unfeasible, the product may be applied and mixed into the soil at any time after planting before emergence of corn. Apply and incorporate 16 2/3 pounds SUTAZINE + 18-6.G per broadcast acre. For band applications, use standard planter attached equipment designed for band application of granular herbicides. Normally, application and incorporation are done in the same operation using equipment that attaches to the back of the applicator, i.e., row wheels, vibro tillers and similar equipment. Incorporate to a depth of 1 to 1 1/2 inches. A rotary hoe can also be used set to cut to a depth of 1 1/2 inches operated at 6 to 8 mph.

NOTE: After-planting application is not recommended for control of nutsedge, Texas panicum, or wild cane. See "BEFORE PLANTING (PREPLANT) APPLICATION" for control of these weeds.

Cultural Practices Following Application

Should weeds develop, a shallow cultivation or rotary hoeing will generally result in better weed control. When cultivating for any reason, it should be shallow, i.e., no more than 1/2 the depth the herbicide was incorporated.

STORAGE AND DISPOSAL

Prohibitions: Do not contaminate water, food, or feed by storage or disposal. Open dumping is prohibited.

Storage: Keep container closed when not in use. Do not store near seeds, fertilizers, or foodstuffs.

Pesticide Disposal: Wastes resulting from the use of this product may be disposed of on site or at an approved waste disposal facility.

Container Disposal: Completely empty bag into application equipment. Then dispose of empty bag in a sanitary landfill or by incineration, or, if allowed by State and local authorities, by burning. If burned, stay out of smoke.

NOTICE TO BUYER AND USER:

Seller warrants that this product conforms to the chemical description on the label and is reasonably fit for the purposes stated on the label when used in accordance with directions under normal conditions of use. This warranty does not extend to the use of this product contrary to label instructions, or under abnormal use conditions, or under conditions not reasonably foreseeable to Seller, and Buyer and User assume the risk of any such use. SELLER DISCLAIMS ALL OTHER WARRANTIES EXPRESSED OR IMPLIED INCLUDING ANY WARRANTY OF FITNESS OR MERCHANTABILITY. SELLER SHALL NOT BE LIABLE FOR CONSEQUENTIAL, SPECIAL OR INDIRECT DAMAGES RESULTING FROM THE USE OR HANDLING OF THIS PRODUCT AND SELLER'S SOLE LIABILITY AND BUYER'S AND USER'S EXCLUSIVE REMEDY SHALL BE LIMITED TO THE REFUND OF THE PURCHASE PRICE.

This product is sold only for uses stated on its label. No express or implied license is granted to use or sell this product under any patent in any country except as specified. Country: United States of America. Patent No. 4,021,224; 4,137,070.

A-4R

The information and specimen labels in this booklet are correct as of the publication date but are subject to change without notice. For current information, contact ICI Americas Inc.

TALON®-G

RODENTICIDE BAIT PACK (MINI–PELLETS)

For Effective Control of Commensal Mice and Rats

Kills warfarin-resistant Norway rats and house mice. Rodents may consume a lethal dose in one feeding with first dead rodents appearing four or five days after treatment begins.

KEEP OUT OF REACH OF CHILDREN

CAUTION

EPA Reg. No. 10182-40

Net Contents: 200/25 gm (0.88 oz) bait packs

This product is protected by US patent numbers 3957824 and 4035505 which are the property of Ward Blenkinsop and Company Limited.

ACTIVE INGREDIENT: Brodifacoum-3-[3-(4'-bromo-[1,1'-biphenyl]-4-yl)-1,2,3,4-tetrahydro-1-naphthalenyl]-4-hydroxy-2H-1-benzopyran-2-one	0.005%
INERT INGREDIENTS:	99.995%
TOTAL	100.000%

PRECAUTIONARY STATEMENTS

HAZARDS TO HUMANS AND DOMESTIC ANIMALS

CAUTION

MAY BE HARMFUL OR FATAL IF SWALLOWED. KEEP AWAY FROM HUMANS, DOMESTIC ANIMALS, AND PETS. WASH HANDS AFTER HANDLING BAIT.

IF BAIT IS EATEN BY HUMANS, CALL A PHYSICIAN AT ONCE. For 24-hour emergency assistance call **ICI Americas Medical Emergency Information Center** 1-800-327-8633.

IF BAIT IS EATEN BY ANIMALS OR PETS, CALL ICI Americas Medical Emergency Information Center 1-800-327-8633 for 24-hour emergency assistance.

In case of significant spill, call CHEMTREC 800/424-9300.

NOTE TO PHYSICIAN OR VETERINARIAN

This product may reduce the clotting ability of the blood and cause hemorrhaging. If poisoning occurs, intramuscular and oral administration of vitamin K_1 are indicated, as in poisoning from overdose of bishydroxy coumarin. For human cases, vitamin K_1 is antidotal at doses of 10-20 mg (*not* mg/kg). For animal cases, vitamin K_1 is antidotal at 2-5 mg/kg. Repeated doses may need to be given up to two weeks (based on monitoring of prothrombin times). In severe cases, blood transfusions may be necessary.

ENVIRONMENTAL HAZARDS

This product is toxic to fish, birds, and wildlife. This product can pose a secondary hazard to birds of prey and mammals. Keep out of any body of water.

DIRECTIONS FOR USE

It is a violation of Federal law to use this product in a manner inconsistent with its labeling.

USE RESTRICTIONS: For control of Norway rats and roof rats in and around homes, industrial, commercial, agricultural and public buildings. TALON may also be used in and around transport vehicles (ships, trains, aircraft) and related port or terminal buildings. Do not use in sewers. This product must be placed in tamper-proof bait boxes or in locations not accessible to children, pets, domestic animals or wildlife. Do not place bait in areas where there is a possibility of contaminating food or surfaces that come in direct contact with food.

Authorized by the USDA for use in Federally inspected meat, poultry and egg processing plants.

SELECTION OF TREATMENT AREAS: Determine areas where rats and/or mice will most likely find and consume the bait. Generally, these areas are along walls, by gnawed openings, in or beside burrows, in corners and concealed places, between floors and walls, or in locations where rodents or their signs have been observed. Remove as much food as possible.

APPLICATION DIRECTIONS

Norway and Roof Rats: Apply 4 to 16 bait packs (usually at intervals of 15 to 30 feet) per placement. Maintain an uninterrupted supply of fresh bait for 10 days or until signs of rat activity cease.

House Mice: Apply 1 bait packet at intervals of 8 to 12 feet per placement; 2 bait packs may be needed at points of very high mouse activity. Maintain an uninterrupted supply of fresh bait for 15 days or until signs of mouse activity cease.

Rats and Mice: Replace contaminated or spoiled bait immediately. Collect and dispose of all dead animals and unconsumed bait according to "Disposal" paragraph. To prevent reinfestation, eliminate food, water, and harborage as much as possible. If reinfestation does occur, repeat treatment. Where a continuous source of infestation is present, establish permanent bait stations and replenish bait as needed.

STORAGE AND DISPOSAL

Do not contaminate water, food or feed by storage or disposal. Do not reuse empty container except for holding additional TALON rodenticide. **Storage:** Store in original container only in a dry place inaccessible to children and pets. **Pesticide Disposal:** Wastes resulting from the use of this product may be disposed of on site or at an approved waste disposal facility. **Container Disposal:** Dispose of bait container in a sanitary landfill or by incineration if allowed by state and local authorities.

NOTICE TO BUYER AND USER

Seller warrants that this product conforms to the chemical description on the label and is reasonably fit for the purposes stated on the label when used in accordance with directions under normal conditions of use. This warranty does not extend to the use of this product contrary to label instructions, or under abnormal use conditions, or under conditions not reasonably foreseeable to Seller, and Buyer and User assume the risk of any such use. SELLER DISCLAIMS ALL OTHER WARRANTIES, EXPRESSED OR IMPLIED, INCLUDING ANY WARRANTY OF FITNESS OR MERCHANTABILITY. SELLER SHALL NOT BE LIABLE FOR CONSEQUENTIAL, SPECIAL OR INDIRECT DAMAGES RESULTING FROM THE USE OR HANDLING OF THIS PRODUCT AND SELLER'S SOLE LIABILITY AND BUYER'S AND USER'S EXCLUSIVE REMEDY SHALL BE LIMITED TO THE REFUND OF THE PURCHASE PRICE.

05-0790-306

The information and specimen labels in this booklet are correct as of the publication date but are subject to change without notice. For current information, contact ICI Americas Inc.

TALON®-G

RODENTICIDE BAIT PACK (PELLETS)

For Effective Control of Commensal Rats

Kills warfarin-resistant Norway rats. Rodents may consume a lethal dose in one feeding with first dead rodents appearing four or five days after treatment begins.

ACTIVE INGREDIENT:
Brodifacoum-
3-[3-(4'-bromo-[1,1'-biphenyl]-4-yl)-1,2,3,4-tetrahydro-1-naphthalenyl]
-4-hydroxy-2H-1-benzopyran-2-one 0.005%
INERT INGREDIENTS: 99.995%

TOTAL 100.000%

EPA Reg. No. 10182-39

This product is protected by US patent numbers 3957824 and 4035505 which are the property of Ward Blenkinsop and Company Limited.

KEEP OUT OF REACH OF CHILDREN
CAUTION

PRECAUTIONARY STATEMENTS

HAZARDS TO HUMANS AND DOMESTIC ANIMALS

CAUTION

MAY BE HARMFUL OR FATAL IF SWALLOWED. KEEP AWAY FROM HUMANS, DOMESTIC ANIMALS, AND PETS. WASH HANDS AFTER HANDLING BAIT.

IF BAIT IS EATEN BY HUMANS, CALL A PHYSICIAN AT ONCE. For 24-hour emergency assistance call **ICI Americas Medical Emergency Information Center** 1-800-327-8633.

IF BAIT IS EATEN BY ANIMALS OR PETS, CALL ICI Americas Medical Emergency Information Center 1-800-327-8633 for 24-hour emergency assistance.

In case of significant spill, call CHEMTREC 800/424-9300.

NOTE TO PHYSICIAN OR VETERINARIAN

This product may reduce the clotting ability of the blood and cause hemorrhaging. If poisoning occurs, intramuscular and oral administration of vitamin K_1 are indicated, as in poisoning from overdose of bishydroxy coumarin. For human cases, vitamin K_1 is antidotal at doses of 10-20 mg (*not* mg/kg). For animal cases, vitamin K_1 is antidotal at 2-5 mg/kg. Repeated doses may need to be given up to two weeks (based on monitoring of prothrombin times). In severe cases, blood transfusions may be necessary.

ENVIRONMENTAL HAZARDS

This product is toxic to fish, birds, and wildlife. This product can pose a secondary hazard to birds of prey and mammals. Keep out of any body of water.

DIRECTIONS FOR USE

It is a violation of Federal law to use this product in a manner inconsistent with its labeling.

USE RESTRICTIONS: For control of Norway rats and roof rats in and around homes, industrial, commercial, agricultural and public buildings. TALON may also be used in and around transport vehicles (ships, trains, aircraft) and related port or terminal buildings. Do not use in sewers. This product must be placed in tamper-proof bait boxes or in locations not accessible to children, pets, domestic animals or wildlife. Do not place bait in areas where there is a possibility of contaminating food or surfaces that come in direct contact with food.

Authorized by the USDA for use in Federally inspected meat, poultry and egg processing plants.

SELECTION OF TREATMENT AREAS: Determine areas where rats and/or mice will most likely find and consume the bait. Generally, these areas are along walls, by gnawed openings, in or beside burrows, in corners and concealed places, between floors and walls, or in locations where rodents or their signs have been observed. Remove as much food as possible.

APPLICATION DIRECTIONS

Norway and Roof Rats: Apply 2 to 8 bait packs (usually at intervals of 15 to 30 feet) per placement. Maintain an uninterrupted supply of fresh bait for 10 days or until signs of rat activity cease.

House Mice: Open pack and apply 1/4 to 1/2 ounce of bait at intervals of 8 to 12 feet per placement. Larger placements (up to 2 ounces) may be needed at points of very high mouse activity. Maintain an uninterrupted supply of fresh bait for 15 days or until signs of mouse activity cease.

Rats and Mice: Replace contaminated or spoiled bait immediately. Collect and dispose of all dead animals and unconsumed bait according to "Disposal" paragraph. To prevent reinfestation, eliminate food, water, and harborage as much as possible. If reinfestation does occur, repeat treatment. Where a continuous source of infestation is present, establish permanent bait stations and replenish bait as needed.

STORAGE AND DISPOSAL

Do not contaminate water, food or feed by storage or disposal. Do not reuse empty container except for holding additional TALON rodenticide. **Storage:** Store in original container only in a dry place inaccessible to children and pets. **Pesticide Disposal:** Wastes resulting from the use of this product may be disposed of on site or at an approved waste disposal facility. **Container Disposal:** Dispose of bait container in a sanitary landfill or by incineration if allowed by state and local authorities.

NOTICE TO BUYER AND USER

Seller warrants that this product conforms to the chemical description on the label and is reasonably fit for the purposes stated on the label when used in accordance with directions under normal conditions of use. This warranty does not extend to the use of this product contrary to label instructions, or under abnormal use conditions, or under conditions not reasonably foreseeable to Seller, and Buyer and User assume the risk of any such use. SELLER DISCLAIMS ALL OTHER WARRANTIES, EXPRESSED OR IMPLIED, INCLUDING ANY WARRANTY OF FITNESS OR MERCHANTABILITY. SELLER SHALL NOT BE LIABLE FOR CONSEQUENTIAL, SPECIAL OR INDIRECT DAMAGES RESULTING FROM THE USE OR HANDLING OF THIS PRODUCT AND SELLER'S SOLE LIABILITY AND BUYER'S AND USER'S EXCLUSIVE REMEDY SHALL BE LIMITED TO THE REFUND OF THE PURCHASE PRICE.

05-0790-305 A-1R

The information and specimen labels in this booklet are correct as of the publication date but are subject to change without notice. For current information, contact ICI Americas Inc.

TALON®-G

RODENTICIDE MINI-PELLETS

For Effective Control of Commensal Mice and Rats

Kills warfarin-resistant Norway Rats and House mice. Rodents may consume a lethal dose in one feeding with first dead rodents appearing four or five days after treatment begins.

KEEP OUT OF REACH OF CHILDREN
CAUTION

EPA Reg. No. 10182-41

This product is protected by US patent numbers 3957824 and 4035505 which are the property of Ward Blenkinsop and Company Limited.

ACTIVE INGREDIENT:
Brodifacoum
3-[3-(4'-bromo-[1,1'-biphenyl]-4-yl)-1,2,3,4-tetrahydro-1-naphthalenyl]-4-hydroxy-2H-1-benzopyran-2-one 0.005%
INERT INGREDIENTS: 99.995%

TOTAL 100.000%

PRECAUTIONARY STATEMENTS

HAZARDS TO HUMANS AND DOMESTIC ANIMALS

CAUTION

MAY BE HARMFUL OR FATAL IF SWALLOWED. KEEP AWAY FROM HUMANS, DOMESTIC ANIMALS, AND PETS. WASH HANDS AFTER HANDLING BAIT.

IF BAIT IS EATEN BY HUMANS, CALL A PHYSICIAN AT ONCE. For 24-hour emergency assistance, call **ICI Americas Medical Emergency Information Center** 1-800-327-8633.

IF BAIT IS EATEN BY ANIMALS OR PETS, call **ICI Americas Medical Emergency Information Center** 1-800-327-8633 for 24-hour emergency assistance. In case of significant spill, call CHEMTREC 800/424-9300.

NOTE TO PHYSICIAN OR VETERINARIAN

This product may reduce the clotting ability of the blood and cause hemorrhaging. If poisoning occurs, intramuscular and oral administration of vitamin K_1 are indicated, as in poisoning from overdose of bishydroxy coumarin. For human cases, vitamin K_1 is antidotal at doses of 10-20 mg (*not* mg/kg). For animal cases, vitamin K_1 is antidotal at 2-5 mg/kg. Repeated doses may need to be given up to two weeks (based on monitoring of prothrombin times). In severe cases, blood transfusions may be necessary.

ENVIRONMENTAL HAZARDS

This product is toxic to fish, birds and wildlife. This product can pose a secondary hazard to birds of prey and mammals. Keep out of any body of water.

DIRECTIONS FOR USE

It is a violation of Federal law to use this product in a manner inconsistent with its labeling.

USE RESTRICTIONS: For control of Norway Rats, Roof rats and House mice in and around homes, industrial, commercial, agricultural and public buildings. TALON may also be used in and around transport vehicles (ships, trains, aircraft) and related port or terminal buildings. Do not use in sewers. This product must be placed in tamper-proof bait boxes or in locations not accessible to children, pets, domestic animals or wildlife. Do not place bait in areas where there is a possibility of contaminating food or surfaces that come in direct contact with food.

Authorized by the USDA for use in federally inspected meat, poultry and egg processing plants.

Selection of Treatment Areas: Determine areas where rats and/or mice will most likely find and consume the bait. Generally, these areas are along walls, by gnawed openings, in or beside burrows, in corners and concealed places, between floors and walls, or in locations where rodents or their signs have been observed. Remove as much food as possible.

APPLICATION DIRECTIONS

Norway and Roof Rats: Apply 4 to 16 ounces of bait (usually at intervals of 15 to 30 feet) per placement. Maintain an uninterrupted supply of fresh bait for 10 days or until signs of rat activity cease.

House Mice: Apply 1/4 to 1/2 ounce of bait at intervals of 8 to 12 feet per placement. Larger placements (up to 2 ounces) may be needed at points of very high mouse activity. Maintain an uninterrupted supply of fresh bait for 15 days or until signs of mouse activity cease.

Rats & Mice: Replace contaminated or spoiled bait immediately. Collect and dispose of all dead animals and unconsumed bait according to "Disposal" paragraph. To prevent reinfestation, eliminate food, water, and harborage as much as possible. If reinfestation does occur, repeat treatment. Where a continuous source of infestation is present, establish permanent bait stations and replenish bait as needed.

STORAGE AND DISPOSAL

Do not contaminate water, food or feed by storage or disposal. Do not reuse empty container except for holding additional TALON rodenticide. **Storage:** Store in original container only in a dry place inaccessible to children and pets. **Pesticide Disposal:** Wastes resulting from the use of this product may be disposed of on site or at an approved waste disposal facility. **Container Disposal:** Dispose of bait container in a sanitary landfill or by incineration if allowed by state and local authorities.

NOTICE TO BUYER AND USER

Seller warrants that this product conforms to the chemical description on the label and is reasonably fit for the purposes stated on the label when used in accordance with directions under normal conditions of use. This warranty does not extend to the use of this product contrary to label instructions, or under abnormal use conditions, or under conditions not reasonably foreseeable to Seller, and Buyer and User assume the risk of any such use. SELLER DISCLAIMS ALL OTHER WARRANTIES, EXPRESSED OR IMPLIED, INCLUDING ANY WARRANTY OF FITNESS OR MERCHANTABILITY. SELLER SHALL NOT BE LIABLE FOR CONSEQUENTIAL, SPECIAL OR INDIRECT DAMAGES RESULTING FROM THE USE OR HANDLING OF THIS PRODUCT AND SELLER'S SOLE LIABILITY AND BUYER'S AND USER'S EXCLUSIVE REMEDY SHALL BE LIMITED TO THE REFUND OF THE PURCHASE PRICE.

05-0790-304

The information and specimen labels in this booklet are correct as of the publication date but are subject to change without notice. For current information, contact ICI Americas Inc.

TALON®-G

RODENTICIDE PELLETS

For Effective Control of Commensal Mice and Rats

Kills Warfarin-Resistant Norway Rats and House Mice. Rodents may Consume a Lethal Dose in One Feeding with First Dead Rodents Appearing Four or Five Days after Treatment Begins.

ACTIVE INGREDIENT:	
Brodifacoum	
3-[3-(4'-bromo-[1,1'-biphenyl]-4-yl)-1,2,3,4-tetrahydro-1-naphthalenyl]-4-hydroxy-2H-1-benzopyran-2-one	0.005%
INERT INGREDIENTS:	99.995%
TOTAL	100.000%

EPA Reg. No. 10182-38

This product is protected by US patent numbers 3957824 and 4035505 which are the property of Ward Blenkinsop and Company Limited.

KEEP OUT OF REACH OF CHILDREN CAUTION

PRECAUTIONARY STATEMENTS

HAZARDS TO HUMANS AND DOMESTIC ANIMALS

CAUTION

MAY BE HARMFUL OR FATAL IF SWALLOWED. KEEP AWAY FROM HUMANS, DOMESTIC ANIMALS, AND PETS. WASH HANDS AFTER HANDLING BAIT.

IF BAIT IS EATEN BY HUMANS, CALL A PHYSICIAN AT ONCE. For 24-hour emergency assistance, call **ICI Americas Inc. Medical Emergency Information Center** 1-800-327-8633.

IF BAIT IS EATEN BY ANIMALS OR PETS, call **ICI Americas Inc. Medical Emergency Information Center** 1-800-327-8633 for 24-hour emergency assistance. In case of significant spill, call CHEMTREC 800/424-9300.

NOTE TO PHYSICIAN OR VETERINARIAN

This product may reduce the clotting ability of the blood and cause hemorrhaging. If poisoning occurs, intramuscular and oral administration of vitamin K_1 are indicated, as in poisoning from overdose of bishydroxy coumarin. For human cases, vitamin K_1 is antidotal at doses of 10-20 mg (*not* mg/kg). For animal cases, vitamin K_1 is antidotal at 2-5 mg/kg. Repeated doses may need to be given up to two weeks (based on monitoring of prothrombin times). In severe cases, blood transfusions may be necessary.

ENVIRONMENTAL HAZARDS

This product is toxic to fish, birds and wildlife. This product can pose a secondary hazard to birds of prey and mammals. Keep out of any body of water.

DIRECTIONS FOR USE

It is a violation of Federal law to use this product in a manner inconsistent with its labeling.

Use Restrictions: For control of Norway Rats, Roof rats and House mice in and around homes, industrial, commercial, agricultural and public buildings. TALON may also be used in and around transport vehicles (ships, trains, aircraft) and related port or terminal buildings. Do not use in sewers. This product must be placed in tamper-proof bait boxes or in locations not accessible to children, pets, domestic animals or wildlife. Do not place bait in areas where there is a possibility of contaminating food or surfaces that come in direct contact with food.

Authorized by the USDA for use in federally inspected meat, poultry and egg processing plants.

Selection of Treatment Areas: Determine areas where rats and/or mice will most likely find and consume the bait. Generally, these areas are along walls, by gnawed openings, in or beside burrows, in corners and concealed places, between floors and walls, or in locations where rodents or their signs have been observed. Remove as much food as possible.

APPLICATION DIRECTIONS

Norway and Roof Rats: Apply 4 to 16 ounces of bait (usually at intervals of 15 to 30 feet) per placement. Maintain an uninterrupted supply of fresh bait for 10 days or until signs of rat activity cease.

House Mice: Apply 1/4 to 1/2 ounce of bait at intervals of 8 to 12 feet per placement. Larger placements (up to 2 ounces) may be needed at points of very high mouse activity. Maintain an uninterrupted supply of fresh bait for 15 days or until signs of mouse activity cease.

Rats & Mice: Replace contaminated or spoiled bait immediately. Collect and dispose of all dead animals and unconsumed bait according to "Disposal" paragraph. To prevent reinfestation, eliminate food, water, and harborage as much as possible. If reinfestation does occur, repeat treatment. Where a continuous source of infestation is present, establish permanent bait stations and replenish bait as needed.

STORAGE AND DISPOSAL

Do not contaminate water, food or feed by storage or disposal. **Storage:** Store in original container only in a dry place inaccessible to children and pets. **Pesticide Disposal:** Wastes resulting from the use of this product may be disposed of on site or at an approved waste disposal facility. **Container Disposal:** Dispose of bait container in a sanitary landfill or by incineration if allowed by state and local authorities.

NOTICE TO BUYER AND USER

Seller warrants that this product conforms to the chemical description on the label and is reasonably fit for the purposes stated on the label when used in accordance with directions under normal conditions of use. This warranty does not extend to the use of this product contrary to label instructions, or under abnormal use conditions, or under conditions not reasonably foreseeable to Seller, and Buyer and User assume the risk of any such use. SELLER DISCLAIMS ALL OTHER WARRANTIES, EXPRESSED OR IMPLIED, INCLUDING ANY WARRANTY OF FITNESS OR MERCHANTABILITY. SELLER SHALL NOT BE LIABLE FOR CONSEQUENTIAL, SPECIAL OR INDIRECT DAMAGES RESULTING FROM THE USE OR HANDLING OF THIS PRODUCT AND SELLER'S SOLE LIABILITY AND BUYER'S AND USER'S EXCLUSIVE REMEDY SHALL BE LIMITED TO THE REFUND OF THE PURCHASE PRICE.

05-0790-303 A-1R

The information and specimen labels in this booklet are correct as of the publication date but are subject to change without notice. For current information, contact ICI Americas Inc.

TILLAM® 6-E

SELECTIVE HERBICIDE

Liquid herbicide for weed control in tobacco, tomatoes and sugar beets

Complete Directions For Use
Use Precautions and Recommendations

ACTIVE INGREDIENT:

S-propyl butylethylthiocarbamate	77.32%
INERT INGREDIENTS:	22.68%
TOTAL	100.00%

Contains 6 Lbs. Active Ingredient Per Gallon
EPA Reg. No. 10182-158

KEEP OUT OF REACH OF CHILDREN

CAUTION

Harmful if Swallowed

STATEMENT OF PRACTICAL TREATMENT

Immediately start the procedures given below and CONTACT A POISON CENTER, A PHYSICIAN OR THE NEAREST HOSPITAL.

If swallowed: Immediately give several glasses of water but *do not* induce vomiting. This product contains hydrocarbon solvent. If vomiting occurs, give fluids again. Have a physician determine if condition of patient will permit induction of vomiting or evacuation of stomach. Do not give anything by mouth to an unconscious or convulsing person.

For skin contact: Flush affected areas with large amounts of running water for a least 15 minutes and get medical attention. Remove and clean contaminated clothing and shoes.

For eye contact: Hold eyelids apart and flush eyes with large amounts of running water for 15 minutes. Get medical attention if irritation occurs.

If inhaled: Remove to fresh air. Get medical attention if respiratory irritation occurs or if breathing becomes difficult.

FOR 24-HOUR EMERGENCY MEDICAL ASSISTANCE CALL 1-800-F-A-S-T-M-E-D (327-8633).
FOR CHEMICAL EMERGENCY: Spill, leak, fire, exposure, or accident call CHEMTREC 1-800-424-9300.

PRECAUTIONARY STATEMENTS

HAZARDS TO HUMANS AND DOMESTIC ANIMALS

CAUTION

Harmful If Swallowed. Avoid contact with skin, eyes and clothing. Avoid inhalation of spray mist. Wear rubber gloves and fresh clothing. Wear goggles to avoid contact with eyes. Wash with soap and water immediately after use or contact. Keep away from heat and flame. Do not contaminate food and feed.

Do not apply this product in such a manner as to directly or through drift, expose workers or other persons. When using the solid set sprinkler method of application, the area being treated must be vacated by unprotected persons.

ENVIRONMENTAL HAZARDS

Keep out of lakes, ponds or streams. Do not contaminate water by cleaning of equipment or disposal of wastes.

Weeds Controlled

Annual Grasses	
Barnyardgrass (Watergrass)	(*Echinochloa* spp.)
Bermudagrass, seedling	*(Cynodon dactylon)*
Crabgrass	(*Digitaria* spp.)
Foxtails-Yellow, Giant, Green	(*Setaria* spp.)
Goosegrass	*(Eleusine indica)*
Millet	(*Setaria* spp.)
Narrow Leaf Signalgrass	*(Brachiaria piligera)*
Wild Oats	*(Avena fatua)*
Annual Broadleaf Weeds:	
Black-eyed-susan; (Sweet Suzy)	*(Rudbeckia serotina)*
Deadnettle (Henbit)	*(Lamium amplexicaule)*
Florida pusley	*(Richardia scabra)*
Hairy nightshade*	*(Solanum villosum)*
Lambsquarters	*(Chenopodium album)*
Nettleleaf goosefoot	*(Chenopodium murale)*
Prostrate pigweed	*(Amaranthus graecizans)*
Purslane; Common	*(Portulaca oleracea)*
Redroot pigweed	*(Amaranthus retroflexus)*
Shepherdspurse	*(Capsella bursa-pastoris)*

The above broadleaf weeds will be controlled only if TILLAM 6-E is applied when conditions are favorable for germination.
*Suppression (partial control)

Perennial Weeds:	
Yellow nutsedge	*(Cyperus esculentus)*
Purple nutsedge	*(Cyperus rotundus)*

Existing stands of Nutsedge must be turned under and chopped up thoroughly before treatment.

GENERAL USE PRECAUTIONS

Read all label directions before using.
TILLAM 6-E should be used only for recommended purposes and at recommended rates (do not overdose).

Applied according to directions and under normal growing conditions, TILLAM 6-E will not harm the treated crop. During germination and early growth, extended periods of unusually cold and wet or hot and dry weather, insect, nematode, or plant disease attack, the use of certain soil applied systemic insecticides, highly saline or alkaline soil conditions, improperly placed fertilizers or soil insecticides may create abnormal conditions that weaken crop seedlings. TILLAM 6-E used under these abnormal conditions could result in crop injury.

DIRECTIONS FOR USE

It is a violation of Federal law to use this product in a manner inconsistent with its labeling.

TILLAM 6-E is a selective herbicide which is mixed (incorporated) into the soil, or applied below the surface of the soil, or applied through solid set sprinklers for control of weeds listed on this label. TILLAM 6-E controls weeds by interfering with normal seed germination and seedling development. It does not control established weeds. All weed growth and crop stubble should be thoroughly worked into the soil before treatment.

SOLID SET SPRINKLERS

USE PRECAUTIONS. Apply this product only through solid set sprinklers irrigation system. Do not apply this product through any other type of irrigation system.

- Crop injury, lack of effectiveness, or illegal pesticide residues in the crop can result from nonuniform distribution of treated water.
- If you have any questions about calibration, you should contact State Extension Service specialists, equipment manufacturers or other experts.
- Do not connect an irrigation system (including greenhouse systems) used for pesticide application to a public water system unless the pesticide label-prescribed safety devices for public water systems are in place.
- A person knowledgeable of the chemigation system and responsible for its operation, or under the supervision of the responsible person, shall shut the system down and make necessary adjustments should the need arise.

USE REQUIREMENTS

- The system must contain a functional check valve, vacuum relief valve, and low pressure drain appropriately located on the irrigation pipeline to prevent water source contamination from back-flow.
- The pesticide injection pipeline must contain a functional, automatic, quick-closing check valve to prevent the flow of fluid back toward the injection pump.
- The pesticide injection pipeline must also contain a functional, normally closed, solenoid-operated valve located on the intake side of the injection pump and connected to the system interlock to prevent fluid from being withdrawn from the supply tank when the irrigation system is either automatically or manually shut down.
- The system must contain functional interlocking controls to automatically shut off the pesticide injection pump when the water pump motor stops.
- The irrigation line or water pump must include a functional pressure switch which will stop the water pump motor when the water pressure decreases to the point where pesticide distribution is adversely affected.
- Systems must use a metering pump, such as a positive displacement injection pump (e.g., diaphragm pump) effectively designed and constructed of materials that are compatible with pesticides and capable of being fitted with a system interlock.
- Do not apply when wind speed favors drift beyond the area intended for treatment.

TILLAM 6-E with Fluid Fertilizer

TILLAM 6-E may be combined with fluid (solution, slurry or suspension) fertilizers. However, physical compatibility with these fertilizers must be determined before combining in the spray tank. See Appendix I giving directions for these combinations. Even though found to be compatible, constant agitation is necessary to keep the TILLAM 6-E evenly mixed with the liquid fertilizer.

Impregnation on Dry Fertilizer

TILLAM 6-E may be impregnated on many dry bulk fertilizers for incorporation prior to transplanting of tobacco to control weeds listed on this label. To insure optimum results, TILLAM 6-E must be uniformly distributed on the fertilizer particles. See Appendix II for special instructions regarding directions for impregnation and use.

Application Directions

The soil should be well-worked and dry enough to permit good soil mixing (incorporation) or sub-surface application. Pour the recommended amount of TILLAM 6-E into the spray tank during the filling operations. Apply in 10 to 50 gallons of water per acre.

For soil incorporated applications, use a properly calibrated, low-pressure, boom-type sprayer having good agitation. For sub-surface applications, soil should be free of trash. Special equipment designed for sub-surface application must be used. Sweeps must be mounted immediately ahead of the planter unit.

In irrigated areas, do not apply TILLAM 6-E before pre-irrigation.

For all band applications, use proportionately less material per acre depending upon row spacing and band width to be treated. Do not use band application on rocky ground because thorough incorporation is not possible.

Incorporation Directions

TILLAM 6-E must be mixed (incorporated) into the soil IMMEDIATELY after application to prevent loss by evaporation of the herbicide. Whenever possible, application and incorporation should be done in the same operation.

Incorporation Before Planting: Use power-driven cultivation equipment set to cut to a depth of two to three inches in light soils and three to four inches in heavy soils OR a tandem disc set to cut to a depth of six inches, followed by a spike-tooth harrow or some other leveling device which extends beyond ends of the disc. For thorough mixing, disc in two different directions (cross disc).

Incorporation at or After Planting: Use power-driven cultivation equipment set to cut to a depth of two to three inches, OR on light soils only, ground-driven tillers (rolling cultivators, rotary hoe, row wheels, et cetera) set to cut to a depth of 1½ inches and operated at 6 to 8 mph. When using ground-driven tillers pre-cultivate if soil is crusted. When incorporating after planting, care must be taken not to disturb the crop seedling.

Sub-Surface Directions

(Tobacco only); Special equipment designed for sub-surface sweep application MUST be used. Sweeps must be mounted immediately ahead of the planter unit.

Covered Sweeps: Set the sweeps to run below the soil surface deep enough to cover the TILLAM 6-E with 5 to 6 inches of soil. Calibrate by measuring the spray band width at the back of the sweep, not the sweep width. NOTE: When applying with sweeps, TILLAM 6-E must be applied deep enough to allow 3-4 inches of soil to remain over the treatment after the transplanting operation.

Bed-Over Method of Application

Apply TILLAM 6-E in a band immediately in front of bedding discs. Set discs so that TILLAM 6-E is covered with 6 inches of soil. Set bedding discs so that treated soil is not folded in. If bed levelers are used at transplanting, beds should be made higher to compensate for leveling, i.e., so that 3 to 4 inches of soil are left above the TILLAM 6-E treatment.

Solid Set Sprinkler Application (Tomatoes only, Western Region): Meter the TILLAM 6-E into the irrigation water during the entire period. Apply in ¼ to ½ acre inch of water or sufficient water to penetrate the soil to a depth of 2 to 4 inches.

Time the TILLAM application to insure that proper penetration of the herbicide corresponds with the end of the irrigation period. Flush the lines then turn the water off promptly. Tailwater (run-off water) from irrigation should be recirculated.

Application of more than recommended quantities of irrigation water may result in decreased product performance by removing the chemical from the zone of effectiveness.

Planting

For maximum weed control benefits, crops should be planted or seeded immediately after application. Do not use a drag behind the planter as it may concentrate TILLAM 6-E over the seed row and cause crop injury.

Cultural Practices Following Application

TILLAM 6-E is not persistent in the soil and susceptible weeds germinating later in the season may not be controlled. Shallow cultivation or approved post-emergence herbicides may be necessary to control susceptible weeds that escape control by TILLAM 6-E as well as those weeds not susceptible to TILLAM 6-E. Do not cultivate deeper than the depth of herbicide incorporation.

When cultivating fields where TILLAM 6-E has been banded in the row, use shields to prevent the movement of untreated soil into the treated row.

RECOMMENDATIONS

Sugar Beets, Pre-Plant: Apply just prior to planting; see DIRECTIONS FOR USE for methods of application and incorporation. Recomended dosages are as follows:

For light textured soils and soils with less than 10% organic matter, apply 2⅔ qts. TILLAM 6-E/acre.

For heavy textured soils and soils with over 10% organic matter, apply 4 qts. TILLAM 6-E/acre.

TOBACCO, FLUE-CURED

Before bedding and transplanting: Apply TILLAM 6-E broadcast at 2⅔ qts. per acre. Incorporate immediately with a tandem disc and bed up immediately prior to transplanting.

or

After bedding but before transplanting: Apply TILLAM 6-E at a broadcast rate of 2⅔ qts. per acre prior to transplanting, reducing amount per acre proportionately depending upon band width and row spacing. Incorporate immediately using a power-driven rotary hoe with variable length blades (i.e., one with short blades on the inside and longer blades on the outside). Bed levelers, if used should be kept shallow as excessive removal of treated soil will destroy weed control.

or

Sub-surface sweeps after bedding but before transplanting: Run at a depth so that when bed is in final shape, 3 to 4 inches of soil will cover the TILLAM 6-E. Apply TILLAM 6-E at 2 qts. per acre, just prior to transplanting. For band treatment, reduce amount per acre proportionately depending upon band width and row spacing.

Tobacco, Light Air Cured

Includes Type 31 (Burley) & Type 32 Tobaccos, Type 41 Cigar Filler. Types 54 and 55 Cigar Binder:

Before transplanting: Apply TILLAM 6-E broadcast at 2⅔ qts. per acre. Incorporate immediately, only with a tandem disc on flat ground followed by transplanting. For band treatment reduce amount per acre proportionately depending upon band width and row spacing. Incorporate immediately, using a power-driven rotary hoe.

TOMATOES

SEE DIRECTIONS FOR USE FOR METHODS OF APPLICATION AND INCORPORATION.

DIRECT SEEDED: (WESTERN REGION ONLY)

Apply TILLAM 6-E just prior to planting.

For light textured soils and soils with less than 10% organic matter, apply 2⅔ qts. TILLAM 6-E/acre.

For heavy textured soils and soils with more than 10% organic matter, apply 4 qts. TILLAM 6-E/acre.

or

Direct Seeded, Layby: (Not pre-plant) Apply TILLAM 6-E as a directed spray to clean cultivated soil. Do not apply within eight (8) days of harvest.

For light textured soils and soils with less than 10% organic matter, apply 2⅔ qts. TILLAM 6-E/acre.

For heavy textured soils and soils with more than 10% organic matter, apply 4 qts. TILLAM 6-E/acre.

or

Direct Seeded, Sprinkler Application

(Western Region only): Apply TILLAM 6-E through solid set sprinklers prior to or immediately after planting, before weeds emerge.

For light textured soils and soils with less than 10% organic matter, apply 2⅔ qts. TILLAM 6-E/acre.

For heavy textured soils and soils with more than 10% organic matter, apply 4 qts. TILLAM 6-E/acre.

or

Plug Planted, Pre-Plant (Western Region only): Prior to plug planting tomatoes, apply and soil incorporate up to 6⅔ quarts of TILLAM 6-E per acre. Use higher rate for hairy nightshade.

Tomato Transplants (Do Not Use Hotcaps)

Before transplanting-Apply TILLAM 6-E prior to transplanting.

or

After transplanting- Apply TILLAM 6-E as a directed spray to clean, cultivated soil. Do not apply within eight (8) days of harvest.

For light textured soils and soils with less than 10% organic matter, apply 2⅔ qts. TILLAM 6-E/acre.

For heavy textured soils and soils with more than 10% organic matter, apply 4 qts. TILLAM 6-E acre.

Tomato Transplants (South Carolina only)

Before bedding and transplanting: Apply TILLAM 6-E at a broadcast rate of 2 qts. per acre just prior to bedding and transplanting, reducing the amount per acre proportionately depending upon band width and row spacing. Use bed-over method of application.

or

After bedding but before transplanting: Apply TILLAM 6-E at a broadcast rate of 2 qts. per acre just prior to transplanting, reducing amount per acre proportionately depending upon band width and row spacing. Apply to the top of the bed and incorporate four to six inches into the bed top.

or

Post-emergence, or after transplanting: Apply TILLAM 6-E at a broadcast rate of 2 qts. per acre reducing amount per acre proportionately depending upon band width and row spacing. Apply in a directed spray to soil around base of tomato plants using bands on either side of the row. Immediately after treatment, sweep out row centers to effectively cover TILLAM 6-E treatment with a layer of soil two inches deep. Do not apply within eight (8) days of harvest.

TILLAM 6-E/DASANIT* Spray Concentrate Tank Mix Combination

*DASANIT is a Registered Trademark of the Parent Company of Farbenfabriken Bayer, GMBH, Leverkusen.

TILLAM 6-E IN COMBINATION WITH DASANIT SPRAY CONCENTRATE controls nematodes, wireworms and weeds controlled by TILLAM 6-E in tobacco (Flue Cured and Light Air Cured).

NOTE: DASANIT is a Restricted Use Pesticide. Refer to the label for DASANIT SPRAY CONCENTRATE for pertinent directions and precautions for use, and restrictions.

Application Directions and Recommendations

The soil should be well worked and dry enough to permit good soil incorporation (mixing into the soil). Apply 2⅔ qts. TILLAM 6-E and 4 qts. DASANIT SPRAY CONCENTRATE in 20 to 50 gallons water per acre. Use a properly calibrated low-pressure boom-type sprayer having good agitation.

Soil Incorporation

The tank mix must be incorporated (mixed) into the soil immediately after application to prevent loss of the herbicide by evaporation. Whenever possible, application and incorporation should be done in the same operation.

Use power-driven cultivation equipment set to cut to a depth of 3" in light soil and 3 to 4" in heavy soil or a tandem disc set to cut to a depth of 4 to 6".

Cross disc at right angles for thorough mixing. The bedding operation should follow the incorporation as soon as is practical. Planting may be done immediately after bedding or at a later date; however, to insure long term weed control, the crop should be planted as soon as possible after application and bedding.

TILLAM 6-E/MOCAP* EC Tank Mix Combinations

*MOCAP is a Registered Trademark of Rhone-Poulenc Inc.

A tank mix of TILLAM 6-E and MOCAP EC can be applied and mixed into the soil before transplanting for control of weeds, wireworms and nematodes in tobacco.

NOTE: MOCAP is a Restricted Use Pesticide. Refer to the label of MOCAP EC for pertinent directions and precautions for use, and restrictions.

Tank Mixing Directions

The TILLAM 6-E/MOCAP EC tank mix combination can be mixed with water and applied as a spray. Pour the recommended amount of TILLAM 6-E directly into the nearly filled spray tank while agitating. Next add the recommended amount of MOCAP EC. Fill the remaining portion of the tank with water; mix thoroughly. Maintain a moderate agitation to provide a uniform mix during application.

Application Directions and Recommendations

The soil should be well-worked and dry enough to permit good soil mixing (incorporation). Apply 2⅔ qts. TILLAM 6-E plus 1 gallon MOCAP EC in 10 to 50 gallons water per acre. Use a properly calibrated low-pressure, boom-type sprayer having good agitation.

Incorporation Directions

The TILLAM 6-E/MOCAP EC tank mix combination must be mixed (incorporated) into the soil immediately after application to prevent loss by evaporation. Whenever possible, application and incorporation should be done in the same operation. Use a tandem disc set to cut to a depth of six inches, followed by a spike-tooth harrow or some other leveling device which extends beyond the ends of the disc. For thorough mixing, disc in two different directions (cross disc).

TILLAM 6-E/PAARLAN* EC Tank Mix Combinations

*PAARLAN is a Registered Trademark of Elanco Products Co., Div. of Eli Lilly and Co.

A tank mix of TILLAM 6-E and PAARLAN EC can be applied and mixed into the soil before transplanting for additional weed control and longer season grass control in tobacco.

NOTE: Refer to the label of PAARLAN EC for pertinent directions and precautions for use, and restrictions.

Tank Mixing Directions

The TILLAM 6-E/PAARLAN EC tank mix combination can be mixed with water and applied as a spray. Pour the recommended amount of TILLAM 6-E directly into the nearly filled spray tank while agitating. Next add the recommended amount of PAARLAN EC. Fill the remaining portion of the tank with water; mix thoroughly. Maintain a moderate agitation to provide a uniform mix during application.

Application Directions and Recommendations

The soil should be well-worked and dry enough to permit good soil mixing (incorporation). Apply 2⅔ qts. TILLAM 6-E plus 2 pints PAARLAN EC in 10 to 50 gallons water per acre. Use a properly calibrated low-pressure, boom-type sprayer having good agitation.

Incorporation Directions

The TILLAM 6-E/PAARLAN EC tank mix combination must be mixed (incorporated) into the soil immediately after application to prevent loss by evaporation. Whenever possible, application and incorporation should be done in the same operation. Use a tandem disc set to cut to a depth of six inches, followed by a spike-tooth harrow or some other leveling device which extends beyond the ends of the disc. For thorough mixing, disc in two different directions (cross disc).

TILLAM 6-E/DEVRINOL® 50-WP Tank Mix Combination Applied Pre-Plant Incorporated

A tank mix combination of TILLAM 6-E and Devrinol 50-WP can be applied and mixed (incorporated) into the soil before planting for weed control in tomatoes and tobacco.

Weeds Controlled

Annual Grasses:	
Barnyardgrass (Watergrass)	*(Echinochloa crus-galli)*
Bermudagrass, seedling	*(Cynodon dactylon)*
Crabgrass	*(Digitaria* spp.*)*
Foxtails (Giant, Green, Yellow)	*(Setaria* spp.*)*
Goosegrass	*(Eleusine indica)*
Millet	*(Setaria* spp.*)*
Narrow Leaf Signalgrass	*(Brachiaria piligera)*
Wild Oats	*(Avena fatua)*
Annual Broadleaf Weeds:	
Black-eyed Susan (Sweet Suzy)	*(Rudbeckia serotina)*
Deadnettle (Henbit)	*(Lamium amplexicaule)*
Florida pusley	*(Richardia scabra)*
Hairy Nightshade*	*(Solanum villosum)*
Lambsquarters	*(Chenopodium album)*
Nettleleaf Goosefoot	*(Chenopodium murale)*
Prostrate Pigweed	*(Amaranthus graecizans)*
Purslane, Common	*(Portulaca oleracea)*
Redroot Pigweed	*(Amaranthus retroflexus)*
Shepherdspurse	*(Capsella bursa-pastoris)*

* Suppression (partial control)

The above broadleaf weeds will be controlled only when the TILLAM/DEVRINOL tank mix is applied when conditions are favorable for germination.

Perennial Weeds:	
Purple Nutsedge	*(Cyperus rotundus)*
Yellow Nutsedge	*(Cyperus esculentus)*

Existing stand of nutsedge must be turned under and chopped up thoroughly before treatment.

Note: Observe all cautions and limitations on labeling of all products used in combination.

Make only one application per season.

In each case where a range of rates is given, the lower rates should be used on light (coarse) textured soils and soils low in organic matter. The higher rates should be used on heavy (fine) textured soils.

After harvest or prior to planting of succeeding crops, a deep moldboard or disc plowing operation must be carried out.

Tank Mixing Directions

The TILLAM 6-E/Devrinol 50-WP tank mix combination can be mixed with water and applied as a spray. Pour the recommended amounts of TILLAM 6-E and Devrinol 50-WP into a spray tank during the filling operations. When using Devrinol 50-WP add first to the spray tank, then add the TILLAM 6-E. Maintain adequate agitation of the tank mix. Apply in 25 to 50 gallons of water per acre.

Application Directions and Recommendations

The soil should be well-worked and dry enough to permit good soil mixing (incorporation).

For soil incorporated applications, use a properly calibrated, low-pressure, boom-type sprayer having good agitation.

In irrigated areas do not apply before pre-irrigation. For all band applications, reduce dosage proportionately depending upon row spacing and band to be treated. Do not use band application on rocky ground because thorough incorporation is not possible.

Incorporation Directions

The herbicides must be mixed (incorporated) into the soil immediately after application to prevent loss by evaporation. Wherever possible, application and incorporation should be done in the same operation.

Incorporation Before Planting

Use power-driven cultivation equipment set to cut to a depth of two to three inches in coarse soils and three to four inches, in heavy soils OR a tandem disc set to cut to a depth of six inches, followed by a spike-tooth harrow or some other leveling device which extends beyond ends of the disc. For thorough mixing, disc in two different directions (cross disc).

Incorporation At Planting

Use power-driven cultivation equipment set to cut to a depth of two to three inches OR, in coarse soils only, ground-driven tillers (rolling cultivators, rotary hoe, row wheels, et cetera) set to cut to a depth of 1½ inches and operated at 6 to 8 mph. When using ground-driven tillers, precultivate if soil is crusted.

Planting

Crops should be planted or seeded immediately after application and incorporation. Do not use a drag behind the planter as it may concentrate the herbicides over the seed row and cause crop injury.

Cultural Practices Following Application

Shallow cultivation may be necessary to control those susceptible weeds that escape control as well as those weeds not susceptible to either herbicide. Do not cultivate deeper than the depth of herbicide incorporation.

When cultivating fields where the tank mix has been banded in the row, use shields to prevent the movement of untreated soil into the treated row.

Recommendations

Tomatoes (Direct seeded [California only] and Transplanted): Apply 2 to 4 pounds DEVRINOL 50-WP plus 2⅔ to 4 qts. TILLAM 6-E per broadcast acre. Use the lower rates on light soils (coarse textured-sandy) and the higher rate on heavy soils (fine textured-clay). Do not use hotcaps on transplanted tomatoes.

Tobacco (Transplanted): Before transplanting, apply 2 pounds DEVRINOL 50-WP plus 2⅔ qts. TILLAM 6-E per broadcast acre.

To avoid injury to crops not specified on the label, do not plant until 12 months after the last DEVRINOL application.

TILLAM 6-E/DYFONATE® 4-EC /DEVRINOL (50-WP or 2-E) Tank Mix Combination

A tank mix combination of TILLAM 6-E, DYFONATE 4-EC and DEVRINOL 50-WP or 2-E can be applied and mixed into the soil before transplanting tomatoes and tobacco, for control of weeds and soil insects. Refer to TILLAM/Devrinol Tank Mix Combination for list of weeds controlled.

NOTE: Dyfonate 4-EC is a Restricted Use Pesticide. Observe all cautions and limitations on labeling of all products used in combinations. Make only one application per season. Do not contaminate water by cleaning of equipment or disposal of wastes.

Directions for Use

The recommended tank mix controls weeds by interfering with normal seed germination and seedling development. It does not control established weeds. All weed growth and crop stubble should be thoroughly worked into the soil before treatment.

For GENERAL USE PRECAUTIONS and DIRECTIONS FOR USE REFER TO TILLAM/DEVRINOL Tank Mix Recommendations for methods of tank mixing, application and incorporation.

Recommendations

Tomatoes (Direct seeded [California only] and Transplanted: Apply 2⅔ to 4 qts. TILLAM 6-E plus 2 to 4 pounds DEVRINOL 50-WP or 2 to 4 qts. DEVRINOL 2-E plus 2 qts. DYFONATE 4-EC for control of symphylans or 4 qts. for control of symphylans and wireworms in 25 to 50 gallons of water per broadcast acre. Use the lower DEVRINOL and TILLAM rates on light soils (coarse texture-sandy) and the higher rates on heavy soils (fine texture-clay). Do not use hotcaps on transplanted tomatoes.

Tobacco (Transplanted): Before transplanting, apply 2⅔ qts. TILLAM 6-E plus 2 pounds DEVRINOL 50-WP or 2 qts. DEVRINOL 2-E plus 2 qts. DYFONATE 4-EC for control of wireworms in 25 to 50 gallons of water per broadcast acre.

To avoid injury to crops not specified on the label, do not plant until 12 months after the last DEVRINOL application.

APPENDIX 1

TILLAM 6-E with Fluid Fertilizers

The following procedure is suggested for determining whether TILLAM 6-E may be combined with a specific fluid fertilizer for spray tank application.

Materials Required:

1. TILLAM 6-E.
2. Fluid fertilizer to be used.
3. Adjuvant for fertilizer tank mix: Compex*, Sponto 168-D*, Uni-Mix*, or equivalent. The adjuvant which provides the best emulsification depends on the specific fertilizer under consideration.
4. Two one-quart, wide-mouth glass jars with lid or stopper.
5. Measuring spoons (a 25 mL pipette or graduated cylinder provides more accurate measurement).
6. Measuring cup, 8 oz. (237 mL).

*Compex, Colloidal Products Incorp., Petaluma, Calif. Sponto 168-D, Retzloff Chemical Company, Houston, Texas, Uni-Mix, Thompson-Hayward Chemical Co., Kansas City, Kansas.

Procedure:

1. Pour a pint (about 473 mL) of the fluid fertilizer into each of the quart jars.
2. Add adjuvant to one of the jars and mix (see rate table).
3. Add the TILLAM 6-E to both jars (see rate table).
4. Close both jars with lid or stopper and mix the contents by turning the jars upside down ten times.
5. Inspect the surface and body of the mixtures—
(A) Immediately after completing the jar inversions.
(B) After allowing the jars to stand quietly for 30 minutes.
(C) And then again after turning the jars upside down 10 times.

If a uniform mix cannot be made, the mixture should not be used. If either mixture remains uniform for 30 minutes, the combination may be used. Should either mixture separate after 30 minutes, but readily remix uniformly with the jar inversions, the mixture can be used if adequate agitation is maintained in the tank. If the mixture with adjuvant is satisfactory, but the one without adjuvant is not, be sure to use the adjuvant in the spray tank. Add the adjuvant first at a rate of 3 pints per 100 gallons of fluid fertilizer; foaming can be minimized by using moderate agitation.

If nondispersible oil, sludge or clumps of solids form in the mixtures, the combinations should not be used.

Rate Table for TILLAM 6-E and Adjuvant with the Fluid Fertilizer**

Gallons of Fluid Fertilizer to be applied per acre	mL or Tsp. of TILLAM 6-E* to be added to 1 pint of fertilizer	
	6-E	
	mL	Tsp.
10	8	1⅔
15	5	1
20	4	¾
25	3	⅔
30	2½	½
40	2	⅖

*Based on field rate of 1 pound active ingredient per acre in the fertilizer volumes indicated. Increase volume proportionately to correspond with intended field rate in terms of pounds active ingredient per acre (e.g., for field rate of 4 pounds actual TILLAM 6-E in 40 gallons fertilizer per acre, add 8 mL or 2 tsp. TILLAM 6-E to each jar for compatibility testing purposes).

**Two (2) milliliters or one-half (½) teaspoon of adjuvant to be added to 1 pint of fluid fertilizer in order to equal the rate of 3 pints of adjuvant per 100 gallons of fluid fertilizer.

APPENDIX II

TILLAM® 6-E Impregnation on Dry Bulk Fertilizers For Use On Tobacco

CAUTION: TILLAM 6-E must not be impregnated on ammonium nitrate, sodium nitrate or potassium nitrate. Such mixtures may cause explosion and fire.

TILLAM 6-E may be impregnated on many dry bulk fertilizers* and applied and incorporated in the soil before planting for the control of grasses and broadleaf weeds in tobacco.

Field results have shown that TILLAM 6-E on bulk dry fertilizers gives weed control equal to TILLAM 6-E applied as a spray in water. However, uniform impregnation of TILLAM 6-E on the dry fertilizer particles and uniform application in the field are necessary to assure good results.

For impregnating TILLAM 6-E on dry fertilizer, use a closed rotary-drum mixer or a similar type of closed blender equipped with suitable spray equipment. The spray nozzle (or nozzles) should be positioned inside of the mixer to provide *uniform* spray coverage of the tumbling fertilizer. The physical properties of fertilizers vary, particulary in liquid absorptive capacity. When absorptivity is sufficient, simple spray impregnation of the fertilizer with TILLAM 6-E provides a satisfactory, dry mixture. If the absorptive capacity is inadequate, use of a highly absorptive powder is required to provide a dry, flowable mixture. Microcel E (Johns-Manville Products Corp.) is the recommended absorbent powder. It should be added separately and uniformly to the prepared TILLAM 6-E fertilizer mixture, in a quantity that is sufficient to provide a suitably flowable mixture. Generally less than 2% by weight of Microcel E is required. The amount of TILLAM 6-E actually required in the manufacture of individual fertilizer mixtures should be determined carefully for each production operation. This is necessary to ensure that the amount of TILLAM 6-E actually contained in the mixture applied to the soil represents the correct rate of use.

All TILLAM 6-E supplementary literature instructions and label recommendations regarding rates per acre, soil incorporation, application, cautions, general use precautions and other directions must be followed.

All individual state regulations relating to bulk dry fertilizer blending, registration, labeling and application are the responsibility of the individual and/or company selling the fertilizer and TILLAM 6-E mixture.

A minimum of 200 pounds and a maximum of 2,000 pounds of approved ingredients* impregnated with TILLAM 6-E at the recommended rate must be applied per acre.

Bulk fertilizer impregnated with TILLAM 6-E should be applied immediately, not stored. It is recommended that all bulk containers be tightly covered while the product is being transported and applied to reduce chances of TILLAM 6-E loss via volatilization.

***Approved Dry Fertilizer Ingredients**

	N	P	K
Ammonium sulfate	21	0	0
Diammonium phosphate	18	46	0
Potassium chloride	0	0	60
Potassium sulfate	0	0	52
Super-phosphate (single)	0	20	0
Super-phosphate (triple)	0	46	0
Urea	45	0	0
Ammonium phosphate sulfate	16	20	0
11-48-0	11	48	0

NOTE: K-Mag and two MAP fertilizers (12.5-50-0, 11-54-0) have been shown to be compatible with TILLAM 6-E and are approved for use.

Rate Chart for Impregnation of Dry Bulk Fertilizers with TILLAM 6-E

Fertilizer Rate Per Acre	TILLAM 6-E Rate Per Acre (2⅔ qts./acre)
200	26⅔ qts./ton
300	17⁴⁄₅ qts./ton
400	13⅓ qts./ton
500	10⅔ qts./ton
600	8⅞ qts./ton
800	6⅔ qts./ton
1000	5⅓ qts./ton
1500	3⅗ qts./ton
2000	2⅔ qts./ton

STORAGE AND DISPOSAL

Prohibitions: Do not contaminate water, food, feed, or fertilizer by storage or disposal. Open dumping is prohibited.

Pesticide Disposal: Rinse spray equipment. Wastes resulting from the use of this product may be disposed of on site or at an approved waste disposal facility.

Container Disposal: Metal—Triple rinse (or equivalent), then offer for recycling, reconditioning, or puncture and dispose of in a sanitary landfill or by other procedures approved by state and local authorities.

Plastic—Triple rinse (or equivalent), then offer for recycling, or reconditioning, or puncture and dispose of in a sanitary landfill or incineration, or, if allowed by State and local authorities, by burning. If burned, stay out of smoke.

NOTICE TO BUYER AND USER:

Seller warrants that this product conforms to the chemical description on the label and is reasonably fit for the purposes stated on the label when used in accordance with directions under normal conditions of use. This warranty does not extend to the use of this product contrary to label instructions, or under abnormal use conditions, or under conditions not reasonably foreseeable to Seller, and Buyer and User assume the risk of any such use. SELLER DISCLAIMS ALL OTHER WARRANTIES EXPRESSED OR IMPLIED INCLUDING ANY WARRANTY OF FITNESS OR MERCHANTABILITY. SELLER SHALL NOT BE LIABLE FOR CONSEQUENTIAL, SPECIAL OR INDIRECT DAMAGES RESULTING FROM THE USE OR HANDLING OF THIS PRODUCT AND SELLER'S SOLE LIABILITY AND BUYER'S AND USER'S EXCLUSIVE REMEDY SHALL BE LIMITED TO THE REFUND OF THE PURCHASE PRICE.

This product is sold only for uses stated on the label. No express or implied license is granted to use or sell this product under any patent in any country except as specified: Country: United States of America.

A-9MR

The information and specimen labels in this booklet are correct as of the publication date but are subject to change without notice. For current information, contact ICI Americas Inc.

TORPEDO™

INSECTICIDE

Only for Sale to, Use and Storage by Professional Pest Control Operators.

ACTIVE INGREDIENT:	
Permethrin (3-Phenoxyphenyl)methyl (±)-*cis, trans*-3-(2,2-dichloroethenyl)-2,2-dimethylcyclopropanecarboxylate*	25.6%
INERT INGREDIENTS	74.4%
TOTAL	100.0%

* *Cis/trans* ratio: Min. 35% (±) *cis* and max. 65% (±) *trans*. TORPEDO contains 2 pounds active ingredient per gallon.

EPA Reg. No. 10182-95
U.S. Patient No. 4,024,163

KEEP OUT OF REACH OF CHILDREN WARNING

PRECAUTIONARY STATEMENTS

HAZARDS TO HUMANS AND DOMESTIC ANIMALS

WARNING

Harmful if swallowed or absorbed through skin. May cause eye irritation.

Avoid contact with eyes, skin or clothing. Avoid breathing vapor or spray mist. Wash thoroughly after handling.

STATEMENT OF PRACTICAL TREATMENT

IF SWALLOWED: Do not induce vomiting. Call a physician. Vomiting should be supervised by a physician or the professional staff because of the possible pulmonary damages via aspiration of the solvent. IF IN EYES: Flush with plenty of water. Get medical attention if irritation persists. IF ON SKIN: Wash with plenty of soap and water.

FOR 24-HOUR EMERGENCY MEDICAL ASSISTANCE CALL 1-800-F-A-S-T-M-E-D (327-8633).
FOR CHEMICAL EMERGENCY: Spill, leak, fire, exposure, or accident call CHEMTREC 1-800-424-9300.

ENVIRONMENTAL HAZARDS

This pesticide is extremely toxic to fish. Use with care when applying in areas adjacent to any body of water. Do not apply directly to water. Do not apply when weather conditions favor drift from treated areas. Do not contaminate water by cleaning of equipment or disposal of wastes. This product is highly toxic to bees exposed to direct treatment or residues on crops or weeds. Do not apply TORPEDO, or allow it to drift to crops or weeds on which bees are actively foraging. Additional information may be obtained from your Cooperative Extension Service. Apply this product only as specified on this label.

PHYSICAL OR CHEMICAL HAZARDS

Do not use or store near heat or open flame.

GENERAL INFORMATION ON THE USE OF THIS PRODUCT

Chemicals for soil treatment are used to establish a barrier against termite attack. The chemical emulsion must be adequately dispersed in the soil to provide a barrier between the wood in the structure and the termite colonies in the soil.

For the effective use of this product, it is necessary that the service technician be familiar with current control practices including trenching, rodding, subslab injection, and low-pressure spray applications. These techniques must be correctly employed to prevent or control infestations by subterranean termite species of *Reticulitermes, Zootermopsis, Heterotermes* and *Coptotermes*. Choice of appropriate procedures should include consideration of such variable factors as the design of the structure, water table, soil type, soil compaction, grade conditions, location and type of domestic water supplied and drainage systems. The biology and behavior of the termite species involved are important factors to be known as well as suspected location of the colony and severity of the infestation within the structure to be protected.

Effective termite control also includes elimination of termite access to moisture by recommending repair of faulty construction grade and/or plumbing. Remove all wood and cellulose containing debris in contact with soil from crawl spaces, porches, and around foundations.

For advice concerning current control practices with relation to the specific local conditions, consult resources in structural pest control and the State regulatory agency.

**DIRECTIONS FOR USE—
SUBTERRANEAN TERMITE CONTROL**

It is a violation of Federal law to use this product in a manner inconsistent with its labeling. This product may not be used against any pests not named on the label. Apply only to establish subsurface termite control barriers specified on product labeling. This labeling must be in the possession of the user at the time of application. Avoid contamination of public and private water supplies by following these precautions:

- Use anti-backflow equipment or procedures to prevent siphonage of pesticide back into water supplies.
- Do not treat soil beneath structures that contain wells or cisterns.
- Extreme care must be taken to avoid runoff. Do not treat soil that is water-saturated or frozen.

Consult State and local specifications for recommended distance of treatment areas from wells. Refer to Federal Housing Administration Specifications for guidance on preconstruction treatments if no State or local government recommendations are available.

After Treatment: Securely plug all holes drilled in construction elements of commonly occupied areas of structures, including unfinished basements, enclosed porches, garages, and workshops.

PRECONSTRUCTION SUBTERRANEAN TERMITE TREATMENT

Effective preconstruction subterranean termite control requires the establishment of an unbroken vertical and/or horizontal chemical barrier between wood in the structure and the termite colonies in the soil.

To meet F.H.A. termite-proofing requirements, follow the latest edition of the Housing and Urban Development (H.U.D.) Minimum Property Standards.

Use a 0.5 to 1.0% emulsion for subterranean termites. Mix 2-4 gallons of TORPEDO in 98 gallons of water. Use the higher rate in hard-to-treat locations or areas of heavy termite pressure. If 0.5% rate is used the property should be inspected annually for possible reinfestation and retreatment if necessary.

Do not apply to any area intended as a plenum air space. Check with the builder or contractor to determine if the design of the structure includes a plenum airspace.

After grading is completed and prior to pouring of the slab, slab-supported or constructed porches, and other critical areas, make the following treatments:

Horizontal Barriers: Before footings are poured, horizontal barriers may be established in footing trenches. Then, after interior grading is completed and prior to the pouring of concrete slabs, horizontal barriers may be established on soil that will be covered by floors, entrance platforms, or porches, and in other critical areas that will be covered by construction. To provide a horizontal barrier, apply the emulsion at the rate of 1 gallon per 10 square feet to fill dirt. If fill is washed gravel or other coarse material, apply at $1^1/_2$ gallons per 10 square feet.

- It is important that the emulsion reaches the soil.
- Applications shall be made with low pressure (less than 50 psi at the nozzle) using a coarse-spray nozzle when establishing horizontal barriers.
- If concrete slabs cannot be poured over soil the same day it has been treated, a water-proof cover such as polyethylene sheeting should be placed over the soil to prevent erosion. This is not necessary if foundation walls have been installed around the treated soil.

Vertical Barriers: After the foundation walls have been poured or built, vertical barriers may be established around the perimeters of floating or supported slabs, around utilities penetrating the slab, and in other critical areas. After the final exterior grading is completed, vertical barriers may be created in back-filled soil against foundation walls. To produce a vertical barrier, apply the emulsion at the rate of 4 gallons per 10 linear feet per foot of depth from grade to the top of the footing. For example, a footing 3 feet deep would require 12 gallons of emulsion per 10 linear feet.

Outside and inside perimeter applications must be made by rodding and/or trenching. When rodding from grade or from the bottom of a shallow trench, rod holes should be spaced in a manner that will allow for application of a continuous chemical barrier. Rod holes should not extend beneath the top of the footings, except when the footing is exposed at or above grade. Special care should be taken to avoid soil washout around the footing.

A trench need not be wider than 6 inches. Rod from the base of a shallow trench to the top of the footings. Low-pressure spray may be used to treat soil which will be replaced in the trench. Mix the emulsion with the soil as it is being replaced in the trench.

Soil should be treated around sewer lines, plumbing, or around any other utility extending from the soil through a slab.

Hollow Masonry Units of the Foundation: In preconstruction situations in which application is not made to soil prior to pouring the footing, treat so as to make a continuous chemical barrier in the voids. Apply the emulsion at the rate of 2 gallons per 10 linear feet. Apply the emulsion so it will reach the footing.

Do not treat in this manner through voids in walls constructed on interior slabs such as basement floors.

Crawl Spaces: For crawl spaces apply at the rate of 4 gallons of emulsion per 10 linear feet per foot of depth from grade to the top of the footing. Application may be made by rodding and/or trenching. If the footing is exposed at or above grade, application should be made with special care to avoid washout around the footing. Treatment should include both sides of foundation and around all piers and pipes extending from the soil. To avoid volatilization into air within the structure, do not make an overall broadcast application to areas intended to be crawl spaces; apply by rodding and/or trenching.

- Rod holes should be spaced (about 1 foot) to provide a continuous chemical barrier.
- Trench need not be wider than 6 inches nor below the foundation. The emulsion should be mixed with the soil as it is being replaced in the trench.

Monolithic Slabs: In the case of a single-pour monolithic slab that does not have a separate foundation or footing, an overall horizontal barrier should be created before the concrete is poured using a rate of 1 gallon of emulsion per 10 square feet. If fill is washed gravel or other coarse material, apply at the rate of $1^1/_2$ gallons per 10 square feet. Critical areas beneath the slab such as utility pipe entries may be treated at the rate of 4 gallons per 10 linear feet around the pipe.

Exterior vertical barriers should be created after the concrete has been poured and final grade established. Apply the emulsion at the rate of 4 gallons per 10 linear feet per foot of depth to the bottom of the concrete.

POSTCONSTRUCTION TREATMENTS

Use a 0.5 to 1.0% emulsion for subterranean termites. Mix 2-4 gallons of TORPEDO in 98 gallons of water. Use the higher rate in hard-to-treat locations or areas of heavy termite pressure. If 0.5% rate is used the property should be inspected annually for possible reinfestation and retreatment, if necessary.

Postconstruction applications shall be made by subslab injection, rodding, and/or trenching using low-pressure spray.

To avoid volatilization into the air within the structure, do not make an overall broadcast application of this product in a crawl space. Rodholes or trenches should not extend below the footing because of the possibility of soil washout by the emulsion.

Do not apply this product to the soil beneath a plenum air space.

Do not apply emulsion until location of heat or air-conditioning ducts, vents, and water and sewer (or plumbing) lines are known and identified. Extreme caution must be taken to avoid contamination of these structural elements and airways. Do not apply this product to soil beneath slabs with subslab or intraslab ducting until ducts are permanently plugged. SURFACE APPLICATION IS PROHIBITED.

Slab-On-Ground: Vertical barriers may be established by subslab injection inside and rodding and/or trenching outside at the rate of 4 gallons of emulsion per 10 linear feet.

Treat the soil from grade to the top of the footing along the outside and, where necessary, along the inside of the foundation perimeter. Treatment may also be required along one side of a partition wall (especially where the wall is connected to the floor by fixtures inserted in the slab) and along cracks, expansion joints, and other critical areas.

Horizontal barriers may be established where necessary by long rodding or by a grid pattern injection using a rate of 1 to $1^1/_2$ gallons of emulsion per 10 square feet depending on fill type and condition.

Drill holes should be spaced in a manner that will allow for application of a continuous chemical barrier.

Where necessary, drill through the foundation walls from the outside and inject the chemical just beneath the slab or along the inside of the foundation.

Along the outside of the foundation walls where shallow foundations exist (1 foot or less), dig a narrow trench approximately 6 inches wide and not below the top of the footing. Apply the emulsion at the rate of 2 gallons per 10 linear feet. As the soil is being replaced into the trench, apply another 2 gallons per 10 linear feet to the backfill. When making soil applications to the foundation extending deeper than 1 foot, follow instructions under: BASEMENTS—OUTSIDE PERIMETER. (See exception for monolithic slabs.)

Hollow Masonry Units of Foundation Walls: Treatment may be made through masonry voids to establish a continuous chemical barrier at the top of the footing. Apply at the rate of 2 gallons of emulsion per 10 linear feet of footing. Where this treatment is necessary, access holes must be drilled *below* the sill plate and should be through a lower mortar joint. Before treatment through basement walls, seal the interior wall and floor expansion joint with mortar, caulk, waterproofing material, or similar impervious sealant. Also, seal openings at the top of the foundation wall. Do not treat in this manner through voids in walls constructed *on* interior slabs such as basement floors.

Basements: Apply the emulsion at the rate of 4 gallons per 10 linear feet per foot of depth from the grade to the top of the footing. For example, a footing 3 feet deep would require 12 gallons of emulsion per 10 linear feet. Application shall be made by subslab injection, trenching and/or rodding.

Inside: Treatment may be required along inside of foundation walls and along one side of interior partition walls (or bearing walls) especially where the wall is connected by fixtures inserted in the floor. Application may also be necessary around sewer pipes, floor drains, conduits, or any cracks in the basement floor. Drill holes should be spaced in a manner that will allow for application of a continuous chemical barrier.

Outside Perimeter: Applications must be made by rodding and/or trenching. When rodding from grade or from the bottom of a shallow trench, rod holes should be spaced in a manner that will allow for application of a continuous chemical barrier. Rod holes should not extend beneath the top of the footings.

A trench need not be wider than 6 inches. Rod from the base of a shallow trench to the top of the footings. Low-pressure spray may be used to treat soil which will be replaced in the trench. Mix the emulsion with the soil as it is being replaced in the trench.

Crawl Spaces: Apply the emulsion at the rate of 4 gallons per 10 linear feet per foot of depth from the grade to the top of the footing. For best results application should be made by rodding and/or trenching.

To avoid volatilization into air within the structure, do not make an overall broadcast application of this product in a crawl space. Do not apply this product to the soil beneath a plenum air space.

Treat both sides of foundation and around all piers and pipes. When rodding from grade or from the bottom of a shallow trench, rod holes should be spaced in a manner that will allow for application of a continuous chemical barrier. Rod holes should not extend beneath the top of the footings.

A trench need not be wider than 6 inches. Rod from the base of a shallow trench to the top of the footings. Low-pressure spray may be used to treat soil which will be replaced in the trench. Mix the emulsion with the soil as it is being replaced in the trench.

- If it is necessary to make an overall barrier under soil in a crawl space, this treatment may only be made by injecting the emulsion several inches below the soil surface.
- It should be recommended that inadequately ventilated crawl spaces be brought into compliance with FHA Minimum Property Standards specifying 1 square foot of ventilated opening per 150 square feet of crawl space area.

Excavation Technique: If treatment must be made in difficult situations such as near wells or cisterns, along faulty foundation walls, and around pipes and utility lines which lead downward from the structure, application may be made in the following manner:

- Trench and remove the soil to be treated onto heavy plastic sheeting or similar liner.
- Treat the soil at the rate of 4 gallons of emulsion per 10 linear feet per foot of depth of the trench. Mix the emulsion thoroughly into the soil taking care to prevent liquid from running off the liner.
- After the treated soil has dried adequately, replace the soil in the trench.

Prior to using this technique near wells or cisterns, consult State, local, or Federal regulatory agencies for information regarding approved treatment practices in your area.

After Treatment: Before leaving the job site, securetly plug all holes drilled in construction elements of commonly occupied areas of structures, including unfinished basements, enclosed porches, garages, and workshops.

RETREATMENT

Retreatment for subterranean termites should be made when there is evidence of reinfestation subsequent to the initial treatment, or there has been a disruption of the chemical barrier in the soil due to construction, excavations, landscaping, etc. Retreatment should be made as a spot application to these areas.

Retreatment may be made to vulnerable areas in accordance with the application techniques described above. This application should be made as a spot treatment to these areas. Routine or annual retreatment of the entire premises should be avoided.

DIRECTIONS FOR USE—TERMITES

(Localized Areas in Structures)

Dosage and Mixing Instructions: TORPEDO is recommended for use as an aqueous emulsion containing 0.5 to 1.0% permethrin. To prepare a 0.5 to 1.0% emulsion, add 2.5 to 5 fl. oz. (75 to 150 ml) of concentrate per gallon of spray.

For control of termites (localized areas of infested wood in structures), apply a 0.5 to 1.0% emulsion to voids and channels in damaged wood and in spaces between wooden members of a structure and between wood and foundations where termite infestation is likely to occur. Application may be made to inaccessible areas by drilling, then injecting the emulsion. Use a sufficient amount of coarse spray to cover the area to the point of runoff. Treatment of localized areas is intended to kill winged reproductive and worker forms of termites in the treated areas and to prevent infestations for a temporary period. This type of application is not intended to be a substitute for soil treatment or mechanical alteration to control subterranean termites.

RECOMMENDATION FOR CONTROL OF CLUSTER FLY

Apply TORPEDO to the outside of the structure. Mix 1 part TORPEDO to 5 parts deodorized lightweight mineral oil; apply to 1,000 square feet of wall area. Use a fogging apparatus which delivers the material in a strong air carrier, producing a small particle size. The apparatus should be held within three feet of the surface being treated. Apply only when air movement is less than 2 miles per hour.

The surface treated should be dry at the time of application. Attics and unoccupied lofts should be treated at the same time and at the same rate.

Generally the north side of structures need not be treated since flies seldom enter from areas not sunlit.

STORAGE AND DISPOSAL

Prohibitions: Do not contaminate water, food or feed by storage or disposal. Open dumping is prohibited. Do not reuse empty container.

Storage: Keep container closed when not in use. Do not store near food or feed. Protect from freezing. In case of spill or leak on floor or paved surfaces, soak up with sand, earth or synthetic absorbent. Remove to chemical waste area.

Pesticide Disposal: Pesticide wastes are toxic. Improper disposal of excess pesticide, spray mixture or rinsate is a violation of Federal law. If these wastes cannot be disposed of by use according to label instructions, contact your State Pesticide or Environmental Control Agency, or the Hazardous Waste representative at the nearest EPA Regional Office for guidance.

Container Disposal: *Metal Containers:* Triple rinse (or equivalent). Then offer for recycling or reconditioning, or puncture and dispose of in a sanitary landfill, or by other procedures approved by state and local authorities. *Plastic Containers:* Triple rinse (or equivalent). Then offer for recycling or reconditioning, or puncture and dispose of in a sanitary landfill, or incineration, or, if allowed by state and local authorities, by burning. If burned, stay out of smoke. *Glass Containers:* Triple rinse (or equivalent). Then dispose of in a sanitary landfill or by other approved state and local procedures.

NOTICE TO BUYER AND USER

Seller warrants that this product conforms to the chemical description on the label and is reasonably fit for the purposes stated on the label when used in accordance with directions under normal conditions of use. This warranty does not extend to the use of this product contrary to label instructions, or under abnormal use conditions, or under conditions not reasonably foreseeable to Seller, and Buyer and User assume the risk of any such use. SELLER DISCLAIMS ALL OTHER WARRANTIES, EXPRESSED OR IMPLIED, INCLUDING ANY WARRANTY OF FITNESS OR MERCHANTABILITY. SELLER SHALL NOT BE LIABLE FOR CONSEQUENTIAL, SPECIAL OR INDIRECT DAMAGES RESULTING FROM THE USE OR HANDLING OF THIS PRODUCT AND SELLER'S SOLE LIABILITY AND BUYER'S AND USER'S EXCLUSIVE REMEDY SHALL BE LIMITED TO THE REFUND OF THE PURCHASE PRICE.

05-0603-014

This is a specimen label. ICI Americas Inc. is not responsible for the accuracy of the information contained herein. As labels are subject to revision, always carefully read and follow the label on the product container.

VAPAM®

Soil Fumigant Solution For All Crops

ACTIVE INGREDIENT:	
Sodium methyldithiocarbamate (anhydrous)	32.7%
INERT INGREDIENTS:	67.3%
TOTAL	100.0%

*Contains 3.18 lbs. active ingredient per gallon

EPA Reg. No. 10182-150

KEEP OUT OF REACH OF CHILDREN

WARNING

STATEMENT OF PRACTICAL TREATMENT

First Aid

Immediately start the procedures given below and CONTACT A POISON CENTER, A PHYSICIAN OR THE NEAREST HOSPITAL. Describe the type and extent of exposure, the victim's symptoms, and follow the advice given.

For skin contact: Flush affected areas with large amounts of running water for 15 minutes and get medical attention. Remove and clean contaminated clothing and shoes.

For eye contact: Immediately flush eyes with large amounts of running water for 15 minutes. Hold the eyelids apart during the flushing to ensure rinsing of the entire surface of the eye and lids with water. Get medical attention immediately.

If inhaled: Remove to fresh air. If breathing is difficult, oxygen may be given with a physicians's advice. If not breathing, clear victim's airway, start mouth-to-mouth artificial respiration, and get medical attention.

If swallowed: Immediately give several glasses of water but do not induce vomiting. If vomiting occurs, give fluids again. Have a physician determine if condition of patient will permit induction of vomiting or evacuation of stomach. Do not give anything by mouth to an unconscious or convulsing person.

FOR 24-HOUR EMERGENCY MEDICAL ASSISTANCE CALL 1-800-F-A-S-T-M-E-D (327-8633).
FOR CHEMICAL EMERGENCY: Spill, leak, fire, exposure, or accident call CHEMTREC 1-800-424-9300.

PRECAUTIONARY STATEMENTS

HAZARDS TO HUMANS AND DOMESTIC ANIMALS

WARNING

Harmful if inhaled or swallowed. Irritating to eyes, nose, throat and skin. Avoid breathing vapor or spray mist. Do not get in eyes, on skin, or clothing. Causes skin irritation. Prolonged contact may cause severe skin irritation. May be fatal if absorbed through skin. In case of contact, immediately remove contaminated clothing. Wash and dry clothing before reuse. Where adequate ventilation is not available, use NIOSH-approved full-face pesticide cannister respirators to reduce exposure. Where exposure potential under the use conditions necessitates a higher level of protection, use a full-face positive-pressure air-supplied respirator.

ENVIRONMENTAL HAZARDS. This product is toxic to fish. Do not apply directly to water. Do not apply where runoff is likely to occur. Do not contaminate water when disposing of equipment washwaters. Apply this product only as specified on the label.

PRODUCT INFORMATION

VAPAM is a water soluble liquid. When applied to properly prepared soil, the liquid is converted into a gaseous fumigant. After sufficient interval of time, the gas dissipates leaving the soil ready for planting. VAPAM is recommended for the control of the following soilborne pests that attack ornamental, food and fibre crops: Weeds and germinating weed seeds [Annual Bluegrass, Bermudagrass, Chickweed, Dandelion, Ragweed, Henbit, Lambsquarters, Amaranthus sp. (Pigweed, Careless weed) Watergrass, Johnsongrass, Nutsedge, Wild morningglory and Purslane] Nematodes, Symphylids, (Garden centipede); and soilborne diseases (Rhizoctonia, Pythium, Phytophthora, Verticillium, Sclerotinia, Oat root fungus and Club root of crucifers).

USE PRECAUTIONS

Keep children and pets out of treated area until dry. Keep off desirable lawns and plants. Do not apply within 3 feet of the drip line of desirable plants, shrubs or trees. Do not use in confined areas without adequate ventilation OR where fumes may enter nearby houses containing growing plants. Do not use in green houses where desirable plants are present. Keep container tightly closed when not in use. Do not store near feed or food.

DIRECTIONS FOR USE

It is a violation of Federal Law to use this product in a manner inconsistent with its labeling.

Before applying VAPAM always cultivate thoroughly area to be treated, breaking up clods and loosening soil deeply and thoroughly. A week before treatment, moisten soil after cultivation to the desired depth; sprinkle or flood irrigate. See section "POTATOES: PACIFIC NORTHWEST AND NORTH CENTRAL STATES-FALL APPLICATION" where no-till stubble, or cover crops exist for specific application directions. Immediately before application, cultivate lightly if the soil has crusted.

To prevent loss from evaporation, use only at times when air temperature is moderate and there is little wind movement. Soil temperature must be from 40°-90° F in the treated zone. Treated zone is defined as the depth of treatment that VAPAM achieves at the time of application. For other conditions, see section, "Cultivation & Planting After Application." Do not apply to the soil surface, as in the sprinkler method, when air temperature is over 90° F or when low humidity or high winds would cause loss of VAPAM before it can be drenched into the soil with additional water. If fumes become unpleasant during treatment, apply more water to seal the fumes into the soil where they should be confined to achieve maximum fumigant benefit.

Where rate ranges for VAPAM are given, use the higher rate if pests (insects, nematodes, etc.) are present in high numbers or if the area to be treated has a history of pest problems.

The activity of VAPAM is increased by the use of a tarp (plastic, paper or fabric) spread loosely over the treated area and secured to prevent removal by wind. Keep covered for a minimum period of 48 hours. Seven days after treatment cultivate area to depth of 2 inches to aerate the soil. Do not seed earlier than 21 days after application when tarping method is used.

APPLICATION IN TANKMIX WITH LIQUID FERTILIZER: VAPAM may be injected in a mixture with liquid fertilizers. Since the composition of liquid fertilizers vary considerably, the physical compatibility of each fertilizer/VAPAM tankmix should be checked by using the following procedure:

Mix a small quantity of VAPAM and liquid fertilizer in a glass container. VAPAM and fertilizer should be mixed in the same ratio as they will be applied to the field (i.e. if 40 gallons of VAPAM and 40 gallons of liquid fertilizer are to be applied per acre, then VAPAM and fertilizer should be mixed in the jar in a 40:40 or 1:1 ratio). Agitate the liquids to attain a complete mixture. If a uniform mix cannot be made, the mixture should not be used. If the mixture remains uniform for 30 minutes, the combination may be used. Should the mixture separate after 30 minutes, but readily remix uniformly with agitation, the mixture can be used if adequate agitation is maintained in the tank.

DO NOT PLACE CAPS ON JAR, AS INCOMPATIBLE MIXES MAY EVOLVE HYDROGEN SULFIDE GAS.

USE PROMPTLY AFTER MIXING WITH WATER OR FERTILIZER. DO NOT ALLOW SOLUTION TO STAND. Flush all equipment with water after each day's use. Disassemble valves and clean carefully.

USE PRECAUTIONS FOR IRRIGATION SYSTEMS

A. Apply this product only through sprinkler including center pivot, lateral move, end tow, side (wheel) roll, traveler, solid set, or hand move; flood (basin); furrow; border; or drip (trickle) irrigation systems. Do not apply this product through any other type of irrigation system.
B. Crop injury, lack of effectiveness, or illegal pesticide residues in the crop can result from nonuniform distribution of treated water.
C. If you have questions about calibration, you should contact State Extension Service specialists, equipment manufacturers or other experts.
D. Do not connect an irrigation system (including greenhouse systems) used for pesticide application to a public water system unless the pesticide label-prescribed safety devices for public water systems are in place.
E. A person knowledgeable of the chemigation system and responsible for its operation, or under the supervision of the responsible person, shall shut the system down and make necessary adjustments should the need arise.

USE PRECAUTIONS FOR FLOOD (BASIN), FURROW AND BORDER IRRIGATION

1. Systems using a gravity flow pesticide dispensing system must meter the pesticide into the water at the head of the field and downstream of a hydraulic discontinuity such as a drop structure or weir box to decrease potential for water source contamination from backflow if water flow stops.
2. Systems utilizing a pressurized water and pesticide injection system must meet the following requirements:
   a. The system must contain a functional check valve, vacuum-relief valve, and low pressure drain appropriately located on the irrigation pipeline to prevent water source contamination from backflow.
   b. The pesticide injection pipeline must contain a functional, automatic, quick-closing check valve to prevent the flow of fluid back toward the injection pump.
   c. The pesticide injection pipeline must also contain a functional, normally closed, solenoid-operated valve located on the intake side of the injection pump and connected to the system interlock to prevent fluid from being withdrawn from the supply tank when the irrigation system is either automatically or manually shut down.
   d. The system must contain functional interlocking controls to automatically shut off the pesticide injection pump when the water pump motor stops.
   e. The irrigation line or water pump must include a functional pressure switch which will stop the water pump motor when the water pressure decreases to the point where pesticide distribution is adversely affected.
   f. Systems must use a metering pump, such as a positive displacement injection pump (e.g., diaphragm pump) effectively designed and constructed of materials that are compatible with pesticides and capable of being fitted with a system interlock.

USE PRECAUTIONS FOR SPRINKLER IRRIGATION

1. The system must contain a functional check valve, vacuum relief valve, and low pressure drain appropriately located on the irrigation pipeline to prevent water source contamination from backflow.
2. The pesticide pipeline must contain a functional, automatic, quick-closing check valve to prevent the flow of fluid back toward the injection pump.
3. The pesticide injection pipeline must also contain a functional, normally closed, solenoid-operated valve located on the intake side of the injection pump and connected to the system interlock to prevent fluid from being withdrawn from the supply tank when the irrigation system is either automatically or manually shut down.
4. The system must contain functional interlocking controls to automatically shut off the pesticide injection pump when the water pump motor stops.

5. The irrigation line or water pump must include a functional pressure switch which will stop the water pump motor when the water pressure decreases to the point where pesticide distribution is adversely affected.
6. Systems must use a metering pump, such as a positive displacement injection pump (e.g., diaphragm pump) effectively designed and constructed of materials that are compatible with pesticides and capable of being fitted with a system interlock.
7. Application of more than recommended quantities of irrigation water may result in decreased product performance by removing the chemical from the zone of effectiveness.
8. Do not apply when wind speed favors drift beyond the area intended for treatment.

USE PRECAUTIONS FOR DRIP (TRICKLE) IRRIGATION

1. **The system must contain a functional check valve, vacuum relief valve and low pressure drain appropriately located on the irrigation pipeline to prevent water source contamination from backflow.**
2. **The pesticide injection pipeline must contain a functional, automatic, quick-closing check valve to prevent the flow of fluid back toward the injection pump.**
3. **The pesticide injection pipeline must also contain a functional, normally closed, solenoid-operated valve located on the intake side of the injection pump and connected to the system interlock to prevent fluid from being withdrawn from the supply tank when the irrigation system is either automatically or manually shut down.**
4. **The system must contain functional interlocking controls to automatically shut off the pesticide injection pump when the water pump motor stops.**
5. **The irrigation line or water pump must include a functional pressure switch which will stop the water pump motor when the water pressure decreases to the point where pesticide distribution is adversely affected.**
6. **Systems must use a metering pump such as a positive displacement injection pump (e.g., diaphragm pump) effectively designed and constructed of materials that are comptatible with pesticides and capable of being fitted with a system interlock.**

CULTIVATION AND PLANTING AFTER APPLICATION

On well drained soils of light to medium texture which are not excessively wet or cold following application, planting may take place 14 to 21 days after treatment. If soils are heavy or especially high in organic matter or remain wet and/or cold (below 60° F) following application of VAPAM, an extended interval should be observed. Where dosages are greater than 100 gal. per acre, wait at least 60 days. On heavy, wet soils, light surface cultivation to break up crusting and promote drying of the soil should be done 5 to 7 days after application. This cultivation may be repeated as necessary. To avoid reinfesting treated soils, cultural practices should be such that untreated soils are not mixed with treated soils.

SPECIAL INSTRUCTIONS: When treating potting soil, or heavier field soils, including soils high in clay or organic matter, it is essentially important that the soils be allowed to aerate and dry thoroughly after using VAPAM. During cold and/or wet weather, frequent shallow cultivations may aid the escape of VAPAM from the soil. If in doubt, transplant a seedling plant and examine for injury before planting crop.

USES, APPLICATION METHODS & RATES

FOR SHALLOW PEST IN SEED BEDS, PLANT BEDS, LAWNS AND OTHER LIMITED AREAS:

SPRINKLING CAN METHOD: Place 1 pt. VAPAM (1½ pt. on very heavy soils or for deep-rooted weeds) in a sprinkling can, fill with water, and sprinkle uniformly over 50 sq. ft. of well-prepared soil. Sprinkle immediately with water until soil is sealed, or tarp for 48 hours.

COMMERCIAL MIXER PROPORTIONER METHOD: Add 1 qt. VAPAM to 3 qt. water in a bucket or other container and apply, through the mixer proportioner, to an area of 100 sq. ft. Sprinkle with water until soil is sealed, or tarp for 48 hours.

SOIL INJECTION: Space injectors 5 in. apart and inject VAPAM 4 in. into well prepared soil. Follow immediately with a roller to smooth and compact the soil surface. Light watering or a tarp after rolling helps prevent gas escape. For seedbeds a dosage of 75 to 100 gal. per acre (1½ pt. to 2 pt. per 100 sq. ft.) is recommended.

ROTARY TILLER: Spray or sprinkle diluted VAPAM immediately in front of tiller. Use 1 qt. VAPAM in 2½ gal. water per each 100 sq. ft. Follow immediately with a roller to smooth and compact the soil surface. Light watering or a tarp after rolling will help prevent gas escape.

Field Application—Where Entire Area Is Being Treated:

SOIL INJECTION: Using injectors spread 5 in. apart or using injector blades inject VAPAM 4 in. deep into well-prepared soil. Follow immediately with a roller to smooth and compact surface. Light watering or a tarp after rolling helps prevent gas escape. For field use, 40 to 100 gal. VAPAM per acre is recommended.

SPRINKLER SYSTEM: Use only those sprinkler systems which give large water droplets to prevent excessive loss. Use 75 to 100 gal. VAPAM per acre. For control of shallow pests (top 1 ft. or less of soil), run sprinklers 5 to 10 min. In next 10 to 20 min. inject into the sprinkler system all VAPAM needed for the area covered. On very light soils keep surface moist by sprinkling for 2 or 3 days. For control of pests deeper than 18 in. in the soil, divide VAPAM into 3 or more equal parts and apply at intervals during the sprinkling period.

CHECK OR FLOOD IRRIGATION: Meter VAPAM at a steady rate into water during irrigation. Use 50 to 100 gal. VAPAM per acre, depending upon the kind of pest and depth desired, in 3 to 18 inches of water per acre.

FIELD APPLICATION TO BEDS OR ROWS:

SOIL INJECTION: VAPAM, at the rate of 75 to 100 gallons per treated acre (1½ to 2 pints per 100 sq. ft.) may be injected into pre-formed plant beds following the directions given above under soil injection. If a wider treated band is desired, space 2 or more injectors (shanks, blades, fertilizer wheels, etc.) at intervals of 5 in. to cover the desired treating width. Roll immediately.

If VAPAM is injected into established plant beds through plastic tarps to terminate growth of a previous crop, and to fumigate the bed in preparation for planting a subsequent crop, the terminated crop should not be used for any food or feed purpose after VAPAM has been applied.

Drip Irrigation: VAPAM must be applied through a drip irrigation system to wet the soil thoroughly in the area being treated.

Meter 25 to-75 gallons VAPAM per treated acre into the drip system during the entire irrigation period. APPLICATION MUST BE CONTINUOUSLY SUPERVISED. THIS IS VERY IMPORTANT: WEED ELIMINATION WILL NOT BE SATISFACTORY IF TOO MUCH WATER IS APPLIED. An adequate concentration of VAPAM must be present at the time of weed seed germination in order to be effective. Further directions for use are as follows:

1. **Ground must be in seed-bed condition, no clods larger than ½" in diameter.**
2. **Beds must be listed, shaped and ready for planting.**
3. **Soil moisture must be at 50% of field capacity in the top 2-3" at time of VAPAM application.**

Tankmix with TILLAM® 6E herbicide (tomatoes only):

A tankmix of VAPAM soil fumigant plus TILLAM 6E herbicide may be used when desired. The mixture must be applied preplant to tomatoes if all directions and precautions pertaining to both VAPAM and TILLAM 6E are followed. Apply through a spray blade, by shank injection, low pressure boom sprayer or (Western Region only) through solid set sprinkler systems.

SOIL COVERING METHODS: (Bed-over methods). VAPAM may be sprayed or dripped onto the soil immediately ahead of bedshaping equipment. Cover the VAPAM with soil to a depth of 3 to 6 inches. The soil should be rolled and compacted immediately. The recommended rate of VAPAM is 50 to 100 gal. per acre, approximately equivalent to 1 to 2 pints per 100 sq. ft. of treated bed.

TREATMENT OF TREE REPLANT SITES: After removing dead or diseased tree and as much of the root system as possible, make a shallow basin over the planting site. Add VAPAM to the stream of water while filling the basin. Use 1 qt. VAPAM per 100 sq. ft. in sufficient water (depending on soil type) to penetrate at least 6 ft. For control of Oak root fungus, use a basin at least 20 x 20 ft. square. Increase dosage to 2 qt. VAPAM per 100 sq. ft. in sufficient water to penetrate to the depth of root system. If water is tanked to the planting site, add VAPAM to the water and mix before filling basin.

TREATMENT OF POTTING SOIL:

A. SPRINKLER METHOD:

1. Spread soil in a smooth layer 4 inches high on concrete or on pretreated soil.
2. Sprinkle VAPAM at rate of 1 pint in 5 gallons of water per 100 sq. ft. of surface area.
3. Layers can be treated one on top of another.
4. Sprinkle top layer with sufficient additional water to seal the surface, or cover with tarp (plastic, kraft paper, etc.)

B. CEMENT MIXER:

1. Add VAPAM to soil mix rate of 1 fl. oz. VAPAM per 2 cu. ft. of soil, in cement or similar mixer. Mix thoroughly.
2. After soil is treated and piled, sprinkle water over entire surface to seal in gas. Or, cover with tarp (plastic, kraft paper, etc.)

C. SHREDDER

1. Dilute VAPAM in sufficient water to obtain even distribution. As soil is ejected from shredder, spray uniformly on soil stream at rate of 1 fl. oz. VAPAM per 2 cu. ft. of soil.
2. After all soil is treated and piled, apply light water seal to entire surface or cover with tarp.

POTATOES: PACIFIC NORTHWEST AND NORTH CENTRAL STATES—Fall Application: To supress nematodes and *Verticillium dahliae* (Early Maturity Disease) in potatoes, use 50 to 100 gallons of VAPAM per treated acre. Inject into the sprinkler system (Refer to USE PRECAUTIONS FOR IRRIGATION SYSTEMS, Part A) all of the VAPAM needed for the area covered and apply in a minimum of 1 acre inch of water. VAPAM may be applied where crop stubble or vegetation exist without prior tillage, provided there is adequate soil penetration of VAPAM.

EARLY MATURITY DISEASES OF POTATOES IN OREGON: Apply 40 gals. VAPAM per acre using thin shank injector rig with shanks spaced 5 inches apart.

NOTE: VAPAM will only control nematodes which are in the treated zone at time of treatment. VAPAM has no soil residual and reinfestation of a field can occur from numerous sources such as deep nematode populations, potato seed pieces, equipment contamination and blowing soil.

TOBACCO PLANT BEDS: Fall applications are recommended wherever possible. Read and follow DIRECTIONS FOR USE carefully. Treatment in the South should generally be made before November 30.

A. TARP METHOD: Prepare the bed 5 to 7 days before application to insure best conditions for weed seed germination and fumigant action of VAPAM. The bed should be free of clods, level and in good tilth. Apply 1 to 1½ gal. of VAPAM in a minimum of 40 gal. of water per 100 sq. yd. Apply uniformly over the entire bed. Cover the bed immediately with a plastic cover. Keep covered no less than one day, but no more than two days. The cover need not be tented, but should be secured to prevent wind from uncovering the treated area. Seven days after date of VAPAM application, loosen the treated soil to a depth of 2 in. Do no seed tobacco earlier than 21 days after VAPAM application.

B. DRENCH METHOD: Apply 2½ gal. VAPAM in 150 to 200 gal. of water per 100 sq. yd. Application may be made with sprinklers, sprayers with nozzles or any suitable equipment. Follow directions given above for seed bed treatment.

SYMPHYLANS CONTROL: Soil should be in good seed bed condition to a depth of 8 to 10 inches. Maintain adequate moisture during Spring season. Treat during July-August when symphylans are in the upper soil surface. Apply 20 gal. VAPAM per treated acre (0.4 pints per 100 sq. ft.) using blade or chisel injectors spaced 5 inches apart. Inject below level of symphylans concentration, usually 6 to 8 inches. Pack soil immediately after application.

PEPPERMINT—Verticillium wilt control. When infestation is limited to small spots in a field, spread can be reduced by treating the soil with 100 gals. VAPAM per treated acre using injector blade or thin shank injector rig with injectors spaced 5 inches apart.

PREVENTION OF ROOT GRAFT TRANSMISSION OF DUTCH ELM AND OAK WILT DISEASE

Immediately after a tree is diagnosed as having Dutch Elm or Oak Wilt disease, isolate the diseased tree from healthy trees with the VAPAM treatment. If a diseased tree is less than 20 feet from a healthy tree or has advanced wilt symptoms, it may be necessary to treat at two sites—one between the diseased and the first healthy-appearing tree and one between the first and the second healthy-appearing trees. This measure is advisable because the casual fungus may have already passed from the diseased to the first healthy-appearing tree before VAPAM was applied.

Use VAPAM diluted one part to three parts water for Dutch Elm disease and diluted one part to ten parts of water for Oak Wilt disease. Drill holes approximately ¾ to 1 inch in diameter, 15 inches deep to 6 to 9 inches apart. Fill each hole with diluted VAPAM to within 2 inches of the soil surface.

Make the line of treatment sufficiently long to kill all roots of the two adjacent trees that are likely to be root-grafted. Apply the chemical slowly and carefully to avoid overflowing the drilled holes, this will reduce grass kill. Tamp each hole closed with the heel. Allow at least two weeks after treatment before removing the diseased tree.

STORAGE AND DISPOSAL

PROHIBITIONS: Do not contaminate water, food or feed by storage or disposal.

STORAGE: Do Not Store Below 0°F. Product crystallizes at lower temperatures. Warm or store at higher temperatures and mix to redisolve crystals and assure uniformity before use.

PESTICIDE DISPOSAL: Wastes resulting from the use of this product may be disposed of on site or at an approved waste disposal facility.

CONTAINER DISPOSAL:

Triple rinse (or equivalent). Then offer for recycling or reconditioning, or puncture and dispose of in a sanitary landfill, or by other procedures approved by State and local authorities.

NOTICE TO BUYER AND USER:

Seller warrants that this product conforms to the chemical description on the label and is reasonably fit for the purposes stated on the label when used in accordance with directions under normal conditions of use. This warranty does not extend to the use of this product contrary to label instructions, or under abnormal use conditions, or under conditions not reasonably foreseeable to Seller, and Buyer and User assume the risk of any such use. SELLER DISCLAIMS ALL OTHER WARRANTIES EXPRESSED OR IMPLIED INCLUDING ANY WARRANTY OF FITNESS OR MERCHANTABILITY. SELLER SHALL NOT BE LIABLE FOR CONSEQUENTIAL, SPECIAL OR INDIRECT DAMAGES RESULTING FROM THE USE OR HANDLING OF THIS PRODUCT AND SELLER'S SOLE LIABILITY AND BUYER'S AND USER'S EXCLUSIVE REMEDY SHALL BE LIMITED TO THE REFUND OF THE PURCHASE PRICE.

This product is sold only for uses stated on its label. No express or implied license is granted to use or sell this product under any patent in any country except as specified: Country: United States of America.

A-4R

The information and specimen labels in this booklet are correct as of the publication date but are subject to change without notice. For current information, contact ICI Americas Inc.

VAPAM®

SOIL FUMIGANT

Soil Fumigant Solution

For Fumigating Garden Soils Before Planting

ACTIVE INGREDIENT:	
Sodium methyldithiocarbamate (Anhydrous)	32.7%
INERT INGREDIENTS:	67.3%
TOTAL	100%

Contains 3.18 lbs. active ingredient per gallon

EPA Reg. No. 10182-229

KEEP OUT OF REACH OF CHILDREN CAUTION

PRECAUTIONARY STATEMENTS

HAZARDS TO HUMANS AND DOMESTIC ANIMALS

CAUTION

Harmful if inhaled or swallowed. Irritating to eyes, nose, throat and skin. Avoid breathing vapor or spray mist. Do not get in eyes, on skin, or clothing. In case of contact, immediately remove contaminated clothing. When applying in enclosed areas, wear a mask or respirator of a type passed by the U.S. Bureau of Mines for Vapam protection. Do not store near food or feed. Keep children and pets out of treated areas.

STATEMENT OF PRACTICAL TREATMENT

First Aid

Immediately start the procedures given below and contact a Poison Center, a physician, or the nearest hospital. Report the type and extent of exposure, describe the victim's symptoms, and follow the advice given.

If on skin: Immediately flush skin with large amount of running water for at least 15 minutes while removing contaminated clothing and shoes. Get medical attention immediately. Wash clothing before reuse.

If in eyes: Immediately flush eyes with large amounts of running water for at least 15 minutes. Hold eyelids apart to ensure rinsing of the entire surface of the eye and lids with water. Get medical attention immediately.

If inhaled: Remove to fresh air. If not breathing, clear the victim's airway and start mouth-to-mouth artificial respiration. If breathing is difficult, give oxygen, preferably with a physician's advice. Get medical attention immediately.

If swallowed: Immediately give several glasses of water but *do not* induce vomiting. If vomiting does occur, give fluids again. Have a physician determine if condition of patient will permit induction of vomiting or evacuation of stomach. Do not give anything by mouth to an unconscious or convulsing person.

FOR 24-HOUR EMERGENCY MEDICAL ASSISTANCE CALL 1-800-F-A-S-T-M-E-D (327-8633).
FOR CHEMICAL EMERGENCY: Spill, leak, fire, exposure, or accident call CHEMTREC 1-800-424-9300.

ENVIRONMENTAL HAZARDS

This product is toxic to fish. Do not apply directly to water or wetlands. Do not contaminate water by cleaning of equipment or disposal of waste. Apply this product only as specified on the label.

PRODUCT INFORMATION

VAPAM is an easy-to-use water solution of a highly effective soil fumigant. It can be used prior to planting a lawn or when preparing a new flower bed or vegetable garden, as an aid in the control of weeds, certain nematodes which attack plants, and common soil-inhabiting plant disease fungi.

When properly applied, VAPAM releases a gaseous fumigant in the soil. This gas later dissipates, leaving the soil ready for planting.

USE PRECAUTIONS

KEEP OFF DESIRABLE LAWNS AND PLANTS. VAPAM may severely injure any plants that it comes in contact with. Therefore, you should read these use precautions carefully. Do not apply within 3 feet of the drip line of plants, shrubs or trees. Do not use in confined areas without adequate ventilation. Do not use in greenhouses where growing plants are present or where fumes may enter nearby houses containing growing plants. After mixing with water, do not allow solution to stand overnight. If fumes from partially treated and water sealed area become unpleasant, sprinkle more water over treated area. **Rinse all equipment with water after each day's use.** Keep children and pets out of treated area. Do not store near food. Keep container tightly closed when not in use. Do not store below 0°F.

DIRECTIONS FOR USE

It is a violation of Federal Law to use this product in a manner inconsistent with its labeling.

For treating seed beds, plant beds, renovating lawns, establishing new lawns, fumigating potting soil.

Before applying VAPAM

A week before applying VAPAM, cultivate the area to be treated, breaking up clods and loosening the soil deeply and thoroughly. Keep soil moist and bring area to final grade and seedbed condition. **This step is essential to good results.**

APPLICATION METHODS AND RATES

Sprinkling Can: Place 1 pint VAPAM (1½ pints on very heavy soils or for deep-rooted weeds) in a sprinkling can, fill with water, stir, and sprinkle uniformly over 50 square feet of well-prepared soil. Sprinkle immediately with water until soil is sealed and no odor of VAPAM can be detected.

Hose-End Sprayers (lawn or fertilizer type): Fill container with 1 pint of VAPAM and spray uniformly on 50 square feet of well-prepared soil. Continue to spray with water until VAPAM odor cannot be detected.

After applying VAPAM

VAPAM should be sealed into the soil by sprinkling after application wetting the soil to a depth of 3-5 inches. Sprinkle lightly so water doesn't puddle. **Keep soil surface wet for 2-3 days after application—then let dry out.** The use of a plastic tarp or other covering over the treated area greatly improves the action of VAPAM. When using a cover, do not wait until the entire area is treated to begin covering but cover each 100 square feet area as it is treated, immediately. Weight the tarp around edges to prevent wind from lifting. Keep covered for two days and then remove. Seven days after applying VAPAM, rake the treated area. Do not rake deeper than 2 inches. If weather is warm and dry, plant after 2 weeks. If weather is cold and wet, rake the soil frequently and do not plant for 4 weeks.

FUMIGATING POTTING SOIL

Treat Your Soil in Wheelbarrow or Garbage Can: Put a 4-inch layer of soil mix in the container and sprinkle surface evenly following the table below. Add another 4-inch layer and repeat sprinkling. Repeat until container is full. Cover with plastic or container lid for 48 hours.

AMOUNT OF VAPAM TO USE FOR EACH 4 INCHES OF SOIL

Diameter of Can	Teaspoons of VAPAM	Water
10"	½ tsp.	4 ozs. (¼ pint)
12"	¾ tsp.	6 ozs.
15"	1¼ tsp.	10 ozs.
20"	2¼ tsp.	18 ozs.

Mix the VAPAM with the water, then sprinkle the soil. For larger areas of potting soil, use VAPAM at a rate of 1 teaspoonful in 8 ozs. (½ pint) of water per square foot. After 5-7 days the treated soil should be shredded or turned or mixed to aerate and help release any VAPAM still in the soil.

The treated soil mixture should then be left for 14 more days before it is put into storage sheds or bins.

NOTE: It is very important that the VAPAM treated soil is thoroughly aerated and dry before being used.

IMPORTANT INFORMATION

If the soils have remained wet or cold (below 60°F at 3" level) do not plant for 4 weeks or longer and aerate soils again to speed release of gases. No crop should be planted until all odor has left the soil. As an added precaution, 10 days after application, plant a few radish seeds which should germinate in about 5 days. If plants are normal, the treated area is safe to plant.

DO:

1. Read the label and follow all manufacturer's directions.
2. Use only when ground temperature 3 inches deep is over 60°F and less than 90°F. In the colder parts of the country, the soil is usually warm enough when the spring flowering shrubs such as forsythia are blooming.
3. VAPAM can be irritating to the skin. We recommend using rubber gloves and rubber boots when applying VAPAM.
4. If VAPAM is splashed on the body, wash immediately, apply soothing ointment and remove contaminated clothing.
5. Clean equipment by rinsing thoroughly in running water.

DO NOT:

1. Do not use VAPAM unless you are prepared to follow all directions completely.
2. Do not apply in greenhouses where desirable plants are growing or in greenhouses connected to other areas where such plants are growing. Fumes from treated soil can injure growing plants.
3. Do not apply indoors without adequate ventilation.

STORAGE AND DISPOSAL

Do not contaminate water, food, or feed by storage or disposal. **Storage:** Do Not Store Below 0°F. Product crystallizes at lower temperatures. Warm or store at higher temperatures and mix to redissolve crystals and assure uniformity before use. **Pesticide Disposal:** Securely wrap original container in several layers of newspaper and discard in trash. **Container Disposal:** Do not reuse container. Rinse thoroughly before discarding in trash.

NOTICE TO BUYER AND USER

Seller warrants that this product conforms to the chemical description on the label and is reasonably fit for the purposes stated on the label when used in accordance with directions under normal conditions of use. This warranty does not extend to the use of this product contrary to label instructions, or under abnormal use conditions, or under conditions not reasonably foreseeable to Seller, and Buyer and User assume the risk of any such use. SELLER DISCLAIMS ALL OTHER WARRANTIES, EXPRESSED OR IMPLIED, INCLUDING ANY WARRANTY OF FITNESS OR MERCHANTABILITY. SELLER SHALL NOT BE LIABLE FOR CONSEQUENTIAL, SPECIAL OR INDIRECT DAMAGES RESULTING FROM THE USE OR HANDLING OF THIS PRODUCT AND SELLER'S SOLE LIABILITY AND BUYER'S AND USER'S EXCLUSIVE REMEDY SHALL BE LIMITED TO THE REFUND OF THE PURCHASE PRICE.

This product is sold only for uses stated on its label. No express or implied license is granted to use or sell this product under any patent in any country except as specified: Country: United States.

A-3R

02-0611-011

This is a specimen label. ICI Americas Inc. is not responsible for the accuracy of the information contained herein. As labels are subject to revision, always carefully read and follow the label on the product container.

VERNAM® 7-E

SELECTIVE HERBICIDE

liquid herbicide for weed control in soybeans and peanuts

ACTIVE INGREDIENT:	
S-propyl dipropylthiocarbamate	88.5%
INERT INGREDIENTS:	11.5%
TOTAL	100.0%

Contains Seven Pounds of Active Ingredient Per Gallon

EPA Reg. No. 10182-221

Complete Directions For Use, Use Precautions and Recommendations

KEEP OUT OF REACH OF CHILDREN

CAUTION

Harmful if Swallowed

STATEMENT OF PRACTICAL TREATMENT

If swallowed: DO NOT induce vomiting, immediately give large quantities of water. If vomiting does occur, give fluids again. Never give anything by mouth to an unconscious person. Call a physician or the nearest Poison Control Center immediately.

For eye contact: Hold eyelids apart and flush with large amounts of running water for at least 15 minutes. Call a physician.

For skin contact: Flush with large amounts of running water for 15 minutes. Remove contaminated clothing and shoes. Get medical attention if skin irritation occurs. Wash clothing before reuse.

If inhaled: Remove to fresh air. Get medical attention if respiratory irritation occurs.

FOR 24-HOUR EMERGENCY MEDICAL ASSISTANCE CALL 1-800-F-A-S-T-M-E-D (327-8633).
FOR CHEMICAL EMERGENCY: Spill, leak, fire, exposure, or accident call CHEMTREC 1-800-424-9300.

PRECAUTIONARY STATEMENTS

HAZARDS TO HUMANS AND DOMESTIC ANIMALS

CAUTION

Harmful if swallowed. Avoid contact with skin, eyes and clothing. Avoid inhalation of spray mist. Wear rubber gloves, goggles and protective clothing. Wash with soap and water immediately after use or contact. Do not contaminate food and feed. Keep away from heat and flame.

ENVIRONMENTAL HAZARDS

Do not apply directly to water. Do not contaminate water by cleaning of equipment or disposal of wastes.

Do not apply this product in such a manner as to directly or through drift, expose workers or other persons. When applied through the sprinkler irrigation method of application, the area being treated must be vacated by unprotected persons.

Weeds Controlled by VERNAM 7-E

VERNAM 7-E will not control established weeds:

Annual Grasses:

Barnyardgrass (watergrass)	*(Echinochloa crus-galli)*
Bermudagrass seedlings**	*(Cynodon dactylon)*
Crabgrass	*(Digitaria* spp.)
Fall Panicum	*(Panicum dichotomiflorum)*
Foxtail millet	*(Setaria italica)*
Giant foxtail	*(Setaria faberi)*
Goosegrass	*(Eleusine indica)*
Green foxtail	*(Setaria viridis)*
Johnsongrass (seedling)	*(Sorghum halepense)*
Wild cane*	*(Sorghum bicolor)*
Yellow foxtail	*(Setaria glauca)*

** Southeastern U.S. Only

* Cultivation is suggested in addition to the VERNAM 7-E treatment in fields with moderate or heavy infestations of wild cane.

Annual Broadleaf Weeds:

Annual morningglory	*(Ipomoea purpurea)*
Balloonvine*	*(Cardiospermun halicacabum)*
Carpetweed	*(Mollugo verticillata)*
Common lambsquarters	*(Chenopodium album)*
Florida purslane (Florida pusley)	*(Richardia scabra)*
Purslane	*(Portulaca oleracea)*
Redroot pigweed	*(Amaranthus retroflexus)*
Sicklepod (coffeeweed)	*(Cassia obtusifolia)*
Smooth pigweed	*(Amaranthus hybridus)*
Velvetleaf (buttonweed)	*(Abutilon theophrasti)*

* In Alabama Only

The above listed broadleaf weeds will be controlled only if treatment is made when conditions are favorable for weed seed germination and growth.

Perennial Grasses:

Bermudagrass*	*(Cynodon dactylon)*
Purple nutsedge	*(Cyperus rotundus)*
Yellow nutsedge	*(Cyperus esculentus)*

* Southeastern U.S. Only

Existing stands of nutsedge must be turned under and chopped up thoroughly before treatment. For nutsedge control VERNAM 7-E must be thoroughly mixed into the soil prior to planting or applied subsurface to the soil with injectors or sweeps. Cultivation is suggested in addition to VERNAM 7-E treatment in fields with moderate to heavy nutsedge infestations.

General Use Precautions

Read entire label before using.

VERNAM 7-E should be used only for recommended purposes and at recommended rates (do not overdose).

Failure to follow label instructions, or use on crops not specifically listed on the label may result in crop injury or destruction.

Applied according to directions and under normal growing conditions, VERNAM 7-E will not harm the treated crop. Malformation of the primary leaves is not unusual. This condition is only temporary and will not affect yields. During germination and early stages of growth, extended periods of unusually cold and wet or hot and dry weather, insect or plant disease attack, the use of certain soil-applied systemic insecticides, improperly placed fertilizers or soil insecticides may create abnormal conditions that weaken crop seedlings. VERNAM 7-E used under these abnormal conditions could result in crop injury.

VERNAM 7-E is recommended for use on mineral soils only.

Do not tank mix VERNAM 7-E with insecticides or fungicides.

Do not contaminate irrigation water or water used for domestic purposes.

In irrigated areas, do not apply VERNAM 7-E before pre-irrigation.

DIRECTIONS FOR USE OF VERNAM 7-E

IT IS A VIOLATION OF FEDERAL LAW TO USE THIS PRODUCT IN A MANNER INCONSISTENT WITH ITS LABELING.

VERNAM 7-E is a selective herbicide which is mixed (incorporated) into the soil or applied below the surface of the soil for control of weeds listed on this label. VERNAM 7-E controls annual weeds as their seeds germinate by interfering with normal germination and seedling development. It does not control established weeds. All weed growth and crop stubble should be thoroughly worked into the soil before treatment.

Application Directions

Pour the recommended amount of VERNAM 7-E into the spray tank during the filling operation. Apply in 10 to 50 gallons of water per acre using a properly calibrated, low-pressure (20 to 40 PSI) sprayer having good agitation. The soil should be well-worked and dry enough to permit good soil mixing (incorporation) or subsurface application. Soil should be as free as possible of trash for subsurface application with sweeps or injectors. Avoid overlaps that will increase dosage above recommended limits.

Fluid Fertilizer Statement

VERNAM 7-E may be combined with fluid (solution, slurry, or suspension) fertilizers. However, physical compatibility with these fluid fertilizers must be determined before combining in the spray tank. See Appendix I for directions regarding these combinations.

Dry Fertilizer Statement

VERNAM 7-E may be impregnated on many dry bulk fertilizers for use *on registered crops.* However, uniform distribution of VERNAM 7-E on fertilizer particles and uniform application are necessary to assure good results. See Appendix II for information and directions for impregnation and use.

Incorporation Directions

VERNAM 7-E must be incorporated into the soil immediately (within minutes) to prevent loss of the herbicide. Whenever possible, application and incorporation should be done in the same operation.

Soil Mixing (Incorporation) Before Planting: The following equipment is commonly used for soil mixing (incorporation) before planting: Power Driven cultivation Equipment (Recommended on all soil types previously primary tilled): Ground speed should be adjusted with all PTO driven cultivation tools to ensure thorough soil mixing.

a. Horizontal action: set to cut 4 to 6 inches deep.

b. Vertical action: set to cut 3 to 4 inches deep.

Tandem Discs: Set to cut to a depth of 4 to 6 inches, operated at 4 to 6 mph, followed by a spike-tooth harrow or some other leveling device which extends beyond the ends of the discs. For more thorough mixing (for control or suppression of deep germinating grasses—wild cane, seedling johnsongrass, and nutsedge), disc in two different directions (cross disc). Prior to the second discing, raise the disc to prevent cutting deeper than 4 to 6 inches.

NOTE: In the Southern Delta of the Mississippi Valley, disc incorporation is not recommended on the heavy (gumbo) clay soils.

Field Cultivators (Recommended for use on coarse textured soils in good tilth): Use 3 to 4 rows of sweeps spaced at 7 inch or less intervals and staggered so that no soil is left unturned, followed by a spike-tooth harrow pulled behind the cultivator. Do not use chisel plows for incorporation. Set the cultivator to cut 4 to 6 inches deep, operate at 5 mph or more. The field cultivator should not be used alone as an incorporation tool where control or suppression of deep-germinating weeds such as nutsedge, johnsongrass seedlings or wild cane is desired. For control of these weeds, use a disc for at least one of the passes over the field.

Soil Mixing (Incorporation) After Planting: Use power-driven cultivation equipment set to cut to a depth of 2 inches, or on coarse textured (light) soils only, ground-driven tillers (rolling cultivators, rotary hoe, row wheels, etc.) set to cut to a depth of 1½ inches and operated at 6 to 8 mph. When incorporating after planting, care must be taken not to disturb the crop seed or seedlings by adjusting tines or blades of incorporation equipment so they are not directly over the drill row.

Subsurface Application Before or at Planting (for peanuts and soybeans only): Special equipment designed for subsurface application MUST be used. Injector units must be rigidly mounted on or IMMEDIATELY ahead or IMMEDIATELY behind the planter unit. Sweeps must be mounted IMMEDIATELY ahead of the planter unit.

(1) **Soil Injection—**Injector shanks must be spaced 2½ to 3 inches apart and mounted in staggered positions to avoid trash buildup. Set shanks to inject VERNAM 7-E 1½ to 3 inches below the soil surface. The width of the band in which weed control is desired will determine the number and spacing of injector shanks required per row. (Example: Four injector shanks spaced 3 inches apart gives a 12-inch band.) A broadcast application can be made by increasing the number of shanks. The two shanks adjacent to the drill row must be 1¼ to 1½ inches on either side of it.

(2) **Covered Sweeps—**Set the sweeps to run 3 to 4 inches below the soil surface. Calibrate by measuring the spray band width at the back edge of the sweep, not the sweep width. For BROADCAST APPLICATIONS, stagger sweeps on double tool bar so that they overlap sufficiently to allow spray bands to meet.

NOTE: When applying with sweeps, VERNAM 7-E must be applied deep enough to allow 3 to 4 inches of soil to remain over the treatment after the planting operation.

PLANTING—SOYBEANS, PEANUTS

Do not plant seed deeper than 2 inches. For VERNAM 7-E application made before planting, seeding can be done immediately, or it may be delayed up to 10 days. If it is necessary to work the soil following application and incorporation, but before planting, do not work the soil deeper than the depth the herbicide was incorporated. To avoid removing VERNAM 7-E from the row (with loss of weed control in the row) do not move or shape soil after treatment, such as by planting in deep furrows.

RECOMMENDATIONS—VERNAM 7-E

These recommendations are given in terms of broadcast (overall) rates of VERNAM 7-E per acre. For band or row application, reduce broadcast rates proportionately, depending upon band width to be treated and row spacing.

Peanuts

VERNAM 7-E may be applied before planting, at planting or immediately after planting. When applying after planting, care must be taken not to disturb the seed. Use 2¹/₃ pints on coarse textured (sandy—light) soils; 3 pints on medium to fine textured (sandy loam to clay loam—medium to heavy) soils.

VERNAM 7-E may be applied through irrigation water after planting before weed growth starts or immediately following clean cultivation. Meter VERNAM 7-E into the irrigation water using a metering device that will introduce a constant flow into the water.

USE PRECAUTIONS FOR IRRIGATION SYSTEMS

A. Apply this product only through sprinkler including center pivot, lateral move, end tow, side (wheel) roll, traveler, soild set, or hand move, flood (basin); furrow; border irrigation systems. Do not apply this product through any type of irrigation system.
B. Crop injury, lack of effectiveness, or illegal pesticide residues in the crop can result from nonuniform distribution of treated water.
C. If you have questions about calibration, you should contact State Extension Service specialists, equipment manufacturers or other experts.
D. Do not connect an irrigation system (including greenhouse systems) used for pesticide application to a public water system unless the pesticide label-prescribed safety devices for public water systems are in place.
E. A person knowledgeable of the chemigation system and responsible for its operation, or under the supervision of the responsible person, shall shut the system down and make necessary adjustments should the need arise.

Furrow: Meter VERNAM 7-E into the water during the entire irrigation period. Do not apply within 45 days or harvest.

USE PRECAUTIONS FOR FLOOD (BASIN), FURROW AND BORDER IRRIGATION

1. System using a gravity flow pesticide dispensing system must meter the pesticide into the water at the head of the field and downstream of a hydraulic discontinuity such as a drop structure or weir box to decrease potential for water source contamination from backflow if water flow stops.
2. Systems utilizing a pressurized water and pesticide injection system must meet the following requirements:
   a. The system must contain a functional check valve, vacuum-relief valve, and low pressure drain appropriately located on the irrigation pipeline to prevent water source contamination from backflow.
   b. The pesticide injection pipeline must contain a functional, automatic, quick-closing check valve to prevent the flow of fluid back toward the injection pump.
   c. The pesticide injection pipeline must also contain a functional, normally closed, solenoid-operated valve located on the intake side of the injection pump and connected to the system interlock to prevent fluid from being withdrawn from the supply tank when the irrigation system is either automatically or manually shut down.
   d. The system must contain functional interlocking controls to automatically shut off the pesticide injection pump when the water pump motor stops.
   e. The irrigation line or water pump must include a functional pressure switch which will stop the water pump motor when the water pressure decreases to the point where pesticide distribution is adversely affected.
   f. Systems must use a metering pump, such as a positive displacement injection pump (e.g., diaphragm pump) effectively designed and constructed of materials that are compatible with pesticides and capable of being fitted with a system interlock.

NOTE: All rates are given on a broadcast basis, per acre. Reduce rates proportionately for hand or strip treatment.

Sprinkler: Meter VERNAM 7-E into the water during the entire period or into sufficient water to penetrate to a depth of 3 to 4 inches. Time this VERNAM application to insure that proper penetration of the herbicide corresponds with the end of the irrigation period. Flush the lines then turn the water off promptly. Tailwater (run-off water) from irrigation should be recirculated. Do not apply within 45 days of harvest.

USE PRECAUTIONS FOR SPRINKLER IRRIGATION

1. The system must contain a functional check valve, vacuum relief valve, and low pressure drain appropiately located on the irrigation pipeline to prevent water source contamination from backflow.
2. The pesticide pipeline must contain a functional, automatic, quick-closing check valve to prevent the flow of fluid back toward the injection pump.
3. The pesticide injection pipeline must also contain a functional, normally closed, solenoid-operated valve located on the intake side of the injection pump and connected to the system interlock to prevent fluid from being withdrawn from the supply tank when the irrigation system is either automatically or manually shut down.
4. The system must contain functional interlocking controls to automatically shut off the pesticide injection pump when the water pump motor stops.
5. The irrigation line or water pump must include a functional pressure switch which will stop the water pump motor when the water pressure decreases to the point where pesticide distribution is adversely affected.
6. Systems must use a metering pump, such as a positive displacement injection pump (e.g., diaphragm pump) effectively designed and constructed of materials that are compatible with pesticides and capable of being fitted with a system interlock.
7. Inject the product with a positive displacement pump into the main line ahead of a right angle turn to insure adequate mixing.
8. Application of more than recommended quantities of irrigation water may result in decreased product performance by removing the chemical from the zone of effectiveness.
9. Do not apply when wind speed favors drift beyond the area intended for treatment, when nozzles do not provide uniform distribution or when lines containing the product must be dismantled and drained.

SOYBEANS

The following recommendations are for soybeans grown in all areas except in the Southeastern States:

VERNAM 7-E may be applied before planting, at planting, or immediately after planting. When applying at or after planting, care must be taken not to disturb the crop seed or seedlings. Use 2¹/₃ pints on coarse textured (sandy—light) soils, 3 pints on medium to fine textured (sandy loam to clay loam—medium to heavy) soils, 3¹/₂ pints on very fine textured (clay—very heavy) soils.

For soybeans grown in the Southeastern States including North Carolina, South Carolina, Georgia, Alabama, Florida and Eastern Tennessee (Tenn. River East):

VERNAM 7-E may be applied before planting, at planting, or immediately after planting. For incorporated applications made before planting do not plant in deep furrows, or use shanks or other devices ahead of planter which bring up untreated soil to the surface. For furrow planted soybeans, apply VERNAM 7-E at or immediately after planting in bands over the row and incorporate with row wheels only. When applying at or after planting, care must be taken not to disturb the crop seed or seedlings. Use 2¹/₃ pints on coarse textured (sandy—light) soils; 3 pints on medium to fine textured (sandy loam to clay loam—medium to heavy) soils.

NOTE: In areas where beds are made prior to planting soybeans, VERNAM 7-E may be applied in a band immediately ahead of the bedding discs, set to cover the VERNAM 7-E with 3 to 4 inches of soil. If bed levelers are used before planting soybeans, make the beds higher to compensate for leveling, so 2 to 3 inches of soil remains over the treatment after the planting operation is done.

Sequential Applications

IMPORTANT: Always consult exact label of each herbicide used for specific recommendations and limitations.

For control of weeds not listed on the VERNAM 7-E label, and for increased control of labeled broadleaf weeds, the following herbicides are suggested as overlay treatments after preplant applications of VERNAM 7-E:

PEANUTS

DYANAP* 3 lbs./gal. liquid

Broadcast rates per acre:
Apply at cracking at the rate of 6 quarts per acre.

ENIDE* 50W + DINITRO (PREMERGE 3 lbs./gal.)

Broadcast rates per acre:
Apply at cracking or up to 7 days after at the rate of 4 lbs. ENIDE 50W and 2 quarts PREMERGE per acre.

LASSO* 4-EC (4 lbs., a.i./gal.) + DINITRO (PREMERGE 3 lbs./gal.)

Broadcast rates per acre:
Apply at cracking at the rate of 2 to 3 quarts of LASSO 4-EC and 2 quarts of PREMERGE per acre.

SOYBEANS

AMIBEN* 2 lbs./gal. liquid

Broadcast rates per acre:
Coarse textured (light) soils—1 Gallon
Medium textured (medium) soils—1¹/₈ Gallons
Fine textured (heavy) soils—1¹/₂ Gallons

BASAGRAN* 4 lbs./gal. liquid

Broadcast rates per acre: Apply ³/₄ to 1 quart per acre. Rates depend upon weeds to be controlled. Consult and follow the recommended rates on the BASAGRAN label.

DYANAP* 3 lbs./gal. liquid

Broadcast rates per acre. Use 6 quarts as a broadcast spray just after the soybeans are planted but before the soybean plants emerge from the soil.

LOROX* 50W

Broadcast rates per acre:

Use 1 to 2½ lbs. LOROX 50W per acre depending on soil type. Use on coarse to medium textured soils. Apply after planting before soybeans emerge. Plant soybeans at least 1¾" deep on flat seed bed only to avoid possible crop injury. Do not use on sand or loamy sand or soil with less than 1% organic matter since crop injury may result.

* AMIBEN chloramben herbicide is a registered trademark of Rhone Poulenc Inc.
* BASAGRAN is a registered trademark of BASF Aktiegesellschaft
* DYANAP is a registered trademark of Uniroyal Chemical, Inc.
* ENIDE is a registered trademark of Tuco, Division of the Upjohn Company
* LASSO is a registered trademark of Monsanto Company
* LOROX is a registered trademark of E.I. Du Pont de Nemours & Co., Inc.
* PREMERGE is a registered trademark of Dow Chemical Company

Tank Mix Directions For Use On Peanuts

VERNAM 7-E may be combined with the compounds listed in this section for tank mix application on peanuts. Recommended rates for each tank mix are listed below. Mixing, spray equipment, soil incorporation, and planting directions for VERNAM 7-E tank mixes on peanuts are listed at the end of the peanut tank mix section.

VERNAM 7-E/BALAN* 1.5 EC For Peanuts

A tank mix combination of Vernam 7-E and BALAN E.C. can be used prior to planting for weed control in peanuts.

* BALAN is a registered trademark of Elanco Products Company

NOTE: Follow all use precautions and warnings that appear on the BALAN 1.5 EC label and on the VERNAM 7-E label. Do not contaminate water by cleaning of equipment or disposal of wastes.

Weeds Controlled

Annual Grasses:	Annual Broadleaves
Annual bluegrass	Balloonvine*
Barnyardgrass (watergrass)	Carpetweed
Crabgrass	Chickweed
Crowfootgrass	Lambsquarters, common
Fall panicum	Florida purslane (pusley)
Field sandbur	Knotweed
Foxtails	Palmer amaranth (carelessweed)
Goosegrass	Pigweed
Junglerice	Purslane
Ryegrass	Redmaids
Johnsongrass seedling	
Texas panicum (buffalograss) (coloradograss)	

* in Alabama Only

Perennial grasses:
Nutsedge

APPLICATION DIRECTIONS

The following directions give broadcast rates of VERNAM 7-E and BALAN 1.5 EC for application prior to planting. Do not apply this combination after planting. BALAN 1.5 EC is not recommended for use on muck or other soil high in organic matter content.

Broadcast Rates Per Acre (Conventional spray and center pivot sprinklers):

Apply 2⅓ pint VERNAM 7-E plus 3 quarts BALAN 1.5-EC on coarse and medium textured (sandy and loam—light and medium) soils.

Apply 3 pints VERNAM 7.E plus 4 quarts BALAN 1.5 EC on fine textured (silt and clay—heavy) soils.

Center Pivot Sprinkler Application: (Limit use to the Southeastern U.S. only.)

Apply VERNAM 7-E/BALAN 1.5-EC with irrigation water before planting or immediately after planting before either peanuts or weeds emerge. It is important that application be done immediately after the last tillage operation since poor results will occur if the weeds have germinated before application of the herbicides. Meter the premixed VERNAM 7-E/BALAN 1.5-EC into the irrigation water during the entire period. Apply in ½ inch of water.

USE PRECAUTIONS:

A. Apply the product only through center pivot irrigation systems containing anti-siphon and check valves which will prevent water source contamination and overflow of the slurry tank and containing interlocking controls between the metering device and the water pump to insure simultaneous shut-off.

B. Inject the product with a positive displacement pump into the main line ahead of a right angle turn to insure adequate mixing.

C. Application of more than recommended quantities of irrigation water may result in decreased product performance by removing the chemical from the zone of effectiveness.

D. Do not apply when wind speed favors drift, when system connections or fittings leak, when nozzles do not provide uniform distribution or when lines containing the product must be dismantled and drained.

VERNAM 7.E/TREFLAN* 4-EC for Peanuts

Selective Herbicide—Tank Mix Directions

For TEXAS and OKLAHOMA ONLY

A tank mix combination of VERNAM 7-E and TREFLAN 4-EC can be used prior to planting for weed control in peanuts. The combination will control a broader spectrum of weeds than either material used alone. It is especially useful where nutsedge and Texas panicum are a problem.

* TREFLAN is a registered Trademark of Elanco Products Company.

Weeds Controlled

Annual Grasses:	Annual Broadleaves:
Annual bluegrass	Balloonvine*
Barnyardgrass (watergrass)	Carpetweed
Brachiaria	Carelessweed
Bromegrass	Chickweed
Cheat	Florida purslane (pusley)
Crabgrass	Goosefoot
Fall panicum	Knotweed
Foxtails (including giant foxtail	Kochia
Goosegrass	Lambsquarters
Junglerice	Pigweeds (spiny, redroot)
Johnsongrass, seedling	Puncturevine
Sandbur	Purslane
Sprangletop	Russian thistle
Stinkgrass	Stinging nettle
Texas panicum (buffalograss) (coloradograss)	

Perennial Grasses:
Nutsedge

* In Alabama Only

APPLICATIONS DIRECTIONS

The following directions give broadcast rates for VERNAM 7-E and TREFLAN 4-EC as a spray tank mix for application prior to planting. Read and follow all directions, use limitations, precautions and warnings that appear on the TREFLAN 4-EC and VERNAM 7-E labels.

Broadcast rates per Acre:

Apply and incorporate before planting 2⅓ pints of VERNAM 7-E and 1 pint TREFLAN 4-EC per acre. For band applications, reduce dosage proportionately, depending on row spacing and band width to be treated.

Use the following directions for the VERNAM 7-E tank mixes on peanuts.

Mixing:

Add the recommended rates of the herbicides to the spray tank during filling and mix thoroughly. Begin adding the herbicides when spray tank is half full. Apply in 10 to 50 gallons of water per acre.

Spray Equipment:

Use a properly calibrated, low pressure, boom-type sprayer having good agitation. Check frequently during application to be sure nozzles are free from clogging and are delivering a uniform spray pattern.

Soil Incorporation:

The tank mix combinations must be mixed (incorporated) into the soil immediately after application. Whenever possible, application and incorporation should be done in the same operation. This can be done by mounting the tanks and boom right on the incorporation rig.

For thorough incorporation, use the following equipment: TANDEM DISC: Set to cut 3 to 6 inches operated at 4-6 mph, followed by a spike-tooth harrow or some other leveling device, which extends beyond the ends of the disc. For thorough mixing, disc in two different directions (cross-disc).

OR: POWER DRIVEN cultivation equipment (tillers, cultivators, hoes) set to cut 2 to 3 inches deep.

Shallow incorporation of less than 2 inches may result in erratic weed control.

Planting:
Peanuts should be planted within 48 hours after the combination has been applied to obtain the maximum period of weed control during the growing season.

TANK MIX DIRECTIONS FOR USE ON SOYBEANS

VERNAM 7-E may be combined with the compounds listed in this section for tank mix applications on soybeans. Recommended rates for each tank mix are listed in the sections below. The mixing, spray equipment, soil incorporation, and planting directions for the VERNAM 7-E tank mixes on soybeans are listed at the end of the soybean tank mix section.

VERNAM 7-E/TREFLAN 4-EC TANK MIX DIRECTIONS FOR SOYBEANS

A tank mix combination of VERNAM 7-E (7 pounds active ingredient per gallon) and TREFLAN 4-EC (4 pounds of active ingredient per gallon) can be used prior to planting for weed control in soybeans.

The combination will control a broader spectrum of weeds than either material used alone. This includes tough weeds such as nutsedge, velvetleaf, foxtails, johnsonsgrass, redroot pigweed and wild cane.

Dependable performance—because the herbicides are mixed into the soil, rainfall is not needed to move them into the weed germinating zone. They are already there to kill grass and broadleaf weeds when they sprout.

WEEDS CONTROLLED:

Annual Grasses:	Annual Broadleaves
Annual bluegrass	Annual morningglory
Barnyardgrass (watergrass)	Balloonvine**
Brachiaria	Carpetweed
Bromegrass	Chickweed
Cheat	Common lambsquarters
Crabgrass	Florida pusley
Fall panicum	Goosefoot
Field sandbur	Knotweed
Foxtails	Kochia
Goosegrass	Palmer amaranth (carelessweed)
Junglerice	Pigweed
Johnsongrass, seedling	Purslane
Sprangletop	Russian thistle
Stinkgrass	Sicklepod (coffeeweed)
Texas panicum (buffalograss) (coloradograss)	Stinging nettle
Wild cane	Velvetleaf (buttonweed)

Perennial Grasses:
Purple nutsedge
Yellow nutsedge
Rhizome johnsongrass*

* Suppression
** In Alabama Only

APPLICATION DIRECTIONS

Read and follow all directions, use limitations, precautions and warnings that appear on the VERNAM 7-E and TREFLAN EC labels. Do not contaminate water by cleaning of equipment or disposal of wastes.

Application Rates

Soil Texture	Broadcast Rates/Acre VERNAM 7-E*	Broadcast Rates/Acre TREFLAN 4-EC*
Coarse (light—sandy soils)	2¹/₃ pints	1 pint
Medium (silt loam & sandy clay soils)	2¹/₃ to 3** pints	1 to 1¹/₂ pints
Fine (heavy—clay soils)	3 to 3¹/₂ pints	1 to 2 pints

* For rhizome johnsongrass suppression use the high rates of TREFLAN 4-EC and VERNAM 7-E.
** For nutsedge, wild cane and velvetleaf control, use the high rate of 3 pints VERNAM 7-E on medium textured soils.

For Rhizome Johnsongrass—Use mold board plow, chisel plow or similar implement to bring rhizomes to the top of the soil. Then follow with a disc set to cut 4-6 inches deep and operate in two different directions at 4-6 mph before application to cut the rhizomes into small (2-3 inch) pieces. This should also destroy any emerged johnsongrass.

Cultivation is suggested in addition to VERNAM 7-E/TREFLAN 4-EC treatment in fields with moderate to heavy infestations of nutsedge, wild cane, velvetleaf, and rhizomes johnsongrass.

VERNAM 7-E/TREFLAN 4-E/METRIBUZIN TANK MIX DIRECTIONS FOR USE ON SOYBEANS

A tank mix of VERNAM 7-E (7 pounds active ingredient per gallon) and TREFLAN 4-EC (4 pounds active ingredient per gallon) and METRIBUZIN 4-L (4 pounds active ingredient per gallon) or METRIBUZIN 50-WP (50% active ingredient by weight) can be used prior to planting for weed control in soybeans.

The combination will control a broader spectrum of weeds than the VERNAM 7-E/TREFLAN 4-EC tank mix.

WEEDS CONTROLLED:

Annual Grasses:	Annual Broadleaves
Annual bluegrass	Balloonvine**
Barnyardgrass (watergrass)	Carpetweed
	Chickweed
Brachiaria	Cocklebur
Bromegrass	Common lambsquarters
Cheat	Common ragweed
Crabgrass	Florida pusley
Fall panicum	Goosefoot
Field sandbur	Knotweed
Foxtails	Kochia
Goosegrass	Palmer amaranth (carelessweed)
Junglerice	
Johnsongrass, seedling	Pennsylvania, smartweed
Sprangletop	Pigweed
Stinkgrass	Prickly sida
Texas panicum (buffalograss) (coloradograss)	Purslane
	Russian thistle
	Sicklepod (coffeeweed)
Wild cane	Stinging nettle
	Velvetleaf (buttonweed)
	Venice mallow

** In Alabama Only

Perennial Grass:
Purple nutsedge
Yellow nutsedge
Rhizome johnsongrass*

* Suppression

APPLICATION DIRECTIONS

Read and follow all directions, use limitations, precautions and warnings that appear on the VERNAM 7-E, TREFLAN 4-EC and appropriate METRIBUZIN (4-L or 50-WP) labels. Do not contaminate water by cleaning of equipment or disposal of wastes.

Application Rates

Soil Texture	Broadcast Rates/Acre VERNAM 7-E*	Broadcast Rates/Acre TREFLAN 4-EC	Broadcast Rates/Acre METRIBUZIN 4L or 50WP
Medium (silt loam & sandy clay soils)	2¹/₃ to 3** pints	1 pint	¹/₂ pint or ¹/₂ pound
Fine (heavy-clay soils)	3 to 3¹/₂ pints	1 pint	¹/₂-³/₄ pint or ¹/₂-³/₄ pound

* For Rhizome johnsongrass suppression use the high rate of VERNAM 7-E.
** For nutsedge, wild cane, and velvetleaf control, use the high rate of 3 pints of VERNAM on medium-textured soils.

Do not use the VERNAM/TREFLAN/METRIBUZIN tank mix on coarse (light sandy) soils.

For Rhizome johnsongrass—use mold board plow, chisel plow or similar implement to bring rhizomes to the top of the soil. Then follow with a disc set to cut 4-6 inches deep and operate to two different directions at 4-6 mph before application to cut the rhizomes into small (2-3 inch) pieces. This should also destroy any emerged johnsongrass.

Cultivation is suggested in addition to VERNAM 7-E/TREFLAN 4-EC/METRIBUZIN (4-L or 50-WP) treatment in fields with moderate to heavy infestations of nutsedge, wild cane velvetleaf, and johnsongrass rhizomes.

VERNAM 7-E/AMIBEN* TANK MIX DIRECTIONS FOR SOYBEANS

A tank mix combination of VERNAM 7-E (7 pounds active ingredient per gallon) and AMIBEN (2 pounds active ingredient per gallon) can be used prior to planting, for weed control in soybeans.

The combination will control a broader spectrum of weeds than either material used alone.

* AMIBEN chloramben herbicide is a registered Trademark of Rhone Poulenc Inc.

WEEDS CONTROLLED:

Annual Grasses:	Annual Broadleaves:
Barnyardgrass (watergrass)	Carpetweed
	Chickweed
Crabgrass	Coffeeweed

Foxtail (giant, green, yellow)	Florida pusley
German millet	Kochia
Goosegrass	Lambsquarters, common
Johnsongrass, seedling	Morningglory, annual
Stinkgrass	Mustard, wild
	Nightshade, black
	Pigweed (redroot, smooth)
	Prickly sida (teaweed)
Pernnial Grasses:	Purslane
Purple nutsedge	Ragweed, common
	Sicklepod
Yellow nutsedge	Smartweed, Pennsylvania
	Spurge, annual
	Velvetleaf (buttonweed)

APPLICATION DIRECTIONS

Read and follow all directions, use limitations, precautions and warnings that appear on the VERNAM 7-E and AMIBEN labels. Do not contaminate water by cleaning of equipment or disposal of wastes.

Application Rates

	Broadcast Rates/Acre	
Soil Texture	VERNAM 7-E	AMIBEN
Medium (silt loam & sandy clay soils)	2⅓ pints	3 quarts

For mixing, spray equipment, soil incorporation, and planting directions, refer to the directions at the end of the VERNAM 7-E tank mix on soybeans section of this label.

VERNAM 7-E/BASALIN* 4-E TANK MIX DIRECTIONS FOR USE ON SOYBEANS

A tank mix combination of VERNAM 7-E (7 pounds active ingredient per gallon) and BASALIN (4 pounds active ingredient per gallon) can be used prior to planting for weed control in soybeans.

The combination will provide better weed control than either herbicide used alone.

* BASALIN is a registered Trademark of BASF Wyandotte Corp.

WEEDS CONTROLLED:

Annual Grasses:	**Annual Broadleaves:**
Barnyardgrass (watergrass)	Carpetweed
Crabgrass	Florida pusley
Foxtail (giant, green, yellow)	Lambsquarters, common
German millet	Morningglory, annual
Goosegrass	Pigweed (redroot, smooth)
Johnsongrass, seedling	Purslane
	Sicklepod
	Velvetleaf (buttonweed)

Perennial Grasses:
Purple nutsedge
Yellow nutsedge

APPLICATION DIRECTIONS

Read and follow all directions, use limitations, precautions and warnings that appear on the VERNAM 7-E and BASALIN 4-E labels. Do not graze treated plant parts to livestock. Do not contaminate water by cleaning of equipment or disposal of wastes.

Application Rates

	Broadcast Rates/Acre	
Soil Texture	VERNAM 7-E	BASALIN 4-E
Coarse (light-sandy soils)	1¾ pints	¾ pint
Medium (silt loam and sandy clay soils)	2⅓ pints	1 pint

For mixing, spray equipment, soil incorporation, and planting directions, refer to the directions at the end of the VERNAM 7-E tank mix on soybeans section of this label.

VERNAM 7-E/LASSO* 4-EC TANK MIX DIRECTIONS FOR USE ON SOYBEANS

A tank mix combination of VERNAM 7-E (7 pounds active ingredient per gallon) and LASSO 4-EC (4 pounds active ingredient per gallon) can be used prior to planting for weed control in soybeans grown in North Central States of Michigan, Indiana, and Illinois. Do not use on sand or loamy sandy soils as injury may occur.

The combination will provide better weed control than either herbicide used alone.

* LASSO is a registered trademark of Monsanto Company.

WEEDS CONTROLLED:

Annual Grasses:	**Annual Broadleaves:**
Barnyardgrass (watergrass)	Carpetweed
Crabgrass	Florida pusley
Fall panicum	Lambsquarters, common
Foxtail (giant, green, yellow)	Morningglory, annual
Goosegrass	Pennsylvania, smartweed
Johnsongrass, seedling	Pigweed (redroot, smooth)
	Purslane
	Sicklepod
	Velvetleaf (buttonweed)

Perennial Grasses:
Purple nutsedge
Yellow nutsedge

APPLICATION DIRECTIONS

Read and follow all directions, use limitations, precautions and warnings that appear on the VERNAM 7-E and LASSO 4-EC labels, Do not contaminate water by cleaning of equipment or disposal of wastes.

Application Rates

	Broadcast Rates/Acre	
Soil Texture	VERNAM 7-E	LASSO 4-EC
Coarse (light-sandy soils)	1¾ pints	2 pints
Medium (soil loam and sandy clay soils)	2⅓ pints**	3 pints
Fine (heavy-clay soils)	3 pints	4 pints

** For nutsedge and velvetleaf control, use the rate of 3 pints of VERNAM 7-E on medium and fine textured (medium and heavy-clay) soils.

For mixing, spray equipment, soil incorporation, and planting directions, refer to the directions at the end of the VERNAM 7-E tank mix on soybean section of this label.

VERNAM 7-E/PROWL* 4-E TANK MIX DIRECTIONS FOR USE ON SOYBEANS

A tank mix combination of VERNAM 7-E (7 pounds active ingredient per gallon) and PROWL 4-E (4 pounds active ingredient per gallon) can be used prior to planting for weed control in soybeans.

The combination will provide better weed control than either herbicide used alone.

* PROWL is a registered trademark of American Cyanamid Company

WEEDS CONTROLLED:

Annual Grasses:	**Annual Broadleaves:**
Barnyardgrass (watergrass)	Carpetweed
Crabgrass	Florida pusley
Foxtail millet	Lambsquarters, common
Foxtail (giant, green, yellow)	Morningglory, annual
Goosegrass	Pigweed (redroot, smooth)
Johnsongrass, seedling	Purslane
	Sicklepod
	Velvetleaf (buttonweed)

Perennial Grasses:
Purple nutsedge
Yellow nutsedge

APPLICATION DIRECTIONS

Read and follow all directions, use limitations, precautions and warnings that appear on the VERNAM 7-E and PROWL 4-E labels. Do not contaminate water by cleaning of equipment or disposal of wastes.

Application Rates

	Broadcast Rates/Acre	
Soil Texture	VERNAM 7-E	PROWL 4-E
Coarse (light-sandy soils)	1¾ pints	¾ pints
Medium (silt loam and sandy clay soils)	2⅓ pints**	1 pint
Fine (heavy—clay soils)	3 pints	1½ pints

** For nutsedge and velvetleaf control, use the rate of 3 pints of VERNAM 7-E on medium textured (medium and heavy—clay) soils.

For mixing, spray equipment, soil incorporation, and planting directions, refer to the directions at the end of the VERNAM 7-E tank mix on soybean section of this label.

VERNAM 7-E/TOLBAN* 4-E TANK MIX DIRECTIONS FOR USE ON SOYBEANS

A tank mix combination of VERNAM 7-E (7 pounds active ingredient per gallon) and TOLBAN 4-E (4 pounds active ingredient per gallon) can be used prior to planting for weed control in soybeans.

The combination will provide better weed control than either herbicide use alone.

* TOLBAN is a registered trademark of Ciba-Geigy Corp.

WEEDS CONTROLLED:

Annual Grasses:
Barnyardgrass (watergrass)
Crabgrass
Foxtail (giant, green, yellow)
German millet
Goosegrass
Johnsongrass, seedling

Annual Broadleaves:
Carpetweed
Florida pusley
Lambsquarters, common
Morningglory, annual
Pigweed (redroot, smooth)
Purslane
Sicklepod
Velvetleaf (buttonweed)

Perennial Grasses:
Purple nutsedge
Yellow nutsedge

APPLICATION DIRECTIONS

Read and follow all directions, use limitations, precautions and warnings that appear on the VERNAM 7-E and TOLBAN 4-E labels. Do not contaminate water by cleaning of equipment or disposal of wastes.

Application Rates

	Broadcast Rates/Acre	
Soil Texture	**VERNAM 7-E**	**TOLBAN 4-E**
Coarse (light-sandy soils)	1¾ pints	¾ pints
Medium (silt loam and sandy clay soils)	2⅓ pints**	1 pint
Fine (heavy—clay soils)	3 pints	1½ pints

** For nutsedge and velvetleaf control, use the rate of 3 pints of VERNAM 7-E on medium textured soils.

For mixing, spray equipment, soil incorporation, and planting directions, refer to the directions at the end of the VERNAM 7-E tank mix on soybean section of this label.

USE THE FOLLOWING DIRECTIONS FOR THE VERNAM 7-E TANK MIXES ON SOYBEANS.

Mixing:
Add the recommended rates of both VERNAM 7-E and TREFLAN 4-EC to the spray tank during filling and mix thoroughly. Begin adding the herbicides when spray tank is half full. Apply in 10-50 gallons of water per acre.

Spray Equipment:
Use a properly calibrated, low pressure, boom-type sprayer having good agitation. Check frequently during application to be sure nozzles are free from clogging and delivering a uniform spray pattern.

DRY FERTILIZER STATEMENT:
VERNAM 7-E/TREFLAN 4-EC tank mix may be impregnated on many dry bulk fertilizers for use on soybeans. However, uniform distribution of VERNAM 7-E/TREFLAN 4-EC on fertilizer particles and uniform application are necessary to assure good results. See Appendix II giving directions for impregnation and use.

Soil Incorporation:
The tank mix combinations with VERNAM 7-E must be mixed (incorporated) into the soil immediately after application. Whenever possible, application and incorporation should be done in the same operation. This can be done by mounting the tanks and boom right on the incorporation rig.

For thorough incorporation, use the following equipment:

Tandem Disc—Set to cut to a depth of 4 inches, followed by a spike-tooth harrow or some other leveling device which extends beyond the ends of the disc. For thorough mixing, disc in two different directions (cross disc).

OR: Power-driven cultivation equipment set to cut to a depth of 2-3 inches.

Planting:
Do not plant seed deeper than 2 inches. Seeding can be done immediately, or it may be delayed up to 10 days. If it is necessary to work the soil following application and incorporation, but before planting, do not work the soil deeper than the depth the herbicide was incorporated i.e., no more than ½ the depth the herbicides were incorporated. To avoid removing the herbicide from the row (with loss of weed control in the row) do not move or shape soil after treatment, such as by planting in deep furrows.

APPENDIX I

PROCEDURE FOR TESTING THE COMPATIBILITY OF VERNAM 7-E AND COMBINATIONS WITH FLUID FERTILIZERS

The following procedure is suggested for determining whether VERNAM 7-E may be combined with a specific fluid fertilizer for spray tank application.

MATERIALS REQUIRED:

1. VERNAM 7-E and/or combination.
2. Fluid fertilizer to be used.
3. Adjuvant for fertilizer tank mix: Compex*, Sponto 168-D*, Unite* or equivalent. The adjuvant which provides the best compatibility depends on the specific fertilizer under consideration.

* Compex is a trademark of Kalo Laboratories, Inc. Kansas City, Missouri; Sponto 168-D is a trademark of Witco Chemical Company, Houston, Texas, Unite is a trademark of Hopkins Agricultural Chemical Company, Madison, Wisconsin.

4. Two one-quart, wide-mouth glass jars with lid.
5. Measuring spoons; a 25 mL pipette or graduated cylinder provides more accurate measurement: 1 tablespoon equals 3 teaspoons, 1 teaspoon equals approximately 5 milliliters.
6. Measuring cup, 8 oz. (237 mL).

PROCEDURE:

1. Pour a pint (about 473 mL) of the fluid fertilizer into each of the quart jars.
2. Add adjuvant to one of the jars and mix (see rate table).
3. Add the VERNAM 7-E to the desired combinations of herbicide to both jars (see rate table for VERNAM 7-E).
4. Close both jars with lid and mix the contents by turning the jars upside down ten times.
5. Inspect the surface and body of the mixtures—
   (A) Immediately after completing the jar inversions.
   (B) After allowing the jars to stand quietly for 30 minutes.
   (C) And then again after turning the jars upside down 10 times.

If a uniform mix cannot be made, the mixture should not be used. If either mixture remains uniform for 30 minutes, the combination may be used. Should either mixture separate after 30 minutes, but readily remix uniformly with the jar inversions, the mixture can be used if adequate agitation is maintained in the tank. If the mixture with adjuvant is satisfactory, but the one without adjuvant is not, be sure to use the adjuvant in the spray tank. Add the adjuvant first at a rate of 3 pints per 100 gallons of fluid fertilizer; foaming can be minimized by using moderate agitation.

If nondispersible oil, sludge or clumps of solids form in the mixtures, the combinations should not be used.

Rate Table For VERNAM 7-E and ADJUVANT**
With The Fluid Fertilizer

Gallons of Fluid Fertilizer To Be Applied Per Acre	mL or Tsp. of VERNAM 7-E* To Be Added To 1 Pint of Fertilizer VERNAM 7-E	
	mL	**Tsp.**
10	7	1⅓
15	5	1
20	4	¾
25	3	⅔
30	2	½
40	2	½

* Based on field rate of 1 pound active ingredient per acre in the fertilizer volumes indicated. Increase volume proportionately to correspond with intended field rate in terms of pounds active ingredient per acre (e.g. for field rate of 4 pounds actual VERNAM 7-E in 40 gallons fertilizer per acre, add 8 mL, or 2 tsp. VERNAM 7-E to each jar for compatibility testing purposes).

** Two (2) milliliters or one-half (½) teaspoon of adjuvant should be added to 1 pint of fluid fertilizer in order to equal the rate of 3 pints of adjuvant per 100 gallons of fluid fertilizer.

APPENDIX II

VERNAM 7-E and VERNAM/TREFLAN 4-EC IMPREGNATION ON DRY BULK FERTILIZERS

CAUTION: VERNAM 7-E alone or in combination with other herbicides must not be impregnated on ammonium nitrate, sodium nitrate or potassium nitrate. Such mixtures may cause explosion and fire.

VERNAM 7-E alone may be impregnated on many* dry bulk fertilizers for use on *registered crops*. VERNAM 7-E in combination with TREFLAN may be impregnated on many* dry bulk fertilizers and applied and incorporated in the soil before planting for the control of *labeled grasses* and broadleaf weeds *in soybeans*.

Field results have shown that VERNAM 7-E and VERNAM/TREFLAN 4-EC on bulk dry fertilizers give weed control equal to VERNAM 7-E or VERNAM 7-E/TREFLAN 4-EC applied as a spray in water or liquid fertilizer. However, uniform impregnation of herbicide solution on the dry fertilizer particles and uniform application in the field are necessary to assure good results.

For impregnating pesticides on dry fertilizers, use a closed rotary-drum mixer or a similar type of closed blender equipped with suitable spray equipment. The spray nozzle (or nozzles) should be positioned inside of the mixer to provide *uniform* spray coverage of the tumbling fertilizer. For the VERNAM 7-E/TREFLAN 4-EC combination, premix in the proper use ratio to prepare a uniform solution for joint spray impregnation. The physical properties of fertilizers vary, particularly in liquid absorptive capacity. When absorptivity is sufficient, simple spray impregnation of the fertilizer with VERNAM 7-E or VERNAM 7-E/TREFLAN 4-EC provides a satisfactory, dry mixture. If the absorptive capacity is inadequate, use of a highly absorptive powder is required to provide a dry, flowable mixture. Microcel E (Johns-Manville Products Corp.) is the recommended absorbent powder. It should be added separately and uniformly to the prepared VERNAM 7-E or VERNAM 7-E/TREFLAN 4-EC fertilizer mixture, in a quantity that is sufficient to provide a suitably flowable mixture. Generally less than 2% by weight of Microcel E is required. The amount of product actually required in the manufacture of individual fertilizer mixture should be determined carefully for each production operation. This is necessary to ensure that the amount of VERNAM 7-E/TREFLAN 4-EC actually contained in the mixture applied to the soil represents the correct rate of use.

All VERNAM 7-E and TREFLAN 4-EC label recommendations regarding rates per acre, soil incorporation, application, cautions, general use precautions and other directions must be followed.

All individual state regulations relating to bulk dry fertilizer blending, registration, labeling and application are the responsibility of the individual and/or company selling the fertilizer and VERNAM 7-E and/or VERNAM/TREFLAN 4-EC mixture.

A minimum of 200 pounds and a maximum of 450 pounds of approved ingredients* impregnated with VERNAM 7-E/TREFLAN 4-EC at the recommended rate must be applied per acre.

Bulk fertilizer impregnated with VERNAM 7-E should be applied immediately, *not stored*. It is recommended that all bulk containers be tightly covered while the product is being transported and applied to reduce chances of VERNAM 7-E loss via volatilization.

*** Approved Dry Fertilizer Ingredients**

	N	P	K
Ammonium Sulfate	21	0	0
Diammonium Phosphate	18	46	0
Potassium Chloride	0	0	60
(Single) Superphosphate	0	20	0
Triple Superphosphate	0	46	0
Urea	45	0	0
Ammonium Phosphate-Sulfate	16	20	0
Potassium Sulfate	0	0	52

NOTE: Sul-PO-MAG, K-MAG, and 11-48-0 have been shown to be compatible with VERNAM 7-E plus TREFLAN 4-EC and are approved for use.

Rate Chart For The Impregnation of Dry Bulk Fertilizers With VERNAM 7-E and VERNAM/TREFLAN 4-EC Added To A Ton of Fertilizer*

Fertilizer Rate		VERNAM 7-E Rate/Acre			TREFLAN 4-EC Rate/Acre	
Per Acre	1 3/4 pts.	2 1/3 pts.	3 pts.	3 1/2 pts.	1 pt.	2 pts.
200 lbs.	8 3/4 qts./ton	11 2/3 qts./ton	15 qts./ton	17 1/2 qts./ton	5 qts./ton	10 qts./ton
250 lbs.	7 qts./ton	9 1/3 qts./ton	12 qts./ton	14 qts./ton	4 qts./ton	8 qts./ton
300 lbs.	5 4/5 qts./ton	7 3/4 qts./ton	10 qts./ton	11 2/3 qts./ton	3 1/3 qts./ton	7 qts./ton
350 lbs.	5 qts./ton	6 2/3 qts./ton	8 3/5 qts./ton	10 qts./ton	2 4/5 qts./ton	6 qts./ton
400 lbs.	4 1/3 qts./ton	5 4/5 qts./ton	7 1/2 qts./ton	8 3/4 qts./ton	2 1/2 qts./ton	5 qts./ton
450 lbs.	3 4/5 qts./ton	5 1/5 qts./ton	6 2/3 qts./ton	7 4/5 qts./ton	2 1/4 qts./ton	4 1/2 qts./ton

* Refer to application rates under "APPLICATION DIRECTIONS".

VERNAM 7-E Typical Physical Data

Pounds/Gallon, 20/20°C	7.88 Approximate (varies with temperature)
Flashpoint:	186°F (tagliabue closed cup)
Viscosity:	Sprayable down to minus 20°F

STORAGE AND DISPOSAL

PROHIBITIONS: Do not contaminate water, food, feed or fertilizer by storage or disposal.

STORAGE: Keep container tightly closed when not in use. Can be stored at temperatures as low as minus 50°F.

PESTICIDE DISPOSAL: Rinse spray equipment. Wastes resulting from the use of this product may be disposed of on site or at an approved waste disposal facility.

CONTAINER DISPOSAL:

METAL: Triple rinse (or equivalent), then offer for recycling or reconditioning, or dispose of in a sanitary landfill, or by procedures approved by state and local authorities.

PLASTIC: Triple rinse (or equivalent), then offer for recycling or reconditioning or puncture and dispose of in a sanitary landfill, or by incineration, or, if allowed by State and local authorities, by burning. If burned, stay out of smoke.

NOTICE TO BUYER AND USER:

Seller warrants that this product conforms to the chemical description on the label and is reasonably fit for the purposes stated on the label when used in accordance with directions under normal conditions of use. This warranty does not extend to the use of this product contrary to label instructions, or under abnormal use conditions, or under conditions not reasonably foreseeable to Seller, and Buyer and User assume the risk of any such use. SELLER DISCLAIMS ALL OTHER WARRANTIES EXPRESSED OR IMPLIED INCLUDING ANY WARRANTY OF FITNESS OR MERCHANTABILITY. SELLER SHALL NOT BE LIABLE FOR CONSEQUENTIAL, SPECIAL OR INDIRECT DAMAGES RESULTING FROM THE USE OR HANDLING OF THIS PRODUCT AND SELLER'S SOLE LIABILITY AND BUYER'S AND USER'S EXCLUSIVE REMEDY SHALL BE LIMITED TO THE REFUND OF THE PURCHASE PRICE.

This product is sold only for uses stated on its label. No express or implied license is granted to use or sell this product under any patent in any Country except as specified: Country: United States of America.

A-14MR

The information and specimen labels in this booklet are correct as of the publication date but are subject to change without notice. For current information, contact ICI Americas Inc.

WEATHERBLOK®

RODENTICIDE

Made with TALON® Rodenticide

Kills Warfarin-Resistant Norway Rats and House Mice. Rodents may Consume a Lethal Dose in One Feeding with First Dead Rodents Appearing Four or Five Days after Treatment Begins.

ACTIVE INGREDIENT:
Brodifacoum
3-[3-(4′-bromo-[1,1′-biphenyl]-4-yl)-1,2,3,4-tetrahydro-1-naphthalenyl]-4-hydroxy-2H-1- benzopyran-2-one 0.005%
INERT INGREDIENTS: 99.995%
TOTAL 100.000%

EPA Reg. No. 10182-48

This product is protected by US patent numbers 3957824 and 4035505 which are the property of Ward Blenkinsop and Company Limited.

KEEP OUT OF REACH OF CHILDREN
CAUTION

PRECAUTIONARY STATEMENTS

HAZARDS TO HUMANS AND DOMESTIC ANIMALS

CAUTION

MAY BE HARMFUL OR FATAL IF SWALLOWED. KEEP AWAY FROM HUMANS, DOMESTIC ANIMALS, AND PETS. WASH HANDS AFTER HANDLING BAIT.

IF BAIT IS EATEN BY HUMANS, CALL A PHYSICIAN AT ONCE. FOR 24-HOUR EMERGENCY MEDICAL ASSISTANCE CALL 1-800-F-A-S-T-M-E-D (327-8633).

IF BAIT IS EATEN BY ANIMALS OR PETS, call **ICI Americas Inc. Medical Emergency Information Center** 1-800-327-8633 for 24-hour emergency assistance.

FOR CHEMICAL EMERGENCY: Spill, leak, fire, exposure, or accident call CHEMTREC 1-800-424-9300.

NOTE TO PHYSICIAN OR VETERINARIAN

This product may reduce the clotting ability of the blood and cause hemorrhaging. If poisoning occurs, intramuscular and oral administration of vitamin K_1 are indicated, as in poisoning from overdose of bishydroxy coumarin. For human cases, vitamin K_1 is antidotal at doses of 10-20 mg (*not* mg/kg). For animal cases, vitamin K_1 is antidotal at 2-5 mg/kg. Repeated doses may need to be given up to two weeks (based on monitoring of prothrombin times). In severe cases, blood transfusions may be necessary.

ENVIRONMENTAL HAZARDS

This product is toxic to fish, birds and wildlife. This product can pose a secondary hazard to birds of prey and mammals. Keep out of any body of water.

DIRECTIONS FOR USE

It is a violation of Federal law to use this product in a manner inconsistent with its labeling.

Use Restrictions: For control of Norway rats, Roof rats and House mice in and around sewers, homes, industrial, commercial, agricultural and public buildings. WEATHER BLOK may also be used in and around transport vehicles (ships, trains, aircraft) and related port or terminal buildings. WEATHER BLOK must be placed in tamper-proof bait boxes or in locations not accessible to children, pets, domestic animals or wildlife such as in burrows which are then closed. Do not place bait in areas where there is a possibility of contaminating food or surfaces that come in direct contact with food.

APPLICATION DIRECTIONS

Norway and Roof Rats: Apply 8-23 WEATHER BLOKS (usually at intervals of 15-30 ft.) per placement. Maintain an uninterrupted supply of fresh bait for 10 days or until signs of rat activity cease. When infestations are limited to overhead areas in or around buildings, tie or fasten WEATHER BLOK securely in protected locations on fences, poles, beams or other overhead areas where signs indicate rodents are active.

For use in sewers, securely attach one end of a wire to WEATHER BLOK and the other end to a stationary structure such as the bottom step of a manhole ladder or to a sewer grate, allowing just enough wire for the block to rest on manhole benching. If benching is not present, suspend block a few inches above the high water line or place WEATHER BLOK on a board supported by opposing steps of the ladder. Securing WEATHER BLOK in this manner will prevent removal by rats or water. Place at least 15 WEATHER BLOKS per manhole. Maintain an uninterrupted supply of fresh bait for at least 10 days or until feeding stops.

House Mice: Apply 1 WEATHER BLOK at intervals of 8-12 feet per placement. Larger placements (2 Bloks) may be needed at points of very high mouse activity. Maintain an uninterrupted supply of fresh bait for 15 days or until signs of mouse activity cease. For use in sewers apply according to above directions for rats.

STORAGE AND DISPOSAL

Do not contaminate water, food or feed by storage or disposal. Do not reuse empty container except for holding additional WEATHER BLOK bait. **Storage:** Store in original container only in a dry place inaccessible to children and pets. **Pesticide Disposal:** Wastes resulting from the use of this product may be disposed of on site or at an approved waste disposal facility. **Container Disposal:** Dispose of bait container in a sanitary landfill or by incineration if allowed by state and local authorities.

NOTICE TO BUYER AND USER

Seller warrants that this product conforms to the chemical description on the label and is reasonably fit for the purposes stated on the label when used in accordance with directions under normal conditions of use. This warranty does not extend to the use of this product contrary to label instructions, or under abnormal use conditions, or under conditions not reasonably foreseeable to Seller, and Buyer and User assume the risk of any such use. SELLER DISCLAIMS ALL OTHER WARRANTIES, EXPRESSED OR IMPLIED, INCLUDING ANY WARRANTY OF FITNESS OR MERCHANTABILITY. SELLER SHALL NOT BE LIABLE FOR CONSEQUENTIAL, SPECIAL OR INDIRECT DAMAGES RESULTING FROM THE USE OR HANDLING OF THIS PRODUCT AND SELLER'S SOLE LIABILITY AND BUYER'S AND USER'S EXCLUSIVE REMEDY SHALL BE LIMITED TO THE REFUND OF THE PURCHASE PRICE.

05-0790-307

The information and specimen labels in this booklet are correct as of the publication date but are subject to change without notice. For current information, contact ICI Americas Inc.

For use in sewers, securely attach one end of a wire to WEATHER BLOCK and the other end to a stationary structure such as the bottom step of a manhole ladder, or to a sewer grate allowing just enough wire for the block to rest on manhole's landing. If landing is not present, suspend block a few inches above the high water line or place WEATHER BLOCK on a hook or support by opposite steps of the ladder. Securing WEATHER BLOCK in this manner will prevent removal by rats or water. Place at least 15 WEATHER BLOCKS per manhole. Maintain an uninterrupted supply of fresh bait for at least 15 days or until feeding stops.

House Mice: Apply 1 WEATHER BLOCK at intervals of 8-12 feet per placement. Larger placements (2 blocks) may be needed at points of very high mouse activity. Maintain an uninterrupted supply of fresh bait for 15 days or until signs of mouse activity cease. For use in sewers apply according to above directions for rats.

STORAGE AND DISPOSAL

Do not contaminate water, food or feed by storage or disposal. Do not reuse empty container except for holding additional WEATHER BLOCK bait. Storage: Store in original container only in a dry place inaccessible to children and pets. Pesticide Disposal: Wastes resulting from the use of this product may be disposed of on site or at an approved waste disposal facility. Container Disposal: Dispose of bait container in a sanitary landfill or by incineration if allowed by state and local authorities.

NOTICE TO BUYER AND USER

Seller warrants that this product conforms to the chemical description on the label and is reasonably fit for the purposes stated on the label when used in accordance with directions under normal conditions of use. This warranty does not extend to the use of this product contrary to label instructions, or under abnormal use conditions, or under conditions not reasonably foreseeable to Seller, and Buyer and User assume the risks of any such use. SELLER DISCLAIMS ALL OTHER WARRANTIES EXPRESSED OR IMPLIED, INCLUDING ANY WARRANTY OF FITNESS OR MERCHANTABILITY. SELLER SHALL NOT BE LIABLE FOR CONSEQUENTIAL, SPECIAL OR INDIRECT DAMAGES RESULTING FROM THE USE OR HANDLING OF THIS PRODUCT AND SELLER'S SOLE LIABILITY AND BUYER'S AND USER'S EXCLUSIVE REMEDY SHALL BE LIMITED TO THE REFUND OF THE PURCHASE PRICE.

85-0790-307

The information and specimen labels in this booklet are correct as of the publication date but are subject to change without notice. For current information, contact ICI Americas Inc.

Mobay Corporation

Agricultural Chemicals Division
Box 4913
Kansas City, MO 64120
(816) 242-2000

BAYLETON®

50% WETTABLE POWDER FUNGICIDE IN WATER SOLUBLE PACKETS

FOR CONTROL OF SPECIFIED DISEASES ON CERTAIN FIELD, FRUIT AND VEGETABLE CROPS

ACTIVE INGREDIENT:
1-(4-Chlorophenoxy)-3,3-dimethyl-1-(1*H*-1,2,4-triazol-1-yl)-2-butanone 50%
INERT INGREDIENTS: 50%

TOTAL 100%

EPA Reg. No. 3125-340 U.S. Patent No. 3,912,752

BAYLETON is a Reg. TM of Bayer AG, Germany
Dithane M-45 is a Reg. TM of Rohm and Haas Company

RECOMMENDED USES

Apples, Cucurbits, Grapes, Pears, Raspberries, Wheat, Barley, Pine Seedlings, Sugar beets, and Grasses grown for seed.

RESTRICTIONS ON WAREHOUSE STORAGE

Protect from rain and moisture. Store in a cool, dry place above 32°F (freezing). Do not store next to herbicides. Do not store near feed or food products.

PACKAGING

2 lb. carton, contains four 8-ounce water soluble packets, 12 per case. Shipping weight 33.5 lbs.

BAYLETON 50% Wettable Powder is a unique fungicide which has the ability to control, as well as prevent, certain important fungus diseases, including rusts, powdery mildew and black rot in wheat, barley, grapes, apples, pineapples, pears, pine seedlings, and grasses grown for seed.

BAYLETON 50% Wettable Powder is absorbed rapidly and works systemically from within the plant.

BAYLETON 50% Wettable Powder mixes readily with water to form a suspension that can be used in all conventional hand- and machine-operated sprayers.

AVISO

PRECAUCION AL USUARIO: Si usted no puede leer o entender inglés, no use este producto hasta que la etiqueta le haya sido explicada ampliamente.

(TO THE USER: If you cannot read or understand English, do not use this product until the label has been fully explained to you.)

DIRECTIONS FOR USE

It is a violation of Federal Law to use this product in a manner inconsistent with its labeling.

Mixing: The enclosed packets are **water soluble.** Do not allow packets to become wet prior to mixing spray solution. Do not handle with wet hands. Reseal outer bag to protect remaining packets.

Ground Application Equipment: The order of mixing is important and should be followed.

1. The spray tank should be filled 1/4 to 1/3 full of water.
2. While recirculating and with the agitator running, add the required number of PVA packets, as determined under "Recommended Applications," into the spray tank as you continue filling the tank to the desired level of water. One packet contains 8 ounces of wettable powder. If it is determined, under "Recommended Applications," that 2 pounds of BAYLETON 50% Wettable Powder should be added to the spray tank, add four 8-ounce packets. The PVA packets should be completely dissolved in 5 to 10 minutes. If an adjuvant is needed, add prior to addition of PVA packets to the spray tank.
3. The addition of other pesticides should be after the PVA packets are completely dissolved, but prior to completion of filling the tank.
4. As with any pesticide, maintain adequate agitation prior to and during spraying.

Aerial Application Equipment: It is important to pre-mix the PVA packets and water before loading into the aerial spray tank. The mixing process should be made according to the same directions for ground application equipment.

CAUTION: Do not use PVA packets in a tank-mix with products that contain Boron or release free chlorine. The resultant reaction of PVA and Boron or free chlorine is a plastic which is not soluble in water or solvents such as diesel oils, kerosene, gasoline or alcohol.

Do not attempt to use the PVA packets directly in diesel oils or summer spray type oils as in ULV or LV uses. PVA packets are water soluble not oil soluble.

NOTICE TO USER: For worker protection during application, wear a hat, long sleeve shirt, and long legged trousers or overalls. Protective clothing should be laundered separately following application.

RECOMMENDED APPLICATIONS

BAYLETON 50% Wettable Powder may be applied in a minimum of 20 gallons of spray solution per acre by ground or in a minimum of 5 gallons of spray solution per acre by aircraft spray equipment except as noted under "Remarks" for each crop. Check equipment calibration frequently. Complete coverage and uniform application are essential for the most effective results, especially when lower spray volumes are applied. If necessary, increase the spray volume per acre for complete crop coverage.

Consult your State Agricultural Experiment Station or Extension Service Specialist or Mobay representative for guidance on the selection of specific dosage rates and timing of applications appropriate for your specific crop and disease. State regulations may contain additional restrictions or requirements. Check the Remarks and Restrictions sections of this label for additional directions.

Do not apply this product through any type of irrigation system.

CROP	DISEASE	RATE OF BAYLETON 50% WETTABLE POWDER		REMARKS (ALSO REFER TO DIRECTIONS FOR USE)
FRUITS		**oz./A**	**oz./100 gal**[1]	
Apples	Cedar apple rust Powdery mildew	2 to 8	1/2 to 2	Make the first application at the green-tip stage and continue applications at 7 to 14 day intervals, or as needed until terminal growth ceases. Use the higher label rate under heavy disease pressure, with extended spray intervals and/or on highly disease susceptible varieties. A maximum of 24 ounces of BAYLETON 50% Wettable Powder may be applied per acre per season. BAYLETON 50% Wettable Powder may be applied up to the day of harvest. Do not graze livestock in treated orchards.
		oz./A	**oz./100 gal**[1]	
Pears	Powdery mildew	2 to 8	1/2 to 2	Make the first application at the budburst stage and continue applications at 7 to 14 day intervals, or as needed. Use the higher label rate under heavy disease pressure, with extended spray intervals, and/or on highly disease susceptible varieties. A maximum of 24 ounces of BAYLETON 50% Wettable Powder may be applied per acre per season. BAYLETON 50% Wettable Powder may be applied up to the day of harvest. Do not graze livestock in treated orchards.
Grapes	Powdery mildew	2 to 6 ozs./A		**GENERAL COMMENTS** Use the higher label rates under heavy disease pressure, with extended spray intervals, and/or on highly disease susceptible varieties. Fields should be observed closely for early disease symptoms when susceptible varieties are planted and/or under prolonged conditions favorable for disease development. A maximum of 18 ounces of BAYLETON 50% WP may be applied per acre per crop season. BAYLETON 50% WP may be applied up to 14 days before harvest. **POWDERY MILDEW:** Make the first application pre-bloom and continue applications at 14 to 21 day intervals, or as needed. **BLACK ROT:**
	Black rot	Protective Schedule 2 to 3 ozs./A Post-infection 3 to 4 ozs./A		

CROP	DISEASE	RATE OF BAYLETON 50% WETTABLE POWDER	REMARKS (ALSO REFER TO DIRECTIONS FOR USE)
			Protective Schedule Make the first application at 10-inch green shoot and continue applications at 7 to 14 day intervals through 5° Brix stage or until veraison (berry coloring) is complete. Under severe disease conditions and/or on highly susceptible varieties, make applications at 7 day intervals. **Post-infection Schedule** During bloom apply within 72 hours after the beginning of an infection period. Do not make spray applications any closer than 7 day intervals. After bloom, continue applications at rates of 2 to 3 ounces of BAYLETON 50% WP at 7 to 14 day intervals, or as needed to maintain control.
Raspberries (California Only)	Powdery mildew	4 ounces	Apply specified dosage per acre in not less than 20 gallons of water using ground equipment. Make first application when disease symptoms first appear and additional applications at 4 to 6 week intervals thereafter as needed. A maximum of 7 total applications may be made in one year with no more than 2 applications within the same 30-day period. Do not harvest within 1 day of application.
SMALL GRAINS	Powdery mildew	2 to 6 ozs./A	Fields should be observed closely for early disease symptoms when susceptible varieties are planted and/or under prolonged conditions favorable for disease development.
Wheat Barley	Rusts (stripe, leaf, and stem) Leaf blotch *(Septoria tritici)* Barley scald	4 to 8 ozs./A	Apply specified dosage per acre in a minimum of 10 gallons of water per acre by ground equipment or 5 gallons of water per acre by air when disease symptoms first appear on leaves or stems. Additional applications should be made if new disease symptoms appear. The total amount of BAYLETON 50% WP should not exceed 16 ozs. per acre per crop season and the last application should not be made within 21 days of harvest. **Powdery mildew:** In areas having severe powdery mildew infection apply 4 to 6 ozs. of BAYLETON 50% WP per acre. **Rusts:** In areas having severe rust infection apply 6 to 8 ozs. of BAYLETON 50% WP per acre. **Leaf blotch:** BAYLETON 50% WP, applied at first sign of infection, will suppress light to moderate leaf blotch infections. For control of severe leaf blotch infections it is recommended that a registered fungicide with good activity against this disease be tank mixed with BAYLETON. **Barley scald:** Apply 4 to 8 ozs. of BAYLETON 50% WP at the first sign of scald symptoms.
GRASSES GROWN FOR SEED Bermudagrass Perennial ryegrass Bluegrass	Rusts *(Puccinia spp.)*	**PREVENTATIVE RATE** 4 to 8 ozs. per acre **CURATIVE RATE** 8 to 16 ozs. per acre	Apply the specified dosage in 20 gallons of water or more per acre with ground sprayers, or in 7 or more gallons of water per acre with aircraft. Thorough coverage is important for optimum disease control. Disease control may be enhanced by the addition of a wetting agent to spray mixtures of BAYLETON 50% WP. A maximum of 2 lbs. of BAYLETON 50% WP may be applied per acre per year. The last application may be made up to 5 days before harvest. Chaff and straw from treated area may be used for feed purposes; however, do not forage or cut green crop or use seed for feed purposes. Regrowth may be grazed in 11 weeks after last application. **RUSTS** **Preventative Schedule** For optimum disease control, apply specified dosage of BAYLETON 50% WP as soon as weather conditions are favorable for rust development. Apply before rust pustules are present, and continue applications at 14 to 21 day intervals depending on disease pressure. **Curative Schedule** Apply the specified dosage if rust pustules are already present. A second application 14 to 21 days later may be necessary. On highly susceptible grass varieties and/or under prolonged conditions favorable for disease development use rates of 12 to 16 oz. per acre and a shorter interval between applications. Rust control is especially important during the period from early head emergence through last flowering. Disease control may not be adequate and yield loss may occur if rust infection is allowed to reach 5 percent or greater before treating. BAYLETON 50% WP will prevent young, developing pustules from producing spores. However, pustules and other symptoms caused by the rust fungus prior to the application of BAYLETON 50% WP will remain visible.
	Powdery mildew	2 to 4 ozs. per acre	**POWDERY MILDEW:** One or two applications at the 2 to 4 oz. rate applied at the first signs of disease and when new powdery mildew symptoms reappear are generally sufficient for disease control. In cases where severe mildew infections are present, use the higher rate.
Pine (Seedlings) (Except California)	Pine rust (Fusiform rust)	4 to 16 ozs.	Begin application prior to infection period and repeat as necessary at 2 to 3 week intervals depending upon disease pressure. Use lower rates in areas of low disease incidence and higher rates in areas of severe disease incidence. A maximum of 64 ozs. of BAYLETON 50% WP may be applied per acre per season. A spreader-sticker is needed to help adhere spray solution to the pine trees.
Sugar beets	Powdery mildew	**BROADCAST** 4 to 16 ozs./A	**GENERAL COMMENTS** Apply the specified dosage per acre in a minimum of 10 gallons of water by ground and 5 gallons of water by air. Under severe disease conditions, use of the higher recommended rate may be necessary. Do not apply more than 16 ozs. of BAYLETON 50% Wettable Powder per acre per season. **BROADCAST APPLICATIONS:** Apply specified dosage per acre as a broadcast foliar spray when disease symptoms first appear. To maintain control or if reinfection occurs, additional applications may be made at 10 to 14 day intervals. Broadcast applications may be made up to 15 days before harvest.

CROP	DISEASE	RATE OF BAYLETON 50% WETTABLE POWDER	REMARKS (ALSO REFER TO DIRECTIONS FOR USE)
		BAND 8 to 16 ozs./A	BAND APPLICATIONS: Apply specified dosage per acre as a spray directed into the leaf whorl in a 2-inch band by turning a flat fan spray nozzle sideways to achieve a narrow band. Band applications may be made up to 30 days before harvest.
Cucurbits [Balsam pear (bitter melon), Chinese waxgourd (Chinese preserving melon), citron melon, cucumber, gherkin, gourds, edible melons including hybrids (including cantaloupe, casaba, crenshaw, honeydew melons, honey balls mango melon, muskmelon, Persian melon), pumpkin, squash, summer squash, winter watermelon, including hybrids]	Powdery mildew *(Erysiphe cichoracearum)* *	2 to 4 ozs. per acre	Apply specified dosage per acre in a minimum of 10 gallons of water by ground and 5 gallons of water by air as a foliar spray when disease symptoms first appear. Repeat at 14 day intervals to maintain control if new disease symptoms appear. Under severe disease conditions use the 4 oz. rate. Applications may be made up to day of harvest. A maximum of 16 ozs. of BAYLETON 50% WP may be applied per acre per season. *BAYLETON 50% WP will not give commercially acceptable control of *Sphaerotheca fuliginea.* 7-12-89

[1] Rates of BAYLETON 50% Wettable Powder are based on a standard of 400 gallons of dilute spray per acre, or the equivalent amount of product per acre in a concentrate spray.

TANK MIXES

The following tank mix recommendations are made for the improved control of specific diseases. Before mixing BAYLETON 50% Wettable Powder with any other fungicide, read and carefully observe the precautionary statements, directions for use, and other information appearing on the EPA registered product labels. The rates given below for Dithane M-45® are for the 80% Wettable Powder formulation.

CROP	CO-APPLIED FUNGICIDE AND RATE	FOR IMPROVED CONTROL OF:	BAYLETON 50% WETTABLE POWDER RATE
Apples	Dithane M-45 @ 1 to 2 lbs./100 gals.	Bitter rot, Black rot, Brown rot, Fly speck, Scab, Sooty blotch	½ to 2 ozs./100 gals.
Barley	Refer to wheat		
Cantaloupe	Refer to melons		
Cucumbers	Refer to melons		
Grapes	Dithane M-45 @ 1½ to 4 lbs./A	Bunch rot Deadarm Downy mildew	2 to 6 ozs./A
Melons	Dithane M-45 @ 2 to 3 lbs./A	Alternaria leaf spot Anthracnose Cercospora leaf spot Downy mildew Gummy stem blight Scab	2 to 4 ozs./A
Pears (Except California)	Refer to apples		
Pumpkins	Refer to melons		
Squash, summer	Refer to melons		
Sugar beets	Dithane M-45 @ 1½ to 2 lbs./A	Cercospora leaf spot	4 to 16 ozs./A
Watermelons	Refer to melons		
Wheat	Dithane M-45 @ 2 lbs./A	Helminthosporium leaf spots Septoria leaf spots (leaf blotch) Septoria glume blotch Tan spot	2 to 8 ozs./A

ROTATIONAL CROPS

Small grains, corn, sorghum, soybeans, beans, peas, and cucurbits may be planted 35 days after the last application of BAYLETON fungicide, however, forage or vines from these crops may not be used for food or feed. Root crops may be planted 120 days after the last application but tops must not be used for food or feed. All crops may be planted 12 months or later after the last application of BAYLETON without any restrictions.

RE-ENTRY STATEMENTS

Do not apply this product in such a manner as to directly or through drift expose workers or other persons. The area being treated must be vacated by unprotected persons. Do not enter treated areas without protective clothing until sprays have dried. Because certain states may require more restrictive re-entry intervals for various crops treated with this product, consult your State Department of Agriculture for further information.

Written or oral warnings must be given to workers who are expected to be in a treated area or in an area about to be treated with this product. Oral warnings should include those statements found under the label heading "Precautionary Statements" and "Re-entry Statements" included on the label. When oral warnings are given, warnings shall be given in a language customarily understood by workers. Oral warnings must be given if there is reason to believe that written warnings cannot be understood by workers. Written warnings must include the following information: WARNING, area treated with BAYLETON 50% Wettable Powder on (date of application). Do not enter without appropriate protective clothing until sprays have dried. **If on skin,** remove contaminated clothing and wash skin immediately with soap and warm water. **If eyes are contaminated,** wash with flowing water for at least 15 minutes. **If swallowed,** vomiting should be induced. (See Precautionary Statements on label).

PRECAUTIONARY STATEMENTS

HAZARDS TO HUMANS AND DOMESTIC ANIMALS

WARNING

Harmful if swallowed, inhaled, or absorbed through the skin. Causes eye irritation. Do not allow to get in eyes, on skin, or on clothing.

Wash hands, arms and face throughly with soap and warm water after use and before eating or smoking. Wash all contaminated clothing with soap and hot water before reuse. Do not contaminate feed or food items. Keep out of reach of children.

STATEMENTS OF PRACTICAL TREATMENT

If swallowed, vomiting should be induced. Administer water freely and induce vomiting by giving one dose (½ oz. or 15 ml.) of syrup of ipecac. If vomiting does not occur within 10 to 20 minutes, administer second dose. If syrup of ipecac is not available, induce vomiting by sticking finger down throat. Repeat until vomit fluid is clear. Never give anything by mouth to an unconscious person. Professional medical assistance should be secured immediately.

If on skin, wash skin immediately with soap and water.

If eyes are contaminated, wash with plenty of water for at least 15 minutes. Get medical attention.

If inhaled, remove to fresh air. If not breathing, give artificial respiration, preferably mouth to mouth. Get medical attention.

SYMPTOMS OF POISONING: The compound does not cause any definite symptoms that would be diagnostic. Poisoning is accompanied by hyperactivity followed by sedation.

NOTE TO PHYSICIAN: No specific antidote. Treat symptomatically.

ENVIRONMENTAL HAZARDS

Do not use on other crops grown for food or forage. Do not apply directly to water. Do not contaminate water by cleaning of equipment or disposal of wastes. Apply this product only as specified on the label.

Do not make applications when weather conditions favor drift from target area.

STORAGE AND DISPOSAL

Pesticide Disposal: Do not contaminate water, food or feed by storage or disposal. Wastes resulting from the use of this product may be disposed of on site or at an approved waste disposal facility.

Container Disposal: Completely empty bag into application equipment. Then dispose of empty bag in a sanitary landfill or by incineration, or if allowed by State and local authorities, by burning. If burned, stay out of smoke.

Store in a cool, dry place and in such a manner as to prevent cross contamination with other pesticides, fertilizers, food and feed. Do not store below freezing (32°F). Exposure to moisture or excessive handling of water soluble packets may cause breakage. Store packets in original container and out of the reach of children, preferably in a locked storage area.

Handle and open container in a manner as to prevent spillage. If the container is leaking or material spilled for any reason or cause, carefully sweep material into a pile. Refer to Precautionary Statements on label for hazards associated with the handling of this material. Do not walk through spilled material. Dispose of pesticide as directed above. In spill or leak incidents, keep unauthorized people away. You may contact the Mobay Emergency Response Team for decontamination procedures or any other assistance that may be necessary. The Mobay Kansas City Emergency Response Telephone No. is 816-242-2582, or contact Chemtrec at 800-424-9300.

IMPORTANT:

This bulletin is not intended to provide adequate information for application. Before using any product, read and carefully observe directions, cautionary statements and other information appearing on the product label. This product is sold subject to the Conditions of Sale set forth on the container label.

READ THE LABEL BEFORE USE

BN 9002

This is a specimen label. Mobay is not responsible for the accuracy of the information contained herein. As labels are subject to revision, always carefully read and follow the label on the product container.

BAYLETON®

50% DRY FLOWABLE FUNGICIDE

FOR CONTROL OF SPECIFIED DISEASES ON CERTAIN FIELD, FRUIT AND VEGETABLE CROPS

ACTIVE INGREDIENT:

1-(4-Chlorophenoxy)-3,3-dimethyl-1-(1*H*-1,2,4- triazol-1-yl)-2-butanone 50%

INERT INGREDIENTS: 50%

TOTAL 100%

EPA Reg. No. 3125-320 U.S. Patent No. 3,912,752

BAYLETON is a Reg. TM of Bayer AG, Germany

Dithane M-45 is a Reg. TM of Rohm and Haas Company

RECOMMENDED USES

Apples, Cucurbits, Grapes, Pears, Pineapple, Raspberries, Wheat, Barley, Pine Seedlings, Pine Seed Nurseries, Sugar Beets, and Grasses grown for seed

RESTRICTIONS ON WAREHOUSE STORAGE

Protect from rain and moisture. Store in a cool, dry place. Do not store next to herbicides. Do not store near feed or food products.

PACKAGING

2 lb. carton, 12 per case. Shipping weight 29 lbs.

BAYLETON 50% Dry Flowable is a unique fungicide which has the ability to control, as well as prevent, certain important fungus diseases, including rusts, powdery mildew and black rot in wheat, barley, grapes, apples, pineapples, pears, pine seedlings, and grasses grown for seed.

BAYLETON 50% Dry Flowable is absorbed rapidly and works systemically from within the plant.

BAYLETON 50% Dry Flowable mixes readily with water to form a suspension that can be used in all convenient hand- and machine- operated sprayers.

AVISO

PRECAUCION AL USUARIO: Si usted no puede leer o entender inglés, no use este producto hasta que la etiqueta le haya sido explicada ampliamente.

(TO THE USER: If you cannot read or understand English, do not use this product until the label has been fully explained to you.)

DIRECTIONS FOR USE

It is a violation of Federal Law to use this product in a manner inconsistent with its labeling.

MIXING

Add the recommended amount of BAYLETON 50% Dry Flowable into the spray tank while filling with water to the desired level. Operate the agitator while mixing. If other materials are added to the spray tank, the BAYLETON 50% Dry Flowable should be thoroughly dispersed prior to the addition of other materials.

COMPATIBILITY

BAYLETON fungicide is physically compatible with many registered insecticides, miticides, and fungicides. To determine the compatibility of BAYLETON fungicide with other products, the following procedure should be followed: Pour the recommended proportions of the products into a suitable container of water, mix thoroughly and allow to stand at least five (5) minutes. If the combination remains mixed or can be remixed readily, the mixture is considered physically compatible. For further information contact your local Mobay representative.

NOTE TO USER

For worker protection during mixing, loading, and during application, wear a hat, long sleeve shirt, and long legged trousers or overalls. In addition, during mixing and loading, wear rubber or neoprene gloves and a dust mask. Protective clothing should be laundered separately following application.

RECOMMENDED APPLICATIONS

BAYLETON 50% Dry Flowable may be applied in a minimum of 20 gallons of spray solution per acre by ground or in a minimum of 5 gallons of spray solution per acre by aircraft spray equipment except as noted under "Remarks" for each crop. Check equipment calibration frequently. Complete coverage and uniform application are essential for the most effective results, especially when lower spray volumes are applied. If necessary, increase the spray volume per acre for complete crop coverage.

Do not apply this product through any type of irrigation system.

Consult your State Agricultural Experiment Station or Extension Service Specialist or Mobay representative for guidance on the selection of specific dosage rates and timing of applications appropriate for your specific crop and disease. State regulations may contain additional restrictions or requirements. Check the Remarks and Restrictions sections of the label for additional directions.

CROP	DISEASE	RATE OF BAYLETON 50% DRY FLOWABLE		REMARKS (ALSO REFER TO DIRECTIONS FOR USE)
		oz./A	oz./100 gal[1]	
FRUITS Apples	Cedar apple rust Powdery mildew	2 to 8	½ to 2	Make the first application at the green-tip stage and continue applications at 7 to 14 day intervals, or as needed until terminal growth ceases. Use the higher label rate under heavy disease pressure, with extended spray intervals, and/or on highly disease susceptible varieties. A maximum of 24 ounces of BAYLETON 50% DF may be applied per acre per season. BAYLETON 50% DF may be applied up to the day of harvest. Do not graze livestock in treated orchards.

CROP	DISEASE	RATE OF BAYLETON 50% DRY FLOWABLE		REMARKS (ALSO REFER TO DIRECTIONS FOR USE)
Pears	Powdery mildew	2 to 8	½ to 2	Make the first application at the budburst stage and continue applications at 7 to 14 day intervals, or as needed. Use the higher label rates under heavy disease pressure, with extended spray intervals, and/or on highly disease susceptible varieties. A maximum of 24 ounces of BAYLETON 50% DF may be applied per acre per season. BAYLETON 50% DF may be applied up to the day of harvest. Do not graze livestock in treated orchards.
Grapes	Powdery mildew	2 to 6 ozs./A		**GENERAL COMMENTS** Use the higher label rates under heavy disease pressure, with extended spray intervals, and/or on highly disease susceptible varieties. Fields should be observed closely for early disease symptoms when susceptible varieties are planted and/or under prolonged conditions favorable for disease development. A maximum of 18 ounces of BAYLETON 50% DF may be applied per acre per crop season. BAYLETON 50% DF may be applied up to 14 days before harvest. **POWDERY MILDEW:** Make the first application pre-bloom and continue applications at 14 to 21 day intervals, or as needed. **BLACK ROT:** **Protective Schedule** Make the first application at 10-inch green shoot and continue applications at 7 to 14 day intervals through 5° Brix stage or until veraison (berry coloring) is complete. Under severe disease conditions and/or on highly susceptible varieties, make applications at 7 day intervals. **Post-infection Schedule** During bloom apply within 72 hours after the beginning of an infection period. Do not make spray applications any closer than 7 day intervals. After bloom, continue applications at rates of 2 to 3 ounces of BAYLETON 50% DF at 7 to 14 day intervals, or as needed to maintain control.
	Black rot	Protective Schedule 2 to 3 ozs./A Post-infection 3 to 4 ozs./A		
Pineapples (Fresh market only)	Fresh fruit rot *(Ceratocystis paradoxa)*	13⅓ oz.		**Post-Harvest Treatment:** Apply specified dosage in 100 gallons of water to fruit after harvest. Fruit should be dipped or sprayed for thorough coverage and allowed to drain. Limit dipping time to not more than 3 minutes.
Rasperries (California Only)	Powdery mildew	4 ounces		Apply specified dosage per acre in not less than 20 gallons of water using ground equipment. Make the first application when disease symptoms first appear and additional applications at 4 to 6 week intervals thereafter as needed. A maximum of 7 total applications may be made in one year with not more than 2 applications within the same 30-day period. Do not harvest within 1 day of application.
SMALL GRAINS Wheat, Barley	Powdery mildew	2 to 6 oz./A		Fields should be observed closely for early disease symptoms when susceptible varieties are planted and/or under prolonged conditions favorable for disease development. Apply specified dosage per acre in a minimum of 10 gallons of water per acre by ground equipment or 5 gallons of water per acre by air when disease symptoms first appear on leaves or stems. Additional applications should be made if new disease symptoms appear. The total amount of BAYLETON 50% DF should not exceed 16 ozs. per acre per crop season and the last application should not be made within 21 days of harvest. **Powdery mildew:** In areas having severe powdery mildew infection apply 4 to 6 ozs. of BAYLETON 50% DF per acre. **Rusts:** In areas having severe rust infection apply 6 to 8 ozs. of BAYLETON 50% DF per acre. **Leaf blotch:** BAYLETON 50% DF, applied at first sign of infection, will suppress light to moderate leaf blotch infections. For control of severe leaf blotch infections it is recommended that a registered fungicide with good activity against this disease be tank mixed with BAYLETON. **Barley scald:** Apply 4 to 8 ozs. of BAYLETON 50% DF at the first sign of scald symptoms.
	Rusts (stripe, leaf, and stem) Leaf blotch *(Septoria tritici)* Barley Scald	4 to 8 oz./A		
GRASSES GROWN FOR SEED Bermudagrass Perennial ryegrass Bluegrass	Rusts *(Puccinia* spp.)	**PREVENTATIVE RATE** 4 to 8 ozs. per acre **CURATIVE RATE** 8 to 16 ozs. per acre		Apply specified dosage in 20 gallons of water or more per acre with ground sprayers, or in 7 or more gallons of water per acre with aircraft. Thorough coverage is important for optimum disease control. Disease control may be enhanced by the addition of a wetting agent to spray mixtures of BAYLETON 50% DF. A maximum of 2 lbs. of BAYLETON 50% DF may be applied per acre per year. The last application may be made up to 5 days before harvest. Chaff and straw from treated area may be used for feed purposes; however, do not forage or cut green crop or use seed for feed purposes. Regrowth may be grazed 11 weeks after last application. **RUSTS** **Preventative Schedule** For optimum disease control, apply specified dosage of BAYLETON 50% DF as soon as weather conditions are favorable for rust development. Apply before rust pustules are present, and continue applications at 14 to 21 day intervals depending on disease pressure. **Curative Schedule** Apply the specified dosage if rust pustules are already present. A second application 14 to 21 days later may be necessary. On highly susceptible grass varieties and/or under prolonged conditions favorable for disease development use rates of 12 to 16 ozs. per acre and a shorter interval between applications. Rust control is especially important during the period from early head emergence through last flowering. Disease control may not be adequate and yield loss may occur if rust infection is allowed to reach 5 percent or greater before treating. BAYLETON 50% DF will prevent young, developing pustules from producing spores. However, pustules and other symptoms caused by the rust fungus prior to the application of BAYLETON 50% DF will remain visible.
	Powdery mildew	2 to 4 ozs. per acre		**POWDERY MILDEW:** One or two applications at the 2 to 4 ozs. rate applied at the first signs of disease and when new powdery mildew symptoms reappear are generally sufficient for disease control. In cases where severe mildew infections are present, use the higher rate.

CROP	DISEASE	RATE OF BAYLETON 50% DRY FLOWABLE	REMARKS (ALSO REFER TO DIRECTIONS FOR USE)
Pine (Seedlings) (Except California)	Pine rust (Fusiform rust)	4 to 16 ozs.	Begin application prior to infection period and repeat as necessary at 2 to 3 week intervals depending upon disease pressure. Use lower rates in areas of low disease incidence and higher rates in areas of severe disease incidence. A maximum of 64 ozs. of BAYLETON 50% DF may be applied per acre per season. A spreader-sticker is needed to help adhere spray solution to the pine trees.
Pine Seed (Nurseries) (Except California)	Fusiform rust (*Cronartium quercuum*)	20 ounces	Mix specified dosage in 100 gals. of water. Cover seeds with the mixture and soak them, at room temperature (75°F), for 24 hours, stirring occasionally. Seed stratification, if necessary, should be completed before treatment with BAYLETON 50% Dry Flowable. Thoroughly air dry all seed before applying any additional treatment or before planting. A spreader sticker may be needed to help adhere BAYLETON to the pine seed. Do not use treated seed for food or feed purposes.
		2 ounces	Apply specified dosage to 50 lbs. of thoroughly wetted pine seeds in a commercial treater or other suitable tumbler apparatus. Allow to mix for at least 10 minutes before applying bird repellent or other seed dressing materials. Thoroughly air dry seed before sowing. Do not use treated seed for food or feed purposes.
Sugar beets	Powdery mildew	**BROADCAST** 4 to 16 ozs./A. **BAND** 8 to 16 ozs./A	**GENERAL COMMENTS** Apply the specified dosage per acre in a minimum of 10 gallons of water by ground and 5 gallons of water by air. Under severe disease conditions, use of the higher recommended rate may be necessary. Do not apply more than 16 ozs. of BAYLETON 50% Dry Flowable per acre per season. **BROADCAST APPLICATIONS:** Apply specified dosage per acre as a broadcast foliar spray when disease symptoms first appear. To maintain control or if reinfection occurs, additional applications may be made at 10 to 14 day intervals. Broadcast applications may be made up to 15 days before harvest. **BAND APPLICATIONS:** Apply specified dosage per acre as a spray directed into the leaf whorl in a 2-inch band by turning a flat fan spray nozzle sideways to achieve a narrow band. Band applications may be made up to 30 days before harvest.
Cucurbits [Balsam pear (bitter melon), Chinese waxgourd (Chinese preserving melon), citron melon, cucumber, gherkin, gourds, edible melons including hybrids (including cantaloupe, casaba, crenshaw, honeydew melons, honey balls mango melon, muskmelon, Persian melon), pumpkin, squash, summer squash, winter watermelon, including hybrids]	Powdery mildew (*Erysiphe cichoracearum*) *	2 to 4 ozs. per acre	Apply specified dosage per acre in a minimum of 10 gallons of water by ground and 5 gallons of water by air as a foliar spray when disease symptoms first appear. Repeat at 14 day intervals to maintain control if new disease symptoms appear. Under severe disease conditions use the 4 oz. rate. Applications may be made up to day of harvest. A maximum of 16 ozs. of BAYLETON 50% DF may be applied per acre per season. *BAYLETON 50% DF will not give commercially acceptable control of *Sphaerotheca fuliginea.* 7-12-89

[1] Rates of BAYLETON 50% Dry Flowable are based on a standard of 400 gallons of dilute spray per acre, or the equivalent amount of product per acre in a concentrate spray.

TANK MIXES: The following tank mix recommendations are made for the improved control of specific diseases. Before mixing BAYLETON 50% Dry Flowable with any other fungicide, read and carefully observe the precautionary statements, directions for use, and other information appearing on the EPA registered product labels. The rates given below for Dithane M-45 fungicide are for the 80% wettable powder formulation.

CROP	CO-APPLIED FUNGICIDES AND RATE	FOR IMPROVED CONTROL OF:		BAYLETON 50% DF RATE
Apples	Dithane M-45@ 1 to 2 lbs./100 gals.	Bitter rot Black rot Brown rot	Fly speck Scab Sooty blotch	½ to 2 ozs./100 gals.
Barley	Refer to wheat			
Cantaloupe Cucumbers	Refer to melons Refer to melons			
Grapes	Dithane M-45 @ 1½ to 4 lbs./A.	Bunch rot	Deadarm Downy mildew	2 to 6 ozs./A.

CROP	CO-APPLIED FUNGICIDES AND RATE	FOR IMPROVED CONTROL OF:		BAYLETON 50% DF RATE
Melons	Dithane M-45 @ 2 to 3 lbs./A.	Alternaria leaf spot Anthracnose Cercospora leaf spot Downy mildew	Gummy stem blight Scab	2 to 4 ozs./A.
Pears (Except California)	Refer to apples			
Pumpkins	Refer to melons			
Squash, summer	Refer to melons			
Sugar beets	Dithane M-45@ 1½ to 2 lbs./A.	Cercospora leaf spot		4 to 16 ozs./A.
Watermelons	Refer to melons			
Wheat	Dithane M-45 @ 2 lbs./A.	Helminthosporium leaf spots Septoria leaf spots (leaf blotch)	Septoria glume blotch Tan spot	2 to 8 ozs./A.

ROTATIONAL CROPS

Small grains, corn, sorghum, soybeans, beans, peas, and cucurbits may be planted 35 days after the last application of BAYLETON fungicide, however forage or vines from these crops may not be used for food or feed. Root crops may be planted 120 days after the last application but tops must not be used for food or feed. All crops may be planted 12 months or later after the last application of BAYLETON without any restrictions.

RE-ENTRY STATEMENTS

Do not apply this product in such a manner as to directly or through drift expose workers or other persons. The area being treated must be vacated by unprotected persons. Do not enter treated areas without protective clothing until sprays have dried. Because certain states may require more restrictive re-entry intervals for various crops treated with this product, consult your State Department of Agriculture for further information.

Written or oral warnings must be given to workers who are expected to be in a treated area or in an area about to be treated with this product. Oral warnings should include those statements found under the label heading "Precautionary Statements" and "Re-entry Statements" included on the label. When oral warnings are given, warnings shall be given in a language customarily understood by workers. Oral warnings must be given if there is reason to believe that written warnings cannot be understood by workers. Written warnings must include the following information: WARNING, area treated with BAYLETON 50% Dry Flowable on (date of application). Do not enter without appropriate protective clothing until sprays have dried. **If on skin,** remove contaminated clothing and wash skin immediately with soap and warm water. **If eyes are contaminated,** wash with flowing water for at least 15 minutes. **If swallowed,** vomiting should be induced. (See Precautionary Statements on label).

PRECAUTIONARY STATEMENTS

HAZARDS TO HUMANS AND DOMESTIC ANIMALS

WARNING

Harmful if swallowed, inhaled, or absorbed through the skin. Causes eye irritation. Do not breathe dust or allow to get in eyes, on skin, or on clothing.

Wash hands, arms and face thoroughly with soap and warm water after use and before eating or smoking. Wash all contaminated clothing with soap and hot water before reuse. Do not contaminate feed or food items. Keep out of reach of children.

STATEMENT OF PRACTICAL TREATMENT

If swallowed, vomiting should be induced. Administer water freely and induce vomiting by giving one dose (½ oz. of 15 ml.) of syrup of ipecac. If vomiting does not occur within 10 to 20 minutes, administer second dose. If syrup of ipecac is not available, induce vomiting by sticking finger down throat. Repeat until vomit fluid is clear. Never give anything by mouth to an unconscious person. Professional medical assistance should be secured immediately.

If on skin, wash skin immediately with soap and water.

If eyes are contaminated, wash with plenty of water for at least 15 minutes. Get medical attention.

If inhaled, remove to fresh air. If not breathing, give artificial respiration, preferably mouth to mouth. Get medical attention.

SYMPTOMS OF POISONING: The compound does not cause any definite symptoms that would be diagnostic. Poisoning is accompanied by hyperactivity followed by sedation.

NOTE TO PHYSICIAN: No specific antidote. Treat symptomatically.

ENVIRONMENTAL HAZARDS

Do not use on other crops grown for food or forage. Keep out of lakes, streams, and ponds. Do not contaminate water by cleaning of equipment or disposal of wastes. Apply this product only as specified on the label.

Do not make applications when weather conditions favor drift from target area.

STORAGE AND DISPOSAL

Pesticide Disposal: Do not contaminate water, food or feed by storage or disposal. Wastes resulting from the use of this product may be disposed of on site or at an approved waste disposal facility.

Container Disposal: Completely empty carton into application equipment. Then dispose of empty carton in a sanitary landfill or by incineration, or if allowed by State and local authorities, by burning. If burned, stay out of smoke.

Store in a cool, dry place and in such a manner as to prevent cross contamination with other pesticides, fertilizers, food, and feed. Store in original container and out of the reach of children, preferably in a locked storage area.

Handle and open container in a manner as to prevent spillage. If the container is leaking or material spilled for any reason or cause, carefully sweep material into a pile. Refer to Precautionary Statements on label for hazards associated with the handling of this material. Do not walk through spilled material. Dispose of pesticide as directed above. In spill or leak incidents, keep unauthorized people away. You may contact the Mobay Emergency Response Team for decontamination procedure or any other assistance that may be necessary. The Mobay Kansas City Emergency Response Telephone No. is 816-242-2582, or contact Chemtrec at 800-424-9300.

IMPORTANT:

This bulletin is not intended to provide adequate information for application. Before using any product, read and carefully observe directions, cautionary statements and other information appearing on the product label. This product is sold subject to the Conditions of Sale set forth on the container label.

READ THE LABEL BEFORE USE

BN 9003

This is a specimen label. Mobay is not responsible for the accuracy of the information contained herein. As labels are subject to revision, always carefully read and follow the label on the product container.

BAYTHROID® 2

EMULSIFIABLE PYRETHROID INSECTICIDE

RESTRICTED USE PESTICIDE

Toxic to Fish and Aquatic Organisms

For retail sale to and use only by Certified Applicators or persons under their direct supervision and only for those uses covered by the Certified Applicator's certification.

FOR CONTROL OF INSECT PESTS ON COTTON

ACTIVE INGREDIENT: Cyano(4-fluoro-3-phenoxyphenyl)methyl 3-(2,2-dichloroethenyl)-2,2-dimethylcyclopropanecarboxylate	25%
INERT INGREDIENTS:	75%
TOTAL	100%

Contains 2 pounds Cyano(4-fluoro-3-phenoxyphenyl) methyl 3-(2,2-dichloroethenyl)-2,2-dimethylcyclopropanecarboxylate per gallon.

EPA Reg. No. 3125-351 EPA Est. 3125-MO-1
U.S. Patent No. 4,218,469

BAYTHROID insecticide is a third-generation synthetic pyrethroid that represents new chemistry to the cotton pyrethroid market. It offers quick knock-down and longer residual control of both boll weevils and Heliothis species at low rates. It also controls lygus, pink bollworms, armyworms, cotton leaf perforators and loopers.

BAYTHROID is a Reg. TM of Bayer AG, Germany

RECOMMENDED USES

Cotton.

RESTRICTIONS ON WAREHOUSE STORAGE

Store in a cool, dry place and away from open flame and extreme heat. Store in such a manner as to prevent cross contamination with other pesticides, fertilizers, food, and feed. Store in original container and out of the reach of children, preferably in a locked storage area.

PACKAGING

½ gal. jug, 8 per case. Shipping weight 38 lbs.

1 gal. jug, 4 per case. Shipping weight 36.7 lbs 15 gal. returnable container.

PELIGRO

PRECAUCION AL USUARIO: Si usted no puede leer o entender inglés, no use este producto hasta que la etiqueta le haya sido explicada ampliamente.

(TO THE USER: If you cannot read or understand English, do not use this product until the label has been fully explained to you.)

DIRECTIONS FOR USE

It is a violation of Federal Law to use this product in a manner inconsistent with its labeling.

BAYTHROID 2 emulsifiable pyrethroid insecticide may be used for control of a broad spectrum of insect pests by contact action. Because of this contact activity, good spray coverage of the crop is needed for the highest level of control.

USE IN CHEMIGATION SYSTEMS

Types of Irrigation Systems

Apply BAYTHROID 2 only through sprinkler, including center pivot, lateral move, side roll, or overhead solid set irrigation systems. Do not apply BAYTHROID 2 through any other type of irrigation system.

GENERAL DIRECTIONS FOR ALL RECOMMENDED TYPES OF IRRIGATION SYSTEMS

Uniform Water Distribution and System Calibration

The irrigation system must provide uniform distribution of treated water. Crop injury, lack of effectiveness, or illegal pesticide residues in the crop can result from nonuniform distribution of treated water.

The system must be calibrated to uniformly apply the rates specified for chemigation application for specific crops. If you have questions about calibration, you should contact State Extension Service specialists, equipment manufacturers or other experts.

Chemigation Monitoring

A person knowledgeable of the chemigation system and responsible for its operation, or under the supervision of the responsible person, shall shut the system down and make necessary adjustments should the need arise.

Drift

Do not apply when wind speed favors drift beyond the area intended for treatment.

Required System Safety Devices

The system must contain a functional check valve, vacuum relief valve, and low pressure drain appropriately located on the irrigation pipeline to prevent water source contamination from backflow.

The pesticide injection pipeline must contain a functional, automatic quick-closing check valve to prevent the flow of fluid back toward the injection pump.

The pesticide injection pipeline must also contain a functional, normally closed, solenoid-operated valve located on the intake side of the injection pump and connected to the system interlock to prevent fluid from being withdrawn from the supply tank when the irrigation system is either automatically or manually shut down.

The system must contain functional interlocking controls to automatically shut off the pesticide injection pump when the water pump motor stops.

The irrigation line or water pump must include a functional pressure switch which will stop the water pump motor when the water pressure decreases to the point where pesticide distribution is adversely affected.

Systems must use a metering pump, such as a positive displacement injection pump (e.g., diaphragm pump) effectively designed and constructed of materials that are compatible with pesticides and capable of being fitted with a system interlock.

Using Water from Public Water Systems

DO NOT APPLY BAYTHROID 2 THROUGH ANY IRRIGATION SYSTEM PHYSICALLY CONNECTED TO A PUBLIC WATER SYSTEM. Public water system means a system for the provision to the public of piped water for human consumption if such system has at least 15 service connections or regularly serves an average of at least 25 individuals daily at least 60 days out of the year.

BAYTHROID 2 may be applied through any of the recommended types of irrigation systems which may be **supplied** by a public water system **only if** the water from the public water system is discharged into a reservoir tank prior to pesticide introduction. There shall be a complete physical break (air gap) between the outlet end of the fill pipe and the top or overflow rim of the reservoir tank of at least twice the inside diameter of the fill pipe. Before beginning chemigation, always make sure that the air gap exists and that there is no blockage of the overflow of the reservoir tank.

Any irrigation system using water supplied from a public water system must also meet the following requirements.

The pesticide injection pipeline must contain a functional, automatic, quick-closing check valve to prevent the flow of fluid back toward the injection pump.

The pesticide injection pipeline must contain a functional, normally closed, solenoid-operated valve located on the intake side of the injection pump and connected to the system interlock to prevent fluid from being withdrawn from the supply tank when the irrigation system is either automatically or manually shut down.

The system must contain functional interlocking controls to automatically shut off the pesticide injection pump when the water pump motor stops or in cases where there is no water pump, when the water pressure decreases to the point where pesticide distribution is adversely affected.

Systems must use a metering pump, such as a positive displacement injection pump (e.g., diaphragm pump) effectively designed and constructed of materials that are compatible with pesticides and capable of being fitted with a system interlock.

Do not apply when wind speed favors drift beyond the area intended for treatment.

Posting Requirements

Posting of areas to be chemigated is required when 1) any part of a treated area is within 300 feet of sensitive areas such as residential areas, labor camps, businesses, day care centers, hospitals, in-patient clinics, nursing homes or any public areas such as schools, parks, playgrounds, or other public facilities not including public roads, or 2) when the chemigated area is open to the public such as golf courses or retail greenhouses.

Posting must conform to the following requirements: Treated areas shall be posted with signs at all usual points of entry and along likely routes of approach from the listed sensitive areas. When there are no usual points of entry, signs must be posted in the corners of the treated areas and in any other location affording maximum visibility to sensitive areas. The printed side of the sign should face away from the treated area towards the sensitive area. The signs shall be printed in English. Signs must be posted prior to application and must remain posted until foliage has dried and soil surface water has disappeared. Signs may remain in place indefinitely as long as they are composed of materials to prevent deterioration and maintain legibility for the duration of the posting period.

All words shall consist of letters at least 2½ inches tall, and all letters and the symbol shall be a color which sharply contrasts with their immediate background. At the top of the sign shall be the words KEEP OUT, followed by an octagonal stop sign symbol at least 8 inches in diameter containing the word STOP. Below the symbol shall be the words PESTICIDES IN IRRIGATION WATER.

Posting required for chemigation does not replace other posting and reentry interval requirements for farmworker safety.

Agitation

For application of BAYTHROID 2 alone, a chemical supply tank is not necessary for premixing since BAYTHROID 2 mixes well with water in the irrigation line. If a chemical supply tank is used for application of

BAYTHROID 2, constant strong mechanical or hydraulic agitation must be maintained in the chemical supply tank during the entire period of application.

Chemical Supply Tank Dilution

If a chemical supply tank is used, you must determine the required amounts of BAYTHROID 2 and water to mix in the tank.

The amount of BAYTHROID 2 needed equals the number of fluid ounces of BAYTHROID 2 to be applied per acre multiplied by the number of acres to be chemigated.

The amount of emulsion needed equals the gallons of emulsion delivered per hour by the injection pump multiplied by the number of hours chemigation will take place.

The amount of water needed equals the amount of emulsion needed minus the amount of BAYTHROID 2 needed.

For example, if you want to apply 1.6 fluid ounces of BAYTHROID 2 per acre to 130 acres in 20 hours and your irrigation pump delivers 15 gallons per hour, you need: 1.6 fluid ounces of BAYTHROID 2 × 130 acres = 208 fluid ounces or 13 pints of BAYTHROID 2. And, you need: 15 gallons per hour × 20 hours = 300 gallons of emulsion, minus 1.625 gallons (13 pints) of BAYTHROID 2 = 298.375 gallons of water.

Cleaning the Chemical Injection System

In order to accurately apply pesticides, the chemical injection system must be kept clean; free from chemical or fertilizer residues and sediments. Refer to your owner's manual or ask your equipment supplier for the cleaning procedure for your injection system.

Flushing the Irrigation System

At the end of the application period, allow time for all lines to flush the pesticides through all nozzles before turning off irrigation water. To ensure the lines are flushed and free of pesticides, a dye indicator may be injected into the lines to mark the end of the application period.

SPRINKLER IRRIGATION SYSTEMS

All directions and requirements listed under the GENERAL DIRECTIONS AND REQUIREMENTS FOR ALL RECOMMENDED TYPES OF IRRIGATION SYSTEMS section of the label must be followed for sprinkler irrigation systems.

In addition, the following directions apply to sprinkler irrigation systems.

Do not apply when wind speed favors drift beyond the area intended for treatment.

It is recommended that nozzles in the immediate area of control panels, chemical supply tanks, pumps and system safety devices be plugged to prevent chemical contamination of these areas.

Center-Pivot and Automatic-Move Linear Systems

Inject the specified dosage per acre continuously for one complete revolution or move of the system. DO NOT USE END GUNS. The system should be run at maximum speed.

Solid Set and Manually Controlled Linear Systems

Injection should be during the last 30 to 60 minutes of a regular irrigation period or as a separate 30 to 60 minute application not associated with a regular irrigation. DO NOT USE END GUNS.

RECOMMENDED APPLICATIONS

CROP	PEST	DOSAGE: POUNDS AI PER ACRE	DOSAGE: BAYTHROID 2 FL. OZS. PER ACRE	REMARKS
Cotton	Beet Armyworms Fall armyworm Thrips Soybean looper	0.05	3.2	Apply specified dosage per acre as needed in sufficient water, at least one gallon, for thorough coverage of foliage. Applications may also be made with ground or aircraft equipment adapted and calibrated for ULV application. BAYTHROID 2 may be mixed with once-refined vegetable oil and applied in a total volume of at least one quart per acre. For application by sprinkler irrigation systems: Apply specified dosage per acre. Follow all directions given under the CHEMIGATION section of the BAYTHROID 2 label.
	Boll weevil Cotton bollworm Tobacco budworm Cabbage looper Cotton aphid Lygus bugs Flea beetles Pink bollworm *Heliothis* (Bollworm/Budworm) Ovicidal	0.025 to 0.05	1.6 to 3.2	Timing of applications should be based on careful scouting and local economic thresholds. A 3 to 5 day application schedule may be required to maintain control of heavy boil weevil infestations. Lower dosage rates are adequate for low to mederate insect pressure, but require careful scouting and may require more frequent applications. Ask your supplier for more detailed information for your area.
	Leafworm Cotton leaf-perforator	0.0125 to 0.025	0.8 to 1.6	For armyworm control, careful scouting and early detection are necessary. Subsequent applications are necessary to maintain control. For cotton aphid, thorough coverage under light infestation pressure is necessary to achieve control. Under heavy infestation pressure a systemic insecticide may be necessary. For maximum contact *Heliothis* ovicidal effect, time application to correspond to peak egg deposition. A total of 32 fluid ounces of BAYTHROID 2 per acre may be applied per crop season. Do not graze treated fields. Applications may be made up to day of harvest.

RATE CONVERSION CHART

Lb. AI/A	Fl. Oz/A	Acres Treated: Per Gallon	Acres Treated: Per Pint
0.0125	0.8	160	20
0.025	1.6	80	10
0.028	1.8	71	9.0
0.033	2.1	61	7.6
0.05	3.2	40	5
0.1	6.4	20	2.5

CROP ROTATION STATEMENT

Do not rotate with small grain crops for a period of one (1) year following application of this active ingredient.

RE-ENTRY STATEMENTS

Do not apply this product in such a manner as to directly or through drift expose workers or other persons. The area being treated must be vacated by unprotected persons. Do not enter treated areas without protective clothing until sprays have dried. Because certain states may require more restrictive re-entry intervals for various crops treated with this product, consult your State Department of Agriculture for further information.

Written or oral warnings must be given to workers who are expected to be in a treated area or in an area about to be treated with this product. Oral warnings should include those statements found under the label heading "Precautionary Statements" and "Re-entry Statements" included on the label. When oral warnings are given, warnings shall be given in a language customarily understood by workers. Oral warnings must be given if there is reason to believe that written warnings cannot be understood by workers. Written warnings must include the following information: DANGER, area treated with BAYTHROID 2 on (date of application). Do not enter without appropriate protective clothing until sprays have dried. **If eyes are contaminated,** wash with flowing water for at least 15 minutes. **If on skin,** remove contaminated clothing and wash skin immediately with soap and warm water. **If swallowed,** do not induce vomiting. Call a physician. (See Precautionary Statements on label.)

PRECAUTIONARY STATEMENTS

HAZARDS TO HUMANS AND DOMESTIC ANIMALS

DANGER

Causes eye damage. May be fatal if inhaled. Harmful if swallowed or absorbed through the skin. Do not get in eyes, on skin or on clothing. Do not breathe vapors, or spray mist.

Do not contaminate feed or food. Keep out of reach of children.

Wear goggles, face shield, or safety glasses. Wash hands, arms, and face thoroughly with soap and warm water after handling. Remove and wash contaminated clothing with soap and hot water before reuse.

SYMPTOMS OF POISONING

In case of poisoning, call a physician immediately. Have patient lie down and keep quiet.

STATEMENTS OF PRACTICAL TREATMENT

If eyes are contaminated, flush with plenty of water. Get medical attention. **If swallowed,** do not induce vomiting. Call a physician. Gastric lavage should be supervised by a physician or professional medical staff because of possible pulmonary damage via aspiration of the solvent.

If on skin, wash skin immediately with soap and warm water. If irritation occurs, get medical attention.

If inhaled, remove to fresh air. If not breathing, give artificial respiration, preferably mouth to mouth. Get medical attention.

TO PHYSICIAN: No specific antidote is available. Treat symptomatically.

ENVIRONMENTAL HAZARDS

This pesticide is extremely toxic to fish. Use with care when applying in areas adjacent to any body of water. Keep out of lakes, streams or ponds. Do not apply when weather conditions favor drift from treated areas. Do not contaminate water when disposing of equipment washwaters. Apply this product only as specified on the label.

This pesticide is highly toxic to bees exposed to direct treatment or residues on crops or weeds. Do not apply BAYTHROID 2 or allow it to drift to crops or weeds on which bees are actively foraging. Additional information may be obtained by consulting your Cooperative Extension Service.

PHYSICAL OR CHEMICAL HAZARDS

Do not use or store near heat or open flame.

STORAGE AND DISPOSAL

Pesticide Disposal: Do not contaminate water, food or feed by storage or disposal.

Pesticide wastes are acutely hazardous. Improper disposal of excess pesticide, spray mixture, or rinsate is a violation of Federal law. If these wastes cannot be disposed of by use according to label instructions, contact your State Pesticide or Environmental Control Agency, or the Hazardous Waste representative at the nearest EPA Regional Office for guidance.

Container Disposal: Triple rinse (or equivalent). Then offer for recycling or reconditioning, or puncture and dispose of in a sanitary landfill, or incineration, or, if allowed by state and local authorities, by burning. If burned, stay out of smoke.

Store in a cool, dry place and away from open flame and extreme heat. Store in such a manner as to prevent cross contamination with other pesticides, fertilizers, food, and feed. Store in original container and out of the reach of children, preferably in a locked storage area.

Handle and open container in a manner as to prevent spillage. If container is leaking invert to prevent leakage. If the container is leaking or material spilled for any reason or cause, carefully dam up spilled material to prevent runoff. Refer to Precautionary Statements on label for hazards associated with the handling of this material. Do not walk through spilled material. Absorb spilled material with absorbing type compounds and dispose of as directed for pesticides above. In spill or leak incidents, keep unauthorized people away. You may contact the Mobay Emergency Response Team for decontamination procedures or any other assistance that may be necessary. The Mobay Kansas City Emergency Response Telephone No. is 816-242-2582, or contact Chemtrec at 800-424-9300.

8-11-89

IMPORTANT:

This bulletin is not intended to provide adequate information for application. Before using the product carefully observe directions, cautionary statements and other information appearing on the product label. This product is sold subject to the Conditions of Sale set forth on the container label.

READ THE LABEL BEFORE USE

BD 9001

This is a specimen label. Mobay is not responsible for the accuracy of the information contained herein. As labels are subject to revision, always carefully read and follow the label on the product container.

BOLSTAR® 6

EMULSIFIABLE INSECTICIDE

RESTRICTED USE PESTICIDE

For retail sale to and use only by Certified Applicators or persons under their direct supervision, and only for those uses covered by the Certified Applicator's certification.

FOR THE CONTROL OF CERTAIN INSECTS ON COTTON AND SOYBEANS

ACTIVE INGREDIENTS:	
O-Ethyl *O*-[4-(methylthio)phenyl] *S*-propyl phosphorodithioate	64%
Aromatic Petroleum Distillate	15%
INERT INGREDIENTS:	21%
TOTAL	100%

Contains 6 lbs. *O*-Ethyl *O*-[4-(methylthio)phenyl]*S*-propyl phosphorodithioate per gallon.

EPA Reg. No. 3125-321 U.S. Patent No. 3,947,529

BOLSTAR and GUTHION are Reg. TMs of Bayer AG, Germany

RECOMMENDED USES

Cotton and Soybeans.

RESTRICTIONS ON WAREHOUSE STORAGE

Do not use, pour, spill or store near heat or open flame.

PACKAGING

5 gallon cans. Shipping weight 51 lbs.

BOLSTAR 6 is an excellent organophosphate insecticide proven effective against pests on cotton and soybeans. On cotton: BOLSTAR 6 controls the *Heliothis* complex, plus many other damaging pests. Through years of grower use, BOLSTAR 6 has delivered excellent control of both resistant and non-resistant tobacco budworms and cotton bollworms. BOLSTAR 6 may be tank mixed with GUTHION® 2L or 2S when GUTHION is used for control of additional insects not in the BOLSTAR 6 label such as boll weevil. BOLSTAR 6 may be applied with ground or aerial equipment with water or refined vegetable oil. On soybeans: BOLSTAR 6 controls a wide range of pests including thrips, leafhoppers, and beet armyworm.

AVISO

PRECAUCION AL USUARIO: Si usted no puede leer o entender inglés, no use este producto hasta que la etiqueta le haya sido explicada ampliamente.

(TO THE USER: If you cannot read or understand English, do not use this product until the label has been fully explained to you.)

DIRECTIONS FOR USE

It is a violation of Federal Law to use this product in a manner inconsistent with its labeling.

BOLSTAR 6 emulsifiable insecticide may be used for the control of susceptible as well as resistant *Heliothis virescens* (tobacco budworm) and *Heliothis zea* (cotton bollworm) on cotton. Control or suppression of additional insects as noted under "RECOMMENDED APPLICATIONS" may also be obtained. BOLSTAR 6 emulsifiable insecticide may be used for the control of certain insects infesting soybeans.

USE IN CHEMIGATION SYSTEMS

Types of Irrigation Systems

Apply BOLSTAR 6 only through sprinkler, including center pivot, lateral move, side roll, or overhead solid set irrigation systems. Do not apply BOLSTAR 6 through any other type of irrigation system.

GENERAL DIRECTIONS FOR ALL RECOMMENDED TYPES OF IRRIGATION SYSTEMS

Uniform Water Distribution and System Calibration

The irrigation system must provide uniform distribution of treated water. Crop injury, lack of effectiveness, or illegal pesticide residues in the crop can result from nonuniform distribution of treated water.

The system must be calibrated to uniformly apply the rates specified for chemigation application for specific crops. If you have questions about calibration, you should contact State Extension Service specialists, equipment manufacturers or other experts.

Chemigation Monitoring

A person knowledgeable of the chemigation system and responsible for its operation, or under the supervision of the responsible person, shall shut the system down and make necessary adjustments should the need arise.

Drift

Do not apply when wind speed favors drift beyond the area intended for treatment.

Required System Safety Devices

The system must contain a functional check valve, vacuum relief valve, and low pressure drain appropriately located on the irrigation pipeline to prevent water source contamination from backflow.

The pesticide injection pipeline must contain a functional, automatic quick-closing check valve to prevent the flow of fluid back toward the injection pump.

The pesticide injection pipeline must also contain a functional, normally closed, solenoid-operated valve located on the intake side of the injection pump and connected to the system interlock to prevent fluid from being withdrawn from the supply tank when the irrigation system is either automatically or manually shut down.

The system must contain functional interlocking controls to automatically shut off the pesticide injection pump when the water pump motor stops.

The irrigation line or water pump must include a functional pressure switch which will stop the water pump motor when the water pressure decreases to the point where pesticide distribution is adversely affected.

Systems must use a metering pump, such as a positive displacement injection pump (e.g., diaphragm pump) effectively designed and constructed of materials that are compatible with pesticides and capable of being fitted with a system interlock.

Using Water from Public Water Systems

DO NOT APPLY BOLSTAR 6 THROUGH ANY IRRIGATION SYSTEM PHYSICALLY CONNECTED TO A PUBLIC WATER SYSTEM. Public water system means a system for the provision to the public of piped water for human consumption if such system has at least 15 service connections or regularly serves an average of at least 25 individuals daily at least 60 days out of the year.

BOLSTAR 6 may be applied through any of the recommended types of irrigation systems which may be **supplied** by a public water system **only if** the water from the public water system is discharged into a reservoir tank prior to pesticide introduction. There shall be a complete physical break (air gap) between the outlet end of the fill pipe and the top or overflow rim of the reservoir tank of at least twice the inside diameter of the fill pipe. Before beginning chemigation, always make sure that the air gap exists and that there is no blockage of the overflow of the reservoir tank.

Any irrigation system using water supplied from a public water system must also meet the following requirements.

The pesticide injection pipeline must contain a functional, automatic, quick-closing check valve to prevent the flow of fluid back toward the injection pump.

The pesticide injection pipeline must contain a functional, normally closed, solenoid-operated valve located on the intake side of the injection pump and connected to the system interlock to prevent fluid from being withdrawn from the supply tank when the irrigation system is either automatically or manually shut down.

The system must contain functional interlocking controls to automatically shut off the pesticide injection pump when the water pump motor stops or in cases where there is no water pump, when the water pressure decreases to the point where pesticide distribution is adversely affected.

Systems must use a metering pump, such as a positive displacement injection pump (e.g., diaphragm pump) effectively designed and constructed of materials that are compatible with pesticides and capable of being fitted with a system interlock.

Do not apply when wind speed favors drift beyond the area intended for treatment.

Agitation

For application of BOLSTAR 6 alone, a chemical supply tank is not necessary for premixing since BOLSTAR 6 mixes well with water in the irrigation line. If a chemical supply tank is used, for application of BOLSTAR 6 alone or in combination with other chemicals, constant strong mechanical or hydraulic agitation must be maintained in the chemical supply tank during the entire period of application.

Chemical Supply Tank Dilution

If a chemical supply tank is used, you must determine the required amounts of BOLSTAR 6 and water to mix in the tank.

The amount of BOLSTAR 6 needed equals the number of pints of BOLSTAR 6 to be applied per acre multiplied by the number of acres to be chemigated.

The amount of emulsion needed equals the gallons of emulsion delivered per hour by the injection pump multiplied by the number of hours chemigation will take place.

The amount of water needed equals the amount of emulsion needed minus the amount of BOLSTAR 6 needed.

For example, if you want to apply 1 pint of BOLSTAR 6 per acre to 130 acres in 20 hours and your injection pump delivers 15 gallons per hour, you need: 1 pint BOLSTAR 6 per acre $\times$ 130 acres = 130 pints or 16.25 gallons of BOLSTAR 6. And, you need: 15 gallons per hour $\times$ 20 hours = 300 gallons of emulsion, minus 16.25 gallons of BOLSTAR 6 = 283.75 gallons of water.

Cleaning the Chemical Injection System

In order to accurately apply pesticides, the chemical injection system must be kept clean; free from chemical or fertilizer residues and sediments. Refer to your owners manual or ask your equipment supplier for the cleaning procedure for your injection system.

Flushing the Irrigation System

At the end of the application period, allow time for all lines to flush the pesticide through all nozzles before turning off irrigation water. To ensure the lines are flushed and free of pesticides, a dye indicator may be injected into the lines to mark the end of the application period.

SPRINKLER IRRIGATION SYSTEMS

All directions and requirements listed under the GENERAL DIRECTIONS AND REQUIREMENTS FOR ALL RECOMMENDED TYPES OF IRRIGATION SYSTEMS section of the label must be followed for sprinkler irrigation systems.

In addition, the following directions apply to sprinkler irrigation systems.

Do not apply when wind speed favors drift beyond the area intended for treatment.

It is recommended that nozzles in the immediate area of control panels, chemical supply tanks, pumps and system safety devices be plugged to prevent chemical contamination of these areas.

Center-Pivot and Automatic-Move Linear Systems

Inject the specified dosage per acre continuously for one complete revolution or move of the system. DO NOT USE END GUNS. The system should be run at maximum speed.

Solid Set and Manually Controlled Linear Systems

Injection should be during the last 30 to 60 minutes of a regular irrigation period or as a separate 30 to 60 minute application not associated with a regular irrigation. DO NOT USE END GUNS.

RECOMMENDED APPLICATIONS

CROP	INSECT	PINTS BOLSTAR 6	REMARKS
FIELD CROPS Cotton	Tobacco budworm, Cotton bollworm, Lygus (nymphs)	2/3 to 2	Apply specified dosage per acre by air or ground equipment alone or in sufficient water or refined vegetable oil for complete coverage. For best results during mid-to-late season or under hot dry conditions use a higher gallonage per acre for better coverage. For example, by air use up to 5 gallons of water per acre. When using oil, one (1) quart of total volume is recommended. Pre-mix material. Oil mixture should be sprayed within 12 hours to avoid compatibility problems. Always run a small jar compatibility test before large scale mixing as each batch of vegetable oil tends to differ greatly in composition. For application by sprinkler irrigation systems: Apply specified dosage per acre. Follow all directions given under the CHEMIGATION section of the label. For **light infestations of larvae,** apply 2/3 to 1 pint (1/2 to 3/4 lb. active) per acre, or for **moderate infestations** apply 1 to 1 1/3 pints (3/4 to 1 lb. active) per acre. Apply as needed with a 5 day interval. **Heavy infestations, or infestations of large larvae:** Apply 1 or 2 applications of the 2 pint (1 1/2 lb. active) rate per acre until infestations are brought under control, then reduce application rates to 2/3 to 1 1/3 pints per acre. Allow at least 5 days between applications. In California and Arizona the minimum use rate is 1 1/3 pints (1 lb. active) per acre, regardless of degree of infestation of tobacco budworm or cotton bollworm. **NOTE:** In areas of frequent rainfall, the higher rates and closer spray intervals (5-day minimum interval required) may be needed for adequate control. In **any case** do not apply more than a total of 8 pints per acre with the last application no closer to harvest than 21 days in California and Arizona, or 14 days in all other states. Do not graze treated fields.
	Fall armyworm, Beet armyworm, Pink bollworm	1 to 2	
	Fleahoppers	2/3	
	Suppression of: Lygus (adults), Whiteflies (except sweet potato whitefly), Spider mites (Carmine and two-spotted)	2/3 to 2	
	Thrips, (except Western flower thrips) Fleahoppers Plant bugs (For early to mid-season use in Southern cotton growing areas)	1/3	

CROP	INSECT	PINTS BOLSTAR 6	REMARKS
			Apply specified dosage per acre by air or ground equipment alone, in water, or in refined vegetable oil as stated above. Apply during early to mid-season when thrips, plant bug or fleahoppers reach economic threshold.
Soybeans (Except California)	Bean thrips, Potato leafhopper, Velvet bean caterpillar	⅓ to 1	Apply specified dosage per acre in sufficient diluent for complete coverage on a 10 to 14 day interval. Apply high rates for heaviest infestations. For application by sprinkler irrigation systems: Apply specified dosage per acre. Follow all directions given under the CHEMIGATION section of the label. **NOTE:** In areas of frequent rainfall the higher rates and closer spray intervals (10-day minimum interval required) may be needed for adequate control. Do not apply more than a total of **4** pints per acre within one season. The last application should be no closer to harvest than **60** days. Grazing or ensilage should be no earlier than 35 days from the last application.
	Plant bugs, *Lygus* spp., Green cloverworm, *Heliothis* spp. (podworm bollworm, corn earworm)	⅔ to 1⅓	
	Beet armyworm, Mexican bean beetle	1 to 1⅓	
	Three-cornered alfalfa hopper (suppression), Mexican bean beetle adults (suppression)		4-18-89

NOTE: BOLSTAR 6 when used on cotton may be tank mixed with GUTHION 2L or GUTHION 2S when GUTHION is used for control of cotton insects at the recommended rates of GUTHION.

Do not apply GUTHION alone within 14 days of harvest of cotton after applying tank mixes of BOLSTAR plus GUTHION. In California and Arizona, GUTHION alone should not be applied within 21 days of harvest of cotton after applying this tank mix.

SPECIAL LOCAL NEED REGISTRATIONS [SECTION 24(c)] FOR BOLSTAR 6*

CROP	INSECT	STATE EPA SLN NO.	DOSAGE	REMARKS
Chrysanthemums (Grown for stock plants)	Beet armyworms	Florida FL 82-0046	2 pints per acre	Apply in 100 gallons of water as a dilute full coverage spray.

* For distribution and use only within the state specified. This Special Local Need registration approved as of date of this publication is, however, subject to cancellation by the state. Contact your nearest Mobay Agricultural Chemicals Division Office for the current status and a copy of the complete individual supplemental Special Local Need labeling. This labeling must be in the possession of the user at time of pesticide application.

RE-ENTRY STATEMENTS

Do not apply this product in such a manner as to directly or through drift expose workers or other persons. The area being treated must be vacated by unprotected persons. Do not enter treated areas without protective clothing until sprays have dried. Because certain states may require more restrictive re-entry intervals for various crops treated with this product, consult your State Department of Agriculture for further information.

Written or oral warnings must be given to workers who are expected to be in a treated area or in an area about to be treated with this product. Oral warnings should include those statements found under the label heading "Precautionary Statements" and "Re-entry Statements" included on the label. When oral warnings are given, warnings shall be given in a language customarily understood by workers. Oral warnings must be given if there is reason to believe that written warnings cannot be understood by workers. Written warnings must include the following information: WARNING, area treated with BOLSTAR 6 on (date of application). Do not enter without appropriate protective clothing until sprays have dried. **If on skin,** remove contaminated clothing and wash skin immediately with soap and warm water. **If eyes are contaminated,** wash with flowing water for at least 15 minutes. **If swallowed,** vomiting should be induced. (See Precautionary Statements on label.)

PRECAUTIONARY STATEMENTS

HAZARDS TO HUMANS AND DOMESTIC ANIMALS

WARNING

CAUSES EYE IRRITATION. MAY BE HARMFUL IF SWALLOWED, INHALED, OR ABSORBED THROUGH THE SKIN.

Do not get in eyes. Do not get on skin. Do not breathe spray mist. Always wash thoroughly with soap and warm water after handling. Wash contaminated clothing with soap and hot water before reuse. Do not contaminate feed or food. Keep out of reach of children.

Should poisoning occur, obtain prompt medical aid. Prolonged exposure will result in cholinesterase depression.

To Physician: The toxic mode of action is to depress cholinesterase. The victim should be treated symptomatically. Atropine sulfate when administered in conjunction with 2-PAM is antidotal.

STATEMENTS OF PRACTICAL TREATMENT

If in eyes, immediately flush eyes with plenty of water for at least 15 minutes. For further treatment, get medical attention.

If swallowed, vomiting should be induced. Administer water freely and induce vomiting by giving one dose (½ ounce or 15 ml.) of syrup of ipecac. If vomiting does not occur within 10 to 20 minutes, administer second dose. If syrup of ipecac is not available, induce vomiting by sticking finger down throat. Repeat until vomit fluid is clear. Never give anything by mouth to an unconscious person. Professional medical assistance should be secured immediately.

If on skin, wash thoroughly with soap and warm water.

ENVIRONMENTAL HAZARDS

This pesticide is toxic to fish and wildlife. Use with care when applying in areas frequented by wildlife or adjacent to any body of water. When BOLSTAR 6 is applied by air at rates of 1½ pints or greater per acre, applications should be made no closer than 100 feet of shorelines of fish bearing waters. Keep out of lakes, streams and ponds. Do not apply when weather conditions favor runoff or drift from target areas. Do not contaminate water by cleaning of equipment or disposal of wastes.

This product is toxic to bees exposed to direct treatment or residues on crops or weeds. Do not apply this product, or allow it to drift to crops or weeds on which bees are foraging. Additional information may be obtained from your Cooperative Extension Service.

Do not use on other crops used for food or forage. Use only according to label directions. Application at rates above those shown may result in illegal crop residues. Do not treat food crops grown in the greenhouse.

Crops to be grown in rotation with cotton should not be planted within 60 days after last application of BOLSTAR 6.

PHYSICAL OR CHEMICAL HAZARDS

Do not use or store near heat or open flame.

STORAGE AND DISPOSAL

Pesticide Disposal: Do not contaminate water, food or feed by storage or disposal. Pesticide wastes are toxic. Improper disposal of excess pesticide, spray mixture, or rinsate is a violation of Federal Law. If these wastes cannot be disposed of by use according to label instructions, contact your State Pesticide or Environmental Control Agency, or the Hazardous Waste representative at the nearest EPA Regional Office for guidance.

Container Disposal: Triple rinse (or equivalent). Then offer for recycling or reconditioning, or puncture and dispose of in a sanitary landfill, or by other procedures approved by State and local authorities.

Store in a cool, dry place, away from open flame and extreme heat, and in such a manner as to prevent cross contamination with other pesticides, fertilizers, food, and feed. Store in original container and out of the reach of children, preferably in a locked storage area.

Handle and open container in a manner as to prevent spillage. If container is leaking invert to prevent leakage. If the container is leaking or material spilled for any reason or cause, carefully dam up spilled material to prevent runoff. Refer to Precautionary Statements on label for hazards associated with the handling of this material. Do not walk through spilled material. Absorb spilled material with absorbing type compounds and dispose of as directed for pesticides above. In spill or leak incidents, keep unauthorized people away. You may contact the Mobay Emergency Response Team for decontamination procedures or any other assistance that may be necessary. The Mobay Kansas City Emergency Response Telephone No. is 816-242-2582, or:

FOR 24-HOUR EMERGENCY MEDICAL ASSISTANCE CALL 1-800-F-A-S-T-M-E-D (327-8633).

FOR CHEMICAL EMERGENCY: Spill, leak, fire, exposure, or accident call **CHEMTREC** 1-800-424-9300

IMPORTANT:

This bulletin is not intended to provide adequate information for application. Before using the product, read and carefully observe directions, cautionary statements and other information appearing on the product label. This product is sold subject to the Conditions of Sale set forth on the container label.

READ THE LABEL BEFORE USE

BS 9001

This is a specimen label. Mobay is not responsible for the accuracy of the information contained herein. As labels are subject to revision, always carefully read and follow the label on the product container.

DEF® 6

EMULSIFIABLE DEFOLIANT

FOR FAST AND EFFECTIVE COTTON DEFOLIATION.

Fast, Clean Leaf Drop For Better Yields of Clean, High-Quality Lint!

ACTIVE INGREDIENT:	
S,S,S-Tributyl phosphorotrithioate	70.5%
INERT INGREDIENTS:	29.5%
TOTAL	100.0%

Contains 6 lbs. *S,S,S*-Tributyl phosphorotrithioate per U.S. gallon.

EPA Reg. No. 3125-282

DEF is a Reg. TM of Bayer AG, Germany

Dropp is a TM of Nor-AM Chemical Corp.

RECOMMENDED USES

For bottom and complete cotton defoliation. May be used in all power-operated ground and aircraft equipment.

RESTRICTIONS ON WAREHOUSE STORAGE

Do not store near fertilizers, seed, insecticides and fungicides. Do not store near heat or open flame. Store in a cool well-ventilated place.

DEF 6 provides superior cotton defoliation. And DEF 6 comes in a bright green can that is easy to identify.

DEF 6 gives fast, clean leaf drop. It works by speeding up the natural defoliation process. DEF stimulates the formation of the abscission layer where the stem joins the stalk, causing leaves to drop cleanly—leaf, stem and all.

And DEF 6 works fast—up to 90 to 100% defoliation within four to seven days. This faster, more complete defoliation induces faster opening of mature bolls, and faster development of later-maturing bolls. With faster-opening bolls, you can pick earlier; before boll rot or Fall rains have the chance to destroy yield and quality of both lint and seed.

With the complete leaf drop, you can pick earlier, with less spindle clogging. Harvesting is more efficient, with less trash and stain that lowers lint quality.

DEF 6 is recommended for complete and/or bottom defoliation. It works with or without the presence of dew or rainfall. DEF 6 is also recommended for defoliation of rank cotton.

AVISO

PRECAUCION AL USUARIO: Si usted no puede leer o entender inglés, no use este producto hasta que la etiqueta le haya sido explicada ampliamente.

(TO THE USER: If you cannot read or understand English, do not use this product until the label has been fully explained to you.)

DIRECTIONS FOR USE

It is a violation of Federal Law to use this product in a manner inconsistent with its labeling.

Under favorable conditions, DEF 6 gives effective defoliation of cotton within four to seven days after application. DEF 6 drops the leaves in a green condition, and does not appear effective until defoliation actually begins. When continued low temperatures prevail at night (below 60°F.), complete defoliation may require nine to fourteen days. DEF 6 may be used whether or not moisture is present on cotton, and the presence or absence of dew or rain, during or following application, does not adversely affect the performance. DEF 6 does not suppress second growth, especially where rainfall follows defoliation. Schedule all applications in accordance with local recommendations. Consult your State Agricultural Experiment Station or Extension Service for specific use information.

Do not apply this product through any type of irrigation system.

COMPLETE DEFOLIATION

PROPER TIMING: DEF 6 should be applied to mature cotton plants when 50% or more of the bolls are open, and 7 to 10 days prior to anticipated picking. Plants are considered suitably mature if the youngest bolls (1) cannot be dented by pressure between thumb and forefinger or (2) cannot easily be cut through with a sharp knife.

THOROUGH COVERAGE: All leaves must be treated to accomplish complete defoliation.

RAINFALL: DEF 6 will provide efficient complete defoliation when properly applied, whether rain or dew is present during or after application, but does not suppress second growth, especially in areas where rainfall follows treatment sufficient to delay picking.

DEFOLIATION AND INHIBITION OF REGROWTH (Except California)

To obtain defoliation, and to provide improved inhibition of regrowth or secondary growth, apply DEF 6 according to label directions, at rates of 1.0 to 1 1/3 pints per acre, as a tank mix combination with 0.066 to 0.1 lb. of Dropp 50WP defoliant and regrowth inhibitor. Best activity will be obtained where the tank mix is applied to mature cotton plants when 60% or more of the bolls are open and where the average mean 24-hour temperature before and after application is above 60°F. Use the lower rates of each product when conditions are favorable for rapid defoliation (high temperatures and high humidity). Adverse conditions, such as low temperature, may require the use of the higher dosages and/or longer times for complete defoliation. A second application of labeled rates of the tank mix combination, or of DEF 6 alone, may be made where necessary.

Do not apply DEF 6 plus Dropp 50% W.P. to immature cotton (less than 60% of bolls open) or use higher than recommended dosages of either product in tank-mix combination, since desiccation and leaf freezing may occur.

Apply as a dilute spray in 5 to 10 gallons of water per acre by air or 10 to 25 gallons of water per acre with ground equipment. Do not apply in oil carrier.

Always follow the mixing suggestions found on the label or on the appropriate adjuvant label.

It is suggested that adjuvants be tested on a limited scale prior to large scale use. Contact your local Mobay representative for additional information.

When applying DEF 6 defoliant and Dropp 50WP defoliant as a tank mix combination, the following order of mixing is recommended: (1) water, (2) Dropp 50WP (according to directions on Dropp label), (3) adjuvant, if used, and (4) DEF 6 after Dropp has completely dispersed.

BOTTOM DEFOLIATION

Much seed cotton is lost due to rot and exposure of the bottom bolls which are the first to mature. An application of DEF 6, using shielded drop nozzles (four 8001 fans or two 1/8 K 1 or 1.5 flooding) to the lower leaves where all the bolls are mature, will cause the treated leaves to drop, exposing the mature bolls to air movement and sunlight. By removing the top 8 to 12 rows of spindles from a picker, the exposed bolls may be harvested. DEF 6 leaves the plant green and flexible so that no damage is done by the picker to the untreated top leaves and immature bolls. Complete defoliation and picking can follow, usually in 3 to 4 weeks. Other additional benefits gained by bottom defoliation, even without picking the exposed bolls, are:

- Reduction of boll rot (up to 1/4 to 1/2 bale per acre).
- Faster rate of opening of lower mature bolls.
- Reduced lodging.
- Higher grade of cotton.
- More cotton from fewer pickings.

DILUTE SPRAYS

MIXING: DEF 6 forms an emulsion when diluted with water, and is suitable for use in all power-operated ground and aircraft sprayers. It is non-corrosive and non-clogging to spray equipment, and does not constitute a fire hazard.

DEF 6 may be added to diesel oil, in accordance with local recommendations, to increase effectiveness, particularly where night temperatures drop to below 60°F., where plants have had insufficient moisture, and on storm-proof varieties of cotton. The use of 3 to 5 gallons of oil per acre is generally required for aircraft spraying, while 5 to 10 gallons is recommended for ground applications. Where oils are used, particular care should be exercised to avoid spray drift to adjacent crops.

DOSAGE: Use specified dosage of DEF 6 per acre in the amount of water or diesel oil necessary to give complete coverage. The quantity of water or oil to be used per acre is dependent somewhat upon the equipment used. In all cases, however, sufficient spray should be applied to treat all leaves. Larger droplets more effectively reach the lower leaves. The gallons of water ordinarily required for adequate coverage are recommended in the table below for both air and ground spray applications. Make sure the spray pattern is that a uniform wetting of leaf surfaces results.

Addition of Adjuvants: DEF or DEF + Dropp may also be applied with an adjuvant (1) blends of vegetable or petroleum-based oils and suitable emulsifiers, or (2) non-ionic surfactants to improve spray coverage. DEF + Dropp tank mix combinations are *not* recommended for use in California. Only products which are exempt from tolerance under 40 CFR 180.1001 may be used. Prior to mixing in the spray tank, it is suggested that a compatibility test be run to determine the compatibility of DEF or DEF + Dropp with the adjuvant to be used.

RECOMMENDED APPLICATIONS

CROP	PINTS DEF 6	GALLONS OF WATER AIR	GALLONS OF WATER GROUND	DIRECTIONS
Cotton	1⅓ to 2	5 to 10	15 to 25	**COMPLETE DEFOLIATION** Apply specified dosage per acre in recommended gallons of water to give thorough coverage to mature cotton when 50% or more of the bolls are open. Applications can be made until first frost, but should not be used within 7 days of picking. For rank cotton, see recommendation below.
	1 to 1½	—	10 to 15	**BOTTOM DEFOLIATION** Apply specified dosage* per acre in recommended gallons of water with spray directed only on the lower part of the plant in the area where only mature bolls are found. Do not use within 7 days of picking. * The bottom defoliation rate should be proportional to the fraction of the plant to be defoliated.
Rank cotton	3	5 to 10	15 to 25	**COMPLETE DEFOLIATION OF RANK COTTON** Apply specified dosage per acre in the recommended amount of water to give thorough coverage of mature cotton when 50% or more of the bolls are open. Schedule 2 applications of the specified dosage two to six weeks apart for complete defoliation of rank cotton. Applications can be made until first frost, but not within 7 days of picking. 8-15-88

SPECIAL LOCAL NEED REGISTRATIONS [SECTION 24(C)] FOR DEF 6*

CROP	USE	STATE EPA SLN NO.	DOSAGE	REMARKS
Cotton & Rank Cotton	Defoliation	Texas TX 81-0045	1⅓ to 3 pints per acre.	Apply in not less than 1½ pints of once refined vegetable oil.

* For distribution and use only within the state specified. This Special Local Need registration approved as of date of this publication is, however, subject to cancellation by the state. Contact your nearest Mobay Agricultural Chemicals Division Office for the current status and a copy of the complete individual supplemental Special Local Need labeling. This labeling must be in the possession of the user at time of pesticide application.

IMPORTANT

Observe all cautions and limitations on labeling of all products used in mixtures. Specifically, when Dropp 50WP is tank mixed with DEF 6 for use on cotton in areas where citrus is grown, observe buffer zone and other restrictions on the Dropp 50WP label and/or in manufacturer's recommendations for Dropp 50WP.

RESTRICTIONS: Do not use on other crops used for food or forage. Avoid spray drift to susceptible plants other than cotton, as this product may injure or defoliate other crops. (Coarse sprays are less likely to drift.)

SPRAY EQUIPMENT CLEANING AND DECONTAMINATION

Do not allow pesticide spray mixtures to dry in spray equipment. Dried pesticide residues may become resuspended and damage other crops if uncleaned spray equipment is used to apply other products during the same or the following year.

Thoroughly clean spray equipment immediately before and after applying DEF 6 and before using spray equipment the following year. Use a cleaner such as "Spic and Span," "Fantastic," or "Formula 409." Wash the spray tank, lines, nozzles and remove dried deposits from exterior surfaces, especially aircraft fabric. This cleaning is particularly important if the spray equipment has been used to apply Dropp 50WP or products containing chlorates.

Immediately after applying DEF 6, remove all unused spray mixture and follow directions on this label for disposal. Thoroughly clean the spray tank, lines, nozzles, and exterior surfaces of spray equipment. Use one of the cleaners listed above. Follow directions on this label for disposal of wash and rinse water.

STORAGE AND DISPOSAL

Pesticide Storage: Do not store near fertilizers, seeds, insecticides, or fungicides.

Pesticide Disposal: Do not contaminate water, food or feed by storage or disposal.

Pesticide wastes are toxic. Improper disposal of excess pesticide, spray mixture, or rinsate is a violation of Federal Law. If these wastes cannot be disposed of by use according to label instructions, contact your State Pesticide or Environmental Control Agency, or the Hazardous Waste representative at the nearest EPA Regional Office for guidance.

Container Disposal: Triple rinse (or equivalent). Then offer for recycling or reconditioning, or puncture and dispose of in a sanitary landfill, or by other procedures approved by State and local authorities.

Store in a cool, dry place, away from open flame and extreme heat, and in such a manner as to prevent cross contamination with other pesticides, fertilizers, food, and feed. Store in original container and out of the reach of children, preferably in a locked storage area.

Handle and open container in a manner as to prevent spillage. If container is leaking invert to prevent leakage. If the container is leaking or material spilled for any reason or cause, carefully dam up spilled material to prevent runoff. Refer to Precautionary Statements on label for hazards associated with the handling of this material. Do not walk through spilled material. Absorb spilled material with absorbing type compounds and dispose of as directed for pesticides above. In spill or leak incidents, keep unauthorized people away. You may contact the Mobay Emergency Response Team for decontamination procedures or any other assistance that may be necessary. The Mobay Kansas City Emergency Response Telephone No. is 816-242-2582, or contact Chemtrec at 800-424-9300.

PRECAUTIONARY STATEMENTS

HAZARDS TO HUMANS AND DOMESTIC ANIMALS

WARNING

May be fatal if swallowed, inhaled, or absorbed through the skin. Do not breathe spray mist. Do not get in eyes, on skin, or on clothing. Wash with soap and warm water after handling and before eating or smoking.

For protection during application wear a hat, long sleeved shirt and long legged trousers. In addition, mixer-loaders must wear rubber or neoprene gloves.

Do not contaminate feed or food. Keep out of reach of children.

STATEMENTS OF PRACTICAL TREATMENT

If swallowed, vomiting should be induced. Administer water freely and induce vomiting by giving one dose (½ oz. or 15 ml.) of syrup of ipecac. If vomiting does not occur within 10 to 20 minutes, administer second dose. If syrup of ipecac is not available, induce vomiting by sticking finger down throat. Repeat until vomit fluid is clear. Never give anything by mouth to an unconscious person. Have patient lie down and keep quiet. Professional medical assistance should be secured immediately.

In case of contact, remove contaminated clothing and flush skin and eyes with plenty of water. Wash contaminated clothing with soap and hot water before reuse. If poisoning occurs, obtain prompt medical aid.

TO PHYSICIAN: This product inhibits cholinesterase resulting in stimulation of the central nervous system, the parasympathetic nervous system and the somatic motor nerves. Poisoning with this product also results in cardiovascular and respiratory symptoms which must be treated as separate pathological entities apart from the cholinergic effects.

Use atropine sulfate to reverse cholinergic symptoms; and maintain a systematic, symptomatic treatment of the cardiovascular and respiratory effects, even though the cholinergic symptoms have ceased. Do not give central nervous system depressants. Watch for pulmonary edema, which may develop in serious cases of poisoning even after 12 hours. The use of oxygen is recommended in case of respiratory distress.

ENVIRONMENTAL HAZARDS

This product is toxic to fish. Keep out of lakes, streams or ponds. Do not apply where runoff is likely to occur. Do not apply when weather conditions favor drift from areas treated. Do not contaminate water by cleaning of equipment or disposal of wastes. Apply this product only as specified on the label.

IMPORTANT:

This bulletin is not intended to provide adequate information for application. Before using the product, read and carefully observe directions, cautionary statements and other information appearing on the product label. This product is sold subject to the Conditions of Sale set forth on the container label.

READ THE LABEL BEFORE USE

DF 9002

This is a specimen label. Mobay is not responsible for the accuracy of the information contained herein. As labels are subject to revision, always carefully read and follow the label on the product container.

DI-SYSTON®

15% GRANULAR SYSTEMIC INSECTICIDE

RESTRICTED USE PESTICIDE

Acute human, oral, dermal, and inhalation toxicity
For retail sale to and use only by Certified Applicators or persons under their direct supervision and only for those uses covered by the Certified Applicator's certification.

BETTER CROP PROTECTION FOR HIGHER CROP YIELDS

ACTIVE INGREDIENT:	
O, O-Diethyl *S*-[2-(ethylthio) ethyl] phosphorodithioate	15%
INERT INGREDIENTS:	85%
TOTAL	100%

EPA Reg. No. 3125-172
DI-SYSTON is a Reg. TM of Bayer AG, Germany

RECOMMENDED USES

Alfalfa	Cotton	Pecans**
Barley	Fruit trees (non-bearing)	Potatoes
Beans	Lettuce	Raspberries***
Broccoli	Ornamental flowers	Sorghum
Brussels sprouts	Ornamental trees and shrubs	Soybeans
Cabbage	Peanuts	Strawberries (propagating plants only)
Cauliflower	Peas	Tobacco
Coffee*		Wheat (Fall)
Clover		

*Puerto Rico only.
**South Central and Southwestern states.
***Nursery stock in Northeast states only.

RESTRICTIONS ON WAREHOUSE STORAGE

Do not store adjacent to or near proximity of foodstuffs, drugs, feeds or any other material intended for consumption by humans or animals. Do not store adjacent to toys, cosmetics, dishes, furniture or clothing. Do not store next to herbicides. Protect from rain and moisture. Store in a cool, well-ventilated place. In case of puncture, use overslip bag and securely close.

PACKAGING

Packaged in 10 pound bags, 5 per case. Shipping weight 51 pounds.

Packaged in 50 pound bags. Shipping weight 52 pounds.

DI-SYSTON 15% Granular systemic insecticide effectively controls insects on alfalfa, cotton, potatoes, sugar beets, ornamentals and many other field and vegetable crops. It gives up to seven weeks protection when applied at planting time as a side-dressing or topically as recommended. DI-SYSTON 15% Granular systemic insecticide provides "built-in" protection because it acts from within the sap stream of the plant. It stops harmful insects before they get started, but does not destroy populations of parasites, predators and pollinators. These beneficial insects continue to live to assist in controlling pests. When DI-SYSTON 15% Granular systemic insecticide is applied at planting time, it is absorbed into the growing plant where it provides many weeks of protection against sucking insects, even season-long protection on some crops. When applied as a side-dressing, DI-SYSTON 15% Granular systemic insecticide moves with soil moisture and is absorbed by the roots.

PELIGRO

PRECAUCION AL USUARIO: Si usted no puede leer o entender inglés, no use este producto hasta que la etiqueta le haya sido explicada ampliamente.

(TO THE USER: If you cannot read or understand English, do not use this product until the label has been fully explained to you.)

DIRECTIONS FOR USE

It is a violation of Federal Law to use this product in a manner inconsistent with its labeling.

DI-SYSTON 15% Granular systemic insecticide enters the plant through the root system. The insecticide, when placed near the roots, is picked up and moves in the sap stream to all parts of the plant, where it kills insects as they feed on the plant.

This formulation of DI-SYSTON 15% Granular is highly versatile, and may be applied by various methods. It can be drilled in the furrow with seed, broadcast or banded on each side of the seed furrow. Foliage broadcast application may be applied by ground or air equipment. Time and method of application and limits for use are discussed in detail for each specific crop under "Remarks" below.

As a preplant treatment for ornamental flowers, shrubs and trees, apply granules evenly to seed bed and mix into soil; seeds or transplants are then placed in the normal manner. Established plants may be treated by distributing granules uniformly under foliage canopy. Work the granules into top soil and water thoroughly. The chemical will leach into the root zone and be taken up by the root system.

RECOMMENDED APPLICATIONS

IMPORTANT: When used on cotton, beans or lettuce at the maximum recommended rates under adverse conditions such as extremely cool or wet or extremely dry weather, DI-SYSTON systemic insecticide may cause some delay in emergence, stunting of seedlings or reduction of stand. Such damage may be more pronounced in light, sandy soils. Plant injury may also occur, or may be enhanced when certain preemergent herbicides are used in conjunction with DI-SYSTON,

CROP	INSECT	DOSAGE DI-SYSTON 15% GRAN.	REMARKS
FIELD CROPS Alfalfa Clover	Aphids Mites	6.7 pounds per acre	Apply only one broadcast treatment but not within 28 days of grazing or cutting for forage. Make only one application per crop season.
Barley	Aphids Mites	6.7 pounds per acre	Apply by drilling or broadcast at planting time, or as a broadcast after emergence. Do not apply more than twice per season, nor within 60 days of harvest for grain. Allow a minimum of 21 days between applications. Do not graze or cut for forage within 30 days of treatment.
Cotton	Aphids Mites Thrips	5 to 8 ozs. per 1000 feet of row (for any row spacing*), or 4 to 6.7 pounds per acre (40 inch row spacing)	**At Planting:** Place the granules in the seed furrow, or in a band on each side of the furrow. **RATE CHART FOR PLANTING TIME APPLICATION** **Heavy Soils** 8 ozs. 15% Gran./1000 ft. of row; 6.7 lbs. 15% Gran./acre **Medium Soils** 6.6 ozs. 15% Gran./1000 ft. of row; 5.3 lbs. 15% Gran./acre **Light Soils** 5 ozs. 15% Gran./1000 ft. of row; 4 lbs. 15% Gran./acre **Post-Plant:** (Irrigated Cotton Only) If an additional treatment is necessary, make one side-dress application, or a single broadcast application over the foliage. Rates up to 13.3 pounds per acre can be applied to first-squaring cotton (90 days before harvest), and rates up to 6.7 pounds per acre can be applied to within 28 days of harvest. Do not apply more than twice per season. Allow a minimum of 21 days between applications. Read "IMPORTANT" statement under "RECOMMENDED APPLICATIONS" for additional information before use on this crop.

CROP	INSECT	DOSAGE DI-SYSTON 15% GRAN.	REMARKS
Cotton (cont'd)	Aphids Mites Thirps	2.5 to 3.3 ounces per 1000 feet of row (for any row spacing*), or 2 to 2.7 pounds per acre (40 inch row spacing)	**Hill Dropped Cotton:** Apply specified dosage with the seed at planting. Use the lower rate where hill spacing is greater than 14 inches apart, or where applications are made to light, sandy soils. Read "IMPORTANT" statement under "RECOMMENDED APPLICATIONS" for additional information before use on this crop.
Peanuts	Aphids Leafhoppers Mites Thirps	7.4 to 14.7 ozs. per 1000 feet of row (for any row spacing*), or 6.7 to 13.3 pounds per acre (36 inch row spacing)	Place the granules in a band on each side of the seed furrow at planting, or as a side-dressing after plants are established. Do not apply directly on the seed. Where necessary, granules may be applied in a band over the row, or as a side-dressing at pegging time. Do not feed treated vines.
Sorghum	Aphids (Greenbugs) (See "NOTE" below)	6 to 8 ozs. per 1000 feet of row (for any row spacing*), or 5 to 6.7 pounds per acre (40 inch row spacing)	**Application at Planting: In-Furrow**—Place the granules in the seed furrow approximately ½ inch above the seed at planting. Do not apply directly on the seed. See "NOTE" below.
		8 ozs. per 1000 feet of row (for any row spacing*), or 6.7 pounds per acre (40 inch row spacing)	**Application at Planting: Band**—Place the granules in a 4 to 6 inch band directly behind the planter shoe, in front of the press wheel. Do not apply directly on the seed. See "NOTE" below. **Post-planting Application: Whorl**—Apply specified dosage by broadcasting into the whorl. Do not apply more than 3 times per season, nor within 30 days of harvest of grain, or 45 days of harvest for forage or fodder. See "NOTE" below.
	NOTE: (1) Post-planting broadcast applications into the whorl may be made following an at-planting application. (2) Due to greenbug resistance in certain areas, DI-SYSTON used alone may not provide satisfactory control. Consult your local agricultural advisor or Extension Service for recommendations.		
Soybeans	Thrips (except flower thrips) Mexican bean beetle	6.7 pounds per acre (36 inch row spacing)	Place the granules in a 4 inch band over the seed furrow behind the planter shoe and in front of the press wheel, or in a band on each side of the seed furrow at planting time. Do not apply directly on the seed.
Tobacco	Aphids Flea beetles	**Transplant Beds:** 1 oz. per 100 square feet of bed	Apply evenly over bed area. Work the granules into the top 2 to 3 inches of soil and seed beds in normal manner. An alternative treatment may be made by broadcasting the granules evenly over the plants after emergence, and watering thoroughly. Do not make more than one application to transplant beds.
		Field Application: 20 to 40 ozs. per 1000 ft. of row (for any row spacing*), or 13.3 to 26.7 lbs. per acre (broadcast, or with 48 inch row spacing)	Apply in a 6 to 12 inch band over the row, or as a broadcast treatment immediately before transplanting. Incorporate the granules by working them into the top 2 to 3 inches of soil. Plant crop in the usual way. Do not make more than 1 field application per crop season, regardless of DI-SYSTON formulation used.
	Southern potato wireworm	13.3 to 26.7 lbs. per acre	When applied as described for "Field Application" above, DI-SYSTON 15% Granular will aid in control of this wireworm.
Wheat (Fall)	Hessian Fly Aphids (Greenbugs) (Oat bird cherry) Grasshoppers	6.7 pounds per acre	Apply by drilling or broadcast at planting time in the Fall. Make only one application, and do not graze or cut for forage within 75 days of treatment. **NOTE:** Due to greenbug resistance in certain areas, DI-SYSTON used alone may not provide satisfactory control. Consult your local agricultural advisor or Extension Service for recommendations.
SEED FIELD CROPS Soybeans	Aphids Leafhoppers Mites Thrips	8 ounces per 1000 feet of row (for any row spacing*), or 6.7 pounds per actre (40 inch row spacing)	Place the granules in a band on each side of the seed furrow at planting or as a side-dressing after plants are established. Do not pasture or use treated crop for feed, food or forage. Do not apply directly to seed.
FRUIT Coffee (Puerto Rico only)	Leafminer	2 to 4 grams per foot of tree height	Distribute granules uniformly under the tree canopy. Do not apply more than twice per year. Make no more than one pre-harvest and one post-harvest application. Do not make the pre-harvest application within 90 days of harvest.
NON-BEARING FRUIT Fruit trees: Apples Apricots Cherries Crab apples Peaches Pears Plums Prunes	Aphids Lace bug Leafhoppers Mites Thrips Whiteflies	2.5 to 27 ozs. per tree, or 2.5 ounces per inch of trunk diameter	Spread granules uniformly from trunk to drip line on all sides. Work into soil, and water thoroughly. The higher dosage is recommended for heavy organic or muck soils. For transplanting, equivalent quantities should be applied into planting hole. Do not apply to trees that will bear fruit during current crop year.
Strawberries (Propagating plants only)	Aphids Mites (except cyclamen mite)	17 to 34.4 ozs. per 1000 ft. of row (for any row spacing*), or 13.3 to 26.7 lbs. per acre (42 inch row spacing)	Apply by side-dressing on each side of the row, or by broadcasting over the row as needed. Do not use fruit from treated plants for food purposes.
Raspberries (Nursery stock in Northeast states only)	Aphids	53.3 lbs. per acre	**Band Application:** Apply specified dosage in a 24 inch band over the row to new or established plantings. Work granules 2 to 3 inches into the soil, using a cultivator or other suitable equipment. A second application of granules may be applied to the soil later in the season. Consult your State Agricultural Experiment Station or Extension Service for proper timing of applications.
NUTS Pecans (South Central and Southwestern states) (see **Note 1**)	Aphids	10 to 20 pounds per acre	Apply by treating 6 foot bands of soil on both sides of the trees. The bands should be located under the main drip area of the trees. Work the granules into the upper 2 to 3 inches of soil. On irrigated pecans, apply water as soon as possible after treatment. Make only 1 application per

CROP	INSECT	DOSAGE DI-SYSTON 15% GRAN.	REMARKS
			season. Early Spring treatment is recommended. Do not harvest nuts within 80 days of treatment. Do not graze grass under treated trees. Do not mix this product with Phosalone.
VEGETABLES Beans	Aphids Leafhoppers Mexican bean beetle (larvae) Mites Thrips	6 to 12 ozs. per 1000 ft. of row (for any row spacing*), or 6.7 to 13.3 lbs. per acre (30 inch row spacing)	For snap or green Lima beans, place the granules in a band on each side of the seed furrow at planting time only. For dry beans, apply as above at planting time or as a side-dressing on each side of the row after plants become established, but not within 60 days of harvest. Do not apply directly on the seed. Do not apply more than once per season. Read "IMPORTANT" statement under "RECOMMENDED APPLICATIONS" for additional information before use on this crop. Do not feed vines or hay.
Brussels sprouts	Aphids Flea beetles Leafhoppers Mites Thrips	7.4 ounces per 1000 ft. of row (for any row spacing*), or 6.7 pounds per acre (36 inch row spacing)	**Transplant Seed Beds:** Apply granules evenly over bed area at the rate of 6.7 pounds per acre. Work into the top 2 to 3 inches of soil and seed beds in the normal manner. An alternative seed bed treatment may be made by broadcasting the granules evenly over the plants after emergence, and watering thoroughly. Do not apply more than one seed bed treatment.
Cauliflower	Aphids Flea beetles Leafhoppers		**Field Applications:** Place the granules in a band on each side of the seed furrow or transplanted row at planting, and where necessary, as a side-dressing after plants become established. Do not apply directly on the seed. Do not apply more than two field applications per season, nor within 30 days of harvest for Brussels sprouts, or 40 days of harvest for cauliflower. Allow a minimum of 21 days between applications.
Broccoli	Aphids Flea beetles Leafhoppers	7.4 ounces per 1000 ft. of row (for any row spacing*), or 6.7 pounds per acre (36 inch row spacing)	**Transplant Seed Beds:** Apply granules evenly over bed area at the rate of 6.7 pounds per acre. Work into the top 2 to 3 inches of soil, and seed beds in normal manner. An alternative seed bed treatment may be made by broadcasting the granules evenly over the plants after emergence, and watering thoroughly. Do not apply more than one seed bed treatment. **Field Applications:** Place the granules as a band on each side of the seed furrow or transplanted row at planting, or as a side-dressing after plants become established. Do not apply directly on the seed. Do not apply more than one field application per season. Do not use within 14 days of harvest for broccoli, or 42 days of harvest for cabbage.
Cabbage (Includes tight heading varieties of Chinese Cabbage)	Aphids Flea beetles		
	Root aphid	11.3 ozs. per 1000 feet of row (for any row spacing*)	
Lettuce	Aphids Leafhoppers Mites Thrips	4 to 8 ozs. per 1000 ft. of row (for any row spacing*), or 6.7 to 13.3 lbs. per acre (20 inch row spacing)	**Spring- or Summer-Seeded Lettuce:** Use the higher rate on heavy organic soils. Place the granules in a band on each side of the row at seeding time only. Do not apply directly on the seed. Do not apply to transplanted lettuce.
Peas	Aphids Leafhoppers	6.7 to 16.7 lbs. per acre	Apply by drilling or broadcasting at planting time, or as a side-dress on each side of the row after emergence. Do not apply more than once per season, nor within 50 days of harvest. Do not feed vines or hay.
Potatoes	Aphids Flea beetles (except tuber flea beetles) Leafhoppers Potato psyllid	15 to 23 ozs. per 1000 ft. of row (for any row spacing*), or 13.5 to 20.5 lbs. per acre (38 inch row spacing)	**Row Treatments:** Place the granules in the seed furrow or in a band on each side of the seed furrow at planting. An additional application may also be made as a side-dressing after plants become established. Do not make more than two applications per season, nor apply within 75 days of harvest. Higher rate to be used in organic soils. See **NOTE 2** below for "foliar applications." Do not mix this product with Phosalone. **NOTE 3:** Colorado Potato Beetle populations in the northeastern states (except Maine), as well as Virginia and North Carolina, have developed resistance to insecticides, including DI-SYSTON. Consult your local Extension Service or MOBAY representative for details.
	Southern potato wireworm Colorado potato beetle (early-season reduction)	20 to 26.7 lbs. per acre	**Broadcast Application:** Apply specified dosage per acre as a pre-plant broadcast application. Work granules into soil to a depth of 2 to 3 inches, using a disc or other suitable equipment. Plant crop in usual manner. A postemergence side-dress may also be made at rates of 15 to 23 ounces per 1000 feet of row (for any row spacing*) after the plants become established. Do not make more than two applications per season, nor apply within 75 days of harvest. See **NOTE 2** below for "foliar applications." Do not mix this product with Phosalone. **NOTE 3:** Resistance to Colorado Potato Beetle has occurred in some areas of the Northeast. Consult your local Extension Service or MOBAY representative for details.
	NOTE 2: East of the Rocky Mountains foliar applications of DI-SYSTON 8 may be made in addition to an **at-planting** soil treatment. Refer to the label for DI-SYSTON 8.		
ORNAMENTALS Flowers (See Note 1)	Aphids Lace bug Leafhoppers Mites Thrips Whiteflies	10.7 ozs. per 150 to 300 feet of row	Apply in a band on each side of the seed furrow, or as a band over the row. Work into soil and water thoroughly. For treatment of Easter lilies and gladiolus, apply the specified dosage in the furrow over the bulbs after they are set, but before covering with soil. Do not make in-furrow applications to garden lily bulbs. The higher dosage is recommended for heavy organic or muck soils.
Shrubs Trees (See Note 1)	Aphids Birch leafminer Camellia scale Elm leaf beetle European elm scale Holly leafminer Lace bug	2.5 to 27 ozs. per tree, or 2.5 ozs. per inch of trunk diameter	Spread granules uniformly from trunk to drip line on all sides. Work into soil and water thoroughly. The higher dosage is recommended for heavy organic or muck soils. For transplanting, equivalent quantities should be applied into planting hole or by dipping seedlings in a puddling clay slurry made up of 1 part by weight of DI-SYSTON 15% Granular in 6 to 14 parts by weight of clay.

CROP	INSECT	DOSAGE DI-SYSTON 15% GRAN.	REMARKS
	Leafhoppers Mimosa webworm Mites Pine tip moth Soft scale (*Coccus hesperidium)* Tea scale Thrips Whiteflies		8-10-88

* ON ROW CROPS WHERE ROW SPACING IS EXTREMELY NARROW, DO NOT PLACE TREATED ZONES (BAND OR FURROW) CLOSER TOGETHER THAN 6 INCHES. **Note 1:** For use by commercial growers or professional applicators only.

SPECIAL LOCAL NEED REGISTRATIONS [SECTION 24(c)] FOR DI-SYSTON 15% GRANULAR*

CROP	INSECT	STATE EPA SLN NO.	DOSAGE	REMARKS
Barley	Hessian fly Aphids **(Greenbugs) (Oat birdcherry) Grasshoppers	Oregon OR 80-0034 Washington WA 85-0036	6.7 pounds per acre	Apply by drilling or broadcast at planting time.
Barley (Irrigated only)	Aphids **(Greenbugs)	Idaho ID 85-0016 Utah UT 89-0001		Broadcast preplant, preemergence or postemergence by air or ground. Irrigated only.
Christmas Trees	Aphids Nantucket Scales Pine Tip Moth Mites	North Carolina NC 88-0001	1 to 1.5 ounces per tree	Apply from trunk to drip line. Work into soil or water thoroughly within 48 hours after application.
Potatoes (sprinkler irrigated only)	Green peach aphids	Oregon OR 79-0042 Washington WA 79-0067	20 pounds per acre	Apply as a postemergence aerial broadcast application.
Potatoes (sprinkler irrigated only)	Aphids	Idaho ID 83-0035 Oregon OR 83-0057 Washington WA 84-0012	20 pounds per acre	Apply suggested rate per acre as a soil application and/or a foliar broadcast treatment.
Wheat	Aphids	California CA 76-0019	6.7 pounds per acre	Apply postemergence as an aerial or ground broadcast application.
	Hessian fly Aphids **(Greenbugs) (Oat birdcherry) Grasshoppers	Oregon OR 80-0034 Washington WA 85-0036	6.7 pounds per acre	Apply by drilling or broadcast at planting time.
Wheat (Irrigated only)	Aphids **(Greenbugs)	Idaho ID 85-0016 Utah UT 89-0001		Broadcast preplant, preemergence or postemergence by air or ground. Irrigated only.
Wheat (Fall) (Spring) Barley	Hessian fly Aphids **(Greenbugs) (Oat birdcherry) Grasshoppers	Montana MT 80-0004	6.7 pounds per acre	Apply by drilling or broadcast at planting time. Make only one application and do not graze or harvest for grain or forage within 60 days of treatment.

*For distribution and use only within the state specified. These Special Local Need registrations approved as of date of this publication are, however, subject to cancellation by the state. Contact your nearest Mobay Agricultural Chemicals Division Office for the current status and a copy of the complete individual supplemental Special Local Need labeling. This labeling must be in the possession of the user at time of pesticide application.
**Due to greenbug resistance in certain areas, DI-SYSTON used alone may not provide satisfactory control. Consult your local agricultural advisor or Extension Service for recommendations.

RESTRICTIONS

In commercial topping of flowers by hand, clean gloves should be worn. Do not use on other crops used for food or forage. Use only according to label directions. Application at rates above those shown may result in illegal crop residues. Do not treat crops grown in the greenhouse.

Do not apply this product with a knapsack or other equipment that is placed on the user's body.

RE-ENTRY STATEMENT FOR FARM WORKERS

Do not apply this product in such a manner as to directly or through drift expose workers or other persons. The area being treated must be vacated by unprotected persons. Do not allow re-entry into treated fields within 24 hours of application unless appropriate protective clothing is worn. Because certain states may require more restrictive re-entry intervals for various crops treated with this product, consult your State Department of Agriculture for further information.

Written or oral warnings must be given to workers who are expected to be in a treated area or in an area about to be treated with this product. Oral warnings should include those statements found under the label heading "Precautionary Statements" and "Re-entry Statements" included on the label. When oral warnings are given, warnings shall be given in a language customarily understood by workers. Oral warnings must be given if there is reason to believe that written warnings cannot be understood by workers. Written warnings must include the following information: DANGER, area treated with DI-SYSTON 15% Granular on (date of application). Do not allow re-entry into treated fields within 24 hours of application unless appropriate protective clothing is worn. **If on skin,** remove contaminated clothing and wash skin immediately with soap and warm water. **If eyes are contaminated,** wash with plenty of flowing water. Get medical attention immediately. **If swallowed,** vomiting should be induced. Get medical attention immediately. (See Precautionary Statements on label.)

STORAGE AND DISPOSAL

Pesticide Disposal: Do not contaminate water, food or feed by storage or disposal.

Pesticide wastes are acutely hazardous. Improper disposal of excess pesticide is a violation of Federal Law. If these wastes cannot be disposed of by use according to label instructions, contact your State Pesticide or Environmental Control Agency, or the Hazardous Waste representative at the nearest EPA Regional Office for guidance.

Container Disposal: Completely empty bag into application equipment. Then dispose of empty bag in a sanitary landfill or by incineration, or, if allowed by State and local authorities, by burning. If burned, stay out of smoke.

Store in a cool, dry place and in such a manner as to prevent cross contamination with other pesticides, fertilizers, food, and feed. Store in original container and out of the reach of children, preferably in a locked storage area.

Handle and open container in a manner as to prevent spillage. If the container is leaking or material spilled for any reason or cause, carefully sweep material into a pile. Refer to Precautionary Statements on label for hazards associated with the handling of this material. Do not walk through spilled material. Dispose of pesticide as directed above. In spill or leak incidents, keep unauthorized people away. You may contact the Mobay Emergency Response Team for decontamination procedures or any other assistance that may be necessary. The Mobay Kansas City Emergency Response Telephone No. is 816-242-2582, or contact Chemtrec at 800-424-9300.

PRECAUTIONARY STATEMENTS

HAZARDS TO HUMANS AND DOMESTIC ANIMALS

Poisonous if swallowed, inhaled, or absorbed through the skin. Rapidly absorbed through the skin. Do not get in eyes or on skin. Do not breathe dust. Wear protective clothing and cotton or natural rubber gloves. If cotton gloves are used, they must be laundered or discarded after each day's use. Do not wear the same gloves for other work. Rubber gloves should be washed with soap and water after each use.

When using DI-SYSTON 15% Granular insecticide in enclosed areas, wear a mask or respirator jointly approved by the Mining Enforcement and Safety Administration (formerly the U.S. Bureau of Mines) and by the National Institute for Occupational Safety and Health under the provisions of 30 CFR Part 11.

Do not contaminate feed or food. Keep out of reach of children. Keep all unprotected persons out of the operating area or vicinity. Wash hands, arms and face thoroughly with soap and warm water before eating or smoking. Wash all contaminated clothing with soap and hot water before reuse.

SYMPTOMS OF POISONING: A sense of "tightness" in the chest. Sweating. Contracted pupils. Stomach pains. Vomiting and diarrhea.

STATEMENTS OF PRACTICAL TREATMENT

TREATMENT: In case of poisoning, call a physician immediately. Have patient lie down and keep quiet. **If swallowed, vomiting should be induced.** Administer water freely and induce vomiting by giving one dose (½ oz. or 15 ml.) of syrup of ipecac. If vomiting does not occur within 10 to 20 minutes, administer second dose. If syrup of ipecac is not available, induce vomiting by sticking finger down throat. Repeat until vomit fluid is clear. Never give anything by mouth to an unconscious person. Professional medical assistance should be secured immediately. **If on skin,** remove contaminated clothing, and wash skin immediately with soap and warm water. Get medical attention immediately. **If eyes are contaminated,** wash with plenty of flowing water. Get medical attention immediately.

NOTE TO PHYSICIAN: Antidote—Administer atropine sulfate in large therapeutic doses. Repeat as necessary to the point of tolerance. 2-PAM is also antidotal, and may be administered in conjunction with atropine.

Compound inhibits cholinesterase, resulting in stimulation of the central nervous system, the parasympathetic nervous system and the somatic motor nerves. Do not give morphine. Watch for pulmonary edema, which may develop in serious cases of poisoning even after 12 hours. At first sign of pulmonary edema, the patient should be placed in an oxygen tent, and treated symptomatically.

ENVIRONMENTAL HAZARDS

This pesticide is toxic to fish and wildlife. Do not apply directly to water or wetlands. Drift and runoff from treated areas may be hazardous to fish in neighboring areas. Do not contaminate water by cleaning of equipment or disposal of wastes. Cover or incorporate spills.

This pesticide is toxic to bees exposed to direct application. Applications should be timed to coincide with periods of minimum bee activity, usually between late evening and early morning.

IMPORTANT:

This bulletin is not intended to provide adequate information for application. Before using the product, read and carefully observe directions, cautionary statements and other information appearing on the product label.

This product is sold subject to the Conditions of Sale set forth on the container label.

READ THE LABEL BEFORE USE

DS 9003

This is a specimen label. Mobay is not responsible for the accuracy of the information contained herein. As labels are subject to revision, always carefully read and follow the label on the product container.

DI-SYSTON® 8

EMULSIFIABLE SYSTEMIC INSECTICIDE

RESTRICTED USE PESTICIDE

Acute human oral, dermal, and inhalation toxicity.
For retail sale to and use only by Certified Applicators or persons under their direct supervision, and only for those uses covered by the Certified Applicator's certification.

8 POUND PER GALLON FORMULATION
FOR EFFECTIVE CROP PROTECTION.

ACTIVE INGREDIENTS:	
Disulfoton *O,O*-Diethyl *S*-[2-(ethylthio)ethyl] phosphorodithioate. .	85%
INERT INGREDIENTS:	15%
TOTAL	100%

Contains 8 lbs. *O,O*-Diethyl *S*-[2-(ethylthio)ethyl] phosphorodithioate per gallon.

EPA Reg. No. 3125-307

DI-SYSTON is a Reg. TM of Bayer AG, Germany
Compex is a Reg. TM of Kalo Laboratories, Inc.

RECOMMENDED USES

Barley	Peas
Beans	Pecans
Broccoli	Potatoes
Brussels sprouts	Sorghum
Cabbage	Strawberries (propagating plants only)
Cauliflower	Tobacco
Corn	Tomatoes
Cotton	Wheat
Lettuce	

RESTRICTIONS ON WAREHOUSE STORAGE

Do not store adjacent to or near proximity of foodstuffs, drugs, feeds or any other material intended for consumption by humans or animals. Do not store adjacent to toys, cosmetics, dishes, furniture or clothing. Do not store next to herbicides.

PACKAGING

5 gallon container. Shipping weight 51.5 lbs. 50 gallon container. Shipping weight 521 lbs.

DI-SYSTON 8 is effective for the control of a variety of harmful insects that attack many field and vegetable crops. DI-SYSTON 8 gives up to seven weeks control of sucking insects. This versatile insecticide can be used as a soil treatment alone, or with liquid fertilizer, and can be sprayed or injected into the furrow at planting time or after emergence. DI-SYSTON 8 enters the plant through the roots, and is translocated through the plant by the sap stream. Mites and sucking insects feeding on the plant pick up a lethal dose of the chemical and die quickly. Also, because it is inside the plant, DI-SYSTON 8 will not destroy beneficial insect populations, nor will it wash off or weather-away.

PELIGRO

PRECAUCION AL USUARIO: Si usted no puede leer o entender inglés, no use este producto hasta que la etiqueta le haya sido explicada ampliamente.

(TO THE USER: If you cannot read or understand English, do not use this product until the label has been fully explained to you.)

DIRECTIONS FOR USE

It is a violation of Federal Law to use this product in a manner inconsistent with its labeling.

DI-SYSTON systemic insecticide enters the plant by absorption through the roots. The insecticide placed near the roots is translocated to all parts of the plant and kills sucking insects feeding on the plant.

The formulation of DI-SYSTON 8 is highly versatile, and offers many forms of application. It can be sprayed or injected into the furrow at planting time or soil-injected as a side-dress treatment at planting time, or after plants become established. Applications may be made in water emulsion, with liquid fertilizer, or undiluted in specially designed equipment for low volume soil injection, but never on crops for which less than 1 quart DI-SYSTON 8 per acre is the recommended dosage. When mixing with certain liquid high-phosphorous fertilizers, the use of Compex® compatibility agent, or other equivalent agents at the rate of 1 quart per 100 gals. of mixture may be needed to assure proper mixing. Time and method of application and limits for use are discussed in detail for each specific crop under "Remarks" below.

RECOMMENDED APPLICATIONS

IMPORTANT: When used on cotton, tomatoes, beans or lettuce at the maximum recommended rates, under adverse conditions, such as extremely cool or wet or extremely dry weather, DI-SYSTON may cause some delay in emergence, stunting of seedlings or reduction of stand. Such damage may be more pronounced in light sandy soils. Plant injury may also occur or may be enhanced when certain preemergent herbicides are used in conjunction with DI-SYSTON.

USE IN CHEMIGATION SYSTEMS

DI-SYSTON 8 may be applied through irrigation systems only on those crops listed under Recommended Applications where application through irrigation systems is recommended.

Types of Irrigation Systems

Apply DI-SYSTON 8 only thorough sprinkler, including center pivot, lateral move, side roll, or overhead solid set irrigation systems. Do not apply DI-SYSTON 8 through any other type of irrigation system.

GENERAL DIRECTIONS FOR ALL RECOMMENDED TYPES OF IRRIGATION SYSTEMS

Uniform Water Distribution and System Calibration

The irrigation system must provide uniform distribution of treated water. Crop injury, lack of effectiveness, or illegal pesticide residues in the crop can result from nonuniform distribution of treated water.

The system must be calibrated to uniformly apply the rates specified. If you have questions about calibration, you should contact State Extension Service specialists, equipment manufacturers or other experts.

Chemigation Monitoring

A person knowledgeable of the chemigation system and responsible for its operation, or under the supervision of the responsible person, shall shut the system down and make necessary adjustments should the need arise.

Drift

Do not apply when wind speed favors drift beyond the area intended for treatment.

Required System Safety Devices

The system must contain a functional check valve, vacuum relief valve, and low pressure drain appropriately located on the irrigation pipeline to prevent water source contamination from backflow.

The pesticide injection pipeline must contain a functional, automatic, quick-closing check valve to prevent the flow of fluid back toward the injection pump.

The pesticide injection pipeline must also contain a functional, normally closed, solenoid-operated valve located on the intake side of the injection pump and connected to the system interlock to prevent fluid from being withdrawn from the supply tank when the irrigation system is either automatically or manually shut down.

The system must contain functional interlocking controls to automatically shut off the pesticide injection pump when the water pump motor stops.

The irrigation line or water pump must include a functional pressure switch which will stop the water pump motor when the water pressure decreases to the point where pesticide distribution is adversely affected.

Systems must use a metering pump, such as a positive displacement injection pump (e.g., diaphragm pump) effectively designed and constructed of materials that are compatible with pesticides and capable of being fitted with a system interlock.

Using Water from Public Water Systems

DO NOT APPLY DI-SYSTON 8 THROUGH ANY IRRIGATION SYSTEM PHYSICALLY CONNECTED TO A PUBLIC WATER SYSTEM. Public water system means a system for the provision to the public of piped water for human consumption if such system has at least 15 service connections or regularly serves an average of at least 25 individuals daily at least 60 days out of the year.

DI-SYSTON 8 may be applied through sprinkler irrigation systems which may be **supplied** by a public water system **only if** the water from the public water system is discharged into a reservoir tank prior to the pesticide introduction. There shall be a complete physical break (air gap) between the outlet end of the fill pipe and to top or overflow rim of the reservoir tank of at least twice the inside diameter of the fill pipe. Before beginning chemigation, always make sure that the air gap exists and that there is no blockage of the overflow of the reservoir tank.

Any irrigation system using water supplied from a public water system must also meet the following requirements.

The pesticide injection pipeline must contain a functional, automatic, quick-closing check valve to prevent the flow of fluid back toward the injection pump.

The pesticide injection pipeline must contain a functional, normally closed, solenoid-operated valve located on the intake side of the injection pump and connected to the system interlock to prevent fluid from being withdrawn from the supply tank when the irrigation system is either automatically or manually shut down.

The system must contain functional interlocking controls to automatically shut off the pesticide injection pump when the water pump motor stops or in cases where there is no water pump, when the water pressure decreases to the point where the pesticide distribution is adversely affected.

Systems must use a metering pump, such as a positive displacement injection pump (e.g., diaphragm pump) effectively designed and constructed of materials that are compatible with pesticides and capable of being fitted with a system interlock.

Do not apply when wind speed favors drift beyond the area intended for treatment.

Posting Requirements

Posting of areas to be chemigated is required when 1) any part of a treated area is within 300 feet of sensitive areas such as residential areas, labor camps, businesses, day care centers, hospitals, in-patient clinics, nursing homes or any public areas such as schools, parks, playgrounds, or other public facilities not including public roads, or 2) when the chemigated area is open to the public such as golf courses or retail greenhouses.

Posting must conform to the following requirements: Treated areas shall be posted with signs at all usual points of entry and along likely routes of approach from the listed sensitive areas. When there are no usual points of entry, signs must be posted in the corners of the treated areas and in any other location affording maximum visibility to sensitive areas. The printed side of the sign should face away from the treated areas towards the sensitive areas. The signs shall be printed in English. Signs must be posted prior to application and must remain posted until foliage has dried and soil surface water has disappeared. Signs may remain in place indefinitely as long as they are composed of materials to prevent deterioration and maintain legibility for the duration of the posting period.

All words shall consist of letters at least 2½ inches tall, and all letters and the symbol shall be a color which sharply contrasts with their immediate background. At the top of the sign shall be the words KEEP OUT, followed by an octagonal stop sign symbol at least 8 inches in diameter containing the word STOP. Below the symbol shall be the words PESTICIDES IN IRRIGATION WATER.

Posting required for chemigation does not replace other posting and reentry interval requirements for farmworker safety.

Agitation

For application of DI-SYSTON 8 alone, a chemical supply tank is not necessary for premixing since DI-SYSTON 8 mixes well with water in the irrigation line. If a chemical supply tank is used, constant strong mechanical or hydraulic agitation must be maintained in the chemical supply tank during the entire period of application.

Chemical Supply Tank Dilution

If a chemical supply tank is used, you must determine the required amounts of DI-SYSTON 8 and water to mix in the tank.

The amount of DI-SYSTON 8 needed equals the number of pints of DI-SYSTON 8 to be applied per acre multiplied by the number of acres to be chemigated.

The amount of emulsion needed equals the gallons of emulsion delivered per hour by the injection pump multiplied by the number of hours chemigation will take place.

The amount of water needed equals the amount of emulsion needed minus the amount of DI-SYSTON 8 needed.

For example, for center-pivot or automatic-move linear systems, if you want to apply 3 pints of DI-SYSTON 8 per acre to 130 acres in 20 hours and your injection pump delivers 15 gallons per hour, you need: 3 pints of DI-SYSTON 8 per acre × 130 acres = 390 pints or 48.75 gallons of DI-SYSTON 8. And, you need: 15 gallons per hour × 20 hours = 300 gallons of emulsion, minus 48.75 gallons of DI-SYSTON 8 = 251.25 gallons of water.

Another example, for solid set and manually controlled linear systems, is if you want to apply 3 pints of DI-SYSTON 8 per acre during the last 60 minutes of a set irrigating 6 acres and your injection pump delivers 20 gallons per hour, you need 3 pints of DI-SYSTON 8 per acre × 6 acres = 18 pints or 2.25 gallons of DI-SYSTON 8 for the 6 acre set. And, you need 20 gallons per hour × 1 hour = 20 gallons of emulsion, minus 2.25 gallons of DI-SYSTON 8 = 17.25 gallons of water for the 6 acre set.

Cleaning the Chemical Injection System

In order to accurately apply pesticides, the chemical injection system must be kept clean; free of chemical or fertilizer residues and sediments. Refer to your owner's manual or ask your equipment supplier for the cleaning procedure for your injection system.

Flushing the Irrigation System

At the end of the application period, allow time for all lines to flush the pesticide through nozzles or emitters before turning off irrigation water. To ensure the lines are flushed and free of pesticides, a dye indicator may be injected into the lines to mark the end of the application period.

EQUIPMENT AREA CONTAMINATION PREVENTION: It is recommended that nozzles in the immediate area of control panels, chemical supply tanks, pumps and system safety devices be plugged to prevent chemical contamination of these areas.

Center-Pivot and Automatic-Move Linear Systems

Inject the specified dosage per acre continuously for one complete revolution or move of the system. DO NOT USE END GUNS. The system should be run at maximum speed.

Solid Set and Manually Controlled Linear Systems

Injection should be during the last 30 to 60 minutes of regular irrigation periods or as a separate 30 to 60 minute application not associated with a regular irrigation. DO NOT USE END GUNS.

CROP	INSECT	DOSAGE—DI-SYSTON 8	REMARKS
FIELD CROPS Barley	Aphids Mites	1 pint per acre	Apply as a soil injection in a water emulsion, or with liquid fertilizer at planting time. Do not apply within 60 days of harvest. Do not graze or cut for forage within 30 days of treatment.
Barley (Foliar Application Spring or Fall)	Aphids (Including greenbugs) (See **NOTE** in Remarks)	1/2 to 1 pint per acre	For application by aircraft or ground equipment: Apply specified dosage per acre in sufficient water to obtain thorough coverage but not less than 1 gallon total volume per acre. Use the lower rates on plants up to tillering (stooling). Higher rates should be used after plants begin to tiller. For application in liquid fertilizer: Mix 1/2 to 1 pint of DI-SYSTON 8 in the amount of liquid fertilizer to be applied per acre. Apply this mixture as a top dressing with suitable ground equipment. Foliar applications may be made following an at-planting application, however, regardless of method of application do not apply more than a total of 2 pints per acre of DI-SYSTON 8 per crop season. Do not make any applications within 30 days of grain harvest. **NOTE:** (1) Do not graze treated fields after any application of DI-SYSTON 8. (2) Due to greenbug resistance in certain areas, DI-SYSTON used alone may not provide satisfactory control. Consult your local agricultural advisor or Extension Service for recommendations.
Corn Field Popcorn	Aphids Leafhoppers Mites Thrips	1.2 fl. ounces per 1000 feet of row for any row spacing* (for 40 inch row spacing use 1 pint per acre)	**Treatment at Planting Time:** As a water emulsion, spray in 4 to 6 inch band directly behind the planter shoe in front of the press wheel, or apply with **liquid fertilizer** in split bands on each side of the corn row. Do not apply directly to the seed. Do not forage or cut treated corn for food, feed or ensilage within 28 days of treatment. **Post-plant Side-dress Treatment:** An alternative treatment for early-season control of aphids, leafhoppers, mites and thrips may be made in a water emulsion, or with liquid fertilizer as a soil-injection on each side of the furrow after emergence. Do not forage or cut treated corn for food, feed or ensilage within 28 days of treatment.
	Aphids Mites	8 to 16 fl. ounces **NOTE:** Apply only one soil treatment and one foliar treatment per crop season.	**Foliar Treatment:** Apply specified dosage per acre as a foliar spray in sufficient water for thorough coverage, but not less than 1 gallon total volume per acre. Do not forage or cut treated corn for food, feed or ensilage within 28 days of treatment.
Cotton	Aphids Mites Thrips	.75 to 1.2 fluid ounces per 1000 feet of row (for any row spacing*), or 9 to 16 fluid ounces per acre (40 inch row spacing)	**Preplant:** As a preplant soil incorporated treatment, DI-SYSTON 8 may be applied in approximately a 16 inch band as a water emulsion either alone or with a preemergent herbicide. Request Bulletin No. DS 6720 for specific recommendations on a herbicide to be used with DI-SYSTON. **Planting:** Apply as a water emulsion, or with liquid fertilizer. Spray in the furrow on either side of the seed, or soil-inject on each side of the furrow. Do not apply directly on the seed. **RATE CHART FOR PREPLANT OR PLANTING TIME APPLICATION** **Heavy Soil:** 1.2 fl. oz./1000 ft. of row, or 1 pt./acre **Medium Soil:** 1 fl. oz./1000 ft. of row, or 12 fl. oz./acre **Light Soil:** .75 fl. oz./1000 ft. of row, or 9 fl. oz./acre **Post-plant:** If an additional treatment is needed, apply in a water emulsion, or with liquid fertilizer as a soil-injection on each side of the row. A single post-plant application at rates up to 2 pints/acre may be applied up to first-squaring (90 days before harvest), or a single post-plant application at rates up to 1 pint/acre may be applied up to 28 days of harvest. Do not apply more than twice per season. Allow a minimum of 21 days between applications. Read "IMPORTANT" statement under "RECOMMENDED APPLICATIONS" for additional information before use on this crop.
	Aphids Thrips	3 to 9 fluid ounces **NOTE:** If foliar applications are made, do not apply soil treatments within the same crop year.	**FOLIAR—Prior to Bloom:** Apply specified dosage per acre as a foliar spray, using aircraft or ground equipment, in sufficient water for thorough coverage, but not less than 1 gallon total volume per acre. Repeat as necessary, but do not apply more than 3 times per season, with the last application no later than bloom.
Sorghum	Aphids (Greenbugs) (See **NOTE** below)	.9 to 1.2 fl. ozs. per 1000 feet of row (for any row spacing*), (for 40 inch row spacing use 12 to 16 fl. ozs. per acre)	**AT-PLANTING APPLICATION: Furrow**—Apply as a water emulsion, or in liquid fertilizer in the seed furrow approximately 1/2 inch above the seed. Do not apply directly on the seed.
		1.2 fl. ozs. per 1000 feet of row (for any row spacing*), (for 40 inch row spacing use 1 pint per acre)	**AT-PLANTING APPLICATION: Band**—Apply as a water emulsion, or in liquid fertilizer in a 4 to 6 inch band directly behind the planter shoe, in front of the press wheel. Do not apply directly on the seed.
		NOTICE: The following side-dress and/or foliar applications may be made in addition to granular or liquid at-planting applications. A **total** of **5 applications** may be made to a single sorghum crop (2 soil applications, granular or liquid, plus 3 foliar applications of rates 8 fluid ounces or less/acre).	

CROP	INSECT	DOSAGE—DI-SYSTON 8	REMARKS
	Aphids (Greenbugs) (See **NOTE** below)	1.2 fl. ozs. per 1000 feet of row (for any row spacing*), (for 40 inch row spacing use 1 pint per acre)	**POST-PLANTING APPLICATION: Side-dress**—Apply as a water emulsion, or with liquid fertilizer injected into the row on both sides of the furrow at any time up to boot stage, but not within 45 days of use for forage or fodder.
	Aphids (Greenbugs) (See **NOTE** below) Sorghum midge (Arkansas, Louisiana, Oklahoma, Texas)	4 to 8 fluid ounces per acre	**POST-PLANTING APPLICATION: Foliar Spray—A maximum of 3 foliar applications may be made of rates 8 fluid ounces or less/acre.** Apply specified dosage per acre in sufficient water to obtain thorough coverage. For aerial application, use not less than 1 gallon of water per acre. Use not less than 5 gallons of water per acre with ground equipment. **For Midge:** Make one application when 50% of the seed heads have emerged from the boot. Make a second application 3 to 5 days later.
	Banks grass mites (in Texas, excluding Trans-Pecos area, and in Oklahoma)	8 fluid ounces per acre	**DAYS FROM LAST APPLICATION TO HARVEST FOR FOLIAR RATES OF 4 TO 8 FLUID OUNCES/ACRE**

APPLICATION SCHEDULE	FOR GRAIN	FOR FORAGE OR FODDER
Any soil **plus any** foliar	34	45
1 or 2 foliars	7	45
3 foliars	34	60

NOTE: Due to greenbug resistance in certain areas, DI-SYSTON used alone may not provide satisfactory control. Consult your local agricultural advisor or Extension Service for recommendations.

CROP	INSECT	DOSAGE—DI-SYSTON 8	REMARKS
Tobacco	Aphids Flea beetles	4 pints (1/2 gallon) per acre	Apply as a water emulsion broadcast spray in sufficient water for thorough coverage. Disk or till the soil to a depth of 2 to 3 inches before setting plants. Do not make more than one field application per crop season, regardless of the formulation of DI-SYSTON used.
Wheat (Foliar application Spring or Fall)	Aphids (Greenbugs) Mites (See **NOTE** in "Remarks")	4 to 12 fluid ounces per acre	**For application by aircraft or ground equipment:** Apply specified dosage per acre in sufficient water to obtain thorough coverage, but not less than 1 gallon total volume per acre. Use the lower rates on plants up to tillering (stooling). Higher rates should be used after plants begin to tiller. **For application in liquid fertilizer:** Mix 6 to 12 fluid ounces of DI-SYSTON 8 in the amount of liquid fertilizer to be applied per acre. Apply this mixture as a top-dressing, with suitable ground equipment. **Fall Applications:** Two applications may be made in the Fall. Allow 30 days between applications. **Spring Applications:** Application at green-up may be made following Fall applications. A second application may be made 30 days following the first Spring application. Do not make **any** applications within 30 days of grain harvest. **NOTE:** (1) Do **not** graze treated fields or cut for forage after **any** application of DI-SYSTON 8. (2) Due to greenbug resistance in certain areas, DI-SYSTON used alone may not provide satisfactory control. Consult your local agricultural advisor or Extension Service for recommendations.
Wheat (Fall)	Hessian fly Aphids (Greenbugs and oat bird cherry—Except California)	0.25 fluid ounces per 1000 feet of row (see chart below) or up to 1 pint per acre	Apply as a soil-injection in water emulsion or with liquid fertilizer at planting time in the Fall. **NOTE:** (1) Do **not** graze treated fields or cut for forage after **any** application of DI-SYSTON 8. (2) Due to greenbug resistance in certain areas, DI-SYSTON used alone may not provide satisfactory control. Consult your local agricultural advisor or Extension Service for recommendations.

Row Spacing	Ounces/Acre	Ounces DI-SYSTON 8 per 1000 feet of row based on 3 inch band
7 or 8 inch rows	15.5	.25
10 inch rows	13.25	.25
12 inch rows	10.75	.25
14 inch rows	9.5	.25
16 inch rows	8.5	.25
18 inch rows	7.0	.25

CROP	INSECT	DOSAGE—DI-SYSTON 8	REMARKS
NON-BEARING FRUIT Strawberries (Propagating plants only)	Aphids Mites (Except cyclamen mite)	2.8 to 5 fluid ounces per 1000 feet of row (for any row spacing*), or 2 to 4 pints per acre (42 inch row spacing)	Apply in a water emulsion, or with liquid fertilizer as a soil injection on each side of the row as needed. Do not use fruit from treated plants for food purposes.
NUTS Pecans (South Central and Southwestern states only)	Aphids	1.5 to 3 pints per acre	Apply by spraying 6 foot bands of soil on both sides of the trees. The bands should be located under the main drip area of the trees. Work the treated soil to a depth of 2 to 3 inches. On irrigated pecans, apply water as soon as possible after treatment. Make only 1 application per season. Early Spring treatment is recommended. Do not harvest nuts within 80 days of treatment. Do not graze grass under treated trees. Do not tank mix this product with Phosalone.
Pecans	Aphids	4 to 6 fluid ounces	**Foliar Applications:** Apply specified dosage in 100 gallons of water as a full-coverage spray, using not more than 400 gallons of finished spray per acre. Repeat as necessary, up to 3 times per season. Do not apply within 30 days of harvest. Do not graze grass under treated trees. Do not tank mix this product with Phosalone.

CROP	INSECT	DOSAGE—DI-SYSTON 8	REMARKS
		¼ to 1 pint	**Application by Aircraft:** Apply specified dosage per acre in sufficient water to obtain thorough coverage, but not less than 5 gallons total volume per acre. Do not apply more than 3 times per season, nor within 30 days of harvest. Do not graze grass under treated trees. Do not tank mix this product with Phosalone.
VEGETABLES Beans	Aphids Leafhoppers Mexican bean beetle (larvae) Mites Thrips	.9 to 1.9 fluid ounces per 1000 feet of row (for any row spacing*), or 1 to 2 pints per acre (30 inch row spacing) (For extremely narrow row spacing, do not exceed the 2 pints per acre rate)	For snap or green Lima beans, apply in a water emulsion, or with liquid fertilizer as a soil injection on each side of the seed furrow at planting time only. For dry beans, apply as above at planting time, or inject as a side-dressing on each side of the row after plants become established, but not within 60 days of harvest. Do not apply directly on the seed. Do not apply more than once per season. Read "IMPORTANT" statement under "RECOMMENDED APPLICATIONS" for additional information before use on this crop.
Brussels sprouts	Aphids Flea beetles Leafhoppers Mites Thrips	1.1 fluid ounces per 1000 feet of row (for any row spacing*), or 1 pint per acre (36 inch row spacing)	**Transplant Seed Beds:** Apply evenly over bed area in a water emulsion as a broadcast spray prior to seeding, at the rate of 1 pint per acre. Work into the top 2 to 3 inches of soil, and plant seed in normal manner. **Field Applications:** Apply specified dosage in a water emulsion, or with liquid fertilizer as an injection on each side of the seed furrow or transplanted row at planting, and where necessary, injected as a side-dressing after plants are established. Do not apply directly on the seed. Do not apply more than two field applications per season, nor within 30 days of harvest for Brussels sprouts, or 40 days of harvest for cauliflower. Allow a minimum of 21 days between applications.
Cauliflower	Aphids Flea beetles Leafhoppers		
Broccoli	Aphids Flea beetles Leafhoppers	1.1 fluid ounces per 1000 feet of row (for any row spacing*), or 1 pint per acre (36 inch row spacing*)	**Transplant Seed Beds:** Apply evenly over bed area in a water emulsion as a broadcast spray prior to seeding, at the rate of 1 pint per acre. Work into the top 2 to 3 inches of soil, and plant seed in normal manner. **Field Applications:** Apply specified dosage in a water emulsion, or with liquid fertilizer as an injection on each side of the seed furrow or transplanted row at planting, or injected as a side-dressing after plants are established. Do not apply directly on the seed. Do not apply more than one field application per season. Do not use within 14 days of harvest for broccoli, or 42 days of harvest for cabbage.
Cabbage (Includes tight heading varieties of Chinese Cabbage)	Aphids Flea beetles		
	Root aphids	1.7 fluid ounces per 1000 feet of row (for any row spacing*)	
Lettuce	Aphids Leafhoppers Mites Thrips	.6 to 1.2 fluid ounces per 1000 feet of row (for any row spacing*), or 1 to 2 pints per acre (20 inch row spacing)	**Spring or Summer Seeded Lettuce:** Apply in a water emulsion, or with liquid fertilizer as an injection on each side of the seed furrow at planting time only. Use the higher rate on heavy organic soils. Do not apply directly on the seed. Do not apply to transplanted lettuce. **Fall or Winter Seeded Lettuce:** Apply specified dosage in a water emulsion, or as a soil-injected side-dressing with liquid fertilizer at thinning time. Do not apply within 60 days of harvest. Read "IMPORTANT" statement under "RECOMMENDED APPLICATIONS" for additional information before use on this crop.
Peas	Aphids Leafhoppers	1 to 2.5 pints per acre	Apply in a water emulsion, or with liquid fertilizer as an in-furrow spray or as an injection on each side of the seed furrow at planting, or injected as a side-dressing after plants are established. Do not apply more than once per season, nor within 50 days of harvest. Do not feed vines or hay.
Potatoes	Aphids	3 to 4 pints per acre	**Soil Application—Preplant Broadcast:** Apply specified dosage per acre as a preplant broadcast application. Incorporate to a depth of 2 to 3 inches. Plant crop in usual manner. Do not make more than two **soil** applications of DI-SYSTON per crop season regardless of method of application or formulation used. Do not tank mix this product with Phosalone.
	Aphids Flea beetles (except tuber flea beetles) Leafhoppers Potato psyllid Southern potato wireworm Colorado potato beetle (early-season reduction)	2.25 to 3.5 fl. oz. per 1000 feet of row (for any row spacing*), or 2 to 3 pints per acre (38 inch row spacing)	**Soil Application—Furrow:** Apply undiluted in specifically designed equipment for low volume soil injection, or apply in a water emulsion, or with liquid fertilizer as an in-furrow spray, or injection on each side of the seed furrow at planting. An **additional** side-dress application may be made as a water emulsion injection after plants become established. Do not make more than two **soil** applications of DI-SYSTON per crop season, regardless of method of application or formulation used. Do not apply second soil application within 75 days of harvest. Higher rate to be used in organic soils. Do not tank mix this product with Phosalone. NOTE: Colorado Potatoe Beetle populations in the northeast states (except Maine), as well as Virginia and North Carolina, have developed resistance to insecticides, including DI-SYSTON. Consult your local Extension Service or MOBAY sales representative for details.
	Green peach aphids (East of the Rocky Mountains only)	6 to 16 fluid ounces per acre	**Foliar applications:** Using aircraft or ground equipment, apply specified dosage per acre in sufficient water to obtain thorough coverage, but not less than 1 gallon total volume per acre. Begin applications when pest first appears. Use the higher rates for heavy infestations. Repeat as necessary, but not more than 3 times per season, nor within 30 days of harvest. Do not tank mix this product with Phosalone.
	Green peach aphid Colorado potato beetle	3 pints per acre	**SPRINKLER IRRIGATION APPLICATION—ONLY IN OREGON, WASHINGTON, IDAHO AND UTAH:** A single application at the rate of 3 pints per acre may be made for control of Green peach aphid and Colorado potato beetle. Do not apply within 60 days of harvest. Application may be made following **at-planting** soil application as recommended on the label or on the label for DI-SYSTON 15% Granular. Follow all directions given under the Chemigation section of the label.

CROP	INSECT	DOSAGE—DI-SYSTON 8	REMARKS
	NOTE: Foliar applications may be made following **at-planting** soil treatment as recommended above, or on the label for DI-SYSTON 15% Granular.		
Tomatoes	Aphids Flea beetles Leafhoppers Leafminers Mites Colorado potato beetle (early season reduction)	1.2 to 3.5 fluid ounces per 1000 feet of row (for any row spacing*), or 1 to 3 pints per acre (38 inch row spacing)	**Transplant Seed Beds:** Apply evenly over bed area in a water emulsion as a broadcast spray prior to seeding, at the rate of 3 pints per acre. Work into the top 2 to 3 inches of soil, and plant seed in normal manner. **Field Applications:** Apply specified dosage in a water emulsion, or with liquid fertilizer as an injection on each side of the seed furrow or transplanted row at planting time. Do not apply directly on the seed. Where 2 field applications are required, use only 1.2 to 2.4 fluid ounces per 1000 feet of row, or 1 to 2 pints per acre, injected on each side of the seed furrow or transplanted row at planting, and as a side-dressing after plants are established. Allow a minimum of 21 days between applications. Do not apply within 30 days of harvest.
	NOTE: Florida Only—Rates up to 10.5 fluid ounces per 1000 feet of row, with a 72-inch row spacing, may be applied to highly alkaline soils as an injection on each side of the seed furrow or transplanted row at planting time only; or where 2 field applications are required, up to 5.25 fluid ounces per 1000 feet of row with a 72-inch row spacing at planting time, and up to 5.25 fluid ounces per 1000 feet of row as a side-dressing after plants become established. Do not apply within 30 days of harvest. Read "IMPORTANT" statement under "RECOMMENDED APPLICATIONS" for additional information before use on this crop. 6-1-89		

* ON ROW CROPS WHERE ROW SPACING IS EXTREMELY NARROW, DO NOT PLACE TREATED ZONES (BAND OR FURROW) CLOSER TOGETHER THAN 6 INCHES.

SPECIAL LOCAL NEED REGISTRATIONS [SECTION 24(c)] FOR DI-SYSTON 8*

CROP	INSECT	STATE EPA SLN NO.	DOSAGE	REMARKS
Asparagus	Asparagus aphids Thrips	Arizona AZ 85-0007	1 pint per acre.	Apply specified dosage to fern stage when infestation appears. A maximum of 3 applications may be made per year. Do not apply within 45 days of harvest.
		California CA 84-0192		Apply specified dosage to fern stage when infestation appears. A maximum of 3 applications may be made per year. Do not apply within 30 days of harvest.
		Oregon OR 84-0032 Washington WA 84-0036		Apply specified dosage to fern stage of asparagus when aphid or thrips infestation appears. A maximum of 3 applications may be made per year. Do not apply within 6 months of harvest.
Asparagus	Asparagus aphids	North Carolina NC 86-0005	1 pint per acre	Apply specified dosage to fern stage when infestation appears. A maximum of 3 applications may be made per year. Do not apply within 5.5 months of harvest.
		Illinois IL 89-0002 Michigan MI 89-0002		Apply specified dosage to fern stage when infestation appears. A maximum of 3 applications per year. Do not apply within 120 days of harvest.
Cotton	Aphids	Texas TX 86-0007	1.6 fluid ounces per acre.	Apply specified dosage as foliar spray by ground or air. A maximum of 2 times per season, but no later than boll opening. Do not apply within 30 days of harvest.
Bermuda grass (Seed crop only)	Thrips Whiteflies Mealybugs	Arizona AZ 79-0044	1 pint per acre.	Application by ground or air.
		California CA 79 0113		
Conservation Reserve Program Acres Seeded to Ground Cover (native and tame grass)	Aphids (Russian Wheat aphid)	Nebraska NE 87-0003 South Dakota SD 87-0007 Wyoming WY 87-0005	½ to 1 pint per acre.	Application by air or ground. Do not graze or harvest treated fields after any application of DI-SYSTON 8.
Grasses grown for seed	Aphids Grasshoppers	South Dakota SD 87-0007	½ to 1 pint per acre.	Apply by air or ground. Do not graze or harvest treated fields for forage. Do not feed seed or straw to livestocks or use straw for livestock bedding.
Lettuce	Lettuce root aphids	California CA 81-0044	1.2 fluid ounce per 1,000 ft. of row (for any row spacing) or 2 pints per acre (20 inch row spacing).	Incorporate by sprinkler injection. Do not use within 60 days of harvest.
Oats	Aphids	Maine ME 88-0001	½ pint per acre	Application by ground or air. Do not apply within 60 days of grain harvest.
Peppers	Aphids	California CA 77-0036	1 qt. per acre	Apply in a water emulsion or with liquid fertilizer, side-dressed after plants are established. Do not apply within 60 days of harvest.
Triticale (Foliar application Spring or Fall)	Aphids (Greenbugs) Mites	Wyoming WY 87-0004	4 to 12 fluid ounces per acre.	Application by ground or air. May be applied in liquid fertilizer.
Wheat (Fall or Spring foliar application)	Aphids Mites	New Mexico NM 88-0001 Oklahoma OK 88-0002	4 to 12 fluid ounces per acre.	FOR USE ONLY ON WHEAT NOT TREATED WITH DI-SYSTON AT PLANTING. Air or ground application. Do not apply within 30 days of grain harvest.

*For distribution and use only within the state specified. These Special Local Need registrations approved as of date of this publication are, however, subject to cancellation by the state. Contact your nearest Mobay Agricultural Chemicals Division Office for the current status and a copy of the complete individual supplemental Special Local Need labeling. This labeling must be in the possession of the user at the time of pesticide application.

RESTRICTIONS

Not for home garden use. Do not use on other crops used for food or forage. Use only according to label directions. Application at rates above those shown may result in illegal crop residues. Do not treat food crops grown in the greenhouse. Do not apply in any manner not specified on the label.

RE-ENTRY STATEMENTS FOR FARMWORKERS

Do not apply this product in such a manner as to directly or through drift expose workers or other persons. The area being treated must be vacated by unprotected persons. Do not allow re-entry into treated fields within 24 hours of application unless appropriate protective clothing is worn. Because certain states may require more restrictive re-entry intervals for various crops treated with this product, consult your State Department of Agriculture for further information.

Written or oral warnings must be given to workers who are expected to be in a treated area or in an area about to be treated with this product. Oral warnings should include those statements found under the label heading "Precautionary Statements" and "Re-entry Statements" included on the label. When oral warnings are given, warnings shall be given in a language customarily understood by workers. Oral warnings must be given if there is reason to believe that written warnings cannot be understood by workers. Written warnings must include the following information: DANGER, area treated with DI-SYSTON 8 on (date of application). Do not allow re-entry into treated fields within 24 hours of application unless appropriate protective clothing is worn. **If on skin,** remove contaminated clothing and wash skin immediately with soap and warm water. **If eyes are contaminated,** wash with flowing water for at least 15 minutes. **If swallowed,** vomiting should be induced. (See Precautionary Statements on label.)

STORAGE AND DISPOSAL

Pesticide Disposal: Do not contaminate water, food or feed by storage or disposal.

Pesticide wastes are acutely hazardous. Improper disposal of excess pesticide, spray mixture, or rinsate is a violation of Federal Law. If these wastes cannot be disposed of by use according to label instructions, contact your State Pesticide or Environmental Control Agency, or the Hazardous Waste representative at the nearest EPA Regional Office for guidance.

Container Disposal: Triple rinse (or equivalent). Then offer for recycling or reconditioning, or puncture and dispose of in a sanitary landfill, or by other procedures approved by State and local authorities.

Store in a cool, dry place and in such a manner as to prevent cross contamination with other pesticides, fertilizers, food, and feed. Store in original container and out of the reach of children, preferably in a locked storage area.

Handle and open container in a manner as to prevent spillage. If container is leaking invert to prevent leakage. If the container is leaking or material spilled for any reason or cause, carefully dam up spilled material to prevent runoff. Refer to Precautionary Statements on label for hazards associated with the handling of this material. Do not walk through spilled material. Absorb spilled material with absorbing type compounds and dispose of as directed for pesticides above. In spill or leak incidents, keep unauthorized people away. You may contact the Mobay Emergency Response Team for decontamination procedures or any other assistance that may be necessary. The Mobay Kansas City Emergency Response Telephone No. is 816-242-2582, or contact Chemtrec at 800-424-9300.

PRECAUTIONARY STATEMENTS

HAZARDS TO HUMANS AND DOMESTIC ANIMALS

Poisonous if swallowed, inhaled or absorbed through the skin. Rapidly absorbed through the skin. Do not get in eyes, on skin or on clothing. Wear protective clothing, natural rubber gloves and goggles.

Do not breathe fumes or spray mist. Wear a pesticide respirator jointly approved by the Mining Enforcement and Safety Administration (formerly the U.S. Bureau of Mines) and by the National Institute for Occupational Safety and Health under the provisions of 30 CFR Part 11.

Do not contaminate feed or food. Keep out of reach of children. Keep all unprotected persons out of the operating area or vicinity. Wash hands, arms and face thoroughly with soap and warm water before eating or smoking.

HANDLING OF CONTAMINATED ARTICLES: Clothing and rags receiving undiluted DI-SYSTON 8 as a result of leaks, spills or mishaps should be removed immediately and not reused. These contaminated articles should be buried in the same manner as the empty container; see container disposal section for directions. (Contaminated articles which should be discarded would include: hats, gloves, aprons, coats, boots, etc. not made of rubber, or coated with rubber or other similar materials). Clothing receiving spray mist or droplets from the mixed, diluted DI-SYSTON 8 should be removed and decontaminated before reuse. This clothing can be decontaminated by machine washing separately from other items with soap or detergent and bleach in hot water twice before reuse. Protective clothing and equipment should be washed down with detergent or soap and bleach in water. Wash water from the cleaning of protective clothing or equipment should not be allowed to run off or otherwise enter water supplies.

SYMPTOMS OF POISONING: A sense of "tightness" in the chest. Sweating. Contracted pupils. Stomach pains. Vomiting and diarrhea.

STATEMENTS OF PRACTICAL TREATMENT

In case of poisoning, call a physician immediately. Have patient lie down and keep quiet. **If swallowed, vomiting should be induced.** Administer water freely and induce vomiting by giving one dose (1/2 oz. or 15 ml.) of syrup of ipecac. If vomiting does not occur within 10 to 20 minutes, administer second dose. If syrup of ipecac is not available, induce vomiting by sticking finger down throat. Repeat until vomit fluid is clear. Never give anything by mouth to an unconscious person. Professional medical assistance should be secured immediately. **If on skin,** remove contaminated clothing and wash skin immediately with soap and warm water. Get medical attention immediately. *If in eyes,* wash with plenty of flowing water. Get medical attention immediately.

NOTE TO PHYSICIAN: Antidote—Administer atropine sulfate in large therapeutic doses. Repeat as necessary to the point of tolerance. 2-PAM is also antidotal, and may be administered but only in conjunction with atropine.

Compound inhibits cholinesterase resulting in stimulation of the central nervous system, the parasympathetic nervous system and the somatic motor nerves. Do not give morphine. Watch for pulmonary edema, which may develop in serious cases of poisoning, even after 12 hours. At first sign of pulmonary edema, the patient should be placed in an oxygen tent and treated symptomatically.

ENVIRONMENTAL HAZARDS

This pesticide is toxic to fish and wildlife. Do not apply directly to water or wetlands. Drift and runoff from treated areas may be hazardous to fish in neighboring areas. Do not contaminate water by disposing of equipment washwaters. Cover or incorporate spills.

This pesticide is toxic to bees exposed to direct application. Applications should be timed to coincide with periods of minimum bee activity, usually between late evening and early morning.

PHYSICAL OR CHEMICAL HAZARDS

Do not use or store near heat or open flame.

IMPORTANT:

This bulletin is not intended to provide adequate information for application. Before using the product, read and carefully observe directions, cautionary statements and other information appearing on the product label. This product is sold subject to the Conditions of Sale set forth on the container label.

READ THE LABEL BEFORE USE

DS 9002

This is a specimen label. Mobay is not responsible for the accuracy of the information contained herein. As labels are subject to revision, always carefully read and follow the label on the product container.

DYLOX®

80% SOLUBLE POWDER INSECTICIDE

A SELECTIVE INSECTICIDE FOR EFFECTIVE CONTROL OF INSECT PESTS ON A VARIETY OF FIELD AND VEGETABLE CROPS

ACTIVE INGREDIENT:	
Dimethyl (2,2,2-trichloro-1-hydroxyethyl)phosphonate	80%
INERT INGREDIENTS:	20%
TOTAL	100%

EPA Reg. No. 3125-184

DYLOX is a Reg. TM of Bayer AG, Germany

RECOMMENDED USES

Alfalfa	Pumpkin
Blueberries	Soybeans (seed)
Clover	Table beets
Corn	Tobacco
Cotton	Tomatoes

RESTRICTIONS ON WAREHOUSE STORAGE

Keep away from space heaters, or walls that sweat. Store below 120°F. Protect from rain and moisture. Store in a cool, dry place. Do not store next to herbicides.

PACKAGING

5 pound carton, 10 per case shipping weight 53 lbs.

50 pound box. Shipping weight 52 lbs.

DYLOX selective insecticide effectively controls a wide variety of insect pests, without appreciably reducing beneficial insect populations, when applied at recommended minimum rates. As a result, DYLOX 80% Soluble Powder supplements natural control factors instead of destroying them. Insect parasites, predators and pollinators continue to perform their good work. DYLOX 80% Soluble Powder dissolves in water, and is recommended for use in all conventional power-operated ground sprayers and aircraft spray equipment.

The unique selective advantage of DYLOX 80% Soluble Powder is lost when the chemical is used in conjunction with, or alternated with pesticides which are non-selective.

PELIGRO

PRECAUCION AL USUARIO: Si usted no puede leer o entender inglés, no use este producto hasta que la etiqueta le haya sido explicada ampliamente.

(TO THE USER: If you cannot read or understand English, do not use this product until the label has been fully explained to you.)

DIRECTIONS FOR USE

It is a violation of Federal Law to use this product in a manner inconsistent with its labeling.

DYLOX insecticide offers the unique advantage of not significantly affecting beneficial insects (parasites, predators and pollinators), especially when applied at the minimum recommended rates per acre. Thus, DYLOX supplements the natural control factors, instead of destroying them. This advantage is lost when DYLOX is used in conjunction with parathion or other non-selective insecticides. Accordingly, for the most effective and economical control, DYLOX should not be used in conjunction with, or alternated in, a spray schedule with pesticides destroying parasites, predators or other beneficial insects. This applies to alfalfa and cotton, in particular. While DYLOX is not recommended for the control of aphids, mites and cabbage looper, some suppression of these pests may result from recommended dosages.

MIXING: DYLOX 80% Soluble Powder insecticide dissolves readily in water, and is suitable for use in all power-operated ground and aircraft sprayers. To dissolve in water, pour the required amount of DYLOX 80% Soluble Powder into full amount of water, and then agitate. Use immediately after mixing. It is recommended that DYLOX 80% Soluble Powder be used in sprayers equipped with 50-mesh screens. If 100-mesh screens are used, some screen clogging may occur.

Do not apply this product through any type of irrigation system.

DOSAGE: Use specified dosage of DYLOX 80% Soluble Powder in the amount of water necessary to give complete coverage of foliage. The type of equipment used will determine the concentration required.

SPRAYING: Complete coverage and thorough application are essential for most effective results. Schedule applications in accordance with local conditions. Consult your State Agricultural Extension Service or Experiment Station for specific timing of applications in your area.

NOTE: Spray droplets from applications of DYLOX 80% Soluble Powder may cause spotting of paint finishes on aluminum and steel siding, automobiles, trucks and tractors. Do not allow spray droplets to contact metal painted surfaces. If accidental exposure does occur, the painted surface should be thoroughly washed immediately after exposure.

RECOMMENDED APPLICATIONS

CROP	INSECT	OUNCES DYLOX 80% S.P.	REMARKS
FIELD CROPS	Alfalfa caterpillar	7½ to 10	Apply specified dosage per acre, using sufficient water for complete coverage, but not less than 1 gallon per acre. Three applications may be made per cutting, with the last application up to the day of cutting.
	Alfalfa webworm	5 to 20	
Alfalfa Clover (including mixed stands with grasses)	Western yellow-striped armyworm	10	
	Beet armyworm Leafhoppers Variegated cutworm	10 to 20	
	Armyworm Alfalfa plant bug Lygus bugs Stink bugs Tarnished plant bug	20	
Corn Field, Sweet, Popcorn	Armyworms Cutworms	10 to 20	Apply specified dosage per acre, using sufficient water for complete coverage, but not less than 1 gallon per acre. Three applications may be made per season. For early application, when plants are 3 to 12 inches high, direct the spray to the lower portions of plants and to the soil around the base of the plants. Later applications may be made as full-coverage spray up to day of harvest.
Cotton	Cotton fleahopper	5 to 20	Apply specified dosage per acre, using sufficient water for complete coverage, but not less than 1 gallon per acre. Do not pasture treated fields within 14 days of application. Do not apply within 7 days of picking. DYLOX application may occasionally cause marginal leaf burn.
	Cotton leafworm Darkling ground beetle Western yellow-striped armyworm	10 to 20	
	Beet armyworm Southern garden leafhopper	20	**NOTE:** For control of light to moderate infestation of cotton fleahoppers, use 5 ounces of DYLOX 80% S.P. per acre. Higher rates may be required if heavy infestations or migrations occur.
	Black fleahopper complex Cotton leaf perforator Leaf roller Lygus bugs Stink bugs	20 to 30	
	Salt-marsh caterpillar	30	
Tobacco	Budworm Hornworm	20	Apply specified dosage per acre, using sufficient water for complete coverage, but not less than 1 gallon per acre. For budworm control, make application directly into bud. Repeat as necessary. Do not treat within 3 days of harvest.
	Green June beetle larvae	10	Apply specified dosage in 100 gallons of water per 100 square yards to uprooted areas of plant beds.

CROP	INSECT	OUNCES DYLOX 80% S.P.	REMARKS
SEED FIELD CROPS Alfalfa Clover	Armyworms Lygus bugs Stink bugs Variegated cutworm	20 to 30	Apply specified dosage per acre, using sufficient water for complete coverage, but not less than 1 gallon per acre. Repeat as necessary, up to 7 days before harvest. Chaff from alfalfa and clover seed crop may be used for feed or forage; however, **do not** cut green crop for these purposes. Do not pasture or use treated soybeans for feed, food, forage or oil purposes.
Soybeans	Armyworms Dipterous leafminers Lygus bugs Stink bugs Variegated cutworm		**NOTE:** Due to lygus bug resistance in certain areas of alfalfa and clover DYLOX used alone may not provide satisfactory control. Consult your local agricultural advisor or extension service for recommendations.
FRUIT Blueberries	Spanworm	16	Apply specified dosage per acre using sufficient water for complete coverage, but not less than 1 gallon per acre. Make a prebloom application as soon as insects appear. A second application may be made during bloom but not less than 7 days after first application. Do not apply after bloom stage.
VEGETABLES Pumpkin	Variegated cutworm	10 to 20	Apply specified dosage per acre, in sufficient water for complete coverage, but not less than 1 gallon per acre. Do not apply more than 3 times per season, nor within 3 days of harvest.
	Squash bug	20	
Table beets	Variegated cutworm	10 to 20	Apply specified dosage per acre, using sufficient water for complete coverage, but not less than 1 gallon per acre. Repeat as necessary. Do not apply within 28 days of harvest. Do not eat tops.
	Alfalfa webworm Beet webworm Dipterous leafminers Salt-marsh caterpillar	20	
	Beet armyworm Lygus bugs	20 to 30	
Tomatoes	Serpentine leafminers Tomato hornworms Dipterous leafminers	20	Apply specified dosage per acre, using sufficient water for complete coverage, but not less than 1 gallon per acre. Repeat as necessary. Do not apply within 21 days of harvest. 8-11-88

PREMISE SPRAY FOR FLY CONTROL: For Fly Control in and around Farm Buildings (Except Dairy Barns and Poultry Houses): Mix 5 pounds of DYLOX 80% Soluble Powder in 40 gallons of water (equivalent to 5 level tablespoonfuls per gallon). Thoroughly wet all areas to be treated by spraying 1 gallon per 500 square feet of surface area, or to the point of runoff. Apply to floors, walls, ceiling, around windows, doors and other areas where flies congregate. Also spray stock pens (after removal of animals), garbage areas and other locations where refuse has collected. Repeat applications every 1 to 2 weeks, or as fly populations increase. For longer residual control, add 1 pound of sugar to each gallon of spray solution.

RESTRICTIONS (Premise Spray): Remove animals before spraying barn. Do not treat portions of buildings where poultry feed. Direct application to walls, floors or other surfaces previously treated with lime, whitewash or other alkaline materials may be ineffective. Do not contaminate feed, water or milk-handling equipment.

Rate/Acre In Ounces Of DYLOX 80% S.P.	Pounds Of DYLOX 80% S.P. To Treat:		
	5 Acres	10 Acres	15 Acres
5	1½	3⅛	4⅔
7½	2⅓	4⅔	7
8	2½	5	7½
10	3⅛	6¼	9⅜
16	5	10	15
20	6¼	12½	18¾
30	9⅜	18¾	28⅛

NOTE: When using a more convenient whole number or fraction of pounds of formulated material, do not exceed maximum registered rate.

SPECIAL LOCAL NEED REGISTRATIONS [SECTION 24(C)] FOR DYLOX 80% SOLUBLE POWDER*

CROP	INSECT	STATE EPA SLN NO.	DOSAGE	REMARKS
Alfalfa	Army cutworm	Kansas KS 76-0002	20 ounces per acre.	Three applications may be made per cutting with last application up to day of cutting.
Blueberries	Gypsy moth	New Jersey NJ 84-0006	1.5 pounds	Apply specified dosage with aerial spray equipment or with ground spray equipment as a broadcast spray just before or during bloom stage when gypsy moth populations reach damaging levels. Do not apply more than one time per season nor within 45 days of harvest.
Cotton	Pink bollworm (adults)	California CA 82-0094	1.25-1.75 pounds per acre.	Apply specified dosage per acre, using sufficient water for complete coverage. Do not apply within 7 days of picking.

*For distribution and use only within the state specified. These Special Local Need registrations approved as of date of this publication are, however, subject to cancellation by the state. Contact your nearest Mobay Agricultural Chemicals Division Office for the current status and a copy of the complete individual supplemental Special Local Need labeling. This labeling must be in the possession of the user at time of pesticide application.

RESTRICTIONS

Do not use on other crops used for food or forage. Use only according to label directions. Application at rates above those shown may result in illegal crop residues. Do not treat food crops grown in the greenhouse.

DO NOT ALLOW DYLOX SPRAY TO DRIFT ONTO SORGHUM. TO DO SO MAY CAUSE BURNING OF THE SORGHUM OR MILO. Keep children and pets off treated areas until spray deposit has dried.

RE-ENTRY STATEMENTS

Do not apply this product in such a manner as to directly or through drift expose workers or other persons. The area being treated must be vacated by unprotected persons. Do not reenter treated fields within 24 hours unless protective clothing is worn. Because certain states may require more restrictive re-entry intervals for various crops treated with this product, consult your State Department of Agriculture for further information.

Written or oral warnings must be given to workers who are expected to be in a treated area or in an area about to be treated with this product. Oral warnings should include those statements found under the label heading "Precautionary Statements" and "Re-entry Statements" included on the label. When oral warnings are given, warnings shall be given in a language customarily understood by workers. Oral warnings must be given if there is reason to believe that written warnings cannot be understood by workers. Written warnings must include the following information: **DANGER,** area treated with DYLOX 80% Soluble Powder on (date of application).

Do not reenter treated fields within 24 hours unless protective clothing is worn. **If on skin,** remove contaminated clothing and wash skin immediately with soap and warm water. **If eyes are contaminated,** wash with flowing water for at least 15 minutes. **If swallowed,** vomiting should be induced. (See Precautionary Statements on label.)

PRECAUTIONARY STATEMENTS

HAZARDS TO HUMANS AND DOMESTIC ANIMALS

DANGER

Corrosive. Causes irreversible eye damage. May be fatal if swallowed. Harmful if inhaled, or absorbed through the skin. Prolonged or frequently repeated skin contact may cause allergic reactions in some people. Do not get in eyes or on skin. Avoid breathing dust and avoid contact with skin. Wash thoroughly with soap and warm water after handling, and before eating, drinking, or using tobacco. Remove contaminated clothing and wash before reuse.

STATEMENTS OF PRACTICAL TREATMENT

If in eyes: Hold eyelids open and flush with plenty of water for 15 minutes. Seek medical attention immediately. **If on skin:** Remove contaminated clothing. Wash affected areas with plenty of soap and water. Get medical attention if irritation develops and persists. If signs of intoxication (poisoning) occur, get medical attention immediately. **If inhaled:** First, remove victim to fresh air or uncontaminated area. If not breathing, give artificial respiration, preferably mouth-to-mouth. Get medical attention as soon as possible. **If swallowed:** Call a physician or Poison Control Center. If medical assistance cannot be given immediately, induce vomiting and get to a hospital. To induce vomiting, administer 1 or 2 glasses of water and stick finger down throat. Repeat until vomitus is clear. Or, if available, administer syrup of ipecac (1/2 oz. or 15 ml). If vomiting does not occur within 5 to 10 minutes, administer a second dose of ipecac. Do not attempt to give anything by mouth or induce vomiting in an unconscious or convulsing person.

NOTE TO PHYSICIAN: ANTIDOTE—Administer atropine sulfate in large therapeutic doses. Repeat as necessary to the point of tolerance. 2-PAM is also antidotal and may be administered in conjunction with atropine.

Compound inhibits cholinesterase resulting in stimulation of the central nervous system, autonomic nervous system, and skeletal muscle. Do not give morphine. Watch for pulmonary edema which may develop in serious cases of poisoning even after 24 to 48 hours. At first sign of pulmonary edema, the patient should be placed in oxygen tent and treated symptomatically.

In case of poisoning, it is also requested that Mobay Corporation, Agricultural Chemicals Division, Kansas City, Missouri, be notified. Telephone 816-242-2582 (week-ends).

ENVIRONMENTAL HAZARDS

PROTECT WILDLIFE!!

This pesticide is toxic to fish and wildlife and is extremely toxic to aquatic invertebrates. Apply the product in strict accordance with label directions. Do not apply when weather conditions favor drift from areas treated. Do not apply to water or wetlands. Do not apply where runoff is likely to occur. Runoff from treated areas may be hazardous to aquatic organisms in neighboring areas. In cleaning of equipment or disposal of wastes, do not contaminate water.

STORAGE AND DISPOSAL

Pesticide Disposal: Do not contaminate water, food or feed by storage or disposal.

Wastes resulting from the use of this product may be disposed of on site or at an approved waste disposal facility.

Container Disposal: Do not use container in connection with food, feed, or drinking water. Completely empty bag into application equipment. Then dispose of empty bag in a sanitary landfill or by incineration, or, if allowed by State and local authorities, by burning. If burned, stay out of smoke.

Store in a cool, dry place and in such a manner as to prevent cross contamination with other pesticides, fertilizers, food, and feed. Store in original container and out of the reach of children, preferably in a locked storage area.

Handle and open container in a manner as to prevent spillage. If the container is leaking or material spilled for any reason or cause, carefully sweep material into a pile. Refer to Precautionary Statements on label for hazards associated with the handling of this material. Do not walk through spilled material. Dispose of pesticide as directed above. In spill or leak incidents, keep unauthorized people away. You may contact the Mobay Emergency Response Team for decontamination procedures or any other assistance that may be necessary. The Mobay Kansas City Emergency Response Telephone No. is 816-242-2582, or contact Chemtrec at 800-424-9300.

IMPORTANT:

This bulletin is not intended to provide adequate information for application. Before using the product, read and carefully observe directions, cautionary statements and other information appearing on the product label. This product is sold subject to the Conditions of Sale set forth on the container label.

READ THE LABEL BEFORE USE

DX 9001

This is a specimen label. Mobay is not responsible for the accuracy of the information contained herein. As labels are subject to revision, always carefully read and follow the label on the product container.

DYRENE®

50% WETTABLE POWDER FOLIAGE FUNGICIDE

FOR CONTROL OF CERTAIN PLANT DISEASES ON CERTAIN CROPS

ACTIVE INGREDIENT:	
4,6-Dichloro-*N*-(2-chlorophenyl)-1,3,5-triazin-2- amine	50%
INERT INGREDIENTS:	50%
TOTAL	100%

EPA Reg. No. 3125-50

DYRENE is a Reg. TM of Mobay Corporation

RECOMMENDED USES

Celery, Onions (Green), Potatoes, Strawberries, and Tomatoes

RESTRICTIONS ON WAREHOUSE STORAGE

Protect from rain and moisture. Keep away from space heaters, and walls that sweat. Do not store next to herbicides. Store in a cool, dry place.

PACKAGING

4 lb. bag, 8 per case. Shipping weight 37 lbs.
75 lb. drum. Shipping weight 85 lbs.

DYRENE foliage fungicide is an organic fungicide for disease control. DYRENE has proved to be very effective in years of commercial usage for the control of the important fungus diseases which attack tomatoes, potatoes, celery, onions (green), and strawberries.

By controlling these destructive fungus diseases, DYRENE promotes increases in crop yields. In addition to its fungicidal effectiveness, DYRENE will not harm foliage, or affect the flavor or odor of treated crops.

PELIGRO

PRECAUCION AL USUARIO: Si usted no puede leer o entender inglés, no use este producto hasta que la etiqueta le haya sido explicada ampliamente.

(TO THE USER: If you cannot read or understand English, do not use this product until the label has been fully explained to you.)

DIRECTIONS FOR USE

It is a violation of Federal Law to use this product in a manner inconsistent with its labeling.

Do not use on seed crops intended to be used for feed or forage. Do not graze treated areas.

Do not reenter treated fields within 24 hours following application of this product.

DYRENE 50% Wettable Powder effectively protects crops from various destructive plant diseases. Complete and thorough coverage of foliage is necessary for satisfactory control.

CROP APPLICATIONS: Apply DYRENE in a protective spray schedule before disease symptoms appear. Repeat if needed at intervals of 7 to 10 days, and during extended periods of wet weather, every 4 or 5 days. Observe time limitations for each use for interval between last application and harvest. Schedule applications in accordance with local conditions. Consult your State Agricultural Experiment Station or Extension Service for specific use information, and for State regulations which may contain additional restrictions or requirements.

Do not apply this product through any type of irrigation system.

RECOMMENDED APPLICATIONS

CROP	DISEASE	POUNDS DYRENE 50% W.P.	REMARKS
FRUIT Strawberries	Gray mold (*Botrytis*) Common leaf spot Leaf scorch	2 to 5	Apply specified dosage using 100 to 200 gallons of water per acre for good coverage.* Additional applications may be made as needed, but do not exceed 30 pounds of DYRENE 50% WP per acre per season. Do not apply within 3 days of harvest.
VEGETABLES Celery	Early blight Late blight *Rhizoctonia* crater rot	2 to 6	Apply specified dosage using 75 to 200 gallons of water per acre for good coverage by ground equipment* or a minimum of 5 gallons of water per acre by air. Additional applications may be made as needed, but do not exceed 60 pounds of DYRENE 50% WP per acre per season. Do not apply within 7 days of harvest. Treated celery should be trimmed.
Onions (green)	*Botrytis leaf blight* Purple blotch (*Alternaria*)	2 to 3	Apply specified dosage using 75 to 100 gallons of water per acre for good coverage.* Additional applications may be made as needed but do not exceed 18 pounds of DYRENE 50% WP per acre per season. Do not apply with 3 days of harvest. Not recommended for use in Florida.
Potatoes	Early blight Late blight	2 to 5	Apply specified dosage using 75 to 175 gallons of water per acre for good coverage by ground equipment* or a minimum of 5 gallons of water per acre by air. Additional applications may be made as needed, but do not exceed 60 pounds of DYRENE 50% WP per acre per season. Do not apply within 3 days of harvest.
Tomatoes	Anthracnose Botrytis gray mold ghost spot Early blight Gray leaf spot Late blight *Septoria* leaf spot	2 to 5	Apply specified dosage using 75 to 175 gallons of water per acre for good coverage by ground equipment* or a minimum of 5 gallons of water per acre by air. Additional applications may be made as needed, but do not exceed 60 pounds of DYRENE 50% WP per acre per season. Do not apply within 3 days harvest. 2-18-88

* This dosage is calculated for conventional hydraulic-type sprayers. When lower volumes of spray are applied per acre, increase the concentration of DYRENE in the spray mixture in order to apply amount of DYRENE per acre equivalent to a full-coverage spray.

SPECIAL LOCAL NEED REGISTRATIONS [SECTION 24(C)] FOR DYRENE 50% WETTABLE POWDER*

CROP	DISEASE	STATE EPA SLN NO.	DOSAGE	REMARKS
Fruits Strawberries Vegetables Celery Onions (Green) Potatoes Tomatoes	Same as Federal label. Black mold added for tomatoes.	California CA 77-0066	Same as Federal label.	This Special Local Need registration is for a product packaged in water soluble packets. (PVA) Aerial applications permitted in not less than 10 gallons of water.
Tomatoes	Anthracnose- *Botrytis* Early blight Gray leaf spot Late blight *Septoria* leaf spot Black mold (*Alternaria tenuis*)	California CA 76-0103	2 to 5 pounds per acre.	Apply specified dosage using 75 to 175 gallons of water per acre for ground application or apply by air in 5 to 15 gallons of water.

* For distribution and use only within the state specified. These Special Local Need registrations approved as of date of this publication are, however, subject to cancellation by the state. Contact your nearest Mobay Agricultural Chemicals Division Office for the current status and a copy of the complete individual supplemental Special Local Need labeling. This labeling must be in the possession of the user at time of pesticide application.

RESTRICTIONS

Do not exceed the maximum dosage. Do not use on other crops used for food or feed. Use only according to label directions. Do not treat food crops grown in the greenhouse.

RE-ENTRY STATEMENTS

Do not apply this product in such a manner as to directly or through drift expose workers or other persons. The area being treated must be vacated by unprotected persons. Do not reenter treated fields within 24 hours following application of this product. Because certain states may require more restrictive re-entry intervals for various crops treated with this product, consult your State Department of Agriculture for further information.

Written or oral warnings must be given to workers who are expected to be in a treated area or in an area about to be treated with this product. Oral warnings should include those statements found under the label heading "Precautionary Statements" and "Re-entry Statements" included on the label. When oral warnings are given, warnings shall be given in a language customarily understood by workers. Oral warnings must be given if there is reason to believe that written warnings cannot be understood by workers. Written warnings must include the following information: DANGER, area treated with DYRENE 50% Wettable Powder on (date of application). Do not reenter treated fields within 24 hours following application of this product. **If in eyes,** hold eyelids open and flush with steady, gentle stream of water for 15 minutes. Get medical attention. **If on skin,** remove contaminated clothing and wash skin immediately with soap and warm water. **If swallowed,** drink large quantities of water.

STORAGE AND DISPOSAL

Do not contaminate water, food or feed by storage or disposal.

Pesticide Disposal: Wastes resulting from the use of this product may be disposed of on site or at an approved waste disposal facility.

Container Disposal: Do not use container in connection with food, feed, or drinking water. Completely empty bag into application equipment. Then dispose of empty bag in a sanitary landfill or by incineration, or, if allowed by State and local authorities, by burning. If burned, stay out of smoke.

Store in a cool, dry place and in such a manner as to prevent cross contamination with other pesticides, fertilizers, food, and feed. Store in original container and out of the reach of children, preferably in a locked storage area.

Handle and open container in a manner as to prevent spillage. If the container is leaking or material spilled for any reason or cause, carefully sweep material into a pile. Refer to Precautionary Statements on label for hazards associated with the handling of this material. Do not walk through spilled material. Dispose of pesticide as directed above. In spill or leak incidents, keep unauthorized people away. You may contact the Mobay Emergency Response Team for decontamination procedures or any other assistance that may be necessary. The Mobay Kansas City Emergency Response Telephone No. is 816-242-2582, or contact Chemtrec at 800-424-9300.

PRECAUTIONARY STATEMENTS

HAZARDS TO HUMANS AND DOMESTIC ANIMALS

DANGER

Causes irreversible eye damage. Harmful if swallowed. Do not get into eyes or on clothing. Avoid contact with skin. Wear goggles, face shield or safety glasses. Wash thoroughly with soap and water after handling. Remove and wash contaminated clothing with soap and hot water before reuse.

NOTE: Protective clothing should be worn during periods of exposure, such as during application, or when contacting treated foliage. DYRENE anilazine may cause an allergic skin reaction in some susceptible persons. This condition should be treated as an allergic dermatitis.

Keep out of reach of children.

STATEMENTS OF PRACTICAL TREATMENT

If in eyes: Hold eyelids open and flush with steady, gentle stream of water for 15 minutes. Get medical attention. **If on skin:** Wash immediately with soap and warm water. If irritation occurs, get medical attention. **If swallowed:** Drink promptly a large quantity of milk, or if milk is not available, large quantities of water. Avoid alcohol. Note to Physician: Probable mucosal damage may contraindicate the use of gastric lavage.

ENVIRONMENTAL HAZARDS

This pesticide is toxic to fish and aquatic invertebrates. Do not apply directly to water or wetlands. Drift and runoff from treated areas may be hazardous to aquatic organisms in neighboring areas. Do not contaminate water by cleaning of equipment or disposal of wastes. Cover or incorporate spills.

IMPORTANT:

This bulletin is not intended to provide adequate information for application. Before using the product, read and carefully observe directions, cautionary statements and other information appearing on the product label. This product is sold subject to the Conditions of Sale set forth on the container label.

Read the label before use.

DY 9002

This is a specimen label. Mobay is not responsible for the accuracy of the information contained herein. As labels are subject to revision, always carefully read and follow the label on the product container.

FURADAN® 3G

INSECTICIDE

RESTRICTED USE PESTICIDE

Due to Avian Toxicity

For retail sale to and application only by Certified Applicators or persons under their direct supervision, and only for those uses covered by Certified Applicator's certification.

FOR CONTROL OF MOSQUITOES AND RICE WATER WEEVIL ON RICE

For Agricultural or Commercial Use Only

ACTIVE INGREDIENT:

Carbofuran*	3%
INERT INGREDIENTS:	97%
TOTAL	100%

*(2,3-dihydro-2,2-dimethyl-7-benzofuranyl methylcarbamate)

EPA Reg. No. 279-2792-3125

FURADAN—FMC trademark

U.S. Patent No. 3,474,171

RECOMMENDED USES

Rice water weevil and mosquitoes in Arkansas, Louisiana, Mississippi, Missouri, and Texas.

RESTRICTIONS ON WAREHOUSE STORAGE

The package will not withstand stacking two pallets high. Do not store next to herbicides. In case of puncture, use overslip bag and securely close.

PACKAGING

50 lb. bag. Shipping weight 51 lbs.

FURADAN insecticide is a carbamate compound designed to control rice water weevils and mosquitoes. It is especially useful in areas reporting resistance of these insects to chlorinated hydrocarbon insecticides. By controlling rice water weevils, FURADAN insecticide stops lodging problems, keeps plants growing normally, without setback or loss of vigor, speeds maturity and permits increased yields.

AVISO

PRECAUCION AL USUARIO: Si usted no puede leer o entender inglés, no use este producto hasta que la etiqueta le haya sido explicada ampliamente.

(TO THE USER: If you cannot read or understand English, do not use this product until the label has been fully explained to you.)

DIRECTIONS FOR USE

It is a violation of Federal Law to use this product in a manner inconsistent with its labeling.

Do not re-enter treated fields less than 24 hours following application. Do not use this product on Long Island, N.Y.

Do not plant with any crop for which carbofuran treatment is not registered for at least 10 months.

Do not rotate with any crop on soil treated at greater than 3.0 pounds active ingredient per acre for at least 10 months.

RICE: In Arkansas, Louisiana, Mississippi, Missouri, and Texas—For control of **rice water weevil and mosquitoes,** apply 15 to 20 pounds of FURADAN 3G per acre 1 day before, or within 21 days after permanent flooding. Do not make more than one application per season. Apply by air or ground equipment. The subsequent use of propanil herbicide may result in crop injury.

For effective larval control of dark rice field mosquito, application must be made within 2 to 4 days after the field has been flooded.

SPECIAL LOCAL NEED REGISTRATIONS [SECTION 24(C)] FOR FURADAN 3G*

CROP	PEST	STATE EPA SLN NO.	DOSAGE	REMARKS
Rice	Rice water weevil Mosquitoes	Arkansas AR 81-0045 Louisiana LA 81-0020 Mississippi MS81-0051	15 to 20 pounds per acre.	One application as needed but not within 60 days of harvest.

* For distribution and use only within the state specified. These Special Local Need registrations approved as of date of this publication are, however, subject to cancellation by the state. Contact your nearest Mobay Agricultural Chemicals Division Office for the current status and a copy of the complete individual supplemental Special Local Need labeling. This labeling must be in the possession of the user at time of pesticide application.

PRECAUTIONARY STATEMENTS

HAZARDS TO HUMANS & (DOMESTIC ANIMALS)

WARNING

Poisonous if swallowed. May be fatal or harmful as a result of skin or eye contact or by breathing dust. Causes cholinesterase inhibition. Warning symptoms of poisoning include weakness, headache, sweating, nausea, vomiting, diarrhea, tightness in chest, blurred vision, pinpoint eye pupils, abnormal flow of saliva, abdominal cramps, and unconsciousness.

Atropine sulfate is antidotal.

Wear long-sleeved clothing and protective gloves when handling. Wash hands and face before eating or smoking. Bathe at the end of the work day. Change clothing daily and wash before reuse.

STATEMENTS OF PRACTICAL TREATMENT

If swallowed, vomiting should be induced. Administer water freely and induce vomiting by giving one dose (1/2 oz. or 15 ml.) of syrup of ipecac. If vomiting does not occur within 10 to 20 minutes, administer second dose. If syrup of ipecac is not available, induce vomiting by sticking finger down throat. Repeat until vomit fluid is clear. Professional medical assistance should be secured immediately. Never give anything by mouth to an unconscious person. Get medical attention. **If in eyes,** immediately flush eyes with plenty of water for at least 15 minutes. **If on skin,** in case of contact, remove contaminated clothing and immediately wash skin with soap and water.

NOTE TO PHYSICIAN: Antidote—FURADAN (carbofuran) is a reversible cholinesterase inhibitor. Do not use oximes such as 2-PAM. Start by giving 2 mg. atropine intramuscularly. According to clinical response, continue until signs of atropinization occur (dry mouth or dilated pupils). If in eye, instill one drop of homatropine.

ENVIRONMENTAL HAZARDS

This pesticide is toxic to fish, birds and other wildlife. Birds feeding on treated areas may be killed. Birds killed by carbofuran pose a hazard to hawks and other birds of prey; bury or otherwise dispose of dead birds to prevent poisoning of other wildlife. Cover or incorporate granules in spill areas. Runoff from treated areas may be hazardous to fish in neighboring areas. Fish may be killed at recommended rates. Do not contaminate wells, wetlands or any body of water by cleaning of equipment or disposal of waste.

Carbofuran is a chemical which can travel (seep or leach) through soil and can contaminate ground water which may be used as drinking water. Carbofuran has been found in ground water as a result of agricultural use. Users are advised not to apply carbofuran where the water table (ground water) is close to the surface and where the soils are very permeable, i.e., well-drained soils such as loamy sands. Your local agricultural agencies can provide further information on the type of soil in your area and the location of ground water.

Note to User: Do not allow this material to drift onto neighboring crops or non-crop areas or use in a manner or at a time other than in accordance with label directions because animal, plant or crop injury, illegal residues or other undesirable results may occur.

STORAGE AND DISPOSAL

Pesticide Storage

Do not store in or around the home.

Keep out of reach of children and animals. Store in original containers only. Store in a cool, dry place and avoid excess heat. Carefully open containers. After partial use, fold and roll back bags, clamp and close tightly. Do not put concentrate or dilute material into food or drink containers. Do not contaminate other pesticides, fertilizers, water, food or feed by storage or disposal.

In case of spill, avoid contact, isolate area and keep out animals and unprotected persons. Confine spills. Call Mobay Kansas City Emergency Response Telephone No. 816-242-2582, or contact Chemtrec at 800-424-9300.

To confine spill: If liquid, dike surrounding area or absorb with sand, cat litter or commercial clay. If dry material, cover to prevent dispersal. Place damaged package in a holding container. Identify contents.

Pesticide Disposal

Pesticide wastes are acutely hazardous. Improper disposal of excess pesticide, spray mixture, or rinsate is a violation of Federal Law. If wastes cannot be disposed of by use according to label instructions, contact your State Pesticide or Environmental Control Agency, or the Hazardous Waste representative at the nearest EPA Regional Office for guidance.

Container Disposal

Completely empty bag into application equipment. Then dispose of empty bag in a sanitary landfill or by incineration, or, if allowed by State and local authorities, by burning. If burned, stay out of smoke.

IMPORTANT:

This bulletin is not intended to provide adequate information for application. Before using the product, read and carefully observe directions, cautionary statements and other information appearing on the product label. This product is sold subject to the Conditions of Sale set forth on the container label.

READ THE LABEL BEFORE USE

FU 9001

This is a specimen label. Mobay is not responsible for the accuracy of the information contained herein. As labels are subject to revision, always carefully read and follow the label on the product container.

FURADAN®

4F INSECTICIDE

RESTRICTED USE PESTICIDE

Due to acute oral and inhalation toxicity.
For retail sale to and application only by Certified Applicators or personnel under their direct supervision.

FOR CONTACT AND SYSTEMIC CONTROL OF PESTS ON CERTAIN FIELD AND VEGETABLE CROPS

ACTIVE INGREDIENT:

*Carbofuran	44.00%
INERT INGREDIENTS:	56.00%
TOTAL	100.00%

*2,3-Dihydro-2,2-dimethyl-7-benzofuranyl methylcarbamate.
This product contains 4 lb. of carbofuran per gallon.

EPA Reg. No. 279-2876-ZC-3125

FURADAN—FMC trademark U.S. Patent No. 3,474,171

RECOMMENDED USES

Alfalfa, Field corn, Sweet corn, Popcorn, Cotton, Pine seedlings, Potatoes, Small grains (wheat, oats, barley), Soybeans, Grapes (California), Strawberries (Washington and Oregon), Sugarcane, Sunflowers*, Tobacco, Ornamentals.
*See label restrictions.

PACKAGING

1 gallon jug, 4 per case. Shipping weight 44 lbs.
2½ gallon jug, 2 per case. Shipping weight 50.63 lbs.
30 gallon drum. Shipping weight 312 pounds.

FURADAN 4F insecticide is a carbamate compound designed to control alfalfa weevil larvae and lygus bugs on alfalfa, corn rootworms on field corn, European corn borer on sweet corn, sugarcane borers and various pests of potatoes, pine seedlings and small grains. It has a long residual life against alfalfa weevil and Egyptian alfalfa weevil, and can be applied earlier than other insecticides to control these larvae.

PELIGRO

PRECAUCION AL USUARIO: Si usted no puede leer o entender inglés, no use este producto hasta que la etiqueta le haya sido explicada ampliamente.

(TO THE USER: If you cannot read or understand English, do not use this product until the label has been fully explained to you.)

DIRECTIONS FOR USE

It is a violation of Federal Law to use this product in a manner inconsistent with its labeling.

RECOMMENDED APPLICATIONS

Do not use this product on Long Island, N.Y.

Do not apply this product in such a manner as to directly or through drift expose workers or other persons. The area being treated must be vacated by unprotected persons.

If prolonged intimate contact with corn and or sorghum foliage will result, do not re-enter treated fields within 14 days of application without wearing proper protective clothing. For all other situations, do not re-enter fields less than 24 hours following application, unless appropriate protective clothing is worn.

Protective clothing means, at least, a hat or other suitable head covering, a long sleeved shirt and long legged trousers or a coverall type garment (all of closely woven fabric covering the body, including the arms and legs), shoes and socks.

Because certain states may require more restrictive re-entry intervals for various crops treated with this product, consult your State Department of Agriculture for further information.

Written or oral warnings must be given to workers who are expected to be in a treated area or in an area about to be treated with this product. Inform workers of areas or fields that may not be entered without specific protective clothing, period of time field must be vacated and appropriate actions to take in case of accidental exposure. An example of such information is given under written warnings. When oral warnings are given, warnings shall be given in a language customarily understood by workers. Oral warnings must be given if there is reason to believe that written warnings cannot be understood by workers. Written warnings must include the following information:

DANGER

Area treated with FURADAN 4F insecticide-nematicide on (insert date). Do not enter without appropriate protective clothing for 24 hours (insert State Department of Agriculture's reentry interval, if more restrictive).

In case of accidental exposure to pesticide spray, wash the skin thoroughly with soap and water. Remove contaminated clothing and wash before reuse. If in eyes, flush with plenty of water. If inhaled, go to an area where the pesticide has not been applied. Get medical attention if needed.

Do not apply this product through any type of irrigation systems.

Do not plant with any crop for which carbofuran treatment is not registered for at least 10 months.

Do not rotate with any crop on soil treated at greater than 3 pounds active ingredient per acre for at least 10 months.

ALFALFA: Alfalfa Weevil Larvae, Egyptian Alfalfa Weevil, Pea Aphid, and in New York State for Snout Beetle control—Apply the amount of FURADAN 4F indicated in the chart when feeding is noticed or when insects appear. Alfalfa Weevil adult—Apply 1 to 2 pints per acre when insects appear. Alfalfa Blotch Leafminer (Northeastern states only) and Potato Leafhopper—Apply 1 to 2 pints per acre when insects appear. Lygus Bugs—Apply 2 pints per acre prior to bloom. Grasshoppers—Apply ¼ to ½ pint per acre when grasshopper feeding is noticed. For control of Blue Alfalfa Aphid (nymphs and wingless adults)—apply FURADAN 4F at ½ to 1 pint per acre when insect feeding is noticed or when insects appear.

Observe the indicated number of days after application before cutting or grazing. Do not apply more than twice per season. Do not apply more than once per cutting. Do not use more than 1 pint per acre in the second application. Apply only to fields planted to pure stands of alfalfa. Do not move bees into alfalfa fields within 7 days of application.

Pints of FURADAN 4F Per Acre	Do not cut or graze within:
1/2	7 days
1	14 days
2	28 days

Minimum gallonage requirements: Ten gallons of finished spray per acre with ground equipment, two gallons per acre with aircraft, except for Blue Alfalfa Aphid use 30 gallons of finished spray per acre with ground equipment, 5 gallons per acre with aircraft.

For waterfowl protection, do not apply on fields in proximity of waterfowl nesting areas and/or on fields where waterfowl are known to repeatedly feed.

COTTON Thrips—Use FURADAN 4F at 2.5 fluid ounces per 1,000 linear feet of row (1 quart per acre with 40 inch row spacing). Apply in the seed furrow at planting. FURADAN 4F may be mixed with water or liquid fertilizer. When FURADAN 4F is used with liquid fertilizers, premix one part of FURADAN 4F with two parts of water. Check physical compatibility before mixing large quanities. Add this premix to the tank of fertilizer along with the rinsings from the premixing container. Maintain agitation in the tank after mixing and during application. Do not mix until ready to use. Do not feed cotton forage.

FIELD CORN, POPCORN, SWEET CORN—At Planting: Corn Rootworms, Flea Beetles, and to aid in the control of first generation European Corn Borer—Use 2.5 fluid ounces of FURADAN 4F per 1,000 linear feet of row (1 quart per acre with 40 inch row spacing). Apply at planting, as a 7 inch band over the row or inject on each side of the row by mixing with water or liquid fertilizers.

Southern Corn Rootworm, Flea Beetles, Seed Corn Maggot, Wireworms, and to aid in the control of first generation European Corn Borer and Armyworm for approximately 4 to 6 weeks after planting—Use 2.5 fluid ounces of FURADAN 4F per 1,000 linear feet of row (1 quart per acre with 40 inch row spacing). Apply at planting directly into the seed furrow.

FURADAN 4F may be mixed with water or liquid fertilizers. If FURADAN 4F is used with liquid fertilizers, premix one part of FURADAN 4F with two parts of water. Check physical compatibility before mixing large quantities. Add this premix to the tank of fertilizer along with the rinsings from the premixing container. Maintain agitation in the tank after mixing and during application. Do not mix until ready to use. Do not feed forage within 30 days of last application.

FIELD CORN, POPCORN—Post Plant: Corn Rootworms (northern, southern and western)—use 2.5 fluid ounces of FURADAN 4F per 1,000 linear feet of row (1 quart per acre with 40 inch row spacing). Apply as a post emergent spray by banding over the row, or by side dressing or basal spraying both sides of the row after corn emerges. Control will generally be improved if the treatment is cultivated into the soil. Do not feed forage within 30 days of last application.

FIELD CORN—Foliar Application: European Corn Borer—Use 1 1/2 to 2 pints of FURADAN 4F per acre as a foliar spray when corn borer eggs begin to hatch. Use the higher rate for heavier pest infestations. For treatment with aerial equipment, apply as broadcast spray using a minimum of 1 gallon of finished spray per acre. For treatment with ground equipment, direct the spray into the corn whorl for first brood and into the ear zone for second brood, using a minimum of 10 gallons of finished spray per acre. Repeat if necessary. Observe all precautions listed below.

Southwestern Corn Borer—Use FURADAN 4F at 1 to 2 pints per acre as a broadcast foliar spray when eggs begin to hatch. Rate used will depend on desired residual activity. If infestation continues, re-treat in 7 days after a 1 pint application and within 14 days after a 2 pint application. Apply in sufficient water for thorough coverage using a minimum of 1 gallon of finished spray per acre with air equipment or 10 gallons of finished spray per acre with ground equipment. Observe all precautions listed below.

Banks Grass Mites (suppression)—FURADAN 4F when applied at 2 pints per acre for the control of European or Southwestern Corn Borers will also suppress Banks Grass Mites.

Grasshoppers—Use FURADAN 4F as a foliar spray at 1/4 to 1/2 pint per acre when insects appear or feeding is noticed.

—Use the 1/4 pint rate under light to moderate population levels (0-14 grasshoppers/sq. yd.).

—Use the 1/2 pint rate under more severe population levels (15 or more grasshoppers/sq. yd.).

Apply in sufficient water for thorough coverage using a minimum of 2 gallons of finished spray per acre with air equipment or 10 gallons of finished spray per acre with ground equipment. Observe all precautions listed below.

Do not make more than two applications per season at the 1 1/2 to 2 pint use rate. Do not make more than four applications per season at the 1 pint use rate. Do not forage, cut or harvest within 30 days of last application. Do not apply on seed corn less than 14 days prior to detasseling or rogueing. If prolonged intimate contact will result, do not re-enter treated field within 14 days of application without wearing proper protective clothing.

SWEET CORN—Foliar Application (MACHINE HARVESTED ONLY): European Corn Borer—For control of second generation borers apply 1 pint of FURADAN 4F per acre. Make first application just prior to first silking and repeat at weekly intervals. Do not make more than four (4) applications per season. Do not apply within seven (7) days of harvest. If prolonged intimate contact will result, do not re-enter treated fields within 14 days of application without wearing proper protective clothing. Do not graze or harvest stalks within 21 days of last application.

Minimum gallonage requirements: 10 gallons of finished spray per acre with ground equipment, 2 gallons per acre with aircraft.

Do not make a foliar application if more than 12 ozs. of FURADAN 10G per 1,000 linear feet of row (10 lbs. per acre with 40" row spacing) or 8 ozs. of FURADAN 15G per 1,000 linear feet of row (6.7 lbs. per acre with 40 inch row spacing) or 2.5 fluid ounces of FURADAN 4F per 1,000 linear feet of row (1 quart per acre with 40 inch row spacing) were used in an at-planting application. At-planting application not permitted in California.

PINE SEEDLINGS: Pales Weevil and Pitch-eating Weevil in pine plantations—Apply a 1% (W/W) active FURADAN clay slurry (see following for preparation) to the roots of pine seedlings prior to transplanting. Treat seedlings by dipping roots or use any other suitable means which allows thorough coating. Keep roots moist until transplanted.

Prepare the slurry as follows: Add 1.6 ounces (2 1/2 tablespoons) of FURADAN 4F to 1/2 gallon of water. Mix thoroughly. Add 2 pounds of pulverized kaolin clay (pH 4.5) to this suspension. Mix thoroughly. This is sufficient to treat the roots of 150 to 200 seedlings.

Adequate ventilation is required for indoor treatment. Wear protective unlined rubber gloves at time of application. Users should avoid contamination of the inside of the gloves. If contamination of the inside of the gloves is suspected, the gloves should be immediately discarded.

POTATOES: Potato Tuberworm (Virginia only); Colorado Potato Beetle, European Corn Borer, Potato Flea Beetle, Potato Leafhopper—Use 1 to 2 pints of FURADAN 4F in sufficient water to treat one acre. Apply when insects first appear, and repeat as necessary to maintain control. Do not make more than 8 foliar applications per season. Do not apply more than 6 pints to the foliage if FURADAN 10 Granules or FURADAN 15G was used at planting. Do not apply more than 2 pints per application. Do not apply within 14 days of harvest. Minimum gallonage requirements: 10 gallons of finished spray per acre with ground equipment and 3 gallons per acre with aircraft.

Do not use this product in Long Island, New York.

For waterfowl protection do not apply:
—immediately before or during furrow irrigation
—on fields in proximity of waterfowl nesting areas
—on fields where waterfowl are known to repeatedly feed

SMALL GRAINS (WHEAT, OATS, BARLEY): Grasshoppers—Use FURADAN 4F at 1/4 to 1/2 pint per acre when insects appear. Cereal Leaf Beetle—Use 1/2 pint per acre when insects appear. Apply before heads emerge from boot. Do not make more than two applications per season. Do not feed treated forage to livestock.

Minimum gallonage requirements: Ten gallons of finished spray per acre with ground equipment, two gallons per acre with aircraft.

For waterfowl protection do not apply on fields in proximity of waterfowl nesting areas and/or on fields where waterfowl are known to repeatedly feed.

SOYBEANS—At Planting Application: Nematodes (Root-Knot, Cyst, Stunt, Ring, Sting, Spiral, Lesion, Lance, Dagger, Stubby-Root)—use 3.75 to 5 fluid ounces per 1,000 linear feet of row (3 to 4 pints per acre with 40-inch row spacing). Apply at planting as a 12-inch surface band. Soybeans may be grazed or cut for forage 30 days or later following an at planting application.

SOYBEANS —Foliar Application: Mexican Bean Beetle—Use FURADAN 4F at 1/2 to 1 pint per acre when insects appear. Grasshoppers—Use FURADAN 4F at 1/4 to 1/2 pint per acre when insects appear. Do not use FURADAN 4F as a foliar application if FURADAN 10G, 15G or 4F was applied to soybeans at planting time. Do not make more than 2 foliar applications per season. Do not apply within 21 days of harvest. Do not graze or feed foliar-treated forage to livestock or cut for silage or hay. Minimum gallonage requirements: 20 gallons of finished spray per acre with ground equipment, 1 1/2 gallons per acre with aircraft.

STRAWBERRIES (For use in Washington and Oregon): Root Weevils—Use FURADAN 4F at 2 to 4 pints per acre (2.6 to 5.1 fluid ounces per 1,000 linear feet of row for 42 inch row spacing). Apply as a 10 to 12 inch band over the row after last harvest but before October 1st. Do not make more than one application per season. Do not apply if berries are present.

SUNFLOWERS (CONFECTIONARY AND OIL)—At Planting: Sunflower Stem Weevil, Sunflower Beetle, Grasshoppers—Use 2.5 to 5.0 fluid ounces of FURADAN 4F per 1000 linear feet of row (1.4 to 2.8 quarts per acre with 30 inch row spacing). Apply directly into the seed furrow -OR- Apply in a 7 inch band. FURADAN 4F may be mixed with water or liquid fertilizer. When FURADAN 4F is used with liquid fertilizers, premix one part of FURADAN 4F with two parts of water. Check physical compatibility before mixing large quantities. Add this premix to the tank of fertilizer along with the rinsings from the premixing container. Maintain agitation in the tank after mixing and during application. Do not mix until ready to use.

Use the higher rate when heavier insect infestations are anticipated.

SUNFLOWERS (CONFECTIONARY AND OIL)—Foliar Application: Sunflower Moth, Banded Sunflower Moth, Stem Weevils, Seed Weevils—Use FURADAN 4F at 1 pint per acre. Grasshoppers—Use FURADAN 4F at 1/4 to 1 pint per acre. Sunflower Beetle—Use FURADAN 4F at 1/4 to 1/2 pint per acre. Apply as a foliar spray when insects appear. When a rate range is indicated, use higher rate for heavier insect infestations. Repeat applications as necessary but not more than four times per season. Apply a minimum of two gallons of finished spray per acre with aircraft and 10 gallons by ground equipment. If prolonged, intimate contact will result, do not re-enter treated field within 14 days of application without wearing proper protective clothing. Do not harvest crop within 28 days of last application.

SUGARCANE: Sugarcane Borer—Apply 1 to 1 1/2 pints FURADAN 4F per acre using ground or aerial equipment. Check sugarcane fields weekly beginning in early June and continuing through August. Make first application only after visible joints form and 5% or more of the plants are infested with young larvae feeding in or under the leaf sheath and which have not bored into the stalks. Repeat whenever field checks indicate the infestation exceeds 5%. Do not apply within 17 days of harvest. Do not use in Hawaii.

GRAPES (For use in California): For control of Nematodes (Root-Knot and Dagger) and Grape Phylloxera. Apply FURADAN 4F insecticide-nematicide at 2 1/2 gallons per acre as a broadcast treatment only, to the soil surface between the vine rows and immediately incorporate by mechanical means. Do not apply within 200 days of harvest. Remove dense weed growth prior to treatment. Do not use on soils of pH 8.0 or greater. Do not make more than one application per crop year. Wear protective clothing including rubber gloves and impervious foot covering at time of application.

TOBACCO—BURLEY: Flea Beetles—Use 1 gallon of FURADAN 4F per acre. Apply as a broadcast spray over the soil surface prior to transplanting and incorporate into the top 3 inches of soil with a suitable device.

TOBACCO—FLUE-CURED: Flea Beetles, Wireworms, and to aid in the control of Budworms and Root-knot Nematodes—Use 1 1/2 gallons of FURADAN 4F per acre—or—use 1 gallon of FURADAN 4F per acre for control of Flea Beetles only. Apply as a broadcast spray over the soil surface prior to forming beds. Incorporate into the top 3 inches of soil. Form beds and plant as usual. This product may induce flecking of the bottom or lower leaves.

ORNAMENTALS—Container Grown (Except California): Root Weevil Larvae—Prepare a stock solution by thoroughly mixing 1 to 2 fluid ounces of FURADAN 4F per 100 gallons of water. Apply as a soil drench in sufficient volume to saturate the entire soil profile within each container. The following guidelines give approximate amounts to use on various sized containers. Make a single application when larvae are present (usually from July to mid-October in outdoor growing areas). Later application may be less effective due to lower temperatures and/or the presence of more mature larvae.

Container Size	Amount of Stock Solution Per Container
6 inches diameter	1 pint
8 inches diameter	1 to 2 pints
10 inches diameter	2 to 4 pints
12 inches diameter	3 to 6 pints
16 inches diameter	4 to 8 pints

Early marginal necrosis or leaf drop may occur on Hydrangea or Birch. Not all species or varieties of ornamentals have been tested. Before treating large numbers of plants of a particular variety, treat a few plants and observe prior to full scale application.

Do not handle treated plants, soil or containers that are wet with spray.

Application of FURADAN 4F through overhead sprinkler equipment is prohibited.

9-9-88

SPECIAL LOCAL NEED REGISTRATIONS [SECTION 24(c)] FOR FURADAN 4F*

CROP	PEST	STATE EPA SLN NO.	DOSAGE	REMARKS
Alfalfa	Spotted alfalfa aphid	Oklahoma OK 82-0003	1 to 2 pints per acre.	Application by ground or air.
Alfalfa (Pure seeded)	Alfalfa blotch leafminer Potato leafhopper Clover root curculio	Pennsylvania PA 78-0012	1 to 2 quarts per acre.	At-planting application.
	Alfalfa blotch leafminer Potato leafhopper Clover root curculio Nematodes	New Jersey NJ 87-0001 New York NY 81-0017		
	Alfalfa blotch leafminer Potato leafhopper	Maryland MD 80-0025 Virginia VA 80-0037		
	Potato leafhopper Clover root curculio	Michigan MI 81-0011		
	Clover root curculio Potato leafhopper Crickets Grasshoppers	North Carolina NC 82-0031	1 to 2 quarts per acre.	At-planting application.
Alfalfa (Pure seeded) (no-till seedings)	Potato leafhopper	Illinois IL 84-0003	1 to 2 quarts per acre.	At-planting application.
Bermuda grass (grown for seed)	Thrips Fulgorids (leafhoppers)	Arizona AZ 77-0017 California CA 80-0168	1 quart per acre in 5 gallons of water by air, or 20 gallons of water by ground application.	Repeat treatment as necessary to maintain control.
Corn (field)	Thripus Fulgorida (leafhoppers) Whiteflies	California CA 78-0190	2 pints per acre.	Imperial and Riverside countries.
	Chinch bug Greenbugs	Kansas KS 81-0027	2 pints per 13,000 feet of row.	In furrow application
	Chinch bugs Aphids Corn rootworms	Louisiana LA 87-0010	2 pints per 13,000 feet of row.	In furrow application.
			1/2 to 1 pint per acre.	Foliar application for chinch bugs and aphids.
	Southwestern corn borer Sugarcane borer	Louisiana LA 79-0031	1 1/2 to 2 pints per acre.	Broadcast foliar spray.
	Southern corn billbug	North Carolina NC 80-0019 South Carolina SC 80-0026	1 to 2 quarts per acre.	Post emergent spray.
Corn (sweet)	European corn borer Corn earworm	Michigan MI 82-0019 Ohio OH 82-0015	1 pint per acre.	Application by air or ground. (MACHINE HARVESTED ONLY)

CROP	PEST	STATE EPA SLN NO.	DOSAGE	REMARKS
Grain sorghum	Greenbug Corn leaf aphid	New Mexico NM 78-0011 Texas TX 78-0027 Oklahoma OK 81-0012	0.5 to 1 pint per acre.	Application by ground or air.
	Chinch bug	Mississippi MS 82-0032 Missouri MO 80-0011		Ground application only.
		Nebraska NE 88-0004	1 pint per acre.	
		Kansas KS 78-0014	1 pint per acre	
		Kansas KS 81-0027	2 pints per 13,000 feet of row.	In furrow application.
	Chinch bug (for early plant protection and stand establishment) Greenbugs	Missouri MO 81-0030	2 pints per 13,000 linear feet of row (40"row spacing).	Apply directly into seed furrow.
	Greenbug	Kansas KS 78-0014 Missouri MO 80-0011	1 pint per acre.	Application by ground or air.
		Nebraska NE 88-0004	½ to 1 pint per acre.	
		Kansas KS 81-0027	2 pints per 13,000 feet of row.	In furrow application.
Grapes	Black vine weevil	Washington WA 82-0075	1 quart per acre.	Ground application after adult weevils have emerged in Spring.
	Nematodes Grape phylloxera (suppression)	California CA 85-0059	Pre-harvest 2 to 3 qts. per acre. Post-harvest 3 to 6 qts. per acre.	Through drip irrigation equipment.
Non-bearing peach and apple trees	Nematodes	West Virginia WV 78-0008	2.5 gallons per acre (10.0 pounds active).	Apply in the Spring when tree growth begins.
Non-bearing peach, nectarine and apple trees		Delaware DE 81-0008 Maryland MD 80-0019 Virginia VA 80-0039	1.5 gallons per acre (6.0 pounds active).	Preplant or post-plant applications.
Non-bearing peach, nectarine, apple, apricot, cherry, and plum trees		Ohio OH 81-0033	1.5 to 2.5 gallons per acre (6.0 to 10.0 pounds per acre).	
Potatoes	Colorado potato beetle Potato leafhopper Potato fleabeetle	Delaware DE 78-0019	Label rates.	By air: minimum 2 gallons finished spray per acre.
Rotation crops (wheat or barley)		Oregon OR 81-0047		May be planted as rotation crops 4 months following the last application of FURADAN 4F or FURADAN 10 Granules provided straw is not harvested or grazed.
Small grains (Wheat, Oats, Barley)	Grasshopper	Montana MT 86-0006 South Dakota SD 86-0005 North Dakota ND 86-0003	¼ pint per acre.	Application by ground or air. Do not apply within 21 days of harvest.
Sorghum	Grasshopper	Nebraska NE 78-0022	¼ to ½ pint per acre.	Application by ground or air.
Sorghum (grain)	Greenbug Corn leaf aphid	Oklahoma OK 81-0012	1 pint per acre.	Application by ground or air.
	Chinch bug		½ to 1 pint per acre.	Ground application only.
Sorghum (grown for grain or forage)	Greenbug Corn leaf aphid Southern corn rootworm Yellow Sugarcane aphid	Texas TX 81-0016	1½ to 2 pints per 13,000 linear feet of row.	Apply in-furrow at-planting.
	Chinch bug		½ to 1 pint in 20 to 30 gallons of water.	Ground application only.
Soybeans	Nematodes	Mississippi MS 82-0017 Missouri MO 82-0017	3 to 4 pints per 13,000 linear feet of row (one acre in 40 inch row spacing).	At planting.
	Mexican bean beetle	Mississippi MS 82-0017 Missouri MO 82-0017	2 to 4 pints per 13,000 linear feet of row (one acre with 40 inch row spacing).	At planting.
Strawberries	Root weevils	California CA 82-0099 New York NY 81-0009 New Jersey NJ 81-0014 Vermont VT 81-0001	4 pints per acre.	Apply after harvest.

CROP	PEST	STATE EPA SLN NO.	DOSAGE	REMARKS
Sugarcane	Diaprepes abbreviatus	Puerto Rico PR 80-0002	1 pint per acre.	Application by ground or air.
	Wireworms	Louisiana LA 87-0011	1 qt. per acre.	Apply in a 15-inch band directly over the planted cane prior to covering.
Tobacco (dark air-cured, dark fire-cured)	Flea beetles	Kentucky KY 79-0012	1 gallon per acre in 20 to 40 gallons of water.	Broadcast spray over soil surface.
Tobacco (dark-fired)		Tennessee TN 81-0026	½ to 1 gallon per acre.	Broadcast spray over soil surface.
Tobacco (flue-cured)	Budworms Hornworms Flea beetles Nematodes (including cyst nematode) Wireworms	Virginia VA 79-0021	1 to 1½ gallons per acre (4 to 6 pounds active).	Broadcast spray over soil surface.
		Virginia VA 78-0021		Tank mix with Ridomil 2E for black shank and blue mold.
	Green peach aphid (aids in control)	South Carolina SC 81-0012		Broadcast over soil surface prior to forming beds.
Tobacco (fire-cured)	Hornworms Flea beetles Blue mold	Virginia VA 81-0021	1 gallon per acre.	Tank mix with Ridomil 2E.
Tobacco (seedleaf)	Flea beetles Hornworms Wireworms Nematodes Aphids (early-season)	Pennsylvania PA 80-0042	1 to 1½ gallons per acre.	Broadcast spray over soil surface.
Tobacco (Maryland type)		Maryland MD 81-0020		

*For distribution and use only within the state specified. These Special Local Need registrations approved as of date of this publication are, however, subject to cancellation by the state. Contact our nearest Mobay Agricultural Chemicals Division Office for the current status and a copy of the complete individual supplemental Special Local Need labeling. This labeling must be in the possession of the user at time of pesticide application.

PRECAUTIONARY STATEMENTS

HAZARDS TO HUMANS AND DOMESTIC ANIMALS

DANGER–POISON

Poisonous if swallowed or inhaled. May be fatal or harmful as a result of skin or eye contact or by breathing spray mist. Causes cholinesterase inhibition. Warning symptoms of poisoning include weakness, headache, sweating, nausea, vomiting, diarrhea, tightness in chest, blurred vision, pinpoint eye pupils, abnormal flow of saliva, abdominal cramps, and unconsciousness. Atropine Sulfate is antidotal.

Wear a pesticide respirator jointly approved by the Mining Enforcement and Safety Administration (formerly the U.S. Bureau of Mines) and by the National Institute for Occupational Safety and Health under the provisions of 30 CFR Part 11. Wear goggles or face shield for eye protection when measuring or mixing. Wear long sleeved clothing and protective gloves when handling. Wash hands and face before eating or smoking. Bathe at the end of the work day. Change clothing daily and wash before reuse.

Keep all unprotected persons out of the operating area or vicinity where there may be danger of drift.

STATEMENTS OF PRACTICAL TREATMENT

If swallowed, vomiting should be induced. Administer water freely and induce vomiting by giving one dose (½ oz. or 15 ml.) of syrup of ipecac. If vomiting does not occur within 10 to 20 minutes, administer second dose. If syrup of ipecac is not available, induce vomiting by sticking finger down throat. Repeat until vomit fluid is clear. Do not induce vomiting or give anything by mouth to an unconscious person. Professional medical assistance should be secured immediately.

If inhaled, remove to fresh air. Call a physician immediately.

If on skin. In case of contact, remove contaminated clothing and immediately wash skin with soap and water.

If in eyes, get medical attention; in the meantime, flush eyes with plenty of water for at least 15 minutes.

ANTIDOTE

Note to Physician: Carbofuran is a reversible cholinesterase inhibitor. Do not use oximes such as 2-PAM. Start by giving 2 mg. atropine intramuscularly. According to clinical response, continue until signs of atropinization occur (dry mouth or dilated pupils). If in eye, instill one drop of homatropine.

ENVIRONMENTAL HAZARDS

This product is toxic to fish, birds and other wildlife. Birds feeding on treated areas may be killed. For waterfowl protection do not apply immediately before or during irrigation or on fields in proximity of waterfowl nesting areas, or on fields where waterfowl are known to repeatedly feed. Drift and runoff from treated areas may be hazardous to fish in neighboring areas. Do not apply directly to water.

Notice: It is a Federal offense to use any pesticide in a manner that results in the death of a member of an endangered species.

The use of FURADAN 4F may pose a hazard to the following Federally designated endangered/threatened species known to be found in certain areas within the named locations: **Attwater's Greater Prairie Chicken—**Texas counties including: Aransas, Austin, Brazoria, Colorado, Galveston, Goliad, Harris, Refugio and Victoria **Aleutian Canada Goose—**California counties including: Colusa, Merced, Stanislaus and Sutter **Kern Primrose Sphinx Moth—**Walker Basin of Kern County, California

This product may not be used in areas where adverse impact on the Federally designated endangered/threatened species noted above is likely. Prior to making applications, the user of this product must determine that no such species are located in or immediately adjacent to the area to be treated. If the user is in doubt whether or not the above named endangered species may be affected, he should contact either the regional U.S. Fish and Wildlife Service office (Endangered Species Specialist) or personnel of the State Fish and Game office.

This product is highly toxic to bees exposed to direct treatment or residues on crops. Do not apply this product or allow it to drift to blooming crops or weeds if bees are visiting the treatment area. Protective information may be obtained from your Cooperative Agricultural Extension Service.

Carbofuran is a chemical which can travel (seep or leach) through soil and can contaminate ground water which may be used as drinking water. Carbofuran has been found in ground water as a result of agricultural use. Users are advised not to apply carbofuran where the water table (ground water) is close to the surface and where the soils are very permeable, i.e., well-drained soils such as loamy sands. Your local agricultural agencies can provide further information on the type of soil in your area and the location of ground water.

STORAGE AND DISPOSAL

Pesticide Storage

Not for use or storage in or around the house.
Do not store below 35°F, (2°C).
Keep out of reach of children and animals. Store in original containers only. Store in a cool, dry place and avoid excess heat. Carefully open containers. After partial use, replace lids and close tightly. Do not put concentrate or dilute material into food or drink containers. Do not contaminate other pesticides, fertilizers, water, food, or feed by storage or disposal.
In case of spill, avoid contact, isolate area and keep out animals and unprotected persons. Confine spills. Call Mobay Kansas City Emergency Response No. (816) 242-2582 or contact Chemtrec at (800) 424-9300.
To confine spill: If liquid, dike surrounding area or absorb with sand, cat litter or commercial clay. If dry material, cover to prevent dispersal. Place damaged package in a holding container. Identify contents.

Pesticide Disposal

Pesticide wastes are acutely hazardous. Improper disposal of excess pesticide, spray mixture, or rinsate is a violation of Federal Law. If these wastes cannot be disposed of by use according to label instructions, contact your State Pesticide or Environmental Control Agency, or the Hazardous Waste representative at the nearest EPA Regional Office for guidance.

Container Disposal

Metal Containers: Triple rinse (or equivalent). Then offer for recycling or reconditioning, or puncture and dispose of in a sanitary landfill, or by other procedures approved by State and local authorities. Do not cut or weld metal containers.

Plastic Containers: Triple rinse (or equivalent). Then offer for recycling or reconditioning, or puncture and dispose of in a sanitary landfill, or incineration, or, if allowed by State and local authorities, by burning. If burned, stay out of smoke.

IMPORTANT:

This bulletin is not intended to provide adequate information for application. Before using the product, read and carefully observe directions, cautionary statements and other information appearing on the product label. This product is sold subject to the Conditions of Sale set forth on the container label.

READ THE LABEL BEFORE USE

FU 9003

This is a specimen label. Mobay is not responsible for the accuracy of the information contained herein. As labels are subject to revision, always carefully read and follow the label on the product container.

GUTHION 2S®

EMULSIFIABLE INSECTICIDE

RESTRICTED USE PESTICIDE

For retail sale to and use only by Certified Applicators or persons under their direct supervision, and only for those uses covered by the Certified Applicator's certification.

FOR CONTROL OF A WIDE VARIETY OF AGRICULTURAL PESTS

ACTIVE INGREDIENTS:

O,O-Dimethyl *S*-[(4-oxo-1,2,3-benzotriazin-3 (4*H*)-yl) methyl] phosphorodithioate	22%
INERT INGREDIENTS:	78%
TOTAL:	100%

Contains 2 lbs. *O,O*-Dimethyl *S* [(4-oxo-1,2,3-benzotriazin-3 (4*H*)-yl) methyl] phosphorodithioate per gallon.

EPA Reg. No. 3125-123.

GUTHION, DEF and BOLSTAR are Reg. TMs of Bayer AG, Germany
Galecron is a trademark of Ciba-Geigy for chlordimeform
Fundel is a trademark of Schering AG Berlin/Bergamen

RECOMMENDED USES

Almonds	Cranberries	Peppers
Apples‡	Cucumbers	Plums***†
Apricots***†	Eggplant	Potatoes
Artichokes	Filberts*	Prunes***†
Barley	Grapes	Quinces‡
Beans (dry and snap)	Loganberries	Raspberries
Blackberries	Melons	Rye
Blackeyed peas	Nectarines***†	Southern pine seed orchards
Blueberries**	Nursery plants	Soybeans
Boysenberries	Oats	Spinach
Broccoli	Onions (green and dry)	Strawberries
Brussels sprouts	Ornamentals	Tomatoes
Cabbage	Pasture grass****	Wheat
Cauliflower	Peaches***†	Walnuts
Celery	Pecans	
Christmas trees		
Citrus fruits		
Cotton		
Crab apples‡		

*Pacific Northwest only
**Eastern and North Central states only
***Eastern U.S.
****States east of the Mississippi River
††West of the Rocky Mountains
‡Washington, Oregon, Idaho and Utah

RESTRICTIONS ON WAREHOUSE STORAGE

Do not store in front of space heaters. Store below 120°F. Do not store adjacent to or near proximity of foodstuffs, drugs, feed or any other material intended for consumption by humans or animals. Do not store adjacent to toys, cosmetics, dishes, furniture or clothing. Do not store next to herbicides. If product freezes-out, it may be reconstituted by warming and agitating. Do not store below 25°F. Protect from heat. Keep away from open flame. Do not heat.

PACKAGING

2 gallon can. Shipping weight: 22 lbs.
5 gallon can. Shipping weight: 52 lbs.
29 gallon drum. Shipping weight: 299 lbs.

GUTHION 2S emulsifiable insecticide has proved to be effective in controlling major insects on a wide variety of fruit and vegetable crops after many years of intensive foreign and domestic tests and years of commercial use.

FRUITS: GUTHION 2S controls major insect pests on a variety of stone fruits, grapes and berries. This insecticide greatly simplifies spray schedules. Its residual activity protects from insect damage on most fruit crops.

VEGETABLES: Registered for use on many vegetables, GUTHION 2S has proved to be effective in controlling potato tuberworm, diamondback moth, Colorado potato beetle and others. As with fruit, GUTHION 2S has residual action.

Liquid concentrate and wettable powder formulations also available.

PELIGRO

PRECAUCION AL USUARIO: Si usted no puede leer o entender inglés, no use este producto hasta que la etiqueta le haya sido explicada ampliamente.

(TO THE USER: If you cannot read or understand English, do not use this product until the label has been fully explained to you.)

DIRECTIONS FOR USE

It is a violation of Federal Law to use this product in a manner inconsistent with its labeling.

MIXING: GUTHION 2S emulsifiable insecticide forms an emulsion when diluted with water, and is suitable for use in all power-operated ground sprayers and aircraft sprayers. To mix with water, pour the required amount of GUTHION 2S into full amount of water, and then agitate. If cold water is to be used, and nozzle or screen plugging is experienced, 16 mesh slotted screens (from Spraying Systems Co.) are recommended.

DOSAGE: Use specified dosage of GUTHION 2S in the amount of water necessary to give complete coverage of foliage. The type of equipment used will determine the concentration required.

SPRAYING: Work to windward. Protect sprayer operators from drift or mist. When low volumes of spray are applied, complete coverage and thorough application are essential for most effective results. Schedule applications in accordance with local conditions. Consult your State Agricultural Experiment Station or Extension Service for specific use information in your area.

CHEMIGATION

GUTHION 2S may be applied through recommended types of irrigation systems to many crops. The REMARKS section for each crop lists the types of applications allowed. If application through irrigation systems is not listed in the REMARKS section for a crop, GUTHION 2S may not be applied to that crop through irrigation systems.

Types of Irrigation Systems

Apply GUTHION 2S only through sprinkler, including center pivot, lateral move, side roll, overhead solid set, or low pressure sprinkler irrigation systems. Do not apply GUTHION 2S through any other type of irrigation systems.

GENERAL DIRECTIONS FOR ALL RECOMMENDED TYPES OF IRRIGATION SYSTEMS

Uniform Water Distribution and System Calibration

The irrigation system must provide uniform distribution of treated water. Crop injury, lack of effectiveness, or illegal pesticide residues in the crop can result from nonuniform distribution of treated water.

The system must be calibrated to uniformly apply the rates specified for chemigation application for specific crops. If you have questions about calibration, you should contact State Extension Service specialists, equipment manufacturers or other experts.

Chemigation Monitoring

A person knowledgeable of the chemigation system and responsible for its operations, or under the supervision of the responsible person, shall shut the system down and make necessary adjustments should the need arise.

Drift

Do not apply when wind speed favors drift beyond the area intended for treatment.

Required System Safety Devices

The system must contain a functional check valve, vacuum relief valve, and low pressure drain appropriately located on the irrigation pipeline to prevent water source contamination from backflow.

The pesticide injection pipeline must contain a functional, automatic, quick-closing check valve to prevent the flow of fluid back toward the injection pump.

The pesticide injection pipeline must also contain a functional, normally closed, solenoid-operated valve located on the intake side of the injection pump and connected to the system interlock to prevent fluid from being withdrawn from the supply tank when the irrigation system is either automatically or manually shut down.

The system must contain functional interlocking controls to automatically shut off the pesticide injection pump when the water pump motor stops.

The irrigation line or water pump must include a functional pressure switch which will stop the water pump motor when the water pressure decreases to the point when pesticide distribution is adversely affected.

Systems must use a metering pump, such as a positive displacement injection pump (e.g., diaphragm pump) effectively designed and constructed of materials that are compatible with pesticides and capable of being fitted with a system interlock.

Using Water from Public Water Systems

DO NOT APPLY GUTHION 2S THROUGH ANY IRRIGATION SYSTEM PHYSICALLY CONNECTED TO A PUBLIC WATER SYSTEM. Public water system means a system for the provision to the public of piped water for human consumption if such system has at least 15 service connections or regularly services an average of at least 25 individuals daily at least 60 days out of the year.

GUTHION 2S may be applied through any of the recommended types of irrigation systems which may be supplied by a public water system only if the water from public water system is discharged into a reservoir tank prior to pesticide introduction. There shall be a complete physical break (air gap) between the outlet end of the fill pipe and the top or overflow rim of the reservoir tank of at least twice the inside diameter of the fill pipe. Before beginning chemigation, always make sure that the air gap exists and that there is no blockage of the overflow or the reservoir tank.

Any irrigation system using water supplied from a public water system must also meet the following requirements:

The pesticide injection pipeline must contain a functional, automatic quick-closing check valve to prevent the flow of fluid back toward the injection pump.

The pesticide injection pipeline must contain a functional, normally closed, solenoid-operated valve located on the intake side of the injection pump and connected to the system interlock to prevent fluid from being withdrawn from the supply tank when the irrigation system is either automatically or manually shut down.

The system must contain functional interlocking controls to automatically shut off the pesticide injection pump when the water pump motor stops or in cases where there is no water pump, when the water pressure decreases to the point where pesticide distribution is adversely affected.

Systems must use a metering pump, such as a positive displacement injection pump (e.g., diaphragm pump) effectively designed and constructed of materials that are compatible with pesticides and capable of being fitted with a system interlock.

Do not apply when wind speed favors drift beyond the area intended for treatment.

Posting Requirements

Posting of areas to be chemigated is required when 1) any part of a treated area is within 300 feet of sensitive areas such as residential areas, labor camps, businesses, day care centers, hospitals, in-patient clinics, nursing homes or any public areas such as schools, parks, playgrounds, or other facilities not including public roads, or 2) when the chemigated area is open to the public such as golf courses or retail greenhouses.

Posting must conform to the following requirements: Treated areas shall be posted with signs at all usual points of entry and along likely routes of approach from the listed sensitive areas. When there are no usual points of entry, signs must be posted in the corners of the treated areas and in any other location affording maximum visibility to sensitive areas. The printed side of the sign should face away from the treated area towards the sensitive areas. The signs shall be printed in English. Signs must be posted prior to application and must remain posted until foliage has dried and soil surface water has disappeared. Signs may remain in place indefinitely as long as they are composed of materials to prevent deterioration and maintain legibility for the duration of the posting period.

All words shall consist of letters at least 2½ inches tall, and all letters and the symbol shall be a color which sharply contrasts with their immediate background. At the top of the sign shall be the words KEEP OUT, followed by an octagonal stop sign symbol at least 8 inches in diameter containing the word STOP. Below the symbol shall be the words PESTICIDES IN IRRIGATION WATER.

Posting required for chemigation does not replace other posting and reentry interval requirements for farmworker safety.

Compatibility

When mixing with other chemicals refer to Compatibility section elsewhere on the label.

Agitation

For application of GUTHION 2S alone, a chemical supply tank is not necessary for premixing since GUTHION 2S mixes well with water in the irrigation line. If a chemical supply tank is used, for application of GUTHION 2S alone or in combination with liquid fertilizer or other chemicals, constant strong mechanical or hydraulic agitation must be maintained in the chemical supply tank during the entire period of application.

Chemical Supply Tank Dilution

If a chemical supply tank is used, you must determine the required amounts of GUTHION 2S to mix in the tank.

The amount of GUTHION 2S needed equals the number of pints of GUTHION 2S to be applied per acre multiplied by the number of acres to be chemigated.

The amount of emulsion needed equals the gallons of emulsion delivered per hour by the injection pump multiplied by the number of hours chemigation will take place.

The amount of water needed equals the amount of emulsion needed minus the amount of GUTHION 2S needed.

For example, if you want to apply 3 pints of GUTHION 2S per acre to 130 acres in 20 hours and your injection pump delivers 15 gallons per hour, you need: 3 pints GUTHION 2S per acre × 130 acres = 390 pints or 48.75 gallons of GUTHION 2S. And, you need: 15 gallons per hour × 20 hours = 300 gallons of emulsion, minus 48.75 gallons of GUTHION 2S = 251.25 gallons of water.

Cleaning the Chemical Injection System

In order to accurately apply pesticides, the chemical injection system must be kept clean; free of chemical or fertilizer residues and sediments. Refer to your owners manual or ask your equipment supplier for cleaning procedure to your injection system.

Flushing the Irrigation System

At the end of the application period, allow time for all lines to flush the pesticide through all nozzles or emitters before turning off irrigation water. To ensure the lines are flushed and free of pesticides, a dye indicator may be injected into the lines to mark the end of the application period.

SPRINKLER IRRIGATION SYSTEMS

All directions and requirements under the GENERAL DIRECTIONS AND REQUIREMENTS FOR ALL RECOMMENDED TYPES OF IRRIGATION SYSTEMS section of the label must be followed for sprinkler irrigation systems.

In addition, the following directions apply to sprinkler irrigation systems:

Do not apply when wind speed favors drift beyond the area intended for treatment.

It is recommended that nozzles in the immediate area of control panels, chemical supply tanks, pumps and system safety devices be plugged to prevent chemical contamination of these areas.

Center-Pivot and Automatic-Move Linear Systems

Inject the specified dosage per acre continuously for one complete revolution or move of the system. DO NOT USE END GUNS. For a foliar application, the system should be run at maximum speed. For a soil application, the system should be run at a slower speed with application in at least ¼ inch of water.

Solid Set and Manually Controlled Linear Systems

For foliar application, injection should be during the last 30 to 60 minutes of regular irrigation period or as a separate 30 to 60 minute application not associated with a regular irrigation. For soil treatment, application should be in at least ¼ inch of water. DO NOT USE END GUNS.

RECOMMENDED APPLICATIONS

CROP	INSECT	PINTS GUTHION 2S	REMARKS
FIELD CROPS	Boll weevil	½ to 1	Apply specified dosage per acre by air or ground equipment in sufficient water for complete coverage but not less than 1 gallon per acre. For application by irrigation systems: Apply specified dosage per acre. Follow all directions given under the CHEMIGATION section of the label. A total of 12 applications per crop season may be made regardless of rate, formulation or method of application used. Applications may be made up to 1 day before hand picking. Cotton may be machine harvested any time after application. GUTHION 2S at rates up to 1 pint per acre may be tank-mixed with Galecron® 4E or Fundal® 4EC for ovicidal control of bollworms and tobacco budworms. For control of the larval stages of resistant and non-resistant bollworm and/or tobacco budworm, BOLSTAR® 6 may be included in tank mixes with GUTHION 2S. When using GUTHION 2S in a tank mix with Galecron 4E or Fundal 4EC, observe all limitations, precautions and rate recommendations which appear on the labels for these products. **See Note below.**
Cotton	Brown cotton leafworm Cotton fleahopper Cotton leafworm Lygus bugs Thrips	1	
	Pink bollworm Rapid plant bug Tarnished plant bug	1 to 2	
	Stink bug	2	
Cotton (Ultra-low-volume spray[1])	Boll weevil	½ to 1	GUTHION 2S may be used undiluted in any ground or aerial spray equipment that has been adapted and calibrated for ultra-low volume spraying. Spray machines must be equipped with accepted low volume devices that will produce droplets within the range of 30 to 100 microns in size. ULV aerial applications should be made at altitudes of 10 to 20 feet. A total of 12 applications may be made per crop season regardless or rate, formulation or method of application used. Application may be made up to 2 days before hand picking. Cotton may be machine harvested any time after application. **Early- and Mid-Season Control:** Apply specified dosage per acre in accordance with local recommendations. **Diapause Weevil Control:** The one-pint-per-acre rate only is recommended for control of diapausing boll weevils. Schedule applications in accordance with local recommendations.
Barley Oats Rye Wheat	Cereal leaf beetle	1½ to 2	Apply specified dosage per acre, by air or ground equipment, in sufficient water for complete coverage, but not less than 1 gallon per acre. Do not apply more than once per season. Do not harvest for food, feed, forage or graze within 30 days of treatment.
Pasture grasses (States east of Mississippi River)	Grasshoppers Meadow spittlebugs (adults)	2 to 3	Apply specified dosage per acre by air or ground equipment in sufficient water for complete coverage but not less than 1 gallon per acre. Do not apply more than once per cutting. Do not graze or harvest within 16 days of application at the 2 pint rate or 21 days at rates above 2 pints.
Soybeans	Aphids Bean leaf beetle Green cloverworm Leafhoppers Leafminers Leaf rollers Stink bugs Velvet bean caterpillar	1½ to 2	Apply specified dosage per acre, by air or ground equipment, in sufficient water to give complete coverage, but not less than 1 gallon per acre. One or 2 applications of rates up to 2 pints per acre may be made up to 45 days of harvest. Do not graze or feed treated vines to livestock.
	Mexican bean beetle	2 to 3	
FRUIT	[1] **NOTE:** It is suggested that when treating fruit during the bloom period, bee keepers should be warned in advance to remove hives a safe distance from orchards to be treated.		
Apples Crab apples Quinces (Washington, Oregon Idaho, and Utah)	Apple maggot Codling moth European apple sawfly Eye-spotted bud moth Fruit tree leaf roller Green fruitworm Orange tortrix Plum curculio Redbanded leaf roller	2 to 4	Apply specified dosage per acre by helicopter or fixed wing aircraft, using equipment capable of applying ultra-low volume sprays[1]. Use Beecomist nozzles with 20 micron sleeves. A total of 4 applications of GUTHION insecticide may be made per crop season regardless of rate, formulation or method of application used. No more than 2 of the 4 applications may be made by aircraft. Allow at least 7 days between applications, and 7 days between last application and harvest.

CROP	INSECT	PINTS GUTHION 2S	REMARKS
Apricots Nectarines Peaches (Eastern U.S.)	Aphids* Cottony peach scale European fruit lecanium scale Forbes scale Lesser peach tree borer Oriental fruit moth Peach tree borer Peach twig borer *Platynota flavedana* leaf roller Plum curculio Redbanded leaf roller San Jose scale* Stink bug Thrips Tarnished plant bug Terrapin scale Walnut scale White peach scale	3½ to 4½	Apply specified dosage per acre as a full coverage spray. Allow at least 14 days between applications. A total of 13½ pints per acre per crop season may be applied to nectarines. A total of 18 pints per acre per crop season may be applied to apricots and peaches. Allow at least 21 days between last application and harvest. For control of peach tree borer, apply 2 or 3 sprays to trunk from ground to scaffold limbs, timed with moth flight. For control of scale apply when crawlers are present. GUTHION 2S is compatible with dormant and summer oils which may be added to peach sprays in accordance with local recommendations.
Apricots Nectarines Peaches (West of the Rocky Mountains)	Lesser peach tree borer Oriental fruit moth Peach tree borer Peach twig borer *Platynota flavedana* leaf roller Plum curculio Redbanded leaf roller Stink bug Thrips Tarnished plant bug	6 to 8	Apply specified dosage per acre. Allow at least 14 days between applications. A total of 13½ pints per acre per crop season may be applied to nectarines. A total of 18 pints per acre per crop season may be applied to apricots and peaches. Allow at least 21 days between last application and harvest. For control of peach tree borer, apply 2 or 3 sprays to trunk from ground to scaffold limbs, timed with moth flight. For control of scale apply when crawlers are present.
Plums Prunes (Eastern U.S.)	Aphids* Codling moth Eye-spotted bud moth Forbes scale Fruit tree leaf roller Lesser peach tree borer Orange tortrix Peach tree borer Peach twig borer Plum curculio Redbanded leaf roller San Jose scale* Stink bug Tarnished plant bug* Tussock moth	3½ to 4½ (1 to 1½ pints per 100 gallons)	Apply specified dosage per acre as a full coverage spray. A total of 13½ pints per acre per crop season may be applied to plums and prunes. Allow at least 10 days between applications. Allow at least 15 days between last application and harvest. For control of peach tree borer, apply 2 or 3 sprays to trunk from ground to scaffold limbs, timed with moth flight. For control of scale apply when crawlers are present. GUTHION 2S is compatible with dormant and summer oils which may be added to prune sprays in accordance with local recommendations.
	American plum borer	6	
Plums Prunes (West of the Rocky Mountains)	Codling moth Eye-spotted bud moth Fruit tree leaf roller Lesser peach tree borer Orange tortrix Peach tree borer Peach twig borer Plum curculio Redbanded leaf roller Stink bug Tarnished plant bug* Tussock moth	4 to 8	Apply specified dosage per acre. A total of 13½ pints per acre per crop season may be applied to plums and prunes. Allow at least 10 days between applications. Allow at least 15 days between last application and harvest. For control of peach tree borer, apply 2 or 3 sprays to trunk from ground to scaffold limbs, timed with moth flight.
	American plum borer	6 to 8	
Blackberries Boysenberries Loganberries Raspberries	Leafhoppers Leaf rollers	1	Apply specified dosage per acre to foliage, using approximately 200 gallons of water for good coverage. Where ground conditions dictate an air application, use specified rate in a minimum of 1 gallon of water per acre. Do not apply within 14 days of harvest. Do not apply more than twice per season.
	Leafminers	1¼	
	Aphids	1¼ to 2	
	Obscure root weevil	2	
	Obscure root weevil	2	For control of root weevils prior to harvest, apply specified dosage per acre to lower portion of canes, and to the soil beneath the plants, using approximately 200 gallons of water. Do not apply more than twice per season. Do not make applications within 3 days of harvest at rates up to 4 pints per acre.
Blueberries (Eastern and North Central states only)	Blueberry maggot Fruitworms Lecanium scale Plum curculio	2 to 3	Apply specified dosage per acre with aerial or ground equipment using sufficient water for good coverage. A total of 3 applications may be made per crop season regardless of rate or formulation used. Allow at least 10 days between applications and at least 7 days between last application and harvest.
Cranberries	Cranberry fruitworm *Sparganothis sulfureana* Tipworm	2 to 4	Apply specified dosage per acre by air using sufficient water for good coverage. For application by irrigation systems: Apply specified dosage per acre. Follow all directions given under the CHEMIGATION section of the label. A total of 3 application may be made per crop season regardless of rate or formulation used.
	Fireworms	4	

CROP	INSECT		PINTS GUTHION 2S	REMARKS
				Allow at least 14 days between applications and at least 21 between last application and harvest.
Citrus fruits	Aphids* Black scale Brown soft scale Chaff scale Citricola scale Citrus mealybug Citrus thrips Cottony-cushion scale European brown snail	Florida red scale Fruit tree leaf rollers Fuller rose beetle Glover scale Orange tortrix Purple scale Snow scale Western tussock moth Whiteflies	8 to 12	Apply specified dosage per acre as a full coverage spray. A single application per year may be applied up to within 7 days of harvest. Where 2 applications are required, allow at least 28 days between the second application and harvest. A total of 2 applications may be applied per fruit year regardless of rate or formulation used. Do not pick fruit or do other work involving contact with the tree (such as pruning) within 7 days after treatment.
	California red scale	Yellow scale	12	
Grapes	Grape berry moth Grape cane girdlers Grape leaf skeletonizer Grape mealybug Leafhoppers* Redbanded leaf rollers Thrips		3 to 4	Apply specified dosage per acre as a full coverage spray. A total of 3 applications may be made per crop season regardless of rate of formulation used. Allow at least 14 days between applications. Minimum dosage specified may be applied up to harvest. For higher rates allow at least 10 days between last application and harvest.
Strawberries	Aphids* Meadow spittlebug Oblique-banded leaf roller Obscure root weevil Omnivorous leaf tier	Pea leaf weevil Small black (grass) weevil Strawberry leaf rollers Whitefly	2	Apply specified dosage per acre as a full coverage spray using sufficient water for good coverage. For application by irrigation systems: Apply specified dosage per acre. Follow all directions given under the CHEMIGATION section of the label. A total of 4 applications may be made per crop season regardless of rate or formulation used. Allow at least 5 days between applications and at least 5 days between last application and harvest.
NUTS Almonds	Navel orange worm Peach twig borer		6 to 8	Apply specified dosage per acre. A total of 2 applications may be made per crop season regardless of rate or formulation used. Allow at least 30 days between applications and at least 60 days between last application and harvest. Do not apply after husks split.
Filberts (Pacific Northwest only)	Apple mealybug Filbert aphid Filbert leaf roller Filbertworm		6 to 8	Apply specified dosage per acre as a full coverage spray. A total of 3 applications may be made per crop season regardless of rate or formulation used. Allow at least 14 days between applications and at least 30 days between last application and harvest. Do not graze livestock in treated groves for 21 days after treatment.
Pecans	Aphids* Fall webworm Hickory shuckworm Leafminers May beetles	Pecan casebearer Southern green stink bug Spittlebug Twig girdlers Walnut caterpillar	6 to 8	Apply specified dosage per acre as a full coverage spray. A total of 3 applications may be made per crop season regardless of rate or formulation used. Allow at least 7 days between applications. Do not apply after shuck split. Livestock may be grazed in treated groves after a 21-day post-treatment interval.
Walnuts	Codling moth Walnut husk fly Navel orangeworm	Filbertworm Red-humped caterpillar	6 to 8	Apply specified dosage per acre as a full coverage spray. A total of 3 applications may be made per crop season regardless of rate or formulation used. Allow at least 14 days between applications. Do not apply after husks split. Livestock may be grazed in treated groves after a 21-day post-treatment interval.
VEGETABLES Artichokes	Plume moth		6	Apply specified dosage per acre in sufficient water for complete coverage. For application by irrigation systems: Apply specified dosage per acre. Follow all directions given under the CHEMIGATION section of the label. A total of 3 applications may be made per crop season regardless of rate, formulation or method of application used.

CROP	INSECT	PINTS GUTHION 2S	REMARKS
			Allow at least 14 days between applications and at least 30 days between last application and harvest. Do not feed or ensile treated forage.
Beans (Snap and dry)	Leaf rollers Mites	1 to 2	Apply specified dosage per acre by air or ground equipment in sufficient water for complete coverage, but not less than 1 gallon per acre. Do not apply to snap beans within 7 days of harvest, nor to dry beans within 30 days of harvest. Do not exceed 4 applications. Do not feed or ensile treated forage.
	Aphids* Bean leaf beetle Green cloverworm Leafhoppers Leafminers Stink bugs Velvet bean caterpillar	1½ to 2	
	Mexican bean beetle Spotted cucumber beetle Striped cucumber beetle Tarnished plant bug Western-striped cucumber beetle	2	
Broccoli Brussels sprouts Cabbage (Includes tight heading varieties of Chinese Cabbage) Cauliflower	Aphids* Cabbage looper Diamondback moth Imported cabbageworm	2 to 3	Apply specified dosage per acre, by air or ground equipment, in sufficient water for complete coverage, but not less than 1 gallon per acre. Do not apply within 7 days of harvest for Brussels sprouts, 15 days of harvest for broccoli and cauliflower, nor within 21 days of harvest for cabbage. Do not exceed more than 3 applications per season.
	Cabbage maggot	½	Mix specified dosage in 50 gallons of water. Apply 4 to 6 ounces of this emulsion per plant immediately after transplanting.
	Cabbage maggot (Transplant fields in California only)	3	Apply specified dosage in 300 to 400 gallons of water per acre as a soil drench in the rows when damage first appears. Additional applications may be necessary. Do not apply within 21 days of harvest.
	Cabbage maggot (Direct seeded fields in California only)	3	Apply specified dosage per acre, in sufficient water for uniform distribution. Mix in the upper 2 inches of soil prior to seeding, or spray in the seed row at planting time. Usually, 2 additional sprays are necessary during the growing season, depending upon time of year and maggot population. Do not apply within 21 days of harvest. Do not exceed more than 3 applications per season.
Blackeyed peas (Southern peas, Crowder peas)	Corn earworm Cowpea curculio	3 to 4	Apply specified dosage per acre, by air or ground equipment, in sufficient water for complete coverage, but not less than 1 gallon per acre. Do not apply more than 4 times per season, nor within 7 days of harvest. Do not use vines for feed or forage, nor pasture treated areas.
	Leafminers Stink bugs	1½ to 2	
Celery	Aphids Leafminers Leafhoppers Spittlebugs Tarnished plant bug	2	Apply specified dosage in 100 gallons of water as a full-coverage spray**, using not more than 200 gallons of finished spray per acre. Do not apply within 14 days of harvest. Do not exceed more than 3 applications per season.
Cucumbers	Spotted cucumber beetle Striped cucumber beetle Western-striped cucumber beetle	2	Apply specified dosage per acre in sufficient water for complete coverage. For application by irrigation systems: Apply specified dosage per acre. Follow all directions given under the CHEMIGATION section of the Label. A total of 3 applications may be made per crop season regardless of rate, formulation or method of application used. Allow at least 7 days between applications and at least 1 day between last application and harvest.
Eggplant	Leafminers	1½ to 2	Apply specified dosage per acre in sufficient water for complete coverage. For application by irrigation systems: Apply specified dosage per acre. Follow all directions given under the CHEMIGATION section of the Label. A total of 3 applications may be made per crop season regardless of rate, formulation or method of application used. Allow at least 7 days between applications. Do not apply after fruit set.
	European corn borer Flea beetles	2	
Onions (Green and dry)	Thrips	2 to 3	Apply specified dosage per acre, by air or ground equipment in sufficient water for complete coverage. For application by irrigation systems: Apply specified dosage per acre. Follow all directions given under the CHEMIGATION section of the label. A total of 3 applications may be made per crop season regardless of rate, formulation or method used. Allow at least 7 days between applications for dry bulb onions. Allow at least 10 days between applications for green onions.

CROP	INSECT	PINTS GUTHION 2S	REMARKS
			For dry onions, allow at least 28 days between last application and harvest. For green onions, allow at least 7 days between last application and harvest.
Peppers	Leafminers	1½ to 2	Apply specified dosage per acre by air or ground equipment, in sufficient water for complete coverage. For application by irrigation systems: Apply specified dosage per acre. Follow all directions given under the CHEMIGATION section of the label. A total of 3 applications may be made per crop season regardless of rate, formulation or method of application used. Allow at least 7 days between applications and 7 days between last application and harvest.
	European corn borer Flea beetles	2	
Potatoes	Colorado potato beetle	1½	Apply specified dosage per acre in sufficient water for complete coverage. For application by irrigation systems: Apply specified dosage per acre. Follow all directions given under the CHEMIGATION section of the label. A total of 3 applications may be made per crop season regardless of rate, formulation or method used. Allow at least 7 days between applications and 7 days between last application and harvest. **NOTE:** Resistance to Colorado potato beetle has occurred in some areas. Consult your local Extension Service or Mobay Sales Representative for details.
	Banded cucumber beetle Leafminers	1½ to 2	
	European corn borer Flea beetle Leafhoppers Spittlebugs Tarnished plant bug	2 to 3	
	Tuberworm	2¼ to 3	
Spinach	Aphids* Leafminers Mites	1½ to 2	Apply specified dosage per acre, by air or ground equipment, in sufficient water for complete coverage, but not less than 1 gallon per acre. Do not apply within 14 days of harvest. Do not exceed more than 3 applications per season.
Tomatoes	Colorado potato beetle	1½	Apply specified dosage per acre by air or ground equipment, in sufficient water for complete coverage. For application by irrigation systems: Apply specified dosage per acre. Follow all directions given under the CHEMIGATION section of the label. A total of 4 applications may be made per crop season regardless of rate, formulation or method of application used. The high rates should be used where heavy infestations of late instar lepidopterous larvae (large worms) and pinworms are present. Allow at least 7 days between applications. Rates of 3 pints or less per acre may be applied up to day of harvest. Rates above 3 pints per acre require an interval of 14 days between last application and harvest. **NOTE:** Resistance of Colorado potato beetle has occurred in some areas. Consult your local Extension Service or Mobay Sales Representative for details.
	Banded cucumber beetle *Drosophila* Green stink bug Leafminers Whitefly	1½ to 2	
	Aphids* European corn borer Flea beetles Grasshoppers Leafhoppers Thrips	2 to 3	
	Tuberworm	2¼ to 3	
	Fruitworm Hornworm Pinworm Yellow-striped armyworm	3 to 6	
MELONS Honeydew melons Muskmelons (cantaloupe) Watermelons Other melons	Leafhoppers Leafminers	1½ to 2	Apply specified dosage per acre in sufficient water to give complete coverage. For application by irrigation systems: Apply specified dosage per acre. Follow all directions given under the CHEMIGATION section of the label. A total of 3 applications may be made per crop season regardless of rate, formulation or method of application used. Allow at least 5 days between applications and at least 7 days between last application and harvest.
	Rindworms Spotted cucumber beetle Striped cucumber beetle Western-striped cucumber beetle	2	
ORNAMENTALS	Aphids* Blackvine weevil *Cerococcus* scale Euonymus scale Juniper scale Lace bugs Leafhoppers Olive scale Oystershell scale *Pulvinaria* scale Thrips	1½ to 2	Apply specified dosage per 100 gallons** of water. Spray all foliage surfaces, including the underside of leaves, for complete coverage. Repeat applications may be made as necessary.
	Brown soft scale Putnam scale	4	
	Black pine leaf scale European elm scale	3 to 4	
Ornamentals, Nursery plants	European pine shoot moth Nantucket pine tip moth	1½ to 3	Apply specified dosage per acre, in sufficient water for good coverage. Time applications to coincide with moth flights.
	Injury to hawthorn or American linden may occur under some conditions. Do not allow children or pets on treated area until material has been washed into soil, and treated area is dry.		

CROP	INSECT	PINTS GUTHION 2S	REMARKS
Christmas Trees	Scale spps. Sawfly spps. European pine shoot moth Eastern pine shoot borer Nantucket pine tip moth	1½ to 3	Apply specified dosage per acre by air or ground equipment in sufficient water to give complete coverage but not less than 1 gallon per acre. Time applications to coincide with suspectible pest development.
Southern pine seed orchards	Coneworm Seedworm	(See Remarks)	Use 6 pints per 100 gallons of water (0.2% dilution) as a full coverage spray**. Use 3 pints per 10 gallons of water (1% dilution) for low volume sprayers. Apply first application within 30 days following conelet closure, followed by 3 to 5 applications (at least 30 days apart). Apply approximately 5 to 10 gallons of the 0.2% dilution with high volume sprayers or approximately 1 to 2 gallons of the 1% dilution with low volume sprayers per tree. Thorough coverage of cones is necessary for maximum control. Apply no more than 6 applications per season. 11-29-88

[1] **NOTE:** This formulation, when used undiluted, may cause spotting of automobile finishes if prolonged exposure is permitted. Do not spray directly over automobiles. If accidental exposure does occur, automobiles should be washed immediately.

* In some areas, these species may have developed resistance to organophosphate insecticides. GUTHION insecticide used alone may not provide satisfactory control in those areas. Consult your local agricultural advisor or cooperative extension service for recommendations.

** This concentration is calculated for conventional hydraulic-type sprayers. When lower volumes of spray per acre are applied with concentrate sprayers, increase the concentration of GUTHION 2S in the spray mixture in order to apply amount of GUTHION 2S per acre equivalent to a full-coverage spray. In Southern Pine Seed orchards where conditions dictate an air application, apply dosage per acre equivalent to a full coverage ground spray in not less than one gallon of water per acre.

SPECIAL LOCAL NEED REGISTRATIONS [SECTION 24(c)] FOR GUTHION 2S*

CROP	INSECT	STATE EPA SLN NO.	DOSAGE	REMARKS
Blueberries	Blueberry maggot	Maine ME 79-0006	1 pint per acre.	Aerial application. Follow state recommendations for timing of sprays.
Celery	Carrot weevil	Ohio OH 81-0018	2 pints per acre.	Do not apply within 14 days of harvest.
Cranberries	Cranberry weevil Cranberry fruitworm *Sparganothis sulfureana* Tipworm	Massachusetts MA 78-0002	2 to 4 pints per acre.	Application by ground or air.
	Fireworms		4 pints per acre.	
Parsley (Root and moss-curled)	Carrot weevil (*Listronotus oregonensis*)	Ohio OH 81-0018	2 pints per acre.	Apply up to 21 days of harvest.
		New Jersey NJ 86-0005	1 quart per acre.	

* For distribution and use only within the state specified. These Special Local Need registrations approved as of date of this publication are, however, subject to cancellation by the state. Contact your nearest Mobay Agricultural Chemicals Division Office for the current status and a copy of the complete individual supplemental Special Local Need labeling. This labeling must be in the possession of the user at time of pesticide application.

RESTRICTIONS

Do not use on other crops used for food or forage. Use only according to label directions. Application at rates above those shown may result in illegal crop residues. Do not graze livestock in treated orchards or groves for 21 days after treatment. Do not treat food crops grown in the greenhouse. Do not apply when weather conditions favor drift from areas treated.

RE-ENTRY STATEMENTS FOR FARM WORKERS

Do not apply this product in such a manner as to directly or through drift expose workers or other persons. The area being treated must be vacated by unprotected persons. Do not enter treated areas for 24 hours after application unless protective clothing is worn, as described in the product labeling. Because certain states may require more restrictive re-entry intervals for various crops treated with this product, consult your State Department of Agriculture for further information.

Written or oral warnings must be given to workers who are expected to be in a treated area or in an area about to be treated with this product. Oral warnings should include those statements found under the label heading "Precautionary Statements" and "Re-entry Statements" included on the label. When oral warnings are given, warnings shall be given in a language customarily understood by workers. Oral warnings must be given if there is reason to believe that written warnings cannot be understood by workers. Written warnings must include the following information: DANGER, area treated with GUTHION 2S on (date of application). Do not enter treated areas for 24 hours after application unless protective clothing is worn, as described in the product labeling. **If on skin,** remove contaminated clothing and wash skin immediately with soap and warm water. **If eyes are contaminated,** wash with flowing water for at least 15 minutes. **If swallowed,** vomiting should be induced. (See Precautionary Statements on label).

ROTATIONAL CROPS

Do not plant root crops other than those with registered azinphos-methyl uses in azinphos-methyl treated soil sooner than 6 months after the last application. Do not plant any other crop other than those with registered azinphos-methyl uses in treated soil sooner than 30 days after last application.

PRECAUTIONARY STATEMENTS

HAZARDS TO HUMANS AND DOMESTIC ANIMALS

DANGER–POISON

Fatal if swallowed, inhaled or absorbed through the skin. Do not get in eyes, on skin or on clothing. Do not breathe vapor or spray mist. Spray operator should work to windward to stay out of drift or mist.

When loading spray tank, or handling the concentrate at any other time, wear protective clothing, natural rubber gloves and goggles. Wear a pesticide respirator jointly approved by the Mining Enforcement and Safety Administration (formerly the U.S. Bureau of Mines) and by the National Institute for Occupational Safety and Health under the provisions of 30 CFR Part 11.

Do not contaminate feed or foodstuffs. Keep out of reach of children and domestic animals. Keep all unprotected persons out of the operating area or vicinity where there may be danger of drift. Wash hands, arms and face thoroughly with soap and warm water before eating or smoking.

HANDLING OF CONTAMINATED ARTICLES: Clothing and rags receiving undiluted GUTHION 2S as a result of leaks, spills or mishaps should be removed immediately, and not reused. These contaminated articles should be buried in the same manner as the empty container; see container disposal section for directions. (Contaminated articles which should be discarded include: hats, gloves, aprons, coats, boots, etc. not made of rubber, or coated with rubber or other similar materials.) Clothing receiving spray mist or droplets from the mixed, diluted GUTHION 2S should be removed and decontaminated before reuse. This clothing can be decontaminated by machine-washing, separately from other items, with soap or detergent and bleach in hot water twice before reuse. Protective clothing and equipment should be washed-down with detergent or soap and bleach in water. Washwater from the cleaning of protective clothing or equipment should not be allowed to runoff or otherwise enter water supplies.

SYMPTOMS OF POISONING: A sense of "tightness" in the chest. Sweating. Contracted pupils. Stomach pains. Vomiting and diarrhea.

STATEMENTS OF PRACTICAL TREATMENT

In case of poisoning, call a physician immediately. Have patient lie down and keep quiet. **If swallowed,** vomiting should be induced. Administer water freely, and induce vomiting by giving one dose (½ oz. or 15 ml.) of syrup of ipecac. If vomiting does not occur within 10 to 20 minutes, administer second dose. If syrup of ipecac is not available, induce vomiting by sticking finger down throat. Repeat until vomit fluid is clear. Never give anything by mouth to an unconscious person. Professional medical assistance should be secured immediately. **If on skin,** remove contaminated clothing, and wash skin immediately with soap and warm water. **If eyes are contaminated,** wash with flowing water for at least 15 minutes.

TO PHYSICIAN: Antidote—Administer atropine sulfate in large therapeutic doses. Repeat as necessary, to the point of tolerance. 2-PAM is also antidotal, and may be administered in conjunction with atropine.

Compound inhibits cholinesterase, resulting in stimulation of the central nervous system, the parasympathetic nervous system and the somatic motor nerves. Do not give morphine. Watch for pulmonary edema, which may develop in serious cases of poisoning, even after 12 hours. At first sign of pulmonary edema, the patient should be placed in an oxygen tent, and treated symptomatically.

ENVIRONMENTAL HAZARDS

This pesticide is extremely toxic to fish and wildlife.

Do not apply directly to water or wetlands (swamps, bogs, marshes, and potholes). Do not contaminate water when disposing of equipment wastewater. Drift and runoff from treated areas may be hazardous to aquatic organisms in neighboring areas.

This product is highly toxic to bees exposed to direct treatment or residues on blooming crops or weeds. Do not apply this product or allow it to drift to blooming crops or weeds if bees are visiting the treatment area. Protective information may be obtained from your Cooperative Agricultural Extension Service.

PHYSICAL OR CHEMICAL HAZARDS

Keep away from open flame. Do not heat.

PROTECTIVE CLOTHING AND WORK SAFETY STATEMENTS

If handled indoors provide mechanical exhaust ventilation.

Keep all unprotected persons, children, livestock, and pets away from treated areas or where there is danger of drift.

Do not rub eyes or mouth with hands. If you feel sick in any way, STOP work and get help right away. See First Aid section of the label.

For questions regarding further treatment, or information about the product, call the National Pesticide Telecommunications Network at 800-858-7378.

HANDLE THE CONCENTRATE ONLY WHEN WEARING THE FOLLOWING PROTECTIVE CLOTHING AND EQUIPMENT:
Wear a protective suit of one or two pieces that covers all parts of the body except the head, hands, and feet. Wear chemical resistant gloves, chemical resistant apron, and chemical resistant shoes, shoe coverings, or boots. Wear goggles or a face shield and a pesticide respirator approved by the National Institute for Occupational Safety and Health under the provision of 30 CFR Part 11.

If handling the concentrate with a closed system, long sleeved shirt and long-legged pants may be substituted for the protective suit and the respirator requirement is waived.

WEAR THE FOLLOWING PROTECTIVE CLOTHING DURING APPLICATION, EQUIPMENT REPAIR, EQUIPMENT CLEANING, DURING REENTRY TO TREATED AREAS, AND DISPOSAL OF THE PESTICIDE:

Wear a protective suit of one or two pieces that covers all parts of the body except the head, hands, and feet. Wear chemical resistant gloves and chemical resistant shoes, shoe coverings, or boots. Wear a chemical resistant hat during airblast application.

During application from an enclosed tractor cab or airplane cockpit, or other suitable vehicle in which the windows are rolled up, long sleeved shirt and long pants may be worn in place of the above protective clothing. Chemical resistant gloves must be available in the cab or cockpit and must be worn while exiting. This clothing is inadequate to protect you during equipment repair, equipment cleaning, reentry, or during pesticide disposal.

IMPORTANT! BEFORE REMOVING GLOVES, WASH THEM WITH SOAP AND WATER. ALWAYS WASH HANDS, FACE, AND ARMS WITH SOAP AND WATER BEFORE SMOKING, DRINKING, EATING OR TOILETING.

After work, take off all clothing and shoes. Shower using soap and water. Do not wear contaminated clothing. Wash protective clothing and protective equipment with soap and water after each use. Respirators must be cleaned and filters replaced according to instructions included with the respirators. Personal clothing worn during use must be laundered separately from household articles. Clothing and protective equipment heavily contaminated or drenched with GUTHION azinphos-methyl must be destroyed according to state and local regulations. HEAVILY CONTAMINATED OR DRENCHED CLOTHING CANNOT BE ADEQUATELY DECONTAMINATED.

During aerial application, human flaggers are prohibited unless in totally enclosed vehicle.

STORAGE AND DISPOSAL

Pesticide Disposal: Do not contaminate water, food or feed by storage or disposal.

Pesticide wastes are acutely hazardous. Improper disposal of excess pesticide, spray mixture, or rinsate is a violation of Federal Law. If these wastes cannot be disposed of by use according to label instructions, contact your State Pesticide or Environmental Control Agency, or the Hazardous Waste representative at the nearest EPA Regional Office for guidance.

Container Disposal: Triple rinse (or equivalent). Then offer for recycling or reconditioning, or puncture and dispose of in a sanitary landfill, or by other procedures approved by State and local authorities.

Store in a cool, dry place and in such a manner as to prevent cross contamination with other pesticides, fertilizers, food, and feed. Do not store below 25°F. Keep away from open flame and extreme heat. Store in original container and out of the reach of children, preferably in a locked storage area.

Handle and open container in a manner as to prevent spillage. If container is leaking invert to prevent leakage. If the container is leaking or material spilled for any reason or cause, carefully dam up spilled material to prevent runoff. Refer to Precautionary Statements on label for hazards associated with the handling of this material. Do not walk through spilled material. Absorb spilled material with absorbing type compounds and dispose of as directed for pesticides above. In spill or leak incidents, keep unauthorized people away. You may contact the Mobay Emergency Response Team for decontamination procedures or any other assistance that may be necessary. The Mobay Kansas City Emergency Response Telephone No. is 816-242-2582, or contact Chemtrec at 800-424-9300.

IMPORTANT:

This bulletin is not intended to provide adequate information for application. Before using the product, read and carefully observe directions, cautionary statements and other information appearing on the product label. This product is sold subject to the Conditions of Sale set forth on the container label.

READ THE LABEL BEFORE USE

G 9003

This is a specimen label. Mobay is not responsible for the accuracy of the information contained herein. As labels are subject to revision, always carefully read and follow the label on the product container.

GUTHION 2L®

EMULSIFIABLE INSECTICIDE

RESTRICTED USE PESTICIDE

Due to Acute Toxicity

For retail sale to and use only by Certified Applicators or persons under their direct supervision, and only for those uses covered by the Certified Applicator's certification.

FOR CONTROL OF A WIDE VARIETY OF AGRICULTURAL PESTS

ACTIVE INGREDIENTS:	
O,O-Dimethyl *S*-[(4-oxo-1,2,3-benzotriazin-3(4*H*)-yl) methyl] phosphorodithioate	22.2%
INERT INGREDIENTS:	77.8%
TOTAL	100.0%

Contains 2 lbs. *O,O*-Dimethyl *S*-[(4-oxo-1,2,3-benzotriazin-3(4*H*)-yl) methyl] phosphorodithioate per gal.

EPA Reg. No. 3125-102

BOLSTAR, GUTHION and DEF are Reg. TMs of Bayer AG, Germany
Galecron is a trademark of Ciba-Geigy for chlordimeform
Fundal is a trademark of Schering AG Berlin/Bergamen

RECOMMENDED USES

Almonds	Cotton	Pecans
Apricots	Cranberries	Peppers
Artichokes	Cucumbers	Plums***†
Barley	Eggplant	Potatoes
Beans (dry and snap)	Filberts*	Prunes***†
Blackberries	Grapes	Raspberries
Blackeyed peas	Loganberries	Rye
Blueberries* *	Melons	Southern pine seed orchards
Boysenberries	Nectarines***†	Soybeans
Broccoli	Nursery plants	Spinach
Brussels sprouts	Oats	Strawberries
Cabbage	Onions (green and dry)	Sugarcane****
Cauliflower	Ornamentals	Tomatoes
Celery	Pasture grass‡	Wheat
Christmas trees	Peaches***†	Walnuts
Citrus fruits		

**** Florida, Louisiana and Texas only
* Pacific Northwest only
** Eastern and North Central states only
‡ States East of the Mississippi River
*** Eastern U.S.
† West of the Rocky Mountains.

RESTRICTIONS ON WAREHOUSE STORAGE

Do not store in front of space heater. Do not store in draft from an open door. Do not store below 45°F. Keep away from open flame. Do not heat. Store below 120°F. Difficult to reconstitute formulation after exposure to cold. Do not store adjacent to, or near proximity of foodstuffs, drugs, feeds or any other material intended for consumption by humans or animals. Do not store adjacent to toys, cosmetics, dishes, furniture or clothing. Do not store next to herbicides. In case of freeze-out, do not sell until reconstituted.

PACKAGING

5 gallon can. Shipping weight: 51 lbs.
50 gallon drum. Shipping weight: 507 lbs.

GUTHION 2L has proved to be effective in controlling insects on a wide variety of fruit, field, and vegetable crops after many years of intensive foreign and domestic tests and years of commercial use.

FRUITS: GUTHION 2L controls major insect pests on a variety of stone fruits, citrus and berries. This insecticide greatly simplifies spray schedules. Its longer residual activity protects from insect damage on most fruit crops.

VEGETABLES: Registered for use on many vegetables, GUTHION 2L has proved to be effective in controlling potato tuberworm, diamondback moth, Colorado potato beetle and others.

COTTON: GUTHION 2L is an effective cotton insecticide. It provides control of the boll weevil, and in areas of weevil resistance, fields treated with GUTHION 2L have consistently shown the lowest weevil populations and damage. For boll weevil control on cotton (conventional or low volume spray), GUTHION 2L at rates up to 1 pint per acre may be tank mixed with Galecron® 4E or Fundal® 4EC for ovicidal control of bollworms and tobacco budworms. GUTHION 2L also gives excellent control of early-season insects, such as thrips, fleahoppers, overwintering boll weevils and plant bugs. GUTHION 2L may be tank mixed in one application with DEF® 6 defoliant.

OTHER SPARY AND WETTABLE POWDER FORMULATIONS ALSO AVAILABLE

PELIGRO

PRECAUCION AL USUARIO: Si usted no puede leer o entender inglés, no use este producto hasta que la etiqueta le haya sido explicada ampliamente.

(TO THE USER: If you cannot read or understand English, do not use this product until the label has been fully explained to you.)

DIRECTIONS FOR USE

It is a violation of Federal Law to use this product in a manner inconsistent with its labeling.

MIXING: GUTHION 2L emulsifiable insecticide forms an emulsion when diluted with water, and is suitable for use in all power-operated ground and aircraft sprayers. To mix with water, pour the required amount of GUTHION 2L into full amount of water, and then agitate. GUTHION 2L may also be applied undiluted as an ultra-low-volume spray with either ground or aircraft equipment that has been adapted and calibrated for ultra-low-volume spraying as described below under "Recommended Applications" for those crops specified.

COMPATIBILITY: GUTHION 2L is physically compatible with pesticides or with liquid fertilizer, first contact your supplier. For further information, contact your local Mobay representative. If your supplier and Mobay representative have no experience with the combination you are considering, you should conduct tests to determine physical compatibility.

DOSAGE: Use specified dosage of GUTHION 2L in the amount of water necessary to give complete coverage of foliage. The type of equipment used will determine the concentration required.

SPRAYING: Work to windward. Protect sprayer operators from drift or mist. When low volumes of spray are applied, complete coverage and thorough application are essential for most effective results. Schedule applications in accordance with local conditions. Consult your State Agricultural Extension Service or Experiment Station for specific use information in your area.

CHEMIGATION

GUTHION 2L may be applied through recommended types of irrigation systems to many crops. The REMARKS section for each crop lists the types of application allowed. If application through irrigation systems is not listed in the REMARKS section for a crop, GUTHION 2L may not be applied to that crop through irrigation systems.

Types of Irrigation Systems

Apply GUTHION 2L only through sprinkler, including center pivot, lateral move, side roll, overhead solid set, or low pressure sprinkler irrigation systems. Do not apply GUTHION 2L through any other types of irrigation systems.

GENERAL DIRECTIONS FOR ALL RECOMMENDED TYPES OF IRRIGATION SYSTEMS

Uniform Water Distribution and System Calibration

The irrigation system must provide uniform distribution of treated water. Crop injury, lack of effectiveness, or illegal pesticide residues in the crop can result from nonuniform distribution of treated water.

The system must be calibrated to uniformly apply the rates specified for chemigation application for specific crops. If you have questions about calibration, you should contact State Extension Service specialist, equipment manufacturers or other experts.

Chemigation Monitoring

A person knowledgeable of the chemigation system and responsible for its operations, or under the supervision of the responsible person, shall shut the system down and make necessary adjustments should the need arise.

Drift

Do not apply when wind speed favors drift beyond the area intended for treatment.

Required System Safety Devices

The system must contain a functional check valve, vacuum relief valve, and low pressure drain appropriately located on the irrigation pipeline to prevent water source contamination from backflow.

The pesticide injection pipeline must contain a functional, automatic, quick-closing check valve to prevent the flow of fluid back toward the injection pump.

The pesticide injection pipeline must also contain a functional, normally closed, solenoid-operated valve located on the intake side of the injection pump and connected to the system interlock to prevent fluid from being withdrawn from the supply tank when the irrigation system is either automatically or manually shut down.

The system must contain functional interlocking controls to automatically shut off the pesticide injection pump when the water pump motor stops.

The irrigation line or water pump must include a functional pressure switch which will stop the water pump motor when the water pressure decreases to the point when pesticide distribution is adversely affected.

Systems must use a metering pump, such as a positive displacement injection pump (e.g., diaphragm pump) effectively designed and constructed of materials that are compatible with pesticides and capable of being fitted with a system interlock.

Using Water from Public Water Systems

DO NOT APPLY GUTHION 2L THROUGH ANY IRRIGATION SYSTEM **PHYSICALLY CONNECTED** TO A PUBLIC WATER SYSTEM. Public water system means a system for the provision to the public of piped water for human consumption if such system has at least 15 service connections or regularly services an average of at least 25 individuals daily at least 60 days out of the year.

GUTHION 2L may be applied through any of the recommended types of irrigation systems which may be **supplied** by a public water system **only if** the water from the public water system is discharged into a reservoir tank prior to pesticide introduction. There shall be a complete physical break (air gap) between the outlet end of the fill pipe and the top or overflow rim of the reservoir tank of at least twice the inside diameter of the fill pipe. Before beginning chemigation, always make sure that the air gap exists and that there is no blockage of the overflow of the reservoir tank.

Any irrigation system using water supplied from a public water system must also meet the following requirements:

The pesticide injection pipeline must contain a functional, automatic quick-closing check valve to prevent the flow of fluid back toward the injection pump.

The pesticide injection pipeline must contain a functional, normally closed, solenoid-operated valve located on the intake side of the injection pump and connected to the system interlock to prevent fluid from being withdrawn from the supply tank when the irrigation system is either automatically or manually shut down.

The system must contain functional interlocking controls to automatically shut off the pesticide injection pump when the water pump motor stops or in cases where there is no water pump, when the water pressure decreases to the point where pesticide distribution is adversely affected.

Systems must use a metering pump, such as a positive displacement injection pump (e.g., diaphragm pump) effectively designed and constructed of materials that are compatible with pesticides and capable of being fitted with a system interlock.

Do not apply when wind speed favors drift beyond the area intended for treatment.

Posting Requirements

Posting of areas to be chemigated is required when 1) any part of a treated area is within 300 feet of sensitive areas such as residential areas, labor camps, businesses, day care centers, hospitals, in-patient clinics, nursing homes or any public areas such as schools, parks, playgrounds, or other public facilities not including public roads, or 2) when the chemigated area is open to the public such as golf courses or retail greenhouses.

Posting must conform to the following requirements: Treated areas shall be posted with signs at all usual points of entry and along likely routes of approach from the listed sensitive areas. When there are no usual points of entry, signs must be posted in the corners of the treated areas and in any other location affording maximum visibility to sensitive areas. The printed side of the sign should face away from the treated area towards the sensitive areas. The signs shall be printed in English. Signs must be posted prior to application and must remain posted until foliage has dried and soil surface water has disappeared. Signs may remain in place indefinitely as long as they are composed of materials to prevent deterioration and maintain legibility for the duration of the posting period.

All words shall consist of letters at least 2½ inches tall, and all letters and the symbol shall be a color which sharply contrasts with their immediate background. At the top of the sign shall be the words KEEP OUT, followed by an octagonal stop sign symbol at least 8 inches in diameter containing the word STOP. Below the symbol shall be the words PESTICIDES IN IRRIGATION WATER.

Posting required for chemigation does not replace other posting and reentry interval requirements for farmworker safety.

Compatibility

When mixing other chemicals refer to Compatibility section elsewhere on the label.

Agitation

For application of GUTHION 2L alone, a chemical supply tank is not necessary for premixing since GUTHION 2L mixes well with water in the irrigation line. If a chemical supply tank is used, for application of GUTHION 2L alone or in combination with liquid fertilizer or other chemicals, constant strong mechanical or hydraulic agitation must be maintained in the chemical supply tank during the entire period of application.

Chemical Supply Tank Dilution

If a chemical tank is used, you must determine the required amounts of GUTHION 2L and water to mix in the tank.

The amount of GUTHION 2L needed equals the number of pints of GUTHION 2L to be applied per acre multiplied by the number of acres to be chemigated.

The amount of emulsion needed equals the gallons of emulsion delivered per hour by the injection pump multiplied by the number of hours chemigation will take place.

The amount of water needed equals the amount of emulsion needed minus the amount of GUTHION 2L needed.

For example, if you want to apply 3 pints of GUTHION 2L per acre to 130 acres in 20 hours and your injection pump delivers 15 gallons per hour, you need: 3 pints GUTHION 2L per acre × 130 acres = 390 pints or 48.75 gallons of GUTHION 2L. And, you need: 15 gallons per hour × 20 hours = 300 gallons of emulsion, minus 48.75 gallons of GUTHION 2L = 251.25 gallons of water.

Cleaning the Chemical Injection System

In order to accurately apply pesticides, the chemical injection system must be kept clean; free of chemical or fertilizer residues and sediments. Refer to your owners manual or ask your equipment supplier for cleaning procedure for your injection system.

Flushing the Irrigation System

At the end of the application period, allow time for all lines to flush the pesticide through all nozzles or emitters before turning off irrigation water. To ensure the lines are flushed and free of pesticides, a dye indicator may be injected into the lines to mark the end of the application period.

SPRINKLER IRRIGATION SYSTEMS

All directions and requirements under the GENERAL DIRECTIONS AND REQUIREMENTS FOR ALL RECOMMENDED TYPES OF IRRIGATION SYSTEMS section of the label must be followed for sprinkler irrigation systems.

In addition, the following directions apply to sprinkler irrigation systems:

Do not apply when wind speed favors drift beyond the area intended for treatment.

It is recommended that nozzles in the immediate area of control panels, chemical supply tanks, pumps and system safety devices be plugged to prevent chemical contamination of these areas

Center-Pivot and Automatic-Move Linear Systems

Inject the specified dosage per acre continuously for one complete revolution or move of the system. DO NOT USE END GUNS. For a foliar application, the system should be run at maximum speed. For a soil application, the system should be run at a slower speed with application in at least ¼ inch of water.

Solid Set and Manually Controlled Linear Systems

For foliar application, injection should be during the last 30 to 60 minutes of regular irrigation period or as a separate 30 to 60 minute application not associated with a regular irrigation. For soil treatment, application should be in at least ¼ inch of water. DO NOT USE END GUNS.

RECOMMENDED APPLICATIONS

CROP	INSECT	PINTS GUTHION 2L	REMARKS
FIELD CROPS	Boll weevil	½ to 1	Apply specified dosage per acre by air or ground equipment in sufficient water for complete coverage but not less than 1 gallon per acre. For application by irrigation systems: Apply specified dosage per acre. Follow all directions given under the CHEMIGATION section of the label. A total of 12 applications per crop season may be made regardless of rate, formulation or method of application used. Cotton receiving late-season applications should not be pastured.
Cotton (Conventional or low-volume spray)	Brown cotton leafworm, Cotton leafworm, Cotton fleahopper, Lygus bugs, Thrips	1	
	Pink bollworm, Tarnished plant bug, Rapid plant bug	1 to 2	GUTHION 2L at rates up to 1 pint per acre may be tank mixed with Galecron 4E or Fundal 4EC for ovicidal control of bollworms and tobacco budworms.
	Stink bug	2	For control of the larval stages of resistant and non-resistant bollworm and/or tobacco budworm, BOLSTAR® 6 may be included in GUTHION 2L tank mixes. When using GUTHION 2L in a tank mix with Galecron 4E or Fundal 4EC, observe all limitations, precautions and rate recommendations which appear on the labels for these products. **See NOTE below.**

Note: GUTHION 2L may be tank mixed in one application with DEF® 6 defoliant. Refer to label for DEF 6 for recommended rates, precautions, and restrictions. Preceding the use of this tank mix, GUTHION 2L may be applied a maximum of 9 times at 1 pint per acre. The maximum rate of DEF 6 in a tank mix combination is 2 pints per acre.

CROP	INSECT	PINTS GUTHION 2L	REMARKS
Cotton (Ultra-low-volume spray)	Boll weevil	½ to 1	GUTHION 2L may be used undiluted in any ground or aerial spray equipment that has been adapted and calibrated for ultra-low-volume spraying. Spray machines must be equipped with accepted low-volume devices that will produce droplets within the range of 30 to 100 microns in size. ULV aerial applications should be made at altitudes of 10 to 20 feet. A total of 12 applications may be made per crop season regardless of rate, formulation or method of application used. Application may be made up to 2 days before hand picking. Cotton may be machine harvested any time after application. Do not graze livestock in treated areas. **See NOTE below.** **Early- and Mid-season Control:** Apply specified dosage per acre in accordance with local recommendations. **Diapause Weevil Control:** The one-pint-per-acre rate only is recommended for control of diapausing boll weevils. Schedule applications in accordance with local recommendations.
Barley Oats Rye Wheat	Cereal leaf beetle	1½ to 2	Apply specified dosage per acre, by air or ground equipment, in sufficient water for complete coverage, but not less than 1 gallon per acre. Do not apply more than once per season. Do not harvest for food, feed, forage or graze within 30 days of treatment.
Pasture grasses (States east of Mississippi River)	Grasshoppers Meadow spittlebugs (adults)	2 to 3	Apply specified dosage per acre by air or ground equipment in sufficient water for complete coverage but not less than 1 gallon per acre. Do not apply more than once per cutting. Do not graze or harvest within 16 days of application at the 2 pint rate or 21 days at rates above 2 pints.
Soybeans	Aphids* Bean leaf beetle Green cloverworm Leafhoppers Leafminers Leaf rollers Stink bugs Velvet bean caterpillar	1½ to 2	Apply specified dosage per acre, by air or ground equipment in sufficient water to give complete coverage, but not less than 1 gallon per acre. One or 2 applications of rates up to 2 pints per acre may be made up to 14 days of harvest. Do not graze or feed treated vines to livestock.
	Mexican bean beetle	2 to 3	
Sugarcane (Conventional spray) (Florida, Louisiana and Texas only)	Sugarcane borer	3	Apply specified dosage per acre by air application, using a minimum of 2 gallons of water per acre. For best results, applications should be made at approximately cane-top level, and not more than 10 feet above cane-top level. A total of 5 applications may be made per season. Allow at least 30 days between the last application and harvest. Bagasse from sugarcane treated with GUTHION 2L may be used for feed of cattle, goats and sheep. Consult your State Agricultural Extension Service or Experiment Station for specifc use information.
Sugarcane (Ultra-low-[1] volume spray) (Florida, Louisiana and Texas only)	Sugarcane borer	3	Apply specified dosage per acre undiluted in any aerial spray equipment that has been adapted and calibrated for ultra-low-volume spraying. Planes must be equipped with accepted low-volume devices that will produce droplets within the range of 30 to 100 microns in size. For best results, applications should be made at approximately cane-top level, and not more than 10 feet above cane-top level. A total of 5 applications may be made per season. Allow at least 30 days between last application and harvest. Bagasse from sugarcane treated with GUTHION 2L may be used for feed of cattle, sheep and goats. Consult your State Agricultural Extension Service or Experiment Station for specific use information.
FRUIT	NOTE: It is suggested that when treating fruit during the bloom period, bee keepers should be warned well in advance to remove hives a safe distance from orchards to be treated.		
Apricots Nectarines Peaches (Eastern U.S.)	Aphids** Cottony peach scale European fruit lecanium scale Forbes scale Lesser peach tree borer Oriental fruit moth Peach tree borer Peach twig borer *Platynota flavedana* leaf roller Plum curculio Redbanded leaf roller San Jose scale* Stink bug Thrips Tarnished plant bug Terrapin scale Walnut scale White peach scale	3½ to 4½	Apply specified dosage per acre as a full coverage spray. Allow at least 14 days between applications. A total of 13½ pints per acre per crop season may be applied to nectarines. A total of 18 pints per acre per crop season may be applied to apricots and peaches. Allow at least 21 days between last application and harvest. For control of peach tree borer, apply 2 to 3 sprays to trunk from ground to scaffold limbs, timed with moth flight. For scale control, apply when crawlers are present. GUTHION 2L is compatible with dormant and summer oils which may be added to peach sprays in accordance with local recommendations.
Apricot Nectarines Peaches (West of the Rocky Mountains)	Lesser peach tree borer Oriental fruit moth Peach twig borer Peach tree borer *Platynota flavedana* leaf roller Plum curculio Redbanded leaf roller Stink bug Thrips Tarnished plant bug	6 to 8	Apply specified dosage per acre. Allow at least 14 days between applications. A total of 13½ pints per acre per crop season may be applied to nectarines. A total of 18 pints per acre per crop season may be applied to apricots and peaches. Allow at least 21 days between last application and harvest. For control of peach tree borer, apply 2 or 3 sprays to trunk from ground to scaffold limb, timed with moth flight.

[1]NOTE: This formulation, when used undiluted, may cause spotting of automobile finishes if prolonged exposure is permitted. Do not spray directly over automobiles. If accidental exposure does occur, automobiles should be washed immediately.

CROP	INSECT	PINTS GUTHION 2L	REMARKS
Plums Prunes (Eastern U.S.)	Aphids* Coding moth Eye-spotted bud moth Forbes scale Fruit tree leaf roller Lesser peach tree borer Orange tortrix Peach tree borer Peach twig borer Plum curculio Redbanded leaf roller San Jose scale* Stink bug Tarnished plant bug* Tussock moth	3½ to 4½ (1 to 1½ pints per 100 gallons	Apply specified dosage per acre as a full coverage spray. A total of 13½ pints per acre per crop season may be applied to plums and prunes. Allow at least 10 days between applications. Allow at least 15 days between last application and harvest. For control of peach tree borer, apply 2 or 3 sprays to trunk from ground to scaffold limbs, timed with moth flight. For control of scale, apply when crawlers are present. GUTHION 2L is compatible with dormant and summer oils which may be added to prune sprays in accordance with local recommendations.
	American plum borer	6	
Plums Prunes (West of the Rocky Mountains)	Codling moth Eye-spotted bud moth Fruit tree leaf roller Lesser peach treeborer Orange tortrix Peach tree borer Peach twig borer Plum curculio Redbanded leaf roller Stink bug Tarnished plant bug* Tussock moth	4 to 8	Apply specified dosage per acre. A total of 13½ pints per acre per crop season may be applied to plums and prunes. Allow at least 10 days between applications. Allow at least 15 days between last application and harvest. For control of peach tree borer, apply 2 or 3 sprays to trunk from ground to scaffold limbs, timed with moth flight.
	American plum borer	6 to 8	
Blackberries Boysenberries Loganberries Raspberries	Leafhoppers Leaf rollers	1	Apply specified dosage per acre to foliage, using approximately 200 gallons of water for good coverage. Where ground conditions dictate an air application of GUTHION 2L, use specified rate in a minimum of 1 gallon of water per acre. Do not apply within 14 days of harvest. Do not apply more than twice per season.
	Leafminers	1¼	
	Aphids*	1¼ to 2	
	Obscure root weevil	2	
	Obscure root weevil	2	For control of root weevils and borers prior to harvest, apply specified dosage per acre to lower portion of canes, and to the soil beneath the plants, using approximately 200 gallons of water. Do not apply more than twice per season. Do not make applications within 3 days of harvest at rates up to 4 pints per acre.
Blueberries (Eastern and North Central states only)	Blueberry maggot Fruitworms Lecanium scale Plum curculio	2 to 3	Apply specified dosage per acre with aerial or ground equipment using sufficient water for good coverage. A total of 3 applications may be made per crop season regardless of rate or formulation used. Allow at least 10 days between applications and at least 7 days between last application and harvest.
Cranberries	Cranberry fruitworm *Sparganothis sulfureana* Tipworm	2 to 4	Apply specified dosage per acre by air using sufficient water for good coverage. For application by irrigation systems: Apply specified dosage per acre. Follow all directions given under the CHEMIGATION seciton of the label. A total of 3 applications may be made per crop season regardless of rate or formulation used. Allow at least 14 days between applications and at least 21 days between last application and harvest.
	Fireworms	4	
Citrus fruits	Aphids* Black scale Brown soft scale Chaff scale Citricola scale Citrus mealybug Citrus thrips Cottony-cushion scale European brown snail Florida red scale Fruit tree leaf rollers Fuller rose beetle Glover scale Orange tortrix Purple scale Snow scale Western tussock moth Whiteflies	8 to 12	Apply specified dosage in 100 gallons of water as a full-coverage spray. A single application per year may be applied up to within 7 days of harvest. Where 2 applications are required, the second spray should not be applied within 28 days of harvest. A total of two applications may be applied per fruit year regardless of rate or formulation. Do not pick fruit, or do other work involving contact with the tree (such as pruning) within 7 days after treatment.
	California red scale Yellow scale	12	
Grapes	Grape berry moth Grape cane girdlers Grape leaf skeletonizer Grape mealybug Leafhoppers* Redbanded leaf roller Thrips	3 to 4	Apply specified dosage per acre as a full coverage spray. A total of 3 applications may be made per crop season regardless of rate or formulation used. Allow at least 14 days between applications. Minimum dosage specified may be applied up to harvest. For higher rates allow at least 7 days between last application and harvest.

CROP	INSECT	PINTS GUTHION 2L	REMARKS
Strawberries	Aphids* Meadow spittlebug Oblique-banded leaf roller Obscure root weevil Omnivorous leaf tier Pea leaf weevil Small black (grass) weevil Strawberry leaf rollers Whitefly	2	Apply specified dosage per acre as a full-coverage spray using sufficient water for good coverage. For application by irrigation systems: Apply specified dosage per acre. Follow all directions given under the CHEMIGATION section of the label. A total of 4 applications may be made per crop season regardless of rate or formulation used. Allow at least 5 days between applications and at least 5 days between last application and harvest.
NUTS Almonds	Peach twig borer Navel orangeworm	6 to 8	Apply specified dosage per acre. A total of 2 applications may be made per crop season regardless of rate or formulation used. Allow at least 30 days between applications and at least 60 days between last application and harvest. Do not apply afer husks split.
Filberts (Pacific Northwest only)	Apple mealybug Filbert aphid Filbert leaf roller Filbertworm	6 to 8	Apply specified dosage per acre as a full coverage spray. A total of 3 applications may be made per crop season regardless of rate or formulation used. Allow at least 14 days between applications and at least 30 days between last application and harvest. Do not graze livestock in treated in groves for 21 days after treatment.
Pecans	Aphids** Fall webworm Hickory shuckworm Leafminers May beetles Pecan casebearer Southern green stink bug Spittlebug Twig girdlers Walnut caterpillar	6 to 8	Apply specified dosage per acre as a full coverage spray. A total of 3 applications may be made per crop season regardless of rate or formulation used. Allow at least 7 days between applications. Do not apply after shuck split. Livestock may be grazed in treated groves after a 21-day post-treatment interval.
Walnuts	Codling moth Filbertworm Navel orangeworm Red-humped caterpillar Walnut husk fly	6 to 8	Apply specified dosage per acre as a full-coverage spray. A total of 3 applications may be made per crop season regardless of rate or formulation used. Allow at least 14 days between applications. Do not apply after husks split. Do not graze livestock in treated groves for 21 days after treatment.
VEGETABLES Artichokes	Plume moth	6	Apply specified dosage per acre in sufficient water for complete coverage. For application by irrigation systems: Apply specified dosage per acre. Follow all directions given under the CHEMIGATION section of the label. A total of 3 applications may be made per crop season regardless of rate, formulation or method of application used. Allow at least 14 days between applications and at least 30 days between last application and harvest. Do not feed or ensile treated forage.
Beans (Snap and Dry)	Leaf rollers Mites	1 to 2	Apply specified dosage per acre by air or ground equipment in sufficient water to give complete coverage, but not less than 1 gallon per acre. Do not apply to snap beans within 7 days of harvest, nor to dry beans within 30 days of harvest. Do not exceed 4 applications. Do not feed or ensile treated forage.
	Aphids* Bean leaf beetle Green cloverworm Leafhoppers Leafminers Stink bugs Velvet bean caterpillar	1½ to 2	
	Mexican bean beetle Spotted cucumber beetle Striped cucumber beetle Tarnished plant bug Western-striped cucumber beetle	2	
Broccoli Brussels sprouts Cabbage (Includes tight heading varieties of Chinese cabbage) Cauliflower	Aphids* Cabbage looper Diamondback moth Imported cabbageworm	2 to 3	Apply specified dosage per acre in sufficient water for complete coverage, but not less than 1 gallon per acre. Do not apply within 7 days of harvest for Brussels sprouts, 15 days of harvest for broccoli and cauliflower, nor within 21 days of harvest for cabbage. Do not exceed more than 3 applications per season.
	Cabbage maggot	½	Mix specified dosage in 50 gallons of water. Apply 4 to 6 ounces of this emulsion per plant immediately after transplanting.
	Cabbage maggot (Transplant fields in California only)	3	Apply specified dosage in 300 to 400 gallons of water per acre as a soil drench in the rows when damage first appears. Additional applications may be necessary. Do not apply within 21 days of harvest. Do not exceed more than 3 applications per season.
	Cabbage maggot (Direct-seeded fields in California only)	3	Apply specified dosage per acre in sufficient water for uniform distribution. Mix in the upper 2 inches of soil prior to seeding, or spray in the seed row at planting time. Usually 2 additional sprays are necessary during the growing season depending upon time of year and maggot population. Do not apply within 21 days of harvest. Do not exceed more than 3 applications per season.
Celery	Aphids* Leafminers Leafhoppers Spittlebugs Tarnished plant bug	2	Apply specified dosage in 100 gallons of water as a full-coverage spray,** using not more than 200 gallons of finished spray per acre. Do not apply within 14 days of harvest. Do not exceed more than 3 applications per season.
Blackeyed peas (Southern peas, Crowder peas)	Corn earworm Cowpea curculio	3 to 4	Apply specified dosage per acre by air, or ground equipment, in sufficient water for complete coverage, but not less than 1 gallon per acre. Do not apply more than 4 times per season, nor within 7 days of harvest. Do not use vines for feed or forage, nor pasture treated areas.
	Leafminers Stink bugs	1½ to 2	

CROP	INSECT	PINTS GUTHION 2L	REMARKS
Cucumbers	Spotted cucumber beetle Striped cucumber beetle Western-striped cucumber beetle	2	Apply specified dosage per acre in sufficient water for complete coverage. For application by irrigation systems: Apply specified dosage per acre. Follow all directions given under the CHEMIGATION section of the label. A total of 3 applications may be made per crop season regardless of rate, formulation or method of application used. Allow at least 7 days between applications and at least 1 day between last application and harvest.
Eggplant	Leafminers	1½ to 2	Apply specified dosage per acre in sufficient water for complete coverage. For application by irrigation systems: Apply specified dosage per acre. Follow all directions given under the CHEMIGATION section of the label. A total of 3 applications may be made per crop season regardless of rate, formulation or method of application used. Allow at least 7 days between applications. Do not apply after fruit set.
	European corn borer Flea beetles	2	
Onions (Green and dry)	Thrips	2 to 3	Apply specified dosage per acre, by air or ground equipment, in sufficient water for complete coverage. For application by irrigation systems: Apply specified dosage per acre. Follow all directions given under the CHEMIGATION section of the label. A total of 3 applications may be made per crop season regardless of rate, formulation or method of application used. Allow at least 7 days between applications for dry bulb onions. Allow at least 10 days between applications for green onions. For dry onions, allow at least 28 days between last application and harvest. For green onions, allow at least 7 days between last application and harvest.
Peppers	Leafminers	1½ to 2	Apply specified dosage per acre, by air or ground equipment, in sufficient water for complete coverage. For application by irrigation systems: Apply specified dosage per acre. Follow all directions given under the CHEMIGATION section of the label. A total of 3 applications may be made per crop season regardless of rate, formulation or method of application used. Allow at least 7 days between applications and 7 days between last application and harvest.
	European corn borer Flea beetles	2	
Potatoes	Colorado potato beetle	1½	Apply specified dosage per acre, in sufficient water for complete coverage. For application by irrigation systems: Apply specified dosage per acre. Follow all directions given under the CHEMIGATION section of the label. A total of 3 applications may be made per crop season regardless of rate, formulation or method of application used. Allow at least 7 days between applications and 7 days between last application and harvest. **NOTE:** Resistance of Colorado potato beetle has occurred in some areas. Consult your Local Extension Service or MOBAY Sales Representative for details.
	Banded cucumber beetle Leafminers	1½ to 2	
	European corn borer Flea beetle Leafhoppers Spittlebugs Tarnished plant bug	2 to 3	
	Tuberworm	2¼ to 3	
Spinach	Aphids* Leafminers Mites	1½ to 2	Apply specified dosage per acre by air or ground equipment in sufficient water for complete coverage, but not less than 1 gallon per acre. Do not apply within 14 days of harvest. Do not exceed more than 3 applications per season.
Tomatoes	Colorado potato beetle	1½	Apply specified dosage per acre, by air or ground equipment in sufficient water for complete coverage. For application by irrigation systems: Apply specified dosage per acre. Follow all directions given under the CHEMIGATION section of the label. A total of 4 applications may be made per crop season regardless of rate, formulation or method of application used. The high rates should be used where heavy infestations of late instar lepidopterous larvae (large worms) and pinworms are present. Allow at least 7 days between applications. Rates of 3 pints per acre or less may be applied up to the day of harvest. Rates above 3 pints require an interval of 14 days between last application and harvest. **NOTE:** Resistance of Colorado potato beetle has occurred in some areas. Consult your Local Extension Service or Mobay Sales Representative for details.
	Banded cucumber beetle *Drosophila* Green stink bug Leafminers Whitefly	1½ to 2	
	Aphids* European corn borer Flea beetles Grasshoppers Leafhoppers Thrips	2 to 3	
	Tuberworm	2¼ to 3	
	Fruitworm Hornworm Pinworm Yellow-striped armyworm	3 to 6	
MELONS Honeydew melons Muskmelons (cantaloupe) Watermelons Other melons	Leafhoppers Leafminers	1½ to 2	Apply specified dosage per acre in sufficient water to give complete coverage. For applications by irrigation systems: Apply specified dosage per acre. Follow all directions given under the CHEMIGATION section of the label. A total of 3 applications may be made per crop season regardless of rate, formulation or method of application used. Allow at least 5 days between applications and at least 7 days between last application and harvest.
	Rindworms Spotted cucumber beetle Striped cucumber beetle Western-striped cucumber beetle	2	

CROP	INSECT	PINTS GUTHION 2L	REMARKS
ORNAMENTALS Ornamentals nursery plants	Aphids* Black vine weevil *Cerococcus* scale Euonymus scale Juniper scale Lace bugs Leafhoppers Olive scale Oystershell scale *Pulvinaria* scale Thrips	1½ to 2	Apply specified dosage per 100 gallons** of water (2 teaspoonfuls per gallon). Spray all foliage surfaces, including the underside of leaves, for complete coverage. Repeat applications may be made as necessary.
	Brown soft scale Putnam scale	4	
	Black pine leaf scale European elm scale	3 to 4	
	European pine shoot moth Nantucket pine tip moth	1½ to 3	Apply specified dosage per acre in sufficient water for good coverage. For application to individual trees, use 1 tablespoonful of GUTHION 2L per gallon of water. Time applications to coincide with moth flights.
Christmas Trees	Scale spp. Sawfly spp. European pine shoot moth Eastern pine shoot borer Nantucket pine tip moth	1½ to 3	Apply specified dosage per acre by air or ground equipment in sufficient water to give complete coverage but not less than 1 gallon per acre. Time applications to coincide with suspectible pest development.
Southern pine seed orchards	Coneworm Seedworm	(See Remarks)	Use 6 pints per 100 gallons of water (0.2% dilution) as a full coverage spray**. Use 3 pints per 10 gallons of water (1% dilution) for low volume sprayers. Apply first application within 30 days following conelet closure, followed by 3 to 5 applications (at least 30 days apart). Apply approximately 5 to 10 gallons of the 0.2% dilution with high volume sprayers or approximately 1 to 2 gallons of the 1% dilution with low volume sprayers per tree. Thorough coverage of cones is necessary for maximum control. 11-29-88

Injury to hawthorn or American linden may occur under some conditions. Do not allow children or pets in treated area until material is dry.

* In some areas, these species may have developed resistance to organophosphate insecticides. GUTHION insecticide used alone may not provide satisfactory control in those areas. Consult your local agricultural advisor or cooperative extension service for recommendations.

** This concentration is calculated for conventional, hydraulic-type sprayers. When lower volumes of spray per acre are applied with concentrate sprayers, increase the concentration of GUTHION 2L in the spray mixture in order to apply the amount of GUTHION 2L per acre equivalent to a full-coverage spray. In Southern Pine Seed Orchards where conditions dictate an air application, apply dosage per acre equivalent to a full coverage ground spray in not less than one gallon of water per acre.

SPECIAL LOCAL NEED REGISTRATIONS [SECTION 24(c)] FOR GUTHION 2L*

CROP	INSECT	STATE EPA SLN NO.	DOSAGE	REMARKS
Cotton	Boll weevil	Louisiana LA 82-0028	½ to 2 pints per acre.	Ultra Low Volume application in vegetable oil.
		Alabama AL 84-0004 South Carolina SC 82-0021 Georgia GA 84-0004 Texas TX 84-0005 Mississippi MS 84-0012 Arizona AZ 88-0020	½ to 1 pint per acre.	

* For distribution and use only within the state specified. These Special Local Need registrations approved as of date of this publication are, however, subject to cancellation by the state. Contact your nearest Mobay Agricultural Chemicals Division Office for the current status and a copy of the complete individual supplemental Special Local Need labeling. This labeling must be in the possession of the user at time of pesticide application.

ROTATIONAL CROPS

Do not plant root crops other than those with registered azinphos-methyl uses in azinphos-methyl treated soil sooner than 6 months after the last application. Do not plant any other crop other than those with registered azinphos-methyl uses in treated soil sooner than 30 days after last application.

RESTRICTIONS

Do not use on other crops used for food or forage. Use only according to label directions. Application at rates above those shown may result in illegal crop residues. Do not graze livestock in treated orchards or groves for 21 days after treatment. Do not treat food crops grown in the greenhouse.

RE-ENTRY STATEMENTS FOR FARM WORKERS

Do not apply this product in such a manner as to directly or through drift expose workers or other persons. The area being treated must be vacated by unprotected persons. Do not enter treated areas for 24 hours after application unless protective clothing is worn, as described in the product labeling. Because certain states may require more restrictive re-entry intervals for various crops treated with this product, consult your State Department of Agriculture for further information.

Written or oral warnings must be given to workers who are expected to be in a treated area or in an area about to be treated with this product. Oral warnings should include those statements found under the label heading "Precautionary Statements" and "Re-entry Statements" included on the label. When oral warnings are given, warnings shall be given in a language customarily understood by workers. Oral warnings must be given if there is reason to believe that written warnings cannot be understood by workers. Written warnings must include the following information: DANGER, area treated with GUTHION 2L on (date of application). Do not enter treated areas for 24 hours after application unless protective clothing is worn as described in the product labeling. **If on skin,** remove contaminated clothing and wash skin immediately with soap and warm water. **If eyes are contaminated,** wash with flowing water for at least 15 minutes. **If swallowed,** vomiting should be induced. (See Precautionary Statements on label).

PRECAUTIONARY STATEMENTS

DANGER–POISON

HAZARDS TO HUMANS AND DOMESTIC ANIMALS

Poisonous if swallowed, inhaled or absorbed through the skin. Do not get in eyes or on skin. Do not breathe fumes or spray mist. Spray operator should work to windward to stay out of drift or mist.

When loading spray tank, or handling the concentrate at any other time, wear protective clothing, natural rubber gloves and goggles. When handling the concentrate, wear a pesticide respirator jointly approved by the Mining Enforcement and Safety Administration (formerly the U.S. Bureau of Mines) and by the National Institute for Occupational Safety and Health under the provisions of 30 CFR Part 11.

Do not contaminate feed or foodstuffs. Keep out of reach of children and domestic animals. Keep all unprotected persons out of the operating area or vicinity where there may be danger of drift. Wash hands, arms and face thoroughly with soap and warm water before eating or smoking.

HANDLING OF CONTAMINATED ARTICLES: Clothing and rags receiving undiluted GUTHION 2L as a result of leaks, spills or mishaps should be removed immediately, and not reused. These contaminated articles should be disposed of in the same manner as the empty container (see container disposal section for directions). Contaminated articles which should be discarded include: hats, gloves, aprons, coats, boots, etc. not made of rubber, or coated with rubber or other similar materials. Clothing receiving spray mist or droplets from the mixed, diluted GUTHION 2L should be removed, and decontaminated before reuse. This clothing can be decontaminated by machine-washing, separately from other items, with soap or detergent and bleach in hot water twice before reuse. Protective clothing and equipment should be washed-down with detergent or soap and bleach in water. Wash-water from the cleaning of protective clothing or equipment should not be allowed to run off, or otherwise enter water supplies.

SYMPTOMS OF POISONING: A sense of "tightness" in the chest. Sweating. Contracted pupils. Stomach pains. Vomiting and diarrhea.

STATEMENTS OF PRACTICAL TREATMENT

In case of poisoning, call a physician immediately. Have patient lie down and keep quiet.

If swallowed, vomiting should be induced. Administer water freely, and induce vomiting by giving one dose (1/2 oz. or 15 ml.) of syrup of ipecac. If vomiting does not occur within 10 to 20 minutes, administer second dose. If syrup of ipecac is not available, induce vomiting by sticking finger down throat. Repeat until vomit fluid is clear. Never give anything by mouth to an unconscious person. Professional medical assistance should be secured immediately.

If on skin, remove contaminated clothing, and wash skin immediately with soap and warm water.

If eyes are contaminated, wash with flowing water for at least 15 minutes.

TO PHYSICIAN: Antidote—Administer atropine sulfate in large therapeutic doses. Repeat as necessary, to the point of tolerance. 2-PAM is also antidotal, and may be administered in conjunction with atropine. Compound inhibits cholinesterase, resulting in stimulation of the central nervous system, the parasympathetic nervous system and the somatic motor nerves. Do not give morphine. Watch for pulmonary edema, which may develop in serious cases of poisoning, even after 12 hours. At first sign of pulmonary edema, the patient should be placed in an oxygen tent, and treated symptomatically.

ENVIRONMENTAL HAZARDS

Do not apply directly to water or wetlands (swamps, bogs, marshes, and potholes). Do not contaminate water by cleaning of equipment or disposal of wastes. Drift and runoff from treated areas may be hazardous to aquatic organisms in neighboring areas.

This product is highly toxic to bees exposed to direct treatment or residues on blooming crops or weeds. Do not apply this product or allow it to drift to blooming crops or weeds if bees are visiting the treatment area. Protective information may be obtained from your Cooperative Agricultural Extension Service.

PHYSICAL OR CHEMICAL HAZARDS

Keep away from open flame. Do not heat.

PROTECTIVE CLOTHING AND WORK SAFETY STATEMENTS

If handled indoors provide mechanical exhaust ventilation.

Keep all unprotected persons, children, livestock, and pets away from treated areas or where there is danger of drift.

Do not rub eyes or mouth with hands. If you feel sick in any way, STOP work and get help right away. See First Aid section of the label.

For questions regarding further treatment, or information about the product, call the National Pesticide Telecommunications Network at 800-858-7378.

HANDLE THE CONCENTRATE ONLY WHEN WEARING THE FOLLOWING PROTECTIVE CLOTHING AND EQUIPMENT:

Wear a protective suit of one or two pieces that covers all parts of the body except the head, hands, and feet. Wear chemical resistant gloves, chemical resistant apron, and chemical resistant shoes, shoe coverings, or boots. Wear goggles or a face shield and a pesticide respirator approved by the National Institute for Occupational Safety and Health under the provision of 30 CFR Part 11.

If handling the concentrate with a closed system, long sleeved shirt and long-legged pants may be substituted for the protective suit and the respirator requirement is waived.

WEAR THE FOLLOWING PROTECTIVE CLOTHING DURING APPLICATION, EQUIPMENT REPAIR, EQUIPMENT CLEANING, DURING REENTRY TO TREATED AREAS, AND DISPOSAL OF THE PESTICIDE:

Wear a protective suit of one or two pieces that covers all parts of the body except the head, hands, and feet. Wear chemical resistant gloves and chemical resistant shoes, shoe coverings, or boots. Wear a chemical resistant hat during airblast application.

During application from an enclosed tractor cab or airplane cockpit, or other suitable vehicle in which the windows are rolled up, long sleeved shirt and long pants may be worn in place of the above protective clothing. Chemical resistant gloves must be available in the cab or cockpit and must be worn while exiting. This clothing is inadequate to protect you during equipment repair, equipment cleaning, reentry, or during pesticide disposal.

IMPORTANT! BEFORE REMOVING GLOVES, WASH THEM WITH SOAP AND WATER. ALWAYS WASH HANDS, FACE, AND ARMS WITH SOAP AND WATER BEFORE SMOKING, DRINKING, EATING OR TOILETING.

After work, take off all clothing and shoes. Shower using soap and water. Do not wear contaminated clothing. Wash protective clothing and protective equipment with soap and water after each use. Respirators must be cleaned and filters replaced according to instructions included with the respirators. Personal clothing worn during use must be laundered separately from household articles. Clothing and protective equipment heavily contaminated or drenched with GUTHION azinphos-methyl must be destroyed according to state and local regulations. HEAVILY CONTAMINATED OR DRENCHED CLOTHING CANNOT BE ADEQUATELY DECONTAMINATED.

During aerial application, human flaggers are prohibited unless in totally enclosed vehicle.

STORAGE AND DISPOSAL

Pesticide Disposal: Do not contaminate water, food or feed by storage or disposal.

Pesticide wastes are acutely hazardous. Improper disposal of excess pesticide, spray mixture, or rinsate is a violation of Federal Law. If these wastes cannot be disposed of by use according to label instructions, contact your State Pesticide or Environmental Control Agency, or the Hazardous Waste representative at the nearest EPA Regional Office for guidance.

Container Disposal: Triple rinse (or equivalent). Then offer for recycling or reconditioning, or puncture and dispose of in a sanitary landfill, or by other procedures approved by State and local authorities.

Store in a cool, dry place and in such a manner as to prevent cross contamination with other pesticides, fertilizers, food, and feed. Keep away from open flame and extreme heat. Do not store below 45°F. Store in original container and out of the reach of children, preferably in a locked storage area.

Handle and open container in a manner as to prevent spillage. If container is leaking invert to prevent leakage. If the container is leaking or material spilled for any reason or cause, carefully dam up spilled material to prevent runoff. Refer to Precautionary Statements on label for hazards associated with the handling of this material. Do not walk through spilled material. Absorb spilled material with absorbing type compounds and dispose of as directed for pesticides above. In spill or leak incidents, keep unauthorized people away. You may contact the Mobay Emergency Response Team for decontamination procedures or any other assistance that may be necessary. The Mobay Kansas City Emergency Response Telephone No. is 816-242-2582, or contact Chemtrec at 800-424-9300.

IMPORTANT:

This bulletin is not intended to provide adequate information for application. Before using the product, read and carefully observe directions, cautionary statements and other information appearing on the product label. This product is sold subject to the Conditions of Sale set forth on the container label.

READ THE LABEL BEFORE USE

G 9004

This is a specimen label. Mobay is not responsible for the accuracy of the information contained herein. As labels are subject to revision, always carefully read and follow the label on the product container.

GUTHION®

35% WETTABLE POWDER CROP INSECTICIDE

FOR EFFECTIVE ECONOMICAL INSECT CONTROL

ACTIVE INGREDIENT:
O,O-Dimethyl *S*-(4-oxo-1,2,3-benzotriazin-3(4H)-yl)methyl phosphorodithioate 35%
INERT INGREDIENTS: 65%
TOTAL 100%

EPA Reg. No. 3125-378
GUTHION is a Reg. TM of Bayer AG, Germany

RECOMMENDED USES

Alfalfa***
Almonds
Apples
Apricots†‡
Artichokes
Barley
Beans
(snap and dry)
Birdsfoot trefoil***
Blackberries
Blackeyed peas
Blueberries**
Boysenberries
Broccoli
Brussels sprouts
Cabbage
Cantaloupe
Cauliflower
Celery
Cherries
Christmas Trees
Citrus fruits
Clover
Crab apples
Cranberries
Cucumbers
Eggplant
Filberts*
Grapes
Honeydew melons
Loganberries
Muskmelons
Nectarines†‡
Nursery plants
Oats
Onions
(green and dry)
Ornamentals
Other melons
Pasture
grasses***
Peaches†‡
Pears
Pecans
Peppers
Plums†‡
Potatoes
Prunes†‡
Quinces
Raspberries
Rye
Southern Pine
Seed Orchards
Soybeans
Spinach
Strawberries
Tomatoes
Walnuts
Watermelons
Wheat

* Pacific Northwest only
** Eastern and North Central states only
*** East of the Mississippi River
† Eastern U.S.
‡ West of the Rocky Mountains

RESTRICTIONS ON WAREHOUSE STORAGE

Store in a cool dry place. Protect from heat. Keep away from open flame. Do not heat.

PACKAGING

3.5 lb. carton, 10 per case. Shipping weight 43.5 lbs.

7 lb. carton, 4 per case.

Wettable powder formulations of GUTHION insecticide have proved effective through many years of extensive commercial use, both foreign and domestic. The 35% Wettable Powder formulation has excellent wetting and suspension properties.

FRUITS: GUTHION 35% WP controls major fruit pests. The use of GUTHION 35% WP greatly simplifies spray schedules. GUTHION 35% WP stays on the job, too. Its residual killing activity gives excellent and continuous control from one cover spray to the next . . . often eliminates "between-cover" applications. A combination of GUTHION 35% WP and dormant oil for delayed-dormant sprays on apples effectively controls early insect pests on this crop. All-season use of GUTHION 35% WP not only eliminates complicated spray schedules, but results in more profitable, finer-finish fruits.

VEGETABLES: Registered for use on many vegetables, GUTHION 35% WP has proved to be exceptionally effective in controlling diamondback moth, Colorado potato beetle and others. As with fruit, GUTHION 35% WP has prolonged residual action . . . does the job and stays on the job.

PELIGRO

PRECAUCION AL USUARIO: Si usted no puede leer o entender inglés, no use este producto hasta que la etiqueta le haya sido explicada ampliamente.

(TO THE USER: If you cannot read or understand English, do not use this product until the label has been fully explained to you.)

DIRECTIONS FOR USE

It is a violation of Federal Law to use this product in a manner inconsistent with its labeling.

MIXING: GUTHION 35% Wettable Powder insecticide mixes easily with water to form a suspension which is readily usable in all conventional ground spray equipment. Mix the required amount of GUTHION 35% Wettable Powder with a small quantity of water. Add this premix through the screen while filling the spray tank or fill the tank to the required level and add the premix. Operate the agitator while mixing.

DOSAGE: Use specified dosage of GUTHION 35% Wettable Powder in the amount of water necessary to give complete coverage of foliage. The type of equipment used will determine the concentration required.

COMPATIBILITY: GUTHION 35% Wettable Powder is physically compatible with many registered pesticides and liquid fertilizers. When considering mixing GUTHION 35% Wettable Powder with other pesticides or with liquid fertilizer, first contact your supplier. For further information, contact your local Mobay representative. If your supplier and Mobay representative have no experience with the combination you are considering, you should conduct tests to determine physical compatibility.

To determine physical compatibility, pour the recommended proportions of each chemical with the same proportion of water as will be present in the chemical supply tank into a suitable container, mix thoroughly and allow to stand for five minutes. If the combination remains mixed, or can be remixed readily, the mixture is considered physically compatible. When mixing wettable powder or dry flowable formulations, add and disperse these first, then add liquid pesticides. Conduct another compatibility test at concentrations which will be present in the irrigation lines. If there is any separation which cannot be remixed readily, Mobay recommends that the combination not be used.

Combinations should be kept agitated and should be applied immediately. Do not allow combinations to sit for prolonged periods in the chemical supply tank or irrigation lines.

SPRAYING: Work to windward. Protect sprayer operators from drift or mist. When low volumes of spray are applied complete coverage and thorough application are essential for most effective results. Schedule applications in accordance with local conditions. Consult your State Agricultural Experiment Station or Extension Service for specific use information in your area.

CHEMIGATION

GUTHION 35% Wettable Powder may be applied through recommended types of irrigation systems to many crops, The REMARKS section for each crop lists the types of applications allowed. If application through irrigation systems is not listed in the REMARKS section for a crop, GUTHION 35% Wettable Powder may not be applied to that crop through irrigation systems.

Do not connect an irrigation system (including greenhouse systems) used for pesticide application to a public water system unless the pesticide label prescribed safety devices for public water systems are in place.

Types of Irrigation Systems

Apply GUTHION 35% Wettable Powder only through sprinkler, including center pivot, lateral move, side roll, overhead solid set, or low pressure sprinkler irrigation systems. Do not apply GUTHION 35% Wettable Powder through any other types of irrigation systems.

GENERAL DIRECTIONS FOR ALL RECOMMENDED TYPES OF IRRIGATION SYSTEMS

Uniform Water Distribution and System Calibration

The irrigation system must provide uniform distribution of treated water. Crop injury, lack of effectiveness, or illegal pesticide residues in the crop can result from nonuniform distribution of treated water.

The system must be calibrated to uniformly apply the rates specified for chemigation application for specific crops. If you have questions about calibration, you should contact State Extension Service specialist, equipment manufacturers or other experts.

Chemigation Monitoring

A person knowledgeable of the chemigation system and responsible for its operations, or under the supervision of the responsible person, shall shut the system down and make necessary adjustments should the need arise.

Drift

Do not apply when wind speed favors drift beyond the area intended for treatment.

Required System Safety Devices

The system must contain a functional check valve, vacuum relief valve, and low pressure drain appropriately located on the irrigation pipeline to prevent water source contamination from backflow.

The pesticide injection pipeline must contain a functional, automatic, quick-closing check valve to prevent the flow of fluid back toward the injection pump.

The pesticide injection pipeline must also contain a functional, normally closed, solenoid-operated valve located on the intake side of the injection pump and connected to the system interlock to prevent fluid from being withdrawn from the supply tank when the irrigation system is either automatically or manually shut down.

The system must contain functional interlocking controls to automatically shut off the pesticide injection pump when the water pump motor stops.

The irrigation line or water pump must include a functional pressure switch which will stop the water pump motor when the water pressure decreases to the point when pesticide distribution is adversely affected.

Systems must use a metering pump, such as a positive displacement injection pump (e.g., diaphragm pump) effectively designed and constructed of materials that are compatible with pesticides and capable of being fitted with a system interlock.

Using Water from Public Water Systems

Public water system means a system for the provision to the public of piped water for human consumption if such system has at least 15 service connections or regularly services an average of at least 25 individuals daily at least 60 days out of the year.

DO NOT APPLY GUTHION 35% WETTABLE POWDER THROUGH ANY IRRIGATION SYSTEM SUPPLIED BY A PUBLIC WATER SYSTEM UNLESS the water from the public water system is discharged into a reservoir tank prior to pesticide introduction. There shall be a complete physical break (air gap) between the outlet end of the fill pipe and the top or overflow rim of the reservoir tank of at least twice the inside diameter of the fill pipe. Before beginning chemigation, always make sure that the air gap exists and that there is no blockage of the overflow of the reservoir tank.

Any irrigation system using water supplied from a public water system must also meet the following requirements:

The pesticide injection pipeline must contain a functional, automatic quick-closing check valve to prevent the flow of fluid back toward the injection pump.

The pesticide injection pipeline must contain a functional, normally closed, solenoid-operated valve located on the intake side of the injection pump and connected to the system interlock to prevent fluid from being withdrawn from the supply tank when the irrigation system is either automatically or manually shut down.

The system must contain functional interlocking controls to automatically shut off the pesticide injection pump when the water pump motor stops or in cases where there is no water pump, when the water pressure decreases to the point where pesticide distribution is adversely affected.

Systems must use a metering pump, such as a positive displacement injection pump (e.g., diaphragm pump) effectively designed and constructed of materials that are compatible with pesticides and capable of being fitted with a system interlock.

Do not apply when wind speed favors drift beyond the area intended for treatment.

Posting Requirements

Posting of areas to be chemigated is required when 1) any part of a treated area is within 300 feet of sensitive areas such as residential areas, labor camps, businesses, day care centers, hospitals, in-patient clinics, nursing homes or any public areas such as schools, parks, playgrounds, or other public facilities not including public roads, or 2) when the chemigated area is open to the public such as golf courses or retail greenhouses.

Posting must conform to the following requirements: Treated areas shall be posted with signs at all usual points of entry and along likely routes of approach from the listed sensitive areas. When there are no usual points of entry, signs must be posted in the corners of the treated areas and in any other location affording maximum visibility to sensitive areas. The printed side of the sign should face away from the treated area towards the sensitive areas. The signs shall be printed in English. Signs must be posted prior to application and must remain posted until foliage has dried and soil surface water has disappeared. Signs may remain in place indefinitely as long as they are composed of materials to prevent deterioration and maintain legibility for the duration of the posting period.

All words shall consist of letters at least $2\frac{1}{2}$ inches tall, and all letters and the symbol shall be a color which sharply contrasts with their immediate background. At the top of the sign shall be the words KEEP OUT, followed by an octagonal stop sign symbol at least 8 inches in diameter containing the word STOP. Below the symbol shall be the words PESTICIDES IN IRRIGATION WATER.

Posting required for chemigation does not replace other posting and reentry interval requirements for farmworker safety.

Compatibility

When mixing with other chemicals refer to Compatibility section elsewhere on the label.

Agitation

For application of GUTHION 35% Wettable Powder, a chemical supply tank is necessary for premixing. For application of GUTHION 35% Wettable Powder alone or in combination with liquid fertilizer or other chemicals, constant strong mechanical or hydraulic agitation must be maintained in the chemical supply tank during the entire period of application.

Chemical Supply Tank Dilution

You must determine the required amounts of GUTHION 35% Wettable Powder to mix in the tank.

The amount of GUTHION 35% Wettable Powder needed equals the number of pounds of GUTHION 35% Wettable Powder to be applied per acre multiplied by the number of acres to be chemigated.

The amount of solution needed equals the gallons of solution delivered per hour by the injection pump multiplied by the number of hours chemigation will take place.

For example, if you want to apply 2 pounds of GUTHION 35% Wettable Powder per acre to 130 acres in 20 hours and your injection pump delivers 15 gallons per hour, you need: 2 pounds GUTHION 35% Wettable Powder per acre $\times$ 130 acres $=$ 260 pounds of GUTHION 35% Wettable Powder. And, you need: 15 gallons per hour $\times$ 20 hours $=$ 300 gallons of suspension $=$ 300 gallons of water.

Cleaning the Chemical Injection System

In order to accurately apply pesticides, the chemical injection system must be kept clean; free of chemical or fertilizer residues and sediments. Refer to your owners manual or ask your equipment supplier for cleaning procedure for your injection system.

Flushing the Irrigation System

At the end of the application period, allow time for all lines to flush the pesticide through all nozzles or emitters before turning off irrigation water. To ensure the lines are flushed and free of pesticides, a dye indicator may be injected into the lines to mark the end of the application period.

Equipment Area Contamination

It is recommended that nozzles in the immediate area of control panels, chemical supply tanks, pumps and system safety devices be plugged to prevent chemical contamination of these areas.

Center-Pivot and Automatic-Move Linear Systems

Inject the specified dosage per acre continuously for one complete revolution or move of the system. DO NOT USE END GUNS. For a foliar application, the system should be run at maximum speed.

Solid Set and Manually Controlled Linear Systems

For foliar application, injection should be during the last 30 to 60 minutes of regular irrigation period or as a separate 30 to 60 minute application not associated with a regular irrigation. DO NOT USE END GUNS.

RECOMMENDED APPLICATIONS

CROP	PEST	POUNDS GUTHION 35% WP	REMARKS
FIELD CROPS Alfalfa Birdsfoot trefoil (East of the Mississippi River Only) Clover	Leafhoppers	$\frac{11}{16}$ to $1\frac{7}{16}$	Apply specified dosage per acre by air or ground equipment. Use at least 10 gallons of water per acre with ground equipment and at least 1 gallon of water per acre for aerial application. It may be necessary to use 20 to 25 gallons of water per acre on heavy growth for control of alfalfa weevil or Egyptian alfalfa weevil with ground equipment. For application by irrigation systems: Apply specified dosage per acre. Follow all directions given under the CHEMIGATION section of the label. Only one application per cutting may be made regardless of rate, formulation or method of application used. Do not apply within 14 days of harvest at the rates of $\frac{11}{16}$ to $1\frac{1}{16}$ pounds per acre or within 16 days of harvest at the rate of $1\frac{7}{16}$ pounds per acre.
	Alfalfa weevil Egyptian alfalfa weevil	$1\frac{1}{16}$ to $1\frac{7}{16}$	
	Alfalfa plant bug Fleahopper Grasshoppers Lygus bugs Spittlebugs	$1\frac{7}{16}$	
Pasture Grasses (States east of the Mississippi River)	Grasshoppers Meadow spittlebugs (adults)	$1\frac{3}{8}$ to $2\frac{1}{8}$	Apply specified dosage per acre using a minimum of 10 gallons of water per acre. Do not apply more than once per cutting. Do not graze or harvest within 16 days of application at the $1\frac{3}{8}$ pound rate or 21 days at rates above $1\frac{3}{8}$ pound.

CROP	PEST	POUNDS GUTHION 35% WP	REMARKS
Barley Oats Rye Wheat	Cereal leaf beetle	1 to $1\frac{3}{8}$	Apply specified dosage per acre in sufficient water for complete coverage. Do not apply more than once per season. Do not harvest for food, feed, forage or graze within 30 days of treatment.
Soybeans	Aphids Bean leaf beetle Green cloverworm Leafhoppers Leaf miners Leaf rollers Stink bugs Velvet bean caterpillar	1 to $1\frac{3}{8}$	Apply specified dosage per acre in sufficient water to give complete coverage. Do not apply within 45 days of harvest. Do not graze or feed treated vines to livestock. Do not exceed more than 3 applications per season.
	Mexican bean beetle	$1\frac{3}{8}$ to $2\frac{1}{8}$	
FRUIT	**NOTE:** It is suggested that when treating fruit during the bloom period, bee keepers should be warned well in advance to remove hives a safe distance from orchards to be treated.		
Apples Crab apples Pears	Aphids* Apple maggot Codling moth European apple sawfly Eye-spotted bud moth Forbes scale Fruit tree leaf roller Green fruitworm* Leafhoppers* Mealybug* Orange tortrix Pear midge Plum curculio Putnam scale Redbanded leaf roller San Jose scale* Stink bug Tarnished plant bug*	$2\frac{13}{16}$ to $4\frac{1}{4}$ ($\frac{11}{16}$ to $1\frac{1}{16}$ pounds per 100 gallons)	Apply specified dosage per acre by air or ground equipment in sufficient water for good coverage. Up to $17\frac{1}{8}$ pounds GUTHION 35% WP may be applied per crop season. Allow at least 7 days between applications and 7 days between last application and harvest. GUTHION 35% WP is compatible with dormant and summer oils which may be added to apple and pear sprays in accordance with local recommendations.
Cherries	Eye-spotted bud moth Forbes scale Fruit flies Fruit tree leaf roller Cherry leaf miner Mineola moth Lesser peach tree borer Plum curculio San Jose scale*	$2\frac{1}{8}$ ($\frac{11}{16}$ pound per 100 gallons)	Apply specified dosage per acre as a full coverage spray. Up to $8\frac{1}{2}$ pounds of GUTHION 35% WP may be applied per acre per crop season. Allow at least 14 days between applications and 15 days between last application and harvest. **California:** Apply post-harvest only.
Apricots Nectarines Peaches (Eastern U.S.)	Aphids* Cottony peach scale European fruit lecanium scale Forbes scale Lesser peach tree borer Oriental fruit moth Peach tree borer Peach twig borer *Platynota flavendana* leaf roller Plum curculio Redbanded leaf roller San Jose scale* Stink bug Thrips Tarnished plant bug Terrapin scale Walnut scale White peach scale	$2\frac{1}{2}$ to $3\frac{3}{16}$ ($\frac{11}{16}$ to $\frac{7}{8}$ pound per 100 gallons)	Apply specified dosage per acre as a full coverage spray. Allow at least 14 days between applications. A total of $9\frac{5}{8}$ pounds per acre per crop season may be applied to nectarines. A total of $12\frac{13}{16}$ pounds per crop season may be applied to apricots and peaches. Allow at least 21 days between last application and harvest. For control of peach tree borer, apply 2 or 3 sprays to trunk from ground to scaffold limbs, timed with moth flight. For control of scale, apply when crawlers are present. GUTHION 35% WP is compatible with dormant and summer oils which may be added to peach sprays in accordance with local recommendations.
Apricots Nectarines Peaches (West of the Rocky Mountains)	Lesser peach tree borer Oriental fruit moth Peach twig borer Peach tree borer *Platynota flavendana* leaf roller Plum curculio Redbanded leaf roller Stink bug Thrips Tarnished plant bug	$4\frac{1}{4}$ to $5\frac{11}{16}$	Apply specified dosage per acre. Allow at least 14 days between applications. A total of $9\frac{5}{8}$ pounds per acre per crop season may be applied to nectarines. A total of $12\frac{13}{16}$ pounds per acre per crop season may be applied to apricots and peaches. Allow at least 21 days between last application and harvest. For control of peach tree borer, apply 2 or 3 sprays to trunk from ground to scaffold limbs, timed with moth flight.
Plums Prunes (Eastern U.S.)	Aphids* Codling moth Eyespotted bud moth Forbes scale Fruit tree leaf roller Lesser peach tree borer Orange tortrix Peach tree borer Peach twig borer Plum curculio Red-banded leaf roller San Jose scale* Stink bug Tarnished plant bug* Tussock moth	$2\frac{1}{2}$ to $3\frac{9}{16}$ ($\frac{11}{16}$ to $\frac{7}{8}$ pound per 100 gallons)	Apply specified dosage per acre as a full coverage spray. A total of $9\frac{5}{8}$ pounds per acre per crop season may be applied to plums and prunes. Allow at least 10 days between applications. Allow at least 15 days between last application and harvest. For control of peach tree borer, apply 2 or 3 sprays to trunk from ground to scaffold limbs, timed with moth flight. For control of scale, apply when crawlers are present. GUTHION 35% WP is compatible with dormant and summer oils which may be added to prune sprays in accordance with local recommendations.
	American plum borer	$4\frac{1}{4}$	

CROP	PEST	POUNDS GUTHION 35% WP	REMARKS
Plums Prunes (West of the Rocky Mountains)	Codling moth Eyespotted bud moth Fruit tree leaf roller Lesser peach tree borer Orange tortrix Peach tree borer Peach twig borer Plum curculio Redbanded leaf roller Stink bug Tarnished plant bug* Tussock moth	$2\frac{13}{16}$ to $5\frac{11}{16}$	Apply specified dosage per acre. A total of $9\frac{5}{8}$ pounds per acre per crop season may be applied to plums and prunes. Allow at least 10 days between applications. Allow at least 15 days between last application and harvest. For control of peach tree borer, apply 2 or 3 sprays to trunk from ground to scaffold limbs, timed with moth flight.
	American plum borer	$4\frac{1}{4}$ to $5\frac{11}{16}$	
Blackberries Boysenberries Loganberries Raspberries	Leafhoppers Leaf rollers	$\frac{5}{8}$	Apply specified dosage per acre to foliage using approximately 200 gallons of water for good coverage. Do not apply more than twice per season nor within 14 days of harvest.
	Leaf miners	$\frac{7}{8}$	
	Aphids	$\frac{7}{8}$ to $1\frac{3}{8}$	
	Obscure root weevil	$1\frac{3}{8}$	
	Obscure root weevil	$1\frac{3}{8}$	For control of root weevils prior to harvest apply specified dosage per acre to lower portion of canes and to the soil beneath the plants using approximately 200 gallons of water. Do not apply more than twice per season. Do not make applications within 3 days of harvest at rates up to $2\frac{7}{8}$ pounds per acre.
Blueberries (Eastern and North Central States only)	Blueberry maggot Fruitworms Lecanium scale Plum curculio	$1\frac{7}{16}$ to $2\frac{1}{8}$	Apply specified dosage per acre with aerial or ground equipment using sufficient water for good coverage. A total of 3 applications may be made per crop season regardless of rate or formulation used. Allow at least 10 days between applications and at least 7 days between last application and harvest. Do not apply later than June 15.
Cranberries	Cranberry fruitworm *Sparganothis sulfureana* Tipworm	$1\frac{7}{16}$ to $2\frac{13}{16}$	Apply specified dosage per acre by air using sufficient water for good coverage. For application by irrigation systems: Apply specified dosage per acre. Follow all directions given under the CHEMIGATION section of the label. A total of 3 applications may be made per crop season regardless of rate or formulation used. Allow at least 14 days between applications and at least 21 days between last application and harvest.
	Fireworms	$2\frac{13}{16}$	
Citrus Fruits	Aphids* Black scale Brown soft scale Chaff scale Citricola scale Citrus mealybug Citrus thrips Cottony-cushion scale European brown snail Florida red scale Fruit tree leaf rollers Fuller rose beetle Glover scale Orange tortrix Purple scale Snow scale Western tussock moth Whiteflies	$5\frac{11}{16}$ to $8\frac{1}{2}$	Apply specified dosage per acre as a full coverage spray. A single application per year may be applied up to within 7 days of harvest. Where 2 applications are required, allow at least 28 days between the second applicaiton and harvest. A total of 2 applications may be applied per fruit year regardless of rate or formulation used. Do not pick fruit or do other work involving contact with the tree (such as pruning) within 7 days of treatment.
	California red scale Yellow scale	$8\frac{1}{2}$	
Grapes	Grape berry moth Grape cane girdlers Grape mealybug Leafhoppers* Red-banded leaf roller Thrips Grape leaf skeletonizer	$2\frac{1}{8}$ to $2\frac{13}{16}$	Apply specified dosage per acre as a full coverage spray. A total of 3 applications may be made per crop season regardless of rate or formulation used. Allow at least 14 days between applications. Minimum dosage specified may be applied up to harvest. With higher rates allow at least 10 days between last application and harvest.
Strawberries	Aphids* Meadow spittlebug Oblique-banded leaf roller Obscure root weevil Omnivorous leaf tier Pea leaf weevil Small black (grass) weevil Strawberry leaf rollers Whitefly	$1\frac{7}{16}$	Apply specified dosage per acre as a full coverage spray using sufficient water for good coverage. For application by irrigation systems: Apply specified dosage per acre. Follow all directions given under the CHEMIGATION section of the label. A total of 4 applications may be made per crop season regardless of rate or formulation used. Allow at least 5 days between applications and at least 5 days between last application and harvest.
NUTS Almonds	Peach twig borer Navel orangeworm	$4\frac{1}{4}$ to $5\frac{11}{16}$	Apply specified dosage per acre.

CROP	PEST	POUNDS GUTHION 35% WP	REMARKS
			A total of 2 applications may be made per crop season regardless of rate or formulation used. Allow at least 30 days between applications and at least 60 days between last application and harvest. Do not apply after husks split.
Filberts (Pacific Northwest only)	Apple mealybug Filbertworm Filbert aphid Filbert leaf roller	$4\frac{1}{4}$ to $5\frac{11}{16}$	Apply specified dosage per acre as a full-coverage spray. A total of 3 applications may be made per crop season regardless of rate or formulation used. Allow at least 14 days between applications and at least 30 days between last application and harvest. Do not graze livestock in treated groves for 21 days after treatment.
Pecans	Aphids* Fall webworm Hickory shuckworm Leaf miners May beetles Pecan casebearer Southern green stink bug Spittlebugs Twig girdlers Walnut caterpillars	$4\frac{1}{4}$ to $5\frac{11}{16}$	Apply specified dosage per acre as a full coverage spray. A total of 3 applications may be made per crop season regardless of rate or formulation used. Allow at least 7 days between applications. Do not apply after shuck split. Livestock may be grazed in treated groves after a 21-day post-treatment interval.
Walnuts	Codling moth Filbertworm Navel orangeworm Walnut husk fly Red-humped caterpillar	$4\frac{1}{4}$ to $5\frac{11}{16}$	Apply specified dosage per acre as a full coverage spray. A total of 3 applications may be made per crop season regardless of rate or formulation used. Allow at least 14 days between applications. Do not apply after husks split. Do not graze livestock in treated groves for 21 days after treatment.
VEGETABLES Artichokes	Plume moth	$4\frac{1}{4}$	Apply specified dosage per acre in sufficient water for complete coverage. For application by irrigation systems: Apply specified dosage per acre. Follow all directions given under the CHEMIGATION section of the label. A total of 3 applications may be made per crop regardless of rate, formulation or method of application used. Allow at least 14 days between applications and at least 30 days between last application and harvest. Do not feed or ensile treated forage.
Beans (Dry & Snap)	Leaf rollers Mites	$\frac{5}{8}$ to $1\frac{3}{8}$	Apply specified dosage per acre in sufficient water for complete coverage. Do not apply to snap beans within 7 days of harvest, nor to dry beans within 30 days of harvest. Do not exceed 4 applications. Do not feed or ensile treated forage.
	Aphids Bean leaf beetle Green cloverworm Leafhoppers Leaf miners Stink bugs Velvet bean caterpillar	1 to $1\frac{3}{8}$	
	Mexican bean beetle Spotted cucumber beetle Striped cucumber beetle Tarnished plant bug Western-striped cucumber beetle	$1\frac{3}{8}$	
Broccoli Brussels sprouts Cabbage (Includes tight heading varieties of Chinese cabbage) Cauliflower	Aphids Cabbage looper Diamondback moth Imported cabbageworm	$1\frac{3}{8}$ to $2\frac{1}{8}$	Apply specified dosage per acre in sufficient water for complete coverage. Do not apply within 7 days of harvest for Brussels sprouts, 15 days of harvest for broccoli and cauliflower, nor within 21 days of harvest for cabbage. Do not exceed more than 3 applications per season.
	Cabbage maggot	$\frac{1}{4}$ to $\frac{1}{2}$	Mix specified dosage in 50 gallons of water. Apply 4 to 6 ounces of this emulsion per plant at or immediately after transplanting.
	Cabbage maggot (Direct-seeded fields in California only)	$2\frac{1}{8}$	Apply specified dosage per acre in sufficient water for uniform distribution. Mix in upper 2 inches of soil prior to seeding, or spray in the seed row at planting time. Usually, 2 additional sprays are necessary during the growing season, depending on time of year and maggot population. Do not apply within 21 days of harvest. Do not exceed more than 3 applications per season.

CROP	PEST	POUNDS GUTHION 35% WP	REMARKS
	Cabbage maggot (Transplant fields in California only)	2 1/8	Apply specified dosage in 300 to 400 gallons of water per acre as a soil drench in rows when damage first appears. Additional applications may be necessary. Do not apply within 21 days of harvest. Do not exceed more than 3 applications per season.
Blackeyed peas (Southern peas, Crowder peas)	Corn earworm Cowpea curculio	2 1/8 to 2 7/8	Apply specified dosage per acre in sufficient water for complete coverage. Do not apply more than 4 times per season nor within 7 days of harvest. Do not use vines for feed or forage nor pasture treated areas.
	Leaf miners Stink bugs	1 to 1 3/8	
Celery	Aphids Leaf miners Leafhoppers Spittlebug Tarnished plant bug	1 3/8	Apply specified dosage in 100 gallons of water as a full-coverage spray** using not more than 200 gallons of finished spray per acre. Do not apply within 14 days of harvest. Do not exceed more than 3 applications per season.
Cucumbers	Spotted cucumber beetle Striped cucumber beetle Western-striped cucumber beetle	1 7/16	Apply specified dosage per acre in sufficient water to give complete coverage. For application by irrigation systems: Apply specified dosage per acre. Follow all directions given under the CHEMIGATION section of the label. A total of 3 applications may be made per crop season regardless of rate, formulation or method of application used. Allow at least 7 days between applications and at least 1 day between last application and harvest.
Eggplant	Leaf miners	1 1/16 to 1 7/16	Apply specified dosage per acre in sufficient water to give complete coverage. For application by irrigation systems: Apply specified dosage per acre. Follow all directions given under the CHEMIGATION section of the label. A total of 3 applications may be made per crop season regardless of rate, formulation or method of application used. Allow at least 7 days between applications. Do not apply after fruit set.
	European corn borer Flea beetles	1 7/16	
Onions (Green & Dry)	Thrips	1 7/16 to 2 1/8	Apply specified dosage per acre by air or ground equipment in sufficient water to give complete coverage. For application by irrigation systems: Apply specified dosage per acre. Follow all directions given under the CHEMIGATION section of the label. A total of 3 applications may be made per crop season regardless of rate, formulation or method of application used. Allow at least 7 days between applications for dry bulb onions. Allow at least 10 days between applications for green onions. For dry onions, allow at least 28 days between last application and harvest. For green onions, allow at least 7 days between last application and harvest.
Peppers	Leaf miners	1 1/16 to 1 7/16	Apply specified dosage per acre by air or ground equipment in sufficient water to give complete coverage. For application by irrigation systems: Apply specified dosage per acre. Follow all directions given under the CHEMIGATION section of the label. A total of 3 applications may be made per crop season regardless of rate, formulation or method of application used. Allow at least 7 days between applications and 7 days between last application and harvest.
	European corn borer Flea beetles	1 7/16	
Potatoes	Colorado potato beetle	1 1/16	Apply specified dosage per acre in sufficient water for complete coverage. For application by irrigation systems: Apply specified dosage per acre. Follow all directions given under the CHEMIGATION section of the label. A total of 3 applications may be made per crop season regardless of rate, formulation or method of application used. Allow at least 7 days between applications and 7 days between application and harvest. **NOTE:** Resistance of Colorado potato beetle has occurred in some areas. Consult your local Extension Service or MOBAY sales representative for details.
	Banded cucumber beetle Leaf miners	1 1/16 to 1 7/16	
	European corn borer Flea beetle Leafhoppers Spittlebugs Tarnished plant bug	1 7/16 to 2 1/8	
	Tuberworm	1 9/16 to 2 1/8	
Spinach	Aphids Leaf miners Mites	1 to 1 3/8	Apply specified dosage per acre in sufficient water for complete coverage. Do not apply within 14 days of harvest. Do not exceed more than 3 applications per season.
Tomatoes	Colorado potato beetle	1 1/16	
	Banded cucumber beetle *Drosophila* Green stink bug Leaf miners Whitefly	1 1/16 to 1 7/16	Apply specified dosage per acre by air or ground equipment in sufficient water for complete coverage. For application by irrigation systems: Apply specified dosage per acre. Follow all directions given under the CHEMIGATION section of the label. A total of 4 applications may be made per crop season regardless of rate, formulation or method of application used. The high rates should be used where heavy infestations of late instar lepidopterous larvae (large worms) and pinworms are present. Rates of 2 1/8 pounds per acre or less may be applied up to day of harvest. Rates above 2 1/8 pounds per acre require an interval of 14 days between application and harvest.

CROP	PEST	POUNDS GUTHION 35% WP	REMARKS
			NOTE: Resistance of Colorado potato beetle has occurred in some areas. Consult your local Extension Service or MOBAY sales representative for details.
	Aphids*, Grasshoppers, European corn borer, Leafhoppers, Flea beetles, Thrips	1 7/16 to 2 1/8	
	Tuberworm	1 9/16 to 2 1/8	
	Fruitworm, Pinworm, Hornworm, Yellow-striped armyworm	2 1/8 to 4 1/4	
MELONS Honeydew melons Muskmelon (cantaloupe) Watermelons Other melons	Leafhoppers Leaf miners	1 1/16 to 1 7/16	Apply specified dosage per acre in sufficient water to give complete coverage. For application by irrigation systems: Apply specified dosage per acre. Follow all directions given under the CHEMIGATION section of the label. A total of 3 applications may be made per crop season regardless of rate, formulation or method of application used.
	Rindworms, Striped cucumber beetle, Spotted cucumber beetle, Western-striped cucumber beetle	1 7/16	Allow at least 5 days between applications and at least 7 days between last application and harvest. 8-8-88
ORNAMENTALS Southern Pine Seed Orchards	Coneworm Seedworm	See Remarks	Use 3.25 pounds per 100 gallons of water (0.2% dilution) as a full coverage spray**. Use 1.6 pounds per 10 gallons of water (1% dilution) for low volume sprayers. Apply first application within 30 days following conelet closure followed by 3 to 5 applications (at least 30 days apart). Apply approximately 5 to 10 gallons of the 0.2% dilution with high volume sprayers or approximately 1 to 2 gallons of the 1% dilution with low volume sprayers per tree. Thorough coverage of cones is necessary for maximum contol.
Ornamentals Nursery Plants	Aphids *Cerococcus* scale Euonymus scale Juniper scale Lace bugs Leafhoppers Mites Olive scale Oystershell scale *Pulvinaria* scale Thrips	1 to 1 3/8	Apply specified dosage per 100 gallons of water (1 1/3teaspoonsuls per gallon). Spray all foliage surfaces including the underside of leaves for complete coverage. For control of black pine leaf, brown soft, European elm, and Putnam scales, use the higher rate. Repeat as necessary.
	Brown soft scale Putnam scale	2 3/4	
	European elm scale Black pine leaf scale	2 1/8 to 2 3/4	
	Cone midge Cone moth	5 1/2 to 11	Apply specified dosage per 100 gallons of water. Time applications to coincide with moth flight when cones are open for pollination. Thorough coverage of cones is necessary for maximum control. Repeat as necessary.
	European pine shoot moth Nantucket pine tip moth	1 to 2 1/8	Apply specified dosage per acre in sufficient water for good coverage. Time application to coincide with moth flights.
	Injury to hawthorn or American linden may occur under some conditions. Do not allow children or pets on treated area until material has been washed into soil and treated area is dry.		
Christmas Trees	Scale spp. Sawfly spp. European pine shoot moth Eastern pine shoot borer Nantucket pine tip moth	1 to 2 1/8	Apply specified dosage per acre by air or groundequipment in sufficient water to give complete coverage but not less than 1 gallon per acre. Time applications to coincide with susceptible pest development. 11-29-88

*In some areas, these species may have developed resistance to organophosphate insecticides. GUTHION insecticide used alone may not provide satisfactory control in those areas. Consult your local agricultural advisor or cooperative extension service for recommendations.

**This concentration is calculated for conventional, hydraulic-type sprayers. When lower volumes of spray are applied per acre with concentrate sprayers, increase the concentration of GUTHION 35% Wettable Powder in the spray mixture in order to apply amount of GUTHION 35% Wettable Powder per acre equivalent to a full coverage spray. In Southern Pine Seed Orchards where conditions dictate an air application, apply dosage per acre equivalent to a full coverage ground spray in not less than one gallon of water per acre.

* In some areas, these species may have developed resistance to organophosphate insecticides. GUTHION insecticide used alone may not provide satisfactory control in those areas. Consult your local agricultural advisor or cooperative extension service for recommendations.

** This concentration is calculated for conventional hydraulic-type sprayers. When lower volumes of spray are applied per acre with concentrate sprayers, the concentration of GUTHION 35% Wettable Powder in the spray mixture must be increased in order to apply the amount of GUTHION 35% Wettable Powder per acre equivalent to a full coverage spray.

Do not allow children or pets in treated area until material is dry.

SPECIAL LOCAL NEED REGISTRATIONS [SECTION 24(c)] FOR GUTHION 35% WETTABLE POWDER*

CROP	INSECT	STATE EPA SLN NO.	DOSAGE	REMARKS
Parsley (Root and moss-curled)	Carrot weevil *Listronotus oregonensis* (Liconte)	New Jersey NJ 89-0008	1 7/16 lbs per acre	A total of 3 applications may be made by ground equipment. Do not apply within 21 days of harvest.
Southern pine seed orchards	Coneworm Seedworm	Mississippi MS88-0004	6.4 lbs. per 100 gals. of water for conventional sprayers or 3.2 lbs. per 10 gals. of water for low volume sprayers	First application within 30 days following conelet closure, followed by 3 to 5 applications at least 30 days apart.

* For distribution and use only within the state specified. These Special Local Need registrations approved as of date of this publication are, however, subject to cancellation by the state. Contact your nearest Mobay Agricultural Chemicals Division Office for the current status and a copy of the complete individual supplement Special Local Need labeling. This labeling must be in the possession of the user at time of pesticide application.

RESTRICTIONS

Do not use on other crops used for food or forage. Use only according to label directions. Application at rates above those shown may result in illegal crop residues. Do not graze livestock in treated orchards or groves for 21 days after treatment. Do not treat food crops grown in the greenhouse.

RE-ENTRY STATEMENTS FOR FARM WORKERS

Do not apply this product in such a manner as to directly or through drift expose workers or other persons. The area being treated must be vacated by unprotected persons. Do not enter treated areas for 24 hours after application unless protective clothing is worn as described in the product labeling. Because certain states may require more restrictive re-entry intervals for various crops treated with this product, consult your State Department of Agriculture for further information.

Written or oral warnings must be given to workers who are expected to be in a treated area or in an area about to be treated with this product. Oral warnings should include those statements found under the label heading "Precautionary Statements" and "Re-entry Statements" included on the label. When oral warnings are given, warnings shall be given in a language customarily understood by workers. Oral warnings must be given if there is reason to believe that written warnings cannot be understood by workers. Written warnings must include the following information: DANGER, area treated with GUTHION 35% Wettable Powder on (date of application). Do not enter treated areas for 24 hours after application unless protective clothing is worn, as described in the product labeling. **If on skin,** remove contaminated clothing and wash skin immediately with soap and warm water. **If eyes are contaminated,** wash with flowing water for at least 15 minutes. **If swallowed,** vomiting should be induced. (See Precautionary Statements on label.)

ROTATIONAL CROPS

Do not plant root crops other than those with registered azinphos-methyl uses in azinphos-methyl treated soil sooner than 6 months after the last application. Do not plant any other crop other than those with registered azinphos-methyl uses in treated soil sooner than 30 days after last application.

PRECAUTIONARY STATEMENTS

HAZARDS TO HUMANS AND DOMESTIC ANIMALS

DANGER–POISON

Fatal if swallowed. Harmful if inhaled or absorbed through the skin. Avoid breathing dust or spray mist. Avoid contact with skin, eyes, or clothing. Spray operator should work to windward to stay out of drift or mist.

When loading spray tank, or handling the powder at any other time, wear protective clothing, natural rubber gloves and goggles. For protection against GUTHION, when handling the powder, wear a pesticide mask or respirator jointly approved by the Mining Enforcement and Safety Administration (formerly the U.S. Bureau of Mines) and by the National Institute for Occupational Safety and Health under the provisions of 30 CFR Part 11.

Do not contaminate feed or foodstuffs. Keep out of reach of children and domestic animals. Keep all unprotected persons out of the operating area or vicinity where there may be danger of drift. Wash thoroughly with soap and warm water after handling. Remove and wash all contaminated clothing with soap and hot water before reuse.

SYMPTOMS OF POISONING: A sense of "tightness" in the chest. Sweating. Contracted pupils. Stomach pains. Vomiting and diarrhea.

STATEMENTS OF PRACTICAL TREATMENT

In case of poisoning, call a physician immediately. Have patient lie down and keep quiet.

If swallowed, vomiting should be induced. Administer water freely and induce vomiting by giving one dose (1/2 oz. or 15 ml.) of syrup of ipecac. If vomiting does not occur within 10 to 20 minutes, administer second dose. If syrup of ipecac is not available, induce vomiting by sticking finger down throat. Repeat until vomit fluid is clear. Never give anything by mouth to an unconscious person. Professional medical assistance should be secured immediately.

If on skin, remove contaminated clothing and wash skin immediately with soap and warm water.

If eyes are contaminated, wash with flowing water for at least 15 minutes.

If inhaled, remove victim to fresh air. If not breathing, give artificial respiration, preferably mouth-to-mouth. Get medical attention.

TO PHYSICIAN: ANTIDOTE—Administer atropine sulfate in large therapeutic doses. Repeat as necessary to the point of tolerance. 2-PAM is also antidotal and may be administered in conjunction with atropine. Compound inhibits cholinesterase resulting in stimulation of the central nervous system, the parasympathetic nervous system, and the somatic motor nerves. Do not give morphine. Watch for pulmonary edema, which may develop in serious cases of poisoning even after 12 hours. At first sign of pulmonary edema, the patient should be placed in an oxygen tent and treated symptomatically.

ENVIRONMENTAL HAZARDS

This pesticide is extremely toxic to fish and wildlife. Do not apply directly to water or wetlands (swamps, bogs, marshes, and potholes). Do not contaminate water by cleaning of equipment or disposal of wastes. Drift and runoff from treated areas may be hazardous to aquatic organisms in neighboring areas.

This product is highly toxic to bees exposed to direct treatment or residues on crops. Protective information may be obtained from your Cooperative Agricultural Extension Service.

PROTECTIVE CLOTHING AND WORK SAFETY STATEMENTS

If handled indoors provide mechanical exhaust ventilation.

Keep all unprotected persons, children, livestock, and pets away from treated areas or where there is danger of drift.

Do not rub eyes or mouth with hands. If you feel sick in any way, STOP work and get help right away. See First Aid section of the label. For questions regarding further treatment, or information about the product, call the National Pesticide Telecommunications Network at 800-858-7378.

HANDLE THE CONCENTRATE ONLY WHEN WEARING THE FOLLOWING PROTECTIVE CLOTHING AND EQUIPMENT:

Wear a protective suit of one or two pieces that covers all parts of the body except the head, hands, and feet. Wear chemical resistant gloves, chemical resistant apron, and chemical resistant shoes, shoe coverings, or boots. Wear goggles or a face shield and a pesticide respirator approved by the National Institute for Occupational Safety and Health under the provision of 30 CFR Part 11.

WEAR THE FOLLOWING PROTECTIVE CLOTHING DURING APPLICATION, EQUIPMENT REPAIR, EQUIPMENT CLEANING, DURING RE-ENTRY TO TREATED AREAS, AND DISPOSAL OF THE PESTICIDE:

Wear a protective suit of one or two pieces that covers all parts of the body except the head, hands, and feet. Wear chemical resistant gloves and chemical resistant shoes, shoe coverings, or boots. Wear a chemical resistant hat during airblast application.

During application from an enclosed tractor cab or airplane cockpit, or other suitable vehicle in which the windows are rolled up, long sleeved shirt and long pants may be worn in place of the above protective clothing. Chemical resistant gloves must be available in the cab or cockpit and must be worn while exiting. This clothing is inadequate to protect you during equipment repair, equipment cleaning, re-entry, or during pesticide disposal.

IMPORTANT! BEFORE REMOVING GLOVES, WASH THEM WITH SOAP AND WATER. ALWAYS WASH HANDS, FACE, AND ARMS WITH SOAP AND WATER BEFORE SMOKING, DRINKING, EATING OR TOILETING.

After work, take off all clothing and shoes. Shower using soap and water. Do not wear contaminated clothing. Wash protective clothing and protective equipment with soap and water after each use. Respirators must be cleaned and filters replaced according to instructions included with the respirators. Personal clothing worn during use must be laundered separately from household articles. Clothing and protective equipment heavily contaminated or drenched with GUTHION azinphos-methyl must be destroyed according to state and local regulations. HEAVILY CONTAMINATED OR DRENCHED CLOTHING CANNOT BE ADEQUATELY DECONTAMINATED.

During aerial application, human flaggers are prohibited unless in totally enclosed vehicle.

STORAGE AND DISPOSAL

Pesticide Disposal: Do not contaminate water, food or feed by storage or disposal.

Pesticide wastes are acutely hazardous. Improper disposal of excess pesticide, spray mixture, or rinsate is a violation of Federal Law. If these wastes cannot be disposed of by use according to label instructions, contact your State Pesticide or Environmental Control Agency, or the Hazardous Waste representative at the nearest EPA Regional Office for guidance.

Container Disposal: Do not use container in connection with food, feed, or drinking water. Completely empty carton into application equipment. Then dispose of empty carton in a sanitary landfill or by incineration, or, if allowed by State and local authorities, by burning. If burned, stay out of smoke.

Store in a cool, dry place and in such a manner as to prevent cross contamination with other pesticides, fertilizers, food, and feed. Store in original container and out of the reach of children, preferably in a locked storage area.

Handle and open container in a manner as to prevent spillage. If the container is leaking or material spilled for any reason or cause, carefully sweep material into a pile. Refer to Precautionary Statements on label for hazards associated with the handling of this material. Do not walk through spilled material. Dispose of pesticide as directed above. In spill or leak incidents, keep unauthorized people away. You may contact the Mobay Emergency Response Team for decontamination procedures or any assistance that may be necessary. The Mobay Kansas City Emergency Response Telephone No. is 816-242-2582, or contact Chemtrec at 800-424-9300.

IMPORTANT: This bulletin is not intended to provide adequate information for application. Before using the product, read and carefully observe directions, cautionary statements and other information appearing on the product label. This product is sold subject to the Conditions of Sale set forth on the container label.

STOP—READ THIS LABEL BEFORE USE

G 9007

This is a specimen label. Mobay is not responsible for the accuracy of the information contained herein. As labels are subject to revision, always carefully read and follow the label on the product container.

GUTHION®

50% WETTABLE POWDER
CROP INSECTICIDE
WATER SOLUBLE PACKETS

FOR EFFECTIVE ECONOMICAL INSECT CONTROL

ACTIVE INGREDIENT:	
O,O-Dimethyl *S*-[(4-oxo-1,2,3-benzotriazin-3(4H)-yl)methyl]phosphorodithioate	50%
INERT INGREDIENTS:	50%
TOTAL	100%

EPA Reg. No. 3125-379

GUTHION is a Reg. TM of Bayer AG, Germany

Wettable powder formulations of GUTHION insecticide have proved effective through many years of extensive commercial use, both foreign and domestic. The 50% Wettable Powder formulation has excellent wetting and suspension properties.

Pre-measured water soluble packets reduce handling and dust and are offered for sale in Arizona and California.

FRUITS: GUTHION 50% WP controls major fruit pests. The use of GUTHION 50% WP greatly simplifies spray schedules. GUTHION 50% WP stays on the job, too. Its residual killing activity gives excellent and continuous control from one cover spray to the next . . . often eliminates "between-cover" applications. A combination of GUTHION 50% WP and dormant oil for delayed-dormant sprays on apples effectively controls early insect pests on this crop. All-season use of GUTHION 50% WP not only eliminates complicated spray schedules, but results in more profitable finer-finish fruits.

VEGETABLES: Registered for use on many vegetables, GUTHION 50% WP has proved to be exceptionally effective in controlling diamondback moth, Colorado potato beetle and others. As with fruit, GUTHION 50% WP has prolonged residual action . . . does the job and stays on the job.

RECOMMENDED USES

Alfalfa**	Cauliflower	Other melons
Almonds	Celery	Peaches***‡
Apples	Cherries	Pears
Apricots***‡	Citrus fruits	Pecans
Artichokes	Clover	Peppers
Barley	Crab apples	Plums***‡
Beans (snap and dry)	Cranberries	Potatoes
Birdsfoot trefoil**	Cucumbers	Prunes***‡
Blackberries	Eggplant	Raspberries
Blackeyed peas	Filberts*	Rye
Blueberries****	Grapes	Soybeans
Boysenberries	Honeydew melons	Spinach
Broccoli	Loganberries	Strawberries
Brussels sprouts	Nectarines***‡	Tomatoes
Cabbage	Oats	Walnuts
Cantaloupe	Onions (green and dry)	Watermelons
		Wheat

*Pacific Northwest only
**East of the Mississippi River
***Eastern U.S.
****Eastern and North Central states only
‡West of the Rocky Mountains

RESTRICTIONS ON WAREHOUSE STORAGE

Keep water soluble packets in this container and store in a cool dry place, but not below freezing (32°F). Protect from heat. Keep away from open flame. Do not heat. Entire inner packets dissolve in water. After opening outer carton, drop the required unopened inner packets into spray tank as directed. Do not excessively handle water soluble packet or expose it to moisture, since this may cause breakage.

PACKAGING

5 lb. carton, contains five 1 lb. water soluble packets, 4 per case, shipping weight 26.5 lbs.

PELIGRO

PRECAUCION AL USUARIO: Si usted no puede leer o entender inglés, no use este producto hasta que la etiqueta le haya sido explicada ampliamente.

(TO THE USER: If you cannot read or understand English, do not use this product until the label has been fully explained to you.)

DIRECTIONS FOR USE

It is a violation of Federal Law to use this product in a manner inconsistent with its labeling.

MIXING: The enclosed packets containing GUTHION 50% Wettable Powder insecticide are **water soluble.** Do not allow packet to become wet prior to adding to the spray tank. Do not handle with wet hands. Reseal outer carton to protect remaining packets.

To prepare the spray mixture, drop the required number of unopened packets, as determined under "Recommended Application", into the spray tank while filling with water to the desired level. Operate the agitator while mixing. Depending on the water temperature and the degree of agitation, the packets should be completely dissolved within approximately 5 minutes from the time they were added to the water.

COMPATIBILITY: GUTHION 50% Wettable Powder is physically compatible with many registered pesticides and liquid fertilizers. When considering mixing GUTHION 50% Wettable Powder with other pesticides or with liquid fertilizer, first contact your supplier. For futher information, contact your local Mobay representative. If your supplier and Mobay representative have no experience with the combination you are considering, you should conduct tests to determine physical compatibility.

To determine physical compatibility, pour the recommended proportions of each chemical with the same proportion of water as will be present in the chemical supply tank into a suitable container, mix thoroughly and allow to stand for five minutes. If the combination remains mixed, or can be remixed readily, the mixture is considered physically compatible. When mixing wettable powder or dry flowable formulations, add and disperse these first, then add liquid pesticides. Conduct another compatibility test at concentrations which will be present in the irrigation lines. If there is any separation which cannot be remixed readily, Mobay recommends that the combination not be used.

Combinations should be kept agitated and should be applied immediately. Do not allow combination to set for prolonged periods in the chemical supply tank or irrigation lines.

DOSAGE: Use specified dosage of GUTHION 50% Wettable Powder in the amount of water necessary to give complete coverage of foliage. Determine the **total amount** of wettable powder to be added to the spray tank based on the rates under "Recommended Applications". For each one pound of wettable powder to be added to the spray tank, use one 1 pound packet. For example, if it is determined that 2 pounds of GUTHION 50% Wettable Powder should be added to the spray tank, add two 1 pound packets. The type of equipment used will determine the concentration required; however, use of these packets is not recommended for making highly concentrated mixtures such as used in aircraft spraying.

SPRAYING: Work to windward. Protect sprayer operators from drift or mist. When low volumes of spray are applied complete coverage and thorough application are essential for most effective results. Schedule applications in accordance with local conditions. Consult your State Agricultural Experiment Station or Extension Service for specific use information in your area.

CHEMIGATION

GUTHION 50% Wettable Powder may be applied through recommended types of irrigation systems to many crops. The REMARKS section for each crop list the types of applications allowed. If application through irrigation systems is not listed in the REMARKS section for a crop, GUTHION 50% Wettable Powder may not be applied to that crop through irrigation systems.

Types of Irrigation Systems

Apply GUTHION 50% Wettable Powder only through sprinkler, including center pivot, lateral move, side roll, overhead solid set, or low pressure sprinkler irrigation systems. Do not apply GUTHION 50% Wettable Powder through any other type of irrigation systems.

GENERAL DIRECTIONS FOR ALL RECOMMENDED TYPES OF IRRIGATION SYSTEMS

Uniform Water Distribution and System Calibration

The irrigation system must provide uniform distribution of treated water. Crop injury, lack of effectiveness, or illegal pesticide residues in the crop can result from nonuniform distribution of treated water.

The system must be calibrated to uniformly apply the rates specified for chemigation application for specific crops. If you have questions about calibration, you should contact State Extension Service specialist, equipment manufacturers or other experts.

Chemigation Monitoring

A person knowledgeable of the chemigation system and responsible for its operations, or under the supervision of the responsible person, shall shut the system down and make necessary adjustments should the need arise.

Drift

Do not apply when wind speed favors drift beyond the area intended for treatment.

Required System Safety Devices

The system must contain a functional check valve, vacuum relief valve, and low pressure drain appropriately located on the irrigation pipeline to prevent water source contamination backflow.

The pesticide injection pipeline must contain a functional, automatic, quick-closing check valve to prevent the flow of fluid back toward the injection pump.

The pesticide injection pipeline must also contain a functional, normally closed, solenoid-operated valve located on the intake side of the injection pump and connected to the system interlock to prevent fluid from being withdrawn from the supply tank when the irrigation system is either automatically or manually shut down.

The system must contain functional interlocking controls to automatically shut off the pesticide injection pump when the water pump motor stops.

The irrigation line or water pump must include a functional pressure switch which will stop the water pump motor when the water pressure decreases to the point when pesticide distribution is adversely affected.

Systems must use a metering pump, such as a positive displacement injection pump (e.g., diaphragm pump) effectively designed and constructed of materials that are compatible with pesticides and capable of being fitted with a system interlock.

Using Water from Public Water Systems

DO NOT APPLY GUTHION 50% WETTABLE POWDER THROUGH ANY IRRIGATION SYSTEM SUPPLIED BY A PHYSICALLY CONNECTED PUBLIC WATER SYSTEM

Public water system means a system for the provision to the public of piped water for human consumption if such system has at least 15 service connections or regularly services an average of at least 25 individuals daily at least 60 days out of the year.

GUTHION 50% Wettable Powder may be applied through any of the recommended types of irrigation systems which may be supplied by a public water system only if the water from the public water system is discharged into a reservoir tank prior to pesticide introduction. There shall be a complete physical break (air gap) between the outlet end of the fill pipe and the top or overflow rim of the reservoir tank of at least twice the inside diameter of the fill pipe. Before beginning chemigation, always make sure that the air gap exists and that there is no blockage of the overflow of the reservoir tank.

Any irrigation system using water supplied from a public water system must also meet the following requirements:

The pesticide injection pipeline must contain a functional, automatic quick-closing check valve to prevent the flow of fluid back toward the injection pump.

The pesticide injection pipeline must contain a functional, normally closed, solenoid-operated valve located on the intake side of the injection pump and connected to the system interlock to prevent fluid from being withdrawn from the supply tank when the irrigation system is either automatically or manually shut down.

The system must contain functional interlocking controls to automatically shut off the pesticide injection pump when the water pump motor stops or in cases where there is no water pump, when the water pressure decreases to the point where pesticide distribution is adversely affected.

Systems must use a metering pump, such as a positive displacement injection pump (e.g., diaphragm pump) effectively designed and constructed of materials that are compatible with pesticides and capable of being fitted with a system interlock.

Do not apply when wind speed favors drift beyond the area intended for treatment.

Posting Requirements

Posting of areas to be chemigated is required when 1) any part of a treated area is within 300 feet of sensitive areas such as residential areas, labor camps, businesses, day care centers, hospitals, in-patient clinics, nursing homes or any public areas such as schools, parks, playgrounds, or other public facilities not including public roads, or 2) when the chemigated area is open to the public such as golf courses or retail greenhouses.

Posting must conform to the following requirements: Treated areas shall be posted with signs at all usual points of entry and along likely routes of approach from the listed sensitive areas. When there are no usual points of entry, signs must be posted in the corners of the treated areas and in any other location affording maximum visibility to sensitive areas. The printed side of the sign should face away from the treated area towards the sensitive areas. The signs shall be printed in English. Signs must be posted prior to application and must remain posted until foliage has dried and soil surface water has disappeared. Signs may remain in place indefinitely as long as they are composed of materials to prevent deterioration and maintain legibility for the duration of the posting period.

All words shall consist of letters at least 2½ inches tall, and all letters and the symbol shall be a color which sharply contrasts with their immediate background. At the top of the sign shall be the words KEEP OUT, followed by an octagonal stop sign symbol at least 8 inches in diameter containing the word STOP. Below the symbol shall be the words PESTICIDES IN IRRIGATION WATER.

Posting required for chemigation does not replace other posting and reentry interval requirements for farmworker safety.

Compatibility

When mixing with other chemicals refer to Compatibility section elsewhere on the label.

Agitation

For application of GUTHION 50% Wettable Powder, a chemical supply tank is necessary for premixing. For application of GUTHION 50% Wettable Powder alone or in combination with liquid fertilizer or other chemicals, constant strong mechanical or hydraulic agitation must be maintained in the chemical supply tank during the entire period of application.

Chemical Supply Tank Dilution

When a chemical supply tank is used, you must determine the required amounts of GUTHION 50% Wettable Powder to mix in the tank.

The amount of GUTHION 50% Wettable Powder alone needed equals the number of pounds of GUTHION 50% Wettable Powder to be applied per acre multiplied by the number of acres to be chemigated.

The amount of solution needed equals the gallons of solution delivered per hour by the injection pump multiplied by the number of hours chemigation will take place.

For example, if you want to apply 2 pounds of GUTHION 50% Wettable Powder per acre to 130 acres in 20 hours and your injection pump delivers 15 gallons per hour, you need: 2 pounds GUTHION 50% Wettable Powder per acre $\times$ 130 acres = 260 pounds of GUTHION 50% Wettable Powder. And, you need: 15 gallons per hour $\times$ 20 hours = 300 gallons of suspension = 300 gallons of water.

Cleaning the Chemical Injection System

In order to accurately apply pesticides, the chemical injection system must be kept clean; free of chemical or fertilizer residues and sediments. Refer to your owners manual or ask your equipment supplier for cleaning procedure for your injection system.

Flushing the Irrigation System

At the end of the application period, allow time for all lines to flush the pesticide through all nozzles or emitters before turning off irrigation water. To ensure the lines are flushed and free of pesticides, a dye indicator may be injected into the lines to mark the end of the application period.

Sprinkler Irrigation Systems

All directions and requirements under the GENERAL DIRECTIONS AND REQUIREMENTS FOR ALL RECOMMENDED TYPES OF IRRIGATION SYSTEMS section of the label must be followed for sprinkler irrigation systems.

In addition, the following directions apply to sprinkler irrigation systems:

Do not apply when wind speed favors drift beyond the area intended for treatment.

It is recommended that nozzles in the immediate area of control panels, chemical supply tanks, pumps and system safety devices be plugged to prevent chemical contamination of these areas.

Center-Pivot and Automatic-Move Linear Systems

Inject the specified dosage per acre continuously for one complete revolution or move of the system. DO NOT USE END GUNS. For a foliar application, the system should be run at maximum speed. For a soil application, the system should be run at a slower speed with application in at least 1/4 inch of water.

Solid Set and Manually Controlled Linear Systems

For foliar application, injection should be during the last 30 to 60 minutes of regular irrigation period **or** as a separate 30 to 60 minute application not associated with a regular irrigation. For soil treatment, application should be in at least 1/4 inch of water. DO NOT USE END GUNS.

RECOMMENDED APPLICATIONS

CROP	INSECT	POUNDS GUTHION 50% WP	REMARKS
FIELD CROPS Alfalfa Birdsfoot trefoil (East of the Mississippi River only) Clover	Leafhoppers Mites	1/2 to 1	Apply specified dosage per acre by air or ground equipment. Use at least 10 gallons of water per acre with ground equipment and at least 1 gallon of water per acre for aerial application. It may be necessary to use 20 to 25 gallons of water per acre on heavy growth for control of alfalfa weevil or Egyptian alfalfa weevil with ground equipment. For application by irrigation systems: Apply specified dosage per acre. Follow all directions given under the CHEMIGATION section of the label. Only one application per cutting may be made regardless of rate, formulation or method of application used. Do not apply within 14 days of harvest at the rates of 11/16 to 1 1/16 pounds per acre or within 16 days of harvest at the rate of 1 7/16 pounds per acre.
	Alfalfa weevil Egyptian alfalfa weevil	3/4 to 1	
	Alfalfa plant bug Fleahopper Grasshoppers Lygus bugs Spittlebugs	1	
Barley Oats Rye Wheat	Cereal leaf beetle	3/4 to 1	Apply specified dosage per acre in sufficient water for complete coverage. Do not apply more than once per season. Do not harvest for food, feed, forage or graze within 30 days of treatment.
Soybeans	Aphids Bean leaf beetle Green cloverworm Leafhoppers Leaf miners Leaf rollers Stink bugs Velvet bean caterpillar	3/4 to 1	Apply specified dosage per acre in sufficient water to give complete coverage. Do not apply within 45 days of harvest. Do not graze or feed treated vines to livestock. Do not exceed more than 3 applications per season.
	Mexican bean beetle	1 to 1 1/2	
FRUIT	NOTE: It is suggested that when treating fruit during the bloom period, bee keepers should be warned well in advance to remove hives a safe distance from orchards to be treated.		
Apples Crab apples Pears	Aphids* Apple maggot Codling moth European apple sawfly Eye-spotted bud moth Forbes scale Fruit tree leaf roller Green fruitworm* Leafhoppers* Mealybug* Orange tortrix Pear midge Plum curculio Putnam scale Redbanded leaf roller San Jose scale* Stink bug Tarnished plant bug*	2 to 3 (1/2 to 3/4 pounds per 100 gallons)	Apply specified dosage per acre by air or ground equipment in sufficient water for good coverage. Up to 12 pounds GUTHION 50% WP may be applied per crop season. Allow at least 7 days between applications and 7 days between last application and harvest. GUTHION 50% WP is compatible with dormant and summer oils which may be added to apple and pear sprays in accordance with local recommendations.
Cherries	Eye-spotted bud moth Forbes scale Fruit flies Fruit tree leaf roller Cherry leaf miner Mineola moth Lesser peach tree borer Plum curculio San Jose scale*	1 1/2 (1/2 pound per 100 gallons)	Apply specified dosage per acre as a full coverage spray. Up to 8 1/2 pounds of GUTHION 50% WP may be applied per acre per crop season. Allow at least 14 days between applications and 15 days between last application and harvest. **California:** Apply post-harvest only.
Apricots Nectarines Peaches (Eastern U.S.)	Aphids* Cottony peach scale European fruit lecanium scale Forbes scale Lesser peach tree borer Oriental fruit moth Peach tree borer Peach twig borer *Platynota flavedana* leaf roller Plum curculio Redbanded leaf roller San Jose scale* Stink bug Thrips Tarnished plant bug Terrapin scale Walnut scale White peach scale	1 3/4 to 2 1/4 (1/2 to 5/8 pound per 100 gallons)	Apply specified dosage per acre as a full coverage spray. Allow at least 14 days between applications. A total of 6 3/4 pounds per acre per crop season may be applied to nectarines. A total of 9 pounds per acre per crop season may be applied to apricots and peaches. Allow at least 21 days between last application and harvest. For control of peach tree borer, apply 2 or 3 sprays to trunk from ground to scaffold limbs, timed with moth flight. For control of scale, apply when crawlers are present. GUTHION 50% WP is compatible with dormant and summer oils which may be added to peach sprays in accordance with local recommendations.

CROP	INSECT		POUNDS GUTHION 50% WP	REMARKS
Apricots Nectarines Peaches (West of the Rocky Mountains)	Lesser peach tree borer Oriental fruit moth Peach twig borer Peach tree borer *Platynota flavedana* leaf roller	Plum curculio Redbanded leaf roller Stink bug Thrips Tarnished plant bug	3 to 4	Apply specified dosage per acre. Allow at least 14 days between applications. A total of 6¾ pounds per acre per crop season may be applied to nectarines. A total of 7 pounds per acre per crop season may be applied to apricots and peaches. Allow at least 21 days between last application and harvest. For control of peach tree borer, apply 2 or 3 sprays to trunk from ground to scaffold limbs, timed with moth flight.
Plums Prunes (Eastern U.S.)	Aphids* Codling moth Eyespotted bud moth Forbes scale Fruit tree leaf roller Lesser peach tree borer Orange tortrix	Peach tree borer Peach twig borer Plum curculio Redbanded leaf roller San Jose scale* Stink bug Tarnished plant bug* Tussock moth	1¾ to 2½ (½ to ⅝pounds per 100 gallons)	Apply specified dosage per acre as a full coverage spray. A total of 6¾ pounds per acre per crop season may be applied to plums and prunes. Allow at least 10 days between applications. Allow at least 15 days between last application and harvest. For control of peach tree borer, apply 2 or 3 sprays to trunk from ground to scaffold limbs, timed with moth flight. For control of scale, apply when crawlers are present. GUTHION 50% WP is compatible with dormant and summer oils which may be added to prune sprays in accordance with local recommendations.
	American plum borer		3	
Plums Prunes (West of the Rocky Mountains)	Codling moth Eye-spotted bud moth Fruit tree leaf roller Lesser peach tree borer Orange tortrix Peach tree borer	Peach twig borer Plum curculio Redbanded leaf roller Stink bug Tarnished plant bug* Tussock moth	2 to 4	Apply specified dosage per acre. A total of 6¾ pounds per acre per crop season may be applied to plums and prunes. Allow at least 10 days between applications. Allow at least 15 days between last application and harvest. For control of peach tree borer, apply 2 or 3 sprays to trunk from ground to scaffold limbs, timed with moth flight.
	American plum borer		3 to 4	
Blackberies Boysenberries Loganberries Raspberries	Leafhoppers	Leaf rollers	½	Apply specified dosage per acre to foliage using approximately 200 gallons of water for good coverage. Do not apply more than twice per season nor within 14 days of harvest.
	Leaf miners		⅝	
	Aphids		⅝ to 1	
	Obscure root weevil		1	For control of root weevils prior to harvest apply specified dosage per acre to lower portion of canes and to the soil beneath the plants using approximately 200 gallons of water. Do not apply more than twice per season. Do not make applications within 3 days of harvest at rates up to 2 pounds per acre.
	Obscure root weevil		1	
Blueberries (Eastern and North Central States only)	Blueberry maggot Fruitworms Lecanium scale Plum curculio		1 to 1½	Apply specified dosage per acre with aerial or ground equipment using sufficient water for good coverage. A total of 3 applications may be made per crop season regardless of rate or formulation used. Allow at least 10 days between applications and at least 7 days between last application and harvest. Do not apply later than June 15.
Cranberries	Cranberry fruitworm *Sparganothis sulfureana* Tipworm		1 to 2	Apply specified dosage per acre by air using sufficient water for good coverage. For application by irrigation systems: Apply specified dosage per acre. Follow all directions given under the CHEMIGATION section of the label. A total of 3 applications may be made per crop season regardless of rate or formulation used. Allow at least 14 days between applications and at least 21 days between last application and harvest.
	Fireworms		2	
Citrus Fruits	Aphids* Black scale Brown soft scale Chaff scale Citricola scale Citrus mealybug Citrus thrips Cottony-cushion snail European brown scale	Florida red scale Fruit tree leaf roller Fuller rose beetle Glover scale Orange tortrix Purple scale Snow scale Western tussock moth Whiteflies	4 to 6	Apply specified dosage per acre as a full coverage spray. A single application per year may be applied up to within 7 days of harvest. Where 2 applications are required, allow at least 28 days between the second application and harvest. A total of 2 applications may be applied per fruit year regardless of rate or formulation used. Do not pick fruit or do other work involving contact with the tree (such as pruning) within 7 days after treatment.
	California red scale	Yellow scale	6	
Grapes	Grape berry moth Grape cane girdlers Grape mealybug Leafhoppers*	Redbanded leaf roller Thrips Grape leaf skeletonizer	1½ to 2	Apply specified dosage per acre as a full coverage spray. A total of 3 applications may be made per crop season regardless of rate or formulation used. Allow at least 14 days between applications. Minimum dosage specified may be applied up to harvest. With higher rates allow at least 10 days between last application and harvest.

CROP	INSECT	POUNDS GUTHION 50% WP	REMARKS
Strawberries	Aphids* Meadow spittlebug Oblique-banded leaf roller Obscure root weevil Omnivorous leaf tier Pea leaf weevil Small black (grass) weevil Strawberry leaf roller Whitefly	1	Apply specified dosage per acre as a full coverage spray using sufficient water for good coverage. For application by irrigation systems: Apply specified dosage per acre. Follow all directions given under the CHEMIGATION section of the label. A total of 4 applications may be made per crop season regardless of rate or formulation used. Allow at least 5 days between applications and at least 5 days between last application and harvest.
NUTS Almonds	Peach twig borer Navel orangeworm	3 to 4	Apply specified dosage per acre by concentrate sprayer or as a full coverage spray using not more than 400 gallons of finished spray per acre. If conditions dictate an aerial spray, apply specified dosage by aircraft in not less than 20 gallons of water per acre. Application is limited to one dormant spray and/or two postbloom sprays per season. Last application may be made up to 28 days before harvest. Allow at least 120 days between the dormant and the first postbloom spray and at least 28 days between the postbloom sprays. Do not graze livestock in treated groves for 21 days after treatment.
Filberts (Pacific Northwest only)	Apple mealybug Filbertworm Filbert aphid Filbert leaf roller	$4\frac{1}{4}$ to $5\frac{11}{16}$	Apply specified dosage per acre as a full coverage spray. A total of 3 applications may be made per crop season regardless of rate or formulation used. Allow at least 14 days between applications and at least 30 days between last application and harvest. Do not graze livestock in treated groves for 21 days after treatment.
Pecans	Aphids* Fall webworm Hickory shuckworm Leafminers May beetles Pecan casebearer Southern green stink bug Spittlebugs Twig girdlers Walnut caterpillars	$4\frac{1}{4}$ to $5\frac{11}{16}$	Apply specified dosage per acre as a full coverage spray. A total of 3 applications may be made per crop season regardless of rate or formulation used. Allow at least 7 days between applications. Do not apply after shuck split. Livestock may be grazed in treated groves after a 21-day post-treatment interval.
Walnuts	Codling moth Filbertworm Navel orangeworm Walnut husk fly Red-humped caterpillar	$4\frac{1}{4}$ to $5\frac{11}{16}$	Apply specified dosage per acre as a full coverage spray. A total of 3 applications may be made per crop season regardless of rate or formulation used. Allow at least 14 days between applications. Do not apply after husks split. Do not graze livestock in treated groves for 21 days after treatment.
VEGETABLES Artichokes	Plume moth	$4\frac{1}{4}$	Apply specified dosage per acre in sufficient water for complete coverage. For application by irrigation systems: Apply specified dosage per acre. Follow all directions given under the CHEMIGATION section of the label. A total of 3 applications may be made per crop season regardless of rate, formulation or method of application used. Allow at least 14 days between applications and at least 30 days between last application and harvest. Do not feed or ensile treated forage.
Beans (Dry and Snap)	Leaf rollers Mites	$\frac{5}{8}$ to $1\frac{3}{8}$	Apply specified dosage per acre in sufficient water for complete coverage. Do not apply to snap beans within 7 days of harvest nor to dry beans within 30 days of harvest. Do not exceed 4 applications. Do not feed or ensile treated forage.
	Aphids Bean leaf beetle Green cloverworm Leafhoppers Leaf miners Stink bugs Velvet bean caterpillar	1 to $1\frac{3}{8}$	
	Mexican bean beetle Spotted cucumber beetle Striped cucumber beetle Tarnished plant bug Western-striped cucumber beetle	$1\frac{3}{8}$	
Broccoli Brussels Sprouts Cabbage (Includes tight heading varieties of Chinese cabbage) Cauliflower	Aphids Cabbage looper Diamondback moth Imported cabbageworm	1 to $1\frac{1}{2}$	Apply specified dosage per acre in sufficient water for complete coverage. Do not apply within 7 days of harvest for Brussels sprouts, 15 days of harvest for broccoli and cauliflower, nor within 21 days of harvest for cabbage. Do not exceed more than 3 applications per season.
	Cabbage maggot	$\frac{1}{4}$ to $\frac{3}{8}$	Mix specified dosage in 50 gallons of water. Apply 4 to 6 ounces of this emulsion per plant at or immediately after transplanting.
	Cabbage maggot (Direct seeded fields in California only)	$1\frac{1}{2}$	Apply specified dosage per acre in sufficient water for uniform distribution. Mix in upper 2 inches of soil prior to seeding or spray in the seed row at planting time. Usually 2 additional sprays are necessary during the growing season, depending on time of year and maggot population. Do not apply within 21 days of harvest. Do not exceed more than 3 applications per season.
	Cabbage maggot (Transplant fields in California only)	$1\frac{1}{2}$	Apply specified dosage in 300 to 400 gallons of water per acre as a soil drench in rows when damage first appears. Additional applications may be necessary. Do not apply within 21 days of harvest. Do not exceed more than 3 applications per season.

CROP	INSECT	POUNDS GUTHION 50% WP	REMARKS
Blackeyed peas (Southern peas, Crowder peas)	Corn earworm Cowpea curculio	1½ to 2	Apply specified dosage per acre in sufficient water for complete coverage. Do not apply more than 4 times per season not within 7 days of harvest. Do not use vines for feed or forage nor pasture treated areas.
	Leaf miners Stink bugs	¾ to 1	
Celery	Aphids Leaf miners Leafhoppers Spittlebug Tarnished plant bug	1	Apply specified dosage in 100 gallons of water as a full coverage spray** using not more than 200 gallons of finished spray per acre. Do not apply within 14 days of harvest. Do not exceed more than 3 applications per season.
Cucumbers	Spotted cucumber beetle Striped cucumber beetle Western-striped cucumber beetle	1	Apply specified dosage per acre in sufficient water for complete coverage. For application by irrigation systems: Apply specified dosage per acre. Follow all directions given under the CHEMIGATION section of the label. A total of 3 applications may be made per crop season regardless of rate, formulation or method of application used. Allow at least 7 days between applications and at least 1 day between last application and harvest.
Eggplant	Leaf miners	¾ to 1	Apply specified dosage per acre in sufficient water to give complete coverage. For application by irrigation systems: Apply specified dosage per acre. Follow all directions given under the CHEMIGATION section of the label. A total of 3 applications may be made per crop season regardless of rate, formulation or method of application used. Allow at least 7 days between applications. Do not apply after fruit set.
	European corn borer Flea beetles	1	
Onions (Green and Dry)	Thrips	1 to 1½	Apply specified dosage per acre by air or ground equipment in sufficient water for complete coverage. For application by irrigation systems: Apply specified dosage per acre. Follow all directions given under the CHEMIGATION section of the label. A total of 3 applications may be made per crop season regardless of rate, formulation or method of application used. Allow at least 7 days between applications for dry bulb onions. Allow at least 10 days between applications for green onions. For dry onions, allow at least 28 days between last application and harvest. For green onions, allow at least 7 days between last application and harvest.
Peppers	Leaf miners	¾ to 1	Apply specified dosage per acre by air or ground equipment in sufficient water for complete coverage. For application by irrigation systems: Apply specified dosage per acre. Follow all directions given under the CHEMIGATION section of the label. A total of 3 applications may be made per crop season regardless of rate, formulation or method of application used. Allow at least 7 days between applications and 7 days between last application and harvest.
	European corn borer Flea beetles	1	
Potatoes	Colorado potato beetle	¾	Apply specified dosage per acre in sufficient water for complete coverage. For application by irrigation systems: Apply specified dosage per acre. Follow all directions given under the CHEMIGATION section of the label. A total of 3 applications may be made per crop season regardless of rate, formulation or method of application used. Allow at least 7 days between applications and 7 days between last application and harvest. NOTE: Resistance of Colorado potato beetle has occurred in some areas. Consult your local Extension Service or MOBAY sales representative for details.
	Banded cucumber beetle Leaf miners	¾ to 1	
	European corn borer Flea beetle Leafhoppers Spittlebugs Tarnished plant bug	1 to 1½	
	Tuberworm	1⅛ to 1½	
Spinach	Aphids Leaf miners Mites	¾ to 1	Apply specified dosage per acre in sufficient water for complete coverage. Do not apply within 14 days of harvest. Do not exceed more than 3 applications per season.
Tomatoes	Colorado potato beetle	¾	Apply specified dosage per acre by air or ground equipment in sufficient water for complete coverage. For application by irrigation systems: Apply specified dosage per acre. Follow all directions given under the CHEMIGATION section of the label. A total of 4 applications may be made per crop season regardless of rate, formulation or method of application used. The high rates should be used where heavy infestations of late instar lepidopterous larvae (large worms) and pinworms are present. Allow at least 7 days between applications. Rates of 1½ pounds per acre or less may be applied up to day of harvest. Rates above 1½ pounds require an interval of 14 days between last application and harvest. NOTE: Resistance of Colorado potato beetle has occurred in some areas. Consult your local Extension Service or MOBAY sales representative for details.
	Banded cucumber beetle *Drosophila* Green stink bug Leaf miners Whitefly	¾ to 1	
	Aphids* European corn borer Flea beetles Grasshoppers Leafhoppers Thrips	1 to 1½	
	Tuberworm	1⅛ to 1½	
	Fruitworm Hornworm Pinworm Yellow-striped armyworm	1½ to 3	

CROP	INSECT	POUNDS GUTHION 50% WP	REMARKS
MELONS Honeydew melons Muskmelon (cantaloupe) Watermelons Other melons	Leafhoppers Leaf miners	¾ to 1	Apply specified dosage per acre in sufficient water to give complete coverage. For application by irrigation systems: Apply specified dosage per acre. Follow all directions given under the CHEMIGATION section of the label. A total of 3 applications may be made per crop season regardless of rate, formulation or method of application used. Allow at least 5 days between applications and at least 7 days between last application and harvest.
	Rindworms Spotted cucumber beetle Striped cucumber beetle Western-striped cucumber beetle	1	
ORNAMENTALS Southern Pine Seed Orchards	Coneworm Seedworm	See Remarks	Use 4 pounds per 100 gallons of water (0.2% dilution) as a full coverage spray**. Use 2.25 pounds per 10 gallons of water (1% dilution) for low volume sprayers. Apply first application within 30 days following conelet closure, followed by 3 to 5 applications (at least 30 days apart). Apply approximately 5 to 10 gallons of the 0.2% dilution with high volume sprayers or approximately 1 to 2 gallons of the 1% dilution with low volume sprayers per tree. Thorough coverage of cones is necessary for maximum control.
Nursery Plants	Aphids *Cerococcus* scale Euonymus scale Juniper scale Lace bugs Leafhoppers Mites Olive scale Oystershell scale *Pulvinaria* scale Thrips	¾ to 1	Apply specified dosage per 100 gallons of water. Spray all foliage surfaces including the underside of leaves for complete coverage. For control of black pine leaf, brown soft, European elm, and Putnam scales, use the higher rate. Repeat as necessary.
	Brown soft scale Putnam scale	2	
	European elm scale Black pine leaf scale	1½ to 2	
	Cone midge Cone moth	4 to 8	Apply specified dosage per 100 gallons of water. Time applications to coincide with moth flight when cones are open for pollination. Thorough coverage of cones is necessary for maximum control. Repeat as necesary.
	European pine shoot moth Nantucket pine tip moth	¾ to 1½	Apply specified dosage per acre in sufficient water for good coverage. Time applications to coincide with moth flights.
	Injury to hawthorn or American linden may occur under some conditions. Do not allow children or pets on treated area until material has been washed into soil and treated area is dry.		
Christmas Trees	Scale spp. Sawfly spp. European pine shoot moth Eastern pine shoot borer Nantucket pine tip moth	¾ to 1½	Apply specified dosage per acre by air or ground equipment in sufficient water to give complete coverage but not less than 1 gallon per acre. Time applications to coincide with susceptible pest development. 9-14-89

* In some areas, these species may have developed resistance to organophosphate insecticides. GUTHION insecticide used alone may not provide satisfactory control in those areas. Consult your local agricultural advisor or cooperative extension service for recommendations.

** This concentration is calculated for conventional hydraulic-type sprayers. When lower volumes of spray are applied per acre with concentrate sprayers increase the concentration of GUTHION 50% Wettable Powder in the spray mixture in order to apply the amount of GUTHION 50% Wettable Powder per acre equivalent to a full coverage spray. In Southern Pine Seed Orchards where conditions dictate an air application, apply dosage per acre equivalent to a full coverage ground spray in not less than one gallon of water per acre.

Do not allow children or pets in treated area until material is dry.

SPECIAL LOCAL NEED REGISTRATIONS [SECTION 24(c)] FOR GUTHION 50% WETTABLE POWDER*

CROP	PEST	STATE EPA SLN NO.	DOSAGE	REMARKS
Cotton	Pink bollworm	California CLA 81-0074	1 to 2 pounds per acre	Application by ground or air.
Pistachios	Navel orangeworm	California CA 79-0149	5 pounds per acre	Apply at onset of hull splitting but prior to 10% hullsplit.
Grapes	Omnivorous leafroller Grape leaffolder	California CA 80-0146	1 to 2 pounds per acre	Up to 3 applications per season.

* For distribution and use only within the state specified. These Special Local Need registrations approved as of date of this publication are, however, subject to cancellation by the state. Contact your nearest Mobay Agricultural Chemicals Division Office for the current status and a copy of the complete individual supplemental Special Local Need labeling. This labeling must be in the possession of the user at time of pesticide application.

RESTRICTIONS

Do not use on other crops used for food or forage. Use only according to label directions. Application at rates above those shown may result in illegal crop residues. Do not graze livestock in treated orchards or groves for 21 days after treatment. Do not treat food crops grown in the greenhouse.

RE-ENTRY STATEMENTS FOR FARM WORKERS

Do not apply this product in such a manner as to directly or through drift expose workers or other persons. The area being treated must be vacated by unprotected persons. Do not enter treated areas for 24 hours after application unless protective clothing is worn, as described in the product labeling. Because certain states may require more restrictive re-entry intervals for various crops treated with this product, consult your State Department of Agriculture for further information.

Written or oral warnings must be given to workers who are expected to be in a treated area or in an area about to be treated with this product. Oral warnings should include those statements found under the label heading "Precautionary Statements" and "Re-entry Statements" included on this label. When oral warnings are given, warnings shall be given in a language customarily understood by workers. Oral warnings must be given if there is reason to believe that written warnings cannot be understood by workers. Written warnings must include the following information: DANGER, area treated with GUTHION 50% Wettable Powder on (date of application). Do not enter treated areas for 24 hours after application unless protective clothing is worn, as described in the product labeling. **If on skin,** remove contaminated clothing and wash skin immediately with soap and warm water. **If eyes are contaminated,** wash with flowing water for at least 15 minutes. **If swallowed,** vomiting should be induced. (See Precautionary Statements on label).

ROTATIONAL CROPS

Do not plant root crops other than those with registered azinphos-methyl uses in azinphos-methyl treated soil sooner than 6 months after the last application. Do not plant any other crop other than those with registered azinphos-methyl uses in treated soil sooner than 30 days after last application.

PRECAUTIONARY STATEMENTS

HAZARDS TO HUMANS AND DOMESTIC ANIMALS

DANGER–POISON

Fatal if swallowed, inhaled or absorbed through the skin. Do not get in eyes or on skin. Do not breath spray mist. Spray operator should work to windward to stay out of drift or mist.

When loading spray tank, wear protective clothing, natural rubber gloves and goggles.

Do not contaminate feed or foodstuffs. Keep out of reach of children and domestic animals. Keep all unprotected persons out of the operating area or vicinity where there may be danger of drift. Wash hands, arms, and face thoroughly with soap and warm water before eating or smoking. Remove and wash all contaminated clothing with soap and hot water before reuse.

SYMPTOMS OF POISONING: A sense of "tightness" in the chest. Sweating. Contracted pupils. Stomach pains. Vomiting and diarrhea.

STATEMENTS OF PRACTICAL TREATMENT

In case of poisoning, call a physician immediately. Have patient lie down and keep quiet.

If swallowed, vomiting should be induced. Administer water freely and induce vomiting by giving one dose (1/2 oz. or 15 ml.) of syrup of ipecac. If vomiting does not occur within 10 to 20 minutes, administer second dose. If syrup of ipecac is not available, induce vomiting by sticking finger down throat. Repeat until vomit fluid is clear. Never give anything by mouth to an unconscious person. Professional medical assistance should be secured immediately.

If on skin, remove contaminated clothing and wash skin immediately with soap and warm water.

If eyes are contaminated, wash with flowing water for at least 15 minutes.

TO PHYSICIAN: ANTIDOTE—Administer atropine sulfate in large therapeutic doses. Repeat as necessary to the point of tolerance. 2-PAM is also antidotal and may be administered in conjunction with atropine.

Compound inhibits cholinesterase resulting in stimulation of the central nervous system, the parasympathetic nervous system, and the somatic motor nerves. Do not give morphine. Watch for pulmonary edema, which may develop in serious cases of poisoning even after 12 hours. At first sign of pulmonary edema, the patient should be placed in an oxygen tent and treated symptomatically.

ENVIRONMENTAL HAZARDS

This pesticide is extremely toxic to fish and wildlife. Do not apply directly to water or wetlands (swamps, bogs, marshes, and potholes). Do not contaminate water by cleaning of equipment or disposal of wastes. Drift and runoff from treated areas may be hazardous to aquatic organisms in neighboring areas.

This product is highly toxic to bees exposed to direct treatment or residues on blooming crops or weeds. Do not apply this product or allow it to drift to blooming crops or weeds if bees are visiting the treatment area. Protective information may be obtained from your Cooperative Agriculture Extension Service.

PROTECTIVE CLOTHING AND WORK SAFETY STATEMENTS

If handled indoors provide mechanical exhaust ventilation.

Keep all unprotected persons, children, livestock, and pets away from treated areas or where there is danger of drift.

Do not rub eyes or mouth with hands. If you feel sick in any way, STOP work and get help right away. See First Aid section of the label.

For questions regarding further treatment, or information about the product, call the National Pesticide Telecommunications Network at 800-858-7378.

HANDLE THE CONCENTRATE ONLY WHEN WEARING THE FOLLOWING PROTECTIVE CLOTHING AND EQUIPMENT:

Wear a protective suit of one or two pieces that covers all parts of the body except the head, hands, and feet. Wear chemical resistant gloves, chemical resistant apron, and chemical resistant shoes, shoe coverings, or boots. Wear goggles or a face shield and a pesticide respirator approved by the National Institute for Occupational Safety and Health under the provision of 30 CFR Part 11.

WEAR THE FOLLOWING PROTECTIVE CLOTHING DURING APPLICATION, EQUIPMENT REPAIR, EQUIPMENT CLEANING, DURING RE-ENTRY TO TREATED AREAS, AND DISPOSAL OF THE PESTICIDE:

Wear a protective suit of one or two pieces that covers all parts of the body except the head, hands, and feet. Wear chemical resistant gloves and chemical resistant shoes, shoe coverings, or boots. Wear a chemical resistant hat during airblast application.

During application from an enclosed tractor cab or airplane cockpit, or other suitable vehicle in which the windows are rolled up, long sleeved shirt and long pants may be worn in place of the above protective clothing. Chemical resistant gloves must be available in the cab or cockpit and must be worn while exiting. This clothing is inadequate to protect you during equipment repair, equipment cleaning, re-entry, or during pesticide disposal.

IMPORTANT! BEFORE REMOVING GLOVES, WASH THEM WITH SOAP AND WATER. ALWAYS WASH HANDS, FACE, AND ARMS WITH SOAP AND WATER BEFORE SMOKING, DRINKING, EATING OR TOILETING.

After work, take off all clothing and shoes. Shower using soap and water. Do not wear contaminated clothing. Wash protective clothing and protective equipment with soap and water after each use. Respirators must be cleaned and filters replaced according to instructions included with the respirators. Personal clothing worn during use must be laundered separately from household articles. Clothing and protective equipment heavily contaminated or drenched with GUTHION azinphos-methyl must be destroyed according to state and local regulations. HEAVILY CONTAMINATED OR DRENCHED CLOTHING CANNOT BE ADEQUATELY DECONTAMINATED.

During aerial application, human flaggers are prohibited unless in totally enclosed vehicle.

STORAGE AND DISPOSAL

Pesticide Disposal: Do not contaminate water, food or feed by storage or disposal.

Pesticide wastes are acutely hazardous. Improper disposal of excess pesticide, spray mixture, or rinsate is a violation of Federal Law. If these wastes cannot be disposed of by use according to label instructions, contact your State Pesticide or Environmental Control Agency, or the Hazardous Waste representative at the nearest EPA Regional Office for guidance.

Container Disposal: Do not use container in connection with food, feed, or drinking water. Completely empty carton into application equipment. Then dispose of empty carton in a sanitary landfill or by incineration, or, if allowed by State and local authorities, by burning. If burned, stay out of smoke.

Store in a cool, dry place and in such a manner as to prevent cross contamination with other pesticides, fertilizers, food, and feed. Do not store below freezing (32° F). Exposure to moisture or excessive handling of water soluble packets may cause breakage. Store packets in original container and out of reach of children, preferably in a locked storage area.

Handle and open container in a manner as to prevent spillage. If the container is leaking or material spilled for any reason or cause, carefully sweep material into a pile. Refer to Precautionary Statements on label for hazards associated with the handling of this material. Do not walk through spilled material. Dispose of pesticide as directed above. In spill or leak incidents, keep unauthorized people away. You may contact the Mobay Emergency Response Team for decontamination procedures or any assistance that may be necessary. The Mobay Kansas City Emergency Response Telephone No. is 816-242-2582, or contact Chemtrec at 800-424-9300.

IMPORTANT:

This bulletin is not intended to provide adequate information for application. Before using the product, read and carefully observe directions, cautionary statements and other information appearing on the product label. This product is sold subject to the Conditions of Sale set forth on the container label.

STOP—READ THE LABEL BEFORE USE

G 9005

This is a specimen label. Mobay is not responsible for the accuracy of the information contained herein. As labels are subject to revision, always carefully read and follow the label on the product container.

MESUROL®

75% WETTABLE POWDER INSECTICIDE—BIRD REPELLENT—MOLLUSCICIDE

RESTRICTED USE PESTICIDE

Due to Avian Toxicity and Fish and Aquatic Organism Toxicity

For retail sale to and use only by certified applicators or persons under their direct supervision and only for those uses covered by the certified applicator's certification.

FOR CONTROL OF CERTAIN INSECTS AND MOLLUSKS, AND FOR BIRD REPELLENCY ON CHERRIES

BEFORE MAKING ANY APPLICATION TO CHERRIES, CONTACT YOUR LOCAL MOBAY REPRESENTATIVE OR STATE AGRICULTURAL AGENT FOR INFORMATION REGARDING THE REGISTRATION STATUS OF MESUROL ON THIS CROP.

ACTIVE INGREDIENT:
methiocarb (4-methylthio-3,5-xylylemethylcarbamate) 75%
INERT INGREDIENTS: 25%

TOTAL 100%

EPA Reg. No. 3125-288

MESUROL is a Reg. TM of Bayer AG, Germany

RECOMMENDED USES

Cherries and Ornamentals

RESTRICTIONS ON WAREHOUSE STORAGE

Do not store in front of space heater. Store below 120°F. Do not store adjacent to, or near foodstuffs, drugs, feeds or any other material intended for consumption by humans or animals. Do not store adjacent to toys, cosmetics, dishes, furniture or clothing. Do not store next to herbicides. Store in a cool dry place.

PACKAGING

5 lb. bag, 6 per case. Shipping weight 36 lbs.

MESUROL 75% Wettable Powder is a carbamate insecticide which effectively controls certain insects on cherries.

The 75% Wettable Powder formulation is Federally registered for use as a bird repellent on cherries. It effectively repels grackles, robins, starlings and other pest birds.

The product may also be used to control snails and slugs and certain other pests on ornamentals.

The product mixes readily with water to form a suspension in all conventional spray equipment.

AVISO

PRECAUCION AL USUARIO: Si usted no puede leer o entender inglés, no use este producto hasta que la etiqueta le haya sido explicada ampliamente.

(TO THE USER. If you cannot read or understand English, do not use this product until the label has been fully explained to you.)

DIRECTIONS FOR USE

It is a violation of Federal Law to use this product in a manner inconsistent with its labeling.

FOR INSECT CONTROL: MESUROL 75% Wettable Powder effectively controls certain specified destructive pests on cherries and ornamentals. Use specified quantities of MESUROL 75% Wettable Powder in sufficient water to give complete coverage of foliage. The type of equipment used will determine the concentration required. Thorough coverage is necessary for maximum insect control.

FOR USE AS A BIRD REPELLENT ON CHERRIES: MESUROL 75% Wettable Powder will aid in the prevention of feeding damage to ripening cherries by certain specified pest birds. Application at rates above those shown may result in illegal crop residues.

MESUROL 75% Wettable Powder mixes readily with water to form a suspension usable in all conventional spray equipment. Maintain adequate agitation in the spray tank to keep the wettable powder in suspension. Do not apply with oil.

Schedule applications in accordance with local conditions.

Do not apply this product through any type of irrigation system.

RECOMMENDED APPLICATIONS

CROP	PEST	DOSAGE MESUROL 75% WP PER ACRE	DOSAGE MESUROL 75% WP PER 100 GALS.	REMARKS
BEARING FRUIT Cherries	Cherry Fruit flies; European red mite; McDaniel mite	1 to 1⅓ pounds	5.3 to 6.9 ounces	For ground application apply specified dosage per 100 gallons of water based on 300 gallons per acre. This concentration is calculated for conventional hydraulic type sprayers. When lower or higher volumes of water are applied, increase or decrease the concentration of MESUROL 75% Wettable Powder in the spray mixture in order to apply amount of MESUROL 75% Wettable Powder per acre equivalent to a full coverage spray. For aerial application apply specified dosage in a minimum of 5 gallons of water per acre. Repeat as necessary with a minimum of 7 days between applications. A maximum of 4 pounds of product may be applied per acre per season regardless of the method of application. Applications may be made up to 14 days prior to harvest. If only 1 application (2 pounds of product maximum) is made in a season, application may be made up to 3 days before harvest. DO NOT GRAZE OR FEED COVER CROPS UNDER TREATED ORCHARDS.
	Plum curculio	1⅓ pounds	6.9 ozs.	
	Repellency of: Robins; Starlings; Finches; Grackles; Sparrows; Bluejays; Cedar Waxwings	2 pounds	10.6 ozs.	
NON-BEARING FRUITS Deciduous fruit trees	Aphids; Mites	½ to 1 lb.		Apply specified dosage per acre in 50 gallons of water. Spray plants and foliage until thoroughly wet. Repeat as necessary applying up to 4 applications per season (8 pounds of product per acre). Do not apply to trees which will bear fruit within 1 year of application.
	Snails; Slugs	2 lbs.		
ORNAMENTALS (Nurseries and Greenhouse only) Flowers; Shrubs; Trees	Aphids; Mites	½ to 1 lb.		Apply specified dosage per acre in 50 gallons of water. Spray plants and foliage until thoroughly wet. Repeat as necessary applying up to 4 applications per season (8 pounds of product per acre). **NOTE:** Do not apply in conjunction with foliar fertilizer applications.
	Snails; Slugs	2 lbs.		4-26-89

PRECAUTIONARY STATEMENTS

HAZARDS TO HUMANS AND DOMESTIC ANIMALS

WARNING

Poisonous if swallowed, inhaled, or absorbed through the skin. Rapidly absorbed through the skin. Do not get in eyes or on skin. Wear protective clothing, natural rubber gloves, and goggles.

Do not breathe dust or spray mist. Wear a mask or respirator similar to those used for protection against organocarbamate insecticides jointly approved by the Mining Enforcement and Safety Administration (formerly the U.S. Bureau of Mines) and by the National Institute for Occupational Safety and Health under the provisions of 30 CFR Part 11.

Do not contaminate feed or food. Keep out of reach of children. Keep all unprotected persons out of the operating area, or vicinity where there may be danger of drift. Wash hands, arms and face thoroughly with soap and warm water before eating or smoking. Wash all contaminated clothing with soap and hot water before reuse.

SYMPTOMS OF POISONING: A sense of "tightness" in the chest. Sweating. Contracted pupils. Stomach pains. Vomiting and diarrhea.

TO PHYSICIAN: Antidote—Administer atropine sulfate in large therapeutic doses. Repeat as necessary to the point of tolerance.

Compound inhibits cholinesterase resulting in stimulation of the central nervous system, the parasympathetic nervous system, and the somatic motor nerves. Do not give morphine. Watch for pulmonary edema, which may develop in serious cases of poisoning even after 12 hours. At first sign of pulmonary edema, the patient should be placed in an oxygen tent and treated symptomatically.

STATEMENTS OF PRACTICAL TREATMENT

If swallowed, vomiting should be induced. Administer water freely, and induce vomiting by giving one dose (½ oz. or 15 ml.) of syrup of ipecac. If vomiting does not occur within 10 to 20 minutes, administer second dose. If syrup of ipecac is not available, induce vomiting by sticking finger down throat. Repeat until vomit fluid is clear. Never give anything by mouth to an unconscious person. Professional medical assistance should be secured immediately. In case of poisoning, call a physician immediately. Have patient lie down and keep quiet.

If on skin, remove contaminated clothing and wash skin immediately with soap and warm water.

If eyes are contaminated, wash with flowing water for at least 15 minutes.

ENVIRONMENTAL HAZARDS

Do not use on other crops used for food or forage. Use only according to label directions. Applications at rates above those shown may result in illegal crop residues.

This pesticide is toxic to fish and very highly toxic to birds. Do not apply directly to water or wetlands (swamps, bogs, marshes, and potholes). Drift and runoff from treated areas may be hazardous to aquatic organisms in neighboring areas. Do not contaminate water by cleaning of equipment, or disposal of wastes.

This product is highly toxic to bees exposed to direct treatment or residues on blooming crops or weeds. Do not apply this product or allow it to drift to blooming crops or weeds if bees are visiting the treatment area.

RE-ENTRY STATEMENTS

Do not apply this product in such a manner as to directly or through drift expose workers or other persons. The area being treated must be vacated by unprotected persons. Do not enter treated areas for 24 hours after application unless protective clothing is worn. Because certain states may require more restrictive re-entry intervals for various crops treated with this product, consult your State Department of Agriculture for further information.

Written or oral warnings must be given to workers who are expected to be in a treated area or in an area about to be treated with this product. Oral warnings should include those statements found under the label heading "Precautionary Statements" and "Re-entry Statements" included on the label. When oral warnings are given, warnings shall be given in a language customarily understood by workers. Oral warnings must be given if there is reason to believe that written warnings cannot be understood by workers. Written warnings must include the following information: WARNING, area treated with MESUROL 75% Wettable Powder on (date of application). Do not enter treated areas for 24 hours after application unless protective clothing is worn. **If on skin,** remove contaminated clothing and wash skin immediately with soap and warm water. **If eyes are contaminated,** wash with flowing water for at least 15 minutes. **If swallowed,** vomiting should be induced. (See Precautionary Statements on label.)

PROTECTIVE CLOTHING AND WORK SAFETY STATEMENTS

REPEATED OR PROLONGED EXPOSURE TO CHOLINESTERASE INHIBITORS SUCH AS ARE CONTAINED IN THIS PRODUCT MAY CAUSE SUSCEPTIBILITY TO ADVERSE EFFECTS FROM VERY SMALL DOSES OF ANY CHOLINESTERASE INHIBITOR.

Keep all unprotected persons, livestock, and pets away from treated areas or where there is a danger of drift.

Do not rub eyes or mouth with hands. If you feel sick in any way, STOP work and get help right away. See First Aid (Practical Treatment) Section of the label.

WEAR THE FOLLOWING PERSONAL PROTECTIVE EQUIPMENT DURING MIXING, LOADING, APPLICATION, REPAIR AND CLEANING OF MIXING/LOADING/APPLICATION EQUIPMENT, DISPOSAL OF THE SPRAY SOLUTION, AND EARLY RE-ENTRY INTO TREATED AREAS BEFORE SPRAYS HAVE DRIED: long sleeve shirt, long pants, shoes, socks, chemical-resistant gloves.

IMPORTANT! Always wash hands, face, and arms with soap and water before smoking, eating, drinking, or toileting.

AFTER WORK: Before removing gloves, wash them with soap and water. Wash hands, face, and arms with soap and water. Launder all clothing worn during use with this product before reusing. Launder this clothing separately from household clothing. Clothing or gloves that have been heavily contaminated or drenched must be destroyed according to state and local regulations. HEAVILY CONTAMINATED OR DRENCHED CLOTHING CANNOT BE ADEQUATELY DECONTAMINATED.

DURING AERIAL APPLICATION, HUMAN FLAGGERS MUST BE IN TOTALLY ENCLOSED VEHICLES.

Aerial applicators are not required to wear chemical resistant gloves if in an enclosed cockpit.

STORAGE AND DISPOSAL

Pesticide Disposal: Do not contaminate water, food or feed by storage or disposal. Wastes resulting from the use of this product may be disposed of on site or at an approved waste disposal facility.

Container Disposal: Do not use container in connection with food, feed, or drinking water. Completely empty bag into application equipment. Then dispose of empty bag in a sanitary landfill or by incineration, or, if allowed by State and local authorities, by burning. If burned, stay out of smoke.

Store in a cool, dry place in such a manner as to prevent cross contamination with other pesticides, fertilizers, food and feed. Store in original container and out of the reach of children, preferably in a locked storage area.

Handle and open container in a manner as to prevent spillage. If the container is leaking or material spilled for any reason or cause, carefully sweep material into a pile. Refer to Precautionary Statements on label for hazards associated with the handling of this material. Do not walk through spilled material. Dispose of pesticide as directed above. In spill or leak incidents, keep unauthorized people away. You may contact the Mobay Emergency Response Team for decontamination procedures or any other assistance that may be necessary. The Mobay Kansas City Emergency Response Telephone No. is 816-242-2582, or contact Chemtrec at 800-424-9300.

IMPORTANT:

This bulletin is not intended to provide adequate information for application. Before using the product, read and carefully observe directions, cautionary statements and other information appearing on the label. This product is sold subject to the Conditions of Sale set forth on the container label.

READ THE LABEL BEFORE USE

ML 9001

This is a specimen label. Mobay is not responsible for the accuracy of the information contained herein. As labels are subject to revision, always carefully read and follow the label on the product container.

METASYSTOX-R™

SPRAY CONCENTRATE SYSTEMIC INSECTICIDE

RESTRICTED USE PESTICIDE

Due to Reproductive Effects

For retail sale to and use only by certified applicators or persons under their direct supervision and only for those uses covered by the certified applicator's certification. Direct supervision for this product is defined as the certified applicator being physically present during mixing, loading, equipment repair, and equipment cleaning. Certified applicators must ensure that all persons involved in these activities under their direct supervision are informed of the precautionary statements.

FOR CONTROL OF APHIDS, MITES AND OTHER PESTS ON A VARIETY OF FOOD CROPS AND FIELD CROPS.

ACTIVE INGREDIENTS:	
S-[2-(Ethylsulfinyl)ethyl] *O,O*-dimethyl phosphorothioate	25%
Aromatic Petroleum Distillate	14%
INERT INGREDIENTS:	61%
TOTAL	100%

Contains 2 lbs. *S*-[2-Ethylsulfinyl)ethyl] *O,O*-dimethyl phosphorothioate per US. gallon.

EPA Reg. No. 3125-111

METASYSTOX-R is a TM of Bayer AG, Germany

RECOMMENDED USES

Alfalfa (seed)	Cucumbers	Peppers
Apples*	Eggplant	Plums*
Apricots*	Filberts†	Prunes*
Beans (snap and lima)	Grapefruit	Pumpkin
	Grapes*	Quinces*
Broccoli	Head lettuce	Safflower
Brussels sprouts	Lemons	Spearmint
Cabbage	Muskmelon (cantaloupe)	Sorghum
Cauliflower		Spanish onions††
Cherries*	Nectarines*	Squash
Citrus*	Oranges	Sugar beets
Clover (seed)	Other melons	Turnips
Cotton	Peaches*	Walnuts
Corn	Pears	Watermelon
Crab apples*	Peppermint	

* Non-bearing only

† Oregon and Washington only

†† West of the Mississippi River only

RESTRICTIONS ON WAREHOUSE STORAGE

Flammable. Keep away from heat and open flame.

PACKAGING

2 gallon can. Shipping wt. 20 lbs.
5 gallon can. Shipping wt. 45 lbs.

METASYSTOX-R Spray Concentrate insecticide is effective for the control of destructive pests that attack certain vegetable, fruit and field crops. It kills these pests by contact and systemic action. METASYSTOX-R Spray Concentrate is absorbed into the foliage, and travels in the sap stream to protect the plant from sucking insects and mites. Weather does not reduce the effectiveness of METASYSTOX-R Spray Concentrate after it has been absorbed into the foliage. The chemical is inside the plant where it cannot wash off or blow away. And, it will not destroy beneficial insect populations after it has been absorbed. METASYSTOX-R Spray Concentrate mixes with water to form an emulsion that is usable in all conventional spray equipment.

AVISO

PRECAUCION AL USUARIO: Si usted no puede leer o entender inglés, no use este producto hasta que la etiqueta le haya sido explicada ampliamente.

(TO THE USER: If you cannot read or understand English, do not use this product until the label has been fully explained to you.)

DIRECTIONS FOR USE

It is a violation of Federal Law to use this product in a manner inconsistent with its labeling.

METASYSTOX-R systemic insecticide penetrates by absorption, and is translocated in the plant. The best time to apply METASYSTOX-R is early morning or evening, especially when crops such as alfalfa and cotton are in bloom. This practice will minimize the possibility of beneficial pollinating insects, including honeybees, from coming into contact with the spray.

MIXING: METASYSTOX-R Spray Concentrate forms an emulsion when diluted with water, and is suitable for use in all power-operated ground and aircraft sprayers. To mix with water, pour the required amount of METASYSTOX-R Spray Concentrate into full amount of water, and then agitate.

Closed systems for mixing and/or loading this product must be capable of removing the pesticide from the shipping container, rinsing the container, and transferring the pesticide and rinsate into mixing tanks and/or application equipment. Protective clothing and equipment, specified in the pesticide labeling, must also be used when mixing and/or loading this pesticide.

DOSAGE: Use specified dosage of METASYSTOX-R Spray Concentrate in the amount of water necessary to give complete coverage of foliage, but in no case use less than one gallon of water per acre. The type of equipment used will determine the concentration required.

SPRAYING: Work to windward. Protect sprayer operators from drift or mist. When low volumes of spray are applied, complete coverage and thorough application are essential for most effective results. Schedule applications in accordance with local conditions. Consult your State Agricultural Experiment Station or Extension Service for specific use information and for state regulations, which may contain additional restrictions or requirements.

USE IN CHEMIGATION SYSTEMS

Types of Irrigation Systems

Apply METASYSTOX-R Spray Concentrate only through sprinkler, including center pivot, lateral move, side roll, overhead solid set or low pressure irrigation systems. Do not apply METASYSTOX-R Spray Concentrate through any other type of irrigation system.

GENERAL DIRECTIONS FOR ALL RECOMMENDED TYPES OF IRRIGATION SYSTEMS

Uniform Water Distribution and System Calibration

The irrigation system must provide uniform distribution of treated water. Crop injury, lack of effectiveness, or illegal pesticide residues in the crop can result from nonuniform distribution of treated water.

The system must be calibrated to uniformly apply the rates specified for chemigation application for specific crops. if you have questions about calibration, you should contact State Extension Service specialists, equipment manufacturers or other experts.

Chemigation Monitoring

A person knowledgeable of the chemigation system and responsible for its operation, or under the supervision of the responsible person, shall shut the system down and make necessary adjustments should the need arise.

Drift

Do not apply when wind speed favors drift beyond the area intended for treatment.

Required System Safety Devices

The system must contain a functional check valve, vacuum relief valve, and low pressure drain appropriately located on the irrigation pipeline to prevent water source contamination from backflow.

The pesticide injection pipeline must contain a functional, automatic quick-closing check valve to prevent the flow of fluid back toward the injection pump.

The pesticide injection pipeline must also contain a functional, normally closed, solenoid-operated valve located on the intake side of the injection pump and connected to the system interlock to prevent fluid from being withdrawn from the supply tank when the irrigation system is either automatically or manually shut down.

The system must contain functional interlocking controls to automatically shut off the pesticide injection pump when the water pump motor stops.

The irrigation line or water pump must include a functional pressure switch which will stop the water pump motor when the water pressure decreases to the point where pesticide distribution is adversely affected.

Systems must use a metering pump, such as a positive displacement injection pump (e.g., diaphragm pump) effectively designed and constructed of matrials that are compatible with pesticides and capable of being fitted with a system interlock.

Using Water from Public Water Systems

DO NOT APPLY METASYSTOX-R SPRAY CONCENTRATE THROUGH ANY IRRIGATION SYSTEM **PHYSICALLY CONNECTED** TO A PUBLIC WATER SYSTEM. Public water system means a system for the provision to the public of piped water for human consumption if such system has at least 15 service connections or regularly serves an average of at least 25 individuals daily at least 60 days out of the year.

METASYSTRO-R Spray Concentrate may be applied through any of the recommended types of irrigation systems which may be **supplied** by a public water system **only if** the water from the public water system is discharged into a reservoir tank prior to pesticide introduction. There shall be a complete physical break (air gap) between the outlet end of the fill pipe and the top or overflow rim of the reservoir tank of at least twice the inside diameter of the fill pipe. Before beginning chemigation, always make sure that the air gap exists and that there is no blockage of the overflow of the reservoir tank.

Any irrigation system using water supplied from a public water system must also meet the following requirements.

The pesticide injection pipeline must contain a functional, automatic, quick-closing check valve to prevent the flow of fluid back toward the injection pump.

The pesticide injection pipeline must contain a functional, normally closed, solenoid-operated valve located on the intake side of the injection pump and connected to the system interlock to prevent fluid from being withdrawn from the supply tank when the irrigation system is either automatically or manually shut down.

The system must contain functional interlocking controls to automatically shut off the pesticide injection pump when the water pump motor stops or in cases where there is no water pump, when the water pressure decreases to the point where pesticide distribution is adversely affected.

Systems must use a metering pump, such as a positive displacement injection pump (e.g., diaphragm pump) effectively designed and constructed of materials that are compatible with pesticides and capable of being fitted with a system interlock.

Do not apply when wind speed favors drift beyond the area intended for treatment.

Agitation

For application of METASYSTOX-R Spray Concentrate alone, a chemical supply tank is not necessary for premixing since METASYSTOX-R Spray Concentrate mixes well with water in the irrigation line. If a chemical supply tank is used, for application of METASYSTOX-R Spray Concentrate alone or in combination with other chemicals, constant strong mechanical or hydraulic agitation must be maintained in the chemical supply tank during the entire period of application.

Chemical Supply Tank Dilution

If a chemical supply tank is used, you must determine the required amounts of METASYSTOX-R Spray Concentrate and water to mix in the tank.

The amount of METASYSTOX-R Spray Concentrate needed equals the number of pints of METASYSTOX-R Spray Concentrate to be applied per acre multiplied by the number of acres to be chemigated.

The amount of emulsion needed equals the gallons of emulsion delivered per hour by the injection pump multiplied by the number of hours chemigation will take place.

The amount of water needed equals the amount of emulsion needed minus the amount of METASYSTOX-R Spray Concentrate needed.

For example, if you want to apply 1 pint of METASYSTOX-R Spray Concentrate per acre to 130 acres in 20 hours and your injection pump delivers 15 gallons per hour, you need: 1 pint METASYSTOX-R Spray Concentrate per acre × 130 acres = 130 pints or 16.25 gallons of METASYSTOX-R Spray Concentrate. And, you need: 15 gallons per hour × 20 hours = 300 gallons of emulsion, minus 16.25 gallons of METASYSTOX-R Spray Concentrate = 283.75 gallons of water.

Cleaning the Chemical Injection System

In order to accurately apply pesticides, the chemical injection system must be kept clean; free from chemical or fertilizer residues and sediments. Refer to your owners manual or ask your equipment supplier for the cleaning procedure for your injection system.

Flushing the Irrigation System

At the end of the appliction period, allow time for all lines to flush the pesticide through all nozzles before turning off irrigation water. To ensure the lines are flushed and free of pesticides, a dye indicator may be injected into the lines to mark the end of the application period.

SPRINKLER IRRIGATION SYSTEMS

All directions and requirements listed under the GENERAL DIRECTIONS AND REQUIREMENTS FOR ALL RECOMMENDED TYPES OF IRRIGATION SYSTEMS section of the label must be followed for sprinkler irrigation systems.

In addition, the following directions apply to sprinkler irrigation systems.

Do not apply when wind speed favors drift beyond the area intended for treatment.

It is recommended that nozzles in the immediate area of control panels, chemical supply tanks, pumps and system safety devices be plugged to prevent chemical contamination of these areas.

Center-Pivot and Automatic-Move Linear Systems

Inject the specified dosage per acre continuously for one complete revolution or move of the system. DO NOT USE END GUNS. The system should be run at maximum speed.

Solid Set and Manually Controlled Linear Systems

Injection should be during the last 30 to 60 minutes of a regular irrigation period or as a separate 30 to 60 minute application not associated with a regular irrigation. DO NOT USE END GUNS.

Low Pressure Irrigation Systems (Mini-sprinkler systems only.)

All directions and requirements listed under the GENERAL DIRECTIONS AND REQUIREMENTS FOR ALL RECOMMENDED TYPES OF IRRIGATION SYSTEMS section of the label must be followed for low pressure Mini-sprinkler irrigation systems.

In addition, the following directions apply to low pressure irrigation systems.

Do not apply when wind speed favors drift beyond the area intended for treatment.

Injection should be during the last 30 to 60 minutes of a regular irrigation period or as a separate 30 to 60 minute application not associated with a regular irrigation.

For systems which do not wet the entire crop area, apply proportionately less than the specified broadcast rate per acre based on the area wetted. For example, if 50% of the crop area is wetted by the irrigation system, apply one half of the specified broadcast rate per acre.

RECOMMENDED APPLICATIONS

CROP	PEST	PINTS MSR S.C.	REMARKS
FIELD CROPS Cotton (Except California)	Aphids Thrips Fleahoppers	½ to 1	Apply specified dosage per acre in sufficient water for complete coverage but not less than one gallon per acre. Do not apply more than 3 times per season nor within 14 days of picking. Do not graze or feed gin trash to dairy or meat animals. For application by sprinkler irrigation systems: Apply specified dosage per acre. Follow all directions given under the CHEMIGATION section of the label.
	Mites	1½ to 2	
Cotton (California only)	Aphids Mites Lygus bugs	1½ to 2	
Corn Field, Sweet, Popcorn	Aphids Corn rootworm beetles Leafhoppers Mites Thrips	1½ to 2	Apply specified dosage per acre in sufficient water for complete coverage but not less than one gallon per acre. For control of corn rootworm beetles apply as needed in accordance with established economic thresholds in your locality. When using one application per season, do not apply within 7 days of harvest of corn, or within 7 days of foraging or harvest of corn fodder. When using two or three applications per season, do not make the last application within 21 days of harvest of corn, or within 21 days of foraging or harvest of corn fodder. Do not apply more than three times per season. For application by sprinkler irrigation systems: Apply specified dosage per acre. Follow all directions given under the CHEMIGATION section of the label.
Peppermint Spearmint	Aphids Mites	3	Apply specified dosage, in at least 20 gallons of water per acre, with ground equipment. For established infestations, make 2 applications 10 to 14 days apart. Do not apply within 14 days of harvest. For application by sprinkler irrigation systems: Apply specified dosage per acre. Follow all directions given under the CHEMIGATION section of the label.
Safflower	Aphids	2	Apply specified dosage per acre, in sufficient water for complete coverage, but not less than 10 gallons per acre. Do not apply more than 3 times per season, nor within 7 days of harvest. For application by sprinkler irrigation systems: Apply specified dosage per acre. Follow all directions given under the CHEMIGATION section of the label.
	Lygus bugs	4	

CROP	PEST	PINTS MSR S.C.	REMARKS
Sorghum	Greenbugs Corn leaf aphids Yellow sugarcane aphid Banks grass mites (Southwest only, including Kansas, Oklahoma, Texas, Colorado, New Mexico and Arizona)	1 to 2	Apply specified dosage per acre, by air or ground equipment, in sufficient water for complete coverage, but not less than one gallon per acre. Use the higher dosage for control of mites. Do not apply more than 3 times per season, nor within 45 days of harvest of grain, or 21 days of grazing or cutting for forage. Do not use on sweet sorghum. **CAUTION:** Since certain sorghum varieties may be sensitive to organophosphate insecticides, METASYSTOX-R Spray Concentrate should be applied to a small area of the field, and observed for a few days, to determine if any crop injury will occur. For application by sprinkler irrigation systems: Apply specified dosage per acre. Follow all directions given under the CHEMIGATION section of the label. **NOTE:** Due to greenbug resistance in certain areas METASYSTOX-R used alone may not provide satisfactory control. Consult your local Agricultural Advisor or Extension Service for recommendations.
Sugar Beets	Aphids Leafhoppers Mites	1½ to 3	Apply specified dosage per acre, in sufficient water for complete coverage, but not less than one gallon per acre. Do not apply more than 6 times per season. Do not harvest beets or use beet tops for feed or forage within 30 days of last treatment. For application by sprinkler irrigation systems: Apply specified dosage per acre. Follow all directions given under the CHEMIGATION section of the labe.
SEED FIELD CROPS Alfalfa Clover	Aphids Leafhoppers Mites Thrips Lygus bugs (California only)	1½ to 2	Apply specified dosage per acre, in sufficient water for complete coverage, but not less than one gallon per acre. Repeat as necessary, up to 21 days before seed harvest. Chaff from seed crop may be used for feed or forage but do not cut green crop for these purposes. For application by sprinkler irrigation systems: Apply specified dosage per acre. Follow all directions given under the CHEMIGATION section of the label.
FRUITS Grapefruit Lemons Oranges	Aphids Texas citrus mite	1 to 1½	Apply specified dosage, in 100 gallons of water, as a full-coverage spray.* Do not apply more than twice per season, nor within 7 days of harvest.
Pears	Aphids Mites	1	Apply specified dosage, in 100 gallons of water, as a full-coverage spray.* Do not apply more than once per season, nor within 30 days of harvest.
NON-BEARING FRUITS Apples Apricots Cherries Crab apples Grapes Nectarines Peaches Plums Prunes Quinces	Aphids Mites	1 to 1½	Apply specified dosage in 100 gallons of water as a full-coverage spray* on non-bearing trees. Repeat as necessary. Do not apply to trees that will bear fruit within the next 12 months. For application by sprinkler irrigation systems: Apply specified dosage per acre. Follow all directions given under the CHEMIGATION section of the label.
Citrus	Aphids Texas citrus mite		
VEGETABLES Beans (snap and lima)	Leafhoppers Mites Lygus bugs (California only)	2	Apply specified dosage in at least 4 gallons of water per acre. Three applications may be made per picking season, with the last application no later than 21 days before harvest. Do not graze or cut treated vines for feed or forage within 21 days of application. For application by sprinkler irrigation systems: Apply specified dosage per acre. Follow all directions given under the CHEMIGATION section of the label.
Broccoli Brussels sprouts Cauliflower	Aphids	1½ to 2	Apply specified dosage per acre in sufficient water for complete coverage, but not less than one gallon per acre. Do not apply more than 3 times per season, nor within 10 days of harvest for brussels sprouts, or 7 days of harvest for broccoli and cauliflower. For application by sprinkler irrigation systems: Apply specified dosage per acre. Follow all directions given under the CHEMIGATION section of the label.
Cabbage (Includes tight heading varieties of Chinese cabbage)	Aphids	1½ to 3	Apply specified dosage per acre in sufficient water for complete coverage, but not less than one gallon per acre. Do not apply more than 3 times per season, nor within 7 days of harvest. For application by sprinkler irrigation systems: Apply specified dosage per acre. Follow all directions given under the CHEMIGATION section of the label.
Cucumbers	Aphids Mites	1½ to 2	Apply specified dosage per acre in sufficient water for complete coverage, but not less than one gallon per acre. Do not apply more than twice per season. For application by sprinkler irrigation systems: Apply specified dosage per acre. Follow all directions given under the CHEMIGATION section of the label.
Eggplant	Aphids Mites	2	Apply specified dosage per acre in sufficient water for complete coverage, but not less than one gallon per acre. Do not apply more than 3 times per season, nor within 7 days of harvest. For application by sprinkler irrigation systems: Apply specified dosage per acre. Follow all directions given under the CHEMIGATION section of the label.
			Apply specified dosage per acre in sufficient water for complete coverage with a minimum of one gallon per acre. A maximum of 3 applications may be made per season. The following preharvest intervals based on number of applications applies only to California's spring and summer crops:

CROP	PEST	PINTS MSR S.C.	REMARKS
Head lettuce	Aphids Mites	1½ to 2	1 application . . . . . 14 days 2 applications . . . . . 21 days 3 applications . . . . . 28 days All crops in Arizona and California's fall and winter crops require a 28 day PHI (regardless of the number of applications). All other states require a 21 day PHI. For application by sprinkler irrigation systems: Apply specified dosage per acre. Follow all directions given under the CHEMIGATION section of the label.
Spanish onions (bulb) (West of the Mississippi River only)	Thrips (Population suppression only)	3	Apply specified dosage per acre in sufficient water for complete coverage, but not less than ten gallons per acre. For best results, make 3 applications at 10 to 14 day intervals with the last application no later than 30 days before harvest. For application by sprinkler irrigation systems: Apply specified dosage per acre. Follow all directions given under the CHEMIGATION section of the label.
Peppers	Aphids	2	Apply specified dosage per acre, in sufficient water for complete coverage, but not less than one gallon per acre. Do not apply more than twice per season. For application by sprinkler irrigation systems: Apply specified dosage per acre. Follow all directions given under the CHEMIGATION section of the label.
Pumpkin Squash (Winter and Summer)	Aphids	1½ to 2	Apply specified dosage per acre in sufficient water for complete coverage, but not less than one gallon per acre. Do not apply more than once per season. Do not apply within 14 days of harvest of pumpkins or Winter squash, nor within 1 day of harvest of Summer squash. For application by sprinkler irrigation systems: Apply specified dosage per acre. Follow all directions given under the CHEMIGATION section of the label.
Turnips	Aphids Mites	1½ to 2	Apply specified dosage per acre in sufficient water for complete coverage, but not less than one gallon per acre. Do not apply more than twice per season. Do not apply within 21 days of harvest if tops are to be used for food, nor within 7 days of harvest if tops are not to be utilized. For application by sprinkler irrigation systems: Apply specified dosage per acre. Follow all directions given under the CHEMIGATION section of the label.
MELONS Muskmelon (Cantaloupe) Other melons	Aphids Mites	1½ to 2	Apply specified dosage per acre in sufficient water for complete coverage, but not less than one gallon per acre. Do not apply more than 3 times per season, nor within 14 days of harvest. For application by sprinkler irrigation systems: Apply specified dosage per acre. Follow all directions given under the CHEMIGATION section of the label.
Watermelons	Aphids Mites	1½ to 2	Apply specified dosage per acre in sufficient water for complete coverage, but not less than one gallon per acre. Do not apply more than twice per season, nor within 7 days of harvest. For application by sprinkler irrigation systems: Apply specified dosage per acre. Follow all directions given under the CHEMIGATION section of the label.
NUTS Filberts (Oregon and Washington only)	Aphids	1 pint per 10 trees (average 4½ in. diameter)	Apply either undiluted or diluted with an equal volume of water. For undiluted application, apply ⅓ fluid ounce (2 teaspoonfuls) of product per inch of trunk diameter. For diluted application apply ⅔ fluid ounce of mixture per inch of trunk diameter. One pint of product is sufficient to treat 10 trees averaging 4½ inches in diameter when applied diluted or undiluted. Apply specified dosage with a paint brush or low pressure sprayer in a band completely around the trunk. On trees with extremely short trunks, it may be necessary to treat bases of scaffold limbs in order to apply the correct dosage. Width of the treated band will vary with trunk diameter. Do not apply more than once per season, nor within 105 days of harvest. Do not use on heavily stressed trees, or on young trees with trunk diameters less than 2 inches.
Walnuts	Aphids Mites	1½	Apply specified dosage per 100 gallons of water as a full-coverage spray.* Do not apply more than once per season, nor within 30 days of harvest. 6-26-89

* This METASYSTOX-R Spray Concentrate dosage is calculated for conventional sprayers. When lower volumes of spray are applied per acre, as with low pressure, low volume, airplane or mist-type equipment, increase the concentration of METASYSTOX-R Spray Concentrate in the spray mixture in order to apply amount of METASYSTOX-R Spray Concentrate per acre equivalent to a full-coverage spray, but in no case use less than one gallon of water per acre.

SPECIAL LOCAL NEED REGISTRATIONS [SECTION 24(c)] FOR METASYSTOX-R SPRAY CONCENTRATE*

CROP	PEST	STATE EPA SLN NO.	DOSAGE	REMARKS
Alfalfa (Seed crop)	Aphids Leafhoppers Mites Thrips Lygus bugs	Idaho ID 78-0001 Oregon OR 77-0067 Nevada NV 77-0014 Washington WA 77-0058	1½ to 2 pints per acre.	Application by ground or air. Lygus bugs is the basis for the SLN registration.
Corn (Field)	Mites	Texas TX 80-0024	1½ to 2 pints per acre	Maximum 3 applications, with a 14 day pre-harvest interval.
Douglas fir seed & breeding orchards	Cone and seed insects	Oregon OR 80-0071 Washington WA 80-0069	3⅛ to 4⅛ gallons per acre.	Application by ground.

CROP	PEST	STATE EPA SLN NO.	DOSAGE	REMARKS
Filberts	Filbert aphid	Oregon OR 80-0079 Washington WA 80-0077	2 pints per acre.	Foliar spray.
Melons	Leafhopper	California CA 82-0104	2 pints per acre.	Application by air at vining stage.
Peppermint	Aphids Mites	Idaho ID 77-0006 Oregon OR 77-0019 Washington WA 77- 0014	3 pints per acre.	Application by ground or air.
Peppers	Leafhopper	California CA 82-0103	2 pints per acre.	Application by air after crop is established. Do not apply more than twice per season.
Spearmint	Aphids Mites	Idaho ID 77-0006 Oregon OR 77-0019 Washington WA 77-0014	3 pints per acre.	Application by ground or air.
Strawberries (Only pre-bloom & post-final harvest)	Strawberry aphid	Oregon OR 79-0067 Washington WA 85-0003	2 pints	Apply by air or ground. Do not apply to fruit. Three applications per year, one application pre-bloom and two post-final harvest.
Christmas Trees	Aphids Mites Thrips	Michigan MI 89-0004	1 to 2 pints per acre	Apply in 5 gallons of water by air and in 20-50 gallons of water by ground.
	Pine Needle Scale	Ohio OH 89-0004	2 pints per acre.	
Field Grown Nursery stock including Christmas Trees	Aphids Mites Scale	Oregon OR 89-8004	1½ to 2 pints per acre.	

* For distribution and use only within the state specified. These Special Local Need registrations approved as of date of this publication are, however, subject to cancellation by the state. Contact your nearest Mobay Agricultural Chemicals Division Office for the current status and a copy of the complete individual supplemental Special Local Need labeling. This labeling must be in the possession of the user at time of pesticide application.

RESTRICTIONS

Do not use on other crops used for food or forage. use only according to label directions. Application at rates above those shown may result in illegal crop residues. Do not treat food crops grown in the greenhouse.

RE-ENTRY STATEMENTS

Do not apply this product in such a manner as to directly or through drift expose workers or other persons. The area being treated must be vacated by unprotected persons. **DO NOT ENTER TREATED AREAS FOR 48 HOURS AFTER APPLICATION UNLESS PROTECTIVE CLOTHING AND EQUIPMENT, AS PRESCRIBED IN THE LABELING, ARE USED.** Because certain states may require more restrictive re-entry intervals for various crops treated with this product, consult your State Department of Agriculture for further information.

Written or oral warnings must be given to workers who are expected to be in a treated area or in an area about to be treated with this product. Oral warnings should include those statements found under the label heading "Precautionary Statements" and "Re-entry Statements" included on the label. When oral warnings are given, warnings shall be given in a language customarily understood by workers. Oral warnings must be given if there is reason to believe that written warnings cannot be understood by workers. Written warnings must include the following information: WARNING, area treated with METASYSTOX-R Spray Concentrate on (date of application). **DO NOT ENTER TREATED AREAS FOR 48 HOURS AFTER APPLICATION UNLESS PROTECTIVE CLOTHING AND EQUIPMENT, AS PRESCRIBED IN THE LABELING, ARE USED.** **If on skin,** remove contaminated clothing and wash skin immediately with soap and warm water. **If eyes are contaminated,** wash with flowing water for at least 15 minutes. **If swallowed,** vomiting should be induced. (See Precautionary Statements on label).

PRECAUTIONARY STATEMENTS

HAZARDS TO HUMANS AND DOMESTIC ANIMALS

WARNING

Causes eye irritation. May be fatal if swallowed, inhaled or absorbed through the skin. Prolonged or frequently repeated skin contact causes allergic reactions in some individuals. Do not get in eyes, on skin, or on clothing. Do not breathe spray mist. Wash thoroughly with soap and warm water after handling. Wash clothing with soap and hot water before reuse.

Avoid contamination of feed or food. Keep out of reach of children.

STATEMENTS OF PRACTICAL TREATMENT

If splashed in eyes. immediately flush eyes with plenty of water for at least 15 minutes. For further treatment, call a physician.

If swallowed, vomiting should be induced. Administer water freely, and induce vomiting by giving one dose (½ oz. or 15 ml.) of syrup of ipecac. If vomiting does not occur within 10 to 20 minutes, administer second dose. If syrup of ipecac is not available, induce vomiting by sticking finger down throat. Repeat until vomit fluid is clear. Never give anything by mouth to an unconscious person. Professional medical assistance should be secured immediately.

If poisoning occurs, obtain prompt medical aid.

NOTE TO PHYSICIAN: May cause cholinesterase depression. Atropine sulfate is antidotal. 2-PAM is also antidotal and may be administered in conjunction with atropine.

ENVIRONMENTAL HAZARDS

This product is toxic to fish and wildlife. Do not apply directly to water or wetlands (swamps, bogs, marshes, and potholes). Do not contaminate water by cleaning of equipment or disposal of wastes. Drift and runoff from treated areas may be hazardous to aquatic organisms in adjacent aquatic sites.

This product is toxic to bees exposed to direct treatment. Do not apply this product while bees are actively visiting the treatment area.

PHYSICAL OR CHEMICAL HAZARDS

Flammable. Keep away from heat and open flame.

PROTECTIVE CLOTHING, EQUIPMENT AND WORK SAFETY STATEMENTS

REPEATED EXPOSURES TO CHOLINESTERASE INHIBITORS SUCH AS ARE CONTAINED IN THIS PRODUCT MAY, WITHOUT WARNING, CAUSE PROLONGED SUSCEPTIBILITY TO VERY SMALL DOSES OF ANY CHOLINESTERASE INHIBITOR.

Keep all unprotected persons, children, livestock, and pets away from treated areas or where there is danger of drift.

Do not rub eyes or mouth with hands. If you feel sick in any way, STOP WORK and get help right away. See First Aid (Practical Treatment) section of label.

If handled indoors, provide mechanical exhaust ventilation.

HANDLE THE CONCENTRATE ONLY WHEN WEARING THE FOLLOWING PROTECTIVE CLOTHING AND EQUIPMENT.

Wear a protective suit of one or two pieces that covers all parts of the body except the head, hands, and feet. Wear chemical resistant gloves, chemical resistant apron, and chemical resistant shoes, shoe coverings, or boots. Wear goggles or a face shield. Wear a pesticide respirator approved by the National Institute for Occupational Safety and Health under the provisions of 30 CFR Part 11.

IF MIXING AND/OR LOADING IN A CLOSED SYSTEM WEAR THE FOLLOWING PROTECTIVE CLOTHING AND EQUIPMENT.

Wear long sleeve shirt and long legged pants. Wear chemical resistant gloves, chemical resistant apron, and chemical resistant shoes, shoe coverings, or boots. Wear goggles or a face shield.

WEAR THE FOLLOWING PROTECTIVE CLOTHING DURING APPLICATION, EQUIPMENT REPAIR, EQUIPMENT CLEANING, AND DURING EARLY REENTRY TO TREATED AREAS.

Wear a protective suit of one or two pieces that covers all parts of the body except the head, hands, and feet. Wear chemical resistant gloves, and chemical resistant boots, shoes, or shoe coverings.

DURING APPLICATION ONLY FROM A TRACTOR WITH A COMPLETELY ENCLOSED CAB OR AERIALLY WITH AN ENCLOSED COCKPIT WEAR THE FOLLOWING PROTECTIVE CLOTHING AND EQUIPMENT.

Wear long sleeve shirt and long legged pants, chemical resistant boots, shoes, or shoe coverings. Clean chemical resistant gloves must be available in the cab/cockpit and must be worn while exiting. This clothing is inadequate to protect you during equipment repair, cleaning or reenty.

IMPORTANT! Before removing gloves, wash them with soap and water. Alway wash hands, face, and arms with soap and water before smoking, eating, drinking or toileting.

AFTER WORK, take off all clothing and shoes. Shower using soap and water. Wear only clean clothes. Do not wear contaminated clothing. Wash protective clothing and protective equipment with soap and water after each use. Respirators must be cleaned and filters replaced according to instructions included with the respirators. Personal and protective clothing worn during use must be laundered separately from household articles. Clothing and protective equipment heavily contaiminated or drenched with oxydemetonmethyl must be destroyed according to state and local regulations. HEAVILY CONTAMINATED OR DRENCHED CLOTHING CANNOT BE ADEQUATELY DECONTAMINATED.

FLAGGING: Human flaggers are prohibited unless in totally enclosed vehicles.

STORAGE AND DISPOSAL

Pesticide Disposal: Do not contaminate water, food or feed by storage or disposal.

Pesticide wastes are toxic. Improper disposal of excess pesticide, spray mixture, or rinsate is a violation of Federal Law. If these wastes cannot be disposed of by use according to label instructions, contact your State Pesticide or Environmental Control Agency, or the Hazardous Waste representative at the nearest EPA Regional Office for guidance.

Container Disposal: Triple rinse (or equivalent). Then offer for recycling or reconditioning, or puncture and dispose of in a sanitary landfill, or by other procedures approved by State and local authorities.

Store in a cool, dry place and in such a manner as to prevent cross contamination with other pesticides, fertilizers, food, and feed. Store away from heat and open flame. Store in original container and out of the reach of children, preferably in a locked storage area.

Handle and open container in a manner as to prevent spillage. If container is leaking invert to prevent leakage. If the container is leaking or material spilled for any reason or cause, carefully dam up spilled material to prevent runoff. Refer to Precautionary Statements on label for hazards associated with the handling of this material. Do not walk through spilled material. Absorb spilled material with absorbing type compounds and dispose of as directed for pesticides above. In spill or leak incidents, keep unauthorized people away. You may contact the Mobay Emergency Response Team for decontamination procedures or any other assistance that may be necessary. The Mobay Kansas City Emergency Response Telephone No. is 816-242-2582, or contact Chemtrec at 800-424-9300.

IMPORTANT:

This bulletin is not intended to provide adequate information for application. Before using the product read and carefully observe directions, cautionary statements and other information appearing on the product label. This product is sold subject to the Conditions of Sale set forth on the container label.

READ THE LABEL BEFORE USE

MS 9001

This is a specimen label. Mobay is not responsible for the accuracy of the information contained herein. As labels are subject to revision, always carefully read and follow the label on the product container.

MONITOR® 4
LIQUID INSECTICIDE

RESTRICTED USE PESTICIDE

Due to Acute Toxicity

For retail sale to and use only by Certified Applicators or persons under their direct supervision, and only for those uses covered by the Certified Applicator's certification.

FOR EFFECTIVE INSECT CONTROL ON COLE CROPS, COTTON, POTATOES, AND SUGAR BEETS

ACTIVE INGREDIENT:	
O,S-Dimethyl phosphoramidothioate*	40%
INERT INGREDIENTS:	60%
TOTAL	100%

Contains 4 lbs. O,S-Dimethyl phosphoramidothioate per gallon.

EPA Reg. No. 3125-280 *U.S. Patent Nos. 3,689,604 and 3,676,555

MONITOR is a Reg. TM of Chevron Chemical Company

RECOMMENDED USES

Cotton, Broccoli, Brussels sprouts, Cabbage, Cauliflower, Celery (Florida only), Potatoes, Sugar beets (Arizona and California only)

RESTRICTIONS ON WAREHOUSE STORAGE

Do not store in front of space heaters. Store below 120°F. Do not store adjacent to or near foodstuffs, drugs, feed or any other material intended for consumption by humans or animals. Do not store adjacent to toys, cosmetics, dishes, furniture or clothing. Do not store next to herbicides. Store in dry place, but not below 15° F. Protect from heat.

PACKAGING

2 gallon can. Shipping weight 24 lbs.
5 gallon can. Shipping weight 55 lbs.

MONITOR 4 liquid insecticide is an organophosphorus compound which controls a wide range of pests by contact and local systemic action. It is effective as a foliar spray against resistant and nonresistant species of aphids, plus cabbage looper, beet armyworm, tuberworm and others. It features excellent cleanup capability, and residual protection, which, supplemented by local systemic activity, continues late into the growing season. MONITOR 4 can also be adapted to a highly efficient preventive program for cole crops and potatoes. Its protection has proved valuable in promoting increased yields. Applied as recommended on cotton, MONITOR 4 provides effective control of several major pests, including mites and thrips. It is compatible with most commonly-used fungicides, and is non-injurious to plants, when used as directed.

PELIGRO

PRECAUCION AL USUARIO: Si usted no puede leer o entender inglés, no use este producto hasta que la etiqueta le haya sido explicada ampliamente.

(TO THE USER: If you cannot read or understand English, do not use this product until the label has been fully explained to you.)

DIRECTIONS FOR USE

It is a violation of Federal Law to use this product in a manner inconsistent with its labeling.

Use the designated amount of product in the following gallons of water per acre, unless otherwise noted:

Ground Application: 25 to 125 gallons of water

Air Application: 3 to 10 gallons of water

Chemigation: 1/8 to 1/4 inch of water

USE IN CHEMIGATION SYSTEMS

Types of Irrigation Systems

Apply MONITOR 4 only through sprinkler, including center pivot, lateral move, side roll, overhead solid set, or low pressure sprinkler irrigation systems. Do not apply MONITOR 4 through any other type of irrigation system.

GENERAL DIRECTIONS FOR ALL RECOMMENDED TYPES OF IRRIGATION SYSTEMS

Uniform Water Distribution and System Calibration

The irrigation system must provide uniform distribution of treated water. Crop injury, lack of effectiveness, or illegal pesticide residues in the crop can result from nonuniform distribution of treated water.

The system must be calibrated to uniformly apply the rates specified for chemigation application for specific crops. If you have questions about calibration, you should contact State Extension Service specialists, equipment manufacturers or other experts.

Chemigation Monitoring

A person knowledgeable of the chemigation system and responsible for its operation, or under the supervision of the responsible person, shall shut the system down and make necessary adjustments should the need arise.

Drift

Do not apply when wind speed favors drift beyond the area intended for treatment.

Required System Safety Devices

The system must contain a functional check valve, vacuum relief valve, and low pressure drain appropriately located on the irrigation pipeline to prevent water source contamination from backflow.

The pesticide injection pipeline must contain a functional, automatic, quick-closing check valve to prevent the flow of fluid back toward the injection pump.

The pesticide injection pipeline must also contain a functional, normally closed, solenoid-operated valve located on the intake side of the injection pump and connected to the system interlock to prevent fluid from being withdrawn from the supply tank when the irrigation system is either automatically or manually shut down.

The system must contain functional interlocking controls to automatically shut off the pesticide injection pump when the water pump motor stops.

The irrigation line or water pump must include a functional pressure switch which will stop the water pump motor when the water pressure decreases to the point where the pesticide distribution is adversely affected.

Systems must use a metering pump, such as as positive displacement injection pump (e.g., diaphragm pump) effectively designed and constructed of materials that are compatible with pesticides and capable of being fitted with a system interlock.

Using Water from Public Water Systems

DO NOT APPLY MONITOR 4 THROUGH ANY IRRIGATION SYSTEM **PHYSICALLY CONNECTED** TO A PUBLIC WATER SYSTEM. Public water system means a system for the provision to the public of piped water for human consumption if such system has at least 15 service connections or regularly serves an average of at least 25 individuals daily at least 60 days out of the year.

MONITOR 4 may be applied through any of the recommended types of irrigation systems which may be **supplied** by a public water system **only if** the water from the public water system is discharged into a reservoir tank prior to pesticide introduction. There shall be a complete physical break (air gap) between the outlet end of the fill pipe and to top or overflow rim of the reservoir tank of at least twice the inside diameter of the fill pipe. Before beginning chemigation, always make sure that the air gap exists and that there is no blockage of the overflow of the reservoir tank.

Any irrigation system using water supplied from a public water system must also meet the following requirements.

The pesticide injection pipeline must contain a functional, automatic, quick-closing check valve to prevent the flow of fluid back toward the injection pump.

The pesticide injection pipeline must contain a functional, normally closed, solenoid-operated valve located on the intake side of the injection pump and connected to the system interlock to prevent fluid from being withdrawn from the supply tank when the irrigation system is either automatically or manually shut down.

The system must contain a functional interlocking control to automatically shut off the pesticide injection pump when the water pump motor stops or in cases where there is no water pump, when the water pressure decreases to the point where the pesticide distribution is adversely affected.

Systems must use a metering pump, such as a positive displacement injection pump (e.g., diaphragm pump) effectively designed and constructed of materials that are compatible with pesticides and capable of being fitted with a system interlock.

Do not apply when wind speed favors drift beyond the area intended for treatment.

Posting Requirements

Posting of areas to be chemigated is required when 1) any part of a treated area is within 300 feet of sensitive areas such as residential areas, labor camps, businesses, day care centers, hospitals, in-patient clinics, nursing homes or any public areas such as schools, parks, playgrounds, or other public facilities not including public roads, or 2) when the chemigated area is open to the public such as golf courses or retail greenhouses.

Posting must conform to the following requirements: Treated areas shall be posted with signs at all usual points of entry and along likely routes of approach from the listed sensitive areas. When there are no usual points of entry, signs must be posted in the corners of the treated areas and in any other location affording maximum visibility to sensitive areas. The printed side of the sign should face away from the treated area towards the sensitive area. The signs shall be printed in English. Signs must be posted prior to application and must remain posted until foliage has dried and soil surface water has disappeared. Signs may remain in place indefinitely as long as they are composed of materials to prevent deterioration and maintain legibility for the duration of the posting period.

All words shall consist of letters at least 2½ inches tall, and all letters and the symbol shall be a color which sharply contrasts with their immediate background. At the top of the sign shall be the words KEEP OUT, followed by an octagonal stop sign symbol at least 8 inches in diameter containing the word STOP. Below the symbol shall be the words PESTICIDES IN IRRIGATION WATER.

Posting required for chemigation does not replace other posting and reentry interval requirements for farmworker safety.

Agitation

For application of MONITOR 4 alone, a chemical supply tank is not necessary for premixing since MONITOR 4 mixes well with water in the irrigation line. If a chemical supply tank is used for application of MONITOR 4, constant strong mechanical or hydraulic agitation must be maintained in the chemical supply tank during the entire period of application.

Chemical Supply Tank Dilution

If a chemical supply tank is used, you must determine the required amounts of MONITOR 4 and water to mix in the tank.

The amount of MONITOR 4 needed equals the number of pints of MONITOR 4 to be applied per acre multiplied by the number of acres to be chemigated.

The amount of emulsion needed equals the gallons of emulsion delivered per hour by the injection pump multiplied by the number of hours chemigation will take place.

The amount of water needed equals the amount of emulsion needed minus the amount of MONITOR 4 needed.

For example, if you want to apply 2 pints of MONITOR 4 per acre to 130 acres in 20 hours and your injection pump delivers 15 gallons per hour, you need: 2 pints of MONITOR 4 per acre × 130 acres = 260 pints or 32.5 gallons of MONITOR 4. And, you need: 15 gallons per hour × 20 hours = 300 gallons of emulsion, minus 32.5 gallons of MONITOR 4 = 267.5 gallons of water.

Cleaning the Chemical Injection System

In order to accurately apply pesticides, the chemical injection system must be kept clean; free of chemical or fertilizer residues and sediments. Refer to your owners manual or ask your equipment supplier for the cleaning procedure for your injection system.

Flushing the Irrigation System

At the end of the application period, allow time for all lines to flush the pesticide through all nozzles or emitters before turning off irrigation water. To ensure the lines are flushed and free of pesticides, a dye indicator may be injected into the lines to mark the end of the application period.

SPRINKLER IRRIGATION SYSTEMS

All directions and requirements listed under the GENERAL DIRECTIONS AND REQUIREMENTS FOR ALL RECOMMENDED TYPES OF IRRIGATION SYSTEMS section of the MONITOR 4 label must be followed for sprinkler irrigation systems.

In addition, the following directions apply to sprinkler irrigation systems.

Do not apply when wind speed favors drift beyond the area intended for treatment.

It is recommended that nozzles in the immediate area of control panels, chemical supply tanks, pumps and system safety devices be plugged to prevent chemical contamination of these areas.

Center-Pivot and Automatic-Move Linear Systems

Inject the specified dosage per acre continuously for one complete revolution or more of the system. DO NOT USE END GUNS. The system should be run at maximum speed.

Solid Set and Manually Controlled Linear Systems

Injection should be during the last 30 to 60 minutes of regular irrigation period or as a separate 30 to 60 minute application not associated with a regular irrigation. DO NOT USE END GUNS.

RECOMMENDED APPLICATIONS

CROP	INSECT	PINTS MONITOR 4	REMARKS
FIELD	Aphids (Early- Thrips season)	⅕ to ⅖ (0.1 to 0.2 lbs. AI/A)	

CROP	INSECT	PINTS MONITOR 4	REMARKS
Cotton	Fleahoppers	⅖ (0.2 lbs. AI/A)	Apply specified dosage per acre as needed up to 50 days before harvest and before bolls open. For whiteflies, fleahoppers and low rates for aphids and thrips, a second application may be needed for heavy infestations. For application by irrigation systems: Apply specified dosage per acre. Follow all directions given under the CHEMIGATION section of the label. Do not feed gin trash to livestock or allow animals to graze on treated areas.
	Whiteflies (Except Sweet potato whitefly)	½ to 1	
	Beet armyworms Cabbage looper Lygus Mites	1 to 2	
	Aphids (Mid-to late-season) Thrips		
	Stink bugs (Arizona and California only)		
VEGETABLES Broccoli Brussels sprouts Cabbage (Includes tight heading varieties of Chinese cabbage) Cauliflower	Cabbage looper Imported cabbageworm Diamondback moth larvae Aphids	1 to 2	**In a Preventive Program:** Beginning when the plants are small, use the 1 pint rate (0.5 pound active) per acre, and apply at 7-day intervals. **For Clean-up of Existing Populations:** Use 1½ to 2 pints (0.75 to 1.0 pound active) per acre. Do not apply to Brussels sprouts later than 14 days before the first harvest, or to cauliflower later than 28 days before first harvest, or to cabbage later than 35 days before first harvest. For broccoli, the dosage rate of 1 pint per acre may be made up to 14 days before first harvest, but do not apply rates above 1 pint per acre within 21 days of first harvest. For application by irrigation systems: Apply specified dosage per acre. Follow all directions given under the CHEMIGATION section of the label.
Celery (Florida only)	Dipterous leafminers	1 to 2 (0.5 to 1.0 lbs. active)	Apply specified dosage per acre with ground equipment only. Up to 10 applications may be made per season at 7 to 10 day intervals. Do not apply within 21 days of harvest. For application by irrigation systems: Apply specified dosage per acre. Follow all directions given under the CHEMIGATION section of the label. All celery must be trimmed in field and no trimming may be fed to livestock. Treated celery must not be dried for celery flakes.
Potatoes	Aphids Potato tuberworm Cabbage looper Colorado potato beetle (Except Long Island, New York) Flea beetles Potato leafhopper Cutworms (Foliage feeding) (Except California) Armyworms (Except California) European corn borer (Except California) Lygus bug (Except California)	1½ to 2	Apply specified dosage (0.75 to 1.0 pound active) per acre. Apply in a 7 to 10 day preventive program, or as necessary. Do not apply later than 14 days before harvest. For application by irrigation systems: Apply specified dosage per acre. Follow all directions given under the CHEMIGATION section of the label. Resistance to Colorado potato beetle has occurred in other areas of New York. Consult your local extension service for details.
Sugar beets (Arizona and California only)	Green peach aphid	1 to 2	Apply specified dosage per acre as a full coverage spray as needed, allowing 14 days between applications. For application by irrigation systems: Apply specified dosage per acre. Follow all directions given under the CHEMIGATION section of the label. Do not make more than 6 applications per crop season and not within 30 days of harvest. Sprays may be directed to obtain full foliar coverage. 9-21-87
	Beet armyworm	2	

SPECIAL LOCAL NEED REGISTRATIONS [SECTION 24(C)] FOR MONITOR 4*

CROP	INSECT	STATE EPA SLN NO.	DOSAGE	REMARKS
Alfalfa (Seed crop)	Lygus Stinkbug Plant Bugs	Arizona AZ 82-0013	1½ to 2 pints per acre.	Application by air.
Cotton	Aphids Beet armyworm Cabbage looper Mites Thrips Lygus	California CA 79-0188	1 to 2 pints per acre.	CA 79-0188, AR 81-0044, MS 81-0055: Do not apply after 65% of the bolls are open. MS 81-0014: Apply as needed up to 50 days before harvest and before bolls open. For application by air apply in 1 to 3 gallons of water.

CROP	INSECT	STATE EPA SLN NO.	DOSAGE	REMARKS
	Whiteflies	Arkansas AR 81-0044 Mississippi MS 81-0055 MS 81-0014	1/2 to 1 pint per acre.	
	Thrips (early-season)	Arkansas AR 81-0044	1/5 to 2/5 pint per acre.	
	Fleahoppers		2/5 pint per acre.	
	Aphids (early-season) Thrips (early-season)	Mississippi MS 81-0055 MS 81-0014	1/5 to 2/5 pint per acre.	
	Fleahoppers		2/5 pint per acre.	
	Beet armyworms Cabbage looper Mites Aphids Thrips (mid-to late-season)	Arkansas AR 81-0044 Mississippi MS 81-0055 MS 81-0014	1 to 2 pints per acre.	
	Western flower thrips	Arkansas AR 87-0007 Louisiana LA 87-0012	3/5 to 2 pints per acre.	Apply up to 50 days before harvest and before bolls open.
Cucumbers	Aphids Pickleworms Melonworm Leafminer	Florida FL 81-0009	1 to 2 pints per acre.	Maximum 12 pints per acre per crop season. Apply up to 7 days of harvest.
Eggplant	Spidermites Aphids Leafminers Lepidopterous larvae	Florida FL 81-0033	1 to 2 pints per acre.	Maximum 14 pints per acre per crop season. Apply up to 7 days of harvest.
Head lettuce	Aphids Cabbage looper Vegetable leafminer	Florida FL 81-0012	1 to 2 pints per acre.	Apply when insects first appear.
Head lettuce (Crisphead only)	Aphids Armyworms Cabbage looper Leafminers Thrips	California CA 80-0186	1 to 2 pints per acre.	Ground application.
Melons	Leafminer Pickleworm Melonworm Rindworm Complex (cabbage looper, tobacco budworm, granulate cutworm)	Florida FL 81-0034	1 to 2 pints per acre.	Apply at 7 day intervals. Do not apply more than 12 pints per acre per crop season.
Melons	Melon aphid Leaf miner Pickleworm Melonworm Rindworm Complex (cabbage looper, corn earworm, tobacco budworm, granulate cutworm) Green stink bug Potato leafhopper Spotted cucumber beetle Two-spotted mite	Texas TX 84-0020	1 to 2 pints per acre.	Apply specified dosage (0.5 to 1.0 pound active) in 20-100 gallons of water per acre at 7 day intervals. Do not apply more than 5 times per season. Applications may be made up to 14 days of harvest for watermelon, cantaloupe and honeydew melons.
	Rindworm Complex (Heliothis, Armyworm, Cutworm, Looper) Leafminer Diabrotica bettle Aphids Leafhopper	California CA 88-0021	1 to 2 pints per acre.	Air or ground application. Maximum 4 applications up to 14 days of harvest.
Peppers	Green peach aphids Leafminers Thrips Flea beetles Aphids	New Mexico NM 82-0008 Texas TX 82-0019 California CA 88-0020	1 pint per acre.	Application by air or ground up to 21 days before harvest.

CROP	INSECT	STATE EPA SLN NO.	DOSAGE	REMARKS
Seed alfalfa	Lygus Thrips Armyworm Spider mite suppression	California CA 83-0064	2 pints per acre.	Apply by ground or air. Do not make more than one pre-bloom application per crop season. Do not buffer spray solution to less than pH 6.0.
Seed Clover	Lygus		2 pints per acre.	**POST-BLOOM:** Apply specified dosage per acre after removing bees from field.
Tomatoes	Aphids Beet armyworm[1] Black cutworm[1] Cabbage looper Colorado potato beetle Fruitworm Hornworm Leafminers Variegated cutworm [1] Not for Indiana SLN	California CA 78-0163 Indiana IN 79-0001 Michigan MI 78-0016 Ohio OH 79-0008 North Carolina NC 89-0007	1½ to 2 pints per acre.	Applications may be made up to 7 days of harvest.
Tomatoes (Fresh fruit only)	Leafminers (adults)	South Carolina SC 78-0007	1½ to 2 pints per acre.	Ground application up to 14 days of harvest.
	Aphids Fruitworm Leafminers Tomato pinworm	Florida FL 80-0046		Application by ground or air, up to 7 days of harvest.

* For distribution and use only within the state specified. These Special Local Need registrations approved as of date of this publication are, however, subject to cancellation by the state. Contact your nearest Mobay Agricultural Chemicals Division Office for the current status and a copy of the complete individual supplemental Special Local Need labeling. This labeling must be in the possession of the user at time of pesticide application.

RESTRICTIONS

Do not apply under conditions involving possible drift to food, forage or other plantings that might be damaged or the crops thereof rendered unfit for sale, use or consumption. Not for use or storage in or around the home.

RE-ENTRY STATEMENTS

Do not apply this product in such a manner as to directly or through drift expose workers or other persons. The area being treated must be vacated by unprotected persons. Do not enter treated areas within 24 hours after application unless appropriate protective clothing is worn. Because certain states may require more restrictive re-entry intervals for various crops treated with this product, consult your State Department of Agriculture for further information.

Written or oral warnings must be given to workers who are expected to be in a treated area or in an area about to be treated with this product. Oral warnings should include those statements found under the label heading "Precautionary Statements" and "Re-entry Statements" included on the label. When oral warnings are given, warnings shall be given in a language customarily understood by workers. Oral warnings must be given if there is reason to believe that written warnings cannot be understood by workers. Written warnings must include the following information: DANGER, area treated with MONITOR 4 on (date of application). Do not enter within 24 hours after application unless appropriate protective clothing is worn. **If on skin,** remove contaminated clothing and wash skin immediately with soap and warm water. **If eyes are contaminated,** wash with flowing water for at least 15 minutes. **If swallowed,** vomiting should be induced. (See Precautionary Statements on label).

PRECAUTIONARY STATEMENTS

HAZARDS TO HUMANS AND DOMESTIC ANIMALS

Keep out of reach of children. Do not inhale. Do not get on skin. Do not take internally.

Poisonous if swallowed, inhaled or absorbed through skin. Rapidly absorbed through skin. Do not get in eyes, on skin or on clothing. Wear a mask or respirator jointly approved by the Mining Enforcement and Safety Administration (formerly the U.S. Bureau of Mines) and by the National Institute for Occupational Safety and Health under the provisions of 30 CFR Part 11. Wear natural rubber gloves, protective clothing and goggles. In case of contact, wash immediately with soap and water. Wash hands, arms and face thoroughly with soap and water before eating or smoking. Keep all unprotected persons out of the operating areas or vicinity where there may be danger of drift. Do not store or transport with feed or food.

NOTE TO PHYSICIAN: Antidote—Atropine is antidotal. 2-PAM is also antidotal, and may be administered in conjunction with atropine.

Handling of Contaminated Articles: Clothing and rags receiving undiluted MONITOR 4 as a result of leaks, spills or mishaps should be removed immediately and not reused. These contaminated articles should be disposed of in the same manner as the empty container; see container disposal section for directions. (Contaminated articles which should be discarded include: hats, gloves, aprons, coats, boots, etc. not made of rubber, or coated with rubber or other similar materials.) Clothing receiving spray mist or droplets from the mixed, diluted MONITOR 4 should be removed and decontaminated before reuse. This clothing can be decontaminated by machine washing, separately from other items, with soap or detergent and bleach in hot water twice before reuse. Protective clothing and equipment should be washed down with detergent or soap and bleach water. Wash water from the cleaning of protective clothing or equipment should not be allowed to run off or otherwise enter water supplies.

STATEMENT OF PRACTICAL TREATMENT

If swallowed, vomiting should be induced. Administer water freely and induce vomiting by giving one dose (½ oz. or 15 ml.) of syrup of ipecac. If vomiting does not occur within 10 to 20 minutes, administer second dose. If syrup of ipecac is not available, induce vomiting by sticking finger down throat. Repeat until vomit fluid is clear. Never give anything by mouth to an unconscious person. Have patient lie down and keep quiet. Professional medical assistance should be secured immediately.

If inhaled, remove victim to fresh air. If not breathing give artificial respiration, preferably mouth to mouth. Get medical assistance immediately.

If on skin. In case of contact, remove contaminated clothing and immediately wash with soap and water.

If in eyes, wash with plenty of water for at least 15 minutes. Get medical attention.

ENVIRONMENTAL HAZARDS

This product is extremely toxic to birds and other wildlife. Birds and other wildlife in treated areas may be killed. Do not apply directly to water bodies or wetlands (e.g., lakes, streams, ponds, canals). Do not contaminate water by cleaning of equipment or disposal of wastes. This product is highly toxic to bees exposed to direct treatment or residues on blooming crops or weeds. Do not apply this product or allow it to drift to blooming crops or weeds if bees are visiting the treatment area.

STORAGE AND DISPOSAL

Pesticide Disposal: Do not contaminate water, food or feed by storage or disposal.

Pesticide wastes are acutely hazardous. Improper disposal of excess pesticide, spray mixture, or rinsate is a violation of Federal Law. If these wastes cannot be disposed of by use according to label instructions, contact your State Pesticide or Environmental Control Agency, or the Hazardous Waste representative at the nearest EPA Regional Office for guidance.

Container Disposal: Metal—Triple rinse (or equivalent). Then offer for recycling or reconditioning, or puncture and dispose of in a sanitary landfill approved by State and local authorities. **Plastic**—Triple rinse (or equivalent). Then offer for recycling or reconditioning, or puncture and dispose of in a sanitary landfill, or incineration, or, if allowed by state and local authorities, by burning. If burned, stay out of smoke.

Store in a cool, dry place and in such a manner as to prevent cross contamination with other pesticides, fertilizers, food, and feed. Store in original container and out of the reach of children, preferably in a locked storage area.

Handle and open container in a manner as to prevent spillage. If container is leaking, invert to prevent leakage. If the container is leaking or material spilled for any reason or cause, carefully dam up spilled material to prevent runoff. Refer to Precautionary Statements on label for hazards associated with the handling of this material. Do not walk through spilled material. Absorb spilled material with absorbing type compounds and dispose of as directed for pesticides above. In spill or leak incidents, keep unauthorized people away. You may contact the Mobay Emergency Response Team for decontamination procedures or any other assistance that may be necessary. The Mobay Kansas City Emergency Response Telephone No. is 816-242-2582, or contact Chemtrec at 800-424-9300.

IMPORTANT:

This bulletin is not intended to provide adequate information for application. Before using the product, read and carefully observe directions, cautionary statements and other information appearing on the product label. This product is sold subject to the Conditions of Sale set forth on the container label.

READ THE LABEL BEFORE USE

MT 9001

This is a specimen label. Mobay is not responsible for the accuracy of the information contained herein. As labels are subject to revision, always carefully read and follow the label on the product container.

MORESTAN®

25% WETTABLE POWDER
MITICIDE—FUNGICIDE—INSECTICIDE

FOR CONTROL OF MITES, POWDERY MILDEW, PEAR PSYLLA AND APHIDS

ACTIVE INGREDIENT:	
6-Methyl-1,3-dithiolo [4,5-*b*]quinoxalin-2-one	25%
INERT INGREDIENTS:	75%
TOTAL	100%

EPA Reg. No. 3125-302

MORESTAN is a Reg. TM of Bayer AG, Germany

RECOMMENDED USES

BEARING FRUIT Apples Apricots Pears Citrus (CA, AZ, TX, FL only)
NON-BEARING FRUIT Apples Apricots Cherries Citrus Nectarines Peaches Pears Plums Prunes

FRUIT Strawberries
OTHER CROPS Macadamia nuts Walnuts

RESTRICTIONS ON WAREHOUSE STORAGE

PACKAGING 5 lb. carton, contains 5 one-pound water soluble packets, 6 per case. Shipping weight 36 lbs.
Protect from rain and moisture. Store in a cool dry place but not below 32°F. Do not store next to herbicides.

MORESTAN 25% Wettable Powder effectively controls several species of mites attacking fruit, macadamia nuts and walnuts. In addition, it protects against powdery mildew on apple trees, controls aphids on walnuts and guards against pear psylla injury.

MORESTAN controls on contact. It is equally effective against mite eggs, nymphs and adults of resistant and non-resistant strains. Its residual activity extends effectiveness.

MORESTAN 25% Wettable Powder mixes readily with water to form a suspension that can be used in conventional air blast sprayers at various concentrations.

Water soluble packets are pre-measured to reduce dust and handling. Offered for sale in Arizona, California, Hawaii, Idaho, Oregon, Utah and Washington.

AVISO

PRECAUCION AL USUARIO: Si usted no puede leer o entender inglés, no use este producto hasta que la etiqueta le haya sido explicada ampliamente.

(TO THE USER: If you cannot read or understand English, do not use this product until the label has been fully explained to you.)

DIRECTIONS FOR USE

It is a violation of Federal Law to use this product in a manner inconsistent with its labeling.

MORESTAN 25% Wettable Powder miticide-fungicide-insecticide is recommended for residual control of aphids, mites, mite eggs and powdery mildew. The formulation mixes readily with water to form a suspension, which is usable in all conventional hand- or machine-operated sprayers. For best results, apply a thorough coverage before mites or mite eggs are present in large numbers, and before mites cause bronzing of the foliage. Apply during good drying conditions (see "RESTRICTIONS"). Consult your State Agricultural Extension Service or Experiment Station for detailed information on timing of applications for most effective control.

MIXING: The enclosed packets containing MORESTAN 25% Wettable Powder are **water soluble.** Do not allow packets to become wet prior to adding to the spray tank. Do not handle with wet hands. Reseal outer carton to protect remaining packets.

To prepare the spray mixture, drop the required number of unopened packets, as determined under "Recommended Applications," into the spray tank while filing with water to the desired level. Operate the agitator while mixing. Depending on the water temperature and the degree of agitation, the packets should be completely dissolved within approximately 5 minutes from the time they were added to the water.

Do not apply this product through any type of irrigation system.

DOSAGE: Use specified dosage of MORESTAN 25% Wettable Powder in the designated amount of water necessary to give complete coverage of foliage. Determine the **total amount** of wettable power to be added to the spray tank based on the rates under "Recommended Applications." **For each 1 pound (16 ounces)** of wettable powder to be added to the spray tank use one 1 pound packet. For example, if it is determined that 2 pounds (32 ounces) of MORESTAN 25% Wettable Powder should be added to the spray tank, add two 1 pound packets. The type of equipment used will determine the concentration required; however, use of these packets is not recommended for making highly concentrated mixtures such as used in aircraft spraying.

RECOMMENDED APPLICATIONS

CROP	PEST	POUNDS MORESTAN 25% WP	REMARKS
BEARING FRUIT Pre-Bloom and Post-Harvest Apples Pears	Powdery mildew (see **NOTE** in Remarks) Mites Mite eggs	½ to 1	Apply specified dosage per 100 gallons of water in pre-bloom sprays (delayed-dormant, pre-pink or pink). Do not apply after first bloom. Additional applications may be made to tree foliage after harvest. **NOTE:** Pre-bloom sprays with MORESTAN 25% W.P. will provide effective control of early-season powdery mildew infections. No additional fungicides are needed for powdery mildew control during the pre-bloom period. Use other recommended fungicides in post-bloom spray programs for seasonal control.
	Pear psylla	1½	
Apricots (Post-harvest only)	Mites (Except cyclamen mite) Mite eggs	4	Apply specified dosage per acre in a minimum of 10 gallons of water per acre by air or 40 gallons of water per acre by ground. For best control thorough coverage of all surfaces is necessary. Make only one application between harvest and up to the time trees break dormancy. Do not graze cover crops.
BEARING CITRUS TREES Citrus (California, Arizona, Texas and Florida only)	Citrus red mite Citrus rust mite Texas citrus mite Mite eggs	5	Apply specified dosage per acre by concentrate or dilute ground spray using 20 to 500 gallons of water per acre. Schedule applications no closer than 90 days apart in Texas and Florida, or 180 days apart in California or Arizona. Allow at least 21 days between last application and harvest. **NOTE:** No more than 1 (one) application may be made to bearing fruit.

CROP	PEST	POUNDS MORESTAN 25% WP	REMARKS
NON-BEARING FRUITS Apples	Powdery mildew Mites Mite eggs	½ to 1	Apply specified dosage per 100 gallons of water.* Spray all foliage surfaces, including the underside of leaves, for complete coverage. Repeat as necessary. For control of powdery mildew, begin preventive applications before disease appears, and apply at 7- to 14-day intervals. During heavy infection periods, make more frequent applications. Do not apply to fruit trees that will bear fruit within one year of application.
Apricots Pears Cherries Plums Nectarines Prunes Peaches	Mites (Except cyclamen mite) Mite eggs	½ to 1	
Citrus	Citrus red mite Citrus rust mite Texas citrus mite Mite eggs	1 to 1½	Apply specified dosage per 100 gallons of water. Spray all surfaces, including the underside of leaves, for complete coverage. Repeat as necessary. Do not apply to trees that will bear fruit within one year of application.
FRUIT Strawberries (Pre-Bloom Only)	Mites (except cyclamen mite) Mite eggs	½ to 1	Apply specified dosage per 100 gallons of water* as a pre-bloom application. Application may be made in propagating beds, and in field plantings during the first year of setting new plants, but not after first bloom. Spray all foliage surfaces, include the underside of leaves for complete coverage. Repeat applications may be made as needed but not after first bloom appears.
NUTS Macadamia nuts	Mites	1 to 2	Apply specified dosage per 100 gallons of water as a full-coverage spray,* using up to 400 gallons of finished spray per acre, depending upon the number and size of trees. Spray all surfaces, including the underside of leaves, for complete coverage. Three applications per season may be made up to the day of harvest.
Walnuts	Aphids Mites Mite eggs	1 to 1½	Apply specified dosage per 100 gallons of water as a full-coverage spray,* using up to 250 gallons of finished spray per acre, depending upon the number and size of trees. Spray all foliage surfaces, including the underside of leaves, for complete coverage. Do not apply more than twice per season, or within 30 days of harvest. 8-18-89

* This concentration is calculated for conventional sprayers. When lower volumes of spray are applied per acre, as with low pressure, low volume, or mist-type equipment, increase the concentration of MORESTAN 25% Wettable Powder in the spray mixture in order to apply amount of MORESTAN 25% Wettable Powder per acre equivalent to a full-coverage spray. Refer also to "Dosage" paragraph.

SPECIAL LOCAL NEED REGISTRATIONS [SECTION 24(c)] FOR MORESTAN 25% WETTABLE POWDER (PVA)*

CROP	PEST	STATE EPA SLN NO.	DOSAGE	REMARKS
Bearing citrus trees	Mites Mite eggs	California CA 79-0059	5 pounds per acre.	By concentrate or dilute ground spray.
Cucurbits (grown for seed)	Powdery Mildew Mites Mite eggs	California CA 87-0010	1.5 pounds	Apply specified dosage per acre by ground or air.

* For distribution and use only within the state specified. These Special Local Need registrations approved as of date of this publication are, however, subject to cancellation by the state. Contact your nearest Mobay Agricultural Chemicals Division Office for the current status and a copy of the complete individual supplemental Special Local Need labeling. This labeling must be in the possession of the user at time of pesticide application.

RESTRICTIONS

Under certain adverse weather conditions, particularly those that create slow drying of spray deposits, injury may occur to foliage of strawberries and deciduous fruit trees having new and/or tender growth.

Do not use summer oils if MORESTAN is used in the cover spray program.

Do not use spreader-sticker materials with MORESTAN.

Do not allow sprays to drift onto other crops, including pastures. Do not graze treated areas.

Do not use on other crops used for food and forage. Use only according to label directions. Applications at rates above those shown may result in illegal crop residues. Do not treat food crops grown in the greenhouse.

RE-ENTRY STATEMENTS

Do not apply this product in such a manner as to directly or through drift expose workers or other persons. The area being treated must be vacated by unprotected persons. Do not enter treated areas without protective clothing until sprays have dried. Because certain states may require more restrictive re-entry intervals for various crops treated with this product, consult your State Department of Agriculture for further information.

Written or oral warnings must be given to workers who are expected to be in a treated area or in an area about to be treated with this product. Oral warnings should include those statements found under the label heading "Precautionary Statements" and "Re-entry Statements" included on the label. When oral warnings are given, warnings shall be given in a language customarily understood by workers. Oral warnings must be given if there is reason to believe that written warnings cannot be understood by workers. Written warnings must include the following information: WARNING, area treated with MORESTAN 25% Wettable Powder on (date of application). Do not enter without appropriate protective clothing until sprays have dried. **If on skin,** remove contaminated clothing and wash skin immediately with soap and warm water. **If eyes are contaminated,** wash with flowing water for at least 15 minutes. **If swallowed,** vomiting should be induced. (See Precautionary Statements on label.)

STORAGE AND DISPOSAL

Pesticide Disposal: Do not contaminate water, food or feed by storage or disposal.

Wastes resulting from the use of this product may be disposed of on site or at an approved waste disposal facility.

Container Disposal: Do not use container in connection with food, feed, or drinking water. Completely empty carton into application equipment. Then dispose of empty carton in a sanitary landfill or by incineration, or, if allowed by State and local authorities, by burning. If burned, stay out of smoke.

Store in a cool, dry place and in such a manner as to prevent cross contamination with other pesticides, fertilizers, food, and feed. Do not store below freezing (30°F). Exposure to moisture or excessive handling of water soluble packets may cause breakage. Store packets in original container and out of the reach of children, preferably in a locked storage area.

Handle and open container in a manner as to prevent spillage. If the container is leaking or material spilled for any reason or cause, carefully sweep material into a pile. Refer to Precautionary Statements on label for hazards associated with the handling of this material. Do not walk through spilled material. Dispose of pesticide as directed above. In spill or leak incidents, keep unauthorized people away. You may contact the Mobay Emergency Response Team for decontamination procedures or any other assistance that may be necessary. The Mobay Kansas City Emergency Response Telephone No. is 816-242-2582, or contact Chemtrec at 800-424-9300.

PRECAUTIONARY STATEMENTS

HAZARDS TO HUMANS AND DOMESTIC ANIMALS

WARNING

Causes substantial but temporary eye injury. May be harmful if inhaled. Do not get in eyes. Avoid contact with skin or clothing. Wear goggles, face shield, or safety glasses when handling the wettable powder. Avoid breathing spray mist. Wash thoroughly with soap and warm water after handling. Remove and wash contaminated clothing with soap and hot water before reuse.

Under conditions of bright sunlight, particularly if wind is present, some persons spraying MORESTAN 25% Wettable Powder experience a skin reaction on exposed parts of the body similar to sunburn. When applying sprays containing more than 1/2 lb. MORESTAN 25% Wettable Powder per 100 gal. water, it is recommended that sufficient clothing be worn to cover most body areas and that a suitable sun-screen lotion such as Uval (Dome Laboratories) or Solbar (Person and Covey) be applied to body areas necessarily left uncovered such as the neck and face.

Do not contaminate feed or food. Keep out of reach of children.

STATEMENTS OF PRACTICAL TREATMENT

If in eyes, hold eyelids open and flush with a steady, gentle stream of water for 15 minutes. Call a physician if irritation persists or develops after flushing. *If swallowed,* administer water freely and induce vomiting by giving one dose (1/2 oz. or 15 ml.) of syrup of ipecac. If vomiting does not occur within 10 to 20 minutes, administer second dose. If syrup of ipecac is not available, induce vomiting by sticking finger down throat. Repeat until vomit fluid is clear. Never give anything by mouth to an unconscious person. Get medical attention immediately.

If on skin, wash skin immediately with soap and warm water. If irritation occurs, get medical attention.

If inhaled, remove to fresh air. If not breathing, give artificial respiration, preferably mouth to mouth. Get medical attention.

TO PHYSICIAN: No specific antidote is available. Treat symptomatically.

ENVIRONMENTAL HAZARDS

MORESTAN is toxic to fish. Do not apply where runoff is likely to occur. In cleaning of equipment or disposal of wastes, do not contaminate streams, lakes, or ponds with this material.

IMPORTANT:

This bulletin is not intended to provide adequate information for application. Before using the product, read and carefully observe directions, cautionary statements and other information appearing on the product label. This product is sold subject to the Conditions of Sale set forth on the container label.

READ THE LABEL BEFORE USE

M 9001

This is a specimen label. Mobay is not responsible for the accuracy of the information contained herein. As labels are subject to revision, always carefully read and follow the label on the product container.

MORESTAN®

25% WETTABLE POWDER
MITICIDE—FUNGICIDE—INSECTICIDE
IN WATER SOLUBLE PACKETS

FOR CONTROL OF MITES, POWDERY MILDEW, PEAR PSYLLA AND APHIDS

ACTIVE INGREDIENT:	
6-Methyl-1,3-dithiolo [4,5-*b*] quinoxalin-2-one	25%
INERT INGREDIENTS:	75%
TOTAL	100%

EPA Reg. No. 3125-302

MORESTAN is a Reg. TM of Bayer AG, Germany

RECOMMENDED USES

BEARING FRUIT	NON-BEARING FRUIT		FRUIT	OTHER CROPS
Apples	Apples	Peaches	Strawberries	Macadamia
Apricots (Post-harvest only)	Apricots	Pears		nuts
Pears	Cherries	Plums		Walnuts
Citrus (CA, AZ, TX, FL only)	Citrus	Prunes		
	Nectarines			

RESTRICTIONS ON WAREHOUSE STORAGE

Protect from rain and moisture. Store in a cool dry place but not below 32°F. Do not store next to herbicides.

PACKAGING

5 lb. carton, contains 5 one-pound water soluble packets, 6 per case. Shipping weight 36 lbs.

MORESTAN 25% Wettable Powder in water soluble packets effectively controls several species of mites attacking fruit, macadamia nuts and walnuts. In addition, it protects against powdery mildew on apple trees, controls aphids on walnuts and guards against pear psylla injury.

MORESTAN controls on contact. It is equally effective against mite eggs, nymphs and adults of resistant and non-resistant strains. Its residual activity extends effectiveness.

MORESTAN 25% Wettable Powder mixes readily with water to form a suspension that can be used in conventional air blast sprayers at various concentrations.

Water soluble packets are pre-measured to reduce dust and handling. Offered for sale in Arizona, California, Hawaii, Idaho, Oregon, Utah and Washington.

AVISO

PRECAUCION AL USUARIO: Si usted no puede leer o entender inglés, no use este producto hasta que la etiqueta le haya sido explicada ampliamente.

(TO THE USER: If you cannot read or understand English, do not use this product until the label has been fully explained to you.)

DIRECTIONS FOR USE

It is a violation of Federal Law to use this product in a manner inconsistent with its labeling.

MORESTAN 25% Wettable Powder miticide-fungicide-insecticide is recommended for residual control of aphids, mites, mite eggs, and powdery mildew. The formulation mixes readily with water to form a suspension which is usable in all conventional hand- or machine- operated sprayers. For best results, apply a thorough coverage before mites or mite eggs are present in large numbers and before mites cause bronzing of the foliage. Apply during good drying conditions. (See Restrictions.) Consult your State Agricultural Extension Service or Experiment Station for detailed information on timing of applications for most effective control.

MIXING: The enclosed packets containing MORESTAN 25% Wettable Powder are **water soluble.** Do not allow packets to become wet prior to adding to the spray tank. Do not handle with wet hands. Reseal outer carton to protect remaining packets. To prepare the spray mixture, drop the required number of unopened packets, as determined under "Recommended Applications" into the spray tank while filling with water to the desired level. Operate the agitator while mixing. Depending on the water temperature and the degree of agitation, the packets should be completely dissolved within approximately 5 minutes from the time they were added to the water.

Do not apply this product through any type of irrigation system.

DOSAGE: Use specified dosage of MORESTAN 25% Wettable Powder in the designated amount of water necessary to give complete coverage of foliage. Determine the **total amount** of wettable powder to be added to the spray tank based on the rates under "Recommended Applications." **For each 1 pound (16 ounces)** of wettable powder to be added to the spray tank use one 1 pound packet. For example, if it is determined that 2 pounds (32 ounces) of MORESTAN 25% Wettable Powder should be added to the spray tank, add two 1 pound packets. The type of equipment used will determine the concentration required; however, use of these packets is not recommended for making highly concentrated mixtures such as used in aircraft spraying.

RECOMMENDED APPLICATIONS

CROP	PEST	POUNDS MORESTAN 25% W.P.	REMARKS
BEARING FRUIT Pre-Bloom and Post-Harvest Apples Pears	Powdery mildew (see NOTE in Remarks) Mites Mite eggs	1/2 to 1	Apply specified dosage per 100 gallons of water in prebloom sprays (delayed-dormant, pre-pink or pink). Do not apply after first bloom. Additional applications may be made to tree foliage after harvest. **Note:** Pre-bloom sprays with MORESTAN 25% W.P. will provide effective control of early-season powdery mildew infections. No additional fungicides are needed for powdery mildew control during the pre-bloom period. Use other recommended fungicides in post-bloom spray programs for seasonal control.
	Pear psylla	1 1/2	

CROP	PEST	POUNDS MORESTAN 25% W.P.	REMARKS
Apricots (Post-harvest only)	Mites (Except cyclamen mite) Mite eggs	4	Apply specified dosage per acre in a minimum of 10 gallons of water per acre by air or 40 gallons of water per acre by ground. For best control thorough coverage of all surfaces is necessary. Make only one application between harvest and up to the time trees break dormancy. Do not graze cover crops.
BEARING CITRUS TREES Citrus (Arizona, California Texas, and Florida only)	Citrus red mite Citrus rust mite Texas citrus mite Mite eggs	5	Apply specified dosage per acre by concentrate or dilute ground spray using 20 to 500 gallons of water per acre. Schedule applications no closer than 90 days apart in Texas and Florida, or 180 days apart in Arizona and California. Allow at least 21 days between last application and harvest. **NOTE:** No more than 1 (one) application may be made to bearing fruit.
NON-BEARING FRUITS Apples	Powdery mildew Mites Mite eggs	½ to 1	Apply specified dosage per 100 gallons of water*. Spray all foliage surfaces, including the underside of leaves, for complete coverage. Repeat as necessary. For control of powdery mildew, begin preventive applications before disease appears, and apply at 7- to 14-day intervals. During heavy infection periods, make more frequent applications. Do not apply to fruit trees that will bear fruit within one year of application.
Apricots Pears Cherries Plums Nectarines Prunes Peaches	Mites (Except cyclamen mite) Mite eggs	½ to 1	
Citrus	Citrus red mite Citrus rust mite Texas citrus mite Mite eggs	1 to 1½	Apply specified dosage per 100 gallons of water. Spray all surfaces, including the underside of leaves, for complete coverage. Repeat as necessary. Do not apply to trees that will bear fruit within one year of application.
FRUIT Strawberries (Pre-Bloom only)	Mites (except cyclamen mite) Mite eggs	½ to 1	Apply specified dosage per 100 gallons of water* as a pre-bloom application. Application may be made in propagating beds, and in field plantings during the first year of setting new plants, but not after first bloom. Spray all foliage surfaces, include the underside of leaves for complete coverage. Repeat applications may be made as needed but not after first bloom appears.
NUTS Macadamia nuts	Mites	1 to 2	Apply specified dosage per 100 gallons of water as a full-coverage spray*, using up to 400 gallons of finished spray per acre, depending upon the number and size of trees. Spray all surfaces, including the underside of leaves, for complete coverage. Three applications per season may be made up to the day of harvest.
Walnuts	Aphids Mites Mite eggs	1 to 1½	Apply specified dosage per 100 gallons of water as a full-coverage spray*, using up to 250 gallons of finished spray per acre, depending upon the number and size of trees. Spray all foliage surfaces, including the underside of leaves, for complete coverage. Do not apply more than twice per season, or within 30 days of harvest. 8-18-89

* This concentration is calculated for conventional sprayers. When lower volumes of spray are applied per acre, as with low pressure, low volume, or mist-type equipment, increase the concentration of MORESTAN 25% Wettable Powder in the spray mixture in order to apply amount of MORESTAN 25% Wettable Powder per acre equivalent to a full-coverage spray. Refer also to **"DOSAGE"** paragraph.

SPECIAL LOCAL NEED REGISTRATIONS [SECTION 24(C)] FOR MORESTAN 25% WETTABLE POWDER (PVA)*

CROP	PEST	STATE EPA SLN NO.	DOSAGE	REMARKS
Bearing citrus trees	Mites Mite eggs	California CA 79-0059	5 pounds per acre.	By concentrate or dilute ground spray.
Cucurbits (grown for seed)	Powdery Mildew Mites Mite eggs	California CA 87-0010	1.5 pounds	Apply specified dosage per acre by ground or air.

* For distribution and use only within the state specified. These Special Local Need registrations approved as of date of this publication are, however, subject to cancellation by the state. Contact your nearest Mobay Agricultural Chemicals Division Office for the current status and a copy of the complete individual supplemental Special Local Need labeling. This labeling must be in the possession of the user at time of pesticide application.

RESTRICTIONS

Under certain adverse weather conditions, particularly those that create slow drying of spray deposits, injury may occur to foliage of strawberries and deciduous fruit trees having new and/or tender growth.

Do not use summer oils if MORESTAN is used in the cover spray program.

Do not use spreader-sticker materials with MORESTAN.

Do not allow sprays to drift onto other crops, including pastures. Do not graze treated areas.

Do not use on other crops used for food and forage. Use only according to label directions. Applications at rates above those shown may result in illegal crop residues. Do not treat food crops grown in the greenhouse.

RE-ENTRY STATEMENTS

Do not apply this product in such a manner as to directly or through drift expose workers or other persons. The area being treated must be vacated by unprotected persons. Do not enter treated areas without protective clothing until sprays have dried. Because certain states may require more restrictive re-entry intervals for various crops treated with this product, consult your State Department of Agriculture for further information.

Written or oral warnings must be given to workers who are expected to be in a treated area or in an area about to be treated with this product. Oral warnings should include those statements found under the label heading "Precautionary Statements" and "Re-entry Statements" included on the label. When oral warnings are given, warnings shall be given in a language customarily understood by workers. Oral warnings must be given if there is reason to believe that written warnings cannot be understood by workers. Written warnings must include the following information: WARNING, area treated with MORESTAN 25% Wettable Powder on (date of application). Do not enter without appropriate protective clothing until sprays have dried. **If on skin,** remove contaminated clothing and wash skin immediately with soap and warm water. **If eyes are contaminated,** wash with flowing water for at least 15 minutes. **If swallowed,** vomiting should be induced. (See Precautionary Statements on label.)

STORAGE AND DISPOSAL

Pesticide Disposal: Do not contaminate water, food or feed by storage or disposal.

Wastes resulting from the use of this product may be disposed of on site or at an approved waste disposal facility.

Container Disposal: Do not use container in connection with food, feed, or drinking water. Completely empty carton into application equipment. Then dispose of empty carton in a sanitary landfill or by incineration, or, if allowed by State and local authorities, by burning. If burned, stay out of smoke.

Store in a cool, dry place and in such a manner as to prevent cross contamination with other pesticides, fertilizers, food, and feed. Do not store below freezing (32°F). Exposure to moisture or excessive handling of water soluble packets may cause breakage. Store packets in original container and out of the reach of children, preferably in a locked storage area.

Handle and open container in a manner as to prevent spillage. If the container is leaking or material spilled for any reason or cause, carefully sweep material into a pile. Refer to Precautionary Statements on label for hazards associated with the handling of this material. Do not walk through spilled material. Dispose of pesticide as directed above. In spill or leak incidents, keep unauthorized people away. You may contact the Mobay Emergency Response Team for decontamination procedures or any other assistance that may be necessary. The Mobay Kansas City Emergency Response Telephone No. is 816-242-2582, or contact Chemtrec at 800-424-9300.

PRECAUTIONARY STATEMENTS

HAZARDS TO HUMANS AND DOMESTIC ANIMALS

WARNING

Causes substantial but temporary eye injury. May be harmful if inhaled. Do not get in eyes. Avoid contact with skin or clothing. Avoid breathing of spray mist. Wash thoroughly with soap and warm water after handling. Remove and wash contaminated clothing with soap and hot water before reuse.

Under conditions of bright sunlight, particularly if wind is present, some persons spraying MORESTAN 25% Wettable Powder experience a skin reaction on exposed parts of the body similar to sunburn. When applying sprays containing more than 1/2 lb. MORESTAN 25% Wettable Powder per 100 gal. of water, it is recommended that sufficient clothing be worn to cover most body areas, and that a suitable sun-screen lotion, such as Uval (Dome Laboratories) or Solbar (Person and Covey) be applied to body areas necessarily left uncovered such as the neck and face.

Do not contaminate feed or food. Keep out of reach of children.

STATEMENTS OF PRACTICAL TREATMENT

If in eyes, hold eyelids open and flush with a steady, gentle stream of water for 15 minutes. Call a physician if irritation persists or develops after flushing.

If swallowed, administer water freely and induce vomiting by giving one dose (1/2 oz. or 15 ml.) of syrup on ipecac. If vomiting does not occur within 10 to 20 minutes, administer second dose. If syrup of ipecac is not available, induce vomiting by sticking finger down throat. Repeat until vomit fluid is clear. Never give anything by mouth to an unconscious person. Get medical attention immediately.

If on skin, wash skin immediately with soap and warm water. If irritation occurs, get medical attention.

If inhaled, remove to fresh air. If not breathing, give artificial respiration, preferably mouth to mouth. Get medical attention.

TO PHYSICIAN: No specific antidote is available. Treat symptomatically.

ENVIRONMENTAL HAZARDS

MORESTAN is toxic to fish. Do not apply where runoff is likely to occur. In cleaning of equipment or disposal of wastes, do not contaminate streams, lakes, or ponds with this material.

IMPORTANT:

This bulletin is not intended to provide adequate information for application. Before using the product, read and carefully observe directions, cautionary statements and other information appearing on the product label. This product is sold subject to the Conditions of Sale set forth on the container label.

READ THE LABEL BEFORE USE

M 9002

This is a specimen label. Mobay is not responsible for the accuracy of the information contained herein. As labels are subject to revision, always carefully read and follow the label on the product container.

NEMACUR®

15% GRANULAR SYSTEMIC INSECTICIDE—NEMATICIDE

RESTRICTED USE PESTICIDE

For sale to and use only by a Certified Applicator for uses authorized by his certification, or by persons under his direct supervision.

For effective control of nematodes in certain field, fruit, and vegetable crops

ACTIVE INGREDIENT:
Ethyl 3-methyl-4-(methylthio)phenyl
(1-methylethyl)phosphoramidate 15%
INERT INGREDIENTS: 85%

TOTAL: 100%

EPA Reg. No. 3125-236

NEMACUR is a Reg. TM of Bayer AG, Germany

RECOMMENDED USES

Brussels sprouts, Cabbage, Citrus, Cotton, Eggplant, Garlic, Non-bearing Strawberry Nursery Stock, Okra, Peanuts.

RESTRICTIONS ON WAREHOUSE STORAGE

Do not store in front of space heater. Do not store adjacent to or near foodstuffs, drugs, feed or any other material intended for consumption by humans or animals. Do not store adjacent to toys, cosmetics, dishes, furniture or clothing. Protect from rain and moisture. Do not store next to herbicides. In cases of puncture, use overslip bag and securely close.

NEMACUR 15% Granular nematicide is an organophosphorus compound which has shown excellent activity against the major genera of nematodes attacking citrus, cotton, peanuts, and vegetable crops.

NEMACUR has long-lasting contact and systemic activity. NEMACUR creates a clean zone around roots and also moves with root systems to keep protecting as roots develop. Used according to directions, a planting time application can provide protection against nematode damage.

PACKAGING

50 lb. bag. Shipping weight: 52 lbs.

PELIGRO

PRECAUCION AL USUARIO: Si usted no puede leer o entender inglés, no use este producto hasta que la etiqueta le haya sido explicada ampliamente.
(TO THE USER: If you cannot read or understand English, do not use this product until the label has been fully explained to you.)

DIRECTIONS FOR USE

It is a violation of Federal Law to use this product in a manner inconsistent with its labeling.

RECOMMENDED APPLICATIONS

Consult your State Agricultural Experiment Station or Extension Service Specialist or Mobay representative for guidance on the selection of specific dosage rates and timing of applications appropriate for your specific crop and pest. State regulations may contain additional restrictions or requirements. Check the "Remarks" and "Restrictions" sections of the label for additional directions.

CROP	PEST	DOSAGE NEMACUR 15% GRANULAR	REMARKS
FIELD CROPS Cotton	Thrips (early-season reduction)	6 to 8 ozs. per 1000 ft. of row for any row spacing. (5 to 6.5 lbs. per acre on 40-inch row)	**In furrow/covering soil:** Apply specified dosage per 1000 ft. of row in furrow or band in covering soil behind the seed drop and in front of covering devices.
	Nematodes Thrips	8 to 18 ozs. per 1000 ft. of row for any row spacing. (6.5 to 14.7 lbs. per acre on 40-inch row spacing)	**In furrow/covering soil:** Apply specified dosage per 1000 ft. of row in furrow or band in covering soil behind the seed drop and in front of covering devices. **Band:** Apply specified dosage per 1000 ft. of row in a 6-12 inch band over the row. Incorporate thoroughly to insure uniform distribution.[1]

Suggested use rates of NEMACUR 15% G. in oz./1000 feet of row based on nematode infestation levels:*

Band width	Low	Moderate	High*
In furrow	8	10	12
In covering soil	8	10	12
6-inch band	8	10	12
8-inch band	10	12	14
10-inch band	11	13	16
12-inch band	12	14	18

CROP	PEST	DOSAGE NEMACUR 15% GRANULAR	REMARKS
			* Use the highest recommended rates in fields with a history of nematode problems. **IMPORTANT:** Do not overlap bands on narrow row crops. Do not feed or graze cotton foliage. Under very heavy thrip pressure additional control measures may be needed.
Peanuts	Nematodes Thrips (early-season reduction)	11 to 18.7 ozs. for any row spacing (or 10 to 17 pounds per acre on 36-inch rows)	Apply specified dosage per 1000 ft. or row as a 12-inch band at planting. Incorporate the granules into the soil. ON NARROW ROW CROPS, DO NOT ALLOW BANDS TO OVERLAP. Use the high rate in fields with high populations of nematodes, in fields where root-knot species other than southern root-knot, *Meloidogyne incognita* are present, or in fields having a history of serious nematode damage. Do not feed or graze green peanut vines or peanut vine hay. Do not hog down treated peanut fields.

[1] The action of some planters during planting operation may provide adequate incorporation. When using planters which do not disturb enough soil for incorporation or under dry conditions, additional incorporation can be achieved with scratchers, paddlewheels, pickwheels, ground driven rototillers or similar equipment behind the planter.

CROP	PEST	DOSAGE NEMACUR 15% GRANULAR	REMARKS
VEGETABLES Cabbage (includes tight heading varieties of Chinese cabbage) (direct seeded and transplanted) (Except cabbage grown for seed) Brussels Sprouts (transplanted) (Except brussels sprouts grown for seed)	Nematodes	9.8 to 18.4 ozs. for any row spacing (or 8.9 to 16.7 lbs. per acre on 36-inch rows	Apply specified dosage per 100 ft. of row as a 12- to 15-inch band. Incorporate the granules into the soil. ON NARROW ROW CROPS, DO NOT ALLOW BANDS TO OVERLAP. **Suggested use rates of NEMACUR 15% G. in ozs./1000 ft. of row based on nematode infestation levels:*** **Band width** — **Low** — **High*** 12-inch band — 9.8 — 14.7 15-inch band — 12.2 — 18.4 * Use the highest recommended rates in fields with a history of nematode problems.
Okra (Except California)	Nematodes	14.7 to 18.4 ozs. (or 13.4 to 16.7 lbs. per acre on 36-inch rows)	Apply specified dosage per 1000 ft. of row as a 12- to 15-inch band at planting. Incorporate the granules into the soil. Use low rate in 12-inch band. Use high rate in 15-inch band. **BAND WIDTH (36-inch row spacing)** **12-inch**: 14.7 oz./1000 ft. of row or 13.4 lbs./acre **15-inch**: 18.4 ox./100 ft. of row or 16.7 lbs./acre ON NARROW ROW CROPS, DO NOT ALLOW BANDS TO OVERLAP.
Non-Bell Peppers (California, Georgia and Puerto Rico Only)	Nematodes	10 to 14.7 ozs. for any row spacing (or 9 to 13.3 lbs. per acre on 36 inch rows)	Apply specified dosage per 1000 ft. of row in a 12-inch band over the row at planting and incorporate the granules into the soil. Use the high rate in fields with high populations of nematodes or in fields having a history of serious nematode damage.
Eggplant	Nematodes	14.7 oz (or 13.4 lbs. per acre on 36-inch rows)	Apply specified dosage per 1000 ft. of row in a 12-inch band over the row at transplanting and immediately incorporate the granules into the soil. ON NARROW ROW CROPS, DO NOT ALLOW BANDS TO OVERLAP.
Garlic	Bulb and Stem Nematode	9.2 to 18.4 ozs./1000 ft. of row for any row spacing (15 to 30 lbs. per acre on 40-inch beds with two seed lines [rows] per bed)	Apply specified dosage in-furrow with the garlic seed (cloves) at planting. For best results, plant the crop within 7 to 10 days after cracking (seed preparation).
FRUIT Citrus (Arizona, California and Texas only) (In California do not apply to Kumquat, Tangelo or Citrus Hybrids)	Nematodes	33½ to 67 lbs.	Apply specified dosage per acre in a band in up to 2 applications per season, not to exceed 67 lbs. of NEMACUR 15% Granular per acre per season. Apply to the soil with ground equipment and incorporate immediately using mechanical incorporation or irrigation. Center the treated band on the tree row using a band width of 50% of the row spacing and covering the feeder root system of the plant. Under dry conditions, follow with overhead or flood irrigation. Do not apply more than 10 lbs. active ingredient (67 lbs of NEMACUR 15% Granular) per acre per season. Do not apply within 30 days of harvest. Do not graze livestock on treated areas. On non-bearing citrus make initial applications when trees have recovered from transplanting shock and have initiated new growth, provided that one or more irrigations have been applied and soil has settled around the roots. Phytotoxicity may occur if applications are made before soil has settled and plants are growing.

CROP	PEST	DOSAGE NEMACUR 15% GRANULAR	REMARKS
Non-bearing Strawberry nursery stock	Nematodes	17 ozs. for any row spacing (or 13.8 pounds per acre on 40-inch rows)	Apply specified dosage in ozs. per 1000 ft. of row in a 12 inch band over the row immediately after transplanting and incorporate with cultivation equipment. Apply a second application in a 12 inch band over the row 8 weeks after transplanting. Incorporate by cultivation or by sprinkler irrigation (1 inch of water) immediately following application. Do not apply more than two applications to non-bearing nursery stock. Use only on strawberry nursery stock that will not be harvested for fruit within 20 months after application. 5-4-89

RESTRICTIONS

Any food crop not specified on this label may be planted into treated areas 120-days after the last application. Any cover crops that are planted during the 120 day period must be plowed under and not grazed.

Do not use on other crops. Use only according to label directions.

Do not apply this product with a knapsack or similar equipment that is placed on the user's body.

PROTECTIVE CLOTHING AND EQUIPMENT

USE ONLY WHEN WEARING THE FOLLOWING PROTECTIVE CLOTHING AND EQUIPMENT DURING MIXING/LOADING, APPLICATION, REPAIR AND CLEANING OF MIXING, LOADING AND APPLICATION EQUIPMENT, DISPOSAL OF THE PESTICIDE, AND EARLY RE-ENTRY INTO TREATED AREAS: Protective suit of one or two pieces covering all parts of the body except head, hands, and feet; chemical-resistant gloves; chemical-resistant shoes (or chemical-resistant shoe covers or chemical-resistant boots); and a NIOSH- or MSHA-approved respirator. In addition, mixer/loaders must wear a chemical resistant apron. During equipment repair and cleaning, the respirator need not be worn. During early re-entry after dust has settled and vapors have dispersed, the respirator need not be worn.

IF MIXING LOADING IS PERFORMED USING A CLOSED SYSTEM, THE FOLLOWING PROTECTIVE CLOTHING AND EQUIPMENT MAY BE WORN AS AN ALTERNATIVE: Long-sleeved shirt and long-legged pants; chemical resistant gloves; chemical resistant apron; shoes and socks.

IF APPLICATION IS PERFORMED USING AN ENCLOSED CAB, THE FOLLOWING PROTECTIVE CLOTHING AND EQUIPMENT MAY BE WORN AS AN ALTERNATIVE: Long-sleeved shirt and long-legged pants; shoes and socks. Chemical resistant gloves must be available in the cab and must be worn when exiting. This clothing is inadequate to protect you during equipment repair or cleaning, re-entry, or pesticide disposal work.

IMPORTANT! If pesticide comes in contact with skin, wash off with soap and water. Always wash hands, face, and arms with soap and water before smoking, eating, drinking, or toileting.

AFTER WORK: Before removing gloves, wash them with soap and water. Take off all work clothes and shoes. Shower using soap and water. Wear clean clothes. Do not reuse contaminated clothing. Personal clothing worn during work must be laundered separately from household articles. Store protective clothing separately from personal clothing. Clean or launder protective clothing after each use. Respirators must be cleaned and filters replaced according to instructions included with the respirator. Protective clothing and equipment that becomes heavily contaminated must be destroyed according to state and local regulations. HEAVILY CONTAMINATED CLOTHING CANNOT BE ADEQUATELY DECONTAMINATED.

RE-ENTRY STATEMENTS

Do not apply this product in such a manner as to directly or through drift expose workers or other persons. The area being treated must be vacated by unprotected persons. Do not re-enter treated area for 48 hours unless appropriate protective clothing is worn. Any person coming in direct contact with treated soil prior to incorporation or irrigation, or with treated wet soil after the initial irrigation, must wear protective clothing and equipment as described in the section "Protective Clothing and Equipment" on the label. Wash thoroughly with soap and water after handling and before eating or smoking.

Because certain states may require more restrictive re-entry intervals for various crops treated with this product, consult your State Department of Agriculture for further information.

Written or oral warnings must be given to workers who are expected to be in a treated area or in an area about to be treated with this product. Oral warnings should include those statements found under the label heading "Precautionary Statements" and "Re-entry Statements" included on the label. When oral warnings are given, warnings shall be given in a language customarily understood by workers. Oral warnings must be given if there is reason to believe that written warnings cannot be understood by workers. Written warnings must include the following information: DANGER, area treated with NEMACUR 15% Granular on (date of application). See "Protective Clothing and Equipment" for use of proper protective clothing. **If on skin,** remove contaminated clothing and wash skin immediately with soap and warm water. **If eyes are contaminated,** wash with flowing water for at least 15 minutes. **If swallowed,** vomiting should be induced. (See Precautionary Statements on label).

STORAGE AND DISPOSAL

Pesticide Disposal: Do not contaminate water, food or feed by storage or disposal.

Pesticide wastes are acutely hazardous. Improper disposal of excess pesticide, or rinsate is a violation of Federal Law. If these wastes cannot be disposed of by use according to label instructions, contact your State Pesticide or Environmental Control Agency, or the Hazardous Waste representative at the nearest EPA Regional Office for guidance.

Container Disposal: Do not use container in connection with food, feed, or drinking water. Completely empty bag into application equipment. Then dispose of empty bag in a sanitary landfill or by incineration, or, if allowed by State and local authorities, by burning. If burned, stay out of smoke.

Store in a cool, dry place and in such a manner as to prevent cross contamination with other pesticides, fertilizers, food, and feed. Store in original container and out of the reach of children, preferably in a locked storage area.

Handle and open container in a manner as to prevent spillage. If the container is leaking or material spilled for any reason or cause, carefully sweep material into a pile. Refer to Precautionary Statements on label for hazards associated with the handling of this material. Do not walk through spilled material. Dispose of pesticide as directed above. In spill or leak incidents, keep unauthorized people away. You may contact the Mobay Emergency Response Team for decontamination procedures or any other assistance that may be necessary. The Mobay Kansas City Emergency Response Telephone No. is 816-242-2582, or contact Chemtrec at 800-424-9300.

PRECAUTIONARY STATEMENTS

HAZARDS TO HUMANS AND DOMESTIC ANIMALS

DANGER–POISON

Fatal if swallowed or absorbed through the skin. Causes eye irritation. May be fatal if inhaled. Do not get in eyes, on skin, or on clothing.

Do not breathe dust. Wear protective clothing, rubber gloves, and goggles or face shield when handling. Wash thoroughly with soap and water after handling and before eating or smoking. Remove and wash contaminated clothing before reuse.

Do not contaminate feed or food. Keep all unprotected persons out of the operating area or vicinity. Wash hands, arms, and face thoroughly with soap and warm water before eating or smoking. Wash all contaminated clothing with soap and hot water before reuse.

Keep out of the reach of children.

Not for use or storage in or around the home.

SYMPTOMS OF POISONING: A sense of "tightness" in the chest. Sweating. Contracted pupils. Stomach pains. Vomiting and diarrhea.

STATEMENTS OF PRACTICAL TREATMENT

In case of poisoning, call a physician immediately. Have patient lie down and keep quiet.

If swallowed, vomiting should be induced. Administer water freely and induce vomiting by giving one dose (1/2 oz. or 15 ml.) of syrup of ipecac. If vomiting does not occur within 10 to 20 minutes, administer second dose. If syrup of ipecac is not available, induce vomiting by sticking finger down throat. Repeat until vomit fluid is clear. Never give anything by mouth to an unconscious person. Professional medical assistance should be secured immediately.

If on skin, remove contaminated clothing and wash skin immediately with soap and warm water.

If eyes are contaminated, wash with flowing water for at least 15 minutes.

If inhaled, remove victim to fresh air. If not breathing, give artificial respiration, preferably mouth-to-mouth.

TO PHYSICIAN: Antidote—Administer atropine sulfate in large therapeutic doses. Repeat as necessary to the point of tolerance. 2-PAM is also antidotal and may be administered in conjunction with atropine.

Compound inhibits cholinesterase resulting in stimulation of the central nervous system, the parasympathetic nervous system and the somatic motor nerves. Do not give morphine. Watch for pulmonary edema, which may develop in serious cases of poisoning even after 12 hours. At fist sign of pulmonary edema, the patient should be placed in an oxygen tent and treated symptomatically.

ENVIRONMENTAL HAZARDS

This pesticide is toxic to fish and wildlife. Birds feeding on treated areas may be killed.

Cover, incorporate, or pick up spilled granules at row ends or turn areas. Runoff from treated areas may be hazardous to aquatic organisms in neighboring areas. Do not apply directly to water or wetlands (swamps, bogs, marshes, and potholes). Do not contaminate water when disposing of equipment washwaters.

IMPORTANT: This bulletin is not intended to provide adequate information for application. Before using the product, read and carefully observe directions, cautionary statements and other information appearing on the label. This product is sold subject to the Conditions of Sale set forth on the container label.

READ THE LABEL BEFORE USE

N 9003

This is a specimen label. Mobay is not responsible for the accuracy of the information contained herein. As labels are subject to revision, always carefully read and follow the label on the product container.

NEMACUR®

15% GRANULAR SYSTEMIC INSECTICIDE-NEMATICIDE

For effective control of nematodes in citrus crops (Except Arizona, California and Texas.)

RESTRICTED USE PESTICIDE

For sale to and use only by a Certified Applicator for uses authorized by his certification, or by persons under his direct supervision.

ACTIVE INGREDIENT:
Ethyl 3-methyl-4-(methylthio)phenyl (1- methylethyl) phosphoramidate 15%
INERT INGREDIENTS: 85%
TOTAL. 100%

EPA Reg. No. 3125-236
NEMACUR is a Reg. TM of Bayer AG, Germany.

RECOMMENDED USES

Citrus (except Arizona, California and Texas)

RESTRICTIONS ON WAREHOUSE STORAGE

Do not store in front of space heater. Do not store adjacent to or near foodstuffs, drugs, feed or any other material intended for consumption by humans or animals. Do not store adjacent to toys, cosmetics, dishes, furniture, or clothing. Protect from rain and moisture. Do not store next to herbicides. In case of puncture, use overslip bag and securely close.

PACKAGING

50 lb. bag. Shipping weight: 52 lbs.

NEMACUR 15% Granular nematicide is an organophosphorus compound that has shown excellent activity against the major genera of nematodes, including burrowing nematodes, which attack citrus crops—oranges, grapefruit, lemons, limes and tangerines.

NEMACUR has long-lasting contact and systemic activity. Used according to directions, it can provide effective control of nematodes that can stunt citrus growth, reduce yields, and lead to general tree decline.

NEMACUR is easy to apply with conventional equipment, and must be incorporated immediately by irrigation or mechanically.

PACKAGING

50 lb. bag. Shipping weight: 52 lbs.

PELIGRO

PRECAUCION AL USUARIO: Si usted no puede leer o entender inglés, no use este producto hasta que la etiqueta le haya sido explicada ampliamente.

(TO THE USER: If you cannot read or understand English, do not use this product until the label has been fully explained to you.)

DIRECTIONS FOR USE

It is a violation of Federal Law to use this product in a manner inconsistent with its labeling.

RECOMMENDED APPLICATIONS

Consult your State Agricultural Experiment Station or Extension Service Specialist or Mobay representative for guidance on the selection of dosage rates and timing of applications appropriate for your specific crop and pest. State regulations may contain additional restrictions or requirements. Check the "Remarks" and "Restrictions" sections of the label for additional directions.

CROP	PEST	DOSAGE NEMACUR 15% G.	REMARKS
Citrus (Except Arizona, California and Texas)	Nematodes	33½ to 67 pounds	Applied the specified dosage per acre in a band in up to 2 applications per season, not to exceed 67 pounds of NEMACUR 15% Granular per acre per season. Apply to the soil with ground equipment and incorporate immediately using mechanical incorporation or irrigation. Center the treated band on the tree row using a band width of 50% of the row spacing and covering the feeder root system of the plant. Under dry conditions, follow with overhead or flood irrigation. Do not apply more than 10 pounds active ingredient (67 pounds of NEMACUR 15% Granular) per acre per season. Do not apply within 30 days of harvest. Do not graze livestock on treated areas. On non-bearing citrus make initial applications when trees have recovered from transplanting shock and have initiated new growth, provided that one or more irrigations have been applied and soil has settled around the roots. Phytotoxicity may occur if applications are made before soil has settled and plants are growing. 10-21-88

RESTRICTIONS

Any food crop not specified on the label may be planted into treated areas 120 days after the last application. Any cover crops that are planted during the 120 day period must be plowed under and not grazed.

Do not use on other crops. Use only according to label directions.

Do not apply this product with a knapsack or similar equipment that is placed on the user's body.

PROTECTIVE CLOTHING AND EQUIPMENT

USE ONLY WHEN WEARING THE FOLLOWING PROTECTIVE CLOTHING AND EQUIPMENT DURING MIXING/LOADING, APPLICATION, REPAIR AND CLEANING OF MIXING, LOADING AND APPLICATION EQUIPMENT, DISPOSAL OF THE PESTICIDE, AND EARLY RE-ENTRY INTO TREATED AREAS: Protective suit of one or two pieces covering all parts of the body except head, hands, and feet; chemical-resistant gloves; chemical-resistant shoes (or chemical-resistant shoe covers or chemical-resistant boots); and a NIOSH- or MSHA-approved respirator. In addition, mixer/loaders must wear a chemical-resistant apron. During equipment repair and cleaning, the respirator need not be worn. During early reentry after dust has settled and vapors have dispersed, the respirator need not be worn.

IF MIXING/LOADING IS PERFORMED USING A CLOSED SYSTEM, THE FOLLOWING PROTECTIVE CLOTHING AND EQUIPMENT MAY BE WORN AS AN ALTERNATIVE: Long-sleeved shirt and long-legged pants; chemical-resistant gloves; chemical-resistant apron, shoes and socks.

IF APPLICATION IS PERFORMED USING AN ENCLOSED CAB, THE FOLLOWING PROTECTIVE CLOTHING AND EQUIPMENT MAY BE WORN AS AN ALTERNATIVE: Long-sleeved shirt and long-legged pants; shoes, and socks. Chemical-resistant gloves must be available in the cab and must be worn when exiting. This clothing is inadequate to protect you during equipment repair or cleaning, reentry, or pesticide disposal work.

IMPORTANT! If pesticide comes in contact with skin, wash off with soap and water. Always wash hands, face, and arms with soap and water before smoking, eating, drinking, or toileting.

AFTER WORK: Before removing gloves, wash them with soap and water. Take off all work clothes and shoes. Shower using soap and water. Wear clean clothes. Do not reuse contaminated clothing. Personal clothing worn during work must be laundered separately from household articles. Store protective clothing separately from personal clothing. Clean or launder protective clothing after each use. Respirators must be cleaned and filters replaced according to instructions included with the respirator. Protective clothing and equipment that becomes heavily contaminated must be destroyed according to state and local regulations. HEAVILY CONTAMINATED CLOTHING CANNOT BE ADEQUATELY DECONTAMINATED.

RE-ENTRY STATEMENTS

Do not apply this product in such a manner as to directly or through drift expose workers or other persons. The area being treated must be vacated by unprotected persons. Do not re-enter treated areas for 48 hours unless appropriate protective clothing is worn. Any person coming in direct contact with treated soil prior to incorporation or irrigation, or with treated wet soil after the initial irrigation, must wear protective clothing and equipment as described in the section "Protective Clothing and Equipment" on the label. Wash thoroughly with soap and water after handling and before eating or smoking.

Because certain states may require more restrictive re-entry intervals for various crops treated with this product, consult your State Department of Agriculture for further information.

Written or oral warnings must be given to workers who are expected to be in a treated area or in an area about to be treated with this product. Oral warnings should include those statements found under the label heading "Precautionary Statements" and "Re-entry Statements" included on the label. When oral warnings are given, warnings shall be given in a language customarily understood by workers. Oral warnings must be given if there is reason to believe that written warnings cannot be understood by workers. Written warnings must include the following information: DANGER, area treated with NEMACUR 15% Granular on (date of application). See "Protective clothing and equipment" for use of proper protective clothing. **If on skin,** remove contaminated clothing and wash skin immediately with soap and warm water. **If eyes are contaminated,** wash with flowing water for at least 15 minutes. **If swallowed, vomiting should be induced.** (See Precautionary Statements on label.)

STORAGE AND DISPOSAL

Pesticide Disposal: Do not contaminate water, food or feed by storage or disposal.

Pesticide wastes are acutely hazardous. Improper disposal of excess pesticide or rinsate is a violation of Federal Law. If these wastes cannot be disposed of by use according to label instructions, contact your State Pesticide or Environmental Control Agency, or the Hazardous Waste representative at the nearest EPA Regional Office for guidance.

Container Disposal: Do not use container in connection with food, feed, or drinking water. Completely empty bag into application equipment. Then dispose of empty bag in a sanitary landfill or by incineration, or, if allowed by State and local authorities, by burining. If burned, stay out of smoke.

Store in a cool, dry place and in such a manner as to prevent cross contamination with other pesticides, fertilizers, food, and feed. Store in original container and out of the reach of children, preferably in a locked storage area.

Handle and open container in a manner as to prevent spillage. If the container is leaking or material spilled for any reason or cause, carefully sweep material into a pile. Refer to Precautionary Statements on label for hazards associated with the handling of this material. Do not walk through spilled material. Dispose of pesticide as directed above. In spill or leak incidents, keep unauthorized people away. You may contact the Mobay Emergency Response Team for decontamination procedures or any other assistance that may be necessary. The Mobay Kansas City Emergency Response Telephone No. is 816-242-2582, or contact Chemtrec at 800-424-9300.

PRECAUTIONARY STATEMENTS

HAZARDS TO HUMANS AND DOMESTIC ANIMALS

DANGER–POISON

Fatal if swallowed or absorbed through the skin. Causes eye irritation. May be fatal if inhaled. Do not get in eyes, on skin, or on clothing. Do not breathe dust. Wear protective clothing, rubber gloves, and goggles or face shield when handling. Wash thoroughly with soap and water after handling and before eating or smoking. Remove and wash contaminated clothing before reuse.

Do not contaminate feed or food. Keep all unprotected persons out of the operating area or vicinity. Wash hands, arms, and face thoroughly with soap and warm water before eating or smoking. Wash all contaminated clothing with soap and hot water before reuse. Keep out of the reach of children.

Not for use or storage in or around the home.

SYMPTOMS OF POISONING: A sense of "tightness" in the chest. Sweating. Contracted pupils. Stomach pains. Vomiting and diarrhea.

STATEMENTS OF PRACTICAL TREATMENT

In case of poisoning, call a physician immediately. Have patient lie down and keep quiet.

If swallowed, vomiting should be induced. Administer water freely, and induce vomiting by giving one dose (1/2 oz. or 15 ml.) of syrup of ipecac. If vomiting does not occur within 10 to 20 minutes, administer second dose. If syrup of ipecac is not available, induce vomiting by sticking finger down throat. Repeat until vomit fluid is clear. Never give anything by mouth to an unconscious person. Professional medical assistance should be secured immediately.

If on skin, remove contaminated clothing and wash skin immediately with soap and warm water.

If eyes are contaminated, wash with flowing water for at least 15 minutes.

If inhaled, remove victim to fresh air. If not breathing give artificial respiration, preferably mouth to mouth.

NOTE TO PHYSICIAN: Antidote—Administer atropine sulfate in large therapeutic doses. Repeat as necessary to the point of tolerance. 2-PAM is also antidotal and may be administered in conjunction with atropine.

Compound inhibits cholinesterase, resulting in stimulation of the central nervous system, the parasympathetic nervous system and the somatic motor nerves. Do not give morphine. Watch for pulmonary edema, which may develop in serious cases of poisoning even after 12 hours. At first sign of pulmonary edema, the patient should be placed in an oxygen tent and treated symptomatically.

ENVIRONMENTAL HAZARDS

This pesticide is toxic to fish and wildlife. Birds feeding on treated areas may be killed.

Cover, incorporate, or pick up spilled granules at row ends or turn areas. Runoff from treated areas may be hazardous to aquatic organisms in neighboring areas. Do not apply directly to water or wetlands (swamps, bogs, marshes, and potholes). Do not contaminate water when disposing of equipment washwaters.

IMPORTANT:

This bulletin is not intended to provide adequate information for application. Before using the product, read and carefully observe directions, cautionary statements and other information appearing on the label. This product is sold subject to the Conditions of Sale set forth on the container label.

READ THE LABEL BEFORE USE

N 9004

This is a specimen label. Mobay is not responsible for the accuracy of the information contained herein. As labels are subject to revision, always carefully read and follow the label on the product container.

NEMACUR 3®

EMULSIFIABLE SYSTEMIC INSECTICIDE—NEMATICIDE

RESTRICTED USE PESTICIDE

For sale to and use only by Certified Applicator authorized by his certification, or by persons under his direct supervision.

FOR EFFECTIVE CONTROL OF NEMATODES IN CERTAIN FIELD AND FRUIT CROPS

ACTIVE INGREDIENT:	
Ethyl 3-methyl-4-(methylthio)phenyl (1-methylethyl) phosphoramidate	35%
INERT INGREDIENTS:	65%
TOTAL	100%

Contains aromatic Petroleum Distillates.

Contains 3 pounds of Ethyl 3-methyl-4-(methylthio)phenyl (1-methylethyl) phosphoramidate per gallon.

EPA Reg. No. 3125-283

NEMACUR is a Reg. TM of Bayer AG, Germany
Treflan is a Reg. TM of Elanco Products Co.

RECOMMENDED USES

Apples, Cherries, Citrus, Cotton, Eggplant, Grapes, Kiwifruit, Nectarine's, Peaches, Peanuts, Pineapples, Raspberries, Tobacco, Nonbearing deciduous fruit and nut trees and commercial tree and vine nurseries.

RESTRICTIONS ON WAREHOUSE STORAGE

Do not store in front of space heater. Store below 120°F, but not below 32°F. Do not store adjacent to, or near foodstuffs, drugs, feed or any other material intended for consumption by humans or animals. Do not store adjacent to toys, cosmetics, dishes, furniture, or clothing. Protect from rain and moisture. Do not store next to herbicides.

PACKAGING

5 gal. drum. Shipping weight: 47 lbs.

NEMACUR 3 emulsifiable nematicide is an organophosphorus compound which has shown excellent activity against the major genera of nematodes attacking cotton, peanuts, grapes, pineapple, tobacco and fruit crops.

NEMACUR 3 has long-lasting contact and systemic activity. NEMACUR 3 creates a clean zone around roots and also moves with the root system to keep protecting as roots develop. Used according to directions, a planting time application can provide protection against nematode damage.

PELIGRO

PRECAUCION AL USUARIO: Si usted no puede leer o entender inglés, no use este producto hasta que la etiqueta le haya sido explicada ampliamente.

(TO THE USER: If you cannot read or understand English, do not use this product until the label has been fully explained to you.)

DIRECTIONS FOR USE

It is a violation of Federal Law to use this product in a manner inconsistent with its labeling.

NEMACUR 3 may be applied by ground equipment or low pressure irrigation (grapes, citrus & non-bearing deciduous fruit & nut trees & commercial tree & vine nurseries). Do not apply by mist sprayer or aerial spray equipment.

Ground application equipment: Unless otherwise specified, apply NEMACUR 3 in a minimum of 20 gallons per **treated** acre. Maintain sufficient agitation during application to insure a uniform mixture. Incorporate NEMACUR 3 immediately after application for best results.

Chemigation:

Apply NEMACUR 3 only through low pressure irrigation equipment (including drip, mini-sprinklers, strip tubing and jets). Do not apply NEMACUR 3 through any type of irrigation system.

The system must contain a functional check valve, vacuum relief valve and low pressure drain appropriately located on the irrigation pipeline to prevent water source contamination from backflow.

The pesticide injection pipeline must contain a functional, automatic, quick-closing check valve to prevent the flow of fluid back toward the injection pump.

The pesticide injection pipeline must also contain a functional, normally closed, solenoid-operated valve located on the intake side of the injection pump and connected to the system interlock to prevent fluid from being withdrawn from the supply tank when the irrigation system is either automatically or manually shut down.

The system must contain functional interlocking controls to automatically shut off the pesticide injection pump when the water pump motor stops.

The irrigation line or water pump must include a functional pressure switch which will stop the water pump motor when the water pressure decreases to the point where pesticide distribution is adversely affected.

Systems must use a metering pump such as a positive displacement injection pump (e.g., diaphragm pump) effectively designed and constructed of materials that are compatible with pesticides and capable of being fitted with a system interlock.

Do not apply when wind speed favors drift beyond the areas intended for treatment.

A pesticide supply tank is not necessary for premixing since NEMACUR 3 mixes well with water in the irrigation line. If a pesticide supply tank is used, add NEMACUR 3 to water in supply tank and maintain sufficient agitation during application to insure a uniform mixture.

Begin irrigation and check emitters to insure all systems are operating normally. Make NEMACUR 3 injection in the first $1/3$ to $1/2$ of the irrigation with an injection time of at least 30 minutes. Maintain the system's operating pressure low enough to prevent fogging and/or misting during application. Inject NEMACUR 3 into the irrigation system after the filter(s). Continue irrigation until system's lines are flushed.

Crop injury, lack of effectiveness, or illegal pesticide residues in the crop can result from nonuniform distribution of treated water.

If you have questions about calibration, you should contact State Extension Service specialists, equipment manufacturers or other experts.

Do not connect an irrigation system (including greenhouse systems) used for pesticide application to a public water system unless the pesticide label-prescribed safety devices for public water systems are in place.

A person knowledgeable of the chemigation system and responsible for its operation, or under the supervision of the responsible person, shall shut the system down and make necessary adjustments should the need arise.

Posting: Posting of areas to be chemigated is required when 1) any part of a treated area is within 300 feet of sensitive areas such as residential areas, labor camps, businesses, day care centers, hospitals, inpatient clinics, nursing homes or any public areas such as schools, parks, playgrounds, or other public facilities not including public roads; or 2) when the chemigated area is open to the public such as golf courses or retail greenhouses.

Posting must conform to the following requirements: Treated areas shall be posted with signs at all usual points of entry and along likely routes of approach from the listed sensitive areas. When there are no usual points of entry, signs must be posted in the corners of the treated areas and in any other location affording maximum visibility to sensitive areas. The printed side of the sign should face away from the treated area towards the sensitive area. The signs shall be printed in English. Signs must be posted prior to application and must remain posted until foliage has dried and soil surface water has disappeared. Signs may remain in place indefinitely as long as they are composed of materials to prevent deterioration and maintain legibility for the duration of the posting period.

All words shall consist of letters at least $2\frac{1}{2}$ inches tall, and all letters and the symbol shall be a color which sharply contrasts with their immediate background. At the top of the sign shall be the words KEEP OUT, followed by an octagonal stop sign symbol at least 8 inches in diameter containing the word STOP. Below the symbol shall be the words PESTICIDES IN IRRIGATION WATER.

Mixing: NEMACUR 3 mixes readily with water. No special premixing should be required if used in spray equipment equipped with a recirculation system or agitation. If the spray mixture is allowed to set overnight, redispersion is readily obtained by agitation or recirculation.

Tank Mixing: NEMACUR 3 can be tank mixed with registered fungicides, herbicides and insecticides. Observe all cautions and limitations on labeling of all products used in tank mixtures.

Compatibility: NEMACUR 3 is physically compatible with many registered pesticides and liquid fertilizers. To determine the compatibility of NEMACUR 3 with other products, the following procedure should be followed: Pour the recommended proportions of the products into a suitable container of water, mix thoroughly and allow to stand at least five (5) minutes. If the combination remains mixed or can be remixed readily, the mixture is considered physically compatible. It is not recommended that combinations not be allowed to set overnight due to possible crystal growth, clabbering, etc. For further information contact your local Mobay representative.

Consult your State Agricultural Experiment Station, or Extension Service Specialist or Mobay representative for guidance on the selection of dosage rates and timing of applications appropriate for your specific crop. State regulations may contain additional restrictions or requirements. Check the remarks and restrictions sections of the product label for additional directions.

RECOMMENDED APPLICATIONS

CROP	PEST	DOSAGE NEMACUR 3	REMARKS
FIELD CROPS Cotton	Thrips (early-season reduction	2.4 to 3.3 fl. ozs. per 1000 ft. of row for any row spacing. (1 to $1\frac{1}{3}$ quarts per acre, based on 40 inch rows.)	**In furrow/covering soil:** Apply specified dosage per 1,000 ft. of row as a spray in furrow or band in covering soil behind the seed drop and in front of covering devices.
	Nematodes Thrips	3.3 to 7.1 fl. ozs. per 1000 ft. of row for any row spacing. ($1\frac{1}{3}$ to 2.9 quarts per acre based on 40 inch rows.)	**In furrow/covering soil:** Apply specified dosage per 1,000 ft. of row as a spray in furrow or band in covering soil behind the seed drop and in front of covering devices. **Band:** Apply specified dosage per 1,000 ft. of row as a spray in a 6 to 12 inch band over the row. Incorporate thoroughly to insure uniform distribution.[1]

Suggested use rates of NEMACUR 3 in fluid oz./1,000 feet of row based on nematode infestation levels:*

Band width	Low	Moderate	High*
In furrow	3.3	4.0	4.8
In covering soil	3.3	4.0	4.8
6 inch band	3.3	4.0	4.8
8 inch band	4.0	4.8	5.7
10 inch band	4.4	5.2	6.4
12 inch band	4.8	5.7	7.1

*Use the highest recommended rates in fields with a history of nematode problems.

CROP	PEST	DOSAGE NEMACUR 3	REMARKS
Cotton (California only)	Nematodes Thrips	9.8 fl. ozs. per 1,000 ft. of row or 4.0 qts. per acre based on 40" rows. Do not exceed 4 qts. per acre regardless of row spacing.	**SOIL INJECTION APPLICATIONS:** Apply specified dosage in a water emulsion or with liquid fertilizer through 2 or more injection shanks centered on the seed row covering an 18" band. Plant crop in usual manner.
Peanuts	Nematodes Thrips (early season reduction)		Apply specified dosage in fl. oz. per 1,000 ft. of row as a water emulsion spray in a 12 inch band over the row at planting. Use sufficient water and incorporate into the soil. ON NARROW ROW CROPS, DO NOT ALLOW BANDS TO OVERLAP.
		4.5 to 7.3 fl. ozs. for any row spacing (or 2 to 3.3 quarts per acre on 36 inch rows).	Use the high rate in fields with high populations of nematodes in fields where root-knot species other than southern root-knot, *Meloidogyne incognita*, are present or in fields having a history of serious nematode damage. Do not feed or graze green peanut vines or peanut vine hay. Do not hog down treated peanut fields.
Tobacco (Not for use on shade-grown tobacco)	Nematodes	$1\frac{1}{3}$ to 2 gallons	Uniformly broadcast the specified dosage in a minimum of 20 gallons of water per acre over the soil surface. Incorporate to a depth of 2 to 4 inches disking or tilling. Use the high rate in fields with high populations of nematodes, in fields where tobacco cyst and/or root-knot species other than southern root-knot *Meloidogyne incognita*, are present or in fields having a history of serious nematode damage. Plant crop in the usual manner.
	Tobacco cyst Nematodes	2 gallons	

IMPORTANT: Do not overlap bands on narrow row crops. Do not feed or graze cotton foliage. Under very heavy thrip pressure, additional control measures may be needed.
[1] The action of some planters during planting operation may provide adequate incorporation. When using planters which do not disturb enough soil for incorporation or under dry conditions, additional incorporation can be achieved with scratchers, paddlewheels, ground driven rototillers or similar equipment behind the planter.

CROP	PEST	DOSAGE NEMACUR 3	REMARKS
FRUIT Apple (Except California) Cherry Nectarine Peach	Nematodes	$1\frac{2}{3}$ to $3\frac{1}{2}$ gallons	**Band:** Apply specified dosage in not less than 10 gallons of solution per acre to the soil with ground equipment and incorporate immediately.* Center the treated band on the tree row using a band width of 50% of row spacing and covering the feeder root system of the plant.
Nectarine Peach	Nematodes	2 quarts to $1\frac{1}{2}$ gallons	**LOW PRESSURE IRRIGATION:** Apply specified dosage per acre to the soil in 1 to 6 applications with at least 14 days between applications using a minimum of $1\frac{1}{2}$ gallons to a maximum of 3 gallons of NEMACUR 3 per acre per season.
	* Incorporate this product into the soil mechanically or supply sufficient irrigation (not to exceed the depth of the root zone). Control of nematodes is best obtained when there is adequate rainfall or irrigation after application. Do not apply within 72 days of harvest of apples, or within 45 days of harvest of peaches or cherries. Do not apply more than $3\frac{1}{2}$ gallons/acre per year per planting site. Do not graze livestock in treated orchards. Do not feed cover crops grown in treated orchards to livestock. **Special Precautions:** Avoid contacting tree foliage with the spray mixture. Do not add oils or surfactants to the spray mixture. Do not allow spray or drift to contact operator. On non-bearing apple, cherry, nectarine and peach make initial applications when trees have recovered from transplanting shock and have initiated new growth, provided that one or more irrigations have been applied and soil has settled around the roots. Phytotoxicity may occur if applications are made before soil has settled and plants are growing.		
Grapes	Nematodes	3 gallons	**Band:** Apply specified dosage in not less than 10 gallons of solution per acre to the soil with ground equipment and incorporate immediately.* Center the treated band on the vine row using a band width of 50% of the row spacing.
		1 quart to 1 gallon	**LOW PRESSURE IRRIGATION:** Apply the specified dosage per acre in one or more applications using a minimum of 1 gallon and a maximum of 2 gallons of NEMACUR 3 per acre per season.
	* Incorporate this product into the soil mechanically or supply sufficient irrigation (not to exceed the depth of the root zone.) **RESTRICTIONS:** On bearing grapes, the last application may be made up to 2 days of harvest. Do not apply more than 3 gallons of NEMACUR 3 (9 pounds active ingredient) per acre in a 50% band or 2 gallons per acre by low pressure irrigation in one season. Do not graze or feed treated crop to livestock. On non-bearing grapes, make initial applications when vines have recovered from transplanting shock and have initiated new growth, provided that one or more irrigations have been applied and soil has settled around the roots. Phytotoxicity may occur if applications are made before soil has settled and plants are growing.		
Kiwifruit (California only)	Nematodes	2 quarts to 1 gallon	**LOW PRESSURE IRRIGATION** Apply specified dosage per acre to the soil in 2 to 6 applications using a minimum of 2 gallons and a maximum of 4 gallons of NEMACUR 3 per acre per year. Start NEMACUR applications when the soil temperature in the spring reaches 55°F or above at the 12-inch depth. To maintain control, an application may be made 5 days after harvest if the soil temperature at the 12-inch depth is 55° or above. A maximum of 3 gallons of NEMACUR 3 may be applied preharvest, and a maximum of 1 gallon may be applied post harvest. Do not apply NEMACUR 3 to newly established kiwifruit vineyards which are less than one year old. Do not apply within 31 days of harvest. Do not graze livestock in treated vineyards. Do not feed cover crops grown in treated vineyards to livestock.
Citrus (Arizona, California and Texas only) (In California do not apply to Kumquat, Tangelo, or Citrus hybrids)	Nematodes	$1\frac{2}{3}$ to $3\frac{1}{2}$ gallons	**BAND:** Apply specified dosage in not less than 10 gallons of solution per acre to the soil with ground equipment and incorporate immediately.* Two applications may be made per season, not to exceed $3\frac{1}{2}$ gallons per acre. Center the treated band on the vine row using a band width of 50% of the row spacing and covering the feeder root system of the plant.

CROP	PEST	DOSAGE NEMACUR 3	REMARKS
		2 quarts to 1½ gallons	**LOW PRESSURE IRRIGATION:** Apply specified dosage per acre to the soil in 1 to 6 applications with at least 14 days between applications using a minimum of 1½ gallons to a maximum of 3 gallons of NEMACUR 3 per acre per season.
	* Incorporate the product into the soil mechanically or supply sufficient irrigation (not to exceed the depth of the root zone). Control of nematodes is best obtained when there is adequate rainfall or irrigation after application. Under dry conditions, follow with irrigation. Do not apply within 30 days of harvest. Do not exceed 10 pounds active ingredient (3⅓ gallons of NEMACUR 3) per acre in a 50% band or 3 gallons of NEMACUR 3 (9 pounds active ingredient) per acre by low pressure irrigation per season. Do not graze livestock in treated areas. On non-bearing citrus make initial applications when trees have recovered from transplanting shock and have initiated new growth, provided that one or more irrigations have been applied and soil has settled around the roots. Phytotoxicity may occur if applications are made before soil has settled and plants are growing.		
Pineapple (Hawaii only)	Nematodes (*Rotylenchulus* and *Meloidogyne* species)	6⅔ gallon	**PREPLANT APPLICATION:** Apply the specified dosage per acre as a water emulsion spray over the entire area to be treated. Incorporate thoroughly to insure uniform distribution. See "NOTE" below. Additional postplant applications may be made to the plant crop at 1 to 3 month intervals by broadcast spray or drip irrigation equipment as indicated below. However, do not apply more than a total of 13⅓ gallons of NEMACUR 3 (40 pounds of active ingredient) per acre per plant crop.
		1⅓ to 8 pints	**POSTPLANT APPLICATION:** **PLANT CROP.** Apply specified dosage in 50 to 250 gallons of water per acre as a broadcast spray or through drip irrigation. Applications may begin immediately after planting. Make additional applications at intervals of 1 to 3 months as needed. Do not apply within 30 days before harvest. See "NOTE" below. Do not apply more than a total of 13⅓ gallons of NEMACUR 3 per acre per plant crop. **RATOON CROP.** Apply specified dosage in 50 to 250 gallons of water per acre as a broadcast spray or through drip irrigation. The first application may be made immediately following crop harvest. Make additional applications at intervals of 1 to 3 months as needed. Do not apply within 30 days before harvest. See "NOTE" below. Do not apply more than a total of 6⅔ gallons of NEMACUR 3 to each ratoon crop. NOTE: Do not use green forage or green fodder (cannery waste, such as cull fruit, fruit skin or shells, crowns, cores and basal ends of fruit may be fed) for animal feed.
Pineapple* (Puerto Rico only)	Nematodes	13⅓ to 26⅔ pints	**POSTPLANT BROADCAST APPLICATION:** **PLANT CROP—**Apply specified dosage (5 to 10 lbs. active ingredient) in 50 to 250 gallons of water per acre as a broadcast spray. Begin applications 1 to 3 months after planting. Make additional applications at intervals of 3 to 6 months as needed. Do not apply within 7½ months before harvest. **FIRST RATOON CROP.** Apply specified dosage in 50 to 250 gallons of water per acre as a broadcast spray. Make the first application immediately following crop harvest. Make additional applications at intervals of 3 to 6 months as needed. Do not apply within 7½ months before harvest.
		Do not apply more than a total of 20 lbs. active ingredient per acre per crop season regardless of the formulation or method of application. Do not use forage or fodder for animal feed. * NEMACUR 3 emulsifiable nematicide has shown excellent activity against the major genera of nematodes infesting pineapple plants. When applied as a broadcast spray after planting, NEMACUR protects young developing root systems. For best results, applications should be made when plants are actively growing.	
Raspberry (Except California)	Nematodes	1 to 2 gallons	**Band:** Apply specified dosage in not less than 10 gallons of solution per acre to the soil with ground equipment and incorporate immediately using mechanical incorporation or irrigation. Center the treated band on the row using a band width of 50% of the row spacing and covering the feeder root system of the plant. Apply during the period of October 1 to December 31 when adequate rainfall can be expected. Do not apply within six months of harvest. Do not apply more than once per year. Use the high rate in fields with high populations of nematodes or in fields having a history of serious nematode damage.
Vegetables Eggplant	Nematodes	5.9 fl. oz. (or 2⅔ quarts per acre on 36-inch rows)	**Band:** Apply specified dosage per 1,000 ft. of row in a 12-inch band over the row at transplanting and immediately incorporate into the soil. ON NARROW ROW CROPS, DO NOT ALLOW BANDS TO OVERLAP.
NON-BEARING* Deciduous fruit & nut trees & commercial tree & vine nurseries	Nematodes	1½ to 3 gallons	**Band:** Apply specified dosage in not less than 10 gallons of solution per acre to the soil with ground equipment and incorporate immediately.** Center the treated band on the tree row using the low rate in a band width of 25% of the row spacing or the high rate in a band width of 50% of the row spacing. Use a minimum band width of "to the drip line".
		1 quart to 1 gallon	**LOW PRESSURE IRRIGATION:** Apply the specified dosage per acre to the soil in one or more applications using a minimum of 1 gallon and a maximum of 2 gallons of NEMACUR 3 per acre per season.

Make initial applications when plants have recovered from transplanting shock and have initiated new growth, provided that one or more irrigations have been applied and soil has settled around the roots. Phytotoxicity may occur if applications are made before soil has settled and plants are established.

Do not exceed 9 pounds active ingredient (3 gallons of NEMACUR 3) per acre in a 50% band or 2 gallons per acre by low pressure irrigation in one season.

*Non-bearing fruit and nut trees are those that will not bear for one year after application, including newly planted and established trees or vines.

Any fruit or nuts that may form on treated trees or vines during this one year period must be destroyed and not used for human or animal consumption.

**Incorporate this product into the soil mechanically or supply sufficient irrigation (not to exceed the depth of the root zone). Control of nematodes is best obtained when there is adequate rainfall or irrigation after application. Under dry conditions, follow with irrigation.

SPECIAL LOCAL NEED REGISTRATIONS [SECTION 24(c)] FOR NEMACUR 3*

CROP	PEST	STATE EPA SLN NO.	DOSAGE	REMARKS
Bulbous iris and Lily bulbs	Nematodes	Oregon OR 80-0063	2 to 4 gallons per acre.	Apply in-furrow at time of planting.
Cotton	Nematodes Thrips (early-season reduction)	New Mexico NM 79-0021 Texas TX 79-0035	2.6 to 5.3 fluid ounces per 1000 feet of row for any row spacing. (1.1 to 2.2 quarts per acre, based on 40-inch rows.)	In a water emulsion or as a tank-mix with Treflan® EC and/or liquid fertilizer.
Iris and narcissus bulbs	Nematodes	Washington WA 76-0034	2 to 4 gallons per acre.	Apply in-furrow at time of planting.
Transplant Cabbage	Nematodes	Florida FL 84-0019	5 fl. oz. per 1,000 ft. of row spacing or 2¼ quarts per acre on 36-inch row spacing.	**Post-plant Drench Application** Apply specified dosage in a 10-inch band directly over the row within 24 hours of transplanting using 62 gallons of water per 1,000 ft. of row, or 800 to 1,000 gallons of water per acre depending on width of row spacing. Restriction: Application must be separate from hand transplanting.

*For distribution and use only within the state specified. These Special Local Need registrations approved as of date of this publication are, however, subject to cancellation by the state. Contact your nearest Mobay Agricultural Chemicals Division Office for the current status and a copy of the complete individual supplemental Special Local Need labeling. This labeling must be in the possession of the user at the time of pesticide application.

RESTRICTIONS

Do not use on other crops. Use only according to label directions. Any food crop not specified on the label may be planted into treated areas 120 days after the last application. Any cover crops that are planted during the 120 day period must be plowed under and not grazed.

PROTECTIVE CLOTHING AND EQUIPMENT

USE ONLY WHEN WEARING THE FOLLOWING PROTECTIVE CLOTHING AND EQUIPMENT DURING MIXING/LOADING, APPLICATION, REPAIR AND CLEANING OF MIXING, LOADING AND APPLICATION EQUIPMENT, DISPOSAL OF THE PESTICIDE, AND EARLY REENTRY INTO TREATED AREAS: Protective suit of one or two pieces covering all parts of the body except head, hands, and feet; chemical-resistant gloves; chemical-resistant shoes (or chemical-resistant shoe covers or chemical-resistant boots); and a NIOSH- or MSHA- approved respirator. In addition, mixer/loaders must wear a chemical-resistant apron. During equipment repair and cleaning, the respirator need not be worn. During early re-entry after sprays have dried and vapors have dispersed, the respirator need not be worn.

IF MIXING/LOADING IS PERFORMED USING A CLOSED SYSTEM, THE FOLLOWING PROTECTIVE CLOTHING AND EQUIPMENT MAY BE WORN AS AN ALTERNATIVE: Long-sleeved shirt and long-legged pants; chemical-resistant gloves; chemical-resistant apron; shoes and socks. Goggles or face shield must be worn if the system is under pressure.

IF APPLICATION IS PERFORMED USING AN ENCLOSED CAB, THE FOLLOWING PROTECTIVE CLOTHING AND EQUIPMENT MAY BE WORN AS AN ALTERNATIVE: Long-sleeved shirt and long-legged pants; shoes, and socks. Chemical-resistant gloves must be available in the cab and must be worn when exiting. This clothing is inadequate to protect you during equipment repair or cleaning, re-entry, or pesticide disposal work.

IMPORTANT! If pesticide comes in contact with skin, wash off with soap and water. Always wash hands, face, and arms with soap and water before smoking, eating, drinking, or toileting.

AFTER WORK: Before removing gloves, wash them with soap and water. Take off all work clothes and shoes. Shower using soap and water. Wear clean clothes. Do not reuse contaminated clothing. Personal clothing worn during work must be laundered separately from household articles. Store protective clothing separately from personal clothing. Clean or launder protective clothing after each use. Respirators must be cleaned and filters replaced according to instructions included with the respirator. Protective clothing and equipment that becomes heavily contaminated or drenched must be destroyed according to state and local regulations. HEAVILY CONTAMINATED OR DRENCHED CLOTHING CANNOT BE ADEQUATELY DECONTAMINATED.

RE-ENTRY STATEMENTS

Do not apply this product in such a manner as to directly or through drift expose workers or other persons. The area being treated must be vacated by unprotected persons.

Do not re-enter treated areas for 48 hours unless appropriate protective clothing is worn. Any person coming in direct contact with treated water, and/or treated soil prior to incorporation or irrigation, or with treated wet soil after the initial irrigation, must wear protective clothing and equipment as described in the section "Protective Clothing and Equipment" on the label. Wash thoroughly with soap and water after handling and before eating or smoking.

Because certain states may require more restrictive re-entry intervals for various crops treated with this product, consult your State Department of Agriculture for further information.

Written or oral warnings must be given to workers who are expected to be in a treated area or in an area about to be treated with this product. Oral warnings should include those statements found under the label headings "Precautionary Statements" and "Re-Entry Statements" included on the label. When oral warnings are given, warnings shall be given in a language customarily understood by workers. Oral warnings must be given if there is reason to believe that written warnings cannot be understood by workers. Written warnings must include the following information: DANGER, area treated with NEMACUR 3 on (date of application). See Protective Clothing and Equipment for use of proper protective clothing. **If on skin,** remove contaminated clothing and wash skin immediately with soap and warm water. **If eyes are contaminated,** wash with flowing water for at least 15 minutes. **If swallowed, vomiting should be induced.** (See Precautionary Statements on label.)

STORAGE AND DISPOSAL

Pesticide Disposal: Do not contaminate water, food or feed by storage or disposal.

Pesticide wastes are acutely hazardous. Improper disposal of excess pesticide, spray mixture, or rinsate is a violation of Federal Law. If these wastes cannot be disposed of by use according to label instructions, contact your State Pesticide or Environmental Control Agency or the Hazardous Waste representative at the nearest EPA Regional Office for guidance.

Container Disposal: Triple rinse (or equivalent). Then offer for recycling or reconditioning, or puncture and dispose of in a sanitary landfill, or by other procedures approved by State and local authorities.

Store in a cool, dry place and in such a manner as to prevent cross contamination with other pesticides, fertilizers, food and feed. Do not store below 32°F. Do not store near heat or open flame. Store in original container and out of the reach of children, preferably in a locked storage area.

Handle and open container in a manner as to prevent spillage. If container is leaking invert to prevent leakage. If the container is leaking or material spilled for any reason or cause, carefully dam up spilled material to prevent runoff. Refer to Precautionary Statements on label for hazards associated with the handling of this material. Do not walk through spilled material. Absorb spilled material with absorbing type compounds and dispose of as directed for pesticides above. In spill or leak incidents, keep unauthorized people away. You may contact the Mobay Emergency Response Team for decontamination procedures or any other assistance that may be necessary. The Mobay Kansas City Emergency Response Telephone No. is 816-242-2582, or contact Chemtrec at 800-424-9300.

PRECAUTIONARY STATEMENTS

HAZARDS TO HUMANS AND DOMESTIC ANIMALS

DANGER–POISON

Fatal if swallowed or absorbed through the skin. Causes irreversible eye damage. May be fatal if inhaled. Do not get in eyes, on skin or on clothing. Wear protective clothing, rubber gloves and goggles or face shield when handling. Wash thoroughly with soap and water after handling and before eating, drinking or using tobacco. Remove contaminated clothing and wash before reuse.

Do not breathe vapor or spray mist. Wear a mask or pesticide respirator jointly approved by the Mine Safety and Health Administration and by the National Institute for Occupational Safety and Health.

Do not contaminate feed or food. Keep out of reach of children. Keep all unprotected persons out of the operating area or vicinity.

HANDLING OF CONTAMINATED ARTICLES: Clothing and rags receiving undiluted NEMACUR 3 as a result of leaks, spills, or mishaps should be removed immediately and not reused. These contaminated articles should be buried in the same manner as the empty container: see container disposal section for directions. (Contaminated articles which should be discarded would include: hats, gloves, aprons, coats, boots, etc. not made of rubber or coated with rubber or other similar materials). Clothing receiving spray mist or droplets from the mixed, diluted NEMACUR 3 should be removed and decontaminated before reuse. This clothing can be decontaminated by machine washing separately from other items with soap or detergent and bleach in hot water twice before reuse. Protective clothing and equipment should be washed down with detergent or soap and bleach in water. Wash water from the cleaning of protective clothing or equipment should not be allowed to run off or otherwise enter water supplies.

SYMPTOMS OF POISONING: A sense of "tightness" in the chest. Sweating. Contracted pupils. Stomach pains. Vomiting and diarrhea.

STATEMENTS OF PRACTICAL TREATMENT

In case of poisoning, call a physician immediately. Have patient lie down and keep quiet.

If swallowed, vomiting should be induced. Administer water freely and induce vomiting by giving one dose (1/2 or 15 ml.) of syrup of ipecac. If vomiting does not occur within 10 to 20 minutes, administer second dose. If syrup of ipecac is not available, induce vomiting by sticking finger down throat. Repeat until vomit fluid is clear. Never give anything by mouth to an unconscious person. Professional medical assistance should be secured immediately.

If on skin, remove contaminated clothing and wash skin immediately with soap and warm water.

If eyes are contaminated, hold eyelids open and flush with a steady, gentle stream of water for at least 15 minutes.

If inhaled, remove victim to fresh air. If not breathing give artificial respiration, preferably mouth to mouth.

NOTE TO PHYSICIAN: Antidote—Administer atropine sulfate in large therapeutic doses. Repeat as necessary to the point of tolerance. 2-PAM is also antidotal and may be administered in conjunction with atropine.

Compound inhibits cholinesterase, resulting in stimulation of the central nervous system, the parasympathetic nervous system, and the somatic motor nerves. Do not give morphine. Watch for pulmonary edema, which may develop in serious cases of poisoning even after 12 hours. At first sign of pulmonary edema, the patient should be placed in an oxygen tent and treated symptomatically.

ENVIRONMENTAL HAZARDS

This pesticide is toxic to fish and wildlife. Drift and runoff from treated areas may be hazardous to aquatic organisms in neighboring areas. Do not apply directly to water or wetlands (swamps, bogs, marshes, and potholes). Do not contaminate water when disposing of equipment washwaters.

Do not use mist sprayers. Use only coarse sprays directed at soil to eliminate spray drift.

Aerial application of this product is prohibited.

IMPORTANT:

This bulletin is not intended to provide adequate information for application. Before using the product, read and carefully observe directions, cautionary statements and other information appearing on the label. This product is sold subject to the Conditions of Sale set forth on the container label.

READ THE LABEL BEFORE USE

N 9001

This is a specimen label. Mobay is not responsible for the accuracy of the information contained herein. As labels are subject to revision, always carefully read and follow the label on the product container.

NEMACUR 3®

EMULSIFIABLE SYSTEMIC INSECTICIDE-NEMATICIDE

RESTRICTED USE PESTICIDE

For sale to and use only by Certified Applicator for uses authorized or persons under his direct supervision.

FOR EFFECTIVE CONTROL OF NEMATODES IN CITRUS CROPS (except Arizona, California, and Texas).

ACTIVE INGREDIENT:	
Ethyl 3-methyl-4-(methylthio)phenyl (1-methylethyl) phosphoramidate	35%
INERT INGREDIENTS:	65%
TOTAL	100%

EPA Reg. No. 3125-283

NEMACUR is a Reg. TM of Bayer AG, Germany

RECOMMENDED USES

Citrus (except Arizona, California, and Texas)

RESTRICTIONS ON WAREHOUSE STORAGE

Do not store in front of space heater. Store below 120°F but not below 32°F. Do not store adjacent to, or near foodstuffs, drugs, feed or any other material intended for consumption by humans or animals. Do not store adjacent to toys, cosmetics, dishes, furniture, or clothing. Protect from rain and moisture. Do not store next to herbicides. In case of puncture, place the container in a recovery drum containing sufficient absorbent material to absorb leakage.

PACKAGING

5 gal. drum. Shipping weight: 47 lbs.

NEMACUR 3 emulsifiable nematicide is an organophosphorus compound that has shown excellent activity against the major genera of nematodes, including burrowing nematodes, which attack citrus—oranges, grapefruit, lemons, limes and tangerines.

NEMACUR has long-lasting contact and systemic activity. Used according to directions, it can provide effective control of nematodes that stunt citrus growth, reduce yields, and lead to general tree decline.

NEMACUR is easy to apply with conventional equipment, broadcast or banded, and must be incorporated immediately by irrigation or mechanically.

PELIGRO

PRECAUCION AL USUARIO: Si usted no puede leer o entender inglés, no use este producto hasta que la etiqueta le haya sido explicada ampliamente.

(TO THE USER: If you cannot read or understand English, do not use this product until the label has been fully explained to you.)

DIRECTIONS FOR USE

It is a violation of Federal Law to use this product in a manner inconsistent with its labeling.

NEMACUR 3 may be applied by ground equipment or low pressure irrigation (Chemigation). Do not apply by mist sprayer or aerial spray equipment.

Ground Application Equipment: Unless otherwise specified, apply NEMACUR 3 in a minimum of 20 gallons of water per **treated** acre. Maintain sufficient agitation during application to insure a uniform mixture. Incorporate NEMACUR 3 immediately after application for best results.

CHEMIGATION

Apply NEMACUR 3 only through low pressure irrigation equipment (including drip, mini-sprinklers, strip tubing and jets). Do not apply NEMACUR 3 through any other type of irrigation system.

The system must contain a functional check valve, vacuum relief valve and low pressure drain appropriately located on the irrigation pipeline to prevent water source contamination from backflow.

The pesticide injection must contain a functional, automatic, quick-closing check valve to prevent the flow of fluid back toward the injection pump.

The pesticide injection pipeline must also contain a functional, normally closed, solenoid-operated valve located on the intake side of the injection pump and connected to the system interlock to prevent fluid from being withdrawn from the supply tank when the irrigation system is either automatically or manually shut down.

The system must contain functional interlocking controls to automatically shut off the pesticide injection pump when the water pump motor stops.

The irrigation line or water pump must include a functional pressure switch which will stop the water pump motor when the water pressure decreases to the point where pesticide distribution is adversely affected.

Systems must use a metering pump such as a positive displacement injection pump (e.g., diaphragm pump) effectively designed and constructed of materials that are compatible with pesticides and capable of being fitted with a system interlock.

Do not apply when wind speed favors drift beyond the areas intended for treatment.

A pesticide supply tank is not necessary for premixing since NEMACUR 3 mixes well with water in the irrigation line. If a pesticide supply tank is used, add NEMACUR 3 to water in supply tank and maintain sufficient agitation during application to insure a uniform mixture.

Begin irrigation and check emitters to insure all systems are operating normally. Make NEMACUR 3 injection in the first 1/3 to 1/2 of the irrigation with an injection time of at least 30 minutes. Maintain the system's operating pressure low enough to prevent fogging and/or misting during application. Inject NEMACUR 3 into the irrigation system after the filter(s). Continue irrigation until the system's lines are flushed.

Crop injury, lack of effectiveness, or illegal pesticide residues in the crop can result from nonuniform distribution of treated water.

If you have questions about calibration, you should contact State Extension Service specialists, equipment manufacturers or other experts.

Do not connect an irrigation system (including greenhouse systems) used for pesticide application to a public water system unless the pesticide label-prescribed safety devices for public water systems are in place.

A person knowledgeable of the chemigation system and responsible for its operation, or under the supervision of the responsible person, shall shut the system down and make necessary adjustments should the need arise.

Posting: Posting of areas to be chemigated is required when 1) any part of a treated area is within 300 feet of sensitive areas such as residential areas, labor camps, businesses, day care centers, hospitals, inpatient clinics, nursing homes or any public areas such as schools, parks, playgrounds, or other public facilities not including public roads; 2) when the chemigated area is open to the public such as golf courses or retail greenhouses.

Posting must conform to the following requirements: Treated areas shall be posted with signs at all usual points of entry and along likely routes of approach from the listed sensitive areas. When there are no usual points of entry, signs must be posted in the corners of treated areas and in any other location affording maximum visibility to sensitive areas. The printed side of the sign should face away from the treated area towards the sensitive area. The signs shall be printed in English. Signs must be posted prior to application and must remain posted until foliage has dried and soil surface water has disappeared. Signs may remain in place indefinitely as long as they are composed of materials to prevent deterioration and maintain legibility for the duration of the posting period.

All words shall consist of letters at least 2 1/2 inches tall, and all letters and the symbol shall be a color which sharply contrasts with their immediate background. At the top of the sign shall be the words KEEP OUT, followed by an octagonal stop sign symbol at least 8 inches in diameter containing the word STOP. Below the symbol shall be the words PESTICIDES IN IRRIGATION WATER.

Mixing: NEMACUR 3 mixes readily with water. No special premixing should be required if used in spray equipment equipped with a recirculation system or agitation. If the spray mixture is allowed to sit overnight, redispersion is readily obtained by agitation or recirculation.

Tank Mixing: NEMACUR 3 can be tank mixed with registered fungicides, herbicides and insecticides. Observe all cautions and limitations on labeling of all products used in tank mixtures.

Compatability: NEMACUR 3 is physically compatible with many registered pesticides and liquid fertilizers. To determine the compatibility of NEMACUR 3 with other products, the following procedure should be followed: Pour the recommended proportions of the products into a suitable container of water, mix thoroughly and allow to stand at least five (5) minutes. If the combination remains mixed or can be re-mixed readily, the mixture is considered physically compatible. It is recommended that combinations not be allowed to set overnight due to possible crystal growth, clabbering, etc. For further information contact your local Mobay representative.

Consult your State Agricultural Experiment Station, or Extension Service Specialist or Mobay representative for guidance on the selection of dosage rates and timimg of applications appropriate for your specific crop. State regulations may contain additional restrictions or requirements. Check the remarks and restrictions sections of the NEMACUR 3 label for additional directions.

RECOMMENDED APPLICATIONS

CROP	PEST	DOSAGE OF NEMACUR 3	REMARKS
FRUIT Citrus (Except Arizona, California, and Texas)	Nematodes	1 2/3 to 3 1/3 gallons	**BAND:** Apply specified dosage in not less than 10 gallons of solution per acre to the soil with ground equipment and incorporate immediately.* Two applications may be made per season, not to exceed 3 1/3 gallons per acre. Center the treated band on the tree row using a band width of 50% of the row spacing and covering the feeder root system of the plant.
		2 quarts to 1 1/2 gallons	**LOW PRESSURE IRRIGATION:** Apply specified dosage per acre to the soil in 1 to 6 applications with at least 14 days between applications using a minimum of 1 1/2 gallons to a maximum of 3 gallons of NEMACUR 3 per acre per season.

* Incorporate this product into the soil mechanically or supply sufficient irrigation (not to exceed the depth of the root zone). Control of nematodes is best obtained when there is adequate rainfall or irrigation after application. Under dry conditions, follow with irrigation.

Do not apply within 30 days of harvest.

Do not exceed 10 pounds active ingredient (3 1/3 gallons of NEMACUR 3) per acre in a 50% band or 3 gallons of NEMACUR 3 (9 pounds active ingredient) per acre by low pressure irrigation per season. Do not graze livestock in treated areas.

On non-bearing citrus make initial application when trees have recovered from transplanting shock and have initiated new growth, provided that one or more irrigations have been applied and soil has settled around the roots. Phytotoxocity may occur if applications are made before soil has settled and plants are growing.

10-21-88

RESTRICTIONS

Do not use on other crops. Use only according to label directions.

Any food crop not specified on the label may be planted into treated areas 120 days after the last application. Any cover crops that are planted during the 120 day period must be plowed under and not grazed.

PROTECTIVE CLOTHING AND EQUIPMENT

USE ONLY WHEN WEARING THE FOLLOWING PROTECTIVE CLOTHING AND EQUIPMENT DURING MIXING/LOADING, APPLICATION, REPAIR AND CLEANING OF MIXING, LOADING AND APPLICATION EQUIPMENT, DISPOSAL OF THE PESTICIDE, AND EARLY RE-ENTRY INTO TREATED AREAS: Protective suit of one or two pieces covering all parts of the body except head, hands, and feet; chemical-resistant gloves; chemical-resistant shoes (or chemical-resistant shoe covers or chemical-resistant boots); and a NIOSH- or MSHA-approved respirator. In addition, mixer/loaders must wear a chemical resistant apron. During equipment repair and cleaning, the respirator need not be worn. During early reentry after sprays have dried and vapors have dispersed, the respirator need not be worn.

IF MIXING/LOADING IS PERFORMED USING A CLOSED SYSTEM, THE FOLLOWING PROTECTIVE CLOTHING AND EQUIPMENT MAY BE WORN AS AN ALTERNATIVE: Long-sleeved shirt and long-legged pants; chemical-resistant gloves; chemical-resistant apron; shoes and socks. Goggles or face shield must be worn if the system is under pressure.

IF APPLICATION IS PERFORMED USING AN ENCLOSED CAB, THE FOLLOWING PROTECTIVE CLOTHING AND EQUIPMENT MAY BE WORN AS AN ALTERNATIVE: Long-sleeved shirt and long-legged pants; shoes and socks. Chemical-resistant gloves must be available in the cab and must be worn when exiting. This clothing is inadequate to protect you during equipment repair or cleaning, re-entry, or pesticide disposal work.

IMPORTANT! If pesticide comes in contact with skin, wash off with soap and water. Always wash hands, face, and arms with soap and water before smoking, eating, drinking, or toileting.

AFTER WORK: Before removing gloves, wash them with soap and water. Take off all work clothes and shoes. Shower using soap and water. Wear clean clothes. Do not reuse contaminated clothing. Personal clothing worn during work must be laundered separately from household articles. Store protective clothing separately from personal clothing. Clean or launder protective clothing after each use. Respirators must be cleaned and filters replaced according to instructions included with the respirator. Protective clothing and equipment that becomes heavily contaminated must be destroyed according to state and local regulations. HEAVILY CONTAMINATED OR DRENCHED CLOTHING CANNOT BE ADEQUATELY DECONTAMINATED.

RE-ENTRY STATEMENTS

Do not apply this product in such a manner as to directly or through drift expose workers or other persons. The area being treated must be vacated by unprotected persons. Do not re-enter treated area for 48 hours unless appropriate protective clothing is worn.

Any person coming in direct contact with treated water, and/or treated soil prior to incorporation or irrigation, or with treated wet soil after the initial irrigation, must wear protective clothing and equipment as described in the sections "Protective Clothing and Equipment" on the label. Wash thoroughly with soap and water after handling and before eating or smoking.

Because certain states may require more restrictive re-entry intervals for various crops treated with this product, consult your State Department of Agriculture for further information.

Written or oral warnings must be given to workers who are expected to be in treated area or in an area about to be treated with this product. Oral warnings should include those statements found under the label heading "Precautionary Statements" and "Re-entry Statements" included on the label. When oral warnings are given, warnings shall be given in a language customarily understood by workers. Oral warnings must be given if there is reason to believe that written warnings cannot be understood by workers. Written warnings must include the following information: DANGER, area treated with NEMACUR 3 on (date of application). See "Re-entry Statements" for use of proper protective clothing. **If on skin,** remove contaminated clothing and wash skin immediately with soap and warm water. **If eyes are contaminated,** wash with flowing water for at least 15 minutes. **If swallowed,** vomiting should be induced. (See Precautionary Statements on label).

STORAGE AND DISPOSAL

Pesticide Disposal: Do not contaminate water, food or feed by storage or disposal.

Pesticide wastes are acutely hazardous. Improper disposal of excess pesticide, spray mixture, or rinsate is a violation of Federal Law. If these wastes cannot be disposed of by use according to label instructions, contact your State Pesticide or Environmental Control Agency, or the Hazardous Waste representative at the nearest EPA Regional Office for guidance.

Container Disposal: Triple rinse (or equivalent). Then offer for recycling or reconditioning, or puncture and dispose of in a sanitary landfill, or by other procedures approved by State and local authorities.

Store in a cool, dry place and in such a manner as to prevent cross contamination with other pesticides, fertilizers, food, and feed. Do not store below 32° F. Do not store near heat or open flame. Store in original container and out of the reach of children, preferably in a locked storage area.

Handle and open container in an manner as to prevent spillage. If container is leaking invert to prevent leakage. If the container is leaking or material spilled for any reason or cause, carefully dam up spilled material to prevent runoff. Refer to Precautionary Statements on label for hazards associated with the handling of this material. Do not walk through spilled material. Absorb spilled material with absorbing type compounds and dispose of as directed for pesticides above. In spill or leak incidents, keep unauthorized people away. You may contact the Mobay Emergency Response Team for decontamination procedures or any other assistance that may be necessary. The Mobay Kansas City Emergency Response Telephone No. is 816-242-2582, or contact Chemtrec at 800-424-9300.

PRECAUTIONARY STATEMENTS

HAZARDS TO HUMANS AND DOMESTIC ANIMALS

DANGER–POISON

Fatal if swallowed or absorbed through the skin. Causes irreversible eye damage. May be fatal if inhaled. Do not get in eyes, on skin, or on clothing. Wear protective clothing, rubber gloves, and goggles or face shield when handling. Wash thoroughly with soap and water after handling and before eating, drinking or using tobacco. Remove contaminated clothing and wash before reuse.

Do not breathe vapors or spray mist. Wear a mask or respirator jointly approved by the Mine Safety and Health Administration and by the National Institute for Occupational Safety and Health.

Do not contaminate feed or food. Keep out of reach of children. Keep all unprotected persons out of the operating area or vicinity.

HANDLING OF CONTAMINATED ARTICLES: Clothing and rags receiving undiluted NEMACUR 3 as a result of leaks, spills, or mishaps should be removed immediately and not reused. These contaminated articles should be buried in the same manner as the empty container: see container disposal section for directions. (Contaminated articles which should be discarded would include: hats, gloves, aprons, coats, boots, etc. not made of rubber, or coated with rubber or other similar materials.) Clothing receiving spray mist or droplets from the mixed, diluted NEMACUR 3 should be removed and decontaminated before reuse. This clothing can be decontaminated by machine washing separately from other items with soap or detergent and bleach in hot water twice before reuse. Protective clothing and equipment should not be allowed to run off or otherwise enter water supplies.

SYMPTOMS OF POISONING: A sense of "tightness" in the chest. Sweating. Contracted pupils. Stomach pains. Vomiting and diarrhea.

STATEMENTS OF PRACTICAL TREATMENT

In case of poisoning, call a physician immediately. Have patient lie down and keep quiet.

If swallowed, vomiting should be induced. Administer water freely, and induce vomiting by giving one dose (1/2 oz. or 15 ml.) of syrup of ipecac. If vomiting does not ocur within 10 to 20 minutes, administer second dose. If syrup of ipecac is not available, induce vomiting by sticking finger down throat. Repeat until vomit fluid is clear. Never give anything by mouth to an unconscious person. Professional medical assistance should be secured immediately.

If on skin, remove contaminated clothing and wash skin immediately with soap and warm water.

If eyes are contaminated, wash with flowing water for at least 15 minutes.

If inhaled, remove victim to fresh air. If not breathing, give artificial respiration, preferably mouth to mouth.

NOTE TO PHYSICIAN: Antidote—Administer atropine sulfate in large therapeutic doses. Repeat as necessary to the point of tolerance. 2-PAM is also antidotal and may be administered in conjunction with atropine.

Compound inhibits cholinesterase, resulting in stimulation of the central nervous system, the parasympathetic nervous system and the somatic motor nerves. Do not give morphine. Watch for pulmonary edema, which may develop in serious cases of poisoning even after 12 hours. At first sign of pulmonary edema, the patient should be placed in an oxygen tent and treated symptomatically.

ENVIRONMENTAL HAZARDS

This pesticide is toxic to fish and other wildlife. Do not apply directly to water. Do not apply when weather conditions favor drift from target area. Do not contaminate water when disposing of equipment washwaters. Do not use mist sprayers. Use only coarse sprays directed at soil to eliminate spray drift. Aerial application of this product is prohibited.

IMPORTANT:

This bulletin is not intended to provide adequate information for application. Before using the product, read and carefully observe directions, cautionary statements and other information appearing on the label. This product is sold subject to the Conditions of Sale set forth on the container label.

READ THE LABEL BEFORE USE

N 9002

This is a specimen label. Mobay is not responsible for the accuracy of the information contained herein. As labels are subject to revision, always carefully read and follow the label on the product container.

SALUTE®

4EC HERBICIDE

FOR CONTROL OF CERTAIN GRASSES AND BROADLEAF WEEDS IN SOYBEANS

ACTIVE INGREDIENTS:

Trifluralin, *alpha, alpha, alpha*-trifluoro-2,6-dinitro-N,N-dipropyl-*p*-toluidine	28%
Metribuzin, 4-amino-6-(1,1-dimethylethyl)-3-(methylthio)-1,2,4-triazin-5(4*H*)-one*	14%
INERT INGREDIENTS:	58%
TOTAL	100%

Contains 2.67 pounds of trifluralin and 1.33 pounds of metribuzin per gallon.

EPA Reg. No. 3125-375 EPA Est. 3125-MO-1

SALUTE is a Reg. TM of Mobay Corporation
SENCOR is a Reg. TM of Bayer AG, Germany
Command and Commence are Reg. TMs of FMC Corporation
Scepter, Squadron and Tri-Scept are Reg. TMs of American Cyanamid Co.
X-77 is a Reg. TM of Chevron Chemical Co.
*U.S. Patent No. 3,671,523

RECOMMENDED USES

SOYBEANS

RESTRICTIONS ON WAREHOUSE STORAGE

Store in a cool, dry place and in such a manner as to prevent cross contamination with other pesticides, fertilizers, food and feed.

PACKAGING

2 1/2 gal. jug, 2 per case. Shipping weight 52.1 lbs.
110 gal. mini-bulk.

SALUTE 4EC is an emulsifiable concentrate preplant herbicide formulated to give you consistent control of a wide spectrum of both grasses and broadleaves commonly found in soybeans.

SALUTE can be applied by ground or aerial equipment and can be used in a SPLIT-SHOT application with SENCOR® herbicide, combined with Command®, Commence®, Scepter®, Squadron®, or Tri-Scept® and can be used with dry or liquid fertilizer.

SALUTE effectively controls such troublesome weeks as: Jimsonweed, common ragweed, sesbania, smartweed, pigweed, prickly sida/teaweed, velvetleaf and bristly starbur.

It also controls annual grasses such as: Barnyardgrass, crabgrass, foxtails, sprangletop, stinkgrass, browntop millet and bluegrass.

AVISO

PRECAUCION AL USUARIO: Si usted no puede leer o entender inglés, no use este producto hasta que la etiqueta le haya sido explicada ampliamente.

(TO THE USER: If you cannot read or understand English, do not use this product until the label has been fully explained to you.)

DIRECTIONS FOR USE

It is a violation of Federal Law to use this product in a manner inconsistent with its labeling.

Do not contaminate food or feed.

GENERAL INFORMATION

MIXING: When using SALUTE 4EC, make sure the sprayer is completely clean, free of rust or corrosion which may occur from winter storage. Examine strainers and screens to be sure the sprayer is clean from previously used pesticides.

Any tank mix containing SALUTE 4EC should be kept minimally agitated and sprayed out immediately, especially under cool weather conditions. Do not allow tank mixes to stand for prolonged periods of time.

When using water that is extremely hard, add 1/2 pint of Ortho X-77® (do not substitute) spreader sticker per 100 gallons of spray mixture. This should be added to the spray tank when at least 1/2 full, but prior to the addition of any herbicide.

The proper mixing procedure for SALUTE 4EC alone:
1. Fill spray tank with the appropriate amount of water.
2. Add recommended rate of SALUTE 4EC.
3. Follow the triple rinse procedure to insure that all product is removed from the container.
4. Begin minimal agitation or recirculation until the herbicide is thoroughly mixed in the tank and spray out immediately.

When mixing SALUTE 4EC with other herbicides:
1. Fill spray tank 2/3 to 3/4 full of water and add other herbicide.
2. Finish filling the tank with the appropriate amount of water and add the recommended rate of SALUTE 4EC.
3. Follow the triple rinse procedure to insure that all product is removed from herbicide containers.
4. Begin minimal agitation or recirculation until the herbicides are thoroughly mixed in the tank and spray out immediately.

APPLICATION: Use a standard low pressure (20 to 40 psi) herbicide boom sprayer equipped with suitable nozzles and screens no finer than 50-mesh in-nozzle and in-line strainers. Agitate thoroughly before and during application with bypass agitation.

Do not apply this product through any type of irrigation system.

Ground Application: Apply the proper rate of SALUTE 4EC in a minimum of 10 to 40 gallons of spray mixture per acre broadcast.

Aerial Application: Where permitted, apply specified rate in a minimum of 2 to 10 gallons of spray mixture per acre. Do not apply aerially when wind speed is greater than 10 mph.

FOR ALL APPLICATIONS OF SALUTE: Sprayer must be accurately calibrated before applying SALUTE 4EC. Check sprayer during application to be sure it is working properly and delivering a uniform spray pattern. As the volume of spray mixture decreases per acre, the importance of accurate calibration and uniform application increases. Avoid over application, misapplication, and boom and spray swath overlapping that will increase spray dosage. (Crop injury may occur as a result.) Avoid spray skips and gaps which allow weeds to grow in untreated soil. Do not apply when weather conditions favor spray drift.

SPRAYER CLEANUP: Spray equipment must be thoroughly cleaned to remove remaining traces of herbicide that might injure other crops to be sprayed. Drain any remaining spray solution of SALUTE 4EC from the spray tank and discard in non-crop areas away from water supplies. Rinse the spray tank and refill with water, adding a heavy-duty detergent at the rate of one cup per 20 gallons of water. Recycle this mixture through the equipment for 5 minutes and spray out. Repeat this procedure twice. Fill the spray tank with clean water, recycle for 5 minutes and spray out. Clean pump and nozzle screens thoroughly. Wash away any spray mixture from the outside of spray tank, nozzles or spray rig. All rinse water must be disposed of in compliance with local, state, and Federal guidelines.

APPLICATION OF SALUTE 4EC IN FLUID FERTILIZERS

SALUTE 4EC may be applied in fluid fertilizer solutions to soybeans by following the appropriate mixing procedures and compatibility check. Compatibility checks should be made for each batch of fluid fertilizer because of the variability of these fertilizers.

COMPATIBILITY CHECK:
1. Premix 2 teaspoonfuls of SALUTE 4EC with 8 teaspoonfuls of water (1:4 ratio) in a quart jar by adding the water first, then SALUTE. Mix thoroughly.
2. Pour 1 pint of fluid fertilizer into the quart jar and shake well.
3. Allow to stand for 5 minutes.

THIS COMPATIBILITY CHECK SHOULD ONLY BE USED WHEN MIXING WITH FLUID FERTILIZERS.

INTERPRETATION OF RESULTS: If the solution in the jar appears to be uniform, without signs of agglomeration, or without a separation of an oily film on top of the fertilizer, the mixture may be used. If not, repeat the compatibility check using twice the amount of water or add a compatibility agent to the water. If separation occurs and can be resuspended by shaking, then application is possible with good agitation in the spray tank.

COMMERCIAL IMPREGNATION AND APPLICATION OF SALUTE 4EC ON DRY BULK FERTILIZER

Dry bulk fertilizer may be impregnated or coated with SALUTE 4EC for application to soybeans. All recommendations, cautions, and special precautions on the label must be followed along with State regulations relating to dry bulk fertilizer blending, impregnating and labeling.

IMPREGNATION: To impregnate, use a system consisting of a belt, conveyor, or closed drum which is used for dry bulk fertilizer blending. Any commonly used fertilizer can be impregnated with SALUTE 4EC except ammonium nitrate, or fertilizers containing ammonium nitrate, potassium nitrate, or sodium nitrate. Do not use on powder limestone.

Apply using a minimum of 200 pounds and a maximum of 450 pounds dry bulk fertilizer per acre. To impregnate or coat the dry bulk fertilizer with SALUTE 4EC, the delivery nozzles must be directed to deliver a fine spray toward the fertilizer for thorough coverage while avoiding spray contact with mixing equipment. Uniform impregnation of SALUTE 4EC to dry bulk fertilizer will vary and if the absorptivity is not adequate, an absorptive powder may be added to produce a dry free-flowing mixture. Micro-Cel E (Johns-Manville Product Corporation) is the recommended absorbent powder. SALUTE 4EC should be pre-mixed with water to form a sprayable mixture. Apply immediately after impregnation.

RATES: Select the recommended rate of SALUTE 4EC per acre from the appropriate section of the label and refer to the formula below to determine the amount of SALUTE 4EC which is to be impregnated on a ton of dry bulk fertilizer based on the amount of fertilizer which will be distributed on one acre.

$$\text{Pints SALUTE 4EC Per Acre} \times \frac{2000}{\text{Lbs. Fertilizer Per Acre}} = \text{Pints SALUTE 4EC Per Ton of Fertilizer}$$

APPLICATION: Uniform application is essential for satisfactory weed control. Accurate calibration of fertilizer application equipment is essential for uniform distribution to the soil surface. The recommended method of application is to apply 1/2 the recommended rate and overlap 50 percent or to double apply by splitting the middles to obtain the best distribution pattern.

If fertilizer materials are excessively dusty, use diesel oil or other suitable additive to reduce dust prior to impregnation as dusty fertilizer will result in poor distribution during application. Crop injury and/or poor weed control may occur where the impregnated fertilizer is not uniformly applied.

SOYBEANS (Except California)

SALUTE 4EC is recommended for preplant incorporated, and the preplant incorporated portion of the SPLIT-SHOT application to control certain weeds in soybeans. All applications can be applied with ground or aerial spray equipment.

Activation: A minimum amount of soil moisture is required to activate SALUTE 4EC.

Grazing and Feeding Treated Vines: Treated vines may be fed to livestock 40 days after the application of SALUTE 4EC.

Rate Ranges: Where a rate range is shown, use a lower rate on soils that are coarse-textured or low in organic matter. Use a higher rate on soils that are relatively fine-textured or high in organic matter.

Replanting: If replanting is necessary in fields previously treated with SALUTE 4EC as directed on the label, the field may be replanted to soybeans. However, a minimum of tillage is recommended. Do not apply a second treatment as injury to soybeans may occur.

SPECIAL PRECAUTIONS

Injury to soybeans may occur when SALUTE 4EC is used under the following conditions:
1. When soils have a calcareous surface area or a pH of 7.5 or higher.
2. Due to the sensitivity of certain soybean varieties, SALUTE 4EC is not recommended for use on sensitive varieties, such as Tracy, Semmes, Altona, Dassel, Vansoy, Coker 102, Coker 156, Portage, Maple Amber,

Govan, NKS-1884, Terra-Vig 505, Terra-Vig 606, Vinton 81, Asgrow 6520, AP 55, AP 71 and Narow. Consult your Mobay representative or your seed supplier for information on the tolerance of newly released soybean varieties prior to use of SALUTE 4EC.
3. When applied in conjunction with soil-applied organic phosphate pesticides.
4. Over application or boom overlapping (may result in stand loss and soil residues).
5. Uneven application or improper incorporation can decrease the level of weed control and/or increase the level of injury.
6. When applied to any soil with less than 0.5% organic matter.
7. Soil incorporation deeper than recommended.
8. When sprayers are not calibrated accurately.
9. When heavy rains occur soon after application, especially in poorly drained areas where water may stand for several days.
10. When soybeans are planted less than 1½" deep.
11. Where atrazine carryover is present.
12. When poor quality soybean seed is used.

SALUTE 4EC ALONE

Preplant Incorporated Application: Incorporate SALUTE 4EC into the top 2 inches of soil within 24 hours of application using a disk, field cultivator, rolling cultivator, or similar implement. Two incorporation passes are recommended. The second incorporation pass should not be deeper than the first. SALUTE 4EC may be applied up to 10 days before planting.

Prepare the soil surface by deep plowing, offset disking or tandem disking prior to the application of SALUTE 4EC. The soil surface should be well prepared and free of clods and trash. Incorporate SALUTE 4EC into the top 2 inches of soil before planting. Variable weed control may result from delayed incorporation, application to a moist, warm soil surface or if the wind velocity is 10 miles per hour or higher. Do not use spike or spring tooth harrows alone for incorporation. Shallow incorporation with implements set to cut less than 2 inches deep may result in erratic weed control.

Incorporation Equipment:
1. Set PTO-driven equipment (tillers, cultivators, hoes) to cut once 2 to 3 inches deep and space rotors to provide a clean sweep of the soil. PTO equipment should not be operated at a speed greater than 4 miles per hour.
2. Set disk to cut 4 to 6 inches deep and operate twice in different directions at 4 to 6 miles per hour.
3. Set mulch treader and other similar disk-type implements to cut 3 to 4 inches deep and operate twice in different directions at 5 to 8 miles per hour.
4. For Coarse and Medium Textured Soils Only: Set field cultivator to cut 3 to 4 inches deep and operate twice at 5 miles per hour. Set rolling cultivator to cut 2 to 4 inches deep and operate twice at 6 to 8 miles per hour. Set bed conditioner (Do-All) to cut 2 to 4 inches deep and operate at 4 to 6 miles per hour unless in flat-planted culture where two passes are required.

For information on applying product in fluid or dry fertilizer refer to the "Application Of SALUTE 4EC in Fluid Fertilizers" or "Commercial Impregnation And Application Of SALUTE 4EC On Dry Bulk Fertilizer" in the label.

Weeds Controlled: SALUTE 4EC, when applied as directed, will control the following weeds.

Annual Grasses:	Annual Broadleaves:	
Barnyardgrass	Bristly Starbur	Purslane
Bluegrass	Carpetweed	Ragweed, Common
Broadleaf Signalgrass	Copperleaf, Hophornbeam	Redweed
Browntop Millet	Florida Beggarweed	Russian Thistle
Crabgrass	Florida Pusley	Sesbania
Crowfootgrass	Galinsoga	Shepherdspurse
Foxtails	Jimsonweed	Smartweed
Goosegrass	Knotweed	Spurred Anoda
Johnsongrass(Seedling)	Kochia	Velvetleaf
Junglerice	Lambsquarters	Venice Mallow
Panicum, Fall	Pigweed	Wild Mustard
Panicum, Texas	Prickly Sida/Teaweed	
Sandbur		
Sprangletop		
Stinkgrass		

SALUTE 4EC will provide suppression (reduce the competition of) cocklebur, red rice, sicklepod and sunflower.

RECOMMENDED APPLICATION
(Broadcast Rates)

	SALUTE 4EC Pints/Acre	
	Organic Matter	
Soil Texture	0.5 to 3%	Over 3%
COARSE[a] (Loamy sand, sandy loam)	1½	1½ to 2
MEDIUM[c] (Loam, silt loam, silt, sandy clay, sandy clay loam)	2¼	2¼ to 2¾
FINE[c] (Silty clay, silty clay loam,[b] clay, clay loam)	2½ to 3	3

[a] Do not use on coarse soils with less than 1% organic matter.
[b] Silty clay loam soils are transitional soils and may be classified as medium-textured soils in some regions of the U.S.
[c] For control of lambsquarters, redroot pigweed, wild mustard and green and yellow foxtails on alkaline (calcareous) soils in Minnesota, Nebraska, South Dakota, and North Dakota only, apply SALUTE 4EC at rates of 2 pints/acre on medium soils and 2 to 2½ pints/acre on fine soils, regardless of soil organic matter percentage (use 2½ pints only where soil pH is less than 7.5 and weed pressure is heavy). For control of other weeds listed on the label, use SALUTE 4EC at full rates recommended in the table above, **but note that crop injury may occur on soils having a calcareous surface area or a pH of 7.5 or higher.**

IN COARSE (LIGHT) SOILS ONLY

(Alabama, Arkansas, Florida, Georgia, Louisiana, Mississippi, Missouri, North Carolina, Oklahoma, South Carolina, Tennessee, Texas, Virginia.)

SALUTE 4EC is recommended for use as a preplant incorporated application in coarse-textured, low organic matter soils in the states listed above for the control of certain weeds in soybeans. Refer to the appropriate sections of the label for specific directions for use, recommendations and restrictions.

MIXING: Refer to "Mixing" under the "General Information" section in the front of the SALUTE 4EC label.

APPLICATION: For specific application information, refer to "Application" under the "General Information" section in the front of the label.

Weeds Controlled: Refer to the "SALUTE 4EC Alone" section of the label for a complete list of weeds controlled.

SALUTE Application

Soil Texture	Organic Matter	SALUTE 4EC Pints/Acre
COARSE SOILS (Sand,[a] loamy sand, sandy loam)	0.5 to 2.0%	1½ to 2¼[b]

[a] Do not use on sand soils with less than 1% organic matter, or on sandy loam or loamy sand soils with less than 0.5% organic matter.
[b] Use the higher rate under heavy weed pressures and/or on soils higher in organic matter.

SPECIAL PRECAUTIONS

For additional precautions, restrictions, limitations and sprayer clean-up information, refer to the appropriate sections of the label.

SPLIT-SHOT APPLICATION

SALUTE 4EC applied preplant incorporated followed by an overlay of SENCOR® herbicide will control more broadleaf and grass weeds in soybeans than when either herbicide is used alone.

Refer to the SENCOR herbicide label and to appropriate sections of the SALUTE 4EC label for directions for soil preparation, herbicide application, incorporation techniques, and additional directions on rates, weeds species controlled, and special precautions.

Application: Apply SALUTE 4EC as a preplant incorporated application up to 21 days before planting at rates listed below. This is followed by SENCOR alone at the indicated rates as a preemergence overlay application after planting but prior to soybean emergence.

On medium and fine textured soils with greater than 2% organic matter, a rate range is recommended for the SENCOR preemergence overlay application. The higher rate should be used (a) in fields with a history of severe broadleaf weed pressure, (b) when the time between preplant incorporated tank mix and preemergence overlay applications approaches the maximum stated above, and/or (c) when the organic matter content of the soil is at the upper end of the indicated range.

Weeds Controlled: In addition to the weeds controlled by SALUTE 4EC alone, SALUTE with SENCOR as a SPLIT-SHOT application will control buffalobur, cocklebur, sicklepod, and sunflower.

SALUTE with SENCOR as a SPLIT-SHOT application will provide suppression (reduce the competition) of red rice.

SPLIT-SHOT APPLICATION
SALUTE 4EC + SENCOR 4
(Broadcast Rates)

	Preplant Incorporated Application	—Followed By—	Preemergence Overlay Application		
	SALUTE 4EC Pints/Acre		SENCOR 4 Pints/Acre Organic Matter		
Soil Texture			0.5 to 2%	2 to 4%	Over 4%
COARSE[a] (Sand, loamy sand, sandy loam)	1½	—followed by—	¼	¼	¼-½

Soil Texture	Preplant Incorporated Application SALUTE 4EC Pints/Acre	—Followed By—	Preemergence Overlay Application SENCOR 4 Pints/Acre Organic Matter 0.5 to 2%	2 to 4%	Over 4%
MEDIUM (Loam, silt loam, sandy clay loam, silt, sandy clay)	2 1/4 2[b]	—followed by—	1/4	1/4-1/2	1/2-3/4[c]
FINE (Silty clay loam,* clay loam, silty clay, clay)	3 2 3/4[b]	—followed by—	1/4	1/4-1/2	1/2-3/4[c]

SPLIT-SHOT APPLICATION
SALUTE 4EC + SENCOR DF
(Broadcast Rates)

Soil Texture	Preplant Incorporated Application SALUTE 4EC Pints/Acre	—Followed By—	Preemergence Overlay Application SENCOR DF Lbs/Acre Organic Matter 0.5 to 2%	2 to 4%	Over 4%
COARSE[a] (Sand, loamy sand, sandy loam)	1 1/2	—followed by—	1/6	1/6	1/6-1/3
MEDIUM (Loam, silt loam, sandy clay loam, silt, sandy clay)	2 1/4 2[b]	—followed by—	1/6	1/6-1/3	1/3-1/2[c]
FINE (Silty clay loam,* clay loam, silty clay, clay)	3 2 3/4[b]	—followed by—	1/6	1/6-1/3	1/3-1/2[c]

*Silty clay loam soils are transitional soils and may be classified as medium-textured soils in some regions of the U.S.

[a] On coarse-textured soils, do not use on sand soil with less than 1% organic matter or on loamy sand or sandy loam soils with less than 0.5% organic matter. On coarse-textured soils with a calcareous surface area or a pH of 7.5 or higher, do not use on sand soils with less than 2% organic matter, or on loamy sand or sandy loam soils with less than 1% organic matter.

[b] Use the lower rate of SALUTE 4EC in the preplant incorporated application on soils having a calcareous surface area or a pH of 7.5 or higher, and in those situations where soils within a field vary extremely in texture or organic matter content.

[c] Reduce this preemergence overlay rate of SENCOR 4 by 1/4 pint/acre (SENCOR DF by 1/6 lb./acre) when using SPLIT-SHOT application on soils with over 4% organic matter and which have a calcareous surface area or a pH of 7.5 or higher.

SALUTE plus Scepter® Tank Mix Application

SALUTE 4EC may be applied with Scepter as a preplant incorporated broadcast application for the control of certain broadleaf weeds and grasses in soybeans. Consult the Scepter label for specific directions on use, recommendations, restrictions and any additional weeds not specified on the label for SALUTE 4EC.

Mixing: Refer to "Mixing" under the "General Information" section in the front of the SALUTE 4EC label.

Application: For specific application information refer to "Application" under the "General Information" section in the front of the SALUTE 4EC label.

Weeds Controlled: In addition to weeds controlled by SALUTE 4EC alone, SALUTE plus Scepter will control the following broadleaf weeds: buffalobur, cocklebur, morningglory (pitted and smallflower), sicklepod, sunflower.

SALUTE 4EC plus Scepter will provide suppression (reduce the competition) of morningglory (ivyleaf and tall) and red rice.

SALUTE plus Scepter Application
(Broadcast Rates)

Soil Texture	SALUTE 4EC (Pints/Acre)	Scepter[b] (Pints/Acre)
COARSE (Loamy sand or sandy loam)	1 1/2	1/3 to 1/2[b]
MEDIUM (Loam, silt loam, silt, sandy clay, sandy clay loam)	2 1/4	1/3 to 1/2[b]
FINE (Silty clay, silty clay loam,* clay, clay loam)	2 1/4 to 3[a]	1/3 to 1/2[b]

*Silty clay loam soils are transitional soils and may be classified as medium-textured soils in some regions of the U.S.

[a] The higher rate is recommended on soils with organic matter greater than 3%.

[b] For maximum control of moderate to heavy infestations of cocklebur and sicklepod, use the higher rate.

Special Precautions

For additional precautions, restrictions, limitations and sprayer clean-up information, refer to the appropriate sections of the SALUTE 4EC and Scepter labels.

Do not use treated vines for feed or forage.

Sequential Application of Scepter Following Salute:
If needed, application of SALUTE alone or in a registered tank mix according to directions on the label, may be followed by an early postemergence application of Scepter for control of cocklebur. Apply 1/6 to 1/3 pint of Scepter in a minimum of 20 gallons of water per acre. Use 1/6 pint of Scepter if cockleburs are less than 3 inches tall or have fewer than 3 leaves and are actively growing. For cockleburs less than 6 inches tall and actively growing use 1/3 pint per acre. Do not use Scepter when soybeans or cockleburs have been subjected to stress conditions such as temperature or moisture extremes. Do not exceed a total of 2/3 pint of Scepter per acre in one season. Wait at least 10 days after application of Scepter before cultivating.

When preparing the spray mixture with Scepter, add 2 pints of nonionic surfactant approved for use on growing crops and containing at least 80% active ingredient per 100 gallons of mixture. Apply crop oil concentrate (COC) at the rate specified on the COC label.

Use Scepter only in the states where it is registered as listed on the product label.

Apply Scepter at least 90 days before harvest of soybeans. Do not graze or feed soybean forage, hay, or straw to livestock.

Refer to the Scepter label for additional cautions and precautions, directions, limitations, and information on environmental hazards and planting of rotational crops.

SALUTE 4EC plus Command® Tank Mix Application

SALUTE 4EC may be applied with Command 4EC as a preplant incorporated broadcast application for the control of certain broadleaf weeds and grasses in soybeans. Consult the Command label for specific directions for use, recommendations and restrictions not specified on the SALUTE label.

Mixing: Refer to "Mixing" under the "General Information" section in the front of the SALUTE 4EC label.

Application: For specific application information, refer to "Application" under the "General Information" section in the front of the SALUTE 4EC label.

Weeds Controlled: In addition to weeds controlled by SALUTE 4EC alone, SALUTE plus Command will provide greater consistency in the control of velvetleaf, especially under heavy infestation.

SALUTE plus Command Application
(Broadcast Rates)

Soil Texture	SALUTE 4EC (Pints/Acre)	Command 4EC (Pints/Acre)
COARSE (Loamy sand or sandy loam)	1 1/2	1/2 to 3/4[a]
MEDIUM (Loam, silt loam, silt, sandy clay, sandy clay loam)	2 1/4	1/2 to 3/4[a]
FINE (Silty clay, silty clay loam,* clay, clay loam)	2 1/4 to 3[a]	1/2 to 3/4[a]

*Silty clay loam soils are transitional soils and may be classified as medium textured soils in some regions of the U.S.

[a] The higher rate is recommended on soils with organic matter greater than 3%.

Special Precautions

Do not plant wheat, oats, barley, rye or alfalfa in the fall or following spring after application as crop injury may occur.
Do not use treated vines for feed or forage.
For additional precautions, restrictions, limitations and sprayer clean-up information, refer to the appropriate sections of the SALUTE 4EC label and the Command 4EC label.
Do not apply where weather conditions favor drift.

SALUTE 4EC plus Commence® Tank Mix Application

SALUTE 4EC may be applied with Commence 5.25 EC as a preplant incorporated broadcast application for the control of certain broadleaf weeds and grasses in soybeans. Consult the Commence label for specific directions for use, recommendations and restrictions not specified on the SALUTE 4EC label.

Mixing: Refer to "Mixing" under the "General Information" section in the front of the SALUTE 4EC label.

Application: For specific application, refer to "Application" under the "General Information" section in the front of the SALUTE 4EC label.

Weeds Controlled: In addition to weeds controlled by SALUTE 4EC alone, SALUTE 4EC plus Commence will provide control of shattercane and greater consistency in the control of velvetleaf, especially under heavy infestation.

SALUTE 4EC plus Commence Application
(Broadcast Rates)

Soil Texture	SALUTE 4EC (Pints/Acre)	Commence 5.25 EC (Pints/Acre)
Soil Texture	SALUTE 4EC (Pints/Acre)	Commence 5.25 EC (Pints/Acre)
COARSE (Loamy sand or sandy loam)	1½	1
MEDIUM (Loam, silt loam, silt, sandy clay, sandy clay loam)	2	1
FINE (Silty clay, silty clay loam,* clay, clay loam)	2¼	1

* Silty clay loam soils are transitional soils and may be classified as medium-textured soils in some regions of the U.S.

Special Precautions

Do not plant wheat, oats, barley, rye or alfalfa in the Fall or following Spring after application as crop injury may occur.
Do not use treated vines for feed or forage.
For additional precautions, restrictions, limitations and sprayer clean-up information, refer to the appropriate sections of the SALUTE 4EC label and the Commence label.

Observe all cautions and limitations on labeling of all products used in mixtures.

Do not apply where weather conditions favor drift.

SALUTE 4EC plus Squadron® Tank Mix Application

SALUTE 4EC may be applied with Squadron 2.33 EC as a preplant incorporated broadcast application for the control of certain broadleaf weeds and grasses in soybeans. Consult the Squadron label for specific directions for use, recommendations, restrictions and any additional weeds not specified on the SALUTE 4EC label.

Mixing: Refer to "Mixing" under the "General Information" section in the front of the SALUTE 4EC label.

Application: For specific application information, refer to "Application" under the "General Information" section in the front of the SALUTE 4EC label.

Weeds Controlled: In addition to weeds controlled by SALUTE 4EC alone, SALUTE 4EC plus Squadron will control the following broadleaf weeds: buffalobur, cocklebur, morningglory (pitted and smallflower), shattercane and sunflower.
SALUTE 4EC plus Squadron will provide suppression (reduce the competition) of morningglory (ivyleaf and tall) and red rice.

SALUTE 4EC plus Squadron Application
(Broadcast Rates)

Soil Texture	SALUTE 4EC (Pints/Acre)	Squadron 2.33 EC (Pints/Acre)
COARSE (Loamy sand or sandy loam)	1½	1½
MEDIUM (Loam, silt loam, silt, sandy clay, sandy clay loam)	2	1½
FINE (Silty clay, silty clay loam,* clay, clay loam)	2¼	1½

* Silty clay loam soils are transitional soils and may be classified as medium-textured soils in some regions of the U.S.

Special Precautions

For additional precautions, restrictions, limitations and sprayer clean-up information, refer to the appropriate sections of the SALUTE 4EC label and the Squadron label.

Observe all cautions and limitations on labeling of all products used in mixtures.

Do not use treated vines for feed or forage.

In coarse soils, crop injury may occur in Fall planted Winter wheat.

SALUTE 4EC plus Tri-Scept® Tank Mix Application

SALUTE 4EC may be applied with Tri-Scept 3EC as a preplant incorporated broadcast application for the control of certain broadleaf weeds and grasses in soybeans. Consult the Tri-Scept label for specific directions for use, recommendations, restrictions and any additional weeds not specified on the label.

Mixing: Refer to "Mixing" under the "General Information" section in the front of the SALUTE 4EC label.

Application: For specific application information, refer to "Application" under the "General Information" section in the front of the SALUTE 4EC label.

Weeds Controlled: In addition to weeds controlled by SALUTE 4EC alone, SALUTE plus Tri-Scept will control the following broadleaf weeds: buffalobur, cocklebur, morningglory (pitted and smallflower), shattercane and sunflower.
SALUTE 4EC plus Tri-Scept will provide suppression (reduce the competition) of morningglory (ivyleaf and tall) and red rice.

SALUTE 4EC plus Tri-Scept Application
(Broadcast Rates)

Soil Texture	SALUTE 4EC (Pints/Acre)	Tri-Scept 3EC (Pints/Acre)
COARSE (Loamy sand or sandy loam)	1½	1
MEDIUM (Loam, silt loam, silt, sandy clay, sandy clay loam)	2	1¼
FINE (Silty clay, silty clay loam,* clay, clay loam)	2¼	1¼

* Silty clay loam soils are transitional soils and may be classified as medium-textured soils in some regions of the U.S.

Special Precautions

For additional precautions, restrictions, limitations and sprayer clean-up information, refer to the appropriate sections of the SALUTE 4EC label and the Tri-Scept label.

Observe all cautions and limitations on labeling of all products used in mixtures.

Do not use treated vines for feed or forage.

In coarse soils, crop injury may occur in Fall planted Winter wheat.

8-15-88

CROP ROTATION RECOMMENDATIONS
Waiting Period After Application of SALUTE 4EC*

4 MONTHS	8 MONTHS	12 MONTHS	18 MONTHS
Alfalfa	Cotton	Other crops	Sugar Beets
Barley	Lentils	not listed	Onions
Asparagus	Peas	(except	And other
Corn	Rice	root crops)	root crops
Forage Grasses			
Potatoes			
Sainfoin			
Soybeans			
Sugarcane			
Tomatoes			
Wheat			

* Cover crops for soil building or erosion control may be planted any time, but do not graze or harvest for food or feed. Stand reductions may occur in some areas.

1. In those portions of Kansas, Nebraska, North Dakota, Oklahoma, South Dakota and Texas where at least 20 inches of irrigation and/or rainfall(total) was used to produce the crop, sorghum or oats should not be planted for 12 months after an application of SALUTE.

2. If less than 20 inches of total water was used to produce the crop, do not plant corn, sorghum, proso millet, or oats for 18 months after an application of SALUTE 4EC. Cool, wet weather conditions during the early stage of growth may increase the possibility of injury to sorghum.

STORAGE AND DISPOSAL

PESTICIDE DISPOSAL: Do not contaminate water, food or feed by storage or disposal.

Wastes resulting from the use of this product may be disposed of on site or at an approved waste disposal facility.

CONTAINER DISPOSAL: Triple rinse (or equivalent). Then offer for recycling or reconditioning, or puncture and dispose of in a sanitary landfill, or incineration, or, if allowed by State and local authorities, by burning. If burned, stay out of smoke.

Store in a cool, dry place and in such a manner as to prevent cross contamination with other pesticides, fertilizers, food, and feed. Store in original container and out of the reach of children, preferably in a locked storage area.

Handle and open container in a manner as to prevent spillage. If container is leaking invert ot prevent leakage. If the container is leaking or material spilled for any reason or cause, carefully dam up spilled material to prevent runoff. Refer to Precautionary Statements on label for hazards associated with the handling of this material. Do not walk through spilled material. Absorb spilled material with absorbing type compounds and dispose of as directed for pesticides above. In spill or leak incidents, keep unauthorized people away. You may contact the Mobay Emergency Response Team for decontamination procedures or any other assistance that may be necessary. The Mobay Kansas City Emergency Response Telephone No. is 816-242-2582, or contact Chemtrec at 800-424-9300.

PRECAUTIONARY STATEMENTS

HAZARDS TO HUMANS AND DOMESTIC ANIMALS

CAUTION

Harmful if swallowed, inhaled or absorbed through the skin. Causes eye irritation. This product may cause allergic skin reaction. Avoid contact with eyes, skin or clothing. Avoid breathing spray mist or vapors. Wear natural rubber or neoprene gloves and protective coveralls to prevent skin contact. Wash immediately with soap and water if skin contamination occurs and after handling. Wash contaminated clothing with soap and hot water before reuse.

Keep out of reach of children.

Obtain prompt medical aid if poisoning should occur.

To Physician: No specific antidote is available. Treat the patient symptomatically.

SYMPTOMS OF POISONING: The compound does not cause any definite symptoms that would be diagnostic. Poisoning is accompanied by breathing difficulties and sedation.

STATEMENTS OF PRACTICAL TREATMENT

If in eyes, immediately flush eyes with plenty of water for at least 15 minutes. For further treatment get medical attention.

If swallowed, call a physician or poison control center. Vomiting should be induced. Administer water freely, and induce vomiting by giving one dose (1/2 oz. or 15 ml.) of syrup of ipecac. If vomiting does not occur within 10 to 20 minutes, administer second dose. If syrup of ipecac is not available, induce vomiting by sticking finger down throat. Repeat until vomit fluid is clear. Do not induce vomiting or give anything by mouth to an unconscious person. Professional medical assistance should be secured immediately.

If on skin, wash thoroughly with soap and warm water.

ENVIRONMENTAL HAZARDS

This pesticide is toxic to fish. Do not apply directly to water or wetlands (swamps, bogs, or marshes). Drift or runoff from treatment areas may be hazardous to aquatic organisms in neighboring aquatic sites. Do not contaminate water by cleaning of equipment or disposal of wastes.

Do not use on crops grown for food, feed or forage unless specified on the label. Do not apply when weather conditions favor drift from treated areas or allow sprays to drift onto adjacent desirable plants. Apply this product only as specified on the label.

Observe all cautions and limitations on labeling of all products used in mixtures.

Ground Water Advisory: Metribuzin is a chemical which can travel (seep or leach) through soil and can contaminate ground water which may be used as drinking water. Metribuzin has been found in ground water as a result of agricultural use. Users are advised not to apply metribuzin where the water table (ground water) is close to the surface and where the soils are very permeable, i.e., well drained soils such as loamy sands. Your local agricultural agencies can provide further information on the type of soil in your area and the location of ground water.

RE-ENTRY STATEMENTS

Do not apply this product in such a manner as to directly or through drift expose workers or other persons. The area being treated must be vacated by unprotected persons. Do not enter treated areas without protective clothing until sprays have dried. Because certain states may require more restrictive re-entry intervals for various crops treated with this product, consult your State Department of Agriculture for further Information.

Written or oral warnings must be given to workers who are expected to be in a treated area or in an area about to be treated with this product. Oral warnings should include those statements found under the label heading "Precautionary Statements" and "Re-entry Statements" included on the label. When oral warnings are given, warnings shall be given in a language customarily understood by workers. Oral warnings must be given if there is reason to believe that written warnings cannot be understood by workers. Written warnings must include the following information: WARNING, area treated with SALUTE 4EC herbicide on (date of application). Do not enter without appropriate protective clothing until sprays have dried. **If on skin,** remove contaminated clothing and wash skin immediately with soap and warm water. **If eyes are contaminated,** wash with flowing water for at least 15 minutes. **If swallowed,** vomiting should be induced. (See Precautionary Statements on label.)

IMPORTANT:

This bulletin is not intended to provide adequate information for application. Before using the product, read and carefully observe directions, cautionary statements and other information appearing on the product label. This product is sold subject to the Conditions of Sale set forth on the container label.

STOP—READ THE LABEL BEFORE USE

SA 9001

This is a specimen label. Mobay is not responsible for the accuracy of the information contained herein. As labels are subject to revision, always carefully read and follow the label on the product container.

SENCOR® DF

75% DRY FLOWABLE HERBICIDE

FOR CONTROL OF CERTAIN GRASSES AND BROADLEAF WEEDS

ACTIVE INGREDIENT:

4-Amino-6-(1,1-dimethylethyl)-3-(methylthio)-1,2,4-triazin-5(4*H*)-one	75%
INERT INGREDIENTS:	25%
TOTAL:	100%

EPA Reg. No. 3125-325 EPA Est. 3125-MO-1

RECOMMENDED USES

Soybeans, Potatoes, Lentils and Dry Field Peas, Established Alfalfa and Sainfoin, Established Asparagus, Carrots, Sugarcane, Tomatoes, Wheat/Fallow/Wheat Rotation, Winter Wheat, Barley, Field Corn, Barley and Wheat/Fallow/Barley and Wheat Rotation.

RESTRICTIONS ON WAREHOUSE STORAGE

Protect from rain and moisture. Store in a cool, dry place.
Do not store in front of a space heater.

PACKAGING

10 lb. jugs, 4 per case.
5 lb. jugs, 8 per case.

SENCOR DF 75% dry flowable herbicide delivers unmatched preemergence broadleaf weed control in soybeans. Also SENCOR DF is now registered for postemergence directed spray applications to soybeans in specified Southern and Southeastern states. It may also be applied with aerial or ground spray equipment for preemergence and/or postemergence weed control in potatoes. SENCOR DF effectively controls such problem broadleaf weeds as velvetleaf, jimsonweed, sesbania (coffeebean, coffeeweed), prickly sida (teaweed), and annual smartweeds, in addition to certain grass weeds.

SENCOR DF is registered for combination applications with Treflan®, Lasso®, Amiben®, Gramoxone® Super, Dual®, Prowl®, Sonalan®, Surflan®, Cannon®, Commence®, Command®, Freedom®, Scepter®, linuron plus Dual or Lasso, and 2,4-D low volatile ester on soybeans. SENCOR DF may be applied to control many weeds in wheat/fallow/wheat rotation, winter wheat, barley, field corn, barley and wheat/fallow/barley and wheat rotation, established alfalfa and sainfoin fields, established asparagus, carrots, sugarcane, and tomatoes.

SENCOR DF herbicide is lightweight and pours cleanly. Concentrated 75% formulation can also reduce handling.

AVISO

PRECAUCION AL USUARIO: Si usted no puede leer o entender inglés, no use este producto hasta que la etiqueta le haya sido explicada ampliamente.

(TO THE USER: If you cannot read or understand English, do not use this product until the label has been fully explained to you.)

DIRECTIONS FOR USE

It is a violation of Federal Law to use this product in a manner inconsistent with its labeling.

GENERAL INFORMATION

Mixing: When using SENCOR DF, make sure the sprayer is completely clean, free of rust or corrosion which occurs from Winter storage. Examine strainers and screens to be sure the sprayer is clean from previously used pesticides.

Any tank mix containing SENCOR DF should be kept agitated and sprayed out immediately. Do not allow tank mixes to stand for prolonged periods of time.

The proper mixing procedure for SENCOR DF alone or in tank mix combinations with other herbicides is:

1. Fill the spray tank 1/4 to 1/3 full with clean water.
2. Add recommended rate of SENCOR DF while recirculating and with agitator running.
3. Follow the triple rinse procedure described under "STORAGE AND DISPOSAL" to insure that all product is removed from the container.
4. Mix thoroughly and add clean water to fill spray tank to desired level.
5. Add the other herbicide to tank last and agitate thoroughly.
6. Continue agitation during application and until sprayer tank is empty.

SOIL TEXTURE: As used on the label, "Coarse Soils" are loamy sand or sandy loam soils. "Medium Soils" are loam, silt loam, silt, sandy clay, or sandy clay loam. "Fine Soils" are silty clay, silty clay loam, clay, or clay loam. Silty clay loam soils are transitional soils and may be classified as medium textured soils in some regions of the U.S.

APPLICATION

Chemigation

SENCOR DF is recommended for application through sprinkler irrigation equipment to potatoes, soybeans, tomatoes, and asparagus as directed on the label. Refer to the crop sections of the label for recommended rates, weeds controlled or suppressed, restrictions, and special precautions.

Apply this product only through sprinkler (including center pivot, lateral move, or solid set) irrigation systems. Do not apply this product through any other type of irrigation system.

Crop injury, lack of effectiveness, or illegal pesticide residues in the crop can result from nonuniform distribution of treated water.

Calibration: (Center Pivot and Self-Propelled Lateral Move Systems)

Sprinkler irrigation systems must be accurately calibrated for application of SENCOR DF. Greater accuracy in calibration (and distribution) will be achieved by injecting a larger volume of a more dilute mixture of product and water per hour. Follow the steps below to calibrate center pivot and lateral move systems:

1. Determine number of minutes required to make one complete revolution while applying 1/4 to 3/4 inch of water per acre.
2. With the system at operating pressure determine the exact number of minutes required to inject one gallon of water.
3. Divide the time required for one revolution (step 1) by the time required to inject one gallon (step 2). This gives total gallons of product-water mixture to be added to nurse tank.
4. Add required amount of water to nurse tank and start the agitation system. Then add sufficient SENCOR DF at the recommended rate (see RECOMMENDED BROADCAST APPLICATIONS) to the nurse tank.

EXAMPLE: If 20 hours (1200 minutes) were required for one revolution and if 2 minutes were required to inject one gallon then a total of 600 gallons of product-water mixture are required ($1200 \div 2 = 600$); to treat 135 acres at 2/3 pound/acre, 90.5 pounds of SENCOR DF are required.

If you have questions about calibration, you should contact State Extension Service specialists, equipment manufacturers or other experts.

Do not connect an irrigation system (including greenhouse systems) used for pesticide application to a public water system unless the pesticide label-prescribed safety devices for public water systems are in place.

A person knowledgeable of the chemigation system and responsible for its operation, or under the supervision of the responsible person, shall shut the system down and make necessary adjustments should the need arise.

The system must contain a functional check valve, vacuum relief valve, and low pressure drain appropriately located on the irrigation pipeline to prevent water source contamination from backflow.

The pesticide injection pipeline must contain a functional, automatic, quick-closing check valve to prevent the flow of fluid back toward the injection pump.

The pesticide injection pipeline must also contain a functional, normally closed, solenoid-operated valve located on the intake side of the injection pump and connected to the system interlock to prevent fluid from being withdrawn from the supply tank when the irrigation system is either automatically or manually shut down.

The system must contain functional interlocking controls to automatically shut off the pesticide injection pump when the water pump motor stops.

The irrigation line or water pump must include a functional pressure switch which will stop the water pump motor when the water pressure decreases to the point where pesticide distribution is adversely affected.

Systems must use a metering pump, such as a positive displacement injection pump (e.g., diaphragm pump) effectively designed and constructed of materials that are compatible with pesticides and capable of being fitted with a system interlock.

Do not apply when wind speed favors drift beyond the area intended for treatment.

Maintain continuous agitation in the injection nurse tanks during the herbicide application, sufficient to keep herbicide in suspension.

Apply specified dosage in 1/4 to 3/4 inch of water (use 1/4 to 1/2 inch of water on sandy soils) per acre as a continuous injection in center pivot and lateral move systems or in the last 15 to 30 minutes of set in permanent solid set sprinkler systems. Application of more than the quantity of irrigation water recommended on the label may result in decreased product performance by removing the chemical from the zone of effectiveness. Where sprinkler distribution patterns do not overlap sufficiently, unacceptable weed control may result. Where sprinkler distribution patterns overlap excessively crop injury may result. Allow sufficient time for pesticide to be flushed through all lines and all nozzles before turning off irrigation water. To insure that lines are flushed and free of remaining pesticide, an indicator dye may be injected into the lines to mark the end of the application period.

Use a minimum of 1 part water to 1 part herbicide for injection. The use of a larger volume of water will insure greater accuracy and more uniform distribution.

Application of SENCOR DF with Herbicide Spray Equipment: Use a standard low pressure (20 to 40 psi.) herbicide boom sprayer equipped with suitable nozzles and screens no finer than 50-mesh in-nozzle and in-line strainers. Agitate thoroughly before and during application with by-pass agitation.

GROUND APPLICATION: Apply the proper rate of SENCOR DF in a minimum of 10 to 40 gallons of spray mixture per acre broadcast.

Banded Application: Use proportionally less SENCOR DF per acre in a band versus a broadcast application. For band application use 1/4 to 1 gallon of spray mix per inch of band width regardless of row spacing.

Examples: To treat a 15-inch band on rows 30 inches apart, use one-half of the broadcast rate of SENCOR DF.
To treat a 14-inch band on rows 42 inches apart, use one-third of the broadcast rate of SENCOR DF.

AERIAL APPLICATION: Where permitted, apply specified rate in a minimum of 2 to 10 gallons of spray mixture per acre. Do not apply aerially when wind speed is greater than 10 mph. **NOTE:** Do not apply aerially when SENCOR DF is tank mixed with Lasso.

For All Applications of SENCOR DF: Sprayer must be accurately calibrated before applying SENCOR DF. Check sprayer during application to be sure it is working properly and delivering a uniform spray pattern. As the volume of spray mixture decreases per acre, the importance of accurate calibration and uniform application increases. Avoid over application, mis-application, and boom and spray swath overlapping that will increase spray dosage. (Crop injury may occur as a result.) Avoid spray skips and gaps which allow weeds to grow in untreated soil. Do not apply when weather conditions favor spray drift and/or when sensitive or cool season crops, such as cole crops, onions, peas, or strawberries are present in adjacent fields or in areas where wheat is growing in coarse textured soils.

SPRAYER CLEANUP: Spray equipment must be thoroughly cleaned to remove remaining traces of herbicide that might injure other crops to be sprayed. Drain any remaining spray solution of SENCOR from the spray tank and discard in non-crop areas away from water supplies. Rinse the spray tank and refill with water, adding a heavy-duty detergent at the rate of one cup per 20 gallons of water. Recycle this mixture through the equipment for 5 minutes and spray out. Repeat this procedure twice. Fill the spray tank with clean water, recycle for 5 minutes and spray out. Clean pump and nozzle screens thoroughly. Wash away any spray mixture from the outside of spray tank, nozzles or spray rig. All rinse water must be disposed of in compliance with local, State, and Federal guidelines.

APPLICATION OF SENCOR DF IN FLUID FERTILIZERS

SENCOR DF may be applied in fluid fertilizer solutions to alfalfa and soybeans by following the appropriate mixing procedures and compatibility check. When using tank mix combinations, be sure all components are compatible.

Compatibility checks of SENCOR DF and tank mix combinations which include SENCOR should be made for each batch of fluid fertilizer because of the variability of these fertilizers.

Compatibility Check:

1. Premix 2 teaspoonfuls of SENCOR DF with 8 teaspoonfuls of water (1:4 ratio) in a quart jar by adding the water first and follow with SENCOR. Mix thoroughly. If a second herbicide is to be used, double the amount of water (1:8 ratio) and add the second herbicide after mixing SENCOR first.
2. Then pour 1 pint of fluid fertilizer into the quart jar and shake well.
3. Allow to stand for 5 minutes.

THIS COMPATIBILITY CHECK SHOULD ONLY BE USED WHEN MIXING WITH FLUID FERTILIZERS.

Interpretation of Results: If the solution in the jar appears to be uniform, without signs of agglomeration, or without a separation of an oily film on top of the fertilizer, the mixture may be used. If not, repeat the compatibility check using twice the amount of water or add a compatibility agent to the water. If separation occurs but the mixture can be resuspended by shaking, then application is possible with good agitation in the spray tank.

Tank Mixing Guidelines:

1. Add the required amount of water and compatibility agent (if required) to the tank. Start agitation system while adding SENCOR DF and follow by adding the fluid fertilizer and agitate.
2. If a second herbicide is to be used, follow as above in 1, but use twice the amount of water. Start agitation and add SENCOR DF and follow by adding the second herbicide, and then continue filling the tank with fluid fertilizer.
3. Maintain continuous agitation to assure uniform spray mixture until the tank is emptied.

COMMERCIAL IMPREGNATION AND APPLICATION OF SENCOR DF ON DRY BULK FERTILIZER

Dry bulk fertilizer may be impregnated or coated with SENCOR DF for application to established alfalfa and to soybeans. All recommendations, cautions, and special precautions on the label must be followed along with State regulations relating to dry bulk fertilizer blending, impregnating and labeling.

Impregnation: To impregnate, use a system consisting of a belt, conveyor, or closed drum which is used for dry bulk fertilizer blending. Any commonly used fertilizer can be impregnated with SENCOR except ammonium nitrate, or fertilizers containing ammonium nitrate, potassium nitrate, or sodium nitrate. Do not use on powder limestone.

Apply using a minimum of 200 pounds dry bulk fertilizer per acre and up to a maximum of 450 pounds per acre. To impregnate or coat the dry bulk fertilizer with SENCOR, the delivery nozzles must be directed to deliver a fine spray toward the fertilizer for thorough coverage while avoiding spray contact with mixing equipment. Uniform impregnation of SENCOR to dry bulk fertilizer will vary and if the absorptivity is not adequate, an absorptive powder may be added to produce a dry, free-flowing mixture. Micro-Cel E (Johns-Manville Product Corporation) is the recommended absorbent powder. SENCOR should be premixed with water to form a sprayable mixture. When another herbicide is used with SENCOR, mix and impregnate immediately.

Apply immediately after impregnation unless experience has shown that impregnated fertilizer can be stored without becoming lumpy and difficult to spread.

Rates: Select the recommended rate of SENCOR DF per acre from the appropriate section of the label and refer to the formula below to determine the amount of SENCOR which is to be impregnated on a ton of dry bulk fertilizer based on the amount of fertilizer which will be distributed on one acre.

$$\text{Pounds SENCOR DF Per Acre} \times \frac{2000}{\text{Lbs. Fertilizer Per Acre}} = \text{Pounds SENCOR DF Per Ton of Fertilizer}$$

APPLICATION: Uniform application is essential for satisfactory weed control. Accurate calibration of fertilizer application equipment is essential for uniform distribution to the soil surface. The recommended method of application is to apply 1/2 the recommended rate and overlap 50 percent or to double apply by splitting the middles to obtain the best distribution pattern.

If fertilizer materials are excessively dusty, use diesel oil or other suitable additive to reduce dust prior to impregnation as dusty fertilizer will result in poor distribution during application. Crop injury and/or poor weed control may occur where the impregnated fertilizer is not uniformly applied.

INCORPORATION AND COMBINATION USES: When SENCOR DF is to be used in combination with another herbicide, follow directions on the label for combinations, rates, crops, incorporation, and special precautions.

SOYBEANS (Except California)

SENCOR herbicide tank mix combinations are recommended for preplant incorporated applications, preemergence surface applications, SPLIT-SHOT and EXTENDED SPLIT-SHOT application. SENCOR DF is also recommended as an overlay application following a preplant incorporated application of a recommended grass herbicide and alone as a preemergence surface application. All these applications can be applied with ground equipment, and some can be applied with aerial spray equipment. In addition, SENCOR DF can be applied as a postemergence directed spray to soybeans in certain states.

Special Precautions (Soybeans)

Injury to soybeans may occur when SENCOR DF is used under the following conditions:

1. When soils have a calcareous surface area or a pH of 7.5 or higher.
2. Due to the sensitivity of certain soybean varieties, SENCOR is not recommended for use on sensitive varieties, such as Tracy, Semmes, Altona, Dassel, Vansoy, Coker 102, Coker 156, Portage, Maple Amber, Govan, NKS-1884, Terra-Vig 505, Terra-Vig 606, Vinton 81, Asgrow 6520, AP 55 and AP 71. Consult your Mobay Representative or your seed supplier for information on the tolerance to SENCOR of newly released soybean varieties, prior to use of SENCOR.
3. When applied in conjunction with soil-applied organic phosphate pesticides.
4. Over application or boom overlapping may result in stand loss and soil residues.
5. Uneven application or improper incorporation can decrease the level of weed control and/or increase the level of injury.
6. When applied to any soil with less than 1/2% organic matter.
7. Soil incorporation deeper than recommended.
8. When sprayers are not calibrated accurately.
9. When heavy rains occur soon after application, especially in poorly drained areas where water may stand for several days.
10. When soybeans are planted less than 1 1/2" deep, particularly in preemergence application.

Activation: A minimum amount of soil moisture is required to activate SENCOR. In areas of low rainfall, preemergence applications to dry soil should be followed with light irrigation of 1/4 acre inch of water. Do not apply heavy irrigation immediately after application. As with many surface-applied herbicides, weed control and crop tolerance may vary with rainfall and/or soil texture.

Grazing and Feeding Treated Vines:

Treated vines may be grazed or fed to livestock 40 days after application when SENCOR DF is applied alone or with Treflan, Dual, Prowl, Lasso, Amiben, Gramoxone, Lasso plus Gramoxone.

Do not use treated vines for feed or forage when SENCOR is applied with Surflan, Surflan plus Gramoxone, Sonalan, linuron plus Lasso, linuron plus Dual.

Rate Ranges:

Where a rate range is shown, use a lower rate on soils that are coarse-textured or low in organic matter. Use a higher rate on soils that are relatively fine-textured or high in organic matter.

Replanting:

If replanting is necessary in fields treated with SENCOR DF as directed on the label, the field may be replanted to soybeans. When replanting, a minimum of tillage is recommended. Do not apply a second treatment as injury to soybeans may occur.

WEEDS CONTROLLED BY SENCOR AND SENCOR HERBICIDE TANK MIX COMBINATIONS

ANNUAL BROADLEAF WEEDS	SENCOR Alone	SENCOR SPLIT-SHOT	SENCOR plus Treflan	SENCOR plus Dual	SENCOR plus Prowl	SENCOR plus Lasso	SENCOR plus Surflan	SENCOR plus Amiben	SENCOR plus Gramoxone	SENCOR plus LASSO and Gramoxone	SENCOR plus Surflan and Gramoxone	EXTENDED SPLIT-SHOT	SENCOR plus Sonalan	SENCOR plus Linuron plus (Lasso or Dual)
Black Nightshade *(Solanum nigrum)*	P	P	P	C	P	C	S	S	P	C	P	C	P	S
Bristly Starbur *(Acanthospermum hispidum)*	C	C	C	C	C	C	C	C	C	C	C	C	C	C
Buffalobur *(Solanum rostratum)*	C	C	P	P	P	P	P	P	P	P	P	C	P	•
Carpetweed *(Mollugo verticillata)*	C	C	C	C	C	C	C	C	C	C	C	C	C	C
Cocklebur *(Xanthium pensylvanicum)*	S	C	S	S	S	S	S	S	C	C	C	C	S	S
Copperleaf, Hophornbeam *(Acalypha ostryaefolia)*	C	C	C	C	C	C	C	C	C	C	C	C	C	C
Florida Beggarweed *(Desmodium tortuosum)*	C	C	C	C	C	C	C	C	C	C	C	C	C	C
Florida Pusley *(Richardia scabra)*	C	C	C	C	C	C	C	C	C	C	C	C	C	C
Galinsoga *(Galinsoga spp.)*	C	C	C	C	C	C	C	C	C	C	C	C	C	C

	SENCOR Alone	SENCOR SPLIT-SHOT	SENCOR plus Treflan	SENCOR plus Dual	SENCOR plus Prowl	SENCOR plus Lasso	SENCOR plus Surflan	SENCOR plus Amiben	SENCOR plus Gramoxone	SENCOR plus LASSO and Gramoxone	SENCOR plus Surflan and Gramoxone	EXTENDED SPLIT-SHOT	SENCOR plus Sonalan	SENCOR plus Linuron plus (Lasso or Dual)
ANNUAL BROADLEAF WEEDS														
Horseweed Marestail *(Conyza canadensis)*	•	•	•	•	•	•	•	•	C	C	C	C	•	•
Jimsonweed *(Datura stramonium)*	C	C	C	C	C	C	C	S	C	C	C	C	C	S
Knotweed *(Polygonum spp.)*	C	C	C	C	C	C	C	C	C	C	C	C	C	C
Kochia *(Kochia scoparia)*	C	C	C	C	C	C	C	C	C	C	C	C	C	C
Lambsquarters *(Chenopodium spp.)*	C	C	C	C	C	C	C	C	C	C	C	C	C	C
Morningglory, Ivyleaf *(Ipomoea hederacea)*	P	P	S	P	P	P	P	P	P	P	P	P	P	P
Morningglory, Pitted *(Ipomoea lacunosa)*	P	P	S	P	P	P	P	P	P	P	P	P	P	P
Morningglory, Smallflower *(Jacquemontia tamnifolia)*	P	P	C	P	P	P	P	P	P	P	P	P	P	P
Morningglory, Tall *(Ipomoea purpurea)*	P	P	S	P	P	P	S	P	P	P	P	P	P	P
Pigweeds *(Amaranthus spp.)*	C	C	C	C	C	C	C	C	C	C	C	C	C	C
Prickly Sida/Teaweed *(Sida spinosa)*	C	C	C	C	C	C	C	C	C	C	C	C	C	C
Purslane *(Portulaca oleracea)*	C	C	C	C	C	C	C	C	C	C	C	C	C	C
Ragweed, Common *(Ambrosia artemisiifolia)*	C	C	C	C	C	C	C	C	C	C	C	C	C	C
Redweed *(Melochia corchorifolia)*	C	C	C	C	C	C	C	C	C	C	C	C	C	C
Russian Thistle *(Salsola kali)*	C	C	C	C	C	C	C	C	C	C	C	C	C	C
Sesbania *(Sesbania spp.)*	C	C	C	C	C	C	C	C	C	C	C	C	C	C
Shepherdspurse *(Capsella bursa-pastoris)*	C	C	C	C	C	C	C	C	C	C	C	C	C	C
Sicklepod *(Cassia obtusifolia)*	S	C	S	C	S	C	S	C	C	C	S	C	S	S
Smartweeds *(Polygonum spp.)*	C	C	C	C	C	C	C	C	C	C	C	C	C	S
Spotted Spurge *(Euphorbia maculata)*	C	C	P	C	P	C	S	P	C	C	C	C	P	•
Spurred Anoda *(Anoda cristata)*	C	C	C	C	C	C	C	C	C	C	C	C	C	•
Sunflower *(Helianthus spp.)*	C	C	S	S	S	S	S	S	S	S	S	C	S	P
Velvetleaf *(Abutilon theophrasti)*	C	C	C	C	C	C	C	C	C	C	C	C	C	C
Venice Mallow *(Hibiscus trionum)*	C	C	C	C	C	C	C	C	C	C	C	C	C	C
Wild Mustards *(Brassica spp.)*	C	C	C	C	C	C	C	C	C	C	C	C	C	C
ANNUAL GRASSES														
Barnyardgrass *(Echinochloa crus-galli)*	S	C	C	C	C	C	C	C	C	C	C	C	C	C
Bluegrass *(Poa annua)*	C	C	C	C	C	C	•	C	C	C	C	C	C	C
Broadleaf Signalgrass *(Brachiaria platyphylla)*	C	C	C	C	C	C	C	C	C	C	C	C	C	•
Browntop Millet *(Panicum ramosum)*	C	C	C	P	C	S	C	P	C	P	C	C	•	•
Crabgrass *(Digitaria spp.)*	C	C	C	C	C	C	C	C	C	C	C	C	C	C

ANNUAL GRASSES

Crowfootgrass *(Dactyloctenium aegyptium)*	C	C	C	C	C	C	C	C	C	C	C	C	•	•
Cupgrass *(Eriochloa gracilis)*	P	C	P	P	P	P	P	P	P	P	P	C	•	•
Foxtails *(Setaria spp.)*	S	C	C	C	C	C	C	C	C	C	C	C	C	C
Goosegrass *(Eleusine indica)*	C	C	C	C	C	C	C	C	C	C	C	C	C	C
Johnsongrass, Seedling *(Sorghum halepense)*	C	C	C	C	C	C	C	P	C	C	C	C	C	•
Junglerice *(Echinochloa colonum)*	C	C	C	C	C	C	•	C	C	C	C	C	C	•
Nutsedge, Yellow *(Cyperus esculentus)*	P	P	P	C	P	C	P	P	P	C	P	C	P	•
Panicum, Fall *(Panicum dichotomiflorum)*	P	C	C	C	C	C	C	P	P	C	P	C	C	C
Panicum, Texas *(Panicum texanum)*	P	C	C	P	C	S	C	P	P	S	P	S	C	•
Red Rice *(Oryza sativa)*	P	C	C	C	P	C	•	P	P	C	P	C	•	•
Sandbur *(Cenchrus spp.)*	P	C	C	P	C	S	•	P	P	C	P	S	•	•
Shattercane *(Sorghum bicolor)*	P	C	C	P	P	P	P	P	P	P	P	P	C	•
Sorghum, Volunteer (Sorghum spp.)	P	C	C	P	P	P	P	P	P	P	P	P	•	P
Sprangletop *(Leptochloa spp.)*	P	C	C	P	P	P	•	P	P	P	P	P	•	P
Stinkgrass *(Eragrostis spp.)*	P	C	C	P	P	P	•	P	P	P	P	P	•	P
Wheat, Volunteer *(Triticum spp.)*	P	P	P	P	P	P	S	P	P	C	C	P	•	P
Witchgrass *(Panicum capillare)*	P	C	C	C	C	C	•	P	P	C	C	C	C	•

C = Control
• = No Information
S = Suppression or Erratic Control (Control may range from poor to excellent)
P = Poor or No Control

SENCOR HERBICIDE ALONE

SENCOR DF Herbicide (Alone) Preemergence Application: The following rates of SENCOR DF may be applied preemergence to soybeans through center pivot or lateral move sprinkler irrigation systems that apply water in a uniform manner. Refer to Chemigation section of the SENCOR DF label for directions. SENCOR DF can be applied broadcast or banded. This application may be made during planting or as a separate operation after planting but before crop emergence. See the "GENERAL INFORMATION" section in the front of the SENCOR DF label.

Do not apply to sand soils; or to sandy loam or loamy sand soils containing less than 2% organic matter. Do not incorporate into soil or apply more than once per season.

POUNDS OF SENCOR DF PER ACRE

SOIL TEXTURE	ORGANIC MATTER		
	Less than 2%	2 to 4%	Over 4%
COARSE SOILS (Sandy loam, loamy sand)	DO NOT USE[3]	1/2	2/3
MEDIUM SOILS[1] (Loam, silt loam, silt, sandy clay, sandy clay loam)	1/2 to 2/3	2/3 to 5/6	5/6 to 1
FINE SOILS[1] (Silty clay, silty clay loam,[2] clay, clay loam)	2/3 to 5/6	5/6 to 1	1 to 1 1/6
MISSISSIPPI DELTA ONLY	1	1 1/6	1 1/3

[1]For control of lambsquarters, redroot pigweed and wild mustard, and for suppression of green, yellow and giant foxtails on alkaline (calcareous) soils in Nebraska, Minnesota, South Dakota and North Dakota only, apply SENCOR DF at rates of 1/3 pound/acre on medium soils and 1/3 to 1/2 pound/acre on fine soils, regardless of soil organic matter percentage (use 1/2 pound only where soil pH is less than 7.5 and weed pressure is heavy). The 1/3 pound/acre rate of SENCOR DF alone can be applied regardless of soil pH. For control of other weeds listed on the label use SENCOR DF at full rates recommended in the table above, **but note that crop injury may occur on soils having a calcareous surface area or a pH of 7.5 or higher.**

[2]Silty clay loam soils are transitional soils and may be classified as medium textured soils in some regions of the U.S.

[3]Refer to the appropriate section of the label for use of SENCOR DF on soybeans in coarse soils with 0.5 to 2.0% organic matter in certain states.

USES OF SENCOR HERBICIDE IN COMBINATIONS WITH OTHER HERBICIDES

SEQUENTIAL APPLICATION OF SCEPTER FOLLOWING SENCOR DF: If needed, application of SENCOR DF alone or in a registered tank mix according to directions on the label, may be followed by an early postemergence application of Scepter for control of cocklebur. Apply 1/6 to 1/3 pint of Scepter in a minimum of 20 gallons of water per acre. Use 1/6 pint of Scepter if cockleburs are less than 3 inches tall or have fewer than 3 leaves and are actively growing. For cockleburs less than 6 inches tall and actively growing, use 1/3 pint per acre. Do not use Scepter when soybeans or cockleburs have been subjected to stress conditions such as temperature or moisture extremes. Do not exceed a total of 2/3 pint of Scepter per acre in one season. Wait at least 10 days after application of Scepter before cultivating.

When preparing the spray mixture with Scepter, add 2 pints of nonionic surfactant approved for use on growing crops and containing at least 80% active ingredient per 100 gallons of mixture. Apply crop oil concentrate (COC) at the rate specified on the COC label.

Use Scepter only in the states where it is registered as listed on the product label.

Apply Scepter at least 90 days before harvest of soybeans. Do not graze or feed soybean forage, hay, or straw to livestock.

Refer to the Scepter label for additional cautions and precautions, directions, limitations, and information on environmental hazards and planting of rotational crops.

SPLIT-SHOT APPLICATION

A preplant incorporated application of SENCOR DF tank mixed with either Treflan, Lasso, Dual, Prowl, or Sonalan and followed by a preemergence surface application of SENCOR alone, after planting but not prior to soybean emergence, will control more broadleaf and grass weeds in soybeans than when either herbicide is used alone.

Refer to the Treflan, Lasso, Dual, Prowl or Sonalan labels, and to appropriate sections of the SENCOR DF label, for directions on soil preparation, herbicide application, incorporation techniques, herbicide rates, weed species controlled, and restrictions for using tank mix combinations of SENCOR. Carefully observe the "Special Precautions" section concerning the use of SENCOR in tank mix combinations on soybeans.

When a SPLIT-SHOT application of SENCOR DF with Prowl, Treflan, or Sonalan is used, the preplant incorporated tank mix may be applied up to 21 days prior to planting soybeans; with Dual or Lasso, the preplant incorporated tank mix may be applied up to 14 days prior to planting.

On medium and fine textured soils with greater than 2% organic matter, a rate range is recommended for the SENCOR DF preemergence overlay application. The higher rate should be used (a) in fields with a history of severe broadleaf weed pressure, (b) when the time between preplant incorporated tank mix and preemergence overlay applications approaches the maximum stated above, and/or (c) when the organic matter content of the soil is at the upper end of the indicated range.

For black nightshade control, refer to the appropriate section of the Lasso, Dual, or Sonalan labels for specific instructions.

SPLIT-SHOT APPLICATION

SOIL TEXTURE[1]	PREPLANT INCORPORATED TANK MIX APPLICATION RATE OF COMBINATION PRODUCT/ACRE	PLUS	RATE OF SENCOR DF POUNDS/ACRE	—FOLLOWED BY—	PREEMERGENCE OVERLAY APPLICATION RATE OF SENCOR DF POUNDS/ACRE ORGANIC MATTER LESS THAN 2%		2 TO 4%		OVER 4%
COARSE (Light) (Sand, loamy sand, sandy loam)	Treflan 1 pt or Lasso 2-2½ qt or Dual 1¼-1½ pt or Prowl 1½ pt or Sonalan 1¼-2 pt	plus	1/3	—FOLLOWED BY—	1/6	or	1/6	or	1/6-1/3
MEDIUM (Loam, silt loam, sandy clay loam, silt, sandy clay)	Treflan 1½ pt or Lasso 2½-3 qt or Dual 1½ pt or Prowl 1½ pt or Sonalan 1¾-2½ pt	plus	1/2 or 1/3[2]	—FOLLOWED BY— —FOLLOWED BY—	1/6 1/3	or or	1/6-1/3 1/3-1/2	or or	1/3-1/2 (1/2-2/3)[3]
FINE (Heavy) (Silty clay loam,* clay loam, silty clay, clay)	Treflan 2 pt or Lasso 2½-3 qt or Dual 2-2½ pt or Prowl 1½-2 pt or Sonalan 2¼-3 pt.	plus	2/3 or 1/2[2]	—FOLLOWED BY— —FOLLOWED BY—	1/6 1/3	or or	1/6-1/3 1/3-1/2	or or	1/3-1/2 (1/2-2/3)[3]

* Silty clay loam soils are transitional soils and may be classified as medium textured soils in some regions of the U.S.

[1]On **coarse textured** soils, do not use on sand soils with less than 1% organic matter, or on loamy sand or sandy loam soils with less than 0.5% organic matter. However, on coarse textured soils **with a calcareous surface area or a pH of 7.5 or higher,** do not use on sand soils with less than 2% organic matter, or on loamy sand or sandy loam soils with less than 1% organic matter.

[2]Use this lower rate of SENCOR DF in the preplant incorporated tank mix **on soils having a calcareous surface area or a pH of 7.5 or higher,** and in those situations where soils within a field vary extremely in texture or organic matter content.

[3]Reduce this preemergence overlay rate of SENCOR DF by 1/6 pound/acre when using SPLIT-SHOT application **on soils with over 4% organic matter and which have a calcareous surface area or a pH of 7.5 or higher.**

EXTENDED SPLIT-SHOT APPLICATION (Includes No-Till, Reduced Till, Ridge Till, Strip Till, Mulch Till)

An early preplant (surface-applied or shallow incorporated) application of SENCOR tank mixed with either Dual or Lasso, followed by a preemergence surface application of SENCOR tank mixed with Dual or Lasso after planting but prior to soybean emergence, will control more broadleaf and grass weeds in soybeans than either herbicide used alone.

An EXTENDED SPLIT-SHOT application will decrease the need for tillage and/or contact herbicides for the control of existing vegetation prior to planting, while providing residual control of weeds after planting.

When an EXTENDED SPLIT-SHOT application of SENCOR DF with Dual or Lasso is used, the preplant tank mix combination may be applied 15 to 30 days prior to planting soybeans. Follow directions on the label accompanying the product for SPLIT-SHOT applications from 0 to 14 days before planting.

Where a rate range is recommended, the higher rates should be used (a) in fields with a history of severe weed pressure, (b) when the time between early preplant tank mix and preemergence overlay applications approaches the maximum 30 days, (c) when the organic matter content of the soil is at the upper end of the indicated range, (d) when heavy crop residues are present on the soil surface, and/or (e) when the early preplant tank mix application is shallow incorporated (e.g. use 2 to 2½ qts Lasso in the early preplant tank mix when surface applied and use 2½ to 3 qts Lasso when the tank mix is to be lightly incorporated).

When weeds exceed to 1 to 1½ inches in height or diameter at application, use a contact herbicide, such as Roundup® or Gramoxone.

Refer to the Dual or Lasso label, and to appropriate sections of the SENCOR DF label for additional information on soil preparation, herbicide application, weeds controlled, precautions, restrictions, limitations and sprayer clean-up.

EXTENDED SPLIT-SHOT APPLICATION

SOIL TEXTURE[1]	EARLY PREPLANT TANK MIX APPLICATION (SURFACE-APPLIED OR SHALLOW INCORPORATED) RATE OF COMBINATION PRODUCT/ACRE	PLUS	RATE OF SENCOR DF POUND/ACRE	FOLLOWED BY	PREEMERGENCE OVERLAY APPLICATION RATE OF COMBINATION PRODUCT/ACRE	PLUS	RATE OF SENCOR DF POUND/ACRE ORGANIC MATTER ½ TO 2%	2 TO 4%	OVER 4%
COARSE (Light) Sand, loamy sand, sandy loam	Dual 1⅓ pt or Lasso 1½-2 qt	plus	1/3-1/2	followed by	Dual 2/3 pt or Lasso 1½ qt	plus	1/6	1/6-1/3	1/3

SOIL TEXTURE[1]	EARLY PREPLANT TANK MIX APPLICATION (SURFACE-APPLIED OR SHALLOW INCORPORATED) RATE OF COMBINATION PRODUCT/ACRE	PLUS	RATE OF SENCOR DF POUND/ACRE	FOLLOWED BY	PREEMERGENCE OVERLAY APPLICATION RATE OF COMBINATION PRODUCT/ACRE	PLUS	RATE OF SENCOR DF POUND/ACRE ORGANIC MATTER ½ TO 2%	2 TO 4%	OVER 4%
MEDIUM Loam, silt loam, sandy clay loam, silt, sandy clay	Dual 1¾ pt or Lasso 2-3 qt	plus	(2)½- ⅔	followed by	Dual ¾ pt or Lasso 1-2 qt	plus	⅓	⅓-½	½-⅔
FINE (Heavy) Silty clay loam,* clay loam, silty clay, clay	Dual 2 pt or Lasso 2-3 qt	plus	(2)⅔-⅚	followed by	Dual 1 pt or Lasso 1-2 qt	plus	⅓	⅓-½	½-⅔

* Silty clay loams soils are transitional soils and may be classified as medium textured soils in some regions of the U.S.

[1] On **coarse textured** soils, do not use on sand soil with less than 1% organic matter. However, on coarse textured soils **with a calcareous surface area or a pH of 7.5 or higher,** do not use on sand soils with less than 2% organic matter, or on loamy sand or sandy loam soils with less than 1% organic matter.

[2] Use the lower rate of SENCOR DF in the early preplant tank mix on **soils having a calcareous surface area or a pH of 7.5 or higher, and in those rare situations where soils within a field vary extremely in texture or organic matter content.**

SENCOR PLUS SONALAN

SENCOR plus Sonalan Overlay Application

SENCOR DF may be applied as a preemergence overlay application following a preplant incorporated application of Sonalan 3 EC. Consult the Sonalan label for specific directions on use, recommendations, restrictions and any additional weeds not specified on the SENCOR DF label.

SENCOR plus Sonalan Tank Mix Application

Incorporate the tank mixture into the top 1 to 2 inches of soil within 21 days before planting according to label directions for Sonalan.

Apply SENCOR DF plus Sonalan preplant incorporated if furrow irrigation is used or when a period of dry weather after application is expected. If soybeans are planted on beds, apply and incorporate the tank mixture after bed formation.

Mixing: Refer to "Mixing" under the "GENERAL INFORMATION" section in the front of the SENCOR DF label.

Application: Sonalan should be uniformly applied and thoroughly mixed into the soil within 2 days after application. For specific application information, refer to the "GENERAL INFORMATION" section in the front of the SENCOR DF label.

BROADCAST RATES

SOIL TEXTURE	SENCOR DF lbs/acre	Sonalan 3 EC pts/acre
COARSE[1] (Sandy loam, loamy sand)	⅓	1¼-2
MEDIUM[3] (Loam, silt loam, silt, sandy clay, sandy clay loam)	½	1¾-2½
FINE[3] (Silty clay, silty clay loam,[2] clay, clay loam)	⅔	2¼-3

[1]Do not use on coarse soils with less than 1% organic matter.

[2]Silty clay loam soils are transitional soils and may be classified as medium textured soils in some regions of the U.S.

[3]For control of lambsquarters, redroot pigweed, wild mustard, and green and yellow foxtails on alkaline (calcareous) soils in Minnesota, Nebraska, South Dakota, and North Dakota only, apply SENCOR DF at rates of ⅓ pound/acre on medium soils and ⅓ to ½ pound/acre on fine soils regardless of soil organic matter percentage (use ½ pound only where soil pH is less than 7.5 and weed pressure is heavy). The ⅓ pound rate of SENCOR DF in tank mix combination with Sonalan can be applied regardless of soil pH. For control of other weeds not listed on the label, use SENCOR DF at full rates recommended in the table above, **but note that crop injury may occur on soils having a calcareous surface area or a pH of 7.5 or higher.**

Special Precautions (SENCOR plus Sonalan)

For additional precautions, restrictions, limitations, incorporation, and sprayer cleanup information, refer to the appropriate sections of the SENCOR DF label and the Sonalan label.

For black nightshade control, refer to the Sonalan label for specific rates and application instructions.

SENCOR PLUS TREFLAN

SENCOR and Treflan Overlay Application

SENCOR DF may be applied as a preemergence broadcast or band overlay application following a preplant incorporated treatment of Treflan. Consult the Treflan label for specific directions for use, recommendations, restrictions and any additional weeds not specified on the SENCOR DF label.

SENCOR plus Treflan Tank Mix Application

A single application of a tank mix combination of SENCOR DF and Treflan EC will control more broadleaf and grass weeds in soybeans than when either herbicide is used alone.

Prepare the soil surface by deep plowing, offset disking or tandem disking prior to the application of the herbicide combination. The soil surface should be well prepared and free of clods and trash.

This SENCOR plus Treflan tank mix combination may be applied and incorporated into the soil up to 10 days before planting.

Mixing: Refer to "Mixing" under the "GENERAL INFORMATION" section in the front of the SENCOR DF label.

Application: For specific application information refer to the "GENERAL INFORMATION" section in the front of the SENCOR DF label.

Apply SENCOR DF plus Treflan to the soil surface and incorporate in the same operation, if possible. Variable weed control may result from delayed incorporation if SENCOR plus Treflan are applied to a wet, warm soil surface or if the wind velocity is 10 miles per hour or higher. Use machinery that mixes SENCOR plus Treflan thoroughly with the soil. Incorporation may be delayed up to 24 hours after application. Shallow incorporation with implements set to cut less than 2 inches deep may result in erratic weed control. Do not use spike or springtooth harrows alone for incorporation.

BROADCAST RATES

SOIL TEXTURE	SENCOR DF Pounds per Acre	Treflan EC Pints per Acre
COARSE[1] (Sandy loam, loamy sand)	⅓	1
MEDIUM (Loam, silt loam, sandy clay, sandy clay loam)[3]	½	1½
FINE (Silty clay, silty clay loam,[2] clay, clay loam)[3]	⅔	2

[1]Do not use on coarse soils with less than 1% organic matter.

[2]Silty clay loam soils are transitional soils and may be classified as medium textured soils in some regions of the U.S.

[3]For control of lambsquarters, redroot pigweed, wild mustard and green and yellow foxtails on alkaline (calcareous) soils in Minnesota, Nebraska, South Dakota, and North Dakota only, apply SENCOR DF at rates of ⅓ pound/acre on medium soils and ⅓ to ½ pound/acre on fine soils, regardless of soil organic matter percentage (use ½ pound only where soil pH is less than 7.5 and weed pressure is heavy). The ⅓ pound rate of SENCOR DF in tank mix combination with Treflan can be applied regardless of soil pH. For control of other weeds listed on the label, use SENCOR DF at full rates recommended in the table above, **but note that crop injury may occur on soils having a calcareous surface area or a pH of 7.5 or higher.**

Incorporation Equipment:

1. Set PTO-driven equipment (tillers, cultivators, hoes) to cut 2 to 3 inches deep and space rotors to provide a clean sweep of the soil. PTO equipment should not be operated at a speed greater than 4 miles per hour.
2. Set disk to cut 4 to 6 inches deep and operate twice in different directions at 4 to 6 miles per hour.
3. Set mulch treader and other similar disk-type implements to cut 3 to 4 inches deep and operate twice in different directions at 5 to 8 miles per hour.

For Coarse and Medium Textured Soils Only

4. Set rolling cultivator to cut 2 to 4 inches deep and operate twice at 6 to 8 miles per hour. Set bed conditioner (Do-All) to cut 2 to 4 inches deep and operate at 4 to 6 miles per hour.

Special Precautions (SENCOR plus Treflan)

Seedling disease, cold weather, excessive moisture, high salt concentration or drought may weaken soybean seedlings and increase the possibility of damage from the tank mix. Do not plant soybeans deeper than 2 inches.

In the Central United States, do not plant sorghum or oats for 12 months where the tank mix has been applied unless 20 inches or more of irrigation and/or rainfall (total) was used to produce the crop. If less than 20 inches total water was used to produce the crop during the year, do not plant either crop for 18 months after the tank mix application. Cool wet weather conditions during the early stage of growth may increase the possibility of injury to sorghum.

For additional precautions, restrictions, limitations and sprayer cleanup information refer to the appropriate section of the SENCOR DF label. Do not use this tank mix combination on soils containing charcoal in Arkansas, Louisiana and Mississippi.

SENCOR PLUS DUAL

SENCOR plus Dual Overlay Application

Apply a preplant incorporated treatment of Dual 8E as directed on that product label for use on soybeans. Follow with a preemergence treatment of SENCOR DF as directed on the label for use on soybeans.

SENCOR plus Dual Tank Mix Applications

Preplant Incorporated Application: Incorporate the tank mixture into the top 2 inches of soil within 14 days before planting using a disk, harrow, rolling cultivator or similar implement.

Apply SENCOR DF plus Dual preplant incorporated if furrow irrigation is used or when a period of dry weather after application is expected. If soybeans are planted on beds, apply and incorporate the tank mixture after bed formation.

Preemergence Application: Dry weather following preemergence application of SENCOR plus Dual tank mixture may reduce effectiveness. If weeds develop, cultivate uniformly with shallow tilling equipment such as a rotary hoe that will not damage soybeans.

Mixing Instructions: Refer to "Mixing" under the "GENERAL INFORMATION" section in the front of the SENCOR DF label. For information on applying SENCOR DF in fluid or dry fertilizer refer to the "Application Of SENCOR DF in Fluid Fertilizers" or "Commercial Impregnation And Application Of SENCOR DF On Dry Bulk Fertilizer" under the "GENERAL INFORMATION" section in the front of the SENCOR DF label.

BROADCAST RATES
SENCOR plus Dual Tank Mix Preemergence Applications

	ORGANIC MATTER			
	0.5% to 3%		Over 3%	
SOIL TEXTURE	SENCOR DF Pounds/Acre	Dual 8E Pints/Acre	SENCOR DF Pounds/Acre	Dual 8E Pints/Acre
COARSE[1] (Loamy sand, Sandy loam)	1/3	1 1/4	1/2	1 1/2
MEDIUM (Loam, silt loam, silt)	1/2	1 1/2	2/3	2
FINE (Silty clay loam,[2] sandy clay loam, silty clay, sandy slay, clay loam, clay)	2/3	2	2/3 to 5/6	2 to 2 1/2
MISSISSIPPI DELTA ONLY (Silty clay, clay)	1	2	1	2 to 2 1/2

[1]Do not use on sand soils. Do not apply SENCOR and Dual overlay or tank mix preemergence on loamy sand with less than 2% organic matter.
[2]Silty clay loam soils are transitional soils and may be classified as medium textured soils in some regions of the U.S.

BROADCAST RATES
SENCOR plus Dual Tank Mix Preplant Incorporated Applications

	ORGANIC MATTER			
	0.5% to 3%		3% or greater	
SOIL TEXTURE	SENCOR DF Pounds/Acre	Dual 8E Pints/Acre	SENCOR DF Pounds/Acre	Dual 8E Pints/Acre
COARSE[1] (Loamy sand, Sandy loam)	1/3	1 1/4	1/3	1 1/2
MEDIUM (Loam, silt loam, silt)	1/2	1 1/2	1/2	2
FINE (Silty clay loam,[2] sandy clay loam, silty clay, sandy clay, clay loam, clay)	2/3	2	2/3	2 to 2 1/2
MISSISSIPPI DELTA ONLY (Silty clay, clay)	2/3 to 5/6	2	2/3 to 5/6	2 to 2 1/2

[1]Do not use on sand soils. Do not apply SENCOR plus Dual tank mix preplant incorporated on sand or loamy sand with less than 2% organic matter or crop injury may occur.
[2]Silty clay loam soils are transitional soils and may be classified as medium textured soils in some regions of the U.S.

Special Precautions (SENCOR and Dual)

For additional precautions, restrictions, limitations and sprayer cleanup information refer to the appropriate sections of the SENCOR DF label and the Dual label.

SENCOR PLUS PROWL

SENCOR plus Prowl Overlay Application

Apply a preplant incorporated treatment of Prowl as directed on that product label for use on soybeans. Follow with a preemergence treatment of SENCOR DF as directed on the SENCOR label for use on soybeans.

SENCOR plus Prowl Tank Mix Application

Preplant Incorporated Application: Prepare the soil by plowing or disking to mix previous crop residues into the soil to a depth of 4 to 6 inches.

For specific application information refer to the "GENERAL INFORMATION" section in the front of the SENCOR DF label.

Incorporate the tank mixture into the top 1 or 2 inches of soil within 7 days after application according to label directions for Prowl. Mechanical incorporation is not required if a rain of one-quarter inch or more occurs within 7 days after application. Soybeans must be planted no later than 7 days after application of the tank mixture.

Preemergence Application: Except for minimum and no-tillage systems, the seed bed should be firm and free of trash and clods. For specific application information refer to the "GENERAL INFORMATION" section in the front of the SENCOR DF label. Do not apply Prowl preemergence north of Interstate 80.

This application must be made after planting and before crop emergence. Do not incorporate.

If cultivation is necessary because of soil crusting, soil compaction or weed germination before rain or irrigation, use shallow tilling equipment such as a rotary hoe that does not damage soybeans.

Mixing Instructions: Refer to "Mixing" under the "GENERAL INFORMATION" section in the front of the SENCOR DF label.

For information on applying SENCOR DF in fluid or dry fertilizer refer to the "Application Of SENCOR DF in Fluid Fertilizers" or "Commercial Impregnation And Application of SENCOR DF On Dry Bulk Fertilizer" under the "GENERAL INFORMATION" section in front of the SENCOR DF label.

SOUTHERN STATES AND EASTERN COASTAL PLAINS

For use only in Alabama, Arkansas, Florida, Georgia, Louisiana, Mississippi, North Carolina, Oklahoma, Southeastern Missouri "Bootheel" Region and Coastal Plains of Delaware*, Maryland*, New Jersey* and Virginia*.

* SENCOR plus Prowl is not recommended for use on soils with less than 2% organic matter in the coastal plain of New Jersey or the Delmarva Peninsula.

BROADCAST RATES
SENCOR plus Prowl Tank Mix Applications

SOIL TEXTURE	SENCOR DF Pounds/Acre	Prowl Pints/Acre
COARSE[1] (Sandy loam, loamy sand)	1/3	1 1/2
MEDIUM (Loam, silt loam, silt, sandy clay, sandy clay loam)	1/2	1 1/2
FINE (Silty clay, silty clay loam,[2] clay, clay loam)	2/3	1 1/2 to 2

[1]Do not use on sand soils. Do not use on loamy sand or sandy loam containing less than 1% organic matter.
[2]Silty clay loam soils are transitional soils and may be classified as medium textured soils in certain regions of the U.S.
Do not use on muck or peat soils.

NORTHEASTERN AND NORTH CENTRAL STATES

For use only in Illinois, Indiana, Iowa, Kansas, Kentucky, Michigan, Minnesota, Nebraska, New York, North Dakota, Ohio, Pennsylvania, South Dakota, Wisconsin and Missouri (except the "Bootheel" Region).

BROADCAST RATES
SENCOR plus Prowl Tank Mix Applications

SOIL TEXTURE	ORGANIC MATTER ½ to 3% SENCOR DF Pounds/Acre	½ to 3% Prowl Pints/Acre	Over 3% SENCOR DF Pounds/Acre	Over 3% Prowl Pints/Acre
COARSE[1] (Sandy loam, loamy sand)	1/3	1	1/2	1 1/2
MEDIUM (Loam, silt loam, sandy clay, sandy clay loam)	1/2	1 1/2 to 2	1/2 to 2/3	1 1/2 to 2
FINE (Silty clay, silty clay loam,[2] clay, clay loam)	1/2 to 2/3	1 1/2 to 2	2/3 to 5/6	2 to 2 1/2

[1]Do not use on sand soils. Do not use on loamy sand or sandy loam containing less than 1% organic matter. Where a range of rates is shown for medium and fine soils, use the higher rate if heavy weed infestations are anticipated.
[2]Silty clay loam soils are transitional soils and may be classified as medium textured soils in certain areas of the U.S.
Do not use on muck or peat soils.

Special Precautions (SENCOR plus Prowl)

Soil incorporation deeper than recommended will reduce weed control and can result in crop injury.

For additional precautions, restrictions, limitations, and sprayer cleanup information, refer to the appropriate sections of the SENCOR DF and Prowl labels.

SENCOR PLUS LASSO

SENCOR plus Lasso Tank Mix Application

PREEMERGENCE: SENCOR DF may be used in a tank mix combination with Lasso as a preemergence band or broadcast application to soybeans in accordance with the specified soil types and dosages recommended.

For specific information regarding spray equipment, dilution rates, mixing, directions for use, methods of application, limitations and restrictions refer to the appropriate section of the SENCOR DF label.

Refer to the Lasso label for pertinent recommendations, directions for use, restrictions and any additional weeds not specified on the SENCOR DF label.

RECOMMENDED APPLICATIONS
SENCOR plus Lasso Tank Mix Preemergence Application (Broadcast Rates)

SOIL TEXTURE	½ to 3% Organic Matter SENCOR DF Lbs./Acre	+	Lasso Qts./Acre	Greater than 3% Organic Matter SENCOR DF Lbs./Acre	+	Lasso Qts./Acre
COARSE[1] (Sandy loam)	1/3	+	1 1/2 to 2	1/2	+	1 1/2 to 2
MEDIUM[2] (Loam, silt loam, silt, sandy clay, sandy clay loam)	1/2	+	1 1/2 to 2	2/3	+	1 1/2 to 2
FINE[2] (Silty clay, silty clay loam,[3] clay, clay loam)	2/3	+	2	2/3 to 5/6	+	2 to 2 1/2
MISSISSIPPI-DELTA ONLY (Silty clay to heavy clay)	SENCOR DF Lbs./Acre 1 1/3	+	Lasso Qts./Acre 2 to 2 1/2			

[1]Do not use SENCOR plus Lasso on sand or loamy sand soils with less than 2% organic matter.
[2]For control of lambsquarters, redroot pigweed, wild mustard and green and yellow foxtails on alkaline (calcareous) soils in Minnesota, Nebraska, South Dakota, and North Dakota only, apply SENCOR DF at rates of 1/3 pound/acre on medium soils and 1/3 to 1/2 pound/acre on fine soils regardless of soil organic matter percentage (use 1/2 pound only where soil pH is less than 7.5 and weed pressure is heavy). The 1/3 pound/acre rate of SENCOR DF in tank mix combination with Lasso can be applied regardless of soil pH. For control of other weeds use SENCOR DF at full rates recommended in the table above, **but note that crop injury may occur on soils having a calcareous surface area or a pH of 7.5 or higher.**
[3]Silty clay loam soils are transitional soils and may be classified as medium textured soils in some regions of the U.S.

DO NOT USE ON MUCK SOILS.

PREPLANT INCORPORATED: For specific application information refer to the "GENERAL INFORMATION" section in the front of the SENCOR DF label.

Apply SENCOR DF plus Lasso preplant incorporated if furrow irrigation is used or when a period of dry weather after application is expected. If soybeans are planted on beds, apply and incorporate the tank mixture after bed formation. Apply within 7 days prior to planting and shallowly incorporate into the upper 1 to 2 inches of soil.

RECOMMENDED APPLICATIONS
SENCOR plus Lasso Tank Mix Preplant Incorporated Applications (Broadcast Rates)

SOIL TEXTURE	SENCOR DF Pounds/Acre	Lasso Quarts/Acre
COARSE[1] (Loamy sand [over 2% organic matter], sandy loam)	1/3	2 to 2 1/2
MEDIUM: (Loam, silt loam, silt)	1/2	2 1/2 to 3
FINE: (Silty clay loam,[2] sandy clay loam, silty clay, sandy clay, clay loam, clay)	2/3	2 1/2 to 3
MISSISSIPPI DELTA ONLY: (Silty clay, clay)	2/3 to 5/6	2 1/2 to 3

[1]Do not use SENCOR plus Lasso on sand or loamy sand soils with less than 2% organic matter.
[2]Silty clay loam soils are transitional soils and may be classified as medium textured soils in some regions of the U.S.

DO NOT USE ON MUCK SOILS.

Special Precautions (SENCOR plus Lasso)

For additional precautions, restrictions, limitations and sprayer cleanup information refer to the appropriate sections of the SENCOR DF and Lasso labels.

SENCOR plus Cannon 3EC Tank Mix Application

SENCOR DF may be tank mixed with Cannon 3EC for preplant incorporated application to control certain weeds in soybeans. Refer to the "GENERAL INFORMATION" section of the SENCOR DF label for information on mixing, application, restrictions, special precautions and weeds controlled by SENCOR DF. See appropriate sections of the Cannon 3EC herbicide label for additional precautionary statements, directions for use, recommendations, restrictions and additional weeds controlled.

RECOMMENDED APPLICATIONS
SENCOR plus Cannon Tank Mix Preplant Incorporated Applications (Broadcast Rates)

SOIL TEXTURE	½ to 3% Organic Matter SENCOR DF Lbs./Acre	+	Cannon 3 EC Qts./Acre	Greater than 3% Organic Matter SENCOR DF Lbs./Acre	+	Cannon 3 EC Qts./Acre
COARSE[1] (Sandy loam)	1/3	+	3	1/2	+	3
MEDIUM[2] (Loam, silt loam, silt, sandy clay, sandy clay loam)	1/2	+	3	2/3	+	3
FINE[2] (Silty clay, silty clay loam,[3] clay, clay loam)	2/3	+	3	2/3 to 5/6	+	3
MISSISSIPPI-DELTA ONLY (Silty clay to heavy clay)	SENCOR DF Lbs./Acre 1 1/3	+	Cannon 3EC Qts./Acre 3			

[1]Do not use SENCOR DF plus Cannon on sand or loamy sand soils with less than 2% organic matter.
[2]For control of lambsquarters, redroot pigweed, wild mustard and green and yellow foxtails on alkaline (calcareous) soils in Minnesota, Nebraska, South Dakota, and North Dakota only, apply SENCOR DF at rates of 1/3 pound/acre on medium soils and 1/3 to 1/2 pound/acre on fine soils regardless of soil organic matter percentage (use 1/2 pound only where soil pH is less than 7.5 and weed pressure is heavy). The 1/3 pound/acre rate of SENCOR DF in tank mix combination with Cannon can be applied regardless of soil pH. For control of other weeds use SENCOR DF at full rates

recommended in the table above, **but note that crop injury may occur on soils having a calcareous surface area or a pH of 7.5 or higher.**

[3]Silty clay loam soils are transitional soils and may be classified as medium textured soils in some regions of the U.S.

DO NOT USE ON MUCK SOILS.

Do not allow sprays to drift on to adjacent desirable plants.

Observe all cautions and limitations on the labeling of all products used in mixtures.

SENCOR PLUS COMMAND TANK MIX APPLICATION

SENCOR DF may be applied in combination with Command 4 EC as a preplant or shallow incorporated application for the control of certain weeds in soybeans. Consult the Command label for specific directions on use, recommendations, restrictions and any additional weeds not specified on the SENCOR DF label.

Mixing: Refer to the "General Information" section in the front of the SENCOR DF label.

Application: SENCOR plus Command 4EC may only be applied with ground equipment as a preplant or shallow incorporated application. SENCOR plus Command should be immediately incorporated into the top 1-3 inches after application unless surface is dry. On dry soils, incorporate into the top 1-3 inches within 3 hours of tank mix application.

Do not apply this tank mix within 1000 feet of towns and subdivisions, commercial vegetable, fruit, nurseries or greenhouse operations.

A minimum of 15 gallons spray volume per acre is recommended with appropriate nozzle types and sizes to produce a coarse spray droplet. The use of an approved agricultural drift reducing additive is recommended for application volumes of 15-40 gallons per acre. The use of an approved agricultural drift reducing additive is required at spray volumes of 10-15 gallons per acre.

NOTE: Off-site movement of Command spray drift or vapors can cause foliar whitening or yellowing of some vegetation. Prior to application of Command, read and strictly follow all precautions and application instructions as set forth in that label.

For additional information on application, refer to the "General Information" section in the front of the SENCOR DF label and the Command label.

Weeds Controlled: Bristly Starbur, Carpetweed, Copperleaf, Florida Beggardweed, Florida Pusley, Galinsoga, Jimsonweed, Knotweed, Lambsquarters, Pigweeds, Prickly Sida/Teaweed, Purslane, Common Ragweed, Redweed, Sesbania, Smartweeds, Spurred Anoda, Velvetleaf, Venice Mallow, Wild Mustards, Barnyardgrass,* Bluegrass, Broadleaf Signalgrass, Crabgrass,* Foxtails (Green, Giant, Yellow,* Robust Purple), Goosegrass, Johnsongrass (seedling),* Fall Panicum,* Texas Panicum and witchgrass.

*Use 2 pts/A Command 4 EC on coarse and medium textured soils with high populations of these weeds.

BROADCAST RATES

SOIL TEXTURE[1]	ORGANIC MATTER			
	0.5 to 3%		Over 3%	
	SENCOR DF lbs/acre	Command 4 EC pts/acre	SENCOR DF lbs/acre	Command 4 EC pts/acre
COARSE[2] (Sandy loam, loamy sand)	1/3	1 1/2-2	1/3	1 1/2-2
MEDIUM (Loam, silt loam, silt, sandy clay, sandy clay loam)	1/3-1/2	1 1/2-2	1/3-1/2	1 1/2-2
FINE (Silty clay, silty clay loam,[3] clay, clay loam)	1/3-1/2	1 1/2-2	1/2-2/3	1 1/2-2

[1]Crop injury may occur on soils having a calcareous surface area or a pH of 7.1 or higher.

[2]Do not use on coarse soils with less than 1% organic matter.

[3]Silty clay loam soils are transitional soils and may be classified as medium textured soils in some regions of the U.S.

Special Precautions (SENCOR DF plus Command)

Do not rotate to wheat, oats, barley, rye, alfalfa or seed corn in the fall of the year of application or in the spring of the following year as crop injury may occur.

Do not apply when weather conditions favor drift.

Do not use treated vines for feed or forage.

Observe all cautions and limitations on labeling of all products used in mixtures.

Do not apply aerially or through irrigation equipment.

SENCOR plus Commence 5.25EC Tank Mix Application

SENCOR DF may be tank mixed with Commence 5.25EC for preplant incorporated application to control certain weeds in soybeans. Refer to the "GENERAL INFORMATION" section of the SENCOR DF label for information on mixing, application, restrictions, special precautions and weeds controlled by SENCOR DF. See appropriate sections of the Commence 5.25EC herbicide label for additional precautionary statements, directions for use, recommendations, restrictions and additional weeds controlled.

RECOMMENDED APPLICATIONS

SENCOR Plus Commence Tank Mix Preplant Incorporated Applications (Broadcast Rates)

SOIL TEXTURE[1]	ORGANIC MATTER			
	1/2 to 3%		Over 3%	
	SENCOR DF lbs/acre	Command 5.25EC pts/acre	SENCOR DF lbs/acre	Commence 5.25EC pts/acre
COARSE[2] (Sandy loam, loamy sand)	1/3	1 1/3-2	1/3	1 1/3-2
MEDIUM (Loam, silt loam, silt, sandy clay, sandy clay loam)	1/3 to 1/2	2 to 2 1/4	1/3 to 1/2	2 to2 1/4
FINE (Silty clay, silty clay loam,[3] clay, clay loam)	1/3 to 1/2	2 2/3	1/2 to 2/3	2 2/3

[1]Crop injury may occur on soils having a calcareous surface area or a pH of 7.1 or higher.

[2]Do not use on coarse soils with less than 1% organic matter.

[3]Silty clay loam soils are transitional soils and may be classified as medium textured soils in some regions of the U.S.

Special Precautions (SENCOR plus Commence)

Do not rotate to wheat, oats, barley, rye, alfalfa or seed corn in the fall of the year of application or in the spring of the following year as crop injury may occur.

Do not apply when weather conditions favor drift.

Do not use treated vines for feed or forage.

Do not apply aerially or through irrigation equipment.

Do not allow sprays to drift on to adjacent desirable plants.

Observe all cautions and limitations on labeling of all products used in mixtures.

SENCOR plus Freedom 3EC Tank Mix Application

SENCOR DF may be tank mixed with Freedom 3EC for preplant incorporated application to control certaim weeds on soybeans. Refer to the "GENERAL INFORMATION" section of the SENCOR DF label for information on mixing, application, restrictions, special precatuions and weeds controlled by SENCOR DF. See appropriate sections of the Freedom 3EC herbicide label for additional precautionary statements, directions for use, recommendations, restrictions and additional weeds controlled.

RECOMMENDED APPLICATIONS

SENCOR plus Freedom Tank Mix Preemergence Application (Broadcast Rates)

SOIL TEXTURE	1/2 to 3% Organic Matter			Greater than 3% Organic Matter		
	SENCOR DF Lbs./Acre	+	Freedom 3EC Qts./Acre	SENCOR DF Lbs./Acre	+	Freedom 3EC Qts./Acre
COARSE[1] (Sandy loam)	1/3	+	2 3/4 to 3 1/2	1/2	+	3 to 3 1/2
MEDIUM[2] (Loam, silt loam, silt, sandy clay, sandy clay loam)	1/2	+	2 3/4 to 3 1/2	2/3	+	3 1/2 to 4
FINE[2] (Silty clay, silty clay loam,[3] clay, clay loam)	2/3	+	3 1/2 to 4	2/3 to 5/6	+	3 1/2 to 4 1/2
MISSISSIPPI-DELTA ONLY (Silty clay to heavy clay)	SENCOR DF Lbs./Acre + Freedom 3EC Qts./Acre: 1 1/3	+	3 1/2 to 4 1/2			

[1]Do not use SENCOR plus Freedom on sand or loamy sand soils with less than 2% organic matter.

[2]For control of lambsquarters, redroot pigweed, wild mustard and green and yellow foxtails on alkaline (calcareous) soils in Minnesota, Nebraska, South Dakota, and North Dakota only, apply SENCOR DF at rates of 1/3 pound/acre on medium soils and 1/3 to 1/2 pound/acre on fine soils regardless of soil organic matter percentage (use 1/2 pound only where soil pH is less than 7.5 and weed pressure is heavy). The 1/3 pound/acre rate of SENCOR DF in tank mix combination with Freedom can be applied regardless of soil pH. For control of other weeds use SENCOR DF at full

rates recommended in the table above, **but note that crop injury may occur on soils having a calcareous surface area or a pH of 7.5 or higher.**

[3]Silty clay loam soils are transitional soils and may be classified as medium textured soils in some regions of the U.S.

DO NOT USE ON MUCK SOILS.

Do not allow sprays to drift on to adjacent desirable plants.

Observe all cautions and limitations on the labeling of all products used in mixtures.

SENCOR plus Command plus a Grass Herbicide Tank Mix Application

SENCOR DF may be applied with Command 4 EC and a grass herbicide (Treflan, Lasso, Dual, Prowl or Sonalan) for the control of certain broadleaf weeds and grasses in soybeans. This combination will provide improved control of heavy infestations of velvetleaf, jimsonweed and common ragweed. Consult the Command, Treflan, Lasso, Dual, Prowl or Sonalan labels for specific directions for use, recommendations, restrictions and additional weeds controlled not specified on the SENCOR DF label.

Mixing: Refer to the "General Information" section in the front of the SENCOR DF label.

Application: For specific application information, refer to the "General Information" section in the front of the SENCOR DF label.

Weeds Controlled: SENCOR plus Command plus Treflan, Lasso, Dual, Prowl or Sonalan will control the following broadleaf weeds and grasses:

Annual Grasses	**Annual Broadleaves**	
Barnyardgrass	Bristly Starbur	Ragweed, Common
Bluegrass	Carpetweed	Redweed
Broadleaf Signalgrass	Cooperleaf, Hophornbeam	Russian Thistle
Browntop Millet	Florida Beggarweed	Sesbania
Crabgrass	Florida Pusley	Sherpherdspurse
Crowfootgrass	Galinsoga	Sicklepod
Foxtails	Jimsonweed	Smartweeds
Goosegrass	Knotweed	Spotted Spurge
Johnsongrass (seedling)	Kochia	Spurred Anoda
Panicum, Fall	Lambsquarters	Velvetleaf
Witchgrass	Pigweed	Venice Mallow
	Prickley Sida/Teaweed	Wild Mustard
	Purslane	

SENCOR and Command plus Treflan, Lasso, Dual, Prowl or Sonalan will suppress (reduce the competition) of cocklebur and sunflower.

Recommended Application
(Broadcast Rates)

A = Treflan (pt/acre) D = Lasso (qt/acre)
B = Dual (pt/acre) E = Sonalan (pt/acre)
C = Prowl (pt/acre)

SOIL TEXTURE[1]	A	B	C	D	E	SENCOR DF (lbs/acre)	Command 4 EC[3] (pts/acre)
COARSE	1	1¼-1½	1½	2-2½	1¼-2	⅓	½ to ¾
MEDIUM	1½	1½-2	1½- 2	2½-3	1¾-2½	⅓ to ½[2]	½ to ¾
FINE	2	2-2½	1½-2½	2½-3	2¼-3	½ to ⅔[2]	½ to ¾

[1]Refer to "Soil Texture" paragraph on the label for specific soil classification. On coarse textured soils **with a calcareous surface area or a pH of 7.5 or higher,** do not use on loamy sand or sandy loam soils with less than 1% organic matter.

[2]The higher rate of SENCOR DF is recommended for the control of sicklepod and hemp sesbania. Use the lower rate of SENCOR DF in the preplant incorporated tank mix **on soils having a calcerous surface area or a pH of 7.5 or higher,** and in those situations where soils within a field vary extremely in texture or organic matter content.

[3]Higher rate is recommended under moderate to heavy weed infestations.

SENCOR PLUS SCEPTER PLUS A GRASS HERBICIDE TANK MIX APPLICATION

SENCOR DF may be applied with Scepter and a grass herbicide (Treflan, Lasso, Dual, Prowl or Sonalan) for the control of certain broadleaf weeds and grasses in soybeans. SENCOR and Scepter plus Treflan or Sonalan may be applied preplant incorporated broadcast. SENCOR and Scepter plus Lasso, Dual or Prowl may be applied preplant incorporated, preemergence broadcast or in a band application. Consult the Scepter, Treflan, Lasso, Dual, Prowl or Sonalan labels for specific directions for use, recommendations, restrictions and additional weeds controlled not specified on the SENCOR DF label.

Mixing: Refer to the "General Information" section in the front of the SENCOR DF label.

Application: For specific application information, refer to the "General Information" section in the front of the SENCOR DF label.

Weeds Controlled: SENCOR plus Scepter plus Treflan, Lasso, Dual, Prowl, or Sonalan will control the following broadleaf weeds and grasses:

Annual Grasses		
Barnyardgrass	Johnsongrass (Seedling)	Coffee Senna
Bluegrass	Panicum, Fall	Copperleaf, Hophornbeam
Broadleaf Signalgrass	Witchgrass	Florida Beggarweed
Browntop Millet	Bristly Starbur	Florida Pulsey
Crabgrass	Buffalobur	Galinsoga
Crowfootgrass	Carpetweed	Jimsonweed
Foxtails	Cocklebur	Knotweed
Goosegrass		
Annual Broadleaves		
Kochia	Purslane	Smartweeds
Lambsquarters	Ragweed, Common	Spotted Spurge
Morningglory, Pitted	Redweed	Spurred Anoda
Morningglory, Smallflower	Russian Thistle	Sunflower
Pigweed	Sebania	Velvetleaf
Prickly Sida/Teaweed	Shepherdspurse	Venice Mallow
	Sicklepod	Wild Mustard

SENCOR and Scepter plus Treflan, Lasso, Dual, Prowl or Sonalan will suppress (reduce the competition) of ivy leaf and tall morningglory, and red rice.

Recommended Application
(Broadcast Rates)

A = Treflan (pt/acre) D = Lasso (qt/acre)
B = Dual (pt/acre) E = Sonalan (pt/acre)
C = Prowl (pt/acre)

SOIL TEXTURE[1]	A	B	C	D	E	SENCOR DF (lbs/acre)	Scepter[3] (pt/acre)
COARSE	1	1¼-1½	1½	2-2½	1¼-2	⅓	⅓ to ½
MEDIUM	1½	1½-2	1½- 2	2½-3	1¾-2½	⅓ to ½[2]	⅓ to ½
FINE	2	2-2½	1½-2½	2½-3	2¼-3	½ to ⅔[2]	⅓ to ½

[1]Refer to "Soil Texture" paragraph on the SENCOR DF label for specific soil classification. On coarse textured soils **with a calcareous surface area or a pH of 7.5 or higher,** do not use on loamy sand or sandy loam soils with less than 1% organic matter.

[2]The higher rate of SENCOR DF is recommended for preemergence tank mix application and for the control of sicklepod and hemp sesbania. Use the lower rate of SENCOR DF in the preplant incorporated tank mix **on soils having a calcareous surface area or a pH of 7.5 or higher,** and in those situations where soils within a field vary extremely in texture or organic matter content.

[3]Higher rate is recommended under moderate to heavy weed infestations.

SENCOR PLUS SURFLAN

SENCOR plus Surflan Tank Mix Application

SENCOR DF may be used in a tank mix combination with Surflan A.S. as a preemergence application to control weeds in soybeans in accordance with the specified soil types and dosages recommended. Consult the Surflan label for specific directions for use, recommendations, restrictions and any additional weeds not specified on the SENCOR DF label.

Mixing Instructions: Refer to the "GENERAL INFORMATION" section in the front of the SENCOR DF label.

Application: Apply within 2 days after planting but before crop emergence.

A one-half inch rain or more or its equivalent in overhead irrigation is needed to activate the herbicide combination. If a one-half inch rain does not occur within 7 days after application, incorporate the herbicide combination into the top one inch of the soil with a ground-driven rotary hoe, rolling cultivator, or similar implement that will not damage soybean plants. Cultivation will control emerging annual weeds and grasses and improve herbicidal effectiveness. If heavy rains pack the soil surface, rotary hoe to break the crust and to aid soybean emergence.

For specific information on application, refer to the "GENERAL INFORMATION" section in the front of the SENCOR DF label.

BROADCAST RATES

SOIL TEXTURE	SENCOR DF Pounds Per Acre	Surflan A.S. Quarts Per Acre
COARSE[1] (Sand, sandy loam, loamy sand)	⅓	½ to ¾
MEDIUM (Loam, silt loam, silt, sandy clay, sandy clay loam)	½	¾ to 1
FINE (Silty clay, silty clay loam,[2] clay, clay loam)	⅔	1 to 1¼
MISSISSIPPI DELTA ONLY	⅔ to ⅚	1 to 1¼

[1]Not recommended on sand with less than 1% organic matter. Do not use on soils containing more than 5% organic matter.

[2]Silty clay loam soils are transitional soils and may be classified as medium textured soils in some regions of the U.S.

Special Precautions (SENCOR plus Surflan)

Do not plant root crops for 12 months following application.

Do not contaminate any body of water nor apply to any area not specified on the label.

For additional precautions, restrictions, limitations and sprayer cleanup information, refer to the appropriate section of the SENCOR DF and Surflan A.S. labels.

SENCOR PLUS AMIBEN

SENCOR plus Amiben Preemergence Tank Mix Application

SENCOR DF may be applied in a tank mix combination with Amiben 2 as a band or broadcast application to soybeans.

SENCOR plus Amiben Preemergence Tank Mix Application
BROADCAST RATES

SOIL TEXTURE	½ to 3% Organic Matter: SENCOR DF Lbs./Acre	+	½ to 3% Organic Matter: Amiben 2 Gals./Acre	Greater than 3% Organic Matter: SENCOR DF Lbs./Acre	+	Greater than 3% Organic Matter: Amiben 2 Gals./Acre
COARSE[1] (Sandy loam, loamy sand)	1/3	+	3/4 to 1	1/2	+	3/4 to 1
MEDIUM (Loam, silt loam, silt, sandy clay, sandy clay loam)	1/2	+	3/4 to 1	2/3	+	1 to 1 1/4
FINE (Silty clay, silty clay loam,[2] clay, clay loam)	2/3	+	1 to 1 1/4	2/3 to 5/6	+	1 to 1 1/4
MISSISSIPPI DELTA ONLY (Silty clay to heavy clay)	SENCOR DF Lbs./Acre + Amiben Gals./Acre: 1 1/3	+	1 1/2			

[1]Do not use on sandy loam or loamy sand soils with less than 2% organic matter. Do not use on sand soils.

[2]Silty clay loam soils are transitional soils and may be classified as medium textured soils in some regions of the U.S.

DO NOT USE ON MUCK SOILS.

Where a rate range is specified, use the higher rate in fields with a history of intense weed pressure. See "Rate Ranges" in the front of the "Soybeans" section on the SENCOR DF label.

For additional precautions and information regarding spray equipment, mixing, application, limitations and restrictions refer to the appropriate section of the SENCOR DF and Amiben 2 labels.

SENCOR PLUS LINURON PLUS (LASSO OR DUAL)

SENCOR plus Linuron plus (Lasso or Dual) Tank Mix Application

SENCOR DF may be applied in combination with linuron 50 WP or 4L and Lasso 4 or Dual 8E as a preemergence application for the control of certain weeds in soybeans. Consult the linuron, Lasso, or Dual labels for specific directions for use, recommendations, restrictions and any additional weeds not specified on the SENCOR DF label.

Mixing: Refer to the "GENERAL INFORMATION" section in the front of the SENCOR DF label.

Application: Applications can be made only with ground spray equipment in accordance with specified soil types and dosages rates. For specific application information, refer to the "GENERAL INFORMATION" section in the front of the SENCOR DF label.

BROADCAST RATES
(0.5 to 3% Organic Matter Only)

SOIL TEXTURE	SENCOR DF Pounds/Acre	linuron 50 WP or 4 L (Pounds/Acre) or (Pints/Acre)	Lasso 4 Quarts/Acre	or	Dual 8E Pints/Acre
COARSE[1] (Sand, loamy sand, sandy loam)	1/6 to 1/4	1/3 to 1/2	3/4 to 1	or	1 to 1 1/4
MEDIUM (Loam, silt loam, silt, sandy clay, sandy clay loam)	1/4 to 1/3	1/2 to 3/4	1 to 1 1/2	or	1 1/4 to 1 1/2
FINE (Silty clay, silty clay loam,[2] clay, clay loam)	1/3 to 1/2	3/4 to 1 1/2	1 1/4 to 2	or	1 1/2 to 2

[1]DO NOT USE SENCOR plus linuron plus (Lasso or Dual) on sand soils with less than 1% organic matter.

[2]Silty clay loam soils are transitional soils and may be classified as medium textured soils in some regions of the U.S.

Special Precautions SENCOR plus linuron plus (Lasso or Dual)

For additional precautions, restrictions, limitations, and sprayer cleanup information, refer to the appropriate sections of the SENCOR DF, linuron, Lasso, or Dual labels.

FOR USE IN COARSE (LIGHT) SOILS IN Alabama, Arkansas, Florida, Georgia, Louisiana, Mississippi, Missouri, North Carolina, Oklahoma, South Carolina, Tennessee, Texas and Virginia.

SENCOR herbicide is recommended alone or in combination with Treflan, Lasso, Dual, Surflan or Amiben for use in coarse-textured, low organic matter soils in the states listed above for the control of certain weeds in soybeans. Refer to the appropriate sections of the SENCOR, Treflan, Lasso, Dual, Surflan or Amiben labels for specific directions for use, recommendations, restrictions and any additional weeds not specified on the SENCOR DF label.

Mixing: Refer to the "GENERAL INFORMATION" section in the front of the SENCOR DF label.

Application: For specific application information, refer to the "GENERAL INFORMATION" section in the front of the SENCOR DF label.

SENCOR (Alone) Preemergence Application

BROADCAST RATES

SOIL TEXTURE	ORGANIC MATTER	SENCOR DF Pounds/Acre
COARSE (LIGHT) SOILS (Sand,[1] loamy sand, sandy loam)	0.5% or above	1/3 to 1/2[2]

[1]Not recommended for use on sand with less than 1% organic matter.

[2]Use the higher rate under heavy weed pressures and/or on soils higher in organic matter.

SENCOR in Combination with Other Herbicides

SENCOR DF is recommended in a tank mix combination with Treflan as a preplant incorporated application or as a preemergence overlay application following a preplant incorporated application of Treflan. SENCOR DF is also recommended for use as a preemergence application in combination with Lasso, Dual, Surflan or Amiben.

BROADCAST RATES

SOIL TEXTURE	ORGANIC MATTER	Combination Product/Acre	Plus SENCOR DF Pounds/Acre
COARSE (LIGHT) SOILS (Sand,[1] loamy sand, sandy loam)	0.5% or above	Preplant Incorporated Treflan 4EC 1 pt	1/3 to 1/2[2]
		Preemergence Lasso 4EC 1 1/2 to 2 qt -or- Dual 8E 1 1/4 to 1 1/2 pt -or- Surflan A.S. 1/2 to 3/4 qt -or- Amiben 2 3/4 to 1 gal	1/3 to 1/2[2]

[1]Not recommended for use on sand with less than 1% organic matter.

[2]Use the higher rate under heavy weed pressures and/or on soils higher in organic matter.

Special Precautions

Do not use on sand soils with less than 1% organic matter, or on sandy loam or loamy sand soils with less than 0.5% organic matter.

For additional precautions, restrictions, limitations and sprayer cleanup information, refer to the appropriate sections of the SENCOR DF, Treflan, Lasso, Dual, Surflan or Amiben labels.

For Contact Action and Residual Control

For Control of Emerged Annual Broadleaf Weeds and Grasses and For Suppression of Emerged Perennials Where Soybeans Will Be Planted Directly Into a Pre-formed Bed, Cover Crop or in Previous Crop Residues as in "Minimum Till" or "No-Till" Crop Production.

SENCOR PLUS GRAMOXONE

SENCOR plus Gramoxone Tank Mix Application:

Apply 1 1/3 to 2 2/3 pints Gramoxone Super plus 1/2 to 1 1/3 pounds SENCOR DF according to the recommendations for SENCOR on the label as a broadcast ground spray before, during or after planting but before emergence of the crop. Soybeans exposed at time of application will be killed. Use the higher rate of Gramoxone for heavier infestations or for taller weeds (4 to 6 inches tall). Apply in 20 to 60 gallons spray mix per acre. Use the higher volume in fields with heavy weed infestations or with large amounts of previous crop residues. Add Ortho X-77® Spreader or similar non-ionic spreader at 1 quart per 100 gallons of diluted spray. This mixture will not control weeds taller than 6 inches.

SENCOR PLUS LASSO AND GRAMOXONE

SENCOR plus Lasso and Gramoxone Tank Mix Application:
Apply 1⅓ to 2⅔ pints Gramoxone Super plus 2 to 2½ quarts Lasso 4 EC plus ⅓ to 1⅓ pounds SENCOR DF as a preemergence broadcast ground spray. Use the higher rate of Gramoxone for weeds 4 to 6 inches tall. Soybeans exposed at time of application will be killed. Refer to the appropriate sections of the SENCOR, Gramoxone and Lasso labels for rates which are dependent upon soil textures, directions, limitations and cautions.

Apply in 20 to 60 gallons spray mix per acre. Use the higher volume in fields with heavy weed infestations or with large amounts of previous crop residues. Add Ortho X-77 Spreader or similar non-ionic spreader at 8 fl. ounces per 100 gallons of diluted spray. Add SENCOR to the spray tank first and mix thoroughly. Then add Lasso followed by Gramoxone and Ortho X-77 Spreader. Constant agitation is recommended at all times. This tank mix will not control shattercane which emerges after treatment or other weeds taller than 6 inches.

SENCOR PLUS SURFLAN AND GRAMOXONE

SENCOR plus Surflan and Gramoxone Tank Mix Application
The combination provides control of weeds and grasses which are emerged at the time of application plus many weeds and grasses which may germinate after planting. Soybeans exposed at time of application will be killed.

Weeds Controlled by Gramoxone: Gramoxone will kill emerged annual broadleaf weeds and grasses and provide top kill of perennials.

When to Apply: Apply the tank mix at planting or within 2 days after planting. Do not spray over top of emerged soybeans.

SENCOR plus Surflan and Gramoxone Preemergence Tank Mix Application

	Broadcast Rates Per Acre		
SOIL TEXTURE	**SENCOR DF**	**Surflan A.S.**	**Gramoxone Super**
COARSE[1]	⅓ pound	¾ quart	1⅓ quarts
MEDIUM	½ pound	1 quart	1⅓ quarts
FINE	⅔ pound	1½ quarts	1⅓ quarts

[1]Not recommended on sand soils with less than 1% organic matter.

Do not use on soils containing more than 5% organic matter.

Mixing: Refer to the "GENERAL INFORMATION" section in the front of the SENCOR DF label.

Add Ortho X-77 Spreader or similar non-ionic spreader at 8 fl. ozs. per 100 gallons of diluted spray mixture.

Application: For specific application information refer to the "GENERAL INFORMATION" section in the front of the SENCOR DF label.

A one-half inch rain or more or its equivalent in overhead irrigation is needed to activate the herbicide combination.

Special Precautions (SENCOR plus Surflan and Gramoxone):
Poor weed control and/or crop injury may result if directions are not followed.

Do not use a rib-type press wheel on your no-till planter or crop injury may result.

For additional precautions, special precautions, restrictions, limitations and sprayer cleanup information, refer to the appropriate section of the SENCOR, Surflan and Gramoxone labels.

SENCOR PLUS 2,4-D Low Volatile Ester Tank Mix Application
SENCOR DF alone or in combination with Dual, Lasso, Surflan or Prowl may be applied with 2,4-D low volatile ester as an early preplant surface application for the control of certain broadleaf weeds and grasses in soybeans in minimum or no-till production. Application is recommended 30 days prior to planting. Apply one pound active ingredient (ai) per acre of 2,4-D low volatile ester with labelled rates of SENCOR DF. Where a grass herbicide is used in the tank mix apply at the rates specified on that product's label.

Refer to the SENCOR DF, 2,4-D low volatile ester, Dual, Lasso, Surflan and Prowl labels for weeds controlled, precautions, restrictions, limitations and sprayer clean-up information.

SOUTHERN AND SOUTHEASTERN STATES ONLY

POSTEMERGENCE DIRECTED SPRAY APPLICATIONS: SENCOR DF can be applied in postemergence directed sprays to soybeans for control of certain weeds which escape preplant or preemergence herbicide applications and for control of additional flushes of weeds that may occur after soybeans have emerged. Postemergence directed sprays of SENCOR DF can be applied to soybeans in addition to a preemergence or preplant application of SENCOR DF according to label directions.

Weeds Controlled: SENCOR DF, applied postemergence to soybeans as a directed spray according to directions on the label, will control the following at rates shown (broadcast basis) when grasses and common ragweed are less than 1 inch tall and other broadleaves are less than 3 inches tall:

⅓ pound/acre		⅓ to ⅔ pound/acre	⅔ pound/acre
Florida Beggarweed *(Desmodium tortuosum)*	Purslane *(Portulaca oleracea)*	Sesbania *(Sesbania spp.)*	Ragweed, Common *(Ambrosia artemisiifolia)*
Pigweeds *(Amaranthus spp.)*	Sicklepod *(Cassia obtusifolia)*	Prickly Sida/Teaweed *(Sida spinosa)*	
Carpetweed *(Mollugo verticillata)*	Velvetleaf *(Abutilon theophrasti)*		
Cocklebur *(Xanthium pensylvanicum)*	Crabgrass *(Digitaria spp.)*		
Dayflower *(Commelina spp.)*			
Mexican weed *(Caperonia castaniifolia)*			

At the rate of ⅔ pound/acre morningglory species *(Ipomoea spp.)* horsenettle, *(Solanum spp.)* Florida pusley, *(Richardia scabra)* spotted spurge *(Euphorbia maculata)* and wild poinsettia *(Euphorbia heterophylla)* are suppressed when SENCOR DF is applied before these weeds are 3 inches tall. The ⅔ pound/acre rate will suppress broadleaf signalgrass *(Brachiaria platyphylla)* up to 1 inch tall.

SENCOR POSTEMERGENCE DIRECTED SPRAY APPLICATIONS
RECOMMENDED APPLICATIONS

CROP	SENCOR DF Pounds Per Acre	REMARKS
Soybeans (Alabama, Arkansas, Florida, Georgia, Kentucky, Louisiana, Mississippi, Missouri, North Carolina, Oklahoma, South Carolina, Tennessee and Texas)	⅓ to ⅔ (broadcast basis)	Apply proper dosage using 10 to 40 gallons of water per acre as a directed spray in a 6 to 8 inch band on each side of the row after soybeans are 8 inches tall and before broadleaf weeds are 3 inches tall and before grasses and common ragweed are 1 inch tall. For best results the spray must cover weed foliage with minimum or no contact of soybean foliage. Add a non-ionic surfactant such as Ortho X-77 to the spray mixture to obtain better wetting of weed leaf surfaces. To determine the correct dosage of SENCOR DF for a band application see "Banded Application" under the "GENERAL INFORMATION" section in the front of the label. If necessary, a second postemergence directed spray application can be made after 7 days. Do not feed or graze green soybean vines. Do not harvest soybeans or use dry soybean vines for feed or forage within 70 days of last application.

SPECIAL PRECAUTIONS (Directed Postemergence)
Do not apply directly to soybeans or serious crop injury will occur. Do not allow spray to contact more than the lower ¼ to ⅓ of soybean plants. Soybean leaves contacted by the spray will be killed.

Do not apply SENCOR DF postemergence to sensitive soybean varieties. See "Special Precautions" in the front of the SENCOR DF label.

To avoid injury to other crops or desirable plants from spray drift, sprayer pressure must not exceed 30 psi and the sprayer must be fitted with nozzles no smaller than 8002 T-Jet (or equivalent). Do not apply under weather conditions which favor drift.

POTATOES

SENCOR herbicide is recommended for use in ground or aircraft spray equipment as a preemergence and/or postemergence application to potatoes. Postemergence applications may be made on russetted or white-skinned potato varieties that are not early maturing. For optimum control of weeds in potatoes, apply SENCOR DF before weeds are 1 inch tall. Refer to "Special Precautions" for potatoes for additional information.

Ground Application: SENCOR DF is recommended for use with ground spray equipment, except air blast sprayers, as a preemergence and/or postemergence application for control of certain grass and broadleaf weeds in potatoes. Apply as a broadcast spray after planting and before crop emergence or after drag-off if this operation is part of the normal cultural practice. Postemergence applications should be applied after crop emergence.

Aerial Application: SENCOR DF is recommended for use in aerial spray equipment as a preemergence and/or postemergence application for control of certain grass and broadleaf weeds in potatoes. Apply as a broadcast spray after planting and before crop emergence or after drag-off if this operation is part of the normal cultural practice. Postemergence applications should be applied after crop emergence.

PREEMERGENCE

Weeds Controlled:

SENCOR DF, applied preemergence to potatoes according to directions, will effectively control:

Broadleaves:		Grasses:
Carpetweed *(Mollugo verticillata)*	Ragweed, Common *(Ambrosia artemisiifolia)*	Broadleaf Signalgrass *(Brachiaria platyphylla)*
Cocklebur *(Xanthium pensylvanicum)*	Shepherdspurse *(Capsella bursa-pastoris)*	Crabgrass *(Digitaria spp.)*
Jimsonweed *(Datura stramonium)*	Sicklepod *(Cassia obtusifolia)*	Foxtails *(Setaria spp.)*
Lambsquarters *(Chenopodium album)*	Smartweed, Pennsylvania *(Polygonum pensylvanicum)*	Johnsongrass, Seedling *(Sorghum halepense)*
Pigweeds *(Amaranthus spp.)*	Wild Mustards *(Brassica spp.)*	Panicum, Fall *(Panicum dichotomiflorum)*

For Control of Wild Mustard (Brassica spp.) in Potatoes (Except California): Apply at the rate of 1/3 to 2/3 pound per acre as a preemergence broadcast spray using aerial or ground spray equipment in accordance with directions on the label. Application should be made at or near ground crack and/or before the mustard is one inch tall. For control of mustard larger than 2 inches in diameter or other weeds listed on the label, a minimum of 2/3 pound per acre must be used. When less than 2/3 pound per acre is used, a postemergence application in accordance with directions on the SENCOR label may be required for control of other weeds or later germinating weed species.

Hard To Control Weeds: Although SENCOR DF applied preemergence may not provide commercially acceptable control in every instance, it will suppress growth of the following weeds and reduce their competition with potato plants.

Broadleaves:	Grasses:
Common Purslane *(Portulaca oleracea)*	Barnyardgrass *(Echinochloa crus-galli)*
	Nutsedge *(Cyperus spp.)*

POSTEMERGENCE

Weeds Controlled:

SENCOR DF, applied early postemergence to potatoes according to directions, will effectively control:

Broadleaves:	Grasses:
Cocklebur *(Xanthium pensylvanicum)*	*Foxtails (Setaria spp.)*
Lambsquarters *(Chenopodium album)*	
Pigweed, Redroot *(Amaranthus retroflexus)*	
Ragweed, Common *(Ambrosia artemisiifolia)*	
Russian Thistle *(Salsola kali)*	
Smartweed, Pennsylvania *(Polygonum pensylvanicum)*	

On medium and fine soils in Idaho, Oregon and Washington, two postemergence applications, each at the rate of 2/3 pound per acre, will control barnyardgrass, *(Echinochloa crus-galli)*, kochia, *(Kochia scoparia)*, sunflower *(Helianthus spp.)*, and Jim Hill mustard *(Sisymbrium altissimum)* in addition to weeds listed above.

Hard To Control Weeds: Although SENCOR DF applied early postemergence may not provide commercially acceptable control in every instance, it will suppress growth of the following weeds and reduce their competition with potato plants.

Broadleaves:	Grasses:
Kochia *(Kochia scoparia)*	*Barnyardgrass (Echinochloa crus-galli)*
Sunflower *(Helianthus spp.)*	

RECOMMENDED BROADCAST APPLICATIONS

CROP	SENCOR DF Pounds/Acre	REMARKS
Potatoes	2/3 to 1 1/3	**PREEMERGENCE:** Apply specified dosage per acre as a broadcast spray to the soil surface after planting and before crop emergence or after drag-off if this operation is part of normal cultural practice*. Distribute the chemical evenly over the area to be treated. Do not incorporate into soil. Use the lower rate on sandy soils.
Potatoes (russetted or white-skinned varieties that are not early maturing)	1/3 to 2/3	**POSTEMERGENCE:** Apply specified dosage per acre as a broadcast spray over the tops of potato plants.* Use rates of 1/3 to 2/3 lb. for control of redroot pigweed and lambsquarters only. Apply the 2/3 lb. rate for control of other weeds listed on the label. Do not apply within 60 days of harvest.
		SPLIT APPLICATIONS: This product may be applied once preemergence and once postemergence as directed above*. Do not apply postemergence within 60 days of harvest.
		IDAHO, OREGON AND WASHINGTON ONLY: Two postemergence applications can be made as broadcast sprays over the tops of potato plants if SENCOR is not applied preemergence. Use 1/3 to 2/3 pound/acre for control of redroot pigweed and lambsquarters only. On coarse (sandy) soils with low organic matter, do not exceed 1/2 pound/acre per application. On medium and heavy soils only, use 2/3 pound/acre for control of other weeds listed on the label. Make the first application early in the season while weeds are still small. Allow at least 14 days before the second application. For optimum control, apply before weeds are 1 inch tall. Do not apply after June 30 if treated land is to be planted to crops other than potatoes. Do not apply within 60 days of harvest.

*Refer to Special Precautions—Potatoes below.

SPRINKLER IRRIGATION APPLICATION TO POTATOES:

SENCOR DF may be applied early postemergence to potatoes through center pivot, permanent solid set, or self-propelled lateral move irrigation systems as directed on the label. Refer to "Chemigation" section of the SENCOR DF label for directions.

RECOMMENDED BROADCAST APPLICATIONS

CROP	SENCOR DF Pounds/Acre	REMARKS
Potatoes (russetted or white-skinned varieties that are not early maturing)	1/3 to 2/3	**POSTEMERGENCE IN SPRINKLER IRRIGATION:** Apply specified dosage in 1/4 to 3/4 inch of water per acre as a continuous injection in center pivot and self-propelled wheel move systems, or in the last 15 to 30 minutes of set in permanent solid set sprinkler systems. On sandy soil, apply in 1/4 to 1/2 inch of water. Use rates of 1/3 to 2/3 pounds for control of redroot pigweed and lambsquarters only. Apply the 2/3 pound rate for control of other weeds listed on the label. Do not apply within 60 days of harvest.

Do not apply when wind speed favors drift, when system connections leak, when nozzles do not provide uniform distribution or when lines containing the product must be dismantled or drained.

SENCOR plus Dual Tank Mix Application (Except California)

SENCOR DF may be applied in a tank mix combination with DUAL 8E as a preemergence broadcast application to potatoes. Apply SENCOR DF at 1/2 to 1 1/3 pounds and Dual 8E at 1.5 to 3 pints per acre according to the respective labels for use of each product alone on potatoes. Refer to appropriate sections of the SENCOR DF and Dual labels for precautionary statements, restrictions, application information and weeds controlled.

SENCOR plus Prowl Tank Mix Application (Except California)

SENCOR DF may be applied in a tank mix combination with Prowl as a preemergence broadcast application to potatoes. Apply SENCOR DF at 2/3 to 1 1/3 pounds and Prowl at 1 to 3 pints per acre according to the respective labels for use of each product alone on potatoes. Refer to the appropriate sections of the SENCOR DF and Prowl labels for precautionary statements, restrictions, application information and weeds controlled.

Special Precautions (Potatoes)

Do Not Use SENCOR DF on Potatoes in Kern County, California.

Do not apply more than $1\frac{1}{3}$ pounds total SENCOR DF per acre in a single crop season regardless of the method of application. **Do not apply within 24 hours of application of other pesticides.** Do not apply postemergence within 3 days after periods of cool, wet or cloudy weather or crop injury may occur. Postemergence applications may cause some chlorosis or minor necrosis.

In addition to early maturing smooth-skinned white, or red-skinned varieties of potatoes, postemergence applications of SENCOR DF are not recommended on Atlantic, Shepody, Chipbelle, Bellchip, and Centennial varieties. Preemergence applications to these varieties may cause crop injury, especially under adverse weather conditions and when higher SENCOR DF rates are used.

Do not apply to sweet potatoes or yams.

Do not plant sensitive crops such as onions, lettuce, cole crops and cucurbits during the next growing season following applications of SENCOR DF or injury may occur.

ALFALFA AND SAINFOIN

SENCOR herbicide is labelled for use in alfalfa and sainfoin in the following areas:

1. Alfalfa and sainfoin (Including mixed stands with grasses) (all areas except California).
2. Alfalfa and sainfoin (Including mixed stands with grasses) (California only).
3. Alfalfa—Tank mix combination with Gramaxone (Colorado, Idaho, Montana, Nevada, Oregon, Utah, Washington, and Wyoming.)
4. Alfalfa—Post-dormant application of SENCOR impregnated on dry fertilizer only.
5. Alfalfa—Non-dormant, non-winter hardy varieties (Arizona only).

SENCOR DF is recommended for use in aerial or ground spray equipment as a broadcast surface application to established crops of alfalfa and sainfoin for the control of certain grass and broadleaf weeds.

APPLICATION: Refer to "GENERAL INFORMATION" in the front of the SENCOR DF label for detailed information on the application of SENCOR DF. For information on applying SENCOR DF in fluid or on dry fertilizer refer to the "Application Of SENCOR DF In Fluid Fertilizers" or "Commercial Impregnation And Application of SENCOR DF ON Dry Bulk Fertilizer" under the "GENERAL INFORMATION" section in the front of the SENCOR DF label.

Special Precautions:

Do not apply SENCOR DF after growth begins in the Spring or before growth ceases in the Fall.

Treat only dormant established alfalfa (except where specified on the SENCOR DF label).

Do not graze or harvest within 28 days after application.

For best weed control, apply SENCOR DF when weeds are less than 2 inches tall or before weed foliage is 2 inches in diameter.

Reduced weed control may occur when extended dry conditions follow application of SENCOR DF.

Crop injury may occur if alfalfa or sainfoin are:

1. Under stress conditions such as diseases, insect infestations, poorly drained soils, drought or Winter injury when SENCOR DF is applied, or
2. When SENCOR DF is applied during the first growing season after seeding or if applied earlier than 12 months after seeding.

ALFALFA AND SAINFOIN
(All Areas except California)

RECOMMENDED BROADCAST APPLICATIONS

CROP	SENCOR DF Pounds/Acre	REMARKS
Alfalfa and Sainfoin (Except California)	$\frac{1}{2}$ to $1\frac{1}{3}$	**ALL AREAS:** (Including Idaho, Oregon, and Washington). Select the proper dosage according to weeds known to be present in the field to be treated. Use the lower rate of SENCOR DF on sandy loam or loamy sand soils in areas other than Idaho, Oregon, and Washington.
	$\frac{1}{3}$ to $\frac{1}{2}$	**IDAHO, OREGON AND WASHINGTON ONLY:** On sandy loam or loamy sand soils apply SENCOR DF at rates of $\frac{1}{3}$ to $\frac{1}{2}$ pound per acre for control of common chickweed only.

FOR USE ON MIXED STANDS OF ALFALFA AND GRASSES:

Rates of $\frac{2}{3}$ to 1 pound of SENCOR DF per acre will provide partial reduction of forage grass stands. These rates may be used to reduce forage grass stands to prevent crowding out of alfalfa. Higher rates will severely reduce forage grass stands.

SENCOR DF should not be used on sand soils. In areas West of the Rocky Mountains, avoid using SENCOR DF on soils with a calcareous surface area, high levels of lime or sodium, or a pH greater than 8.2.

Weeds Controlled:

$\frac{1}{3}$ to $\frac{1}{2}$ pound SENCOR DF/acre (Idaho, Oregon, and Washington Only)
Chickweed, common *(Cerastium vulgatum)*

$\frac{1}{2}$ to $\frac{2}{3}$ pound SENCOR DF/acre
Cheat *(Bromus secalinus)*
Downy Brome *(Bromus tectorum)*
Japanese Brome *(Bromus japonicus)*
Rescuegrass *(Bromus catharticus)*

$\frac{2}{3}$ to $1\frac{1}{3}$ pounds SENCOR DF/acre

Broadleaves:
Fleabane, Rough *(Erigeron strigosus)*
Flixweed *(Descurainia sophia)*
Henbit *(Lamium amplexicaule)*
Kochia *(Kochia scoparia)*
Lambsquarters, Common *(Chenopodium album)*
Marestail (Horseweed) *(Hippuris vulgaris)*
Meadow Salsify *(Tragopogon pratensis)*
Mustard, Blue *(Chorispora tenella)*
Mustard, Jim Hill (tumble) *(Sisymbrium altissimum)*
Mustard, Tansy *(Descurainia pinnata)*
Pepperweed, Virginia *(Lepidium virginicum)*
Pennycress, Field *(Thlaspi arvense)*
Pigweed, Redroot *(Amaranthus retroflexus)*
Prickly Lettuce *(Lactuca serriola)*
Shepherdspurse *(Capsella bursa-pastoris)*
White Cockle *(Melandrium album)*
Wild Buckwheat *(Polygonum convolvulus)*
Yellow Rocket *(Barbarea vulgaris)*

Grasses:
Foxtail, Green *(Setaria viridis)*
Little Barley *(Hordeum pusillum)*
Smooth Brome *(Bromus inermis)*
Wild Oats *(Avena fatua)*

$1\frac{1}{3}$ pounds SENCOR DF/acre

Broadleaves:
Chickweed, Mousear *(Cerastium vulgatum)*
Dandelion *(Taraxacum officinale)*
Ragweed, Common *(Ambrosia artemisiifolia)*

Grasses:
Barnyardgrass *(Echinochloa crus-galli)*
Bluegrass *(Poa annua)*
Foxtail Barley *(Hordeum jubatum)*

Weeds Partially Controlled: At the rate of $1\frac{1}{3}$ pounds/acre SENCOR DF may be used to reduce the competition from curly dock *(Rumex crispus)*. At $\frac{2}{3}$ to $1\frac{1}{3}$ pounds/acre, SENCOR DF may be used to reduce the competition of German moss or knawel *(Scleanthus annus)*.

ALFALFA AND SAINFOIN (California Only)
(Including Mixed Stands With Grasses)

SENCOR DF is recommended for use in aerial or ground spray equipment as a broadcast surface application to dormant established crops of alfalfa and sainfoin.

APPLICATION: SENCOR DF is recommended for use in aerial or ground spray equipment as a broadcast surface application to dormant established crops of alfalfa and sainfoin for control of certain grass and broadleaf weeds. Do not apply SENCOR DF after growth begins in the Spring or before growth ceases in the Fall. Do not apply to either alfalfa or sainfoin during the first growing season after seeding.

For information on applying SENCOR DF in fluid fertilizer solutions to alfalfa, refer to the appropriate section of the SENCOR DF label.

For information on commercial impregnation and application of SENCOR DF on dry bulk fertilizer, refer to the appropriate section of the SENCOR DF label.

Weeds Controlled:

$\frac{1}{2}$ to $\frac{2}{3}$ pounds SENCOR DF/acre

Cheatgrass (downy brome)

⅔ to 1⅓ pounds SENCOR DF/acre

Broadleaves:		Grasses:
Chickweed, Common *(Cerastium vulgatum)*	Mustard, Tansy *(Descurainia pinnata)*	Smooth Brome *(Bromus inermis)*
Flixweed *(Descurainia sophia)*	Pepperweed, Virginia *(Lepidium virginicum)*	Wild Oats *(Avena fatua)*
Henbit *(Lamium amplexicaule)*	Shepherdspurse *(Capsella bursa-pastoris)*	
Kochia *(Kochia scoparia)*	White Cockle *(Melandrium album)*	
Meadow Salsify *(Tragopogon pratensis)*	Wild Buckwheat *(Polygonum convolvulus)*	
Mustard, Blue *(Chlorispora tenella)*	Yellow Rocket *(Barbarea vulgaris)*	

1⅓ pounds SENCOR DF/Acre

Broadleaves:		Grasses:
Dandelion *(Taraxacum officinale)*	Barnyardgrass *(Echinochloa crus-galli)*	Foxtail Barley *(Hordeum jubatum)*
	Bluegrass *(Poa annua)*	

RECOMMENDED BROADCAST APPLICATIONS

CROP	SENCOR DF Pounds/Acre	REMARKS
Alfalfa and Sainfoin (California only)	½ to 1⅓	Select the proper dosage according to weeds known to be present in the field to be treated. Apply specified dosage in 20 to 40 gallons of water per acre with ground spray equipment or 3 to 10 gallons of water per acre with aerial spray equipment fitted with nozzles suitable for broadcast applications of herbicides. Treat only dormant established crops of alfalfa and sainfoin. Injury may occur to alfalfa if SENCOR DF is applied earlier than 12 months after seeding. Do not apply after Spring growth begins or before growth ceases in the Fall. Do not graze or harvest within 28 days after application. At the 1⅓ pound/acre rate, SENCOR DF may be used for suppression of curly dock.

FOR USE ON MIXED STANDS OF ALFALFA AND GRASSES: Rates of ⅔ to 1⅓ pounds of SENCOR DF per acre will provide partial reduction of forage grass stands. These rates may be used to reduce forage grass stands to prevent crowding out of alfalfa. Higher rates will severely reduce forage grass stands.

Do not apply with aerial spray equipment when wind speed is greater than 10 mph. Do not apply when weather conditions favor spray drift and/or when sensitive cool season crops, such as cole crops, onions, peas, or strawberries, are present in adjacent fields. Applications should not be made when weather conditions favor spray drift, especially in areas where wheat is growing on coarse textured soils in adjacent fields, or injury may occur.

ALFALFA

SENCOR plus Gramoxone Tank Mix

(Colorado, Idaho, Montana, Nevada, Oregon, Utah, Washington and Wyoming)

APPLICATION: SENCOR DF plus Gramoxone tank mix application is recommended for use, during the dormant season, in aerial or ground spray equipment as a broadcast surface application to established (at least 1 year old) alfalfa for the control of certain grass and broadleaf weeds. Do not apply SENCOR/Gramoxone tank mix to regrowth (after grazing or cutting) that is more than 2 inches tall. Apply once per season. Do not apply following cuttings during growing season. Use a minimum of 10 gallons of water per acre with aerial spray equipment and a minimum of 20 gallons of water per acre with ground spray equipment. Add Ortho X-77 Spreader (non-ionic) at 1 to 2 pints per 100 gallons of spray solution.

WEEDS CONTROLLED: SENCOR DF (½ to 1 pound/acre) plus Gramoxone Super (2 to 2⅔ pints/acre) tank mix application will control established weeds. Gramoxone controls weeds by contact activity.

Bluegrass	Chickweed
Henbit	Downy Brome
Cheat	Japanese Brome
Rescuegrass	Shepherdspurse

⅓ to ½ pound of SENCOR per acre may be used for common chickweed only on sandy soils.
Use SENCOR DF at ⅔ 1 pound/acre for control of the following weeds:

Ryegrass	Kochia	Prickly Lettuce
Sow Thistle	Meadow Salsify	Little Barley
Groundsel	Pepperweed	Redroot Pigweed
Smooth Brome	Jim Hill Mustard	Common Lambsquarters
Wild Oats	Green Foxtail	Marestail (Horseweed)
Blue Mustard	Yellow Rocket	Field Pennycress
Tansy Mustard	White Cockle	Rough Fleabane
Flixweed	Wild Buckwheat	

RECOMMENDED BROADCAST APPLICATIONS

DOSAGE (Pounds or Pints/Acre)	REMARKS
SENCOR DF ½ to 1 pound plus Gramoxone Super 2 to 2⅔ pints	Apply specified dosages of SENCOR and Gramoxone in at least 10 gallons of water per acre with aerial equipment, or at least 20 gallons of water per acre with ground equipment. Do not apply this tank mix to alfalfa regrowth (after grazing or cutting) if more than 2 inches tall. For best weed control, apply when broadleaf weeds and grasses are 1-6 inches tall and are actively growing. Care should be taken to avoid overlaps. Use lower rates of SENCOR on sandy loam or loamy sand soils. Reduced weed control may occur when extended dry conditions follow application of SENCOR. Crop injury may occur if alfalfa is under stress conditions such as diseases, insect infestations, drought or winter injury or if SENCOR is applied to alfalfa earlier than 12 months after seeding.

FOR USE ON MIXED STANDS OF ALFALFA AND GRASSES: Rates of ⅔ to 1 pound of SENCOR DF per acre will provide partial reduction of forage grass stands. These rates may be used to reduce forage grass stands to prevent crowding out of alfalfa.
Do not graze or harvest within 42 days after application.

In areas west of the Rockies, avoid the use of SENCOR on soils with calcareous surface, soils with high levels of lime or sodium, and with a pH greater than 8.2.

Do not apply when weather conditions favor spray drift. Aerial application should not be made when wind speed is greater than 10 mph.

Do not use on sand soil.

Refer to the Gramoxone label for additional directions, weed species controlled and precautions.

ALFALFA

Post Dormant Application of SENCOR Impregnated on Dry Fertilizer Only

SENCOR DF may be applied after dormancy has broken, but prior to three inches of new alfalfa shoot growth, only when impregnated on dry fertilizer in Connecticut, Illinois, Indiana, Iowa, Kansas, Kentucky, Michigan, Minnesota, Nebraska, New York, Ohio, Pennsylvania, South Dakota and Wisconsin. Apply at rates of 1 to 1⅓ pounds per acre as directed on the label for application during dormancy. Apply only when alfalfa foliage is dry or crop injury may occur. When using this application method, do not harvest or graze treated alfalfa for 60 days after application.

ALFALFA

Non-Dormant, Non-Winter Hardy Varieties

(Arizona only)

SENCOR is recommended as a broadcast surface application to established crops of non-dormant alfalfa varieties for preemergence and postemergence control of certain winter annual weeds following either a Fall or Winter sheep grazing/green-chop harvest.

Weeds Controlled:

Mouse Barley	Shepherdspurse
Nettleleaf Goosefoot	Littleseed Canarygrass
Prickly Lettuce	Lambsquarters
Field Pepperweed	Little Mallow (Cheeseweed) (Malva)
Silversheath Knotweed	
London Rocket (Mustard)	Spiny Sowthistle

RECOMMENDED BROADCAST APPLICATIONS

CROP	SENCOR DF Pounds/Acre	REMARKS
Alfalfa Non-Dormant Non-Winter Hardy Varieties	1/2 to 2/3	Apply specified dosage by aerial or ground spray equipment in 7 to 40 gallons of water per acre. Treat established alfalfa stubble after Fall or Winter sheep grazing or green-chop harvest and prior to the time regrowth is 2" tall. Alfalfa foliage present at time of application can exhibit yellowing. Injury may occur to alfalfa in areas of high salt concentration where the crop is stunted and/or has a poorly developed root system, or if alfalfa is under stressed growing conditions such as diseases, insect infestations, or drought. For most effective postemergence weed control, treatment should be made before weeds are 2" tall or before leaf rosettes are 2" wide. For maximum control, rainfall (1/4" or more) or irrigation is necessary within 30 days of treatment, however, do not flood irrigate within 2 days after treatment. Use 1/2 pound SENCOR on sand soil when only mustard, goosefoot, lambsquarters, or canarygrass are the weeds to be controlled. Do not apply earlier than 6 months after seeding. Do not graze or harvest within 28 days after application.

Special Precautions:

Maintain continuous mechanical agitation in the spray tank to insure a uniform spray mixture.

Do not apply with aerial spray equipment when wind speed is greater than 10 mph. Do not apply when weather conditions favor spray drift and/or when sensitive cool season crops, such as cole crops, onions, peas or strawberries are present in adjacent fields. Applications should not be made when weather conditions favor spray drift especially in areas where wheat is growing on coarse textured soils in adjacent fields, or injury may occur.

ASPARAGUS (Established)

SENCOR DF is recommended for use in aerial or ground spray equipment or sprinkler irrigation (center pivot, lateral move, or solid set) systems as a single preemergence broadcast application or as a split application consisting of a preemergence broadcast application followed by a post harvest broadcast application. Refer to the "GENERAL INFORMATION" section of the SENCOR DF label for direction.

Weeds Controlled:

SENCOR DF, applied to established asparagus according to directions, will effectively control:

Broadleaves:		Grasses:
Chickweed, Common *(Stellaria media)*	(Ragweed, Common *(Ambrosia artemisiifolia)*	Crabgrass *(Digitaria spp.)*
Jimsonweed *(Datura stramonium)*	Smartweed, Pennsylvania *(Polygonum pensylvanicum)*	Foxtails *(Setaria spp.)*
Lambsquarters *(Chenopodium album)*	Sorrel, Red *(Rumex acetosella)*	Sandbur, Field *(Cenchrus pauciflorus)*
Pigweed, Redroot *(Amaranthus retroflexus)*	Velvetleaf *(Abutilon theophrasti)*	

RECOMMENDED BROADCAST APPLICATIONS

CROP	SENCOR DF Pounds/Acre	REMARKS
Asparagus	1 1/3 to 2 2/3	**PREEMERGENCE APPLICATION ONLY:** Make a single surface application in early Spring before asparagus spears or ferns emerge. If the field is to be disked, apply SENCOR DF after disking but before the crop emerges. Use the lower rate for control of the broadleaf weeds listed above. Use the higher rate in fields with a history of severe infestations of grasses and for maximum residual control. Do not apply within 14 days of harvest.
	2/3 to 1 1/3 preemergence plus 1 1/3 to 2 post harvest	**SPLIT APPLICATION PREEMERGENCE AND POST HARVEST** **Preemergence Application:** Apply before asparagus spears or ferns emerge. If the field is to be disked, apply after disking but prior to crop emergence. Do not apply within 14 days of harvest. **Post Harvest Application:** Apply after last harvest of the season but prior to emergence. The lower combination rates may be used for control of common ragweed, lambsquarters, redroot pigweed, and red sorrel. Use the higher combination rates for other weeds listed or in fields with severe grass infestations or for maximum post harvest control of emerged weeds.

IMPORTANT: The total amount of SENCOR DF applied in one crop season may not exceed 2 2/3 pounds per acre.

Special Precautions (Asparagus)

Do not use on newly seeded asparagus nor on young plants during the first growing season after setting crowns.

DO NOT APPLY POST HARVEST APPLICATIONS UNTIL AFTER THE LAST HARVEST OF SPEARS.

CARROTS

Special Conditions of Sale Provision for Use on Carrots: The following directions for use were developed under the direction of IR-4 (government minor crops use program). As such, the testing was done independently from the testing program of Mobay Corporation. Buyer is advised that Mobay makes no assurances regarding satisfaction with the product and all risks of crop injury or product performance are assumed by the Buyer.

Apply SENCOR DF herbicide with ground equipment as specified below under "Recommended Applications". For effective control of broadleaf weeds with postemergence applications, apply SENCOR before weeds are 1 inch in height or diameter. Thorough spray coverage is essential for adequate weed control.

Do not use air blast or other high pressure spray equipment to make postemergence applications of SENCOR. Refer to the appropriate section of the SENCOR DF label for additional information regarding spray equipment dilution rates, mixing, spray cleanup, restrictions, container disposal and cautions.

Refer to "MIXING" under the "GENERAL INFORMATION" section in the front of the SENCOR DF label.

For specific application information see "General Information" and "Application" sections in the front of the SENCOR DF label.

Weeds controlled: SENCOR DF applied to carrots according to directions will effectively control:

Carpetweed *(Mullugo verticillata)*	Pigweed, Redroot *(Amaranthus retroflexus)*
Galinsoga *(Galinsoga parviflora)*	Pigweed, Smooth *(Amaranthus hybridus)*
Horseweed *(Conyza canadensis)*	Prickly Lettuce *(Lactuca serriola)*
Lambsquarters, Common *(Chenopodium album)*	Sheperdspurse *(Capsella bursa-pastoris)*
Mustard, wild *(Sinapis arvensis)*	Pineappleweed *(Matricaria matricarioides)*

RECOMMENDED BROADCAST APPLICATIONS

CROP	SENCOR DF Pounds/Acre	REMARKS
Carrots (except California)	1/3	Apply specified dosage per acre as a broadcast spray over the tops of carrot plants. Application should be made after carrots have formed 5-6 true leaves but before weeds are 1 inch in height or diameter. If needed, a second application may be made after an interval of at least 3 weeks. Applications may be made up to 60 days of harvest. Do not apply to carrots grown for seed.

IMPORTANT: The total amount of SENCOR DF applied in one crop season must not exceed 2/3 pound per acre.

Special Precautions:

Do not apply within 3 days after periods of cool, wet or cloudy weather or crop injury will occur.

Do not apply SENCOR within 3 days of any other chemical unless specified on the SENCOR DF label.

Do not apply on very hot days or excessive crop injury will result.

Do not apply until carrots have at least 5-6 true leaves. Earlier application will result in excessive crop damage.

Crop injury or delayed maturity may result from applications of SENCOR if carrots are growing under stress conditions such as periods of drought or cool, wet and cloudy weather preceding application.

Following an application of SENCOR DF, chlorosis (yellowing) and burning of the leaf tissue may occur.

For newly introduced varieties of carrots with unknown tolerance to SENCOR, treat only a small area to determine if SENCOR can be used without injury to the crop.

FIELD CORN

(Iowa, Kansas, Missouri, Nebraska, and South Dakota Only)

SENCOR DF herbicide is recommended for the added effect on the control of certain broadleaf weeds in corn when applied preemergence in a tank mix combination with labeled rates of Lasso plus atrazine, Lasso plus Bladex®, Dual plus atrazine, or Dual plus Bladex. Refer to the appropriate product labels for weeds controlled by these combinations.

SENCOR DF in each of the tank mix combinations will control the following weeds:

Lambsquarters *(Chenopodium album)*	Pigweed, Redroot (carelessweed) *(Amaranthus retroflexus)*	Smartweed, Pennsylvania *(Polygonum pensylvanicum)*	Velvetleaf *(Abutilon theophrasti)*

MIXING: Select proper dosages based on soil type as shown below. Fill the spray tank $1/4$ to $1/3$ full with clean water and add SENCOR DF to the spray tank first. Mix thoroughly and then add atrazine or Bladex to the spray tank. Add water to fill spray tank to desired level. Mix thoroughly and then add Lasso or Dual to the spray tank. Maintain sufficient agitation during mixing and spraying operations to insure a uniform spray mixture.

APPLICATION: For specific application information see the "GENERAL INFORMATION" section in the front of the SENCOR DF label.

RECOMMENDED BROADCAST APPLICATIONS
RATES PER ACRE:

SOIL TEXTURE	SENCOR DF	Lasso 4 EC	or Dual 8 EC	Atrazine 4L or 80 WP	or Bladex 4L or 80 WP
MEDIUM (Loam, silt loam, silt, sandy clay, sandy clay loam)	$1/3$ lb.	1.5 to 2.0 qts.	1.5 to 2.0 pts.	0.75 to 1.5 qts. of 4L or 0.95 to 1.9 lbs. of 80 WP	1.25 to 2.5 qts. of 4L or 1.5 to 3.1 lbs. of 80 WP
FINE (Silty clay, silty clay loam,[1] clay, clay loam)	$1/3$ lb.	2.0 to 2.5 qts.	2.0 to 2.5 pts.	0.75 to 1.5 qts. of 4L or 0.95 to 1.9 lbs. of 80 WP	1.5 to 3.0 qts. of 4L or 1.75 to 3.75 lbs. of 80 WP

[1] Silty clay loam soils are transitional soils and may be classified as medium textured soils in some regions of the U.S.

Follow the Lasso, Dual, Bladex, or atrazine labels for use rates on specific soil textures and organic matter content. Not recommended for use on muck soils as reduced weed control may result.

Special Precautions: (Field Corn)

Not for use on coarse textured soils or soils containing less than 2% organic matter content.

Do not use on soils having a calcareous surface or high in lime or sodium and/or a pH of 7.0 or higher.

Corn seed should be planted $1\,1/2$ inches deep.

Use in hybrid seed corn production fields only if both parents are known to be tolerant to SENCOR DF and other herbicides used.

Observe all cautions and limitations on labeling of all products used in mixtures.

LENTILS AND DRY FIELD PEAS

(Idaho, East of the Cascades in Oregon & Washington, and Montana)

SENCOR herbicide is recommended as a preemergence and postemergence application for the suppression of certain broadleaf weeds in lentils and dry field peas.

WEEDS SUPPRESSED*

Common Chickweed**	Lambsquarters
Dog Fennel	Shepherdspurse**
Field Pennycress	Wild Mustard
Henbit**	

* Suppression is a reduction in weed size and growth compared to a non-treated area in the same field.

** Preemergence application only.

PREEMERGENCE APPLICATION: A single preemergence application at $1/3$ to $1/2$ lb. of SENCOR DF per acre in 20 to 40 gallons of water per acre with ground spray equipment may be made per crop year. Apply SENCOR before or after planting, BUT BEFORE CROP GERMINATES. Thorough incorporation, either by rainfall or by mechanical means, is essential for weed suppression. Under dry conditions, incorporate SENCOR DF into the top 1 to 2 inches of soil with spike harrows, or similar shallow incorporation equipment, then cross harrow to insure uniform soil incorporation. Where soil surface is moist at the time of application and rain follows before weed emergence, a broadcast application should provide adequate weed suppression.

Use the higher rate on fine-textured soils (high in clay or organic matter) and in fields with a history of high weed populations.

POSTEMERGENCE APPLICATION: Use $1/6$ to $1/4$ lb. of SENCOR DF per acre on **lentils** and **spring peas.** On **winter peas,** use $1/4$ to $1/3$ lb. of SENCOR DF per acre. For suppression of dog fennel, use $1/3$ lb. SENCOR DF per acre. Apply specified dosage in 20 to 40 gallons of water per acre and not over 40 psi with ground spray equipment as a broadcast spray when weeds are small (less than 2" in height or diameter) and before crop is 6 inches tall.

One postemergence application may be made per season.

Temporary chlorosis of crop may occur. There is an added risk of crop injury if a postemergence application is made following a previous metribuzin application.

Do not apply over very moist soils or wet crop foliage. Do not apply postemergence applications within 3 days after periods of cool, wet, or cloudy weather or crop injury may occur.

Do not apply within 24 hours of treatment with other pesticides.

SPECIAL PRECAUTIONS (all applications):

Crop injury may result if crop is under stress conditions caused by cold weather, low fertility, disease or insect damage.

Crop injury may also result if application is followed by heavy rain.

Do not use on coarse-textured soils, sandy soils or any soil with less than 1.5% organic matter.

Do not apply to "Estin" lentils.

Do not use on clay knobs or poorly covered subsoils.

Do not apply on shallow seedings less than 2" deep (preemergence only).

Do not apply within 50 days of harvest of peas, or within 75 days of harvest of lentils. Do not graze or feed treated vines to livestock within 40 days after application.

Maintain continuous spray tank agitation to keep material in suspension. Avoid overlapping and shut off spray booms while turning, slowing or stopping, or crop injury will occur.

NOTE: This treatment may cause some chlorosis or minor necrosis. Because lentil and pea varieties may vary in their susceptibility to SENCOR DF, determining crop tolerance prior to adoption as a field scale practice is suggested to prevent possible injury.

For additional precautions, restrictions, limitations, and sprayer clean-up information refer to the appropriate sections of the SENCOR DF label.

SUGARCANE (Hawaii only)

SENCOR, a selective herbicide, is effective as a preemergence and an early postemergence broadcast application for control of certain grass and broadleaf weeds. When applied as a spot treatment, it also provides excellent control of perennial grasses and broadleaves.

Ground Application: SENCOR DF should be mixed by filling the spray tank half full of clean water. Then add the recommended amount of SENCOR DF to suit the total tank capacity and the rate of application per acre (preferably 25 to 35 gallons per acre). Complete filling the tank and maintain sufficient agitation during mixing and spraying to insure a uniform spray mixture.

Aerial Application: SENCOR DF is recommended for use in aerial spray equipment as a preemergence or postemergence application to irrigated sugarcane. Aerial spray equipment should be calibrated to apply the proper amount of SENCOR DF in 7 to 10 gallons of spray mixture per acre.

SENCOR DF applied preemergence or postemergence to the sugarcane as a broadcast spray or spot treatment will effectively control the following when weeds are less than 3 inches in height.

Weeds Controlled in Irrigated and Non-irrigated Sugarcane:

Broadleaves:

- **Amaranth, Spiny** *(Amaranthus spinosus)*
- Euphorbia, Wild *(Euphorbia spp.)*
- Fireweed *(Erechtites hieracifolius)*
- Floras paintbrush *(Emilia sonochifolia)*
- Spurge, Garden *(Euphorbia hirta)*
- Spurge, Graceful *(Euphorbia glomerifera)*

Grasses:

- Crabgrass *(Digitaria spp.)*
- Guineagrass *(Panicum maximum)*
- Plushgrass *(Chloris radiata)*
- *Ricegrass* *(Oryzopsis hymenoides)*
- Wiregrass *(Eleusine indica)*

Weeds Controlled in Irrigated Sugarcane Only:

Broadleaves:

- Amaranth Spleen *(Amaranthus dubius)*
- Haole Koa *(Leucaena leucocephala)*
- *Hialoa* *(Waltheria americana)*
- Hilahila *(Mimosa pudica)*
- Purslane, Common *(Portulaca oleracea)*
- *Rattlepod* *(Crotalaria spectabilis)*

Grasses:

- Alexandergrass *(Brachiaria plantaginea)*
- Bristly foxtail *(Setaria verticillata)*

Weeds Controlled in Non-irrigated Sugarcane Only:

Broadleaves:

- Ageratum *(Ageratum conyzoides)*
- Richardia *(Richardia brasiliensis)*
- Tarweed *(Cuphea carthagenesis)*

RECOMMENDED BROADCAST APPLICATIONS

CROP	SENCOR DF Pounds/Acre	REMARKS
Sugarcane (Hawaii only)	2⅔ to 5⅓ (non-irrigated) 5⅓ to 8 (irrigated)	**PREEMERGENCE:** (Irrigated and non-irrigated sugarcane) Apply specified dosage per acre as a broadcast spray to the soil surface. Applications should be made within two weeks after planting prior to cane emergence or shortly after emergence (spike stage). OR **EARLY POSTEMERGENCE:** (Irrigated and non-irrigated sugarcane) Apply specified dosage per acre as a broadcast spray over the cane. Application may be delayed as long as 4 to 6 weeks after planting provided weeds are less than 3 inches in height.
	2⅔ to 5⅓	OR **POSTEMERGENCE:** Apply specified dosage per acre as a broadcast spray to control weeds prior to "close in" time when cane shades out the weed growth.
	3⅓ to 6⅔	**SPOT TREATMENT:** Apply specified dosage in 30 to 50 gallons of finished spray per acre. Spot treatments may be used to control weeds in missed areas, corners of fields, or areas of hard to control weeds.

NOTE: Do not apply more than 10⅔ pounds of SENCOR DF (8 lb. active ingredient)/acre per crop cycle regardless of the method of application. The last application may be made up to 17 months of harvest.

SPECIAL PRECAUTION:

Do not use treated foliage for feed or forage.

SUGARCANE (Louisiana and Texas Only)

Preemergence and postemergence applications of SENCOR DF with aerial or ground spray equipment are recommended for control of the following weeds in sugarcane in Louisiana and Texas:

Broadleaves:

- Amaranth, Spiny *(Amaranthus spinosus)*
- Bindweed, Field *(Convolvulus arvensis)*
- Chickweed *(Cerastium vulgatum)*
- Henbit *(Lamium amplexicaule)*
- Lambsquarters *(Chenopodium album)*
- *London rocket* *(Sisymbrium irio)*
- Marestail *(Conyza canadensis)*
- Mustard, Wild *(Brassica kaber)*
- Pigweeds *(Amaranthus spp.)*
- Purslane *(Portulaca oleracea)*
- Sowthistle *(Sonchus spp.)*

Grasses:

- Broadleaf Signalgrass *(Brachiaria platyphylla)*
- Crabgrass *(Digitaria spp.)*
- Foxtails *(Setaria spp.)*
- Johnsongrass, Seedling *(Sorghum halepense)*
- Oats, Winter *(Avena spp.)*

RECOMMENDED APPLICATIONS

CROP	SENCOR DF Pounds/Acre	REMARKS
Sugarcane (Louisiana and Texas only)	2 to 4	**BROADCAST:** Apply specified dosage per acre using 20 to 30 gallons of water with ground equipment or 5 gallons of water with aircraft spray equipment. Apply as a broadcast spray during the Fall after planting or to the stubble after harvest. Make a second application early in the Spring.
	1 to 2	**BAND:** Apply specified dosage in 10 to 20 gallons of water per acre in a 30 to 36 inch band over the row during the Fall after planting or to the stubble after harvest. Make a second application early in the Spring.

Special Precautions: (Louisiana and Texas only)

Do not use treated foliage for feed or forage.

Use the higher rate on heavy clay soil and soil with a high percentage of organic matter. If necessary, a third application may be made in late Spring at layby. Do not apply within 60 days of harvest.

SUGARCANE (Florida Only)

Postemergence over-the-top or directed spray applications of SENCOR DF are recommended for the control of the following weeds in sugarcane in Florida.

Broadleaves:

- Amaranth, Spiny (seedling) *(Amaranthus spinosus)*
- Butterweed *(Cressleaf groundsel)* *(Senecio glabellus)*
- Cudweed *(Gnaphalium spp)*
- Purslane *(Portulaca oleracea)*

Grasses:

- *Crabgrass, large *(Digitaria sanguinalis)*
- Foxtail, bristlegrass *(Setaria magna)*
- Goosegrass *(Eleusine indica)*
- Panicum, broadleaf *(Panicum adspersum)*
- Signalgrass, Broadleaf *(Brachiaria platyphylla)*

* Best control is achieved when applications are made when this weed is less than 4" in diameter.

Recommended Broadcast Applications

CROP	SENCOR DF Pounds/Acre	REMARKS
Sugarcane	1⅓ to 2⅔	**GROUND APPLICATION:** SENCOR DF may be used in one or two applications with a minimum of 14 days between each application. Apply when weeds are less than 6 inches tall in 10 to 40 gallons of spray mixture per acre. Do not use more than 2⅔ pounds per acre in a single growing season. **POSTEMERGENCE BROADCAST OR BAND:** Apply over the top of stubble or plant cane while sugarcane is less than 14 inches tall. **POSTEMERGENCE DIRECTED SPRAY:** Apply to sugarcane that is a minimum of 14 inches tall and before row closing.

	1⅓ to 2	**AERIAL APPLICATION:** Apply when weeds are less than 4 inches tall in 5 to 10 gallons of spray mixture per acre. Apply to stubble or plant cane while the sugarcane is less than 14 inches tall.

SENCOR PLUS Atrazine TANK MIX: SENCOR DF may be used with atrazine as a preemergence or postemergence (before row closing) application to sugarcane. Rates for SENCOR DF are 1 to 2⅔ pounds/acre and atrazine 80% WP (4L) are 2½ to 5 lb/acre (2 to 4 quarts/acre). For additional information on precautions, instructions, limitations, application, and weeds controlled, refer to the SENCOR DF label and the atrazine label.

Special Precautions: (Florida only)

Do not use on sand soils.

Spray contact with sugarcane foliage may result in minor leaf margin chlorosis and/or necrosis.

Do not apply within 60 days of harvest.

Do not use treated crop for feed or forage.

Avoid spray overlaps or variations in application speed that may result in insufficient or excessive rates of application.

TOMATOES

Apply SENCOR herbicide with ground equipment to seeded and transplanted tomatoes as specified under "Recommended Applications" on the label.

For effective control of grasses and broadleaf weeds with postemergence applications, apply SENCOR DF before weeds are 1 inch tall. Thorough spray coverage on weed foliage is essential for adequate control with postemergence applications.

Do not use air blast or other high pressure spray equipment to make postemergence applications of SENCOR DF. Refer to the appropriate section of the label for additional information regarding spray equipment, dilution rates, mixing, sprayer cleanup, restrictions, container disposal and cautions.

Refer to "Mixing" under the "GENERAL INFORMATION" section in the front of the label.

For specific application information see under the "GENERAL INFORMATION" section in the front of the SENCOR DF label.

Weeds Controlled:

Preplant Incorporated Applications—Transplant Tomatoes Only

Broadleaves:		Grasses:
Galinsoga *(Galinsoga spp.)*	*Pigweed, Redroot *(Amaranthus retroflexus)*	*Goosegrass *(Eleusine indica)*
Lambsquarters *(Chenopodium album)*	*Purslane, Common *(Portulaca oleracea)*	

Preplant incorporated applications applied as directed will suppress foxtails, panicums and barnyardgrass.

SENCOR plus Treflan Tank Mix

This tank mix combination applied preplant incorporated as directed on the label will control the weeds listed above plus those weeds listed on the Treflan label.

POSTEMERGENCE APPLICATIONS—ESTABLISHED TOMATOES

For effective control of weeds with postemergence applications, apply SENCOR DF before weeds are 1 inch tall.

Broadcast Sprays—⅓ to ½ lb. of SENCOR DF/acre

Directed Sprays—⅔ to 1⅓ lbs. of SENCOR DF/acre

Broadleaves:		Grasses:
Carpetweed *(Mollugo verticillata)*	Pigweeds *(Amaranthus spp.)*	*Foxtail, Yellow *(Setaria glauca)*
Fumitory *(Fumaria officinalis)*	Purslane *(Portulaca oleracea)*	Goosegrass *(Eleusine indica)*
Galinsoga *(Galinsoga spp.)*	*Ragweed, Common *(Ambrosia artemisiifolia)*	(Plus weeds listed under "Broadcast Sprays")
*Jimsonweed *Datura stramonium*	Smartweed, Pennsylvania *(Polygonum pensylvanicum)*	
*Ladysthumb *(Polygonum persicaria)*	Toadflax *(Linaria spp.)*	
Lambsquarters *(Chenopodium album)*	*Velvetleaf *(Abutilon theophrasti)*	
Mustard, Wild *(Brassica kaber)*		

* For optimum control of these weeds, use the highest rate recommended on the label for the type of application to be made. Repeat postemergence applications may be needed for best control.

Postemergence applications as directed on the label will suppress barnyardgrass and crabgrass when these weeds are less than 1 inch tall.

RECOMMENDED BROADCAST APPLICATIONS

CROP	SENCOR DF *Pounds/Acre	REMARKS
Tomatoes	⅓ to ⅔	**PREPLANT INCORPORATED—TRANSPLANT TOMATOES ONLY (EXCEPT CALIFORNIA):** Apply specified dosage in 10 to 40 gallons of water per acre as a broadcast spray to the soil surface immediately before transplanting. Incorporate to a depth of 2 to 4 inches with equipment capable of uniformly mixing the chemical into the soil. This application may be made alone or in a tank mix combination with Treflan EC. Refer to the Treflan label for specific rate of application and for additional precautions and restrictions for tomatoes.
	⅓ to ½	**POSTEMERGENCE BROADCAST SPRAY—ESTABLISHED TOMATOES:** Apply specified dosage in 20 to 75 gallons of water per acre as a broadcast spray or apply in ¼ to ¾ inch of water (¼ to ½ inch of water on sandy soils) per acre as a continuous injection in center pivot and lateral move systems, or in the last 15 to 30 minutes of set in permanent solid set sprinkler systems over the plants in single or multiple applications with a minimum of 14 days between treatments. For transplanted tomatoes do not apply within 14 days of transplanting. Transplants **must** have recovered from transplant shock and new growth evident. Do not apply to established tomatoes within 24 hours of application of other pesticides. Do not tank mix with other pesticides. (See "Special Precautions" below.)
	⅔ to 1⅓	**POSTEMERGENCE DIRECTED SPRAY—ESTABLISHED TOMATOES:** Apply specified dosage in 20 to 75 gallons of water per acre as a directed spray in single or multiple applications with a minimum of 14 days between treatments. Avoid contacting tomato foliage with spray. This method of treatment is recommended for use in fields with a history of severe weed pressure or in fields infested with hard to control weeds. For transplanted tomatoes, do not apply within 14 days of transplanting. Transplants **must** have recovered from transplant shock and new growth evident. Do not apply to established tomatoes within 24 hours of application of other pesticides. (See "Special Precautions" below.) When banding see the appropriate section in the front of the label.

* Use the higher rate in fields with a history of severe weed pressure and for maximum residual weed control.

Special Precautions (Tomatoes)

Do not apply more than a total of 1⅓ pounds SENCOR DF per crop season.

Do not apply the total amount of 1⅓ pounds SENCOR DF within a time span of less than 35 days, except in the case of directed sprays.

Allow at least 14 days between applications, regardless of dosage or method of application.

Do not apply within 7 days of harvest.

Do not apply within 3 days after periods of cool, wet or cloudy weather, or crop injury will occur.

Do not use hot caps on tomatoes within 7 days before or at any time after application of SENCOR DF.

Do not treat seeded tomatoes until plants have reached the 5 to 6 leaf stage.

Crop injury or delayed maturity may result from broadcast or directed spray applications if tomatoes are growing under stress conditions such as periods of drought or cool, wet and cloudy weather preceding application.

For newly introduced tomato varieties with unknown tolerance to SENCOR, treat only a small area to determine if SENCOR DF can be used without injury to the crop.

DO NOT USE SENCOR DF ON TOMATOES IN KERN COUNTY, CALIFORNIA.

CEREALS (Barley and Wheat)

For the control of certain grass and broadleaf weeds, SENCOR herbicide is recommended as a single postemergence application in barley (Winter and Spring) and Winter wheat (see the appropriate section of the label for geographic limitations).

SENCOR DF may be used in cereals (barley and wheat) as follows:

1. Barley (Spring and Winter)—Only East of the Cascade Mountains in Idaho, Oregon, Montana, Utah, and Washington.
2. Winter wheat in Idaho, Oregon, Montana, Utah & Washington.
3. Winterwheat in Idaho, Montana, Nevada, Oregon, Utah, and Washington—Early Application (2-lead stage).
4. Dryland (Non-irrigated) Winter wheat in Kansas, Oklahoma & Texas.
5. Dryland (Non-irrigated) Winter wheat only (Arkansas, Louisiana, Mississippi & Tennessee).
6. Dryland (Non-Irrigated) Winter Wheat only (Georgia).

Mixing: See the "Mixing" section under "GENERAL INFORMATION" in the front of the label for specific mixing procedures. Do not tank mix SENCOR DF with pesticides other than those specified below. When tank mixing, carefully follow the instructions on the label. Refer to the other product label registered for the specific use intended for directions, rates, and weeds controlled.

Application: Refer to the "GENERAL INFORMATION" section in the front of the label for detailed information on the application of SENCOR DF.

Apply SENCOR DF when wheat stand is uniform in height. For best results apply SENCOR DF when weeds are small (less than 1 to 1½ inches in diameter or height). **In the Fall** apply after wheat has a minimum of 3 tillers and 4 secondary roots at least 2 inches in length, but before Winter dormancy and wheat jointing occurs. **In the Spring** apply after wheat has recovered from Winter dormancy and after wheat has fully tillered and has developed secondary roots at least 2 inches in length, but before jointing.

Apply SENCOR DF only once per crop season to wheat or barley.

Grazing Treated Wheat

Wheat fields treated with SENCOR DF may be grazed 14 days following application.

Special Precautions (Cereals)

Crop injury may occur on barley or wheat when SENCOR DF is used under the following conditions:

1. When applied while the crop is under stress such as Winter kill, frost damage, disease, drought or excessive moisture, or when these conditions follow the application.
2. When applied in combination with fluid fertilizer.
3. When applications are made prior to the growth stage specified on the label.
4. When soils are high in lime or sodium, have a pH greater than 7.7, or are calcareous soils.
5. When proper application instructions are not followed.

Weed control may not be observed for 2 to 4 weeks after application; under very dry conditions, weed control may not be apparent for 4 to 6 weeks. Moisture (at least ½") is needed within 2 to 3 weeks after application to move SENCOR into the weed root zone. Lack of adequate moisture after application may result in poor control.

Do not exceed the rates specified on the label.

BARLEY (Spring and Winter)

FOR USE ONLY EAST OF THE CASCADE MOUNTAINS IN IDAHO, OREGON, UTAH, MONTANA AND WASHINGTON ON THE VARIETIES COMPANA, HANNCHEN, HECTOR, HUDSON, LUTHER, KAMIAK, KLAGES, OLYMPIC, PIROLINE AND STEPTOE.

A single postemergence application of SENCOR DF is recommended for control of certain grass and broadleaf weeds in Spring and Winter barley.

RECOMMENDED BROADCAST APPLICATIONS

SOIL TEXTURE	SENCOR DF Pounds/Acre % Organic Matter[1] 1 to 2	Over 2	REMARKS
COARSE (Sandy loam, loamy sand)	⅓	⅓ to ½	Barley should be seeded to obtain a uniform crop stand at application time. Apply after barley has fully tillered and developed secondary roots but before jointing and before weeds are 2 inches tall or have 2 inch diameter rosettes. Do not graze or harvest treated barley for feed before crop maturity.
MEDIUM (Loam, silt loam, silt, sandy clay, sandy clay loam)	⅓ to ½	½	
FINE (Silty clay, silty clay loam,[2] clay, clay loam)	½	½ to ⅔	

[1] Do not use on any soils containing less than 1% organic matter.

[2] Silty clay loam soils are transitional soils and may be classified as medium textured soils in some regions of the U.S.

Weeds Controlled in Barley:

Broadleaves:		Grasses:
Chickweed, Common *(Stellaria media)*	*Mustard, Blue or Purple *(Chorispora tenella)*	*Cheatgrass *(Bromus secalinus)*
Fiddleneck, Tarweed *(Amsinckia lycopsoides)*	Mustard, Jim Hill *(Sisymbrium altissimum)*	*Downy Brome *(Bromus tectorum)*
Henbit *(Lamium amplexicaule)*	*Mustard, Wild *(Brassica kaber)*	*Foxtails *(Setaria spp.)*
*Kochi *(Kochia scoparia)*	Pigweeds *(Amaranthus spp.)*	*Ryegrass, Annual *(Lolium multiflorum)*
Lambsquarters *(Chenopodium album)*		*Windgrass *(Apera sp.)*

* Use the highest rate recommended for soil texture and percent organic matter for the control of these weeds.

Hard to Control Weed:

Wild Oats *(Avena fatua)*

NOTE: Control of wild oats may be variable, ranging from poor to excellent depending on timing of application, stage of growth of wild oats at time of application, temperatures and soil moisture conditions following treatment.

SENCOR in Combination with Other Herbicides

SENCOR DF may be applied as directed above or may be tank mixed with 2,4-D, Banvel®, Bronate®, Buctril®, Finesse® or Igran®. Refer to the other product labels registered for this use for additional directions, rates and weed species controlled.

Observe all cautions and limitations on labelling of all products used in mixtures.

WINTER WHEAT

FOR USE ON WINTER WHEAT IN IDAHO, OREGON, UTAH AND WASHINGTON ON THE VARIETIES DAWS, GAINES, HATTON, HILL 81, HYSLOP, LUKE, McDERMID, MORO NuGAINES, PAHA, PECK, SPRAGUE, STEPHENS, WANSER, FARO, TYEE, AND WESTON AND IN MONTANA ON THE VARIETIES CENTURK, CHEYENNE AND WINALTA.

A single postemergence application of SENCOR DF is recommended for control of certain grass and broadleaf weeds in Winter wheat.

RECOMMENDED BROADCAST APPLICATIONS[1]

SOIL TEXTURE	SENCOR DF Pounds/Acre 0.75% to 1% Organic Matter[2]	1% to 2% Organic Matter	More than 2% Organic Matter
COARSE (Loamy sand, sandy loam)	⅓ to ⅖	⅓ to ⅖	⅓ to ⅖
MEDIUM (Loam, silt loam, silt, sandy clay, sandy clay loam)	⅓ to ⅖	⅖ to ½	⅖ to ⅗
FINE (Silty clay, silty clay loam,[3] clay, clay loam)	⅓ to ½	½ to ⅔	½ to ⅔

[1] Do not use on gravelly or thinly covered or exposed subsoil areas (clay knobs) as crop injury may occur.

[2] Do not use on any soils containing less than 0.75% organic matter.

[3] Silty clay loam soils are transitional soils and may be classified as medium textured soils in some regions of the U.S.

Weeds Controlled in Wheat:

Broadleaves:

Annual Polemonium (Jacob's Ladder) *(Polemonium micranthum)*
Chickweed, Common *(Stellaria media)*
Conical Catchfly *(Silene conica)*
Corncockle *(Agrostemma githago)*
Fiddleneck, Tarweed *(Amsinckia lycopsoides)*
Gromwell *(Lithospermum officinale)*
Henbit *(Lamium amplexicaule)*
Ivyleaf Speedwell *(Veronica hederifolia)*
Knotweed, Prostrate *(Polygonum aviculare)*
*Kochia *(Kochia scoparia)*
Lambsquarters *(Chenopodium album)*
Madwort (Catchweed) *(Asperugo procumbens)*
Mayweed (Dog Fennel) (Eupatorium capillifoium)
Miners Lettuce *(Claytonia perfoliata)*
*Mustard, Blue or Purple *(Chorispora tenella)*
*Mustard, Jim Hill (tumble) *(Sisymbrium altissimum)*
*Mustard, Wild *(Brassica kaber)*
Pennycress, Field *(Thlaspi arvense)*
Pineappleweed *(Matricaria matricarioides)*
Prickly Lettuce *(Lactuca serriola)*
Pigweeds *(Amaranthus spp.)*
Redstem Filaree *(Erodium cicutarium)*
*Russian Thistle *(Salsola iberica)*
Shepherdspurse *(Capsella bursa-pastoris)*

Grasses:

*Blackgrass (Black Twitchgrass) *(Alopecurus myosuroides)*
*Bluegrass, bublous *(Poa Bulbosa)*
*Cheatgrass *(Bromus secalinus)*
*Downy Brome *(Bromus tectorum)*
*Foxtails *(Setaria spp.)*
*Rescuegrass *(Bromus catharticus)*
*Ryegrass, Italian *(Lolium multiflorum)*
*Wild Barley *(Hordeum leporinum)*
*Windgrass *(Apera spica-venti)*

* Suppression only may occur.

For suppression of these weeds, use the highest rate recommended for soil texture and percent organic matter. Suppression is a reduction in weed size and growth compared to a non-treated area in the same field.

Hard-To-Control Weeds:

Broadleaves:

Cowcockle *(Vaccaria pyamidata)*
Mustard, Tansy *(Descurainia pinnata)*
Wild Buckwheat *(Polygonum convolvulus)*

Grasses:

Ripgut Brome *(Bromus diandrus)*
Wild Oats *(Avena fatua)*

NOTE: Control of these weeds may be variable, ranging from poor to excellent depending on timing of application, stage of growth at the time of application, temperatures, and soil moisture conditions following treatment.

SENCOR In Combination With Other Herbicides

SENCOR DF may be applied as a tank-mix with Glean®, 2,4-D, MCPA, Igran, Banvel, Bronate or Buctril. Refer to the other product labels registered for this use for additional directions, rates and weed species controlled. Observe all cautions and limitations on labeling of all products used in mixtures.

IRRIGATED WHEAT/BARLEY

Follow timing directions for dryland wheat/barley found at the beginning of this Cereals section. After application of SENCOR DF, do not apply more than 1/2" of water for the first irrigation; the maximum amount for each additional irrigation should not exceed 1".

SENCOR plus Igran Tank Mix Application

Apply SENCOR DF at 1/3 to 1/2 pound per acre and Igran 80 WP at 1/2 to 3/4 pound per acre according to label directions. The rate of each product depends on soil type. Do not use Igran on sand or loamy sand soils. Use lower rate of Igran on sandy loam soils.

Broadleaf Weed Control Only:

Where only broadleaf control is needed, SENCOR DF applied at the rates listed below in a tank mix combination with Brominal, Brominal Plus, ME4 Brominal, 3 + 3 Brominal, Bronate or Buctril can be used to control wild buckwheat, in addition to all the weeds listed on these labels.

RECOMMENDED BROADCAST APPLICATIONS

SENCOR DF Pounds/Acre	REMARKS
1/4 to 1/3	Apply SENCOR DF in a tank mix combination with Buctril or Bronate. User higher rate of SENCOR DF for control of larger broadleaf weeds that are 2 inches in diameter or height. Apply before weeds are greater than 3 inches in diameter or height. Not for use on sand soils.

Special Precautions:

Refer to the "General Mixing and Application," and "Precautions" in the front of the Cereal section.

Temporary chlorosis or injury may occur after application of SENCOR DF to wheat, especially if frost or snow occurs within a few days after application. Heavy rainfall (1 inch or more) within 4 days following application may result in crop injury.

WINTER WHEAT

FOR USE IN IDAHO, MONTANA, NEVADA, OREGON, UTAH AND WASHINGTON ON THE VARIETES CENTURK, CHEYENNE, DAWS, FARO, GAINES, HATTON, HILL 81, HYSLOP, LUKE, McDERMID, MORO, NuGAINES, PAHA, PECK, SPRAGUE, STEPHENS, TYEE, WANSER, WESTON AND WINALTA.

SENCOR DF 75% dry flowable herbicide, in a tank mix combination with one of the herbicide products specified below, may be applied to wheat after the crop has reached the 2-leaf stage for control of the following broadleaf weeds:

Common Chickweed	*(Stellaria media)*
Henbit	*(Lamium amplexicaule)*
Prickly Lettuce	*(Lactuca serriola)*
Blue Mustard	*(Chorispora tenella)*
Field Pennycress	*(Thlaspi arvense)*
Tarweed Fiddleneck	*(Amsinckia lycopsoides)*

RECOMMENDED APPLICATIONS

CROP	SENCOR DF Pounds/Acre	REMARKS
Winter Wheat	1/8 to 3/16 (2 to 3 ounces)	Use only in a tank mix combination for broadleaf weed control*. The specified dosage per acre may be added to other herbicides registered for use on wheat after the crop had reached the 2-leaf stage, but not during Winter dormancy. See the appropriate sections of the SENCOR DF label and the label for other herbicide products that are tank mixed with SENCOR DF 75% dry flowable herbicide for directions, application rates, mixing procedures, special precautions and restrictions before using.

* May be used with: Bronate, Buctril, Banvel, 2,4-D, MCPA, Glean, Ally, Finesse, or Harmony.
Observe all cautions and limitations on labeling of all products used in mixtures.

DRYLAND (NON-IRRIGATED) WINTER WHEAT

FOR USE ONLY ON DRYLAND WINTER WHEAT IN KANSAS, OKLAHOMA AND TEXAS ON THESE VARIETIES: GARST 64, HAWK, LARNED, NEWTON, PIONEER 2157, STALLION, TAM W101, TAM 105, VICTORY, AND WRANGLER.

Weeds Controlled in Wheat:

Broadleaves:

Annual Polemonium (Jacob's Ladder) *(Polemonium micranthum)*
Chickweed, Common *(Stellaria media)*
Fiddleneck, Tarweed *(Amsinckia lycopsoides)*
Henbit *(Lamium amplexicaule)*
*Kochia *(Kochia scoparia)*
Lambsquarters *(Chenopodium album)*
*Mustard, Blue *(Chorispora tenalla)*
Mustard, Jim Hill (Sisymbrium altissimum)
*Mustard, Wild *(Brassica kaber)*
Pigweeds *(Amaranthus spp.)*

Grasses:

*Cheatgrass *(Bromus secalinus)*
*Downy Brome *(Bromus tectorum)*
*Foxtails *(Setaria spp.)*
*Rescuegrass *(Bromus catharticus)*

* For control of these weeds use the highest rate recommended for soil texture and percent organic matter.

Hard to Control Weeds:* Italian ryegrass *(Lolium multiflorum)*, *Wild Oats *(Avena fatua)*, and *Wild Buckwheat *(Polygonum convolvulus)*.

NOTE: Control of these weeds may be variable, ranging from poor to excellent depending on timing of application, stage of growth at time of application, temperatures and soil moisture conditions following treatment.

RECOMMENDED BROADCAST APPLICATIONS

SOIL TEXTURE	SENCOR DF Pounds/Acre — % Organic Matter[1] — Fall: 1 to 2	Fall: Over 2	Spring: 1 to 2	Spring: Over 2	REMARKS
COARSE (Sandy, loam, loamy sand)	$1/3$	$1/3$	$1/3$	$1/3$ to $1/2$	Apply specified dosage per acre with aerial or ground spray equipment as a broadcast postemergence spray. For Fall or Spring application, refer to "Application" in the beginning of this Cereals section for information on timing of application and growth restrictions.
MEDIUM (Loam, silt loam, silt, sandy clay, sandy clay loam)	$1/3$ to $1/2$	$1/2$	$1/2$ to $2/3$	$1/2$ to $5/6$	
FINE (Silty clay, silty clay loam,[2] clay, clay loam)	$1/2$	$1/2$ to $2/3$	$1/2$ to $5/6$	$2/3$ to 1	

[1] Do not use on any soils containing less than 1% organic matter.
[2] Silty clay loam soils are transitional soils and may be classified as medium textured soils in some regions of the U.S.

SENCOR in Combination with Other Herbicides

SENCOR DF may be applied as directed above or may be tank mixed with 2,4-D, Banvel, Bronate, Brominal, or Buctril. Refer to the other product labels registered for additional directions, rates, and weed species controlled.

DRYLAND (NON-IRRIGATED) WINTER WHEAT ONLY

(Arkansas, Louisiana, Mississippi and Tennessee)

Weeds Controlled:

Broadleaves:

- Bittercress (*Cardamine spp.*)
- *Buttercup (*Ranunculus spp.*)
- Carolina Geranium (*Geranium carolinianum*)
- Chickweed, Common (*Stellaria media*)
- Chickweed, Mousear (*Cerastium vulgatum*)
- Cutleaf Evening Primrose (*Oenothera Iaciniata*)
- Henbit (*Lamium amplexicaule*)
- *Kochia (*Kochia scoparia*)
- Lambsquarters, Common (*Chenopodium album*)
- Mayweed (Dog Fennel) (*Eupatorium capillifolium*)
- Mustard, Wild (*Brassica kaber*)
- Pepperweed, Virginia (*Lepidium virginicum*)
- Pigweeds (*Amaranthus spp.*)
- Shepherdspurse (*Capsella bursa-pastoris*)
- Whitlowgrass, Spring *Draba Verna*
- Wild Turnip (*Brassica rapa*)

Grasses:

- Bluegrass, Annual (*Poa annua*)
- *Cheatgrass (*Bromus secalinus*)
- *Downy Brome (*Bromus tectorum*)
- Little Barley (*Hordeum pusillum*)
- *Rescuegrass (*Bromus catharticus*)
- *Ryegrass, Italian (*Lolium multiflorum*)

*For control of these weeds on fine soils, use SENCOR DF at $2/3$ pound/acre.

NOTE: Control of Wild Buckwheat (*Polygonum convolvulus*) and Winter Vetch (*Vicia spp.*) may be variable with SENCOR DF, ranging from poor to excellent depending on growth at time of application, temperatures and moisture conditions following application.

Recommended for use only on the following wheat varieties:

Abe, Arthur 71, Caldwell, Coker 747, Coker 762, Coker 68-15, Doublecrop, Hunter, McNair 1003, McNair 1813, Nelson, Pike, Southern Belle and Tyler.

RECOMMENDED BROADCAST APPLICATIONS

SOIL TEXTURE	SENCOR DF[1] Pounds/Acre	REMARKS
COARSE (Sandy loam, loamy sand)	$1/3$	Applied specified dosage per acre as a broadcast postemergence spray. For wheat plants that are coming out of Winter dormancy, allow at least two weeks for plants to recover before application. If wheat has been grazed during the Winter, allow at least two weeks for recovery before application. Refer to the beginning of this Cereals section for information on timing and growth stages for application.
MEDIUM (Loam, silt loam, silt, sandy clay, sandy clay loam)	$1/2$	
FINE (Silty clay, silty clay loam,[2] clay, clay loam)	$1/2$-$2/3$[3]	

[1] Do not use on coarse soils containing less than 1% organic matter or on medium or fine textured soils containing less than 0.5% organic matter.
[2] Silty clay loam soils are transitional soils and may be classified as medium textured soils in some regions of the U.S.
[3] Use $2/3$ pound/acre for control of weed species designated with an asterisk under "Weeds Controlled."

SENCOR in Combination with Other Herbicides:

SENCOR DF may be applied as directed above and may be tank mixed with 2,4-D, Banvel, Bronate, or Buctril. Refer to the other registered product labels for additional directions, rates, and weed species controlled.

Special Precautions

Temporary chlorosis may occur after application of SENCOR DF to wheat, especially if frost occurs.

DRYLAND (NON-IRRIGATED) WINTER WHEAT ONLY

(Georgia)

Weeds Controlled:

Broadleaves:

- Bittercress (*Cardamine spp.*)
- Buttercup (*Ranunculus spp.*)
- Carolina Geranium (*Geranium carolinianum*)
- Chickweed, Common (*Stellaria media*)
- Chickweed, Mousear (*Cerastium vulgatum*)
- Cutleaf Evening Primrose (*Oenothera laciniata*)
- Henbit (*Lamium amplexicaule*)
- Kochia (*Kochia scoparia*)
- Lambsquarters, Common (*Chenopodium album*)
- Mayweed (Dog Fennel) (*Eupatorium capillifolium*)
- Mustard, Wild (*Brassica kaber*)
- Pepperweed, Virginia (*Lepidium virginicum*)
- Pigweeds (*Amaranthus spp.*)
- Sheperdspurse (*Capsella bursa-pastoris*)
- Whitlowgrass, Spring (*Draba Verna*)
- Wild Turnip (*Brassica rapa*)
- Wild Radish (*Raphanus raphanistrum*)

Grasses:

- Bluegrass, Annual (*Poa annua*)
- Cheatgrass *Bromus secalinus*)
- Downy Brome (*Bromus tectorum*)
- Little Barley (*Hordeum pusillum*)
- Rescuegrass (*Bromus catharticus*)
- Ryegrass, Italian (*Lolium multiflorum*)

NOTE: Control of Wild Buckwheat (*Polygonum convolvulus*) and Winter Vetch (*Vicia spp.*) may be variable with SENCOR DF, ranging from poor to excellent depending on growth at time of application, temperatures and moisture conditions following application.

Recommended for use only on the following wheat varieties:

Abe, Arthur 71, Caldwell, Coker 747, Coker 762, Coker 797, Coker 68-15, Doublecrop, Florida 302, Hunter, McNair 1003, McNair 1813, Nelson, Oasis, Omega 78, Pike, Southern Belle, Stacy and Tyler.

RECOMMENDED BROADCAST APPLICATIONS

SOIL TEXTURE	SENCOR DF Pounds/Acre	REMARKS
COARSE (Sandy loam, loamy sand)	$1/3$	Apply specified dosage per acre as a broadcast postemergence spray. For wheat plants that are coming out of Winter dormancy, allow at least two weeks for plants to recover before application. If wheat has been grazed during the Winter, allow at least two weeks for recovery before application. Do not apply before 75 days after planting. Apply only after wheat has tillers, but before November 15 in the Piedmont area and northern part of the state, and before December 1 in the Coastal Plain area.
MEDIUM Loam, silt loam, silt, sandy clay, sandy clay loam)	$1/2$	

Do not use on Coarse soils containing less than 1% organic matter or on Medium textured soils containing less than 0.5% organic matter.

SENCOR in Combination with Other Herbicides:

SENCOR DF may be applied as directed above and may be tank mixed with 2,4-D, Banvel, Bronate or Buctril. Refer to the other registered product labels for additional directions, rates and weed species controlled.

Special Precautions

Temporary chlorosis may occur after application of SENCOR DF to wheat, especially if frost occurs.

FOR WEED CONTROL IN A WHEAT/FALLOW/WHEAT ROTATION (Idaho, Oregon, Utah and Washington Only)

SENCOR DF may be applied to provide weed control during the fallow period after wheat harvest or in the Spring before Winter wheat is planted. Winter wheat can be seeded 4 months (120 days) after Spring application. Mechanical tillage or the application of a contact herbicide may be required to control weeds germinating prior to seeding of Winter wheat. Best results will be obtained where straw and chaff are evenly distributed across the field.

For specific application information see under the "GENERAL INFORMATION" section in the front of the SENCOR DF label.

Where weed growth is present at application time, SENCOR DF should be applied with Gramoxone or other contact herbicides. Refer to the other product labels registered for additional directions, rates and weed species controlled.

Weeds Controlled:

Broadleaves:		Grasses:
Chickweed, Common (*Stellaria media*)	Mustard, Tansy (*Descurainia pinnata*)	Cheatgrass (*Bromus secalinus*)
Henbit (*Lamium amplexicaule*)	Mustard, Treacle (*Eyrsimum repandum*)	Downy Brome (*Bromus tectorum*)
*Kochia (*Kochia scoparia*)	Mustard, Wild (*Brassica kaber*)	*Wheat, Volunteer (*Triticum spp.*)
Lambsquarters (*Chenopodium album*)	Pennycress, Field (Fanweed) (*Thlaspi arvense*)	
Mustard, Blue or Purple (*Chorispora tenella*)	Pigweeds (*Amaranthus spp.*)	
Mustard, Jim Hill (*Sisymbrium altissimum*)	*Russian Thistle (*Salsola iberica*)	
	*Wild Sunflower (*Helianthus spp.*)	

* **NOTE:** Since control of these weeds may be variable depending on moisture following application, the higher labelled rate is recommended.

After Harvest Application (Fall Fallow): SENCOR DF may be applied to wheat stubble after harvest in the Fall. Apply $\frac{2}{3}$ to $\frac{5}{6}$ pound per acre broadcast before weeds emerge. Use higher rate for longer weed control or for weeds designated as requiring the higher rate for control. Rainfall ($\frac{1}{2}$ inch or more) is necessary for herbicide activation.

Do not plant crops in treated areas for at least 10 months following Fall applications.

SENCOR DF may be applied to $\frac{2}{3}$ to $\frac{5}{6}$ pound per acre as directed above for a Fall application. If other vegetation is present at the time of application use a contact herbicide.

Spring Application (Summer Fallow): SENCOR DF may be applied to wheat stubble in the Spring. Apply $\frac{1}{2}$ to $\frac{2}{3}$ pound per acre broadcast before weeds emerge in the Spring. Use higher rate for longer weed control or weeds designated as requiring higher rate for control. Rainfall ($\frac{1}{2}$ inch or more) is necessary for herbicide activation.

Precautions and Restrictions:

Do not graze treated fields.

Do not plant Spring seeded cereals following Fall fallow applications of SENCOR.

Where SENCOR was applied in the Fall, do not apply SENCOR in the Spring.

FOR WEED CONTROL IN A FALLOW ROTATION WITH BARLEY AND WHEAT (Colorado, Kansas, Montana, Nebraska and Wyoming Only)

SENCOR DF may be applied to provide weed control during the fallow period after wheat or barley harvest or in the Spring before planting of Winter wheat or barley. Mechanical tillage or the application of a contact herbicide may be required to control weeds germinating prior to seeding of Winter wheat or barley.

Refer to "Mixing" in the "GENERAL INFORMATION" section in the front of the label.

For specific application information see the "GENERAL INFORMATION" section in the front of the SENCOR DF label.

Where weed growth is present at application time, SENCOR DF should be applied with Gramoxone, Roundup, or other contact herbicide. Refer to the other product labels registered for additional directions, rates, and weed species controlled. Do not plant crops in treated areas, earlier than 10 months following Fall applications.

Weeds Controlled:

Broadleaves:		Grasses:
Chickweed, Common (*Stellaria media*)	Mustard, Tansy (*Descuraini pinnata*)	Cheatgrass (*Bromus secalinus*)
Cowcockle (*Vaccaria pyramidata*)	Mustard, Treacle (*Eyrsimum repandum*)	Downy Brome (*Bromus tectorum*)
Henbit (*Lamium amplexicaule*)	Mustard, Wild (*Brassica kaber*)	Foxtail, Green (*Setaria viridis*)
*Kochia (*Kochia scoparia*)	Pennycress, Field (Fanweed) (*Thlaspi arvense*)	*Wheat, Volunteer (*Triticum spp.*)
Lambsquarters (*Chenopodium album*)	Pigweeds (*Amaranthus spp.*)	*Wild Oats (*Avena fatua*)
Mustard, Blue or Purple (*Chorispora tenella*)	*Russian Thistle (*Salsola iberica*)	
Mustard, Jim Hill (*Sisymbrium altissimum*)	*Wild Sunflower (*Helianthus spp.*)	

* **NOTE:** Since control of these weeds may vary depending on moisture following application, use the higher rate recommended below.

AFTER HARVEST APPLICATION (Fall Fallow)—SENCOR DF may be applied to the stubble after harvest in the Fall. Apply $\frac{5}{6}$ to 1 pound per acre broadcast before weeds emerge. Use the higher rate for longer weed control or for weeds designated as requiring the higher rate for control. Rainfall ($\frac{1}{2}$ inch or more) is necessary for herbicide activation.

SPRING APPLICATION (Summer Fallow)—SENCOR DF may be applied to the stubble in the Spring. Apply $\frac{1}{2}$ to $\frac{2}{3}$ pound per acre broadcast before weeds emerge in the Spring. Use the higher rate for longer weed control or weeds designated as requiring the higher rate for control. Rainfall ($\frac{1}{2}$ inch or more) is necessary for herbicide activation. Wheat or barley can be seeded 120 days after Spring application.

Precautions and Restrictions

Do not graze treated fields.

Do not plant Spring seeded cereals following Fall applications for fallow.

Where SENCOR was applied in the Fall, do not apply SENCOR in the Spring.

SPECIAL LOCAL NEED REGISTRATIONS [SECTION 24(c)] FOR SENCOR DF HERBICIDE*

CROP	PEST	STATE EPA SLN NO.	DOSAGE	REMARKS
Alfalfa	Various broadleaf weeds and grasses	Washington WA 85-0005 Oregon OR-85-0020	$\frac{1}{2}$ to $1\frac{1}{3}$ pounds	Application by ground or air. Do not graze or harvest within 28 days of application.
Barley (Spring)	Various weeds	Alaska AK 83-0003	$\frac{1}{6}$ to $\frac{1}{3}$ pound per acre.	Postemergence ground application.
Lentils and Peas	Various broadleaf weeds and grasses	Idaho ID 87-0005 Oregon OR 87-0002 Washington Wa 87-0008	See label rates.	Preemergence and postemergence before crop germinates.
Potatoes	Weeds	Idaho ID 81-0046 Oregon	See label rates.	Tank mix with Eptam 7-E.

SPECIAL LOCAL NEED REGISTRATIONS [SECTION 24(c)] FOR SENCOR DF HERBICIDE*

CROP	PEST	STATE EPA SLN NO.	DOSAGE	REMARKS
		OR 81-0040 Washington WA 86-0015		
Potatoes	Various weeds	California CA 83-0034 Idaho ID 87-0017 Washington WA 86-0015	See label rates.	Postemergence application to russetted or while skinned varieties that are not early maturing.
Soybeans	Weeds	Michigan MI 81-0012	1/3 to 1/2 pound per acre.	For use on alkaline (calcareous) soils
	Red rice	Louisiana LA 81-0007	See label rates.	Overlay application following Teflan PPI or tank mixed with Lasso.
	Various broadleaf weeds and grasses	North Carolina NC 86-006	2/3 to 1 1/2 pounds per acre.	Application alone or as a tank mix on blackland soils.
Sugarcane (Fallow)	Various grass and broadleaf weeds	Louisiana LA 84-0018	1 1/3 to 2 pounds per acre.	Broadcast (Summer fallow). Application by ground or air in late May or early June.
Tomatoes (chemical fallow)	Common chickweed London rocket Shepherdspurse Wild mustard Lambsquarters Redroot pigweed Purslane	California CA 79-0234	2/3 to 1 pound per acre.	Broadcast by air or ground at planting. Not for use in Kern County, California.
Corn (Fall fallow)	Common chickweed London rocket Sheperdspurse Wild mustard Fiddleneck Malva	California CA 84-0007	2/3 pound per acre.	Broadcast (Fall fallow) application by air or ground. Do not apply through irrigation system.
Wheat (Winter)	Ivyleaf speedwell	Oregon OR 81-0013	1/3 pound per acre.	Postemergence. Application by ground or air.

* For distribution and use only within the state specified. These Special Local Need registrations approved as of date of this publication are, however, subject to cancellation by the state. Contact your nearest Mobay Agricultural Chemicals Division Office for the current status and a copy of the complete individual supplemental Special Local Need labeling. This labeling must be in the possession of the user at time of pesticide application.

CROP ROTATION RECOMMENDATIONS

Waiting Period After SENCOR Herbicide Application[1]

4 Months		8 Months	12 Months	18 Months
Alfalfa	Sainfoin	Barley	Other Crops not listed (except root crops)	Sugar Beets
Barley[2]	Soybeans	Cotton		Onions
Asparagus	Sugarcane	Lentils		And other root crops
Corn	Tomatoes	Peas		
Forage Grasses	Wheat[2]	Rice		
Potatoes		Wheat		

[1] Cover crops for soil building or erosion control may be planted any time, but do not graze or harvest for food or feed. Stand reductions may occur in some areas.

[2] Following peas, lentils or soybeans.

PRECAUTIONARY STATEMENTS

HAZARDS TO HUMANS AND DOMESTIC ANIMALS

CAUTION

May be harmful if swallowed or inhaled. Avoid contact with eyes, skin or clothing. Avoid breathing spray mist. Wash thoroughly with soap and warm water after handling. Wash contaminated clothing with soap and hot water before reuse.

Do not contaminate feed or food. Keep out of reach of children.

Obtain prompt medical aid if poisoning should occur.

To Physician: Treat the patient symptomatically.

Symptoms of Poisoning: The compound does not cause any definite symptoms that would be diagnostic. Poisoning is accompanied by breathing difficulties and sedation.

STATEMENT OF PRACTICAL TREATMENT

If in eyes, immediately flush eyes with plenty of water for at least 15 minutes. For further treatment get medical attention.

If swallowed, vomiting should be induced. Administer water freely, and induce vomiting by giving one dose (1/2 oz. or 15 ml.) of syrup of ipecac. If vomiting does not occur within 10 to 20 minutes, administer second dose. If syrup of ipecac is not available, induce vomiting by sticking finger down throat. Repeat until vomit fluid is clear. Never give anything by mouth to an unconscious person. Professional medical assistance should be secured immediately.

If on skin, wash thoroughly with soap and warm water.

ENVIRONMENTAL HAZARDS

Do not use on other crops grown for food or forage. Do not apply directly to water or wetlands. Do not apply when weather conditions favor drift from areas treated. Do not contaminate water by cleaning of equipment or disposal of wastes. Apply this product only as specified on the label.

Do not allow sprays to drift on to adjacent desirable plants.

Observe all cautions and limitations on labeling of all products used in mixtures.

GROUND WATER ADVISORY. Metribuzin is a chemical which can travel (seep or leach) through soil and can contaminate ground water which may be used as drinking water. Metribuzin has been found in ground water as a result of agricultural use. Users are advised not to apply metribuzin where the water table (ground water) is close to the surface, and where the soils are very permeable, i.e., well drained soils such as loamy sands. Your local agricultural agencies can provide further information on the type of soil in your area and the location of ground water.

RE-ENTRY STATEMENTS

Do not apply this product in such a manner as to directly or through drift expose workers or other persons. The area being treated must be vacated by unprotected persons. Do not enter treated areas without protective clothing until sprays have dried. Because certain states may require more restrictive re-entry intervals for various crops treated with this product, consult your State Department of Agriculture for further information.

Written or oral warnings must be given to workers who are expected to be in a treated area or in an area about to be treated with this product. Oral warnings should include those statements found under the label heading "Precautionary Statements" and "Re-entry Statements" included on the label. When oral warnings are given, warnings shall be given in a language customarily understood by workers. Oral warnings must be given if there is reason to believe that written warnings cannot be understood by workers. Written warnings must include the following information: WARNING, area treated with SENCOR DF herbicide on (date of application). Do not enter without appropriate protective clothing until sprays have dried. **If on skin,** remove contaminated clothing and wash skin immediately with soap and warm water. **If eyes are contaminated,** wash with flowing water for at least 15 minutes. **If swallowed, vomiting should be induced.** (See Precautionary Statements on label).

STORAGE AND DISPOSAL

Pesticide Disposal: Do not contaminate water, food or feed by storage or disposal. Wastes resulting from the use of this product may be disposed of on site or at an approved waste disposal facility.

Container Disposal: Triple rinse (or equivalent). Then offer for recycling or reconditioning, or puncture and dispose of in a sanitary landfill, or incineration, or, if allowed by State and local authorities, by burning. If burned, stay out of smoke.

Store in a cool, dry place and in such a manner as to prevent cross contamination with other pesticides, fertilizers, food, and feed. Store in original container and out of the reach of children, preferably in a locked storage area.

Handle and open container in a manner as to prevent spillage. If the container is leaking or material spilled for any reason or cause, carefully sweep spilled material into a pile. Refer to Precautionary Statements on label for hazards associated with the handling of this material. Do not walk through spilled material. Dispose of pesticide as directed above. In spill or leak incidents, keep unauthorized people away. You may contact the Mobay Emergencey Response Team for decontamination procedures or any other assistance that may be necessary. The Mobay Kansas City Emergency Response Telephone No. is 816-242-2582, or contact Chemtrec at 800-424-9300.

SENCOR is a Reg. TM of Bayer AG, Germany
Amiben and Brominal are Reg. TMs of Union Carbide Agricultural Products Co., Inc.
Banvel is a Reg. TM of Velsicol Chemical Corp.
Bladex is a Reg. TM of Shell Chemical Co.
Bronate and Buctril are Reg. TMs of Rhone-Poulenc Chemical Co.
Command is a Reg. TM of FMC Corp.
Dual and Igran are Reg. TMs of Ciba-Geigy.
Glean is a Reg. TM of E.I. duPont de Nemours and Co.
Gramoxone is a Reg. TM of ICI Americas Inc.
Lasso and Roundup are Reg. TMs of Monsanto Co.
Ortho X-77 is a Reg. TM of Chevron Chemical Co.
Prowl and Scepter are Reg. TMs of American Cyanamid Co.
Surflan, Sonalan and Treflan are Reg. TMs of Elanco Products Co.

IMPORTANT:

This bulletin is not intended to provide adequate information for application. Before using any product, read and carefully observe directions, cautionary statements, and other information appearing on the product label. This product is sold subject to the Conditions of Sale set forth on the container label.

READ THE LABEL BEFORE USE

This is a specimen label. Mobay is not responsible for the accuracy of the information contained herein. As labels are subject to revision, always carefully read and follow the label on the product container.

SENCOR® 4

FLOWABLE HERBICIDE

FOR CONTROL OF CERTAIN GRASSES AND BROADLEAF WEEDS

ACTIVE INGREDIENT: 4-Amino-6-(1,1-dimethylethyl)-3-(methylthio)-1,2,4-triazin-5(4*H*)-one	41%
INERT INGREDIENTS:	59%
TOTAL:	100%

Contains 4 pounds of 4-Amino-6-(1,1-dimethylethyl)-3-(methylthio)-1,2,4-triazin-5 (4*H*)-one per gallon.

EPA Reg. No. 3125-314 EPA Est. 3125-MO-1

RECOMMENDED USES

Soybeans, Potatoes, Lentils, Dry Field Peas, Established Alfalfa and Sainfoin, Established Asparagus, Carrots, Tomatoes, Barley, Field Corn, Winter Wheat, Wheat/Fallow/Wheat Rotation, Barley and Wheat/Fallow/Barley and Wheat Rotation, and Sugarcane

RESTRICTIONS ON WAREHOUSE STORAGE

Protect from rain and moisture.
Store in a cool, dry place.
Do not store in front of a space heater.

SENCOR 4 flowable herbicide delivers unmatched preemergence broadleaf weed control in soybeans. SENCOR 4 may be used in ground or aircraft spray equipment as a preemergence and/or postemergence application to potatoes. SENCOR 4 also may be applied to control many weeds in barley, winter wheat, field corn, in established alfalfa and sainfoin, and carrots, asparagus. It effectively controls such problem broadleaf weeds as velvetleaf, jimsonweed, sesbania, prickly sida/teaweed, sicklepod and annual smartweeds. It also controls certain grasses. SENCOR 4 is registered for combination applications with Cannon®, Command®, Commence®, Treflan®, Lasso®, Amiben®, Gramoxone® Super, Surflan®, Dual®, Prowl®, Roundup®, and Sonalan® on soybeans.

AVISO

PRECAUCION AL USUARIO: Si usted no puede leer o entender inglés, no use este producto hasta que la etiqueta le haya sido explicada ampliamente.

(TO THE USER: If you cannot read or understand English, do not use this product until the label has been fully explained to you.)

DIRECTIONS FOR USE

It is a violation of Federal Law to use this product in a manner inconsistent with its labeling.

GENERAL INFORMATION

Mixing: When using SENCOR 4, make sure the sprayer is completely clean, free of rust or corrosion which occurs from Winter storage. Examine strainers and screens to be sure the sprayer is clean from previously used pesticides.

Any tank mix containing SENCOR 4 should be kept agitated and sprayed out immediately. Do not allow tank mixes to stand for prolonged periods of time.

The proper mixing procedure for SENCOR 4 alone or in tank mix combinations with other herbicides is:

1. Fill the spray tank 1/4 to 1/3 full with clean water.
2. Add recommended rate of SENCOR 4 while recirculating and with agitator running.
3. Follow the triple rinse procedure described under "STORAGE AND DISPOSAL" to insure that all product is removed from the container.
4. Mix thoroughly and add clean water to fill spray tank to desired level.
5. Add the other herbicide to tank last and agitate thoroughly.
6. Continue agitation during application and until sprayer tank is empty.

SOIL TEXTURE: As used on the label, "Coarse soils" are loamy sand or sandy loam soils. "Medium soils" are loam, silt loam, silt, sandy clay, or sandy clay loam. "Fine soils" are silty clay, silty clay loam, clay, or clay loam. Silty clay loam soils are transitional soils and may be classified as medium textured soils in some regions of the U.S.

CHEMIGATION

SENCOR 4 is recommended for application through sprinkler irrigation equipment to potatoes, soybeans, tomatoes, and asparagus as directed on the label. Refer to the crop sections of the label for recommended rates, weeds controlled or suppressed, restrictions, and special precautions.

Apply this product only through sprinkler (including center pivot, lateral move, or solid set) irrigation systems. Do not apply this product through any other type of irrigation system.

Crop injury, lack of effectiveness, or illegal pesticide residues in the crop can result from nonuniform distribution of treated water.

Calibration: (Center Pivot and Self-Propelled Lateral Move Systems) Sprinkler irrigation systems must be accurately calibrated for application of SENCOR 4. Greater accuracy in calibration (and distribution) will be achieved by injecting a larger volume of a more dilute mixture of product and water per hour. Follow the steps below to calibrate center pivot and lateral move systems:

1. Determine number of minutes required to make one complete revolution while applying 1/4 to 3/4 inch of water per acre.
2. With the system at operating pressure determine the exact number of minutes required to inject one gallon of water.
3. Divide the time required for one revolution (step 1) by the time required to inject one gallon (step 2). This gives total gallons of product-water mixture to be added to nurse tank.
4. Add required amount of water to nurse tank and start the agitation system. Then add sufficient SENCOR 4 at the recommended rate (see RECOMMENDED BROADCAST APPLICATIONS) to the nurse tank.

EXAMPLE: If 20 hours (1200 minutes) were required for one revolution and if 2 minutes were required to inject one gallon then a total of 600 gallons of product-water mixture are required ($1200 \div 2 = 600$); to treat 135 acres at 1 pint/acre, 135 pints (16 gallons and 7 pints) of SENCOR 4 are required.

If you have questions about calibration, you should contact State Extension Service specialists, equipment manufacturers or other experts.

Do not connect an irrigation system (including greenhouse systems) used for pesticide application to a public water system unless the pesticide label-prescribed safety devices for public water systems are in place.

A person knowledgeable of the chemigation system and responsible for its operation, or under the supervision of the responsible person, shall shut the system down and make necessary adjustments should the need arise.

The system must contain a functional check valve, vacuum relief valve, and low pressure drain appropriately located on the irrigation pipeline to prevent water source contamination from backflow.

The pesticide injection pipeline must contain a functional, automatic, quick-closing check valve to prevent the flow of fluid back toward the injection pump.

The pesticide injection pipeline must also contain a functional, normally closed, solenoid-operated valve located on the intake side of the injection pump and connected to the system interlock to prevent fluid from being withdrawn from the supply tank when the irrigation system is either automatically or manually shut down.

The system must contain functional interlocking controls to automatically shut off the pesticide injection pump when the water pump motor stops.

The irrigation line or water pump must include a functional pressure switch which will stop the water pump motor when the water pressure decreases to the point where pesticide distribution is adversely affected.

Systems must use a metering pump, such as a positive displacement injection pump (e.g., diaphragm pump) effectively designed and constructed of materials that are compatible with pesticides and capable of being fitted with a system interlock.

Do not apply when wind speed favors drift beyond the area intended for treatment.

Maintain continuous agitation in the injection nurse tanks during the herbicide application, sufficient to keep herbicide in suspension.

Apply specified dosage in 1/4 to 3/4 inch of water (use 1/4 to 1/2 inch of water on sandy soils) per acre as a continuous injection in center pivot and lateral move systems or in the last 15 to 30 minutes of set in permanent solid set sprinkler systems. Application of more than the quantity of irrigation water recommended on the label may result in decreased product performance by removing the chemical from the zone of effectiveness. Where sprinkler distribution patterns do not overlap sufficiently, unacceptable weed control may result. Where sprinkler distribution patterns overlap excessively crop injury may result. Allow sufficient time for pesticide to be flushed through all lines and all nozzles before turning off irrigation water. To insure that lines are flushed and free of remaining pesticide, an indicator dye may be injected into the lines to mark the end of the application period.

Use a minimum of 1 part water to 1 part herbicide for injection. The use of a larger volume of water will insure greater accuracy and more uniform distribution.

Application of SENCOR 4 with Herbicide Spray Equipment: Use a standard low pressure (20 to 40 psi.) herbicide boom sprayer equipped with suitable nozzles and screens no finer than 50-mesh in-nozzle and in-line strainers. Agitate thoroughly before and during application with bypass agitation.

GROUND APPLICATION: Apply the proper rate of SENCOR 4 in a minimum of 10 to 40 gallons of spray mixture per acre broadcast.

Banded Application: Use proportionally less SENCOR 4 per acre in a band versus a broadcast application. For band application use 1/4 to 1 gallon of spray mix per inch of band width regardless of row spacing.

EXAMPLES: To treat a 15-inch band on rows 30 inches apart, use one-half of the broadcast rate of SENCOR 4.

To treat a 14-inch band on rows 42 inches apart, use one-third of the broadcast rate of SENCOR 4.

AERIAL APPLICATION: Where permitted, apply specified rate in a minimum of 2 to 10 gallons of spray mixture per acre. Do not apply aerially when wind speed is greater than 10 mph. **NOTE:** Do not apply aerially when SENCOR is tank mixed with Lasso.

For All Applications of SENCOR 4: Sprayer must be accurately calibrated before applying SENCOR 4. Check sprayer during application to be sure it is working properly and delivering a uniform spray pattern. As the volume of spray mixture decreases per acre, the importance of accurate calibration and uniform application increases. Avoid over application, misapplication, and boom and spray swath overlapping that will increase spray dosage. (Crop injury may occur as a result.) Avoid spray skips and gaps which allow weeds to grow in untreated soil. Do not apply when weather conditions favor spray drift and/or when sensitive or cool season crops, such as cole crops, onions, peas, or strawberries are present in adjacent fields or in areas where wheat is growing in coarse textured soils.

SPRAYER CLEANUP:
Spray equipment must be thoroughly cleaned to remove remaining traces of herbicide that might injure other crops to be sprayed. Drain any remaining spray solution of SENCOR 4 from the spray tank and discard in non-crop areas away from water supplies. Rinse the spray tank and refill with water, adding a heavy-duty detergent at the rate of one cup per 20 gallons of water. Recycle this mixture through the equipment for 5 minutes and spray out. Repeat this procedure twice. Fill the spray tank with clean water, recycle for 5 minutes and spray out. Clean pump and nozzle screens thoroughly. Wash away any spray mixture from the outside of spray tank, nozzle or spray rig. All rinse water must be disposed of in compliance with local, State, and Federal guidelines.

APPLICATION OF SENCOR 4 IN FLUID FERTILIZERS

SENCOR 4 may be applied in fluid fertilizer solutions to alfalfa and soybeans by following the appropriate mixing procedures and compatibility check. When using tank mix combinations, be sure all components are compatible.

Compatibility checks of SENCOR 4 and tank mix combinations which include SENCOR should be made for each batch of fluid fertilizer because of the variability of these fertilizers.

Compatibility Check:
1. Premix 2 teaspoonfuls of SENCOR 4 with 8 teaspoonfuls of water (1:4 ratio) in a quart jar by adding the water first and followed with SENCOR. Mix thoroughly. If a second herbicide is to be used, double the amount of water (1:8 ratio) and add the second herbicide after mixing SENCOR first.
2. Then pour 1 pint of fluid fertilizer into the quart jar and shake well.
3. Allow to stand for 5 minutes.

THIS COMPATIBILITY CHECK SHOULD ONLY BE USED WHEN MIXING WITH FLUID FERTILIZERS.

Interpretation of Results:
If the solution in the jar appears to be uniform, without signs of agglomeration, or without a separation of an oily film on top of the fertilizer, the mixture may be used. If not, repeat the compatibility check using twice the amount of water or add a compatibility agent to the water. If separation occurs and can be resuspended by shaking, then application is possible with good agitation in the spray tank.

Tank Mixing Guidelines:
1. Add the required amount of water and compatibility agent (if required) to the tank. Start agitation system while adding SENCOR 4 and follow by adding the fluid fertilizer and agitate.
2. If a second herbicide is to be used, follow as above in 1, but use twice the amount of water. Start agitation and add SENCOR 4 and follow by adding the second herbicide, and then continue filling the tank with fluid fertilizer.
3. Maintain continuous agitation to assure uniform spray mixture until the tank is emptied.

COMMERCIAL IMPREGNATION AND APPLICATION OF SENCOR 4 ON DRY BULK FERTILIZER

Dry bulk fertilizer may be impregnated or coated with SENCOR 4 for application to established alfalfa and to soybeans. All recommendations, cautions, and special precautions on the label must be followed along with State regulations relating to dry bulk fertilizer blending, impregnating and labeling.

Impregnation: To impregnate, use a system consisting of a belt, conveyor, or closed drum which is used for dry bulk fertilizer blending. Any commonly used fertilizer can be impregnated with SENCOR except ammonium nitrate, or fertilizers containing ammonium nitrate, potassium nitrate, or sodium nitrate. Do not use on powder limestone.

Apply using a minimum of 200 pounds dry bulk fertilizer per acre and up to a maximum of 450 pounds per acre. To impregnate or coat the dry bulk fertilizer with SENCOR, the delivery nozzles must be directed to deliver a fine spray toward the fertilizer for thorough coverage while avoiding spray contact with mixing equipment. Uniform impregnation of SENCOR to dry bulk fertilizer will vary and if the absorptivity is not adequate, an absorptive powder may be added to produce a dry free-flowing mixture. Micro-Cel E (Johns-Manville Product Corporation) is the recommended absorbent powder. SENCOR should be pre-mixed with water to form a sprayable mixture. When another herbicide is used with SENCOR, mix and impregnate immediately.

Apply immediately after impregnation unless experience has shown that impregnated fertilizer can be stored without becoming lumpy and difficult to spread.

Rates: Select the recommended rate of SENCOR 4 per acre from the appropriate section of the label and refer to the formula below to determine the amount of SENCOR which is to be impregnated on a ton of dry bulk fertilizer based on the amount of fertilizer which will be distributed on one acre.

$$\text{Pints SENCOR 4 Per Acre} \times \frac{2000}{\text{Lbs. Fertilizer Per Acre}} = \text{Pints SENCOR 4 Per Ton of Fertilizer}$$

Application: Uniform application is essential for satisfactory weed control. Accurate calibration of fertilizer application equipment is essential for uniform distribution to the soil surface. The recommended method of application is to apply 1/2 the recommended rate and overlap 50 percent or to double apply by splitting the middles to obtain the best distribution pattern.

If fertilizer materials are excessively dusty, use diesel oil or other suitable additive to reduce dust prior to impregnation as dusty fertilizer will result in poor distribution during application. Crop injury and/or poor weed control may occur where the impregnated fertilizer is not uniformly applied.

INCORPORATION AND COMBINATION USES: When SENCOR 4 is to be used in combination with another herbicide, follow directions on the label for combinations, rates, crops, incorporation, and special precautions.

SOYBEANS (Except California)

SENCOR herbicide tank mix combinations are recommended for preplant incorporated applications, preemergence surface applications, SPLIT-SHOT application and EXTENDED SPLIT-SHOT application. SEN-

COR is also recommended as an overlay application following a preplant incorporated application of a recommended grass herbicide and alone as a preemergence surface application. All these applications can be applied with ground equipment, and some can be applied with aerial spray equipment. In addition, SENCOR can be applied as a postemergence directed spray to soybeans in certain states.

Special Precautions (Soybeans)

Injury to soybeans may occur when SENCOR 4 is used under the following conditions:

1. When soils have a calcareous surface area or a pH of 7.5 or higher.
2. Due to the sensitivity of certain soybean varieties, SENCOR is not recommended for use on sensitive varieties, such as Tracy, Semmes, Altona, Vansoy, Coker 102, Coker 156, Portage, Maple Amber, Govan, NKS-1884, Terra-Vig 505, Terra-Vig 606, Vinton 81, Asgrow 6520, AP 55 and AP 71. Consult your Mobay representative or your seed supplier for information on the tolerance to SENCOR of newly released soybean varieties, prior to use of SENCOR.
3. When applied in conjunction with soil-applied organic phosphate pesticides.
4. Over application or boom overlapping may result in stand loss and soil residues.
5. Uneven application or improper incorporation can decrease the level of weed control and/or increase the level of injury.
6. When applied to any soil with less than ½% organic matter.
7. Soil incorporation deeper than recommended.
8. When sprayers are not calibrated accurately.
9. When heavy rains occur soon after application, especially in poorly drained areas where water may stand for several days.
10. When soybeans are planted less than 1½" deep, particularly in preemergence application.

Activation: A minimum amount of soil moisture is required to activate SENCOR. In areas of low rainfall, preemergence applications to dry soil should be followed with light irrigation of ¼ acre inch of water. Do not apply heavy irrigation immediately after application. As with many surface-applied herbicides, weed control and crop tolerance may vary with rainfall and/or soil texture.

Grazing and Feeding Treated Vines: Treated vines may be grazed or fed to livestock 40 days after application when SENCOR 4 is applied alone or with Treflan,® Dual,® Prowl,® Lasso,® Amiben,® Gramoxone®, Lasso plus Gramoxone.

Do not use treated vines for feed or forage when SENCOR is applied with Surflan,® Surflan plus Gramoxone, Sonalan,® linuron plus Lasso, linuron plus Dual.

Rate Ranges: Where a rate range is shown, use a lower rate on soils that are coarse-textured or low in organic matter. Use a higher rate on soils that are relatively fine-textured or high in organic matter.

Replanting: If replanting is necessary in fields treated with SENCOR 4 as directed on the label, the field may be replanted to soybeans. When replanting, a minimum of tillage is recommended. Do not apply a second treatment as injury to soybeans may occur.

WEEDS CONTROLLED BY SENCOR AND SENCOR HERBICIDE TANK MIX COMBINATIONS

ANNUAL BROADLEAF WEEDS	SENCOR Alone	SENCOR SPLIT-SHOT	SENCOR plus Treflan	SENCOR plus Dual	SENCOR plus Prowl	SENCOR plus Lasso	SENCOR plus Surflan	SENCOR plus Amiben	SENCOR plus Gramoxone	SENCOR plus LASSO and Gramoxone	SENCOR plus Surflan and Gramoxone	EXTENDED SPLIT-SHOT	SENCOR plus Sonalan	SENCOR plus Linuron plus (Lasso or Dual)
Black Nightshade (*Solanum nigrum)*	P	P	P	C	P	C	S	S	P	C	P	C	P	S
Bristly Starbur (*Acanthospermum hispidum)*	C	C	C	C	C	C	C	C	C	C	C	C	C	C
Buffalobur (*Solanum rostratum)*	C	C	P	P	P	P	P	P	P	P	P	C	P	•
Carpetweed (*Mollugo verticillata)*	C	C	C	C	C	C	C	C	C	C	C	C	C	C
Cocklebur (*Xanthium pensylvanicum)*	S	C	S	S	S	S	S	S	C	C	C	C	S	S
Copperleaf, Hophornbeam (*Acalypha ostryaefolia)*	C	C	C	C	C	C	C	C	C	C	C	C	C	C
Florida Beggarweed (*Desmodium tortuosum)*	C	C	C	C	C	C	C	C	C	C	C	C	C	C
Florida Pusley (*Richardia scabra)*	C	C	C	C	C	C	C	C	C	C	C	C	C	C
Galinsoga (*Galinsoga spp.)*	C	C	C	C	C	C	C	C	C	C	C	C	C	C
Horseweed Marestail (*Conyza canadensis)*	•	•	•	•	•	•	•	•	•	C	C	C	•	•
Jimsonweed (*Datura stramonium)*	C	C	C	C	C	C	C	S	C	C	C	C	C	S
Knotweed (*Polygonum spp.)*	C	C	C	C	C	C	C	C	C	C	C	C	C	C
Kochia (*Kochia scoparia)*	C	C	C	C	C	C	C	C	C	C	C	C	C	C
Lambsquarters (*Chenopodium spp.)*	C	C	C	C	C	C	C	C	C	C	C	C	C	C
Morningglory, Ivyleaf (*Ipomoea hederacea)*	P	P	S	P	P	P	P	P	P	P	P	P	P	P
Morningglory, Pitted (*Ipomoea lacunosa)*	P	P	S	P	P	P	P	P	P	P	P	P	P	P
Morningglory, Smallflower (*Jacquemontia tamnifolia)*	P	P	C	P	P	P	P	P	P	P	P	P	P	P

	SENCOR Alone	SENCOR SPLIT-SHOT	SENCOR plus Treflan	SENCOR plus Dual	SENCOR plus Prowl	SENCOR plus Lasso	SENCOR plus Surflan	SENCOR plus Amiben	SENCOR plus Gramoxone	SENCOR plus LASSO and Gramoxone	SENCOR plus Surflan and Gramoxone	EXTENDED SPLIT-SHOT	SENCOR plus Sonalan	SENCOR plus Linuron plus (Lasso or Dual)
ANNUAL BROADLEAF WEEDS														
Morningglory, Tall *(Ipomoea purpurea)*	P	P	S	P	P	P	S	P	P	P	P	P	P	P
Pigweeds *(Amaranthus spp.)*	C	C	C	C	C	C	C	C	C	C	C	C	C	C
Prickly Sida/Teaweed *(Sida spinosa)*	C	C	C	C	C	C	C	C	C	C	C	C	C	C
Purslane *(Portulaca oleracea)*	C	C	C	C	C	C	C	C	C	C	C	C	C	C
Ragweed, Common *(Ambrosia artemisiifolia)*	C	C	C	C	C	C	C	C	C	C	C	C	C	C
Redweed *(Melochia corchorifolia)*	C	C	C	C	C	C	C	C	C	C	C	C	C	C
Russian Thistle *(Salsola kali)*	C	C	C	C	C	C	C	C	C	C	C	C	C	C
Sesbania *(Sesbania spp.)*	C	C	C	C	C	C	C	C	C	C	C	C	C	C
Shepherdspurse *(Capsella bursa-pastoris)*	C	C	C	C	C	C	C	C	C	C	C	C	C	C
Sicklepod *(Cassia obtusifolia)*	S	C	S	C	S	C	S	C	C	C	S	C	S	S
Smartweeds *(Polygonum spp.)*	C	C	C	C	C	C	C	C	C	C	C	C	C	S
Spotted Spurge *(Euphorbia maculata)*	C	C	P	C	P	C	S	P	C	C	C	C	P	•
Spurred Anoda *(Anoda cristata)*	C	C	C	C	C	C	C	C	C	C	C	C	C	•
Sunflower *(Helianthus spp.)*	C	C	S	S	S	S	S	S	S	S	S	C	S	P
Velvetleaf *(Abutilon theophrasti)*	C	C	C	C	C	C	C	C	C	C	C	C	C	C
Venice Mallow *(Hibiscus trionum)*	C	C	C	C	C	C	C	C	C	C	C	C	C	C
Wild Mustards *(Brassica spp.)*	C	C	C	C	C	C	C	C	C	C	C	C	C	C
ANNUAL GRASSES														
Barnyardgrass *(Echinochloa crus-galli)*	S	C	C	C	C	C	C	C	C	C	C	C	C	C
Bluegrass *(Poa annua)*	C	C	C	C	C	C	•	C	C	C	C	C	C	C
Broadleaf Signalgrass *(Brachiaria platyphylla)*	C	C	C	C	C	C	C	C	C	C	C	C	C	•
Browntop Millet *(Panicum ramosum)*	C	C	C	P	C	S	C	P	C	P	C	C	•	•
Crabgrass *(Digitaria spp.)*	C	C	C	C	C	C	C	C	C	C	C	C	C	C
Crowfootgrass *(Dactyloctenium aegyptium)*	C	C	C	C	C	C	C	C	C	C	C	C	•	•
Cupgrass *(Eriochloa gracilis)*	P	C	P	P	P	P	P	P	P	P	P	C	•	•
Foxtails *(Setaria spp.)*	S	C	C	C	C	C	C	C	C	C	C	C	C	C
Goosegrass *(Eleusine indica)*	C	C	C	C	C	C	C	C	C	C	C	C	C	C
Johnsongrass, Seedling *(Sorghum halepense)*	C	C	C	C	C	C	C	P	C	C	C	C	C	•
Junglerice *(Echinochloa colonum)*	C	C	C	C	C	C	•	C	C	C	C	C	C	•
Nutsedge, Yellow *(Cyperus esculentus)*	P	P	P	C	P	C	P	P	P	C	P	C	P	•
Panicum, Fall *(Panicum dichotomiflorum)*	P	C	C	C	C	C	C	P	P	C	P	C	C	C

ANNUAL BROADLEAF WEEDS	SENCOR Alone	SENCOR SPLIT-SHOT	SENCOR plus Treflan	SENCOR plus Dual	SENCOR plus Prowl	SENCOR plus Lasso	SENCOR plus Surflan	SENCOR plus Amiben	SENCOR plus Gramoxone	SENCOR plus LASSO and Gramoxone	SENCOR plus Surflan and Gramoxone	EXTENDED SPLIT-SHOT	SENCOR plus Sonalan	SENCOR plus Linuron plus (Lasso or Dual)
Panicum, Texas (*Panicum texanum*)	P	C	C	P	C	S	C	P	P	S	P	S	C	•
Red Rice (*Oryza sativa*)	P	C	C	C	P	C	•	P	P	C	P	C	•	•
Sandbur (*Cenchrus spp.*)	P	C	C	P	C	S	•	P	P	C	P	S	•	•
Shattercane (*Sorghum bicolor*)	P	C	C	P	P	P	P	P	P	P	P	P	C	•
Sorghum, Volunteer (Sorghum spp.)	P	C	C	P	P	P	P	P	P	P	P	P	•	P
Sprangletop (*Leptochloa spp.*)	P	C	C	P	P	P	•	P	P	P	P	P	•	P
Stinkgrass (*Eragrostis spp.*)	P	C	C	P	P	P	•	P	P	P	P	P	•	P
Wheat, Volunteer (*Triticum spp.*)	P	P	P	P	P	P	S	P	P	C	C	P	•	P
Witchgrass (*Panicum capillare*)	P	C	C	C	C	C	•	P	P	C	C	C	C	•

C = Control
• = No Information
S = Suppression or Erratic Control (Control may range from poor to excellent)
P = Poor or No Control

SENCOR Herbicide Alone

SENCOR Herbicide (Alone) Preemergence Application: The following rates of SENCOR 4 may be applied preemergence to soybeans through center pivot or lateral move sprinkler irrigation systems that apply water in a uniform manner. Refer to "Chemigation" section of the SENCOR 4 label for directions. SENCOR 4 can be applied broadcast or banded. This application may be made during planting or as a separate operation after planting but before crop emergence.

See the "Application of SENCOR with Herbicide Spray Equipment" section under "GENERAL INFORMATION" in the front of the label.

Do not apply to sand soils; or to sandy loam or loamy sand soils containing less than 2% organic matter. Do not incorporate into soil or apply more than once per season.

Pints Of SENCOR 4 Per Acre

Soil Texture	Organic Matter: Less than 2%	2 to 4%	Over 4%
COARSE SOILS (Sandy loam, loamy sand)	DO NOT USE[3]	3/4	1
MEDIUM SOILS[1] (Loam, silt loam, silt, sandy clay, sandy clay loam)	3/4 to 1	1 to 1 1/4	1 1/4 to 1 1/2
FINE SOILS[1] (Silty clay, silty clay loam,[2] clay, clay loam)	1 to 1 1/4	1 1/4 to 1 1/2	1 1/2 to 1 3/4
MISSISSIPPI DELTA ONLY	1 1/2	1 3/4	2

[1] For control of lambsquarters, redroot pigweed and wild mustard, and for suppression of green, yellow and giant foxtails on alkaline (calcareous) soils in Nebraska, Minnesota, South Dakota and North Dakota only, apply SENCOR 4 at rates of 1/2 pint/acre on medium soils and 1/2 to 3/4 pint/acre on fine soils regardless of soil organic matter percentage (use 3/4 pint only where soil pH is less than 7.5 and weed pressure is heavy). The 1/2 pint/acre rate of SENCOR 4 alone can be applied regardless of soil pH. For control of other weeds listed on the label use SENCOR 4 at full rates recommended in the table above, **but note that crop injury may occur on soils having a calcareous surface area or a pH of 7.5 or higher.**

[2] Silty clay loam soils are transitional soils and may be classified as medium textured soils in some regions of the U.S.

[3] Refer to the appropriate section of the label for use of SENCOR 4 on soybeans in coarse soils with 0.5 to 2.0% organic matter in certain states.

USES OF SENCOR HERBICIDE IN COMBINATION WITH OTHER HERBICIDES

SEQUENTIAL APPLICATION OF SCEPTER® FOLLOWING SENCOR 4

If needed, application of SENCOR 4 alone or in a registered tank mix according to directions on the label, may be followed by an early postemergence application of Scepter for control of cocklebur. Apply 1/6-1/3 pint of Scepter in a minimum of 20 gallons of water per acre. Use 1/6 pint of Scepter if cockleburs are less than 3 inches tall or have fewer than 3 leaves and are actively growing. For cockleburs less than 6 inches tall and actively growing use 1/3 pint per acre. Do not use Scepter when soybeans or cockleburs have been subjected to stress conditions such as temperature or moisture extremes. Do not exceed a total of 2/3 pint of Scepter per acre in one season. Wait at least 10 days after application of Scepter before cultivating.

When preparing the spray mixture with Scepter, add 2 pints of nonionic surfactant approved for use on growing crops and containing at least 80% active ingredient per 100 gallons of mixture. Apply crop oil concentrate (COC) at the rate specified on the COC label.

Use Scepter only in the states where it is registered as listed on the product label.

Apply Scepter at least 90 days before harvest of soybeans. Do not graze or feed soybean forage, hay, or straw to livestock.

Refer to the Scepter label for additional cautions and precautions, directions, limitations, and information on environmental hazards and planting of rotational crops.

SPLIT-SHOT APPLICATION

A preplant incorporated application of SENCOR 4 tank mixed with either Treflan, Lasso, Dual, Prowl, or Sonalan and followed by a preemergence surface application of SENCOR alone, after planting but prior to soybean emergence, will control more broadleaf and grass weeds in soybeans than when either herbicide is used alone.

Refer to the Treflan, Lasso, Dual, Prowl, or Sonalan labels and to appropriate sections of the SENCOR 4 label for directions on soil preparation, herbicide application, incorporation techniques, herbicide rates, weed species controlled, and restrictions for using tank mix combinations of SENCOR. Carefully observe the "Special Precautions" section concerning the use of SENCOR in tank mix combinations on soybeans.

When a SPLIT-SHOT application of SENCOR 4 with Prowl, Treflan, or Sonalan is used, the preplant incorporated tank mix may be applied up to 21 days prior to planting soybeans; with Dual or Lasso, the preplant incorporated tank mix may be applied up to 14 days prior to planting.

On medium and fine textured soils with greater than 2% organic matter, a rate range is recommended for the SENCOR 4 preemergence overlay application. The higher rate should be used (a) in fields with a history of severe broadleaf weed pressure, (b) when the time between preplant incorporated tank mix and preemergence overlay applications approaches the maximum stated above, and/or (c) when the organic matter content of the soil is at the upper end of the indicated range.

For black nightshade control, refer to the appropriate section of the Lasso, Dual, or Sonalan labels for specific instructions.

SPLIT-SHOT APPLICATION

	PREPLANT INCORPORATED TANK MIX APPLICATION			—FOLLOWED BY—	PREEMERGENCE OVERLAY APPLICATION				
	RATE OF COMBINATION PRODUCT/ACRE	PLUS	RATE OF SENCOR 4 PINT/ACRE		RATE OF SENCOR 4 PINT/ACRE ORGANIC MATTER				
SOIL TEXTURE[1]					LESS THAN 2%		2 TO 4%		OVER 4%
COARSE (Light) (Sand, loamy sand, sandy loam)	Treflan 1 pt. or Lasso 2-2½ qt. Dual 1¼-1½ pt. or Prowl 1½ pt. or Sonalan 1¼-2 pt.	plus—	½	—FOLLOWED BY—	¼	or	¼	or	¼-½
MEDIUM (Loam, silt loam, sandy clay loam, silt, sandy clay)	Treflan 1½ pt. or Lasso 2½-3 qt. or Dual 1½ pt. or Prowl 1½ pt. or Sonalan 1¾-2½ pt.	plus—	¾	—FOLLOWED BY—	¼	or	¼-½	or	½-¾
			or ½[2]	—FOLLOWED BY—	½	or	½-¾	or	(¾-1)[3]
FINE (Heavy) (Silty clay loam,* clay loam, silty clay, clay)	Treflan 2 pt. or Lasso 2½-3 qt. or Dual 2-2½ pt. or Prowl 1½-2 pt. or Sonalan 2¼-3 pt.	plus—	1.0	—FOLLOWED BY—	¼	or	¼-½	or	½-¾
			or ¾[2]	—FOLLOWED BY—	½	or	½-¾	or	(¾-1)[3]

* Silty clay loam soils are transitional soils and may be classified as medium textured soils in some regions of the U.S.

[1] On **coarse textured** soils, do not use on sand soils with less than 1% organic matter, or on loamy sand or sandy loam soils with less than 0.5% organic matter. However, on coarse textured soils **with a calcareous surface area or a pH of 7.5 or higher,** do not use on sand soils with less than 2% organic matter, or on loamy sand or sandy loam soils with less than 1% organic matter.

[2] Use this lower rate of SENCOR 4 in the preplant incorporated mix **on soils having a calcareous surface area or a pH of 7.5 or higher,** and in those situations where soils within a field vary extremely in texture or organic matter content.

[3] Reduce this preemergence overlay rate of SENCOR 4 by ¼ pint/acre when using SPLIT-SHOT application **on soils with over 4% organic matter and which have a calcareous surface area or a pH of 7.5 or higher.**

EXTENDED SPLIT-SHOT APPLICATION

(Includes No-Till, Reduced Till, Ridge Till, Strip Till, Mulch Till)

An early preplant (surface-applied or shallow incorporated) application of SENCOR 4 tank mixed with either Dual or Lasso, followed by a preemergence surface application of SENCOR tank mixed with Dual or Lasso after planting but prior to soybean emergence, will control more broadleaf and grass weeds in soybeans than either herbicide used alone.

An EXTENDED SPLIT-SHOT application will decrease the need for tillage and/or contact herbicides for the control of existing vegetation prior to planting, while providing residual control of weeds after planting.

When an EXTENDED SPLIT-SHOT application of SENCOR 4 with Dual or Lasso is used, the preplant tank mix combination may be applied 15 to 30 days prior to planting soybeans. Follow directions on the label accompanying the product for SPLIT-SHOT applications from 0-14 days before planting.

Where a rate range is recommended, the higher rates should be used (a) in fields with a history of severe weed pressure, (b) when the time between early preplant tank mix and preemergence overlay applications approaches the maximum 30 days, (c) when the organic matter content of the soil is at the upper end of the indicated range, (d) when heavy crop residues are present on the soil surface, and/or (e) when the early preplant tank mix application is shallow incorporated (e.g. use 2 to 2½ qts Lasso in the early preplant tank mix when surface applied and use 2½ to 3 qts Lasso when the tank mix is to be lightly incorporated).

When weeds exceed 1 to 1½ inches in height or diameter at application, use a contact herbicide, such as Roundup® or Gramoxone.

Refer to the Dual or Lasso label, and to appropriate sections of the SENCOR 4 label for additional information on soil preparation, herbicide application, weeds controlled, precautions, restrictions, limitations, and sprayer clean-up.

EXTENDED SPLIT-SHOT APPLICATION

EARLY PREPLANT TANK MIX APPLICATION (SURFACE-APPLIED OR SHALLOW INCORPORATED)				FOLLOWED BY	PREEMERGENCE OVERLAY APPLICATION				
SOIL TEXTURE(1)	RATE OF COMBINATION PRODUCT/ ACRE	PLUS	RATE OF SENCOR 4 PINT/ ACRE		RATE OF COMBINATION PRODUCT/ ACRE	PLUS	RATE OF SENCOR 4 PINT/ACRE ORGANIC MATTER ½ TO 2%	2 TO 4%	OVER 4%
COARSE (light) Sand, loamy sand, sandy loam	Dual 1⅓ pt or Lasso 1½-2 qt	plus	½-¾	followed by	Dual ⅔ pt or Lasso 1½ qt	plus	¼	¼-½	½
MEDIUM Loam, silt loam, sandy clay loam, silt sandy clay	Dual 1¾ pt or Lasso 2-3 qt	plus	(2)¾-1	followed by	Dual ¾ pt or Lasso 1-2 qt	plus	½	½-¾	¾-1

EARLY PREPLANT TANK MIX APPLICATION (SURFACE-APPLIED OR SHALLOW INCORPORATED)				FOLLOWED BY	PREEMERGENCE OVERLAY APPLICATION				
SOIL TEXTURE(1)	RATE OF COMBINATION PRODUCT/ ACRE	PLUS	RATE OF SENCOR 4 PINT/ ACRE		RATE OF COMBINATION PRODUCT/ ACRE	PLUS	RATE OF SENCOR 4 PINT/ACRE ORGANIC MATTER ½ TO 2%	2 TO 4%	OVER 4%
FINE (Heavy) Silty clay loam,* clay loam, silty clay, clay	Dual 2 pt or Lasso 2-3 qt	plus	(2)1-1¼	followed by	Dual 1 pt or Lasso 1-2 qt	plus	½	½-¾	¾-1

* Silty clay loam soils are transitional soils and may be classified as medium textured soils in some regions of the U.S.

(1) On **coarse textured** soils, do not use on sand soil with less than 1% organic matter. However, on coarse textured soils **with a calcareous surface area or a pH of 7.5 or higher,** do not use on sand soils with less than 2% organic matter, or on loamy sand or sandy loam soils with less than 1% organic matter.

(2) Use the lower rate of SENCOR 4 in the early preplant tank mix **on soils having a calcareous surface area or a pH of 7.5 or higher,** and in those rare situations where soils within a field vary extremely in texture or organic matter content.

SENCOR PLUS SONALAN

SENCOR plus Sonalan Overlay Application

SENCOR 4 may be applied as a preemergence overlay application following a preplant incorporated application on Sonalan 3 EC. Consult the Sonalan label for specific directions on use, recommendations, restrictions and any additional weeds not specified on the SENCOR label.

SENCOR plus Sonalan Tank Mix Application

Incorporate the tank mixture into the top 1 to 2 inches of soil within 21 days before planting according to label directions for Sonalan.

Apply SENCOR 4 plus Sonalan preplant incorporated if furrow irrigation is used or when a period of dry weather after application is expected. If soybeans are planted on beds, apply and incorporate the tank mixture after bed formation.

Mixing: Refer to the "General Information" section in the front of the SENCOR 4 label.

Application: Sonalan should be uniformly applied and throughly mixed into the soil within 2 days after application. For specific application information, refer to "Application" under the "General Information" section in the front of the SENCOR 4 label.

BROADCAST RATES

Soil Texture	SENCOR 4 pts/acre	Sonalan 3 EC pts/acre
COARSE[1] (Sandy loam, loamy sand)	½	1¼-2
MEDIUM[3] (Loam, silt loam, silt, sandy clay, sandy clay loam)	¾	1¾-2½
FINE[3] (Silty clay, silty clay loam,[2] clay, clay loam)	1	2¼-3

[1] Do not use on coarse soils with less than 1% organic matter.

[2] Silty clay loam soils are transitional soils and may be classified as medium textured soils in some regions of the U.S.

[3] For control of lambsquarters, redroot pigweed, wild mustard, and green and yellow foxtails on alkaline (calcareous) soils in Minnesota, Nebraska, South Dakota, and North Dakota only, apply SENCOR 4 at rates of ½ pint/acre on medium soils and ½ to ¾ pint/acre on fine soils regardless of soil organic matter percentage (use ¾ pint only where soil pH is less than 7.5 and weed pressure is heavy). The ½ pint rate of SENCOR 4 in tank mix combination with Sonalan can be applied regardless of soil pH. For control of other weeds not listed on the label, use SENCOR 4 at full rates recommended in the table above, **but note that crop injury may occur on soils having a calcareous surface area or a pH of 7.5 or higher.**

Special Precautions (SENCOR plus Sonalan)

For additional precautions, restrictions, limitations, incorporation, and sprayer clean-up information, refer to the appropriate sections of the SENCOR 4 label and the Sonalan label.

For black nightshade control, refer to the Sonalan label for specific rates and application instructions.

SENCOR PLUS TREFLAN

SENCOR plus Treflan Overlay Application

SENCOR 4 may be applied as a preemergence broadcast or band overlay application following a preplant incorporated treatment of Treflan. Consult the Treflan label for specific directions for use, recommendations, restrictions and any additional weeds not specified on the SENCOR label.

SENCOR plus Treflan Tank Mix Application

A single application of a tank mix combination of SENCOR 4 and Treflan EC will control more broadleaf and grass weeds in soybeans than when either herbicide is used alone.

Prepare the soil surface by deep plowing, offset disking or tandem disking prior to the application of the herbicide combination. The soil surface should be well prepared and free of clods and trash.

This SENCOR plus Treflan tank mix combination may be applied and incorporated into the soil up to 10 days before planting.

Mixing: Refer to the "GENERAL INFORMATION" section in the front of the SENCOR 4 label.

Application: For specific application information refer to the "GENERAL INFORMATION" section in the front of the SENCOR 4 label.

Apply SENCOR 4 plus Treflan to the soil surface and incorporate in the same operation, if possible. Variable weed control may result from delayed incorporation if SENCOR plus Treflan are applied to a wet, warm soil surface or if the wind velocity is 10 miles per hour or higher. Use machinery that mixes SENCOR plus Treflan thoroughly with the soil. Incorporation may be delayed up to 24 hours after application. Shallow incorporation with implements set to cut less than 2 inches deep may result in erratic weed control. Do not use spike or spring-tooth harrows alone for incorporation.

BROADCAST RATES

Soil Texture	SENCOR 4 Pints per Acre	Treflan EC Pints per Acre
COARSE[1] (Sandy loam, loamy sand)	½	1
MEDIUM (Loam, silt loam, silt, sandy clay, sandy clay loam)[3]	¾	1½
FINE (Silty clay, silty clay loam[2], clay, clay loam)[3]	1	2

[1] Do not use on coarse soils with less than 1% organic matter.

[2] Silty clay loam soils are transitional soils and may be classified as medium textured soils in some regions of the U.S.

[3] For control of lambsquarters, redroot pigweed, wild mustard and green and yellow foxtails on alkaline (calcareous) soils in Minnesota, Nebraska, South Dakota, and North Dakota only, apply SENCOR 4 at rates of ½ pint/acre on medium soils and ½ to ¾ pint/acre on fine soils, regardless of soil organic matter percentage (use ¾ pint only where soil pH is less than 7.5 and weed pressure is heavy). The ½ pint rate of SENCOR 4 in tank mix combination with Treflan can be applied regardless of soil pH. For control of other weeds listed on the label, use SENCOR 4 at full rates recommended in the table above, **but note that crop injury may occur on soils having a calcareous surface area or a pH of 7.5 or higher.**

Incorporation Equipment:

1. Set PTO-driven equipment (tillers, cultivators, hoes) to cut 2 to 3 inches deep and space rotors to provide a clean sweep of the soil. PTO equipment should not be operated at a speed greater than 4 miles per hour.
2. Set disk to cut 4 to 6 inches deep and operate twice in different directions at 4 to 6 miles per hour.
3. Set mulch treader and other similar disk-type implements to cut 3 to 4 inches deep and operate twice in different directions at 5 to 8 miles per hour.

For Coarse and Medium Textured Soils Only

4. Set rolling cultivator to cut 2 to 4 inches deep and operate twice at 6 to 8 miles per hour. Set bed conditioner (Do-All) to cut 2 to 4 inches deep and operate at 4 to 6 miles per hour.

Special Precautions (SENCOR plus Treflan)

Seedling disease, cold weather, excessive moisture, high salt concentration or drought may weaken soybean seedlings and increase possibility of damage from the tank mix. Do not plant soybeans deeper than 2 inches.

In the Central United States, do not plant sorghum or oats for 12 months where the tank mix has been applied unless 20 inches or more of irrigation and/or rainfall (total) was used to produce the crop. If less than 20 inches total water was used to produce the crop during the year, do not plant either crop for 18 months after the tank mix application. Cool wet weather conditions during the early stage of growth may increase the possibility of injury to sorghum.

For additional precautions, restrictions, limitations and sprayer cleanup information refer to the appropriate section of the label. Do not use this tank mix combination on soils containing charcoal in Arkansas, Louisiana and Mississippi.

SENCOR PLUS DUAL

SENCOR plus Dual Overlay Application

Apply a preplant incorporated treatment of Dual 8E as directed on that product label for use on soybeans. Follow with a preemergence treatment of SENCOR 4 as directed on the label for use on soybeans.

SENCOR plus Dual Tank Mix Applications

Preplant Incorporated Application: Incorporate the tank mixture into the top 2 inches of soil within 14 days before planting using a disk, harrow, rolling cultivator or similar implement.

Apply SENCOR 4 plus Dual preplant incorporated if furrow irrigation is used or when a period of dry weather after application is expected. If soybeans are planted on beds, apply and incorporate the tank mixture after bed formation.

Preemergence Application: Dry weather following preemergence application of SENCOR plus Dual tank mixture may reduce effectiveness. If weeds develop, cultivate uniformly with shallow tilling equipment such as rotary hoe that will not damage soybeans.

Mixing Instructions: Refer to the "GENERAL INFORMATION" section in the front of the SENCOR 4 label.

For information on applying SENCOR in fluid or dry fertilizer refer to the "Application Of SENCOR 4 In Fluid Fertilizers" or "Commercial Impregnation And Application OF SENCOR 4 On Dry Bulk Fertilizer" under the "GENERAL INFORMATION" section in the front of the SENCOR 4 label.

Broadcast Rates
SENCOR plus Dual Tank Mix Preemergence Applications

SOIL TEXTURE	Organic Matter			
	0.5% to 3%		Over 3%	
	SENCOR 4 Pints/Acre	Dual 8E Pints/Acre	SENCOR 4 Pints/Acre	Dual 8E Pints/Acre
COARSE[1] (Loamy sand, sandy loam)	1/2	1 1/4	3/4	1 1/2
MEDIUM (Loam, silt loam, silt)	3/4	1 1/2	1	2
FINE (Silty clay loam,[2] sandy clay loam, silty clay, sandy clay, clay loam, clay)	1	2	1 to 1 1/4	2 to 2 1/2
MISSISSIPPI DELTA ONLY (Silty clay, clay)	1 1/2	2	1 1/2	2 to 2 1/2

[1] Do not use on sand soils. Do not apply SENCOR 4 and Dual overlay or tank mix preemergence on loamy sand with less than 2% organic matter.

[2] Silty clay loam soils are transitional soils and may be classified as medium textured soils in some regions of the U.S.

Broadcast Rates
SENCOR plus Dual Tank Mix Preplant Incorporated Applications

SOIL TEXTURE	Organic Matter			
	0.5% to less than 3%		3% or greater	
	SENCOR 4 Pints/Acre	Dual 8E Pints/Acre	SENCOR 4 Pints/Acre	Dual 8E Pints/Acre
COARSE[1] (Loamy sand, sandy loam)	1/2	1 1/4	1/2	1 1/2
MEDIUM (Loam, silt loam, silt)	3/4	1 1/2	3/4	2
FINE (Silty clay loam,[2] sandy clay loam, silty clay, sandy clay, clay loam, clay)	1	2	1	2 to 2 1/2
MISSISSIPPI DELTA ONLY (Silty clay, clay)	1 to 1 1/4	2	1 to 1 1/4	2 to 2 1/2

[1] Do not use on sand soils. Do not apply SENCOR 4 plus Dual tank mix preplant incorporated on sand or loamy sand with less than 2% organic matter or crop injury may occur.

[2] Silty clay loam soils are transitional soils and may be classified as medium textured soils in some regions of the U.S.

Special Precautions (SENCOR and Dual)

For additional precautions, restrictions, limitations and sprayer cleanup information refer to the appropriate sections of the SENCOR label and the Dual label.

SENCOR PLUS PROWL

SENCOR plus Prowl Overlay Application

Apply a preplant incorporated treatment of Prowl as directed on that product label for use on soybeans. Follow with a preemergence treatment of SENCOR 4 as directed on the label for use on soybeans.

SENCOR plus Prowl Tank Mix Application

Preplant Incorporated Application: Prepare the soil by plowing or disking to mix previous crop residues into the soil to a depth of 4 to 6 inches.

For specific application information refer to "Application" under the "GENERAL INFORMATION" section in the front of the SENCOR 4 label.

Incorporate the tank mixture into the top 1 or 2 inches of soil within 7 days after application according to label directions for Prowl. Mechanical incorporation is not required if a rain of one-quarter inch or more occurs within 7 days after application. Soybeans must be planted no later than 7 days after application of the tank mixture.

Preemergence Application: Except for minimum and no-tillage systems, the seed bed should be firm and free of trash and clods.

For specific application information refer to the "GENERAL INFORMATION" section in the front of the SENCOR 4 label. Do not apply Prowl preemergence north of Interstate 80.

This application must be made after planting and before crop emergence. Do not incorporate.

If cultivation is necessary because of soil crusting, soil compaction or weed germination before rain or irrigation, use shallow tilling equipment such as a rotary hoe that does not damage soybeans.

Mixing Instructions: Refer to the "GENERAL INFORMATION" section in the front of the SENCOR 4 label.

For information on applying SENCOR 4 in fluid or dry fertilizer refer to the "Application Of SENCOR 4 In Fluid Fertilizers" or "Commercial Impregnation And Application Of SENCOR 4 On Dry Bulk Fertilizer" under the "GENERAL INFORMATION" section in the front of the SENCOR 4 label.

SOUTHERN STATES AND EASTERN COASTAL PLAINS

For use only in Alabama, Arkansas, Florida, Georgia, Louisiana, Mississippi, North Carolina, Oklahoma, Southeastern Missouri "Bootheel" Region, and Coastal Plains of Delaware*, Maryland*, New Jersey * and Virginia*.

*SENCOR plus Prowl is not recommended for use on soils with less than 2% organic matter in the coastal plain of New Jersey or the Delmarva Peninsula.

Broadcast Rates
SENCOR plus Prowl Tank Mix Applications

SOIL TEXTURE	SENCOR 4 Pints/Acre	Prowl Pints/Acre
COARSE[1] (Sandy loam, loamy sand)	1/2	1 1/2
MEDIUM (Loam, silt loam, silt, sandy clay, sandy clay loam)	3/4	1 1/2
FINE (Silty clay, silty clay loam,[2] clay, clay loam)	1	1 1/2 to 2

[1] Do not use on sand soils. Do not use on loamy sand or sandy loam containing less than 1% organic matter.

[2] Silty clay loam soils are transitional soils and may be classified as medium textured soils in certain regions of the U.S.

Do not use on muck or peat soils.

NORTHEASTERN AND NORTH CENTRAL STATES

For use only in Illinois, Indiana, Iowa, Kansas, Kentucky, Michigan, Minnesota, Nebraska, New York, North Dakota, Ohio, Pennsylvania, South Dakota, Wisconsin and Missouri (except the "Bootheel" Region).

Broadcast Rates
SENCOR plus Prowl Tank Mix Applications

SOIL TEXTURE	Organic Matter			
	1/2 to 3%		Over 3%	
	SENCOR 4 Pints/Acre	Prowl Pints/Acre	SENCOR 4 Pints/Acre	Prowl Pints/Acre
COARSE[1] (Sandy loam, loamy sand)	1/2	1	3/4	1 1/2

SOIL TEXTURE	Organic Matter ½ to 3%: SENCOR 4 Pints/ Acre	Prowl Pints/Acre	Organic Matter Over 3%: SENCOR 4 Pints/Acre	Prowl Pints/Acre
MEDIUM (Loam, silt loam, sandy clay, sandy clay loam)	¾	1½ to 2	¾ to 1	1½ to 2
FINE (Silty clay, silty clay loam,[2] clay, clay loam)	¾ to 1	1½ to 2	1 to 1¼	2 to 2½

[1] Do not use on sand soils. Do not use on loamy sand or sandy loam containing less than 1% organic matter. Where a range of rates is shown for medium and fine soils, use the higher rate if heavy weed infestations are anticipated.

[2] Silty clay loam soils are transitional soils and may be classified as medium textured soils in certain areas of the U.S.

Do not use on muck or peat soils.

Special Precautions (SENCOR plus Prowl)

Soil incorporation deeper than recommended will reduce weed control and can result in crop injury.

For additional precautions, restrictions, limitations, and sprayer cleanup information, refer to the appropriate sections of the SENCOR 4 and Prowl labels.

SENCOR PLUS LASSO

SENCOR plus Lasso Tank Mix Application

PREEMERGENCE: SENCOR 4 may be used in a tank mix combination with Lasso as a preemergence band or broadcast application to soybeans in accordance with the specified soil types and dosages recommended.

For specific information regarding spray equipment, dilution rates, mixing, directions for use, methods of application. limitations and restrictions refer to the appropriate section of the SENCOR 4 label.

Refer to the Lasso label for pertinent recommendations, directions for use, restrictions and any additional weeds not specified on the SENCOR 4 label.

Recommended Applications
SENCOR plus Lasso Tank Mix Preemergence Application (Broadcast Rates)

SOIL TEXTURE	½ to 3% Organic Matter: SENCOR 4 Pts./Acre	+	Lasso/ Qts./Acre	Greater than 3% Organic Matter: SENCOR 4 Pts./Acre	+	Lasso Qts./Acre
COARSE[1] (Sandy Loam)	½	+	1½ to 2	¾	+	1½ to 2
MEDIUM[2] (Loam, silt loam, silt, sandy clay, sandy clay loam)	¾	+	1½ to 2	1	+	1½ to 2
FINE[2] (Silty clay, silty clay loam,[3] clay, clay loam)	1	+	2	1 to 1¼	+	2 to 2½
MISSISSIPPI DELTA ONLY (Silty clay to heavy clay)	SENCOR 4 Pts./Acre + Lasso Qts./Acre: 2 + 2 to 2½					

[1] Do not use SENCOR plus Lasso on sand or loamy sand soils with less than 2% organic matter.

[2] For control of lambsquarters, redroot pigweed, wild mustard and green and yellow foxtails on alkaline (calcareous) soils in Minnesota, Nebraska, South Dakota, and North Dakota only, apply SENCOR 4 at rates of ½ pint/acre on medium soils and ½ to ¾ pint/acre on fine soils regardless of soil organic matter percentage (use ¾ pint only where soil pH is less than 7.5 and weed pressure is heavy). The ½ pint/acre rate of SENCOR 4 in tank mix combination with Lasso can be applied regardless of soil pH. For control of other weeds use SENCOR 4 at full rates recommended in the table above, **but note that crop injury may occur on soils having a calcareous surface area or a pH of 7.5 or higher.**

[3] Silty clay loam soils are transitional soils and may be classified as medium textured soils in some regions of the U.S.

DO NOT USE ON MUCK SOILS.

PREPLANT INCORPORATED: For specific application information refer to the "GENERAL INFORMATION" section in the front of the SENCOR 4 label.

Apply SENCOR 4 plus Lasso preplant incorporated if furrow irrigation is used or when a period of dry weather after application is expected. If soybeans are planted on beds, apply and incorporate the tank mixture after bed formation. Apply within 7 days prior to planting and shallowly incorporate into the upper 1 to 2 inches of soil.

Recommended Applications
SENCOR plus Lasso Tank Mix Preplant Incorporated Applications (Broadcast Rates)

Soil Texture	SENCOR 4 Pints/ Acre	Lasso Quarts/Acre
COARSE[1] (Loamy sand [over 2% organic matter], sandy loam)	½	2 to 2½
MEDIUM (Loam, silty loam, silt)	¾	2½ to 3
FINE (Silty clay loam,[2] sandy clay loam, silty clay, sandy clay, clay loam, clay)	1	2½ to 3
MISSISSIPPI DELTA ONLY: (Silty clay, clay)	1 to 1¼	2½ to 3

[1] Do not use SENCOR plus Lasso on sand or loamy sand soils with less than 2% organic matter.

[2] Silty clay loam soils are transitional soils and may be classified as medium textured soils in some regions of the U.S.

DO NOT USE ON MUCK SOILS.

Special Precautions (SENCOR plus Lasso)

For additional precautions, restrictions, limitations and sprayer cleanup information refer to the appropriate sections of the SENCOR 4 and Lasso labels.

SENCOR plus Cannon 3 EC Tank Mix Application

SENCOR 4 may be tank mixed with Cannon 3 EC for preplant incorporated application to control certain weeds in soybeans. Refer to the "GENERAL INFORMATION" section of the SENCOR 4 label for information on mixing, application, restrictions, special precautions and weeds controlled by SENCOR 4. See appropriate sections of the Cannon 3 EC herbicide label for additional precautionary statements, directions for use, recommendations and additional weeds controlled.

RECOMMENDED APPLICATIONS
SENCOR plus Cannon Tank Mix Preplant Incorporated Application (Broadcast Rates)

SOIL TEXTURE	½ to 3% Organic Matter: SENCOR 4 Pts./Acre	+	Cannon 3 EC Qts./Acre-	Greater than 3% Organic Matter: SENCOR 4 Pts./Acre	+	Cannon 3 EC/ Qts./Acre-
COARSE[1] (Sandy Loam)	½	+	3	¾	+	3
MEDIUM[2] (Loam, silt loam, silt, sandy clay, sandy clay loam)	¾	+	3	1	+	3
FINE[2] (Silty clay, silty clay loam,[3] clay, clay loam)	1	+	3	1 to 1¼	+	3
MISSISSIPPI DELTA ONLY (Silty clay to heavy clay)	SENCOR 4 Pts./Acre + Cannon 3 EC Qts./Acre: 2 + 3					

[1] Do not use SENCOR plus Cannon on sand or loamy sand soils with less than 2% organic matter.

[2] For control of lambsquarters, redroot pigweed, wild mustard, green and yellow foxtails on alkaline (calcareous) soils in Minnesota, Nebraska, South Dakota, and North Dakota only, apply SENCOR 4 at rates of ½ pint/acre on medium soils and ½ to ¾ pint/acre on fine soils regardless of soil organic matter percentage (use ¾ pint only where soil pH is less than 7.5 and weed pressure is heavy). The ½ pint/acre of SENCOR 4 in tank mix combination with Cannon can be applied regardless of soil pH. For control of other weeds, use SENCOR 4 at full rates recommended in the table above, **but note that crop injury may occur on soils having a calcareous surface area or a pH of 7.5 or higher.**

[3] Silty clay loam soils are transitional soils and may be classified as medium textured soils in some regions of the U.S.

DO NOT USE ON MUCK SOILS.

Do not allow sprays to drift on to adjacent desirable plants.

Observe all cautions and limitations on the labeling of all products used in mixtures.

SENCOR plus Commence 5.25 EC Tank Mix Application

SENCOR 4 may be tank mixed with Commence 5.25 EC for preplant incorporated application to control certain weeds in soybeans. Refer to the "GENERAL INFORMATION" section of the SENCOR 4 label for information on mixing, application, restrictions, special precautions and weeds controlled by SENCOR 4. See appropriate sections of the Commence 5.25 EC herbicide label for additional precautionary statements, directions for use, recommendations and additional weeds controlled.

RECOMMENDED APPLICATIONS
SENCOR plus Commence—Tank Mix Preplant Incorporated Application
(Broadcast Rates)

SOIL TEXTURE[1]	Organic Matter			
	1/2 to 3%		Over 3%	
	SENCOR 4 Pts/Acre	Commence 5.25 EC Pts/Acre	SENCOR 4 Pts/Acre	Commence 5.25 EC Pts/Acre
COARSE[1] (Sandy loam, loamy sand)	1/2	1 1/3 to 2	1/2	1 1/3 to 2
MEDIUM (Loam, silt loam, silt, sandy clay, sandy clay loam)	1/2 to 3/4	2 to 2 1/4	1/2 to 3/4	2 to 2 1/4
FINE (Silty clay, silty clay loam,[3] clay, clay loam)	1/2 to 3/4	2 2/3	3/4 to 1	2 2/3

[1] Crop injury may occur on soils having a calcareous surface area or a pH of 7.1 of higher.
[2] Do not use on coarse soils with less than 1% organic matter.
[3] Silty clay loam soils are transitional soils and may be classified as medium textured soils in some regions of the U.S.

Special Precautions (SENCOR plus Commence)

Do not rotate to wheat, oats, barley, rye, alfalfa or seed corn in the fall of the year of application or in the spring of the following year as crop injury may occur.
Do not apply when weather conditions favor drift.
Do not use treated vines for feed or forage.
Do not apply aerially or through irrigation equipment.
Do not allow sprays to drift on to adjacent desirable plants.
Observe all cautions and limitations on the labeling of all products used in mixtures.

SENCOR PLUS COMMAND® TANK MIX APPLICATION

SENCOR 4 may be applied in combination with Command 4 EC as a preplant or shallow incorporated application for the control of certain weeds in soybeans. Consult the Command label for specific directions on use, recommendations, restrictions and any additional weeds not specified on the SENCOR 4 label.

Mixing: Refer to the "General Information" section in the front of the SENCOR 4 label.

Application: SENCOR 4 plus Command 4 EC may only be applied with ground equipment as a preplant or shallow incorporated application. SENCOR plus Command should be immediately incorporated into the top 1-3 inches after application unless surface is dry. On dry soils, incorporate into the top 1-3 inches within 3 hours of tank mix application.

Do not apply this tank mix within 1000 feet of towns and subdivisions, commercial vegetable, fruit, nurseries or greenhouse operations.

A minimum of 15 gallons per acre spray volume is recommended with appropriate nozzle types and sizes to produce a coarse spray droplet. The use of an approved agricultural drift reducing additive is recommended for application volumes of 15-40 gallons per acre. The use of an approved agricultural drift reducing additive is required at spray volumes of 10-15 gallons per acre.

NOTE: Off-site movement of Command spray drift or vapors can cause foliar whitening or yellowing of some vegetation. Prior to application of Command, read and strictly follow all precautions and application instructions as set forth in that label.

For additional information on application, refer to the "General Information" section in the front of the SENCOR 4 label and the Command 4 EC label.

Weeds Controlled: Bristly Starbur, Carpetweed, Copperleaf, Florida Beggarweed, Florida Pusley, Galinsoga, Jimsonweed, Knotweed, Lambsquarters, Pigweeds, Prickly Sida/Teaweed, Purslane, Common Ragweed, Redweed, Sesbania, Smartweeds, Spurred Anoda, Velvetleaf, Venice Mallow, Wild Mustards, Barnyardgrass*, Bluegrass, Broadleaf Signalgrass, Crabgrass*, Foxtails (Green, Giant, Yellow*, Robust Purple), Goosegrass, Johnsongrass (seedling)*, Fall Panicum*, Texas Panicum, and Witchgrass.
*Use 2 pts/A Command 4 EC on coarse and medium textured soils with high populations of these weeds.

RECOMMENDED APPLICATIONS
(Broadcast Rates)

Soil Texture[1]	Organic Matter			
	0.5 to 3%		Over 3%	
	SENCOR 4 (pts/acre)	Command 4 E (pts/acre)	SENCOR 4 (pts/acre)	Command 4 EC (pts/acre)
COARSE[2] (Sandy loam, loamy sand)	1/2	1 1/2 - 2	1/2	1 1/2 - 2
MEDIUM (Loam, silt loam, silt, sandy clay, sandy clay loam)	1/2 - 3/4	1 1/2 - 2	1/2 - 3/4	1 1/2 - 2
FINE Silty clay, silty clay loam,[3] clay, clay loam)	1/2 - 3/4	1 1/2 - 2	3/4 - 1	1 1/2 - 2

[1] Crop injury may occur on soils having a calcareous surface area or a pH of 7.1 or higher.
[2] Do not use on coarse soils with less than 1% organic matter.
[3] Silty clay loam soils are transitional soils and may be classified as medium textured soils in some regions of the U.S.

Special Precautions (SENCOR plus Command)

Do not rotate to wheat, oats, barley, rye, alfalfa or seed corn in the fall of the year of application or in the spring of the following year as crop injury may occur.

Do not apply when weather conditions favor drift.

Do not use treated vines for feed or forage.

Observe all cautions and limitations on labeling of all products used in mixtures.

Do not apply aerially or through irrigation equipment.

SENCOR plus Freedom 3 EC Tank Mix Application

SENCOR 4 may be tank mixed with Freedom 3 EC for preplant incorporated application to control certain weeds in soybeans. Refer to the "GENERAL INFORMATION" Section of the SENCOR 4 label for information on mixing, application, restrictions, special precautions and weeds controlled by SENCOR 4. See appropriate sections of the Freedom 3 EC herbicide label for additional precautionary statements, directions for use, recommendations and additional weeds controlled.

RECOMMENDED APPLICATIONS
SENCOR plus Freedom Tank Mix Preplant Incorporated Application
(Broadcast Rates)

SOIL TEXTURE	1/2 to 3% Organic Matter			Greater than 3% Organic Matter		
	SENCOR 4 Pts./Acre	+	Freedom 3 EC Qts./Acre-	SENCOR 4 Pts./Acre	+	Freedom 3 EC/ Qts./Acre
COARSE[1] (Sandy Loam)	1/2	+	2 3/4 to 3 1/2	3/4	+	3 to 3 1/2
MEDIUM[2] (Loam, silt loam, silt, sandy clay, sandy clay loam)	3/4	+	2 3/4 to 3 1/2	1	+	3 1/2 to 4
FINE[2] (Silty clay, silty clay loam,[3] clay, clay loam)	1	+	3 1/2 to 4	1 to 1 1/4	+	3 1/2 to 4 1/2
MISSISSIPPI DELTA ONLY (Silty clay to heavy clay)	SENCOR 4 Pts./Acre + Freedom 3 EC Qts./Acre 2 + 3 1/2 to 4 1/2					

[1] Do not use SENCOR plus Freedom on sand or loamy sand soils with less than 2% organic matter.
[2] For control of lambsquarters, redroot pigweed, wild mustard, green and yellow foxtails on alkaline (calcareous) soils in Minnesota, Nebraska, South Dakota, and North Dakota only, apply SENCOR 4 at rates of 1/2 pint/acre on medium soils and 1/2 to 3/4 pint/acre on fine soils regardless of soil organic matter percentage (use 3/4 pint only where soil pH is less than 7.5 and weed pressure is heavy). The 1/2 pint/acre of SENCOR 4 in tank mix combination with Freedom can be applied regardless of soil pH. For control of other weeds, use SENCOR 4 at full rates recommended in the table above, **but note that crop injury may occur on soils having a calcareous surface area or a pH of 7.5 or higher.**
[3] Silty clay loam soils are transitional soils and may be classified as medium textured soils in some regions of the U.S.

DO NOT USE ON MUCK SOILS.

Do not allow sprays to drift on to adjacent desirable plants.

Observe all cautions and limitations on the labeling of all products used in mixtures.

SENCOR Plus Command Plus a Grass Herbicide Tank Mix Application

SENCOR 4 may be applied with Command 4 EC and a grass herbicide (Treflan, Lasso, Dual, Prowl or Sonalan) for the control of certain broadleaf weeds and grasses in soybeans. This combination will provide improved control of heavy infestations of velvetleaf, jimsonweed and common ragweed. SENCOR and Command plus a grass herbicide may be applied preplant incorporated broadcast. Consult the Command, Treflan, Lasso, Dual, Prowl or Sonalan labels for specific directions for use, recommendations, restrictions and additional weeds controlled not specified on the SENCOR 4 label.

Mixing: Refer to the "General Information" section in the front of the SENCOR 4 label.

Application: For specific application information, refer to the "General Information" section in the front of the SENCOR 4 label.

Weeds Controlled:

SENCOR plus Command plus Treflan, Lasso, Dual, Prowl or Sonalan will control the following broadleaf weeds and grasses:

Annual Grasses	Annual Broadleaves	
Barnyardgrass	Bristly Starbur	Ragweed, Common
Bluegrass	Carpetweed	Redweed
Broadleaf Signalgrass	Copperleaf, Hophornbeam	Russian Thistle
Browntop Millet	Florida Beggarweed	Sesbania
Crabgrass	Florida Pusley	Shepherdspurse
Crowfootgrass	Galinsoga	Sicklepod
Foxtails	Jimsonweed	Smartweeds
Goosegrass	Knotweed	Spotted Spurge
Johnsongrass (seedling)	Kochia	Spurred Anoda
Panicum, Fall	Lambsquarters	Velvetleaf
Witchgrass	Pigweed	Venice Mallow
	Prickly Sida/Teaweed	Wild Mustards
	Purslane	

SENCOR and Command plus Treflan, Lasso, Dual, Prowl or Sonalan will provide suppression (reduce the competition) of cocklebur and sunflower.

Recommended Application (Broadcast Rates)

A = Treflan (pt/acre)
B = Dual (pt/acre)
C = Prowl (pt/acre)
D = Lasso (qt/acre)
E = Sonalan (pt/acre)

Soil[1] Texture	A	B	C	D	E	SENCOR 4 (pts/acre)	Command 4 EC[3] (pts/acre)
Coarse	1	1¼-1½	1½	2-2½	1¼-2	½	½ to ¾
Medium	1½	1½- 2	1½-2	2½-3	1¾-2½	½ to ¾[2]	½ to ¾
Fine	2	2-2½	1½-2½	2½-3	2¼-3	¾ to 1[2]	½ to ¾

[1] Refer to "Soil Texture" paragraph on the SENCOR 4 label for specific soil classification. On coarse textured soils **with a calcareous surface area or a pH of 7.5 or higher,** do not use on loamy sand or sandy loam soils with less than 1% organic matter.

[2] The higher rate of SENCOR 4 is recommended for the control of sicklepod and hemp sesbania. Use the lower rate of SENCOR in the preplant incorporated tank mix **on soils having a calcareous surface area or a pH of 7.5 or higher,** and in those situations where soils within a field vary extremely in texture or organic matter content.

[3] Higher rate is recommended under moderate to heavy weed infestations.

SENCOR Plus Scepter Plus a Grass Herbicide Tank Mix Application

SENCOR 4 may be applied with Scepter and a grass herbicide (Treflan, Lasso, Dual, Prowl or Sonalan) for the control of certain broadleaf weeds and grasses in soybeans. SENCOR and Scepter plus Treflan or Sonalan may be applied preplant incorporated broadcast. SENCOR and Scepter plus Lasso, Dual or Prowl may be applied preplant incorporated, preemergence broadcast or in a band application. Consult the Scepter, Treflan, Lasso, Dual, Prowl or Sonalan labels for specific directions for use, recommendations, restrictions and additional weeds controlled not specified on the SENCOR 4 label.

Mixing: Refer to the "General Information" section in the front of the SENCOR 4 label.

Application: For specific application information, refer to the "General Information" section in the front of the SENCOR 4 label.

Weeds Controlled:

SENCOR plus Scepter plus Treflan, Lasso, Dual, Prowl or Sonalan will control the following broadleaf weeds and grasses:

Annual Grasses	Annual Broadleaves	
Barnyardgrass	Bristly Starbur	Pigweed
Bluegrass	Buffalobur	Prickly Sida/Teaweed
Broadleaf Signalgrass	Carpetweed	Purslane
Browntop Millet	Cocklebur	Ragweed, Common
Crabgrass	Coffee Senna	Redweed
Crowfootgrass	Copperleaf, Hophornbeam	Russian Thistle
Foxtails	Florida Beggarweed	Sesbania
Goosegrass	Florida Pusley	Shepherdspurse
Johnsongrass (seedling)	Galinsoga	Sicklepod
Panicum, Fall	Jimsonweed	Smartweeds
Witchgrass	Knotweed	Spotted Spurge
	Kochia	Spurred Anoda
	Lambsquarters	Sunflower
	Morningglory, Pitted	Velvetleaf
	Morningglory, Smallflower	Venice Mallow
		Wild Mustards

SENCOR and Scepter plus Treflan, Lasso, Dual, Prowl or Sonalan will provide suppression (reduce the competition) of ivyleaf and tall morningglory, and red rice.

Recommended Application (Broadcast Rates)

A = Treflan (pt/acre)
B = Dual (pt/acre)
C = Prowl (pt/acre)
D = Lasso (qt/acre)
E = Sonalan (pt/acre)

Soil[1] Texture	A	B	C	D	E	SENCOR 4 (pts/acre)	Scepter[3] (pts/acre)
Coarse	1	1¼-1½	1½	2-2½	1¼-2	½	⅓ to ½
Medium	1½	1½-2	1½-2	2½-3	1¾-2½	½ to ¾[2]	⅓ to ½
Fine	2	2-2½	1½-2½	2½-3	2¼-3	¾ to 1[2]	⅓ to ½

[1] Refer to "Soil Texture" for specific soil classification. On coarse textured soils **with a calcareous surface area or a pH of 7.5 or higher,** do not use on loamy sand or sandy loam soils with less than 1% organic matter.

[2] The higher rate of SENCOR 4 is recommended for preemergence tank mix application and for the control of sicklepod and hemp sesbania. Use the lower rate of SENCOR 4 in the preplant incorporated tank mix **on soils having a calcareous surface area or a pH of 7.5 or higher,** and in those situations where soils within a field vary extremely in texture or organic matter content.

[3] Higher rate is recommended under moderate to heavy weed infestations.

SENCOR PLUS SURFLAN

SENCOR plus Surflan Tank Mix Application

SENCOR 4 may be used in a tank mix combination with Surflan A.S. as a preemergence application to control weeds in soybeans in accordance with the specified soil types and dosages recommended. Consult the Surflan label for specific directions for use, recommendations, restrictions and any additional weeds not specified on the SENCOR 4 label.

Mixing Instructions: Refer to the "GENERAL INFORMATION" section in the front of the SENCOR label.

Application: Apply within 2 days after planting but before crop emergence.

A one-half inch rain or more or its equivalent in overhead irrigation is needed to activate the herbicide combination. If a one-half inch rain does not occur within 7 days after application, incorporate the herbicide combination into the top one inch of the soil with a ground-driven rotary hoe, rolling cultivator, or similar implement that will not damage soybean plants. Cultivation will control emerging annual weeds and grasses and improve herbicide effectiveness. If heavy rains pack the soil surface, rotary hoe to break the crust and to aid soybean emergence.

For specific information on application, refer to the "GENERAL INFORMATION" section in the front of the SENCOR 4 label.

Broadcast Rates

Soil Texture	SENCOR 4 Pints Per Acre	Surflan A.S. Quarts Per Acre
COARSE[1](Sand, sandy loam, loamy sand)	½	½ to ¾
MEDIUM (Loam, silt loam, silt, sandy clay, sandy clay loam)	¾	¾ to 1
FINE (Silty clay, silty clay loam,[2] clay, clay loam)	1	1 to 1¼
MISSISSIPPI DELTA ONLY	1 to 1¼	1 to 1¼

[1] Not recommended on sand with less than 1% organic matter. Do not use on soils containing more than 5% organic matter.

[2] Silty clay loam soils are transitional soils and may be classified as medium textured soils in some regions of the U.S.

Special Precautions (SENCOR plus Surflan)

Do not plant root crops for 12 months following application.

Do not contaminate any body of water nor apply to any area not specified on the label.

For additional precautions, restrictions, limitations and sprayer clean-up information, refer to the appropriate section of the SENCOR 4 label and the Surflan A.S. label.

SENCOR PLUS AMIBEN

SENCOR plus Amiben Preemergence Tank Mix Application

SENCOR 4 may be applied in a tank mix combination with Amiben 2 as a band or broadcast application to soybeans.

SENCOR plus Amiben Preemergence Tank Mix Application
Broadcast Rates

Soil Texture	½ to 3% Organic Matter			Greater than 3% Organic Matter		
	SENCOR 4 Pts./Acre	+	Amiben 2 Gal./Acre	SENCOR 4 Pts./Acre	+	Amiben 2 Gal./Acre
COARSE[1] (Sandy loam, loamy sand)	½	+	¾ to 1	¾	+	¾ to 1
MEDIUM (Loam, silt loam, silt, sandy clay, sandy clay loam)	¾	+	¾ to 1	1	+	1 to 1¼
FINE (Silty clay, silty clay loam,[2] clay, clay loam)	1	+	1 to 1¼	1 to 1¼	+	1 to 1¼
MISSISSIPPI DELTA ONLY (Silty clay to heavy clay)	SENCOR 4 Pts./Acre + Amiben Gal./Acre 2 + 1½					

[1] Do not use on sandy loam or loamy sand soils with less than 2% organic matter. Do not use on sand soils.

[2] Silty clay loam soils are transitional soils and may be classified as medium textured soils in some regions of the U.S.

Do not use on muck soils.

Where a rate range is specified, use the higher rate in fields with a history of intense weed pressure. See "Rate Ranges" in the front of the "Soybeans" section on the SENCOR 4 label.

For additional precautions and information regarding spray equipment, mixing, application, limitations and restrictions, refer to the appropriate section of the SENCOR 4 and Amiben 2 labels.

SENCOR PLUS LINURON PLUS (Lasso or Dual)

SENCOR plus linuron plus (Lasso or Dual) Tank Mix Application

SENCOR 4 may be applied in combination with linuron 50 WP or 4 L and Lasso 4 or Dual 8 E as a preemergence application for the control of certain weeds in soybeans. Consult the linuron, Lasso, or Dual labels for specific directions for use, recommendations, restrictions and any additional weeds not specified on the SENCOR 4 label.

Mixing: Refer to the "GENERAL INFORMATION" section in the front of the SENCOR 4 label.

Application: Applications can be made only with ground spray equipment in accordance with specified soil types and dosage rates. For specific application information, refer to the "GENERAL INFORMATION" section in the front of the label.

Broadcast Rates
(0.5 to 3% Organic Matter Only)

Soil Texture	SENCOR 4 Pts/Acre	linuron 50 WP or 4 L (Lbs/Acre) or (Pts/Acre)	Lasso 4 Qts/Acre	or	Dual 8 E Pts/Acre
COARSE[1] (Sand, loamy sand, sandy loam)	¼ to ⅜	⅓ to ½	¾ to 1	or	1 to 1¼
MEDIUM (Loam, silt loam, silt, sandy clay, sandy clay loam)	⅜ to ½	½ to ¾	1 to 1½	or	1¼ to 1½
FINE (Silty clay, silty clay loam,[2] clay, clay loam)	½ to ¾	¾ to 1½	1¼ to 2	or	1½ to 2

[1] Do not use SENCOR plus linuron plus (Lasso or Dual) on sand soils with less than 1% organic matter.

[2] Silty clay loam soils are transitional soils and may be classified as medium textured soils in some regions of the U.S.

Special Precautions SENCOR plus linuron plus (Lasso or Dual)

For additional precautions, restrictions, limitations, and sprayer cleanup information, refer to the appropriate sections of the SENCOR 4 label and the linuron label and the Lasso or Dual labels.

FOR USE IN COARSE (LIGHT) SOILS in Alabama, Arkansas, Florida, Georgia, Louisiana, Mississippi, Missouri, North Carolina, Oklahoma, South Carolina, Tennessee, Texas, Virginia.

SENCOR 4 herbicide is recommended alone or in combination with Treflan, Lasso, Dual, Surflan or Amiben for use in coarse-textured, low organic matter soils in the states listed above for the control of certain weeds in soybeans. Refer to the appropriate sections of the SENCOR, Treflan, Lasso, Dual, Surflan or Amiben labels for specific directions on use, recommendations, restrictions and any additional weeds not specified on the SENCOR 4 label.

Mixing: Refer to the "GENERAL INFORMATION" section in the front of the SENCOR 4 label.

Application: For specific application information refer to the "GENERAL INFORMATION" section in the front of the SENCOR 4 label.

SENCOR (Alone) Preemergence Application

Broadcast Rates

Soil Texture	Organic Matter	SENCOR 4 Pts/Acre
Coarse (Light) Soils Sand,[1] loamy sand, sandy loam	0.5% or above	½ to ¾[2]

[1] Not recommended for use on sand with less than 1% organic matter.

[2] Use the higher rate under heavy weed pressures and/or on soils higher in organic matter.

SENCOR in Combination with Other Herbicides

SENCOR 4 is recommended in a tank mix combination with Treflan as a preplant incorporated application or as a preemergence overlay application following a preplant incorporated application of Treflan. SENCOR 4 is also recommended for use as a preemergence application in combination with Lasso, Dual, Surflan or Amiben.

Broadcast Rates

Soil Texture	Organic Matter	Combination Product/Acre	Plus- SENCOR 4 Pts/Acre
COARSE (LIGHT) SOILS Sand,[1] loamy sand, sandy loam	0.5% or above	**Preplant Incorporated** Treflan 4EC 1 pt –or– **Preemergence** Lasso 4 EC 1½ to 2 qts –or– Dual 8EC 1¼ to 1½ pts –or– Surflan A.S. ½ to ¾ qt –or– Amiben 2¾ to 1 gal	½ to ¾[2] ½ to ¾[2]

[1] Not recommended for use on sand with less than 1% organic matter.

[2] Use the higher rate under heavy weed pressures and/or on soils higher in organic matter.

Special Precautions

Do not use on sand soils with less than 1% organic matter, or on sandy loam or loamy sand soils with less than 0.5% organic matter.

For additional precautions, restrictions, limitations and sprayer cleanup information, refer to the appropriate sections of the SENCOR 4, Treflan, Lasso, Dual, Surflan or Amiben labels.

FOR CONTACT ACTION AND RESIDUAL CONTROL

For Control of Emerged Annual Broadleaf Weeds and Grasses and For Suppression of Emerged Perennials Where Soybeans Will Be Planted Directly Into a Pre-formed Bed, Cover Crop or in Previous Crop Residues as in "Minimum Till" or "No-Till" Crop Production.

SENCOR plus Gramoxone

SENCOR plus Gramoxone Tank Mix Application

Apply 1⅓ to 2⅔ pints of Gramoxone Super plus ¾ to 2 pints SENCOR 4 according to the recommendations for SENCOR on the label as a broadcast ground spray before, during or after planting, but before emergence of the crop. Soybeans exposed at time of application will be killed. Use the higher rate of Gramoxone for heavier infestations or for taller weeds (4 to 6 inches tall). Apply in 20 to 60 gallons spray mix per acre. Use the higher volume in fields with heavy weed infestations or with large amounts of previous crop residues. Add Ortho X-77® Spreader or similar non-ionic spreader at 1 quart per 100 gallons of diluted spray. This mixture will not control weeds taller than 6 inches.

SENCOR plus Lasso and Gramoxone

SENCOR plus Lasso and Gramoxone Tank Mix Application

Apply 1⅓ to 2⅔ pints of Gramoxone Super plus 2 to 2½ quarts Lasso 4 EC plus ½ to 2 pints SENCOR 4 as a preemergence broadcast ground spray. Use the higher rate of Gramoxone for weeds 4 to 6 inches tall. Soybeans exposed at time of application will be killed. Refer to the appropriate sections of the SENCOR, Gramoxone and Lasso labels for rates which are dependent upon soil textures, directions, limitations and cautions.

Apply in 20 to 60 gallons spray mix per acre. Use the higher volume in fields with heavy weed infestations or with large amounts or previous crop residues. Add Ortho X-77 Spreader or similar non-ionic spreader at 8 fl. ounces per 100 gallons of diluted spray. Add SENCOR to the spray tank first and mix thoroughly. Then add Lasso followed by Gramoxone and X-77 Spreader. Constant agitation is recommended at all times. This tank mix will not control shattercane which emerges after treatment or other weeds taller than 6 inches.

SENCOR plus Surflan and Gramoxone

SENCOR plus Surflan and Gramoxone Tank Mix Application

The combination provides control of weeds and grasses which are emerged at the time of application plus many weeds and grasses which may germinate after planting. Soybeans exposed at time of application will be killed.

Weeds Controlled by Gramoxone: Gramoxone will kill emerged annual broadleaf weeds and grasses and provide top kill of perennials.

When to Apply: Apply the tank mix at planting or within 2 days after planting. Do not spray over top of emerged soybeans.

SENCOR plus Surflan and Gramoxone Preemergence Tank Mix Application

Soil Texture	Broadcast Rates Per Acre		
	SENCOR 4	Surflan A.S.	Gramoxone Super
COARSE[1]	1/2 pint	3/4 quart	1 1/3 quarts
MEDIUM	3/4 pint	1 quart	1 1/3 quarts
FINE	1 pint	1 1/2 quarts	1 1/3 quarts

[1] Not recommended on sand soils with less than 1% organic matter.
Do not use on soils containing more than 5% organic matter.

Mixing: Refer to the "GENERAL INFORMATION" section in the front of the SENCOR 4 label. Add Ortho X-77 Spreader or similar non-ionic spreader at 8 fl. ozs. per 100 gallons of diluted spray mixture.

Application: For specific application information refer to the "GENERAL INFORMATION" section in the front of the SENCOR 4 label.

A one-half inch rain or more or its equivalent in overhead irrigation is needed to activate the herbicide combination.

Special Precautions (SENCOR plus Surflan and Gramoxone):
Poor weed control and/or crop injury may result if directions are not followed.

Do not use a rib-type press wheel on your no-till planter or crop injury may result.

For additional precautions, special precautions, restrictions, limitations and sprayer cleanup information, refer to the appropriate section of the SENCOR 4, Surflan and Gramoxone labels.

SENCOR Plus 2,4-D Low Volatile Ester

SENCOR 4 alone or in combination with Dual, Lasso, Surflan or Prowl may be applied with 2,4-D low volatile ester as an early preplant surface application for the control of certain broadleaf weeds and grasses in soybeans in minimum or no-till production. Apply one pound active ingredient (ai) per acre of 2,4-D low volatile ester with labelled rates of SENCOR 4. Where a grass herbicide is used in the tank mix, apply at the rates specified on that product's label. Application is recommended 30 days prior to planting.

Refer to the SENCOR 4, 2,4-D low volatile ester, Dual, Lasso, Surflan and Prowl labels for weeds controlled, precautions, restrictions, limitations and sprayer clean-up information.

SOUTHERN AND SOUTHEASTERN STATES ONLY

POSTEMERGENCE DIRECTED SPRAY APPLICATIONS:

SENCOR 4 can be applied in postemergence directed sprays to soybeans for control of certain weeds which escape preplant or preemergence herbicide applications and for control of additional flushes of weeds that may occur after soybeans have emerged. Postemergence directed sprays of SENCOR 4 can be applied to soybeans in addition to a preemergence or preplant application of SENCOR 4 according to label directions.

Weeds Controlled:
SENCOR 4, applied postemergence to soybeans as a directed spray according to directions on the label, will control the following at rates shown (broadcast basis) when grasses and common ragweed are less than 1 inch tall and other broadleaves are less than 3 inches tall:

1/2 pint/acre

Florida Beggarweed *(Desmodium tortuosum)*	Mexicanweed *(Caperonia castaniifolia)*
Pigweeds *(Amaranthus spp.)*	Purslane *(Portulaca oleracea)*
Carpetweed *(Mollugo verticillata)*	Sicklepod *(Cassia obtusifolia)*
Cocklebur *(Xanthium pensylvanicum)*	Velvetleaf *(Abutilon theophrasti)*
Dayflower *(Commelina spp.)*	Crabgrass *(Digitaria spp.)*

1/2 to 1 pint/acre	1 pint/acre
Sesbania *(Sesbania spp.)*	Ragweed, Common *(Ambrosia artemisiifolia)*
Prickly sida/Teaweed *(Sida spinosa)*	

At the rate of 1 pint/acre morningglory species, *(Ipomoea spp.)* horsenettle, *(Solanum spp.)* Florida pusley, *(Richardia scabra)* spotted spurge *(Euphorbia maculata)* and wild poinsettia *(Euphorbia heterophylla)* are suppressed when SENCOR 4 is applied before these weeds are 3 inches tall. The 1 pint/acre rate will suppress broadleaf signalgrass *(Brachiaria platyphylla)* up to 1 inch tall.

SENCOR Postemergence Directed Spray Applications
Recommended Applications

Crop	SENCOR 4 Pints Per Acre	Remarks
Soybeans (Alabama, Arkansas, Florida, Georgia, Kentucky, Louisiana, Mississippi, Missouri, North Carolina, Oklahoma, South Carolina, Tennessee and Texas)	1/2 to 1 (broadcast basis)	Apply proper dosage using 10 to 40 gallons of water per acre as a directed spray in a 6 to 8 inch band on each side of the row after soybeans are 8 inches tall and before broadleaf weeds are 3 inches tall and before grasses and common ragweed are 1 inch tall. For best results the spray must cover weed foliage with minimum or no contact with soybean foliage. Add a non-ionic surfactant such as Ortho X-77® to the spray mixture to obtain better wetting of weed leaf surfaces. To determine the correct dosage of SENCOR 4 for a band application see "Banded Application" under "Application" in the "GENERAL INFORMATION" section in the front of the label. If necessary, a second postemergence directed spray application can be made after 7 days. Do not feed or graze green soybean vines. Do not harvest soybeans or use dry soybean vines for feed or forage within 70 days of last application.

SPECIAL PRECAUTIONS (Directed Postemergence):
Do not apply directly to soybeans or serious crop injury will occur. Do not allow spray to contact more than the lower 1/4-1/3 of soybean plants. Soybean leaves contacted by the spray will be killed.

Do not apply SENCOR 4 postemergence to sensitive soybean varieties. See "Special Precautions" in front of the SENCOR 4 label.

To avoid injury to other crops or desirable plants from spray drift, sprayer pressure must not exceed 30 psi and the sprayer must be fitted with nozzles no smaller than 8002 T-Jet (or equivalent). Do not apply under weather conditions which favor drift.

POTATOES

SENCOR herbicide is recommended for use in ground or aircraft spray equipment as a preemergence and/or postemergence application to potatoes. Postemergence applications may be made only on russeted or white-skinned potato varieties that are not early maturing. For optimum control of weeds in potatoes, apply SENCOR 4 before weeds are 1 inch tall. Refer to "Special Precautions" for potatoes for additional information.

Ground Application: SENCOR 4 is recommended for use with ground spray equipment, except air blast sprayers, as a preemergence and/or postemergence application for control of certain grass and broadleaf weeds in potatoes. Apply as a broadcast spray after planting and before crop emergence or after drag-off if this operation is part of the normal cultural practice. Postemergence applications should be applied after crop emergence.

Aerial Application: SENCOR 4 is recommended for use in aerial spray equipment as a preemergence and/or postemergence application for control of certain grass and broadleaf weeds in potatoes. Apply as a broadcast spray after planting and before crop emergence or after drag-off if this operation is part of the normal cultural practice. Postemergence applications should be applied after crop emergence.

PREEMERGENCE

Weeds Controlled:
SENCOR 4, applied preemergence to potatoes according to directions, will effectively control:

Broadleaves:		Grasses:
Carpetweed *(Mollugo verticillata)*	Ragweed, Common *(Ambrosia artemisiifolia)*	Broadleaf Signalgrass *(Brachiaria platyphylla)*
Cocklebur *(Xanthium pensylvanicum)*	Shepherdspurse *(Capsella bursa-pastoris)*	Crabgrass *(Digitaria spp.)*
		Foxtails *(Setaria spp.)*
Jimsonweed *(Datura stramonium)*	Sicklepod *(Cassia obtusifolia)*	Johnsongrass, Seedling *(Sorghum halepense)*
Lambsquarters *(Chenopodium album)*	Smartweed, Pennsylvania *(Polygonum pensylvanicum)*	Panicum, Fall *(Panicum dichotomiflorum)*
Pigweeds *(Amaranthus spp.)*	Wild Mustards *(Brassica spp.)*	

For Control of Wild Mustard (Brassica spp.) in Potatoes (Except California)—Apply at the rate of 1/2 to 1 pint per acre as a preemergence broadcast spray using aerial or ground spray equipment in accordance with directions on the label. Application should be made at or near ground crack and/or before the mustard is one inch tall. For control of mustard larger than 2 inches in diameter or other weeds listed on the label, a minimum of 1 pint per acre must be used. When less than 1 pint per acre is used, a postemergence application in accordance with directions on the SENCOR 4 label may be required for control of other weeds or later germinating weed species.

Hard To Control Weeds: Although SENCOR 4 applied preemergence may not provide commercially acceptable control in every instance, it will suppress growth of the following weeds and reduce their competition with potato plants.

Broadleaves:	**Grasses:**
Common Purslane *(Portulaca oleracea)*	Barnyardgrass *(Echinochloa crus-galli)* Nutsedge *(Cyperus spp.)*

POSTEMERGENCE

Weeds Controlled:

SENCOR 4, applied early postemergence to potatoes according to directions, will effectively control:

Broadleaves:		**Grasses:**
Cocklebur *(Xanthium pensylvanicum)* Lambsquarters *(Chenopodium album)* Pigweed, Redroot *(Amaranthus retroflexus)*	Ragweed, Common *(Ambrosia artemisiifolia)* Russian Thistle *(Salsola kali)* Smartweed, Pennsylvania *(Polygonum pensylvanicum)*	Foxtails *(Setaria spp.)*

On medium and fine soils in Idaho, Oregon and Washington, two postemergence applications, each at the rate of 1 pint per acre, will control barnyardgrass *(Echinochloa crus-galli)*, kochia *(Kochia scoparia)*, sunflower *(Helianthus spp.)*, and Jim Hill mustard *(Sisymbrium altissimum)* in addition to weeds listed above.

Hard to Control Weeds: Although SENCOR 4 applied early postemergence may not provide commercially acceptable control in every instance, it will suppress growth of the following weeds and reduce their competition with potato plants.

Broadleaves:	**Grasses:**
Kochia *(Kochia scoparia)* Sunflower *(Helianthus spp.)*	Barnyardgrass *(Echinochloa crus-galli)*

Recommended Broadcast Applications

Crop	SENCOR 4 Pints Per Acre	Remarks
Potatoes	1 to 2	**PREEMERGENCE:** Apply specified dosage per acre as a broadcast spray to the soil surface after planting and before crop emergence or after drag-off if this operation is part of normal cultural practice.* Distribute the chemical evenly over the area to be treated. Do not incorporate into soil. Use the lower rate on sandy soils.
Potatoes (russetted or white-skinned varieties that are not early maturing)	1/2 to 1	**POSTEMERGENCE:** Apply specified dosage per acre as a broadcast spray over the tops of potato plants.* Use rates of 1/2 to 1 pint for control of redroot pigweed and lambsquarters only. Apply the 1 pt. rate for control of other weeds listed on the label. Do not apply within 60 days of harvest.
		SPLIT APPLICATIONS: This product may be applied once preemergence and once postemergence as directed above.* Do not apply postemergence within 60 days of harvest.
		IDAHO, OREGON AND WASHINGTON ONLY: Two postemergence applications can be made as broadcast sprays over the tops of potato plants if SENCOR 4 is not applied preemergence. Use 1/2 to 1 pint/acre for control of redroot pigweed and lambsquarters only. On coarse (sandy) soils with low organic matter, do not exceed 3/4 pint/acre per application. On medium and heavy soils only, use 1 pint/acre for control of other weeds listed on the label. Make the first application early in the season while weeds are still small. Allow at least 14 days before the second application. For optimum control, apply before weeds are 1 inch tall. Do not apply after June 30 if treated land is to be planted to crops other than potatoes. Do not apply within 60 days of harvest.

*Refer to Special Precautions—Potatoes below.

SPRINKLER IRRIGATION APPLICATION TO POTATOES:

SENCOR 4 may be applied early postemergence to potatoes through center pivot, permanent solid set, or self-propelled lateral-move irrigation systems as directed on the label. Refer to "Chemigation" section of the SENCOR 4 label for directions.

Recommended Broadcast Applications

Crop	SENCOR 4 Pints/Acre	Remarks
Potatoes (russetted or white-skinned varieties that are not early maturing)	1/2 to 1	**POSTEMERGENCE IN SPRINKLER IRRIGATION:** Apply specified dosage in 1/4 to 3/4 inch of water per acre as a continuous injection in center pivot and self-propelled wheel move systems or in the last 15 to 30 minutes of set in permanent solid set sprinkler systems. On sandy soil, apply 1/4 to 1/2 to 1 pint for control of redroot pigweed and lambsquarters only. Apply the 1 pint rate for control of other weeds listed on the label. Do not apply within 60 days of harvest.

Do not apply when wind speed favors drift, when system connections leak, when nozzles do not provide uniform distribution or when lines containing the product must be dismantled or drained.

SENCOR plus Dual Tank Mix Application (Except California)

SENCOR 4 may be applied in a tank mix combination with Dual 8E as a preemergence broadcast application to potatoes. Apply SENCOR 4 at 0.75 to 2 pints and Dual 8E at 1.5 to 3 pints per acre according to the respective labels for use of each product alone on potatoes. Refer to appropriate section of the SENCOR 4 and Dual labels for precautionary statements, restrictions, application information and weeds controlled.

SENCOR plus Prowl Tank Mix Application (Except California)

SENCOR 4 may be applied in a tank mix combination with Prowl as a preemergence broadcast application to potatoes. Apply SENCOR 4 at 1 to 2 pints and Prowl at 1 to 3 pints per acre according to the respective labels for use of each product alone on potatoes. Refer to the appropriate sections of the SENCOR 4 and Prowl labels for precautionary statements, restrictions, application information and weeds controlled.

Special Precautions (Potatoes)

Do Not Use SENCOR 4 on Potatoes in Kern County, California.

Do not apply more than 2 pints total SENCOR 4 per acre in a single crop season regardless of the method of application. **Do not apply within 24 hours of application of other pesticides.** Do not apply postemergence within 3 days after periods of cool, wet or cloudy weather or crop injury may occur. Postemergence applications may cause some chlorosis or minor necrosis.

In addition to early maturing smooth-skinned white, or red-skinned varieties of potatoes, postemergence applications of SENCOR 4 are not recommended on Atlantic, Shepody, Chipbelle, Bellchip, and Centennial varieties. Preemergence applications to these varieties may cause crop injury, especially under adverse weather conditions and when higher SENCOR 4 rates are used.

Do not apply to sweet potatoes or yams.

Do not plant sensitive crops such as onions, lettuce, cole crops and cucurbits during the next growing season following applications of SENCOR 4 or injury may occur.

ALFALFA AND SAINFOIN

SENCOR herbicide is labelled for use in alfalfa and sainfoin in the following areas:

1. Alfalfa and sainfoin (including mixed stands with grasses) (all areas except California).
2. Alfalfa and sainfoin (including mixed stands with grasses) (California only).
3. Alfalfa—Tank mix combination with Gramoxone (Colorado, Idaho, Montana, Nevada, Oregon, Utah, Washington, and Wyoming).
4. Alfalfa—Post-dormant application of SENCOR impregnated on dry fertilizer only.

SENCOR 4 is recommended for use in aerial or ground spray equipment as a broadcast surface application to established crops of alfalfa and sainfoin for the control of certain grass and broadleaf weeds.

APPLICATION: Refer to "GENERAL INFORMATION" in the front of the SENCOR 4 label for detailed information on the application of SENCOR 4. For information on applying SENCOR 4 in fluid or on dry fertilizer refer to the "Application Of SENCOR 4 In Fluid Fertilizers" or "Commercial Impregnation And Application Of SENCOR 4 On Dry Bulk Fertilizer" under the "GENERAL INFORMATION" section in the front of the SENCOR 4 label.

Special Precautions:

Do not apply SENCOR 4 after growth begins in the Spring or before growth ceases in the Fall.

Treat only dormant established alfalfa (except where specified on the SENCOR 4 label).

Do not graze or harvest within 28 days after application.

For best weed control, apply SENCOR 4 when weeds are less than 2 inches tall or before weed foliage is 2 inches in diameter.

Reduced weed control may occur when extended dry conditions follow application of SENCOR 4.

Crop injury may occur if alfalfa or sainfoin are:

1. Under stress conditions such as diseases, insect infestations, poorly drained soils, drought or Winter injury when SENCOR 4 is applied, or
2. When SENCOR 4 is applied during the first growing season after seeding or if applied earlier than 12 months after seeding.

ALFALFA AND SAINFOIN (All Areas except California)

Recommended Broadcast Applications

Crop	SENCOR 4 Pints/Acre	Remarks
Alfalfa and Sainfoin (Except California)	¾ to 2	**ALL AREAS:** (Including Idaho, Oregon, and Washington) Select the proper dosage according to weeds known to be present in field to be treated. Use the lower rate of SENCOR on sandy loam or loamy sand soils in areas other than Idaho, Oregon, and Washington.
	½ to ¾	**IDAHO, OREGON AND WASHINGTON ONLY:** On sandy loam or loamy sand soils apply SENCOR 4 at rates of ½ to ¾ pint per acre for control of common chickweed only

FOR USE ON MIXED STANDS OF ALFALFA AND GRASSES:

Rates of 1 to 1½ pints of SENCOR 4 per acre will provide partial reduction of forage grass stands. These rates may be used to reduce forage grass stands to prevent crowding out of alfalfa. Higher rates will severely reduce forage grass stands.

SENCOR 4 should not be used on sand soils. In areas West of the Rocky Mountains, avoid using SENCOR 4 on soils with a calcareous surface area, high levels of lime or sodium, or a pH greater than 8.2.

Weeds Controlled:

½ to ¾ pint SENCOR 4/acre (Idaho, Oregon, and Washington Only)
Chickweed, Common *(Cerastium vulgatum)*

¾ to 1 pint SENCOR 4/acre
Cheat *(Bromus secalinus)*
Downy Brome *(Bromus tectorum)*
Japanese Brome *(Bromus japonicus)*
Rescuegrass *(Bromus catharticus)*

1 to 2 pints SENCOR 4/acre

Broadleaves:		Grasses:
Fleabane, Rough *(Erigeron strigosus)*	Mustard, Tansy *(Descurainia pinnata)*	Foxtail, Green *(Setaria viridis)*
Flixweed *(Descurainia sophia)*	Pepperweed *(Lepidium virginicum)*	Little Barley *(Hordeum pusillum)*
Henbit *(Lamium amplexicaule)*	Pennycress, Field *(Thlaspi Arvense)*	Smooth Brome *(Bromus inermis)*
Kochia *(Kochia scoparia)*	Pigweed, Redroot *(Amananthus retroflexus*	Wild Oats *(Avena fatua)*
Lambsquarters, Common *(Chenopodium album)*	Prickly Lettuce *(Lactuca serriola)*	
Marestail (Horseweed) *(Hippuris vulgaris)*	Shepherdspurse *(Capsella bursa-pastoris)*	
Meadow Salsify *(Tragopogon pratensis)*	White Cockle *(Melandrium album)*	
Mustard, Blue *(Chorispora tenella)*	Wild Buckwheat *(Polygonum convolvulus)*	
Mustard, Jim Hill (tumble) *(Sisymbrium altissimum)*	Yellow Rocket *(Barbarea vulgaris)*	

2 pints SENCOR 4/acre

Broadleaves:	Grasses:
Chickweed, Mousear *(Cerastium vulgatum)*	Barnyardgrass *(Echinochloa crus-galli)*
Dandelion *(Taraxacum officinale)*	Bluegrass *(Poa annua)*
Ragweed, Common *(Ambrosia artemisiifolia)*	Foxtail Barley *(Hordeum jubatum)*

Weeds Partially Controlled: At the rate of 2 pints/acre SENCOR 4 may be used to reduce the competition from curly dock *(Rumex crispus)*. At 1 to 2 pints/acre, SENCOR 4 may be used to reduce the competition of German moss or Knawel *(Scleanthus annus)*.

ALFALFA AND SAINFOIN (California Only) (Including Mixed Stands With Grasses)

SENCOR 4 is recommended for use in aerial or ground spray equipment as a broadcast surface application to dormant established crops of alfalfa and sainfoin.

APPLICATION: SENCOR 4 is recommended for use in aerial or ground spray equipment as a broadcast surface application to dormant established crops of alfalfa and sainfoin for control of certain grass and broadleaf weeds. Do not apply SENCOR 4 after growth begins in the Spring or before growth ceases in the Fall. Do not apply to either alfalfa or sainfoin during the first growing season after seeding.

For information on applying SENCOR 4 in fluid fertilizer solutions to alfalfa, refer to the appropriate section of the SENCOR 4 label.

For information on commercial impregnation and application of SENCOR 4 on dry bulk fertilizer, refer to the appropriate section of the SENCOR 4 label.

Weeds Controlled:

¾ to 1 pint SENCOR 4/acre
Cheatgrass (downy brome)

1 to 2 pints SENCOR 4/acre

Broadleaves:		Grasses:
Chickweed, Common *(Cerastium vulgatum)*	Mustard, Tansy *(Descurainia pinnata)*	Smooth Brome *(Bromus inermis)*
Flixweed *(Descurainia sophia)*	Pepperweed, Virginia *(Lepidium virginicum)*	Wild Oats *(Avena fatua)*
Henbit *(Lamium amplexicaule)*	Shepherdspurse *(Capsella bursa-pastoris)*	
Kochia *(Kochia scoparia)*	White Cockle *(Melandrium album)*	
Meadow Salsify *(Tragopogon pratensis)*	Wild Buckwheat *(Polygonum convolvulus)*	
Mustard, Blue *(Chlorispora tenella)*	Yellow Rocket *(Barbarea vulgaris)*	

2 Pints SENCOR 4/acre

Broadleaves:	Grasses:	
Dandelion *(Taraxacum officinale)*	Barnyardgrass *(Echinochloa crus-galli)*	Foxtail Barley *(Hordeum jubatum)*

Bluegrass
(Poa annua)

Recommended Broadcast Applications

Crop	SENCOR 4 Pints/Acre	Remarks
Alfalfa and Sainfoin (California only)	3/4 to 2	Select the proper dosage according to weeds known to be present in the field to be treated. Apply specified dosage in 20 to 40 gallons of water per acre with ground spray equipment or 3 to 10 gallons of water per acre with aerial spray equipment fitted with nozzles suitable for broadcast applications of herbicides. Treat only dormant established crops of alfalfa and sainfoin. Injury may occur to alfalfa if SENCOR is applied earlier than 12 months after seeding. Do not apply after Spring growth begins or before growth ceases in the Fall. Do not graze or harvest within 28 days after application. At the 2 pint/acre rate, SENCOR 4 may be used for suppression of curly dock.

FOR USE ON MIXED STANDS OF ALFALFA AND GRASSES: Rates of 1 to 1½ pints of SENCOR 4 per acre will provide partial reduction of forage grass stands. These rates may be used to reduce forage grass stands to prevent crowding out of alfalfa. Higher rates will severely reduce forage grass stands.

Do not apply with aerial spray equipment when wind speed is greater than 10 mph. Do not apply when weather conditions favor spray drift and/or when sensitive cool season crops such as cole crops, onions, peas, or strawberries, are present in adjacent fields. Applications should not be made when weather conditions favor spray drift, especially in areas where wheat is growing on coarse textured soils in adjacent fields, or injury may occur.

ALFALFA Tank Mix with Gramoxone

(Colorado, Idaho, Montana, Nevada, Oregon, Utah, Washington, and Wyoming)

APPLICATION: SENCOR 4 plus Gramoxone tank mix application is recommended for use, during the dormant season, in aerial or ground spray equipment as a broadcast surface application to established (at least 1 year old) alfalfa for the control of certain grass and broadleaf weeds. Do not apply SENCOR/Gramoxone tank mix to regrowth (after grazing or cutting) that is more than 2 inches tall. Apply once per season. Do not apply following cuttings during growing season. Use a minimum of 10 gallons of water per acre with aerial spray equipment and a minimum of 20 gallons of water per acre with ground spray equipment. Add Ortho X-77 Spreader (nonionic) at 1 to 2 pints per 100 gallons of spray solution.

WEEDS CONTROLLED: SENCOR 4 (3/4 1½ pint/acre) plus Gramoxone Super (2 to 2⅔ pints/acre) tank mix application will control established weeds. Gramoxone controls weeds by contact activity.

Bluegrass	Chickweed
Henbit	Downy Brome
Cheat	Japanese Brome
Rescuegrass	Shepherdspurse

½ to 3/4 pint of SENCOR 4 per acre may be used for common chickweed only on sandy soils. Use SENCOR 4 at 1 to 1½ pints/acre for control of the following weeds:

Ryegrass	Kochia	Prickly Lettuce
Sow Thistle	Meadow Salisify	Little Barley
Groundsel	Pepperweed	Redroot Pigweed
Smooth Brome	Jim Hill Mustard	Common Lambsquarters
Wild Oats	Green Foxtail	Marestail (Horseweed)
Blue Mustard	Yellow Rocket	Field Pennycress
Tansy Mustard	White Cockle	Rough Fleabane
Flixweed	Wild Buckwheat	

RECOMMENDED APPLICATIONS

DOSAGE (Pints/Acre)	REMARKS
SENCOR 4 ½ to 1½ pints plus Gramoxone Super 2 to 2⅔ pints	Apply specified dosages of SENCOR and Gramoxone in at least 10 gallons of water per acre with aerial equipment or at least 20 gallons of water per acre with ground equipment. Do not apply this tank mix to alfalfa regrowth (after grazing or cutting) if more than 2 inches tall. For best weed control, apply when broadleaf weeds and grasses are 1-6 inches tall and are actively growing. Care should be taken to avoid overlaps. Use lower rates of SENCOR on sandy loam or loamy sand soils. Reduced weed control may occur when extended dry conditions follow application of SENCOR. Crop injury may occur if alfalfa is under stress conditions such as diseases, insect infestations, drought or winter injury or if SENCOR is applied to alfalfa earlier than 12 months after seeding.

FOR USE ON MIXED STANDS OF ALFALFA and GRASSES: Rates of 1 to 1½ pints of SENCOR 4 per acre will provide partial reduction of forage grass stands. These rates may be used to reduce forage grass stands to prevent crowding out of alfalfa.
Do not graze or harvest within 42 days after application.
In areas west of the Rockies, avoid the use of SENCOR on soils with calcareous surface, soils with high levels of lime or sodium, and with a pH greater than 8.2.
Do not apply when weather conditions favor spray drift. Aerial application should not be made when wind speed is greater than 10 mph.
Do not use on sand soil.
Refer to the Gramoxone label for additional directions, weed species controlled and precautions.

ALFALFA

Post Dormant Application of SENCOR Impregnated on Dry Fertilizer Only—SENCOR 4 may be applied after dormancy has broken, but prior to three inches of new alfalfa shoot growth, only when impregnated on dry fertilizer in Connecticut, Illinois, Indiana, Iowa, Kansas, Kentucky, Michigan, Minnesota, Nebraska, New York, Ohio, Pennsylvania, South Dakota and Wisconsin. Apply at rates of 1½ to 2 pints per acre as directed on the label for application during dormancy. Apply only when alfalfa foliage is dry or crop injury may occur. When using this application method, do not harvest or graze treated alfalfa for 60 days after application.

ASPARAGUS (Established)

SENCOR 4 is recommended for use in aerial or ground spray equipment or sprinkler irrigation (center pivot, lateral move, or solid set) systems as a single preemergence broadcast application or as a split application consisting of a preemergence broadcast application followed by a post harvest broadcast application.

Refer to "Chemigation" and "Application of SENCOR 4 with Herbicide Spray Equipment" sections of the SENCOR 4 label for directions.

Weeds Controlled:
SENCOR 4, applied to established asparagus according to directions, will effectively control:

Broadleaves:		Grasses:
Chickweed, Common *(Stellaria media)*	Ragweed, Common *(Ambrosia artemisiifolia*	Crabgrass *(Digitaria spp.)*
Jimsonweed *(Datura stramonium)*	Smartweed, Pennsylvania *(Polygonum pensylvanicum)*	Foxtails *(Setaria spp.)*
Lambsquarters *(Chenopodium album*	Sorrel, Red *(Rumex acetosella)*	Sandbur, Field *(Cenchrus pauciflorus)*
Pigweed, Redroot *(Amaranthus retroflexus)*	Velvetleaf *(Abutilon theophrasti)*	

Recommended Broadcast Applications

Crop	SENCOR 4 Pints/Acre	Remarks
Asparagus	2 to 4	**PREEMERGENCE APPLICATION ONLY:** Make a single surface application in early Spring before asparagus spears or ferns emerge. If the field is to be disked, apply SENCOR 4 after disking but before the crop emerges. Use the lower rate for control of the broadleaf weeds listed above. Use the higher rate in fields with a history of severe infestations of grasses and for maximum residual control. Do not apply within 14 days of harvest.
	1 to 2 preemergence plus 2 to 3 post harvest	**SPLIT APPLICATION PREEMERGENCE AND POST HARVEST** **Preemergence Application:** Apply before asparagus spears or ferns emerge. If the field is to be disked, apply after disking but prior to crop emergence. Do not apply within 14

		days of harvest. **Post Harvest Application:** Apply after last harvest of the season but prior to emergence. The lower combination rates may be used for control of common ragweed, lambsquarters, redroot pigweed, and red sorrel. Use the higher combination rates for other weeds listed or in fields with severe grass infestations or for maximum post harvest control of emerged weeds.

IMPORTANT: The total amount of SENCOR 4 applied in one crop season may not exceed 4 pints per acre.

Special Precautions (Asparagus)

Do not use on newly seeded asparagus nor on young plants during the first growing season after setting crowns.

DO NOT APPLY POST HARVEST APPLICATIONS UNTIL AFTER THE LAST HARVEST OF SPEARS.

CARROTS
(Except California)

Special Conditions of Sale Provision for Use on Carrots: The following directions for use were developed under the direction of IR-4 (government minor crops use program). As such the testing was done independently from the testing program of Mobay Corporation. Buyer is advised that Mobay makes no assurances regarding satisfaction with the product and all risks of crop injury or product performance are assumed by the Buyer.

Apply SENCOR herbicide with ground equipment as specified below under "Recommended Applications." For effective control of broadleaf weeds with postemergence applications, apply SENCOR before weeds are 1 inch in height or diameter. Thorough spray coverage is essential for adequate weed control.

Do not use air blast or other high pressure spray equipment to make postemergence applications of SENCOR. Refer to the appropriate section of the SENCOR 4 label for additional information regarding spray equipment, dilution rates, mixing, sprayer cleanup, restrictions, container disposal and cautions.

Refer to "Mixing" under the "General Information" section in the front of the SENCOR 4 label.

For specific application information see "General Information" and "Application" sections in the front of the SENCOR 4 label. Weeds controlled: SENCOR 4 applied to carrots according to directions will effectively control:

Carpetweed *(Mullugo verticillata)*	Pigweed, Redroot *(Amaranthus retroflexus)*
Galinsoga *(Galinsoga parviflora)*	Pigweed, Smooth *(Amaranthus hybridus)*
Horseweed *(Conyza canadensis)*	Prickly Lettuce *(Lactuca serriola)*
Lambsquarters, Common *(Chenopodium album)*	Shepherdspurse *(Capsella bursa-pastoris)*
Mustard, Wild *(Sinapis arvensis)*	Pineappleweed *(Matricaria matricarioides)*

RECOMMENDED APPLICATIONS

CROP	SENCOR 4 Pints/Acre	REMARKS
Carrots (except California)	½	Apply specified dosage per acre as a broadcast spray over the tops of carrot plants. Application should be made after carrots have formed 5-6 true leaves but before weeds are 1 inch in height or diameter. If needed, a second application may be made after an interval of at least 3 weeks. Applications may be made up to 60 days of harvest. Do not apply to carrots grown for seed.

IMPORTANT: The total amount of SENCOR 4 applied in one crop must not exceed **one** pint per acre.

Special Precautions:

Do not apply within 3 days after periods of cool, wet or cloudy weather or crop injury will occur.

Do not apply SENCOR within 3 days of any other chemical unless specified on the SENCOR 4 label.

Do not apply on very hot days or excessive crop injury will result.

Do not apply until carrots have at least 5-6 true leaves. Earlier applications will result in excessive crop damage.

Crop injury or delay maturity may result from applications of SENCOR if carrots are growing under stress conditions such as periods of drought or cool, wet and cloudy weather preceding application.

Following an application of SENCOR 4, chlorosis (yellowing) and burning of the leaf tissue may occur.

For newly introduced varieties of carrots with unknown tolerance to SENCOR, treat only a small area to determine if SENCOR can be used without injury to the crop.

FIELD CORN
(Iowa, Kansas, Missouri, Nebraska, and South Dakota Only)

SENCOR 4 is recommended for the added effect on the control of certain broadleaf weeds in corn when applied preemergence in a tank mix combination with labeled rates of Lasso plus atrazine, Lasso plus Bladex,® Dual plus atrazine, or Dual plus Bladex. Refer to the appropriate product labels for weeds controlled by these combinations.

SENCOR 4 in each of the tank mix combinations will control the following weeds:

Lambsquarters *(Chenopodium album)*	Smartweed, Pennsylvania *(Polygonum pensylvanicum)*
Pigweed, Redroot (carelessweed) *(Amaranthus retroflexus)*	Velvetleaf *(Abutilon theophrasti)*

Mixing: Select proper dosages based on soil type as shown below. Fill the spray tank ¼ to ⅓ full with clean water and add SENCOR 4 to the spray tank first. Mix thoroughly and then add atrazine or Bladex to the spray tank. Add water to fill spray tank to desired level. Mix thoroughly and then add Lasso or Dual to the spray tank. Maintain sufficient agitation during mixing and spraying operations to insure a uniform spray mixture.

Application: For specific application information see the "GENERAL INFORMATION" section in the front of the SENCOR 4 label.

Recommended Broadcast Applications

Rates Per Acre:

Soil Texture	SENCOR 4	Lasso 4 EC	or Dual 8 E	Atrazine 4L or 80 WP	or Bladex 4L or 80 WP
MEDIUM (Loam, silt loam, silt, sandy clay, sandy clay loam)	½ pt.	1.5 to 2.0 qts.	1.5 to 2.0 pts.	0.75 to 1.5 qts. or 4L or 0.95 to 1.9 lbs. of 80 WP	1.25 to 2.5 qts. of 4L or 1.5 to 3.1 lbs. of 80 WP
FINE (Silty clay, silty clay loam,[1] clay, clay loam)	½ pt.	2.0 to 2.5 qts.	2.0 to 2.5 pts.	0.75 to 1.5 qts. of 4L or 0.95 to 1.9 lbs. of 80 WP	1.5 to 3.0 qts. of 4L or 1.75 to 3.75 lbs. of 80 WP

[1] Silty clay loam soils are transitional soils and may be classified as medium textured soils in some regions of the U.S.

Follow the Lasso, Dual, Bladex, or atrazine labels for use rates on specific soil textures and organic matter content. Not recommended for use on muck soils as reduced weed control may result.

Special Precautions: (Field Corn)

Not for use on coarse textured soils or soils containing less than 2% organic matter content.

Do not use on soils having a calcareous surface or high in lime or sodium and/or a pH of 7.0 or higher.

Corn seed should be planted 1½ inches deep.

Use in hybrid seed corn production fields only if both parents are known to be tolerant to SENCOR 4 and other herbicides used.

Observe all cautions and limitations on labeling of all products used in mixtures.

LENTILS AND DRY FIELD PEAS

(Idaho, East of the Cascades in Oregon & Washington, and Montana)

SENCOR herbicide is recommended as a preemergence and postemergence application for the suppression of certain broadleaf weeds in lentils and dry field peas.

WEEDS SUPPRESSED*

Common Chickweed**	Lambsquarters
Dog Fennel	Shepherdspurse**
Field Pennycress	Wild Mustard
Henbit**	

*Suppression is a reduction in weed size and growth compared to a non-treated area in the same field.

**Preemergence application only.

PREEMERGENCE APPLICATION

A single preemergence application at ½ to ¾ pt. of SENCOR 4 in 20 to 40 gallons of water per acre with ground spray equipment may be made per crop year. Apply SENCOR 4 before or after planting. BUT BEFORE

CROP GERMINATES. Thorough incorporation, either by rainfall or by mechanical means, is essential for weed suppression. Under dry conditions, incorporate SENCOR 4 into the top 1 to 2 inches of soil with spike harrows, or similar shallow incorporation equipment then cross harrow to insure uniform soil incorporation. Where soil surface is moist at the time of application and rain follows before weed emergence, a broadcast application should provide adequate weed suppression. Use the higher rate on fine-textured soils (high in clay or organic matter) and in fields with a history of high weed populations.

POSTEMERGENCE APPLICATION:

Use 1/4 to 3/8 pt. of SENCOR 4 per acre on **lentils** and **spring peas.** On **winter peas,** use 3/8 to 1/2 pt. of SENCOR 4 per acre. For suppression of dog fennel, use 1/2 pt. SENCOR 4 per acre. Apply specified dosage in 20 to 40 gallons of water per acre and not over 40 psi with ground spray equipment as a broadcast spray when weeds are small (less than 2" in height or diameter) and before crop is 6 inches tall.

One postemergence application may be made per season.

Temporary chlorosis of crop may occur. There is an added risk of crop injury if a postemergence application is made following a previous metribuzin application.

Do not apply over very moist soils or wet crop foliage. Do not apply postemergence applications within 3 days after periods of cool, wet, or cloudy weather or crop injury may occur.

Do not apply within 24 hours of treatment with other pesticides.

SPECIAL PRECAUTIONS (all applications):

Crop injury may result if crop is under stress conditions caused by cold weather, low fertility, disease or insect damage.

Crop injury may also result if application is followed by heavy rain.

Do not use on coarse-textured soils, sandy soils or any soil with less than 1.5% organic matter.

Do not apply to "Estin" lentils.

Do not use on clay knobs or poorly covered subsoils.

Do not apply on shallow seedings less than 2" deep (preemergence only).

Do not apply within 50 days of harvest of peas, or within 75 days of harvest of lentils. Do not graze or feed treated vines to livestock within 40 days after application.

Maintain continuous spray tank agitation to keep material in suspension. Avoid overlapping and shut off spray booms while turning, slowing or stopping, or crop injury will occur.

NOTE: This treatment may cause some chlorosis or minor necrosis. Because lentil and pea varieties may vary in their susceptibility to SENCOR 4, determining crop tolerance prior to adoption as a field scale practice is suggested to prevent possible injury.

For additional precautions, restrictions, limitations, and sprayer clean-up information refer to the appropriate sections of the SENCOR 4 label.

SUGARCANE (Hawaii only)

SENCOR, a selective herbicide, is effective as a preemergence and an early postemergence broadcast application for control of certain grass and broadleaf weeds. When applied as a spot treatment, it also provides excellent control of perennial grasses and broadleaves.

Ground Application: SENCOR 4 should be mixed by filling the spray tank half full of clean water. Then add the recommended amount of SENCOR 4 to suit the total tank capacity and the rate of application per acre (preferably 25 to 35 gallons per acre). Complete filling the tank and maintain sufficient agitation during mixing and spraying to insure a uniform spray mixture.

Aerial Application: SENCOR 4 is recommended for use in aerial spray equipment as a preemergence or postemergence application to irrigated sugarcane. Aerial spray equipment should be calibrated to apply the proper amount of SENCOR 4 in 7 to 10 gallons of spray mixture per acre.

SENCOR 4 applied preemergence or postemergence to the sugarcane as a broadcast spray or spot treatment will effectively control the following when weeds are less than 3 inches in height.

Weeds Controlled in Irrigated and Non-irrigated Sugarcane:

Broadleaves:

- Amaranth, Spiny *(Amaranthus spinosus)*
- Euphorbia, Wild *(Euphorbia spp.)*
- Fireweed *(Erechtites hieracifolius)*
- Floras Paintbrush *(Emilia sonochifolia)*
- Spurge, Garden *(Euphorbia hirta)*
- Spurge, Graceful *(Euphorbia glomerifera)*

Grasses:

- Crabgrass *(Digitaria spp.)*
- Guineagrass *(Panicum maximum)*
- Plushgrass *(Chloris radiata)*
- Ricegrass *(Oryzopsis hymenoides)*
- Wiregrass *(Eleusine indica)*

Weeds Controlled in Irrigated Sugarcane Only:

Broadleaves:

- Amaranth, Spleen *(Amaranthus dubius)*
- Haole Koa *(Leucaena leucocephala)*
- Hialoa *(Waltheria americana)*
- Hilahila *(Mimosa pudica)*
- Purslane, Common *(Portulaca oleracea)*
- Rattlepod *(Crotalaria spectabilis)*

Grasses:

- Alexandergrass *(Brachiaria plantaginea)*
- Bristly Foxtail *(Setaria verticillata)*

Weeds Controlled in Non-irrigated Sugarcane Only:

Broadleaves:

- Ageratum *(Ageratum conyzoides)*
- Richardia *(Richardia brasiliensis)*
- Tarweed *(Cuphea carthagenesis)*

Recommended Broadcast Applications

Crop	SENCOR 4 Pints/Acre	Remarks
Sugarcane (Hawaii only)	4 to 8 (1/2 to 1 gal.) (non-irrigated)	PREEMERGENCE: (Irrigated and non-irrigated sugarcane) Apply specified dosage per acre as a broadcast spray to the soil surface. Applications should be made within two weeks after planting prior to cane emergence or shortly after emergence (spike stage).
		OR
	8 to 12 (1 to 1 1/2 gal.) (irrigated)	EARLY POSTEMERGENCE: (Irrigated and non-irrigated sugarcane) Apply specified dosage per acre as a broadcast spray over the cane. Application may be delayed as long as 4 to 6 weeks after planting provided weeds are less than 3 inches in height.
		OR
	4 to 8 (1/2 to 1 gal.)	POSTEMERGENCE: Apply specified dosage per acre as a broadcast spray to control weeds prior to "close in" time when cane shades out the weed growth.
	5 to 10 (1 1/8 to 1 1/4 gal.)	SPOT TREATMENT: Apply specified dosage in 30 to 50 gallons of finished spray per acre. Spot treatments may be used to control weeds in missed areas, corners of fields, or areas of hard to control weeds.

NOTE: Do not apply more than 16 pints (2 gal.) of SENCOR 4 (8 lb. active ingredient)/acre per crop cycle regardless of the method of application. The last application may be made up to 17 months of harvest.

Special Precaution: Do not use treated foliage for feed or forage.

SUGARCANE (Louisiana and Texas Only)

Preemergence and postemergence applications of SENCOR 4 with aerial or ground spray equipment are recommended for control of the following weeds in sugarcane in Louisiana and Texas:

Broadleaves:

- Amaranth, Spiny *(Amaranthus spinosus)*
- Bindweed, Field *(Convolvulus arvensis)*
- Chickweed *(Cerastium vulgatum)*
- Henbit *(Lamium amplexicaule)*
- Lambsquarters *(Chenopodium album)*
- *London Rocket* *(Sisymbrium irio)*
- Marestail *(Conyza canadensis)*
- Mustard, Wild *(Brassica kaber)*
- Pigweeds *(Amaranthus spp.)*
- Purslane *(Portulaca oleracea)*
- Sowthistle *(Sonchus spp.)*

Grasses:

- Broadleaf Signalgrass *(Brachiaria platyphylla)*
- Crabgrass *(Digitaria spp.)*
- Foxtails *(Setaria spp.)*
- Johnsongrass, Seeding *(Sorghum halepense)*
- Oats, Winter *(Avena spp.)*

Recommended Broadcast Applications

Crop	SENCOR 4 Pints/Acre	Remarks
Sugarcane (Louisiana and Texas only)	3 to 6	**BROADCAST:** Apply specified dosage per acre using 20 to 30 gallons of water with ground equipment or 5 gallons of water with aircraft spray equipment. Apply as a broadcast spray

		during the Fall after planting or to the stubble after harvest. Make a second application early in the Spring.
	$1\frac{1}{2}$ to 3	**BAND:** Apply specified dosage in 10 to 20 gallons of water per acre in a 30 to 36 inch band over the row during the Fall after planting or to the stubble after harvest. Make a second application early in the Spring.

Special Precautions: (Louisiana and Texas only)

Do not use treated foliage for feed or forage.

Use the higher rate on heavy clay soil and soil with a high percentage of organic matter. If necessary, a third application may be made in late Spring at layby. Do not apply within 60 days of harvest.

SUGARCANE (Florida Only)

Postemergence over-the-top or directed spray applications of SENCOR 4 are recommended for control of the following weeds in sugarcane in Florida.

Broadleaves:	Grasses:
Amaranth, Spiny (seedling) *(Amaranthus spinosus)*	*Crabgrass, large *(Digitaris sanguinalis)*
Butterweed *(Cressleaf groundsel)* *(Senecio glabellus)*	Foxtail, Bristlegrass *(Setaria magna)*
Cudweed (*Gnaphalium* spp)	Goosegrass *(Eleusine indica)*
Purslane *(Portulaca oleracea)*	Panicum, Broadleaf *(Panicum adspersum)*
	Signalgrass, Broadleaf *(Brachiaria platyphylla)*

Recommended Broadcast Applications

Crop	SENCOR 4 Pints/Acre	Remarks
Sugarcane	2 to 4	**GROUND APPLICATION:** SENCOR 4 may be used in one or two applications with a minimum of 14 days between each application. Apply when weeds are less than 6 inches tall in 10 to 40 gallons of spray mixture per acre. Do not use more than 4 pints per acre in a single growing season. **POSTEMERGENCE BROADCAST OR BAND:** Apply over the top of stubble or plant cane while sugarcane is less than 14 inches tall. **POSTEMERGENCE DIRECTED SPRAY:** Apply to sugarcane that is a minimum of 14 inches tall and before row closing.
	2 to 3	**AERIAL APPLICATION:** Apply when weeds are less than 4 inches tall in 5 to 10 gallons of spray mixture per acre. Appy to stubble or plant cane while the sugarcane is less than 14 inches tall.

SENCOR PLUS Atrazine TANK MIX: SENCOR 4 may be used with atrazine as a preemergence or postemergence (before row closing) application to sugarcane. Rates for SENCOR 4 are $1\frac{1}{2}$ to 4 pints/acre and atrazine 80% WP (4L) are $2\frac{1}{2}$ to 5 lb/acre (2 to 4 quarts/acre). For additional information on precautions, instructions, limitations, application, and weeds controlled, refer to the SENCOR 4 label and the atrazine label.

Special Precautions: (Florida only)

Do not use on sand soils.

Spray contact with sugarcane foliage may result in minor leaf margin chlorosis and/or necrosis.

Do not apply within 60 days of harvest.

Do not use treated crop for feed or forage.

Avoid spray overlaps or variations in application speed that may result in insufficient or excessive rates of application.

TOMATOES

Apply SENCOR herbicide with ground equipment to seeded and transplanted tomatoes as specified under "Recommended Applications" on the label.

For effective control of grasses and broadleaf weeds with postemergence applications, apply SENCOR 4 before weeds are 1 inch tall. Thorough spray coverage on weed foliage is essential for adequate control with postemergence applications.

Do not use air blast or other high pressure spray equipment to make postemergence applications of SENCOR 4. Refer to the appropriate section of the label for additional information regarding spray equipment, dilution rates, mixing, sprayer cleanup, restrictions, container disposal and cautions.

Refer to the "GENERAL INFORMATION" section in the front of the label.

For specific application information see the "GENERAL INFORMATION" section in the front of the SENCOR 4 label.

Weeds Controlled:

PREPLANT INCORPORATED APPLICATIONS—TRANSPLANT TOMATOES ONLY

Broadleaves:		Grasses:
Galinsoga *(Galinsoga spp.)*	*Pigweed, Redroot *(Amaranthus retroflexus)*	*Goosegrass *(Eleusine indica)*
Lambsquarters *(Chenopodium album)*	*Purslane, Common *(Portulaca oleracea)*	

Preplant incorporated applications applied as directed will suppress foxtails, panicums and barnyardgrass.

SENCOR/Treflan Tank Mix

This tank mix combination applied preplant incorporated as directed on the label will control the weeds listed above plus those weeds listed on the Treflan label.

POSTEMERGENCE APPLICATIONS—ESTABLISHED TOMATOES

For effective control of weeds with postemergence applications, apply SENCOR 4 before weeds are 1 inch tall.

Broadcast Sprays—½ to ¾ pt. of SENCOR 4/acre

Directed Sprays 1 to 2 pts. of SENCOR 4/acre

Broadleaves:		Grasses:
Carpetweed *(Mollugo verticillata)*	Pigweeds *(Amaranthus spp.)*	*Foxtail, Yellow *(Setaria glauca)*
Fumitory *(Fumaria officinalis)*	Purslane *(Portulaca oleracea)*	Goosegrass *(Eleusine indica)*
Galinsoga *(Galinsoga spp.)*	*Ragweed, Common *(Ambrosia artemisiifolia)*	*(Plus weeds listed under "Broadcast Sprays")*
*Jimsonweed *(Datura stramonium)*	*Smartweed, Pennsylvania *(Polygonum pensylvanicum)*	
*Ladysthumb *(Polygonum persicaria*	Toadflax *(Linaria spp.)*	
Lambsquarters *(Chenopodium album)*	*Velvetleaf *(Abutilon theophrasti)*	
Mustard, Wild *(Brassica kaber)*		

*For optimum control of these weeds, use the highest rate recommended on the label for the type of application to be made. Repeat postemergence applications may be needed for best control.

Postemergence applications as directed on the label will suppress barnyardgrass and crabgrass when these weeds are less than 1 inch tall.

Recommended Broadcast Applications

Crop	SENCOR 4 *Pints/Acre	Remarks
	½ to 1	**PREPLANT INCORPORATED—TRANSPLANT TOMATOES ONLY (EXCEPT CALIFORNIA):** Apply specified dosage in 10 to 40 gallons of water per acre as a broadcast spray to the soil surface immediately before transplanting. Incorporate to a depth of 2 to 4 inches with equipment capable of uniformly mixing the chemical into the soil. This application may be made alone or in a tank mix combination with Treflan EC. Refer to the Treflan label for specific rate of application and for additional precautions and restrictions for tomatoes.

Crop	SENCOR 4 *Pints/Acre	Remarks
Tomatoes	1/2 to 3/4	**POSTEMERGENCE BROADCAST SPRAY—ESTABLISHED TOMATOES:** Apply specified dosage in 20 to 75 gallons of water per acre as a broadcast spray or apply in 1/4 to 3/4 inch of water (1/4 to 1/2 inch of water on sandy soils) per acre as a continuous injection in center pivot and lateral move systems or in the last 15 to 30 minutes of set in permanent solid set sprinkler systems over the plants in single or multiple applications with a minimum of 14 days between treatments. For transplanted tomatoes do not apply within 14 days of transplanting. Transplants **must** have recovered from transplant shock and new growth evident. Do not apply to established tomatoes within 24 hours of application of other pesticides. Do not tank mix with other pesticides. (See "Special Precautions" below.)
	1 to 2	**POSTEMERGENCE DIRECTED SPRAY—ESTABLISHED TOMATOES:** Apply specified dosage in 20 to 75 gallons of water per acre as a directed spray in single or multiple applications with a minimum of 14 days between treatments. Avoid contacting tomato foliage with spray. This method of treatment is recommended for use in fields with a history of severe weed pressure or in fields infested with hard to control weeds. For transplanted tomatoes, do not apply within 14 days of transplanting. Transplants **must** have recovered from transplant shock and new growth evident. Do not apply to established tomatoes within 24 hours of application of other pesticides. (See "Special Precautions" below.) When banding see the appropriate section in the front of the label.

*Use the higher rate in fields with a history of severe weed pressure and for maximum residual weed control.

Special Precautions (Tomatoes)

Do not apply more than a total of 2 pts. SENCOR 4 per crop season.

Do not apply the total amount of 2 pts. SENCOR within a time span of less than 35 days, except in the case of directed sprays.

Allow at least 14 days between applications regardless of dosage or method of application.

Do not apply within 7 days of harvest.

Do not apply within 3 days after periods of cool, wet or cloudy weather, or crop injury will occur.

Do not use hot caps on tomatoes within 7 days before or at any time after application of SENCOR 4.

Do not treat seeded tomatoes until plants have reached the 5 to 6 leaf stage.

Crop injury or delayed maturity may result from broadcast or directed spray applications if tomatoes are growing under stress conditions such as periods of drought or cool, wet and cloudy weather preceding application.

For newly introduced tomato varieties with unknown tolerance to SENCOR, treat only a small area to determine if SENCOR 4 can be used without injury to the crop.

DO NOT USE SENCOR 4 ON TOMATOES IN KERN COUNTY, CALIFORNIA.

CEREALS (Barley and Wheat)

For the control of certain grass and broadleaf weeds, SENCOR herbicide is recommended as a single postemergence application in barley (Winter and Spring) and Winter wheat (see the appropriate section of the label for geographic limitations).

SENCOR 4 may be used in cereals (barley and wheat) as follows:

1. Barley (Spring and Winter)—Only East of the Cascade Mountains in Idaho, Oregon, Montana & Washington.
2. Winter wheat in Idaho, Oregon, Montana, Utah & Washington.
3. Winter wheat in Idaho, Montana, Nevada, Oregon, Utah & Washington—Early Application (2-leaf stage).
4. Dryland (Non-irrigated) Winter wheat in Kansas, Oklahoma & Texas.
5. Dryland (Non-irrigated) Winter Wheat only (Arkansas, Louisiana, Mississippi & Tennessee).
6. Dryland (Non-irrigated) Winter wheat only (Georgia).

Mixing: See the "GENERAL INFORMATION" section in the front of the label for specific mixing procedures. Do not tank mix SENCOR 4 with pesticides other than those specified below. When tank mixing, carefully follow the instructions on the label. Refer to the other product labels registered for the specific use intended for directions, rates, and weeds controlled.

Application: Refer to the "GENERAL INFORMATION" section in the front of the label for detailed information on the application of SENCOR 4.

Apply SENCOR 4 when wheat stand is uniform in height. For best results apply SENCOR 4 when weeds are small (less than 1 to 1 1/2 inches in diameter or height). **In the Fall** apply after wheat has a minimum of 3 tillers and 4 secondary roots at least 2 inches in length, but before Winter dormancy and wheat jointing occurs. **In the Spring** apply after wheat has recovered from Winter dormancy and after wheat has fully tillered and has developed secondary roots at least 2 inches in length, but before jointing.

Apply SENCOR 4 only once per crop season to wheat or barley.

Grazing Treated Wheat: Wheat fields treated with SENCOR 4 may be grazed 14 days following application.

Special Precautions (Cereals)

Crop injury may occur on barley or wheat when SENCOR 4 is used under the following conditions:

1. When applied while the crop is under stress such as Winter kill, frost damage, disease, drought or excessive moisture, or when these conditions follow the application.
2. When applied in combination with fluid fertilizer.
3. When applications are made prior to the growth stage specified on the label.
4. When soils are high in lime or sodium, have a pH greater than 7.7, or are calcareous soils.
5. When proper application instructions are not followed.

Weed control may not be observed for 2 to 4 weeks after application; under very dry conditions, weed control may not be apparent for 4 to 6 weeks. Moisture (at least 1/2") is needed within 2 to 3 weeks after application to move SENCOR into the weed root zone. Lack of adequate moisture after application may result in poor control.

Do not exceed the rates specified on the label.

BARLEY (Spring and Winter)

FOR USE ONLY EAST OF THE CASCADE MOUNTAINS IN IDAHO, OREGON, UTAH, MONTANA AND WASHINGTON ON THE VARIETIES BOYER, COMPANA, HANNCHEN, HECTOR, HESK, HUDSON, LUTHER, KAMIAK, KLAGES, OLYMPIC, PIROLINE AND STEPTOE.

A single postemergence application of SENCOR 4 is recommended for control of certain grass and broadleaf weeds in Spring and Winter barley.

Recommended Broadcast Applications

Soil Texture	SENCOR 4 Pints/Acre, % Organic Matter[1]: 1 to 2%	Over 2%	Remarks
COARSE (Sandy loam, loamy sand)	1/2	1/2 to 3/4	Barley should be seeded to obtain a uniform crop stand at application time. Apply after barley has fully tillered and developed secondary roots but before jointing and before weeds are 2 inches tall or have 2 inch diameter rosettes. Do not graze or harvest treated barley for feed before crop maturity.
MEDIUM (Loam, silt loam, silt, sandy clay, sandy clay loam)	1/2 to 3/4	3/4	
FINE (Silty clay, silty clay loam,[2] clay, clay loam)	3/4	3/4 to 1	

[1]Do not use on any soils containing less than 1% organic matter.
[2]Silty clay loam soils are transitional soils and may be classified as medium textured soils in some regions of the U.S.

Weeds Controlled in Barley:

Broadleaves:	Grasses:
Chickweed, Common *(Stellaria media)*	*Cheatgrass *(Bromus secalinus)*
Fiddleneck, Tarweed *(Amsinckia lycopsoides)*	*Downy Brome *(Bromus tectorum)*
Henbit *(Lamium amplexicaule)*	*Foxtails *(Setaria spp.)*
*Kochia *(Kochia scoparia)*	*Ryegrass, Annual *(Lolium multiflorum)*
Lambsquarters *(Chenopodium album)*	*Windgrass (*Apera* spp.)
*Mustard, Blue or Purple *(Chorispora tenella)*	
*Mustard, Jim Hill	

(*Sisymbrium altissimum*)
*Mustard, Wild
(*Brassica kaber*)
Pigweeds
(*Amaranthus spp.*)

*Use the highest rate recommended for soil texture and percent organic matter for the control of these weeds.

Hard to Control Weed:
Wild Oats (*Avena fatua*).

NOTE: Control of wild oats may be variable ranging from poor to excellent depending on timing of application, stage of growth of wild oats at time of application, temperatures and soil moisture conditions following treatment.

SENCOR in Combination with Other Herbicides
SENCOR 4 may be applied as directed above or may be tank mixed with 2,4-D, Banvel®, Bronate®, Buctril®, Finesse®, or Igran®. Refer to the other product labels registered for this use for additional directions, rates, and weed species controlled.

WINTER WHEAT

FOR USE ON WINTER WHEAT IN IDAHO, OREGON, UTAH AND WASHINGTON ON THE VARIETIES DAWS, GAINES, HATTON, HILL 81, HYSLOP, LUKE, McDERMID, MORO, NuGAINES, PAHA, PECK, SPRAGUE, STEPHENS, WANSER, FARO, TYEE, AND WESTON AND IN MONTANA ON THE VARIETIES CENTURK, CHEYENNE AND WINALTA.

A single postemergence application of SENCOR 4 is recommended for control of certain grass and broadleaf weeds in Winter wheat.

Recommended Broadcast Applications[1]

Soil Texture	SENCOR 4 Pints/Acre		
	0.75% to 1% Organic Matter[2]	1% to 2% Organic Matter	More than 2% Organic Matter
COARSE (Loamy sand, sandy loam)	0.5 to 0.6	0.5 to 0.6	0.5 to 0.6
MEDIUM (Loam, silt loam, silt, sandy clay, sandy clay loam)	0.5 to 0.6	0.6 to 0.75	0.6 to 0.9
FINE (Silty clay, silty clay loam,[3] clay, clay loam)	0.5 to 0.75	0.75 to 1.0	0.75 to 1.0

[1]Do not use on gravelly or thinly-covered or exposed subsoil areas (clay knobs) as crop injury may occur.
[2]Do not use on any soils containing less than 0.75% organic matter.
[3]Silty clay loam soils are transitional soils and may be classified as medium textured soils in some regions of the U.S.

Weeds Controlled in Wheat:

Broadleaves:

Annual Polemonium (Jacob's Ladder) (*Polemonium micranthum*)
Chickweed, Common (*Stellaria media*)
Conical Catchfly (*Silene conica*)
Corncockle (*Agrostemma githago*)
Fiddleneck, Tarweed (*Amsinckia lycopsoides*)
Gromwell (*Lithospermum officinale*)
Henbit (*Lamium amplexicaule*)
Ivyleaf Speedwell (*Veronica hederifolia*)
Knotweed, Prostrate (*Polygonum aviculare*)
Mayweed, (dog fennel) (*Eupatorium capillifoium*)
Miners Lettuce (*Claytonia perfoliata*)
*Mustard, blue or purple (*Chorispora tenella*)
*Mustard, Jim Hill (tumble) *Sisymbrium altissimum*)
*Mustard, Wild (*Brassica kaber*)
Pennycress, Field (*Thlaspi arvense*)
Pineappleweed (*Matricaria matricarioides*)
Prickly Lettuce (*Latuca serriola*)
Pigweeds (*Amaranthus spp.*)
*Kochia (*Kochia scoparia*)
Lambsquarters (*Chenopodium album*)
Madwort (Catchweed) (*Asperugo procumbens*)
Redstem Filaree (*Erodium cicutarium*)
*Russian Thistle (*Salsola iberica*)
Shepherdspurse (*Capsella bursa-pastoris*)

Grasses:

*Blackgrass (Black twitchgrass) (*Alopecurus myosuroides*)
*Bluegrass, bulbous (*Poa bulbosa*)
*Cheatgrass (*Bromus secalinus*)
*Downy Brome (*Bromus tectorum*)
*Foxtails (*Setaria spp.*)
*Rescuegrass (*Bromus catharticus*)
*Ryegrass, Italian (*Lolium multiflorum*)
*Wild Barley (*Hordeum leporinum*)
*Windgrass (*Apera spica-venti*)

*Suppression only may occur.

For suppression of these weeds, use the highest rate recommended for soil texture and percent organic matter. Suppression is a reduction in weed size and growth compared to a non-treated area in the same field.

Hard-To-Control Weeds:

Broadleaves:

Cowcockle (*Vaccaria pyamidata*)
Mustard, Tansy (*Descurainia pinnata*)
Wild Buckwheat (*Polygonum convolvulus*)

Grasses:

Ripgut Brome (*Bromus diandrus*)
Wild Oats (*Avena fatua*)

NOTE: Control of these weeds may be variable, ranging from poor to excellent depending on timing of application, stage of growth at the time of application, temperatures, and soil moisture conditions following treatment.

SENCOR in Combination With Other Herbicides
SENCOR 4 may be applied as a tank mix with Finesse, Glean®, 2,4-D, MCPA, Igran®, Banvel, Bronate or Buctril®. Refer to the other product labels registered for this use for additional directions, rates and weed species controlled. Observe all cautions and limitations on labeling of all products used in mixtures.

IRRIGATED WHEAT/BARLEY
Follow timing directions for dryland wheat/barley found at the beginning of this cereals section. After application of SENCOR 4, do not apply more than 1/2 inch of water for the first irrigation; the maximum amount for each additional irrigation should not exceed 1 inch.

SENCOR plus Igran Tank Mix Application
Apply SENCOR 4 at 1/2 to 3/4 pint per acre and Igran 80 WP at 1/2 to 3/4 pound per acre according to label directions. The rate of each product depends on soil type. Do not use Igran on sand or loamy sand soils. Use lower rate of Igran on sandy loam soils.

BROADLEAF WEED CONTROL ONLY:
Where only broadleaf control is needed, SENCOR 4 applied at the rates listed below in a tank mix combination with Bronate, or Buctril can be used to control wild buckwheat, in addition to all the weeds listed on these labels.

Recommended Broadcast Applications

SENCOR 4 Pints/Acre	Remarks
3/8 to 1/2	Apply SENCOR 4 in a tank mix combination with Buctril or Bronate. Use higher rate of SENCOR for control of larger broadleaf weeds that are 2 inches in diameter or height. Apply before weeds are greater than 3 inches in diameter or height. Not for use on sand soils.

Special Precautions: Refer to the "General Mixing and Application," and "Precautions" in the front of the Cereal section.

Temporary chlorosis or injury may occur after application of SENCOR 4 to wheat, especially if frost or snow occurs within a few days after application. Heavy rainfall (1 inch or more) within 4 days following application may result in crop injury.

WINTER WHEAT
FOR USE IN IDAHO, MONTANA, NEVADA, OREGON, UTAH AND WASHINGTON ON THE VARIETIES CENTURK, CHEYENNE, DAWS, FARO, GAINES, HATTON, HILL 81, HYSLOP, LUKE, McDERMID, MORO, NuGAINES, PAHA, PECK, SPRAGUE, STEPHENS, TYEE WANSER, WESTON AND WINALTA

SENCOR 4 flowable herbicide in a tank mix combination with one of the herbicide products specified below, may be applied to wheat after the crop has reached the 2-leaf stage for control of the following broadleaf weeds:

Common Chickweed	(*Stellaria media*)
Henbit	(*Lamium amplexicaule*)
Prickly Lettuce	(*Lactuca serriola*)
Blue Mustard	(*Chorispora tenella*)
Field Pennycress	(*Thlaspi arvense*)
Tarweed Fiddleneck	(*Amsinckia lycopsoides*)

RECOMMENDED APPLICATIONS

Crop	SENCOR 4 Fluid Ounces Per Acre	Remarks
Winter Wheat	3 to 4 1/2	Use only in a tank mix combination for broadleaf weed control*. The specified dosage per acre may be added to other herbicides registered for use on wheat after the crop has reached the 2-leaf stage, but not during winter dormancy. See the appropriate sections of the label and the label for other herbicide products that are tank mixed with SENCOR 4 flowable herbicide for directions, application rates, mixing procedures, special precautions and restrictions before using.

* May be used with Bronate, Buctril, Banvel, 2,4-D, MCPA, Glean, Ally®, Finesse, or Harmony®.

Observe all cautions and limitations on labeling of all products used in mixtures.

DRYLAND (NON-IRRIGATED) WINTER WHEAT

FOR USE ONLY ON DRYLAND WINTER WHEAT IN KANSAS, OKLAHOMA AND TEXAS ON THESE VARIETIES: TAM W-101, TAM-105, GARST 64, HAWK, LARNED, NEWTON, PIONEER 2157, STALLION, VICTORY, AND WRANGLER

Weeds Controlled In Wheat:

Broadleaves:

- Annual Polemonium (Jacob's Ladder) *(Polemonium micranthum)*
- Chickweed, Common *(Stellaria media)*
- Fiddleneck, Tarweed *(Amsinckia lycopsoides)*
- Henbit *(Lamium amplexicaule)*
- *Kochia *(Kochia scoparia)*
- Lambsquarters *(Chenopodium album)*
- Mustard, Blue *(Chorispora tenella)*
- *Mustard, Jim Hill *(Sisymbrium altissimum)*
- *Mustard, Wild *(Brassica kaber)*
- Pigweeds *(Amaranthus spp.)*

Grasses:

- *Cheatgrass *(Bromus tectorum)*
- *Downy Brome *(Bromus tectorum)*
- *Foxtails *(Setaria spp.)*
- *Rescuegrass *(Bromus catharticus)*

*For control of these weeds, use the highest rate recommended for soil texture and percent organic matter.

Hard to Control Weeds: *Italian ryegrass, *(Lolium multiflorum)*, *Wild Oats *(Avena fatua)*, and *Wild Buckwheat *(Polygonum convolvulus)*.

NOTE: Control of these weeds may be variable, ranging from poor to excellent depending on timing of application, stage of growth at time of application, temperatures and soil moisture conditions following treatment.

Recommended Broadcast Applications

Soil Texture	SENCOR 4 Pint/Acre — % Organic Matter[1] — Fall: 1 to 2	Fall: Over 2	Spring: 1 to 2	Spring: Over 2	Remarks
COARSE (Sandy loam, loamy sand)	1/2	1/2	1/2	1/2 to 3/4	Apply specified dosage per acre with aerial or ground spray equipment as a broadcast postemergence spray.
MEDIUM (Loam, silt loam, silt, sandy clay, sandy clay loam)	1/2 to 3/4	3/4	3/4 to 1	3/4 to 1 1/4	For Fall or Spring application, refer to "Application" in the beginning of the Cereals section for information on timing of application and growth restrictions.
FINE (Silty clay, silty clay loam,[2] clay, clay loam)	3/4	3/4 to 1	3/4 to 1 1/4	1 to 1 1/2	

[1] Do not use on any soils containing less than 1% organic matter.

[2] Silty clay loam soils are transitional soils and may be classified as medium textured soils in some regions of the U.S.

SENCOR in Combination with Other Herbicides

SENCOR 4 may be applied as directed above or may be tank mixed with 2,4-D, Banvel, Bronate, Brominal, or Buctril. Refer to the other product labels registered for additional directions, rates, and weed species controlled.

DRYLAND (NON-IRRIGATED) WINTER WHEAT ONLY
(Arkansas, Louisiana, Mississippi and Tennessee)

Weeds Controlled:

Broadleaves:

- Bittercress *(Cardamine spp.)*
- *Buttercup *(Ranunculus spp.)*
- Carolina Geranium *(Geranium carolinianum)*
- Chickweed, Common *(Stellaria media)*
- Chickweed, Mousear *(Cerastium vulgatum)*
- Cutleaf Evening Primrose *(Oenothera laciniata)*
- Henbit *(Lamium amplexicaule)*
- *Kochia *(Kochia scoparia)*
- Lambsquarters, Common *(Chenopodium album)*
- Mayweed (Dog Fennel) *(Eupatorium capillifolium)*
- Mustard, Wild *(Brassica kaber)*
- Pepperweed, Virginia *(Lepidium virginicum)*
- Pigweeds *(Amaranthus spp.)*
- Shepherdspurse *(Capsella bursa-pastoris)*
- Whitlowgrass, Spring *(Draba verna)*
- Wild Turnip *(Brassica rapa)*

Grasses:

- Bluegrass, Annual *(Poa annua)*
- *Cheatgrass *(Bromus secalinus)*
- *Downy Brome *(Bromus tectorum)*
- Little Barley *(Hordeum pusillum)*
- *Rescuegrass *(Bromus catharticus)*
- *Ryegrass, Italian *(Lolium multiflorum)*

*For control of these weeds on fine soils, use SENCOR 4 at 1 pint/acre.

NOTE: Control of Wild Buckwheat *(Polygonum convolvulus)* and Winter Vetch *(Vicia spp.)* may be variable with SENCOR 4, ranging from poor to excellent depending on growth at time of application, temperatures and moisture conditions following application.

Recommended for use only on the following wheat varieties:
Abe, Arthur 71, Caldwell, Coker 747, Coker 762, Coker 68-15, Double-crop, Hunter, McNair 1003, McNair 1813, Nelson, Pike, Southern Belle & Tyler.

Recommended Broadcast Applications

Soil Texture	SENCOR 4[1] Pints/Acre	Remarks
COARSE (Sandy loam, loamy sand)	1/2	Apply specified dosage per acre as a broadcast postemergence spray. For wheat plants that are coming out of Winter dormancy, allow at least two weeks for plants to recover before application. If wheat has been grazed during the Winter, allow at least two weeks for recovery before application. Refer to the beginning of this Cereals section for information on timing and growth stages for application.
MEDIUM (Loam, silt loam, sandy clay, sandy clay loam)	3/4	
FINE (Silty clay, silty clay loam,[2] clay, clay loam)	3/4 to 1[3]	

[1] Do not use on coarse soils containing less than 1% organic matter or on medium or fine textured soils containing less than 0.5% organic matter.

[2] Silty clay loam soils are transitional soils and may be classified as medium textured soils in some regions of the U.S.

[3] Use 1 pint/acre for control of weed species designated with an asterisk under "Weeds Controlled."

SENCOR in Combination with Other Herbicides:

SENCOR 4 may be applied as directed above and may be tank mixed with 2,4-D, Banvel, Bronate, or Buctril. Refer to the other registered product labels for additional directions, rates, and weed species controlled.

Special Precautions

Temporary chlorosis may occur after application of SENCOR 4 to wheat, especially if frost occurs.

DRYLAND (NON-IRRIGATED) WINTER WHEAT ONLY (Georgia)

Weeds Controlled:

Broadleaves:

Bittercress *(Cardamine spp.)*
Buttercup *(Ranunculus spp.)*
Carolina Geranium *(Geranium carolinianum)*
Chickweed, Common *(Stellaria media)*
Chickweed, Mousear *(Cerastium vulgatum)*
Cutleaf Evening Primrose *(Oenothera laciniata)*
Henbit *Lamium amplexicaule)*
Kochia *(Kochia scoparia)*
Lambsquarters, Common *(Chenopodium album)*
Mayweed (Dog Fennel) *(Eupatorium capillifolium)*
Mustard, Wild *(Brassica kaber)*
Pepperweed, Virginia *(Lepidium virginicum)*
Pigweeds *(Amaranthus spp.)*
Shepherdspurse *(Capsella bursa-pastrois)*
Whitlowgrass, Spring *(Draba Verna)*
Wild Turnip *(Brassica rapa)*
Wild Radish *(Raphanus raphanistrum)*

Grasses:

Bluegrass, Annual *(Poa annua)*
Cheatgrass *(Bromus secalinus)*
Downy Brome *(Bromus tectorum)*
Little Barley *(Hordeum pusillum)*
Rescuegrass *(Bromus catharticus)*
Ryegrass, Italian *(Lolium multiflorum)*

NOTE: Control of Wild Buckwheat *(Polygonum convolvulus)* and Winter Vetch *(Vicia spp.)* may be variable with SENCOR 4, ranging from poor to excellent depending on growth at time of application, temperatures and moisture conditions following application.
Recommended for use only on the following wheat varieties:
Abe, Arthur 71, Caldwell, Coker 747, Coker 762, Coker 797, Coker 68-15, Doublecrop, Florida 302, Hunter, McNair 1003, McNair 1813, Nelson, Oasis, Omega 78, Pike, Southern Belle & Tyler.

Recommended Broadcast Applications

Soil Texture	SENCOR 4 Pints/Acre	Remarks
COARSE (Sandy loam, loamy sand)	½	Apply specified dosage per acre as a broadcast postemergence spray. For wheat plants that are coming out of Winter dormancy, allow at least two weeks for plants to recover before application. If wheat has been grazed during the winter, allow at least two weeks for recovery before application. Do not apply before 75 days after planting. Apply only after wheat has 4 tillers, but before jointing. Wheat must be planted before November 15 in the Piedmont area and northern part of the state, and before December 1 in the Coastal Plain area.
MEDIUM (Loam, silt loam, silt, sandy clay, sandy clay loam)	¾	

Do not use on coarse soils containing less than 1% organic matter or on Medium Fine textured soils containing less than 0.5% organic matter.

SENCOR in Combination with Other Herbicides:

SENCOR 4 may be applied as directed above and may be tank mixed with 2,4-D, Banvel, Bronate, or Buctril. Refer to the other registered product labels for additional directions, rates, and weed species controlled.

Special Precautions

Temporary chlorosis may occur after application of SENCOR 4 to wheat, especially if frost occurs.

FOR WEED CONTROL IN A WHEAT/FALLOW/WHEAT ROTATION (Idaho, Oregon, Utah and Washington Only)

SENCOR 4 may be applied to provide weed control during the fallow period after wheat harvest or in the Spring before Winter wheat is planted. Winter wheat can be seeded 4 months (120 days) after Spring application. Mechanical tillage or the application of a contact herbicide may be required to control weeds germinating prior to seeding of Winter wheat. Best results will be obtained where straw and chaff are evenly distributed across the field.

For specific application information see the "GENERAL INFORMATION" section in the front of the SENCOR 4 label.

Where weed growth is present at application time, SENCOR 4 should be applied with Gramoxone or other contact herbicide. Refer to the other product labels registered for additional directions, rates, and weed species controlled.

Weeds Controlled:

Broadleaves:

Chickweed, Common *(Stellaria media)*
Henbit *(Lamium amplexicaule)*
*Kochia *(Kochia scoparia)*
Lambsquarters *(Chenopodium album)*
Mustard, Blue or Purple *(Chorispora tenella)*
Mustard, Jim Hill *(Sisymbrium altissimum)*
Mustard, Tansy *(Descuainia pinnata)*
Mustard, Treacle *(Eyrsimum repandum)*
Mustard, Wild *(Brassica kaber)*
Pennycress, Field (Fanweed) *(Thlaspi arvense)*
Pigweeds *(Amaranthus spp.)*
*Russian Thistle *(Salsola iberica)*
*Wild Sunflower *(Helianthus spp.)*

Grasses:

Cheatgrass *(Bromus secalinus)*
Downy Brome *(Bromus tectorum)*
*Wheat, Volunteer *(Triticum spp.)*

***NOTE:** Since control of these weeds may be variable depending on moisture following application, the higher labelled rate is recommended.

After Harvest Application (Fall Fallow): SENCOR 4 may be applied to wheat stubble after harvest in the Fall. Apply 1 to 1¼ pints per acre broadcast before weeds emerge. Use higher rate for longer weed control or for weeds designated as requiring the higher rate for control. Rainfall (½ inch or more) is necessary for herbicide activation.

Do not plant crops in treated areas for at least 10 months following Fall applications.

SENCOR 4 may be applied at 1 to 1¼ pints per acre as directed above for a Fall application. If other vegetation is present at the time of application use a contact herbicide.

Spring Application (Summer Fallow): SENCOR 4 may be applied to wheat stubble in the Spring. Apply ¾ to 1 pint per acre broadcast before weeds emerge in the Spring. Use higher rate for longer weed control or weeds designated as requiring higher rate for control Rainfall (½ inch or more) is necessary for herbicide activation.

Precautions and Restrictions:

Do not graze treated fields.

Do not plant Spring seeded cereals following Fall fallow applications of SENCOR.

Where SENCOR was applied in the Fall, do not apply SENCOR in the Spring.

FOR WEED CONTROL IN A FALLOW ROTATION WITH BARLEY AND WHEAT (Colorado, Kansas, Montana, Nebraska and Wyoming only)

SENCOR 4 may be applied to provide weed control during the fallow period after wheat or barley harvest or in the Spring before planting of Winter wheat or barley, Mechanical tillage or the application of a contact herbicide may be required to control weeds germinating prior to seeding of Winter wheat or barley.

Refer to "Mixing" in the "GENERAL INFORMATION" section in the front of the label.

For specific application information see "GENERAL INFORMATION" section in the front of the SENCOR 4 label.

Where weed growth is present at application time, SENCOR 4 should be applied with Gramoxone, Roundup®, or other contact herbicide. Refer to the other product labels registered for additional directions, rates, and weed species controlled. Do not plant crops in treated areas earlier than 10 months following Fall applications.

WEEDS CONTROLLED:

Broadleaves:

Chickweed, Common (*Stellaria media*)
Cowcockle (*Vaccaria pyramidata*)
Henbit (*Lamium amplexicaule*)
*Kochia (*Kochia scoparia*)
Lambsquarters (*Chenopodium album*)
Mustard, Blue or Purple (*Chorispora tenella*)
Mustard, Jim Hill (*Sisymbrium altissimum*)
Mustard Tansy (*Descurainia pinnata*)
Mustard, Treacle (*Eyrsimum repandum*)
Mustard, Wild (*Brassica kaber*)
Pennycress, Field (Fanweed) (*Thlaspi arvense*)
Pigweeds (*Amaranthus spp.*)
Russian Thistle (*Salsola iberica*)
Wild Sunflower (*Helianthus spp.*)

Grasses:

Cheatgrass (*Bromus secalinus*)
Downy Brome (*Bromus tectorum*)
*Foxtail, Green (*Setaria viridis*)
*Wheat, Volunteer (*Triticum spp.*)
*Wild Oats (*Avena fatua*)

***NOTE:** Since control of these weeds may vary depending on moisture following application, use the higher rate recommended below.

AFTER HARVEST APPLICATION (Fall Fallow)—SENCOR 4 may be applied to the stubble after harvest in the Fall. Apply 1¼ to 1½ pints per acre broadcast before weeds emerge. Use the higher rate for longer weed control or for weeds designated as requiring the higher rate for control. Rainfall (½ inch or more) is necessary for herbicide activation.

SPRING APPLICATION (Summer Fallow)—SENCOR 4 may be applied to wheat stubble in the Spring. Apply ¾ to 1 pint per acre broadcast before weeds emerge in the Spring. Use the higher rate for longer weed control or weeds designated as requiring the higher rate for control. Rainfall (½ inch or more) is necessary for herbicide activation. Wheat or barley can be seeded 120 days after Spring application.

Precautions and Restrictions:

Do not graze treated fields.

Do not plant Spring seeded cereals following Fall applications fallow.

Where SENCOR was applied in the Fall, do not apply SENCOR in the Spring.

3-20-89

SPECIAL LOCAL NEED REGISTRATIONS [SECTION 24(c)] FOR SENCOR 4*

CROP	PEST	STATE EPA SLN NO.	DOSAGE	REMARKS
Alfalfa	Various broadleaf weeds and grasses	Washington WA 85-0006 Oregon OR 85-0019	¾ to 2 pts.	Application by ground or air. Do not graze or harvest within 28 days of application.
Barley (Spring)	Various weeds	Alaska AK 83-0002	¼ to ½ pint/acre.	Postemergence ground application.
Lentils and Peas	Various broadleaf weeds and grasses	Idaho ID 87-0004 Oregon OR 87-0001 Washington WA 87-0007	See label rates.	Preemergence and postemergence before crop germinates.
Lentils and dry field peas	Suppression of certain broadleaf weeds	Idaho ID 80-0020	½ to ¾ pint per acre.	Apply before or after planting but before crop emergence. Northern Idaho only
Potatoes	Weeds	Idaho ID 81-0045 Oregon OR 81-0039 Washington WA 81-0041	Label rates.	Tank mix with Eptam 7-E
	Various Weeds	Idaho ID 87-0016 Woodington WA 86-0016	See label rates	Postemergence application to russetted varieties of potatoes that are not early maturing.
	Various broadleaf and grass weeds including quackgrass	Maine ME 79-0009	1 to 2 pints per acre.	May be tank mixed with 7 to 10 pounds DOWPON M as a preemergence broadcast spray.
Soybeans	Red rice	Louisiana LA 79-0009	See SENCOR Lasso and Treflan labels for specific dosage.	Apply as a preemergence broadcast or band overlay.
	Various weeds	Michigan MI 79-0021	½ to ¾ pints per acre.	Use on alkaline soils.
Soybeans	Various broadleaf weeds and grasses	North Carolina NC 86-0007	1 to 2 pints per acre.	Application alone or as a tank mix on blackened soils.
		Alabama AL 81-0026 Louisiana LA 79-0019 Missouri MO 79-0012 Tennessee TN 85-0005	½ to 1 pint per acre.	Use according to label directions. Do not apply through irrigation system.
Sugarcane	Various grass and broadleaf weeds	Louisiana LA 84-0017	2 to 3 pints	BROADCAST (Summer Fallow): Apply specified dosage per acre using 20 to 30 gallons of water with ground equipment or 5 gallons of water with aircraft spray equipment. Apply in late May to early June after spring tillage and rows have been rebuilt. Use the lower rate if the application is made less than 5 weeks prior to planting. Apply to a soil surface which is free of weeds, trash and clods.
Tomatoes (Chemical fallow)	Various weeds	California CA 79-0235	1 to 1.5 pints per acre	Application by ground or air. Not for use in Kern County.

CROP	PEST	STATE EPA SLN NO.	DOSAGE	REMARKS
Wheat (Dryland Winter only)	Various grass and broadleaf weeds	Missouri MO 84-0004	½ to 1½ pints per acre.	On Caldwell, Centurk 78, Coker 747, Hart, Newton, Pike, Pioneer 876, Pioneer 578, TAM-105, and TAM W-101 wheat variety.

* For distribution and use only within the state specified. These Special Local Need registrations approved as of date of this publication are, however, subject to cancellation by the state. Contact your nearest Mobay Agricultural Chemicals Division Office for the current status and a copy of the complete individual supplemental Special Local Need labeling. This labeling must be in the possession of the user at time of pesticide application.

SPECIAL LOCAL NEED REGISTRATIONS [SECTION 24(c)] FOR SENCOR 4*

CROP	PEST	STATE EPA SLN NO.	DOSAGE	REMARKS
Tomatoes (Chemical fallow)	Various weeds	California CA 79-0235	1 to 1.5 pints per acre.	Application by ground or air. Not for use in Kern County.
Wheat (Dryland Winter only)	Various grass and broadleaf weeds	Missouri MO 84-0004	½ to 1½ pints per acre.	On Caldwell, Centurk 78, Coker 747, Hart, Newton, Pike, Pioneer 876, Pioneer 578, TAM-105, and TAM W-101 wheat variety.

* For distribution and use only within the state specified. These Special Local Need registrations approved as of date of this publication are, however, subject to cancellation by the state. Contact your nearest Mobay Agricultural Chemicals Division Office for the current status and a copy of the complete individual supplemental Special Local Need labeling. This labeling must be in the possession of the user at time of pesticide application.

CROP ROTATION RECOMMENDATIONS

Waiting Period After SENCOR Herbicide Application[1]

4 Months	8 Months	12 Months	18 Months
Alfalfa	Barley	Other Crops not listed (except root crops)	Sugar Beets
Barley[2]	Cotton		Onions
Asparagus	Lentils		And other root crops
Corn	Peas		
Forage Grasses	Rice		
Potatoes	Wheat		
Sainfoin			
Soybeans			
Sugarcane			
Tomatoes			
Wheat[2]			

[1] Cover crops for soil building or erosion control may be planted any time, but do not graze or harvest for food or feed. Stand reductions may occur in some areas.
[2] Following peas, lentils or soybeans.

PRECAUTIONARY STATEMENTS

HAZARDS TO HUMANS AND DOMESTIC ANIMALS

CAUTION

May be harmful if swallowed or inhaled. Avoid contact with eyes, skin or clothing. Avoid breathing spray mist. Wash thoroughly with soap and warm water after handling. Wash contaminated clothing with soap and hot water before reuse.

Do not contaminate feed or food. Keep out of reach of children.

Obtain prompt medical aid if poisoning should occur.

To Physician: Treat the patient symptomatically.

Symptoms of Poisoning: The compound does not cause any definite symptoms that would be diagnostic. Poisoning is accompanied by breathing difficulties and sedation.

STATEMENT OF PRACTICAL TREATMENT

If in eyes, immediately flush eyes with plenty of water for at least 15 minutes. For further treatment get medical attention.

If swallowed, vomiting should be induced. Administer water freely, and induce vomiting by giving one dose (½ oz. or 15 ml.) of syrup of ipecac. If vomiting does not occur within 10 to 20 minutes, administer second dose. If syrup of ipecac is not available, induce vomiting by sticking finger down throat. Repeat until vomit fluid is clear. Never give anything by mouth to an unconscious person. Professional medical assistance should be secured immediately.

If on skin, wash thoroughly with soap and warm water.

ENVIRONMENTAL HAZARDS

Do not use on other crops grown for food or forage. Do not apply directly to water or wetlands. Do not apply when weather conditions favor drift from areas treated. Do not contaminate water by cleaning of equipment or disposal of wastes. Apply this product only as specified on the label.

Do not allow sprays to drift onto adjacent desirable plants.

Observe all cautions and limitations on labeling of all products used in mixtures.

GROUND WATER ADVISORY. Metribuzin is a chemical which can travel (seep or leach) through soil and can contaminate ground water which may be used as drinking water. Metribuzin has been found in ground water as a result of agricultural use. Users are advised not to apply metribuzin where the water table (ground water) is close to the surface, and where the soils are very permeable, i.e., well drained soils such as loamy sands. Your local agricultural agencies can provide further information on the type of soil in your area and the location of ground water.

RE-ENTRY STATEMENTS

Do not apply this product in such a manner as to directly or through drift expose workers or other persons. The area being treated must be vacated by unprotected persons. Do not enter treated areas without protective clothing until sprays have dried. Because certain states may require more restrictive re-entry intervals for various crops treated with this product, consult your State Department of Agriculture for further information.

Written or oral warnings must be given to workers who are expected to be in a treated area or in an area about to be treated with this product. Oral warnings should include those statements found under the label heading "Precautionary Statements" and "Re-entry Statements" included on the SENCOR 4 label. When oral warnings are given, warnings shall be given in a language customarily understood by workers. Oral warnings must be given if there is reason to believe that written warnings cannot be understood by workers. Written warnings must include the following information: WARNING, area treated with SENCOR 4 flowable herbicide on (date of application). Do not enter without appropriate protective clothing until sprays have dried. **If on skin,** remove contaminated clothing and wash skin immediately with soap and warm water. **If eyes are contaminated,** wash with flowing water for at least 15 minutes. **If swallowed, vomiting should be induced.** (See Precautionary Statements on label).

STORAGE AND DISPOSAL

Pesticide Disposal: Do not contaminate water, food or feed by storage or disposal.

Wastes resulting from the use of this product may be disposed of on site or at an approved waste disposal facility.

Container Disposal: Triple rinse (or equivalent). Then offer for recycling or reconditioning, or puncture and dispose of in a sanitary landfill, or incineration, or, if allowed by State and local authorities, by burning. If burned, stay out of smoke.

Store in a cool, dry place and in such a manner as to prevent cross contamination with other pesticides, fertilizers, food, and feed. Store in original container and out of the reach of children, preferably in a locked storage area.

Handle and open container in a manner as to prevent spillage. If container is leaking invert to prevent leakage. If the container is leaking or material spilled for any reason or cause, carefully dam up spilled material to prevent runoff. Refer to Precautionary Statements on label for hazards associated with the handling of this material. Do not walk through spilled material. Absorb spilled material with absorbing type compounds and dispose of as directed for pesticides above. In spill or leak incidents, keep unauthorized people away. You may contact the Mobay Emergency Response Team for decontamination procedures or any other assistance that may be necessary. The Mobay Kansas City Emergency Response Telephone No. is 816-242-2582, or contact Chemtrec at 800-424-9300.

SENCOR is a Reg. TM of Bayer AG, Germany
Amiben is a Reg. TM of Union Carbide Agricultural Products Co., Inc.
Banvel is a Reg. TM of Velsicol Chemical Corp.
Bladex is a Reg. TM of Shell Chemical Co.
Bronate and Buctril are Reg. TMs of Rhone-Poulenc Chemical Co.
Command and Commence are Reg. TMs of FMC Corporation
Dual and Igran are Reg. TMs of Ciba-Geigy
Glean, Ally, Finesse, and Harmony are Reg. TMs of E. I. duPont de Nemours and Co.
Gramoxone is a Reg. TM of ICI Americas, Inc.
Common, Freedom, Lasso and Roundup are Reg. TMs of Monsanto Co.
Prowl and Scepter are Reg. TMs of American Cyanamid Co.

Surflan, Sonalan and Treflan are Reg. TMs of Elanco Products Co.
X-77 is a Reg. TM of Chevron Chemical Co.

IMPORTANT:
This bulletin is not intended to provide adequate information for application. Before using any product, read and carefully observe directions, cautionary statements and other information appearing on the product label. This product is sold subject to the Conditions of Sale set forth on the container label.

READ THE LABEL BEFORE USE

SN 9002

This is a specimen label. Mobay is not responsible for the accuracy of the information contained herein. As labels are subject to revision, always carefully read and follow the label on the product container.

TURBO® 8EC

HERBICIDE

FOR CONTROL OF CERTAIN GRASSES AND BROADLEAF WEEDS IN SOYBEANS AND POTATOES

ACTIVE INGREDIENTS:
Metolachlor, 2-chloro-*N*-(2-ethyl-6-methyl-phenyl)-*N*-(2-methoxy-1-methylethyl) acetamide* 70%
Metribuzin, 4-amino-6-(1,1-dimethylethyl)-3-(methylthio)-1,2,4-triazin-5 (4*H*)-one** 15%
INERT INGREDIENTS: 15%
TOTAL: 100%

Contains 6.55 pounds of metolachlor and 1.45 pounds of metribuzin per gallon.

EPA Reg. No. 3125-366 EPA Est. 3125-MO-1

TURBO is a Reg. TM of Mobay Corporation
SENCOR is a Reg. TM of Bayer AG, Germany
Command is a Reg. TM of FMC Corporation
Scepter is a Reg. TM of American Cyanamid Co.
Gramoxone is a Reg. TM of ICI Americas, Inc.
Roundup is a Reg. TM of Monsanto Co.
X-77 is a Reg. TM of Chevron Chemical Co.

*U.S. Patent No. 3,937,730
**U.S. Patent No. 3,671,523

RECOMMENDED USES

Soybeans, Potatoes

RESTRICTIONS ON WAREHOUSE STORAGE

Store in a cool, dry place and in such a manner as to prevent cross contamination with other pesticides, fertilizers, food and feed.

PACKAGING

2½ gal. jug, 2 per case. 110 gal. mini-bulk

TURBO 8EC is an emulsifiable concentrate preplant or preemergence herbicide for soybeans and potatoes. TURBO 8EC effectively controls a broad spectrum of both grasses and broadleaves and can be applied with aerial or ground equipment as a preplant incorporated, preemergence, SPLIT-SHOT or EXTENDED SPLIT-SHOT application to soybeans, or preemergence to potatoes. TURBO 8EC can effectively control such annual grasses as: foxtails, crabgrass, barnyardgrass, and fall panicum. In addition, it also controls annual broadleaves like: black nightshade, cocklebur, jimsonweed, lambsquarters, pigweed, prickly sida, ragweed, sesbania, sicklepod, smartweed and velvetleaf. For control of additional broadleaf weed species, TURBO 8EC may be used on soybeans in combination with a registered broadleaf herbicide or with Roundup® or Gramoxone® for reduced-till or no-till systems.

AVISO

PRECAUCION AL USUARIO: Si usted no puede leer o entender inglés, no use este producto hasta que la etiqueta le haya sido explicada ampliamente.

(TO THE USER: If you cannot read or understand English, do not use this product until the label has been fully explained to you.)

DIRECTIONS FOR USE

It is a violation of Federal Law to use this product in a manner inconsistent with its labeling.

Do not contaminate feed or food.

GENERAL INFORMATION

Mixing: When using TURBO 8EC, make sure the sprayer is completely clean, free of rust or corrosion which occurs from Winter storage. Examine strainers and screens to be sure the sprayer is clean from previously used pesticides.

Any tank mix containing TURBO 8EC should be kept agitated and sprayed out immediately. Do not allow tank mixes to stand for prolonged periods of time.

The proper mixing procedure for TURBO 8EC alone or in tank mix combinations with other herbicides is:

Emulsifiable Concentrates:
1. Fill the spray tank ¼ to ⅓ full with clean water.
2. Add the recommended rate of TURBO 8EC while recirculating and with agitator running.
3. Follow the triple rinse procedure described under "STORAGE AND DISPOSAL" to ensure that all product is removed from the container.
4. Mix thoroughly and add clean water to fill spray tank to desired level.
5. Add the other emulsifiable concentrate herbicide to tank last and agitate thoroughly.
6. Continue agitation during application and until sprayer tank is empty.

Other Products: If the tank mix combination is to be made using a wettable powder, liquid solution, liquid flowable, or water dispersable granules, TURBO 8EC should be added last after the other herbicide has been thoroughly dispersed.

Application: Use a standard low pressure (20 to 40 psi.) herbicide boom sprayer equipped with suitable nozzles and screens no finer than 50-mesh in-nozzle and in-line strainers. Agitate thoroughly before and during application with bypass agitation.

Do not apply this product through any type of irrigation system.

Ground Application: Apply the proper rate of TURBO 8EC in a minimum of 10 to 40 gallons of spray mixture per acre broadcast. For banded application, use proportionally less TURBO 8EC per acre in a band versus a broadcast application. For band application, use ¼ to 1 gallon of spray mix per inch of band width regardless of row spacing.

Examples: To treat a 15-inch band on rows 30 inches apart, use one-half of the broadcast rate of TURBO 8EC. To treat a 14-inch band on rows 42 inches apart, use one-third of the broadcast rate of TURBO 8EC.

Aerial Application: Where permitted, apply specified rate in a minimum of 2 to 10 gallons of spray mixture per acre. Do not apply aerially when wind speed is greater than 10 mph.

FOR ALL APPLICATIONS OF TURBO 8EC: Sprayer must be accurately calibrated before applying TURBO 8EC. Check sprayer during application to be sure it is working properly and delivering a uniform spray pattern. As the volume of spray mixture decreases per acre, the importance of accurate calibration and uniform application increases. Avoid over application, misapplication, and boom and spray swath overlapping that will increase spray dosage. (Crop injury may occur as a result.) Avoid spray skips and gaps which allow weeds to grow in untreated soil. Do not apply when weather conditions favor spray drift and/or when sensitive or cool season crops, such as cole crops, onions, peas, or strawberries are present in adjacent fields or in areas where wheat is growing in coarse textured soils.

SPRAYER CLEANUP

Spray equipment must be thoroughly cleaned to remove remaining traces of herbicide that might injure other crops to be sprayed. Drain any remaining spray solution of TURBO 8EC from the spray tank and discard in non-crop areas away from water supplies. Rinse the spray tank and refill with water, adding a heavy-duty detergent at the rate of one cup per 20 gallons of water. Recycle this mixture through the equipment for 5 minutes and spray out. Repeat this procedure twice. Fill the spray tank with clean water, recycle for 5 minutes and spray out. Clean pump and nozzle screens thoroughly. Wash away any spray mixture from the outside of spray tank, nozzles or spray rig. All rinse water must be disposed of in compliance with local, state, and Federal guidelines.

APPLICATION OF TURBO 8EC IN FLUID FERTILIZERS

TURBO 8EC may be applied in fluid fertilizer solutions to soybeans by following the appropriate mixing procedures and compatibility check. Compatibility checks of TURBO 8EC should be made for each batch of fluid fertilizer because of the variability of these fertilizers.

COMPATIBILITY CHECK:
1. Premix 2 teaspoonfuls of TURBO 8EC with 8 teaspoonfuls of water (1:4 ratio) in a quart jar by adding the water first and follow with TURBO 8EC. Mix thoroughly.
2. Then pour 1 pint of fluid fertilizer into the quart jar and shake well.
3. Allow to stand for 5 minutes.

THIS COMPATIBILITY CHECK SHOULD ONLY BE USED WHEN MIXING WITH FLUID FERTILIZERS.

Interpretation of results: If the solution in the jar appears to be uniform, without signs of agglomeration, or without a separation of an oily film on top of the fertilizer, the mixture may be used. If not, repeat the compatibility check using twice the amount of water or add a compatibility agent to the water. If separation occurs and can be resuspended by shaking, then application is possible with good agitation in the spray tank.

COMMERCIAL IMPREGNATION AND APPLICATION OF TURBO 8EC ON DRY BULK FERTILIZER

Dry bulk fertilizer may be impregnated or coated with TURBO 8EC for application to soybeans. All recommendations, cautions, and special precautions on the label must be followed along with State regulations relating to dry bulk fertilizer blending, impregnating and labeling.

Impregnation: To impregnate, use a system consisting of a belt, conveyor, or closed drum which is used for dry bulk fertilizer blending. Any commonly used fertilizer can be impregnated with TURBO 8EC except ammonium nitrate, or fertilizers containing ammonium nitrate, potassium nitrate, or sodium nitrate. Do not use on powder limestone.

Apply using a minimum of 200 pounds dry bulk fertilizer per acre and up to a maximum of 450 pounds per acre. To impregnate or coat the dry bulk fertilizer with TURBO 8EC, the delivery nozzles must be directed to deliver a fine spray toward the fertilizer for thorough coverage while avoiding spray contact with mixing equipment. Uniform impregnation of TURBO 8EC to dry bulk fertilizer will vary and if the absorptivity is not adequate, an absorptive powder may be added to produce a dry free-flowing mixture. Micro-Cel E (Johns-Manville Product Corporation) is the recommended absorbent powder. TURBO 8EC should be pre-mixed with water to form a sprayable mixture. Apply immediately after impregnation unless experience has shown that impregnated fertilizer can be stored without becoming lumpy and difficult to spread.

Rates: Select the recommended rate of TURBO 8EC per acre from the appropriate section of the label and refer to the formula below to determine the amount of TURBO 8EC which is to be impregnated on a ton of dry bulk fertilizer based on the amount of fertilizer which will be distributed on one acre.

$$\text{Pints TURBO 8EC Per Acre} \times \frac{2000}{\text{Lbs. Fertilizer Per Acre}} = \text{Pint TURBO 8EC Per Ton of Fertilizer}$$

Application: Uniform application is essential for satisfactory weed control. Accurate calibration of fertilizer application equipment is essential for uniform distribution to the soil surface. The recommended method of application is to apply $\frac{1}{2}$ the recommended rate and overlap 50 percent or to double apply by splitting the middles to obtain the best distribution pattern.

If fertilizer materials are excessively dusty, use diesel oil or other suitable additive to reduce dust prior to impregnation as dusty fertilizer will result in poor distribution during application. Crop injury and/or poor weed control may occur where the impregnated fertilizer is not uniformly applied.

SOYBEANS (Except California)

TURBO 8EC is recommended for preplant or shallow incorporated, preemergence, SPLIT-SHOT and EXTENDED SPLIT-SHOT application to control certain weeds in soybeans. All applications where permitted can be applied with ground or aerial spray equipment.

Activation: A minimum amount of soil moisture is required to activate TURBO 8EC. In areas of low rainfall, preemergence applications to dry soil should be followed with light irrigation of $\frac{1}{4}$ acre inch of water. Do not apply heavy irrigation immediately after application. As with many surface-applied herbicides, weed control and crop tolerance may vary with rainfall and/or soil texture.

Grazing and Feeding Treated Vines: Treated vines may be grazed or fed to livestock 40 days after the application of TURBO 8EC.

Rate Ranges: Where a rate range is shown, use a lower rate on soils that are coarse-textured or low in organic matter. Use higher rate on soils that are relatively fine-textured or high in organic matter.

Replanting: If replanting is necessary in fields previously treated with TURBO 8EC as directed on the label, the field may be replanted to soybeans. However, a minimum of tillage is recommended. Do not apply a second treatment as injury to soybeans may occur.

SPECIAL PRECAUTIONS (SOYBEANS)

Injury to soybeans may occur when TURBO 8EC is used under the following conditions:

1. When soils have a calcareous surface area or a pH of 7.5 or higher.
2. Due to the sensitivity of certain soybean varieties, TURBO 8EC is not recommended for use on sensitive varieties, such as Tracy, Semmes, Altona, Dassel, Vansoy, Coker 102, Coker 156, Portage, Maple Amber, Govan, NKS-1884, Terra-Vig 505, Terra-Vig 606, Vinton 81, Asgrow 6520, AP 55, AP 71, and Narow. Consult your Mobay representative or your seed supplier for information on the tolerance to TURBO 8EC on newly released soybean varieties prior to use of TURBO 8EC.
3. When applied in conjunction with soil-applied organic phosphate pesticides.
4. Over application or boom overlapping may result in stand loss and soil residues.
5. Uneven application or improper incorporation can decrease the level of weed control and/or increase the level of injury.
6. When applied to any soil with less than $\frac{1}{2}$% organic matter.
7. Soil incorporation deeper than recommended.
8. When sprayers are not calibrated accurately.
9. When heavy rains occur soon after application, especially in poorly drained areas where water may stand for several days.
10. When soybeans are planted less than $1\frac{1}{2}$" deep, particularly when TURBO 8EC is applied preemergence.
11. Where atrazine carryover is present.
12. When using poor quality soybean seed.

TURBO 8EC ALONE

Preplant Incorporated Application: Incorporate TURBO 8EC into the top 2 inches of soil within 14 days before planting using a disk, field cultivator, rolling cultivator, or similar implement. Except when shallow mixing, two incorporation passes are recommended. Apply TURBO 8EC preplant incorporated if furrow irrigation is used or when a period of dry weather after application is expected. If soybeans are planted on beds, apply and incorporate the tank mixture after bed formation.

Preemergence Application: Dry weather following preemergence application of TURBO 8EC may reduce effectiveness. If weeds develop, cultivate uniformly with shallow tilling equipment such as a rotary hoe that will not damage soybeans.

For information on applying product in fluid or dry fertilizer refer to the "Application OF TURBO 8EC In Fluid Fertilizers" or "Commercial Impregnation And Application Of TURBO 8EC On Dry Bulk Fertilizer" in the label.

Weeds Controlled: TURBO 8EC, when applied as directed, will control the following weeds.

Annual Grasses:	**Annual Broadleaves:**	
Barnyardgrass	Black Nightshade	Purslane
Bluegrass	Bristly Starbur	Ragweed, Common
Broadleaf Signalgrass	Carpetweed	Redweed
Crabgrass	Copperleaf, Hophornbeam	Russian Thistle
Crowfootgrass	Florida Beggarweed	Sesbania
Cupgrass, Prairie	Florida Pusley	Shepherdspurse
Cupgrass, Southwestern	Galinsoga	Sicklepod
Foxtails	Jimsonweed	Smartweeds
Goosegrass	Knotweed	Spotted Spurge
Junglerice	Kochia	Spurred Anoda
Panicum, Fall	Lambsquarters	Velvetleaf
Red Rice	Pigweed	Venice Mallow
Witchgrass	Prickly Sida/Teaweed	Wild Mustards

TURBO 8EC will provide suppression or erratic control (ranging from poor to excellent) of cocklebur, seedling Johnsongrass, hairy nightshade, yellow nutsedge, Texas panicum, sandbur, shattercane, volunteer sorghum and sunflower.

RECOMMENDED APPLICATIONS

Broadcast Rates

Soil Texture	TURBO 8EC Pints/Acre	
	Organic Matter 0.5 to 3%	Organic Matter Over 3%
COARSE[a] (Loamy sand, sandy loam)	$1\frac{1}{2}$[c]	$1\frac{3}{4}$-2[b]
MEDIUM (Loam, silt loam, silt, sandy clay, sandy clay loam)	2-$2\frac{1}{4}$	$2\frac{1}{4}$-$2\frac{3}{4}$[b]
FINE (Silty clay, silty clay loam,* clay, clay loam)	$2\frac{3}{4}$	$2\frac{3}{4}$-$3\frac{1}{2}$[b]

* Silty clay loam soils are transitional soils and may be classified as medium-textured soils in some regions of the U.S.

[a] Do not use on sand soils. On coarse-textured soils, do not use on loamy sand soils with less than 2% organic matter.

[b] For preplant incorporated application, use the lower rate range.

[c] For Southern and Southeastern states. See section on IN COARSE (LIGHT) SOILS ONLY.

IN COARSE (LIGHT) SOILS ONLY

(Alabama, Arkansas, Florida, Georgia, Louisiana, Mississippi, Missouri, North Carolina, Oklahoma, South Carolina, Tennessee, Texas, Virginia.)

TURBO 8EC is recommended for use as a preplant incorporated and preemergence application in coarse-textured, low organic matter soils in the states listed above for the control of certain weeds in soybeans. Refer to the appropriate sections of the label for specific directions on use, recommendations and restrictions.

Mixing: Refer to "Mixing" under the "General Information" section in the front of the label.

Application: For specific application information, refer to "Application" under the "General Information" section in the front of the label.

Weeds Controlled: Refer to the "TURBO 8EC Alone" section in the label for a complete list of weeds controlled.

TURBO 8EC Preemergence Application

Soil Texture	Organic Matter	TURBO 8EC Pints/Acre
COARSE SOILS (Sand[a], loamy sand, sandy loam)	0.5 or above	1½-2½[b]

[a] Not recommended for use on sand with less than 1% organic matter.
[b] Use the higher rate under heavy weed pressures and/or on soils higher in organic matter. Soybean varieties with a high level of metribuzin tolerance are recommended when using the high rate. For maximum control of sicklepod, use a preemergence application.

SPECIAL PRECAUTIONS:

For additional precautions, restrictions, limitations and sprayer clean-up information, refer to the appropriate sections of the label.

TURBO 8EC plus Overlay Application of SENCOR® Herbicide

TURBO 8EC applied preplant or shallow incorporated followed by an overlay application of SENCOR herbicide will control more broadleaf and grass weeds in soybeans than when either herbicide is used alone.

Refer to the SENCOR herbicide label and to appropriate sections of the TURBO 8EC label for directions, for soil preparation, herbicide application, incorporation techniques, and for additional directions on rates, weeds species controlled, and special precautions.

Applications: Apply TURBO 8EC as a preplant or shallow incorporated application up to 14 days before planting at rates listed below. This is followed by SENCOR herbicide alone at the indicated rates as preemergence overlay application after planting but prior to soybean emergence.

On medium and fine textured soils with greater than 2% organic matter, a rate range is recommended for the preemergence overlay application of SENCOR herbicide. The higher rate should be used (a) in fields with a history of severe broadleaf weed pressure, (b) when the time between preplant incorporated tank mix and preemergence overlay applications approaches the maximum 14 days as stated above, and/or (c) when the organic matter content of the soil is at the upper end of the indicated range.

Weeds Controlled: In addition to the weeds controlled by TURBO 8EC alone, TURBO 8EC herbicide with SENCOR herbicide will control buffalobur, cocklebur and sunflower.

TURBO 8EC Plus SENCOR 4 Application
(Broadcast Rates)

SOIL TEXTURE	Preplant Incorporated Application TURBO 8EC pts/acre	—FOLLOWED BY—	Preemergence Overlay Application SENCOR 4 (pts/acre) ORGANIC MATTER 0.5 to 2%	2 to 4%	Over 4%
COARSE[a] Sand, loamy sand, sandy loam)	1½	—followed by—	¼	¼	¼-½
MEDIUM (Loam, silt loam, sandy clay loam, silt, sandy clay)	2	—followed by—	¼	¼-½	½-¾
FINE (Silty clay loam*, clay loam, silty clay, clay)	2¾	—followed by—	¼	¼-½	½-¾

TURBO 8EC Plus SENCOR DF Application
(Broadcast Rates)

SOIL TEXTURE	Preplant Incorporated Application TURBO 8EC pts/acre	—FOLLOWED BY—	Preemergence Overlay Application SENCOR DF (lbs/acre) ORGANIC MATTER 0.5 to 2%	2 to 4%	Over 4%
COARSE[a] (Sand, loamy sand, sandy loam)	1½	—followed by—	1/6	1/6	1/6-1/3
MEDIUM (Loam, silt loam, sandy clay loam, silt, sandy clay)	2	—followed by—	1/6	1/6-1/3	1/3-½
FINE (Silty clay loam*, clay loam, silty clay, clay)	2¾	—followed by—	1/6	1/6-1/3	1/3-½

* Silty clay loam soils are transitional soils and may be classified as medium-textured soils in some regions of the U.S.
[a] Do not use on coarse-textured soil with less than 1% organic matter, a calcareous surface area or a pH of 7.5 or higher.

SPECIAL PRECAUTIONS:

For additional precautions, restrictions, limitations and sprayer clean-up information, refer to the appropriate sections of the TURBO 8EC label and the SENCOR herbicide label.

SPLIT-SHOT APPLICATION

TURBO 8EC applied preplant or shallow incorporated followed by a preemergence application of TURBO 8EC will control more broadleaf and grass weeds in soybeans than when either herbicide is used alone.

Refer to the appropriate sections of the label for directions for soil preparation, herbicide application, incorporation techniques, and for additional directions on rates, weeds species controlled, and special precautions.

Applications: Apply TURBO 8EC as a preplant or shallow incorporated application up to 14 days before planting at rates listed below. This is followed by TURBO 8EC at the indicated rates as a preemergence overlay application after planting but prior to soybean emergence.

Where a rate range is recommended for TURBO 8EC, the higher rate should be used (a) in fields with a history of severe broadleaf weed pressure, (b) when the time between preplant incorporated and preemergence overlay applications approaches the maximum 14 days as stated above, and/or (c) when the organic matter content of the soil is over 3 percent.

Weeds Controlled: In addition to the weeds controlled by TURBO 8EC alone, SPLIT-SHOT application of TURBO 8EC will control buffalobur, cocklebur and sunflower.

TURBO SPLIT-SHOT APPLICATION
(Broadcast Rates)

SOIL TEXTURE	Preplant Incorporated Application TURBO 8EC (pts/acre)	—FOLLOWED BY—	Preemergence Overlay Application TURBO 8EC (pts/acre)
COARSE[a] (Sand, loamy sand, sandy loam)	1-1½	—followed by—	¾-1
MEDIUM (Loam, silt loam, sandy clay loam, silt, sandy clay)	1¾-2	—followed by—	¾-1
FINE (Silty clay loam*, clay loam, silty clay, clay)	2-2½[b]	—followed by—	¾-1[b]

* Silty clay loam soils are transitional soils and may be classified as medium-textured soils in some regions of the U.S.
[a] Do not use on coarse-textured soil with less than 1% organic matter, a calcareous surface area or a pH of 7.5 or higher.
[b] Total not to exceed 3 pts/acre on soils with organic matter less than 3 percent.

SPECIAL PRECAUTIONS:

For additional precautions, restrictions, limitations and sprayer clean-up information, refer to the appropriate sections of the TURBO 8EC label.

TURBO 8EC plus Scepter® Tank Mix Application

TURBO 8EC may be applied with Scepter as a preplant incorporated, preemergence broadcast or band application for the control of certain broadleaf weeds and grasses in soybeans. Consult the Scepter label for specific directions on use, recommendations, restrictions, and any additional weeds not specified on the TURBO 8EC label.

Mixing: Refer to "Mixing" under the "General Information" section in front of the label.

Application: For specific application information, refer to "Application" under the "General Information" section in the front of the label.

Weeds Controlled: In addition to weeds controlled by TURBO 8EC alone, TURBO 8EC plus Scepter will control the following annual broadleaf weeds:

Buffalobur	Morningglory, smallflower
Cocklebur	Sicklepod
Morningglory, pitted	Sunflower

TURBO 8EC plus Scepter will provide suppression (reduce the competition) of ivyleaf and tall morningglory, and giant ragweed.

TURBO 8EC plus Scepter Application
(Broadcast Rates)

SOIL TEXTURE	TURBO 8EC[a] (pts/acre)	Scepter[b] (pts/acre)
COARSE (Loamy sand or sandy loam)	1½[c]	1/3-½

SOIL TEXTURE	TURBO 8EC[a] (pts/acre)	Scepter[b] (pts/acre)
MEDIUM (Loam, silt loam, silt, sandy clay, sandy clay loam)	1³/₄-2¹/₄	¹/₃-¹/₂
FINE (Silty clay, silty clay loam*, clay, clay loam)	2¹/₄-2³/₄	¹/₃-¹/₂

* Silty clay loam soils are transitional soils and may be classified as medium-textured soils in some regions of the U.S.

[a] Higher rate is recommended on soils with organic matter greater than 3%.

[b] For preemergence application, use the higher rate. For maximum control of moderate to heavy infestations of cocklebur, giant ragweed and sicklepod, use the higher rate and a preplant incorporated application.

[c] For Southern and Southeastern States in coarse soils, see the "IN COARSE (LIGHT) SOILS ONLY" section of the label for TURBO 8EC rates.

SPECIAL PRECAUTIONS:

For additional precautions, restrictions, limitations and sprayer clean-up information, refer to the appropriate sections of the TURBO 8EC and the Scepter labels.

Do not use vines treated with Scepter for feed or forage.

SEQUENTIAL APPLICATION OF SCEPTER FOLLOWING TURBO:

If needed, application of TURBO 8EC alone or in a registered tank mix, according to directions on the label, may be followed by an early postemergence application of Scepter for control of cocklebur. Apply ¹/₆ to ¹/₃ pint of Scepter in a minimum of 20 gallons of water per acre. Use ¹/₆ pint of Scepter if cockleburs are less than 3 inches tall or have fewer than 3 leaves and are actively growing. For cockleburs less than 6 inches tall and actively growing, use ¹/₃ pint per acre. Do not use Scepter when soybeans or cockleburs have been subjected to stress conditions such as temperature or moisture extremes. Do not exceed a total of ²/₃ pint of Scepter per acre in one season. Wait at least 10 days after application of Scepter before cultivating.

When preparing spray mixture with Scepter, add 2 pints of nonionic surfactant approved for use on growing crops and containing at least 80% active ingredient per 100 gallons of mixture. Apply crop oil concentrate (COC) at the rate specified on the COC label.

Use Scepter only in the states where it is registered as listed on the product label.

Apply Scepter at least 90 days before harvest of soybeans. Do not graze or feed soybean forage, hay, or straw to livestock.

Refer to the Scepter label for additional cautions and precautions, directions, limitations, and information on environmental hazards and planting of rotational crops.

TURBO 8EC plus Command® 4EC Tank Mix Application

TURBO 8EC may be applied with Command 4EC as a preplant or shallow incorporated broadcast application for the control of certain broadleaf weeds and grasses in soybeans. Consult the Command 4EC label for specific directions for use, recommendations and restrictions not specified on the TURBO 8EC label.

Mixing: Refer to "Mixing" under the "General Information" section in front of the label.

Application: For specific application information, refer to "Application" under the "General Information" section in the front of the label.

Weeds Controlled: In addition to weeds controlled by TURBO 8EC alone, TURBO 8EC plus Command 4EC will provide improved control of heavy infestations of velvetleaf, jimsonweed and common ragweed.

TURBO 8EC plus Command 4EC Application (Broadcast Rates)

SOIL TEXTURE	TURBO 8EC[a] (pts/acre)	Command 4EC (pts/acre)
COARSE (Loamy sand or sandy loam)	1¹/₂-1³/₄[b]	¹/₂-³/₄
MEDIUM (Loam, silt loam, silt, sandy clay, sandy clay loam)	1³/₄-2¹/₄	¹/₂-³/₄
FINE (Silty clay, silty clay loam*, clay, clay loam)	2¹/₄-2³/₄	¹/₂-³/₄

* Silty clay loam soils are transitional soils and may be classified as medium-textured soils in some regions of the U.S.

[a] Higher rate is recommended on soils with organic matter greater than 3%.

[b] For Southern and Southeastern States in coarse soils, see the "IN COARSE (LIGHT) SOILS ONLY" section of the label for TURBO 8EC rates.

SPECIAL PRECAUTIONS:

Do not plant wheat, oats, barley, rye or alfalfa in the fall or following spring after application as crop injury may occur.

Do not use vines treated with Command 4EC for feed or forage.

For additional precautions, restrictions, limitations and sprayer clean-up information, refer to the appropriate sections of the TURBO 8EC label and the Command 4EC label.

Do not apply where weather conditions favor drift.

TURBO 8EC IN REDUCED-TILLAGE OR NO-TILL SYSTEM

Application: TURBO 8EC may be applied in combination with Gramoxone® Super Roundup® or 2,4-D low volatile ester for the control of annual grasses and broadleaf weeds and the suppression of emerged perennial weeds where soybeans are directly seeded into a stale seedbed, cover crop, or in previous crop residues.

Special Precautions: Poor weed control and/or crop injury may result if directions are not followed. Do not use a rib-type press wheel on your no-till planter or crop injury may result. For additional precautions, restrictions, limitations, and sprayer clean-up information, refer to the appropriate sections of the Gramoxone Super, Roundup, 2,4-D low volatile ester and the TURBO 8EC label.

TURBO 8EC plus Gramoxone Application: Apply 1¹/₃ to 2²/₃ pints of Gramoxone with TURBO 8EC at the rates listed below as a preemergence broadcast ground spray. Use the higher rate of Gramoxone for weeds 4 to 6 inches tall. Soybeans exposed at the time of application will be killed. Apply in 20 to 60 gallons spray mix per acre. Use the higher volume in fields with heavy weed infestation or with large amounts of previous crop residues. When a surfactant is needed, add Ortho X-77® Spreader or similar non-ionic spreader at 8 fl. ounces per 100 gallons of diluted spray. Refer to the Gramoxone label for rates, directions, limitations, and precautions.

TURBO 8EC plus Roundup Application: Apply 1 to 1.5 quarts of Roundup with TURBO 8EC at the rates listed below as a preemergence broadcast ground spray. Use the higher rates of Roundup for weeds greater than 6 inches tall. Refer to the Roundup label for additional rate information, directions, limitations and precautions.

TURBO 8EC plus 2,4-D low volatile ester Application: Apply 1.0 lb. active ingredient (ai) per acre of 2,4-D low volatile ester with TURBO 8EC at the rates listed below as an early preplant surface application. Application is recommended 30 days prior to planting. Refer to the 2,4-D low volatile ester label for additional rate information, directions, limitations and precautions.

Weeds Controlled: TURBO 8EC herbicide combinations will control the weeds listed in the "TURBO 8EC" Alone section. Refer to the Gramoxone, Roundup or 2,4-D low volatile ester labels for additional weeds controlled.

NO-TILL OR REDUCED-TILL APPLICATION (Broadcast Rates)

SOIL TEXTURE	TURBO 8EC Pints/Acre[a]
COARSE[b] (Loamy sand, sandy loam)	1¹/₂-2¹/₄
MEDIUM (Loam, silt loam, silt, sandy clay, sandy clay loam)	2¹/₄-3
FINE (Silty clay, silty clay loam[c], clay, clay loam)	3-3³/₄

[a] Use low rate range for low trash level or soils with less than 3% organic matter. Use the higher rate range for high trash level or soils with greater than 3% organic matter.

[b] Do not use on sand soils. On coarse-textured soils, do not use on loamy sand soils with less than 2% organic matter.

[c] Silty clay loam soils are transitional soils and may be classified as medium-textured soils in some regions of the U.S.

EXTENDED SPLIT-SHOT APPLICATION

An early preplant (surface-applied or shallow incorporated) application of TURBO 8EC, followed by a preemergence surface application of TURBO 8EC after planting but prior to soybean emergence, will provide more consistent control of broadleaf and grass weeds than a single application.

An EXTENDED SPLIT-SHOT application will decrease the need for tillage and/or contact herbicides for the control of existing vegetation prior to planting, while providing residual control of weeds after planting.

Application: An early preplant application may be made 15 to 30 days prior to planting soybeans. Follow this application with a preemergence overlay application after planting but before crop emergence. Follow directions on the label accompanying the product for SPLIT-SHOT applications from 0 to 14 days before planting.

Where a rate range is recommended, the higher rates should be used (a) in fields with a history of severe weed pressure, (b) when the time between early preplant and preemergence overlay applications approaches the maximum 30 days, (c) when the organic matter content of the soil is over 3 percent and/or (d) when heavy crop residues are present on the soil surface.

When weeds exceed 1 to 1¹/₂ inches in height or diameter at application, use a contact herbicide, such as Roundup or Gramoxone.

Weeds Controlled: In addition to weeds controlled by TURBO 8EC alone, the EXTENDED SPLIT-SHOT application will control buffalobur, cocklebur and sunflower.

EXTENDED SPLIT-SHOT APPLICATION
(Broadcast Rates)

SOIL TEXTURE	Early Preplant Application TURBO 8EC (pts/acre)	—FOLLOWED BY—	Preemergence Overlay Application TURBO 8EC (pts/acre)
COARSE[a] (Sand, loamy sand, sandy loam)	1½-2	—followed by—	¾-1¼
MEDIUM (Loam, silt loam, sandy clay loam, silt, sandy clay)	1¾-2¼	—followed by—	1-1½
FINE (Silty clay loam*, clay loam, silty clay, clay)	2-2½[b]	—followed by—	1¼-1¾[b]

* Silty clay loam soils are transitional soils and may be classified as medium-textured soils in some regions of the U.S.

[a] On coarse-textured soils, do not use on sand soils with less than 1% organic matter. However, on coarse-textured soils with a calcareous surface area or a pH of 7.5 or higher, do not use on sand soils with less than 2% organic matter, or on loamy sand or sandy loam soils with less than 1% organic matter.

[b] Total not to exceed 3¾ pts of TURBO 8EC/acre per use season.

POTATOES (Except California)

TURBO 8EC is recommended as a preemergence broadcast spray for the control of certain broadleaf weeds and grasses in potatoes.

Application: Apply with ground (except air blast sprayers) or aerial spray equipment as a preemergence broadcast application. Apply after planting but before crop emergence or after drag-off if this operation is part of the normal cultural practice.

Weeds Controlled: TURBO 8EC, applied preemergence to potatoes according to directions, will effectively control:

Annual Grasses	Annual Broadleaves:	
Barnyardgrass	Black Nightshade	Velvetleaf
Broadleaf Signalgrass	Carpetweed	Venice Mallow
Crabgrass	Galinsoga	Wild Mustards
Crowfootgrass	Jimsonweed	
Cupgrass, Prairie	Lambsquarter	
Cupgrass, Southwestern	Pigweeds	
Foxtails	Prickly Sida	
Goosegrass	Ragweed, Common	
Junglerice	Russian Thistle	
Nutsedge, Yellow	Sesbania	
Panicum, Fall	Shepherdspurse	
Red Rice	Sicklepod	
Witchgrass	Smartweed, Pennsylvania	

TURBO 8EC will suppress (reduce the competition of) cocklebur, common purslane, hairy nightshade, Johnsongrass (seedling), Kochia, sandbur, and volunteer sorghum.

RECOMMENDED APPLICATIONS
(Broadcast Preemergence Rates)
TURBO 8EC Pints/Acre

SOIL MIXTURE	ORGANIC MATTER 0.5 to 3%	ORGANIC MATTER Over 3%
COARSE ([a]Loamy sand, sandy loam)	2-2½	2½-3
MEDIUM/FINE (Loam, silt loam, silt, sandy clay, sandy clay loam, silty clay, silty clay loam, clay, clay loam)	3-3½	3½-4

[a] Do not use on sand soils.
Silty clay loam soils are transitional soils and may be classified as medium textured soils in some regions of the U.S.

SPECIAL PRECAUTIONS:

Do not apply within 24 hours of application of other pesticides.
Do not use TURBO 8EC at or after potatoes have emerged.
Applications to Atlantic, Shepody, Chip Bell, Bell Chip and Centennial varieties may cause crop injury, especially under adverse weather conditions and when higher label rates are used.
Do not apply to sweet potatoes or to yams.
Do not plant sensitive crops such as onions, lettuce, sugar beets, cole crops and cucurbits during the next growing season following applications of TURBO 8EC or injury may occur.
Do not harvest within 60 days after application.
Do not use on muck or peat soils.

Postemergence Sequential Applications with SENCOR:
TURBO 8EC may be followed by a postemergence application of ½ to 1 pint/acre of SENCOR 4 or ⅓ to ⅔ pounds/acre of SENCOR DF. If TURBO 8EC is used on medium/fine soils with greater than 3% organic matter, use ½ pint of SENCOR 4 or ⅓ pound of SENCOR DF. For additional precautions, restrictions and limitations, refer to the label for SENCOR 4 or SENCOR DF.

CROP ROTATION RECOMMENDATIONS[a] 10-1-88

Waiting Period After Application of TURBO[b]

4½ Months	8 Months	12 Months	18 Months
Winter Barley Winter Wheat	Corn Cotton Peas Potatoes Rice Soybeans Spring Barley Spring Wheat	Alfalfa Asparagus Forage Grasses Lentils Sainfoin Sugar Cane Tomatoes Other Crops not listed (except root crops)	Onions Sugar Beets And Other Root Crops

[a] Cover crops for soil building or erosion control may be planted any time, but do not graze or harvest for food or feed. Stand reductions may occur in some areas.

[b] Crop rotation recommendations do not include the tank mix applications of Scepter or Command plus TURBO. Refer to the label of the other product for additional restrictions.

PRECAUTIONARY STATEMENTS

HAZARDS TO HUMANS AND DOMESTIC ANIMALS

CAUTION

May be harmful if swallowed, inhaled or absorbed through the skin. This product may cause allergic skin reactions. Wear protective clothing (coveralls and gloves) while handling or using this product. Avoid contact with eyes, skin or clothing. Avoid breathing spray mist. Wash thoroughly with soap and warm water after handling. Wash contaminated clothing with soap and hot water before reuse.

Keep out of reach of children.

Obtain prompt medical aid if poisoning should occur.

TO PHYSICIAN: No specific antidote is available. Treat the patient symptomatically.

SYMPTOMS OF POISONING: The compound does not cause any definite symptoms that would be diagnostic. Poisoning is accompanied by breathing difficulties and sedation.

STATEMENTS OF PRACTICAL TREATMENT

If in eyes, immediately flush eyes with plenty of water for at least 15 minutes. For further treatment get medical attention.

If swallowed, call a physician or poison control center. Vomiting should be induced. Administer water freely, and induce vomiting by giving one dose (½ oz. or 15 ml.) of syrup of ipecac. If vomiting does not occur within 10 to 20 minutes, administer second dose. If syrup of ipecac is not available, induce vomiting by sticking finger down throat. Repeat until vomit fluid is clear. Do not induce vomiting or give anything by mouth to an unconscious person. Professional medical assistance should be secured immediately.

If on skin, wash thoroughly with soap and warm water.

ENVIRONMENTAL HAZARDS

Do not use on other crops grown for food or forage. Do not apply directly to water or wetlands (swamps, bogs, marshes and potholes). Do not contaminate water by cleaning of equipment or disposal of wastes. Do not apply when weather conditions favor drift from areas treated. Apply this product only as specified on the label.

Do not allow sprays to drift on to adjacent desirable plants.

Observe all cautions and limitations on labeling of all products used in mixtures.

Ground Water and Surface Water Advisories
Metolachlor has been identified in limited sampling of ground water and there is the possibility that it may leach through the soil to ground water, especially where soils are coarse and ground water is near the surface. Following application and during rainfall events that cause runoff, metolachlor may reach surface water bodies including streams, rivers and reservoirs.

Care must be taken when using this product to prevent back siphoning into wells, spills or improper disposal of excess pesticide, spray mixtures or rinsates.

Check valves or antisiphoning devices must be used on all mixing equipment.

Metribuzin is a chemical which can travel (seep or leach) through soil and can contaminate ground water which may be used as drinking water.

Metribuzin has been found in ground water as a result of agricultural use. Users are advised not to apply metribuzin where the water table (ground water) is close to the surface and where the soils are very permeable, i.e., well drained soils such as loamy sands. Your local agricultural agencies can provide further information on the type of soil in your area and the location of ground water.

STORAGE AND DISPOSAL

PESTICIDE DISPOSAL: Do not contaminate water, food or feed by storage or disposal.
Wastes resulting from the use of this product may be disposed of on site or at an approved waste disposal facility.

CONTAINER DISPOSAL: Triple rinse (or equivalent). Then offer for recycling or reconditioning, or puncture and dispose of in a sanitary landfill, or incineration, or, if allowed by State and local authorities, by burning. If burned, stay out of smoke.

Store in a cool, dry place and in such a manner as to prevent cross contamination with other pesticides, fertilizers, food and feed. Store in original container and out of the reach of children, preferably in a locked storage area.

Handle and open container in a manner as to prevent spillage. If container is leaking invert to prevent leakage. If the container is leaking or material spilled for any reason or cause, carefully dam up spilled material to prevent runoff. Refer to Precautionary Statements on label for hazards associated with the handling of this material. Do not walk through spilled material. Absorb spilled material with absorbing type compounds and dispose of as directed for pesticides above. In spill or leak incidents, keep unauthorized people away. You may contact the Mobay Emergency Response Team for decontamination procedures or any other assistance that may be necessary. The Mobay Kansas City Emergency Response Telephone No. is 816-242-2582, or contact Chemtrec at 800-424-9300.

RE-ENTRY STATEMENTS

Do not apply this product in such a manner as to directly or through drift expose workers or other persons. The area being treated must be vacated by unprotected persons. Do not enter treated areas without protective clothing until sprays have dried. Because certain states may require more restrictive re-entry intervals for various crops treated with this product, consult your State Department of Agriculture for further information.

Written or oral warnings must be given to workers who are expected to be in a treated area or in an area about to be treated with this product. Oral warnings should include those statements found under the label heading. "Precautionary Statements" and "Re-entry Statements" included on the label. When oral warnings are given, warnings shall be given in a language customarily understood by workers. Oral warnings must be given if there is reason to believe that written warnings cannot be understood by workers. Written warnings must include the following information: WARNING, area treated with TURBO 8EC herbicide on (date of application). Do not enter without appropriate protective clothing until sprays have dried. **If on skin,** remove contaminated clothing and wash skin immediately with soap and warm water. **If eyes are contaminated,** wash with flowing water for at least 15 minutes. **If swallowed,** vomiting should be induced. (See Precautionary Statements on label.)

IMPORTANT:

This bulletin is not intended to provide adequate information for application. Before using any product, read and carefully observe directions, cautionary statements and other information appearing on the product label. This product is sold subject to the Conditions of Sale set forth on the container label.

READ THE LABEL BEFORE USE

TU 9001

This is a specimen label. Mobay is not responsible for the accuracy of the information contained herein. As labels are subject to revision, always carefully read and follow the label on the product container.

Monsanto Company

Agricultural Products
800 North Lindbergh Boulevard
St. Louis, MO 63167
(314) 694-1000

BRONCO®
HERBICIDE

RESTRICTED USE PESTICIDE

due to oncogenicity. For retail sale to and use only by Certified Applicators or persons under their direct supervision and only for those uses covered by the Certified Applicator's certification.

The use of this product may be hazardous to your health. This product contains alachlor, which has been determined to cause tumors in laboratory animals. In case of an emergency involving this product or for user safety information on this product, call collect, day or night (314) 694-4000.

For weed control in Corn, Soybeans and Grain Sorghum (Milo)

SHAKE WELL BEFORE USING

COMPLETE DIRECTIONS FOR USE

AVOID CONTACT WITH DESIRABLE PLANTS AND TREES SINCE SEVERE INJURY OR DESTRUCTION MAY RESULT.

ACTIVE INGREDIENTS:	
Alachlor	27.6%
Isopropylamine salt of glyphosate	14.8%
INERT INGREDIENTS	57.6%
TOTAL	100.0%

Contains 2.6 pounds of 2-chloro-2′-6′diethyl-N-(methoxymethyl) acetanilide and 1.4 pounds* of the isopropylamine salt of N-(phosphonomethyl) glycine per U.S. gallon. *Equivalent to 1.04 pounds per U.S. gallon of the acid, glyphosate.

EPA Reg. No. 524-341-AA

U.S. Pat. No. 3,799,758 covers use.

Other patents pending.

1988-1 817.37-000.37/53

Read the entire label before using this product.

Use only according to label instructions.

Read "LIMIT OF WARRANTY AND LIABILITY" before buying or using. If terms are not acceptable, return at once unopened.

LIMIT OF WARRANTY AND LIABILITY

This company warrants that this product conforms to the chemical description on the label and is reasonably fit for the purposes set forth in the Complete Directions for Use label booklet ("Directions") when used in accordance with those Directions under the conditions described therein. NO OTHER EXPRESS OR IMPLIED WARRANTY OF FITNESS FOR PARTICULAR PURPOSE OR MERCHANTABILITY IS MADE. This warranty is also subject to the conditions and limitations stated herein.

Buyer and all users shall promptly notify this company of any claims whether based in contract, negligence, strict liability, other tort or otherwise.

Buyer and all users are responsible for all loss or damage from use or handling which results from conditions beyond the control of this company, including but not limited to, incompatibility with products other than those set forth in the Directions, unusual weather, weather conditions which are outside the range considered normal at the application site and for the time period when the product is applied, as well as weather conditions which are outside the application ranges set forth in the Directions, application in any manner not explicitly set forth in the Directions, moisture conditions outside the moisture range specified in the Directions, or the presence of products other than those set forth in the Directions in or on the soil or crop.

THE EXCLUSIVE REMEDY OF THE USER OR BUYER, AND THE LIMIT OF THE LIABILITY OF THIS COMPANY OR ANY OTHER SELLER FOR ANY AND ALL LOSSES, INJURIES OR DAMAGES RESULTING FROM THE USE OR HANDLING OF THIS PRODUCT (INCLUDING CLAIMS BASED IN CONTRACT, NEGLIGENCE, STRICT LIABILITY, OTHER TORT OR OTHERWISE) SHALL BE THE PURCHASE PRICE PAID BY THE USER OR BUYER FOR THE QUANTITY OF THIS PRODUCT INVOLVED, OR, AT THE ELECTION OF THIS COMPANY OR ANY OTHER SELLER, THE REPLACEMENT OF SUCH QUANTITY OR, IF NOT ACQUIRED BY PURCHASE, REPLACEMENT OF SUCH QUANTITY. IN NO EVENT SHALL THIS COMPANY OR ANY OTHER SELLER BE LIABLE FOR ANY INCIDENTAL OR CONSEQUENTIAL DAMAGES.

The buyer and all users are deemed to have accepted the terms of this LIMIT OF WARRANTY AND LIABILITY which may not be varied by any verbal or written agreement.

PRECAUTIONARY STATEMENTS

HAZARDS TO HUMANS AND DOMESTIC ANIMALS

KEEP OUT OF REACH OF CHILDREN

DANGER!

CAUSES EYE BURNS AND SKIN IRRITATION.

HARMFUL IF SWALLOWED.

MAY CAUSE ALLERGIC SKIN REACTION.

Do not get in eyes, on skin, or on clothing.

Wear goggles or face shield, rubber gloves, long trousers, long-sleeved shirt or jacket of tightly woven material along with boots high enough to cover ankles when transferring and mixing, and when adjusting, repairing or cleaning equipment. Wear rubber boots when pouring from open containers. Clothing which comes in contact with Bronco must be washed before reuse. Clothing or other materials which become drenched with the concentrated pesticide must be disposed of in a sanitary landfill, by incineration or, if allowed by state and local authorities, by burning. If burned, stay out of smoke.

Wash thoroughly with soap and water after handling.

FIRST AID: IF IN EYES, immediately flush with plenty of water for at least 15 minutes. Call a physician.

IF ON SKIN: immediately flush with plenty of water while removing contaminated clothing. Call a physician. Wash clothing before reuse. Sensitized persons should avoid further contact and reuse of clothing.

IF SWALLOWED, induce vomiting immediately by giving two glasses of water and sticking finger down throat. Call a physican. NEVER GIVE ANYTHING BY MOUTH TO AN UNCONSCIOUS PERSON.

In case of emergency involving this product, Call Collect, day or night, (314) 694-4000.

ENVIRONMENTAL HAZARDS

This chemical has been identified in limited ground water sampling and there is the possibility that it can leach through the soil to ground water, especially where soils are coarse and ground water is near the surface. Following application, and during rainfall events that cause runoff, this chemical may reach surface water bodies including streams, rivers and reservoirs.

Care must be taken when using this product to prevent back siphoning into wells, spills or improper disposal of excess pesticides, spray mixtures or rinsates.

Check valves or antisiphoning devices must be used on all mixing and/or irrigation equipment. Disposal of excess pesticide, spray mixtures, or rinsate should be according to label use instructions or according to the state pesticide or environmental control agency or the hazardous waste representative at the nearest EPA regional office.

Do not apply directly to water or wetland (swamps, bogs, marshes or potholes). Do not contaminate water by cleaning of equipment or disposal of wastes.

PHYSICAL OR CHEMICAL HAZARDS

COMBUSTIBLE.

Do not use or store near heat or flame.

Use only with adequate ventilation.

In case of:

FIRE, use water spray, foam, dry chemical or CO_2,

SPILL OR LEAK, flush area with water spray.

Spray solutions of this product should be mixed, stored and applied only in stainless steel, aluminum, fiberglass, plastic and plastic-lined steel containers.

DO NOT MIX, STORE OR APPLY THIS PRODUCT OR SPRAY SOLUTIONS OF THIS PRODUCT IN GALVANIZED STEEL OR UNLINED STEEL (EXCEPT STAINLESS STEEL) CONTAINERS OR SPRAY TANKS. This product or spray solutions of this product react with such containers and tanks to produce hydrogen gas which may form a highly combustible gas mixture. This gas mixture could flash or explode, causing serious personal injury, if ignited by open flame, spark, welder's torch, lighted cigarette or other ignition source.

STORAGE AND DISPOSAL

Do not contaminate water, foodstuffs, seed or feed by storage and disposal.

STORAGE:
Store above 40°F. to keep product in solution.

Below 40°F. crystals may form and settle to the bottom. If crystals form, place in a warm room and shake the container frequently until crystals dissolve.

Keep container closed to prevent spills and contamination.

For bulk containers, see the container label for alternate storage information.

DISPOSAL:
Wastes of this pesticide may cause eye burns and skin irritation and may be dangerous. Improper disposal of excess pesticide, spray mixtures, or rinsate is a violation of Federal law. If these wastes cannot be disposed of according to label use instructions, contact your State Pesticide or Environmental Control Agency, or the Hazardous Waste representative at the nearest EPA Regional Office for guidance.

Empty container retains vapor and product residue. Observe all labeled safeguards until container is cleaned, reconditioned or destroyed.

(See the individual container label for disposal information.)

GENERAL INFORMATION

Bronco, applied alone or in tank mixtures, is recommended for the preemergence and postemergence control of many annual grasses and broadleaf weeds in soybeans and corn grown under NO-TILL or other minimum tillage practices. This product will also provide reduced competition from certain emerged perennial weeds.

MOISTURE CONDITIONS

Rainfall or sprinkler irrigation within 6 hours after application may reduce control of emerged weeds.

Moisture in the form of rainfall or irrigation is required after application to move this product into the weed germination zone for preemergence weed control. The amount of moisture required depends on the soil moisture content at the time of application, soil type, and plant residues. Performance is improved when 1/3 to 3/4 inch of moisture is received within one week after application and prior to weed emergence. High intensity or excessive rainfall or irrigation after application may reduce preemergence weed control on coarse soils.

READ AND CAREFULLY OBSERVE PRECAUTIONARY STATEMENTS AND ALL OTHER INFORMATION APPEARING ON THE LABELING OF ALL PRODUCTS USED IN MIXTURES.

ATTENTION

AVOID DRIFT, EXTREME CARE MUST BE USED WHEN APPLYING THIS PRODUCT TO PREVENT INJURY TO DESIRABLE PLANTS, TREES AND CROPS.

Do not allow the herbicide solution to mist, drip, drift, or splash onto desirable vegetation since minute quantities of this herbicide can cause severe damage or destruction to crops, plants, or other areas on which treatment was not intended. The likelihood of injury occurring from the use of this product is greatest when winds are gusty or in excess of 5 miles per hour or when other conditions, including lesser wind velocities, will allow spray drift to occur. When spraying avoid combinations of pressure and nozzle type that will result in splatter or fine particles (mist) which are likely to drift. AVOID APPLYING AT EXCESSIVE SPEED OR PRESSURE.

DO NOT MIX THIS PRODUCT WITH LIQUID FERTILIZERS, HERBICIDE OILS OR ANY MATERIALS OTHER THAN WATER, EXCEPT THOSE LISTED ON THIS LABEL.

NOTE: Use of this product in any manner not consistent with this label may result in injury to persons, animals or crops, or other unintended consequences.

DO NOT APPLY THIS PRODUCT TO EMERGED CORN OR SOYBEANS.

DO NOT FEED FORAGE FROM OR GRAZE TREATED AREAS WITHIN 8 WEEKS AFTER APPLICATION.

Flush sprayer with clean water after use.

SOIL TEXTURE

Applicators should evaluate soil and weed conditions carefully to assure that they choose the lowest effective label rate.

The recommended use rates of this product and the other herbicides labeled for use in tank mixtures with this product generally vary with soil texture. Unless soil texture is specifically named, rate tables throughout this label refer to only three soil texture groups: COARSE, MEDIUM and FINE. The following is a complete listing of soil textures included in each of these three soil texture groups:

Soil Texture Group	Soil Texture
COARSE	sand, loamy sand, sandy loam
MEDIUM	loam, silt loam, silt, sandy clay loam
FINE	silty clay loam, clay loam, sandy clay, silty clay, clay

Refer to the above table to determine the corresponding soil texture group for the soil to be treated.

WEEDS CONTROLLED

When applied as directed under conditions described, this product alone or in tank mixtures with other products listed will control or reduce competition from the weeds indicated. Refer to the "DIRECTIONS FOR USE" section to determine rates for each crop.

Reduced control of emerged weeds may occur when weeds are under poor growing conditions such as from drought stress, when application follows recent mowing or when excessing amounts of dust cover the plants.

NOTE: For additional information on weeds controlled by this product applied alone or in tank mixtures with other products in DOUBLE CROP systems, consult the "Double Crop" section of this label.

This product applied alone or in tank mixtures with other products listed will CONTROL the following EMERGED WEEDS:

ANNUAL GRASSES

Bluegrass (annual)
Poa annua
Brome (downy)
Bromus tectorum
Corn (volunteer)
Zea mays
Crabgrass
Digitaria ischaemum
Digitaria sanguinalis
Foxtail (giant)
Setaria faberi
Foxtail (green)
Setaria viridis
Foxtail (yellow)
Setaria lutescens
Panicum (fall)*
Panicum dichotomiflorum
Sandbow (field)
Cenchrus spp.
Shattercane
Sorghum bicolor
Wheat (volunteer)
Triticum aestivum

ANNUAL BROADLEAVES

Fleabane
Erigeron spp.
Kochia
Kochia scoparia
Lambsquarters
Chenopodium album
Lettuce (prickly)
Lactuca serriola
Marestail, Horseweed
Conyza canadensis
Pigweed, Carelessweed (redroot)
Amaranthus retroflexus
Pigweed (smooth)
Amaranthus hybridus
Ragweed (common)
Ambrosia artemisifolia
Ragweed (giant)
Ambrosia trifida
Smartweed
Polygonum pensylvanicum
Sunflowers
Helianthus annuus
Thistle (Russian)
Salsola kali
Velvetleaf, Buttonweed
Abutilon theophrasti

* Five (5) quarts of this product are required for satisfactory postemergence control of fall panicum.

This product applied alone or in tank mixtures with other products listed will REDUCE COMPETITION from the following EMERGED WEEDS:

PERENNIAL GRASSES

Bluegrass (Kentucky)
Poa spp.
Bromegrass (smooth)
Bromus inermis
Fescue
Festuca spp.
Muhly (wirestem)
Muhlenbergia frondosa
Johnsongrass
Sorghum halepense
Orchardgrass
Dactylis glomerata
Quackgrass
Agropyron repens
Ryegrass (perennial)
Lolium perenne
Timothy
Phleum pratense

PERENNIAL BROADLEAVES

Alfalfa
Medicago sativa
Clover (red)
Trifolium pratense
Clover (white)
Trifolium repens
Dock (curly)
Rumex crispus
Milkweed
Asclepias spp.
Mullein (common)
Verbascum thapsus
Smartweed (swamp)
Polygonum coccineum
Thistle (Canada)
Cirsium arvense

This product applied alone or in tank mixture with other products will provide PREEMERGENCE CONTROL of the following weeds:

ANNUAL GRASSES

Barnyardgrass
Echinochloa crus-galli
Crabgrass
Digitaria ischaemum
Digitaria sanguinalis
Foxtail (giant)
Setaria faberi
Foxtail (green)
Setaria viridis
Foxtail (yellow)
Setaria lutescens
Goosegrass
Elusine indica
Panicum (fall)
Panicum dichotomiflorum
Signalgrass (broadleaf)
Brachiaria spp.
Witchgrass
Panicum capillare

ANNUAL BROADLEAVES

When applied as directed under conditions described, Bronco alone or in tank mixtures with products listed in the following table will provide pre-emergence control or reduced competition from the additional grass and broadleaf weeds as indicated. Refer to the crop on which treatment is intended in the "DIRECTIONS FOR USE" section of this label to determine the rate of Bronco alone and of tank mixtures recommended for use in each crop.

NOTE: C = Control   R = Reduced Competition

BRONCO®	Tank Mixtures					
	Atrazine	Bladex™	Lorox™	Princep™	Sencor/™	Lexone™
Carpetweed Mollugo verticillata	C	C	C	C	C	C
Cocklebur Xanthium pensylvanicum	•	R	R	R	R	R
Jimsonweed Datura stramonium	•	C	R	R	R	R
Kochia Kochia scoparia	•	C	•	•	•	•
Lambsquarters Chenopodium album	R	C	C	C	C	C
Morningglory (annual) Ipomeoea purpurea	•	R	R	R	R	R
Mustard Brassica spp.	•	C	C	C	C	C
Nightshade (black) Solanum nigrum	C*	C	C	C	C	C*
Nightshade (hairy) Solanum sarachoides	C*	C	C	C	C	C*
Pigweed, Carelessweed (redroot) Amaranthus retroflexus	C	C	C	C	C	C
Pigweed (smooth) Amaranthus hybridus	C	C	C	C	C	C
Purslane Portulaca oleracea	C	C	C	C	C	C
Pusley (Florida) Richardia scabra	C	C	C	C	C	C
Ragweed (common) Ambrosia artemisiifolia	R	C	C	C	C	C
Ragweed (giant) Ambrosia trifida	•	C	R	R	R	R
Sandbur (field) Cenchrus spp.	•	C	•	•	•	•
Sida (prickly), teaweed Sida spinosa	R	C	C	C	C	C
Smartweed Polygonum pensylvanicum Polygonum persicaria	R	C	C	C	C	C
Velvetleaf, Buttonweed Abutilon theoprasti	•	R	R	R	R	R

* Use the maximum labeled rates to achieve control.
™ Bladex is a registered trademark of the Shell Chemical Company.
™ Lorox and Lexone are trademarks of E.I. duPont de Nemours and Company.
™ Princep is a registered trademark of the Ciba-Geigy Corporation.
™ Sencor is a trademark of the parent company of Farbenfabriken Bayer GmbH, Leverkusen.

REFER TO SPECIFIC PRODUCT LABELS FOR ADDITIONAL WEEDS CONTROLLED, CROP ROTATION RESTRICTIONS AND CAUTIONARY STATEMENTS.

APPLICATION INFORMATION

SHAKE CONTAINER VIGOROUSLY PRIOR TO MIXING AND USE.

Do not apply this product through any type of irrigation system.

Broadcast Treatment—Apply this product and labeled tank mixture products in 10 to 30 gallons of water, or 10 to 50 gallons of liquid nitrogen (28 percent or 32 percent solutions only) per acre, before, during or after planting, but PRIOR TO CROP EMERGENCE. Application of this product in fluid fertilizer should only be used for treatment of annual weeds less than 6 inches tall. Spray coverage should be uniform and complete. Under conditions of dense stubble or plant residue or very high weed populations, higher rates within the recommended ranges may be needed for best results.

DO NOT APPLY BY AIR.

MIX ONLY WITH CLEAN WATER.

Ammonium Sulfate—The addition of ammonium sulfate in the spray solution may increase the performance of Bronco and Bronco tank mixtures on emerged annual weeds under adverse growing conditions. When using ammonium sulfate, add 2 percent dry ammonium sulfate by weight or 17 pounds per 100 gallons of water. Ammonium sulfate should be added to the water in the spray tank and completely dissolved prior to adding herbicide or surfactant. Do not mix ammonium sulfate in fluid fertilizer solutions.

If ammonium sulfate is added directly to the spray tank, add slowly with agitation. Adding too quickly may clog outlet lines. Nozzle tip plugging may result from the use of low-quality ammonium sulfate. To determine quality, perform a jar test by adding ⅓ cup of ammonium sulfate to 1 gallon of water and agitate for 1 minute. If undissolved sediment is observed, predissolve the ammonium sulfate in water and filter prior to addition to the spray tank.

Following use of ammonium sulfate, thoroughly rinse the spray system with clean water to reduce corrosion.

Bronco tank mixtures with ammonium sulfate are not approved in the state of California.

Surfactants—Nonionic surfactants which are labeled for use with herbicides may be used with Bronco. Do not reduce rates of Bronco when adding surfactant. Use 0.5 percent surfactant concentration (2 quarts per 100 gallons of spray solution) when using surfactants which contain at least 50 percent active ingredient or a 1 percent surfactant concentration (4 quarts per 100 gallons of spray solution) for those surfactants containing less than 50 percent active ingredient. Read and carefully observe surfactant cautionary statements and other information appearing on the surfactant label.

MIXING, SPRAYING AND HANDLING INSTRUCTIONS

NOTE

The use of a mechanical transfer system is required to be used by all mixer/loaders and/or applicators who treat 300 acres or more annually with pesticides containing alachlor as an active ingredient.

Direct contact or exposure to Bronco or spray mixtures of Bronco should be minimized. The following instructions for transfer, mixing, cleaning or repairing equipment should be followed in order to minimize this exposure. Review the protective clothing requirements as listed in the Precautionary Statements section of this label (front panel) and do not use this product until you have the necessary protection clothing.

2.5 Gallon Containers

Open pouring from these containers can result in exposure from splashing or spilling. Special care in lifting and pouring are strongly recommended.

Bulk Containers

Open pouring from these containers can result in exposure from splashing or spilling and is not recommended. This product should be transferred from these containers to the mix or spray tank using pumps or transfer probes. The probe or pump should not be removed from the container or disconnected until the container is emptied and rinsed. Use the pump or probe system to rinse the empty container and transfer the rinsate directly to the mix or spray tank.

Equipment Cleaning & Repair

Cleaning and repair of transfer systems and application equipment is a source of exposure to this product. Care should be taken to minimize exposure during cleaning and repair of transfer systems and application equipment. Whenever possible, these systems or equipment should be rinsed before being cleaned or repaired. When repairs must be made during transfer or application, the equipment should be shutdown, and special care taken to avoid contact with the pesticide.

Always predetermine the compatibility of Bronco with other labeled herbicides by mixing small proportional quantities in advance. NOTE: Reduced results may occur if muddy water is used, such as water from ponds or ditches.

Mix this product or labeled tank mixtures with clean water as follows:

1. Place a 20 to 35-mesh screen or wetting basket over filling port.
2. Through the screen, fill the sprayer tank one-half full with water.
3. Add tank mix components. If a wettable powder is used, make a slurry with water, and add it SLOWLY through the screen into the tank. Continue agitation.
   If flowable herbicides are used, premix one part flowable with one part water. Add diluted mixture SLOWLY through the screen into the tank.

4. Premix one part of Bronco with one part water. Add diluted mixture SLOWLY through the screen into the tank.
5. Complete filling the sprayer tank with water. Remove hose from tank immediately to avoid siphoning back into the water source. Maintain good agitation at all times until the contents of the tank are sprayed.

NOTE: If spray mixture is allowed to settle at any time, thorough agitation is required to resuspend the mixture before spraying is resumed. Keep bypass line on or near bottom of tank to minimize foaming. Screen size in nozzle or line strainers should be no finer than 50 mesh. Carefully select proper nozzle to avoid spraying a fine mist. Check for even distribution of spray droplets. For best results with ground application equipment, use flat fan or whirl-chamber nozzle. To reduce loss of chemical due to drift of a fine mist, apply at pressure less than 40 psi.

DIRECTIONS FOR USE

It is a violation of Federal Law to use this product in any manner inconsistent with this labeling.

SHAKE WELL BEFORE USING

NO-TILL AND MINIMUM TILLAGE SYSTEMS CORN, SOYBEANS AND GRAIN SORGHUM (MILO)

When applied as directed under the conditions described, this product or labeled tank mixtures will control many emerged annual weeds, suppress many emerged perennial weeds and give preemergence control of many annual weeds when corn, soybeans or grain sorghum will be planted directly into a cover crop, or in previous crop residues. This product or labeled tank mixtures will not control regrowth from perennial weeds or unemerged perennial weeds.

For use on grain sorghum, apply Bronco only on fields where the grain sorghum seed has been properly treated with a seed protectant containing the active ingredient, flurazole.

For mixing instructions, see the "MIXING AND SPRAYING INSTRUCTIONS" section of this label.

The higher recommended rates listed for each soil texture grouping should be used when:

- PREVIOUS CROP RESIDUES ARE EXTREMELY DENSE
- EMERGED WEEDS ARE OVER 6 INCHES TALL
- HEAVY INFESTATIONS OF ANNUAL GRASSES OR BROADLEAVES ARE EXPECTED
- SOILS CONTAIN MORE THAN 3 PERCENT ORGANIC MATTER

For postemergence weed control in double crop systems, small grain stubble should be cut to a height of 10 to 12 inches, leaving enough weed leaf surface for adequate herbicide coverage and uptake. Consult the "Double Crop" section of this label for additional information and specific use rates.

BRONCO USE RATES

The following table provides the recommended rates for this product in no-till and minimum tillage systems in CORN, SOYBEANS and GRAIN SORGHUM. For recommended Bronco rates on medium and fine textured soils in DOUBLE CROP systems, consult the "Double Crop" section of this label.

Soil Texture Group	Broadcast Rate of BRONCO® (Qts./A.)
Coarse	3 to 5
Medium	4 to 5
Fine	4 to 5

DOUBLE CROP SYSTEMS SOYBEANS AND GRAIN SORGHUM (MILO)

Bronco or Bronco plus a labeled tank mixture will control the weed species listed below in medium and fine texture soils at the 3 to 5 quarts per acre rate. For additional information for weeds controlled on coarse textured soils, consult the "WEEDS CONTROLLED" section of this label.

Annual Grasses	Annual Broadleaves*
Bluegrass (annual)	Lambsquarters
Brome (downy)	Marestail**
Corn (volunteer)	Pigweed (redroot)
Crabgrass	Pigweed (smooth)
Foxtail (giant)	Ragweed (common)
Foxtail (green)	
Foxtail (yellow)	
Wheat (volunteer)	

* A tank mixture of Bronco plus Lorox or Lexone or Sencor is recommended to improve control of emerged labeled weeds in double crop soybeans. Tank mixtures of Bronco plus atrazine are recommended to improve weed control in double crop grain sorghum. Consult the soybean and grain sorghum tank mixture sections of this label for recommended rates.

** A Bronco rate of 3 quarts per acre plus tank mixture with Lorox or Lexone or Sencor will control light infestations of this broadleaf weed. When medium to heavy infestations exist use the 4 to 5 quart per acre rate of Bronco rate alone or in tank mixture with Lorox or Lexone or Sencor. Consult the tank mixtures section of this label for recommended use rates.

When using the 3 quart per acre rate of Bronco or Bronco tank mixtures, flat fan nozzles or low volume flood jet nozzles and application volumes of 20 gallons per acre or less are recommended for optimum performance. Refer to the "GENERAL INFORMATION" section of this label for additional information.

The higher Bronco rates (4 to 5 quarts per acre) should be used in double crop systems when:

- Weeds not listed under Double Crop Systems are present (Note: A 5 quart per acre Bronco rate must be used to control emerged fall panicum).
- Weeds are drought stressed.
- Heavy infestations of annual grasses or broadleaves are present.

TANK MIXTURES

BRONCO® plus Roundup® and/or Lasso®

In cases of heavy weed infestation or hard-to-control weed species found on this label, Roundup OR Lasso herbicides OR both can be added to the recommended treatment rates to obtain improved control under the following methods of application:

Postemergence	Roundup
Preemergence	Lasso
Postemergence and Preemergence	Roundup and Lasso

Refer to the respective label booklets for Roundup and Lasso for specific recommendations regarding the timing of application.

The following table shows the maximum amounts of Roundup OR Lasso herbicides, OR both, that can be added to specific treatment rates of Bronco.

Recommended Tank Mixture Rates (Qts./A.)		
BRONCO®	Roundup® and/or	Lasso®
3	2	2.2
3.5	1.8	1.7
4	1.6	1.4
4.5	1.4	1.1
5	1.25	0.75

TANK MIXTURE—CORN

BRONCO® plus ATRAZINE OR BRONCO plus BLADEX OR BRONCO plus PRINCEP

Apply BRONCO® alone or in tank mixture with atrazine, OR Bladex OR Princep herbicides for control of grasses and broadleaf weeds listed in the "WEEDS CONTROLLED" section of this label when CORN is planted directly into a cover crop or in previous crop residues.

Refer to specific product labels for crop rotation restrictions and cautionary statements of all products used in these tank mixtures.

See the following table for recommended rates of atrazine, OR Bladex OR Princep in this tank mixture applied on various soil types. To determine the recommended rates for Bronco refer to the rate table for this product.

Soil Texture Group*	Broadcast Rates (Qts./A)** Atrazine 4L	Bladex 4L	Princep 4L
Coarse—Sandy Loam Only	1 to 1.2	1 to 1.6	1 to 1.2
Medium	1.2 to 1.6	1.2 to 1.6	1.2 to 1.6
Fine	1.6 to 2	1.6 to 2.2	1.6 to 2

* Refer to the "SOIL TEXTURE" section of this label to determine the corresponding soil texture group for the soil to be treated.

** When using 4L formulations use the rates listed.

- Use equivalent rates when using 80W or 90 percent dry flowable formulations. One quart of the 4L formulation of these products equals 1.25 pounds of the 80W formulation. One quart of atrazine 4L or Princep 4L equals 1.1 pounds of the 90 percent dry flowable formulations of these products.
- Use the higher rates when recommended for each product listed when soils contain more than 3 percent organic matter.

TANK MIXTURES CORN

BRONCO® plus ATRAZINE plus PRINCEP OR BRONCO® plus ATRAZINE plus BLADEX

For additional preemergence control of heavy weed infestation and/or hard-to-control weeds, a three-way tank mixture of Bronco herbicide plus Atrazine plus Princep OR Bronco plus Atrazine plus Bladex may be applied. Refer to the following rate tables and the rate table for Bronco to determine the recommended rates for these tank mixtures on various soil types.

Broadcast Rates (Qts./A.)**

Soil Texture Group*	Atrazine 4L	+ Princep 4L
Coarse	0.75 to 1	0.75 to 1
Medium	1 to 1.25	1 to 1.5
Fine	1 to 1.25	1 to 1.5
	Atrazine 4L	**+ Bladex 4L**
Coarse	0.75 or 1	1 to 1.25
Medium	1 to 1.25	1.25 to 2
Fine	1 to 1.25	1.25 to 2

* Refer to the "SOIL TEXTURE" section of this label to determine the corresponding soil texture group for the soil to be treated.
** When using 4L formulation use the rates listed.

- Use equivalent rates when using 80W or 90 percent dry flowable formulations. One quart of the 4L formulation of these products equals 1.25 pounds of the 80W formulation. One quart of atrazine 4L or Princep 4L equals 1.1 pounds of the 90 percent dry flowable formulations of these products.
- USE THE LOWER RATES RECOMMENDED FOR EACH PRODUCT LISTED WHEN SOILS CONTAIN LESS THAN 3 PERCENT ORGANIC MATTER.

NOTE: PLANT ONLY CORN, PEANUTS, SORGHUM OR SOYBEANS* the year following use of this mixture.

*There is a possibility of crop injury due to carryover of atrazine if soybeans are planted the following year. DO NOT plant soybeans the year following use of these tank mixtures on furrow irrigated corn. DO NOT graze treated areas or feed treated forage to livestock for 21 days following application of these tank mixtures.

NOTE: Land treated with Princep should not be planted to any crop except corn for one year following treatment as crop injury may occur.

TANK MIXTURES FULL SEASON AND DOUBLE CROP SOYBEANS

BRONCO® plus LOROX
OR
BRONCO plus LEXONE OR SENCOR

Apply Bronco herbicide alone or in tank mixtures with Lorox, OR Lexone, OR Sencor herbicides for control of grasses and broadleaf weeds listed in the "WEEDS CONTROLLED" section of this label when SOYBEANS are planted directly into a cover crop, stale seedbed, or in previous residues such as wheat stubble.

Refer to specific product labels for crop rotation restrictions and cautionary statements of all products used in these tank mixtures.

See the following table for recommended rates of Lorox, OR Lexone OR Sencor in tank mixtures applied on various soil types. To determine rates for Bronco refer to the rate table for this product.

Broadcast Rates (Qts./A)

Soil Texture Group*	Lorox L**†	Lexone 4L or Sencor 4F†
Coarse—Sandy		
Loam Only	0.5 to 0.75	0.25 to 0.375
Medium	0.75 to 1	0.375 to 0.5
Fine	1 to 1.5	0.5 to 0.75

* Refer to the "SOIL TEXTURE" section of this label to determine the corresponding soil texture group for the soil to be treated.
** Use the higher rate of Lorox L in the recommended ranges on soils containing more than 3 percent organic matter.
† Use rates listed in this label when using Lorox L, Lexone 4L or Sencor 4F. Use equivalent rates when using the 50 WP or DF formulations of these products. One pint of Lorox L, Lexone 4L or Sencor 4F equals 1 pound of the corresponding 50 WP formulation or ⅔ pound of the corresponding DF formulation.

THE FOLLOWING INFORMATION APPLIES ONLY TO USE OF LEXONE OR SENCOR.

On the silty clay or heavy clay soils of the Mississippi Delta, use 0.75 to 1 quart of Lexone or Sencor per acre.

Use the higher rate of Lexone or Sencor in the recommended ranges on soils with greater than 2 percent organic matter.

Do not use on muck soils.

Do not apply on alkaline soils with a ph higher than 7.4.

Crop injury may occur if any atrazine was applied on the soil the year before use of this Lexone or Sencor tank mixtures.

DO NOT REPLANT CROPS OTHER THAN SOYBEANS FOR 120 DAYS AFTER APPLICATION.

TANK MIXTURES FULL SEASON AND DOUBLE CROP GRAIN SORGHUM (MILO)

BRONCO® plus ATRAZINE

Apply Bronco alone or in tank mixture with atrazine herbicide for control of grasses and broadleaf weeds listed in the "WEEDS CONTROLLED" section of this label when grain sorghum is planted directly into a cover crop or in previous crop residues.

Refer to specific product labels for crop rotation restrictions and cautionary statements of all products used in these tank mixtures.

See the following for recommended rates of atrazine in this tank mixture applied on various soil types. To determine rates for Bronco refer to the rate table for this product.

Soil Texture Group*	Broadcast Rates (Qts./A) Atrazine 4L
Coarse (Except Sand)	0.75 to 1
Medium	1 to 1.25
Fine	1.2 to 1.5

* Refer to the "SOIL TEXTURE" section of this label to determine the corresponding soil texture group for the soil to be treated.
** When using 4L formulations use the rates listed.

- Use equivalent rates when using 80W or 90 percent dry flowable formulations. One quart of the 4L formulation of this product equals 1.25 pounds of the 80W formulation. One quart of atrazine 4L equals 1.1 pounds of the 90 percent dry flowable formulation of this product.
- Use the higher rates when recommended for each product listed when soils contain more than 3 percent organic matter.

®Roundup and ®Lasso are registered trademarks of Monsanto Company.
1988-1 817.37-000.37/53

In case of an emergency involving this product, Call Collect, day or night, (314) 694-4000.

BUCKLE®

GRANULAR HERBICIDE

Granular selective herbicide for spring preplant application to control wild oats and pigeongrass in Barley and for fall preplant application to control wild oats and pigeongrass in Durum Wheat and Barley.

ACTIVE INGREDIENT:
S (2,3,3-Trichloroallyl) diisopropyl-thiocarbamate 10%
α, α, α trifluoro-2,6-dinitro-N,N-dipropyl-p-toluidine 3%
INERT INGREDIENTS: 87%
TOTAL 100%

EPA Reg. No. 524-375 EPA Est. 524-IA-1

Read the entire label before using this product.

Use only according to label instructions.

REFORMULATION OR REPACKAGING IS PROHIBITED.

Read "LIMIT OF WARRANTY AND LIABILITY" before buying or using. If items are not acceptable, return at once unopened.

PRECAUTIONARY STATEMENTS

HAZARDS TO HUMANS AND DOMESTIC ANIMALS

KEEP OUT OF REACH OF CHILDREN

CAUTION!

MAY CAUSE EYE IRRITATION

Avoid contact with skin, eyes, or clothing.

Wash thoroughly with soap and water after handling.

FIRST AID: IF IN EYES, immediately flush with plenty of water at least 15 minutes. Call a physician.

In case of an emergency involving this product, Call Collect, day or night, (314) 694-4000.

ENVIRONMENTAL HAZARDS

This pesticide is toxic to fish. Do not apply directly to water or wetlands (swamps, bogs, or marshes). Drift or runoff from treatment areas may be hazardous to aquatic organisms in neighboring aquatic sites. Do not contaminate water by cleaning of equipment or disposal of wastes.

STORAGE AND DISPOSAL

Do not contaminate water, foodstuffs, seed or feed by storage or disposal.

STORAGE:

Keep bag closed to prevent spills and contamination.

DISPOSAL:

Wastes resulting from the use of this product that cannot be used or chemically reprocessed should be disposed of in a landfill approved for pesticide disposal or buried on site in a safe place away from water supplies. All disposal should be in accordance with applicable Federal, state or local procedures.

Emptied container retains vapor and product residue. Observe all labeled safeguards until container is destroyed.

Completely empty bag into application equipment. Then dispose of empty bag in a sanitary landfill or by incineration, or, if allowed by state and local authorities, by burning. If burned, stay out of smoke.

If you have any questions concerning this label or the proper application or use of Buckle, contact your local Monsanto Sales Representative.

DIRECTIONS FOR USE

It is a violation of Federal law to use this product in any manner inconsistent with its labeling.

This product is recommended for spring application prior to planting barley, and for fall application prior to planting durum wheat and barley to control wild oats, green foxtail (Setaria viridis) and yellow foxtail (Setaria lutescens). Green and yellow foxtail are commonly referred to as pigeongrass.

Application equipment must be properly calibrated: application of too much herbicide may injure the crop; application of too little may result in poor weed control. Recommended rates should be followed in order to:

1 Avoid crop injury

2. Avoid herbicide residues at harvest

3. Control weeds

Use of this product not consistent with this label may result in injury to persons, animals or crops, or other unintended consequences. Keep container closed to prevent spills and contamination.

Application to a field which is wet, lumpy, rough or ridged will result in reduced weed control and promote crop thinning. Incorporation must be completed within 24 hours after application and before germination of the weeds. If weeds have emerged prior to treating and/or planting, they must be controlled. All deep tillage by cultivators or double disc implements must be completed prior to application. Do not plow following application of this herbicide.

When using this herbicide, a strip should be left untreated for proof of results. Weed control may be evaluated also by removing a surface inch or two of soil at the time of germination to inspect the number of weed seedlings killed before emergence.

Weeds are usually killed before emergence, but occasionally and particularly under dry conditions, plants may reach the 3 to 4 leaf stage before they die.

Under conditions of prolonged high temperature at the time of germination or extreme drought in the spring, this product may not maintain the usual high standard of weed control.

ATTENTION

DELAYED EMERGENCE, STAND REDUCTION, STUNTING AND YIELD LOSS MAY RESULT DUE TO COLD OR WET CONDITIONS, IRRIGATION DURING GERMINATION OR EMERGENCE, DEEP PLANTING, SOIL CRUSTING, DISEASE, INSECTS, POOR SEED QUALITY, OR SELECTION OF A VARIETY GENETICALLY SUSCEPTIBLE TO STRESS. THESE ARE CONDITIONS BEYOND THE CONTROL OF MONSANTO, AND THE GROWER SHOULD CONSULT THE STATE UNIVERSITY EXTENSION SPECIALIST OR SEED PRODUCER OR SUPPLIER FOR LOCAL VARIETAL INFORMATION AND RECOMMENDED TILLAGE AND PLANTING PRACTICES AND DATES.

NOTE: Slight stand reductions do not normally affect yield.

NOTE: Do not graze livestock on treated crops.

Do not apply BUCKLE® for durum wheat on soil that has been planted to soybeans or sugar beets and treated with trifluralin the previous year.

Sugar beets, red beets, or spinach should not be planted for 14 months after application of this product.

If less than 20 inches of total water were used to produce the crop, do not plant any of the above-mentioned crops for 20 months after application of this product. Cool, wet conditions during the early stage of growth may increase the possibility of injury to sorghum.

Do not use this product on soils containing more than 10% organic matter or on muck soils.

Do not apply this product to soils which are subject to prolonged periods of flooding.

Sorghum, proso millet, corn or oats should not be planted for 16 months after application of this product.

Certain vegetable crops should not be planted within 5 months following application of this product. Consult a Monsanto representative for those vegetable crops which can or cannot be planted safely.

Use in any manner not consistent with label directions can result in serious injury to the crop.

HOW TO APPLY AND CALIBRATE EQUIPMENT

This granular herbicide must be applied through a specially designed ground applicator capable of applying small quantities of granules evenly.

It is important that the applicator be calibrated properly to deliver the desired amount of this product to avoid applying too little, or too much, material. To give even distribution, scatter plates (similar to those used for applying granules in a band) must be attached to each delivery tube or outlet in such a manner to give overall coverage. To calibrate, attach a collector pan, or strong, vented plastic bag, over each spreader plate or delivery tube. Operate over normal terrain to be treated at 4 to 5 miles per hour. Collect the granules from all outputs after covering the desired distance.

For example, if a 14-foot applicator is being used, stake off a distance of 324 feet in the field to be treated (distance equal to 1/10 acre treated). After attaching a bag, or collection pan, to each outlet, collect the granules while driving the desired distance. Check to see that each outlet disperses the same amount of granules. Combine all samples and weigh. For the above set of conditions, the quantity of granules that should be collected for the following recommended rates for broadcast treatment are:

	Recommended Rate Per Acre	
	10 lbs	12.5 lbs.
Proper amount to be collected	1.0 lb	1.25 lbs

If more or less than the desired quantity is collected, adjust opening accordingly and again collect the granules while driving the staked-off distance. Continue this procedure until the proper amount is delivered.

FIELD PREPARATION

Before applying this product, be sure the soil is in good working condition. All deep tillage by cultivation, or double disc implements, must be completed prior to application of this herbicide. If stubble ground is being treated, one or two passes with a field cultivator may be required before application and incorporation.

INCORPORATION DIRECTIONS

Fall Applications

When incorporating with a single pass in the fall, a second incorporation must be performed in the spring during seedbed preparation. If no spring work is anticipated, then both incorporations should be done in the fall.

Spring Applications

Apply granules and incorporate with equipment such as a culti-harrow or duckfoot cultivator. A second incorporation at right angles should provide best results.

General Incorporation Guidelines

For proper incorporation of BUCKLE, set incorporation implement to work the soil no deeper than 3 or 4 inches. Do not use disc implements for incorporation.

On summer fallow ground which is loose and free of lumps and trash, incorporate this product using equipment similar to a field cultivator, culti-harrow or a duckfoot with a rod weeder attachment.

For applications to fields of standing stubble, fields may be worked once or twice with a field cultivator or chisel plow. Apply granules and incorporate with a field cultivator or culti-harrow.

If soil must be ridged after incorporating to prevent soil erosion by high winds, the depth of ridging should be kept to a minimum.

In preparing the herbicide treated area for seeding in the spring, care must be taken to avoid working the soil any deeper than fall tillage.

APPLICATION RATES

Apply BUCKLE in the spring prior to planting barley, or in the fall prior to planting barley or durum wheat at the rates listed in the Application Rate Table.

Application Time	Barley	Durum Wheat
Spring	10-12.5 lb/A	
Fall	10-12.5 lb/A	10-12.5 lb/A

In the fall, apply within 3 weeks of normal freezeup or until snow cover occurs.

Apply higher rates on heavy weed infestations within the recommended range.

LIMIT OF WARRANTY AND LIABILITY

This Company warrants that this product conforms to the chemical description on the label and is reasonably fit for the purposes set forth in the Complete Directions for Use ("Directions") when used in accordance with those Directions under the conditions described therein. NO OTHER EXPRESS WARRANTY OR IMPLIED WARRANTY OF FITNESS FOR PARTICULAR PURPOSE OR MERCHANTABILITY IS MADE. This warranty is also subject to the conditions and limitations stated herein.

Buyer and all users shall promptly notify this Company of any claims whether based in contract, negligence, strict liability, other tort or otherwise.

Buyer and all users are responsible for all loss or damage from use or handling which results from conditions beyond the control of this Company, including, but not limited to, incompatibility with products other than those set forth in the Directions, unusual weather, weather conditions which are outside the range considered normal at the application site and for the time period when the product is applied, as well as weather conditions which are outside the application ranges set forth in the Directions, application in any manner not explicitly set forth in the Directions, moisture conditions outside the moisture range specified in the Directions, or the presence of products other than those set forth in the Directions in or on the soil, crop or treated vegetation.

THE EXCLUSIVE REMEDY OF THE USER OR BUYER, AND THE LIMIT OF THE LIABILITY OF THIS COMPANY OR ANY OTHER SELLER FOR ANY AND ALL LOSSES, INJURIES OR DAMAGE RESULTING FROM THE USE OR HANDLING OF THIS PRODUCT (INCLUDING CLAIMS BASED IN CONTRACT, NEGLIGENCE, STRICT LIABILITY, OTHER TORT OR OTHERWISE) SHALL BE THE PURCHASE PRICE PAID BY THE USER OR BUYER FOR THE QUANTITY OF THIS PRODUCT INVOLVED, OR, AT THE ELECTION OF THIS COMPANY OR ANY OTHER SELLER THE REPLACEMENT OF SUCH QUANTITY, OR, IF NOT ACQUIRED BY PURCHASE, REPLACEMENT OF SUCH QUANTITY. IN NO EVENT SHALL THIS COMPANY OR ANY OTHER SELLER BE LIABLE FOR ANY INCIDENTAL CONSEQUENTIAL OR SPECIAL DAMAGES.

Buyer and all users are deemed to have accepted the terms of this LIMIT OF WARRANTY AND LIABILITY which may not be varied by any verbal or written agreement.

* Registered trademark of Monsanto Company

843.62-000 03CG

In case of an emergency involving this product Call Collect, day or night. (314) 694-4000.

This sample label is current as of January 1, 1989. The product descriptions and recommendations provided in this sample label are for background information only. Always refer to the label on the product container before using Monsanto or any other agrichemical product.

This specimen label is intended for use only as a guide in providing general information regarding the use of this product. Monsanto is not responsible for the accuracy of the information contained herein. As labels are subject to revision, always carefully read and follow the label on the product container.

BULLET

HERBICIDE

Selective herbicide for preemergence weed control in corn (all types).

RESTRICTED USE PESTICIDE

due to oncogenicity. For retail sale to and use only by Certified Applicators or persons under their direct supervision and only for those uses covered by the Certified Applicator's certification.

The use of this product may be hazardous to your health. This product contains alachlor, which has been determined to cause tumors in laboratory animals.

In case of an emergency involving this product or for user safety information on this product, Call Collect, day or night (314) 694-4000.

SHAKE WELL BEFORE USING

Complete Directions For Use

ACTIVE INGREDIENTS:*	
Alachlor,[2-chloro-2′,6′-diethy-N-(methoxymethyl) acetanilide]	25.2%
Atrazine,[2-chloro-4-(ethylamino)-6 -(isopropylamino)s-triazine]	14.3%
Related atrazine compounds	0.8%
INERT INGREDIENTS:	59.7%
TOTAL	100.0%

*Contains 300 grams/litre or 2.5 pounds/gallon of alachlor and 180 grams/litre or 1.5 pounds/gallon of atrazine and related compounds.

EPA Reg. No. 524-418

Bullet® is a registered trademark of Monsanto Company.

1988-1 843.25-000.01/CG

Read the entire label before using this product.

Use only according to label instructions.

Read "LIMIT OF WARRANTY AND LIABILITY" before buying or using. If terms are not acceptable, return at once unopened.

REFORMULATION IS PROHIBITED

LIMIT OF WARRANTY AND LIABILITY

This Company warrants that this product conforms to the chemical description on the label and is reasonably fit for the purposes set forth in the Complete Directions for Use label booklet ("Directions") when used in accordance with those Directions under the conditions described therein. NO OTHER EXPRESS WARRANTY OR IMPLIED WARRANTY OF FITNESS FOR PARTICULAR PURPOSE OR MERCHANTABILITY IS MADE. This warranty is also subject to the conditions and limitations stated herein.

Buyer and all users shall promptly notify this Company of any claims whether based in contract, negligence, strict liability, other tort or otherwise.

Buyer and all users are responsible for all loss or damage from use or handling which results from conditions beyond the control of this Company, including, but not limited to, incompatibility with products other than those set forth in the Directions, unusual weather, weather conditions which are outside the range considered normal at the application site and for the time period when the product is applied, as well as weather conditions which are outside the application ranges set forth in the Directions, application in any manner not explicity set forth in the Directions, moisture conditions outside the moisture range specified in the Directions, or the presence of products other than those set forth in the Directions in or on the soil, crop or treated vegetation.

THE EXCLUSIVE REMEDY OF THE USER OR BUYER, AND THE LIMIT OF THE LIABILITY OF THIS COMPANY OR ANY OTHER SELLER FOR ANY AND ALL LOSSES, INJURIES OR DAMAGE RESULTING FROM THE USE OR HANDLING OF THIS PRODUCT (INCLUDING CLAIMS BASED IN CONTRACT, NEGLIGENCE, STRICT LIABILITY, OTHER TORT OR OTHERWISE) SHALL BE THE PURCHASE PRICE PAID BY THE USER OR BUYER FOR THE QUANTITY OF THIS PRODUCT INVOLVED, OR, AT THE ELECTION OF THIS COMPANY OR ANY OTHER SELLER, THE REPLACEMENT OF SUCH QUANTITY, OR, IF NOT ACQUIRED BY PURCHASE, REPLACEMENT OF SUCH QUANTITY. IN NO EVENT SHALL THIS COMPANY OR ANY OTHER SELLER BE LIABLE FOR ANY INCIDENTAL, CONSEQUENTIAL OR SPECIAL DAMAGES.

Buyer and all users are deemed to have accepted the terms of this LIMIT OF WARRANTY AND LIABILITY which may not be varied by any verbal or written agreement.

PRECAUTIONARY STATEMENTS

HAZARDS TO HUMANS AND DOMESTIC ANIMALS

CAUTION!

HARMFUL IF INHALED.

Avoid breathing vapor or spray mist.

Seek medical attention if breathing difficulty develops.

Wear goggles or face shield, rubber gloves, long trousers, long-sleeved shirt or jacket of tightly woven material along with boots high enough to cover ankles when transferring and mixing, and when adjusting, repairing or cleaning equipment. Wear rubber boots when pouring from open containers or when reentry is made into fields following application of this product through center pivot irrigation systems and the field is still wet. Clothing which comes in contact with this product must be washed before reuse. Clothing or other materials which become drenched with the concentrated pesticide must be disposed of in sanitary landfill, by incineration or, if allowed by state and local authorities, by burning. If burned, stay out of smoke.

Wash thoroughly with soap and water after handling.

STATEMENT OF PRACTICAL TREATMENT

IF INHALED: Remove victim to fresh air. If not breathing, give artificial respiration, preferably mouth-to-mouth. Get medical attention.

In case of emergency involving this product, Call Collect, day or night, (314) 694-4000.

ENVIRONMENTAL HAZARDS

This chemical has been identified in limited ground water sampling and there is the possibility that it can leach through the soil to ground water, especially where soils are coarse and ground water is near the surface. Following application, and during rainfall events that cause runnoff, this chemical may reach surface water bodies including streams, rivers and reservoirs.

Care must be taken when using this product to prevent back-siphoning into wells, spills or improper disposal of excess pesticide, spray mixtures or rinsates.

Check valves or antisiphoning devices must be used on all mixing and/or irrigation equipment. Disposal of excess pesticide, spray mixtures or rinsate should be according to label use instructions or according to the state pesticide or environmental control agency or the hazardous waste representative at the nearest EPA regional office.

This product is toxic to fish. Do not apply directly to water or wetlands. Do not contaminate water when disposing of equipment washwaters.

PHYSICAL OR CHEMICAL HAZARDS

In case of: SPILL or LEAK, soak up, remove and keep away from humans and domestic animals.

DIRECTIONS FOR USE

It is a violation of Federal law to use this product in any manner inconsistent with its labeling.

STORAGE AND DISPOSAL

Do Not contaminate water, foodstuffs, feed or seed by storage or disposal.

STORAGE:

STORE ABOVE –4°F TO KEEP PRODUCT FROM FREEZING. If frozen, thaw before use.

DISPOSAL:

Wastes resulting from the use of this product that cannot be used or chemically reprocessed should be disposed of in a landfill approved for pesticide disposal or in accordance with applicable Federal, state or local procedures.

Emptied container retains product residue. Observe all labeled safeguards until container is cleaned, reconditioned or destroyed.

(See individual container label for disposal information).

GENERAL INFORMATION

This product is recommended for the control of many annual grasses and broadleaf weeds. Except for minimum and conservation tillage systems, the seed bed should be fine, firm and free of clods and trash.

Performance is improved when moisture occurs within one week after application. The amount required depends primarily upon the amount of soil moisture in the weed germination zone at the time of application, soil type and organic matter content, but 1/3 to 3/4 inch is generally sufficient for performance of this product. High intensity or excessive rainfall, or excessive irrigation, after application may reduce control.

Read and carefully observe the cautionary statements and all other information appearing on this label and on the labels of Gramoxone™ Super, Bladex™ and Princep™ herbicides before using in combination with this product.

Use of this product in any manner not consistent with this label may result in injury to persons, animals or crops, or other unintended consequences. Keep container closed to prevent spills and contamination.

Do not apply when conditions favor drift.

Flush sprayer with clean water after use.

Do not graze treated area or feed treated forage to livestock for 21 days following application.

NOTE: Use this product for weed control in corn only. PLANT ONLY CORN, PEANUTS, SORGHUM (MILO) OR SOYBEANS THE YEAR FOLLOWING THE USE OF THIS MIXTURE. IF SOYBEANS ARE TO BE PLANTED, THERE IS A POSSIBILITY OF CROP INJURY DUE TO CARRYOVER OF ATRAZINE. Do not plant soybeans in areas where furrow irrigation is practiced.

™Bladex is a trademark of E.I. du Pont de Nemours and Company.
™Gramoxone is a trademark of ICI Americas Inc.
™Princep is a trademark if Ciba Geigy Corporation.

SOIL TEXTURE

Applicators should evaluate soil and weed conditions carefully to assure that they choose the lowest effective label rate.

The recommended use rates of this product and the other herbicides labeled for use in tank mixtures with this product generally vary with soil texture. Unless soil texture is specifically named, rate tables throughout this label refer to only three soil textural groups: coarse, medium and fine. The following is a complete listing of soil textures included in each of these three soil textural groups:

SOIL TEXTURE GROUP	SOIL TEXTURE
COARSE:	sand, loamy sand, sandy loam
MEDIUM:	loam, silt loam, silt, sandy clay loam
FINE:	silty clay loam, clay loam, sandy clay, silty clay, clay

Refer to the above table to determine the corresponding soil textural group for the soil to be treated.

MIXING, SPRAYING AND HANDLING INSTRUCTIONS

NOTE

The use of a mechanical transfer system is required to be used by all mixer/loaders and/or applicators who treat 300 acres or more annually with pesticides containing alachlor as an active ingredient.

Direct contact or exposure to this product or spray mixtures of this product should be minimized. The following instructions for transfer, mixing, cleaning or repairing equipment should be followed in order to minimize this exposure. Review the protective clothing requirements as listed in the Precautionary Statements section of this label (front panel) and do not use this product until you have the necessary protective clothing.

2.5 Gallon Containers

Open pouring from these containers can result in exposure from splashing or spilling. Special care in lifting and pouring are strongly recommended.

Bulk Containers

Open pouring from these containers can result in exposure from splashing or spilling and is not recommended. This product should be transferred from these containers to the mix or spray tank using pumps or transfer probes. The probe or pump should not be removed from the container or disconnected until the container is emptied and rinsed. Use the pump or probe system to rinse the empty container and transfer the rinsate directly to the mix or spray tank.

Equipment Cleaning & Repair

Cleaning and repair of transfer systems and application equipment is a source of exposure to this product. Care should be taken to minimize exposure during cleaning and repair of transfer systems and application equipment. Whenever possible, these systems or equipment should be rinsed before being cleaned or repaired. When repairs must be made during transfer or application, the equipment should be shutdown, and special care taken to avoid contact with the pesticide.

Always predetermine the compatibility of this product with water carrier or sprayable fluid fertilizer or other labeled herbicides by mixing small proportional quantities in advance.

Mix this product with the appropriate carrier as follows:

1. Place a 20 to 35 mesh screen or wetting basket over the filling port.
2. Through the screen, fill the sprayer tank one-half full with the carrier.
3. If a compatibility agent is necessary to improve mixing or to prevent the formation of undesirable and unsprayable gels or precipitates, while agitating, add the compatibility agent to the carrier already in the tank.
   Use only compatibility agents cleared by the Federal Food and Drug Administration for this use. Follow the directions for use on the compatibility agent label. Check for adequate agitation.
4. Premix one part of this product with one part water. Add the diluted mixture SLOWLY through the screen into the tank. When necessary, this product may be diluted with a fertilizer carrier instead of water, after compatibility has been confirmed.
5. With minimum tillage systems, add the required amounts of Gramoxone Super near the end of the filling process.
6. Complete filling the sprayer tank with carrier. Remove the hose from the tank immediately to avoid siphoning back into the carrier source. Maintain good agitation until the contents of the tank are sprayed. NOTE: If spray mixture is allowed to settle at any time, thorough agitation is required to resuspend the mixture before spraying is resumed.

Keep by-pass line on or near the bottom of the tank to minimize foaming. Screen size in nozzle or line strainers should be no finer than 50 mesh. Carefully select the correct nozzle to avoid spraying a fine mist. For best results, use flat fan or flood jet nozzles. Check for even distribution of spray droplets. To reduce loss of the chemical due to drift of a fine mist, apply at pressures from 20 to 40 psi.

STANDARD SPRAYABLE FLUID FERTILIZER COMPATIBILITY TEST

Herbicides may not always mix evenly throughout a sprayable fluid fertilizer or the components may separate too quickly to make their combined use of practical value. This may be due to certain characteristics of the different fluid fertilizers. A simple test using small quantities of the components is suggested to provide compatibility potential.

The test follows:

A. Materials Required for A Compatibility Test

1. Two one-quart jars with lid or stopper (marked "with" and "without").
2. TEAspoons (for a more exacting test, a five to ten milliliter (ml.) pipette or graduated cylinder is desirable.
3. A minimum of one quart of the sprayable fluid fertilizer to be tested.
4. The herbicide(s) to be mixed.
5. A compatibility agent (the purpose of the adjuvant is to keep the fertilizer and herbicide(s) in suspension, if this assistance is needed).

B. Procedure

1. Add one pint of the sprayable fluid fertilizer that will be used, or other herbicide carrier, to each jar marked "with"and "without".

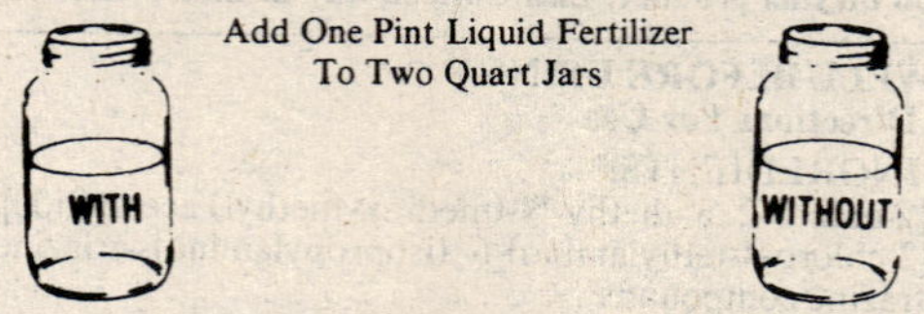

2. To the jar marked "with", add 1/4 TEAspoon or 1.2 milliters of a suitable compatibility agent; shake gently for five to ten seconds to mix (1/4 TEAspoon is the equivalent of two pints per 100 gallons of liquid fertilizer).

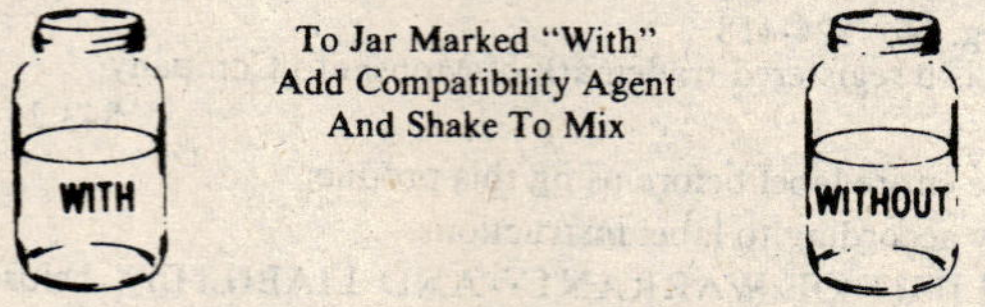

3. To each jar add the appropriate amount of herbicide(s). If more than one is used, add them separately with the wettable powders added first, flowables second and liquid last. Shake gently five to ten seconds after each addition.

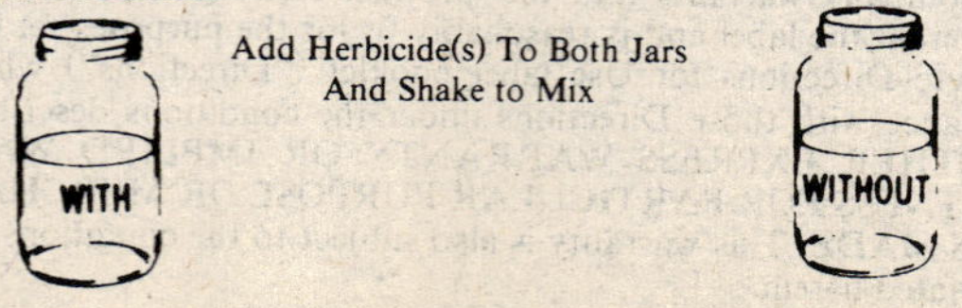

Herbicide	Rate/a		Amount to Be Added Per Pint of Sprayable Fluid Fertilizer (Assuming Volume is 25 gal/a) Level TEAspoons		Milliters
This Product	3 qts	=	2.9	or	14.2
	3.5 qts	=	3.4	or	16.6
	4 qts	=	3.8	or	19
	4.5 qts	=	4.3	or	21.4
	5 qts	=	4.8	or	23.8

This compatibility test is designed for 25 gallons of spray solution per acre with the maximum labeled rate of herbicide. For changes in spray volume or herbicide rate, make appropriate changes in the ingredients of the test. Regardless of spray volume, the amount of compatibility agent should be equal to two or three pints (two pints = 1/4 TEAspoon or 1.2 milliters, three pints = 3/8 TEAspoon or 1.8 milliters per pint of sprayable fluid fertilizer) per 100 gallons of liquid fertilizer.

C. Observations and Decisions

1. If the herbicide(s) and the sprayable fluid fertilizer are compatible.
2. If a compatibility agent is necessary.

Five minutes after the final addition and mixing, observe both jars for the formation of large flakes, sludge, gels or other precipitates. Observe if the herbicide(s) cannot be physically mixed with the liquid fertilizer but remains as small oily particles in the solution.

If incompatibility in any form described above occurs in the jar "with" the compatibility agent added, the liquid fertilizer and herbicide(s) should not be used together in the same spray tank.

If incompatibility as described above occurs in the jar "without" the adjuvant but not in the jar "with" adjuvant, the use of a compatibility agent is recommended.

Both jars should be allowed to stand and be observed periodically for one-half hour. If the separate layers of liquid fertilizer and additives can be resuspended by shaking, commercial application is possible. Wettable powders will either settle to the bottom of the tank or jar, or float to the top, depending upon the density of the fertilizers.

If the herbicide(s) is compatible with fluid fertilizer in the foregoing test without using a compatibility agent, fluid fertilizer may be used for the premixing. If it is not compatible, the herbicide(s) should be premixed with water before adding to the spray tank.

APPLICATION WITH DRY BULK FERTILIZER

The herbicide-ferlitizer impregnation process must be conducted only at commercial fertilizer or chemical dealerships adequately equipped for this procedure.

Dry bulk fertilizer may be impregnated with this product for use on corn. This product must be applied with 200 to 450 pounds of dry bulk fertilizer per acre and shallowly incorporated within 7 days prior to planting. The correct amount of herbicide must be applied per acre in the blend as recommended in this label for the crop, weed and soil type treated. Refer to the table for broadcast rate per acre in the "Surface Application or Preplant Incorporation" section of this label to determine the recommended broadcast rate per acre for the herbicide treatment to be applied.

The following table provides a reference to determine the amount of LIQUID herbicide to be mixed per ton of dry bulk fertilizer for a range of herbicide recommendations for fertilizer rates per acre:

		RECOMMENDED QUARTS LIQUID HERBICIDE/ACRE			
Fertilizer Rate (lb/acre)	Acres Covered (per ton)	3	3.5 (quarts of herbicide/ ton dry bulk fertilizer)	4	4.5
200	10	30	35	40	45
250	8	24	28	32	36
300	6.7	20	23.3	26.6	30
350	5.7	17.1	20	22.8	25.7
400	5	15	17.5	20	22.5
450	4.5	13.3	15.6	17.8	20

To determine the amount of herbicide needed for rates not included in the preceding table, use the following formula:

$$\frac{\text{Recommended herbicide rate (lb or qt)/acre} \times 2000}{\text{pounds fertilizer/acre}} = \text{Pounds or quarts of herbicide per ton of dry bulk fertilizer}$$

Mix and blend the dry fertilizer and herbicide mixture in a closed rotary-drum-type mixer allowing sufficient time to ensure uniform coverage. Use at least one ton of dry fertilizer per mixing operation. Inject the herbicide into the drum over a minimum of a 2-minute period and allow at least 2 additional minutes mixing time to ensure uniformity. The nozzle used to spray the herbicide treatment must be placed inside the mixer to provide uniform spray coverage of the tumbling fertilizer.

If the dry fertilizer used has inadequate adsorptive capacity, use a highly adsorptive powder such as Microcel™ E to provide a free-flowing mixture. The adsorptive powder must be added separately and uniformly to the prepared herbicide-fertilizer mixture. Generally 2 percent by weight of Microcel E is required. The following is a partial list of dry fertilizers which may be impregnated with this product.

Ammonium sulfate	21-0-0
Ammonium phosphate-sulphate	16-20-0
Diammonium phosphate	18-46-0
Potassium chloride	0-0-60
Potassium sulfate	0-0-52
*Urea	46-0-0

*Some ureas may be phytotoxic when applied on corn. Use only ureas known to be safe to corn.

NOTE: DO NOT impregnate tank mixtures of this product with other herbicides or fertilizers containing ammonium nitrate, potassium nitrate or sodium nitrate.

™Microcel is a trademark of Manville Products Corporation.

APPLICATION

Spread the herbicide-fertilizer mixture with a calibrated applicator. Be certain the material is applied uniformly to the soil surface. When using spin spreaders, a 100 percent overlap is recommended to ensure even distribution. Spin applicators should be calibrated to apply one-half the desired rate per acre and application overlapped by doubling back to cover one-half of the previous swath. Dribble spreaders must be calibrated with the herbicide-fertilizer mixture since the flow pattern for the herbicide-coated fertilizer will generally be slower than straight dry fertilizer.

WEEDS CONTROLLED

Applicators should evaluate soil and weed conditions carefully to assure that they choose the lowest effective label rate.

When applied as directed this product will CONTROL the following weeds:

ANNUAL GRASSES

Barnyardgrass *Echinochloa crus-galli*	Panicum, browntop *Panicum fasiculatum*
Crabgrass *Digitaria ischaemum* *Digitaria sanguinalis*	Panicum, fall *Panicum dichotomiflorium*
Foxtail, giant *Setaria faberi*	Sandbur, Grassbur* *Cenchrus incertus*
Foxtail, green *Setaria viridis*	Signalgrass, broadleaf *Brachiaria spp.*
Foxtail, yellow *Setaria lutescens*	Sprangletop, red *Leptochloa filiformis*
Goosegrass *Eleusine indica*	Witchgrass *Panicum capillare*

*Apply 4 to 5 quarts of this product per acre. Preemergence surface applications should be applied to a freshly prepared seedbed. Rotary hoe or initiate overhead irrigation (.033 to 0.55 inch) within 24 hours of application for best control.

ANNUAL BROADLEAVES

Beggarweed, Florida *Desmodium tortuosum*	Nightshade, black *Solanum nigrum*
Carpetweed *Mollugo verticillata*	Nightshade, hairy *Solanum sarrachoides*
Cocklebur, common* *Xanthium strumarium*	Pigweed, Carelessweed *Amaranthus spp.*
Ground cherry, annual *Physalis spp.*	Purslane *Portulaca oleracea*
Jimsonweed *Datura stramonium*	Pusley, Florida *Richardia scabra*
Kochia *Kochia scoparia*	Ragweed, common *Ambrosia artemisifolia*
Lambsquarters *Chenopodium album*	Sida, prickly; Teaweed *Sida spinosa*
Marestail, Horseweed *Conyza canadensis*	Smartweed *Polygonum pensylvanicum*
Morningglory, annual* *Ipomoea purpurea*	*Polygonium persicaria*
Mustard *Brassica spp.*	Velvetleaf, Buttonweed* *Abutilon theophrasti*

Use a minimum of 4 to 5 quarts of this product to control these weeds. Use the higher rate in areas of heavy weed infestation, coarse-textured soils or medium and fine-textured soils with organic matter content greater than 3 percent.

SEDGE

Nutsedge, yellow*
Cyperus esculentus

*Use 5 quarts per acre of this product on medium and fine textured soils and preplant incorporate only for control of this weed. When applied premergence to coarse-textured soils with less than 3 percent organic matter at rates of 4 to 5 quarts per acre, this product will provide reduced competition from this weed.

When applied preplant incorporated or immediately after planting and within 5 days of last tillage, this product at a rate of 4 to 5 quarts per acre on a broadcast basis will reduce competition from the following HARD-TO-CONTROL weeds:

ANNUAL GRASSES	
Johnsongrass, seedling *Sorghum halepense*	Shattercane, wild cane *Sorghum bicolor*
Rice, red *Oryza sativa*	Sorghum, volunteer *Sorghum bicolor*
ANNUAL BROADLEAF	
Sicklepod *Cassia obtusifolia*	

APPLICATION SYSTEMS

Ground Application—Apply as a broadcast treatment in 10 or more gallons of water or sprayable fluid fertilizer per acre. Do not apply during periods of gusty winds or winds in excess of 15 MPH.

Band Treatment—Apply a broadcast equivalent rate and volume per acre. To determine these:

$$\frac{\text{Band with in inches}}{\text{Row width in inches}} \times \text{Broadcast RATE per acre} = \text{Band RATE per acre}$$

$$\frac{\text{Band width in inches}}{\text{row width in inches}} \times \text{Broadcast VOLUME per acre} = \text{Band VOLUME per acre}$$

Mount spray equipment on the planter with nozzles behind press wheel and apply this product at planting time.

Aerial—Fixed wing; helicopter—Apply the recommended rates of this product or approved tank mixtures in 3 to 10 gallons of water per acre. AVOID DRIFT—DO NOT APPLY DURING INVERSION CONDITIONS, WHEN WINDS ARE GUSTY, OR UNDER ANY OTHER CONDITION WHICH WILL ALLOW DRIFT. Coarse sprays are less likely to drift; therefore, do not use nozzles or nozzle configurations which dispense spray or fine spray droplets. Do not angle nozzles forward into the airstream and do not increase spray volume by increasing nozzle pressure. Ensure uniform application. To avoid streaked, uneven or overlapped application, use mechanical marking systems only; HUMAN FLAGGERS PROHIBITED!

CHEMIGATION

Apply this product and labeled tank mixtures only through center pivot irrigation systems. Do not apply this product through any other type of irrigation system.

Crop injury, lack of effectiveness, or illegal pesticide residues in the crop can result from nonuniform distribution of treated water. Use only in systems that apply uniformly.

If you have questions about calibration, you should contact State Extension Service specialists, equipment manufactures or other experts.

Do not chemigate through systems connected to a public water system.

A person knowledgeable of the chemigation system and responsible for its operation, or under the supervision of the responsible person, shall shut the system down and make necessary adjustments should the need arise. Do not apply when system connections or fittings leak, when nozzles do not provide uniform distribution or when the line containing the product must be dismantled and drained.

The system must contain a functional check valve, vacuum relief valve, and low pressure drain appropriately located on the irrigation pipeline to prevent water source contamination from back flow.

The pesticide injection pipeline must contain a functional, automatic, quick-closing check valve to prevent the flow of fluid back toward the injection pump. The pesticide injection pipeline must also contain a functional, normally closed, solenoid operated valve located on the intake side of the injection pump and connected to the system interlock to prevent fluid from being withdrawn from the supply tank when the irrigation system is either automatically or manually shut down.

The system must contain functional interlocking controls to automatically shut off the pesticide injection pump when the water pump motor stops.

The irrigation line or water pump must include a functional pressure switch which will stop the water pump motor when the water pressure decreases to the point where pesticide distribution is adversely affected.

Systems must use a metering pump, such as a positive displacement injection pump (e.g., diaphragm pump) effectively designed and constructed of materials that are compatible with pesticides and capable of being fitted with a systems interlock. Pumps, injection equipment, agitation equipment, hoses and connections between supply tank and the point of injection must be constructed of materials which are resistant to Lasso.

Do not apply when wind speed favors drift beyond the area intended for treatment.

Meter this product into the center pivot irrigation system after planting and before crop emergence. Herbicide application should be made in 1/3 to 3/4 inch of water per acre. Do not apply in more than 3/4 inch of water per acre under any conditions or reduced performance may occur. On very sandy soils (more than 60 percent sand and less than 1 per cent organic matter), apply this product in a maximum of 1/2 inch water per acre. Good agitation must be maintained during the entire application period. Flush the system with water when application is complete. Refer to the "MIXING, SPRAYING, AND HANDLING INSTRUCTIONS" section of this label for mixing procedures.

NOTE: Applications can be made by injection through center pivot irrigation equipment only when specified for the crop to be treated in the "DIRECTIONS FOR USE" section of this label. Do not apply this product or any labeled tank mixture through center pivot irrigation unless the treatment is specifically recommended in this label for the crop to be treated.

CULTIVATION INFORMATION

If annual grass or broadleaf weed seedlings emerge, rotary hoe immediately to improve weed control.

If cultivaiton is necessary because of soil crusting or compaction, tillage should be shallow to minimize dilution of this herbicide.

Except for the above, as long as the grasses and weeds are controlled, delay cultivation as long as possible.

If band application is used, avoid throwing soil during cultivation onto the band before weeds emerge in the band.

CORN

Surface Application—Apply this product in 10 or more gallons of water per acre any time from immediately after planting until weeds or grasses reach the 2-leaf stage and the crop is no more than 5 inches tall. Some leaf burn to the crop may occasionally occur, but this should not affect subsequent crop growth or yield. This production can also be applied with sprayable fluid fertilizer in 10 or more gallons per acre. Applications with sprayable fluid fertilizer must be made prior to crop emergence. Performance may be improved when application is made prior to weed emergence. Applications made beyond the 2-leaf stage of weed growth will not provide satisfactory weed control. On sweet corn, use this product before crop or weeds emerge. See the following table for recommended broadcast rates per acre for surface applications.

Preplant Incorporation—Broadcast spray this product in 10 or more gallons of water or sprayable fluid fertilizer and incorporate thoroughly into the top 2 inches of soil within 7 days prior to planting. When using one-pass incorporation (Surface Blend™) the equipment used must provide uniform and shallow blending to avoid streaking the treatment. To ensure shallow incorporation of 2 inches or less, pull-type tillage equipment should be set to work the soil NO DEEPER THAN 4 INCHES.

This product is diluted when incorporated into the soil; therefore, SHALLOW AND UNIFORM incorporation of the herbicide treatment is essential.

™Surface Blend is a trademark of Monsanto Company.

See the following table for the recommended broadcast rates per acre for preplant incorporated applications:

Surface Application of Preplant Incorporated Application Rates

Soil	Broadcast Rate Per Acre (Quarts)	
Textural Group*	Less Than 3% organic matter	3% or more organic matter
Coarse**	2.5	3
Medium	3	3.75
Fine	3.75	3.75 to 4.5

Use up to 5 quarts of this product per acre to control cocklebur, annual morningglory, velvetleaf or heavy grass infestations or to reduce competition from the hard-to-control weeds listed on this label.

NOTE: To CONTROL yellow nutsedge, apply 5 quarts of this product per acre on medium and fine textured soils and preplant incorporate only.

*Refer to the "SOIL TEXTURE" section of this label to determine the corresponding soil textural group for the soil to be treated.

**When applied through center pivot irrigation or under sprinkler irrigation systems on coarse soils, use a minimum of 3 quarts of this product per acre. DO NOT exceed 1/3 to 3/4 inch of water per acre with the first irrigation as reduced effectiveness may result.

BLADEX plus THIS PRODUCT (TANK MIXTURE)

Apply this tank mixture in water or sprayable fluid fertilizer solutions for control of annual grasses and broadleaf weeds listed in the "WEEDS CONTROLLED" section of this label. Refer to the rate table for this product and to the Bladex rate table below for either surface application or preplant incorporation.

Surface Application or Preplant Incorporated Application Rates

Soil	Broadcast Rate Per Acre (Quarts) Bladex	
Textural Group	Less Than 3% organic matter	3% or more organic matter
Coarse*	1	1 to 1.25

Soil Textural Group	Broadcast Rate Per Acre (Quarts) Bladex	
	Less Than 3% organic matter	3% or more organic matter
Medium	1 to 1.25	1.5 to 2
Fine	1.25 to 1.5	1.5 to 2

*Do not use this mixture on sand or loamy sand soils with less than 2 percent organic matter.

EARLY PREPLANT SYSTEMS

For use in no-till and other conservation tillage systems in DELAWARE, ILLINOIS, INDIANA, IOWA, KANSAS, KENTUCKY, MARYLAND, MINNESOTA, MISSOURI, NEBRASKA, NORTH CAROLINA, OHIO, SOUTH DAKOTA, VIRGINIA, AND WISCONSIN.

This product, when applied in a single application (alone or in a tank-mix combination with Bladex or Princep), split application or as a sequential application to Bladex or Princep in early preplant programs, will provide preemergence control or reduced competition of listed annual grasses and broadleaf weeds.

If emerged weeds are present at the time of treatment, Gramoxone Super or 2,4-D should be added to the product according to the directions for use on their respective product labels. If unsatisfactory weed control occurs (due to excessively dry or excessively wet conditions) following the earlier application, a postemergence application of an appropriate labeled grass and/or broadleaf weed herbicide may be used. If a postemergence treatment includes the herbicide used early preplant, do not exceed the labeled rate for corn on a given soil texture. Observe all precautions and limitations on the labels for Bladex, Princep, Gramoxone Super, 2,4,-D and other postemergence herbicides before use of these products.

RECOMMENDED RATE AND TIMING OF APPLICATION

Single application of Bullet herbicide—Single application treatments should be made less than 30 days before planting but prior to weed emergence. Single application treatments should be restricted to areas north of Interstate 70 when west of the Mississippi River and north of Interstate 64 when east of the Mississippi River, where cool spring weather minimizes or slows biodegration of herbicides.

See the following rate table for recommended broadcast rates per acre for single application.

Soil Textural Group	Broadcast Rate Per Acre (Quarts)** BULLET®
Coarse*	3.5 to 4
Medium	3.75 to 4.75
Fine	4.75 to 6

The single application treatments of Bullet herbicide may also be tank-mixed with 1 to 1.25 quarts per acre of Princep or 1.5 to 2 quarts per acre of Bladex to provide improved control of fall panicum and crabgrass.

*Single applications on coarse soils should not be made more than 2 weeks prior to planting.

**Use the lower rate where light weed infestation is expected and the higher rate where heavy weed infestation is expected.

Split application of Bullet herbicide—Apply 60 percent of the recommended rate as a split treatment prior to weed emergence and no more than 45 days prior to planting, and the remaining 40 percent at or immediately following planting.

FOR CONTROL OF FALL PANICUM, CRABGRASS OR BROADLEAF SIGNALGRASS, USE THE SEQUENTIAL APPLICATION OF BULLET HERBICIDE FOLLOWING BLADEX OR PRINCEP.

See the following table for recommended broadcast rates per acre for split application.

Soil Textural Group	Broadcast Rate Per Acre (Quarts)** BULLET®
Coarse*	3.5 to 4.5
Medium	3.75 to 4.75
Fine	4.75 to 6

*Use the lower rate where light weed infestation is expected and the higher rate where heavy weed infestation is expected.

Sequential applicaiton of Bullet herbicide—The sequential program of Bullet herbicide following Bladex or Princep should be utilized for the control of fall panicum, crabgrass or broadleaf signalgrass. Apply 1.5 to 2 quarts per acre of Bladex or 1 to 1.25 quarts per acre of Princep prior to weed emergence and no more than 45 days prior to planting. At or immediately following planting, apply the recommended rate of Bullet herbicide.

For recommended rates of this product following applications of Bladex or Princep, see the following sequential application table.

Soil Textural Group	Broadcast Rate Per Acre (Quarts)** BULLET®
Coarse*	3 to 4
Medium	3.5 to 4.5
Fine	3.75 to 4.75

*Use the lower rate where light weed infestation is expected and the higher rate where heavy weed infestation is expected.

When using Bladex 80W or Bladex 90 DF, use equivalent rates. One quart of Bladex 4L equals 1.25 pounds of Bladex 80W and 1.1 pounds of Bladex 90 DF. When using Princep 80W or Princep 90, use equivalent rates. One quart of Princep 4L equals 1.25 pounds of Princep 80W and 1.1 pounds of Princep 90.

DO NOT use Bladex on coarse-textured soils or soils with less than 1 percent organic matter.

NOTE: Soil treated with Princep should not be planted to any crop except corn for one year following treatment, as crop injury may occur. After harvest of the treated crop, plow and thoroughly till the soil in the fall or spring to minimize possible injury to spring seeded rotational crops.

MINIMUM TILLAGE SYSTEMS

TANK MIXTURES OF THIS PRODUCT plus GRAMOXONE SUPER

When applied as directed under the conditions described, these tank mixtures control many emerged annual weeds, suppress many emerged perennial weeds and give preemergence control of many annual weeds when corn will be planted directly into a cover crop, established sod or in previous crop residues. These tank mixtures will not control regrowth from perennial weeds.

Refer to specific product labels for crop rotation restrictions and cautionary statements of all products used in the tank mixture.

For mixing instructions, see the "MIXING, SPRAYING AND HANDLING INSTRUCTIONS" section of this label.

Apply the tank mixture of this product with Gramoxone Super in 20 to 60 gallons of water per acre immediately before, during or after planting but BEFORE CROP EMERGENCE. As previous crop residue or vegetation increases, spray volume should be increased within the recommended range to insure complete coverage.

For recommended rates of this product with tank mixtures with Gramoxone Super see the following minimum tillage rate table.

Soil Textural Group	Broadcast Rate Per Acre (Quarts)** BULLET®
Coarse	3
Medium	3.75
Fine	3.75 to 4.5

*Use up to 5 quarts of this product per acre to control cocklebur, annual morningglory, velvetleaf or heavy grass infestations, or to reduce competition from the hard-to-control weeds listed on this label.

*Refer to the "SOIL TEXTURE" section of this label to determine the corresponding textural group for the soil to be treated.

ATTENTION

AVOID DRIFT—EXTREME CARE MUST BE USED WHEN APPLYING THESE TANK MIXTURES TO PREVENT INJURY TO DESIRABLE PLANTS AND CROPS. Do not allow spray mist to drift since even minute quantities of spray can cause severe damage or destruction to nearby crops, plants or other areas on which treatment is not intended. Do not apply when winds are gusty or in excess of 5 miles per hour or when other conditions, including lesser wind velocities will allow drift to occur. When spraying avoid combinations of pressure and nozzle type that will result in fine particles (mist) which are more likely to drift.

For weeds controlled preemergence, see the "WEEDS CONTROLLED" section of this label. For control of emerged weeds, see the "WEEDS CONTROLLED" sections of the labels for Gramoxone Super. REFER TO THE GRAMOXONE SUPER LABEL FOR PRECAUTIONARY STATEMENTS.

GRAMOXONE SUPER plus THIS PRODUCT (TANK MIXTURE)

When used as directed, Gramoxone Super in a tank mixture with this product controls many emerged annual weeds and suppresses many emerged perennial weeds. Apply 1.5 to 2.5 pints of Gramoxone Super (0.25 to 0.5 pound active) with the amount of this product recommended in the "MINIMUM TILLAGE SYSTEMS" section of this label per acre immediately before, during or after planting but BEFORE CROP EMERGENCE. Use the 2.5 pint rate if weeds are 4 to 6 inches tall. This mixture may not control weeds taller than 6 inches. Add a nonionic spreader surfactant (approved for use on crops) containing at least 75 percent surface active agent at 8 fluid ounces per 100 gallons of diluted spray solution.

LASSO® and/or ATRAZINE plus THIS PRODUCT (TANK MIXTURE)

For hard-to-control weeds, additional amounts of Lasso herbicide and/or atrazine may be added to the recommended treatment rates for this product to provide improved control. For more consistent control of common cocklebur, annual morningglory or velvetleaf, additional atrazine may be applied so the total atrazine rate is at least 1.5 quarts per acre on medium soil with less than 3 percent organic matter, and 1.5 to 2 quarts on medium and fine soils with 3 percent or greater organic matter content. The following table shows the amounts of Lasso herbicide and/or atrazine that can be added to specific treatment rates of this product.

Recommended Rate (quarts)	Additional Product (quarts)	
BULLET®	LASSO®	ATRAZINE (maximum)
2.5	1.25 to 2.25	1.1
3	1 to 2	0.9
3.5	0.75 to 1.75	0.7
4	0.5 to 1.5	0.5

®Lasso is a registered trademark of Monsanto Company.

In case of an emergency involving this product Call Collect, day or night (314) 694-4000.

This sample label is current as of January 1, 1989. The product descriptions and recommendations provided in this sample label are for background information only. Always refer to the label on the product container before using Monsanto or any other agrichemical product.

CANNON®
HERBICIDE

RESTRICTED USE PESTICIDE

due to oncogenicity. For retail sale to and use only by Certified Applicators or persons under their direct supervision and only for those uses covered by the Certified Applicator's certification.

The use of this product may be hazardous to your health. This product contains alachlor which has been determined to cause tumors in laboratory animals.

In case of an emergency involving this product or for user safety information on this product, call collect, day or night (314) 694-4000.

Emulsifiable herbicide for weed control in soybeans and dry edible beans (all types).

ACTIVE INGREDIENTS:*
Alachlor 29.5%
Trifluralin 5.9%
INERT INGREDIENTS: 64.6%
TOTAL 100.0%

Contains petroleum distillates.

* Contains 300 grams/litre or 2.5 pounds/gallon of 2-chloro-2',6'-diethyl-N-(methoxymethyl)acetanilide and 60 grams/litre or 0.5 pound/gallon of α, α, α-trifluoro-2,6-dinitro-N,N-dipropyl-p-toluidine.

U.S. Pat. No. 3,547,620 EPA Reg. No. 524-412

Complete Directions for Use

®CANNON is a registered trademark of Monsanto Company.

1988-2 843.29-000.07/53

Read the entire label before using this product.

Use only according to label instructions.

Read "LIMIT OF WARRANTY AND LIABILITY" before buying or using. If terms are not acceptable, return at once unopened.

REFORMULATION IS PROHIBITED.

LIMIT OF WARRANTY AND LIABILITY

This Company warrants that this product conforms to the chemical description on the label and is reasonably fit for the purposes set forth in the Complete Directions for Use label booklet ("Directions") when used in accordance with those Directions under the conditions described therein. NO OTHER EXPRESS WARRANTY OR IMPLIED WARRANTY OF FITNESS FOR PARTICULAR PURPOSE OR MERCHANTABILITY IS MADE. This warranty is also subject to the conditions and limitations stated herein.

Buyer and all users shall promptly notify this Company of any claims whether based on contract, negligence, strict liability, other tort or otherwise.

Buyer and all users are responsible for all loss or damage from or handling which results from conditions beyond the control of this Company, including, but not limited to, incompatibility with products other than those set forth in the Directions, unusual weather, weather conditions which are outside the range considered normal at the application site and for the time period when the product is applied, as well as weather conditions which are outside the application ranges set forth in the Directions, application in any manner not explicitly set forth in the Directions, moisture conditions outside the moisture range specified in the Directions, or the presence of products other than those set forth in the Directions in or on the soil, crop or treated vegetation.

THE EXCLUSIVE REMEDY OF THE USER OR BUYER, AND THE LIMIT OF THE LIABILITY OF THIS COMPANY OR ANY OTHER SELLER FOR ANY AND ALL LOSSES, INJURIES OR DAMAGES RESULTING FROM THE USE OR HANDLING OF THIS PRODUCT (INCLUDING CLAIMS BASED IN CONTRACT, NEGLIGENCE, STRICT LIABILITY, OR OTHER TORT OR OTHERWISE) SHALL BE THE PURCHASE PRICE PAID BY THE USER OR BUYER FOR THE QUANTITY OF THIS PRODUCT INVOLVED, OR, AT THE ELECTION OF THIS COMPANY OR ANY OTHER SELLER, THE REPLACEMENT OF SUCH QUANTITY, OR, IF NOT ACQUIRED BY PURCHASE, REPLACEMENT OF SUCH QUANTITY. IN NO EVENT SHALL THIS COMPANY OR ANY OTHER SELLER BE LIABLE FOR ANY INCIDENTAL, CONSEQUENTIAL OR SPECIAL DAMAGES.

Buyer and all users are deemed to have accepted the terms of this LIMIT OF WARRANTY AND LIABILITY which may not be varied by any verbal or written agreement.

PRECAUTIONARY STATEMENTS

HAZARDS TO HUMANS AND DOMESTIC ANIMALS

KEEP OUT OF REACH OF CHILDREN

WARNING!

CAUSES SUBSTANTIAL BUT TEMPORARY EYE INJURY.

HARMFUL IF SWALLOWED.

MAY CAUSE ALLERGIC SKIN REACTION.

Do not get in eyes, on skin or on clothing.

Wear goggles or face shield, rubber gloves, long trousers, long-sleeved shirt or jacket of tightly woven material along with boots high enough to cover ankles when transferring and mixing, and when adjusting, repairing or cleaning equipment. Wear rubber boots when pouring from open containers. Clothing which comes in contact with this product must be washed before reuse. Clothing or other materials which become drenched with the concentrated pesticide must be disposed of in a sanitary landfill, by incineration or, if allowed by state and local authorities, by burning. If burned, stay out of smoke.

Wash thoroughly with soap and water after handling.

FIRST AID: IF IN EYES, immediately flush with plenty of water for at least 15 minutes. Call a physician.

IF ON SKIN, immediately flush with plenty of water while removing contaminated clothing. As soon as soap is available, wash thoroughly with soap and water. Wash clothing before reuse. Sensitized persons should avoid further contact and reuse of contaminated clothing.

IF SWALLOWED, do **NOT** induce vomiting. Give large quantities of water. Call a physician. NEVER GIVE ANYTHING BY MOUTH TO AN UNCONSCIOUS PERSON.

In case of emergency involving this product,
Call Collect, day or night, (314) 694-4000.

ENVIRONMENTAL HAZARDS

This product contains alachlor which has been identified in limited ground water sampling and there is the possibility that it can leach through the soil to ground water, especially where soils are coarse and ground water is near the surface. Following application, and during rainfall events that cause runoff, this chemical may reach surface water bodies including streams, rivers, and reservoirs.

Care must be taken when using this product to prevent back-siphoning into wells, spills or improper disposal of excess pesticide, spray mixtures or rinsates.

Check valves or antisiphoning devices must be used on all mixing equipment. Disposal of excess pesticide, spray mixtures, or rinsate should be according to label use instructions or according to the state pesticide or environmental control agency or the hazardous waste representative at the nearest EPA regional office.

Do not apply directly to water or wetlands. Do not contaminate water when disposing of equipment washwaters.

DIRECTIONS FOR USE

It is a violation of Federal law to use this product in any manner inconsistent with its labeling.

STORAGE AND DISPOSAL

Do not contaminate water, foodstuffs, seed or feed by storage or disposal.

STORAGE:

STORE ABOVE 32° F TO KEEP PRODUCT IN SOLUTION.

Below 32° F crystals may form and settle to the bottom. If crystals form, place in a warm room and roll and/or shake the container frequently to redissolve before using. For bulk containers, see the container label for alternate storage information.

DISPOSAL: Wastes of this pesticide may cause eye irritation and may be dangerous. Improper disposal of excess pesticide, spray mixtures, or rinsate is a violation of Federal law. If these wastes cannot be disposed of according to label use instructions, contact your State Pesticide or Environmental Control Agency, or the Hazardous Waste representative at the nearest EPA Regional Office for guidance.

Emptied container retains vapor and product residue. Observe all labeled safeguards until container is cleaned, reconditioned or destroyed. DO NOT CUT OR WELD ON OR NEAR THIS CONTAINER.

(See the individual container label for disposal information.)

PRECAUTIONARY STATEMENTS

CANNON® herbicide is recommended for the control and to reduce competition of annual grasses and broadleaf weeds listed in the "WEEDS CONTROLLED" section of this label. Apply this product and shallowly incorporate prior to planting to blend the herbicide into the upper 1 to 2 inches of soil.

Performance is often improved when sufficient rainfall or irrigation occurs to moisten the soil within the weed germination zone. If adequate moisture is not received prior to weed emergence, then shallow cultivation is recommended. Refer to the "APPLICATION AND TIMING METHODS" section of this label for details.

Read and carefully observe the precautionary statements and all other information appearing on the labels of all products used in mixtures and sequential treatments.

Do not flood irrigate following application of this product or tank mixtures.

Do not apply when conditions favor drift.

Use of this product in any manner not consistent with this label may result in injury to persons, animals or crops, or other unintended consequences. Keep container closed to prevent spills and contamintion.

Drain and thoroughly flush sprayer with clean water after use and also prior to utilizing spray equipment for treating crops other than soybeans or dry edible beans.

Do not graze treated area or feed treated soybean forage hay or straw to livestock following application.

®CANNON is a registered trademark of Monsanto Company.

SOIL TEXTURE

Applicators should evaluate soil and weed conditions carefully to assure that they choose the lowest effective label rate.

The recommended use rates of this product and the other herbicides labeled for use in tank mixtures with this product generally vary with soil texture. Unless soil texture is specifically named, rate tables throughout this label refer to only three soil textural groups: coarse, medium and fine. The following is a complete listing of soil textures included in each of these three soil textural groups:

Soil Textural Group	Soil Texture
COARSE:	sand, loamy sand, sandy loam
MEDIUM:	loam, silt loam, silt, sandy clay loam
FINE:	silty clay loam, clay loam, sandy clay, silty clay, clay

Refer to the above table to determine the corresponding soil textural group for the soil to be treated.

MIXING, SPRAYING AND HANDLING INSTRUCTIONS

NOTE

The use of a mechanical transfer system is required to be used by all mixer/loaders and/or applicators who treat 300 acres or more annually with pesticides containing alachor as an active ingredient.

Direct contact or exposure to this product or spray mixtures of this product should be minimized. The following instructions for transfer, mixing, cleaning or repairing equipment should be followed in order to minimize this exposure. Review the protective clothing requirements as listed in the "PRECAUTIONARY STATEMENTS" section of this label (front panel) and do not use this product until you are wearing the necessary protective clothing.

2.5 Gallon Containers

Open pouring from these containers can result in exposure from splashing or spilling. Special care in lifting and pouring is strongly recommended.

Bulk Containers

This product should be transferred from these containers to the mix or spray tank using pumps or transfer probes. The probe or pump should not be removed from the container or disconnected until the container is emptied and rinsed. Use the pump or probe system to rinse the empty container and transfer the rinsate directly to the mix or spray tank.

Equipment Cleaning and Repair

Cleaning and repair of transfer systems or application equipment is a source of exposure to this product. Care should be taken to minimize exposure during cleaning and repair of transfer systems and application equipment. Whenever possible, these systems or this equipment should be rinsed before being cleaned or repaired. When repairs must be made during transfer or application, the equipment must be shut down and special care should be taken to avoid contact with the pesticide.

Always predetermine the compatibility of this product or labeled tank mixtures of this product with water carrier or sprayable fluid fertilizer carrier by mixing small proportional quantities in advance. See the "STANDARD SPRAYABLE FLUID FERTILIZER COMPATIBILITY TEST" section in this label to determine the compatibility of this product and the labeled tank mixtures recommended for use with sprayable fluid fertilizer carrier. For soybeans, only CANNON and tank mixtures of CANNON herbicide may be applied in sprayable fluid fertilizers.

Mix this product or labeled tank mixtures of this product with the appropriate carrier as follows:

1. Place a 20 to 35 mesh screen or wetting basket over filling port.
2. Through the screen, fill the sprayer tank one-half full with the appropriate carrier.
3. If a compatibility agent is necessary to improve mixing, or to prevent the formation of undesirable and unsprayable gels or precipitates, while agitating add it to the carrier already in the tank. Use only compatibility agents cleared by FDA for this use. Read and follow all directions for use, cautionary statements and all other information appearing on the selected compatibility agent label. Check for adequate agitation.
4. If a wettable powder or dry flowable formulation is used, make a slurry with water and add it slowly through the screen into the tank. Continue agitation.
5. If a flowable formulation is used, add slowly through the screen into the tank. Mixing and compatibility may be improved when flowable is premixed one part flowable with one part water and added to the tank in diluted form.
6. Add CANNON slowly through the screen into the tank. Mixing and compatibility may be improved when this product is prediluted with two parts of water and added to the tank in diluted form.
7. Complete filling the sprayer tank with carrier. Remove hose from tank immediately after filling to avoid siphoning back into the carrier source.

Maintain good agitation at all times until the contents of the tank are sprayed. NOTE: If spray mixture is allowed to settle at any time, thorough agitation is required to resuspend the mixture before spraying is resumed.

Keep by-pass line on or near bottom of tank to minimize foaming. Screen size in nozzle or line strainers should be no finer than 50 mesh. Carefully select the correct nozzle to avoid spraying a fine mist. Check for even distribution of spray droplets. To reduce loss of chemical due to drift of a fine mist, apply at nozzle pressures below 40 psi.

STANDARD SPRAYABLE FLUID FERTILIZER COMPATIBILITY TEST

Herbicides may not always mix evenly throughout a sprayable fluid fertilizer or the components may separate too quickly to make their combined use of practical value. This may be due to certain characteristics of the different fluid fertilizers. A simple test using small quantities of the components is suggested to determine compatibility potential.

The test follows:

A. Materials Required For A Compatibility Test

1. Two one-quart jars with lid or stopper (marked "with" and "without").
2. TEAspoon (for a more exacting test, a five to ten milliliter (ml) pipette or graduated cylinder is desirable).
3. Sprayable fluid fertilizer to be tested.
4. The herbicides to be mixed.
5. A compatibility agent (the purpose of the adjuvant is to help keep the fertilizer and crop protection chemical in suspension, if this assistance is needed).

B. Procedure

1. Add one pint of the sprayable fluid fertilizer that will be used or other herbicide carrier to each jar marked "with" and "without."

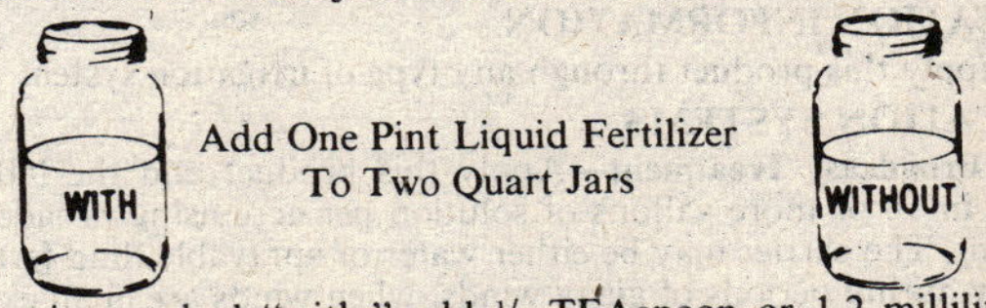

2. To the jar marked "with," add 1/4 TEAspoon or 1.2 milliliters of a suitable compatibility agent, shake gently for five to ten seconds to mix (1/4 TEAspoon in one pint is the equivalent of two pints per 100 gallons of liquid fertilizer).

To Jar Marked "With" Add Compatibility Agent And Shake To Mix

3. To each jar add the appropriate amount of herbicide(s). If more than one is used, add them separately with the wettable powders or dry flowables added first, flowables second and liquid last. Shake gently five to ten seconds after each addition.

Add Herbicide(s) To Both Jars And Shake To Mix

Herbicide	Amount To Be Added Per Pint Of Sprayable Fluid Fertilizer (Assuming Volume is 25 Gal/A) Rate/A		Level TEAspoons
Wettable	1 lb	=	1.5
Powders	2 lb	=	3
or	3 lb	=	4.5
Dry Flowables	4 lb	=	6
	5 lb	=	7.5

Herbicide	Rate/A		Level TEAspoons		ML
Emulsifiable	1 pint	=	0.5	or	2.4
Concentrates or	1 quart	=	1	or	4.7
Flowables or	2 quarts	=	2	or	9.5
Liquids or	3 quarts	=	3	or	14.2
Solutions	1 gallon	=	4	or	19.0
	5 quarts	=	5	or	23.8

The compatibility test is designed for 25 gallons of spray per acre with the maximum labeled rate of herbicide. For changes in spray volume or herbicide rate, make appropriate changes in the ingredients of the test.

Regardless of spray volume, the amount of compatibility agent should be equal to two or three pints (two pints = 1/4 TEAspoon or 1.2 milliliters, three pints = 3/8 TEAspoon or 1.8 milliliters per pint of sprayable fluid fertilizer) per 100 gallons of liquid fertilizer.

C. Observations and Decisions

1. If the herbicide(s) and the sprayable fluid fertilizer are compatible.
2. If a compatibility agent is necessary.

Five minutes after the final addition and mixing, observe both jars for the formation of large flakes, sludge, gels or other precipitates. Observe if the herbicide(s) cannot be physically mixed with the liquid fertilizer but remains as small oily particles in the solution.

If incompatibility in any form described above occurs in the jar "with" the compatibility agent added, the liquid fertilizer and the herbicide(s) should not be used together in the same spray tank.

If compatibility as described above occurs in the jar "without" the adjuvant but not in the jar "with" adjuvant, the use of a compatibility adjuvant is recommended.

Both jars should be allowed to stand and be observed periodically for one-half hour. If the separate layers of liquid fertilizer and additives can be resuspended by shaking, commercial application is possible. An emulsifiable concentrate normally will go to the top after standing, wettable powders will either settle to the bottom of the tank or jar, or float to the top, depending upon the density of the fertilizers.

If the herbicide(s) is compatible with fluid fertilizer in the foregoing test without using a compatibility agent, fluid fertilizer may be used for premixing. If it is not compatible without the compatibility agent, the herbicide(s) should be premixed with water before adding to the spray tank.

APPLICATION INFORMATION

Do not apply this product through any type of irrigation system.

APPLICATION SYSTEMS

Ground Broadcast Treatment—Apply this product and the labeled tank mixtures in 10 or more gallons of solution per acre using broadcast boom equipment. The carrier may be either water or sprayable fluid fertilizer. Do not apply during periods of gusty winds, when winds are in an excess of 15 mph or when other conditions favoring drift exist.

Aerial—Fixed wing helicopter—Apply the recommended rates of this product or approved tank mixtures in 3 to 10 gallons of water per acre. AVOID DRIFT—DO NOT APPLY DURING INVERSION CONDITIONS, WHEN WINDS ARE GUSTY, OR UNDER ANY OTHER CONDITION WHICH WILL ALLOW DRIFT. Coarse sprays are less likely to drift; therefore, do not use nozzles or nozzle configurations which dispense spray or fine spray droplets. Do not angle nozzles forward into the airstream and do not increase spray volume by increasing nozzle pressure. Ensure uniform application. To avoid streaked, uneven or overlapped application, use mechanical marking systems only. HUMAN FLAGGERS PROHIBITED!

Application with Dry Bulk Fertilizer—The herbicide/fertilizer impregnation process must be completed only at commercial fertilizer dealerships equipped for this procedure.

Dry bulk fertilizer may be impregnated with this product or the tank mixtures of this product plus Lexone™ or Sencor™. CANNON herbicide and these tank mixtures must be applied with 200 to 450 pounds of dry bulk fertilizer per acre and shallowly incorporated within 7 days prior to planting. The correct amount of herbicide must be applied per acre in the blend recommended in this label for the crop selected and weed species to be controlled.

The following table provides a reference to determine the amount of LIQUID herbicide to be mixed per ton of dry bulk fertilizer for a range of herbicide recommendations and fertilizer rates per acre.

Fertilizer Rate (lb/acre)	Acres Covered (per ton)	Recommended Quarts Liquid Herbicide/Acre 3	3.5 (quarts of herbicide/ton dry bulk fertilizer)	4
200	10	30	35	40
250	8	24	28	32
300	6.7	20	23.3	26.6
350	5.7	17.1	20	22.8
400	5	15	17.5	20
450	4.5	13.3	15.6	17.8

To determine the amount of herbicide needed for rates not included in the preceding table, use the following formula:

$$\frac{\text{Recommended herbicide rate (lb or qt)/acre} \times 2000}{\text{pounds fertilizer/acre}} = \text{pounds or quarts of herbicide per ton of dry bulk fertilizer}$$

Mix and blend the dry fertilizer and herbicide mixture in a closed rotary drum-type mixer, allowing sufficient time to ensure uniform coverage. Use at least one ton of dry fertilizer per mixing operation. Inject the herbicide into the drum over a minimum of a 2-minute period and allow at least 2 additional minutes mixing time to ensure uniformity. The nozzle used to spray the herbicide treatment must be placed inside the mixer to provide uniform spray coverage of the tumbling fertilizer.

If the dry fertilizer used has inadequate adsorptive capacity, use a highly adsorptive material such as Agsorb™, NP-79™, or Microcel™ E to provide a free-flowing mixture. The adsorptive material and dry fertilizer should be thoroughly mixed prior to adding liquid herbicide. Generally the drying material weight is no more than 2 to 4 percent of the total weight of the dry fertilizer. The following table provides a list of dry fertilizers which may be impregnated with CANNON or tank mixtures of CANNON with other herbicides.

™ Agsorb is a trademark of the Oil Dri Corporation.
™ Lexone is a trademark of E. I. du Pont de Nemours and Company.
™ Microcel is a trademark of Manville Products Corporation.
™ NP-79 is a trademark of Eagle Pitcher Industries.
™ Sencor is a trademark of the parent company of Farbenfabriken Bayer Gmbh, Leverkusen.

Fertilizer	CANNON®	CANNON + Lexone Sencor
Ammonium sulfate (21-0-0)	Yes	Yes
Ammonium phosphate-sulfate (16-20-0)	Yes	Yes
Diammonium phosphate (18-46-0)	Yes	Yes
Potassium chloride (0-0-60)	Yes	Yes
Single super-phosphate (0-20-0)	Yes	No
Treble super-phosphate (0-46-0)	Yes	No
Urea (46-0-0)	Yes	Yes

NOTE: DO NOT impregnate this product or tank mixtures of this product with other herbicides on fertilizers containing ammonium nitrate, potassium nitrate or sodium nitrate.

Spread the dry fertilizer-herbicide mixture uniformly with an adequately calibrated applicator. Fertilizers impregnated with this product or tank mixtures of this product with other herbicides must be spread uniformly.

Non-uniform spreading of the dry fertilizer/herbicide mixture may result in unsatisfactory weed control or crop injury.

APPLICATION TIMING AND METHODS

Preplant Incorporation—This product and the labeled tank mixtures may be mixed into the soil using shallow incorporation equipment any time within 7 days prior to planting. Apply the recommended treatment rate to the soil surface as a broadcast application and shallowly incorporate into the upper 2 inches of soil within 24 hours after application. If weeds emerge after treatment, rotary hoe or shallowly cultivate immediately to improve performance.

1-Pass Incorporation (Surface Blend™): Shallowly incorporate (Surface Blend) the treatment into the upper 1 to 2 inches of the soil. Equipment should be operated at manufacturer's designated speed for incorporation to ensure adequate mixing and distribution of the herbicide treatment in the soil. Equipment design, including any drag attachments, must be adequate to avoid soil ridging which may result in streaked or reduced weed control. Soil conditions, including moisture content and crop residue levels, must be suitable to allow thorough and uniform mixing with the equipment used for 1-pass incorporation.

Contact Monsanto Company for additional information regarding recommended equipment and methods for Surface Blend application.

2-Pass Incorporation: When 2-pass incorporation is used, shallowly incorporate the herbicide treatment into the upper 1 to 2 inches of the soil with equipment set to work the soil NO DEEPER THAN 4 INCHES. The second pass must be made at an angle and no deeper than the first pass to ensure proper distribution of the herbicide treatment in the soil.

Cultivation Information—Delay cultivation after application for as long as possible unless weeds or grasses emerge. Shallowly cultivate or rotary hoe immediately if weeds or grasses emerge. If cultivation is necessary because of soil crusting or compaction, set equipment shallow and minimize lateral soil movement to avoid dilution or displacement of the herbicide treatment.

™ Surface Blend is a Trademark of the Monsanto Company.

WEEDS CONTROLLED

Applicators should evaluate soil and weed conditions carefully to assure that they choose the lowest effective label rate. When applied as directed under conditions described, annual grasses and broadleaf weeds are CONTROLLED by this product.

CANNON®

ANNUAL GRASSES	ANNUAL BROADLEAVES
Barnyardgrass *Echinochloa crus-galli*	**Carpetweed** *Mollugo verticilata*
Crabgrass *Digitaria ischaemum* *Digitaria sanguinalis*	**Galinsoga** *Galinsoga spp.*
Foxtail, giant *Setaria faberi*	**Kochia** *Kochia scoparia*
Foxtail: green, robust purple, robust white *Setaria viridis*	**Groundcherry, cutleaf** *Physalis angulata*
Foxtail, yellow *Setaria lutescens*	**Nightshade: black*, eastern black*** *Solanum nigrum* *Solanum ptycanthum*
Goosegrass *Eleusine indica*	**Nightshade, hairy*** *Solanum sarrachoides*
Panicum, fall *Panicum dichotomiflorum*	**Pigweed, Carelessweed** *Amaranthus spp.*
Signalgrass *Brachiaris platyphylla*	**Purslane** *Portulaca oleracea*
Witchgrass *Panicum capillare*	**Pusley, Florida** *Richardia scabra*

* Use a minimum of 4 quarts of this product per acre alone or in tank mixture for control of these weeds.

When applied preplant incorporated, 4 quarts of this product per acre applied alone or in a labeled tank mixture REDUCES COMPETITION from the following hard-to-control annual grasses and broadleaf weeds.

ANNUAL GRASSES	ANNUAL BROADLEAVES
Cupgrass, woolly *Eriochloa villosa*	**Beggarweed, Florida** *Desmodium tortuosum*
Johnsongrass, seedling *Sorghum halepense*	**Lambsquarters** *Chenopodium album*
Sandbur, Grassbur *Cenchrus incertus*	**Ragweed, common** *Ambrosia artemisiifolia*
	Sida, prickly; Teaweed *Sida spinosa*
	Smartweed *Polygonum pensylvanicum* *Polygonum persicaria*

When applied as directed under conditions described, tank mixtures of this product with products listed in the following table will control or reduce competition from the additional broadleaf weeds as indicated. Refer to "DIRECTIONS FOR USE" section of this label to determine the tank mixtures and rates recommended for use in each crop.

NOTE: C = Controlled **R = Reduced Competition**

	Canopy™	Command™	Preview™	Scepter™	Sencor/ Lexone
Beggarweed *Desmodium tortuosum*	C	•	R	R	•
Cocklebur *Xanthium strumarium*	C	•	C	C	R
Coffeeweed, hemp sesbania *Sesbania exaltata*	C	•	C	•	•
Copperleaf, hophornbeam *Acalypha spp.*	C	•	C	•	•
Groundcherry, annual *Physalis spp.*	C	•	C	C	C
Jimsonweed *Datura stramonium*	C	•	C	C	R
Lambsquarters *Chenopodium album*	C	C	C	C	C
Morningglory:					
Tall *Ipomea purpurea*	C	•	R	R	•
Pitted *Ipomea lacunosa*	C	•	R	C	•
Ivyleaf* *Ipomea hederacea*	C	•	R	R	•
Entire leaf Ipomea hederacea var. integriuscula	C	•	R	R	•
Smallflower *Jacquemontra tamnifolia*	C	•	•	C	•
Mustard *Brassica ssp.*	C	C	C	C	C
Ragweed, common *Ambrosia artemisiifolia*	C	C	C	C	C
Sicklepod *Cassia obtusifolia*	C	•	•	R	R
Sida, prickly; Teaweed *Sida spinosa*	C	C	C	C	•
Smartweed *Polygonum pensylvanicum* *Polygonum persicaria*	C	R	C	C	C
Velvetleaf, Buttonweed *Abutilon theophrasti*	C	C	C	C	R
Waterhemp *Amaranthus tuberculatus*	•	•	C	C	•

™ Canopy is a trademark of E. I. du Pont de Nemours and Company.
™ Command is a trademark of FMC Corporation.
™ Preview is a trademark of E. I. du Pont de Nemours and Company.
™ Scepter is a trademark of American Cyanamid Company.

SOYBEANS

CANNON®

Apply this product in water solution or sprayable fluid fertilizer at 4 quarts per acre. Refer to the "WEEDS CONTROLLED" section of this label to determine specific weed species.

NOTE: For control of annual grasses only, use 3 quarts per acre.

TANK MIXTURES

CANNON® plus COMMAND

Apply 4 quarts of this product per acre, plus 3/4 pint of Command per acre in water or sprayable fluid fertilizer. Refer to the "WEEDS CONTROLLED" section of this label to determine specific weed species.

NOTE: Observe all precautions and limitations on the Command Label.

DO NOT APPLY BY AIR.

CANNON® plus LEXONE or SENCOR

Apply this tank mixture in water or sprayable fluid fertilizer. Refer to the "WEEDS CONTROLLED" section of this label to determine specific weed species.

Soil* Textural Group	Broadcast Rates Per Acre	
	CANNON® (quarts)	Lexone 4L** or Sencor 4F (pint)
Coarse	3	0.5
Medium	3	0.75
Fine	3	1

* Refer to the "SOIL TEXTURE" section of the label to determine the corresponding textural group of the soil to be treated.

** Use the rates listed in this label when using Lexone 4L or Sencor 4F. Use the equivalent rates when using the 50 WP or DF formulations of these products. One pint of Lexone 4L or Sencor 4F equals one pound of the corresponding 50 WP formulation or 2/3 pound of the corresponding DF formulations.

NOTE: Observe all precautions and limitations on the Lexone and Sencor labels (i.e., soil type, organic matter, soil pH).

CANNON® plus PREVIEW

Apply this tank mixture in water or sprayable fluid fertilizer. Refer to the "WEEDS CONTROLLED" section of this label to determine specific weed species.

Soil Textural Group*	Broadcast Rates Per Acre		
		½ to 3% Organic Matter	3% or More Organic Matter
	CANNON® (quarts)	Preview (ounces)	Preview (ounces)
Medium	3	7	8
Fine	3	8	9 to 10

* Refer to the "SOIL TEXTURE" section of this label to determine the corresponding textural group of the soil to be treated.

NOTE: Observe all precautions and limitations on the Preview label (i.e., soil type, organic matter, soil pH, recropping interval, crop rotations).

DO NOT APPLY BY AIR.

CANNON® plus CANOPY

Apply this tank mixture in water or sprayable fluid fertilizer. Refer to the "WEEDS CONTROLLED" section of this label to determine specific weed species.

Soil Textural Group*	Broadcast Rates Per Acre	
	CANNON® (quarts)	Canopy (ounces)
Medium	3	8 to 12
Fine	3	10 to 14

* Refer to the "SOIL TEXTURE" section of this label to determine the corresponding textural group of the soil to be treated.

** Use the higher rate of Canopy in areas of heavy weed infestation.

NOTE: Observe all precautions and limitations on the Canopy label (i.e., soil type, organic matter, soil pH, recropping interval, crop rotations).

DO NOT APPLY BY AIR.

CANNON® plus SCEPTER

Apply 2-1/2 quarts of this product plus 2/3 pint of Scepter per acre in water or sprayable fluid fertilizer. Refer to the "WEEDS CONTROLLED" section of this label to determine specific weed species.

Apply 2-1/2 quarts of this product plus 1/2 pint of Scepter per acre for control of common lambsquarter, cocklebur, and the weeds listed as controlled in the CANNON section of this label.

NOTE: Observe all precautions and limitations on the Scepter label (i.e., crop rotation).

BASAGRAN™ SEQUENTIAL TREATMENT

Soybeans Only

Basagran may be applied using 3/4 to 1-1/2 quarts per acre following applications of CANNON herbicide or tank mixes of CANNON plus Command, CANNON plus Lexone or Sencor, CANNON plus Preview, CANNON plus Canopy, or CANNON plus Scepter for the control of emerged weeds such as cocklebur, velvetleaf, and other weed species listed on the label for Basagran. Read and follow the directions for use appearing on the label for Basagran.

NOTE: Observe all precautions and limitations on the Basagran label.

™ Basagran is a trademark of BASF Corporation.

SEQUENTIAL POSTEMERGENCE TREATMENT

Soybeans Only

SCEPTER

Scepter may be applied using a 1/3 pint per acre following applications of CANNON herbicide, or tank mixes of CANNON plus Command, or CANNON plus Lexone or Sencor, for the control of emerged cocklebur and pigweed when less than twelve inches tall. Wait at least ten days before cultivating after a postemergence treatment of Scepter. Read and follow the postemergence directions for use appearing on the label for Scepter.

This product may be used in the following states: Alabama, Arkansas, Delaware, Florida, Georgia, Illinois, Indiana, Iowa, eastern Kansas, Kentucky, Louisiana, southern Michigan, Mississippi, eastern Nebraska, New Jersey, North Carolina, Ohio, Oklahoma, South Carolina, Tennessee, Texas, Virginia and southern Wisconsin.

NOTE: Observe all precautions and limitations on the Scepter label (i.e., crop rotation, harvest intervals and grazing restrictions).

DRY EDIBLE BEANS (ALL TYPES)

CANNON®

FOR USE ONLY WEST OF THE MISSISSIPPI RIVER, EXCEPT IN KERN COUNTY, CALIFORNIA.

Apply 4 quarts of this product per acre. Refer to the "WEED CONTROLLED" section of this label to determine specific weed species.

In case of emergency involving this product, Call Collect, day or night, (314) 694-4000.

FALLOW MASTER™

HERBICIDE

For control or suppression of emerged weeds in fallow and reduced tillage small grain systems in the states of Colorado, Idaho, Kansas, Montana, Nebraska, New Mexico, North Dakota, Oklahoma, Oregon, South Dakota, Texas, Utah, Washington, and Wyoming.

ACTIVE INGREDIENTS:*

Glyphosate [N-phosphonomethyl glycine] as the isopropylamine salt	16.5%
Dicamba (3,6 dichloro-o-anisic acid) as the isopropylamine salt	7.0%
INERT INGREDIENTS:	76.5%
TOTAL	100.0%

* Contains 180 grams per litre or 1.5 lbs per U.S. gallon of the active ingredient, glyphosate, as its isopropylamine salt and 72 grams per litre or 0.6 lb per U.S. gallon of the active ingredient, dicamba as its isopropylamine salt. Equivalent to 132 grams per litre or 1.1 lbs per U.S. gallon of the acid glyphosate and 60 grams per litre or 0.5 lb per U.S. gallon of the acid dicamba.

The glyphosate component of Fallow Master herbicide is protected by U.S. Patent No. 3,799,758 and U.S. Patent No. 4,405,531. Other patents are pending.

No license granted under any non U.S. patent.

Complete Directions for Use

AVOID CONTACT WITH FOLIAGE OF CROP OR OTHER DESIRABLE VEGETATION SINCE SEVERE INJURY OR DESTRUCTION MAY RESULT.

EPA Reg. No. 524-390

™ Fallow Master is a trademark of, and used under license from, Sandoz, Ltd.

1988-1 886.65-000.07/CG

Read the entire label before using this product.

Use only according to label instructions.

Read "LIMIT OF WARRANTY AND LIABILITY" before buying or using. If terms are not acceptable, return at once unopened.

REFORMULATION OR REPACKAGING IS PROHIBITED.

LIMIT OF WARRANTY AND LIABILITY

This Company warrants that this product conforms to the chemical description on the label and is reasonably fit for the purposes set forth in the Complete Directions for Use label booklet ("Directions") when used in accordance with those Directions under the conditions described therein. NO OTHER EXPRESS WARRANTY OR IMPLIED WARRANTY OF FITNESS FOR PARTICULAR PURPOSE OR MERCHANTABILITY IS MADE. This warranty is also subject to the conditions and limitations stated herein.

Buyer and all users shall promptly notify this Company of any claims whether based in contract, negligence, strict liability, other tort or otherwise.

Buyer and all users are responsible for all loss or damage from use or handling which results from conditions beyond the control of this Company, including, but not limited to, incompatibility with products other than those set forth in the Directions, application to or contact with desirable vegetation, unusual weather, weather conditions which are outside the range considered normal at the application site and for the time period when the product is applied, as well as weather conditions which are outside the application ranges set forth in the Directions, application in any manner not explicitly set forth in the Directions, moisture conditions outside the moisture range specified in the Directions, or the presence of products other than those set forth in the Directions in or on the soil, crop or treated vegetation.

THE EXCLUSIVE REMEDY OF THE USER OR BUYER, AND THE LIMIT OF THE LIABILITY OF THIS COMPANY OR ANY OTHER SELLER FOR ANY AND ALL LOSSES, INJURIES OR DAMAGES RESULTING FROM THE USE OR HANDLING OF THIS PRODUCT (INCLUDING CLAIMS BASED IN CONTRACT, NEGLIGENCE, STRICT LIABILITY, OTHER TORT OR OTHERWISE) SHALL BE THE PURCHASE PRICE PAID BY THE USER OR BUYER FOR THE QUANTITY OF THIS PRODUCT INVOLVED, OR, AT THE ELECTION OF THIS COMPANY OR ANY OTHER SELLER, THE REPLACEMENT OF SUCH QUANTITY OR, IF NOT ACQUIRED BY PURCHASE, REPLACEMENT OF SUCH QUANTITY. IN NO EVENT SHALL THIS COMPANY OR ANY OTHER SELLER BE LIABLE FOR ANY INCIDENTAL, CONSEQUENTIAL OR SPECIAL DAMAGES.

Buyer and all users are deemed to have accepted the terms of this LIMIT OF WARRANTY AND LIABILITY which may not be varied by any verbal or written agreement.

PRECAUTIONARY STATEMENTS

HAZARDS TO HUMANS AND DOMESTIC ANIMALS

KEEP OUT OF REACH OF CHILDREN

DANGER!

CAUSES EYE BURNS.

HARMFUL IF SWALLOWED OR INHALED.

MAY CAUSE ALLERGIC SKIN REACTION.

Do not get in eyes, on skin or on clothing.

Wear goggles or face shield and rubber gloves in transferring and mixing and when adjusting, repairing or cleaning equipment.

Avoid breathing vapor or spray mist.

Wash thoroughly with soap and water after handling.

FIRST AID: IF IN EYES, immediately flush with plenty of water for at least 15 minutes. Call a physician.

IF ON SKIN, immediately flush with plenty of water while removing contaminated clothing. As soon as soap is available, wash skin thoroughly with soap and water. If irritation persists, call a physician. Wash clothing before reuse. Sensitized persons should avoid further contact and reuse of contaminated clothing.

IF INHALED, remove individual to fresh air. Seek medical attention if breathing difficulty develops.

IF SWALLOWED, this product will cause gastrointestinal tract irritation. Immediately dilute by swallowing water or milk. Call a physician. NEVER GIVE ANYTHING BY MOUTH TO AN UNCONSCIOUS PERSON.

In case of an emergency involving this product, Call Collect, day or night, (314) 694-4000.

ENVIRONMENTAL HAZARDS

Do not apply directly to water or wetlands (swamps, bogs, marshes or potholes). Do not contaminate water when disposing of washwaters.

PHYSICAL OR CHEMICAL HAZARDS

Spray solutions of this product should be mixed, stored and applied only in stainless steel, aluminum, fiberglass, plastic and plastic-lined steel containers.

DO NOT MIX, STORE OR APPLY THIS PRODUCT OR SPRAY SOLUTIONS OF THIS PRODUCT IN GALVANIZED STEEL OR UNLINED STEEL (EXCEPT STAINLESS STEEL) CONTAINERS OR SPRAY TANKS. This product or spray solutions of this product react with such containers and tanks to produce hydrogen gas which may form a highly combustible gas mixture. This gas mixture could flash or explode, causing serious personal injury, if ignited by open flame, spark, welder's torch, lighted cigarette or other ignition source.

GENERAL INFORMATION

Fallow Master is a postemergence herbicide for control or suppression of emerged weeds in fallow and reduced tillage systems. This product should be applied 15 days prior to planting of wheat, barley, oats, or sorghum (grain or forage).

This product enters the plant through the foliage and moves throughout the plant. Visual effects of control are a gradual wilting or yellowing of the plant, which advances to complete browning of aboveground growth and deterioration of affected underground plant parts. Visible symptoms will usually develop on labeled weeds within 2 to 4 days after application, but may not occur for 7 or more days. Extremely cool or cloudy weather following treatment may slow activity of this product and delay the visual effects of control.

Keep people and domestic animals off treated areas until foliage has dried.

APPLICATION PRECAUTIONS AND WARNINGS

Do not apply this product through any type of irrigation system.

- Delay planting of wheat, barley, oats, or sorghum (grain or forage) for 15 days after applications of this product.
- Do not plant any crop other than wheat, barley, oats, corn or sorghum (grain or forage) for 3 months after application or until this product has disappeared from the soil.
- This product is recommended for control of emerged weeds prior to establishment of labeled crops. Large amounts of green or decaying vegetation left standing or incorporated into the seedbed may enhance the development of disease in newly planted crops. This may result in poor emergence and/or stands, especially under cool and/or wet conditions.
- Do not feed or forage treated vegetation from treated areas within 8 weeks after application.
- Spraying early to control young weeds before dense stands develop, or light cultivation to assist weed decay, will favor preparation of suitable seed beds.
- THIS PRODUCT MAY CAUSE INJURY TO DESIRABLE TREES, PLANTS OR CROPS, ESPECIALLY SENSITIVE BROADLEAF PLANTS SUCH AS BEANS, COTTON, FLOWERS, FRUIT TREES, GRAPES, ORNAMENTALS, PEAS, POTATOES, SOYBEANS, SUNFLOWERS, TOMATOES AND OTHER BROADLEAF PLANTS IF BROUGHT INTO CONTACT WITH THEIR ROOTS, STEMS OR FOLIAGE. THESE PLANTS ARE MOST SENSITIVE DURING THEIR ACTIVE GROWTH AND DEVELOPMENT STAGE.
- Do not use aerial equipment to apply this product when sensitive crops and plants are growing in the vicinity of area to be treated.
- Applications should be made only when there is no hazard from spray drift since very small quantities of spray, which may not be visible, may severely injure susceptible crops or desirable vegetation.
- The likelihood of injury occurring to adjacent crops from the use of this product is greatest when winds are gusty or in excess of 5 miles per hour or when other conditions including lesser wind velocities will favor spray drift.
- Movement of this product on soil particles during windstorms may cause damage to susceptible plants that are contacted. This hazard is reduced if rainfall occurs shortly after application.
- Buyer and all users are responsible for all loss or damage in connection with the use or handling of mixtures of this herbicide or other materials that are not expressly recommended in this labeling. Mixing this product with herbicides or other materials not recommended on this label may result in reduced performance. Use of this product in any manner not consistent with this label may result in injury to persons, animals or crops, or other unintended consequences.
- Fallow Master is subject to all state and county regulations for dicamba.

DIRECTIONS FOR USE

It is a violation of Federal law to use this product in any manner inconsistent with its labeling.

STORAGE AND DISPOSAL

Do not contaminate water, food, or feed by storage or disposal. Do not store near fertilizers, seeds, insecticides or fungicides.

STORAGE:
STORE ABOVE 40°F. to keep product in solution. If crystals form, place in a warm room (72°F.), allow the product to reach room temperature and roll or shake periodically until crystals have dissolved. Keep container closed to prevent spills and contamination.

For bulk containers, see the container label for alternate storage information.

DISPOSAL:
Wastes resulting from the use of this product that cannot be used or chemically reprocessed should be disposed of in a landfill approved for pesticide

disposal or in accordance with applicable Federal, state or local procedures.

See the individual container label for disposal information.

TIMING OF APPLICATION

This product should be applied postemergence to vigorously growing weeds when they have reached the recommended size given in the "RECOMMENDED RATES AND WEEDS CONTROLLED" section of this label. Application should be delayed until maximum emergence of the target weeds, but before weeds exceed the maximum size recommended. For annual weeds, allow 1 day after treatment before tillage.

Reduced control may result if applications are made during poor growing conditions such as drought stress, disease or insect damage or if weeds have been mowed, grazed or cut. Heavy dust on foliage or an overstory canopy covering targeted weeds may also reduce control.

Rainfall or irrigation occurring within 6 hours after application may reduce effectiveness. Heavy rainfall or irrigation within 2 hours after application may wash this product off the foliage and a repeat application may be required.

RECOMMENDED RATES AND WEEDS CONTROLLED

For best results, apply this product after most weed seeds have germinated but before seedhead formation in grasses or flower bud formation in broadleaves.

When applied as directed, this product will provide control or suppression of the grass and broadleaf weed species listed below. Rates recommended are for maximum weed height at treatment time.

	FALLOW MASTER™	
Annual Weed Species	Rate per Acre	Maximum Height
Foxtail, green *Setaria viridis*	22 oz	12″
Barley *Hordeum vulgare* **Brome, downy*** *Bromus tectorum* **Cheat*** *Bromus secalinus* **Foxtail** *Setaria spp.* **Kochia*** *Kochia scoparia* **Lettuce, prickly*** *Lactuca serriola*	32 oz	6″
Lambsquarters *Chenopodium album* **Mustard, tansy** *Descurainia pinnata* **Mustard, tumble** *Sisymbrium altissimum* **Pigweed** *Amaranthus spp.* **Sandbur, field** *Cenchrus spp.* **Stinkgrass** *Eragrestis cilianensis* **Wheat** *Triticum aestivum* **Witchgrass** *Panicum capillare*	32 oz	12″
Barnyardgrass *Echinochloa crus-galli* **Buffalobur** *Solanum rostratum* **Goatgrass** *Aegilops cylindrica* **Mustard, blue** *Chorispora tenella* **Panicum, fall** *Panicum dichotomiflorum*	44 oz	6″
Oats, wild *Avena fatua* **Thistle, Russian** *Salsola kali*	44 oz	12″

* For control of heavy infestations or dense, overwintered stands, use 44 ounces.

AID TO TILLAGE

This product used in combination with preplant and conventional fallow tillage practices will provide control of downy brome, cheat, volunteer wheat, tansy mustard and foxtails.

Apply 22 fluid ounces of this product in 3 to 10 gallons of water per acre to weeds that are actively growing. Treat when weeds are less than 6 inches in height. Application must be followed by conventional tillage practices before regrowth of the treated plant occurs. Allow at least 1 day after application before tillage.

POSTHARVEST APPLICATIONS

This product will provide control of weeds following grain harvest. Weeds should be allowed to regrow after damage incurred during harvesting operations, and to recover from environmental stress, before application of this product. Weeds should be treated prior to the heading stage of annual grasses and before broadleaf weeds exceed 24 inches in height. Ammonium sulfate will improve performance on annual weeds under stress conditions.

Weeds controlled at the 32 ounce-per-acre rate include: downy brome, green foxtail, stinkgrass and volunteer wheat.

Weeds controlled at the 44 ounce-per-acre rate include kochia, lambsquarters, mustard, pigweed and Russian thistle.

Weeds controlled at the 52 ounce per acre rate include: barnyardgrass, sandbur, witchgrass, yellow foxtail, and prickly lettuce.

FALLOW MASTER plus ATRAZINE or BLADEX™.

For use only in Colorado, Kansas, Nebraska, Oklahoma, Oregon, South Dakota and Washington.

Tank mixtures of FALLOW MASTER plus atrazine or Bladex will provide postemergence control of listed annual weeds in fallow and reduced tillage systems. In addition, these tank mixtures will provide soil residual control of weeds listed on the atrazine and Bladex labels.

™ Bladex is a trademark of E. I. du Pont de Nemours and Company.

ANNUAL WEEDS CONTROLLED

Barnyardgrass *Echinochloa crus-galli*	**Pigweed, redroot** *Amaranthus retroflexus*
Brome, downy *Bromus tectorum*	**Sandbur, field** *Cenchrus spp.*
Foxtail, green *Setaria viridis*	**Stinkgrass** *Eragrostis cilianensis*
Kochia *Kochia scoparia*	**Thistle, Russian** *Salsola kali*
Lambsquarters *Chenopodium album*	**Wheat** *Triticum aestivum*
Lettuce. prickly *Lactuca serriola*	**Witchgrass** *Panicum capillare*
Mustard, tansy *Descurainia pinnata*	

Apply 32 ounces of this product plus 1 pound or less of the active ingredient, atrazine, per acre. Use 44 ounces of this product plus 2 pounds or less of the active ingredient, atrazine, per acre. Use 52 ounces of this product plus 1 to 3 pounds of the actrive ingredient, atrazine, per acre. Use 44 ounces of this product plus 2.4 to 4 pounds of the active ingredient, cyanazine, in Bladex per acre. In Oregon and Washington, do not exceed 1 pound of the active ingredient, atrazine, per acre. Barnyardgrass will be suppressed at the 52 ounce per acre rate and will require up to 70 ounces per acre for control. The addition of ammonium sulfate is recommended to increase the performance of Fallow Master herbicide plus atrazine or Bladex tank mixtures.

Consult the atrazine and Bladex labels (90 DF, 80W, 4L) for use rates, soil type, planting, cropping and other restrictions, as well as other precautionary statements and use according to the most restrictive label.

These tank mixtures may be applied with ground or aerial equipment. See the "APPLICATION EQUIPMENT" section for instructions.

For ground application, apply in 3 to 10 gallons of water per acre.

For aerial application, apply in 3 to 5 gallons of water per acre.

AMMONIUM SULFATE

The addition of 2 percent dry ammonium sulfate by weight (or liquid equivalent) or 17 pounds per 100 gallons of water may increase the performance of this product on annual weeds. The improvement in performance may be apparent where environmental stress is a concern. Low-quality ammonium sulfate may contain material that will not readily dissolve which could result in nozzle tip plugging. To determine quality, perform a jar test by adding 1/3 cup of ammonium sulfate to 1 gallon of water and agitate for 1 minute. If undissolved sediment is observed, predissolve the ammonium sulfate in water and filter prior to addition to the spray tank. If ammonium sulfate is added directly to the spray tank, add slowly with agitation. Adding too quickly may clog outlet line. ENSURE THAT AMMONIUM SULFATE IS COMPLETELY DISSOLVED IN THE SPRAY TANK BEFORE ADDING THIS PRODUCT. THOROUGHLY RINSE THE SPRAY SYSTEM WITH CLEAN WATER AFTER USE TO REDUCE CORROSION.

NOTE: When using ammonium sulfate, compatibility problems may occur at carrier volumes below 5 GPA.

APPLICATION EQUIPMENT

This product may be applied using either ground or aerial spray equipment. Use extreme care to avoid misting or drifting of herbicide solution onto foliage, green stems or fruit of desirable crops, trees, or plants during both growing and dormant periods since even very small quantities of spray can cause severe plant injury.

GROUND APPLICATION: Apply recommended rates of this product in 3 to 10 gallons of water per acre as a broadcast spray. For optimum spray distribution and coverage, use flat fan or low-volume flood nozzles. When using flood nozzles, space them no more than 40 inches apart and ensure double overlap of spray pattern. Refer to the manufacturer's recommendations for correct pressure and nozzle height above the target canopy. Avoid pressure and nozzles which produce fine droplets or mist.

Use appropriate marking devices to ensure uniform spray coverage and best results from this product.

AERIAL APPLICATION: Apply the recommended rates of this product in 3 to 5 gallons of water per acre as a broadcast spray. DO NOT APPLY DURING INVERSION CONDITIONS, WHEN WINDS ARE GUSTY, OR WHEN OTHER CONDITIONS WILL FAVOR DRIFT. DRIFT MAY CAUSE DAMAGE TO ANY VEGETATION CONTACTED TO WHICH TREATMENT WAS NOT INTENDED. TO PREVENT INJURY TO ADJACENT DESIRABLE VEGETATION, APPROPRIATE BUFFER ZONES MUST BE MAINTAINED.

Coarse sprays are less likely to drift; therefore, do not use nozzles or nozzle configurations which dispense spray as fine spray droplets. Do not angle nozzles forward into the airstream and do not increase spray volume by increasing nozzle pressure.

MIXING INSTRUCTIONS

Fill the spray tank to about 3/4 of the desired volume with clean water. Add the recommended amount of this product, then complete the filling process while maintaining agitation. Remove the hose from the mixing tank immediately after filling to avoid siphoning back into the carrier source. During mixing and application, foaming of the spray solution may occur. To prevent or minimize foam, terminate by-pass and return lines at the tank bottom and/or use an agriculturally approved antifoam or defoaming agent.

Additional surfactant is not recommended for this formulation.

NOTE: Reduced results may occur if water containing soil is used, such as water from ponds and unlined ditches.

SPRAYER CLEANUP

The steps listed below are suggested for thorough cleaning of spray equipment following applications of this product. Failure to clean the sprayer thoroughly may result in injury to desirable vegetation subsequently sprayed with the equipment.

1. Hose down thoroughly the inside as well as outside surfaces of equipment while filling the spray tank half full with water. Flush by operating the sprayer until the system is purged of the rinse water.

2. Fill the tank with water while adding 1 quart of household ammonia for every 25 gallons of water. Operate the pump to circulate the ammonia solution through the boom and nozzles. Let the solution stand for several hours, preferably overnight.

3. Flush the solution out of the spray tank through the boom.

4. Remove the nozzles and screens and flush the system with two full tanks of water.

Thoroughly wash aircraft, especially landing gear, after each day of spraying to remove residues of this product accumulated during spraying or from spills. PROLONGED EXPOSURE OF THIS PRODUCT TO UNCOATED STEEL SURFACES MAY RESULT IN CORROSION AND POSSIBLE FAILURE OF THE PART. LANDING GEAR ARE MOST SUSCEPTIBLE. The maintenance of an organic coating (paint) which meets aerospace specification MIL-C-38413 may prevent corrosion.

886.65-000.07/CG

In case of an emergency involving this product, Call Collect, day or night, (314) 694-4000.

This sample label is current as of January 1, 1989. The product descriptions and recommendations provided in this sample label are for background information only. Always refer to the label on the product container before using Monsanto or any other agrichemical product.

This specimen label is intended for use only as a guide in providing general information regarding the use of this product. Monsanto is not responsible for the accuracy of the information contained herein. As labels are subject to revision, always carefully read and follow the label on the product container.

FAR-GO®

HERBICIDE

Emulsifiable herbicide for control of wild oats in winter wheat, spring and durum wheat, barley, green peas, field dried peas and lentils.

ACTIVE INGREDIENT:	
*S-(2,3,3-trichloroallyl)-diisopropylthiocarbamate	46.3%
INERT INGREDIENTS:	53.7%
TOTAL	100.0%

Contains petroleum distillates.

*Contains 4 pounds S-(2,3,3-trichloroallyl)-diisopropylthiocarbamate per gallon.

FAR-GO® is a registered trademark of Monsanto Company.

EPA Reg. No. 524-145-AA

1988-1 815.15-000.65/CG

Complete Directions for Use

Read the entire label before using this product.

Use only according to label instructions.

Read "LIMIT OF WARRANTY AND LIABILITY" before buying or using. If terms are not acceptable, return at once unopened.

REFORMULATION OR REPACKAGING IS PROHIBITED.

LIMIT OF WARRANTY AND LIABILITY

This Company warrants that this product conforms to the chemical description on the label and is reasonably fit for the purposes set forth in the Complete Directions for Use label booklet ("Directions") when used in accordance with those Directions under the conditions described therein. NO OTHER EXPRESS OR IMPLIED WARRANTY OF FITNESS FOR PARTICULAR PURPOSE OR MERCHANTABILITY IS MADE. This warranty is also subject to the conditions and limitations stated herein.

Buyer and all users shall promptly notify this Company of any claims whether based in contract, negligence, strict liability, other tort or otherwise.

Buyer and all users are responsible for all loss or damage from use or handling which results from conditions beyond the control of this Company, including but not limited to, incompatibility with products other than those set forth in the Directions, unusual weather, weather conditions which are outside the range considered normal at the application site and for the time period when the product is applied, as well as weather conditions which are outside the application ranges set forth in the Directions, application in any manner not explicitly set forth in the Directions, moisture conditions outside the moisture range specified in the Directions, or the presence of products other than those set forth in the Directions in or on the soil, crop or treated vegetation.

THE EXCLUSIVE REMEDY OF THE USER OR BUYER, AND THE LIMIT OF THE LIABILITY OF THIS COMPANY OR ANY OTHER SELLER FOR ANY AND ALL LOSSES, INJURIES OR DAMAGES RESULTING FROM THE USE OR HANDLING OF THIS PRODUCT (INCLUDING CLAIMS BASED IN CONTRACT, NEGLIGENCE, STRICT LIABILITY, OTHER TORT OR OTHERWISE) SHALL BE THE PURCHASE PRICE PAID BY THE USER OR BUYER FOR THE QUANTITY OF THIS PRODUCT INVOLVED, OR, AT THE ELECTION OF THIS COMPANY OR ANY OTHER SELLER, THE REPLACEMENT OF SUCH QUANTITY, OR, IF NOT ACQUIRED BY PURCHASE, REPLACEMENT OF SUCH QUANTITY. IN NO EVENT SHALL THIS COMPANY OR ANY OTHER SELLER BE LIABLE FOR ANY INCIDENTAL OR CONSEQUENTIAL OR SPECIAL DAMAGES.

The Buyer and all users are deemed to have accepted the terms of this LIMIT OF WARRANTY AND LIABILITY which may not be varied by any verbal or written agreement.

PRECAUTIONARY STATEMENTS

HAZARDS TO HUMANS AND DOMESTIC ANIMALS

KEEP OUT OF REACH OF CHILDREN

WARNING!

CAUSES SKIN AND EYE IRRITATION.

HARMFUL IF SWALLOWED.

Do not get in eyes, on skin or on clothing.

Avoid breathing vapors or spray mist.

Wash thoroughly with soap and water after handling.

To avoid contact with skin or eyes, wear rubber gloves and goggles when handling or spraying.

FIRST AID: IF IN EYES, immediately flush with plenty of water for at least 15 minutes. Call a physician.

IF ON SKIN, wash with plenty of soap and water. Remove contaminated clothing. Wash clothing before reuse. If irritation persists, get medical attention.

IF SWALLOWED, do NOT induce vomiting. Give large quantities of water. Call a physician immediately. NEVER GIVE ANYTHING BY MOUTH TO AN UNCONSCIOUS PERSON.

In case of an emergency involving this product, Call Collect, day or night, (314) 694-4000.

ENVIRONMENTAL HAZARDS

Avoid direct application to any body of water. Do not contaminate water by disposal of waste or cleaning of equipment.

PHYSICAL OR CHEMICAL HAZARDS

COMBUSTIBLE.

Do not use, pour, spill or store near heat or flame.

Use only with adequate ventilation.

In case of:
FIRE, use water spray, foam, dry chemical or CO_2.
SPILL or LEAK, flush area with water spray.

STORAGE AND DISPOSAL

Do not contaminate water, foodstuffs, seed or feed by storage or disposal.

STORAGE:

STORE ABOVE 32°F TO KEEP FROM FREEZING.

Freezing will result in crystals which settle to the bottom. If allowed to freeze, place in a warm room (72°F.) and roll and shake the container frequently for several days to redissolve before using. For bulk containers, see the container label for alternate storage information.

Keep this container closed to prevent spills, evaporation and contamination.

DISPOSAL:

Wastes of this pesticide may cause eye and skin irritation and may be dangerous. Improper disposal of excess pesticide, spray mixtures, or rinsate is a violation of Federal law. If these wastes cannot be disposed of by use according to label instructions, contact your State Pesticide or Environmental Control Agency, or the Hazardous Waste representative at the nearest EPA Regional Office for guidance.

Empty container retains vapor and product residue. Observe all labeled safeguards until container is cleaned, reconditioned or destroyed.

(See individual container label for disposal information.)

GENERAL INFORMATION

This herbicide, or any labeled tank mixture of this herbicide, should be sprayed on the soil and incorporated before weeds germinate. If weeds, including wild oats, have emerged prior to treating and/or planting, they must be controlled.

Before applying this herbicide or any labeled tank mixture of this herbicide, the soil should be in good working condition. This applies particularly to clay soils. Avoid application to a field left in a ridged condition. Application to a field which is wet, lumpy, rough or ridged, will result in reduced wild oat control and increased crop thinning.

This product or any labeled tank mixture of this herbicide must be incorporated into the soil immediately after spraying.

Seeding may be done either before or after spraying and incorporation, depending upon the crop that is to be sown. See the "Directions for Use" section of this label. If seeding is delayed, shallow reworking of the treated area before seeding will not destroy the effects of this herbicide.

Weed control may be evaluated by removing a surface inch of soil at the time of germination to inspect the number of wild oats that were killed before emergence.

This herbicide is primarily absorbed by wild oat shoots from the treated layer of the soil. Wild oats are usually controlled before they emerge through the soil. Occasionally, and particularly under dry conditions and cold soil temperatures, some wild oats may emerge and reach the 3 to 4-leaf stage before dying. Under conditions of prolonged high temperature at the time of wild oat germination or drought in the spring, this product may not provide the usual high standard of wild oat control.

ATTENTION

DELAYED EMERGENCE, STAND REDUCTION, STUNTING AND YIELD LOSS MAY RESULT DUE TO COLD OR WET CONDITIONS, IRRIGATION DURING GERMINATION OR EMERGENCE, DEEP PLANTING, SOIL CRUSTING, DISEASE, INSECTS, POOR SEED QUALITY OR SELECTION OF A VARIETY GENETICALLY SUSCEPTIBLE TO STRESS. THESE ARE CONDITIONS BEYOND THE CONTROL OF MONSANTO, AND THE GROWER SHOULD CONSULT THE STATE UNIVERSITY EXTENSION SPECIALIST OR SEED PRODUCER OR SUPPLIER FOR LOCAL VARIETAL INFORMATION AND RECOMMENDED TILLAGE AND PLANTING PRACTICES AND DATES.

Domestic oats should not be seeded where Far-Go was used the previous year.

Do not graze livestock on treated crops.

IN THE STATE OF MONTANA, do not use this herbicide on fields to be seeded to hard red spring wheats with press drills, if the field is or will be irrigated in the current growing season.

Do not use postplant incorporation with hoe drills. This practice will result in unacceptable seeding depth and an excessive concentration of this herbicide in the seed zone.

Use of this product in any manner not consistent with this label may result in injury to persons, animals or crops, or other unintended consequences.

STANDARD SPRAYABLE FLUID FERTILIZER COMPATIBILITY TEST

Herbicides may not always mix evenly throughout a sprayable fluid fertilizer or the compounds may separate too quickly to make their combined use of practical value. This may be due to certain characteristics of the different fluid fertilizers. A simple test using small quantities of the components is suggested to provide compatibility potential.

The test follows:

A. Materials Required For A Compatibility Test

1. Two one-quart jars with lid or stopper (marked "with" and "without").
2. Teaspoons (for a more exacting test, a five to ten milliliter (ml.) pipette or graduated cylinder is desirable).
3. A minimum of one quart of the sprayable fluid fertilizer to be tested.
4. The herbicide to be mixed.
5. A compatibility agent (The purpose of the adjuvant is to help keep the fertilizer and crop protection chemical in suspension, if this assistance is needed).

B. Procedure

1. Add one pint of the sprayable fluid fertilizer that will be used or other herbicide carrier to each jar marked "with" and "without".

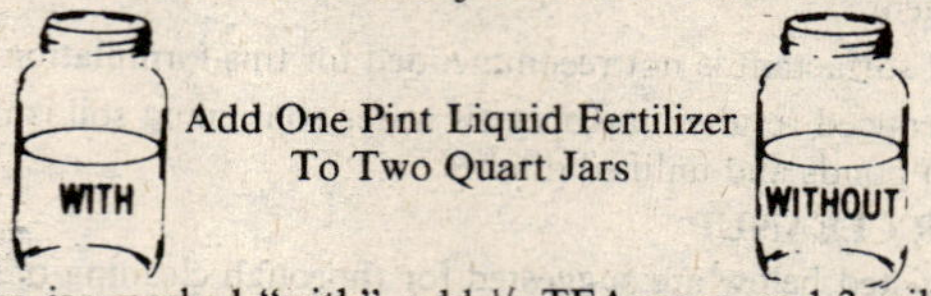

2. To the jar marked "with", add 1/4 TEAspoon or 1.2 milliliters of a suitable compatibility agent; shake gently for five to ten seconds to mix. (1/4 TEAspoon is the equivalent of two pints per 100 gallons of liquid fertilizer.)

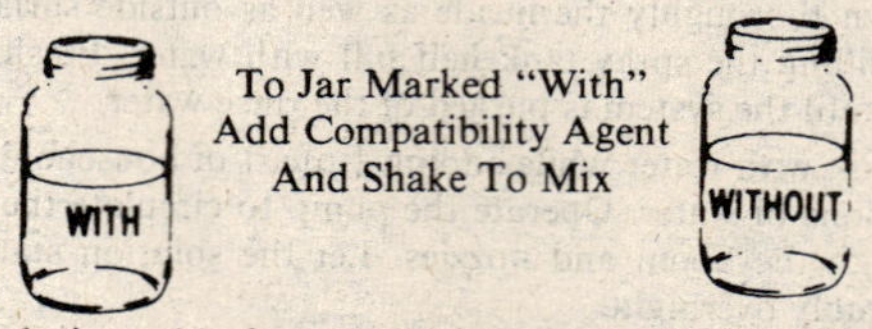

3. To each jar add the appropriate amount of herbicide(s). Shake gently for five to ten seconds after each addition.

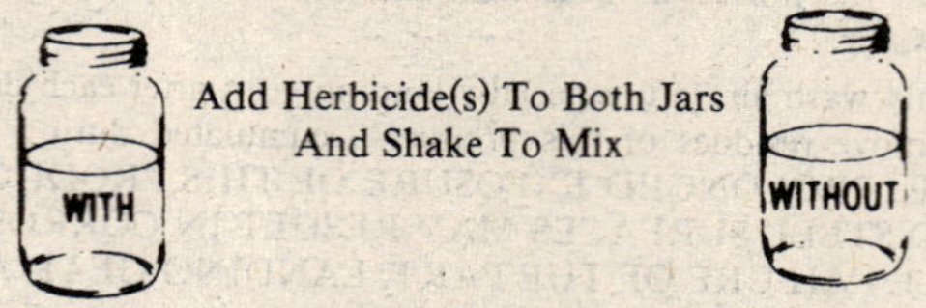

			Amount To Be Added Per Pint Of Sprayable Fluid Fertilizer (Assuming Volume is 25 gal/A)		
Herbicide	Rate/A		Level TEAspoons		Milliliters
FAR-GO®	1 qt.	=	0.9	or	4.7
	1.25 qt.	=	1.2	or	5.9

This compatibility test is designed for 25 gallons of spray per acre with the maximum labeled rate of herbicide. For changes in spray volume, make appropriate changes in the ingredients of the test. Regardless of spray volume, the amount of compatibility agent should be equal to two or three pints (two pints = 1/4 TEAspoon or 1.2 milliliters, three pints = 3/8 TEAspoon or 1.8 milliliters per pint of sprayable fluid fertilizer) per 100 gallons of liquid fertilizer.

C. Observations and Decisions

1. If the herbicide(s) and the sprayable fluid fertilizer are compatible.
2. If a compatibility agent is necessary.

Five minutes after the final addition and mixing, observe both jars for the formation of large flakes, sludge, gels or other precipitates. Observe if the herbicide(s) cannot be physically mixed with the liquid fertilizer but remains as small oily particles in the solution.

If incompatibility in any form described above occurs in the jar "with" the compatibility agent added, the liquid fertilizer and the herbicide(s) should not be used together in the same spray tank.

If incompatibility as described above occurs in the jar "without" the adjuvant but not in the jar "with" adjuvant, the use of a compatibility adjuvant is recommended.

Both jars should be allowed to stand and to be observed periodically for one-half hour. If the separate layers of liquid fertilizer and additives can be resuspended by shaking, commercial application is possible. This product normally will go to the top after standing.

If the herbicide(s) is compatible with fluid fertilizer in the foregoing test without having to use a compatibility agent, fluid fertilizer may be used for the premixing. If it is not compatible, the herbicide(s) should be premixed with water before adding to the spray tank.

MIXING INSTRUCTIONS

Always predetermine the compatibility of this herbicide or labeled tank mix of this herbicide with water carrier or sprayable fluid fertilizer carrier by mixing small proportional quantities in advance. See the "Standard Sprayable Fluid Compatibility Test" section of the label for instructions as to how to determine the compatibility of this herbicide with sprayable fluid fertilizer.

NOTE: This herbicide can be applied in sprayable fluid fertilizer in the spring to barley, spring and durum wheat only.

Mix Far-Go or labeled tank mix of Far-Go with the appropriate carrier as follows:

1. Fill the tank approximately three-quarters full with water. Check for adequate agitation.
2. Carefully add the proper amount of herbicide(s) into the tank while mixing.
3. Fill the remaining portion of the tank with water.
4. Maintain good agitation at all times until the contents of the tank are sprayed.
5. Mix thoroughly again if the solution has been left in the tank for any length of time.
6. The sprayer by-pass valve should be in operation before and during application.

APPLICATION INFORMATION

Do not apply this product through any type of irrigation system.

Ground Broadcast Treatment—Apply this product in 10 or more gallons of water or sprayable fluid fertilizer per acre using broadcast boom equipment. Higher volumes will usually result in more uniform spraying and less evaporation.

Band Treatment—Apply a broadcast equivalent rate and volume per acre. To determine these:

$$\frac{\text{Band width in inches}}{\text{Row width in inches}} \times \text{Broadcast RATE per acre} = \text{Band RATE per acre}$$

$$\frac{\text{Band width in inches}}{\text{Row width in inches}} \times \text{Broadcast VOLUME per acre} = \text{Band VOLUME per acre}$$

Before spraying, check all hose connections to see that they are properly secured. Spraying equipment must be properly calibrated. Adjust the height of the spray boom in the field to ensure an even distribution of the chemical on the soil. Pressure should not be above 30 p.s.i. and just high enough to provide a uniform spray pattern. Boomless sprayers should not be used because the distribution of this product will be too uneven.

A field marker or chain should be attached to both ends of the sprayer boom to serve as a marker, thus avoiding overlapping or untreated strips. Overlapping will double the recommended rate and crop injury may result.

Spraying during winds above 10 miles per hour may result in a loss of herbicide below the amount required for good wild oat control.

WEEDS CONTROLLED WITH FAR-GO®

When applied as directed under conditions described on this label, Far-Go CONTROLS the following weeds:

Oats (wild) Avena fatua	**Ryegrass (annual)** Lolium multiflorum (Oregon only)

DIRECTIONS FOR USE

It is a violation of Federal Law to use this product in any manner inconsistent with its labeling.

FIELD PREPARATION

Before applying this herbicide or any labeled tank mixture, the soil should be in good working condition. All deep tillage must be completed prior to application of this herbicide. If plowed ground is being treated, a single cultivation is recommended prior to spraying. If stubble ground with a heavy trash cover is being treated, or if the soil is wet or lumpy (making proper mixing difficult), two cultivations before spraying are recommended.

HOW TO INCORPORATE

For proper incorporation of Far-Go, set incorporation implement to work the soil no deeper than 3 to 4 inches. Incorporation should be performed immediately after herbicide application. Do not use disk implements for incorporation.

CROPPING SYSTEMS

FAR-GO®

FALL TREATMENT: Winter Wheat and Winter Barley

Crop	FAR-GO® Rate (Qts/A)
Winter Wheat* and Winter Barley* Apply just before seeding to soon after seeding before wild oats germinate.	1.25
Winter Wheat and Winter Barley Apply just before seeding.	1.5**

* DO NOT use on clay knobs.

** For use only within Colorado, Kansas, Nebraska, Oklahoma and Texas to control Bromus species (*B. tectorum, B. japonicus, and B. secalinus*), or on fallow ground within Montana for suppression of cheatgrass (*B. tectorum*).

For applications to winter wheat and winter barley after seeding, apply and shallowly incorporate with a spike tooth, or springtooth harrow followed with a second incorporation at right angles. When incorporating after seeding winter wheat and winter barley, adjust cultivation depth so as not to disturb seed.

FALL TREATMENT: Barley, Durum Wheat, and Spring Wheat

Crop	FAR-GO® (Qts/A)
Spring or Durum Wheat	1
Barley	1.25

Apply within 3 weeks of soil freezeup, since breakdown of this product is minimal at soil temperatures of 40°F. or less, and very limited herbicide will be lost.

For applications to provide wild oat control in spring planted spring wheat, durum wheat, and barley, apply on summer fallow ground which is free of lumps and trash, and single-pass incorporate with equipment similar to a culti-harrow or a duckfoot cultivator multiweeder, or springtooth harrow.

For application to stubble ground with heavy trash cover, fields should be worked with a field cultivator or chisel plow to remove as much surface trash as possible. When a chisel plow is used it should be followed by a field cultivator before application. Apply this herbicide and single-pass incorporate with a field cultivator or culti-harrow.

When incorporating with a single pass in the fall, a second incorporation must be performed in the spring during seedbed preparation. If no spring work is anticipated, then both incorporations should be done in the fall.

If soil must be ridged after incorporating to prevent soil erosion by high winds, the depth of ridging should be kept to a minimum. Deep ridging may decrease the effectiveness of the herbicide or result in crop injury in the spring.

In preparing the herbicide-treated area for seeding in the spring, care must be taken to avoid working the soil any deeper then 3 to 4 inches.

SPRING TREATMENT: Spring Wheat, Barley, Green Peas, Field Dried Peas, and Lentils

Before Seeding Applications

Crop	FAR-GO® Rate (Qts/A)
Spring and Durum Wheat	1
Barley—Apply just before seeding.	1.25
Green Peas*	1.25
Field Dried Peas*	1.25
Lentils—Apply up to 3 weeks before seeding.	1.25

* Leaf crinkling and delayed maturity of peas may occur, particularly on clay points in the Northwest, but this is usually more than offset by a reduction of wild oat competition and increased yields.

Before Seeding Incorporation: For applications to stubble ground, work fields twice with a disk, field cultivator or chisel plow to provide soil in a good working condition and to remove surface trash. Apply and incorporate with equipment such as a culti-harrow, springtooth cultivator or duckfoot cultivator. A second incorporation at right angles should provide best results.

SPRING TREATMENT: Barley, Green Peas, Field Dried Peas, Lentils, Durum Wheat and Spring Wheat

After Seeding Incorporation

Crop	FAR-GO® Rate (Qts/A)
Spring and Durum Wheat	1
Barley	1.25
Green Pea*	1.25
Field Dried Peas*	1.25
Lentils	1.25

* Leaf crinkling and delayed maturity of peas may occur, particularly on clay points in the Northwest, but this is usually more than offset by a reduction of wild oat competition and increased yields.

Apply after seeding before crop emergence and before wheat and barley kernel sprouts exceed 1/2 inch in length and before green peas, field dried peas, and lentils shoot sprouts exceed 1/4 inch in length.

After Seeding Incorporation: Apply immediately after seeding and shallowly incorporate with equipment such as a flex multiweeder or harrow, and follow with a second incorporation at a right angle. Adjust incorporation equipment to a depth so as not to disturb the seed.

FAR-GO® PLUS TREFLAN™ (TANK MIXTURE)

Barley and Wheat

When applied as directed under conditions described on this label, this tank mixture CONTROLS these additional weeds:

Foxtail (green)	Foxtail (yellow)
Setaria viridis	Setaria lutescens

Green and yellow foxtail are commonly referred to as pigeon grass.

™ Treflan is a trademark of Elanco Products Co.

SPRING TREATMENT: Barley, Spring Wheat, and Durum Wheat

After Seeding Incorporation

Crop	FAR-GO® (Qts/A)	+	Treflan (Qts/A)
Spring Wheat, Durum Wheat and Barley	1		0.5

After Seeding Incorporation: Apply immediately after seeding and shallowly incorporate with equipment such as a flex multiweeder or harrow, and follow with a second incorporation at a right angle. Adjust incorporation equipment to a depth so as not to disturb the seed.

NOTE: Do not apply the Far-Go herbicide plus Treflan tank mix to soils which are subject to prolonged periods of flooding or excessive irrigation.

If less than 25 inches of total water is used to produce the crop, do not plant sorghum or oats for 18 months after application.

Refer to Treflan label for crop rotation restrictions and cautionary statements.

ANNUAL CANARYGRASS SEED PRODUCTION

FALL TREATMENT: Apply the herbicide at a rate of 1 quart per acre within 3 weeks of soil freezeup. The seeding depth of the crop should be 1 to 2 inches.

SPRING TREATMENT: Incorporate this product according to the "Before Seeding Incorporation" instructions in the "Spring Treatment" section of this label. Apply the herbicide at a rate of 1 quart per acre just before seeding. The seeding depth of the crop should be 1 to 2 inches.

FERTILIZER IMPREGNATION WITH FAR-GO®

FOR USE IN ALL STATES EXCEPT CALIFORNIA

The application of Far-Go impregnated on fertilizer may be made as follows:

Crops: Winter Wheat, Spring Wheat, Durum Wheat and Barley
Time of Application and Rate:

- Fall Applied and Fall Incorporated
  - Spring wheat and barley
    1.25 Qts/A of Far-Go (The 1.5 qt. rate may be used only on fine textured soil.)
  - Winter wheat
    1.25 Qts/A of FAR-GO
    Preplant Incorporated
- Spring Applied and Preplant Incorporated
  - Spring barley only
    1.25 Qts/A of FAR-GO
  - Spring and durum wheat
    1 Qt/A of FAR-GO
    Before Seeding Incorporation: Apply and immediately incorporate with equipment such as a culti-harrow, springtooth cultivator or duckfoot cultivator. A second incorporation at right angles should provide best results. Do not use other PPI herbicides in conjunction with this herbicide treatment.

APPLICATION WITH DRY BULK FERTILIZER

FOR USE IN ALL STATES EXCEPT CALIFORNIA

The herbicide/fertilizer impregnation process must be done at commercial fertilizer or chemical dealerships that are properly equipped for this procedure.

Dry bulk fertilizer may be impregnated with FAR-GO herbicide. FAR-GO must be impregnated on a minimum of 150 pounds per acre of dry fertilizer. Applications of dry bulk fertilizer impregnated with FAR-GO herbicide may be made in the fall or spring, but only preplant incorporated treatments to barley, spring wheat, and durum wheat may be made in the spring. Spring planted wheat or barley may be seeded into fall FAR-GO herbicide applications in accordance with label directions. Applications should be made as soon as possible after blending.

The following table provides a reference to determine the amount of liquid herbicide to be mixed per ton of dry bulk fertilizer.

Dry Fertilizer (Lbs/Acre)	Acres Covered (Per Ton)	FAR-GO® Herbicide Rate (Quarts Per Acre) 1.25	1.50
		(Quarts Per Ton)	
150	13.3	16.5	20.0
160	12.5	15.8	18.8
170	11.7	14.8	17.5
180	11.1	14.0	16.8
190	10.5	13.0	15.8
200	10.0	12.5	15.0
210	9.5	12.0	14.3
220	9.1	11.5	13.5
230	8.7	11.0	13.0
240	8.3	10.5	12.5
250	8.0	10.0	12.0
260	7.7	9.6	11.6
270	7.4	9.3	11.1

Mix and blend the dry fertilizer and herbicide in a closed rotary drum-type mixer allowing sufficient time to ensure uniform coverage. The nozzle used to spray the herbicide treatment must be placed inside the mixer to provide uniform spray coverage of the tumbling fertilizer.

Spread the fertilizer/herbicide mixture with a properly calibrated applicator. Be certain the material is applied uniformly to the soil surface. When using fan spreaders, a 100 percent overlap is recommended. Fan spreaders should be calibrated to put out one-half the desired rate per acre and application overlapped by doubling back to cover one-half of the previous swath. Dribble spreaders must be calibrated with the fertilizer/herbicide mixture, since the flow pattern will generally be slower than straight dry fertilizer. Nonuniform spreading of the fertilizer/herbicide mixture may result in unsatisfactory weed control or crop injury.

Follow normal Far-Go herbicide incorporation procedures to incorporate the fertilizer/herbicide mixture.

A partial list of fertilizers which may be impregnated with Far-Go are:

Diammonium phosphate	18-46-0
Treble super-phosphate	0-46-0
Urea	46-0-0
Ammonium nitrate	33-0-0
Ammonium sulfate	21-0-0
Potassium chloride	0-0-60
Potassium sulfate	0-0-52

WARNING: FAR-GO CAN BE IMPREGNATED ON ANY COMMONLY USED DRY FERTILIZER EXCEPT THOSE CONTAINING POTASSIUM NITRATE OR SODIUM NITRATE. WHEN MIXING FAR-GO WITH AMMONIUM NITRATE, IT IS IMPERATIVE THAT FAR-GO BE EVENLY MIXED, THAT CONCENTRATIONS ABOVE 5 PERCENT BY WEIGHT OF FAR-GO BE AVOIDED, AND THAT THE MIXTURE BE UTILIZED WITHIN 24 HOURS OF MIXING WITH IMMEDIATE APPLICATION RECOMMENDED. CONCENTRATIONS IN EXCESS OF 5 PERCENT BY WEIGHT OF FAR-GO WITH AMMONIUM NITRATE OF LONG-TERM STORAGE OF FAR-GO AND AMMONIUM NITRATE COULD RESULT IN EXCESSIVE CONCENTRATIONS OF FAR-GO IN THE MIXTURE DUE TO DRAINING OF FAR-GO. THESE EXCESSIVE CONCENTRATIONS OF FAR-GO MAY CAUSE PERSONAL INJURY AND/OR PROPERTY DAMAGE FROM FIRE AND/OR EXPLOSION IF IGNITED OR DETONATED. EXCESSIVE CONCENTRATIONS OF FAR-GO IN THE MIXTURE MAY ALSO CAUSE CROP INJURY. AMMONIUM NITRATE IS AN OXIDIZING AGENT AND ALL REGULATIONS RELATING TO ITS STORAGE, BLENDING, LABELING, OR TRANSPORTATION ARE THE RESPONSIBILITY OF

THE ENTITY MIXING AND/OR SELLING THE FERTILIZER/ FAR-GO HERBICIDE MIXTURE.

If you have any questions concerning this Label Pamphlet or the proper application or use of FAR-GO herbicide, contact your local Monsanto Sales Representative.

In case of an emergency involving this product, Call Collect, day or night, (314) 694-4000.

This sample label is current as of January 1, 1989. The product descriptions and recommendations provided in this sample label are for background information only. Always refer to the label on the product container before using Monsanto or any other agrichemical product.

This specimen label is intended for use only as a guide in providing general information regarding the use of this product. Monsanto is not responsible for the accuracy of the information contained herein. As labels are subject to revision, always carefully read and follow the label on the product container.

FAR-GO® GRANULAR

HERBICIDE

Granular selective herbicide for fall or spring application to control wild oats in winter wheat, spring and durum wheat, barley, green peas, field dried peas and lentils, and for fall application to suppress *Bromus species (B. tectorum, B. secalinus and B. japonicus) in winter wheat and in winter barley.*

ACTIVE INGREDIENT:	
S-(2,3,3-Trichloroallyl)diisopropylthiocarbamate	10%
INERT INGREDIENTS:	90%
TOTAL	100%

EPA Reg. No. 524-292-AA

1988-1 843.53-000.43CG

®FAR-GO is a registered trademark of Monsanto Company.

CAUTION

KEEP OUT OF REACH OF CHILDREN

COMPLETE DIRECTIONS FOR USE

In case of an emergency involving this product, Call Collect, day or night, (314) 694-4000.

Read the entire label before using this product.

Use only according to label instructions.

Read "LIMIT OF WARRANTY AND LIABILITY" before buying or using. If terms are not acceptable, return at once unopened.

REFORMULATION OR REPACKAGING IS PROHIBITED.

LIMIT OF WARRANTY AND LIABILITY

This company warrants that this product conforms to the chemical description on the label and is reasonably fit for the purposes set forth in the Complete Directions for Use label booklet ("Directions"), when used in accordance with those Directions under the conditions described therein. NO OTHER EXPRESS OR IMPLIED WARRANTY OF FITNESS FOR PARTICULAR PURPOSE OR MERCHANTABILITY IS MADE. This warranty is also subject to the conditions and limitations stated herein.

Buyer and all users shall promptly notify this Company of any claims whether based in contract, negligence, strict liability, other tort or otherwise.

Buyer and all users are responsible for all loss or damage from use or handling which results from conditions beyond the control of this Company, including, but not limited to, incompatibility with products other than those set forth in the Directions, unusual weather, weather conditions which are outside the range considered normal at the application site and for the time period when the product is applied as well as weather conditions which are outside the application ranges set forth in the Directions, application in any manner not explicitly set forth in the Directions, moisture conditions outside the moisture range specified in the Directions, or the presence of products other than those set forth in the Directions in or on the soil, crop or treated vegetation.

THE EXCLUSIVE REMEDY OF THE USER OR BUYER, AND THE LIMIT OF THE LIABILITY OF THIS COMPANY OR ANY OTHER SELLER FOR ANY AND ALL LOSSES, INJURIES OR DAMAGES RESULTING FROM THE USE OR HANDLING OF THIS PRODUCT (INCUDING CLAIMS BASED IN CONTRACT, NEGLIGENCE, STRICT LIABILITY, OTHER TORT OR OTHERWISE) SHALL BE THE PURCHASE PRICE PAID BY THE USER OR BUYER FOR THE QUANTITY OF THIS PRODUCT INVOLVED, OR, AT THE ELECTION OF THIS COMPANY OR ANY OTHER SELLER, THE REPLACEMENT OF SUCH QUANTITY OR, IF NOT ACQUIRED BY PURCHASE, REPLACEMENT OF SUCH QUANTITY. IN NO EVENT SHALL THIS COMPANY OR ANY OTHER SELLER BE LIABLE FOR ANY INCIDENTAL, CONSEQUENTIAL OR SPECIAL DAMAGES.

The buyer and all users are deemed to have accepted the terms of this LIMIT OF WARRANTY AND LIABILITY which may not be varied by any verbal or written agreement.

PRECAUTIONARY STATEMENTS

HAZARDS TO HUMANS AND DOMESTIC ANIMALS

Keep out of reach of children.

CAUTION!

MAY CAUSE EYE IRRITATION.

Avoid contact with skin, eyes or clothing.

Wash thoroughly with soap and water after handling.

FIRST AID: IF IN EYES, immediately flush with plenty of water for at least 15 minutes. Call a physician.

In case of emergency involving this product, Call Collect, day or night, (314) 694-4000.

ENVIRONMENTAL HAZARDS

Avoid direct application to any body of water. Do not contaminate water by disposal of waste or cleaning of equipment.

DIRECTIONS FOR USE

It is a violation of Federal Law to use this product in any manner inconsistent with its labeling.

STORAGE AND DISPOSAL

Do not contaminate water, foodstuffs, feed or seed by storage and disposal.

STORAGE

Keep bag closed to prevent spills and contamination.

DISPOSAL

Wastes resulting from the use of this product that cannot be used or chemically reprocessed should be disposed of in a landfill approved for pesticide disposal or buried on site in a safe place away from water supplies. All disposal should be in accordance with applicable Federal, state or local procedures.

Emptied container retains vapor and product residue. Observe all labeled safeguards until container is destroyed.

Completely empty bag into application equipment. Then dispose of empty bag in a sanitary landfill or by incineration, or, if allowed by state and local authorities, by burning. If burned, stay out of smoke.

This herbicide is recommended for wild oat control in barley, peas, lentils, durum, spring and winter wheat only, and for suppression of downy brome *(Bromus tectorum)*, cheat *(Bromus secarinus)* and Japanese brome *(Bromus japonicus)* in winter wheat and winter barley.

Other crops should not be treated with this product because injury may occur. For barley, durum and spring wheat, this product may be applied on the soil either in the fall or in the spring before wild oats germinate. For lentils and peas, this product may only be applied in the spring before wild oats germinate. For winter wheat, this product may only be applied in the fall before wild oats germinate.

Application equipment must be properly calibrated: application of too much herbicide may injure the crop; application of too little may result in poor wild oat control. Recommended rates should be followed in order to:

1. Avoid crop injury
2. Avoid crop residue at harvest
3. Control wild oats
4. Suppress *Bromus* species

Application to a field which is wet, lumpy, rough or ridged will result in reduced wild oat control and promote crop thinning. Incorporation must be completed within 48 hours after application and before germination of the wild oats. If weeds, including wild oats, have emerged prior to treating and/or planting, they must be controlled. All deep tillage by cultivators or double disc implements must be completed prior to application. Do not plow following application of this herbicide.

Seeding may be done either before or after application, depending upon the crop that is to be sown. If seeding is delayed, shallow reworking of the treated area before seeding will not destroy the effects of the chemical.

When using this herbicide, a strip should be left untreated for proof of results. Weed control may be evaluated also by removing a surface inch or two of the soil at the time of germination to inspect the number of wild oats that were killed before emergence.

Wild oats are ususally killed before emergence, but occasionally, and particularly under dry conditions, plants may reach the 3 to 4-leaf stage before they die.

Under conditions of prolonged high temperature at the time of germination or extreme drought in the spring, this product may not maintain the usual high standard of wild oat control.

ATTENTION

DELAYED EMERGENCE, STAND REDUTION, STUNTING AND YIELD LOSS MAY RESULT DUE TO COLD OR WET CONDITIONS, IRRIGATION DURING GERMINATION OR EMERGENCE, DEEP PLANTING, SOIL CRUSTING, DISEASE, INSECTS, OR SELECTION OF A VARIETY GENETICALLY SUSCEPTIBLE TO STRESS. THESE ARE CONDITIONS BEYOND THE CONTROL OF MONSANTO, AND THE GROWER SHOULD CONSULT THE STATE UNIVERSITY EXTENTION SPECIALIST OR SEED PRODUCER OR SUPPLIER FOR LOCAL VARIETAL INFORMATION AND RECOMMENDED TILLAGE AND PLANTING PRACTICES AND DATES.

Domestic oats should not be seeded where Far-Go was used the previous year.

Do not graze livestock on treated crops.

IN THE STATE OF MONTANA, do not use this herbicide on fields to be seeded to hard red spring wheat with press drills, if the field is or will be irrigated in the current growing season.

Do not use postplant incorporation with hoe drills. This practice will result in unacceptable seeding depth and an excessive concentration of this herbicide in the seed zone.

Use of this product not consistent with this label may result in injury to persons, animals or crops, or other unintended consequences. Keep container closed to prevent spills and contamination.

HOW TO APPLY AND CALIBRATE EQUIPMENT

This granular herbicide must be applied through a specially designed ground applicator or airplane capable of of applying small quantities of granules evenly.

GROUND EQUIPMENT: It is important that the applicator be calibrated properly to deliver the desired amount of this product to avoid applying too little, or too much, material. To give even distribution, scatter plates (similar to those used for applying granules in a band) must be attached to each delivery tube or outlet in such a manner to give overall coverage. To calibrate, attach a collector pan, or strong plastic bag, over each spreader plate or delivery tube. Operate over normal terrain to be treated at 4 to 5 miles per hour. Collect the granules from all outputs after covering the desired distance.

For example: If a 14-foot applicator is being used, stake off a distance of 324 feet in the field to be treated (distance equal to 1/10 acre treated). After attaching a bag, or collection pan, to each outlet, collect the granules while driving the desired distance. Check to see that each outlet disperses the same amount of granules. Combine all samples and weigh. For the above set of conditions, the quantity of granules that should be collected for the following recommended rates for broadcast treatment are:

	Recommended Rate Per Acre		
	10 lbs.	12.5 lbs.	15 lbs.
Proper amount to be collected	1.0 lbs.	1.25 lbs.	1.5 lbs.

If more or less than the desired quantity is collected, adjust opening accordingly and again collect the granules while driving the staked-off distance. Continue this procedure until the proper amount is delivered.

AIRPLANE: For aerial applications, attachments designed for applying low volumes of granules must be used. In order to ensure uniform aerial application, it is recommended that the field distribution pattern is checked, and any necessary gate and vent modifications are made to ensure an even pattern distribution. In order to ensure uniform application and to avoid overlapping and possible crop injury, it is recommended that two flagmen, one at each end of the field, be used.

FIELD PREPARATION

Before applying this product, be sure the soil is in good working condition. All deep tillage by cultivation, or double disc implements, must be completed prior to application of this herbicide. If stubble ground is being treated, one or two passes with a field cultivator may be required before application and incorporation.

INCORPORATED APPLICATIONS

When summer fallow, or plowed ground is being treated, a field cultivator will provide adequate incorporation. For proper incorporation of Far-Go, set incorporation implement to work the soil no deeper than 3 to 4 inches. Do not use disc implements for incorporation.

FALL—On summer fallow ground which is loose and free of lumps and trash, incorporate this product using equipment similar to a culti-harrow or a duckfoot with rod weeder attachment.

For applications to fields or standing stubble, fields may be worked once or twice with a field cultivator or chisel plow. Apply granules and incorporate with a field cultivator or culti-harrow.

If soil must be ridged after incorporating, to prevent soil erosion by high winds, the depth of ridging should be kept to a minimum.

In preparing the herbicide treated area for seeding in the spring, care must be taken to avoid working the soil any deeper than fall tillage.

When incorporating with a single pass in the fall, a second incorporation must be preformed in the spring during seedbed preparation. If no spring work is anticipated, then both incorporations should be done in the fall.

For applications to winter wheat after seeding, apply and shallowly incorporate with a spike tooth, or spring tooth harrow set shallow as not to disturb wheat seed.

SPRING—Before Seeding Incorporation: For applications to fields of standing stubble, work fields once or twice with a disc, field cultivator or chisel plow, to provide soil in a good working condition. Apply granules and incorporate with equipment such as a culti-harrow or duckfoot cultivator. A second incorporation at right angles should provide best results.

FOR SPRING AND DURUM WHEAT AND BARLEY IN MONTANA ONLY. For suppression of Persian darnel (Lolium *temulentum*), apply this product before seeding and shallowly incorporate.

After Seeding Incorporation: Apply granules immediately after seeding and shallow incorporate at right angles with equipment such as a flex multi-weeder or harrow. Adjust incorporation equipment to a depth so as not to disturb the seed.

When to Use Fall Treatment Incorporated		
Crop	**Rate (Lbs./A)**	**When to Apply**
Spring and Durum Wheat and Barley	12.5 to 15	*Within 3 weeks of normal freezeup or until snow cover occurs
Winter Wheat and Winter Barley**	12.5 to 15	Before or after seeding.

Apply lower rates on light soils and apply higher rates on heavy soils.

Incorporate within 48 hours. For methods, see "FIELD PREPARATION" and "INCORPORATED APPLICATIONS" sections.

* Average soil temperature at the 2-inch depth must be 40°F. or less.

** For suppression of downy brome (Bromus *tectorum*), cheat (Bromus *secalinus*) and japanese brome (Bromus *japonicus*) apply 15 pounds per acre prior to planting and shallow incorporate. In Colorado, Kansas, Nebraska, Oklahoma and South Dakota plant with hoedrills only.

Spring Treatment Incorporated		
Crop	**Rate (Lbs./A)**	**When to Apply**
Spring and Durum Wheat	10.0 to 12.5	After seeding before wild oats germinate.
Spring and Durum Wheat	10.0 to 12.5*	Before seeding and before wild oats germinate.
Barley	12.5 to 15.0	Just before or immediately after seeding, before wild oats germinate.
Lentils	12.5 to 15.0	Just before or immediately after seeding, before wild oats germinate.
Peas	12.5 to 15.0	Just before or immediately after seeding, before wild oats germinate.

Apply lower rates on light soils and apply higher rates on heavy soils.

Incorporate within 48 hours. For methods, see "FIELD PREPARATION" and "INCORPORATED APPLICATIONS" sections.

* For press drills use the 10.0 lb. rate. For HOEDRILLS use the 12.5 lb. rate.

Best spring and durum wheat results will be obtained with at least a 3-day delay between herbicide incorporation and seeding.

SURFACE APPLICATION WITH DELAYED INCORPORATION

USE ONLY IN IDAHO, MINNESOTA, MONTANA, NORTH DAKOTA AND UTAH

This product may be applied as a surface application with a delayed incorporation prior to planting spring wheat, durum wheat, and barley at a rate of 15 pounds per acre for all soil types.

Surface applications of this product which will be followed by a delayed incorporation in the spring may be made beginning 3 weeks prior to soil freeze-up in the fall and ending before spring thaw. (Average soil temperature at the 2-inch depth must be 40° F or less).

Surface applications must be incorporated in the spring; two passes are recommended. If surface applications are made because soil conditions do not permit incorporation and these conditions change, making incorporation possible, incorporate even if several weeks after application.

This surface application may be made in standing stubble or to fields with surface residues. Surface applications should not be made to fields covered with snow or with excessive crop residue, which will not allow granule contact with the soil.

Surface applications with delayed incorporation may not provide the same level of wild oat control as fall incorporated applications. Wild oat control resulting from delayed incorporation may be reduced if drought or abnormally warm temperatures occur between application and crop emergence in the spring. To achieve the most reliable wild oat control, this product should be incorporated within 48 hours of application as described in the "FIELD PREPARATIONS" and "INCORPORATED APPLICATIONS" sections of this label.

NO-TILL APPLICATION FOR WILD OAT CONTROL IN SPRING PLANTED CROPS IN MONTANA ONLY

This product may be applied prior to planting no-till spring wheat, durum wheat, and barley at a rate of 15 pounds per acre for all soil types.

No-till applications of this product may be made 3 weeks prior to soil freeze-up and ending before spring thaw. (Average soil temperature at the 2-inch depth must be 40° F or less).

No-till applications may be made in standing stubble or to fields with surface residue. No-till applications should not be made under conditions which will not allow granule contact with the soil.

No-till applications are to be made in reduced tillage/no-till systems only. If this product is used in any other system, it must be applied and incorporated as described elsewhere in this label.

FOR FALL PLANTED CROPS

FOR USE IN COLORADO, KANSAS, NEBRASKA, OKLAHOMA AND SOUTH DAKOTA — This product can be surface-applied prior to seeding of winter wheat without incorporation when seeding is done with a hoedrill. Apply 15 pounds per acre for suppression of downy brome (Bromus *techorum*), cheat (Bromus *secalinus*) and Japanese brome (Bromus *japonicus*).

FOR USE IN IDAHO, MONTANA, OREGON, UTAH and WASHINGTON — Apply this product at a rate of 15 pounds per acre as a surface application ahead of a Yielder™ type no-till drill to obtain suppression of wild oats and downy brome in winter wheat or barley. A flextine harrow attached to the drill or used shallowly following planting at a right angle to the drill rows may improve herbicide performance.

Yielder™ is a trademark of Yielder Drill, Inc.

843.53-000.43/CG

If you have any questions concerning this Label Pamphlet or the proper application or use of Far-Go herbicide, contact your local Monsanto Sales Representative.

In case of an emergency involving this product, Call Collect, day or night, (314) 694-4000.

This sample label is current as of January 1, 1989. The product descriptions and recommendations provided in this sample label are for background information only. Always refer to the label on the product container before using Monsanto or any other agrichemical product.

This specimen label is intended for use only as a guide in providing general information regarding the use of this product. Monsanto is not responsible for the accuracy of the information contained herein. As labels are subject to revision, always carefully read and follow the label on the product container.

FREEDOM

HERBICIDE

Emulsifiable herbicide for weed control in soybeans only.

RESTRICTED USE PESTICIDE

due to oncogenicity. For retail sale to and use only by Certified Applicators or persons under their direct supervision and only for those uses covered by the Certified Applicator's certification.

The use of this product may be hazardous to your health. This product contains alachlor which has been determined to cause tumors in laboratory animals.

In case of an emergency involving this product or for user safety information on this product, call collect, day or night (314) 694-4000.

ACTIVE INGREDIENTS:*	
Alachlor	31.7%
Trifluralin	3.9%
INERT INGREDIENTS:	64.4%
TOTAL	100.0%

Contains petroleum distillates.

*Contains 320 grams/litre or 2.67 pounds/gallon of 2-chloro-2', 6'-diethyl-N-(methoxymethyl) acetanilide and 40 grams/litre or 0.33 pound/gallon of (α, α, α- trifluoro-2, 6-dinitro-N, N-dipropyl p-toluidine).

EPA Reg. No. 524-422

Complete Directions for Use

™FREEDOM is a trademark of Monsanto Company.

1989-1

843.31-000.01/CG

Read the entire label before using this product.

Use only according to label instructions.

Read "LIMIT OF WARRANTY AND LIABILITY" before buying or using. If terms are not acceptable, return at once unopened.

REFORMULATION IS PROHIBITED.

LIMIT OF WARRANTY AND LIABILITY

This Company warrants that this product conforms to the chemical description on the label and is reasonably fit for the purposes set forth in the Complete Directions for Use label booklet ("Directions") when used in accordance with those Directions under the conditions described therein. NO OTHER EXPRESS WARRANTY OR IMPLIED WARRANTY OF FITNESS FOR PARTICULAR PURPOSE OR MERCHANTABILITY IS MADE. This warranty is also subject to the conditions and limitations stated herein.

Buyer and all users shall promplty notify this Company of any claims whether based on contract, negligence, strict liability, other tort or otherwise.

Buyer and all users are responsible for all loss or damage from use or handling which results from conditions beyond the control of this Company, including, but not limited to, incompatibility with products other than those set forth in the Directions, unusual weather, weather conditions which are outside the range considered normal at the application site and for the time period when the product is applied, as well as weather conditions which are outside the application ranges set forth in the Directions, application in any manner not explicitly set forth in the Directions, moisture conditions outside the moisture range specified in the Directions, or the presence of products other than those set forth in the Directions in or on the soil, crop or treated vegetation.

THE EXCLUSIVE REMEDY OF THE USER OR BUYER, AND THE LIMIT OF THE LIABILITY OF THIS COMPANY OR ANY OTHER SELLER FOR ANY AND ALL LOSSES, INJURIES OR DAMAGES RESULTING FROM THE USE OR HANDLING OF THIS PRODUCT (INCLUDING CLAIMS BASED IN CONTRACT, NEGLIGENCE, STRICT LIABILITY, OTHER TORT OR OTHERWISE) SHALL BE THE PURCHASE PRICE PAID BY THE USER OR BUYER FOR THE QUANTITY OF THIS PRODUCT INVOLVED, OR, AT THE ELECTION OF THIS COMPANY OR ANY OTHER SELLER, THE REPLACEMENT OF SUCH QUANTITY, OR, IF NOT ACQUIRED BY PURCHASE, REPLACEMENT OF SUCH QUANTITY. IN NO EVENT SHALL THIS COMPANY OR ANY OTHER SELLER BE LIABLE FOR ANY INCIDENTAL, CONSEQUENTIAL OR SPECIAL DAMAGES.

Buyer and all users are deemed to have accepted the terms of this LIMIT OF WARRANTY AND LIABILITY which may not be varied by any verbal or written agreement.

PRECAUTIONARY STATEMENTS

HAZARDS TO HUMANS AND DOMESTIC ANIMALS

KEEP OUT OF REACH OF CHILDREN

WARNING!

CAUSES SUBSTANTIAL BUT TEMPORARY EYE INJURY

CAUSES SKIN IRRITATION.

HARMFUL IF SWALLOWED.

MAY CAUSE RESPIRATORY TRACT IRRITATION.

MAY CAUSE ALLERGIC SKIN REACTION.

Do not get in eyes, on skin or on clothing.

Avoid breathing vapors or spray mist.

Wear goggles or face shield, rubber gloves, long trousers, long-sleeved shirt or jacket of tightly woven material along with boots high enough to cover ankles when transferring and mixing, and when adjusting, repairing or cleaning equipment. Wear rubber boots when pouring from open containers. Clothing which comes in contact with this product must be washed before reuse. Clothing or other materials which become drenched with the concentrated pesticide must be disposed of in a sanitary landfill, by incineration or, if allowed by state and local authorities, by burning. If burned, stay out of smoke. Wash thoroughly with soap and water after handling.

FIRST AID: IF IN EYES, immediately flush with plenty of water for at least 15 minutes. Get medical attention.

IF ON SKIN, immediately flush with plenty of water while removing contaminated clothing. As soon as soap is available, wash thoroughly with soap and water. Wash clothing before reuse. Sensitized persons should avoid further contact and reuse of contaminated clothing. Get medical attention.

IF SWALLOWED, drink promptly a large quantity of milk, egg whites, gelatin solution, or if these are not available, drink large quantities of water. Avoid alcohol. Get medical attention. NEVER GIVE ANYTHING BY MOUTH TO AN UNCONSCIOUS PERSON.

IF INHALED, remove individual to fresh air. Get medical attention if breathing difficulty develops.

In case of an emergency involving this product, Call Collect, day or night, (314) 694-4000.

ENVIRONMENTAL HAZARDS

This product contains alachlor which has been identified in limited ground water sampling and there is the possibility that it can leach through the soil to ground water, especially where soils are coarse and ground water is near the surface. Following application, and during rainfall events that cause runoff, this chemical may reach surface water bodies including streams, rivers, and reservoirs.

Care must be taken when using this product to prevent back-siphoning into wells, spills or improper disposal of excess pesticide, spray mixtures or rinsates.

Check valves or antisiphoning devices must be used on all mixing equipment. Disposal of excess pesticide, spray mixtures, or rinsate should be according to label use instructions or according to the state pesticide or environmental control agency or the hazardous waste representative at the nearest EPA regional office.

Do not apply directly to water or wetlands. Do not contaminate water when disposing of equipment washwaters.

DIRECTIONS FOR USE

It is a violation of Federal law to use this product in any manner inconsistent with its labeling.

STORAGE AND DISPOSAL

Do not contaminate water, foodstuffs, feed or seed by storage or disposal.

STORAGE:

STORE ABOVE 32°F TO KEEP PRODUCT IN SOLUTION.

Below 32°F crystals may form and settle to the bottom. If crystals form, place in a warm room and roll and/or shake the container frequently to redissolve before using. For bulk containers, see the container label for alternate storage information.

DISPOSAL:

Wastes of this pesticide may cause eye irritation and may be dangerous. Improper disposal of excess pesticide, spray mixtures, or rinsate is a violation of Federal law. If these wastes cannot be disposed of according to label use instructions, contact your State Pesticide or Enviornmental Control Agency, or the Hazardous Waste representative at the nearest EPA Regional Office for guidance.

Emptied container retains vapor and product residue. Observe all labeled safeguards until container is cleaned, reconditioned or destroyed. DO NOT CUT OR WELD ON OR NEAR THIS CONTAINER.

(See the individual container label for disposal information.)

GENERAL INFORMATION

This product is recommended for the control and to reduce competition of annual grasses and broadleaf weeds listed in the "WEEDS CONTROLLED" section of this label. Apply this product and shallowly incorporate prior to planting to blend the herbicide into the upper 1 to 2 inches of soil.

Performance is often improved when sufficient rainfall or irrigation occurs to moisten the soil within the weed germination zone. If adequate moisture is not received prior to weed emergence, then shallow cultivation is recommended. Refer to the "APPLICATION AND TIMING METHODS" section of this label for details.

Read and carefully observe the precautionary statements and all other information appearing on the labels of all products used in mixtures and sequential treatments.

Do not flood irrigate following application of this product or tank mixtures.

Do not apply when conditions favor drift. Use of this product in any manner not consistent with this label may result in injury to persons, animals or crops, or other unintended consequences. Keep container closed to prevent spills and contamination.

Drain and thoroughly flush sprayer with clean water after use and also prior to utilizing spray equipment for treating crops other than soybeans.

Do not graze treated area or feed treated soybean forage, hay or straw to livestock following application.

SOIL TEXTURE

Applicators should evaluate soil and weed conditions carefully to assure that they choose the lowest effective label rate.

The recommended use rates of this product and the other herbicides labeled for use in tank mixtures with this product generally vary with soil texture. Unless soil texture is specifically named, rate tables throughout this label refer to only three soil textural gropus: coarse, medium and fine. The following is a complete listing of soil textures included in each of these three soil textural groups:

Soil Textural Group	Soil Texture
COARSE	sand, loamy sand, sandy loam
MEDIUM	loam, silt loam, silt, sandy clay loam
FINE	silty clay loam, clay loam, sandy clay, silty clay, clay

Refer to the above table to determine the corresponding soil textural group for the soil to be treated.

MIXING, SPRAYING AND HANDLING INSTRUCTIONS

NOTE

The use of a mechanical transfer system is required to be used by all mixer/loaders and/or applicators who treat 300 acres or more annually with pesticides containing alachlor as an active ingredient.

Direct contact or exposure to this product or spray mixtures of this product should be minimized. The following instructions for transfer, mixing, cleaning or repairing equipment should be followed in order to minimize this exposure. Review the protective clothing requirements as listed in the "PRECAUTIONARY STATEMENTS" section of this label (front panel) and do not use this product until you are wearing the necessary protective clothing.

2.5 Gallon Containers

Open pouring from these containers can result in exposure from splashing or spilling. Special care in lifting and pouring is strongly recommended.

Bulk Containers

This product should be transferred from these containers to the mix or spray tank using pumps or transfer probes. The probe or pump should not be removed from the container or disconnected until the container is emptied and rinsed. Use the pump or probe system to rinse the empty container and transfer the rinsate directly to the mix or spray tank.

Equipment Cleaning and Repair

Cleaning and repair of transfer systems or application equipment is a source of exposure to this product. Care should be taken to minimize exposure during cleaning and repair of transfer systems and application equipment. Whenever possible, these systems or this equipment should be rinsed before being cleaned or repaired. When repairs must be made during transfer or application, the equipment must be shut down and special care should be taken to avoid contact with the pesticide.

Always predetermine the compatibility of this product or labeled tank mixtures of this product with water carrier or sprayable fluid fertilizer carrier by mixing small proportional quantities in advance. See the "STANDARD SPRAYABLE FLUID FERTILIZER COMPATIBILITY TEST" section in this label to determine the compatibility of this product and the labeled tank mixtures recommended for use with sprayable fluid fertilizer carrier.

Mix this product or labeled tank mixture of this product with the appropriate carrier as follows:

1. Place a 20 to 35 mesh screen or wetting basked over filling port.
2. Through the screen, fill the sprayer tank one-half full with the appropriate carrier.
3. If a compatibility agent is necessary to improve mixing, or to prevent the formation of undesirable and unsprayable gels or precipitates, while agitating, add it to the carrier already in the tank. Use only compatibility agents cleared by FDA for this use. Read and follow all directions for use, cautionary statements and all other information appearing on the selected compatibility agent label. Check for adequate agitation.
4. If a wettable powder or dry flowable formulation is used, make a slurry with water and add it slowly through the screen into the tank. Continue agitation.
5. If a flowable formulation is used, add slowly through the screen into the tank. Mixing and compatibility may be improved when flowable is premixed one part flowable with one part water and added to the tank in diluted form.
6. Add Freedom herbicide slowly thorugh the screen into the tank. Mixing and compatibility may be improved when this product is prediluted with two parts of water and added to the tank in diluted form.
7. Complete filling the sprayer tank with carrier. Remove hose from tank immediately after filling to avoid siphoning back into the carrier source.

Maintain good agitation at all times until the contents of the tank are sprayed. NOTE: If spray mixture is allowed to settle at any time, thorough agitation is required to resuspend the mixture before spraying is resumed.

Keep by-pass line on or near bottom of tank to minimize foaming. Screen size in nozzle or line strainers should be no finer than 50 mesh. Carefully select the correct nozzle to avoid spraying a fine mist. Check for even distribution of spray droplets. To reduce loss of chemical due to drift of a fine mist, apply at nozzle pressures below 40 psi.

STANDARD SPRAYABLE FLUID FERTILIZER COMPATIBILITY TEST

Herbicides may not always mix evenly throughout a sprayable fluid fertilizer or the components may separate too quickly to make their combined use of practical value. This may be due to certain characteristics of the different fluid fertilizers. A simple test using small quantities of the components is suggested to determine compatibility potential.

The test follows:

A. Materials Required For A Compatibility Test

1. Two one-quart jars with lid or stopper (marked "with" and "without").
2. TEAspoon (for a more exacting test, a five to ten milliter (ml) pipette or graduated cylinder is desirable).
3. Sprayable fluid fertilizer to be tested.
4. The herbicides to be mixed.
5. A compatibility agent (the purpose of the adjuvant is to help keep the fertilizer and crop protection chemical in suspension, if this assistance is needed).

B. Procedure

1. Add one pint of the sprayable fluid fertilizer that will be used or other herbicide carrier to each jar marked "with" and "without".

Add One Pint Liquid Fertilizer
To Two Quart Jars

2. To the jar marked "with", add 1/4 TEAspoon or 1.2 milliters of a suitable compatibility agent, shake gently for five to ten seconds to mix (1/4 TEAspoon in one pint is the equivalent of two pints per 100 gallons of liquid fertilizer).

To Jar Marked "With"
Add Compatibility Agent
And Shake To Mix

3. To each jar add the appropriate amount of herbicide(s) listed on the following tables. If more than one is used, add them separately with the wettable powders or dry flowables added first, flowables second and liquid last. Shake gently five to ten seconds after each addition.

Add Herbicide(s) To Both Jars
And Shake To Mix

Herbicide	Amount To Be Added Per Pint Of Sprayable Fluid Fertilizer (Assuming Volume is 25 gal/a) Rate/A		Level TEAspoons
Wettable Powders or Dry Flowables	1 lb	=	1.5
	2 lb	=	3
	3 lb	=	4.5
	4 lb	=	6
	5 lb	=	7.5

Herbicide	Rate/A		Level TEAspoons		Milliters
Emulsifiable Concentrates or Flowables or Liquids or Solutions	1 pint	=	0.5	or	2.4
	1 quart	=	1	or	4.7
	2 quarts	=	2	or	9.5
	3 quarts	=	3	or	14.2
	1 gallon	=	4	or	19.0
	5 quarts	=	5	or	23.8

The compatibility test is designed for 25 gallons of spray per acre. For changes in spray volume of herbicide rate, make appropriate changes in the ingredients of the test. Regardless of spray volume, the amount of compatibility agent should be equal to two or three pints (two pints = 1/4 TEAspoon or 1.2 milliters, three pints = 3/8 TEAspoon or 1.8 milliters per pint of sprayable fluid fertilizer) per 100 gallons of liquid fertilizer.

C. Observations and Decisions

1. If the herbicide(s) and the sprayable fluid fertilizer are compatible.
2. If a compatibility agent is necessary.

Five minutes after the final addition and mixing, observe both jars for the formation of large flakes, sludge, gels or other precipitates. Observe if the herbicide(s) cannot be physically mixed with the liquid fertilizer but remains as small oily particles in the solution.

If incompatibility in any form described above occurs in the jar "with" the compatibility agent added, the liquid fertilizer and the herbicide(s) should not be used together in the same spray tank.

If incompatibility as described above occurs in the jar "without" the adjuvant but not in the jar "with" adjuvant, the use of a compatibility adjuvant is recommended.

Both jars should be allowed to stand and be observed periodically for one-half hour. If the separate layers of liquid fertilizer and additives can be resuspended by shaking, commercial application is possible. An emulsifiable concentrate normally will go to the top after standing, wettable powders will either settle to the bottom of the tank or jar, or float to the top, depending upon the density of the fertilizers.

If the herbicide(s) is compatible with fluid fertilizer in the foregoing test without using a compatibility agent, fluid fertilzer may be used for premixing. If it is not compatible without the compatibility agent, the herbicide(s) should be premixed with water before adding to the spray tank.

APPLICATION INFORMATION

Do not apply this product through any type of irrigation system.

APPLICATION SYSTEMS

Ground Broadcast Treatment—Apply this product and the labeled tank mixtures in 10 or more gallons of solution per acre using broadcast boom equipment. The carrier may be either water or sprayable fluid fertilizer. Do not apply during periods of gusty winds, when winds are in excess of 15 mph or when other conditions favoring drift exist.

Aerial—Fixed wing; helicopter—Apply the recommended rates of this product or approved tank mixtures in 3 to 10 gallons of water per acre. AVOID DRIFT—DO NOT APPLY DURING INVERSION CONDITIONS, WHEN WINDS ARE GUSTY, OR UNDER ANY OTHER CONDITION WHICH WILL ALLOW DRIFT. Coarse sprays are less likely to drift; therefore, do not use nozzles or nozzle configurations which dispense spray or fine spray droplets. Do not angle nozzles forward into the airstream and do not increase spray volume by increasing nozzle pressure. Ensure uniform application. To avoid streaked, uneven or overlapped application, use mechanical marking systems only. HUMAN FLAGGERS PROHIBITED!

Application with Dry Bulk Fertilizer—The herbicide/fertilizer impregnation process must be completed only at commercial fertilizer dealerships equipped for this procedure.

Dry bulk fertilizer may be impregnated with this product or the tank mixtures of this product plus Lexone™ or Sencor™. FREEDOM herbicide and these tank mixtures must be applied with 200 to 450 pounds of dry bulk fertilizer per acre and shallowly incorporated within 7 days prior to planting. The correct amount of herbicide must be applied per acre in the blend recommended in this label for the crop selected and weed species to be controlled.

The following table provides a reference to determine the amount of LIQUID herbicide to be mixed per ton of dry bulk fertilizer for a range of herbicide recommendations and fertilizer rates per acre.

Fertilizer Rate (lb/a)	Acres Covered (per ton)	RECOMMENDED QUARTS LIQUID HERBICIDE/ACRE 3	3.5 (quarts of herbicide/ton dry bulk fertilizer)	4
200	10	30	35	40
250	8	24	28	32
300	6.7	20	23.3	26.6
350	5.7	17.1	20	22.8
400	5	15	17.5	20
450	4.5	13.3	15.6	17.8

To determine the amount of herbicide needed for rates not included in the preceding table, use the following formula:

$$\frac{\text{Recommended herbicide rate (lb or qt/acre} \times 2000)}{\text{pounds fertilizer/acre}} = \text{pounds or quarts of herbicide per ton of dry bulk fertilizer}$$

Mix and blend the dry fertilizer and herbicide mixture in a closed rotary drum-type mixer, allowing sufficient time to ensure uniform coverage. Use at least one ton of dry fertilizer per mixing operation. Inject the herbicide into the drum over a minimum of a 2-minute period and allow at least 2 additional minutes mixing time to ensure uniformity. The nozzle used to spray the herbicide treatment must be placed inside the mixer to provide uniform spray coverage of the tumbling fertilizer.

If the dry fertilizer used has inadequate adsorptive capacity, use a highly adsorptive material such as Agsorb™, NP-79™, or Microcel™ E to provide a free-flowing mixture. The adsorptive material and dry fertilizer should be thoroughly mixed prior to adding liquid herbicide. Generally the drying material weight is no more than 2 to 4 percent of the total weight of the dry fertilizer. The following table provides a list of dry fertilizers which may be impregnated with this product or tank mixtures of this product with other herbicides.

™Agsorb is a trademark of Oil Dri Corporation.
™Lexone is a trademark of E.I. du Pont de Nemours and Company.
™Microcel is a trademark of Manville Products Corporation.
™NP-79 is a trademark of Eagle Pitcher Industries.
™Sencor is a trademark of the parent company of Farbenfabriken Bayer Gmbh.

Fertilizer	FREEDOM™	FREEDOM + Lexone or Sencor
Ammonium sulfate (21-0-0)	Yes	Yes
Ammonium phosphate-sulfate (16-20-0)	Yes	Yes
Diammonium phosphate		

Fertilizer	FREEDOM™	FREEDOM + Lexone or Sencor
(18-46-0)	Yes	Yes
Potassium chloride (0-0-60)	Yes	Yes
Single super-phosphate (0-20-0)	Yes	Yes
Treble super-phosphate (0-46-0)	Yes	Yes
Urea (46-0-0)	Yes	Yes

NOTE: DO NOT impregnate this product or tank mixtures of this product with other herbicides on fertilizers containing ammonium nitrate, potassium nitrate or sodium nitrate.

Spread the dry fertilizer-herbicide mixture uniformly with an adequately calibrated applicator. Fertilzers impregnated with this product or tank mixtures of this product with other herbicides must be spread uniformly. Non-uniform spreading of the dry fertilizer/herbicide mixture may result in unsatisfactory weed control or crop injury.

APPLICATION TIMING AND METHODS

Preplant Incorporation—This product and the labeled tank mixtures may be mixed into the soil using shallow incorporation equipment any time within 7 days prior to planting. Apply the recommended treatement rate to the soil surface as a broadcast application and shallowly incorporate into the upper 2 inches of soil within 24 hours after application. If weeds emerge after treatment, rotary hoe or shallowly cultivate immediately to improve performance.

1-Pass Incorporation (Surface Blend™): Shallowly incorporate (Surface Blend) the treatment into the upper 1 to 2 inches of the soil. Equipment should be operated at manufacturer's designated speed for incorporation to ensure adequate mixing and distribution of the herbicide treatment in the soil. Equipment design, including any drag attachments, must be adequate to avoid soil ridging which may result in streaked or reduced weed control. Soil conditions, including moisture content and crop residue levels, must be suitable to allow thorough and uniform mixing with the equipment used for 1-pass incorporation.

Contact Monsanto Company for additional information regarding recommended equipment and methods for Surface Blend application.

2-Pass Incorporation: When 2-pass incorporation is used, shallowly incorporate the herbicide treatment into the upper 1 to 2 inches of the soil with equipment set to work the soil NO DEEPER THAN 4 INCHES. The second pass must be made at an angle and no deeper than the first pass to ensure uniform distribution of the herbicide treatment in the soil.

Cultivation Information—Delay cultivation after application for as long as possible unless weeds or grasses emerge. Shallowly cultivate or rotary hoe immediately if weeds or grasses emerge. If cultivation is necessary because of soil crusting or compaction, set equipment shallow and minimize lateral soil movement to avoid dilution or displacement of the herbicide treatment.

™Surface Blend is a trademark of Monsanto Company

WEEDS CONTROLLED

Applicators should evaluate soil and weed conditions carefully to assure that they choose the lowest effective label rate. When applied as directed under conditions described, this product and tank mixtures of this product will control or reduce competition from the weeds listed.

C = Control R = Reduced Competition

	FREEDOM™	FREEDOM + Canopy™	FREEDOM + Command™	FREEDOM + Preview™	FREEDOM + Scepter™	FREEDOM + Sencor™ /Lexone™
ANNUAL GRASSES						
Barnyardgrass	C	C	C	C	C	C
Crabgrass: hairy, smooth	C	C	C	C	C	C
Cupgrass, woolly[1]	R	R	R	R	R	R
Foxtail: giant, green, robust purple, robust white, yellow	C	C	C	C	C	C
Goosegrass	C	C	C	C	C	C
Johnsongrass, seedling[2]	R	R	R	R	C	R
Panicum: browntop, fall	C	C	C	C	C	C
Panicum, Texas[3]	R	R	R	R	R	R
Rice, red[4]	C	C	C	C	C	C
Sandbur, Grassbur[2]	R	R	R	R	R	R
Shattercane, wild cane[1]	R	R	R	R	R	R
Signalgrass, broadleaf	C	C	C	C	C	C
Sprangletop, red	C	C	C	C	C	C
Witchgrass	C	C	C	C	C	C
ANNUAL BROADLEAVES						
Beggarweed, Florida[2]	R	C	•	R	•	C
Carpetweed	C	C	C	C	C	C
Cocklebur	•	C	•	C	C	R
Coffeeweed: hemp sesbania	•	C	•	C	•	•
Copperleaf, hophornbeam	•	C	•	C	•	•
Galinsoga	C	C	C	C	C	C
Groundcherry, annual	•	C	•	C	C	C
Groundcherry, cutleaf[5]	C	C	C	C	C	C
Jimsonweed	•	C	C	C	C	R
Kochia	•	C	C	C	C	C
Lambsquarters[5]	R	C	C	C	C	C
Morningglory:						
Tall, ivyleaf, entireleaf	•	C	•	R	R	•
Pitted	•	C	•	R	C	•
Smallflower	R	C	•	•	C	•
Mustard	•	C	•	C	C	•
Nightshade:						
eastern black[4], hairy[4]	C	C	C	C	C	C
Pigweed, Carelessweed	C	C	C	C	C	C
Purslane	C	C	C	C	C	C
Pusley, Florida	C	C	C	C	C	C
Ragweed, common[5]	R	C	C	C	C	C
Ragweed, giant	•	R	•	R	R	•
Sicklepod	•	C	•	•	R	R
Sida, prickly; Teaweed[5]	R	C	C	C	C	•
Smartweed[5]	R	C	R	C	C	C
Starbur, bristly	R	C	•	•	C	C
Sunflower, common	•	R	•	R	C	•
Velvetleaf; Buttonweed	•	C	C	C	C	R
Waterhemp	C	C	C	C	C	C
SEDGE						
Nutsedge, yellow[4]	C	C	C	C	C	C

™Canopy and Preview are trademarks of E.I. du Pont de Nemours and Company.
™Command is a trademark of FMC Corporation.
™Scepter is a trademark of American Cyanamid Company.

[1]Apply 5 to 5.5 quarts of this product per acre alone or in tank mixture on this weed.
[2]Apply 4 to 5 quarts of this product per acre alone or in tank mixture on this weed.
[3]Apply 4.5 quarts of this product per acre, followed by sprinkler irrigation if rainfall is not expected.
[4]Apply 4.5 to 5.5 quarts of this product per acre alone or in tank mixture on this weed. When tank mixed with Scepter, use 3 to 4 quarts of this product per acre.
[5]Apply 4 to 5 quarts of this product per acre when used alone on this weed.

For scientific names for the weeds listed above, refer to the appendix at the end of this booklet.

SOYBEANS

FREEDOM™

Apply this product in water or sprayable fluid fertilizer solution for control of yellow nutsedge and the annual grasses and broadleaf weeds listed in the "WEEDS CONTROLLED" section of this label.

APPROVED APPLICATION SYSTEMS

Ground—Broadcast boom
Aerial—Fixed wing; helicopter
Dry Bulk Fertilizer Impregnation

Reference: The "APPLICATION SYSTEMS" section of this label provides detailed information on the application of this product using the system selected.

RECOMMENDED RATES: Refer to the following table for the recommended rates for this product. Applications which are not consistent with recommendations in this label may result in unsatisfactory weed control or crop injury.

Soil Textural Group*	Broadcast Rate per Acre (Quarts)**	
	Less than 3% organic matter	3% or more organic matter
Coarse	3.5 to 4	3.5 to 4
Medium	3.5 to 4.5	3.5 to 4.5
Fine	4 to 4.5	4.5 to 5

*Refer to the "SOIL TEXTURE" section of this label to determine the corresponding textural group for the soil to be treated.
**Use the higher rate in the recommended range in areas of heavy weed infestations. Use a minimum of 4.5 quarts per acre on coarse textured soils for control of yellow nutsedge or red rice. Rates should be increased to a maximum of 5.5 quarts on medium and fine soils when yellow nutsedge or red rice is present.

FREEDOM™ plus CANOPY TANK MIXTURE

Apply this tank mixture in water or sprayable fluid fertilizer solution for control of yellow nutsedge and the annual grasses and broadleaf weeds listed in the "WEEDS CONTROLLED" section of this label.

NOTE: Observe all precautions and limitations on the labels for Freedom and Canopy herbicides, including precautions on soil pH, sensitive soybean varieties, minimum recropping interval and rotational guidelines.

DO NOT plant rice or peanuts within 18 months after application of this tank mixture.

APPROVED APPLICATION SYSTEM

Ground—Broadcast boom

Reference: The "APPLICATION SYSTEMS" section of the label provides detailed information on the application of this tank mixture using the system selected.

RECOMMENDED RATES: Refer to the following table for the recommended broadcast treatment rates for this tank mixture. Applications which are not consistent with recommendations in this label may result in unsatisfactory weed control or crop injury.

	Broadcast Rates per Acre**		
	½ to 3% organic matter	3% or more organic matter	
Soil Textural Group*	FREEDOM™ (quarts)	FREEDOM (quarts)	Canopy (ounces)
Coarse-Sandy Loam Only	2.75 to 3.5	3 to 3.5	6
Medium	2.75 to 3.5	3.5 to 4	8
Fine	3.5 to 4	3.5 to 4.5	10

*Refer to the "SOIL TEXTURE" section of this label to determine the corresponding textural group for the soil to be treated. DO NOT APPLY TO SOIL WITH LESS THAN ½% organic matter.
**Use the higher rate of FREEDOM herbicide in the recommended ranges in areas of heavy grass infestations, heavy crop residue or when broadleaf signalgrass is present. See "WEEDS CONTROLLED" section of this label for specific rate recommendations on seedling johnsongrass, red rice and yellow nutsedge. Use the higher rate of Canopy in the recommended ranges on soils with higher organic matter or heavy pressure from large deep germinating weed seeds.

FREEDOM™ plus COMMAND TANK MIXTURE

Apply this tank mixture in water or sprayable fluid fertilizer solution for control of yellow nutsedge and the annual grasses and broadleaf weeds listed in the "WEEDS CONTROLLED" section of this label.

NOTE: Observe all precautions and limitations on the labels for FREEDOM and Command herbicides before using this tank mixture.

APPROVED APPLICATION SYSTEMS

Ground—Broadcast boom

Reference: The "APPLICATION SYSTEMS" section of this label provides detailed information on the application of this tank mixture using the system selected.

RECOMMENDED RATES: Apply 3 to 3.5 quarts of FREEDOM herbicide plus 1 pint of Command 4E per acre. In areas of heavy pigweed and lambsquarter infestations, use 3.5 to 4 quarts of FREEDOM per acre in the tank mixture.

FREEDOM™ plus LEXONE or FREEDOM plus SENCOR TANK MIXTURES

Apply this tank mixture in water or sprayable fluid fertilizer solution for control of yellow nutsedge and the annual grasses and broadleaf weeds listed in the "WEEDS CONTROLLED" section of this label.

NOTE: DO NOT use on muck soils or on alkaline soils with greater than 7.4 pH. Crop injury may result if atrazine was applied the year preceding use of this tank mixture.

Observe all precautions and limitations on the labels for FREEDOM, Lexone and Sencor herbicides before using this tank mixture.

DO NOT REPLANT CROPS OTHER THAN SOYBEANS FOR 120 DAYS AFTER APPLICATION OF THESE TANK MIXTURES.

APPROVED APPLICATION SYSTEMS

Ground—Broadcast boom

Aerial—Fixed wing: helicopter

Dry Bulk Fertilizer Impregnation

Reference: The "APPLICATION SYSTEMS" section of this label provides detailed information on the application of this tank mixture using the system selected.

RECOMMENDED RATES: Refer to the following table for the recommended broadcast treatment rates for this tank mixture. Applications which are not consistent with recommendations in this label may result in unsatisfactory weed control or crop injury.

	Broadcast Rates per Acre**			
	Less than 3% organic matter		3% or more organic matter	
Soil Textural Group*	FREEDOM™ + (quarts)	Lexone 4L** or Sencor 4L** (pint)	FREEDOM + (quarts)	Lexone 4L** or Sencor 4L** (pint)
Coarse-Sandy Loam Only	2.75 to 3.5	.05	3 to 3.5	0.75
Medium	2.75 to 3.5	0.5 to 0.75	3.5 to 4	0.75
Fine	3.5 to 4	0.75 to 1	3.5 to 4.5	1

*Refer to the "SOIL TEXTURE" section of this label to determine the corresponding soil textural group for the soil to be treated.
**Use the rates listed in this label when using Lexone 4L or Sencor 4F. Use the equivalent rates when using the 50WP or DF formulations of these products. One pint of Lexone 4L or Sencor 4F equals one pound of the corresponding 50WP formulation or ⅔ pound of the corresponding DF formulation.

NOTE: DO NOT use on a sand or loamy sand soil. Use 1.5 pints of Lexone 4L or Sencor 4F per acre in this tank mixture when applying on heavy clay soils of the Mississippi Delta. Use a minimum of 4.5 to 5.5 quarts of FREEDOM herbicide in this tank mixture on medium or fine textured soils to control black or hairy nightshade.

For improved control of heavy velvetleaf infestations, add ½ pint of Command herbicide per acre to this tank mixture.

Observe all precautions and limitations on the labels for Lexone, Sencor and Command before using these tank mixtures.

FREEDOM™ plus PREVIEW TANK MIXTURE

Apply this tank mixture in water or sprayable fluid fertilizer solution for control of yellow nutsedge and the annual grasses and broadleaf weeds listed in the "WEEDS CONTROLLED" section of this label.

NOTE: Observe all precautions and limitations on the labels for FREEDOM and Preview, including precautions on soil pH, sensitive soybean varieties, minimum recropping interval and rotational guidelines.

APPROVED APPLICATION SYSTEMS

Ground—Broadcast boom

Reference: The "APPLICATION SYSTEMS" section of the label provides detailed information on the application of this tank mixture using the system selected.

RECOMMENDED RATES: Refer to the following table for the recommended broadcast treatment rates for this tank mixture. Applications which are not consistent with recommendations in this label may result in unsatisfactory weed control or crop injury.

	Broadcast Rates per Acre**			
	½ to 3% organic matter		3% or more organic matter	
Soil Textural Group*	FREEDOM™ + (quarts)	Preview (ounces)	FREEDOM + (quarts)	Preview (ounces)
Coarse-Sandy Loam Only	2.75 to 3.5	6	3 to 3.5	7
Medium	2.75 to 3.5	7	3.5 to 4	8
Fine	3.5 to 4	8	3.5 to 4.5	9

*Refer to the "SOIL TEXTURE" section of this label to determine the corresponding soil textural group for the soil to be treated. DO NOT APPLY TO SOIL WITH LESS THAN ½% organic matter.
**Use the higher rate of FREEDOM herbicide in the recommended ranges in areas of heavy grass infestations, heavy crop residue or when broadleaf signalgrass is present. See "WEEDS CONTROLLED" section of this label for specific rate recommendations on seedling johnsongrass, red rice and yellow nutsedge.

FREEDOM™ plus plus SCEPTER TANK MIXTURE

Apply this tank mixture in water or sprayable fluid fertilizer solution for control of yellow nutsedge and the annual grasses and broadleaf weeds listed in the "WEEDS CONTROLLED" section of this label.

Note: Observe all precautions and limitations on the labels for FREEDOM and Scepter herbicides before using this tank mixture including precautions on minimum recropping interval and rotational guidelines.

APPROVED APPLICATION SYSTEMS

Ground—Broadcast boom

Aerial—Fixed wing; helicopter

Reference: The "APPLICATION SYSTEMS" section of this label provides detailed information on the application of this tank mixture using the system selected.

RECOMMENDED RATES: Refer to the following table for the recommended broadcast treatment rates for this tank mixture. Applications which are not consistent with recommendations in this label may result in unsatisfactory weed control or crop injury. See the "WEEDS CONTROLLED" section of this label for recommended rates and application methods for control of certain weeds.

Soil Textural Group*	Broadcast Rates per Acre**		
	Less than 3% organic matter	3% or more organic matter	
	FREEDOM™ (quarts)	FREEDOM (quarts)	Scepter*** (pints)
Coarse	2.5 to 3	2.5 to 3	½ to ⅔
Medium	2.5 to 3	2.75 to 3.25	½ to ⅔
Fine	2.75 to 3.25	3 to 3.5	½ to ⅔

*Refer to the "SOIL TEXTURE" section of this label to determine the corresponding soil textural group for the soil to be treated.

**Use the higher rate of FREEDOM herbicide in the recommended ranges in areas of heavy grass infestations, heavy crop residue or when the following weeds are present: fall panicum, barnyardgrass, broadleaf signalgrass, seedling johnsongrass, red rice or yellow nutsedge.

***Use the higher rates of Scepter in the recommended ranges when red rice, sicklepod, Florida beggarweed or velvetleaf are present. For improved control of heavy velvetleaf infestations, add ½ pint of Command 4E herbicide per acre to this tank mixture.

Observe all precautions and limitations on the labels for Scepter and Command before using these tank mixtures.

APPENDIX

LIST OF COMMON AND SCIENTIFIC NAME OF ANNUAL GRASSES AND BROADLEAF WEEDS.

ANNUAL GRASSES

Barnyardgrass
Echinochloa crus-galli
Crabgrass:
Digitaria schaemum
Digitaria sanguinalis
Cupgrass, wooly
Eriochloa villosa
Foxtail, giant
Setaria faberi
Foxtail, green robust purple, robust white
Setaria viridis
Foxtail, yellow
Setaria lutescens
Goosegrass
Eleusine indica
Johnsongrass
Sorghum halepense
Panicum, browntop
Panicum fasciculatum
Panicum fall
Panicum dichotomiflorum
Panicum, Texas
Panicum texanum
Rice, red
Orvza sativa
Sandbur, Grassbur
Cenchrus incertus
Shattercane, wild cane
Sorghum bicolor
Signalgrass, broadleaf
Brachiaria platyphylla
Sprangletop, red
Leptochloa filiformis
Witchgrass
Panicum capillare

ANNUAL BROADLEAVES

Beggarweed, Florida
Desmodium tortuosum
Carpetweed
Mollugo verticillata
Cocklebur
Xanthium strumarium
Coffeeweed
Hemp sesbania
Sesbania exaltata
Copperleaf, hophornbeam
Acalypha spp.
Galinsoga
Galinsoga spp.
Groundcherry, annual
Physalis spp.
Groundcherry, cutleaf
Physalis angulata
Jimsonweed
Datura stramonium
Kochia
Kochia scoparia
Lambsquarters
Chenopodium album
Morningglory:
Tall
Ipomea purpurea
Pitted
Ipomea lacunosa
Ivyleaf
Ipomea hederacea
Entireleaf
Ipomea hederacea var. intergriuscula
Mustard
Brassica spp.
Nightshade, black
Solanum nigrum
Nightshade, hairy
Solanum sarrachoides
Pigweed, Carelessweed
Amaranthus spp.
Purslane
Portulaca oleracea
Pusley, Florida
Richardia scabra
Ragweed common
Ambrosia artemisifolia
Ragweed, giant
Ambrosia trifida
Sicklepod
Cassia obtusifolia
Sida. prickly; Teaweed
Sida spinosa
Smartweed
Polygonum pensylvanicum
Polygonum persicaria
Starbur, bristly
Acanthospermum hispidum
Sunflower, commonn
Helianthus annuus
Velvetleaf, Buttonweed
Abutilon theophrasti
Waterhemp
Amaranthus tuberculatos
Smallflower
Jacquemontia tamnifolia

SEDGE

Nutsedge, yellow
Cyperus esculentus

STATE REGISTRATIONS

FREEDOM™ herbicide has been approved by the U.S. Environmental Protection Agency for the crop, uses and sites listed in this label. This product is registered for use in the following states:

Alabama	Louisiana
Arkansas	Mississippi
Florida	Missouri
Georgia	Nebraska
Illinois	South Carolina
Kansas	Tennessee
Kentucky	Texas

In case of emergency involving this product, Call Collect, day or night, (314) 694-4000.

843.31-001.01/CG

This sample label is current as of January 1, 1989. The product descriptions and recommendations provided in this sample label are for background information only. Always refer to the label on the product container before using Monsanto or any other agrichemical product.

HONCHO®

HERBICIDE

FOR USE IN ALABAMA, ARIZONA, ARKANSAS, DELAWARE, FLORIDA, GEORGIA, LOUISIANA, MARYLAND, MISSISSIPPI, NEVADA, NEW MEXICO, NORTH CAROLINA, OREGON, SOUTH CAROLINA, TENNESSEE, TEXAS, VIRGINIA, AND WASHINGTON.

Complete Directions for Use

AVOID CONTACT WITH FOLIAGE, GREEN STEMS, OR FRUIT OF CROPS, DESIRABLE PLANTS AND TREES, SINCE SEVERE INJURY OR DESTRUCTION MAY RESULT.

®HONCHO is a registered trademark of Monsanto Company.

ACTIVE INGREDIENT:

*Isopropylamine salt of glyphosate	41.5%
INERT INGREDIENTS:	58.5%
TOTAL	100.0%

*Contains 480 grams per litre or 4 pounds of the active ingredient isopropylamine salt of N-(phosphonomethyl) glycine per U.S. gallon. Equivalent to 356 grams per litre or 3 pounds per U.S. gallon of the acid, glyphosate.

Product is protected by U.S. Pat. No. 3,799,758 and U.S. Pat. No. 4,405,531. Other patents pending.

No license granted under any non-U.S. patent.

EPA Reg. No. 524-326

1988-2 892.67-000.08/53

Read the entire label before using this product.

Use only according to label instructions.

Read "LIMIT OF WARRANTY AND LIABILITY" before buying or using. If terms are not acceptable, return at once unopened.

REFORMULATION OR REPACKAGING IS PROHIBITED.

LIMIT OF WARRANTY AND LIABILITY

This Company warrants that this product conforms to the chemical description on the label and is reasonably fit for the purposes set forth in the Complete Directions for Use label booklet ("Directions") when used in accordance with those Directions under the conditions described therein. NO OTHER EXPRESS WARRANTY OR IMPLIED WARRANTY OF FITNESS FOR PARTICULAR PURPOSE OR MERCHANTABILITY IS MADE. This warranty is also subject to the conditions and limitations stated herein.

Buyer and all users shall promptly notify this Company of any claims whether based in contract, negligence, strict liability, other tort or otherwise.

Buyer and all users are responsible for all loss or damage from use or handling which results from conditions beyond the control of this Company, including, but not limited to, incompatibility with products other than those set forth in the Directions, application to or contact with desirable vegetation, unusual weather, weather conditions which are outside the range considered normal at the application site and for the time period when the product is applied, as well as weather conditions which are outside the application ranges set forth in the Directions, application in any manner not explicitly set forth in the Directions, moisture conditions outside the moisture range specified in the Directions, or the presence of products other than those set forth in the Directions in or on the soil, crop or treated vegetation.

THE EXCLUSIVE REMEDY OF THE USER OR BUYER, AND THE LIMIT OF THE LIABILITY OF THIS COMPANY OR ANY OTHER SELLER FOR ANY AND ALL LOSSES, INJURIES OR DAMAGES RESULTING FROM THE USE OR HANDLING OF THIS PRODUCT (INCLUDING CLAIMS BASED IN CONTRACT, NEGLIGENCE, STRICT LIABILITY, OTHER TORT OR OTHERWISE) SHALL BE THE PURCHASE PRICE PAID BY THE USER OR BUYER FOR THE QUANTITY OF THIS PRODUCT INVOLVED, OR, AT THE ELECTION OF THIS COMPANY OR ANY OTHER SELLER, THE REPLACEMENT OF SUCH QUANTITY, OR, IF NOT ACQUIRED BY PURCHASE, REPLACEMENT OF SUCH QUANTITY. IN NO EVENT SHALL THIS COMPANY OR ANY OTHER SELLER BE LIABLE FOR ANY INCIDENTAL, CONSEQUENTIAL OR SPECIAL DAMAGES.

Buyer and all users are deemed to have accepted the terms of this LIMIT OF WARRANTY AND LIABILITY which may not be varied by any verbal or written agreement.

PRECAUTIONARY STATEMENTS

HAZARDS TO HUMANS AND DOMESTIC ANIMALS

KEEP OUT OF REACH OF CHILDREN

CAUTION!

MAY CAUSE EYE IRRITATION.

Avoid contact with eyes, skin or clothing.

Wash thoroughly with soap and water after handling.

FIRST AID: IF IN EYES, immediately flush with plenty of water for at least 15 minutes. Call a physician.

IF ON SKIN, immediately flush with plenty of water. Wash clothing before reuse.

In case of an emergency involving this product, Call Collect, day or night, (314) 694-4000.

ENVIRONMENTAL HAZARDS

Do not apply directly to water or wetland (swamps, bogs, marshes or potholes). Do not contaminate water when disposing of equipment washwaters.

PHYSICAL OR CHEMICAL HAZARDS

Spray solutions of this product should be mixed, stored and applied only in stainless steel, aluminum, fiberglass, plastic and plastic-lined steel containers.

DO NOT MIX, STORE OR APPLY THIS PRODUCT OR SPRAY SOLUTIONS OF THIS PRODUCT IN GALVANIZED STEEL OR UNLINED STEEL (EXCEPT STAINLESS STEEL) CONTAINERS OR SPRAY TANKS. This product or spray solutions of this product react with such containers and tanks to produce hydrogen gas which may form a highly combustible gas mixture. This gas mixture could flash or explode, causing serious personal injury, if ignited by open flame, spark, welder's torch, lighted cigarette or other ignition source.

GENERAL INFORMATION

DO NOT APPLY THIS PRODUCT USING AERIAL SPRAY EQUIPMENT EXCEPT UNDER CONDITIONS AS SPECIFIED WITHIN THIS LABEL.

This product, a water soluble liquid, mixes readily with water to be applied as a foliar spray for the control or destruction of most herbaceous plants. It may be applied through most standard industrial or field-type sprayers after dilution and thorough mixing with water in accordance with label instructions.

This product moves through the plant from the point of foliage contact to and into the root system. Visible effects on most annual weeds occur within 2 to 4 days, but on most perennial weeds may not occur for 7 days or more. Extremely cool or cloudy weather following treatment may slow down activity of this product and delay visual effects of control. Visible effects are a gradual wilting and yellowing of the plant which advances to complete browning of aboveground growth and deterioration of underground plant parts.

Unless otherwise specified on this label, delay application until vegetation has emerged and reached the stages described for control of such vegetation under the "Weeds Controlled" section of this label. Unemerged plants arising from unattached underground rhizomes or root stocks of perennials will not be affected by the spray and will continue to grow. For this reason, best control of most perennial weeds is obtained when treatment is made at late growth stages approaching maturity.

Always use the higher rate of this product per acre within the recommended range when (1) weed growth is heavy or dense, or (2) weeds are growing in an undisturbed (noncultivated) area.

Do not treat weeds under poor growing conditions such as drought stress, disease or insect damage, as reduced weed control may result. Reduced results may also occur when treating weeds heavily covered with dust.

Reduced control may result when applications are made to annual or perennial weeds that have been mowed, grazed, or cut, and have not been allowed to regrow to the recommended stage for treatment.

Rainfall or irrigation occurring within 6 hours after application may reduce effectiveness. Heavy rainfall or irrigation within 2 hours after application may wash the chemical off the foliage and a repeat treatment may be required.

This product does not provide residual weed control. For subsequent residual weed control, follow a label-approved herbicide program. Read and carefully observe the cautionary statements and all other information appearing on the labels of all herbicides used.

Buyer and all users are responsible for all loss or damage in connection with the use or handling of mixtures of Honcho with herbicides or other materials that are not expressly recommended in this labeling. Mixing this product with herbicides or other materials not recommended on this label may result in reduced performance.

For best results, spray coverage should be uniform and complete. Do not spray weed foliage to the point of runoff.

Keep people and pets off treated areas until spray solution has dried to prevent transfer of this product onto desirable vegetation.

ATTENTION

DOMESTIC ANIMALS: This product is considered to be relatively nontoxic to dogs and other domestic animals; however, ingestion of this product or large amounts of freshly sprayed vegetation may result in temporary gastrointestinal irritation (vomiting, diarrhea, colic, etc.). If such symptoms are observed, provide the animal with plenty of fluids to prevent dehydration. Call a veterinarian if symptoms persist for more than 24 hours.

AVOID DRIFT. EXTREME CARE MUST BE USED WHEN APPLYING THIS PRODUCT TO PREVENT INJURY TO DESIRABLE PLANTS AND CROPS.

Do not allow the herbicide solution to mist, drip, drift, or splash onto desirable vegetation, since minute quantities of this herbicide can cause severe damage or destruction to the crop, plants, or other areas on which treatment was not intended. The likelihood of injury occurring from the use of this product is greatest when winds are gusty or in excess of 5 miles per hour or when other conditions, including lesser wind velocities, will allow spray drift to occur. When spraying, avoid combinations of pressure and nozzle type that will result in splatter or fine particles (mist) which are likely to drift. AVOID APPLYING AT EXCESSIVE SPEED OR PRESSURE.

NOTE: Use of this product in any manner not consistent with this label may result in injury to persons, animals or crops, or other unintended consequences. Keep container closed to prevent spills and contamination.

MIXING, ADDITIVES AND APPLICATION INSTRUCTIONS

APPLY THESE SPRAY SOLUTIONS IN PROPERLY MAINTAINED AND CALIBRATED EQUIPMENT CAPABLE OF DELIVERING DESIRED VOLUMES. DO NOT APPLY UNDER WIND OR OTHER CONDITIONS WHICH ALLOW DRIFT TO OCCUR. HAND GUN APPLICATIONS SHOULD BE PROPERLY DIRECTED TO AVOID SPRAYING DESIRABLE PLANTS. NOTE: REDUCED RESULTS MAY OCCUR IF WATER CONTAINING SOIL IS USED, SUCH AS WATER FROM PONDS AND UNLINED DITCHES.

MIXING

This product requires use with a nonionic surfactant. When using this product, mix 4 to 8 quarts (1 to 2 percent solution) of a nonionic surfactant per 100 gallons of spray solution. Use the 8-quart rate for any surfactant containing less than 50 percent active ingredient. Read and carefully observe surfactant cautionary statements and other information appearing on the surfactant label.

This product mixes readily with water. Mix spray solutions of this product as follows: Fill the mixing or spray tank with the required amount of water. Add the proper amount of this product (see "Directions for Use" and "Weeds Controlled" sections of this label) and nonionic surfactant near the end of the filling process and mix well. Remove hose from tank immediately after filling to avoid siphoning back into the carrier source. During mixing and application, foaming of the spray solution may occur. To prevent or minimize foam, avoid the use of mechanical agitators, place the filling hose below the surface of the spray solution, terminate by-pass and return lines at the bottom of the tank and, if needed, use an approved antifoam or defoaming agent.

CLEAN SPRAYER AND PARTS IMMEDIATELY AFTER USING THIS PRODUCT BY THOROUGHLY FLUSHING WITH WATER AND DISPOSE OF RINSATE ACCORDING TO LABELED USE OR DISPOSAL INSTRUCTIONS.

TANK MIXTURES

Always predetermine the compatability of labeled tank mixtures of this product with water carrier by mixing small proportional quantities in advance.

Mix labeled tank mixtures of this product with water as follows:

1. Place a 20 to 35 mesh screen or wetting basket over filling port.
2. Through the screen, fill the sprayer tank one-half full with water and start agitation.
3. If a wettable powder is used, make a slurry with the water carrier, and add it SLOWLY through the screen into the tank. Continue agitation.
4. If a flowable formulation is used, premix one part flowable with one part water. Add diluted mixture SLOWLY through the screen into the tank. Continue agitation.
5. If an emulsifiable concentrate formulation is used, premix one part emulsifiable concentrate with two parts water. Add diluted mixture SLOWLY through the screen into the tank. Continue agitation.
6. Continue filling the sprayer tank with water and add the required amount of this product near the end of the filling.
7. Add nonionic surfactant to the spray tank before completing the filling process.
8. Add individual formulations to the spray tank as follows: wettable powder, flowable, emulsifiable concentrate, drift control additive, water soluble liquid followed by surfactant.

Maintain good agitation at all times until the contents of the tank are sprayed. If the spray mixture is allowed to settle, thorough agitation is required to resuspend the mixture before spraying is resumed.

Keep by-pass line on or near bottom of tank to minimize foaming. Screen size in nozzle or line strainers should be no finer than 50 mesh. Carefully select proper nozzle to avoid spraying a fine mist. For best results with conventional ground application equipment, use flat fan nozzles.

Clean sprayer and parts immediately after using this product by thoroughly flushing with water.

ADDITIVES

AMMONIUM SULFATE

The addition of 2 percent dry ammonium sulfate by weight or 17 pounds per 100 gallons of water may increase the performance of this product and this product plus 2,4-D amine or Banvel™ tank mixtures on annual and perennial weeds. The improvement in performance may be apparent where environmental stress is a concern. Low-quality ammonium sulfate may contain material that will not readily dissolve which could result in nozzle tip plugging. To determine quality, perform a jar test by adding 1/3 cup of ammonium sulfate to 1 gallon of water and agitate for 1 minute. If undissolved sediment is observed, predissolve the ammonium sulfate in water and filter prior to addition to the spray tank. If ammonium sulfate is added directly to the spray tank, add slowly with agitation. Adding too quickly may clog outlet lines. Ensure that ammonium sulfate is completely dissolved in the spray tank before adding herbicides or surfactant. Thoroughly rinse the spray system with clean water after use to reduce corrosion.

NOTE: The use of ammonium sulfate as an additive does not preclude the need for surfactant. Do not use herbicide rates lower than recommended in this label.

™ Banvel is a trademark of Sandoz, Inc.

COLORANTS OR DYES

Agriculturally approved colorants or marking dyes may be added to this product. Colorants or dyes used in spray solutions of this product may reduce performance, especially at lower rates or dilutions. Use colorants or dyes according to manufacturer's recommendations.

APPLICATION EQUIPMENT AND TECHNIQUES

Do not apply this product through any type of irrigation system.

This product may be applied with the following application equipment:

Aerial—Fixed Wing and Helicopter

Broadcast Spray

Controlled Droplet Applicator (CDA)—Hand-held or boom-mounted applicators which produce a spray consisting of a narrow range of droplet sizes.

Hand-held and High-Volume Spray Equipment—Knapsack and backpack sprayers, pump-up pressure sprayers, handguns, handwands, lances and other hand-held spray equipment used to direct the spray onto weed foliage and vehicle mounted high-volume spray equipment for spray-to-wet applications.

Selective Equipment—Shielded sprayers

See the appropriate part of this section for specific rates of application and instructions.

AERIAL EQUIPMENT

Use the recommended rates of this product in 3 to 15 gallons of water per acre unless otherwise specified on this label. See the "Weeds Controlled" section of this label for specific rates. Unless otherwise specified, do not exceed the one quart rate per acre of this product. **(Aerial applications of this product may be made in annual cropping conventional tillage systems, preharvest. Refer to the individual use area sections of this label for recommended volumes and application rates.)** FOR AERIAL APPLICATION IN CALIFORNIA, REFER TO THE FEDERAL LABEL FOR AERIAL APPLICATIONS IN THAT STATE FOR SPECIFIC INSTRUCTIONS, RESTRICTIONS AND REQUIREMENTS.

Avoid direct application to any body of water.

AVOID DRIFT—DO NOT APPLY DURING INVERSION CONDITIONS, WHEN WINDS ARE GUSTY, OR UNDER ANY OTHER CONDITION WHICH WILL ALLOW DRIFT. DRIFT MAY CAUSE DAMAGE TO ANY VEGETATION CONTACTED TO WHICH TREATMENT IS NOT INTENDED. TO PREVENT INJURY TO ADJACENT DESIRABLE VEGETATION, APPROPRIATE BUFFER ZONES MUST BE MAINTAINED.

Coarse sprays are less likely to drift; therefore, do not use nozzles or nozzle configurations which dispense spray as fine spray droplets. Do not angle nozzles forward into the airstream and do not increase spray volume by increasing nozzle pressure.

Drift control additives may be used. When a drift control additive is used, read and carefully observe the cautionary statements and all other information appearing on the additive label.

Ensure uniform application—To avoid streaked, uneven or overlapped application, use appropriate marking devices.

Thoroughly wash aircraft, especially landing gear, after each day of spraying to remove residues of this product accumulated during spraying or from spills. PROLONGED EXPOSURE OF THIS PRODUCT TO UNCOATED STEEL SURFACES MAY RESULT IN CORROSION AND POSSIBLE FAILURE OF THE PART. LANDING GEAR ARE MOST SUSCEPTIBLE. The maintenance of an organic coating (paint) which meets aerospace specification MIL-C-38413 may prevent corrosion.

BROADCAST EQUIPMENT

For control of annual or perennial weeds listed on this label using broadcast equipment—Use the recommended rates of this product in 3 to 40 gallons of water per acre as a broadcast spray unless otherwise specified on this label. See the "Weeds Controlled" section of this label for specific rates. As density of weeds increases, spray gallonage should be increased within the recommended range to ensure complete coverage. Carefully select proper nozzle to avoid spraying a fine mist. For best results with ground application equipment, use flat fan nozzles. Check for even distribution of spray droplets.

CONTROLLED DROPLET APPLICATION (CDA)

The rate of this product applied per acre by vehicle-mounted CDA equipment must not be less than the amount recommended in this label when applied by conventional broadcast equipment. For vehicle-mounted CDA equipment, apply 3 to 15 gallons of water per acre.

For the control of labeled annual weeds with hand-held CDA units, apply a 20 percent solution of this product plus a 5 percent solution of nonionic surfactant at a flow rate of 2 fluid ounces per minute and a walking speed of 1.5 MPH (1 quart per acre). For the control of labeled perennial weeds, apply a 20 to 40 percent solution of this product plus a 5 to 10 percent solution of nonionic surfactant at a flow rate of 2 fluid ounces per minute and a walking speed of 0.75 MPH (2 to 4 quarts per acre).

Controlled droplet application equipment produces a spray pattern which is not easily visible. Extreme care must be exercised to avoid spray or drift contacting the foliage or any other green tissue of desirable vegetation as damage or destruction may result.

HAND-HELD and HIGH-VOLUME EQUIPMENT

Use coarse sprays only.

Mix this product in clean water and apply to foliage of vegetation to be controlled. For applications made on a spray-to-wet basis, spray coverage should be uniform and complete. Do not spray to the point of runoff.

For control of annual weeds listed on this label, apply 0.5 percent solution of this product plus nonionic surfactant to weeds less than 6 inches in height or runner length. Apply prior to seedhead formation in grasses or bud formation in broadleaves. Allow three or more days before tillage or mowing.

For annual weeds over 6 inches tall, or unless otherwise specified, use a 1 percent solution. For best results, use a 2 percent solution on harder-to-control perennials, such as bermudagrass, dock, field bindweed, hemp dogbane, milkweed and Canada thistle.

Less than complete coverage of weeds may result from the use of spray equipment designed for motorized spot treatments. Where less than complete coverage of annual weeds occurs, use a 5 percent solution. Do not reduce recommended concentrations of this product when adding surfactant.

Prepare the desired volume of spray solution by mixing the amount of this product in water as shown in the following table:

Spray Solution

Desired Volume	Amount of HONCHO®				
	½%	1%	1½%	2%	5%
1 gallon	⅔ oz	1⅓ oz	2 oz	2⅔ oz	6½ oz
25 gallons	1 pt	1 qt	1½ qt	2 qt	5 qt
100 gallons	2 qt	1 gal	1½ gal	2 gal	5 gal

2 tablespoons = 1 ounce

Add surfactant at a rate of 1 to 2 oz. per gallon of spray mix.

For use in knapsack sprayers, it is suggested that the proper amount of this product be mixed with water in a larger container. Fill sprayer with the mixed solution and add the correct amount of surfactant.

For hand-held CONTROLLED DROPLET APPLICATORS, see the "Controlled Droplet Applicator (CDA)" section of this label.

SELECTIVE EQUIPMENT

This product may be applied through a shielded applicator after dilution and thorough mixing with water to listed weeds growing in any noncrop site specified on this label and only when specifically recommended in cropping systems.

A shielded applicator directs the herbicide solution onto weeds while shielding desirable vegetation from the herbicide.

AVOID CONTACT WITH DESIRABLE VEGETATION.

Contact of the herbicide solution with desirable vegetation may result in damage or destruction. Droplets, mist, foam, or splatter of the herbicide solution settling on desirable vegetation may result in discoloration, stunting, or destruction.

Weeds not contacted by the herbicide solution will not be affected. This may occur in dense clumps, severe infestations, or when the height of the weeds varies so that not all weeds are contacted. In these instances, repeat treatment may be necessary.

See the "Weeds Controlled" section of this label for recommended stage of growth for specific weeds.

NOTE

- Maintain equipment in good operating condition. Avoid leakage or dripping onto desirable vegetation.
- Adjust height of applicator to insure adequate contact with weeds.
- Keep shields on shielded applicators adjusted to protect desirable vegetation.
- DO NOT operate equipment at ground speeds greater than 5 mph. Weed control may be affected by speed of application equipment. As weed density increases, reduce equipment ground speed to insure good coverage of weeds.

SHIELDED APPLICATORS

When applied as directed under conditions described for shielded applicators, this product will control those weeds listed in the "Weeds Controlled" section of this label.

Shielded applicators which apply the herbicide solution as a spray band should be calibrated on a broadcast equivalent rate and volume basis. To determine these:

$$\frac{\text{Band width in inches}}{\text{Row width in inches}} \times \begin{array}{c}\text{Herbicide}\\\text{broadcast}\\\text{RATE}\\\text{per acre}\end{array} = \begin{array}{c}\text{Herbicide}\\\text{band RATE}\\\text{per acre}\end{array}$$

$$\frac{\text{Band width in inches}}{\text{Row width in inches}} \times \begin{array}{c}\text{Broadcast}\\\text{VOLUME of}\\\text{solution}\\\text{per acre}\end{array} = \begin{array}{c}\text{Band VOLUME}\\\text{of solution}\\\text{per acre}\end{array}$$

Use nozzles that provide uniform coverage within the treated area. EXTREME CARE MUST BE EXERCISED TO AVOID CONTACT WITH DESIRABLE VEGETATION.

For specific rates of application and instructions for control of various annual and perennial weeds, see the "Weeds Controlled" section of this label.

WEEDS CONTROLLED

This herbicide controls many annual and perennial grasses and broadleaf weeds.

ANNUAL WEEDS

- Apply to actively growing grass and broadleaf weeds.
- Allow at least 3 days after treatment before tillage.
- For maximum agronomic benefit, apply when weeds are 6 inches or less in height.
- To prevent seed production, applications should be made prior to seedhead formation.
- This product does not provide residual control; therefore, delay application until maximum weed emergence. Repeat treatments may be necessary to control later germinating weeds.

LOW-VOLUME BROADCAST APPLICATION (LOW-RATE TECHNOLOGY)

When applied as directed under the conditions described, this product will control the weeds listed below when:

1. Water carrier volumes are 3 to 10 gallons per acre for ground applications and 3 to 5 gallons per acre for aerial applications. (See the "Aerial Equipment" section of this label for approved sites.)

NOTE

- The addition of 2 percent dry ammonium sulfate by weight or 17 pounds per 100 gallons of water may increase the performance of this product on annual weeds. The improvement in performance may be apparent where environmental stress is a concern. Refer to the "Mixing, Additives and Application Instructions" section of this label.
- Do not tank-mix with soil-residual herbicides when using these rates unless otherwise specified.
- For weeds that have been mowed, grazed, or cut, allow regrowth to occur prior to treatment.
- Refer to the "Tank Mixtures" portion of this section for control of additional broadleaf weeds.

Weed Species	Maximum Height	Rate Per Acre**
Foxtail *Setaria spp.*	12"	8 oz.
Barnyardgrass *Echinochloa crus-galli*	6"	12 oz.
Bluegrass, annual *Poa annua*	(0 to 4" (4 to 6"	16 oz.[1]) 24 oz.[1])
Brome, downy* *Bromus tectorum*		
Mustard, blue *Chorispora tenella*		
Mustard, tansy *Descurainia pinnata*		
Mustard, tumble *Sisymbrium altissimum*		
Mustard, wild *Sinapis arvensis*		
Spurry, umbrella *Holosteum umbellatum*		
Barley *Hordeum vulgare*	12"	
Rye *Secale cereale*		
Sandbur, field *Cenchrus spp.*		
Shattercane *Sorghum bicolor umbellatum*		
Stinkgrass *Eragrostis cilianensis*		
Wheat *Triticum aestivum*	18"	
Morningglory *Ipomea spp.*	2"	16 oz.
Sicklepod *Cassia obtusifolia*		
Bluegrass, bulbous *Poa bulbosa*	6"	
Cheat *Bromus secalinus*		
Chickweed, mouseear *Cerastium vulgatum*		
Corn *Zea mays*		
Groundsel, common *Senecio vulgaris*		
Horseweed/Marestail *Conyza canadensis*		
Lambsquarters, common *Chenopodium album*		
Pennycress, field **Fanweed** *Thlaspi arvense*		
Rocket, London *Sisymbrium irio*		
Ryegrass, Italian *Lolium multiflorum*		
Shepherdspurse *Capsella bursa-pastoris*		
Buttercup *Ranunculus spp.*	12"	
Cocklebur *Xanthium strumarium*		
Crabgrass *Digitaria spp.*		
Dwarfdandelion *Krigia cespitosa*		

Weed Species	Maximum Height	Rate Per Acre**
Falseflax, smallseed *Camelina microcarpa*		
Foxtail, Carolina *Alopecurus carolinianus*		
Johnsongrass, seedling *Sorghum halepense*		
Lambsquarters, common *Chenopodium album*		
Oats, wild *Avena fatua*		
Panicum, fall *Panicum dichotomiflorum*		
Panicum, Texas *Panicum texanum*		
Pigweed, redroot *Amaranthus retroflexus*		
Pigweed, smooth *Amaranthus hybridus*		
Witchgrass *Panicum capillare*		
Sicklepod *Cassia obtusifolia*	3 to 4"	24 oz.
Signalgrass, broadleaf *Brachiaria platyphylla*	4"	
Horseweed/Marestail *Conyz canadensis* **Lambsquarters, common** *Chenopodium album*	7 to 12"	
Teaweed *Sida spinosa*	4"	32 oz.
Rice, red *Oryza jativa*	6"	
Geranium, Carolina *Geranium carolinianum* **Goosegrass** *Eleusine indica* **Primrose, cutleaf evening** *Oenothera laciniate* **Pusley, Florida** *Richardia scabra*	12"	
Sicklepod *Cassia obtusifolia* **Spanishneedles** *Bidens bipinnata*	5 to 12"	
Filaree *Erodium spp.*	12"	48 oz.

[1] Use these rates in Arkansas, Mississippi, Louisiana, and Texas for preplant treatments.

* For control in no-till systems, use 16 fluid ounces per acre.

** For those rates less than 32 ounces per acre, this product at rates up to 32 ounces per acre may be used where heavy weed densities exist.

TANK MIXTURES

HONCHO® plus BANVEL plus NONIONIC SURFACTANT

HONCHO plus 2,4-D AMINE plus NONIONIC SURFACTANT

DO NOT APPLY BANVEL OR 2,4-D AMINE TANK MIXTURES BY AIR IN CALIFORNIA.

These tank mixtures are recommended for use in fallow and reduced tillage areas only. Follow use directions as given in this "Low-Volume Broadcast Application" section.

Honcho plus Banvel or 2,4-D amine will control the annual grasses and broadleaf weeds previously listed for this product alone at the indicated heights (except 8 ounces per acre applications), plus the following broadleaf weeds. For those weeds previously listed at 8 ounces of this product herbicide alone per acre, use the 12-ounce rate in these tank mixtures.

NOTE: Refer to the specific product labels for crop rotation restrictions and cautionary statements of all products used in tank mixtures. Some crop injury may occur if Banvel is applied within 45 days of planting. The addition of Banvel in a mixture with this product may provide short-term residual control of selected weed species.

Apply 12 to 16 ounces of this product plus 0.25 lb. a.i. of Banvel or 0.5 pound a.e. of 2,4-D amine per acre to control dense populations of the following annual broadleaf weeds when less than 6 inches in height:

Kochia* *Kochia scoparia*	**Pigweed, redroot** *Amaranthus retroflexus*
Lambsquarters *Chenopodium album*	**Thistle, Russian** *Salsola kali*
Lettuce, prickly *Lactuca serriola*	

* Controlled with Banvel tank mixture only.

HIGH-VOLUME BROADCAST APPLICATIONS

When applied as directed under the conditions described, this product will control the weeds listed below when water carrier volumes are 10 to 40 gallons per acre for ground applications.

Use 1 quart of this product per acre if weeds are less than 6 inches tall. If weeds are over 6 inches tall, use 1.5 quarts of this product per acre. If weeds have been mowed, grazed, or cut, allow adequate time for new growth to recommended stages prior to treatment. These rates will also provide control of weeds listed in the "Low-Volume Broadcast Application" section.

Weed Species	
Balsamapple* *Momordica charantia*	**Panicum** *Panicum spp.*
Bassia, fivehook *Bassia hyssopifolia*	**Ragweed, common** *Ambrosia artemisiifolia*
Brome *Bromus spp.*	**Ragweed, giant** *Ambrosia trifida*
Fiddleneck *Amsinckia spp.*	**Smartweed, Pennsylvania** *Polygonum pensylvanicum*
Flaxleaf, hairy *Conyza bonariensis*	**Sowthistle, annual** *Sonchus oleraceus*
Fleabane *Erigeron spp.*	**Sunflower** *Helianthus annuus*
Kochia *Kochia scoparia*	**Thistle, Russian** *Salsola kali*
Lettuce, prickly *Lactuca serriola*	**Velvetleaf** *Abutilon theophrasti*

* Apply with hand-held equipment only.

PERENNIAL WEEDS

Apply this product as follows to control or destroy most perennial weeds:

NOTE: If weeds have been mowed or tilled, do not treat until plants have resumed active growth and are at the recommended stages.

Repeat treatments may be necessary to control weeds regenerating from underground parts or seed. Repeat treatments must be made prior to crop emergence.

The addition of 2 percent dry ammonium sulfate by weight or 17 pounds per 100 gallons of water may increase the performance of this herbicide on perennial weeds. The improvement in performance may be apparent where environmental stress is a concern. Refer to the "Mixing, Additives and Application Instructions" section of this label.

When applied as recommended under the conditions described, this product WILL CONTROL the following PERENNIAL WEEDS:

Alfalfa *Medicago sativa*	**Horsenettle** *Solanum carolinense*
Alligatorweed* *Alternanthera philoxeroides*	**Horseradish** *Armoracia rusticana*
Artichoke, Jerusalem *Helianthus tuberosus*	**Johnsongrass** *Sorghum halepense*
Bahiagrass *Paspalum notatum*	**Kikuyugrass** *Pennisetum clandestinum*
Bentgrass *Agrostis spp.*	**Knapweed** *Centaurea repens*
Bermudagrass *Cynodon dactylon*	**Lantana** *Lantana camara*
Bindweed, field *Convolvulus arvensis*	**Milkweed** *Asclepias spp.*
Bluegrass, Kentucky *Poa spp.*	**Mullein, common** *Verbascum thapsus*
Blueweed, Texas *Helianthus ciliaris*	**Napiergrass** *Pennisetum purpureum*

Brackenfern *Pteridium aquilinum*	**Nightshade, silverleaf** *Solanum elaeagnifolium*
Bromegrass, smooth *Bromus inermis*	**Nutsedge; purple, yellow** *Cyperus rotundus* *Cyperus esculentus*
Bursage, woollyleaf *Franseria tomentosa*	**Orchardgrass** *Dactylis glomerata*
Canarygrass, reed *Phalaris arundinacea*	**Pampas** *Cortaderia jubata*
Cattail *Typha spp.*	**Paragrass** *Brachiaria mutica*
Clover, red *Trifolium pratense*	**Ryegrass, perennial** *Lolium perenne*
Clover, white *Trifolium repens*	**Smartweed, swamp** *Polygonum coccineum*
Cogongrass *Imperat cylindric*	**Sweet potato, wild*** *Ipomoea pandurata*
Dallisgrass *Paspalum dilatatum*	**Thistle, Canada** *Cirsium arvense*
Dandelion *Taraxacum officinale*	**Timothy** *Phleum pratense*
Dock, curly *Rumex crispus*	**Torpedograss*** *Panicum repens*
Dogbane, hemp *Apocynum cannabinum*	**Vaseygrass** *Paspalum urvillei*
Fescues *Festuca spp.*	**Wheatgrass, western** *Agropyron smithii*
Fescue, tall *Festuca arundinacea*	
Guineagrass *Pancium maximum*	

* Partial control

Alfalfa—Apply 1 quart of this product in 3 to 10 gallons of water per acre. Make application after last hay cutting in the fall. Allow alfalfa to regrow to a height of 6 to 8 inches or more prior to treatment. Applications should be followed with deep tillage at least 7 days after treatment, but before soil freeze-up.

Alligatorweed—Apply 4 quarts of this product per acre or apply a $1^1/_2$ percent solution with hand-held equipment to provide partial control. Apply when most of the plants are in bloom. Repeat applications will be required to maintain such control.

Bentgrass—For suppression in grass seed production areas. For ground applications only, apply 1.5 quarts of this product in 10 to 20 gallons of water per acre. Ensure entire crown area has resumed growth prior to a fall application. Bentgrass should have at least 3 inches of growth. Tillage prior to treatment should be avoided. Tillage 7 to 10 days after application is recommended for best results. Failure to use tillage after treatment may result in unacceptable control.

Bermudagrass—For control, apply 5 quarts of this product per acre and, for partial control, apply 3 quarts per acre. Treat when bermudagrass is actively growing and seedheads are present. Retreatment may be necessary to maintain control. Allow 7 or more days after application before tillage.

Bindweed, field/Blueweed, Texas—Apply 4 to 5 quarts of this product per acre west of the Mississippi River and 3 to 4 quarts per acre east of the Mississippi River. Apply when weed is actively growing and is at or beyond full bloom. Do not treat when weed is under drought stress, as good soil moisture is necessary for active growth. For best results, apply in late summer or fall. Fall treatments must be applied before a killing frost. Allow 7 or more days after application before tillage.

Bindweed, field (suppression on irrigated land where annual tillage is performed)—Apply 1 quart of this product in 3 to 10 gallons of water per acre. Apply to actively growing bindweed that has reached a length of 12 inches or greater. Allow maximum weed emergence and runner growth. Do not treat when weed is under drought stress, as good soil moisture is necessary for active growth. Allow 3 or more days after application before tillage.

Bluegrass, Kentucky/Bromegrass, smooth/Orchardgrass—Apply 2 quarts of this product in 10 to 40 gallons of water per acre when the grasses are actively growing and most plants have reached boot-to-early seedhead stage of development. For partial control in pasture or hay crop renovation, apply 1 to $1^1/_2$ quarts of this product in 5 to 10 gallons of water per acre.

Apply to actively growing plants when most have reached 4 to 12 inches in height. Allow 7 or more days after application before tillage.

Brackenfern—Apply 3 to 4 quarts of this product per acre as a broadcast spray or as a 1 to $1^1/_2$ percent solution with hand-held equipment. Apply to fully expanded fronds which are at least 18 inches long.

Bursage, woollyleaf—For control, apply 2 quarts of this product plus 1 pint of Banvel per acre. For partial control, apply 1 quart of this product plus 1 pint of Banvel per acre. Apply 3 to 20 gallons of water per acre. Apply when plants are producing new active growth which has been initiated by moisture for at least 2 weeks and when plants are at or beyond flowering.

Canarygrass, reed/Timothy/Wheatgrass, western—Apply 2 to 3 quarts of this product per acre. For best results, apply to actively growing plants when most have reached the boot-to-head stage of growth. Allow 7 or more days after application before tillage.

Cogongrass—Apply 3 to 5 quarts of this product in 10 to 40 gallons of water per acre. Apply when cogongrass is at least 18 inches tall and actively growing in later summer or fall. Allow 7 or more days after application before tillage or mowing. Due to uneven stages of growth and the dense nature of vegetation preventing good spray coverage, repeat treatments may be necessary to maintain control.

Dogbane, hemp/Horseradish/Knapweed—Apply 4 quarts of this product per acre. Apply when actively growing and when most weeds have reached the late bud to flower stage of growth. Following crop harvest or mowing, allow weeds to regrow to a mature stage prior to treatment. For best results, apply in late summer or fall. Allow 7 or more days after application before tillage.

Fescue, tall—Apply 3 quarts of this product in 10 to 40 gallons of water per acre to actively growing plants when most have reached boot-to-early seedhead stage of development.

Fall applications only—Apply 1 quart of this product in 3 to 10 gallons of water per acre. Apply to fescue in the fall when actively growing and plants have 6 to 12 inches of new growth. Allow 7 or more days after application before tillage. A sequential application of 1 pint per acre of this product will improve long-term control and control seedlings germinating after fall treatments or the following spring.

Guineagrass—Apply 3 quarts of this product per acre or use a 1 percent solution with hand-held equipment. Apply to actively growing guineagrass when most have reached at least the 7-leaf stage of growth. Ensure thorough coverage when using hand-held equipment. Allow 7 or more days after application before tillage.

Johnsongrass/Ryegrass, perennial—Apply 1 to 3 quarts of this product per acre. In annual cropping systems, apply 1 to 2 quarts of this product per acre. Apply 1 quart of this product in 3 to 10 gallons of water per acre. Use 2 quarts of this product when applying 10 to 40 gallons of water per acre. In noncrop or areas where annual tillage (no-till) is not performed, apply 2 to 3 quarts of this product in 10 to 40 gallons of water per acre. For best results, apply to actively growing plants when most have reached the boot-to-head stage of growth. Allow 7 or more days after application before tillage. Do not tank-mix with residual herbicides when using the 1 quart per acre rate.

For burndown of johnsongrass, apply 1 pint per acre in 3 to 10 gallons of water per acre before the plants reach a height of 12 inches. For this use, allow at least 3 days after treatment before tillage.

Spot Treatment (partial control or suppression)—Apply a 1 percent solution of this product when johnsongrass is 12 to 18 inches in height. Coverage should be uniform and complete.

Kikuyugrass—Apply 2 to 3 quarts of this product per acre. Spray when most kikuyugrass is at least 8 inches in height (3 or 4-leaf stage of growth) and actively growing. Allow 3 or more days after application before tillage.

Lantana—Apply this product as a 1 to $1^1/_4$ percent solution using hand-held equipment only. Apply to actively growing lantana at or beyond the bloom stage of growth. Use the higher application rate for plants that have reached the woody stage of growth. Allow 7 or more days after application before tillage.

Milkweed, common—Apply 3 quarts of this product per acre. Apply when actively growing and most of the milkweed has reached the late bud to flower stage of growth. Following small grain harvest or mowing, allow milkweed to regrow to a mature stage prior to treatment. Allow 7 or more days after application before tillage.

Nightshade silverlead—For control, apply 2 quarts of this product in 3 to 10 gallons of water per acre. Applications should be made when at least 60 percent of plants have berries. Fall treatments must be applied before a killing frost. Allow 7 or more days after application before tillage. Do not treat when weed is under drought stress, as good soil moisture is necessary for active growth.

Nutsedge; purple, yellow—Apply 3 quarts of this product per acre as a broadcast spray, or apply a 1 percent solution from hand-held equipment to control existing nutsedge plants and immature nutlets attached to treated plants. Treat when plants are in flower or when new nutlets can be found at rhizome tips. Nutlets which have not germinated will not be controlled and may germinate following treatment. Repeat treatments will be required for long-term control. Wait 7 days after treatment before tillage. Tillage will stimulate nutlet germination.

Apply 1 pint to 1 quart of this product per acre in 3 to 40 gallons of water per acre to provide partial control of existing nutsedge plants. Treat when plants have 3 to 5 leaves (less than 6 inches). Repeat treatments at this rate will be required to control subsequent emerging plants. Wait 7 days after treatment before tillage or mowing.

Sweet Potato, wild—Apply this product as a 2 percent solution using hand-held equipment. Apply to actively growing weeds that are at or beyond the bloom stage of growth. Repeat applications will be required. Allow the plant to reach the recommended stage of growth before retreatment. Allow 7 or more days before tillage.

Pampas—Apply this product as a 1½ to 2 percent solution using hand-held equipment. Apply to plants that are actively growing at or beyond the boot stage of growth. Thorough coverage is necessary for best control.

Thistle, Canada—Apply 2 to 3 quarts of this product per acre. Apply to actively growing thistles when most are at or beyond the bud stage of growth. After harvest, mowing or tillage in the late summer or fall, allow at least 4 weeks for initiation of active growth and rosette development prior to the application of this product. Fall treatments must be applied before a killing frost. Allow 3 or more days after application before tillage.

For suppression of Canada thistle, apply 1 quart of this product in 3 to 10 gallons of water per acre in the late summer or fall after harvest, mowing or tillage. Allow rosette regrowth to a minimum of 6 inches in diameter before treating. Applications can be made as long as leaves are still green and plants are actively growing at the time of application. Allow 3 or more days after application before tillage.

Torpedograss—Apply 4 to 5 quarts of this product per acre to provide partial control of torpedograss. Apply to actively growing torpedograss when most plants are at or beyond the seedhead stage of growth. Repeat applications will be required to maintain control. Fall treatments must be applied before frost. Allow 7 or more days after application before tillage.

Other perennials listed on this label—Apply 3 to 5 quarts of this product per acre. Apply when actively growing and most have reached early head or early bud stage of growth. Allow 7 or more days after application before tillage.

TANK MIXTURES

When applied as directed, under the conditions described, Honcho plus Banvel or 2,4-D amine will suppress or control the following perennial broadleaf weed:

Bindweed, field
Convolvulus arvensis

For suppression, apply 16 ounces of this product plus 0.5 lb a.e. of 2,4-D amine or 0.25 lb a.i. of Banvel in 3 to 10 gallons of water per acre for ground applications and 3 to 5 gallons of water per acre for aerial applications. Applications should be delayed until maximum emergence has occurred and when vines are between 6 to 18 inches in length. These tank mixtures may be applied using aerial application equipment (except in California) in fallow and reduced tillage systems only.

For control, apply 2 quarts of this product plus 0.5 pound a.i. of Banvel in 10 to 20 gallons of water per acre. At these rates, apply using ground application equipment only.

Refer to the specific product labels for crop rotation restrictions and cautionary statements of all products in tank mixture.

(Do not apply tank mixtures of this product plus 2,4-D or Banvel by air in California)

Irrigated Agricultural Lands

Apply 1 to 2 quarts of this product plus 1 quart of 2,4-D amine in 10 to 20 gallons of water per acre with ground equipment only for partial control (suppression) of field bindweed. Applications should be made following harvest or in fall fallow ground when bindweed is actively growing and the majority of runners are 12 inches or more in length. Use at least one irrigation to promote active bindweed growth.

Dogbane, hemp
Apocynum cannabium

For suppression, apply 16 ounces of this product plus 16 ounces of 2,4-D amine in 3 to 10 gallons of water per acre for ground applications, and 3 to 5 gallons of water per acre for aerial applications (except in California). Delay application until maximum emergence of dogbane has occurred. For control, apply 4 quarts of this product in 3 to 5 gallons of water per acre using ground equipment only.

Smartweed, swamp
Polygonum coccienum

For control, apply 16 ounces of this product plus 0.5 pound active ingredient of 2,4-D amine in 3 to 10 gallons of water per acre in the late summer or fall. Apply when plants are actively growing and most have reached the early bud stage of growth. Allow 7 or more days after application before tillage.

WOODY BRUSH AND TREES

When applied as recommended under the conditions described, this product CONTROLS or PARTIALLY CONTROLS the following woody brush, plants and trees:

Alder
Alnus spp.
Ash*
Fraxinus spp.
Aspen, quaking
Populus tremuloides
Bearmat (Bearclover)
Chamebatia foliolosa
Birch
Betula spp.
Blackberry
Rubus spp.
Broom:
French
Cytisus monspessulanus
Scotch
Cytisus scoparius
Buckwheat, California*
Eriogonum fasciculatum
Cascara*
Rhamnus purshiana
Catsclaw*
Acacia greggi
Ceanothus:
Ceanothus spp.
Chamis:
Adenostoma fasciculatum
Cherry
Bitter
Prunus emarginata
Black
Prunus serotina
Pia
Prunus pensylvanica
Coyote brush
Baccharis consanguinea
Creeper, Virginia*
Parthenocissus quinquefolia
Dewberry
Rubus trivialis
Elderberry
Sambucus spp.
Elm*
Ulmus spp.
Eucalyptus, bluegum
Eucalyptus glotulus
Hasardia*
Haplopappus squamosus
Hawthorn
Crataegus spp.
Hazel
Corylus spp.
Honeysuckle
Lonicera spp.
Kudza
Pueraria lobata
Locust, black*
Robinia pseudoacacia
Modrone*
Arbutus menziesii
Manzanita
Arctostaphylos spp.
Maple:
Red**
Acer rubrum
Sugar
Acer saccharum
Vine*
Acer circinatum
Monkey Flower*
Mimulus guttatus
Oak:
Black*
Quercus velutina
Northern Pin
Quercus palustris
Post
Quercus stellata
Red
Quercus rubra
Southern Red
Quercus falcata
White*
Quercus alba
Persimmon*
Diospyros spp.
Poison Ivy
Rhus radicans
Poison Oak
Rhus toxicodendron
Poplar*, yellow
Liriodendron tulipifera
Raspberry
Rubus spp.
Rosa multiflora
Rosa multiflora
Sage, black
Salvia mellifera
Sagebrush, California
Artemisia californica
Salmonberry
Rubus spectabilis
Sassafras
Sassafras aibidum
Sourwood
Oxydendrum arboreum
Sumac:
Poison*
Rhus vernix
Smooth*
Rhus glabra
Winged*
Rhus copallina
Sweetgum
Liquidambar styraciflua
Swordfern*
Polystichum munitum
Tallowtree, Chinese
Sapium sebiferum
Tan Oak
Lithocarpus densiflorus
Thimbleberry
Rubus parviflorus
Tobacco, tree*
Nicotiana glauca
Trumpetcreeper
Campsis radicans
Willow
Salix spp.

---

* Partial control
** See below for control or partial control instructions.

NOTE: If brush has been mowed or tilled or trees have been cut, do not treat until regrowth has reached the recommended stages of growth.

Apply this product when plants are actively growing, and unless otherwise directed, after full leaf expansion. Use the higher rate for larger plants and/or dense areas of growth. On vines, use the higher rate for plants that have reached the woody stage of growth. Best results are obtained when application is made in late summer or fall after fruit formation.

In arid areas, best results are obtained when application is made in the spring to early summer when brush species are at high moisture content and are flowering.

Ensure thorough coverage when using hand-held equipment. Symptoms may not appear prior to frost or senescence with fall treatments.

Allow 7 or more days after application before tillage, mowing or removal. Repeat treatments may be necessary to control plants regenerating from underground parts or seed. Some autumn colors on undesirable deciduous species are acceptable provided no major leaf drop has occurred. Reduced performance may result if fall treatments are made following a frost.

See "Directions for Use", and "Mixing, Additives and Application Instructions" sections of this label for labeled uses and specific application instructions.

Apply this product as follows to control or partially control the following woody brush and trees.

Alder/Dewberry/Honeysuckle/Post Oak/Raspberry—For control, apply 3 to 4 quarts per acre of this product as a broadcast spray or as a 1 to 1½ percent solution with hand-held equipment.

Aspen, quaking/Cherry, bitter, black, pin/Hawthorn/Oak, southern red/ Sweetgum/Trumpetcreeper—For control, apply 2 to 3 quarts of this product per acre as a broadcast spray or as a 1 to 1½ percent solution with hand-held equipment.

Birch/Elderberry/Hazel/Salmonberry/Thimbleberry—For control, apply 2 quarts per acre of this product as a broadcast spray or as a 1 percent solution with hand-held equipment.

Blackberry—For control, apply 3 to 4 quarts per acre of this product as a broadcast spray, or 1 to 1½ percent solution with hand-held equipment. Make application after plants have reached full leaf maturity. Best results are obtained when applications are made in late summer or fall. After berries have set or dropped in late fall, blackberry can be controlled by applying a ¾ percent solution of this product with hand-held equipment. For control of blackberries after leaf drop and until killing frost or as long as stems are green, apply 3 to 4 quarts of this product in 10 to 40 gallons of water per acre.

Broom, French, Scotch—For control, apply a 1½ to 2 percent solution with hand-held equipment.

Buckwheat, California/Hasardia/Monkey Flower/Tobacco, tree — For partial control of these species, apply a 1 to 2 percent solution of this product as a foliar spray with hand-held equipment. Thorough coverage of foliage is necessary for best results.

Catsclaw—For partial control, apply as a 1 to 1½ percent solution with hand-held equipment.

Coyote Brush—For control, apply a 1½ to 2 percent solution with hand-held equipment when at least 50 percent of the new leaves are fully developed.

Eucalyptus, bluegum—For control of eucalyptus resprouts, apply a 2 percent solution of this product with hand-held equipment when resprouts are 6 to 12 feet tall. Ensure complete coverage. Apply when plants are growing actively. Avoid application to drought-stressed plants. For control of eucalyptus trees 2 inches or less in diameter, cut trees as close to the soil surface as desired. Apply a 50 to 100 percent solution of this product to the freshly cut surface immediately after cutting. Delay in applying this product may result in reduced performance.

Kudzu—For control, apply 4 quarts of this product per acre as a broadcast spray or as a 2 percent solution with hand-held equipment. Repeat applications will be required to maintain control.

Maple, red**—For control, apply as a 1 to 1½ percent solution with hand-held equipment when at least 50 percent of the new leaves are fully developed. For partial control, apply 2 to 4 quarts of this product per acre as a broadcast spray.

Maple, sugar/Oak, northern pin/Oak, red—For control, apply as a 1 to 1½ percent solution with hand-held equipment when at least 50 percent of the new leaves are fully developed.

Poison Ivy/Poison Oak—For control, apply 4 to 5 quarts of this product per acre as a broadcast spray or as a 2 percent solution with hand-held equipment. Repeat applications may be required to maintain control. Fall treatments must be applied before leaves lose green color.

Rose, multiflora—For control, apply 2 quarts of this product per acre as a broadcast spray or as a 1 percent solution with hand-held equipment. Treatments should be made prior to leaf deterioration by leaf-feeding insects.

Sage, black/Sagebrush, California/Chamise/Tallowtree, Chinese — For control of these species, apply a 1 percent solution of this product as a foliar spray with hand-held equipment. Thorough coverage of foliage is necessary for best results.

Willow—For control, apply 3 quarts of this product per acre as a broadcast spray or as a 1 percent solution with hand-held equipment.

Other Woody Brush and Trees listed on this label*—For partial control, apply 2 to 4 quarts of this product per acre as a broadcast spray or as a 1 to 1½ percent solution with hand-held equipment.

DIRECTIONS FOR USE

It is a violation of Federal law to use this product in any manner inconsistent with its labeling.

STORAGE AND DISPOSAL

Do not contaminate water, foodstuffs, seed or feed by storage or disposal.

STORAGE:

STORE ABOVE 10°F (-12°C) TO KEEP PRODUCT FROM CRYSTALLIZING.

Crystals will settle to the bottom. If allowed to crystallize, place in a warm room 68°F(20°C) for several days to redissolve and shake or roll to mix well before using.

DISPOSAL:

Wastes resulting from the use of this product that cannot be used or chemically reprocessed should be disposed of in a landfill approved for pesticide disposal or in accordance with applicable Federal, state or local procedures.

Emptied container retains vapor and product residue. Observe all label safeguards until container is destroyed.

Do not reuse container. Triple rinse container. Then puncture and dispose of in a sanitary landfill, or by incineration, or, if allowed by state and local authorities, by burning. If burned, stay out of smoke.

See "General Information" and "Mixing, Additives and Application Instructions" sections of this label for essential product performance information and the following sections for specific recommended uses.

EXTREME CARE MUST BE EXERCISED TO AVOID CONTACT OF SPRAY WITH FOLIAGE OF DESIRABLE TURFGRASSES, TREES, SHRUBS, OR OTHER DESIRABLE VEGETATION, SINCE SEVERE DAMAGE OR DESTRUCTION MAY RESULT.

NOTE: If spraying areas adjacent to desirable plants, use a shield made of cardboard, sheet metal or plyboard while spraying to help prevent spray from contacting foliage of desirable plants.

Repeat treatments may be necessary to control weeds regenerating from underground parts or seeds.

This product does not provide residual weed control. For subsequent weed control, follow a label-approved herbicide program.

Read and carefully observe the cautionary statements and all other information appearing on the labels of all herbicides used.

FARMSTEAD WEED CONTROL

When applied as directed for "Noncrop Uses" under the conditions described, this product controls undesirable vegetation listed on this label around farmstead building foundations, along and in fences, shelterbelts, and for general nonselective farmstead weed control.

For specific rates of application and instructions for control of various annual and perennial weeds, see the "Weeds Controlled" section of this label.

GRASSES FOR SEED PRODUCTION

PREPLANT AND RENOVATION

When applied as directed under the conditions described, this product controls most existing vegetation prior to the planting or renovation of grass seed production areas.

For specific rates of application and instructions for control of various annual and perennial weeds, and woody brush and trees, see the "Weeds Controlled" section of this label.

For maximum control of existing vegetation, delay planting to determine if any regrowth from escaped underground plant parts occurs. Where repeat treatments are necessary, sufficient regrowth must be attained prior to application. For warm-season grasses, such as bermudagrass, summer or fall applications provide best control.

DO NOT DISTURB SOIL OR UNDERGROUND PLANT PARTS BEFORE TREATMENT. Tillage or renovation techniques such as vertical mowing, coring or slicing should be delayed for 7 days after application to allow proper translocation into underground plant parts.

GRASSES FOR SEED PRODUCTION

Apply this product to actively growing weeds at the stages of growth given in the "Weeds Controlled" section of this label prior to planting or renovation of turf or forage grass areas grown for seed production.

DO NOT feed or graze treated areas within 8 weeks after application.

CROPPING SYSTEMS

See "General Information" and "Mixing Additives and Application Instructions" sections of this label for essential product performance information.

See the following Cropping System sections for specific recommended uses.

EXTREME CARE MUST BE EXERCISED TO AVOID CONTACT OF SPRAY WITH FOLIAGE, GREEN STEMS OR FRUIT OF DESIRABLE CROPS, PLANTS, TREES OR OTHER DESIRABLE VEGETATION SINCE SEVERE DAMAGE OR DESTRUCTION MAY RESULT.

Repeat treatments may be necessary to control weeds regenerating from underground parts or seed. Except as otherwise specified on this label, repeat treatments must be made before the crop emerges in accordance with the instructions of this label.

Except as otherwise specified in a crop section of this label, the combined total of all treatments must not exceed 8 quarts per acre of this product per year.

Do not plant subsequent crops other than those on the label for 30 days following application.

For other cropping systems, except following spot treatment in forage grasses and legumes, do not harvest or feed treated crops for 8 weeks after application. Allow 14 days following spot treatment or selective equipment use before grazing domestic livestock or harvesting forage grasses and legumes.

When applied as directed for "Cropping Systems", under the conditions described, this product controls annual and perennial weeds listed on this label, prior to the emergence of direct seeded crops or prior to transplanting of crops listed on this label.

ALFALFA*	LENTILS
ARTICHOKE (Jerusalem)	LETTUCE
	LOGANBERRY
ASPARAGUS*	MELONS***
BARLEY*	MUSTARD GREENS
BEANS (All)	OATS*
BEET GREENS	OKRA
BEETS (Red, Sugar)	OLALLIEBERRY
BLACKBERRY	ONION
BLUEBERRY	PEANUTS
BOYSENBERRY	PEAS (All)
BROCCOLI	PEPPER***
CABBAGE	PINEAPPLE††
CARROT	POTATO (Irish, Sweet)
CAULIFLOWER	PUMPKIN***
CELERY	RADISH
CHICORY	RASPBERRY (Black, Red)
CORN (All)*	
COTTON*	RICE**
CUCUMBER***	RUTABAGA
CURRANT	SORGHUM (Milo)
DEWBERRY	SOYBEANS*
EGGPLANT***	SPINACH
ELDERBERRY	SQUASH*** (Summer, Winter)
FORAGE GRASSES*	
FORAGE LEGUMES*	TOMATILO***
GARLIC***	TOMATOES***†
GOOSEBERRY	TURNIPS
GOURDS***	WATERCRESS***
HORSERADISH	WATERMELON***
HUCKLEBERRY	WHEAT*
KALE	

* Spot treatments may be applied in these crops.

** Do not treat rice fields or levees when the fields contain flood water.

*** Apply only prior to planting. Allow at least 3 days between application and planting.

† Use is restricted to direct seeded crops only.

†† Do not feed or graze treated pineapple forage following application.

Spot Treatment (Only those crops with "*" can be spot treated)—Applications in growing crops must be made prior to heading of small grains and milo, initial pod set in soybeans, silking of corn, and boll opening on cotton.

For forage grasses and forage legumes, see "Spot Treatment" in the "Pastures" section of "Cropping Systems" in this label.

For dilution and rates of application using boom or hand-held equipment, see the "Mixing, Additives and Application Instructions" and "Weeds Controlled" sections of this label.

NOTE: FOR FORAGE GRASSES AND FORAGE LEGUMES, NO MORE THAN ONE-TENTH OF ANY ACRE SHOULD BE TREATED AT ONE TIME. FOR ALL OTHER CROPS, DO NOT TREAT MORE THAN 10 PERCENT OF THE TOTAL FIELD AREA TO BE HARVESTED.

THE CROP RECEIVING SPRAY IN TREATED AREA WILL BE KILLED. TAKE CARE TO AVOID DRIFT OR SPRAY OUTSIDE TARGET AREA FOR THE SAME REASON.

Selective Equipment—This product may be applied through shielded applicators. Shielded applicators may be used in tree crops and grapes.

See the "Selective Equipment" part of the "Application Equipment and Techniques" section of this label for information on proper use and calibration of this equipment.

Allow at least the following time intervals between application and harvest:

Apples, Avocado, Cherry, Citrus, Grapes, Pear	14 days
Rutabagas	14 days
Stone Fruit	17 days
Nut Crops	21 days

ASPARAGUS

When applied as directed for "Cropping Systems", under the conditions described, this product controls weeds listed on this label in asparagus.

For specific rates of application and instructions for control of various annual and perennial weeds, see the "Weeds Controlled" section of this label.

Prior to Crop Emergence—Apply this product prior to planting or emergence of spears in an established crop for the control of emerged labeled annual and perennial weeds. DO NOT APPLY WITHIN A WEEK BEFORE THE FIRST SPEARS EMERGE.

Spot Treatment—Apply this product immediately after cutting, but prior to the emergence of new spears. Do not treat more than 10 percent of the total field to be harvested. Do not harvest within 5 days of treatment.

Postharvest—Apply this product after the last harvest and all spears have been removed. If spears are allowed to regrow, delay application until ferns have developed. Delayed treatments should be applied as directed or shielded spray in order to avoid contact of the spray with ferns, stems or spears. Direct contact of the spray with the asparagus may result in serious crop injury.

NOTE: Select and use recommended types of spray equipment for postemergence postharvest applications. A directed spray is any application where the spray pattern is aligned in such a way as to avoid direct contact of the spray with the crop. A shielded spray is any application where a physical barrier is positioned and maintained between the spray and the crop to prevent contact of the spray with the crop.

BERRIES AND SMALL FRUIT

Apply as a preplant broadcast application, or as a directed spray.

See "General Information" and "Mixing, Additives and Application Instructions" sections of this label for essential product performance information.

See the "Selective Equipment" part of the "Application Equipment and Techniques" section of this label for information on recommended use and calibration of this equipment.

Do not permit herbicide solution to contact desirable vegetation, including green shoots, canes or foliage.

FALLOW AND REDUCED TILLAGE SYSTEMS

Use this product in fallow and reduced tillage systems for control of annual weeds prior to emergence of crops listed in this label. Refer to the "Weeds Controlled" section of this label for specific rates and instructions. This product may be applied using ground or aerial spray equipment. See the "Application Equipment and Techniques" section of this label for instructions.

TANK MIXTURES

HONCHO® plus BANVEL plus NONIONIC SURFACTANT

HONCHO plus 2,4-D AMINE plus NONIONIC SURFACTANT

HONCHO plus GOAL™ plus NONIONIC SURFACTANT

DO NOT APPLY BANVEL, OR 2,4-D AMINE TANK MIXTURES BY AIR IN CALIFORNIA.

The addition of Banvel in a mixture with this product may provide short-term residual control of selected weed species. Some crop injury may occur if Banvel is applied within 45 days of planting. Refer to the Banvel and 2,4-D labels for cropping restrictions and other use instructions.

HONCHO® PLUS GOAL TANK MIXTURES

This product alone or in tank mixtures with Goal plus 0.5 to 1 percent nonionic surfactant by total spray volume will provide control of those weeds listed below.

Make applications when weeds are actively growing and at the recommended stages of growth. Avoid spraying when weeds are subject to moisture stress, when dust is on the foliage or when straw canopy covers the weeds.

HONCHO® 12 oz/acre		HONCHO 16 oz/acre	
Wheat	18"*	Annual grasses at left plus:	
Barley	12"	Ryegrass annual	6"
Bluegrass annual	6"	Chickweed	6"
Barnyardgrass	6"	Groundsel	6"
Rye	6"	Marestail	6"
		Rocket, London	6"
		Shepherdspurse	6"
		Crabgrass	12"
		Johnsongrass, seedling	12"
		Lambsquarters	12"
		Oats, wild	12"
		Pigweed, redroot	12"
		Mustards	12"
HONCHO 12 oz/acre + GOAL 2 to 4 oz/acre**		**HONCHO 16 oz/acre + GOAL** 2 to 4 oz/acre**	
Annual grasses above plus:		Annual weeds above plus:	

HONCHO® 12 oz/acre		HONCHO 16 oz/acre	
Cheeseweed	3"	Cheeseweed	6"
Chickweed	3"	Groundsel	6"
Groundsel	3"	Chickweed	6"
Rocket, London	6"	Rocket London	12"
Shepherdspurse	6"	Shepherdspurse	12"

* Maximum height or length in inches.

** Use the higher rate when weeds approach maximum recommended height or stands are dense.

These recommended tank mixtures may be applied using ground or aerial spray equipment. Refer to the "Weeds Controlled" section of this label for specific rates and instructions.

™ Goal is a trademark of Rhom and Haas Company.

PASTURES

Apply this product prior to planting forage grasses and legumes.

Pasture or Hay Crop Renovation—When applied as a broadcast spray, this product controls the annual and perennial weeds listed in this label prior to planting forage grasses or legumes. Remove domestic livestock before application and wait 8 weeks after application before grazing or harvesting.

Spot Treatment—When applied as a spot treatment as recommended, this product controls annual and perennial weeds listed in this label which are growing in pastures, forage grasses and forage legumes composed of bahiagrass, bermudagrass, bluegrass, brome, fescue, orchardgrass, ryegrass, timothy, wheatgrass, alfalfa or clover.

For spot treatment, apply in areas where the movement of domestic livestock can be controlled. No more than one-tenth of any acre should be treated at one time. Further applications may be made in the same area at 30 day intervals. Remove domestic livestock before application and wait 14 days after application before grazing livestock or harvesting.

MINIMUM TILLAGE SYSTEMS
TANK MIXTURES
CORN AND SOYBEANS

When applied as recommended under the conditions described, these tank mixtures control many emerged weeds, and give preemergence control of many annual weeds when corn or soybeans will be planted directly into a cover crop, established sod, or in previous crop residues.

Refer to specific product labels for crop rotation restrictions and cautionary statements of all products used in these tank mixtures. For mixing instructions, see the "Mixing, Additives and Application Instructions" section of this label.

CORN

HONCHO® plus LASSO®
HONCHO plus LASSO plus ATRAZINE
HONCHO plus LASSO plus BLADEX™
HONCHO plus LASSO plus SIMAZINE
HONCHO plus ATRAZINE plus SIMAZINE

SOYBEANS

HONCHO® plus LASSO®
HONCHO plus LASSO plus LINURON
HONCHO plus LASSO plus LEXONE™
HONCHO plus LASSO plus SENCOR™

Apply these tank mixtures in 10 to 40 gallons of water per acre after planting or during planting in such manner that the planter does not disturb the treated soil. Do not apply these mixtures after crop emergence.

REDUCED CONTROL MAY RESULT IF THIS PRODUCT IS USED IN TANK MIXTURES CONTAINING FLUID FERTILIZERS.

CONTROL OF EMERGED WEEDS

Annual Weeds—Apply to actively growing grasses and broadleaf weeds. Use 1 quart of this product per acre in these tank mixtures if weeds are less than 6 inches tall. If weeds are over 6 inches tall, apply 1.5 quarts of this product per acre. For emerged annual weeds controlled, see the "Weeds Controlled" section of this label.

Perennial Weeds—At normal application rates in minimum tillage systems, perennial weeds may not be at the proper stage of growth for control. See the "General Information" section of this label for the proper stage of growth for perennial weeds.

Use of 2 to 4 quarts of this product per acre in these tank mixtures, under these conditions provides top kill and reduces competition from many emerged perennial grass and broadleaf weeds. For emerged perennial weeds controlled, see the "Weeds Controlled" section of this label.

To obtain the desired stage of growth, it may be necessary to apply this product alone in the late summer or fall and then follow with a label-approved, seedling weed control program at planting.

NOTE: When using these tank mixtures, do not exceed 4 quarts of this product per acre.

USE OF THESE TANK MIXTURES FOR BERMUDAGRASS OR JOHNSONGRASS CONTROL IN MINIMUM TILLAGE SYSTEMS IS NOT RECOMMENDED. For bermudagrass control, follow the instructions under the "Control of Perennial Weeds" section of this label and then use a label-approved, seedling weed-control program in a minimum tillage or conventional tillage system. For johnsongrass control, follow the instructions under the "Control of Perennial Weeds" portion in the "Weeds Controlled" section of this label, and then use a label-approved seedling weed-control program with conventional tillage.

The following are labeled ranges for the recommended preemergence herbicides:

Product	Rates/Acre (Quarts)	Rates/Acre (Pounds)
Lasso	2 to 4	•
Atrazine 4L	1 to 2	•
Atrazine 80WP	•	1.25 to 2.3
Atrazine 90DF	•	1.1 to 2
Bladex 4L	1 to 2.2	•
Bladex 80WP	•	1.25 to 2.75
Simazine 4L	1 to 2.2	•
Simazine 80WP	•	1.25 to 2.75
Linuron 4L	0.5 to 1.5	•
Linuron 50WP	•	1 to 3
Lexone 4L or Sencor 4F	0.25 to 0.75	•
Lexone or Sencor 50 WP	•	0.5 to 1.55
Lexone or Sencor 75DF	•	0.33 to 1

Read and carefully observe the label claims, cautionary statements, rates and all other information on the labels of all products. Use according to the most restrictive label directions for each product in the mixture.

®Lasso is a registered trademark of Monsanto Company.

™ Bladex and Lexone are trademarks of E.I. du Pont de Nemours and Company.

™ Sencor is a trademark of the parent company of Farbenfabriken Bayer GmbH, Leverkusen.

PREHARVEST APPLICATIONS

When applied as directed under the conditions described, this product controls annual and perennial weeds listed on this label prior to the harvest of COTTON.

Weed Control—For specific rates of application and instructions for control of various annual and perennial weeds, see the "Weeds Controlled" section of this label.

Broadcast Applications—This product may be applied using either aerial or ground spray equipment. For ground application with broadcast equipment, apply this product in 10 to 40 gallons of water per acre. For aerial applications, apply this product in 3 to 15 gallons of water per acre.

To control johnsongrass using multiple nozzle-directed spray equipment, apply 1 quart of this product in 10 to 20 gallons of water per acre. Ensure complete coverage.

For partial control of field bindweed, apply 1 quart of this product in 3 to 40 gallons of water per acre. Apply when bindweed is actively growing and 12 inches or greater in length. Reduced performance may result if bindweed is under drought stress.

TANK MIXTURES

HONCHO® plus DEF™ 6
HONCHO plus FOLEX™

When applied as recommended under the conditions described, these tank mixtures control annual and perennial weeds listed on this label prior to the harvest of cotton. For application guidelines, precautions and defoliant rates, refer to the supplemental label for cotton harvest aid.

Timing of Application—Apply this product or these tank mixtures for preharvest weed control after 60 percent of the cotton bolls have opened.

NOTE: DO NOT APPLY TO CROPS GROWN FOR SEED.

Allow a minimum of 7 days between application and harvest.

Do not feed or graze treated cottonseed forage or hay following preharvest applications.

™Def is a trademark of Mobay Chemical Corporation.

™Folex is a trademark of Rhone-Poulenc, Inc.

TREE AND VINE CROPS

This product is recommended for weed control in established groves, vineyards, or orchards, or for site preparation prior to transplanting crops listed in this section. Applications may be made with boom equipment, CDA, shielded sprayers, hand-held and high-volume wands, lances, or orchard guns, or with wiper applicator equipment, except as directed in this section. See the "Application Equipment and Techniques" section of this label for specific information on use of equipment.

When applying this product alone, refer to the "Weeds Controlled" section of this label for recommended rates to be used.

NOTE

Repeat treatments may be necessary to control weeds originating from underground parts of untreated weeds or from seeds. This product does not provide residual weed control. For subsequent weed control, follow a program using residual herbicides or use repeated applications of this product. Do not apply more than 10.6 quarts of this product per acre per year.

EXTREME CARE MUST BE EXERCISED TO AVOID CONTACT OF HERBICIDE SOLUTION, SPRAY, DRIFT, OR MIST WITH FOLIAGE OR GREEN BARK OF TRUNK, BRANCHES, SUCKERS, FRUIT, OR OTHER PARTS OF TREES OR VINES. CONTACT OF THIS PRODUCT WITH OTHER THAN MATURED BROWN BARK CAN RESULT IN SERIOUS CROP DAMAGE.

AVOID PAINTING OUT STUMPS WITH THIS PRODUCT AS INJURY RESULTING FROM ROOT GRAFTING MAY OCCUR IN ADJACENT TREES.

Reduced control may result when applications are made to annual or perennial weeds that have been mowed, grazed or cut and have not been allowed to regrow to the recommended stage for treatment.

For specific rates of applications and instructions, see the "Weeds Controlled" section of this label.

TANK MIXTURES TREE AND VINE CROPS

When applied as a tank mixture, this product provides control of the emerged annual weeds and partial control of the emerged perennial weeds listed in this label. The following residual herbicides will provide preemergence control of those weeds listed in the individual product labels.

HONCHO® plus GOAL™ 1.6E
HONCHO plus KARMEX™ WP
HONCHO plus KROVAR™ I
HONCHO plus KROVAR™ II
HONCHO plus SIMAZINE, CALIBER™ 90
HONCHO plus SIMAZINE 4L
HONCHO plus SIMAZINE 80W
HONCHO plus SOLICAM™ 80WP
HONCHO plus SURFLAN™ AS
HONCHO plus SURFLAN 75W
HONCHO plus SIMAZINE
(80W, or 4L, or CALIBER 90)
plus SURFLAN (AS or 75W)
HONCHO plus GOAL (1.6E) plus
SURFLAN (AS or 75W)
HONCHO plus GOAL (1.6E) plus SIMAZINE
(80W, or 4L, or CALIBER 90)*
HONCHO plus GOAL (1.6 E) plus
SURFLAN (AS or 75W) plus
SIMAZINE (80W, 4L, or CALIBER 90)

*See the Honcho plus Goal plus simazine tank mixture section below for specific use instructions.

Do not apply these tank mixtures in Puerto Rico.

CONTROL OF EMERGED WEEDS

Annual Weeds—Apply 1 quart per acre of this product in these tank mixtures when weeds are less than 12 inches tall. For weeds greater than 12 inches tall, apply 1.5 quarts per acre.

Perennial Weeds—For partial control of perennial weeds using these tank mixtures, apply 2 to 5 quarts per acre of this product. Follow the recommendations in the "Weeds Controlled" section of this label for stage of growth and rate of application for specific perennial weeds.

PREEMERGENCE WEED CONTROL

The following are the labeled rates for the recommended residual herbicides:

Product	Rate/Acre
Goal 1.6E	5 qts.
Karmex WP	2 to 8 lbs.
Krovar I	2 to 8 lbs.
Krovar II	2 to 8 lbs.
Simazine, Caliber 90	1.1 to 10.6 lbs.
Simazine, 80W	1.25 to 12 lbs.
Simazine 4L	2 to 9.6 qts.
Solicam 80WP	2.5 to 5 lbs.
Surflan AS	2 to 4 qts.
Surflan 75W	2.66 to 5.33 lbs

NOTE: These residual herbicides may provide postemergence activity on certain annual weed species.

Refer to the individual product labels for specific crops, rates, geographical restrictions and precautionary statements.

Read and carefully observe the label claims, cautionary statements, rates and all other information on the labels of all products.

For specific rates of application and instructions, see the "Weeds Controlled" section of this label.

HONCHO plus GOAL

This product alone and this product plus Goal tank mixtures will control or suppress annual weeds listed below.

Apply recommended rates of this product or this product plus Goal tank mixtures in 3 to 10 gallons of water per acre. Apply when weeds are actively growing and less than 6 inches in height. If weeds are under drought stress, irrigate prior to application.

RECOMMENDED RATES AND WEEDS CONTROLLED
C = Control S = Suppression

Weed Species	Honcho® (fl oz/acre) 8	12	16	Honcho + Goal (fl oz/acre) 12 + (4 to 16)
Bluegrass, annual	C	C	C	C
Barley	C	C	C	C
Cheeseweed (3")	S	S	S	C
Rocket, London	S	S	C	C
Shepherdspurse	S	S	C	C
Chickweed	S	C	C	C
Fleabane, flaxleaf	S	C	C	C
Red Maids	S	C	C	C
				16 + (4 to 16)
Barnyardgrass	S	C	C	C
Cheeseweed (3 to 6")	S	S	S	C
Crabgrass	S	S	C	C
Groundsel, common	S	S	C	C
Junglerice	S	S	C	C
Lambsquarters	S	S	C	C
Marestail	S	S	C	C
Pigweed, redroot	S	S	C	C
Ryegrass, common	S	S	C	C
Sowthistle, annual	S	S	C	C
Purslane, common	S	S	C	C
Nettle, stinging	S	S	C	C
Filaree	S	S	S	S

HONCHO® plus GOAL plus SIMAZINE
(California only)

This product may be tank-mixed with Goal plus simazine for broad-spectrum postemergence and preemergence control. Refer to the following table for approved crops and rate ranges for each product in this tank mix. Refer to the individual product labels for weeds controlled, geographical restrictions, precautionary statements and specific use rates.

Crop	Rate Ranges* HONCHO® (quarts)	+	GOAL 1.6E (quarts)	+	SIMAZINE 80W** (pounds)
Grapes	1 to 1.5		1.25 to 5		2.5 to 5
Almonds	1 to 1.5		1.25 to 5		1.25 to 2.5

Use equivalent rate of Caliber 90 or simazine 4L.
*Rate dependent on weeds present.
**Rate dependent on soil type and weeds present.

HONCHO® plus GOAL plus SIMAZINE/SURFLAN

This product plus Goal, applied alone, or as a three-way or four-way tank mix with simazine and/or Surflan will provide postemergence control of those weeds listed below when applied as recommended.

Refer to individual product labels for preemergence rates, weeds controlled, precautionary statements and other important information.

Apply these tank mixtures in 3 to 40 gallons of water per acre.

WEEDS CONTROLLED

Weed Species	Maximum Height (inches)	Honcho® + Goal (quarts/acre)	
Barley, wild	**4**	**0.5 to 1**	**+ 0.5 to 5**
Bluegrass, annual	**4**	**0.5 to 1**	**+ 0.5 to 5**
Cheeseweed, common	**4**	**0.5 to 1**	**+ 0.5 to 5**
Chickweed, common	**4**	**0.5 to 1**	**+ 0.5 to 5**
Flaxleaf, fleabane	**4**	**0.5 to 1**	**+ 0.5 to 5**
Groundsel, common	**4**	**0.5 to 1**	**+ 0.5 to 5**
Rocket, London	**4**	**0.5 to 1**	**+ 0.5 to 5**
Marestail	**4**	**0.5 to 1**	**+ 0.5 to 5**
Pineappleweed	**4**	**0.5 to 1**	**+ 0.5 to 5**
Shepherdspurse	**4**	**0.5 to 1**	**+ 0.5 to 5**
Sowsthistle, common	**4**	**0.5 to 1**	**+ 0.5 to 5**
Cheeseweed, common	**12**	**1 to 1.5**	**+ 1 to 5**
Nettle, stinging	**12**	**1 to 1.5**	**+ 1 to 5**
Filaree	**12**	**1 to 1.5**	**+ 1 to 5**

PERENNIAL GRASS SUPPRESSION ORCHARD FLOORS

When applied as directed, this product will suppress vegetative growth as indicated below.

Bahiagrass

This product will provide significant inhibition of seedhead emergence and will suppress vegetative growth for a period of approximately 45 days with a single application and approximately 120 days with sequential applications. Apply this product 1 to 2 weeks after full green-up or after mowing to a uniform height of 3 to 4 inches. Applications must be made prior to seedhead emergence. Apply 6 fluid ounces of this product in 10 to 25 gallons of water per acre.

Sequential applications of this product may be made at approximately 45-day intervals to extend the period of seedhead and vegetative growth suppression. For continued seedhead suppression, sequential applications must be made prior to seedhead emergence. Apply no more than 2 sequential applications per year. As a first sequential application, apply 4 fluid ounces of this product per acre. A second sequential application of 2 to 4 fluid per acre may be made approximately 45 days after the last application.

Bermudagrass

For burndown, apply 1 to 2 quarts of this product in 3 to 20 gallons of water per acre. Use 1 quart in 3 to 20 gallons of water per acre east of the Rocky Mountains. Use 1 to 2 quarts of this product in 3 to 10 gallons of water per acre west of the Rocky Mountains. Use this treatment only if reduction of the bermudagrass stand can be tolerated. When burndown is required prior to harvest, allow at least 21 days to ensure sufficient time for burndown to occur.

Suppression only—(east of the Rocky Mountains). Apply 6 to 16 fluid ounces of this product in 3 to 20 gallons of water per acre no sooner than 1 to 2 weeks after full green-up. Mowing prior to application may occur provided a minimum height of 3 inches is maintained. Rates of 6 to 10 fluid ounces should be used where a lesser degree of suppression is desired. Sequential applications may be made when regrowth occurs and bermudagrass injury and stand reduction can be tolerated.

Suppression only—(west of the Rocky Mountains). Apply 16 fluid ounces of this product in 3 to 10 gallons of water per acre to bermudagrass up to 6 inches in height and no sooner than 1 to 2 weeks after full green-up. Mowing prior to application may occur provided a minimum height of 3 inches is maintained. Sequential applications may be made when regrowth occurs and bermudagrass injury and stand reduction can be tolerated.

Cool-Season Grass Covers

For suppression of tall fescue, fine fescue, orchardgrass and quackgrass, apply 8 fluid ounces of this product in 10 to 20 gallons of water per acre. For best suppression, add ammonium sulfate to the spray solution at a rate of 2 percent by weight or 17 pounds per 100 gallons of spray solution.

For suppression of Kentucky bluegrass covers, apply 6 fluid ounces of this product per acre. Do not add ammonium sulfate.

For best results, mow cool-season grass cover in the spring to even the height and apply the recommended rate of this product 3 to 4 days after mowing. Avoid treating cool-season grass covers under poor growing conditions, such as drought stress (drip irrigation), disease or insect damage.

LOW-VOLUME APPLICATION (Florida and Texas)

For burndown or control of listed perennial weeds, apply the recommended rate of this product plus 0.5 to 1 percent nonionic surfactant by total spray volume in 3 to 20 gallons of water per acre. Where weed foliage is dense, use 10 to 20 gallons of water per acre. Apply when weeds are actively growing and at the growth stages listed in the "Perennial Weeds Controlled" section of this label. If perennial weeds are mowed, allow weeds to regrow to the recommended stage of growth.

S = Suppression, B = Burndown, C = Control, PC = Partial control

Weed Species	HONCHO® Rate per Acre 1 qt	2 qt	3 qt	5 qt
Bermudagrass	B	•	PC	C
Guineagrass				
Texas and Florida Ridge	B	C	C	C
Florida Flatwoods	•	B	C	C
Paragrass	B	C	C	C
Torpedograss	S	•	PC	C

™Caliber is a trademark of Ciba-Geigy Corporation.
™Goal is a trademark of Rohm and Haas Company.
™Karmex and Krovar are trademarks of E.I. du Pont de Nemours and Company.
™Solicam is a trademark of Sandoz, Inc.
™Surflan is a trademark of Elanco Products Company.

TREE CROPS

Citrus*: citron, grapefruit, kumquat, lemon, lime, orange, tangelo, tangerine.

Nuts:** almond, filbert, macadamia, pecan, pistachio, walnut.

Pome Fruit*: apple, pear.

Stone Fruit*:** apricots, cherries, nectarines, olives, peaches, plums/prunes.

For cherries, any application equipment listed in this section may be used in all states.

For citron and olives, apply as a directed spray only.

Any application equipment listed in this section may be used in apricots, nectarines, peaches, and plums/prunes growing in Arizona, California, Colorado, Idaho, Kansas, Kentucky, New Jersey, North Dakota, Oklahoma, Oregon, Texas, Utah, and Washington, except for peaches grown in the states specified in the following paragraph. In all other states use wiper equipment only.

For PEACHES grown in Alabama, Arkansas, Florida, Georgia, Louisiana, Mississippi, North Carolina, South Carolina, and Tennessee only, apply with a shielded boom sprayer or shielded wiper applicator which prevents any contact of this product with the foliage or bark of trees. Apply no later than 90 days after first bloom. Applications made after this time may result in severe damage. Remove suckers and low hanging limbs at least 10 days prior to application. Avoid applications near trees with recent pruning wounds or other mechanical injury. Apply only near trees which have been planted in the orchard for 2 or more years. EXTREME CARE MUST BE TAKEN TO ENSURE NO PART OF THE PEACH TREE IS CONTACTED.

Tropical Fruit: acerola, avocado*, banana (plantains), coffee****, figs*, guava, mango*, papaya, tea. Allow a minimum of 1 day between last application and harvest of guava and papaya. In coffee and banana, delay applications 3 months after transplanting to allow the new coffee or banana plant to become established.

NOTE:

*Allow a minimum of 14 days between last application and harvest.
**Allow a minimum of 21 days between last application and harvest.
***Allow a minimum of 17 days between last application and harvest.
***Allow a minimum of 28 days between last application and harvest.

VINE CROPS

Kiwi Fruit

Grapes: Any variety of table, wine, or raisin grape may be treated with any equipment listed in this section.

In the northeast and Great Lakes regions, applications must be made prior to the end of bloom stage of grapes to avoid injury.

NOTE:

Applications should not be made when green shoots, canes, or foliage are in the spray zone.

Allow a minimum of 14 days between last application and harvest.

In case of an emergency involving this product, Call Collect, day or night, (314) 694-4000.

This speciman label is intended for use only as a guide in providing general information regarding the use of this product. Monsanto is not responsible for the accuracy of the information contained herein. As labels are subject to revision, always carefully read and follow the label on the product container.

LANDMASTER® BW

BINDWEED HERBICIDE

For suppression of emerged field bindweed and control or suppression of annual weeds in fallow and reduced tillage small grains systems.

AVOID CONTACT WITH FOLIAGE OF CROP OR OTHER DESIRABLE VEGETATION SINCE SEVERE INJURY OR DESTRUCTION MAY RESULT.

ACTIVE INGREDIENTS:*	
Glyphosate [N-(phosphonomethyl) glycine] as the isopropylamine salt	12.9%
2,4-D (2,4-dichlorophenoxyacetic acid) as the isopropylamine salt	20.6%
INERT INGREDIENTS:	66.5%
TOTAL	100.0%

*Contains 144 grams per litre or 1.2 lbs per U.S. gallon of the active ingredient, glyphosate, as its isopropylamine salt and 227 grams per litre or 1.9 lbs per U.S. gallon of the active ingredient, 2,4-D, as its isopropylamine salt. Equivalent to 108 grams per litre or 0.9 lb per U.S. gallon of the acid, glyphosate and 182 grams per litre or 1.5 lbs per U.S. gallon of the acid, 2,4-D.

The glyphosate component of LANDMASTER® BW herbicide is protected by U.S. Patent No. 3,799,758 and U.S. Patent No. 4,405,531. Other patents are pending.

No license granted under any non-U.S. patent.

Complete Directions for Use

EPA Reg. No. 524-351

® LANDMASTER is a registered trademark of Monsanto Company.

1988-2 897.14-000.37/CG

Read the entire label before using product.

Use only according to label instructions.

Read "LIMIT OF WARRANTY AND LIABILITY" before buying or using. If terms are not acceptable, return at once unopened.

REFORMULATION OR REPACKAGING IS PROHIBITED.

LIMIT OF WARRANTY AND LIABILITY

This Company warrants that this product conforms to the chemical description on the label and is reasonably fit for the purposes set forth in the Complete Directions for Use label booklet ("Directions") when used in accordance with those Directions under the conditions described therein. NO OTHER EXPRESS WARRANTY OR IMPLIED WARRANTY OF FITNESS FOR PARTICULAR PURPOSE OR MERCHANTABILITY IS MADE. This warranty is also subject to the conditions and limitations stated herein.

Buyer and all users shall promptly notify this Company of any claims whether based in contract, negligence, strict liability, other tort or otherwise.

Buyer and all users are responsible for all loss or damage from use or handling which results from conditions beyond the control of this Company, including, but not limited to, incompatibility with products other than those set forth in the Directions, application to or contact with desirable vegetation, unusual weather, weather conditions which are outside the range considered normal at the application site and for the time period when the product is applied, as well as weather conditions which are outside the application ranges set forth in the Directions, application in any manner not explicitly set forth in the Directions, moisture conditions outside the moisture range specified in the Directions, or the presence of products other than those set forth in the Directions in or on the soil, crop or treated vegetation.

THE EXCLUSIVE REMEDY OF THE USER OR BUYER, AND THE LIMIT OF THE LIABILITY OF THIS COMPANY OR ANY OTHER SELLER FOR ANY AND ALL LOSSES, INJURIES OR DAMAGES RESULTING FROM THE USE OR HANDLING OF THIS PRODUCT (INCLUDING CLAIMS BASED IN CONTRACT, NEGLIGENCE, STRICT LIABILITY, OTHER TORT OR OTHERWISE) SHALL BE THE PURCHASE PRICE PAID BY THE USER OR BUYER FOR THE QUANTITY OF THIS PRODUCT INVOLVED, OR, AT THE ELECTION OF THIS COMPANY OR ANY OTHER SELLER, THE REPLACEMENT OF SUCH QUANTITY, OR, IF NOT ACQUIRED BY PURCHASE, REPLACEMENT OF SUCH QUANTITY. IN NO EVENT SHALL THIS COMPANY OR ANY OTHER SELLER BE LIABLE FOR ANY INCIDENTAL, CONSEQUENTIAL OR SPECIAL DAMAGES.

Buyer and all users are deemed to have accepted the terms of this LIMIT OF WARRANTY AND LIABILITY which may not be varied by any verbal or written agreement.

PRECAUTIONARY STATEMENTS

HAZARDS TO HUMANS AND DOMESTIC ANIMALS

KEEP OUT OF REACH OF CHILDREN

DANGER!

CAUSES EYE BURNS.

HARMFUL IF SWALLOWED.

MAY CAUSE ALLERGIC SKIN REACTION.

Do not get in eyes, on skin, or on clothing.

Wear goggles or face shield and rubber gloves in transferring and mixing, and when adjusting, repairing or cleaning equipment.

Wash thoroughly with soap and water after handling.

FIRST AID: IF IN EYES, immediately flush with plenty of water for at least 15 minutes. Call a physician.

IF ON SKIN, immediately flush with plenty of water for at least 15 minutes while removing contaminated clothing. Wash clothing before reuse. Sensitized persons should avoid further contact and reuse of contaminated clothing.

IF SWALLOWED, induce vomiting immediately by either giving two glasses of water and sticking finger down throat or, when available, administering syrup of ipecac as prescribed on its container. Call a physician. NEVER GIVE ANYTHING BY MOUTH TO AN UNCONSCIOUS PERSON.

In case of an emergency involving this product, Call Collect, day or night, (314) 694-4000.

ENVIRONMENTAL HAZARDS

Do not apply directly to water or wetland (swamps, bogs, marshes or potholes). Do not contaminate water by cleaning of equipment or disposal of wastes.

PHYSICAL OR CHEMICAL HAZARDS

Spray solutions of this product should be mixed, stored and applied only in stainless steel, aluminum, fiberglass, plastic and plastic-lined steel containers.

DO NOT MIX, STORE OR APPLY THIS PRODUCT OR SPRAY SOLUTIONS OF THIS PRODUCT IN GALVANIZED STEEL OR UNLINED STEEL (EXCEPT STAINLESS STEEL) CONTAINERS OR SPRAY TANKS. This product or spray solutions of this product react with such containers and tanks to produce hydrogen gas which may form a highly combustible gas mixture. This gas mixture could flash or explode, causing serious personal injury, if ignited by open flame, spark, welder's torch, lighted cigarette or other ignition source.

GENERAL INFORMATION

LANDMASTER BW is a postemergence herbicide for control or suppression of emerged weeds in fallow and reduced tillage systems, prior to planting or emergence of wheat, barley, corn, oats, rye or sorghum (grain or forage).

This product enters the plant through the foliage and moves throughout the plant. Visual effects of control are a gradual wilting or yellowing of the plant, which advances to complete browning of aboveground growth and deterioration of affected underground plant parts. Visible symptoms will usually develop within 2 to 4 days after application, but may not occur for 7 or more days. Extremely cool or cloudy weather following treatment may slow activity of this product and delay the visual effects of control.

APPLICATION PRECAUTIONS AND WARNINGS

- Do not plant any crop other than wheat, barley, corn, oats, rye or sorghum (grain or forage) for 3 months after treatment or until this product has disappeared from the soil.
- Applications of this product after planting and prior to crop emergence may cause crop injury if rainfall or environmental conditions delaying crop emergence are experienced.
- This product is recommended for control of emerged weeds prior to establishment of labeled crops. Large amounts of green or decaying vegetation left standing or incorporated into the seedbed may enhance the development of disease in newly planted crops. This may result in poor emergence and/or stands, especially under cool and/or wet conditions.
- Spraying early to control young weeds before dense stands develop or light cultivation to assist weed decay will favor preparation of suitable seedbeds.
- In reduced tillage and no-till systems, ensure good seed to soil contact and proper seeding depth.
- Do not feed or forage treated vegetation from treated areas within 8 weeks after application.
- DO NOT APPLY IN THE VICINITY OF 2,4-D SENSITIVE CROPS SUCH AS COTTON, GRAPES, TOMATOES AND OTHER DESIRABLE VEGETATION.
- Applications should be made only when there is no hazard from spray drift since very small quantities of spray, which may not be visible, may severely injure susceptible crops or desirable vegetation.
- The likelihood of injury occurring to adjacent crops from the use of this product is greatest when winds are gusty or in excess of 5 miles per hour or when other conditions including lesser wind velocities will favor spray drift.
- Movement of this product on soil particles during windstorms may cause damage to susceptible plants that are contacted. This hazard is reduced if rainfall occurs shortly after application.
- Buyer and all users are responsible for all loss or damage in connection with the use or handling of mixtur s of this herbicide or other materials that are not expressly recommended in this labeling. Mixing this product with herbicides or other materials not recommended on this label may result in reduced performance. Use of this product in any manner not consistent with this label may result in injury to persons, animals or crops, or other unintended consequences.

LANDMASTER BW is subject to all state and county regulations for 2,4-D amine.

DIRECTIONS FOR USE

It is a violation of Federal law to use this product in any manner inconsistent with its labeling.

STORAGE AND DISPOSAL

Do not contaminate water, food, or feed by storage or disposal. Do not store near fertilizers, seeds, insecticides or fungicides.

STORAGE:

STORE ABOVE 40°F. to keep product in solution. If crystals form, place in a warm room (72°F.), allow the product to reach room temperature and roll or shake periodically until crystals have dissolved. Keep container closed to prevent spills and contamination.

For bulk containers, see the container label for alternate storage information.

DISPOSAL:

Wastes of this pesticide may cause eye burns and may be dangerous. Improper disposal of excess pesticide, spray mixtures, or rinsate is a violation of Federal law. If these wastes cannot be disposed of according to label use instructions, contact your State Pesticide or Environmental Control Agency, or the Hazardous Waste representative at the nearest EPA Regional Office for guidance.

Emptied container retains vapor and product residue. Observe all labeled safeguards until container is cleaned, reconditioned or destroyed.

(See the individual container label for disposal information.)

TIMING OF APPLICATION

This product should be applied postemergence to vigorously growing weeds when they have reached the recommended size given in the "RECOMMENDED RATES AND WEEDS CONTROLLED" section of this label. Application should be delayed until maximum emergence of the target weeds, but before weeds exceed the maximum size recommended. For annual weeds, allow 1 day after treatment before tillage. For field bindweed, allow at least 7 days after treatment before tillage.

Reduced control may result if treatments are made during poor growing conditions such as drought stress, disease or insect damage or if weeds have been mowed, grazed or cut. Heavy dust on foliage or an overstory canopy covering targeted weeds may also reduce control.

Rainfall or irrigation occurring within 6 hours after application may reduce effectiveness. Heavy rainfall or irrigation within 2 hours after application will wash this product off the foliage and a repeat treatment will be required.

RECOMMENDED RATES AND WEEDS CONTROLLED

For best results, apply this product after most weed seeds have germinated but before seedhead formation in grasses or flower bud formation in broadleaves.

When applied as directed, this product will provide control or suppression of the grass and broadleaf weed species listed below. Rates recommended are for maximum weed height at treatment time.

Perennial Weed Species	Rate Per Acre	LANDMASTER® BW Length of Vines/ Treatment Height
Bindweed, field** *Convolvulus arvensis*	54 oz (Suppression only)	6 to 18"

Annual Weed Species	Rate Per Acre	Maximum Height
Foxtail, green *Setaria viridis*	27 oz	12"
Barley *Hordeum vulgare* **Brome, downy*** *Bromus tectorum* **Cheat*** *Bromus secalinus* **Foxtail** *Setaria spp.* **Kochia*** *Kochia scoparia* **Lettuce, prickly*** *Lactuca serriola* **Oats, wild** *Avena fatua* **Puncturevine** *Tribulus terrestris* **Purslane, common** *Portulaca oleracea*	40 oz	6"
Cocklebur *Xanthium strumarium* **Lambsquarters** *Chenopodium album* **Mustard, tansy** *Descurainia pinnata* **Mustard, tumble** *Sisymbrium altissimum* **Pigweed, redroot** *Amaranthus retroflexus* **Pigweed, smooth** *Amaranthus hybridus* **Rye** *Secale cereale* **Stinkgrass** *Eragrostis cilianensis* **Thistle, Russian** *Salsola kali* **Wheat** *Triticum aestivum* **Witchgrass** *Panicum capillare*	40 oz	12"
Barnyardgrass *Echinochloa crus-galli* **Buffalobur** *Solanum rostratum* **Goatgrass** *Aegilops cylindrica* **Mustard, blue** *Chorispora tenella* **Panicum, fall** *Panicum dichotomiflorum*	54 oz	6"
Oats, wild *Avena fatua* **Witchgrass** *Panicum capillare*	54 oz	12"

* For improved control in no-till systems or heavy infestations or overwintered stands, use 54 ounces. For best results on light kochia infestations, treat after the plant has passed through the woolly stage of growth and is 3 to 6 inches in height. When treating medium to heavy infestations or plants that are in the woolly stage (1 to 3 inches in height), add 2 fluid ounces of Banvel™ to the recommended rate of LANDMASTER BW. Refer to the Banvel label for planting, cropping and other restrictions. Follow all precautions on the Banvel label.

** LANDMASTER BW may also be used as a 1 percent solution on a spray-to-wet basis for annual weed control and field bindweed suppression. Spray coverage should be uniform and complete. Do not spray to the point of runoff.

Prepare the desired volume of spray by mixing the amount of LANDMASTER BW in clean water as shown in the following table:

Spray Solutions

Desired Volume	Amount of LANDMASTER® BW
1 gallon	1⅓ ounces
25 gallons	1 quart
100 gallons	1 gallon
2 tablespoons = 1 ounce	

™ Banvel is a trademark of Sandoz, Inc.

POSTHARVEST APPLICATIONS

This product will provide control of weeds following grain harvest. Weeds should be allowed to regrow after damage incurred during harvest operations and to recover from environmental stress before application of this product. Weeds should be treated prior to the heading stage of annual grasses and before broadleaf weeds exceed 24 inches in height. Ammonium sulfate will improve performance on annual weeds under stress conditions.

Weeds controlled at the 40 ounce-per-acre rate include downy brome, green foxtail, stinkgrass, and volunteer wheat.

Weeds controlled at the 54 ounce-per-acre rate include field bindweed, kochia, lambsquarters, mustard, pigweed and Russian thistle.

Weeds controlled at the 64 ounce-per-acre rate include barnyardgrass, sandbur, witchgrass, yellow foxtail and prickly lettuce.

FARMSTEADS, DITCHES AND FENCEROWS

When applied as directed this product will control downy brome, bulbous bluegrass, kochia, tumble mustard, tansy mustard and prickly lettuce, and provide suppression of crested wheatgrass, smooth bromegrass, and field bindweed.

Use a 1 percent solution of this product for spray-to-wet, spot spraying applications. Spray coverage should be uniform and complete. Do not spray to the point of runoff. Treat when weeds are small, actively growing and free of dust. Use 108 ounces of this product in 5-10 gallons of water per acre for broadcast boom applications.

AID TO TILLAGE

LANDMASTER BW herbicide used in conjunction with preplant and conventional fallow tillage practices will provide control of downy brome, cheat, volunteer wheat, tansy mustard, and foxtail. Apply 27 fluid ounces of this product in 3 to 10 gallons of water per acre to weeds that are actively growing. Treat when weeds are less than 6 inches in height. Application must be followed by conventional tillage practices before regrowth of the treated plant occurs. Allow at least 1 day after application before tillage.

AMMONIUM SULFATE

The addition of 2 percent dry ammonium sulfate by weight (or liquid equivalent) or 17 pounds per 100 gallons of water may increase the performance of LANDMASTER BW and LANDMASTER BW herbicide tank mixtures on annual weeds. The improvement in performance may be apparent where environmental stress is a concern. Low-quality ammonium sulfate may contain material that will not readily dissolve which could result in nozzle tip plugging. To determine quality, perform a jar test by adding ⅓ cup of ammonium sulfate to 1 gallon of water and agitate for 1 minute. If undissolved sediment is observed, predissolve the ammonium sulfate in water and filter prior to addition to the spray tank. If ammonium sulfate is added directly to the spray tank, add slowly with agitation. Adding too quickly may clog outlet line. ENSURE THAT AMMONIUM SULFATE IS COMPLETELY DISSOLVED IN THE SPRAY TANK BEFORE ADDING HERBICIDES. THOROUGHLY RINSE THE SPRAY SYSTEM WITH CLEAN WATER AFTER USE TO REDUCE CORROSION.

NOTE: Compatibility problems may occur at carrier volumes below 5 GPA.

APPLICATION EQUIPMENT

Do not apply this product through any type of irrigation system.

This product may be applied using either ground, aerial spray or hand-held equipment. Use extreme care to avoid misting or drifting of herbicide solution onto foliage, green stems or fruit of desirable crops, trees, or plants during both growing and dormant periods since even very small quantities of spray can cause severe plant injury.

GROUND APPLICATION: Apply recommended rates of this product in 3 to 10 gallons of water per acre as a broadcast spray. For optimum spray distribution and coverage, use flat fan or low volume flood nozzles. When using flood nozzles, space them no more than 40 inches apart and ensure double overlap of spray pattern. Refer to the manufacturer's recommendations for correct pressure and nozzle height above the target canopy. Avoid pressure and nozzles which produce fine droplets or mist.

Use appropriate marking devices to ensure uniform spray coverage and best results from LANDMASTER BW herbicide.

HAND-HELD and HIGH-VOLUME EQUIPMENT (use coarse sprays only): Mix this product in clean water and apply to foliage of vegetation to be controlled. For applications made on a spray-to-wet basis, spray coverage should be uniform and complete. Do not spray to the point of runoff. Use a 1 percent solution for annual weed control and field bindweed suppression.

AERIAL APPLICATION: Apply the recommended rates of this product in 3 to 5 gallons of water per acre as a broadcast spray. DO NOT APPLY DURING INVERSION CONDITIONS, WHEN WINDS ARE GUSTY, OR WHEN OTHER CONDITIONS WILL FAVOR DRIFT. DRIFT MAY CAUSE DAMAGE TO ANY VEGETATION CONTACTED TO WHICH TREATMENT WAS NOT INTENDED. TO PREVENT INJURY TO ADJACENT DESIRABLE VEGETATION, APPROPRIATE BUFFER ZONES MUST BE MAINTAINED.

Coarse sprays are less likely to drift, therefore, do not use nozzles or nozzle configurations which dispense spray as fine spray droplets. Do not angle nozzles forward into the airstream and do not increase spray volume by increasing nozzle pressure.

MIXING INSTRUCTIONS

LANDMASTER® BW ALONE

Fill the spray tank to about 3/4 of the desired volume with clean water. Add the recommended amount of LANDMASTER BW herbicide, then complete the filling process while maintaining agitation. Remove the hose from the mix tank immediately after filling to avoid siphoning back into the carrier source. During mixing and application, foaming of the spray solution may occur. To prevent or minimize foam, terminate by-pass and return lines at the tank bottom and/or use an agriculturally approved antifoam or defoaming agent.

Additional surfactant is not necessary for this formulation.

NOTE: Reduced results may occur if water containing soil is used, such as water from ponds and unlined ditches.

TANK MIXTURES

Always predetermine the compatibility of labeled tank mixtures of this herbicide with water carrier by mixing small proportional quantities in advance.

Mix labeled tank mixtures of Landmaster BW herbicide with water as follows:

1. Place a 20 to 35 mesh screen or wetting basket over filling port.
2. Through the screen, fill the sprayer tank one-half full with water and start agitation.
3. If a wettable powder is used, make a slurry with the water carrier, and add it SLOWLY through the screen into the tank. Continue agitation.
4. If a flowable formulation is used, premix one part flowable with one part water. Add diluted mixture SLOWLY through the screen into the tank. Continue agitation.
5. If an emulsifiable concentrate formulation is used, premix one part emulsifiable concentrate with two parts water. Add diluted mixture slowly through the screen into the tank. Continue agitation.
6. Continue filling the sprayer tank with water and add the required amount of Landmaster BW herbicide near the end of the filling.
7. Add individual formulations to the spray tank as follows: wettable powder, flowable, emulsifiable concentrate, drift control additive, water soluble liquid.

Maintain good agitation at all times until the contents of the tank are sprayed. If the spray mixture is allowed to settle, thorough agitation is required to resuspend the mixture before spraying is resumed. Keep bypass line on or near bottom of tank to minimize foaming. Screen size in nozzle should be no finer than 100 mesh and in-line strainers should be no finer than 50 mesh. Carefully select proper nozzle to avoid spraying a fine mist. For best results with conventional ground application equipment, use flat fan nozzles.

Check label of all products used in tank mix for cleaning instructions. Clean as per the hardest material to remove.

SPRAYER CLEANUP

CLEAN THE ENTIRE SPRAYER AFTER APPLICATION OF THIS PRODUCT. Failure to clean the sprayer thoroughly may result in injury to desirable crops which are subsequently sprayed.

First, add clean water to the tank and thoroughly rinse the entire sprayer system. Secondly, fill the tank with water and ammonia. Add 1 quart of household ammonia per 25 gallons of water. Pump enough solution through the hoses, boom and nozzles to fill these parts completely. Then fill the tank, close and leave for 24 hours before draining and rinsing thoroughly with water.

Application or use of other agriculture chemicals with the equipment used for this product may result in injury to desirable vegetation.

Thoroughly wash aircraft, especially landing gear, after each day of spraying to remove residues of this product accumulated during spraying or from spills. PROLONGED EXPOSURE OF THIS PRODUCT TO UNCOATED STEEL SURFACES MAY RESULT IN CORROSION AND POSSIBLE FAILURE OF THE PART. LANDING GEAR ARE MOST SUSCEPTIBLE. The maintenance of an organic coating (paint) which meets aerospace specification MIL-C-38413 may prevent corrosion.

897.14-000.37/CG

This sample label is current as of January 1, 1989. The product descriptions and recommendations provided in this sample label are for background information only. Always refer to the label on the product container before using Monsanto or any other agrichemical product.

In case of an emergency involving this product, Call Collect, day or night, (314) 694-4000.

This specimen label is intended for use only as a guide in providing general information regarding the use of this product. Monsanto is not responsible for the accuracy of the information contained herein. As labels are subject to revision, always carefully read and follow the label on the product container.

LANDMASTER® II

HERBICIDE

For control or suppression of emerged weeds in fallow and reduced tillage small grain systems.

ACTIVE INGREDIENTS:*

Glyphosate [N-phosphonomethyl glycine] as the isopropylamine salt	13.3%
2,4-D (2,4-dichlorophenoxyacetic acid) as the isopropylamine salt	11.1%
INERT INGREDIENTS:	75.6%
TOTAL	100.0%

*Contains 144 grams per litre or 1.2 lbs per U.S. gallon of the active ingredient, glyphosate, as its isopropylamine salt and 120 grams per litre or 1 lb per U.S. gallon of the active ingredient, 2,4-D, as its isopropylamine salt. Equivalent to 108 grams per litre or 0.9 lb per U.S. gallon of the acid, glyphosate and 96 grams per litre or 0.8 lb per U.S. gallon of the acid, 2,4-D.

The glyphosate component of LANDMASTER II herbicide is protected by U.S. Patent No. 3,799,758, and U.S. Patent No. 4,405,531, Other patents are pending.

No license granted under any non-U.S. patent.

Complete Directions for Use

AVOID CONTACT WITH FOLIAGE OF CROP OR OTHER DESIRABLE VEGETATION SINCE SEVERE INJURY OR DESTRUCTION MAY RESULT.

EPA Reg. No. 524-376

®LANDMASTER is a registered trademark of Monsanto Company.

1988-2 886.14-000.20/CG

Read the entire label before using product.

Use only according to label instructions.

Read "LIMIT OF WARRANTY AND LIABILITY" before buying or using. If terms are not acceptable, return at once unopened.

REFORMULATION IS PROHIBITED.

LIMIT OF WARRANTY AND LIABILITY

This Company warrants that this product conforms to the chemical description on the label and is reasonably fit for the purposes set forth in the Complete Directions for Use label booklet ("Directions") when used in accordance with those Directions under the conditions described therein. NO OTHER EXPRESS WARRANTY OR IMPLIED WARRANTY OF FITNESS FOR PARTICULAR PURPOSE OR MERCHANTABILITY IS MADE. This warranty is also subject to conditions and limitations stated herein.

Buyer and all users shall promptly notify this Company of any claims whether based in contract, negligence, strict liability, other tort or otherwise.

Buyer and all users are responsible for all loss or damage from use or handling which results from conditions beyond the control of this Company, including, but not limited to, incompatibility with products other than those set forth in the Directions, application to or contact with desirable vegetation, unusual weather, weather conditions which are outside the range considered normal at the application site and for the time period when the product is applied, as well as weather conditions which are outside the application ranges set forth in the Directions, application in any manner not explicitly set forth in the Directions, moisture conditions outside the moisture range specified in the Directions, or the presence of products other than those set forth in the Directions in or on the soil, crop or treated vegetation.

THE EXCLUSIVE REMEDY OF THE USER OR BUYER, AND THE LIMIT OF THE LIABILITY OF THIS COMPANY OR ANY OTHER SELLER FOR ANY AND ALL LOSSES, INJURIES OR DAMAGES RESULTING FROM THE USE OR HANDLING OF THIS PRODUCT (INCLUDING CLAIMS BASED IN CONTRACT, NEGLIGENCE, STRICT LIABILITY, OTHER TORT OR OTHERWISE) SHALL BE THE PURCHASE PRICE PAID BY THE USER OR BUYER FOR THE QUANTITY OF THIS PRODUCT INVOLVED, OR, AT THE ELECTION OF THIS COMPANY OR ANY OTHER SELLER, THE REPLACEMENT OF SUCH QUANTITY OR, IF NOT ACQUIRED BY PURCHASE, REPLACEMENT OF SUCH QUANTITY. IN NO EVENT SHALL THIS COMPANY OR ANY OTHER SELLER BE LIABLE FOR ANY INCIDENTAL, CONSEQUENTIAL, OR SPECIAL DAMAGES.

The Buyer and all users are deemed to have accepted the terms of this LIMIT OF WARRANTY AND LIABILITY which may not be varied by any verbal or written agreement.

PRECAUTIONARY STATEMENTS

HAZARDS TO HUMANS AND DOMESTIC ANIMALS

KEEP OUT OF REACH OF CHILDREN

DANGER!

CAUSES EYE BURNS.

HARMFUL IF SWALLOWED OR INHALED.

Do not get in eyes, on skin or on clothing.

Wear goggles or face shield and rubber gloves in transferring and mixing and when adjusting, repairing or cleaning equipment.

Avoid breathing vapors or spray mist.

Wash thoroughly with soap and water after handling.

FIRST AID: IF IN EYES, immediately flush with plenty of water for at least 15 minutes. Call a physician. Remove material from skin and clothing.

IF INHALED, remove individual to fresh air. Seek medical attention if breathing difficulty develops.

IF SWALLOWED, induce vomiting immediately by either giving two glasses of water and sticking finger down throat or, when available, administering syrup of ipecac as prescribed on its container. Call a physician. NEVER GIVE ANYTHING BY MOUTH TO AN UNCONSCIOUS PERSON.

In case of an emergency involving this product, Call Collect, day or night, (314) 694-4000.

ENVIRONMENTAL HAZARDS

Do not apply directly to water or wetland (swamps, bogs, marshes or potholes). Do not contaminate water when disposing of equipment washwaters.

PHYSICAL OR CHEMICAL HAZARDS

Spray solutions of this product should be mixed, stored and applied only in stainless steel, aluminum, fiberglass, plastic and plastic-lined steel containers.

DO NOT MIX, STORE OR APPLY THIS PRODUCT OR SPRAY SOLUTIONS OF THIS PRODUCT IN GALVANIZED STEEL OR UNLINED STEEL (EXCEPT STAINLESS STEEL) CONTAINERS OR SPRAY TANKS. This product or spray solutions of this product react with such containers and tanks to produce hydrogen gas which may form a highly combustible gas mixture. This gas mixture could flash or explode, causing serious personal injury, if ignited by open flame, spark, welder's torch, lighted cigarette or other ignition source.

GENERAL INFORMATION

LANDMASTER II is a postemergence herbicide for control or suppression of emerged weeds in fallow and reduced tillage systems, prior to planting or emergence of wheat, barley, corn, oats, rye or sorghum (grain or forage).

This product enters the plant through the foliage and moves throughout the plant. Visual effects of control are a gradual wilting or yellowing of the plant, which advances to complete browning of above-ground growth and deterioration of affected underground plant parts. Visible symptoms will usually develop on labeled weeds within 2 to 4 days after application, but may not occur for 7 or more days. Extremely cool or cloudy weather following treatment may slow activity of this product and delay the visual effects of control.

APPLICATION PRECAUTIONS AND WARNINGS

- Do not plant any crop other than corn, wheat, barley, oats, rye or sorghum (grain or forage) for 3 months after treatment or until this product has disappeared from the soil.
- Applications of this product after planting and prior to crop emergence may cause crop injury if rainfall or environmental conditions delaying crop emergence are experienced.
- This product is recommended for control of emerged weeds prior to establishment of labeled crops. Large amounts of green or decaying vegetation left standing or incorporated into the seedbed may enhance the development of disease in newly planted crops. This may result in poor emergence and/or stands, especially under cool and/or wet conditions.
- Spraying early to control young weeds before dense stands develop or light cultivation to assist weed decay will favor preparation of suitable seedbeds.
- In reduced tillage and no-till systems, ensure good seed to soil contact and proper seeding depth.
- Do not feed or forage treated vegetation from treated areas within 8 weeks after application.
- DO NOT APPLY IN THE VICINITY OF 2,4-D SENSITIVE CROPS SUCH AS COTTON, GRAPES, TOMATOES AND OTHER DESIRABLE VEGETATION.
- Applications should be made only when there is no hazard from spray drift since very small quantities of spray, which may not be visible, may severely injure susceptible crops or desirable vegetation.
- The likelihood of injury occurring to adjacent crops from the use of this product is greatest when winds are gusty or in excess of 5 miles per hour, or when other conditions including lesser wind velocities will favor spray drift.
- Movement of this product on soil particles during windstorms may cause damage to susceptible plants that are contacted. This hazard is reduced if rainfall occurs shortly after application.
- Buyer and all users are responsible for all loss or damage in connection with the use or handling of mixtures of this herbicide or other materials that are not expressly recommended in this labeling. Mixing this product with herbicides or other materials not recommended on this label may result in reduced performance. Use of this product in any manner not consistent with this label may result in injury to persons, animals or crops, or other unintended consequences.

LANDMASTER II is subject to all state and county regulations for 2,4-D amine.

DIRECTIONS FOR USE

It is a violation of Federal law to use this product in any manner inconsistent with its labeling.

STORAGE AND DISPOSAL

Do not contaminate water, food, or feed by storage or disposal. Do not store near fertilizers, seeds, insecticides or fungicides.

STORAGE:

STORE ABOVE 40°F to keep product in solution. If crystals form, place in a warm room (72°F), allow the product to reach room temperature and roll or shake periodically until crystals have dissolved. Keep container closed to prevent spills and contamination.

For bulk containers, see the container label for alternate storage information.

DISPOSAL:

Wastes of this pesticide may cause eye burns and may be dangerous. Improper disposal of excess pesticide, spray mixtures, or rinsate is a violation of Federal law. If these wastes cannot be disposed of by use according to label instructions, contact your State Pesticide or Environmental Control Agency, or the Hazardous Waste representative at the nearest EPA Regional Office for guidance.

Empty container retains vapor and product residue. Observe all labeled safeguards until container is cleaned, reconditioned or destroyed.

(See the individual container label for disposal information.)

TIMING OF APPLICATION

This product should be applied postemergence to vigorously growing weeds when they have reached the recommended size given in the "RECOMMENDED RATES AND WEEDS CONTROLLED" section of this label. Application should be delayed until maximum emergence of the target weeds, but before weeds exceed the maximum size recommended. For annual weeds, allow 1 day after treatment before tillage. For field bindweed, allow at least 7 days after treatment before tillage.

Reduced control may result if treatments are made during poor growing conditions such as drought stress, disease or insect damage or if weeds have been mowed, grazed or cut. Heavy dust on foliage or an overstory canopy covering targeted weeds may also reduce control.

Rainfall or irrigation occurring within 6 hours after application may reduce effectiveness. Heavy rainfall or irrigation within 2 hours after application will wash this product off the foliage and a repeat treatment will be required.

•For field bindweed and quackgrass, do not till between harvest and fall applications or in fall or spring prior to spring application.

RECOMMENDED RATES AND WEEDS CONTROLLED

For best results, apply this product after most weed seeds have germinated but before seedhead formation in grasses or flower bud formation in broadleaves.

When applied as directed, LANDMASTER II herbicide will provide control or suppression of the grass and broadleaf weed species listed below. Rates recommended are for maximum weed height at treatment time.

Annual Weed Species	LANDMASTER® II Rate Per Acre	LANDMASTER® II Maximum Height
Foxtail, green *Setaria viridis*	27 oz	12″
Barley *Hordeum vulgare*; Brome, downy* *Bromus tectorum*; Cheat* *Bromus secalinus*; Foxtail *Setaria spp.*; Kochia* *Kochia scoparia*; Lettuce, prickly* *Lactuca serriola*; Mustard, tumble *Sisymbrium altissimum*; Oats, wild *Avena fatua*; Puncturevine *Tribulus terrestris*; Purslane, common *Portulaca oleracea*	40 oz	6″
Cocklebur *Xanthium pensylvanicum*; Lambsquarters *Chenopodium album*; Mustard, tansy *Descurainia pinnata*; Pigweed, redroot *Amaranthus retroflexus*; Pigweed, smooth *Amaranthus hybridus*; Rye *Secale cerale*; Stinkgrass *Eragrostis cilianensis*; Thistle, Russian *Salsola kali*; Wheat *Triticum aestivum*; Witchgrass *Panicum capillare*	40 oz	12″
Barnyardgrass *Echinochloa crus-galli*; Buffalobur *Solanum rostratum*; Goatgrass *Aegilops cylindrica*; Mustard (blue) *Chorispora tenella*; Panicum (fall) *Panicum dichotomiflorum*	54 oz	6″
Oats, wild *Avena fatua*; Sandbur, field *Cendrus spp.*	54 oz.	12″

Perennial Weed Species	Rate Per Acre	Length of Vines/ Treatment Height
Bindweed (field)** *Convolvulus arvensis*	64 to 80 oz (Suppression only)	6 to 18″

*For improved control in no-till systems or heavy infestations or overwintered stands, use 54 ounces. For best results on light kochia infestations, treat after the plant has passed through the woolly stage of growth and is 3 to 6 inches in height. When treating medium to heavy infestations or plants that are in the woolly stage (1 to 3 inches in height), add 2 fluid ounces of Banvel™ to the recommended rate of LANDMASTER II. The addition of ammonium sulfate is recommended for use with LANDMASTER II plus Banvel tank mixes. Refer to the "AMMONIUM SULFATE" section for complete instructions. Refer to the Banvel label for planting, cropping and other restrictions. Follow all precautions on the Banvel label.

**Use the higher recommended rates of this product for dense, well-established stands of field bindweed and for longer-term suppression. LANDMASTER II may be used as a 1 percent solution on a spray-to-wet basis for annual weed control and field bindweed suppression. Spray coverage should be uniform and complete. Do not spray to the point of runoff.

Prepare the desired volume of spray solution by mixing the amount of LANDMASTER II in clean water as shown in the following table:

SPRAY SOLUTIONS

Desired Volume	Amount of LANDMASTER® II
1 gallon	1⅓ ounces
25 gallons	1 quart
100 gallons	1 gallon
2 tablespoons = 1 ounce	

™ Banvel is a trademark of Sandoz, Inc.

LANDMASTER II plus ATRAZINE or BLADEX™ TANK MIXTURE

For use only in Colorado, Kansas, Nebraska, Oklahoma, Oregon, South Dakota and Washington.

Tank mixtures of LANDMASTER II plus atrazine or Bladex will provide postemergence control of listed annual weeds in fallow and reduced tillage systems. In addition, these tank mixtures will provide soil residual control of weeds listed on the atrazine and Bladex labels.

™Bladex is a trademark of E. I. du Pont de Nemours and Company.

ANNUAL WEEDS CONTROLLED

Barnyardgrass *Echinochloa crus-galli*	Pigweed, redroot *Amaranthus retroflexus*
Brome, downy *Bromus tectorum*	Sandbur, field *Cenchrus spp.*
Foxtail, green *Setaria virdis*	Stinkgrass *Eragrostis cilianensis*
Kochia *Kochia scoparia*	Thistle, Russian *Salsola kali*
Lambsquarters *Chenopodium album*	Wheat *Triticum aestivum*
Lettuce, prickly *Lactuca serriola*	Witchgrass *Panicum cappillare*
Mustard, tansy *Descurainia pinnata*	

Apply 40 ounces of this product plus 1 pound or less of the active ingredient, atrazine, per acre. Use 54 ounces of this product plus 2 pounds or less of the active ingredient, atrazine, per acre.

Use 64 ounces of this product plus 1 to 3 pounds of the active ingredient, atrazine, per acre. Use 54 ounces of this product with 2.4 to 4 pounds of the active ingredient, cyanazine, in Bladex per acre. In Oregon and Washington, do not exceed 1 pound of the active ingredient, atrazine, per acre. Barnyardgrass will be suppressed at the 64 ounce per acre rate and will require up to 86 ounces per acre for control. The addition of ammonium sulfate is recommended to increase the performance of this product plus atrazine or Bladex tank mixtures.

Consult the atrazine and Bladex labels (90 DF, 80W, 4L) for use rates, soil type, planting, cropping and other restrictions, as well as other precautionary statements and use according to the most restrictive label.

These tank mixtures may be applied with ground or aerial equipment. See the "APPLICATION EQUIPMENT" section for instructions.

POSTHARVEST APPLICATIONS

This product will provide control of weeds following grain harvest. Weeds should be allowed to regrow after damage incurred during harvest operations and to recover from environmental stress before application of this product. Weeds should be treated prior to the heading stage of annual grasses and before broadleaf weeds exceed 24 inches in height. Ammonium sulfate will improve performance on annual weeds under stress conditions.

Weeds controlled at the 40 ounce-per-acre rate include downy brome, green foxtail, stinkgrass, volunteer wheat.

Weeds controlled at the 54 ounce-per-acre rate include kochia, lambsquarters, mustard, pigweed and Russian thistle.

Weeds controlled at the 64 ounce-per-acre rate include barnyardgrass, sandbur, witchgrass, yellow foxtail and prickly lettuce.

AID TO TILLAGE

LANDMASTER II herbicide used in conjunction with preplant and conventional fallow tillage practices will provide control of downy brome, cheat, volunteer wheat, tansy mustard, and foxtails. Apply 27 fluid ounces of this product in 3 to 10 gallons of water per acre to weeds that are actively growing. Treat when weeds are less than 6 inches in height. Application must be followed by conventional tillage practices before regrowth of the treated plant occurs. Allow at least 1 day after application before tillage.

AMMONIUM SULFATE

The addition of 2 percent dry ammonium sulfate by weight (or liquid equivalent) or 17 pounds per 100 gallons of water may increase the performance of LANDMASTER II and LANDMASTER II herbicide tank mixtures on annual weeds. The improvement in performance may be apparent where environmental stress is a concern. Low-quality ammonium sulfate may contain material that will not readily dissolve which could result in nozzle tip plugging. To determine quality, perform a jar test by adding 1/3 cup of ammonium sulfate to 1 gallon of water and agitate for 1 minute. If undissolved sediment is observed, predissolve the ammonium sulfate in water and filter prior to addition to the spray tank. If ammonium sulfate is added directly to the spray tank, add slowly with agitation. Adding too quickly may clog outlet line. ENSURE THAT AMMONIUM SULFATE IS COMPLETELY DISSOLVED IN THE SPRAY TANK BEFORE ADDING HERBICIDES. THOROUGHLY RINSE THE SPRAY SYSTEM WITH CLEAN WATER AFTER USE TO REDUCE CORROSION.

NOTE: Compatibility problems may occur at carrier volumes below 5 GPA.

APPLICATION EQUIPMENT

Do not apply this product through any type of irrigation system.

This product may be applied using either ground or aerial spray equipment. Use extreme care to avoid misting or drifting of herbicide solution onto foliage, green stems or fruit of desirable crops, trees, or plants during both growing and dormant periods, since even very small quantities of spray can cause severe plant injury.

GROUND APPLICATION: Apply recommended rates of this product in 3 to 10 gallons of water per acre as a broadcast spray. For optimum spray distribution and coverage, use flat fan or low-volume flood nozzles. When using flood nozzles, space them no more than 40 inches apart and ensure double overlap of spray pattern. Refer to the manufacturer's recommendations for correct pressure and nozzle height above the target canopy. Avoid pressure and nozzles which produce fine droplets or mist.

Use appropriate marking devices to ensure uniform spray coverage and best results from LANDMASTER II herbicide.

HAND-HELD AND HIGH-VOLUME EQUIPMENT (use coarse sprays only): Mix this product in clean water and apply to foliage of vegetation to be controlled. For applications made on a spray-to-wet basis, spray coverage should be uniform and complete. Do not spray to the point of runoff. Use a 1 percent solution for field bindweed.

AERIAL APPLICATION: Apply the recommended rates of this product in 3 to 5 gallons of water per acre as a broadcast spray. DO NOT APPLY DURING INVERSION CONDITIONS. WHEN WINDS ARE GUSTY, OR WHEN OTHER CONDITIONS WILL FAVOR DRIFT. DRIFT MAY CAUSE DAMAGE TO ANY VEGETATION CONTACTED TO WHICH TREATMENT WAS NOT INTENDED. TO PREVENT INJURY TO ADJACENT DESIRABLE VEGETATION. APPROPRIATE BUFFER ZONES MUST BE MAINTAINED.

Coarse sprays are less likely to drift; therefore, do not use nozzles or nozzle configurations which dispense spray as fine spray droplet. Do not angle nozzles forward into the airstream and do not increase spray volume by increasing nozzle pressure.

MIXING INSTRUCTIONS

LANDMASTER® II ALONE

Fill the spray tank to about 3/4 of the desired volume with clean water. Add the recommended amount of LANDMASTER II herbicide, then complete the filling process while maintaining agitation. Remove the hose from the mix tank immediately after filling to avoid siphoning back into the carrier source. During mixing and application, foaming of the spray solution may occur. To prevent or minimize foam, terminate by-pass and return lines at the tank bottom and/or use an agriculturally approved antifoam or defoaming agent.

Additional surfactant is not recommended for this formulation.

NOTE: Reduced results may occur if water containing soil is used, such as water from ponds and unlined ditches.

TANK MIXTURES

Always predetermine the compatibility of labeled tank mixtures of this herbicide with water carrier by mixing small proportional quantities in advance.

Mix labeled tank mixtures of LANDMASTER II herbicide with water as follows:

1. Place a 20 to 35 mesh screen or wetting basket over filling port.
2. Through the screen, fill the sprayer tank one-half full with water and start agitation.
3. If a wettable powder is used, make a slurry with the water carrier, and add it SLOWLY through the screen into the tank. Continue agitation.
4. If a flowable formulation is used, premix one part flowable with one part water. Add diluted mixture SLOWLY through the screen into the tank. Continue agitation.
5. If an emulsifiable concentrate formulation is used, premix one part emulsifiable concentrate with two parts water. Add diluted mixture slowly through the screen into the tank. Continue agitaton.
6. Continue filling the sprayer tank with water and add the required amount of LANDMASTER II herbicide near the end of the filling.
7. Add individual formulations to the spray tank as follows: wettable powder, flowable, emulsifiable concentrate, drift control additive and water soluble liquid.

Maintain good agitation at all times until the contents of the tank are sprayed. If the spray mixture is allowed to settle, thorough agitation is required to resuspend the mixture before spraying is resumed.

Keep by-pass line on or near bottom of tank to minimize foaming. Screen size in nozzle should be no finer than 100 mesh and in-line strainers should be no finer than 50 mesh. Carefully select proper nozzle to avoid spraying a fine mist. For best results with conventional ground application equipment, use flat fan nozzles.

Check label of all products used in tank mix for cleaning instructions. Clean as per the hardest material to remove.

SPRAYER CLEANUP

CLEAN THE ENTIRE SPRAYER AFTER APPLICATION OF THIS PRODUCT. Failure to clean the sprayer thoroughly may result in injury to desirable crops which are subsequently sprayed. First, add clean water to the tank and thoroughly rinse the entire sprayer system. Secondly, fill the tank with water and ammonia. Add 1 quart of household ammonia per 25 gallons of water. Pump enough solution through the hoses, boom and nozzles to fill these parts completely. Then fill the tank, close and leave for 24 hours before draining and rinsing thoroughly with water.

Application or use of other agricultural chemicals with the equipment used for this product may result in injury to desirable vegetation.

Thoroughly wash aircraft, especially landing gear, after each day of spraying to remove residues of this product accumulated during spraying or from spills. PROLONGED EXPOSURE OF THIS PRODUCT TO UNCOATED STEEL SURFACES MAY RESULT IN CORROSION AND POSSIBLE FAILURE OF THE PART. LANDING GEAR ARE MOST SUSCEPTIBLE. The maintenance of an organic coating (paint) which meets aerospace specification MIL-C-38413 may prevent corrosion.

886.14-000.20/53

In case of an emergency involving this product, Call Collect, day or night, (314) 694-4000.

This sample label is current as of January 1, 1989. The product descriptions and recommendations provided in this sample label are for background information only. Always refer to the label on the product container before using Monsanto or any other agrichemical product.

This specimen label is intended for use only as a guide in providing general information regarding the use of this product. Monsanto is not responsible for the accuracy of the information contained herein. As labels are subject to revision, always carefully read and follow the label on the product container.

LARIAT®

FLOWABLE HERBICIDE

RESTRICTED USE PESTICIDE

due to oncogenicity. For retail sale to and use only by Certified Applicators or persons under their direct supervision and only for those uses covered by the Certified Applicator's certification.

The use of this product may be hazardous to your health. This product contains alachlor, which has been determined to cause tumors in laboratory animals.

In case of an emergency involving this product or for user safety information on this product, call collect, day or night (314) 694-4000.

Selective herbicide for preemergence weed control in corn (all types), and grain sorghum (milo).

ACTIVE INGREDIENTS:*

Alachlor, (2-chloro-2'6'-diethyl-N-(methoxymethyl). acetanilide)	27.2%
Atrazine, (2-chloro-4-(ethylamino)-6-(isopropylamino)-s-triazine)	15.5%
Related atrazine compounds	0.8%
INERT INGREDIENTS	56.5%
TOTAL	100.0%

*Contains 300 grams/litre or 2.5 pounds/gallon of alachlor and 180 grams/litre or 1.5 pounds/gallon of atrazine and related compounds.

SHAKE WELL BEFORE USING

Complete Directions For Use

®Lariat is a registered trademark of Monsanto Company.

EPA Reg. No. 524-329-AA 1988-1 843.89-000.24/53

Read the entire label before using this product.

Use only according to label instructions.

Read "LIMIT OF WARRANTY AND LIABILITY" before buying or using. If terms are not acceptable, return at once unopened.

REFORMULATION IS PROHIBITED.

LIMIT OF WARRANTY AND LIABILITY

This Company warrants that this product conforms to the chemical description on the label and is reasonably fit for the purposes set forth in the Complete Directions for Use label booklet ("Directions") when used in accordance with those Directions under the conditions described therein. NO OTHER EXPRESS WARRANTY OR IMPLIED WARRANTY OF FITNESS FOR PARTICULAR PURPOSE OF MERCHANTABILITY IS MADE. This warranty is also subject to the conditions and limitations stated herein.

Buyer and all users shall promptly notify this Company of any claims whether based in contract negligence, strict liability, other tort or otherwise.

Buyer and all users are responsible for all loss or damage from use or handling which results from conditions beyond the control of this Company, including, but not limited to, incompatibility with products other than those set forth in the Directions, unusual weather, weather conditions which are outside the range considered normal at the application site and for the time period when the product is applied, as well as weather conditions which are outside the application ranges set forth in the Directions, application in any manner not explicitly set forth in the Directions, moisture conditions outside the moisture range specified in the Directions, or the presence of products other than those set forth in the Directions in or on the soil, crop or treated vegetation.

THE EXCLUSIVE REMEDY OF THE USER OR BUYER, AND THE LIMIT OF THE LIABILITY OF THIS COMPANY OR ANY OTHER SELLER FOR ANY AND ALL LOSSES, INJURIES OR DAMAGES RESULTING FROM THE USE OR HANDLING OF THIS PRODUCT (INCLUDING CLAIMS BASED IN CONTRACT, NEGLIGENCE, STRICT LIABILITY, OTHER TORT OR OTHERWISE) SHALL BE THE PURCHASE PRICE PAID BY THE USER OR BUYER FOR THE QUANTITY OF THIS PRODUCT INVOLVED, OR, AT THE ELECTION OF THIS COMPANY OR ANY OTHER SELLER. THE REPLACEMENT OF SUCH QUANTITY OR, IF NOT ACQUIRED BY PURCHASE. REPLACEMENT OF SUCH QUANTITY. IN NO EVENT SHALL THIS COMPANY OR ANY OTHER SELLER BE LIABLE FOR ANY INCIDENTAL, CONSEQUENTIAL OR SPECIAL DAMAGES.

Buyer and all users are deemed to have accepted the terms of this LIMIT OF WARRANTY AND LIABILITY which may not be varied by any verbal or written agreement.

PRECAUTIONARY STATEMENTS

HAZARDS TO HUMANS AND DOMESTIC ANIMALS

KEEP OUT OF REACH OF CHILDREN

WARNING!

HARMFUL IF SWALLOWED OR ABSORBED THROUGH SKIN.

CAUSES EYE AND SKIN IRRITATION.

MAY CAUSE ALLERGIC SKIN REACTION.

Do not get in eyes, on skin or on clothing.

Wear goggles or face shield, rubber gloves, long trousers, long-sleeved shirt or jacket of tightly woven material along with boots high enough to cover ankles when transferring and mixing, and when adjusting, repairing or cleaning equipment. Wear rubber boots when pouring from open containers or when reentry is made into fields following application of this product through center pivot irrigation systems and the field is still wet. Clothing which comes in contact with this product must be washed before reuse. Clothing or other materials which become drenched with the concentrated pesticide must be disposed of in a sanitary landfill, by incineration or, if allowed by state and local authorities, by burning. If burned, stay out of smoke.

Wash thoroughly with soap and water after handling.

FIRST AID: IF IN EYES, immediately flush with plenty of water for at least 15 minutes. Call a physician.

IF ON SKIN, immediately flush with plenty of water while removing contaminated clothing. Call a physician. Wash clothing before reuse. Sensitized persons should avoid further contact and reuse of contaminated clothing.

IF SWALLOWED, induce vomiting immediately by giving two glasses of water and sticking finger down throat. Call a physician. NEVER GIVE ANYTHING BY MOUTH TO AN UNCONSCIOUS PERSON.

In case of an emergency involving this product, Call Collect, day or night, (314) 694-4000.

ENVIRONMENTAL HAZARDS

This chemical has been identified in limited ground water sampling and there is the possibility that it can leach through the soil to ground water, especially where soils are coarse and ground water is near the surface. Following application, and during rainfall events that cause runoff, this chemical may reach surface water bodies including streams, rivers and reservoirs.

Care must be taken when using this product to prevent back siphoning into wells, spills or improper disposal of excess pesticide, spray mixtures or rinsates.

Check valves or antisiphoning devices must be used on all mixing and/or irrigation equipment. Disposal of excess pesticide, spray mixtures or rinsate should be according to label use instructions or according to the state pesticide or environmental control agency or the hazardous waste representative at the nearest EPA regional office.

This product is toxic to fish. Do not apply directly to water or wetlands. Do not contaminate water by cleaning of equipment or disposal of wastes.

PHYSICAL OR CHEMICAL HAZARDS

COMBUSTIBLE.

Do not use or store near heat or flame.

Use only with adequate ventilation.

In case of:

FIRE, use water spray, foam, dry chemical or CO_2

SPILL or LEAK, soak up, remove and keep away from humans and domestic animals.

STORAGE AND DISPOSAL

Do not contaminate water, foodstuffs, seed or feed by storage or disposal.

STORAGE:

STORE ABOVE 40°F.

Below 32°F this product may freeze. If frozen, place in a warm room and roll and/or shake the container frequently to redissolve before using. For bulk containers, see the container label for alternate storage information.

DISPOSAL:

Wastes of this pesticide may cause eye and skin irritation and may be dangerous. Improper disposal of excess pesticide, spray mixtures, or rinsate is a violation of Federal law. If these wastes cannot be disposed of according to label use instructions, contact your State Pesticide or Environmental Control Agency, or the Hazardous Waste representative at the nearest EPA Regional Office for guidance.

Emptied container retains vapor and product residue. Observe all labeled safeguards until container is cleaned, reconditioned or destroyed.

(See the individual container label for disposal information).

GENERAL INFORMATION

This product is recommended for the control of may annual grasses and broadleaf weeds. Except for minimum and conservation tillage systems, the seed bed should be fine, firm and free of clods and trash.

Performance is improved when moisture occurs within one week after application. The amount required depends primarily upon the amount of soil moisture in the weed germination zone at the time of application, soil type and organic matter content, but 1/3 to 3/4 inch is generally sufficient for performance of this product. High intensity or excessive rainfall, or excessive irrigation, after application may reduce control.

Read and carefully observe the cautionary statements and all other information appearing on this label and on the labels of Gramoxone ™ Super and Roundup® herbicides before using in combination with this product.

Use of this product in any manner not consistent with this label may result in injury to persons, animals or crops, or other unintended consequences. Keep container closed to prevent spills and contamination.

Do not apply when conditions favor drift.

Flush sprayer with clean water after use.

Do not graze treated area or feed treated forage to livestock for 21 days following application.

NOTE: Use this product for weed control in corn and grain sorghum (milo) only. PLANT ONLY CORN, PEANUTS, SORGHUM (MILO)

OR SOYBEANS THE YEAR FOLLOWING THE USE OF THIS MIXTURE. IF SOYBEANS ARE TO BE PLANTED, THERE IS A POSSIBILITY OF CROP INJURY DUE TO CARRYOVER OF ATRAZINE. Do not plant soybeans in areas where furrow irrigation is practiced. On grain sorghum (milo) grown in Texas, use only on the fine textured soils of the Gulf Coast and Blacklands. Do not exceed 3.25 quarts per acre of this product as possible crop injury may result due to atrazine.

®Trademark of Monsanto Company.

™ Gramoxone Super is a trademark of ICI America Inc.

SOIL TEXTURE

Applicators should evaluate soil and weed conditions carefully to assure that they choose the lowest effective label rate.

The recommended use rates of this product and the other herbicides labeled for use in tank mixtures with this product generally vary with soil texture. Unless soil texture is specifically named, rate tables throughout this label refer to only three soil texture groups: coarse, medium and fine. The following is a complete listing of soil textures included in each of these three soil texture groups:

Soil Texture Group	Soil Texture
COURSE:	sand, loamy sand, sandy loam
MEDIUM:	loam, silt loam, silt, sandy clay loam
FINE:	silty clay loam, clay loam, sandy clay, silty clay, clay

Refer to the above table to determine the corresponding soil texture group for the soil to be treated.

MIXING, SPRAYING AND HANDLING INSTRUCTIONS

NOTE

The use of a mechanical transfer system is required to be used by all mixer/loaders and/or applicators who treat 300 acres or more annually with pesticides containing alachlor as an active ingredient.

Direct contact or exposure to this product or spray mixtures of this product should be minimized. The following instructions for transfer, mixing, cleaning or repairing equipment should be followed in order to minimize this exposure. Review the protective clothing requirements as listed in the Precautionary Statements section of this label (front panel) and do not use this product until you have the necessary protective clothing.

2.5 Gallon Containers

Open pouring from these containers can result in exposure from splashing or spilling. Special care in lifting and pouring are strongly recommended.

Bulk Containers

Open pouring from these containers can result in exposure from splashing or spilling and is not recommended. This product should be transferred from these containers to the mix or spray tank using pumps or transfer probes. The probe or pump should not be removed from the container or disconnected until the container is emptied and rinsed. Use the pump or probe system to rinse the empty container and transfer the rinsate directly to the mix or spray tank.

Equipment Cleaning & Repair

Cleaning and repair of transfer systems and application equipment is a source of exposure to this product. Care should be taken to minimize exposure during cleaning and repair of transfer systems and application equipment. Whenever possible, these systems or equipment should be rinsed before being cleaned or repaired. When repairs must be made during transfer or application, the equipment should be shut down, and special care taken to avoid contact with the pesticide.

Always predeterime the compatibility of this product with water carrier or sprayable fluid fertilizer or other labeled herbicides by mixing small proportional quantities in advance.

Mix this product with appropriate carrier as follows:

1. Place a 20 to 35 mesh screen or wetting basket over filling port.
2. Through the screen, fill the sprayer tank one-half full with the carrier.
3. If a compatibility agent is necessary to improve mixing or to prevent the formation of undesirable and unsprayable gels or precipitates, while agitating add it to the carrier already in the tank.

Use only compatibility agents, cleared by the Federal Food and Drug Administration for this use. Follow the directions for use of the compatibility agent label. Check for adequate agitation.

4. Premix one part of this product with one part water. Add diluted mixture SLOWLY through the screen into the tank. When necessary, this product may be diluted with a fertilizer carrier instead of water, after compatibility has been confirmed.
5. With minimum tillage systems, add the required amounts of Roundup or Gramoxone Super near the end of the filling process.
6. Complete filling the sprayer tank with carrier. Remove hose from tank immediately to avoid siphoning back into the carrier source. Maintain good agitation at all times until the contents of the tank are sprayed.

NOTE: If spray mixture is allowed to settle at any time, thorough agitation is required to resuspend the mixture before spraying is resumed.

Keep by-pass line on or near the bottom of tank to minimize foaming. Screen size in nozzle or line strainers should be no finer than 50 mesh. Carefully select proper nozzle to avoid spraying a fine mist. For best results, use flat fan or flood jet nozzles. Check for even distribution of spray droplets. To reduce loss ot the chemical due to drift of a fine mist, apply at pressures from 20 to 40 psi.

STANDARD SPRAYABLE FLUID FERTILIZER COMPATIBILITY TEST

Herbicides may not always mix evenly throughout a sprayable fluid fertilizer or the components may separate too quickly to make their combined use of practical value. This may be due to certain characteristics of the different fluid fertilizers. A simple test using small quantities of the components is suggested to provide compatibility potential.

The test follows:

A. Materials Required For A Compatibility Test

1. Two one-quart jars with lid or stopper (marked "with" and "without").
2. TEAspoons (for a more exacting test, a five to ten milliliter (ml.) pipette or graduated cylinder is desirable.
3. A minimum of one quart of the sprayable fluid fertilizer to be tested.
4. The herbicide(s) to be mixed.
5. A compatibility agent (the purpose of the adjuvant is to help keep the fertilizer and herbicide(s) in suspension, if this assistance is needed).

B. Procedure

1. Add one point of the sprayable fluid fertilizer and will be used or other herbicide carrier to each jar marked "with " and "without".

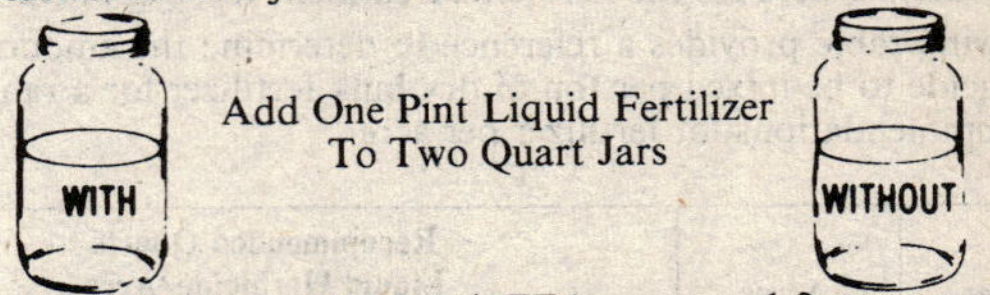

2. To the jar marked "with ", add 1/4 TEAspoon or 1.2 milliliters of a suitable compatibility agent; shake gently for five to ten seconds to mix. (1/4 TEAspoon is the equivalent of two pints per 100 gallons of liquid fertilizer.)

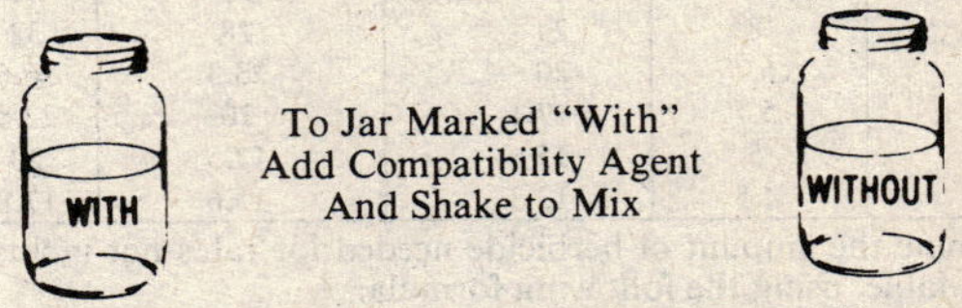

3. To each jar add the appropriate amount of herbicide(s). If more than one is used, add them separately with the wettable powders added first, flowables second and liquid last. Shake gently five to ten seconds after each addition.

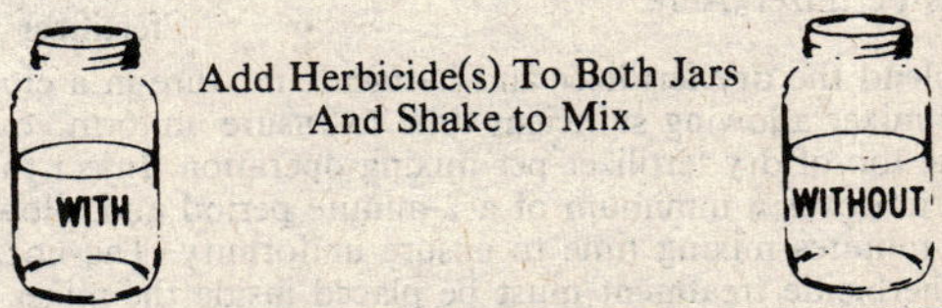

	Amount to Be Added Per Pint of Sprayable Fluid Fertilizer (Assuming Volune is 25 Gal/A)			
Herbicide	Rate/A	Level TEAspoons		Milliliters
This Product	3 qts =	2.9	or	14.2
	3.5 qts =	3.4	or	16.6
	4 qts =	3.8	or	19
	4.5 qts =	4.3	or	21.4
	5 qts =	4.8	or	23.8

This compatibility test is designed for 25 gallons of spray per acre with the maximum labeled rate of herbicide. For changes in spray volume or herbicide rate, make appropriate changes in the ingredients of the test. Regardless of spray volume, the amount of compatibility agent should be equal to two or three pints (two pints = 1/4 TEAspoon or 1.2 milliliters, three pints = 3/8 TEAspoon or 1.8 milliliters per pint of sprayable fluid fertilizer) per 100 gallons of liquid fertilizer.

C. Observations and Decisions

1. If the herbicide(s) and the sprayable fluid fertilizer are compatible.
2. If a compatibility agent is necessary.

Five minutes after the final addition and mixing, observe both jars for the formation of large flakes, sludge, gels or other precipitates. Observe if the herbicide(s) cannot be physically mixed with the liquid fertilizer but remains as small oily particles in the solution.

If incompatibility in any form described above occurs in the jar "with" the compatibility agent added, the liquid fertilizer and herbicide(s) should not be used together in the same spray tank.

If incompatibility as described above occurs in the jar "without" the adjuvant but not in the jar "with" adjuvant, the use of a compatibility adjuvant is recommended.

Both jars should be allowed to stand and be observed periodically for one-half hour. If the separate layers of liquid fertilizer and additives can be resuspended by shaking, commercial application is possible. Wettable powders will either settle to the bottom of the tank or jar, or float to the top, depending upon the density of the fertilizers.

If the herbicide(s) is compatible with fluid fertilizer in the foregoing test without having to use a compatibility agent, fluid fertilizer may be used for the premixing. If it is not compatible, the herbicide(s) should be premixed with water before adding to the spray tank.

APPLICATION WITH DRY BULK FERTILIZER

The herbicide/fertilizer impregnation process must be conducted only at commercial fertilizer or chemical dealerships properly equipped for this procedure.

Dry bulk fertilizer may be impregnated with this product for use on corn. This product must be applied with 200 or 450 pounds of dry bulk fertilizer per acre and shallowly incorporated within 7 days prior to planting. The proper amount of herbicide must be applied per acre in the blend as recommended in this label for the crop, weed and soil type treated. Refer to the table for broadcast rate per acre in the "Surface Application or Preplant Incorporated" section of this label to determine the recommended broadcast rate per acre for the herbicide treatment to be applied.

The following table provides a reference to determine the amount of LIQUID herbicide to be mixed per ton of dry bulk fertilizer for a range of herbicide recommendations for fertilizer per acre:

Fertilizer Rate (Lb./Acre)	Acres Covered (Per Ton)	Recommended Quarts Liquid Herbicide/Acre			
		3	3.5	4	4.5
		(Quarts of Herbicide/Ton Dry Bulk Fertilizer)			
200	10	30	35	40	45
250	8	24	28	32	36
300	6.7	20	23.3	26.6	30
350	5.7	17.1	20	22.8	25.7
400	5	15	17.5	20	22.5
450	4.5	13.3	15.6	17.8	20

To determine the amount of herbicide needed for rates not included in the preceding table, using the following formula:

$$\frac{\text{Recommended Herbicide Rate (lb. or qt.)/Acre} \times 2000}{\text{Pounds Fertilizer/Acre}} = \text{Pounds or Quarts of herbicide per ton of dry bulk fertilizer}$$

Mix and blend the dry fertilizer and herbicide mixture in a closed rotary-drum-type mixer allowing sufficient time to ensure uniform coverage. Use at least one ton of dry fertilizer per mixing operation. Inject the herbicide into the drum over a minimum of a 2-minute period and allow at least 2 additional minutes mixing time to ensure uniformity. The nozzle used to spray the herbicide treatment must be placed inside the mixer to provide uniform spray coverage of the tumbling fertilizer.

If the dry fertilizer used has inadequate adsorptive capacity, use a highly adsorptive powder such as Microcel ™ E to provide a free-flowing mixture. The adsorptive powder must be added separately and uniformly to the prepared herbicide- fertilizer mixture. Generally 2 percent by weight of Microcel E is required. The following is a partial list of dry fertilizers which may be impregnated with this product.

Ammonium sulfate	21-0-0
Ammonium phosphate-sulfate	16-20-0
Diammonium phosphate	18-46-0
Potassium chloride	0-0-60
Potassium sulfate	0-0-52
*Urea	46-0-0

* Some ureas may be phytotoxic when applied on corn.

Use only ureas known to be safe to corn.

NOTE: Do not impregnate tank mixtures of this product with other herbicides or fertilizers containing ammonium nitrate, potassium nitrate or sodium nitrate.

™ Microcel is a trademark of Manville Products Corporation.

APPLICATION

Spread the fertilizer/chemical mixture with a properly calibrated applicator. Be certain the material is applied uniformly to the soil surface. When using spin spreaders, a 100 percent overlap is recommended to ensure even distribution. Spin applicators should be calibrated to apply one-half the desired rate per acre and application overlapped by doubling back to cover one-half of the previous swath. Dribble spreaders must be calibrated with the fertilizer/chemical mixture since the flow pattern for the herbicide-coated fertilizer will generally be slower than straight dry fertilizer.

WEEDS CONTROLLED

Applicators should evaluate soil and weed conditions carefully to assure that they choose the lowest effective label rate.

When applied as directed, this product will CONTROL the following weeds:

ANNUAL GRASSES	ANNUAL BROADLEAVES
Barnyardgrass *Echinochloa crus-galli*	Carpetweed *Mollugo verticillata*
Crabgrass Digitaria *ischaemum*, *Digitaria sanguinalis*	Jimsonweed *Datura stramonium*
Foxtail, giant *Setaria faberi*	Kochia *Kochia scoparia*
Foxtail, green *Setaria virdis*	Lambsquarters *Chenopodium album*
Foxtail, yellow *Setaria lutescens*	Mustard *Brassica spp.*
Goosegrass *Eleusine indica*	Nightshade, black *Solanum nigrum*
Panicum, browntop *Panicum fasciculatum*	Nightshade, hairy *Solanum sarrachoides*
Panicum, fall *Panicum dichotomiflorum*	Pigweed, Carelessweed *Amaranthus spp.*
Signalgrass, broadleaf *Brachiaria spp.*	Purslane *Portulaca oleracea*
Sprangletop, red *Leptochloa filiformis*	Pusley, Florida *Richardia scabra*
Witchgrass *Panicum capillare*	Ragweed, common *Ambrosia artemisiifolia*
SEDGE	Smartweed *Polygonum pensylvanicum*
Nutsedge, yellow* *Cyperus esculentus*	*Polygonum persicaria*

* Use 5 quarts per acre of this product on medium and fine-textured soils and preplant incorporate only for control of this weed.

When applied preplant incorporated or immediately after planting and within 5 days of last tillage, this product at a rate of 4 to 5 quarts per acre on a broadcast basis will reduce competition from the following HARD-TO-CONTROL weeds:

ANNUAL GRASSES	ANNUAL BROADLEAVES
Johnsongrass, seedling *Sorghum halepense*	Beggarweed, Flordia *Desmodium tortuosum*
Rice, red *Oryza sativa*	Cocklebur *Xanthium strumarium*
Sandbur, Grassbur *Cenchrus incertus*	Morningglory, annual *Ipomoea purpurea*
SEDGE	Sida, spiny, Teaweed
Nutsedge, yellow *Cyperus esculentus*	Velvetleaf, Buttonweed *Abutilon theophrasti*

DIRECTIONS FOR USE

It is violation of Federal law to use this product in any manner inconsistent with its labeling.

Aerial—Fixed wing; helicopter. Apply the recommended rates of this product or approved tank mixtures in 3 to 10 gallons of water per acre. AVOID DRIFT . DO NOT APPLY DURING INVERSION CONDITIONS, WHEN WINDS ARE GUSTY, OR UNDER ANY OTHER CONDITION WHICH WILL ALLOW DRIFT. Coarse sprays are less likely to drift, therefore do not use nozzles or nozzle configurations which dispense spray or fine spray droplets. Do not angle nozzles forwards into the airstream and do not increase spray volume by increasing nozzle pressure. Ensure uniform application. To avoid streaked, uneven or overlapped application, use mechanical marking systems only. HUMAN FLAGGERS PROHIBITED!

GROUND APPLICATION

Apply as a broadcast treatment in 10 or more gallons of water or sprayable fluid fertilizer per acre. Do not apply during periods of gusty winds or winds in excess of 15 MPH.

Band Treatment—Apply a broadcast equivalent rate and volume per acre to determine these:

$$\frac{\text{Band Width in inches}}{\text{Row width in inches}} \times \text{Broadcast RATE per acre} = \text{Band RATE per acre}$$

Band Width in inches / Row width in inches	×	Broadcast VOLUME per acre	=	Band VOLUME per acre

Mount spray equipment on the planter with nozzles behind press wheel and apply this product at planting time.

CHEMIGATION

Apply this product and labeled tank mixtures only through center pivot irrigation systems. Do not apply this product through any other type of irrigation system.

Crop injury, lack of effectiveness, or illegal pesticide residues in the crop can result from nonuniform distribution of treated water. Use only in systems that apply uniformly.

If you have questions about calibration, you should contact State Extension Service specialists, equipment manufacturers or other experts.

Do not chemigate through systems connected to a public water system.

A person knowledgeable of the chemigation system and responsible for its operation, or under the supervision of the responsible person, shall shut the system down and make necessary adjustments should the need arise. Do not apply when system connections or fittings leak, when nozzles do not provide uniform distibution or when the line containing the product must be dismantled and drained.

The system must contain a functional check valve, vacuum relief valve, and low pressure drain appropriately located on the irrigation pipeline to prevent water source contamination form back flow.

The pesticide injection pipeline must contain a functional, automatic, quick-closing check valve to prevent the flow of fluid back toward the injection pump.

The pesticide injection pipeline must also contain a functional, normally closed, solenoid operated valve located on the intake side of the injection pump and connected to the system interlock to prevent fluid from being withdrawn from the supply tank when the irrigation system is either automatically or manually shut down.

The system must contain functional interlocking controls to automatically shut off the pesticide injection pump when the water pump motor stops.

The irrigation line or water pump must include a functional pressure switch which will stop the water pump motor when the water pressure decreases to the point where pesticide distribution is adversely affected. System must use a metering pump, such as a positive displacement injection pump (e.g. diaphragm pump) effectively designed and constructed of materials that are compatible with pesticides and capable of being fitted with a system interlock, Pumps, infection equipment, agitation equipment, hoses and connections between supply tank and the point of injection must be constructed of materials which are resistant to Lasso.

Do not apply when wind speed favors drift beyond the area intended for treatment.

Meter this product into the center pivot irrigation system after planting and before crop emergence. Herbicide application should be made in 1/3 to 3/4 inch of water per acre. Do not apply in more than 3/4 inch of water per acre under any conditions or reduced performance may occur. On very sandy soils (more than 60% sand and less than 1% organic matter), apply this product in a maximum of 1/2 inch water per acre. Good agitation must be maintained during the entire application period. Flush the system with water when application is complete. Refer to the "MIXING, SPRAYING AND HANDLING INSTRUCTIONS" section of the label for mixing procedures.

NOTE: Applications can be made by injection through center pivot irrigation equipment only when specified for the crop to be treated in the "DIRECTIONS FOR USE" section of this label. Do not apply this product or any labeled tank mixture through center pivot irrigation unless the treatment is specifically recommended in this label for the crop to be treated.

CULTIVATION INFORMATION

If annual grass or broadleaf weed seedlings emerge, rotary hoe immediately to improve weed control.

If cultivation is necessary because of soil crusting or compaction, tillage should be shallow to minimize dilution of this herbicide.

Except for the above, as long as the grasses and weeds are controlled, delay cultivation as long as possible.

If band application is used, avoid throwing soil during cultivation onto the band before weeds emerge in the band.

CORN

Surface Application—Apply this product in 10 or more gallons of water per acre any time from immediately after planting until weeds or grasses reach the 2-leaf stage and the crop is no more than 5 inches tall. Some leaf burn to the crop may occasionally occur, but this should not affect subsequent crop growth or yield. This product can also be applied with sprayable fluid fertilizer in 10 or more gallons per acre. Applications with sprayable fluid fertilizer **must be made prior to crop emergence.** Performance may be improved when application is made prior to weed emergence. Applications made beyond the 2-leaf stage of weed growth will not provide satisfactory weed control. On sweet corn, use this product before crop or weeds emerge. See the following table for recommended broadcast rates per acre for surface applications.

Preplant Incorporation—Broadcast spray this product in 10 or more gallons of water or sprayable fluid fertilizer and incorporate thoroughly into the top 2 inches of soil within 7 days prior to planting. When using one-pass incorporation (Surface Blend™), the equipment used must provide uniform and shallow blending to avoid streaking the treatment. To insure shallow incorporation of 2 inches or less, pull-type tillage equipment should be set to work the soil NO DEEPER THAN 4 INCHES.

This product is diluted when incorporated into the soil. Therefore, SHALLOW AND UNIFORM incorporation of the herbicide treatment is essential.

™ Surface Blend is a trademark of Monsanto Company.

See the following table for recommended broadcast rates per acre for preplant incorporated applications:

Surface Application or Preplant Incorporation Application Rates

Soil Texture Group*	Broadcast Rate Per Acre (Quarts) Less Than 3% Organic Matter	3% Or More Organic Matter
Coarse**	2.5	3
Medium	3	3.75
Fine	3.75	3.75 to 4.5

Use up to 5 quarts of this product per acre to control heavy grass infestations where present, or to reduce competition form the hard-to-control weeds listed on this label.

NOTE: To CONTROL yellow nutsedge, apply 5 quarts of this product per acre on medium and fine-textured soils and preplant incorporate only.

*Refer to the "SOIL TEXTURE" section of this label to determine the corresponding texture group for the soil to be treated.

**When applied through center pivot irrigation or under sprinkler irrigation systems on coarse soils, use a minimum of 3 quarts of this product per acre. DO NOT exceed 1/3 to 3/4 inch of water per acre with the first irrigation as reduced effectiveness may result.

BLADEX™ PLUS THIS PRODUCT (TANK MIXTURE)

Apply this tank mixture in water or sprayable fluid fertilizer solutions for control of annual grasses and broadleaf weeds listed in the "WEEDS CONTROLLED" section of this label. Refer to the rate table for this product and to the Bladex rate table below for either surface application or preplant incorporation.

Surface Application or Preplant Incorporated Application Rates

Soil Texture Group	Broadcast Rate Per Acre (Quarts) Bladex: Less Than 3% Organic Matter	3% Or More Organic Matter
Coarse*	1	1 to 1.25
Medium	1 to 1.25	1.5 to 2
Fine	1.25 to 1.5	1.5 to 2

* Do not use this mixture on sand or loamy sand soils with less than 2% organic matter.

* Bladex is a trademark of E.I. du Pont de Nemours and Company.

MINIMUM TILLAGE SYSTEMS

TANK MIXTURES OF THIS PRODUCT PLUS ROUNDUP® OR GRAMOXONE SUPER

When applied as directed under the conditions described, these tank mixtures control many emerged annual weeds, suppress many emerged perennial weeds and give preemergence control of may annual weeds when corn will be planted directly into a cover crop, established sod or in previous crop residues. These tank mixtures will not control regrowth from perennial weeds.

DO NOT APPLY BY AIR.

Refer to specific product labels for crop rotation restrictions and cautionary statements of all products used in the tank mixture.

For mixing instructions, see the "MIXING, SPRAYING AND HANDLING INSTRUCTIONS" section of this label.

Apply the tank mixture of this product with Roundup in 10 to 40 gallons of water, or the tank mixture of this product with Gramoxone Super in 20 to 60 gallons of water per acre immediately before, during or after planting but BEFORE CROP EMERGENCE. As previous crop residue or vegetation increases, spray gallonage should be increased within the recommended range to insure complete coverage.

For recommended rates of this product with tank mixtures of Roundup or Gramoxone Super see the following minimum tillage rate table.

Soil Texture Group*	Broadcast Rate Per Acre (Quarts)
	Lariat®
Coarse	3
Medium	3.75
Fine	3.75 to 4.5

Use up to 5 quarts of this product per acre to control heavy grass infestations where present, or to reduce competition from the hard-to-control weeds listed on this label.

*Refer to the "SOIL TEXTURE" section of this label to determine the corresponding texture group for the soil to be treated.

ATTENTION

AVOID DRIFT—EXTREME CARE MUST BE USED WHEN APPLYING THESE TANK MIXTURES TO PREVENT INJURY TO DESIRABLE PLANTS AND CROPS. Do not allow spray mist to drift since even minute quantities of spray can cause severe damage or destruction to nearby crops, plants or other areas on which treatment is not intended. Do not apply when winds are gusty or in excess of 5 miles per hour or when other conditions, including lesser wind velocities, will allow drift to occur. When spraying, avoid combinations of pressure and nozzle type that will result in fine particles (mist) which are more likely to drift.

For weeds controlled preemergence, see the "WEEDS CONROLLED" section of this label. For control of emerged weeds, see the "WEEDS CONTROLLED" sections of the labels for Roundup or Gramoxone Super.

PLEASE NOTE THAT GRAMOXONE SUPER IS A HIGHLY TOXIC SUBSTANCE WHICH CAN BE FATAL IF MISUSED. READ THE ENTIRE GRAMOXONE SUPER LABEL AND USE ONLY AS DIRECTED.

ROUNDUP® PLUS THIS PRODUCT (TANK MIXTURE)

Annual Weeds—Use 1.5 quarts of Roundup herbicide per acre with the amount of this product recommended in the "MINIMUM TILLAGE SYSTEMS" section of this label.

Perennial Weeds—At normal application dates in minimum tillage systems, perennial weeds may not be at the proper stage of growth for control. See the "GENERAL INFORMATION" section of the label booklet for Roundup for the proper stage of growth for perennial weeds. Use of 2 to 4 quarts of Roundup herbicide in a tank mixture with the amount of this product recommended in the rate table of this label under these conditions provides top kill and reduces competition from many emerged perennial grass and broadleaf weeds. To obtain control, follow label recommendations for stage of growth and rate of application for specific perennial weeds. To obtain the desired stage of growth, it may be necessary to apply Roundup herbicide alone in the late summer or fall and then follow with this product at planting.

NOTE: When using this tank mixture, do not exceed 4 quarts of Roundup herbicide per acre.

USE OF THIS TANK MIXTURE FOR BERMUDAGRASS OR JOHNSONGRASS CONTROL IS NOT RECOMMENDED.

GRAMOXONE SUPER PLUS THIS PRODUCT (TANK MIXTURE)

When used as directed, Gramoxone Super in tank mixture with this product controls many emerged annual weeds and suppresses many emerged perennial weeds. Apply 1.5 to 2.5 pints of Gramoxone Super (0.25 to 0.5 lb active) with the amount of this product recommended in the "MINIMUM TILLAGE SYSTEMS" section of this label per acre immediately before, during or after planting but BEFORE CROP EMERGENCE. Use the 2.5 pint rate if weeds are 4 to 6 inches tall. This mixture may not conrol weeds taller than 6 inches. Add a nonionic spreader surfactant (approved for use on crops) containing at least 75 percent surface active agent at 8 fluid ounces per 100 gallons of diluted spray.

GRAIN SORGHUM (MILO)

Early preplant surface, preplant incorporated and preemergence surface applications must be made ONLY to grain sorghum (milo) planted with seed that has been properly treated with the Screen® seed protectant. Postemergence surface applications can be made to grain sorghum (milo) planted with protected or unprotected seed (grain sorghum seed not treated with a seed protectant).

NOTE: IN TEXAS, USE ONLY ON THE FINE TEXTURED SOILS OF THE GULF COAST AND BLACKLANDS.

Preemergence Surface Application—Apply this product in 10 or more gallons of water or sprayable fluid fertilizer per acre after planting, before the crop and weed emergence, and within 5 days of last preplant tillage.

®Screen is a registered trademark of Monsanto Company.

See the following table for recommended broadcast rates per acre for surface application.

NOTE: Applications made to grain sorghum (milo) growing on alkali soils or where cuts, fills or erosion have exposed calcareous or alkali subsoils may result in crop injury.

Preplant Incorporation—Broadcast spray this product in 10 or more gallons of water or sprayable fluid fertilizer and incorporate thoroughly into the top 1 to 2 inches of soil within 7 days prior to planting. When using one-pass incorporation (Surface Blend™), the equipment used must provide uniform and shallow blending to avoid streaking. To insure shallow incorporation of 2 inches or less, pull-type tillage equipment should be set to work the soil NO DEEPER THAN 4 INCHES.

This product is diluted when incorporated into the soil therefore, SHALLOW AND UNIFORM incorporation of the herbicide treatment is essential.

See the following table for recommended broadcast rates per acre for preplant incorporation applications:

Surface Application or Preplant Incorporation Application Rates

Soil Texture Group*	Broadcast Rate Per Acre (Quarts)	
	Less Than 1.5% Organic Matter	1.5% Or More Organic Matter
Coarse**	2.5	2.75
Medium	2.75	2.75 to 3.75
Fine	3	3 to 4

Use up to 5 quarts of this product per acre on medium or fine-textured soils, to control heavy grass infestation where present, or to reduce competition from the hard-to-control weeds listed on the label.

NOTE: When applied through center pivot irrigation or under sprinkler irrigation systems on medium or fine textured soils only, use 4 to 5 quarts of this product. Do not exceed 3/4 inch of water per acre with the first irrigaiton as reduced effectiveness may result.

*Refer to the "SOIL TEXTURE" section of this label to determine the corresponding texture group for the soil to be treated.

**DO NOT use this product for preplant incorporated application in coarse soils.

Postemergence Surface—Apply this product in 10 or more gallons of water per acre after crop emergence but before the weeds or grasses reach the 2-leaf stage and before the cop is 5 inches tall to reduce competition from weeds listed as controlled on this label. DO NOT APPLY IN SPRAYABLE FLUID FERTILIZER AFTER CROP EMERGENCE as crop injury may result. Some leaf burn to the crop may occasionally occur when applied early postemergence to grain sorghum (milo), but this should not affect subsequent growth or yield. Weeds which are emerged at time of application are not controlled by this product. If weeds are emerged at application, shallowly cultivate immediately to improve control.

See the "Surface Application or Preplant Incorporated Application Rates" for grain sorghum (milo) for recommended broadcast rates per acre for postemergence surface applications.

Early Preplant Surface Application—Apply 3.5 to 4 quarts of this product per acre in 10 or more gallons of water or sprayable fertilizer as a single application within 14 days prior to planting. When making split applications, apply 3.5 quarts of this product within 30 days prior to planting, followed by 1 to 1.5 quarts of this product per acre at planting. If emerged weeds are present at the time of the early preplant application or at-planting application, a labeled contact herbicide or tillage should be used. The at-planting application should be made before crop or weed emergence. Minimize movement of treated soil out of the row or untreated soil to the surface during planting or weed control may be reduced. Do not use on coarse or fine-textured soils.

LASSO® and/or ATRAZINE PLUS THIS PRODUCT (TANK MIXTURE)

For hard-to-control weeds, additional amounts of Lasso and/or atrazine may be added to the recommended treatment rates for this product. The following table shows the amounts of Lasso and/or atrazine that can be added to specific treatment rates of this product.

Recommended Rate (Quarts)	Additional Product (Quarts)	
Lariat®	Lasso®	Atrazine (Maximum)
2.5	1.25 to 2.25	1.1
3	1 to 2	0.9
3.5	0.75 to 1.75	0.7
4	0.5 to 1.5	0.5

In case of emergency involving this product. Call Collect, day or night, (314) 694-4000.

 843.89-000.24/53

LASSO®
HERBICIDE

Emulsifiable herbicide for weed control in corn, soybeans, peanuts, grain sorghum (milo), and crops listed.

RESTRICTED USE PESTICIDE

due to oncogenicity. For retail sale to and use only by Certified Applicators or persons under their direct supervision and only for those uses covered by the Certified Applicator's certification.

The use of this product may be hazardous to your health. This product contains alachlor, which has been determined to cause tumors in laboratory animals.

In case of an emergency involving this product or for user safety information on this product, call collect, day or night (314) 694-4000.

Complete Directions for Use

ACTIVE INGREDIENT:	
*Alachlor	45.1%
INERT INGREDIENTS:	54.9%
TOTAL	100.0%

Contains petroleum distillates

*Contains 480 grams/litre or 4 pounds/gallon of 2-chloro-2',6'-diethyl-N-(methoxymethyl)acetanilide.

U.S. Pat No. 3,547,620 EPA Reg. No. 524-314-2A

For use of this product in California see page 1682.

1989-1 843.41-002.05/53

Read the entire label before using this product.

Use only according to label instructions.

Read "LIMIT OF WARRANTY AND LIABILITY" before buying or using. If terms are not acceptable, return at once unopened.

LIMIT OF WARRANTY AND LIABILITY

This Company warrants that this product conforms to the chemical description on the label and is reasonably fit for the purposes set forth in the Complete Directions for Use label booklet ("Directions") when used in accordance with those Directions under the conditions described therein. NO OTHER EXPRESS WARRANTY OR IMPLIED WARRANTY OF FITNESS FOR PARTICULAR PURPOSE OF MERCHANTABILITY IS MADE. This warranty is also subject to the conditions and limitations stated herein.

Buyer and all users shall promptly notify this Company of any claims whether based in contract, negligence, strict liability, other tort or otherwise.

Buyer and all users are responsible for all loss or damage from use or handling which results from conditions beyond the control of this Company, including, but not limited to, incompatibility with products other than those set forth in the Directions, unusual weather, weather conditions which are outside the range considered normal at the application site and for the time period when the product is applied, as well as weather conditions which are outside the application ranges set forth in the Directions, application in any manner not explicitly set forth in the Directions, moisture conditions outside the moisture range specified in the Directions, or the presence of products other than those set forth in the Directions in or on the soil, crop or treated vegetation.

THE EXCLUSIVE REMEDY OF THE USER OR BUYER, AND THE LIMIT OF THE LIABILITY OF THIS COMPANY OR ANY OTHER SELLER FOR ANY AND ALL LOSSES, INJURIES OR DAMAGES RESULTING FROM THE USE OR HANDLING OF THIS PRODUCT (INCLUDING CLAIMS BASED IN CONTRACT, NEGLIGENCE, STRICT LIABILITY, OTHER TORT OR OTHERWISE) SHALL BE THE PURCHASE PRICE PAID BY THE USER OR BUYER FOR THE QUANTITY OF THIS PRODUCT INVOLVED, OR, AT THE ELECTION OF THIS COMPANY OR ANY OTHER SELLER, THE REPLACEMENT OF SUCH QUANTITY, OR, IF NOT ACQUIRED BY PURCHASE, REPLACEMENT OF SUCH QUANTITY. IN NO EVENT SHALL THIS COMPANY OR ANY OTHER SELLER BE LIABLE FOR ANY INCIDENTAL, CONSEQUENTIAL OR SPECIAL DAMAGES.

Buyer and all users are deemed to have accepted the terms of this LIMIT OF WARRANTY AND LIABILITY which may not be varied by any verbal or written agreement.

PRECAUTIONARY STATEMENTS

HAZARDS TO HUMANS AND DOMESTIC ANIMALS

DANGER!

KEEP OUT OF REACH OF CHILDREN

CAUSES EYE BURNS AND SKIN IRRITATION.

MAY CAUSE RESPIRATORY TRACT IRRITATION.

HARMFUL IF SWALLOWED.

MAY CAUSE ALLERGIC SKIN REACTION.

Do not get in eyes, on skin or on clothing.

Avoid breathing vapor or spray mist.

Wear goggles or face shield, rubber gloves, long trousers, long-sleeved shirt or jacket of tightly woven material along with boots high enough to cover ankles when transferring and mixing, and when adjusting, repairing or cleaning equipment. Wear rubber boots when pouring from open containers or when reentry is made into fields following application of this product through center pivot irrigation systems and the field is still wet. Clothing which comes in contact with LASSO must be washed before reuse. Clothing or other materials which become drenched with the concentrated pesticide must be disposed of in a sanitary landfill, by incineration or, if allowed by state and local authorities, by burning. If burned, stay out of smoke.

Wash thoroughly with soap and water after handling.

FIRST AID: IF IN EYES, immediately flush with plenty of water for at least 15 minutes. Call a physician.

IF ON SKIN, immediately flush with plenty of water while removing contaminated clothing. As soon as soap is available, wash thoroughly with soap and water. Wash clothing before reuse. Sensitized persons should avoid further contact and reuse of contaminated clothing. Get medical attention.

IF SWALLOWED, Do NOT induce vomiting. Give large quantities of water. Get medical attention. NEVER GIVE ANYTHING BY MOUTH TO AN UNCONSCIOUS PERSON.

IF INHALED, remove to fresh air.

Get medical attention if breathing difficulty develops.

In case of an emergency involving this product, Call Collect, day or night, (314) 694-4000.

ENVIRONMENTAL HAZARDS

This chemical has been identified in limited ground water sampling and there is the possibility that it can leach through the soil to ground water, especially where soils are coarse and ground water is near the surface. Following application, and during rainfall events that cause runoff, this chemical may reach surface water bodies including streams, rivers and reservoirs.

Care must be taken when using this product to prevent back siphoning into wells, spills or improper disposal of excess pesticide, spray mixtures or rinsates.

Check valves or antisiphoning devices must be used on all mixing and/or irrigation equipment. Disposal of excess pesticide, spray mixtures or rinsate should be according to label use instructions or according to the state pesticide or environmental control agency or the hazardous waste representative at the nearest EPA regional office.

Do not apply directly to water or wetlands. Do not contaminate water when disposing of equipment washwaters.

PHYSICAL OR CHEMICAL HAZARDS

COMBUSTIBLE.

Do not use or store near heat or flame.

Use only with adequate ventilation.

In case of:

FIRE, use water spray, foam, dry chemical or CO_2.

SPILL or LEAK, flush area with water spray.

DIRECTIONS FOR USE

It is a violation of Federal law to use this product in any manner inconsistent with its labeling.

STORAGE AND DISPOSAL

Do not contaminate water, foodstuffs, seed or feed by storage or disposal.

STORAGE:

STORE ABOVE 32°F TO KEEP PRODUCT IN SOLUTION.

Below 32°F crystals may form and settle to the bottom. If crystals form, place in a warm room and roll and/or shake the container frequently to redissolve before using. For bulk containers, see the container label for alternate storage information.

DISPOSAL:

Wastes of this pesticide may cause eye burns and skin irritation and may be dangerous. Improper disposal of excess pesticide, spray mixtures, or rinsate is a violation of Federal law. If these wastes cannot be disposed of according to label use instructions, contact your State Pesticide or Environmental Control Agency, or the Hazardous Waste representative at the nearest EPA Regional Office for guidance.

Emptied container retains vapor and product residue. Observe all labeled safeguards until container is cleaned, reconditioned or destroyed.

(See the individual container label for disposal information.)

GENERAL INFORMATION

LASSO® herbicide is recommended for control of yellow nutsedge and the annual grasses and broadleaf weeds listed in the "WEEDS CON-

TROLLED" section of this label. This product may be applied either as a surface application after planting or shallowly incorporated prior to planting to blend the herbicide treatment into the upper 1 to 2 inches of soil. Except for minimum or conservation tillage systems, the seedbed should be fine, firm and free of clods and trash.

Read and carefully observe precautionary statements and all other information appearing on the labeling of all products used in mixtures and sequential treatments.

Do not flood irrigate following application of this product or tank mixtures. This product will not control emerged seedlings.

Do not apply when conditions favor drift.

Use of this product not consistent with this label may result in injury to persons, animals or crops, or other unintended consequences. Keep container closed to prevent spills and contamination.

Flush sprayer with clean water after use.

® LASSO is a registered trademark of Monsanto Company.

SOIL TEXTURE

Applicators should evaluate soil and weed conditions carefully to assure that they choose the lowest effective label rate.

The recommended use rates of this product and the other herbicides labeled for use in tank mixtures with this product generally vary with soil texture. Unless soil texture is specifically named, rate tables throughout this label refer to only three soil texture groups: coarse, medium and fine. The following is a complete listing of soil textures included in each of these three soil texture groups:

Soil Textural Group	Soil Texture
COARSE:	sand, loamy sand, sandy loam
MEDIUM:	loam, silt loam, silt, sandy clay loam
FINE:	silty clay loam, clay loam, sandy clay, silty clay, clay

Refer to the above table to determine the corresponding soil textural group for the soil to be treated.

MIXING, SPRAYING AND HANDLING INSTRUCTIONS

NOTE

The use of a mechanical transfer system is required to be used by all mixer/loaders and/or applicators who treat 300 acres or more annually with pesticides containing alachlor as an active ingredient.

Direct contact or exposure to LASSO or spray mixtures of LASSO should be minimized. The following instructions for transfer, mixing, cleaning or repairing equipment should be followed in order to minimize this exposure. Review the protective clothing requirements as listed in the Precautionary Statements section of this label and do not use this product until you have the necessary protective clothing.

2.5 Gallon Containers

Open pouring from these containers can result in exposure from splashing or spilling. Special care in lifting and pouring are strongly recommended.

55 Gallon and Bulk Containers

Open pouring from these containers can result in exposure from splashing or spilling and is not recommended. This product should be transferred from these containers to the mix or spray tank using pumps or transfer probes. The probe or pump should not be removed from the container or disconnected until the container is emptied and rinsed. Use the pump or probe system to rinse the empty container and transfer the rinsate directly to the mix or spray tank.

Equipment Cleaning & Repair

Cleaning and repair of transfer systems and application equipment is a source of exposure to this product. Care should be taken to minimize exposure during cleaning and repair of transfer systems and application equipment. Whenever possible, these systems or equipment should be rinsed before being cleaned or repaired.

When repairs must be made during transfer or application, the equipment should be shutdown, and special care taken to avoid contact with the pesticide.

Always predetermine the compatibility of LASSO or labeled tank mixtures of LASSO with water carrier or sprayable fluid fertilizer carrier by mixing small proportional quantities in advance. See the "STANDARD SPRAYABLE FLUID FERTILIZER COMPATIBILITY TEST" section in this label to determine the compatibility of this product and the labeled tank mixtures recommended for use with sprayable fluid fertilizer carrier. NOTE: LASSO and all tank mixtures listed for corn and soybeans in the "DIRECTIONS FOR USE" section may be applied in sprayable fluid fertilizers. DO NOT apply LASSO or labeled tank mixtures of LASSO in sprayable fluid fertilizers on any other crops.

Mix LASSO or labeled tank mixture of LASSO with the appropriate carrier as follows:

1. Place a 20 to 35 mesh screen or wetting basket over filling port.
2. Through the screen, fill the sprayer tank one-half full with the appropriate carrier.
3. If a compatibility agent is necessary to improve mixing or to prevent the formation of undesirable and unsprayable gels or precipitates, while agitating add it to the carrier already in the tank. Use only compatibility agents, cleared by FDA for this use. Read and follow all directions for use, cautionary statements and all other information appearing on the selected compatibility agent label. Check for adequate agitation.
4. If a wettable powder or dry flowable formulation is used, make a slurry with water and add it slowly through the screen into the tank. Continue agitation.
5. If a flowable formulation is used, add slowly through screen into the tank. Mixing and compatibility may be improved when flowable is premixed one part flowable with one part water and added to the tank in diluted form.
6. Add LASSO slowly through the screen into the tank. Mixing and compatibility may be improved when this product is prediluted with two parts of water and added to the tank in diluted form.
7. Complete filling the sprayer tank with carrier. If Roundup herbicide or Gramoxone™ Super is used, add the required amount near the end of the filling process. Remove hose from tank immediately after filling to avoid siphoning back into the carrier source.

Maintain good agitation at all times until the contents of the tank are sprayed. NOTE: If spray mixture is allowed to settle at any time thorough agitation is required to resuspend the mixture before spraying is resumed.

Keep by-pass line on or near bottom of tank to minimize foaming. Screen size in nozzle or line strainers should be no finer than 50 mesh. Carefully select proper nozzle to avoid spraying a fine mist. Check for even distribution of spray droplets. To reduce loss of the chemical due to drift of a fine mist, apply at nozzle pressures below 40 psi.

® Roundup is a registered trademark of Monsanto Company.

™ Gramoxone is a trademark of ICI Americas Inc.

STANDARD SPRAYABLE FLUID FERTILIZER COMPATIBILITY TEST

Herbicides may not always mix evenly throughout a sprayable fluid fertilizer or the components may separate too quickly to make their combined use of practical value. This may be due to certain characteristics of the different fluid fertilizers. A simple test using small quantities of the components is suggested to provide compatibility potential.

The test follows:

A. Materials Required For A Compatibility Test

1. Two one-quart jars with lid or stopper (marked "with" and "without").
2. TEAspoons (for a more exacting test, a five to ten milliliter (ml) pipette or graduated cylinder is desirable).
3. Sprayable fluid fertilizer to be tested.
4. The herbicide chemicals to be mixed.
5. A compatibility agent (the purpose of the adjuvant is to help keep the fertilizer and crop protection chemical in suspension, if this assistance is needed.)

B. Procedure

1. Add one pint of the sprayable fluid fertilizer that will be used or other herbicide carrier to each jar marked "with" and "without."

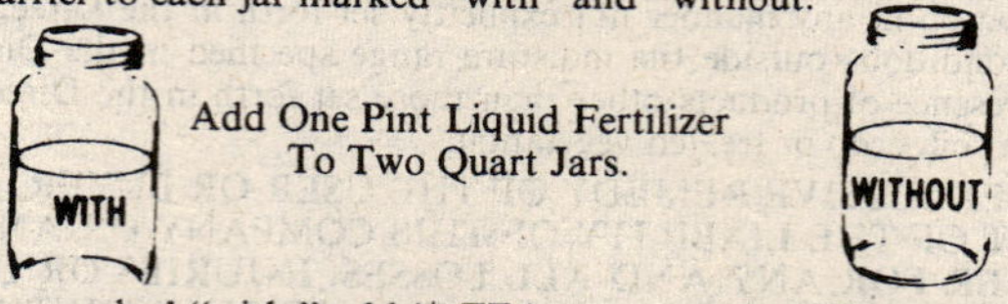

2. To the jar marked "with," add 1/4 TEAspoon or 1.2 milliliters of a suitable compatibility agent; shake gently for five to ten seconds to mix (1/4 TEAspoon in one pint is the equivalent of two pints per 100 gallons of liquid fertilizer).

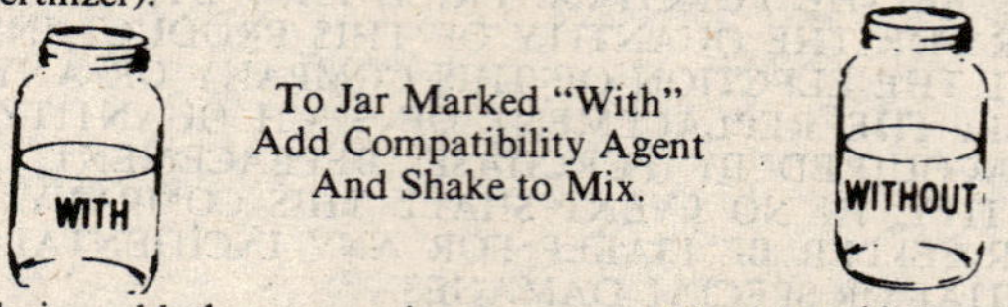

3. To each jar add the appropriate amount of herbicide(s) listed in the following tables. If more than one is used, add them separately with the wettable powders or dry flowables added first, flowables second and liquid last. Shake gently five to ten seconds after each addition.

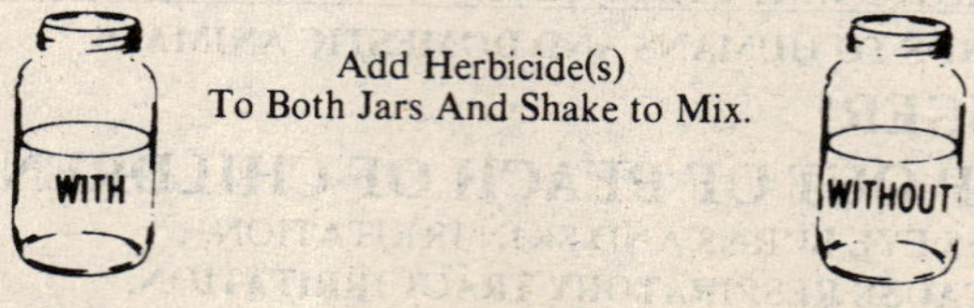

Herbicide	Rate/A		Amount to Be Added Per Pint Of Sprayable Fluid Fertilizer (Assuming Volume is 25 gal/A) Level TEAspoons
Wettable	1 lb.	=	1.5
Powders	2 lb.	=	3
or	3 lb.	=	4.5
Dry Flowables	4 lb.	=	6
	5 lb.	=	7.5

Herbicide	Rate/A		Level TEAspoons		Milliliters
Emulsifiable	1 pint	=	0.5	or	2.4
Concentrates or	1 quart	=	1	or	4.7
Flowables or	2 quarts	=	2	or	9.5
Liquids or	3 quarts	=	3	or	14.2
Solutions	1 gallon	=	4	or	19.0
	5 quarts	=	5	or	23.8

This compatability test is designed for 25 gallons of spray per acre. For changes in spray volume or herbicide rate, make appropriate changes in the ingredients of the test. Regardless of spray volume, the amount of compatibility agent should be equal to two or three pints (two pints = $^1/_4$ TEAspoon or 1.2 milliliters, three pints = $^3/_8$ TEAspoon or 1.8 milliliters per pint of sprayable fluid fertilizer) per 100 gallons of liquid fertilizer.

C. Observations and Decisions

1. If the herbicide(s) and the sprayable fluid fertilizer are compatible.
2. If a compatibility agent is necessary.

Five minutes after the final addition and mixing, observe both jars for the formation of large flakes, sludge, gels or other precipitates. Observe if the herbicide(s) cannot be physically mixed with the liquid fertilizer but remains as small oily particles in the solution.

If incompatibility in any form described above occurs in the jar "with" the compatibility agent added, the liquid fertilizer and the herbicide(s) should not be used together in the same spray tank.

If incompatibility as described above occurs in the jar "without" the adjuvant but not in the jar "with" adjuvant, the use of a compatibility adjuvant is recommended.

Both jars should be allowed to stand and be observed periodically for one-half hour. If the separate layers of liquid fertilizer and additives can be resuspended by shaking, commercial application is possible. An emulsifiable concentrate normally will go to the top after standing; wettable powders will either settle to the bottom of the tank or jar, or float to the top, depending upon the density of the fertilizers.

If the herbicide(s) is compatible with fluid fertilizer in the foregoing test without having to use a compatibility agent, fluid fertilizer may be used for the premixing. If it is not compatible without the compatibility agent, the herbicide(s) should be premixed with water before adding to the spray tank.

APPLICATION INFORMATION

APPLICATION SYSTEMS

Ground Broadcast Treatment—Apply this product and the labeled tank mixtures in 10 or more gallons of solution per acre using broadcast boom equipment. The carrier may be either water or sprayable fluid fertilizer as specified for the crop to be treated in the "DIRECTIONS FOR USE" section of this label. Do not apply during periods of gusty winds, when winds are in excess of 15 mph or when other conditions favoring drift exist.

Band Treatment—Apply a broadcast equivalent rate and volume per acre. To determine these:

$$\frac{\text{Bank width in inches}}{\text{Row width in inches}} \times \text{Broadcast RATE per acre} = \text{Band RATE per acre}$$

$$\frac{\text{Band width in inches}}{\text{Row width in inches}} \times \text{Broadcast VOLUME per acre} = \text{Band VOLUME per acre}$$

CHEMIGATION

Apply this product and labeled tank mixtures only through center pivot irrigation systems. Do not apply this product through any other type of irrigation system.

Crop injury, lack of effectiveness, or illegal pesticide residues in the crop can result from nonuniform distribution of treated water. Use only in systems that apply uniformly.

If you have questions about calibration, you should contact State Extension Service specialists, equipment manufacturers or other experts.

Do not chemigate through systems connected to a public water system.

A person knowledgeable of the chemigation system and responsible for its operation, or under the supervision of the responsible person, shall shut the system down and make necessary adjustments should the need arise. Do not apply when system connections or fittings leak, when nozzles do not provide uniform distribution or when the line containing the product must be dismantled and drained.

The system must contain a functional check valve, vacuum relief valve, and low pressure drain appropriately located on the irrigation pipeline to prevent water source contamination from back flow.

The pesticide injection pipeline must contain a functional, automatic, quick-closing check valve to prevent the flow of fluid back toward the injection pump.

The pesticide injection pipeline must also contain a functional, normally closed, solenoid operated valve located on the intake side of the injection pump and connected to the system interlock to prevent fluid from being withdrawn from the supply tank when the irrigation system is either automatically or manually shut down.

The system must contain functional interlocking controls to automatically shut off the pesticide injection pump when the water pump motor stops.

The irrigation line or water pump must include a functional pressure switch which will stop the water pump motor when the water pressure decreases to the point where pesticide distribution is adversely affected.

Systems must use a metering pump, such as a positive displacement injection pump (e.g., diaphragm pump) effectively designed and constructed of materials that are compatible with pesticides and capable of being fitted with a systems interlock. Pumps, injection equipment, agitation equipment, hoses and connections between *supply* tank and the point of injection must be constructed of materials which are resistant to LASSO.

Do not apply when wind speed favors drift beyond the area intended for treatment.

Meter LASSO herbicide or a labeled tank mixture of LASSO into the center pivot irrigation system after planting and before crop emergence. Herbicide application should be made in $^1/_3$ to $^3/_4$ inch of water per acre. Do not apply in more than $^3/_4$ inch of water per acre under any conditions or reduced performance may occur. On very sandy soils (more than 60% sand and less than 1% organic matter), apply these herbicides in a maximum of $^1/_2$ inch water per acre. Good agitation must be maintained during the entire application period. Flush the system with water when application is complete. Refer to the "MIXING, SPRAYING AND HANDLING INSTRUCTIONS" section of the label for mixing procedures.

NOTE: Applications can be made by injection through center pivot irrigation equipment only when specified for the crop to be treated in the "DIRECTIONS FOR USE" section of this label. Do not apply this product or any labeled tank mixture through center pivot irrigation unless the treatment is specifically recommended in this label for the crop to be treated.

Posting of areas to be chemigated is required when any part of a treated area is within 300 feet of sensitive areas such as residential areas, labor camps, businesses, day care centers, hospitals, in-patient clinics, nursing homes or any public area such as schools, parks, playgrounds or other public facilities not including public roads.

Posting must conform to the following requirement. Treated areas shall be posted with signs at all usual points of entry and along likely routes of approach from the listed sensitive areas. When there are not usual points of entry, signs must be posted in the corners of the treated areas and in any other location affording maximum visibility to sensitive areas. The printed side of the sign should face away from the treated area towards the sensitive area. Signs must be posted prior to application and must remain posted until foliage has dried and soil surface water has disappeared. Posting signs may be obtained from your local farm chemical dealer.

Aerial — Fixed wing helicopter — Apply the recommended rates of this product or approved tank mixtures in 3 to 10 gallons of water per acre. AVOID DRIFT — DO NOT APPLY DURING INVERSION CONDITIONS, WHEN WINDS ARE GUSTY, OR UNDER ANY OTHER CONDITION WHICH WILL ALLOW DRIFT. Coarse sprays are less likely to drift, therefore do not use nozzles or nozzle configurations which dispense spray or fine spray droplets. Do not angle nozzles forward into the airstream and do not increase spray volume by increasing nozzle pressure. Ensure uniform application. To avoid streaked, uneven or overlapped application, use mechanical marking systems only, HUMAN FLAGGERS PROHIBITED!

Application with Dry Bulk Fertilizer—The herbicide-fertilizer impregnation process must be completed only at commercial fertilizer or chemical dealerships properly equipped for this procedure.

Dry bulk fertilizer may be impregnated with LASSO or the tank mixtures of LASSO plus atrazine or LASSO plus Bladex on corn; and LASSO or the tank mixtures of LASSO plus Lexone or LASSO plus Sencor on soybeans. LASSO and these tank mixtures must be applied with 200 to 450 pounds of dry bulk fertilizer per acre and shallowly incorporated within 7 days prior to planting. The herbicide must be applied as recommended in this label for the crop, weed and soil type. Refer to the table for broadcast rate per acre in the appropriate crop section of this label to determine the recommended rate per acre for the herbicide treatment to be applied.

The following table provides a reference to determine the amount of LIQUID herbicide to be mixed per ton of dry bulk fertilizer for a range of herbicide recommendations for fertilizer rates per acre:

Fertilizer Rate (Lb/Acre)	Acres Covered (Per Ton)	Recommended Quarts Liquid Herbicide/Acre (Quarts of Herbicide/Ton Dry Bulk Fertilizer)			
		1.5	2	2.5	3
200	10	15	20	25	30
250	8	12	16	20	24
300	6.7	10	13.3	16.7	20
350	5.7	8.6	11.4	14.3	17.1
400	5	7.5	10	12	15
450	4.5	6.7	8.9	11.1	13.3

To determine the amount of herbicide needed for rates not included in the preceding table, use the following formula:

$$\frac{\text{Recommended herbicide rate (lb. or qt)/acre} \times 2000}{\text{pounds fertilizer/acre}} = \text{Pounds or quarts of herbicide per tone of dry bulk fertilizer}$$

Mix and blend the dry fertilizer and herbicide mixture in a closed rotary drum-type mixer allowing sufficient time to ensure uniform coverage. Use at least one ton of dry fertilizer per mixing operation. Inject the herbicide into the drum over a minimum of a 2-minute period and allow at least 2 additional minutes mixing time to ensure uniformity. The nozzle used to spray the herbicide treatment must be placed inside the mixer to provide uniform spray coverage of the tumbling fertilizer.

If the dry fertilizer used has inadequate adsorptive capacity, use a higher adsorptive material such as Agsorb™, NP-79™, Microcel™ E, or LASSO® II granules to provide a freeflowing mixture. The adsorptive material and dry fertilizer should be thoroughly mixed prior to adding liquid herbicide. Follow the guidelines for quantities to add of nonherbicidal adsorptive material.

LASSO II granules can be used as an adsorptive agent with complete fertilizer blends of 350 pounds or greater per acre. This procedure is not recommended for straight urea applications. See the table below for the correct rates of LASSO II granules and LASSO per acre. If the total amount of LASSO or LASSO liquid herbicide in the tank mixture to be impregnated exceeds 3.25 quarts per acre additional nonherbicidal absorptive material may be necessary to ensure a free flowing mixture.

Pounds Plus Quarts of Herbicide/Acre To Impregnate on Dry Bulk Fertilizer. Recommended Total LASSO® Broadcast Rate/Acre (Quarts)					
3		3.5		4	
LASSO II (Lbs)	LASSO (Quarts)	LASSO II (Lbs)	LASSO (Quarts)	LASSO II (Lbs)	LASSO (Quarts)
5	2.25	5	2.75	5	3.25

™ Agsorb is a trademark of Oil Dri Corporation.
™ Microcel is a trademark of Manville Products Corporation.
™ NP-79 is a trademark of Eagle Picher Industries.

The following table provides a partial list of dry fertilizers which may be impregnated with LASSO or tank mixtures of LASSO with other herbicides:

Fertilizer	LASSO®	LASSO + atrazine	LASSO + Bladex	LASSO + Lexone or Sencor
Ammonium sulfate (21-0-0)	Yes	Yes	Yes	Yes
Ammonium phosphate-sulfate (16-20-0)	Yes	Yes	Yes	Yes
Diammonium phosphate (18-46-0)	Yes	Yes	Yes	Yes
Potassium chloride (0-0-60)	Yes	Yes	Yes	Yes
Potassium sulfate (0-0-52)	Yes	Yes	Yes	Yes
Single super-phosphate (0-20-0)	Yes	No	No	Yes
Treble super-phosphate (0-46-0)	Yes	No	No	Yes
*Urea (46-0-0)	Yes	Yes	Yes	Yes

* Some ureas may be phytotoxic when applied on corn. Use only ureas known to be safe to corn. Blends suitable for use in soybeans, such as 0-26-26 are also approved for use with LASSO and the tank mixtures listed.

NOTE: DO NOT impregnate LASSO or tank mixtures of LASSO with other herbicides on fertilizers containing ammonium nitrate, potassium nitrate or sodium nitrate.

Spread the herbicide-dry fertilizer mixture uniformly with a properly calibrated applicator. Fertilizers impregnated with LASSO or tank mixtures of LASSO with other herbicides must be spread at half-rate and overlapped 100 percent to obtain full rate and uniform distribution. Non-uniform spreading of the fertilizer-herbicide mixture may result in unsatisfactory weed control or crop injury.

™ Bladex and Lexone are trademarks of E. I. duPont de Nemours and Company.
™ Sencor is a trademark of the parent company of Farbenfabricken Bayer GmbH.

APPLICATION TIMING AND METHODS

NOTE: Unless specifically recommended, only one treatment of LASSO may be applied to a labeled crop.

Preemergence Surface Application—This product and all labeled tank mixtures except LASSO plus Command™, LASSO plus Eptam™ and LASSO plus Treflan™ may be applied to the soil surface after planting and prior to either crop or weed emergence. Apply within 5 days of last preplant tillage. If weeds emerge after treatment, or if treatment is applied more than 5 days after last preplant tillage, rotary hoe or shallowly cultivate immediately to improve performance. Precipitation or overhead sprinkler irrigation is required after application to move the herbicide treatment into the weed germination zone. The amount of precipitation or overhead sprinkler irrigation required depends on existing soil mixture, soil type and percent organic matter content, but 1/3 to 3/4 inch is normally adequate. Performance is improved when moisture is received within 7 days after application and prior to weed emergence. High intensity or excessive rainfall or excessive irrigation after application may reduce control.

Postemergence Surface Application—This product and certain tank mixtures may be applied to specific crops at the ground cracking stage of the crop or after crop emergence. Refer to the specific treatment and crop intended in the "DIRECTIONS FOR USE" section of the label to determine if applications at cracking or postemergence to the crop are recommended and to determine the proper weed and crop growth stage limitations. Precipitation or overhead sprinkler irrigation is required after application to move the herbicide treatment into the weed germination zone to control unemerged weeds. The amount of precipitation or irrigation required depends on existing soil moisture, soil type and percent organic matter content, but 1/3 to 3/4 inch is normally adequate. If weeds emerge after treatment, rotary hoe or shallowly cultivate immediately to improve performance.

Preplant Incorporation—This product and many of the labeled tank mixtures may be mixed into the soil using shallow incorporation equipment any time within 7 days prior to planting. Refer to the crop to be treated in the "DIRECTIONS FOR USE" section of this label to determine if preplant incorporation is recommended. Apply the recommended treatment rate to the soil surface as a broadcast application. Precipitation or irrigation is required to bring incorporated herbicide treatments into contact with germinating weed seedlings. The amount of precipitation or irrigation required depends on existing soil moisture, soil type and percent organic matter content, but 1/3 to 3/4 inch is normally adequate. If weeds emerge after treatment, rotary hoe or shallowly cultivate immediately to improve performance.

NOTE: Do not incorporate unless specifically recommended in this label for the crop and herbicide treatment selected. Incorporation of this product and its labeled tank mixtures on coarse soil types may reduce length of weed control due to leaching with rainfall or irrigation. Incorporation is not recommended on coarse soils in the southeastern states.

1-Pass Incorporation (Surface Blend™): Shallowly incorporate (Surface Blend) the treatment into the upper 1 to 2 inches of the soil. Equipment should be operated at manufacturer's designed speed for incorporation to ensure adequate mixing and distribution of the herbicide treatment in the soil. Equipment design including any drag attachments must be adequate to avoid soil ridging which may result in streaked or reduced weed control. Soil conditions, including moisture content and crop residue levels, must be suitable to allow thorough and uniform mixing with the equipment used for 1-pass incorporation.

Contact Monsanto Company for additional information regarding recommended equipment and methods for Surface Blend application.

2-Pass Incorporation: When 2-pass incorporation is used, shallowly incorporate the herbicide treatment into the upper 1 to 2 inches of the soil with equipment set to work the soil NO DEEPER THAN 4 INCHES. The second pass must be made at an angle to and no deeper than the first pass to ensure proper distribution of the herbicide treatment in the soil.

Cultivation Information—Delay cultivation after application for as long as possible unless weeds or grasses emerge. Shallowly cultivate or rotary hoe immediately if weeds or grasses emerge. If cultivation is necessary because of soil crusting or compaction, set equipment shallow and minimize lateral soil movement to avoid dilution or displacement of the herbicide treatment. Only after weeds emerge in the treated band, set cultivator to throw soil into the row covering the band. On peanuts, do not throw soil into the row.

™ Command is a trademark of FMC Corporation.
™ Eptam is a trademark of ICI Americas Inc.
™ Surface Blend is a trademark of Monsanto Company.
™ Treflan is a trademark of Elanco Products Company.

WEEDS CONTROLLED

Applicators should evaluate soil and weed conditions carefully to assure that they choose the lowest effective label rate.

When applied as directed under conditions described, this product and tank mixtures of this product will control or reduce competition from the weeds listed. Refer to the crop on which treatment is intended to determine the rates and tank mixtures recommended.

For the scientific names of the weeds listed below, refer to the appendix at the end of this label.

NOTE: C = Control R = Reduced Competition

ANNUAL GRASSES

		LASSO® plus						
	LASSO®	Amiben™	atrazine	Banvel™	Bladex™	Canopy™	Command™	Gemini™
Barnyardgrass	C	C	C	C	C	C	C	C
Crabgrass	C	C	C	C	C	C	C	C
Cupgrass, woolly[1]	R	R	R	•	R	R	R	•
Foxtail: giant, green, robust purple, robust white, yellow	C	C	C	C	C	C	C	C
Goosegrass	C	C	C	C	C	C	C	C
Johnsongrass, seedling[2,8]	R	R	R	R	R	R	R	R
Millet, proso[3]	C	•	C	•	C	•	•	•
Panicum: browntop, fall	C	C	C	C	C	C	C	C
Panicum, Texas[4]	R	R	R	R	R	R	R	R
Rice, red[5]	C	C	C	•	C	C	C	•
Sandbur, Grassbur[2,6]	R	•	C	•	•	•	•	•
Shattercane, wild cane[1]	R	R	R	•	R	R	R	•
Signalgrass, broadleaf	C	C	C	C	C	C	C	C
Sprangletop, red	C	C	C	C	C	C	C	C
Witchgrass	C	C	C	C	C	C	C	C

ANNUAL BROADLEAVES

		LASSO® plus						
	LASSO®	Amiben™	atrazine	Banvel™	Bladex™	Canopy™	Command™	Gemini™
Beggarweed, Florida[2]	R	C	C	•	C	C	•	C
Carpetweed	C	C	C	C	C	C	C	C
Cocklebur	•	•	R	R	R	C	•	C
Coffeeweed, Hemp sesbania	•	•	•	•	•	C	•	C
Copperleaf, hophornbeam	•	•	•	•	•	C	•	•
Galinsoga	C	C	C	C	C	C	C	C
Groundcherry, annual	•	•	C	•	•	C	•	C
Groundcherry, cutleaf[7]	C	C	C	C	C	C	C	C
Jimsonweed	•	•	C	•	•	C	C	C
Kochia	•	•	C	•	•	•	•	•
Lambsquarters[2]	R	C	C	C	C	C	C	C
Morningglory: Tall[8]	•	•	R	R	R	C	•	C
Pitted ivyleaf[8], entireleaf[8]	•	•	•	•	•	C	•	C
Smallflower	R	R	C	•	C	C	•	C
Mustard	•	R	C	C	C	C	•	C
Nightshade: black[7]; hairy[7]	C	C	C	C	C	C	C	C
Pigweed, Carelessweed	C	C	C	C	C	C	C	C
Purslane	C	C	C	C	C	C	C	C
Pusley, Florida	C	C	C	C	C	C	C	C
Ragweed, common[2]	R	C	C	C	C	C	C	C
Ragweed, giant	•	•	•	•	•	R	•	R
Sicklepod	•	•	C	•	C	C	•	R
Sida, prickly; Teaweed[2]	R	•	•	•	•	C	C	C
Smartweed[2]	R	C	C	C	C	C	R	C
Starbur, bristly	R	R	C	•	C	C	•	•
Sunflower, common	•	•	•	•	•	R	•	•
Velvetleaf, Buttonweed[8]	•	R	R	R	R	C	C	C
Waterhemp	C	C	C	C	C	C	C	C
SEDGE								
Nutsedge, yellow[5]	C	C	C	•	C	C	C	•

ANNUAL GRASSES

	LASSO® plus						
	linuron	Lorox™ Plus	Preview™	propazine	Scepter™	Sencor™	Treflan™
Barnyardgrass	C	C	C	C	C	C	C
Crabgrass	C	C	C	C	C	C	C
Cupgrass, woolly[1]	•	•	R	•	R	R	R
Foxtail: giant, green, robust purple, robust white, yellow	C	C	C	C	C	C	C
Goosegrass	C	C	C	C	C	C	C
Johnsongrass, seedling[2,8]	R	R	R	R	C	R	R
Millet, proso[3]	•	•	•	•	•	•	•
Panicum: browntop, fall	C	C	C	C	C	C	C
Panicum, Texas[4]	R	R	R	R	R	R	R
Rice, red[5]	•	•	C	C	C	C	C
Sandbur, Grassbur[2,6]	•	•	•	•	•	•	•
Shattercane, wild cane[1]	•	•	R	R	R	R	R
Signalgrass, broadleaf	C	C	C	C	C	C	C
Sprangletop	C	C	C	C	C	C	C
Witchgrass	C	C	C	C	C	C	C

ANNUAL BROADLEAVES

	LASSO® plus						
	linuron	Lorox™ Plus	Preview™	propazine	Scepter™	Sencor™	Treflan™
Beggarweed, Florida[2]	•	•	R	•	•	R	•
Carpetweed	C	C	C	C	C	C	C
Cocklebur	•	C	C	R	C	R	•
Coffeeweed	•	•	C	•	•	•	•
Cooperleaf, hophornbeam	•	•	C	•	•	•	•
Galinsoga	C	C	C	C	C	C	C
Groundcherry, annual	C	C	C	C	C	C	•
Cutleaf[7]	C	C	C	C	C	C	C
Jimsonweed	•	C	C	C	C	R	•
Kochia	•	•	•	C	•	•	C
Lambsquarters[2]	C	C	C	C	C	C	•
Morningglory: Tall[8]	•	R	R	R	R	•	•
Pitted	•	R	R	•	C	•	•
Ivyleaf[8], entireleaf[8]	•	R	R	•	R	•	•
Smallflower	•	•	•	•	C	•	•
Mustard	C	C	C	C	C	•	•
Nightshade: black[7]; hairy[7]	C	C	C	C	C	C	C
Pigweed, Carelessweed	C	C	C	C	C	C	C
Purslane	C	C	C	C	C	C	C
Pusley, Florida	C	C	C	C	C	C	C
Ragweed, common[2]	C	C	C	C	C	C	•
Ragweed, giant	•	R	R	•	R	•	•
Sicklepod	•	•	•	C	R	R	•
Sida, prickly; Teaweed[2]	C	C	C	•	C	•	•
Smartweed[2]	C	C	C	C	C	C	•
Starbur, bristly	•	•	•	•	C	C	•
Sunflower, common	•	•	R	•	C	•	•
Velvetleaf, Buttonweed[8]	R	C	C	R	C	R	•
Waterhemp	C	C	C	C	C	C	C
SEDGE							
Nutsedge, yellow[5]	•	•	C	C	C	C	C

[1] Use 4 quarts per acre of this product applied alone or in tank-mix combinations. Preplant incorporate only to reduce competition from this weed. In areas of heavy weed infestations, apply an additional 2 quarts per acre of LASSO any time from planting until corn is 5 inches tall or any time

before soybeans exceed unifoliate stage (first two true leaves). This sequential treatment must be made prior to emergence of the weed.

[2] For reduced competition from this weed using LASSO alone, use 2.5 to 4 quarts of LASSO per acre.

[3] Plant corn into freshly prepared seedbed. Apply 4 quarts of LASSO per acre preemergence surface immediately after corn planting. Apply an additional 2 quarts of LASSO per acre delayed preemergence 6 to 10 days after planting, but before wild proso millet and corn emerge. If weather conditions prevent the timely application of the delayed preemergence application, apply a labeled postemergence herbicide.

[4] Use a minimum of 3 quarts of LASSO per acre and apply through center pivot irrigation or under sprinkler irrigation systems. Only LASSO and LASSO plus atrazine tank mixture can be applied through center pivot irrigation systems. Only LASSO and LASSO plus atrazine tank mixture can be applied through center pivot irrigation systems.

[5] Use 3 to 4 quarts of LASSO per acre applied alone or in tank mixture and apply preplant incorporated only for control on medium and fine-textured soils. When tank-mixed with Scepter, use 2 to 3 quarts of LASSO per acre and preplant incorporate only. On coarse textured soils with less than 3 percent organic matter, use 3 to 4 quarts of LASSO per acre as a preemergence surface application to reduce competition from yellow nutsedge.

[6] Apply 2.5 to 3 quarts of LASSO plus 1 to 2 quarts of atrazine 4L per acre. Preemergence surface application should be applied to a freshly prepared seedbed. For best control, rotary hoe or initiate overhead irrigation (0.33 to 0.5 inch) within 24 hours of application.

[7] Use a minimum of 2.5 quarts of LASSO per acre on coarse soils, or 3 to 4 quarts of LASSO per acre on medium and fine soils when used alone or in tank mixture with Amiben, Banvel, Canopy, Command, Preview, Sencor/Lexone or Treflan for control of this weed.

[8] For best results on these species when using LASSO plus Scepter tank mixture, use preplant incorporated treatments.

NOTE: The 3-way tank mixture of LASSO plus atrazine plus Bladex controls or reduces competition as specified for the respective weeds listed under the LASSO plus atrazine and LASSO plus Bladex recommendations

The 3-way tank mixture of LASSO plus Amiben, plus Lexone or Sencor controls or reduces competition as specified for the respective weeds listed under the LASSO plus Amiben and LASSO plus Lexone, or LASSO plus Sencor recommendations.

Certain tank mixtures of LASSO and broadleaf herbicides may be combined with applications of Roundup® herbicide or Gramoxone Super for control of many emerged weeds prior to corn or soybean emergence. Refer to the Roundup or Gramoxone Super label for a list of emerged weeds controlled by tank mixtures with these products.

TM Amiben is a trademark of Rhone-Poulenc.
TM Banvel is a trademark of Sandoz, Inc.
TM Basagran is a trademark of BASF Aktiengelsellschaft Ludwigshafen am Rhein.
TM Canopy, Gemini, Lorox and Preview are trademarks of E.I. du Pont de Nemours and Company.
TM Scepter is a trademark of American Cyanamid Corporation.

CROPS

CORN
(All Types)

LASSO®

Apply this product in water or sprayable fluid fertilizer solution for control of yellow nutsedge and the annual grasses and broadleaf weeds listed in the "WEEDS CONTROLLED" section of this label.

APPROVED APPLICATION SYSTEMS

Ground—Broadcast boom; banded

Aerial—Fixed wing; helicopter

Center Pivot Irrigation

Dry Bulk Fertilizer Impregnation

Reference: The "APPLICATION SYSTEMS" section of this label provides detailed information on the application of this product using the system selected.

APPROVED APPLICATION METHODS

Preplant Incorporated—Apply this product within 7 days prior to planting and shallowly incorporate or Surface Blend into the upper 1 to 2 inches of soil. Irrigation within 10 days following planting may improve weed control.

Preemergence Surface—Apply this product after planting, before crop and weed emergence and within 5 days after last preplant tillage operation.

Postemergence Surface—This product may be applied after crop emergence to reduce competition from weeds listed as controlled on this label. Apply before corn exceeds 5 inches in height.

Minimize herbicide contact with the corn foliage since leaf burn may occur. Subsquent growth or yield should not be affected.

Weeds which are emerged at time of application are not controlled by this product. If weeds are emerged at application, shallowly cultivate or rotary hoe immediately to improve performance. DO NOT apply in sprayable fluid fertilizer after crop emergence.

NOTE: The highest levels of weed control will be obtained when applications are made prior to the emergence of the crop and weeds.

Reference: The "APPLICATION TIMING AND METHODS" section of this label provides detailed information and procedures for the application timing and method selected.

Sequential Treatment—This product may be applied as two sequential treatments for improved control of hard-to-control weeds. The combined total application rate for the two treatments must not exceed 8 quarts per acre per year. The second treatment must be applied before the corn exceeds 5 inches in height.

RECOMMENDED RATES: Refer to the following tables for the recommended broadcast treatment rates for this product. Applications which are not consistent with recommendations in this label may result in unsatisfactory weed control or crop injury.

Surface Application Rates

Soil Textural Group*	Broadcast Rate Per Acre (Quarts)**	
	Less Than 3% Organic Matter	3% or More Organic Matter
Coarse***	2 to 2.25	2 to 2.25
Medium	2 to 2.75	2 to 2.75
Fine	2 to 2.75	2.5 to 3.25

* Refer to the "SOIL TEXTURE" section of this label to determine the corresponding textural group for the soil to be treated.

** Use the higher rate in the recommended range in areas of heavy weed infestations. Use a minimum of 2.5 quarts of this product per acre on coarse soils and 3 to 4 quarts on medium or fine textured soils to control black or hairy nightshade.

*** When applied through center pivot irrigation or under sprinkler irrigation systems on coarse soils, use a minimum of 3 quarts per acre of this product.

NOTE: For postemergence surface treatment—When applied following LASSO or tank mixtures of LASSO, the total amount of LASSO applied alone or in combination must not exceed 8 quarts per acre per year. Do not apply in sprayable fluid fertilizer.

On coarse soils with 10% or more organic matter use 4 to 6 quarts of LASSO herbicide per acre. On muck or peat soils use 6 to 8 quarts per acre. Use the higher rate of LASSO herbicide in the recommended ranges in areas of heavy weed infestations.

Preplant Incorporated Application Rates

Soil Textural Group*	Broadcast Rate Per Acre (Quarts)**	
	Less Than 3% Organic Matter	3% or More Organic Matter
Coarse***	2.5	2.5
Medium	2.5 to 3	2.5 to 3
Fine	2.5 to 3	3 to 3.5

* Refer to the "SOIL TEXTURE" section of this label to determine the corresponding textural group for the soil to be treated.

** Use the higher rate in the recommended range in areas of heavy weed infestations. Use 2.5 to 3.5 quarts per acre where furrow irrigation is used. Use a minimum of 3 quarts on coarse textured soils for control of yellow nutsedge or red rice. Rates may be increased to a maximum of 4 quarts on any soil type when heavy infestations of yellow nutsedge or red rice are present.

On coarse soils with 10% or more organic matter use 4 to 6 quarts of LASSO herbicide per acre. On muck or peat soils use 6 to 8 quarts per acre. Use the higher rate of LASSO herbicide in the recommended ranges in areas of heavy weed infestations.

LASSO® PLUS ATRAZINE
(TANK MIXTURE)

Apply this tank mixture in water or sprayable fluid fertilizer solutions for control of yellow nutsedge and the annual grasses and broadleaf weeds listed in the "WEEDS CONTROLLED" section of this label.

APPROVED APPLICATION SYSTEMS

Ground—Broadcast boom; banded

Aerial—Fixed wing; helicopter

Center Pivot Irrigation

Dry Bulk Fertilizer Impregnation

Reference: The "APPLICATION SYSTEMS" section of this label provides detailed information on the application of this tank mixture using the system selected.

APPROVED APPLICATION METHODS

Preplant Incorporated—Apply this tank mixture within 7 days prior to planting and shallowly incorporate or Surface Blend into the upper 1 to 2 inches of soil.

Preemergence Surface—Apply this tank mixture after planting, before crop and weed emergence and within 5 days after last preplant tillage operation.

Postemergence Surface—Apply this tank mixture after crop emergence until weeds reach the 2-leaf stage and the crop is no more than 5 inches tall. Applications made beyond the 2-leaf stage will not provide satisfactory weed control. Some leaf burn may occasionally occur when applied early postemergence to corn, but subsequent growth or yield should not be affected.

DO NOT apply in sprayable fluid fertilizer after crop emergence since crop injury may result.

Reference: The "APPLICATION TIMING AND METHODS" section of this label provides detailed information and procedures for the application timing and method selected.

NOTE: PLANT ONLY CORN, PEANUTS, SORGHUM OR SOYBEANS* the year following use of this mixture.

*There is a possibility of crop injury due to carryover of atrazine if soybeans are planted the following year. DO NOT plant soybeans the year following use of this tank mixture on furrow irrigated corn. DO NOT graze treated area or feed treated forage to livestock for 21 days following application of this tank mixture.

RECOMMENDED RATES: Refer to the following tables for the recommended broadcast treatment rates for this tank mixture. Applications which are not consistent with recommendations in this label may result in unsatifactory weed control or crop injury.

Surface Application Rates

Soil Textural Group*	Broadcast Rate Per Acre (Quarts)**			
	Less Than 3% Organic Matter		3% or More Organic Matter	
	LASSO® +	Atrazine 4L***	LASSO +	Atrazine 4L***
Coarse	1.5 to 2	0.75 to 2	1.5 to 2	1 to 2
Medium	1.5 to 2	1 to 2	2 to 2.25	1.25 to 2
Fine	2 to 2.25	1.25 to 2	2 to 2.75	1.25 to 2

* Refer to the "SOIL TEXTURE" section of this label to determine the corresponding textural group for the soil to be treated.

** Use the higher rates in the recommended ranges in areas of heavy weed infestations.

*** Use rates listed in this label when using atrazine 4L. Use equivalent rates when using atrazine 80W or 90% dry flowable formulations. One quart of atrazine 4L equals 1.25 pounds of atrazine 80W or 1.1 pounds of atrazine 90% dry flowable.

NOTE: When applied through center pivot irrigation or under sprinkler irrigation systems on coarse soils, use 1 quart per acre of atrazine 4L and a minimum of 2 quarts per acre of LASSO in this tank mixture.

When applied following LASSO or tank mixtures of LASSO, the total amount of LASSO applied alone or in combination must not exceed 8 quarts per acre per year. The total amount of atrazine 4L must not exceed 3.2 quarts per acre per year. Do not apply in sprayable fluid fertilizer. Minimize herbicide contact with the corn foliage since leaf burn may occur. Subsequent growth or yield should not be affected. Do not graze or feed treated forage to livestock for 21 days following application.

Preplant Incorporated Application Rates

Soil Textural Group*	Broadcast Rate Per Acre (Quarts)**			
	Less Than 3% Organic Matter		3% or More Organic Matter	
	LASSO® +	Atrazine 4L***	LASSO +	Atrazine 4L***
Coarse	1.75 to 2.25	1.25 to 2	1.75 to 2.25	1.25 to 2
Medium	1.75 to 2.25	1.25 to 2	2.25 to 2.5	1.5 to 2
Fine	2.25 to 2.5	1.5 to 2	2.25 to 2.75	1.5 to 2

* Refer to the "SOIL TEXTURE" section of this label to determine the corresponding textural group for the soil to be treated.

** Use the higher rates in the recommended ranges in areas of heavy weed infestations.

*** Use rates listed in this label when using atrazine 4L. Use equivalent rates when using atrazine 80W or 90% dry flowable formulations. One quart of atrazine 4L equals 1.25 pounds of atrazine 80W or 1.1 pounds of atrazine 90% dry flowable.

LASSO® plus BANVEL (TANK MIXTURE)

Field and Silage Corn Only

For use in ILLINOIS, IOWA, MINNESOTA and WISCONSIN on level or flat-planted field corn on fine textured (silty clay loam, clay loam, sandy clay, silty clay or clay) soils with more than 4% organic matter.

Apply in water or sprayable fluid fertilizer solutions for control of the annual grasses and broadleaf weeds listed in the "WEEDS CONTROLLED" section of this label.

NOTE: Use on coarse or medium textured soils or on fine textured soils with 4% or less organic matter may result in crop injury and/or destruction.

APPROVED APPLICATION SYSTEMS

Ground—Broadcast boom; banded

NOTE: DO NOT apply by air or injection through center pivot irrigation systems.

Reference: The "APPLICATION SYSTEMS" section of this label provides detailed information on the proper application of this tank mixture using ground spray equipment.

APPROVED APPLICATION METHODS

Preemergence Surface—Apply this tank mixture after planting, before crop and weeds emerge and within 5 days of last preplant tillage operation. Corn seeds must be planted 1½ inches or deeper beneath the soil surface. Direct chemical contact with corn seed must be avoided since crop injury may result. Apply far enough behind planter equipment to avoid any incorporation by the planter wheel or other covering device. If corn seeds are planted less than 1½ inches beneath the soil surface, delay application until corn has spiked.

Postemergence Surface—This tank mixture may be applied after crop emergence until weeds reach the 2-leaf stage and the crop is no more than 3 inches tall. Applications made beyond the 2-leaf stage will not provide satisfactory grass control. If rain does not occur within 3 days after application, rotary hoe immediately to control emerged weeds.

Some leaf burn may occasionally occur when applied early postemergence to corn, but this should not affect subsequent growth or yield. DO NOT apply in sprayable fluid fertilizer after crop emergence.

Reference: The "APPLICATION TIMING AND METHODS" section of this label provides detailed information and procedures for the application timing and method selected.

NOTE: PREVENT DRIFT TO SOYBEANS OR OTHER DESIRABLE PLANTS. Do not use on furrow irrigated corn, or when corn is planted at the bottom of a furrow, utilizing lister, till or other similar planting methods. DO NOT incorporate prior to planting or corn emergence. If it is necessary to drag for leveling or rotary hoe to break soil crust, DO NOT disturb the soil more than ½ inch deep.

RECOMMENDED RATES: Refer to the following table for the recommended broadcast treatment rates for this tank mixture. Applications which are not consistent with recommendations in this label may result in unsatifactory weed control or crop injury and/or destruction.

Soil Textural Group	Broadcast Rate Per Acre (Quarts)	
	More than 4% organic matter	
	LASSO®	Banvel
Fine—silty clay loam through clay	2.5	0.5

LASSO plus BLADEX (TANK MIXTURE)

Apply this tank mixture in water or sprayable fluid fertilizer solutions for control of yellow nutsedge, and the annual grasses and broadleaf weeds listed in the "WEEDS CONTROLLED" section of this label.

APPROVED APPLICATION SYSTEMS

Ground—Broadcast boom; banded

Aerial—Fixed wing; helicopter

Dry Bulk Fertilizer Impregnation

Reference: The "APPLICATION SYSTEMS" section of this lable provides information on the proper application of this tank mixture using the system selected.

APPROVED APPLICATION METHODS

Preplant Incorporated—Apply this tank mixture within 7 days prior to planting and shallowly incorporate or Surface Blend into the upper 1 to 2 inches of soil.

Preemergence Surface—Apply this tank mixture after planting, before crop and weed emergence and within 5 days after last preplant tillage operation.

Reference: The "APPLICATION TIMING AND METHODS" section of this label provides detailed information and procedures for the application timing and method selected.

RECOMMENDED RATES: Refer to the following tables for the recommended broadcast treatment rates for this tank mixture. Applications which are not consistent with recommendations in this label may result in unsatifactory weed control or crop injury.

Surface Application Rates

Soil Textural Group*	Broadcast Rate Per Acre (Quarts)			
	1% to 3% Organic Matter		3% or more Organic Matter	
	LASSO® +	Bladex 4L***	LASSO +	Bladex 4L***
Coarse**	1.5 to 2	1 to 1.2	1.5 to 2	1.2 to 1.6
Medium	1.5 to 2	1.2 to 1.4	2 to 2.25	1.6 to 2
Fine	2 to 2.25	1.4 to 1.8	2 to 2.75	1.8 to 2.2

* Refer to the "SOIL TEXTURE" section of this label to determine the corresponding textural group for the soil to be treated.

** On coarse textured soils containing less than 2% organic matter, use LASSO plus Bladex only on sandy loam. Do not use on sand and loamy sand with less than 2% organic matter.

*** Use rates listed in this label using Bladex 4L. Use equivalent rates when using Bladex 80W. One quart of Bladex 4L equals 1.25 lbs of Bladex 80W.

Preplant Incorporated Application Rates

Soil Textural Group*	Broadcast Rate Per Acre (Quarts)			
	1% to 3% Organic Matter		3% or More Organic Matter	
	LASSO® +	Bladex 4L***	LASSO +	Bladex 4L***
Coarse**	1.75 to 2.25	1.2 to 1.4	1.75 to 2.25	1.4 to 1.8
Medium	1.75 to 2.25	1.4 to 1.6	2.25 to 2.5	1.8 to 2.2
Fine	2.25 to 2.5	1.6 to 2	2.25 to 2.75	2 to 2.4

NOTE: See "Incorporation" instructions.

* Refer to the "SOIL TEXTURE" section of this label to determine the corresponding textural group for the soil to be treated.

** On coarse textured soils containing less than 2% organic matter, use LASSO plus Bladex only on sandy loam. Do not use on sand and loamy sand with less than 2% organic matter.

*** Use rates listed in this label when using Bladex 4L. Use equivalent rates when using Bladex 80W. One quart of Bladex 4L equal 1.25 pounds of Bladex 80W.

LASSO® plus PRINCEP™ (TANK MIXTURE)

Apply this tank mixture in water or sprayable fluid fertilizer solutions for control of yellow nutsedge and the annual grasses and broadleaf weeds listed in the "WEEDS CONTROLLED" section of this label.

APPROVED APPLICATION SYSTEMS

Ground—Broadcast boom, banded

Aerial—Fixed wing: helicoper

Dry Bulk Fertilizer Impregnation

Reference: THE "APPLICATION SYSTEMS" section of this label provides information on the application of this tank mixture using the system selected.

APPROVED APPLICATION METHODS

Preplant Incorporated—Apply this tank mixture within 7 days prior to planting and shallowly incorporate or Surface Blend into the upper 1 to 2 inches of soil.

Preemergence Surface—Apply this tank mixture after planting, before crop and weed emergence and within 5 days after last preplant tillage operation.

Reference: the "APPLICATION TIMING AND METHODS" section of this label provides detailed information and procedures for the application timing and method selected.

NOTE: LAND TREATED WITH PRINCEP SHOULD NOT BE PLANTED TO ANY CROP EXCEPT CORN FOR ONE YEAR FOLLOWING TREATMENT AS CROP INJURY MAY OCCUR. AFTER HARVEST OF TREATED CROP, PLOW AND THOROUGHLY TILL THE SOIL IN THE FALL OR SPRING TO MINIMIZE POSSIBLE INJURY TO SPRING SEEDED ROTATIONAL CROPS.

RECOMMENDED RATES: Refer to the following tables for the recommended broadcast treatment rates for this tank mixture. Applications which are not consistent with recommendations in this label may result in unsatisfactory weed control or crop injury.

Surface Application Rates

Soil Textural Group*	Broadcast Rate Per Acre (Quarts)**			
	Less than 3% organic matter		3% or more organic matter	
	LASSO® +	Princep 4L***	LASSO +	Princep 4L***
Coarse—sandy loam only	1.5 to 2	0.75 to 2	1.5 to 2	1 to 2
Medium	1.25 to 2	1 to 2	2 to 2.25	1.25 to 2
Fine	2 to 2.25	1.25 to 2	2 to 2.75	1.25 to 2

* Refer to the "SOIL TEXTURE" section of this label to determine the corresponding soil textural group for the soil to be treated.

** Use the higher rates in the recommended ranges in acres of heavy weed infestations.

*** Use rates listed in this label when using Princep 4L. Use equivalent rates when using Princep 80W or 90% dry flowable formulations. One quart of the 4L formulation equals 1.25 pounds of the 80W formulation and 1.1 pounds of the 90% dry flowable formulations.

Preplant Incorporated Application Rates

Soil Textural Group*	Broadcast Rate Per Acre (Quarts)**			
	Less than 3% organic matter		3% or more organic matter	
	LASSO® +	Princep 4L***	LASSO +	Princep 4L***
Coarse—sandy loam only	1.75 to 2.25	1.25 to 2	1.75 to 2.25	1.25 to 2
Medium	1.75 to 2.25	1.25 to 2	2.25 to 2.5	1.5 to 2
Fine	2.25 to 2.5	1.5 to 2	2.25 to 2.75	1.5 to 2

* Refer to the "SOIL TEXTURE" section of this label to determine the corresponding soil textural group for the soil to be treated.

** Use the higher rates in the recommended ranges in acres of heavy weed infestations.

*** Use rates listed in this label when using Princep 4L. Use equivalent rates when using Princep 80W or 90% dry flowable formulations. One quart of the 4L formulation equals 1.25 pounds of the 80W formulation and 1.1 pounds of the 90% dry flowable formulations.

LASSO® plus ATRAZINE PLUS BLADEX (TANK MIXTURE)

Apply this tank mixture in water or sprayable fluid fertilizer solutions for control of yellow nutsedge and the annual grasses and broadleaf weeds listed in the "WEEDS CONTROLLED" section of this label.

APPROVED APPLICATION SYSTEMS

Ground—Broadcast boom; banded

Aerial—Fixed wing; helicopter

Reference: The "APPLICATION SYSTEMS" section of this label provides detailed information on the proper application of this tank mixture using the system selected.

APPROVED APPLICATION METHODS

Preemergence Surface—Apply this tank mixture after planting, before crop and weed emergence and within 5 days after last preplant tillage operation.

Preplant Incorporated—Apply this tank mixture within 7 days prior to planting and shallowly incorporate or Surface Blend into the upper 1 to 2 inches of soil.

Reference: The "APPLICATION TIMING AND METHODS" section of this label provides detailed information and procedures for the application timing and method selected.

NOTE: PLANT ONLY CORN, PEANUTS, SORGHUM OR SOYBEANS* the year following use of this mixture.

*There is a possibility of crop injury due to carryover of atrazine if soybeans are planted the following year. DO NOT plant soybeans the year following use of this tank mixture on furrow irrigated corn. DO NOT graze treated area or feed treated forage to livestock for 21 days following application of this tank mixture.

RECOMMENDED RATES: Refer to the following tables for the recommended broadcast treatment rates for this tank mixture. Applications which are not consistent with recommendations in this label may result in unsatisfactory weed control or crop injury.

Surface Application Rates

Soil Textural Group*	Broadcast Rate Per Acre (Quarts)		
	Less Than 3% Organic Matter		
	LASSO +	atrazine + 4L**	Bladex 4L***
Coarse****	1.5 to 2	0.75	0.5 to 1
Medium	1.5 to 2	0.75 to 1	1 to 1.25
Fine	2 to 2.25	1 to 1.25	1 to 1.5

Soil Textural Group*	Broadcast Rate Per Acre (Quarts)		
	3% or More Organic Matter		
	LASSO +	atrazine + 4L**	Bladex 4L***
Coarse****	1.5 to 2	0.75 to 1	1 to 1.25
Medium	2 to 2.25	1 to 1.25	1 to 2
Fine	2 to 2.75	1.25 to 1.5	1 to 2

* Refer to the "SOIL TEXTURE" section of this label to determine the corresponding textural group for the soil to be treated.

** Use rates listed in this label when using atrazine 4L. Use equivalent rates when using atrazine 80W or 90% dry flowable formulations. One quart of atrazine 4L equals 1.25 pounds of atrazine 80W or 1.1 pounds of atrazine 90% dry flowable.

*** Use rates listed in this label using Bladex 4L. Use equivalent rates when using Bladex 80W. One quart of Bladex 4L equals 1.25 pounds of Bladex 80W.

**** Do not use this mixture on sand or loamy sand soils with less than 2% organic matter.

NOTE: Use the higher rates in the recommended ranges in areas of heavy weed infestations.

Preplant Incorporated Application Rates

Soil Textural Group*	Broadcast Rate Per Acre (Quarts)		
	Less Than 3% Organic Matter		
	LASSO +	atrazine + 4L**	Bladex 4L***
Coarse****	1.75 to 2.25	1	0.5 to 1
Medium	1.75 to 2.25	1 to 1.25	1 to 1.25
Fine	2.25 to 2.5	1.25 to 1.5	1 to 1.5

Soil Textural Group*	Broadcast Rate Per Acre (Quarts)		
	3% or More Organic Matter		
	LASSO +	atrazine + 4L**	Bladex 4L***
Coarse****	1.75 to 2.25	1 to 1.25	1 to 1.25
Medium	2.25 to 2.5	1.25 to 1.5	1 to 2

Soil Textural Group*	Broadcast Rate Per Acre (Quarts) 3% or More Organic Matter LASSO +	atrazine + 4L**	Bladex 4L***
Fine	2.25 to 2.75	1.5 to 1.75	1 to 2

* Refer to the "SOIL TEXTURE" section of this label to determine the corresponding soil textural group for the soil to be treated.

** Use rates listed in this label when using atrazine 4L. Use equivalent rates when using atrazine 80W or 90% dry flowable formulations. One quart of atrazine 4L equals 1.25 pounds of atrazine 80W or 1.1 pounds of atrazine 90% dry flowable.

*** Use rates listed in this label using Bladex 4L. Use equivalent rates when using Bladex 80W. One quart of Bladex 4L equals 1.25 pounds of Bladex 80W.

**** Do not use this mixture on sand or loamy sand soils with less than 2% organic matter.

NOTE: Use the higher rates in the recommended ranges in areas of heavy weed infestations.

LASSO® plus ATRAZINE plus PROWL™ (TANK MIXTURE)

Apply this tank mixture in water or sprayable fluid fertilizer solutions for control of triazine resistant lambsquarter and pigweed in addition to yellow nutsedge and the annual grasses and broadleaf weeds listed in the "WEEDS CONTROLLED" section of this label.

APPROVED APPLICATION SYSTEMS

Ground—Broadcast boom; banded.

Aerial—Fixed wing; helicopter.

Reference: The "APPLICATION SYSTEMS" section of this label provides information on the application of this tank mixture using the system selected.

APPROVED APPLICATION METHOD

Preemergence Surface—Apply this tank mixture after planting, before crop and weed emergence and within 5 days after last preplant tillage operation.

DO NOT preplant incorporate this tank mixture as serious crop injury can result.

Reference: The "APPLICATION TIMING AND METHODS" section of this label provides detailed information and procedures for the application timing and method selected.

NOTE: PLANT ONLY CORN, PEANUTS, SORGHUM OR SOYBEANS* the year following use of this mixture.

* There is a possibility of crop injury due to carryover of atrazine if soybeans are planted the following year. DO NOT plant soybeans the year following use of this tank mixture on furrow irrigated corn. DO NOT graze treated area or feed treated forage to livestock for 21 days following application of this tank mixture.

RECOMMENDED RATES: Refer to the following tables for the recommended broadcast treatment rates for this tank mixture. Applications which are not consistent with recommendations in this label may result in unsatisfactory weed control or crop injury.

Soil Textural Group*	Broadcast Rate Per Acre (Quarts) Less Than 3% Organic Matter LASSO +	atrazine 4L** +	Prowl
Coarse	1.5 to 2	0.75 to 2	0.75
Medium	1.5 to 2	1 to 2	0.75 to 1
Fine	2 to 2.25	1.25 to 2	1

Soil Textural Group*	Broadcast Rate Per Acre (Quarts) 3% or More Organic Matter LASSO +	atrazine + 4L** +	Prowl
Coarse	1.5 to 2	1 to 2	0.75 to 1
Medium	2 to 2.25	1.25 to 2	1
Fine	2 to 2.75	1.25 to 2	1

* Refer to the "SOIL TEXTURE" section of this label to determine the corresponding soil textural group for the soil to be treated.

** Use rates listed in this label when using atrazine 4L. Use equivalent rates when using atrazine 80W or 90% dry flowable formulations. One quart of atrazine 4L equals 1.25 pounds of atrazine 80W or 1.1 pounds of atrazine 90% dry flowable.

NOTE: Use the higher rates in the recommended ranges in areas of heavy weed infestations.

™Prowl is a trademark of American Cyanamid Corporation.

LASSO® plus BLADEX plus PROWL™ (TANK MIXTURE)

Apply this tank mixture in water or sprayable fluid fertilizer solutions for control of triazine resistant lambsquarter and pigweed in addition to yellow nutsedge and the annual grasses and broadleaf weeds listed in the "WEEDS CONTROLLED" section of this label.

APPROVED APPLICATION SYSTEMS

Ground—Broadcast boom; banded.

Aerial—Fixed wing; helicopter.

Reference: The "APPLICATION SYSTEMS" section of this label provides information on the application of this tank mixture using the system selected.

APPROVED APPLICATION METHOD

Preemergence Surface—Apply this tank mixture after planting, before crop and weed emergence and within 5 days after last preplant tillage operation.

DO NOT preplant incorporate this tank mixture as serious crop injury can result.

Reference: The "APPLICATION TIMING AND METHODS" section of this label provides detailed information and procedures for the application timing and method selected.

RECOMMENDED RATES: Refer to the following tables for the recommended broadcast treatment rates for this tank mixture. Applications which are not consistent with recommendations in this label may result in unsatisfactory weed control or crop injury.

Soil Textural Group*	Broadcast Rate Per Acre (Quarts) 1% to 3% Organic Matter LASSO +	Bladex + 4L** +	Prowl
Coarse**	1.5 to 2	1 to 1.2	0.75
Medium	1.5 to 2	1.2 to 1.4	0.75 to 1
Fine	2 to 2.25	1.4 to 1.8	1

Soil Textural Group*	Broadcast Rate Per Acre (Quarts) 3% or More Organic Matter LASSO +	Bladex,7 4L** +	Prowl
Coarse**	1.5 to 2	1.2 to 1.6	0.75 to 1
Medium	2 to 2.25	1.6 to 2	1
Fine	2 to 2.75	1.8 to 2.2	1

* Refer to the "SOIL TEXTURE" section of this label to determine the corresponding soil textural group for the soil to be treated.

** Do not use this tank mixture on sand or loamy sand soils with less than 2% organic matter.

*** Use rates listed in this label using Bladex 4L. Use equivalent rates when using Bladex 80W. One quart of Bladex 4L equals 1.25 pounds of Bladex 80W.

NOTE: Use the higher rates in the recommended ranges in areas of heavy weed infestations.

LASSO® plus ATRAZINE plus LINURON (TANK MIXTURE)

FOR USE ONLY IN DELAWARE, MARYLAND, NEW JERSEY, NEW YORK, PENNSYLVANIA, VIRGINIA AND WEST VIRGINIA.

Apply this tank mixture in water or sprayable fluid fertilizer solutions for control of triazine resistant lambsquarter and pigweed in addition to yellow nutsedge and the annual grasses and broadleaf weeds listed in the "WEEDS CONTROLLED" section of this label.

APPROVED APPLICATION SYSTEMS

Ground—Broadcast boom; banded.

Reference: The "APPLICATION SYSTEMS" section of this label provides information on the application of this tank mixture using the system selected.

APPROVED APPLICATION METHOD

Preemergence Surface—Apply this tank mixture after planting, before crop and weed emergence and within 5 days after last preplant tillage operation.

Reference: The "APPLICATION TIMING AND METHODS" section of this label provides detailed information and procedures for the application timing and method selected.

NOTE: PLANT ONLY CORN, PEANUTS, SORGHUM OR SOYBEANS* the year following use of this mixture.

* There is a possibility of crop injury due to carryover of atrazine if soybeans are planted the following year. DO NOT plant soybeans the year following use of this tank mixture on furrow irrigated corn. DO NOT graze treated area or feed treated forage to livestock for 21 days following application of this tank mixture.

RECOMMENDED RATES: Refer to the following tables for the recommended broadcast treatment rates for this tank mixture. Applications which are not consistent with recommendations in this label may result in unsatisfactory weed control or crop injury.

Soil Textural Group*	Broadcast Rate Per Acre (Quarts) Less Than 3% Organic Matter		
	LASSO +	atrazine 4L** +	linuron 4L***
Coarse—sandy loam only	1.5 to 2	0.75 to 2	0.33
Medium	1.5 to 2	1 to 2	0.5
Fine	2 to 2.25	1.25 to 2	0.5

Soil Textural Group*	Broadcast Rate Per Acre (Quarts) 3% or More Organic Matter		
	LASSO +	atrazine 4L** +	linuron 4L***
Coarse—sandy loam only	1.5 to 2	1 to 2	0.5
Medium	2 to 2.25	1.25 to 2	0.5
Fine	2 to 2.75	1.25 to 2	0.5

* Refer to the "SOIL TEXTURE" section of this label to determine the corresponding soil textural group for the soil to be treated.

** Use rates listed in this label when using atrazine 4L. Use equivalent rates when using atrazine 80W or 90% dry flowable formulations. One quart of atrazine 4L equals 1.25 pounds of atrazine 80W or 1.1 pounds of atrazine 90% dry flowable.

*** Use the rates listed in this label when using linuron 4L. Use equivalent rates when using linuron 50WP or 50% dry flowable formulations. One quart of linuron 4L equals 2 pounds of linuron 50WP or 2 pounds of linuron 50% dry flowable.

NOTE: Use the higher rates in the recommended ranges in areas of heavy weed infestations.

Conservation or Minimum Tillage Systems

TANK MIXTURES

AT-PLANTING APPLICATIONS

When applied as directed under the conditions described, these tank mixtures control many emerged annual weeds, suppress many emerged perennial weeds and give premergence control of many annual grasses and weeds when corn will be planted directly into a cover crop, established sod or in previous crop residues. These tank mixtures will not control regrowth from perennial weeds.

DO NOT APPLY BY AIR.

Refer to specific product labels for crop rotation restrictions and cautionary statements of all products used in the tank mixture.

For mixing instructions, see the "MIXING AND SPRAYING INSTRUCTIONS" section of this label.

LASSO® plus ROUNDUP®
or
LASSO plus GRAMOXONE SUPER
or
LASSO plus ATRAZINE plus ROUNDUP
or
LASSO plus ATRAZINE plus GRAMOXONE SUPER
or
LASSO plus BLADEX plus ROUNDUP
or
LASSO plus BLADEX plus GRAMOXONE SUPER
or
LASSO plus PRINCEP™ plus ROUNDUP
or
LASSO plus PRINCEP plus GRAMOXONE SUPER
or
LASSO plus ATRAZINE plus BLADEX plus ROUNDUP
or
LASSO plus ATRAZINE plus BLADEX plus GRAMOXONE SUPER
or
LASSO plus ATRAZINE plus PRINCEP plus ROUNDUP
or
LASSO plus ATRAZINE plus PRINCEP plus GRAMOXONE SUPER

Apply these tank mixtures with Roundup in 10 to 40 gallons of water, or the tank mixtures with Gramoxone Super in 20 to 60 gallons of water spray solution per acre immediately before, during or after planting, but BEFORE CROP EMERGENCE. As density of stubble, crop residue or weeds increases, spray gallonage and rate should be increased within the recommended ranges to insure complete coverage. In the absence of emerged vegetation, delete the Roundup or Gramoxone Super portion of these tank mixtures.

™ Princep is a trademark of Ciba-Geigy Corporation.

CONTROL OR SUPPRESSION OF EMERGED WEEDS

ROUNDUP®

Annual Weeds—Apply 1 quart of Roundup herbicide per acre in these tank mixtures if weeds are less than 6 inches tall, and use 1.5 quarts of Roundup if weeds are more than 6 inches tall for control of many emerged annual grasses and broadleaf weeds. For emerged annual weeds controlled, see the "WEEDS CONTROLLED" section of the label for Roundup.

Perennial Weeds—At normal application dates in minimum tillage systems, perennial weeds may not be at the proper stage of growth for control. See the "GENERAL INFORMATION" section of the label for Roundup for the proper stage of growth for perennial weeds. Use of 2 to 4 quarts of Roundup per acre in the above mixtures under these conditions provides top kill and reduces competition from many emerged perennial grasses and broadleaf weeds.

For emerged perennial weeds controlled see the "WEEDS CONTROLLED" section of the label for Roundup. To obtain control of perennial weeds, follow recommendations on the label for Roundup for stage of growth and rate of application.

USE OF THIS MIXTURE FOR BERMUDAGRASS OR JOHNSONGRASS CONTROL IS NOT RECOMMENDED.

NOTE: When using these tank mixtures, do not exceed 4 quarts of Roundup herbicide per acre.

GRAMOXONE SUPER

When used as directed, Gramoxone Super in labeled tank mixtures controls many emerged annual weeds and suppresses many emerged perennial weeds.

Broadcast Treatment—Apply 1.5 to 2.5 pints of Gramoxone Super (0.25 to 0.5 lb. active) in these tank mixtures immediately before, during or after planting but BEFORE CROP EMERGENCE. Use the 2.5 pints rate if any weeds are 4 to 6 inches tall. This mixture may not control weeds taller than 6 inches. As density of stubble, crop residue or weeds increases, spray gallonage should be increased within the recommended range for complete coverage. Add a nonionic spreader surfactant (approved for use on crops) containing at least 75 percent surfactant active agent at 8 ounces per 100 gallons of diluted spray. REFER TO THE GRAMOXONE SUPER LABEL FOR PRECAUTIONARY STATEMENTS.

PREEMERGENCE WEED CONTROL

LASSO®

For weeds controlled preemergence, see the sections of this label on LASSO. See the following table for recommended rates of LASSO in this tank mixture on various soil types:

Soil Textural Group*	BROADCAST RATE PER ACRE (QUARTS) LASSO®
Coarse	2.5
Medium	3
Fine	3 to 3.5

* Refer to the "SOIL TEXTURE" section of the label to determine the corresponding textural group for the soil to be treated.

Use the higher rate of LASSO in the recommended ranges in areas of heavy grass infestation, or when organic matter content is 3 percent or more.

LASSO® plus ATRAZINE

For weeds controlled preemergence, see the sections of this label for LASSO and LASSO plus atrazine.

See the following table for recommended rates of LASSO plus atrazine in this tank mixture on various soil types.

Soil Textural Group*	Broadcast Rate Per Acre (Quarts)	
	LASSO® +	Atrazine 4L**
Coarse—sandy loam only	1.75 to 2	1 to 2
Medium	2 to 2.5	1.2 to 2
Fine	2.5 to 3	1.6 to 2

* Refer to the "SOIL TEXTURE" section of this label to determine the corresponding textural group for the soil to be treated.

** Use rates listed in this label when using atrazine 4L. Use equivalent rates when using atrazine 80W or 90% dry flowable formulations. One quart of atrazine 4L equals 1.25 pounds of atrazine 80W or 1.1 pounds of atrazine 90% dry flowable.

Use the higher rate of LASSO in the recommended ranges in areas of heavy grass infestation or when fall panicum or crabgrass will be present.

Use the higher rate of atrazine in the recommended ranges on soils with greater than 3% organic matter.

NOTE: PLANT ONLY CORN, PEANUTS, SORGHUM OR SOYBEANS† the year following use of this mixture.

† There is a possibility of crop injury due to carryover of atrazine if soybeans are planted the following year. DO NOT plant soybeans the year following use of this tank mixture on furrow irrigated corn.

DO NOT graze treated area or feed treated forage to livestock for 21 days following application of this tank mixture.

LASSO® plus BLADEX

For weeds controlled preemergence, see the "WEEDS CONTROLLED" section of the label for LASSO and this tank mixture.

See the following table for recommended rates of LASSO plus Bladex in this tank mixture on various soil types.

Soil Textural Group*	Broadcast Rate Per Acre (Quarts)	
	LASSO® +	Bladex 4L**
Coarse—sandy loam only	1.75 to 2	1 to 1.6
Medium	2 to 2.5	1.2 to 1.6
Fine	2.5 to 3	1.6 to 2.2

* Refer to the "SOIL TEXTURE" section of this label to determine the corresponding textural group for the soil to be treated.

** When using Bladex 80W use equivalent rates. One quart of Bladex 4L equals 1.25 pounds of Bladex 80W.

Use the higher rate of LASSO in the recommended ranges in areas of heavy grass infestation or when fall panicum or crabgrass will be present.

Use the higher rate of Bladex in the recommended ranges on soils with greater than 3% organic matter.

NOTE: Do not use this mixture on sand or loamy sand soils with less than 2% organic matter.

LASSO® plus PRINCEP

For weeds controlled preemergence see the "WEEDS CONTROLLED" section of the labels for LASSO and Princep herbicides.

See the following table for recommended rates of LASSO plus Princep in this tank mixture on various soil types.

Soil Textural Group*	Broadcast Rate Per Acre (Quarts)	
	LASSO® +	PRINCEP 4L**
Coarse—sandy loam only	1.75 to 2	1 to 1.2
Medium	2 to 2.5	1.2 to 1.6
Fine	2.5 to 3	1.6 to 2.2

* Refer to the "SOIL TEXTURE" section of this label to determine the corresponding soil textural group for the soil to be treated.

** When using Princep 80W use equivalent rates. One quart of Princep 4L equals 1.25 pounds of Princep 80W.

Use the higher rate of LASSO in the recommended ranges in areas of heavy grass infestation or when fall panicum or crabgrass will be present.

Use the higher rate of Princep in the recommended ranges on soils with greater than 3% organic matter.

NOTE: LAND TREATED WITH PRINCEP SHOULD NOT BE PLANTED TO ANY CROP EXCEPT CORN FOR ONE YEAR FOLLOWING TREATMENT AS CROP INJURY MAY OCCUR. AFTER HARVEST OF TREATED CROP, PLOW AND THOROUGHLY TILL THE SOIL IN THE FALL OR SPRING TO MINIMIZE POSSIBLE INJURY TO SPRING SEEDED ROTATIONAL CROPS.

■ **LASSO® plus ATRAZINE plus BLADEX** ■

This tank mixture may be applied for additional preemergence control of heavy weed infestation and/or hard-to-control weeds listed in the "WEEDS CONTROLLED" section of the label. Refer to the following rate table to determine the recommended rates for this tank mixture on various soil types.

Soil Textural Group*	Broadcast Rate Per Acre (Quarts)		
	LASSO +	atrazine 4L** +	Bladex 4L***
Coarse****	1.75 to 2	0.75 to 1	1 to 1.25
Medium	2 to 2.5	1 to 1.25	1 to 2
Fine	2.25 to 3	1 to 1.25	1 to 2

* Refer to the "SOIL TEXTURE" section of this label to determine the corresponding soil textural group for the soil to be treated.

** Use rates listed in this label when using atrazine 4L. Use equivalent rates when using atrazine 80W or 90% dry flowable formulations. One quart of atrazine 4L equals 1.25 pounds of atrazine 80W or 1.1 pounds of atrazine 90% dry flowable.

*** Use rates listed in this label when using Bladex 4L. Use equivalent rates when using Bladex 80W. One quart of Bladex 4L equals 1.25 pounds of Bladex 80W.

**** DO NOT use this mixture on sand or loamy sand soils with less than 2% organic matter.

Use the higher rates in the recommended ranges in area of heavy weed infestations.

NOTE: PLANT ONLY corn, peanuts, sorghum or soybeans* the year following use of this mixture.

* There is a possibility of crop injury to carryover of atrazine if soybeans are planted the following year. DO NOT plant soybeans the year following use of this tank mixture on furrow irrigated corn.

DO NOT graze treated area or feed treated forage to livestock for 21 days following application of this tank mixture.

LASSO® plus ATRAZINE plus PRINCEP

For additional preemergence control of heavy weed infestation and/or hard-to-control weeds, a three-way tank mixture of LASSO herbicide plus atrazine plus Princep may be applied. Refer to the following rate table to determine the recommended rates for this tank mixture on various soil types.

Soil Textural Group*	Broadcast Rate Per Acre (Quarts)**		
	LASSO +	Atrazine 4L*** +	Princep 4L***
Coarse	1.75 to 2	0.75 to 1	0.75 to 1
Medium	2 to 2.5	1 to 1.25	1 to 1.5
Fine	2.5 to 3	1 to 1.25	1 to 1.5

* Refer to the "SOIL TEXTURE" section of this label to determine the corresponding soil textural group for the soil to be treated.

** When using 4L formulations use the rates listed. Use equivalent rates when using 80W or 90% dry flowable formulations. One quart of the 4L formulation of these products equals 1.25 pounds of 80W formulation. One quart of atrazine 4L or Princep 4L equals 1.1 pounds of the 90% dry flowable formulations of these products.

Use the higher rates of atrazine and Princep in the recommended ranges on soils with greater than 3% organic matter.

In the areas of heavy grass infestation, use the higher rate of LASSO in the recommended ranges when fall panicum, crabgrass, or soils with greater than 3% organic matter will be present.

NOTE: LAND TREATED WITH PRINCEP SHOULD NOT BE PLANTED TO ANY CROP EXCEPT CORN FOR ONE YEAR FOLLOWING TREATMENT AS CROP INJURY MAY OCCUR. AFTER HARVEST OF TREATED CROP, PLOW AND THOROUGHLY TILL THE SOIL IN THE FALL OR SPRING TO MINIMIZE POSSIBLE INJURY TO SPRING SEEDED ROTATIONAL CROPS. PLANT ONLY CORN, PEANUTS, SORGHUM OR SOYBEANS† THE YEAR FOLLOWING USE OF THIS MIXTURE.

† There is a possibility of crop injury due to carryover of atrazine if soybeans are planted the following year.

DO NOT plant soybeans the year following use of this tank mixture on furrowed irrigated corn.

DO NOT graze treated area or feed treated forage to livestock for 21 days following application of this tank mixture.

EARLY PREPLANT APPLICATION

LASSO® plus ATRAZINE (TANK MIXTURE)

For use in no-till and other conservation tillage systems in DELAWARE, ILLINOIS, INDIANA, IOWA, KANSAS, KENTUCKY, MARYLAND, MINNESOTA, MISSOURI, NEBRASKA, NORTH CAROLINA, OHIO, SOUTH DAKOTA, VIRGINIA, AND WISCONSIN.

This tank mixture, when applied in a single application (alone or in a 3-way combination with Bladex or Princep), split application or as a sequential application to Bladex or Princep in early preplant programs, will provide preemergence control or reduced competition of listed annual grasses and broadleaf weeds.

If emerged weeds are present at the time of treatment, Roundup®, Gramoxone Super or 2,4-D should be added to the product according to the directions for use on their respective product labels. If unsatisfactory weed control occurs (due to excessively dry or excessively wet conditions) following the earlier application, a postemergence application of a appropriate labelled grass and/or broadleaf weed herbicide may be used. If a postemergence treatment includes the herbicide used early preplant, do not exceed the labelled rate for corn on a given soil texture. Observe all precautions and limitations on the labels for LASSO, atrazine, Bladex, Princep, Roundup, Gramoxone Super, 2,4-D and other postemergence herbicides before use of these products.

RECOMMENDED RATE AND TIMING OF APPLICATION:

Single application—Application of this tank mixture should be made less than 30 days before planting but prior to weed emergence. This application should be restricted to areas north of Interstate 70 when west of the Mississippi River and north of Interstate 64 when east of the Mississippi River where cool spring weather minimizes or slows biodegradation of herbicides.

See the following table for recommended rates.

Soil Textural Group*	Broadcast Rate Per Acre (Quarts)**	
	LASSO® +	atrazine 4L
Coarse*	2.25 to 2.5	1.25 to 1.5
Medium	2.5 to 3	1.5 to 2

Soil Textural Group*	Broadcast Rate Per Acre (Quarts)**	
	LASSO® +	atrazine 4L
Fine	3 to 3.75	1.75 to 2.25

This tank mixture may also be mixed in a 3-way combination with 1 to 1.25 quarts per acre of Princep or 1.5 to 2 quarts per acre of Bladex to provide improved control of fall panicum and crabgrass.

* Applications on coarse soils should not be made more than 2 weeks prior to planting.

** Use the lower rate where light weed infestation is expected and the higher rate where heavy weed infestation is expected.

Split application—Apply 60 percent of the recommended rate as a split treatment prior to weed emergence and no more than 45 days prior to planting and the remaining 40 percent at or immediately following planting.

FOR CONTROL OF FALL PANICUM, CRABGRASS OR BROADLEAF SIGNALGRASS, USE THE SEQUENTIAL APPLICATION OF THIS TANK MIXTURE FOLLOWING BLADEX OR PRINCEP.

See the following table for recommended rates.

Soil Textural Group*	Broadcast Rate Per Acre (Quarts)*	
	LASSO® +	atrazine 4L
Coarse	2.25 to 2.75	1.25 to 1.75
Medium	2.5 to 3	1.5 to 2
Fine	3 to 3.75	1.75 to 2.25

* Use the lower rate where light weed infestation is expected and the higher rate where heavy weed infestation is expected.

Sequential application—Application of this tank mixture following Bladex or Princep should be utilized for the control of fall panicum, crabgrass or broadleaf signalgrass. Apply 1.5 to 2 quarts per acre of Bladex or 1 to 1.25 quarts per acre of Princep prior to weed emergence and no more than 45 days prior to planting. At or immediately following planting apply the recommended rate of this tank mixture.

Following applications of Bladex or Princep see the following table for recommended rates.

Soil Textural Group*	Broadcast Rate Per Acre (Quarts)*	
	LASSO® +	atrazine 4L
Coarse	1.75 to 2.5	1 to 1.5
Medium	2.25 to 2.75	1.25 to 1.75
Fine	2.5 to 3	1.5 to 2

* Use the lower rate where light weed infestation is expected and the higher rate where heavy weed infestation is expected.

Use rates listed in this label when using the 4L formulation of atrazine, Bladex or Princep. Use equivalent rates of the 80W or 90% dry flowable formulations of atrazine, Bladex or Princep. One quart of the 4L formulation of atrazine, Bladex or Princep equals 1.25 pounds of the 80W and 1.1 pounds of the 90% dry flowable formulations of atrazine, Bladex or Princep.

DO NOT use Bladex on coarse textured soils or soils of less than 1 percent organic matter.

Plant only corn, peanuts, sorghum (milo) or soybeans the year following the use of this program. There is a possibility of crop injury due to carryover of atrazine is soybeans are planted the following year.

NOTE: Soil treated with Princep should not be planted to any crop except corn for one year following treatment as crop injury may occur. After harvest of the treated crop, plow and thoroughly till the soil in the fall or spring to minimize possible injury to spring seeded rotational crops.

GRAIN SORGHUM (MILO)

LASSO®

Apply this product in water or sprayable fluid fertilizer solution for control of yellow nutsedge and the annual grasses and broadleaf weeds listed in the "WEEDS CONTROLLED" section of this label.

Preplant incorporated and preemergence surface applications must be made ONLY to grain sorghum planted with seed that has been properly treated with Screen® seed protectant. Postemergence surface applications can be made to grain sorghum planted with protected or unprotected seed (grain sorghum seed not treated with a seed protectant).

® Screen is a registered trademark of Monsanto Company.

APPROVED APPLICTION SYSTEMS

Center Pivot Irrigation

Aerial — Fixed wing; helicopter

Ground—Broadcast boom; banded

Reference: The "APPLICATION SYSTEMS" section of this label provides detailed information on the application of this product using the system selected.

APPROVED APPLICATION METHODS

Preplant Incorporated—Apply this product within 7 days prior to planting and shallowly incorporate or Surface Blend into the upper 1 to 2 inches of soil. Irrigation within 10 days following planting may improve weed control.

Preemergence Surface—Apply this product after planting, before crop and weed emergence, and within 5 days after last preplant tillage operation.

Postemergence Surface—Apply this product in water after crop emergence but before the weeds or grasses reach the 2-leaf stage and before the crop is 5 inches tall to reduce competition from weeds listed as controlled on this label. DO NOT APPLY IN SPRAYABLE FLUID FERTILIZER AFTER CROP EMERGENCE. Some leaf burn to the crop may occasionally occur when applied early postemergence to grain sorghum, but this should not affect subsequent growth or yield. Weeds which are emerged at time of application are not controlled by this product. If weeds are emerged at application, shallowly cultivate immediately to improve control.

Reference: The "APPLICATION TIMING AND METHODS" sections of this label provides detailed information and procedures for the application timing and method selected.

RECOMMENDED RATES: Refer to the following table for the recommended broadcast treatment rates for this product. Applications which are not consistent with recommendations in this label may result in unsatisfactory weed control or crop injury.

Surface Application Rates

Soil Textural Group*	Broadcast Rate Per Acre (Quarts)**
	LASSO®
Coarse	1.5 to 2
Medium	2 to 2.25
Fine	2 to 2.5

* Refer to the "SOIL TEXTURE" section of this label to determine the corresponding textural group for the soil to be treated.

** Use the higher rates in the recommended ranges in areas of heavy weed infestation.

NOTE: When applied through CENTER PIVOT IRRIGATION or under sprinkler irrigation systems, use a minimum of 3 quarts per acre of this product on coarse soils.

Preplant Incorporated Application Rates

Soil Textural Group*	Broadcast Rate Per Acre (Quarts)**
	LASSO®
Coarse	2 to 2.5
Medium	2.5 to 2.75
Fine	2.5 to 3

* Refer to the "SOIL TEXTURE" section of this label to determine the corresponding texture group for the soil to be treated.

** Rates may be increased to a maximum of 4 quarts on any soil type when heavy infestation of yellow nutsedge is present.

Postemergence Surface Application Rates—See the "Preemergence Surface" application rates for grain sorghum for recommended broadcast rates per acre.

LASSO® plus ATRAZINE (TANK MIXTURE)

Apply this tank mixture in water or sprayable fluid fertilizer solutions for control of yellow nutsedge and the annual grasses and broadleaf weeds listed in the "WEEDS CONTROLLED" section of this label.

Early preplant surface, preplant incorporated and preemergence surface applications must be made ONLY to grain sorghum planted with seed that has been properly treated with Screen® seed protectant. Postemergence surface applications can be made to grain sorghum planted with protected or unprotected seed (grain sorghum seed not treated with a seed protectant).

NOTE: In Texas, use only on the fine textured soils of the Gulf Coast.

APPROVED APPLICATION SYSTEMS

Center Pivot Irrigation

Aerial — Fixed wing; helicopter

Ground—Broadcast boom; banded

Reference: The "APPLICATION SYSTEMS" section of this label provides detailed information on the proper application of this tank mixture using the system selected.

APPROVED APPLICATION METHODS

Preplant Incorporated—Apply this tank mixture within 7 days prior to planting and shallowly incorporate or Surface Blend into the upper 1 to 2 inches of soil.

Preemergence Surface—Apply this tank mixture after planting, before crop and weed emergence and within 5 days after last preplant tillage operation.

Postemergence Surface—Apply this tank mixture in water after crop emergence but before the weeds or grasses reach the 2-leaf stage and before the crop is 5 inches tall to reduce competition from weeds listed as controlled on this label. DO NOT APPLY IN SPRAYABLE FLUID FER-

TILIZER AFTER CROP EMERGENCE. Some leaf burn to the crop may occasionally occur when applied early postemergence to grain sorghum, but this should not affect subsequent growth or yield. Weeds which are emerged at the time of application are not controlled by this tank mixture. If weeds are emerged at application, shallowly cultivate immediately to improve control.

Early Preplant Surface—Apply this tank mixture in water or sprayable fluid fertilizer as a single application within 30 days prior to planting. When making split applications, apply this tank mixture within 45 days prior to planting, followed by an additional application of LASSO at planting. If emerged weeds are present at the time of the early preplant application or at-planting application, a labeled contact herbicide or tillage should be used. The at-planting application should be made before crop or weeds emerge. Minimize movement of treated soil out of the row or untreated soil to the surface during planting or weed control may be reduced. Do not use in coarse or fine textured soils.

Reference: The "APPLICATION TIMING AND METHODS" section of this label provides detailed information and procedures for the application timing and method selected.

NOTE: PLANT ONLY CORN, PEANUTS, SORGHUM OR SOYBEANS* the year following use of this mixture.

* There is a possibility of crop injury due to carry-over of atrazine if soybeans are planted the following year. DO NOT plant soybeans the year following use of this tank mixture on furrow irrigated milo. On grain sorghum grown in Texas, use only on the fine textured soils of the Gulf Coast and Blacklands. The atrazine 4L rate should not exceed 1.25 quarts per acre as possible crop injury may result due to atrazine. DO NOT graze treated area or feed treated forage to livestock for 21 days following application of this tank mixture.

RECOMMENDED RATES: Refer to the following tables for the recommended broadcast treatment rates for this tank mixture. Applications which are not consistent with recommendations in this label may result in unsatisfactory weed control or crop injury.

Surface Application Rates

Soil Textural Group*	Broadcast Rate Per Acre (Quarts)**			
	Less Than 3% Organic Matter		3% or More Organic Matter	
	LASSO® +	Atrazine 4L***	LASSO +	Atrazine 4L***
Coarse (except sand)	1.5	0.75 to 1	1.5 to 1.75	0.75 to 1
Medium	1.5 to 1.75	1 to 1.25	1.5 to 2	1 to 1.5
Fine	1.5 to 2	1.25 to 1.5	1.75 to 2.25	1.25 to 1.75

* Refer to the "SOIL TEXTURE" section of this label to determine the corresponding textural group for the soil to be treated.

** Use the higher rates in the recommended ranges in areas of heavy weed infestations, or for fields under irrigation.

*** Use rates listed in this label when using atrazine 4L. Use equivalent rates when using atrazine 80W or 90% dry flowable formulations. One quart of atrazine 4L equals 1.25 pounds of atrazine 80W or 1.1 pounds of atrazine 90% dry flowable.

NOTE: When applied through center pivot irrigation or under sprinkler irrigation systems use a minimum of 2 quarts per acre of LASSO and 1 quart per acre of atrazine 4L in this tank mixture on coarse soils.

PREPLANT INCORPORATED APPLICATION RATES

Soil Textural Group*	Broadcast Rate Per Acre (Quarts)**			
	Less Than 1.5% Organic Matter		1.5% or More Organic Matter	
	LASSO® +	atrazine 4L***	LASSO +	atrazine 4L***
Medium	1.5 to 2	1 to 1.25	1.75 to 2	1 to 1.5
Fine	1.75 to 2	1 to 1.5	2 to 2.5	1.25 to 1.75

* Refer to the "SOIL TEXTURE" section of this label to determine the corresponding textural group for the soil to be treated.

NOTE: Use 3 to 4 quarts of LASSO per acre for control of yellow nutsedge or to reduce competition from the hard-to-control weeds listed on this label.

Postemergence Surface Application Rates—See the "Preemergence Surface" application rates for recommended broadcast rates per acre.

Early Preplant Application Rates

Soil Textural Group*	Broadcast Rate Per Acre (Quarts)**		
	Early Application		At-planting Application
	LASSO® +	Atrazine 4L	LASSO
Single Application			
Coarse (except sand)	2.25 to 3	1 to 1.25	•
Medium	2.75 to 3	1.5 to 2	•
Split Application			
Coarse (except sand)	1.75 to 2	1 to 1.5	0.75 to 1
Medium	1.75 to 2	1.5 to 2	0.75 to 1

* Refer to the "SOIL TEXTURE" section of this label to determine the corresponding textural group for the soil to be treated.

** Use the higher rate in the recommended range in areas of heavy weed infestation.

LASSO®plus PROPAZINE (TANK MIXTURE)

Apply this tank mixture in water or sprayable fluid fertilizer solution for control of yellow nutsedge and the annual grasses and broadleaf weeds listed in the "WEEDS CONTROLLED" section of this label. Apply only to grain sorghum seed that has been properly treated with a protectant containing the active ingredient, flurazole.

APPROVED APPLICATION SYSTEMS

Ground—Broadcast boom; banded

Aerial — Fixed wing; helicopter

Reference: The "APPLICATION SYSTEMS" section of this label provides detailed information on the proper application of this tank mixture using the equipment selected.

APPROVED APPLICATION METHOD

Preemergence Surface—Apply this tank mixture after planting, before crop and weed emergence and within 5 days after last tillage operation.

Reference: The "APPLICATION TIMING AND METHODS" section of this label provides detailed information and procedures for the application method and timing selected.

NOTE: Applications of this tank mixture to alkali soils where cuts, fills, or erosion have exposed calcareous of alkali subsoils may result in crop injury from propazine. Do not plant crops other than corn or soybeans for 18 months following application of this tank mixture due to possibility of crop injury from propazine carryover.

RECOMMENDED RATES: Refer to the following tables for the recommended broadcast treatment rates for this tank mixture. Applications which are not consistent with recommendations in this label may result in unsatisfactory weed control or crop injury.

Surface Application Rates

Soil Textural Group*	Broadcast Rate Per Acre (Quarts)**	
	LASSO® +	Propazine 4L***
Coarse	1.5 to 1.75	0.75 to 1****
Medium	1.75 to 2	1
Fine	2 to 2.25	1

* Refer to the "SOIL TEXTURE" section of this label to determine the corresponding textural group for the soil to be treated.

** Use the higher rates in the recommended ranges in areas of heavy weed infestation, for fields under irrigation, or where soil organic matter is greater than 3 percent.

*** Use rates listed in this label when using propazine 4L. Use equivalent rates when using propazine 80W or 90% dry flowable formulations. One quart of propazine 4L equals 1.25 pounds of propazine 80W or 1.1 pounds of propazine 90% dry flowable.

**** In coarse soils with less than 0.75% organic matter, use 0.4 quart of propazine 4L per acre. In coarse soils with 0.75% to 1% organic matter, use 0.6 quart propazine per acre. When using these rates, the level and duration of weed control may be lessened if either an irrigation or a rainfall in excess of 1 inch occurs after application.

CONSERVATION OR MINIMUM TILLAGE SYSTEMS TANK MIXTURES AT-PLANTING APPLICATIONS

Preemergence surface applications must be made **ONLY** to grain sorghum planted with seed that has been properly treated with Screen® seed protectant.

Weed applied as directed under the conditions described, these tank mixtures control many emerged annual weeds, suppress many emerged perennial weeds and give premergence control of many annual grasses and weeds when grain sorghum (milo) will be planted directly into a cover crop, established sod or in previous crop residues. These tank mixtures will not control regrowth from perennial weeds.

DO NOT APPLY BY AIR.

Refer to specific labels for crop rotation restriction and cautionary statements of all products used in the tank mixture.

For mixing instructions, see the "MIXING, SPRAYING AND HANDLING INSTRUCTIONS" section of this label.

LASSO plus ROUNDUP®
or
LASSO plus GRAMOXONE SUPER
or
LASSO plus ATRAZINE plus ROUNDUP
or
LASSO plus ATRAZINE plus GRAMOXONE SUPER

Apply these tank mixtures with Roundup in 10 to 40 gallons of water, or the tank mixtures with Gramoxone Super in 20 to 60 gallons of spray solution per acre immediately before, during or after planting, but BEFORE CROP EMERGENCE. As density of stubble, crop residue or weeds increases, spray gallonage and rate should be increased within the recommended ranges to insure complete coverage. In the absence of emerged vegetation, delete the Roundup or Gramoxone Super portion of these tank mixtures.

CONTROL OR SUPPRESSION OF EMERGED WEEDS

Roundup®

Annual Weeds—Apply 1 quart of Roundup herbicide per acre in these tank mixtures if weeds are less than 6 inches tall, and use 1.5 quarts of Roundup if weeds are more than 6 inches tall for control of many emerged annual grasses and broadleaf weeds. For emerged annual weeds controlled, see the "WEEDS CONTROLLED" section of the label for Roundup.

Perennial Weeds—At normal application dates in minimum tillage systems, perennial weeds may not be at the proper stage of growth for control. See the "GENERAL INFORMATION" section of the label for Roundup for the proper stage of growth for perennial weeds. Use of 2 to 4 quarts of Roundup per acre in the above mixtures under these conditions provides top kill and reduces competition from many emerged perennial grasses and broadleaf weeds.

For emerged perennial weeds controlled, see the "WEEDS CONTROLLED" section of the label for Roundup. To obtain control of perennial weeds, follow the recommendations on the label for Roundup for stage of growth and rate of application.

USE OF THIS MIXTURE FOR BERMUDAGRASS OR JOHNSONGRASS CONTROL IS NOT RECOMMENDED.

NOTE: When using these tank mixtures, do not exceed 4 quarts of Roundup herbicide per acre.

GRAMOXONE SUPER

When used as directed, Gramoxone Super in these tank mixtures controls many emerged annual weeds and suppresses many emerged perennial weeds.

Broadcast Treatment—Apply 1.5 to 2.5 pints of Gramoxone Super per acre in these tank mixtures immediately before, during or after planting, but BEFORE CROP EMERGENCE. Use the 2.5-pint rate if any weeds are 4 to 6 inches tall. These mixtures may not control weeds taller than 6 inches. As density of stubble, crop residue or weeds increases, spray gallonage should be increased within the recommended range for complete coverage. Add a nonionic spreader surfactant (approved for use on crops) containing at least 75% surfactant active agent at 8 fluid ounces per 100 gallons of diluted spray. REFER TO THE GRAMOXONE SUPER LABEL FOR PRECAUTIONARY STATEMENTS.

PREMERGENCE WEED CONTROL

■ **LASSO®** ■

For weeds controlled preemergence, see the "WEEDS CONTROLLED" section of this label. See the following table for rcommended rates of LASSO in this tank mixture on various soil types:

Soil Textural Group*	Broadcast Rate Per Acre (Quarts) LASSO®
Coarse	2 to 2.5
Medium	2.5 to 2.75
Fine	2.5 to 3

* Refer to the "SOIL TEXTURE" section of the label to determine the corresponding soil textural group for the soil to be treated.

Use the higher rate of LASSO in the recommended ranges in areas of heavy grass infestation, or when organic matter content is 3 percent or more.

■ **LASSO® plus ATRAZINE** ■

For weeds controlled preemergence, see the "WEEDS CONTROLLED" section of this label.

See the following table for recommended rates of LASSO plus atrazine in this tank mixture on various soil types.

Soil Textural Group*	Broadcast Rate Per Acre (Quarts) LASSO®	+ atrazine 4L**
Coarse—sandy loam only	1.5 to 2	0.75 to 1
Medium	1.75 to 2.25	1 to 1.25
Fine	2 to 2.5	1.25 to 1.5

* Refer to the "SOIL TEXTURE" section of this label to determine the corresponding soil textural group for the soil to be treated.

** Use rates listed in this label when using atrazine 4L. Use equivalent rates when using atrazine 80W or 90% dry flowable formulations. One quart of atrazine 4L equals 1.25 pounds of atrazine 80W or 1.1 pounds of atrazine 90% dry flowable.

Use the higher rate of LASSO in the recommended ranges in areas of heavy grass infestation or when fall panicum or crabgrass will be present.

Use the higher rate of atrazine in the recommended ranges on soils with greater than 3 percent organic matter.

NOTE: PLANT ONLY CORN, PEANUTS, SORGHUM OR SOYBEANS the year following use of this mixture.

There is a possibility of crop injury due to carryover of atrazine if soybeans are planted the following year. DO NOT plant soybeans the year following use of this tank mixture on furrow irrigated corn.

DO NOT graze treated areas or feed treated forage to livestock for 21 days following application of this tank mixture.

SOYBEANS

LASSO®

Apply this product in water or sprayable fluid fertilizer solution for control of yellow nutsedge, and the annual grasses and broadleaf weeds listed in the "WEEDS CONTROLLED" section of this label.

APPROVED APPLICATION SYSTEMS

Ground—Broadcast boom; banded

Aerial—Fixed wing; helicopter

Center Pivot Irrigation

Dry Bulk Fertilizer Impregnation

Reference: The "APPLICATION SYSTEMS" section of this label provides detailed information on the application of this product using the system selected.

APPROVED APPLICATION METHODS

Preplant Incorporated—Apply this product within 7 days prior to planting and shallowly incorporate or Surface Blend into the upper 1 to 2 inches of soil.

Preemergence Surface—Apply this product after planting, before crop and weed emergence and within 5 days after last preplant tillage operation.

Postemergence Surface—This product may be applied after crop emergence to reduce competition from weeds listed as controlled on this label. Apply before soybeans exceed the unifoliate stage (first two true leaves).

Weeds which are emerged at time of application are not controlled by this product. If weeds are emerged at application, shallowly cultivate or rotary hoe immediately to improve performance. DO NOT apply in sprayable fluid fertilizer after crop emergence.

NOTE: The highest levels of weed control will be obtained when applications are made prior to the emergence of the crop and weeds.

DO NOT graze treated area or feed treated forage to livestock following postemergence application of this product.

Reference: The "APPLICATION TIMING AND METHODS" section of this label provides detailed information and procedures for the application timing and method selected.

Sequential Treatment—This product may be applied as two sequential treatments for improved control of hard-to-control weeds. The combined total application rate for the two treatments must not exceed 8 quarts per acre per year. The second treatment must be applied before soybeans exceed unifoliate stage (first two true leaves).

RECOMMENDED RATES: Refer to the following tables for the recommended broadcast treatment rates for this product. Applications which are not consistent with recommendations in this label may result in unsatisfactory weed control or crop injury.

Surface Application Rates

Soil Textural Group*	Broadcast Rate Per Acre (Quarts)** Less Than 3% Organic Matter	3% or More Organic Matter
Coarse***	2 to 2.25	2 to 2.25
Medium	2 to 2.75	2 to 2.75
Fine	2 to 2.75	2.5 to 3.25

* Refer to the "SOIL TEXTURE" section of this label to determine the corresponding textural group for the soil to be treated.

** Use the higher rate in the recommended range in areas of heavy weed infestations. Use a minimum of 2.5 quarts of this product per acre on coarse soils and 3 to 4 quarts on medium or fine textured soils to control black or hairy nightshade.

*** When applied through center pivot irrigation or under sprinkler irrigation systems on coarse soils, use a minimum of 3 quarts per acre of this product.

On coarse soils with 10% more organic matter use 4 quarts of LASSO herbicide per acre.

Preplant Incorporated Application Rates

Soil Textural Group*	Broadcast Rate Per Acre (Quarts)** Less Than 3% Organic Matter	3% or More Organic Matter
Coarse	2.5	2.5
Medium	2.5 to 3	2.5 to 3
Fine	2.5 to 3	3 to 3.5

* Refer to the "SOIL TEXTURE" section of this label to determine the corresponding textural group for the soil to be treated.

** Use the higher rate in the recommended range in areas of heavy weed infestations. Use a minimum of 3 quarts per acre on coarse textured soils for control of yellow nutsedge or red rice. Rates may be increased to a maximum of 4 quarts on any soil type when heavy infestations of yellow nutsedge or red rice are present.

On coarse soils with 10% or more organic matter use 4 quarts of LASSO herbicide per acre.

LASSO® plus AMIBEN (TANK MIXTURE)

Apply this tank mixture in water or sprayable fluid fertilizer solution for control of the annual grasses and broadleaf weeds listed in the "WEEDS CONTROLLED" section of this label.

DO NOT use this tank mixture on sand or loamy sand soils as crop injury may result.

APPROVED APPLICATION SYSTEMS

Ground—Broadcast boom; banded
Aerial—Fixed wing; helicopter

Reference: The "APPLICATION SYSTEMS" section of this label provides detailed information on the application of this tank mixture using the system selected.

APPROVED APPLICATION METHODS

Preplant Incorporated—Apply this tank mixture within 7 days prior to planting and shallowly incorporate or Surface Blend into the upper 1 to 2 inches of soil.

Preemergence Surface—Apply after planting, before crop and weed emergence and within 5 days after last preplant tillage operation.

Reference: The "APPLICATION TIMING AND METHODS" section of this label provides detailed information and procedures for the application timing and method selected.

RECOMMENDED RATES:

Preplant Incorporated—Apply 2.5 to 3 quarts of LASSO plus 4 quarts of Amiben per acre.

NOTE: Use the 3 quart per acre rate of LASSO in areas of heavy weed infestation. Use 3 to 4 quarts of LASSO when black or hairy nightshade, yellow nutsedge or red rice is present.

Preemergence Surface—Apply 2 quarts LASSO plus 4 quarts Amiben per acre.

NOTE: Use a minimum of 2.5 quarts of LASSO on coarse soils and 3 to 4 quarts in this tank mixture on medium or fine textured soils to control black or hairy nightshade.

LASSO plus CANOPY (TANK MIXTURE)

Apply this tank mixture in water or sprayable fluid fertilizer solution for control of annual grasses and broadleaf weeds in the "WEEDS CONTROLLED" section of this label.

NOTE: Observe all precautions and limitations on the LASSO and Canopy labels including precautions on soil pH, sensitive soybean varieties, minimum recropping interval and rotational guidelines.

DO NOT plant rice or peanuts within 18 months after application of this tank mixture.

APPROVED APPLICATION SYSTEM

Ground—Broadcast boom

Reference: The "APPLICATION SYSTEMS" section of the label provides detailed information on the proper application of this tank mixture using the system selected.

APPROVED APPLICATION METHODS

Preplant Incorporated—Apply this tank mixture within 7 days prior to planting and shallowly incorporate or Surface Blend into the upper 1 to 2 inches of soil. Preplant incorporation may cause slight soybean injury early, but soybeans should recover with no subsequent effect on growth or yield.

Preemergence Surface—Apply this tank mixture after planting, before crop and weed emergence and within 5 days after the last preplant tillage operation.

Reference: The "APPLICATION TIMING AND METHODS" section of this label provides detailed information and procedures for application timing and method selected.

RECOMMENDED RATES: Refer to the following tables for the recommended broadcast treatment rates for this tank mixture. Applications which are not consistent with recommendations in this label may result in unsatisfactory weed control or crop injury.

Surface Application Rates

Soil Textural Group*	Broadcast Rates Per Acre**		
	½ to 3% Organic Matter	3% or More Organic Matter	
	LASSO® (quarts)	LASSO (quarts)	Canopy (ounces)
Coarse—sandy loam only	1.75 to 2.25	1.75 to 2.25	6
Medium	1.75 to 2.25	2 to 2.5	8
Fine	2 to 2.5	2.25 to 2.75	10

* Refer to the "SOIL TEXTURE" section of this label to determine the corresponding soil textural group for the soil to be treated. DO NOT APPLY TO SOIL WITH LESS THAN ½% organic matter.

** Use the higher rate of LASSO in the recommended ranges in areas of heavy grass infestations, heavy crop residue or when broadleaf signalgrass is present.

Preplant Incorporated Application Rate

Soil Textural Group	Broadcast Rates Per Acre**		
	½ to 3% Organic Matter	3% or More Organic Matter	
	LASSO® (quarts)	LASSO (quarts)	Canopy (ounces)
Coarse—sandy loam only	2 to 2.5	2 to 2.5	6
Medium	2 to 2.5	2.25 to 2.75	8
Fine	2.25 to 2.75	2.5 to 3	10

* Refers to the "SOIL TEXTURE" section of this label to determine the corresponding soil textural group of the soil to be tested. DO NOT APPLY TO SOIL WITH LESS THAN ½% organic matter.

** Use the higher rate of LASSO in the recommended ranges in areas of heavy grass infestations, heavy crop residue or when broadleaf signalgrass is present. See the "WEEDS CONTROLLED" section of this label for specific rate recommendations on seedling johnsongrass, red rice and yellow nutsedge.

LASSO® plus COMMAND (TANK MIXTURE)

Apply this tank mixture in water or sprayable fluid fertilizer solution for control of yellow nutsedge and the annual grasses and broadleaf weeds listed in the "WEEDS CONTROLLED" section of this label.

NOTE: Observe all precautions and limitations on the LASSO and Command labels before use of this tank mixture.

APPROVED APPLICATION SYSTEMS

Ground—Broadcast boom; banded

Reference: The "APPLICATION SYSTEMS" section of this label provides detailed information on the proper application of this tank mixture using the system selected.

APPROVED APPLICATION METHODS

Preplant Incorporated—Apply this tank mixture within 7 days prior to planting and shallowly incorporate or Surface Blend into the upper 1 to 2 inches of soil.

Reference: The "APPLICATION TIMING AND METHODS" section of this label provides detailed information and procedures for application timing and method selected.

RECOMMENDED RATES: Apply 2 to 2.5 quarts of LASSO plus 1 pint of Command 4E per acre (⅔ pint of Command 6E). In areas of heavy pigweed and lambsquarter infestations, use 2.5 to 3 quarts of LASSO per acre in the tank mixture.

LASSO® plus GEMINI (TANK MIXTURE)

Apply this tank mixture in water or sprayable fluid fertilizer solution for control of annual grasses and broadleaf weeds in the "WEEDS CONTROLLED" section of this label.

Note: Observe all precautions and limitations on the LASSO and Gemini labels including precautions on soil pH, minimum recropping interval and rotational guidelines.

DO NOT plant rice or peanuts within 18 months after application of this tank mixture.

APPROVED APPLICATION SYSTEMS

Ground—Broadcast boom; banded

Reference: The "APPLICATION SYSTEMS" section of this label provides detailed information on the proper application of this tank mixture using the system selected.

APPROVED APPLICATION METHODS

Preemergence Surface—Apply this tank mixture after planting, before crop and weed emergence and within 5 days after last preplant tillage operation.

Reference: The "APPLICATION TIMING AND METHODS" section of this label provides detailed information and procedures for application timing and method selected.

RECOMMENDED RATES: Refer to the following tables for the recommended broadcast treatment rates for this tank mixture. Applications which are not consistent with recommendations in this label may result in unsatisfactory weed control or crop injury.

Surface Application Rates

Soil Textural Group*	Broadcast Rate Per Acre**		
	½ to 3% Organic Matter	3% or More Organic Matter	
	LASSO® (quarts)	LASSO (quarts)	Gemini (ounces)
Coarse—sandy loam only	1.75 to 2.25	1.75 to 2.25	12 to 16
Medium	1.75 to 2.25	2 to 2.5	16 to 20
Fine	2 to 2.5	2.25 to 2.75	20 to 24

* Refer to the "SOIL TEXTURE" section of this label to determine the corresponding soil textural group for the soil to be treated.

** Use the higher rate of LASSO in the recommended ranges in areas of heavy grass infestation, heavy crop residue or when broadleaf signalgrass is present. Use the higher rate of Gemini in the recommended ranges on soils with higher organic matter or heavy pressure from large deep germinating weed seeds.

LASSO® plus LEXONE
or
LASSO plus SENCOR
(TANK MIXTURE)

Apply this tank mixture in water or sprayable fluid fertilizer solution for control of yellow nutsedge and the annual grasses and broadleaf weeds listed in the "WEEDS CONTROLLED" section of this label.

NOTE: DO NOT use on muck soils or on alkaline soils with greater than 7.4 pH. Crop injury may result if atrazine was applied the year preceding use of this tank mixture.

DO NOT REPLANT CROPS OTHER THAN SOYBEANS FOR 120 DAYS AFTER APPLICATION OF THESE TANK MIXTURES.

APPROVED APPLICATION SYSTEMS

Ground—Broadcast boom; banded
Aerial—Fixed wing; helicopter

Dry Bulk Fertilizer Impregnation

Reference: The "APPLICATION SYSTEMS" section of this label provides detailed information on the application of this tank mixture using the system selected.

APPROVED APPLICATION METHODS

Preplant Incorporated—Apply this tank mixture within 7 days prior to planting and shallowly incorporate or Surface Blend into the upper 1 to 2 inches of soil. Preplant incorporation may cause slight soybean injury early, but soybeans should recover with no subsequent effect on growth or yield.

Preemergence Surface—Apply this tank mixture after planting, before crop and weed emergence and within 5 days after last preplant tillage operation.

Reference: The "APPLICATION TIMING AND METHODS" section of this label provides detailed information and procedures for application timing and method selected.

RECOMMENDED RATES: Refer to the following tables for the recommended broadcast treatment rates for this tank mixture. Applications which are not consistent with recommendations in this label may result in unsatisfactory weed control or crop injury.

Surface Application Rates

Soil Textural Group	Broadcast Rate Per Acre			
	Less Than 3% Organic Matter		3% or More Organic Matter	
	LASSO® + (Quarts)	Lexone 4L** or Sencor 4L** (Pint)	LASSO + (Quarts)	Lexone 4L** or Sencor 4L** (Pint)
Coarse—sandy loam only	1.75 to 2.25	0.5	1.75 to 2.25	0.75
Medium	1.75 to 2.25	0.5 to 0.75	2 to 2.5	0.75
Fine	2 to 2.5	0.75 to 1	2.25 to 2.75	1

* Refer to the "SOIL TEXTURE" section of this label to determine the corresponding soil textural group for the soil to be treated.
** Use the rates listed in this label when using Lexone 4L or Sencor 4F. Use the equivalent rates when using the 50WP or DF formulations of these products. One pint of Lexone 4L or Sencor 4F equals one pound of the corresponding 50WP formulation or ⅔ pound of the corresponding DF formulation.

NOTE: DO NOT use on sand or loamy sand soil. Use 1.5 pints of Lexone 4L or Sencor 4F per acre in this tank mixture when applying on heavy clay soil of the Mississippi Delta. Use a minimum of 2.5 quarts of LASSO on coarse soils and 3 to 4 quarts in this tank mixture on medium or fine textured soils to control black or hairy nightshade.

Preplant Incorporated Application Rates

Soil Textural Group*	Broadcast Rate Per Acre (Quarts)***			
	Less Than 3% Organic Matter		3% or More Organic Matter	
	LASSO® (quarts)	Lexone 4L** or Sencor 4F** (pint)	LASSO (quarts)	Lexone 4L** or Sencor 4F** (pint)
Coarse—sandy loam only	2 to 2.5	0.5	2 to 2.5	0.75
Medium	2 to 2.5	0.5 to 0.75	2.25 to 2.75	0.75
Fine	2.25 to 2.75	0.75 to 1	2.5 to 3	1

* Refer to the "SOIL TEXTURE" section of this label to determine the corresponding soil textural group for the soil to be treated.
** Use the rates listed in this label when using Lexone 4L or Sencor 4F. Use the equivalent rates when using the 50WP or DF formulations of these products. One pint of Lexone 4L or Sencor 4F equals one pound of the corresponding 50WP formulation or ⅔ pound of the corresponding DF formulation.
*** Use the 3 quart per acre rate of LASSO in areas of heavy weed infestation. Use 3 to 4 quarts of LASSO when black or hairy nightshade, yellow nutsedge or red rice is present.

NOTE: Do not use on coarse textured soils. Do not use on medium or fine textured soils with less than 1% organic matter. These mixtures, preplant incorporated, may cause slight soybean injury early, but the soybeans should recover and subsequent growth or yield should not be affected.

LASSO® plus AMIBEN plus LEXONE or
SENCOR (TANK MIXTURES)

Apply this tank mixture in water solution for control of the annual grasses and broadleaf weeds listed in the "WEEDS CONTROLLED" section of this label.

NOTE: DO NOT use on sand, loamy sand, or on soils with less than 1% organic matter. DO NOT use the foliage from soybeans treated with LASSO plus Amiben plus Lexone or Sencor for feed or forage.

APPROVED APPLICATION SYSTEMS

Ground—Broadcast boom

Aerial—Fixed wing; helicopter

Reference: The "APPLICATION SYSTEMS" section of this label provides detailed information on the proper application of this tank mixture using the system selected.

APPROVED APPLICATION METHODS

Preplant Incorporation—Apply this tank mixture within 7 days prior to planting and shallowly incorporate or Surface Blend into the upper 1 to 2 inches of soil. Preplant incorporation may cause slight soybean injury early but soybeans should recover with no subsequent effect on growth or yield.

Preemergence Surface—Apply this tank mixture after planting, before crop and weed emergence and within 5 days after last preplant tillage operation.

Reference: The "APPLICATION TIMING AND METHODS" section of this label provides detailed information and procedures for application timing methods selected.

RECOMMENDED RATES: Refer to the following tables for recommended broadcast treatment rates. Applications which are not consistent with recommendations in this label may result in unsatisfactory weed control or crop injury.

Soil Textural Group*	BROADCAST RATE PER ACRE		
	Less Than 3% Organic Matter		
	LASSO® (quarts)	Amiben (quarts)	Lexone 4L** Sencor 4F (pint)
Coarse—sandy loam only	1.75 to 2.25	3	0.5
Medium	1.75 to 2.25	3 to 4	0.5
Fine	2 to 2.5	4 to 5	0.75

Soil Textural Group*	Broadcast Rate Per Acre		
	3% or More Organic Matter		
	LASSO® (quarts)	Amiben (quarts)	Lexone 4L** Sencor 4F (pint)
Coarse—sandy loam only	1.75 to 2.25	3 to 4	0.75
Medium	2 to 2.5	4 to 5	0.5 to 0.75
Fine	2.25 to 2.75	4 to 6	0.75 to 1

* Refer to the "SOIL TEXTURE" section of this label to determine the corresponding soil textural group for the soil to be treated. On alkaline (calcareous) soils with a pH of 7.4 or greater, apply Lexone 4L or Sencor 4F at 0.5 pint/A on medium soils and 0.5 to 0.75 pint or corresponding equivalent of WP or DF formulations on fine soils. Use the higher rates of LASSO in the recommended ranges in areas of heavy weed infestations and for preplant incorporation applications. Use a minimum of 2.5 quarts of LASSO on coarse soils and 3 to 4 quarts in this tank mixture on medium or fine soils, when black or hairy nightshade are present. Use 3 to 4 quarts of LASSO per acre preplant incorporated for control of yellow nutsedge.
** Use rates listed in this label when using Lexone 4L or Sencor 4F. Use equivalent rates when using the 50WP or DF formulations of these products. One pint of Lexone 4L or Sencor 4F equals 1 pound of the corresponding 50WP formulation or ⅔ pound of the corresponding DF formulation.

LASSO® plus Linuron
(TANK MIXTURE)

Apply this tank mixture in water or sprayable fluid fertilizer solution for control of the annual grasses and broadleaf weeds listed in the "WEEDS CONTROLLED" section of this label.

NOTE: Plant seed at least 1¾ inches deep and DO NOT use on sand or loamy sand or on soils with less than 1% organic matter as crop injury may occur. DO NOT spray over the top of emerged soybeans.

APPROVED APPLICATION SYSTEMS

Ground—Broadcast boom; banded

Reference: The "APPLICATION SYSTEMS" section of this label provides detailed information on the proper application of this tank mixture using the system selected.

APPROVED APPLICATION METHODS

Preemergence Surface—Apply after planting, before crop and weed emergence and within 5 days after last preplant tillage operation.
Reference: The "APPLICATION TIMING AND METHODS" section of this label provides detailed information and procedures for application timing and method selected.

RECOMMENDED RATES: Refer to the following table for the recommended broadcast treatment rates for this tank mixture. Applications which are not consistent with recommendations in this label may result in unsatisfactory weed control or crop injury.

Surface Application Rates

Soil Textural Group*	Broadcast Rate Per Acre (Quarts)			
	Less Than 3% Organic Matter		3% or More Organic Matter	
	LASSO® +	linuron** 4L***	LASSO® +	linuron** 4L***
Coarse—sandy loam only	1.75 to 2.25	0.5	1.75 to 2.25	0.5 to 0.75
Medium	1.75 to 2.25	0.5 to 0.75	2 to 2.5	0.75 to 1
Fine	2 to 2.5	1	2.25 to 2.75	1.25 to 1.5

* Refer to the "SOIL TEXTURE" section of this label to determine the corresponding soil textural group for the soil to be treated.
** Use the rates listed on this label when using linuron 4L. Use equivalent rates when using linuron 50WP or 50 percent dry flowable formulations. One quart of linuron 4L equals 2 pounds of linuron 50WP or 2 pounds of linuron 50 percent dry flowable.

LASSO® plus LOROX PLUS (TANK MIXTURE)

Apply this tank mixture in water or sprayable fluid fertilizer solution for control of annual grasses and broadleaf weeds listed in the "WEEDS CONTROLLED" section of this label.

NOTE: Observe all precautions and limitations on the labels for LASSO and Lorox Plus, including precautions on soil pH, minimum recropping interval and rotation guidelines.

APPROVED APPLICATION SYSTEMS

Ground—Broadcast boom; banded.

Reference: The "APPLICATION SYSTEM" section of this label provides detailed information on the proper application of this tank mixture using the system selected.

APPROVED APPLICATION METHODS

Preemergence Surface—Apply this tank mixture after planting, before crop and weed emergence and within 5 days after last preplant tillage operation.

Reference: The "APPLICATION TIMING AND METHODS" section of this label provides detailed information and procedures for application timing and method selected.

RECOMMENDED RATES: Refer to the following table for the recommended broadcast treatment rates for this tank mixture. Application which are not consistent with recommendations in this label may result in unsatisfactory weed control or crop injury.

Surface Application Rates

Soil Textural Group*	Broadcast Rate Per Acre**		
	½ to 3% Organic Matter	3% or More Organic Matter	
	LASSO® (quarts)	LASSO (quarts)	Lorox Plus (ounces)
Coarse—sandy loam only	1.75 to 2.25	1.75 to 2.25	12 to 14
Medium	1.75 to 2.25	2 to 2.5	14 to 16
Fine	2 to 2.5	2.25 to 2.75	16 to 18

* Refer to the "SOIL TEXTURE" section of this label to determine the corresponding soil textural group for the soil to be treated.
** Use the higher rate of LASSO in the recommended ranges in areas of heavy grass infestations, heavy crop residue or when broadleaf signalgrass is present. Use the higher rate of Lorox Plus in the recommended ranges of soils with higher organic matter or heavy pressure from large deep germinating weed seeds.

LASSO® plus PREVIEW (TANK MIXTURE)

Apply this tank mixture in water or sprayable fluid fertilizer solution for control of annual grasses and broadleaf weeds listed in the "WEEDS CONTROLLED" section of this label.

NOTE: Observe all precautions and limitations on the labels for LASSO and Preview, including precautions on soil pH, sensitive soybean varieties, minimum recropping interval and rotational guidelines.

APPROVED APPLICATION SYSTEMS

Ground—Broadcast boom

Reference: The "APPLICATION SYSTEMS" section of the label provides detailed information on the proper application of this tank mixture using the system selected.

APPROVED APPLICATION METHODS

Preplant Incorporated—Apply this tank mixture within 7 days prior to planting and shallowly incorporate or Surface Blend into the upper 1 to 2 inches of soil. Preplant incorporation may cause slight soybean injury early, but soybeans should recover with no subsequent effect on growth or yield.

Preemergence Surface—Apply this tank mixture after planting, before crop and weed emergence and with 5 days after last preplant tillage operation.

Reference: The "APPLICATION TIMING AND METHODS" section of this label provides detailed information and procedures for application timing and method selected.

RECOMMENDED RATES: Refer to the following tables for the recommended broadcast treatment rates for this tank mixture. Applications which are not consistent with recommendations in this label may result in unsatisfactory weed control or crop injury.

Surface Application Rates

Soil Textural Group*	Broadcast Rate Per Acre			
	½ to 3% Organic Matter		3% or More Organic Matter	
	LASSO® (Quarts)	Preview (Ounces)	LASSO (Quarts)	Preview (Ounces)
Coarse—Sandy loam only	1.75 to 2.25	6	1.75 to 2.25	7
Medium	1.75 to 2.25	7	2 to 2.5	8
Fine	2 to 2.5	8	2.25 to 2.75	9

* Refer to the "SOIL TEXTURE" section of this label to determine the corresponding soil textural group for the soil to be treated. DO NOT APPLY TO SOIL WITH LESS THAN ½% organic matter.
** Use the higher rate of LASSO in the recommended ranges in areas of heavy weed infestations, heavy crop residue or when broadleaf signalgrass is present.

Preplant Incorporated Application Rates

Soil Textural Group*	Broadcast Rate Per Acre**			
	½ to 3% Organic Matter		3% or More Organic Matter	
	LASSO® (Quarts)	Preview (Ounces)	LASSO (Quarts)	Preview (Ounces)
Coarse—sandy loam only	2 to 2.5	6	2 to 2.5	7
Medium	2 to 2.5	7	2.25 to 2.75	8
Fine	2.25 to 2.75	8	2.5 to 3	9

* Refer to the "SOIL TEXTURE" section of this label to determine the corresponding soil textural group for the soil to be treated. DO NOT APPLY TO SOIL WITH LESS THAN ½% organic matter.
** Use the higher rate of LASSO in the recommended ranges in areas of heavy grass infestations, heavy crop residue or when broadleaf signalgrass is present. See the "WEEDS CONTROLLED" section of this label for specific rate recommendations on seedling johnsongrass, red rice and yellow nutsedge.

LASSO® plus SCEPTER (TANK MIXTURE)

Apply this tank mixture in water or sprayable fluid fertilizer solution for control of yellow nutsedge and the annual grasses and broadleaf weeds listed in the "WEEDS CONTROLLED" section of this label.

Note: Observe all precautions and limitations on the LASSO and Scepter labels before use of this tank mixture including precautions on minimum recropping interval and rotational guidelines.
DO NOT apply this tank mixture in sprayable fluid fertilizer.

APPROVED APPLICATION SYSTEMS

Ground—Broadcast boom; banded

Aerial — Fixed wing; helicopter

Reference: The "APPLICATION SYSTEMS" section of this label provides detailed information on the proper application of this tank mixture using the system selected.

APPROVED APPLICATION METHODS

Preplant Incorporated—Apply this tank mixture within 7 days prior to planting and shallowly incorporate or Surface Blend into the upper 1 to 2 inches of soil.

Preemergence Surface—Apply this tank mixture after planting, before crop and weed emergence and within 5 days after last preplant tillage operation.

Reference: The "APPLICATION TIMING AND METHODS" section of this label provides detailed information and procedures for application timing and method selected.

RECOMMENDED RATES: Refer to the following table for the recommended broadcast treatment rates for this tank mixture. Applications which are not consistent with recommendations in this label may result in unsatisfactory weed control or crop injury. See the "WEEDS CONTROLLED" section of this label for recommended rates and application methods for control of certain weeds.

Soil Textural Group*	Broadcast Rate Per Acre**		
	Less than 3% organic matter	3% or more organic matter	
	LASSO® (Quarts)	LASSO (Quarts)	Scepter*** (Pints)
Coarse	1.5 to 2	1.5 to 2	1/2 to 2/3
Medium	1.5 to 2	1.75 to 2.25	1/2 to 2/3
Fine	1.75 to 2.25	2 to 2.5	1/2 to 2/3

* Refer to the "SOIL TEXTURE" section of this label to determine the corresponding soil textural group for the soil to be treated.

** Use the higher rate of LASSO in the recommended ranges in areas of heavy grass infestations, heavy crop residue or when the following weeds are present: fall panicum, barnyardgrass, broadleaf signalgrass, seedling johnsongrass, red rice or yellow nutsedge.

*** Use the higher rate of Scepter in the recommended range when red rice, sicklepod, beggarweed or velvetleaf are present.

For improved control of heavy velvetleaf infestations, add 1/2 pint of Command 4E herbicide per acre to this tank mixture and preplant incorporate. Observe all precautions and limitations on the Scepter and Command labels before use of this tank mixture.

LASSO® plus TREFLAN
(TANK MIXTURE)

Apply this tank mixture in water or sprayable fluid fertilizer solution for control of the annual grasses and broadleaf weeds listed in the "WEEDS CONTROLLED" section of this label.

APPROVED APPLICATION SYSTEMS

Ground—Broadcast boom; banded

Reference: The "APPLICATION SYSTEMS" section of the label provides detailed information on the proper application of this tank mixture using the system selected.

APPROVED APPLICATION METHODS

Preplant Incorporated—Apply this tank mixture within 7 days prior to planting and shallowly incorporate or Surface Blend into the upper 1 to 2 inches of soil within eight (8) hours after application.

Reference: The "APPLICATION TIMING AND METHODS" section of this label provides detailed information and procedures for the application timing and method selected.

RECOMMENDED RATES:

Preplant Incorporated—Apply 2.5 to 3 quarts of LASSO plus 0.5 quarts of Treflan per acre. Applications which are not consistent with the recommendation in this label may result in unsatisfactory weed control or crop injury.

Conservation or Minimum
Tillage Systems
TANK MIXTURES

SOYBEANS ONLY

When applied as directed under the conditions described, these tank mixtures control many emerged annual weeds, suppress many emerged perennial weeds and give preemergence control of many annual grasses and weeds when soybeans will be planted directly into a cover crop, stale seedbed, or in previous crop residues such as wheat stubble.

These tank mixtures will not control regrowth from perennial weeds.

DO NOT APPLY BY AIR.

For mixing instructions, see the "MIXING, SPRAYING AND HANDLING INSTRUCTIONS" section of this label.

LASSO® plus ROUNDUP®
or
LASSO plus GRAMOXONE SUPER
or
LASSO plus CANOPY plus ROUNDUP
or
LASSO plus CANOPY plus GRAMOXONE SUPER
or
LASSO plus GEMINI plus ROUNDUP
or
LASSO plus GEMINI plus GRAMOXONE SUPER
or
LASSO plus LINURON plus ROUNDUP
or
LASSO plus LINURON plus GRAMOXONE SUPER
or
LASSO plus LOROX PLUS plus ROUNDUP
or
LASSO plus LOROX PLUS plus GRAMOXONE SUPER
or
LASSO plus PREVIEW plus ROUNDUP
or
LASSO plus PREVIEW plus GRAMOXONE SUPER
or
LASSO plus SCEPTER plus ROUNDUP
or
LASSO plus SCEPTER plus GRAMOXONE SUPER
or
LASSO plus LEXONE plus ROUNDUP
or
LASSO plus LEXONE plus GRAMOXONE SUPER
or
LASSO plus SENCOR plus ROUNDUP
or
LASSO plus SENCOR plus GRAMOXONE SUPER

Apply these tank mixtures with Roundup in 10 to 40 gallons of water, or the tank mixtures with Gramoxone Super in 20 to 60 gallons of water spray solution per acre immediately before, during or after planting, but BEFORE CROP EMERGENCE. As density of stubble, crop residue or weeds increases, spray gallonage and rate should be increased within the recommended ranges to insure complete coverage. In the absence of emerged vegetation delete the Roundup or Gramoxone Super portion of these mixes.

CONTROL OR SUPPRESSION OF EMERGED WEEDS

ROUNDUP®

Annual Weeds—Apply 1 quart of Roundup per acre in these tank mixtures if weeds are less than 6 inches tall, and use 1.5 quarts of Roundup if weeds are more than 6 inches tall for control of many emerged annual grasses and broadleaf weeds. For emerged annual weeds controlled, see the "WEEDS CONTROLLED" section of the label for Roundup.

Perennial Weeds—At normal application dates in minimum tillage systems, perennial weeds may not be at the proper stage of growth for control. See the "GENERAL INFORMATION" section of the label for Roundup for the proper stage of growth for perennial weeds. Use of 2 to 4 quarts of Roundup per acre in this tank mixture under these conditions provides top kill and reduces competition from many emerged perennial grasses and broadleaf weeds. For perennial weeds controlled, see the "WEEDS CONTROLLED" section of the label for Roundup.

To obtain control of perennial weeds, follow recommendations on the label for Roundup for stage of growth and rate of application.

USE OF THIS TANK MIXTURE FOR BERMUDAGRASS OR JOHNSONGRASS CONTROL IS NOT RECOMMENDED.

NOTE: When using these tank mixtures, do not exceed 4 quarts of Roundup herbicide per acre.

GRAMOXONE SUPER

When used as directed, Gramoxone Super in tank mixture with labeled LASSO tank mixtures controls many emerged annual weeds and suppresses many emerged perennial weeds.

Broadcast Treatment—Apply 1.5 to 2.5 pints of Gramoxone Super (0.25 to 0.5 lb active) in this tank mixture immediately before, during or after planting but BEFORE CROP EMERGENCE. Use the 2.5 pint rate if any weeds are 4 to 6 inches tall. This mixture may not control weeds taller than 6 inches. As density of stubble, crop residue or weeds increases, spray gallonage should be increased within the recommended range for complete coverage. Add a nonionic spreader surfactant (approved for use in crops) containing at least 75 percent surface active agent at 8 fluid ounces per 100 gallons of diluted spray. REFER TO THE LABEL FOR GRAMOXONE SUPER FOR PRECAUTIONARY STATEMENTS.

PREEMERGENCE WEED CONTROL

LASSO®

For weeds controlled preemergence, see the "WEEDS CONTROLLED" section of this label.

See the following table for recommended rates of LASSO on various soil types.

Soil Textural Group*	Broadcast Rate Per Acre (Quarts)
	LASSO®
Coarse	2.5
Medium	3
Fine	3 to 3.5

* Refer to the "SOIL TEXTURE" section of the label to determine the corresponding soil textural group for the soil to be treated.

Use the higher rate in the recommended ranges in areas of heavy grass infestation, or when organic matter content is 3 percent or more.

LASSO® plus CANOPY

For weeds controlled preemergence see the "WEEDS CONTROLLED" section of this label for LASSO and LASSO plus Canopy.

NOTE: Observe all precautions and limitations on the LASSO and Canopy labels including precautions on soil pH, sensitive soybean varieties and rotational crop restrictions.

See the following table for recommended rates of LASSO and Canopy in this tank mixture.

Soil Textural Group*	Broadcast Rate Per Acre**	
	LASSO® (Quarts) +	CANOPY (Ounces)
Course—sandy loam only	1.75 to 2.25	6
Medium	2 to 2.5	8
Fine	2.5 to 3	10

* Refer to the "SOIL TEXTURE" section of this label to determine the corresponding soil textural group for the soil to be treated.

** Use the higher rate of LASSO and Canopy in areas of heavy weed infestation.

LASSO® plus GEMINI

For weed controlled preemergence see the "WEEDS CONTROLLED" section of this label for LASSO plus Gemini.

Note: Observe all precautions and limitations on the LASSO and Gemini labels including precautions on soil pH and rotational crop restrictions.

See the following table for recommended rates of LASSO and Gemini in this tank mixture.

Soil Textural Group*	Broadcast Rate Per Acre**	
	LASSO® (quarts) +	Gemini (ounces)
Coarse—sandy loam only	1.75 to 2.25	12 to 16
Medium	2 to 2.5	16 to 20
Fine	2.5 to 3	20 to 24

* Refer to the "SOIL TEXTURE" section of this label to determine the corresponding soil textural group for the soil to be treated.

** Use the higher rate of LASSO and Gemini in areas of heavy weed infestation.

LASSO® plus LEXONE
or
LASSO plus SENCOR

For weeds controlled preemergence, see the "WEEDS CONTROLLED" section of this label for LASSO and LASSO plus Lexone 4L or Sencor 4F.

See the following table for recommended rates of LASSO plus Lexone 4L or LASSO plus Sencor 4F in this tank mixture on various soil types.

Soil Textural Group*	Broadcast Rate Per Acre**	
	LASSO® (quarts) +	LEXONE 4L** or SENCOR 4F** (pints)
Coarse—sandy loam only	1.75 to 2.25	0.5 to 0.75
Medium	2 to 2.5	0.75 to 1
Fine	2.5 to 3	1 to 1.5***

* Refer to the "SOIL TEXTURE" section of this label to determine the corresponding soil textural group for the soil to be treated.

** Use the rates listed in this label when using Lexone 4L or Sencor 4F. Use the equivalent rates when using the 50WP or DF formulations of these products. One pint of Lexone 4L or Sencor 4F equals one pound of the corresponding 50WP formulation or ⅔ pound of the corresponding DF formulation.

*** On the heavy clay soil of the Mississippi Delta, use 1.5 pints of Lexone 4L or Sencor 4F per acre.

Use the higher rate of LASSO in the recommended ranges in areas of heavy grass infestations or when fall panicum or crabgrass will be present.

Use the higher rate of Lexone or Sencor in the recommended ranges on soils with greater than 3% organic matter.

DO NOT use this mixture on sand or loamy sand soils as crop injury from Lexone or Sencor may occur.

DO NOT use on muck soils.

DO NOT apply on alkaline soils with a pH greater than 7.4.

Crop injury may occur if atrazine was applied the year before use of this tank mixture.

DO NOT REPLANT CROPS OTHER THAN SOYBEANS FOR 120 DAYS AFTER APPLICATION OF THESE TANK MIXTURES.

LASSO® plus LINURON

For weeds controlled preemergence, see the "WEEDS CONTROLLED" section of this label for LASSO and LASSO plus linuron.

See the following table for recommended rates of LASSO and linuron in this tank mixture.

Soil Textural Group*	Broadcast Rate Per Acre (Quarts)	
	LASSO® +	Linuron 4L**
Coarse—sandy loam only	1.75 to 2.25	0.5 to 0.75
Medium	2 to 2.5	0.75 to 1
Fine	2.5 to 3	1 to 1.5

* Refer to the "SOIL TEXTURE" section of the label to determine the corresponding soil textural group for the soil to be treated.

** Use the rates listed on this label when using linuron 4L. Use equivalent rates when using linuron 50WP or 50% dry flowable formulation. One quart of linuron 4L equals 2 pounds of linuron 50WP or 2 pounds of linuron 50% dry flowable.

Use the higher rate of LASSO in the recommended ranges in areas of heavy grass infestation or when fall panicum or crabgrass will be present.

Use the higher rate of linuron in the recommended ranges on soils with greater than 3% organic matter.

DO NOT use this tank mixture on sand or loamy sand or on soil with less than 1% organic matter as crop injury from linuron may occur.

LASSO® plus LOROX PLUS

For weeds controlled preemergence see the "WEEDS CONTROLLED" section of this label for LASSO and LASSO plus Lorox Plus.

NOTE: Observe all precautions and limitations on the labels for LASSO and Lorox Plus, including precautions on soil pH and rotational crop restrictions.

See the following table for recommended rates of LASSO and Lorox Plus in this tank mixture.

Soil Textural Group*	Broadcast Rate Per Acre**	
	LASSO® (Quarts) +	Lorox Plus (Ounces)
Coarse—sandy loam only	1.75 to 2.25	12 to 14
Medium	2 to 2.5	14 to 16
Fine	2.5 to 3	16 to 18

* Refer to the "SOIL TEXTURE" section of this label to determine the corresponding soil textural group for the soil to be treated.

** Use the higher rate of LASSO and Lorox Plus in areas of heavy weed infestation.

LASSO® plus PREVIEW

For weed controlled preemergence see the "WEEDS CONTROLLED" section of this label for LASSO and LASSO plus Preview.

NOTE: Observe all precautions and limitations on the labels for LASSO and Preview, including precautions on soil pH, sensitive soybean varieties and rotational crop restrictions.

See the following table for recommended rates of LASSO and Preview in this tank mixture.

Soil Textural Group*	Broadcast Rate Per Acre**	
	LASSO® (Quarts) +	Preview (Ounces)
Coarse—sandy loam only	1.75 to 2.25	6
Medium	2 to 2.5	7
Fine	2.5 to 3	8

* Refer to the "SOIL TEXTURE" section of this label to determine the corresponding soil textural group for the soil to be treated.

** Use the higher rate of LASSO and Preview in areas of heavy weed infestation.

LASSO® plus SCEPTER

For weeds controlled preemergence, see the "WEEDS CONTROLLED" section of this label for LASSO and LASSO plus Scepter.

Note: Observe all precautions and limitations on the LASSO and Scepter labels before use of this tank mixture including minimum recropping interval and rotational guidelines.

See the following table for recommended rates of LASSO and Scepter in this tank mixture.

Soil Textural Group*	Broadcast Rate Per Acre**	
	LASSO® (Quarts)	Scepter (Pints)
Coarse	1.75 to 2.25	⅔
Medium	2 to 2.5	⅔
Fine	2.5 to 3	⅔

* Refer to the "SOIL TEXTURE" section of this label to determine the corresponding textural group for the soil to be treated.

** Use the higher rate of LASSO in areas of heavy weed infestation.

PEANUTS

LASSO®

Apply this product in water solution for control of yellow nutsedge and the annual grasses and broadleaf weeds listed in the "WEEDS CONTROLLED" section of this label.

APPROVED APPLICATION SYSTEMS

Ground—Broadcast boom; banded

Aerial — Fixed wing; helicopter

Center Pivot Irrigation

Reference: The "APPLICATION SYSTEMS" section of this label provides detailed information on the proper application of this product using the system selected.

APPROVED APPLICATION METHODS

Preplant Incorporated*—Apply within 7 days prior to planting and shallowly incorporate or Surface Blend into the upper 1 to 2 inches of soil. For control of Texas panicum, this application should be made as a tank mixture with a labeled rate of Balan EC™, Prowl™ or Sonolan DF™ herbicide.

Postplant Incorporated*—Apply after planting and shallowly incorporate or Surface Blend into the upper 1 to 2 inches of soil prior to peanut germination. Incorporation depth after planting must be kept above the seed to avoid mechanical injury.

Preemergence Surface—Apply after planting, before crop and weed emergence and within 5 days after last preplant tillage operation.

Ground-crack Surface — Apply this product at the ground-crack stage of peanuts.

Reference: The "APPLICATION TIMING AND METHODS" section of this label provides detailed information and procedures for application timing and method selected.

* Incorporation on coarse soils may result in reduced length of weed control due to leaching with rainfall or irrigation. For improved control of yellow nutsedge and labeled broadleaf weeds, a labeled sequential ground-crack herbicide application should be made.

NOTE: Do not use treated forage for animal fodder.

™ Balan and Sonolan are trademarks of Elanco Products Company.

RECOMMENDED RATES: Refer to the following table for recommended broadcast treatment rates for this product. Applications which are not consistent with recommendations in this label may result in unsatisfactory weed control or crop injury.

Application Method	LASSO®* (Quarts)
Preplant or Postplant Incorporated	3 to 4
Preemergence or Ground Crack Surface	3 to 4

* Use the higher rate in the recommended range if heavy weed infestations are present.

DRY BEANS

Dry Beans (All Types)

FOR USE ONLY WEST OF THE MISSISSIPPI RIVER, EXCEPT IN KERN COUNTY, CALIFORNIA.

LASSO®

Apply this product in water solution for control of yellow nutsedge and the annual grasses and broadleaf weeds listed in the "WEEDS CONTROLLED" section of this label.

NOTE: DO NOT APPLY ON DRY BEANS AFTER PLANTING AS CROP INJURY MAY OCCUR. This product may delay crop maturity and/or reduce yield if cold, wet soil conditions occur after planting.

APPROVED APPLICATION SYSTEMS

Ground—Broadcast boom

Reference: The "APPLICATION SYSTEMS" section of this label provides detailed information on the proper application of this product using the system selected.

APPROVED APPLICATION METHODS

Preplant Incorporated—Apply this product within 7 days prior to planting and shallowly incorporate or Surface Blend into the upper 1 to 2 inches of soil.

Reference: The "APPLICATION TIMING AND METHODS" section of the label provides detailed information and procedures for application timing and method selected.

RECOMMENDED RATES: Apply 2.5 to 3 quarts of this product per acre and shallowly incorporate prior to planting. Use the higher rate in areas of heavy weed infestations or when black or hairy nightshade or yellow nutsedge are present. Applications which are not consistent with recommendations in this label may result in unsatisfactory weed control or crop injury.

LASSO® FOLLOWED BY AMIBEN

Dry Beans (All Types)

FOR USE IN MICHIGAN ONLY

Apply Amiben as a sequential preemergence application to LASSO preplant incorporated.

APPROVED APPLICATION SYSTEMS

Ground—Broadcast boom

Reference: The "APPLICATION SYSTEMS" section of this label provides detailed information on the application of these products using the system selected.

APPROVED APPLICATION METHODS

LASSO®

Preplant Incorporated—Apply this product within 7 days prior to planting and shallowly incorporate or Surface Blend into the upper 1 to 2 inches

Reference: The "APPLICATION, TIMING and METHODS" section of this label provides detailed information and procedures for application timing and method selected.

AMIBEN

Preemergence Surface—Apply Amiben after planting and before crop and weed emergence. Read and carefully follow all use instructions and precautionary statements listed on the label.

Reference: The "APPLICATION, TIMING AND METHODS" section of this label provides detailed information and procedures for application timing and method selected.

RECOMMENDED RATES:

LASSO—Apply 2 quarts of this product per acre and shallowly incorporate or Surface Blend prior to planting.

NOTE: DO NOT USE ON SAND OR LOAMY SANDS. DO NOT APPLY ON DRY BEANS AFTER PLANTING AS CROP INJURY MAY OCCUR. This product may delay crop maturity and/or reduce yield if cold, wet soil conditions occur after planting.

AMIBEN—Apply 4 quarters of this product per acre.

NOTE: When LASSO followed by Amiben is used on soils with 6% or more organic matter, weed control may be reduced.

LASSO® plus EPTAM™ (TANK MIXTURE)

Dry Beans (All Types)

FOR USE ONLY IN MICHIGAN AND WEST OF THE MISSISSIPPI RIVER, EXCEPT IN CALIFORNIA

Apply this tank mixture in water solution for control of yellow nutsedge and the annual grasses and broadleaf weeds listed in the "WEEDS CONTROLLED" section of this label.

NOTE: Read the label for Eptam thoroughly for rotational crop restrictions. This tank mixture may delay crop maturity and/or reduce yield if cold, wet soil conditions occur after planting.

APPROVED APPLICATION SYSTEMS

Ground—Broadcast boom

Reference: The "APPLICATION SYSTEMS" section of this label provides information on the application of this product using the system selected.

APPROVED APPLICATION METHODS

Preplant Incorporated—Apply this tank mixture within 7 days prior to planting and shallowly incorporate into the upper 2 inches of soil immediately.

Reference: The "APPLICATION TIMING AND METHODS" section of this label provides detailed information and procedures for application timing and method selected.

RECOMMENDED RATES: Apply 2 to 3 quarts of LASSO plus 2 to 3 pints of Eptam 7-E per acre. Use the higher recommended rate of LASSO for this tank mixture when heavy weed infestations, black or hairy nightshade or yellow nutsedge are present. USE ONLY THE 2 QUART RATE OF LASSO IN MICHIGAN. Applications which are not consistent with recommendations in this label may result in unsatisfactory weed control or crop injury.

™ Eptam is a trademark of the Stauffer Chemical Company.

LASSO® plus TREFLAN (TANK MIXTURE)

Dry Beans (All Types)

FOR USE ONLY WEST OF THE MISSISSIPPI RIVER, EXCEPT IN KERN COUNTY, CALIFORNIA

Apply this tank mixture in water solution for control of yellow nutsedge and the annual grasses and broadleaf weeds listed in the "WEEDS CONTROLLED" section of this label.

NOTE: Read the label for Treflan thoroughly for rotational crop restrictions. This tank mixture may delay crop maturity and/or reduce yield if cold, wet soil conditions occur after planting.

APPROVED APPLICATION SYSTEMS

Ground—Broadcast boom

Reference: The "APPLICATION SYSTEMS" section of this label provides detailed information on the application of this product using the system selected.

APPROVED APPLICATION METHODS

Preplant Incorporated—Apply this tank mixture within 7 days prior to planting and shallowly incorporate into the upper 2 inches of soil within 8 hours after application.

Reference: The "APPLICATION TIMING AND METHODS" section of this label provides detailed information and procedures for application timing and method selected.

RECOMMENDED RATES: Apply 2.5 to 3 quarts of this product plus 1/2 quart of Treflan per acre. Use the higher rate of this product for this tank mixture in areas of heavy weed infestations or when black or hairy nightshade or yellow nutsedge are present. Applications which are not consistent with recommendations in this label may result in unsatisfactory weed control or crop injury.

LASSO®
Red Kidney Bean Types Only
FOR USE IN ILLINOIS, INDIANA AND WISCONSIN ONLY

Apply this product in water solution for control of yellow nutsedge and the annual grasses and broadleaf weeds listed in the "WEEDS CONTROLLED" section of this label.

NOTE: This product may delay crop maturity and/or reduce yield if cold, wet soil conditions occur after planting.

NOTE: Only one application of LASSO herbicide may be made per cropping season.

APPROVED APPLICATION SYSTEMS

Ground—Broadcast boom; banded

Reference: The "APPLICATION SYSTEMS" section of this label provides detailed information on the application of this tank mixture using the system selected.

APPROVED APPLICATION METHODS

Preplant Incorporated—Apply this product within 7 days prior to planting and shallowly incorporate or Surface Blend into the upper 1 to 2 inches of soil.

Preemergence Surface—Apply this product after planting and before crop and weed emergence and within 5 days after last preplant tillage operation.

Reference: The "APPLICATION AND METHODS" section of this label provides detailed information and procedures for application timing and method selected.

RECOMMENDED RATES: Apply 2.5 to 3 quarts of this product per acre. Use the higher recommended rate when heavy weed infestations, black or hairy nightshade or yellow nutsedge are present. Preplant incorporate only for control of yellow nutsedge.

Lima Beans (Green)

FOR USE IN ALL STATES, EXCEPT IN KERN COUNTY, CALIFORNIA

LASSO®

Apply this product in water solution for control of yellow nutsedge and the annual grasses and broadleaf weeds listed in the "WEEDS CONTROLLED" section of this label.

NOTE: This product may delay crop maturity and or reduce yield if cold, wet soil conditions occur after planting.

APPROVED APPLICATION SYSTEMS

Ground—Broadcast boom

Reference: The "APPLICATION SYSTEMS" section of this label provides detailed information on the application of this product using the system selected.

APPROVED APPLICATION METHODS

Preplant Incorporated—Apply this product within 7 days prior to planting and shallowly incorporate or Surface Blend into the upper 1 to 2 inches of soil.

Preemergence Surface—Apply this product after planting and before crop and weed emergence and within 5 days after last preplant tillage operation.

Reference: The "APPLICATION TIMING AND METHODS" section of this label provides detailed information and procedures for application timing and method selected.

RECOMMENDED RATES: Refer to the following table for recommended broadcast treatment rates.

Preplant Incorporated Application Rates

Soil Textural Group*	Broadcast Rate Per Acre (Quarts) LASSO®
Coarse	2
Medium and Fine**	2.5 to 3

*Refer to the "SOIL TEXTURE" section of this label to determine the corresponding soil textural group for the soil to be treated.
**Use the higher rate in the recommended range in areas of heavy weed infestation or when black or hairy nightshade are present.

RECOMMENDED RATES: Refer to the following table for recommended broadcast treatment rates.

Surface Application Rates

Soil Textural Group*	Broadcast Rate Per Acre (Quarts) LASSO®
Coarse	1.5 to 2
Medium and Fine**	2.5 to 3

*Refer to the "SOIL TEXTURE" section of this label to determine the corresponding soil textural group for the soil to be treated.
**Use the higher rate in the recommended range in areas of heavy weed infestation or when black or hairy nightshade are present.

NOTE: Preemergence applications are not recommended for use in Illinois or Wisconsin as crop injury may occur.

MUNG BEANS

LASSO®

FOR USE IN OKLAHOMA ONLY

Apply this product in water solution for control of the annual grasses and broadleaf weeds listed in the "WEEDS CONTROLLED" section of this label.

APPROVED APPLICATION SYSTEMS

Ground—Broadcast boom; banded

Reference: The "APPLICATION SYSTEMS" section of this label provides detailed information on the application of product using the system selected.

APPROVED APPLICATION METHODS

Preemergence Surface—Apply this product after planting, before crop and weed emergence and within 5 days after last preplant tillage operation.

Reference: The "APPLICATION TIMING AND METHODS" section of this label provides detailed information and procedures for application timing and method selected.

RECOMMENDED RATES: Refer to the following table for recommended broadcast rates.

Surface Application Rates

Soil Textural Group	Broadcast Rate Per Acre (Quarts) LASSO®
Coarse*	1.5
Medium and Fine	2

*DO NOT USE ON SAND AS CROP INJURY MAY OCCUR.

SUNFLOWERS

LASSO®

Apply this product in water or sprayable fluid fertilizer solution for control of yellow nutsedge, and the annual grasses and broadleaf weeds listed in the "WEEDS CONTROLLED" section of this label.

APPROVED APPLICATION SYSTEMS

Ground—Broadcast boom; banded

Aerial — Fixed wing; helicopter

Reference: The "APPLICATION SYSTEMS" section of this label provides detailed information on the proper application of this product using the system selected.

APPROVED APPLICATION METHODS

Preplant Incorporated—Apply this product within 7 days prior to planting and shallowly incorporate or Surface Blend into the upper 1 to 2 inches of soil.

Preemergence Surface—Apply this product after planting before crop and weed emergence and within 5 days after preplant tillage operation.

Reference: The "APPLICATION TIMING AND METHODS" section of this label provides detailed information and procedures for the application timing and method selected.

NOTE: DO NOT graze treated area or feed treated forage to livestock following application of this product.

RECOMMENDED RATES: Refer to the following tables for the recommended broadcast treatment rates for this product. Applications which are not consistent with recommendations in this label may result in unsatisfactory weed control or crop injury.

Surface Application Rates

Soil Textural Group*	Broadcast Rate Per Acre (Quarts) LASSO®
Coarse	2
Medium	2.5 to 3
Fine	3 to 4

* Refer to the "SOIL TEXTURE" section of this label to determine the corresponding soil textural group for the soil to be treated.

Preplant Incorporated Application Rates

Soil Textural Group*	Broadcast Rate Per Acre (Quarts) LASSO®
Coarse	2.5
Medium	3 to 3.5
Fine	3.5 to 4

* Refer to the "SOIL TEXTURE" section of this label to determine the corresponding soil textural group for the soil to be treated.

LASSO® plus AMIBEN (TANK MIXTURE)

Apply this tank mixture in water solution for control of annual grasses and broadleaf weeds listed in the "WEEDS CONTROLLED" section of this label.

APPROVED APPLICATION SYSTEMS

Ground—Broadcast boom; banded

Aerial — Fixed wing; helicopter

Reference: The "APPLICATION SYSTEMS" section of this label provides detailed information on the proper application of this tank mixture using the system selected.

APPROVED APPLICATION METHODS

Preplant Incorporated —Apply this tank mixture within 7 days prior to planting and shallowly incorporate or Surface Blend into the upper 1 to 2 inches of soil.

Preemergence Surface—Apply after planting, before crop and weed emergence and within 5 days after last preplant tillage operation.

Reference: The "APPLICATION TIMING AND METHODS" section of this label provides detailed information and procedures for the application timing and method selected.

NOTE: DO NOT graze treated area or feed treated forage to livestock following application of this tank mixture.

RECOMMENDED RATES:

Preplant Incorporated—Apply 2.5 to 3 quarts of LASSO plus 4 quarts of Amiben per acre.

Preemergence Surface—Refer to the following table for the recommended broadcast treatment rates for LASSO plus Amiben. Applications which are not consistent with recommendations in this label may result in unsatisfactory weed control or crop injury.

Surface Application Rates

Soil Textural Group*	Broadcast Rate Per Acre (Quarts) LASSO® +	Amiben
Coarse	2	4
Medium	2 to 2.5	4
Fine	2.5 to 3	4

*Refer to the "SOIL TEXTURE" section of this label to determine the corresponding soil textural group for the soil to be treated.

COTTON

LASSO®

FOR USE ONLY IN TEXAS, OKLAHOMA, LOUISIANA, NEW MEXICO, ARIZONA AND CALIFORNIA

Apply this product in water solution for control of the annual grasses and broadleaf weeds listed in the "WEEDS CONTROLLED" section of this label.

NOTE: DO NOT use this product on sandy or loamy sand soils. This treatment may cause stunting and/or reduce yield if seedling disease, cold weather, excessive moisture or drought stress occurs after application. If the crop must be replanted into treated soil, the seedbed should be tilled lightly prior to replanting. DO NOT plant in furrows (use only in flat planted and cotton planted on top of beds).

NOTE: Only one application of LASSO herbicide may be made per cropping season.

APPROVED APPLICATION SYSTEMS

Ground—Broadcast boom; banded

Reference: The "APPLICATION SYSTEMS" section of this label provides detailed information on the proper application of this product using the system selected.

NOTE: If a band treatment is used then the spray nozzles must be located behind the drags or the covering mechanism of the planter to avoid the hazard of cotton injury and reduced yield.

APPROVED APPLICATION METHODS

Preplant Incorporated (Except Louisiana)—Apply this product within 7 days prior to planting and shallowly incorporate into the upper 1 to 2 inches of soil.

Preemergence Surface—Apply this product after planting, before crop or weed emergence and within 5 days after last preplant tillage operation.

Reference: The "APPLICATION TIMING AND METHODS" section of this label provides detailed information and procedures for application timing and method selected.

RECOMMENDED RATES: Refer to the following tables for the recommended broadcast treatment for this product.

Surface Application Rates*

Soil Textural Group**	Broadcast Rate Per Acre (Quarts) LASSO®
Coarse—sandy loam only	2
Medium	2.5
Fine	3

*FOR LOUISIANA ONLY, apply 1 quart of this product per acre on sandy loam, 1.5 quarts per acre on medium soils, and 2 quarts per acre on fine soils.

NOTE: FOR PLANTING DATES AND AGRONOMIC PRACTICES, REFER TO THE LOCAL USE GUIDE FOR LOUISIANA.

**Refer to the "SOIL TEXTURE" section of this label to determine the corresponding soil textural group for the soil to be treated.

Preplant Incorporated Application Rates*

Soil Textural Group**	Broadcast Rate Per Acre (Quarts) LASSO®
Coarse—sandy loam only	2.5 to 3
Medium	3 to 3.5
Fine	3.5 to 4

*DO NOT USE PREPLANT INCORPORATION IN LOUISIANA.

**Refer to the "SOIL TEXTURE" section of this label to determine the corresponding soil textural group for the soil to be treated.

WOODY ORNAMENTALS

LASSO®

FOR USE ON JUNIPER (JUNIPERUS SPP.) AND YEW (TAXUS SPP.)

Apply this product in water solution for control of the annual grasses and broadleaf weeds listed in the "WEEDS CONTROLLED" section of this label.

NOTE: Topical application or contact of this product with green foilage of the woody ornamental may cause injury and is not recommended. In case green foliage is accidentally sprayed, overhead irrigation within 1 hour after application will minimize injury symptoms (i.e. leaf tip burn, discoloration, distortion, etc.). This product is recommended for this use for application by certified, professional applicators only.

DO NOT USE THIS PRODUCT ON SEEDBEDS, UNROOTED CUTTINGS OR PRIOR TO TRANSPLANTING. DO NOT USE THIS PRODUCT IN GREENHOUSES OR OTHER ENCLOSED STRUCTURES, APPLIED THROUGH IRRIGATION SYSTEMS OR MIXED WITH SPRAYABLE FLUID FERTILIZERS. APPLICATIONS DURING PERIODS WHEN TEMPERATURE EXCEEDS 90°F MAY CAUSE INJURY TO ORNAMENTALS.

APPROVED APPLICATION SYSTEMS

Ground—Broadcast boom; banded

Reference: The "APPLICATION SYSTEMS" section of this label provides detailed information on the application of this tank mixture using the system selected.

APPROVED APPLICATION METHODS

Post-Transplant Directed—Apply this product as a directed spray after transplanting or to established plantings prior to weed emergence.

Reference: The "APPLICATION TIMING AND METHODS" section of this label provides detailed information and procedures for application timing and method selected.

RECOMMENDED RATES: Apply 4 quarts of this product per acre as a broadcast treatment. Applications which are not consistent with the recommendations in this label may result in unsatisfactory weed control or injury to ornamental plantings.

NOTE: Repeat applications of this product may be required for sustained weed control. DO NOT exceed 3 applications per year or retreat within 21 days.

APPENDIX

LIST OF COMMON AND SCIENTIFIC NAME OF ANNUAL GRASSES AND BROADLEAF WEEDS.

ANNUAL GRASSES

Barnyardgrass
Echinochloa crus-galli
Crabgrass
Digitaria ischaemum
Digitaria sanguinalis
Cupgrass, woolly
Eriochloa villosa
Foxtail, giant
Setaria faberi
Foxtail, green, robust purple, robust white
Setaria viridis
Foxtail, yellow
Setaria lutescens
Goosegrass
Eleusine indica
Johnsongrass, seedling
Sorghum halepense
Millet, proso
Panicum miliaceum
Panicum, browntop
Panicum fasciculatum
Panicum, fall
Panicum dichotomiflorum
Panicum, Texas
Panicum texanum
Rice, red
Oryza sativa
Sandbur, Grassbur
Cenchrus incertus
Shattercane, wild cane
Sorghum bicolor
Signalgrass, broadleaf
Brachiaria platyphylla
Sprangletop, red
Leptochloa filiformis
Witchgrass
Panicum capillare

ANNUAL BROADLEAVES

Beggarweed, Florida
Desmodium tortuosum
Carpetweed
Mollugo verticillata
Cocklebur
Xanthium strumarium
Coffeeweed,
Hemp sesbania
Sesbania exaltata
Copperleaf, hophornbeam
Acalypha spp.
Galinsoga
Galinsoga spp.
Groundcherry, annual
Physalis spp.
Groundcherry, cutleaf
Physalis angulata
Jimsonweed
Datura stramonium
Kochia
Kochia scoparia
Lambsquarters
Chenopodium album
Morningglory:
Tall
Ipomea purpurea
Pitted
Ipomea lacunosa
Ivyleaf
Ipomea hederacea
Entireleaf
Ipomea hederacea var. intergriuscula
Smallflower
Jacquemontia tamnifolia
Mustard
Brassica spp.
Nightshade, black
Solanum nigrum
Nightshade, hairy
Solanum sarrachoides
Pigweed, Carlessweed
Amaranthus spp.
Purslane
Portulaca oleracea
Pusley, Florida
Richardia scabra
Ragweed, common
Ambrosia artemisiifolia
Ragweed, giant
Ambrosia trifida
Sicklepod
Cassia obtusifolia
Sida, prickly; Teaweed
Sida spinosa
Smartweed
Polygonum pensylvanicum
Polygonum persicaria
Starbur, bristly
Acanthospermum hispidum
Sunflower, common
Helianthus annuus
Velvetleaf, Buttonweed
Abutilon theophrasti
Waterhemp
Amaranthus tuberculatos

SEDGE

Nutsedge, yellow
Cyperus esculentus

CALIFORNIA

LASSO® herbicide has been approved by the U.S. Environmental Protection Agency for the uses, crops and sites listed on this label. Approval of the items listed below is pending under the State of California registration requirements.

These use conditions, crops and sites may not be treated with this product in California until approval is received:

- Use of LASSO II in dry fertilizer impregnation applications.
- LASSO early postemergence on grain sorghum (milo).
- LASSO plus atrazine mix early postemergence on grain sorghum (milo).
- LASSO plus atrazine early preplant on grain sorghum (milo).
- Tank mixtures with Canopy, Command, Gemini, Lorox Plus, Preview, Sceptor and Treflan on soybeans.
- Tank mixtures with Gramoxone Super on corn or soybeans.
- LASSO plus atrazine early preplant on grain sorghum (milo).
- Control of *Ipomoea hederacea.*
- Control of *Ipomoea hederacea var.*
- Control of *Ipomoea integriuscula.*
- Control of *Ipomoea lacunosa.*
- Control of *Jacquemontia tamnifolia.*
- Control of *Leptochloa filiformus.*
- Control of *Panicum fasciculatum.*
- Aerial application

In case of an emergency involving this product, Call Collect, day or night, (314) 694-4000.

This sample label is current as of January 1, 1989. The product descriptions and recommendations provided in this sample label are for background information only. Always refer to the label on the product container before using Monsanto or any other agrichemical product.

LASSO® II

HERBICIDE

Selective Herbicide For Preemergence Weed Control In Corn, Soybeans, Peanuts, Sunflowers And Woody Ornamentals.

RESTRICTED USE PESTICIDE

due to oncogenicity. For retail sale to and use only by Certified Applicators or persons under their direct supervision and only for those uses covered by the Certified Applicator's certification.

The use of this product may be hazardous to your health. This product contains alachlor, which has been determined to cause tumors in laboratory animals.

In case of an emergency involving this product or for user safety information on this product, call collect, day or night (314) 694-4000.

ACTIVE INGREDIENT:	
Alachlor:2-chloro-2′, 6′-diethyl-N- (methoxymethy)acetanilide. . . .	15%
INERT INGREDIENTS:	85%
TOTAL.	100%

U.S.Patent 3,547,620

EPA Reg. No. 524-296-AA

Complete Directions for Use.

1988-2 843.77-000.36/CG

Read the entire label before using this product.

Use only according to label instructions.

Read "LIMIT OF WARRANTY AND LIABILITY" before buying or using. If terms are not acceptable, return at once unopened.

LIMIT OF WARRANTY AND LIABILITY

This company warrants that this product conforms to the chemical description on the label and is reasonably fit for the purposes set forth in the Complete Directions for Use Label Booklet "Directions" when used in accordance with those directions under the conditions described therein. NO OTHER EXPRESS OR IMPLIED WARRANTY OF FITNESS FOR PARTICULAR PURPOSE OR MERCHANTABILITY IS MADE. This warranty is also subject to the conditions and limitations stated herein.

Buyer and all users shall promptly notify this company of any claims whether based on contract, negligence, strict liability, other tort or otherwise.

Buyer and all users are responsible for all loss or damage from use or handling which results from conditions beyond the control of this Company, including, but not limited to, incompatibility with products other than those set forth in the Directions, unusual weather, weather conditions which are outside the range considered normal at the application site and for the time period when the product is applied, as well as weather conditions which are outside the application ranges set forth in the Directions, application in any manner not explicitly set forth in the Directions, moisture conditions outside the moisture range specified in the Directions, or the presence of products other than those set forth in the Directions in or on the soil, crop or treated vegetation.

THE EXCLUSIVE REMEDY OF THE USER OR BUYER, AND THE LIMIT OF THE LIABILITY OF THIS COMPANY OR ANY OTHER

SELLER FOR ANY AND ALL LOSSES, INJURIES OR DAMAGES RESULTING FROM THE USE OR HANDLING OF THIS PRODUCT (INCLUDING CLAIMS BASED IN CONTRACT, NEGLIGENCE, STRICT LIABILITY, OTHER TORT OR OTHERWISE) SHALL BE THE PURCHASE PRICE PAID BY THE USER OR BUYER FOR THE QUANTITY OF THIS PRODUCT INVOLVED, OR, AT THE ELECTION OF THIS COMPANY OR ANY OTHER SELLER, THE REPLACEMENT OF SUCH QUANTITY OR, IF NOT ACQUIRED BY PURCHASE, REPLACEMENT OF SUCH QUANTITY. IN NO EVENT SHALL THIS COMPANY OR ANY OTHER SELLER BE LIABLE FOR ANY INCIDENTAL OR CONSEQUENTIAL DAMAGES.

The buyer and all users are deemed to have accepted the terms of this LIMIT OF WARRANTY AND LIABILITY which may not be varied by any verbal or written agreement.

In case of an emergency involving this product, Call Collect, day or night, (314) 694-4000.

PRECAUTIONARY STATEMENTS

HAZARDS TO HUMANS AND DOMESTIC ANIMALS

KEEP OUT OF REACH OF CHILDREN

WARNING!

IRRITATING TO EYES.

MAY CAUSE ALLERGIC SKIN REACTION.

Do not get in eyes, on skin or on clothing.

Avoid breathing dust.

Do not take internally.

Wash thoroughly with soap and water after handling.

Wear rubber gloves, long trousers, long sleeved shirt or jacket of tightly woven material along with boots high enough to cover ankles when loading and when adjusting, repairing or cleaning equipment. Clothing which comes in contact with Lasso must be washed before reuse.

FIRST AID: IF IN EYES, immediately flush with plenty of water for at least 15 minutes. Call a physician.

IF ON SKIN, immediately flush with plenty of water. Remove contaminated clothing. If irritation persists, get medical attention.

In case of emergency involving this product, Call Collect, day or night, (314) 694-4000.

ENVIRONMENTAL HAZARDS

This chemical has been identified in limited ground water sampling and there is the possibility that it can leach through the soil to ground water, especially where soils are coarse and ground water is near the surface. Following application, and during rainfall events that cause runoff, this chemical may reach surface water bodies including streams, rivers and reservoirs.

Do not apply directly to water or wetlands. Do not contaminate water by cleaning of equipment or disposal of wastes.

STORAGE AND DISPOSAL

Do not contaminate water, foodstuffs, seed or feed by storage and disposal.

Wastes resulting from the use of this product that cannot be used or chemically reprocessed should be disposed of in a landfill approved for pesticide disposal or in accordance with applicable Federal, state or local procedures.

Completely empty bag into application equipment. Then dispose of empty bag in a sanitary landfill or by incineration, or, if allowed by state and local authorities, by burning. If burned, stay out of smoke.

APPLICATION EQUIPMENT, CALIBRATION AND HANDLING INSTRUCTIONS

NOTE

Direct contact or exposure to Lasso should be minimized. The following instructions for loading and repairing equipment should be followed in order to minimize this exposure. Review the protective clothing requirements as listed in the Precautionary Statements section of this label and do not use this product until you have the necessary protective clothing.

Equipment Cleaning and Repair

Cleaning and repair of application equipment is a source of exposure to this product. Care should be taken to minimize exposure during cleaning and repair of application equipment. Whenever possible, this equipment should be rinsed before being cleaned or repaired. When repairs must be made during application, special care should be taken to avoid contact with the pesticide.

Broadcast Treatment—Use broadcast applicators that can apply these granules evenly.

Band Treatment (Behind Press Wheel only) — Use applicators designed for this purpose. Calculate the amount of granules per acre needed for band treatment as follows:

$$\frac{\text{Band width in inches}}{\text{Row width in inches}} \times \text{Broadcast Rate per acre} = \text{Band Rate per acre}$$

For example: If 26 pounds per acre are recommended for broadcast treatment and you wish to treat a band of 14 inches wide in crop rows spaced 38 inches apart:

$$\frac{14}{38} \times 26 = \text{9.6 pounds of granules needed for the band treatment}$$

NOTE: For application of this product do not rely on settings used for other granular herbicides.

Calibration—The following settings are only guidelines for calibration; they do not allow for equipment variation or wear. Check settings after initial calibration so that the proper rate is applied.

	Granular Applicator Setting			
	Broadcast Rate Per Acre			
	16 Pounds		20 Pounds	
Applicator	4 mph	5 mph	4 mph	5 mph
Allis Chalmers	2	2	3	3
Gandy (bolt-on)	12	13	13	15
(snap-on)	12	13	14	15
I.H. before 1981	1/9	2/2	2/2	2/6
1981—	1/9	2/1	2/1	2/4
John Deere—				
metal hopper	1/27	1/31	1/31	2/1
plastic hopper.				
−10,20 etc. gauge	13	15	15	19
−5,15 etc. gauge	11	13	13	15
Noble-hopper bottom hole •	10	13	13	15
Noble and White-bottom hole ■	7	8	9	10
Buffalo				
Fluted-row feed	2.0	2.0	2.5	2.5
Plateless planter	8	10	10	13

	Granular Applicator Setting			
	Broadcast Rate Per Acre			
	23 Pounds		26 Pounds	
Applicator	4 mph	5 mph	4 mph	5 mph
Allis Chalmers	4	4	4	4
Gandy (bolt-on)	14	17	16	18
(snap-on)	15	16	16	17
I.H. before 1981	2/4	2/9	2/6	3/5
1981 -	2/3	2/6	2/5	2/8
John Deere -				
metal hopper	1/33	2/4	2/2	2/6
plastic hopper				
−10,20 etc gauge	17	21	19	24
−5,15 etc. gauge	14	16	15	17
Noble-hopper bottom hole •	14	17	16	19
Noble and White bottom hole ■	10	12	11	13
Buffalo				
Fluted-row feed	3.0	3.0	3.5	3.5
Plateless planter	12	15	13	18

It is important that the applicator be calibrated properly to deliver the desired amount of this product. To calibrate, set the applicator and attach a bag over the spreader plate or remove the delivery tube and attach a plastic nursing bottle over the spout. Operate at a constant speed on terrain to be planted. Four to 5 mph is suggested. For a 14-inch band, collect the granules for 500 linear feet (583 sq. ft. area) and measure the VOLUME or WEIGHT collected. Tap plastic nursing bottle or other measuring container gently to settle the granules when taking a reading.

Refer to the following table to determine the broadcast rate per acre for the amount collected. If necessary adjust the application equipment so that each spout dispenses the proper amount.

Broadcast Rate Per Acre	Volume Collected in 500 linear feet for a 14 inch band	Weight Collected in 500 linear feet for a 14 inch band
16 lb.	4.4 fl. oz.	3.4 oz.
20 lb.	5.6 fl. oz.	4.3 oz.
23 lb.	6.4 fl. oz.	4.9 oz.
26 lb.	7.2 fl. oz.	5.6 oz.

Broadcast Rate Per Acre	Volume Collected in 500 linear feet for a 14 inch band	Weight Collected in 500 linear feet for a 14 inch band

For other band widths calculate the calibration distance as follows:

$$\frac{14 \text{ inches}}{\text{Band width in inches}} \times 500 = \text{Calibration distance in linear feet.}$$

CULTIVATION INFORMATION

If grass or weed seedlings emerge before it rains rotary hoeing will improve performance.

If cultivation is necessary because of soil crusting or compaction, tillage should be shallow to minimize dilution of this herbicide.

Except for the above, as long as the grasses and weeds are controlled, delay cultivation as long as possible.

Only after weeds emerge in the treated band, set cultivator to throw soil into the row covering the band. On peanuts, do not throw soil into the row.

WEED CONTROL

When applied as directed under the conditions described, this product CONTROLS the following weeds:

ANNUAL GRASSES

Barnyardgrass Echinochloa crus-galli
Crabgrass Digitaria sanguinalis
Foxtail (giant) Setaria faberi
Foxtail (green) Setaria viridis
Foxtail (yellow) Setaria lutescens
Goosegrass Eleusine indica
Panicum (fall) Panicum dichtomiflorum
Signalgrass (broadleaf) Brachiaria spp.
Witchgrass Panicum

ANNUAL BROADLEAVES

Carpetweed Mollugo verticillata
Pigweed, Carelessweed Amaranthus spp.
Purslane Portulaca oleracea
Pusley (Florida) Richardia scabria

When applied immediately after planting, under the conditions described, this product at a rate of 16 to 26 pounds per acre on a broadcast basis REDUCES COMPETITION from the following hard to control weeds:

SEDGES

Nutsedge (yellow) Cyperus esculentus

ANNUAL BROADLEAVES

Beggarweed (Florida) Desmondium tortuosum
Lambsquarters Chenopodium album
Nightshade (black) Solanum nigrum
Ragweed (common) Ambrosia artemisiifolia
Smartweed Polygonum pensylvanicum, Polygonum persicaria
Sida (spiny), Teaweed Sida spinosa

This product WILL NOT CONTROL Grass, Field Bindweed, Bull Nettle, Canada Thistle, Rhizome Johnsongrass, Morningglory and Quackgrass.

DIRECTIONS FOR USE

It is a violation of Federal law to use this product in any manner inconsistent with its labeling.

This product is recommended for preemergence control of most annual grasses and certain broadleaf weeds. It should be applied to the soil surface prior to crop or weed emergence. The soil surface should be freshly worked and free of weed growth. The seed bed should be fine, firm and free of clods and trash.

Better results are obtained when moisture occurs within one week after application. High intensity or excessive rainfall, or excessive irrigation, after application may reduce control.

Use of this product not consistent with this label may result in injury to persons, animals or crops, or other unintended consequences. Keep bag closed to prevent spills and contamination.

FIELD CORN, HYBRID SEED CORN, SILAGE CORN, SWEET CORN AND SOYBEANS

Applicators should evaluate soil and weed conditions carefully to assure that they choose the lowest effective label rate.

For control of listed annual grasses and broadleaf weeds apply after planting and before crop or weeds emerge. For band treatment apply granules behind the press wheel. See table below for recommended rates on various soil textures.

Soil Texture	Broadcast Rate per Acre: Less Than 3% Organic Matter	More than 3% Organic Matter
MEDIUM		
Loams	16 lb	20 lb
HEAVY		
Silty Clay Loam thru Clay	20 lb	26 lb

For Lister or Deep Furrow Planted Corn

Broadcast Treatment—Apply at the rate of 16 pounds per acre.

Band Treatment (Behind Press Wheel)—For a 10 inch band treatment over the row (40 inch spacing) apply 4 pounds per acre. For a 7 inch band treatment over the row (40 inch spacing), apply 2.8 pounds per acre.

If heavy rains follow application, weed control on the top and sides of the furrow may not be as good as that obtained in the bottom of the furrow.

Where deep lister planting is practiced, only partial weed control can be expected on steep furrow sides. It is suggested that a 7 inch band be applied at the bottom of the furrow. Better results can be expected where the granules lie on an even flat surface which is not subject to shifting by wind or rain.

For Furrow Irrigated Corn and Soybeans

This product must be incorporated into the soil by harrow, finger weeder or rotary hoe. Shallow incorporation of at least 1/2 inch but not more than 2 inches is effective for weed control. Incorporation deeper than 2 inches may result in unsatisfactory weed control. Incorporation depth must be kept above the seed. Soil moisture at application must be sufficient for crop and weed emergence. Irrigation within 10 days following planting may improve weed control. See the table above for recommended rates.

PEANUTS

Broadcast—Apply 20 pounds of this product per acre from immediately after planting and until either crop or weeds emerge. NOTE: Hard-to-control weeds require 23 to 26 pounds per acre.

NOTE: Do not use treated forage for animal fodder.

SUNFLOWERS

For control of listed annual grasses and broadleaf weeds apply after planting and before crops or weeds emerge. For band treatment apply granules behind the press wheel.

Broadcast Treatment—Apply at the rate of 20 to 26 pounds per acre.

Band Treatment (Behind Press Wheel) — For a 10 inch band treatment over the row (40 inch spacing) apply 5 to 6.5 pounds per acre. For a 7 inch band treatment over the row (40 inch spacing), apply 3.5 to 4.6 pounds per acre. Use the higher rate in areas of heavy weed infestations, to reduce competition from hard-to-control weeds listed in this label, or on soil with 3 percent or more organic matter.

WOODY ORNAMENTALS

Field and Linear Grown Ornamentals

Use in **Cotoneaster** (Cotoneaster spp.), **Crabapple** (Malus spp.), **Dogwood** (Cornus spp.), **Euonymus** (Euonymus spp. shrub spp. only), **Holly** (Ilex spp.), **Juniper** (Juniperus spp.), and **Yew** (Taxus spp.).

CONTAINER GROWN ORNAMENTALS

Use in Holly (Ilex spp.) and Juniper (Juniperus spp.).

DO NOT USE IN GREENHOUSES OR OTHER ENCLOSED STRUCTURES.

NOT RECOMMENDED FOR DOMESTIC APPLICATION EXCEPT BY CERTIFIED APPLICATORS.

When applied as directed under the conditions described, this product controls the weeds listed on this label in newly planted or established field and container grown woody ornamentals.

DO NOT USE THIS PRODUCT ON SEEDBEDS, UNROOTED CUTTINGS OR PRIOR TO TRANSPLANTING. DO NOT APPLY THIS PRODUCT TO PLANTS WHOSE FOLIAGE IS MOIST OR WET AS GRANULES MAY STICK TO THIS FOLIAGE AND CROP INJURY MAY RESULT. OVERHEAD IRRIGATION APPLIED WITHIN ONE HOUR AFTER APPLICATION WILL WASH GRANULES OFF THE FOLIAGE AND INJURY WILL BE MINIMIZED.

Broadcast Treatment—Apply this product at a broadcast rate of 27 pounds per acre anytime after transplanting and before weeds emerge or after existing weeds have been removed. See the "Weed Control" and "Cultivation Information" sections of this label for further information. Repeat applications of this product may be required for sustained long-term weed control. Do not exceed three (3) applications of this product per year.

NOTE: Plant injury symptoms (leaf tip burn, discoloration, distortion, etc.) may occur with applications of this product to actively growing woody ornamentals. Do not retreat within twenty-one (21) days.

In case of emergency involving this product, call Collect, day or night, (314) 694-4000.

This sample label is current as of January 1, 1989. The product descriptions and recommendations provided in this sample label are for background information only. Always refer to the label on the product container before using Monsanto or any other agrichemical product.

LASSO® MICRO-TECH

HERBICIDE

Herbicide for weed control in corn, soybeans and crops listed.

RESTRICTED USE PESTICIDE

due to oncogenicity. For retail sale to and use only by Certified Applicators or persons under their direct supervision and only for those uses covered by the Certified Applicator's certification.

The use of this product may be hazardous to your health. This product contains alachlor, which has been determined to cause tumors in laboratory animals.

In case of an emergency involving this product or for user safety information on this product, call collect day or night (314) 694-4000.

ACTIVE INGREDIENTS:*	
Alachlor, [2-chloro-2',6'-diethyl-N-(methoxymethyl)acetanilide]	41.5%
INERT INGREDIENTS:	58.5%
TOTAL	100.0%

*Contains 480 grams/litre or 4 pounds/gallon.

Product protected by U.S. Patent No. 4,280,833 and 4,417,916.

Complete Directions for Use

*LASSO® MICRO-TECH® is referred to throughout these instructions as "LASSO."

EPA Reg. No. 524-344-ZA

1989-1 843.11-000.62/53

Read the entire label before using this product.

Use only according to label instructions.

Read "LIMIT OF WARRANTY AND LIABILITY" before buying or using. If terms are not acceptable, return at once unopened.

LIMIT OF WARRANTY AND LIABILITY

This Company warrants that this product conforms to the chemical description on the label and is reasonably fit for the purposes set forth in the Complete Directions for Use label booklet ("Directions") when used in accordance with those Directions under the conditions described therein. NO OTHER EXPRESS WARRANTY OR IMPLIED WARRANTY OF FITNESS FOR PARTICULAR PURPOSE OF MERCHANTABILITY IS MADE. This warranty is also subject to the conditions and limitations stated herein.

Buyer and all users shall promptly notify this Company of any claims whether based in contract, negligence, strict liability, other tort or otherwise.

Buyer and all users are responsible for all loss or damage from use or handling which results from conditions beyond the control of this Company, including, but not limited to, incompatibility with products other than those set forth in the Directions, unusual weather, weather conditions which are outside the range considered normal at the application site and for the time period when the product is applied, as well as weather conditions which are outside the application ranges set forth in the Directions, application in any manner not explicitly set forth in the Directions, moisture conditions outside the moisture range specified in the Directions, or the presence of products other than those set forth in the Directions in or on the soil, crop or treated vegetation.

THE EXCLUSIVE REMEDY OF THE USER OR BUYER, AND THE LIMIT OF THE LIABILITY OF THIS COMPANY OR ANY OTHER SELLER FOR ANY AND ALL LOSSES, INJURIES OR DAMAGES RESULTING FROM THE USE OR HANDLING OF THIS PRODUCT (INCLUDING CLAIMS BASED IN CONTRACT, NEGLIGENCE, STRICT LIABILITY, OTHER TORT OR OTHERWISE) SHALL BE THE PURCHASE PRICE PAID BY THE USER OR BUYER FOR THE QUANTITY OF THIS PRODUCT INVOLVED, OR, AT THE ELECTION OF THIS COMPANY OR ANY OTHER SELLER, THE REPLACEMENT OF SUCH QUANTITY OR, IF NOT ACQUIRED BY PURCHASE, REPLACEMENT OF SUCH QUANTITY. IN NO EVENT SHALL THIS COMPANY OR ANY OTHER SELLER BE LIABLE FOR ANY INCIDENTAL, CONSEQUENTIAL OR SPECIAL DAMAGES.

Buyer and all users are deemed to have accepted the terms of this LIMIT OF WARRANTY AND LIABILITY which may not be varied by any verbal or written agreement.

PRECAUTIONARY STATEMENTS

HAZARDS TO HUMANS AND DOMESTIC ANIMALS

KEEP OUT OF REACH OF CHILDREN

CAUTION!

MAY CAUSE EYE AND SKIN IRRITATION.

Avoid contact with eyes, skin and clothing.

Wear goggles or face shield, rubber gloves, long trousers, long sleeved shirt or jacket of tightly woven material along with boots high enough to cover ankles when transferring and mixing, and when adjusting, repairing or cleaning equipment. Wear rubber boots when pouring from open containers or when reentry is made into field following application of this product through center pivot irrigation systems and the field is still wet. Clothing which comes in contact with LASSO must be washed before reuse. Clothing or other materials which become drenched with the concentrated pesticide must be disposed of in a sanitary landfill, by incineration or, if allowed by state and local authorities, by burning. If burned, stay out of smoke.

Wash thoroughly with soap and water after handling.

FIRST AID: IF ON SKIN, immediately flush with plenty of water while removing contaminated clothing. As soon as soap is available, wash thoroughly with soap and water. Wash clothing before reuse. Sensitized persons should avoid further contact and reuse of contaminated clothing.

In case of an emergency involving this product, Call Collect, day or night, (314) 694-4000.

ENVIRONMENTAL HAZARDS

This chemical has been identified in limited ground water sampling and there is the possibility that it can leach through to soil to ground water, especially where soils are coarse and ground water is near the surface. Following application, and during rainfall events that cause runoff, this chemical may reach surface water bodies including streams, rivers and reservoirs.

Care must be taken when using this product to prevent back siphoning into wells, spills or improper disposal of excess pesticide, spray mixtures or rinsates.

Check valves or antisiphoning devices must be used on all mixing and/or irrigation equipment. Disposal of excess pesticide, spray mixtures, or rinsate should be according to label use instructions or according to the state pesticide or environmental control agency or the hazardous waste representative at the nearest EPA regional office.

Do not apply directly to water or wetlands. Do not contaminate water when disposing of equipment washwaters.

PHYSICAL OR CHEMICAL HAZARDS

In case of:

SPILL or LEAK, soak up, remove and keep away from humans and domestic animals.

DIRECTIONS FOR USE

It is a violation of federal law to use this product in any manner inconsistent with its labeling.

STORAGE AND DISPOSAL

Do not contaminate water, foodstuffs, seed or feed by storage and disposal.

STORAGE:

STORE ABOVE -4°F. TO KEEP PRODUCT FROM FREEZING. If frozen, thaw before use.

DISPOSAL:

Wastes resulting from the use of this product that cannot be used or chemically reprocessed should be disposed of in a landfill approved for pesticide disposal or in accordance with applicable Federal, state or local procedures.

Emptied container retains product residue. Observe all labeled safeguards until container is cleaned, reconditioned or destroyed.

(See individual container label for disposal information.)

GENERAL INFORMATION

LASSO® herbicide is recommended for control of yellow nutsedge and the annual grasses and broadleaf weeds listed in the "WEEDS CONTROLLED" section of this label. This product may be applied either as a surface application after planting or shallowly incorporated prior to planting to blend the herbicide treatment into the upper 1 to 2 inches of soil. Except for minimum or conservation tillage systems, the seedbed should be fine, firm and free of clods and trash.

®LASSO is a registered trademark of Monsanto Company.

Read and carefully observe precautionary statements and all other information appearing on the labeling of all products used in mixtures and sequential treatments.

Do not flood irrigate following application of this product or tank mixtures. This product will not control emerged seedlings.

Do not apply when conditions favor drift.

Use of this product not consistent with this label may result in injury to persons, animals or crops, or other unintended consequences. Keep container closed to prevent spills and contamination.

Flush sprayer with clean water after use.

®LASSO is a registered trademark of Monsanto Company.

SOIL TEXTURE

Applicators should evaluate soil and weed conditions carefully to assure that they choose the lowest effective label rate.

The recommended use rates of this product and the other herbicides labeled for use in tank mixtures with this product generally vary with soil texture. Unless soil texture is specifically named, rate tables throughout this label refer to only three soil texture groups: coarse, medium and fine. The following is a complete listing of soil textures included in each of these three soil texture groups:

Soil Textural Group	Soil Texture
COARSE:	sand, loamy sand, sandy loam
MEDIUM:	loam, silt loam, silt, sandy clay loam
FINE:	silty clay loam, clay loam, sandy clay, silty clay, clay

Refer to the above table to determine the corresponding soil textural group for the soil to be treated.

MIXING, SPRAYING AND HANDLING INSTRUCTIONS

NOTE

The use of a mechanical transfer system is required to be used by all mixer/loaders and/or applicators who treat 300 acres or more annually with pesticides containing alachlor as an active ingredient.

Direct contact or exposure to LASSO or spray mixtures of LASSO should be minimized. The following instructions for transfer, mixing, cleaning or repairing equipment should be followed in order to minimize this exposure. Review the protective clothing requirements as listed in the Precautionary Statements section of this label and do not use this product until you have the necessary protective clothing.

2.5 Gallon Containers

Open pouring from these containers can result in exposure from splashing or spilling. Special care in lifting and pouring are strongly recommended.

55 Gallon and Bulk Containers

Open pouring from these containers can result in exposure from splashing or spilling and is not recommended. This product should be transferred from these containers to the mix or spray tank using pumps or transfer probes. The probe or pump should not be removed from the container or disconnected until the container is emptied and rinsed. Use the pump or probe system to rinse the empty container and transfer the rinsate directly to the mix or spray tank.

Equipment Cleaning & Repair

Cleaning and repair of transfer systems and application equipment is a source of exposure to this product. Care should be taken to minimize exposure during cleaning and repair of transfer systems and application equipment. Whenever possible, these systems or equipment should be rinsed before being cleaned or repaired. When repairs must be made during transfer or application, the equipment should be shutdown, and special care taken to avoid contact with the pesticide.

Always predetermine the compatibility of LASSO or labeled tank mixtures of LASSO with water carrier or sprayable fluid fertilizer carrier by mixing small proportional quantities in advance. See the "STANDARD SPRAYABLE FLUID FERTILIZER COMPATIBILITY TEST" section in this label to determine the compatibility of this product and the labeled tank mixtures recommended for use with sprayable fluid fertilizer carrier.

NOTE: LASSO and all tank mixtures listed for corn and soybeans in the "DIRECTIONS FOR USE" section may be applied in sprayable fluid fertilizers. DO NOT apply LASSO or labeled tank mixtures of LASSO in sprayable fluid fertilizer on any other crops.

Mix LASSO or labeled tank mixture of LASSO, EXCEPT LASSO plus TREFLAN, with the appropriate carrier as follows:

1. Place a 20 to 35 mesh screen or wetting basket over filling port.
2. Through the screen, fill the sprayer tank one-half full with the appropriate carrier.
3. If a compatibility agent is necessary to improve mixing or to prevent the formation of undesirable and unsprayable gels or precipitates, while agitating add it to the carrier already in the tank. Use only compatibility agents, cleared by FDA for this use. Read and follow all directions for use, cautionary statements and all other information appearing on the selected compatibility agent label. Check for adequate agitation.
4. If a wettable powder or dry flowable formulation is used, make a slurry with water and add it slowly through the screen into the tank. Continue agitation.
5. If a flowable formulation is used, add slowly through screen into the tank. Mixing and compatibility may be improved when flowable is premixed one part flowable with one part water and added to the tank in diluted form.
6. Add LASSO slowly through the screen into the tank. Mixing and compatibility may be improved when this product is prediluted with two parts of water and added to the tank in diluted form.
7. Complete filling the sprayer tank with carrier. If Gramoxone™ Super is used, add the required amount near the end of the filling process. Remove hose from tank immediately after filling to avoid siphoning back into the carrier source.

Mix LASSO plus Treflan with water as follows:

1. Place a 20 to 35 mesh screen or wetting basket over filling port.
2. Through the screen, fill the sprayer tank one-half full with the appropriate carrier.
3. Add the Treflan while continuing agitation.
4. Add the LASSO next while continuing agitation.
5. Complete filling the tank with water.

Maintain good agitation at all times until the contents of the tank are sprayed. NOTE: If spray mixture is allowed to settle at any time thorough agitation is required to resuspend the mixture before spraying is resumed.

Keep by-pass line on or near bottom of tank to minimize foaming. Screen size in nozzle or line strainers should be no finer than 50 mesh. Carefully select proper nozzle to avoid spraying a fine mist. Check for even distribution of spray droplets. To reduce loss of the chemical due to drift of a fine mist, apply at nozzle pressures below 40 psi.

™Gramoxone is a trademark of ICI Americas Inc.

™Treflan is a trademark of Elanco Products Company.

STANDARD SPRAYABLE FLUID FERTILIZER COMPATIBILITY TEST

Herbicides may not always mix evenly throughout a sprayable fluid fertilizer or the components may separate too quickly to make their combined use of practical value. This may be due to certain characteristics of the different fluid fertilizers. A simple test using small quantities of the components is suggested to provide compatibility potential.

The test follows:

A. Materials Required For A Compatibility Test

1. Two one-quart jars with lid or stopper (marked "with" and "without").
2. TEAspoons (for a more exacting test, a five to ten milliliter (ml.) pipette or graduated cylinder is desirable).
3. Sprayable fluid fertilizer to be tested.
4. The herbicide chemicals to be mixed.
5. A compatibility agent (the purpose of the adjuvant is to help keep the fertilizer and crop protection chemical in suspension, if this assistance is needed.)

B. Procedure

1. Add one pint of the sprayable fluid fertilizer that will be used or other herbicide carrier to each jar marked "with" and "without."

Add One Pint Liquid Fertilizer To Two Quart Jars.

2. To the jar marked "with," add 1/4 TEAspoon or 1.2 milliters of a suitable compatibility agent; shake gently for five to ten seconds to mix. (1/4 TEAspoon in one pint is the equivalent of two pints per 100 gallons of liquid fertilizer.)

To Jar Marked "With" Add Compatibility Agent And Shake to Mix.

3. To each jar add the appropriate amount of herbicide(s) listed in the following tables. If more than one is used, add them separately with the wettable powders or dry flowables added first, flowables second and liquid last. Shake gently five to ten seconds after each addition.

Add Herbicide(s) To Both Jars And Shake to Mix.

Herbicide	Rate/a		Amount to Be Added Per Pint Of Sprayable Fluid Fertilizer (Assuming Volume is 25 gal/A) Level Teaspoons
Wettable	1 lb.	=	1.5
Powders	2 lb.	=	3
or	3 lb.	=	4.5
Dry Flowables	4 lb.	=	6
	5 lb.	=	7.5

Herbicide	Rate/a		Level Teaspoons		Milliters
Emulsifiable	1 pint	=	0.5	or	2.4
Concentrates or	1 quart	=	1	or	4.7
Flowables or	2 quarts	=	2	or	9.5
Liquids or	3 quarts	=	3	or	14.2
Solutions	1 gallon	=	4	or	19.0
	5 quarts	=	5	or	23.8

This compatibility test is designed for 25 gallons of spray per acre. For changes in spray volume or herbicide concentration, make appropriate changes in the ingredients of the test. Regardless of spray volume, the amount of compatibility agent should be equal to two or three pints (two pints = $^1/_4$ TEAspoon or 1.2 milliliters, three pints = $^3/_8$ TEAspoon or 1.8 milliliters per pint of sprayable fluid fertilizer) per 100 gallons of liquid fertilizer.

C. Observations and Decisions

1. If the herbicide(s) and the sprayable fluid fertilizer are compatible.
2. If a compatibility agent is necessary.

Five minutes after the final addition and mixing, observe both jars for the formation of large flakes, sludge, gels or other precipitates. Observe if the herbicide(s) cannot be physically mixed with the liquid fertilizer but remains as small oily particles in the solution.

If incompatibility in any form described above occurs in the jar "with" the compatibility agent added, the liquid fertilizer and the herbicide(s) should not be used together in the same spray tank.

If incompatibility as described above occurs in the jar "without" the adjuvant but not in the jar "with" adjuvant, the use of a compatibility adjuvant is recommended.

Both jars should be allowed to stand and be observed periodically for one-half hour. If the separate layers of liquid fertilizer and additives can be resuspended by shaking, commercial application is possible. An emulsifiable concentrate normally will go to the top after standing; wettable powders will either settle to the bottom of the tank or jar, or float to the top, depending upon the density of the fertilizers.

If the herbicide(s) is compatible with fluid fertilizer in the foregoing test without having to use a compatibility agent, fluid fertilizer may be used for the premixing. If it is not compatible without the compatibility agent, the herbicide(s) should be premixed with water before adding to the spray tank.

APPLICATION INFORMATION

APPLICATION SYSTEMS

Ground Broadcast Treatment—Apply this product and the labeled tank mixtures in 15 or more gallons of solution per acre using broadcast boom equipment. The carrier may be either water or sprayable fluid fertilizer as specified for the crop to be treated in the "DIRECTIONS FOR USE" section of this label. Do not apply during periods of gusty winds, when winds are in excess of 15 mph or when other conditions favoring drift exist.

Band Treatment—Apply a broadcast equivalent rate and volume per acre. To determine these:

$$\frac{\text{Band width in inches}}{\text{Row width in inches}} \times \text{Broadcast RATE per acre} = \text{Band RATE per acre}$$

$$\frac{\text{Band width in inches}}{\text{Row width in inches}} \times \text{Broadcast VOLUME per acre} = \text{Band VOLUME per acre}$$

CHEMIGATION

Apply this product and labeled tank mixtures only through center pivot irrigation systems. Do not apply this product through any other type of irrigation system.

Crop injury, lack of effectiveness, or illegal pesticide residues in the crop can result from nonuniform distribution of treated water. Use only in systems that apply uniformly.

If you have questions about calibration, you should contact State Extension Service specialists, equipment manufacturers or other experts.

Do not chemigate through systems connected to a public water system.

A person knowledgeable of the chemigation system and responsible for its operation, or under the supervision of the responsible person, shall shut the system down and make necessary adjustments should the need arise. Do not apply when system connections or fittings leak, when nozzles do not provide uniform distribution or when the line containing the product must be dismantled and drained.

The system must contain a functional check valve, vacuum relief valve, and low pressure drain appropriately located on the irrigation pipeline to prevent water source contamination from back flow.

The pesticide injection pipeline must contain a functional, automatic, quick-closing check valve to prevent the flow of fluid back toward the injection pump.

The pesticide injection pipeline must also contain a functional, normally closed, solenoid operated valve located on the intake side of the injection pump and connected to the system interlock to prevent fluid from being withdrawn from the supply tank when the irrigation system is either automatically or manually shut down.

The system must contain functional interlocking controls to automatically shut off the pesticide injection pump when the water pump motor stops.

The irrigation line or water pump must include a functional pressure switch which will stop the water pump motor when the water pressure decreases to the point where pesticide distribution is adversely affected.

Systems must use a metering pump, such as a positive displacement injection pump (e.g., diaphragm pump) effectively designed and constructed of materials that are compatible with pesticides and capable of being fitted with a systems interlock. Pumps, injection equipment, agitation equipment, hoses and connections between *supply* tank and the point of injection must be constructed of materials which are resistant to LASSO.

Do not apply when wind speed favors drift beyond the area intended for treatment.

Meter LASSO herbicide or a labeled tank mixture of LASSO into the center pivot irrigation system after planting and before crop emergence. Herbicide application should be made in $^1/_3$ and $^3/_4$ inch of water per acre. Do not apply in more than $^3/_4$ inch of water per acre under any conditions or reduced performance may occur. On very sandy soils (more than 60% sand and less than 1% organic matter), apply these herbicides in a maximum of $^1/_2$ inch water per acre. Good agitation must be maintained during the entire application period. Flush the system with water when application is complete. Refer to the "MIXING, SPRAYING AND HANDLING INSTRUCTIONS" section of the label for mixing procedures.

NOTE: Applications can be made by injection through center pivot irrigation equipment only when specified for the crop to be treated in the "DIRECTIONS FOR USE" section of this label. Do not apply this product or any labeled tank mixture through center pivot irrigation unless the treatment is specifically recommended in this label for the crop to be treated.

Aerial — Fixed wing; helicopter — Apply the recommended rates of this product or approved tank mixtures in 3 to 10 gallons of water per acre. AVOID DRIFT — DO NOT APPLY DURING INVERSION CONDITIONS, WHEN WINDS ARE GUSTY, OR UNDER ANY OTHER CONDITION WHICH WILL ALLOW DRIFT. Coarse sprays are less likely to drift; therefore, do not use nozzles or nozzle configurations which dispense spray or fine spray droplets. Do not angle nozzles forward into the airstream and do not increase spray volume by increasing nozzle pressure. Ensure uniform application. To avoid streaked, uneven or overlapped application, use mechanical marking systems only, HUMAN FLAGGERS PROHIBITED!

Application with Dry Bulk Fertilizer—The herbicide-fertilizer impregnation process must be done only at commercial fertilizer or chemical dealerships properly equipped for this procedure.

Dry bulk fertilizer may be impregnated with LASSO or the tank mixtures of LASSO plus Atrazine or LASSO plus Bladex on corn; and LASSO or the tank mixtures of LASSO plus Lexone or Sencor on soybeans. LASSO and these tank mixtures must be applied with 200 to 450 pounds of dry bulk fertilizer per acre and shallowly incorporated within 7 days prior to planting. The herbicide must be applied as recommended in this label for the crop, weed and soil type. Refer to the table for broadcast rate per acre in the appropriate crop section of this label to determine the recommended rate per acre for the herbicide treatment to be applied.

The following table provides a reference to determine the amount of LIQUID herbicide to be mixed per ton of dry bulk fertilizer for a range of herbicide recommendations for fertilizer rates per acre:

		Recommended Quarts Liquid Herbicide/Acre			
		1.5	2	2.5	3
Fertilizer Rate (Lb./Acre)	Acres Covered (Per Ton)	(Quarts of Herbicide/Ton Dry Bulk Fertilizer)			
200	10	15	20	25	30
250	8	12	16	20	24
300	6.7	10	13.3	16.7	20
350	5.7	8.6	11.4	14.3	17.1
400	5	7.5	10	12	15

	Recommended Quarts Liquid Herbicide/Acre				
		1.5	2	2.5	3
Fertilizer Rate (Lb./Acre)	Acres Covered (Per Ton)	(Quarts of Herbicide/Ton Dry Bulk Fertilizer)			
450	4.5	6.7	8.9	11.1	13.3

To determine the amount of herbicide needed for rates not included in the preceding table, use the following formula:

$$\frac{\text{Recommended herbicide rate (lb. or qt.)/acre} \times 2000}{\text{pounds fertilizer/acre}} = \text{pounds or quarts of herbicide per ton of dry bulk fertilizer}$$

Mix and blend the dry fertilizer and herbicide mixture in a close rotary-drum type mixture allowing sufficient time to ensure uniform coverage. Use at least one ton of dry fertilizer per mixing operation. Inject the herbicide into the drum over a minimum of a 2-minute period and allow at least 2 additional minutes mixing time to ensure uniformity. The nozzle used to spray the herbicide treatment must be placed inside the mixer to provide uniform spray coverage of the tumbling fertilizer.

If the fertilizer used has inadequate adsorptive capacity, use a higher adsorptive material such as Agsorb™, NP-79™ or Microcel™ E to provide a free-flowing mixture. The adsorptive material and dry fertilizer should be thoroughly mixed prior to adding the herbicide. The following table provides a partial list of dry fertilizers which may be impregnated with LASSO or tank mixtures of LASSO with other herbicides:

Fertilizer	LASSO®	LASSO + atrazine	LASSO + Bladex™	LASSO + Lexone™ or Sencor™
Ammonium sulfate (21-0-0)	Yes	Yes	Yes	Yes
Ammonium phosphate-sulfate (16-20-0)	Yes	Yes	Yes	Yes
Diammonium phosphate (18-46-0)	Yes	Yes	Yes	Yes
Potassium chloride (0-0-60)	Yes	Yes	Yes	Yes
Potassium sulfate (0-0-52)	Yes	Yes	Yes	Yes
Single super-phosphate (0-20-0)	Yes	No	No	Yes
Treble super-phosphate (0-46-0)	Yes	No	No	Yes
*Urea (46-0-0)	Yes	Yes	Yes	Yes

*Blends suitable for use in soybeans, such as 0-26-26 are also approved for use with LASSO and the tank mixtures listed.

NOTE: DO NOT impregnate LASSO or tank mixtures of LASSO with other herbicides on fertilizers containing ammonium nitrate, potassium nitrate or sodium nitrate.

Spread the herbicide-dry fertilizer mixture uniformly with a properly calibrated applicator. Fertilizers impregnated with LASSO or tank mixtures of LASSO with other herbicides must be spread at half-rate and overlapped 100 percent to obtain full rate and uniform distribution. Non-uniform spreading of the fertilizer-herbicide mixture may result in unsatisfactory weed control or crop injury.

™Agsorb is a trademark of Oil Dri Corporation.
™Bladex and Lexone are trademarks of the E.I. du Pont de Nemours and Company.
™ Microcel is a trademark of Manville Products Corporation.
™NP-79 is a trademark of Eagle Pitcher Industries.
™Sencor is a trademark of the parent company of Farbenfabricken Bayer GmbH.

APPLICATION TIMING AND METHODS

NOTE: Unless specifically recommended, only one treatment of LASSO may be applied to a labeled crop.

Preemergence Surface Application—This product and all labeled tank mixtures except LASSO plus Command™, LASSO plus Eptam™ and LASSO plus Treflan may be applied to the soil surface after planting and prior to either crop or weed emergence. Apply within 5 days of last preplant tillage. If weeds emerge after treatment, or if treatment is applied more than 5 days after last preplant tillage, rotary hoe or shallowly cultivate immediately to improve performance.

Precipitation or overhead sprinkler irrigation is required after application to move the herbicide treatment into the weed germination zone. The amount of precipitation or overhead sprinkler irrigation required depends on existing soil moisture, soil type and percent organic matter content, but 1/3 to 3/4 inches is normally adequate. Performance is improved when moisture is received within 7 days after application and prior to weed emergence. High intensity or excessive rainfall or excessive irrigation after application may reduce control.

Preplant Incorporation—This product and many of the labeled tank mixtures may be mixed into the soil, using shallow incorporation equipment, any time within 7 days prior to planting. Refer to the crop to be treated in the "DIRECTIONS FOR USE" section of this label to determine if preplant incorporation is recommended. Apply the recommended treatment rate to the soil surface as a broadcast application. Precipitation or irrigation is required to bring incorporated herbicide treatments into contact with germinating weed seedlings. The amount of precipitation or irrigation required depends on existing soil moisture, soil type and percent organic matter content, but 1/3 to 3/4 inch is normally adequate. If weeds emerge after treatment, rotary hoe or shallowly cultivate immediately to improve performance.

NOTE: Do not incorporate unless specifically recommended in this label for the crop and herbicide treatment selected. Incorporation of this product and its labeled tank mixtures on coarse soil types may reduce length of weed control due to leaching with rainfall or irrigation. Incorporation is not recommended on coarse soils in the southeastern states.

1-Pass Incorporation (Surface Blend™): Shallowly incorporate (Surface Blend) the treatment into the upper 1 to 2 inches of the soil. Equipment should be operated at manufacturer's designed speed for incorporation to ensure adequate mixing and distribution of the herbicide treatment in the soil. Equipment design, including any drag attachments, must be adequate to avoid soil ridging which may result in streaked or reduced weed control. Soil conditions, including moisture content and crop residue levels, must be suitable to allow thorough and uniform mixing with the equipment used for 1-pass incorporation.

Contact Monsanto Company for additional information regarding recommended equipment and methods for Surface Blend application.

2-Pass Incorporation: When 2-pass incorporation is used, shallowly incorporate the herbicide treatment into the upper 1 to 2 inches of the soil with equipment set to work the soil NO DEEPER THAN 4 INCHES. The second pass must be made at an angle to and no deeper than the first pass to ensure proper distribution of the herbicide treatment in the soil.

Cultivation Information—Delay cultivation after application for as long as possible unless weeds or grasses emerge. Shallowly cultivate or rotary hoe immediately if weeds or grasses emerge. If cultivation is necessary because of soil crusting or compaction, set equipment shallow and minimize lateral soil movement to avoid dilution or displacement of the herbicide treatment. If a band application is used and weeds have emerged in the treated band, set cultivator to throw soil into the row, covering the band.

™Surface Blend is a trademark of Monsanto Company.
™Command is a trademark of FMC Corporation.
™Eptam is a trademark of ICI Americas Inc.

WEEDS CONTROLLED

Applicators should evaluate soil and weed conditions carefully to assure that they choose the lowest effective label rate.

When applied as directed under conditions described, this product and tank mixtures of this product will control or reduce competition from the weeds listed. Refer to the crop on which treatment is intended to determine the rates and tank mixtures recommended.

For the scientific names to the weeds listed below, refer to the appendix at the end of this booklet.

NOTE: C = Control R = Reduced Competition

ANNUAL GRASSES

		LASSO® plus						
	LASSO®	Amiben™	atrazine	Banvel™	Bladex®	Canopy™	Command™	Gemini™
Barnyardgrass	C	C	C	C	C	C	C	C
Crabgrass	C	C	C	C	C	C	C	C
Cupgrass, wooly[1]	R	R	R	•	R	R	R	•
Foxtail: giant, green, robust purple, robust white, yellow	C	C	C	C	C	C	C	C
Goosegrass	C	C	C	C	C	C	C	C
Johnsongrass, seedling[2,8]	R	R	R	R	R	R	R	R
Millet, proso[3]	C	•	C	•	C	•	•	•
Panicum; browntop, fall	C	C	C	C	C	C	C	C
Panicum, Texas[4]	R	R	R	R	R	R	R	R
Rice, red[5]	C	C	C	•	C	C	C	•
Sandbur, Grassbur[2,6]	R	•	C	•	•	•	•	•
Shattercane, wild cane[1]	R	R	R	•	R	R	R	•
Signalgrass, broadleaf	C	C	C	C	C	C	C	C
Sprangletop, red	C	C	C	C	C	C	C	C
Witchgrass	C	C	C	C	C	C	C	C

ANNUAL BROADLEAVES

	LASSO®	LASSO® plus Amiben™	atrazine	Banvel™	Bladex™	Canopy™	Command™	Gemini™
Beggarweed, Florida[2]	R	C	C	•	C	C	•	C
Carpetweed	C	C	C	C	C	C	C	C
Cocklebur	•	•	R	R	R	C	•	C
Coffeeweed, Hemp sesbania	•	•	•	•	•	C	•	C
Copperleaf, hophornbeam	•	•	•	•	•	C	•	•
Galinsoga	C	C	C	C	C	C	C	C
Groundcherry, annual	•	•	C	•	•	C	•	C
Groundcherry, cutleaf[7]	C	C	C	C	C	C	C	C
Jimsonweed	•	•	C	•	•	C	C	C
Kochia	•	•	C	•	•	•	•	•
Lambsquarters[2]	R	C	C	C	C	C	C	C
Morningglory: Tall[8]	•	•	R	R	R	C	•	C
Pitted, Ivyleaf[8], entireleaf[8]	•	•	•	•	•	C	•	C
Smallflower	R	R	C	•	C	C	•	C
Mustard	•	R	C	C	C	C	•	C
Nightshade: black[7]; hairy[7]	C	C	C	C	C	C	C	C
Pigweed, Carlessweed	C	C	C	C	C	C	C	C
Purslane	C	C	C	C	C	C	C	C
Pusley, Florida	C	C	C	C	C	C	C	C
Ragweed, common[2]	R	C	C	C	C	C	C	C
Ragweed, giant	•	•	•	•	•	R	•	R
Sicklepod	•	•	C	•	C	C	•	R
Sida, prickly, Teaweed[2]	R	•	•	•	•	C	C	C
Smartweed[2]	R	C	C	C	C	C	R	C
Starbur, bristly	R	R	C	•	C	C	•	•
Sunflower, common	•	•	•	•	•	R	•	•
Velvetleaf, Buttonweed[1]	•	R	R	R	R	C	C	C
Waterhemp	C	C	C	C	C	C	C	C
SEDGE								
Nutsedge, yellow[5]	C	C	C	•	C	C	C	•

NOTE: C = Control **R = Reduced Competition**

ANNUAL GRASSES

	LASSO® plus linuron	Lorox™	Preview	propazine	Scepter™	Sencor™	Treflan™
Barnyardgrass	C	C	C	C	C	C	C
Crabgrass	C	C	C	C	C	C	C
Cupgrass, wooly[1]	•	•	R	•	R	R	R
Foxtail; giant, green, robust purple, robust white; yellow	C	C	C	C	C	C	C
Goosegrass	C	C	C	C	C	C	C
Johnsongrass, seeding[2,8]	R	R	R	R	C	R	R
Millet, proso[3]	•	•	•	•	•	•	•
Panicum: browntop, fall	C	C	C	C	C	C	C
Pancium, Texas[4]	R	R	R	R	R	R	R
Rice, red[5]	•	•	C	C	C	C	C
Sandbur, Grassbur[2,6]	•	•	•	•	•	•	•
Shattercane, wild cane[1]	•	•	R	R	R	R	R
Signalgrass, broadleaf	C	C	C	C	C	C	C
Sprangletop, red	C	C	C	C	C	C	C
Witchgrass	C	C	C	C	C	C	C

ANNUAL BROADLEAVES

	LASSO® plus linuron	Lorox™ Plus	Preview™	propazine	Scepter™	Sencor™	Treflan™
Beggarweed, Florida[2]	•	•	R	•	•	R	•
Carpetweed	C	C	C	C	C	C	C
Cocklebur	•	C	C	R	C	R	•
Coffeeweed	•	•	C	•	•	•	•
Cooperleaf, hophornbeam	•	•	C	•	•	•	•
Galinsoga	C	C	C	C	C	C	C
Groundcherry, annual	C	C	C	C	C	C	•
Groundcherry, cutleaf[7]	C	C	C	C	C	C	C
Jimsonweed	•	C	C	C	C	R	•
Kochia	•	•	•	C	•	•	C
Lambsquarters[2]	C	C	C	C	C	C	•
Morningglory: Tall[8]	•	R	R	R	R	•	•
Pitted	•	R	R	•	C	•	•
Ivyleaf[8]; entireleaf[8]	•	R	R	•	R	•	•
Smallflower	•	•	•	•	C	•	•
Mustard	C	C	C	C	C	•	•
Nightshade: black[7]; hairy[7]	C	C	C	C	C	C	C
Pigweed, Carelessweed	C	C	C	C	C	C	C
Purslane	C	C	C	C	C	C	C
Pusley, Florida	C	C	C	C	C	C	C
Ragweed, common[2]	C	C	C	C	C	C	•
Ragweed, giant	•	R	R	•	R	•	•
Sicklepod	•	•	•	C	R	R	•
Sida, prickly; Teaweed[2]	C	C	C	•	C	•	•
Smartweed[2]	C	C	C	C	C	C	•
Starbur, bristly	•	•	•	•	C	C	•
Sunflower, common	•	•	R	•	C	•	•
Velvetleaf, Buttonweed[8]	R	C	C	R	C	R	•
Waterhemp	C	C	C	C	C	C	C
SEDGE							
Nutsedge, yellow[5]	•	•	C	C	C	C	C

[1] Use 4 quarts per acre of this product applied alone or in tank-mix combinations. Preplant incorporate only to reduce competition from this weed.

[2] For reduced competition from this weed with LASSO alone, use 2.5 to 4 quarts of LASSO per acre.

[3] Plant corn into freshly prepared seedbed. Apply 4 quarts of LASSO per acre preemergence surface immediately after corn planting for reduced competition from this weed. Improved control may be obtained by applying a labeled postemergence herbicide.

[4] Use a minimum of 3 quarts of LASSO per acre and apply through center pivot irrigation or under sprinkler irrigation systems. Only LASSO and LASSO plus atrazine tank mixture can be applied through center pivot irrigation systems.

[5] Use 3 to 4 quarts of LASSO per acre applied alone or in tank mixture and apply preplant incorporated only for control on medium and fine-textured soils. When tank mixed with Scepter, use 2 to 3 quarts of LASSO per acre and preplant incorporate only. On coarse-textured soils with less than 3% organic matter, use 3 to 4 quarts of LASSO per acre as a preemergence surface application to reduce competition from yellow nutsedge.

[6] Apply 2.5 to 3 quarts of LASSO plus 1 to 2 quarts of atrazine 4L per acre. Preemergence surface applications should be applied to a freshly prepared seedbed. For best control, rotary hoe or initiate overhead irrigation (0.33 to 0.5 inch) within 24 hours of application.

[7] Use a minimum of 2.5 quarts of LASSO per acre on coarse soils, or 3 to 4 quarts of LASSO per acre on medium and fine soils when used alone or in tank mixture with Amiben, Banvel, Canopy, Command, Preview, Sencor/Lexone or Treflan for control of this weed.

[8] For best results on these species when using LASSO plus Scepter tank mixture, use preplant incorporated treatments.

NOTE: The 3-way tank mixture of LASSO plus atrazine plus Bladex controls or reduces competition as specified for the respective weeds listed under the LASSO plus atrazine and LASSO plus Bladex recommendations.

The 3-way tank mixture of LASSO plus Amiben plus Lexone or Sencor controls or reduces competition as specified for the respective weeds listed under the LASSO plus Amiben and LASSO plus Lexone or LASSO plus Sencor recommendations.

Certain tank mixtures of LASSO and broadleaf herbicides may be combined with application of Gramoxone Super for control of many emerged weeds prior to corn or soybean emergence. Refer to the Gramoxone Super label for a list of emerged weeds controlled by tank mixtures with these products.

™ Amiben is a trademark of Rhone-Poulenc.

™ Banvel is a trademark of Sandoz, Inc.

™ Canopy, Gemini, Lorox and Preview are trademarks of E. I. du Pont de Nemours and Company.

™ Scepter is a trademark of American Cyanamid Corporation.

CROPS

CORN (All Types)

LASSO®

Apply this product in water or sprayable fluid fertilizer solution for control of yellow nutsedge and the annual grasses and broadleaf weeds listed in the "WEEDS CONTROLLED" section of this label.

APPROVED APPLICATION SYSTEMS

Ground—Broadcast boom; banded

Aerial—Fixed wing; helicopter

Center Pivot Irrigation

Dry Bulk Fertilizer Impregnation

Reference: The "APPLICATION SYSTEMS" section of this label provides detailed information on the proper application of this product using the system selected.

APPROVED APPLICATION METHODS

Preplant Incorporated—Apply this product within 7 days prior to planting and shallowly incorporate or Surface Blend into the upper 1 to 2 inches of soil. Irrigation within 10 days following planting may improve weed control.

Preemergence Surface—Apply this product after planting, before crop and weed emergence and within 5 days after last preplant tillage operation.

DO NOT apply this product postemergence, sequential or lay-by.

DO NOT apply any alachlor containing product sequential to this product.

Reference: The "APPLICATION TIMING AND METHODS" section of this label provides detailed information and procedures for the application timing and method selected.

RECOMMENDED RATES: Refer to the following tables for the recommended broadcast treatment rates for this product. Applications which are not consistent with recommendations in this label may result in unsatisfactory weed control or crop injury.

Surface Application Rates

Soil Textural Group*	Broadcast Rate Per Acre (Quarts)**	
	Less Than 3% Organic Matter	**3% or More Organic Matter**
Coarse***	2 to 2.25	2 to 2.25
Medium	2 to 2.75	2 to 2.75
Fine	2 to 2.75	2.5 to 3.25

* Refer to the "SOIL TEXTURE" section of this label to determine the corresponding soil textural group for the soil to be treated.

** Use the higher rate in the recommended range in areas of heavy weed infestations. Use a minimum of 2.5 quarts of this product per acre on coarse soils and 3 to 4 quarts on medium or fine textured soils to control black or hairy nightshade.

*** When applied through center pivot irrigation or under sprinkler irrigation systems on coarse soils, use a minimum of 3 quarts per acre of this product.

Preplant Incorporated Application Rates

Soil Textural Group*	Broadcast Rate Per Acre (Quarts)**	
	Less Than 3% Organic Matter	**3% or More Organic Matter**
Coarse***	2.5	2.5
Medium	2.5 to 3	2.5 to 3
Fine	2.5 to 3	3 to 3.5

* Refer to the "SOIL TEXTURE" section of this label to determine the corresponding soil textural group for the soil to be treated.

** Use the higher rate in the recommended range in areas of heavy weed infestations. Use 2.5 to 3.5 quarts per acre where furrow irrigation is used. Use a minimum of 3 quarts on coarse textured soils for control of yellow nutsedge or red rice. Rates may be increased to a maximum of 4 quarts on any soil type when heavy infestations of yellow nutsedge or red rice are present.

LASSO® PLUS ATRAZINE
(TANK MIXTURE)

Apply this tank mixture in water or sprayable fluid fertilizer solutions for control of yellow nutsedge and the annual grasses and broadleaf weeds listed in the "WEEDS CONTROLLED" section of this label.

APPROVED APPLICATION SYSTEMS

Ground—Broadcast boom; banded

Aerial—Fixed wing; helicopter

Center Pivot Irrigation

Dry Bulk Fertilizer Impregnation

Reference: The "APPLICATION SYSTEMS" section of this label provides detailed information on the proper application of this tank mixture using the system selected.

APPROVED APPLICATION METHODS

Preplant Incorporated—Apply this tank mixture within 7 days prior to planting and shallowly incorporate or Surface Blend into the upper 1 to 2 inches of soil.

Preemergence Surface—Apply this tank mixture after planting, before crop and weed emergence and within 5 days after last preplant tillage operation.

Reference: The "APPLICATION TIMING AND METHODS" section of this label provides detailed information and procedures for the application timing and method selected.

NOTE: PLANT ONLY CORN, PEANUTS, SORGHUM OR SOYBEANS* the year following use of this mixture.

* There is a possibility of crop injury due to carryover of atrazine if soybeans are planted the following year. DO NOT plant soybeans the year following use of this tank mixture on furrow irrigated corn. DO NOT graze treated area or feed treated forage to livestock for 21 days following application of this tank mixture.

RECOMMENDED RATES: Refer to the following tables for the recommended broadcast treatment rates for this tank mixture. Applications which are not consistent with recommendations in this label may result in unsatisfactory weed control or crop injury.

Surface Application Rates

Soil Textural Group*	Broadcast Rate Per Acre (Quarts)**			
	Less Than 3% Organic Matter		**3% or More Organic Matter**	
	LASSO®	**+ Atrazine 4L***	**LASSO**	**+ Atrazine 4L***
Coarse	1.5 to 2	0.75 to 2	1.5 to 2	1 to 2
Medium	1.5 to 2	1 to 2	2 to 2.25	1.25 to 2
Fine	2 to 2.25	1.25 to 2	2 to 2.75	1.25 to 2

* Refer to the "SOIL TEXTURE" section of this label to determine the corresponding soil textural group for the soil to be treated.

** Use the higher rates in the recommended ranges in areas of heavy weed infestations.

*** Use rates listed in this label when using atrazine 4L. Use equivalent rates when using atrazine 80W or 90% dry flowable formulations. One quart of atrazine 4L equals 1.25 pounds of atrazine 80W or 1.1 pounds of atrazine 90% dry flowable.

NOTE: When applied through center pivot irrigation or under sprinkler irrigation systems on coarse soils, use 1 quart per acre of atrazine 4L and a minimum of 2 quarts per acre of LASSO in this tank mixture.

Preplant Incorporated Application Rates

Soil Textural Group*	Broadcast Rate Per Acre (Quarts)**			
	Less Than 3% Organic Matter		**3% or More Organic Matter**	
	LASSO®	**+ Atrazine 4L***	**LASSO**	**+ Atrazine 4L***
Coarse	1.75 to 2.25	1.25 to 2	1.75 to 2.25	1.25 to 2
Medium	1.75 to 2.25	1.25 to 2	2.25 to 2.5	1.5 to 2
Fine	2.25 to 2.5	1.5 to 2	2.25 to 2.75	1.5 to 2

* Refer to the "SOIL TEXTURE" section of this label to determine the corresponding soil textural group for the soil to be treated.

** Use the higher rates in the recommended ranges in areas of heavy weed infestations.

*** Use rates listed in this label when using atrazine 4L. Use equivalent rates when using atrazine 80W or 90% dry flowable formulations. One quart of atrazine 4L equals 1.25 pounds of atrazine 80W or 1.1 pounds of atrazine 90% dry flowable.

LASSO® plus BANVEL
(TANK MIXTURE)
Field and Silage Corn Only

For use in ILLINOIS, IOWA, MINNESOTA and WISCONSIN on level or flat-planted field corn on fine textured (silty clay loam, clay loam, sandy clay, silty clay or clay) soils with more than 4 percent organic matter.

Apply in water or sprayable fluid fertilizer solutions for control of the annual grasses and broadleaf weeds listed in the "WEEDS CONTROLLED" section of this label.

NOTE: Use on coarse or medium textured soils or on fine textured soils with 4 percent or less organic matter may result in crop injury and/or destruction.

APPROVED APPLICATION SYSTEMS

Ground—Broadcast boom; banded

NOTE: DO NOT apply by air or injection through center pivot irrigation systems.

Reference: The "APPLICATION SYSTEMS" section of this label provides detailed information on the application of this tank mixture using ground spray equipment.

APPROVED APPLICATION METHODS

Preemergence Surface—Apply this tank mixture after planting, before crop and weeds emerge and within 5 days of last preplant tillage operation. Corn seeds must be planted 1½ inches or deeper beneath the soil surface. Direct chemical contact with corn seed must be avoided since crop injury may result. Apply far enough behind planter equipment to avoid any incorporation by the planter wheel or other covering device. If corn seeds are planted less than 1½ inches beneath the soil surface, delay application until corn has spiked.

Reference: The "APPLICATION TIMING AND METHODS" section of this label provides detailed information and procedures for the application timing and method selected.

NOTE: PREVENT DRIFT TO SOYBEANS OR OTHER DESIRABLE PLANTS. Do not use on furrow irrigated corn, or when corn is planted at the bottom of a furrow, utilizing lister, till or other similar planting methods. DO NOT incorporate prior to planting or corn emergence. If it is necessary to drag for leveling or rotary hoe to break soil crust, DO NOT disturb the soil more than ½ inch deep.

RECOMMENDED RATES: Refer to the following table for the recommended broadcast treatment rates for this tank mixture. Applications which are not consistent with recommendations in this label may result in unsatisfactory weed control or crop injury and/or destruction.

Soil Textural Group	Broadcast Rate Per Acre (Quarts)	
	More Than 4% Organic Matter	
	LASSO®	**Banvel**
Fine—silty clay loam through clay	2.5	0.5

LASSO® plus Bladex (TANK MIXTURE)

Apply this tank mixture in water or sprayable fluid fertilizer solutions for control of yellow nutsedge, and the annual grasses and broadleaf weeds listed in the "WEEDS CONTROLLED" section of this label.

APPROVED APPLICATION SYSTEMS

Ground—Broadcast boom; banded

Aerial—Fixed wing; helicopter

Dry Bulk Fertilizer Impregnation

Reference: The "APPLICATION SYSTEMS" section of this label provides information on the proper application of this tank mixture using the system selected.

APPROVED APPLICATION METHODS

Preplant Incorporated—Apply this tank mixture within 7 days prior to planting and shallowly incorporate or Surface Blend into the upper 1 to 2 inches of soil.

Preemergence Surface—Apply this tank mixture after planting, before crop and weed emergence and within 5 days after last preplant tillage operation.

Reference: The "APPLICATION TIMING AND METHODS" section of this label provides detailed information and procedures for the application timing and method selected.

RECOMMENDED RATES: Refer to the following tables for the recommended broadcast treatment rates for this tank mixture. Applications which are not consistent with recommendations in this label may result in unsatisfactory weed control or crop injury.

Surface Application Rates

Soil Textural Group*	Broadcast Rate Per Acre (Quarts)			
	1% to 3% Organic Matter		3% or More Organic Matter	
	LASSO® +	Bladex 4L***	LASSO +	Bladex 4L***
Coarse**	1.5 to 2	1 to 1.2	1.5 to 2	1.2 to 1.6
Medium	1.5 to 2	1.2 to 1.4	2 to 2.25	1.6 to 2
Fine	2 to 2.25	1.4 to 1.8	2 to 2.75	1.8 to 2.2

* Refer to the "SOIL TEXTURE" section of this label to determine the corresponding soil textural group for the soil to be treated.

** On coarse textured soils containing less than 2% organic matter, use LASSO plus Bladex only on sandy loam. Do not use on sand and loamy sand with less than 2% organic matter.

*** Use rates listed in this label using Bladex 4L. Use equivalent rates when using Bladex 80W. One quart of Bladex 4L equals 1.25 lbs of Bladex 80W.

Preplant Incorporated Application Rates

Soil Textural Group*	Broadcast Rate Per Acre (Quarts)			
	1% to 3% Organic Matter		3% or More Organic Matter	
	LASSO® +	Bladex 4L***	LASSO +	Bladex 4L***
Coarse**	1.75 to 2.25	1.2 to 1.4	1.75 to 2.25	1.4 to 1.8
Medium	1.75 to 2.25	1.4 to 1.6	2.25 to 2.5	1.8 to 2.2
Fine	2.25 to 2.5	1.6 to 2	2.25 to 2.75	2 to 2.4

NOTE: See "Incorporation" instructions.

* Refer to the "SOIL TEXTURE" section of this label to determine the corresponding soil textural group for the soil to be treated.

** On coarse textured soils containing less than 2% organic matter, use LASSO plus Bladex only on sandy loam. Do not use on sand and loamy sand with less than 2% organic matter.

*** Use rates listed in this label when using Bladex 4L. Use equivalent rates when using Bladex 80W. One quart of Bladex 4L equal 1.25 pounds of Bladex 80W.

LASSO® plus ATRAZINE PLUS BLADEX (TANK MIXTURE)

Apply this tank mixture in water or sprayable fluid fertilizer solutions for control of yellow nutsedge and the annual grasses and broadleaf weeds listed in the "WEEDS CONTROLLED" section of this label.

APPROVED APPLICATION SYSTEMS

Ground—Broadcast boom; banded

Aerial—Fixed wing; helicopter

Reference: The "APPLICATION SYSTEMS" section of this label provides detailed information on the proper application of this tank mixture using the system selected.

APPROVED APPLICATION METHODS

Preemergence Surface—Apply this tank mixture after planting, before crop and weed emergence and within 5 days after last preplant tillage operation.

Preplant Incorporated—Apply this tank mixture within 7 days prior to planting and shallowly incorporate or Surface Blend into the upper 1 to 2 inches of soil.

Reference: The "APPLICATION TIMING AND METHODS" section of this label provides detailed information and procedures for the application timing and method selected.

NOTE: PLANT ONLY CORN, PEANUTS, SORGHUM OR SOYBEANS* the year following use of this mixture.

* There is a possibility of crop injury due to carryover of atrazine if soybeans are planted the following year. DO NOT plant soybeans the year following use of this tank mixture on furrow irrigated corn. DO NOT graze treated area or feed treated forage to livestock for 21 days following application of this tank mixture.

RECOMMENDED RATES: Refer to the following tables for the recommended broadcast treatment rates for this tank mixture. Applications which are not consistent with recommendations in this label may result in unsatisfactory weed control or crop injury.

Surface Application Rates

Soil Textural Group*	Broadcast Rate Per Acre (Quarts)		
	Less Than 3% Organic Matter		
	LASSO® +	atrazine 4L**	+ Bladex 4L***
Coarse****	1.5 to 2	0.75	0.5 to 1
Medium	1.5 to 2	0.75 to 1	1 to 1.25
Fine	2 to 2.25	1 to 1.25	1 to 1.5

Soil Textural Group*	Broadcast Rate Per Acre (Quarts)		
	3% Or More Organic Matter		
	LASSO +	atrazine 4L**	+ Bladex 4L***
Coarse****	1.5 to 2	0.75 to 1	1 to 1.25
Medium	2 to 2.25	1 to 1.25	1 to 2
Fine	2 to 2.75	1.25 to 1.5	1 to 2

* Refer to the "SOIL TEXTURE" section of this label to determine the corresponding soil textural group for the soil to be treated.

** Use rates listed in this label when using atrazine 4L. Use equivalent rates when using atrazine 80W or 90% dry flowable formulations. One quart of atrazine 4L equals 1.25 pounds of atrazine 80W or 1.1 pounds of atrazine 90% dry flowable.

*** Use rates listed in this label using Bladex 4L. Use equivalent rates when using Bladex 80W. One quart of Bladex 4L equals 1.25 pounds of Bladex 80W.

**** Do not use this mixture on sand or loamy sand soils with less than 2% organic matter.

NOTE: Use the higher rates in the recommended ranges in areas of heavy weed infestations.

Preplant application section of Lasso & STR & Bladex on pg 40 & 41 omitted.

LASSO® plus ATRAZINE plus PROWL™ (TANK MIXTURE)

Apply this tank mixture in water or sprayable fluid fertilizer solutions for control of triazine resistant lambsquarter and pigweed in addition to yellow nutsedge and the annual grasses and broadleaf weeds listed in the "WEEDS CONTROLLED" section of this label.

APPROVED APPLICATION SYSTEMS

Ground—Broadcast boom; banded

Aerial—Fixed wing; helicopter

Reference: The "APPLICATION SYSTEMS" section of this label provides information on the application of this tank mixture using the system selected.

APPROVED APPLICATION METHOD

Preemergence Surface—Apply this tank mixture after planting, before crop and weed emergence and within 5 days after last preplant tillage operation.

DO NOT preplant incorporate this tank mixture as serious crop injury can result.

Reference: The "APPLICATION TIMING AND METHODS" section of this label provides detailed information and procedures for the application timing and method selected.

NOTE: PLANT ONLY CORN, PEANUTS, SORGHUM OR SOYBEANS* the year following use of this mixture.

*There is a possibility of crop injury due to carryover of atrazine if soybeans are planted the following year. DO NOT plant soybeans the year following use of this tank mixture on furrow irrigated corn. DO NOT graze treated area or feed treated forage to livestock for 21 days following application of this tank mixture.

RECOMMENDED RATES—Refer to the following tables for the recommended broadcast treatment rates for this tank mixture. Applications which are not consistent with recommendations in this label may result in unsatisfactory weed control or crop injury.

Soil Textural Group*	Broadcast Rate Per Acre (Quarts)		
	Less than 3% Organic matter		
	LASSO® +	atrazine 4L** +	Prowl
Coarse	1.5 to 2	0.75 to 2	0.75
Medium	1.5 to 2	1 to 2	0.75 to 1
Fine	2 to 2.25	1.25 to 2	1

Soil Textural Group*	Broadcast Rate Per Acre (Quarts)		
	3% Or More Organic Matter		
	LASSO +	atrazine 4L** +	Prowl
Coarse	1.5 to 2	1 to 2	0.75 to 1
Medium	2 to 2.25	1.25 to 2	1
Fine	2 to 2.75	1.25 to 2	1

* Refer to the "SOIL TEXTURE" section of this label to determine the corresponding soil textural group for the soil to be treated.

** Use rates listed in this label when using atrazine 4L. Use equivalent rates when using atrazine 80W or 90% dry flowable formulations. One quart of atrazine 4L equals 1.25 pounds of atrazine 80W or 1.1 pounds of atrazine 90% dry flowable.

NOTE: Use the higher rates in the recommended ranges in areas of heavy weed infestations.

™ **Prowl is a trademark of American Cynamid Corporation.**

LASSO® plus BLADEX plus PROWL (TANK MIXTURE)

Apply this tank mixture in water or sprayable fluid fertilizer solutions for control of triazine resistant lambsquarter and pigweed in addition to yellow nutsedge and the annual grasses and broadleaf weeds listed in the "WEEDS CONTROLLED" section of this label.

APPROVED APPLICATION SYSTEMS

Ground—Broadcast boom; banded

Aerial—Fixed wing; helicopter

Reference: The "APPLICATION SYSTEMS" section of this label provides information on the application of this tank mixture using the system selected.

APPROVED APPLICATION METHOD

Preemergence Surface—Apply this tank mixture after planting, before crop and weed emergence and within 5 days after last preplant tillage operation.

DO NOT preplant incorporate this tank mixture as serious crop injury can result.

Reference: The "APPLICATION TIMING AND METHODS" section of this label provides detailed information and procedures for the application timing and method selected.

RECOMMENDED RATES: Refer to the following tables for the recommended broadcast treatment rates for this tank mixture. Applications which are not consistent with recommendations in this label may result in unsatisfactory weed control or crop injury.

Soil Textural Group*	Broadcast Rate Per Acre (Quarts)		
	1% to 3% Organic matter		
	LASSO®	Bladex + 4L*** +	Prowl
Coarse**	1.5 to 2	1 to 1.2	0.75
Medium	1.5 to 2	1.2 to 1.4	0.75 to 1
Fine	2 to 2.25	1.4 to 1.8	1

Soil Textural Group*	Broadcast Rate Per Acre (Quarts)		
	3% Or More Organic Matter		
	LASSO	Bladex + 4L*** +	Prowl
Coarse**	1.5 to 2	1.2 to 1.6	0.75 to 1
Medium	2 to 2.25	1.6 to 2	1
Fine	2 to 2.75	1.8 to 22	1

* Refer to the "SOIL TEXTURE" section of this label to determine the corresponding soil textural group for the soil to be treated.

** Do not use this tank mixture on sand or loamy sand soils with less than 2% organic matter.

*** Use rates listed in this label when using Bladex 4L. Use equivalent rates when using Bladex 80W. One quart of Bladex 4L equals 1.25 lbs. of Bladex 80W.

NOTE: Use the higher rates in the recommended ranges in areas of heavy weed infestations.

LASSO® plus ATRAZINE plus LINURON (TANK MIXTURE)

FOR USE ONLY IN DELAWARE, MARYLAND, NEW JERSEY, NEW YORK, PENNSYLVANIA, VIRGINIA AND WEST VIRGINIA

Apply this tank mixture in water or sprayable fluid fertilizer solutions for control of triazine resistant lambsquarter and pigweed in addition to yellow nutsedge and the annual grasses and broadleaf weeds listed in the "WEEDS CONTROLLED" section of this label.

APPROVED APPLICATION SYSTEMS

Ground—Broadcast boom; banded

Reference: the "APPLICATION SYSTEMS" section of this label provides information on the application of this tank mixture using the system selected.

APPROVED APPLICATION METHOD

Preemergence Surface—Apply this tank mixture after planting, before crop and weed emergence and within 5 days after last preplant tillage operation.

Reference: The "APPLICATION TIMING AND METHODS" section of this label provides detailed information and procedures for the application timing and method selected.

NOTE: PLANT ONLY CORN, PEANUTS, SORGHUM, OR SOYBEANS* the year following use of this mixture.

*There is a possibility of crop injury due to carryover of atrazine if soybeans are planted the following year. DO NOT plant soybeans the year following use of this tank mixture on furrow irrigated corn. DO NOT graze treated area or feed treated forage to livestock for 21 days following application of this tank mixture.

RECOMMENDED RATES: Refer to the following tables for the recommended broadcast treatment rates for this tank mixture. Applications which are not consistent with recommendations in this label may result in unsatisfactory weed control of crop injury.

Soil Textural Group*	Broadcast Rate Per Acre (Quarts)		
	Less than 3% Organic matter		
	LASSO® +	atrazine 4L** +	Linuron 4L***
Coarse—sandy loam only	1.5 to 2	0.75 to 2	0.33
Medium	1.5 to 2	1 to 2	0.5
Fine	2 to 2.25	1.25 to 2	0.5

Soil Textural Group*	Broadcast Rate Per Acre (Quarts)		
	3% Or More Organic Matter		
	LASSO +	atrazine 4L** +	Linuron 4L***
Coarse—sandy loam only	1.5 to 2	1 to 2	0.5
Medium	2 to 2.25	1.25 to 2	0.5
Fine	2 to 2.75	1.25 to 2	0.5

* Refer to the "SOIL TEXTURE" section of this label to determine the corresponding soil textural group for the soil to be treated.

** Use rates listed in this label when using atrazine 4L. Use equivalent rates when using atrazine 80W or 90% dry flowable formulations. One quart of atrazine 4L equals 1.25 pounds of atrazine 80W or 1.1 pounds of atrazine 90% dry flowable.

*** Use the rates listed in this label when using linuron 4L. Use equivalent rates when using linuron 50WP or 50% dry flowable formulations. One quart of linuorn 4L equals 2 pounds of linuron 50WP or 2 pounds of linuron 50% dry flowable.

NOTE: Use the higher rates in the recommended ranges in areas of heavy weed infestations.

CONSERVATION OR MINIMUM TILLAGE SYSTEMS TANK MIXTURES

AT-PLANTING APPLICATION

When applied as directed under the conditions described, these tank mixtures control many emerged annual weeds, suppress many emerged perennial weeds and give preemergence control of many annual grasses and weeds when corn will be planted directly into a cover crop, established sod or in previous crop residues. These tank mixtures will not control regrowth from perennial weeds.

DO NOT APPLY BY AIR.

For mixing instructions, see the "MIXING, SPRAYING AND HANDLING INSTRUCTIONS" section of this label.

■ **LASSO® plus GRAMOXONE SUPER** ■

or

■ **LASSO plus ATRAZINE plus GRAMOXONE SUPER** ■

or

■ **LASSO plus BLADEX plus GRAMOXONE SUPER** ■

or

■ **LASSO plus PRINCEP™ plus GRAMOXONE SUPER** ■

or

■ **LASSO plus ATRAZINE plus BLADEX plus GRAMOXONE SUPER** ■

or

■ **LASSO plus ATRAZINE plus PRINCEP plus GRAMOXONE SUPER** ■

Apply these tank mixtures with Gramoxone Super in 20 to 60 gallons of water spray solution per acre immediately before, during or after planting, but BEFORE CROP EMERGENCE. As density of stubble, crop residue or weeds increases, spray gallonage and rate should be increased within the recommended ranges to insure complete coverage. In the absence of emerged vegetation, delete the Gramoxone Super portion of these tank mixtures.

TMPrincep is a trademark of Ciba-Geigy Corporation.

CONTROL OR SUPPRESSION OF EMERGED WEEDS ■

GRAMOXONE SUPER ■

When used as directed, Gramoxone Super in tank mixture with LASSO and atrazine controls many emerged annual weeds and suppresses many emerged perennial weeds.

Broadcast Treatment—Apply 1.5 to 2.5 pints of Gramoxone Super in these tank mixture immediately before, during or after planting but BEFORE CROP EMERGENCE. Use the 2.5 pints rate if any weeds are 4 to 6 inches tall. This mixture may not control weeds taller than 6 inches. As density of stubble, crop residue or weeds increases, spray gallonage should be increased within the recommended range for complete coverage. Add a nonionic spreader surfactant (approved for use on crops) containing at least 75 percent surfactant active agent at 8 ounces per 100 gallons of diluted spray. REFER TO THE GRAMOXONE SUPER LABEL FOR PRECAUTIONARY STATEMENTS.

PREEMERGENCE WEED CONTROL ■

LASSO®■

For weeds controlled preemergence, see the sections of this label on LASSO. See the following table for recommended rates of LASSO in this tank mixture on various soil types:

Soil Textural Group*	Broadcast Rate Per Acre (Quarts) LASSO®
Coarse	2.5
Medium	3
Fine	3 to 3.5

* Refer to the "SOIL TEXTURE" section of the label to determine the corresponding soil textural group for the soil to be treated.

Use the higher rate of LASSO in the recommended ranges in areas of heavy grass infestations, or when organic matter content is 3 percent or more.

LASSO® plus ATRAZINE ■

For weeds controlled preemergence, see the sections of this label for LASSO and LASSO plus atrazine.

See the following table for recommended rates of LASSO plus atrazine in this tank mixture on various soil types.

Soil Textural Group*	Broadcast Rate Per Acre (Quarts)	
	LASSO® +	Atrazine 4L**
Coarse—sandy loam only	1.75 to 2	1 to 2
Medium	2 to 2.5	1.2 to 2
Fine	2.5 to 3	1.6 to 2

* Refer to the "SOIL TEXTURE" section of this label to determine the corresponding soil textural group for the soil to be treated.

** Use rates listed in this label when using atrazine 4L. Use equivalent rates when using atrazine 80W or 90% dry flowable formulations. One quart of atrazine 4L equals 1.25 pounds of atrazine 80W or 1.1 pounds of atrazine 90% dry flowable.

Use the higher rate of LASSO in the recommended ranges in areas of heavy grass infestation or when fall panicum or crabgrass will be present.

Use the higher rate of atrazine in the recommended ranges on soils with greater than 3% organic matter.

NOTE: PLANT ONLY CORN, PEANUTS, SORGHUM OR SOYBEANS the year following use of this mixture.

There is a possibility of crop injury due to carryover of atrazine if soybeans are planted the following year. DO NOT plant soybeans the year following use of this tank mixture on furrow irrigated corn.

DO NOT graze treated area or feed treated forage to livestock for 21 days following application of this tank mixture. ■

LASSO® plus BLADEX ■

For weeds controlled preemergence, see the "WEEDS CONTROLLED" section of the label for LASSO and this tank mixture.

See the following table for recommended rates of LASSO plus Bladex in this tank mixture with Roundup on various soil types.

Soil Textural Group*	Broadcast Rate Per Acre (Quarts)	
	LASSO® +	Bladex 4L**
Coarse-sandy loam only	1.75 to 2	1 to 1.6
Medium	2 to 2.5	1.2 to 1.6
Fine	2.5 to 3	1.6 to 2.2

* Refer to the "SOIL TEXTURE" section of this label to determine the corresponding soil textural group for the soil to be treated.

** When using Bladex 80W use equivalent rates. One quart of Bladex 4L equals 1.25 pounds of Bladex 80W.

Use the higher rate of LASSO in the recommended ranges in areas of heavy grass infestation or when fall panicum or crabgrass will be present.

Use the higher rate of Bladex in the recommended ranges on soils with greater than 3% organic matter.

NOTE: Do not use this mixture on sand or loamy sand soils with less than 2% organic matter.

LASSO® plus PRINCEP

For weeds controlled preemergence see the "WEEDS CONTROLLED" section of the labels for LASSO and Princep herbicides.

See the following table for recommended rates of LASSO plus Princep in this tank mixture on various soil types.

Soil Textural Group*	Broadcast Rate Per Acre (Quarts)	
	LASSO® +	Princep 4L**
Coarse-sandy loam only	1.75 to 2	1 to 1.2
Medium	2 to 2.5	1.2 to 1.6
Fine	2.5 to 3	1.6 to 2.2

* Refer to the "SOIL TEXTURE" section of this label to determine the corresponding soil textural group for the soil to be treated.

** When using Princep 80W use equivalent rates. One quart of Princep 4L equals 1.25 pounds of Princep 80W.

Use the higher rate of LASSO in the recommended ranges in areas of heavy grass infestation or when fall panicum or crabgrass will be present.

Use the higher rate of Princep in the recommended ranges on soils with greater than 3% organic matter.

NOTE: LAND TREATED WITH PRINCEP SHOULD NOT BE PLANTED TO ANY CROP EXCEPT CORN FOR ONE YEAR FOLLOWING TREATMENT AS CROP INJURY MAY OCCUR. AFTER HARVEST OF TREATED CROP, PLOW AND THOROUGHLY TILL THE SOIL IN THE FALL OR SPRING TO MINIMIZE POSSIBLE INJURY TO SPRING SEEDED ROTATIONAL CROPS. ■

LASSO® plus ATRAZINE plus BLADEX■

This tank mixture may be applied for additional preemergence control of heavy weed infestation and/or hard-to-control weeds listed in the "WEEDS CONTROLLED" section of this label. Refer to the following rate table to determine the recommended rates for this tank mixtures on various soil types.

Soil Textural Group*	Broadcast Rate Per Acre (Quarts)**		
	LASSO® +	atrazine 4L** +	Bladex 4L***
Coarse****	1.75 to 2	0.75 to 1	1 to 1.25
Medium	2 to 2.5	1 to 1.25	1 to 2
Fine	2.25 to 3	1 to 1.25	1 to 2

* Refer to the "SOIL TEXTURE" section of this label to determine the corresponding soil textural group for the soil to be treated.

** Use rates listed in this label when using atrazine 4L. Use equivalent rates when using atrazine 80W or 90% dry flowable formulations. One quart of atrazine 4L equals 1.25 pounds of atrazine 80W or 1.1 pounds of atrazine 90% dry flowable.

*** Use rates listed in this label when using Bladex 4L. Use equivalent rates when using Bladex 80W. One quart of Bladex 4L equals 1.25 pounds of Bladex 80W.

**** DO NOT use this mixture on sand or loamy sand soils with less than 2% organic matter.

Use the higher rates in the recommended ranges in area of heavy weed infestations.

NOTE: PLANT ONLY corn, peanuts, sorghum or soybeans* the year following use of this mixture.

*There is a possibility of crop injury due to carryover of atrazine if soybeans are planted the following year. DO NOT plant soybeans the year following use of this tank mixture on furrow irrigated corn.

DO NOT graze treated area or feed treated forage to livestock for 21 days following application of this tank mixture. ■

LASSO® plus ATRAZINE plus PRINCEP■

For additional preemergence control of heavy weed infestation and/or hard-to-control weeds, a three-way tank mixture of LASSO herbicide plus atrazine plus Princep may be applied. Refer to the following rate table to determine the recommended rates for this tank mixture on various soil types.

Soil Textural Group*	Broadcast Rate Per Acre (Quarts)**		
	LASSO® +	atrazine 4L*** +	Princep 4L***
Coarse	1.75 to 2	0.75 to 1	0.75 to 1
Medium	2 to 2.5	1 to 1.25	1 to 1.5
Fine	2 to 3	1 to 1.25	1 to 1.5

* Refer to the "SOIL TEXTURE" section of this label to determine the corresponding soil textural group for the soil to be treated.

** When using 4L formulations use the rates listed. Use equivalent rates when using 80W or 90% dry flowable formulations. One quart of the 4L formulation of these products equals 1.25 pounds of the 80W formulation. One quart of atrazine 4L or Princep 4L equals 1.1 pounds of the 90% dry flowable formulations of these products.

Use the higher rates of atrazine and Princep in the recommended ranges on soils with greater than 3% organic matter.

In the areas of heavy grass infestation, use the higher rate of LASSO in the recommended ranges when fall panicum, crabgrass, or soils with greater than 3% organic matter will be present.

NOTE: LAND TREATED WITH PRINCEP SHOULD NOT BE PLANTED TO ANY CROP EXCEPT CORN FOR ONE YEAR FOLLOWING TREATMENT AS CROP INJURY MAY OCCUR. AFTER HARVEST OF TREATED CROP, PLOW AND THOROUGHLY TILL THE SOIL IN THE FALL OR SPRING TO MINIMIZE POSSIBLE INJURY TO SPRING SEEDED ROTATIONAL CROPS. PLANT ONLY CORN, PEANUTS, SORGHUM OR SOYBEANS† THE YEAR FOLLOWING USE OF THIS MIXTURE.

† There is a possibility of crop injury due to carryover of atrazine if soybeans are planted the following year.

DO NOT plant soybeans the year following use of this tank mixture on furrowed irrigated corn.

DO NOT GRAZE TREATED AREA OR FEED TREATED FORAGE TO LIVESTOCK FOR 21 days following application of this tank mixture.

EARLY PREPLANT APPLICATION

LASSO® plus ATRAZINE (TANK MIXTURE)

For use in no-till and other conservation tillage systems in DELAWARE, ILLINOIS, INDIANA, IOWA, KANSAS, KENTUCKY, MARYLAND, MINNESOTA, MISSOURI, NEBRASKA, NORTH CAROLINA, OHIO, SOUTH DAKOTA, VIRGINIA, AND WISCONSIN.

This tank mixture, when applied in a single application (alone or in a 3-way combination with Bladex or Princep), split application or as a sequential application to Bladex or Princep in early preplant programs, will provide preemergence control or reduced competition of listed annual grasses and broadleaf weeds.

If emerged weeds are present at the time of treatment Gramoxone Super or 2,4-D should be added to the product according to the directions for use on their respective product labels. If unsatisfactory weed control occurs (due to excessively dry or excessively wet conditions) following the earlier application, a postemergence application of an appropriate labelled grass and/or broadleaf weed herbicide may be used. If a postemergence treatment includes the herbicide used early preplant, do not exceed the labelled rate for corn on a given soil texture. Observe all precautions and limitations on the labels for LASSO, atrazine, Bladex, Princep, Gramoxone Super, 2,4-D and other post-emergence herbicides before use of these products.

RECOMMENDED RATE AND TIMING OF APPLICATION

Single application—Application of this tank mixture should be made less than 30 days before planting but prior to weed emergence. This application should be restricted to areas north of Interstate 70 when west of the Mississippi River and north of Interstate 64 when east of the Mississippi River where cool spring weather minimizes or slows biodegradation of herbicides.

See the following table for recommended rates.

Soil Textural Group	Broadcast Rate Per Acre (Quarts)	
	LASSO® +	atrazine 4L
Coarse*	2.25 to 2.5	1.25 to 1.5
Medium	2.5 to 3	1.5 to 2
Fine	3 to 3.75	1.75 to 2.25

This tank mixture may also be mixed in a 3-way combination with 1 to 1.25 quarts per acre of Princep or 1.5 to 2 quarts per acre of Bladex to provide improved control of fall pancium and crabgrass.

* Applications on coarse soils should *not* be made more than 2 weeks prior to planting.

** Use the lower rate where light weed infestation is expected and the higher rate where heavy weed infestation is expected.

Split application—Apply 60 percent of the recommended rate as a split treatment prior to weed emergence and no more than 45 days prior to planting and the remaining 40 percent at or immediately following planting.

FOR CONTROL OF FALL PANCIUM, CRABGRASS OR BROADLEAF SIGNALGRASS USE THE SEQUENTIAL APPLICATION OF THIS TANK MIXTURE FOLLOWING BLADEX OR PRINCEP.

See the following table for recommended rates.

Soil Textural Group	Broadcast Rate Per Acre (Quarts)	
	LASSO® +	atrazine 4L
Coarse	2.25 to 2.75	1.25 to 1.75
Medium	2.5 to 3	1.5 to 2
Fine	3 to 3.75	1.75 to 2.25

*Use the lower rate where light weed infestation is expected and the higher rate where heavy weed infestation is expected.

Sequential application—Application of this tank mixture following Bladex or Princep should be utilized for the control of fall panicum, crabgrass or broadleaf signalgrass. Apply 1.5 to 2 quarts per acre of Bladex or 1 to 1.25 quarts per acre of Princep prior to weed emergence and no more than 45 days prior to planting. At or immediately following planting apply the recommended rate of this tank mixture.

Following applications of Bladex or Princep see the following table for recommended rates.

Soil Textural Group	Broadcast Rate Per Acre (Quarts)	
	LASSO® +	atrazine 4L
Coarse	1.75 to 2.5	1 to 1.5
Medium	2.25 to 2.75	1.25 to 1.75
Fine	2.5 to 3	1.5 to 2

* Use the lower rate where light weed infestaton is expected and the higher rate where heavy weed infestation is expected.

Use rates listed in this label when using the 4L forumlation of atrazine, Bladex or Princep. Use equivalent rates of the 80W or 90% dry flowable formulations of atrazine, Bladex or Princep. One quart of the 4L formulation of atrazine, Bladex or Princep equals 1.25 pounds of the 80W and 1.1 pounds of the 90% dry flowable formulations of atrazine, Bladex or Princep.

DO NOT use Bladex on coarse textured soils or soils of less than 1 percent organic matter.

Plant only corn, peanuts, sorghum (milo) or soybeans the year following the use of this program. There is a possibility of crop injury due to carryover of atrazine if soybeans are planted the following year.

NOTE: Soil treated with Princep should not be planted to any crop excpet corn for one year following treatment as crop injury may occur. After harvest of the treated crop, plow and thoroughly till the soil in the fall or spring to minimize possible injury to spring seeded rotational crops.

SOYBEANS

LASSO®

Apply one treatment of this product in water or sprayable fluid fertilizer solution for control of yellow nutsedge and the annual grasses and broadleaf weeds listed in the "WEEDS CONTROLLED" section of this label.

APPROVED APPLICATION SYSTEMS

Ground—Broadcast boom; banded

Aerial—Fixed wing; helicopter

Center Pivot Irrigation

Dry Bulk Fertilizer Impregnation

Reference: The "APPLICATION SYSTEMS" section of this label provides detailed information on the proper application of this product using the system selected.

APPROVED APPLICATION METHODS

Preplant Incorporated—Apply this product within 7 days prior to planting and shallowly incorporate or Surface Blend into the upper 1 to 2 inches of soil.

Preemergence Surface—Apply this product after planting, before crop and weed emergence and within 5 days after last preplant tillage operation.

Reference: The "APPLICATION TIMING AND METHODS" section of this label provides detailed information and procedures for the application timing and method selected.

RECOMMENDED RATES: Refer to the following tables for the recommended broadcast treatment rates for this product. Applications which are not consistent with recommendations in this label may result in unsatisfactory weed control or crop injury.

Surface Application Rates

Soil Textural Group*	Broadcast Rate Per Acre (Quarts)**	
	Less Than 3% Organic Matter	3% or More Organic Matter
Coarse***	2.25	2 to 2.25
Medium	2 to 2.75	2 to 2.75
Fine	2 to 2.75	2.5 to 3.25

* Refer to the "SOIL TEXTURE" section of this label to determine the corresponding soil textural group for the soil to be treated.

** Use the higher rate in the recommended range in areas of heavy weed infestations. Use a minimum of 2.5 quarts of this product per acre on coarse soils and 3 to 4 quarts on medium or fine-textured soils to control black or hairy nightshade.

*** When applied through center pivot irrigation or under sprinkler irrigation systems on coarse soils, use a minimum of 3 quarts per acre of this product.

On coarse soils with 10% or more organic matter, use 4 quarts of LASSO herbicide per acre.

Preplant Incorporated Application Rates

Soil Textural Group*	Broadcast Rate Per Acre (Quarts)**	
	Less Than 3% Organic Matter	3% or More Organic Matter
Coarse***	2.5	2.5
Medium	2.5 to 3	2.5 to 3
Fine	2.75 to 3	3 to 3.5

* Refer to the "SOIL TEXTURE" section of this label to determine the corresponding soil textural group for the soil to be treated.
** Use the higher rate in the recommended range in areas of heavy weed infestation. Use a minimum of 3 quarts per acre on coarse-textured soils for control of yellow nutsedge or red rice. Rates may be increased to a maximum of 4 quarts on any soil type when heavy infestations of yellow nutsedge or red rice are present.
*** On coarse soils with 10% or more organic matter, use 4 quarts of LASSO herbicide per acre.

NOTE: Do not feed soybean hay from treated crop.

LASSO® plus AMIBEN (TANK MIXTURE)

Apply one treatment of this tank mixture in water or sprayable fluid fertilizer solution for control of the annual grasses and broadleaf weeds listed in the "WEEDS CONTROLLED" section of this label.

DO NOT use this tank mixture on sand or loamy sand soils as crop injury may result.

APPROVED APPLICATION SYSTEMS

Ground—Broadcast boom; banded

Aerial—Fixed wing; helicopter

Reference: The "APPLICATION SYSTEMS" section of this label provides detailed information on the application of this tank mixture using the system selected.

APPROVED APPLICATION METHODS

Preplant Incorporated—Apply this tank mixture within 7 days prior to planting and shallowly incorporate or Surface Blend into the upper 1 to 2 inches of soil.

Preemergence Surface—Apply after planting, before crop and weed emergence and within 5 days after last preplant tillage operation.

Reference: The "APPLICATION TIMING AND METHODS" section of this label provides detailed information and procedures for the application timing and method selected.

RECOMMENDED RATES:

Preplant Incorporated—Apply 2.5 to 3 quarts of LASSO plus 4 quarts of Amiben per acre.

NOTE: Use the 3 quart per acre rate of LASSO in areas of heavy weed infestation. Use 3 to 4 quarts of LASSO when black or hairy nightshade, yellow nutsedge or red rice is present.

Preemergence Surface—Apply 2 quarts of LASSO plus 4 quarts of Amiben per acre.

NOTE: Use a minimum of 2.5 quarts of LASSO on coarse soils and 3 to 4 quarts in this tank mixture on medium or fine textured soils to control black or hairy nightshade.

NOTE: Do not feed soybean hay from treated crop.

LASSO® plus CANOPY (TANK MIXTURE)

Apply this tank mixture in water or sprayable fluid fertilizer solution for control of annual grasses and broadleaf weeds in the "WEEDS CONTROLLED" section of this label.

NOTE: Observe all precautions and limitations on the LASSO and Canopy labels, including precautions on soil pH, sensitive soybean varieties, minimum recropping interval and rotational guidelines.

DO NOT plant rice or peanuts within 18 months after application of this tank mixture.

APPROVED APPLICATION SYSTEMS

Ground—Broadcast boom

Reference: The "APPLICATION SYSTEMS" section of this label provides detailed information on the proper application of this tank mixture using the system selected.

APPROVED APPLICATION METHODS

Preplant Incorporated—Apply this tank mixture within 7 days prior to planting and shallowly incorporate or Surface Blend into the upper 1 to 2 inches of soil. Preplant incorporation may cause slight soybean injury early, but soybeans should recover with no subsequent effect on growth or yield.

Preemergence Surface—Apply this tank mixture after planting, before crop and weed emergence and within 5 days after last preplant tillage operation.

Reference: The "APPLICATION TIMING AND METHODS" section of this label provides detailed information and procedures for application timing and method selected.

RECOMMENDED RATES: Refer to the following tables for the recommended broadcast treatment rates for this tank mixture. Applications which are not consistent with recommendations in this label may result in unsatisfactory weed control or crop injury.

Surface Application Rates

Soil Textural Group*	Broadcast Rate Per Acre**		
	½ To 3% Organic Matter	3% or More Organic Matter	
	LASSO® (Quarts)	LASSO (Quarts)	Canopy (Ounces)
Coarse—sandy loam only	1.75 to 2.25	1.75 to 2.25	6
Medium	1.75 to 2.25	2 to 2.5	8
Fine	2 to 2.5	2.25 to 2.75	10

* Refer to the "SOIL TEXTURE" section of this label to determine the corresponding soil textural group for the soil to be treated. DO NOT APPLY TO SOIL WITH LESS THAN ½ % organic matter.
** Use the higher rate of LASSO in the recommended ranges in areas of heavy grass infestations, heavy crop residue or when broadleaf signalgrass is present.

Preplant Incorporated Application Rates

Textural Group*	Broadcast Rate Per Acre**		
	½ to 3% Organic Matter	3% or More Organic Matter	
	LASSO® (Quarts)	LASSO (Quarts)	Canopy (Ounces)
Coarse—sandy loam only	2 to 2.5	2 to 2.5	6
Medium	2 to 2.5	2.25 to 2.75	8
Fine	2.25 to 2.75	2.5 to 3	10

* Refer to the "SOIL TEXTURE" section of this label to determine the corresponding soil textural group for the soil to be treated. DO NOT APPLY TO SOIL WITH LESS THAN ½ % organic matter.
** Use the higher rate of LASSO in the recommended ranges in areas of heavy grass infestations, heavy crop residue or when broadleaf signalgrass is present. See "WEEDS CONTROLLED" section of this label for specific rate recommendations on seedling johnsongrass, red rice and yellow nutsedge.

LASSO® plus COMMAND (TANK MIXTURE)

Apply this tank mixture in water or sprayable fluid fertilizer solution for control of yellow nutsedge and the annual grasses and broadleaf weeds listed in the "WEEDS CONTROLLED" section of this label.

NOTE: Observe all precautions and limitations on the LASSO and Command labels before use of this tank mixture.

APPROVED APPLICATION SYSTEMS

Ground—Broadcast boom; banded

Reference: The "APPLICATION SYSTEMS" section of this label provides detailed information on the application of this tank mixture using the system selected.

APPROVED APPLICATION METHODS

Preplant Incorporated—Apply this tank mixture within 7 days prior to planting and shallowly incorporate or Surface Blend into the upper 1 to 2 inches of soil.

Reference: The "APPLICATION TIMING AND METHODS" section of this label provided detailed information and procedures for application timing and method selected.

RECOMMENDED RATES: Apply 2 to 2.5 quarts of LASSO plus 1 pint of Command 4E per acre (⅔ pint of Command 6E). In areas of heavy pigweed and lambsquarter infestations, use 2.5 to 3 quarts of LASSO per acre in the tank mixture.

LASSO® plus GEMINI (TANK MIXTURE)

Apply this tank mixture in water or sprayable fluid fertilizer solution for control of annual grasses and broadleaf weeds in the "WEEDS CONTROLLED" section of this label.

NOTE: Observe all precautions and limitations on the LASSO and Gemini labels, including precautions on soil pH, minimum recropping interval and rotational guidelines.

DO NOT plant rice or peanuts within 18 months after application of this tank mixture.

APPROVED APPLICATIONS SYSTEMS

Ground—Broadcast boom; banded

Reference: The "APPLICATION SYSTEMS" section of this label provides detailed information on the application of this tank mixture using the system selected.

APPROVED APPLICATION METHODS

Preemergence Surface—Apply this tank mixture after planting, before crop and weed emergence and within 5 days after last preplant tillage operation.

Reference: The "APPLICATION TIMING AND METHODS" section of this label provides detailed information and procedures for application timing and methods selected.

RECOMMENDED RATES: Refer to the following tables for the recommended broadcast treatment rates for this tank mixture. Applications which are not consistent with recommendations in this label may result in unsatisfactory weed control or crop injury.

Surface Application Rates

Soil Textural Group*	Broadcast Rate Per Acre: ½ To 3% Organic Matter — LASSO® (Quarts)	+	Broadcast Rate Per Acre: 3% or More Organic Matter — LASSO (Quarts)	+	Gemini (Ounces)
Coarse—sandy loam only	1.75 to 2.25		1.75 to 2.25		12 to 16
Medium	1.75 to 2.25		2 to 2.5		16 to 20
Fine	2 to 2.5		2.25 to 2.65		20 to 24

* Refer to the "SOIL TEXTURE" section of this label to determine the corresponding soil textural group for the soil to be treated.
** Use the higher rate of LASSO in the recommended ranges in areas or heavy grass infestations, heavy crop residue or when broadleaf signalgrass is present. Use the higher rate of Gemini in the recommended ranges on soils with higher organic matter or heavy pressure from large deep germinaying weed seeds.

LASSO® plus LEXONE or LASSO plus SENCOR (TANK MIXTURE)

Apply one treatment of this tank mixture in water or sprayable fluid fertilizer solution for control of yellow nutsedge and the annual grasses and broadleaf weeds listed in the "WEEDS CONTROLLED" section of this label.

NOTE: DO NOT use on muck soils or on alkaline soils with greater than 7.4 pH. Crop injury may result if atrazine was applied the year preceding use of this tank mixture.

DO NOT REPLANT CROPS OTHER THAN SOYBEANS FOR 120 DAYS AFTER APPLICATION OF THESE TANK MIXTURES.

APPROVED APPLICATION SYSTEMS

Ground—Broadcast boom; banded

Aerial—Fixed wing; helicopter

Dry Bulk Fertilizer Impregnation

Reference: The "APPLICATION SYSTEMS" section of this label provides detailed information on the application of this tank mixture using the system selected.

APPROVED APPLICATION METHODS

Preplant Incorporated—Apply this tank mixture within 7 days prior to planting and shallowly incorporate or Surface Blend into the upper 1 to 2 inches of soil. Preplant incorporation may cause slight soybean injury early, but soybeans should recover with no subsequent effect on growth or yield.

Preemergence Surface—Apply this tank mixture after planting, before crop and weed emergence and within 5 days after last preplant tillage operation.

Reference: The "APPLICATION TIMING AND METHODS" section of this label provides detailed information and procedures for application timing and method selected.

RECOMMENDED RATES: Refer to the following tables for the recommended broadcast treatment rates for this tank mixture. Applications which are not consistent with recommendations in this label may result in unsatisfactory weed control or crop injury.

Surface Application Rates

Soil Textural Group*	Less Than 3% Organic Matter — LASSO® (Quarts)	+ Lexone 4L** Or Sencor 4F** (Pints)	3% or More Organic Matter — LASSO (Quarts)	+ Lexone 4L** or Sencor 4F** (Pints)
Coarse-sandy loam only	1.75 to 2.25	0.5	1.75 to 2.25	0.75
Medium	1.75 to 2.25	0.5 to 0.75	2 to 2.5	0.75
Fine	2 to 2.5	0.75 to 1	2.25 to 2.75	1

(Broadcast Rate Per Acre**)

* Refer to the "SOIL TEXTURE" section of this label to determine the corresponding soil textural group for the soil to be treated.
** Use the rates listed in this label when using Lexone 4L or Sencor 4F. Use the equivalent rates when using the 50 WP or DF formulations of these products. One pint of Lexone 4L or Sencor 4F equals one pound of the corresponding 50 WP formulation or ⅔ pound of the corresponding DF formulation.

NOTE: DO NOT use on sand or loamy sand soil. Use 1.5 to 2 pints of Lexone 4L or Sencor 4F per acre in this tank mixture when applying on heavy clay soils of the Mississippi Delta. Use a minimum of 2.5 quarts of LASSO on coarse soils and 3 to 4 quarts in this tank mixture on medium or fine-textured soils to control black or hairy nightshade.

Preplant Incorporated Application Rates

Soil Textural Group*	Less Than 3% Organic Matter — LASSO® (Quarts)	Lexone 4L** Or Sencor 4F** (Pint)	3% or More Organic Matter — LASSO (Quarts)	Lexone 4L** or Sencor 4F** (Pint)
Coarse-sandy loam only	2 to 2.5	0.5	2 to 2.25	0.75
Medium	2 to 2.25	0.5 to 0.75	2.25 to 2.75	0.75
Fine	2.25 to 2.75	0.75 to 1	2.5 to 3	1

(Broadcast Rate Per Acre (Quarts)**)

* Refer to the "SOIL TEXTURE" section of this label to determine the corresponding soil textural group for the soil to be treated.
** Use the rates listed in this label when using Lexone 4L or Sencor 4F. Use the equivalent rates when using the 50 WP or DF formulations of these products. One pint of Lexone 4L or Sencor 4F equals one pound of the corresponding 50 WP formulation or ⅔ pound of the corresponding DF formulations.
*** Use the 3 quart per acre rate of LASSO in areas of heavy weed infestation. Use 3 to 4 quarts of LASSO when black or hairy nightshade, yellow nutsedge or red rice is present.

NOTE: Do not use on coarse-textured soils. Do not use on medium or fine-textured soils with less than 1% organic matter. These mixtures, preplant incorporated, may cause slight soybean injury early, but the soybeans should recover and subsequent growth or yield should not be affected.

NOTE: Do not feed soybean hay from treated crop.

LASSO® plus AMIBEN plus LEXONE or SENCOR (TANK MIXTURE)

Apply this tank mixture in water solution for control of the annual grasses and broadleaf weeds listed in the "WEEDS CONTROLLED" section of this label.

NOTE: DO NOT use on sand, loamy sand, or on soils with less than 1% organic matter. DO NOT use the foilage from soybeans treated with LASSO plus Amiben plus Lexone or Sencor for feed or forage.

APPROVED APPLICATION SYSTEMS

Ground—Broadcast boom

Aerial—Fixed wing; helicopter

Reference: The "APPLICATION SYSTEMS" section of this label provides detailed information on the application of this tank mixture using the system selected.

APPROVED APPLICATION METHODS

Preplant Incorporation—Apply this tank mixture within 7 days prior to planting and shallowly incorporate or Surface Blend into the upper 1 to 2 inches of soil. Preplant incorporation may cause slight soybean injury early but soybeans should recover with no subsequent effect on growth or yield.

Preemergence Surface—Apply this tank mixture after planting, before crop and weed emergence and within 5 days after last preplant tillage operation.

Reference: The "APPLICATION TIMING AND METHODS" section of this label provides detailed information and procedures for application timing methods selected.

RECOMMENDED RATES: Refer to the following tables for the recommended broadcast treatment rates. Applications which are not consistent with recommendations in this label may result in unsatisfactory weed control or crop injury.

Soil Textural Group*	Broadcast Rate Per Acre, Less than 3% Organic matter — LASSO® (quarts)	Amiben (quarts)	Lexone 4L** Sencor 4F (pint)
Coarse—sandy loam only	1.75 to 2.25	3	0.5
Medium	1.75 to 2.25	3 to 4	0.5
Fine	2 to 2.5	4 to 5	0.75

Soil Textural Group*	Broadcast Rate Per Acre, 3% Or More Organic Matter — LASSO® (quarts)	Amiben (quarts)	Lexone 4L** Sencor 4L (pint)
Coarse—sandy loam only	1.75 to 2.25	3 to 4	0.75
Medium	2 to 2.5	4 to 5	0.5 to 0.75
Fine	2.25 to 2.75	4 to 6	0.75 to 1

* Refer to the "SOIL TEXTURE" section of this label to determine the corresponding soil textural group for the soil to be treated. On alkaline (calcareous) soils with

a pH of 7.4 or greater, apply Lexone 4L or Sencor 4F at 0.5 pint/A on medium soils and 0.5 to 0.75 pint or corresponding equivalent of WP or DF formulations on fine soils. Use the higher rates of LASSO in the recommended ranges in areas of heavy weed infestations and for preplant incorporation applications. Use a minimum of 2.5 quarts of LASSO on coarse soils and 3 to 4 quarts in this tank mixture on medium or fine soils, when black or hairy nightshade are present. Use 3 to 4 quarts of LASSO per acre preplant incorporated for control of yellow nutsedge.

** Use rates listed in this label when using Lexone 4L or Sencor 4F. Use equivalent rates when using the 50WP or DF formulations of these products. One pint of Lexone 4L or Sencor 4F equals 1 pound of the corresponding 50WP formulation or 2/3 pound of the corresponding DF fomulation.

LASSO® plus LINURON (TANK MIXTURE)

Apply one treatment of this tank mixture in water or sprayable fluid fertilizer solution for control of the annual grasses and broadleaf weeds listed in the "WEEDS CONTROLLED" section of this label.

NOTE: Plant seed at least 1¾ inches deep and DO NOT use on sand or loamy sand or on soils with less than 1% organic matter as crop injury may occur. DO NOT spray over the top of emerged soybeans.

APPROVED APPLICATION SYSTEMS

Ground—Broadcast boom; banded

Reference: The "APPLICATION SYSTEMS" section of this label provides detailed information on the proper application of this tank mixture using the system selected.

APPROVED APPLICATION METHODS

Preemergence Surface—Apply after planting, before crop and weed emergence and within 5 days after last preplant tillage operation.

Reference: The "APPLICATION TIMING AND METHODS" section of this label provides detailed information and procedures for application timing and method selected.

RECOMMENDED RATES: Refer to the following tables for the recommended broadcast treatment rates for this tank mixture. Applications which are not consistent with recommendations in this label may result in unsatisfactory weed control or crop injury.

Surface Application Rates

Soil Textural Group*	Broadcast Rate Per Acre (Quarts): 1 to 3% Organic Matter: LASSO® +	Linuron 4L**	3% or More Organic Matter: LASSO +	Linuron 4L**
Coarse—sandy loam only	1.75 to 2.25	0.5	1.75 to 2.25	0.5 to 0.75
Medium	1.75 to 2.25	0.5 to 0.75	2 to 2.5	0.75 to 1
Fine	2 to 2.5	1	2.25 to 2.75	1.25 to 1.5

* Refer to the "SOIL TEXTURE" section of this label to determine the corresponding soil textural group for the soil to be treated.

** Use the rates listed when using linuron 4L. Use equivalent rates when using linuron 50WP or 50% dry flowable formulations. One quart of linuron 4L equals 2 pounds of linuron 50WP or 2 pounds of linuron 50% dry flowable.

NOTE: Do not feed soybean hay from treated crop.

LASSO® plus LOROX PLUS (TANK MIXTURE)

Apply this tank mixture in water or sprayable fluid fertilizer solution for control of annual grasses and broadleaf weeds listed in the "WEEDS CONTROLLED" section of this label.

Note: Observe all precautions and limitations on the labels for LASSO and Lorox Plus, including precautions on soil pH, minimum recropping interval and rotational guidelines.

APPROVED APPLICATION SYSTEMS

Ground—Broadcast boom; banded

Aerial—Fixed wing; helicopter

Reference: The "APPLICATION SYSTEMS" section of this label provides detailed information on the proper application of this tank mixture using the system selected.

APPROVED APPLICATION METHODS

Preemergence Surface—Apply this tank mixture after planting, before crop and weed emergence and within 5 days after last preplant tillage operation.

Reference: The "APPLICATION TIMING AND METHODS" section of this label provides detailed information and procedures for application timing and method selected.

RECOMMENDED RATES: Refer to the following tables for the recommended broadcast treatment rates for this tank mixture. Applications which are not consistent with recommendations in this label may result in unsatisfactory weed control or crop injury.

Surface Application Rates

Soil Textural Group*	Broadcast Rate Per Acre: ½ to 3% Organic Matter: LASSO® (Quarts) +	3% or More Organic Matter: LASSO (quarts) +	Lexone Plus (Ounces)
Coarse—sandy loam only	1.75 to 2.25	1.75 to 22.5	12 to 14
Medium	1.75 to 2.25	2 2.25	14 to 16
Fine	2 to 2.5	2.25 to 2.75	16 to 18

* Refer to the "SOIL TEXTURE" section of this label to determine the corresponding soil textural group for the soil to be treated.

** Use the higher rate of LASSO in the recommended ranges in areas of heavy grass infestations, heavy crop residue or when broadleaf signalgrass is present. Use the higher rate of Lorox Plus in the recomended ranges on soils with higher organic matter or heavy pressure from large deep germinating weed seeds.

LASSO® plus PREVIEW (TANK MIXTURE)

Apply this tank mixture in water or sprayable fluid fertilizer solution for control of annual grasses and broadleaf weeds listed in the "WEEDS CONTROLLED" section of this label.

Note: Observe all precautions and limitations on the labels for LASSO and Preview, including precautions on soil pH, sensitive soybean varieties, minimum recropping interval and rotational guidelines.

APPROVED APPLICATION SYSTEMS

Ground—Broadcast boom

Reference: The "APPLICATION SYSTEMS" section of this label provides detailed information on the application of this tank mixture using the system selected.

APPROVED APPLICATION METHODS

Preplant Incorporated—Apply this tank mixture within 7 days prior to planting and shallowly incorporate or Surface Blend into the upper 1 to 2 inches of soil. Preplant incorporation may cause slight soybean injury early, but soybeans should recover with no subsequent affect on growth or yield.

Preemergence Surface—Apply this tank mixture after planting, before crop and weed emergence and within 5 days after last preplant tillage operation.

Reference: The "APPLICATION TIMING AND METHODS" section of this label provides detailed information and procedures for application timing and method selected.

RECOMMENDED RATES: Refer to the following tables for the recommended broadcast treatment rates for this tank mixture. Applications which are not consistent with recommendations in this label may result in unsatisfactory weed control or crop injury.

Surface Application Rates

Soil Textural Group*	Broadcast Rate Per Acre**: ½ to 3% Organic Matter: LASSO® (Quarts) +	Preview (Ounces)	3% or More Organic Matter: LASSO (Quarts) +	Preview (Ounces)
Coarse-sandy-loam only	1.75 to 2.25	6	1.75 to 2.25	7
Medium	1.75 to 2.25	7	2 to 2.5	8
Fine	2 to 2.5	8	2.25 to 2.75	9

* Refer to the "SOIL TEXTURE" section of this label to determine the corresponding soil textural group for the soil to be treated. DO NOT APPLY TO SOIL WITH LESS THAN ½% organic matter.

** Use the higher rate of LASSO in the recommended ranges in areas of heavy grass infestations, heavy crop residue or when broadleaf signalgrass is present.

Preplant Incorporated Application Rates

Soil Textural Group*	Broadcast Rate Per Acre**: ½ To 3% Organic Matter: LASSO® (Quarts) +	Preview (Ounces)	3% or More Organic Matter: LASSO (Quarts) +	Preview (Ounces)
Coarse—sandy loam only	2 to 2.5	6	2 to 2.5	7
Medium	2 to 2.5	7	2.25 to 2.75	8
Fine	2.25 to 2.75	8	2.5 to 3	9

* Refer to the "SOIL TEXTURE" section of this label to determine the corresponding soil textural group for the soil to be treated. DO NOT APPLY TO SOIL WITH LESS THAN ½% organic matter.

** Use the higher rate of LASSO in the recommended ranges in areas of heavy grass infestation, heavy crop residue or when broadleaf signalgrass is present. See "WEEDS CONTROLLED" section of this label for specific rate recommendations on seedling johnsongrass, red rice and yellow nutsedge.

LASSO® plus SCEPTER (TANK MIXTURE)

Apply this tank mixture in water or sprayable fluid fertilizer solution for control of yellow nutsedge and the annual grasses and broadleaf weeds listed in the "WEEDS CONTROLLED" section of this label.

NOTE: Observe all precautions and limitations on the LASSO and Scepter labels before use of this tank mixture, including precautions on minimum recropping interval and rotational guidelines.

DO NOT apply this tank mixture in sprayable fluid fertilizer.

APPROVED APPLICATION SYSTEMS

Ground—Broadcast boom; banded

Reference: The "APPLICATION SYSTEMS" section of this label provides detailed information on the application of this tank mixture using the system selected.

APPROVED APPLICATION METHODS

Preplant Incorporated—Apply this tank mixture within 7 days prior to planting and shallowly incorporate or Surface Blend into the upper 1 to 2 inches of soil.

Preemergence Surface—Apply this tank mixture after planting, before crop and weed emergence and within 5 days after last preplant tillage operation.

Reference: The "APPLICATION TIMING AND METHODS" section of this label provides detailed information and procedures for application timing and method selected.

RECOMMENDED RATES: Refer to the following table for the recommended broadcast treatment rates for this tank mixture. Applications which are not consistent with recommendations in this label may result in unsatisfactory weed control or crop injury. See the "WEEDS CONTROLLED" section of this label for recommended rates and application methods for control of certain weeds.

Soil Textural Group*	Broadcast Rate Per Acre**		
	Less than 3% organic matter	3% of More organic matter	
	LASSO® (Quarts)	LASSO (Quarts)	Scepter*** (Pint)
Coarse	1.5 to 2	1.5 to 2	1/2 to 2/3
Medium	1.5 to 2	1.75 to 2.25	1/2 to 2/3
Fine	1.75 to 2.25	2 to 2.5	1/2 to 2/3

* Refer to the "SOIL TEXTURE" section of this label to determine the corresponding soil textural group for the soil to be treated.

** Use the higher rate of LASSO in the recommended ranges in areas of heavy grass infestations, heavy crop residue or when the following weeds are present: fall panicum, barnyardgrass, broadleaf signalgrass, seedling johnsongrass, red rice or yellow nutsedge.

***Use the higher rate of Scepter in the recommended range when red rice, sickleod, beggarweed or velvetleaf are present.

For improved control of heavy velvetleaf infestations, add 1/2 pint of Command 4E herbicide per acre to this tank mixture and preplant incorporate. Observe all precautions and limitations on the Scepter and Command labels before use of this tank mixture.

LASSO® plus TREFLAN (TANK MIXTURE)

Apply this tank mixture in water solution for control of the annual grasses and broadleaf weeds listed in the "WEEDS CONTROLLED" section of this label.

APPROVED APPLICATION SYSTEMS

Ground—Broadcast boom; banded

Reference: The "APPLICATION SYSTEMS" section of this label provides detailed information on the proper application of this tank mixture using the system selected.

APPROVED APPLICATION METHODS

Preplant Incorporated—Apply this tank mixture within 7 days prior to planting and shallowly incorporate or Surface Blend into the upper 1 to 2 inches of soil within eight (8) hours after application.

Reference: The "APPLICATION TIMING AND METHODS" section of this label provides detailed information and procedures for the application timing and method selected.

RECOMMENDED RATES:

Preplant Incorporated—Apply 2.5 to 3 quarts of LASSO plus 0.5 quarts of Treflan per acre. Applications which are not consistent with the recommendation in this label may result in unsatisfactory weed control or crop injury.

CONSERVATION OR MINIMUM TILLAGE SYSTEMS TANK MIXTURES

SOYBEANS ONLY

When applied as directed under the conditions described, these tank mixtures control many emerged annual weeds, suppress many emerged perennial weeds and give preemergence control of many annual grasses and weeds when soybeans will be planted directly into a cover crop, stale seedbed, or in previous crop residues such as wheat stubble.

These tank mixtures will not control regrowth from perennial weeds.

DO NOT APPLY BY AIR.

For mixing instructions, see the "MIXING, SPRAYING AND HANDLING INSTRUCTIONS" section of this label.

LASSO® plus GRAMOXONE SUPER

LASSO plus CANOPY plus GRAMOXONE SUPER

LASSO plus GEMINI plus GRAMOXONE SUPER

LASSO plus LEXONE plus GRAMOXONE SUPER

LASSO plus LINURON plus GRAMOXONE SUPER

LASSO plus LOROX PLUS plus GRAMOXONE SUPER

LASSO plus PREVIEW plus GRAMOXONE SUPER

LASSO plus SCEPTER plus GRAMOXONE SUPER

LASSO plus SENCOR plus GRAMOXONE SUPER

Apply one treatment of these mixtures in 20 to 60 gallons of water per acre immediately before, during or after planting, but BEFORE CROP EMERGENCE. As density of stubble, crop residue or weeds increase, spray gallonage and rate should be increased within the recommended ranges to insure complete coverage. In the absence of emerged vegetation delete the Gramoxone Super portion of these mixes.

CONTROL OR SUPPRESSION OF EMERGED WEEDS

GRAMOXONE SUPER

When used as directed, Gramoxone Super in tank mixture with labeled LASSO tank mixtures controls many emerged annual weeds and suppresses many emerged perennial weeds.

Broadcast Treatment—Apply 1.5 to 2.5 pints of Gramoxone Super (0.25 to 0.5 lb. active) in this tank mixture immediately before, during or after planting but BEFORE CROP EMERGENCE. Use the 2.5 pints rate if any weeds are 4 to 6 inches tall. This mixture may not control weeds taller than 6 inches. As density of stubble, crop residue or weeds increase, spray gallonage should be increased within the recommended range for complete coverage. Add a nonionic spreader surfactant (approved for use on crops) containing at least 75 percent surface active agent. REFER TO THE LABEL FOR GRAMOXONE SUPER FOR PRECAUTIONARY STATEMENTS.

■LASSO® ■

For weeds controlled preemergence, see the section of this label on LASSO. See the following table for recommended rates of LASSO in this tank mixture on various soil types.

Soil Textural Group*	Broadcast Rate Per Acre (Quarts)
	LASSO®
Coarse	2.5
Medium	3
Fine	3 to 3.5

*Refer to the "SOIL TEXTURE" section of the label to determine the corresponding soil textural group for the soil to be treated.

Use the higher rate of LASSO in the recommended ranges in areas of heavy grass infestations, or when organic matter content is 3 percent or more.

LASSO® plus CANOPY

For weeds controlled preemergence, see the "WEEDS CONTROLLED" section of this label for LASSO and LASSO plus Canopy.

NOTE: Observe all precautions and limitations on the LASSO and Canopy labels, including precautions on soil pH, sensitive soybean varieties and rotational crop restrictions.

See the following table for recommended rates of LASSO and Canopy in this tank mixture.

Soil Textural Group*	Broadcast Rate Per Acre**	
	LASSO® + (Quarts)	Canopy (Ounces)
Coarse—sandy loam only	1.75 to 2.25	6
Medium	2 to 2.5	8
Fine	2.5 to 3	10

* Refer to the "SOIL TEXTURE" section of this label to determine the corresponding soil textural group for the soil to be treated.

** Use the higher rate of LASSO and Canopy in areas of heavy weed infestation.

LASSO® plus GEMINI

For weeds controlled preemergence, see the "WEEDS CONTROLLED" section of this label for LASSO plus Gemini.

NOTE: Observe all precautions and limitations on the LASSO and Gemini labels, including precautions on soil pH and rotational crop restrictions.

See the following table for recommended rates of LASSO and Gemini in this tank mixture.

Soil Textural Group*	Broadcast Rate Per Acre**	
	LASSO® (Quarts) +	Gemini (Ounces)
Coarse—sandy loam only	1.75 to 2.25	12 to 16
Medium	2 to 2.5	16 to 20
Fine	2.5 to 3	20 to 24

* Refer to the "SOIL TEXTURE" section of this label to determine the corresponding soil textural group for the soil to be treated.
** Use the higher rate of LASSO and Gemini in areas of heavy weed infestation.

LASSO® plus LEXONE
or
LASSO plus SENCOR

For weeds controlled preemergence, see the "WEEDS CONTROLLED" section of this label for LASSO and LASSO plus Lexone 4L or Sencor 4F.

See the following table for recommended rates of LASSO plus Lexone 4L, or LASSO plus Sencor 4F in this tank mixture on various soil types.

Soil Textural Group*	Broadcast Rate Per Acre	
	LASSO® (quarts) +	Lexone 4L** or Sencor 4F** (pints)
Coarse—sandy loam only	1.75 to 2.25	0.5 to 0.75
Medium	2 to 2.5	0.75 to 1
Fine	2.5 to 3	1 to 1.5***

* Refer to the "SOIL TEXTURE" section of this label to determine the corresponding soil textural group for the soil to be treated.
** Use the rates listed in this label when using Lexone 4L or Sencor 4F. Use the equivalent rates when using the 50WP or DF formulations of these products. One pint of Lexone 4L or Sencor 4F equals one pound of the corresponding 50WP formulation or ⅔ pound of the corresponding DF formulation.
*** On the heavy clay soils of the Mississippi Delta, use 1.5 to 2 pints of Lexone 4L or Sencor 4F per acre.

Use the higher rate of LASSO in the recommended ranges in areas of heavy grass infestations of when fall panicum or crabgrass will be present.

Use the higher rate of Lexone or Sencor in the recommended ranges on soils with greater than 3% organic matter.

DO NOT use this mixture on sand or loamy sand soils as crop injury from Lexone or Sencor may occur.

DO NOT use on muck soils.

DO NOT apply on alkaline soils with a pH greater than 7.4.

Crop injury may occur if atrazine was applied the year before use of this tank mixture.

DO NOT REPLANT CROPS OTHER THAN SOYBEANS FOR 120 DAYS AFTER APPLICATION OF THESE TANK MIXTURES.

PREEMERGENCE WEED CONTROL

LASSO® plus LINURON

For weeds controlled preemergence, see the "WEEDS CONTROLLED" section of this label for LASSO and LASSO plus linuron.

See the following table for recommended rates of LASSO and linuron 4L in this tank mixture.

Soil Textural Group*	Broadcast Rate Per Acre (Quarts)	
	LASSO® +	linuron 4L**
Coarse—sandy loam only	1.75 to 2.25	0.5 to 0.75
Medium	2 to 2.5	0.75 to 1
Fine	2.5 to 3	1 to 1.5

* Refer to the "SOIL TEXTURE" section of this label to determine the corresponding soil textural group for the soil to be treated.
** Use the rates listed when using linuron 4L. Use equivalent rates when using linuron 50WP or 50 percent dry flowable formulations. One quart of linuron 4L equals 2 pounds of linuron 50WP or 2 pounds of linuron 50 percent dry flowable.

Use the higher rate of LASSO in the recommended ranges in areas of heavy grass infestation or when fall panicum or crabgrass will be present.

Use the higher rate of linuron 4L in the recommended ranges on soils with greater than 3% organic matter.

DO NOT use this tank mixture on sand or loamy sand or on soil with less than 1% organic matter as crop injury from linuron may occur.

NOTE: Do not feed soybean hay from treated crop.

LASSO® plus LOROX PLUS

For weeds controlled preemergence, see the "WEEDS CONTROLLED" section of this label for LASSO and LASSO plus Lorox Plus.

NOTE: Observe all precautions and limitations on the labels for LASSO and Lorox Plus, including precautions on soil pH and rotational crop restrictions.

See the following table for recommended rates of LASSO and Lorox Plus in this tank mixture.

Soil Textural Group*	Broadcast Rate Per Acre**	
	LASSO® (Quarts) +	Lorox Plus (Ounces)
Coarse—sandy loam only	1.75 to 2.25	12 to 14
Medium	2 to 2.5	14 to 16
Fine	2.5 to 3	16 to 18

* Refer to the "SOIL TEXTURE" section of this label to determine the corresponding soil textural group for the soil to be treated.
** Use the higher rate of LASSO and Lorox Plus in areas of heavy weed infestation.

LASSO® plus PREVIEW

For weeds controlled preemergence, see the "WEEDS CONTROLLED" section of this label for LASSO and LASSO plus Preview.

NOTE: Observe all precautions and limitations on the labels for LASSO and Preview, including precautions on soil pH, sensitive soybean varieties and rotational crop restrictions.

See the following table for recommended rates of LASSO and Preview in this tank mixture.

Soil Textural Group*	Broadcast Rate Per Acre**	
	LASSO® (Quarts) +	Preview (Ounces)
Coarse—sandy loam only	1.75 to 2.25	6
Medium	2 to 2.5	7
Fine	2.5 to 3	8

* Refer to the "SOIL TEXTURE" section of this label to determine the corresponding soil textural group for the soil to be treated.
** Use the higher rate of LASSO and Preview in areas of heavy weed infestation.

LASSO® plus SCEPTER

For weeds controlled preemergence, see the "WEEDS CONTROLLED" section of this label for LASSO and LASSO plus Scepter.

NOTE: Observe all precautions and limitations on the LASSO and Scepter labels before use of this tank mixture, including minimum recropping interval and rotational guidelines.

See the following table for recommended rates of LASSO and Scepter in this tank mixture.

Soil Textural Group*	Broadcast Rate Per Acre**	
	LASSO® (Quarts)	Scepter (Pint)
Coarse	1.75 to 2.25	⅔
Medium	2 to 2.5	⅔
Fine	2.5 to 3	⅔

*Refer to the "SOIL TEXTURE" section of this label to determine the corresponding textural group for the soil to be treated.
**Use the higher rate of LASSO in areas of heavy weed infestations.

DRY BEANS

Dry Beans (All Types)

FOR USE ONLY WEST OF THE MISSISSIPPI RIVER, EXCEPT IN CALIFORNIA.

LASSO®

Apply one treatment of this product in water solution for control of yellow nutsedge and the annual grasses and broadleaf weeds listed in the "WEEDS CONTROLLED" section of this label.

NOTE: DO NOT APPLY ON DRY BEANS AFTER PLANTING AS CROP INJURY MAY OCCUR. This product may delay crop maturity and/or reduce yield if cold, wet soil conditions occur after planting.

APPROVED APPLICATION SYSTEMS

Ground—Broadcast boom

Reference: The "APPLICATION SYSTEMS" section of this label provides detailed information on the application of this product using the system selected.

APPROVED APPLICATION METHODS

Preplant Incorporated—Apply this product within 7 days prior to planting and shallowly incorporate or Surface Blend into the upper 1 to 2 inches of soil.

Reference: The "APPLICATION TIMING AND METHODS" section of this label provides detailed information and procedures for application timing and method selected.

RECOMMENDED RATES: Apply 2.5 to 3 quarts of this product per acre and shallowly incorporate prior to planting. Use the higher rate in areas of heavy weed infestations or when black or hairy nightshade or yellow nutsedge are present. Applications which are not consistent with recommendations in this label may result in unsatisfactory weed control or crop injury.

LASSO® FOLLOWED BY AMIBEN

Dry Beans (All Types)

FOR USE IN MICHIGAN ONLY

Apply Amiben as a sequential preemergence application to LASSO preplant incorporated.

APPROVED APPLICATION SYSTEMS

Ground—Broadcast boom

Reference: The "APPLICATION SYSTEMS" section of this label provides detailed information on the application of these products using the system selected.

APPROVED APPLICATION METHODS

LASSO®

Preplant Incorporated—Apply this product within 7 days prior to planting and shallowly incorporate or Surface Blend into the upper 1 to 2 inches of soil.

Reference: The "APPLICATION, TIMING AND METHODS" section of this label provides detailed information and procedures for application timing and method selected.

AMIBEN

Preemergence Surface—Apply Amiben or after planting and before crop and weed emergence. Read and carefully follow all use instructions and precautionary statements listed on the respective label.

Reference: The "APPLICATION, TIMING AND METHODS" section of this label provides detailed information and procedures for application timing and method selected.

RECOMMENDED RATES:

LASSO—Apply 2 quarts of this product per acre and shallowly incorporate or Surface Blend prior to planting.

NOTE: DO NOT USE ON SAND OR LOAMY SANDS. DO NOT APPLY ON DRY BEANS AFTER PLANTING AS CROP INJURY MAY OCCUR. This product may delay crop maturity and/or reduce yield if cold, wet soil conditions occur after planting.

AMIBEN—Apply 4 quarts of this product per acre.

NOTE: When LASSO followed by Amiben is used on soils with 6% or more organic matter, weed control may be reduced.

LASSO® plus EPTAM™

(TANK MIXTURE)

Dry Beans (All Types)

FOR USE ONLY IN MICHIGAN AND WEST OF THE MISSISSIPPI RIVER, EXCEPT IN CALIFORNIA

Apply this tank mixture in water solution for control of yellow nutsedge and the annual grasses and broadleaf weeds listed in the "WEEDS CONTROLLED" section of this label.

NOTE: Read the label for Eptam thoroughly for rotational crop restrictions. This tank mixture may delay crop maturity and/or reducing yield of corn, wet soil conditions occur after planting.

APPROVED APPLICATION SYSTEMS

Ground—Broadcast boom

Reference: The "APPLICATION SYSTEMS" section of this label provides detailed information on the application of this tank mixture using the system selected.

APPROVED APPLICATION METHODS

Preplant Incorporated—Apply this tank mixture within 7 days prior to planting and shallowly incorporate into the upper 2 inches of soil immediately.

Reference: The "APPLICATION TIMING AND METHODS" section of this label provides detailed information and procedures for application timing and method selected.

RECOMMENDED RATES: Apply 2 to 3 quarts of LASSO plus 2 to 3 pints of Eptam 7-E per acre. Use the higher recommended rate of LASSO for this tank mixture when heavy weed infestations, black or hairy nightshade or yellow nutsedge are present. USE ONLY THE 2 QUART RATE OF LASSO IN MICHIGAN. Applications which are not consistent with recommendations in this label may result in unsatisfactory weed control or crop injury.

™ Eptam is a trademark of ICI Americas Inc.

LASSO® plus TREFLAN

(TANK MIXTURE)

Dry Beans (All Types)

FOR USE ONLY WEST OF THE MISSISSIPPI RIVER, EXCEPT IN CALIFORNIA

Apply one treatment of this tank mixture in water solution for control of yellow nutsedge and the annual grasses and broadleaf weeds listed in the "WEEDS CONTROLLED" section of this label.

NOTE: Read the label for Treflan thoroughly for rotational crop restrictions. This tank mixture may delay crop maturity and/or reduce yield if cold, wet soil conditions occur after planting.

APPROVED APPLICATION SYSTEMS

Ground—Broadcast Boom

Reference: The "APPLICATION SYSTEMS" section of this label provides detailed information on the application of this tank mixture using the system selected.

APPROVED APPLICATION METHODS

Preplant Incorporated—Apply this tank mixture within 7 days prior to planting and shallowly incorporate into the upper 2 inches of soil within 8 hours after application.

Reference: The "APPLICATION TIMING AND METHODS" section of this label provides detailed information and procedures for application timing and method selected.

RECOMMENDED RATES: Apply 2.5 to 3 quarts of this produce plus 1/2 quart of Treflan per acre. Use the higher rate of this product for this tank mixture in areas of heavy weed infestations or when black or hairy nightshade or yellow nutsedge are present. Applications which are not consistent with recommendations in this label may result in unsatisfactory weed control or crop injury.

LASSO®

Red Kidney Bean Type Only

FOR USE IN ILLINOIS AND WISCONSIN ONLY

Apply one treatment of this product in water solution for control of yellow nutsedge and the annual grasses and broadleaf weeds listed in the "WEEDS CONTROLLED" section of this label.

NOTE: This product may delay crop maturity and/or reduce yield if cold, wet soil conditions occur after planting.

NOTE: Only one application of LASSO herbicide may be made per cropping season.

APPROVED APPLICATION SYSTEMS

Ground—Broadcast boom; banded

Reference: The "APPLICATION SYSTEMS" section of this label provides detailed information on the application of this tank mixture using the system selected.

APPROVED APPLICATION METHODS

Preplant Incorporated—Apply this product within 7 days prior to planting and shallowly incorporate or Surface Blend into the upper 1 to 2 inches of soil.

Preemergence Surface—Apply this product after planting, before crop and weed emergence, and within 5 days after last preplant tillage operation.

Reference: The "APPLICATION TIMING AND METHODS" section of this label provides detailed information and procedures for application timing and method selected.

RECOMMENDED RATES: Apply 2.5 to 3 quarts of this product per acre. Use the higher recommended rate when heavy weed infestations, black or hairy nightshade or yellow nutsedge are present. Preplant incorporate only for control of yellow nutsedge.

LIMA BEANS (GREEN)

FOR USE IN ALL STATES, EXCEPT IN CALIFORNIA

LASSO®

Apply one treatment of this product in water solution for control of yellow nutsedge and the annual grasses and broadleaf weeds listed in the "WEEDS CONTROLLED" section of this label.

NOTE: This product may delay crop maturity and/or reduce yield if cold, wet soil conditions occur after planting.

APPROVED APPLICATION SYSTEMS

Ground—Broadcast boom

Reference: The "APPLICATION SYSTEMS" section of this label provides detailed information on the proper application of this product using broadcast boom equipment.

APPROVED APPLICATION METHODS

Preplant Incorporated—Apply this product within 7 days prior to planting and shallowly incorporate or Surface Blend into the upper 1 to 2 inches of soil.

Preemergence Surface—Apply this product after planting and before crop and weed emergence and within 5 days after last preplant tillage operation.

Reference: The "APPLICATION TIMING AND METHODS" section of this label provides detailed information and procedures for application timing and method selected.

RECOMMENDED RATES: Refer to the following table for recommended broadcast treatment rates.

Surface Application Rates

Soil Textural Group*	Broadcast Rate per Acre (Quarts) LASSO®
Coarse	1.5 to 2
Medium and Fine**	2.5 to 3

* Refer to the "SOIL TEXTURE" section of this label to determine the corresponding soil textural group for the soil to be treated.
** Use the higher rates in the recommended range in areas of heavy weed infestations or when black or hairy nightshade are present.

NOTE: Preemergence applications are not recommended for use in Illinois or Wisconsin as crop injury may occur.

Preplant Incorporated Application Rates

Soil Textural Group*	Broadcast Rate Per Acre (Quarts) LASSO®
Coarse	2
Medium and Fine**	2.5 to 3

* Refer to the "SOIL TEXTURE" section of this label to determine the corresponding soil textural group for the soil to be treated.
** Use the higher rates in the recommended ranges in areas of heavy weed infestations, or when black or hairy nightshade or yellow nutsedge are present.

MUNG BEANS

LASSO®

FOR USE IN OKLAHOMA ONLY

Apply one treatment of this product in water solution for control of the annual grasses and broadleaf weeds listed in the "WEEDS CONTROLLED" section of this label.

APPROVED APPLICATION SYSTEMS

Ground—Broadcast boom; banded

Reference: The "APPLICATION SYSTEMS" section of this label provides detailed information on the application of product using the system selected.

APPROVED APPLICATION METHODS

Preemergence Surface—Apply this product after planting and before crop and weed emergence and within 5 days after last preplant tillage operation.

Reference: The "APPLICATION TIMING AND METHODS" section of this label provides detailed information and procedures for application timing and method selected.

RECOMMENDED RATES: Refer to the following table for recommended broadcast treatment rates.

Surface Application Rates

Soil Textural	Broadcast Rate Per Acre (Quarts) LASSO®
Coarse*	1.5
Medium and Fine	2

* DO NOT USE ON SAND AS CROP INJURY MAY OCCUR.

WOODY ORNAMENTALS

LASSO®

FOR USE ON JUNIPER (JUNIPERUS SPP.) AND YEW (TAXUS SPP.)

Apply one treatment of this product in water solution for control of the annual grasses and broadleaf weeds listed in the "WEEDS CONTROLLED" section of this label.

NOTE: Topical application or contact of this product with green foliage of the woody ornamental may cause injury and is not recommended. In case green foliage is accidentally sprayed, overhead irrigation within 1 hour after application will minimize injury symptoms (i.e. leaf tip burn, discoloration, distortion, etc.). This product is recommended for this use for application by certified, professional applicators only.

DO NOT USE THIS PRODUCT ON SEEDBEDS, UNROOTED CUTTINGS OR PRIOR TO TRANSPLANTING. DO NOT USE THIS PRODUCT IN GREENHOUSES OR OTHER ENCLOSED STRUCTURES, APPLIED THROUGH IRRIGATION SYSTEMS OR MIXED WITH SPRAYABLE FLUID FERTILIZERS. APPLICATIONS DURING PERIODS WHEN TEMPERATURE EXCEEDS 90°F MAY CAUSE INJURY TO ORNAMENTALS.

APPROVED APPLICATION SYSTEMS

Ground—Broadcast boom; banded

Reference: The "APPLICATION SYSTEMS" section of this label provides detailed information on the application of this tank mixture using the system selected.

APPROVED APPLICATION METHODS

Post Transplant Directed—Apply this product as a directed spray after transplanting or to established plantings prior to weed emergence.

Reference: The "APPLICATION TIMING AND METHODS" section of this label provides detailed information and procedures for application timing and method selected.

RECOMMENDED RATES: Apply 4 quarts of this product per acre as a broadcast treatment. Applications which are not consistent with the recommendations in this label may result in unsatisfactory weed control or injury to ornamental plantings.

NOTE: Repeat applications of this product may be required for sustained weed control. DO NOT exceed 3 applications per year or retreat within 21 days.

APPENDIX

LIST OF COMMON AND SCIENTIFIC NAME OF ANNUAL GRASSES AND BROADLEAF WEEDS.

ANNUAL GRASSES

Barnyardgrass
Echinochloa crus-galli
Crabgrass
Digitaria ischaemum
Digitaria sanguinalis
Cupgrass, woolly
Eriochloa villosa
Foxtail, giant
Setaria faberi
Foxtail, green, robust purple, robust white
Setaria viridis
Foxtail, yellow
Setaria lutescens
Goosegrass
Eleusine indica
Johnsongrass, seedling
Sorghum halepense
Millet, proso
Panicum miliaceum
Panicum, browntop
Panicum fasciculatum
Panicum, fall
Panicum dichotomiflorum
Panicum, Texas
Panicum texanum
Rice, red
Oryza sativa
Sandbur, Grassbur
Cenchrus incertus
Shattercane, wild cane
Sorghum bicolor
Signalgrass, broadleaf
Brachiaria platyphylla
Sprangletop, red
Leptochloa filiformis
Witchgrass
Panicum capillare

ANNUAL BROADLEAVES

Beggarweed, Florida
Desmodium tortuosum
Carpetweed
Mollugo verticillata
Cocklebur
Xanthium strumarium
Coffeeweed,
Hemp sesbania
Sesbania exaltata
Copperleaf, hophornbeam
Acalypha spp.
Galinsoga
Galinsoga spp.
Groundcherry, annual
Physalis spp.
Groundcherry, cutleaf
Physalis angulata
Jimsonweed
Datura stramonium
Kochia
Kochia scoparia
Lambsquarters
Chenopodium album
Morningglory:
Tall
Ipomea purpurea
Pitted
Ipomea lacunosa
Ivyleaf
Ipomea hederacea
Entireleaf
Ipomea hederacea var. intergriuscula
Smallflower
Jacquemontia tamnifolia
Mustard
Brassica spp.
Nightshade, black
Solanum nigrum
Nightshade, hairy
Solanum sarrachoides
Pigweed, Carelessweed
Amaranthus spp.
Purslane
Portulaca oleracea
Pusley, Florida
Richardia scabra
Ragweed, common
Ambrosia artemisiifolia
Ragweed, giant
Ambrosia trifida
Sicklepod
Cassia obtusifolia
Sida, prickly; Teaweed
Sida spinosa
Smartweed
Polygonum pensylvanicum
Polygonum persicaria
Starbur, bristly
Acanthospermum hispidum
Sunflower, common
Helianthus annuus
Velvetleaf, Buttonweed
Abutilon theophrasti
Waterhemp
Amaranthus tuberculatos

SEDGE

Nutsedge, yellow
Cyperus esculentus

843.11-000.62/53 EPA Reg. No. 524-344-ZA

In case of an emergency involving this product, Call Collect, day or night, (314) 694-4000.

PONDMASTER™

AQUATIC HERBICIDE

Controls Emerged Weeds and Brush in Farm Ponds and Aquatic Sites
Complete Directions for Use

ACTIVE INGREDIENT:

*Isopropylamine salt of glyphosate	41.0%
INERT INGREDIENTS:	59.0%
TOTAL	100.0%

*Contains 480 grams per litre or 4 pounds of the active ingredient isopropylamine salt of N-(phosphonomethyl)glycine per U.S. gallon. Equivalent to 356 grams per litre or 3 pounds per U.S. gallon of the acid, glyphosate.

®Registered trademark of Monsanto Company

EPA Reg. No. 524-308

1988-1 892.36-000.01/CG

Store in secured place in original container.

Read the entire label before using this product.

Use only according to label instructions.

NOTICE: Buyer assumes all responsibility for safety and use not in accordance with directions. If these terms are not acceptable, return at once unopened.

AVOID CONTACT WITH FOLIAGE, GREEN STEMS OR FRUIT OF CROPS, DESIRABLE PLANTS AND TREES, SINCE SEVERE INJURY OR DESTRUCTION MAY RESULT.

PRECAUTIONARY STATEMENTS

HAZARDS TO HUMANS AND DOMESTIC ANIMALS

KEEP OUT OF REACH OF CHILDREN

WARNING!

CAUSES EYE IRRITATION.

HARMFUL IF SWALLOWED OR ABSORBED THROUGH SKIN.

Do not get in eyes, on skin or on clothing.

Wash thoroughly with soap and water after handling.

FIRST AID: IF IN EYES, immediately flush with plenty of water for at least 15 minutes. Call a physician.

IF ON SKIN, flush with plenty of water while removing contaminated clothing. Wash clothing before reuse.

IF SWALLOWED, this product will cause gastrointestinal tract irritation. Immediately dilute by swallowing water or milk. Call a physician.

KEEP PEOPLE AND PETS OFF TREATED AREAS UNTIL SPRAY SOLUTION HAS DRIED.

DOMESTIC ANIMALS: This product is considered to be relatively nontoxic to dogs and other domestic animals. However, ingestion of this product or large amounts of freshly sprayed vegetation may result in temporary gastrointestinal irritation (vomiting, diarrhea, colic, etc.). If such symptoms are observed, provide the animal with plenty of fluids to prevent dehydration. Call a veterinarian if symptoms persist for more than 24 hours.

In case of an emergency involving this product, Call Collect, day or night, (314) 694-4000.

ENVIRONMENTAL HAZARDS

Do not contaminate water by cleaning of equipment or disposal of wastes.

Treatment of aquatic weeds can result in oxygen loss from decomposition of dead plants. This loss can cause fish suffocation.

PHYSICAL OR CHEMICAL HAZARDS

Mix, store or apply this product ONLY in stainless steel, aluminum, fiberglass, plastic and plastic-lined steel containers. DO NOT MIX, STORE OR APPLY THIS PRODUCT OR SPRAY SOLUTIONS OF THIS PRODUCT IN GALVANIZED STEEL OR UNLINED STEEL (EXCEPT STAINLESS STEEL) CONTAINERS OR SPRAY TANKS. Use of this product in such containers could result in the formation of an explosive hydrogen gas mixture which could flash or explode if ignited by open flames or any other ignition source.

DIRECTIONS FOR USE

It is a violation of Federal law to use this product in any manner inconsistent with its labeling.

Pondmaster is a nonselective herbicide that is applied as a foliar spray for the control or destruction of most emerged aquatic grasses, weeds and brush in farm ponds, irrigation ditches and other aquatic sites.

This product is absorbed by the leaves and moves throughout the stem and roots to control the entire plant. Visible symptoms may require a week or more to appear. Symptoms are a gradual wilting and yellowing of the sprayed plant followed by deterioration of both shoots and roots. This product has no herbicidal activity in the soil or water and will not wash or leach to affect nearby vegetation.

This product controls emerged weeds and brush species in and around all bodies of fresh and brackish water which may be flowing, nonflowing or transient. This includes lakes, rivers, streams, ponds, seeps, irrigation and drainage ditches, canals, reservoirs and similar sites.

This product does not control plants which are either completely submerged or have a majority of their foliage under water. Treated water may be used for irrigation, fishing, swimming, boating or other recreational and domestic purposes after application.

DO NOT APPLY THIS PRODUCT WITHIN ONE-HALF MILE UPSTREAM OF A POTABLE (DRINKING) WATER INTAKE IN FLOWING WATER (I.E., RIVER, STREAM, ETC.) OR WITHIN ONE-HALF MILE OF A POTABLE WATER INTAKE IN A STANDING BODY OF WATER SUCH AS A LAKE, POND OR RESERVOIR.

DO NOT APPLY THIS PRODUCT ON RICE LEVEES WHEN FLOOD WATER IS PRESENT.

DO NOT APPLY IN ESTUARIES.

DO NOT APPLY THROUGH ANY TYPE OF IRRIGATION SYSTEM.

Consult local state fish and game agency and water control authorities before applying this product to public water. **Permits may be required to treat such water.**

THIS PRODUCT CAN INJURE OR DESTROY ALL VEGETATION CONTACTED. AVOID SPRAY DRIFT CONTACT WITH DESIRABLE LAWN GRASSES, FLOWERS, VEGETABLES, SHRUBS OR TREES. DO NOT CONTACT GREEN BARK OF TREES OR SHRUBS. IF DESIRABLE VEGETATION IS CONTACTED, WASH IMMEDIATELY WITH WATER.

STORAGE AND DISPOSAL

Do not contaminate water, foodstuffs, feed or seed by storage or disposal.

STORAGE:

Store this product in its original container and in a secured storage area. In case of SPILL or LEAK, soak up with paper towels and discard in trash.

DISPOSAL:

To dispose of unused product, securely wrap original container in several layers of newspaper and discard in trash.

Empty container retains vapor and product residue. Observe all labeled safeguards until container is discarded.

Do not reuse container. Rinse thoroughly before discarding in trash.

WEEDS CONTROLLED

When applied as recommended under the conditions described, this product controls many common weeds, including the following:

ANNUALS

Barnyardgrass	Pennycress, field
Bluegrass, annual	Pigweed, redroot; smooth
Chickweed	Ragweed, common; giant
Crabgrass	Ryegrass, annual
Foxtail	Sandbur, field
Lambsquarters, common	Shepherdspurse
Lettuce, prickly	Sowthistle
Mustard, Tansy	

PERENNIALS

Alligatorweed	Loosestrife, purple
Bahiagrass	Lotus, American
Bermudagrass	Nutsedge, purple; yellow
Bindweed, field	Orchardgrass
Bluegrass, Kentucky	Phragmites
Bromegrass, smooth	Quackgrass
Canarygrass, reed	Ryegrass, perennial
Cattail	Smartweed, swamp
Clover, red; white	Spatterdock
Cutgrass, giant	Thistle, Canada
Dallisgrass	Torpedograss
Dandelion	Tules, common
Fescue spp.	Vaseygrass
Fescue, tall	Waterhyacinth
Guineagrass	Waterlettuce
Johnsongrass	Waterprimrose

BRUSH AND TREES

Alder	Oak

PERENNIALS

Blackberry	Poison Ivy
Cherry	Raspberry
Dewberry	Rose, multiflora
Elderberry	Saltbush
Florida holly	Tallowtree, Chinese
Honeysuckle	Trumpetcreeper
Kudzu	Waxmyrtle, southern
Maple	Willow

MIXING AND SPRAYING

Hand-Held Sprayer—Add 2-2/3 ounces (1 ounce = 2 tbsp.) of this product to 1 gallon of clean water.

Ensure uniform and complete spray coverage, but do not spray plants to the point that the spray solution runs off. Use coarse spray. Do not spray during windy conditions, as spray drift may cause damage to any vegetation contacted.

Use pump-up type or hand-trigger type sprayers. USE OF HOSE-END SPRAYERS OR SPRINKLE TYPE DEVICES MAY RESULT IN POOR AND/OR ERRATIC RESULTS.

WHEN TO APPLY

Apply to vigorously growing plants. Apply to annual weeds anytime after emergence. For best results on perennial weeds, apply after seedheads, flowers or berries appear. On woody brush or trees, apply after leaves have fully expanded. Applications made at less than the recommended stage of growth may result in reduced control. Where mowing, clipping or tilling has occurred, allow weeds to regrow to the recommended growth stage before spraying this product.

Rainfall or watering occurring within 6 hours after application may reduce effectiveness. Heavy rainfall or irrigation within 2 hours after application may wash the product off the foliage and a repeat treatment may be required. Delay disturbing treated areas for at least 7 days after application. This allows time for this product to move within the plant for the most effective results.

Floating mats of vegetation may require retreatment. If wash-off of sprayed foliage by wave-action occurs within 6 hours of application, then retreatment may be required. Do not retreat within 24 hours following initial treatment. Make applications while traveling upstream to prevent concentration of this product in water.

CLEANING EQUIPMENT

Triple-rinse spray tank and flush all sprayer components with water to remove residues of this product. After thorough cleaning, equipment may be used to apply other products.

Product protected by U.S. Pat. Nos. 3,799,758 and 4,405,531. Other patents are pending.

 892-36-000-01/53

In case of an emergency involving this product, Call Collect, day or night, (314) 694-4000.

This sample label is current as of April 1, 1988. The product descriptions and recommendations provided in this sample label are for background information only. Always refer to the label on the product container before using Monsanto or any other agrichemical product.

This specimen label is intended for use only as a guide in providing general information regarding the use of this product. Monsanto is not responsible for the accuracy of the information contained herein. As labels are subject to revision, always carefully read and follow the label on the product container.

RAMROD®

FLOWABLE HERBICIDE

Selective herbicide for preemergence weed control in field corn, hybrid seed corn, silage corn, sweet corn, grain sorghum (milo), processing green peas, processing pumpkins, and flax.

ACTIVE INGREDIENTS:	
* Propachlor, (2-chloro-N-isopropylacetanilide)	42.0%
INERT INGREDIENTS:	58.0%
TOTAL	100.0%

* Contains 480 grams/liter or 4 pounds/gallon of propachlor.

® Ramrod is a registered trademark of Monsanto Company

Complete Directions for Use.

SHAKE WELL BEFORE USING

EPA Reg. No. 524-331-AA 1988-1 848.21-000.19/CG

Read the entire label before using this product.

Use only according to label instructions.

Read "LIMIT OF WARRANTY AND LIABILITY" before buying or using. If terms are not acceptable, return at once unopened.

REFORMULATION OR REPACKAGING IS PROHIBITED.

LIMIT OF WARRANTY AND LIABILITY

This company warrants that this product conforms to the chemical description on the label and is reasonably fit for the purposes set forth in the Complete Directions for Use label booklet ("Directions") when used in accordance with those Directions under the conditions described therein. NO OTHER EXPRESS OR IMPLIED WARRANTY OF FITNESS FOR PARTICULAR PURPOSE OR MERCHANTABILITY IS MADE. This warranty is also subject to the conditions and limitations stated herein.

Buyer and all users shall promptly notify this company of any claims whether based in contract, negligence, strict liability, other tort or otherwise.

Buyer and all users are responsible for all loss or damage from use or handling which results from conditions beyond the control of this company, including, but not limited to, incompatability with products other than those set forth in the Directions, unusual weather, weather conditions which are outside the range considered normal at the application site and for the time period when the product is applied as well as weather conditions which are outside the application ranges set forth in the Directions, application in any manner not explicitly set forth in the Directions, moisture conditions outside the moisture range specified in the Directions, or the presence of products other than those set forth in the Directions in or on the soil, crop or treated vegetation.

THE EXCLUSIVE REMEDY OF THE USER OR BUYER, AND THE LIMIT OF THE LIABILITY OF THIS COMPANY OR ANY OTHER SELLER FOR ANY AND ALL LOSSES, INJURIES OR DAMAGES RESULTING FROM THE USE OR HANDLING OF THIS PRODUCT (INCLUDING CLAIMS BASED IN CONTRACT, NEGLIGENCE, STRICT LIABILITY, OTHER TORT OR OTHERWISE) SHALL BE THE PURCHASE PRICE PAID BY THE USER OR BUYER FOR THE QUANTITY OF THIS PRODUCT INVOLVED, OR, AT THE ELECTION OF THIS COMPANY OR ANY OTHER SELLER, THE REPLACEMENT OF SUCH QUANTITY, OR, IF NOT ACQUIRED BY PURCHASE, REPLACEMENT OF SUCH QUANTITY. IN NO EVENT SHALL THIS COMPANY OR ANY OTHER SELLER BE LIABLE FOR ANY INCIDENTAL, CONSEQUENTIAL OR SPECIAL DAMAGES.

Buyer and all users are deemed to have accepted the terms of this LIMIT OF WARRANTY AND LIABILITY which may not be varied by any verbal or written agreement.

PRECAUTIONARY STATEMENTS

HAZARDS TO HUMANS AND DOMESTIC ANIMALS

KEEP OUT OF REACH OF CHILDREN

WARNING!

HARMFUL IF SWALLOWED OR ABSORBED THROUGH SKIN.

CAUSES EYE AND SKIN IRRITATION.

MAY CAUSE ALLERGIC SKIN REACTION.

Do not get in eyes, on skin, or on clothing.

Keep container closed.

While handling, wear rubber gloves.

Wash thoroughly after handling.

FIRST AID: IN CASE OF CONTACT, immediately flush eyes or skin with plenty of water for at least 15 minutes while removing contaminated clothing. For eyes, call a physician. Wash clothing before reuse.

IF SWALLOWED, drink 1 or 2 glasses of water and induce vomiting by touching back of throat with finger or blunt object. Do not induce vomiting or give anything by mouth to an unconscious person. Get medical attention.

In case of emergency involving this product, call collect, day or night, (314) 694-4000.

ENVIRONMENTAL HAZARD

This product is toxic to fish and wildlife. Keep out of lakes, streams, or ponds. Do not apply where runoff is likely to occur. Do not apply when weather conditions favor drift from areas treated.

GENERAL INFORMATION

This product is recommended for control of many annual grasses and certain broadleaf weeds. For best results it should be applied to the soil surface before crop or weeds emerge. The seedbed should be fine, firm and free of clods and trash; application should be made within 5 days after the last tillage for weed control. Do not apply when conditions favor drift. Moisture, either as rain or irrigation, is required after application to activate this product. On coarse soils one-third inch of rainfall, and on medium and fine textured and/or high organic soils one-third to three-fourths inch of rainfall is required depending upon original soil moisture. Better results are obtained when moisture occurs within 10 days after application.

Generally lower amounts of rainfall are required when soil moisture is high at time of planting.

NOTE: Use of this product in any manner not consistent with this label may result in injury to persons, animals or crops, or other unintended consequences. Keep container closed to prevent spills and contamination.

Flush sprayer with clean water after use.

Do not use this product or any of its labeled tank mixtures in fields where flood irrigation is to be practiced.

DO NOT apply in sprayable fluid fertilizer after crop emergence since crop injury may result.

Do not graze or feed sorghum (milo) forage, silage or fodder (stubble) from treated fields to dairy animals.

Read and carefully observe the cautionary statements and all other information appearing on labeling of all products used in mixtures.

MIXING AND SPRAYING INSTRUCTIONS

Always predetermine the compatibility of this product or its labeled tank mix with water carrier or sprayable fluid fertilizer carrier by mixing small proportional quantities in advance. See the Standard Sprayable Fluid Fertilizer Compatibility Test section on this label.

Mix this product or its labeled tank mix with the appropriate carrier as follows:

1. Place a 20 to 35 mesh screen or wetting basket over the filling port.
2. Through the screen, fill the sprayer tank one-half full with the appropriate carrier.
3. If a compatibility agent is necessary to improve mixing or to prevent the formation of undesirable and unsprayable gels or precipitates, while agitating add it to the carrier already in the tank. Use only compatibility agents cleared by Federal Food and Drug Administration for this use. Read and follow all directions for use, cautionary statements and all other information appearing on the selected compatibility agent. Check for adequate agitation.
4. Add the required amount of this product. Continue agitation.
5. If a labeled tank mix is being prepared, premix the other product in water. If a wettable powder is used, make a slurry. If flowable atrazine, propazine or Modown are used, premix one part flowable with one part water. Add to tank and continue agitation.
6. Complete filling the sprayer tank with carrier. Remove hose from tank immediately after filling to avoid siphoning back into carrier source. NOTE: Maintain good agitation at all times until the contents of the tank are sprayed, otherwise, settling which can lead to clogging may occur. Close tank valves and drain all lines when spraying is delayed.

Keep by-pass line on or near bottom of tank to minimize foaming. Screen size in nozzle or line strainers should be no finer than 50 mesh. Carefully select proper nozzle to avoid spraying a fine mist. For best results, use flat fan or flood jet nozzle. Check for even distribution of spray droplets. To reduce loss of the chemical due to drift of a fine mist, apply at pressures less than 40 psi.

STANDARD SPRAYABLE FLUID FERTILIZER COMPATIBILITY TEST

Herbicides may not always mix evenly throughout a sprayable fluid fertilizer or the components may separate too quickly to make their combined use of practical value. This may be due to certain characteristics of the different fluid fertilizers. A simple test using small quantities of the components is suggested to provide compatibility potential.

A. Materials Required For A Compatibility Test

1. Two one-quart jars with lid or stopper (marked "with" and "without").
2. Teaspoons (for a more exacting test, a five to ten milliliter (ml.) pipette or graduated cylinder is desirable).
3. A minimum of one quart of the sprayable fluid fertilizer to be tested.
4. The herbicide(s) to be mixed.
5. A compatability agent (the purpose of the adjuvant is to help keep the fertilizer and herbicide(s) in suspension, if this assistance is needed).

B. Procedure

1. Add one pint of the sprayable fluid fertilizer that will be used or other herbicide carrier to each jar marked "with" and "without".

Add One Pint Liquid
Fertilizer To Two
Quart Jars

2. To the jar marked "with", add ¼ TEAspoon or 1.2 milliliters of a suitable compatibility agent; shake gently for five to ten seconds to mix. (¼ TEAspoon is the equivalent of two pints per 100 gallons of liquid fertilizer.)

To Jar Marked "With"
Add Compatibility Agent
And Shake To Mix

3. To each jar add the appropriate amount of herbicide(s). If more than one is used, add them separately with the wettable powders added first, flowables second and liquid last. Shake gently five to ten seconds after each addition.

Add Herbicide(s) To
Both Jars And Shake
To Mix

Herbicide	Rate/A		Amount To Be Added Per Pint Of Sprayable Fluid Fertilizer (Assuming Volume is 25 gal/A) Level TEAspoons		Milliliters
This Product	2.5 qts	=	2.4	or	11.8
	3.0 qts	=	2.9	or	14.2
	3.5 qts	=	3.4	or	16.6
	4.0 qts	=	3.8	or	19.0
	4.5 qts	=	4.3	or	21.4
	5.0 qts	=	4.8	or	23.8
	5.5 qts	=	5.3	or	26.2
	6.0 qts	=	5.8	or	29.6

This compatibility test is designed for 25 gallons of spray per acre with the maximum labeled rate of herbicide. For changes in spray volume or herbicide concentration, make appropriate changes in the ingredients of the test. Regardless of spray volume, the amount of compatibility agent should be equal to two or three pints (two pints = ¼ TEAspoon or 1.2 milliliters, three pints = ⅜ TEAspoon or 1.8 milliliters per pint of sprayable fluid fertilizer) per 100 gallons of liquid fertilizer.

C. Observations And Decisions

1. If the herbicide(s) and the sprayable fluid fertilizer are compatible.
2. If a compatibility agent is necessary.

Five minutes after the final addition and mixing, observe both jars for the formation of large flakes, sludge, gels or other precipitates. Observe if the herbicide(s) cannot be physically mixed with the liquid fertilizer but remains as small oily particles in the solution.

If incompatibility in any form described above occurs in the jar "with" the compatibility agent added, the liquid fertilizer and the herbicide(s) should not be used together in the same spray tank.

If incompatibility as described above occurs in the jar "without" the adjuvant but not in the jar "with" adjuvant, the use of a compatibility adjuvant is recommended.

Both jars should be allowed to stand and be observed periodically for one-half hour. If the separate layers of liquid fertilizer and additives can be resuspended by shaking, commercial application is possible. Wettable powders will either settle to the bottom of the tank or jar, or float to the top, depending upon the density of the fertilizers.

If the herbicide(s) is compatible with fluid fertilizer in the foregoing test without having to use a compatibility agent, fluid fertilizer may be used for the premixing. If it is not compatible, the herbicide(s) should be premixed with water before adding to the spray tank.

APPLICATION INFORMATION

GROUND APPLICATION

Do not apply this product through any type of irrigation system.

Broadcast Treatment—Apply in 15 or more gallons of water or sprayable fluid fertilizer per acre according to this label. Do not apply during periods of gusty winds or winds in excess of 15 mph.

Band Treatment—Apply a broadcast equivalent rate and volume per acre. To determine these:

$$\frac{\text{Band width in inches}}{\text{Row width in inches}} \times \text{Broadcast RATE per acre} = \text{Band RATE per acre}$$

$$\frac{\text{Band width in inches}}{\text{Row width in inches}} \times \text{Broadcast VOLUME per acre} = \text{Band VOLUME per acre}$$

Mount spray equipment on the planter with nozzles behind press wheel and apply this product at planting time.

NOTE: For band application nozzle placement, refer to the nozzle manufacturer's specification as a guide for determining proper nozzle height.

AERIAL APPLICATION

Only Ramrod Flowable or the labeled tank mixtures of Ramrod Flowable plus atrazine and Ramrod Flowable plus propazine may be applied by aerial spray equipment.

AVOID DRIFT—Do not apply when winds are gusty or in excess of 5 miles per hour or when other conditions, including lesser wind velocities, will allow drift to occur. Any resulting drift can damage nearby crops, vegetation or other areas on which treatment is not intended, and can cause application to be streaked or uneven, reducing the effectiveness of this product. In order to ensure uniform aerial application and to avoid overlapping and possible crop injury, it is recommended that 2 flagmen be used, one at each end of the field. Apply this product in 5 or more gallons of water per acre.

Coarse sprays are less likely to drift; therefore, do not use nozzles or nozzle configurations which dispense spray as fine spray droplets. Do not angle nozzles forward into the airstream and do not increase spray volume by increasing nozzle pressure.

CULTIVATION INFORMATION

As long as weeds are controlled, delay cultivation as long as possible.

If weeds seedlings emerge, shallow cultivate or rotary hoe immediately to improve weed control.

If cultivation or rotary hoeing is necessary because of soil crusting or compaction, tillage should be shallow to minimize dilution and displacement of this herbicide or tank mixture.

WEED CONTROL WITH RAMROD FLOWABLE

When applied as directed, under the conditions described, this product CONTROLS the following weeds:

Barnyardgrass Echinoichloa crus-galli	**Goosegrass** Eleusine indica
Carpetweed Mollugo verticillata	**Panicum (fall)** Panicum dichotomiflorum
Crabgrass Digitaria spp.	**Pigweed** Amaranthus spp.
Foxtail (giant) Setaria faberi	**Purslane** Portulaca oleracea
Foxtail (green) Setaria viridis	**Pusley (Florida)** Richardia scabra
Foxtail (yellow) Setaria lutescens	**Ryegrass (annual)** Lolium spp

When applied immediately after planting, under the conditions described, this product at a rate of 5 to 6 quarts per acre on a broadcast basis REDUCES COMPETITION from the following hard-to-control weeds:

Buckwheat (wild) Polygonum convolvulus	**Ragweed (common)** Ambrosia artemisiifolia
Groundsel Senecio vulgaris	**Signalgrass (broadleaf)** Brachiaria platyphylla
Lambsquarters Chenopodium album	**Smartweed, Ladysthumb** Polygonum persicaria

DIRECTIONS FOR USE

It is a violation of Federal Law to use this product in any manner inconsistent with its labeling.

STORAGE AND DISPOSAL

Do not contaminate water, foodstuffs, seed or feed by storage or disposal.

STORAGE:

STORE ABOVE 20°F. AND BELOW 100°F. TO KEEP PRODUCT IN SUSPENSION.

Below 20°F. the product may freeze. If frozen, mix well to resuspend when temperature is above 40°F.

Above 100°F. the product may become thick and unsprayable.

DISPOSAL:

Wastes resulting from the use of this product that cannot be used or chemically reprocessed should be disposed of in a landfill approved for pesticide disposal or in accordance with applicable Federal, state or local procedures.

(See the individual container label for disposal information.)

SHAKE WELL BEFORE USING

RAMROD FLOWABLE

FIELD CORN, HYBRID SEED CORN, SILAGE CORN AND SWEET CORN

Preemergence Surface Treatment — For control of listed annual grasses and broadleaf weeds, apply this product in 15 or more gallons of water or sprayable fluid fertilizer per acre on the soil surface any time from immediately after planting and before the crop or weeds emerge. See the following table for recommended rates. In areas of heavy grass or broadleaf weed infestations or when listed hard-to-control weeds are present, use the higher rate in the recommended ranges.

Soil Texture	Broadcast Rate in Quarts Per Acre: Less than 3% organic matter	3% or more organic matter
COARSE Sand thru Sandy Loam	4 to 4.5	4.5 to 5
MEDIUM Loam thru Sandy Clay Loam	4.5 to 5	5.25 to 5.5
FINE Silty Clay Loam thru Clay	5.25 to 5.5	5.5 to 6

For Lister or Deep Furrow Planted Corn — Apply this product at the broadcast rate of 5.25 to 6 quarts per acre in 15 or more gallons of water. If heavy rains follow application, the treatment on the top and sides of the furrow may not provide as good weed control as that obtained in the bottom of the furrow.

GRAIN SORGHUM (MILO)

Preemergence Surface Treatment — For control of listed annual grasses and broadleaf weeds apply 4 to 5 quarts of this product in 15 or more gallons of water or sprayable fluid fertilizer solutions per acre any time from immediately after planting and before the crop or weeds emerge. The higher rate should be used on high organic matter soils (greater than 3%) or in areas of heavy grass and broadleaf weed infestations or when listed hard-to-control weeds are present.

For Lister or Deep Furrow Planted Milo — Apply this product at the broadcast rate of 5 quarts per acre in 15 or more gallons of water. If heavy rains follow application, the treatment on the top and sides of the furrow may not provide as good weed and grass control as that obtained in the bottom of the furrow.

PEAS, GREEN (FOR PROCESSING)

Preemergence Surface Treatment — Apply 4 to 5 quarts in 15 or more gallons of water per acre immediately after planting or prior to emergence of crop or weeds.

PUMPKINS (FOR PROCESSING ONLY)

Preemergence Surface Treatment — For control of listed annual grasses and broadleaf weeds, apply this product in 15 or more gallons of water per acre on the soil surface any time from immediately after planting and before the crop or weeds emerge. See the following table for recommended rates. In areas of heavy grass or broadleaf weed infestations or when listed hard-to-control weeds are present, use the higher rate in the recommended ranges.

Soil Texture*	Broadcast Rate in Quarts per Acre: 1 to 3% organic matter	More than 3% organic matter
COARSE Sand thru Sandy Loam	4 to 4.5	4.5 to 5
MEDIUM Loam thru Sandy Clay Loam	4.5 to 5	5.25 to 5.5
FINE Silty Clay Loam thru Clay	5.25 to 5.5	5.5 to 6

*Do not apply to soil with less than 1% organic matter.

NOTE: Apply this product only once per growing season. Do not apply within 90 days of harvest.

DO NOT apply to "jack-o'-lantern" type pumpkins.

FLAX

Preemergence Surfact Treatment — For control of listed grasses and broadleaf weeds, apply 4 quarts of this product in 15 or more gallons of water or liquid fertilizer per acre. This product can be applied on the soil surface any time from immediately after planting and before the crop or weed emerge.

RAMROD FLOWABLE PLUS ATRAZINE (Tank Mixture)

FIELD CORN, HYBRID SEED CORN, SILAGE CORN, SWEET CORN, AND GRAIN SORGHUM

NOTE FOR GRAIN SORGHUM:

IN TEXAS USE ONLY ON THE FINE TEXTURE SOILS OF THE GULF COAST.

When applied as directed under the conditions described, this tank mixture CONTROLS these additional annual broadleaf weeds.

Kochia Kochia scoparia	**Nightshade (hairy)** Solanum sarachoides
Mustard	**Smartweed (Pennsylvania)**

Brassica spp.	Polygonum
Nightshade (black)	pensylvanicum
Solanum nigrum	

When applied as directed under the conditions described, this tank mixture REDUCES COMPETITION from these additional hard-to- control weeds.

Cocklebur	**Morningglory (annual)**
Xanthium	Ipomoea purpurea
strumanium	**Signalgrass (broadleaf)**
Jimsonweed	Brachiaria platyphylla
Datura stramonium	**Velvetleaf, Buttonweed**
	Abutilon theophrasti

Preemergence Surface Treatment — For control of listed annual grasses and broadleaf weeds, apply this tank mixture in 15 or more gallons of water for sprayable fluid fertilizer per acre on the soil surface any time from immediately after planting and before either the corp or weeds emerge. See the following table for recommended rates. In areas of heavy grass or broadleaf weed infestations or when listed hard-to- control weeds are present, use the higher rate in recommended range.

Soil Texture	Broadcast Rate in Pounds Per Acre: Less than 3% organic matter		3% or more organic matter	
	Ramrod Flowable + (Quarts)	atrazine* 4L (Quarts)	Ramrod Flowable + (Quarts)	atrazine* 4L (Quarts)
COARSE** Sand thru Sandy Loam	2.5	1.0	3	1.2
MEDIUM Loam thru Sandy Clay Loam	3	1.2	3.4	1.4
FINE Silty Clay Loam thru Clay	3	1.2 to 1.4	3.4 to 4	1.4 to 1.6

*Use rates listed in this label when using atrazine 4L. Use equivalent rates when using atrazine 80W or 90% dry flowable formulations. One quart of atrazine 4L equals 1.25 pounds of atrazine 80W or 1.1 pounds of atrazine 90% dry flowable.
**Do not use on grain sorghum on coarse texture soils with less than 3% organic matter.

PLANT ONLY CORN, SORGHUM OR SOYBEANS THE YEAR FOLLOWING THE USE OF THIS TANK MIXTURE. IF SOYBEANS ARE TO BE PLANTED, THERE IS A POSSIBILITY OF CROP INJURY DUE TO CARRYOVER OF ATRAZINE. Do not plant soybeans in areas where furrow irrigation is practiced. Applications made on alkali soils or where cuts, fills or erosion have exposed calcareous or alkali subsoils may result in crop injury to milo.

Emerged Weed Control — For Corn Only — This tank mixture can be applied until weeds reach the 2-leaf stage, but prior to corn emergence. Applications beyond the 2-leaf stage will not provide satisfactory results.

Better results will be obtained if rainfall occurs within 5 to 7 days after application. If rainfall does not occur, use shallow cultivation to remove emerged weeds. Avoid cultivation methods which will displace the treated soil layer.

See the rate table in the Preemergence Treatment section of this label for the recommended rate for the soil texture and organic matter content to be treated.

For Lister or Deep Furrow Planted Corn — Apply this tank mixture at the higher broadcast rate listed in the rate table in the preemergence section of this label for the soil texture and organic matter content to be treated.

If heavy rains follow application, the treatment on the top and sides of the furrow may not provide as good weed control as that obtained in the bottom of the furrow.

RAMROD FLOWABLE PLUS PROPAZINE (Tank Mixture)

GRAIN SORGHUM (MILO)

When applied as directed under the conditions described, this tank mixture CONTROLS these additional annual broadleaf weeds:

Kochia	**Lambsquarters**
Kochia scoparia	Chenopodium album

When applied as directed under conditions described, this tank mixture REDUCES COMPETITION from this additional annual broadleaf weed:

Morningglory (annual)
Ipomoea purpurea

Preemergence Surface Treatment — For control of listed annual grasses and broadleaf weeds, apply this tank mixture in 15 or more gallons of water per acre anytime after planting and before either crop or weeds emerge. Application must be made within 5 days of last tillage operation for best results. See the following table for recommended rates:

Soil Texture	Broadcast Rate per Acre Quarts: Ramrod® Flowable +	Propazine 4L More than 1% Organic matter
COARSE Sand thru Sandy Loam	2.5 to 3	.8 to 1*
MEDIUM THRU FINE Loam thru Clay	3	1

*Use the higher rates within the recommended range for this tank mixture when heavy weed infestations exist, harder-to-control species are present, or soil organic matter is greater than 3 percent.
** In coarse soil types with 1% or less organic matter, use the following rates of propazine 80W in the tank mixture:

Coarse Soils Percent Organic Matter	Quarts Propazine 4L per acre***
Less than .75%	.4
.75% to 1%	.6

***Use rates listed in this label when using propazine 4L. Use equivalent rates when using atrazine 80W or 90% dry flowable formulations. One quart 4L equals 1.25 pounds of 80W or 1.1 pounds of 90% dry flowable.

NOTE: Applications of this tank mixture to alkali soils where cuts, fills, or erosions have exposed calcareous or alkali subsoils may result in crop injury. Do not plant crops other than corn or sorghum for 18 months following application of this tank mixture due to possibility of crop injury from propazine carryover.

Whenever using reduced rates of propazine recommended for coarse soils with less than 1% organic matter, duration of weed control may be reduced. Weed control may be reduced if either an irrigation or a rainfall in excess of one inch occurs after application.

All applicable directions, restrictions and precautions on the container labels for Ramrod Flowable herbicide and propazine are to be followed.

RAMROD FLOWABLE PLUS MODOWN™ 4F FLOWABLE (Tank Mixture)

GRAIN SORGHUM (MILO)

When applied as directed under the conditions described, this tank mixture CONTROLS these additional annual broadleaf weeds:

Kochia	**Lambsquarters**
Kochia scoparia	Chenopodium album

When applied as directed under the conditions described, this tank mixture REDUCES COMPETITION from these additional annual broadleaf weeds:

Jimsonweed	**Sida (prickly), Teaweed**
Datura stramonium	Sida spinosa
Morningglory (annual)	**Velvetleaf, Buttonweed**
Ipomoea purpurea	Abutilon theophrasti

Preemergence Surface Treatment — For control of listed annual grasses and broadleaf weeds, apply this tank mixture in 15 or more gallons of water per acre any time after planting and before either crop or weeds emerge. Application must be made within 5 days of last tillage operation for best results. See the following table for recommended rates:

Soil Texture	Broadcast Rate in Quarts per Acre*: Ramrod® Flowable +	Modown 4F Flowable
COARSE Sand thru Sandy Loam	3	1.25
MEDIUM and FINE Loam thru Clay	3 to 4	1.25 to 1.5

*Use the higher rate within the recommended range for this tank mixture when heavy weed infestations exist, harder-to-control species are present, or soil organic matter is greater than 3 percent.
TM Trademark of Mobil Chemical Company

NOTE: Do not apply this tank mixture after milo has begun to crack the soil surface just prior to emergence. Rainfall which splashes treated soil onto the crop at time of emergence may result in crop injury from Modown. Do not rotate to crops other than corn, soybeans, grain sorghums, small grains or rice prior to one year following treatment with this tank mixture.

All applicable directions, restrictions and precautions on the container labels for Ramrod Flowable herbicide and Modown herbicide are to be followed.

848.21-000.19/CG

In case of an emergency involving this product, Call Collect, day or night, (314) 694-4000

This sample is current as of January 1, 1989. The product descriptions and recommendations provided in this sample label are for background information only. Always refer to the label on the product container before using Monsanto or any other agrichemical product.

RAMROD® AND ATRAZINE

FLOWABLE HERBICIDE MIXTURE

Selective herbicide for preemergence weed control in grain sorghum (milo), field corn, hybrid seed corn, silage corn and sweet corn.

Complete Directions for Use.

SHAKE WELL BEFORE USING

ACTIVE INGREDIENTS:	
Propachlor, (2-chloro-N-isopropylacetanilide)	31.5%
Atrazine, [2-Chloro-4-(ethylamino)-6-(isopropylamino)-s- triazine]	10.0%
Related atrazine compounds	0.5%
INERT INGREDIENTS:	58.0%
TOTAL	100.0%

Contains 360 grams/litre or 3 pounds/gallon of propachlor and 120 grams/litre or 1 pound/gallon of atrazine and related atrazine compounds.

®Ramrod is a registered trademark of Monsanto Company

EPA Reg. No. 524-328-AA 1988-1 848.29-000.43/CG

Read the entire label before using this product.

Use only according to label instructions.

Read "LIMIT OF WARRANTY AND LIABILITY" before buying or using. If terms are not acceptable, return at once unopened.

REFORMULATION OR REPACKAGING IS PROHIBITED.

LIMIT OF WARRANTY AND LIABILITY

This Company warrants that this product conforms to the chemical description on the label and is reasonably fit for the purposes set forth in the Complete Directions for Use label booklet ("Directions") when used in accordance with those Directions under the conditions described therein. NO OTHER EXPRESS WARRANTY OR IMPLIED WARRANTY OF FITNESS FOR PARTICULAR PURPOSE OR MERCHANTABILITY IS MADE. This warranty is also subject to the conditions and limitations stated herein.

Buyer and all users shall promptly notify this company of any claims whether based in contract, negligence, strict liability, other tort or otherwise.

Buyer and all users are responsible for all loss or damage from use or handling which results from conditions beyond the control of this company, including, but not limited to, incompatibility with products other than those set forth in the Directions, unusual weather, weather conditions which are outside the range considered normal at the application site and for the time period when the product is applied, as well as weather conditions which are outside the application ranges set forth in the Directions, application in any manner not explicitly set forth in the Directions, moisture conditions outside the moisture range specified in the Directions, or the presence of products other than those set forth in the Directions in or on the soil, crop or treated vegetation.

THE EXCLUSIVE REMEDY OF THE USER OR BUYER, AND THE LIMIT OF THE LIABILITY OF THIS COMPANY OR ANY OTHER SELLER FOR ANY AND ALL LOSSES, INJURIES OR DAMAGES RESULTING FROM THE USE OR HANDLING OF THIS PRODUCT (INCLUDING CLAIMS BASED IN CONTRACT, NEGLIGENCE, STRICT LIABILITY, OTHER TORT OR OTHERWISE) SHALL BE THE PURCHASE PRICE PAID BY THE USER OR BUYER FOR THE QUANTITY OF THIS PRODUCT INVOLVED, OR, AT THE ELECTION OF THIS COMPANY OR ANY OTHER SELLER, THE REPLACEMENT OF SUCH QUANTITY, OR, IF NOT ACQUIRED BY PURCHASE, REPLACEMENT OF SUCH QUANTITY. IN NO EVENT SHALL THIS COMPANY OR ANY OTHER SELLER BE LIABLE FOR ANY INCIDENTAL OR CONSEQUENTIAL DAMAGES.

Buyer and all users are deemed to have accepted the terms of this LIMIT OF WARRANTY AND LIABILITY which may not be varied by any verbal or written agreement.

PRECAUTIONARY STATEMENTS

HAZARDS TO HUMANS AND DOMESTIC ANIMALS

KEEP OUT OF REACH OF CHILDREN

WARNING!

HARMFUL IF SWALLOWED OR ABSORBED THROUGH SKIN.

CAUSES EYE AND SKIN IRRITATION.

MAY CAUSE ALLERGIC SKIN REACTION.

Do not get in eyes, on skin, or on clothing.

Keep container closed.

While handling, wear rubber gloves.

Wash thoroughly after handling.

FIRST AID: IN CASE OF CONTACT, immediately flush eyes or skin with plenty of water for at least 15 minutes while removing contaminated clothing. For eyes, call a physician. Wash clothing before reuse.

IF SWALLOWED, drink 1 or 2 glasses of water and induce vomiting by touching back of throat with finger or blunt object. Do not induce vomiting or give anything by mouth to an unconscious person. Get medical attention.

In case of emergency involving this product, call collect, day or night, (314) 694-4000

ENVIRONMENTAL HAZARD

This product is toxic to fish and wildlife. Keep out of lakes, streams, or ponds. Do not apply where runoff is likely to occur. Do not apply when weather conditions favor drift from areas treated.

Apply this product only as specified on this label.

GENERAL INFORMATION

This herbicide mixture is recommended for control of many annual grasses and certain broadcast weeds. For best results it should be applied to the soil surface before crop or weeds emerge. The seedbed should be fine, firm and free of clods and trash; application should be made within 5 days after the last tillage for weed control. Do not apply when conditions favor drift. Some moisture, either as rain or irrigation, is required soon after application to activate this herbicide mixture. On coarse texture soil one-third inch of rainfall, and on medium and fine textured and/or high organic soils one-third to three-fourths inch of rainfall is required depending upon original soil moisture. Better results are obtained when moisture occurs within 10 days after application. Generally lower amounts of rainfall are required where soil moisture is high at time of planting.

NOTE: Use of this product in any manner not consistent with this label may result in injury to persons, animals or crops, or other unintended consequences. Keep container closed to prevent spills and contamination.

Flush sprayer with clean water after use.

Do not use this product in fields where flood irrigation is to be practiced.

DO NOT apply in sprayable fluid fertilizer after crop emergence since crop injury may result.

Do not graze or feed sorghum (milo) forage, silage or fodder (stubble) from treated fields to dairy animals.

PLANT ONLY CORN, SORGHUM OR SOYBEANS THE YEAR FOLLOWING THE USE OF THIS TANK MIXTURE. IF SOYBEANS ARE TO BE PLANTED, THERE IS A POSSIBILITY OF CROP INJURY DUE TO CARRYOVER OF ATRAZINE. Do not plant soybeans in areas where furrow irrigation is practiced.

In Texas, on grain sorghum, use only on the fine textured soils of the Gulf Coast.

Applications made to sorghum growing on alkali soils or where cuts, fills or erosions have exposed calcareous or alkali subsoils may result in crop injury.

MIXING AND SPRAYING INSTRUCTIONS

Always predetermine the compatibility of this product with water carrier or sprayable fluid fertilizer carrier by mixing small proportional quantities in advance. See the Standard Sprayable Fluid Fertilizer Compatibility Test section on this label.

Mix this product or its labelled tank mix with the appropriate carrier as follows:

1. Place a 20 to 35 mesh screen or wetting basket over the filling port.
2. Through the screen, fill the sprayer tank one-half full with the appropriate carrier.
3. If a compatibility agent is necessary to improve mixing or to prevent the formation of undesirable and unsprayable gels or precipitates, while agitating add it to the carrier already in the tank. Use only compatibility agents cleared by Federal Food and Drug Administration for this use. Read and follow all directions for use, cautionary statements and all other informa-

tion appearing on the selected compatibility agent. Check for adequate agitation.

4. Add the required amount of this product. Continue agitation.

5. Complete filling the sprayer tank with carrier. Remove hose from tank immediately after filling to avoid siphoning back into carrier source. Maintain good agitation at all times until the contents of the tank are sprayed, otherwise, settling which can lead to clogging may occur. Close tank valves and drain all lines when spraying is delayed.

Keep by-pass line on or near bottom of tank to minimize foaming. Screen size in nozzle or line strainers should be no finer than 50 mesh. Carefully select proper nozzle to avoid spraying a fine mist. For best results, use flat fan or flood jet nozzles. Check for even distribution of spray droplets. To reduce loss of the chemical due to drift of a fine mist, apply at pressures less than 40 psi.

STANDARD SPRAYABLE FLUID FERTILIZER COMPATIBILITY TEST

Herbicides may not always mix evenly throughout a sprayable fluid fertilizer or the components may separate too quickly to make their combined use of practical value. This may be due to certain characteristics of the different fluid fertilizers. A simple test using small quantities of the components is suggested to provide compatibility potential.

The test follows:

A. Materials Required For A Compatibility Test

1. Two one-quart jars with lid or stopper (marked "with" and "without").
2. Teaspoons (for a more exacting test, a five to ten mililiter (ml.) pipette or graduated cylinder is desirable).
3. A minimum of one quart of the sprayable fluid fertilizer to be tested.
4. The herbicide(s) to be mixed.
5. A compatability agent (the purpose of the adjuvant is to help keep the fertilizer and herbicide(s) in suspension, if this assistance is needed).

B. Procedure

1. Add one point of the sprayable fluid fertilizer that will be used or other herbicide carrier to each jar marked "with" and "without".

Add One Pint Liquid Fertilizer
To Two Quart Jars.

2. To the jar marked "with", add 1/4 TEAspoon or 1.2 milliliters of a suitable compatibility agent; shake gently for five to ten seconds to mix. (1/4 TEAspoon is the equivalent of two pints per 100 gallons of liquid fertilizer.)

To Jar Marked "with"
Add Compatibility Agent
And Shake To Mix.

3. To each jar add the appropriate amount of this porduct. Shake gently for five to ten seconds.

Add Herbicide(s) to Both Jars
And Shake To Mix

Herbicide	Rate/a	Amount To Be Added Per Pint Of Sprayable Fluid Fertilizer (Assuming Volume is 25 gal/A): Level TEA-spoons		Milliliters
This Product				
	3.5 qts =	3.4	or	16.6
	4.0 qts =	3.8	or	19.0
	4.5 qts =	4.3	or	21.4
	5.0 qts =	4.8	or	23.8
	5.5 qts =	5.3	or	26.2

This compatibility test is designed for 25 gallons of spray per acre with the maximum labeled rate of herbicide. For changes in spray volume or herbicide concentration, make appropriate changes in the ingredients of the test. Regardless of spray volume, the amount of compatibility agent should be equal to two or three pints (two pints = 1/4 TEAspoon or 1.2 mililiters, three pints = 3/8 TEAspoon or 1.8 mililiters per pint of sprayable fluid fertilizer) per 100 gallons of liquid fertilizer.

C. Observations And Decisions

1. If the herbicide(s) and the sprayable fluid fertilizer are compatible.
2. If a compatibily agent is necessary.

Five minutes after the final addition and mixing, observe both jars for the formation of large flakes, sludge, gels or other precipitates. Observe if the herbicides cannot be physically mixed with the liquid fertilizer but remains as small oily particles in the solution.

If incompatibility in any form described above occurs in the jar "with" the compatibility agent added, the liquid fertilizer and the herbicide(s) should not be used together in the same spray tank.

If incompatibility as described above occurs in the jar "without" the adjuvant but not in the jar "with" adjuvant, the use of a compatibility adjuvand is recommended.

Both jars should be allowed to stand and be observed periodically for one-half hour. If the separate layers of liquid fertilizer and additives can be resuspended by shaking, commercial application is possible. Wettable powders will either settle to the bottom of the tank or jar, or float to the top, depending upon the density of the fertilizers.

If this product is compatible with fluid fertilizer in the foregoing test without having to use a compatibility agent, fluid fertilizer may be used for the premixing. If it is not compatible, the herbicide(s) should be premixed with water before adding to the spray tank.

APPLICATION INFORMATION

GROUND APPLICATION

Do not apply this product through any type of irrigation system.

Broadcast Treatment—Apply in 15 or more gallons of water or sprayable fluid fertilizer per acre according to this label. Do not apply during periods of gusty winds or winds in excess of 15 mph.

Band Treatment — Apply a broadcast equivalent rate and volume per acre. To determine these:

$$\frac{\text{Band width in inches}}{\text{Row width in inches}} \times \text{Broadcast RATE per acre} = \text{Band RATE per acre}$$

$$\frac{\text{Band width in inches}}{\text{Row width in inches}} \times \text{Broadcast VOLUME per acre} = \text{Band VOLUME per acre}$$

Mount spray equipment on the planter with nozzles behind press wheel and apply this product at planting time.

NOTE: For band application nozzle placement, refer to the nozzle manufacturer's specification as a guide for determining proper nozzle height.

AERIAL APPLICATION

AVOID DRIFT — Do not apply when winds are gusty or in excess of 5 miles per hour or when other conditions, including lesser wind velocities, will allow drift to occur. Any resulting drift can damage nearby crops, vegetation or other areas on which treatment is not intended, and can cause application to be streaked or uneven, reducing the effectiveness of this product. In order to ensure uniform aerial application and to avoid overlapping and possible crop injury, it is recommended that 2 flagmen be used, one at each end of the field. Apply this product in 5 or more gallons of water per acre.

Coarse sprays are less likely to drift; therefore, do not use nozzles or nozzle configurations which dispense spray as fine spray droplets. Do not angle nozzles forward into the airstream and do not increase spray volume by increasing nozzle pressure.

CULTIVATION INFORMATION

As long as weeds are controlled, delay cultivation as long as possible.

If weed seedlings emerge, shallow cultivate or rotary hoe to improve weed control.

If cultivation or rotary hoeing is necessary because of soil crusting or compaction, tillage should be shallow to minimize dilution and displacement of this herbicide or tank mixture.

WEEDS CONTROL

When applied as directed, under the conditions described, this product CONTROLS the following weeds:

Barnyardgrass, Watergrass *Echinochloa crus-galli*	**Nightshade (black)** *Solanum nigrum*
Buckwheat (wild) *Polygonum convolvulus*	**Nightshade (hairy)** *Solanum sarachoides*
Carpetweed *Mollugo verticillata*	**Panicum (fall)** *Panicum dichotomiflorum*
Crabgrass *Digitaria spp.*	**Pigweed** *Amaranthus spp.*
Foxtail (giant) *Setaria faberi*	**Purslane** *Portulaca oleracea*
Foxtail (green) *Setaria viridis*	**Pusley (Florida)** *Richardia scabra*
Foxtail (yellow) *Setaria lutescens*	**Ragweed** *Ambrosia artemisiifolia*
Goosegrass *Eleusine indica*	**Ryegrass (annual)** *Lolium spp.*
Groundsel *Senecio vulgaris*	**Smartweed** *Polygonum pensylvanicum* *Polygonum persicaria*
Kochia *Kochia scoparia*	
Lambsquarters *Chenopodium album*	
Mustard *Brassica spp.*	

When applied immediately after planting, under the conditions described, this product at a rate of 4 to 5.5 quarts per acre REDUCES COMPETITION from the following hard to control weeds:

Buttonweed, Velvetleaf *Abutilon theophrasti*	**Jimsonweed** *Datura stramonium*
Cocklebur *Xanthium strumarium*	**Morningglory (annual)** *Ipomoea purpurea*
	Signalgrass (broadleaf) *Brachiaria platyphylla*

DIRECTIONS FOR USE

It is a violation of Federal Law to use this product in any manner inconsistent with its labeling.

STORAGE AND DISPOSAL

Do not contaminate water, foodstuffs, seed or feed by storage or disposal.

STORAGE:

STORE ABOVE 20°F. AND BELOW 100°F. TO KEEP PRODUCT IN SUSPENSION.

Below 20°F. the product may freeze. If frozen, mix well to resuspend when temperature is above 40°F.

Above 100°F. the product may become thick and unsprayable.

DISPOSAL:

Wastes resulting from the use of this product that cannot be used or chemically reprocessed should be disposed of in a landfill approved for pesticide disposal or in accordance with applicable Federal, state or local procedures.

(See the individual container label for disposal information.)

SHAKE WELL BEFORE USING

FOR WEED CONTROL IN GRAIN SORGHUM (MILO), FIELD CORN, HYBRID SEED CORN, SILAGE CORN AND SWEET CORN.

Preemergence Surface Treatment — For control of listed annual grasses and broadleaf weeds, apply this product in 15 or more gallons of water or sprayable fluid fertilizer per acre on the soil surface any time from immediately after planting and before the crop or weeds emerge. See the following table for recommended rates. In areas of heavy grass or broadleaf weed infestations or when listed hard-to-control weeds are present, use the higher rate in the recommended ranges.

Soil Texture	Broadcast Rate in Quarts Per Acre	
	Less than 3% organic matter	3% or more organic matter
COARSE*		
Sand thru Sandy Loam	3.5	4
MEDIUM		
Loam thru Sandy Clay Loam	4	4.5
FINE		
Silty Clay Loam thru Clay	4 to 4.5	4.5 to 5.5

* Do not use on grain sorghum on coarse texture soils with less than 3% organic matter.

Note for Corn only — This product can be applied until weeds reach the 2-leaf stage but prior to corn emergence. Applications beyond the 2-leaf stage will not provide satisfactory results. Better results will be obtained if rainfall occurs within 5 to 7 days after application. If rainfall does not occur, use shallow cultivation with a rolling cultivator or similar implement to remove emerged weeds. Avoid cultivation methods which will displace the treated section of this label for the recommended rate for the soil texture and organic matter content to be treated.

For Lister or Deep Furrow Planted Corn — Apply this product at the higher broadcast rate listed in the rate table in the preemergence section of this label for the soil texture and organic matter content to be treated.

If heavy rains follow application, the treatment on the top and sides of the furrow may not provide as good weed control as that obtained in the bottom of the furrow.

848.29-000.43/CG

In case of an emergency involving this product, Call Collect, day or night, (314) 694-4000.

This sample label is current as of January 1, 1989. The product descriptions and recommendations provided in this sample label are for background information only. Always refer to the label on the product container before using Monsanto or any other agrichemical product.

RAMROD® 20 Granular

SELECTIVE HERBICIDE

For pre-emergence weed control in field corn, hybrid seed corn, silage corn, sweet corn, grain sorghum (milo); and soybeans (for seed only).

ACTIVE INGREDIENTS:	
2-chloro-N-isoprophylacetanilide	20%
INERT INGREDIENTS:	80%
TOTAL	100%

EPA Reg. No. 524-152-AA

KEEP OUT OF REACH OF CHILDREN

READ "LIMIT OF WARRANTY AND LIABILITY" BEFFORE BUYING OR USING. IF TERMS ARE NOT ACCEPTABLE RETURN AT ONCE UNOPENED.

WARNING!

IRRITATING TO SKIN, EYES, NOSE AND THROAT. MAY CAUSE ALLERGIC SKIN REACTION.

In case of an emergency involving this product, Call Collect, day or night, (314) 694-4000.

®Registered Trademark of Monsanto Company.

Avoid contact with eyes and skin.
While handling, wear freshly laundered cotton or clean rubber gloves.
Avoid breathing dust.
Do not take internally.
In case of contact, flush skin or eyes with plenty of water; for eyes, get medical attention.
Wash thoroughly after handling.
Launder clothing before re-use.
Avoid contamination of seed, feed, and foodstuffs.

Read the entire label before using this product.

Use only according to label instructions.

Reformulation or repackaging is prohibited.

LIMIT OF WARRANTY AND LIABILITY

This Company warrants that this product conforms to the chemical description on the label and is reasonably fit for the purposes set forth in the Complete Directions for Use label booklet ("Directions"), when used in accordance with those Directions under the conditions described therein. NO OTHER EXPRESS OR IMPLIED WARRANTY OF FITNESS FOR PARTICULAR PURPOSE OR MERCHANTABILITY IS MADE. This warranty is also subject to the conditions and limitations stated herein.

Buyer and all users shall promptly notify this company of any claims whether based in contract, negligence, strict liability, other tort or otherwise.

Buyer and all users are responsible for all loss or damage from use or handling which results from conditions beyond the control of this company, including but not limited to incompatibility with products other than those set forth in the Directions, unusual weather, weather conditions which are outside the range considered normal at the application site and for the time period when the product is applied, as well as weather conditions which are outside the application ranges set forth in the Directions, application in any manner not explicitly set forth in the Directions, moisture conditions outside the moisture range specified in the Directions, or the presence of products other than those set forth in the Directions in or on the soil, crop or treated vegetation.

THE EXCLUSIVE REMEDY OF THE USER OR BUYER, AND THE LIMIT OF THE LIABILITY OF THIS COMPANY OR ANY OTHER SELLER FOR ANY AND ALL LOSSES, INJURIES OR DAMAGES RESULTING FROM THE USE OR HANDLING OF THIS PRODUCT (INCLUDING CLAIMS BASED IN CONTRACT, NEGLIGENCE, STRICT LIABILITY, OTHER TORT OR OTHERWISE) SHALL BE THE PURCHASE PRICE PAID BY THE USER OR BUYER FOR THE QUANTITY OF THIS PRODUCT INVOLVED, OR, AT THE ELECTION OF THIS COMPANY OR ANY OTHER SELLER, THE REPLACEMENT OF SUCH QUANTITY OR, IF NOT ACQUIRED BY PURCHASE, REPLACEMENT OF SUCH QUANTITY. IN NO EVENT SHALL THIS COMPANY OR ANY OTHER SELLER BE LIABLE FOR ANY INCIDENTAL, CONSEQUENTIAL OR SPECIAL DAMAGES.

The buyer and all users are deemed to have accepted the terms of this LIMIT OF WARRANTY AND LIABILITY which may not be varied by any verbal or written agreement.

GENERAL INFORMATION

Ramrod 20 is recommended for use as a pre-emergence herbicide to control most annual grasses and certain broadleaf weeds. It should be applied to the soil surface prior to weed or crop emergence. The soil surface should be freshly worked and free of weed top growth. The seed bed should be fine, firm and free of clods and trash.

Some moisture either as rain or irrigation is required soon after application to activate Ramrod 20. On light sandy loam soil one-third inch of rainfall and on heavy clay and or high organic soils one-third to three-fourths inch of rainfall is required depending upon original soil moisture. Best results are obtained when moisture occurs within 10 days after application. Generally lower amounts of rainfall are required where soil moisture is high at time of planting.

Ramrod 20 with lesser amounts of rainfall in some instances will allow the grass seedlings to emerge, but they will be severely stunted and with a subsequent cultivation they can be covered up.

If soil crusting becomes a problem and hoeing is necessary, hoe in the same direction the rows are planted.

Where deep lister planting is practiced only partial weed control can be expected on steep furrow sides. It is suggested that a 7 inch band be applied at the bottom of the furrow. Best results can be expected where the granules lie on an even flat surface not subject to shifting by wind or rain.

With broadcast treatment, as long as Ramrod 20 is controlling the grasses and weeds, there is no need for cultivation. For band treated fields, when first cultivation is made, set cultivator to throw soil into the row covering the band.

NOTE: Rates of application given on this label are general. Consult your Agricultural Experiment Station or Extension Service Weed Specialists concerning specific usage, dosages and methods of application.

Use of this product in any manner not consistent with this label may result in injury to persons, animals or crops, or other unintended consequences. Keep container closed to prevent spills and contamination. Do not contaminate fertilizers, insecticides, fungicides and seed with Ramrod.

This product is toxic to fish. Keep out of lakes, ponds and streams. Do not contaminate water by cleaning of equipment or disposal of wastes. Apply this product only as specified on this label.

Do not reuse container. Destroy when empty.

WEEDS CONTROLLED

Annual Ryegrass
Lolium spp.
Barnyardgrass (Watergrass)
Echinochloa crus-galli
Carpetweed
Mollugo verticillata
Crabgrass
Digitaria spp.
Fall Panicum
Panicum dichotomiflorum
Florida Pusley
Richardia scabra
Giant Foxtail (Wild Millet)
Setaria faberii
Green Foxtail
Setaria viridis
Yellow Foxtail
Setaria glauca
Goosegrass
Eleusine indica
Groundsel
Senecia vulgaris
Lambsquarters
Chenopodium album
Pigweed
Amaranthus spp.
Purslane
Portulaca oleracea
Ragweed
Ambrosia artemisiifolia
Sandbur
Cenchrus spp.
Shepherdspurse*
Capsella bursa-pastoris
Smartweed*
Polygonum persicaria only
Wild Buckwheat*
Polygonum convolvulus

*At 25 to 30 pounds per acre granular Ramrod 20.

Granular Ramrod 20 WILL NOT CONTROL Johnsongrass, Canada thistle, field bindweed, quackgrass, morningglory, cocklebur and bull nettle.

CALIBRATION AND APPLICATION INSTRUCTIONS

It is important that the applicator be calibrated properly to deliver the desired amount of Ramrod 20 to avoid applying too little or too much material. To calibrate, attach a cloth bag or strong plastic bag over the spreader plate or remove the delivery tube and attach a plastic nursing bottle over the spout. Operating over normal terrain to be planted, at planting speed, collect the granules for five hundred (500) linear feet and measure the VOLUME ounces collected. Use the following table to determine the correct volume of granules required for a 14-inch band at a given broadcast application rate.

Lbs. Per Acre To Be Applied **(Broadcast Rate)**	Volume To Be Collected **(In 500 Feet)**
20	6¼ oz.
25	7¾ oz.
30	9½ oz.

Tap plastic nursing bottle or other measuring container gently to settle the granules when taking a reading. Adjust the application equipment so that each spout dispenses the proper amount.

For broadcast (over-all treatment)–use broadcast applicators or fertilizer spreaders that can apply small amounts of granules evenly.

For band treatment—use applicators designed for this purpose. Calculate the amount of granules per acre needed for band treatment as follows.

$$\frac{\text{Band Width in Inches}}{\text{Inches Between Crop Rows}} \times \text{Recommended Over-all Rate} = \text{Pounds of Ramrod 20 needed per acre for band treatment}$$

For example: If 25 pounds per acre are recommended for over-all treatment and you wish to treat a band of 14 inches wide in crop rows spaced 38 inches apart.

$$\frac{14}{38} \times 25 = \text{9.2 pounds of granules needed per acre of crop for the band treatment}$$

For application of granular Ramrod 20 do not rely on settings used for other granular herbicides.

DIRECTIONS FOR USE

It is a violation of Federal law to use this product in any manner inconsistent with it's labeling.

STORAGE AND DISPOSAL

Do not contaminate water, foodstuffs, feed or seed by storage or disposal.

DISPOSAL:

Wastes resulting from the use of this product that cannot be used or chemically reprocessed should be disposed of in a landfill approved for pesticide disposal or in accordance with applicable Federal, state or local procedures.

Completely empty bag into application equipment. Then dispose of empty bag in a sanitary landfill or by incineration, or, if allowed by state and local authorities, by burning. If burned, stay out of smoke.

FIELD CORN, HYBRID SEED CORN, SILAGE CORN AND SWEET CORN.

Broadcast (over-all treatment)—Apply 20 to 30 pounds of granular Ramrod 20 per acre on soil surface after planting and prior to emergence of either crop or weeds. See table below for recommended rates on various soil textures and levels of organic matter. In areas of heavy grass or broadleaf weed infestations use the higher rate shown for your soil texture.

Soil Texture	Broadcast Rates per Acre	
	Less than 3% organic matter	More than 3% organic matter
Sand to Sandy Loam	20 lbs.	25 lbs.
Silt Loam	22 lbs.	27 lbs.
Clay Loam to Clay	25 lbs.	30 lbs.

Band Treatment (Behind Press Wheel)—Apply granular Ramrod 20 on soil surface after planting and prior to emergence of either crop or weeds. Use the higher broadcast rate for high organic soils, or in areas of heavy grass or broadleaf weed infestations. With the broadcast rate for your soil texture, the following table gives the pounds per acre needed for your row spacing.

BAND TREATMENT

Broadcast Rate/Acre	20 lbs.	22 lbs.	25 lbs.	27 lbs.	30 lbs.
	14" Band on Various Row Spacings				
Row Spacings	lb./acre	lb./acre	lb./acre	lb./acre	lb./acre
40	7.0	7.7	8.8	9.5	10.5
38	7.4	8.1	9.2	9.9	11.0
36	7.8	8.6	9.7	10.5	11.7
30	9.3	10.3	11.7	12.6	14.0

FOR LISTER OR DEEP FURROW PLANTED CORN

Broadcast (over-all treatment)—Apply granular Ramrod 20 at the broadcast rate of 25 to 30 pounds per acre.

Band Treatment (Behind Press Wheel)—For a 7 inch band treatment over the row (38 inch spacing), apply 4.6 to 5.5 pounds of granular Ramrod 20 per acre.

If heavy rains follow application, the treatment on the top and sides of the furrow may not provide as good weed and grass control as that obtained in the bottom of the furrow.

FOR FURROW IRRIGATED CORN

Granular Ramrod 20 must be incorporated into the soil by harrow, finger weeder or rotary hoe for weed control on furrow irrigated land. Shallow incorporation of at least one-half inch but not more than 2 inches is most effective for weed control. Incorporation deeper than 2 inches will result in unsatisfactory weed control. If applied on very dry soil, irrigation is necessary to activate granular Ramrod 20. See the table above for recommended rates on various soil textures and levels of organic matter.

GRAIN SORGHUM (milo)
SOYBEANS (to be grown for seed only)

Broadcast (over-all treatment)—Apply 20 to 25 pounds of granular Ramrod 20 per acre on soil surface after planting and prior to emergence of either crop or weeds. In areas of heavy grass or broadleaf weed infestations and on high organic matter soils, use 25 pounds per acre.

Band Treatment (Behind Press Wheel)—Apply granular Ramrod 20 on the soil surface after planting and prior to emergence of either crop or weeds. Use the higher rate for high organic soil or in areas of heavy grass or broadleaf weed infestations.

BAND TREATMENTS

Broadcast Rate/Acre	20 lbs.	22 lbs.	25 lbs.
	14" Band on Various Row Spacings		
Row Spacings	lb./acre	lb./acre	lb./acre
40	7.0	7.7	8.8
38	7.4	8.1	9.2
36	7.8	8.6	9.7
30	9.3	10.3	11.7

NOTE: Do not graze or feed sorghum (milo) forage or silage from treated fields to dairy animals.

Do not use soybean seed for food, feed or oil purposes.

Do not graze or feed soybean forage from treated area.

®Registered trademark of Monsanto Company

This sample label is current as of January 1, 1989. The product descriptions and recommendations provided in this sample label are for background information only. Always refer to the label on the product container before using Monsanto or any other agrichemical product.

EPA Est. 524-1A-1 843.30-000.29M.

RANGER®

HERBICIDE

For control of quackgrass and listed weeds in annual cropping systems, pastures and sods

AVOID CONTACT WITH FOLIAGE OF CROP OR OTHER DESIRABLE VEGETATION, SINCE SEVERE INJURY OR DESTRUCTION MAY RESULT.

Complete Directions for Use

ACTIVE INGREDIENT:
* Isopropylamine salt of glyphosate 28.6%
INERT INGREDIENTS: 71.4%
TOTAL 100.0%

* Contains 324 grams per litre or 2.7 pounds of the active ingredient isopropylamine salt of N-(phosphonomethyl) glycine per U.S. gallon. Equivalent to 240 grams per litre or 2 pounds per U.S. gallon of the acid, glyphosate.

Product protected by U.S. Pat. Nos. 3,799,758 and 4,405,531. Other patents pending.

No license granted under any non-U.S. patent.

EPA Reg. No. 524-382

® RANGER is a registered trademark of Monsanto Company.
1989-1 892.20-000.24/53

Read the entire label before using this product.

Use only according to label instructions.

Read "LIMIT OF WARRANTY AND LIABILITY" before buying or using. If terms are not acceptable, return at once unopened.

REFORMULATION OR REPACKAGING IS PROHIBITED.

LIMIT OF WARRANTY AND LIABILITY

This Company warrants that this product conforms to the chemical description on the label and is reasonably fit for the purposes set forth in the Complete Directions for Use label booklet ("Directions") when used in accordance with those Directions under the conditions described therein. NO OTHER EXPRESS WARRANTY OR IMPLIED WARRANTY OF FITNESS FOR PARTICULAR PURPOSE OR MERCHANTABILITY IS MADE. This warranty is also subject to the conditions and limitations stated herein.

Buyer and all users shall promptly notify this Company of any claims whether based in contract, negligence, strict liability, other tort or otherwise.

Buyer and all users are responsible for all loss or damage from use or handling which results from conditions beyond the control of this Company, including, but not limited to, incompatibility with products other than those set forth in the Directions, application to or contact with desirable vegetation, unusual weather, weather conditions which are outside the range considered normal at the application site and for the time period when the product is applied, as well as weather conditions which are outside the application ranges set forth in the Directions, application in any manner not explicitly set forth in the Directions, moisture conditions outside the moisture range specified in the Directions, or the presence of products other than those set forth in the Directions in or on the soil, crop or treated vegetation.

THE EXCLUSIVE REMEDY OF THE USER OR BUYER, AND THE LIMIT OF THE LIABILITY OF THIS COMPANY OR ANY OTHER SELLER FOR ANY AND ALL LOSSES, INJURIES OR DAMAGES RESULTING FROM THE USE OR HANDLING OF THIS PRODUCT (INCLUDING CLAIMS BASED IN CONTRACT, NEGLIGENCE, STRICT LIABILITY, OTHER TORT OR OTHERWISE) SHALL BE THE PURCHASE PRICE PAID BY THE USER OR BUYER FOR THE QUANTITY OF THIS PRODUCT INVOLVED, OR, AT THE ELECTION OF THIS COMPANY OR ANY OTHER SELLER, THE REPLACEMENT OF SUCH QUANTITY, OR, IF NOT ACQUIRED BY PURCHASE, REPLACEMENT OF SUCH QUANTITY. IN NO EVENT SHALL THIS COMPANY OR ANY OTHER SELLER BE LIABLE FOR ANY INCIDENTAL, CONSEQUENTIAL OR SPECIAL DAMAGES.

Buyer and all users are deemed to have accepted the terms of this LIMIT OF WARRANTY AND LIABILITY which may not be varied by any verbal or written agreement.

PRECAUTIONARY STATEMENTS

HAZARDS TO HUMANS AND DOMESTIC ANIMALS

KEEP OUT OF REACH OF CHILDREN

DANGER!

CAUSES EYE BURNS.

HARMFUL IF SWALLOWED OR INHALED.

Do not get in eyes, on skin or on clothing.

Wear goggles or face shield and rubber gloves in transferring and mixing and when adjusting, repairing or cleaning equipment.

Avoid breathing vapor or spray mist.

Wash thoroughly with soap and water after handling.

FIRST AID: IF IN EYES, immediately flush with plenty of water for at least 15 minutes. Call a physician.

IF INHALED, remove individual to fresh air. Seek medical attention if breathing difficulty develops.

IF SWALLOWED, this product will cause gastrointestinal tract irritation. Immediately dilute by swallowing water or milk. Call a physician.

In case of an emergency involving this product, Call Collect, day or night, (314) 694-4000.

ENVIRONMENTAL HAZARDS

Do not apply directly to water or wetland (swamps, bogs, marshes or potholes). Do not contaminate water by cleaning of equipment or disposal of wastes.

PHYSICAL OR CHEMICAL HAZARDS

Spray solutions of this product should be mixed, stored and applied only in stainless steel, aluminum, fiberglass, plastic and plastic-lined steel containers.

DO NOT MIX, STORE OR APPLY THIS PRODUCT OR SPRAY SOLUTIONS OF THIS PRODUCT IN GALVANIZED STEEL OR UNLINED STEEL (EXCEPT STAINLESS STEEL) CONTAINERS OR SPRAY TANKS. This product or spray solutions of this product react with such containers and tanks to produce hydrogen gas which may form a highly combustible gas mixture. This gas mixture could flash or explode, causing serious personal injury, if ignited by open flame, spark, welder's torch, lighted cigarette or other ignition source.

GENERAL INFORMATION

RANGER is a postemergence herbicide used in agricultural systems for control of quackgrass and other weeds listed on this label.

This product enters the plant through the foliage and moves throughout the plant. Visual effects of control are a gradual wilting or yellowing of the plant, which advances to complete browning of above-ground growth and deterioration of affected underground plant parts. Visible symptoms will usually develop on labeled weeds within 2 to 4 days after application, but may not occur for 7 or more days. Extremely cool or cloudy weather following treatment may slow activity of this product and delay the visual effects of control.

This product does not provide residual weed control. For subsequent residual weed control, follow a label-approved herbicide program. Read and carefully observe the cautionary statements and all other information appearing on the label of all herbicides used.

APPLICATION PRECAUTIONS AND WARNINGS

- Apply only when there is no hazard from spray drift since very small quantities of spray which may not be visible may severely injure susceptible crops or desirable vegetation.
- The likelihood of injury occurring to adjacent crops from the use of this product is greatest when winds are gusty or in excess of 5 miles per hour or when other conditions including lesser wind velocities will favor spray drift.
- Do not feed or forage vegetation from treated areas within 8 weeks after application.
- Buyer and all users are responsible for all loss or damage in connection with the use or handling of mixtures of this product with herbicides or other materials that are not expressly recommended in this labeling. Mixing this product with herbicides or other materials not recommended on this label may result in reduced performance. Use of this product in any manner not consistent with this label may result in injury to persons, animals, or crops, or other unintended consequences.

DIRECTIONS FOR USE

It is a violation of Federal law to use this product in any manner inconsistent with its labeling.

STORAGE AND DISPOSAL

Do not contaminate water, foodstuffs, feed or seed by storage or disposal.

STORAGE:

At temperatures below 50°F, the viscosity of this product may increase to a point that pouring and mixing could be difficult.

DISPOSAL:

Wastes resulting from the use of this product that cannot be used or chemically reprocessed should be disposed of in a landfill approved for pesticide disposal or in accordance with applicable Federal, state or local procedures.

Emptied container retains vapor and product residue. Observe all labeled safeguards until container is destroyed.

Do not reuse container. Triple rinse container. Then puncture and dispose of in a sanitary landfill, or by incineration, or, if allowed by state or local authorities, by burning. If burned, stay out of smoke.

Timing of Application

This product should be applied postemergence to vigorously growing weeds when they have reached the recommended size given in the "WEEDS CONTROLLED" section of this label. Application should be delayed until maximum emergence of the target weeds.

For quackgrass control in the fall, do not till between harvest and application. For quackgrass control in the spring, do not till in the fall or spring prior to application. A tillage operation following application may give improved quackgrass control. Allow 3 or more days after application before tillage.

When applied as directed under the conditions described, this product controls the weeds listed on this label, prior to the emergence of the following crops.

Alfalfa	Forage legumes
Barley	Oats
Corn (all)	Wheat
Forage grasses	

WEEDS CONTROLLED

For control of quackgrass and associated annual weeds, apply 3 pints per acre of this product in recommended water volumes when the plants have reached the 3 to 4-leaf stage and are at least 6 to 8 inches high. In sod situations, this product may be tank mixed with 2,4-D or Banvel™ for forage legume control. (Refer to the specific product labels for use rates, crop rotation restrictions and cautionary statements.)

When applied as directed under the conditions described, this product controls the following annual and perennial weeds listed below:

RECOMMENDED RATES AND WEEDS CONTROLLED

Annual Weed Species	Maximum Height	Rate Per Acre
Foxtail Setaria spp.	12"	12 oz.
Barnyardgrass Echinochloa crus-galli	6"	18 oz.
Bluegrass, annual Poa annua		
Brome, downy* Bromus tectorum		
Mustard, blue Chorispora tenella		
Mustard, tansy Descurainia pinnata		
Mustard, tumble Sisymbrium altissimum		
Mustard, wild Sinapis arvensis		
Spurry, umbrella Holsteum umbellatum	6"	18 oz.
Barley Hordeum vulgare	12"	
Rye Secale cereale		
Sandbur, field Cenchrus spp.		
Shattercane Sorghum bicolor		
Stinkgrass Eragrostis cilianensis		
Wheat Triticum aestivum	18"	
Morningglory Ipomea spp.	3"	24 oz.
Bluegrass, bulbous Poa bulbosa	6"	
Cheat Bromus secalinus		
Chickweed, mouseear Cerastium vulgatum		
Corn, volunteer Zea mays		
Groundsel, common Senecio vulgaris		
Horseweed/Marestail Conyza canadensis		

Annual Weed Species	Maximum Height	Rate Per Acre
Pennycress, field **Fanweed** Thlaspi arvense **Rocket, London** Sisymbrium irio **Ryegrass, Italian** Lolium multiflorum **Shepherdspurse** Capsella bursa-pastoris		
Buttercup Ranunculus spp. **Cocklebur** Xanthium pensylvanicum **Crabgrass** Digitaria spp. **Dwarfdandelion** Krigia cespitosa **Falseflax, smallseed** Camelina microcarpa **Lambsquarters, common** Chenopodium album **Oats, wild** Avena fatua **Panicum, fall** Panicum dichotomiflorum **Panicum, Texas** Panicum texanum **Pigweed, redroot** Amaranthus retroflexus **Pigweed, smooth** Amaranthus hybridus **Witchgrass** Panicum capillare	12"	

* For control in no-till systems, use 24 fluid ounces per acre.

Perennial Weed Species	Growth Stage	Rate Per Acre
Fescue (tall)* Festuca arundinacea	4 to 12"	3 pts.
Muhly, wirestem Muhlenbergia frondosa	8" or more	3 pts.
Orchardgrass Dactylis glomerata	4 to 12"	3 pts.
Quackgrass Agropyron repens	3 to 4 leaves	3 pts.

* Fall applications only.
TM Banvel is a trademark of Sandoz Inc.

APPLICATION EQUIPMENT

Do not apply this product through any type of irrigation system.

This product may be applied using either ground or aerial spray equipment. Use extreme care to avoid misting or drifting of herbicide solution onto foliage, green stems or fruit of desirable crops, trees, or plants during both growing and dormant periods, since even small quantities of spray can cause severe plant injury.

Ground Application: Apply recommended rates of this product in 5 to 30 gallons of water per acre as a broadcast spray. For optimum spray distribution and coverage, use flat fan or low-volume flood nozzles. When using flood nozzles, space them no more than 40 inches apart and ensure double overlap of spray pattern. Refer to the manufacturer's recommendations for correct pressure and nozzle height above the target canopy. Avoid pressure and nozzle combinations which produce fine droplets or mist susceptible to drift.

Use appropriate marking devices to ensure uniform spray coverage and best results from this product.

Aerial Application: Apply the recommended rates of this product in 3 to 15 gallons of water per acre as a broadcast spray. DO NOT APPLY DURING INVERSION CONDITIONS, WHEN WINDS ARE GUSTY, OR WHEN OTHER CONDITIONS WILL FAVOR DRIFT. DRIFT MAY CAUSE DAMAGE TO ANY VEGETATION CONTACTED TO WHICH TREATMENT WAS NOT INTENDED. TO PREVENT INJURY TO ADJACENT DESIRABLE VEGETATION, APPROPRIATE BUFFER ZONES MUST BE MAINTAINED.

Coarse sprays are less likely to drift; therefore, do not use nozzles or nozzle configurations which dispense spray as fine spray droplets. Do not angle nozzles forward into airstream and do not increase spray volume by increasing nozzle pressure.

Drift control additives may be used when making aerial applications. When a drift control additive is used, read and carefully observe the cautionary statements and all other information appearing on the additive label.

MIXING INSTRUCTIONS

Fill the spray tank to about 3/4 of the desired volume with clean water. Add the recommended amount of this product, then complete the filling process while maintaining agitation. Remove the hose from the mix tank immediately after filling to avoid siphoning back into the carrier source. During mixing and application, foaming of the spray solution may occur. To prevent or minimize foam, terminate by-pass and return lines at the tank bottom and/or use an agriculturally approved anti-foam or defoaming agent.

Mixing this product with herbicides or other materials not recommended on this label may result in reduced performance. Do not add additional surfactant to the spray solution when using this product.

NOTE: Reduced results may occur if water containing soil is used, such as water from ponds and unlined ditches.

When mixing in water below 50°F without vigorous agitation, this product should be premixed with water in a ratio of two parts water to one part this product.

SPRAYER CLEANUP

Clean sprayer and parts immediately after using this product by thoroughly flushing with water.

Thoroughly wash aircraft, especially landing gear, after each day of spraying to remove residues of this product accumulated during spraying or from spills. PROLONGED EXPOSURE OF THIS PRODUCT TO UNCOATED STEEL SURFACES MAY RESULT IN CORROSION AND POSSIBLE FAILURE OF THE PART. LANDING GEAR ARE THE MOST SUSCEPTIBLE. The maintenance of an organic coating (paint) which meets aerospace specification MIL-C-38413 may prevent corrosion.

 892.20-000.24/53

In case of emergency involving this product, Call Collect, day or night, (314) 694-4000.

This specimen label is intended for use only as a guide in providing general information regarding the use of this product. Monsanto is not responsible for the accuracy of the information contained herein. As labels are subject to revision, always carefully read and follow the label on the product container.

ROUNDUP®
HERBICIDE

Complete Directions For Use

AVOID CONTACT WITH FOLIAGE, GREEN STEMS, OR FRUIT OF CROPS, DESIRABLE PLANTS AND TREES, SINCE SEVERE INJURY OR DESTRUCTION MAY RESULT.

This product has been approved for use in California except as stated otherwise on page 1731.

ACTIVE INGREDIENT:
*Isopropylamine salt of glyphosate 41.0%
INERT INGREDIENTS: 59.0%
TOTAL 100.0%

*Contains 480 grams per litre or 4 pounds of the active ingredient isopropylamine salt of N-(phosphonomethyl) glycine per U.S. gallon. Equivalent to 356 grams per litre or 3 pounds per U.S. gallon of the acid, glyphosate.

EPA Reg. No. 524-308-AA
Product protected by U.S. Pat. No. 3,799,758 and U.S. Pat. No. 4,405,531. Other patents pending.

In case of an emergency involving this product, Call Collect, day or night. (314)694-4000.

1989–1 897.10–004.85/53

Read each of these sections of this label for essential product performance information.

Read the entire label before using this product.

Use only according to label instructions.

Read "LIMIT OF WARRANTY AND LIABILITY" before buying or using. If terms are not acceptable, return at once unopened.

LIMIT OF WARRANTY AND LIABILITY

(Not applicable to consumer applications applied by the homeowner for noncommercial purposes as permitted by the supplemental labeling for one-quart containers.)

This Company warrants that this product conforms to the chemical description on the label and is reasonably fit for the purposes set forth in the Complete Directions for Use label booklet ("Directions") when used in accordance with those Directions under the conditions described therein. NO OTHER EXPRESS WARRANTY OR IMPLIED WARRANTY OF FITNESS FOR PARTICULAR PURPOSE OR MERCHANTABIL-

ITY IS MADE. This warranty is also subject to the conditions and limitations stated herein.

Buyer and all users shall promptly notify this Company of any claims whether based in contract, negligence, strict liability, other tort or otherwise.

Buyer and all users are responsbile for all loss or damage from use or handling which results from conditions beyond the control of this Company, including but not limited to, incompatibility with products other than those set forth in the Directions, application to or contact with desirable vegetation, unusual weather, weather conditions which are outside the range considered normal at the application site and for the time period when the product is applied, as well as weather conditions which are outside the application ranges set forth in the Directions, application in any manner not explicitly set forth in the Directions, moisture conditions outside the moisture range specified in the Directions, or the presence of products other than those set forth in the Directions in or on the soil, crop or treated vegetation.

THE EXCLUSIVE REMEDY OF THE USER OR BUYER, AND THE LIMIT OF THE LIABILITY OF THIS COMPANY OR ANY OTHER SELLER FOR ANY AND ALL LOSSES, INJURIES OR DAMAGES RESULTING FROM THE USE OR HANDLING OF THIS PRODUCT (INCLUDING CLAIMS BASED IN CONTRACT, NEGLIGENCE, STRICT LIABILITY, OTHER TORT OR OTHERWISE) SHALL BE THE PURCHASE PRICE PAID BY THE USER OR BUYER FOR THE QUANTITY OF THIS PRODUCT INVOLVED, OR, AT THE ELECTION OF THIS COMPANY OR ANY OTHER SELLER, THE REPLACEMENT OF SUCH QUANTITY OR, IF NOT ACQUIRED BY PURCHASE, REPLACEMENT OF SUCH QUANTITY. IN NO EVENT SHALL THIS COMPANY OR ANY OTHER SELLER BE LIABLE FOR ANY INCIDENTAL, CONSEQUENTIAL OR SPECIAL DAMAGES.

Buyer and all users are deemed to have accepted the terms of this LIMIT OF WARRANTY AND LIABILITY which may not be varied by any verbal or written agreement.

PRECAUTIONARY STATEMENTS

HAZARDS TO HUMANS AND DOMESTIC ANIMALS

KEEP OUT OF REACH OF CHILDREN

WARNING!

CAUSES EYE IRRITATION
HARMFUL IS SWALLOWED
MAY CAUSE SKIN IRRITATION.

Do not get in eyes, on skin or on clothing.

Wash thoroughly after handling.

FIRST AID: IF IN EYES, immediately flush with plenty of water for at least 15 minutes. Call a physician.

IF ON SKIN, immediately flush with plenty of water. Remove contaminated clothing. Wash clothing before reuse.

IF SWALLOWED, this product will cause gatrointestinal tract irritation. Immediately dilute by swallowing water or milk. Call a physician.

In case of an emergency involving this product, Call Collect, day day or night, (314) 694–4000.

ENVIRONMENTAL HAZARDS

Do not apply directly to water or wetland (swamps, bogs, marshes or potholes). Do not contaminate water by cleaning of equipment or disposal of wastes.

PHYSICAL OR CHEMICAL HAZARDS

Spray solutions of this product should be mixed, stored and applied only in stainless steel, aluminum, fiberglass, plastic and plastic-lined steel containers.

DO NOT MIX, STORE OR APPLY THIS PRODUCT OR SPRAY SOLUTIONS OF THIS PRODUCT IN GALVANIZED STEEL OR UNLINED STEEL (EXCEPT STAINLESS STEEL) CONTAINERS OR SPRAY TANKS. This product or spray solutions of this product react with such containers and tanks to produce hydrogen gas which may form a highly combustible gas mixture. This gas mixture could flash or explode, causing serious personal injury if ignited by open flame, spark, welder's torch, lighted cigarette or other ignition source.

GENERAL INFORMATION

DO NOT APPLY THIS PRODUCT USING AERIAL SPRAY EQUIPMENT EXCEPT UNDER CONDITIONS AS SPECIFIED WITHIN THIS LABEL.

ROUNDUP® herbicide, a water soluble liquid mixes readily with water to be applied as a foliar spray for the control or destruction of most herbaceous plants. It may be applied through most standard industrial or field-type sprayers after dilution and thorough mixing with water in accordance with label instructions.

This product moves through the plant from the point of foliage contact to and into the root system. Visible effects on most annual weeds occur within 2 to 4 days, but on most perennial weeds may not occur for 7 days or more. Extremely cool or cloudy weather following treatment may slow down activity of this product and delay visual effects of control. Visible effects are a gradual wilting and yellowing of the plant which advances to complete browning of above ground growth and deterioration of underground plant parts.

Unless otherwise specified on this label, delay application until vegetation has emerged and reached the stages described for control of such vegetation under the "Weeds Controlled" section of this label. Unemerged plants arising from unattached underground rhizomes or root stocks of perennials will not be affected by the spray and will continue to grow. For this reason, best control of most perennial weeds is obtained when treatment is made at late growth stages approaching maturity.

Always use the higher rate of this product per acre within the recommended range when (1) weed growth is heavy or dense, or (2) weeds are growing in an undisturbed (noncultivated) area.

Do not treat weeds under poor growing conditions such as drought stress, disease or insect damage, as reduced weed control may result. Reduced results may also occur when treating weeds heavily covered with dust.

Reduced control may result when applications are made to annual or perennial weeds that have been mowed, grazed, or cut, and have not been allowed to regrow to the recommended stage for treatment.

Rainfall or irrigation occurring within 6 hours after application may reduce effectiveness. Heavy rainfall or irrigation within 2 hours after application may wash the chemical off the foliage and a repeat treatment may be required.

ROUNDUP herbicide does not provide residual weed control. For subsequent residual weed control, follow a label-approved herbicide program. Read and carefully observe the cautionary statements and all other information appearing on the labels of all herbicides used.

Buyer and all users are responsible for all loss or damage in connection with the use or handling of mixtures of ROUNDUP with herbicides or other materials that are not expressly recommended in this labeling. Mixing this product with herbicides or other materials not recommended on this label may result in reduced performance.

For best results, spray coverage should be uniform and complete. Do not spray weed foliage to the point of runoff.

Keep people and pets off treated areas until spray solution has dried.

DOMESTIC ANIMALS: This product is considered to be relatively nontoxic to dogs and other domestic animals. However, ingestion of this product or large amounts of freshly sprayed vegetation may result in temporary gastrointestinal irritation (vomiting, diarrhea, colic, etc.). If such symptoms are observed, provide the animal with plenty of fluids to prevent dehydration. Call a veterinarian if symptoms persist for more than 24 hours.

® Registered trademark of Monsanto Company

ATTENTION

AVOID DRIFT, EXTREME CARE MUST BE USED WHEN APPLYING THIS PRODUCT TO PREVENT INJURY TO DESIRABLE PLANTS AND CROPS.

Do not allow the herbicide solution to mist, drip, drift, or splash onto desirable vegetation since minute quantities of this herbicide can cause severe damage or destruction to the crop, plants, or other areas on which treatment was not intended. The likelihood of injury occurring from the use of this product is greatest when winds are gusty or in excess of 5 miles per hour or when other conditions, including lesser wind velocities, will allow spray drift to occur. When spraying, avoid combinations of pressure and nozzle type that will result in splatter or fine particles (mist) which are likely to drift. AVOID APPLYING AT EXCESSIVE SPEED OR PRESSURE.

NOTE: Use of this product in any manner not consistent with this label may result in injury to persons, animals or crops, or other unintended consequences. Keep container closed to prevent spills and contamination.

MIXING, ADDITIVES AND APPLICATION INSTRUCTIONS

APPLY THESE SPRAY SOLUTIONS IN PROPERLY MAINTAINED AND CALIBRATED EQUIPMENT CAPABLE OF DELIVERING DESIRED VOLUMES. DO NOT APPLY UNDER WIND OR OTHER CONDITIONS WHICH ALLOW DRIFT TO OCCUR. HAND GUN APPLICATIONS SHOULD BE PROPERLY DIRECTED TO AVOID SPRAYING DESIRABLE PLANTS. NOTE: REDUCED RESULTS MAY OCCUR IF WATER CONTAINING SOIL IS USED, such as WATER FROM PONDS AND UNLINED DITCHES.

MIXING

This product mixes readily with water. Mix spray solutions of this product as follows: Fill the mixing or spray tank with the required amount of water. Add the proper amount of this product (see "Directions for Use" and "Weeds Controlled" sections of this label) near the end of the filling process and mix well. Remove hose from tank immediately after filling to avoid siphoning back into the carrier source. During mixing and application foaming of the spray solution may occur. To prevent or minimize foam, avoid the use of mechanical agitators, place the filling hose below the surface of the spray solution, terminate by-pass and return lines at the bottom of the tank and if needed use an approved anti-foam or defoaming agent.

TANK MIXTURES

Always predetermine the compatibility of labeled tank mixtures of this herbicide with water carrier by mixing small proportional quantities in advance.

Mix labeled tank mixtures of ROUNDUP herbicide with water as follows:

1. Place a 20 to 35 mesh screen or wetting basket over filling port.
2. Through the screen, fill the sprayer tank one-half full with water and start agitation.
3. If a wettable powder is used, make a slurry with the water carrier, and add it SLOWLY through the screen into the tank. Continue agitation.
4. If a flowable formulation is used, pre-mix one part flowable with one part water. Add diluted mixture SLOWLY through the screen into the tank. Continue agitation.
5. If an emulsifiable concentrate formulation is used, pre-mix one part emulsifiable concentrate with two parts water. Add diluted mixture slowly through the screen into the tank. Continue agitation.
6. Continue filling the sprayer tank with water and add the required amount of ROUNDUP herbicide near the end of the filling.
7. Where nonionic surfactant is recommended, add this to the spray tank before completing the filling process.
8. Add individual formulations to the spray tank as follows: wettable powder, flowable, emulsifiable concentrate, drift control additive, water soluble liquid followed by surfactant.

Maintain good agitation at all times until the contents of the tank are sprayed. If the spray mixture is allowed to settle, thorough agitation is required to resuspend the mixture before spraying is resumed.

Keep by-pass line on or near bottom of tank to minimize foaming. Screen size in nozzle or line strainers should be no finer than 50 mesh. Carefully select proper nozzle to avoid spraying a fine mist. For best results with conventional ground application equipment, use flat fan nozzles.

Clean sprayer and parts immediately after using this product by thoroughly flushing with water.

ADDITIVES

SURFACTANTS

Nonionic surfactants which are labeled for use with herbicides may be used. Do not reduce rates of this herbicide when adding surfactant. When adding additional surfactant or label instructions require the use of additional surfactant, use 0.5 percent surfactant concentration (2 quarts per 100 gallons of spray solution) when using surfactants which contain at least 50 percent active ingredient or a 1 percent surfactant concentration (4 quarts per 100 gallons of spray solution) for those surfactants containing less than 50 percent active ingredient. Read and carefully observe surfactant cautionary statements and other information appearing on the surfactant label.

AMMONIUM SULFATE

The addition of 2 percent dry ammonium sufate by weight or 17 pounds per 100 gallons of water may increase the performance of ROUNDUP and ROUNDUP plus 2,4-D amine or Banvel™ tank mixtures on annual and perennial weeds. The improvement in performance may be apparent where environmental stress is a concern. Low-quality ammonium sulfate may contain material that will not readily dissolve which could result in nozzle tip plugging. To determine quality, perform a jar test by adding 1/3 cup of ammonium sulfate to 1 gallon of water and agitate for 1 minute. If undissolved sediment is observed, predissolve the ammonium sulfate in water and filter prior to addition to the spray tank. If ammonium sulfate is added directly to the spray tank, add slowly with agitation. Adding too quickly may clog outlet lines. Ensure that ammonium sulfate is completely dissolved in the spray tank before adding herbicides or surfactant. Thoroughly rinse the spray system with clean water after use to reduce corrosion.

NOTE: The use of ammonium sulfate as an additive does not preclude the need for additional surfactant. Do not use herbicide rates lower than recommended in this label.

™ Banvel is a trademark of Sandoz, Inc.

COLORANTS OR DYES

Agriculturally approved colorants or marking dyes may be added to this product. Colorants or dyes used in spray solutions of this product may reduce performance, especially at lower rates or dilutions. Use colorants or dyes according to the manufacturer's recommendations.

APPLICATION EQUIPMENT AND TECHNIQUES

Do not apply this product through any type of irrigation system.

This product may be applied with the following application equipment:

Aerial—Fixed Wing and Helicopter

Broadcast Spray

Controlled Droplet Applicator (CDA)—Hand-held or boommounted applicators which produce a spray consisting of a narrow range of droplet sizes.

Hand-Held and High-Volume Spray Equipment—Knapsack and backpack sprayers, pump-up pressure sprayers, handguns, handwands, lances and other hand-held spray equipment used to direct the spray onto weed foliage and vehicle mounted high-volume spray equipment for spray-to-wet applications.

Selective equipment—Recirculating sprayers, shielded sprayers and wiper applicators.

See the appropriate part of this section for specific rates of application and instructions.

AERIAL EQUIPMENT

Use the recommended rates of this herbicide in 3 to 15 gallons of water per acre unless otherwise specified on this label. See the "Weeds Controlled" section of this label for specific rates. Unless otherwise specified, do not exceed the one-quart rate per acre of this product.

(Aerial applications of this product may be made in annual cropping conventional tillage systems, fallow and reduced tillage systems, preharvest, silvicultural sites, and rights-of-way. Refer to the individual use area sections of this label for recommended volumes and application rates.) FOR AERIAL APPLICATION IN CALIFORNIA, REFER TO THE FEDERAL SUPPLEMENTAL LABEL FOR AERIAL APPLICATIONS IN THAT STATE FOR SPECIFIC INSTRUCTIONS, RESTRICTIONS AND REQUIREMENTS.

Avoid direct application to any body of water.

AVOID DRIFT—DO NOT APPLY DURING INVERSION CONDITIONS, WHEN WINDS ARE GUSTY, OR UNDER ANY OTHER CONDITION WHICH WILL ALLOW DRIFT. DRIFT MAY CAUSE DAMAGE TO ANY VEGETATION CONTACTED TO WHICH TREATMENT IS NOT INTENDED. TO PREVENT INJURY TO ADJACENT DESIRABLE VEGETATION, APPROPRIATE BUFFER ZONES MUST BE MAINTAINED.

Coarse sprays are less likely to drift, therefore, do not use nozzles or nozzle configurations which dispense spray as fine spray droplets. Do not angle nozzles forward into the airstream and do not increase spray volume by increasing nozzle pressure.

Drift control additives may be used. When a drift control additive is used, read and carefully observe the cautionary statements and all other information appearing on the additive label.

Ensure uniform application—To avoid streaked, uneven or overlapped application, use appropriate marking devices.

Thoroughly wash aircraft, especially landing gear after each day of spraying to remove residues of this product accumulated during spraying or from spills. PROLONGED EXPOSURE OF THIS PRODUCT TO UNCOATED STEEL SURFACES MAY RESULT IN CORROSION AND POSSIBLE FAILURE OF THE PART. LANDING GEAR ARE MOST SUSCEPTIBLE. The maintenance of an organic coating (paint) which meets aerospace specifications MIL-C-38413 may prevent corrosion.

BROADCAST EQUIPMENT

For control of annual or perennial weeds listed on this label using broadcast equipment—Use the recommended rates of this product in 3 to 40 gallons of water per acre as a broadcast spray unless otherwise specified on this label. See the "Weeds Controlled" section of this label for specific rates. As density of weeds increases, spray gallonage should be increased within the recommended range to ensure complete coverage. Carefully select proper nozzle to avoid spraying a fine mist. For best results with ground application equipment, use flat fan nozzles. Check for even distribution of spray droplets.

CONTROLLED DROPLET APPLICATION (CDA)

The rate of this product applied per acre by vehicle-mounted CDA equipment must not be less than the amount recommended in this label when applied by conventional broadcast equipment. For vehicle-mounted CDA equipment apply 3 to 15 gallons of water per acre.

For the control of labeled annual weeds with hand-held CDA units, apply a 20 percent solution of this product at a flow rate of 2 fluid ounces per minute and a walking speed of 1.5 MPH (1 quart per acre). For the control of labeled perennial weeds, apply a 20 to 40 percent solution of this product at a flow rate of 2 fluid ounces per minute and a walking speed of 0.75 MPH (2 to 4 quarts per acre).

Controlled droplet application equipment produces a spray pattern which is not easily visible. Extreme care must be exercised to avoid spray or drift contacting the foliage or any other green tissue of desirable vegetation as damage or destruction may result.

HAND-HELD and HIGH-VOLUME EQUIPMENT

Use coarse sprays only.

Mix this product in clean water and apply to foliage of vegetation to be controlled. For applications made on a spray-to-wet basis, spray coverage should be uniform and complete. Do not spray to the point of runoff.

For control of annual weeds listed on this label, apply a 0.5 percent solution of this product plus nonionic surfactant to weeds less than 6 inches in height or runner length. Apply prior to seedhead formation in grass or bud formation in broadleaves. Allow three or more days before tillage or mowing.

For annual weeds over 6 inches tall, or when not using additional surfactant, or unless otherwise specified, use a 1 percent solution. For best results, use a 2 percent solution on harder-to-control perennials, such as bermudagrass, dock, field bindweed, hemp dogbane, milkweed and Canada thistle.

Less than complete coverage of weeds may result from the use of spray equipment designed for motorized spot treatments. Where less than complete coverage of annual weeds occurs, use a 5 percent solution. Do not reduce recommended concentrations of ROUNDUP when adding surfactant.

Prepare the desired volume of spray solution by mixing the amount of this product in water as shown in the following table:

Spray Solution

Desired Volume	Amount of ROUNDUP® ½%	1%	1½%	2%	5%
1 gallon	⅔ oz	1⅓ oz	2 oz	2⅔ oz	6½ oz
25 gallons	1 pt	1 qt	1½ qt	2 qt	5 qt
100 gallons	2 qt	1 gal	1½ gal	2 gal	5 gal
2 tablespoons = 1 ounce					

For use in knapsack sprayers, it is suggested that the proper amount of this product be mixed with water in a larger container. Fill sprayer with the mixed solution.

For hand-held WIPER APPLICATORS see the "Selective Equipment" section and for hand-held CONTROLLED DROPLET APPLICATORS see the "Controlled Droplet Applicator (CDA)" section of this label.

SELECTIVE EQUIPMENT

This product may be applied through a recirculating spray system, a shielded applicator, or a wiper applicator after dilution and thorough mixing with water to listed weeds growing in any noncrop site specified on this label and only when specifically recommended in cropping systems.

A recirculating spray system directs the spray solution onto weeds growing above desirable vegetation, while spray solution not intercepted by weeds is collected and returned to the spray tank for reuse.

A shielded applicator directs the herbicide solution onto weeds while shielding desirable vegetation from the herbicide.

A wiper applicator applies the herbicide solution onto weeds by rubbing the weed with an absorbent material containing the herbicide solution.

AVOID CONTACT WITH DESIRABLE VEGETATION

Contact of the herbicide solution with the desirable vegetation may result in damage or destruction. Applicators used above desired vegetation should be adjusted so that the lowest spray stream or wiper contact point is at least two inches above the desirable vegetation. Droplets, mist, foam, or splatter of the herbicide solution setting on desirable vegetation may result in discoloration, stunting, or destruction.

Applications made above the crops should be made when the weeds are a minimum of 6 inches above the desirable vegetation. Better results may be obtained when more of the weed is exposed to the herbicide solution. Weeds not contacted by the herbicide solution will not be affected. This may occur in dense clumps, severe infestations, or when the height of the weeds varies so that not all weeds are contacted. In these instances, repeat treatment may be necessary.

See the "Weeds Controlled" section of this label for recommended stage of growth for specific weeds.

NOTE

- Maintain equipment in good operating condition. Avoid leakage or dripping onto desirable vegetation.
- Adjust height of applicator to insure adequate contact with weeds.
- Keep nozzle tips and wiping surfaces clean.
- Keep spray patterns aligned into recovery chamber of the recirculating sprayer.
- Keep shields on shielded applicators adjusted to protect desirable vegetation.
- Maintain recommended roller RPM on roller applicators while in use.
- Keep wiper material at proper degree of saturation with herbicide solution.
- DO NOT use wiper equipment when weeds are wet.
- DO NOT operate equipment at ground speeds greater than 5 mph. Weed control may be affected by speed of application equipment. As weed density increases, reduce equipment ground speed to insure good coverage of weeds.
- Be aware that on sloping ground the herbicide solution may migrate, causing dripping on the lower end and drying on the upper end of a wiper applicator.
- Variation in equipment design may affect weed control. With wiper applicators, the wiping material and its orientation must allow delivery of sufficient quantities of the recommended herbicide solution directly to the weed.
- Care must be taken with all types of wipers to ensure that the absorbent material does not become oversaturated, causing the herbicide to drip on desirable vegetation.
- Mix only the amount of solution to be used during a one-day period, as reduced activity may result from use of leftover solutions. With all equipment, drain and clean sprayer and wiper parts immediately after using this product by thoroughly flushing with water.

RECIRCULATING SPRAYERS

Recirculating spray calibration is made on the basis of ground speed and delivery volume. Two procedures can be used to calibrate: (1) determine the discharge being delivered per minute, then operate at the designated ground speed, or (2) select the desired ground speed and then adjust the sprayer to deliver the recommended volume per minute (this may require nozzle changes). Use the appropriate table below.

Do not operate at nozzle pressure above 20 PSI.

Table 1. Use this table when calibrating box or row-type recirculating sprayers. Box or row-type sprayer calibration is based on the total discharge collected per row. Use only straight stream or 15° fan-type nozzles.

*Volume Per Minute Per Row	
MPH	Ounces
2	25 to 35
3	38 to 51
4	51 to 68
5	65 to 86

*NOTE: Be certain the amount collected is for all spray streams treating one row.

Table 2. Use this table when calibrating broadcast type recirculating sprayers. Broadcast recirculating sprayer calibration is based on the discharge collected per minute from one nozzle on a 20-inch spacing.

Volume Per Minute Per Nozzle	
MPH	Ounces
2	7 to 9
3	10 to 13
4	13 to 18
5	16 to 22

When applied as recommended under the conditions described for recirculating sprayers, this product will control the following weeds growing a minimum of 6 inches above desirable vegetation.

Perennial Broadleaf Weeds—To SUPPRESS the following weeds, mix in a ratio of 4 quarts of this product in 20 gallons of water and apply as directed.

Dogbane, hemp *Apocynum cannabinum*
Milkweed *Asclepias syriaca*

Perennial Grasses and Annual Broadleaf Weeds—To control the following weeds, mix a ratio of 3 quarts of this product in 20 gallons of water and apply as directed.

Cocklebur *Xanthum strumarium*
Pigweed, redroot *Amaranthus retroflexus*
Johnsongrass *Sorghum halepense*
Sunflower *Helianthus annuss*

Annual Grasses—To control the following weeds, mix in a ratio of 2 quarts of this product in 20 gallons of water and apply as directed:

Corn *Zea mays*
Shattercane *Sorgnum bicolor*

SHIELDED APPLICATORS

When applied as directed under conditions described for shielded applicators, this product will control those weeds listed in the "Weeds Controlled" section of this label.

Shielded applicators which apply the herbicide solution as a spray band should be calibrated on a broadcast equivalent rate and volume basis. To determine these:

$$\frac{\text{Band width in inches}}{\text{Row width in inches}} \times \text{Herbicide Broadcast RATE per acre} = \text{Herbicide Band RATE per acre}$$

$$\frac{\text{Band width in inches}}{\text{Row width in inches}} \times \text{Broadcast VOLUME of solution per acre} = \text{Band VOLUME of solution per acre}$$

Use nozzles that provide uniform coverage within the treated area. EXTREME CARE MUST BE EXERCISED TO AVOID CONTACT WITH DESIRABLE VEGETATION.

For specific rates of application and instructions for control of various annual and perennial weeds, see the "Weeds Controlled" section of this label.

WIPER APPLICATORS

Wiper applicators include either roller or wick devices which physically wipe appropriate concentrations or amounts of this product directly onto the weed. Equipment must be designed, maintained, and operated to prevent the herbicide solution from contacting desirable vegetation. Operate this equipment at ground speeds no greater than 5 mph. Performance may be improved by reducing speed in areas of heavy weed infestations to ensure adequate wiper saturation. Better results may be obtained if two applications are made in opposite directions.

Do not add surfactant to the herbicide solution.

For Roller Applicators—Mix 1 gallon of this product in enough water to prepare 10 gallons of herbicide solution (10 percent solution). Apply this solution to perennial weeds or annual broadleaf weeds listed in this "Wiper Applicators" section.

Mix 1 gallon of this product in enough water to provide 20 gallons of herbicide solution (5 percent solution) Apply this solution to annual grasses listed in this "Wiper Applicators" section.

Roller speed should be maintained at 40 to 60 RPM.

For Wick or Wiper Applicators—Mix 1 gallon of this product in 2 gallons of water to prepare a 33 percent solution. Apply this solution to weeds listed in this "Wiper Applicators" section.

In severe infestations, reduce equipment ground speed to ensure that adequate amounts of this product are wiped on the weeds. A second treatment in the opposite direction may be beneficial.

Do not permit herbicide solution to contact desirable vegetation.

When applied as recommended under the conditions described for "Wiper Applicators" this product CONTROLS the following weeds:

Annual Grasses

Corn
Zea Mays
Panicum, Texas
Panicum texanum
Rye, common
Secale cereale
Shattercane
Sorghum bicolor

Annual Broadleaves

Sicklepod
Cassia obtusifolia
Spanishneedles
Bidens bipinnata
Starbur, bristly
Acanthospermum hispidum

When applied as recommended under the conditions described for "Wiper Applicators" this product SUPPRESSES the following weeds:

Annual Broadleaves

Beggarweed, Florida
Desmodium tortuosum
Dogfennel
Eupatorium capiliflorium
Pigweed, redroot
Amaranthus retroflexus
Ragweed, common
Ambrosia artemisiifolia
Ragweed, giant
Ambrosia trifida
Sunflower
Helianthus annuus
Thistle, musk
Carduus nutans
Velvetleaf
Abutilon theophrasti

Perennial Grasses

Bermudagrass
Cynodon dactylon
Guineagrass
Panicum maximum
Johnsongrass
Sorghum halepense
Smutgrass
Sporobolus poiretii
Vaseygrass
Paspalum urvillei

Perennial Broadleaves

Dogbane, hemp
Apocynum cannabinum
Milkweed
Ascelepias syriaca
Nightshade, silverleaf
Solanum elaeagnifolium
Thistle, Canada
Cirsium arvense

WEEDS CONTROLLED

This herbicide controls many annual and perennial grasses and broadleaf weeds.

ANNUAL WEEDS

- Apply to actively growing grass and broadleaf weeds.
- Allow at least 3 days after treatment before tillage.
- For maximum agronomic benefit, apply when weeds are 6 inches or less in height.
- To prevent seed production, applications should be made prior to seedhead formation.
- This product does not provide residual control; therefore, delay application until maximum weed emergence. Repeat treatments may be necessary to control later germinating weeds.

LOW-VOLUME BROADCAST APPLICATION (LOW-RATE TECHNOLOGY)

When applied as directed under the conditions described, this product will control the weeds listed below when:

1. Water carrier volumes are 3 to 10 gallons per acre for ground applications and 3 to 5 gallons per acre for aerial applications. (See the "Aerial Application" section of this label for approved sites.)
2. A nonionic surfactant is added at 0.5 to 1 percent by total spray volume. Use 0.5 percent surfactant concentration when using surfactants which contain at least 50 percent active ingredient or a 1 percent surfactant concentration for those surfactants containing less than 50 percent active ingredient.

NOTE

- The addition of 2 percent dry ammonium sulfate by weight or 17 pounds per 100 gallons of water may increase the performance of this herbicide on annual weeds. The improvement in performance may be apparent where environmental stress is a concern. Refer to the "Mixing, Additives and Application Instructions" section of this label.
- Do not tank-mix with soil residual herbicides when using these rates unless otherwise specified.
- For weeds that have been mowed, grazed, or cut, allow regrowth to occur prior to treatment.
- Refer to the "Tank Mixtures" portion of this section for control of additional broadleaf weeds.

Weed Species	Maximum Height	Rate Per Acre**
Foxtail *Setaria spp.*	12″	8 oz.
Barnyardgrass *Echinochloa crus-galli*	6″ (0 to 4″) (4 to 6″)	12 oz. 16 oz.[1]) 24 oz.[1])
Bluegrass, annual *Poa annua*		
Brome, downy* *Bromus tectorum*		
Mustard, blue *Chorispora tenella*		
Mustard, tansy *Descurainia pinnata*		
Mustard, tumble *Sisymbrium altissimum*		
Mustard, wild *Sinapis arvensis*		
Spurry, umbrella *Holosteum umbellatum*		
Barley *Hordeum vulgare*	12″	
Rye *Secale cereale*		
Sandbur, field *Cenchrus spp.*		
Shattercane *Sorghum bicolor*		
Stinkgrass *Eragrostis cilianensis*		
Wheat *Triticum aestivum*	18″	
Morningglory *Ipomea spp.*	2″	16 oz.
Sicklepod *Cassia obtusifolia*		
Bluegrass, bulbous *Poa bulbosa*	6″	
Cheat *Bromus secalinus*		
Chickweed, mouseear *Cerastium vulgatum*		
Corn *Zea mays*		
Groundsel, common *Senecio vulgaris*		
Horseweed/Marestail *Conyza canadensis*		
Lambsquarters, common *Chenopodium album*		
Pennycress, field **Fanweed** *Thlaspi arvense*		
Rocket, London *Sisymbrium irio*		
Ryegrass, Italian *Lolium multiflorum*		
Shepherdspurse *Capsella bursa-pastoris*		
Buttercup *Ranunculus spp.*	12″	
Cocklebur *Xanthium strumarium*		
Crabgrass *Digitaria spp.*		

Weed Species	Maximum Height	Rate Per Acre**
Dwarfdandelion *Krigia cepitosa*		
Falseflax, smallseed *Camelina microcarpa*		
Foxtail, Carolina *Alopecurus carolinianus*		
Johnsongrass, seedling *Sorghum halepense*		
Oats, wild *Avena fatua*		
Panicum, fall *Panicum dichotomiflorum*		
Panicum, Texas *Panicum texanum*		
Pigweed, redroot *Amaranthus retroflexus*		
Pigweed, smooth *Amaranthus hybridus*		
Witchgrass *Panicum capillare*		
Sicklepod *Cassia obtusifolia*	3 to 4"	24 oz.
Signalgrass, broadleaf *Brachiaria platyphylla*	4"	
Horseweed/Marestail *Conyza canadensis*	7 to 12"	
Lambsquarters, common *Chenopodium album*		
Teaweed *Sida spinosa*	4"	32 oz.
Rice, red *Oryza sativa*	6"	
Geranium, Carolina *Geranium carolinianum*	12"	
Goosegrass *Eleusine indica*		
Primrose, cutleaf evening *Oenothera laciniate*		
Pusley, Florida *Richardia scabra*		
Sicklepod *Cassia obtusifolia*	5 to 12"	
Spanishneedles *Bidens bipinnata*		
Filaree *Erodium spp.*	12"	48 oz.

[1] Use these rates in Alabama, Arkansas, Mississippi, Missouri, Louisiana and Texas for preplant treatments.

* For control in no-till systems, use 16 fluid ounces per acre.

** For those rates less than 32 ounces per acre, ROUNDUP at rates up to 32 ounces per acre may be used where heavy weed densities exist.

TANK MIXTURES

ROUNDUP® plus BANVEL plus NONIONIC SURFACTANT
ROUNDUP plus 2,4-D AMINE plus NONIONIC SURFACTANT

DO NOT APPLY BANVEL OR 2,4-D AMINE TANK MIXTURES BY AIR IN CALIFORNIA.

These tank mixtures are recommended for use in fallow and reduced tillage areas only. Follow use directions as given in the "Low-volume Broadcast Application" section.

ROUNDUP plus Banvel or 2,4-D amine will control the annual grasses and broadleaf weeds previously listed for ROUNDUP alone at the indicated heights (except 8 ounces per acre applications), plus the following broadleaf weeds. For those weeds previously listed at 8 ounces of ROUNDUP herbicide alone per acre, use the 12-ounce rate in these tank mixtures.

NOTE: Refer to the specific product labels for crop rotation restrictions and cautionary statements of all products used in tank mixtures. Some crop injury may occur if Banvel is applied within 45 days of planting. The addition of Banvel in a mixture with this product may provide short-term residual control of selected weed species.

Apply 12 to 16 ounces of this product plus 0.25 lb. a.i. of Banvel or 0.5 pound a.e. of 2,4-D amine, plus 0.5 to 1 percent nonionic surfactant by total spray volume per acre to control dense populations of the following annual broadleaf weeds when less than 6 inches in height.

Kochia* *Kochia scoparia*	**Pigweed, redroot** *Amaranthus retroflexus*
Lambsquarters *Chenopodium album*	**Thistle, Russian** *Salsola kali*
Lettuce, prickly *Lactuca serriola*	

* Controlled with Banvel tank mixture only.

HIGH-VOLUME BROADCAST APPLICATIONS

When applied as directed under the conditions described, this product will control the weeds listed below when water carrier volumes are 10 to 40 gallons per acre for ground applications.

Use 1 quart of this product per acre if weeds are less than 6 inches tall. If weeds are over 6 inches tall, use 1.5 quarts of this product per acre. If weeds have been mowed, grazed, or cut allow adequate time for new growth to recommended stages prior to treatment. These rates will also provide control of weeds listed in the "Low-Volume Broadcast Application" section.

Weed Species

Balsamapple* *Momordica charantia*	**Panicum** *Panicum spp.*
Bassia, fivehook *Bassia hyssopifolia*	**Ragweed, common** *Ambrosia artemisiifolia*
Brome *Bromus spp.*	**Ragweed, giant** *Ambrosia trifida*
Fiddleneck *Amsinckia spp.*	**Smartweed, Pennsylvania** *Polygonum pensylvanicum*
Flaxleaf Fleabane *Conyza bonariensis*	**Sowthistle, annual** *Sonchus oleraceus*
Fleabane *Erigeron spp.*	**Sunflower** *Helianthus annuus*
Kochia *Kochia scoparia*	**Thistle, Russian** *Salsola kali*
Lettuce, prickly *Lactuca serriola*	**Velvetleaf** *Abutilon theophrasti*

* Apply with hand-held equipment only.

PERENNIAL WEEDS

Apply this product as follows to control or destroy most perennial weeds:

NOTE: If weeds have been mowed or tilled, do not treat until plants have resumed active growth and are at the recommended stages.

Repeat treatments may be necessary to control weeds regenerating from underground parts or seed. Repeat treatments must be made prior to crop emergence.

The addition of 2 percent dry ammonium sulfate by weight or 17 pounds per 100 gallons of water may increase the performance of this herbicide on perennial weeds. The improvement in performance may be apparent where environmental stress is a concern. Refer to the "Mixing, Additives and Application Instructions" section of this label.

When applied as recommended under the conditions described, this product WILL CONTROL the following PERENNIAL WEEDS:

Alfalfa *Medicago sativa*	**Horseradish** *Armoracia rusticana*
Alligatorweed* *Alternanthera philoxeroides*	**Johnsongrass** *Sorghum halepense*
Artichoke, Jerusalem *Helianthus tuberosus*	**Kikuyugrass** *Pennisetum clandestinum*
Bahiagrass *Paspalum notatum*	**Knapweed** *Centaurea repens*
Bentgrass *Agrostis spp.*	**Lantana** *Lantana camara*
Bermudagrass *Cynodon dactylon*	**Milkweed** *Asclepias spp.*
Bindweed (field) *Convolvulus arvensis*	**Muhly, wirestem** *Muhlenbergia frondonsa*
Bluegrass (Kentucky) *Poa spp.*	**Mullein, common** *Verbascum thapsus*
Blueweed (Texas) *Helianthus ciliaris*	**Napiergrass** *Pennisetum purpureum*
Brackenfern *Pteridium aquilinum*	**Nightshade, silverleaf** *Solanum elaeagnifolium*
Bromegrass (smooth) *Bromus inermis*	**Nutsedge, purple, yellow** *Cyperus rotundus, Cyperus esculentus*
Bursage (woollyleaf) *Franseria tomentosa*	**Orchardgrass** *Dactylis glomerata*
Canarygrass (reed) *Phalaris arundinacea*	**Pampas** *Cortaderia jubata*
Cattail *Typha spp.*	**Paragrass** *Brachiaria mutica*
Clover (red) *Trifolium pratense*	**Phragmites*** *Phragmites spp.*
Clover (white) *Trifolium repens*	**Quackgrass** *Agropyron repens*
Cogongrass *Imperata cylindrica*	**Ryegrass, perennial** *Lolium perenne*

Dallisgrass
Paspalum dilatatum
Dandelion
Taraxacum officinale
Dock, curly
Rumex crispus
Dogbane, hemp
Apocynum cannabinum
Fescues
Festuca spp.
Fescue, tall
Festuca arundinacea
Guineagrass
Panicum maximum
Horsenettle
Solanum carolinense
Smartweed, swamp
Polygonum coccineum
Sweet potato, wild*
Ipomoea pandurata
Thistle, Canada
Cirsium arvense
Timothy
Phleum pratense
Torpedograss*
Panicum repens
Vaseygrass
Paspalum urvilei
Wheatgrass, western
Agropyron smithii

* Partial Control

See "Directions for Use" and "Mixing, Additives and Application Instructions" sections of this label for labeled uses and specific application instructions.

Alfalfa—Apply 1 quart of this product per acre plus 0.5 to 1 percent nonionic surfactant by total spray volume in 3 to 10 gallons of water per acre. Make application after the last hay cutting in the fall. Allow alfalfa to regrow to a height of 6 to 8 inches or more prior to treatment. Applications should be followed with deep tillage at least 7 days after treatment, but before soil freeze-up.

Alligatorweed—Apply 4 quarts of this product per acre or apply a 1½ percent solution with hand-held equipment to provide partial control. Apply when most of the plants are in bloom. Repeat applications will be required to maintain such control.

Bentgrass—For suppression in grass seed production areas. For ground applications only, apply 1.5 quarts of this product plus 0.5 to 1 percent nonionic surfactant by total spray volume in 10 to 20 gallons of water per acre. Ensure entire crown area has resumed growth prior to a fall application. Bentgrass should be actively growing and have at least 3 inches of growth. Tillage prior to treatment should be avoided. Tillage 7 to 10 days after application is recommended for best results. Failure to use tillage after treatment may result in unacceptable control.

Bermudagrass—For control, apply 5 quarts of this product per acre and for partial control, apply 3 quarts per acre. Treat when bermudagrass is actively growing and seedheads are present. Retreatment may be necessary to maintain control. Allow 7 or more days after application before tillage.

Bindweed, field/Blueweed, Texas—Apply 4 to 5 quarts of this product per acre west of the Mississippi River and 3 to 4 quarts per acre east of the Mississippi River. Apply when weed is actively growing and is at or beyond full bloom. Do not treat when weed is under drought stress as good soil moisture is necessary for active growth. New leaf development indicates active growth. For best results, apply in late summer or fall. Fall treatments must be applied before a killing frost. Allow 7 or more days after application before tillage.

Bindweed, field (suppression irrigated lands where annual tillage is performed): Apply 1 quart of this product plus 0.5 to 1 percent nonionic surfactant by total spray volume in 3 to 10 gallons of water per acre. Apply to actively growing bindweed that has reached a length of 12 inches or greater. Allow maximum weed emergence and runner growth. Do not treat when weed is under drought stress as good soil moisture is necessary for active growth. Allow 3 or more days after application before tillage.

Bluegrass, Kentucky/Bromegrass, smooth/Orchardgrass—Apply 2 quarts of this product in 10 to 40 gallons of water per acre when the grasses are actively growing and most plants have reached boot-to-early seedhead stage of development. For partial control in pasture or hay crop renovation apply 1 to 1½ quarts of this product plus 0.5 to 1 percent nonionic surfactant by total spray volume in 3 to 10 gallons of water per acre. Apply to actively growing plants when most have reached 4 to 12 inches in height. Allow 7 or more days after application before tillage.

Orchardgrass (sods going to no-till corn)—Apply 1 to 1.5 quarts of this product per acre plus 0.5 to 1 percent nonionic surfactant by total spray volume in 3 to 10 gallons of water per acre. Apply to orchardgrass that is a minimum of 12 inches tall for spring applications and 6 inches tall for fall applications. Allow at least three days following application before planting. A sequential application of 3.75 to 4.5 quarts of Lariat® herbicide, or equivalent, will be necessary for optimum results. Lariat should be applied within 3 to 10 days following spring applications to prevent annual weed growth.

* Lariat is a registered trademark of Monsanto Company.

Brackenfern—Apply 3 to 4 quarts of this product per acre as a broadcast spray or as a 1 to 1½ percent solution with hand-held equipment. Apply to fully expanded fronds which are at least 18 inches long.

Bursage, woollyleaf—For control apply 2 quarts of this product plus 1 pint of Banvel per acre. For partial control apply 1 quart of this product plus 1 pint of Banvel per acre. Add 0.5 to 1 percent nonionic surfactant by total spray volume and apply 3 to 20 gallons of water per acre. Apply when plants are producing new active growth which has been initiated by moisture for at least 2 weeks and when plants are at or beyond flowering.

Canarygrass, reed/Timothy/Wheatgrass, western—Apply 2 to 3 quarts of this product per acre. For best results, apply to actively growing plants when most have reached the boot-to-head stage of growth. Allow 7 or more days after application before tillage.

Cogongrass—Apply 3 to 5 quarts of this product plus 0.5 to 1 percent nonionic surfactant in 10 to 40 gallons of water per acre. Apply when Cogongrass is at least 18 inches tall and actively growing in late summer or fall. Allow 7 or more days after application before tillage or mowing. Due to uneven stages of growth and the dense nature of vegetation preventing good spray coverage, repeat treatments may be necessary to maintain control.

Dogbane, hemp/Knapweed/Horseradish—Apply 4 quarts of this product per acre. Apply when actively growing and when most weeds have reached the late bud to flower stage of growth. Following crop harvest or mowing, allow weeds to regrow to a mature stage prior to treatment. For best results, apply in late summer or fall. Allow 7 or more days after application before tillage.

Fescue, tall—Apply 3 quarts of this product in 10 to 40 gallons of water per acre to actively growing plants when most have reached boot-to-early seedhead stage of development.

Fall applications only—Apply 1 quart of this product plus 0.5 to 1 percent nonionic surfactant by total spray volume in 3 to 10 gallons of water per acre. Apply to fescue in the fall when actively growing and plants have 6 to 12 inches of new growth. Allow 7 or more days after application before tillage. A sequential application of 1 pint per acre of this product plus nonionic surfactant will improve long-term control and control seedlings germinating after fall treatments or the following spring.

Guineagrass—Apply 3 quarts of this product per acre or use a 1 percent solution with hand-held equipment. Apply to actively growing guineagrass when most has reached at least the 7-leaf stage of growth. Ensure thorough coverage when using hand-held equipment. Allow 7 or more days after application before tillage.

Johnsonson/Ryegrass, perennial—Apply 1 to 3 quarts of this product per acre. In annual cropping systems apply 1 to 2 quarts of this product per acre. Apply 1 quart of this product plus 0.5 to 1 percent nonionic surfactant by total spray volume in 3 to 10 gallons of water per acre. Use 2 quarts of this product when applying 10 to 40 gallons of water per acre. In noncrop or areas where annual tillage (no-till), is not performed, apply 2 to 3 quarts of this product in 10 to 40 gallons of water per acre. For best results, apply to actively growing plants when most have reached the boot-to-head stage of growth or in the fall prior to frost. Allow 7 or more days after application before tillage. Do not tank-mix with residual herbicides when using the 1 quart per acre rate.

For burndown of Johnsongrass, apply 1 pint per acre plus 0.5 to 1 percent nonionic surfactant in 3 to 10 gallons of water per acre before the plants each reach a height of 12 inches. For this use, allow at least 3 days after treatment before tillage.

Spot Treatment (partial control or suppression)—Apply a 1 percent solution of this product plus 0.5 to 1 percent nonionic surfactant by total spray volume when johnsongrass is 12 to 18 inches in height. Coverage should be uniform and complete.

Kikuyugrass—Apply 2 to 3 quarts of this product per acre. Spray when most kikuyugrass is at least 8 inches in height (3 or 4-leaf stage of growth) and actively growing. Allow 3 or more days after application before tillage.

Lantana—Apply this product as a 1 to 1¼ percent solution using hand-held equipment only. Apply to actively growing lantana at or beyond the bloom stage of growth. Use the higher application rate for plants that have reached the woody stage of growth. Allow 7 or more days after application before tillage.

Milkweed, common—Apply 3 quarts of this product per acre. Apply when actively growing and most of the milkweed has reached the late bud to flower stage of growth. Following small grain harvest or mowing, allow milkweed to regrow to a mature stage prior to treatment. Allow 7 or more days after application before tillage.

Muhly, wirestem—Apply 1 to 2 quarts of this product per acre. Use 1 quart of this product plus 0.5 to 1 percent nonionic surfactant by total spray volume in 3 to 10 gallons of water per acre. Use 2 quarts of this product when applying 10 to 40 gallons of water per acre or in pasture, sod, or noncrop areas. Spray when the wirestem muhly is 8 inches or more in height and actively growing. Do not till between harvest and fall applications or in the fall or spring prior to spring applications. Allow 3 or more days after application before tillage. This product will not provide residual control of wirestem muhly from seeds which germinate after a ROUNDUP herbicide application. Do not tank mix with residual herbicides when using the 1 quart per acre rate.

Nightshade, silverleaf—For control, apply 2 quarts of this product plus 0.5 to 1 percent nonionic surfactant by total spray volume in 3 to 10 gallons of water per acre. Applications should be made when at least 60 percent of the plants have berries. Fall treatments must be applied before a killing frost. Allow 7 or more days after application before tillage. Do not treat when weed is under drought stress as good soil moisture is necessary for active growth.

Nutsedge: purple, yellow—Apply 3 quarts of this product per acre as a broadcast spray, or apply a 1 percent solution from hand-held equipment to control existing nutsedge plants and immature nutlets attached to treated plants. Treat when plants are in flower or when new nutlets can be found at rhizome tips. Nutlets which have not germinated will not be controlled and may germinate following treatment. Repeat treatments will be required for long-term control. Wait 7 days after treatment before tillage. Tillage will stimulate nutlet germination. Apply 1 pint to 1 quart of this product per acre, plus 0.5 to 1 percent nonionic surfactant in 3 to 40 gallons of water per acre to provide suppression or partial control of existing nutsedge plants. Treat when plants have 3 to 5 leaves (less than 6 inches). Repeat treatments will be required to control subsequent emerging plants. Wait 7 days after treatment before tillage or mowing.

Pampas—Apply this product as a 1½ to 2 percent solution using hand-held equipment. Apply to plants that are actively growing at or beyond the boot stage of growth. Thorough coverage is necessary for best control.

Phragmites—For partial control of phragmites in Florida and the counties of other states bordering the Gulf of Mexico, apply 5 quarts per acre as a broadcast spray or apply a 2 percent solution from hand-held equipment. In other areas of the U.S., apply 3 quarts per acre as a broadcast spray or apply a 1 percent solution from hand-held equipment for partial control. For best results, treat during late summer or fall months or when plants are actively growing and in full bloom. Treatment before or after this stage may lead to reduced control. Due to the dense nature of the vegetation, which may prevent good spray coverage or uneven stages of growth, repeat treatments may be necessary to maintain control. Visual control symptom will be slow to develop.

Quackgrass—In Annual Cropping Systems, or in Pastures and Sods Followed by Deep Tillage: Apply 1 to 2 quarts of this product per acre. For the one quart rate, apply 0.5 to 1 percent nonionic surfactant by total spray volume in 3 to 10 gallons of water per acre. For the 2 quart rate, apply in 10 to 40 gallons of water per acre. Do not tank mix with residual herbicides when using the 1 quart rate. Spray when quackgrass is 8 to 12 inches in height and actively growing. Do not till between harvest and fall applications or in fall or spring prior to spring application. Allow 3 or more days after application before tillage. In pastures or sods, for best results use a moldboard plow.

Quackgrass—Pasture or Sod or Other Noncrop Areas Where Deep Tillage is Not Planned Following Application: Apply 2 to 3 quarts in 10 to 40 gallons of water per acre. Spray when the quackgrass is greater than 8 inches tall and actively growing. Do not till between harvest and fall application or in fall or spring prior to spring application. Allow 3 or more days after application before tillage.

Sweet Potato, wild—Apply this product as a 2 percent solution using hand-held equipment. Apply to actively growing weeds that are at or beyond the bloom stage of growth. Repeat applications will be required. Allow the plant to reach the recommended stage of growth before retreatment. Allow 7 or more days before tillage.

Thistle, Canada—Apply 2 to 3 quarts of this product per acre. Apply to actively growing thistles when most are at or beyond the bud stage of growth. After harvest, mowing or tillage in the late summer or fall, allow at least 4 weeks for initiation of active growth and rosette development prior to the application of this product. Fall treatments must be applied before a killing frost. Allow 3 or more days after application before tillage.

For suppression of Canada thistle, apply 1 quart per acre of this product plus 0.5 to 1 percent nonionic surfactant by total spray volume in 3 to 10 gallons of water per acre in the late summer or fall after harvest, mowing or tillage. Allow rosette regrowth to a minimum of 6 inches in diameter before treating. Applications can be made as long as leaves are still green and plants are actively growing at the time of application. Allow 3 or more days after application before tillage.

Torpedograss—Apply 4 to 5 quarts of this product per acre to provide partial control of torpedograss. Apply to actively growing torpedograss when most plants are at or beyond the seedhead stage of growth. Repeat applications will be required to maintain control. Fall treatments must be applied before frost. Allow 7 or more days after application before tillage.

Other perennials listed on this label—Apply 3 to 5 quarts of this product per acre. Apply when actively growing and most have reached early head or early bud stage of growth. Allow 7 or more days after application before tillage.

TANK MIXTURES

When applied as directed, under the conditions described, ROUNDUP plus Banvel or 2,4-D amine will suppress or control the following perennial broadleaf weed:

Bindweed, field
Convolvulus arvensis

For suppression, apply 16 ounces of this product plus 0.5 lb. a.e. of 2,4-D amine or 0.25 lb. a.i. of Banvel, plus 0.5 to 1 percent nonionic surfactant by total spray volume in 3 to 10 gallons of water per acre for ground applications and 3 to 5 gallons of water per acre for aerial applications. Applications should be delayed until maximum emergence has occurred and when vines are between 6 to 18 inches in length. These tank mixtures may be applied using aerial application equipment (except in California) in fallow and reduced tillage systems only.

For control, apply 2 quarts of this product plus 0.5 pound a.i. of Banvel in 10 to 20 gallons of water per acre. At these rates, apply using ground application equipment only.

Irrigated Agricultural Lands

Apply to 1 to 2 quarts of this product plus 1 quart per acre of 2,4-D amine in 10 to 20 gallons of water per acre with ground equipment only for partial control (suppression) of field bindweed. Applications should be made following harvest or in fall fallow ground when the bindweed is actively growing and the majority of runners are 12 inches or more in length. The use of at least one irrigation will promote active bindweed growth.

Dogbane, hemp
Apocynum cannabium

For suppression, apply 16 ounces of this product plus 16 ounces of 2,4-D amine plus 0.5 to 1 percent nonionic surfactant by total spray volume in 3 to 10 gallons of water per acre for ground applications and 3 to 5 gallons of water per acre for aerial applications. Delay applications until maximum emergence of dogbane has occurred. For control, apply 4 quarts of this product in 3 to 20 gallons of water using ground application equipment only.

Smartweed, Swamp
Polygonum coccineum

For control, apply 16 ounces of this product plus 0.5 pounds active ingredient of 2,4-D amine plus 0.5 to 1 percent nonionic surfactant by total volume in 3 to 10 gallons of water per acre in the late summer or fall. Apply when plants are actively growing and most have reached the early bud stage of growth. Allow 7 or more days after application before tillage.

Refer to the specific product labels for crop rotation restrictions and cautionary statements of all products in tank mixture.

Woody Brush and Trees

When applied as recommended under the conditions described, this product CONTROLS or PARTIALLY CONTROLS the following woody brush plants and trees

Alder
Alnus spp.
Ash*
Fraxinus spp.
Aspen, quaking
Populus tremuloides
Bearmat (Bearclover)
Chamaebatia foliosa
Birch
Betula spp.
Blackberry
Rubus spp.
Broom: French
Cytisus monspessulanus
Scotch
Cytisus scoparius
Buckwheat, California*
Eriogonum fasciculatum
Cascara*
Rhamnus purshiana
Catsclaw*
Acacia greggi
Ceanothus*
Ceanothus spp.
Chamise
Adenostoma fasciculatum
Cherry: Bitter
Prunus emarginata
Black
Prunus serotina
Pin
Prunus pensylvanica
Coyote brush
Baccharis consanguinea
Creeper, Virginia*
Parthenocissus quinquefolia
Dewberry
Rubus trivialis
Elderberry
Sambucus spp.
Elm*
Ulmus spp.
Eucalyptus, Bluegum
Eucalyptus glotulus
Hasardia*
Haplopappus squamosus
Hawthorn
Crataegus spp.
Hazel
Corylus spp.
Honeysuckle
Monkey Flower*
Mimulus guttatus
Oak: Black*
Quercus velutina
Northern Pin
Quercus palustris
Post
Quercus stellata
Red
Quercus rubra
Southern Red
Quercus falcata
White*
Quercus alba
Persimmon*
Diospyros spp.
Poison Ivy
Rhus radicans
Poison Oak
Rhus toxicodendron
Poplar*, yellow
Liriodendron tulipifera
Raspberry
Rubus spp.
Rose, multiflora
Rosa multiflora
Sage, black
Salvia mellifera
Sagebrush, California
Artemisia californica
Salmonberry
Rubus spectabilis
Sassafras
Sassafras aibidum
Sourwood
Oxydendrum arboreum
Sumac: Poison*
Rhus vernix
Smooth*
Rhus glabra
Winged*
Rhus copallina
Sweetgum
Liquidambar styraciflua
Swordfern*
Polystichum munitum
Tallowtree, Chinese
Sapium sebiferum
Tan Oak
Lithocarpus densiflorus
Thimbleberry

Woody Brush and Trees

Lonicera spp.	*Rubus parviflorus*
Kudzu	**Tobacco, tree***
Pueraria lobata	*Nicotiana glauca*
Locust, black*	**Trumpetcreeper**
Robinia pseudoacacia	*Campsis radicans*
Madrone	**Willow**
Arbutus menziesii	*Salix spp.*
Manzanita	
Arctostaphylos spp.	
Maple: Red**	
Acer rubrum	
Sugar	
Acer saccharum	
Vine*	
Acer circinatum	

* Partial control

** See below for control or partial control instructions.

NOTE: If brush has been mowed or tilled or trees have been cut, do not treat until regrowth has reached the recommended stages of growth.

Apply this product when plants are actively growing, and unless otherwise directed, after full leaf expansion. Use the higher rate for larger plants and/or dense areas of growth. On vines, use the higher rate for plants that have reached the woody stage of growth. Best results are obtained when application is made in late summer or fall after fruit formation.

In arid areas, best results are obtained when application is made in the spring to early summer when brush species are at high moisture content and are flowering.

Ensure thorough coverage when using hand-held equipment. Symptoms may not appear prior to frost or senescence with fall treatments.

Allow 7 or more days after application before tillage, mowing or removal. Repeat treatments may be necessary to control plants regenerating from underground parts or seed. Some autumn colors on undesirable deciduous species are acceptable provided no major leaf drop has occurred. Reduced performance may result if fall treatments are made following a frost.

See "Directions for Use," and "Mixing, Additives and Application Instructions" sections of this label for labeled uses and specific application instructions.

Apply this product as follows to control or partially control the following woody brush and trees.

Alder/Dewberry/Honeysuckle/Post Oak/Raspberry—For control apply 3 to 4 quarts per acre of this product as a broadcast spray or as a 1 to 1½ percent solution with hand-held equipment.

Aspen, quaking/Cherry: bitter, black, pin/Hawthorn/Oak, southern red/Sweetgum/Trumpetcreeper—For control apply 2 to 3 quarts of this product per acre as a broadcast spray or as a 1 to 1½ percent solution with hand-held equipment.

Birch/Elderberry/Hazel/Salmonberry/Thimbleberry—For control, apply 2 quarts per acre of this product as a broadcast spray or as a 1 percent solution with hand-held equipment.

Blackberry—For control, apply 3 to 4 quarts per acre of this product as a broadcast spray, or 1 to 1½ percent solution with hand-held equipment. Make application after plants have reached full leaf maturity. Best results are obtained when applications are made in late summer or fall. After berries have set or dropped in late fall, blackberry can be controlled by applying a ¾ percent solution of this product plus 0.5 to 1 percent nonionic surfactant by total spray volume with hand-held equipment. For control of blackberries after leaf drop and until killing frost or as long as stems are green, apply 3 to 4 quarts of this product in 10 to 40 gallons of water per acre.

Broom: French, Scotch—For control, apply a 1½ to 2 percent solution with hand-held equipment.

Buckwheat, California/Hasardia/Monkey Flower/Tobacco, tree—For partial control of these species, apply a 1 to 2 percent solution of this product as a foliar spray with hand-held equipment. Thorough coverage of foliage is necessary for best results.

Catsclaw—For partial control, apply as a 1 to 1½ percent solution with hand-held equipment.

Coyote Brush—For control, apply a 1½ to 2 percent solution with hand-held equipment when at least 50 percent of the new leaves are fully developed.

Eucalyptus/Bluegum—For control of eucalyptus resprouts, apply a 2 percent solution of this product with hand-held equipment when resprouts are 6 to 12 feet tall. Ensure complete coverage. Apply when plants are growing actively. Avoid application to drought-stressed plants.

Kudzu—For control, apply 4 quarts of this product per acre as a broadcast spray or as a 2 percent solution with hand-held equipment. Repeat applications will be required to maintain control.

Maple, red**—For control, apply as a 1 to 1½ percent solution with hand-held equipment when at least 50 percent of the new leaves are fully developed. For partial control, apply 2 to 4 quarts of this product per acre as a broadcast spray.

Maple, sugar/Oak, northern pin/Oak red—For control, apply as a 1 to 1½ percent solution with hand-held equipment when at least 50 percent of the new leaves are fully developed.

Poison Ivy/Poison Oak—For control, apply 4 to 5 quarts of this product per acre as a broadcast spray or as a 2 percent solution with hand-held equipment. Repeat applications may be required to maintain control. Fall treatments must be applied before leaves lose green color.

Rose, multiflora—For control, apply 2 quarts of this product per acre as a broadcast spray or as a 1 percent solution with hand-held equipment. Treatments should be made prior to leaf deterioration by leaf-feeding insects.

Sage, black/Sagebrush, California/Chamise/Tallowtree, Chinese—For control of these species, apply a 1 percent solution of this product as a foliar spray with hand-held equipment. Thorough coverage of foliage is necessary for best results.

Willow—For control, apply 3 quarts of this product per acre as a broadcast spray or as a 1 percent solution with hand-held equipment.

Other Woody Brush and Trees listed on this label—For partial control, apply 2 to 4 quarts of this product per acre as a broadcast spray or as a 1 to 2 percent solution with hand-held equipment.

DIRECTIONS FOR USE

It is a violation of Federal law to use this product in any manner inconsistent with its labeling.

STORAGE AND DISPOSAL

Do not contaminate water, foodstuffs, seed or feed by storage or disposal.

See container label for STORAGE AND DISPOSAL instructions.

NONCROP USES

See "General Information" and "Mixing, Additives and Application Instructions" sections of this label for essential product performance information and the following "Noncrop" sections for specific recommended uses.

EXTREME CARE MUST BE EXERCISED TO AVOID CONTACT OF SPRAY WITH FOLIAGE OF DESIRABLE TURFGRASSES, TREES, SHRUBS, OR OTHER DESIRABLE VEGETATION SINCE SEVERE DAMAGE OR DESTRUCTION MAY RESULT.

NOTE: If spraying areas adjacent to desirable plants, use a shield made of cardboard, sheet metal or plyboard while spraying to help prevent spray from contacting foliage of desirable plants.

Repeat treatments may be necessary to control weeds regenerating from underground parts or seeds.

ROUNDUP herbicide does not provide residual weed control. For subsequent weed control, follow a label-approved herbicide program.

Read and carefully observe the cautionary statements and all other information appearing on the labels of all herbicides used.

INDUSTRIAL, RECREATIONAL AND PUBLIC AREAS

When applied as directed for "Noncrop uses," under conditions described, this product controls annual and perennial weeds listed on this label growing in areas such as airports, ditch banks, dry ditches, dry canals, fencerows, golf courses, highways, industrial plant sites, lumber yards, parking areas, parks, petroleum tank farms and pumping installations, pipelines, power and telephone rights-of-way, railroads, roadsides, schools, storage areas, other public areas and similar industrial or noncrop areas.

For specific rates of application and instructions for control of various annual and perennial weeds and woody brush and trees, see the "Weeds Controlled" section of this label.

This product may be applied with recirculating sprayers, shielded applicators, or wiper applicators in any noncrop site specified on this label. See the "Selective Equipment part of "Application Equipment and Techniques" section of this label for information on proper use and calibration of this equipment.

TANK MIXTURES FOR INDUSTRIAL SITES AND FORESTRY SITE PREPARATIONS

ROUNDUP® plus OUST™

Use on industrial sites including airports, industrial plants, lumberyards, petroleum tank farms, pumping stations, pipelines, railroads, roadsides, storage areas or other similar sites where bare ground is desired.

This tank mixture may also be used as a site preparation treatment for sites to be planted to jack pine, loblolly pine, red pine, slash pine and Virginia pine.

When applied as directed for "Noncrop Uses" under the conditions described, this product plus Oust provides control of annual weeds listed in the "Weeds Controlled" section of the label for this product and Oust, and control or partial control of the perennial weeds listed below.

Apply 1 to 2 quarts of this product with 2 to 4 ounces of Oust in 10 to 40 gallons of spray solution per acre as a broadcast spray to actively growing weeds.

This mixture may be applied by aerial equipment in site prep operations. When applied by air, use the recommended rates in 5 to 15 gallons of spray solution per acre.

For control of annual weeds, use the lower rates of these products.

For control of the listed perennial weeds, use the higher rates of both products. For partial control, use the lower rates.

Bahiagrass *Paspalum notatum*	**Johnsongrass**** *Sorghum halepense*
Bermudagrass* *Cynodon dactylon*	**Poorjoe**** *Diodia teres*
Broomsedge *Andropogon virginicus*	**Quackgrass** *Agropyron repens*
Dock, curly *Rumex crispus*	**Trumpetcreeper*** *Campsis radicans*
Dogfennel *Eupatorium capilliforium*	**Vaseygrass** *Paspalum urvillei*
Fescue, tall *Festuca arundinacea*	**Vervain, blue** *Verbena hastata*

* Suppression at the higher rates only.
** Control at the lower rates.

Read and carefully observe the cautionary statements and all other information appearing on the labels of all herbicides used.

™ Oust is a trademark of E.I. du Pont de Nemours and Company.

TANK MIXTURES NONCROP SITES

When applied as a tank mixture, this product provides control of the emerged annual weeds and partial control of the emerged perennial weeds listed in this label. When applied as a tank mixture, the following residual herbicides will provide preemergence control of the weeds listed in the individual product labels.

ROUNDUP plus DIURON
ROUNDUP® plus KROVAR™ I
ROUNDUP plus KROVAR II
ROUNDUP plus RONSTAR™ 50WP
ROUNDUP plus SIMAZINE, CALIBER™ 90
ROUNDUP plus SIMAZINE 4L
ROUNDUP plus SIMAZINE 80W
ROUNDUP plus SURFLAN™ 75W
ROUNDUP plus SURFLAN AS

When tank mixing with residual herbicides, add an agriculturally approved nonionic surfactant at 0.5 to 1 percent by volume of spray solution. See the "Mixing, Additives and Application Instructions" section of this label before preparing these tank mixtures.

Read and carefully observe the label claims, cautionary statements, recommended use rates and all other information on the labels of all products used in these tank mixtures. Use according to the most restrictive label directions for each product in the mixture.

CONTROL OF EMERGED WEEDS

Annual Weeds—Apply 1 quart per acre of this product in these tank mixtures when weeds are less than 6 inches tall and 1½ quarts per acre when weeds are more than 6 inches tall.

Perennial Weeds—For partial control of perennial weeds using these tank mixtures, apply 2 to 5 quarts per acre of this product. Follow the recommendations in the "Weeds Controlled" section of this label for stage of growth and rate of application for specific perennial weeds.

PREEMERGENCE WEED CONTROL

The following are the labeled rates for the recommended residual herbicides for noncrop uses:

Product	Rate/Acre
Diuron Karmex™ 50WP	2 to 8 lbs
Krovar I	4 to 6 lbs
Krovar II	2 to 5 lbs
Ronstar 50WP	4 to 8 lbs
Simazine, Caliber 90	5.3 to 11.1 lbs
Simazine 80W	6 to 12½ lbs
Simazine 4L	4.8 to 10 qts
Surflan 75W	2⅔ to 5⅓ lbs
Surflan AS	2 to 4 qts

Refer to the individual product labels for specific noncrop sites, rates, carrier volumes and precautionary statements.

Mix only the quantity of spray solution which can be used during the same day. Do not allow these tank mixtures to stand overnight as this may result in reduced weed control.

APPLY THESE TANK MIXTURES THROUGH CONVENTIONAL BROADCAST EQUIPMENT ONLY.

™ Caliber is a trademark of Ciba-Geigy Corporation.

™ Karmex and ™Krovar are trademarks of E.I. du Pont de Nemours and Company.
™ Ronstar is a trademark of Rhone-Poulenc, Inc.
™ Surflan is a trademark of Elanco Products Company

FARMSTEAD WEED CONTROL

When applied as directed for "Noncrop Uses," under conditions described, this product controls undesirable vegetation listed on this label around farmstead building foundations, along and in fences, shelterbelts, and for general nonselective farmstead weed control.

For specific rates of application and instructions for control of various annual and perennial weeds, see the "Weeds Controlled" section of this label.

ORNAMENTALS

THIS PRODUCT IS NOT RECOMMENDED FOR USE AS AN OVER-THE-TOP BROADCAST SPRAY IN ORNAMENTALS.

When applied as directed for the conditions described for "Noncrop Uses," this product controls undesirable vegetation listed on this label prior to planting ornamentals, in established ornamentals, and within and around greenhouses and shadehouses.

For specific rates of application and instructions for control of various annual and perennial weeds, see the "Weeds Controlled" section of this label.

Where repeat applications are necessary, do not exceed 10.6 quarts of this product per acre per year.

Site Preparation—Following preplant applications of this product, any ornamental species may be planted. Precautions should be taken to protect nontarget plants during site preparation applications.

Greenhouse/Shadehouse Use—This product may be used to control weeds listed on this label which are growing in greenhouses. Desirable vegetation must not be present during application and air circulation fans must be turned off.

Postdirected Spray—Use as a directed spray toward the base of established woody ornamentals species listed below.

Arborvitae *Thuja spp.*	**Lilac** *Syringa spp.*
Azalea *Rhododendron spp.*	**Magnolia** *Magnolia spp.*
Boxwood *Buxus spp.*	**Maple** *Acer spp.*
Crabapple *Malus spp.*	**Oak** *Quercus spp.*
Euonymus *Euonymus spp.*	**Privet** *Ligustrum spp.*
Fir *Abies spp.* *Pseudotsuga spp.*	**Pine** *Pinus spp.*
Hollies *Ilex spp.*	**Spruce** *Picea spp.*
	Yew *Taxus spp.*

SILVICULTURAL SITES, RIGHTS-OF-WAY and CHRISTMAS TREES

NOTE: NOT RECOMMENDED FOR USE AS AN OVER-THE-TOP BROADCAST SPRAY IN SILVICULTURAL NURSERIES, OR CHRISTMAS TREE PLANTATIONS.

When applied as directed for "Noncrop Uses" under conditions described, this product controls undesirable vegetation listed on this label. This product also suppresses or controls undesirable vegetation listed on this label when applied at recommended rates for release of established coniferous species listed on this label.

For specific rates of application and instructions for control of various brush, annual and perennial weeds, see the "Weeds Controlled" section of this label. For specific rates of application for release of listed coniferous species, see the "Conifer Release" part of this section of the label.

Where repeat applications are necessary, do not exceed 10.6 quarts of this product per acre per year.

Aerial Application—This product may be applied using aerial spray equipment for silvicultural site preparation, conifer release and rights-of-way treatments. See the "Application Equipment and Techniques" part of the "Mixing, Additives and Application Instructions" section of this label for information on how to properly spray this product by air.

DO NOT APPLY THIS PRODUCT BY AIR TO RIGHTS-OF-WAY SITES IN THE STATE OF CALIFORNIA.

In order to reduce the aerial application drift hazard to aquatic sites*, to nontarget sites, or any site containing desirable vegetation, always maintain appropriate buffer zones. A buffer zone of the following minimum distances should be maintained:

- Helicopters using a Microfoil™ boom, a Thru-Valve™ boom (TVB-45), or equivalent drift control systems, should maintain at least a 50-foot buffer zone.

- When using other aerial equipment:

1. Maintain at least a 75-foot buffer zone for applications using 2 quarts or less per acre of this product.
2. Maintain at least a 125-foot buffer zone for application using more than 2 quarts per acre of this product.
3. Maintain at least a 400-foot buffer zone for applications on rights-of-way when applied from 75 feet or more above ground level.

These distances should be increased if conditions favoring drift exist.

* Aquatic sites include all lakes, ponds, and streams used for significant domestic purposes or angling.

™ Microfoil is a trademark of Union Carbide Agricultural Products, Inc.

™ Thru-Valve is a trademark of Waldrum Specialties, Inc.

SITE PREPARATION

Following preplant applications of this product, any silvicultural species may be planted.

POSTDIRECTED SPRAY

In established silvicultural sites, use as a spray on the foliage of undesirable vegetation. Care must be exercised to avoid contact of spray, drift or mist with foliage or green bark of desirable species.

CONIFER RELEASE

For release, apply only where conifers have been established for more than one year. Vegetation should not be disturbed prior to treatment or until visual symptoms appear after treatment. Symptoms of treatment are slow to appear, especially in woody species treated in late fall. Injury may occur to conifers treated for release, especially where spray patterns overlap or the higher rates are applied or when applications are made during periods of active conifer growth. **Do not use additional surfactant with conifer release applications.**

Applications must be made after formation of final conifer resting buds in the fall or prior to initial bud swelling in spring. Some autumn colors on undesirable deciduous species are acceptable provided no major leaf drop has occurred. Use the following rates for conifer release to control or partially control the weeds listed in the "Weeds Controlled" section of this label.

For release of the following conifer species:

Douglas fir *Pseudotsuga menziesii*	**Pines*** *Pinus spp.*
Fir *Abies spp.*	**Spruce** *Picea spp.*
Hemlock *Tsuga spp.*	

* Includes all species except eastern white pine, loblolly pine or slash pine.

Apply 1½ to 2 quarts of this product per acre except in Washington and Oregon, west of the crest of the Cascade Mountains. For spring treatments west of the crest of the Cascade Mountains, apply 1 quart of this product per acre before conifer bud swell for control of annual weeds. For fall treatments in Washington and Oregon, west of the crest of the Cascade Mountains, apply 1 to 1½ quarts of this product per acre before any major leaf drop of deciduous species.

For release of western hemlock apply 1 quart of this product per acre.

For release of the following conifer species:

Loblolly pine *Pinus taeda*	**Slash pine** *Pinus elliottii*
	Eastern white pine *Pinus strobus*

Late Season Application—Apply 1½ to 2 quarts of this product in a minimum of 5 gallons of spray solution per acre during early autumn. Applications made prior to September 1 or when conditions are conducive to rapid growth of conifers will result in potential for increased injury in the form of tip and/or needle burn. Injury may decrease with later applications. Some autumn colors are acceptable at time of application. Apply prior to frost or leaf drop of undesirable plants. Applications made according to label direction will release loblolly pine, eastern white pine and slash pine by reducing competition from the following species:

Ash *Fraxinus spp.*	**Persimmon** *Diospyros spp.*
Cherry:	**Poplar, yellow** *Liriodendron tulipfera*
Black *Prunus serotina*	**Sassafras** *Sassafras albidum*
Pin *Prunus pensylvanica*	**Sourwood** *Oxydendrum arboreum*
Elm *Ulmus spp.*	**Sumac:**
Hawthorn *Crataegus spp.*	**Poison** *Rhus vernix*
Locust, black *Robina pseudoacacia*	**Smooth** *Rhus glabra*
Maple, red *Acer rubra*	**Winged** *Rhus copallina*
Oak:	**Sweetgum** *Liquidambar styraciflua*
Black *Quercus velutina*	
Post *Quercus stellata*	
Southern Red *Quercus falcata*	
White *Quercus alba*	

Apply only to those sites where woody brush and trees listed in this label constitute the majority of the undesirable species.

ROUNDUP plus Oust Tank Mixtures for Conifer Release from Herbaceous Weeds

To release **loblolly pines** from herbaceous weeds, tank mixtures of this product with Oust will provide control of annual weeds listed in the "Weeds Controlled" section of this and the Oust label, and partial control of the perennial weeds listed below.

Apply 16 to 24 fluid ounces of ROUNDUP with 2 to 4 ounces of Oust in 10 to 30 gallons of spray solution per acre. Make application to actively growing weeds as a broadcast spray over the top of the young loblolly pines.

This tank mixture may be applied using aerial equipment. When applying by air, use the recommended rate in 5 to 15 gallons of spray solution per acre.

For control of annual weeds below 12 inches in height (or runner length on annual vines) use the low rates of both products. Use the higher rates of both products when annual weeds are in more advanced stages of growth and approaching flower or seed formation.

Use the higher rates of both products for partial control of the following perennial weeds. Use the lower rates for suppression of growth.

Bahiagrass *Paspalum notatum*	**Fescue, tall** *Festuca arundinacea*
Broomsedge *Andropogon virginicus*	**Johnsongrass**** *Sorghum halepense*
Dock, curly *Rumex crispus*	**Poorjoe**** *Diodia teres*
Dogfennel *Eupatorium capilliforium*	**Trumpetcreeper*** *Campsis radicans*
	Vaseygrass *Paspalum urvillei*
	Vervain, blue *Verbena hastata*

* Suppression at the higher rates only.

** Control at the higher rates.

Pine damage may occur or can be accentuated if treatment takes place when young trees are under stress from drought, flood water, insects or disease.

Read and observe the cautionary statements and all other information appearing on the label of all herbicides used.

NOTE TO USER

This product must not be used in areas where adverse inpact on Federally designated endangered, threatened plant or aquatic species is likely.

Prior to making applications, the user of this product must determine that no such species are located in or immediately adjacent to the area to be treated.

CUT STUMP TREATMENTS

Woody vegetation may be controlled by treating freshly cut stumps of trees and resprouts with this product.

Apply this product using suitable equipment to ensure coverage of the entire cambium. Cut vegetation close to the soil surface. Apply a 50 to 100 percent solution of this product to the freshly cut surface immediately after cutting. Delays in application may result in reduced performance. For best results, applications should be made during periods of active growth and full leaf expansion.

When used according to directions for cut stump application, this product will CONTROL, PARTIALLY CONTROL or SUPPRESS many types of woody brush and tree species, some of which are listed below:

Alder *Alnus spp.*	**Saltcedar** *Tamarisk spp.*
Eucalyptus/Bluegum *Eucalyptus glotulus*	**Sweetgum** *Liquidambar styraciflua*
Madrone *Arbutus menziesii*	**Tan Oak** *Lithocarpus densiflorus*
Oak *Quercus spp.*	**Willow** *Salix spp.*
Reed, giant *Arundo donax*	

INJECTION AND FRILL APPLICATIONS

Woody vegetation may be controlled by injection or frill application of this product. Apply this product using suitable equipment which must penetrate into the living tissue. Apply the equivalent of 1 ml of this product per each 2 to 3 inches of trunk diameter. This is best achieved by applying 50 to 100 percent concentration of this material either to a continuous frill around the tree or as cuts evenly spaced around the tree below all branches. As tree diameter increases in size, better results are achieved by applying diluted material to a continuous frill or more closely spaced cuttings. Avoid application techniques that allow runoff to occur from frill or cut areas in species that exude sap freely after frills or cutting. In species such as this, make frill or cut at an oblique angle so as to produce a cupping effect and use undiluted material. For best results, application should be made during periods of active growth and after full leaf expansion.

This treatment WILL CONTROL the following woody species:

Oak *Quercus spp.*	**Sweetgum** *Liquidambar styraciflua*
Poplar *Populus spp.*	**Sycamore** *Platanus occidentalis*

This treatment WILL SUPPRESS the following woody species:

Black gum *Nyssa sylvatica*	**Hickory** *Carya spp.*
Dogwood *Cornus spp.*	**Maple, red** *Acer rubrum*

TURFGRASSES AND GRASSES FOR SEED PRODUCTION

PREPLANT AND RENOVATION

When applied as directed for "Noncrop Uses," under conditions described, this product controls most existing vegetation prior to the planting or renovation of either turfgrasses or grass seed production areas.

For specific rates of application and instructions for control of various annual and perennial weeds, and woody brush and trees, see the "Weeds Controlled" section of this label.

For maximum control of existing vegetation, delay planting to determine if any regrowth from escaped underground plant parts occurs. Where repeat treatments are necessary, sufficient regrowth must be attained prior to application. For warm-season grasses, such as bermudagrass, summer or fall application provide best control.

DO NOT DISTURB SOIL OR UNDERGROUND PLANT PARTS BEFORE TREATMENT. Tillage or renovation techniques such as vertical mowing, coring or slicing should be delayed for 7 days after application to allow proper translocation into underground plant parts.

TURFGRASSES

Where existing vegetation is growing in a field or unmowed situation, apply this product to actively growing weeds at the stages of growth given in the "Weeds Controlled" section of this label.

Where existing vegetation is growing under mowed turfgrass management, apply this product after omitting at least one regular mowing to allow sufficient growth for good interception of the spray.

Desirable turfgrasses may be planted following the above procedures.

GRASSES FOR SEED PRODUCTION

Apply this product to actively growing weeds at the stages of growth given in the "Weeds Controlled" section of this label prior to planting or renovation of turf or forage grass areas grown for seed producton.

DO NOT feed or graze treated areas within 8 weeks after application.

ANNUAL WEED CONTROL IN DORMANT BERMUDAGRASS AND BAHIAGRASS TURF

When applied as directed for "Noncrop Uses," this product will provide control or suppression of many winter annual weeds and tall fescue for effective release of dormant bermudagrass and bahiagrass turf. Refer to the rate table for ROUNDUP alone under the "Release of Bermudagrass and Bahiagrass" section of this label for recommended rates and volumes on the species to be suppressed or controlled. Treat only when turf is dormant and prior to spring greenup. Spot treatments or broadcast applications of the product in excess of 16 ounces per acre may result in injury or delayed greenup in highly maintained turfgrass areas; i.e., golf courses, lawns, etc. DO NOT APPLY TANK MIXTURES of ROUNDUP plus Oust in highly maintained turfgrass areas.

RELEASE OF BERMUDAGRASS OR BAHIAGRASS

NOTE: Use only in areas where bermudagrass or bahiagrass are desirable ground covers and where some temporary injury or discoloration can be tolerated. Use ROUNDUP plus Oust tank mixtures only on railroads, highways, utility plant sites, or other right-of-way areas.

When applied as directed for "Noncrop Uses," this product will provide control or suppression of many winter annual weeds and tall fescue for effective release of dormant bermudagrass or bahiagrass. This product may be tank-mixed with Oust as recommended for residual control. Make applications to dormant bermudagrass or bahiagrass. ROUNDUP plus Oust tank mixture may delay greenup. **To avoid delays in greenup and minimize injury, do not add more than 1 ounce per acre of Oust on bermudagrass or more than ½ ounce per acre on bahiagrass, or treat when these grasses are in a semi-dormant condition.**

For best results on winter annuals, treat when plants are in an early growth stage (below 6 inches in height) after most have germinated. For best results on tall fescue, treat when fescue is in or beyond the 4 to 6-leaf stage.

WEEDS CONTROLLED

Rate recommendations for control or suppression of winter annuals and tall fescue are listed below:

Apply the recommended rates of this product alone or as a tank mixture in 10 to 25 gallons of water, plus 0.5 to 1 percent nonionic surfactant by total spray volume per acre.

For the best recommendation for the mixture of weeds within your geographic areas, contact your Monsanto sales representative.

Weeds Controlled or Suppressed with ROUNDUP® Alone*

NOTE: C = Control
S = Suppression

Weed Species	ROUNDUP® Oz/Acre 8	12	16	24	32	64
Barley, little *Hordeum pusilium*	S	C	C	C	C	C
Bedstraw, catchweed *Galium aparine*	S	C	C	C	C	C
Bluegrass, annual *Poa annua*	S	C	C	C	C	C
Chervil *Chaerophyllum tainturieri*	S	C	C	C	C	C
Chickweed, common *Stellaria media*	S	C	C	C	C	C
Clover, crimson *Trifolium incarnatum*	•	S	S	C	C	C
Clover, large hop *Trifolium campestre*	•	S	S	C	C	C
Fescue, tall *Festuca arundinaceae*	•	•	•	•	S	S
Geranium, Carolina *Geranium carolinianum*	•	•	S	S	C	C
Henbit *Lamium amplexicaule*	•	S	C	C	C	C
Ryegrass						
Italian *Lolium multiflorum*	•	•	S	C	C	C
Speedwell, corn *Veronica arvensis*	S	C	C	C	C	C
Vetch, common *Vicia sativa*	•	•	S	C	C	C

* *These rates apply only to sites where an established competitive turf is present.*

Weeds Controlled or Suppressed with ROUNDUP® Plus Oust*

NOTE: C = Control
S = Suppression

ROUNDUP® + Oust (Oz/Acre)

Weed Species	ROUNDUP + Oust	8 + ¼	12 + ¼	12 + ½	16 + ¼	16 + ½	12 + 1	16 + 1
Barley, little *Hordeum pusilium*		C	C	C	C	C	C	C
Bedstraw, catchweed *Galium aparine*		C	C	C	C	C	C	C
Bluegrass, annual *Poa annua*		S	C	C	C	C	C	C

Weed Species	ROUNDUP + Oust	8 + ¼	12 + ¼	12 + ½	16 + ¼	16 + ½	12 + 1	16 + 1
Chervil *Chaerophyllum tainturieri*		C	C	C	C	C	C	C
Chickweed, common *Stellaria media*		S	C	C	C	C	C	C
Clover, crimson *Trifolium incarnatum*		S	S	S	S	C	C	C
Clover, large hop *Trifolium campestre*		•	•	S	S	S	C	C
Fescue, tall *Festuca arundinaceae*		•	•	•	•	•	S	S
Geranium, Carolina *Geranium carolinianum*		•	S	S	C	C	C	C
Henbit *Lamium amplexicaule*		•	S	C	C	C	C	C
Ryegrass, Italian *Lolium multiflorum*		•	S	S	C	C	C	C
Speedwell, corn *Veronica arvensis*		S	C	C	C	C	C	C
Vetch, common *Vicia sativa*		C	C	C	C	C	C	C

* These rates or mixtures of rates apply only to sites where an established competitive turf is present.

RELEASE OF ACTIVELY GROWING BERMUDAGRASS

When applied as directed, this product will aid in the release of bermudagrass by providing control of annual species listed in the "Weeds Controlled" section of this and the Oust label, and suppression or partial control of certain perennial weeds.

For control or suppression of those annual species listed on this label, use 1 to 3 pints of this product as a broadcast spray in 10 to 25 gallons of spray solution per acre. Use the lower rate when treating annual weeds below 6 inches in height (or length of runner in annual vines). Use higher rates as size of plants increases or as they approach flower or seedhead formation.

Use the higher rate of this product for partial control of the following perennial species. Use the lower rates for suppression of growth. For best results, see the "Weeds Controlled" section of this label for proper stage of growth.

Bahiagrass *Paspalum notatum*	**Johnsongrass**** *Sorghum halepense*
Bluestem, silver *Andropogon saccharoides*	**Trumpetcreeper*** *Campsis radicans*
Fescue, tall *Festuca arundinacea*	**Vaseygrass** *Paspalum urvillei*

* Suppression at higher rates only.
** Control at the higher rates.

This product may be tank-mixed with Oust. If tank-mixed, use no more than 1 to 2 pints per acre of ROUNDUP herbicide with 1 to 2 ounces of Oust per acre.

Use the lower rates of both mixtures to control annual weeds below 6 inches in height (or runner length in annual vines) that are listed in the "Weeds Controlled" section of this booklet and the Oust label. Use the higher rates as annual weeds increase in size and approach the flower or seedhead stages.

Use the higher rates of this product to provide partial control of the following perennial weeds. Use the lower rates for suppression of growth.

Bahiagrass *Paspalum notatum*	**Johnsongrass*** *Sorghum halepense*
Bluestem, silver *Andropogon saccharoides*	**Poorjoe**** *Diodia teres*
Broomsedge *Andropogon virginicus*	**Trumpetcreeper*** *Campsis radicans*
Dock, curly *Rumex crispus*	**Vaseygrass** *Paspalum urvillei*
Dogfennel *Eupatorium capilliforium*	**Vervain, blue** *Verbena hastata*
Fescue, tall *Festuca arundinacea*	

* Suppression at higher rates only.
** Control at the higher rates.

Use only on well-established bermudagrass. Bermudagrass injury may result from the treatment but regrowth will occur under moist conditions. Repeat applications in the same season are not recommended, since severe injury may result.

Read and carefully observe the cautionary statements and all other information appearing on the labels of all herbicides used.

BAHIAGRASS SEEDHEAD AND VEGETATIVE SUPPRESSION

When applied as directed in the indicated noncrop areas (roadsides, airports, golf course roughs, and plant sites) this product will provide significant inhibition of seedhead emergence and will suppress vegetative growth for a period of approximately 45 days with single applications and approximately 120 days with sequential applications.

Apply ROUNDUP herbicide 1 to 2 weeks after full greenup of bahiagrass or after the bahiagrass has been mowed to a uniform height of 3 to 4 inches. Applications must be made prior to seedhead emergence. Apply 6 fluid ounces per acre of this product plus 0.5 to 1 percent nonionic surfactant by total spay volume in 10 to 25 gallons of water per acre.

Sequential applications of this product plus 0.5 to 1 percent nonionic surfactant by total spray volume may be made at approximately 45 day intervals to extend the period of seedhead and vegetative growth suppression. For continued seedhead suppression, sequential applications must be made prior to seedhead emergence. Apply no more than 2 sequential applications per year. As a first sequential application, apply 4 fluid ounces of this product per acre plus nonionic surfactant. A second sequential application of 2 to 4 fluid ounces per acre plus nonionic surfactant may be made approximately 45 days after the last application.

A tank mixture of ROUNDUP plus Oust **may be applied only on roadsides** for seedhead inhibition and vegetative suppression. Apply 6 fluid ounces per acre of this product plus 0.25 ounce per acre of Oust, plus 0.5 to 1 percent nonionic surfactant by total spray volume 1 to 2 weeks following an initial spring mowing. When using ROUNDUP plus Oust for suppression of bahiagrass, make only one application per year.

CROPPING SYSTEMS

See "General Information" and "Mixing, Additives and Application Instructions" sections of this label for essential product performance information.

See the following "Cropping Systems" sections for specific recommended uses.

EXTREME CARE MUST BE EXERCISED TO AVOID CONTACT OF SPRAY WITH FOLIAGE, GREEN STEMS OR FRUIT OF DESIRABLE CROPS, PLANTS, TREES OR OTHER DESIRABLE VEGETATION SINCE SEVERE DAMAGE OR DESTRUCTION MAY RESULT.

Repeat treatments may be necessary to control weeds regenerating from underground parts or seed. Except as otherwise specified on this label, repeat treatments must be made before the crop emerges in accordance with the instructions of this label.

Except as otherwise specified in a crop section of this label, the combined total of all treatments must not exceed 8 quarts per acre of this product per year.

Do not plant subsequent crops other than those on the label for 30 days following application.

For other cropping systems, except following spot treatment in forage grasses and legumes, do not harvest or feed treated crops for 8 weeks after application. Allow 14 days following spot treatment or selective equipment use before grazing domestic livestock or harvesting forage grasses and legumes.

When applied as directed for "Cropping Systems," under the conditions described, this product controls annual and perennial weeds listed on this label, prior to the emergence of direct seeded crops or prior to transplanting of crops listed on this label.

ALFALFA*	KALE
ARTICHOKE, JERUSALEM	LENTILS
	LETTUCE
ASPARAGUS*	
BARLEY*	LOGANBERRY
BEANS (All)	MELONS***
BEET GREENS	MUSTARD GREENS
BEETS (Red, Sugar)	OATS*
BLACKBERRY	OKRA
BLUEBERRY	OLALLIEBERRY
BOYSENBERRY	ONION
BROCCOLI	PEANUTS
CABBAGE	PEAS (All)
CARROT	PEPPER***
CAULIFLOWER	PINEAPPLE****
CELERY	POTATO (Irish, Sweet)

CHICORY	PUMPKIN***
CORN (All)*	RADISH
COTTON*	RASPBERRY (Black, Red)
CRANBERRY	RICE**
CUCUMBER***	RUTABAGA
CURRANT	SORGHUM (Milo)
DEWBERRY	SOYBEANS*
EGGPLANT***	SPINACH
ELDERBERRY	SQUASH*** (Summer, Winter)
FORAGE GRASSES*	TOMATILO***
FORAGE LEGUMES*	TOMATOES***†
GARLIC***	TURNIPS
GOOSEBERRY	WATERMELON***
GOURDS***	WATERCRESS***
HORSERADISH	WHEAT*
HUCKLEBERRY	

* Spot treatments may be applied in these crops.
** Do not treat rice fields or levees when the fields contain flood water.
*** Apply only prior to planting. Allow at least 3 days between application and planting.
**** Do not feed or graze treated pineapple forage following application.
† Use is restricted to direct seeded crops only.

Spot Treatment (Only those crops with "*" can be spot treated.)—Applications in growing crops must be made prior to heading of small grains and milo, initial pod set in soybeans, silking of corn, boll opening on cotton.

For forage grasses and forage legumes see "Spot Treatment" in the "Pastures" section of "Cropping Systems" in this label.

For dilution and rates of application using boom or hand-held equipment, see "Mixing, Additives and Application Instructions" and "Weeds Controlled" sections of this label.

NOTE: FOR FORAGE GRASSES AND FORAGE LEGUMES. NO MORE THAN ONE-TENTH OF ANY ACRE SHOULD BE TREATED AT ONE TIME. FOR ALL OTHER CROPS, DO NOT TREAT MORE THAN 10 PERCENT OF THE TOTAL FIELD AREA TO BE HARVESTED.

THE CROP RECEIVING SPRAY IN TREATED AREA WILL BE KILLED. TAKE CARE TO AVOID DRIFT OR SPRAY OUTSIDE TARGET AREA FOR THE SAME REASON.

Selective Equipment—This product may be applied through recirculating sprayers, shielded applicators, or wiper applicators in cotton and soybeans. Shielded and wiper applicators may also be used in tree crops and grapes. Wiper applicators may be used in rutabagas, forage grasses and forage legumes, including pasture sites and sorghum milo.

See the "Selective Equipment" part of the "Application Equipment and Techniques" section of this label for information on proper use and calibration of this equipment.

Allow at least the following time intervals between application and harvest:

Cotton, Soybeans	7 days
Apples, Avocado, Cherry, Citrus, Grapes, Pear, Rutabagas	14 days
Stone Fruit	17 days
Nut Crops	21 days
Sorghum (milo)*	40 days

* Do not use roller applicators. Do not feed or graze treated milo fodder. Do not ensile treated vegetation.

ASPARAGUS

When applied as directed for "Cropping Systems" under the conditions described, this product controls weeds listed on this label in asparagus.

For specific rates of applications and instructions for control of various annual and perennial weeds, see the "Weeds Controlled" section of this label.

Prior to Crop Emergence—Apply this product prior to crop emergence for the control of emerged labeled annual and perennial weeds. DO NOT APPLY WITHIN A WEEK BEFORE THE FIRST SPEARS EMERGE.

Spot Treatment—Apply this product immediately after cutting, but prior to the emergence of new spears. Do not treat more than 10 percent of the total field area to be harvested. Do not harvest within 5 days of treatment.

Postharvest—Apply this product after the last harvest and all spears have been removed. If spears are allowed to regrow, delay application until ferns have developed. Delayed treatments should be applied as directed or shielded spray in order to avoid contact of the spray with ferns, stems or spears. Direct contact of the spray with the asparagus may result in serious crop injury.

NOTE: Select and use recommended types of spray equipment for post-emergence post harvest applications. A directed spray is any application where the spray pattern is aligned in such a way as to avoid direct contact of the spray with the crop. A shielded spray is any application where a physical barrier is positioned and maintained between the spray and the crop to prevent contact of spray with the crop.

BERRIES AND SMALL FRUITS

For cranberries, apply after fruit set and no later than 30 days before harvest.

For other berries, apply as a preplant broadcast application, or as a directed spray or wiper application post-planting.

Wiper applicators may be used in cranberries in accordance with instructions in this section.

See "General Information" and "Mixing, Additives and Application Instructions" sections of this label for essential product performance information.

See the "Selective Equipment" part of the "Application Equipment and Techniques" section of this label for information on recommended use and calibration of this equipment.

For Wick or other Wiper Applicators—Mix 1 gallon of this product in 4 gallons of water to prepare a 20 percent solution. Apply the solution to emerged weeds. Apply after cranberry fruit set and no later than 30 days before harvest.

In severe infestations, reduce equipment ground speed to insure that adequate amounts of this product are wiped on the weeds. A second treatment in the opposite direction may be beneficial.

Do not permit herbicide solution to contact desirable vegetation, including green shoots, canes, or foliage.

FALLOW AND REDUCED TILLAGE SYSTEMS

FOR AERIAL APPLICATION IN CALIFORNIA, REFER TO SUPPLEMENTAL LABEL

Use this product in fallow and reduced tillage systems for control of annual weeds prior to emergence of crops listed in this label. Refer to the "Weeds Controlled" section of this label for specific rates and instructions. This product may be applied using ground or aerial spray equipment. See the "Application Equipment and Techniques" section of this label for instructions.

TANK MIXTURES
ROUNDUP® plus BANVEL
plus NONIONIC SURFACTANT
ROUNDUP plus 2,4-D AMINE
plus NONIONIC SURFACTANT
ROUNDUP plus GOAL™
plus NONIONIC SURFACTANT

DO NOT APPLY BANVEL, GLEAN, OR 2,4-D AMINE TANK MIXTURES BY AIR IN CALIFORNIA.

The additions of Banvel in a mixture with this product may provide short-term residual control of selected weed species. Some crop injury may occur if Banvel is applied within 45 days of planting. Refer to the Banvel and 2, 4-D labels for cropping restrictions and other use instructions.

ROUNDUP® plus Goal Tank Mixtures

This product alone or in tank mixtures with Goal plus 0.5 to 1 percent nonionic surfactant by total spray volume will provide control of those weeds listed below.

Make applications when weeds are actively growing and at the recommended stages of growth. Avoid spraying when weeds are subject to moisture stress, when dust is on the foliage or when straw canopy covers the weeds.

ROUNDUP® 12 oz/acre		ROUNDUP 16 oz/acre	
Wheat	18"*	Annual grasses at left plus:	
Barley	12"	Ryegrass, annual	6"
Bluegrass, annual	6"	Chickweed	6"
Barnyardgrass	6"	Groundsel	6"
Rye	6"	Marestail	6"
		Rocket, London	6"
		Shepherdspurse	6"
		Crabgrass	12"
		Johnsongrass, seedling	12"
		Lambsquarters	12"
		Oats, wild	12"
		Pigweed, redroot	12"
		Mustards	12"

ROUNDUP 12 oz/acre + Goal** 2 to 4 oz/acre		ROUNDUP 16 oz/acre + Goal** 2 to 4 oz/acre	
Annual grasses above plus:		Annual weeds above plus	
Cheeseweed, common	3"	Cheeseweed, common	6"

Chickweed	3"	Groundsel	6"
Groundsel	3"	Chickweed	12"
Rocket, London	6"	Rocket, London	12"
Shepherdspurse	6"	Shepherdspurse	12"

* Maximum height or length in inches.

** Use the higher rate when weeds approach maximum recommended height or stands are dense.

These recommended tank mixtures may be applied using ground or aerial spray equipment. Refer to the "Weeds Controlled" section of this label for specific rates and instructions.

™Glean is a trademark of E. I du Pont de Nemours and Company

™Goal is a trademark of Rohm and Haas Company.

AID TO TILLAGE

This product, when used in conjunction with preplant tillage practices, will provide control of downy brome, cheat, volunteer wheat, tansy mustard and foxtail. Apply 8 fluid ounces of this product plus 0.5 to 1 percent nonionic surfactant by total spray volume in 3 to 10 gallons of water per acre. Make applications when weeds are actively growing and before they are 6 inches in height. Application must be followed by conventional tillage practices no later than 15 days after treatment and before regrowth occurs. Allow at least 1 day after application before tillage. Tank mixtures with residual herbicides may result in reduced performance.

PASTURES

Apply this product prior to planting forage grasses and legumes.

Pasture or Hay Crop Renovation—When applied as a broadcast spray this product controls the annual and perennial weeds listed in this label prior to planting forage grasses or legumes. Remove domestic livestock before application and wait 8 weeks after application before grazing or harvesting.

Spot Treatment—When applied as a spot treatment as recommended, this product controls annual and perennial weeds listed in this label which are growing in pastures, forage grasses and forage legumes composed of bahiagrass, bermudagrass, bluegrass, brome, fescue, orchardgrass, ryegrass, timothy, wheatgrass, alflfa or clover.

Wiper Application—When applied as directed this product controls or supresses the weeds listed under "Wiper Applicators" in the "Selective Equipment" section of this label. For spot treatment and wiper application, apply in areas where the movement of domestic livestock can be controlled. No more than one-tenth of any acre should be treated at one time. Further applications may be made in the same area at 30 day intervals. Remove domestic livestock before application and wait 14 days after application before grazing livestock or harvesting.

SUGARCANE

When applied as directed for "Cropping Systems" under the condition described, this product controls those emerged annual and perennial weeds listed on this label growing in or around sugarcane or in fields to be planted to sugarcane. This product will also control undesirable sugarcane.

NOTE: Where repeat treatments are necessary, do not exceed a total of 10.6 quarts of this product per acre per year. Do not apply to vegetation in or around ditches, canals or ponds containing water to be used for irrigation.

Broadcast Treatment—Apply this product in 10 to 40 gallons of water per acre on emerged weeds growing in fields to be planted to sugarcane.

For specific rates of application and instructions for control of various annual and perennial weeds, see the "Weeds Controlled" section of this label.

For removal of last stubble or ratoon cane, apply 4 to 5 quarts of this product in 10 to 40 gallons of water per acre to new growth having at least 7 or more new leaves. Allow 7 or more days after application before tillage.

Spot Treatment in or Around Sugarcane Fields—For dilution and rates of application using hand-held equipment, see "Mixing, Additives and Application Instructions" and "Weeds Controlled" sections of this label.

For control of volunteer or diseased sugarcane, make a 1 percent solution of this product in water and spray to wet the foliage of vegetation to be controlled.

NOTE: When spraying volunteer or diseased sugarcane, the plants should have at least 7 new leaves.

Avoid spray contact with healthy cane plants since severe damage or destruction may result.

Do not feed or graze treated sugarcane forage following application.

MINIMUM TILLAGE SYSTEMS

Tank Mixtures

CORN AND SOYBEANS

When applied as recommended under the conditions described, these tank mixtures control many emerged weeds, and give preemergence control of many annual weeds when corn or soybeans will be planted directly into a cover crop, established sod, or in previous crop residues.

Refer to specific product labels for crop rotation restrictions and cautionary statements of all products used in these tank mixtures. For mixing instructions, see the "Mixing, Additives and Application Instructions" section of this label.

CORN
ROUNDUP® plus LASSO®
ROUNDUP plus LASSO plus ATRAZINE
ROUNDUP plus LASSO plus BLADEX™
ROUNDUP plus LASSO plus SIMAZINE
ROUNDUP plus ATRAZINE plus SIMAZINE
SOYBEANS
ROUNDUP® plus LASSO®
ROUNDUP plus LASSO plus LINURON
ROUNDUP plus LASSO plus LEXONE™
ROUNDUP plus LASSO plus SENCOR™

Apply these tank mixtures in 10 to 40 gallons of water per acre after planting or during planting in such manner that the planter does not disturb the treated soil. Do not apply these mixtures after crop emergence.

REDUCED CONTROL MAY RESULT IF THIS PRODUCT IS USED IN TANK MIXTURES CONTAINING FLUID FERTILIZERS.

CONTROL OF EMERGED WEEDS

Annual Weeds—Apply to actively growing grasses and broadleaf weeds. Use 1 quart of ROUNDUP herbicide per acre in these tank mixtures if weeds are less than 6 inches tall. If weeds are over 6 inches tall, apply 1.5 quarts of this product per acre. For emerged annual weeds controlled, see the "Weeds Controlled" section of this label.

Perennial Weeds—At normal application rates in minimum tillage systems, perennial weeds may not be at the proper stage of growth for control. See the "General Information" section of this label for the proper stage of growth for perennial weeds.

Use of 2 to 4 quarts of ROUNDUP herbicide per acre in these tank mixtures, under these conditions provides top kill and reduces competition from many emerged perennial grass and broadleaf weeds. For emerged perennial weeds controlled, see the "Weeds Controlled" section of this label.

To obtain the desired stage of growth, it may be necessary to apply ROUNDUP herbicide alone in the late summer or fall and then follow with a label-approved, seedling weed control program at planting.

NOTE: When using these tank mixtures, do not exceed 4 quarts of ROUNDUP herbicide per acre.

USE OF THESE TANK MIXTURES FOR BERMUDAGRASS OR JOHNSONGRASS CONTROL IN MINIMUM TILLAGE SYSTEMS IS NOT RECOMMENDED. For bermudagrass control, follow the instructions under "Control of Perennial Weeds" section of this label and then use a label-approved, seedling weed-control program in a minimum tillage or conventional tillage system. For Johnsongrass control, follow the instructions under "Control of Perennial Weeds" section of the label, and then use a label-approved seedling weed-control program with conventional tillage.

The following are labeled ranges for the recommended preemergence herbicides:

Product	Rates/Acre (Quarts)	Rates/Acre (Pounds)
Lasso	2 to 4	•
Atrazine 4L	1 to 2	•
Atrazine 80WP	•	1.25 to 2.3
Atrazine 90DF	•	1.1 to 2
Bladex 4L	1 to 2.2	•
Bladex 80WP	•	1.25 to 2.75
Simazine 4L	1 to 2.2	•
Simazine 80WP	•	1.25 to 2.75
Linuron 4L	0.5 to 1.5	•
Linuron 50WP	•	1 to 3
Lexone 4L or Sencor 4F	0.25 to 0.75	•
Lexone or Sencor 50WP	•	0.5 to 1.55
Lexone or Sencor 75DF	•	0.33 to 1

Read and carefully observe the label claims, cautionary statements, rates and all other information on the labels of all products. Use according to the most restrictive label directions for each product in the mixture.

® Lasso is a registered trademark of Monsanto Company.

™ Bladex and ™Lexone are trademarks of E. I. du Pont de Nemours and Company.

™ Sencor is a trademark of the parent company of Farbenfabriken Bayer GmbH, Leverkusen.

PREHARVEST APPLICATIONS

When applied as directed under the conditions described, this product controls annual and perennial weeds listed on this label prior to the harvest of COTTON.

Weed Control—For specific rates of application and instructions for control of various annual and perennial weeds, see the "Weeds Controlled" section of this label.

Broadcast Applications—This product may be applied using either aerial or ground spray equipment. For ground applications with broadcast equipment, apply this product in 10 to 40 gallons of water per acre. For aerial applications, apply this product in 3 to 15 gallons of water per acre.

To control Johnsongrass using multiple nozzle-directed spray equipment, apply 1 quart of this product plus 0.5 to 1 percent nonionic surfactant by total spray volume in 10 to 20 gallons of water per acre. Ensure complete coverage.

For partial control of field bindweed, apply one quart of this product plus 0.5 to 1 percent nonionic surfactant by total spray volume in 3 to 40 gallons of water per acre. Apply when bindweed is actively growing and 12 inches or greater in length. Reduced performance may result if bindweed is under drought stress.

Tank Mixtures

ROUNDUP® plus Def™ 6
ROUNDUP plus Folex™

When applied as recommended under the conditions described, these tank mixtures control annual and perennial weeds listed on this label prior to the harvest of cotton. For application guidelines, precautions and defoliant rates, refer to the supplemental label for cotton harvest aid.

Timing of Application—Apply this product or these tank mixtures for preharvest weed control after 60 percent of the cotton bolls have opened.

NOTE: DO NOT APPLY TO CROPS GROWN FOR SEED.

Allow a minimum of 7 days between application and harvest.

Do not feed or graze treated cottonseed forage or hay following preharvest applications.

™Def is a trademark of Mobay Chemical Corporation.
™Folex is a trademark of Rhone-Poulenc, Inc.

TREE AND VINE CROPS

This product is recommended for weed control in established groves, vineyards, or orchards, or for site preparation prior to transplanting crops listed in this section. Applications may be made with boom equipment. CDA, shielded sprayers, hand-held and high-volume wands, lances, or orchard guns, or with wiper applicator equipment, except as directed in this section. See the "Application Equipment and Techniques" section of this label for specific information on use of equipment.

When applying ROUNDUP alone, refer to the "Weeds Controlled" section for recommended rates to be used.

NOTE

Repeat treatments may be necessary to control weeds originating from underground parts of untreated weeds or from seeds. This product does not provide residual weed control. For subsequent weed control, follow a program using residual herbicides or use repeated applications of this product. Do not apply more than 10.6 quarts of this product per acre per year.

EXTREME CARE MUST BE EXERCISED TO AVOID CONTACT OF HERBICIDE SOLUTION, SPRAY, DRIFT, OR MIST WITH FOLIAGE OR GREEN BARK OF TRUNK, BRANCHES, SUCKERS, FRUIT, OR OTHER PARTS OF TREES OR VINES. CONTACT OF THIS PRODUCT WITH OTHER THAN MATURED BROWN BARK CAN RESULT IN SERIOUS CROP DAMAGE.

AVOID PAINTING OUT STUMPS WITH THIS PRODUCT AS INJURY RESULTING FROM ROOT GRAFTING MAY OCCUR IN ADJACENT TREES.

Reduced control may result when applications are made to annual or perennial weeds that have been mowed, grazed or cut and have not been allowed to regrow to the recommended stage for treatment.

For specific rates of applications and instructions, see the "Weeds Controlled" section of this label.

TANK MIXTURES TREE AND VINE CROPS

When applied as a tank mixture, this product provides control of the emerged annual weeds and partial control of the emerged perennial weeds listed in this label. The following residual herbicides will provide preemergence control of those weeds listed in the individual product labels.

ROUNDUP® plus GOAL™ 1.6E
ROUNDUP plus KARMEX™ WP
ROUNDUP plus KROVAR™ I
ROUNDUP plus KROVAR™ II
ROUNDUP plus SIMAZINE, CALIBER™ 90
ROUNDUP plus SIMAZINE 4L
ROUNDUP plus SIMAZINE 80W
ROUNDUP plus SOLICAM™ 80 WP
ROUNDUP plus SURFLAN™ AS
ROUNDUP plus SURFLAN 75W
ROUNDUP plus SIMAZINE (80W, or 4L, or CALIBER 90) plus SURFLAN (AS or 75W)
ROUNDUP plus GOAL (1.6E) plus SURFLAN (AS or 75W)
ROUNDUP plus GOAL (1.6E) plus SIMAZINE (80W, or 4L or CALIBER 90)*
ROUNDUP plus GOAL (1.6E) plus SURFLAN (AS or 75W) plus SIMAZINE (80W, 4L, or CALIBER 90)

* See the ROUNDUP plus Goal plus simazine tank mixture section below for specific use instructions.

When tank-mixing with residual herbicides, add an agriculturally approved nonionic surfactant.

Do not apply these tank mixtures in Puerto Rico.

CONTROL OF EMERGED WEEDS

Annual Weeds—Apply 1 quart per acre of this product in these tank mixtures, when weeds are less than 12 inches tall. For weeds greater than 12 inches tall, apply 1.5 quarts per acre.

Perennial Weeds—For partial control of perennial weeds using these tank mixtures, apply 2 to 5 quarts per acre of this product. Follow the recommendations in the "Weeds Controlled" section of this label for stage of growth and rate of application for specific perennial weeds.

PREEMERGENCE WEED CONTROL

The following are the labeled rates for the recommended residual herbicides:

Product	Rate/Acre
Goal 1.6E	1.25 to 5 qts.
Karmex WP	2 to 8 lbs.
Krovar I	2 to 8 lbs.
Krovar II	2 to 8 lbs.
Simazine, Caliber 90	1.1 to 10.6 lbs.
Simazine, 80W	1¼ to 12 lbs.
Simazine 4L	2 to 9.6 qts.
Solicam 80WP	2½ to 5 lbs.
Surflan AS	2 to 4 qts.
Surflan 75W	2⅔ to 5⅓ lbs.

NOTE: These residual herbicides may provide postemergence activity on certain annual weed species.
Refer to the individual product labels for specific crops, rates, geographical restrictions and precautionary statements.

Read and carefully observe the label claims, cautionary statements, rates and all other information on the labels of all products.

For specific rates of application and instructions see the "Weeds Controlled" section of this label.

ROUNDUP® Plus GOAL

ROUNDUP alone and ROUNDUP plus Goal tank mixtures will control or suppress annual weeds listed below.

Apply the recommended rates of ROUNDUP or ROUNDUP plus Goal tank mixtures plus 0.5 to 1 percent nonionic surfactant by spray volume in 3 to 10 gallons of water per acre. Apply when weeds are actively growing and less than 6 inches in height. If weeds are under drought stress, irrigate prior to application.

Recommended Rates and Weeds Controlled
C = Control S = Suppression

Weeds Species	ROUNDUP rate (fl oz/acre) 8	12	16	ROUNDUP + Goal (fl oz/acre) 16 + 4 to 16
Bluegrass, annual	C	C	C	C
Barley	C	C	C	C
Cheeseweed, common (3")	S	S	S	C
Rocket, London	S	S	C	C
Shepherdspurse	S	S	C	C
Chickweed	S	C	C	C
Fleabane, flaxleaf	S	C	C	C
Red Maids	S	C	C	C
				16 + 4 to 16
Barnyardgrass	S	C	C	C
Cheeseweed, common (3-6")	S	S	S	C
Crabgrass	S	S	C	C
Groundsel, common	S	S	C	C
Junglerice	S	S	C	C
Lambsquarters	S	S	C	C
Marestail	S	S	C	C
Pigweed, redroot	S	S	C	C
Ryegrass, common	S	S	C	C
Sowthistle, annual	S	S	C	C
Purslane, common	S	S	C	C
Nettle, stinging	S	S	C	C
Filaree	S	S	S	S

ROUNDUP® plus Goal plus SIMAZINE
(California only)

ROUNDUP may be tank-mixed with Goal plus simazine for broad-spectrum postemergence and preemergence control. Refer to the following table for approved crops and rate ranges for each product in this tank mix. Refer to the individual product labels for weeds controlled, geographical restrictions, precautionary statements and specific use rates.

	Rate Ranges*				
Crops	ROUNDUP® (quarts)	+	Goal 1.6E (quarts)	+	Simazine 80W** (pounds)
Grapes	1 to 1.5		1.25 to 5		2.5 to 5
Almond	1 to 1.5		1.25 to 5		1.25 to 2.5

Use equivalent rate of Caliber 90 or simazine 4L
*Rate dependent on weeds present.
**Rate dependent on soil type and weeds present.

ROUNDUP® plus GOAL plus SIMAZINE/SURFLAN

This product plus Goal, applied alone or as a three-way or four-way with simazine and/or Surflan, will provide postemergence control of those weeds listed below when applied as recommended.

Refer to individual product labels for preemergence rates, weeds controlled, precautionary statements and other important information.

Apply these tank mixtures in 3 to 40 gallons of water. Add 0.5 to 1 percent nonionic surfactant by total spray volume to the spray solution.

Weeds Controlled

Weed Species	Maximum Height (inches)	ROUNDUP® + Goal (quarts/acre)
Barley, wild	4	0.5 to 1 + 0.5 to 5
Bluegrass, annual	4	0.5 to 1 + 0.5 to 5
Cheeseweed, common	4	0.5 to 1 + 0.5 to 5
Chickweed, common	4	0.5 to 1 + 0.5 to 5
Flaxleaf, fleabane	4	0.5 to 1 + 0.5 to 5
Groundsel, common	4	0.5 to 1 + 0.5 to 5
Marestail	4	0.5 to 1 + 0.5 to 5
Pineappleweed	4	0.5 to 1 + 0.5 to 5
Rocket, London	4	0.5 to 1 + 0.5 to 5
Shepherdspurse	4	0.5 to 1 + 0.5 to 5
Sowthistle, annual	4	0.5 to 1 + 0.5 to 5
Cheeseweed	12	1 to 1.5 + 1 to 5
Nettle, stinging	12	1 to 1.5 + 1 to 5
Filaree	12	1 to 1.5 + 1 to 5

TM Goal is a trademark of Rohm and Haas Company.
TM Solicam is a trademark of Sandoz, Inc.

PERENNIAL GRASS SUPPRESSION ORCHARD FLOORS

When applied as directed this product will suppress vegetative growth as indicated below.

Bahiagrass

This product will provide significant inhibition of seedhead emergence and will suppress vegetative growth for a period of approximately 45 days with a single application and approximately 120 days with sequential applications. Apply this product 1 to 2 weeks after full green-up or after mowing to a uniform height of 3 to 4 inches. Applications must be made prior to seedhead emergence. Apply 6 fluid ounces of this product plus 0.5 to 1 percent nonionic surfactant by total spray volume in 10 to 25 gallons of water per acre.

Sequential applications of this product plus nonionic surfactant may be made at approximately 45-day intervals to extend the period of seedhead and vegetative growth suppression. For continued seedhead suppression, sequential applications must be made prior to seedhead emergence. Apply no more than 2 sequential applications per years. As a first sequential application, apply 4 fluid ounces of this product plus nonionic surfactant. A second sequential application of 2 to 4 fluid ounces may be made approximately 45 days after the last application.

Bermudagrass

For burndown, apply 1 to 2 quarts of this product plus 0.5 to 1 percent nonionic surfactant by total spray volume in 3 to 20 gallons of water per acre. Use 1 quart of this product in 3 to 20 gallons of water per acre east of the Rocky Mountains. Use 1 to 2 quarts of this product in 3 to 10 gallons of water per acre west of the Rocky Mountains. Use this treatment only if reduction of the bermudagrass stand can be tolerated. When burndown is required prior to harvest, allow at least 21 days to ensure sufficient time for burndown to occur.

Suppression only (east of the Rocky Mountains)—Apply 6 to 16 fluid ounces plus 0.5 to 1 percent nonionic surfactant by total spray volume in 3 to 20 gallons of water per acre no sooner than 1 to 2 weeks after full green-up. Mowing prior to application may occur provided a minimum height of 3 inches is maintained. Rates of 6 to 10 fluid ounces plus nonionic surfactant should be used in shaded conditions or where a lesser degree of suppression is desired. Sequential applications may be made when regrowth occurs and bermudagrass injury and stand reduction can be tolerated.

Suppression only (west of the Rocky Mountains)—Apply 16 fluid ounces of this product plus 0.5 to 1 percent nonionic surfactant by total spray volume in 3 to 10 gallons of water per acre up to 6 inches in height and no sooner than 1 to 2 weeks after full green-up. Mowing prior to application may occur provided a minimum height of 3 inches is maintained. Sequential applications may be made when regrowth occurs and bermudagrass injury and stand reduction can be tolerated.

Cool Season Grass Covers

For suppression of tall fescue, fine fescue, orchardgrass and quackgrass, apply 8 fluid ounces of this product plus 0.5 to 1 percent nonionic surfactant by total spray volume in 10 to 20 gallons of water per acre. For best suppression, add ammonium sulfate to the spray solution at a rate of 2 percent by weight or 17 pounds per 100 gallons of spray solution.

For suppression of Kentucky bluegrass covers, apply 6 fluid ounces of this product plus nonionic surfactant. Do not add ammonium sulfate.

For best results, mow cool-season grass covers in the spring to even their heights and apply the recommended rate of this product 3 to 4 days after mowing. Avoid treating cool season grass covers under poor growing conditions, such as drought stress (drip irrigation), disease or insect damage.

LOW VOLUME APPLICATION (FLORIDA AND TEXAS)

For burndown or control of listed perennial weeds, apply the recommended rates of this product plus 0.5 to 1 percent nonionic surfactant by total spray volume in 3 to 20 gallons of water per acre. Where weed foliage is dense, use 10 to 20 gallons of water per acre.

Apply when weeds are actively growing and at the growth stages listed in the "Perennial Weeds Controlled" section of this label. If perennial weeds are mowed, allow weeds to regrow to the recommended stage of growth.

S = Suppression; B = Burndown; PC = Partial control; C = Control

Weed Species	ROUNDUP Rate Per Acre 1 qt	2 qts	3 qts	5 qts
Bermudagrass	B	•	PC	C
Guineagrass	B	C	C	C
Texas and Florida Ridge				
Florida Flatwoods	•	B	C	C
Paragrass	B	C	C	C
Torpedograss	S	•	PC	C

TREE CROPS

Citrus*: citron, grapefruit, kumquat, lemon, lime, orange, tangelo, tangerine.

Nuts**: almond, filbert, macadamia, pecan, pistachio, walnut.

Pome Fruit*: apple, pear.

Stone Fruit***: apricots, cherries, nectarines, olives, peaches, plums/prunes.

For cherries, any application equipment listed in this section may be used in all states.

For citron and olives, apply as a directed spray only.

Any application equipment listed in this section may be used in apricots, nectarines, peaches, and plum/prunes growing in Arizona, California, Colorado, Idaho, Kansas, Kentucky, New Jersery, North Dakota, Oklahoma, Oregon, Texas, Utah, and Washingtion, except for peaches grown in the states specified in the following paragraph. In all other states use wiper equipment only.

For PEACHES grown in Alabama, Arkansas, Florida, Georgia, Louisiana, Mississippi, North Carolina, South Carolina, and Tennessee only, apply with a shielded boom sprayer or shielded wiper applicator which prevents any contact of this product with the foliage or bark of trees. Apply no later than 90 days after first bloom. Applications made after this time may result in severe damage. Remove suckers and low hanging limbs at least 10 days prior to application. Avoid applications near trees with recent pruning wounds or other mechanical injury. Apply only near trees which have been planted in the orchard for 2 or more years. EXTREME CARE MUST BE TAKEN TO ENSURE NO PART OF THE PEACH TREE IS CONTACTED.

Tropical Fruit: acerola, avocado*, banana (plantains), coffee****, figs*, guava, mango*, papaya, tea. Allow a minimum of 1 day between last application and harvest of guava and papaya. In coffee and banana, delay applications 3 months after transplanting to allow the new coffee or banana plant to become established.

NOTE

* Allow a minimum of 14 days between last application and harvest.
** Allow a minimum of 21 days between last application and harvest of these crops.
*** Allow a minimum of 17 days between last application and harvest.
**** Allow a minimum of 28 days between last application and harvest.

VINE CROPS

Kiwi Fruit*

Grapes: *Any variety of table, wine, or raisin grape may be treated with any equipment listed in this section.

In the northeast and Great Lakes regions, applications must be made prior to the end of bloom stage of grapes to avoid injury.

*Applications should not be made when green shoots, canes, or foliage are in the spray zone.

*Allow a minimum of 14 days between last application and harvest.

CALIFORNIA

ROUNDUP® herbicide has been approved by the U.S. Environmental Protection Agency for the uses, crops and sites listed in this label and by California under label designation 1988-1. Approval of the items listed below is pending under the State of California registration requirements. With the exception of these items, this booklet contains the material approved by California in label 1988-1.

These use conditions, crops and sites may not be treated with this product in California until approval is received:

- Aerial application of ROUNDUP plus Oust at the rate of 5 to 15 gallons of spray solution per acre for conifer release.
- Prior to planting or transplanting of srawberries.
- Tank Mixes with Larat and Roundup.

EPA Reg. No. 524-3038-AA
897.10-004.85/53

This product has been approved for use in California except as stated otherwise on page ________.

In case of an emergency involving this product. Call Collect, day or night, (314) 694-4000.

This specimen label is intended for use only as a guide in providing general information regarding the use of this product. Monsanto is not responsible for the accuracy of the information contained herein. As labels are subject to revision, always carefully read and follow the label on the product container.

ROUNDUP® RT SHUTTLE 100

REDUCED TILLAGE HERBICIDE

For control of annual and perennial weeds in reduced tillage systems.

ACTIVE INGREDIENT:	
*Isopropylamine salt of glyphosate	41.0%
INERT INGREDIENTS:	59.0%
TOTAL	100.0%

* Contains 480 grams per litre or 4 pounds of the active ingredient isopropylamine salt of N-(phosphonomethyl) glycine per U.S. gallon. Equivalent to 356 grams per litre or 3 pounds per U.S. gallon of the acid, glyphosate.

Product is protected by
U.S. Pat. No. 3,799,758 and U.S. Pat. No. 4,405,531.
Other patents pending.

EPA Reg. No. 524-308

ROUNDUP® is a registered trademark of Monsanto Company

1988-2 897.15-000.04/CG

Complete Directions for Use

AVOID CONTACT WITH FOLIAGE, GREEN STEMS, OR FRUIT OF CROPS, DESIRABLE PLANTS AND TREES, SINCE SEVERE INJURY OR DESTRUCTION MAY RESULT.

Read the entire label before using this product.

Use only according to label instructions.

Read "LIMIT OF WARRANTY AND LIABILITY" before buying or using. If terms are not acceptable, return at once unopened.

REFORMULATION OR REPACKAGING IS PROHIBITED.

LIMIT OF WARRANTY AND LIABILITY

This Company warrants that this product conforms to the chemical description on the label and is reasonably fit for the purposes set forth in the Complete Directions for Use label booklet ("Directions") when used in accordance with those Directions under the conditions described therein. NO OTHER EXPRESS WARRANTY OR IMPLIED WARRANTY OF FITNESS FOR PARTICULAR PURPOSE OR MERCHANTABILITY IS MADE. This warranty is also subject to the conditions and limitations stated herein.

Buyer and all users shall promptly notify this Company of any claims whether based in contract, negligence, strict liability, other tort or otherwise.

Buyer and all users are responsible for all loss or damage from use or handling which results from conditions beyond the control of this Company, including, but not limited to, incompatibility with products other than those set forth in the Directions, application to or contact with desirable vegetation, unusual weather, weather conditions which are outside the range considered normal at the application site and for the time period when the product is applied, as well as weather conditions which are outside the application ranges set forth in the Directions, application in any manner not explicitly set forth in the Directions, moisture conditions outside the moisture range specified in the Directions, or the presence of products other than those set forth in the Directions in or on the soil, crop or treated vegetation.

THE EXCLUSIVE REMEDY OF THE USER OR BUYER, AND THE LIMIT OF THE LIABILITY OF THIS COMPANY OR ANY OTHER SELLER FOR ANY AND ALL LOSSES, INJURIES OR DAMAGES RESULTING FROM THE USE OR HANDLING OF THIS PRODUCT (INCLUDING CLAIMS BASED IN CONTRACT, NEGLIGENCE, STRICT LIABILITY, OTHER TORT OR OTHERWISE) SHALL BE THE PURCHASE PRICE PAID BY THE USER OR BUYER FOR THE QUANTITY OF THIS PRODUCT INVOLVED, OR, AT THE ELECTION OF THIS COMPANY OR ANY OTHER SELLER, THE REPLACEMENT OF SUCH QUANTITY, OR, IF NOT ACQUIRED BY PURCHASE, REPLACEMENT OF SUCH QUANTITY. IN NO EVENT SHALL THIS COMPANY OR ANY OTHER SELLER BE LIABLE FOR ANY INCIDENTAL, CONSEQUENTIAL OR SPECIAL DAMAGES.

Buyer and all users are deemed to have accepted the terms of this LIMIT OF WARRANTY AND LIABILITY which may not be varied by any verbal or written agreement.

PRECAUTIONARY STATEMENTS

HAZARDS TO HUMANS AND DOMESTIC ANIMALS

KEEP OUT OF REACH OF CHILDREN

WARNING!

CAUSES EYE IRRITATION.
HARMFUL IF SWALLOWED.
MAY CAUSE SKIN IRRITATION.

Do not get in eyes, on skin or on clothing.

Wash thoroughly with soap and water after handling.

FIRST AID: IF IN EYES, immediately flush with plenty of water for at least 15 minutes. Call a physician.

IF ON SKIN, immediately flush with plenty of water. Remove contaminated clothing. Wash clothing before reuse.

IF SWALLOWED, this product will cause gastrointestinal tract irritation. Immediately dilute by swallowing water or milk. Call a physician.

In case of an emergency involving this product, Call Collect, day or night, (314) 694-4000.

ENVIRONMENTAL HAZARDS

Do not apply directly to water or wetland (swamps, bogs, marshes or potholes). Do not contaminate water by cleaning of equipment or disposal of wastes.

PHYSICAL OR CHEMICAL HAZARDS

Spray solutions of this product should be mixed, stored and applied only in stainless steel, aluminum, fiberglass, plastic and plastic-lined steel containers.

DO NOT MIX, STORE OR APPLY THIS PRODUCT OR SPRAY SOLUTIONS OF THIS PRODUCT IN GALVANIZED STEEL OR UNLINED STEEL (EXCEPT STAINLESS STEEL) CONTAINERS OR SPRAY TANKS. This product or spray solutions of this product react with such containers and tanks to produce hydrogen gas which may form a highly combustible gas mixture. This gas mixture could flash or explode, causing serious personal injury, if ignited by open flame, spark, welder's torch, lighted cigarette or other ignition source.

GENERAL INFORMATION

DO NOT APPLY THIS PRODUCT USING AERIAL SPRAY EQUIPMENT EXCEPT UNDER CONDITIONS AS SPECIFIED WITHIN THIS LABEL.

ROUNDUP® RT herbicide, a water soluble liquid, mixes readily with water to be applied as a foliar spray for the control or destruction of most herbaceous plants. It may be applied through most standard industrial or field-type sprayers after dilution and thorough mixing with water in accordance with label instructions.

This product moves through the plant from the point of foliage contact to and into the root system. Visible effects on most annual weeds occur within 2 to 4 days but on most perennial weeds may not occur for 7 days or more. Extremely cool or cloudy weather following treatment may slow down activity of this product and delay visual effects of control. Visible effects are a gradual wilting and yellowing of the plant which advances to complete browning of aboveground growth and deterioration of underground plant parts.

Unless otherwise specified on this label, delay application until vegetation has emerged and reached the stages described for control of such vegetation under the "WEEDS CONTROLLED" section of this label. Unemerged plants arising from unattached underground rhizomes or root stocks of perennials will not be affected by the spray and will continue to grow. For this reason, best control of most perennial weeds is obtained when treatment is made at late growth stages approaching maturity.

Always use the higher rate of this product per acre within the recommended range when (1) weed growth is heavy or dense, or (2) weeds are growing in an undisturbed (noncultivated) area.

Do not treat weeds under poor growing conditions such as drought stress, disease or insect damage, as reduced weed control may result. Reduced results may also occur when treating weeds heavily covered with dust.

This product is recommended for control of emerged weeds prior to establishment of labeled crops. Large amounts of green or decaying vegetation left standing or incorporated into the seedbed may enhance the development of disease in newly planted crops. This may result in poor emergence and/or stands, especially under cool and/or wet conditions.

Spraying early to control young weeds before dense stands develop or light cultivation to assist weed decay will favor preparation of suitable seedbeds.

In reduced tillage and no-till systems, ensure good seed to soil contact and proper seeding depth.

Reduced control may result when applications are made to annual or perennial weeds that have been mowed, grazed, or cut, and have not been allowed to regrow to the recommended stage for treatment.

Rainfall or irrigation occurring within 6 hours after application may reduce effectiveness. Heavy rainfall or irrigation within 2 hours after application may wash the chemical off the foliage and a repeat treatment may be required.

This product does not provide residual weed control. For subsequent residual weed control, follow a label-approved herbicide program. Read and carefully observe the cautionary statements and all other information appearing on the labels of all herbicides used.

Buyer and all users are responsible for all loss or damage in connection with the use or handling of mixtures of this product with herbicides or other materials that are not expressly recommended in this labeling. Mixing this product with herbicides or other materials not recommended on this label may result in reduced performance.

For best results, spray coverage should be uniform and complete. Do not spray weed foliage to the point of runoff.

Keep people and pets off treated areas until spray solution has dried to prevent transfer of this product onto desirable vegetation.

® Registered trademark of Monsanto Company

ATTENTION: AVOID DRIFT EXTREME CARE MUST BE USED WHEN APPLYING THIS PRODUCT TO PREVENT INJURY TO DESIRABLE PLANTS AND CROPS.

Do not allow the herbicide solution to mist, drip, drift, or splash onto desirable vegetation since minute quantities of this herbicide can cause severe damage or destruction to the crop, plants, or other areas on which treatment was not intended. The likelihood of injury occurring from the use of this product is greatest when winds are gusty or in excess of 5 miles per hour or when other conditions, including lesser wind velocities, will allow spray drift to occur. When spraying, avoid combinations of pressure and nozzle type that will result in splatter or fine particles (mist) which are likely to drift. AVOID APPLYING AT EXCESSIVE SPEED OR PRESSURE.

NOTE: Use of this product in any manner not consistent with this label may result in injury to persons, animals, crops or other unintended consequences. Keep container closed to prevent spills and contamination.

MIXING, ADDITIVES AND APPLICATION INSTRUCTIONS

APPLY THESE SPRAY SOLUTIONS IN PROPERLY MAINTAINED AND CALIBRATED EQUIPMENT CAPABLE OF DELIVERING DESIRED VOLUMES. DO NOT APPLY UNDER WIND OR OTHER CONDITIONS WHICH ALLOW DRIFT TO OCCUR. HAND GUN APPLICATIONS SHOULD BE PROPERLY DIRECTED TO AVOID SPRAYING DESIRABLE PLANTS. NOTE: REDUCED RESULTS MAY OCCUR IF WATER CONTAINING SOIL IS USED, such as WATER FROM PONDS AND UNLINED DITCHES.

MIXING

This product mixes readily with water. Mix spray solutions of this product as follows: Fill the mixing or spray tank with the required amount of water. Add the proper amount of this product (see "DIRECTIONS FOR USE" and "WEEDS CONTROLLED" sections of this label) near the end of the filling process and mix well.

Remove hose from tank immediately after filling to avoid siphoning back into the carrier source. During mixing and application foaming of the spray solution may occur. To prevent or minimize foam, avoid the use of mechanical agitators, place the filling hose below the surface of the spray solution, terminate by-pass and return lines at the bottom of the tank and, if needed, use an approved anti-foam or defoaming agent.

TANK MIXTURES

Always predetermine the compatibility of labeled tank mixtures of this herbicide with water carrier by mixing small proportional quantities in advance.

Mix labeled tank mixtures of this product with water as follows:

1. Place a 20 to 35 mesh screen or wetting basket over filling port.
2. Through the screen, fill the sprayer tank one-half full with water and start agitation.
3. If a wettable powder is used, make a slurry with the water carrier, and add it SLOWLY through the screen into the tank. Continue agitation.
4. If a flowable formulation is used, premix one part flowable with one part water. Add diluted mixture SLOWLY through the screen into the tank. Continue agitation.
5. If an emulsifiable concentrate formulation is used, premix one part emulsifiable concentrate with two parts water. Add diluted mixture slowly through the screen into the tank. Continue agitation.
6. Continue filling the sprayer tank with water and add the required amount of this product near the end of the filling.
7. Where nonionic surfactant is recommended, add this to the spray tank before completing the filling process.
8. Add individual formulations to the spray tank as follows: wettable powder, flowable, emulsifiable concentrate, drift control additive, water soluble liquid followed by surfactant.

Maintain good agitation at all times until the contents of the tank are sprayed. If the spray mixture is allowed to settle, thorough agitation is required to resuspend the mixture before spraying is resumed.

Keep by-pass line on or near bottom of tank to minimize foaming. Screen size in nozzle or line strainers should be no finer than 50 mesh. Carefully select proper nozzle to avoid spraying a fine mist. For best results with conventional ground application equipment, use flat fan nozzles.

Clean sprayer and parts immediately after using this product by thoroughly flushing with water.

ADDITIVES

SURFACTANTS

Nonionic surfactants which are labeled for use with herbicides may be used. Do not reduce rates of this herbicide when adding surfactant. When adding additional surfactant or when label instructions require use of additional surfactants, use 0.5 percent surfactant concentration (2 quarts per 100 gallons of spray solution) when using surfactants which contain at least 50 percent active ingredient or a 1 percent surfactant concentration (4 quarts per 100 gallons of spray solution) for those surfactants containing less than 50 percent active ingredient. Read and carefully observe surfactant cautionary statements and other information appearing on the surfactant label.

AMMONIUM SULFATE

The addition of 2 percent dry ammonium sulfate by weight or 17 pounds per 100 gallons of water may increase the performance of this product and this product plus 2,4-D amine or Banvel™ tank mixtures on annual weeds. The improvement in performance may be apparent where environmental stress is a concern. Low-quality ammonium sulfate may contain material that will not readily dissolve which could result in nozzle tip plugging. To determine quality, perform a jar test by adding 1/3 cup of ammonium sulfate to 1 gallon of water and agitate for 1 minute. If undissolved sediment is observed, predissolve the ammonium sulfate in water and filter prior to addition to the spray tank. If ammonium sulfate is added directly to the spray tank, add slowly with agitation. Adding too quickly may clog outlet lines. Ensure that ammonium sulfate is completely dissolved in the spray tank before adding herbicides or surfactant. Thoroughly rinse the spray system with clean water after use to reduce corrosion.

NOTE: The use of ammonium sulfate as an additive does not preclude the need for additional surfactant. Do not use herbicide rates lower than recommended in this label.

™Banvel is a trademark of Sandoz, Inc.

APPLICATION EQUIPMENT AND TECHNIQUES

Do not apply this product through any type of irrigation system.

This product may be applied using either ground or aerial spray equipment. Use extreme care to avoid misting or drifting of herbicide solution onto foliage, green stems or fruit of desirable crops, trees, or plants during both growing and dormant periods, since even very small quantities of spray can cause severe plant injury.

GROUND APPLICATION: Apply recommended rates of this product in 3 to 10 gallons of water per acre as a broadcast spray. For optimum spray distribution and coverage, use flat fan or low-volume nozzles. When using flood nozzles, space them no more than 40 inches apart and ensure double overlap of spray pattern. Refer to the manufacturer's recommendations for

correct pressure and nozzle height above the target canopy. Avoid pressure and nozzles which produce fine droplets or mist.
Use appropriate marking devices to ensure uniform spray coverage and best results from the herbicide.

AERIAL APPLICATION: Apply the recommended rates of this product in 3 to 5 gallons of water per acre as a broadcast spray. DO NOT APPLY DURING INVERSION CONDITIONS, WHEN WINDS ARE GUSTY, OR UNDER ANY OTHER CONDITION WHICH WILL ALLOW DRIFT. DRIFT MAY CAUSE DAMAGE TO ANY VEGETATION CONTACTED TO WHICH TREATMENT WAS NOT INTENDED. TO PREVENT INJURY TO ADJACENT DESIRABLE VEGETATION, APPROPRIATE BUFFER ZONES MUST BE MAINTAINED.

Coarse sprays are less likely to drift; therefore, do not use nozzles or nozzle configurations which dispense spray as fine spray droplets. Do not angle nozzles forward into the airstream and do not increase spray volume by increasing nozzle pressure.

SPRAYER CLEANUP

CLEAN THE ENTIRE SPRAYER AFTER APPLICATION OF THIS PRODUCT. Failure to clean the sprayer thoroughly may result in injury to desirable crops which are subsequently sprayed. First, add clean water to the tank and thoroughly rinse the entire sprayer system. Secondly, fill the tank with water and ammonia. Add 1 quart of household ammonia per 25 gallons of water. Pump enough solution through the hoses, boom and nozzles to fill these parts completely. Then fill the tank, close and leave for 24 hours before draining and rinsing thoroughly with water.

Application or use of other agricultural chemicals with the equipment used for this product may result in injury to desirable vegetation.

Thoroughly wash aircraft, especially landing gear, after each day of spraying to remove residues of this product accumulated during spraying or from spills. PROLONGED EXPOSURE OF THIS PRODUCT TO UNCOATED STEEL SURFACES MAY RESULT IN CORROSION AND POSSIBLE FAILURE OF THE PART. LANDING GEAR ARE MOST SUSCEPTIBLE. The maintenance of an organic coating (paint) which meets aerospace specification MIL-C-38413 may prevent corrosion.

WEEDS CONTROLLED

This herbicide controls many annual and perennial grasses and broadleaf weeds.

ANNUAL WEEDS

- Apply to actively growing grass and broadleaf weeds.
- Allow at least 3 days after treatment before tillage.
- For maximum agronomic benefit, apply when weeds are 6 inches or less in height.
- To prevent seed production, applications should be made prior to seedhead formation.
- This product does not provide residual control; therefore, delay application until maximum weed emergence. Repeat treatments may be necessary to control later germinating weeds.
- Do not use these rates for weeds that have been mowed, grazed or cut, unless weed regrowth has occurred and weeds are vigorously growing.
- Refer to the "TANK MIXTURES" portion of this section for control of additional broadleaf weeds.

Recommended Rates and Weeds Controlled

Annual Weed Species	Maximum Height	Rate Per Acre**
Foxtail *Setaria spp.*	12"	8 oz.
Barnyardgrass *Echinochloa crus-galli*	6"	12 oz.
Bluegrass, annual *Poa annua*		
Brome, downy* *Bromus tectorum*		
Mustard, blue *Chorispora tenella*		
Mustard, tansy *Descurainia pinnata*		
Mustard, tumble *Sisymbrium altissimum*		
Mustard, wild *Sinapsis arvensis*		
Spurry, Umbrella *Holosteum umbellatum*		
Barley *Hordeum vulgare*	12"	
Rye *Secale cereale*		
Sandbur, field *Cenchrus spp.*		
Shattercane *Sorghum bicolor*		
Stinkgrass *Eragrostis cilianensis*		
Wheat *Triticum aestivum*	18"	12 oz.
Morningglory *Ipomea spp.*	3"	16 oz.
Bluegrass, bulbous *Poa bulbosa*	6"	16 oz.
Cheat *Bromus secalinus*		
Chickweed, mouseear *Cerastium vulgatum*		
Corn, volunteer *Zea mays*		
Groundsel, common *Senecio vulgaris*		
Horseweed/Marestail *Conyza canadensis*		
Pennycress, field; Fanweed *Thlaspi arvense*		
Rocket, London *Sisymbrium irio*		
Ryegrass, Italian *Lolium multiflorum*		
Shepherdspurse *Capsella bursa-pastoris*		
Buttercup *Ranunculus spp.*	12"	16 oz.
Cocklebur *Xanthium strumarium*		
Crabgrass *Digitaria spp.*		
Dwarfdandelion *Krigia cespitosa*		
Falseflax, smallseed *Camelina microcarpa*		
Foxtail, Carolina *Alopercurus carolinianus*		
Johnsongrass, seedling *Sorghum halepense*		
Lambsquarters, common *Chenopodium album*		
Oats, wild *Avena fatua*		
Panicum, fall *Panicum dichotomiflorum*		
Panicum, Texas *Panicum texanum*		
Pigweed, redroot *Amaranthus retroflexus*		
Pigweed, smooth *Amaranthus hybridus*		
Witchgrass *Panicum capillare*		
Signalgrass, broadleaf *Brachiaria platyphylla*	6"	24 oz.
Teaweed *Sida spinosa*	3"	32 oz.
Eveningprimrose, cutleaf *Oenothera laciniate*	12"	32 oz.
Geranium, Carolina *Geranium carolianianum*		
Goosegrass *Eleusine indica*		
Pusley, Florida *Richardia scabra*		
Sicklepod *Cassia obtusifolia*		
Spanishneedles *Bidens bipinnata*		
Filaree *Erodium spp.*	12"	48 oz.

*For control in no-till systems, use 16 fluid ounces per acre.
**For those rates less than 32 ounces per acre, this product at rates up to 32 ounces per acre may be used where heavy weed densities exist.

DIRECTIONS FOR USE

It is a violation of Federal law to use this product in any manner inconsistent with its labeling.

STORAGE AND DISPOSAL

Do not contaminate water, foodstuffs, seed or feed by storage or disposal.

DISPOSAL:

Wastes resulting from the use of this product that cannot be used or chemically reprocessed should be disposed of in a landfill approved for pesticide disposal or in accordance with applicable Federal, state or local procedures.

Emptied container retains vapor and product residue. Observe all labeled safeguards until container is cleaned, reconditioned or destroyed.

Do not reuse container. Return emptied container per the Monsanto container return program. If not returned, triple rinse container, then puncture and dispose of in a sanitary landfill, or by incineration, or, if allowed by state and local authorities, by burning. If burned, stay out of smoke.

FALLOW AND REDUCED TILLAGE SYSTEMS

Use this product in fallow and reduced tillage systems for control of annual weeds prior to emergence of crops listed in this label. Refer to the "WEEDS CONTROLLED" section of this label for specific rates and instructions. This product may be applied using ground or aerial spray equipment. See the "APPLICATION EQUIPMENT AND TECHNIQUES" section of this label for instructions.

TANK MIXTURES

ROUNDUP® RT plus BANVEL

ROUNDUP RT plus 2,4-D AMINE

ROUNDUP RT plus

ATRAZINE or BLADEX™

DO NOT APPLY BANVEL OR 2,4-D AMINE TANK MIXTURES BY AIR IN CALIFORNIA

The additions of Banvel in a mixture with this product may provide short-term residual control of selected weed species. Some crop injury may occur if Banvel is applied within 45 days of planting. Refer to the Banvel and 2,4-D labels for cropping restrictions and other use instructions. These recommended tank mixtures may be applied using ground or aerial spray equipment. Refer to the "WEEDS CONTROLLED" section of this label for specific rates and instructions.

This product plus Banvel or 2,4-D amine will control the annual grasses and broadleaf weeds previously listed for this product alone at the indicated heights (except 8-ounces-per-acre applications), plus the following broadleaf weeds. For those weeds previously listed at 8 ounces of this product alone per acre, use the 12-ounce rate in these tank mixtures.

NOTE: Refer to the specific product labels for crop rotation restrictions and cautionary statements of all products used in tank mixtures. Some crop injury may occur if Banvel is applied within 45 days of planting. The addition of Banvel in a mixture with this product may provide short-term residual control of selected weed species.

Apply 12 to 16 ounces of this product plus 0.25 pound a.i. of Banvel, or 0.5 pound a.e. of 2,4-D amine plus 0.5 to 1 percent nonionic surfactant by total spray volume per acre to control dense populations of the following annual broadleaf weeds when less than 6 inches in height:

Kochia* *Kochia scoparia*	**Pigweed, Redroot** *Amaranthus retroflexus*
Lambsquarters *Chenopodium album*	**Thistle, Russian** *Salsola kali*
Lettuce, prickly *Lactuca serriola*	

*Controlled with Banvel tank mixture only.

This product plus atrazine or Bladex plus nonionic surfactant tank mixture provides postemergence control of listed annual weeds in fallow and reduced tillage systems. In addition, this tank mixture will provide soil residual control of weeds listed on the atrazine and Bladex label.

For use only in Colorado, Kansas, Nebraska, Oklahoma, Oregon, South Dakota, and Washington.

ANNUAL WEEDS CONTROLLED

Barnyardgrass *Echinochloa crus-galli*	**Mustard, tansy** *Descurainia pinnata*
Brome, downy *Bromus tectorum*	**Pigweed** *Amaranthus* spp.
Foxtail, green *Setaria viridis*	**Sandbur, field** *Cenchrus* spp.
Kochia *Kochia scoparia*	**Stinkgrass** *Eragrostis cilianensis*
Lambsquarters *Chenopodium album*	**Thistle, Russian** *Salsola kali*
Lettuce, prickly *Lactuca serriola*	**Wheat** *Triticium aestivum*
	Witchgrass *Panicum capillare*

Apply 12 ounces of this product with 1 pound or less of the active ingredient atrazine per acre or 16 ounces of this product with 2 pounds or less of the active ingredient atrazine per acre or 20 ounces of this product with 1 to 3 pounds of the active ingredient atrazine per acre plus 0.5 to 1.0 percent nonionic surfactant by total spray volume. Apply 16 ounces of this product with 2.4 to 4 pounds of the active ingredient cyanazine in Bladex per acre plus 0.5 to 1 percent nonionic surfactant. Barnyardgrass will be suppressed at the 20 ounce per acre rate and will require up to 26 ounces per acre for control. In Oregon and Washington, do not exceed 1 pound of the active ingredient atrazine per acre. The addition of 2 percent dry ammonium sulfate by weight (or liquid equivalent) or 17 pounds per 100 gallons of water is recommended to increase the performance of this product plus atrazine or Bladex plus nonionic surfactant tank mixtures.

Consult the atrazine or Bladex labels (90 DF, 4L, 80W) for use rates, soil type, planting, cropping and other restrictions, as well as other precautionary statements and use according to the most restrictive label.

These tank mixtures may be applied with ground or aerial equipment. See the "APPLICATION EQUIPMENT AND TECHNIQUES" section for instructions.

For ground application, apply in 3 to 10 gallons of water per acre.

For aerial application, apply 3 to 5 gallons of water per acre.

POSTHARVEST APPLICATIONS

This product will provide control of existing weeds following grain harvest. Weeds should be allowed to regrow after damage incurred during harvest operations, and to recover from environmental stress, before application of this product. Weeds should be treated prior to the heading stage of annual grasses and before broadleaf weeds exceed 24 inches in height. Nonionic surfactant should be added at 0.5 to 1 percent by total spray volume. Use of ammonium sulfate is recommended and may improve performance on annual weeds.

Apply 12 ounces of this product per acre plus nonionic surfactant to control downy brome, green foxtail, stinkgrass and volunteer wheat.

To control kochia, lambsquarters, mustard, pigweed and Russian thistle, apply 16 ounces of this product plus 0.25 pounds a.i. of Banvel or 0.5 pounds a.e. of 2,4-D amine plus nonionic surfactant per acre.

Apply 20 ounces of this product per acre plus nonionic surfactant to control barnyardgrass, sandbur, and witchgrass. Addition of 0.25 pound a.i. of Banvel or 0.5 pound a.e. of 2,4-D amine plus nonionic surfactant to 20 ounces of this product will control prickly lettuce.

® Bladex is a trademark of E. I. DuPont De Nemours and Company.

AID TO TILLAGE

This product used in conjunction with preplant and conventional fallow tillage practices will provide control of downy brome, volunteer wheat, tansy mustard and foxtail.

Apply 8 fluid ounces per acre (1/2 pint) of this product plus 0.5 to 1 percent nonionic surfactant by total spray volume as a broadcast spray in 3 to 10 gallons of water per acre.

For ground application, apply in 3 to 10 gallons of water per acre. For aerial applications, apply in 3 to 5 gallons of water per acre. Do not mix with residual herbicides.

Application of this product should be made when weeds are actively growing and before weeds are greater than 6 inches tall. Application must be followed by conventional tillage practices before regrowth of the treated plants occurs. Allow at least one day after application before tillage. Tillage must occur no later than 15 days following application.

This product does not provide residual weed control. Repeat treatments may be necessary to control subsequent germinating weeds.

Do not treat weeds under poor growing conditions such as drought stress, or disease or insect damage as reduced weed control may result. Reduced results also may occur when treating weeds heavily covered with dust, or when straw canopy covers the weeds.

PERENNIAL WEEDS

Apply this product as follows to control or destroy most perennial weeds:

NOTE: If weeds have been mowed or tilled, do not treat until plants have resumed active growth and are at the recommended stages. Repeat treatments may be necessary to control weeds regenerating from underground parts or seed. Repeat treatments must be made prior to crop emergence. The addition of 2 percent dry ammonium sulfate by weight or 17 pounds per 100 gallons of water may increase the performance of this herbicide on perennial weeds. The improvement in performance may be apparent where environmental stress is a concern. Refer to the "MIXING, ADDITIVES AND APPLICATION INSTRUCTIONS" section of this label.

When applied as recommended under the conditions described, this product will CONTROL the following perennial weeds:

Alfalfa *Medicago sativa*	**Quackgrass** *Agropyron repens*
Johnsongrass *Sorghum halepense*	**Ryegrass, perennial** *Lolium perenne*
Nightshade, silverleaf *Solanum elaeagnifolium*	**Thistle, Canada** *Cirsium arvense*

Johnsongrass/Ryegrass, perennial—Apply 1 to 3 quarts of this product per acre. In annual cropping systems, apply 1 to 2 quarts of this product per acre. Apply 1 quart of this product plus 0.5 to 1 percent nonionic surfactant by total spray volume in 5 to 10 gallons of water per acre. Use 2 quarts of this product when applying 10 to 40 gallons of water per acre. In noncrop or areas where annual tillage (no-till), is not performed, apply 2 to 3 quarts of this product in 10 to 40 gallons of water per acre. For best results, apply to actively growing plants when most have reached the boot-to-head stage of growth. Allow 7 or more days after application before tillage. Do not tank-mix with residual herbicides when using the 1 quart per acre rate. See "DIRECTIONS FOR USE" and "MIXING, ADDITIVES AND APPLICATION INSTRUCTIONS" sections of this label for labeled uses and specific application instructions.

For burndown of contacted vegetation, apply 1 pint per acre plus 0.5 to 1 percent nonionic surfactant in 5 to 10 gallons of water per acre before the plants reach a height of 12 inches. For this use, allow at least 3 days after treatment before tillage.

Quackgrass—In Annual Cropping Systems or in Pastures and Sods Followed by Deep Tillage: Apply 1 to 2 quarts of this product per acre. For the one-quart rate, apply 0.5 to 1 percent nonionic surfactant by total spray volume in 5 to 10 gallons of water per acre. For the 2-quart rate, apply in 10 to 40 gallons of water per acre. Do not tank-mix with residual herbicides when using the 1 quart rate. Spray when quackgrass is 8 to 12 inches high and actively growing. Do not till between harvest and fall applications or in the fall or spring prior to spring application. Allow 3 or more days after application before tillage. In pastures or sods, for best results, use a moldboard plow.

Quackgrass—Pasture or Sod or Other Noncrop Areas Where Deep Tillage is Not Planned Following Application: Apply 2 to 3 quarts in 10 to 40 gallons of water per acre. Spray when the quackgrass is greater than 8 inches tall and actively growing. Do not till between harvest and fall application or in the fall or spring prior to spring application. Allow 3 or more days after application before tillage.

See "DIRECTIONS FOR USE" and "MIXING, ADDITIVES, and APPLICATIONS INSTRUCTIONS" sections of this label for labeled uses and specific application instructions.

Thistle, Canada—Apply 2 to 3 quarts of this product per acre. Apply to actively growing thistles when most are at or beyond the bud stage of growth. After harvest, mowing or tillage in the late summer or fall, allow at least 4 weeks for initiation of active growth and rosette development prior to the application of this product. Fall treatments must be applied before a killing frost. Allow 3 or more days after application before tillage. See "DIRECTIONS FOR USE" and "MIXING, ADDITIVES AND APPLICATION INSTRUCTIONS" sections of this label for labeled uses and specific application instructions.

Other perennials listed on this label—Apply 3 to 5 quarts of this product per acre. Apply when actively growing and when most have reached early head or early bud stage of growth. Allow 7 or more days after application before tillage.

See "DIRECTIONS FOR USE" and "MIXING, ADDITIVES, AND APPLICATION INSTRUCTIONS" sections of this label for labeled uses and specific application instructions.

TANK MIXTURES

When applied as directed under the conditions described. this product plus Banvel or 2,4-D amine will SUPPRESS or CONTROL the following perennial broadleaf weed:

Bindweed, field
Convolvulus arvensis

For suppression, apply 16 ounces of this product plus 0.5 lb a.e. of 2,4-D amine or 0.25 lb a.i. of Banvel, plus 0.5 to 1 percent nonionic surfactant by total spray volume in 3 to 10 gallons of water per acre for ground applications and 3 to 5 gallons of water per acre for aerial applications. Applications should be delayed until maximum emergence has occurred and when vines are between 6 to 18 inches in length. These tank mixtures may be applied using aerial application equipment in fallow and reduced tillage systems only.

For control, apply 2 quarts of this product plus 0.5 pound a.i. of Banvel in 3 to 20 gallons of water per acre. At these rates, apply using ground application equipment only.

CROPPING SYSTEMS

See "GENERAL INFORMATION" and "APPLICATION EQUIPMENT AND TECHNIQUES" sections of this label for essential product performance information.

Repeat treatments may be necessary to control weeds regenerating from underground parts or seed. Except as otherwise specified on this label, repeat treatments must be made before the crop emerges, in accordance with the instructions of this label.

Except as otherwise specified in a crop section of this label, the combined total of all treatments must not exceed 8 quarts per acre of this product per year.

Do not plant subsequent crops other than those on this label for one year following application.

For other cropping systems, except following spot treatment in forage grasses and legumes, do not harvest or feed treated crops for 8 weeks after application. Allow 14 days following spot treatment or selective equipment use before grazing domestic livestock or harvesting forage grasses and legumes.

This product controls annual and perennial weeds listed on this label, prior to the emergence of these crops.

ALFALFA*	LENTILS
BARLEY*	OATS*
BEETS (Red, Sugar)	PEAS (All)
POTATO (Irish, sweet)	CORN (All)
COTTON*	FORAGE GRASSES*
FORAGE LEGUMES*	SORGHUM (Milo)
SOYBEANS*	WHEAT

*Spot treatments may be applied in these crops.

Spot Treatment (Only those crops with "*" can be spot treated)—Applications in growing crops must be made prior to heading of small grains and milo, initial pod set in soybeans, silking of corn, and boll opening on cotton.

For dilution and rates of application using boom or hand-held equipment, see the "APPLICATION EQUIPMENT AND TECHNIQUES" and "WEEDS CONTROLLED" section of this label.

897.15-000. (041CG) 1988-2

In case of an emergency involving this product Call Collect, day or night (314) 694-4000.

This sample label is current as of January 1, 1989. The product descriptions and recommendations provided in this sample label are for background information only. Always refer to the label on the product container before using Monsanto or any other agrichemical product.

This specimen label is intended for use only as a guide in providing general information regarding the use of this product. Monsanto is not responsible for the accuracy of the information contained herein. As labels are subject to revision, always carefully read and follow the label on the product container.

SHOWDOWN™

MAXIMUM LOAD GRANULE HERBICIDE

Granular selective herbicide for fall application to control wild oats in spring and durum wheat and barley.

ACTIVE INGREDIENT:
S-(2,3,3-Trichloroallyl)-
diisopropylthiocarbamate. 20.0%
INERT INGREDIENTS: 80.0%
TOTAL. 100.0%

EPA Reg. No. 524-416 843.55-000.03/53

Read the entire label before using this product.

Use only according to label instructions.

Read "LIMIT OF WARRANTY AND LIABILITY" before buying or using. If terms are not acceptable, return at once unopened.

REFORMULATION OR REPACKAGING IS PROHIBITED.

LIMIT OF WARRANTY AND LIABILITY

This Company warrants that this product conforms to the chemical description on the label and is reasonably fit for the purposes set forth in the Complete Directions for Use label booklet ("Directions") when used in accordance with those Directions under the conditions described therein. NO OTHER EXPRESS WARRANTY OR IMPLIED WARRANTY OF FITNESS FOR PARTICULAR PURPOSE OR MERCHANTABILITY IS MADE. This warranty is also subject to the conditions and limitations stated herein. Buyer and all users shall promptly notify this Company of any claims whether based in contract, negligence, strict liability, other tort or otherwise.

Buyer and all users are responsible for all loss or damage from use or handling which results from conditions beyond the control of this Company, including but not limited to, incompatibility with products other than those set forth in the Directions, unusual weather, weather conditions which are outside the range considered normal at the application site and for the time period when the product is applied, as well as weather conditions which are outside the application ranges set forth in the Directions, application in any manner not explicitly set forth in the Directions, moisture conditions outside the moisture range specified in the Directions, or the presence of products other than those set forth in the Directions in or on the soil, crop or treated vegetation.

THE EXCLUSIVE REMEDY OF THE USER OR BUYER, AND THE LIMIT OF THE LIABILITY OF THIS COMPANY OR ANY OTHER SELLER FOR ANY AND ALL LOSSES, INJURIES OR DAMAGES RESULTING FROM THE USE OR HANDLING OF THIS PRODUCT (INCLUDING CLAIMS BASED IN CONTRACT, NEGLIGENCE, STRICT LIABILITY, OTHER TORT OR OTHERWISE) SHALL BE THE PURCHASE PRICE PAID BY THE USER OR BUYER FOR THE QUANTITY OF THIS PRODUCT INVOLVED, OR, AT THE ELECTION OF THIS COMPANY OR ANY OTHER SELLER, THE REPLACEMENT OF SUCH QUANTITY, OR, IF NOT ACQUIRED BY PURCHASE, REPLACEMENT OF SUCH QUANTITY. IN NO EVENT SHALL THIS COMPANY OR ANY OTHER SELLER BE LIABLE FOR ANY INCIDENTAL, CONSEQUENTIAL OR SPECIAL DAMAGES.

Buyer and all users are deemed to have accepted the terms of this LIMIT OF WARRANTY AND LIABILITY which may not be varied by any verbal or written agreement.

PRECAUTIONARY STATEMENTS

HAZARDS TO HUMANS AND DOMESTIC ANIMALS

CAUTION!

HARMFUL IF SWALLOWED.
CAUSES EYE IRRITATION.

Avoid contact with skin, eyes or clothing.

Wash thoroughly with soap and water after handling.

FIRST AID: IF IN EYES, flush with plenty of water for at least 15 minutes. Call a physician if irritation persists.

IF SWALLOWED, call a physician or Poison Control Center. Drink 1 or 2 glasses of water and induce vomiting by touching back of throat with finger. Do not induce vomiting or give anything by mouth to an unconscious person.

In case of an emergency involing this product, Call Collect, day or night, (314) 694-4000.

ENVIRONMENTAL HAZARDS

Avoid direct application to any body of water. Do not contaminate water when disposing of equipment washwaters.

DIRECTIONS FOR USE

It is a violation of Federal law to use this product in any manner inconsistent with its labeling.

STORAGE AND DISPOSAL

Do not contaminate water, foodstuffs, feed or seed by storage or disposal.

STORAGE:

Keep bag closed to prevent spills and contamination.

DISPOSAL:

Wastes resulting from the use of this product that cannot be used or chemically reprocessed should be disposed of in a landfill approved for pesticide disposal or in accordance with applicable Federal, state or local procedures.

Emptied container retains vapor and product residue. Observe all labeled safeguards until container is destroyed.

Completely empty container into application equipment. Then dispose of empty container in a sanitary landfill or by incineration, or, if allowed by state and local authorities, by burning. If burned, stay out of smoke.

This herbicide is recommended for wild oat control in barley, durum and spring wheat only. Other crops should not be treated with this product because injury may occur. For barley, durum and spring wheat, this product may be applied on the soil in the fall before wild oats germinate.

Application equipment must be properly calibrated: application of too much herbicide may injure the crop; application of too little may result in poor wild oat control. Recommended rates should be followed in order to:

1. Avoid crop injury
2. Avoid crop residue at harvest
3. Control wild oats

Application to a field which is wet, lumpy, rough or ridged will result in reduced wild oat control and promote crop thinning. Incorporation must be completed within 48 hours after application and before germination of the wild oats. If weeds, including wild oats, have emerged prior to treating and/or planting, they must be controlled. All deep tillage by cultivators or double disc implements must be completed prior to application. Do not plow following application of this herbicide.

Seeding may be done either before or after application, depending upon the crop that is to be sown. If seeding is delayed, shallow reworking of the treated area before seeding will not destroy the effects of the chemical.

When using this herbicide, a strip should be left untreated for proof of results. Weed control may be evaluated also by removing a surface inch or two of the soil at the time of germination to inspect the number of wild oats that were killed before emergence.

Wild oats are usually killed before emergence, but occasionally, and particularly under dry conditions, plants may reach the 3 to 4-leaf stage before they die.

Under conditions of prolonged high temperature at the time of germination, or extreme drought in the spring, this product may not maintain the usual high standard of wild oat control.

ATTENTION

DELAYED EMERGENCE, STAND REDUCTION, STUNTING AND YIELD LOSS MAY RESULT DUE TO COLD OR WET CONDITIONS, IRRIGATION DURING GERMINATION OR EMERGENCE, DEEP PLANTING, SOIL CRUSTING, DISEASE, INSECTS, OR SELECTION OF A VARIETY GENETICALLY SUSCEPTIBLE TO STRESS. THESE ARE CONDITIONS BEYOND THE CONTROL OF MONSANTO, AND THE GROWER SHOULD CONSULT THE STATE UNIVERSITY EXTENSION SPECIALIST OR SEED PRODUCER OR SUPPLIER FOR LOCAL VARIETAL INFORMATION AND RECOMMENDED TILLAGE AND PLANTING PRACTICES AND DATES.

Domestic oats should not be seeded where this product was used the previous year.

Do not graze livestock on treated crops.

IN THE STATE OF MONTANA, do not use this herbicide on fields to be seeded to hard red spring wheat with press drills, if the field is or will be irrigated in the current growing season.

Do not use postplant incorporation with hoe drills. This practice will result in unacceptable seeding depth and an excessive concentration of this herbicide in the seed zone.

Use of this product not consistent with this label may result in injury to persons, animals or crops, or other unintended consequences. Keep container closed to prevent spills and contamination.

HOW TO APPLY AND CALIBRATE EQUIPMENT

This granular herbicide must be applied through a specially designed ground applicator capable of applying small quantities of granules evenly.

GROUND EQUIPMENT: It is important that the applicator be calibrated properly to deliver the desired amount of this product to avoid applying too little, or too much, material. To give even distribution, scatter plates (similar to those used for applying granules in a band) must be attached to each delivery tube or outlet in such a manner to give overall coverage. To calibrate, attach a collector pan, or strong plastic bag, over each spreader plate or delivery tube. Operate over normal terrain to be treated at 4 to 5 miles per hour. Collect the granules from all outputs after covering the desired distance.

For example: If a 14-foot applicator is being used, stake off a distance of 324 feet in the field to be treated (distance equal to 1/10 acre treated). After attaching a bag, or collection pan, to each outlet, collect the granules while driving the desired distance. Check to see that each outlet disperses the same amount of granules. Combine all samples and weigh. For the above set of conditions, the quantity of granules that should be collected for the following recommended rates of broadcast treatment are:

	RECOMMENDED RATE PER ACRE	
	6.3 lbs	7.5 lbs
Proper amount to be collected	0.63 lb	0.75 lb

If more or less than the desired quantity is collected, adjust opening accordingly and again collect the granules while driving the staked-off distance. Continue this procedure until the proper amount is delivered.

FIELD PREPARATION

Before applying this product, be sure the soil is in good working condition. All deep tillage by cultivation, or double disc implements, must be completed prior to application of this herbicide. If stubble ground is being treated, one or two passes with a field cultivator may be required before application and incorporation.

INCORPORATED APPLICATIONS

When summer fallow or plowed ground is being treated, a field cultivator will provide adequate incorporation. For proper incorporation of this product, set incorporation implement to work the soil no deeper than 3 to 4 inches. Do not use disc implements for incorporation.

FALL—On summer fallow ground which is loose and free of lumps and trash, incorporate this product using equipment similar to a culti-harrow or a duckfoot with rod weeder attachment.
For applications to fields or standing stubble, fields may be worked once or twice with a field cultivator or chisel plow. Apply granules and incorporate with a field cultivator or culti-harrow.

If soil must be ridged after incorporating, to prevent soil erosion by high winds, the depth of ridging should be kept to a minimum.

In preparing the herbicide treated area for seeding in the spring, care must be taken to avoid working the soil any deeper than fall tillage.

When incorporating with a single pass in the fall, a second incorporation must be performed in the spring during seedbed preparation. If no spring work is anticipated, then both incorporations should be done in the fall.

WHEN TO USE
FALL TREATMENT INCORPORATED

CROP	RATE (lbs/a)	WHEN TO APPLY
Spring and Durum Wheat, and Barley	6.3 to 7.5	*Within 3 weeks of normal freeze-up or until snow cover occurs.
Winter Wheat and Winter Barley	6.3 to 7.5	Before or after seeding.

Apply lower rates on light soils and apply higher rates on heavy soils.

Incorporate within 48 hours. For methods, see "FIELD PREPARATION" and "INCORPORATED APPLICATIONS" sections.

* Average soil temperature at the 2 inch depth must be 40°F. or less.

SURFACE APPLICATION WITH DELAYED INCORPORATION

USE ONLY IN MINNESOTA, MONTANA AND NORTH DAKOTA

This product may be applied as a surface application with a delayed incorporation prior to planting spring wheat, durum wheat, and barley at a rate of 7.5 pounds per acre for all soil types.

Surface applications of this product which will be followed by a delayed incorporation in the spring may be made beginning 3 weeks prior to soil freeze-up in the fall and ending before spring thaw. (Average soil temperature at the 2-inch depth must be 40°F or less).

Surface applications must be incorporated in the spring: two passes are recommended. If surface applications are made because soil conditions do not permit incorporation and these conditions change, making incorporation possible, incorporate even if several weeks after application.

This surface application may be made in standing stubble or to fields with surface residues. Surface applications should not be made to fields covered with snow or with excessive crop residue, which will not allow granule contact with the soil.

Surface applications with delayed incorporation may not provide the same level of wild oat control as fall incorporated applications. Wild oat control resulting from delayed incorporation may be reduced if drought or abnormally warm temperatures occur between application and crop emergence in the spring. To achieve the most reliable wild oat control, this product should be incorporated within 48 hours of application as described in the "FIELD PREPARATIONS" and "INCORPORATED APPLICATIONS" sections of this label.

843.55-000.03/53

In case of an emergency involving this product, Call Collect, day or night, (314) 694-4000.

This specimen label is intended for use only as a guide in providing general information regarding the use of this product. Monsanto is not responsible for the accuracy of the information contained herein. As labels are subject to revision, always carefully read and follow the label on the product container.

NOR-AM Chemical Company

3509 Silverside Road
P.O. Box 7495
Wilmington, DE 19803
(302) 575-2000

ANTOR® 4ES

EMULSIFIABLE SOLUTION

Herbicide for Selective Weed Control in Sugar Beets, Red Beets (Intended for Processing or Grown for Seed Only), Spinach and Bermudagrass (Grown for Seed Only)

FOR AGRICULTURAL USE ONLY

ACTIVE INGREDIENT:	Percent by Weight
Diethatyl ethyl: [N-(chloroacetyl)-N-(2,6-diethylphenyl) glycine, ethyl ester]	43.3%
INERT INGREDIENTS:	56.7%
TOTAL	100.0%

This product contains 4 lbs. active ingredient per gallon.

EPA Reg. No. 45639-54 U.S. Patent No.: 3,780,090

PRECAUCION AL USUARIO: Si usted no lee ingles, no use este producto hasta que la etiqueta haya sido explicada ampliamente.

KEEP OUT OF REACH OF CHILDREN

WARNING — AVISO

STATEMENT OF PRACTICAL TREATMENT

IF SWALLOWED—DO NOT INDUCE VOMITING. CALL A PHYSICIAN IMMEDIATELY. SEE "NOTE TO PHYSICIAN."

IF IN EYES—FLUSH WITH PLENTY OF WATER. GET MEDICAL ATTENTION.

IF ON THE SKIN—WASH SKIN WITH SOAP AND PLENTY OF WATER.

NOTE TO PHYSICIAN—CONTAINS MONOCHLOROTOLUENE SOLVENT.

PRECAUTIONARY STATEMENTS

HAZARDS TO HUMANS AND DOMESTIC ANIMALS

WARNING

The active ingredient in ANTOR® may cause skin sensitization reactions in certain individuals. Wear coveralls and gloves while handling or using this product. Causes substantial but temporary eye injury. Harmful if swallowed. Do not get in eyes or on clothing. Wear goggles, face shield, or safety glasses when handling. Wash thoroughly with soap and water after handling. Avoid breathing of spray mist. Avoid contamination of food.

ENVIRONMENTAL HAZARDS

Do not apply directly to any body of water.

Do not apply when weather conditions favor drift from areas treated.

Do not contaminate water by cleaning of equipment or disposal of wastes.

PHYSICAL OR CHEMICAL HAZARDS

Do not use, pour, or store near heat or open flame.

IN CASE OF FIRE, LEAKY OR DAMAGED CONTAINERS, OR OTHER EMERGENCY, REPORT AT ONCE BY TOLL-FREE TELEPHONE TO: 800-424-9300.

STORAGE AND DISPOSAL

STORAGE: Protect ANTOR® 4ES from freezing. Open dumping is prohibited.

PESTICIDE DISPOSAL: Do not contaminate water, food, or feed by storage or disposal. Pesticide wastes are toxic. Improper disposal of excess pesticides, spray mixture, or rinsate is a violation of Federal Law. If these wastes cannot be disposed of according to label instructions, contact your State Pesticide or Environmental Control Agency, or the Hazardous Waste representative at the nearest EPA Regional Office for guidance.

CONTAINER DISPOSAL: Triple rinse (or equivalent). Then offer for recycling or reconditioning, or puncture and dispose of in a sanitary landfill, or by other procedures approved by state and local authorities.

Do not reuse empty container.

GENERAL USE PRECAUTIONS: Unusually dry, windy weather following application of ANTOR 4ES or tank-mixes, which dries the upper soil layer, may reduce effectiveness.

If crop is lost due to climatic or soil conditions following application of ANTOR 4ES or tank-mixes, do not re-treat field with ANTOR 4ES or tank-mixes within the same season.

Do not allow spray mixture to stand in tank overnight. Flush and drain equipment after each day's use.

Where ANTOR 4ES is used in tank-mix with TCA, do not use treated sugar beet tops for feed or forage.

Do not mix ANTOR 4ES or tank-mixes with liquid fertilizers.

ROTATIONAL CROPS: Crops, except wheat, may be rotated at 3 months post-application. Wheat may be planted 9 months after application.

DIRECTIONS FOR USE

It is a violation of Federal Law to use this product in a manner inconsistent with its labeling.

Read entire Directions for Use and Disclaimer of Warranties before using this product.

Do not apply this product in combination with insecticides or fungicides.

Do not apply this product in such a manner as to directly or through drift expose workers or other persons. The area being treated must be vacated by unprotected persons.

When ANTOR® 4ES is applied preemergence, the following statements apply:

REENTRY STATEMENT

Do not enter treated areas without protective clothing until sprays have dried.

Because certain states may require more restrictive reentry intervals for various crops treated with this product, consult your State Department of Agriculture for further information.

Written or oral warnings must be given to workers who are expected to be in a treated area or in an area about to be treated with this product. (Indicate specific oral warnings which inform workers of areas or fields that may not be entered without specific protective clothing, period of time field must be vacated, and appropriate actions to take in case of accidental exposure.) When oral warnings are given, warnings shall be given in a language customarily understood by workers. Oral warnings must be given if there is a reason to believe that written warnings cannot be understood by workers. Written warnings must include the following information:

WARNING: Area treated with ANTOR® 4ES herbicide on (date of application). Do not enter without appropriate protective clothing until spray has dried. In case of accidental exposure, see Statement of Practical Treatment on the label.

GENERAL INFORMATION

ANTOR® 4ES is a selective soil-applied herbicide for use in sugar beets, red beets, spinach, and Bermudagrass. It can be applied either preplant incorporated or preemergence. The seedbed should be firm, fine and free of clods and trash. When applied preemergence, ANTOR 4ES requires rainfall or irrigation prior to weed germination for optimum weed control. Similarly, when applied preplant incorporated, moisture in the form of rainfall, irrigation or existing soil moisture must be available to activate the product.

When dry weather conditions prevail or furrow irrigation is used, preplant incorporation of ANTOR 4ES is recommended for best results.

SPRAY EQUIPMENT: Apply ANTOR® 4ES alone or in tank-mixes using standard low pressure (20–50 psi) spray equipment. The equipment should be carefully calibrated before use and checked frequently during application to ensure that it is functioning properly. Uniformly apply the recommended rates. Use a spray volume of 10 to 50 gallons per acre broadcast basis. Use proportional amounts for band spraying. See section on Band Applications for calculations.

MIXING THE SPRAY: For ANTOR 4ES alone, add 1/4 to 1/2 the total amount of water to the tank. Then add the required amount of ANTOR and the remaining quantity of water with the bypass agitator running. When tank-mixing ANTOR 4ES with wettable powders, mix the proper amount of wettable powder formulation into half the required amount of water in the spray tank with the agitator running. Add the required amount of ANTOR 4ES and then add the remaining required amount of water. Before tank-mixing with other herbicides, refer to the specific product labels for additional mixing directions.

BAND APPLICATION: To determine the amount of ANTOR 4ES or tank-mixed emulsifiable concentrates needed per acre by band application, divide band width in inches by row width and multiply the answer by the recommended broadcast rate in quarts. To determine the pounds of wettable powder needed per acre when tank-mixed with ANTOR 4ES, divide band width in inches by row width in inches and multiply the answer by the recommended broadcast rate in pounds.

APPLICATION THROUGH IRRIGATION SYSTEMS (CHEMIGATION):

Apply this product only through sprinkler including center pivot, lateral move, end tow, side (wheel) roll, traveler, big gun, solid set, or hand move;

flood (basin); furrow; or border irrigation system(s). Do not apply this product through any other type of irrigation system.

Crop injury, lack of effectiveness, or illegal pesticide residues in the crop can result from non-uniform distribution of treated water.

If you have questions about calibration, you should contact State Extension Service specialists, equipment manufacturers or other experts.

Do not connect an irrigation system (including greenhouse systems) used for pesticide application to a public water system unless the pesticide label-prescribed safety devices for public water systems are in place.

A person knowledgeable of the chemigation system and responsible for its operation, or under the supervision of the responsible person, shall shut the system down and make necessary adjustments should the need arise.

WEED SPECIES CONTROLLED—SUGAR BEETS

ANNUAL GRASS WEEDS

Barnyardgrass (watergrass) *Echinochloa crus-galli*
Yellow foxtail *Setaria glauca*
Green foxtail *Setaria viridis*
Giant foxtail *Setaria faberi*
Foxtail millet *Setaria italica*
Wild oats *Avena fatua*
Canarygrass *Phalaris canariensis*
Johnsongrass (seedling) *Sorghum halepense*
Large crabgrass *Digitaria sanguinalis*
Italian ryegrass (annual) *Lolium multiflorum*
Broadleaf signalgrass *Brachiaria platyphylla*
Witchgrass *Panicum capillare*
Junglerice *Echinochloa colonum*
Fall panicum *Panicum dichotomiflorum*

ANNUAL BROADLEAF WEEDS

Redroot pigweed *Amaranthus retroflexus*
Prostrate pigweed *Amaranthus blitoides*
Black nightshade *Solanum nigrum*
Annual sowthistle *Sonchus oleraceus*
Groundcherry *Physalis longifolia*
Shepherdspurse *Capsella bursa-pastoris*

SUGAR BEETS

ANTOR® 4ES USED ALONE—ALL AREAS

SUGAR BEETS—PREPLANT INCORPORATED: In arid regions where furrow irrigation is practiced or where the rainfall pattern is erratic, soil incorporation of this product is required to provide optimum weed control.

Apply ANTOR® 4ES to the soil surface broadcast or as a band treatment preplant or at the time of planting and incorporate. Use the recommended rates in Dosage Table 1.

Incorporation immediately after application is not required but must be done prior to weed germination. Use a hooded-power or ground-driven rotary tiller, rolling cultivator or similar equipment properly adjusted to uniformly incorporate ANTOR 4ES to a depth of 1 to 2 inches. Deeper incorporation may reduce effectiveness.

Where sugar beets are grown in beds, apply ANTOR 4ES after bedding and incorporate. Irrigate until tops of beds are thoroughly wetted. Incorporation is recommended in the Red River Valley of Minnesota and North Dakota where rainfall is erratic or marginal.

SUGAR BEETS—PREEMERGENCE: Preemergence application of ANTOR 4ES is recommended where sprinkler irrigation is used for crop germination and where rainfall can be depended on to activate the product.

Apply ANTOR 4ES to the soil surface, either broadcast or as a band treatment at time of planting or shortly after, but prior to weed germination. One-half inch of rainfall or irrigation is usually adequate for activation. It is important that the top 2 inches of soil be thoroughly wetted. Refer to Dosage Table 1 for recommended rates.

Dosage Table 1: Sugar Beets—ANTOR 4ES® Alone—All Areas

Soil Texture	Broadcast Rate of ANTOR 4ES Per Acre(1)
Coarse: Loamy sands and sandy loams	3 qts.
Medium/Fine: Loams, silt loams, clay loams and clays containing less than 5% organic matter	4 to 6 qts.(2)
Fine: Clay loams and clays containing more than 5% organic matter	6 qts.

(1) See section on Band Application for calculation of proportional amounts.
(2) Use higher rates on heavy clay soils or where black nightshade or groundcherry are expected to be a problem.

NOTE: TO AVOID EXCESSIVE PHYTOTOXICITY TO FALL PLANTED SUGAR BEETS SOUTH OF THE TEHACHAPI MOUNTAINS IN CALIFORNIA WHEN TEMPERATURES ARE IN EXCESS OF 100°F, apply ANTOR at the rate of 2 to 2½ lbs. active ingredient (2 to 2½ quarts ANTOR 4ES) per acre broadcast equivalent. A maximum rate of 2½ to 3½ quarts ANTOR 4ES should be used when temperatures are below 100°F. Power incorporation of ANTOR 4ES lessens the chance of phytotoxicity when applied at the higher rate.

SUGAR BEETS—ANTOR® 4ES + PYRAMIN®(*) FL TANK-MIX

(California, Michigan, Ohio and Northwest Intermountain States only)

A tank-mix of ANTOR 4ES with Pyramin FL may be used in specified geographic areas for a broader spectrum of weed control. In addition to the weeds controlled by ANTOR 4ES alone, this mixture controls:

Common ragweed *Ambrosia artemisiifolia*
Wild mustard *Brassica kaber*
Pennsylvania smartweed *Polygonum pensylvanicum*
Common lambsquarters *Chenopodium album*
Common purslane *Portulaca oleracea*
Nettleleaf goosefoot *Chenopodium murale*
Henbit *Lamium amplexicaule*
Fanweed (stinkweed) *Thlaspi arvense*

Do not use this mixture under conditions where Pyramin FL alone is not recommended. Before use, read the Pyramin FL label for additional information and precautions.

(*) Pyramin® is a registered trademark of BASF Aktiengesellschaft.

CALIFORNIA: For preplant incorporation with furrow irrigation, apply the tank-mix to the soil surface prior to, or at time of, planting and incorporate to a depth of 1 to 2 inches using a rotary type tiller. Use the recommended rates in Dosage Table 2. Irrigate until tops of beds are thoroughly wetted.

For preemergence application with sprinkler irrigation, apply the tank-mix to the soil surface at time of planting or shortly after, but prior to weed germination, using the rates recommended in Dosage Table 2. Irrigate prior to crop and weed germination. Follow directions and precautions on the Pyramin FL label for use preemergence with sprinkler irrigation. Do not exceed ¾ inch of sprinkler irrigation per set until the sugar beets have two true leaves.

NORTHWEST, INTERMOUNTAIN STATES: For use preemergence under sprinkler irrigation only. Apply tank-mix to the soil surface at the time of planting or shortly after, but prior to weed germination, using the rates recommended in Dosage Table 2. Irrigate prior to crop and weed germination. Follow directions and precautions on the Pyramin FL label for use preemergence with sprinkler irrigation. Do not exceed ¾ inch of sprinkler irrigation per set until the sugar beets have two true leaves.

MICHIGAN AND OHIO: For use preemergence without irrigation, apply the tank-mix to the soil surface at the time of planting or shortly after, but prior to weed germination using rates recommended in Dosage Table 2.

Dosage Table 2: Sugar Beets—ANTOR® 4ES + Pyramin FL Tank-Mix

Area	Soil Texture	Broadcast Rate Per Acre(1) ANTOR 4ES (qts.)	Pyramin FL (qts.)
California, Northwest, Intermountain States, Michigan and Ohio	**Coarse:** Sandy loams only	2	2.5
	Medium: Loams, silt loams and clay loams containing less than 3% organic matter	2 to 3(2)	3
	Fine: Clay loams and clays containing more than 3% organic matter	3	3

(1) See section on Band Applications for calculation of proportional amounts.
(2) Use higher rate on clay loam soils.

SUGAR BEETS—ANTOR® 4ES + TCA TANK-MIX (Minnesota and North Dakota Only): Tank-mixes of ANTOR 4ES + TCA may be used in the Red River Valley of North Dakota and Minnesota for improved control of foxtails (*Setaria* spp.).

This tank-mix may be applied either preemergence or preplant incorporated; however, shallow (up to 2") incorporation will ensure more consistent weed control where rainfall is erratic or marginal.

See Dosage Table 3 for recommended rates.

Dosage Table 3: Sugar Beets—ANTOR® 4ES + TCA Tank-Mix

Soil Texture	Broadcast Rate Per Acre(1) ANTOR 4ES (qts.)	TCA 4.76 lb ai/gal(2)	TCA 79.3% WSP(3)
Coarse: Loamy sands, and sandy loams	NOT RECOMMENDED		

Dosage Table 3: Sugar Beets—ANTOR® 4ES + TCA Tank-Mix

Soil Texture	Broadcast Rate Per Acre(1)		
	ANTOR 4ES (qts.)	TCA 4.76 lb ai/gal(2)	TCA 79.3% WSP(3)
Medium/Fine: Loams, silt loams, clay loams and clays containing less than 5% organic matter	4 to 6(4)	10 pts. or	7.5 lbs.
Fine: Clay loams and clays containing more than 5% organic matter	6	10 pts. or	7.5 lbs.

(1) See section on Band Applications for calculations of proportional amounts.
(2) Solution.
(3) Water Soluble Powder.
(4) Use higher rates on heavy clay soils.

SUGAR BEETS—ANTOR® 4ES + EPTAM®(1) TANK-MIX OR SEQUENCE (Minnesota and North Dakota Only): The tank-mix or sequence of ANTOR 4ES and Eptam 7E may be used in the Red River Valley of Minnesota and North Dakota for improved control of:

Green foxtail *Setaria viridis*
Yellow foxtail *Setaria glauca*
Wild oats *Avena fatua*

In addition, this tank mix will reduce competition from the following weeds:

Kochia *Kochia scoparia*
Common lambsquarters *Chenopodium album*
Wild buckwheat *Polygonum convolvulus*

Tank-mixes of ANTOR 4ES + EPTAM 7E may be used in the spring. Apply preplant incorporated with incorporation of 2 to 4 inches.

See Dosage Table 4 for recommended rates.

(1) Eptam® is a registered trademark of ICI Americas, Inc..

Dosage Table 4: Sugar Beets—ANTOR® 4ES + Eptam® 7E Tank-Mix (Spring Applications Only)

Soil Texture	Broadcast Rate Per Acre(1)	
	ANTOR 4ES (qts.)	Eptam 7E (pts.)
Coarse: Loamy sands, sandy loams, or soils containing less than 4% organic matter	NOT RECOMMENDED	
Medium/Fine: Loams, silt loams, clay loams and clays containing more than 4% organic matter	4	2.25

(1) See section on Band Applications for calculations of proportional amounts.

SUGAR BEETS—EPTAM® 7E or 10G (FALL APPLIED) + ANTOR® 4ES (SPRING APPLIED): Fall applications of Eptam 7E or 10G at labeled rates may be followed by spring applications of ANTOR 4ES at 4 qts. per acre on a broadcast basis.

Note: ANTOR/Eptam tank-mixes or sequences are not recommended for use on soils containing less than 4% organic matter.

Do not use this mixture under conditions where Eptam alone is not recommended. Before use, read the Eptam label for additional information and precautions.

SUGAR BEETS—SEQUENTIAL TREATMENT WITH BETANEX® OR BETAMIX®—ALL AREAS: A sequential treatment of ANTOR® 4ES applied preplant incorporated or preemergence at recommended rates followed by a postemergence treatment of BETANEX or BETAMIX will control the weeds listed under the Weed Species Controlled section, and the additional weeds:

Kochia *Kochia scoparia*
Common lambsquarters *Chenopodium album*
Common ragweed *Ambrosia artemisiifolia*
Wild buckwheat *Polygonum convolvulus*
Common chickweed *Stellaria media*
Common purslane *Portulaca oleracea*
Londonrocket *Sisymbrium irio*
Nettleleaf goosefoot *Chenopodium murale*

The postemergence treatment should be applied according to the Directions for Use on the BETANEX and BETAMIX labels.

Read and follow the precautionary statements and all other information appearing on the BETANEX and BETAMIX labels.

SUGAR BEETS—APPLICATION OF ANTOR® 4ES VIA SPRINKLER SYSTEM INJECTION: In arid regions where overhead sprinkler irrigation is used to germinate the sugar beet crop, ANTOR 4ES may be injected through sprinkler irrigation systems.

SPRINKLER CHEMIGATION

IMPORTANT: Read all instructions under "Application Through Irrigation Systems (Chemigation)" given in the GENERAL INFORMATION section of this label.

The system must contain a functional check valve, vacuum relief valve, and low pressure drain appropriately located on the irrigation pipeline to prevent water source contamination from backflow.

The pesticide injection pipeline must contain a functional, automatic, quick-closing check valve to prevent the flow of fluid back toward the injection pump.

The pesticide injection pipeline must also contain a functional, normally closed, solenoid-operated valve located on the intake side of the injection pump and connected to the system interlock to prevent fluid from being withdrawn from the supply tank when the irrigation system is either automatically or manually shut down.

The system must contain functional interlocking controls to automatically shut off the pesticide injection pump when the water pump motor stops.

The irrigation line or water pump must include a functional pressure switch which will stop the water pump motor when the water pressure decreases to the point where pesticide distribution is adversely affected.

Systems must use a metering pump, such as a positive displacement injection pump (e.g., diaphragm pump) effectively designed and constructed of materials that are compatible with pesticides and capable of being fitted with a system interlock.

Do not apply when wind speed favors drift beyond the area intended for treatment.

Diluted ANTOR should be metered into the central pivot irrigation system at the rates recommended in Dosage Table 5. Applications should be made after seeding and before emergence of sugar beets or weeds. Apply ANTOR® 4ES in 1/3 to 3/4 inch of water per acre.

Agitation of the diluted ANTOR in the chemical source (or slurry) tank should be maintained during the entire application period.

Inject ANTOR 4ES with a positive displacement pump into the main water line ahead of a right-angle turn to insure adequate mixing. Do not apply ANTOR 4ES in greater than the recommended quantities of irrigation water per acre or decreased activity will result.

Do not use ANTOR 4ES in irrigation systems where connections or fittings leak, or where nozzles do not give uniform distribution. Flush the irrigation system completely with water upon completion of the application to remove ANTOR 4ES from the line before dismantling or draining. When sprinkler distribution patterns do not overlap, unacceptable weed control may result. Where overlap is excessive, crop injury may result.

Dosage Table 5: Sugar Beets—ANTOR® 4ES Applied Via Sprinkler Systems

Soil Texture	Broadcast Rate Per Acre ANTOR 4ES (qts.)
Coarse: Loamy sands and sandy loams	3
Medium/Fine: Loams, silt loams, clay loams and clays	4

SUGAR BEETS—ANTOR® 4ES + NORTRON® TANK-MIX (California, Colorado, Montana, Nebraska, Texas and Wyoming only): Tank-mixes of ANTOR 4ES with NORTRON EC can be used in specified geographic areas for control of the additional weeds:

Kochia *Kochia scoparia*
Common lambsquarters *Chenopodium album*
Russian thistle *Salsola kali*
Volunteer barley *Hordeum* sp.
Volunteer wheat *Triticum* sp.
Common purslane *Portulaca oleracea*
Common chickweed *Stellaria media*
Ladysthumb *Polygonum persicaria*
Pennsylvania smartweed *Polygonum pensylvanicum*

ANTOR® 4ES + NORTRON® tank-mixes may be applied preplant incorporated where furrow irrigation is practiced using shallow incorporation implements such as rolling cultivators or harrows, or using power tillers set to a depth of 1–2 inches.

Preemergence applications of ANTOR 4ES + NORTRON tank-mixes may be used when application is followed by sprinkler irrigation prior to weed germination and crop emergence, or where rainfall can be expected to occur within 7–10 days after application.

ANTOR 4ES may also be utilized in sequential applications with NORTRON where a broadcast preemergence treatment of ANTOR 4ES is followed by a band overlay of NORTRON preemergence. The same dosage rates should be used as in the tank-mix. Refer to Dosage Table 6 for recommended rates.

Before use, read the NORTRON EC label for additional information and precautions.

Dosage Table 6: Sugar Beets—ANTOR® 4ES + NORTRON® Tank Mix—(California, Colorado, Montana, Nebraska, Texas and Wyoming)

Soil Texture	Broadcast Rates Per Acre[1]	
	ANTOR 4ES (qts.)	NORTRON 1.5 EC (qts.)
Coarse: Loamy sands and sandy loams	1.5	4
Medium/Fine: Loams and silt loams	2	5.3
Fine: Clay loams and clays	3 to 4[2]	6.6

[1] See section on Band Applications for calculations of proportional amounts.
[2] Use higher rates on heavy clay soils.

SUGAR BEETS—ANTOR® 4ES + NORTRON® + PYRAMIN® FL TANK-MIX (Michigan only): A tank-mix of ANTOR 4ES + NORTRON EC + Pyramin FL may be used in Michigan for control of the additional weeds:

Common ragweed *Ambrosia artemisiifolia*
Wild mustard *Brassica kaber*
Pennsylvania smartweed *Polygonum pensylvanicum*
Common lambsquarters *Chenopodium album*
Common purslane *Portulaca oleracea*
Nettleleaf goosefoot *Chenopodium murale*
Common chickweed *Stellaria media*
Henbit *Lamium amplexicaule*
Fanweed (stinkweed) *Thlaspi arvense*
Kochia *Kochia scoparia*
Russian thistle *Salsola kali*
Volunteer barley *Hordeum* sp.
Volunteer wheat *Tritcum* sp.
Ladysthumb *Polygonum persicaria*

For preemergence use without irrigation, apply the tank-mix to the soil surface at the time of planting or shortly after, but prior to weed germination. Use rates recommended in Dosage Table 7.

Do not use this mixture under conditions where Pyramin alone is not recommended. Before use, read the Pyramin FL and NORTRON EC labels for additional information and precautions.

Dosage Table 7: Sugar Beets—ANTOR® 4ES + NORTRON® + Pyramin—(Michigan)

Soil Texture	Broadcast Rates Per Acre		
	ANTOR 4ES (qts.)	NORTRON EC (qts.)	Pyramin FL (qts.)
Coarse: Sandy loams only	1.5	4	2.25
Medium: Loams, silt loams and clay loams containing less than 3% organic matter	2	5.3	2.25
Fine: Clay loams and clays containing more than 3% organic matter	3	5.3	2.25

SUGAR BEETS—ANTOR® 4ES + RO-NEET 6E®[1] TANK-MIX (Idaho, Montana, Eastern Oregon, Wyoming and McKenzie and Williams Counties of Western North Dakota only): A tank mix of ANTOR 4ES + RO-NEET 6E may be used in Idaho, Montana, Eastern Oregon, Wyoming and Williams and McKenzie counties of Western North Dakota for broader spectrum weed and grass control including lambsquarters and nightshade.

[1] RO-NEET® is a registered trademark of ICI Americas, Inc.

See Dosage Table 8 for recommended rates.

Dosage Table 8: Sugar Beets—ANTOR® 4ES + RO-NEET—Tank Mix—(Idaho, Montana, Western North Dakota, Eastern Oregon and Wyoming)

Soil Texture	Broadcast Rates Per Acre[1]	
	ANTOR 4ES (pts.)	Ro-Neet 6E (pts.)
Coarse: Gravelly coarse sand	2.5	1.67
Medium Coarse: Loamy sands and sandy loams	3	2
Medium Fine: Loams and silt loams	3.5	2.33
Fine: Clay loams and clays	4	2.67

[1] Rates in this table are broadcast rates. In areas where bedding-up is practiced, rates should be reduced by one-third (1/3).

WEED SPECIES CONTROLLED: RED BEETS AND SPINACH

ANNUAL GRASS WEEDS

Barnyardgrass (watergrass) *Echinochloa crus-galli*
Yellow foxtail *Setaria glauca*
Green foxtail *Setaria viridis*
Giant foxtail *Setaria faberi*
Foxtail millet *Setaria italica*
Wild oats *Avena fatua*
Canarygrass *Phalaris canariensis*
Johnsongrass (seedling) *Sorghum halepense*
Large crabgrass *Digitaria sanguinalis*
Italian ryegrass (annual) *Lolium multiflorum*
Broadleaf signalgrass *Brachiaria platyphylla*
Witchgrass *Panicum capillare*
Junglerice *Echinochloa colonum*
Fall panicum *Panicum dichotomiflorum*

ANNUAL BROADLEAF WEEDS

Redroot pigweed *Amaranthus retroflexus*
Prostrate pigweed *Amaranthus blitoides*
Spiny pigweed *Amaranthus spinosus*
Annual sowthistle *Sonchus oleraceus*
Sibara *Sibara virginica*
Henbit *Lamium amplexicaule*
Smallflower galinsoga *Galinsoga parviflora*
Hairy galinsoga *Galinsoga ciliata*
Groundcherry *Physalis longifolia*
Shepherdspurse *Capsella bursa-pastoris*
Pineappleweed *Matricaria matricanoides*

METHOD OF APPLICATION: RED BEETS, SPINACH

SOILS: Avoid application of ANTOR 4ES to red beets or spinach growing in soils with a pH of 7.2 or higher as crop injury may occur.

Preplant Incorporated

In arid regions where furrow irrigation is practiced or where the rainfall pattern is erratic, soil incorporation of ANTOR® 4ES is required to provide optimum weed control.

Incorporation immediately after treatment is not required but must be done prior to weed germination. Use a hooded-power or ground-driven rotary tiller, rolling cultivator, or similar equipment properly adjusted to uniformly incorporate ANTOR 4ES to a maximum depth of 1 to 2 inches. **Deeper incorporation may reduce effectiveness.**

Preemergence

Preemergence application of ANTOR 4ES is recommended where sprinkler irrigation is used for crop germination and where rainfall can be depended on to activate the product.

Apply ANTOR 4ES to the soil at time of planting or shortly after, but prior to weed germination. One-half inch of rainfall or irrigation is usually adequate for activation. It is important that the top 2 inches of soil be thoroughly wetted.

NOTE: Excessive soil moisture resulting from furrow irrigation, frequent sprinkler irrigation or abundant rainfall will cause dilution of ANTOR 4ES in the surface soil layer and may reduce effectiveness of the product. If soil moisture content is low, it is recommended that the field be irrigated prior to ANTOR application. If needed, a follow-up sprinkler irrigation of no more than one-half inch (1/2") of water per acre following ANTOR application may be undertaken to activate and incorporate the herbicide into the soil. Use of furrow irrigation is *not* recommended following ANTOR 4ES application to achieve incorporation of the herbicide.

RED BEETS (Intended for Processing or Grown for Seed only)

Do not use ANTOR 4ES or ANTOR 4ES tank mixes on red beets within 75 days of harvest. Apply ANTOR 4ES preplant incorporated or preemergence using rates recommended in Dosage Table 9.

Dosage Table 9: Red Beets—ANTOR 4ES® Alone

Soil Texture	Broadcast Rate of ANTOR 4ES Per Acre[1]
Coarse to Medium/Fine: Loamy sands and sandy loams loams, silt loams, clay loams and clays containing less than 5% organic matter	3–4 qts.
Fine: Clay loams and clays containing more than 5% organic matter	4–6 qts.

[1] Within each soil texture category use the higher rate on the finer textured soils and/or where shepherdspurse or groundcherry is expected to be a problem.

RED BEETS—ANTOR® 4ES + PYRAMIN®(*) FL TANK MIX

A tank mix of ANTOR® 4ES with Pyramin® FL may be used in red beets to control a broader spectrum of weeds. In addition to the weeds controlled by ANTOR® 4ES alone, this mixture controls:

Common ragweed *Ambrosia artemisiifolia*
Wild mustard *Sinapsis arvensis*
Pennsylvania smartweed *Polygonum pensylvanicum*
Fanweed (stinkweed) *Thlaspi arvense*
Common lambsquarters *Chenopodium album*
Common purslane *Portulaca oleracea*
Nettleleaf goosefoot *Chenopodium murale*

Apply this tank-mix preemergence using rates recommended in Dosage Table 10. Do not use this mixture under conditions where Pyramin FL alone is not recommended. Before use, read the Pyramin FL label for additional information and precautions.

Dosage Table 10: Red Beets—ANTOR® 4ES + Pyramin FL Tank Mix

Soil Texture	Broadcast Rates Per Acre	
	ANTOR 4ES (qts.)	Pyramin FL (qts.)
Coarse: Sandy loams only	2	3
Medium/Fine: Loams, silt loams, and clay loams containing less than 3% organic matter	2 to 3(¹)	3
Fine: Clay loams and clays containing more than 3% organic matter	3	3

(¹) Use the higher rate on clay loam soils.

RED BEETS—ANTOR® 4ES + RO-NEET®(**) 6E TANK MIX

(Wisconsin and Minnesota only)

A tank mix of ANTOR 4ES with RO-NEET 6E may be used in Wisconsin and Minnesota to control a broader spectrum of weeds including lambsquarters and nightshade. Before use, read the RO-NEET label for additional information and precautions. Incorporate as specified on the RO-NEET label.

Dosage Table 11: Red Beets—ANTOR® 4ES + RO-NEET 6E Tank-Mix (Wisconsin and Minnesota)

Soil Texture	Broadcast Rates Per Acre	
	ANTOR® 4ES (qts.)	Ro-Neet® 6E (qts.)
Coarse: Gravelly coarse sand (Less than 2% organic matter)	1.5	1.0
Medium/Coarse: Loamy sands and sandy loams (2–3% organic matter)	2.5	1.6
Medium/Fine: Loams and silt loams (3–5% organic matter)	3	2
Fine: Clay loams and clays (greater than 5% organic matter)	4	2

(*) Pyramin® is a registered trademark of BASF Aktiengesellschaft.
(**) RO-NEET is a registered trademark of ICI Americas, Inc.

SPINACH

Apply ANTOR® 4ES preplant incorporated or preemergence using rates recommended in Dosage Table 12.

Dosage Table 12: Spinach—ANTOR® 4ES Alone

Soil Texture	Broadcast Rate of ANTOR® 4ES Per Acre(¹)(²)
Coarse: Loamy sands and sandy loams	3–4 qts.
Medium/Fine: Loams, silt loams, clay loams and clays containing less than 5% organic matter	4 qts.

(¹) Within each soil texture category use the higher rate on the finer-textured soils and/or where shepherdspurse or groundcherry is expected to be a problem.
(²) The use of ANTOR® 4ES on clay loams or clays containing more than 5% organic matter has not been fully investigated. Do not use ANTOR® 4ES on these soils.

USE PRECAUTIONS: SPINACH AND RED BEETS

To avoid possible crop injury, do not apply ANTOR 4ES to late-planted or second-crop spinach or to spinach or red beets after June 15 of the growing season in the following counties in Texas and New Mexico:

TEXAS		NEW MEXICO
Gaines	Andrews	Curry
Dawson	Palmer	Roosevelt
Bailey	Deaf Smith	Lea
Crosby	Lamb	
Lubbock	Hockey	
Lynn	Borden	
Terry	Garza	
Cochran		

BERMUDAGRASS Weeds Species Controlled

(Grown for Seed only)

ANTOR® 4ES is an effective preemergence herbicide for use in established Bermudagrass grown for seed *(Cynodon dactylon)*. ANTOR 4ES controls germinating weeds only after shoot and root absorption. It should be applied in the irrigation water (water run) prior to weed germination.

ANNUAL GRASS WEEDS

Sprangletop *Leptochloa spp.*
Watergrass *Echinochloa oryzoides*
Canarygrass *Phalaris spp.*

ANNUAL BROADLEAF WEEDS

Redroot pigweed *Amaranthus retroflexus*

USE PRECAUTIONS: BERMUDAGRASS

Do not apply ANTOR 4ES more than three times per season.

Apply ANTOR 4ES to Bermudagrass which has been established for a minimum of 4 months. If applied earlier, crop injury may occur.

Take steps to insure that tail water does not reach other crops.

RESTRICTION

Do not graze or feed treated crop, forage, or hay to livestock.

The treated crop is to be used only for seed production.

ENVIRONMENTAL HAZARDS

Do not apply directly to any body of water except as specified on this label.

POSTING REQUIREMENTS

Areas being treated with pesticides applied through irrigation systems must be posted at the usual entrances to the area, at bulletin boards where workers assemble and where the treated area lies adjacent to roads, residential areas, or any other areas such as schools, parks or other public facilities. Posted signs must be legible from 25 feet away and be printed in English and any other appropriate language such as Spanish. Signs must be posted prior to application, during application and for at least one week after application, but may remain in place indefinitely. The sign must indicate the following:

NOTICE

Pesticides are applied in irrigation water in this field.

Do not enter this field when the irrigation system is operating.

Do not drink, bathe in or play in water or in furrows, puddles, ponds, canals, or ditches associated with this irrigation system.

APPLICATION RATE

Spring Application:

For early season weed control meter 2.0–4.0 quarts per acre of ANTOR 4ES into the irrigation water of established Bermudagrass prior to weed emergence. Use the higher rate for extended residual control, heavy weed pressure, or on heavy soil types. Refer to the flow chart and sample calculation for additional information.

Spring and Summer Application:

For control of weeds in both the spring and early summer months meter ANTOR 4ES at a rate of 2.0 quarts per acre in the first spring irrigation followed by a second application in 6–8 weeks of 2.0 quarts per acre applied into the irrigation water. Refer to the flow chart and sample calculation for additional information.

Fall Application:

For control of cool season winter weeds, meter 2.0–4.0 quarts per acre of ANTOR 4ES into the irrigation water prior to weed emergence. Application timing may coincide with the last irrigation of the year. Use the higher rate for extended residual control, heavy weed pressure or on heavy soil types. Refer to the flow chart and sample calculation for additional information.

Application made after weed emergence may result in unacceptable weed control.

APPLICATION METHOD

FLOOD (BASIN), FURROW AND BORDER CHEMIGATION: IMPORTANT: Read all instructions under "Application Through Irrigation Systems (Chemigation)" given in the GENERAL INFORMATION Section of this label.

Systems using a gravity flow pesticide dispensing system must meter the pesticide into the water at the head of the field and downstream of a hydraulic discontinuity such as a drop structure or weir box to decrease potential for water source contamination from backflow if water flow stops.

Systems utilizing a pressurized water and pesticide injection system must meet the following requirements:

a. The system must contain a functional check valve, vacuum relief valve, and low pressure drain appropriately located on the irrigation pipeline to prevent water source contamination from backflow.
b. The pesticide injection pipeline must contain a functional, automatic, quick-closing check valve to prevent the flow of fluid back toward the injection pump.
c. The pesticide injection pipeline must also contain a functional, normally closed, solenoid-operated valve located on the intake side of the injection pump and connected to the system interlock to prevent fluid from being withdrawn from the supply tank when the irrigation system is either automatically or manually shut down.
d. The system must contain functional interlocking controls to automatically shut off the pesticide injection pump when the water pump motor stops.
e. The irrigation line of water pump must include a functional pressure switch which will stop the water pump motor when the water pressure decreases to the point where pesticide distribution is adversely affected.
f. Systems must use a metering pump, such as a postive displacement injection pump (e.g., diaphragm pump) effectively designed and constructed of materials that are compatible with pesticides and capable of being fitted with a system interlock.

Type of Device:

The constant flow device is recommended for metering ANTOR 4ES into the irrigation water for water-run applications of border type flood irrigation. This device is not recommended for other types of irrigation systems.

The necessary hardware is a length of pipe to reach from the chemical drum to just above the water level, a petcock valve, screen to prevent particles from clogging the flow orifice and a coupling to hold the orifice disc in place. Once the flow has begun by opening the petcock, open the venting petcock above the T coupler. This forms a constant head siphon, the flow rate governed by the orifice aperture.

Set Up and Calibration:

Set the ANTOR 4ES drum on planks over the ditch with the constant flow device pointed downward. Place the drum behind a gate or near some structure where there is turbulence for mixing of the herbicide. Start the herbicide flow to commence with the start of the irrigation cycle calibrated to finish with the end of the irrigation period. To calibrate, estimate the length of time of the irrigation and the number of acres to be treated. Calculate the flow rate for the amount of ANTOR 4ES to be applied and check the flow chart for the appropriate size orifice. (Refer to the flow rate chart and sample calculation.) Select the appropriate orifice and measure the flow of herbicide for 1 minute in the measuring cup. Compare this to your calculations and adjust the orifice accordingly so that the correct amount is metered during the irrigation cycle. Tail water (run-off) from flood irrigation should be recirculated or controlled so that it does not leave the field.

Flow Rates for ANTOR® 4ES Using Various Tee Jet Orifices*

Tee Jet Orifice	CC Per Minute	Ounces Per Minute	Quarts Per Hour
.015	10	.34	.64
.016	11	.37	.69
.018	13	.44	.83
.020	17	.57	1.07
.022	22	.74	1.39
.024	24	.81	1.52
.026	30	1.01	1.89
.028	37	1.25	2.34
.030	42	1.42	2.66
.032	47	1.59	2.98
.034	54	1.83	3.43
.035	59	2.00	3.75
.037	61	2.06	3.86
.040	73	2.47	4.63
.041	78	2.64	4.95
.043	84	2.84	5.33
.045	95	3.21	6.02
.046	98	3.31	6.21
.048	103	3.48	6.53
.049	113	3.82	7.16
.051	122	4.13	7.74
.052	126	4.26	7.99
.054	136	4.60	8.63
.055	144	4.89	9.17
.057	153	5.17	9.69
.059	168	5.68	10.65
.061	173	5.85	10.97
.063	184	6.22	11.66
.065	190	6.42	12.04
.067	207	7.00	13.13
.070	238	8.05	15.09
.075	270	9.13	17.12
.080	296	10.01	18.77
.086	339	11.46	21.49
.093	384	12.99	24.36
.103	470	15.89	29.79

* Figures were taken at 60°F and are approximate. Occasionally check to measure flow in the field to make sure you have the correct orifice and because flow rates vary with temperatures.

Sample Calculation

The following sample calculation is appropriate for ANTOR® 4ES applied as water-run with a constant flow device in bermudagrass grown for seed.

1. Size of block to be watered in one irrigation setting.
   eg. **10 acres**
2. ANTOR 4ES rate per acre.
   eq. **2 quarts**
3. Total amount of ANTOR 4ES to be applied to 10 acre block.
   eg. 2 qts/ANTOR 4ES × 10 acres = **20 quarts**
4. Amount of time to irrigate the 10 acre block.
   eg. **3 hours**
5. Amount of ANTOR 4ES applied per hour.
   eg. 20 qts. ÷ 3 hours = **6.67 quarts/hr.**
6. Ounces of ANTOR 4ES applied per hour.
   eg. 6.67 qts/hr × 32 oz/qt = **213.4 oz/hr.**
7. Ounces of ANTOR 4ES applied per minute.
   eg. 213.4 oz/hr ÷ 60 min/hr = **3.56 oz./min.**
8. Cubic centimeters (CC) of ANTOR 4ES per minute.
   eg. 3.56 oz/min × 29.6 ml/oz = **105.37 ml/min.**
9. Check flow chart for flow rate.
   eg. 105.37 ml/min. ≃ **orifice No. 0.48**
10. Always calibrate the orifice before beginning because of differences in orifice flow rate.

IMPORTANT: READ BEFORE USE

By using this product, user accepts the following conditions, warranty, disclaimer of warranties and limitations of liability.

CONDITIONS: The directions for use of this product are believed to be adequate and should be followed carefully. However, because of extreme weather and soil conditions, manner of use and other factors beyond NOR-AM Chemical Company's control, it is impossible for NOR-AM to eliminate all risks associated with the use of this product. As a result, crop injury or ineffectiveness is always possible.

DISCLAIMER OF WARRANTIES: THERE ARE NO WARRANTIES, EXPRESS OR IMPLIED, OF MERCHANTABILITY OR OF FITNESS FOR A PARTICULAR PURPOSE OR OTHERWISE, WHICH EXTEND BEYOND THE STATEMENTS MADE ON THIS LABEL. No agent of NOR-AM Chemical Company is authorized to make any warranties beyond those contained herein or to modify the warranties contained herein. UNDER NO CIRCUMSTANCES SHALL NOR-AM CHEMICAL COMPANY BE LIABLE FOR ANY INCIDENTAL OR CONSEQUENTIAL DAMAGES RESULTING FROM FAILURE TO USE THIS PRODUCT IN ACCORDANCE WITH LABEL DIRECTIONS.

LIMITATIONS OF LIABILITY: In no event shall NOR-AM Chemical Company's liability for damages arising out of the use of this product exceed the purchase price of the product used.

518-1B-AES-15M-Rev 9/89

This is a specimen label. NOR-AM Chemical Company is not responsible for the accuracy of the information contained herein. As labels are subject to revision, always carefully read and follow the label on the product container.

APOLLO®

OVICIDE MITICIDE

A Suspension Concentrate (SC) Formulation for Mite Control on Pears

ACTIVE INGREDIENT:	Percent by Weight
3,6-bis (2-chlorophenyl)-1,2,4,5-tetrazine	42%
INERT INGREDIENTS:	58%
TOTAL	100%

Contains 1 pound of active ingredient per quart.

EPA Registration No. 45639-135

KEEP OUT OF REACH OF CHILDREN

CAUTION

STATEMENT OF PRACTICAL TREATMENT

IF IN EYES: Flush eyes with plenty of water. Call a physician if irritation persists.

IF ON SKIN: Wash with plenty of soap and water. Get medical attention if irritation persists.

PRECAUTIONARY STATEMENTS

HAZARDS TO HUMANS AND DOMESTIC ANIMALS

CAUTION

Avoid contact with skin, eyes or clothing. In case of skin contact, carefully remove contaminated clothing and wash affected area with soap and water. In case of eye contact, flush eyes with plenty of water. Get medical attention if irritation persists.

Wash hands and exposed skin before eating, drinking or smoking and after spraying or handling. Wash all contaminated clothing thoroughly before reuse.

ENVIRONMENTAL HAZARDS

Do not apply directly to water. Do not apply when weather conditions favor drift from treated area. Do not contaminate water when disposing of equipment wash waters.

IN CASE OF FIRE, LEAKY OR DAMAGED CONTAINERS, OR OTHER EMERGENCY, REPORT AT ONCE BY TOLL-FREE TELEPHONE TO: 800-424-9300.

IMPORTANT—Read the entire Directions for Use and Disclaimer of Warranties before using this product.

STORAGE AND DISPOSAL

Prohibitions: Do not contaminate water, food or feed by storage or disposal.

Storage: Store in original container, tightly closed, in a safe place.

Pesticide Disposal: Wastes resulting from the use of this product may be disposed of on site or at an approved waste disposal facility.

Container Disposal: Triple rinse (or equivalent). Then offer for recycling or reconditioning, or puncture and dispose of in a sanitary landfill, or incineration, or if allowed by state and local authorities, by burning. If burned, stay out of smoke.

DIRECTIONS FOR USE

It is a violation of federal law to use this product in a manner inconsistent with its labeling.

GENERAL INFORMATION

APOLLO® SC is a specific miticide which acts as an ovicide and on the young motile stages. APOLLO SC is relatively non-toxic to beneficial insects and mites. It has good residual activity on European red mites, two-spotted spider mites, McDaniel mites and yellow spider mites when applied according to label directions.

TARGET SPECIES

European red mite	*Panonychus ulmi*
Two-spotted spider mite	*Tetranychus urticae*
McDaniel mite	*Tetranychus mcdanieli*
Yellow spider mite	*Eotetranychus carpini boreallis*

USE ON PEARS

APOLLO SC may be applied by ground using either dilute or concentrate spray equipment. Do not apply by air.

Concentrate Application (Less than 400 gallons per acre): Apply 4–8 ounces of APOLLO SC per acre.

Dilute Application (400 gallons per acre or more): Use 1–2 ounces of APOLLO SC per 100 gallons of spray. Do not apply less than 4 ounces of APOLLO SC per acre.

Recommendations for Use

Because APOLLO® SC acts primarily against the egg and larval stages, apply APOLLO SC at the first sign of mite activity. If the mite population is approaching damaging levels or if pear rust mite is present, combine APOLLO SC with an appropriate adulticide such as CARZOL® SP. When tank mixing APOLLO SC with an adulticide, use label rates for both products.

IMPORTANT—Do not apply within 21 days of harvest. Do not graze treated areas. Do not apply this product through any type of irrigation system.

DO NOT APPLY MORE THAN ONE APPLICATION OF APOLLO SC PER SEASON.

IMPORTANT: READ BEFORE USE

By using this product, user accepts the following conditions, warranty, disclaimer of warranties and limitations of liability.

CONDITIONS: The directions for use of this product are believed to be adequate and should be followed carefully. However, because of extreme weather and soil conditions, manner of use and other factors beyond NOR-AM Chemical Company's control, it is impossible for NOR-AM to eliminate all risks associated with the use of this product. As a result, crop injury or ineffectiveness is always possible.

DISCLAIMER OF WARRANTIES: THERE ARE NO WARRANTIES, EXPRESS OR IMPLIED, OF MECHANTABILITY OR OF FITNESS FOR A PARTICULAR PURPOSE OR OTHERWISE, WHICH EXTEND BEYOND THE STATEMENTS MADE ON THIS LABEL. No agent of NOR-AM Chemical Company is authorized to make any warranties beyond those contained herein or to modify the warranties contained herein. UNDER NO CIRCUMSTANCES SHALL NOR-AM CHEMICAL COMPANY BE LIABLE FOR ANY INCIDENTAL OR CONSEQUENTIAL DAMAGES RESULTING FROM FAILURE TO USE THIS PRODUCT IN ACCORDANCE WITH LABEL DIRECTIONS.

LIMITATIONS OF LIABILITY: In no event shall NOR-AM Chemical Company's liability for damages arising out of the use of this product exceed the purchase price of the product used.

530-A-1B-5M-3/89

This is a specimen label. NOR-AM Chemical Company is not responsible for the accuracy of the information contained herein. As labels are subject to revision, always carefully read and follow the label on the product container.

BETAMIX®

SUGAR BEET HERBICIDE

Postemergence Herbicide for Control of Weeds in Sugar Beets

FOR AGRICULTURAL USE ONLY

ACTIVE INGREDIENTS:	
Phenmedipham*	8.0%
Desmedipham**	8.0%
INERT INGREDIENTS:	84.0%
TOTAL	100.0%

Contains 1.3 lbs. active ingredient per gallon.

This product contains the toxic inert ingredient isophorone.

*3-methoxycarbonylaminophenyl-3-methylcarbanilate

**Ethyl *m*-hydroxycarbanilate carbanilate (ester)

EPA Reg. No. 45639-87 U.S Patent No. 3,404,975

U.S. Patent No. 3,692,820

PRECAUCION AL USUARIO: Si usted no lee ingles, no use este producto hasta que la etiqueta haya sido explicada ampliamente.

KEEP OUT OF REACH OF CHILDREN

WARNING —AVISO

STATEMENT OF PRACTICAL TREATMENT

IF SWALLOWED: CALL A PHYSICIAN OR POISON CONTROL CENTER. DO NOT INDUCE VOMITING.

NOTE TO PHYSICIAN: EMPTY STOMACH CONTENTS BY GASTRIC LAVAGE. AVOID ASPIRATION.

IF IN EYES: FLUSH WITH PLENTY OF WATER. GET MEDICAL ATTENTION.

IF ON SKIN: WASH IMMEDIATELY WITH PLENTY OF SOAP AND WATER.

DO NOT REUSE THIS CONTAINER. DESTROY WHEN EMPTY.

PRECAUTIONARY STATEMENTS

HAZARDS TO HUMANS

WARNING

HARMFUL IF SWALLOWED. MAY PRODUCE SEVERE IRRITATION OF EYES AND IRRITATION OF SKIN. AVOID BREATHING SPRAY MIST. AVOID CONTACT WITH SKIN AND EYES.

Avoid contamination of feed and foodstuffs.

ENVIRONMENTAL HAZARDS

This pesticide is toxic to fish and aquatic organisms. Do not apply directly to water or wetlands (swamps, bogs, marshes, and potholes). Drift and runoff from treated areas may be hazardous to fish and aquatic organisms in adjacent aquatic sites. Do not contaminate water by cleaning of equipment or disposal of wastes.

PHYSICAL OR CHEMICAL HAZARDS

DO NOT USE OR STORE NEAR HEAT OR OPEN FLAME.

STORAGE AND DISPOSAL

STORAGE: Store in original container and keep closed. Store in a cool, dry place. Do not use or store near heat or open flame.

PESTICIDE DISPOSAL: Do not contaminate water, food or feed by storage or disposal. Pesticide wastes are toxic. Improper disposal of excess pesticide, spray mixture, or rinsate is a violation of Federal Law. If these wastes cannot be disposed of by use according to label instructions, contact your State Pesticide or Environmental Control Agency, or the Hazardous Waste representative at the nearest EPA Regional Office for guidance.

CONTAINER DISPOSAL: Triple rinse (or equivalent). Then offer for recycling or reconditioning, or puncture and dispose of in a sanitary landfill, or by other procedures approved by state and local authorities.

IN CASE OF FIRE, LEAKY OR DAMAGED CONTAINERS, OR OTHER EMERGENCY, REPORT AT ONCE BY TOLL-FREE TELEPHONE TO: 800-424-9300.

GENERAL INFORMATION

BETAMIX® is a selective postemergence herbicide for use in sugar beets (including beets grown for seeds): Annual sowthistle (*Sonchus oleraceus*), Black nightshade (*Solanum nigrum*), Coast fiddleneck (*Amsinckia intermedia*), Common chickweed (*Stellaria media*), Common lambsquarters (*Chenopodium album*), Common ragweed (*Ambrosia artemisiifolia*), Green foxtail (*Setaria viridis*), Groundcherry (*Physalis lanceifolia*), Kochia* (*Kochia scoparia*), London rocket (*Sisymbrium irio*), Nettleleaf goosefoot (*Chenopodium murale*), Pigeon grass (yellow foxtail) (*Setaria glauca*), Prostrate pigweed (*Amaranthus gracizans*), Purslane (*Portulaca oleraceus*), Redroot pigweed (*Amaranthus retroflexus*), Shepherdspurse (*Capsella bursa-pastoris*), Wild buckwheat (*Polygonum convolvulus*), Wild mustard (*Brassica kaber*).

* Spray Kochia while in the rosette stage, less than 1" in diameter. For best results, spray when the weeds are at the two-leaf stage.

USE PRECAUTIONS: DO NOT APPLY BETAMIX TO SUGAR BEETS LATER THAN 90 DAYS PRIOR TO HARVEST.

DO NOT EXCEED A TOTAL OF 12 PINTS BETAMIX PER ACRE PER SEASON.

DO NOT APPLY THIS PRODUCT THROUGH ANY TYPE OF IRRIGATION SYSTEM.

BETAMIX MAY CAUSE BEET INJURY IF THE CROP IS UNDER STRESS FROM ONE OR MORE OF THE FOLLOWING CONDITIONS:

- Rapid climatic changes from cool, overcast days, to hot (85° or over), bright days.
- Windy conditions or drought.
- Use of a preplant or postemergence herbicide or other chemicals.
- Insect or disease injury.
- Close cultivation.

If stress conditions are present, delay application in order to give plants a chance to recover.

If extreme weather conditions are of short duration, delay spraying until the end of such a period. If BETAMIX applications must be made on days with extreme temperature and/or brightness, delay spraying until evening.

DO NOT OVERTREAT: The use of higher than recommended rates may cause beet injury.

Do not spray while dew is present.

Rainfall within 6 hours of spraying may reduce weed kill. Do not allow spray drift to contact adjacent crops which may be injured by spray drift.

IMPORTANT: BETAMIX may cause temporary growth retardation and/or chlorosis or tipburn on sugar beets. Beets usually resume normal growth within 10 days.

DIRECTIONS FOR USE

It is a violation of Federal Law to use this product in a manner inconsistent with its labeling. Read the entire Directions for Use before using this product.

REENTRY STATEMENT

Do not apply this product in such a manner as to directly or through drift expose workers or other persons. The area being treated must be vacated by unprotected persons.

Reentry into treated areas is prohibited until leaves are completely dry.

Do not enter treated areas without protective clothing.

Because certain states may require more restrictive reentry intervals for various crops treated with this product, consult your State Department of Agriculture for further information.

Written or oral warnings must be given to workers who are expected to be in a treated area or in an area about to be treated with this product. (Indicate specific oral warnings which inform workers of areas or fields that may not be entered without specific protective clothing, period of time field must be vacated and appropriate actions to take in case of accidental exposure.) When oral warnings are given, warnings shall be given in a language customarily understood by workers. Oral warnings must be given if there is reason to believe that written warnings cannot be understood by workers. Written warnings must include the following information:

WARNING: Area treated with BETAMIX® on (date of application). Do not enter without appropriate protective clothing until leaves are completely dry. In case of accidental exposure see Statement of Practical Treatment on the label.

When used as directed, BETAMIX is selective in beets past the two-true leaf stage. Beets may be severely injured if treated before the two-true leaf stage. The stage of growth of the weeds is very important for satisfactory control. For best results, spray when weeds are at the two-true leaf stage. Best results are obtained when the weeds are actively growing and are not under water or heat stress.

Mixing the Spray: Make sure the Sprayer is CLEAN.

BETAMIX emulsifiable concentrate formulation contains sufficient wetting agents for optimum coverage. Do not add additional wetting agents or other spray adjuvants. Add sufficient water to fill the lines. Then add the desired amount of BETAMIX and the remaining quantity of water with the bypass agitator running. Bypass agitation is sufficient. Mechanical agitation is not necessary. Only use freshly prepared spray emulsions.

Always spray immediately after preparing the spray solution. Prepare only enough spray solution to last less than four hours.

RATE OF APPLICATION: By Ground: Apply BETAMIX at the rate of 4½ to 7½ pints in 20 to 50 gallons of water broadcast basis. For band application, see dosage chart.

By Air: Apply BETAMIX at the rate of 4½ to 7½ pints per acre using 5 to 15 gallons of spray per acre. Use the 7½-pint rate only on well-established beets which are not under stress.

In order to avoid phytotoxic spray drift to non-target crops during application of BETAMIX, the following buffer zones should be observed:

Cotton, Potatoes, Sunflowers, Sorghum, Wheat	50 feet
Blackeye beans, Cabbage, Flax	100 feet
Lettuce, Rape, Tomatoes	300 feet

DO NOT APPLY WHEN WIND SPEED IS OVER 10 MILES PER HOUR. AVOID APPLICATIONS WHEN CONDITIONS FAVOR DRIFT.

To avoid excessive phytotoxicity to fall-planted sugar beets in Arizona and south of the Tehachapi Mountains in California when temperatures are above 85°F, apply BETAMIX at the rate of 4 pints per acre (broadcast equivalent). Evening applications are recommended.

For further information, contact your County Agricultural Agent, Farm Advisor or NOR-AM Chemical Company.

REPEAT APPLICATION OF BETAMIX: For control of later germinating weeds, make a second application of BETAMIX. Use 4½ to 6 pints of BETAMIX. Allow at least 7 days between first and second applications. Apply when beets have at least 4 leaves. For best results, use the higher rate and spray when weeds are at the two-true leaf stage. Apply lower rates when the beets are under stress as explained on the left panel.

SPLIT (LOW RATE) APPLICATIONS: (ALL SUGAR BEET AREAS EXCEPT CALIFORNIA): Split (low rate) applications of BETAMIX at 2 to 3 pints per acre broadcast equivalent in 5 to 10 gallons of water may be applied to sugar beets at any stage of growth when necessary to control early germinating, difficult to control weeds. The spray must be applied when the earliest emerging weeds have reached cotyledon size.

Any weeds which are not completely controlled by the first treatment will usually be checked and controlled by repeat applications. The following application should be made 5 to 7 days after the preceding application or when another flush of weeds germinates. If the second application is delayed, conventional treatment as described above will be necessary.

Apply the 4½ to 7½ pint rates only to sugar beets past the two-true leaf stage.

SPLIT APPLICATIONS: CALIFORNIA:

Split applications of BETAMIX at 2½ to 6 pints per acre broadcast equivalent may be made when sugar beets have at least two true leaves. Use the lower rate when temperatures are above 85°F or the crop is under stress. Allow 4 to 7 days between applications.

DOSAGE CHART FOR BAND APPLICATION

Broadcast Equivalent	Band Width	Band Rate — Row Spacing 22"	28"	32"
2 pints/acre	5"	7.3 fl. oz.	5.7 fl. oz.	5.3 fl. oz.
	7"	10.1 fl. oz.	8.0 fl. oz.	7.5 fl. oz.
3 pints/acre	5"	10.9 fl. oz.	8.5 fl. oz.	8.0 fl. oz.
	7"	15.2 fl. oz.	12.0 fl. oz.	11.2 fl. oz.
4½ pints/acre	5"	16.4 fl. oz.	12.9 fl. oz.	12.1 fl. oz.
	7"	22.9 fl. oz.	18.0 fl. oz.	16.9 fl. oz.
6 pints/acre	5"	21.8 fl. oz.	17.1 fl. oz.	16.0 fl. oz.
	7"	30.5 fl. oz.	24.0 fl. oz.	22.5 fl. oz.
7½ pints/acre	5"	27.3 fl. oz.	21.4 fl. oz.	20.1 fl. oz.
	7"	38.2 fl. oz.	30.0 fl. oz.	28.1 fl. oz.

IMPORTANT: READ BEFORE USE

By using this product, user accepts the following conditions, warranty, disclaimer of warranties and limitations of liability.

CONDITIONS: The directions for use of this product are believed to be adequate and should be followed carefully. However, because of extreme weather and soil conditions, manner of use and other factors beyond NOR-AM Chemical Company's control, it is impossible for NOR-AM to eliminate all risks associated with the use of this product. As a result, crop injury or ineffectiveness is always possible.

DISCLAIMER OF WARRANTIES: THERE ARE NO WARRANTIES, EXPRESS OR IMPLIED, OF MERCHANTABILITY OR OF FITNESS FOR A PARTICULAR PURPOSE OR OTHERWISE, WHICH EXTEND BEYOND THE STATEMENTS MADE ON THIS LABEL. No agent of NOR-AM Chemical Company is authorized to make any warranties beyond those contained herein or to modify the warranties contained herein. UNDER NO CIRCUMSTANCES SHALL NOR-AM CHEMICAL COMPANY BE LIABLE FOR ANY INCIDENTAL OR CONSEQUENTIAL DAMAGES RESULTING FROM FAILURE TO USE THIS PRODUCT IN ACCORDANCE WITH LABEL DIRECTIONS.

LIMITATIONS OF LIABILITY: In no event shall NOR-AM Chemical Company's liability for damages arising out of the use of this product exceed the purchase price of the product used.

518-IB-BTM-7.5M-Rev. 5/89

This is a specimen label. NOR-AM Chemical Company is not responsible for the accuracy of the information contained herein. As labels are subject to revision, always carefully read and follow the label on the product container.

BETANEX®

SUGAR BEET HERBICIDE

Postemergence Herbicide for Control of Redroot Pigweed and Other Weeds in Sugar Beets

FOR AGRICULTURAL USE ONLY

ACTIVE INGREDIENT:
Desmedipham* 16.0%
INERT INGREDIENTS: 84.0%
TOTAL 100.0%

Contains 1.3 lbs. active ingredient per gallon.
*Ethyl m-hydroxycarbanilate carbanilate (ester)
This product contains the toxic inert ingredient isophorone.

EPA REG. NO. 45639-86 — U.S. Patent No. 3,404,975; U.S. Patent No. 3,898,075

PRECAUCION AL USUARIO: Si usted no lee ingles, no use este producto hasta que la etiqueta haya sido explicada ampliamente.

KEEP OUT OF REACH OF CHILDREN

WARNING —AVISO

STATEMENT OF PRACTICAL TREATMENT

IF SWALLOWED: CALL A PHYSICIAN OR POISON CONTROL CENTER. DO NOT INDUCE VOMITING.

NOTE TO PHYSICIAN: EMPTY STOMACH CONTENTS BY GASTRIC LAVAGE. AVOID ASPIRATION.

IF IN EYES: FLUSH WITH PLENTY OF WATER. GET MEDICAL ATTENTION.

IF ON SKIN: WASH IMMEDIATELY WITH PLENTY OF SOAP AND WATER.

DO NOT REUSE THIS CONTAINER. DESTROY WHEN EMPTY.

PRECAUTIONARY STATEMENTS

WARNING

HAZARDS TO HUMANS

HARMFUL IF SWALLOWED. MAY PRODUCE SEVERE IRRITATION OF EYES AND IRRITATION OF SKIN. AVOID BREATHING SPRAY MIST. AVOID CONTACT WITH SKIN AND EYES.

AVOID CONTAMINATION OF FEED AND FOODSTUFFS.

ENVIRONMENTAL HAZARDS—THIS PRODUCT IS TOXIC TO FISH. KEEP OUT OF LAKES, STREAMS, OR PONDS. DO NOT APPLY WHERE RUNOFF IS LIKELY TO OCCUR. DO NOT APPLY WHEN WEATHER CONDITIONS FAVOR DRIFT FROM AREAS TREATED. DO NOT CONTAMINATE WATER WHEN DISPOSING OF EQUIPMENT WASHWATERS.

PHYSICAL OR CHEMICAL HAZARDS—DO NOT USE OR STORE NEAR HEAT OR OPEN FLAME.

USE PRECAUTIONS

DO NOT APPLY BETANEX TO SUGAR BEETS LATER THAN 90 DAYS PRIOR TO HARVEST.

DO NOT EXCEED A TOTAL OF 12 PINTS BETANEX PER ACRE PER SEASON.

DO NOT APPLY THIS PRODUCT THROUGH ANY TYPE OF IRRIGATION SYSTEM.

BETANEX MAY CAUSE BEET INJURY IF THE CROP IS UNDER STRESS FROM ONE OR MORE OF THE FOLLOWING CONDITIONS:

Rapid climatic changes from cool, overcast days, to hot (85° or over), bright days.

Windy conditions or drought.

Use of a preplant or postemergence herbicide or other chemicals.

Insect or disease injury.

Close cultivation.

If stress conditions are present, delay application in order to give plants a chance to recover.

If extreme weather conditions are of short duration, delay spraying until the end of such a period.

DO NOT OVERTREAT: The use of higher than recommended rates may cause beet injury.

Do not spray while dew is present.

Rainfall within 6 hours of spraying may reduce weed kill. Do not allow spray drift to contact adjacent crops which may be injured by spray drift.

Important: BETANEX may cause temporary growth retardation and/or chlorosis or tip-burn on sugar beets. Beets usually resume normal growth within 10 days.

STORAGE AND DISPOSAL

Storage: Store in original container and keep closed. Store in a cool, dry place. Do not use or store near heat or open flame.

Pesticide Disposal: Do not contaminate water, food or feed by storage or disposal. Pesticide wastes are toxic. Improper disposal of excess pesticide, spray mixture, or rinsate is a violation of Federal Law. If these wastes cannot be disposed of by use according to label instructions, contact your State Pesticide or Environmental Control Agency, or the Hazardous Waste representative at the nearest EPA Regional Office for guidance.

Container Disposal: Triple rinse (or equivalent). Then offer for recycling or reconditioning, or puncture and dispose of in a sanitary landfill, or by other procedures approved by state and local authorities.

IN CASE OF FIRE, LEAKY OR DAMAGED CONTAINERS, OR OTHER EMERGENCY, REPORT AT ONCE BY TOLL-FREE TELEPHONE TO: 800-424-9300.

GENERAL INFORMATION

BETANEX® is a selective postemergence herbicide for use in sugar beets for control of the following weeds: Annual sowthistle (*Sonchus oleraceus*), Black nightshade (*Solanum nigrum*), Coast fiddleneck (*Amsinckia intermedia*), Common chickweed (*Stellaria media*), Common lambsquarters (*Chenopodium album*), Common ragweed (*Ambrosia artemisliifolia*), Groundcherry (*Physalis lanceifolia*), London rocket (*Sisymbrium irio*), Nettleleaf goosefoot (*Chenopodium murale*), Prostrate pigweed (*Amaranthus gracizans*), Purslane (*Portulaca oleracas*), Redroot pigweed (*Amaranthus retroflexus*), Shepherdspurse (*Capsella bursa-pastoris*), Wild buckwheat (*Polygonum convolvulus*), Wild mustard (*Brassica kaber*).

DIRECTIONS FOR USE

It is a violation of Federal Law to use this product in a manner inconsistent with its labeling. Read the entire Directions for Use before using this product.

REENTRY STATEMENT

Do not apply this product in such a manner as to directly or through drift expose workers or other persons. The area being treated must be vacated by unprotected persons.

Reentry into treated areas is prohibited until leaves are completely dry.

Do not enter treated areas without protective clothing.

Because certain states may require restrictive reentry intervals for various crops treated with this product, consult your State Department of Agriculture for further information.

Written or oral warnings must be given to workers who are expected to be in a treated area or in an area about to be treated with this product. (Indicate specific oral warnings which inform workers of areas or fields that may not be entered without specific protective clothing, period of time field must be vacated and appropriate actions to take in case of accidental exposure.) When oral warnings are given, warnings shall be given in a language customarily understood by workers. Oral warnings must be given if there is reason to believe that written warnings cannot be understood by workers. Written warnings must include the following information:

WARNING. Area treated with BETANEX® on (date of application). Do not enter without appropriate protective clothing until leaves are completely dry. In case of accidental exposure see Statement of Practical Treatment on the label.

When used as directed, BETANEX is selective in sugar beets past the two-true leaf stage. Beets may be severely injured if treated before the two-true leaf stage. For best results, spray when the weeds are at the two-true leaf stage. Best results are obtained when the weeds are actively growing and are not under water or heat stress.

Mixing the Spray: Make sure the Sprayer is CLEAN.

BETANEX emulsifiable concentrate formulation contains sufficient wetting agents for optimum coverage. Do not add additional wetting agents or other spray adjuvants. Add sufficient water to fill the lines. Then add the desired amount of BETANEX and the remaining quantity of water with the bypass agitator running. Bypass agitation is sufficient. Mechanical agitation is not necessary. Only use freshly prepared spray emulsions.

RATE OF APPLICATION

By Ground: Apply BETANEX at the rate of 4½ to 7½ pints in 20 to 50 gallons of water broadcast basis. For band application, see dosage chart.

By Air: Apply BETANEX at the rate of 4½ to 7½ pints per acre using 5 to 15 gallons of spray per acre.

Use the 7½-pint rate only on well-established beets which are not under stress.

In order to avoid phytotoxic spray drift to non-target crops during application of BETANEX the following buffer zones should be observed:

Cotton, Potatoes, Sunflowers, Sorghum, Wheat	50 feet
Blackeye Beans, Cabbage, Flax	100 feet
Lettuce, Rape, Tomatoes	300 feet

DO NOT APPLY WHEN WIND SPEED IS OVER 10 MILES PER HOUR. AVOID APPLICATIONS WHEN CONDITIONS FAVOR DRIFT.

To avoid excessive phytotoxicity to fall-planted sugar beets in Arizona and south of the Tehachapi Mountains in California when temperatures are above 85°F, apply BETANEX at the rate of 4 pints per acre (broadcast equivalent). Evening applications are recommended.

For further information, contact your County Agricultural Agent, Farm Advisor or NOR-AM Chemical Company.

Repeat Application of BETANEX®: For control of later germinating weeds, make a second application of BETANEX. Use 4½ to 6 pints of BETANEX. Allow at least 7 days between first and second applications. Apply when beets have at least 4 leaves. For best results, use the higher rate and spray when weeds are at the two-true leaf stage. Apply lower rates when the beets are under stress as explained on the left panel.

Split (Low Rate) Applications: (All Sugar Beet Areas Except California): Split (low rate) applications of BETANEX at 2 to 3 pints per acre broadcast equivalent in 5 to 10 gallons of water may be applied to sugar beets at any stage of growth when necessary to control early germinating, difficult-to-control weeds. The spray must be applied when the earliest emerging weeds have reached cotyledon size.

Any weeds which are not completely controlled by the first treatment will usually be checked and controlled by repeat applications. The following application should be made 5 to 7 days after the preceding application or when another flush of weeds germinates. If the second application is delayed, conventional treatment as described above will be necessary.

Apply the 4½ to 7½ pint rates only to sugar beets past the two-true leaf stage.

DOSAGE CHART FOR BAND APPLICATION

		Band Rate—Row Spacing		
Broadcast Equivalent	**Band Width**	**22″**	**28″**	**32″**
2 pints/acre	5″	7.3 fl. oz.	5.7 fl. oz.	5.3 fl. oz.
	7″	10.1 fl. oz.	8.0 fl. oz.	7.5 fl.oz.
3 pints/acre	5″	10.9 fl. oz.	8.5 fl. oz.	8.0 fl. oz.
	7″	15.2 fl. oz.	12.0 fl. oz.	11.2 fl. oz.
4½ pints/acre	5″	16.4 fl. oz.	12.9 fl. oz.	12.1 fl. oz.
	7″	22.9 fl. oz.	18.0 fl. oz.	16.9 fl. oz.
6 pints/acre	5″	21.8 fl. oz.	17.1 fl. oz.	16.0 fl. oz.
	7″	30.5 fl. oz.	24.0 fl. oz.	22.5 fl. oz.
7½ pints/acre	5″	27.3 fl. oz.	21.4 fl. oz.	20.1 fl. oz.
	7″	38.2 fl. oz.	30.0 fl. oz.	28.1 fl. oz.

IMPORTANT: READ BEFORE USE

By using this product, user accepts the following conditions, warranty, disclaimer of warranties and limitations of liability.

CONDITIONS: The directions for use of this product are believed to be adequate and should be followed carefully. However, because of extreme weather and soil conditions, manner of use and other factors beyond NOR-AM Chemical Company's control, it is impossible for NOR-AM to eliminate all risks associated with the use of this product. As a result, crop injury or ineffectiveness is always possible.

DISCLAIMER OF WARRANTIES: THERE ARE NO WARRANTIES, EXPRESS OR IMPLIED, OF MERCHANTABILITY OR OF FITNESS FOR A PARTICULAR PURPOSE OR OTHERWISE, WHICH EXTEND BEYOND THE STATEMENTS MADE ON THIS LABEL. No agent of NOR-AM Chemical Company is authorized to make any warranties beyond those contained herein or to modify the warranties contained herein. UNDER NO CIRCUMSTANCES SHALL NOR-AM CHEMICAL COMPANY BE LIABLE FOR ANY INCIDENTAL OR CONSEQUENTIAL DAMAGES RESULTING FROM FAILURE TO USE THIS PRODUCT IN ACCORDANCE WITH LABEL DIRECTIONS.

LIMITATIONS OF LIABILITY: In no event shall NOR-AM Chemical Company's liability for damages arising out of the use of this product exceed the purchase price of the product used.

518-IB-BTN-5M-Rev. 10/88

This is a specimen label. NOR-AM Chemical Company is not responsible for the accuracy of the information contained herein. As labels are subject to revision, always carefully read and follow the label on the product container.

BOTEC®

PEANUT SEED PROTECTANT

ACTIVE INGREDIENTS:

2,6-Dichloro-4-nitroaniline*	30.0% w/w
Captan (N-trichloromethylthio-4-cyclohexene-1,2-dicarboximide)	29.3% w/w
Captan Related Derivatives	0.7% w/w
INERT INGREDIENTS:	40.0% w/w
TOTAL	100.0% w/w

*Botran®

This product contains the toxic inert ingredient Rhodamine B.

EPA Reg. No. 45639-121

KEEP OUT OF REACH OF CHILDREN

DANGER

Causes irreversible eye damage. Harmful if swallowed or inhaled. May cause allergic skin reactions. Do not get in eyes. Wear goggles or face shield when handling. Avoid contact with skin and clothing. Remove and separately launder clothing before reuse.

STATEMENT OF PRACTICAL TREATMENT

IF IN EYES: FLUSH WITH PLENTY OF WATER. CALL A PHYSICIAN.

IF SWALLOWED: DRINK PROMPTLY A LARGE QUANTITY OF MILK, EGG WHITES, GELATIN SOLUTIONS, OR IF THESE ARE NOT AVAILABLE, DRINK LARGE QUANTITIES OF WATER. AVOID ALCOHOL.

IF INHALED: REMOVE VICTIM TO FRESH AIR. IF NOT BREATHING GIVE ARTIFICIAL RESPIRATION, PREFERABLY MOUTH-TO-MOUTH. GET MEDICAL ATTENTION.

PRECAUTIONARY STATEMENTS

DANGER

Causes irrreversible eye damage. Harmful if swallowed or inhaled. May cause allergic skin reactions. Do not get in eyes. Wear goggles or face shield when handling. Avoid contact with skin and clothing. Remove and separately launder clothing before reuse.

ENVIRONMENTAL HAZARDS

This pesticide is toxic to fish. Do not contaminate water when disposing of equipment washwaters. In the event that treated seeds are spilled and cannot be salvaged, completely cover them with soil.

GENERAL INFORMATION

BOTEC® Peanut Seed Protectant controls seed rots caused by Aspergillus and Rhizopus species.

DIRECTIONS FOR USE

It is a violation of Federal law to use this product in a manner inconsistent with its labeling.

BOTEC Peanut Seed Protectant is formulated for dry treatment of seed. Apply with any type of dust treater which will insure complete coverage of the seed. Use at the rate of 4 to 6 ounces per 100 pounds of seed. When used as directed, this product colors seeds red. Do Not use treated seed for food, feed or oil purposes. Bags with treated seed should bear a tag or label cautioning against their use for food, feed or oil purposes, as well as reuse for packing feed or food stuffs. Dispose of excess treated seed in accordance with approved Federal, State or local procedures under the Resource Conservation and Recovery Act.

USE PRECAUTIONS

Wear goggles or face shield when handling. Wash thoroughly with soap and water after handling. Remove contaminated clothing and separately launder clothing before reuse.

Mixers/loaders when mixing and loading this product must wear a long sleeved shirt and long pants or a coverall that covers all parts of the body except the head, hands and feet. Wear chemical resistant gloves, goggles or a face shield and a chemically resistant apron.

AFTER WORK: Take off all clothes and shoes. Shower using soap and water. Do not wear contaminated clothing. Wash protective clothing and equipment with soap and water after each use. Personal clothing worn during use must be laundered separately from household articles. Clothing and protective equipment heavily contaminated or drenched with captan must be destroyed according to state and local regulations.

A dust mask must be worn when mixing and loading this product.

STORAGE AND DISPOSAL

Do not reuse empty container.
Do not contaminate water, food or feed by storage or disposal.

STORAGE: Store in original container and keep closed. Store in a cool, dry place.

PESTICIDE DISPOSAL—Do not contaminate water, food or feed by storage or disposal. Pesticide wastes are acutely hazardous. Improper disposal of excess pesticide, spray mixture, or rinsate is a violation of Federal Law. If these wastes cannot be disposed of by use according to label instructions, contact your State Pesticide or Environmental Control Agency or the Hazardous Waste Representative at the nearest EPA Regional Office for guidance.

CONTAINER DISPOSAL: Completely empty liner by shaking and tapping sides and bottom to loosen clinging particles. Empty residue into application equipment. Then dispose of liner in a sanitary landfill or by incineration if allowed by State and local authorities. Dispose of drum in the same manner.

IN CASE OF FIRE, LEAKY OR DAMAGED CONTAINERS, OR OTHER EMERGENCY, REPORT AT ONCE BY TOLL-FREE TELEPHONE TO: 800-424-9300.

IMPORTANT: READ BEFORE USE

By using this product, user accepts the following conditions, warranty, disclaimer of warranties and limitations of liability.

CONDITIONS: The directions for use of this product are believed to be adequate and should be followed carefully. However, because of extreme weather and soil conditions, manner of use and other factors beyond NOR-AM Chemical Company's control, it is impossible for NOR-AM to eliminate all risks associated with the use of this product. As a result, crop injury or ineffectiveness is always possible.

DISCLAIMER OF WARRANTIES: THERE ARE NO WARRANTIES, EXPRESS OR IMPLIED, OF MERCHANTABILITY OR OF FITNESS FOR A PARTICULAR PURPOSE OR OTHERWISE, WHICH EXTEND BEYOND THE STATEMENTS MADE ON THIS LABEL. No agent of NOR-AM Chemical Company is authorized to make any warranties beyond those contained herein or to modify the warranties contained herein. UNDER NO CIRCUMSTANCES SHALL NOR-AM CHEMICAL COMPANY BE LIABLE FOR ANY INCIDENTAL OR CONSEQUENTIAL DAMAGES RESULTING FROM FAILURE TO USE THIS PRODUCT IN ACCORDANCE WITH LABEL DIRECTIONS.

LIMITATIONS OF LIABILITY: In no event shall NOR-AM Chemical Company's liability for damages arising out of the use of this product exceed the purchase price of the product used.

533-SL-BC-3M-Rev. 10/88

This is a specimen label. NOR-AM Chemical Company is not responsible for the accuracy of the information contained herein. As labels are subject to revision, always carefully read and follow the label on the product container.

BOTRAN® 75W

FUNGICIDE

For: Apricot, Carrot, Celery, Cucumber, Endive, Garlic, Grape, Lettuce, Nectarine, Onion, Peach, Plum and Prune, Potato, Rhubarb, Snap Bean, Sweet Cherry, Sweet Potato, Tomato and Ornamentals.

See folder in bottom flap of bag for usage directions.

ACTIVE INGREDIENT	
2,6-Dichloro-4-nitroaniline	75% w/w
INERT INGREDIENTS	25% w/w
TOTAL	100% w/w

EPA Reg. No. 45639-110

PRECAUCION AL USUARIO: Si usted no lee ingles, no use este producto hasta que la etiqueta haya sido explicada ampliamente.

KEEP OUT OF REACH OF CHILDREN

CAUTION —PRECAUCION

STATEMENT OF PRACTICAL TREATMENT

IF ON SKIN: REMOVE CONTAMINATED CLOTHING AND WASH WITH SOAP AND WATER.
IF IN EYES: FLUSH WITH WATER. IF EYE IRRITATION PERSISTS, CALL A PHYSICIAN.
IF SWALLOWED: DRINK WARM WATER AND INDUCE VOMITING BY TOUCHING BACK OF THROAT WITH FINGER AND/OR BY ADMINISTERING SYRUP OF IPECAC. FOLLOW WITH MILK AND A MILD CATHARTIC (MILK OF MAGNESIA). CALL A PHYSICIAN OR POISON CONTROL CENTER IMMEDIATELY. NEVER GIVE FLUIDS OR INDUCE VOMITING IF PATIENT IS UNCONSCIOUS OR HAVING CONVULSIONS.

PRECAUTIONARY STATEMENTS

HAZARDS TO HUMANS AND DOMESTIC ANIMALS

CAUTION

HARMFUL IF SWALLOWED. AVOID INHALATION OF DUST OR SPRAY MIST. AVOID CONTACT WITH SKIN AND EYES.

ENVIRONMENTAL HAZARDS

Toxic to fish. Do not apply directly to water or wetlands. Do not apply where runoff is likely to occur. Do not apply when weather conditions favor drift from areas treated. Do not contaminate water when disposing of equipment washwaters.

STORAGE AND DISPOSAL

Do not contaminate water, food or feed by storage or disposal.

STORAGE: Store in original container and keep closed. Store in a cool, dry place.

PESTICIDE DISPOSAL: Wastes resulting from the use of this product may be disposed of on site or at an approved waste disposal facility.

CONTAINER DISPOSAL: Completely empty bag into application equipment. Then dispose of empty bag in a sanitary landfill or by incineration, or, if allowed by State and local authorities, by burning. If burned, stay out of smoke.

DO NOT REUSE EMPTY CONTAINER. In case of fire, leaky or damaged containers, or other emergency, report at once by toll-free telephone to: (800) 424-9300.

DIRECTIONS FOR USE

It is a violation of Federal law to use this product in a manner inconsistent with its labeling.

Do not apply this product in such a manner as to directly or through drift expose workers or other persons. The area being treated must be vacated by unprotected persons.

Do not apply this product through any type of irrigation system except as described in the Special Local Need registrations use directions.

EQUIVALENTS: *One and one-third pounds as found in Directions for Use equals one 4 pound bag in 300 gallons.* One (1) level tablespoonful per gallon equals one (1) pound per 100 gallons.

COMPATIBILITY: BOTRAN 75W is compatible with wettable powder formulations of captan, zineb, Guthion, parathion, carbaryl, and endosulfan. Normally BOTRAN 75W can be mixed with other fungicides, insecticides, or defoliants formulated as wettable powders. It is suggested that such combinations be used on a small number of plants before treating large areas. However, combination sprays of BOTRAN 75W and miscible oil formulations of insecticides, particularly organic phosphorous compounds may cause plant injury.

REENTRY STATEMENT

Do not enter treated areas without protective clothing until sprays have dried. Because certain states may require more restrictive reentry intervals for various crops treated with this product, consult your State Department of Agriculture for further information.

Written or oral warnings must be given to workers who are expected to be in a treated area or in an area about to be treated with this product. When oral warnings are given, warnings shall be given in a language customarily understood by workers. Oral warnings must be given if there is reason to believe that written warnings cannot be understood by workers.

DIRECTIONS FOR USE ON EDIBLE CROPS

CROP	DISEASE	DOSAGE[1] LB PER 100 GAL. OF WATER	USE
Apricot	Fruit decay *Monilinia* Brown Rot *Rhizopus* rot	1 1/3	Allow 10 days and 1 day prior to harvest. Allow 1 day between application and harvest.
	Postharvest decay *Rhizopus stolonifer* *Monilinia* rot (Brown rot)	1	Apply as combination of fruit wax plus BOTRAN 75W plus Benlate 50W (0.5 lb. per 100 gal. of water) during packing line operations. Mix the combination in a tank equipped with mechanical agitation. Mechanically agitate combination during use. Apply the BOTRAN-Benlate-wax combination to the fruit at the rate of 35–45 gal/hr. Flat-fan or flood jet nozzles should be used. Collect spray runoff in a drain tray and discard. Care should be taken to prevent plugging of nozzles.
Note: Do not treat fruit packed in containers which do not allow adequate drainage. R. arrhizus control in apricots may only be in the range of 60% or less.			
Carrot (Storage only)	Post-harvest decay *Sclerotinia sclerotiorum* (dip)	1	Dip carrots for 10 seconds in well agitated suspension. Do not rinse after treatment. Add 1/2 lb. BOTRAN 75W to 100 gallons of dip suspension after 500 bushels are treated. After each 1000 bushels treated, drain and flush tank. Refill with fresh dip suspension.
Cucumber (greenhouse)	White mold *Sclerotinia sclerotiorum*	1 1/3	Apply to diseased areas of plants. Additional applications may be required after 14 days.
Grape	Bunch rot *Botrytis* rot Storage rot	1 1/3	BOTRAN 75W may be applied at onset of bloom to soon after shatter. Use 2 2/3 lb/acre of BOTRAN 75W in 50 to 400 gallons of water per acre for control of Botrytis bunch rot (*Botrytis cinerea*). Up to 5 1/3 pounds of BOTRAN 75W may be applied per season. Use BOTRAN dust after cluster tightening (see BOTRAN dust labels for complete directions for use). When combining BOTRAN 75W with gibberellin, use with last gibberellin application.
Note: Use limited to grapes grown west of the Rocky Mountains only.			
Peach and Nectarine	Blossom blight *Monilinia*	1 1/3	Apply at pink bud and full bloom.
	Fruit decay *Monilinia* Brown rot *Rhizopus* rot	1 1/3	Apply 18 days, 10 days and 1 day prior to harvest. Allow 1 day between application and harvest.
Note: When combining with captan, use 1 to 1 1/3 lb. BOTRAN 75W to 1–2 lbs captan 50W or equivalent per 100 gallons of water as recommended.			
Peach (only)	Post-harvest decay *Rhizopus* rot *Botrytis* rot	1	**Spray or dip**—Immediately after picking, spray or dip peaches with suspension of BOTRAN 75W. Do not recirculate used spray suspension. **Processing line operations**—BOTRAN 75W is most effective when applied during or after the defuzzing process, either by brushes, spray, or in the hydrocooler. **Brushes**—Apply BOTRAN 75W to the brush with traversing feeder nozzle capable of metering approximately 10 gallons of BOTRAN 75W suspension per hour with the valve completely open. Agitate the spray suspension. Unused suspension that drips from the brush should be caught in a drain tray and discarded. **Spray**—Direct a mist spray down upon the rotating peaches. Use pressures between 20-25 psi with delivery system of approximately 45 gallons per hour. TJ80 nozzles are suggested. Agitate the spray suspension. Collect spray runoff in a drain tray and discard. **Hydrocooler**—As volume decreases, make up to original volume by adding BOTRAN 75W suspension of 1 Lb per 100 gal. Drain and flush daily.
	Note: Do not treat peaches packed in containers which do not allow adequate drainage.		
	Post-harvest decay *Rhizopus* rot *Botrytis* rot *Monilinia* rot (*Brown rot*)	2 2/3	For freezing and canning only. Immediately after picking dip peaches in suspension of BOTRAN 75W. Continuous agitation is necessary. As volume decreases, make up to original volume by adding BOTRAN 75W suspension of 2 2/3 lb per 100 gal. of water. Drain and flush system daily.
WARNING: Peaches treated with this rate of BOTRAN will have residues exceeding the established tolerance of BOTRAN on peaches. Do not use this fruit for the fresh fruit market. Processing for freezing and canning removes all excess residue.			
Nectarine (only)	Post-harvest decay *Rhizopus* rot *Monilinia* rot (brown rot)	2 2/3	Apply only with fruit wax. Mix the BOTRAN with wax in a tank equipped with mechanical agitation. The BOTRAN-wax combination should be mechanically agitated during use: by-pass agitation will not keep the BOTRAN in suspension. Fruit should be sprayed during packing line operations. Apply the BOTRAN-wax combination to the fruit at the rate of 30-35 gal./hour at 18-30 psi. Flat-fan or flood jet nozzles should be used. Collect spray run-off in a drain tray and discard. Care should be taken to prevent plugging of nozzles.
NOTE: If wax containing OPP is used, the concentration of OPP in the final mixture should not exceed 0.25%.			
Plum and Prune	Blossom blight *Monilinia* Brown rot	1 1/3	Apply at popcorn and full bloom.
Plum	Post-harvest decay *Rhizopus* rot *Monilinia* rot (Brown rot)		Apply only with fruit wax. Mix BOTRAN 75W with wax in a tank equipped with mechanical agitation. The BOTRAN-wax combination should be mechanically agitated during use: by-pass agitation will not keep the BOTRAN in suspension. Use flat-fan or flood jet nozzles to uniformly spray fruit during packing line operations. Care should be taken to prevent plugging of nozzles. Collect spray run-off in a drain tray and discard.
		2 2/3	For use in conventional applicators (30-50 gal/hr), apply 100 gallons of the diluted BOTRAN-wax combination to approximately 50,000 pounds of fruit.
		10-12	For use in low volume applicators (5-8 gal/hr), apply 100 gallons of the BOTRAN-wax combination to approximately 500,000 pounds of fruit.
Rhubarb (hot house)	*Botrytis* leaf rot	1 1/3	Begin applications when first buds emerge from crowns. Repeat at weekly intervals. Do not use later than 3 days before harvest.

DIRECTIONS FOR USE ON EDIBLE CROPS

CROP	DISEASE	DOSAGE[1] LB PER 100 GAL. OF WATER	USE
Sweet Cherry	Blossom blight *Monilinia*	1 1/3	Apply at popcorn, bloom, full bloom and petal fall.
	Fruit decay *Monilinia* Brown rot *Rhizopus* rot	1 1/3	Apply 10 days and 1 day prior to harvest.
	Post-harvest decay *Rhizopus* rot	1 1/3	Spray cherries, while sorting, with a suspension of BOTRAN 75W. Do not recirculate used spray suspension.
Note: When combining with captan, use 1 to 1 1/3 lb BOTRAN 75W to 1–2 lbs captan 50W or equivalent per 100 gallons of water as recommended.			
Tomato	Stem canker *Botrytis* (greenhouse)	1	Spray stem of plant from ground level up to a height of 18 to 24 inches. Repeat weekly or as necessary. Seedlings or newly set transplants may be injured by drenching.
Sweet potato Post-harvest	Post-harvest decay *Rhizopus* rot (Spray)	1	Spray, using 8001 fan-shaped nozzle at 50 to 70 psi immediately after washing while the potatoes are conveyed on chain rollers or belts. Do not rinse after treatment. In recirculating systems, as volume decreases, make up to original volume with suspension equivalent to 1 lb BOTRAN 75W per 100 gal. of water. Where black rot is a problem, do not recirculate used spray suspension.
	(Dip)	1	Dip roots for 10 to 15 seconds in well agitated suspension of BOTRAN 75W. Do not rinse after treatment. Add 1/2 lb BOTRAN 75W to 100 gallons of treating suspension after 500 bushels are treated. After each 1000 bushels treated, drain and flush tank. Refill with fresh dip suspension.
Plantbed	Sclerotial blight *Sclerotium rolfsii*		*Seed Dip. Dip seed sweet potatoes 10 to 15 seconds in well agitated suspension containing 1 lb BOTRAN 75W in 7.5 gal of water. Drain sweet potatoes and bed promptly. Prepare fresh BOTRAN 75W suspension daily.
Seed Dip	Scurf *Monilochaetes infuscans*	*	**OR** **Bed Spray. Use 3 to 3 3/4 lb of BOTRAN 75W in 14 gal of water per 1,000 square feet of plant-bed. Spray or sprinkle BOTRAN 75W suspension over bedded sweet potatoes before covering them with soil.
Bed Spray		**	**Caution:** Do not plant tomatoes as follow-up crop.
Celery	Pink rot *Sclerotinia sclerotiorum*	2	Begin application when disease is anticipated or approximately 10 weeks before harvest. Repeat at 7-day intervals in summer or 14-day intervals in fall and winter. Do not apply within 7 days of harvest.
		5 1/3	**OR** If early disease is not likely, make a single application of 5 1/3 lbs of BOTRAN 75W per acre 4 to 8 weeks before harvest. Use sprayer with drop nozzle boom to direct spray (100 gal per acre) to base of plants and adjacent soil.
Potato	*Botrytis* blight White mold *Sclerotinia sclerotiorum*	2	Apply 100 gallons per acre beginning at layby. Continue at 10- to 14-day intervals. Do not apply within 14 days of harvest. Do not feed treated potatoes to livestock. LIMIT USE TO NORTH CENTRAL STATES ONLY.
Lettuce-Field (Leaf and Head)	Drop or Watery Soft Rot *Sclerotinia sclerotiorum* *S. minor*	2-3 1/3	Apply BOTRAN 75W at 2 lbs. per acre preemergence in 20-50 gallons of water in a 4-6 inch band over the transplanted or seeded row. Make a second application at 3 1/3 lbs. per acre immediately after thinning in 50-100 gallons of water. Do not use more than 5 1/3 lbs. BOTRAN 75W per acre per season. - OR-
		5 1/3	Apply immediately after thinning. Use 50-100 gallons of water per acre. Do not apply this product again after this single application.
	Botrytis gray mold rot *Botrytis cinerea*	2 2/3	Apply prior to thinning (2 to 6 leaf stage) and repeat immediately after thinning. Do not apply within 14 days of harvest. Use sprayer with drop nozzle to direct spray (50-100 gallons water per acre) to lower portion of plants and surrounding soil.
Lettuce-Greenhouse (Leaf only)	Botrytis gray mold rot *Botrytis cinerea*	2 2/3	Apply 100 gallons of spray mixture per acre 7 days following transplanting. Repeat application when plants are half mature. Do not apply to wilted plants, seedlings, or later than 14 days before harvest.
Note: Some leaf bronzing may be noted soon after application, but the effect is of a temporary nature.			
Endive (escarole) (field)	*Botrytis* gray mold rot *Botrytis cinerea*	2 2/3	Apply 100 gallons of spray mixture per acre 7 days following transplanting. Repeat application when plants are half mature. Do not apply to wilted plants, seedlings, or later than 14 days before harvest. **OR**
		2 2/3	Apply BOTRAN 75W spray at thinning and repeat application 7 days later. Do not apply within 14 days of harvest. Use sprayer with drop nozzle boom to direct spray (100 gal per acre) to lower portion of plants and surrounding soil.
Note: Some leaf bronzing may be noted soon after application, but the effect is of a temporary nature. Do not apply more than two applications of BOTRAN 75W per season.			
Onion and Garlic	White rot *Sclerotium*	32-40*** (broadcast, or 43,560 linear feet of furrow)	**Direct seeding—**mix (broadcast-incorporate) BOTRAN 75W in top 1 1/2 inches of soil 1-2 weeks before onions are seeded. Use restricted to Eastern Oregon, Eastern Washington and Western Idaho. **Planting sets or buds—In Furrow:** Use low pressure (30 to 40 psi) spray equipment. Use 50-mesh or coarser screens. Center the front nozzle on the furrow so that spray strikes the soil around the sets or buds and the rear nozzle sprays the covering soil as it fills the furrow. **Broadcast:** Use 100 gallons per acre for broadcast treatment. Do not exceed 40 lbs per acre broadcast.
Note: Plow to 8 inches and cross-disc BOTRAN 75W treated area prior to seeding sensitive follow-up crops to prevent injury (lettuce). Do not plant spinach as follow-up crop in treated soil. *** Use 6–10 lb per acre in North Central States.			
Snap Bean	White mold *Sclerotinia sclerotiorum* *(Sclerotiniose)*	4 (pole varieties) 3 (bush *varieties)*	Begin applications when disease is anticipated and continue at 7-day intervals during periods favorable to development of disease. Use 100 gallons (pole varieties) and 75 gallons (bush-varieties) per acre. Do not apply within 2 days of any harvest. Do not feed treated foliage to livestock.

DIRECTIONS FOR USE ON ORNAMENTAL CROPS

CROP	DISEASE	DOSAGE[1] LB PER 100 GAL. OF WATER	USE
Rose, Hydrangea	Blight *Botrytis*	1⅓	**Field**—Apply BOTRAN 75W prior to shipment or storage. **Stored**—Spray storage area thoroughly; spray or dip plants; repeat application during long storage periods.
Chrysanthemum, Geranium, Rose	Blight *Botrytis*	½ to ¾	**Stock cuttings or greenhouse plants**—Begin when disease is anticipated or first appears. Spray foliage and flowers at 7- to 14-day intervals. During prolonged damp, cool periods, apply every 5 to 7 days.
Gladiolus	Dry rot *Stromatinia* (soil)	20	Spray or dust into furrow at planting time, so that treated soil surrounds corms. 20 lb will treat one acre in furrows 1 ft. wide spaced 3 ft apart (14,520 linear ft. of furrow).
	Blight *Botrytis*	¾	Apply spray mixture with enough pressure to force fungicide between spikes. Apply to bunched cut flower spikes after harvest and before shipment or storage.

[1] Use suggested amount of BOTRAN 75W per 100 gallons water (unless otherwise specified) in full coverage sprays in conventional sprayers (not to exceed a maximum of 500 gallons per acre) or equivalent amount in concentrate or aerial equipment.

SPECIAL LOCAL NEED REGISTRATIONS

POTATOES

Control of White Mold *(Sclerotinia sclerotiorum)* with BOTRAN 75W

EPA SLN Numbers: CA-790042 (Amended 7/3/84), ID-800030, OR-800013, WA-860026, and NV-860006

FOR USE IN: CALIFORNIA, IDAHO, OREGON, WASHINGTON AND NEVADA.

For control of White mold *(Sclerotinia sclerotiorum)*, use from **2 to 6 pounds of BOTRAN 75W Fungicide per acre** in the first application. Begin treatment when vines are within 6 to 8 inches from closing of rows just prior to layby. If disease persists, subsequent applications of 2 pounds per acre may be applied at 7 to 10 day intervals: do not exceed 10 pounds per season. Do not apply within 20 days of harvest. Do not feed treated potatoes to livestock.

APPLICATION WITH THE FOLLOWING EQUIPMENT:

1. Sprinkler Irrigation:
See "Application Through Sprinkler Irrigation Equipment (Potatoes and Carrots)" below for specific instructions.

2. Ground Equipment:
Use 100 gallons per acre of water.

3. Aerial Equipment:
Use at least 10 gallons per acre as the minimum spray volume.

CARROTS

Control of White Mold *(Sclerotinia sclerotiorum)* with BOTRAN® 75W

EPA SLN NUMBER: CA-860002

FOR USE IN CALIFORNIA ONLY

LOCATION: Sandy soils located in Kern County.

CROP/SITE/COMMODITY: Carrots.

TARGET PEST/PROBLEM: *Sclerotinia sclerotiorum* (white mold).

DOSAGE: Apply 3 to 6 lbs. of product per acre (2.54–4.50 lbs. active ingredient).

DILUTION RATE: Not applicable.

METHOD OF APPLICATION: Ground application through sprinkler irrigation equipment **only.** See "Application Through Sprinkler Irrigation Equipment (Potatoes and Carrots)" below for specific use instructions.

FREQUENCY/TIMING OF APPLICATION: Begin treatment just prior to row closure or when disease is anticipated. Maximum of 9 lbs. of product can be applied per growing season. Make one application of 6 lbs. followed by 3 lbs. **OR** three applications at 3 lbs. each.

FIELD REENTRY AFTER APPLICATION: Do not enter treated areas without protective clothing until sprays have dried.

PREHARVEST INTERVAL: 21 days.

OTHER REQUIREMENTS: (1) Do not feed treated carrots or carrot tops to livestock; and (2) Carrot tops and Crop residues must be disced under within two weeks following completion of harvest.

Valid until withdrawn, suspended or cancelled by the EPA, the Kern County Agricultural Commissioner, NOR-AM Chemical Company, or the Department of Food and Agriculture.

Authorization must be obtained from the county agricultural commissioner prior to this use. This does not constitute a recommendation of the Department of Food and Agriculture and will not prevent quarantine action if illegal residues are found on or in the crop. Neither the Department nor the county agricultural commissioner, the manufacturer or formulator makes any warranty of merchantability, fitness of purpose, or otherwise, expressed or implied, concerning the use of a pesticide in accordance with these provisions. The user and/or grower acknowledges the preceding disclaimer and accepts liability for any possible damage resulting from this use. Do not use in mixture with other pesticides unless provided for in the labeling. Trial on a small area to check out unanticipated problems is suggested.

24(c) Registrant: Kern County Agricultural Commissioner.

APPLICATION THROUGH SPRINKLER IRRIGATION EQUIPMENT (POTATOES AND CARROTS)

Apply this product only through sprinkler including center pivot, solid-set, or portable (wheel move) irrigation systems.

Do not apply this product through any other type of irrigation system.

Crop injury, lack of effectiveness, or illegal pesticide residues in the crop can result from non-uniform distribution of treated water.

If you have questions about calibration, you should contact State Extension Service specialists, equipment manufacturers or other experts.

Do not connect an irrigation system (including greenhouse systems) used for pesticide application to a public water system unless the pesticide label–prescribed safety devices for public water systems are in place.

A person knowledgeable of the chemigation system and responsible for its operation, or under the supervision of the responsible person, shall shut the system down and make necessary adjustments should the need arise.

The system must contain a functional check valve, vacuum relief valve, and low pressure drain appropriately located on the irrigation pipeline to prevent water source contamination from backflow.

The pesticide injection pipeline must contain a functional, automatic, quick-closing check valve to prevent the flow of fluid back toward the injection pump.

The pesticide injection pipeline must also contain a functional, normally closed, solenoid-operated valve located on the intake side of the injection pump and connected to the system interlock to prevent fluid from being withdrawn from the supply tank when the irrigation system is either automatically or manually shut down.

The system must contain functional interlocking controls to automatically shut off the pesticide injection pump when the water pump motor stops.

The irrigation line or water pump must include a functional pressure switch which will stop the water pump motor when the water pressure decreases to the point where pesticide distribution is adversely affected.

Systems must use a metering pump, such as a positive displacement injection pump (e.g., diaphragm pump) effectively designed and constructed of materials that are compatible with pesticides and capable of being fitted with a system interlock.

Do not apply when wind speed favors drift beyond the area intended for treatment.

CALIBRATION AND APPLICATION TECHNIQUES WHEN USING CENTER-PIVOT, SOLID-SET, OR PORTABLE (WHEEL MOVE) SPRINKLER IRRIGATION EQUIPMENT:

A. Center-pivot Irrigation Equipment. Operate system and injection equipment at normal pressures recommended by the manufacturer of injection equipment used. Fill tank of injection equipment with water. Operate system for one complete circle, measuring time required, amount of water injected, and acreage contained in circle. Mix recommended amount of BOTRAN 75W Fungicide for acreage to be covered into same amount of water used during calibration and inject into system continuously for one revolution. Shut off injection equipment after one revolution, but continue to operate irrigation system until BOTRAN 75W Fungicide has been cleared from last sprinkler head.

B. Solid-set and Portable (wheel move) Irrigation Equipment. Determine acreage covered by sprinkler. Fill tank of injection equipment with water and adjust flow to use contents over a 10 to 30 minute period. Mix desired amount of BOTRAN 75W Fungicide for acreage to be covered into quantity of water used during calibration and operate entire system at normal pressures recommended by the manufacturer of injection equipment used for amount of time established during calibration. BOTRAN 75W Fungicide should be injected at the end of the irrigation cycle or as a separate application. Stop injection equipment after treatment is completed and continue to operate irrigation system until BOTRAN 75W Fungicide has been cleared from last sprinkler head.

DO NOT run irrigation system without safety valves or other devices to prevent back-siphoning of BOTRAN 75W Fungicide into water source. Irrigation water treated with BOTRAN 75W Fungicide should be maintained on the treated area until the water is absorbed into the soil. The tank containing BOTRAN 75W Fungicide should be connected to the suction side of the irrigation pump or other pressurized equipment attached to the irrigation line.

Agitation of the diluted BOTRAN in the chemical source (or slurry) tank should be maintained during the entire application period.

GREENHOUSE SEED POTATOES

Control of Botrytis blight and white mold *(Sclerotinia sclerotiorum)* with BOTRAN® 75W

EPA SLN NUMBER: CA-860042

FOR USE IN CALIFORNIA ONLY

LOCATION: Santa Cruz County.

CROP/SITE/COMMODITY: Greenhouse Grown Seed Potatoes or Transplants.

TARGET PEST/PROBLEM: Botrytis blight, white mold.

DOSAGE: Apply two level tablespoons of product per gallon of water.

DILUTION RATE: Not applicable.

METHOD OF APPLICATION: Handsprayer. Spray vegetation to the point of drip.

FREQUENCY/TIMING OF APPLICATION: Apply at 7–10 day intervals as necessary. Maximum of 9 applications per crop.

FIELD REENTRY AFTER APPLICATION: Do not enter treated area until spray residue has dried.

PREHARVEST INTERVAL: Not applicable.

OTHER REQUIREMENTS: Treated potatoes shall not be used for human or livestock consumption.

Valid until withdrawn, suspended or cancelled by the EPA, the NOR-AM Chemical Company, the Santa Cruz County Agricultural Commissioner, or the Department of Food and Agriculture.

Authorization must be obtained from the county agricultural commissioner prior to this use. This does not constitute a recommendation of the Department of Food and Agriculture and will not prevent quarantine action if illegal residues are found on or in the crop. Neither the Department nor the county agricultural commissioner, the manufacturer or formulator makes any warranty of merchantability, fitness of purpose, or otherwise, expressed or implied, concerning the use of a pesticide in accordance with these provisions. The user and/or grower acknowledges the preceding disclaimer and accepts liability for any possible damage resulting from this use. Do not use in mixture with other pesticides unless provided for in the labeling.

Trial on a small area to check out unanticipated problems is suggested.

24(c) Registrant: Santa Cruz County Agricultural Commissioner.

FENNEL

Control of White Mold *(Sclerotinia sclerotiorum)* with BOTRAN 75W

EPA SLN Number: CA-800185

FOR USE IN CALIFORNIA

Dosage: 5¹/₃ lbs. per 100 gallons of water *or* 2 lbs. per 100 gallons of water if early disease develops.

Dilution Rate: 100 gallons of water per acre.

Method of Application: Ground. Use sprayer with drop nozzle boom to direct spray to base of plants and adjacent soil.

Frequency/Timing of Application: Use the 5¹/₃ lb. rate if early disease is not likely. Make a single application per acre 4 to 8 weeks before harvest. If the 2 lb. rate is used, begin application when disease is anticipated or approximately 10 weeks before harvest and repeat at 7 day intervals in summer or 14 day intervals in fall and winter.

Field Reentry After Application: When spray dries.

Preharvest Interval: 7 days.

Other Requirements: None.

Valid until withdrawn, suspended or cancelled by the U.S. Environmental Protection Agency or NOR-AM Chemical Company. All applicable directions, restrictions, and precautions on the EPA-registered label(s) are to be followed.

CONTAINER, BAREROOT, AND GREENHOUSE GROWN CONIFER NURSERY STOCK

Control of *Botrytis* with BOTRAN 75W

EPA SLN Numbers: CA-800176, ID-820006, MT-820001, OR-800076, and WA-800065

FOR USE IN: CALIFORNIA, IDAHO, MONTANA, OREGON AND WASHINGTON

BOTRAN 75W may be applied to conifer seedlings of Douglasfir, hemlock, noble fir, Sitka spruce, Western red cedar and Redwood to prevent or control blight *(Botrytis).*

Use 1.3 pounds of BOTRAN 75W per 100 gallons of water as foliar spray. Begin treatment when disease is anticipated or first appears. Spray foliage until thoroughly wet. Repeat treatment at 10–14 day intervals. Rate may be increased to 2.6 pounds of BOTRAN 75W to control especially heavy *Botrytis* outbreaks. Alternating BOTRAN 75W with other fungicides EPA-registered for use on these sites may be done in order to minimize buildup of tolerance of *Botrytis* to any one fungicide.

IMPORTANT: READ BEFORE USE

By using this product, user accepts the following conditions, warranty, disclaimer of warranties and limitations of liability.

CONDITIONS: The directions for use of this product are believed to be adequate and should be followed carefully. However, because of extreme weather and soil conditions, manner of use and other factors beyond NOR-AM Chemical Company's control, it is impossible for NOR-AM to eliminate all risks associated with the use of this product. As a result, crop injury or ineffectiveness is always possible.

DISCLAIMER OF WARRANTIES: THERE ARE NO WARRANTIES, EXPRESS OR IMPLIED, OF MERCHANTABILITY OR OF FITNESS FOR A PARTICULAR PURPOSE OR OTHERWISE, WHICH EXTEND BEYOND THE STATEMENTS MADE ON THIS LABEL. No agent of NOR-AM Chemical Company is authorized to make any warranties beyond those contained herein or to modify the warranties contained herein. UNDER NO CIRCUMSTANCES SHALL NOR-AM CHEMICAL COMPANY BE LIABLE FOR ANY INCIDENTAL OR CONSEQUENTIAL DAMAGES RESULTING FROM FAILURE TO USE THIS PRODUCT IN ACCORDANCE WITH LABEL DIRECTIONS.

LIMITATIONS OF LIABILITY: In no event shall NOR-AM Chemical Company's liability for damages arising out of the use of this product exceed the purchase price of the product used.

532-BN-IB-5M-Rev. 9/89

This is a specimen label. NOR-AM Chemical Company is not responsible for the accuracy of the information contained herein. As labels are subject to revision, always carefully read and follow the label on the product container.

CARZOL® SP

MITICIDE/INSECTICIDE

FOR AGRICULTURAL USE ONLY

ACTIVE INGREDIENT:	Percent by Weight
Formetanate hydrochloride*	92%
*N,N-dimethyl-N'[3-[[(methylamino)carbonyl]oxy]phenyl] methanimidamide monohydrochloride	
INERT INGREDIENTS	8%
TOTAL	100%

EPA Reg. No. 45639-74 U.S. Pat. Nos. 3,336,186; Re 27,347; 3,542,853

PRECAUCION AL USUARIO: Si usted no lee ingles, no use este producto hasta que la etiqueta haya sido explicada ampliamente.

KEEP OUT OF REACH OF CHILDREN

DANGER–POISON

PELIGRO—VENENO

STATEMENT OF PRACTICAL TREATMENT

IF SWALLOWED—DRINK 1 OR 2 GLASSES OF WATER AND INDUCE VOMITING BY TOUCHING BACK OF THROAT WITH FINGER. DO NOT INDUCE VOMITING OR GIVE ANYTHING BY MOUTH TO AN UNCONSCIOUS PERSON. GET MEDICAL ATTENTION.

IF ON SKIN—WASH IMMEDIATELY WITH SOAP AND WATER.

IF IN EYES—FLUSH WITH WATER FOR AT LEAST 15 MINUTES AND GET MEDICAL ATTENTION.

NOTICE TO PHYSICIAN—CARZOL® SP IS A CARBAMATE WHICH ACTS AS A CHOLINESTERASE INHIBITOR. SPECIFIC ANTIDOTE—ATROPINE. DO NOT USE 2-PAM.

DO NOT REUSE THIS BOX

PRECAUTIONARY STATEMENTS

DANGER

HAZARDS TO HUMANS AND DOMESTIC ANIMALS

Fatal if swallowed. Do not breathe spray mist. Can cause irreversible eye damage. May cause skin sensitization. Do not get in eyes, on skin or on clothing. Wear goggles, face shield or safety glasses. Wear a pesticide respirator approved jointly by the Mining Enforcement and Safety Administration (formerly U.S. Bureau of Mines) and by the National Institute for Occupational Safety and Health under the provisions of 30 CFR Part II. Wash thoroughly with soap and water after handling and before eating or smoking. Remove contaminated clothing and wash before reuse.

Do not apply this product in such a manner as to directly or through drift expose workers or other persons. The area being treated must be vacated by unprotected persons.

ENVIRONMENTAL HAZARDS

This pesticide is toxic to wildlife. Do not apply directly to water or wetlands. Do not contaminate water by cleaning of equipment or disposal of wastes. Cover or incorporate spills. This product is toxic to bees exposed to direct treatment or residues on blooming crops or weeds. Do not apply this product if bees are visiting the treatment area. Do not allow animals to graze in orchard areas treated with CARZOL SP.

STORAGE AND DISPOSAL

PESTICIDE DISPOSAL—

Do not contaminate water, food, or feed by storage or disposal. Open dumping is prohibited.

Pesticide wastes are acutely hazardous. Improper disposal of excess pesticide, spray mixture, or rinsate is a violation of Federal Law. If these wastes cannot be disposed of by use according to label instructions, contact your State Pesticide or Environmental Control Agency, or the Hazardous Waste representative at the nearest EPA Regional Office for guidance.

CONTAINER DISPOSAL—

Completely empty box into application equipment. Then dispose of empty box in a sanitary landfill or by incineration, or, if allowed by State and local authorities, by burning. If burned, stay out of smoke.

IN CASE OF FIRE, LEAKY OR DAMAGED CONTAINERS, OR OTHER EMERGENCY, REPORT AT ONCE BY TOLL-FREE TELEPHONE TO: 800-424-9300.

DIRECTIONS FOR USE

It is a violation of Federal law to use this product in a manner inconsistent with its labeling.

REENTRY STATEMENT

Do not enter treated areas for 24 hours unless protective clothing is worn.

Because certain states may require more restrictive reentry intervals for various crops treated with this product, consult your State Department of Agriculture for further information.

Written or oral warnings must be given to workers who are expected to be in a treated area or in an area about to be treated with this product. (Indicate specific oral warnings which inform workers of areas or fields that may not be entered without specific protective clothing, period of time field must be vacated and appropriate actions to take in case of accidental exposure.) When oral warnings are given, warnings shall be given in a language customarily understood by workers. Oral warnings must be given if there is reason to believe that written warnings cannot be understood by workers. Written warnings must include the following information:

DANGER: Area treated with CARZOL SP on (date of application). Do not enter without appropriate protective clothing for 24 hours or until leaves are completely dry. In case of accidental exposure see Statement of Practical Treatment on the label.

CARZOL SP is completely soluble in water. Preparation of a premix of slurry before adding to the spray tank is not necessary. Add the desired amount of CARZOL SP directly to the spray tank 1/2 full of water with the agitator running. Then add the remainder of the water.

NOTE: CARZOL SP is not stable in alkaline water. Do not prepare the spray mixture until just before application. Do not prepare more spray mixture than will be used within 4 hours.

USE PRECAUTIONS

To avoid excessive residues, observe the following limitations:

- CITRUS: Do not apply more than 5 lbs. CARZOL SP per acre in a 12-month period. Allow 30 days between applications.
  NOTE: If brown soft scale is present in the grove, a suitable scallicide should be applied after the CARZOL SP application.
- APPLES, PEARS: Do not apply more than 4 lbs. CARZOL SP per acre in one crop season.
- PEACHES: Do not apply more than 2 1/2 lbs. CARZOL SP per acre in one crop season.
- NECTARINES: Do not apply more than 4 lbs. CARZOL SP per acre in one crop season.
- PLUMS, PRUNES: Do not apply more than 3 lbs. CARZOL SP per acre in one crop season.
- ALFALFA (Seed Crop Only): Use this product only on pure stands of alfalfa to be harvested for seed. Do not allow spray to drift onto adjacent crops. Do not spray while bees are visiting the treatment area. Do not allow livestock to graze treated fields. Do not use treated crops for livestock feed. Crop residues are not to be fed to livestock or baled for such use. Fields treated with CARZOL SP must be cleared after harvest by burning or by disking into the soil. CARZOL SP does not control aphids. If aphids appear, add an aphicide to the spray.
- Do not apply this product through any type of irrigation system.

To avoid excessive residues observe the pre-harvest intervals shown in the table.
For other use limitations see Use Precautions.

CROP	PEST	APPLICATION
ORANGES, LEMONS, TANGERINES, TANGELOS AND GRAPEFRUIT (Arizona only)	THRIPS	Apply 1/2 to 1 1/2 lbs. of CARZOL SP per acre as outside coverage spray when thrips appear. For aerial application, use a minimum of 10 gallons of water per acre. Apply as needed up to a limit of 5 lbs. CARZOL SP per acre per 12 month period. Allow 7 days before harvest.
ORANGES, LEMONS TANGERINES, TANGELOS AND GRAPEFRUIT (California only)	THRIPS	Apply 1/2 to 1 1/2 lbs. of CARZOL SP per acre as outside coverage spray when thrips appear. For aerial application, use a minimum of 10 gallons of water per acre. Apply no more than twice per season. Allow 7 days before harvest.
ORANGES, LEMONS, TANGERINES AND GRAPEFRUIT (Florida)	RUST MITE APOPKA WEEVIL	Use 1–2 oz. CARZOL SP per 100 gallons of water when mites appear. For full cover dilute spray, apply 500 to 1500 gallons of water per acre depending on the size of the trees. For concentrate spray, apply 1 lb. CARZOL SP per acre. Repeat application as needed. Allow 7 days before harvest.
ORANGES, LEMONS, TANGERINES AND GRAPEFRUIT (Texas)	RUST MITE	Use 2–4 oz. CARZOL SP per 100 gallons of water when mites appear. For full cover dilute spray, apply 500 gallons of water per acre. Repeat application as needed. Allow 7 days before harvest.
APPLES	TENTIFORM LEAF MINER	Use 4–8 oz. CARZOL SP per 100 gallons of water at pink and petal fall. For full cover dilute spray, apply 400 to 800 gallons of water per acre depending on the size of the trees. For concentrate spray, apply 1–4 lbs. CARZOL SP per acre in at least 50 gallons of water. Allow 7 days before harvest.
APPLES	WHITE APPLE LEAFHOPPER	Use 2–4 oz. CARZOL SP per 100 gallons when leafhoppers appear. For full cover dilute spray, apply 400 gallons of water per acre. For concentrate spray, apply 1/2–1 lb. CARZOL SP per acre in at least 50 gallons of water. Repeat application as needed. Allow 7 days before harvest.

CROP	PEST	APPLICATION
APPLES AND PEARS	EUROPEAN RED MITE, TWO-SPOTTED MITE	Use 4–8 oz. CARZOL SP per 100 gallons of water when mites appear. For full cover dilute spray, apply 400 to 800 gallons of water per acre depending on the size of the trees and severity of the mite infestation. For concentrate spray, apply 1–4 lbs. CARZOL SP per acre in at least 50 gallons of water. Repeat application as needed. Allow 7 days before harvest.
APPLES (Also see below for Arizona only)	THRIPS	Use 4–5 oz. CARZOL SP per 100 gallons of water at bloom and repeat as necessary through petal fall. For full cover, dilute spray, apply 400 gallons of water per acre. For concentrate spray or dilute application of less than 400 gallons, apply 1 to 1¼ lbs. CARZOL SP per acre in a minimum of 50 gallons of water per acre. Apply at night or while bees are not working in blooms. Allow 7 days before harvest.
PEARS	PEAR RUST MITE	Use 4 oz. CARZOL SP per 100 gallons of water. Apply post bloom to harvest at first indication of rust mite infestation. For full cover dilute spray, apply 400–800 gallons of water per acre depending on the size of the trees. For concentrate spray, apply 1–2 lbs. CARZOL SP per acre in at least 50 gallons of water. Repeat application as needed. Allow 7 days before harvest.
PEACHES AND NECTARINES	EUROPEAN RED MITE, McDANIEL MITE, TWO-SPOTTED MITE	Use 4 oz. CARZOL SP per 100 gallons of water when mites appear. For full cover dilute spray, apply 400 to 500 gallons of water per acre depending on the size of the trees. For concentrate spray, apply 1–1¼ lbs. CARZOL SP per acre in at least 50 gallons of water. Repeat application as needed. Allow 21 days before harvest for peaches, 14 days for nectarines.
PEACHES	LYGUS BUGS, STINK BUGS	Use 4 oz. CARZOL SP per 100 gallons of water at first sign of infestation. For full cover dilute spray, apply 400–500 gallons of water per acre. For concentrate spray, apply 1–1¼ lbs. CARZOL SP per acre. Repeat application as needed. Allow 21 days before harvest.
NECTARINES	THRIPS	Use 4–6 oz. CARZOL SP per 100 gallons of water when thrips infest blossoms through petal fall. For full cover dilute spray, apply 400 to 500 gallons of water per acre depending on the size of the trees. For concentrate or aerial spray, apply 1–2 lbs. CARZOL SP per acre. Apply at night or while bees are not working in blooms. Repeat application as needed. Allow 14 days before harvest.
PLUMS AND PRUNES	EUROPEAN RED MITE, McDANIEL MITE, TWO-SPOTTED MITE	Use 4 oz. CARZOL SP per 100 gallons of water when mites appear. For full cover dilute spray, apply 400 to 600 gallons of water per acre depending on the size of the trees. For concentrate spray, apply 1–1½ lbs. CARZOL SP per acre in at least 50 gallons of water Repeat application as needed. Allow 7 days before harvest.
ALFALFA (seed crop only in California, Oregon, Washington, Idaho and Nevada. Use pre-bloom only on alfalfa in all areas except California.)	LYGUS BUGS	Use ½–1 lb. CARZOL SP per acre. Apply when lygus bugs reach damaging numbers. Repeat application as needed. Consult your State Agricultural Experiment Station or State Extension Service Specialist for proper timing.
	CONSPERSE STINK BUGS (California only)	Use 1 lb. CARZOL SP per acre for knockdown of a heavy nymphal population. A lygus control program will control stink bugs. When applied by air use a minimum of 5 gallons per acre. When applied by ground use a minimum of 20 gallons per acre.
	TWO-SPOTTED MITE	Use ½–1 lb. CARZOL SP per acre. Apply when mites appear. When applied by air use a minimum of 10 gallons of spray per acre. When applied by ground use a minimum of 20 gallons per acre. Repeat application as needed.

SPECIAL LOCAL NEED REGISTRATION—EPA SLN NO. AZ-880009

APPLES (Arizona only)	THRIPS	CARZOL SP is used for the control of thrips on apples by aerial application at rates of 1 to 1¼ lbs. CARZOL SP per acre with a dilution rate of 10 gallons of water per acre minimum. Apply 2-4 times at 4-6 day intervals between bloom and petal fall. Do not apply more than 4 lbs. CARZOL SP per acre in one crop season. Apply at night or while bees are not working in blooms. Do not re-enter treated areas until leaves are completely dry or until 24 hours after application. Allow 7 days before harvest.

SPECIAL LOCAL NEED REGISTRATION—EPA SLN NO. AZ-890009

ORANGES, LEMONS TANGERINES, TANGELOS AND GRAPEFRUIT (Arizona only)	THRIPS	Apply ½ to 1½ lbs. of CARZOL SP per acre as outside coverage spray when thrips appear. For aerial applications, use a minimum of 10 gallons of water per acre. Apply as needed up to a limit of 5 lbs. CARZOL SP per acre per 12 month period. Allow 7 days before harvest. Do not reenter treated areas for 48 hours unless protective clothing is worn. Allow 14 days between applications.

IMPORTANT: READ BEFORE USE

By using this product, user accepts the following conditions, warranty, disclaimer of warranties and limitations of liability.

CONDITIONS: The directions for use of this product are believed to be adequate and should be followed carefully. However, because of extreme weather and soil conditions, manner of use and other factors beyond NOR-AM Chemical Company's control, it is impossible for NOR-AM to eliminate all risks associated with the use of this product. As a result, crop injury or ineffectiveness is always possible.

DISCLAIMER OF WARRANTIES: THERE ARE NO WARRANTIES, EXPRESS OR IMPLIED, OF MERCHANTABILITY OR OF FITNESS FOR A PARTICULAR PURPOSE OR OTHERWISE, WHICH EXTEND BEYOND THE STATEMENTS MADE ON THIS LABEL. No agent of NOR-AM Chemical Company is authorized to make any warranties beyond those contained herein or to modify the warranties contained herein. UNDER NO CIRCUMSTANCES SHALL NOR-AM CHEMICAL COMPANY BE LIABLE FOR ANY INCIDENTAL OR CONSEQUENTIAL DAMAGES RESULTING FROM FAILURE TO USE THIS PRODUCT IN ACCORDANCE WITH LABEL DIRECTIONS.

LIMITATIONS OF LIABILITY: In no event shall NOR-AM Chemical Company's liability for damages arising out of the use of this product exceed the purchase price of the product used.

525-1B-C-10M-Rev. 9/89

This is a specimen label. NOR-AM Chemical Company is not responsible for the accuracy of the information contained herein. As labels are subject to revision, always carefully read and follow the label on the product container.

DROPP® 50WP

COTTON DEFOLIANT

Packed in Water-Soluble Bags

FOR AGRICULTURAL USE ONLY

ACTIVE INGREDIENT:

Thidiazuron*	50%
*N-phenyl-N'-1,2,3-thiadiazol-5-ylurea	
INERT INGREDIENTS:	50%
TOTAL	100%

EPA Reg. No. 45639-89 U.S. Patent Nos. 4,294,605; 3,883,567

PRECAUCION AL USUARIO: Si usted no lee ingles, no use este producto hasta que la etiqueta haya sido explicada ampliamente.

KEEP OUT OF REACH OF CHILDREN

CAUTION —CAUCION

STATEMENT OF PRACTICAL TREATMENT

IF SWALLOWED: CALL A PHYSICIAN OR POISON CONTROL CENTER. DRINK 1 OR 2 GLASSES OF WATER AND INDUCE VOMITING BY TOUCHING BACK OF THROAT WITH FINGER. DO NOT INDUCE VOMITING OR GIVE ANYTHING BY MOUTH TO AN UNCONSCIOUS PERSON.

IF INHALED: REMOVE PERSON TO FRESH AIR.

IF IN EYES: FLUSH WITH PLENTY OF WATER FOR FIFTEEN MINUTES.

IF ON SKIN: WASH WITH PLENTY OF SOAP AND WATER.

THIS BOX CONTAINS DROPP 50 WP IN WATER SOLUBLE BAGS. INNER BAGS DISSOLVE IN WATER AND CONTENTS DISPERSE. AFTER OPENING OUTER BOX, OPEN FOIL LINER, IMMEDIATELY DUMP ENTIRE UNOPENED INNER BAGS INTO THE MIX TANK. DO NOT HANDLE SOLUBLE BAGS OR EXPOSE THEM TO MOISTURE, SINCE THIS MAY CAUSE BREAKAGE.

PRECAUTIONARY STATEMENTS

HAZARDS TO HUMANS AND DOMESTIC ANIMALS

Harmful if Swallowed, Inhaled, or Absorbed Through the Skin. Avoid Breathing Spray Mist. Avoid Contact with Skin, Eyes, and Clothing. Do Not Contaminate Food or Feedstuffs.

Do not apply this product in such a manner as to directly or through drift expose workers or other persons. The area being treated must be vacated by unprotected persons.

ENVIRONMENTAL HAZARDS

Do not apply directly to water or wetlands (swamps, bogs, marshes and potholes). Do not contaminate water when disposing of equipment washwaters. Do not apply when weather conditions favor drift from target area.

STORAGE AND DISPOSAL

STORAGE: Do not store box under wet conditions. Handle box carefully when stored at temperatures of less than 50°F to avoid breakage of soluble bags.

PESTICIDE DISPOSAL: Do not contaminate water, food or feed by storage or disposal. Wastes resulting from the use of this product may be disposed of on site or at an approved waste disposal facility.

CONTAINER DISPOSAL: Completely empty box into application equipment. Then dispose of empty box in a sanitary landfill or by incineration, or, if allowed by State and local authorities, by burning. If burned, stay out of smoke.

Do not reuse empty container.

IN CASE OF FIRE, LEAKY OR DAMAGED CONTAINERS, OR OTHER EMERGENCY, REPORT AT ONCE BY TOLL-FREE TELEPHONE TO: 800-424-9300.

GENERAL INFORMATION

DROPP® 50WP cotton defoliant is used for removal of leaves prior to harvesting. DROPP disperses easily in water and can be used in any ground and aircraft equipment designed for application of wettable powders. It may take several days before the effect of DROPP becomes noticeable. Adverse conditions, such as low temperatures, may require higher dosages and/or longer times for more complete defoliation. DROPP inhibits regrowth after defoliation and thus reduces staining of lint during harvesting and ginning.

USE PRECAUTIONS:

Do not apply this product through any type of irrigation system.

DO NOT ALLOW SPRAY DRIFT TO CONTACT TREES OR CROPS OTHER THAN MATURE COTTON OR COTTON WHICH YOU DESIRE TO DEFOLIATE.

Rainfall within 24 hours after application will reduce the effectiveness of DROPP. Use only freshly prepared sprays.

Do not store spray mixture overnight. Do not feed foliage from treated cotton plants or gin trash to livestock.

Do not plant small grain, sorghum, corn or root crops earlier than two weeks after application of DROPP. Do not plant legumes, alfalfa or leafy vegetables earlier than two months after application of DROPP. Do not use immature crops for food or feed.

DIRECTIONS FOR USE

It is a violation of Federal law to use this product in a manner inconsistent with its labeling.

REENTRY STATEMENT:

Reentry into treated areas is prohibited until leaves are completely dry or until 24 hours after application.

Do not enter treated areas without protective clothing.

Because certain states may require more restrictive reentry intervals for various crops treated with this product, consult your State Department of Agriculture for further information.

Written or oral warnings must be given to workers who are expected to be in a treated area or in an area about to be treated with this product. (Indicate specific oral warnings which inform workers of areas or fields that may not be entered without specific protective clothing, period of time field must be vacated and appropriate actions to take in case of accidental exposure.)

When oral warnings are given, warnings shall be given in a language customarily understood by workers. Oral warnings must be given if there is reason to believe that written warnings cannot be understood by workers. Written warnings must include the following information:

CAUTION: Area treated with DROPP 50WP on (date of application). Do not enter without appropriate protective clothing for 24 hours or until leaves are completely dry. In case of accidental exposure see Statement of Practical Treatment on the label.

TIME OF APPLICATION: Apply DROPP 50WP cotton defoliant only to mature cotton plants when 60–70% or more of the bolls are open, and the remaining bolls are determined to be mature.

A boll is mature when it is too hard to be dented when squeezed between thumb and fingers, too hard to be sliced with a sharp knife, and when the seed coat becomes light brown in color.

Apply DROPP 50WP at least 5 days prior to anticipated harvest.

NOTE: Good defoliation with DROPP is dependent upon three factors:

1. High air temperature
2. High humidity
3. High moisture content in leaf of cotton plant

Do not use DROPP 50WP alone or in combinations when nighttime temperatures are expected to fall below 60°F.

Performance of DROPP 50WP alone or in tank mixes is variable under low temperatures. Ideally, nighttime temperatures 2–3 days following application should be above 60°F or total defoliation can be affected. Use of DROPP 50WP on drought-stressed cotton (low leaf moisture, thick cuticle, etc.) may result in less than satisfactory defoliation.

USE OF ADJUVANTS (CROP OILS): The use of adjuvants such as petroleum-based crop oils or penetrating oils have been shown to improve performance during low nighttime temperatures (60–65°F) or when defoliating drought-stressed cotton. Refer to specific adjuvant label for rate recommendations.

CAUTION: The addition of adjuvants can cause desiccation and/or leaf freezing during periods of high temperatures. Compounds that desiccate leaf tissue are not recommended.

MIXING INSTRUCTIONS: Fill the spray tank with one-half of the total amount of water to be used. After opening each box to be used, open its foil liner. Then, with the agitator running, immediately dump the entire contents of the required number of boxes of soluble bags into the mix tank and fill the tank with the additional quantity of water required. Use entire contents of each box.

TANK MIX ADJUVANTS: When using tank mixes of DROPP 50WP and organophosphates, the use of a surfactant or compatibility agent is recommended. Prior to mixing in the spray tank, a small scale compatability test should be conducted.

Use only products which are exempt from tolerance under 40 CFR 180.1001.

Always follow the mixing instructions on the label of the appropriate adjuvant.

APPLICATION: Use the specified dosage of DROPP 50WP cotton defoliant in sufficient amount of water to give complete coverage of the foliage with uniform wetting of leaf surfaces. Agitate the spray mix during application.

DOSAGE: One box (5 × 1 lb. bags) of DROPP treats 12.5–25 acres (0.2–0.4 lb. DROPP per acre). Apply in 10–25 gallons of spray per acre by ground equipment and 2–10 gallons per acre by air. Use the higher rates during periods of low temperatures. (60-65°)

Two applications not exceeding a total of 1 box per 8.3 acres (0.6 lbs./acre) may be necessary to defoliate rank cotton.

BOTTOM DEFOLIATION FOR USE IN ARIZONA ONLY: DROPP 50WP may be used as an aid to improve air circulation and help reduce boll rot caused by moisture.

Apply DROPP 50WP at .1 lbs. formulated product per acre in 10–25 gallons of water per acre.

Application to **bottom** 1/3 of plant **only** recommended.

Timing of application should be based upon the latest date when ground application equipment can enter the field without crop damage.

PRE-CONDITIONING FOR USE IN ARIZONA AND CALIFORNIA ONLY: DROPP 50WP may be used as a pre-conditioner to enhance the activity of a defoliant application.

Apply DROPP 50WP at 0.1– 0.2 lbs. formulated product per acre. Apply in 10–25 gallons of water per acre by ground application or 2–10 gallons of water per acre by aerial application.

Timing of application is recommended to be 7–14 days prior to the use of another defoliant. Refer to the second product label prior to use for complete recommendations.

TANK MIX OF DROPP® 50WP PLUS PREP[1]

The tank mix of DROPP 50WP + Prep is recommended to improve overall defoliation, and as an aid in accelerating the opening of mature unopened cotton bolls. Best activity will be obtained where the tank mix is applied to mature cotton plants with 60% open bolls and the remaining bolls are determined to be mature (See Time Of Application section of this label). Do not apply tank mix before sufficient unopened bolls have matured to produce the desired cotton yield. Do not apply tank mix if nighttime temperatures are expected to fall below 60°F; unsatisfactory defoliation may result. Certified cotton seed producers should determine boll and seed maturity prior to treatment.

Apply DROPP 50WP at a rate of 0.1– 0.2 lbs. formulated product plus Prep at 1.0–2.0 lbs. active ingredient per acre.

Higher rate of DROPP should be used where excessive regrowth is anticipated, or during cooler temperatures.

Apply as a dilute spray as follows:

Application	Gal. Water/Acre
Aerial	3–10
Ground	10 –50

TANK MIX OF DROPP® 50WP PLUS DEF 6[2] OR DROPP 50WP PLUS FOLEX 6 EC[3] FOR USE IN LA., MS., ARK., CA., AZ., ALA., GA., NC., SC., TN. ONLY: The tank mix of DROPP 50WP and DEF 6 or FOLEX 6 EC is recommended to improve defoliation and inhibit regrowth under less than ideal conditions. Less than ideal conditions are those when the nighttime temperatures are expected to be 60°F on the date of application **and** for three days following application.

Best activity will be obtained when the tank mix is applied to mature cotton plants with 60% or more open bolls.

RATE OF APPLICATION

DROPP 50WP + DEF 6 = Use .1–.2 lbs DROPP 50WP + .5–1.0 pints DEF 6

DROPP 50WP + FOLEX 6 EC = Use .1–.2 lbs DROPP 50WP + 0.5 –1.0 pints FOLEX 6 EC

Higher rate of DROPP 50WP should be used where excessive regrowth is anticipated.

Maximum rates of DROPP and phosphate defoliants applied in a tank mix during very high air temperatures can cause desiccation.

Use lower rate of DEF 6 or FOLEX 6 EC under ideal nighttime temperature conditions (65°F); higher rate during less than ideal or adverse conditions.

Apply as a dilute spray in 5–10 gallons of water per acre by air or 10 –25 gallons of water per acre with ground equipment.

When applying DEF 6 or FOLEX 6 EC defoliant plus DROPP 50WP as a tank mix, the following mixing sequence is recommended:

1) Add required water to spray tank, agitate
2) Add DROPP 50WP according to label directions (see **mixing instructions**)
3) Add DEF 6 or FOLEX 6 EC according to rate desired **after** the DROPP 50WP has completely dispersed.

A second application of the labeled tank mix or of DROPP alone may be made where necessary.

DO NOT USE TANK MIX OF DROPP 50WP PLUS DEF 6 OR FOLEX 6EC IN THE RIO GRANDE VALLEY COUNTIES OF STARR, HIDALGO, WILLACY, AND CAMERON.

IMPORTANT: CLEAN-OUT INSTRUCTIONS

Tank mixes of DROPP 50WP with organophosphates may form a residue in application equipment. **While still fresh and moist,** this residue is easily removed by flushing the entire system with a commercial tank cleaner. We have recently found that a tank cleaner such as **Cleaner Concentrate**** manufactured by Valco Chemical Division of Harlingen, Texas, is effective for removing these fresh residues.

Do not allow the spray solution to dry in the application equipment. **Immediately following application,** clean all equipment (mix tanks, pumps, transfer lines, application tanks, sumps, booms and nozzles and all related equipment) thoroughly with cleaner and water.

Should small quantities of DROPP 50WP remain in inadequately cleaned tanks, they may be released during subsequent applications and may cause damage to crops. NOR-AM Chemical Company accepts no liability for damage to crops due to inadequately cleaned tanks.

** In limited trials, this cleaner has shown promise in cleaning dried tank residues.

The procedure for removing dried residues requires allowing the diluted cleaning solution to stand in equipment filled to capacity, for 7 days followed by thorough flushing.

1 Prep is a trademark of Rhone-Poulenc, Inc.
2 DEF 6 is a registered trademark of Mobay Chemical Corporation, Agricultural Chemical Division
3 FOLEX 6 EC is a registered trademark of Rhone-Poulenc Inc., Agrochemical Division.

IMPORTANT: READ BEFORE USE

By using this product, user accepts the following conditions, warranty, disclaimer of warranties and limitations of liability.

CONDITIONS: The directions for use of this product are believed to be adequate and should be followed carefully. However, because of extreme weather and soil conditions, manner of use and other factors beyond NOR-AM Chemical Company's control, it is impossible for NOR-AM to eliminate all risks associated with the use of this product. As a result, crop injury or ineffectiveness is always possible.

DISCLAIMER OF WARRANTIES: THERE ARE NO WARRANTIES, EXPRESS OR IMPLIED, OF MERCHANTABILITY OR OF FITNESS FOR A PARTICULAR PURPOSE OR OTHERWISE, WHICH EXTEND BEYOND THE STATEMENTS MADE ON THIS LABEL. No agent of NOR-AM Chemical Company is authorized to make any warranties beyond those contained herein or modify the warranties contained herein. UNDER NO CIRCUMSTANCES SHALL NOR-AM CHEMICAL COMPANY BE LIABLE FOR ANY INCIDENTAL OR CONSEQUENTIAL DAMAGES RESULTING FROM FAILURE TO USE THIS PRODUCT IN ACCORDANCE WITH LABEL DIRECTIONS.

LIMITATIONS OF LIABILITY: In no event shall NOR-AM Chemical Company's liability for damages arising out of the use of this product exceed the purchase price of the product used.

527-IB-DWP-10M-Rev. 7/89

This is a specimen label. NOR-AM Chemical Company is not responsible for the accuracy of the information contained herein. As labels are subject to revision, always carefully read and follow the label on the product container.

MITAC® WP

50% WETTABLE POWDER INSECTICIDE FOR PEARS

ACTIVE INGREDIENT:	Percent by Weight
Amitraz (N'-(2,4-dimethylphenyl)-N-[[(2,4-dimethylphenyl) imino]methyl]-N-methylmethanimidamide)	50%
INERT INGREDIENTS:	50%
TOTAL	100%

This product contains the toxic inert ingredient paraformaldehyde.

EPA Reg. No. 45639-61 U.S. Patent Nos. 3781355; 3864497

PRECAUCION AL USUARIO: Si usted no lee ingles, no use este producto hasta que la etiqueta haya sido explicada ampliamente.

KEEP OUT OF REACH OF CHILDREN

WARNING — AVISO

STATEMENT OF PRACTICAL TREATMENT

IF IN EYES: FLUSH EYES WITH CLEAN WATER. GET MEDICAL ATTENTION IF IRRITATION PERSISTS.

IF SWALLOWED: DRINK PROMPTLY A LARGE QUANTITY OF MILK, EGG WHITES, GELATIN SOLUTION, OR IF THESE ARE NOT AVAILABLE, DRINK LARGE QUANTITIES OF WATER. AVOID ALCOHOL.

IF ON SKIN: WASH HANDS WITH SOAP AND WATER EACH TIME BEFORE EATING, DRINKING, OR SMOKING. AT THE END OF THE WORKING DAY, BATHE ENTIRE BODY WITH SOAP AND PLENTY OF WATER.

ANTIDOTE: NONE KNOWN.

PRECAUTIONARY STATEMENTS

HAZARDS TO HUMANS

WARNING

CAUSES EYE IRRITATION. HARMFUL IF SWALLOWED. REPEATED SKIN CONTACT MAY CAUSE AN ALLEGIC REACTION. DO NOT GET IN EYES. WEAR GOGGLES OR FACE SHIELD. WASH THOROUGHLY AFTER HANDLING. AVOID BREATHING DUST OR SPRAY MIST.

PROTECTIVE CLOTHING & EQUIPMENT: During mixing/loading or application wear a protective suit which has long sleeves and long pants. Wear chemical resistant gloves, a hat, boots and goggles or face shield. A helmet with visor may be substituted for the hat and goggles during aerial application. Mixer/loaders should also wear a chemical resistant apron when handling the concentrated product. Wash thoroughly with soap and water after handling and before eating, urinating, or smoking. Remove and wash clothing before reuse. Clothing should be laundered separately from household articles. Replace gloves frequently. Clothing which has been drenched and used gloves should be disposed of in accordance with state or local regulations. Instead of clothing and equipment specified above, the applicator can use an enclosed tractor cab or cockpit with properly filtered air supply.

ENVIRONMENTAL HAZARDS

This product is toxic to fish. Do not apply directly to water. Drift and run-off from treated areas may be hazardous to fish in adjacent sites. Do not contaminate water by cleaning of equipment or disposal of wastes. Apply this pesticide only as specified on this label.

Do not apply this product through any type of irrigation system.

STORAGE AND DISPOSAL

STORAGE: Store in original container and keep closed. Store in a cool, dry place.

PESTICIDE DISPOSAL: Do not contaminate water, food, or feed by storage or disposal. Wastes resulting from the use of this product may be disposed of on site or at an approved waste disposal facility.

Do not reuse empty container.

CONTAINER DISPOSAL: Completely empty bag into application equipment. Then dispose of empty bag in a sanitary landfill or by incineration, or, if allowed by State and local authorities, by burning. If burned, stay out of smoke.

IN CASE OF FIRE, LEAKY OR DAMAGED CONTAINERS, OR OTHER EMERGENCY, REPORT AT ONCE BY TOLL-FREE TELEPHONE TO: 800-424-9300.

GENERAL INFORMATION

MITAC® WP is an insecticide for use on pears. When applied as directed, MITAC WP will control pear psylla. MITAC WP can be applied by air or ground equipment, as dilute or concentrate sprays. The rate of MITAC WP per acre is based on thorough coverage with a dilute spray of 400 gallons from a conventional sprayer.

When using concentrate sprayers, the amount of water is reduced but the amount of MITAC WP per acre should remain the same as when using dilute sprays.

COMPATIBILITY: MITAC WP is compatible with most commonly used pear insecticides, miticides and fungicides; however, do not combine MITAC WP with detergents, sulfur, Bordeau mixture, NAA or other highly alkaline materials as a tank mix.

DO NOT GRAZE TREATED AREAS.

PEAR PSYLLA: Apply 1½ to 3 pounds MITAC WP per acre. Where full dilute coverage requires less than 400 gallons per acre, use 6 to 12 ounces per 100 gallons.

For prebloom control of overwintering adults, apply MITAC WP after eggs have moved into the female psylla oviduct, but prior to heavy egg laying. For best results, MITAC WP should be applied when daily maximum temperatures exceed 50°F. Consult your local University Extension Service for further information concerning best timing for control of overwintering adult psylla.

If weather or orchard conditions do not permit spraying before heavy egg laying, spray after the majority of the first generation eggs have hatched.

For postbloom control, apply MITAC WP as needed. Best control will be obtained when the majority of pear psylla are in the adult or young nymphal stages of development.

Rate should not exceed 3 pounds of MITAC WP per acre per application. Do not apply within 7 days of harvest. Do not apply more than 18 pounds per acre of MITAC WP during the use season.

DIRECTIONS FOR USE

It is a violation of Federal Law to use this product in a manner inconsistent with its labeling.

Do not apply this product in such a manner as to directly or through drift expose workers or other persons. The area being treated must be vacated by unprotected persons.

REENTRY STATEMENT

Do not re-enter treated areas for 24 hours without protective clothing.

Because certain states may require more restrictive reentry intervals for various crops treated with this product, consult your State Department of Agriculture for further information.

Written or oral warnings must be given to workers who are expected to be in a treated area or in an area about to be treated with this product. (Indicate specific oral warnings which inform workers of areas or fields that may not be entered without specific protective clothing, period of time field must be vacated and appropriate actions to take in case of accidental exposure.) When oral warnings are given, warnings shall be given in a language customarily understood by workers. Oral warnings must be given if there is reason to believe that written warnings cannot be understood by workers. Written warnings must include the following information:

WARNING: Area treated with MITAC® WP on (date of application). Do not enter without appropriate protective clothing for 24 hours. In case of accidental exposure see Statement of Practical Treatment on the label.

IMPORTANT: READ BEFORE USE

By using this product, user accepts the following conditions, warranty, disclaimer of warranties and limitations of liability.

CONDITIONS: The directions for use of this product are believed to be adequate and should be followed carefully. However, because of extreme weather and soil conditions, manner of use and other factors beyond NOR-AM Chemical Company's control, it is impossible for NOR-AM to eliminate all risks associated with the use of this product. As a result, crop injury or ineffectiveness is always possible.

DISCLAIMER OF WARRANTIES: THERE ARE NO WARRANTIES, EXPRESS OR IMPLIED, OF MERCHANTABILITY OR OF FITNESS FOR A PARTICULAR PURPOSE OR OTHERWISE, WHICH EXTEND BEYOND THE STATEMENTS MADE ON THIS LABEL. No agent of NOR-AM Chemical Company is authorized to make any warranties beyond those contained herein or modify the warranties contained herein. UNDER NO CIRCUMSTANCES SHALL NOR-AM CHEMICAL COMPANY BE LIABLE FOR ANY INCIDENTAL OR CONSEQUENTIAL DAMAGES RESULTING FROM FAILURE TO USE THIS PRODUCT IN ACCORDANCE WITH LABEL DIRECTIONS.

LIMITATIONS OF LIABILITY: In no event shall NOR-AM Chemical Company's liability for damages arising out of the use of this product exceed the purchase price of the product used.

750-IB-MWP-3M-Rev. 9/89

This is a specimen label. NOR-AM Chemical Co. is not responsible for the accuracy of the information contained herein. As labels are subject to revision, always carefully read and follow the label on the product container.

NORTRON® EC

SUGAR BEET HERBICIDE

For selective control of weeds in sugar beets

GRASS SEED HERBICIDE

For Selective Control of Weeds in Certain Grass Seed Crops in Western Oregon and Western Washington

FOR AGRICULTURAL USE ONLY

ACTIVE INGREDIENT:	Percent by Weight
Ethofumesate: (2-ethoxy–2,3-dihydro-3,3-dimethyl-5-benzofuranyl methanesulfonate*)	19%
INERT INGREDIENTS:	81%
TOTAL	100%

This product contains 1.5 lbs. active ingredient per gallon.

EPA Reg. No. 45639-5

*Protected by U.S. Patent No. 3,689,507 and 4,072,495

PRECAUCION AL USUARIO: Si usted no lee ingles, no use este producto hasta que la etiqueta haya sido explicada ampliamente.

KEEP OUT OF REACH OF CHILDREN

DANGER —PELIGRO

CORROSIVE, CAUSES EYE DAMAGE.
WEAR GOGGLES OR FACE SHIELD WHEN HANDLING.

STATEMENT OF PRACTICAL TREATMENT

IF SWALLOWED—DRINK PROMPTLY A LARGE QUANTITY OF MILK, EGG WHITES, GELATIN SOLUTION OR IF THESE ARE NOT AVAILABLE, DRINK LARGE QUANTITIES OF WATER. AVOID ALCOHOL. CALL A PHYSICIAN.

IF ON SKIN—FLUSH WITH PLENTY OF WATER FOR AT LEAST 15 MINUTES.

IF IN EYES—IMMEDIATELY FLUSH WITH PLENTY OF WATER FOR AT LEAST 15 MINUTES. CALL A PHYSICIAN.

PRECAUTIONARY STATEMENTS

HAZARDS TO HUMANS AND DOMESTIC ANIMALS

DANGER

Corrosive. Causes eye damage. Do not get in eyes. Wear goggles or face shield when handling. Harmful if absorbed through skin. Avoid contact with skin.

FIRST AID—In case of contact, immediately flush eyes or skin with plenty of water for at least 15 minutes. For eyes, call a physician. Get medical attention if skin irritation persists. Remove and wash contaminated clothing before reuse.

ENVIRONMENTAL HAZARDS

Do not apply directly to water. Do not apply when weather conditions favor drift from treated area.

Do not contaminate water by cleaning of equipment or disposal of wastes.

PHYSICAL OR CHEMICAL HAZARDS

DO NOT USE, POUR, SPILL OR STORE NEAR HEAT OR OPEN FLAME.

IN CASE OF FIRE, LEAKY OR DAMAGED CONTAINERS, OR OTHER EMERGENCY, REPORT AT ONCE BY TOLL-FREE TELEPHONE TO: 800-424-9300.

GENERAL USE PRECAUTIONS

NORTRON® EC or tank-mixes should be used for recommended purposes and at recommended rates only. (DO NOT OVERTREAT.)

Do not graze livestock on treated crops.

Do not apply this product through any type of irrigation system.

If crop is lost due to climatic or soil conditions following application of NORTRON EC or tank-mixes, do not plant crops other than sugar beets or ryegrass in treated land during the same season. Do not retreat field with NORTRON EC. If fields are replanted to sugar beets, reseed into treated band.

Do not rotate with any crops other than sugar beets or ryegrass for 12 months after application. Thorough tillage, including moldboard plowing, should precede the planting of crops other than sugar beets or ryegrass. Do not use NORTRON EC on muck or peat soils.

Do not allow spray mixture to stand in tank overnight. Flush and drain spray equipment after each day's use.

NORTRON EC may soften PVC hosing if not drained.

Store unused spray mixture in tightly-sealed containers and protect from frost.

Do not use NORTRON EC in water having a temperature below 40°F., as crystallization of spray mixture may occur in the nozzles and strainers.

This label must be in the possession of the user at the time of pesticide application.

STORAGE AND DISPOSAL

STORAGE: Store NORTRON EC in a cool place above 34°F.

PESTICIDE DISPOSAL: Do not contaminate water, food or feed by storage or disposal. Pesticide wastes are acutely hazardous. Improper disposal of excess pesticide, spray mixture, or rinsate is a violation of Federal Law. If these wastes cannot be disposed of by use according to label instructions, contact your State Pesticide or Environmental Control Agency, or the Hazardous Waste representative at the nearest EPA Regional Office for guidance.

CONTAINER DISPOSAL: Triple rinse (or equivalent). Then offer for recycling or reconditioning, or puncture and dispose of in a sanitary landfill, or by other procedures approved by state and local authorities.

DO NOT REUSE EMPTY CONTAINER.

GENERAL INFORMATION

WEED SPECIES CONTROLLED—SUGAR BEETS

NORTRON® EC is a selective herbicide for use in sugar beets for the control of the weed species listed below. It provides effective control of these weeds for up to 10 weeks following application.

WEED SPECIES CONTROLLED

Annual Broadleaf Weeds

Black nightshade (*Solanum nigrum*), Common chickweed (*Stellaria media*) Common lambsquarters (*Chenopodium album*), Common purslane (*Portulaca oleracea*), Kochia (*Kochia scoparia*), Ladysthumb (*Polygonum persicaria*), Pennsylvania smartweed (*Polygonum pensylvanicum*), Redroot pigweed (*Amaranthus retroflexus*), Russian thistle (*Salsola kali var. tenuifolia*), Wild buckwheat (*Polygonum convolvulus*).

Annual Grass Weeds

Annual bluegrass (*Poa annua*), Barnyardgrass* (*Echinochloa crus-galli*), Canarygrass (*Phalaris canariensis*), Green foxtail (*Setaria viridis*), Large crabgrass (*Digitaria sanguinalis*), Volunteer barley (*Hordeum* sp.), Volunteer wheat (*Triticum* sp.), Wild oats** (*Avena fatua*), Yellow foxtail (*Setaria glauca*).

*Control of barnyardgrass may be reduced with the NORTRON® EC + Pyramin because of the lower rate of NORTRON EC recommended.

**Control of wild oats has been inconsistent in Minnesota and North Dakota.

NORTRON EC alone will also reduce competition from these HARD-TO-CONTROL weeds:

Annual Sowthistle (*Sonchus oleraceus*), Puncturevine (*Tribulus terrestris*), Sherpherdspurse (*Capsella bursa-pastoris*), Purple nutsedge (*Cyperus rotundus*), Yellow nutsedge (*Cyperus esculentus*).

WEED SPECIES CONTROLLED—GRASS SEED

NORTRON EC is a selective herbicide for use in ryegrass, tall fescue, and bentgrass seed crops in Western Oregon and Western Washington. It effectively controls or reduces competition from those weed species listed below. NORTRON EC may be applied preemergence to new seedings of annual ryegrass or postemergence to established stands of perennial ryegrass, tall fescue, or bentgrass. Application to bentgrass is restricted to plantings which have been established for one year or longer. Soil should be moist at time of application. NORTRON EC is less effective when applied to dry soil. Rainfall shortly after application is necessary for activation. Make only one application per season, either preemergence or postemergence, not both.

WEEDS CONTROLLED: Annual bluegrass (*Poa annua*), Rattail fescue (*Festuca myuros*), Seedling volunteer wheat (*Triticum* spp.), Seedling volunteer barley (*Hordeum* spp.), Soft chess (*Bromus mollis*), Seedling Wild oats (*Avena fatua*), Downy brome (*Bromus tectorum*), Common chickweed (*Stellaria media*), Common vetch (*Vicia sativa*), Common velvetgrass (*Holcus lanatus*), Mannagrass (*Glyceria* spp.).

Note: NORTRON EC does not control volunteer ryegrass. Other registered herbicides will be required to control volunteer annual or perennial ryegrass seedlings.

DIRECTIONS FOR USE

It is a violation of federal law to use this product in a manner inconsistent with its labeling.

Read entire Directions for Use and Disclaimer of Warranties before using this product.

Do not apply this product in such a manner as to directly or through drift expose workers or other persons. The area being treated must be vacated by unprotected persons.

When NORTRON® EC is applied preemergence, or postemergence, the following statements apply:

REENTRY STATEMENT

Do not enter treated areas without protective clothing until sprays have dried.

Because certain states may require more restrictive reentry intervals for various crops treated with this product, consult your State Department of Agriculture for further information.

Written or oral warnings must be given to workers who are expected to be in a treated area or in an area about to be treated with this product. (Indicate specific oral warnings which inform workers of areas or fields that may not be entered without specific protective clothing, period of time field must be vacated, and appropriate actions to take in case of accidental exposure.) When oral warnings are given, warnings shall be given in a language customarily understood by workers. Oral warnings must be given if there is reason to believe that written warnings cannot be understood by workers. Written warnings must include the following information:

DANGER: Area treated with NORTRON® EC herbicide on (date of application). Do not enter without appropriate protective clothing until spray has dried. In case of accidental exposure, see Statement of Practical Treatment on the label.

NORTRON EC WITH LIQUID FERTILIZERS: NORTRON EC and tank mixes may be combined with liquid (solution, slurry or suspension) fertilizers. However, physical compatibility with these fertilizers should be determined before combining in the spray tank. See Appendix I for directions for testing combinations. Constant agitation in the spray tank is recommended when combining NORTRON EC with compatible liquid fertilizers.

RYEGRASS, TALL FESCUE, AND BENTGRASS SEED CROPS

(For use in Western Oregon and Western Washington only):

NORTRON EC may cause stunting and stand reduction of newly seeded perennial ryegrass and tall fescue, if the crop is planted late in the fall and subjected to adverse climatic conditions or pesticides which restrict normal growth.

If vegetative matter or stover from previous crop was burned, sufficient rainfall or overhead irrigation must have occurred to dissipate the charcoal residue remaining after burning prior to NORTRON EC treatment. Failure to allow for dissipation of charcoal residue may result in reduced weed control.

Spray equipment: Use a fixed boom power sprayer properly calibrated to a constant speed and rate of delivery. Do not use smaller than 50 mesh strainer. Avoid overlapping of spray swath. Shut off boom while starting, turning or stopping to avoid overlapping. Apply in 10 to 50 gallons of water per acre at low pressure (20–50 psi).

Soil preparation: A firm, fine and level seedbed free of trash and vegetative matter will provide best results from preemergence applications. Large clods can reduce effectiveness of NORTRON EC. All existing vegetative growth should be thoroughly worked into the soil before treatment.

NEW SEEDINGS OF ANNUAL OR PERENNIAL RYEGRASS:

Before weed emergence: Apply NORTRON EC after seeding and prior to weed emergence. For best results apply to moist soil. Apply 1/2 to 1 gallon per acre. Use the lower rate for control of annual bluegrass and common chickweed. For control of rattail fescue and other weeds listed use 3/4 to 1 gallon per acre.

After weed emergence: Apply NORTRON EC at earliest possible weed growth stage but not later than the 4-leaf stage. Rattail fescue, which is more difficult to control, must be treated no later than the 2-leaf stage. Apply 3/4 to 1 1/4 gallons per acre. Use the highest rate where rattail fescue is present and where weed infestation is heavy.

NEW SEEDINGS OF FALL-PLANTED PERENNIAL RYEGRASS AND TALL FESCUE TREATED WITH DIURON PLUS CHARCOAL:

Timing of application: Apply NORTRON EC following crop emergence and after sufficient rainfall and/or overhead irrigation has occurred to dissipate the charcoal band (approximately 4 inches). Use dosage rates listed in table below. Surface debris may result in reduced weed control. Failure to allow for complete dissipation of the charcoal band may result in reduced weed control within the crop row. For best results, apply NORTRON EC to a moist soil surface.

Before using diuron, read the diuron label for rate recommendations, timing of applications, directions for use and precautionary statements. Do not exceed maximum dosage rates for either herbicide.

Dosage Table		
Crop	Rate Per Acre	Remarks
Perennial ryegrass and Tall fescue	1/2 to 1 gal	For effective control, annual bluegrass must be treated before the 4 leaf stage, rattail fescue, wild oats and volunteer wheat must be treated before the 2 leaf stage. Use the lower rate for control of annual bluegrass and common chickweed, use the higher rate for control of rattail fescue, wild oats and other weeds listed.

ESTABLISHED STANDS OF PERENNIAL RYEGRASS AND TALL FESCUE:

Before weed emergence: Apply NORTRON EC at 3/4 to 1 gallon per acre prior to weed emergence. Use higher rate where rattail fescue is expected to be a problem. For best results, apply to moist soil. Crop residue and debris will reduce effectiveness of treatment and should be removed or destroyed.

After weed emergence: Apply NORTRON EC at earliest possible weed growth stage but not later than the 4-leaf stage. Rattail fescue, which is more difficult to control, must be treated no later than the 2-leaf stage. Apply 3/4 to 1 gallon per acre. Use the higher rate where rattail fescue is present. Where weed pressure is very heavy and rattail fescue is at the maximum stage of growth for treating, a rate of 1 1/4 gallons of NORTRON EC is recommended.

ESTABLISHED STANDS OF BENTGRASS: Apply only to well-established stands which have been seeded for not less than 12 months. Straw from previous crop must be removed or destroyed. Failure to do so may result in reduced weed control.

Before weed emergence: Apply NORTRON EC at 1/2 to 1 gallon per acre prior to weed emergence. Use higher rate where rattail fescue is expected to be a problem. For best results, apply to moist soil.

After weed emergence: Apply NORTRON EC at earliest possible weed growth stage but no later than the 4-leaf stage. Rattail fescue, which is more difficult to control, must be treated no later than the 2-leaf stage. Apply at the rate of 1/2 to 1 gallon per acre. Use higher rate when rattail fescue is a problem. Do not apply more than 1 gallon of NORTRON EC per acre on bentgrass.

SUGAR BEETS

Apply tank-mixes only in specified regions or states and in accordance with directions on label.

PREPLANT INCORPORATED AND PREEMERGENCE APPLICATIONS

SOIL PREPARATION: The soil should be prepared according to good agricultural practices. Large clods can reduce the effectiveness of NORTRON® EC and tank-mixes. All existing vegetative growth should be thoroughly worked into the soil before treatment.

SPRAY EQUIPMENT: Apply NORTRON EC alone or in tank-mixes to the soil using standard low pressure (20–50 psi) spray equipment. Spray equipment should be carefully calibrated before use and checked frequently during application to see that it is functioning properly. Do not use smaller than 50-mesh strainer. Uniformly apply the recommended rates of NORTRON EC or NORTRON EC tank-mixes in 10–60 gallons of water per acre on a broadcast basis. Avoid overlaps since crop injury may result. When applying NORTRON EC or tank-mixes in a band, check to make certain that the band width is accurate for the dosage rate being applied.

The spray tank and lines should be thoroughly cleaned and rinsed prior to using NORTRON EC.

INCORPORATION EQUIPMENT: Where soil incorporation is recommended use a hooded-power or ground-driven rotary tiller, rolling cultivator or similar equipment properly adjusted to uniformly incorporate NORTRON® EC or tank-mixes to a depth of 1 to 2 inches. Deeper incorporation may reduce effectiveness. Do not apply NORTRON EC or tank-mixes through soil injector shanks. Incorporation should be accomplished prior to planting. If done after planting, proper precautions should be taken to avoid damaging or moving the crop seed. See below for Layering Application.

LAYERING APPLICATION:

Spring: Form beds with appropriate bedding equipment. Pre-irrigate field if necessary. Remove bed top with suitable de-ridging machinery to provide a minimum width of 10" across the top of the bed. Apply NORTRON EC in a band at the recommended rate indicated in the appropriate regional dosage table and cover the treated band with 1 inch of soil using ditchers or discs equipment. Shape the bed with roller shaper and irrigate until the tops of the beds are thoroughly wetted. Irrigate from furrows on both sides of the row.

Fall: This method of application can be used when spring moisture is marginal or where irrigation water is not available at planting time. Fall bedding utilizes the winter-accumulated moisture to enhance activation of the herbicide and to aid in germination of the sugar beet crop.

Prepare the field (as for planting; plow, pack, and float, etc.), in the fall, usually late September or October. Apply NORTRON EC in a band to the soil surface at the recommended rate indicated in the appropriate regional dosage table. Be sure that the soil surface to be treated is free from trash and vegetation.

Cover the treated bands with soil and form beds or ridges using ditchers or discs. In the spring when the soil is sufficiently dry to be worked, de-ridge the beds down to within 1/2" to 1" of the treated layer using suitable equipment such as the Kirchner bedder or Oregon Northslope harrow. When de-ridging, maintain the original bedding guidance system by using a bull tongue chisel, slide guides or similar equipment. This will ensure that the planter will follow in the treated band. Plant sugar beets in the de-ridged area when the soil conditions allow.

MIXING DIRECTIONS: When mixing NORTRON® EC in the spray tank with Pyramin or TCA, add NORTRON EC first. Agitate spray solution thoroughly and continuously. See Pyramin label for additional mixing directions.

GENERAL APPLICATION: Sugar beets grown under rainfall: Apply NORTRON EC alone or in a tank-mix preemergence at time of planting or shortly after, but prior to weed germination. NORTRON EC or tank-mix does not require mechanical soil incorporation provided that sufficient rainfall occurs shortly following application to activate the chemical. One-half inch of rainfall is usually adequate for activation. In areas where rainfall can be marginal for activation, such as the Red River Valley (Minnesota and North Dakota), it is recommended that NORTRON EC or the tank-mix be applied before or at the time of planting and incorporated into the soil.

Sugar beets grown under furrow irrigation: Apply NORTRON® EC alone or in a tank-mix to the soil surface preplant or at time of planting, but prior to weed germination, and incorporate into the soil. Where sugar beets are grown in beds, apply NORTRON EC or tank-mix after bedding and incorporate. Since NORTRON EC or tank-mix must have moisture to control weeds effectively, irrigate until tops of beds are thoroughly wetted.

Sugar beets grown under sprinkler irrigation: Apply NORTRON EC alone or in tank-mix preemergence at time of planting or shortly after, and irrigate prior to crop and weed germination. Repeat irrigation as necessary to maintain good moisture in upper soil layer. Apply at least one-half inch of water during first irrigation. Do not mechanically incorporate NORTRON EC or tank-mix into the soil under sprinkler irrigation.

CULTURAL PRACTICES FOLLOWING APPLICATION: When properly applied, NORTRON EC alone or in a tank-mix will provide up to 10 weeks control of susceptible weed species. When cultivating fields in which NORTRON EC or tank-mixes have been banded, care should be exercised to minimize the movement of untreated soil into the treated band. Where a broadcast application has been made, do not cultivate deeper than two inches, as this reduces the effectiveness of NORTRON EC or tank-mixes.

NORTRON® EC Alone
Dosage Table 1 (All Regions Except North Dakota and Minnesota):

Soil Texture	RATE OF NORTRON EC PER ACRE(1)			
		7-Inch Band Width(2)		
	Broadcast	22″ Row	28″ Row	30″ Row
Coarse Textured Soils: Sands, loamy sands and sandy loams	3/4 to 1 1/4 Gallons	2 to 3 1/4 Pints	1 1/2 to 2 1/2 Pints	1 1/4 2 1/4 Pints
Medium Texture Soils: Silt loams, clay loams which contain less than 3% organic matter	1 1/4 to 2 Gallons	3 1/4 to 5 Pints	2 1/2 to 4 Pints	2 1/4 to 3 3/4 Pints
Fine Textured Soils: Clay loams which contain more than 3% organic matter and clays	2 to 2 1/2 Gallons	5 to 6 1/4 Pints	4 to 5 Pints	3 3/4 to 4 3/4 Pints

(1) Use the higher rate within each soil texture category on the finer texture soils and/or where Kochia, barnyardgrass or black nightshade are expected to be a problem.
(2) For other band or row widths, adjust the rate in proportion to the area actually treated.

Dosage Table 2 (North Dakota and Minnesota only):

Soil Texture		NORTRON® EC/Acre
		7-Inch Band Width(1)
	Broadcast	22″ Row
Coarse Textured Soils: Sandy loams only	2 Gallons	5 Pints
Medium Textured Soils: Silt loams and clay loams	2 Gallons	5 Pints
Fine Textured Soils: Heavy clays	2 1/2 Gallons	6 1/4 Pints

(1) For other band or row widths, adjust the rate in proportion to the area actually treated

NORTRON® EC + ANTOR® 4ES (tank-mix)

CALIFORNIA, COLORADO, MONTANA, NEBRASKA, TEXAS, AND WYOMING ONLY:

This tank-mix controls these additional weeds:

Italian ryegrass (annual) (*Lolium multi-florum*), Annual sowthistle (*Sonchus oleraceus*), Fall panicum (*Panicum dichotomiflorum*), Foxtail millet (*Setaria italica*), Groundcherry (*Physalis longifolia*), Junglerice (*Echinochloa colonum*), Prostrate pigweed (*Amaranthus blitoides*), Shepherdspurse (*Capsella bursa-pastoris*), Witchgrass (*Panicum capillare*).

NORTRON® EC + ANTOR® 4ES tank-mix may be applied preplant incorporated where furrow irrigation is practiced using shallow incorporation implements such as rolling cultivators or harrows, or using power tillers set to a depth of 1–2 inches.

Preemergence applications of NORTRON EC + ANTOR ES tank-mix may be used when application is followed by sprinkler irrigation prior to weed germination and crop emergence, or where rainfall can be expected to occur within 7–10 days after application.

ANTOR 4ES may also be utilized in sequential applications with NORTRON EC where a broadcast preemergence treatment of ANTOR 4ES is followed by a band overlay of NORTRON EC preemergence. The same dosage rates should be used as in the tank-mix.

Refer to the Dosage Table 3 for recommended rates.

Before use, read the ANTOR 4ES label for additional information and precautions.

Dosage Table 3 (California, Colorado, Montana, Nebraska, Texas, and Wyoming only):

Soil Texture	ANTOR® 4ES Per Acre(1)	NORTRON® EC Per Acre(1)
Coarse Loamy sands and sandy loams	1.5 Quarts	4 Quarts
Medium: Loams and silt loams	2 Quarts	5.3 Quarts
Fine Clay loams and clays	3 to 4 Quarts (2)	6.6 Quarts

(1) See section on Band Applications for calculations of proportional amounts.
(2) Use higher rate on heavy clay soils.

NORTRON EC + PYRAMIN®* (tank-mix)

This tank-mix controls these additional broadleaf weeds: Annual sowthistle (*Sonchus oleraceus*), Black mustard (*Brassica nigra*), Buffalobur (*Solanum rostratum*), Coast fiddleneck (*Amsinckia intermedia*), Common groundsel (*Senecio vulgaris*). Common ragweed (*Ambrosia artemisiifolia*), Cutleaf nightshade (*Solanum triflorum*), Groundcherry (*Physalis* spp.), Henbit (*Lamium amplexicaule*), Nettleleaf goosefoot (*Chenopodium murale*), Prickly lettuce (Lactuca serriola), Prostrate knotweed (*Polygonum aviculare*), Shepherdspurse (*Capsella bursa-pastoris*).

CALIFORNIA (winter-grown sugar beets only): Under sprinkler irrigation or where natural rainfall is adequate, apply this tank-mix preemergence. See Pyramin label for precautions regarding application of sprinkler irrigation. Where furrow irrigation is to be used, apply preplant incorporated. Prepare seedbed or form beds for planting. Then use rotary tiller type of incorporation and incorporate not more than 2 inches deep. Plant sugar beets and irrigate. Sub-irrigate until tops of beds are thoroughly wetted. Refer to Dosage Table 4 for recommended rates.

Dosage Table 4 (California, winter-grown sugar beets only):

Soil Texture	NORTRON® EC Per Acre		PYRAMIN® FL Per Acre	
	Broadcast (For calibration purposes only)	10-Inch Band Width(1)(2) 30″ Row	Broadcast (For calibration purposes only)	10-Inch Band Width(1) 30″ Row
Coarse Textured Soils: Sands, loamy sands and sandy loams	NOT RECOMMENDED			
Medium Textured Soils: Silt loams, clay loams which contain less than 3% organic matter	1 to 1 1/4 Gallons	2 2/3 to 3 1/3 Pints	3 Quarts	1 Quart
Fine Textured Soils: Clay loams which contain more than 3% organic matter and clays	1 1/3 to 1 3/4 Gallons	3 1/2 to 4 1/2 Pints	3 Quarts	1 Quart

(1) For other band or row widths, adjust rates in proportion to the area actually treated. Do not apply this mixture broadcast.
(2) Use the higher rate of NORTRON® EC within each soil texture category on the finer-textured soils and/or where volunteer barley or wheat are expected to be a problem.

* Pyramin® is a registered trademark of BASF Aktiengesellschaft

PRECAUTIONS: Do not exceed 3/4 inch of sprinkler irrigation per set until the beets have two true leaves. Do not use the tank-mix under conditions where Pyramin alone is not recommended. Before use, read the Pyramin label for additional information and precautions.

CENTRAL AND EASTERN STATES, INCLUDING MICHIGAN AND OHIO: This tank-mix controls the additional weed, common ragweed, and other broadleaf weeds specified in the weed table. Apply preemergence at the time of planting or shortly after, but before weed germination using recommended rates listed in Dosage Table 5. Do not mechanically incorporate the herbicides into the soil as crop injury may result. Do not use this tank-mix where Pyramin alone is not recommended. Before use, read Pyramin label for additional information and precautions.

Dosage Table 5 (Central and Eastern States only):

Soil Texture	NORTRON® EC/Acre			PYRAMIN® FL/Acre		
		7-Inch Band Width(1)			7-Inch Band Width(1)	
	Broad-cast	22" Row	28" Row	Broad-cast	22" Row	28" Row
Coarse Textured Soils: Sandy loams only	1 Gallon	2 1/2 Pints	2 Pints	2.25 Quarts	.75 Quart	.5 Quart
Medium Textured Soils: Silt and clay loams which contain less than 3% organic matter	1 1/3 Gallons	3 1/3 Pints	2 2/3 Pints	3 Quarts	1 Quart	.75 Quart
Fine Textured Soils: Clay loams which contain more than 3% organic matter and clays	1 2/3 Gallons	4 1/4 Pints	3 1/3 Pints	3 Quarts	1 Quart	.75 Quart

(1) For other band or row widths, adjust the rate in proportion to the area actually treated.

NORTRON® EC + ANTOR® 4ES + PYRAMIN® (tank-mix)

MICHIGAN ONLY: A tank-mix of NORTRON EC + ANTOR 4ES + Pyramin may be used in Michigan for control of the additional weeds:

Common ragweed (*Ambrosia artemisiifolia*), Wild mustard (*Brassica kaber*), Pennsylvania smartweed (*Polygonum pensylvanicum*), Common lambsquarters (*Chenopodium album*), Common purslane (*Portulaca oleracea*), Nettleleaf goosefoot (*Chenopodium murale*), Common chickweed (*Stellaria media*), Henbit (*Lamium amplexicaule*), Fanweed (stinkweed) (*Thlaspi arvense*), Kochia (*Kochia scoparia*), Russian thistle (*Salsola kali*), Volunteer barley (*Hordeum* sp.), Volunteer wheat (*Triticum* sp.), ladysthumb (*Polygonum persicaria*).

For preemergence use without irrigation, apply the tank-mix to the soil surface at the time of planting or shortly after, but prior to weed germination. Use rates recommended in Dosage Table 6.

Do not use this mixture under conditions where Pyramin alone is not recommended. Before use, read the Pyramin and ANTOR 4ES labels for additional information and precautions.

Dosage Table 6 (Michigan only):

Soil Texture	Broadcast Rates Per Acre		
	Nortron® EC (qts.)	Antor® 4ES (qts.)	Pyramin® FL (qts.)
Coarse Textured Soils: Sandy loams only	4	1.5	2.25
Medium Textured Soils: Loams, silt loams and clay loams containing less than 3% organic matter	5.3	2	2.25
Fine Textured Soils: Clay loams and clays containing more than 3% organic matter	5.3	3	2.25

NORTRON EC + TCA (tank-mix)**

CENTRAL AND EASTERN STATES, INCLUDING MICHIGAN AND OHIO: The control of certain grass weeds, particularly yellow foxtail and barnyardgrass, can be enhanced by the use of this tank-mix.

Apply premergence at time of planting or shortly after but prior to weed germination. Apply at rates recommended in Dosage Table 7. Before use, read TCA label for additional information and precautions.

** TCA is a registered trademark of Hopkins Chemical Company

Dosage Table 7 (Central and Eastern States, including Michigan and Ohio):

Soil Texture	Nortron® EC/Acre		TCA/Acre(1)	
		7-Inch Band(2)		7-Inch Band(2)
	Broadcast	28" Row	Broadcast	28" Row
Coarse Textured Soils: Sandy loams only	1 Gallon	2 Pints	3 Quarts	1½ Pints
Medium Textured Soils: Silt and clay loams which contain less than 3% organic matter	1⅓ Gallons	2⅔ Pints	1 Gallon	1 Quart
Fine Textured Soils: Clay loams which contain more than 3% organic matter and clays	2 Gallons	4 Pints	1 Gallon	1 Quart

(1) Based on commercial liquid formulation containing the equivalent of 6 pounds of dry 90% sodium TCA per gallon.

(2) For other band or row widths, adjust the rate in proportion to the area actually treated.

EPTAM®* (fall-applied) + NORTRON® EC (spring-applied)

NORTH DAKOTA AND MINNESOTA ONLY (sequential treatment): The control of certain weeds, particularly yellow foxtail and redroot pigweed, can be improved by the sequential treatment of Eptam 7E applied in the fall followed by a spring application of NORTRON EC. Furthermore, the addition of NORTRON EC provides up to 10 weeks' control of susceptible weeds.

Apply Eptam 7E in the fall in accordance with directions on the label.

Apply NORTRON EC in the spring to the soil surface preplant or at the time of planting and incorporate into the soil. Apply at recommended rates listed in Dosage Table 8. This sequential treatment is not recommended for use on coarse-textured soils.

Do not use in areas or under conditions where Eptam 7E is not recommended. Before use, read Eptam 7E label for additional information and precautions.

* Eptam® is a registered trademark of ICI Americas Inc.

Dosage Table 8 (North Dakota and Minnesota only):

Soil Texture	NORTRON® EC Per Acre	
		7-Inch Band Width(1)
	Broadcast	22" Row
Coarse Textured Soils: Sands, loamy sands and sandy loams	NOT RECOMMENDED	
Medium Textured Soils: Silt loams and clay loams	2 Gallons	5 Pints
Fine Textured Soils: Heavy clays	2 Gallons	5 Pints

(1) For other band or row widths, adjust the rate in proportion to the area actually treated.

POSTEMERGENCE APPLICATION

(For use in California, Michigan, Ohio, Minnesota, and North Dakota only):

GENERAL INFORMATION

The tank-mixes of NORTRON® EC plus BETANEX® and NORTRON EC plus BETAMIX® applied postemergence broaden and enhance the control of weeds.

Furthermore, control of susceptible weeds which may germinate after treatment can be obtained provided overhead moisture is applied to activate NORTRON EC. The choice of tank-mixes is dependent upon weed species present.

NORTRON EC alone is not recommended for postemergence use.

Timing of Application: Apply the tank-mix NORTRON EC + BETANEX or NORTRON EC + BETAMIX when sugar beets have at least 4 fully expanded true leaves. Apply at rates recommended in the Dosage Tables. Use the higher rate of NORTRON EC where increased residual weed control is desired. Where Eptam has been applied preplant (fall or spring-applied), do not apply NORTRON EC + BETANEX or BETAMIX tank-mix before the sugar beets have 6 expanded true leaves. Sugar beets may be killed or severely injured if treated at earlier stages of crop growth. See Postemergence Use Precautions for additional information on proper use.

Mixing the Spray: NORTRON® EC, BETANEX® and BETAMIX® are formulated as emulsifiable concentrates and are easily mixed. Add NORTRON EC to the water in the spray tank followed by BETANEX or BETAMIX while agitating the spray solution thoroughly. Refer to the BETANEX or BETAMIX labels for additional precautions and information on mixing.

Spray Equipment: Apply the mixture using standard low pressure (20–60 psi) spray equipment. Ensure that the sprayer is thoroughly clean. Spray equipment should be carefully calibrated before use and checked frequently during application to see that it is functioning properly. Uniformly apply the recommended rate in 20–60 gallons of water per acre on a broadcast basis. Avoid overlaps, since crop injury may result. When applying in a band, check to make certain that the band width is accurate for the dosage rate being applied. Do not use strainer smaller than 50-mesh.

Moisture Following Application/Residual Weed Control: Rainfall or sprinkler irrigation within 6 hours of spraying may reduce weed control; however, moisture after this period of time is advantageous for moving NORTRON® EC into the 1–2" layer of soil where it can be absorbed by the roots of sprayed and germinating weeds to provide optimum control. One-half inch or more of sprinkler irrigation is required to activate NORTRON EC on most soil types.

Residual control of weeds is dependent upon soil moisture conditions, rate of herbicide used, and soil texture. The activity of NORTRON EC in the soil is reduced as the soil texture becomes finer and organic matter increases.

CAUTION:

MINNESOTA AND NORTH DAKOTA—Do not apply to crop which has been previously treated with NORTRON EC.

CALIFORNIA, MICHIGAN AND OHIO—Following a preemergence treatment of NORTRON EC, do not apply NORTRON EC postemergence where more than 2 gallons were applied preemergence. No more than a total of 2¾ gallons of NORTRON EC should be applied in a single growing season. See Postemergence Use Precautions for additional information on proper use.

NORTRON® EC + BETANEX® (tank-mix)

MICHIGAN, OHIO, MINNESOTA, AND NORTH DAKOTA ONLY: NORTRON EC applied postemergence in a tank-mix with BETANEX broadens and enhances the control of troublesome weeds. Furthermore, preemergence control of susceptible weeds which may germinate following treatment can be obtained provided overhead moisture is sufficient to activate NORTRON EC.

WEED SPECIES CONTROLLED:

Group I: Weeds controlled up to the 6-leaf stage.

Redroot pigweed (*Amaranthus retroflexus*), Wild mustard (*Brassica kaber*), Nettleleaf goosefoot (*Chenopodium murale*), London rocket (*Sisymbrium irio*), Common lambsquarters (*Chenopodium album*).

Group II: Weeds controlled up to the 4-leaf stage.

Common chickweed (*Stellaria media*), Black nightshade (*Solanum nigrum*), Common ragweed (*Ambrosia artemisiifolia*), Shepherdspurse (*Capsella bursa-pastoris*), Groundcherry (*Physalis lanceifolia*), Pennsylvania smartweed (*Polygonum pensylvanicum*), Ladysthumb (*Polygonum persicaria*).

Group III: Weeds controlled up to the 2-leaf stage.

Annual sowthistle (*Sonchus oleraceus*), Common purslane (*Portulaca oleracea*), Wild buckwheat (*Polygonum convolvulus*), Kochia*(*Kochia scoparia*)

* Spray kochia while in the rosette stage, less than 1" in diameter.

Dosage Table 9 (Michigan, Ohio, Minnesota and North Dakota only):									
Rate of NORTRON® EC Per Acre(1)					Rate of BETANEX® Per Acre(1)				
		Row Spacing					Row Spacing		
Broadcast Rate	Band(2) Width (in)	22"	28"	30"	Broadcast Rate	Band(2) Width (in)	22"	28"	30"
3/4-1 Gallons	7	2-2 2/3 Pints	1 1/2-2 Pints	1 1/3-1 3/4 Pints	4 1/2-6 Pints	7	1 1/2-2 Pints	1 1/8-1 1/2 Pints	1-1 1/3 Pints

(1) Use the higher rate on larger weeds and sugar beets.

(2) For other band or row widths, adjust rates in proportion to the area actually treated.

NORTRON® EC + BETAMIX® (tank-mix)

CALIFORNIA, MICHIGAN, OHIO, MINNESOTA AND NORTH DAKOTA ONLY: NORTRON EC applied postemergence in a tank-mix with BETAMIX broadens and enhances the control of troublesome weeds. Furthermore, preemergence control of susceptible weeds, which may germinate following treatment, can be obtained provided overhead moisture is sufficient to activate NORTRON EC.

WEED SPECIES CONTROLLED:

Group I: Weeds controlled up to the 6-leaf stage.

Redroot pigweed (*Amaranthus retroflexus*), Wild mustard (*Brassica kaber*), Nettleleaf goosefoot (*Chenopodium murale*), London rocket (*Sisymbrium irio*), Common lambsquarters (*Chenopodium album*).

Group II: Weeds controlled up to the 4-leaf stage.

Common chickweed (*Stellaria media*), Black nightshade (*Solanum nigrum*), Common ragweed (*Ambrosia artemisiifolia*), Shepherdspurse (*Capsella bursa-pastoris*), Groundcherry (*Physalis lanceifolia*), Pennsylvania smartweed (*Polygonum pensylvanicum*), Ladysthumb (*Polygonum persicaria*), Prostrate pigweed (*Amaranthus blitoides*), Coast fiddleneck (*Amsinckia intermedia*).

Group III: Weeds controlled up to the 2-leaf stage.

Annual sowthistle (*Sonchus oleraceus*), Kochia* (*Kochia scoparia*), Common purslane** (*Portulaca oleracea*), Prostrate knotweed (*Polygonum aviculare*), Wild buckwheat (*Polygonum convolvulus*), Green foxtail*** (*Setaria viridis*), Yellow foxtail (pigeongrass)*** (*Setaria glauca*), Annual bluegrass*** (*Poa annua*), Canarygrass*** (*Phalaris canariensis*).

* Spray kochia while in the rosette stage, less than 1" in diameter.

** Group II weed in California.

*** For best control, overhead moisture required.

Dosage Table 10 (California, Michigan, Ohio, Minnesota and North Dakota only):									
Rate of NORTRON® EC Per Acre(1)					Rate of BETAMIX® Per Acre(1)				
		Row Spacing					Row Spacing		
Broadcast Rate	Band(2) Width (in)	22"	28"	30"	Broadcast Rate	Band(2) Width (in)	22"	28"	30"
3/4-1 Gallons	7	2-2 2/3 Pints	1 1/2-2 Pints	1 1/3-1 3/4 Pints	4 1/2-6 Pints	7	1 1/2-2 Pints	1-1 1/2 Pints	1 1/3 Pints

(1) Use the higher rate on larger weeds and sugar beets.

(2) For other band or row widths, adjust rates in proportion to the area actually treated.

PREPLANT AND PREEMERGENCE USE PRECAUTIONS:

NORTRON® EC applied alone or in tank-mixes according to label directions and under normal growing conditions may cause temporary leaf fusion, distortion and stunting. Crop injury may occur during early growth when crop is stressed due to herbicide residue carryover, highly saline or alkaline soils, unusually cold and wet weather or improperly placed fertilizers or soil insecticides.

Unusually dry, windy weather following application of NORTRON EC, which dries the upper soil layer, may reduce effectiveness.

Where NORTRON EC is used in a tank-mix with TCA, do not use treated sugar beet tops for feed or forage.

POSTEMERGENCE USE PRECAUTIONS: Make only one application of NORTRON® EC + BETANEX® or BETAMIX® tank-mix during each growing season.

Do not apply NORTRON EC + BETANEX or BETAMIX tank-mix to sugar beets later than 90 days prior to harvest.

Crop Planting Precautions: If crop is lost due to unfavorable growth conditions following treatment, do not replant with crops other than sugar beets in treated land during the same season. If fields are replanted to sugar beets, reseed into treated band. Do not retreat field with NORTRON EC in the same season.

NORTRON EC + BETANEX OR BETAMIX MAY CAUSE SUGAR BEET INJURY OR STAND LOSS IF THE CROP IS UNDER STRESS FROM ONE OR MORE OF THE FOLLOWING CONDITIONS.

Rapid climatic changes from cool, overcast days, to hot (80°F or over), bright days. When the air temperature is, or is likely to be, above 80°F on the day of spraying, application should be made in the evening when the temperature is lower.
Frost within seven days following treatment;
Windy or drought conditions;
Use of a preplant or preemergence herbicide or other chemicals;
Insect or disease injury;
Close cultivation.

If stress conditions are present, delay application until crop has recovered.

DO NOT OVERTREAT: The use of higher-than-recommended rates may cause beet injury.

Do not spray while dew is present.

Rainfall or sprinkler irrigation within 6 hours of application may reduce weed kill.

Do not allow spray drift to contact adjacent crops which may be injured by spray drift.

IMPORTANT: This tank-mix may cause temporary growth retardation and/or chlorosis or tip-burn on sugar beets. Sugar beets usually resume normal growth within 10 days.

APPENDIX I

NORTRON EC WITH LIQUID FERTILIZERS

The following procedure is suggested for evaluation of physical compatibility of NORTRON EC in mixtures with liquid fertilizers for spray tank application.

MATERIAL REQUIRED:

1. NORTRON EC - components of tank-mixes if intended for use
2. Liquid fertilizer to be used
3. Adjuvant for fertilizer tank mix: Compex* or E-Z Mix*.
4. Two (or more) one quart, wide mouth containers with lids or stoppers
5. Measuring spoons - (25 ml pipette or graduated cylinder provides more accurate measurement)
6. Measuring cup, 8 fl. oz. (237 ml)

* Compex - Kalo Laboratories, Inc., Kansas City, MO

* E-Z Mix - United Agri-Products, Greeley, CO

PROCEDURE:

1. Pour one pint (473 ml) of the liquid fertilizer into each of the quart containers.
2. Add adjuvant(s) to one or more of the containers and mix; follow label directions of adjuvant.
3. Add the NORTRON EC and tank mix components to the containers (see rate table below).
4. Close the containers with lids or stoppers and mix contents by inverting the containers ten times.
5. Inspect the surface and body of the mixture:
   a) immediately after mixing,
   b) after allowing mixtures to stand quietly for 30 minutes,
   c) immediately after mixing again (invert the containers ten more times.

If a uniform mixture does not occur, the spray tank mixture should not be used. If any of the mixtures remain uniform for 30 minutes, that mixture may be used in spray tank applications. Should any of the mixtures separate after 30 minutes but remix readily into a uniform mixture with inversion of the container, the mixture may be used provided that adequate agitation is maintained in the spray tank. If a NORTRON EC + fertilizer mixture utilizing an adjuvant is satisfactory, but the one without adjuvant is not, be sure to use the adjuvant in the spray tank at the rate recommended on the label, which was used in this test.

If nondispersable oil, sludge or clumps of solids form in the mixtures, those combinations should not be used for spray tank application.

RATE TABLE FOR NORTRON EC MIXTURES WITH LIQUID FERTILIZERS

Gal. of Liquid Fertilizer to be Applied Per Acre	*ml or tsp. of NORTRON EC to be added to 1 pint of fertilizer	
	ml	tsp
20	47	9.5
30	32	6.5
40	24	5.0
50	19	4.0
60	16	3.25

* Based on field use rate of 3.0 lb ai/acre (2 gal/acre) in the fertilizer volumes indicated. Adjust amount of NORTON EC added proportionately to correspond with intended field use rate recommended on NORTRON EC label for soil type. Add the proportionate amount of tank mix component (e.g., ANTOR 4ES, Pyramin) if intended for use, based on volume of NORTRON EC used in the test.

IMPORTANT: READ BEFORE USE

By using this product, user accepts the following conditions, warranty, disclaimer of warranties and limitations of liability.

CONDITIONS: The directions for use of this product are believed to be adequate and should be followed carefully. However, because of extreme weather and soil conditions, manner of use and other factors beyond NOR-AM Chemical Company's control, it is impossible for NOR-AM to eliminate all risks associated with the use of this product. As a result, crop injury or ineffectiveness is always possible.

DISCLAIMER OF WARRANTIES: THERE ARE NO WARRANTIES, EXPRESS OR IMPLIED, OF MERCHANTABILITY OR FITNESS FOR A PARTICULAR PURPOSE OR OTHERWISE, WHICH EXTEND BEYOND THE STATEMENTS MADE ON THIS LABEL. No agent of NOR-AM Chemical Company is authorized to make any warranties beyond those contained herein or to modify the warranties contained herein. UNDER NO CIRCUMSTANCES SHALL NOR-AM CHEMICAL COMPANY BE LIABLE FOR ANY INCIDENTAL OR CONSEQUENTIAL DAMAGES RESULTING FROM FAILURE TO USE THIS PRODUCT IN ACCORDANCE WITH LABEL DIRECTIONS.

LIMITATIONS OF LIABILITY: In no event shall NOR-AM Chemical Company's liability for damages arising out of the use of this product exceed the purchase price of the product used.

518-IB-NEC-5M-Rev. 5/89

This is a specimen label. NOR-AM Chemical Company is not responsible for the information contained herein. As labels are subject to revision, always carefully read and follow the label on the product container.

ROTATE® 2 1/2 G

INSECTICIDE

For Use on Non-Bearing Nut, Citrus and other Fruit Trees
DO NOT USE ON FOOD CROPS

ACTIVE INGREDIENTS:	By Weight
Bendiocarb: (2,2-dimethyl-1,3 benzodioxol-4-yl methylcarbamate*)	2.5%
INERT INGREDIENTS:	97.5%
TOTAL	100.0%

EPA Reg. No. 45639-100 *Protected by U.S. Patent No. 3,736,338

KEEP OUT OF REACH OF CHILDREN

CAUTION

STATEMENT OF PRACTICAL TREATMENT

IF SWALLOWED: DRINK 1 OR 2 GLASSES OF WATER AND INDUCE VOMITING BY TOUCHING BACK OF THROAT WITH FINGER. CALL PHYSICIAN OR POISON CONTROL CENTER. DO NOT INDUCE VOMITING OR GIVE ANYTHING BY MOUTH TO AN UNCONSCIOUS PERSON. GET MEDICAL ATTENTION.

IF INHALED: REMOVE VICTIM TO FRESH AIR.

IF ON SKIN: WASH WITH PLENTY OF SOAP AND WATER. GET MEDICAL ATTENTION IF IRRITATION PERSISTS.

IF IN EYES: FLUSH WITH PLENTY OF WATER.

NOTE TO PHYSICIAN: BENDIOCARB IS A MODERATE REVERSIBLE CHOLINESTERASE INHIBITOR. INTRAVENOUS INJECTION OF ATROPINE IS ANTIDOTAL.

PRECAUTIONARY STATEMENTS

HAZARDS TO HUMANS AND DOMESTIC ANIMALS

CAUTION

Harmful if swallowed. Do not take internally. Do not breathe dust. Avoid contact with skin, eyes, or clothing. May be absorbed through the skin. Wear long sleeves. Change clothing daily. Wash hands thoroughly after handling. Keep children and pets off treated area during irrigation and until area has dried. Avoid contamination of feed and foodstuffs.

ENVIRONMENTAL HAZARDS

This pesticide is toxic to fish and wildlife. Do not apply directly to water or wetlands (swamps, bogs, marshes, and potholes). Runoff from treated areas may be hazardous to aquatic organisms in neighboring areas. Do not contaminate water by cleaning of equipment or disposal of waste. Collect, cover or incorporate granules spilled on soil surface.

This product is highly toxic to bees exposed to direct treatment of residues on plants. Protective information may be obtained from your Cooperative Agricultural Extension Service.

IMPORTANT: Read the entire Directions for Use and Disclaimer of Warranties before using this product.

IN CASE OF FIRE, LEAKY OR DAMAGED CONTAINERS, OR OTHER EMERGENCY, REPORT AT ONCE BY TOLL-FREE TELEPHONE TO: 800-424-9300.

DIRECTIONS FOR USE

It is a violation of federal law to use this product in a manner inconsistent with its labeling.

Store this product in its original, closed container in a dry, locked area away from children and domestic animals. Do not reuse empty container.

PESTICIDE DISPOSAL: Securely wrap original container in several layers of newspaper and discard in trash.

CONTAINER DISPOSAL: Do not reuse bag. Discard bag in trash.

GENERAL INFORMATION

ROTATE® 2½ G is a broad spectrum residual insecticide for the control of certain insects on non-bearing nut, citrus and other fruit trees.

APPLICATION EQUIPMENT: ROTATE® 2½ G is a granular formulation. Apply with a fertilizer spreader or equivalent equipment as uniformly as possible.

NON-BEARING NUT, CITRUS AND OTHER FRUIT TREES: (Trees not expected to bear nuts or fruit for 12 months following last application of ROTATE® 2½ G).

Applications should be made using the rates shown below:

Pest	Dosage of Product	Application Procedure
Ants	1 oz. per tree or ant mound (110–200 oz./acre)	Apply around base of trees, transplants, resets or around ant mound. Repeat as necessary.

IMPORTANT: READ BEFORE USE

By using this product, user accepts the following conditions, warranty, disclaimer of warranties and limitations of liability.

CONDITIONS: The directions for use of this product are believed to be adequate and should be followed carefully. However, because of extreme weather and soil conditions, manner of use and other factors beyond NOR-AM Chemical Company's control, it is impossible for NOR-AM to eliminate all risks associated with the use of this product. As a result, crop injury or ineffectiveness is always possible.

DISCLAIMER OF WARRANTIES: THERE ARE NO WARRANTIES, EXPRESS OR IMPLIED OF MERCHANTABILITY OR OF FITNESS FOR A PARTICULAR PURPOSE OR OTHERWISE, WHICH EXTEND BEYOND THE STATEMENTS MADE ON THIS LABEL. No agent of NOR-AM Chemical Company is authorized to make any warranties beyond those contained herein or modify the warranties contained herein. UNDER NO CIRCUMSTANCES SHALL NOR-AM CHEMICAL COMPANY BE LIABLE FOR ANY INCIDENTAL OR CONSEQUENTIAL DAMAGES RESULTING FROM FAILURE TO USE THIS PRODUCT IN ACCORDANCE WITH LABEL DIRECTIONS.

LIMITATIONS OF LIABILITY: In no event shall NOR-AM Chemical Company's liability for damages arising out of the use of this product exceed the purchase price of the product used.

725-IB-RT 2½ G-5M-Rev. 10/88

This is a specimen label. NOR-AM Chemical Company is not responsible for the information contained herein. As labels are subject to revision, always carefully read and follow the label on the product container.

VORLEX®

SOIL FUMIGANT

RESTRICTED USE PESTICIDE

For retail sale to and use only by Certified Applicators or persons under their direct supervision and only for those uses covered by the Certified Applicator's Certification.

FOR AGRICULTURAL USE ONLY

ACTIVE INGREDIENTS:	
1,3-dichloropropene and other related chlorinated C_3 hydrocarbons	40.0%
Methylisothiocyanate	20.0%
INERT INGREDIENTS:	40.0%
TOTAL	100.0%

Formula Weight: 8.6 lbs/Gallon

EPA Reg. No. 45639-90 U.S. PAT. Nos. 3,205,129; 3,113,908; 3,240,666

PRECAUCION AL USUARIO: Si usted no lee ingles, no use este producto hasta que la etiqueta haya sido explicada ampliamente.

KEEP OUT OF REACH OF CHILDREN

DANGER–POISON

PELIGRO—VENENO

STATEMENT OF PRACTICAL TREATMENT

IF INHALED: REMOVE TO FRESH AIR. IF NOT BREATHING, GIVE ARTIFICIAL RESPIRATION, PREFERABLY MOUTH-TO-MOUTH. IF BREATHING IS DIFFICULT, GIVE OXYGEN. CALL A PHYSICIAN.

IF ON SKIN: IMMEDIATELY FLUSH SKIN WITH PLENTY OF WATER FOR AT LEAST 15 MINUTES WHILE REMOVING CONTAMINATED CLOTHING AND SHOES. CALL A PHYSICIAN. IF WATER IS NOT IMMEDIATELY AVAILABLE, REMOVE EXCESS CHEMICAL FROM SKIN WITH ABSORBENT MATERIAL SUCH AS TOWEL OR DRY SOIL, THEN PROCEED AT ONCE TO LOCATION WHERE WATER IS AVAILABLE AND THOROUGHLY WASH CONTAMINATED SKIN WITH PLENTY OF WATER. CALL A PHYSICIAN.

IF IN EYES: IMMEDIATELY FLUSH EYES WITH PLENTY OF WATER OR A 5% SOLUTION OF SODIUM BICARBONATE FOR AT LEAST 15 MINUTES; CALL A PHYSICIAN. CONTINUE TREATMENT WITH OPHTHALMIC HYDROCORTISONE OINTMENT.

IF SWALLOWED: IMMEDIATELY CALL A PHYSICIAN OR POISON CONTROL CENTER. DO NOT INDUCE VOMITING. NOTE TO PHYSICIAN: EMPTY STOMACH CONTENTS BY GASTRIC LAVAGE.

DO NOT REUSE ORIGINAL CONTAINERS UNLESS RECONDITIONED

PRECAUTIONARY STATEMENTS

HAZARDS TO HUMANS AND DOMESTIC ANIMALS

DANGER. CAUSES SEVERE EYE DAMAGE. MAY BE FATAL IF INHALED, ABSORBED THROUGH SKIN, OR SWALLOWED. DO NOT GET IN EYES, ON SKIN OR ON CLOTHING. DO NOT TAKE INTERNALLY. DO NOT BREATHE VAPOR. USE ONLY WITH ADEQUATE VENTILATION.

Causes burns of skin and eyes. May produce kidney and liver damage upon prolonged exposure. Wear chemical worker goggles, face shield or safety glasses. Wash thoroughly with soap and water after handling, and before eating and smoking.

GENERAL DIRECTIONS

NOTE TO USERS: CANCER HAZARD WARNING STATEMENT

The use of this product may be hazardous to your health. This product contains 1,3-dichloropropene which has been determined to cause tumors in laboratory animals. Risks can be reduced by closely following the use directions and precautions, and by wearing protective clothing specified elsewhere on this label.

HANDLING PRECAUTIONS

DO NOT GET ON SKIN—NOTE TO USERS:

Required clothing and equipment during the handling and application of VORLEX:

One-piece coveralls which have long sleeves and long pants constructed of laminated fabric as specified in the USDA/EPA Guide for Commercial Applicators.

Liquid-proof hat such as a plastic hard hat with a plastic sweat band.

Heavy duty liquid-proof (neoprene/synthetic) work gloves and boots.

Any article worn while handling VORLEX must be washed before reusing. Immediately remove all clothing which has been drenched or has otherwise absorbed VORLEX from any spill. Dispose of contaminated clothing in a sanitary landfill by incineration, or if allowed by state and local authorities, by burning. If burned, stay out of smoke.

Never wear protective gear having odor of VORLEX. Wash and aerate all protective gear thoroughly after use until all odor is gone.

DO NOT INHALE VAPORS—NIOSH- or MSHA-approved respiratory protection should be worn when VORLEX liquid soil fumigant is exposed to the atmosphere or when conducting operations which vent to the atmosphere. A NIOSH- or MSHA-approved half-faced respirator with chemical worker's goggles or full-face respirators shall be used during small spills, repairs, calibrations, transfers, sampling and when working in poorly ventilated areas. When in use canisters or cartridges shall be replaced daily or sooner if specified by manufacturer or at first sign of odor breakthrough, whichever comes first. NIOSH-approved cartridges, such as a WILLSON-R21 will be adequate for short-term situations such as listed above.

Where very high concentrations of vapors might be expected (such as large spills in poorly ventilated areas) a self-contained or air-supplied respirator should be used.

DO NOT GET IN EYES—Wear eye protection such as chemical worker's goggles, safety glasses or full-face respirator when handling VORLEX.

DO NOT SWALLOW—Do not use the mouth to siphon VORLEX from containers or to blow out clogged lines, nozzles, etc.

FIELD REENTRY—Workers entering the treated area within 72 hours after application of VORLEX must wear protective clothing.

For Chemical Emergency—Spill, Leak, Fire, Exposure Or Accident—Call 800-424-9300 Day Or Night.

STORAGE, SHIPMENT AND DISPOSAL

STORAGE: Store in tightly-closed original container in a cool place away from dwellings. Do not allow contamination of seeds, plants, fertilizers, or other pesticide chemicals. Do not contaminate food, feedstuffs, drugs, or domestic water supplies. In outside storage, store drums (make sure that bungs are tightly closed) on sides to avoid accumulation of rain water in top or bottom of recessed areas.

SHIPMENT: Do not ship or store with food, feeds, drugs, or clothing.

PESTICIDE DISPOSAL: Do not contaminate water, food or feed by storage or disposal. Pesticide wastes are toxic. Improper disposal of excess pesticide, spray mixture, or rinsate is a violation of Federal Law. If these wastes cannot be disposed of by use according to label instructions, contact your State Pesticide or Environmental Control Agency, or the Hazardous Waste Representative at the nearest EPA Regional Office for guidance.

CONTAINER DISPOSAL: Remove bungs in a well-ventilated area, protected from precipitation, to clear drums of volatile residues, and offer for recycling or reconditioning. Do not puncture, cut, or weld empty drums until thoroughly ventilated and all volatiles are removed (to prevent explosion and/or burning), then puncture, crush, and dispose of in a State-approved sanitary landfill. Replace bungs before shipping. Other State-approved empty container disposal procedures may also be used.

DO NOT REUSE ORIGINAL CONTAINERS UNLESS RECONDITIONED

GENERAL INFORMATION

VORLEX Soil Fumigant is used as a preplanting treatment in both field and greenhouse soils.

In the soil, when used as directed, VORLEX vaporizes and diffuses, helping to control plant parasitic nematodes and harmful fungi causing damping-off, root-rot and wilt. When used as directed, VORLEX also helps to control wireworms and symphylans. When used as directed, VORLEX helps control weeds such as pigweed, dandelion, lambsquarters, bermudagrass, purslane, crabgrass, quackgrass, spurge, fleabane, sedge, chickweed, ragweed, mustard, Poa spp., smartweed, wild radish, dodder, velvetleaf and shepherdspurse.

PHYSICAL OR CHEMICAL HAZARDS

FLAMMABLE

DO NOT USE, POUR, SPILL, OR STORE NEAR HEAT OR OPEN FLAME. DO NOT CUT OR WELD CONTAINER. FIRE OR EXPLOSION MAY RESULT EVEN IF CONTAINER IS EMPTY.

ENVIRONMENTAL HAZARDS

This product is toxic to fish and wildlife. Do not spill or empty into streams, ponds or any other body of water. Do not contaminate water or wetlands by cleaning of equipment, disposal of wastes, or direct application. In case of spills, properly dispose of contaminated materials.

DIRECTIONS FOR USE

It is a violation of Federal Law to use this product in a manner inconsistent with its labeling.

USE PRECAUTIONS: Do not apply this product through any type of irrigation system.

As with most soil fumigants, application of VORLEX may temporarily raise the level of ammonia nitrogen and soluble salts in the soil which may result in plant injury. This is most likely to occur when heavy rates of fertilizer or fumigant are applied to soils that are either cold, wet, acid or high in organic matter.

Do not add fertilizers containing ammonia or ammonium salts or undecomposed manure to the soil at or near the time of fumigation. Use only fertilizers containing nitrates until the soil temperature is above 65°F and the crop is well established.

Do not allow liquid or vapors to come in contact with lawn and desirable plants; do not apply on areas beneath which roots of desirable plants extend.

Do not use in greenhouses which contain living plants.

Do not fumigate if the soil temperature at 3 inch depths is over 90°F.

Avoid reinfestation of treated soil. Do not contaminate treated soil with untreated soil. Do not use transplants or tools which could carry soil-borne pests from infested land. Clean equipment carefully before using.

VORLEX may be corrosive under certain conditions. Flush all applicators with fuel oil or kerosene immediately after use. DO NOT USE WATER. Do not use containers or equipment of aluminum, magnesium or their alloys. Close containers tightly.

TIME OF APPLICATION: Treatment can be made at any time the soil is workable; fall or winter application is desirable for crops which are planted early in the spring.

Soil temperature: VORLEX is active in soils as cold as 35°F except against soil insects.

PREPARATION OF SOIL: FOR OPTIMUM CONTROL WITH VORLEX FOLLOW THESE STEPS CAREFULLY. (1) Work crop remains into the soil so that they are well decomposed before treatment. (2) In shrub and tree replant sites remove as much of the root system as possible. (3) Break up all clods and loosen soil thoroughly by cultivation one week before treatment. Deep tillage is recommended, particularly for deep-rooted crops. Always work soil deeper than intended application depth. (4) Keep soil moist from the surface down through the root zone and maintain in this condition for one week before fumigation. DO NOT FUMIGATE DRY SOILS.

APPLICATION: Use injection type applicators.

Broadcast treatment: Set shanks 6" to 8" apart; for band treatment, set shanks 8" apart. Inject at a depth of 6" to 8"[1].

Row treatment: Adjust injection chisels to same spacing as planter. Seal the treated strips by bedding or listing. Always inject the fumigant at least 6–8" below the final soil (planting) surface. Before planting, remove the tops of the beds, being careful that none of this soil falls over on the fumigated band.

DOSAGE

Pest	Soil Type	Overall (Broadcast) Application		Row or Band Application	
		Per Acre	Per 100 sq. ft.	Per Chisel	Per 100 Linear Ft.
Nematodes ..	Mineral	7-15 gal.	2 1/8-4 1/2 fl.oz.	40-87 ml.[6]	1.4-3.0 fl.oz.[6]
Nematodes ..	Organic[2] Muck	10-25 gal.	3-7 1/2 fl. oz.	58-146 ml.[6]	2-5 fl.oz.[6]
Symphylans[3]	—	10-15 gal.	3-4 1/2 fl. oz.	—	—
Weeds[1], fungi, nematodes and soil insects (wire worms)	Mineral	25-50 gal.[4]	7 1/2-15 fl. oz.[4]	145-290 ml.[4,5]	5-10 fl. oz.[4,5]
Weeds[1], fungi, nematodes and soil insects (wire worms)	Organic[2] Muck	40-60 gal.	3/4-1 1/8 pint	232-348 ml.[5]	8-12 fl. oz.[5]

1. For best weed control use the higher dosages indicated. Inject at 4-inch depth and cover with a tarp for 4 days.
2. If VORLEX is used in heavy clay soils, an extended aeration period may be required.
3. To obtain best symphylan control remove all residues prior to fumigation.
4. For citrus and other orchard trees where deep penetration is desirable, use 60–70 gallons of VORLEX per acre (1 1/8–1 3/8 pints per 100 sq. ft.)
5. Row or band treatment does not give effective soil insect control.
6. Row fumigation may not effectively control nematodes on root crops where infestation is severe.

IMMEDIATELY after application, drag the soil to remove chisel marks, then firm with a smooth roller. If additional water is required, apply only the amount necessary to wet the soil to a depth of one-half inch. If a tarp seal is used, lower dosages may be effective.

In soils 70°F or higher at 6" depth, special attention should be given to sealing the surface of the soil. Tarping gives the best seal.

FUMIGATION PERIOD: After application at soil temperatures of 60°F or above, keep soil moist and undisturbed for a minimum of 4 days; 7 days is preferable. Use a shorter fumigation period only if it has been shown effective.

Colder soils require a longer fumigation period. If these conditions occur, allow at least 1 to 3 weeks for the fumigation period depending upon the temperature.

In fall or winter application the fumigation period should extend until spring. The soil should be undisturbed during this period.

AERATION PERIOD: For soil temperatures above 60°F (at 6" depth), aerate the soil at the end of the fumigation period by cultivation, e.g. discing. Repeated cultivation will aid release of the fumigant and is especially desirable if heavy rains occur during the fumigation period. If soil crusting occurs, it retards aeration and must be broken up. Repeated cultivation may be necessary to prevent soil crusting.

Aerate until the fumigant has completely escaped from the soil. This is usually accomplished by allowing 1 week between cultivation (end of fumigation period) and planting for each 12 gallons per acre of VORLEX.

For soil temperatures below 60°F (at 6" depth), longer aeration periods than those mentioned above will be required. Repeated deep cultivation is especially important when heavy rains or very wet soil accompany low soil temperatures, since these conditions retard aeration.

DO NOT PLANT WHEN THE ODOR OF THE FUMIGANT IS STILL DETECTABLE.

IMPORTANT: READ BEFORE USE

By using this product, user accepts the following conditions, warranty, disclaimer of warranties and limitations of liability.

CONDITIONS: The directions for use of this product are believed to be adequate and should be followed carefully. However, because of extreme weather and soil conditions, manner of use and other factors beyond NOR-AM Chemical Company's control, it is impossible for NOR-AM to eliminate all risks associated with the use of this product. As a result, crop injury or ineffectiveness is always possible.

DISCLAIMER OF WARRANTIES: THERE ARE NO WARRANTIES, EXPRESS OR IMPLIED, OF MERCHANTABILITY OR OF FITNESS FOR A PARTICULAR PURPOSE OR OTHERWISE, WHICH EXTEND BEYOND THE STATEMENTS MADE ON THIS LABEL. No agent of NOR-AM Chemical Company is authorized to make any warranties beyond those contained herein or to modify the warranties contained herein. UNDER NO CIRCUMSTANCES SHALL NOR-AM CHEMICAL COMPANY BE LIABLE FOR ANY INCIDENTAL OR CONSEQUENTIAL DAMAGES RESULTING FROM FAILURE TO USE THIS PRODUCT IN ACCORDANCE WITH LABEL DIRECTIONS.

LIMITATIONS OF LIABILITY: In no event shall NOR-AM Chemical Company's liability for damages arising out of the use of this product exceed the purchase price of the product used.

528-IB-V-10M-Rev. 12/88

Pennwalt Corporation

Agchem Division
Three Parkway
Philadelphia, PA 19102
(215) 587-7000

ACCELERATE®

A HARVEST AID FOR COTTON

Contains 0.52 lbs. Endothall* per gallon

ACTIVE INGREDIENT:

Mono (N,N-dimethylalkylamine)** salt of endothall*	15.9%
INERT INGREDIENTS:	84.1%
TOTAL	100.0%

*7-oxabicyclo[2.2.1]heptane-2,3-dicarboxylic acid equivalent 5.5%
**Alkyl groups as in fatty acids of coconut oil

EPA Registration No. 4581-284 EPA Establishment No. 4581-TX-1

KEEP OUT OF REACH OF CHILDREN

PELIGRO DANGER

PRECAUCION AL USUARIO: Si usted no lee ingles, no use este producto hasta que la etiqueta haya sido explicado ampliamente.

STATEMENT OF PRACTICAL TREATMENT

IF IN EYES, immediately flush with plenty of water for at least 15 minutes. Call a physician.

IF ON SKIN, immediately flush with plenty of water for at least 15 minutes. Remove and wash contaminated clothing before reuse.

IF SWALLOWED, drink promptly a large quantity of milk, egg whites, gelatin solution or if these are not available, drink large quantities of water. Avoid alcohol. Call a physician immediately.

NOTE TO PHYSICIAN: Probable mucosal damage may contraindicate the use of gastric lavage. Measures against circulatory shock, respiratory depression, and convulsion may be needed.

PRECAUTIONARY STATEMENTS

HAZARDS TO HUMANS (AND DOMESTIC ANIMALS)

DANGER

CORROSIVE, CAUSES IRREVERSIBLE EYE DAMAGE. CAUSES SKIN IRRITATION. MAY BE FATAL IF ABSORBED THROUGH SKIN. HARMFUL IF SWALLOWED OR INHALED. DO NOT GET IN EYES, ON SKIN OR ON CLOTHING. WEAR PROTECTIVE CLOTHING, RUBBER GLOVES AND GOGGLES OR FACE SHIELD WHEN HANDLING. Wash thoroughly with soap and water after handling and before eating or smoking. Avoid breathing spray mist. Remove contaminated clothing and wash before reuse.

ENVIRONMENTAL HAZARDS

Avoid contact with or drift to other crops, trees, or plants as injury may occur.

This product is toxic to fish and wildlife. Keep out of lakes, streams or ponds. Do not contaminate water when disposing of equipment washwaters.

GENERAL INFORMATION

ACCELERATE is a liquid concentrate that when tank mixed with sodium chlorate type (containing suitable fire suppressants) or organic phosphate cotton defoliants causes a faster and more dependable leaf drop prior to hand picking or machine harvesting. The combination of ACCELERATE and sodium chlorate type or organic phosphate defoliants usually drops leaves in a green condition in approximately one-half the time of sodium chlorate type or organic phosphate defoliants without ACCELERATE. Under favorable defoliation conditions, satisfactory leaf drop will require 3-4 days; whereas, under adverse conditions, such as low temperatures, toughened plants or actively growing plants, it may be extended to 6-7 days. ACCELERATE is non-corrosive to metal and will not damage metal parts normally used in applying spray chemicals. Recommended rates of ACCELERATE used alone without sodium chlorate type or organic phosphate defoliants will not give satisfactory defoliation. Wash out sprayer immediately after use.

DIRECTIONS FOR USE

It is a violation of Federal law to use this product in a manner inconsistent with its labeling.

Do not apply this product through any type of irrigation system.

RE-ENTRY STATEMENT

Do not apply this product in such a manner as to directly or through drift expose workers or other persons. The area being treated must be vacated by unprotected persons.

Do not enter treated areas without protective clothing until sprays have dried.

Because certain states may require more restrictive reentry intervals for various crops treated with this product, consult your State Department of Agriculture for further information.

Written or oral warnings must be given to workers who are expected to be in an area about to be treated with this product. In case of accidental exposure, workers should wash thoroughly with soap and water and contaminated clothing should be removed and washed before reuse. If symptoms of skin or eye irritation occur call a physician. When oral warnings are given, warnings shall be given in a language customarily understood by workers. Oral warnings shall be given if there is reason to believe that written warnings cannot be understood by workers. Written warnings must include the following information: "DANGER. Area treated with ACCELERATE Cotton Harvest Aid on (date of application). Do not enter without protective clothing until spray has dried. In case of accidental exposure, workers should wash thoroughly with soap and water and contaminated clothing should be removed and washed before reuse. If symptoms of skin or eye irritation occur call a physician."

Under normal conditions, apply the combination of ACCELERATE and sodium chlorate type or organic phosphate defoliants when fiber quality of the top bolls will not be damaged by loss of the top leaves which feed these bolls (top bolls should be firm to thumb pressure or difficult to cut with a sharp knife). The combination of ACCELERATE and sodium chlorate type or organic phosphate defoliants will not damage the bolls directly or cause them to open prematurely. Ground machines should have at least 3 nozzles per row including one overhead nozzle and have pressure and speed adjusted to give fine droplets yet maintain the correct spray volume of 15 to 25 gallons per acre. For aerial application rates of 5-12 gallons of total spray should be used. Aerial applicators should fly as closely as possible to the plant tops and avoid application during the excessive temperatures of midday and when winds may cause drift to adjacent fields. If spraying is interrupted, be sure to start agitation before spraying is resumed.

Apply specified spray dosage per acre in recommended gallons of water to give thorough coverage to mature cotton when 50% or more of bolls are open. Applications can be made until frost.

APPLICATION RATES

Observe all cautions and limitations on labeling of all products used in mixtures.

ACCELERATE WITH ORGANIC PHOSPHATE DEFOLIANTS:

Under ideal defoliation conditions tank mix with water 1 pint of organic phosphate defoliant with 1½ pints of ACCELERATE per acre. Under adverse defoliation conditions, use 1½ pints of ACCELERATE with an increased rate, up to the maximum label rate, of organic phosphate defoliant. Always add ACCELERATE to organic phosphate defoliants previously tank mixed with water. Do not mix together without water. Be sure mixing tank is clean because residues of insecticides may cause curdling.

ACCELERATE WITH SODIUM CHLORATE TYPE DEFOLIANTS:

Under ideal defoliation conditions, tank mix 1 pint of ACCELERATE with label recommended rate of sodium chlorate type defoliants per acre. Under adverse defoliation conditions increase rate of ACCELERATE to 1½ pints per acre.

STORAGE AND DISPOSAL

Do not contaminate water, food, or feed by storage or disposal.

Storage Instructions: Store in the original container. Do not store in a manner where cross-contamination with other pesticides, fertilizers, food or feed could occur. In the event of spillage during handling or storage, absorb with sand or other inert material and dispose of absorbent in accordance with the Pesticide Disposal Instructions listed below.

Pesticide Disposal Instructions: Pesticide wastes are acutely hazardous. Improper disposal of excess pesticide, spray mixture, or rinsate is a violation of Federal law. If these wastes cannot be disposed of by use according to label instructions, contact your State Pesticide or Environmental Control Agency, or the Hazardous Waste representative at the nearest EPA Regional Office for guidance.

Container Disposal Instructions: Triple rinse (or equivalent). Then offer for recycling or reconditioning, or puncture and dispose of in a sanitary landfill, or by incineration, or, if allowed by state and local authorities, by burning. If burned, stay out of smoke.

COMPATIBILITY: ACCELERATE Cotton Harvest Aid is compatible with Arsenic Acid, Bolls-Eye®, DEF®, Dropp®, Folex®, Gramoxone®, Harvade®, Paraquat, Prep® and Sodium Chlorate and may be applied in sequence or as a tank mixture. Good agitation in the spray tank is essential and a tank mixture should not be allowed to stand without agitation for more than 5 to 10 minutes. ACCELERATE should be added last to avoid excessive foaming. Read and observe all appropriate label use directions and precautions for the defoliants used.

Bolls-Eye® is a trademark of Vertac Chemical Corp.; DEF® is a trademark of Mobay Chemical Corp.; Dropp® is a trademark of NOR-AM Chemical Co.; Folex® is a trademark of Rhone-Poulenc Inc.; Gramoxone® is a trademark of ICI Plant Protection Division; Harvade® is a trademark of Uniroyal Chemical; Prep® is a trademark of Union Carbide Agricultural Products Co. Inc.

WARRANTY AND DISCLAIMER

Pennwalt Corporation warrants that this material conforms to the chemical description on the label and is reasonably fit for the purposes referred to in the Directions for Use, subject to the risks referred to therein. PENNWALT MAKES NO OTHER EXPRESS OR IMPLIED WARRANTY OF FITNESS OR MERCHANTABILITY OR ANY OTHER EXPRESS OR IMPLIED WARRANTY. IN NO CASE SHALL PENNWALT OR SELLER BE LIABLE FOR CONSEQUENTIAL, SPECIAL OR INDIRECT DAMAGES RESULTING FROM THE USE OR HANDLING OF THIS PRODUCT INCLUDING, BUT NOT LIMITED TO, LOSS OF PROFITS, BUSINESS REPUTATION, OR CUSTOMERS; LABOR COST; OR OTHER EXPENSES INCURRED IN PLANTING OR HARVESTING.

Pennwalt and seller offer this product and the buyer and user accept it subject to the foregoing conditions of sale and warranty which may be varied only by agreement in writing signed by a duly authorized representative of Pennwalt.

5-C201S-01 D8 (4/88)

AQUATHOL® K

AQUATIC HERBICIDE

ACTIVE INGREDIENT:

Dipotassium salt of endothall*	40.3%
INERT INGREDIENTS:	59.7%
TOTAL	100.0%

*7-oxabicyclo [2.2.1]heptane-2,3-dicarboxylic acid equivalent 28.6%
Contains per gallon 4.23 lbs. dipotassium endothall
(equivalent to 3.0 lbs. endothal acid)

EPA Registration No. 4581-204 EPA Establishment No. 4581-TX-1

DANGER–POISON

KEEP OUT OF REACH OF CHILDREN

STATEMENT OF PRACTICAL TREATMENT

IF SWALLOWED, drink promptly a large quantity of milk, egg whites, gelatin solution or if these are not available, drink large quantities of water. Avoid alcohol. Call a physician immediately.

IF ON SKIN, immediately flush with plenty of water for at least 15 minutes. Remove and wash contaminated clothing before reuse.

IF IN EYES, immediately flush with plenty of water for at least 15 minutes. Call a physician.

NOTE TO PHYSICIAN: Probably mucosal damage may contraindicate the use of gastric lavage. Measures against circulatory shock, respiratory depression and convulsion may be needed.

GENERAL INFORMATION

AQUATHOL K is a liquid concentrate soluble in water which is effective against a broad range of aquatic plants with a margin of safety to fish.

Dosage rates indicated for the application of AQUATHOL K are measured in "Parts Per Million" (ppm) of dipotassium endothall. Only 0.5 to 5.0 ppm are generally required for aquatic weed control, whereas some fish species are tolerant to approximately 100 ppm or over.

DIRECTIONS FOR USE

It is a violation of Federal law to use this product in a manner inconsistent with its labeling.

AQUATIC WEEDS CONTROLLED AND DOSAGE RATE CHARTS

AQUATHOL K is recommended for the control of the following aquatic weeds in irrigation and drainage canals, ponds and lakes at the rates indicated. Since the active ingredient is water soluble and tends to diffuse from the area treated, select the dosage rate applicable to the area to be treated. Use the lower rate in each range of rates where the growth is young and growing and/or where the weed stand is not heavy. Marginal treatments of large bodies of water require higher rates as indicated.

HOW TO APPLY:

AQUATHOL K is a contact killer; consequently, do not apply before weeds are present. Application as early as possible after weeds are present is recommended to permit use of lower application rates. However, for best results water temperature should be at 65°F or above. If an entire pond is treated at one time, or if the dissolved oxygen level is low at time of application, decay of weeds may remove enough oxygen from the water, causing fish to suffocate. Water containing very heavy vegetation should be treated in sections to prevent suffocation of fish. Sections should be treated 5-7 days apart. Carefully measure size and depth of area to be treated and determine amount of AQUATHOL K to apply from chart. For best results apply on a calm day where there is little wave action.

AQUATHOL K should be sprayed on the water or injected below the water surface and should be distributed as evenly as possible. It may be applied as it comes from the container or diluted with water depending on the equipment. Some dilution will give better distribution.

In instances where the nuisance to be controlled is an exposed surface problem (i.e., some of the broad-leaved pond weeds) it is important to get good contact coverage utilizing the highest concentration (least water dilution) compatible with the type of equipment used so that even distribution is achieved.

Necessary approval and/or permits should be obtained in states where required.

COMMON NAME	LATIN NAME	ENTIRE POND OR LARGE AREA TREATMENT	SPOT OR LAKE MARGIN TREATMENT
Bass Weed	Potamogeton amplifolius	2.0-3.0 ppm	3.0-4.0 ppm
Bur Reed	Sparganium spp.	3.0-4.0 ppm	4.0-5.0 ppm
Coontail	Ceratophyllum spp.	1.0-2.0 ppm	2.0-3.0 ppm
Hydrilla	Hydrilla verticillata	2.0-3.0 ppm	3.0-4.0 ppm
Milfoil	Myriophyllum spp.	2.0-3.0 ppm	3.0-4.0 ppm
Pondweed			
Bushy	Najas spp.	0.5-1.5 ppm	2.0-3.0 ppm
Curly-Leaf	Potamogeton crispus	0.5-1.5 ppm	2.0-3.0 ppm
Flat-Stem	Potamogeton natans	1.0-2.0 ppm	2.0-3.0 ppm
Horned	Zannichellia spp.	1.0-2.0 ppm	2.0-3.0 ppm
Sago	Potamogeton pectinatus	1.0-2.0 ppm	2.0-3.0 ppm
______	Potamogeton americanus	2.0-3.0 ppm	3.0-4.0 ppm
______	Potamogeton diversifolius	1.0-2.0 ppm	2.0-3.0 ppm
______	Potamogeton filiformis	2.0-3.0 ppm	3.0-4.0 ppm
______	Potamogeton pusillus	1.0-2.0 ppm	2.0-3.0 ppm
Water Star Grass	Heteranthera spp.	2.0-3.0 ppm	3.0-4.0 ppm

RATE OF APPLICATION—LAKES AND PONDS

The following chart indicates the total quantity of material to be applied.

APPROXIMATE GALLONS OF AQUATHOL K FOR ONE ACRE (208′ × 208′) TREATMENT

	DOSAGE IN GALLONS FOR VARIOUS CONCENTRATIONS IN PPM						
DEPTH	0.5 ppm	1.0 ppm	1.5 ppm	2.0 ppm	3.0 ppm	4.0 ppm	5.0 ppm
1 ft.	0.3	0.6	1.0	1.3	1.9	2.6	3.2
2 ft.	0.6	1.3	1.9	2.6	3.8	5.1	6.4
4 ft.	1.3	2.6	3.8	5.1	7.7	10.2	12.8
6 ft.	1.9	3.8	5.8	7.6	11.5	15.3	19.2

RATE OF APPLICATION—IRRIGATION AND DRAINAGE CANALS* *

The following indicates the total quantity of material to be applied.

GALLONS OF AQUATHOL K REQUIRED TO TREAT 1 MILE OF CANAL 1 FOOT DEEP*

	WIDTH OF CANAL IN FEET			
PPM	**5**	**10**	**15**	**20**
1.0 ppm	0.4	0.75	1.2	1.5
2.0 ppm	0.75	1.5	2.3	3.0
3.0 ppm	1.2	2.3	3.5	4.5
4.0 ppm	1.5	3.0	4.5	6.0
5.0 ppm	2.0	3.8	5.7	7.5

The minimum contact time with weeds for optimum results should be 2 hours.
*For deeper water, adjust rate accordingly.
**Not for this use in California.

PRECAUTIONARY STATEMENTS

HAZARDS TO HUMANS (AND DOMESTIC ANIMALS)

DANGER

FATAL IF ABSORBED THROUGH SKIN. MAY BE FATAL IF SWALLOWED. HARMFUL IF INHALED. CORROSIVE, CAUSES IRREVERSIBLE EYE DAMAGE. DO NOT GET IN EYES, ON SKIN OR ON CLOTHING. WEAR PROTECTIVE CLOTHING, RUBBER GLOVES, AND GOGGLES OR FACE SHIELD WHEN HANDLING. Wash thoroughly with soap and water after handling and before eating or smoking. Remove contaminated clothing and wash before reuse. Avoid breathing spray mist.

ENVIRONMENTAL HAZARDS

Avoid contact with or drift to other crops or plants as injury may result. Wash out spray equipment with water after each operation.

Do not use fish from treated water for food or feed within 3 days of treatment. Do not use treated water for watering livestock, for preparing agricultural sprays for food crops, for irrigation, or for domestic purposes within the following periods:

Up to 0.5 ppm dipotassium salt
(0.35 ppm acid equivalent)—7 days after application

Up to 4.25 ppm dipotassium salt
(3.0 ppm acid equivalent)—14 days after application

Up to 5.0 ppm dipotaqssium salt
(3.5 ppm acid equivalent)—25 days after application

NOTE: Areas treated with AQUATHOL K may be used for swimming twenty-four hours after treatment.

Treated water can be used for sprinkling bent grass immediately.

STORAGE AND DISPOSAL

Do not contaminate water, food, or feed by storage or disposal.

Storage Instructions: Store in the original container. Do not store in a manner where cross-contamination with other pesticides, fertilizers, food or feed could occur. Storage at temperatures below 32°F may result in the product freezing or crystallizing. Should this occur the product must be warmed to 50°F or higher and thoroughly agitated. In the event of a spillage during handling or storage, absorb with sand or other inert material and dispose of absorbant in accordance with the Pesticide Disposal Instructions listed below.

Pesticide Disposal Instructions: Pesticide wastes are acutely hazardous. Improper disposal of excess pesticide, spray mixture, or rinsate is a violation of Federal law. If these wastes cannot be disposed of by use according to label instructions, contact your State Pesticide or Environmental Control Agency, or the Hazardous Waste representative at the nearest EPA Regional Office for guidance.

Container Disposal Instructions: Triple rinse (or equivalent). Then offer for recycling or reconditioning, or puncture and dispose of in a sanitary landfill, or incineration, or, if allowed by state and local authorities, by burning. If burned, stay out of smoke.

WARRANTY AND DISCLAIMER

Pennwalt Corporation warrants that this material conforms to the chemical description on the label and is reasonably fit for the purposes referred to in the Directions for Use, subject to the risks referred to therein. PENNWALT MAKES NO OTHER EXPRESS OR IMPLIED WARRANTY. IN NO CASE SHALL PENNWALT OR SELLER BE LIABLE FOR CONSEQUENTIAL, SPECIAL OR INDIRECT DAMAGES RESULTING FROM THE USE OR HANDLING OF THIS PRODUCT INCLUDING, BUT NOT LIMITED TO, LOSS OF PROFITS, BUSINESS REPUTATION, OR CUSTOMERS; LABOR COST; OR OTHER EXPENSES INCURRED IN PLANTING OR HARVESTING.

Pennwalt and seller offer this product and the buyer and user accept it subject to the foregoing conditions of sale and warranty which may be varied only by agreement in writing signed by a duly authorized representative of Pennwalt.

5-C102S-01 H4 (8/84)

AQUATHOL®

GRANULAR AQUATIC HERBICIDE

ACTIVE INGREDIENT:

Dipotassium salt of endothall*	10.1%
INERT INGREDIENTS:	89.9%
TOTAL	100.0%

* 7-oxabicyclo [2.2.1]heptane-2,3-dicarboxylic acid equivalent 7.2%

EPA Registration No. 4581-201 EPA Establishment No. 228-IL-1

KEEP OUT OF REACH OF CHILDREN

DANGER

STATEMENT OF PRACTICAL TREATMENT

IF IN EYES, hold eyelids open and flush with a steady, gentle stream of water for 15 minutes. Call a physician.

IF SWALLOWED, drink promptly a large quantity of milk, egg whites, gelatin solution or if these are not available, drink large quantities of water. Avoid alcohol. Call a physician immediately.

IF ON SKIN, immediately flush with plenty of water for at least 15 minutes. Remove and wash contaminated clothing before reuse.

NOTE TO PHYSICIAN: Probable mucosal damage may contraindicate the use of gastric lavge. Measures against circulatory shock, respiratory depression and convulsion may be needed.

PRECAUTIONARY STATEMENTS

HAZARDS TO HUMANS (AND DOMESTIC ANIMALS)

DANGER

CORROSIVE. CAUSES IRREVERSIBLE EYE DAMAGE. HARMFUL IF SWALLOWED, ABSORBED THROUGH SKIN OR INHALED. AVOID CONTACT WITH SKIN OR CLOTHING. DO NOT GET IN EYES. WEAR GOGGLES OR FACE SHIELD WHEN HANDLING. Avoid breathing dust. Wash thoroughly with soap and water after handling. Remove contaminated clothing and wash before reuse.

ENVIRONMENTAL HAZARDS

Avoid contact with or drift to other crops or plants as injury may result.

Do not use treated water for irrigation or for agricultural sprays on food crops or for domestic purposes within 7 days of treatment.

Do not use fish from treated water for food or feed within 3 days of treatment.

GENERAL INFORMATION

AQUATHOL GRANULAR is a granular aquatic herbicide for use in ponds and lakes which, under field test conditions has shown to be effective against a broad range of aquatic plants with a margin of safety to fish. Dosage rates indicated for the applications of AQUATHOL GRANULAR are measured in "Parts Per Million" (ppm). 1 ppm as a dosage rate means that there would be 1 part of AQUATHOL GRANULAR'S active ingredient in 1,000,000 parts of water. Only 1/2 to 5 ppm are generally required for aquatic weed control, whereas some fish species are tolerant to approximately 100 ppm or over. For best results treat areas of one acre or more and/or margins of at least 100 feet in large bodies of water.

Thoroughly clean application equipment immediately after use.

NOTE: Areas treated with AQUATHOL GRANULAR may be used for swimming 24 hours after treatment.

DIRECTIONS FOR USE

It is a violation of Federal law to use this product in a manner inconsistent with its labeling.

HOW TO APPLY:

AQUATHOL GRANULAR is a contact killer, consequently, do not apply before weeds are present. Application as early as possible after weeds are present is recommended to permit use of lower application rates. However, for best results water temperature should be at a minimum of 65°F. If an entire pond is treated at one time, or if the dissolved oxygen level is low at the time of application, decay of weeds may remove enough oxygen from the water causing fish to suffocate. Water containing very heavy vegetation should be treated in sections to prevent suffocation of fish. Sections should be treated 5-7 days apart. Carefully measure size and depth of area to be treated and determine proper amount of AQUATHOL GRANULAR to apply from chart. For best results apply on a calm day where there is little wave action.

Scatter AQUATHOL GRANULAR as evenly as possible over treated areas. A cyclone seeder is useful for this purpose. In instances where the nuisance to be controlled is an exposed surface problem (i.e., some of the broadleaved pond weeds) it is important to get good contact coverage of the problem.

Necessary approval and/or permits should be obtained in states where required.

HOW TO DETERMINE DOSAGE RATE
(Active Ingredient)

AQUATHOL GRANULAR is recommended for the control of the following aquatic weeds at the rates indicated. Since AQUATHOL GRANULAR's active ingredient is water soluble and tends to diffuse from the area treated, select the dosage rate applicable to the area to be treated. Use the lower rate in each range of rates when the growth is young and growing and/or where the weed stand is not heavy. Marginal treatments of large bodies of water require highest rates as indicated.

WEEDS CONTROLLED AND AQUATHOL GRANULAR DOSAGE RATE CHART

COMMON NAME	LATIN NAME	ENTIRE POND OR LARGE AREA TREATMENT	SPOT OR LAKE MARGIN TREATMENT
Bass Weed	Potamogeton amplifolius	2.0-3.0 ppm	3.0-4.0 ppm
Bur Reed	Aparganium spp.	3.0-4.0 ppm	4.0-5.0 ppm
Coontail	Ceratophyllum spp.	1.0-2.0 ppm	2.0-3.0 ppm
Milfoil	Myriophyllum spp.	2.0-3.0 ppm	3.0-4.0 ppm
Pondweed			
Bushy	Najas spp.	0.5-1.5 ppm	2.0-3.0 ppm
Curly-Leaf	Potamogeton crispus	0.5-1.5 ppm	2.0-3.0 ppm
Flat-Stem	Potamogeton zosteriformis	2.0-3.0 ppm	3.0-4.0 ppm
Floating-Leaf	Potamogeton natans	1.0-2.0 ppm	2.0-3.0 ppm
Horned	Zannichellia spp.	1.0-2.0 ppm	2.0-3.0 ppm
Sago	Potamogeton pectinatus	1.0-2.0 ppm	2.0-3.0 ppm
	Potamogeton nodosus	2.0-3.0 ppm	2.0-3.0 ppm
	Potamogeton diversifolius	1.0-2.0 ppm	2.0-3.0 ppm
	Potamogeton filiformis	2.0-3.0 ppm	3.0-4.0 ppm
	Potamogeton pusillus	1.0-2.0 ppm	2.0-3.0 ppm
Water Star Grass	Heteranthera spp.	2.0-3.0 ppm	3.0-4.0 ppm

HOW TO DETERMINE QUANTITY TO BE APPLIED

The following charts indicate the total quantity of material to be applied for certain size areas.

APPROXIMATE POUNDS OF AQUATHOL GRANULAR FOR ONE ACRE*
TREATMENT DOSAGE IN POUNDS FOR VARIOUS CONCENTRATIONS IN PPM

DEPTH	0.5 ppm	1.0 ppm	1.5 ppm	2.0 ppm	2.5 ppm	3.0 ppm	4.0 ppm	5.0 ppm
1 Ft. Deep	13 lbs.	27 lbs.	40 lbs.	54 lbs.	67 lbs.	81 lbs.	108 lbs.	135 lbs.
2 Ft. Deep	27 lbs.	54 lbs.	81 lbs.	108 lbs.	134 lbs.	161 lbs.	215 lbs.	269 lbs.
3 Ft. Deep	40 lbs.	81 lbs.	121 lbs.	161 lbs.	202 lbs.	242 lbs.	323 lbs.	403 lbs.
4 Ft. Deep	54 lbs.	108 lbs.	161 lbs.	215 lbs.	269 lbs.	323 lbs.	430 lbs.	538 lbs.
5 Ft. Deep	67 lbs.	135 lbs.	202 lbs.	269 lbs.	336 lbs.	403 lbs.	538 lbs.	673 lbs.
6 Ft. Deep	81 lbs.	161 lbs.	242 lbs.	323 lbs.	403 lbs.	484 lbs.	646 lbs.	807 lbs.

FOR 1000 SQUARE FEET TREATMENT
DOSAGE IN POUNDS FOR VARIOUS CONCENTRATIONS IN PPM

DEPTH	0.5 ppm	1.0 ppm	1.5 ppm	2.0 ppm	2.5 ppm	3.0 ppm	4.0 ppm	5.0 ppm
1 Ft. Deep	0.3 lbs.	0.6 lbs.	0.9 lbs.	1.2 lbs.	1.5 lbs.	1.9 lbs.	2.5 lbs.	3.1 lbs.
2 Ft. Deep	0.6 lbs.	1.2 lbs.	1.9 lbs.	2.5 lbs.	3.1 lbs.	3.7 lbs.	4.9 lbs.	6.2 lbs.
3 Ft. Deep	0.9 lbs.	1.9 lbs.	2.8 lbs.	3.7 lbs.	4.6 lbs.	5.6 lbs.	7.4 lbs.	9.3 lbs.
4 Ft. Deep	1.2 lbs.	2.5 lbs.	3.7 lbs.	4.9 lbs.	6.2 lbs.	7.4 lbs.	9.9 lbs.	12.4 lbs.
5 Ft. Deep	1.5 lbs.	3.1 lbs.	4.6 lbs.	6.2 lbs.	7.7 lbs.	9.3 lbs.	12.4 lbs.	15.4 lbs.
6 Ft. Deep	1.9 lbs.	3.7 lbs.	5.6 lbs.	7.4 lbs.	9.3 lbs.	11.1 lbs.	14.8 lbs.	18.5 lbs.

*One acre equals approximately 208′ × 208′

Where the area to be treated is greater than those listed in the charts proceed as follows:

a. Compute the approximate surface acreage
b. Compute the average depth
c. Multiply a. by b. to determine the total number of acre/feet
d. Multiply the pounds required at the 1 foot depth under the rate to be used by the number of acre/feet to determine total quantity to be used.

STORAGE AND DISPOSAL

Do not contaminate water, food, or feed by storage or disposal.

Storage Instructions: Store in the original container, preferably in a locked storage area. Do not store in a manner where cross-contamination with other pesticides, fertilizers, food or feed could occur. If spilled during storage or handling sweep up spillage and dispose of in accordance with the Pesticide Disposal Instructions listed below.

Pesticide Disposal Instructions: Pesticide wastes are acutely hazardous. Improper disposal of excess pesticide or rinsate is a violation of Federal law. If these wastes cannot be disposed of by use according to label instructions, contact your State Pesticide or Environmental Control Agency, or the Hazardous Waste representative at the nearest EPA Regional Office for guidance.

Container Disposal Instructions: Triple rinse (or equivalent). Then offer for recycling or reconditioning, or puncture and dispose of in a sanitary landfill, or by incineration, or, if allowed by state and local authorities, by burning. If burned, stay out of smoke.

WARRANTY AND DISCLAIMER

Pennwalt Corporation warrants that this material conforms to the chemical description on the label and is reasonably fit for the purposes referred to in the Directions for Use, subject to the risks referred to therein. PENNWALT MAKES NO OTHER EXPRESS OR IMPLIED WARRANTY OF FITNESS OR MERCHANTABILITY OR ANY OTHER EXPRESS OR IMPLIED WARRANTY. IN NO CASE SHALL PENNWALT OR SELLER BE LIABLE FOR CONSEQUENTIAL, SPECIAL OR INDIRECT DAMAGES RESULTING FROM THE USE OR HANDLING OF THIS PRODUCT INCLUDING, BUT NOT LIMITED TO, LOSS OF PROFITS, BUSINESS REPUTATION, OR CUSTOMERS; LABOR COST; OR OTHER EXPENSES INCURRED IN PLANTING OR HARVESTING.

Pennwalt and seller offer this product and the buyer and user accept it subject to the foregoing conditions of sale and warranty which may be varied only by agreement in writing signed by a duly authorized representative of Pennwalt.

5-F103S-02 A6 (2/85)

DES-I-CATE®

A POTATO VINE KILLER AND A HARVEST AID FOR ALFALFA AND CLOVER

Contains 0.52 lbs. endothall* per gallon

ACTIVE INGREDIENT:

Mono (N,N-dimethylalkylamine)** salt of endothall*	15.9%
INERT INGREDIENTS:	84.1%
TOTAL	100.0%

*7-oxabicyclo[2.2.1]heptane-2,3-dicarboxylic acid equivalent 5.5%
**Alkyl groups as in fatty acids of coconut oil

EPA Registration No. 4581-206   EPA Establishment No. 4581-TX-1

KEEP OUT OF REACH OF CHILDREN

PELIGRO   DANGER

PRECAUCION AL USUARIO: Si usted no lee ingles, no use este producto hasta que la etiqueta haya sido explicado ampliamente.

STATEMENT OF PRACTICAL TREATMENT

IF IN EYES, immediately flush with plenty of water for at least 15 minutes. Call a physician.

IF ON SKIN, immediately flush with plenty of water for at least 15 minutes. Remove and wash contaminated clothing before reuse.

IF SWALLOWED, drink promptly a large quantity of milk, egg whites, gelatin solution or if these are not available, drink large quantities of water. Avoid alcohol. Call a physician immediately.

NOTE TO PHYSICIAN: Probable mucosal damage may contraindicate the use of gastric lavage. Measures against circulatory shock, respiratory depression, and convulsion may be needed.

PRECAUTIONARY STATEMENTS

HAZARDS TO HUMANS (AND DOMESTIC ANIMALS)

DANGER

CORROSIVE. CAUSES IRREVERSIBLE EYE DAMAGE. CAUSES SKIN IRRITATION. MAY BE FATAL IF ABSORBED THROUGH SKIN. HARMFUL IF SWALLOWED OR INHALED. DO NOT GET IN EYES, ON SKIN OR ON CLOTHING. WEAR PROTECTIVE CLOTHING, RUBBER GLOVES AND GOGGLES OR FACE SHIELD WHEN HANDLING. Wash thoroughly with soap and water after handling and before eating or smoking. Avoid breathing spray mist. Remove contaminated clothing and wash before reuse.

ENVIRONMENTAL HAZARDS

Avoid contact with or drift to other crops, trees, or plants as injury may occur. To avoid drift use coarse sprays and do not apply in windy weather.

This product is toxic to fish and wildlife. Keep out of lakes, streams or ponds. Do not contaminate water by cleaning of equipment, or disposal of wastes.

Do not feed treated forage to dairy animals being finished for slaughter. Do not use on forage to be sold commercially or to be shipped interstate. Do not graze treated areas.

GENERAL INFORMATION

DES-I-CATE is a water soluble formulation of endothall containing a special inorganic salt and an amine for use as a defoliant or desiccant for drying out, hardening, and removing green plant growth from certain seed crops prior to harvesting to permit direct combining, such as alfalfa and clover.

It is also a contact-type herbicide for killing potato vines. In addition to facilitating harvesting, DES-I-CATE assists in setting potato skins.

Use the recommended rate of DES-I-CATE with water as a carrier. DES-I-CATE disperses readily in water and remains uniformly in solution. It is suggested to fill the tank with water first and then add DES-I-CATE since there is a possibility of foaming. In case of excessive foaming add small amount of antifoam to spray tank.

In aerial as well as ground applications, complete coverage of plants with DES-I-CATE in the liquid state is essential for best results. Therefore, large spray droplet size should be used and application should be made preferably during periods of highest relative humidity (early morning or evening) if daytime temperatures are expected to reach 80° or higher. Wash out sprayer immediately after use.

DIRECTIONS FOR USE

It is a violation of Federal law to use this product in a manner inconsistent with its labeling.

Do not apply this product through any type of irrigation system.

RE-ENTRY STATEMENT

Do not apply this product in such a manner as to directly or through drift expose workers or other persons. The area being treated must be vacated by unprotected persons.

Do not enter treated areas without protective clothing until sprays have dried.

Because certain states may require more restrictive reentry intervals for various crops treated with this product, consult your State Department of Agriculture for further information.

Written or oral warnings must be given to workers who are expected to be in an area about to be treated with this product. In case of accidental exposure, workers should wash thoroughly with soap and water and contaminated clothing should be removed and washed before reuse. If symptoms of skin or eye irritation occur call a physician. When oral warnings are given, warnings shall be given in a language customarily understood by workers. Oral warnings shall be given if there is reason to believe that written warnings cannot be understood by workers. Written warnings must include the following information: "DANGER. Area treated with DES-I-CATE Harvest Aid on (date of application). Do not enter without protective clothing until spray has dried. In case of accidental exposure, workers should wash thoroughly with soap and water and contaminated clothing should be removed and washed before reuse. If symptoms of skin or eye irritation occur call a physician."

POTATOES

Apply to potato vines 10 to 14 days prior to harvest, using the recommended rate of DES-I-CATE thoroughly mixed with water as follows:

For Aerial Applications: Light to medium vine growth—Apply 1½ to 2 gallons of DES-I-CATE in 5 to 10 gallons of total spray per acre.

Heavy vine growth—Apply 2 gallons of DES-I-CATE in 5 to 10 gallons total spray per acre. Use of 10 gallons total spray will improve coverage on heavy vine growth. For exceptionally heavy vines it is sometimes advisable to apply one gallon of DES-I-CATE in 5 gallons of spray in one direction followed immediately by a similar amount applied at right angles to the first set of swaths.

For Ground Applications: Light to medium vine growth—Apply 1½ to 2 gallons of DES-I-CATE per acre in 20 to 100 gallons of total spray. Use higher rate during cool, cloudy weather.

Heavy vine growth—Apply 2 gallons of DES-I-CATE per acre in 20 to 100 gallons of total spray. For Northwest conditions, it is suggested that total spray in ground equipment not exceed 40 gallons per acre.

NOTE: The addition of 3 to 5 gallons per acre of diesel fuel or 1 pint of paraffin base herbicidal oil for each 20 gallons of total spray may increase speed and overall vine kill. No wetting agent or emulsifier is required since DES-I-CATE will normally emulsify diesel fuel or other herbicidal oil. Do not use high rate, diesel fuel or other oils under conditions favorable for rapid vine kill such as low soil moisture or high temperatures since stem-end discolorations may occur. When using diesel fuel, always add the oil last to the DES-I-CATE and water to avoid the possibility of forming an invert emulsion.

ALFALFA AND CLOVER

(Seed Crops Only—Seed from treated fields should be used for planting purposes only. Do not use seed for food, feed, or oil purposes.)

Alfalfa and Clovers (Crimson C., Alsike C., Ladino C., Red C., Sweet C. and White C.).

For Aerial Applications: The following rates are recommended:

Medium stands: Use 1¼ gallons of DES-I-CATE in 8½ gallons of water per acre in one application.

Heavy stands: Use ¾ to 1¼ gallons of DES-I-CATE in 8½ to 9 gallons of water per acre and repeat application in 3 to 5 days; or use 1½ gallons of DES-I-CATE in 8 gallons of water per acre in one application.

For Ground Application: Use the same rates of DES-I-CATE per acre, but increase the water to make a total spray of 15 to 20 gallons per acre.

When to combine is dependent on the weather but usually the desiccated alfalfa or clover is ready to combine 5 to 10 days after treatment.

NOTE: Seed crops should not be sprayed until the seed is mature and the plants have lost moisture normally following the end of the irrigation season. Spraying before the seed has reached maximum dry weight may result in yield reduction. Seed crops are normally ready for combining 5 to 10 days after treatment. For best results combine the seed crop as soon as it is ready for harvest.

STORAGE AND DISPOSAL

Do not contaminate water, food, or feed by storage or disposal.

Storage Instructions: Store in the original container. Do not store in a manner where cross-contamination with other pesticides, fertilizers, food or feed could occur. In the event of spillage during handling or storage, absorb with sand or other inert material and dispose of absorbent in accordance with the Pesticide Disposal Instructions listed below.

Pesticide Disposal Instructions: Pesticide wastes are acutely hazardous. Improper disposal of excess pesticide, spray mixture, or rinsate is a violation of Federal law. If these wastes cannot be disposed of by use according to label instructions, contact your State Pesticide or Environmental Control Agency, or the Hazardous Waste representative at the nearest EPA Regional Office for guidance.

Container Disposal Instructions: Triple rinse (or equivalent). Then offer for recycling or reconditioning, or puncture and dispose of in a sanitary landfill, or by incineration, or, if allowed by state and local authorities, by burning. If burned, stay out of smoke.

WARRANTY AND DISCLAIMER

Pennwalt Corporation warrants that this material conforms to the chemical description on the label and is reasonably fit for the purposes referred to in the Directions for Use, subject to the risks referred to therein. PENNWALT MAKES NO OTHER EXPRESS OR IMPLIED WARRANTY OF FITNESS OR MERCHANTABILITY OR ANY OTHER EXPRESS OR IMPLIED WARRANTY. IN NO CASE SHALL PENNWALT OR SELLER BE LIABLE FOR CONSEQUENTIAL, SPECIAL OR INDIRECT DAMAGES RESULTING FROM THE USE OR HANDLING OF THIS PRODUCT INCLUDING, BUT NOT LIMITED TO, LOSS OF PROFITS, BUSINESS REPUTATION, OR CUSTOMERS; LABOR COST; OR OTHER EXPENSES INCURRED IN PLANTING OR HARVESTING.

Pennwalt and seller offer this product and the buyer and user accept it subject to the foregoing conditions of sale and warranty which may be varied only by agreement in writing signed by a duly authorized representative of Pennwalt.

5-C204S-01 D8 (4/88)

SUPPLEMENTAL LABELING

FOR USE IN HOPS SUCKER SUPPRESSION
NOT FOR USE IN CALIFORNIA

USE	RATE/ACRE	REMARKS
To suppress hops sucker growth	0.5 - 1.0 lb. ai	Apply once or twice per season. Make first application when main bines are 2 - 3 feet tall. Subsequent application may be made 7 - 14 days later when hops suckers are 1 - 2 feet long. Spray only basal portion of plant. Avoid spraying main bine within 1 - 2 feet of growing point. Apply in sufficient water for adequate coverage (15 - 40 gpa). Do not apply within 85 days of harvest. Apply with ground equipment only.

Observe all applicable precautions and limitations on the EPA-registered label for this product.

DESICCANT L-10®

HARVEST AID FOR COTTON

RESTRICTED USE PESTICIDE DUE TO ACUTE TOXICITY AND ONCOGENICITY

For retail sale to and use only by certified applicators or persons under their direct supervision, and only for those uses covered by the certified applicator's certification.

ACTIVE INGREDIENT:
Arsenic Acid (H_3AsO_4) 75%
INERT INGREDIENTS: 25%
TOTAL 100.0%

Total arsenic, all in water soluble form, expressed as metallic 39.6%
Weight of solution per gallon at 68°F . . . . 15.7 lb.

EPA Reg. NO. 4581-231 EPA Est. NO. 4581-TX-1

Do not apply this product through any type of irrigation system.

KEEP OUT OF REACH OF CHILDREN

DANGER–POISON

PRECAUCION AL USUARIO: Si usted no lee ingles, no use este producto hasta que la etiqueta haya sido explicado ampliamente.

IF SWALLOWED: Call a physician or Poison Control Center. Drink 1 to 2 glasses of water and induce vomiting by touching back of throat with finger. Do not induce vomiting or give anything by mouth to an unconscious person.

IF ON SKIN: Wash with plenty of soap and water. Get medical attention.

IF INHALED: Remove victim to fresh air. If not breathing, give artificial respiration, preferably mouth-to-mouth. Get medical attention.

IF IN EYES. Hold eyelids open and flush with a steady, gentle stream of water for 15 minutes. Get medical attention.

PRECAUTIONARY STATEMENTS

HAZARDS TO HUMANS (AND ANIMAL SAFETY)

DANGER

CONTAINS INORGANIC ARSENIC
CANCER HAZARD
HARMFUL IF INHALED OR SWALLOWED
USE ONLY WITH ADEQUATE VENTILATION OR RESPIRATORY PROTECTION
FATAL IF SWALLOWED, ABSORBED THROUGH SKIN, OR INHALED. CORROSIVE. CAUSES IRREVERSIBLE EYE DAMAGE AND SKIN BURNS. Do not get in eyes, on skin, or on clothing. Do not breathe spray mist. Wear protective clothing and rubber gloves. Wear a mask or pesticide air purifying respirator equipped with a high efficiency filter jointly approved by the Mining Enforcement and Safety Administration, and the National Institute for Occupational Safety and Health. Wear goggles or face shield when handling. Wash thoroughly with soap and water after handling and before eating or smoking. Remove contaminated clothing and wash before reuse.

ENVIRONMENTAL HAZARDS

Keep children, pets, livestock and poultry away from treated areas. Avoid contamination of food and water to be consumed by humans and animals. Since DESICCANT L-10 is injurious to all plant foliage, avoid contact with or drift to desirable plants or crops. Special precautions should be taken to avoid air drift to areas and buildings occupied by humans, animals or poultry.

This product is toxic to wildlife. Do not contaminate streams, lakes, ponds, woodlands or similar non-crop areas. This product is highly toxic to bees exposed to direct treatment or residues on crops. Birds feeding on treated areas may be killed. Do not apply where runoff is likely to occur. Do not contaminate water by cleaning of equipment or disposal of wastes. Protective information may be obtained from your Cooperative Agricultural Extension Service.

See below for endangered species restrictions.

STORAGE AND DISPOSAL

Do not contaminate water, food, or feed by storage or disposal.

Storage Instructions: Store in the original container.

DESICCANT L-10 has acidic properties and reacts with fabric, galvanized metals, black iron, and certain other metals. Do not use galvanized metal containers for storage as highly toxic arsine gas may be formed.

Do not store in a manner where cross-contamination with other pesticides, fertilizers, food or feed could occur. In the event of spillage during handling or storage, absorb with sand or other inert material and dispose of absorbent in accordance with the Pesticide Disposal Instructions listed below.

Pesticide Disposal Instructions: Pesticide wastes are acutely hazardous. Improper disposal of excess pesticide, spray mixture, or rinsate is a violation of Federal law. If these wastes cannot be disposed of by use according to label instructions, contact your State Pesticide or Environmental Control agency, or the Hazardous Waste representative at the nearest EPA Regional Office for guidance.

Container Disposal Instructions: Triple rinse (or equivalent). Then offer for recycling or reconditioning, or puncture and dispose of in a sanitary landfill, or by incineration, or, if allowed by state and local authorities, by burning. If burned, stay out of smoke.

GENERAL INFORMATION

DESICCANT L-10 is a liquid concentrate containing arsenic acid and is especially effective as a cotton desiccant. DESICCANT L-10 is recommended particularly for use on cotton which is to be stripped. Cotton which has been properly desiccated with DESICCANT L-10 is suitable for machine picking or stripping as normally there will be an absence of green leaves with little or no regrowth. Do not tank mix DESICCANT L-10 with other defoliants or desiccants.

DESICCANT L-10 has acidic properties and reacts with fabric, galvanized metal, black iron, and certain other metals. Any metallic equipment or nozzles used in application of DESICCANT L-10 should be made of stainless steel as there is little or no reaction with this metal at field strength dilution. Do not use galvanized metal containers or spraying equipment as highly toxic arsine gas may be formed. To reduce possible corrosion to metals and deterioration of fabric and other parts, all equipment including airplanes which have been in contact with or exposed to DESICCANT L-10, or its solution, should be thoroughly washed with water immediately after use.

DIRECTIONS FOR USE

It is a violation of Federal Law to use this product in a manner inconsistent with its labeling.

SAFETY PRECAUTIONS

When mixing/loading or otherwise handling the concentrate wear midforearm to elbow length chemical resistant gloves, chemical resistant shoes or boots, goggles or face shield, and a protective suit which has long sleeves and long pants. The protective suit must be worn over normal work clothes.

When applying the diluted spray solution wear a protective suit which has long sleeves and long pants, chemical resistant gloves, and chemical resistant shoes or boots. The protective suit must be worn over normal work clothes.

When applying the dilute spray from an enclosed aircraft cockpit or enclosed tractor cab wear an uncontaminated long sleeve shirt and long pants. Any article of clothing worn while applying the product must be cleaned before reusing. Launder workclothes separate from household articles. Clothing which has been drenched or heavily contaminated should be disposed of in accordance with state or local regulations.

RE-ENTRY STATEMENT

Do not apply this product in such a manner as to directly or through drift expose workers or other persons. The area being treated must be vacated by unprotected persons.

Do not enter treated areas without protective clothing until sprays have dried.

Because certain states may require more restrictive reentry intervals for various crops treated with this product, consult your State Department of Agriculture for further information.

Written or oral warnings must be given to workers who are expected to be in an area about to be treated with this product. In case of accidental exposure, workers should wash thoroughly with soap and water and contaminated clothing should be removed and washed before reuse. If symptoms of skin or eye irritation occur call a physician. When oral warnings are given, warnings shall be given in a language customarily understood by workers. Oral warnings shall be given if there is reason to believe that written warnings cannot be understood by workers. Written warnings must include the following information: "DANGER. Area treated with DESICCANT L-10 Harvest Aid for Cotton on (date of application). Do not enter without protective clothing until spray has dried. In case of accidental exposure, workers should wash thoroughly with soap and water and contaminated clothing should be removed and washed before reuse. If symptoms of skin or eye irritation occur call a physician."

APPLY DESICCANT L-10 at the rate of 1½ quarts per acre using sufficient water to make 4 to 10 gallons of spray. For best results, thorough uniform coverage is necessary. Application should be made when air is calm and weather is clear and hot. Apply after the crop has matured, as earlier application may reduce yield. Do not apply within 4 days of harvest. Do not apply more than 1½ quarts per acre to avoid excessive residues of arsenic acid on raw cottonseed.

RESTRICTIONS

This pesticide is toxic to wildlife. Do not apply directly to water or wetlands (swamps, bogs, marshes and potholes). Do not contaminate water by cleaning of equipment or disposal of wastes.

Do not graze or use treated plants or gin trash for food or forage.

DESICCANT L-10 is injurious to all plant foliage.

Pesticide wastes are acutely hazardous. Improper disposal of excess pesticide, spray mixture or rinsate is a violation of Federal law. If these wastes cannot be disposed of by use according to label instructions, contact your State Pesticide or Environmental Control Agency or the Harardous Waste representative at the nearest EPA Regional office for guidance.

ENDANGERED SPECIES RESTRICTIONS

The use of any pesticide in a manner that may kill or otherwise harm endangered or threatened species or adversely modify their habitat is a violation of federal laws. The use of this product is controlled to prevent death or harm to endangered or threatened species that occur in the following counties or elsewhere in their range.

Before using this pesticide in the following counties, you must obtain the Cropland Endangered Species Bulletin (EPA/ES-CROP). The use of this pesticide is prohibited in these counties unless specified otherwise in the Bulletin. The EPA Bulletin is available from either your County Agricultural Extension Agent, the Endangered Species Specialist in your State Wildlife Agency Headquarters or the appropriate Regional Office of the U.S. Environmental Protection Agency (EPA). This Bulletin must be reviewed prior to pesticide use and retained.

Texas Counties	Endangered Species
Fort Bend, Refugio, Victoria	Attwater's Greater Prairie Chicken

WARRANTY AND DISCLAIMER

Pennwalt Corporation warrants that this material conforms to the chemical description on the label and is reasonably fit for the purposes referred to in the Directions for Use, subject to the risks referred to therein. PENNWALT MAKES NO OTHER EXPRESS OR IMPLIED WARRANTY OF FITNESS OR MERCHANTABILITY OR ANY OTHER EXPRESS OR IMPLIED WARRANTY. IN NO CASE SHALL PENNWALT OR SELLER BE LIABLE FOR CONSEQUENTIAL, SPECIAL OR INDIRECT DAMAGES RESULTING FROM THE USE OR HANDLING OF THIS PRODUCT INCLUDING, BUT NOT LIMITED TO, LOSS OF PROFITS, BUSINESS REPUTATION, OR CUSTOMERS; LABOR COST; OR OTHER EXPENSES INCURRED IN PLANTING OR HARVESTING.

Pennwalt and seller offer this product and the buyer and user accept it subject to the foregoing conditions of sale and warranty which may be varied only by agreement in writing signed by a duly authorized representative of Pennwalt.

5-C203S-01 G7 (6/87)

ENDOTHAL

TURF HERBICIDE

1 PINT TREATS 10,000 SQUARE FEET

ACTIVE INGREDIENT:

Disodium salt of endothall*	19.2%
INERT INGREDIENTS:	80.8%
TOTAL	100.0%

*7-oxabicyclo[2.2.1]heptane-2,3-dicarboxylic acid equivalent 15.5%
Contains 1.8 lbs. disodium endothall per gallon
(equivalent to 1.46 lbs. endothal acid)

EPA Registration No. 4581-79 EPA Establishment No. 228-IL-1

KEEP OUT OF REACH OF CHILDREN

DANGER

STATEMENT OF PRACTICAL TREATMENT

IF IN EYES, hold eyelids open and flush with a steady, gentle stream of water for 15 minutes.

IF SWALLOWED, drink promptly a large quantity of milk, egg whites, gelatin solution or if these are not available, drink large quantities of water. Avoid alcohol. Call a physician immediately.

IF ON SKIN, immediately flush with plenty of water for at least 15 minutes. Remove and wash contaminated clothing before reuse.

NOTE TO PHYSICIAN: Probable mucosal damage may contraindicate the use of gastric lavage. Measures against circulatory shock, respiratory depression and convulsion may be needed.

PRECAUTIONARY STATEMENTS

HAZARDS TO HUMANS AND DOMESTIC ANIMALS

DANGER

MAY BE FATAL IF SWALLOWED OR ABSORBED THROUGH SKIN. CORROSIVE. CAUSES IRREVERSIBLE EYE DAMAGE. DO NOT GET IN EYES OR ON SKIN OR CLOTHING. WEAR PROTECTIVE CLOTHING, RUBBER GLOVES AND GOGGLES OR FACE SHIELD WHEN HANDLING. Avoid breathing spray mist. Avoid contact with skin or clothing. Wash thoroughly with soap and water after handling and before eating or smoking. Remove contaminated clothing and wash before reuse.

ENVIRONMENTAL HAZARDS

Avoid contact with or drift to trees or other plants, as injury may occur. To avoid drift, do not apply in windy weather. Do not contaminate water when disposing of equipment washwaters.

This product is toxic to wildlife.

DIRECTIONS FOR USE

It is a violation of Federal law to use this product in a manner inconsistent with its labeling.

Do not apply this product through any type of irrigation system.

ENDOTHAL TURF HERBICIDE is an aqueous solution of the disodium salt of endothall for use as a selective spray for lawns to control black medic, burclover, cheat and other annual bromes, chickweed (in the South), cranesbill and filaree, dichondra (pony foot), goathead, henbit, knotweed, lespedeza, little barley, oxalis, rescuegrass, ryegrass, sweetclover, veronica (speedwell), vetch, white clover and for suppression of *Poa annua* in bent and blue grass turf. For control of certain other broadleaved weeds, 2,4-D in accordance with manufacturer's directions, may be added to the sprayer with ENDOTHAL TURF HERBICIDE.

Under certain conditions some temporary turf browning may occur to St. Augustine, Bermuda and other turf grasses, especially when temperatures are over 80°F and the weather is dry. Apply only on well established, vigorously growing turf. Do not use on putting greens or Red fescue turfs.

HOW TO MIX AND APPLY: Mix 1½ ounces (3 tablespoonfuls) of ENDOTHAL TURF HERBICIDE in 1 to 2 gallons of water and apply evenly to 1,000 square feet or for larger areas mix one quart in 20 to 50 gallons of water applying evenly as a fine mist to one acre. For best results, make 2 applications at 2-3 week intervals. Apply only when moisture and temperature conditions are favorable for good turf growth. For *Poa annua* control, use 3 early spring applications at 3/4 ounce (1 1/2 tablespoonfuls) per 1,000 square feet. Apply evenly as a fine mist; do not use more than recommended amounts.

STORAGE AND DISPOSAL

Do not contaminate water, food, or feed by storage or disposal.

Storage Instructions: Store in the original container. Do not store in a manner where cross-contamination with other pesticides, fertilizers, food or feed could occur. Storage at temperatures below 32°F may result in the product freezing or crystalizing. Should this occur the product must be warmed to 50°F or higher and thoroughly agitated. In the event of a spillage during handling or storage, absorb with sand or other inert material and dispose of absorbant in accordance with the Pesticide Disposal Instructions listed below.

Pesticide Disposal Instructions: Pesticide wastes are acutely hazardous. Improper disposal of excess pesticide, spray mixture, or rinsate is a violation of Federal law. If these wastes cannot be disposed of by use according to label instructions, contact your State Pesticide or Environmental Control Agency, or the Hazardous Waste representative at the nearest EPA Regional Office for guidance.

Container Disposal Instructions: Triple rinse (or equivalent). Then offer for recycling or reconditioning, or puncture and dispose of in a sanitary landfill, or by incineration, or, if allowed by state and local authorities, by burning. If burned, stay out of smoke.

WARRANTY AND DISCLAIMER

Pennwalt Corporation warrants that this material conforms to the chemical description on the label and is reasonably fit for the purposes referred to in the Directions For Use, subject to the risks referred to therein. PENNWALT MAKES NO OTHER EXPRESS OR IMPLIED WARRANTY OF FITNESS OR MERCHANTABILITY OR ANY OTHER EXPRESS OR IMPLIED WARRANTY. IN NO CASE SHALL PENNWALT OR SELLER BE LIABLE FOR CONSEQUENTIAL, SPECIAL OR INDIRECT DAMAGES RESULTING FROM THE USE OR HANDLING OF THIS PRODUCT INCLUDING, BUT NOT LIMITED TO, LOSS OF PROFITS, BUSINESS REPUTATION, OR CUSTOMERS; LABOR COST; OR OTHER EXPENSES INCURRED IN PLANTING OR HARVESTING.

Pennwalt and seller offer this product and the buyer and user accept it subject to the foregoing conditions of sale and warranty which may be varied only by agreement in writing signed by a duly authorized representative of Pennwalt.

5-F501S-01 D8 (4/88)

HERBICIDE 273

SUGAR BEET HERBICIDE

ACTIVE INGREDIENT:
Dipotassium salt of endothall* 40.3%
INERT INGREDIENTS: 59.7%
TOTAL 100.0%

*7-oxabicyclo[2.2.1]heptane-2, 3-dicarboxylic acid equivalent 28.6%
This product contains 4.23 lbs. dipotassium endothall per gallon (equivalent to 3 lbs. endothall)

EPA Registration No. 4581-223 EPA Establishment No. 228-IL-1

DANGER–POISON

KEEP OUT OF REACH OF CHILDREN

PRECAUCION AL USUARIO: Si usted no lee ingles, no use este producto hasta que la etiqueta haya sido explicado ampliamente.

STATEMENT OF PRACTICAL TREATMENT

If swallowed, drink promptly a large quantity of milk, egg whites, gelatin solution or, if these are not available, drink large quantities of water. Avoid alcohol. Call a physician immediately.

If on skin, immediately flush with plenty of water for at least 15 minutes after removing contaminated clothing and shoes.

If in eyes, immediately flush with plenty of water for at least 15 minutes. Call a physician.

NOTE TO PHYSICIAN: Probable mucosal damage may contraindicate the use of gastric lavage. Measures against circulatory shock, respiratory depression, and convulsion may be needed.

GENERAL INFORMATION

HERBICIDE 273 is a liquid formulation which mixes readily with water for use in the control of cheatgrass, redroot or pigweed, Kochia, foxtails, burclover, blueweed (henbit), Texas blueweed, volunteer barley, wild buckwheat, barnyardgrass, bullgrass, annual bluegrass, ragweed, purslane, smartweed, wild carrot, volunteer sunflowers, and shepherdspurse in sugar beets.

Mix the recommended amounts of HERBICIDE 273 listed in the tables below in 2 to 15 gallons of water per acre for aerial application and in 10 to 40 gallons of water per acre for ground equipment.

This material is known to be phytotoxic to a wide variety of plants, especially strawberries. For aerial application a 50-foot buffer zone downwind from the wing edge is recommended to avoid injury to desirable plants. Apply when there is a positive wind of less than 10 mph. A height of 6 to 10 feet over the crop is recommended. Keep nozzle pressure below 40 psi.

DIRECTIONS FOR USE

It is a violation of Federal law to use this product in a manner inconsistent with its labeling.

Do not apply this product through any type of irrigation system.

RE-ENTRY STATEMENT

Do not apply this product in such a manner as to directly or through drift expose workers or other persons. The area being treated must be vacated by unprotected persons.

Do not enter treated areas without protective clothing until sprays have dried.

Because certain states may require more restrictive reentry intervals for various crops treated with this product, consult your State Department of Agriculture for further information.

Written or oral warnings must be given to workers who are expected to be in an area about to be treated with this product. In case of accidental exposure, workers should wash thoroughly with soap and water and contaminated clothing should be removed and washed before reuse. If symptoms of skin or eye irritation occur call a physician. When oral warnings are given, warnings shall be given in a language customarily understood by workers. Oral warnings shall be given if there is reason to believe that written warnings cannot be understood by workers. Written warnings must include the following information: "DANGER. Area treated with HERBICIDE 273 Sugar Beet Herbicide on (date of application). Do not enter without protective clothing until spray has dried. In case of accidental exposure, workers should wash thoroughly with soap and water and contaminated clothing should be removed and washed before reuse. If symptoms of skin or eye irritation occur call a physician."

PRE-EMERGENCE APPLICATIONS

Field tests have shown that best pre-emergence results are obtained when HERBICIDE 273 is sprayed onto the seed bed during planting in a manner that spray would be incorporated into the soil not exceeding 1 1/2 inches deep. High moisture content of soil is essential for best weed control. Apply before sugar beets emerge from the soil.

HERBICIDE 273 Per Acre*

Broadcast Rates	Band Rates—Pints			
	Band Width	Row Spacing 22″	28″	32″
1.0-2.2 Gallons/Acre	5″	2.0-4.0	1.5-3.0	1.25-3.0
	7″	2.5-5.5	2.0-4.5	1.75-4.0
	10″	3.5-8.0	3.0-6.5	2.5 -5.5

*NOTE: For fine soils, use the lower rate. For use on the soils of Eastern Colorado and Western Nebraska, apply 1.5 to 3 pints HERBICIDE 273 for 7 inch band application on rows 22 inches wide. Adjust rate proportionately for other band widths and row spacings.

POST-EMERGENCE APPLICATIONS

Make application when the sugar beets have 2 to 4 true leaves (do not count the cotyledonary leaves, which are the first 2 leaves to appear). Under certain weather conditions, some temporary marginal leaf burn may occur to the sugar beets, especially at the higher rates, but recovery is normally rapid. For best results, apply when the temperature is between 60°F and 80°F. Do not apply later than 60 days after sugar beet emergence.

HERBICIDE 273 Per Acre*

Broadcast Rates	Band Rates—Fluid Ounces			
	Band Width	Row Spacing 22″	28″	32″
1-2 Quarts/Acre	5″	7-15	6-11	5-10
	7″	10-20	8-16	7-14
	10″	15-29	11-23	10-20

*NOTE: Use the lower rates of HERBICIDE 273 when the sugar beets have only 2 true leaves, or when spraying very susceptible weeds such as smartweed or wild buckwheat at early growth stage (2 to 4 true leaves).

TANK MIXES OF HERBICIDE 273 AND SELECTED HERBICIDES

FOR POST-EMERGENCE USE ON SUGAR BEETS ONLY

Follow all directions and precautions that appear on both product labels. In addition to the weeds previously mentioned, tank mix combinations of HERBICIDE 273 plus one of the herbicides listed below will control a broader spectrum of annual broadleaf weeds and seedling grasses than either herbicide used alone. For best results, apply when sugar beets are at the 4 to 6 true leaf stage, and when the weeds are actively growing and are not under water or heat stress. Since leaf burn may occur, do not apply tank mixes before sugar beets have 4 expanded true leaves.

Mixing Directions—Fill sprayer tank half full with water. Start agitation with a mechanical or good hydraulic agitator. While agitating, add the required amount of HERBICIDE 273 followed by required amount of one of the other herbicides listed below. Add the remaining quantity of water with the bypass agitator running. Mix only the amount of spray that will be used immediately.

HERBICIDE 273 PLUS BETANAL®—Controls seedlings of annual broadleaf weeds, such as knotweed, lambsquarters, nightshade, mustard, London rocket, goosefoot, ground cherry, fiddleneck, chickweed, and green and yellow foxtail. Do not add any wetting agents or other spray adjuvants. Do not apply later than 90 days prior to harvest.

HERBICIDE 273 PLUS BETANEX®—Controls seedlings of annual broadleaf weeds, such as lambsquarters, mustard, nightshade, goosefoot, and London rocket. Do not add any wetting agents or other spray adjuvants. Do not apply later than 90 days prior to harvest.

HERBICIDE 273 PLUS DOWPON® M—Controls seedlings of annual broadleaf weeds and grasses, such as foxtails, crabgrass, panicums, wild oats, bermudagrass, bluegrass, Johnsongrass, maidencane, and quackgrass.

HERBICIDE 273 PLUS PYRAMIN W®—Controls seedlings of annual broadleaf weeds, such as lambsquarters, mustard, fanweed, nightshade, nettleleaf, and goosefoot. Use 50 mesh screens in the line and Tee-jet #8004E nozzles or larger, or their equivalent.

Rates of Application—See Example Below

Post-Emergence Herbicide	Broadcast Rates Per Acre	Band Rates Per Acre Pints/Pounds			
		Band Width	Row Spacing 22″	28″	32″
HERBICIDE 273	2-3 pts	5″	0.5-0.7	0.4-0.6	0.3-0.4
		7″	0.7-1.0	0.5-0.8	0.5-0.7
		10″	0.9-1.3	0.7-1.1	0.6-0.9
BETANAL	6-7.5 pts	5″	1.4-1.7	1.0-1.2	0.9-1.1
		7″	2.0-2.5	1.5-1.9	1.3-1.6
		10″	2.7-3.4	2.1-2.6	1.8-2.2
BETANEX	6-7.5 pts	5″	1.4-1.7	1.0-1.2	0.9-1.1
		7″	2.0-2.5	1.5-1.9	1.3-1.6
		10″	2.7-3.4	2.1-2.6	1.8-2.2
DOWPON M (Dalapon)	2-3 lbs.	5″	0.5-0.7	0.3-0.5	0.3-0.5
		7″	0.7-1.0	0.5-0.8	0.5-0.7
		10″	0.9-1.4	0.7-1.1	0.6-0.9
PYRAMIN W	4-5 lbs	5″	0.9-1.1	0.7-0.9	0.6-0.7
		7″	1.3-1.6	1.0-1.3	0.8-1.0
		10″	1.8-2.3	1.4-1.9	1.2-1.5

NOTE: Use the low rates for fine soils or when the temperature exceeds 80°F.

Example: **BROADCAST TREATMENT**—Tank mix 2 to 3 pints of HERBICIDE 273 and 2 to 3 pounds of DOWPON M per acre.
BAND TREATMENT—For a 10″ band and 32″ row spacing tank mix 0.6 to 0.9 pint of HERBICIDE 273 and 0.6 to 0.9 pound of DOWPON M per acre.

®BETANAL is a registered trademark of Schering A.G. Berlin/Bergkamen
®BETANEX is a registered trademark of Schering A.G. Berlin/Bergkamen
®DOWPON M is a registered trademark of Dow Chemical Company
®PYRAMIN W is a registered trademark of BASF Wyandotte Corporation

PRECAUTIONARY STATEMENTS

HAZARDS TO HUMANS (AND DOMESTIC ANIMALS)

DANGER

FATAL IF ABSORBED THROUGH SKIN. MAY BE FATAL IF SWALLOWED. HARMFUL IF INHALED. CORROSIVE, CAUSES IRREVERSIBLE EYE DAMAGE. DO NOT GET IN EYES, ON SKIN OR ON CLOTHING. WEAR PROTECTIVE CLOTHING, RUBBER GLOVES, AND GOGGLES OR FACE SHIELD WHEN HANDLING. Wash thoroughly with soap and water after handling and before eating or smoking. Remove contaminated clothing and wash before reuse. Avoid breathing spray mist.

ENVIRONMENTAL HAZARDS

This product is toxic to wildlife. Do not apply directly to water. Do not contaminate water by disposal of equipment washwaters. Avoid contact with or drift to other crops or plants as injury may result.

STORAGE AND DISPOSAL

Do not contaminate water, food, or feed by storage or disposal.

Storage Instructions: Store in the original container. Do not store in a manner where cross-contamination with other pesticides, fertilizers, food or feed could occur. Storage at temperatures below 32°F may result in the product freezing or crystalizing. Should this occur the product must be warmed to 50°F or higher and thoroughly agitated. In the event of a spillage during handling or storage, absorb with sand or other inert material and dispose of absorbant in accordance with the Pesticide Disposal Instructions listed below.

Pesticide Disposal Instructions: Pesticide wastes are acutely hazardous. Improper disposal of excess pesticide, spray mixture, or rinsate is a violation of Federal law. If these wastes cannot be disposed of by use according to label instructions, contact your State Pesticide or Environmental Control Agency, or the Hazardous Waste representative at the nearest EPA Regional Office for guidance.

Container Disposal Instructions: Triple rinse (or equivalent). Then offer for recycling or reconditioning, or puncture and dispose of in a sanitary landfill, or by incineration, or, if allowed by state and local authorities, by burning. If burned, stay out of smoke.

WARRANTY AND DISCLAIMER

Pennwalt Corporation warrants that this material conforms to the chemical description on the label and is reasonably fit for the purposes referred to in the Directions for Use, subject to the risks referred to therein. PENNWALT MAKES NO OTHER EXPRESS OR IMPLIED WARRANTY OF FITNESS OR MERCHANTABILITY OR ANY OTHER EXPRESS OR IMPLIED WARRANTY. IN NO CASE SHALL PENNWALT OR SELLER BE LIABLE FOR CONSEQUENTIAL, SPECIAL OR INDIRECT DAMAGES RESULTING FROM THE USE OR HANDLING OF THIS PRODUCT INCLUDING, BUT NOT LIMITED TO, LOSS OF PROFITS, BUSINESS REPUTATION, OR CUSTOMERS; LABOR COST; OR OTHER EXPENSES INCURRED IN PLANTING OR HARVESTING.

Pennwalt and seller offer this product and the buyer and user accept it subject to the foregoing conditions of sale and warranty which may be varied only by agreement in writing signed by a duly authorized representative of Pennwalt.

5-F504S-01 D8 (4/88)

HYDROTHOL® 191

AQUATIC ALGICIDE AND HERBICIDE

ACTIVE INGREDIENT:	
Mono(N,N-dimethylalkylamine**) salt of endothall*.	53.0%
INERT INGREDIENTS:	47.0%
TOTAL.	100.0%

Contains 2 lb. endothall* per gallon

*7-oxabicyclo [2.2.1] heptane-2,3-dicarboxylic acid equivalent 23.36%
**Alkyl groups as derived from Coconut Oil

EPA Registration No. 4581-174 EPA Establishment No. 4581-TX-1

KEEP OUT OF REACH OF CHILDREN

DANGER–POISON

STATEMENT OF PRACTICAL TREATMENT

IF SWALLOWED, drink promptly a large quantity of milk, egg whites, gelatin solution or if these are not available, drink large quantities of water. Avoid alcohol. Call a physician immediately.

IF ON SKIN, immediately flush with plenty of water for at least 15 minutes. Remove and wash contaminated clothing before reuse.

IF IN EYES, immediately flush with plenty of water for at least 15 minutes. Call a physician.

NOTE TO PHYSICIAN: Probable mucosal damage may contraindicate the use of gastric lavage. Measures against circulatory shock, respiratory depression and convulsion may be needed.

GENERAL INFORMATION

HYDROTHOL 191 is a liquid concentrate soluble in water and is a highly effective aquatic herbicide and algicide for use in irrigation and drainage canals, lakes and ponds to control the following weeds and algae: Najas, Elodea, Coontail, Potamogeton, Milfoil sp., Zannichellia, Vallisneria,

Cladophora, Pithophora, Spirogyra, and Chara, when weeds are actively growing. Dosage rates indicated are measured in "Parts per Million" (ppm). 1.0 ppm as a dosage rate means that there would be 1 part endothall (acid) in 1,000,000 parts of water.

Necessary approval and/or permits should be obtained in States where required. Consult state water or conservation authorities before applying to public waters or to ponds, canals or streams which flow into public waters.

DIRECTIONS FOR USE

It is a violation of Federal law to use this product in a manner inconsistent with its labeling.

IRRIGATION AND DRAINAGE CANALS

Apply HYDROTHOL 191 as a surface spray with uniform coverage or inject under the surface of the water. The minimum contact time with weeds for optimum results should be two hours.

HEAVY INFESTATION—Apply at rate of 4.0 to 6.75 gal. per acre foot (equivalent to 3.0-5.0 ppm). See table below.

MODERATE OR LIGHT INFESTATIONS—Apply at rate of 1.4 to 2.7 gal. per acre foot (equivalent to 1.0-2.0 ppm). See table below.

GALLONS OF HYDROTHOL 191 REQUIRED TO TREAT 1 MILE OF DITCH, 1 FOOT DEEP

	WIDTH OF DITCH OR CANAL IN FEET									
ppm	4	6	8	10	12	14	16	18	20	22
1.0 ppm	0.7	1.0	1.3	1.7	1.9	2.3	2.6	2.9	3.3	3.6
2.0 ppm	1.3	2.0	2.6	3.3	3.9	4.6	5.2	5.9	6.5	7.2
3.0 ppm	2.0	2.8	3.9	4.9	5.9	6.8	7.8	8.8	9.8	10.8
4.0 ppm	2.6	3.9	5.2	6.6	7.8	9.1	10.4	11.8	13.1	14.4
5.0 ppm	3.2	4.9	6.5	8.2	9.7	11.5	13.1	14.7	16.4	18.0

Note: 1.1 pt. HYDROTHOL 191 per acre foot equals 0.1 ppm of endothall (acid).

LAKES AND PONDS

ALGAE CONTROL: Cladophora, Pithophora, Spirogyra, and Chara can be controlled with applications of 0.6 to 2.2 pints per acre foot (0.05 to 0.2 ppm) applied as a uniform surface spray or injected under water surface. Pithophora may require 0.2 ppm or higher. Dosages may be increased to 0.4 to 1.1 gals (0.3 to 0.8 ppm) where greater longevity of control is desired and for marginal treatments. Repeat treatments when algal growth reappears.

SUBMERGED AQUATICS: Due to the toxicity to fish, the use of HYDROTHOL 191 for submerged aquatic weeds is suggested only by commercial applicators on a marginal or sectional rather than overall type treatment. Use 0.7 to 3.4 gals. per acre foot (0.5 to 2.5 ppm) on a marginal treatment either sprayed on or injected below the surface. Use dosages over 1.0 ppm on very narrow margins or in areas where some fish kill is not objectionable. Do not treat more than 1/10 of the lake or pond at one time with doses in excess of 1.0 ppm.

APPROXIMATE QUANTITY OF HYDROTHOL 191 FOR ONE ACRE* TREATMENT

	Dosages for Various Concentrations In ppm								
DEPTH	0.05	0.1	0.2	0.3	0.5	0.8	1.0	2.0	2.5
1 ft.	0.6 pts	1.1 pts	2.2 pts	.4 gal	.7 gal	1.1 gal	1.4 gal	2.7 gal	3.4 gal
2 ft.	1.1 pts	2.2 pts	4.3 pts	.8 gal	1.4 gal	2.2 gal	2.8 gal	5.4 gal	6.8 gal
4 ft.	2.2 pts	4.3 pts	8.6 pts	1.6 gal	2.7 gal	4.4 gal	5.4 gal	10.9 gal	13.6 gal
6 ft.	3.3 pts	6.5 pts	12.9 pts	2.5 gal	4.1 gal	6.5 gal	8.2 gal	16.3 gal	20.4 gal

*One acre equals approximately 208′ × 208′

PRECAUTIONARY STATEMENTS

HAZARDS TO HUMANS (AND DOMESTIC ANIMALS)

DANGER

FATAL IF ABSORBED THROUGH SKIN. MAY BE FATAL IF SWALLOWED. CORROSIVE, CAUSES IRREVERSIBLE EYE DAMAGE AND SKIN BURNS. HARMFUL IF INHALED. DO NOT GET IN EYES, ON SKIN OR ON CLOTHING. WEAR PROTECTIVE CLOTHING, RUBBER GLOVES, AND GOGGLES OR FACE SHIELD WHEN HANDLING. Wash thoroughly with soap and water after handling and before eating or smoking. Remove contaminated clothing and wash before reuse. Avoid breathing spray mist.

ENVIRONMENTAL HAZARDS

Fish will be killed by dosages in excess of 0.3 ppm.
Do not use where fish are important resources.

Avoid contact with or drift to desirable plants or crops as injury may result.

Clean out application equipment after each operation.

Do not use fish from treated water for food or feed within three days after treatment.

Do not use treated water for watering livestock, for preparing agricultural sprays for food crops, for irrigation, or for domestic purposes within the following periods:

Up to 0.3 ppm— 7 days after application
Up to 3.0 ppm—14 days after application
Up to 5.0 ppm—25 days after application

STORAGE AND DISPOSAL

Do not contaminate water, food, or feed by storage or disposal.

Storage Instructions: Store in the original container. Do not store in a manner where cross-contamination with other pesticides, fertilizers, food or feed could occur. In the event of spillage during handling or storage, absorb with sand or other inert material and dispose of absorbant in accordance with the Pesticide Disposal Instructions listed below.

Pesticide Disposal Instructions: Pesticide wastes are acutely hazardous. Improper disposal of excess pesticide, spray mixture, or rinsate is a violation of Federal law. If these wastes cannot be disposed of by use according to label instructions, contact your State Pesticide or Environmental Control Agency, or the Hazardous Waste representative at the nearest EPA Regional Office for guidance.

Container Disposal Instructions: Triple rinse (or equivalent). Then offer for recycling or reconditioning, or puncture and dispose of in a sanitary landfill, or by incineration, or, if allowed by state and local authorities, by burning. If burned, stay out of smoke.

WARRANTY AND DISCLAIMER

Pennwalt Corporation warrants that this material conforms to the chemical description on the label and is reasonably fit for the purposes referred to in the Directions for Use, subject to the risks referred to therein. PENNWALT MAKES NO OTHER EXPRESS OR IMPLIED WARRANTY OF FITNESS OR MERCHANTABILITY OR ANY OTHER EXPRESS OR IMPLIED WARRANTY. IN NO CASE SHALL PENNWALT OR SELLER BE LIABLE FOR CONSEQUENTIAL, SPECIAL OR INDIRECT DAMAGES RESULTING FROM THE USE OR HANDLING OF THIS PRODUCT INCLUDING, BUT NOT LIMITED TO, LOSS OF PROFITS, BUSINESS REPUTATION, OR CUSTOMERS; LABOR COST; OR OTHER EXPENSES INCURRED IN PLANTING OR HARVESTING.

Pennwalt and seller offer this product and the buyer and user accept it subject to the foregoing conditions of sale and warranty which may be varied only by agreement in writing signed by a duly authorized representative of Pennwalt.

5-C110S-01 H8(7/88)

HYDROTHOL® 191

GRANULAR AQUATIC ALGICIDE AND HERBICIDE

ACTIVE INGREDIENT:

Mono(N,N-dimethylalkylamine*) salt of endothall**	11.2%
INERT INGREDIENTS:	88.8%
TOTAL	100.0%

*Alkyl groups as derived from Coconut Oil
**7-oxabicyclo [2.2.1] heptane-2,3-dicarboxylic acid equivalent 5%

EPA Registration No. 4581-172 EPA Establishment No. 228-IL-1

KEEP OUT OF REACH OF CHILDREN

DANGER

STATEMENT OF PRACTICAL TREATMENT

IF IN EYES, hold eyelids open and flush with a steady, gentle stream of water for 15 minutes. Call a physician.

IF SWALLOWED, drink promptly a large quantity of milk, egg whites, gelatin solution or if these are not available, drink large quantities of water. Avoid alcohol. Call a physician immediately.

IF ON SKIN, immediately flush with plenty of water for at least 15 minutes. Remove and wash contaminated clothing before reuse.

NOTE TO PHYSICIAN: Probable mucosal damage may contraindicate the use of gastric lavage. Measures against circulatory shock, respiratory depression and convulsion may be needed.

PRECAUTIONARY STATEMENTS

HAZARDS TO HUMANS AND DOMESTIC ANIMALS

DANGER

CORROSIVE. CAUSES IRREVERSIBLE EYE DAMAGE. HARMFUL IF SWALLOWED, ABSORBED THROUGH SKIN OR INHALED. AVOID CONTACT WITH SKIN OR CLOTHING. DO NOT GET IN EYES. WEAR GOGGLES OR FACE SHIELD WHEN HANDLING. Avoid breathing dust. Wash thoroughly with soap and water after handling. Remove contaminated clothing and wash before reuse.

ENVIRONMENTAL HAZARDS

Avoid contact with or drift to other crops or plants as injury may result.

Fish will be killed by dosages in excess of 0.3 ppm. Do not use where fish are important resources. Do not use fish from treated water for food or feed within three days after treatment.

Do not use treated water for watering livestock, for preparing agricultural sprays for food crops, for irrigation or for domestic purposes within the following periods:

Up to 0.3 ppm— 7 days after application

Up to 3.0 ppm—14 days after application

Up to 5.0 ppm—25 days after application

GENERAL INFORMATION

HYDROTHOL 191 Granular is a highly effective algicide and aquatic weed killer for use in irrigation and drainage canals, lakes and ponds to control the following weeds and algae: Najas, Elodea, Coontail, Potamogeton, Milfoil sp., Zannichellia, Vallisneria, Cladophora, Spirogyra, Pithophora and Chara, when weeds are actively growing. Dosage rates indicated are measured in "Parts per Million" (ppm). 1 ppm as a dosage rate means that there would be 1 part endothall (acid) in 1,000,000 parts of water. Thoroughly clean application equipment immediately after use.

Necessary approval and/or permits should be obtained in states where required. Consult state water or conservation authorities before applying to public waters or to ponds, canals or streams which flow into public waters.

DIRECTIONS FOR USE

It is a violation of Federal law to use this product in a manner inconsistent with its labeling.

IRRIGATION AND DRAINAGE CANALS

HEAVY INFESTATIONS—Evenly spread 160 to 270 lbs. HYDROTHOL 191 Granular per acre foot of water (3.0 to 5.0 ppm) with suitable aerial or ground equipment.

MODERATE OR LIGHT INFESTATIONS—Use 55 to 110 lbs. per acre foot HYDROTHOL 191 Granular (1.0 to 2.0 ppm) applied evenly.

Pounds of HYDROTHOL 191 Granular required per mile of canal or swath, 1 ft. deep.					
	Width of Canal, Ditch or Swath				
ppm	10	15	20	30	40
0.5	33	49	65	95	130
1.0	65	98	130	190	260
2.0	130	195	260	380	520
3.0	195	294	390	570	780
4.0	260	392	520	760	1040
5.0	325	490	650	950	1300

Note: 5.5 lbs. per acre foot equals 0.1 ppm of Endothall (acid).

LAKES AND PONDS

ALGAE CONTROL—Cladophora, Pithophora, Spirogyra, and Chara can be controlled by even applications of 3 lbs. to 11 lbs. HYDROTHOL 191 Granular per acre foot (0.05 to 0.2 ppm) applied by suitable air or water equipment. HYDROTHOL 191 Granular is especially effective on bottom growth control. Pithophora may require 0.2 ppm or higher. Dosages may be increased to 27-82 lbs. of HYDROTHOL 191 Granular (0.5 to 1.5 ppm) where greater longevity of control is desired and for marginal treatments. For floating mats HYDROTHOL 191 Granular should be scattered as evenly as possible over the mat. Repeat treatments when algal growth reappears.

SUBMERGED AQUATICS—Due to the toxicity to fish, the use of HYDROTHOL 191 Granular for submerged aquatics is suggested only by commercial applicators on a marginal or sectional treatment basis. Use 27 to 136 lbs. HYDROTHOL 191 Granular per acre foot (0.5 to 2.5 ppm) scattered evenly as possible. Use dosages over 1.0 ppm on narrow margins or in areas where some fish kill is not objectionable.

Do not treat more than 1/10 of the lake or pond at one time with doses in excess of 1.0 ppm. Approximate pounds of HYDROTHOL 191 Granular for one acre* treatment at various depths and ppm concentrations.

	PPM Concentrations						
Depth	0.05 ppm	0.1 ppm	0.2 ppm	0.5 ppm	1.0 ppm	1.5 ppm	2.5 ppm
1 ft.	3	5	11	27	54	82	136
2 ft.	5	11	22	54	109	163	272
4 ft.	11	22	44	109	217	326	544
6 ft.	16	33	65	163	326	489	815

* One acre equals approximately 208′ × 208′.

FOR USE ON RICE IN CALIFORNIA ONLY

HYDROTHOL 191 GRANULAR is a formulation of a submersed aquatic weed killer for use in rice fields to control Southern Naiad *(Najas guadalupensis)*, American Pondweed *(Potamogeton nodosus)*, Horned Pondweed *(Zannichellia palustris)*, Sago Pondweed *(Potamogeton pectinatus)*, and Coontail *(Ceratophyllum demersum)*.

TIME OF APPLICATION

Apply 25 to 60 days after sowing, when aquatic weeds are present. Do not apply before rice emerges above surface of water. Do not apply after rice starts heading. Do not apply HYDROTHOL 191 GRANULAR more than once yearly.

RATE OF APPLICATION

The ideal average depth of water at time of application is 6 to 8 inches; however, based on depth, use rates as follows:

Depth of Water in Inches	Lbs. HYDROTHOL 191 GRANULAR Per Acre	ppm Acid Equivalent
4-6	40	2.2-1.5
6-8	50	1.9-1.4
8-10	60	1.7-1.4

APPLICATION EQUIPMENT

Apply with ground or air equipment designed for uniform distribution of granules. Clean out application equipment after each use.

Necessary approval and/or permits should be obtained if required. Consult State Fish and Game agencies before applying this product.

Do not release water from flooded fields within 10 days of application. Carefully read regular product label and follow appropriate precautions and directions.

STORAGE AND DISPOSAL

Do not contaminate water, food, or feed by storage or disposal.

Storage Instructions: Store in the original container, preferably in a locked storage area. Do not store in a manner where cross-contamination with other pesticides, fertilizers, food or feed could occur. If spilled during storage or handling sweep up spillage and dispose of in accordance with the Pesticide Disposal Instructions listed below.

Pesticide Disposal Instructions: Pesticide wastes are acutely hazardous. Improper disposal of excess pesticide or rinsate is a violation of Federal law. If these wastes cannot be disposed of by use according to label instructions, contact your State Pesticide or Environmental Control Agency, or the Hazardous Waste representative at the nearest EPA Regional Office for guidance.

Container Disposal Instructions: Triple rinse (or equivalent). Then offer for recycling or reconditioning, or puncture and dispose of in a sanitary landfill, or, by incineration, or, if allowed by state and local authorities, by burning. If burned, stay out of smoke.

WARRANTY AND DISCLAIMER

Pennwalt Corporation warrants that this material conforms to the chemical description on the label and is reasonably fit for the purposes referred to in the Directions for Use, subject to the risks referred to therein. PENNWALT MAKES NO OTHER EXPRESS OR IMPLIED WARRANTY OF FITNESS OR MERCHANTABILITY OR ANY OTHER EXPRESS OR IMPLIED WARRANTY. IN NO CASE SHALL PENNWALT OR SELLER BE LIABLE FOR CONSEQUENTIAL, SPECIAL OR INDIRECT DAMAGES RESULTING FROM THE USE OR HANDLING OF THIS PRODUCT INCLUDING, BUT NOT LIMITED TO, LOSS OF PROFITS, BUSINESS REPUTATION, OR CUSTOMERS; LABOR COSTS; OR OTHER EXPENSES INCURRED IN PLANTING OR HARVESTING.

Pennwalt and seller offer this product and the buyer and user accept it subject to the foregoing conditions of sale and warranty which may be varied only by agreement in writing signed by a duly authorized representative of Pennwalt.

5-F111S-01 H8 (7/88)

KNOX OUT® 2 FM

FLOWABLE MICROENCAPSULATED INSECTICIDE

SHAKE WELL BEFORE USING

ACTIVE INGREDIENTS:
Diazinon® (0,0-Diethyl 0-[6 methyl-2-(1- methylethyl)-4-pyrimidinyl]phosphorothioate) 23.0%
INERT INGREDIENTS: 77.0%

TOTAL 100.0%

Contains 2 pounds Diazinon per gallon
U.S. Patents 3,429,827 and 3,577,515

EPA Registration No. 4581-335 EPA Establishment No. 4581-TX-1

KEEP OUT OF REACH OF CHILDREN

CAUTION

HARMFUL IF SWALLOWED, INHALED, OR ABSORBED THROUGH SKIN

GENERAL INFORMATION

KNOX OUT 2FM is a flowable microencapsulated formulation containing 2 pounds Diazinon per gallon.

It is to be used for residual pest control in and around buildings including, but not limited to, homes, apartments, stores, hospitals (non-patient room areas), nursing homes, interiorscapes, manufacturing plants, warehouses, schools, restaurants, hotels, motels, food processing plants, and on vessels, aircraft, boats, ships, rail cars, buses, trucks, and trailers.

KNOX OUT 2FM can be sprayed on any surface which will not be damaged or stained by water. Under conditions of proper application, KNOX OUT 2FM will not leave a visible deposit. However, if unevenly applied or sprayed to run-off, a visible deposit may occur on some dark surfaces and may be wiped off with a wet cloth or sponge.

Due to its controlled-release nature, KNOX OUT 2FM gives excellent residual kill of the pests listed below; however, effective knockdown may take up to 24 hours after application.

DIRECTIONS FOR USE

It is a violation of Federal law to use this product in a manner inconsistent with its labeling.

Do not apply this product through any type of irrigation system.

Vacate patient rooms and school rooms before treatment. Allow spray deposits to dry thoroughly before the room is reoccupied.

Thoroughly wash out spray equipment before using KNOX OUT 2FM. Shake concentrate well before diluting. When diluting, first add approximately half the water to the spray tank, then add the required quantity of KNOX OUT 2FM, and finally add the rest of the water. Occasionally agitate spray during use to assure even coverage. Shake or re-agitate sprayer before use if application is interrupted.

Do not use this formulation with oil.

IF SPRAY SCREENS ARE USED, THEY SHOULD BE 50 MESH OR LARGER.

Unless otherwise directed, dilute KNOX OUT 2FM with water to make a 0.5% to 1% spray** as indicated below:

Amount of Finished Spray	Amount* of KNOX OUT 2FM to use for: 0.5% Spray**	1% Spray**
1 Pint	0.3 fl. oz.	0.6 fl. oz.
2 Pints	0.6 fl. oz.	1.2 fl. oz.
4 Pints	1.2 fl. oz.	2.5 fl. oz.
1 Gallon	2.5 fl. oz.	5.0 fl. oz.
2 Gallons	5.0 fl. oz.	10.0 fl. oz.
3 Gallons	7.5 fl. oz.	15.0 fl. oz.
10 Gallons	25.0 fl. oz. (1.6 pints)	50.0 fl. oz. (1.6 quarts)

*1 Tablespoon = 0.5 fl. oz.
**Active ingredient basis

Cockroaches and Silverfish—Make spot applications of KNOX OUT 2FM as a 0.5% to 1% fine spray or by paint brush to drawers, shelves, around water pipes, storage areas, closets, behind and under cabinets, refrigerators, sinks, stoves and in similar areas where these insects hide.

Repeat treatments of KNOX OUT 2FM should be made as needed, but generally will not be necessary more than once every 60 days. Use the low rate for light infestations or monthly maintenance sprays, and the higher concentration for heavy infestations, initial clean out, or where insecticide resistance is suspected.

Ants—Make spot applications of 0.5% to 1% KNOX OUT 2FM as a spray to baseboards, window frames, wall voids, crawl spaces, cracks and crevices, and similar hiding or habitation areas. Apply to trails, around doors and windows, and other areas where these pests may crawl or enter the premises. Repeat application as necessary.

Carpenter Ants—Make spot applications of 0.5% to 1% KNOX OUT 2FM to any trails, around doors and windows, wall voids, crawl spaces, and any other places where ants enter premises. Where possible, treat ant nest or infested wood directly. Repeat application as necessary.

Carpet Beetles—Make applications of 0.5% to 1% KNOX OUT 2FM as a spray on and under edges of floor coverings, under carpets and furniture, and in closets or other localities where these insects are found. Entire carpets may be treated. When treating entire carpets see directions under **Fleas and Ticks.** Repeat application as necessary.

Crickets—Make spot applications of 0.5% to 1% of KNOX OUT 2FM as a spray to baseboards, behind and under cabinets and appliances, and other hiding places. Areas around doors and water pipes should be treated to help prevent crickets from entering the premises. Repeat application as necessary.

Flies—Make spot applications of 1% KNOX OUT 2FM as a spray to window frames, screens, dumpsters, trash cans, and similar areas where these pests may congregate. Repeat application as necessary.

Fleas and Ticks—Thoroughly apply KNOX OUT 2FM as a 1% fine spray to pet beds and resting quarters, adjacent cracks and crevices, baseboards, carpets and floor coverings, between and under cushions of upholstered furniture, and other areas where these pests may be present. Entire carpets may be treated. Applications to most surfaces should be as light as possible, approximately 2 quarts finished spray per 800 square feet. It may be necessary to use a higher volume of spray, approximately 3 to 4 quarts per 800 square feet, on long, dense shag carpets. Old bedding of pets should be replaced with clean, fresh bedding after treatment. Animals should not be sprayed with this product.

Perimeter Treatment—Residual spray for control of Ants, Carpenter Ants, Cockroaches, Crickets, Fleas, Silverfish, and Ticks:

Apply a 0.5% to 1% KNOX OUT 2FM spray to exterior surfaces of building, porches, screens, window frames, eaves, patios, garages, and other areas where these pests congregate or enter premises.

To help prevent infestations of buildings by the above pests, treat a band of soil 6 to 10 feet wide around and adjacent to the building. Also, treat the building foundation to a height of 2 to 3 feet where pests are active and may find entrance. KNOX OUT 2FM may be applied in compressed air equipment or power spray equipment. Apply as a coarse spray to thoroughly and uniformly wet the band area. A rate of 4 to 6 ounces of KNOX OUT 2FM per 1,000 square feet is desirable.

Food Handling Establishments—Places other than private residences in which food is held, processed, prepared, packaged, or served. Applications are limited to spot and/or crack and crevice treatments only.

Make spot and/or crack and crevice applications of KNOX OUT 2FM using a 0.5% to 1% spray in and around both food and non-food locations, treating localized areas where cockroaches, silverfish, ants, and flies have been seen or are suspected of hiding or entering. Do not use as a space spray or allow spray to contact food, food processing surfaces, or utensils. Make applications using low pressure.

Crack and crevice applications should be made using equipment capable of delivering a pin stream or of applying small amounts of insecticide directly into or between equipment and floors, openings leading to voids in walls, floors, and ceilings, and similar locations where these pests may hide or enter.

Spot treatments should not cover areas of more than 2 square feet and not more than 20% of the floor surface should be treated.

This product is authorized for use in inedible product areas of Federal Inspected Meat and Poultry Plants.

Stored Product Pests—For control of exposed stages of Sawtoothed Grain Beetle, Confused Flour Beetle, Red Flour Beetle, Rice Weevil, Grain Weevil, Warehouse Beetle:

Make spot and/or crack and crevice treatment of 0.5% to 1% KNOX OUT 2FM to localized areas including shelves, cupboards, walls and floors, where food products suspected of being infested are stored. Food should be removed or covered during treatment. This treatment is supplemental to good housekeeping, since it will have no effect on insects breeding within packaging. Infested food packaging should be destroyed.

Use In Greenhouses And Interiorscapes—For control of Aphids, Fungus Gnat Larvae, Leafminers, Mealybugs, (Citrus, Stem and Root), Mites, Scale Insects and Thrips:

Apply KNOX OUT 2FM to the ornamentals listed below at a rate of 0.5 to 1.0 fl. oz. per gallon of water (3-6 pints/100 gallons).

Treat when insects appear. Use higher rate for heavy insect infestations. Repeat applications as necessary.

ORNAMENTALS

African Violets	Flat Topped Sedum	Petunia
Agapanthus	Floss Flower	Photinia
Ageratum	Friendship Plant	Pineapple Guava
Aluminum Plant	Gardenia	Pittosporum

Alyssum	Gazinia	Podocarpus
Arrowhead	Geranium	Primrose
Azalea	Golddust	Privet
Begonia	Honeysuckle	Pyracantha
Bird's Nest Fern	Hydrangea	Raphiolepsis
Boston Fern	Ilex	Rhododendron
Bottlebrush	Impatiens	Rose
Cactus	Ivy	Salvia
Calendula	Jade Plant (Crassula)	Sansevieria
Camelia	Japanese Arelia	Scarlet Sage
Celosia Jewel Box	Juniper	Snake Plant
Chrysanthemum	Leucothoe	Snapdragon
Cockscomb	Lilac	Spider Plant
Coleus	Malus	Star Jasmine
Cotoneaster	Maranta	Strawflower
Crape Myrtle	Marigold	Tahitian Bridal Veil
Croton	Nandia	Velvet Plant (Gynura)
Daisy	Oleander	Verbena
Dianthus	Orchid	Virvia
Dogwood	Osmanthus	Wandering Jew
Escalonia	Pachysandra	Wax Plant
Euonymus	Periwinkle	Zinnia

Do not use on Bella Palm, Dracaena, Hoya, Maidenhand Fern, Neath, Poinsettia, and Prayer Plant.

KNOX OUT 2FM has demonstrated excellent plant safety; however, all varieties of the plants listed above have not been tested. When treating large numbers of plants of a single variety, spray a few plants and observe for phytotoxicity prior to full scale application.

KNOX OUT 2FM may not control some organophosphate resistant ornamental plant pests.

Use In and Around Livestock Housing—For control of Flies in Barns and Animal Sleeping Quarters (except Dairy Barns, Milk Rooms and Poultry Houses):

Apply a 1% KNOX OUT 2FM spray to ceilings and walls of livestock sheds, calf barns, hog barns, loading sheds, and other farm buildings except dairy barns, milkrooms and poultry houses. Areas where flies congregate, such as garbage dumps and corrals, should be sprayed thoroughly. Repeat applications as fly populations warrant.

Remove animals from buildings prior to treatment and keep animals out until spray has dried.

Do not contaminate milk, feed, drinking water, or apply to animals or poultry.

Yellowjackets—Use in Baits for Control of Scavenger Yellowjackets *Vespula pensylvanica, V. vulgaris)* in the 11 Contiguous Western States Only (AZ, CA, CO, ID, MT, NM, NV, OR, UT, WA, WY).

Consult the Supplemental Label available in these states.

PRECAUTIONARY STATEMENTS

HAZARDS TO HUMANS AND DOMESTIC ANIMALS

CAUTION

STATEMENT OF PRACTICAL TREATMENT

IF SWALLOWED—Drink 1 or 2 glasses of water and induce vomiting by touching back of throat with finger or blunt object. Do not induce vomiting or give anything by mouth to an unconscious person.

IF ON SKIN—Remove contaminated clothing immediately and wash skin with soap and water.

IF IN EYES—Wash thoroughly with water for at least 15 minutes. Call a physician if symptoms occur.

NOTE TO PHYSICIAN: The active ingredient of KNOX OUT 2FM is an organophosphate cholinesterase inhibitor. Atropine is antidotal only if symptoms of cholinesterase inhibition are present. Morphine is contraindicated.

HARMFUL IF SWALLOWED, INHALED OR ABSORBED THROUGH SKIN. Avoid contact with skin, eyes or clothing. Avoid breathing spray mist. Wash thoroughly with soap and water after handling. Remove contaminated clothing and wash before reuse. Keep out of reach of children.

Do not apply to humans, their clothing or bedding. Do not allow children or pets to contact treated surfaces until spray has dried. Do not contaminate food or use on household pets. Cover fish tanks prior to application.

Avoid contamination of food, feedstuffs, or water supply. Do not contaminate food preparation surfaces, kitchen utensils, dishes, or food storage containers.

Do not spray where electrical short circuits might result, such as wall outlets, conduits, etc.

ENVIRONMENTAL HAZARDS

This product is toxic to fish, birds and other wildlife. Do not apply directly to water and wetlands. Do not contaminate water when disposing of equipment washwaters.

This product is highly toxic to bees.

STORAGE AND DISPOSAL

Do not contaminate water, food or feed by storage or disposal.

STORAGE: Store in original container only, preferably in a locked storage area. Do not store in a manner where cross contamination with other pesticides, fertilizer, food or feed could occur. If spilled during storage or handling, soak up spillage with absorbant material and dispose of in accordance with Pesticide Disposal instructions listed below.

PESTICIDE DISPOSAL: Wastes resulting from the use of this product may be disposed of on site or at an approved waste disposal facility.

CONTAINER DISPOSAL: Triple rinse container, then offer for recycling or reconditioning, or puncture and dispose of in a sanitary landfill, or by incineration, or, if allowed by state and local authorities, by burning. If burned, stay out of smoke.

®Diazinon is a registered trademark of Ciba-Geigy Corporation

WARRANTY AND DISCLAIMER

Pennwalt Corporation warrants that this material conforms to the chemical description on the label and is reasonably fit for the purposes referred to in the Directions for Use, subject to the risks referred to therein. PENNWALT MAKES NO OTHER EXPRESS OR IMPLIED WARRANTY OF FITNESS OR MERCHANTABILITY OR ANY OTHER EXPRESS OR IMPLIED WARRANTY. IN NO CASE SHALL PENNWALT OR SELLER BE LIABLE FOR CONSEQUENTIAL, SPECIAL OR INDIRECT DAMAGES RESULTING FROM THE USE OR HANDLING OF THIS PRODUCT INCLUDING, BUT NOT LIMITED TO, LOSS OF PROFITS, BUSINESS REPUTATION, OR CUSTOMERS; LABOR COST; OR OTHER EXPENSES INCURRED IN PLANTING OR HARVESTING.

Pennwalt and seller offer this product and the buyer and user accept it subject to the foregoing conditions of sale and warranty which may be varied only by agreement in writing signed by a duly authorized representative of Pennwalt.

5-Z411S-01 E8 (5/88)

KRYOCIDE®

INSECTICIDE

ACTIVE INGREDIENT:

Cryolite: sodium aluminofluoride (Fluorine—not less than 50%)	96.0%
INERT INGREDIENTS:	4.0%
TOTAL	100.0%

EPA Registration No. 4581-116 EPA Establishment No. 33770-JP-01

CAUTION

KEEP OUT OF REACH OF CHILDREN

See Specific Caution Statements On Panel Below

PRECAUTIONARY STATEMENTS

HAZARDS TO HUMANS AND DOMESTIC ANIMALS

CAUTION

HARMFUL IF SWALLOWED, ABSORBED THROUGH SKIN, OR INHALED. CAUSES MODERATE EYE IRRITATION. Avoid contact with skin, eyes, or clothing. Avoid breathing spray mist or dust. Wash thoroughly with soap and water after handling. Remove contaminated clothing and wash before reuse.

Do not apply this product in such a manner as to directly or through drift expose workers or other persons. The area being treated must be vacated by unprotected persons.

ENVIRONMENTAL HAZARDS

This product is toxic to fish. Keep out of any body of water. Do not apply where runoff is likely to occur. Do not apply directly to water or wetlands (i.e. swamps, bogs, marshes and potholes). Do not contaminate water when disposing of equipment washwaters.

GENERAL INFORMATION

KRYOCIDE is a fine powder and suitable for spray or dust application by ground equipment or air. A thorough application should be made, covering the underside of the leaves as well as the top if possible.

DIRECTIONS FOR USE

It is a violation of Federal law to use this product in a manner inconsistent with its labeling.

Do not apply this product through any type of irrigation system.

Do not enter treated areas without protective clothing until spray has dried or dust has settled. Because certain states may require more restrictive reentry intervals for various crops treated with this product, consult your State Department of Agriculture for further information.

Written or oral warnings must be given to workers who are expected to be in a treated area or in an area about to be treated with this product. When oral warnings are given, such warnings shall be given in a language customarily understood by the workers. Oral warnings must be given if there is reason to believe that written warnings cannot be understood by the workers. Written warnings must include the following information: " 'CAUTION. Area treated with KRYOCIDE® Insecticide on *(date of application)* Do not enter without protective clothing until sprays have dried or dusts have settled.' "

KRYOCIDE® is recommended for control of the insects on the crops listed below.

NOTE: Do not use KRYOCIDE in combination with lime or compounds containing free lime. Do not apply KRYOCIDE immediately before a harvest.

Remove excess residues on edible portions of fruit and vegetables by washing, brushing or other effective means.

POUNDS PER ACRE

CROP	INSECTS	RATE LBS/A SPRAY	RATE LBS/A DUST	REMARKS
Broccoli Cabbage Cauliflower Collards Mustard	Cabbage Looper Climbing Cutworms Corn Earworm Diabrotica Beetles (including cucumber beetles) Diamondback Moth Caterpillar Flea Beetles Imported Cabbageworm Yellowstriped Armyworm	25-50	25-50	Do not apply after edible parts start to form on broccoli; or after heads form on cabbage; or after seedling stage on cauliflower, collards or mustard.
Cantaloupe Cucumber Squash Watermelon	Cabbage Looper Diabrotica Beetles (including cucumber beetles) Flea Beetles Melonworm Pickleworm	15-25	25-50	Remove excess residues on edible portions by washing, brushing or other effective means.
Citrus	Citrus Cutworm Fruittree Leafroller Fuller Rose Beetle Garden Tortrix Halcocera Katydids Orange Tortrix Orangedog Variegated Cutworm	25-50	25-50	Remove excess residues on edible portions of fruit by washing, brushing, or other effective means.
Grapes	Cutworms Grape Leaffolder Omnivorous Leafroller Orange Tortrix Western Grapeleaf Skeletonizer	6-8 by ground in 25-200 gal. water 8 by air (helicopter) in 20 gal. water	10-20	Apply just as fruit starts to form. When needed to control later broods, make a second application after the fruit is 1/4" or larger in diameter. Do not use as a dust on grapes after fruit starts to form. Do not make more than 2 applications per year. Do not apply later than 30 days before harvest. For application to grapes grown for table use after the fruit is 1/4" or larger in diameter, use of concentrate sprays will reduce visible residue spots which may result from the use of dilute applications at that stage.
Lettuce	Cabbage Looper Corn Earworm Tobacco Budworm	15-30	15-30	Do not apply to lettuce after seedling stage.
Peppers	Hornworms Pepper Weevil	25-50	25-50	Remove excess residues on edible portions by washing, brushing, or other effective means.
Tomatoes	Blister Beetles Cabbage Looper Flea Beetles Fruitworms Hornworms Tomato Pinworm	15-30	15-30	Apply when plants are first set out and repeat as needed. Remove excess residues on edible portions by washing, brushing, or other effective means.
Ornamentals Shade Trees	Codling Moth Flea Weevil Fuller Rose Beetle Gypsy Moth Caterpillar Katydids Plum Curculio	25-50	25-50	

STORAGE AND DISPOSAL

Do not contaminate water, food, or feed by storage or disposal.

Storage Instructions: Store in the original container in a dry area. Do not store in a manner where cross-contamination with other pesticides, fertilizers, food or feed could occur. If spilled during storage or handling sweep up spillage and dispose of in accordance with the Pesticide Disposal Instructions listed below.

Pesticide Disposal Instructions: Wastes resulting from the use of this product may be disposed of on site or at an approved waste disposal facility.

Container Disposal Instructions: Completely empty bag into application equipment. Then dispose of empty bag in a sanitary landfill or by incineration, or, if allowed by State and local authorities, by burning. If burned, stay out of smoke.

WARRANTY AND DISCLAIMER

Pennwalt Corporation warrants that this material conforms to the chemical description on the label and is reasonably fit for the purposes referred to in the Directions for Use, subject to the risks referred to therein. PENNWALT MAKES NO OTHER EXPRESS OR IMPLIED WARRANTY OF FITNESS OR MERCHANTABILITY OR ANY OTHER EXPRESS OR IMPLIED WARRANTY. IN NO CASE SHALL PENNWALT OR SELLER BE LIABLE FOR CONSEQUENTIAL, SPECIAL OR INDIRECT DAMAGES RESULTING FROM THE USE OR HANDLING OF THIS PRODUCT INCLUDING, BUT NOT LIMITED TO, LOSS OF PROFITS, BUSINESS REPUTATION, OR CUSTOMERS; LABOR COST; OR OTHER EXPENSES INCURRED IN PLANTING OR HARVESTING.

Pennwalt and seller offer this product and the buyer and user accept it subject to the foregoing conditions of sale and warranty which may be varied only by agreement in writing signed by a duly authorized representative of Pennwalt.

5-E617G-01 H8 (7/88)

MANEB 80

FUNGICIDE

ACTIVE INGREDIENT:

Maneb (Manganese ethylenebisdithiocarbamate)	80.0%
(Equivalent to manganese metallic 16.5%)	
INERT INGREDIENTS:	20.0%
TOTAL	100.0%

EPA Reg No. 4581-255 EPA Est No. 33910-HL-01

KEEP OUT OF REACH OF CHILDREN

CAUTION

STATEMENT OF PRACTICAL TREATMENT

IF SWALLOWED, induce vomiting immediately by giving two glasses of water and sticking finger down throat. Call a physician. Never give anything by mouth to an unconscious person.

IF IN EYES, Immediately flush eyes with plenty of water for at least 15 minutes. Get medical attention if irritation persists.

IF ON SKIN, immediately flush skin with plenty of water. Get medical attention if irritation persists.

See the following pages for General Information, Directions for Use, and Chemigation Instructions. Do not apply this product through any irrigation system unless these instructions are followed.

PRECAUTIONARY STATEMENTS

HAZARDS TO HUMANS (AND DOMESTIC ANIMALS)

CAUTION

HARMFUL IF SWALLOWED, ABSORBED THROUGH SKIN, OR INHALED. CAUSES MODERATE EYE IRRITATION. Avoid contact with skin, eyes, or clothing. Avoid breathing spray mist or dust. Wash thoroughly with soap and water after handling. Remove contaminated clothing and wash before reuse.

Do not apply this product in such a manner as to directly or through drift expose workers or other persons. The area being treated must be vacated by unprotected persons.

ENVIRONMENTAL HAZARDS

This product is toxic to fish. Drift and runoff from treated areas may be hazardous to aquatic organisms in neighboring areas. Do not apply directly to water or wetlands (swamps, bogs, marshes or potholes). Do not apply where runoff is likely to occur. Do not contaminate water when disposing of equipment washwaters.

GENERAL INFORMATION

MANEB 80 may be used as a spray for the control of many fungus plant diseases. It mixes readily with water, suspends well, and adheres to sprayed surfaces with a minimum of runoff.

DIRECTIONS FOR USE

It is a violation of Federal law to use this product in a manner inconsistent with its labeling.

WORKER SAFETY RULES

Do not enter treated areas within 24 hours after application unless wearing the protective clothing listed below. Because certain states may require more restrictive reentry intervals for various crops treated with this product, consult your State Department of Agriculture for further information.

After sprays have dried, do not enter or allow entry into treated areas or areas where there is a danger of drift until the 24-hour reentry interval has expired unless wearing the personal protective equipment listed on this label.

Keep all unprotected persons, children, livestock, and pets away from treated area or where there is danger of drift.

Do not rub eyes or mouth with hands. See First Aid (Practical Treatment Section).

PERSONAL PROTECTIVE EQUIPMENT

HANDLERS (MIXERS, LOADERS, AND APPLICATORS) AND EARLY REENTRY WORKERS MUST WEAR THE FOLLOWING PROTECTIVE CLOTHING AND EQUIPMENT: a long-sleeved shirt and long pants or a coverall; chemical resistant gloves, shoes, socks, and goggles or a face shield. During mixing and loading, a chemical resistant apron must also be worn.

During application from a tractor with a completely enclosed cab with positive pressure filtration, or aerially with an enclosed cockpit, a long-sleeved shirt and long pants must be worn in place of the above protective clothing. Chemical resistant gloves must be available in the cab or cockpit and worn while exiting.

IMPORTANT! Before removing gloves, wash them with soap and water. Always wash hands, face, and arms with soap and water before eating, smoking or drinking. Always wash hands and arms with soap and water before using the toilet.

After work take off all clothes and shoes. Shower using soap and water. Wear only clean clothes. Do not use contaminated clothing. Wash protective clothing and protective equipment with soap and water after each use. Personal clothing worn during use must be laundered separately from household articles. Clothing and protective equipment drenched with maneb must be destroyed according to state and local regulations.

DRENCHED CLOTHING CANNOT BE ADEQUATELY DECONTAMINATED.

During aerial application, human flaggers are prohibited unless in totally enclosed vehicles.

REENTRY STATEMENT

Written or oral warnings must be given to workers who are expected to be in a treated area to be in a treated area or in an area about to be treated with this product. When oral warnings are given, such warnings shall be given in a language customarily understood by the workers. Oral warnings must be given if there is reason to believe that written warnings cannot be understood by the workers. Written warnings must include the following information: " 'CAUTION. Area treated with MANEB 80 Fungicide on *(date of application)*. Do not enter without protective clothing within 24 hours after application.' "

INSTRUCTIONS FOR APPLICATION

MANEB 80 is recommended for control of the diseases on the crops listed below.

SPRAYS:

Place MANEB 80 into suspension prior to adding an adjuvant to the spray mixture. Read and carefully observe the precautionary statements and all other information appearing on product labels prior to spray preparation.

FOR GROUND APPLICATIONS use at the rate indicated in sufficient water for thorough coverage, minimum 10 gallons. Increased volume of water may be necessary as foliage density increases.

To prepare mixture, sift in the required amount of product when the agitator paddles or the jet or recirculating pipe from pump are covered. Fill the tank with water when the product is thoroughly mixed. Three to four level teaspoonfuls of MANEB 80 per one gallon of water is equivalent to 1½ to 2 pounds per 100 gallons.

FOR AIR APPLICATION use at the rate indicated in sufficient water for thorough coverage, at least 3 gallons for field crops. Increased volume of water may be necessary as foliage density increases.

CROP	DISEASES	RATE LBS./ACRE	REMARKS
Corn (Sweet)	Helminthosporium Leaf Blight	1½	Begin application when first signs of disease appear and apply every 7 to 10 days while the disease is present (under severe conditions, shorten intervals to 4 to 5 days). Add a recommended surfactant or spreader/sticker to the spray tank. Do not feed treated forage to livestock.
Potatoes	Early and Late Blight	1½-2	Begin when plants are 2 to 6 inches high. Repeat at 5 to 10 day intervals, more often and using the higher rate under severe disease pressure.
Potatoes (Seed Pieces) Not for use in California	Fusarium Seed Piece Decay	1¼ lb/ 10 gal	Seed piece treatment only. Dip whole or cut tubers. Spread in cool place if held before planting. Do not use treated seed pieces for food, feed, or oil purposes.
Sugar Beets	Cercospora Leaf Spot	3	Apply when disease threatens repeat at 7 to 10 day intervals for 3 to 5 applications. Do not apply within 14 days of harvest. Do not feed treated tops to livestock.

CROP	DISEASES	RATE LBS./ACRE	REMARKS
Almonds	Brown Rot Blossom and Twig Blight Scab Shot-hole Fungus (Leaf Blight)	6-8	Apply at popcorn, full-bloom, and petal fall or every 7 to 10 days if bloom is staggered. Omit petal fall spray if only brown rot is present. Use higher rate under severe disease pressure. If applied after petal fall, do not feed hulls to dairy animals or animals being finished for slaughter. Do not apply later than 5 weeks after petal fall.
Grasses (Ornamentals— Grown for Replanting Seed Purposes Only) Not for use in California	Leaf, Stem and Stripe Rust	2-3	Begin when rust pustules are first seen. Repeat at 7 to 14 day intervals. Use the higher rate under severe disease pressure. Do not feed treated grass in livestock.
Roses	Cercospora Leaf Spot (Texas) Rust (California) Black Spot	1.0-1.5	Spray when first leaves unfold; repeat weekly as needed.

CHEMIGATION

NOT FOR THIS USE IN CALIFORNIA
GENERAL INSTRUCTIONS

Apply this product only through sprinkler irrigation systems including center pivot, lateral move, end tow, side (wheel) roll, traveler, big gun, solid set, or hand move. Do not apply this product through any other type of irrigation system.

Crop injury, lack of effectiveness, or illegal pesticide residues in the crop can result from nonuniform distribution of treated water.

If you have any questions about calibration, you should contact State Extension Service specialists, equipment manufacturers or other experts.

A person knowledgeable of the chemigation system and responsible for its operation, or under the supervision of the responsible person, shall shut the system down and make necessary adjustments should the need arise.

Do not connect chemigation system to any public water system. Public water system means a system for the provision of piped water for human consumption if such system has at least 15 service connections or regularly serves an average of at least 25 individuals daily at least 60 days out of the year.

SYSTEM REQUIREMENTS

Systems utilizing a pressurized water and pesticide injection system must meet the following requirements:

The system must contain a functional check valve, vacuum relief valve, and low pressure drain appropriately located on the irrigation pipeline to prevent water source contamination from backflow.

The pesticide injection pipeline must contain a functional, automatic, quick-closing check valve to prevent the flow of fluid back toward the injection pump.

The pesticide injection pipeline must also contain a functional, normally closed, solenoid-operated valve located on the intake side of the injection pump and connected to the system interlock to prevent fluid from being withdrawn from the supply tank when the irrigation system is either automatically or manually shut down.

The system must contain functional interlocking controls to automatically shut off the pesticide injection pump when the water pump motor stops.

The irrigation line or water pump must include a functional pressure switch which will stop the water pump motor when the water pressure decreases to the point where pesticide distribution is adversely affected.

Systems must use a metering pump, such as a positive displacement injection pump (e.g., diaphragm pump) effectively designed and constructed of materials that are compatible with pesticides and capable of being fitted with a system interlock.

INSTRUCTIONS FOR SPRINKLER (OVERHEAD) IRRIGATION:

Observe the requirements in the System Requirements section above.

Do not apply when wind speed favors drift beyond the area intended for treatment.

Apply MANEB 80 only through systems containing anti-syphon and check valves designed to prevent water source contamination or overflow of the mix tank and containing interlocking controls between the metering device and the water pump to insure simultaneous shut-off.

Maintain a gentle continuous agitation in mix tank during mixing and application to assure a uniform suspension.

Greater accuracy in calibration and distribution will be achieved by injecting a larger volume of a more dilute suspension per unit time.

Application of more than recommended quantities of irrigation water per acre may result in decreased product performance.

Do not apply when wind speed favors drift, when system connections or fittings leak, when nozzles do not provide uniform distribution or when lines containing the product cannot be flushed and must be dismantled and drained. In a center pivot system, block the nozzle set nearest the well/pivot/injection unit to prevent spray being applied to this area.

Where sprinkler distribution patterns do not overlap sufficiently, unacceptable disease control may result.

Allow sufficient time for pesticide to be flushed through all lines and all nozzles before turning off irrigation water.

MANEB 80 may be applied in conjunction with chemically neutral liquid fertilizers. Application in conjunction with highly alkaline fertilizers, such as aqueous ammonia, may cause a degradation of the pesticide, resulting in reduced performance and should be avoided.

Check local restrictions and requirements regarding sprinkler irrigation applications, as they may vary from state to state.

SPRAY PREPARATION:

Remove scale, pesticide residues, and other foreign matter from the chemical tank and entire injector system. Flush with clean water.

Prepare a suspension of MANEB 80 in a mix tank. Fill the tank with 1/2 or 3/4 the desired amount of water. Start mechanical or hydraulic agitation. Slowly add the required amount of MANEB 80 and then the remaining volume of water.

APPLICATION INSTRUCTIONS:

Set sprinkler system to deliver 0.1 to 1.25 inches of water per acre. Volumes of water higher than this may reduce efficacy. Start sprinkler and then uniformly inject the suspension of MANEB 80 into the irrigation water line so as to deliver the desired rate per acre. The suspension of MANEB 80 should be injected with a positive displacement pump into the main line ahead of a right angle turn to insure adequate mixing.

NOTE: When treatment with MANEB 80 has been completed, do not irrigate the treated area for 24 to 48 hours to prevent washing the chemical off the crop.

STORAGE AND DISPOSAL

Do not contaminate water, food, or feed by storage or disposal.

Storage Instructions: Store in the original container in a dry area. If allowed to become wet the product will deteriorate and represent a fire hazard. Keep away from sources of ignition (e.g. sparks and open flame). Close bag when not in use. Do not store in manner where cross-contamination with other pesticides, fertilizers, food or feed could occur. If spilled during storage or handling sweep up spillage and dispose of in accordance with the Pesticide Disposal Instructions listed below.

Pesticide Disposal Instructions: Wastes resulting from the use of this product may be disposed of on site or at an approved waste disposal facility.

Container Disposal Instructions: Completely empty bag into application equipment. Then dispose of empty bag in a sanitary landfill or by incineration, or, if allowed by State and local authorities, by burning. If burned, stay out of smoke.

WARRANTY AND DISCLAIMER

Pennwalt Corporation warrants that this material conforms to the chemical description on the label and is reasonably fit for the purposes referred to in the Directions for Use, subject to the risks referred to therein. PENNWALT MAKES NO OTHER EXPRESS OR IMPLIED WARRANTY OF FITNESS OR MERCHANTABILITY OR ANY OTHER EXPRESS OR IMPLIED WARRANTY. IN NO CASE SHALL PENNWALT OR SELLER BE LIABLE FOR CONSEQUENTIAL, SPECIAL OR INDIRECT DAMAGES RESULTING FROM THE USE OR HANDLING OF THIS PRODUCT INCLUDING, BUT NOT LIMITED TO, LOSS OF PROFITS, BUSINESS REPUTATION, OR CUSTOMERS: LABOR COST, OR OTHER EXPENSES INCURRED IN PLANTING OR HARVESTING.

Pennwalt and seller offer this product and the buyer and user accept it subject to the foregoing conditions of sale and warranty which may be varied only by agreement in writing signed by a duly authorized representative of Pennwalt.

5-N311S-01 (9/88) J9

MANEB PLUS ZINC F4

FUNGICIDE

SHAKE WELL BEFORE USING
NOT FOR USE IN CALIFORNIA

ACTIVE INGREDIENT:
Maneb (Manganese Ethylenebisdithiocarbamate) 37%
(Equivalent to manganese metallic 7.6%)
INERT INGREDIENTS: 63%
TOTAL 100%

This product contains 4 pounds Maneb per gallon.

EPA Reg No. 4581-359 EPA Establishment No. 4581-TX-1

KEEP OUT OF REACH OF CHILDREN

CAUTION

STATEMENT OF PRACTICAL TREATMENT

IF SWALLOWED, call a physician or Poison Control Center. Drink 1 or 2 glasses of water and induce vomiting by touching back of throat with finger. Do not induce vomiting or give anything by mouth to an unconscious person.

IF IN EYES, flush eyes with plenty of water. Call a physician if irritation persists.

IF ON SKIN, wash with plenty of soap and water. Get medical attention if irritation persists.

Do not apply this product through any irrigation system unless these instructions are followed.

PRECAUTIONARY STATEMENTS

HAZARDS TO HUMANS (AND DOMESTIC ANIMALS)
CAUTION

HARMFUL IF SWALLOWED, ABSORBED THROUGH SKIN, OR INHALED. CAUSES MODERATE EYE IRRITATION. Avoid contact with skin, eyes, or clothing. Avoid breathing mist. Wash thoroughly with soap and water after handling. Remove contaminated clothing and wash before reuse. Wear protective clothing (long pants, long sleeved shirt, impermeable gloves, hat and boots) during mixing and loading.

Do not apply this product in such a manner as to directly or through drift expose workers or other persons. The area being treated must be vacated by unprotected persons.

ENVIRONMENTAL HAZARDS

This product is toxic to fish. Do not apply to water. Do not apply where runoff is likely to occur. Do not contaminate water when disposing of equipment washwaters.

PHYSICAL OR CHEMICAL HAZARDS

Keep away from fire and sparks. Store in a cool dry place.

GENERAL INFORMATION

DIRECTIONS FOR USE

It is a violation of Federal law to use this product in a manner inconsistent with its labeling.

Do not use this product in such a manner as to directly or through drift expose workers or other persons. The area being treated must be vacated by unprotected persons.

Do not enter treated areas within 24 hours of application unless wearing the protective clothing listed below. Because certain states may require more restrictive reentry intervals for various crops treated with this product, consult your State Department of Agriculture for further information.

After sprays have dried, do not enter or allow entry into treated areas where there is a danger of drift until the 24-hour reentry interval has expired unless wearing the personal protective equipment listed on this label.

Keep all unprotected persons, children, livestock, and pets away from treated area or where there is a danger of drift.

Do not rub eyes or mouth with hands. See First Aid (Practical Treatment Section).

PERSONAL PROTECTIVE EQUIPMENT

HANDLERS (MIXERS, LOADERS, AND APPLICATORS) AND EARLY REENTRY WORKERS MUST WEAR THE FOLLOWING PROTECTIVE CLOTHING ABD EQUIPMENT: a long-sleeved shirt and long pants or a coverall; chemical resistant gloves; shoes, socks, and goggles or a face shield. During mixing and loading, a chemical resistant apron must also be worn.

During application from a tractor with a completely enclosed cab with positive pressure filtration, or aerially with an enclosed cockpit, a long-sleeved shirt and long pants must be worn in place of the above protective clothing. Chemical resistant gloves must be available in the cab or cockpit and worn while exiting.

IMPORTANT! Before removing gloves, wash them with soap and water. Always wash hands, face, and arms with soap and water before eating, smoking or drinking. Always wash hands and arms with soap and water before using the toilet.

After work take off all clothes and shoes. Shower using soap and water. Wear only clean clothes. Do not use contaminated clothing. Wash protective clothing and protective equipment with soap and water after each use. Personal clothing worn during use must be laundered separately from household articles. Clothing and protective equipment drenched with maneb must be destroyed according to state and local regulations.

DRENCHED CLOTHING CANNOT BE ADEQUATELY DECONTAMINATED

During aerial application, human flaggers are prohibited unless in totally enclosed vehicles.

REENTRY STATEMENT

Written or oral warnings must be given to workers who are expected to be in a treated area or in an area about to be treated with this product. When oral warnings are given, such warnings shall be given in a language customarily understood by the workers. Oral warnings must be given if there is reason to believe that written warnings cannot be understood by the workers. Written warnings must include the following information "'CAUTION. Area treated with MANEB PLUS ZINC F4 Fungicide on (date of application). Do not enter without protective clothing within 24 hours after application.'"

DIRECTIONS FOR USE AS A SPRAY

FOR GROUND APPLICATIONS use at the rate indicated in sufficient water for thorough coverage, minimum of 10 gallons. Increased volume of water may be necessary as foliage density increases.

FOR AIR APPLICATION use at the rate indicated in sufficient water for thorough coverage, at least 3 gallons. Increased volume of water may be necessary as foliage density increases.

FRUIT AND NUT CROP USES

Crop	Diseases	Rate Pt/Acre	Remarks
Almonds	Brown Rot Blossom and Twig Blight Scab Shot-hole Fungus (Leaf Blight)	9½-12¾	Apply at popcorn, full bloom, and petal fall or every 7 to 10 days if bloom is staggered. Omit petal fall spray if only brown rot is present. If applied after petal fall, do not feed hulls to dairy animals or animals being finished for slaughter. Use the higher rate under severe disease pressure. Do not apply later than 5 weeks after petal fall.

VEGETABLE CROP USES

Crop	Diseases	Rate Pt/Acre	Remarks
Corn (Sweet)	Helminthosporium Blight	2½	Begin application when first signs of disease appear and apply every 7 to 10 days while the disease is present (under severe conditions, shorten intervals to 4 to 5 days). Add a recommended surfactant or spreader/sticker to the spray tank. Do not feed treated forage to livestock.
Potatoes	Early and Late Blights	2½-3¼	Begin when plants are 2 to 6 inches high. Repeat at 5- to 10-day intervals, more often under severe disease pressure.
Potatoes (Seed Pieces)	Fusarium Seed Piece Decay	1.0 qt/10 gal	Seed piece treatment only. Dip whole or cut tubers. Spread in cool place if held before planting. Do not use treated seed pieces for food, feed, or oil purposes.

AGRICULTURAL FIELD CROP USES

Crop	Diseases	Rate Pt/100 Gal	Remarks
Grasses (Ornamental—Grown for Replanting Seed Purposes Only)	Leaf, Stem and Stripe Rust	3½-4¾	Begin when rust pustules are first seen. Repeat at 7- to 14-day intervals. Use higher rate under severe disease pressure. Do not feed treated grass to livestock.

Crop	Diseases	Rate Pt/100 Gal	Remarks
Sugar Beets	Leaf Spot *(Cercospora)*	4³/₄	Begin when disease threatens. Repeat at 7- to 10-day intervals for 3 to 5 applications. Do not apply within 14 days of harvest. Do not feed treated tops to livestock.

AGRICULTURAL ORNAMENTAL CROP USES

Crop	Diseases	Rate Qt/100 Gal	Remarks
Carnations	Alternaria Leaf Spot Anthracnose Botrytis Blight	0.8-1.2	Begin when new growth starts. Repeat weekly.
Chrysanthemums	Asochyta Ray Blight Botrytis Petal Spot Septoria Leaf Spot	12.8 fl oz PLUS Captan 50W ³/₄ lbs per 100 gal water	Apply twice weekly during the blooming season for Petal Spot and Ray Blight. For Septoria Leaf Spot apply weekly throughout the season.
Dahlias	Alternaria Leaf Spot Botrytis Blight	0.8-1.2	Begin when new growth starts. Repeat weekly.
Dogwood	Anthracnose	0.8-1.2	Begin when buds open. Repeat when bracts fall 4 weeks later and in later summer.
Gladiolus	Botrytis Blight Curvularia and Stemphylium Leaf Spots	1.2	Begin when flower spikes are developing. Repeat 2 to 3 times at weekly intervals.
Grasses (Nongrazed Areas)	Brown Patch	4.8 fl oz in sufficient water/1000 sq ft	Begin at first sign of disease. Repeat at 7- to 14-day intervals.
	Dollar Spot	9.6-12.8 fl oz in sufficient water/1000 sq ft	Apply as for Brown Patch, above.
	Melting-Out	4.8-6.4 fl oz in sufficient water/1000 sq ft	Apply as for Brown Patch, above.
	Leaf, Stem and Stripe Rusts	2.4	Begin when rust pustules are first seen. Repeat at 7- to 14-day intervals. Do not graze treated areas. Do not feed clippings to livestock.
Lillies	Botrytis Blight	0.8-1.2	Begin with new growth. Repeat weekly.
Pansies	Anthracnose	0.8-1.2	Begin with new growth. Repeat weekly.
Peonies	Alternaria Leaf Spot Botrytis Blight Phytophthora Blight	0.8-1.2	Apply to foliage and soil in early spring and early fall and 7- to 10-day intervals during the growing season.
Roses	Black Spot Cercospora Leaf Spot (Texas) Rust (California)	0.8-1.2	Begin when first leaves unfold. Repeat at 7- to 10-day intervals.
Snapdragons	Rust	0.8-1.2	Begin with emergence. Repeat weekly.
Zinnias	Alternaria Blight Leaf Spot Botrytis	0.8-1.2	Begin with emergence. Repeat weekly.

CHEMIGATION

NOT FOR THIS USE IN CALIFORNIA
GENERAL INSTRUCTIONS

Apply this product only through sprinkler irrigation systems including center pivot, lateral move, end tow, side (wheel) roll, traveler, big gun, solid set, or hand move. Do not apply this product through any other type of irrigation system.

Crop injury, lack of effectiveness, or illegal pesticide residues in the crop can result from nonuniform distribution of treated water.

If you have any questions about calibration, you should contact State Extension Service specialists, equipment manufacturers or other experts.

A person knowledgeable of the chemigation system and responsible for its operation, or under the supervision of the responsible person, shall shut the system down and make necessary adjustments should the need arise.

Do not connect chemigation system to any public water system. Public water system means a system for the provision of piped water for human consumption if such system has at least 15 service connections or regularly serves an average of at least 25 individuals daily at least 60 days out of the year.

SYSTEM REQUIREMENTS

Systems utilizing a pressurized water and pesticide injection system must meet the following requirements:

The system must contain a functional check valve, vacuum relief valve, and low pressure drain appropriately located on the irrigation pipeline to prevent water source contamination from backflow.

The pesticide injection pipeline must contain a functional, automatic, quick-closing check valve to prevent the flow of fluid back toward the injection pump.

The pesticide injection pipeline must also contain a functional, normally closed, solenoid-operated valve located on the intake side of the injection pump and connected to the system interlock to prevent fluid from being withdrawn from the supply tank when the irrigation system is either automatically or manually shut down.

The system must contain functional interlocking controls to automatically shut off the pesticide injection pump when the water pump motor stops.

The irrigation line or water pump must include a functional pressure switch which will stop the water pump motor when the water pressure decreases to the point where pesticide distribution is adversely affected.

Systems must use a metering pump, such as a positive displacement injection pump (e.g., diaphragm pump) effectively designed and constructed of materials that are compatible with pesticides and capable of being fitted with a system interlock.

INSTRUCTIONS FOR SPRINKLER (OVERHEAD) IRRIGATION:

Observe the requirements in the System Requirements section above.

Do not apply when wind speed favors drift beyond the area intended for treatment.

Apply MANEB PLUS ZINC F4 only through systems containing anti-syphon and check valves designed to prevent water source contamination or overflow of the mix tank and containing interlocking controls between the metering device and the water pump to insure simultaneous shut-off.

Maintain a gentle continuous agitation in mix tank during mixing and application to assure a uniform suspension.

Greater accuracy in calibration and distribution will be achieved by injecting a larger volume of a more dilute suspension per unit time.

Application of more than recommended quantities of irrigation water per acre may result in decreased product performance.

Do not apply when wind speed favors drift, when system connections or fittings leak, when nozzles do not provide uniform distribution or when lines containing the product cannot be flushed and must be dismantled and drained. In a center pivot system, block the nozzle set nearest the well/pivot/injection unit to prevent spray being applied to this area.

Where sprinkler distribution patterns do not overlap sufficiently, unacceptable disease control may result.

Allow sufficient time for pesticide to be flushed through all lines and all nozzles before turning off irrigation water.

MANEB PLUS ZINC F4 may be applied in conjunction with chemically neutral liquid fertilizers. Application in conjunction with highly alkaline fertilizers, such as aqueous ammonia, may cause a degradation of the pesticide, resulting in reduced performance and should be avoided.

Check local restrictions and requirements regarding sprinkler irrigation applications, as they may vary from state to state.

SPRAY PREPARATION

Remove scale, pesticide residues, and other foreign matter from the chemical tank and entire injector system. Flush with clean water.

Prepare a suspension of MANEB PLUS ZINC F4 in a mix tank. Fill the tank with ¹/₂ or ³/₄ the desired amount of water. Start mechanical or hydraulic agitation. Slowly add the required amount of MANEB PLUS ZINC F4 and then the remaining volume of water.

APPLICATION INSTRUCTIONS

Set sprinkler system to deliver 0.1 to 1.25 inches of water per acre. Volumes of water higher than this may reduce efficacy. Start sprinkler and then uniformly inject the suspension of MANEB PLUS ZINC F4 into the irrigation water line so as to deliver the desired rate per acre. The suspension of MANEB PLUS ZINC F4 should be injected with a positive displacement pump into the main line ahead of a right angle turn to insure adequate mixing.

NOTE: When treatment with MANEB PLUS ZINC F4 has been completed, do not irrigate the treated area for 24 to 48 hours to prevent washing the chemical off the crop.

STORAGE AND DISPOSAL

Do not contaminate water, food or feed by storage or disposal.

Storage Instructions. Store in the original container. Keep away from fire and sparks. Store in a cool place, but not below 32°F. Do not allow product to become overheated in storage as this may bring on chemical changes which will impair the fungicide effectiveness of MANEB PLUS ZINC F4 Fungicide. Keep container closed when not in use. Do not store in a manner where cross-contamination with other pesticides, fertilizers, food or feed could occur. If spilled during storage or handling, absorb with sand or other inert material and dispose of absorbant in accordance with the Pesticide Disposal Instructions listed below.

Pesticide Disposal Instructions: Wastes resulting from the use of this product may be disposed of on site or at an approved waste disposal facility.

Container Disposal Instructions: Triple rinse (or equivalent). Then offer for recycling or reconditioning, or puncture and dispose of in a sanitary landfill, or by incineration, or, if allowed by state and local authorities, by burning. If burned, stay out of smoke.

WARRANTY AND DISCLAIMER

Pennwalt Corporation warrants that this material conforms to the chemical description on the label and is reasonably fit for the purposes referred to in the Directions for Use, subject to the risks referred to therein. PENNWALT MAKES NO OTHER EXPRESS OR IMPLIED WARRANTY OF FITNESS OR MERCHANTABILITY OR ANY OTHER EXPRESS OR IMPLIED WARRANTY. IN NO CASE SHALL PENNWALT OR SELLER BE LIABLE FOR CONSEQUENTIAL, SPECIAL OR INDIRECT DAMAGES RESULTING FROM THE USE OR HANDLING OF THIS PRODUCT INCLUDING, BUT NOT LIMITED TO, LOSS OF PROFITS, BUSINESS REPUTATION, OR CUSTOMERS; LABOR COST; OR OTHER EXPENSES INCURRED IN PLANTING OR HARVESTING.

Pennwalt and seller offer this product and the buyer and user accept it subject to the foregoing conditions of sale and warranty which may be varied only by agreement in writing signed by a duly authorized representative of Pennwalt.

5–Z324D–01 (9/89) J9

MICROTHIOL® Special

Micronized Wettable Sulfur

ACTIVE INGREDIENT:

Sulfur	80.0%
INERT INGREDIENTS	20.0%
TOTAL	100.0%

EPA Reg. No. 4581-373 EPA Est. No. 44420-FR-1

KEEP OUT OF REACH OF CHILDREN

CAUTION

STATEMENT OF PRACTICAL TREATMENT

IF ON SKIN, wash with soap and water.

IF IN EYES, immediately flush eyes with plenty of water for at least 15 minutes. Get medical attention if irritation persists.

IF SWALLOWED, give one or two glasses of water and induce vomiting by sticking finger down throat. Never give anything by mouth to an unconscious person.

PRECAUTIONARY STATEMENTS

HAZARDS TO HUMANS (AND DOMESTIC ANIMALS)

CAUTION

May cause irritation of eyes. Harmful if swallowed, inhaled, or absorbed through skin. Avoid breathing dust or spray mist. Avoid contact with skin, eyes, or clothing. Wash thoroughly with soap and water after handling. Remove contaminated clothing and wash before reuse.

ENVIRONMENTAL HAZARDS

Do not apply directly to water or wetlands (swamps, bogs, marshes and potholes). Do not contaminate water by disposal of equipment washwater. Do not apply where runoff is likely to occur. Do not apply when weather conditions favor drift from treated areas.

DIRECTIONS FOR USE

It is a violation of Federal Law to use this product in a manner inconsistent with its labelling.
Do not apply this product through any type of irrigation system.

REENTRY STATEMENT

Do not enter treated areas without protective clothing within 24 hours after application. Protective clothing means, at least, a hat or other suitable head covering, a long-sleeved shirt and long legged trousers, shoes and socks. Because certain states may require more restrictive reentry intervals for various crops treated with this product, consult your State Department of Agriculture for futher information.

Written or oral warnings must be given to workers who are expected to be in a treated area or in an area about to be treated with this product. When oral warnings are given, such warnings shall be given in a language customarily understood by the workers. Oral warnings must be given when there is reason to believe that written warnings cannot be understood by the workers. Written warnings must include the following information: "CAUTION, Area treated with MICROTHIOL SPECIAL on (date of application)."

APPLICATION INSTRUCTIONS

Apply through any type of handheld, knapsack, mechanical or pneumatic spraying equipment.
MIXING: Pour the required amount of MICROTHIOL SPECIAL into the indicated amount of water while stirring. MICROTHIOL SPECIAL disperses instantly.

Dilute Application:

Ground: Specified rate in 20 to 60 gallons of water per acre.
Orchard: Specified rate in 100 to 800 gallons of water per acre.

Concentrated Application:

Ground: Specified rate in 5 to 10 gallons of water per acre.
Orchard: Specified rate in 20 to 100 gallons of water per acre.

Aerial Application:

Ground: Specified Rate in 3 to 20 gallons of water per acre.
Orchard: Specified rate in 10 to 20 gallons of water per acre.

COMPATIBILITY

MICROTHIOL SPECIAL is completely compatible with Bordeaux mixture, copper fungicides, liquid fertilizers and all organic fungicides, insecticides, acaricides and herbicides.

USE INSTRUCTIONS

Some crops may be damaged by sulfur under certain climatic conditions. Do not use on any crop unless sulfur has been shown to be safe in your locality. Do not allow spray to drift onto sulfur-sensitive crops such as apricots, cranberries, spinach and Anjou pears. Sulfur may burn foliage when temperature is high. Do not apply at such times. Do not use within two weeks of an oil spray treatment. For citrus, do not apply within 21 days of an oil spray. When growing crops for processing, consult the processor before applying sulfur.

GRAPES: For control of powdery mildew, bud mite, blister mite and red spider mite.
Mildew: Apply throughout the season at 14 day intervals according to the stage of development of the vegetation and the intensity of the attack.

Before flowering: Apply a maximum of 5 lbs per acre.
After flowering: Apply a maximum of 6 lbs per acre.
Mites: Apply a maximum of 10 lbs per acre at budburst, making sure to wet thoroughly.
Note: Concord and other labrusca type grapes may be injured.

SMALL FRUITS AND BERRIES:

RASPBERRIES: For control of powdery mildew. Apply a maximum of 15 lbs per acre each week from first bloom to fruit set.

STRAWBERRIES: For control of powdery mildew and red spider mite. Apply a maximum of 10 lbs per acre at early leaf stage and continue as necessary.

BLACKBERRIES, BOYSENBERRIES, DEWBERRIES, LOGANBERRIES, BLUEBERRIES, GOOSEBERRIES, HUCKLEBERRIES AND CURRANTS: For control of powdery mildew. Apply a maximum of 15 lbs per acre before blossom and continue at 10 day intervals as necessary.

ORCHARDS:

APPLES AND PEARS: For control of powdery mildew, scab, blister mite and red spider mite. Before flowering: Apply a maximum of 30 lbs per acre. After flowering: Apply a maximum of 15 lbs per acre.

CITRUS: For control of red spider mite, flat mite, rust mite and silver mite. Apply a maximum of 20 lbs per acre as necessary.

PEACHES, PLUMS, CHERRIES, ALMONDS, NECTARINES, PRUNES: For control of powdery mildew (except on almonds), brown rot, leaf spot, silver mite, flat mite and red spider mite. Apply a maximum of 30 lbs per acre at bloom and repeat as necessary at intervals of 12 days.

AVOCADOS: For control of brown mite. Apply a maximum of 30 lbs per acre as needed.

MANGOES: For control of powdery mildew. Apply a maximum of 30 lbs per acre before flowering and continue at intervals of 20 days.

WALNUTS: For control of pacific mite, almond mite, twospotted mite and red spider mite. Apply a maximum of 20 lbs per acre as required.

PEANUTS: For control of powdery mildew, leaf spot and red spider mite. Apply a maximum of 5 lbs per acre at early leaf stage and repeat at 18 day intervals.

SUGAR BEETS: For control of powdery mildew and red spider mite. Apply 4 lbs per acre as required at 18 days intervals.

COTTON: For control of red spider mite. Apply a maximum of 5 lbs per acre as necessary.

CEREALS: Corn, wheat, barley, oats, rye, sorghum: For control of red spider mite, twospotted mite, pacific mite and atlantic mite. Apply a maximum of 15 lbs per acre when disease first appears and continue as necessary.

VEGETABLES:

BEANS: For control of leaf spot and powdery mildew. Apply a maximum of 7 lbs per acre at early leaf stage and repeat every 14 days as necessary.

SOYBEANS: For control of leaf spot and powdery mildew. Apply a maximum of 12.5 lbs per acre at early leaf stage and repeat every 14 days as necessary.

CARROTS: For control of powdery mildew and Petrobia mite. Apply 10 lbs per acre at early leaf stage and repeat every 14 days as necessary.

BROCCOLI, BRUSSELS SPROUTS, CABBAGE, CAULIFLOWER, COLLARDS, KALE, MUSTARD GREENS, RUTABAGAS, PEPPERS, TURNIPS: For control of powdery mildew and red spider mite. Apply 5 lbs per acre at early leaf stage and repeat every 14 days as necessary.

ONIONS AND GARLIC: For control of powdery mildew and Petrobia mite. Apply 6 lbs per acre when disease first appears and repeat as necessary.

TOMATOES: For control of russet mite. Apply a maximum of 3 lbs per acre as necessary. Thorough coverage is required.

POTATOES (Not for this Use in California). For control of powdery mildew and red spider mite. Apply at early leaf stage and repeat every 14 days as necessary, using 5 lbs. per acre.

CUCURBITS: Melons, cucumbers, squash: For control of powdery mildew. Apply a maximum of 25 lbs per acre when disease first appears and continue as necessary.

ORNAMENTALS AND ROSES: For control of powdery mildew and leaf spot. Apply a maximum of 5.6 lbs in 100 gallons of water when disease first appears and repeat at 10 day intervals as required.

NOTE: Consult State Agricultural Station or State Extension Service Specialist for rate recommended for your local area.

STORAGE AND DISPOSAL

STORAGE INSTRUCTIONS: Store in a closed package in a cool, dry place. Do not store near flammable materials. Do not store in a manner where cross-contamination with other pesticides, fertilizers, food or feed could occur. If spilled during storage or handling sweep up spillage and dispose of in accordance with the Pesticide Disposal Instructions below.

PESTICIDE DISPOSAL INSTRUCTIONS: Wastes resulting from the use of this produce may be disposed of on site or at an approved waste disposal facility.

CONTAINER DISPOSAL INSTRUCTIONS: Completely empty bag by shaking and tapping sides and bottom to loosen clinging particles. Empty residue into application equipment, then dispose of bag in a sanitary landfill, by incineration, or, if allowed by State and local authorities, by burning. If burned, stay out of smoke.

WARRANTY AND DISCLAIMER

Pennwalt Corporation warrants that this material conforms to the chemical description on the label and is reasonably fit for the purposes referred to in the Directions for Use, subject to the risks referred to therein. PENNWALT MAKES NO OTHER EXPRESS OR IMPLIED WARRANTY OR FITNESS OR MERCHANTABILITY OR ANY OTHER EXPRESS OR IMPLIED WARRANTY. IN NO CASE SHALL PENNWALT OR SELLER BE LIABLE FOR CONSEQUENTIAL, SPECIAL OR INDIRECT DAMAGES RESULTING FROM THE USE OR HANDLING OF THIS PRODUCT INCLUDING, BUT NOT LIMITED TO, LOSS OF PROFITS, BUSINESS REPUTATION, OR CUSTOMERS: LABOR COST: OR OTHER EXPENSES INCURRED IN PLANTING OR HARVESTING.

Pennwalt and seller offer this product and the buyer and user accept it subject to the foregoing conditions of sale and warranty which may be varied only by agreement in writing signed by a duly authorized representative of Pennwalt.

5-U619S-01 K8 (11/88)

PENNCAP-M®

MICROENCAPSULATED INSECTICIDE

RESTRICTED USE PESTICIDE

Due to very high acute toxicity to humans and birds.

For retail sale to and use only by certified applicators or persons under their direct supervision and only for those uses covered by the certified applicator's certification. Direct supervision for this product is defined as the certified applicator being physically present during application, mixing, loading, repair and cleaning of application equipment. Commercial certified applicators must also ensure that all persons involved in these activities are informed of the precautionary statements.

ACTIVE INGREDIENTS:	
O,O-Dimethyl O-p-nitrophenyl phosphorothioate*	20.9%
Related Isomers	1.1%
Xylene Base Aromatic Solvent	4.9%
INERT INGREDIENTS:	73.1%
TOTAL	100.0%

* Methyl Parathion (Contains 2 pounds Methyl Parathion and Related Isomers per Gallon) U.S. Patent Nos. 3,429,827; 3,577,515; 3,959,464

EPA Registration No. 4581-292 EPA Establishment No. 4581-TX-1

KEEP OUT OF REACH OF CHILDREN

AVISO WARNING

PRECAUCION AL USUARIO: Si usted no lee ingles, no use este producto hasta que le etiqueta haya sido explicado ampliamente.

STATEMENT OF PRACTICAL TREATMENT

IF SWALLOWED, induce vomiting immediately by giving two glasses of water and sticking finger down throat. Call a physician. Never give anything by mouth to an unconscious person.

IF ON SKIN, immediately wash with soap and water and flush with plenty of water.

IF IN EYES, immediately flush eyes with plenty of water for at least 15 minutes.

See below for Precautionary and First Aid statements and Note to Physician.

PRECAUTIONARY STATEMENTS

HAZARDS TO HUMANS (AND DOMESTIC ANIMALS)

MAY BE FATAL IF SWALLOWED. HARMFUL IF ABSORBED THROUGH SKIN OR INHALED.

Wash thoroughly with soap and water after handling and before eating or smoking. Avoid contact with skin, eyes, or clothing. Avoid breathing vapors and spray mist. Remove contaminated clothing and wash before reuse. Use in ventilated areas since accumulation of xylene vapors may be hazardous. Use with adequate ventilation. Avoid prolonged breathing of vapor.

FIRST AID: Call a physician in all cases of suspected poisoning. It is suggested that a suitably equipped local medical facility be identified prior to using this product.

Warning symptoms of organophosphate poisoning include weakness, headache, tightness in chest, blurred vision, non-reactive pin-point pupils, salivation, sweating, nausea, vomiting, diarrhea and abdominal cramps. If these symptoms appear transport the patient immediately to the nearest hospital. Keep patient prone and quiet. Start artificial respiration immediately if patient is not breathing.

Note to Physician: The active ingredient of PENNCAP-M is an organophosphate cholinesterase inhibitor. Atropine is antidotal. 2-PAM is also antidotal and may be administered in conjunction with atropine. Morphine is contraindicated.

ENVIRONMENTAL HAZARDS

This pesticide is highly toxic to aquatic invertebrates and wildlife. Birds in treated areas may be killed. Shrimp and other aquatic organisms may be killed at recommended application rates. Do not apply directly to water or wetlands (swamps, bogs, marshes and potholes). Runoff and drift from target areas may be hazardous to aquatic organisms in adjacent aquatic sites. Do not contaminate water by cleaning of equipment or disposal of wastes. Do not apply when weather conditions favor drift or runoff from target areas.

THIS PRODUCT IS HIGHLY TOXIC TO BEES exposed to direct treatment or residues on crops or blooming weeds. Do not apply this product or allow it to drift to blooming crops or weeds if bees are visiting the areas to be treated. Refer to and follow the more specific and applicable precautions in the directions for certain crops. Protective information may be obtained from your Cooperative Extension Service.

STORAGE AND DISPOSAL

Do not contaminate water, food, or feed by storage or disposal.

Storage Instructions: Store in the original container. Do not store in a manner where cross-contamination with other pesticides, fertilizers, food or feed could occur. If spilled during storage or handling, absorb with sand or other inert material and dispose of absorbant in accordance with the Pesticide Disposal Instructions listed below.

Pesticide Disposal Instructions: Pesticide wastes are acutely hazardous. Improper disposal of excess pesticide, spray mixture, or rinsate is a violation of Federal Law. If these wastes cannot be disposed of by use according to label instructions, contact your State Pesticide or Environmental Control Agency, or the Hazardous Waste representative at the nearest EPA Regional Office for guidance.

Container Disposal Instructions: Triple rinse (or equivalent). Then offer for recycling or reconditioning, or puncture and dispose of in a sanitary landfill, or by incineration, or, if allowed by state and local authorities, by burning. If burned, stay out of smoke.

DIRECTIONS FOR USE

It is a violation of Federal law to use this product in a manner inconsistent with its labeling.

PENNCAP-M is a flowable formulation consisting of water suspension of polymeric-type microcapsules which contain methyl parathion. It is applied by air or ground equipment by diluting it with an amount of water suitable for the specific crop and type of spraying equipment. For air application, use at least 2 gallons total spray per acre unless specified otherwise.

First add approximately half of the water to the spray tank, then add the required quantity of PENNCAP-M with good agitation, and finally add the rest of the water. This procedure should be followed to insure proper mixing of PENNCAP-M.

Mixtures of PENNCAP-M with emulsifiable concentrates, organic solvents, or some surfactants may cause increased toxicity and shorter residual effectiveness. This effect is more pronounced with concentrated sprays and extended periods between mixing and application.

DO NOT USE SCREEN OR NOZZLES FINER THAN 50 MESH. STIR OR SHAKE BEFORE USING.

PENNCAP M will not control organophosphate resistant insects.

WORK SAFETY RULES

The National Pesticide Telecommunication Network is available for recommendations regarding poisoning management, emergency treatment, and other information regarding the toxicity of methyl parathion. The toll free number for the National Pesticide Telecommunication Network is 1-800-858-7378.

If handled indoors provide mechanical exhaust ventilation. Keep all unprotected persons, children, livestock, and pets away from treated area or where there is danger of drift. Do not rub eyes or mouth with hands. If you feel sick in any way, STOP work and get help right away. See First Aid (Practical Treatment) section on main container label.

HANDLE THE CONCENTRATE ONLY WHEN WEARING THE FOLLOWING PROTECTIVE CLOTHING AND EQUIPMENT: Wear a protective suit of one or two pieces that covers all parts of the body except the head, hands, and feet. Wear chemical resistant gloves, chemical resistant apron, and chemical resistant shoes, shoe coverings, or boots. Wear goggles or a face shield. Wear a pesticide respirator approved by the National Institute for Occupational Safety and Health under the provisions of 30 CFR part II.

USE ONLY WHEN WEARING THE FOLLOWING PROTECTIVE CLOTHING AND EQUIPMENT DURING MIXING, LOADING, APPLICATION, REPAIR AND CLEANING OF APPLICATION EQUIPMENT, DISPOSAL OF PESTICIDE AND EARLY REENTRY INTO TREATED AREAS: protective suit of one or two pieces covering all parts of the body except the head, hands, and feet; chemical resistant gloves; chemical resistant shoes (or chemical resistant shoe coverings or chemical resistant boots); goggles or face shield; hood or wide brimmed hat; NIOSH approved respirator.

IF MIXING/LOADING IS PERFORMED USING A CLOSED SYSTEM, THE FOLLOWING PROTECTIVE CLOTHING AND EQUIPMENT MAY BE WORN AS AN ALTERNATIVE: long-sleeve shirt; long-legged pants; chemical resistant gloves; chemical resistant apron; shoes and socks. Goggles or face shield must be worn when the system is under pressure. All other protective clothing and equipment required for use with open systems must be available nearby.

IF APPLICATION IS PERFORMED USING AN ENCLOSED CAB OR COCKPIT, THE FOLLOWING PROTECTIVE CLOTHING AND EQUIPMENT MAY BE WORN AS AN ALTERNATIVE: long-sleeve shirt and long-legged pants; shoes and socks. Chemical resistant gloves must be available in the cab or cockpit and must be worn during entry to and exit from the application vehicle. All other protective clothing and equipment required for use during application must be available in the cab and must be worn when exiting the cab into treated area. When used for this purpose, contaminated clothing may not be brought back into the cab unless in an enclosure such as a plastic bag. REMEMBER—THIS CLOTHING IS INADEQUATE TO PROTECT YOU DURING REPAIR AND CLEANING OF APPLICATION EQUIPMENT AND EARLY REENTRY TO TREATED AREAS! REFER TO PROTECTIVE CLOTHING AND EQUIPMENT REQUIREMENTS ABOVE.

IMPORTANT! If pesticide comes into contact with skin, wash off with soap and water and contact a physician immediately. Always wash hands, face, and arms with soap and water before smoking, eating, drinking, or going to the bathroom.

AFTER WORK: before removing gloves, wash them with soap and water. Take off all work clothes and shoes. Shower using soap and water. Wear only clean clothes when leaving job—do not wear contaminated clothing. Personal clothing worn during work must be stored and laundered separately from protective clothing and household articles. Store protective clothing separately from personal clothing. Clean or launder protective clothing after each use. Respirators must be cleaned and filters replaced according to instructions included with the respirators. Protective clothing and protective equipment heavily contaminated or drenched with methyl parathion must be destroyed according to state and local regulations. HEAVILY CONTAMINATED OR DRENCHED CLOTHING CANNOT BE ADEQUATELY DECONTAMINATED. DURING AERIAL APPLICATION, HUMAN FLAGGERS ARE PROHIBITED.

RE-ENTRY STATEMENT

Do not apply this product in such a manner as to directly or through drift exposure workers or other persons. The area being treated must be vacated by unprotected persons.

Reentry into treated fields is prohibited for 48 hours after application, unless all protective clothing and equipment required for early reentry as described under Work Safety Rules are worn.

Written or oral warnings must be given to workers who are expected to be in an area about to be treated with this product. In case of accidental exposure, workers should wash thoroughly with soap and water and contaminated clothing should be removed and washed before reuse. When oral warnings are given, warnings shall be given in a language customarily understood by workers. Oral warnings shall be given if there is reason to believe that written warnings cannot be understood by workers. Written warnings must include the following information. "WARNING. Area treated with PENNCAP-M® Insecticide on (date of application). Do not enter without appropriate protective clothing within 48 hours after application. In case of accidental exposure, workers should wash thoroughly with soap and water and contaminated clothing should be removed and washed before reuse."

FIELD CROPS

Crop	Insects	Pints/ Acre	Remarks
Alfalfa	Blue Alfalfa Aphid	1½-2	Do not spray alfalfa during bloom to avoid injury to bees. Do not apply to alfalfa grown for seed within 7 days prior to bloom or while bloom is present in the treatment area. Do not apply within 15 days of harvest or grazing. Do not apply if the weeds in the treatment area are in bloom.
	Grasshoppers*	1-3	
	Alfalfa Blotch Leafminer Alfalfa Plant Bug Alfalfa Weevil (larvae and adults) Green Cloverworm Potato Leafhopper Pea Aphid Spittlebugs Tarnished Plant Bug	2-3	
	Cutworms Egyptian Alfalfa Weevil	2-4	
Corn (Field)	Corn Rootworm Adults (for protection of silks)	1-2	First Generation European Corn Borer: Whorl Application—2 pints per acre by ground Broadcast Application—4 pints per acre. Second Generation European Corn Borer: Broadcast Application—2 to 4 pints per acre. Use the low rate for light infestations. When environmental conditions favor extended moth flights of the second generation of European Corn Borer, two correctly timed applications of PENNCAP-M at the lowest recommended rate may provide improved control over a single application at a higher rate.** Apply PENNCAP-M at 2 pints/acre at the time of the initial
	Grasshoppers*	1-3	
	Aphids Flea Beetles True Armyworm**	2-3	
	Corn Earworm European Corn Borer Southwestern Corn Borer Western Bean Cutworm	2-4	
	Black Cutworm	4	

Crop	Insects	Pints/ Acre	Remarks
			hatch of the second generation. Repeat 10-14 days later. For low volume aerial applications to control Corn Rootworm Adults, Aphids, European Corn Borer and Western Bean Cutworm use the indicated rates of PENNCAP-M at a minimum of ½ gallon total spray per acre.** Application of undiluted PENNCAP-M is not recommended. Equipment that has been specially adapted and calibrated for low volume should be used. Best control will be achieved using spray devices that will produce droplets within the size range of 100 to 200 microns. The following conditions must be met: 1. Equipment must be adjusted to distribute spray uniformly over the spray swath; 2. Wind conditions and other factors such as temperature and humidity are such that the spray is deposited on the target; 3. Local regulations do not prohibit low volume aerial sprays. For most efficient European Corn Borer and Corn Earworm control, use a spray schedule suitably adapted to the area. For best Black Cutworm control, apply PENNCAP-M under moist soil conditions. To avoid injury to bees, do not apply during pollen shed if bees are visiting the areas to be treated during their normal foraging hours. Do not harvest, cut for forage, or graze within 12 days of application.
Corn (Sweet)	Corn Rootworm Adults (for protection of silks)	1-2	First Generation European Corn Borer: Whorl Application—2 pints per acre by ground Broadcast Application—4 pints per acre. Second Generation European Corn Borer: Broadcast Application—2 to 4 pints per acre. Repeat applications as necessary to maintain control. For most efficient European Corn Borer and Corn Earworm control, use a spray schedule suitably adapted to the area. For best Black Cutworm control, apply PENNCAP-M under moist soil conditions. To avoid injury to bees, do not apply during pollen shed if bees are visiting the areas to be treated during their normal foraging hours. Do not apply later than 3 days before harvest. Do not cut for forage or graze within 12 days of application.
	Grasshoppers*	1-3	
	Aphids Flea Beetles True Armyworm**	2-3	
	Corn Earworm European Corn Borer Sap Beetles	2-4	
	Black Cutworm	4	
Cotton	Fleahoppers Plant Bugs (including **Lygus** spp.) Thrips**	½- 1	Do not apply within 7 days of hand picking or harvest.
	Boll Weevil	1-3	
	Bollworm Tobacco Budworm	4-6	

Rangeland, Pasture, Grass grown for seed production, and Roadside Areas	Grasshoppers*	1-3	Do not apply within 15 days of harvest or grazing.
	Thrips**	2	
	True Armyworm**	2-3	
Rice	Stink Bugs	1-2	Do not apply within 14 days of application of propanil.
	Grasshoppers*	1-3	Do not apply within 15 days of harvest.
Soybeans	Grasshoppers*	1-3	Do not make more than 2 applications per season. Do not apply within 20 days of harvest or grazing.
	Bean Leaf Beetle Green Cloverworm Mexican Bean Beetle Potato Leafhopper Silverspotted Skipper Stink Bugs Threecornered Alfalfa Hopper Thrips** Velvetbean Caterpillar	2-3	
	Corn Earworm (Podworm) Japanese Beetle**	3-4	
Tobacco	Grasshoppers*	1-3	Do not apply within 5 days of priming. Avoid contact with plant juices when priming or cutting tobacco.
	Aphids Tobacco Hornworm Vegetable Weevil	2-4	
	Flea Beetle Tobacco Budworm	4-6	
Wheat, Oats and Barley	Aphids (including Greenbug)	1-2	Do not apply within 15 days of harvest or grazing.
	Grasshoppers*	1-3	
	True Armyworm**	2-3	

FRUIT CROPS

Crop	Insects	Pints/ Acre	Remarks
Apples and Pears	Apple Maggot Apple and Thorn Skeletonizer Colding Moth Fruit Tree Leafroller** Green Fruitworm Lesser Appleworm Oblique-banded Leafroller** Oriental Fruit Moth Oyster Shell Scale* Pandemis Leafroller** Pear Plant Bug (pears only) Plum Curculio Red-banded Leafroller** San Jose Scale Tarnished Plant Bug Tufted Apple Budmoth Variegated Leafroller** White-marked Tussock Moth Woolly Apple Aphid* *	4-8 OR 1-2 pints/100 gals. of dilute spray	Under heavy insect pressure, use the higher dosage; consult your local agricultural experiment station for proper timing of sprays. Do not apply more than 8 pints per acre per treatment. Do not tank-mix PENNCAP-M with streptomycin since injury to fruit and foliage may result. Do not apply to D'Anjou pears as a concentrate or semi-concentrate spray since fruit injury is likely to occur. Do not make initial application before total petal fall. Do not apply when a substantial number of weeds or cover crops in orchards are in bloom and bees are actively visiting the area to be treated. Do not apply with 14 days of harvest.

Crop	Insects	Pints/ Acre	Remarks
Cherries	Black Cherry Aphid** Fruit Flies (including Cherry Fruit Fly) Green Fruitworms Leafrollers** Plum Curculio	3-6 OR 1-2 pints/ 100 gals. of dilute spray	For additional control during post-harvest period, treat interplanted trees, all unpicked fruit, and pollinators. Do not apply more than 6 pints per acre per treatment. Do not make initial application before total petal fall. Do not apply when a substantial number of weeds or cover crops in orchards are in bloom and bees are actively visiting the area to be treated. Do not graze or feed cover crops from treated orchards. Do not apply within 14 days of harvest.
Grapes	Grape Aphids Grape Blossom Midge Grape Leafhopper (East of the Rocky Mountains) Potato Leafhopper Red-banded Leafroller**	4 OR 2 pints/ 100 gals. of dilute spray	Do not apply more than 12 pints per acre per treatment. Do not apply within 14 days of harvest.
	Grape Berry Moth Japanese Beetle**	4-8 OR 2-4 pints/ 100 gals. of dilute spray	
	Rose Chafer	6-8 OR 3-4 pints/ 100 gals. of dilute spray	
	Grape Leafhopper (West of the Rocky Mountains)	8-12 OR 4-6 pints/ 100 gals. of dilute spray	
Nectarines, Peaches, Plums, and Prunes	Lesser Peachtree Borer Peachtree Borer	4½-6 OR 1½-2 pints/ 100 gals. of dilute spray	Do not apply more than 8 pints per acre per treatment. Do not apply more than 2½ gallons per acre per year to nectarines. Do not make initial application before total petal fall. Do not apply when a substantial number of weeds or cover crops in orchards are in bloom and bees are actively visiting the area to be treated. Do not graze or feed cover crops from treated orchards. Do not apply later than 14 days before harvest.
	Catfacing Insects Oriental Fruit Moth Plum Curculio	3-6 OR 1-2 pints/ 100 gals. of dilute spray	
	Peach Twig Borer	8	Do not apply in less than 200 gallons per acre. Make dormant or delayed dormant applications only. May be mixed with suitable spray oil. Do not allow tank mixes of PENNCAP-M and oil to stand but use immediately. For precautions relating to the protection of pollinators and the grazing or feeding of cover crops, see above.
Nectarines and Peaches (Arizona and California only)	Insects as above	As above	Do not use more than 2½ gallons per acre on peaches from January 1 to harvest. Do not use more than 8 pints per acre per treatment when using dilute sprays on nectarines. Do not apply more than 2½ gallons per acre per year to nectarines. For precautions relating to the protection of pollinators and the grazing or feeding of cover crops, see above. Do not apply within 21 days of harvest.

VEGETABLE CROPS

Crop	Insects	Pints/ Acre	Remarks
Artichokes	Artichoke Plume Moth	4	Do not apply within 7 days of harvest.
Beans (Dry, Southern Peas)	Aphids** Cucumber Beetles (**Diabrotica** spp.) Green Cloverworm Leafhoppers Lygus Bugs Mexican Bean Beetle	2	Beans—To avoid injury to bees, do not apply during the period from 7 days prior to first bloom through peak bloom. After this time, PENNCAP-M may be applied to beans as long as bees are not actively visiting the area. Southern Peas—Start application when blooms are first observed and repeat at 3 to 5 day intervals. Do not apply within 15 days of harvest.
	Cowpea Curculio European Corn Borer	2-4	
	Corn Earworm	4	
Succulent Beans (except California)	Aphids** Cucumber Beetles (**Diabrotica** spp.) Green Cloverworm Leafhoppers Lygus Bugs Mexican Bean Beetle	2	Do not make more than 3 applications per season. Do not apply within 3 days of harvest.
	Cowpea Curculio European Corn Borer	2-4	
	Corn Earworm	4	
Succulent Beans (California)			For succulent beans in California, follow directions under Beans (Dry, Southern Peas) above, same insects and rates.
Cabbage	Diamondback Moth** Flea Beetles	2-4	Do not apply within 21 days of harvest.
	Imported Cabbageworms	4-6	
Onions	Thrips	2	Do not apply when onions are blooming to avoid injury to bees. Do not apply within 15 days of harvest.
Peas (White Blooming)	Pea Aphid Pea Leaf Weevil Pea Weevil	2	Do not apply more than 4 pints per acre per year. Do not apply if the crop or weeds in the treatment area are in bloom. Do not apply within 10 days of harvest for pods. Do not harvest for forage or graze within 15 days of application.
Peas (Austrian Winter)	Pea Aphid Pea Leaf Weevil Pea Weevil	2	See Peas (White Blooming). Do not apply within 7 days of bloom. Do not apply if the crop or weeds in the treatment area are in bloom.
Potatoes	Grasshoppers*	1-3	Do not apply within 5 days of harvest.
	Cutworms European Corn Borer Flea Beetles Potato Leafhopper Tarnished Plant Bug	2-4	
	Colorado Potato Beetle	2-6	

Crop	Insects	Pints/ Acre	Remarks
Tomatoes	Flea Beetles Leafminers	2-4	Do not apply within 15 days of harvest.
	Colorado Potato Beetle Hornworms Tomato Fruitworm Tomato Pinworm	4	

* For grasshoppers, use higher rates if the majority of insects are large or weather is cool.

** Not for this use in California.

ULV AERIAL APPLICATION IN WATER TO COTTON NOT FOR THIS USE IN CALIFORNIA

APPLICATION INSTRUCTIONS

Use spray equipment that has been specifically adapted and calibrated for ultra-low-volume spraying of PENNCAP-M. The following conditions are to be met:

1. Equipment must be adjusted to distribute the spray uniformly over the spray swath.
2. Do not apply when wind conditions favor drift from treated areas.

Spray equipment should be cleaned as soon as possible after completion of spraying. Strainers should be cleaned daily. Do not allow PENNCAP-M to dry in or on application equipment before cleaning.

Crop	Insect	Rates	Remarks
Cotton	Boll Weevil	PENNCAP-M 1-2 pints Plus an equal amount of water 1-2 pints	Begin applications when damaging levels of boll weevils appear. Repeat applications as necessary to maintain insect control. Add a material for **Heliothis** control to the tank mix when bollworms and/or budworms (**Heliothis** spp.) are present. Use the rate recommended on the label for the specific material used. Cotton may be machine harvested any time after application.

TANK MIXES WITH PENNCAP-M

TANK MIX FOR USE ON APPLES AND PEARS

(All Apple and Pear Growing Areas Except California)

Insects	Rates	Remarks
Apple Maggot Apple and Thorn Skeletonizer Codling Moth Fruit Tree Leafroller Green Fruitworm Lesser Appleworm Oblique-banded Leafroller Oriental Fruit Moth Oyster Shell Scale Pandemis Leafroller Pear Plant Bug (pears only) Plum Curculio Red-banded Leafroller San Jose Scale Tarnished Plant Bug Tufted Apple Budmoth Variegated Leafroller White-marked Tussock Moth Woolly Apple Aphid	PENNCAP-M 2-4 pints/acre Plus azinphos-methyl 50W 1 lb./acre OR PENNCAP-M ½- 1 pint Plus azinphos-methyl 50W ¼ lb./100 gallons for dilute spray	Do not apply more than 4 pints PENNCAP-M plus 1 pound azinphos-methyl 50W per acre per treatment. Do not apply within 7 days of harvest.

Observe all limitations, precautions, and rate recommendations which appear on the labels for these products.

TANK MIX FOR USE ON APPLES AND PEARS

(All Apple and Pear Growing Areas Except California)

Insects	Rates	Remarks
Apple Maggot Apple and Thorn Skeletonizer Codling Moth Fruit Tree Leafroller Green Apple Aphid Green Fruitworm Lesser Appleworm Oblique-banded Leafroller Oriental Fruit Moth Oyster Shell Scale Pandemis Leafroller Pear Plant Bug (pears only) Plum Curculio Red-banded Leafroller San Jose Scale Tarnished Plant Bug Tentiform Leafminer Tufted Apple Budmoth Variegated Leafroller White Apple Leafhopper White-marked Tussock Moth Woolly Apple Aphid	PENNCAP-M 2-4 pints/acre Plus one of the following: ZOLONE ® 3EC* 2-6 pints/acre ZOLONE 25W 4-8 lbs./acre OR PENNCAP-M ½- 1 pint Plus one of the following: ZOLONE 3EC ½- 1½ pints ZOLONE 25W 1-2 lbs./100 gallons for dilute spray	Do not apply more than 4 pints PENNCAP-M plus 6 pints Zolone 3EC or 8 pounds Zolone 25W per acre per treatment. Do not apply within 14 days of harvest.

Observe all limitations, precautions, and rate recommendations which appear on the labels for these products.

® ZOLONE is a registered trademark of Rhone-Poulenc Chemical Company.

TANK MIX FOR USE ON APPLES

(All Apple Growing Areas Except California)

Insects	Rates	Remarks
Apple Maggot Apple and Thorn Skeletonizer Codling Moth Fruit Tree Leafroller Green Apple Aphid Green Fruitworm Lesser Appleworm Oblique-banded Leafroller Oriental Fruit Moth Oyster Shell Scale Pandemis Leafroller Plum Curculio Red-banded Leafroller San Jose Scale Tarnished Plant Bug Tentiform Leafminer Tufted Apple Budmoth Variegated Leafroller White Apple Leafhopper White-mârked Tussock Moth Woolly Apple Aphid	PENNCAP-M 2-4 pints/acre Plus one of the following: methomyl 1.8L 2-4 pints/acre methomyl 90SP ½-1 lb./acre OR PENNCAP-M ½-1 pint Plus one of the following: methomyl 1.8L ½-1 pint methomyl 90SP 2-4 ounces/100 gallons for dilute spray	Do not apply more than 4 pints PENNCAP-M plus 4 pints methomyl 1.8L or 1 pound methomyl 90SP per acre per treatment. Do not apply within 8 days of harvest.

Observe all limitations, precautions, and rate recommendations which appear on the labels for these products.

TANK MIX FOR USE ON APPLES
(All Apple Growing Areas East of the Rocky Mountains)

Insects	Rates	Remarks
Apple Maggot Apple and Thorn Skeletonizer Codling Moth Fruit Tree Leafroller Green Apple Aphid Green Fruitworms Lesser Appleworm Oblique-banded Leafroller Oriental Fruit Moth Plum Curculio Red-banded Leafroller San Jose Scale Tentiform Leafminer Tufted Apple Budmoth Variegated Leafroller White Apple Leafhopper Woolly Apple Aphid	PENNCAP-M 2-4 pints/acre Plus THIODAN ® 50WP 2/lbs. acre OR for dilute sprays PENNCAP-M 1/2-1 pint Plus THIODAN 50WP 1/2 lb./100 gallons of spray mixture	Under heavy pressure from maggots, fruitworms, leafrollers, curculio, scale and budmoth use the higher dosage of PENNCAP-M. Do not apply within 21 days of harvest.

Observe all limitations, precautions, and rate recommendations which appear on the labels for these products.
® THIODAN is a registered trademark of Hoechst AG.

TANK MIX FOR USE ON COTTON
(All Cotton Growing Areas Except California)

Insects	Rates/Acre	Remarks
Boll Weevil Bollworm Tobacco Budworm	PENNCAP-M 2-3 pints Plus one of the following: GALECRON ® SP FUNDAL ® SP ** 1/8-1/4 pound GALECRON 4E FUNDAL 4E 1/8-1/4 pint	Always prepare fresh mixtures just prior to use. Do not apply within 21 days of harvest. Do not feed foliage from treated cotton plants or gin trash to livestock.

Observe all limitations, precautions, and rate recommendations which appear on the labels for these products.
* GALECRON is a registered trademark of Ciba-Geigy.
** FUNDAL is a registered trademark of Nor-AM Chemical Co.

TANK MIX FOR USE ON COTTON
(All Cotton Growing Areas Except California)

Insects	Rates/Acre	Remarks
Armyworms Boll Weevil Bollworm Cabbage Looper Plant Bugs Tobacco Budworm Whiteflies	PENNCAP-M 1-3 pints Plus ORTHENE ® Insecticide* 1/2-1 pound	Do not apply within 21 days of harvest.

Observe all limitations, precautions, and rate recommendations which appear on the labels for these products.
*ORTHENE is a registered trademark of Chevron Chemical Company.

TANK MIX FOR USE ON COTTON
(All Cotton Growing Areas Except California)

Insects	Rates/Acre	Remarks
Boll Weevil Bollworm Tobacco Budworm	PENNCAP-M 1-2 pints Plus AMBUSH ® Insecticide* 1/5-4/5 pint	Low rates indicated should be used under low insect pressure or in a season long program under close scouting supervision. Do not apply within 14 days of harvest.

Observe all limitations, precautions, and rate recommendations which appear on the labels for these products.
* AMBUSH is a registered trademark of ICI Americas, Inc.

TANK MIX FOR USE ON COTTON
(All Cotton Growing Areas Except California)

Insects	Rates/Acre	Remarks
Boll Weevil Bollworm Tobacco Budworm	PENNCAP-M 1-2 pints Plus PYDRIN ® Insecticide* 1/6- 2/3 pint	Low rates indicated should be used under low insect pressure or in season long program under close scouting supervision. Do not apply within 14 days of harvest.

Observe all limitations, precautions, and rate recommendations which appear on the labels for these products.
* PYDRIN is a registered trademark of Shell Chemical Company.

CHEMIGATION

NOT FOR THIS USE IN CALIFORNIA
GENERAL INSTRUCTIONS

Apply this product only through sprinkler irrigation systems including center pivot, lateral move, end tow, side (wheel) roll, traveler, big gun, solid set, or hand move. Do not apply this product through any other type of irrigation system.

Crop injury, lack of effectiveness, or illegal pesticide residues in the crop can result from nonuniform distribution of treated water.

If you have any questions about calibration, you should contact State Extension Service specialists, equipment manufacturers or other experts.

A person knowledgeable of the chemigation system and responsible for its operation, or under the supervision of the responsible person, shall shut the system down and make necessary adjustments should the need arise.

Do not connect chemigation system to any public water system. Public water system means a system for the provision of piped water for human consumption if such system has at least 15 service connections or regularly serves an average of at least 25 individuals daily at least 60 days out of the year.

SYSTEM REQUIREMENTS

Systems utilizing a pressurized water and pesticide injection system must meet the following requirements:

The system must contain a functional check valve, vacuum relief valve, and low pressure drain appropriately located on the irrigation pipeline to prevent water source contamination from backflow.

The pesticide injection pipeline must contain a functional, automatic, quick-closing check valve to prevent the flow of fluid back toward the injection pump.

The pesticide injection pipeline must also contain a functional, normally closed, solenoid-operated valve located on the intake side of the injection pump and connected to the system interlock to prevent fluid from being withdrawn from the supply tank when the irrigation system is either automatically or manually shut down.

The system must contain functional interlocking controls to automatically shut off the pesticide injection pump when the water pump motor stops.

The irrigation line or water pump must include a functional pressure switch which will stop the water pump motor when the water pressure decreases to the point where pesticide distribution is adversely affected.

Systems must use a metering pump, such as a positive displacement injection pump (e.g., diaphragm pump) effectively designed and constructed of materials that are compatible with pesticides and capable of being fitted with a system interlock.

INSTRUCTIONS FOR SPRINKLER (OVERHEAD) IRRIGATION:

Observe the requirements in the System Requirements section above.

Apply PENNCAP-M Insecticide only through systems containing anti-syphon and check valves designed to prevent water source contamination or overflow of the mix tank and containing interlocking controls between the metering device and the water pump to insure simultaneous shut-off.

Maintain a gentle continuous agitation in mix tank during mixing and application to assure a uniform suspension.

Greater accuracy in calibration and distribution will be achieved by injecting a larger volume of a more dilute suspension per unit time.

Application of more than recommended quantities of irrigation water per acre may result in decreased product performance.

Do not apply when wind speed favors drift, when system connections or fittings leak, when nozzles do not provide uniform distribution or when lines containing the product cannot be flushed and must be dismantled and drained. In a center pivot system, block the nozzle set nearest the well/pivot/injection unit to prevent spray being applied to this area. All screens should be 50 mesh or larger.

Where sprinkler distribution patterns do not overlap sufficiently, unacceptable insect control may result.

Allow sufficient time for pesticide to be flushed through all lines and all nozzles before turning off irrigation water.

PENNCAP-M may be applied in conjunction with chemically neutral liquid fertilizers. Application in conjunction with highly alkaline fertilizers, such as aqueous ammonia, may cause a degradation of the pesticide, resulting in reduced performance and should be avoided.

Check local restrictions and requirements regarding sprinkler irrigation applications, as they may vary from state to state.

SPRAY PREPARATION:

Remove scale, pesticide residues, and other foreign matter from the chemical tank and entire injector system. Flush with clean water.

Prepare a suspension of PENNCAP-M Insecticide in a mix tank. Fill the tank with 1/2 or 3/4 of the desired amount of water. Start mechanical or hydraulic agitation. Slowly add the required amount of PENNCAP-M and then the remaining volume of water.

APPLICATION INSTRUCTIONS:

Set sprinkler system to deliver 0.1 to 1.25 inches of water per acre. Volumes of water higher than this may reduce efficacy. Start sprinkler and then uniformly inject the suspension of PENNCAP-M into the irrigation water line so as to deliver the desired rate per acre. The suspension of PENNCAP-M should be injected with a positive displacement pump into the main line ahead of a right angle turn to insure adequate mixing.

NOTE: When treatment with PENNCAP-M has been completed, do not irrigate the treated area for 24 to 48 hours to prevent washing the chemical off the crop.

WARRANTY AND DISCLAIMER

Pennwalt Corporation warrants that this material conforms to the chemical description on the label and is reasonably fit for the purposes referred to in the Directions for Use, subject to the risks referred to therein. PENNWALT MAKES NO OTHER EXPRESS OR IMPLIED WARRANTY OF FITNESS OR MERCHANTABILITY OR ANY OTHER EXPRESS OR IMPLIED WARRANTY. IN NO CASE SHALL PENNWALT OR SELLER BE LIABLE FOR CONSEQUENTIAL, SPECIAL OR INDIRECT DAMAGES RESULTING FROM THE USE OR HANDLING OF THIS PRODUCT INCLUDING, BUT NOT LIMITED TO, LOSS OF PROFITS, BUSINESS REPUTATION, OR CUSTOMERS; LABOR COST; OR OTHER EXPENSES INCURRED IN PLANTING OR HARVESTING.

Pennwalt and seller offer this product and the buyer and user accept it subject to the foregoing conditions of sale and warranty which may be varied only by agreement in writing signed by a duly authorized representative of Pennwalt.

5-Z401S-01 B8 (11/87)

PENNCOZEB®

FUNGICIDE

An 80% Coordination Product of Manganese and Zinc and Ethylenebisdithiocarbamate

ACTIVE INGREDIENT:

A coordination product of zinc ion and manganese ethylene bisdithiocarbamate 80%

Manganese++ 16%

Zinc++ 2%

Ethylenebisdithiocarbamate ion ($C_4H_6N_2S_4$) 62%

INERT INGREDIENTS: 20%

TOTAL 100%

EPA Registration No. 4581-358 EPA Establishment No. 33910-HL-01

KEEP OUT OF REACH OF CHILDREN

CAUTION

STATEMENT OF PRACTICAL TREATMENT

IF SWALLOWED, call a physician or Poison Control Center. Drink 1 or 2 glasses of water and induce vomiting by touching back of throat with finger. Do not induce vomiting or give anything by mouth to an unconscious person.

IF ON SKIN, wash with plenty of soap and water. Get medical attention.

IF IN EYES, flush eyes with plenty of water. Call a physician if irritation persists.

DIRECTIONS FOR USE

It is a violation of Federal law to use this product in a manner inconsistent with its labeling.

Do not enter treated areas without protective clothing within 24 hours after application. Because certain states may require more restrictive reentry intervals for various crops treated with this product, consult your State Department of Agriculture for further information.

Written or oral warnings must be given to workers who are expected to be in a treated area or in an area about to be treated with this product. When oral warnings are given, such warnings shall be given in a language customarily understood by the workers. Oral warnings must be given if there is reason to believe that written warnings cannot be understood by the workers. Written warnings must include the following information " 'CAUTION. Area treated with PENNCOZEB Fungicide on (*date of application*)' "

After sprays have dried/dusts have settled do not enter or allow entry into treated areas until the 24-hour reentry interval has expired unless wearing the personal protective equipment listed below.

WORKER SAFETY RULES

Keep all unprotected persons, children, livestock, and pets away from treated area or where there is danger of drift.

Do not rub eyes or mouth with hands. See Statement of Practical Treatment.

PERSONAL PROTECTIVE EQUIPMENT—For Mixers, Loaders, Applicators and Early Reentry Workers.

HANDLE THIS PRODUCT ONLY WHEN WEARING THE FOLLOWING PROTECTIVE CLOTHING AND EQUIPMENT: a long-sleeve shirt and long pants or a coverall that covers all parts of the body except the head, hands, and feet; chemical resistant gloves; shoes, socks, and goggles or a face shield. During mixing and loading, a chemical resistant apron must also be worn.

During application from a tractor with a completely enclosed cab with positive pressure filtration, or aerially with an enclosed cockpit, a long-sleeve shirt and long pants may be worn in place of the above protective clothing. Chemical resistant gloves must be available in the cab or cockpit and worn while exiting.

IMPORTANT! Before removing gloves, wash them with soap and water. Always wash hands, face, and arms with soap and water before eating, smoking or drinking. Always wash hands and arms with soap and water before using the toilet.

After work take off all clothes and shoes. Shower using soap and water. Wear only clean clothes. Do not use contaminated clothing. Wash protective clothing and protective equipment with soap and water after each use. Personal clothing worn during use must be laundered separately from household articles. Clothing and protective equipment heavily contaminated or drenched with PENNCOZEB must be destroyed according to state and local regulations. HEAVILY CONTAMINATED OR DRENCHED CLOTHING CANNOT BE ADEQUATELY DECONTAMINATED.

During aerial application, human flaggers are prohibited unless in totally enclosed vehicles.

PENNCOZEB is a wettable powder containing a coordination product of zinc ion and manganese ethylenebisdithiocarbamate, and is recommended for use as a spray or dust for the control of many important plant diseases.

AS A SPRAY, (Ground or aerial equipment) apply at the rate shown. Use sufficient water to provide thorough coverage, usually 20-100 gallons per acre for ground equipment and 5-8 gallons per acre for aircraft. Use a Spreader Sticker at rate of 4 fl. oz. per 100 gallons spray if needed. Add product slowly to water in the spray tank with agitation, or premix thoroughly in separate holding tank for concentrate or aircraft sprayers. Continuous agitation is required to keep the product in suspension.

AS A DUST, apply product at the rate shown, using proper amount of dust containing 6% to 10% PENNCOZEB (4.8% to 8% active). Dust may be prepared by diluting and thoroughly mixing product with pyrophylite or other neutral diluent; commonly used insecticides may displace an equivalent amount of diluent. Use dust mixtures as soon as possible after preparation.

FIELD CROPS

Crops	Diseases	Rate Lbs/ Acre	Remarks
Sugar Beets	Cercospora Leaf Spot	1 1/2-2	Begin applications when disease first threatens and repeat at 7- to 10-day intervals. Do not apply within 14 days of harvest.
Peanuts	Cercospora Leaf Spot	1-2	Begin applications when disease first appears and repeat at 10- to 14-day intervals, or at 7-day intervals during humid weather. If peanut vine hay is to be used for livestock feed, do not use more than 1 1/2 pounds per acre per application. Do not apply within 14 days of harvest.

Crops	Diseases	Rate Lbs/ Acre	Remarks
Cotton (Southwest only)	Rust	$1^1/_4$-2	Begin applications when rust first appears in the area and repeat at 10- to 14-day intervals. Do not apply after bolls open; do not graze treated areas or feed gin trash to livestock.
Wheat	Helminthosporium Leaf Spots Septoria Leaf Spots Septoria Glume Blotch Leaf Rust Tan Spot	2	Start application at onset of disease or when plants are in tillering to jointing stage and repeat at 7- to 10-day intervals. Do not make more than 3 applications in one season. Do not apply within 26 days of harvest. Do not graze livestock in treated areas within 26 days after application.

VEGETABLE CROPS

Crops	Diseases	Rate Lbs/ Acre	Remarks
Potatoes	Early and Late Blights	1-2	Begin applications when plants are 4-6 inches high, and repeat at 7- to 10-day intervals (more frequently under severe disease conditions) throughout the season.
Potato Tuber Treatment	Fusarium Seed Piece Decay		Dip whole or cut seed pieces in mixture of $1^1/_4$ lbs. of product per 50 gallons of water, drain and plant immediately. If seed pieces are to be held before planting, spread out to dry in a cool place. For dust treatment, thoroughly coat with dust containing 10% PENNCOZEB (8% active) at a rate of 1 lb. per 100 lbs. of seed. Do not use treated seed pieces for food or feed purposes.
Tomatoes	Early and Late Blights Gray Leaf Spot Gray Leaf Mold Anthracnose	$1^1/_2$-3	Begin applications when seedlings emerge and repeat at 7-day intervals throughout the season. Do not apply within 5 days of harvest.
Corn: Sweet and Popcorn	Helminthosporium Leaf Blight	$1^1/_2$	Apply PENNCOZEB in sufficient water for thorough coverage. Start applications when disease first appears and repeat at 4- to 7-day intervals. Do not apply within 7 days of harvest. Do not feed treated forage (plant stalk and leaves) to livestock.
Onions (Dry Bulb)	Botrytis Leaf Blight Downy Mildew Neck Rot Purple Blotch	2-3	Begin applications when diseases are first reported in the area and repeat at 7-day intervals throughout the season. Do not apply within 7 days of harvest; do not apply to exposed bulbs.
Asparagus	Rust	2	Begin applications when rust first appears and repeat at 10-day intervals as needed. Use only on ferns after spears have been harvested.

FRUIT CROPS

Crops	Diseases	Rate Lbs/ Acre	Remarks
Cranberries	Fruit Rot	3-6	Begin applications at mid-bloom and repeat at 7- to 10-day intervals. Do not apply within 30 days of harvest.
Grapes	Black Rot Bunch Rot Dead Arm Downy Mildew	$1^1/_2$-4	Apply when shoots are $^1/_2$ to $1^1/_2$ inches long, when shoots are 3-5 inches long and when shoots are 8-10 inches long. Continue at 10- to 14-day intervals, but do not apply within 66 days of harvest. In California, do not apply after fruit set.

CHEMIGATION

NOT FOR THIS USE IN CALIFORNIA
GENERAL INSTRUCTIONS

Apply this product only through sprinkler irrigation systems including center pivot, lateral move, end tow, side (wheel) roll, traveler, big gun, solid set, or hand move. Do not apply this product through any other type of irrigation system.

Crop injury, lack of effectiveness, or illegal pesticide residues in the crop can result from nonuniform distribution of treated water.

If you have any questions about calibration, you should contact State Extension Service specialists, equipment manufacturers or other experts.

A person knowledgeable of the chemigation system and responsible for its operation, or under the supervision of the responsible person, shall shut the system down and make necessary adjustments should the need arise.

Do not connect chemigation system to any public water system. Public water system means a system for the provision of piped water for human consumption if such system has at least 15 service connections or regularly serves an average of at least 25 individuals daily at least 60 days out of the year.

SYSTEM REQUIREMENTS

Systems utilizing a pressurized water and pesticide injection system must meet the following requirements:

The system must contain a functional check valve, vacuum relief valve, and low pressure drain appropriately located on the irrigation pipeline to prevent water source contamination from backflow.

The pesticide injection pipeline must contain a functional, automatic, quick-closing check valve to prevent the flow of fluid back toward the injection pump.

The pesticide injection pipeline must also contain a functional, normally closed, solenoid-operated valve located on the intake side of the injection pump and connected to the system interlock to prevent fluid from being withdrawn from the supply tank when the irrigation system is either automatically or manually shut down.

The system must contain functional interlocking controls to automatically shut off the pesticide injection pump when the water pump motor stops.

The irrigation line or water pump must include a functional pressure switch which will stop the water pump motor when the water pressure decreases to the point where pesticide distribution is adversely affected.

Systems must use a metering pump, such as a positive displacement injection pump (e.g., diaphragm pump) effectively designed and constructed of materials that are compatible with pesticides and capable of being fitted with a system interlock.

INSTRUCTIONS FOR SPRINKLER (OVERHEAD) IRRIGATION:

Observe the requirements in the System Requirements section above.

Do not apply when wind speed favors drift beyond the area intended for treatment.

Apply PENNCOZEB only through systems containing anti-syphon and check valves designed to prevent water source contamination or overflow of the mix tank and containing interlocking controls between the metering device and the water pump to insure simultaneous shut-off.

Maintain a gentle continuous agitation in mix tank during mixing and application to assure a uniform suspension.

Greater accuracy in calibration and distribution will be achieved by injecting a larger volume of a more dilute suspension per unit time.

Application of more than recommended quantities of irrigation water per acre may result in decreased product performance.

Do not apply when wind speed favors drift, when system connections or fittings leak, when nozzles do not provide uniform distribution or when lines containing the product cannot be flushed and must be dismantled and drained. In a center pivot system, block the nozzle set nearest the well/pivot/injection unit to prevent spray being applied to this area.

Where sprinkler distribution patterns do not overlap sufficiently, unacceptable disease control may result.

Allow sufficient time for pesticide to be flushed through all lines and all nozzles before turning off irrigation water.

PENNCOZEB may be applied in conjunction with chemically neutral liquid fertilizers. Application in conjunction with highly alkaline fertilizers, such as aqueous ammonia, may cause a degradation of the pesticide, resulting in reduced performance and should be avoided.

Check local restrictions and requirements regarding sprinkler irrigation applications, as they may vary from state to state.

SPRAY PREPARATION:

Remove scale, pesticide residues, and other foreign matter from the chemical tank and entire injector system. Flush with clean water.

Prepare a suspension of PENNCOZEB in a mix tank. Fill the tank with 1/2 or 3/4 the desired amount of water. Start mechanical or hydraulic agitation. Slowly add the required amount of PENNCOZEB and then the remaining volume of water.

APPLICATION INSTRUCTIONS:

Set sprinkler system to deliver 0.1 to 1.25 inches of water per acre. Volumes of water higher than this may reduce efficacy. Start sprinkler and then uniformly inject the suspension of PENNCOZEB into the irrigation water line so as to deliver the desired rate per acre. The suspension of PENNCOZEB should be injected with a positive displacement pump into the main line ahead of a right angle turn to insure adequate mixing.

NOTE: When treatment with PENNCOZEB has been completed, do not irrigate the treated area for 24 to 48 hours to prevent washing the chemical off the crop.

PRECAUTIONARY STATEMENTS

HAZARDS TO HUMANS (AND DOMESTIC ANIMALS)
CAUTION

May cause irritation of eyes. Harmful if swallowed, inhaled, or absorbed through skin. Avoid contact with eyes, skin, or clothing, Wash thoroughly with soap and water after handling.

Protective clothing (long pants, long sleeved shirt, gloves, hat and boots) must be worn during mixing and loading.

Do not apply this product in such a manner as to directly or through drift expose workers or other persons. The area being treated must be vacated by unprotected persons.

ENVIRONMENTAL HAZARDS

This pesticide is toxic to fish. Drift and runoff from treated areas may be hazardous to aquatic organisms in neighboring areas. Do not apply directly to water or wetlands (swamps, bogs, marshes, and potholes), except as specified for the labeled use on cranberries. Do not contaminate water when disposing of equipment washwaters.

CHEMICAL HAZARDS

Do not allow to become wet during storage. This may lead to certain chemical changes which will reduce the effectiveness of the product as a fungicide and create vapors which may be flammable. Keep container closed when not in use.

STORAGE AND DISPOSAL

STORAGE INSTRUCTIONS: Store in the original container in a dry area. If allowed to become wet the product will deteriorate and represent a fire hazard. Keep away from sources of ignition (e.g. sparks and open flame). Close bag when not in use. Do not store in a manner where cross-contamination with other pesticides, fertilizers, food or feed could occur. If spilled during storage or handling sweep up spillage and dispose of in accordance with the Pesticide Disposal Instructions listed below.

PESTICIDE DISPOSAL INSTRUCTIONS: Wastes resulting from the use of this product may be disposed of on site or at an approved waste disposal facility.

CONTAINER DISPOSAL INSTRUCTIONS: Completely empty bag into application equipment. Then dispose of bag in a sanitary landfill, by incineration, or if allowed by State and local authorities, by burning. If burned, stay out of smoke.

WARRANTY AND DISCLAIMER

Pennwalt Corporation warrants that this material conforms to the chemical description on the label and is reasonably fit for the purposes referred to in the Directions for Use, subject to the risks referred to therein. PENNWALT MAKES NO OTHER EXPRESS OR IMPLIED WARRANTY OF FITNESS OR MERCHANTABILITY OR ANY OTHER EXPRESS OR IMPLIED WARRANTY. IN NO CASE SHALL PENNWALT OR SELLER BE LIABLE FOR CONSEQUENTIAL, SPECIAL OR INDIRECT DAMAGES RESULTING FROM THE USE OR HANDLING OF THIS PRODUCT INCLUDING, BUT NOT LIMITED TO, LOSS OF PROFITS, BUSINESS REPUTATION, OR CUSTOMERS; LABOR COST; OR OTHER EXPENSES INCURRED IN PLANTING OR HARVESTING.

Pennwalt and seller offer this product and the buyer and user accept it subject to the foregoing conditions of sale and warranty which may be varied only by agreement in writing signed by a duly authorized representative of Pennwalt.

5-N315S-01 (9/89) J9

PENNCOZEB® DF

75% DRY FLOWABLE FUNGICIDE
NOT FOR USE IN CALIFORNIA

A Coordination Product of Manganese and Zinc and Ethylenebisdithiocarbamate

ACTIVE INGREDIENT:

A coordination product of zinc ion and manganese ethylenebisdithiocarbamate	75.0
Manganese + +	15.0%
Zinc + +	1.9%
Ethylenebisdithiocarbamate ion ($C_4H_6N_2S_4$)	58.1%
INERT INGREDIENTS	25.0
TOTAL	100.0%

EPA Registration No. 4581-370 EPA Establishment No. 33910-HL-01

KEEP OUT OF REACH OF CHILDREN
CAUTION

STATEMENT OF PRACTICAL TREATMENT

IF SWALLOWED, call a physician or Poison Control Center. Drink 1 or 2 glasses of water and induce vomiting by touching back of throat with finger. Do not induce vomiting or give anything by mouth to an unconscious person.

IF ON SKIN, wash with plenty of soap and water. Get medical attention.

IF IN EYES, flush eyes with plenty of water. Call a physician if irritation persists.

See Inside Panel for Additional Precautionary Statements.

For Directions For Use and Chemigation Instructions see Following Pages.

PRECAUTIONARY STATEMENTS

Hazards To Humans (And Domestic Animals)
CAUTION

May cause irritation of eyes. Harmful if swallowed, inhaled, or absorbed through skin. Avoid contact with eyes, skin, or clothing. Wash thoroughly with soap and water after handling.

ENVIRONMENTAL HAZARDS

This pesticide is toxic to fish. Drift and runoff from treated areas may be hazardous to aquatic organisms in neighboring areas. Do not apply directly to water or wetlands (swamps, bogs, marshes, and potholes), except as specified for the labeled use on cranberries. Do not contaminate water when disposing of equipment washwaters.

CHEMICAL HAZARDS

Do not allow to become wet during storage. This may lead to certain chemical changes which will reduce the effectiveness of the product as a fungicide and create vapors which may be flammable. Keep container closed when not in use.

DIRECTIONS FOR USE

It is a violation of Federal law to use this product in a manner inconsistent with its labeling.

Do not enter treated areas without protective clothing within 24 hours after application. Because certain states may require more restrictive reentry intervals for various cops treated with this product, consult your State Department of Agriculture for further information.

Written or oral warnings must be given to workers who are expected to be in a treated area or in an area about to be treated with this product. When oral warnings are given, such warnings shall be given in a language customarily understood by the workers. Oral warnings must be given if there is reason to believe that written warnings cannot be understood by the workers. Written warnings must include the following information " 'CAUTION. Area treated with PENNCOZEB 75DF Fungicide on (*date of application*).' "

After sprays have dried/dusts have settled do not enter or allow entry into treated areas until the 24-hour reentry interval has expired unless wearing the personal protective equipment listed below.

WORKER SAFETY RULES

Keep all unprotected persons, children, livestock, and pets away from treated area or where there is danger of drift.

Do not rub eyes or mouth with hands. See Statement of Practical Treatment.

PERSONAL PROTECTIVE EQUIPMENT—For Mixers, Loaders, Applicators and Early Reentry Workers.

HANDLE THIS PRODUCT ONLY WHEN WEARING THE FOLLOWING PROTECTIVE CLOTHING AND EQUIPMENT: a long-sleeve shirt and long pants or a coverall that covers all parts of the body except the head, hands, and feet; chemical resistant gloves; shoes, socks, and goggles or a face shield. During mixing and loading, a chemical resistant apron must also be worn.

During application from a tractor with a completely enclosed cab with positive pressure filtration, or aerially with an enclosed cockpit, a long-sleeve shirt and long pants may be worn in place of the above protective clothing. Chemical resistant gloves must be available in the cab or cockpit and worn while exiting.

IMPORTANT! Before removing gloves, wash them with soap and water. Always wash hands, face, and arms with soap and water before eating, smoking or drinking. Always wash hands and arms with soap and water before using the toilet.

After work take off all clothes and shoes. Shower using soap and water. Wear only clean clothes. Do not use contaminated clothing. Wash protective clothing and protective equipment with soap and water after each use. Personal clothing worn during use must be laundered separately from household articles. Clothing and protective equipment heavily contaminated or drenched with PENNCOZEB must be destroyed according to state and local regulations. HEAVILY CONTAMINATED OR DRENCHED CLOTHING CANNOT BE ADEQUATELY DECONTAMINATED.

During aerial application, human flaggers are prohibited unless in totally enclosed vehicles.

PENNCOZEB 75DF is a dry flowable fungicide containing a coordination product of zinc ion and manganese ethylenebisdithiocarbamate, and is recommended for use as a spray for the control of many important plant diseases.

AS A SPRAY, (Ground or aerial equipment) apply at the rate shown. Use sufficient water to provide thorough coverage, usually 20-100 gallons per acre for ground equipment and 5-8 gallons per acre for aircraft. Use a Spreader Sticker at rate of 4 fl. oz. per 100 gallons spray if needed. Add product slowly to water in the spray tank with agitation, or premix thoroughly in separate holding tank for concentrate or aircraft sprayers. Continuous agitation is required to keep the product in suspension.

NOTE: Where a range of rates is given, use the higher rate under conditions of severe disease pressure.

FIELD CROPS

CROPS	DISEASES	RATE LBS/ ACRE	REMARKS
Barley, Oats, Rye	Helminthosporium Leaf Blight Septoria Leaf Spot	2	Begin applications at tillering to jointing stage and repeat at 7- to 10-day intervals. Do not make more than three applications; do not apply within 26 days of harvest; do not graze treated area within 26 days after application.
Sugar Beets	Cercospora Leaf Spot	1½-2	Begin applications when disease first threatens and repeat at 7- to 10-day intervals. Do not apply within 14 days of harvest.
Peanuts	Cercospora Leaf Spot	1-2	Begin applications when disease first appears and repeat at 10- to 14-day intervals, or at 7-day intervals during humid weather. If peanut vine hay is to be used for livestock feed, do not use more than 1½ pounds per acre per application. Do not apply within 14 days of harvest.
Cotton (Southwest only)	Rust	1¼-2	Begin applications when rust first appears in the area and repeat at 10- to 14-day intervals. Do not apply after bolls open; do not graze treated areas or feed gin trash to livestock.
Wheat	Helminthosporium Leaf Spots Septoria Leaf Spots Septoria Glume Blotch Leaf Rust Tan Spot	2	Start application at onset of disease or when plants are in tillering to jointing stage and repeat at 7- to 10-day intervals. Do not make more than 3 applications in one season. Do not apply within 26 days of harvest. Do not graze livestock in treated areas within 26 days after application.

VEGETABLE CROPS

CROPS	DISEASES	RATE LBS/ ACRE	REMARKS
Potatoes	Early and Late Blights	1-2	Begin applications when plants are 4-6 inches high, and repeat at 7- to 10-day intervals (more frequently under severe disease conditions) throughout the season.
Potato Tuber Treatment	Fusarium Seed Piece Decay		Dip whole or cut seed pieces in mixture of 1¼ lbs. of product per 50 gallons of water, drain and plant immediately. If seed pieces are to be held before planting, spread out to dry in a cool place. Do not use treated seed pieces for food or feed purposes.
Tomatoes	Early and Late Blights Gray Leaf Spot Gray Leaf Mold Anthracnose	1½-3	Begin applications when seedlings emerge and repeat at 7-day intervals throughout the season. Do not apply within 5 days of harvest.
Cucumbers, Melons, Summer Squash	Anthracnose Alternaria Leaf Spot Cercospora Leaf Spot Downy Mildew Gummy Stem Blight Scab	2-3	Begin applications at the 2-leaf stage and repeat at 5- to 7-day intervals. Do not apply within 5 days of harvest.
Corn: Sweet and Popcorn	Helminthosporium Leaf Blight	1½	Apply PENNCOZEB in sufficient water for thorough coverage. Start applications when disease first appears and repeat at 4- to 7-day intervals. Do not apply within 7 days of harvest. Do not feed treated forage (plant stalk and leaves) to livestock.
Corn: Field	Helminthosporium Leaf Blight	1½	Apply PENNCOZEB in sufficient water for thorough coverage. Start applications when disease first appears and repeat at 4- to 7-day intervals. Do not apply within 40 days of harvest.
Onions (Dry Bulb)	Botrytis Leaf Blight Downy Mildew Neck Rot Purple Blotch	2-3	Begin applications when diseases are first reported in the area and repeat at 7-day intervals thoughout the season. Do not apply within 7 days of harvest; do not apply to exposed bulbs.
Celery, Fennel	Early and Late Blights	2	Begin applications in plant beds as soon as plants emerge; repeat at 3- to 5-day intervals in plant bed, and at 7-day intervals after plants are set in field. Do not apply within 14 days of harvest.
Carrots	Alternaria Leaf Spot Cercospora Leaf Spot	1½-2	Begin applications when disease first appears and repeat at 7- to 10-day intervals. Do not apply within 7 days of harvest. Do not use treated tops for food or feed purposes.
Asparagus	Rust	2	Begin applications when rust first appears and repeat at 10-day intervals as needed. Use only on ferns after spears have been harvested.

FRUIT CROPS

CROPS	DISEASES	RATE LBS/ ACRE	REMARKS
Apples, Pears, Crabapples, Quince	Bitter Rot Black Rot Brown Rot Cedarapple Rust Fly Speck Sooty Blotch Scab	1-2 per 100 gal finished spray	Apply up to 10 lbs. of product per acre per application (for low volume sprays, adjust accordingly). Apply at 7- to 14-day intervals throughout the season. For scab—begin at green tip or delayed dormant stage; for cedarapple rust, at blossom time. For other diseases, at first cover. Add Spreader Sticker to spray mixture for last 3 cover sprays.
			FOR APPLES: Do not apply within 30 days of harvest in Northern states

CROPS	DISEASES	RATE LBS/ ACRE	REMARKS
			or 21 days in Southern states. (Southern area applies to Arkansas, Delaware, Georgia, Illinois, Indiana, Kansas, Kentucky, Maryland, Missouri, New Jersey, North Carolina, Ohio, Pennsylvania, South Carolina, Tennessee, Virginia, and West Virginia. Northern states applies to New England, New York, Michigan, Far West and other states not included under "Southern States"). Do not graze livestock in treated areas.
			FOR PEARS, CRABAPPLES, AND QUINCE: Do not apply within 15 days of harvest and do not graze livestock in treated areas.
Cranberries	Fruit Rot	3-6	Begin applications at mid-bloom and repeat at 7- to 10-day intervals. Do not apply within 30 days of harvest.
Grapes	Black Rot Bunch Rot Dead Arm Downy Mildew	$1\frac{1}{2}$-4	Apply when shoots are $\frac{1}{2}$ to $1\frac{1}{2}$ inches long, when shoots are 3-5 inches long and when shoots are 8-10 inches long. Continue at 10- to 14-day intervals, but do not apply within 66 days of harvest. In California, do not apply after fruit set.
Papayas	Anthracnose Phytophthora Fruit Rot	2-$2\frac{1}{2}$	Add Spreader Sticker to spray mixture. Begin applications at flowering time; repeat at 14-day intervals or at 7-day intervals under severe disease conditions. Treat crown and blossom area.

CHEMIGATION

NOT FOR USE IN CALIFORNIA

GENERAL INSTRUCTIONS

Apply this product only through sprinkler irrigation systems including center pivot, lateral move, end tow, side (wheel) roll, traveler, big gun, solid set, or hand move. Do not apply this product through any other type of irrigation system.

Crop injury, lack of effectiveness, or illegal pesticide residues in the crop can result from nonuniform distribution of treated water.

If you have any questions about calibration, you should contact State Extension Service specialists, equipment manufacturers or other experts.

A person knowledgeable of the chemigation system and responsible for its operation, or under the supervision of the responsible person, shall shut the system down and make necessary adjustments should the need arise.

Do not connect chemigation system to any public water system. Public water system means a system for the provision of piped water for human consumption if such system has at least 15 service connections or regularly serves an average of at least 25 individuals daily at least 60 days out of the year.

SYSTEM REQUIREMENTS

Systems utilizing a pressurized water and pesticide injection system must meet the following requirements:

The system must contain a functional check valve, vacuum relief valve, and low pressure drain appropriately located on the irrigation pipeline to prevent water source contamination from backflow.

The pesticide injection pipeline must contain a functional, automatic, quick-closing check valve to prevent the flow of fluid back toward the injection pump.

The pesticide injection pipeline must also contain a functional, normally closed, solenoid-operated valve located on the intake side of the injection pump and connected to the system interlock to prevent fluid from being withdrawn from the supply tank when the irrigation system is either automatically or manually shut down.

The system must contain functional interlocking controls to automatically shut off the pesticide injection pump when the water pump motor stops.

The irrigation line or water pump must include a functional pressure switch which will stop the water pump motor when the water pressure decreases to the point where pesticide distribution is adversely affected.

Systems must use a metered pump, such as a positive displacement injection pump (e.g., diaphragm pump) effectively designed and constructed of materials that are compatible with pesticides and capable of being fitted with a system interlock.

INSTRUCTIONS FOR SPRINKLER (OVERHEAD) IRRIGATION:

Observe the requirements in the System Requirements section above.

Do not apply when wind speed favors drift beyond the area intended for treatment.

Apply PENNCOZEB only through systems containing anti-syphon and check valves designed to prevent water source contamination or overflow of the mix tank and containing interlocking controls between the metering device and the water pump to insure simultaneous shut-off.

Maintain a gentle continuous agitation in mix tank during mixing and application to assure a uniform suspension.

Greater accuracy in calibration and distribution will be achieved by injecting a larger volume of a more dilute suspension per unit time.

Application of more than recommended quantities of irrigation water per acre may result in decreased product performance.

Do not apply when wind speed favors drift, when system connections or fittings leak, when nozzles do not provide uniform distribution or when lines containing the product cannot be flushed and must be dismantled and drained. In a center pivot system, block the nozzle set nearest the well/pivot/injection unit to prevent spray being applied to this area.

Where sprinkler distribution patterns do not overlap sufficiently, unacceptable disease control may result.

Allow sufficient time for pesticide to be flushed through all lines and all nozzles before turning off irrigation water.

PENNCOZEB may be applied in conjunction with chemically neutral liquid fertilizers. Application in conjunction with highly alkaline fertilizers, such as aqueous ammonia, may cause a degradation of the pesticide, resulting in reduced performance and should be avoided.

Check local restrictions and requirements regarding sprinkler irrigation applications, as they may vary from state to state.

SPRAY PREPARATION:

Remove scale, pesticide residues, and other foreign matter from the chemical tank and entire injector system. Flush with clean water.

Prepare a suspension of PENNCOZEB in a mix tank. Fill the tank with $\frac{1}{2}$ or $\frac{3}{4}$ the desired amount of water. Start mechanical or hydraulic agitation. Slowly add the required amount of PENNCOZEB and then the remaining volume of water.

APPLICATION INSTRUCTIONS:

Set sprinkler system to deliver 0.1 to 1.25 inches of water per acre. Volumes of water higher than this may reduce efficacy. Start sprinkler and then uniformly inject the suspension of PENNCOZEB into the irrigation water line so as to deliver the desired rate per acre. The suspension of PENNCOZEB should be injected with a positive displacement pump into the main line ahead of a right angle turn to insure adequate mixing.

NOTE: When treatment with PENNCOZEB has been completed, do not irrigate the treated area for 24 to 48 hours to prevent washing the chemical off the crop.

STORAGE AND DISPOSAL

STORAGE INSTRUCTIONS: Store in the original container in a dry area. If allowed to become wet the product will deteriorate and represent a fire hazard. Keep away from sources of ignition (e.g., sparks and open flame). Close bag when not in use. Do not store in a manner where cross-contamination with other pesticides, fertilizers, food or feed could occur. If spilled during storage or handling sweep up spillage and dispose of in accordance with the Pesticide Disposal Instructions listed below.

PESTICIDE DISPOSAL INSTRUCTIONS: Wastes resulting from the use of this product may be disposed of on site or at an approved waste disposal facility.

CONTAINER DISPOSAL INSTRUCTIONS: Completely empty bag into application equipment. Then dispose of bag in a sanitary landfill, by incineration, or if allowed by State and local authorities, by burning. If burned, stay out of smoke.

WARRANTY AND DISCLAIMER

Pennwalt Corporation warrants that this material conforms to the chemical description on the label and is reasonably fit for the purposes referred to in the Directions for Use, subject to the risks referred to therein. PENNWALT MAKES NO OTHER EXPRESS OR IMPLIED WARRANTY OF FITNESS OR MERCHANTABILITY OR ANY OTHER EXPRESS OR IMPLIED WARRANTY. IN NO CASE SHALL PENNWALT OR SELLER BE LIABLE FOR CONSEQUENTIAL, SPECIAL OR INDIRECT DAMAGES RESULTING FROM THE USE OR HANDLING OF THIS PRODUCT INCLUDING, BUT NOT LIMITED TO, LOSS OF PROFITS, BUSINESS REPUTATION, OR CUSTOMERS; LABOR COST; OR OTHER EXPENSES INCURRED IN PLANTING OR HARVESTING.

Pennwalt and seller offer this product and the buyer and user accept it subject to the foregoing conditions of sale and warranty which may be varied only by agreement in writing signed by a duly authorized representative of Pennwalt.

Made and Printed in Holland 5-Z318S-01 3/89 (C9)

PENNCOZEB®

TURF AND ORNAMENTALS FUNGICIDE

An 80% Coordination Product of Manganese and Zinc and Ethylenebisdithiocarbamate

ACTIVE INGREDIENT:
A coordination product of zinc ion and manganese bisdithiocarbamate 80%
Manganese++ 16%
Zinc++ 2%
Ethylenebisdithiocarbamate ion 62%
($C_4H_6N_2S_4$)
INERT INGREDIENTS: 20%
TOTAL. 100%

EPA Registration No. 4581-357 EPA Establishment No. 33910-HL-01

KEEP OUT OF REACH OF CHILDREN

CAUTION

STATEMENT OF PRACTICAL TREATMENT

IF SWALLOWED, call a physician or Poison Control Center. Drink 1 or 2 glasses of water and induce vomiting by touching back of throat with finger. Do not induce vomiting or give anything by mouth to an unconscious person.

IF ON SKIN, wash with plenty of soap and water. Get medical attention.

IF IN EYES, flush eyes with plenty of water. Call a physician if irritation persists.

PRECAUTIONARY STATEMENTS

HAZARDS TO HUMANS (AND DOMESTIC ANIMALS)

CAUTION

May cause irritation of eyes. Harmful if swallowed, inhaled, or absorbed through skin. Avoid contact with eyes, skin, or clothing. Wash thoroughly with soap and water after handling.

Protective clothing (long pants, long sleeved shirt, gloves, hat and boots) must be worn during mixing and loading.

Do not apply this product in such a manner as to directly or through drift expose workers or other persons. The area being treated must be vacated by unprotected persons.

ENVIRONMENTAL HAZARDS

This product is toxic to fish. Keep out of any body of water. Do not apply where runoff is likely to occur. Do not contaminate water by disposal of equipment washwaters. Do not apply when weather conditions favor drift from treated areas.

CHEMICAL HAZARDS

Do not allow to become wet during storage. This may lead to certain chemical changes which will reduce the effectiveness of the product as a fungicide and create vapors which may be flammable. Keep container closed when not in use.

REENTRY STATEMENT

Do not enter treated areas without protective clothing within 24 hours after application. Because certain states may require more restrictive reentry intervals for various crops treated with this product, consult your State Department of Agriculture for further information.

Written or oral warnings must be given to workers who are expected to be in a treated area or in an area about to be treated with this product. When oral warnings are given, such warnings shall be given in a language customarily understood by the workers. Oral warnings must be given if there is reason to believe that written warnings cannot be understood by the workers. Written warnings must include the following information " 'CAUTION. Area treated with PENNCOZEB TURF AND ORNAMENTALS FUNGICIDE on (*date of application*).' "

After sprays have dried dusts have settled do not enter or allow entry into treated areas until the 24-hour reentry interval has expired unless wearing the personal protective equipment listed below.

WORKER SAFETY RULES

Keep all unprotected persons, children, livestock, and pets away from treated area or where there is danger of drift.

Do not rub eyes or mouth with hands. See Statement of Practical Treatment.

PERSONAL PROTECTIVE EQUIPMENT—For Mixers, Loaders, Applicators and Early Reentry Workers.

HANDLE THIS PRODUCT ONLY WHEN WEARING THE FOLLOWING PROTECTIVE CLOTHING AND EQUIPMENT: a long-sleeve shirt and long pants or a coverall that covers all parts of the body except the head, hands, and feet; chemical resistant gloves, shoes, socks, and goggles or a face shield. During mixing and loading, a chemical resistant apron must also be worn.

During application from a tractor with a completely enclosed cab with positive pressure filtration, or aerially with an enclosed cockpit, a long-sleeve shirt and long pants may be worn in place of the above protective clothing. Chemical resistant gloves must be available in the cab or cockpit and worn while exiting.

IMPORTANT! Before removing gloves, wash them with soap and water. Always wash hands, face, and arms with soap and water before eating, smoking or drinking. Always wash hands and arms with soap and water before using toilet.

After work take off all clothes and shoes. Shower using soap and water. Wear only clean clothes. Do not use contaminated clothing. Wash protective clothing and protective equipment with soap and water after each use. Personal clothing worn during use must be laundered separately from household articles. Clothing and protective equipment heavily contaminated or drenched with PENNCOZEB must be destroyed according to state and local regulations. HEAVILY CONTAMINATED OR DRENCHED CLOTHING CANNOT BE ADEQUATELY DECONTAMINATED.

During aerial application, human flaggers are prohibited unless in totally enclosed vehicles.

DIRECTIONS FOR USE

It is a violation of Federal law to use this product in a manner inconsistent with its labeling.

PENNCOZEB TURF AND ORNAMENTALS FUNGICIDE is a wettable powder containing a coordination product of zinc ion and manganese ethylenebisdithiocarbamate, and is recommended for use as a spray or dust for the control of many important plant diseases. When used according to directions, it provides very high fungicidal activity and can be safely used on both turf grasses and certain ornamentals. PENNCOZEB provides control of most common turf grass diseases and it is also effective in controlling major fungus diseases of certain ornamentals.

Diseases of turf grass and ornamentals can attack suddenly and unexpectedly causing severe damage and may even result in total loss of large areas of valuable turf grass and ornamental plants. The use of a regular protective spray program will minimize the risk of disease damage and can generally be accomplished with lower rates and less frequent fungicide applications. Once diseases have become established, higher rates of fungicide and more frequent applications are required to bring them under control. Follow a regular protective program for maximum product performance.

To mix, slowly sift PENNCOZEB TURF AND ORNAMENTALS FUNGICIDE into spray tank while it is being filled and the agitator is running. Add insecticide and nutritional sprays last. PENNCOZEB TURF AND ORNAMENTALS FUNGICIDE is compatible with most commonly used pesticides. Read and observe the precautionary statements and other information appearing on product labels used in mixtures.

AS A SPRAY, (Ground or aerial equipment) apply at the rate shown. Use sufficient water to provide thorough coverage.

AS A DUST, apply product at the rate shown, using proper amount of dust containing 6% to 10% PENNCOZEB (4.8% to 8% active). Dust may be prepared by diluting and thoroughly mixing product with pyrophylite or other neutral diluent; commonly used insecticides may displace an equivalent amount of diluent. Use dust mixtures as soon as possible after preparation.

CROP	DISEASES	RATE OF PENNCOZEB PER APPLICATION OZ./1000 SQ. FT.	REMARKS
TURF			Refer to General Use Instructions except when more specific directions are given for individual crops or diseases.
GENERAL USE INSTRUCTIONS			Start applications when grass greens up in the spring or when disease first appears and repeat at 7 to 14 day intervals or until disease threat is past. When conditions are especially favorable for disease development, apply maximum fungicide use rate on a 7 day spray schedule. Apply in sufficient water for adequate coverage. Do not graze treated areas. Do not feed clippings to livestock.

CROP	DISEASES	RATE OF PENNCOZEB PER APPLICATION OZ./1000 SQ. FT.	REMARKS
	Copper Spot Red Thread Fusarium Blight *(F. Roseum)* Slime Molds	4-8	
	Helminthosporium Melting Out Rhizoctonia Brown Patch Rust	4	Apply on a 7 day schedule for Rhizoctonia.
	Pythium Blight	8	Apply at 5 day intervals or more frequently if conditions are especially favorable for disease development.
	Dollar Spot	6-8	A cadmium fungicide may be added to supplement Dollar Spot control under severe disease conditions.
	Algae	6	
	Fusarium Snow Mold	6-8	Apply at 2 week intervals during winter.

CROP	DISEASES	REMARKS
ORNAMENTALS		Refer to General Use Instructions except when more specific directions are given for individual crops or diseases.
GENERAL USE INSTRUCTIONS		Apply the equivalent of 1½ lbs. PENNCOZEB TURF AND ORNAMENTALS FUNGICIDE per 100 gallons of water (1½ level tablespoons per gallon) in full coverage spray. To improve performance, add 2 to 4 oz. spreader-sticker per 100 gallons of spray. Begin spraying when plants are well leafed out or at first sign of disease and apply at 7 to 10 day intervals throughout the season, or follow State Extension Service recommendations for disease control on the following ornamental plants:
Azaleas	Petal Blight	Spray 2 to 3 times a week while flowers are opening. Direct spray into flowers and spray ground under bushes thoroughly.
Camelias	Petal Blight	Refer to azaleas.
Carnation	Leaf Spot	
Chrysanthemum	Botrytis Petal Spot	Apply at 1 to 2 lbs. per 100 gallons or 1½ level tablespoons per gallon in full coverage spray. Apply twice weekly during the blooming season.
Crabapple (ornamental)	Scab Cedar-Apple Rust	
Dahlia	Botrytis Blight	
Dogwood, Flowering	Anthracnose	Apply when buds begin to open, when bracts have fallen, 4 weeks later, and again in late summer after flower buds for next season are formed.
Gladiolus	Curvularia Leaf Spot Botrytis Blossom Blight	On flower spikes, use at ¾ lb. per 100 gallons. Make weekly applications starting before diseases appear and increase to 2 or 3 applications per week during periods of heavy disease and during rainy weather.
Holly	Purple Spot	
Honeysuckle	Herpobasidium Blight	
Iris	Leaf Spot	
Lily	Botrytis Blight	
Pachysandra	Volutella Blight	Use a drenching spray of 2 lbs. per 50 gallons of water per 5,000 sq. ft. of bed area. Start application at first sign of disease and make at least 5 applications at 10 to 14 day intervals.
Pansy	Anthracnose	
Peony	Phythophthora Blight Botrytis Blight	Apply in early spring and early fall, drenching soil around plants as well as the foliage. Promptly destroy all infected plant parts.
Rhododendron	Petal Blight	Refer to azalea.
Roses	Black Spot	
Snapdragon	Rust	
Tulip	Botrytis Blight (fire)	
Zinnia	Leaf Blight	

CHEMIGATION

NOT FOR THIS USE IN CALIFORNIA
GENERAL INSTRUCTIONS

Apply this product only through sprinkler irrigation systems including center pivot, lateral move, end tow, side (wheel) roll, traveler, big gun, solid set, or hand move. Do not apply this product through any other type of irrigation system.

Crop injury, lack of effectiveness, or illegal pesticide residues in the crop can result from nonuniform distribution of treated water. If you have any questions about calibration, you should contact State Extension Service specialists, equipment manufacturers or other experts.

A person knowledgeable of the chemigation system and responsible for its operation, or under the supervision of the responsible person, shall shut the system down and make necessary adjustments should the need arise.

Do not connect chemigation system to any public water system. Public water system means a system for the provision of piped water for human consumption if such system has at least 15 service connections or regularly serves an average of at least 25 individuals daily at least 60 days out of the year.

SYSTEM REQUIREMENTS

Systems utilizing a pressurized water and pesticide injection system must meet the following requirements:

The system must contain a functional check valve, vacuum relief valve, and low pressure drain appropriately located on the irrigation pipeline to prevent water source contamination from backflow.

The pesticide injection pipeline must contain a functional, automatic, quick-closing check valve to prevent the flow of fluid back toward the injection pump.

The pesticide injection pipeline must also contain a functional, normally closed, solenoid-operated valve located on the intake side of the injection pump and connected to the system interlock to prevent fluid from being withdrawn from the supply tank when the irrigation system is either automatically or manually shut down.

The system must contain functional interlocking controls to automatically shut off the pesticide injection pump when the water pump motor stops.

The irrigation line or water pump must include a functional pressure switch which will stop the water pump motor when the water pressure decreases to the point where pesticide distribution is adversely affected.

Systems must use a metering pump, such as a positive displacement injection pump (e.g., diaphragm pump) effectively designed and constructed of materials that are compatible with pesticides and capable of being fitted with a system interlock.

INSTRUCTIONS FOR SPRINKLER (OVERHEAD) IRRIGATION:

Observe the requirements in the System Requirements section above.

Do not apply when wind speed favors drift beyond the area intended for treatment.

Apply PENNCOZEB only through systems containing anti-syphon and check valves designed to prevent water source contamination or overflow of the mix tank and containing interlocking controls between the metering device and the water pump to insure simultaneous shut-off.

Maintain a gentle continuous agitation in mix tank during mixing and application to assure a uniform suspension.

Greater accuracy in calibration and distribution will be achieved by injecting a larger volume of a more dilute suspension per unit time.

Application of more than recommended quantities of irrigation water per acre may result in decreased product performance.

Do not apply when wind speed favors drift, when system connections or fittings leak, when nozzles do not provide uniform distribution or when lines containing the product cannot be flushed and must be dismantled and drained. In a center pivot system, block the nozzle set nearest the well pivot injection unit to prevent spray being applied to this area.

Where sprinkler distribution patterns do not overlap sufficiently, unacceptable disease control may result.

Allow sufficient time for pesticide to be flushed through all lines and all nozzles before turning off irrigation water.

PENNCOZEB may be applied in conjunction with chemically neutral liquid fertilizers. Application in conjunction with highly alkaline fertilizers, such as aqueous ammonia, may cause a degradation of the pesticide, resulting in reduced performance and should be avoided.

Check local restrictions and requirements regarding sprinkler irrigation applications, as they may vary from state to state.

SPRAY PREPARATION:

Remove scale, pesticide residues, and other foreign matter from the chemical tank and entire injector system. Flush with clean water.

Prepare a suspension of PENNCOZEB in a mix tank. Fill the tank with 1/2 or 3/4 the desired amount of water. Start mechanical or hydraulic agitation.

Slowly add the required amount of PENNCOZEB and then the remaining volume of water.

APPLICATION INSTRUCTIONS:

Set sprinkler system to deliver 0.1 to 1.25 inches of water per acre. Volumes of water higher than this may reduce efficacy. Start sprinkler and then uniformly inject the suspension of PENNCOZEB into the irrigation water line so as to deliver the desired rate per acre. The suspension of PENNCOZEB should be injected with a positive displacement pump into the main line ahead of a right angle turn to insure adequate mixing.

NOTE: When treatment with PENNCOZEB has been completed, do not irrigate the treated area for 24 to 48 hours to prevent washing the chemical off the crop.

STORAGE AND DISPOSAL

Do not contaminate water, food or feed by storage or disposal.

STORAGE INSTRUCTIONS: Store in the original container in a dry area. If allowed to become wet the product will deteriorate and represent a fire hazard. Keep away from sources of ignition (e.g. sparks and open flame). Close bag when not in use. Do not store in a manner where cross-contamination with other pesticides, fertilizers, food or feed could occur. If spilled during storage or handling sweep up spillage and dispose of in accordance with the Pesticide Disposal Instructions listed below.

PESTICIDE DISPOSAL INSTRUCTIONS: Wastes resulting from the use of this product may be disposed of on site or at an approved waste disposal facility.

CONTAINER DISPOSAL INSTRUCTIONS: Completely empty bag into application equipment. Then dispose of bag in a sanitary landfill, by incineration, or if allowed by State and local authorities, by burning. If burned, stay out of smoke.

WARRANTY AND DISCLAIMER

Pennwalt Corporation warrants that this material conforms to the chemical description on the label and is reasonably fit for the purposes referred to in the Directions for Use, subject to the risks referred to therein. PENNWALT MAKES NO OTHER EXPRESS OR IMPLIED WARRANTY OF FITNESS OR MERCHANTABILITY OR ANY OTHER EXPRESS OR IMPLIED WARRANTY. IN NO CASE SHALL PENNWALT OR SELLER BE LIABLE FOR CONSEQUENTIAL, SPECIAL OR INDIRECT DAMAGES RESULTING FROM THE USE OR HANDLING OF THIS PRODUCT INCLUDING, BUT NOT LIMITED TO, LOSS OF PROFITS, BUSINESS REPUTATION, OR CUSTOMERS; LABOR COST; OR OTHER EXPENSES INCURRED IN PLANTING OR HARVESTING.

Pennwalt and seller offer this product and the buyer and user accept it subject to the foregoing conditions of sale and warranty which may be varied only by Agreement in writing signed by a duly authorized representative of Pennwalt.

5-N316S-01 L6 (8 85)

TOPSIN® M

THIOPHANATE-METHYL FUNGICIDE

70% Wettable Powder

ACTIVE INGREDIENT:	
Thiophanate-methyl (dimethyl[(1,2-phenylene)-bis(iminocarbonothioyl)] bis[carbamate])*	70%
INERT INGREDIENTS:	30%
TOTAL	100%

*Also known as dimethyl 4,4'-o-phenylenebis[3-thioallophanate]

®TOPSIN is a registered trademark of Nippon Soda Company, Ltd., and is licensed to Pennwalt Corporation, and is covered by one or more of the following U.S. Patents: 3,769,308; 3,856,847; 4,020,095; and 4,029,813.

EPA Registration No. 4581-322 EPA Establishment No. 39578-TX-1

KEEP OUT OF REACH OF CHILDREN

CAUTION

HAZARDS TO HUMANS (AND DOMESTIC ANIMALS)

HARMFUL IF SWALLOWED, ABSORBED THROUGH SKIN, OR INHALED. CAUSES MODERATE EYE IRRITATION.

IF SWALLOWED, induce vomiting immediately by giving two glasses of water and sticking finger down throat. Call a physician. Never give anything by mouth to an unconscious person.

IF IN EYES, immediately flush eyes with plenty of water for at least 15 minutes. Get medical attention if irritation persists.

IF ON SKIN, immediately flush skin with plenty of water. Get medical attention if irritation persists.

Avoid contact with skin, eyes, or clothing. Avoid breathing dust or spray mist. Wash thoroughly with soap and water after handling. Remove contaminated clothing and wash before reuse.

Do not apply this product in such a manner as to directly or through drift expose workers or other persons. The area being treated must be vacated by unprotected persons.

ENVIRONMENTAL HAZARDS

This pesticide is toxic to fish. Do not apply directly to water or wetlands. Do not apply where runoff is likely to occur. Do not contaminate water when disposing of equipment washwater.

GENERAL INFORMATION AND DIRECTIONS FOR USE

GENERAL INFORMATION

Apply TOPSIN M with ground or aerial equipment, using sufficient volume of spray to provide thorough coverage. Add required amount of TOPSIN M to partially filled tank agitated by mechanical or hydraulic means, and then add remaining required amount of water. Continuous agitation is required to keep the material in suspension. Do not tank mix with highly alkaline pesticides, such as Bordeaux mixture or lime sulfur. No claim of compatability with other pesticides is implied. Use the higher rate under conditions of severe disease pressure. Also, see local State Extension Service recommendations for application schedules.

IMPORTANT: If, after using TOPSIN M as recommended, treatment is not effective, a tolerant strain of fungi may be present. Consult your State Agricultural Experiment Station or your State Agricultural Extension Service for advice on the prompt use of some other suitable fungicide.

DIRECTIONS FOR USE

It is a violation of Federal law to use this product in a manner inconsistent with its labeling.

Chemigation instructions follow tank mix instructions. Do not apply through any irrigation system unless these instructions are followed.

Do not enter treated areas without protective clothing until spray has dried. Because certain states may require more restrictive reentry intervals for various crops treated with this product, consult your State Department of Agriculture for further information.

Written or oral warnings must be given to workers who are expected to be in a treated area or in an area about to be treated with this product. When oral warnings are given, such warnings shall be given in a language customarily understood by the workers. Oral warnings must be given if there is reason to believe that written warnings cannot be understood by the workers. Written warnings must include the following information: "CAUTION. Area treated with TOPSIN M 70W Thiophanate-methyl Fungicide on *(date of application)*." Do not enter without protective clothing until sprays have dried.

Crop	Diseases	Lbs/ Acre	Lbs/ 100 Gal	Remarks
Almonds	Brown Rot Blossom Blight (**Monilinia** sp.)	1½-2		Apply at early pink bud and early petal fall.
Apples (pre-harvest)	Apple Scab (**Venturia** sp.) Bitter Rot (**Glomerella** sp.) Flyspeck (**Schizothyrium** sp.) Powdery Mildew (**Podosphaera** sp.) Sooty Blotch (**Gloeodes** sp.) Black Rot (**Physalospora** sp.)	1-1½	¼-⅜	Apply at 5- to 10-day intervals from green tip through petal fall; continue at 7- to 14-day intervals in cover sprays. NOTE: For control of apple diseases in the Southeastern States of Florida, Georgia, South Carolina, North Carolina and Alabama, use 3 to 4 pounds TOPSIN M per acre.
Apples (post-harvest)	Postharvest Fruit Rots (**Botrytis** sp.) (**Gloeosporium** sp.) (**Penicillium** spp.)		½-1 (420-840 ppm a.i.)	Dip or spray fruit thoroughly as soon as possible after harvest. Do not dip for more than 2 minutes. Double the concentration of TOPSIN M when incorporated in wax and applied in flow-through sprays to the fruit.
Beans	Gray Mold (**Botrytis** sp.) White Mold (**Sclerotinia** sp.)	1½-2 **OR** 1-1½	1-1⅓ **OR** ⅔-1	Apply once at 50% to 70% of full bloom. **OR** Apply twice with the first application at 10% to 30% of full bloom and a second application at 4 to 7 days later or at peak bloom. Do not apply to snap or dry beans within 14 days of harvest or to lima beans within 28 days of harvest.
Celery	Early Blight (**Cercospora** sp.) Late Blight (**Septoria** sp.)	½		Apply to the base of the plants on a 14- to 21-day schedule beginning about 2 weeks after transplanting. **OR** Apply as a foliar spray every 7 to 14 days beginning about 2 weeks after transplanting. Do not apply within 7 days of harvest. Do not make more than 4 applications per season.
Cucurbits (Cucumbers, melons, pumpkins, summer and winter squash) (Not for this use in California)	Anthracnose Gummy Stem Blight Powdery Mildew Target Spot (Cucumbers)	¼-½ for ground applications OR ½ for aerial applications		Begin applications when plants begin to run or when disease first appears, and repeat at 7- to 14-day intervals as needed. For Target Spot, use 7-day intervals as needed.
Onions	White Rot (**Sclerotium** sp.)	½-⅓ lb 1000 ft. of row (12 inch row spacing)		Apply TOPSIN M by spraying directly into the open furrow at the time of planting seed, sets or bulbs. Not for this use through any type of irrigation system.
Peanuts	Leaf Spot (**Cercospora** spp.)	½		Begin applications 35 days after planting or when disease first appears and repeat at 14- to 21-day intervals as needed. Use the 14-day interval under severe disease pressure. Do not apply within 14 days of harvest.
Pecans	Brown Spot (**Cercospora** sp.) Downy Spot (**Mycosphaerella** sp.) Powdery Mildew (**Microsphaera** sp.) Scab (**Fusicladium** sp.)	½-1		Use the higher rate for trees over 30 feet tall and for aerial applications in Arkansas, Louisiana, Mississippi, Oklahoma, and Texas. Begin applications when first leaves are showing and repeat at 3- to 4-week intervals until shuck split. Do not apply after shuck split.
Soybeans	Anthracnose (**Colletotrichum** sp.) Brown Spot (**Septoria** sp.) Frogeye Leaf Spot (**Cercospora** sp.) Pod and Stem Blight (**Diaporthe** sp. and the imperfect stage, **Phomopsis** sp.) Purple Seed Stain (**Cercospora** sp.)	½-1		Apply from full bloom to when pods are ⅛" to ¼" in length. Make a second application 14 to 21 days later. Do not make the second application later than 14 days after pods average ¼" in length or when beans form in the pod. Use the high rate under severe disease pressure. FOR SEED BEANS ONLY—For seed quality make a single application at the high rate when beans form in the pod. Do not make more than 2 applications per year. Do not graze or feed treated vines or hay to livestock.
Stone Fruits (pre-harvest) —Apricots	Brown Rot Blossom Blight (**Monilinia** sp.) Fruit Brown Rot (**Monilinia** sp.)	1½	½	Apply at early bloom (red bud). Make a second application at full bloom. In addition, for Fruit Brown Rot, apply 1 or 2 sprays starting 3 weeks before harvest. If needed under severe disease pressure, apply additional sprays at 10- to 14-day intervals between full bloom and final pre-harvest sprays. Do not apply within 1 day of harvest.
—Cherries	Brown Rot Blossom Blight (**Monilinia** sp.) Fruit Brown Rot (**Monilinia** sp.)	1½	½	Apply at early bloom (early popcorn). Make a second application at full bloom. In addition, for Fruit Brown Rot, apply 1 or 2 sprays starting 3 weeks before harvest. If needed under severe disease pressure, apply additional sprays at 10- to 14-day intervals between full bloom and final preharvest sprays. Do not apply within 1 day of harvest.

Crop	Diseases	Lbs/ Acre	Lbs/ 100 Gal	Remarks
	Cherry Leaf Spot (**Coccomyces** sp.)	1 1/8-1 1/2	3/8-1/2	Apply at petal fall or before, when leaves first unfold, and at first, second and third cover at 10- to 14-day intervals and one spray 14 to 21 days after harvest. Do not apply within 1 day of harvest.
	Powdery Mildew (**Podosphaera** sp. and **Sphaerotheca** sp.)	1 1/2 **PLUS** 1 1/8-1 1/2	1/2 **PLUS** 3/8-1/2	Apply at early bloom (early popcorn). Make a second application at full bloom. **PLUS** Apply at shuck fall and first cover. Do not apply within 1 day of harvest.
—Nectarines	Brown Rot Blossom Blight (**Monilinia** sp.) Fruit Brown Rot (**Monilinia** sp.)	1 1/2	1/2	Apply at early bloom (pink bud). Make a second application at full bloom if conditions favor disease development. In addition, for Fruit Brown Rot, apply 1 or 2 sprays starting 3 weeks before harvest. If needed under severe disease pressure, apply additional sprays at 10- to 14-day intervals between full bloom and final preharvest sprays. Do not apply within 1 day of harvest.
—Peaches	Brown Rot Blossom Blight (**Monilinia** sp.) Fruit Brown Rot (**Monilinia** sp.)	1 1/2-2 1/4	1/2-3/4	Apply at early bloom (pink bud). Make a second application at full bloom if conditions favor disease development. In addition, for Fruit Brown Rot, apply 1 or 2 sprays starting 3 weeks before harvest. If needed under severe disease pressure, apply additional sprays at 10- to 14-day intervals between full bloom and final preharvest sprays. Do not apply within 1 week of harvest.
	Peach Scab (**Cladosporium** sp.)	1 1/2-2 1/4 **PLUS** 1 1/8-1 1/2	1/2-3/4 **PLUS** 3/8-1/2	Apply at early bloom (pink bud). Make a second application at full bloom if conditions favor disease development. **PLUS** Apply at shuck split and at first cover sprays. Do not apply within 1 day of harvest.
—Plums and Prunes	Brown Rot Blossom Blight (**Monilinia** sp.) Fruit Brown Rot (**Monilinia** sp.)	1 1/2	1/2	Apply at early bloom (green tip). Make a second application at full bloom. In addition, for Fruit Brown Rot, apply 1 or 2 sprays starting 3 weeks before harvest. If needed under severe disease pressure, apply additional sprays at 10- to 14-day intervals between full bloom and final preharvest sprays. Do not apply within 1 day of harvest.
	Black Knot (**Dibotryon** sp.)	1 1/2	1/2	Apply at pre-bloom, petal fall, and at first, second, and third cover sprays at 10- to 14-day intervals. Do not apply within 1 day of harvest.
	Leaf Spot (**Coccomyces** sp.)	1 1/2	1/2	Apply at petal fall, shuck, and at first, second, and third cover sprays at 10- to 14-day intervals and 1 spray 14 to 21 days after harvest. Do not apply within 1 day of harvest.
Stone Fruits (post-harvest) —Apricots —Cherries —Nectarines —Peaches —Plums —Prunes	Postharvest Fruit Rots Fruit Brown Rot (**Monilinia** sp.)		3/4-1 (630-840 ppm a.i.)	Dip or spray fruit thoroughly as soon as possible after harvest. Maximum postharvest treatment time is 1 min. NOTE: TOPSIN M will not control fruit rots caused by **Rhizopus** spp. and **Alternaria** spp.
Strawberries	Fruit Rot (**Botrytis** sp.) Leaf Blight (**Dendrophoma** sp.) Leaf Scorch (**Diplocarpon** sp.)	3/4-1		Begin applications at early bloom and continue at 7- to 10-day intervals. Use the higher rate East of the Rocky Mountains and under conditions of severe disease pressure. Do not apply within 1 day of harvest.
Sugar Beets	Leaf Spot (**Cercospora** sp.)	3/8-1/2		Begin applications when disease first appears and repeat at 14- to 21-day intervals as needed. Do not apply within 21 days of harvest.
Sugar Cane (Hawaii only)	Pineapple Disease (**Thielaviopsis** sp. or **Ceratocystis** sp.)		1/2 (1:1,600).	Apply to cut seed pieces as a cold or hot dip. Cold Dip—Immerse seed pieces to provide thorough coverage; remove and allow to drain
			1/4 (1:3.200)	Hot Dip—Maintain temperature of dip at 52°C and soak seed pieces for 20 minutes; remove and allow to drain. Do not use treated seed pieces for food, feed or fodder. Do not graze or feed treated forage to livestock.
Fall-Seeded Wheat (Idaho, Oregon & Washington only)	**Pseudocercosporella** sp. (Foot Rot, Strawbreaker, Eye Spot)	1 lb. (0.7 lb. ai/A)		Apply TOPSIN M at the rate indicated in a single application by air or ground after wheat is tillered but before stem elongation has begun. Use sufficient water to obtain thorough coverage. Do not make more than one application per season. Do not cut for hay within 90 days of application. Do not allow livestock to graze in treated areas before harvest.

TANK MIX FOR USE ON APPLES
(Not for use in California)

Diseases	Lbs/ Acre	Lbs/ 100 Gal	Remarks
Scab (**Venturia** sp.) Bitter Rot (**Glomerella** sp.) Black Rot (**Physalospora** sp.) Flyspeck (**Schizothyrium** sp.) Powdery Mildew (**Podosphaera** sp.) Sooty Blotch (**Gloeodes** sp.)	TOPSIN M 1/2-3/4 + Captan 50WP 3-4 OR Captan 80WP 2-2 1/2	TOPSIN M 1/8-3/16 + Captan 50WP 3/4-1 OR Captan 80WP 1/2-5/8	Apply at 1/2" green tip and repeat at 7- to 14-day intervals or as needed through the cover sprays. Use the high rate of TOPSIN M for varieties more susceptible to powdery mildew. If an application is missed during an infection period, apply the higher rates as soon as possible after the infection period in order to deactivate scab and to prevent further infection. NOTE: Spray injury may result if Captan is used with, immediately before, or closely following an oil spray.

Observe the more restrictive cautions and limitations on labeling of products used in mixtures.

TANK MIX FOR USE ON APPLES
(Not for use in California)

Diseases	Lbs/ Acre	Lbs/ 100 Gal	Remarks
Scab (**Venturia** sp.) Bitter Rot (**Glomerella** sp.) Black Rot (**Physalospora** sp.) Flyspeck (**Schizothyrium** sp.) Powdery Mildew (**Podosphaera** sp.) Sooty Blotch (**Gloeodes** sp.) Brown Rot (**Monilinia** sp.) Cedar Apple Rust (**Gymnosporangium** sp.) Quince Rust (**Gymnosporangium** sp.)	TOPSIN M 1/2-3/4 + Mancozeb 80WP 3-4	TOPSIN M 1/8-3/16 + Mancozeb 80WP 3/4-1	Apply at 1/2" green tip and repeat at 7- to 14-day intervals or as needed through the cover sprays. One quart spray oil may be added per 100 gallons. Add oil to the spray tank last. Use the high rate of TOPSIN M and add spray oil to the spray mixture for varieties more susceptible to powdery mildew, and for scab if an application is missed during an infection period. (Apply as soon as possible after the infection period in order to deactivate scab and to prevent further infection.) Do not apply within 30 days of harvest. Do not graze treated areas.

Observe the more restrictive cautions and limitations on labeling of products used in mixtures.

TANK MIX FOR USE ON APPLES
(Not for use in California)

Diseases	Lbs/ Acre	Lbs/ 100 Gal	Remarks
Scab (**Venturia** sp.) Bitter Rot (**Glomerella** sp.) Black Rot (**Physalospora** sp.) Flyspeck (**Schizothyrium** sp.) Powdery Mildew (**Podosphaera** sp.) Sooty Blotch (**Gloeodes** sp.) Cedar Apple Rust (**Gymnosporangium** sp.) Bull's Eye Rot (**Gloeosporium** sp.)	TOPSIN M 1/2-3/4 + Maneb 80WP 6-8	TOPSIN M 1/8-3/16 + Maneb 80WP 1 1/2-2	Begin applications at green tip and repeat at pink bud and at 7- to 14-day intervals or as needed through the cover sprays. Use the higher rates and closer interval under severe disease conditions. With Golden Delicious apples, adverse effects on fruit finish and color may result. Use the lower rates of TOPSIN M when finish and color are of primary importance. Do not apply within 30 days of harvest. Do not graze livestock in treated areas.

Observe the more restrictive cautions and limitations on labeling of products used in mixtures.

TANK MIX FOR USE ON APPLES
(Not for use in California)

Diseases	Lbs/ Acre	Lbs/ 100 Gal	Remarks
Scab (**Venturia** sp.) Bitter Rot (**Glomerella** sp.) Black Rot (**Physalospora** sp.) Flyspeck (**Schizothyrium** sp.) Powdery Mildew (**Podosphaera** sp.) Sooty Blotch (**Gloeodes** sp.)	TOPSIN M 1/2-3/4 + Polyram 80WP 3-4	TOPSIN M 1/8-3/16 + Polyram 80WP 3/4-1	Begin applications at green tip and repeat applications as needed through the cover sprays. Do not apply within 30 days of harvest. Do not graze livestock in treated areas.

Observe the more restrictive cautions and limitations on labeling of products used in mixtures.

TANK MIX FOR USE ON APPLES
(Not for use in California)

Diseases	Lbs/ Acre	Lbs/ 100 Gal	Remarks
Scab (**Venturia** sp.) Bitter Rot (**Glomerella** sp.) Black Rot (**Physalospora** sp.) Flyspeck (**Schizothyrium** sp.) Powdery Mildew (**Podosphaera** sp.) Sooty Blotch (**Gloeodes** sp.)	TOPSIN M 1/2-3/4 + Sulfur 12-30 lbs a.i.	TOPSIN M 1/8-3/16 + Sulfur 3-7 1/2 lbs a.i.	Begin applications at green tip and repeat at pink bud and at 7- to 14 day intervals or as needed through the cover sprays. Use the higher rates and closer interval under severe disease conditions.

Observe the more restrictive cautions and limitations on labeling of products used in mixtures.

TANK MIX FOR USE ON PEANUTS
(Not for use in California)

Diseases	Lbs/ Acre	Remarks
Leaf Spot (**Cercospora** sp.) Rust (**Puccinia** sp.) Web Blotch (**Ascochyta** sp.)	TOPSIN M 1/4 + Mancozeb 80WP 1-2	Begin application 35 to 40 days after planting or when disease first appears. Repeat at the following intervals: for **Cercospora** leaf spot at 10 to 14 days; for Rust at 7 to 10 days; and for **Ascochyta** web blotch at 7 to 14 days. Spray oil may be added at the rate of 1 pint to 1 quart per acre. Do not graze or feed treated vines, hay, or hulls to livestock. Do not apply within 14 days of harvest.

Observe the more restrictive cautions and limitations on labeling of products used in mixtures.

TANK MIX FOR USE ON STONE FRUITS
(Apricots, Cherries, Nectarines, Peaches, Plums and Prunes)
(Not for use in California)

Diseases	Lbs/ Acre	Lbs/ 100 Gal	Remarks
Brown Rot Blossom Blight (**Monilinia** spp.) Fruit Brown Rot (**Monilinia** spp.) Peach Scab (**Cladosporium** sp.) Powdery Mildew (**Podosphaera** sp. and **Sphaerotheca** sp.) Cherry Leaf Spot (**Coccomyces** sp.) Russet Scab Black Knot (**Dibotryon** sp.)	TOPSIN M 3/4-1 + Captan 50WP 3-6 **OR** Captan 80WP 2-3 3/4	TOPSIN M 1/4-1/3 + Captan 50WP 1-2 **OR** Captan 80WP 5/8-1 1/4	BROWN ROT BLOSSOM BLIGHT—Apply at early bloom stages before infection occurs (apricots—red bud; cherries—early popcorn; nectarines & peaches—pink bud; plums & prunes—green tip). If blossoming is prolonged or conditions favorable for disease persist, make a second application 14 days later. FRUIT BROWN ROT—After blossom blight sprays, make a preharvest application any time from 3 weeks before harvest to 1 day before harvest. Make a second application if conditions favorable for disease persist or harvest is prolonged. PEACH SCAB, POWDERY MILDEW—Use same schedule as for Brown Rot

Diseases	Lbs/ Acre	Lbs/ 100 Gal	Remarks
			Blossom Blight and continue at 10- to 14-day intervals to 1 day before harvest. Make an additional application 2 to 3 weeks after harvest. CHERRY LEAF SPOT—Use same schedule as for Brown Rot Blossom Blight and continue at 10- to 14-day intervals to 1 day before harvest. Make an additional application 2 to 3 weeks after harvest. RUSSET SCAB (PLUMS & PRUNES)—Use same schedule as for Brown Rot Blossom Blight, include full coverage spray at full bloom. BLACK KNOT—Apply at first, second, and third cover sprays at 10- to 14-day intervals. Do not apply within 1 day of harvest. NOTE: This tank mixture will not control bacterial blast, but will supplement a dormant program for peach leaf curl and shot-hole (**Coryneum** blight); refer to other fungicide labels for dormant period control information. Do not use this tank mixture with spray oils.

Observe the more restrictive cautions and limitations on labeling of products used in mixtures.

TANK MIX FOR USE ON STONE FRUITS
(Apricots, Cherries, Nectarines, Peaches, Plums and Prunes)
(Not for use in California)

Diseases	Lbs/ Acre	Lbs/ 100 Gal	Remarks
Brown Rot Blossom Blight (**Monilinia** sp.) Fruit Brown Rot (**Monilinia** sp.) Powdery Mildew (**Podosphaera** sp. and **Sphaerotheca** sp.) Leaf Spot (**Coccomyces** sp.) Shot-hole (**Coryneum** sp.) Rust (**Tranzschelia** sp.) Peach Canker Peach Scab (**Cladosporium** sp.) Black Knot (**Dibotryon** sp.)	TOPSIN M 3/4-1 1/2 + Sulfur 12-18 lbs a.i.	TOPSIN M 1/4-1/2 + Sulfur 4-6 lbs a.i.	BROWN ROT BLOSSOM BLIGHT—Make application at early bloom & again at full bloom. FRUIT BROWN ROT—In addition to bloom sprays, apply 1 or 2 sprays starting 3 weeks before harvest. If needed under severe disease pressure, apply additional sprays at 10- to 14-day intervals between full bloom and final preharvest sprays. POWDERY MILDEW—In addition to bloom sprays, make applications at shuck fall and first cover. LEAF SPOT—Make application when leaves first unfold and first, second, and third cover at 10- to 14-day intervals and 1 spray 14 to 21 days after harvest. SHOT-HOLE, RUST, PEACH CANKER & PEACH SCAB—Apply at pink bloom, petal fall, shuck split, shuck fall, and at 10- to 14-day intervals as long as necessary. BLACK KNOT—Apply at first, second and third cover sprays at 10- to 14-day intervals. Do not apply within 1 day of harvest.

Observe the more restrictive cautions and limitations on labeling of products used in mixtures.

TANK MIX FOR USE ON STONE FRUITS
(Apricots, Cherries, Nectarines and Peaches)
(Not for use in California)

Diseases	Crop	Lbs/ Acre	Lbs/ 100 Gal	Remarks
Brown Rot Blossom Blight (**Monilinia** sp.) Fruit Brown Rot (**Monilinia** sp.) Powdery Mildew (**Podosphaera** sp. and **Sphaerotheca** sp.) Cherry Leaf Spot (**Coccomyces** sp.) Shot-hole (**Coryneum** sp.) Rust (**Tranzschelia** sp.) Peach Canker Peach Scab (**Cladosporium** sp.)	Apricots and Nectarines	TOPSIN M 3/4-1 1/2 + Ziram 3-4 1/2 lbs a.i.	TOPSIN M 1/4-1/2 + Ziram 1-1 1/2 lbs a.i.	BROWN ROT BLOSSOM BLIGHT—Make application at early bloom and again at full bloom. FRUIT BROWN ROT—In addition to bloom sprays, apply 1 or 2 sprays starting 3 weeks before harvest. If needed under severe disease pressure, apply additional sprays at 10- to 14-day intervals between full bloom and final pre-harvest sprays. POWDERY MILDEW—In addition to bloom sprays, make applications at shuck fall and first cover. CHERRY LEAF SPOT—Make application when leaves first unfold & first, second, and third cover at 10- to 14-day intervals and 1 spray 14-21 days after harvest. SHOT-HOLE, RUST, PEACH CANKER & PEACH SCAB—Apply at pink, bloom, petal fall, shuck split, shuck fall, and at 10- to 14-day intervals as long as necessary. For apricots, do not apply after early cover periods. For cherries, do not apply within 7 days of harvest. For nectarines & peaches, last application should be made no later than at petal fall.
	Cherries	TOPSIN M 3/4-1 1/2 + Ziram 3 3/8 lbs a.i.	TOPSIN M 1/4-1/2 + Ziram 1 1/8 lbs a.i.	
	Peaches	TOPSIN M 3/4-1 1/2 + Ziram 3-4 1/2 lbs a.i.	TOPSIN M 1/4-1/2 + Ziram 1-1 1/2 lbs a.i.	

Observe the more restrictive cautions and limitations on labeling of products used in mixtures.

TANK MIX FOR USE ON STONE FRUITS
(Apricots, Nectarines, and Peaches)
(Not for use in California)

Diseases	Lbs/ Acre	Lbs/ 100 Gal	Remarks
Brown Rot Blossom Blight (**Monilinia** sp.) Fruit Brown Rot (**Monilinia** sp.) Peach Scab (**Cladosporium** sp.) Powdery Mildew (**Podosphaera** sp. and **Sphaerotheca** sp.)	TOPSIN M 1 1/2 + Maneb 80WP 6-8	TOPSIN M 1/2 + Maneb 80WP 2-2 2/3	BROWN ROT BLOSSOM BLIGHT—Apply at early bloom stages before infection occurs (apricots—red bud; nectarines & peaches—pink bud). For this application only, TOPSIN M plus Maneb 80WP may be used in combination with superior-type spray oil. If blossoming is prolonged or conditions favorable for disease persist, make a second application 14 days later.

Diseases	Lbs/ Acre	Lbs/ 100 Gal	Remarks
			FRUIT BROWN ROT—After blossom blight sprays, make an additional application 3 weeks before harvest. PEACH SCAB, POWDERY MILDEW—Use same schedule as for Brown Rot Blossom Blight plus applications at shuck split, shuck fall, and 14 days later. Do not apply within 14 days of harvest. NOTE: This tank mixture will not control bacterial blast, but will supplement a dormant program for peach leaf curl and shot-hole (**Coryneum** blight); refer to other fungicide labels for dormant period control information.

Observe the more restrictive cautions and limitations on labeling of products used in mixtures.

CHEMIGATION

NOT FOR THIS USE IN CALIFORNIA
GENERAL INSTRUCTIONS

Apply this product only through sprinkler irrigation systems including center pivot, lateral move, end tow, side (wheel) roll, traveler, big gun, solid set, or hand move. Do not apply this product through any other type of irrigation system.

Crop injury, lack of effectiveness, or illegal pesticide residues in the crop can result from nonuniform distribution of treated water.

If you have any questions about calibration, you should contact State Extension Service specialists, equipment manufacturers or other experts.

A person knowledgeable of the chemigation system and responsible for its operation, or under the supervision of the responsible person, shall shut the system down and make necessary adjustments should the need arise.

Do not connect chemigation system to any public water system. Public water system means a system for the provision of piped water for human consumption if such system has at least 15 service connections or regularly serves an average of at least 25 individuals daily at least 60 days out of the year.

SYSTEM REQUIREMENTS

Systems utilizing a pressurized water and pesticide injection system must meet the following requirements:

The system must contain a functional check valve, vacuum relief valve, and low pressure drain appropriately located on the irrigation pipeline to prevent water source contamination from backflow.

The pesticide injection pipeline must contain a functional, automatic, quick-closing check valve to prevent the flow of fluid back toward the injection pump.

The pesticide injection pipeline must also contain a functional, normally closed, solenoid-operated valve located on the intake side of the injection pump and connected to the system interlock to prevent fluid from being withdrawn from the supply tank when the irrigation system is either automatically or manually shut down.

The system must contain functional interlocking controls to automatically shut off the pesticide injection pump when the water pump motor stops.

The irrigation line or water pump must include a functional pressure switch which will stop the water pump motor when the water pressure decreases to the point where pesticide distribution is adversely affected.

Systems must use a metering pump, such as a positive displacement injection pump (e.g., diaphragm pump) effectively designed and constructed of materials that are compatible with pesticides and capable of being fitted with a system interlock.

INSTRUCTIONS FOR SPRINKLER (OVERHEAD) IRRIGATION:

Observe the requirements in the System Requirements section above.

Do not apply when wind speed favors drift beyond the area intended for treatment.

Apply TOPSIN M only through systems containing anti-syphon and check valves designed to prevent water source contamination or overflow of the mix tank and containing interlocking controls between the metering device and the water pump to insure simultaneous shut-off.

Maintain a gentle continuous agitation in mix tank during mixing and application to assure a uniform suspension.

Greater accuracy in calibration and distribution will be achieved by injecting a larger volume of a more dilute suspension per unit time.

Application of more than recommended quantities of irrigation water per acre may result in decreased product performance.

Do not apply when wind speed favors drift, when system connections or fittings leak, when nozzles do not provide uniform distribution or when lines containing the product cannot be flushed and must be dismantled and drained. In a center pivot system, block the nozzle set nearest the well/pivot/injection unit to prevent spray being applied to this area.

Where sprinkler distribution patterns do not overlap sufficiently, unacceptable disease control may result.

Allow sufficient time for pesticide to be flushed through all lines and all nozzles before turning off irrigation water.

TOPSIN M may be applied in conjunction with chemically neutral liquid fertilizers. Application in conjunction with highly alkaine fertilizers, such as aqueous ammonia, may cause a degradation of the pesticide, resulting in reduced performance and should be avoided.

Check local restrictions and requirements regarding sprinkler irrigation applications, as they may vary from state to state.

SPRAY PREPARATION:

Remove scale, pesticide residues, and other foreign matter from the chemical tank and entire injector system. Flush with clean water.

Prepare a suspension of TOPSIN M in a mix tank. Fill the tank with 1/2 or 3/4 the desired amount of water. Start mechanical or hydraulic agitation. Slowly add the required amount of TOPSIN M and then the remaining volume of water.

APPLICATION INSTRUCTIONS:

Set sprinkler system to deliver 0.1 to 1.25 inches of water per acre. Volumes of water higher than this may reduce efficacy. Start sprinkler and then uniformly inject the suspension of TOPSIN M into the irrigation water line so as to deliver the desired rate per acre. The suspension of TOPSIN M should be injected with a positive displacement pump into the main line ahead of a right angle turn to insure adequate mixing.

NOTE: When treatment with TOPSIN M has been completed, do not irrigate the treated area for 24 to 48 hours to prevent washing the chemical off the crop.

STORAGE AND DISPOSAL

Do not contaminate water, food, or feed by storage or disposal.

Storage Instructions: Store in the original container in a dry area. Do not store in a manner where cross-contamination with other pesticides, fertilizers, food or feed could occur. If spilled during storage or handling, sweep up spillage and dispose of in accordance with the Pesticide Disposal Instructions listed below.

Pesticide Disposal Instructions: Wastes resulting from the use of this product may be disposed of on site or at an approved waste disposal facility.

Container Disposal Instructions: After bag has been emptied, dispose of empty bag in a sanitary landfill or by incineration, or if allowed by State and local authorities, by burning. If burned, stay out of smoke.

WARRANTY AND DISCLAIMER

Pennwalt Corporation warrants that this material conforms to the chemical description on the label and is reasonably fit for the purposes referred to in the Directions for Use, subject to the risks referred to therein. PENNWALT MAKES NO OTHER EXPRESS OR IMPLIED WARRANTY OF FITNESS OR MERCHANTABILITY OR ANY OTHER EXPRESS OR IMPLIED WARRANTY. IN NO CASE SHALL PENNWALT OR SELLER BE LIABLE FOR CONSEQUENTIAL, SPECIAL OR INDIRECT DAMAGES RESULTING FROM THE USE OR HANDLING OF THIS PRODUCT INCLUDING, BUT NOT LIMITED TO, LOSS OF PROFITS, BUSINESS REPUTATION, OR CUSTOMERS, LABOR COST, OR OTHER EXPENSES INCURRED IN PLANTING OR HARVESTING.

Pennwalt and seller offer this product and the buyer and user accept it subject to the foregoing conditions of sale and warranty which may be varied only by agreement in writing signed by a duly authorized representative of Pennwalt.

5-A301S-01 D8 (4/88)

TOPSIN® M

TURF AND ORNAMENTALS FUNGICIDE
70% WETTABLE POWDER

NOT FOR USE IN CALIFORNIA

ACTIVE INGREDIENT:

Thiophanate-methyl - dimethyl[(1,2-phenylene)bis (iminocarbonothioyl)] bis[carbamate]*	70%
INERT INGREDIENTS:	30%
TOTAL	100%

*Also known as dimethyl 4,4-o-phenylenebis[3-thioallophanate]

*TOPSIN is a registered trademark Nippon Soda Company, Ltd. and is licensed to Pennwalt Corporation, and is covered by one or more of the following U.S. Patents 3,769,308; 3,856,847; 4,020,095; and 4,029,813.

EPA Registration No. 4581-288 EPA Establishment No. 4581-TX-1

KEEP OUT OF REACH OF CHILDREN

CAUTION

STATEMENT OF PRACTICAL TREATMENT

IF SWALLOWED, induce vomiting immediately by giving two glasses of water and sticking finger down throat. Call physician. Never give anything by mouth to an unconscious person.

See Side Panel for Additional Precautionary Statements

IF IN EYES, immediately flush eyes with plenty of water for at least 15 minutes. Get medical attention if irritation persists.

IF ON SKIN, immediately flush skin with plenty of water. Get medical attention if irritation persists.

NOTE: FOR GENERAL INFORMATION AND DIRECTIONS FOR USE REFER TO SIDE PANEL.

PRECAUTIONARY STATEMENTS

HAZARDS TO HUMANS (AND DOMESTIC ANIMALS)

HARMFUL IF SWALLOWED, ABSORBED THROUGH SKIN, OR INHALED. CAUSES MODERATE EYE IRRITATION.

Avoid contact with skin, eyes, or clothing. Avoid breathing dust or spray mist. Wash thoroughly with soap and water after handling. Remove contaminated clothing and wash before reuse.

Do not apply this product in such a manner as to directly or through drift expose workers or other persons. The area being treated must be vacated by unprotected persons.

ENVIRONMENTAL HAZARDS

This pesticide is toxic to fish. Do not apply directly to water. Do not apply where runoff is likely to occur. Do not contaminate water by disposal of equipment washwaters.

GENERAL INFORMATION

Apply TOPSIN M with ground or aerial equipment, using sufficient volume of spray to provide thorough coverage. Add required amount of TOPSIN M to partially filled tank agitated by mechanical or hydraulic means and then add remaining required amount of water. Continuous agitation is required to keep the material in suspension. Do not tank mix with copper-containing materials or with highly alkaline pesticides, such as Bordeaux mixture or lime sulfur. No claim of compatability with other pesticides is implied. Use the higher rates under conditions of severe disease pressure. Also, see State Extension Service recommendations for application schedules.

IMPORTANT: If, after using TOPSIN M as recommended, treatment is not effective, a tolerant strain of fungi may be present (consult your local Pennwalt representative). Consult your state Agricultural Extension Service for advice on the prompt use of some other suitable fungicide.

DIRECTIONS FOR USE

It is a violation of Federal law to use this product in a manner inconsistent with its labeling.

TURF

For use on golf course greens, tees, and fairways of Bentgrasses, Bluegrasses, Bermudagrasses, Fescues, Ryegrasses, St. Augustine grasses or their mixtures. TOPSIN M is not phytotoxic to any of the above mentioned grasses when used in accordance with label. TOPSIN M is to be used for the prevention and control of the below mentioned diseases (provide both curative and protective action).

Use spray mixture the same day it is prepared. Spray uniformly over the area to be treated. Apply after mowing or avoid mowing for twelve hours after application.

Apply recommended amounts in sufficient water to obtain thorough coverage, usually five gallons per 1,000 square feet of turf area. When treating golf greens, always treat aprons and approaches. Use the highest recommended rate under conditions of severe disease pressure.

FUSARIUM BLIGHT **(Fusarium roseum)**—Apply TOPSIN M at 4 to 8 ounces per 1,000 square feet. Water into the root zone by irrigating with one inch of water immediately after application. Make two applications at 10 to 14-day intervals, beginning applications when the disease first appears.

DOLLAR SPOT **(Sclerotinia homoeocarpa)**, COPPER SPOT **(Gloeocerospora sorghi)**—Apply TOPSIN M at 3/4 to 1 1/2 ounces per 1,000 square feet beginning when the disease first appears and continue at 10 to 14-day intervals through the season or as needed.

LARGE BROWN PATCH **(Rhizoctonia solani)**—Apply TOPSIN M at 3/4 to 1 1/2 ounces per 1,000 square feet beginning when the disease first appears and continue at 10 to 14-day intervals through the season or as needed. When conditions are unusually favorable for development of disease, reduce interval to 5 to 7 days.

STRIPE SMUT **(Ustilago striiformis)**—Apply TOPSIN M at 4 to 8 ounces per 1,000 square feet. Water into the root zone by irrigating with one inch of water immediately after application. Make two applications at 14 to 21-day intervals. Begin applications when the disease first appears, either in the spring or fall.

ORNAMENTALS, ROSES, FLOWERS, SHADE TREES (FIELD AND GREENHOUSE)

FOLIAR SPRAY: Begin applications when disease first appears and repeat at 10 to 14-day intervals throughout the growing season; shorten interval during humid, rainy weather.

Use at the following rates:

Use 1/2 pound per 100 gallons (1 tablespoon per 2 gallons) for Powdery Mildew and Botrytis Gray Mold.

Use 1 pound per 100 gallons for Anthracnose (for shade trees and woody ornamentals, begin at bud break and make 2 or 3 additional applications at 10 to 14-day intervals): Black Spot of roses: Cercospora, Entomosporium, Ramularia, and Septoria Leaf Spots; Ascochyta and Phomopsis Blights; Didymellina Leaf Spots of iris; Corynespora Leaf Spot of ligustrum; Ovulinia of azalea and rhododendron (begin as flowers open); Scab of pyracantha and flowering crab.

Addition of a surfactant to the spray mixture improves distribution of the spray on hard-to-wet plants such as roses.

For aerial application, use 1/2 to 1 pound per acre.

DRENCH TREATMENT: Botrytis, Fusarium, Rhizoctonia, and Sclerotinia stem, crown, and root rots on herbaceous annuals, perennials, and bedding plants; Cylindrocladium and Thielaviopsis rots on woody ornamentals such as azaleas, rhododendrons, conifers, and poinsettias—Use 1 pound per 100 gallons; apply as a drench or heavy spray (1 to 2 pints per square foot) after transplanting into propagation beds or containers. Repeat at 2 to 4-week intervals during periods favorable for disease. TOPSIN M does not control **Pythium** spp. or **Pytophthora** spp.

PREPLANT DIP TREATMENT:

Plants or Cuttings—For diseases listed under Drench Treatment, use 1 pound per 100 gallons of water; immerse plants or cuttings for 10 to 15 minutes; remove and allow to drain.

Bulbs (Easter Lily, Tulip, Gladiolus, Daffodil, Iris) Fusarium and Penicillium Rots—Use 1 2/3 pounds per 100 gallons of water (2 tablespoonfuls per gallon). Soak cleaned bulbs for 15 to 30 minutes in warm dip (80 to 85°F) preferably within 48 hours after digging. Dry bulbs after treatment. If bulbs are for forcing, treat bulbs that have been heat-cured.

CHEMIGATION

NOT FOR THIS USE IN CALIFORNIA

GENERAL INSTRUCTIONS

Apply this product only through sprinkler irrigation systems including center pivot, lateral move, end tow, side (wheel) roll, traveler, big gun, solid set, or hand move. Do not apply this product through any other type of irrigation system.

Crop injury, lack of effectiveness, or illegal pesticide residues in the crop can result from nonuniform distribution of treated water.

If you have any questions about calibration, you should contact State Extension Service specialists, equipment manufacturers or other experts.

A person knowledgeable of the chemigation system and responsible for its operation, or under the supervision of the responsible person, shall shut the system down and make necessary adjustments should the need arise.

Do not connect chemigation system to any public water system. Public water system means a system for the provision of piped water for human consumption if such system has at least 15 service connections or regularly serves an average of at least 25 individuals daily at least 60 days out of the year.

SYSTEM REQUIREMENTS

Systems utilizing a pressurized water and pesticide injection system must meet the following requirements:

The system must contain a functional check valve, vacuum relief valve, and low pressure drain appropriately located on the irrigation pipeline to prevent water source contamination from backflow.

The pesticide injection pipeline must contain a functional, automatic, quick-closing check valve to prevent the flow of fluid back toward the injection pump.

The pesticide injection pipeline must also contain a functional, normally closed, solenoid-operated valve located on the intake side of the injection pump and connected to the system interlock to prevent fluid from being withdrawn from the supply tank when the irrigation system is either automatically or manually shut down.

The system must contain functional interlocking controls to automatically shut off the pesticide injection pump when the water pump motor stops.

The irrigation line or water pump must include a functional pressure switch which will stop the water pump motor when the water pressure decreases to the point where pesticide distribution is adversely affected.

Systems must use a metering pump, such as a positive displacement injection pump (e.g. diaphragm pump) effectively designed and constructed of materials that are compatible with pesticides and capable of being fitted with a system interlock.

INSTRUCTIONS FOR SPRINKLER (OVERHEAD) IRRIGATION:

Observe the requirements in the System Requirements section.

Do not apply when wind speed favors drift beyond the area intended for treatment.

Apply TOPSIN M only through systems containing anti-syphon and check valves designed to prevent water source contamination or overflow of the mix tank and containing interlocking controls between the metering device and the water pump to insure simultaneous shut-off.

Maintain a gentle continous agitation in mix tank during mixing and application to assure a uniform suspension.

Greater accuracy in calibration and distribution will be achieved by injecting a larger volume of a more dilute suspension per unit time.

Application of more than recommended quantities of irrigation water per acre may result in decreased product performance.

Do not apply when wind speed favors drift, when system connections or fittings leak, when nozzles do not provide uniform distribution or when lines containing the product cannot be flushed and must be dismantled and drained. In a center pivot system block the nozzle set nearest the well pivot injection unit to prevent spray being applied to this area.

Where sprinkler distribution patterns do not overlap sufficiently, unacceptable disease control may result. Allow sufficient time for pesticide to be flushed through all lines and all nozzles before turning off irrigation water.

TOPSIN M may be applied in conjunction with chemically neutral liquid fertilizers. Application in conjunction with highly alkaline fertilizers such as a aqueous ammonia, may cause a degradation of the pesticide, resulting in reduced performance and should be avoided.

Check local restrictions and requirements regarding sprinkler irrigation application, as they may vary from state to state.

SPRAY PREPARATION

Remove scale, pesticide residues, and other foreign matter from the chemical tank and entire injector system. Flush with clean water.

Prepare a suspension of TOPSIN M in a mix tank. Fill the tank with 1/2 or 3/4 the desired amount of water. Start mechanical or hydraulic agitation. Slowly add the required amount of TOPSIN M and then the remaining volume of water.

APPLICATION INSTRUCTIONS

Set sprinkler system to deliver 0.1 to 1.25 inches of water per acre. Volumes of water higher than this may reduce efficacy. Start sprinkler and then uniformly inject the suspension of TOPSIN M into the irrigation water line so as to deliver the desired rate per acre. The suspension of TOPSIN M should be injected with a positive displacement pump into the main line ahead of a right angle turn to insure adequate mixing.

Note: When treatment with TOPSIN M has been completed, do not irrigate the treated area for 24 to 48 hours to prevent washing the chemical off the crop.

STORAGE AND DISPOSAL

Do not contaminate water, food, or feed by storage or disposal.

Storage Instructions: Store in the original container in a dry area. Do not store in a manner where cross-contamination with other pesticides, fertilizers, food or feed could occur. If spilled during storage or handling, sweep up spillage and dispose of in accordance with the Pesticide Disposal Instructions listed below.

Pesticide Disposal Instructions: Wastes resulting from the use of this product may be disposed of on site or at an approved waste disposal facility.

Container Disposal Instructions: After bag has been emptied, dispose of empty bag in a sanitary landfill or by incineration, of if allowed by State and local authorities, by burning. If burned, stay out of smoke.

WARRANTY AND DISCLAIMER

Pennwalt Corporation warrants that this material conforms to the chemical description on the label and is reasonably fit for the purposes referred to in the Directions for Use, subject to the risks referred to therein. PENNWALT MAKES NO OTHER EXPRESS OR IMPLIED WARRANTY OF FITNESS OR MERCHANTABILITY OR ANY OTHER EXPRESS OR IMPLIED WARRANTY. IN NO CASE SHALL PENNWALT OR SELLER BE LIABLE FOR CONSEQUENTIAL, SPECIAL OR INDIRECT DAMAGES RESULTING FROM THE USE OR HANDLING OF THIS PRODUCT INCLUDING, BUT NOT LIMITED TO, LOSS OF PROFITS, BUSINESS REPUTATION, OR CUSTOMERS; LABOR COST; OR OTHER EXPENSES INCURRED IN PLANTING OR HARVESTING.

Pennwalt and seller offer this product and the buyer and user accept it subject to the foregoing conditions of sale and warranty which may be varied only by agreement in writing signed by a duly authorized representative of Pennwalt.

TOPSIN® M 4.5F

THIOPHANATE-METHYL FUNGICIDE

SHAKE WELL BEFORE USING

4.5 Pounds per Gallon Flowable

ACTIVE INGREDIENT:	
Thiophanate-methyl (dimethyl[(1,2-phenylene)-bis(iminocarbonothioyl)]bis[carbamate])*	46.2%
INERT INGREDIENTS:	53.8%
TOTAL	100.0%

EPA Registration No. 4581-352 EPA Establishment No. 4581-TX-1

*Also known as dimethyl 4,4'-o-phenylenebis[3-thioallophanate]

®TOPSIN is a registered trademark of Nippon Soda Company, Ltd., and is licensed to Pennwalt Corporation. Covered by one or more of the following U.S. patents: 3,769,308; 3,856,847; 4,020,095; and 4,029,813.

KEEP OUT OF REACH OF CHILDREN

CAUTION

STATEMENT OF PRACTICAL TREATMENT

IF SWALLOWED, induce vomiting immediately by giving two glasses of water and sticking finger down throat. Call a physician. Never give anything by mouth to an unconscious person.

IF IN EYES, immediately flush eyes with plenty of water for at least 15 minutes. Get medical attention if irritation persists.

IF ON SKIN, immediately flush skin with plenty of water.

GENERAL INFORMATION

Apply TOPSIN M with ground or aerial equipment, using sufficient volume of spray to provide thorough coverage. Add required amount of TOPSIN M to partially filled tank agitated by mechanical or hydraulic means, and then add remaining required amount of water. Continuous agitation is required to keep the material in suspension. Do not tank mix with highly alkaline pesticides, such as Bordeaux mixture or lime sulfur. No claim of compatibility with other pesticides is implied. Use the higher rate under conditions of severe disease pressure. Also, see local State Extension Service recommendations for application schedules.

IMPORTANT: If, after using TOPSIN M as recommended, treatment is not effective, a tolerant strain of fungi may be present. Consult your State Agricultural Experiment Station or your State Agricultural Extension Service for advice on the prompt use of some other suitable fungicide.

DIRECTIONS FOR USE

It is a violation of Federal law to use this product in a manner inconsistent with its labeling.

Do not enter treated areas without protective clothing until spray has dried. Because certain states may require more restrictive reentry intervals for various crops treated with this product, consult your State Department of Agriculture for further information.

Written or oral warnings must be given to workers who are expected to be in a treated area or in an area about to be treated with this product. When oral warnings are given, such warnings shall be given in a language customarily understood by the workers. Oral warnings must be given if there is reason to believe that written warnings cannot be understood by the workers. Written warnings must include the following information: "CAUTION. Area treated with TOPSIN M 4.5F Thiophanate-methyl Fungicide on *(date of application)*. Do not enter without protective clothing until sprays have dried."

CHEMIGATION

NOT FOR THIS USE IN CALIFORNIA

GENERAL INSTRUCTIONS

Apply this product only through sprinkler irrigation systems including center pivot, lateral move, end tow, side (wheel) roll, traveler, big gun, solid set, or hand move. Do not apply this product through any other type of irrigation system.

Crop injury, lack of effectiveness, or illegal pesticide residues in the crop can result from nonuniform distribution of treated water.

If you have any questions about calibration, you should contact State Extension Service specialists, equipment manufacturers or other experts.

A person knowledgeable of the chemigation system and responsible for its operation, or under the supervision of the responsible person, shall shut the system down and make necessary adjustments should the need arise.

Do not connect chemigation system to any public water system. Public water system means a system for the provision of piped water for human consumption if such system has at least 15 service connections or regularly serves an average of at least 25 individuals daily at least 60 days out of the year.

SYSTEM REQUIREMENTS

Systems utilizing a pressurized water and pesticide injection system must meet the following requirements:

The system must contain a functional check valve, vacuum relief valve, and low pressure drain appropriately located on the irrigation pipeline to prevent water source contamination from backflow.

The pesticide injection pipeline must contain a functional, automatic, quick-closing check valve to prevent the flow of fluid back toward the injection pump.

The pesticide injection pipeline must also contain a functional, normally closed, solenoid-operated valve located on the intake side of the injection pump and connected to the system interlock to prevent fluid from being withdrawn from the supply tank when the irrigation system is either automatically or manually shut down.

The system must contain functional interlocking controls to automatically shut off the pesticide injection pump when the water pump motor stops.

The irrigation line or water pump must include a functional pressure switch which will stop the water pump motor when the water pressure decreases to the point where pesticide distribution is adversely affected.

Systems must use a metering pump, such as a positive displacement injection pump (e.g., diaphragm pump) effectively designed and constructed of materials that are compatible with pesticides and capable of being fitted with a system interlock.

INSTRUCTIONS FOR SPRINKLER (OVERHEAD) IRRIGATION:

Observe the requirements in the System Requirements section above.

Do not apply when wind speed favors drift beyond the area intended for treatment.

Apply TOPSIN M 4.5F only through systems containing anti-syphon and check valves designed to prevent water source contamination or overflow of the mix tank and containing interlocking controls between the metering device and the water pump to insure simultaneous shut-off.

Maintain a gentle continuous agitation in mix tank during mixing and application to assure a uniform suspension.

Greater accuracy in calibration and distribution will be achieved by injecting a larger volume of a more dilute suspension per unit time.

Application of more than recommended quantities of irrigation water per acre may result in decreased product performance.

Do not apply when wind speed favors drift, when system connections or fittings leak, when nozzles do not provide uniform distribution or when lines containing the product cannot be flushed and must be dismantled and drained. In a center pivot system, block the nozzle set nearest the well pivot injection unit to prevent spray being applied to this area.

Where sprinkler distribution patterns do not overlap sufficiently, unacceptable disease control may result.

Allow sufficient time for pesticide to be flushed through all lines and all nozzles before turning off irrigation water.

TOPSIN M 4.5F may be applied in conjunction with chemically neutral liquid fertilizers. Application in conjunction with highly alkaline fertilizers, such as aqueous ammonia, may cause a degradation of the pesticide, resulting in reduced performance and should be avoided.

Check local restrictions and requirements regarding sprinkler irrigation applications, as they may vary from state to state.

SPRAY PREPARATION

Remove scale, pesticide residues, and other foreign matter from the chemical tank and entire injector system. Flush with clean water.

Prepare a suspension of TOPSIN M 4.5F in a mix tank. Fill the tank with 1/2 or 3/4 the desired amount of water. Start mechanical or hydraulic agitation. Slowly add the required amount of TOPSIN M 4.5F and then the remaining volume of water.

APPLICATION INSTRUCTIONS:

Set sprinkler system to deliver 0.1 to 1.25 inches of water per acre. Volumes of water higher than this may reduce efficacy. Start sprinkler and then uniformly inject the suspension of TOPSIN M 4.5F into the irrigation water line so as to deliver the desired rate per acre. The suspension of TOPSIN M 4.5F should be injected with a positive displacement pump into the main line ahead of a right angle turn to insure adequate mixing.

NOTE: When treatment with TOPSIN M 4.5F has been completed, do not irrigate the treated area for 24 to 48 hours to prevent washing the chemical off the crop.

Crop	Diseases	Fl Oz/A	Fl Oz 100 Gal	Remarks
Almonds	Brown Rot Blossom Blight (**Monilinia** sp.)	30-40		Apply at early pink bud and early petal fall.
Apples (pre-harvest)	Apple Scab (**Venturia** sp.) Bitter Rot (**Glomerella** sp.) Flyspeck (**Schizothyrium** sp.) Powdery Mildew (**Podosphaera** sp.) Sooty Blotch (**Gloeodes** sp.) Black Rot (**Physalospora** sp.)	20-30	5-7½	Apply at 5- to 10-day intervals from green tip through petal fall; continue at 7- to 14-day intervals in cover sprays. NOTE For control of apple diseases in the Southeastern States of Florida, Georgia, South Carolina, North Carolina and Alabama, use 60-80 fl oz. TOPSIN M per acre.
Beans	Gray Mold (**Botrytis** sp.) White Mold (**Sclerotinia** sp.)	30-40 **OR** 20-30	20-27 **OR** 13-20	Apply once at 50% to 70% of full bloom. **OR** Apply twice with the first application at 10% to 30% of full bloom and a second application at 4 to 7 days later or at peak bloom. Do not apply to snap or dry beans within 14 days of harvest or to lima beans within 28 days of harvest.
Celery	Early Blight (**Cercospora** sp.) Late Blight (**Septoria** sp.)	10		Apply to the base of the plants on a 14- to 21-day schedule beginning about 2 weeks after transplanting. **OR** Apply as a foliar spray every 7 to 14 days beginning about 2 weeks after transplanting. Do not apply within 7 days of harvest. Do not make more than 4 applications per season.
Peanuts	Leaf Spot (**Cercospora** spp.)	10		Begin applications 35 days after planting or when disease first appears and repeat at 14- to 21-day intervals as needed. Use the 14-day interval under severe disease pressure. Do not apply within 14 days of harvest.
Pecans	Brown Spot (**Cercospora** sp.) Downy Spot (**Mycosphaerella** sp.) Powdery Mildew (**Micosphaera** sp.) Scab (**Fusicladium** sp.)	10-20		Use the higher rate for trees over 30 feet tall and for aerial applications in Arkansas, Louisiana, Mississippi, Oklahoma, and Texas. Begin applications when first leaves are showing and repeat at 3- to 4-week intervals until shuck split. Do not apply after shuck split.
Soybeans	Anthracnose (**Colletotrichum** sp.) Brown Spot (**Septoria** sp.) Frogeye Leaf Spot (**Cercospora** sp.) Pod and Stem Blight (**Diaporthe** sp. and the imperfect stage. **Phomopsis** sp.) Purple Seed Stain (**Cercospora** sp.)	10-20		Apply from full bloom to when pods are 1/8" to 1/4" in length. Make a second application 14 to 21 days later. Do not make the second application later than 14 days after pods average 1/4" in length or when beans form in the pod. Use the high rate under severe disease pressure. FOR SEED BEANS ONLY—For seed quality make a single application at the high rate when beans form in the pod.

Crop	Diseases	Fl Oz/A	Fl Oz 100 Gal	Remarks
				Do not make more than 2 applications per year. Do not graze or feed treated vines or hay to livestock.
Stone Fruits (pre-harvest) —Apricots	Brown Rot Blossom Blight (**Monilinia** sp.) Fruit Brown Rot (**Monilinia** sp.)	30	10	Apply at early bloom (red bud). Make a second application at full bloom. In addition, for Fruit Brown Rot, apply 1 or 2 sprays starting 3 weeks before harvest. If needed under severe disease pressure, apply additional sprays at 10- to 14-day intervals between full bloom and final pre-harvest sprays. Do not apply within 1 day of harvest.
—Cherries	Brown Rot Blossom Blight (**Monilinia** sp.) Fruit Brown Rot (**Monilinia** sp.)	30	10	Apply at early bloom (early popcorn). Make a second application at full bloom. In addition, for Fruit Brown Rot, apply 1 or 2 sprays starting 3 weeks before harvest. If needed under severe disease pressure, apply additional sprays at 10- to 14-day intervals between full bloom and final preharvest sprays Do not apply within 1 day of harvest.
	Cherry Leaf Spot (**Coccomyces** sp.)	24-30	8-10	Apply at petal fall or before, when leaves first unfold, and at first, second and third cover at 10- to 14-day intervals and one spray 14 to 21 days after harvest. Do not apply within 1 day of harvest.
	Powdery Mildew (**Podosphaera** sp. and **Sphaerotheca** sp.)	30 **PLUS** 24-30	10 **PLUS** 8-10	Apply at early bloom (early popcorn). Make a second application at full bloom. **PLUS** Apply at shuck fall and first cover. Do not apply within 1 day of harvest.
-Nectarines	Brown Rot Blossom Blight (**Monilinia** sp.) Fruit Brown Rot (**Monilinia** sp.)	30	10	Apply at early bloom (pink bud). Make a second application at full bloom if conditions favor disease development. In addition, for Fruit Brown Rot, apply 1 or 2 sprays starting 3 weeks before harvest. If needed under severe disease pressure, apply additional sprays at 10- to 14-day intervals between full bloom and final preharvest sprays. Do not apply within 1 day of harvest.
—Peaches	Brown Rot Blossom Blight (**Monilinia** sp.) Fruit Brown Rot (**Monilinia** sp.)	30-45	10-15	Apply at early bloom (pink bud). Make a second application at full bloom if conditions favor disease development.

				In addition for Fruit Brown Rot, apply 1 or 2 sprays starting 3 weeks before harvest. If needed under severe disease pressure, apply additional sprays at 10- to 14-day intervals between full bloom and final preharvest sprays Do not apply within 1 day of harvest.
	Peach Scab (**Cladosporium** sp.)	30-45 **PLUS** 24-30	10-15 **PLUS** 8-10	Apply at early bloom (pink bud). Make a second application at full bloom if conditions favor disease development. **PLUS** Apply at shuck split and at first cover sprays Do not apply within 1 day of harvest
—Plums and Prunes	Brown Rot Blossom Blight (**Monilinia** sp.) Fruit Brown Rot (**Monilinia** sp.)	30	10	Apply at early bloom (green tip). Make a second application at full bloom. In addition for Fruit Brown Rot, apply 1 or 2 sprays starting 3 weeks before harvest. If needed under severe disease pressure, apply additional sprays at 10- to 14-day intervals between full bloom and final preharvest sprays. Do not apply within 1 day of harvest.
	Black Knot (**Dibotryon** sp.)	30	10	Apply at pre-bloom, petal fall, and at first, second, and third cover sprays at 10- to 14-day intervals. Do not apply within 1 day of harvest.
	Leaf Spot (**Coccomyces** sp.)	30	10	Apply at petal fall, shuck, and at first, second and third cover sprays at 10- to 14-day intervals and 1 spray 14 to 21 days after harvest. Do not apply within 1 day of harvest.
Straw-berries	Fruit Rot (**Botrytis** sp.) Leaf Blight (**Dendrophoma** sp.) Leaf Scorch (**Diplocarpon** sp.)	15-20		Begin applications at early bloom and continue at 7- to 10-day intervals. Use the higher rate East of the Rocky Mountains and under conditions of severe disease pressure. Do not apply within 1 day of harvest.
Sugar Beets	Leaf Spot (**Cercospora** sp.)	8-10		Begin applications when disease first appears and repeat at 14- to 21-day intervals as needed. Do not apply within 21 days of harvest.
Sugar Cane (Hawaii only)	Pineapple Disease (**Thielaviopsis** sp. or **Ceratocystis** sp.)		10 (1:1,600)	Apply to cut seed pieces as a cold or hot dip. Cold Dip—Immerse seed pieces to provide thorough coverage; remove and allow to drain.
			5 (1:3,200)	Hot Dip—Maintain temperature of dip at 52°C and soak seed pieces for 20 minutes; remove and allow to drain. Do not use treated seed pieces for food, feed or fodder. Do not graze or feed treated forage to livestock.

FOR USE ON FALL-SEEDED WHEAT IN IDAHO, OREGON AND WASHINGTON ONLY

(Not for use in California)

Diseases	Fl Oz/A	Fl Oz/100 Gal	Remarks
Pseudocercosporella sp. (Foot Rot, Strawbreaker, Eye Spot)	20		Apply TOPSIN M at the rate indicated in a single application by air or ground after wheat is tillered but before stem elongation has begun. Use sufficient water to obtain thorough coverage. Do not make more than one application per season. Do not cut for hay within 90 days of application. Do not allow livestock to graze in treated areas before harvest.

TANK MIX FOR USE ON APPLES

(Not for use in California)

Diseases	Rate/ Acre	Rate/ 100 Gal	Remarks
Scab (**Venturia** sp.) Bitter Rot (**Glomerella** sp.) Black Rot (**Physalospora** sp.) Flyspeck (**Schizothyrium** sp.) Powdery Mildew (**Podosphaera** sp.) Sooty Blotch (**Gloeodes** sp.) Brown Rot (**Monilinia** sp.) Cedar Apple Rust (**Gymnosporangium** sp.) Quince Rust (**Gymnosporangium** sp.)	TOPSIN M 10-16 fl oz + Mancozeb 80WP 3-4 lbs	TOPSIN M 2½-4 fl oz + Mancozeb 80WP ¾-1 lb	Apply at ½" green tip and repeat at 7- to 14-day intervals or as needed through the cover sprays. One quart spray oil may be added per 100 gallons. Add oil to the spray tank last. Use the high rate of TOPSIN M and add spray oil to the spray mixture for varieties more susceptible to powdery mildew, and for scab if an application is missed during an infection period (apply as soon as possible after infection period in order to deactivate scab and to prevent further infection). Do not apply within 30 days of harvest. Do not graze treated areas.

Observe the more restrictive cautions and limitations on labeling of products used in mixtures

TANK MIX FOR USE ON APPLES

(Not for use in California)

Diseases	Rate/ Acre	Rate/100 Gal	Remarks
Scab (**Venturia** sp.) Bitter Rot (**Glomerella** sp.) Black Rot (**Physalospora** sp.) Flyspeck (**Schizothyrium** sp.) Powdery Mildew (**Podosphaera** sp.) Sooty Blotch (**Gloeodes** sp.)	TOPSIN M 10-16 fl oz + Captan 50WP 3-4 lbs OR Captan 80WP 2-2½ lbs	TOPSIN M 2½-4 fl oz + Captan 50WP ¾-1 lb OR Captan 80WP ½-⅝ lb	Apply at ½" green tip and repeat at 7- to 14-day intervals or as needed through the cover sprays. Use the high rate of TOPSIN M for varieties more susceptible to powdery mildew. If an application is missed during an infection period, apply the higher rates as soon as possible after the infection period in order to deactivate scab and to prevent further infection. NOTE: Spray injury may result if Captan is used with immediately before, or closely following an oil spray.

Observe the more restrictive cautions and limitations on labeling of products used in mixtures.

TANK MIX FOR USE ON APPLES

(Not for use in California)

Diseases	Rate/ Acre	Rate/ 100 Gal	Remarks
Scab (**Venturia** sp.) Bitter Rot (**Glomerella** sp.) Black Rot (**Physalospora** sp.) Flyspeck (**Schizothyrium** sp.) Powdery Mildew (**Podosphaera** sp.) Sooty Blotch (**Gloeodes** sp.) Cedar Apple Rust (**Gymnosporangium** sp.) Bull's Eye Rot (**Gloeosporium** sp.)	TOPSIN M 10-16 fl oz + Maneb 80WP 6-8 lbs	TOPSIN M 2½-4 fl oz + Maneb 80WP 1½-2 lbs	Begin applications at green tip and repeat at pink bud and at 7- to 14-day intervals or as needed through the cover sprays. Use the higher rates and closer interval under severe disease conditions. With Golden Delicious apples, adverse effects on fruit finish and color may result. Use the lower rates of TOPSIN M when finish and color are of primary importance. Do not apply within 30 days of harvest. Do not graze livestock in treated areas.

Observe the more restrictive cautions and limitations on labeling of products used in mixtures.

TANK MIX FOR USE ON APPLES

(Not for use in California)

Diseases	Rate/ Acre	Rate/100 Gal	Remarks
Scab (**Venturia** sp.) Bitter Rot (**Glomerella** sp.) Black Rot (**Physalospora** sp.) Flyspeck (**Schizothyrium** sp.) Powdery Mildew (**Podosphaera** sp.) Sooty Blotch (**Gloeodes** sp.)	TOPSIN M 10-16 fl oz + Polyram 80WP 3-4 lbs	TOPSIN M 2½-4 fl oz + Polyram 80WP ¾-1 lb	Begin applications at green tip and repeat applications as needed through the cover sprays Do not apply within 30 days of harvest. Do not graze livestock in treated areas.

Observe the more restrictive cautions and limitations on labeling of products used in mixtures.

TANK MIX FOR USE ON APPLES

(Not for use in California)

Diseases	Rate/ Acre	Rate/100 Gal	Remarks
Scab (**Venturia** sp.) Bitter Rot (**Glomerella** sp.) Black Rot (**Physalospora** sp.) Flyspeck (**Schizothyrium** sp.) Powdery Mildew (**Podosphaera** sp.) Sooty Blotch (**Gloeodes** sp.)	TOPSIN M 10-16 fl oz + Sulfur 12-30 lbs a.i.	TOPSIN M 2½-4 fl oz + Sulfur 3-7½ lbs a.i.	Begin applications at green tip and repeat at pink bud and at 7- to 14-day intervals or as needed through the cover sprays. Use the higher rates and closer interval under severe disease conditions.

Observe the more restrictive cautions and limitations on labeling of products used in mixtures.

TANK MIX FOR USE ON STONE FRUITS

(Apricots, Cherries, Nectarines, Peaches, Plums and Prunes)
(Not for use in California)

Diseases	Rate/ Acre	Rate/100 Gal	Remarks
Brown Rot Blossom Blight (**Monilinia** spp.) Fruit Brown Rot (**Monilinia** spp.) Peach Scab (**Cladosporium** sp.) Powdery Mildew (**Podosphaera** sp. and **Sphaerotheca** sp.) Cherry Leaf Spot (**Coccomyces** sp.) Russet Scab Black Knot (**Dibotryon** sp.)	TOPSIN M 15-20 fl oz + Captan 50WP 3-6 lbs **OR** Captan 80WP 2-3¾ lbs	TOPSIN M 5-6½ fl oz + Captan 50WP 1-2 lbs **OR** Captan 80WP ⅝-1¼ lbs	BROWN ROT BLOSSOM BLIGHT—Apply at early bloom stages before infection occurs (apricots—red bud; cherries—early popcorn; nectarines & peaches—pink bud; plums & prunes—green tip). If blossoming is prolonged or conditions favorable for disease persist, make a second application 14 days later. FRUIT BROWN ROT—After blossom blight sprays, make a preharvest application any time from 3 weeks before harvest to 1 day before harvest. Make a second application if conditions favorable for disease persist or harvest is prolonged. PEACH SCAB, POWDERY MILDEW—Use same schedule as for Brown Rot Blossom Blight and continue at 10- to 14-day intervals to 1 day before harvest. Make an additional application 2 to 3 weeks after harvest. CHERRY LEAF SPOT—Use same schedule as for Brown Rot Blossom Blight and continue at 10- to 14-day intervals to 1 day before harvest. Make an additional application 2 to 3 weeks after harvest. RUSSET SCAB (PLUMS & PRUNES)—Use same schedule as for Brown Rot Blossom Blight; include full coverage spray at full bloom. BLACK KNOT—Apply at first, second, and third cover sprays at 10- to 14-day intervals. Do not apply within 1 day of harvest. NOTE: This tank mixture will not control bacterial blast, but will supplement a dormant program for peach leaf curl and shot-hole (**Coryneum** blight); refer to other fungicide labels for dormant period control information. Do not use this tank mixture with spray oils.

Observe the more restrictive cautions and limitations on labeling of products used in mixtures.

TANK MIX FOR USE ON STONE FRUITS

(Apricots, Cherries, Nectarines, Peaches, Plums and Prunes)
(Not for use in California)

Diseases	Rate/ Acre	Rate/100 Gal	Remarks
Brown Rot Blossom Blight (**Monilinia** sp.) Fruit Brown Rot (**Monilinia** sp.) Powdery Mildew (**Podosphaera** sp. and **Sphaerotheca** sp.) Leaf Spot (**Coccomyces** sp.) Shot-hole (**Coryneum** sp.) Rust (**Tranzschellia** sp.) Peach Canker Peach Scab (**Cladosporium** sp.) Black Knot (**Dibotryon** sp.)	TOPSIN M 15-30 fl oz + Sulfur 12-18 lbs a.i.	TOPSIN M 5-10 fl oz + Sulfur 4-6 lbs a.i.	BROWN ROT BLOSSOM BLIGHT—Make application at early bloom & again at full bloom. FRUIT BROWN ROT—In addition to bloom sprays, apply 1 or 2 sprays starting 3 weeks before harvest. If needed under severe disease pressure, apply additional sprays at 10- to 14-day intervals between full bloom and final preharvest sprays. POWDERY MILDEW—In addition to bloom sprays, make applications at shuck fall and first cover. LEAF SPOT—Make application when leaves first unfold and first, second, and third cover at 10- to 14-day intervals and 1 spray 14 to 21 days after harvest. SHOT-HOLE, RUST, PEACH CANKER & PEACH SCAB—Apply at pink, bloom, petal fall, shuck split, shuck fall, and at 10- to 14-day intervals as long as necessary. BLACK KNOT—Apply at first, second, and third cover sprays at 10- to 14-day intervals. Do not apply within 1 day of harvest.

Observe the more restrictive cautions and limitations on labeling of products used in mixtures.

TANK MIX FOR USE ON STONE FRUITS

(Apricots, Cherries, Nectarines, and Peaches)
(Not for use in California)

Diseases	Crop	Rate/ Acre	Rate/100 Gal	Remarks
Brown Rot Blossom Blight (**Monilinia** sp.) Fruit Brown Rot (**Monilinia** sp.) Powdery Mildew (**Podosphaera** sp. and **Sphaerotheca** sp.) Cherry Leaf Spot (**Coccomyces** sp.) Shot-hole (**Coryneum** sp.) Rust (**Tranzschellia** sp.) Peach Canker Peach Scab (**Cladosporium** sp.)	Apricots and Nectarines	TOPSIN M 15-30 fl oz + Ziram 3-4½ lbs a.i.	TOPSIN M 5-10 fl oz + Ziram 1-1½ lbs a.i.	BROWN ROT BLOSSOM BLIGHT—Make application at early bloom and again at full bloom. FRUIT BROWN ROT—In addition to bloom sprays, apply 1 or 2 sprays starting 3 weeks before harvest. If needed under severe disease pressure, apply additional sprays at 10- to 14-day intervals between full bloom and final preharvest sprays. POWDERY MILDEW—In addition to bloom sprays, make applications at shuck fall and first cover. CHERRY LEAF SPOT—Make application when leaves first unfold & at first, second, and third cover at 10- to 14-day intervals and 1 spray 14-21 days after harvest. SHOT-HOLE, RUST, PEACH CANKER & PEACH SCAB—Apply at pink, bloom, petal fall, shuck split, shuck fall, and at 10- to 14-day intervals as long as necessary. For apricots, do not apply after early cover periods. For cherries, do not apply within 7 days of harvest. For nectarines & peaches, last application should be made no later than at petal fall.
	Cherries	TOPSIN M 15-30 fl oz + Ziram 3⅜ lbs a.i.	TOPSIN M 5-10 fl oz + Ziram 1⅛ lbs a.i.	
	Peaches	TOPSIN M 15-30 fl oz + Ziram 3-4½ lbs a.i.	TOPSIN M 5-10 fl oz + Ziram 1-1½ lbs a.i.	

Observe the more restrictive cautions and limitations on labeling of products used in mixtures.

TANK MIX FOR USE ON STONE FRUITS
(Apricots, Nectarines and Peaches)
(Not for use in California)

Diseases	Rate/ Acre	Rate/100 Gal	Remarks
Brown Rot Blossom Blight (**Monilinia** sp.) Fruit Brown Rot (**Monilinia** sp.) Peach Scab (**Cladosporium** sp.) Powdery Mildew (**Podosphaera** sp. and **Sphaerotheca** sp.)	TOPSIN M 30 fl oz + Maneb 80WP 6-8 lbs	TOPSIN M 10 fl oz + Maneb 80WP 2-2⅔ lbs	BROWN ROT BLOSSOM BLIGHT—Apply at early bloom stages before infection occurs (apricots—red bud, nectarines & peaches—pink bud) For this application only, TOPSIN M plus Maneb 80WP may be used in combination with superior-type spray oil. If blossoming is prolonged or conditions favorable for disease persist, make a second application 14 days later. FRUIT BROWN ROT—After blossom blight sprays, make an additional application 3 weeks before harvest. PEACH SCAB POWDERY MILDEW—Use same schedule as for Brown Rot Blossom Blight plus applications at shuck split, shuck fall, and 14 days later. Do not apply within 14 days of harvest. NOTE: This tank mixture will not control bacterial blast, but will supplement a dormant program for peach leaf curl and shot-hole (**Coryneum** blight); refer to other fungicide labels for dormant period control information.

Observe the more restrictive cautions and limitations on labeling of products used in mixtures.

TANK MIX FOR USE ON PEANUTS
(Not for use in California)

Diseases	Rate/ Acre	Remarks
Leaf Spot (**Cercospora** sp.) Rust (**Puccinia** sp.) Web Blotch (**Ascochyta** sp.)	TOPSIN M 5 fl oz + Mancozeb 80WP 1-2 lbs	Begin application 35 to 40 days after planting or when disease first appears. Repeat at the following intervals: for **Cercospora** leaf spot at 10 to 14 days; for Rust at 7 to 10 days; and for **Ascochyta** web blotch at 7 to 14 days. Spray oil may be added at the rate of 1 pint to 1 quart per acre. Do not graze or feed treated vines, hay or hulls to livestock. Do not apply within 14 days of harvest.

Observe the more restrictive cautions and limitations on labeling of products used in mixtures.

PRECAUTIONARY STATEMENTS

HAZARDS TO HUMANS (AND DOMESTIC ANIMALS)
CAUTION

HARMFUL IF SWALLOWED, ABSORBED THROUGH SKIN, OR INHALED. CAUSES MODERATE EYE IRRITATION.

Avoid contact with skin, eyes, or clothing. Avoid breathing spray mist. Wash thoroughly with soap and water after handling. Remove contaminated clothing and wash before reuse.

Do not apply this product in such a manner as to directly or through drift expose workers or other persons. The area being treated must be vacated by unprotected persons.

ENVIRONMENTAL HAZARDS

This pesticide is toxic to fish. Do not apply directly to water. Do not apply where runoff is likely to occur. Do not contaminate water by cleaning of equipment or disposal of wastes.

STORAGE AND DISPOSAL

Do not contaminate water, food, or feed by storage or disposal.

Storage Instructions: Store in the original container. Do not store in a manner where cross-contamination with other pesticides, fertilizers, food or feed could occur. If spilled during storage or handling, absorb with sand or other inert material and dispose of absorbant in accordance with the Pesticide Disposal Instructions listed below.

Pesticide Disposal Instructions: Wastes resulting from the use of this product may be disposed of on site or at an approved waste disposal facility.

Container Disposal Instructions: Triple rinse (or equivalent). Then offer for recycling or reconditioning, or puncture and dispose of in a sanitary landfill, or by incineration, or, if allowed by state and local authorities, by burning. If burned, stay out of smoke.

WARRANTY AND DISCLAIMER

Pennwalt Corporation warrants that this material conforms to the chemical description on the label and is reasonably fit for the purposes referred to in the Directions for Use, subject to the risks referred to therein. PENNWALT MAKES NO OTHER EXPRESS OR IMPLIED WARRANTY OF FITNESS OR MERCHANTABILITY OR ANY OTHER EXPRESS OR IMPLIED WARRANTY. IN NO CASE SHALL PENNWALT OR SELLER BE LIABLE FOR CONSEQUENTIAL. SPECIAL OR INDIRECT DAMAGES RESULTING FROM THE USE OR HANDLING OF THIS PRODUCT INCLUDING. BUT NOT LIMITED TO LOSS OF PROFITS, BUSINESS REPUTATION, OR CUSTOMERS, LABOR COST, OR OTHER EXPENSES INCURRED IN PLANTING OR HARVESTING

Pennwalt and seller offer this product and the buyer and user accept it subject to the foregoing conditions of sale and warranty which may be varied only by agreement in writing signed by a duly authorized representative of Pennwalt.

5-Z304S-01 D8 (4/88)

TOPSIN® M 4.5F

TURF AND ORNAMENTALS FUNGICIDE

NOT FOR USE IN CALIFORNIA
SHAKE WELL BEFORE USING
4.5 Pounds per Gallon Flowable

ACTIVE INGREDIENT:
Thiophanate-methyl (dimethyl[(1,2-phenylene)-bis(iminocarbonothioyl)]bis[carbamate])* 46.2%
INERT INGREDIENTS: 53.8%
TOTAL 100.0%

*Also known as dimethyl 4,4'-o-phenylenebis[3-thioallophanate]
*TOPSIN is a registered trademark of Nippon Soda Company, Ltd., and is licensed to Pennwalt Corporation. Covered by one or more of the following U.S. Patents: 3,769,308; 3,856,847; 4,020,095; and 4,029,813.

EPA Registration No. 4581-352 EPA Establishment No. 4581-TX-1

KEEP OUT OF REACH OF CHILDREN
CAUTION

STATEMENT OF PRACTICAL TREATMENT

IF SWALLOWED, induce vomiting immediately by giving two glasses of water and sticking finger down throat. Call a physician. Never give anything by mouth to an unconscious person.

IF IN EYES, immediately flush eyes with plenty of water for at least 15 minutes. Get medical attention if irritation persists.

IF ON SKIN, immediately flush skin with plenty of water.
See back panel for Precautionary Statements

PRECAUTIONARY STATEMENTS

HAZARDS TO HUMANS (AND DOMESTIC ANIMALS)

HARMFUL IF SWALLOWED, ABSORBED THROUGH SKIN, OR INHALED. CAUSES MODERATE EYE IRRITATION.

Avoid contact with skin, eyes, or clothing. Avoid breathing spray mist. Wash thoroughly with soap and water after handling. Remove contaminated clothing and wash before reuse.

Do not apply this product in such a manner as to directly or through drift expose workers or other persons. The area being treated must be vacated by unprotected persons.

ENVIRONMENTAL HAZARDS

This pesticide is toxic to fish. Do not apply directly to water. Do not apply where runoff is likely to occur. Do not contaminate water by disposal of equipment washwaters.

GENERAL INFORMATION

Apply TOPSIN M with ground or aerial equipment, using sufficient volume of spray to provide thorough coverage. Add required amount of TOPSIN M to partially filled tank agitated by mechanical or hydraulic means and then add remaining required amount of water. Continuous agitation is required to keep the material in suspension. Do not tank mix with copper-containing materials or with highly alkaline pesticides, such as Bordeaux mixture or lime sulfur. No claim of compatability with other pesticides is implied. Use the higher rates under conditions of severe disease pressure. Also, see local State Extension Service recommendations for application schedules.

IMPORTANT: If, after using TOPSIN M as recommended, treatment is not effective, a tolerant strain of fungi may be present (consult your local

Pennwalt representative). Consult your State Agricultural Experiment Station or your State Agricultural Extension Service for advice on the prompt use of some other suitable fungicide.

DIRECTIONS FOR USE

It is a violation of Federal law to use this product in a manner inconsistent with its labeling.

TURF

For use on golf course greens, tees, and fairways of Bentgrasses, Bluegrasses, Bermudagrasses, Fescues, Ryegrasses, St. Augustine grasses or their mixtures. TOPSIN M is not phytotoxic to any of the above mentioned grasses when used in accordance with the label. TOPSIN M is to be used for the prevention and control of the below mentioned diseases (provides both curative and protective action).

Use spray mixture the same day it is prepared. Spray uniformly over the area to be treated. Apply after mowing or avoid mowing for twelve hours after application.

Apply recommended amounts in sufficient water to obtain thorough coverage, usually five gallons per 1,000 square feet of turf area. When treating golf greens, always treat aprons and approaches. Use the highest recommended rate under conditions of severe disease pressure.

FUSARIUM BLIGHT (**Fusarium roseum**)—Apply TOPSIN M at 5 to 10 fluid ounces per 1,000 square feet. Water into the root zone by irrigating with one inch of water immediately after application. Make two applications at 10 to 14-day intervals, beginning applications when the disease first appears.

DOLLAR SPOT (**Sclerotinia homoeocarpa**), COPPER SPOT (**Gloeocercospora sorghi**)—Apply TOPSIN M at 1 to 2 fluid ounces per 1,000 square feet beginning when the disease first appears. Continue at 10 to 14-day intervals through the season or as needed.

LARGE BROWN PATCH (**Rhizoctonia solani**)—Apply TOPSIN M at 1 to 2 fluid ounces per 1,000 square feet beginning when the disease first appears and continue at 10 to 14-day intervals through the season or as needed. When conditions are unusually favorable for development of disease, reduce interval to 5 to 7 days.

STRIPE SMUT (**Ustilago striiformis**)—Apply TOPSIN M at 5 to 10 fluid ounces per 1,000 square feet. Water into the root zone by irrigating with one inch of water immediately after application. Make two applications at 14 to 21-day intervals. Begin applications when the disease first appears, either in the spring or fall.

ORNAMENTALS, ROSES, FLOWERS, SHADE TREES (FIELD AND GREENHOUSE)

FOLIAR SPRAY: Begin applications when disease first appears and repeat at 10 to 14-day intervals throughout the growing season; shorten interval during humid, rainy weather.

Use at the following rates:
Use 10 fluid ounces per 100 gallons (1 teaspoonful per 2 gallons) for Powdery Mildew and Botrytis Gray Mold.

Use 20 fluid ounces per 100 gallons for Anthracnose (for shade trees and woody ornamentals, begin at bud break and make 2 or 3 additional applications at 10 to 14-day intervals). Black Spot of roses; Cercospora, Entomosporium, Ramularia, and Septoria Leaf Spots; Ascochyta and Phomopsis Blights; Didymellina Leaf Spots of iris; Corynespora Leaf Spot of ligustrum; Ovulinia of azalea and rhododendron (begin as flowers open); Scab of pyracantha and flowering crab.

Addition of a surfactant to the spray mixture improves distribution of the spray on hard-to-wet plants such as roses.

For aerial application, use 10-20 fluid ounces per acre.

DRENCH TREATMENT: Botrytis, Fusarium, Rhizoctonia, and Sclerotinia stem, crown, and root rots on herbaceous annuals, perennials, and bedding plants; Cylindrocladium and Thielaviopsis rots on woody ornamentals such as azaleas, rhododendrons, conifers, and poinsettias—Use 20 fluid ounces per 100 gallons; apply as a drench or heavy spray (1 to 2 pints per square foot) after transplanting into propagation beds or containers. Repeat at 2 to 4-week intervals during periods favorable for disease. TOPSIN M does not control **Pythium** spp. or **Phytophthora** spp.

PREPLANT DIP TREATMENT:

Plants or Cuttings—For diseases listed under Drench Treatment, use 20 fluid ounces per 100 gallons of water; immerse plants or cuttings for 10 to 15 minutes; remove and allow to drain.

Bulbs (Easter Lily, Tulip, Gladiolus, Daffodil, Iris) Fusarium and Penicillium Rots—Use 33 fluid ounces per 100 gallons of water (2 teaspoonfuls per gallon). Soak cleaned bulbs for 15 to 30 minutes in warm dip (80° to 85°F) preferably within 48 hours after digging. Dry bulbs after treatment. If bulbs are for forcing, treat bulbs that have been heat-cured.

CHEMIGATION

NOT FOR THIS USE IN CALIFORNIA
GENERAL INSTRUCTIONS

Apply this product only through sprinkler irrigation systems including center pivot, lateral move, end tow, side (wheel) roll, traveler, big gun, solid set, or hand move. Do not apply this product through any other type of irrigation system.

Crop injury, lack of effectiveness, or illegal pesticide residues in the crop can result from nonuniform distribution of treated water.

If you have any questions about calibration, you should contact State Extension Service specialists, equipment manufacturers or other experts.

A person knowledgeable of the chemigation system and responsible for its operation, or under the supervision of the responsible person, shall shut the system down and make necessary adjustments should the need arise.

Do not connect chemigation system to any public water system. Public water system means a system for the provision of piped water for human consumption if such system has at least 15 service connections or regularly serves an average of at least 25 individuals daily at least 60 days out of the year.

SYSTEM REQUIREMENTS

Systems utilizing a pressurized water and pesticide injection system must meet the following requirements.

The system must contain a functional check valve, vacuum relief valve and low pressure drain appropriately located on the irrigation pipeline to prevent water source contamination from backflow.

The pesticide injection pipeline must contain a functional, automatic, quick-closing check valve to prevent the flow of fluid back toward the injection pump.

The pesticide injection pipeline must also contain a functional, normally closed, solenoid-operated valve located on the intake side of the injection pump and connected to the system interlock to prevent fluid from being withdrawn from the supply tank when the irrigation system is either automatically or manually shut down.

The system must contain functional interlocking controls to automatically shut off the pesticide injection pump when the water pump motor stops.

The irrigation line or water pump must include a functional pressure switch which will stop the water pump motor when the water pressure decreases to the point where pesticide distribution is adversely affected.

Systems must use a metering pump, such as a positive displacement injection pump (e.g., diaphragm pump) effectively designed and constructed of materials that are compatible with pesticides and capable or being fitted with a system interlock.

INSTRUCTIONS FOR SPRINKLER (OVERHEAD) IRRIGATION:

Observe the requirements in the System Requirements section above.

Do not apply when wind speed favors drift beyond the area intended for treatment.

Apply TOPSIN M only through systems containing anti-syphon and check valves designed to prevent water source contamination or overflow of the mix tank and containing interlocking controls between the metering device and the water pump to insure simultaneous shut-off.

Maintain a gentle continuous agitation in mix tank during mixing and application to assure a uniform suspension.

Greater accuracy in calibration and distribution will be achieved by injecting a larger volume of a more dilute suspension per unit time.

Application of more than recommended quantities of irrigation water per acre may result in decreased product performance.

Do not apply when wind speed favors drift, when system connections or fittings leak, when nozzles do not provide uniform distribution or when lines containing the product cannot be flushed and must be dismantled and drained in a center pivot system, block the nozzle set nearest the well pivot injection unit to prevent spray being applied to this area.

Where sprinkler distribution patterns do not overlap sufficiently, unacceptable disease control may result.

Allow sufficient time for pesticide to be flushed through all lines and all nozzles before turning off irrigation water.

TOPSIN M may be applied in conjunction with chemically neutral liquid fertilizers. Application in conjunction with highly alkaline fertilizers, such as aqueous ammonia, may cause a degradation of the pesticide, resulting in reduced performance and should be avoided.

Check local restrictions and requirements regarding sprinkler irrigation applications, as they may vary from state to state.

SPRAY PREPARATION

Remove scale, pesticide residues, and other foreign matter from the chemical tank and entire injector system. Flush with clean water.

Prepare a suspension of TOPSIN M in a mix tank. Fill the tank with 1/2 or 3/4 the desired amount of water. Start mechanical or hydraulic agitation. Slowly add the required amount of TOPSIN M and then the remaining volume of water.

APPLICATION INSTRUCTIONS

Set sprinkler system to deliver 0.1 to 1.25 inches of water per acre. Volumes of water higher than this may reduce efficacy. Start sprinkler and then uniformly inject the suspension of TOPSIN M into the irrigation water line so as to deliver the desired rate per acre. The suspension of TOPSIN M should be injected with a positive displacement pump into the main line ahead of a right angle turn to insure adequate mixing.

NOTE: When treatment with TOPSIN M has been completed, do not irrigate the treated area for 24 to 48 hours to prevent washing the chemical off the crop.

STORAGE AND DISPOSAL

Do not contaminate water, food, or feed by storage or disposal.

Storage Instructions: Store in the original container in a dry area. Do not store in a manner where cross-contamination with other pesticides, fertilizers, food or feed could occur. If spilled during storage or handling, sweep up spillage and dispose of in accordance with the Pesticide Disposal Instructions listed below.

Pesticide Disposal Instructions: Wastes resulting from the use of this product may be disposed of on site or at an approved waste disposal facility.

Container Disposal Instructions: After bag has been emptied, dispose of empty bag in a sanitary landfill or by incineration, of if allowed by State and local authorities, by burning. If burned, stay out of smoke.

WARRANTY AND DISCLAIMER

Pennwalt Corporation warrants that this material conforms to the chemical description on the label and is reasonably fit for the purposes referred to in the Directions for Use, subject to the risks referred to therein PENNWALT MAKES NO OTHER EXPRESS OR IMPLIED WARRANTY OF FITNESS OR MERCHANTABILITY OR ANY OTHER EXPRESS OR IMPLIED WARRANTY. IN NO CASE SHALL PENNWALT OR SELLER BE LIABLE FOR CONSEQUENTIAL, SPECIAL OR INDIRECT DAMAGES RESULTING FROM THE USE OR HANDLING OF THIS PRODUCT INCLUDING, BUT NOT LIMITED TO, LOSS OF PROFITS, BUSINESS REPUTATION, OR CUSTOMERS, LABOR COST, OR OTHER EXPENSES INCURRED IN PLANTING OR HARVESTING.

Pennwalt and seller offer this product and the buyer and user accept it subject to the foregoing conditions of sale and warranty which may be varied only by agreement in writing signed by a duly authorized representative of Pennwalt.

ZIRAM F-4

FUNGICIDE

ACTIVE INGREDIENT:

Ziram (Zinc dimethyldithiocarbamate)	39.2%
(Zinc expressed as metallic . . . 8.4%)	
INERT INGREDIENTS:	60.8%
TOTAL	100.0%

This product contains 4 pounds Ziram per gallon.
1 Gallon Ziram F-4 is equivalent to 5¼ pounds Ziram W-76

EPA Registration No. 4581-230 EPA Establishment No. 4581-TX-1

KEEP OUT OF REACH OF CHILDREN

AVISO WARNING

PRECAUCION AL USUARIO:
Si usted no lee ingles, no use este producto hasta que le etiqueta haya sido explicado ampliamente.

STATEMENT OF PRACTICAL TREATMENT

In case of contact with skin, wash with soap and water; for eyes, flush with plenty of water for 15 minutes and get medical attention.

PRECAUTIONARY STATEMENTS

HAZARDS TO HUMANS AND DOMESTIC ANIMALS

CAUSES EYE AND SKIN IRRITATION. DO NOT GET IN EYES, ON SKIN, OR ON CLOTHING. HARMFUL IF SWALLOWED, INHALED, OR ABSORBED THROUGH SKIN. Avoid inhalation of spray mist. Wash contaminated clothing before reuse.

ENVIRONMENTAL HAZARDS

This product is toxic to fish. Use with care when applying in areas adjacent to any body of water. Keep out of lakes, streams, or ponds. Do not apply when weather conditions favor drift from target area. Do not contaminate water when disposing of equipment washwaters.

GENERAL INFORMATION

ZIRAM F-4 is a liquid paste formulation effective against many fungus diseases. It has excellent adhesive properties and is safe on a wide range of plants. It is compatible with Parathion, Malathion, and many commonly used pesticides.

DIRECTIONS FOR USE

It is a violation of Federal law to use this product in a manner inconsistent with its labeling.

Do not apply this product through any type of irrigation system.

For dilute application, use ZIRAM F-4 at recommended rate per 100 gallons of spray unless otherwise stated. Three teaspoonfuls per gallon of water is equivalent to 3 pints per 100 gallons of water.

For concentrate applications, by air, use ZIRAM F-4 in 20 to 30 gallons of spray per acre; by ground, use in a minimum of 40 gallons of spray per acre.

Use the high rate under severe disease pressure or when conditions promote rapid disease development.

RE-ENTRY STATEMENT

Do not apply this product in such a manner as to directly or through drift expose workers or other persons. The area being treated must be vacated by unprotected persons.

Do not enter treated areas until spray has dried unless protective clothing is worn.

Because certain states may require more restrictive reentry intervals for various crops treated with this product, consult your State Department of Agriculture for further information.

Written or oral warnings must be given to workers who are expected to be in an area about to be treated with this product. In case of accidental exposure, workers should wash thoroughly with soap and water and contaminated clothing should be removed and washed before reuse. When oral warnings are given, warnings shall be given in a language customarily understood by workers. Oral warnings must be given if there is reason to believe that written warnings cannot be understood by workers. Written warnings must include the following information. "WARNING. Area treated with ZIRAM F-4 Fungicide on (date of application). Do not enter without appropriate protective clothing until spray has dried. In case of accidental exposure, workers should wash thoroughly with soap and water and contaminated clothing should be removed and washed before reuse."

FRUIT AND NUT CROPS

CROP	PTS/ 100 GALLONS	PTS/A	DISEASE	DIRECTIONS	LIMITATIONS (see footnotes)
Almonds	2-4	12-18	Shot-hole (Fungus)	Apply at popcorn, full-bloom, petal fall, and 5 weeks later.	1,2
			Brown Rot (Blossom and Twig Blight)	Apply at popcorn and full bloom.	
			Leaf Blight Scab	Apply at petal fall and 5 weeks later.	
Apples	2	11-15	Apple Scab (West Coast Only)	Apply at pre-bloom, calyx, and cover sprays. In Washington, apply in first cover spray and in early September on Delicious variety and late September on Winesaps as a protectant spray only.	3, 4
			Bull's Eye Rot (Pacific Northwest Only)	Apply in the first cover spray and in a pre-harvest spray before fall rains begin. Repeat as necessary, especially if excessive rainfall occurs.	
Apricots†	3	12-15	Brown Rot	Apply in pre-bloom, blossom, and petal fall stages and in early cover sprays.	5
			Shot-hole (Fungus)	Apply at pre-bloom, early bloom, and full bloom. Under severe disease conditions, or when weather is extremely wet, make additional applications as often as every 5 days until full bloom.	
	3-4			For post-harvest application in the fall before heavy rains.	6
Cherries†	2		Brown Rot	Apply at popcorn, full bloom, petal fall, and as necessary in cover applications.	9

CROP	PTS/ 100 GALLONS	PTS/A	DISEASE	DIRECTIONS	LIMITATIONS (see footnotes)
			Leaf Spot	Apply at petal fall, shuck fall, and about 2 weeks after shuck fall.	
Nectarines, Peaches	2-4	12-15	Peach Leaf Curl Shot-hole (Fungus)	Apply at dormant and popcorn stages. For West Coast use on peaches for leaf curl, apply just before fall rains or in November and repeat in 3 months while trees are still dormant and again in popcorn stage.	12, 13
	2-3		Brown Rot	Apply at pink bud, 25% to 75% of full bloom, petal fall, and cover stages.	
Pears	2	11-15	Pear Scab (West Coast Only)	Apply in pre-bloom, calyx, and cover stages.	3, 4
			Bull's Eye Rot (Pacific Northwest Only)	Apply pre-harvest prior to fall rains. Repeat if necessary, especially if excessive rainfall occurs.	
Pecans†	3		Scab	Begain applications when leaves are 1/4 to 1/2 inch long. Repeat after pollination is complete and up to 5 cover sprays at 3 to 4-week intervals through August.	14

ORNAMENTAL CROPS

Crop	Pts/ 100 Gallons	Disease	Directions
Asters	2 pts/12 gals	Rust	Begin application in early season and continue at 7-day intervals until blossoms form.
Carnations,* Marigolds, Morning Glory, Snapdragons*	2 1/2	Rust *See below for additional disease control	Apply at 7 to 10-day intervals as needed throughout growing season.
Carnations, Chrysanthe-mums, Dahlias,* Gladiolus, Snapdragons	2 1/2	Leaf Spot *See below for additional disease control	Apply at 7 to 10-day intervals as needed throughout growing season.
Azaleas, Camellias	3-4	Petal Blight	Begin application with first bloom. Repeat at 7 to 10-day intervals through bloom. Apply as a drench to mulch and litter beneath and around bushes.
Dahlias, Delphinium, Golden Glow, Hollyhocks, Phlox	2 1/2	Powdery Mildew	Apply at 7-day intervals throughout growing season.
Roses	2 1/2	Black Spot Powdery Mildew	Begin application when leaves unfold. Repeat at 7 to 10-day intervals throughout season
Pine Seedlings (nursery beds)	4	Fusiform Rust	Begin application at time of emergence. Repeat at 3 to 5-day intervals to mid-June.

Limitations

1. 12 pints/acre—Do not apply later than 5 weeks after petal fall or
2. 25 pints/acre—Do not apply later than petal fall
3. 18 pints/acre—No time limitation or
4. 24 pints/acre—Do not apply within 7 days of harvest
5. 20 pints/acre—Do not apply after early cover period
6. 22 pints/acre—Post-harvest application
7. NA
8. NA
9. 8 pints/acre—Do not apply within 7 days of harvest
10. NA
11. NA
12. 22 pints/acre—Last application at petal fall or
13. 15 pints/acre—Remove excess residues by wiping or brushing
14. 30 pints/acre—Do not apply later than August

†NOT FOR THIS USE IN CALIFORNIA

STORAGE AND DISPOSAL

Do not contaminate water, food, or feed by storage or disposal.

Storage Instructions: Store in the original container. Store in frost-free room. Store in the shade. Do not store in a manner where cross-contamination with other pesticides, fertilizers, food or feed could occur. If spilled during storage or handling, absorb with sand or other inert material and dispose of absorbant in accordance with the Pesticide Disposal Instructions listed below.

Pesticide Disposal Instructions: Pesticide wastes are acutely hazardous. Improper disposal of excess pesticide, spray mixture, or rinsate is a violation of Federal Law. If these wastes cannot be disposed of by use according to label instructions, contact your State Pesticide or Environmental Control Agency, or the Hazardous Waste representative at the nearest EPA Regional Office for guidance.

Container Disposal Instructions: Triple rinse (or equivalent). Then offer for recycling or reconditioning, or puncture and dispose of in a sanitary landfill, or by incineration, or, if allowed by state and local authorities, by burning. If burned, stay out of smoke.

WARRANTY AND DISCLAIMER

Pennwalt Corporation warrants that this material conforms to the chemical description on the label and is reasonably fit for the purposes referred to in the Directions for Use, subject to the risks referred to therein. PENNWALT MAKES NO OTHER EXPRESS OR IMPLIED WARRANTY OF FITNESS OR MERCHANTABILITY OR ANY OTHER EXPRESS OR IMPLIED WARRANTY. IN NO CASE SHALL PENNWALT OR SELLER BE LIABLE FOR CONSEQUENTIAL, SPECIAL, OR INDIRECT DAMAGES RESULTING FROM THE USE OR HANDLING OF THIS PRODUCT INCLUDING, BUT NOT LIMITED TO, LOSS OF PROFITS, BUSINESS REPUTATION, OR CUSTOMERS; LABOR COST; OR OTHER EXPENSES INCURRED IN PLANTING OR HARVESTING.

Pennwalt and seller offer this product and the buyer and user accept it subject to the foregoing conditions of sale and warranty which may be varied only by agreement in writing signed by a duly authorized representative of Pennwalt.

5-C312S-01 E8 (5/88)

ZIRAM W-76

FUNGICIDE

ACTIVE INGREDIENT:

Ziram (Zinc dimethyldithiocarbamate). 76%

(Zinc expressed as metallic . . . 16.25%)

INERT INGREDIENTS: 24%

TOTAL. 100%

EPA Registration No. 4581-140 EPA Establishment No. 33910-HI-01

KEEP OUT OF REACH OF CHILDREN

PELIGRO DANGER

PRECAUCION AL USARIO: Si usted no lee ingles, no use este producto hasta que le etiqueta haya sido explicado ampliamente.

STATEMENT OF PRACTICAL TREATMENT

IF IN EYES: Hold eyelids open and flush with a steady, gentle stream of water for 15 minutes. Get medical attention.

IF INHALED: Remove victim to fresh air. If not breathing give artificial respiration, preferably mouth-to mouth. Get medical attention.

IF SWALLOWED: Call a physician or Poison Control Center. Drink 1 or 2 glasses of water and induce vomiting by touching back of throat with finger. Do not induce vomiting or give anything by mouth to an unconscious person.

IF ON SKIN: Wash with plenty of soap and water. Get medical attention.

PRECAUTIONARY STATEMENTS

HAZARDS TO HUMANS (AND DOMESTIC ANIMALS)

DANGER

CORROSIVE. CAUSES IRREVERSIBLE EYE DAMAGE. MAY BE FATAL IF INHALED. HARMFUL IF SWALLOWED OR ABSORBED THROUGH SKIN. Do not get in eyes or on clothing. Wear goggles, face shield, or safety glasses when handling. Do not breathe dust or spray mist. Wear a mask or pesticide respirator jointly approved by the Mining Enforcement and Safety Administration, and the National Institute for Occupational Safety and Health. Avoid contact with skin or clothing. Wash thoroughly with soap and water after handling. Remove contaminated clothing and wash before reuse.

ENVIRONMENTAL HAZARDS

This product is toxic to fish. Use care when applying in areas adjacent to any body of water. Keep out of lakes, streams, or ponds. Do not apply when weather conditions favor drift from target area. Do not contaminate water when disposing of equipment washwaters.

PHYSICAL OR CHEMICAL HAZARDS

Keep away from fire and sparks.

GENERAL INFORMATION

ZIRAM W-76 is effective against many fungus diseases. It has excellent adhesive properties and is safe on a wide range of plants. It is compatible with Parathion, Malathion, and many commonly used pesticides. Wash the required amount of ZIRAM W-76 through the filter screen while filling the sprayer with water, or make into a paste with water and pour into the spray tank while filling.

DIRECTIONS FOR USE

It is a violation of Federal law to use this product in a manner inconsistent with its labeling.

Do not apply this product through any type of irrigation system.

For dilute application, use ZIRAM W-76 at recommended rate per 100 gallons of spray unless otherwise stated. Three level teaspoonfuls per gallon of water is equivalent to one pound per 100 gallons of water.

For concentrate applications, by air, use ZIRAM W-76 in 20 to 30 gallons of spray per acre; by ground, use in a minimum of 40 gallons of spray per acre.

Use the high rate under severe disease pressure or when conditions promote rapid disease development.

Do not apply this product in such a manner as to directly or through drift expose workers or other persons. The area being treated must be vacated by unprotected persons.

Do not enter treated areas without protective clothing until sprays have dried.

Because certain states may require more restrictive reentry intervals for various crops treated with this product, consult your state Department of Agriculture for further information.

Written or oral warnings must be given to workers who are expected to be in an area about to be treated with this product. In case of accidental exposure, workers should wash thoroughly with soap and water and contaminated clothing should be removed and washed before reuse. If symptoms of skin or eye irritation occur call a physician. When oral warnings are given, warnings shall be given in a language customarily understood by workers. Oral warnings shall be given if there is reason to believe that written warnings cannot be understood by workers. Written warnings must include the following information. "DANGER. Area treated with ZIRAM W-76 Fungicide on (date of application). Do not enter without protective clothing until spray has dried. In case of accidental exposure, workers should wash thoroughly with soap and water and contaminated clothing should be removed and washed before reuse. If symptoms of skin or eye irritation occur call a physician."

FRUIT AND NUT CROPS

Crop	Lbs/100 Gallons	Lbs/A	Disease	Directions	Limitations (see footnotes)
Almonds	1½-2 + 4 oz. Spreader-Sticker	8-12	Shot-hole (Fungus)	Apply at popcorn, full-bloom, petal fall, and 5 weeks later	1,2
			Brown Rot (Blossom and Twig Blight)	Apply at popcorn and full bloom	
			Leaf Blight Scab	Apply at petal fall and 5 weeks later	
Apples†	1-1½	7½-10	Apple Scab (West Coast Only)	Apply at pre-bloom, calyx, and cover sprays. In Washington, apply in first cover spray and in early September on Delicious variety and late September on Winesaps as a protectant spray only.	3,4
			Bull's-Eye Rot (Pacific North-west Only)	Apply in the first cover spray and in a pre-harvest spray before fall rains begin. Repeat as necessary, especially if excessive rainfall occurs.	
Apricots	1½	8-10	Brown Rot	Apply in pre-bloom blossom, and petal fall stages and in early cover sprays.	5
	1½ + 4 oz. Spreader-Sticker		Shot-hole (Fungus)	Apply at pre-bloom, early bloom, and full bloom. Under severe disease conditions, or when weather is extremely wet, make additional applications as often as every 5 days until full bloom.	
	1½-2			For post-harvest application in the fall before heavy rains.	6
Cherries†	1½		Brown Rot	Apply at popcorn, full bloom, petal fall, and as necessary in cover applications.	9
			Leaf Spot	Apply at petal fall, shuck fall, and about 2 weeks after shuck fall.	
Nectarines, Peaches	1½-3	8-10	Peach Leaf Curl Shot-hole (Fungus)	Apply at dormant and popcorn stages. For West Coast use on peaches for leaf curl, apply just before fall rains or in November and repeat in 3 months while trees are still dormant and again in popcorn stage.	12,13
	1½-3		Brown Rot	Apply at pink bud, 25% to 75% of full bloom, petal fall, and cover stages.	
Pears	1-1½	7½-10	Pear Scab (West Coast Only)	Apply in pre-bloom, calyx, and cover stages	3,4

Crop	Lbs/100 Gallons	Lbs/A	Disease	Directions	Limitations (see footnotes)
			Bull's Eye Rot (Pacific North-west Only)	Apply pre-harvest prior to fall rains. Repeat if necessary, especially if excessive rainfall occurs.	
Pecans†	2		Scab	Begin applications when leaves are 1/4 to 1/2 inch long. Repeat after pollination is complete and up to 5 cover sprays at 3-4 week intervals through August.	14

ORNAMENTAL CROPS†

Crop	Lbs/100 Gallons	Disease	Directions
Asters	1 lb/12 gals	Rust	Begin application in early season and continue at 7-day intervals until blossoms form.
Carnations,* Marigolds, Morning Glory, Snapdragons*	1⅔	Rust *See below for additional disease control	Apply at 7 to 10-day intervals as needed throughout growing season.
Carnations, Chrysanthemums, Dahlias,* Gladiolus, Snapdragons*	1⅔	Leaf Spot *See below for additional disease control	Apply at 7 to 10-day intervals as needed throughout growing season.
Azaleas, Camellias	1½-2	Petal Blight	Begin application with first bloom. Repeat at 7 to 10-day intervals through bloom. Apply as a drench to mulch and litter beneath and around bushes.
Dahlias, Delphinium, Golden Glow, Hollyhocks, Phlox	1⅔	Powdery Mildew	Apply at 7-day intervals throughout growing season.
Roses	1⅔	Black Spot Powdery Mildew	Begin application when leaves unfold. Repeat at 7 to 10-day intervals throughout season.

Pine Seedlings (nursery beds)	2	Fusiform Rust	Begin application at time of emergence. Repeat at 3 to 5-day intervals to mid-June.

† NOT FOR THIS USE IN CALIFORNIA

Limitations

1. 8.0 lbs/acre—Do not apply later than 5 weeks after petal fall or
2. 16.0 lbs/acre—Do not apply later than petal fall
3. 11.5 lbs/acre—No time limitation or
4. 15.5 lbs/acre—Do not apply within 7 days of harvest
5. 10.0 lbs/acre—Do not apply after early cover period
6. 15.0 lbs/acre—Post harvest application
7. N/A
8. N/A
9. 5.0 lbs/acre—Do not apply within 7 days of harvest
10. N/A
11. N/A
12. 14.5 lbs/acre—Last application at petal fall or
13. 10.0 lbs/acre—Remove excess residues by wiping or brushing
14. 19.0 lbs/acre—Do not apply later than August

STORAGE AND DISPOSAL

Do not contaminate water, food, or feed by storage or disposal.

Storage Instructions: Store in the original container in a dry area. Keep away from sources of ignition (e.g. sparks and open flame). Close bag when not in use. Do not store in a manner where cross-contamination with other pesticides, fertilizers, food or feed could occur. If spilled during storage or handling sweep up spillage and dispose of in accordance with the Pesticide Disposal instructions listed below.

Pesticide Disposal Instructions: Pesticide wastes are acutely hazardous. Improper disposal of excess pesticide, spray mixture, or rinsate is a violation of Federal Law. If these wastes cannot be disposed of by use according to label instructions, contact your State Pesticide or Environmental Control Agency, or the Hazardous Waste representative at the nearest EPA Regional Office for guidance.

Container Disposal Instructions: Completely empty bag into application equipment. Then dispose of empty bag in a sanitary landfill or by incineration, or, if allowed by State and local authorities, by burning. If burned, stay out of smoke.

WARRANTY AND DISCLAIMER

Pennwalt Corporation warrants that this material conforms to the chemical description on the label and is reasonably fit for the purposes referred to in the Directions for Use, subject to the risks referred to therein. PENNWALT MAKES NO OTHER EXPRESS OR IMPLIED WARRANTY OF FITNESS OR MERCHANTABILITY OR ANY OTHER EXPRESS OR IMPLIED WARRANTY. IN NO CASE SHALL PENNWALT OR SELLER BE LIABLE FOR CONSEQUENTIAL, SPECIAL, OR INDIRECT DAMAGES RESULTING FROM THE USE OR HANDLING OF THIS PRODUCT INCLUDING, BUT NOT LIMITED TO, LOSS OF PROFITS, BUSINESS REPUTATION, OR CUSTOMERS; LABOR COST; OR OTHER EXPENSES INCURRED IN PLANTING OR HARVESTING.

Pennwalt and seller offer this product and the buyer and user accept it subject to the foregoing conditions of sale and warranty which may be varied only by agreement in writing signed by a duly authorized representative of Pennwalt.

5-N314S-01 E8 (5/88)

Rhone-Poulenc Ag Company

P.O. Box 12014
2 T. W. Alexander Drive
Research Triangle Park, NC 27709
(919) 549-2000

ALIETTE®

FUNGICIDE

ACTIVE INGREDIENT:
Aluminum tris (O-ethyl phosphonate) 80.0%
INERT INGREDIENT: 20.0%

Total 100.0%

EPA Reg. No. 264-467 EPA EST. NO.

KEEP OUT OF REACH OF CHILDREN

CAUTION PELIGRO

PRECAUCION AL USUARIO: Si usted no lee ingles, no use este producto hasta que la etiqueta le haya sido explicada ampliamente.

STATEMENT OF PRACTICAL TREATMENT

IF IN EYES: Hold eyelids open and flush with a steady, gentle stream of water for 15 minutes. Get medical attention if irritation persists.

IF ON SKIN: Wash with plenty of soap and water. Get medical attention if irritation persists.

IF INHALED: Remove victim to fresh air. If not breathing, give artificial respiration, preferably mouth to mouth. Get medical attention.

For PRODUCT USE Information Call 1-800-334- 9745.
For MEDICAL and TRANSPORTATION Emergencies ONLY Call 24 Hours A Day 1-800-334-7577.

PRECAUTIONARY STATEMENTS

CAUTION

HAZARDS TO HUMANS AND DOMESTIC ANIMALS

Harmful if absorbed through the skin or inhaled. Avoid contact with skin, eyes, or clothing. Wash thoroughly with soap and water after handling.

STATEMENT OF PRACTICAL TREATMENT

IF IN EYES: Hold eyelids open and flush with a steady, gentle stream of water for 15 minutes.

IF ON SKIN: Wash with plenty of soap and water. Get medical attention if irritation persists.

IF INHALED: Remove victim to fresh air. If not breathing, give artificial respiration, preferably mouth to mouth. Get medical attention.

ENVIRONMENTAL HAZARDS

This product is toxic to aquatic and estuarine invertabrates. Drift and runoff may be hazardous to aquatic and estuarine organisms in neighboring areas.

Do not apply directly to water or wetlands (swamps, bogs, marshes and potholes). Do not contaminate water when disposing of equipment washwaters.

DIRECTIONS FOR USE

IT IS A VIOLATION OF FEDERAL LAW TO USE THIS PRODUCT IN A MANNER INCONSISTENT WITH ITS LABELING.

REENTRY STATEMENT

Do not apply this product in such a manner as to directly or through drift expose workers or other persons. The area being treated must be vacated by unprotected persons. Do not enter treated areas without protective clothing until sprays have dried.

GENERAL INFORMATION

NOTE TO USER: During mixing and applying this product, wear full protective clothing (long pants, long sleeve shirt, water resistant gloves and either face shield or goggles).

Do not contaminate water when disposing of equipment washwaters.

DIRECTIONS THROUGH SPRINKLER IRRIGATION SYSTEMS

Use of ALIETTE® Fungicide through Chemigation is not allowed in California.

Apply this product only through sprinkler irrigation systems including microjet, drip, solid set and center pivot. Do not apply this product through any other type of irrigation system.

SPRAY PREPARATION: Remove scale, pesticide residues, and other foreign matter from the chemical tank and entire injector system. Flush with clean water.

APPLICATION INSTRUCTIONS: First prepare a suspension of ALIETTE® in a mix tank. Fill tank with 1/2 to 3/4 the desired amount of water. Start mechanical or hydraulic agitation. Add the required amount of ALIETTE®, and then the remaining volume of water. Then set sprinkler to deliver 0.1 to 0.3 inch of water per acre. Start sprinkler and uniformly inject the suspension of ALIETTE® into the irrigation water line so as to deliver the desired rate per acre. The suspension of ALIETTE® should be injected with a positive displacement pump into the main line ahead of a right angle turn to insure adequate mixing. If you should have any other questions about calibration, you should contact State Extension Service specialists, equipment manufacturers or other experts.

NOTE

For Microjet and Drip Irrigations Systems

When treatment with ALIETTE® has been completed, further field irrigation over the treated area should be avoided for 24 to 48 hours.

For Solid Set and Center Pivot Irrigation Systems

When treatment with ALIETTE® has been completed, further field irrigation over the treated area should be avoided until foliage is dry to prevent washing the chemical off the crop.

GENERAL PRECAUTIONS FOR APPLICATIONS THROUGH SPRINKLER IRRIGATION SYSTEMS

Maintain continuous agitation in mix tank during mixing and application to assure a uniform suspension.

The system must contain a functional check valve, vacuum relief valve, and low pressure drain appropriately located on the irrigation pipeline to prevent water source contamination from backflow. The pesticide injection pipeline must contain a functional, automatic, quick-closing check valve to prevent the flow of fluid back toward the injection pump. The pesticide injection pipeline must also contain a functional, normally closed solenoid-operated valve located on the intake side of the injection pump and connected to the system interlock to prevent fluid from being withdrawn from the supply tank when the irrigation system is either automatically or manually shutdown. The system must contain functional interlocking controls to automatically shut off the pesticide injection pump when the water pump motor stops. The irrigation line or water pump must include a functional pressure switch which will stop the water pump motor when the water pressure decreases to the point where pesticide distribution is adversely affected. Systems must use a metering pump, such as a positive displacement injection pump (e.g., diaphragm pump) effectively designed and constructed of materials that are compatible with pesticides and capable of being fitted with a system interlock.

Do not apply when wind speed favors drift, when system connection or fittings leak, when nozzles do not provide uniform distribution or when lines containing the product must be dismantled and drained.

Crop injury, lack of effectiveness, or illegal pesticide residues in the crop may result from nonuniform distribution of treated water.

Allow sufficient time for pesticide to be flushed through all lines and all nozzles before turning off irrigation water. A person knowledgeable of the chemigation system and responsible for its operation shall shut the system down and make necessary adjustments should the need arise. Do not connect an irrigation system (including greenhouse systems) used for pesticide application to a public water system unless the label-prescribed safety devices for public water supplies are in place.

PINEAPPLE

When used in conjunction with good cultural management practices, ALIETTE® is effective in controlling Heart Rot *(Phytophthora parasitica)* on pineapples. The dip treatment is critical to insure early protection from Phytophthora for good stands. Foliar sprays should be applied to established plants, when aggravating environmental conditions (i.e. excessive rainfall) occur or are anticipated.

ALIETTE® should be applied on pineapples in accordance with the table below:

Disease	Application	Rate Lbs./Acre	Application Program
Heart Rot *(Phytophthora parasitica)*	Dip	2.5*	Apply as a pre-plant dip immediately prior to planting.
	Foliar	3.75**	A maximum of 6 sprays may be applied at 3 month intervals. Do not apply within 3 months of harvest.

* 100 gallons is intended to treat the number of slips required to plant one acre.

** Apply in sufficient water for thorough coverage. Do not exceed 400 gallons per acre.

CITRUS BEARING

ALIETTE® applied in conjunction with good cultural management practices will provide effective control of Phytophthora foot rot, gummosis, root rot and fruit brown rot caused by *Phytophthora sp.* Applications of ALIETTE® should being when conditions favor disease development.

ALIETTE® should be applied on citrus in accordance with the table below:

Disease	Application Method	Rate Lbs./ Product/A	Application Program
Phytophthora Foot & Root Rot	Foliar	5.0	Apply as a foliar spray to wet when conditions favor disease development (i.e. Feb-April, May-June, Sept.-Oct.)
Brown Rot	Foliar	5.0	Apply as a foliar spray to wet when conditions favor disease development (i.e. November-March)
Phytophthora Foot Rot or Gummosis	Trunk Spray or Paint	2.5-5.0*	Mix the desired amount of ALIETE with 5 gallons of water and apply to the trunk lesion in a sufficient volume to thoroughly wet the entire lesion. If no lesion is present the application should be made from the soil line up the trunk approximately 2 feet. *Under severe disease conditions (i.e. trunk lesions) the higher rate should be used.

ALIETTE® 80WP Fungicide may be applied up to four (4) times per season. Any combination of the above methods of application can be used, however, *do not exceed a total of four (4) applications or 20 lbs/Acre per year.* Do not apply ALIETTE® 80WP within 90 days of harvest.
Do not graze or feed forage from treated groves to livestock.

NON-BEARING CITRUS

(Trees that will not produce marketable fruit for 12 months after the last application)

Aliette applied in conjunction with good cultural management practices will provide effective control of Phytophthora foot rot, gummosis and root rot caused by *Phytophthora sp.*

ALIETTE® should be applied to non-bearing citrus in accordance with the table below:

Disease	Application Method	Rate lbs product	Application Program
Phytophthora Foot & Root Rot	Foliar	5.0/100gal.	Apply as a foliar spray to wet when conditions favor disease development (i.e. March-April, May-June, Sept.-Oct.)
	Trunk Spray or Paint	2.5-5.0*/5 gal.	Mix the desired amount of ALIETTE with 5 gallons of water and apply to the trunk lesion in a sufficient volume to thoroughly wet the entire lesion. If no lesion is present, the application should be made from the soil line up the trunk approximately two (2) feet. *Under severe disease conditions (i.e. Trunk lesions) the higher rate should be used. Nursery tree resets and new plantings should be treated at thed time of planting.

Ground application should not exceed 100 gallons per acre application. ALIETTE® 80WP Fungicide may be applied up to four (4) times per year. Any combination of the above mentioned methods of application can be used, however, *do not exceed a total of four (4) applications or 20 lbs./ Acre per year.*

NON-BEARING POME FRUIT i.e Apples and Pears

(Trees that will not produce marketable fruit for 12 months after the last ALIETTE application.)

ALIETTE applied in conjunction with good cultural management practices will provide effective control of collar and root rot caused by *Phytophthora sp.*

ALIETTE should be applied to nursery and field grown non-bearing pome trees in accordance with the table below:

Disease	Application Method	Rate lbs product/ 100 gals	Application Program
Phytophthora Collar and Root Rot	Foliar	2.5 - 5.0*	Apply as a foliar spray programs of ALIETTE® on a 30-60 day interval when conditions favor disease development. Thorough coverage is required. Do not apply more than 5.0 lbs of ALIETTE® 80WP per acre per application. Do not exceed 20 lbs of ALIETTE® 80WP per acre per season. Nursery tree resets and new plantings should be treated after leaf emergence.

*Under moderate disease pressure apply ALIETTE® 80WP, 3 or 4 times at 5.0 lbs/100 gallons on a 60 day spray interval or 6-8 applications at 2.5 lbs/100 gallons on a 30 day interval.

NON-BEARING STONE FRUIT: Apricots, Cherries, Nectarines, Peaches & Plums NON-BEARING ALMONDS (Trees that will not produce marketable fruit for 12 months after the last ALIETTE® application.)

ALIETTE® applied in conjunction with good cultural management practices will provide effective control of collar and root rot caused by *Phytophthora* sp.

ALIETTE® should be applied to nursery and field grown non-bearing stone fruit and non-bearing almonds in accordance with the table below:

Disease	Application Method	Rate lbs product/ 100 gals.	Application Program
Phytophthora Collar and Root Rot	Foliar	5.0	Apply a foliar spray program of ALIETTE® on a 60 day interval when conditions favor disease development. Thorough coverage is required. Do not apply more than 5.0 lbs of ALIETTE® per acre per application. Nursery tree resets and new plantings should be treated after leaf emergence.
Almond Pruning wound canker	Paint or Spray	3.0 - 5.0*	Apply as a paint or spray to the pruning wound area in a sufficient volume to thoroughly wet the entire wound surface. Under severe disease conditions (i.e. active canker lesion), the higher rate should be used.

* For small volume application, mix 0.5 - 0.8 oz. (3 - 4.5 level tablespoons) product per gallon of water. This is equivalent to 3 - 5 lbs per 100 gallons. Gallonage should not exceed 100 gallons per acer per application. ALIETTE® 80W may be applied up to four (4) times per year. Any combination of the above methods of application can be used, but not to exceed a total of four (4) applications or 20 lbs per acer per year.

NON-BEARING AVOCADOS

(Trees that will not produce marketable fruit for 12 months after the last Aliette application.)

When used in conjunction with good cultural management practices, ALIETTE® is effective in controlling avocado root rot *(Phytophthora cinnamomi).* ALIETTE® applications should begin at transplanting or the start of the growing season.

ALIETTE® may be used as a pre-plant drench treatment and/or in a foliar spray program in accordance with the details in the table below:

Disease	Application	Rate	Application Program
Phytophthora Root Rot *(Phytophthora cinnamomi)*	Drench	Use at concentration of 5.0 oz./10 gallons	Apply 1 quart of solution to the pot or sleeve of each tree 2-3 days prior to transplanting.
	Foliar	5.0 lbs/100 gallons	Begin foliar applications at the start of the growing season. A maximum of 4 foliar sprays per year may be applied at 60 day intervals. The foliar spray should be applied to wet. DO NOT EXCEED 100 GALLONS PER ACRE.

ASPARAGUS (FOR USE IN CALIFORNIA ONLY)

Disease	Rate Lbs. Product/Acre	Application Program
Asparagus spear slime and crown rot *(Phytophthora sp.)*	5.0	Apply ALIETTE® once per season. ALIETTE® should be applied as a foliar spray to fully expanded asparagus ferns. Do not apply to ferns that are beginning to senesce. Thorough coverage is required, not to exceed 50 gallons per acre. ALIETTE® should not be applied within 110 days of harvest.

GINSING (FOR USE IN WISCONSIN ONLY)

Disease	Rate Lbs. Product/100 Gallons	Application Program
Phytophthora foliar and root rot *(Phytophthora cactorum)*	5.0	Make the first application of ALIETTE® as conditions become favorable for disease development in the spring. Continue applications on a seven (7) day interval as long as conditions favor disease development. Do not exceed more than nine (9) applications of ALIETTE® per season or apply within 31 days of harvest. Mix in a minimum of 100 gallons of water per acre.

STORAGE AND DISPOSAL

Do not contaminate water, food or feed by storage or disposal.

PESTICIDE DISPOSAL:

Pesticide wastes are acutely hazardous. Improper disposal of excess pesticide, spray mixture, or rinsate is a violation of Federal Law. If these wastes cannot be disposed of by use according to label instructions, contact your State Pesticide or Environmental Control Agency, or the Hazardous Waste representative at the nearest EPA Regional Office for guidance.

CONTAINER DISPOSAL:

Completely empty bag into application equipment. Then dispose of empty bag in a sanitary landfill or by incineration, or, if allowed by State and local authorities, by burning. If burned, stay out of smoke.

LIMITED WARRANTY AND DISCLAIMER

The manufacturer warrants (a) that this product conforms to the chemical description on the label; (b) that this product is reasonably fit for the purposes set forth in the directions for use when it is used in accordance with such directions; and (c) that the directions, warning and other statements on this label are based upon responsible experts' evaluation of reasonable tests of effectiveness, of toxicity to laboratory animals and to plants, and of residues on food crops and upon reports of field experience. Tests have not been made on all varieties or in all states or under all conditions. THE MANUFACTURER NEITHER MAKES NOR INTENDS, NOR DOES IT AUTHORIZE ANY AGENT OR REPRESENTATIVE TO MAKE, ANY OTHER WARRANTIES, EXPRESS OR IMPLIED, AND IT EXPRESSLY EXCLUDES AND DISCLAIMS ALL IMPLIED WARRANTIES OF MERCHANTABILITY OR FITNESS FOR A PARTICULAR PURPOSE.

THIS WARRANTY DOES NOT EXTEND TO, AND THE BUYER SHALL BE SOLELY RESPONSIBLE FOR, ANY AND ALL LOSS OR DAMAGE WHICH RESULTS FROM THE USE OF THIS PRODUCT IN ANY MANNER WHICH IS INCONSISTENT WITH THE LABEL DIRECTIONS, WARNINGS OR CAUTIONS.

BUYER'S EXCLUSIVE REMEDY AND MANUFACTURER'S OR SELLER'S EXCLUSIVE LIABILITY FOR ANY AND ALL CLAIMS, LOSSES, DAMAGES, OR INJURIES RESULTING FROM THE USE OR HANDLING OF THIS PRODUCT, WHETHER OR NOT BASED IN CONTRACT, NEGLIGENCE, STRICT LIABILITY IN TORT OR OTHERWISE SHALL BE LIMITED, AT THE MANUFACTURER'S OPTION, TO REPLACEMENT OF OR THE REPAYMENT OF THE PURCHASE PRICE FOR, THE QUANTITY OF PRODUCT WITH RESPECT TO WHICH DAMAGES ARE CLAIMED. IN NO EVENT SHALL THE MANUFACTURER OR SELLER BE LIABLE FOR SPECIAL, INDIRECT OR CONSEQUENTIAL DAMAGES RESULTING FROM THE USE OR HANDLING OF THIS PRODUCT.

NOTICE TO BUYER

Purchase of this material does not confer any rights under patents governing this product or the use thereof in countries outside of the United States.

AG 90032

THIS SPECIMEN LABEL IS INTENDED FOR USE ONLY AS A GUIDE IN PROVIDING GENERAL INFORMATION REGARDING THE DIRECTIONS, WARNINGS AND CAUTIONS ASSOCIATED WITH THE USE OF THIS PRODUCT. AS WITH ANY AGRICULTURAL CHEMICAL, ALWAYS FOLLOW THE LABEL INSTRUCTIONS ON THE PACKAGE BEFORE USING.

AMIBEN® GRANULAR

brand CHLORAMBEN HERBICIDE

Controls Annual Broadleaf Weeds and Annual Grasses in Soybeans and Other Crops

ACTIVE INGREDIENTS:
Ammonium Salt of Chloramben
(3-amino, 2,5-dichlorobenzoic acid)* 9.7%
Ammonium salts or related aminodichlorobenzoic acids 1.1%
INERT INGREDIENTS: 89.2%
TOTAL: 100.0%

*Total aminodichlorobenzoic acids 10.0% by weight. Chloramben equivalent 9.0% by weight. Related aminodichlorobenzoic acid equivalent 1.0% by weight.

EPA REG. NO. 264-167 EPA EST. NO. 264-IA-1

KEEP OUT OF REACH OF CHILDREN

CAUTION

For PRODUCT USE Information Call 1-800-334-9745.
For MEDICAL and TRANSPORTATION Emergencies ONLY Call 24 Hours A Day 1-800-334-7577

PRECAUTIONARY STATEMENTS

CAUTION

HAZARDS TO HUMANS AND DOMESTIC ANIMALS

Harmful if swallowed. Avoid contact with skin, eyes, clothing.

STATEMENT OF PRACTICAL TREATMENT

IN CASE OF EYE CONTACT: rinse thoroughly with water and get medical attention.

IN CASE OF SKIN CONTACT: wash immediately with soap and water and remove contaminated clothing.

ENVIRONMENTAL HAZARDS

Do not contaminate water used for irrigation or domestic purposes by cleaning of equipment or disposal of waste.

GENERAL INFORMATION

Thoroughly clean hoppers and hoses of granular applicators before using to apply AMIBEN® Granular Chloramben Herbicide.

Do not graze treated fields of sunflowers or use these plants as livestock feed.

AMIBEN® Granular Chloramben Herbicide has many desirable features: Outstanding control of annual broadleaf weeds and annual grasses. Excellent crop tolerance. No harmful residue in soil, no change in normal crop rotations.

One of the most consistent preemergence herbicides available. Tested and recommended by state colleges and extension agents. Proven effective on millions of acres of soybeans.

HOW AMIBEN® GRANULAR CHLORAMBEN HERBICIDE WORKS

AMIBEN® Granular Chloramben Herbicide is most effective applied at planting time, when spreading and planting are done in the same operation. Subsequent moisture moves the chemical into the soil where germinating weed seeds take it up and are controlled.

AMIBEN® Granular remains active and effective for several weeks, long enough generally for crops to fill middles and shade out late season weed growth.

AMIBEN® Granular controls the following important Annual Grass and Broadleaf Weeds:

barnyardgrass (water grass)	Echinochloa crusgalli
carpetweed	Mollugo verticillata
chickweed, annual	Stellaria media
coffeeweed (sesbania)	Sesbania exaltata
crabgrass, hairy (large)	Digitaria sanguinalis
crabgrass, smooth	Digitaria ischaemum
fall panicgrass	Panicum dichotomiflorum
Florida beggarweed	Desmodium tortuosum
foxtail, giant	Setaria faberii
foxtail, green	Setaria viridis
foxtail, yellow	Setaria glauca

goosegrass (silver crabgrass)	Eleusine indica
johnsongrass-seedling	Sorghum halepense
kochia	Kochia scoparia
lambsquarter, common	Chenopodium album
mustard, wild	Brassica kaber
nightshade, black	Solanum nigrum
pigweeds	Amaranthus sp.
prickly sida (teaweed)	Sida spinosa
purslane (pusley)	Portulaca oleracea
ragweed, common	Ambrosia artemisiifolia
Russian thistle	Salsola kali
smartweed, Pennsylvania	Polygonum pennsylvanicum
spurge, annual	Euphorbia maculata
stinkgrass	Eragrostis cilianesis
velvetleaf (buttonweed)	Abutilon theophrasti
Venice mallow	Hibiscus trionum

Velvetleaf, jimsonweed, buffalobur, wild oats, and cocklebur often germinate in soil at depths below the chemical penetration. Even though populations of these weeds are reduced under normal rainfall conditions, weed seeds which germinate outside the chemical zone are not affected.

For control of cocklebur and suppression of annual morning glory, follow AMIBEN® Granular Chloramben Herbicide treatment with an early, directed, postemergence application of BUTYRAC® 200 Broadleaf Herbicide on soybeans.

Sprouts coming from roots of established perennial weeds such as horsenettle, hedge bindweed, nutsedge (nutgrass), milkweed, dogbane, quackgrass, and johnsongrass are not controlled by any surface preemergence chemical treatment, including AMIBEN® Granular.

Band or broadcast applications of AMIBEN® Granular may be made for broadleaf weed control over the top of preplant soil-incorporated grass herbicides.

GENERAL INFORMATION

DRYBEANS

Environmental conditions and cultural practices can combine to create conditions favorable for rapid, early, development of severe levels of soil-borne diseases in drybeans. The potential for these diseases causing drybean yield reductions must be evaluated when determining where and how drybeans will be grown.

Environmental conditions favoring rapid disease development and resulting injury to drybeans are:

(1) Cool air and soil temperatures during and soon after drybean emergence; (2) Heavy rainfall which prolongs saturated soil conditions; and (3) Soil types such as Kawkawlin Loam which are poorly drained.

These soils are subject to compaction and prolonged water saturation. While severe levels of soil disease are commonly associated with saturated soil conditions, the severity of disease infestations can also be increased by drought conditions which are unfavorable for rapid drybean root growth.

Cultural practices favoring rapid disease development causing injury to drybeans are:

(1) High levels of disease inoculum from plant residues of soybeans or drybeans grown in the field during prior years; (2) Use of disease susceptible varieties; (3) Planting drybeans in low lying fields with poorly drained or compacted soils; (4) Failure to use a 3-year rotation with corn or small grains to reduce inoculum buildup in the field; and (5) High rates of soil applied herbicides which may cause additional stress.

Under conditions favoring excessive root disease development, as outlined above, high rates of AMIBEN® Granular Chloramben Herbicide, as with other soil applied herbicides, may increase the potential for yield reductions caused by high levels of root disease.

DIRECTIONS (For Use On Drybeans)

To minimize potential for disease-induced yield reductions, do not use more than 15 lbs. of AMIBEN® Granular per broadcast equivalent acre, either alone, or as an overlay treatment with other approved drybean herbicides.

Under conditions favoring excessive disease development do not apply AMIBEN® Granular as a surface preemergence treatment.

Follow the label directions for incorporation of AMIBEN® Granular. When the 15 lbs. per acre rate of AMIBEN® Granular is used under heavy weed populations, reduced weed control is to be expected.

SOYBEANS, DRY BEANS (including white, navy, kidney, pinto, lima), PEANUTS (Oklahoma, North Texas), SUNFLOWERS

Plant and treat the same operation. Apply at the rate of 20 to 30 pounds AMIBEN® Granular Chloramben Herbicide per broadcast acre. The higher rate should be used on clay loams, clay, or high organic (3 to 8%) mineral soils.

When Band Applying:

Mount the spreader behind the last packer or row wheel so that planting and spreading are done in the same operation. Proper calibration and maintenance of desired spreader height above the ground are very important. Weeds growing in the middle between the treated rows should be cultivated as soon as necessary. Use care while cultivating to avoid throwing fresh soil on treated bands.

If sufficient rain does not fall within 7 days after spreading AMIBEN® Granular but there is enough soil moisture to germinate weeds and grasses, a light cultivation with a rotary hoe or similar tool will up-root these small broadleaf weeds and grasses. The shallow mixing of AMIBEN® Granular in the surface soil will not interfere with the action of AMIBEN® Granular when rains come.

If heavy rain packs the soil surface, rotary hoe the field to aid bean emergence. Do not disrupt the chemical band with deep cultivation.

For dry beans grown in low rainfall areas, incorporate AMIBEN® Granular into the top 1 to 2 inches of soil with a disk or harrow. Plant beans immediately after incorporation.

AMIBEN® Granualar Chloramben Herbicide split application to soybeans: Apply 15 pounds of AMIBEN® Granular per acre as a broadcast treatment just prior to planting. Incorporate AMIBEN® Granular by shallow cultivation. The same day, if possible, plant beans and apply AMIBEN® Granular in a band over the row at a rate equivalent to 10 to 15 pounds AMIBEN® Granular per broadcast acre.

PROGRAMMED WEED CONTROL FOR SOYBEANS AND SUNFLOWERS

AMIBEN® Granular Chloramben Herbicide applied over preplant "TREFLAN"

AMIBEN® Granular Chloramben Herbicide may be applied in a band over the soybean row at planting time in fields where Treflan has been applied as a preplant soil incorporated herbicide. Apply AMIBEN® Granular on a band at a rate equivalent to 20 to 30 pounds per broadcast acre. Use the higher rate of AMIBEN® Granular herbicide where heavy populations of smartweed, velvetleaf, ragweed, wild mustard or black nightshade are expected. Refer to Treflan label for exact directions, for rate of application, method of incorporation and timing for your area. Observe all directions and precautions on the AMIBEN® Granular Chloramben Herbicide label.

CORN

Use only on heavy soils in Missouri, Minnesota, Iowa, Ohio, Illinois, Kansas, Nebraska and Indiana. *If rains and extended cold weather occur after planting, corn injury may result.* Plant corn at least 2 to 2½ inches deep. Do not use AMIBEN® Granular on corn grown in light textured sandy soils. Apply 20 pounds of AMIBEN® Granular Chloramben Herbicide per acre. For best results, plant and treat at the same time.

SWEET POTATOES

Apply AMIBEN® Granular Chloramben Herbicide at the rate of 40 pounds per broadcast acre at planting time. For the first few cultivations after treatment, set sweeps so that soil in treated bands is not disturbed. For the final cultivation, set sweeps closer to pull soil as high as possible over the sweet potato row.

ESTABLISHED TOMATOES AND PEPPERS (from transplants or seed)

Use 30 to 40 pounds AMIBEN® Granular Chloramben Herbicide per acre as an overall treatment. Apply AMIBEN® Granular once during the season, using the timing recommended below for your soil type and cultural practice. *APPLY AMIBEN® Granular Chloramben Herbicide only when tomato and pepper foliage is dry.*

Times to Apply: (Use only one.)

A. **Transplants on Medium to Heavy Soils.**
On medium to heavy soils, apply AMIBEN® Granular after plants are established (recovered from wilting that follows transplanting). Do not use AMIBEN® Granular at this time for tomatoes or peppers growing on light sandy soil as injury may result.

B. **Transplants on Light Sandy Soils.**
On light textured sandy soils cultivate or hoe to control weeds after transplanting. Apply AMIBEN® Granular immediately after the last cultivation and before weeds emerge. Weeds in the crop row should be removed prior to application.

C. **Direct Seeded Tomatoes or Peppers.**
On direct seeded tomatoes or peppers, do not apply until plants have reached the 5 to 6 leaf stage. Cultivate before application of AMIBEN® Granular Chloramben Herbicide to remove existing weeds. Weeds in the crop row should be removed prior to application of AMIBEN® Granular. *Apply AMIBEN® Granular Chloramben Herbicide only when tomato or pepper foliage is dry.*

SEEDED SQUASH AND PUMPKIN

Treat at planting time or as soon after planting as possible, before weeds and crop emerge. Mount the spreader on the planter so that spreading and seeding are carried out in the same operation. Apply at a rate 30 pounds AMIBEN® Granular Chloramben Herbicide per broadcast acre on light soils; and 40 pounds per broadcast acre on heavy or high organic soils. If band applications are used, use the table below to determine the rate of AMIBEN® Granular to use per banded acre on your soil type. If band applications are used, use normal cultivations after squash or pumpkin emerge to control weeds growing in the untreated strips between the rows.

Pounds AMIBEN® Granular Chloramben Herbicide Per Banded Acre

Row Width	Band Width	Light Soils	High Organic Soils
72 inch	14 width	5.9 lbs	7.8 lbs
76 inch	14 width	5.6 lbs	7.4 lbs
80 inch	14 width	5.3 lbs	7.0 lbs

After squash and pumpkin emerge, use normal cultivations to control weeds growing in the untreated strips between the rows.

LIMA BEANS - Overall Treatment

Apply 20 to 40 pounds of AMIBEN® Granular per broadcast acre. Treat at planting time or as soon after planting as possible, before weeds and lima beans emerge.

SEEDLING ASPARAGUS - Overall Treatment

Apply 30 pounds of AMIBEN® granular per acre as soon as possible after asparagus is planted and before seeds emerge. Irrigation with 1/2 inch of water following application is suggested to produce maximum weed control.

Band Treatments
Pounds Amiben® Granular Chloramben Herbicide Per Banded Acre

Row Width	Band Width	@ 20 lb/A*	@ 25 lb/A*	@ 30 lb/A*	@ 40 lb/A*
21 in	10 in	9 lbs 8 oz	11 lbs 14 oz	14 lbs 5 oz	19 lbs
28 in	12 in	8 lbs 10 oz	10 lbs 11 oz	12 lbs 14 oz	17 lbs 2 oz
	14 in	10 lbs	12 lbs 8 oz	15 lbs	20 lbs
30 in	12 in	8 lbs	10 lbs	12 lbs	16 lbs
	14 in	9 lbs 5 oz	11 lbs 11 oz	14 lbs	18 lbs 11 oz
36 in	12 in	6 lbs 11 oz	8 lbs 5 oz	10 lbs	13 lbs 5 oz
	14 in	7 lbs 13 oz	9 lbs 11 oz	11 lbs 11 oz	15 lbs 10 oz
38 in	12 in	6 lbs 5 oz	7 lbs 14 oz	9 lbs 8 oz	12 lbs 10 oz
	14 in	7 lbs 6 oz	9 lbs 3 oz	11 lbs	14 lbs 11 oz
40 in	12 in	6 lbs	7 lbs 8 oz	9 lbs	12 lbs
	14 in	7 lbs	8 lbs 13 oz	10 lbs 8 oz	14 lbs

* Pounds AMIBEN® Granular Chloramben Herbicide per broadcast acre

TO DETERMINE THE AMOUNT OF AMIBEN® GRANULAR TO PURCHASE, FOLLOW THESE 4 STEPS AND USE THE ABOVE TABLE.

1. Select row width (Example 28 inch)
2. Select band width (Example 24 inch)
3. Select broadcast rate * that fits your weed problems (**Soybean example**-use 30 lb/A where Velvetleaf and Black nightshade are problems).
4. **Formula:**
   15 lbs.per acre × number of acres equals Amount to Purchase.

HOW TO USE THE AMIBEN® GRANULAR EQUIPMENT CHARTS: Steps (1). Identify your granular equipment. Note differences between "old" and "new" equipment. (2). Decide ground speed within a range of 2 to 10 mph. (3). Decide broadcast rate within a range of 20 to 30 lbs. per broadcast acre. Note that each rate range is in a column "below" the miles per hour. (**Example**): if you have the new Gandy, and if you decide 5 mph and a broadcast rate of 30 lbs. AMIBEN® Granular per acre, your gauge setting is 20.

OLD NOBLE
(round opening in bottom of plastic hopper)

	MILES PER HOUR								
Gage setting	2	3	4	5	6	7	8	9	10
6	17.4								
7	20.3								
8	23.3								
9	26.4								
10	29.6	19.8							
11	32.4	21.6	16.2						
12		23.4	17.6						
13		25.3	19.0						
14		27.2	20.4						
15		29.1	21.8						
16		31.0	23.3						
17			24.8	19.8					
18			26.3	21.0					
19			27.6	22.1					
20			29.5	23.6	19.6				
21			31.2	24.9	20.8				
22				26.3	21.9				
23				27.7	23.1	19.8			
24				29.1	24.3	20.8			
25				30.7	25.6	21.9	19.2		
26					26.9	23.1	20.2		
27					24.2	21.2	18.9		
28					29.6	25.4	22.2	19.8	
29					31.0	26.6	23.2	20.7	
30						27.7	24.3	21.6	19.4
31						29.4	25.7	22.9	20.6
32						31.1	27.2	24.2	21.8
33							28.7	25.5	23.0
34							30.2	26.8	24.2
35								28.4	25.6
36								30.1	27.1
37									28.9
38									30.6

NEW NOBLE
(snap-on bottom)

	MILES PER HOUR								
Gage setting	2	3	4	5	6	7	8	9	10
4	18.3								
5	21.5								
6	24.8								
7	28.0	18.7							
8	31.3	20.9							
9		23.0							
10		25.2	18.9						
11		28.7	21.5						
12		32.2	24.2	19.3					
13			26.8	21.5					
14			29.5	23.6	19.6				
15			32.1	25.7	21.4				
16				27.8	23.2	19.9			
17				29.9	24.9	21.4	18.7		
18				32.0	26.7	22.9	20.0		
19					28.5	24.4	21.3	19.0	
20					30.2	25.9	22.7	20.1	
21						27.4	24.0	21.3	19.2
22						28.9	25.3	22.5	20.2
23						30.4	26.6	23.7	21.3
24							27.9	24.8	22.4
25							29.3	26.0	23.4
26							30.6	27.2	24.5
27								28.4	25.5
28								29.5	26.6
29								30.7	27.6
30									28.7
31									29.7
32									30.8

OLD GANDY
(bolt on bottom)

	MILES PER HOUR								
Gage setting	2	3	4	5	6	7	8	9	10
10	20.0								
11	25.0								
12	30.0	20.0							
13		23.5	17.6						
14		27.5	20.6						
15		31.5	23.6	18.9					
16			26.6	21.3					
17			29.9	23.9	19.9				
18			33.1	26.5	22.1	18.9			
19				29.2	24.3	20.9	18.3		
20				32.1	26.7	22.9	20.1		
21					29.4	25.2	22.1	19.6	
22					32.5	27.9	24.4	21.7	19.5
23						30.5	26.7	23.7	21.4
24							29.3	26.1	23.5
25							32.0	28.4	25.6
26								31.0	27.9
27									30.4

NEW GANDY
(snap-on bottom)

	MILES PER HOUR								
Gage setting	2	3	4	5	6	7	8	9	10
10	17.0								
11	20.8								
12	24.8								
13	29.3	19.5							
14	34.8	23.2	17.4						
15		27.5	20.6						
16		31.8	23.9	19.1					
17			27.4	21.9	18.3				

18			31.1	24.9	20.8				
19				28.0	23.3	20.0			
20				30.9	25.7	22.1	19.3		
21					28.3	24.3	21.3	18.9	
22					31.3	26.9	23.5	20.9	18.8
23						29.8	26.1	23.2	20.9
24						32.6	28.6	25.4	22.9
25							31.2	27.7	25.0
26								30.5	27.5
27									30.0

OLD JOHN DEERE
(metal hoppers, 3 range gage)

	MILES PER HOUR								
Gage setting	2	3	4	5	6	7	8	9	10
1/22	18.7								
1/23	21.5								
1/24	24.3								
1/25	27.0								
1/26	29.8	19.9							
1/27	32.5	21.7							
1/28		23.5							
1/29		25.4	19.0						
1/30		27.2	20.4						
1/31		27.2	20.4						
1/32		27.2	20.4						
1/33		27.2	20.4						
2/1		28.6	21.4						
2/2		29.8	22.3						
2/3		31.0	23.2						
2/4			24.1						
2/5			25.0	20.0					
2/6			25.9	20.7					
2/7			26.8	21.5					
2/8			27.7	22.2					
2/9			28.6	22.9					
2/10			29.5	23.6	19.7				
2/11			30.4	24.3	20.3				
2/12				25.0	20.9				
2/13				25.7	21.5				
2/14				26.5	22.1				
2/15				27.2	22.6				
2/16				27.6	23.0				
2/17				28.0	23.4	20.0			
2/18				28.5	23.7	20.3			
2/19				28.9	24.1	20.6			
2/20				29.3	24.4	20.9			
2/21				29.7	24.8	21.2			
2/22				30.2	25.1	21.5			
2/23					25.5	21.8			
2/24					25.8	22.1			
2/25					26.2	22.5			
2/26					26.5	22.8	19.9		
2/27					26.9	23.1	20.2		
2/28					27.3	23.4	20.4		
2/29					27.6	23.4	20.7		
2/30					28.0	24.0	21.0		
2/31					28.8	24.7	21.6		
2/32					29.6	25.4	22.2	19.7	
2/33					30.4	26.1	22.8	20.3	
3/1						27.9	24.4	21.7	19.5
3/2						28.9	25.3	22.5	20.2
3/3						29.9	26.2	23.3	21.0
3/4						31.0	27.1	24.1	21.7
3/5							28.1	24.9	22.4
3/6							28.9	25.7	23.1
3/7							29.8	26.5	23.9
3/8							30.8	27.3	24.6
3/9								28.1	25.3
3/10								29.0	26.1
3/11								29.8	26.8
3/12								30.6	27.5
3/13									28.2
3/14									29.0
3/15									29.7
3/16									30.4

NEW JOHN DEERE
(plastic hoppers 10-20-30 gage)

	MILES PER HOUR								
Gage setting	2	3	4	5	6	7	8	9	10
8	19.3								
9	22.4								
10	25.5								
11	28.0	18.7							
12	30.5	20.4							
13		22.0							
14		23.7							
15		25.4	19.1						
16		27.1	20.3						
17		28.8	21.6						
18		30.5	22.9						
19			24.1	19.3					
20			25.4	20.3					
21			26.9	21.6					
22			28.5	22.8	19.0				
23			30.3	24.2	20.2				
24				25.6	21.3				
25				27.0	22.5	19.3			
26				28.4	23.7	20.3			
27				30.1	25.1	21.5			
28					26.5	22.7	19.9		
29					27.9	23.9	21.0		
30					29.4	25.2	22.0		
31					31.3	26.8	23.4	18.8	
32						28.8	25.2	20.2	
33						31.1	27.2	21.8	
34							29.4	23.5	
35							31.6	25.3	
36								27.0	
37								29.3	
38								31.8	

NEWEST JOHN DEERE
(plastic hoppers 5-15-20 gage)

	MILES PER HOUR								
Gage setting	2	3	4	5	6	7	8	9	10
7	16.0								
8	20.3								
9	24.6								
10	28.9	19.3							
11	33.2	22.1							
12		25.0	18.8						
13		28.3	21.2						
14		31.5	23.6	18.9					
15			26.1	20.9					
16			28.6	22.9	19.1				
17			31.2	25.0	20.8				
18				27.0	22.5	19.3			
19				29.5	24.6	21.1			
20				32.0	26.7	22.9	20.0		
21					28.8	24.7	21.6	19.2	
22					30.9	26.5	23.2	20.6	18.6
23						28.7	25.1	22.3	20.1
24						30.9	27.0	24.0	21.6
25							29.2	25.9	23.4
26							31.4	27.9	25.1
27								29.8	26.9
28								31.8	28.6
29									30.7

OLD INTERNATIONAL HARVESTER (prior to 1981)

	MILES PER HOUR								
Gage setting	2	3	4	5	6	7	8	9	10
1/4	17.1								
1/5	20.7								
1/6	24.2								
1/7	27.8	18.5							
1/8	31.4	20.9							
1/9		23.3							
2/0		25.7	19.3						
2/1		27.7	20.8						
2/2		29.7	22.2						
2/3		31.7	23.7	19.0					
2/4			25.2	20.2					
2/5			26.7	21.4					
2/6			28.2	22.6					
2/7			29.7	23.8	19.8				
2/8			31.2	24.6	20.8				
2/9				26.1	21.8				
3/0				27.3	22.8	19.5			
3/1				28.5	23.8	20.4			
3/2				29.7	24.8	21.2			
3/3				30.9	25.8	22.1	19.3		
3/4					26.8	22.9	20.1		
3/5					27.8	23.8	20.8		
3/6					28.7	24.6	21.6		
3/7					29.7	25.5	22.3	19.8	
3/8					30.7	26.3	23.1	20.5	
3/9						27.2	23.8	21.2	
4/0						28.0	24.5	21.8	19.6
4/1						29.0	25.4	22.6	20.3
4/2						30.0	26.3	23.3	21.0
4/3							27.1	24.1	21.7
4/4							28.0	24.9	22.4
4/5							28.8	25.6	23.1
4/6							29.7	26.4	23.7
4/7							30.5	27.1	24.4
4/8								27.9	25.1
4/9								28.7	25.8
5/0								29.4	26.5
5/1								30.2	27.2
5/2									27.9
5/3									28.5
5/4									29.2
5/5									29.9
5/6									30.6

NEW INTERNATIONAL HARVESTER (1981 and later)

	MILES PER HOUR								
Gage setting	2	3	4	5	6	7	8	9	10
1/5	16.4								
1/6	20.5								
1/7	24.5								
1/8	28.9	19.3							
1/9	32.6	21.8							
2/0		24.5							
2/1		26.5	19.9						
2/2		28.5	21.4						
2/3		30.5	22.9						
2/4			24.4	19.5					
2/5			25.9	20.7					
2/6			27.4	21.9					
2/7			28.9	23.1	19.3				
2/8			30.4	24.3	20.3				
2/9				25.5	21.3				
3/0				26.7	22.3				
3/1				27.9	23.3	19.9			
3/2				29.1	24.3	20.8			
3/3				30.3	25.3	21.7			
3/4					26.3	22.5	19.7		
3/5					27.3	23.4	20.5		
3/6					28.3	24.2	21.2		
3/7					29.3	25.1	21.9	19.5	
3/8					30.3	26.0	22.7	20.2	
3/9						26.8	23.5	20.9	
4/0						27.7	24.2	21.5	
4/1						28.5	25.0	22.2	20.0
4/2						29.4	25.7	22.9	20.6
4/3						30.3	26.5	23.5	21.2
4/4							27.2	24.2	21.8
4/5							28.0	24.9	22.4
4/6							28.7	25.5	23.0
4/7							29.5	26.2	23.6
4/8							30.2	26.9	24.2
4/9								27.5	24.8
5/0								28.2	25.4
5/1								28.9	26.0
5/2								29.5	26.6
5/3								30.2	27.2
5/4									27.8
5/5									28.4
5/6									29.0
5/7									29.6
5/8									30.2

ALLIS CHALMERS (1981 and later)

	MILES PER HOUR	
Gage setting	Rate is uniform for all speeds	#/Br. A
3		18.1
4		22.2
5		29.0
6		34.7

STORAGE AND DISPOSAL

STORAGE

Do not store near fertilizers, seed, insecticides, or fungicides. Do not contaminate water, food, or feed by storage or disposal.

PESTICIDE DISPOSAL

Open dumping is prohibited. Wastes resulting from the use of this product may be disposed of on site or at an approved waste disposal facility.

CONTAINER DISPOSAL

Do not reuse empty container. (Triple rinse or equivalent), then dispose of empty container in a sanitary landfill or by incineration, or if allowed by state and local authorities, by burning. If burned, stay out of smoke.

Consult Federal, state, or local disposal authorities for approved alternative procedures such as limited open burning.

LIMITED WARRANTY AND DISCLAIMER

The manfacturer warrants (a) that this product conforms to the chemical description on the label, (b) that this product is reasonably fit for the purposes set forth in the directions for use when it is used in accordance with such directions, and (c) that the directions, warning, and other statements on this label are based upon responsible experts' evaluation of reasonable tests of effectiveness, of toxicity to laboratory animals and to plants, and of residues on food crops, and upon reports of field experience. Tests have not been made on all varieties or in all states or under all conditions. THE MANUFACTURER NEITHER MAKES NOR INTENDS, NOR DOES IT AUTHORIZE ANY AGENT OR REPRESENTATIVE TO MAKE ANY OTHER WARRANTIES, EXPRESS OR IMPLIED, AND IT EXPRESSLY EXCLUDES AND DISCLAIMS ALL IMPLIED WARRANTIES OF MERCHANTABILITY OR FITNESS FOR A PARTICULAR PURPOSE.

This warranty does not extend to, and the Buyer shall be solely responsible for, any and all loss or damage which results from the use of this product in any manner which is inconsistent with the label directions, warnings, or cautions.

BUYER'S EXCLUSIVE REMEDY AND MANUFACTURER'S OR SELLER'S EXCLUSIVE LIABILITY FOR ANY AND ALL CLAIMS, LOSSES, DAMAGES, OR INJURIES RESULTING FROM THE USE OR HANDLING OF THIS PRODUCT, WHETHER OR NOT BASED ON CONTRACT, NEGLIGENCE, STRICT LIABILITY IN TORT OR OTHERWISE SHALL BE LIMITED, AT THE MANUFACTURER'S OPTION, TO REPLACEMENT OF, OR THE REPAYMENT OF THE PURPOSE PRICE FOR, THE QUANTITY OF PRODUCT WITH RESPECT TO WHICH DAMAGES ARE CLAIMED. IN NO EVENT SHALL MANUFACTURER OR SELLER BE LIABLE FOR SPECIAL, INDIRECT OR CONSEQUENTIAL DAMAGES RESULTING FROM THE USE OR HANDLING OF THIS PRODUCT.

NOTICE TO BUYER

Purchase of this material does not confer any rights under patents governing this product or the use thereof in countries outside of the United States.

AMIBEN is a registered trademark of Rhône-Poulenc

Form No. AG90000

THIS SPECIMEN LABEL IS INTENDED FOR USE ONLY AS A GUIDE IN PROVIDING GENERAL INFORMATION REGARDING THE DIRECTIONS, WARNINGS AND CAUTIONS ASSOC-

ATED WITH THE USE OF THIS PRODUCT. AS WITH ANY AGRICULTURAL CHEMICAL, ALWAYS FOLLOW THE LABEL INSTRUCTIONS ON THE PACKAGE BEFORE USING.

AMIBEN® DS

CHLORAMBEN HERBICIDE

Controls Annual Broadleaf Weeds and Annual Grasses in Soybeans and Other Crops

ACTIVE INGREDIENTS:	
Sodium salt of chloramben (3-amino-2,5-dichlorobenzoic acid)*	83%
INERT INGREDIENTS	17%
TOTAL:	100%

*Chloramben acid equivalent: 75% by weight

KEEP OUT OF REACH OF CHILDREN

CAUTION

E.P.A. Reg. No. 264-305 E.P.A. Est. No. 264-GA-01

For PRODUCT USE Information Call 1-800-334-9745
For MEDICAL and TRANSPORTATION Emergencies ONLY Call 24 Hours A Day 1-800-334-7577

AMIBEN® DS CHLORAMBEN HERBICIDE

Controls Annual Broadleaf Weeds and Annual Grasses in Soybeans and Other Crops

Outstanding control of annual broadleaf weeds and grasses.

Excellent crop tolerance.

No harmful residue in soil—no changes in normal crop rotations. Completely water soluble.

Easy to mix and handle.

One of the most consistent preemergence herbicides available.

Tested and recommended by state colleges and extension agents.

Proven effective on millions of acres of soybeans.

Only the following recommendations referenced in this label are registered for use in California.

CROPS

- Asparagus, seedling
- Drybeans (White, Navy, Kidney, Pinto)
- Lima beans
- Pumpkin
- Squash
- Sunflowers
- Sweet Potatoes
- Soybeans
  - AMIBEN® DS with fertilizer
  - AMIBEN® DS preemergence or preplant incorporated
  - AMIBEN® DS and cultivation
  - AMIBEN® DS plus Lasso
  - AMIBEN® DS plus Treflan
  - AMIBEN® DS plus Lorox
  - AMIBEN® DS plus Prowl

AMIBEN® DS is not registered for control of Venice mallow (Hibiscus trionum) in California.

PRECAUTIONARY STATEMENTS

CAUTION

HAZARDS TO HUMANS AND DOMESTIC ANIMALS

Harmful if swallowed. Avoid contact with skin, eyes or clothing. Avoid breathing spray mist.

STATEMENT OF PRACTICAL TREATMENT

IF SWALLOWED: call a physician or Poison Control Center. Drink 1 to 2 glasses of water and induce vomiting by touching back of throat with finger. Do not induce vomiting or give anything by mouth to an unconscious person.

IN CASE OF SKIN CONTACT: wash immediately with soap and water and remove contaminated clothing and wash before reuse.

In case of eye contact flush with plenty of water. Get medical attention.

IF INHALED: remove victim to fresh air.

ENVIRONMENTAL HAZARDS

Do not apply directly to water. Do not contaminate water used for irrigation or domestic purposes by cleaning of equipment or disposal of waste.

GENERAL INFORMATION

AMIBEN® DS Chloramben Herbicide is effective applied preplant incorporated, preemergence and early postemergence. When used preemergence, it is most effective applied the same day as planting. Rain, irrigation, water, or mechanical incorporation moves AMIBEN® DS into the soil where weed seeds sprout.

Uniform spray application over the soil surface is important. About 10-20 gallons of water or sprayable fluid fertilizer for ground and about 3-5 gallons for air application is enough spray solution for good coverage. See specific sections of this label for any special spray volume requirements.

AMIBEN® DS Chloramben Herbicide remains active and effective for several weeks, long enough generally for crops to fill middles and shade out the late season weed growth. Performance is improved when moisture is received within 5 days after application and prior to weed emergence. On coarse and medium soils, one third inch of rainfall or overhead irrigation is sufficient to move AMIBEN® DS Herbicide into the seed germination zone. On fine soils, one half inch of rainfall or overhead irrigation is needed. High intensity or excessive rainfall or excessive irrigation after application may reduce control by moving the product below the weed seed germination zone.

In coarse textured soils with less than 1% organic matter heavy rains may move AMIBEN® DS Herbicide below the weed seed germination zone and produce erratic weed control. Weed control may be reduced on peak or muck soils (over 8% organic matter).

AMIBEN® DS Chloramben Herbicide controls the following important Annual Grass and Broadleaf Weeds:

barnyardgrass (water grass)	*Echinochloa crusgalli*
carpetweed	*Mullugo verticillata*
chickweed, annual	*Stellaria media*
chickweed, mouseear	*Cerastium vulgatum*
coffeeweed (sesbania)	*Sesbania exaltata*
corn spurry	*Spergola arvensis*
crabgrass, hairy (large)	*Digitaria sanguinalis*
crabgrass, smooth	*Digitaria ischaemum*
fall panicgrass	*Panicum dichotomiflorum*
Florida beggarweed	*Desmodium tortuosum*
foxtail, giant	*Setaria faberii*
foxtail, green	*Setaria viridis*
foxtail, yellow	*Setaria glauca*
goosegrass (silver crabgrass)	*Eleusine indica*
groundsel	*Senecio vulgaris*
johnsongrass-seedling	*Sorghum halepense*
kochia	*Kochia scoparia*
lambsquarters, common	*Chenopodium album*
lovegrass, Mediterranean	*Eragrostis barrelieri*
mustard, wild	*Brassica kaber*
nightshade, black	*Solanum nigrum*
nightshade, Eastern black	*Solanum ptycanthum*
nightshade, hairy	*Solanum sarachoides*
pigweed, prostrate	*Amaranthus biltoides*
pigweed, smooth	*Amaranthus hybridus*
pigweed, redroot	*Amaranthus retroflexus*
prickly sida (teaweed)	*Sida spinosa*
puncturevine	*Tribulus terrestris*
purslane (pusley)	*Portulaca oleracea*
ragweed, common	*Ambrosia artemisiifolia*
Russian thistle	*Salsola kali*
shepherds purse	*Capsella bursa-pastoris*
smartweed, Pennsylvania	*Polygonum pensylvanicum*
sowthistle, annual	*Sonchus oleraceus*
spurge, annual	*Euphorbia maculata*
stinkgrass	*Eragrostis cilianensis*
tall waterhemp	*Amaranthus tuberculatos*
velvetleaf (buttonweed)	*Abutilon theophrasti*
Venice mallow	*Hibiscus trionum*
wild proso millet	*Panicum miliaceum*

Band application of AMIBEN® DS may be made for broadleaf weed control over the top of preplant soil incorporated grass killers.

Deep germinating annual weeds and sprouts from established perennial weeds are not controlled. Sprouts coming from roots of established perennial weeds such as horsenettle, hedge bindweed, nutsedge (nutgrass), milkweed, dogbane, quackgrass, and johnsongrass are not controlled by any selective, surface preemergence chemical treatment including AMIBEN® DS.

DIRECTIONS FOR USE

IT IS A VIOLATION OF FEDERAL LAW TO USE THIS PRODUCT IN A MANNER INCONSISTENT WITH ITS LABELING.

Clean sprayers thoroughly before using to apply AMIBEN® DS CHLORAMBEN HERBICIDE. To clean sprayer after using AMIBEN® DS, rinse all parts of sprayer with water several times.

Avoid spray drift to adjacent crops as injury may result. Coarse sprays are less likely to drift.

DIRECTIONS FOR INCORPORATION OF AMIBEN® DS CHLORAMBEN HERBICIDE

AMIBEN® DS Chloramben Herbicide applied alone, and several AMIBEN® DS tank mixtures as specified in the crops section of this label, may be incorporated into the top 1 to 3 inches of the final seedbed before planting. When preplant incorporation is recommended in the crops section of this label the following equipment and operating procedures should be used. Be sure to use machinery that mixes AMIBEN® DS thoroughly with the soil. Incorporation with implements set to cut less than 2 inches deep may result in erratic weed control.

Recommended Equipment Includes:

1. Two passes with a disc set to cut 4 inches deep at 4 to 6 mph with the second pass running at an angle to the first. A tandem or double-disc operated one time does not provide adequate incorporation.
2. One pass with a field cultivator set to cut 3 to 4 inches deep and operated at 6 mph or more. The field cultivator should have 3 to 4 rows of sweeps, spaced at intervals of 7 inches or less and staggered so that no soil is left unturned. Chisel points should not be used. Do not set cultivator deeper than 4 inches. The field cultivator used alone or in combination with the double-disc will provide effective incorporation. If used in combination with a double-disc the spray boom should be mounted behind the disc blades but ahead of the cultivator sweeps.
3. One pass with a rolling cultivator set to cut 2 to 3 inches deep and operated at 6 to 8 mph. Rolling cultivators are adequate for use on coarse and medium textured soils only.
4. One pass with a mulch treader or other similar disc type implement set to cut 2 to 3 inches deep and operated at 5 to 8 MPH.
5. PTO-driven equipment (tillers, cultivators, hoes) set to cut 2 to 3 inches deep with rotors spaced to provide a clean sweep of the soil operated one time. PTO-driven equipment should not be operated at a speed greater than 4 mph.

MIXING INSTRUCTIONS

AMIBEN® DS Chloramben Herbicide in water or fluid fertilizer: first fill the spray tank 1/4 to 1/2 full with clean water. Start agitation and add AMIBEN® DS while adding the remaining required water or fluid fertilizer.

AMIBEN® DS Chloramben Herbicide Plus Emulsifiable Concentrates in Water: first fill tank 1/4 to 1/2 full with clean water, start agitation and add AMIBEN® DS. Mix thoroughly and add water until tank is 3/4 full, continue agitation and add the emulsifiable concentrate material. Fill tank with remaining required water and continue agitation during spraying until tank is empty. Agitation should sweep bottom contents of the tank up into the main body of the spray mixture for thorough mixing. If a compatibility agent is required it should be thoroughly mixed with the water before adding the emulsifiable concentrate. To prevent foaming, avoid stirring or splashing air into the mixture by placing the end of the fill pipe below the surface of the water in the spray tank.

AMIBEN® DS Chloramben Herbicide Plus Wettable Powders or Flowables: Fill tank 1/2 full with clean water, start sprayer agitation and add the recommended amount of the flowable or wettable powder. Mix thoroughly and add AMIBEN® DS while filling tank with remaining required water. Continue agitation during entire spraying operation until tank has been emptied. Agitation should sweep bottom contents of the tank up into the main body of the spray mixture for thorough mixing and uniform suspension.

DIRECTIONS FOR USING AMIBEN® DS CHLORAMBEN HERBICIDE THROUGH OVERHEAD IRRIGATION SYSTEMS

AMIBEN® DS Chloramben Herbicide may be applied as a preemergence treatment through center pivot, lateral move, side (wheel) roll, solid set or hand move irrigation systems only. AMIBEN® DS may be applied alone, or in tank mixtures with any of the other herbicides on this label which are registered for this type of application. Do not apply this product through any other type of irrigation system. Use the broadcast rates per acre for each product as specified elsewhere on this label for each of the tank mixtures, and follow all use precautions below and on respective labels for any other products applied in this manner.

Specific Requirements for Chemigation Application

1. The system must contain a functional check valve, vacuum relief valve, and low pressure drain appropriately located on the irrigation pipeline to prevent water source contamination from backflow.
2. The pesticide injection pipeline must contain a functional, automatic, quick-closing check valve to prevent the flow of fluid back toward the injection pump.
3. The pesticide injection pipeline must also contain a functional, normally closed, solenoid-operated valve located on the intake side of the injection pump and connected to the system interlock to prevent fluid from being withdrawn from the supply tank when the irrigation system is either automatically or manually shut down.
4. The system must contain functional interlocking controls to automatically shut off the pesticide injection pump when the water pump motor stops.
5. The irrigation line or water pump must include a functional pressure switch which will stop the water pump motor when the water pressure decreases to the point where pesticide distribution is adversely affected.
6. Systems must use a metering pump, such as a positive displacement injection pump (e.g., diaphragm pump) effectively designed and constructed of materials that are compatible with pesticides and capable of being fitted with a system interlock.
7. Do not apply when wind speed favors drift beyond the area intended for treatment.
8. Agitation is recommended in the pesticide supply tank when applying the AMIBEN® DS Chloramben Herbicide.
9. AMIBEN® DS Chloramben Herbicide should be applied continuously for the duration of the water application with center pivot and continuous lateral move systems. Application of AMIBEN® DS should be made during the last 30 to 45 minutes of the irrigation set with other overhead sprinkler systems.
10. Set the sprinkler system to deliver approximately 0.5 inch or less of water per acre. Application of more than 0.5 inch of water per acre may reduce weed control by moving the herbicide below the weed germination zone.
11. Remove scale, pesticide residues and other foreign matter from the supply tank and entire injector system. Flush with clean water.
12. If the herbicide(s) is diluted in the supply tank, fill the tank with half of the water amount desired, add the herbicide(s) and then add remaining water amount with agitation.
13. Start the sprinklers and then inject AMIBEN DS Chloramben Herbicide into the irrigation line. AMIBEN® DS Chloramben Herbicide should be injected with a positive displacement pump into the main line at least 8 feet ahead of a right angle turn to insure adequate mixing.

Refer to the AMIBEN® DS Chloramben Herbicide label for detailed information on application rates and timings.

Chemigation User Precautions

Application of more than 0.5 inch/acre of irrigation water may result in decreased product performance on certain soils.

Do not apply when conditions favor drift, when system connections or fittings leak, or when nozzles do not provide uniform distribution.

Allow sufficient time for pesticide to be flushed through all the lines and nozzles before turning off irrigation water.

Crop injury, lack of effectiveness, or illegal pesticides residues in the crop can result from non-uniform distribution of treated water.

Do not connect an irrigation system used for pesticide application to a public water system.

If you have questions about calibration, you should contact a State Extension Service specialists, equipment manufactures or other experts.

A person knowledgeable of the chemigation system and responsible for its operation, or under the supervision of the responsible person shall shut the system down and make necessary adjustments should the need arise.

DIRECTIONS FOR USING AMIBEN® DS CHLORAMBEN HERBICIDE IN CONSERVATION OR REDUCED TILLAGE SYSTEMS

When applied as directed below, AMIBEN® DS Chloramben Herbicide tank mixtures will control many emerged annual weeds, suppress many emerged perennial weeds and will give preemergence control of many annual grasses and broadleaf weeds in conservation tillage or reduced tillage systems.

When used in tillage systems with practices which eliminate all existing vegetation before planting, AMIBEN® DS Chloramben Herbicide may be used alone or in tank mixture with any of the other products on this label which are labeled for use in reduced tillage operations. AMIBEN® DS and AMIBEN® DS combinations with other preemergence or preplant incorporated herbicides will *not* control emerged weeds. The presence of previous crop residue (trash) which is uniformly distributed on the soil surface will not affect the performance of AMIBEN® DS, providing sufficient rainfall (0.375 to 0.5 inch) is received before weeds begin to germinate. IF WEEDS HAVE BEGUN TO EMERGE AT TIME OF APPLICATION FOLLOW THE DIRECTIONS BELOW.

When used in tillage systems where emerged weeds are present at planting, either Roundup, or Paraquat plus Ortho X-77 spreader should be added to the AMIBEN® DS tank mixture. Refer to the respective Roundup, Paraquat and Ortho X77 labels for crops labeled, weeds controlled, use rates, and specific precautions to be taken. Use a minimum of 15 to 40 gallons of water per acre and apply the tank mixture during or after planting, but *BEFORE CROP EMERGENCE.* As density of stubble, crop residue or weeds increase, spray gallonage and rate should be increased, within recommended ranges, to assure good coverage.

MIXING INSTRUCTIONS FOR USE OF AMIBEN® DS WITH SPRAYABLE FLUID FERTILIZERS.

In most cases, AMIBEN® DS Chloramben Herbicide can be mixed and applied with liquid nitrogen, blended clear liquids, or suspension type fertilizers.

Be sure the spray tank is clean and uncontaminated. First fill the spray tank 1/4 to 1/2 full with clean water. Start agitation and add AMIBEN® DS Chloramben Herbicide. Add required fluid fertilizer and water to fill tank. After the tank is full, no further agitation is necessary.

The compatibility of tank mixtures of AMIBEN® DS with other herbicides in combination with liquid fertilizers should be evaluated on a small scale before mixing spray tank quantities.

DIRECTIONS FOR AMIBEN® DS CHLORAMBEN HERBICIDE IMPREGNATION ON DRY FERTILIZER

Determine rate of fertilizer and AMIBEN® DS Chloramben Herbicide needed per acre. Higher rates of AMIBEN® DS should be used when heavy infestations of velvetleaf, black nightshade, mustard or Pennsylvania smartweed are expected.

Determine the number of gallons of AMIBEN® DS Solution needed per ton of fertilizer from Table 1 using the appropriate rates of fertilizer and AMIBEN® DS. Example: At a fertilizer rate of 350 lbs/A and an AMIBEN DS Chloramben Herbicide rate per acre of 3.0 lbs it would take 3.5 gallons of AMIBEN® DS solution per ton of fertilizer.

To prepare the AMIBEN® DS Solution mix two 12 pound bags of AMIBEN® DS in 5 gallons of water. It is necessary to maintain good agitation to keep AMIBEN® DS in suspension.

Apply up to 5 gallons of the AMIBEN® DS Solution to each ton of fertilizer. Provide sufficient mixing during impregnation to assure even distribution of all fertilizer particles.

Table 1
Gallons Of AMIBEN® DS Solution[1] Per Ton Of Fertilizer

Fertilizer	AMIBEN DS® Rates Per Acre		
Lbs/A	2.4 Lbs	3.0 Lbs	3.6 Lbs
Lbs			
150	*	*	*
200	5.0 Gal.	*	*
250	4.0 Gal.	5.0 Gal.	*
300	3.3 Gal.	4.2 Gal.	5.0 Gal.
350	3.0 Gal.	3.5 Gal.	4.3 Gal.
400	2.5 Gal.	3.1 Gal.	3.8 Gal.
450	2.3 Gal.	2.8 Gal.	3.3 Gal.
500	2.0 Gal.	2.5 Gal.	3.0 Gal.

[1] Solution = 24 lbs AMIBEN® DS Chloramben Herbicide Per 5 Gal. Water.
* DO NOT USE = Fertilizer will become too wet resulting in poor handling and spreading.

Note: For fertilizer rates not in Table 1 apply proportional amounts of the standard AMIBEN® DS solution per ton of fertilizer from the table above.

Example: For 375 lb fertilizer per acre and 2.4 lb AMIBEN® DS per acre apply 2.75 gallons of AMIBEN® DS solution per ton of fertilizer.

DIRECTIONS FOR USE ON SOYBEANS

AMIBEN® DS CHLORAMBEN HERBICIDE PREEMERGENCE OR PREPLANT INCORPORATED

AMIBEN® DS Chloramben Herbicide may be used as a preemergence or preplant incorporated herbicide. AMIBEN® DS may be applied and incorporated at planting or several days prior to planting. Uneven spray application or improper soil incorporation of AMIBEN® DS can produce erratic weed control. Incorporate chemical in the top 0 to 2 inch level as most weeds sprout in this zone. Refer to the "Directions for Incorporation" section of this label for approved incorporation equipment and operating procedures.

In coarse textured soils heavy rains on incorporated AMIBEN® DS Chloramben Herbicide may move it below the weed seed germination zone and produce erratic weed control.

Make all broadcast treatments according to the following table

Soil Texture	AMIBEN® DS Lbs/A
Loamy sand, sandy loam	2.4 to 3.0
Loam, silt loam, silt	3.0 to 3.6
Silty clay loam, silty clay, sandy clay, clay loam, clay	3.6

Use the higher rate of AMIBEN® DS when heavy populations of black nightshade, velvetleaf, wild mustard, common ragweed or Pennsylvania smartweed are expected. To insure good spray coverage use 10 to 20 gallons of water per acre with ground equipment or a minimum of 5 gallons of water per acre by aircraft.

When band spraying, mount the spray nozzle on the last packer or row wheel so that planting and spraying are done in the same operation. Proper calibration and maintenance of desired nozzle height above the ground are very important. Weeds growing in the middles between the treated rows should be cultivated as soon as necessary.

Use care while cultivating to avoid throwing fresh soil on treated bands.

If sufficient rain does not fall within 5 days after spraying AMIBEN® DS Herbicide, but there is enough soil moisture to germinate weeds and grasses, a light cultivation with a rotary hoe or similar tool will up-root these small broadleaf weeds and grasses. The shallow mixing of AMIBEN® DS in the surface soil will not interfere with the action of AMIBEN® DS when rains come.

If heavy rain packs the soil surface, rotary hoe the field to aid bean emergence. Do not disrupt the chemical band with deep cultivation.

AMIBEN® DS CHLORAMBEN HERBICIDE AND "COMMAND" FOR SOYBEANS

A tank mixture of AMIBEN® DS Chloramben Herbicide and Command 4-EC may be applied as a shallow (1-2 inch) preplant incorporated broadcast tank mix prior to planting, or AMIBEN® DS herbicide may be applied as a preemergence overlay treatment following a preplant incorporated application of Command 4-EC.

When applied according to directions given below the tank mixture or the sequential treatment of Command 4-EC and AMIBEN® DS will control all weeds listed on the respective labels for AMIBEN® DS and for Command 4-EC including velvetleaf, annual nightshade, Jimsonweed, common ragweed, Pennsylvania smartweed, barnyardgrass and foxtails.

Apply AMIBEN® DS at the rate of 1.8 to 2.4 lbs per broadcast acre and Command 4-EC at the rate of 1.5 to 2.0 pints per broadcast acre. Use the higher rate of AMIBEN® DS on finer soils or when heavy populations of pigweed, annual nightshade, ragweed, smartweed, barnyardgrass or foxtail are expected. Use the higher rate of Command 4-EC on finer soils or when heavy populations of velvetleaf, Jimsonweed or ragweed are expected.

When applying mixtures of AMIBEN® DS Chloramben Herbicide plus Command, observe all cautions and limitations appearing on the labels of each of the products. Observe application precautions, rotational crop/grazing and feeding restrictions and replanting instructions on the Command 4-EC label.

SPECIAL PRECAUTION

Off-site movement of spray drift or vapors of Command herbicide can cause foliar whitening or yellowing of some plants. Prior to making applications, read and strictly follow all precautions and application instructions on the Command 4-EC label.

AMIBEN® DS CHLORAMBEN HERBICIDE AND "LASSO" FOR SOYBEANS

A tank mixture of AMIBEN® DS Chloramben Herbicide and Lasso may be applied as a preplant soil incorporated herbicide treatment. AMIBEN® DS Herbicide may be applied as a preemergence broadcast tank mix treatment with Lasso at time of planting. Following a preplant soil incorporated application of Lasso, AMIBEN® DS may be applied preemergence at time of planting.

A tank mixture of the sequential treatment of AMIBEN® DS Chloramben Herbicide and Lasso will control the following annual weeds: barnyardgrass, crabgrass, foxtail (green, yellow, giant), fall panicum, lambsquarters, pigweed, smartweed, ragweed, black nightshade, velvetleaf and wild mustard, as well as all other weeds listed on the AMIBEN® DS and Lasso labels.

For uniform application of the tank mixture of AMIBEN® DS Chloramben Herbicide and Lasso use 10 to 20 gallons of water per acre for ground application. To prepare the tank mixture refer to the "Mixing Instructions" portion of this label. DO NOT mix AMIBEN® DS Herbicide and Lasso concentrates before the addition of water as an emulsion will not be formed.

Apply AMIBEN® DS Chloramben Herbicide and Lasso preplant incorporated or preemergence using the appropriate rates from the table below. For preplant incorporated treatments apply the tank mixture to the soil and incorporate into the top two inches of soil as directed in the "Directions for Incorporation" portion of this label. For preemergence treatments, apply the tank mixture during planting (behind the planter). For sequential applications apply Lasso alone preplant incorporated as specified on the Lasso label. Follow with a preemergence application of AMIBEN® DS Herbicide during planting (behind the planter).

With preemergence treatments rain is necessary to move the chemicals into the soil and make them available to control germinating weed seeds. If no rainfall occurs within 5 days following application, a shallow incorporation no deeper than 2 inches is recommended. High intensity or excessive rainfall after application may reduce control.

Make all broadcast treatments according to the following table:

Soil Texture	AMIBEN® DS Lbs/A	Lasso Qts/A
Loamy sand, sandy loam	2.4	1.5 to 2.0
Loam, silt loam, silt	2.4 to 3.0	2.0 to 2.5
Silty clay loam, silty clay, sandy clay, clay loam, clay	3.0 to 3.6	2.5 to 3.0

Use the higher rate of AMIBEN® DS when heavy populations of black nightshade, velvetleaf, mustard or smartweed are expected. Use the higher rate of Lasso when treatment is incorporated or when high populations of black nightshade, witchgrass or foxtail are expected.

For band treatments divide the band width by the row width and multiply by the appropriate rates indicated above to determine the amount of chemical to use per acre.

Read thoroughly the precautionary statements and all other information appearing on labels for both Lasso and AMIBEN® DS Chloramben Herbicide.

AMIBEN® DS CHLORAMBEN HERBICIDE TANK MIXTURE WITH "LASSO" AND "SENCOR" OR "LEXONE" FOR SOYBEANS

GENERAL INFORMATION

A tank mixture of AMIBEN® DS Chloramben Herbicide plus Lasso plus Sencor or Lexone when applied in accordance with the following directions, either preplant soil incorporated or preemergence, effectively controls most weeds listed on the current AMIBEN® DS, Lasso, and Lexone WP/4L/DF or Sencor WP/4/DF labels.

DIRECTIONS FOR USE

Except as directed in the "Conservation or Reduced Tillage" portion of this label, the seedbed should be free of existing weeds before application. Surface trash (plant residue from previous crop) should be uniformly distributed, and the seedbed should be free of clods. The tank mixture should be applied in from 10 to 20 gallons of water per acre using any properly calibrated low pressure sprayer equipped with herbicide tips and in-line strained screens no finer than 50 mesh. To prepare the tank mixture refer to the "Mixing Instructions" portion of this label. Apply the entire contents of the spray tank within 4 hours of initial mixing.

Preplant Soil Incorporated Use:

Incorporation should place the AMIBEN® DS Chloramben Herbicide plus Lasso plus Sencor or Lexone into the top 1 to 3 inches of the final seedbed. Refer to the "Directions for Incorporation" portion of this label for recommended incorporation equipment and operating procedures.

Apply the tank mixture using the table below to determine the proper rate of each chemical for your soil type.

Preemergence Use:

The tank mixture of AMIBEN® DS Chloramben Herbicide plus Lasso plus Lexone or Sencor may be applied as a preemergence broadcast treatment at time of planting. Use the table below to determine the proper rate of each chemical for your soil type.

Make all broadcast treatments according to the following table:

Soil Texture	AMIBEN® DS Lbs/A	LASSO 4E Qts/A	SENCOR 50WP or LEXONE WP Lbs/A	SENCOR 4 or LEXONE 4L Pts/A	SENCOR DF or LEXONE DF Lbs/A
Coarse (loamy sand, sandy loam)	2.4	1.5 to 2.0	0.5	0.5	0.33
Medium (loam, silt loam, silt)	2.4 to 3.0	2.0 to 2.5	0.5 to 0.75*	0.5 to 0.75*	0.33 to 0.5*
Fine (clay, loam, silty clay, sandy clay, clay loam, silty clay loam)	3.0 to 3.6	2.5 to 3.0	0.75 to 1.0*	0.75 to 1.0*	0.5 to 0.67*

* On Clarion Webster soils in Minnesota and Iowa or on similar alkaline (calcareous) soils a pH of 7.5 or above, apply Sencor WP or Lexone WP at 0.5 pound/A, Sencor 4 or Lexone 4L at 0.5 pint/A, or Lexone DF or Sencor DF at 0.33 pound/A on medium soils. On fine soils use 0.5 to 0.75 pound/A Sencor WP or Lexone WP, 0.5 to 0.75 pint Sencor 4 or Lexone 4L or 0.33 to 0.5 pound/A Lexone DF or Sencor DF. Use the higher rate only where soil pH is less than 7.5 and where pressure is heavy.

Use the higher rate of AMIBEN® DS where heavy stands of Pennsylvania smartweed, velvetleaf, ragweed, wild mustard, or black nightshade are expected. Use the higher rate of Lasso when treatments are incorporated or where heavy stands of black nightshade, foxtail, witchgrass, panicum, or yellow nutsedge are expected. Use the high rate of Sencor or Lexone where heavy stands of Florida pusley, sesbania, velvetleaf or Jimsonweed are expected except when soil pH is over 7.5.

CULTIVATION AFTER PLANTING

Soil treated with AMIBEN® DS Chloramben Herbicide plus Lasso plus Sencor or Lexone may be shallow-cultivated, rotary-hoed, or hand-hoed without reducing the weed control activity of the tank mixture. With preplant soil incorporated treatments, do not cultivate deeper than the AMIBEN® DS plus Lasso plus Sencor or Lexone treated layer of soil since this may bring untreated soil to the surface and poor weed control may result. If sufficient rain does not fall within 5 days after preemergence applications of the tank mixture, but there is enough soil moisture to germinate weeds and grasses, a light cultivation with a rotary hoe or similar tool will uproot these small broadleaf weeds and grasses.

SPECIAL PRECAUTIONS

Applied according to directions and under normal growing conditions, the tank mix of AMIBEN® DS plus Lasso plus Sencor or Lexone will not harm the treated crop. Over-application may result in crop injury. Uneven application or improper soil incorporation of AMIBEN® DS plus Lasso plus Sencor or Lexone can result in erratic weed control or crop injury. Seedling disease, cold weather, deep planting, excessive moisture, soil pH over 7.5, high salt concentration, or drought may weaken crop seedlings and increase the possibility of damage from the tank mixtures. Under these conditions, delayed crop development or reduced yields may result.

Do not use on sand or on medium to fine soils with less than 1% organic matter. This mixture applied preplant incorporated may cause slight soybean injury early, but the soybeans should recover and the subsequent growth or yields should not be affected.

Do not use the foliage from soybeans treated with the tank mix of AMIBEN® DS and Lasso and Sencor or AMIBEN® DS and Lasso and Lexone for feed or forage.

Caution: Observe all cautions and limitations on labeling of all products used in mixtures.

AMIBEN® DS AND "LASSO" AND "LOROX" FOR SOYBEANS

A tank mixture of AMIBEN® DS Chloramben Herbicide and Lasso and Lorox L or Lorox 50W may be applied as a preemergence broadcast tank mix treatment at time of planting.

A tank mixture of AMIBEN® DS Herbicide and Lasso and Lorox will control the following annual weeds: barnyardgrass, carpetweed, crabgrass, foxtail (green, yellow, giant), fall panicum, lambsquarters, pigweed, smartweed, ragweed, black nightshade, velvetleaf and wild mustard, as well as all other weeds listed on the AMIBEN® DS, Lasso and Lorox labels. Velvetleaf and cocklebur often germinate in soil at depths below the chemical penetration. Even though populations of these weeds are reduced under normal rainfall conditions, weed seeds which germinate outside the chemical zone are not affected.

Except as directed in the "Conservation or Reduced Tillage" portion of this label, the seedbed should be free of existing weeds before application. Surface trash (plant residue from previous crop) should be uniformly distributed, and the seedbed should be free of clods. For uniform application of the tank mixture of AMIBEN® DS Herbicide and Lasso and Lorox use 10 to 20 gallons of water per acre for ground application. To prepare the tank mixture refer to the "Mixing Instructions" portion of this label. DO NOT mix AMIBEN® DS Herbicide and Lasso and Lorox concentrates before the addition of water as an emulsion will not be formed.

Apply AMIBEN® DS and Lasso and Lorox as a preemergence treatment using the appropriate rates from the table below. Apply the tank mixture during planting (behind the planter).

Rain is necessary to move the chemicals into the soil and make them available to control germinating weed seeds. If no rainfall occurs within 5 days following application, a shallow incorporation no deeper than 2 inches is recommended. High intensity or excessive rainfall after application may reduce control.

Make all broadcast treatments according to the following table:

3% or Less Organic Matter

Soil Texture	AMIBEN ® DS Lbs/A	Lasso Qts/A	Lorox 50W or Lorox L Lbs or pts/A
Loamy sand, sandy loam	2.4	1.5 to 2.0	0.67 to 1.5
Loam, silt loam, silt	2.4 to 3.0	2.0 to 2.5	1.0 to 1.67
Silty clay loam, silty clay, sandy clay, clay loam, clay	3.0 to 3.6	2.5 to 3.0	1.25 to 2.0

Greater than 3% Organic Matter

Soil Texture	AMIBEN ® DS Lbs/A	Lasso Qts/A	Lorox 50W or Lorox L Lbs or pts/A
Loamy sand, sandy loam	2.4	1.5 to 2.0	1.5 to 2.0
Loam, silt loam, silt	2.4 to 3.0	2.0 to 2.5	1.67 to 2.5
Silty clay loam, silty clay, sandy clay, clay loam, clay	3.0 to 3.6	2.5 to 3.0	2.0 to 3.0

Use the higher rate of AMIBEN® DS when heavy populations of black nightshade, velvetleaf, mustard or smartweed are expected. Use the higher rate of Lasso when high populations of black nightshade, witchgrass or foxtail are expected. Use the higher rate of Lorox on soils with organic matter contents which are toward the high end of the ranges given in the table.

For band treatments divide the band width by the row width and multiply by the appropriate rates indicated above to determine the amount of chemical to use per acre.

Plant seeds at least 1.75 inches deep and do not treat if beans have begun to emerge.

Read thoroughly the precautionary statements and all other information appearing on labels for Lasso, Lorox L or Lorox 50W and AMIBEN® DS.

AMIBEN® DS CHLORAMBEN HERBICIDE AND "DUAL" FOR SOYBEANS

A tank mixture of AMIBEN® DS Chloramben Herbicide and Dual 8E may be applied as a preplant soil incorporated herbicide treatment. AMIBEN® DS Herbicide may be applied as a preemergence broadcast tank mix treatment with Dual 8E at time of planting. Following a preplant soil incorporated application of Dual 8E, AMIBEN® DS Herbicide may be applied preemergence at time of planting.

When applied according to label directions, tank mixtures of AMIBEN® DS and Dual will control all weeds controlled by AMIBEN® DS alone or by Dual alone including fall panicum, foxtail millet, red rice, signalgrass, witchgrass, yellow nutsedge and Florida pusley. Refer to the general information section of this label and the Dual label for additional weeds controlled by AMIBEN® DS and Dual 8E.

For uniform application of the tank mixture of AMIBEN® DS and Dual, use 10 to 20 gallons of water with ground equipment. To prepare the tank mixture, refer to the "Mixing Instructions" portion of this label.

Apply AMIBEN® DS and Dual preplant incorporated or preemergence using the appropriate rates from the table below. For preplant incorporated treatments apply the tank mixture to the soil and incorporate into the top two inches of soil following the procedures in the "Directions for Incorporation" portion of this label. Use a preplant incorporate application if furrow irrigation is used or when a period of dry weather is expected. If soybeans are planted on beds, apply and incorporate the tank mixture after bed formation. For preemergence treatments, apply the tank mixture during planting (behind the planter) or after planting before weeds or soybeans emerge. For sequential applications, apply Dual 8E alone preplant incorporated as specified on the Dual label. Follow with a preemergence application of AMIBEN® DS Herbicide during planting (behind the planter).

Make all broadcast treatments according to the following table

Soil Texture	AMIBEN® DS Lbs/A	Dual 8E Pts/A
Loamy sand, sandy loam	2.4	1.5 to 2.0
Loam, silt loam, silt	2.4 to 3.0	2.0 to 2.5
Silty clay loam, silty clay, sandy clay, clay loam, clay	3.0 to 3.6	2.5 to 3.0

Do not use on muck or peat soils.

Use the higher rate of AMIBEN® DS when heavy populations of black nightshade, velvetleaf, mustard or smartweed are expected. Use the higher rate of Dual when treatment is incorporated, when organic matter is over 3%, or when high populations of black nightshade, witchgrass, panicum, foxtail, or yellow nutsedge are expected.

Before applying this combination read thoroughly all precautionary statements and all other information appearing on or attached to the containers of AMIBEN® DS Herbicide and Dual 8E.

AMIBEN® DS CHLORAMBEN HERBICIDE TANK MIXTURE WITH "DUAL" AND "SENCOR" OR "LEXONE" FOR SOYBEANS

GENERAL INFORMATION

A tank mixture of AMIBEN® DS Chloramben Herbicide plus Dual 8E plus Sencor or Lexone when applied in accordance with the following directions, either preplant soil incorporated or preemergence, effectively controls most weeds listed on the current AMIBEN® DS, Dual 8E, and Lexone WP/4L/DF or Sencor WP/4/DF labels.

DIRECTIONS FOR USE

Except as directed in the "Conservation or Reduced Tillage" portion of this label, the seedbed should be free of existing weeds before application. Surface trash (plant residue from previous crop) should be uniformly distributed, and the seedbed should be free of clods. The mixture should be applied in from 10 to 20 gallons of water per acre using any properly calibrated low pressure sprayer equipped with herbicide tips and in-line strained screens no finer than 50 mesh. To prepare the tank mixture refer to the "Mixing Instructions" portion of this label. Apply the entire contents of the spray tank within 4 hours of initial mixing.

Preplant Soil Incorporated Use:

Incorporation should place the AMIBEN® DS plus Dual 8E plus Sencor or Lexone into the top 1 to 3 inches of the final seedbed as directed in the "Directions for Incorporation" portion of this label. Apply the tank mixture using the table below to determine the proper rate of each chemical for your soil type.

Preemergence Use:

The tank mixtures of AMIBEN® DS plus Dual 8E plus Lexone or Sencor may be applied as a preemergence broadcast treatment at time of planting. Use the table below to determine the proper rate of each chemical for your soil type.

Make all broadcast treatments according to the following table

Soil Texture	AMIBEN® DS Lbs/A	DUAL 8E Pts/A	SENCOR 50WP or LEXONE WP Lbs/A	SENCOR 4 or LEXONE 4L Pts/A	SENCOR DF or LEXONE DF Lbs/A
Coarse (loamy sand, sandy loam)	2.4	1.5 to 2.0	0.5	0.5	0.33
Medium (loam, silt loam, silt)	2.4 to 3.0	2.0 to 2.5	0.5 to 0.75*	0.5 to 0.75*	0.33 to 0.5*
Fine (clay, clay loam, silty clay, sandy clay, sandy clay loam, silty clay loam)	3.0 to 3.6	2.5 to 3.0	0.75 to 1.0*	0.75 to 1.0*	0.5 to 0.67*

* On Clarion Webster soils in Minnesota and Iowa or on similar alkaline (calcareous) soils a pH of 7.5 or above, apply Sencor WP or Lexone WP at 0.5 pound/A, Sencor 4 or Lexone 4L at 0.5 pint/A, or Lexone DF or Sencor DF at 0.33 pound/A on medium soils. On fine soils use 0.5 to 0.75 pound/A Sencor WP or Lexone WP, 0.5 to 0.75 pint Sencor 4 or Lexone 4L or 0.33 to 0.5 pound/A Lexone DF or Sencor DF. Use the higher rate only where soil pH is less than 7.5 and where weed pressure is heavy.

Use the higher rate of AMIBEN® DS where heavy stands of Pennsylvania smartweed, velvetleaf, ragweed, wild mustard, or black nightshade are expected. Use the higher rate of Dual 8E when treatments are incorporated, when organic matter is greater than 3%, or where heavy stands of black nightshade, foxtail, witchgrass, panicum, or yellow nutsedge are expected. Use the high rate of Sencor or Lexone where heavy stands of Florida pusley, sesbania, velvetleaf or Jimsonweed are expected except when soil pH is over 7.5.

CULTIVATION AFTER PLANTING

Soil treated with AMIBEN® DS plus Dual plus Sencor or Lexone may be shallow-cultivated, rotary-hoed, or hand-hoed without reducing the weed control activity of the tank mixture. With preplant soil incorporated treatments, do not cultivate deeper than the AMIBEN® DS plus Dual plus Sencor or Lexone treated layer of soil since this may bring untreated soil to the surface and poor weed control may result. If sufficient rain does not fall within 5 days after preemergence applications of the tank mixture, but there is enough soil moisture to germinate weeds and grasses, a light cultivation with a rotary hoe or similar tool will uproot these small broadleaf weeds and grasses.

SPECIAL PRECAUTIONS

Applied according to directions and under normal growing conditions, the tank mix of AMIBEN® DS plus Dual plus Sencor or Lexone will not harm the treated crop. Over-application may result in crop injury. Uneven application or improper soil incorporation of AMIBEN® DS plus Dual plus Sencor or Lexone can result in erratic weed control or crop injury. Seedling disease, cold weather, deep planting, excessive moisture, soil pH over 7.5, high salt concentration, or drought may weaken crop seedlings and increase the possibility of damage from the tank mixture. Under these conditions, delayed crop development or reduced yields may result.

Do not use on sand or on medium to fine soils with less than 1% organic matter. This mixture applied preplant incorporated may cause slight soybean injury early, but the soybeans should recover and the subsequent growth or yields should not be affected.

Do not use the foliage from soybeans treated with the tank mix of AMIBEN® DS and Dual and Sencor or AMIBEN® DS and Dual and Lexone for feed or forage.

Caution: Observe all cautions and limitations on labeling of all products used in mixtures.

AMIBEN® DS CHLORAMBEN HERBICIDE AND "DUAL" AND "LOROX" FOR SOYBEANS

A tank mixture of AMIBEN® DS Chloramben Herbicide and Dual and Lorox L or Lorox 50W may be applied as a preemergence broadcast tank mix treatment at time of planting.

A tank mixture of AMIBEN® DS Herbicide and Dual and Lorox will control the following annual weeds: barnyardgrass, carpetweed, crabgrass, foxtail (green, yellow, giant), fall panicum, lambsquarters, pigweed, smartweed, ragweed, black nightshade, velvetleaf and wild mustard, as well as all other weeds listed on the AMIBEN® DS, Dual and Lorox labels. Velvetleaf and cocklebur often germinate in soil at depths below the chemical penetration. Even though populations of these weeds are reduced under normal rainfall conditions, weed seeds which germinate outside the chemical zone are not affected.

Except as directed in the "Conservation or Reduced Tillage" portion of this label, the seedbed should be free of existing weeds before application. Surface trash (plant residue from previous crop) should be uniformly distributed, and the seedbed should be free of clods. For uniform application of the tank mixture of AMIBEN® DS Herbicide and Dual and Lorox use 10 to 20 gallons of water per acre for ground application. To prepare the tank mixture refer to the "Mixing Instructions" portion of this label.

Apply AMIBEN® DS and Dual and Lorox as a preemergence treatment using the appropriate rates from the table below. Apply the tank mixture during planting (behind the planter).

Rain is necessary to move the chemicals into the soil and make them available to control germinating weed seeds. If no rainfall occurs within 5 days following application, a shallow incorporation no deeper than 2 inches is recommended. High intensity or excessive rainfall after application may reduce control.

Make all broadcast treatments according to the following table:

3% or less Organic Matter

Soil Texture	AMIBEN® DS Lbs/A	Dual Pts/A	Lorox 50W or Lorox L Lbs or pts/A
Loamy sand, sandy loam	2.4	1.5 to 2.0	0.67 to 1.5
Loam, silt loam, silt	2.4 to 3.0	2.0 to 2.5	1.0 to 1.67
Silty clay loam, silty clay, sandy clay, clay loam, clay	3.0 to 3.6	2.5 to 3.0	1.25 to 2.0

Greater than 3% Organic Matter

Soil Texture	AMIBEN® DS Lbs/A	Dual Pts/A	Lorox 50W or Lorox L Lbs or pts/A
Loamy sand, sandy loam	2.4	1.5 to 2.0	1.5 to 2.0
Loam, silt loam, silt	2.4 to 3.0	2.0 to 2.5	1.67 to 2.5
Silty clay loam, silty clay, sandy clay, clay loam, clay	3.0 to 3.6	2.5 to 3.0	2.0 to 3.0

Use the higher rate of AMIBEN® DS when heavy populations of black nightshade, velvetleaf, mustard or smartweed are expected. Use the higher rate of Dual when high populations of black nightshade, witchgrass or foxtail are expected. Use the higher rate of Lorox on soils with organic matter contents which are toward the high end of the ranges given in the table.

For band treatments divide the band width by the row width and multiply by the appropriate rates indicated above to determine the amount of chemical to use per acre.

Plant seeds at least 1.75 inches deep and do not treat if beans have begun to emerge.

Read thoroughly the precautionary statements and all other information appearing on labels for Dual, Lorox L or Lorox 50W and AMIBEN® DS.

AMIBEN® DS CHLORAMBEN HERBICIDE AND "TREFLAN" FOR SOYBEANS

AMIBEN® DS Chloramben Herbicide may be applied in a band over the soybean row at planting time in fields where Treflan has been applied as a preplant soil incorporated herbicide. Or, AMIBEN® DS Herbicide may be applied several days prior to planting as a broadcast tank mix with Treflan. The tank mixture should be used as a spring preplant soil incorporated treatment. The combination of AMIBEN® DS Herbicide with Treflan improves broadleaf weed control of species such as Pennsylvania smartweed, velvetleaf and ragweed. For broadcast treatments, incorporate chemicals immediately and thoroughly to an approximate depth of 2 inches with a disc, field cultivator or similar tool set to cut a depth of 4 to 6 inches. Refer to the Treflan label for additional incorporation information.

Make all broadcast treatments according to the following table

Soil Texture	AMIBEN® DS Lbs/A	TREFLAN Pts/A
Loamy sand, sandy loam	2.4	1
Loam, silt loam, silt	2.4 to 2.0	1.5
Silty clay loam, silty clay, sandy clay, clay loam, clay	3.0 to 3.6	2

Do not use on muck or peat soils.

Use the higher rate of AMIBEN® DS when heavy populations of black nightshade, velvetleaf, mustard or smartweed are expected.

Before applying this combination, read thoroughly all precautionary statements and all other information appearing on or attached to the containers of AMIBEN® Herbicide and Treflan.

AMIBEN® DS CHLORAMBEN HERBICIDE TANK MIXTURES WITH "TREFLAN" AND "SENCOR" OR "LEXONE" FOR SOYBEANS

GENERAL INFORMATION

A tank mix of AMIBEN® DS Chloramben Herbicide plus Treflan plus Sencor or Lexone when applied preplant soil incorporated as recommended herein for use on soybeans, effectively controls the weeds listed on the current AMIBEN® DS, Treflan and Sencor WP/4/DF or Lexone WP/4L/DF labels.

DIRECTIONS FOR USE

Except as directed in the "Conservation or Reduced Tillage" portion of this label, the seedbed should be free of existing weeds before application. Surface trash (plant residue from previous crop) should be uniformly distributed, and the seedbed should be free of clods.

To prepare the tank mixture refer to the "Mixing Instructions" portion of this label. Apply the entire contents of the spray tank within 4 hours of initial mixing. Apply the tank mixture in from 10 to 20 gallons of water per acre using any properly calibrated low pressure sprayer equipped with herbicide tips and in-line strained screens no finer than 50 mesh.

For best results AMIBEN® DS plus Treflan plus Sencor or Lexone should be incorporated as soon as possible after application. The three way tank mixture must be incorporated within 24 hours after application. If AMIBEN® DS plus Treflan plus Sencor or Lexone is applied to a wet, warm soil surface or if the wind velocity is 10 mph or higher, variable weed control may result from delaying incorporation beyond 24 hours.

Incorporation should place the AMIBEN® DS Herbicide plus Treflan plus Sencor or Lexone into the top 2 to 3 inches of the final seedbed. Refer to the incorporation portion of the Treflan label for approved incorporation equipment and operating procedures.

CULTIVATION AFTER PLANTING

Soil treated with AMIBEN® DS plus Treflan plus Sencor or Lexone may be shallow-cultivated, rotary-hoed or hand-hoed without reducing the weed control activity of the tank mixture. Do not cultivate deeper than the AMIBEN® DS plus Treflan plus Sencor or Lexone treated layer of soil since this may bring untreated soil to the surface and poor weed control may result.

Make all broadcast treatments according to the following table

Soil Texture	AMIBEN® DS Lbs/A	Treflan 4EC Pts/A	Sencor 50WP or Lexone WP lbs/A	Sencor 4 or Lexone 4L Pts/A	Sencor DF or Lexone DF lbs/A
Coarse (sand, loamy sand, sandy loam)[a]	2.4	1.0	0.5	0.5	0.33
Medium (loam, silt loam, silt)	2.4 to 3.0*	1.5	0.5 to 0.75**	0.5 to 0.75**	0.33 to 0.5**
Fine (clay, clay loam, silty clay, sandy clay, sandy clay loam, silty clay loam)	3.0 to 3.6*	2.0	0.75 to 1.0**	0.75 to 1.0**	0.5 to 0.67**

[a] Do not use Sencor or Lexone on coarse soils with less than 1% organic matter.

* Use the higher rate of AMIBEN® DS Chloramben Herbicide when velvetleaf or black nightshade is a problem.

** On Clarion Webster soils in Minnesota and Iowa or on similar alkaline (calcareous) soils with a pH of 7.5 or above apply Sencor WP or Lexone WP at 0.5 pounds/A, Sencor 4 or Lexone 4L at 0.5 pint/A or Lexone DF or Sencor DF at 0.33 pound/A on medium soils and at 0.5 to 0.75 pound/A Sencor WP or Lexone WP, 0.5 to 0.75 pint/A Sencor 4 or Lexone 4L or 0.33 to 0.5 pound/A Lexone DF or Sencor DF on fine soils. Use the higher rate only where soil pH is less than 7.5 and where weed pressure is heavy.

SPECIAL PRECAUTIONS

Applied according to directions and under normal growing conditions, the AMIBEN® DS and Treflan and Sencor or AMIBEN® DS and Treflan and Lexone tank mix will not harm the treated crop. Over-application may result in crop injury. Uneven application or improper soil incorporation of the AMIBEN® DS and Treflan and Sencor or AMIBEN® DS and Treflan and Lexone tank mix can result in erratic weed control or crop injury. Seedling disease, cold weather, deep planting, excessive moisture, soil pH over 7.5, high salt concentration, or drought may weaken crop seedlings and increase possibility of damage from the tank mixtures. Under these conditions, delayed crop development or reduced yields may result.

Caution: Observe all cautions and limitations on labeling of all products used in mixtures.

Do not use the foliage from soybeans treated with the AMIBEN® DS and Treflan and Sencor or AMIBEN® DS and Treflan and Lexone tank mix for feed or forage.

AMIBEN® DS CHLORAMBEN HERBICIDE AND "SONALAN" FOR SOYBEANS

AMIBEN® DS Chloramben Herbicide may be applied as a preemergence band or broadcast treatment in soybeans at planting time in fields where Sonalan has been applied as a preplant soil incorporated herbicide. Or AMIBEN® DS Herbicide may be applied prior to planting as a broadcast tank mix with Sonalan. The tank mixture should be used as a spring preplant soil incorporated treatment. The combination of AMIBEN® DS Herbicide with Sonalan improves broadleaf weed control of species such as Pennsylvania smartweed, velvetleaf, common ragweed, black nightshade and wild mustard. For broadcast treatments, incorporate chemicals immediately and thoroughly to an approximate depth of 2 inches as described on the Sonalan label.

Make all broadcast treatments according to the following table:

Soil Texture	AMIBEN® DS Lbs/A	Sonalan Pts/A
Loamy sand, sandy loam	2.4	1.25 to 2.0
Loam, silt loam, silt	2.4 to 3.0	1.75 to 2.5
Silty clay loam, silty clay, sandy clay, clay loam, clay	3.0 to 3.6	2.25 to 3.0

Use the higher rate of AMIBEN® DS where a heavy population of smartweed, velvetleaf, ragweed, wild mustard or black nightshade are expected. Use the higher rate of Sonalan on finer textured soils and on soils with organic matter content over 3%.

AMIBEN® DS CHLORAMBEN HERBICIDE AND "PROWL" FOR SOYBEANS

A tank mixture of AMIBEN® DS Chloramben Herbicide and Prowl may be applied as a preplant soil incorporated herbicide treatment or AMIBEN® DS Herbicide may be applied as a preemergence broadcast treatment at planting time in soybean fields where Prowl has been applied as a preplant soil incorporated herbicide treatment. On soybeans grown south of Interstate 80 the tank mixture of AMIBEN® DS plus Prowl may also be applied as a preemergence broadcast treatment. The combination may be used only in Illinois, Indiana, Iowa, Kansas, Kentucky, Michigan, Minnesota, Nebraska, New York, North Dakota, Ohio, Pennsylvania, South Dakota, Wisconsin and Missouri except in the "Bootheel Region".

When applied according to directions, tank mixtures of AMIBEN® DS Herbicide and Prowl herbicide either preplant incorporated or preemergence, or the sequential treatment of AMIBEN® DS preemergence over Prowl preplant incorporated will control fall panicum, signalgrass, Venice mallow and velvetleaf in addition to all other weeds listed on the AMIBEN® DS and Prowl labels.

For uniform application of the tank mixture of AMIBEN® DS and Prowl, use 10 to 20 gallons of water per acre with ground equipment or 3 to 5 or more gallons of water per acre by aircraft. To prepare the tank mixture refer to the "Mixing Instructions" portion of this label.

Except as directed in the "Conservation or Reduced Tillage" portion of this label, the seedbed should be free of existing weeds before application. Surface trash (plant residue from previous crop) should be uniformly distributed, and the seedbed should be free of clods. When using the tank mixture preplant soil incorporated or when using Prowl preplant incorporated followed by a preemergence application of AMIBEN® DS the herbicides should be applied within 7 days prior to planting. When applied as a preplant soil incorporated treatment the tank mixture of AMIBEN® DS plus Prowl should be thoroughly incorporated into the top 0 to 2 inches of soil using equipment and techniques described elsewhere on this label and on the Prowl label. When using the tank mixture preemergence, treatments should be made at planting or within 2 days after planting before weeds and soybeans emerge.

Make all broadcast treatments according to the following table

Soil Texture	AMIBEN® DS Lbs/A	Prowl Pts/A
Loamy sand, sandy loam	2.4	1.0 to 2.0
Loam, silt loam, silt	2.4 to 3.0	1.5 to 2.5
Silty clay loam, silty clay, sandy clay, clay loam, clay	3.0 to 3.6	2.0 to 3.0

Use the higher rate of AMIBEN® DS when heavy populations of black nightshade, velvetleaf, wild mustard, common ragweed or Pennsylvania smartweed are expected. Use the higher rate of Prowl when treatments are incorporated on soils with more than 3% organic matter, or where heavy stands of foxtail, witchgrass or panicum are expected. Do not use on peat or muck soils.

If sufficient rain does not fall within 5 days after preemergence treatment, but there is enough soil moisture to germinate weeds and soybeans, a light cultivation with a rotary hoe or similar tool will uproot these small weeds. The shallow mixing of the herbicides into the soil surface will not interfere with their action when rains come. If soybean crop loss occurs due to weather conditions, soybeans may be replanted. If replanting is necessary, DO NOT rework the soil.

Before applying this combination, read thoroughly the precautionary statements and all other information appearing on or attached to the containers of both the Prowl and AMIBEN® DS herbicides.

AMIBEN® DS CHLORAMBEN HERBICIDE TANK MIXTURE WITH "PROWL" AND "SENCOR" OR "LEXONE" FOR SOYBEANS

GENERAL INFORMATION

A tank mixture of AMIBEN® DS Chloramben Herbicide plus Prowl plus Sencor or Lexone may be applied either preplant soil incorporated or preemergence in soybeans grown in the northeastern and north central states (Illinois, Indiana, Iowa, Kansas, Kentucky, Michigan, Minnesota, Nebraska, New York, North Dakota, Ohio, Pennsylvania, South Dakota, Wisconsin, Missouri (except the "Bootheel Region"), all areas of Delaware, Maryland, New Jersey, and Virginia (except the Coastal Plains of these states). Do not make preemergence applications of Prowl north of Interstate 80. When applied according to the following directions the tank mixture of AMIBEN® DS plus Prowl plus Sencor or Lexone effectively controls the weeds listed on the current AMIBEN®, Prowl and Sencor WP/4/DF or Lexone WP/4L/DF labels.

DIRECTIONS FOR USE

Except as directed in the "Conservation or Reduced Tillage" portion of this label, the seedbed should be free of existing weeds before application. Surface trash (plant residue from previous crop) should be uniformly distributed, and the seedbed should be free of clods. Uniformly apply the recommended AMIBEN® DS plus Prowl plus Sencor or Lexone tank mixture in 10 to 20 gallons of water per acre with ground equipment or in 3 to 5 or more gallons of water per acre by aircraft using herbicide nozzle tips and in line strained screens no finer than 50 mesh. To prepare the tank mixture refer to the "Mixing Instructions" portion of this label. Apply the entire contents of the spray tank within 4 hours of initial mixing.

PREPLANT INCORPORATED USE

The tank mixture of AMIBEN® DS plus Prowl plus Sencor or Lexone must be incorporated into the soil and soybeans must be planted within 7 days after application. Uneven spray application or improper soil incorporation can produce erratic weed control. The tank mixture should be thoroughly incorporated into the top 1 to 2 inches of soil using equipment and techniques described elsewhere on this label.

PREEMERGENCE USE

Apply the tank mixture of AMIBEN® DS plus Prowl plus Sencor or Lexone as a broadcast treatment at time of planting. This treatment is most effective when adequate rainfall or overhead irrigation is received within 5 days after application. If sufficient rain does not fall within 5 days after spraying the tank mixture of AMIBEN® DS plus Prowl plus Sencor or Lexone, but there is enough soil moisture to germinate weeds and grasses a light cultivation with a rotary hoe or similar tool will up-root these small broadleaf weeds and grasses. The shallow mixing of the tank mixture in the surface soil will improve consistency of performance and reduce dependency on rainfall to activate the herbicides, but will not interfere with their action when rains come.

APPLICATION RATES

Make all broadcast treatments according to the following table

Soil Texture	AMIBEN® DS Lbs/A	Prowl Pts/A	Sencor 50WP or Lexone WP Lbs/A	Sencor 4 or Lexone 4L Pts/A	Sencor DF or Lexone DF Lbs/A
Coarse (loamy sand, sandy loam)	2.4	1.5 to 2.0	0.5	0.5	0.33
Medium (loam, silt loam, silt)	2.4 to 3.0	1.5 to 2.5	0.5 to 0.75*	0.5 to 0.75*	0.33 to 0.5*
Fine (clay, clay loam, silty clay, sandy clay, sandy clay loam, silty clay loam)	3.0 to 3.6	2.0 to 3.0	0.75 to 1.0*	0.75 to 1.0*	0.5 to 0.67*

* On Clarion Webster soils in Minnesota and Iowa or on similar alkaline (calcareous) soils with a pH of 7.5 or above apply Sencor WP or Lexone WP at 0.5 pound/A, Sencor 4 or Lexone 4L at 0.5 pint/A, or Lexone DF or Sencor DF at 0.33 pound/A on medium soils. On fine soils use 0.5 to 0.75 pound/A Sencor WP or Lexone WP, 0.5 to 0.75 pint Sencor 4 or Lexone 4L or 0.33 to 0.5 pound/A Lexone DF or Sencor DF. Use the higher rate only where soil pH is less than 7.5 and where pressure is heavy.

Use the higher rate of AMIBEN® DS where heavy stands of Pennsylvania smartweed, velvetleaf, ragweed, wild mustard, or black nightshade are expected. Use the higher rate of Prowl when treatments are incorporated on soils with more than 3% organic matter, or where heavy stands of foxtail, witchgrass or panicum are expected. Use the high rate of Sencor or Lexone where heavy stands of Florida pusley, sesbania, velvetleaf or Jimsonweed are expected except when soil pH is over 7.5.

SPECIAL PRECAUTIONS

Soybean seed should be planted at least 1 ½ inches below soil surface. Soil incorporation deeper than 2 inches will reduce weed control and can result in crop injury. Seedling disease, cold weather, deep planting, excessive soil moisture, soil pH over 7.5, high salt concentration, or drought may weaken crop seedlings and increase possibility of damage from the tank mixtures. Under these conditions delayed crop development or reduced yields may result.

Do not use on sensitive soybean varieties such as Tracy, Semmes, Altona, Vansoy or Coker 102.

Do not use the foliage from soybeans treated with the AMIBEN® DS plus Prowl plus Sencor or Lexone tank mix for feed or forage.

Caution: Observe all cautions and limitations on labeling of all products used in mixtures.

AMIBEN® DS CHLORAMBEN HERBICIDE AND METRIBUZIN ("SENCOR" OR "LEXONE") TANK MIXTURE FOR SOYBEANS

AMIBEN® DS Chlorambem Herbicide and Metribuzin may be applied in a tank mixture combination as a preemergence band or broadcast application. Use standard spray equipment, and 10 to 20 gallons of spray solution per broadcast acre. To prepare the tank mixture refer to the "Mixing Instructions" portion of this label.

This tank mixture controls Florida pusley, sesbania (coffeebean or coffeeweed), shepherdspurse and sicklepod in addition to those weeds controlled by AMIBEN® DS Herbicide. Performance on cocklebur, fall panicum and black nightshade may be erratic depending on soil temperature, depth of weed seeds in soil and amount of moisture. Control may be improved with timely cultivation.

Make all broadcast treatments according to the following table

3% or Less Organic Matter

Soil Texture	AMIBEN® DS * Lbs/A	Sencor 50W or Lexone WP Lbs/A	Sencor 4 or Lexone 4L Pts/A	Sencor DF or Lexone DF Lbs/A
Loamy sand to sandy loam (greater than 2% organic matter only)	2.4	0.5	0.5	0.33
Loam to sandy clay loam	2.4 to 3.0	0.5	0.75	0.5
Silty clay loam to clay	3.0 to 3.6	1.0	1.0	0.67

Greater than 3% Organic Matter

Soil Texture	AMIBEN® DS Lbs/A	Sencor 50W or Lexone WP Lbs/A	Sencor 4 or Lexone 4L Pts/A	Sencor DF or Lexone DF Lbs/A
Loamy sand to sandy loam (greater than 2% organic matter only)	2.4	0.75	0.75	0.5
Loam to sandy clay loam	2.4 to 3.0	1.0	1.0	0.67
Silty clay loam to clay	3.0 to 3.6	1.0	1.0	0.67

In the Mississippi Delta only, on silty clay to heavy clay textured soil use 3.6 pounds of AMIBEN® DS plus Lexone 50 WP - 1.5 lb., Lexone DF - 1 lb. or 2 lbs. of Sencor 50 WP, 2 pts. of Sencor 4, or 1 lb. of Sencor DF. Do not use on muck soils. Use the high rates when heavy weed stands are expected.

For band application rates divide Band width by the Row spacing and multiply by rate/acre selected from the chart above.

SPECIAL PRECAUTIONS

Injury may occur to soybeans following application of Metribuzin on soils having a calcareous surface area or a pH of 7.5 or higher.

Sensitive soybean varieties such as Tracy, Semmes, Altona, Vansoy or Coker 102, should not be treated with metribuzin or injury may result.

If initial planting does not produce a satisfactory stand, fields treated with AMIBEN® DS Herbicide and Metribuzin tank mixtures may be replanted to soybeans. If replanting is necessary, do not rework soil. Do not apply a second treatment over the first treatment as injury to soybeans may occur. Read Sencor or Lexone labels and literature carefully to understand conditions to insure crop safety. For additional precautions and information concerning spray equipment, mixing directions, and restrictions refer to the appropriate sections of the AMIBEN® DS Herbicide and the Lexone/Sencor labels. Do not use vines treated with the AMIBEN® DS, Metribuzin tank mix for feed or forage.

AMIBEN® DS CHLORAMBEN HERBICIDE AND "LOROX" TANK MIXTURE FOR SOYBEANS

A tank mixture of AMIBEN® DS Herbicide and Lorox 50W or Lorox L will control annual weeds such as barnyardgrass, carpetweed, crabgrass, fall panicum, foxtail, lambsquarters, pigweed, ragweed and Pennsylvania smartweed. Velvetleaf and cocklebur often germinate in soil at depths below the chemical penetration. Even though populations of these weeds are reduced under normal rainfall conditions, weed seeds which germinate outside the chemical zone are not affected.

Treatments should be made at planting time as a simple preemergence application using 10 to 20 gallons of water per acre. For air application use 3 to 5 gallons per acre. To prepare the tank mixture refer to the "Mixing Instructions" portion of this label.

Plant seeds at least 1.75 inches deep and do not treat if beans have begun to emerge.

Make all broadcast treatments according to the following table

	3% or Less Organic Matter		Greater than 3% Organic Matter	
Soil Texture	AMIBEN® DS Lbs/A	Lorox 50W or Lorox L Lbs or pts/A	AMIBEN® DS Lbs/A	Lorox 50W or Lorox L Lbs or pts/A
Sandy Loam	2.4	0.67 to 1.5	2.4	1.5 to 2.0
Silt Loam	3.0	1.0 to 1.67	3.0	1.67 to 2.5
Clay Loam	3.6	1.25 to 2.0	3.6	2.0 to 3.0

For band treatments, use proportionally less AMIBEN® DS Chloramben Herbicide and Lorox. Before applying this tank mixture as a broadcast or band treatment, read thoroughly the precautionary statements and all other information appearing on both product labels.

AMIBEN® DS CHLORAMBEN HERBICIDE AND "BASALIN" FOR SOYBEANS

AMIBEN® DS Chloramben Herbicide may be applied over soybeans at planting time in fields where Basalin herbicide has been applied as a preplant soil incorporated herbicide. Or, AMIBEN® DS Herbicide may be applied as a preplant incorporated tank mix with Basalin.

When applied according to direction, tank mixtures of AMIBEN® DS Herbicide and Basalin or the sequential treatment will control fall panicum, signalgrass, and common purslane in addition to those weeds controlled with AMIBEN® DS Herbicide alone.

For uniform application of the tank mixture of AMIBEN® DS and Basalin use 10 to 20 gallons of water per acre with ground equipment. To prepare the tank mixture refer to the "Mixing Instructions" portion of this label. The tank mixture may be applied and incorporated at planting or several days prior to planting.

When using Basalin herbicide followed by preemergence application of AMIBEN® DS, Basalin may be applied as many as 6 to 8 weeks before planting or up to planting. Except as directed in the "Conservation or Reduced Tillage" portion of this label, the seedbed should be free of existing weeds before application. Surface trash (plant residue from previous crop) should be uniformly distributed, and the seedbed should be free of clods. Basalin is to be incorporated into the soil within 8 hours of application prior to planting. Follow the incorporation instructions on the Basalin label. Read all instructions and cautions on both the AMIBEN® DS and Basalin labels.

Apply AMIBEN® DS Herbicide at a rate of 2.4 to 3.6 pounds per broadcast acre. Apply Basalin herbicide at 1.5 to 3 pints per broadcast acre. Use the higher rate of the rate range on heavier, finer textured soils or on soils with over 4% organic matter. Do not use on peat or muck soils.

Before applying this combination read thoroughly the precautionary statements and all other information appearing on or attached to the containers of both Basalin and AMIBEN® DS herbicides.

AMIBEN® DS CHLORAMBEN HERBICIDE TANK MIXTURES WITH "BASALIN" AND "SENCOR" OR "LEXONE" FOR SOYBEANS

GENERAL INFORMATION

A tank mix of AMIBEN® DS Chloramben Herbicide plus Basalin plus Sencor or Lexone when applied preplant soil incorporated as recommended herein for use on soybeans, effectively controls the weeds listed on the current AMIBEN® DS, Basalin and Sencor WP/4/DF or Lexone WP/4L/DF labels.

DIRECTIONS FOR USE

Except as directed in the "Conservation or Reduced Tillage" portion of this label, the seedbed should be free of existing weeds before application. Surface trash (plant residue from previous crop) should be uniformly distributed, and the seedbed should be free of clods.

To prepare the tank mixture refer to the "Mixing Instructions" portion of this label. Apply the entire contents of the spray tank within 4 hours of initial mixing. Apply the tank mixture in from 10 to 20 gallons of water per acre using any properly calibrated low pressure sprayer equipped with herbicide tips and in-line strained screens no finer than 50 mesh.

For best results AMIBEN® DS plus Basalin plus Sencor or Lexone should be incorporated as soon as possible after application. The three way tank mixture must be incorporated within 24 hours after application. If AMIBEN® DS plus Basalin plus Sencor or Lexone is applied to a wet, warm soil surface or if the wind velocity is 10 mph or higher, variable weed control may result from delaying incorporation beyond 24 hours.

Incorporation should place the AMIBEN® DS Herbicide plus Basalin plus Sencor or Lexone into the top 2 to 3 inches of the final seedbed. Refer to the incorporation portion of the Basalin label for approved incorporation equipment and operating procedures.

CULTIVATION AFTER PLANTING

Soil treated with AMIBEN® DS plus Basalin plus Sencor or Lexone may be shallow-cultivated, rotary-hoed or hand-hoed without reducing the weed control activity of the tank mixture. Do not cultivate deeper than the

AMIBEN® DS plus Basalin plus Sencor or Lexone treated layer of soil since this may bring untreated soil to the surface and poor weed control may result.

Make all broadcast treatments according to the following table

Soil Texture	AMIBEN® DS Lbs/A	Basalin Pts/A	Sencor 50WP or Lexone WP lbs/A	Sencor 4 or Lexone 4L Pts/A	Sencor DF or Lexone DF lbs/A
Coarse (sand, loamy sand, sandy loam) [a]	2.4	1.0	0.5	0.5	0.33
Medium (loam, silt loam, silt)	2.4 to 3.0*	1.5	0.5 to 0.75**	0.5 to 0.75**	0.33 to 0.5**
Fine (clay, clay loam, silty clay, sandy clay, sandy clay loam, silty clay loam)	3.0 to 3.6*	2.0	0.75**	0.75**	0.5**

[a] Do not use Sencor or Lexone on coarse soils with less than 1% organic matter.

* Use the higher rate of AMIBEN® DS Chloramben Herbicide when velvetleaf or black nightshade is a problem.

** On Clarion Webster soils in Minnesota and Iowa or on similar alkaline (calcareous) soils with a pH of 7.5 or above apply Sencor WP or Lexone WP at 0.5 pound/A, Sencor 4 or Lexone 4L at 0.5 pint/A or Lexone DF or Sencor DF at 0.33 pound/A on medium soils and at 0.5 to 0.75 pound/A Sencor WP or Lexone WP, 0.5 to 0.75 pint/A Sencor 4 or Lexone 4L or 0.33 to 0.5 pound/A Lexone DF or Sencor DF on fine soils. Use the higher rate only where soil pH is less than 7.5 and where weed pressure is heavy.

SPECIAL PRECAUTIONS

Applied according to directions and under normal growing conditions, the AMIBEN® DS and Basalin and Sencor or AMIBEN® DS and Basalin and Lexone tank mix will not harm the treated crop. Over-application may result in crop injury. Uneven application or improper soil incorporation of the AMIBEN® DS and Basalin and Sencor or AMIBEN® DS and Basalin and Lexone tank mix can result in erratic weed control or crop injury. Seedling disease, cold weather, deep planting, excessive moisture, soil pH over 7.5, high salt concentration, or drought may weaken crop seedlings and increase possibility of damage from the tank mixtures. Under these conditions, delayed crop development or reduced yields may result.

Caution: Observe all cautions and limitations on labeling of all products used in mixtures.

Do not use the foliage from soybeans treated with the AMIBEN® DS and Basalin and Sencor or AMIBEN® DS and Basalin and Lexone tank mix for feed or forage.

AMIBEN® DS CHLORAMBEN HERBICIDE AND "REWARD" FOR SOYBEANS

GENERAL INFORMATION

A tank mixture of AMIBEN® DS Chloramben Herbicide and Reward can be used prior to planting for weed control in soybeans or AMIBEN® DS may be applied over soybeans at planting time when Reward has been applied as a preplant soil incorporated herbicide.

The combination will control a broader spectrum of weeds than either material alone.

WEEDS CONTROLLED

Nutsedge (yellow & purple)	Velvetleaf - buttonweed
Barnyardgrass (watergrass)	Stinkgrass
Crabgrass (hairy & smooth)	Carpetweed
Foxtail (giant, green, yellow)	Chickweed
German millet	Coffeeweed
Goosegrass	Florida pusley
Johnsongrass (seedling)	Kochia
Mustard	Lambsquarters (common)
Nightshade	Morningglory (annual)
Prickly sida - teaweed	Pigweed (redroot & smooth)
Ragweed (common)	Purslane
Smartweed (Pennsylvania)	Sicklepod
	Spurge (annual)

DIRECTIONS FOR USE

Reward must be incorporated into the soil immediately (within minutes) to prevent loss of the herbicide. Whenever possible application and incorporation of Reward or a tank mixture of AMIBEN® DS and Reward should be done in the same operation. Refer to the Reward label for equipment and operating instructions to best incorporate the tank mixture.

Sequential Applications:

Before planting, apply Reward at the rates listed in the table below for your soil type. Incorporate into the soil as described above and on the Reward label. At time of planting, apply AMIBEN® DS Herbicide at the rates listed in the table below as a broadcast treatment.

Tank Mix Application:

On medium textured soils a tank mixture of AMIBEN® DS and Reward may be applied and incorporated prior to planting. Use the rates of AMIBEN® DS and Reward listed in the table below. Follow the incorporation directions described elsewhere on this label. To prepare the tank mixture refer to the "Mixing Instructions" portion of this label. Apply in 10 to 20 gallons of water per acre.

Make all broadcast treatments according to the following table

Soil Type	AMIBEN® DS Lbs/Acre	Reward Pts/Acre
Coarse (sandy-light)	2.4	2.5
Medium to Fine (sandy loam to clay loam)	3.0	3.0
Fine to Very Fine (clay)	3.6	3.5

Read and follow all directions, use limitations, precautions and warnings that appear on the AMIBEN® DS and Reward labels.

AMIBEN® DS CHLORAMBEN HERBICIDE TANK MIX WITH "SURFLAN" FOR WEED CONTROL IN SOYBEANS

GENERAL INFORMATION

The AMIBEN® DS Chloramben Herbicide and Surflan tank mixture effectively controls all annual grass and broadleaf weeds listed on the Surflan and AMIBEN® DS Herbicide labels including: chickweed, coffeeweed, kochia, wild mustard, black nightshade, prickly sida (teaweed), common ragweed, Russian thistle, Pennsylvania smartweed, annual spurge, stinkgrass and velvetleaf (buttonweed).

AMIBEN® DS and Surflan tank mixtures may be applied at time of planting, using a properly calibrated, low-pressure herbicide sprayer that will provide uniform soil coverage. Agitate thoroughly before and during application. Follow normal AMIBEN® DS and Surflan procedures for soil preparation. Do not spray over top of emerged soybeans.

DIRECTIONS FOR USE

Make all broadcast application according to the following table

Soil Texture	AMIBEN® DS Lbs/A	Surflan 4AS Qts/A	Surflan 75W Lbs/A
Coarse Soils	2.4 to 3.6	0.5 to 0.75	0.67 to 1.0
Medium Soils	2.4 to 3.6	0.75 to 1.0	1 to 1.33
Fine Soils	2.4 to 3.6	1.0 to 1.25	1.33 to 1.67

The higher rate of AMIBEN® DS Herbicide should be used on clay loams, clay or high organic (3 to 5%) mineral soils. Do not use the AMIBEN® DS and Surflan tank mix on soils containing more than 5% organic matter.

SPECIAL PRECAUTION

Poor weed control and/or crop injury may result if directions are not followed. Over-application may result in crop injury and/or soil residue of Surflan. Do not use treated vines for feed or forage. Observe all precautions on both the AMIBEN® DS and Surflan labels.

AMIBEN® DS CHLORAMBEN HERBICIDE PROGRAMMED CONTROL OF WILD PROSO MILLET IN SOYBEANS

When used as directed below, a weed control program of sequential AMIBEN® DS Chloramben Herbicide applications will control wild proso millet in addition to all other weed species listed on the AMIBEN® DS label.

DIRECTIONS FOR USE

1 to 4 days before planting apply a preplant incorporated tank mixture of 2.4 to 3.6 pounds AMIBEN® DS plus the full labeled rate of another grass control material such as Treflan, Basalin, Prowl, Lasso or Dual as directed elsewhere on this label. After soybean emergence, but before emergence of wild proso millet seedlings make a second broadcast application of 2.4 to 3.6 pounds AMIBEN® DS per acre. This application should be made when soybeans are in the cotyledon to 2nd trifoliate growth stage, but before emergence of wild proso millet seedlings. Under normal conditions this will be 10 to 21 days after planting. If wild proso millet or other weeds have begun to emerge use a rotary hoe to remove emerged seedling weeds.

Do not feed straw or green fodder from soybeans treated with AMIBEN® DS postemergence. Observe caution statements on this and all other herbicide labels.

AMIBEN® DS CHLORAMBEN HERBICIDE POSTEMERGENCE ON SOYBEANS

APPLICATION EQUIPMENT

BROADCAST APPLICATIONS

Good coverage is essential for maximum weed control. Use 10 to 20 gallons of water per acre by ground or 3 to 5 gallons of water per acre by air. Wide spaced, low pressure, flood type nozzles do not give adequate coverage. Nozzle pressure should be at least 30 pounds per square inch. Width of swath must not exceed length of boom.

BAND APPLICATIONS

Soil uptake of AMIBEN® DS Chloramben Herbicide from postemergence applications is important. For this reason AMIBEN® DS Herbicide may be applied in a postemergence band ONLY if the following precautions are taken: Spray nozzles must be directed to spray both the emerged weeds and the soil surface. On postemergence band sprayers with multiple nozzles the side nozzles should be directed toward the ground at approximately a 45 degree angle to apply a band of spray on the soil surface equal in width to one half the soybean row spacing. Both side nozzles should spray the same soil surface area; one from each side of the row. When these directions are followed the broadcast rates given below should be reduced by one half to give the same rate of active ingredient in the band that would be achieved from a broadcast application at the full rate. Use sufficient water for good coverage.

AMIBEN® DS CHLORAMBEN HERBICIDE AND "CROP OIL" FOR POSTEMERGENCE WEED CONTROL IN SOYBEANS

For programmed weed control apply a preemergence or preplant incorporated grass herbicide at planting followed by a later application of AMIBEN® DS plus crop oil concentrate when soybeans are in the cracking to 4th trifoliate stage of growth. Apply a tank mixture of 3.6 pounds of AMIBEN® DS plus 1 quart per acre of crop oil concentrate to control or suppress 1 to 8 inch velvetleaf and 1 to 3 inch Pennsylvania smartweed, common ragweed and pigweed, causing weeds to either die or cease growth and become noncompetitive with the crop. Application at this timing provides both postemergence activity on the above species and extended preemergence activity on those late germinating annual grass and broadleaf weeds listed in the general information section of the AMIBEN® DS label which have not yet germinated at the time of treatment. Do not apply later than 33 days after planting. Make applications in 10 to 20 gallons water per acre by ground or 3 to 5 gallons water per acre by air. Good coverage is essential for maximum weed control. Wide spaced, low pressure, flood type nozzles do not give satisfactory coverage. Nozzle pressure should be at least 30 pounds per square inch. Width of swath must not exceed length of boom. Under good soil moisture conditions this treatment will control small grassy weeds (1 to 2 leaf) if rainfall occurs within 48 hours of application.

Caution: Do not use soybean straw or green fodder which has been treated postemergence with AMIBEN® DS for livestock feed. Do not apply later than 33 days after planting.

Do not apply more than 3.6 pounds per acre of AMIBEN® DS alone or in combination as a postemergence treatment. If a preemergence or preplant soil incorporated application of AMIBEN® DS is made, do not apply more than a total of 7.2 pounds of AMIBEN® DS per acre per growing season (3.6 pounds preemergence plus 3.6 pounds postemergence).

AMIBEN® DS CHLORAMBEN HERBICIDE AND CULTIVATION POSTEMERGENCE ON SOYBEANS

If heavy rainfall has prevented application of preplant incorporated or preemergence herbicides, AMIBEN® DS Herbicide may be applied when soybeans are in the cracking to second trifoliate stage of growth at the rate of 3.0 to 3.6 pounds AMIBEN® DS Herbicide per acre. Grass and broadleaf weeds which have emerged **must** be removed. To remove these germinated weeds, use shallow cultivation with a tool such as a rotary hoe before or after AMIBEN® DS Herbicide has been applied. Avoid ridging the soil if a cultivator is used after AMIBEN® DS has been applied. Use the higher rate of fine textured soils or when heavy weed populations are expected.

Caution: Do not use soybean straw or green fodder which has been treated postemergence with AMIBEN® DS for livestock feed. Do not apply later than 33 days after planting.

Do not apply more than 3.6 pounds per acre of AMIBEN® DS alone or in combination as a preplant soil incorporated or preemergence treatment. If a postemergence application of AMIBEN® DS is made, do not apply more than a total of 7.2 pounds of AMIBEN® DS per acre per growing season (3.6 pounds preemergence plus 3.6 pounds postemergence).

AMIBEN® DS CHLORAMBEN HERBICIDE and "BLAZER" FOR POSTEMERGENCE WEED CONTROL IN SOYBEANS

GENERAL INFORMATION

When applied to young actively growing broadleaf weed seedlings the combination of AMIBEN® DS Chloramben Herbicide plus Blazer will control all weeds listed on the Blazer label including cocklebur, jimsonweed, morningglory, Pennsylvania smartweed, pigweed, Eastern black nightshade, and common ragweed and will also provide preemergence control of those grass and broadleaf weeds listed in the general information section of the AMIBEN® DS label which have not yet germinated at time of treatment. While the combination of AMIBEN® DS plus Blazer will provide suppression of small (2 to 3 leaf) foxtail and fall panicum it must not be used as a basic grass control program and should only be applied to fields which have received a preemergence or preplant incorporated grass herbicide treatment as directed on the AMIBEN® DS label. In addition to providing preemergence control of late germinating weeds the combination of AMIBEN® DS plus Blazer will provide control of velvetleaf and increased control of Pennsylvania smartweed, and pigweed which are approaching the maximum weed size listed on the Blazer label.

USE DIRECTIONS

Apply a tank mixture of 3.0 to 3.6 pounds AMIBEN® DS plus 1.5 to 2 pints Blazer when broadleaf weeds have 2 to 4 leaves as directed on the Blazer label (usually 14 to 21 days after planting). When the lower rate is used, 1 quart per acre of crop oil concentrate must be added as directed below and on the Blazer label. The 1 pint per acre rate of Blazer will not provide adequate control of morningglory or nightshade. Use the higher rate on morningglory and nightshade and on large weeds. Use 10 to 20 gallons of water per acre with ground equipment, or 5 to 10 gallons of water per acre by air to insure good coverage. Wide spaced, low pressure, flood type nozzles do not give satisfactory coverage. Nozzle pressure should be at least 30 pounds per square inch. Width of swath must not exceed length of boom. Do not make applications later than 33 days after planting. Refer to the Blazer label for specific directions regarding the addition of a surfactant or crop oil. Follow these same directions when using the tank mixture of AMIBEN® plus Blazer to determine when to add surfactant or crop oil except that 1 quart per acre of crop oil concentrate must be added to provide acceptable control of velvetleaf.

Caution: Do not use soybean straw or green fodder which has been treated postemergence with AMIBEN® DS plus Blazer for livestock feed. Read and observe all directions and cautions on labels for Blazer and AMIBEN®. Do not apply later than 33 days after planting.

AMIBEN® DS CHLORAMBEN HERBICIDE AND "BUTYRAC®" 200 FOR POSTEMERGENCE WEED CONTROL IN SOYBEANS

GENERAL INFORMATION

When applied as directed below a tank mixture of AMIBEN® DS Chloramben Herbicide plus Butyrac® 200 will control or suppress pigweed, velvetleaf, cocklebur and Pennsylvania smartweed causing them to either die or become noncompetitive with the crop. In addition, this tank mixture will provide preemergence control of those grass and broadleaf weeds listed in the general information section of the AMIBEN® DS label which have not yet germinated at time of treatment. The tank mixture of AMIBEN® DS plus Butyrac® 200 will not control emerged grassy weeds and should only be used on fields which have received a preemergence or preplant incorporated grass control herbicide. This combination may cause temporary twisting, curling or malformation of soybean leaves and stems. Soybeans will usually outgrow these effects with no reduction in yield. Do not apply to fields which are under stress from drought, cool temperature, or disease as a reduction in yield can be expected. Good coverage is essential for maximum weed control. Wide spaced, low pressure, flood type nozzles do not give satisfactory coverage. Nozzle pressure should be at least 30 pounds per square inch. Width of swath must not exceed length of boom.

USE DIRECTIONS

Apply a tank mixture of 3.0 to 3.6 pounds AMIBEN® DS plus 2 fluid ounces Butyrac® 200 in 10 to 20 gallons of water per acre by ground equipment when soybeans have 2 to 4 trifoliate leaves but before weeds are 4 inches tall. Use the higher rate when weeds are over 2 inches tall. Do not make application later than 33 days after planting. When weed pressure is heavy, addition of 1 pint per acre of crop oil concentrate has improved weed control, but may cause more soybean malformation as described above.

Caution: Do not use soybean straw or green fodder which has been treated postemergence with AMIBEN® DS plus Butyrac® 200 for livestock feed. Read and observe directions and cautions on labels for Butyrac® 200 and AMIBEN®. Do not make applications later than 33 days after planting.

AMIBEN® DS CHLORAMBEN HERBICIDE PLUS "ALANAP" FOR POSTEMERGENCE WEED CONTROL IN SOYBEANS

GENERAL INFORMATION

When applied according to the directions below, a tank mixture of AMIBEN® DS Chloramben Herbicide plus Alanap will control or suppress velvetleaf, pigweed, cocklebur, and Pennsylvania smartweed causing them to either die or become non-competitive with the crop. While the combination of AMIBEN® DS plus Alanap often provides control or suppression of small (2 to 3 leaf) annual grasses it should only be applied to fields which have received a preemergence or preplant incorporated application of a grass control herbicide. The tank mixture of AMIBEN® DS plus Alanap may cause temporary stunting of soybeans. Under normal conditions the soybeans will recover and yields will not be affected.

USE DIRECTIONS

Appy a tank mixture of 3.0 to 3.6 pounds AMIBEN® DS plus 2 to 3 quarts Alanap in 10 to 20 gallons per acre of water by ground equipment or in 5 or more gallons per acre of water by aerial equipment. To assure good coverage use a minimum of 30 pounds per square inch of nozzle pressure. Apply when soybeans have 3 to 6 trifoliate leaves, but no later than 33 days after planting. Use the higher rates when weeds are larger (over 4 inches). Addition of 1 quart per acre of a surfactant or crop oil concentrate has shown increased control of these large weeds.

Caution: Do not use soybean straw or green fodder which has been treated postemergence with AMIBEN® DS plus Alanap for livestock feed. Read and observe all directions and cautions on labels for Alanap and

AMIBEN® DS. Do not make applications later than 33 days after planting.

DIRECTIONS FOR USE ON DRYBEANS (INCLUDING WHITE, NAVY, KIDNEY, PINTO, LIMA AND ADZUKI BEANS).

GENERAL INFORMATION

Environmental conditions and cultural practices can combine to create conditions favorable for rapid, early development of severe levels of soil-borne diseased in drybeans. The potential for these diseases causing drybean yield reductions must be evaluated when determining where and how drybeans will be grown.

Environmental conditions favoring rapid disease development and root injury to drybeans are: 1) Cool air and soil temperatures during and soon after drybean emergence, 2) Heavy rainfall which prolongs saturated soil conditions, and 3) Soil types such as Kawkawlin loam which are poorly drained. These soils are subject to compaction and prolonged water saturation. While severe levels of soil diseases are commonly associated with saturated soil conditions, the severity of disease infestations can also be increased by droughty conditions which are unfavorable for rapid drybean root growth.

Cultural practices favoring rapid disease development causing injury to drybeans are: 1) High levels of disease inoculum from plant residues of soybeans or drybeans grown in the field during prior years, 2) Use of disease susceptible varieties, 3) Planting drybeans in low lying fields with poorly drained or compacted soils, 4) Failure to use a 3 year rotation with corn or small grains to reduce inoculum buildup in the field and 5) High rates of soil applied herbicides which may cause additional stress.

Under conditions favoring excessive root disease development, as outlined above, high rates of AMIBEN® DS, as with other soil applied herbicides, may increase the potential for yield reductions caused by high levels of root disease. To minimize potential for disease induced yield reductions, as outline above, do not use more than 1.8 lb. AMIBEN® DS per broadcast equivalent acre, either alone, or in any of the several tank mixtures on the label. Under conditions favoring excessive disease development, do not apply AMIBEN® DS as a surface preemergence treatment. Follow the label directions for incorporation of AMIBEN® DS used either alone or in combination with any of the other product listed on the label. When the 1.8 lb. per acre rate of AMIBEN® DS is used under heavy weed populations reduced weed control is to be expected.

AMIBEN® DS CHLORAMBEN HERBICIDE PREEMERGENCE OR PREPLANT INCORPORATED

AMIBEN® DS Chloramben Herbicide may be used on drybeans as a preemergence or preplant incorporated herbicide. AMIBEN® DS may be applied and incorporated at planting or several days prior to planting. Uneven spray application or improper soil incorporation of AMIBEN® DS can produce erratic weed control. Refer to the "Directions for Incorporation" portion of this label for approved incorporation equipment and recommended operating procedures.

Make all broadcast treatments according to the following table

Soil Texture	AMIBEN® DS Lbs/A
Loamy sand, sandy loam	2.4 to 3.0
Loam, silt loam, silt	3.0 to 3.6
Silty clay loam, silty clay, sandy clay, clay loam, clay	3 to 6

Use the higher rate of AMIBEN® DS when heavy populations of black nightshade, velvetleaf, wild mustard, common ragweed or Pennsylvania smartweed are expected. To insure good spray coverage use 10 to 20 gallons of water per acre with ground equipment or a minimum of 5 gallons of water per acre by aircraft.

When band spraying, mount the spray nozzle on the packer or row wheel so that planting and spraying are done in the same operation. Proper calibration and maintenance of desired nozzle height above the ground are very important. Weeds growing in the middles between the treated rows should be cultivated as soon as necessary.

Use care while cultivating to avoid throwing fresh soil on treated bands.

If sufficient rain does not fall within 5 days after spraying AMIBEN® DS Herbicide as a preemergence treatment, but there is enough soil moisture to germinate weeds and grasses, a light cultivation with a rotary hoe or similar tool will up-root these small broadleaf weeds and grasses. The shallow mixing of AMIBEN® DS in the surface soil will not interfere with the action of AMIBEN® DS when rains come.

If heavy rain packs the soil surface, rotary hoe in the field to aid bean emergence. Do not disrupt the chemical band with deep cultivation. In coarse textured soils heavy rains on incorporated AMIBEN® DS Herbicide may move it below the weed seed germination zone and produce erratic weed control.

DO NOT apply more than 3.6 pounds per acre of AMIBEN® DS Herbicide to the same field of drybeans in the same season.

AMIBEN® DS CHLORAMBEN HERBICIDE AND "TREFLAN" FOR DRYBEANS

AMIBEN® DS Chloramben Herbicide may be applied as a preemergence band or broadcast treatment in drybeans at planting time in fields where Treflan has been applied as a preplant soil incorporated herbicide. Or AMIBEN® DS Herbicide may be applied several days prior to planting as a broadcast tank mix with Treflan. The tank mixture should be used as a spring preplant soil incorporated treatment. The combination of AMIBEN® DS Herbicide with Treflan improves broadleaf weed control of species such as Pennsylvania smartweed, velvetleaf and ragweed. For broadcast treatments, incorporate chemicals immediately and thoroughly to an approximate depth of 2 inches as described on the Treflan label.

Make all broadcast treatments according to the following table

Soil Texture	AMIBEN® DS Lbs/A	TREFLAN pts/A
Loam sand, sandy loam	2.4	1
Loam, silt loam, silt	2.4 to 2.0	1.5
Silty clay loam, silty clay, sandy clay, clay loam, clay	3.0 to 3.6	2

Do not use on muck or peat soils.

Use the higher rate of AMIBEN® DS when heavy populations of black nightshade, velvetleaf, mustard or smartweed are expected.

Before applying this combination read thoroughly all precautionary statements and all other information appearing on or attached to the containers of AMIBEN® DS Herbicide and Treflan.

AMIBEN® DS CHLORAMBEN HERBICIDE AND "EPTAM" FOR DRYBEANS

AMIBEN® DS Chloramben Herbicide may be applied as a preplant soil incorporated tank mixture with Eptam Herbicide. Or AMIBEN® DS may be applied preemergence over drybeans at planting time where Eptam has been applied as preplant soil incorporated treatment. When applied preemergence, effectiveness depends on adequate rainfall after treatment. When applied according to directions, this combination will control annual broadleaves including black nightshade, ragweed, Pennsylvania smartweed, wild mustard and annual grasses.

DIRECTIONS FOR USE

Apply AMIBEN® DS Herbicide either preplant soil incorporated or preemergence at a rate of 2.4 to 3.6 pounds per acre. Apply Eptam 7E preplant soil incorporated at a rate of 2.5 to 3.5 pints per acre. Use the higher rates on fine textured soils or where weed pressure is heavy.

The AMIBEN® DS plus Eptam tank mix should be incorporated immediately after application with a field cultivator, tandem disc, mulch treader, rolling cultivator or power driven tiller as described on the Eptam label. Thoroughly mix in the top 2-3 inches of the soil surface by incorporating twice in different directions. Refer to the Eptam label for further information on incorporation procedures.

Rainfall is not as critical for activation of AMIBEN® DS Herbicide when it is applied preplant incorporated as when surface applied.

Apply in 10 to 20 gallons of water per acre. To pepare the spray mixture refer to the "Mixing Instructions" portion of this label. Do not mix the concentrates before adding to water.

If sufficient rain does not occur within 5 days after spraying AMIBEN® DS preemergence, a light cultivation or incorporation with a rotary hoe or similar tool is recommended.

Do not use Eptam or an AMIBEN® DS plus Eptam tank mixture on adzuki beans, cowpeas (blackeye peas, blackeye beans), soybeans, lima beans or other flat podded beans except Romano. Under abnormal weather conditions stunting may occur on Gratiot, Michilite, Sanilac, Seafarer, and Seaway varieties. Do not use on muck or charcoal soils. Read and observe all directions and cautions on labels for both AMIBEN® DS and Eptam.

AMIBEN® DS CHLORAMBEN HERBICIDE PLUS "EPTAM" AND "TREFLAN" FOR DRYBEANS

AMIBEN® DS Chloramben Herbicide may be applied as a preplant soil incorporated tank mixture with Eptam and Treflan herbicides. Or AMIBEN® DS may be applied prememergence over drybeans at planting time where Eptam has been applied as preplant soil incorporated treatment. When applied preemergence, effectiveness depends on adequate rainfall after treatment. When applied according to directions, this combination will control annual broadleaves including black nightshade, ragweed, Pennsylvania smartweed, wild mustard and annual grasses.

DIRECTIONS FOR USE

Apply AMIBEN® DS Herbicide either preplant soil incorporated or preemergence at a rate of 2.4 to 3.6 pounds per acre. Apply Eptam preplant soil incorporated at a rate of 2.5 pints per acre and Treflan preplant soil incorporated at a rate of 1 pint per acre. Use the higher rate of AMIBEN® DS on fine textured soils or where weed pressure is heavy.

The AMIBEN® DS plus Eptam plus Treflan or the Eptam plus Treflan tank mixes should be incorporated immediately after application following the equipment and operating procedures described on the Eptam and Treflan labels.

Rainfall is not as critical for activation of AMIBEN® DS Herbicide when it is applied preplant incorporated as when surface applied.

Apply the tank mixture in 10 to 20 gallons of water per acre. To prepare the spray mixture refer to the "Mixing Instructions" portion of this label. Do not mix the concentrates before adding to water.

If sufficient rain does not occur within 5 days after spraying AMIBEN® DS preemergence, light cultivation or incorporation with a rotary hoe or similar tool is recommended.

Do not use Eptam or an AMIBEN® DS plus Eptam plus Treflan tank mixture on adzuki beans, cowpeas (blackeye peas, blackeye beans), soybeans, lima beans or other flat podded beans except Romano. Under abnormal weather conditions stunting may occur on Gratiot, Michilite, Sanilac, Seafarer, and Seaway varieties. Do not use on muck or charcoal soils. Read and observe all directions and cautions on labels for AMIBEN® DS, Eptam and Treflan.

AMIBEN® DS CHLORAMBEN HERBICIDE AND "SONALAN" FOR DRYBEANS

AMIBEN® DS Chloramben Herbicide may be applied as a preemergence band or broadcast treatment in drybeans at planting time in fields where Sonalan has been applied as a preplant soil incorporated herbicide. Or, AMIBEN® DS Herbicide may be applied prior to planting as a broadcast tank mix with Sonalan. The tank mixture should be used as a spring preplant soil incorporated treatment. The combination of AMIBEN® DS Herbicide with Sonalan improves broadleaf weed control of species such as Pennsylvania smartweed, velvetleaf, common ragweed, black nightshade and wild mustard. For broadcast treatments, incorporate chemicals immediately and thoroughly to an approximate depth of 2 inches as described on the Sonalan label.

Make all broadcast treatments according to the following table

Soil Texture	AMIBEN® DS Lbs/A	Sonalan Pts/A
Loamy sand, sandy loam	2.4	1.25 to 2.0
Loam, silt loam, silt	2.4 to 3.0	1.75 to 2.5
Silty clay loam, silty clay, sandy clay, clay loam, clay	3.0 to 3.6	2.25 to 3.0

Use the higher rate of AMIBEN® DS where a heavy population of smartweed, velvetleaf, ragweed, wild mustard or black nightshade are expected. Use the higher rate of Sonalan on finer textured soils and on soils with organic matter content over 3%.

AMIBEN® DS CHLORAMBEN HERBICIDE AND "PROWL" FOR DRYBEANS (except Adzuki beans)

A tank mixture of AMIBEN® DS Chloramben Herbicide and Prowl may be applied as a preplant soil incorporated herbicide treatment or AMIBEN® DS Herbicide may be applied as a preemergence broadcast treatment at planting time in drybean fields where Prowl has been applied as a preplant soil incorporated herbicide treatment.

When applied according to directions, tank mixtures of AMIBEN® DS Herbicide and Prowl herbicide either preplant incorporated or the sequential treatment of AMIBEN® DS preemergence over Prowl preplant incorporated will control fall panicum, signalgrass, Venice mallow and velvetleaf in addition to all other weeds listed on the AMIBEN® DS and Prowl labels.

For uniform application of the tank mixture of AMIBEN® DS and Prowl, use 10 to 20 gallons of water per acre with ground equipment or 3 to 5 or more gallons of water per acre by aircraft. To prepare the tank mixture refer to the "Mixing Instructions" portion of this label.

Except as directed in the "Conservation or Reduced Tillage" portion of this label, the seedbed should be free of existing weeds before application. Surface trash (plant residue from previous crop) should be uniformly distributed, and the seedbed should be free of clods. When using the tank mixture preplant soil incorporated or when using Prowl preplant incorporated followed by a preemergence application of AMIBEN® DS the herbicides should be applied within 7 days prior to planting. When applied as a preplant soil incorporated treatment the tank mixture of AMIBEN® DS plus Prowl should be thoroughly incorporated into the top 0 to 2 inches of soil using equipment and techniques described elsewhere on this label and on the Prowl label. When using the sequential treatment AMIBEN® DS should be applied at planting.

Make all broadcast treatments according to the following table

Soil Texture	AMIBEN® DS Lbs/A	Prowl Pts/A
Loamy sand, sandy loam	2.4	1.0 to 2.0
Loam, silt loam, silt	2.4 to 3.0	1.5 to 2.5
Silty clay loam, silty clay, sandy clay, clay loam, clay	3.0 to 3.6	2.0 to 3.0

Use the higher rate of AMIBEN® DS when heavy populations of black nightshade, velvetleaf, wild mustard, common ragweed or Pennsylvania smartweed are expected. Use the higher rate of Prowl when treatments are incorporated on soils with more than 3% organic matter, or where heavy stands of foxtail, witchgrass or panicum are expected. Do not use on peat or muck soils.

If sufficient rain does not fall within 5 days after preemergence AMIBEN® DS treatment, but there is enough soil moisture to germinate weeds and soybeans a light cultivation with a rotary hoe or similar tool will uproot these small weeds. The shallow mixing of the herbicides into the soil surface will not interfere with their action when rains come. If drybean crop loss occurs due to weather conditions, drybeans may be replanted. If replanting is necessary, DO NOT rework the soil.

Before applying this combination, read thoroughly the precautionary statements and all other information appearing on or attached to the containers of both the Prowl and AMIBEN® DS herbicides.

AMIBEN® DS CHLORAMBEN HERBICIDE AND "PROWL" AND "EPTAM 7E" FOR DRYBEANS (except Adzuki beans)

A tank mixture of AMIBEN® DS Chloramben Herbicide and Prowl and Eptam 7E may be applied as a preplant soil incorporated herbicide treatment or AMIBEN® DS Herbicide may be applied as a preemergence broadcast treatment at planting time in drybean fields where Prowl and Eptam 7E have been applied as a preplant soil incorporated herbicide treatment.

When applied according to directions, tank mixtures of AMIBEN® DS Herbicide and Prowl Herbicide and Eptam 7E either preplant incorporated or the sequential treatment of AMIBEN® DS preemergence over Prowl and Eptam 7E preplant incorporated will control fall panicum, signalgrass, Venice mallow, nightshade and velvetleaf in addition to all other weeds listed on the AMIBEN® DS, Prowl and Eptam 7E labels.

For uniform application of the tank mixture of AMIBEN® DS, Prowl and Eptam 7E, use 10 to 20 gallons of water per acre with ground equipment or 3 to 5 or more gallons of water per acre by aircraft. To prepare the tank mixture refer to the "Mixing Instructions" portion of this label.

Except as directed in the "Conservation or Reduced Tillage" portion of this label, the seedbed should be free of existing weeds before application. Surface trash (plant residue from previous crop) should be uniformly distributed, and the seedbed should be free of clods. When using the tank mixture preplant soil incorporated or when using Prowl and Eptam preplant incorporated followed by a preemergence application of AMIBEN® DS plus Prowl and Eptam 7E should be thoroughly incorporated into the top 0 to 2 inches of soil using equipment and techniques described elsewhere on this label and on the Prowl and Eptam 7E label. When using the sequential treatment AMIBEN® DS should be applied at time of planting.

Make all broadcast treatments according to the following table

Soil Texture	AMIBEN® DS Lbs/A	Prowl Pts/A	Eptam 7E Pts/A
Loamy sand, sandy loam	2.4	1.0 to 2.0	2.5
Loam, silt loam, silt	2.4 to 3.0	1.5 to 2.5	2.5 to 3.0
Silty clay loam, silty clay, sandy clay loam, clay	3.0 to 3.6	2.0 to 3.0	3.0 to 3.5

Use the higher rate of AMIBEN® DS when heavy populations of black nightshade, velvetleaf, wild mustard, common ragweed or Pennsylvania smartweed are expected. Use the higher rate of Prowl when treatments are incorporated on soils with more than 3% organic matter, or where heavy stands of foxtail, witchgrass or panicum are expected. Use the higher rate of Eptam 7E on finer textured soils or where weed pressure is heavy. Do not use on peat or muck soils.

If sufficient rain does not fall within 5 days after preemergence AMIBEN® DS treatment, but there is enough soil moisture to germinate weeds and soybeans a light cultivation with a rotary hoe or similar tool will uproot these small weeds. The shallow mixing of the herbicides into the soil surface will not interfere with their action when rains come. If drybean crop loss occurs due to weather conditions, drybeans may be replanted. If replanting is necessary, DO NOT rework the soil.

Before applying this combination, read thoroughly the cautionary statements and all other information appearing on or attached to the containers of Prowl, Eptam 7E and AMIBEN® DS herbicides.

AMIBEN® DS CHLORAMBEN HERBICIDE AND "DUAL" FOR DRYBEANS

A tank mixture of AMIBEN® DS Chloramben Herbicide and Dual 8E may be applied as a preplant soil incorporated herbicide treatment. AMIBEN® DS Herbicide may be applied as a preemergence broadcast tank mix treatment with Dual 8E at time of planting. Following a preplant soil incorporated application of Dual 8E, AMIBEN® DS Herbicide may be applied preemergence at time of planting.

When applied according to label directions, tank mixtures of AMIBEN® DS Herbicide and Dual will control all weeds controlled by AMIBEN® DS alone or by Dual alone including fall panicum, foxtail millet, red rice, signalgrass, witchgrass, yellow nutsedge and black nightshade. Refer to the general information section of this label and the Dual label for additional weeds controlled by AMIBEN® DS and Dual 8E.

For uniform application of the tank mixture of AMIBEN® DS and Dual, use 10 to 20 gallons of water with ground equipment. To prepare the tank mixture, refer to the "Mixing Instructions" portion of this label.

Apply AMIBEN® DS and Dual preplant incorporated or preemergence using the appropriate rates from the table below. For preplant incorporated treatments apply the tank mixture to the soil and incorporate into the top two inches of soil just before planting following the procedures in the "Directions for Incorporation" portion of this label. Use a preplant incorporated application if furrow irrigation is used or when a period of dry weather is expected. If drybeans are planted on beds apply and incorporate the tank mixture after bed formation. For preemergence treatments, apply the tank mixture during planting (behind the planter) or after planting before weeds or drybeans emerge. For sequential applications apply Dual 8E alone preplant incorporated and follow with preemergence application of AMIBEN® DS Herbicide during planting (behind the planter) or after planting before weeds or drybeans emerge.

Make all broadcast treatments according to the following table

Soil Texture	AMIBEN® DS Lbs/A	Dual 8E Pts/A
Loamy sand, sandy loam	2.4	1.5
Loam, silt loam, silt	2.4 to 3.0	1.5 to 2.0
Silty clay loam, silty clay, sandy clay, clay loam, clay	3.0 to 3.6	2.0 to 2.5

Do not use on muck or peat soils.

Use the higher rate of AMIBEN® DS when heavy populations of black nightshade, velvetleaf, mustard or smartweed are expected. Use the higher rate of Dual when treatment is incorporated, when organic matter is over 3%, or when high populations of black nightshade, witchgrass, panicum, foxtail, or yellow nutsedge are expected.

Before applying this combination read thoroughly all precautionary statements and all other information appearing on or attached to the containers of AMIBEN® DS Herbicide and Dual 8E.

AMIBEN® DS CHLORAMBEN HERBICIDE AND "LASSO" FOR DRYBEANS

AMIBEN® DS Chloramben Herbicide may be applied as a preemergence broadcast treatment at time of planting in fields where Lasso has been applied as a preplant soil incorporated treatment or a tank mixture of AMIBEN® DS and Lasso may be applied as a preplant soil incorporated treatment.

When applied according to label directions the tank mixture or the sequential treatment of AMIBEN® DS plus Lasso will control all weeds listed on both the AMIBEN® DS and Lasso labels including nightshade, ragweed, Pennsylvania smartweed, lambsquarters, foxtails and barnyardgrass when applied within 7 days prior to planting.

To prepare the tank mixture refer to the "Mixing Instructions" portion of this label. DO NOT mix AMIBEN® DS Herbicide and Lasso concentrates before the addition of water as an emulsion will not be formed.

When used as a tank mixture preplant soil incorporated the AMIBEN® DS Herbicide and Lasso must be uniformly incorporated within 7 days prior to planting, no deeper than the top 2 inches of soil. Refer to the incorporation directions given elsewhere on this label and the Lasso label. Use the rates indicated in the table below.

When AMIBEN® DS is applied following a preplant soil incorporated application of Lasso follow the incorporation directions on the Lasso label. Apply as a broadcast treatment in 10 to 20 gallons of water per acre at the rates indicated in the table below.

Make all broadcast treatments according to the following table

Soil Texture	AMIBEN® DS LBS/A	Lasso QTS/A
Loamy sand, sandy loam	DO NOT USE	DO NOT USE
Loam, silt loam, silt	2.4 to 3.0	1.5 to 2.0
Silty clay loam, silty clay, sandy clay, clay loam, clay	3.0 to 3.6	2.0 to 2.5

Do not use on muck or peat soils

Before applying the tank mixture of AMIBEN® DS plus Lasso or AMIBEN® DS preemergence following Lasso preplant incorporated, read all directions and cautions on both the AMIBEN® DS and Lasso labels.

DIRECTIONS FOR USE ON SUNFLOWERS

Do not graze treated field of sunflowers or use these plants as livestock feed.

AMIBEN® DS PREEMERGENCE OR PREPLANT INCORPORATED

AMIBEN® DS Chloramben Herbicide may be used on sunflowers as a preemergence or preplant incorporated herbicide. AMIBEN® DS may be applied and incorporated at planting or several days prior to planting. Uneven spray application or improper soil incorporation of AMIBEN® DS can produce erratic weed control. Incorporate the chemical in the top 0 to 2 inch level as most weeds sprout in this zone. Refer to the "Directions for Incorporation" portion of this label for equipment and operating procedures to be used for proper incorporation.

Make all broadcast treatments according to the following table

Soil Texture	AMIBEN® DS Lbs/A
Loamy sand, sandy loam	2.4 to 3.0
Loam, silt loam, silt	3.0 to 3.6
Silty clay loam, silty clay, sandy clay, clay loam, clay	3.6

Use the higher rate of AMIBEN® DS when heavy populations of black nightshade, velvetleaf, wild mustard, common ragweed or Pennsylvania smartweed are expected. To insure good spray coverage use 10 to 20 gallons of water per acre with ground equipment or a minimum of 5 gallons of water per acre by aircraft.

When band spraying, mount the spray nozzle on the packer or row wheel so that planting and spraying are done in the same operation. Proper calibration and maintenance of desired nozzle height above the ground are very important. Weeds growing in the middles between the treated rows should be cultivated as soon as necessary.

Use care while cultivating to avoid throwing fresh soil on treated bands.

If sufficient rain does not fall within 5 days after spraying AMIBEN® DS Herbicide as a preemergence treatment, but there is enough soil moisture to germinate weeds and grasses, a light cultivation with a rotary hoe or similar tool will up-root these small broadleaf weeds and grasses. The shallow mixing of AMIBEN® DS in the surface soil will not interfere with action of AMIBEN® DS when rains come.

In coarse textured soils heavy rains on incorporated AMIBEN® DS Herbicide may move it below the weed seed germination zone and produce erratic weed control.

If heavy rain packs the soil surface, rotary hoe in the field to aid sunflower emergence. Do not disrupt the chemical band with deep cultivation.

DO NOT apply more than 3.6 pounds per acre of AMIBEN® DS Herbicide to the same field of sunflowers in the same season.

AMIBEN® DS CHLORAMBEN HERBICIDE AND "PROWL" FOR SUNFLOWERS

A tank mixture of AMIBEN® DS Chloramben Herbicide and Prowl may be applied as a preplant soil incorporated herbicide treatment or AMIBEN® DS Herbicide may be applied as a preemergence broadcast treatment at planting time in sunflowers fields where Prowl has been applied as a preplant soil incorporated herbicide treatment. The tank mixture of AMIBEN® DS plus Prowl may also be applied as a preemergence broadcast treatment.

When applied according to directions, tank mixtures of AMIBEN® DS Herbicide and Prowl herbicide either preplant incorporated or preemergence, or the sequential treatment of AMIBEN® DS preemergence over Prowl preplant incorporated will control fall panicum, signalgrass, Venice mallow and velvetleaf in addition to all other weeds listed on the AMIBEN® DS and Prowl labels.

For uniform application of the tank mixture of AMIBEN® DS and Prowl, use 10 to 20 gallons of water per acre with ground equipment or 3 to 5 or more gallons of water per acre by aircraft. To prepare the tank mixture refer to the "Mixing Instructions" portion of this label.

Except as directed in the "Conservation or Reduced Tillage" portion of this label, the seedbed should be free of existing weeds before application. Surface trash (plant residue from previous crop) should be uniformly distributed, and the seedbed should be free of clods. When using the tank mixture preplant soil incorporated or when using Prowl preplant incorporated followed by a preemergence application of AMIBEN® DS the herbicides should be applied within 7 days prior to planting. When applied as a preplant soil incorporated treatment the tank mixture of AMIBEN® DS plus Prowl should be thoroughly incorporated into the top 0 to 2 inches of soil using equipment and techniques described elsewhere on this label and on the Prowl label. When using the tank mixture preemergence, treatments should be made at planting or within 2 days after planting before weeds and sunflowers emerge.

Make all broadcast treatments according to the following table

Soil Texture	AMIBEN® DS Lbs/A	Prowl Pts/A
Loamy sand, sandy loam	2.4	1.0 to 2.0
Loam, silt loam, silt	2.4 to 3.0	1.5 to 2.5
Silty clay loam, silty clay, sandy clay, clay loam, clay	3.0 to 3.6	2.0 to 3.0

Use the higher rate of AMIBEN® DS when heavy populations of black nightshade, velvetleaf, wild mustard, common ragweed or Pennsylvania smartweed are expected. Use the higher rate of Prowl when treatments are incorporated on soils with more than 3% organic matter, or where heavy stands of foxtail, witchgrass or panicum are expected. Do not use on peat or muck soils.

If sufficient rain does not fall within 5 days after preemergence treatment, but there is enough soil moisture to germinate weeds and sunflowers a light cultivation with a rotary hoe or similar tool will uproot these small weeds. The shallow mixing of the herbicides into the soil surface will not interfere with their action when rains come. If sunflower crop loss occurs due to weather conditions, sunflowers or soybeans may be replanted. If replanting is necessary, DO NOT rework the soil.

Before applying this combination, read thoroughly the precautionary statements and all other information appearing on or attached to the containers of both the Prowl and AMIBEN® DS herbicides.

AMIBEN® DS CHLORAMBEN HERBICIDE AND "TREFLAN" FOR WEED CONTROL IN SUNFLOWERS

GENERAL INFORMATION

AMIBEN® DS Chloramben Herbicide may be applied in a band or broadcast over sunflowers at planting time in fields where Treflan has been applied as a preplant soil incorporated herbicide, or AMIBEN® DS may be applied several days prior to planting as a broadcast tank mix with Treflan. The tank mixture should be used as a spring preplant soil incorporated treatment.

The combination of AMIBEN® DS Herbicide with Treflan improves broadleaf weed control of species such as mustard, Pennsylvania smartweed, velvetleaf and ragweed.

For broadcast treatments, incorporate chemicals immediately and thoroughly in the top 2 to 3 inches as most weeds sprout in this zone. Refer to the Treflan label for equipment and operating directions for incorporation of the tank mixture.

In coarse textured soils heavy rains on incorporated AMIBEN® DS may move it below the weed seed germination zone and cause erratic weed control.

If sufficient rain does not fall within 5 days after a preemergence application of AMIBEN® DS, but there is enough soil moisture to germinate weeds and grasses, a light cultivation with a rotary hoe or similar tool will uproot these small broadleaf weeds and grasses. The shallow mixing of AMIBEN® DS Herbicide in the surface soil will not interfere with the action of AMIBEN® DS when rains come.

DIRECTIONS FOR USE

Apply AMIBEN® DS Herbicide and Treflan at the rates indicated in the table below. If mustard, common ragweed or black nightshade are present, use higher rates of AMIBEN® DS for best control. Do not use on muck or charcoal soils. Read and observe all directions and precautions on the AMIBEN® DS Chloramben and Treflan labels.

Make all broadcast treatments according to the following table

Soil Texture	AMIBEN® DS Lbs/A	Treflan 4EC Pts/A
Coarse	2.4	1.0
Medium	2.4 to 3.6*	1.5
Fine	2.4 to 3.6*	2.0

* Use the 3.6 pound rate if mustard, common ragweed or black nightshade is present for best control.

AMIBEN® DS CHLORAMBEN HERBICIDE AND "EPTAM" FOR SUNFLOWERS

For use only in the states of Minnesota, North Dakota and South Dakota.

AMIBEN® DS Chloramben Herbicide may be applied as a preplant soil incorporated tank mixture with Eptam Herbicide. Or AMIBEN® DS may be applied preemergence over sunflowers at planting time where Eptam has been applied as preplant soil incorporated treatment. When applied preemergence, effectiveness depends on adequate rainfall after treatment. When applied according to directions, this combination will control annual broadleaves including black nightshade, ragweed, Pennsylvania smartweed, wild mustard and annual grasses.

DIRECTIONS FOR USE

Apply AMIBEN® DS Herbicide either preplant soil incorporated or preemergence at a rate of 2.4 to 3.6 pounds per acre. Apply Eptam 7E preplant soil incorporated at a rate of 2.5 to 3.5 pints per acre. Use the higher rates on fine textured or high organic matter soils, or where weed pressure is heavy.

The AMIBEN® DS plus Eptam tank mix should be incorporated immediately after application with a field cultivator, tandem disc, mulch treader, rolling cultivator or power driven tiller as described on the Eptam label. Thoroughly mix in the top 2-3" of the soil surface by incorporating twice in different directions. Refer to the Eptam label for further information on incorporation procedures.

Rainfall is not as critical for activation of AMIBEN® DS Herbicide when it is applied preplant incorporated as when surface applied.

Apply in 10 to 20 gallons of water per acre. To prepare the spray mixture refer to the "Mixing Instructions" portion of this label. Do not mix the concentrates before adding to water.

If sufficient rain does not occur within 5 days after spraying AMIBEN® DS preemergence, a light cultivation or incorporation with a rotary hoe or similar tool is recommended.

AMIBEN® DS CHLORAMBEN HERBICIDE AND "LASSO" FOR SUNFLOWERS

A tank mixture of AMIBEN® DS Chloramben Herbicide and Lasso may be applied as a preplant soil incorporated herbicide treatment. AMIBEN® DS Herbicide may be applied as a preemergence broadcast tank mix treatment with Lasso at time of planting. Following a preplant soil incorporated application of Lasso, AMIBEN® DS may be applied preemergence at time of planting.

For uniform application of the tank mixture of AMIBEN® DS Herbicide and Lasso use 10 to 20 gallons of water per acre for ground application. To prepare the tank mixture refer to the "Mixing Instructions" portion of this label. DO NOT mix AMIBEN® DS Herbicide and Lasso concentrates before the addition of water as an emulsion will not be formed.

Apply AMIBEN® DS and Lasso preplant incorporated or preemergence using the appropriate rates from the table below. For preplant incorporated treatments apply the tank mixture to the soil and incorporate into the top 2 inches of soil as directed in the "Directions for Incorporation" portion of this label. For preemergence treatments, apply the tank mixture during planting (behind the planter) or after planting before weeds or sunflowers emerge. For sequential applications apply Lasso alone preplant incorporated as specified on the Lasso label. Follow with a preemergence application of AMIBEN® DS Herbicide during planting (behind the planter).

With preemergence treatments rain is necessary to move the chemicals into the soil and make them available to control germinating weed seeds. If no rainfall occurs within 5 days following application, a shallow incorporation no deeper than 2 inches is recommended. High intensity or excessive rainfall after application may reduce control.

Make all broadcast treatments according to the following table

Soil Texture	AMIBEN® DS Lbs/A	Lasso Qts/A
Loamy sand, sandy loam	2.4	1.5 to 2.0
Loam, silt loam, silt	2.4 to 3.0	2.0 to 2.5
Silty clay loam, silty clay, sandy clay, clay loam, clay	3.0 to 3.6	2.5 to 3.0

Use the higher rate of AMIBEN® DS when heavy populations of black nightshade, velvetleaf, mustard or smartweed are expected. Use the higher rate of Lasso when treatment is incorporated or when high populations of Black nightshade, witchgrass or foxtail are expected.

Read thoroughly the cautionary statements and all other information appearing on labels for both Lasso and AMIBEN® DS.

DIRECTIONS FOR USE ON PEANUTS

AMIBEN® DS CRACKING APPLICATIONS ON PEANUTS

AMIBEN® DS Chloramben Herbicide controls many annual grass and broadleaf weeds in peanuts when applied as a cracking treatment. It is especially effective in controlling Florida beggarweed *(Desmodium tortuosum)*.

DIRECTIONS FOR USE

Apply AMIBEN® DS at a rate of 2.4 to 3.6 pounds per broadcast acre. Apply as a ground cracking treatment up to the time when the oldest peanut plants reach the 2 leaf stage. Use a minimum of 5 gallons of water per acre for air applications or 10 gallons of water per acre for ground applications. If weeds have begun to emerge at time of application addition of 1 quart per acre of crop oil concentrate is recommended to improve control of these weeds.

DO NOT apply more than 3.6 pounds per acre of AMIBEN® DS Herbicide to the same field of peanuts in the same season.

FOR PEANUTS IN OKLAHOMA AND NORTH TEXAS

At time of planting apply 2.4 to 3.6 pounds AMIBEN® DS Chloramben Herbicide per broadcast acre. The higher rate should be used on clay loams, clay or high organic (3 to 8%) mineral soils. Use enough water for good spray coverage.

When band spraying, mount the spray nozzle behind the last packer or row wheel so that planting and spraying are done in the same operation. Proper calibration and maintenance of desired nozzle height above the ground are very important. Weeds growing in the middles between the treated rows should be cultivated as soon as necessary.

Use care while cultivating to avoid throwing fresh soil on treated bands.

If sufficient rain does not fall within 5 days after spraying AMIBEN® DS Herbicide, but there is enough soil moisture to germinate weeds and grasses, a light cultivation with a rotary hoe or similar tool will up-root these small broadleaf weeds and grasses. The shallow mixing of AMIBEN® DS in the surface soil will not interfere with the action of AMIBEN® DS when rains come.

DO NOT apply more than 3.6 pounds per acre of AMIBEN® DS Herbicide to the same field of peanuts in the same season.

DIRECTIONS FOR USE ON SWEET POTATOES

Apply AMIBEN® DS Chloramben Herbicide at the rate of 4.8 pounds per broadcast acre at planting time. For the first few cultivations after treatment, set sweeps so that soil in treated bands is not disturbed. For the final cultivation, set sweeps closer to pull soil as high as possible over the sweet potato row and treated band. For best results, plant and treat at the same time.

DIRECTIONS FOR USE ON SQUASH AND PUMPKIN

Treat preplant incorporated or preemergence at planting time or as soon after planting as possible, before weeds and crops emerge. Follow incorporation instructions in the appropriate section of this label.

In coarse textured soils, heavy rains on incorporated AMIBEN® DS Chloramben Herbicide may move it below the weed seed germination zone and produce erratic weed control.

For preemergence application, the sprayer may be mounted on the planter so that spraying and seeding are carried out in the same operation. Apply 3.6 pounds of AMIBEN® DS on lighter soils and 4.8 pounds of AMIBEN® DS on heavier or high organic soils per broadcast acre in at least 10 to 20 gallons of water. If band applications are used, calculate the rate by dividing the band width by the row width and multiplying by 3.6 or 4.8 pounds of AMIBEN® DS per broadcast acre whichever is appropriate for your soil type. After squash or pumpkin emerge, use normal cultivations to control weeds that grow in strips unprotected by any chemical.

BAND TREATMENTS

Row Width	Band Width	Lbs AMIBEN® DS Per Band Acre	
		Lighter Soils	Heavier or High Organic Soils
72 inch	14 inch	0.72	0.96
	20 inch	1.02	1.32
76 inch	14 inch	0.66	0.90
	20 inch	0.96	1.26
80 inch	14 inch	0.66	0.84
	20 inch	0.90	1.20

DIRECTIONS FOR USE ON SEEDLING ASPARAGUS

Apply 3.6 pounds of AMIBEN® DS Chloramben Herbicide per broadcast acre in about 10 to 20 gallons of water as soon as possible after asparagus seed is planted and before weeds emerge. Irrigation with about 1/2 inch of water after applying AMIBEN® DS is suggested to produce maximum weed control.

DIRECTIONS FOR USE ON LIMA BEANS

Apply 2.4 to 4.8 pounds of AMIBEN® DS Chloramben Herbicide per broadcast acre in about 10 to 20 gallons of water.

Treat preplant incorporated or preemergence at planting time or as soon after planting as possible, before weeds and lima beans emerge. Follow incorporation instructions under the appropriate section of this label.

In coarse textured soils, heavy rains on incorporated AMIBEN® DS Herbicide may move it below the weed seed germination zone and produce erratic weed control.

If band applications are used, calculate the rate by dividing the band width by the row width and multiplying by 2.4, 3.0, 3.6, or 4.8 pounds of AMIBEN® DS per broadcast acre whichever is appropriate for your weed problem and soil type. Mount band sprayer on planter so that spraying and seeding are carried out in the same operation. Make final adjustment on proper nozzle height when actually spraying in the field.

DIRECTIONS FOR USE ON SNAPBEANS (GREEN AND YELLOW)

Apply AMIBEN® DS at the rate of 2.4 to 3.0 pounds par broadcast acre as a preemergence treatment in sufficient water for good spray coverage (10 to 20 gallons of water per acre). Use the higher rate of AMIBEN® DS on heavy or high organic mineral soils or where Black Nightshade is a problem. Use the lower rate on loam soils. Do not use on sandy soils low in clay content or with less than 1% organic matter. Attach the applicator to the planter so that planting and spraying are done in the same operation. Rain or sprinkler irrigation carries AMIBEN® DS into the soil where weed seeds sprout. If weeds begin to emerge after AMIBEN® DS application, but prior to rainfall or irrigation, a light cultivation is recommended to up-root the small broadleaf weeds and grasses. The shallow mixing of AMIBEN® DS in the soil surface will not interfere with its weed killing action when rain comes.

WARNING: Stunting of beans has been observed when rains wash the chemical into contact with the germinating seed or the root system of young plants. Under good growing conditions plants normally recover and yields are generally not affected.

Heavy or prolonged rainfall may move AMIBEN® DS below the weed seed germination zone and produce erratic weed control.

AMIBEN® DS CHLORAMBEN HERBICIDE AND "DUAL" FOR SNAPBEANS (GREEN AND YELLOW)

AMIBEN® DS Herbicide may be applied as a preemergence broadcast tank mix treatment with Dual 8E at time of planting. Following a preplant soil incorporated application of Dual 8E, AMIBEN® DS Herbicide may be applied preemergence at time of planting.

When applied according to label directions, tank mixtures of AMIBEN® DS Herbicide and Dual will control all weeds controlled by AMIBEN® DS alone or by Dual alone including fall panicum, foxtail millet, red rice, signalgrass, witchgrass, yellow nutsedge and black nightshade. Refer to the general information section of this label and the Dual label for additional weeds controlled by AMIBEN® DS and Dual 8E.

For uniform application of the tank mixture of AMIBEN® DS and Dual, use 10 to 20 gallons of water with ground equipment. To prepare the tank mixture, refer to the "Mixing Instructions" portion of this label.

Apply AMIBEN® DS and Dual using the appropriate rates from the table below. For preemergence treatments, apply the tank mixture during planting (behind the planter) or after planting before weeds or snapbeans emerge. For sequential applications apply Dual 8E alone preplant incorporated and follow with a preemergence application of AMIBEN® DS Herbicide during planting (behind the planter).

Make all broadcast treatments according to the following table

Soil Texture	AMIBEN® DS Lbs/A	Dual 8E Pts/A
Loamy sand, sandy loam	DO NOT USE	1.5 to 2.0
Loam, silt loam, silt	2.4 to 3.0	2.0 to 2.5
Silty clay loam, silty clay, sandy clay, clay loam, clay	2.4 to 3.0	2.0 to 3.0

Do not use on muck or peat soils.

Use the higher rate of AMIBEN® DS when heavy populations of black nightshade, velvetleaf, mustard or smartweed are expected. Use the higher rate of Dual when treatment is incorporated, when organic matter is over 3%, or when high populations of black nightshade, witchgrass, panicum, foxtail, or yellow nutsedge are expected.

Before applying this combination read thoroughly all precautionary statements and all other information appearing on or attached to the containers of AMIBEN® DS Herbicide and Dual 8E.

DIRECTIONS FOR USE ON FIELD SEEDED TOMATOES PROTECTED WITH ACTIVATED CARBON

AMIBEN® DS Chloramben Herbicide may be applied as a broadcast preemergence herbicide treatment on field seeded tomatoes which have been protected by the use of an activated carbon plus vermiculite anticrustant system as described below. Do not use AMIBEN® DS on field seeded tomatoes without the use of the activated carbon plus vermiculite protection system as crop injury will result.

Thoroughy mix 1 pound of activated carbon with each cubic foot of number 2 or number 3 horticultural grade vermiculite. Apply 1 cubic foot of this mixture per 600 feet of seeded row (15 cubic feet per acre) by filling the seed furrow with the mixture. Use a positive feed applicator to deliver the activated carbon plus vermiculite anticrustant mixture to the seed furrow directly ahead of the planter press wheel.

Immediately after seeding and before weeds or tomatoes emerge, apply 2.4 to 3.6 pounds per acre of AMIBEN® DS in 10 to 40 gallons of water per acre as a broadcast treatment. Use only on medium or heavy soils. Use the higher rate of AMIBEN® DS on heavier soils or where high populations of black nightshade, velvetleaf, mustard or smartweed are expected.

The use of the activated carbon plus vermiculite anticrustant system in the seed furrow will protect the tomato seedling from the phytotoxic effects of AMIBEN® DS. When used in this manner AMIBEN® DS *will not control* weed seeds germinating within the vermiculite treated row.

DIRECTIONS FOR USE ON FIELD SEEDED CUCUMBERS PROTECTED WITH ACTIVATED CARBON

AMIBEN® DS Chloramben Herbicide may be applied as a broadcast preemergence herbicide treatment on field seeded cucumbers which have been protected by the use of an activated carbon plus vermiculite anticrustant system as described below. Do not use AMIBEN® DS on field seeded cucumbers without the use of the activated carbon plus vermiculite protection system as crop injury will result.

Thoroughly mix 1 pound of activated carbon with each cubic foot of number 2 or number 3 horticultural grade vermiculite. Apply 1 cubic foot of this mixture per 600 feet of seeded row (15 cubic feet per acre) by filling the seed furrow with the mixture. Use a positive feed applicator to deliver the activated carbon plus vermiculite anticrustant mixture to the seed furrow directly ahead of the planter press wheel.

Immediately after seeding and before weeds or cucumbers emerge, apply 2.4 to 3.6 pounds per acre of AMIBEN® DS in 10 to 40 gallons of water per acre as a broadcast treatment. Use only on medium or heavy soils. Use the higher rate of AMIBEN® DS on heavier soils or where high populations of black nightshade, velvetleaf, mustard or smartweed are expected.

The use of the activated carbon plus vermiculite anticrustant system in the seed furrow will protect the cucumber seedling from the phytotoxic effects of AMIBEN® DS. When used in this manner AMIBEN® DS *will not control* weed seeds germinating within the vermiculite treated row.

DIRECTIONS FOR USE ON FIELD SEEDED CANTALOUPE PROTECTED WITH ACTIVATED CARBON

AMIBEN® DS Chloramben Herbicide may be applied as a broadcast preemergence herbicide treatment on field seeded cantaloupe which have been protected by the use of an activated carbon plus vermiculite anti-crustant system as described below. Do not use AMIBEN® DS on field seeded cantaloupe without the use of the activated carbon plus vermiculite protection system as crop injury will result.

Thoroughly mix 1 pound of activated carbon with each cubic foot of number 2 or number 3 horticultural grade vermiculite. Apply 1 cubic foot of this mixture per 600 feet of seeded row (15 cubic feet per acre) by filling the seed furrow with the mixture. Use a positive feed applicator to deliver the activated carbon plus vermiculite anticrustant mixture to the seed furrow directly ahead of the planter press wheel.

Immediately after seeding and before weeds or cantaloupe emerge, apply 2.4 to 3.6 pounds per acre of AMIBEN® DS in 10 to 40 gallons of water per acre as a broadcast treatment. Use only on medium or heavy soils. Use the higher rate of AMIBEN® DS on heavier soils or where high populations of black nightshade, velvetleaf, mustard or smartweed are expected.

The use of the activated carbon plus vermiculite anticrustant system in the seed furrow will protect the cantaloupe seedling from the phytotoxic effects of AMIBEN® DS. When used in this manner AMIBEN® DS will not *control* weed seeds germinating within the vermiculite treated row.

DIRECTIONS FOR USE ON CORN

Use only on heavy soils in Missouri, Minnesota, Iowa, Ohio, Illinois, Kansas, Nebraska and Indiana. If rains and extended cold weather occur after planting, corn injury may result.

Apply 2.4 pounds of AMIBEN® DS Chloramben Herbicide per acre in at least 10 to 20 gallons of water. Plant and treat at the same time or as soon after planting as possible before weeds and crop emerge.

Plant corn at least 2 to 2½ inches deep. Do not use AMIBEN® DS on corn grown in light textured sandy soils.

AMIBEN® DS CHLORAMBEN HERBICIDE AND "ATRAZINE" 80W TANK MIXTURE

Application: Apply a tank mixture of AMIBEN® DS Chloramben Herbicide and Atrazine 80W during or immediately after planting. A light incorporation with a rotary hoe or similar equipment may be desirable, especially if there is no rainfall soon after application. Do not use AMIBEN® DS and Atrazine for weed control in crops other than corn. Observe all directions on the Atrazine label, especially those relating to crop rotation.

Use only in the states of Missouri, Minnesota, Iowa, Ohio, Illinois, Kansas, Nebraska and Indiana.

Amounts of use: For silty loam soils, apply 1.2 pounds of AMIBEN® DS and 1.25 pounds of Atrazine. For heavier soils, apply 1.2 pounds of AMIBEN® DS and 2.5 pounds of Atrazine or 1.8 pounds of AMIBEN® DS and 1.87 pounds of Atrazine. This mixture is not recommended for use on very light soils such as sands, or sandy loams.

STORAGE AND DISPOSAL

Do not store near fertilizers, seed, insecticides, or fungicides. Open dumping is prohibited.

PESTICIDE DISPOSAL

Do not contaminate water, food, or feed by storage or disposal. Wastes resulting from the use of this product may be disposed of on site or at an approved waste disposal facility.

CONTAINER DISPOSAL

Do not reuse empty container. Completely empty bag into application equipment. Then dispose of empty bag in a sanitary landfill or by incineration, or if allowed by state and local authorities, by burning. If burned, stay out of smoke.

Consult Federal, state or local disposal authorities for approved alternative procedures such as limited open burning.

LIMITED WARRANTY AND DISCLAIMER

The manufacturer warrants (a) that this product conforms to the chemical description on the label; (b) that this product is reasonably fit for the purposes set forth in the directions for use when it is used in accordance with such directions; and (c) that the directions, warnings and other statements on this label are based upon responsible experts' evaluation of reasonable tests of effectiveness, of toxicity to laboratory animals and to plants, and of residues on food crops, and upon reports of field experience. Tests have not been made on all varieties or in all states or under all conditions.

THE MANUFACTURER NEITHER MAKES NOR INTENDS, NOR DOES IT AUTHORIZE ANY AGENT OR REPRESENTATIVE TO MAKE, ANY OTHER WARRANTIES, EXPRESS OR IMPLIED, AND IT EXPRESSLY EXCLUDES AND DISCLAIMS ALL IMPLIED WARRANTIES OR MERCHANTABILITY OR FITNESS FOR A PARTICULAR PURPOSE.

THIS WARRANTY DOES NOT EXTEND TO, AND THE BUYER SHALL BE SOLELY RESPONSIBLE FOR, ANY AND ALL LOSS OR DAMAGE WHICH RESULTS FROM THE USE OF THIS PRODUCT IN A MANNER WHICH IS INCONSISTENT WITH THE LABEL DIRECTIONS, WARNINGS OR CAUTIONS.

BUYER'S EXCLUSIVE REMEDY AND MANUFACTURER'S OR SELLER'S EXCLUSIVE LIABILITY FOR ANY AND ALL CLAIMS, LOSSES, DAMAGES, OR INJURIES RESULTING FROM THE USE OR HANDLING OF THIS PRODUCT, WHETHER OR NOT BASED ON CONTRACT, NEGLIGENCE, STRICT LIABILITY IN TORT OR OTHERWISE SHALL BE LIMITED, AT THE MANUFACTURER'S OPTION, TO REPLACEMENT OF, OR THE REPAYMENT OF THE PURCHASE PRICE FOR, THE QUANTITY OF PRODUCT WITH RESPECT TO WHICH DAMAGES ARE CLAIMED. IN NO EVENT SHALL MANUFACTURER OR SELLER BE LIABLE FOR SPECIAL, INDIRECT OR CONSEQUENTIAL DAMAGES RESULTING FROM THE USE OR HANDLING OF THIS PRODUCT.

NOTICE TO BUYER

Purchase of this material does not confer any rights under patents governing this product or the use thereof in countries outside of the United States.

Blazer is a registered trademark of Rohm and Haas Company.
Treflan and Surflan are registered trademarks of Elanco Products Company.
Lasso and Roundup are registered trademarks of Monsanto Company.
Dual is a registered trademark of Ciba-Geigy Corporation.
Sencor is a registered trademark of Farbenfabrik Bayer GmbH, Leverkusen.
Lorox and Lexone are registered trademarks of E.I. duPont de Nemours and Company.
Prowl is a registered trademark of American Cyanamid Company.
Alanap is a registered trademark of Uniroyal, Inc.
Reward and Eptam are registered trademarks of Stauffer Chemical Company.
Basalin is a registered trademark of BASF Wyandotte Corporation.

AMIBEN® DS and BUTYRAC® 200 are trademarks of Rhône-Poulenc AG90001

THIS SPECIMEN LABEL IS INTENDED FOR USE ONLY AS A GUIDE IN PROVIDING GENERAL INFORMATION REGARDING THE DIRECTIONS, WARNINGS AND CAUTIONS ASSOCIATED WITH THE USE OF THIS PRODUCT. AS WITH ANY AGRICULTURAL CHEMICAL, ALWAYS FOLLOW THE LABEL INSTRUCTIONS ON THE PACKAGE BEFORE USING.

AMID-THIN® W

For thinning apples and pears

ACTIVE INGREDIENT:	
1-Naphthaleneacetamide	8.4%
INERT INGREDIENTS:	91.6%
TOTAL	100.0%

E.P.A. Reg. No. 264-137 AA E.P.A. Est. No. 57062-OH-001

KEEP OUT OF REACH OF CHILDREN

CAUTION

For PRODUCT USE Information Call 1-800-334-9745
For MEDICAL And TRANSPORTATION Emergencies ONLY Call 24 Hours A Day 1-800-334-7577

PRECAUTIONARY STATEMENTS

HAZARDS TO HUMANS AND DOMESTIC ANIMALS

CAUTION

Harmful if swallowed. Avoid contact with skin, eyes, and clothing. Avoid breathing spray mist.

STATEMENT OF PRACTICAL TREATMENT

In case of eye contact flush with water for at least 15 minutes. In case of skin contact wash with soap and water. Wash contaminated clothing before reuse.

ENVIRONMENTAL HAZARDS

Do not apply directly to water or wetlands. Do not contaminate water when disposing of equipment washwaters. Do not contaminate irrigation ditches or water used for irrigation or domestic purposes. Do not apply when weather conditions favor drift from treated areas or use in a greenhouse.

DIRECTIONS FOR USE

It is a violation of Federal law to use this product in a manner inconsistent with its labeling.

AMID-THIN® W contains naphthaleneacetamide and is used to thin many varieties of apples as a petal fall spray.

Do not apply AMID-THIN® W through any type of irrigation system.

GENERAL WORKER PROTECTION STATEMENT

Do not apply this product in such a manner as to directly or through drift expose workers or other persons. The area treated must be vacated by unprotected persons. Do not enter treated areas without protective clothing until sprays have dried. Because certain states may require more restrictive reentry intervals for various crops treated with this product, consult your State Department of Agriculture for further information.

Written or oral warnings must be given to workers who are expected to be in a treated area or in an area about to be treated with this product. Advise workers to stay out of fields during application and until sprays have dried. Regular long-sleeved work clothing should be worn when working in treated fields. See PRECAUTIONARY STATEMENTS; STATEMENT OF PRACTICAL TREATMENT for information on accidental exposures. When oral warnings are given, warnings shall be given in a language customarily understood by workers. Oral warnings must be given if there is reason to believe that written warnings cannot be understood by workers. Written warnings must include the following information: appropriate signal word (CAUTION), area treated with AMID-THIN® W, date of application, appropriate protective clothing, and reentry interval (i.e., until sprays have dried).

DIRECTIONS FOR THINNING APPLES

Many factors affect fruit set including: tree vigor, provisions for cross-pollination, weather conditions and bee activity. Other important considerations are last year's fruit set level and current year's food reserve status which is affected by crop load and foliage vigor in the preceeding year.

Limit the use of the higher concentrations given in the table to vigorous trees having ample provisions for pollination and good fruit set. Use the lower concentrations on weaker, less vigorous trees with less food reserves. One of the outstanding advantages of AMID-THIN® W from the standpoint of over thinning is its wide range of safety at the concentrations suggested.

It is important that early summer varieties are sprayed no later than petal fall since applications two to three weeks later often given insufficient thinning and may result in the development of inferior (pigmy) fruit.

APPLE THINNING TABLE - PPM TO USE

This table is offered only as a guide

I. Application at petal fall in normal years 1	
Yellow Transparent, Williams Early Red, Olenburg (Dutchess), Red Astrachan	25-40 ppm
Red Delicious	15-50 ppm
Yellow Newton	40-50 ppm
II. Application at 2-2½ weeks after full bloom 2*	
Early McIntosh, Wealthy, Golden Delicious, Baldwin, Rome Beauty, Jonathan, Stayman	15-50 ppm
Grimes Golden, Gravenstein, York	40-50 ppm
Cortland, Macoun, Northern Spy	40 ppm
Rhode Island Greening	25-40 ppm

Important: Do not apply AMID-THIN® W to Red Delicious apples grown in the Pacific Northwest as "pigmy" fruit may develop.

FRUITONE® N is labeled for use on Red Delicious apples.

1. Petal Fall: Approximately 90% to 95% of petals off.
2. Full Bloom: 90% of the blossoms open with petals beginning to fall from the king flowers.

* Spraying could be delayed to 3 weeks for the 2 to 2½ weeks after full bloom varieties if unusually cool weather has slowed growth.

Weather conditions, amount of bloom and other factors that make for good set will vary in different localities. Contact your local extension pomologists as to time of application and rates to be used for your specific thinning problems.

DIRECTIONS FOR THINNING BARTLETT AND BOSC PEARS

Spray AMID-THIN® W at a rate of 10 to 50 ppm at petal fall or within 5 to 7 days after petal fall. Use up to 400 gallons of spray solution per acre (2 lb. per acre AMID-THIN® W). Applying AMID-THIN® W in this manner reduces the time and hand labor required for supplemental thinning of pears to the desired level.

Spray Preparation Chart

(Gallons of spray/Acre)	Amount of AMID-THIN® W 75	100	150	300
Rate (ppm)	lb. (oz.)	lb.	lb.	lb.
15	0.11 (1.8)	0.15	0.23	0.45
25	0.19 (3.0)	0.25	0.38	0.75
40	0.30 (4.8)	0.40	0.60	1.20
50	0.38 (6.1)	0.50	0.75	1.50

DIRECTIONS FOR MIXING

Prepare sufficient spray mixture to ensure thorough coverage of trees. When using AMID-THIN® W, fill the tank with water and then add AMID-THIN® W with agitator running and keep under constant agitation during spray operations. To ensure proper mixing of AMID-THIN® W, allow a few minutes of agitation after material has been added to water before beginning spray operation.

STORAGE AND DISPOSAL

STORAGE

Do not store near fertilizers, seeds, insecticides, or fungicides.

PESTICIDE DISPOSAL

Do not contaminate water, food, or feed by storage or disposal. Open dumping is prohibited.

Wastes resulting from the use of this product may be disposed of on site or at an approved waste disposal facility.

CONTAINER DISPOSAL

Do not re-use empty container. Completely empty container into application equipment. Then dispose of container in a sanitary landfill or by incineration, or, if allowed by State and local authorities, by burning. If container is burned, stay out of smoke.

LIMITED WARRANTY AND DISCLAIMER

The manufacturer warrants that this product conforms to the chemical description on the label; that this product is reasonably fit for the purposes set forth in the directions for use when it is used in accordance with such directions; and that the directions, warnings and other statements on this label are based upon responsible experts' evaluation of reasonable tests of effectiveness, of toxicity to laboratory animals and to plants, and of residues on food crops, and upon reports of field experience. Test have not been made on all varieties or in all states or under all conditions. THE MANUFACTURER NEITHER MAKES NOR INTENDS, NOR DOES IT AUTHORIZE ANY AGENT OR REPRESENTATIVE TO MAKE ANY OTHER WARRANTIES, EXPRESS OR IMPLIED, AND IT EXPRESSLY EXCLUDES AND DISCLAIMS ALL IMPLIED WARRANTIES OF MERCHANTABILITY OR FITNESS FOR A PARTICULAR PURPOSE.

THIS WARRANTY DOES NOT EXTEND TO, AND THE BUYER SHALL BE SOLELY RESPONSIBLE FOR, ANY AND ALL LOSS OR DAMAGE WHICH RESULTS FROM THE USE OF THIS PRODUCT IN ANY MANNER WHICH IS INCONSISTENT WITH THE LABEL DIRECTIONS, WARNINGS, OR CAUTIONS.

BUYER'S EXCLUSIVE REMEDY AND MANUFACTURER'S OR SELLER'S EXCLUSIVE LIABILITY FOR ANY AND ALL CLAIMS, LOSSES, DAMAGES OR INJURIES RESULTING FROM THE USE OR HANDLING OF THIS PRODUCT, WHETHER OR NOT BASED IN CONTRACT, NEGLIGENCE, STRICT LIABILITY IN TORT OR OTHERWISE, SHALL BE LIMITED, AT THE MANUFACTURER'S OPTION, TO REPLACEMENT OF OR THE REPAYMENT OF THE PURCHASE PRICE FOR, THE QUANTITY OF PRODUCT WITH RESPECT TO WHICH DAMAGES ARE CLAIMED, IN NO EVENT SHALL MANUFACTURER OR SELLER BE LIABLE FOR SPECIAL INDIRECT OR CONSEQUENTIAL DAMAGES RESULTING FROM THE USE OR HANDLING OF THIS PRODUCT.

NOTICE TO BUYER

Purchase of this material does not confer any rights under patents governing this product or the use thereof in countries outside of the United States.

AMID-THIN is a registered trademark of Rhône-Poulenc.

AG-90035

THIS SPECIMEN LABEL IS INTENDED FOR USE ONLY AS A GUIDE IN PROVIDING GENERAL INFORMATION REGARDING THE DIRECTIONS, WARNINGS AND CAUTIONS ASSOCIATED WITH THE USE OF THIS PRODUCT. AS WITH ANY AGRICULTURAL CHEMICAL, ALWAYS FOLLOW THE LABEL INSTRUCTIONS ON THE PACKAGE BEFORE USING.

AMITROL T™ brand

LIQUID HERBICIDE

RESTRICTED USE PESTICIDE

FOR RETAIL SALE TO AND APPLICATION ONLY BY CERTIFIED APPLICATORS OR PERSONNEL UNDER THEIR DIRECT SUPERVISION.

Controls quackgrass, Canada thistle and other tough annual and perennial weeds.

ACTIVE INGREDIENT:

Amitrole (3-amino-1,2,4-triazole)*	21.6%
INERT INGREDIENTS:	78.4%
TOTAL:	100.0%

*Equivalent to 2 lbs. per gallon.

KEEP OUT OF REACH OF CHILDREN

CAUTION

EPA Reg. No. 264-135-ZA EPA Est. No.

FOR PRODUCT USE INFORMATION CALL 1-800-334-9745

For MEDICAL AND TRANSPORTATION Emergencies ONLY Call 24 HOURS A DAY 1-800-334-7577

PRECAUTIONARY STATEMENTS

CAUTION

HAZARDS TO HUMANS AND DOMESTIC ANIMALS

HARMFUL IF SWALLOWED.
Avoid contact with eyes, skin and clothing.

The use of this product may be hazardous to your health. This product contains Amitrole which has been determined to cause cancer in laboratory animals. Wear a respirator during mixing and loading of this product.

Wear lightweight, waterproof clothing (jumpsuit or coverall and boots or shoes, gloves, and a wide-brimmed plastic hard hat) when applying this product to control dense, tall vegetation. Workers applying this product in other situations must wear lightweight waterproof gloves and boots (or shoes).

STATEMENT OF PRACTICAL TREATMENT

IF SWALLOWED: Induce vomiting immediately. Get medical attention.

IF ON SKIN: Wash thoroughly with soap and water. Remove contaminated clothing and wash before reuse.

IF IN EYES: Flush eyes with plenty of water.

NOTE TO PHYSICIAN: This product may affect thyroid function. Overexposure can be determined by monitoring thyroid function. If worker is under treatment for hypo- or hyperthyroidism, he should be removed from spray crew until the euthyroid state is achieved.

ENVIRONMENTAL HAZARDS

Do not allow spray or spray drift to contaminate edible crops or water intended for irrigation, drinking or domestic purposes as no chemical residue in crops is permitted. Spray drift from this product may injure cotton, tomatoes, lawns, ornamentals and other desirable vegetation. Coarse sprays are less likely to drift.

Do not use spray equipment for other purposes.

Do not use in a greenhouse.

Do not allow spray solution to remain in a sprayer any longer than necessary. Rinse all spray equipment with clean water immediately after each use to prevent corrosion of metal parts.

Do not allow livestock to graze or feed in treated non-crop areas.

Do not apply directly to water or wetlands. Do not contaminate water by cleaning of equipment or disposal of wastes.

DIRECTIONS FOR USE

IT IS A VIOLATION OF FEDERAL LAW TO USE THIS PRODUCT IN A MANNER INCONSISTENT WITH ITS LABELING.

KNAPSACK SPRAYERS - Spray weeds thoroughly, wetting all leaves and stems to ground line. Mix 1/2 cup (4 fluid ounces) of AMITROL-T™ brand liquid herbicide per gallon of water. One gallon of spray solution is usually enough to cover 1 square rod (272 sq. ft.) infested with weeds. In intermountain states or other dryland areas, mix 1/4 cup of AMITROL-T™ per gallon of water and apply 2 gallons of spray solution per square rod.

POWER SPRAYERS - Apply enough of the spray mixture to wet thoroughly all leaves and stems to ground line.

Woody Plants and Vines

	Gallons AMITROL-T™	Gallons Water	When to Apply
Poison ivy-poison oak	1	100	Treat from the time foliage is fully developed until plants begin to go dormant. For effective control, all leaves, stems and suckers must be thoroughly wet to ground line.
Honeysuckle*-Kudzu	2	100	
Salmon berry-Western dewberry (blackberry)-Big leaf maple	2	100	
Wild cherry-Ash-Locust-Sumac	1 1/2	100	*If regrowth of honeysuckle appears, retreat in August or the following Spring.

WEEDS IN NON-CROP AREAS SUCH AS RAILROADS, ROADSIDES, FENCE ROWS AND HARDWOOD NURSERIES

WEEDS	SPOT TREATMENTS[1] SPRAY MIXTURE: GALLONS AMITROL-T™	GALLONS WATER	ACRE TREATMENTS[2] AMOUNTS PER ACRE: GALLONS AMITROL-T™	MIN. GALS. WATER	WHERE TO APPLY
CONTROL: annual grasses and broadleaf weeds (foxtail, ryegrass, cheatgrass, barnyard grass, annual bluegrass, wild hemp or marijuana, pigweed, kochia, sunflower, wild barley)	1/2	50	1	100	Spray when most weeds are 3 to 4 inches tall.
CONTROL annual grasses, annual and perennial broadleaf weeds (whitetop, Canada thistle, volunteer alfalfa, milkweed) SUPPRESS perennial grasses (ripgut grass, reed canary grass, quackgrass).	1	50	2	100	Spray when most weeds are 6 to 10 inches tall.
Canada thistle sowthistle	1	50	2	40	Spray when most thistle are and in the bud to bloom stage. Do not mow treated plants in the Far West, treat during the bloom stage
quackgrass (couch, witch)	1	50	2	40	Spray when vigorous young growth is 6 to 8 inches tall.
whitetop (hoary cress)	1	50	2	40	Spray in spring when weeds are in the bud to bloom stage. Do not mow treated plants.
leafy spurge (Western states)	1	50	2	40	
milkweed	1	50	1	50	
bermudagrass	2	50	4	50	Treat from mid-summer to early fall when shoots are growing vigorously. Spot treat any green regrowth with a solution of 1 gal. AMITROL-T™ per 50 gals. of water.
nutsedge (nut grass)	1	50	2	50	Spray when shoots are 4 to 5 inches tall and growing actively.
horsenettle	1	50	2	100	In East and Midwest, spray when horsenettle is in the bud to bloom stage. In the Far West treat from mid-summer to early frost.
horsetail rush	1 1/2	50	3	100	Spray when 6 to 8 inches tall and growing actively.
dock	1/2	50	1	50	Apply anytime when dock is actively growing preferably before development of bloomstalk.
wild chrysanthemum in hardwood nurseries	2	50	4	100	Apply as directed spray on young weed growth 4 to 6 inches tall. Keep spray off foliage of hardwood nursery stock.

[1] A second treatment may be necessary to control weeds not thoroughly sprayed or weeds coming from dormant seeds and nuts or old roots. Apply when regrowth is young and growing actively.

[2] INTERMOUNTAIN STATES:
Canada thistle-use 4 gals. AMITROL-T™ in 100 gals. water per acre. All other weeds-use the rates of AMITROL-T™ per acre as listed in the preceding chart, applied in a minimum of 100 gals. of water per acre.

STORAGE AND DISPOSAL

STORAGE
Do not contaminate water, food, or feed by storage or disposal.

Do not store near fertilizers, seeds, insecticides or fungicides.

PESTICIDE DISPOSAL
Open dumping is prohibited.
Pesticide wastes are toxic. Improper disposal of excess pesticide, spray mixture or rinsate is a violation of Federal law. If these wastes cannot be disposed by use according to label instructions, contact your State Pesticide or Environmental Control Agency or the Hazardous Waste Representative at the nearest EPA Regional Office for guidance.

CONTAINER DISPOSAL
Do not reuse container.
Triple rinse or equivalent. Then offer for recycling or reconditioning or puncture and dispose of in a sanitary landfill or by incineration, or if allowed by state and local authorities, by burning. If burned, stay out of smoke.

LIMITED WARRANTY AND DISCLAIMER
The manufacturer warrants (a) that this product conforms to the chemical description on the label; (b) that this product is reasonably fit for the purposes set forth in the directions for use when it is used in accordance with such directions; and (c) that the directions, warnings and other statements on this label are based upon responsible experts' evaluation of reasonable tests of effectiveness, of toxicity to laboratory animals and upon reports of field experience. Tests have not been made on all varieties or in all states or under all conditions. THE MANUFACTURER NEITHER MAKES NOR INTENDS, NOR DOES IT AUTHORIZE ANY AGENT OR REPRESENTATIVE TO MAKE, ANY OTHER WARRANTIES, EXPRESS OR IMPLIED, AND IT EXPRESSLY EXCLUDES AND DISCLAIMS ALL IMPLIED WARRANTIES OF MERCHANTABILITY OR FITNESS FOR A PARTICULAR PURPOSE.

THIS WARRANTY DOES NOT EXTEND TO, AND THE BUYER SHALL BE SOLELY RESPONSIBLE FOR, ANY AND ALL LOSS OR DAMAGE WHICH RESULTS FROM THE USE OF THIS PRODUCT IN ANY MANNER WHICH IS INCONSISTENT WITH THE LABEL DIRECTIONS, WARNINGS, OR CAUTIONS.

BUYER'S EXCLUSIVE REMEDY AND MANUFACTURER'S OR SELLER'S EXCLUSIVE LIABILITY FOR ANY AND ALL CLAIMS, LOSSES, DAMAGES, OR INJURIES RESULTING FROM THE USE OR HANDLING OF THIS PRODUCT, WHETHER OR NOT BASED IN CONTRACT, NEGLIGENCE, STRICT LIABILITY IN TORT OR OTHERWISE, SHALL BE LIMITED, AT THE MANUFACTURER'S OPTION, TO REPLACEMENT OF, OR THE REPAYMENT OF THE PURCHASE PRICE FOR, THE QUANTITY OF PRODUCT WITH RESPECT TO WHICH DAMAGES ARE CLAIMED. IN NO EVENT SHALL MANUFACTURER OR SELLER BE LIABLE FOR SPECIAL, INDIRECT OR CONSEQUENTIAL DAMAGES RESULTING FROM THE USE OR HANDLING OF THIS PRODUCT.

NOTICE TO BUYER
Purchase of this material does not confer any rights under patents governing this product or the use thereof in countries outside of the United States.

AMITROL is a trademark of Rhône-Poulenc

AG90002

EPA Approval 8/89

THIS SPECIMEN LABEL IS INTENDED FOR USE ONLY AS A GUIDE IN PROVIDING GENERAL INFORMATION REGARDING THE DIRECTIONS, WARNINGS AND CAUTIONS ASSOCIATED WITH THE USE OF THIS PRODUCT. AS WITH ANY AGRICULTURAL CHEMICAL, ALWAYS FOLLOW THE LABEL INSTRUCTIONS ON THE PACKAGE BEFORE USING.

AMIZOL® brand
INDUSTRIAL HERBICIDE

RESTRICTED USE PESTICIDE
FOR RETAIL SALE TO AND APPLICATION ONLY BY CERTIFIED APPLICATORS OR PERSONNEL UNDER THEIR DIRECT SUPERVISION.

For general weed control in industrial and similar noncrop areas. Use alone or in combination with other weed control herbicides having long residual action.

ACTIVE INGREDIENT:	
Amitrole (3-amino-1,2,4-triazole)	90%
INERT INGREDIENTS:	10%
TOTAL:	100%

E.P.A. Reg. No. 264-119-AA E.P.A. Est. No.

KEEP OUT OF REACH OF CHILDREN
CAUTION

FOR MEDICAL And TRANSPORTATION Emergencies ONLY call 24 HOURS A DAY 1-800-334-7577
FOR PRODUCT USE INFORMATION CALL 800-334-9745

PRECAUTIONARY STATEMENTS

CAUTION
HAZARD TO HUMANS AND DOMESTIC ANIMALS
HARMFUL IF SWALLOWED.
Avoid contact with eyes, skin and clothing.

The use of this product may be hazardous to your health. This product contains amitrole which has been determined to cause cancer in laboratory animals. Wear a respirator during mixing and loading of this product.

Wear lightweight, waterproof clothing (jumpsuit or coverall and boots or shoes, gloves, and a wide-brimmed plastic hard hat) when applying this product to control dense, tall vegetation. Workers applying this product in other situations must wear lightweight waterproof gloves and boots (or shoes).

STATEMENT OF PRACTICAL TREATMENT
IF SWALLOWED: Induce vomiting immediately. Get medical attention.

IF ON SKIN: Wash thoroughly with soap and water. Remove contaminated clothing and wash before reuse.

IF IN EYES: Flush eyes with plenty of water.

NOTE PHYSICIAN This product may effect thyroid function. Overexposure can be determined by monitoring thyroid function. If worker is under treatment for hypo- or hyperthyroidism, he should be removed from spray crew until the euthyroid state is achieved.

ENVIRONMENTAL HAZARDS
Do not apply directly to water or wetlands.

Do not contaminate water by cleaning of equipment or disposal of wastes.

Do not allow spray or spray drift to contaminate edible crops or water intended for irrigation, drinking or other domestic purposes.

Do not allow livestock to graze or feed in treated non-crop areas.

Avoid spray drift as this product may injure susceptible plants such as cotton, tomatoes, and some ornamentals.

Coarse sprays are less likely to drift.

Do not use same sprayer for other purposes unless thoroughly cleaned. To clean sprayer following use of AMIZOL®, rinse all parts of sprayer with water several times.

Do not use in a greenhouse.

GENERAL DIRECTIONS
By combining AMIZOL® Industrial Herbicide with residual-type weed control herbicide, you make a mixture that provides a rapid knockdown of existing vegetation, better control of deep rooted perennial weeds, and prevents most weed growth for one or more seasons.

WHEN TO USE
Apply AMIZOL® Industrial Herbicide combinations anytime weeds are green and actively growing. It is not necessary to apply these combinations before weeds come up.

Weeds treated after they have developed strong stems may remain standing for the rest of the season, even though they are dead. These standing weeds are unsightly and a fire hazard. Therefore, for most effective control of tall weeds, cut and remove the weeds, if possible, then spray the regrowth. For best results, treat when soil moisture is adequate for rapid weed growth.

MIXING SPRAYS
Use enough water to thoroughly wet weed foliage - 50 to 100 gallons per acre. In highway spraying, a strip 8 feet wide and 1 mile long is about 1 acre. Dissolve the required amount of AMIZOL® in water, then add the other product. Wettable powders require constant agitation while mixing and spraying.

AMIZOL® Industrial Herbicide or combinations of AMIZOL® with Telvar, Karmex, Hyvar, Krovar, Tandex, Aatrex and Princep are non-corrosive, non-flammable and non-volatile. AMIZOL® combinations with Sodium TCA are non-flammable and non-volatile, but sprayers should be washed out immediately after each use to prevent corrosion.

Some of the Weeds Controlled

alfalfa	foxtail	purslane
bluegrasses	goosegrass	quackgrass
Canada thistle	groundsel	ragweed
carpetweed	hempnettle	shepherdspurse
catch fly	jimsonweed	smartweed
chickweed	knotweed	sowthistle
chrysanthemum weed (mugwort)	kochia	stinkweed
	lambsquarters	velvetgrass
cocklebur	mustard	white cockle
crabgrass	nightshade	wild buckwheat
dandelion	pigweed	wild oats
downy brome (cheat)	plantain	witchgrass
	poison ivy	yellow rocket

DIRECTIONS FOR USE

IT IS A VIOLATION OF FEDERAL LAW TO USE THIS PRODUCT IN A MANNER INCONSISTENT WITH ITS LABELING.

I. Use AMIZOL® brand Industrial Herbicide alone:	Amounts per Acre	
	AMIZOL®	Water
a. To control annual grass and broadleaf weeds where residual control is not required, spray when most weeds are 3-4 inches tall.	2 lbs.	100 gal.
b. To control annual grasses, annual and perennial broadleaf weeds and suppress perennial grasses, spray when most weeds are 6-10 inches tall.	4 lbs.	100 gal.
c. Poison ivy, poison oak: treat after leaves are fully developed in the spring until leaves start to turn color in the fall.	2 lbs.	100 gal.

II. Use AMIZOL® brand Industrial Herbicide combinations with Princep 80W:	Amounts per Acre	
	AMIZOL®	Princep
a. For seasonal control of most weeds where safety to nearby desirable trees, shrubs and ornamentals is important.	4 lbs.	12½ lbs.
b. For economical control of easy-to-control (annual) weeds; for areas with short growing season.	2 lbs.	6¼ lbs.
c. For grass and broadleaf weed control in nurseries; ornamental plantings in median strips. Use a directional spray.	1 lb.	3¼ lbs.

CAUTION: Lilac, privet, honeysuckle, barberry and boxwood are susceptible to this mixture and may be injured if contacted by spray.

Where johnsongrass or bermudagrass is a problem, a mixture of AMIZOL® brand Industrial Herbicide and Telvar or Karmex is recommended.

III. Use AMIZOL® brand Industrial Herbicide combinations with Karmex or Telvar:	Amounts per Acre	
	AMIZOL®	KARMEX or Telvar
a. For seasonal control of most weeds.	4 lbs.	20 lbs.
b. For economical control of easy-to-control (annual) weeds; for areas with short growing seasons.	2 lbs.	10 lbs.

CAUTION: MIXTURES CONTAINING TELVAR OR KARMEX SHOULD NOT BE APPLIED WHERE MATERIAL WILL WASH OR LEACH INTO THE ROOT ZONE OF VALUABLE TREES OR SHRUBS, SINCE SEVERE DAMAGE MAY RESULT. We recommend using Karmex in high rainfall areas or where Karmex has given best results in the past. Use Telvar in areas of proportionately lower rainfall or where Telvar has given best results in the past.

IV. Use AMIZOL® brand Industrial Herbicide combinations with Sodium TCA 95%:	Amounts per Acre	
	AMIZOL®	Sodium TCA
a. For general control of most grass and broadleaf weeds.	5 lbs.	50 lbs.
b. For control of johnsongrass and bermudagrass, or heavy stands of other grasses.	5 lbs.	100 lbs.

NOTE: Use at least 1 gallon of water for each pound of Sodium TCA applied. Repeat applications will be necessary where annual weeds sprout following earlier applications. The combination of 5 lbs. AMIZOL® brand Industrial Herbicide plus 20 lbs. of Radapon is preferred in low rainfall areas.

V. AMIZOL® brand INDUSTRIAL HERBICIDE IN SPECIAL COMBINATIONS

To control existing vegetation and prevent reinvasion of weeds on railroad ballast and berm areas, highway shoulders and industrial areas, mix 1 to 4 lbs. AMIZOL® brand Industrial Herbicide per acre with one of the following residual herbicides:

AMOUNTS TO USE PER ACRE

	Where Mixtures May Be Used (X)				
	Arid areas or Short Growing Seasons	Humid areas or Full Growing Seasons	Railroad Ballast & Berm	Highway Shoulders	Industrial Areas
Aatrex® 80W (atrazine)	5-10 lbs.	20 lbs.	—	X	X
Hyvar® (bromacil)	4 lbs.	10 lbs.	X	—	X
Karmex® (diuron)	5-10 lbs.	20 lbs.	X	—	X
Krovar® (Bromacil + diuron)	3-6 lbs.	10 lbs.	X	—	—
Princep® 80W (simazine)	5-10 lbs.	20 lbs.	X	—	X
Tandex® (karbutilate)	4 lbs.	10 lbs.	—	—	X
Telvar® (monuron)	5-10 lbs.	20 lbs.	X	—	—

For faster knockdown of annual and perennial broadleaf weeds, add 1 to 4 qts. of 2,4-D ester or amine per acre to the mixtue of Hyvar X or Krovar with AMIZOL®.

PRECAUTION: Mixtures containing Telvar, Karmex, Aatrex, Tandex, Krovar or Hyver X, should not be applied where material will wash or leach into the root zone of valuable trees or shrubs, since severe damage may result.

AMIZOL® brand INDUSTRIAL HERBICIDE FOR WOODY PLANTS

Use AMIZOL® brand Industrial Herbicide in a Highway right-of-way maintenance program as a special tool for cleaning up spot infestations of the species listed below. Most of these are killed to ground line by foliage sprays of 2,4-D, but resprout badly the next year. AMIZOL® is a good translocator and kills most of the roots as well as the tops. AMIZOL® is specific on the species listed. It should not be expected to kill other species.

Use the standard foliage stem method for high volume sprays along fencerows, highway rights-of-way. Spray all foliage and stems of plants from the time foliage is fully developed until plants begin to go dormant. For effective control, all leaves, stems and suckers must be thoroughly wet to ground line.

Amounts to Use

Species	Pounds AMIZOL® per 100 gallons of water	
salmonberry	4	
western dewberry (blackberry)	4	
bigleaf maple	4	Add 4 ounces of spreader sticker per 100 gallons of water.
honeysuckle	4	
kudzu	4	
locust	3 to 4	
ash	3 to 4	
sumac	3 to 4	

USE AMIZOL® brand INDUSTRIAL HERBICIDE COMBINATIONS FOR GENERAL WEED CONTROL

I AMIZOL® alone
II AMIZOL® and simazine (Princep 80W)
III AMIZOL® and monuron or diuron (Telvar or Karmex)
IV AMIZOL® and Sodium TCA 95%
V AMIZOL® in special combinations

Where To Use	Recommended Combinations (in order of preference)
Highways	
guard rails	II, IV, I
bridge abutments	II, III, IV, V
shoulders	V, II, III, IV, I
center strips	II, III, IV, I, V
embankments	I, IV, II, III, V
Railroads	
yards	V, III, II, IV
around buildings	V, II, III
switches	V, III, II
signal equipment	V, III, II
loading areas	V, III, II
ballast, berm	V
Public Utilities	
power substations	II, III, V
pole yards	II, III, V
parking areas	II, III, V
transmission towers	II, III, V
fence lines	II, III, IV
woody plants	I
Industrial Areas	
parking lots	II, III, IV
tank farms	V, III, II
lumber yards	V, II, III
open storage	V, III, II
fence lines	II, III, IV, V
drive-in theaters	II, III, IV, V
Nurseries	
liners	II
ornamentals	II

STORAGE AND DISPOSAL

STORAGE

Do not store near fertilizers, seeds, insecticides, or fungicides.

Do not contaminate water, food, or feed by storage or disposal.

PESTICIDE DISPOSAL

Pesticide wastes are toxic. Improper disposal of excess pesticide, spray mixture, or rinsate is a violation of Federal law. If these wastes cannot be disposed of by use according to label instructions, contact your State Pesticide or Environmental Control Agency or the Hazardous Waste representative at the nearest EPA Regional Office for guidance.

CONTAINER DISPOSAL

Completely empty liner by shaking and tapping sides and bottom to loosen clinging particles. Empty residue into application equipment. Then dispose of liner in sanitary landfill or by incineration if allowed by state or local authorities. If drum cannot be reused, dispose of in the same manner.

LIMITED WARRANTY AND DISCLAIMER

The manufacturer warrants (a) that this product conforms to the chemical description on the label; (b) that this product is reasonably fit for the purposes set forth in the direction for use when it is used in accordance with such directions; (c) and that the directions, warnings and other statements on this label are based upon responsible experts' evaluation of reasonable tests of effectiveness of toxicity to laboratory animals and to plants, and upon reports of field experience. Tests have not been made on all varieties or in all states or under all conditions. THE MANUFACTURER NEITHER MAKES NOR INTENDS, NOR DOES IT AUTHORIZE ANY AGENT OR REPRESENTATIVE TO MAKE ANY OTHER WARRANTIES, EXPRESS OR IMPLIED, AND IT EXPRESSLY EXCLUDES AND DISCLAIMS ALL IMPLIED WARRANTIES OF MERCHANTABILITY OR FITNESS FOR A PARTICULAR PURPOSE.

THIS WARRANTY DOES NOT EXTEND TO, AND THE BUYER SHALL BE SOLELY RESPONSIBLE FOR, ANY AND ALL LOSS OR DAMAGE WHICH RESULTS FROM THE USE OF THIS PRODUCT IN ANY MANNER WHICH IS INCONSISTENT WITH THE LABEL DIRECTIONS, WARNINGS, OR CAUTIONS.

BUYER'S EXCLUSIVE REMEDY AND MANUFACTURER'S OR SELLER'S EXCLUSIVE LIABILITY FOR ANY AND ALL CLAIMS, LOSSES, DAMAGES OR INJURIES RESULTING FROM THE USE OR HANDLING OF THIS PRODUCT, WHETHER OR NOT BASED IN CONTRACT, NEGLIGENCE, STRICT LIABILITY IN TORT OR OTHERWISE, SHALL BE LIMITED, AT THE MANUFACTURER'S OPTION, TO REPLACEMENT OF, OR THE REPAYMENT OF THE PURCHASE PRICE FOR, THE QUANTITY OF PRODUCT WITH RESPECT TO WHICH DAMAGES ARE CLAIMED. IN NO EVENT SHALL MANUFACTURER OR SELLER BE LIABLE FOR SPECIAL, INDIRECT OR CONSEQUENTIAL DAMAGES RESULTING FROM THE USE OR HANDLING OF THIS PRODUCT.

NOTICE TO BUYER

Purchase of this material does not confer any rights under patents governing this product or the use thereof in countries outside of the United States.

AMIZOL is a registered trademark of Rhône-Poulenc

AG 90003

EPA Approval: 1/13/87

THIS SPECIMEN LABEL IS INTENDED FOR USE ONLY AS A GUIDE IN PROVIDING GENERAL INFORMATION REGARDING THE DIRECTIONS, WARNINGS AND CAUTIONS ASSOCIATED WITH THE USE OF THIS PRODUCT. AS WITH ANY AGRICULTURAL CHEMICAL, ALWAYS FOLLOW THE LABEL INSTRUCTIONS ON THE PACKAGE BEFORE USING.

AQUA-KLEEN®

GRANULAR AQUATIC HERBICIDE

For controlling water weeds

ACTIVE INGREDIENT:
*2,4-Dichlorophenoxyacetic acid, butoxyethyl ester 27.6%
INERT INGREDIENTS: 72.4%
TOTAL 100.0%

*2,4-Dichlorophenoxyacetic acid equivalent 19% by weight. Isomer specific by AOAC method No. 6.D01-5

EPA Reg. No. 264-109 AA   EPA Est. No. 264-PA-01

KEEP OUT OF REACH OF CHILDREN
CAUTION

For PRODUCT USE Information Call 1-800-334-9745
For MEDICAL And TRANSPORTATION Emergencies ONLY Call 24 hours A Day 1-800-334-7577

PRECAUTIONARY STATEMENTS

CAUTION

HAZARDS TO HUMANS AND DOMESTIC ANIMALS

Harmful if swallowed or absorbed through skin. Causes slight eye irritation. Avoid contact with skin, eyes, or clothing. Wash thoroughly with soap and water after handling.

STATEMENT OF PRACTICAL TREATMENT

IF SWALLOWED: Call a physician or Poison Control center. If patient is conscious and alert, give 2–3 glasses of water or milk to drink. If available, give one tablespoon of Syrup of Ipecac to induce vomiting. If vomiting has not occurred in 20 minutes, the same dose of Syrup of Ipecac may be repeated one additional time. Alternatively, induce vomiting by touching back of throat with finger. Do not induce vomiting or give anything by mouth to an unconscious person. Get Medical Attention.

IF ON SKIN: Wash with plently of soap and water, if available, while removing contaminated clothing. Wash clothing before reuse.

IF IN EYES: Hold eyelids open and flush with a steady, gentle stream of water for at least 15 minutes. Get medical attention if irritation persists.

NOTE TO PHYSICIAN

No specific antidote is available. All treatments should be based on observed signs and symptoms of distress in the patient. Overexposure to materials other than this product may have occurred.

ENVIRONMENTAL HAZARDS

This product is toxic to fish. Drift or runoff may adversely affect fish and non-target plants. Do not apply to water except as specified on this label. Do not contaminate water when disposing of equipment washwaters.

Do not use in or near a greenhouse. Clean spreader equipment thoroughly before using it for any other purpose. Vapors from this product may injure susceptible plants in the immediate vicinity. Avoid drift of dust to susceptible plants.

GENERAL INFORMATION

AQUA-KLEEN® Aquatic Herbicide contains 2,4-D formulated as the low-volatile butoxyethyl ester. This has been formulated on special heat treated attaclay granules that resist rapid decomposition in water. AQUA-KLEEN® sinks quickly to lake or pond bottoms and releases the weed killing chemical in the critical root zone area.

OXYGEN RATIO

Fish breathe oxygen in the water and a water-oxygen ratio must be maintained. Decaying weeds use up oxygen, but during the period when AQUA-KLEEN® Aquatic Herbicide should be used, the weed mass is fairly sparse and the weed decomposition rate is slow enough so that the water-oxygen ratio is not disturbed by treating the entire area at one time.

If treatments must be applied later in the season when the weed mass is dense and repeat treatments are needed, spread granules in lanes, leaving buffer strips which can then be treated when vegetation in treated lanes has disintegrated. During the growing season, weeds decompose in a 2 to 3 week period following treatment.

Buffer lanes should be 50 to 100 feet wide. Treated lanes should be as wide as the buffer strips.

Buffer lanes should be 50 to 100 feet wide. Treated lanes should be as wide as the buffer strips.

PERMIT TO USE CHEMICALS IN WATER

In many states, permits are required to control weeds by chemical means in public water. If permits are required, they may be obtained from the Chief, Fish Division, State Department of Conservation, or the State Department of Public Health.

DIRECTIONS FOR USE

IT IS A VIOLATION OF FEDERAL LAW TO USE THIS PRODUCT IN A MANNER INCONSISTENT WITH ITS LABELING.

WHEN TO APPLY

For best results, spread AQUA-KLEEN® Aquatic Herbicide in the spring and early summer, during the time weeds start to grow. If desired, this timing can be checked by sampling the lake bottom in areas heavily infested with weeds the year before.

If treatments are delayed until weeds form a dense mat or reach the surface, two treatments may be necessary. Make the second treatment when weeds show signs of recovery, but no later than mid-August.

Occasionally, a second application will be necessary if heavy regrowth occurs or weeds move in from untreated areas.

HOW TO APPLY

FOR LARGE AREAS: Use a fertilizer spreader or mechanical seeder such as the Gerber or Gandy. Before spreading any chemical, calibrate your method of application to be sure of spreading the proper amount. When using boats and power equipment, you must determine the proper combination of (1) boat speed, (2) rate of delivery from the spreader, and (3) width of swath covered by the granules.

FOR SMALL AREAS - AROUND DOCKS OR ISOLATED PATCHES OF WEEDS: Use a portable spreader such as the Cyclone seeder. Estimate or measure out the area you want to treat. Weigh out the amount of material needed and spread this uniformly over the area. More uniform coverage is obtained by dividing the required amount in two and covering the area twice, applying the second half at right angles to the first.

Use the following formula to calibrate your spreader's delivery in pounds of AQUA-KLEEN® Aquatic Herbicide per minute.

miles per hour × spreader width × pounds per acre ÷ 495
= pounds per minute

Example: To apply 100 pounds of AQUA-KLEEN® per acre using a spreader that covers a 20 foot swatch from a boat traveling at 4 miles per hour, set the spreader to deliver 16 pounds of AQUA-KLEEN® granules per minute.

4 mph × 20 feet × 100 Lbs/A ÷ 495 = 16 Lbs/min.

When spreading AQUA-KLEEN® from a boat, we suggest setting out guide posts at frequent intervals marking out the area. This helps to steer a straight course.

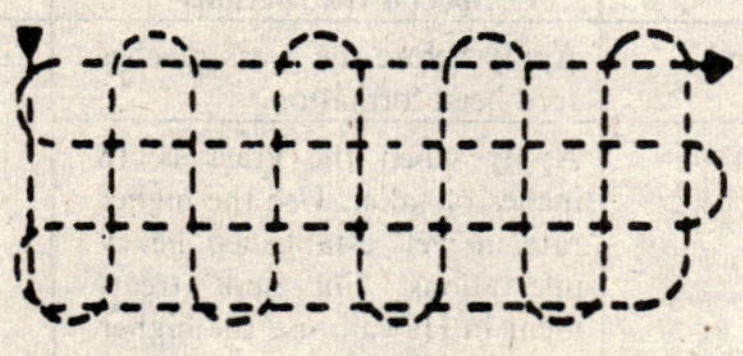

AMOUNTS OF USE

Rates of application vary with the resistance of weed species to the chemical, density of weed mass at time of treatment, water depth, and rate of water flow through the treated area. Use the higher rate for dense weeds, when water is more than 8 feet deep and where there is a large volume turnover.

		AQUA-KLEEN®	
		Pounds Per Acre	Pounds Per 2000 Sq. Ft.
Susceptible Weeds		100	5
Water milfoil	(Myriophyllum spp.)		
Water stargrass	(Heteranthera dubia)		
Slightly To Moderately Resistant Weeds		150 - 200	7½- 10
Bladderwort	(Urticularia spp.)		
White water lilly	(Nymphaea spp.)		
Yellow water lily or spatterdock*	(Nuphar spp.)		
Water shield	(Brasenia spp.)		
Water chestnut	(Trapa natans)		
Coontail*	(Ceratophyllum demersum)		

* Repeat treatments may be needed

This product is designed to control the weeds listed on the label. Control of other weeds may not be satisfactory.

Generally, weeds are difficult to control in lakes where water replacement comes from bottom springs.

WATER pH

Lake water that is extremely acid or alkaline may influence the effectiveness of AQUA-KLEEN® Aquatic Herbicide. A pH on the acid side (pH 6.0 or below) generally favors the action of the chemical while a pH on the alkaline side (pH 8.0 or above) may reduce the action. If regrowth occurs within a period of 6 to 8 weeks, a second application may be needed.

STORAGE AND DISPOSAL

STORAGE

Do not contaminate water, food or feed by storage or disposal.
Do not store near fertilizers, seeds, insecticides or fungicides.

PESTICIDE DISPOSAL

Pesticide wastes are toxic. Improper disposal of excess pesticide, spray mixture, or rinsate is a violation of Federal Law. If these wastes cannot be disposed of by use according to label instructions, contact your State Pesticide or Environmental Control Agency or the Hazardous Waste representative at the nearest EPA Regional Office for guidance.

CONTAINER DISPOSAL

Do not reuse empty bag. Completely empty bag into application equipment. Then dispose of empty bag in a sanitary landfill or by incineration, or if allowed by State and local authorities, by burning. If bag is burned, stay out of smoke.

LIMITED WARRANTY AND DISCLAIMER

The manufacturer warrants (a) that this product conforms to the chemical description on the label; (b) that this product is reasonably fit for the purposes set forth in the directions for use when it is used in accordance with such directions; and (c) that the direction, warnings and other statements on this label are based upon responsible experts' evaluation of reasonable tests of effectiveness of toxicity to laboratory animals and to plants, and of residues on food crops and upon reports of field experience. Tests have not been made on all varieties or in all states or under all conditions. THE MANUFACTURER NEITHER MAKES NOR INTENDS, NOR DOES IT AUTHORIZE ANY AGENT OR REPRESENTATIVE TO MAKE, ANY OTHER WARRANTIES, EXPRESS OR IMPLIED, AND IT EXPRESSLY EXCLUDES AND DISCLAIMS ALL IMPLIED WARRANTIES OF MERCHANTABILITY OR FITNESS FOR A PARTICULAR PURPOSE.

THIS WARRANTY DOES NOT EXTEND TO, AND THE BUYER SHALL BE SOLELY RESPONSIBLE FOR, ANY AND ALL LOSS OR DAMAGE WHICH RESULTS FROM THE USE OF THIS PRODUCT IN ANY MANNER WHICH IS INCONSISTENT WITH THE LABEL DIRECTIONS, WARNINGS, OR CAUTIONS.

BUYER'S EXCLUSIVE REMEDY AND MANUFACTURER'S OR SELLER'S EXCLUSIVE LIABILITY FOR ANY AND ALL CLAIMS, LOSSES, DAMAGES OR INJURIES RESULTING FROM THE USE OR HANDLING OF THIS PRODUCT, WHETHER OR NOT BASED IN CONTRACT, NEGLIGENCE, STRICT LIABILITY IN TORT OR OTHERWISE SHALL BE LIMITED, AT THE MANUFACTURER'S OPTION, TO REPLACEMENT OF, OR THE REPAYMENT OF THE PURCHASE PRICE FOR, THE QUANTITY OF PRODUCT WITH RESPECT TO WHICH DAMAGES ARE CLAIMED. IN NO EVENT SHALL MANUFACTURER OR SELLER BE LIABLE FOR SPECIAL, INDIRECT OR CONSEQUENTIAL DAMAGES RESULTING FROM THE USE OR HANDLING OF THIS PRODUCT.

NOTICE TO BUYER

Purchase of this material does not confer any rights under patents governing this product or the use thereof in countries outside of the United States.

AQUA-KLEEN® is a registered trademark of RHÔNE-POULENC.

AG 90005

EPA Approval: 3/12/80

THIS SPECIMEN LABEL IS INTENDED FOR USE ONLY AS A GUIDE IN PROVIDING GENERAL INFORMATION REGARD-

ING THE DIRECTIONS, WARNINGS AND CAUTIONS ASSOCIATED WITH THE USE OF THIS PRODUCT. AS WITH ANY AGRICULTURAL CHEMICAL, ALWAYS FOLLOW THE LABEL INSTRUCTIONS ON THE PACKAGE BEFORE USING.

ASULOX®

HERBICIDE

For Postemergent Weed Control in Sugarcane, Turf, Ornamentals, Christmas Tree Plantings, and Non-Cropland.

ACTIVE INGREDIENT:
Sodium salt of asulam (methyl sulfanily carbamate)* 36.2%
INERT INGREDIENTS: 63.8%

* Equivalent to 33.1% asulam or not less than 3.34 lbs. per gallon.

KEEP OUT OF REACH OF CHILDREN

CAUTION

EPA Reg. No. 264-447 EPA Est. No. 572-42-FL-001

For PRODUCT USE Information Call 1-800-334-9745
For MEDICAL And TRANSPORTATION Emergencies ONLY Call 24 Hours A Day 1-800-334-7577

PRECAUTIONARY STATEMENTS

CAUTION

HAZARD TO HUMANS AND DOMESTIC ANIMALS

Avoid contact with skin, eyes, or clothing. Wash thoroughly after handling.

STATEMENT OF PRACTICAL TREATMENT

IF ON SKIN: Wash with plenty of soap and water. If irritation persists, get medical attention.

IF IN EYES: Flush with water for 15 minutes and get medical attention if irritation persists.

ENVIRONMENTAL HAZARDS

Do not apply directly to water or wetlands (swamps, bogs, lagoons, marshes or potholes). Do not clean equipment or dispose of equipment washwater in a manner that will contaminate resources. Do not apply when weather conditions favor drift from treated area. Do not contaminate water by cleaning of equipment or disposal of wastes.

DIRECTIONS FOR USE

IT IS A VIOLATION OF FEDERAL LAW TO USE THIS PRODUCT IN A MANNER INCONSISTENT WITH ITS LABELING.
Do not apply ASULOX® Herbicide through any type of irrigation systems.

SUGARCANE

ASULOX® Herbicide can be applied to either plant cane or cane grown from stubble. Apply ASULOX® as a water mix spray for ground applications. Use 15 to 100 gallons of water per acre, depending on local practice. For aerial application, ASULOX® Herbicide should be mixed in 3 to 5 gallons of water per acre, except in Hawaii, where 5 to 10 gallons of water per acre should be used.

Addition of an adijuvant cleared for use on growing crops to the ASULOX® Herbicide water mix spray will improve weed control when environmental conditions are not optimal. Use either a non-ionic surfactant containing a minimum of 80% active ingredient at the rate of 1 to 2 quarts per 100 gallons (0.25% to 0.5% V/V) of water mix spray or a crop oil concentrate containing 80 to 85% paraffin based petroleum oil and 15 to 20% non-ionic surfactant at the rate of 4 quarts per 100 gallons (1% V/V) of water mix spray.

The rates of ASULOX® Herbicide given below are for broadcast applications. For banded application reduce the rate proportionally to the width of the band according to the following formula:

$$\frac{\text{Band width (inches)}}{\text{Row width (inches)}} \times \text{Broadcast rate} = \text{Band rate/acre.}$$

For spot treatments use 6 to 12 pints of ASULOX® in 100 gallons of water and apply in an amount not to exceed 50 gallons of total spray solution per acre. Do not exceed 8 pints per acre treatment.

ASULOX® Herbicide should be used when the weeds are actively growing.

Do not plant crops other than sugarcane in treated area. Cover crops may be planted if plowed under and not grazed.
Do not treat sugarcane within 90 days of harvest.

Do not graze or feed sugarcane fodder and forage to livestock.
Do not rotate with any crop which is not registered for use with Asulam for 1 year following the last application of this chemical.

SUGARCANE

For Single Application Per Growing Season:

Weed Species	Special Instructions	Rate
Itchgrass or Raoulgrass *(Rottboellia exaltata)*	Apply when the grass is 12 inches tall or less (addition of surfactant is necessary).	8 Pints/Acre
Johnsongrass *(Sorghum halepense)*	Apply when the grass is between 18 and 24 inches tall.	
Paragrass or Californiagrass *(Brachiaria mutica or Panicum purpurascens)*	Apply when the grass is 6 to 8 inches tall or less	
Crabgrass *(Digitaria spp.)*	If treatment is made before the grass reaches seed head formation then the lower rate should be used. If the grass is in early seed head formation then the higher rate should be used.	6 to 8 Pints/Acre
Alexandergrass *(Brachiaria plantaginea)* Foxtail *(Setaria spp.)* Goosegrass *(Eleusine indica)* Broadleaf Panicum *(Panicum adspersum)* Barnyardgrass *(Echinochloa crusgalli)*	If treatment is made when the grass is 6 to 8 inches tall or less, then the lower rate should be used. If the grass is greater than 8 inches tall then the higher rate should be used.	

SUGARCANE

Two Applications Per Growing Season

This may be required when initial weed infestations are heavy or reinfestation from seed occurs. Two applications may also be used when treating weed species which germinate at different times during one growing season.

Weed Species	Special Instructions	First Application	Second Application
Crabgrass *(Digitaria spp.)*	At each application the grass should be treated before seed head formation.	6 to 8 Pints/Acre	6 to 8 Pints/Acre
Itchgrass or Raoulgrass *(Rottboellia exaltata)*	At each application the grass should be 12 inches tall or less. (Addition of surfactant is necessary.)	8 Pints/Acre	8 Pints/Acre
Johnsongrass *(Sorghum halepense)*	At each application the grass should be between 18 and 24 inches tall.	8 Pints/Acres	4 to 8 Pints/Acre

NON—CROPLAND

ASULOX® Herbicide may be used as a postemergent treatment to control weeds on non-cropland areas such as:

Boundary fences	Pipeline rights-of-way
Ditchbanks	Railroad rights-of-way and yards
Fence rows	Storage areas and industrial plant sites
Highway and roadside rights-of-way	Utility rights-of-way and yards
Lumberyards	Warehouse lots

A surfactant may be added to the spray solution at 0.25% by volume. (Use an approved non-ionic surfactant.)
Apply ASULOX® as a water-mix spray for ground applications using 20 to 100 gallons of solution per acre, depending on local practice, to control the following weed species:

Weed Species	Special Instructions	Rate
Crabgrass *(Digitaria spp.)*	Apply before the grass reaches seed head formation.	1 to 2 Gal./Acre
Johnsongrass *(Sorghum halepense)*	Apply when the grass is 18 inches or taller. Use the higher rate in well established heavy infestations. For spot treatment in Hawaii, use the higher rate in 100 gallons of solution and apply an amount not to exceed 50 gallons of total solution per acre.	

Weed Species	Special Instructions	Rate
Paragrass or Californiagrass *(Brachiaria mutica = Panicum purpurascens)*	Apply before the grass reaches seed head formation. For spot treatment in Hawaii, use the same rate in 100 gallons of solution and apply an amount not to exceed 50 gallons of total solution per acre.	
Western Bracken *(Pteridium aquilinum var. pubescens)*	Apply when the fern is in full frond.	7 to 8 Pints/Acre

CHRISTMAS TREE PLANTINGS

ASULOX® Herbicide may be used as a postemergent treatment in Christmas Tree Plantings where Douglas Fir, Grand Fir, Nobel Fir or Scotch Pine are grown. Do not graze or feed foliage from treated areas to livestock.

ASULOX® Herbicide should be applied as a water mix spray. For ground application, use a minimum of 20 gallons of solution per acre; and for aerial application, use a minimum of 10 gallons of solution per acre. Do not use a wetting agent with ASULOX® Herbicide.

Weed Species	Special Instructions	Rate
Western Bracken *(Pteridium aquilinum var. pubescens)*	Apply after bud break and hardening or firming of new tree growth. Bracken should be in full frond prior to treatment.	1 Gal./Acre

TURF

ASULOX® Herbicide can be applied on St. Augustinegrass and Tifway 419 Bermudagrass turf. Apply one application per season postemergence to the weeds listed below. Use 20 to 50 gallons of water per acre in the spray solution.

Turf Species	Weed Species	Rate
St. Augustinegrass Tifway 419 Bermudagrass	Bullgrass *(Paspalum supinum)* Crabgrass *(Digitaria sp.)* Goosegrass *(Eleusine indica)* Sandbur *(Cenchrus sp.)*	5 Pints/Acre

Do not use a surfactant. Do not apply to turf which is under stress or freshly mowed.

ORNAMENTALS

ASULOX® Herbicide can be applied as a single, postemergent, broadcast application on the following ornamentals:

Junipers	Yews
Juniperus andorra	*Taxus cuspidata*
Juniperus chinensis	*Taxus media*
Juniperus conferta	*Pondocarpus macrophyllus*
Juniperus horizontalis	
Juniperus litoralis	
Juniperus sabina	

Treatment should be made with a minimum of 20 gallons of water per acre. Do not use a surfactant.

Weed Species		Special Instructions	Rate
Barnyardgrass	*(Echinochloa crus-galli)*	Apply when the weeds are between the stages of early seedling and early seed head formation.	1 Gal/Acre
Crabgrass	*(Digitaria sp.)*		
Fall Panicum	*(Panicum dichotomiflorum)*		
Foxtails	*(Setaria sp.)*		
Goosegrass	*(Eleusine indica)*		
Horseweed (marestail)	*(Conyza canadensis)*		

Local conditions may affect the use of this chemical. Consult State Agricultural Extension or Experiment Station weed specialists for specific recommendations for local weed problems and for information on possible lower dosages.

STORAGE AND DISPOSAL

Do not contaminate water, food, or feed by storage or disposal. Open dumping is prohibited. Store at temperatures above 20° F.

PESTICIDE DISPOSAL: Wastes resulting from the use of this product may be disposed of on site or at an approved waste disposal facility.

CONTAINER DISPOSAL: Triple rinse or equivalent. Puncture and dispose of in a sanitary landfill or incineration, or, if allowed by state and local authorities, by burning. If burned, stay out of smoke.

LIMITED WARRANTY AND DISCLAIMER

The manufacturer warrants (a) that this product conforms to the chemical description on the label; (b) that this product is reasonably fit for the purposes set forth in the directions for use when it is used in accordance with such directions; and (c) that the directions, warnings and other statements on this label are based upon responsible experts evaluation of reasonable tests of effectiveness of toxicity to laboratory animals and to plants, and of residues on food crops, and upon reports of field experience. Tests have not been made on all varieties or in all states or under all conditions. THE MANUFACTURER NEITHER MAKES NOR INTENDS, NOR DOES IT AUTHORIZE ANY AGENT OR REPRESENTATIVE TO MAKE, ANY OTHER WARRANTIES, EXPRESS OR IMPLIED, AND IT EXPRESSLY EXCLUDES AND DISCLAIMS ALL IMPLIED WARRANTIES OF MERCHANTABILITY OR FITNESS FOR A PARTICULAR PURPOSE.

THIS WARRANTY DOES NOT EXTEND TO, AND THE BUYER SHALL BE SOLELY RESPONSIBLE FOR, ANY AND ALL LOSS OR DAMAGE WHICH RESULTS FROM THE USE OF THIS PRODUCT IN ANY MANNER WHICH IS INCONSISTENT WITH THE LABEL DIRECTIONS, WARNINGS OR CAUTIONS.

BUYER'S EXCLUSIVE REMEDY AND MANUFACTURER'S OR SELLER'S EXCLUSIVE LIABILITY FOR ANY AND ALL CLAIMS, LOSSES, DAMAGES OR INJURIES RESULTING FROM THE USE OR HANDLING OF THIS PRODUCT, WHETHER OR NOT BASED IN CONTRACT, NEGLIGENCE, STRICT LIABILITY IN TORT OR OTHERWISE, SHALL BE LIMITED, AT THE MANUFACTURER'S OPTION, TO REPLACEMENT, OF OR THE REPAYMENT OF THE PURCHASE PRICE FOR, THE QUANTITY OF PRODUCT WITH RESPECT TO WHICH DAMAGES ARE CLAIMED. IN NO EVENT SHALL MANUFACTURER OR SELLER BE LIABLE FOR SPECIAL, INDIRECT OR CONSEQUENTIAL DAMAGES RESULTING FROM THE USE OR HANDLING OF THIS PRODUCT.

NOTICE TO BUYER

Purchase of this material does not confer any rights under patents governing this product or the use thereof in countries outside of the United States.

AG90006

ASULOX is a registered trademark of Rhône-Poulenc © 1989 Rhône-Poulenc Ag Company

 EPA Approval: 10/27/88

THIS SPECIMEN LABEL IS INTENDED FOR USE ONLY AS A GUIDE IN PROVIDING GENERAL INFORMATION REGARDING THE DIRECTIONS, WARNINGS AND CAUTIONS ASSOCIATED WITH THE USE OF THIS PRODUCT. AS WITH ANY AGRICULTURAL CHEMICAL, ALWAYS FOLLOW THE LABEL INSTRUCTIONS ON THE PACKAGE BEFORE USING.

BRONATE®

HERBICIDE

RESTRICTED USE PESTICIDE

For retail sale to and use only by Certified Applicators or persons directly under their supervision and only for those uses covered by the Certified Applicator's certification. See precautionary statements for reasons this product is classified Restricted Use.

FOR CONTROL OF CERTAIN BROADLEAF WEEDS IN WHEAT, BARLEY, OATS, AND RYE.

ACTIVE INGREDIENT:

Octanoic acid ester of bromoxynil*(3,5-dibromo-4-hydroxybenzonitrile) 31.7%

Isooctyl ester of 2-methyl-chlorophenoxyacetic acid** 34.0%

INERT INGREDIENTS: 34.3%

*Bromoxynil octanoate equivalent to 21.8% of bromoxynil or not less than 2.0 pounds of bromoxynil per gallon.

**Equivalent to 21.8% 2-methyl-chlorophenoxyacetic acid or not less than 2.0 pounds MCPA acid per gallon.

EPA Reg. No. 264-438 EPA Est. No.

KEEP OUT OF REACH OF CHILDREN

WARNING

For PRODUCT USE Information Call 1-800-334-9745
For MEDICAL And TRANSPORTATION Emergencies ONLY Call 24 Hours A Day 1-800-334-7577

PRECAUTIONARY STATEMENTS

HAZARDS TO HUMANS AND DOMESTIC ANIMALS

WARNING - RESTRICTED USE

Use of this product may be hazardous to your health. Exposure to this product has been shown to cause birth defects in laboratory animals. Women of childbearing age should be particularly careful when handling this product to avoid ingestion & skin contact.

Harmful if swallowed, absorbed through skin, or inhaled. Avoid contact with skin, eyes or clothing. Avoid breathing spray mist.

Users of this product are encouraged to participate in Rhône-Poulenc's educational program dealing with potential hazards and safer use of this product. Further information on this program can be obtained by contacting Rhônc-Poulenc Ag Company at 1-800-334-9745.

ALL PERSONS HANDLING THE CONCENTRATE MUST OBSERVE THE FOLLOWING REQUIREMENTS.

Wear clean cotton (or cloth) coveralls that cover all parts of the body except the head, hands, and feet. The coveralls must be worn over a long sleeve shirt and long pants. Wear clean nitrile goves, a chemical resistant apron, goggles or face shield, and chemical resistant shoes, shoe coverings, or boots. If you use a mechanical transfer system for all mixing and loading operations, use of a chemical resistant apron is optional.

If you will handle a total of 30 gallons or more of this product per day, you must use a mechanical transfer system for all mixing and loading operations. If this product is packaged in a 30 gallon drum, you must use a mechanical transfer system which terminates in a drip-free hard coupling which may be used only with a spray or mix tank which has been fitted with a compatible coupling. If you do not presently own or have access to a mechanical transfer system with this type of coupling, contact your dealer for information on how to obtain such a system or to modify your present system. When using a mechanical transfer system, do not remove or disconnect the pump or probe from the container until the container has been emptied and rinsed. The pump or probe system must be used to rinse the empty container and to transfer the rinsate directly to the mixing or spray tank.

ALL PERSONS APPLYING THIS PRODUCT, OR REPAIRING OR CLEANING EQUIPMENT USED WITH THIS PRODUCT, MUST OBSERVE THE FOLLOWING REQUIREMENTS.

When applying this product from a tractor (unless the tractor has a completely enclosed cab), or when repairing or cleaning equipment used with the product, you must wear clean nitrile gloves and clean cotton (or cloth) coveralls that cover all parts of the body except the head, hands, and feet. Coveralls must be worn over a long sleeve shirt and long pants.

All applicators (except for pilots) and all persons repairing or cleaning equipment used with this product must wear chemical resistant shores, shoe coverings, or boots.

Application from a tractor with a completely enclosed cab or aerial application is required whenever this product is applied to 180 or more acres in a day. To avoid contamination, coveralls and gloves worn when handling the concentrate must be removed prior to entering an enclosed cab or cockpit. When applying from a tractor with an enclosed cab, clean coveralls and clean nitrile gloves must be kept inside the cab, and must be worn when exiting the cab to perform in-field maintenance or repair.

To reduce exposure to residues, wash the spray rig, tractor, and all other equipment used to handle or apply this product with water daily or before using the equipment for any other purpose.

APPLICATION BY CHEMIGATION must be done by fixed pipe, overhead sprinkler systems or hand moved pipe. If hand moved pipe is used for chemigation, the pipe must not be handled in any way until 24 hours after chemigation has been completed and residues have been flushed from the system. When applying by chemigation, no person may enter the application site unless in an enclosed vehicle.

DURING AERIAL APPLICATION, human flaggers are prohibited unless in enclosed vehicles. Aerial application is prohibited within 300 feet of residential areas (e.g., homes, schools, hospitals, shopping areas, etc.)

GENERAL REQUIREMENTS FOR ALL USE

IMPORTANT! Before removing gloves or starting a new work operation, rinse the outside of the gloves thoroughly with water. Always remove gloves and wash your hands and face with soap and water before smoking, eating, drinking, or toileting.

AFTER WORK. Gloves must be discarded at the end of each day. Take off all clothes and shoes. Shower with soap and water. Do not wear or re-use contaminated clothing. Wash protective equipment with water after each use. Protective clothing worn during use must be laundered separately from other household articles.

STATEMENT OF PRACTICAL TREATMENT

IF SWALLOWED: Get medical attention. Do not induce vomiting, contains petroleum distillates. Do not give anything by mouth to an unconscious person.

IF ON SKIN: Wash with plenty of soap and water. Get medical attention if irritation persists.

IF IN EYES: Flush with water for 15 minutes. Get medical attention.

ENVIRONMENTAL HAZARDS

This pesticide is toxic to wildlife and fish. Use with care when applying to areas frequented by wildlife or adjacent to any body of water. Do not apply directly to water. Do not apply when weather conditions favor drift from target areas. Do not contaminate water when disposing of equipment washwaters.

PHYSICAL OR CHEMICAL HAZARDS

Do not use or store near heat or open flame.

IT IS A VIOLATION OF FEDERAL LAW TO USE THIS PRODUCT IN A MANNER INCONSISTENT WITH ITS LABELING.

GENERAL INFORMATION

BRIONATE® Herbicide is a selective postemergence product for control of broadcast weeds in small grains. To broaden the spectrum of control, BRONATE® Herbicide may be tankmixed with Glean®, Avenge®, or MCPA. To ensure maximum safety and weed control, for all cautions and limitations on the label and the labels of products used in a tank mixture with BRONATE® Herbicide.

GENERAL WEED LIST

Postemergence application of BRONATE® Herbicide will control the following weeds when sprayed in the seedling stage. Maximum weed stage of growth is listed under BRONATE RECOMMENDATIONS.

MOST SUSCEPTIBLE BROADLEAF WEEDS

Annual sowthistle	(Sonchus oleraceus)
Black mustard	(Brassica nigra)
Black nightshade	(Solanum nigrum)
Common cocklebur	(Xanthium pensylvanicum)
Common lambsquarters	(Chenopodium album)
Common tarweed	(Hemizonia congesta)
Cow cockle	(Saponaria vaccaria)
Eastern black nightshade	(Solanum ptycanthum)
Coast fiddleneck	(Amsinckia intermedia)
Field pennycress	(Thlaspi arvense)
Green smartweed	(Polygonum scabrum)
Hairy nightshade	(Solanum sarachoides)
Jimsonweed	(Datura stramonium)
Ladysthumb	(Polygonum persicaria)
London rocket	(Sisymbrium irio)
Marshelder	(Iva xanthifolia)
Pennsylvania smartweed	(Polygonum pensylvanicum)
Pepperweed spp.	(Lepidium app.)
Redroot pigweed	(Amaranthus retroflexus)
Russian thistle	(Salsola kali)
Shepherdspurse	(Capsella bursa-pastoris)
Silverleaf nightshade	(Solanum elaeagnifolium)
Smooth pigweed	(Amaranthus hybridus)
Spiny pigweed	(Amaranthus spinosus)
[1]Sunflower	(Helianthus annuus)
Tall Waterhemp	(Amaranthus tuberculatus)
Tartary buckwheat	(Fagopyrum tataricum)
Tumble mustard	(Sisymbrium altissimum)
Wild buckwheat	(Polygonum convolvulus)
Wild mustard	(Brassica kaber)
Yellow rocket	(Barbarea vulgaris)

[1] For control of sunflower, delay application until first sunflower seedlings emerging are 4 inches in height.

SUSCEPTIBLE BROADLEAF WEEDS

Blue (purple) mustard	(Chlorispora tenella)
Common groundsel	(Senecio vulgaris)
Common ragweed	(Ambrosia artemisiifolia)
Corn chamomile	(Anthemis arvensis)
Corn gromwell	(Lithospermum arvense)
Fumitory	(Fumaria officinalis)
Giant ragweed	(Ambrosia trifida)
Hemp sesbania	(Sesbania exaltata)
Henbit	(Lamium amplexicaule)
Ivyleaf morningglory	(Ipomoea hederacea)
Knawel	(Scleranthus annuus)
Kochia	(Kochia scoparia)
Mayweed	(Anthemis cotula)
Prostrate knotweed	(Polygonum aviculare)
Tall morningglory	(Ipomoea purpurea)
Tansy mustard	(Descurainia pinnata)
Tarweed	(Hemizonia spp.)
Velvetleaf	(Abutilon theophrasti)
Wild radish	(Raphanus raphanistrum)

Weeds germinating after spraying will not be controlled.

WEED SUPPRESSION

Canada thistle (Cirsium arvense)

BRONATE® Herbicide applied at 1½ pints per acre provides burn down of top growth. Regrowth may occur. Make applications when Canada thistle is 8 inches tall to the bud stage.

WHEAT, BARLEY, OATS, AND RYE

BRONATE® Herbicide may be applied to wheat and barley from the 3 leaf stage up to the boot stage. Applications should be made to weeds soon after emergence for best weed control and to prevent competition to the crop. Weed control may be reduced when weeds are stressed from lack of moisture and cold temperatures. Avoid spray drift to non-target areas. Do not graze treated fields for 30 days following application.

BRONATE® Herbicide can be applied to wheat and barley alone or in a tank mixture with Glean®, Avenge®, or MCPA to broaden the spectrum of weed control. BRONATE® can also be applied to oats and rye alone or in a tank mixture with MCPA. To ensure maximum safety and weed control, following all cautions and limitations on this label and the labels of products used in tank mixtures with BRONATE® Herbicide.

MIXING, LOADING AND HANDLING DIRECTIONS

2.5 Gallons Containers

It is strongly recommended that special care be taken in mixing and loading this product. Hands should be placed on the container in such a way as to avoid possible drip or splash. Correct procedures for mixing and loading are provided in Rhône-Poulenc Ag Company's Educational Program.

30 Gallon and Bulk Containers

If you will handle a total of 30 gallons or more of this product per day, you must use a mechanical transfer system for all mixing and loading operations. If this product is packaged in a 30 gallon drum, you must use a mechanical transfer system which terminates in a drip-free hard coupling which may be used only with a spray or mix tank which has been fitted with a compatible coupling. If you do not presently own or have access to a mechanical transfer system with this type of coupling, contact your dealer for information on how to obtain such a system or to modify your present system. When using a mechanical transfer system, do not remove or disconnect the pump or probe from the container until the container has been emptied and rinsed. The pump or probe system must be used to rinse the empty container and to transfer the rinsate directly to the mixing or spray tank.

BRONATE® HERBICIDE ALONE: Fill the spray tank ½ to ¾ full with clean water. Begin agitation and add the recommended amount of BRONATE®. Add water to the spray tank to the desired level. Maintain sufficient agitation to ensure a uniform spray mixture during application.

TANK MIXTURES: Fill the spray tank ½ to ¾ full with clean water and begin agitation. If tank mixing with Glean®, add the recommended amount of Glean® first. After Glean® is thoroughly mixed with water, add the recommended amount of BRONATE® Herbicide and add water to the spray tank to the desired level. If tank mixing with Avenge® or MCPA, add the recommended amount of BRONATE® Herbicide first. After BRONATE® is thoroughly mixed with water, add the recommended amount of the other herbicide to the spray tank and add water to the spray tank to the desired level. Maintain sufficient agitation while mixing and during application to ensure a uniform spray mixture.

APPLICATION PROCEDURES

BRONATE® Herbicide alone or in a tank mixture with Glean®, Avenge®, or MCPA can be applied by both ground and aerial equipment.

GROUND APPLICATION

Use a standard herbicide boom sprayer that provides uniform and accurate application. Sprayer should be equipped with flat fan nozzles spaced a maximum of 20 inches apart and screens no finer than 50 mesh in the nozzle tips and in-line strainers.

Applications of BRONATE® Herbicide used alone and in tank mixtures should be used in a minimum spray volume of 10 gallons per acre with a minimum spray pressure of 30 psi at the nozzle tips.

AERIAL APPLICATION

During aerial application, human flaggers are prohibited unless in enclosed vehicles. Aerial application is prohibited within 300 feet of residential areas (e.g., homes, schools, hospitals, shipping areas, etc.)

BRONATE® Herbicide and tank mixtures with BRONATE® can be applied by aerial equipment. Use a minimum of 5 gallons of spray volume per acre using diaphragm-type nozzles that produce cone or fan spray patterns with a maximum spray pressure of 40 psi at the nozzle tips. Nozzles should be located no farther than ¾ the distance from the center of the aircraft to the end of the wing or rotor. Nozzles should be oriented at an angle between straight down and straight back.

The aircraft should discharge the spray a maximum of 10 feet above the crop. To minimize drift, do not apply during periods of gusty winds or when wind exceeds 5 mph.

SPRAYABLE LIQUID FERTILIZERS

BRONATE® Herbicide can be applied to small grains using sprayable liquid fertilizer as the carrier. Predetermine the compatibility with the liquid fertilizer by mixing small proportional quantities in advance. Agitation must be maintained during filling and application operations to ensure that BRONATE® is evenly mixed with the fertilizer.

RECOMMENDED USES

The recommended rate of BRONATE® Herbicide used alone or in a tank mixture with another herbicide is determined by the crop (fall or spring seeded), geographic area, application equipment (ground or aerial application), weed species, and stage of growth of weeds at time of application.

BRONATE® HERBICIDE RECOMMENDATIONS

WHEAT AND BARLEY
(Ground And Aerial Application)

PRODUCT	CROP/GEOGRAPHICAL AREA	RATE	TIMING OF APPLICATION: CROP	TIMING OF APPLICATION: WEEDS
BRONATE®	Fall seeded wheat and barley throughout the United States and spring seeded wheat and barley in Idaho, Oregon, and Washington.	1 Pint/A	Apply to wheat and barley after the 3 leaf stage but before the crop reaches the boot stage.	MOST SUSCEPTIBLE BROADLEAF WEEDS Apply to weeds up to the 8 leaf stage or 4 inches in height, whichever comes first. If weed forms rosette, apply before weeds exceed 2 inches in diameter.
		1½ Pints/A		SUSCEPTIBLE BROADLEAF WEEDS Apply to weeds up to the 4 leaf stage or 2 inches in height, whichever comes first. If weed forms rosette, apply before weeds exceed 1 inch in diameter.
	Spring seeded wheat and barley except Idaho, Oregon and Washington.	1 Pint/A		MOST SUSCEPTIBLE AND SUSCEPTIBLE BROADLEAF WEEDS Apply to weeds that do not exceed the 8 leaf stage or 4 inches in height, whichever comes first. If weed forms rosette, apply before weeds exceed 2 inches in diameter. Apply to kochia up to 2 inches in height.
		1½ Pints/A		Apply to control kochia that is 2 to 4 inches in height.

BRONATE® HERBICIDE TANK MIXTURE RECOMMENDATIONS

PRODUCT	CROP/GEOGRAPHICAL AREA	RATE	TIMING OF APPLICATION CROP	TIMING OF APPLICATION WEEDS
BRONATE® + Glean® (Tankmix)	Fall seeded wheat and barley throughout the United States and spring seeded wheat and barley in Idaho, Oregon and Washington.	3/4 Pint/A + 1/6-1/3 Ounce/A	Apply to wheat and barley after the 2 leaf stage but before the crop reaches the boot stage.	MOST SUSCEPTIBLE BROADLEAF WEEDS Apply to weeds up to the 8 leaf stage or 4 inches in height, whichever comes first. If weed forms rosette, apply before weeds exceed 2 inches in diameter.
		1 Pint/A + 1/6-1/3 Ounce/A		SUSCEPTIBLE BROADLEAF WEEDS Apply to weeds up to the 4 leaf stage or 2 inches in height, whichever comes first. If weed forms rosette, apply before weeds exceed 1 inch in diameter.
	Spring seeded wheat and barley except Idaho, Oregon, and Washington.	1 Pint/A + 1/6-1/3 Ounce/A		MOST SUSCEPTIBLE AND SUSCEPTIBLE BROADLEAF WEEDS Apply to weeds that do not exceed the 8 leaf stage or 4 inches in height, whichever comes first. If weed forms rosette, apply before weeds exceed 2 inches in diameter. Apply to kochia up to 2 inches in height.
		1 1/2 Pints/A + 1/6-1/3 Ounce/A		Apply to control kochia that is 2 to 4 inches in height.

A tankmix application of BRONATE® Herbicide and Glean® broadens the spectrum of weed control. See Glean label for listing of additional susceptible weed species and crop rotation instructions.

Glean® is a trademark of E.I. Dupont de Nemours & Company, Inc.

BRONATE® HERBICIDE AND AVENGE

A tank mixture of BRONATE® Herbicide + Avenge can be used to control wild oats in addition to the broadleaf weeds controlled by BRONATE®. The rate and application timing for weed control with BRONATE® Herbicide in the tank mixture is the same as when BRONATE® is used alone. See BRONATE® HERBICIDE RECOMMENDATIONS.

Apply BRONATE® Herbicide + Avenge tank mixtures when the majority of the wild oats are in the 3 to 5 leaf stage of growth. This frequently coincides with barley in the 3 to 7 leaf stage, spring seeded wheat in the 5 to 6 leaf stage, and fall seed wheat in the 4 leaf to tillered stage of growth. See Avenge label for varietal restrictions and recommended Avenge rates as related to wild oat density.

Avenge® is a trademark of American Cyanamid Company.

BRONATE® HERBICIDE AND MCPA

A tank mixture of BRONATE® Herbicide at the recommended rate and MCPA at 1/4 lb. ai/A is recommended when the weed infestations of MCPA susceptible weeds are heavy or the weeds are at the maximum recommended stage of growth for BRONATE®. The tank mixture will control the weeds listed on the MCPA label in addition to the weeds listed on the BRONATE® Herbicide label. The rate of BRONATE® Herbicide in the tank mixture is the same as when BRONATE® is used alone. See BRONATE® HERBICIDE RECOMMENDATIONS.

A tank mixture of BRONATE® Herbicide + MCPA can be applied after the 3 leaf stage but before the crop reaches the boot stage.

OATS AND RYE (Ground And Aerial Application)

BRONATE® Herbicide alone and in a tank mixture with MCPA can be applied to oats and rye.

Recommendations for BRONATE® Herbicide alone and in a tank mixture with MCPA for use on oats and rye are the same as for wheat and barley.

BRONATE® HERBICIDE WEED CONTROL IN COVER CROPS ON CONSERVATION RESERVE ACRES

DIRECTIONS FOR USE

CROPS: Crested wheatgrasses (all), tall wheatgrass, western wheatgrass, bluebunch wheatgrass, intermediate wheatgrass, sheep fescue, Russian wildrye, and Canby bluegrass.

TARGET PEST: BRONATE® Herbicide will control all broadleaf weeds mentioned in the General Weed List found on the package label. Follow all use directions on this label and on the BRONATE® Herbicide package label to ensure product performance, and crop safety.

WHEN TO APPLY

BRONATE® Herbicide may be applied with standard ground boom and aerial hydraulic spray systems to forage crops listed here anytime after the 2 to 3 leaf stage, but before weeds exceed recommended stages of growth for treatment. Refer to USE RECOMMENDATIONS below for further instructions.

HOW TO APPLY

Ground Application: Use a standard herbicide boom sprayer that provides uniform and accurate application. Sprayer should be equipped with flat fan nozzles spaced a maximum of 20 inches apart and screens no finer than 50 mesh in the nozzle tips and in-line strainers. Apply in a minimum spray volume of 10 gallons per acre with minimum spray pressure of 30 psi at the nozzle tips.

Aerial Application: The sprayer should be equipped to provide a uniform and accurate application. Use a minimum spray volume of 5 gallons per acre using diaphragm-type nozzles that produce cone or fan spray patterns with a maximum spray pressure of 40 psi at the nozzle tips. Nozzles should be located no further than 3/4 the distance from the center of the aircraft to the end of the wing or rotor. Nozzles should be oriented at an angle between straight down and straight back. The aircraft should discharge the spray a maximum of 10 feet above the crop. To minimize drift, do not apply during periods of gusty winds or when wind exceeds 5 mph.

In order to ensure maximum safety and weed control, follow label recommendations on this label and all the cautions and limitations on the package labels of the tank mixture components.

PRODUCT	BRONATE® HERBICIDE WEED CONTROL IN COVER CROPS ON CONSERVATION RESERVE ACRES — USE RECOMMENDATIONS TIMING OF APPLICATION — CROP	WEED[1]
1 1/2 Pints/A	Apply anytime in the fall or spring after the crop has reached the 2 to 3 leaf stage.	MOST SUSCEPTIBLE AND SUSCEPTIBLE BROADLEAF WEEDS Apply to weeds up to the 8 leaf stage or 4 inches in height, whichever comes first. If weed forms rosette, apply before weeds exceed 2 inches in diameter.

[1] Refer to General Weed List on the BRONATE® Herbicide label for a complete list of SUSCEPTIBLE WEEDS.

USE RESTRICTIONS

Do not allow livestock to graze treated fields.

Do not feed treated grasses to livestock.

Mix with clean water only, do not use spray adjuvants or fluid fertilizers as crop injury may occur.

Do not apply to alfalfa.

BRONATE® AND METRIBUZEN HERBICIDE (SENCOR®/LEXONE®) TANK MIX

DIRECTIONS FOR USE CROP

Fall seeded wheat and barley, and spring seeded barley. Use only in the states of Idaho, Oregon, and Washington.

TARGET PEST

Henbit, ivyleaf speed well, mayweed and tansy mustard up to the 4 leaf stage or 2 inches in height or diameter whichever comes first, in addition to the weeds controlled by BRONATE®. See package label for list of weeds controlled by BRONATE®.

RATE

Apply BRONATE® at 1 pint per acre and metribuzen (Sencor/Lexone) at 1/8 - 3/16 lb ai/A.

APPLICATION

Apply in the spring after wheat and barley has started to grow and has a well established secondary root system with at least 3 to 4 tillers, but before the forming of joints in the stem.

GROUND APPLICATION: Use a standard boom sprayer that provides uniform and accurate application. Sprayer should be equipped with flat fan nozzles spaced a maximum of 20 inches apart and screens no finer than 50 mesh in the nozzle tips and in-line strainers. Apply in a minimum spray volume of 10 gallons per acre with minimum spray pressure of 30 psi at the nozzle tips.

AERIAL APPLICATION: The sprayer should be equipped to provide a uniform and accurate application. Use a minimum spray volume of 5 gallons per acre using diaphragm - type nozzles that produce cone or fan spray patterns with a maximum spray pressure of 40 psi at the nozzle tips. Nozzles should be located no further than 3/4 the distance from the center of the aircraft to the end of the wing or rotor. Nozzles should be oriented at an angle between straight down and straight back. The aircraft should discharge the spray a maximum of 10 feet above the crop. To minimize drift, do not apply during periods of gusty winds or when wind exceeds 5 mph.

In order to ensure maximum safety and weed control, follow label recommendations on this label and all the cautions and limitations on the package labels of the tank mixture components. Sencor is a registered trademark of the parent company of Farbenfabriken Bayer Gmbh, Leverkusen. Lexone is a registered trademark of E.I. Du Pont de Nemours Company, Inc.

AUTOMATED SPRINKLER IRRIGATION SYSTEMS

If hand-moved pipe is used for chemigation, the pipe must not be handled in any way until 24 hours after chemigation has been completed and residues have been flushed from the system. When applying by chemigation, no person may enter the application site unless in an enclosed vehicle.

Directions For Use: Fall and spring seeded wheat and barley.

Use only in the states of Colorado, Idaho, Kansas, Montana, Nebraska, Nevada, Oregon, South Dakota, Washington, and Wyoming.

Refer to the package label for MOST SUSCEPTIBLE AND SUSCEPTIBLE WEED species.

BRONATE® Herbicide may be applied through automated sprinkler irrigation systems to wheat and barley after the 3 leaf stage, but before the crop reaches the boot stage.

Use 2 pints per acre in 0.25 to 0.5 acre-inch of water. **BRONATE® at 2 pints/A must be applied with a mechanical transfer system only. See MIXING, LOADING AND HANDLING INSTRUCTIONS section for complete details. Apply to weeds up to the 4 leaf stage or 2 inches in height, whichever comes first. If weed forms rosette, apply before weeds exceed 1 inch in diameter.**

AUTOMATED SPRINKLER IRRIGATION SYSTEMS

Apply BRONATE® Herbicide through sprinkler including center pivot, lateral move, side (wheel) roll, solid set or hand move irrigation systems only. Do not apply this product through any other type of irrigation system.

SPECIFIC REQUIREMENTS FOR APPLICATION THROUGH AUTOMATED SPRINKLER IRRIGATION SYSTEMS.

1. The system must contain a functional check valve, vacuum relief valve, and low pressure drain appropriately located on the irrigation pipeline to prevent water source contamination from backflow.
2. The pesticide injection pipeline must contain a functional, automatic quick-closing check valve to prevent the flow of fluid back toward the injection pump.
3. The pesticide injection pipeline must also contain a functional, normally closed, solenoid-operated valve located on the intake side of the injection pump and connected to the system interlock to prevent fluid from being withdrawn from the supply tank when the irrigation system is either automatically or manually shut down.
4. The system must contain functional interlocking controls to automatically shut off the pesticide injection pump when the water pump motor stops.
5. The irrigation line or water pump must include a functional pressure switch which will stop the water pump motor when the water pressure decreases to the point where pesticide distribution is adversely affected.
6. Systems must use a metering pump, such as a positive displacement injection pump (e.g., diaphragm pump) effectively designed and constructed of materials that are compatible with pesticides and capable of being fitted with a system interlock.
7. Do not apply when wind speed favors drift beyond the area intended for treatment.
8. Agitation is recommended in the pesticide supply tank when applying the BRONATE® Herbicide.
9. BRONATE® Herbicide should be applied continuously for the duration of the water application with center pivot and continuous lateral move systems. Application of BRONATE® should be made during the last 30 to 45 minutes of the irrigation set with other overhead sprinkler system.
10. For best performance, set the sprinkler system to deliver approximately 0.5 inch or less of water per acre.
11. Remove scale, pesticide residues and other foreign matter from the supply tank and entire injector system. Flush with clean water.
12. If BRONATE® Herbicide is diluted in the supply tank, fill the tank with half of the water amount desired, add the BRONATE® and then add remaining water amount with agitation.
13. Start the sprinklers and then inject BRONATE® Herbicide into the irrigation line. BRONATE® should be injected with a positive displacement pump into the main line at least 8 feet ahead of a right angle turn to ensure adequate mixing. Refer to the BRONATE® Herbicide label for detailed information on application rates and timings.

CHEMIGATION USER PRECAUTIONS

Application of more than 0.5 inch/acre of irrigation water may result in decreased product performance on certain soils.

Do not apply when conditions favor drift, when system connections or fittings leak, or when nozzles do not provide uniform distribution.

Allow sufficient time for pesticide to be flushed through all the lines and nozzles before turning off irrigation water.

Crop injury, lack of effectiveness, or illegal pesticide residues in the crop can result from non-uniform distribution of treated water.

Do not connect an irrigation system used for pesticide application to a public water system.

If you have questions about calibration, you should contact State Extension Service Specialists, equipment manufactures or other experts.

A person knowledgeable of the chemigation system and responsible for its operation, or under the supervision of the responsible person, shall shut the system down and make necessary adjustments should the need arise.

STORAGE AND DISPOSAL

STORAGE

Do not contaminate water, food or feed by storage or disposal. Store at temperatures above 3°F. If allowed to freeze, remix before using.

PESTICIDE DISPOSAL

Wastes resulting from the use of this product may be disposed of on site or at an approved waste disposal facility.

CONTAINER DISPOSAL

Triple rinse (or equivalent). Then offer for recycling or reconditioning, or puncture and dispose of in a sanitary landfill, or incineration or if allowed by state and local authorities, by burning. If burned, stay out of smoke.

LIMITED WARRANTY AND DISCLAIMER

The manufacturer warrants that this product conforms to the chemical description on the label; that this product is reasonably fit for the purposes set forth in the directions for use when it is used in accordance with such directions; that the directions, warnings, and other statements on this label are based upon responsible experts' evaluation of reasonable tests of effectiveness, of toxicity to laboratory animals and to plants, and of residues on food crops, and upon reports of field experience. Tests have not been made on all varieties or in all states or under all conditions. THE MANUFACTURER NEITHER MAKES NOR INTENDS, NOR DOES IT AUTHORIZE ANY AGENT OR REPRESENTATIVE TO MAKE, ANY OTHER WARRANTIES, EXPRESS OR IMPLIED, AND IT EXPRESSLY EXCLUDES AND DISCLAIMS ALL IMPLIED WARRANTIES OF MERCHANTABILITY OR FITNESS FOR A PARTICULAR PURPOSE.

THIS WARRANTY DOES NOT EXTEND TO, AND THE BUYER SHALL BE SOLELY RESPONSIBLE FOR, ANY AND LOSS OR DAMAGE WHICH RESULTS FROM THE USE OF THIS PRODUCT IN ANY MANNER WHICH IS INCONSISTENT WITH THE LABEL DIRECTIONS, WARNINGS OR CAUTIONS.

BUYER'S EXCLUSIVE REMEDY AND MANUFACTURER'S OR SELLER'S EXCLUSIVE LIABILITY FOR ANY AND ALL CLAIMS, LOSSES, DAMAGES, OR INJURIES RESULTING FROM THE USE OR HANDLING OF THIS PRODUCT, WHETHER OR NOT BASED IN CONTRACT, NEGLIGENCE, STRICT LIABILITY IN TORT OR OTHERWISE, SHALL BE LIMITED, AT THE MANUFACTURER'S OPTION, TO REPLACEMENT OF, OR THE REPAYMENT OF THE PURCHASE PRICE FOR, THE QUANTITY OF PRODUCT WITH RESPECT TO WHICH DAMAGES ARE CLAIMED. IN NO EVENT SHALL MANUFACTURER OR SELLER BE LIABLE FOR SPECIAL, INDIRECT OR CONSEQUENTIAL DAMAGES RESULTING FROM THE USE OR HANDLING OF THIS PRODUCT.

NOTICE TO BUYER

Purchase of this material does not confer any rights under patents governing this product or the use thereof in countries outside of the United States.

BRONATE is a registered trademark of Rhône-Poulenc

AG90007

THIS SPECIMEN LABEL IS INTENDED FOR USE ONLY AS A GUIDE IN PROVIDING GENERAL INFORMATION REGARDING THE DIRECTIONS, WARNINGS AND CAUTIONS ASSOCIATED WITH THE USE OF THIS PRODUCT. AS WITH ANY AGRICULTURAL CHEMICAL, ALWAYS FOLLOW THE LABEL INSTRUCTIONS ON THE PACKAGE BEFORE USING.

BUCTRIL®
HERBICIDE

RESTRICTED USE PESTICIDE

For retail sale to and use only by Certified Applicators or persons directly under their supervision and only for those uses covered by the Certified Applicator's certification. See precautionary statements for reasons this product is classified Restricted Use.

For the control of certain broadleaf weeds in small grains, seedling alfalfa, corn, sorghum, flax, garlic, onion, annual canarygrass and mint.

ACTIVE INGREDIENT:
Octanoic acid ester of bromoxynil* (3,5-dibromo-4-hydroxybenzonitrile) 33.4%
INERT INGREDIENTS: 66.6%
TOTAL 100.0%

* Bromoxynil octanoate equivalent to 22.9% of bromoxynil or not less than 2.0 pounds of bromoxynil per gallon.

E.P.A. Reg. No. 264-437 E.P.A. Est. No. (See last page)

KEEP OUT OF REACH OF CHILDREN
AVISO WARNING

PRECAUTION AL USUARIO: Si usted no lee ingles, no use este producto hasta que le etiqueta haya sido explicado ampliamente

For PRODUCT USE Information Call 1-800-334-9745.
For MEDICAL And TRANSPORTATION Emergencies ONLY Call 24 Hours A Day 1-800-334-7577.

PRECAUTIONARY STATEMENTS

HAZARD TO HUMANS AND DOMESTIC ANIMALS

WARNING- RESTRICTED USE

Use of this product may be hazardous to your health. Exposure to this product has been shown to cause birth defects in laboratory animals. Women of childbearing age should be particularly careful when handling this product to avoid ingestion and skin contact.
Harmful if swallowed or absorbed through skin or inhaled. Avoid contact with skin, eyes or clothing. Avoid breathing spray mist.

Users of this product are encouraged to participate in Rhône-Poulenc's educational program dealing with potential hazards and safer use of this product. Further information on this program can be obtained by contacting Rhône-Poulenc Ag Company at 1-800-334-9745.

ALL PERSONS HANDLING THE CONCENTRATE MUST OBSERVE THE FOLLOWING REQUIREMENTS.

Wear clean cotton (or cloth) coveralls that cover all parts of the body except the head, hands, and feet. The coveralls must be worn over a long sleeve shirt and long pants. Wear clean nitrile gloves, a chemical resistant apron, goggles or face shield, and chemical resistant shoes, shoe coverings, or boots. If you use a mechanical transfer system for all mixing and loading operations, use of a chemical resistant apron is optional.

If you handle a total of 30 gallons or more of this product per day, you must use a mechanical transfer system for all mixing and loading operations. If this product is packaged in a 30 gallon drum, you must use a mechanical transfer system which terminates in a drip-free hard coupling which may be used only with a spray or mix tank which has been fitted with a compatible coupling. If you do not presently own or have access to a mechanical transfer system with this type of coupling, contact your dealer for information on how to obtain such a system or to modify your present system. When using a mechanical transfer system, do not remove or disconnect the pump or probe from the container until the container has been emptied and rinsed. The pump or probe system must be used to rinse the empty container and to transfer the rinsate directly to the mixing or spray tank.

ALL PERSONS APPLYING THIS PRODUCT, OR REPAIRING OR CLEANING EQUIPMENT USED WITH THIS PRODUCT, MUST OBSERVE THE FOLLOWING REQUIREMENTS.

When applying this product from a tractor (unless the tractor has a completely enclosed cab), or when repairing or cleaning equipment used with the product, you must wear clean nitrile gloves and clean cotton (or cloth) coveralls that cover all parts of the body except the head, hands, and feet. Coveralls must be worn over a long sleeve shirt and long pants.

All applicators (except for pilots) and all persons repairing or cleaning equipment used with this product must wear chemical resistant shoes, shoe coverings, or boots.

Application from a tractor with a completely enclosed cab or aerial application is required whenever this product is applied to 180 or more acres in a day. To avoid contamination, coveralls and gloves worn when handling the concentrate must be removed prior to entering an enclosed cab or cockpit. When applying from a tractor with an enclosed cab, clean coveralls and clean nitrile gloves must be kept inside the cab, and must be worn when exiting the cab to perform in-field maintenance or repair.

To reduce exposure to residues, wash the spray rig, tractor, and all other equipment used to handle or apply this product with water daily or before using the equipment for any other purpose.

APPLICATION BY CHEMIGATION must be done by fixed pipe, overhead sprinkler systems or hand moved pipe. If hand moved pipe is used for chemigation, the pipe must not be handled in any way until 24 hours after chemigation has been completed and residues have been flushed from the system. When applying by chemigation, no person may enter the application site unless in an enclosed vehicle.

DURING AERIAL APPLICATION, human flaggers are prohibited unless in enclosed vehicles. Aerial application is prohibited within 300 feet of residential areas (e.g., homes, schools, hospitals, shopping areas, etc.)

GENERAL REQUIREMENTS FOR ALL USE

IMPORTANT! Before removing gloves or starting a new work operation, rinse the outside of the gloves thoroughly with water. Always remove gloves and wash your hands and face with soap and water before smoking, eating, drinking, or toileting.

AFTER WORK. Gloves must be discarded at the end of each day. Take off all clothes and shoes. Shower with soap and water. Do not wear or re-use contaminated clothing. Wash protective equipment with water after each use. Protective clothing worn during use must be laundered separately from other household articles.

STATEMENT OF PRACTICAL TREATMENT

IF SWALLOWED: Get medical attention. Do not induce vomiting, contains petroleum distillates. Do not give anything by mouth to an unconscious person.

IF on SKIN: Wash with plenty of soap and water. Get medical attention if irritation persists.

IF in EYES: Flush with water for 15 minutes. Get medical attention.

ENVIRONMENTAL HAZARDS

This pesticide is toxic to wildlife and fish. Use with care when applying to areas frequented by wildlife or adjacent to any body of water. Do not apply directly to water. Do not apply when weather conditions favor drift from target areas. Do not contaminate water when disposing of equipment washwaters.

PHYSICAL AND CHEMICAL HAZARDS

Do not use or store near heat or open flame.

DIRECTIONS FOR USE

IT IS A VIOLATION OF FEDERAL LAW TO USE THIS PRODUCT IN A MANNER INCONSISTENT WITH ITS LABELING.

REENTRY STATEMENT

Do not apply this product in such a manner as to directly or through drift expose workers or other persons. The area being treated must be vacated by unprotected persons.

Do not enter treated areas without protective clothing until sprays have dried.

Because certain states may require more restrictive reentry intervals for various crops treated with this product, consult your State Department of Agriculture for further information.

Written or Oral warnings must be given to workers who are expected to be in a treated area or in an area about to be treated with this product. Oral warnings must be given if there is reason to believe that written warnings cannot be understood by workers and given in a language customarily understood by the workers. Oral and written warnings must include the **Statement of Practical Treatment** and **Reentry Statements** as they appear on the label.

GENERAL INFORMATION

BUCTRIL® is a selective postemergence herbicide for control of important broadleaf weeds infesting small grains (wheat, barley, oats, rye, triticale), seedling alfalfa, field corn, popcorn, grain sorghum, flax, onion, garlic, annual canarygrass and mint. Optimum weed control is obtained when BUCTRIL® is applied to actively growing weed seedlings. BUCTRIL® is primarily a contact herbicide, therefore thorough coverage of the weed seedlings is essential for optimum control.

BUCTRIL® is formulated as an emulsifiable concentrate of octanoic acid ester of bromoxynil containing the equivalent of 2 pounds of bromoxynil per gallon. To broaden the spectrum of control in wheat and barley, BUCTRIL® may be tankmixed with Glean®, Finesse®, Avenge®, MCPA, 2,4-D or Hoelon®. In oats and rye, BUCTRIL® may be tankmixed with MCPA or 2,4-D. In field corn and popcorn, BUCTRIL® can be tank mixed with atrazine, Banvel®, or 2,4-D. BUCTRIL® may also be tank-mixed with atrazine in grain sorghum. See individual crop recommendations for proper chemical rates to be used. To ensure maximum crop safety and weed control, follow all cautions and limitations on this label and the labels of products used in the tank mixture with BUCTRIL®.

MIXING, LOADING AND HANDLING INSTRUCTIONS

2.5 Gallon Containers

It is strongly recommended that special care to taken in mixing and loading this product. Hands should be placed on the container in such a way as to avoid possible drip or splash. Correct procedures for mixing and loading are provided in Rhône-Poulenc Ag Company's Educational Program.

30 Gallon and Bulk Containers.

If you will handle a total of 30 gallons or more of this product per day, you must use a mechanical transfer system for all mixing and loading operations. If this product is packaged in a 30 gallon drum, you must use a mechanical transfer system which terminates in a drip-free hard coupling which may be used only with a spray or mix tank which has been fitted with a compatible coupling. If you do not presently own or have access to a mechanical transfer system with this type of coupling, contact your dealer for information on how to obtain such a system or to modify your present system. When using a mechanical transfer system, do not remove or disconnect the pump or probe from the container until the container has been emptied and rinsed. The pump or probe system must be used to rinse the empty container and to transfer the rinsate directly to the mixing or spray tank.

BUCTRIL® **Alone:** Fill the spray tank ½ to ¾ full with clean water. Begin agitation and add the recommended amount of BUCTRIL®. Add water to the spray tank to the desired level. Maintain sufficient agitation to ensure a uniform spray mixture during application.

Tank Mixtures: Fill the spray tank ½ to ¾ full with clean water and begin agitation. If tankmixing with either Glean or Atrazine, add the recommended amount of Glean or Atrazine first. After the herbicide is thoroughly mixed with water add the recommended amount of BUCTRIL® and add water to the spray tank to the desired level.

Maintain sufficient agitation while mixing and during application to ensure a uniform spray mixture. If spray mixture is allowed to remain without agitation for short periods of time, be sure to agitate until uniformly mixed before application.

APPLICATION PROCEDURES

BUCTRIL® alone and in a tank mixture with Glean, Avenge, MCPA, 2,4-D, Hoelon, Finesse and Atrazine can be applied by both ground and aerial equipment. In addition BUCTRIL® and BUCTRIL® + Atrazine can be applied by automated sprinkler irrigation systems. The following provides recommended methods of application for each crop.

Type of Application Equipment

Crop	Ground	Aerial	Automated Sprinkler Irrigation
SMALL GRAINS	X	X	X
SEEDLING ALFALFA	X	X	X
FIELD AND POPCORN	X	X	X
GRAIN SORGHUM	X	X	X
FLAX	X	X	
GARLIC	X	X	X
ONION	X		X
ANNUAL CANARYGRASS	X		
MINT	X		
GRASS GROWN FOR SEED OR SOD PRODUCTION	X		

(X) indicates recommended application use

GROUND APPLICATION

Use a standard herbicide boom sprayer that provides uniform and accurate application. Sprayer should be equipped with screens no finer than 50 mesh in the nozzle tips and in-line strainers. Specific application instructions are described under each crop heading.

AERIAL APPLICATION

During aerial application, human flaggers are prohibited unless in enclosed vehicles. Aerial application is prohibited within 800 feet of residential areas (e.g. homes, schools, hospitals, shopping areas, etc.) To obtain adequate spray coverage, use a minimum of 5 gallons of spray volume per acre using diaphragm-type nozzles that produce cone or fan spray patterns with a maximum spray pressure of 40 psi at the nozzle tips. Nozzles should be located no farther than ¾ the distance from the center of the aircraft to the end of the wing or rotor. Nozzles should be oriented at an angle between straight down and straight back.

The aircraft should discharge the spray a maximum of 10 feet above the crop. To minimize drift, do not apply during periods of gusty winds or when wind exceeds 5 mph.

AUTOMATED SPRINKLER IRRIGATION SYSTEMS

BUCTRIL® Herbicide can be applied through overhead sprinkler irrigation systems to small grains, seedling alfalfa, field and popcorn, grain sorghum, onions and garlic.

Apply BUCTRIL® Herbicide through sprinkler including center pivot, lateral move, side (wheel) roll, solid set or hand move irrigation systems only. If hand moved pipe is used for chemigation, the pipe must not be handled in any way for 24 hours after chemigation has been completed and residues have been flushed from the system. When applying by chemigation, no person may enter the application site unless in an enclosed vehicle. Do not apply this product through any other type of irrigation system.

Specific Requirements For Application Through Automated Sprinkler Irrigation System

1. The system must contain a functional check valve, vacuum relief valve, and low pressure drain appropriately located on the irrigation pipeline to prevent water source contamination from backflow.
2. The pesticide injection pipeline must contain a functional, automatic, quick-closing check valve to prevent the flow of fluid back toward the injection pump.
3. The pesticide injection pipeline must also contain a functional, normally closed, solenoid-operated valve located on the intake side of the injection pump and connected to the system interlock to prevent fluid from being withdrawn from the supply tank when the irrigation system is either automatically or manually shut down.
4. The system must contain functional interlocking controls to automatically shut off the pesticide injection pump when the water pump motor stops.
5. The irrigation line or water pump must include a functional pressure switch which will stop the water pump motor when the water pressure decreases to the point where pesticide distribution is adversely affected.
6. Systems must use a metering pump, such as a positive displacement injection pump (e.g., diaphragm pump) effectively designed and constructed of materials that are compatible with pesticides and capable of being fitted with a system interlock.
7. Do not apply when wind speed favors drift beyond the area intended for treatment.
8. Agitation is recommended in the pesticide supply tank when applying the BUCTRIL® Herbicide.
9. BUCTRIL® Herbicide should be applied continuously for the duration of the water application with center pivot and continuous lateral move systems. Application of BUCTRIL® Herbicide should be made during the last 30-45 minutes of the irrigation set with other overhead sprinkler systems.
10. For best performance, set the sprinkler system to deliver approximately 0.5 inch or less of water per acre.
11. Remove scale, pesticide residues and other foreign matter from the supply tank and entire injector system. Flush with clean water.
12. If BUCTRIL® Herbicide is diluted in the supply tank, fill the tank with half of the water amount desired, add the BUCTRIL® and then add remaining water amount with agitation.
13. Start the sprinklers and then inject BUCTRIL® Herbicide into the irrigation line. BUCTRIL® should be injected with a positive displacement pump into the main line at least 8 feet ahead of a right angle turn to insure adequate mixing. Refer to the BUCTRIL® Herbicide label for detailed information on application rates and timings.

Chemigation User Precautions

Application of more than 0.5 inch/acre of irrigation water may result in decreased product performance on certain soils.

Do not apply when conditions favor drift, when system connections or fittings leak, or when nozzles do not provide uniform distribution.

Allow sufficient time for pesticide to be flushed through all the lines and nozzles before turning off irrigation water.

Crop injury, lack of effectiveness, or illegal pesticide residues in the crop can result from non-uniform distribution of treated water.

Do not connect an irrigation system used for pesticide application to a public water system.

If you have questions about calibration, you should contact State Extension Service specialists, equipment manufactures or other experts.

A person knowledgeable of the chemigation system and responsible for its operations, or under the supervision of the responsible person, shall shut the system down and make necessary adjustments should the need arise.

GENERAL WEED LIST

Postemergence application of BUCTRIL® Herbicide will control the following weeds when sprayed in the seedling stage. Maximum weed stage of growth is listed under RECOMMENDED USES for each crop.

Most Susceptible Broadleaf Weed Species

Annual Sowthistle *(Sonchus oleraceus)*
Black Nightshade *(Solanum nigrum)*
Blue Mustard *(Chorispora tenella)*
Coast Fiddleneck *(Amsinckia intermedia)*
Common Cocklebur *(Xanthium pensylvanicum)*
Common Lambsquarters *(Chenopodium album)*
Common Tarweed *(Hemizonia congesta)*
Eastern Black Nightshade *(Solanum ptycanthum)*
Field Pennycress *(Thlaspi arvense)*
Green Smartweed *(Polygonum scabrum)*
Hairy Nightshade *(Solanum sarachoides)*
Jimsonweed *(Datura stramonium)*
Ladysthumb *(Polygonum persicaria)*
Pennsylvania Smartweed *(Polygonum pensylvanicum)*
Pepperweed spp. (annual) *(Lepidium* spp.*)*
Shepherdspurse *(Capsella bursa-pastoris)*
Silverleaf Nightshade *(Solanum elaeagnifolium)*
Tartary Buckwheat *(Fagopyrum tatoricum)*
Sunflower *(Helianthus annus)*[1]
Wild Buckwheat *(Polygonum convolvulus)*

(1) For control of sunflower, delay application until first emerging sunflower seedlings are 4 inches in height.

Susceptible Broadleaf Weed Species

Buffalobur *(solanum rostratum)*
Common Groundsel *(Senecio vulgaris)*
Common ragweed *(Ambrosia artemisiifolia)*
Corn Chamomile *(Anthemis arvensis)*
Corn Gromwell *(Lithospermum arvense)*
Cow Cockle *(Saponaria vaccaria)*
Giant Ragweed *(Ambrosia trifida)*
Hemp Sesbania *(Sesbania exaltata)*
Ivyleaf morningglory *(pomoea hederacea)*
Knawel *(Scleranthus annus)*
Kochia *(Kochia scoparia)*[2]
London Rocket *(Sisymbrium irio)*
Mayweed *(Anthemis cotula)*
Prostrate Knotweed *(Polygonum aviculare)*
Redroot Pigweed *(Amaranthus retroflexus)*[2]
Russian Thistle *(Salsola kali)*
Spiny Pigweed *(Amaranthus spinosus)*[2]
Tall Morningglory *(pomoea purpurea)*
Tall Waterhemp *(Amaranthus tuberculatus)*[2]
Tumble mustard *(Sisymbrium altissimum)*
Velvetleaf *(Abutilon theophrasti)*
Venice Mallow *(Hibiscus trionum)*
Wild Mustard *(Brassica kaber)*
Wild Radish *(Raphanus raphanistrum)*

(2) For effective control, these weeds should not exceed the 4 leaf stage or 2 inches in height, whichever comes first.

WEED SUPPRESSION

BUCTRIL® suppresses the growth of Canada Thistle *(Cirsium arvense)* by burning down top growth. Regrowth may occur.

COMPATIBILITY OF INSECTICIDES WITH BUCTRIL®

The following foliar insecticides are compatible with BUCTRIL® Herbicide as tank mixtures.

Insecticide

Common Name	Trade Name	Formulation
Diazinon	Various	Emulsifiable Concentrate
Dimethoate	Various	Emulsifiable Concentrate
Trichlorfon	Dylox	Soluble Powder
Carbofuran	Furadan	Flowable
Chlorpyrifos	Lorsban	Emulsifiable Concentrate
Malathion	Various	Emulsifiable Concentrate
Oxydemeton-methyl	Metasystox-R	Sprayable Concentrate
Permethrin	Pounce	Emulsifiable Concentrate
Fenvalerate	Pydrin	Emulsifiable Concentrate
Carbaryl	SEVIN	Sprayable wettable powder
Demeton	Sy-tox	Emulsifiable Concentrate

When mixing an insecticide with BUCTRIL®, fill the spray tank 1/2 to 3/4 full with clean water and begin agitation. If tank mixing with either Dylox, Furadan or SEVIN®, add the recommended amount of insecticide first. After the insecticide is thoroughly mixed with water add the recommended amount of BUCTRIL® Herbicide and add water to the spray tank to the desired level.

If tank mixing with the other insecticides listed above, add the recommended amount of BUCTRIL® first. After BUCTRIL® is thoroughly mixed with water, add the recommended amount of insecticide to the spray tank and add water to the spray tank to the desired level.

Maintain sufficient agitation while mixing and during application to ensure a uniform spray mixture. If spray mixture is allowed to remain without agitation for short periods of time, be sure to agitate until uniformly mixed before application.

FOLLOW ALL CAUTIONS AND LIMITATIONS ON THE LABELS OF PRODUCTS USED IN TANK MIXTURES WITH BUCTRIL®.

Dylox®, Metasystox® and Systox® are Registered Trademarks of the Parent Company of Farbenfabriken Bayer GmbH, Leverkusen.
Furadan® is a Registered Trademark of FMC Corporation.
Lorsban® is a Trademark of the Dow Chemical Company.
Pounce® is a Registered Trademark of FMC Corporation.
Pydrin® is a Registered Trademark of Shell Chemical Company.
SEVIN® is the Registered Trademark of Rhône-Poulenc for carbaryl insecticide.

WHEAT, BARLEY, OATS, RYE AND TRITICALE

BUCTRIL® can be applied to small grains from emergence up to the boot stage. Applications should be made to weeds soon after emergence for best weed control and to prevent competition to the crop. Reduced weed control may occur when weeds are stressed from lack of moisture or cold temperatures. Avoid spray drift to nontarget areas. Do not graze treated fields for 30 days following application.

BUCTRIL® can be applied to wheat and barley alone or in a tank mixture with Glean, Avenge, MCPA, 2,4-D, Finesse or Hoelon to broaden the spectrum of weed control. BUCTRIL® can also be applied to oats and rye alone or in a tank mixture with MCPA or 2,4-D. To ensure maximum crop safety and weed control, follow all cautions and limitations on this label and the labels of products used in tank mixtures with BUCTRIL®.

APPLICATION PROCEDURES

Ground Application

BUCTRIL® used alone and in tank mixtures can be applied by ground equipment using a minimum spray volume of 10 gallons per acre. Use flat fan nozzles spaced a maximum of 20 inches apart with a minimum spray pressure of 30 psi at the nozzle tips. Other nozzle types may not provide adequate coverage of the weeds to ensure optimum weed control.

Aerial Application

BUCTRIL® and tank mixtures with BUCTRIL® can be applied by aerial equipment using a minimum spray volume of 5 gallons per acre with a maximum spray pressure of 40 psi at the nozzle tips. Refer to AERIAL APPLICATION under the GENERAL INFORMATION section for complete use directions.

AUTOMATED SPRINKLER IRRIGATION SYSTEMS

Refer to GENERAL INFORMATION section for complete use directions and precautions.

SPRAYABLE LIQUID FERTILIZERS

BUCTRIL® can be applied to small grains using sprayable liquid fertilizer as the carrier. Predetermine the compatibility with the liquid fertilizer by mixing small proportional quantities in advance. Agitation must be maintained during filling and application operations to ensure that BUCTRIL® is evenly mixed with the fertilizer. Leaf burn may occur when BUCTRIL® is applied with liquid fertilizer, but new leaves are not adversely affected.

RECOMMENDED USES

The recommended rate of BUCTRIL® used alone or in a tank mixture with another herbicide is determined by the crop (winter or spring seeded), geographic area, application equipment (ground or aerial weed species and stage of growth of weeds at time of application.

Recommendations are as follows:

WHEAT AND BARLEY BUCTRIL® RECOMMENDATIONS

PRODUCT	CROP/GEOGRAPHICAL AREA	RATE	TIMING OF APPLICATION: CROP	TIMING OF APPLICATION: WEED
BUCTRIL®	Fall seeded wheat and barley throughout the United States and spring seeded wheat and barley in Idaho, Oregon and Washington.	1½ Pints/A	Apply to wheat and from emergence to the boot stage.	Most Susceptible and Susceptible Broadleaf Weeds: Apply to weeds up to the 4 leaf stage or 2 inches in height whichever comes first. If weed forms rosette, apply before weeds exceed 1 inch in diameter.
	Spring seeded wheat and barley except Idaho, Oregon and Washington	1 Pint/A	Apply to wheat and from emergence to the boot stage.	Most Susceptible and Susceptible Broadleaf Weeds: Apply BUCTRIL® at 1 pint/A to weeds that do not exceed the 4 leaf stage or 2 inches in height whichever comes first. If weed forms rosette, apply before weeds exceed 1 inch in diameter.
		1½ Pints/A		Use BUCTRIL® at 1½ pints/A for control of kochia that is 2 to 4 inches in height and pigweed *(Amaranthus* spp*)* that does not exceed the 4 leaf stage or 2 inches in height whichever comes first.

BUCTRIL® TANK MIXTURE RECOMMENDATIONS

PRODUCT	CROP/GEOGRAPHICAL AREA	RATE	TIMING OF APPLICATION: CROP	TIMING OF APPLICATION: WEED
BUCTRIL® + GLEAN (Tankmix)	Fall seeded wheat and barley throughout the United States and spring seeded wheat and barley in Idaho, Oregon and Washington.	¾ Pint/A + 1/6 - 1/3 oz/A	Apply to wheat and barley in the fall or spring at any time after the crop is in the 2 to 3 leaf stage, but before the boot stage.	Most Susceptible Broadleaf Weeds: Apply to weeds up to the 8 leaf stage or 4 inches in height whichever comes first. If weed forms rosette, apply before weeds exceeds 2 inches in diameter.
		1 Pint/A + 1/6 - 1/3 oz/A		Susceptible Broadleaf Weeds: Apply to weeds up to the 4 leaf stage or 2 inches in height whichever comes first. If weed forms rosette, apply before weeds exceed 1 inch in diameter.
	Spring seeded wheat and barley except Idaho, Oregon Washington.	¾ Pint/A + 1/6 - 1/3 oz/A		Most Susceptible and Susceptible Broadleaf Weeds: Apply BUCTRIL® + GLEAN at ¾ pint/A + 1/6 - 1/3 ounces/A to weeds that do not exceed the 4 leaf stage or 2 inches in height whichever comes first. If weed forms rosette, apply before weeds exceed 1 inch in diameter.
		1½ Pints/A + 1/6 - 1/3 oz/A		Use BUCTRIL® + GLEAN at 1½ pints/A + 1/6 - 1/3 ounces/A for control of kochia that is 2 to 4 inches in height and pigweed *(Amaranthus* spp*)* that does not exceed the 4 leaf stage or 2 inches in height whichever comes first.

A tankmix application of BUCTRIL® and GLEAN broadens the spectrum of weed control. See GLEAN label for listing of additional susceptible weed species and crop rotation instructions.

GLEAN® is a registered trademark of E.I. DuPont de Nemours & Co., Inc.

BUCTRIL® AND AVENGE

A tank mixture of BUCTRIL® + AVENGE can be used to control wild oats in addition to the broadleaf weeds controlled by BUCTRIL®. The rate and application timing for broadleaf weed control with BUCTRIL® in the tank mixture is the same as when BUCTRIL® is used alone. See BUCTRIL® RECOMMENDATIONS.

Apply BUCTRIL® + AVENGE tank mixtures when the majority of the wild oats are in the 3 to 5 leaf stage of growth. This frequently coincides with barley in the 2 to 7 leaf stage, spring seeded wheat in the 5 to 6 leaf stage, and fall seeded wheat in the 4 leaf to tillered stage of growth.

See Avenge label for varietal restrictions and recommended Avenge rates as related to wild oat density.

AVENGE is a trademark of American Cyanamid Company.

BUCTRIL® AND MCPA or 2,4-D

A tank mixture of BUCTRIL® at the recommended rate and MCPA or 2,4-D at ¼ to ½ pound of active ingredient per acre is used to broaden the spectrum of broadleaf weed control. The tank mixture will control the weeds listed on the MCPA or 2,4-D labels in addition to the weeds listed on the BUCTRIL® label. The rate of BUCTRIL® in the tank mixture is the same as when BUCTRIL® is used alone. See BUCTRIL® RECOMMENDATIONS.

BUCTRIL® + MCPA tank mixtures can be applied after the 3 leaf stage but before the crop reaches the boot stage. BUCTRIL® + 2,4-D tank mixtures can be applied after the crop has tillered, but before the forming joints in the stem.

BUCTRIL® Herbicide and Hoelon

A tank mixture of BUCTRIL® + Hoelon can be used to control annual grasses in addition to the broadleaf weeds controlled by BUCTRIL®. The rate and application timing for broadleaf weed control with BUCTRIL® in the tank mixture is the same as when BUCTRIL® is used alone. See BUCTRIL® RECOMMENDATIONS.

BUCTRIL® + Hoelon tank mixtures can be applied to fall seeded wheat and spring seeded wheat and barley up to the jointing stage of growth. Annual grasses should be in the 1 to 3 leaf stage of growth; up to the 4 leaf stage of wheat. See Hoelon label to determine recommended Hoelon rate in relation to the annual grassy weed stage of growth.

Hoelon is a trademark of Hoechst AG.

BUCTRIL® Herbicide AND FINESSE

Use only in the states of Idaho, Oregon, and Washington.

CROP

Fall and Spring seeded Wheat and Barley

TARGET PESTS

All weeds listed on the BUCTRIL® label and the Finesse product label.

BUCTRIL® RECOMMENDATIONS

Apply BUCTRIL® at ¾ to 1½ pints/A and Finesse at 1/6 to 1/3 ounce/A.

Apply to wheat and barley in the fall or spring at any time after the crop is in the 2-3 leaf stage, but before the boot stage.

Ground Application: Use a standard herbicide boom sprayer that provides uniform and accurate application. Sprayer should be equipped with flat fan nozzles spaced a maximum of 20 inches apart and screens no finer than 50 mesh in the nozzle tips and in-line strainers. Apply a minimum spray volume of 10 gallons per acre with minimum spray pressure of 30 psi at the nozzle tips.

Aerial Application: The sprayer should be equipped to provide a uniform and accurate application. Use a minimum spray volume of 5 gallons per acre using diaphragm type nozzles that produce cone or fan spray patterns with a maximum spray pressure of 40 psi at the nozzles tips. Nozzles should be located no further than 3/4 the distance from the center of the aircraft to the end of the wing or rotor. Nozzles should be oriented at an angle between straight down and straight back. The aircraft should discharge the spray a maximum of 10 feet above the crop. To minimize drift, do not apply during periods of gusty winds or when wind exceeds 5 MPH.

In order to ensure maximum safety and weed control, follow label recommendations on this label and all the cautions and limitations on the package labels of the tank mixture components.

Finesse is a registered trademark of E.I. DuPont de Nemours & Co. (Inc.), Wilmington, DE.

OATS, RYE AND TRITICALE

BUCTRIL® used alone or in a tank mixture with MCPA or 2,4-D can be applied to oats and rye. BUCTRIL® should be used alone when applied to triticale. Use the same rates and application timing as recommended for wheat and barley. Refer to previous section for WHEAT AND BARLEY.

WHEAT, BARLEY, OATS, RYE AND TRITICALE
(Automated Sprinkler Irrigation Application)

BUCTRIL® can be applied through automated sprinkler irrigation systems to fall and spring seeded wheat, barley, oats, rye and triticale from emergence to the boot stage. Apply BUCTRIL® at 2 pints/A in 1/4 to 1/2 acre-inch of water. BUCTRIL® Herbicide at 2 pints/A must be applied with a mechanical transfer system only. See MIXING, LOADING and HANDLING INSTRUCTIONS section for complete details. See GENERAL INFORMATION section for complete use directions and precautions. Refer to GENERAL WEED LIST in GENERAL INFORMATION section for list of MOST SUSCEPTIBLE BROADLEAF WEEDS and SUSCEPTIBLE BROADLEAF WEEDS to postemergence applications of BUCTRIL® Herbicide.

MOST SUSCEPTIBLE BROADLEAF WEEDS

Apply to weeds up to the 8 leaf stage or 4 inches in height whichever comes first. If weed forms rosette apply before weeds exceed 2 inches in diameter.

SUSCEPTIBLE BROADLEAF WEEDS

Apply to weeds up to 4 leaf stage or 2 inches in height whichever comes first. If weed forms rosette, apply before the weeds exceed 1 inch in diameter.

WEED SIZE IS CRITICAL WHEN BUCTRIL® HERBICIDE IS APPLIED THROUGH AUTOMATED SPRINKLER IRRIGATION SYSTEMS.

WEEDS SHOULD NOT EXCEED THE MAXIMUM INDICATED STAGE OF GROWTH.

DO NOT APPLY BUCTRIL® HERBICIDE THROUGH AUTOMATED SPRINKLER IRRIGATION SYSTEMS IN CALIFORNIA.

SEEDLING ALFALFA (Not for use on seedling Alfalfa in California)

Apply BUCTRIL® in the spring when weed seedlings are actively growing. For optimum control apply BUCTRIL® in sufficient spray volume to ensure good coverage of the weeds.

BUCTRIL® Herbicide Recommendations

BUCTRIL® can be applied to seedling alfalfa with a minimum of 2 trifoliate leaves.

Apply BUCTRIL® at 1 pint/A to control the MOST SUSCEPTIBLE BROADLEAF WEEDS and 1 1/2 pints/A for control of SUSCEPTIBLE BROADLEAF WEEDS (See GENERAL WEED LIST). Weeds should not exceed the 4 leaf stage or 2 inches in height, whichever comes first, or before rosettes are 1.5 inches in diameter. For improved control of kochia and pigweed (sp), tansy mustard, and henbit, BUCTRIL® at 1 pint/A can applied in tank mixture with 2,4-DB at 0.5 lb. ai/A. BUCTRIL ® will not adequately control over-wintered pennycress, henbit, and mustards.

For ground application, use a minimum of 20 gallons of spray volume per acre. Use flat fan nozzles spaced a maximum of 20 inches apart with a minimum spray pressure of 30 psi at the nozzles tips.

For aerial application, use a minimum of 5 gallons of spray volume per acre with a maximum of 40 psi at the nozzle tips. Refer to AERIAL APPLICATION under GENERAL INFORMATION section of the BUCTRIL® Herbicide label for complete use directions.

For use through automated sprinkler irrigation systems, BUCTRIL® can be applied at 2 pints per acre only. BUCTRIL® Herbicide at 2 pints/A must be applied with a mechanical transfer system only. See MIXING, LOADING, AND HANDLING INSTRUCTIONS section for complete details. See General Information section for complete use directions and precautions.

RESTRICTIONS AND PRECAUTIONS

Crop leaf burn can occur following BUCTRIL® application. Warm, humid conditions may enhance leaf burn. New crop growth will not be affected. Alfalfa yield should not be reduced although total biomass tonnage may decrease compared to a weedy field due to weed removal.

Do not treat when temperatures exceed 70°F at and 3 days following application because unacceptable crop injury may occur.

Do not apply when alfalfa is under moisture, temperature, insect or disease stress or has been stressed by other pesticide carryover or application.

Do not add a surfactant or crop oil because increased alfalfa injury will occur.

Do not cut for feed or graze spring treated alfalfa within 30 days following treatment.

Do not cut for feed or graze fall or winter treated alfalfa until spring, at least 60 days following treatment.

Total BUCTRIL® Herbicide applied to a seedling alfalfa crop should not exceed 2 pints/A.

Follow all restrictions and precautions on the 2,4-DB label when a BUCTRIL® + 2,4-DB tank mixture is used.

FIELD CORN, POPCORN AND GRAIN SORGHUM

BUCTRIL® is applied as a broadcast application in field corn, popcorn or grain sorghum growth for grain, seed or silage under conventional minimum tillage, ridge tillage or no-till systems. For optimum control spray seedling weeds that are actively growing and not stressed from lack of moisture or low temperatures. Avoid spray drift to non-target areas.

A tank mixture of BUCTRIL® + ATRAZINE may be used in field corn, popcorn or grain sorghum to broaden the spectrum of control and provide residual control of germinating seedlings. In field corn, BUCTRIL® may also be used in a tank mixture with Banvel or 2,4-D.

When environmental conditions are quite cool and cloudy or hot and humid, some leaf burn may occur on the corn or grain sorghum foliage. The crop rapidly outgrows this condition and new growth is unaffected. To ensure maximum crop safety and weed control, follow all precautions and limitations on this label and the labels of products used in tank mixtures with BUCTRIL®.

APPLICATION PROCEDURES

GROUND TREATMENT

BUCTRIL® used alone and in tank mixtures with Atrazine or 2,4-D can be applied with ground equipment using a minimum of 10 gallons of spray volume per acre. Use flat fan nozzles spaced a maximum of 20 inches apart with a minimum spray pressure of 30 psi at the nozzle tips. Other nozzle types may not provide adequate coverage of the weeds to ensure optimum control. When weed infestations are heavy, use of a higher spray volume and spray pressure will be helpful to ensure uniform weed coverage. When corn or grain sorghum is large enough to interfere with the spray pattern, drop nozzles should be used to obtain uniform coverage of the weeds.

When BUCTRIL® is used in a tank mixture with Banvel, use a minimum of 20 gallons of spray volume per acre. Use flat fan nozzles spaced a maximum of 20 inches apart with a spray pressure less than 20 psi at the nozzle tips. DO NOT APPLY TANK MIXTURES OF BUCTRIL® & 2,4-D or BUCTRIL® & BANVEL WITH AERIAL EQUIPMENT.

AERIAL EQUIPMENT

BUCTRIL® used alone or in tank mixtures with Atrazine can be applied by aerial equipment. Use a minimum of 5 gallons of spray volume per acre with a maximum spray pressure of 40 psi at the nozzle tips. Refer to AERIAL APPLICATION under GENERAL INFORMATION section for complete use directions.

AUTOMATED SPRINKLER IRRIGATION SYSTEMS

Refer to GENERAL INFORMATION section for complete use directions and precautions.

BUCTRIL® RECOMMENDATIONS

Postemergence applications of BUCTRIL® can be applied to field corn, popcorn and grain sorghum at 1 and 1 1/2 pints/A. A second application is recommended if a new flush of weeds occurs following the first application. The total cumulative rate should not exceed 3 pints/A per season. Application can be made to corn up to tassel emergence, but do not apply within 30 days of ensilage. Application may be made to grain sorghum up to the boot stage but do not apply within 30 days of ensilage. Refer to APPLICATION RATE TABLE for weeds controlled.

Application Rate Table for Corn and Grain Sorghum

Weed Species[1]	1 Pint/A		1½ Pints/A	
When determining leaf stage, count all leaves except cotylendonary leaves	Max. Leaf Stage	Max. Weed Height (inches)	Max. Leaf Stage	Max. Weed Height (inches)
Black Nightshade *(Solanum nigrum)*	6	6	6	6
Buffalobur *(Solanum rostratum)*	4	2	6	4
Commom Cocklebur *(Xanthium pensylvanicum)*	6	8	8	10
Common Lambsquarter *(Chenopodium album)*	-	6	-	8
Common Ragweed *(Ambrosia artemisiifolia)*	6	4	8	6

Weed Species[1] When determining leaf stage, count all leaves except cotylendonary leaves	1 Pint/A Max. Leaf Stage	1 Pint/A Max. Weed Height (inches)	1½ Pints/A Max. Leaf Stage	1½ Pints/A Max. Weed Height (inches)
Eastern Black Nightshade *(Solanum ptycanthum)*	6	6	6	6
Giant Ragweed *(Ambrosia trifida)*	6	4	6	6
Hemp Sesbania *(Sesbania exaltata)*	-	-	4	4
Ivyleaf Morningglory *(Ipomoea hederacea)*	3	3	4	4
Jimsonweed *(Datura stramonium)*	4	4	6	6
Ladysthumb *(Polygonum persicaria)*	4	4	6	6
Pennsylvania Smartweed *(Polygonum pensylvanicum)*	4	4	6	6
Redroot Pigweed *(Amaranthus retroflexus)*	-	-	4	2
Spiny Pigweed *(Amaranthus spinosus)*	-	-	4	2
Sunflower *(Helianthus annus)*	4	6	6	8
Tall Morningglory *(Ipomoea purpurea)*	3	3	4	4
Tall Waterhemp *(Amaranthus tuberculatus)*	-	-	4	2
Velvetleaf *(Abutilon theophrasti)*	4	3	6	5
Venice Mallow *(Hibiscus trionum)*	-	-	4	2
Wild Buckwheat *(Polygonum convolvulus)*	4	6	6	8
Wild Mustard *(Brassica kaber)*	-	-	4	4
Weed Suppressed[2]				
Canada Thistle *(Cirsium arvense)*	Not Recommended		8 inch to bud stage	

[1] When determining leaf statge, count all leaves except cotyledonary leaves.
[2] BUCTRIL® suppresses the growth by burning down of top growth. Regrowth may occur.

PREPLANT THROUGH PREEMERGENCE APPLICATION

BUCTRIL® Herbicide can be applied before planting up until just prior to field corn, popcorn or grain sorghum emergence to control actively growing weed seedlings.

BUCTRIL® Herbicide Recommendations

Product	Rate[1]	Crop	Weed[2]
BUCTRIL®	1 to 1½ Pints/A	Apply to field corn, popcorn and grain sorghum before planting time up until just prior to crop emergence.	See APPLICATION RATE TABLE for list of weeds and corresponding stage of growth that are controlled by BUCTRIL® or BUCTRIL® & Atrazine at recommended rates of application.
BUCTRIL® + Atrazine[3,4]	1 to 1½ Pints/A + ½ to 1⅕ Lbs. ai/A		

[1] When preplant through preemergence treatment is followed with a BUCTRIL® postemergence application, the cumulative rate should not exceed 3 pints/A per season.
[2] For control of additional weeds not listed in the APPLICATION RATE TABLE for field corn, popcorn and grain sorghum, see GENERAL WEED LIST. Use BUCTRIL® at 1 pint/A for control of MOST SUSCEPTIBLE WEEDS and 1½ pints/A for control of SUSCEPTIBLE WEEDS. Apply BUCTRIL® to weeds that do not exceed the 4 leaf stage or 2 inches in height, whichever comes first. If weed forms rosette, apply before weed exceeds 1 inch in diameter.
[3] Refer to ATRAZINE CONVERSION TABLE under the corn section for conversion of active ingredient rates of Atrazine formulation rates.
[4] Follow all cautions and limitations on the product labels used in tank mixtures with BUCTRIL®.

BUCTRIL® TANK MIXTURE RECOMMENDATIONS

BUCTRIL® and ATRAZINE: Postemergence applications of BUCTRIL® at ¾, 1 and 1½ pints/A can be applied to field corn, popcorn and grain sorghum using the following use directions. Refer to APPLICATION RATE TABLE for weeds controlled.

APPLICATION RATE TABLE FOR BUCTRIL® & ATRAZINE TANK MIXTURES

WEED SPECIES[1]	BUCTRIL® & ATRAZINE RATE (TANK MIX)											
	¾ PINT/A + ½ LB AI/A		¾ PINT/A + 1⅕ LB AI/A		1 PINT/A + ½ LB AI/A		1 PINT/A + 1⅕ LB AI/A		1½ PINTS/A + ½ LB AI/A		1½ PINTS/A + 1⅕ LB AI/A	
When determining leaf stage, count all leaves except cotylendonary leaves	MAX. LEAF STAGE	MAX. WEED HEIGHT	MAX. LEAF STAGE	MAX. WEED HEIGHT	MAX LEAF STAGE	MAX. WEED HEIGHT	MAX. LEAF STAGE	MAX. WEED HEIGHT	MAX. LEAF STAGE	MAX. WEED HEIGHT	MAX. LEAF STAGE	MAX. WEED HEIGHT
Black Nightshade *(Solanum nigrum)*	4	4	4	4	6	6	6	6	6	6	6	6
Buffalobur *(Solanum rostratum)*	4	4	4	4	6	4	6	4	6	4	6	4
Common Cocklebur *(Xanthium pensylvanicum)*	6	8	8	10	8	10	10	12	10	12	10	12
Common Lambsquarters *(Chenopodium album)*	-	6	-	10	-	10	-	12	-	12	-	12
Common Ragweed *(Ambrosia artemisiifolia)*	6	4	8	6	8	6	8	6	8	6	8	6
Eastern Black Nightshade *(Solanum ptycanthum)*	4	4	4	4	6	6	6	6	6	6	6	6
Giant Ragweed *(Ambrosia trifida)*	4	4	4	4	6	6	6	6	6	6	6	6
Hemp Sesbania *(Sesbania exaltata)*	4	4	4	4	4	4	4	4	4	4	4	4
Ivyleaf Morningglory *(Ipomoea hederacea)*	3	3	4	4	4	4	4	4	4	4	4	4
Jimsonweed *(Datura stramonium)*	4	4	4	4	6	6	6	6	6	6	6	6
Ladysthumb *(Polygonum persicaria)*	4	4	4	4	6	6	8	8	8	8	8	8
Pennsylvania Smartweed *(Polygonum pensylvanicum)*	4	4	4	4	6	6	8	8	8	8	8	8
Prickly Sida *(Sida spinosa)*	-	-	6	2	4	1	6	2	4	1	6	2
Puncturevine *(Tribulus terrestris)*	-	-	-	-	-	-	6	4	6	4	6	4
Redroot Pigweed[3] *(Amaranthus retroflexus)*	4	2	8	6	6	4	8	6	6	4	8	6
Smooth Pigweed[3] *(Amaranthus hybridus)*	4	2	6	4	4	2	6	4	6	4	6	4
Spiny Pigweed[3] *(Amaranthus spinosus)*	4	2	8	6	6	4	8	6	6	4	8	6
Sunflower *(Helianthus annus)*	6	8	8	10	8	10	10	12	10	12	10	12
Tall Morningglory *(Ipomoea purpurea)*	3	3	4	4	4	4	4	4	4	4	4	4
Tall Waterhemp[3] *(Amaranthus tuberculatus)*	4	2	8	6	6	4	8	6	6	4	8	6
Velvetleaf *(Abutilon theophrasti)*	4	3	4	3	6	5	6	5	8	6	8	6
Venice Mallow *(Hibiscus trionum)*	4	2	4	2	4	2	4	2	4	2	4	2
Wild Buckwheat *(Polygonum convolvulus)*	6	8	8	10	8	10	10	12	10	12	10	12
Wild Mustard *(Sinapis arvensis)*	4	4	4	4	4	4	4	4	4	4	4	4
WEEDS SUPPRESSED[2]												

WEED SPECIES[1]	BUCTRIL® & ATRAZINE RATE (TANK MIX)											
	3/4 PINT/A + 1/2 LB AI/A		3/4 PINT/A + 1 1/5 LB AI/A		1 PINT/A + 1/2 LB AI/A		1 PINT/A + 1 1/5 LB AI/A		1 1/2 PINTS/A + 1/2 LB AI/A		1 1/2 PINTS/A + 1 1/5 LB AI/A	
When determining leaf stage, count all leaves except cotylendonary leaves	MAX. LEAF STAGE	MAX. WEED HEIGHT	MAX. LEAF STAGE	MAX. WEED HEIGHT	MAX LEAF STAGE	MAX. WEED HEIGHT	MAX. LEAF STAGE	MAX. WEED HEIGHT	MAX. LEAF STAGE	MAX. WEED HEIGHT	MAX. LEAF STAGE	MAX. WEED HEIGHT
Canada Thistle *(Cirsium arvense))*	Not Recommended		Not Recommended			8"bud		8" bud		8" bud		8" bud

[1] When determining leaf stage, count all leaves except cotylendonary leaves.
[2] Selected rates of BUCTRIL® + Atrazine tank mixtures suppress the growth by burning down of top growth. Regrowth may occur.
[3] If pigweeds *(Armaranthus sp.)* present in the field to be treated have been identified as triazine resistant biotypes, use BUCTRIL® at 1 1/2 pints/A in a tank mixture with atrazine at 1/2 or 1 1/5 lb. ai/A. Applications should be made when pigweeds do not exceed the 4-leaf stage and 2 inches in height.

FIELD CORN AND POPCORN BUCTRIL® RECOMMENDATIONS

PRODUCT	RATE	TIMING OF APPLICATION CROP	WEED[1]
BUCTRIL®	1 Pint/A	Apply to corn between the 3-leaf stage and prior to tassel emergence. Use a maximum of 2 applications per season.	See APPLICATION Rate Table for list of weeds and corresponding stage of growth that are controlled by BUCTRIL at recommended rates of application.
	1 1/2 Pint/A	Apply to corn between the 4-leaf stag and prior to tassel emergence. Use a maximum of 2 applications per season.	

[1] For control of additional weeds not listed in the APPLICATION RATE TABLE for field corn, popcorn and grain sorghum, see GENERAL WEED LIST: Use BUCTRIL at 1 pint/A for control of MOST SUSCEPTIBLE WEEDS and 1 1/2 pints/A for control of SUSCEPTIBLE WEEDS. Apply BUCTRIL® to weeds that do not exceed the 4-leaf stage or 2 inches in height, whichever comes first. If weed forms rosette, apply before weed exceeds 1 inch in diameter.

FIELD CORN AND POPCORN BUCTRIL® TANKMIX RECOMMENDATIONS

PRODUCT	RATE	TIMING OF APPLICATION CROP	WEED[1]
BUCTRIL + Atrazine[1,2]	1 Pint/A + 1/2-1 1/5 lb ai/A	Apply to corn between the 3-leaf stage and before corn is 30 inches tall. Make up to 2 applications per season.	See Application Rate Table for list of weeds and corresponding stage of growth that are controlled by BUCTRIL & Atrazine tank mixtures at recommended rates of application.
BUCTRIL + Atrazine[1,2]	1 Pint/A + 1/2-1 1/5 lb ai/A	Apply to corn between the 4-leaf stage and before corn is 30 inches tall. Make up to 2 applications per season.	These tank mixtures also provide some residual control of germinating weed seedings

[1] Refer to ATRAZINE CONVERSION TABLE under the corn section for conversion of active ingredient rates of Atrazine for formulation rates.
[2] Follow all precautions and limitations on the labels of products used in tank mixtures with BUCTRIL®.

ATRAZINE CONVERSION TABLE[1]

ATRAZINE FORMULATION	ATRAZINE RATE POUNDS OF ACTIVE INGREDIENT PER ACRE	ATRAZINE FORMULATION RATE PER ACRE
Atrazine 4L	1/2	1 Pint
	1 1/5	2 2/5 Pints
Atrazine 80WP	1/2	5/8 Pound
	1 1/5	1 1/2 Pounds
Aatrex Nine-O	1/2	3/5 Pound
	1 1/5	1 1/3 Pounds

[1] Follow all precautions and limitations on the labels of products used in tank mixture with BUCTRIL®.

FIELD CORN (Ground Application Only) BUCTRIL® TANKMIX RECOMMENDATIONS

PRODUCT	RATE	TIMING OF APPLICATION CROP	WEED
BUCTRIL® + Banvel[1] (tankmix)	1 Pint/A + 1/4 lb ai/A	Apply to corn between the 3-leaf stage and before corn is 36 inches tall or 15 days before tassel emergence, whichever comes first.	All weeds controlled by BUCTRIL® at recommended rates of application plus Canada thistle when applied from the 8-inch to bud stage and field bindweed up to the mid bloom stage.
	1 1/2 Pints/A + 1/4 lb ai/A	Apply to corn between the 4-leaf stage and before corn is 36 inches tall or 15 days before tassel emergence, whichever comes first.	

[1] SPECIAL PRECAUTIONS FOR BUCTRIL® & BANVEL TANK MIXTURES: BUCTRIL® & Banvel Tank mixtures should be applied in a minimum of 20 gallons of spray volume per acre with a spray pressure below 20 psi at the nozzle tips.

DO NOT USE A TANK MIXTURE OF BUCTRIL® & BANVEL UNLESS ALL CAUTIONS AND LIMITATIONS ON THE BANVEL LABEL ARE FOLLOWED.

FIELD CORN
(Ground Application Only)
BUCTRIL® TANKMIX RECOMMENDATIONS

PRODUCT	RATE	TIMING OF APPLICATION CROP	WEED
BUCTRIL® Herbicide + 2,4-D[1] (Tankmix)	1 Pint/A + 1/4 lb ai/A	Apply to corn between the 3 leaf stage and prior to tassel emergence. When corn exceeds 8 inches in height, drop nozzles must be used to apply BUCTRIL® Herbicide & 2,4-D tank mixtures as a directed spray.	All weeds controlled by BUCTRIL® Herbicide at recommended rates of application plus control of wild mustard that is larger than the 4-leaf stage or 4-inches in height, whichever comes first.
	1 1/2 Pints/A + 1/4 lb. ai/A	Apply to corn between the 4 leaf stage corn exceeds 8 inches in height, drop nozzles must be used to apply BUCTRIL® Herbicide & 2,4-D tank mixtures as a directed spray.	

2,4-D CONVERSION TABLE[1]

ATRAZINE FORMULATION	2,4-D RATE POUNDS OF ACTIVE INGREDIENT PER ACRE	ATRAZINE FORMULATION RATE PER ACRE
2,4-D 4 lb/gallon	1/4	1/2 Pint
2,4-D 6 lb/gallon	1/4	1/3 Pint

SPECIAL PRECAUTIONS FOR BUCTRIL® + 2,4-D TANK MIXTURE: Postemergence applications of 2,4-D causes brittleness to corn. Winds or cultivations may cause breakage while corn is brittle. Follow all cautions and limitations on the labels of products used in tank mixtures with BUCTRIL®.

FIELD CORN AND POPCORN (Automated Sprinkler Irrigation Application)

BUCTRIL® can be applied through automated sprinkler irrigation systems to field and popcorn between the 3-leaf stage and before corn is 30″ tall. Apply BUCTRIL® at 1.5 pints/A + Atrazine at 1/2 to 1 1/5 lb. ai/A in 1/4 to 1/2 acre-inch of water. See GENERAL INFORMATION section for complete use directions and precautions. Refer to GENERAL WEED LIST in the GENERAL INFORMATION section for list of MOST SUSCEPTIBLE BROADLEAF WEEDS and SUSCEPTIBLE BROADLEAF WEEDS to postemergence applications BUCTRIL®.

Most Susceptible Broadleaf Weeds

Apply to weeds up to the 8-leaf stage or 4 inches in height whichever comes first. If weed forms rosette apply before weeds exceeds 2 inches in diameter.

Susceptible Broadleaf Weeds

Apply to weeds up to the 4-leaf stage or 2 inches in height whichever comes first. If weed forms rosette apply before weeds exceed 1 inch in diameter.

WEED SIZE IS CRITICAL WHEN BUCTRIL® IS APPLIED THROUGH AUTOMATED SPRINKLER IRRIGATION SYSTEMS. WEEDS SHOULD NOT EXCEED THE MAXIMUM INDICATED STAGE OF GROWTH.

Do not apply BUCTRIL® through automated sprinkler irrigation systems in California.

GRAIN SORGHUM BUCTRIL® RECOMMENDATIONS

PRODUCT	RATE	TIMING OF APPLICATION CROP	WEED[1]
BUCTRIL®	1 Pint/A	Apply to grain sorghum between the 3-leaf stage up to the boot stage Use a maximum of 2 applications per season.	See Application Rate Table for list of weeds and corresponding stages of growth that are controlled by BUCTRIL® at recommended rates of application.
	1 1/2 Pints/A	Apply to grain sorghum between the 4-leaf stage up to the boot stage. Use a maximum of 2 applications per season.	

[1] For control of additional weeds not listed in the APPLICATION RATE TABLE for field corn, popcorn and grain sorghum, see GENERAL WEED LIST. Use BUCTRIL® a + 1 pint/A for control of MOST SUSCEPTIBLE WEEDS and 1 1/2 pints/A for control of SUSCEPTIBLE WEEDS. Apply BUCTRIL® to weeds that do not exceed the 4-leaf stage or inches in height, whichever comes first. If weed forms rosette, apply before weed exceeds 1 inch in diameter.

GRAIN SORGHUM BUCTRIL TANK MIX RECOMMENDATIONS

PRODUCT	RATE	TIMING OF APPLICATION CROP	WEED
BUCTRIL + ATRAZINE[1,2] (tank mix)	1/4 To 1 Pint/A + 1/2-1 1/5 lb ai/A	Apply to grain sorghum between the 3-leaf stage up to the boot stage Use a maximum of 2 applications per season	See Application Rate Table for list of weeds and corresponding stages of growth that are controlled by BUCTRIL® & Atrazine tank mixtures at recommended rates of application. These tank mixtures also provide some residual control of germinating weed seedings.
	1 1/2 Pints/A + 1/2-1 1/5 lb ai/A	Apply to grain sorghum between the 4-leaf stage up to the boot stage. Use a maximum of 2 applications per season.	

[1] Refer to ATRAZINE CONVERSION TABLE under the corn section for conversion of active ingredient rates of atrazine to formulation rates.

[2] Follow all cautions and limitations on the labels of products used in tank mixtures with BUCTRIL.

GRAIN SORGHUM (Automated Sprinkler Irrigation Application)

BUCTRIL® can be applied through automated sprinkler irrigation systems to grain sorghum from the 3-leaf up to the boot stage. Apply BUCTRIL® at 1 1/2 pints/A + Atrazine at 1/2 to 1 1/5 lbs. ai/A in 1/4 to 1/2 acre-inch of water. See GENERAL INFORMATION section for complete use directions and precautions. Refer to GENERAL WEED LIST in the GENERAL INFORMATION section for list of MOST SUSCEPTIBLE BROADLEAF WEEDS and SUSCEPTIBLE BROADLEAF WEEDS.

MOST SUSCEPTIBLE BROADLEAF WEEDS

Apply to weeds up to the 8 leaf stage or 4 inches in height whichever comes first. If weed forms rosette apply before weeds exceed 2 inches in diameter.

SUSCEPTIBLE BROADLEAF WEEDS

Apply to weeds up to the leaf stage or 2 inches in height whichever comes first. If weed forms rosette apply before weeds exceed 1 inch in diameter. WEED SIZE IS CRITICAL WHEN BUCTRIL® IS APPLIED THROUGH AUTOMATED SPRINKLER IRRIGATION SYSTEMS. WEEDS SHOULD NOT EXCEED THE MAXIMUM INDICATED STAGE OF GROWTH.

Do not apply BUCTRIL® through automated sprinkler systems in California.

SPECIAL USE DIRECTIONS FOR CONTROL OF LARGE BROADLEAF WEEDS

Sequential applications of BUCTRIL® or BUCTRIL® tank mixtures may be necessary for control of broadleaf weeds that exceed the maximum stage of growth recommended with a single application of BUCTRIL® or BUCTRIL® tank mixture. Thorough coverage of the weeds is essential to obtain optimum weed control.

COMMON COCKLEBUR, COMMON LAMBSQUARTERS, SUNFLOWER

For control of common cocklebur and common lambsquarters up to 14 inches in height and sunflower up to 18 inches in height, use a postemergence application of BUCTRIL® at 1 pint/A. Make a second application of BUCTRIL® at the same rate 7 to 10 days later.

VELVETLEAF

For control of velvet leaf up to 14 inches in height, use postemergence application of BUCTRIL® at 1 ½ pints/A or BUCTRIL® + Atrazine tank mixture at 1 pint/A + 1 ⅕ lb. ai/A. Make a second application of BUCTRIL® at 1 pint/A 7 to 10 days later.

RESTRICTIONS AND LIMITATIONS

BUCTRIL® does not control grasses. Therefore, it is recommended that a suitable grass control program be used to provide any required grass control.

Addition of a spray additive, or mixture with liquid fertilizers may cause excessive crop injury.

Do not apply BUCTRIL® if heavy rainfall is expected within 1 hour of application.

Seed corn producers should consult the respective seed corn company regarding tolerance of certain seed production inbred lines to BUCTRIL.

Do not plant rotational crops until the following season.

Do not apply within 30 days of ensilage.

The total cumulative rate should not exceed 3 pints/A per season.

FLAX

Apply BUCTRIL® when weed seedlings are actively growing. For optimum control apply BUCTRIL® in sufficient spray volume to ensure good coverage of the weeds. Do not apply to flax under humid conditions and when air temperature is above 85°F as injury may occur. Do not add spray additive with BUCTRIL.

BUCTRIL RECOMMENDATIONS

BUCTRIL® at 1 pint/A should be applied postemergence to flax that is 2 to 8 inches in height. Do not apply BUCTRIL® to flax during or after the bud stage.

BUCTRIL® at 1 pint/A will control the MOST SUSCEPTIBLE and SUSCEPTIBLE BROADLEAF WEEDS (See GENERAL WEED LIST) when applied at the recommended weed stage of growth. Weeds should not exceed the 4 leaf stage or 2 inches in height, whichever comes first. If weed forms rosette, apply before weeds exceed 1 inch in diameter.

For ground application, use a minimum of 10 gallons of spray volume per acre. Use flat fan nozzles spaced a maximum of 20 inches apart with a minimum spray pressure of 30 psi at the nozzle tips. Other nozzle types may not provide adequate coverage of weeds to ensure optimum weed control.

For aerial application, use a minimum of 5 gallons of spray volume per acre with a maximum spray pressure of 40 psi at the nozzle tips. Refer to AERIAL APPLICATION under GENERAL INFORMATION section for complete use directions.

GARLIC

Apply BUCTRIL® when weed seedlings are actively growing. For optimum control apply BUCTRIL® in sufficient spray volume to ensure good coverage of the weeds.

BUCTRIL® RECOMMENDATIONS

BUCTRIL® can be applied after the crop emerges but before garlic is 12 inches in height.

For control of MOST SUSCEPTIBLE AND SUSCEPTIBLE BROADLEAF WEEDS (See GENERAL WEED LIST) apply BUCTRIL® at 1½ pints/A to weeds up to the 4 leaf stage or 2 inches in height, whichever comes first. If weed forms rosette, apply before weeds exceed 1 inch in diameter.

For ground application, use a minimum of 20 gallons of spray volume per acre. Use flat fan nozzles spaced a maximum of 20 inches apart with a minimum spray pressure of 30 psi at the nozzle tips. Other nozzle types may not provide adequate coverage of the weeds to ensure optimum weed control.

For aerial application, use a minimum of 5 gallons of spray volume per acre with a maximum spray pressure of 40 psi at the nozzle tips. Refer to AERIAL APPLICATION under GENERAL INFORMATION section for complete use directions.

Do not apply BUCTRIL® within 112 days of anticipated harvest.

ONIONS

Apply BUCTRIL® at 1 to 1½ pints per acre in at least 50-70 gallons of water per acre by ground equipment or chemigation only. Apply when onions have 2-5 true leaves. Water volume is important. CONCENTRATED SPRAYS KILL ONIONS. Thorough and uniform coverage is necessary for good weed control. Do not treat thrips or sand damaged crops. DO NOT ADD SURFACTANTS. DO NOT APPLY WITH AERIAL EQUIPMENT.

See GENERAL WEED LIST on the BUCTRIL® package label for weeds controlled. Use BUCTRIL® at 1 pint/A for control of MOST SUSCEPTIBLE WEEDS and 1½ pints/A for control of SUSCEPTIBLE WEEDS. Application should be made when weeds do not exceed the 4 leaf stage or 2 inches in height, whichever comes first. If weed forms rosette, apply before weeds exceed 1 inch in diameter.

In onion producing areas, certain environmental conditions reduce development of waxy coating on the onion leaves, thus increasing the possibility of injury. Dry soil, dry onion foliage, high light intensity, low humidity, and high temperatures tend to increase the waxy coating on onion leaves and thus reducing chances for injury. It is essential that the soil and onion foliage be dry at the time of application. Humidity should be low and dew should be off the plants.

Applications should not be made 24 hours before or after cultivation, irrigation, or rain, longer than 24 hours if possible. It is best to spray on a bright, sunny day after dew has completely dried from onion foliage.

Do not use BUCTRIL® Herbicide on onions grown under low light intensity, in areas such as Oregon, west of the Cascades.

The sensitivity of onions varies with the variety and environmental conditions. Therefore, even if all the label directions are followed, BUCTRIL® application still may cause injury to onions under certain circumstances.

ANNUAL CANARYGRASS

Apply BUCTRIL® and MCPA tank mixtures when weed seedlings are actively growing. For optimum control applications should be made in a sufficient spray volume to ensure good coverage of the weeds.

BUCTRIL® RECOMMENDATIONS

BUCTRIL® at 1 to 1½ pints/A tankmixed with MCPA at ¼ to ½ pound active ingredient per acre should be applied to annual canarygrass with a minimum of 3 leaves. BUCTRIL® and MCPA tank mixtures will control the MOST SUSCEPTIBLE and SUSCEPTIBLE BROADLEAF WEEDS (SEE GENERAL WEED LIST) when applied at the recommended stage of growth. Weeds should not exceed the 4 leaf stage or 2 inches in height, whichever comes first. If weed forms rosette, apply before weeds exceed 1 inch in diameter.

Applications should be made in a minimum of 10 gallons of spray volume per acre. Use flat fan nozzles spaced a maximum of 20 inches apart with a minimum spray pressure of 30 psi at the nozzle tips. Other nozzle types may not provide adequate coverage of the weeds to ensure optimum weed control.

PRECAUTIONS

Do not apply when annual canarygrass is under stress from lack of water, or when the crop forms a canopy over the weeds.

Do not graze livestock in treated annual canarygrass fields. Do not harvest within 75 days after application.

PEPPERMINT AND SPEARMINT (ESTABLISHED ONLY) BUCTRIL® RECOMMENDATIONS

BUCTRIL® should be applied only on established peppermint and spearmint which has been harvested for at least one year prior to treatment. BUCTRIL® should be applied after emergence of the broadleaf weeds but before they have more than 4 to 6 leaves. See GENERAL WEED LIST for weeds controlled by BUCTRIL®.

Applications should be made at the rate of 1 to 1½ pints/A. Applications should not be made when air temperatures exceed or are expected to exceed 70°F within 5 days after application. A second application may be made if needed, but the cumulative rate of BUCTRIL® should not exceed 6 pints/A during a single growing season. Do not apply within 70 days of harvest.

All applications of BUCTRIL® to mint should be in a minimum of 10 gallons of spray volume per acre. Use flat fan nozzles spaced a maximum of 20 inches apart with a minimum spray pressure of 30 psi at the nozzle tips. Other nozzle types may not provide adequate coverage of the weeds to ensure optimum weed control.

PRECAUTIONS

BUCTRIL® may cause temporary stunting and leaf chlorosis when applied to growing mint. Do not apply to mint growing under adverse conditions such as insect, nematode and disease infestations, high soil salt content, drought, excessive moisture or mint suffering from winter injury.

ESTABLISHED AND NEWLY PLANTED GRASSES FOR SOD OR SEED PRODUCTION

Postemergence applications of BUCTRIL® controls many seedling broadleaf weeds commonly found in grasses grown for seed. See GENERAL WEED LIST for weeds controlled by BUCTRIL®. Established grasses tolerant to BUCTRIL® include bentgrasses, Kentucky Bluegrass, Fescues, Ryegrass, Bermudagrass, St. Augustinegrass and Zoyiagrass. BUCTRIL® may also be used on seedling grasses such as Merion, Park, Delta, or common Kentucky Bluegrasses, Pennlawn, Chewings, Illahee or Alta Fescues, Orchard grass, Highland, Seaside, or Astoria Bentgrasses, perennial Ryegrasses, Bahiagrass and newly sprigged Zoyiagrass planted for seed or sod production, or forage grasses planted for seed production only.

APPLICATION PROCEDURES

Apply BUCTRIL® at 1½ (0.6 fl oz per 1000 sq. ft.) in 10 to 100 gallons of spray volume per acre. When using a spray volume from 10 to 40 gpa, use flat fan nozzles spaced a maximum of 20 inches apart with a minimum spray pressure of 30 psi at the nozzles tips. When using higher spray volumes, flood jet nozzles may be used with a minimum spray pressure of 30 psi at the nozzle tips.

BUCTRIL® HERBICIDE WEED CONTROL IN COVER CROPS ON CONSERVATION RESERVE ACRES

DIRECTIONS FOR USE

CROPS

Crested wheatgrasses (all), tall wheatgrass, western wheatgrass, bluebunch wheatgrass, intermediate wheatgrass, perennial ryegrass, fescue, bentgrasses, Kentucky bluegrass, Canby bluegrass, bahiagrass, Russian wildrye, seedling alfalfa.

TARGET WEEDS

BUCTRIL® will control broadleaf weeds mentioned in the GENERAL WEED LIST found on the package label. Follow all use directions on this label and on the BUCTRIL® package label to ensure product performance and crop safety.

WHEN TO APPLY

BUCTRIL® may be applied with standard ground boom and aerial hydraulic spray systems to forage crops listed here anytime after the 2-3 leaf stage, but before weeds exceed recommended stages of growth for treatment. Seedling alfalfa may be treated anytime after the 3-trifoliate leaf stage.

HOW TO APPLY

Ground Application: Use a standard herbicide boom sprayer that provides uniform and accurate application. Sprayer should be equipped with flat fan nozzles spaced a maximum of 20 inches apart and screens no finer than 50 mesh in the nozzle tips and in-line strainers. Apply in a minimum spray volume of 10 gallons per acre with minimum spray pressure of 30 psi at the nozzle tips.

Aerial Application: The sprayer should be equipped to provide a uniform and accurate application. Use a minimum spray volume of 5 gallons per acre using diaphragm-type nozzle that produce cone or fan spray patterns with a maximum spray pressure of 40 psi at the distance from the center of the aircraft to the end of the wing or rotor. Nozzles should be oriented at an angle between straight down and straight back. The aircraft should discharge the spray a maximum of 10 feet above the crop. To minimize drift, do not apply during periods of gusty winds or when wind exceeds 5 MPH.

In order to ensure maximum safety and weed control, follow label recommendations on this label and all the precautions and limitations on the package labels of the tank mixture components.

USE RECOMMENDATIONS

BUCTRIL®
Timing of Application

Rate	Crop	Weed[1]
BUCTRIL® 1 ½ Pints/A	Apply anytime in the fall or spring after grasses have reached the 2-3 leaf stage or when alfalfa has reached the 3-trifoliate leaf stage	Most Susceptible And Susceptible Broadleaf Weeds: Apply to weeds up to the 4 leaf stage or 2 inches in height whichever comes first. If weed forms rosette, apply before weeds exceed 1 inch in diameter.

[1] Refer to the general weed list on the BUCTRIL® package label for a complete list of susceptible weeds.

BUCTRIL® AND MCPA—GRASS CROPS ONLY—Do not use on alfalfa.

A tank mixture of BUCTRIL® at the recommended rate + MCPA at ¼ to ½ pound of active ingredient per acre is used to broaden the spectrum of broadleaf weed control. The tank mixture will control the weeds listed on the MCPA label, in addition to the weeds listed on the BUCTRIL® label. The rate of BUCTRIL® in the tank mixture is the same as when BUCTRIL® is used alone. See BUCTRIL® RECOMMENDATIONS.

BUCTRIL® + MCPA tank mixtures can be applied after the 3 leaf stage but before the crop reached the boot stage. Apply only to the following grass species: Crested wheatgrasses, tall wheatgrass, blue bunch wheatgrass, intermediate wheatgrasses, sheep fescue, canby bluegrass, and Russian wildrye. Do not apply to alfalfa.

USE RESTRICTIONS

Do not allow livestock to graze treated fields.

Do not feed treated grasses or alfalfa to livestock.

Do not apply BUCTRIL® + MCPA tank-mixtures to alfalfa.

Mix with clean water only. Do not use spray additives or fluid fertilizers as crop injury may occur.

Do not apply BUCTRIL® to alfalfa when air temperature exceed 80°F as crop injury may occur.

STORAGE AND DISPOSAL

STORAGE

Do not contaminate water, food or feed by storage or disposal.
Do not store near fertilizers or seeds. Store at temperatures above 3°F. If allowed to freeze, remix before using.

PESTICIDE DISPOSAL

Wastes resulting from the use of this product may be disposed of on site or at an approved waste disposal facility.

CONTAINER DISPOSAL

Triple rinse or equivalent. Then offer for recycling or reconditioning, or puncture and dispose of in a sanitary landfill, or by incineration or if allowed by state and local authorities, by burning. If burned, stay out of smoke.

LIMITED WARRANTY AND DISCLAIMER

The manufacturer warrants (a) that this product conforms to the chemical description on the label; (b) that this product is reasonably fit for the purpose set forth in the directions for use when it is used in accordance with such directions; and (c) that the directions, warnings and other statements on this label are based upon responsible experts evaluation of reasonable tests of effectiveness of toxicity to laboratory animals and to plants, and of residues on food crops and upon reports of field experience. Tests have not been made on all varieties or in all states or under all conditions. THE MANUFACTURER NEITHER MAKES NOR INTENDS, NOR DOES IT AUTHORIZE ANY AGENT OR REPRESENTATIVE TO MAKE, ANY OTHER WARRANTIES EXPRESS OR IMPLIED AND IT EXPRESSLY EXCLUDES AND DISCLAIMS ALL IMPLIED WARRANTIES OF MERCHANTABILITY OR FITNESS FOR A PARTICULAR PURPOSE.

THIS WARRANTY DOES NOT EXTEND TO AND THE BUYER SHALL BE SOLELY RESPONSIBLE FOR ANY AND ALL LOSS OR DAMAGE WHICH RESULTS FROM THE USE OF THIS PRODUCT IN ANY MANNER WHICH IS INCONSISTENT WITH THE LABEL DIRECTIONS, WARNINGS OR CAUTIONS.

BUYER'S EXCLUSIVE REMEDY AND MANUFACTURER'S EXCLUSIVE LIABILITY FOR ANY AND ALL CLAIMS LOSSES DAMAGE OR INJURIES RESULTING FROM THE USE OR HANDLING OF THIS PRODUCT, WHETHER OR NOT BASED IN CONTRACT, NEGLIGENCE, STRICT LIABILITY IN TORT OR OTHERWISE, SHALL BE LIMITED AT THE MANUFACTURER'S OPTION TO REPLACEMENT OF OR THE REPAYMENT OF THE PURCHASE PRICE FOR THE QUANTITY OF PRODUCT WITH RESPECT TO WHICH DAMAGES ARE CLAIMED. IN NO EVENT SHALL MANUFACTURER OR SELLER BE LIABLE FOR SPECIAL, INDIRECT OR CONSEQUENTIAL DAMAGES RESULTING FROM THE USE OR HANDLING OF THIS PRODUCT.

NOTICE TO BUYER

Purchase of this material does not confer any rights under patents governing this product or the use thereof in countries outside of the United States.

BUCTRIL is a trademark of Rhône-Poulenc

AG 90008

THIS SPECIMEN LABEL IS INTENDED FOR USE ONLY AS A GUIDE IN PROVIDING GENERAL INFORMATION REGARDING THE DIRECTIONS, WARNINGS AND CAUTIONS ASSOCIATED WITH THE USE OF THIS PRODUCT. AS WITH ANY AGRICULTURAL CHEMICAL, ALWAYS FOLLOW THE LABEL INSTRUCTIONS ON THE PACKAGE BEFORE USING.

BUCTRIL® + ATRAZINE
HERBICIDE

RESTRICTED USE PESTICIDE

FOR RETAIL SALE TO AND USE ONLY BY CERTIFIED APPLICATORS OR PERSONS DIRECTLY UNDER THEIR SUPERVISION AND ONLY FOR THOSE USES COVERED BY THE CERTIFIED APPLICATOR'S CERTIFICATION. SEE PRECAUTIONARY STATEMENTS FOR REASONS THIS PRODUCT IS CLASSIFIED RESTRICTED USE.

Postemergent herbicide for control of certain broadleaf weeds in corn and sorghum.

ACTIVE INGREDIENT:	
Octanoic acid ester of bromoxynil* (3,5-dibromo-4-hydroxybenzonitrile)	15.74%
Atrazine** (2-chloro-4-ethylamino-6-isopropylamino-S-triazine)	21.62%
INERT INGREDIENTS:	62.64%
TOTAL	100.00%

*Product contains bromoxynil octanoate equivalent to 10.81% of bromoxynil or 1.0 pound of bromoxynil per gallon.
**Product contains 2.0 pounds of atrazine per gallon.

Shake Well Before Using

EPA Reg. No. 264-477 EPA Est. No. 264-MO-02

For PRODUCT USE Information Call 1-800-334-9745
For MEDICAL And TRANSPORTATION Emergencies ONLY Call 24 Hours A Day 1-800-334-7577

KEEP OUT OF REACH OF CHILDREN
CAUTION

SHAKE WELL BEFORE USING

PRECAUTIONARY STATEMENTS

HAZARDS TO HUMANS AND DOMESTIC ANIMALS
CAUTION–RESTRICTED USE

Use of this product may be hazardous to your health. Exposure to this product has been shown to cause birth defects in laboratory animals. Women of childbearing age should be particularly careful when handling this product to avoid ingestion and skin contact.
Harmful if swallowed or absorbed through skin or inhaled. Avoid contact with skin, eyes or clothing. Avoid breathing spray mist.

Users of this product are encouraged to participate in Rhone-Poulenc's educational program dealing with potential hazards and safer use of this product. Further information on this program can be obtained by contacting Rhone-Poulenc Ag Company at 1-800-334-9745.

ALL PERSONS HANDLING THE CONCENTRATE MUST OBSERVE THE FOLLOWING REQUIREMENTS.

Wear clean cotton (or cloth) coveralls that cover all parts of the body except the head, hands, and feet. The coveralls must be worn over a long sleeve shirt and long pants. Wear clean nitrile gloves, a chemical resistant apron, goggles or face shield, and chemical resistant shoes, shoe coverings, or boots. If you use a mechanical transfer system for all mixing and loading operations, use of a chemical resistant apron is optional.

If you will handle a total of 30 gallons or more of this product per day, you must use a mechanical transfer system for all mixing and loading operations. If this product is packaged in a 30 gallon drum, you must use a mechanical transfer system which terminates in a drip-free hard coupling which may be used only with a spray or mix tank which has been fitted with a compatible coupling. If you do not presently own or have access to a mechanical transfer system with this type of coupling, contact your dealer for information on how to obtain such a system or to modify your present system. When using a mechanical transfer system, do not remove or disconnect the pump or probe from the container until the container has been emptied and rinsed. The pump or probe system must be used to rinse the empty container and to transfer the rinsate directly to the mixing or spray tank.

ALL PERSONS APPLYING THIS PRODUCT, OR REPAIRING OR CLEANING EQUIPMENT USED WITH THIS PRODUCT, MUST OBSERVE THE FOLLOWING REQUIREMENTS.

When applying this product from a tractor (unless the tractor has a completely enclosed cab), or when repairing or cleaning equipment used with the product, you must wear clean nitrile gloves and clean cotton (or cloth) coveralls that cover all parts of the body except the head, hands, and feet. Coveralls must be worn over a long sleeve shirt and long pants.

All applicators (except for pilots) and all persons repairing or cleaning equipment used with this product must wear chemical resistant shoes, shoe coverings, or boots.

Application from a tractor with a completely enclosed cab or aerial application is required whenever this product is applied to 180 or more acres in a day. To avoid contamination, coveralls and gloves worn when handling the concentrate must be removed prior to entering an enclosed cab or cockpit. When applying from a tractor with an enclosed cab, clean coveralls and clean nitrile gloves must be kept inside the cab, and must be worn when exiting the cab to perform in-field maintenance or repair.

To reduce exposure to residues, wash the spray rig, tractor, and all other equipment used to handle or apply this product with water daily or before using the equipment for any other purpose.

APPLICATION BY CHEMIGATION must be done by fixed pipe, overhead sprinkler systems or hand moved pipe. If hand moved pipe is used for chemigation, the pipe must not be handled in any way until 24 hours after chemigation has been completed and residues have been flushed from the system. When applying by chemigation, no person may enter the application site unless in an enclosed vehicle.

DURING AERIAL APPLICATION, human flaggers are prohibited unless in enclosed vehicles. Aerial application is prohibited within 300 feet of residential areas (e.g., homes, schools, hospitals, shopping areas, etc.)

GENERAL REQUIREMENTS FOR ALL USE

IMPORTANT! Before removing gloves or starting a new work operation, rinse the outside of the gloves thoroughly with water. Always remove gloves and wash your hands and face with soap and water before smoking, eating, drinking, or toileting.

AFTER WORK. Gloves must be discarded at the end of each day. Take off all clothes and shoes. Shower with soap and water. Do not wear or re-use contaminated clothing. Wash protective equipment with water after each use. Protective clothing worn during use must be laundered separately from other household articles.

STATEMENT OF PRACTICAL TREATMENT

IF SWALLOWED: Get medical attention. Do not induce vomiting, contains petroleum distillates. Do not give anything by mouth to an unconscious person.

IF ON SKIN: Wash with plenty of soap and water. Get medical attention if irritation persists.

IF IN EYES: Flush with water for 15 minutes. Get medical attention.

ENVIRONMENTAL HAZARDS

This pesticide is toxic to wildlife and fish. Use with care when applying to areas frequented by wildlife or adjacent to any body of water. Do not apply directly to water. Do not apply when weather conditions favor drift from target areas. Do not contaminate water by cleaning of equipment or disposal of equipment washwaters.

PHYSICAL AND CHEMICAL HAZARDS

Do not use or store near heat or open flame.

GENERAL INFORMATION

BUCTRIL® + Atrazine Herbicide is a selective postemergence product for control of important broadleaf weeds infesting field corn, popcorn and grain sorghum grown for grains, seed or silage under conventional, minimum tillage, ridge tillage or no-till systems. The recommended rates of BUCTRIL® + Atrazine Herbicide pre-mix flowable herbicide are 1 1/2, 2 and 3 pints/A. Each pint of BUCTRIL® + Atrazine Herbicide contains the equivalent of 1/2 pint of BUCTRIL® Herbicide and 1/4 pound of Atrazine.

BUCTRIL® + Atrazine Herbicide controls weeds primarily by contact action, therefore, thorough coverage of the weeds is essential for maximum control. Weeds should be actively growing and not stressed from lack of moisture or low temperatures before and at time of application. BUCTRIL® + Atrazine Herbicide also provides some residual control of germinating seedlings. When environmental conditions are quite cool and cloudy or very hot and humid, some leaf burn may occur on corn or grain sorghum. The crop rapidly outgrows this condition and new growth is unaffected. To ensure maximum crop safety and weed control, follow all cautions and limitations on this label.

DIRECTIONS FOR USE

It is a violation of Federal Law to use this product in a manner inconsistent with its labeling.

REENTRY STATEMENT

Do not apply this product in such a manner as to directly or through drift expose workers or other persons. The area being treated must be vacated by unprotected persons. Do not enter treated areas without protective clothing until sprays have dried.

Because certain states may require more restrictive reentry intervals for various crops treated with this product, consult your State Department of Agriculture for further information.

Written or oral warnings must be given to workers who are expected to be in a treated area or in an area about to be treated with this product. Oral warnings must be given if there is reason to believe that written warnings cannot be understood by workers and given in a language customarily understood by the workers.

Oral and written warnings must include the Statement of Practical Treatment and Reentry Statements as they appear on the label.

MIXING, LOADING AND HANDLING INSTRUCTIONS

2.5 Gallon Containers

It is strongly recommended that special care be taken in mixing and loading this product. Hands should be placed on the container in such a way as to avoid possible drip or splash. Correct procedures for mixing and loading are provided in Rhone-Poulenc Ag Company's Education Program.

30 Gallon and Bulk Containers

If you will handle a total of 30 galloons or more of this product per day, you must use a mechanical transfer system for all mixing and loading operations. If this product is packaged in a 30 gallon drum, you must use a mechanical transfer system which terminates in a drip-free hard coupling which may be used only with a spray or mix tank which has been fitted with a compatible coupling. If you do not presently own or have access to a mechanical transfer system with this type of coupling, contact your dealer for information on how to obtain such a system or to modify your present system. When using a mechanical transfer system, do not remove or disconnect the pump or probe from the container until the container has been emptied and rinsed. The pump or probe system must be used to rinse the empty container and to transfer the rinsate directly to the mixing or spray tank.

Fill the spray tank 1/2 to 3/4 full with clean water. Begin agitation and add the recommended amount of BUCTRIL® + Atrazine Herbicide. Add water to the spray tank to the desired level. Maintain sufficient agitation to ensure a uniform spray mixture during application. If spray mixture is allowed to remain without agitation for even a short period of time, be sure to agitate until uniformly mixed before application.

APPLICATION PROCEDURES

GROUND APPLICATIONS

Use a standard herbicide boom sprayer that provides uniform and accurate application. Sprayer should be equipped with screens no finer than 50 mesh in the nozzle tips and in-line strainers. Use flat fan nozzles spaced a maximum of 20 inches apart and a minimum of 10 gallons of spray volume per acre with a minimum spray pressure of 30 psi at the nozzle tips. Other nozzle types may not provide adequate coverage of the weeds to ensure optimum control. When weed infestations are heavy, use of a higher spray volume and spray pressure will be helpful to ensure uniform coverage. When corn is large enough to interfere with the spray pattern, drop nozzles should be used to obtain uniform coverage of the weeds.

AERIAL APPLICATION

During aerial application, human flaggers are prohibited unless in enclosed vehicles. Aerial application is prohibited within 300 feet of residential areas (e.g. homes, schools, hospitals, shoping areas, etc.)

Use a minimum of 5 gallons of spray volume per acre using diaphragm-type nozzles that produce cone or fan spray patterns with a maximum spray pressure of 40 psi at the nozzle tips. Nozzles should be located no farther than 3/4 distance from the center of the aircraft to the end of the wing rotor. Nozzles should be oriented at an angle between straight down and straight back.

The aircraft should discharge the spray a maximum of 10 feet above the crop. To minimize drift, do not apply during periods of gusty winds or when wind exceeds 5 mph.

AUTOMATED SPRINKLER IRRIGATION SYSTEMS

If hand-moved pipe is used for chemigation, the pipe must not be handled in any way until 24 hours after chemigation has been completed and residues have been flushed from the system. When applying by chemigation, no person may enter the application site unless in an enclosed vehicle.

BUCTRIL® +Atrazine Herbicide at 3 pts/A can be applied through overhead sprinkler irrigation systems to field corn, popcorn, and grain sorghum.

Apply BUCTRIL® + Atrazine Herbicide through sprinkler including center pivot, lateral move, side (wheel) roll, solid set or hand move irrigation systems only. Do not apply this product through any other type of irrigation system.

SPECIFIC REQUIREMENTS FOR APPLICATION THROUGH AUTOMATED SPRINKLER IRRIGATION SYSTEMS.

1. The system must contain a functional check valve, vacuum relief valve, and low pressure drain appropriately located on the irrigation pipeline to prevent water source contamination from backflow.
2. The pesticide injection pipeline must contain a functional, automatic quick-closing check valve to prevent the flow of fluid back toward the injection pump.
3. The pesticide injection pipeline must also contain a functional, normally closed, solenoid-operated valve located on the intake side of the injection pump and connected to the system interlock to prevent fluid from being withdrawn from the supply tank when the irrigation system is either automatically or manually shut down.
4. The system must contain functional interlocking controls to automatically shut off the pesticide injection pump when the water pump motor stops.
5. The irrigation line or water pump must include a functional pressure switch which will stop the water pump motor when the water pressure decreases to the point where pesticide distribution is adversely effected.
6. Systems must use a metering pump, such as a positive displacement injection pump (e.g., diaphragm pump) effectively designed and constructed of materials that are compatible with pesticides and capable of being fitted with a system interlock.
7. Do not apply when wind speed favors drift beyond the area intended for treatment.
8. Agitation is recommended in the pesticide supply tank when applying the BUCTRIL® Herbicide + Atrazine.
9. BUCTRIL® Herbicide + Atrazine should be applied continuously for the duration of the water application with center pivot and continuous lateral move systems. Application of BUCTRIL® + Atrazine should be made during the last 30 to 45 minutes of the irrigation set with other overhead sprinkler system.
10. For best performance, set the sprinkler system to deliver approximately 0.5 inch or less of water per acre.
11. Remove scale, pesticide residues and other foreign matter from the supply tank and entire injector system. Flush with clean water.
12. If BUCTRIL® + Atrazine Herbicide is diluted in the supply tank, fill the tank with half of the water amount desired, add the BUCTRIL® + Atrazine and then add remaining water amount with agitation.
13. Start the sprinklers and then inject BUCTRIL® + Atrazine Herbicide into the irrigation line. BUCTRIL® + Atrazine should be injected with a positive displacement pump into the main line at least 8 feet ahead of a right angle turn to ensure adequate mixing. Refer to the BUCTRIL® + Atrazine Herbicide label for detailed information on application rates and timings.

CHEMIGATION USER PRECAUTIONS

Application of more than 0.5 inch/acre of irrigation water may result in decreased product performance on certain soils.

Do not apply when conditions favor drift, when system connections or fittings leak, or when nozzles do not rpovide uniform distribution.

Allow sufficient time for pesticide to be flushed through all the lines and nozzles before turning off irrigation water.

Crop injury, lack of effectiveness, or illegal pesticide residues in the crop can result from non-uniform distribution of treated water.

Do not connect an irrigation system used for pesticide application to a public water system.

If you have questions about calibration, you should contact State Extension Service Specialists, equipment manufactures or other experts.

A person knowledgeable of the chemigation system and responsible for its operation, or under the supervision of the responsible person, shall shut the system down and make necessary adjustments should the need arise.

BUCTRIL® + ATRAZINE HERBICIDE RECOMMENDATIONS

Postemergence applications of BUCTRIL® + Atrazine Herbicide at 1 1/2, 2 and 3 pints/A can be applied to field corn, popcorn and grain sorghum using the following use directions. Refer to APPLICATION RATE TABLE for weeds controlled.

APPLICATION RATE TABLE FOR CORN AND GRAIN SORGHUM

WEED SPECIES When determining leaf stage, count all leaves except cotyledonary leaves		BUCTRIL® + ATRAZINE RATE					
		1 1/2 PINTS/A		2 PINTS/A		3 PINTS/A	
		MAXIMUM LEAF (INCHES)	MAXIMUM WEED HEIGHT (INCHES)	MAXIMUM LEAF (INCHES)	MAXIMUM WEED HEIGHT (INCHES)	MAXIMUM LEAF (INCHES)	MAXIMUM WEED HEIGHT (INCHES)
Black Nightshade	Solanum nigrum	4	4	6	6	6	6
Buffalobur	Solanum rostratum	4	4	6	4	6	4
Common Cocklebur	Xanthium pensylvanicum	6	8	8	10	10	12
Common Lambsquarters	Chenopodium album	—	6	—	10	—	12
Common Ragweed	Ambrosia artemisiifolia	6	4	8	6	8	6
Eastern Black Nightshade	Solanum ptycanthum	4	4	6	6	6	6

WEED SPECIES When determining leaf stage, count all leaves except cotyledonary leaves		BUCTRIL® + ATRAZINE RATE					
		1½ PINTS/A		2 PINTS/A		3 PINTS/A	
		MAXIMUM LEAF (INCHES)	MAXIMUM WEED HEIGHT (INCHES)	MAXIMUM LEAF (INCHES)	MAXIMUM WEED HEIGHT (INCHES)	MAXIMUM LEAF (INCHES)	MAXIMUM WEED HEIGHT (INCHES)
Entireleaf Morningglory	Ipomoea hedevacea	—	—	4	3	4	3
Giant Ragweed	Ambrosia trifida	4	4	6	6	6	6
Hemp Sesbania	Sesbania exaltata	4	4	4	4	4	4
Ivyleaf Morningglory	Ipomoea hederacea	3	3	4	4	4	4
Jimsonweed	Datura stramonium	4	4	6	6	6	6
Ladysthumb	Polygonum persicaria	4	4	6	6	8	8
Palmleaf Morningglory	Ipomoea wrightii	—	—	4	3	4	3
Pennsylvania Smartweed	Polygonum pensylvanicum	4	4	6	6	8	8
Prickly Sida	Sida spinosa	—	—	4	1	6	2
Puncturevine	Tribulus terrestris	—	—	—	—	6	—
Purple Morningglory	Ipomoea muricata	—	—	2	3	2	3
Redroot Pigweed[1/]	Amaranthus retroflexus	4	2	6	4	8	6
Small Flower Morningglory	Jacquemontia tamnifolia	—	—	4	3	4	3
Smooth Pigweed[1/]	Amaranthus hybridus	4	2	4	2	6	4
Spiny Pigweed[1/]	Amaranthus spinosus	4	2	6	4	8	6
Sunflower	Helianthus annuus	6	8	8	10	10	12
Tall Morningglory	Ipomoea purpurea	3	3	4	4	4	4
Tall Waterhemp[1/]	Amaranthus tuberculatus	4	2	6	4	8	6
Velvetleaf	Abutilon theophrasti	4	3	6	5	8	6
Venice mallow	Hibiscus trionum	4	2	4	2	4	2
Wild Buckwheat	Polygonum convolvulus	6	8	8	10	10	12
Wild Mustard	Brassica kaber	4	4	4	4	4	4

WEED SUPPRESSED

Canada Thistle Cirsium arvense BUCTRIL® Herbicide + Atrazine suppresses the growth by burning down of top growth. Regrowth may occur.	NOT RECOMMENDED	8″ TO BUD STAGE	8″ TO BUD STAGE

[1/] If pigweeds *(Amaranthus spp.)* present in the field to be treated have been identified as triazine resistant biotypes, use BUCTRIL® + Atrazine Herbicide only at 3 pints/A. Applications should be made when pigweeds do not exceed the 4-leaf stage and 2 inches in height.

FIELD CORN AND POPCORN (Ground and Aerial Applications)

BUCTRIL® + ATRAZINE HERBICIDE RECOMMENDATIONS

		TIMING OF APPLICATION	
PRODUCT	RATE	CROP	WEED
BUCTRIL® + Atrazine	1½ Pints/A	Apply to corn from the 3-leaf stage to the 8-leaf stage, but before corn is 30 inches tall. Make up to 2 applications per season.	See APPLICATION RATE TABLE for list of weeds and corresponding maximum stage of growth that are controlled by BUCTRIL® Herbicide + Atrazine at 1½ pints/A.
	2 Pints/A	Apply to corn from the 3-leaf stage to the 8-leaf stage, but before corn is 30 inches tall. Make up to 2 applications per season.	See APPLICATION RATE TABLE for list of weeds and corresponding maximum stage of growth that are controlled by BUCTRIL® + Atrazine Herbicide at 2 pints/A.
	3 Pints/A	Apply to corn from the 4-leaf stage to the 8-leaf stage, but before corn is 30 inches tall. Make up to 2 applications per season.	See APPLICATION RATE TABLE for list of weeds and corresponding maximum stage of growth that are controlled by BUCTRIL® Herbicide + Atrazine at 3 pints/A.

GRAIN SORGHUM
(Ground and Aerial Applications)

BUCTRIL® + ATRAZINE Herbicide RECOMMENDATIONS

		TIMING OF APPLICATION	
PRODUCT	RATE	CROP	WEED
BUCTRIL® + Atrazine	1½ Pints/A	Apply to grain sorghum from the 2-leaf stage to the 10-leaf stage, not to exceed 10 inches in height. Make up to 2 applications per season.	See APPLICATION RATE TABLE for list of weeds and corresponding maximum stage of growth that are controlled by BUCTRIL® + Atrazine Herbicide at 1½ pints/A.
	2 Pints/A	Apply to grain sorghum from the 3-leaf stage to the 10-leaf stage, not to exceed 10 inches in height. Make up to 2 applications per season.	See APPLICATION RATE TABLE for list of weeds and corresponding maximum stage of growth that are controlled by BUCTRIL® + Atrazine Herbicide at 2 pints/A.

PRODUCT	RATE	TIMING OF APPLICATION CROP	WEED
	3 Pints/A	Apply to grain sorghum from the 4-leaf stage to the 10-leaf stage, not to exceed 10 inches in height. Make up to 2 applications per season.	See APPLICATION RATE TABLE for list of weeds and corresponding maximum stage of growth that are controlled by BUCTRIL® + Atrazine Herbicide at 3 pints/A.

RESTRICTIONS AND LIMITATIONS

Postemergence applications of BUCTRIL® + Atrazine Herbicide will not adequately control grasses. Therefore, it is recommended that a suitable grass control program be used to provide any required grass control.

Do not add a spray additive, or mix with liquid fertilizers because excessive crop injury may occur.

Do not apply BUCTRIL® + Atrazine Herbicide if heavy rainfall is expected within 1 hour of application.

Seed corn producers should consult the respective seed corn company regarding tolerance of seed production inbred lines to BUCTRIL® + Atrazine Herbicide.

Do not cut crop for feed or graze within 30 days after application.

A second application of BUCTRIL® + Atrazine Herbicide may be applied if a new flush of weeds occurs following the first application. The total cumulative rate should not exceed 6 pints/A per season.

Do not use BUCTRIL® + Atrazine Herbicide on any crop other than stated on this label.

Application to grain sorghum growing under stress caused by minor element deficiency or to grain sorghum growing on highly calcareous soil may result in crop injury.

CROP ROTATION (Field corn, popcorn and grain sorghum)

Do not rotate to any crop except field corn, popcorn or grain sorghum until the following year, or injury to the rotated crop may occur.

If BUCTRIL® Herbicide + Atrazine is applied after mid-June, do not rotate with crops other than field corn, popcorn or grain sorghum the next year, or injury to the rotated crop may occur.

In the High Plains and intermountain areas of the West where rainfall is sparse and erratic or where irrigation is required, use only when field corn, popcorn or grain sorghum is grown the following year or when a crop of untreated field corn, popcorn or grain sorghum is to precede other rotational crops.

Following harvest of grain sorghum, plow (moldboard or disk plow) and thoroughly till soil in fall or spring to minimize crop injury to spring-seeded rotational crops.

Injury may occur to soybeans planted the year following application on soils having a calcareous surface layer.

Do not plant sugar beets, tobacco, vegetables (including dry beans), spring-seeded small grains, or small seeded legumes and grasses the year following application, or crop injury may occur.

STORAGE AND DISPOSAL

STORAGE:
Do not contaminate water, food or feed by storage or disposal.
Do not store near fertilizers or seeds. Store at temperatures above 32°F. If exposed to freezing temperatures, store at temperatures above 55°F for 24 hours or until completely thawed. Shake well before using.

PESTICIDE DISPOSAL: Wastes resulting from the use of this product may be disposed of on site or at an approved waste disposal facility.

CONTAINER DISPOSAL: Triple rinse (or equivalent). Then puncture and dispose of in a sanitary landfill, or incineration, or if allowed by State and Local authorities, by burning. If burned stay out of smoke.

Metal
Triple rinse (or equivalent). Then offer for recycling or reconditioning, or puncture and dispose of in a sanitary landfill, or incineration, or if allowed by State and Local authorities.

LIMITED WARRANTY AND DISCLAIMER

The manufacturer warrants that this product conforms to the chemical description on the label; that this product is reasonably fit for the purposes set forth in the directions for use when it is used in accordance with such directions; that the directions, warnings and other statements on this label are based upon responsible experts' evaluation of reasonable tests of effectiveness, of toxicity to laboratory animals and to plants, and of residues on food crops, and upon reports of field experience. Tests have not been made on all varieties or in all states or under all conditions. THE MANUFACTURER NEITHER MAKES NOR INTENDS, NOR DOES IT AUTHORIZE ANY AGENT OR REPRESENTATIVE TO MAKE, ANY OTHER WARRANTIES, EXPRESS OR IMPLIED, AND IT EXPRESSLY EXCLUDES AND DISCLAIMS ALL IMPLIED WARRANTIES OF MERCHANTABILITY OR FITNESS FOR A PARTICULAR PURPOSE.

THIS WARRANTY DOES NOT EXTEND TO, AND THE BUYER SHALL BE SOLELY RESPONSIBLE FOR, ANY AND ALL LOSS OR DAMAGE WHICH RESULTS FROM THE USE OF THIS PRODUCT IN ANY MANNER WHICH IS INCONSISTENT WITH THE LABEL DIRECTIONS, WARNINGS OR CAUTIONS.

BUYER'S EXCLUSIVE REMEDY AND MANUFACTURER'S OR SELLER'S EXCLUSIVE LIABILITY FOR ANY AND ALL CLAIMS, LOSSES, DAMAGES OR INJURIES RESULTING FROM THE USE OR HANDLING OF THIS PRODUCT, WHETHER OR NOT BASED IN CONTRACT, NEGLIGENCE, STRICT LIABILITY IN TORT OR OTHERWISE, SHALL BE LIMITED, AT THE MANUFACTURER'S OPTION, TO REPLACEMENT OF, OR THE REPAYMENT OF THE PURCHASE PRICE FOR, THE QUANTITY OF PRODUCT WITH RESPECT TO WHICH DAMAGES ARE CLAIMED. IN NO EVENT SHALL MANUFACTURER OR SELLER BE LIABLE FOR SPECIAL, INDIRECT OR CONSEQUENTIAL DAMAGES RESULTING FROM THE USE OR HANDLING OF THIS PRODUCT.

NOTICE TO BUYER

Purchase of this material does not confer any rights under patents governing this product or the use thereof in countries outside of the United States.

BUCTRIL® is a registered trademark of Rhône-Poulenc Ag Company

AG 90009

THIS SPECIMEN LABEL IS INTENDED FOR USE ONLY AS A GUIDE IN PROVIDING GENERAL INFORMATION REGARDING THE DIRECTIONS, WARNINGS AND CAUTIONS ASSOCIATED WITH THE USE OF THIS PRODUCT. AS WITH ANY AGRICULTURAL CHEMICAL, ALWAYS FOLLOW THE LABEL INSTRUCTIONS ON THE PACKAGE BEFORE USING.

BUTYRAC® 175

Controls Cocklebur and Other Broadleaf Weeds in Soybeans and Peanuts

ACTIVE INGREDIENT:

4-(2,4-Dichlorophenoxy) butyric acid, dimethylamine salt	23.0%
INERT INGREDIENTS:	77.0%
TOTAL	100.0%

*4-(2,4-Dichlorophenoxy) butyric acid equivalent 19.5% by weight or 1¾ pounds per gallon.

EPA Reg. No. 264-164 EPA Est. No. 264-MO-01

KEEP OUT OF REACH OF CHILDREN
CAUTION

For PRODUCT USE Information Call 1-800-334-9745
For MEDICAL and TRANSPORTATION Emergencies ONLY Call 24 Hours A Day 1-800-334-7577

PRECAUTIONARY STATEMENTS

CAUTION

HAZARDS TO HUMANS AND DOMESTIC ANIMALS

Harmful if swallowed. Avoid contact with skin, eyes or clothing.

STATEMENT OF PRACTICAL TREATMENT

IF SWALLOWED: If patient is conscious and alert, give 2-3glasses of water to drink. Do not induce vomiting. GET MEDICAL ATTENTION.

IF IN EYES: Hold eyelids open and flush with a steady gentle stream of water for at least 15 minutes. GET MEDICAL ATTENTION, preferably an ophthalmologist.

IF IN SKIN: Wash skin with plenty of soap and water, if available, while removing contaminated clothing. Wash clothing separately before reuse.

NOTE TO PHYSICIAN

No specific antidote is available. Treat symptomatically. Overexposure to materials other than this product may have occurred.

ENVIRONMENTAL HAZARDS

Do not apply directly to water or wetlands (swamps, bogs, marshes and pot holes). Do not contaminate water when disposing of equipment washwaters. Avoid spray drift to susceptible plants such as cotton, beans, tomatoes and ornamentals, as this product may cause injury. Coarse sprays are less likely to drift. Do not use same spray equipment for other purposes.

Under very high temperature, vapors from this product may injure susceptible plants in the immediate vicinity.

Do not use in a greenhouse.

DIRECTIONS FOR USE

IT IS A VIOLATION OF FEDERAL LAW TO USE THIS PRODUCT IN A MANNER INCONSISTENT WITH ITS LABELING.

GENERAL INFORMATION

A small amount of 2,4-Dd or MCPA, mixed with BUTYRAC® 175 can cause serious injury to soybeans. A sprayer previously used to apply these chemicals must be thoroughly cleaned before applying BUTYRAC® 175.

Do not apply BUTYRAC® through any type of irrigation system.

This material is not recommended as a preemergence treatment. Do not use on sweet clover.

SOYBEANS

OVER-ALL APPLICATIONS: For control of cocklebur and morningglory.

Use 0.8 to 1.0 pint of BUTYRAC® 175 per acre (equivalent to 0.175 to 0.22 pounds active 2,4-DB acid per acre). It is very important to calibrate sprayers before application to be sure you are applying the right amount.

Apply BUTYRAC® 175 from 7 to 10 days before bloom up to midbloom, when soybeans are about knee-high and growing actively. Soybean foliage should be dark green indicating nodulation and nitrification are already underway. Do not apply over-all sprays earlier than this stage of growth.

IMPORTANT—MIDWEST: Cocklebur must form a protective canopy over the bean plants before spraying or injury to the soybeans may result. Application after mid-bloom may affect flowering and reduce soybean yields.

Apply as a broadcast spray. Use a boom sprayer with flat fan-type nozzles. Space nozzles so that one nozzle is over each row and over each middle. Adjust the height of the boom above the crop to give complete coverage of all weeds. Use low pressure (20-40 psi). Do not use boom jet nozzles.

For a time, soybeans will show some effects from the chemical in the form of temporary twisting or curling of the stems and leaves. However, these effects disappear in about 3 to 5 days. Cocklebur will die out in from 3 days to 3 weeks.

DIRECTED BAND APPLICATIONS

For control of cocklebur and morningglory, apply the equivalent of 0.8 to 1.0 pint of BUTYRAC® 175 per broadcast acre (equivalent to 0.175 to 0.22 pounds active 2,4-DB acid per acre). Use sufficient water for good coverage, at least 10 gallons per acre.

Treat when beans are at least 8 inches tall. Adjust nozzle height to allow spray to contact no more than the bottom one-third of soybean plants. Do not allow spray or spray drift to contact growing terminals of beans, as excessive crop injury will result.

For best results treat when the cocklebur and morningglory are no more than 3 inches tall. All leaves and growing terminals of weeds should be contacted by the spray.

For control of weeds one inch or less in height and/or suppression weeds greater than about one inch in height such as velvetleaf, jimsonweed, common ragweed and lambsquarters, apply 1.6 to 1.8 pints of BUTYRAC® 175 per acre (equivalent to 0.35 to 0.4 lb. active 2,4-DB acid per acre). Do not use more than two applications per season. Stunting of the soybeans may be observed when two applications of the higher rates are used for weed control.

To insure proper placement of spray on lower one-third of soybean plants, apply with sprayer nozzles mounted on skid shoes, oiling shoes, or cultivators with gauge wheels. Do not mount on booms with drop nozzles, or on cultivators without gauge wheels. Use fan type nozzle tips #8001 or larger, or their equivalent. Nozzle pressure should be less than 40 psi to reduce spray drift.

IMPORTANT

Drought stressed beans should not be sprayed.

Do not harvest beans within 60 days after application.

Use either the over-all or the directed application. Do not use both applications on the same field in the same season.

Do not use this product on soybeans that show symptoms of Phytophthora root rot disease.

Do not add any wetting agents or detergents to the spray solution.

Be sure sprayer is clean before using BUTYRAC® 175 in it.

DIRECTED BAND APPLICATIONS OF MIXTURES OF BUTYRAC® 175 AND LOROX 50W*

For control of cocklebur, morningglory, sesbania, velvetleaf, lambsquarters, prickly sida (teaweed), jimsonweed, ragweed and crabgrass (2 leaf), apply 1.0 pint of BUTYRAC® 175 plus 1 lb. Lorox per broadcast acre (equivalent to 0.22 pounds active 2,4-DB and 1/2 lb. Linuron active per acre). Use a proportional amount of the two chemicals on the band (for example, on a 12 inch band in 36 inch rows use 1/3 pt of BUTYRAC® 175 and 1/3 lb. of Lorox per acre). Use an #8002 T-Jet (or equivalent) or larger with a minimum of 25 gals. per broadcast acre and spray pressure of 20 to 25 psi, to avoid spray drift. One pint of surfactant such as DuPont WK per 2.5 gallons of spray mix may be added. Continuous agitation in the spray tank is required to keep the material in suspension. Agitate by mechanical or hydraulic means; if by-pass or return line is used, it should terminate at the bottom of the tank to minimize foaming. Openings in nozzle screens should be equal to or larger than 50 mesh.

Apply directed spray when beans are at least 8 inches tall and when weeds do not exceed 2 inches in height. Adjust nozzle height to spray the lower one-third of the soybean plant. Apply with sprayer nozzles mounted on skid shoes, oiling shoes, or on cultivators with gauge wheels. Do not mount on booms with drop nozzles or on cultivators without gauge wheels. If a new flush of weeds occurs after initial treatment, make a second application but do not apply within 60 days of harvest or make more than 2 applications per growing season. Do not allow spray or spray drift to contact growing terminals of beans, as excessive crop injury will result. Do not feed soybean forage or hay to livestock from fields treated with this mixture. Harvested soybeans may be used for feed or oil purposes.

NOTE: Do not use on sands, loamy sands or gravelly soils or on exposed subsoils. Do not use on soils containing less than ½% organic matter as crop injury may result.

* Lorox 50W is a registered trademark of E.I. DuPont de Nemours & Co (Inc.)

PEANUTS

BUTRYAC® 175 has given control or suppression of the following weeds depending on the size of the weed.

Common Name	Botanical Name	Suppression only if larger than:
Pigweed or careless weed	*Amaranthus sp.*	3 inches
morningglory	*Ipomea sp.*	3 foot
cocklebur	*Xanthium sp.*	3 foot
velvetleaf	*Abutilon theophrasti*	1 inch
Virginia Copperleaf	*Acalypha virginica*	1 inch
teaweed	*Sida spinosa*	1 inch
common ragweed	*Ambrosia artemisiifolia*	1 inch
jimsonweed	*Datura stramonium*	1½ inch
devils claw	*Proboscidea louisianica*	2 inches
goatweed	*Croton lindheimeri*	1 inch
lambsquarters	*Chenopodium album*	1 inch
silverleaf nightshade	*Sodanum elaeagnifolium*	4 inches

For best results apply BUTYRAC® 175 when the weeds are one to two inches in size.

For ground application: Use 10 to 30 gallons of spray solution per acre. **For air application:** Use 5 gallons of spray solution per acre.

Georgia, Virginia, North Carolina, Alabama, Florida, Arkansas

Apply 0.9 pints to 1.1 pints (equivalent to 0.2 to 0.25 pounds active 2,4-DB acid) of BUTYRAC® 175 per acre on peanuts 2 to 12 weeks after planting. For late germinating cocklebur and morningglory, apply a second application about 3 weeks after the first application.

In the **Virginia-Carolina** area, do not apply later than 45 days before harvest.

Oklahoma, Texas, New Mexico

Apply 0.9 to 1.8 pints (0.2 to 0.4 pounds active 2,4-DB acid) of BUTYRAC® 175 per acre. Use the low rate on morningglory and cocklebur up to 12 inches in size. Use the higher rate on other weeds. BUTYRAC® 175 may be applied twice during the season. For best suppression of Sida, space the treatments 14 days apart. The second application should not be made later than 100 days after planting or 30 days before harvest.

AMOUNT OF BUTYRAC® 175 TO USE

Acreage to Spray	Rate Desired @0.2lb/A Active	Rate Desired @0.4lb/A Active
1	0.9 pt. or 14.6 oz.	1.8 pt. or 29.0 oz.
8.75	1 gal.	2 gal.

Do not feed treated peanut vines and peanut hay to livestock. Do not apply BUTYRAC® 175 if peanuts are under stress from drought as injury to peanuts may occur.

Do not use on sweet clover.

STORAGE AND DISPOSAL

STORAGE

Do not store near fertilizers, seeds, insecticides, or fungicides.

PESTICIDE DISPOSAL

Do not contaminate water, food, or feed by storage or disposal. Wastes resulting from the use of this product may be disposed of on site or at an approved waste disposal facility.

CONTAINER DISPOSAL

Do not reuse empty container. Triple rinse or equivalent. Then offer for recycling, or reconditioning or puncture and dispose of in a sanitary landfill or incineration or if allowed by state and local authorities, by burning. If burned, stay out of smoke.

LIMITED WARRANTY AND DISCLAIMER

The manufacturer warrants that this product conforms to the chemical description on the label; that this product is reasonably fit for the purposes set forth in the directions for use when it is used in accordance with such directions; and that the directions, warnings and other statements on this label are based upon responsible experts' evaluation of reasonable tests of effectiveness, of toxicity to laboratory animals and to plants, and of residues on food crops, and upon reports of field experience. Tests have not been made on all varieties or in all states or under all conditions. THE MANUFACTURER NEITHER MAKES NOR INTENDS, NOR DOES IT AUTHORIZE ANY AGENT OR REPRESENTATIVE TO MAKE, ANY OTHER WARRANTIES, EXPRESS OR IMPLIED, AND IT EXPRESSLY EXCLUDES AND DISCLAIMS ALL IMPLIED WARRANTIES OF MERCHANTABILITY OR FITNESS FOR A PARTICULAR PURPOSE.

THIS WARRANTY DOES NOT EXTEND TO, AND THE BUYER SHALL BE SOLELY RESPONSIBLE FOR, ANY AND ALL LOSS OR DAMAGE WHICH RESULTS FROM THE USE OF THIS PRODUCT IN ANY MANNER WHICH IS INCONSISTENT WITH THE LABEL DIRECTIONS, WARNINGS OR CAUTIONS.

BUYER'S EXCLUSIVE REMEDY AND MANUFACTURER'S OR SELLER'S EXCLUSIVE LIABILITY FOR ANY AND ALL CLAIMS, LOSSES, DAMAGES, OR INJURIES RESULTING FROM THE USE OR HANDLING OF THIS PRODUCT, WHETHER OR NOT BASED IN CONTRACT, NEGLIGENCE, STRICT LIABILITY IN TORT OR OTHERWISE, SHALL BE LIMITED, AT THE MANUFACTURER'S OPTION TO REPLACEMENT OF, OR THE REPAYMENT OF THE PURCHASE PRICE FOR, THE QUANTITY OF PRODUCT WITH RESPECT TO WHICH DAMAGES ARE CLAIMED. IN NO EVENT SHALL MANUFACTURER OR SELLER BE LIABLE FOR SPECIAL, INDIRECT OR CONSEQUENTIAL DAMAGES RESULTING FROM THE USE OR HANDLING OF THIS PRODUCT.

NOTICE TO BUYER

Purchase of this material does not confer any rights under patents governing this product or the use thereof in countries outside of the United States.

BUTYRAC is a trademark of Rhône-Poulenc

AG 90011

EPA approval: 5/20/74

THIS SPECIMEN LABEL IS INTENDED FOR USE ONLY AS A GUIDE IN PROVIDING GENERAL INFORMATION REGARDING THE DIRECTIONS, WARNINGS AND CAUTIONS ASSOCIATED WITH THE USE OF THIS PRODUCT. AS WITH ANY AGRICULTURAL CHEMICAL, ALWAYS FOLLOW THE LABEL INSTRUCTIONS ON THE PACKAGE BEFORE USING.

BUTYRAC® 200

BROADLEAF HERBICIDE

Controls Broadleaf Weeds in Soybeans, Peanuts, Alfalfa, Trefoil and Clovers.

ACTIVE INGREDIENT:	
4-(2,4-Dichlorophenoxy)butyric acid, dimethylamine salt*	25.9%
INERT INGREDIENTS:	74.1%
TOTAL	100.0%

*4-(2,4-Dichlorophenoxy)butyric acid equivalent 22.0% by weight or 2 pounds per gallon.

EPA Reg. No. 264-105ZA EPA Est. No. 264-MO-01

KEEP OUT OF REACH OF CHILDREN

DANGER PELIGRO

PRECAUCION AL USUARIO: Si usted no lee ingles, no use este producto hasta que la informacion en la etiqueta le ha sido explicada ampliamente.

PRACTICAL TREATMENT

IF IN EYES: FLUSH EYES WITH LARGE AMOUNT OF WATER AND GET MEDICAL ATTENTION.

For PRODUCT USE Information Call 1-800-334-9745
For MEDICAL AND TRANSPORTATION Emergencies ONLY Call 24 Hours A Day 1-800-334-7577

PRECAUTIONARY STATEMENTS

DANGER

HAZARD TO HUMANS AND DOMESTIC ANIMALS

Corrosive. Causes eye damage. Harmful if swallowed. May be fatal if absorbed through the skin. Do not get in eyes, on skin or on clothing. Avoid breathing spray mist. Wear goggles or face shield when handling concentrate. In case of contact immediately flush eyes or skin with plenty of water. Get medical attention if irritation persists. Remove and wash contaminated clothing before reuse.

STATEMENT OF PRACTICAL TREATMENT

IF SWALLOWED: If patient is conscious and alert, give 2-3 glasses of water to drink. Do not induce vomiting. GET MEDICAL ATTENTION.

IF ON SKIN: Immediately wash skin with plenty of soap and water, if available, while removing contaminated clothing and shoes. Shoes and clothing contaminated by substantial spillage of concentrated product should be discarded in a manner which limits further exposure. Otherwise, wash clothing separately before reuse. GET MEDICAL ATTENTION.

IF INHALED: Remove victim to fresh air. If not breathing, give artificial respiration. Administer oxygen if necessary. GET MEDICAL ATTENTION.

IF IN EYES: Hold eyelids open and flush with a steady, gentle stream of water for at least 15 minutes. GET MEDICAL ATTENTION, PREFERABLY AN OPHTHALMOLOGIST.

NOTE TO PHYSICIAN No specific antidote is available.

All treatments should be based on observed signs and symptoms of distress in the patient. Overexposure to materials other than this product may have occurred.

ENVIRONMENTAL HAZARDS

This product is toxic to fish. Do not apply directly to water or wetlands (swamps, bogs, marshes and potholes). Do not contaminate water when disposing of equipment washwaters. Do not contaminate irrigation ditches or water used for irrigation or domestic purposes. Do not use in or near a greenhouse.

PROTECTION STATEMENTS

Do not apply BUTYRAC® 200 Broadleaf Herbicide directly to or otherwise permit it to come in contact with cotton, okra, grapes, tomatoes, fruit trees, vegetables, flowers, or other desirable crop or ornamental plants. Do not permit spray mist to drift onto susceptible plants since very small quantities of the BUTYRAC® 200 Broadleaf Herbicide can cause severe injury during the growing or dormant periods. Use coarse sprays to minimize drift. Do not apply with nozzles that produce fine spray droplets.

Drift from ground application may be reduced by: (1) keeping the spray boom as near to the crop as possible in order to obtain complete coverage; (2) by applying 10 gallons or more of spray per acre; (3) by using no more than 20 pounds of pressure at the nozzle tips; and (4) by not spraying when wind exceeds 5 miles per hour.

Drift from aerial application may be reduced by: (1) applying as near to the target as possible to obtain adequate coverage; (2) by applying 5 or more gallons of spray per acre; (3) by using 20 pounds pressure or less at the nozzle tips; (4) by using nozzles which produce a coarse spray pattern; (5) by spraying when there is no possibility for a temperature inversion at time of spraying.

Applications by aircraft, ground rig and hand sprayers should be carried out only when there is no hazard from spray drift.

DIRECTIONS FOR USE

IT IS A VIOLATION OF FEDERAL LAW TO USE THIS PRODUCT IN A MANNER INCONSISTENT WITH ITS LABELING.

GENERAL INFORMATION

Spray tank residues of 2,4-D or MCPA mixed with Butyrac® 200 Broadleaf Herbicide can cause serious crop or ornamental plant injury. A sprayer previously used to apply these chemicals must be thoroughly cleaned with alkali and water before applying BUTYRAC® 200.

Evaluate tank mixtures not listed on this label for compatibility and crop safety on a small area before applying to the entire field.

BUTYRAC® 200 may form an insoluble precipitate in very hard water. If you expect to mix BUTYRAC® 200 with very hard water, test compatibility by mixing a small amount of BUTYRAC® 200 in the proposed dilution ratios, shake and observe. A compatibility agent approved for use on growing crops such as UNITE or E-Z MIX may be tested to reduce precipitation. Whenever hard water is used to dilute BUTYRAC® 200, spray immediately and do not allow spray mixture to sit overnight.

Do not apply BUTYRAC® 200 Broadleaf Herbicide through any type of irrigation system.

WEED CONTROL IN SOYBEANS

OVER-THE-TOP APPLICATION

(For control of cocklebурs and annual morningglories in Southern states.) Entire leaf or ivyleaf morningglory *(Ipomea hederacea)* is most responsive. Activity on other morningglory species such as pitted or palmleaf may range from suppression to control depending on weed size, rate used and environmental conditions. In general, actively growing, smaller weeds are most responsive.

IMPORTANT: Over-the-top application can reduce yields.

MIDWEST: When BUTYRAC® 200 is applied alone at the 0.7 to 0.9 pint rates, use it only as a directed band application in IA, IL, IN, KS, KY (except the KY Purchase area), MO (except the MO Bootheel), MI,

MN, NE, ND, OH, SD and WI. See directions for BAND APPLICATION.

OTHER SOYBEAN PRODUCING AREAS: Limit over-the-top application to those cultivars and conditions known to provide adequate tolerance.

Apply 0.7 to 0.9 pint (11.2 to 14.4 fluid ounces) of BUTYRAC® 200 Broadleaf Herbicide per acre as a broadcast spray. It is very important to calibrate sprayers before application to be sure the right amount is applied.

Apply BUTYRAC® 200 Broadleaf Herbicide from 7 to 10 days before bloom up to mid-bloom, when soybeans are about knee-high and growing actively. Soybean foliage should be dark green indicating that nodulation and nitrification are already underway. Do not apply over-the-top sprays at these rates earlier than this stage of growth. Application after mid-bloom may affect flowering and reduce soybean yields.

Use a boom sprayer with hollow cone type nozzles. Adjust the height of the boom so nozzles are above the crop to give complete coverage of all weeds. Use low pressure (20-40 psi). Do not use flood type nozzles operating with extremely low pressures as coverage will not be adequate.

For a time, soybeans will show some temporary effects from the chemical in the form of twisted or curled stems and leaves. However, these effects disappear in about 3 to 5 days. Yield may be reduced but is not usually affected.

Soybean herbicide tolerance particularly during flowering can be affected by variety and environmental conditions. If uncertain about the tolerance of Butyrac® 200 over-the-top application, check with local agricultural extension personnel or test first on small area. Allow 3 to 21 days for cocklebur control.

DIRECTED BAND APPLICATION:

To control cockleburs and annual morningglories when soybeans are 8 or more inches tall apply the equivalent of 0.7 to 0.9 pint (11.2 to 14.4 fluid ounces) of BUTYRAC® 200 Broadleaf Herbicide per broadcast acre as a directed band treatment. For best results, spray when the cockleburs and morningglories are no more than 3 inches tall.

To control other broadleaf weeds one inch or less in height and/or suppress weeds including velvetleaf, jimsonweed, common ragweed and lambsquarters taller than about one inch, apply as a directed treatment of 1.4 to 1.6 pints (22.4 to 25.6 fluid ounces) of BUTYRAC® 200 Broadleaf Herbicide per broadcast acre. Use enough water for good coverage, at least 10 gallons per acre. Apply no more than two treatments per season. Soybean plants may be stunted when two applications of the higher rate are used for weed control.

Apply with sprayer nozzles mounted on skid shoes, oiling shoes or cultivators with gauge wheels to insure proper placement of spray on only the lower one-third of the soybean plants. *Do not mount nozzles on booms with drop pipes or on cultivators without gauge wheels.* Use flat fan type nozzles, 8001 or larger or the equivalent. *Do not allow spray to contact growing terminals of bean plants, as excessive crop injury will result.* Nozzle pressure should be less than 40 psi to control spray drift.

USE PRECAUTIONS:

IMPORTANT: Beans stressed by drought or other influences should not be sprayed. Do not use this product on soybeans that show symptoms of disease such as Phytophthora root rot.

Do not apply under weather conditions which favor drift. Do not allow to contact other crops or desirable plants.

Be sure sprayer is clean before applying BUTYRAC® 200 Broadleaf Herbicide.

Do not graze or feed soybean hay within 60 days after application of Butyrac 200 tank mix application.

Do not harvest beans within 60 days after spray application.

Do not treat soybeans with a tank mixture of BUTYRAC 200 Broadleaf Herbicide and SEVIN® Carbaryl Insecticide as severe injury may result.

TANK MIXTURES OF BUTYRAC® 200 BROADLEAF HERBICIDE AND OTHER HERBICIDES

Applying tank mixtures of low rates of BUTYRAC® 200 Broadleaf Herbicide with other soybean herbicides as directed below can reduce competition from early weed populations and can improve weed control or control mid- to late-season weed populations, thus minimizing the likelihood of yield reduction in soybean fields with heavy broadleaf weed infestations. However, treating soybeans under stress (as from drought or disease such as Phytophthora root rot) or in any other manner not directed on this label can cause crop injury and yield reduction. The advantage of treating weeds with an appropriate tank mixture should be balanced against any potential for loss. Base your decision on an evaluation of each factor. Follow the local recommendations of your state, Cooperative Extension Service, or other agricultural weed control authority. Tank mixtures of BUTYRAC® 200 with other herbicidees for use on soybeans are not registered for use in California.

BROADCAST OVERALL TREATMENT

BUTYRAC® 200 BROADLEAF HERBICIDE PLUS ALANAP®-L FOR SOYBEANS

For postemergence control of mid-season cockleburs, giant ragweed, wild mustard, volunteer sunflower.

For best results use 3 ounces of BUTYRAC® 200 plus 2 quarts of Alanap-L on weeds that are 10-12 inches tall. For weeds taller than 12 inches, use 4 ounces of BUTYRAC® 200 and 3 quarts of Alanap-L. BUTYRAC® 200 + Alanap-L used according to directions will control or suppress growth of above weeds.

GROUND APPLICATION

Spray 3 to 4 ounces of BUTYRAC® 200 Broadleaf Herbicide with 2 to 3 quarts of Alanap-L in 10 to 20 gallons of water per acre (broadcast rate) on soybeans that are about 18 inches tall (7 to 10 days before bloom through mid-bloom).

The spray boom should be positioned 18-24" above the tops of the soybean plants or weeds and be equipped with hollow cone nozzles. Maintain a sufficient spray pressure (approximately 40 psi) to insure thorough coverage of all weeds.

AERIAL APPLICATION

Apply 3 to 4 ounces of BUTYRAC® 200 Broadleaf Herbicide with 2 quarts of Alanap-L on weeds that are 10-12 inches tall such as common cocklebur, giant ragweed, wild mustard or sunflower. However, if these weeds are taller than 12 inches use 4 ounces of BUTYRAC® 200 with 3 quarts of Alanap-L.

If applied according to directions BUTYRAC® 200 Herbicide with Alanap-L will control or suppress growth of above weeds, reducing their competition with the soybean crop.

NOTE: Applications of this tank mix should not be made when weather conditions favor drift. Do not apply this tank mix if rain is expected within 6 hours after application. Observe all precautions and limitations on the labels for BUTYRAC® 200 and Alanap-L.

BUTYRAC® 200 BROADLEAF HERBICIDE PLUS BASAGRAN® HERBICIDE FOR SOYBEANS.

For improved postemergence control of annual morningglories and cocklebur.

To the label rate of Basagran Herbicide selected according to the other weed species present, add 2 fluid ounces per acre of BUTYRAC® 200 Broadleaf Herbicide for annual morningglories and cockleburs.

Broadcast the tank mix of BUTYRAC® 200 Broadleaf Herbicide and Basagran Herbicide when weeds are actively growing and before annual morningglory vines are a maximum of 6 inches long in the South (AL, AR, FL, GA, LA, MS, NC, OK, SC, TN, TX, VA) or a maximum of 10 inches long in all other states. This is normally up to the time when soybeans have two expanded trifoliate leaves. If application is delayed until weeds are larger than maximum size stated on the Basagran herbicide label, control will be inadequate.

Use a minimum of 20 gallons of total spray mixture per acre with ground equipment and 40 psi pressure with flat fan or hollow cone nozzles spaced 20 inches apart.

Fill the spray tank half way with water, add the recommended amount of BUTYRAC® 200 and Basagran Herbicides while the agitator is running, then finish filling.

The tank mix is effective partly through contact action, so weeds must be thoroughly covered with spray. Canopies of large crop and weed leaves shelter smaller weeds and prevent adeqate spray coverage. Soybeans tolerate the tank mix of BUTYRAC® 200 Broadleaf Herbicide and Basagran; however, under certain conditions soybean leaves may burn, crinkle and bronze.

NOTE: Observe all precautions and limitations on the labels for BUTYRAC® 200 and Basagran.

BUTYRAC® 200 BROADLEAF HERBICIDE PLUS BLAZER® 2L or TACKLE® HERBICIDE FOR SOYBEANS

For controlling morningglories, cockleburs, common ragweed, jimsonweed and redroot pigweed when weed size exceeds that specified on the Blazer 2L or TACKLE® label.

Mix 2 fluid ounces of BUTYRAC® 200 Broadleaf Herbicide with 1.5 to 2 pints of Blazer 2L or 2 pints of TACKLE® Herbicide for each acre being treated.

For optimum control, apply the tank misture of BUTYRAC® 200 Herbicide and Blazer 2L or TACKLE® to weeds growing actively, before they are 12 inches tall or long. Applications at later stages will result in partial control or suppression. The use of this tank mix will cause soybean foliage injury and may reduce yields. Control with this mixture may decrease with increasing weed size, weed density or heavy soybean canopy, due to poor spray coverage. Treatment at last stages will only partially control or merely suppress large weeds.

For best results, the tank mixture of BUTYRAC® 200 Broadleaf Herbicide and Blazer 2L or TACKLE® Herbicide should be applied with ground equipment. Always use at least 20 gallons of water per acre to assure maximum coverage of weeds partially covered by soybean foliage. In extremely weedy fields, increasing the spray volume above 20 gallons of water will usually assist in activity. When using this tank mixture, nozzle pressures must not exceed 40 psi to avoid excessive spray drift. The addition of surfactant is not recommended since hormonal crop response will be intensified.

NOTE: Observe all precautions and limitations on the labels for BUTYRAC® 200, TACKLE® and Blazer 2L Herbicides.

BUTYRAC® 200 BROADLEAF HERBICIDE PLUS REFLEX 2LC HERBICIDE FOR SOYBEANS

BUTYRAC® 200 Broadleaf Herbicide may be applied in a tank mixture with Reflex 2LC for improved postemergence control of morningglory (annual, ivyleaf, tall); common cocklebur; and ragweed (common, giant).

Use 2-3 ounces of BUTYRAC® 200 with the appropriate label rate of Reflex 2LC. Refer to the Reflex label for details on regarding geographical recommendation, rotation limitations, application stage and weed species controlled. Follow all use precautions and restrictions for both BUTYRAC® 200 and Reflex.

DIRECTED TREATMENT

BUTYRAC® 200 BROADLEAF HERBICIDE PLUS LOROX® L ON SOYBEANS

For directed postemergence control of morningglory, sicklepod, teaweed, cocklebur, velvetleaf, sesbania and other problem weeds.

Apply BUTYRAC® 200 Broadleaf Herbicide plus Lorox L as a directed spray to cover weed foliage with minimum contact of the soybean plant. Do not spray higher than 3 inches on the soybean stem or crop injury may result. Do not spray overtop of soybean plants.

For soybeans at least 8 inches high and when weeds do not exceed 4 inches in height, apply 13 fluid ounces of BUTYRAC® 200 Broadleaf Herbicide with 1 pint of Lorox L per acre. A second application may be made if needed, but do not make more than two applications per season.

For directed broadcast application: use a single flood-type spray nozzle (K series or equivalent) per middle mounted on an oiling shoe or gauge wheel. For directed band treatment: use two nozzles per row mounted on oiling shoes or gauge wheels, one on each side of row. To avoid spray drift, which may cause crop injury, do not exceed nozzle pressure of 25 psi or use nozzle tips smaller than 8002 TeeJet (or equivalent) and do not spray when it is windy. For best results, use a preemergence treatment or cultivation to control early weed growth and to increase the differential between height of soybeans and weeds.

NOTE: Do not use on sands, loamy sands or gravelly soils, on exposed subsoils, or on soils with less than ½% organic matter. Do not apply within 60 days of harvest. Do not feed soybean forage or hay to livestock from fields treated postemergence. Harvested soybeans may be used for food, feed, or oil purposes.

Observe all precautions and limitations on labels for BUTYRAC® 200 and Lorox L.

BUTYRAC® 200 BROADLEAF HERBICIDE PLUS SENCOR® FOR SOYBEANS

For controlling weeds that escape preplant or preemergence herbicides or for new flushes of weeds germinating after soybeans have emerged.

Apply as a directed spray in a 6 to 8 inch band on each side of the row after soybeans are 8 inches tall and before broadleaf weeds are 3 inches tall and grasses and common ragweed are 1 inch tall.

Add 0.9 pint of BUTYRAC® 200 Broadleaf Herbicide to ⅓-⅔ pound Sencor DF per broadcast acre, depending on the weed species. Use the lowest rate for Florida beggarweed, redroot pigweed (carelessweed), carpetweed, common cocklebur, dayflower, Mexicanweed, common purslane, sicklepod (coffeeweed), velvetleaf, and large and smooth crabgrass.

Use ½ to ⅔ pound Sencor DF per broadcast area for hemp sesbania, prickly sida (spiny sida or teaweed), and ⅔ pound per acre for burcucumber, common ragweed, morningglory species, wild gourd, and wild poinsettia. At the rate of ⅔ pounds Sencor DF per broadcast acre, horsenettle, Florida purslane (Florida pusley) and spotted spurge are suppressed when Sencor is applied before these weeds are 3 inches tall. This rate will suppress broadleaf signalgrass up to 1 inch tall. Equivalent rates of Sencor 4 may be used.

To prepare the mixture, fill the sprayer tank half full of water and add the required amount of Sencor. Agitate thoroughly, then add the required amount of BUTYRAC® 200 Broadleaf Herbicide and add more water to fill sprayer tank to desired level.

For maximum contact activity plus residual weed control, a directed broadcast application between the rows is recommended. For best results, the spray must cover weed foliage with minimum or no contact of soybean foliage. Add Ortho® X-77 non-ionic spreader or similar surfactant/oil combination to the spray mixture to wet the weed leaf surfaces better.

Do not apply Sencor directly to soybeans or serious crop injury will occur. Soybean leaves contacted by the spray will be killed. Limit spray to only the lower ¼ to ⅓ of soybean plants. Soybean leaves contacted by the spray will be killed.

Do not apply Sencor postemergence to Tracy, Semmes, Altona, Vansoy, or Coker 102, Coker 156, Portage, Maple Amber, Govan, NKS-1884, Terra-Vig 505, Terra-Vig 606, Vinton 81, Asgrow 6520, Agri Pro 55 and Agri Pro 71 soybeans.

Observe all precautions and limitations on the labels for BUTYRAC® 200 and Sencor.

WEED CONTROL IN PEANUTS

BUTYRAC® 200 BROADLEAF HERBICIDE WILL CONTROL OR SUPPRESS THE FOLLOWING WEEDS DEPENDING ON WEED SIZE:

Common Name	Botanical Name	Maximum Size Controlled*
Pigweed or carelessweed	*Amaranthus* spp.	3 inches
Morningglory	*Ipomoea* spp.	3 feet
Cocklebur	*Xanthium* spp.	3 feet
Velvetleaf	*Abultilon theophrasti*	1 inch
Virginia copperleaf	*Acalypha virginica*	1 inch
Teaweed	*Sida spinosa*	1 inch
Common ragweed	*Ambrosia artemisiifolia*	1 inch
Jimsonweed	*Datura stramonium*	1.5 inches
Devilsclaw	*Proboscidea louisianica*	2 inches
Goatweed	*Croton lindheimeri*	1 inch
Lambsquarters	*Chenopodium album*	1 inch
Silverleaf nightshade	*Solanum elaeagnifolium*	4 inches

* Growth of larger weeds will be only suppressed.

For best results apply BUTYRAC® 200 Broadleaf Herbicide to peanuts when the weeds are 1 to 2 inches in size.

FOR GROUND APPLICATION: Use 10 to 30 gallons of spray solution per acre.

FOR AIR APPLICATION: Use a minimum of 5 gallons of spray solution per acre.

Alabama, Arkansas, Florida, Georgia, Louisiana, North Carolina, South Carolina, Tennessee

Apply 0.8 pint to 1.0 pint (12.8 to 16 fluid onces) of BUTYRAC® 200 Broadleaf Herbicide per acre on peanuts 2 to 12 weeks after planting. For late germinating cockleburs and morningglories, make a second application about 3 weeks after the first one.

In the Virginia Carolina area, do not apply later than 45 days before harvest.

Oklahoma, Texas, New Mexico

Apply 0.8 to 1.6 pint (12.8 to 25.6 fluid ounces) of BUTYRAC® 200 Broadleaf Herbicide per acre. Use the low rate on morningglories and cockleburs up to 12 inches in size. Use the higher rate on other weeds as listed above. BUTYRAC® 200 Broadleaf Herbicide may be applied twice during the season. For best suppression of Sida make applications 14 days apart. The second application should be made not more than 100 days after planting or 30 days before harvest.

Do not feed treated peanut vines or peanut hay to livestock. Do not apply BUTYRAC® 200 Broadleaf Herbicide if peanut plants are under stress from drought as injury may occur.

WEED CONTROL IN SEEDLING ALFALFA, BIRDSFOOT TREFOIL, ALSIKE, LADINO OR RED CLOVER

Best results will be obtained by applying BUTYRAC® 200 Broadleaf Herbicide to spring- or fall-seeded legumes after their emergence when seedling weeds are small, no more than 2-3 inches high or rosettes less than 2 inches across and not bolting. Later application to more mature weeds or to new growth from old established weeds will not be satisfactory.

Weed and crop seedlings should be growing strongly for greatest selectivity. Control is difficult under cold or droughty conditions. Rainfall within 7-10 days after treatment can cause injury. For irrigated crops, apply BUTYRAC® 200 Broadleaf Herbicide as soon as possible after irrigation. Delay the next irrigation for at least 7 to 10 days after spraying.

WEED CONTROL IN ESTABLISHED ALFALFA

Certain weeds will emerge in the fall and over-winter in the rosette stage. Best control of these weeds will result from postemergence treatment in late fall or early winter rather than in spring. Treat when weeds are no more than 2-3 inches high or rosettes less than 2 inches across. Follow state recommendations for timing and rates to be used under local conditions.

Rainfall within 10 days after treatment can cause injury.

Overhead sprinkler irrigation should be delayed as long as possible (10 days or more) following application of BUTYRAC® 200 Broadleaf Herbicide to avoid washing the chemical into the root zone. In established alfalfa there may be some twisting of stems and malformations of leaves. This condition is usually outgrown.

Note: See additional use precautions below.

BUTYRAC® 200 BROADLEAF HERBICIDE DIRECTIONS FOR USE IN SEEDLING AND ESTABLISHED ALFALFA AND IN SEEDLING BIRDSFOOT TREFOIL, ALSIKE, LADINO OR RED CLOVER

SEEDLING BROADLEAF WEEDS CONTROLLED:

Common lambsquarters, redroot pigweed, field pennycress, wild mustard, common ragweed, cocklebur, yellow rocket, Russian thistle, annual morningglory species:

Less than 1 inch high	1 to 2 quarts per acre
1 to 3 inches high	2 to 3 quarts per acre
Smartweed, curled dock:	
Up to 3 inches high	3 quarts per acre

This material does not adequately control over-wintered pennycress and mustards. Use the higher rate in dry, low humidity areas.

Apply BUTYRAC® 200 Broadleaf Herbicide in 10 or more gallons of water per acre when using ground equipment and in at least 5 gallons per acre with aerial equipment. The volume of spray applied should provide good coverage.

USE PRECAUTIONS

1. Do not graze established alfalfa or feed straw or hay from treated crops to livestock within 30 days after application. Do not graze or feed seedling alfalfa, clover or birdsfoot trefoil within 60 days after application.
2. Do not use on sweet clover, peas or other legumes not mentioned, nor on established clovers grown for seed.
3. Do not apply when crop is stressed from lack of moisture.
4. Do not spray when daytime temperatures are expected to exceed 90° F within the next two or three days or are likely to fall below 40° F during or shortly after treatment.
5. Do not add any wetting agents or detergents to the spray solution.

WEED CONTROL IN CONSERVATION RESERVE PROGRAM AREAS

BUTYRAC® 200 Broadleaf Herbicide can be used to control seedling broadleaf weeds in Conservation Reserve Program (CRP) areas, with cover crop stands of grasses (minimum 6 leaf stage or tillering) seedling or established alfalfa and seedling birdsfoot trefoil, alsike, ladino or red clover. If established alfalfa is in the CRP cover crop treated with BUTYRAC® 200, some temporary twisting of stems and malformation of leaves may occur.

Rainfall within 10 days after treatment can cause injury to legumes in CRP areas.

Overhead sprinkler irrigation should be delayed as long as possible (10 days or more) following application of BUTYRAC® 200 to avoid washing the chemical into the root zone and causing legume injury.

BUTYRAC® 200 BROADLEAF HERBICIDE DIRECTIONS FOR USE IN CRP AREAS CONTAINING GRASSES (MINIMUM 6 LEAF STAGE & TILLERING) AND SEEDLING OR ESTABLISHED ALFALFA AND SEEDLING BIRDSFOOT TREFOIL, ALSIKE, LADINO OR RED CLOVER

Note: See additional use precautions below.

SEEDLING BROADLEAF WEEDS CONTROLLED:

Common lambsquarters, redroot pigweed, field pennycress, wild mustard, common ragweed, cocklebur, yellow rocket, Russian thistle, annual morningglory species:

Less than 1 inch high	1 to 2 quarts per acre
1 to 3 inches high	2 to 3 quarts per acre
Smartweed, curly dock:	
Up to 3 inches high	3 quarts per acre

This material does not adequately control over-wintered field pennycress and mustards.

Use the higher rate in dry, low humidity areas.

Apply BUTYRAC® 200 Broadleaf Herbicide in 10 or more gallons of water per acre when using ground equipment and in at least 5 gallons per acre with aerial equipment. The volume of spray applied should provide good coverage.

Note: See additional use precautions below.

USE PRECAUTIONS

1. Do not graze or harvest treated CRP areas.
2. Do not use on sweet clover, peas or other legumes not mentioned.
3. Do not apply when cover crop is stressed from lack of moisture.
4. Do not spray when daytime temperatures are expected to exceed 90°F, within the next two or three days or, are likely to fall below 40°F during or shortly after treatment as excessive legume injury may result.
5. Do not add any wetting agents or detergents to the spray solution.

STORAGE AND DISPOSAL

STORAGE

Do not store near fertilizers, seeds, insecticides, or fungicides.

PESTICIDE DISPOSAL

Do not contaminate water, food or feed by storage or disposal.
Pesticide wastes are acutely hazardous. Improper disposal of excess pesticide, spray mixture, or rinsate is a violation of Federal law. If these wastes cannot be disposed of by use according to label instructions, contact your State Pesticide or Environmental Control Agency, or the Hazardous Waste representative at the nearest EPA Regional Office for guidance.

CONTAINER DISPOSAL

Triple rinse (or equivalent). Then offer for recycling or reconditioning, or puncture and dispose of in a sanitary landfill, or incineration, or, if allowed by state and local authorities, by burning. If burned, stay out of smoke.

Local conditions may affect the use of herbicides. Consult your State Agricultural Experiment Station, Farm Advisors, or Extension Weed Specialists for advice in selecting treatment from this label to best fit local conditions.

Be sure that use of this product conforms to all applicable laws, rules, and regulations. Certain states have restrictions pertaining to application distances from susceptible crops. The applicator should become familiar with these laws, rules, or regulations and follow them exactly.

LIMITED WARRANTY AND DISCLAIMER

The manufacturer warrants that this product conforms to the chemical description on the label; that this product is reasonably fit for the purposes set forth in the directions for use when it is used in accordance with such directions; and that the directions, warnings and other statements on this label are based upon responsible experts' evaluation of reasonable tests of effectiveness, or toxicity to laboratory animals and to plants, and of residues on food crops, and upon reports of field experience. Tests have not been made on all varieties or in all states or under all conditions. THE MANUFACTURER NEITHER MAKES NOR INTENDS, NOR DOES IT AUTHORIZE ANY AGENT OR REPRESENTATIVE TO MAKE, ANY OTHER WARRANTIES, EXPRESS OR IMPLIED, AND IT EXPRESSLY EXCLUDES AND DISCLAIMS ALL IMPLIED WARRANTIES OF MERCHANTABILITY OR FITNESS FOR A PARTICULAR PURPOSE.

THIS WARRANTY DOES NOT EXTEND TO, AND THE BUYER SHALL BE SOLELY RESPONSIBLE FOR, ANY AND ALL LOSS OR DAMAGE WHICH RESULTS FROM THE USE OF THIS PRODUCT IN ANY MANNER WHICH IS INCONSISTENT WITH THE LABEL DIRECTIONS, WARNINGS OR CAUTIONS.

BUYER'S EXCLUSIVE REMEDY AND MANUFACTURER'S OR SELLER'S EXCLUSIVE LIABILITY FOR ANY AND ALL CLAIMS, LOSSES, DAMAGES, OR INJURIES RESULTING FROM THE USE OR HANDLING OF THIS PRODUCT, WHETHER OR NOT BASED IN CONTRACT, NEGLIGENCE, STRICT LIABILITY IN TORT OR OTHERWISE, SHALL BE LIMITED, AT THE MANUFACTURER'S OPTION, TO REPLACEMENT OF, OR THE REPAYMENT OF THE PURCHASE PRICE FOR, THE QUANTITY OF PRODUCT WITH RESPECT TO WHICH DAMAGES ARE CLAIMED. IN NO EVENT SHALL MANUFACTURER OR SELLER BE LIABLE FOR SPECIAL, INDIRECT OR CONSEQUENTIAL DAMAGES RESULTING FROM THE USE OR HANDLING OF THIS PRODUCT.

NOTICE TO BUYER

Purchase of this material does not confer any rights under patents governing this product or the use thereof in countries outside of the United States.

AG 90010

EPA Approval 9/24/87

THIS SPECIMEN LABEL IS INTENDED FOR USE ONLY AS A GUIDE IN PROVIDING GENERAL INFORMATION REGARDING THE DIRECTIONS, WARNINGS AND CAUTIONS ASSOCIATED WITH THE USE OF THIS PRODUCT. AS WITH ANY AGRICULTURAL CHEMICAL, ALWAYS FOLLOW THE LABEL INSTRUCTIONS ON THE PACKAGE BEFORE USING.

CERONE®

PLANT REGULATOR

For Reducing Lodging in Barley and Wheat

For Agricultural or Commercial Use Only. Not for Residential Use.

ACTIVE INGREDIENT:

Ethephon (2-chloroethyl) phosphonic acid*	39.9%
INERT INGREDIENT:	60.1%
TOTAL	100.0%

*Contains four pounds ethephon per gallon

EPA Reg. No. 264-377 EPA EST NO. 264-PA-01

KEEP OUT OF REACH OF CHILDREN

DANGER

PRECAUCION AL USUARIO: Si usted no lee ingles, no use este producto hasta que la informacion en la etiqueta le haya sido explicado ampliamente.

For PRODUCT USE Information Call 1-800-334-9745
For MEDICAL And TRANSPORTATION Emergencies ONLY Call 24 Hours A Day 1-800-334-7577

The use of this product for a variety of plant growth regulation uses is protected by various United States and foreign patents including U.S. Patents 3,879,188 and 4,240,819. No license is granted to use this product in countries other than the United States or for any use not contemplated by this label. Liability for patent infringement may result from use or sale of this product outside the United States.

PRECAUTIONARY STATEMENTS

DANGER

HAZARDS TO HUMANS AND DOMESTIC ANIMALS

Corrosive. Causes irreversible eye damage. Causes skin irritation. Do not get in eyes, on skin or on clothing. Wear full-length trousers, long-sleeved shirt, protective gloves, and goggles or face shield when handling.

Harmful or fatal if swallowed. Avoid contamination of food. Avoid inhalation of spray mist. May cause respiratory irritation. Change to clean work clothes daily. Wash thoroughly with soap and water after handling and before eating, drinking or using tobacco. Remove and wash contaminated clothing before reuse. Keep away from domestic animals.

Call a physician in cases of suspected poisoning. If poisoning is suspected in animals, call a veterinarian.

STATEMENT OF PRACTICAL TREATMENT

IF ON SKIN: Flush immediately with plenty of water for at least 15 minutes. Remove contaminated clothing and shoes.

IF INHALED: Stop spraying and move to uncontaminated area. Call a physician.

IF IN EYES: Flush immediately with plenty of water for at least 15 minutes and call a physician.

IF SWALLOWED: Do not induce vomiting. Promptly drink a large quantity of milk, egg whites, gelatin solution or if these are not available, drink large quantities of water. Avoid alcohol. Call a physician immediately.

NOTE TO PHYSICIAN

There is no specific antidote. Treatment should be directed at the control of symptoms and clinical conditions. This material possesses the characteristics of a strong acid, and may cause mucosal damage if swallowed. Do not induce vomiting. Gastric lavage should be undertaken with care if necessary. Appropriate conventional treatment for circulatory shock, respiratory depression and convulsions may be needed.

ENVIRONMENTAL HAZARDS

Do not apply directly to water or wetlands (swamps, bogs, marshes, and potholes). Do not contaminate water when disposing of equipment washwaters. Do not contaminate water used for irrigation or domestic purposes. Avoid spray drift to nearby crops as this product will cause modifications in plant growth. Plant injury or reduced yields may result.

Do not plant another crop within 30 days after treatment.

IMPORTANT

Do not use this product for purposes other than those listed on the label.

GENERAL INFORMATION

CERONE® Plant Regulator is intended for use in barley and wheat to reduce lodging. CERONE® should not be used in non-lodging situations or where losses due to lodging may be negligible.

Lodged crops increase harvest costs, reduce harvest efficiency and often decrease recoverable grain yields. Lodging also contributes to uneven maturity, higher moisture content and loss of grain quality. Lodging reduction can improve harvest efficiency and grain recovery.

The primary effect of CERONE® Plant Regulator is increased straw strength which increases the plant's resistance to lodging. CERONE® may also shorten the plant depending on variety, environmental conditions and the rate used. CERONE® also reduces "necking" in barley and the resulting yield decreases due to shattering and head loss prior to harvest.

Since CERONE® Plant Regulator treatment improves lodging resistance, barley and wheat crops can be managed for maximum economic yield. Variety selection, seeding, fertility, weed, disease and insect control are important considerations. Consult your local extension service for recommendations for your area.

USE PRECAUTIONS

Do not apply CERONE® Plant Regulator through any type of irrigation system.

Yield or quality decreases may occur if CERONE® is not properly used. CERONE® Plant Regulator should be used only on vigorously growing grain crops that are likely to lodge.

Management practices aimed at maximizing economic yields can produce conditions conducive to disease development. If disease is present such as rust, septoria, or mildew, CERONE® treatment may enhance the severity of the disease. Growers planning to use CERONE® Plant Regulator should plant disease-resistant varieties or monitor their crop for disease and use effective fungicides.

Do not apply CERONE® Plant Regulator to crops already lodged, or under stress from insect or disease damage. For moderate insect or disease pressure, use appropriate control measures alone or in a tank mix with CERONE® as recommended on this label.

Do not apply CERONE® if lack of moisture or other conditions make lodging unlikely as application under such conditions may result in yield loss.

Do not apply CERONE® Plant Regulator to crops under moisture stress or under temperature stress or on saline soils as this may result in yield loss.

Do not apply CERONE® if daytime temperatures are expected to exceed 85°F in non-irrigated or 90°F in irrigated crops, if temperatures are expected to fall below 35°F, or if wide fluctations in temperatures are expected within 5 days of application.

CERONE® may delay heading approximately 1 to 2 days and harvest maturity approximately 1 to 4 days. Extreme temperatures (below 35°F or greater than 85°F in non-irrigated and 90°F in irrigated crops) within 5 days after application may result in further maturity delay.

For CERONE® to be effective the temperature following application should be at least 60°F.

Do not apply CERONE® to crops seeded later than local recommendation. This is especially important in small grain production areas with short growing seasons where possible maturity delay may increase harvest difficulties.

CERONE® treatment may produce undesirable secondary tillers in some spring barley varieties especially under moisture or temperature stress. Treatment is not recommended for Azure barley, or Tyler wheat.

Mix only the amount of spray you expect to use each day. Do not allow mixed solution to stand overnight. Do not add adjuvants, surfactants or wetting agents to CERONE® Plant Regulator other than those recommended on this label.

Do not tank mix with herbicides or nitrogen solutions.

Avoid spray overlap which may result in an excess application rate and yield decrease.

Do not apply CERONE® Plant Regulator if rain is expected within 4 hours.

Allow 7 days following application for the beginning of lodging reduction activity.

DO NOT GRAZE OR FORAGE TREATED CROP OR CUT FOR HAY OR SILAGE. Mature straw at normal harvest may be consumed by animals.

TREATMENT DECISION GUIDE

It is important to inspect fields immediately prior to application to determine if lodging is expected. Use CERONE® only under the following conditions:

1. The crop is expected to lodge and result in significant loss of recoverable yield, grain quality, and/or harvest efficiency.
2. The crop is not under stress from disease or insect damage.
3. Soil moisture or irrigation is adequate to prevent crop stress for 7 days following application.
4. Temperature fluctuations (below 35°F or above 85°F nonirrigated and 90°F irrigated) are not expected for 5 days after application.
5. Applicant can be made between growth stages 8 to 10.

DIRECTIONS FOR USE

IT IS A VIOLATION OF FEDERAL LAW TO USE THIS PRODUCT IN A MANNER INCONSISTENT WITH ITS LABELING.

SPRAY PREPARATION

Add 1/2 to 3/4 of the required amount of water to the spray tank and begin agitation. Add the required amount of CERONE® Plant Regulator, followed by the remaining amount of water. Prepare only as much spray solution as can be used on the day of mixing. Do not allow spray solution to stand overnight.

Do not spill the concentrated product on spray equipment, or any airplane parts. ANY SPILLS SHOULD BE RINSED IMMEDIATELY WITH PLENTY OF WATER AS CERONE® PLANT REGULATOR IS CORROSIVE. Use of a nurse tank is highly recommended for avoiding possible spills of concentrated formulation on spray equipment or airplane parts.

EQUIPMENT CLEANING

Because of the acidic nature of this product, prolonged exposure to spray deposit will damage acrylic plastics, certain paints, and metals. Rinse thoroughly with a detergent and water all exposed acrylic plastic-type materials (e.g., aircraft windshields), and painted surfaces *within an hour* after exposure to spray deposits.

At the end of each day, rinse thoroughly with a detergent and water the spray system and all the metal parts of the aircraft exposed to the spray deposits.

COMPATIBILITY

CERONE® Plant Regulator can be tank-mixed with approved cereal fungicides including: Tilt®, Bayleton®, and mancozeb (Dithan® M-45, Marzate® 200). Triton CS-7 can be added to mancozeb fungicide as specified on mancozeb product labels. Other spray adjuvants are not recommended. Adequate spray tank agitation is recommended when using fungicide tank mixtures. Do not tank mix with herbicides or nitrogen solutions.

NOTE: The CERONE® rate is limited to 0.5 pint per acre in tank mixes with Tilt on non-irrigated barley and wheat.

APPLICATION TIMING

Apply when the flag leaf is just visible to boot stage, but before the awns have emerged or the sheath has split (Feekes-Large Scale 8-10, Zadok's Code 37-45). Do not allow spray solution to contact exposed heads, as damage and reduced yield may result. Inspect fields carefully to determine that application can be made at the proper stage.

APPLICATION

In some areas late seeded barley may be subject to extreme hot weather during anthesis which can result in yield loss. In this situation do not use CERONE® as yield losses may be increased. For this reason do not apply after July 1 in Minnesota and North Dakota.

Do not apply to crops on non-irrigated saline soils. Saline soils can increase plant stress particularly under low moisture or low fertility conditions. In non-irrigated areas, CERONE® should be used only on fields with salinity levels less than 1.0 mmho (ECe, millimhos/cm @ 25°C).
Do not apply to crops without good subsoil moisture at boot stage.

GROUND: CERONE® Plant Regulator should be applied in a minimum of 7 gallons of water per acre when applied by conventional ground equipment. With conventional ground equipment the use of flat fan nozzles is recommended. A minimum of 5 gallons of water per acre may be applied when using controlled droplet application (CDA) or air foil type equipment. Sprayer boom height must be far enough above the crop canopy to allow a uniform spray pattern. Sprayer speed should not exceed 10 mph to insure adequate spray coverage.

AIR: CERONE® Plant Regulator should be applied in a minimum of 3 gallons of water per acre by air.

USE RATES

For CERONE® use rates follow the directions below for non-irrigated and irrigated barley and wheat crops. Generally where moderate lodging pressure is expected 0.5 pint per acre will provide adequate lodging reduction. And in most situations where heavy to severe lodging pressure is expected, 0.5 to 0.75 pint will effectively reduce lodging.

The 1.0 pint rate should be limited to situations where lodging is expected to result in a high loss of yield potential; for example: 1) an irrigated crop with unusually severe lodging pressure, 2) an exceptionally tall lodging prone variety, or 3) a cereal type such as durum known to be subject to severe lodging.

It is important to remember that management practices and environmental conditions will greatly effect the amount of lodging pressure and rate of CERONE® needed.

Some varieties may be more responsive or sensitive to CERONE® than others. On such varieties the CERONE® rate should be limited to 0.5 pint per acre. Consult your state extension specialist or Rhone-Poulenc representative for more information.

Do not apply more than a total of 1.0 pint (0.5 lb. active) of CERONE® Plant Regulator per acre per year.

Do not harvest wheat or barley within 40 days of last CERONE® application.

NON-IRRIGATED BARLEY AND WHEAT

Do not apply CERONE® Plant Regulator to non-irrigated barley and wheat in the states of: Arizona, California (except in the Sacramento Valley and west of the coastal range); Colorado, Idaho, Montana, New Mexico, Nevada, North Dakota (except the Red River Valley); Oregon (except west of the Cascade Range); South Dakota, Texas, Utah, Washington (except west of the Cascade Range); Wyoming, or where summer fallow is required to produce a crop. This includes those areas where expected annual precipitation is less than 20 inches.

CERONE® Plant Regulator is recommended in other non-irrigated areas of the U.S. and the following Western non-irrigated areas: In the Sacramento Valley and west of the Coastal Range in California; West of the Cascade Range in Washington and Oregon; and in the Red River Valley of North Dakota when soil moisture is adequate to prevent crop stress. This includes those areas where expected annual precipitation is greater than 20 inches.

IMPORTANT

Limit the CERONE® rate to 0.5 pint/acre when used in combination with Tilt® in non-irrigated wheat and barley.

Limit the CERONE® rate to 0.5 pint/acre when temperatures are 80 to 85°F.

Do not use CERONE® if daytime temperatures for the period 5 days after application are expected to exceed 85°F.

IRRIGATED BARLEY AND WHEAT

Prior to treatment with CERONE®, irrigation is recommended to avoid stress. Irrigation may be resumed 24 hours after treatment and should be resumed if conditions are hot and dry and continued until grain head filling is completed.

Moisture and/or heat stress during anthesis and grain filling periods have been shown to cause loss of significant yield potential. With CERONE® use, avoiding plant stress during these growth periods is essential in obtaining grain yield and quality.

IMPORTANT: Limit the CERONE® rate to 0.5 pint/acre when temperatures are 85 to 90°F. Do not use CERONE® if daytime temperatures for the period 5 days following application are expected to exceed 90°F.

SPRING AND WINTER BARLEY

Use 0.5 pint where moderate lodging is expected, 0.5 to 0.75 pint if heavy lodging pressure is expected. Some tall varieties with vigorous growth may require 1 pint per acre.

SPRING WHEATS

For most spring wheats and most situations, use 0.5 pint per acre. Use 0.75 pint where severe lodging pressure is expected.

DURUM WHEATS

Use 0.5 pint per acre in moderate lodging situations including semi-dwarf durum varieties. Under heavy pressure 0.75 pint may be required. Use 0.75 to 1.0 pint where severe lodging is expected. In severe lodging situations these rates may not be adequate to control lodging in some tall durum wheats such as "Vic".

WINTER WHEATS (hard red, hard white, soft red, soft white)

Use 0.5 pint for moderate lodging pressure, 0.5 to 0.75 for heavy lodging pressure, and 0.75 to 1.0 pint if severe lodging pressure is expected. In severe lodging situations these rates may not be adequate to control lodging in some tall straw varieties such as "Roughrider" and "Agassiz".

CERONE® Plant Regulator Rate Guide For Barley And Wheat
Read Detailed Directions In This Label First

Situation	Expected Lodging Pressure		
	Moderate	Heavy	Severe
	(CERONE® Pints/Acre)		
Barley (spring & winter)	1/2	1/2 to 3/4	3/4 to 1
Winter Wheat	1/2	1/2 to 3/4	3/4 to 1
Most Spring Wheats	1/2	1/2	3/4
Durum Wheats	1/2	1/2 to 3/4	3/4 to 1
High Temperature[1] Or Sensitive Variety	1/2	1/2	1/2

NOTE: [1] Non-irrigated--80 to 85°F limit rate to 1/2 pint/acre and above 85°F do not treat.
Irrigated----85 to 90°F limit rate to 1/2 pint/acre and avove 90°F do not treat.

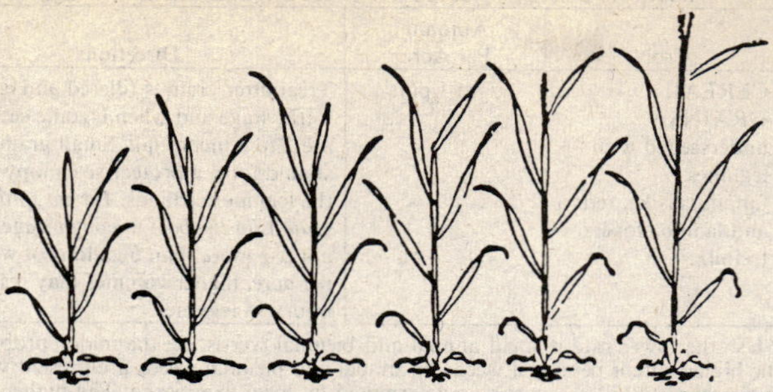

GROWTH DESCRIPTION	2nd node detectable	Flag leaf just visible	Flag leaf ligule visible	Boot swollen	First visible	3/4 of inflorescence completed
Feekes—Large Scale	7	8	9	10	10.1	10.4
Zodok's Code	32	37	39	45	50	57
Recommended CERONE Treatment Time						

STORAGE AND DISPOSAL

Do not contaminate water, food, or feed by storage or disposal.

If container is broken or contents have spilled, follow all precautions indicated above and clean up immediately. Before cleaning up, put on full-length trousers, long-sleeved shirt, protective gloves and goggles or face-shield. Soak up spill with absorbent media such as sand, earth or other suitable material and dispose of waste at an approved waste disposal facility.

PESTICIDE DISPOSAL

Pesticide wastes are acutely hazardous. Improper disposal of excess pesticide, spray mixture, or rinsate is a violation of Federal law. If these wastes cannot be disposed of by use according to label instructions, contact your State Pesticide or Environmental Control Agency or the Hazardous Waste representative at the nearest EPA Regional Office for guidance.

CONTAINER DISPOSAL

Do not reuse empty container. Triple rinse or equivalent. Then offer for recycling or reconditioning or puncture and dispose of in a sanitary landfill, or by incineration, or if allowed by state and local authorities, by burning. If burned, stay out of smoke.

LIMITED WARRANTY AND DISCLAIMER

The manufacturer warrants (a) that this product conforms to the chemical description on the label; (b) that this product is reasonably fit for the purposes set forth in the direction for use when it is used in accordance with such directions; and (c) that the directions, warnings and other statements on this label are based upon responsible experts' evaluation of reasonable tests of effectiveness, of toxicity to laboratory animals and to plants, and of residues on food crops and upon reports of field experience. Tests have not been made on all varieties or in all states or under all conditions. THE MANUFACTURER NEITHER MAKES NOR INTENDS, NOR DOES IT AUTHORIZE ANY AGENT OR REPRESENTATIVE TO MAKE, ANY OTHER WARRANTIES, EXPRESS OR IMPLIED, AND IT EXPRESSLY EXCLUDES AND DISCLAIMS ALL IMPLIED WARRANTIES OF MERCHANTABILITY OR FITNESS FOR A PARTICULAR PURPOSE.

THIS WARRANTY DOES NOT EXTEND TO, AND THE BUYER SHALL BE SOLELY RESPONSIBLE FOR, ANY AND ALL LOSS OR DAMAGE WHICH RESULTS FROM THE USE OF THIS PRODUCT IN ANY MANNER WHICH IS INCONSISTENT WITH THE LABEL DIRECTIONS, WARNINGS OR CAUTIONS.

BUYER'S EXCLUSIVE REMEDY AND MANUFACTURER'S OR SELLER'S EXCLUSIVE LIABILITY FOR ANY AND ALL CLAIMS, LOSSES, DAMAGES OR INJURIES RESULTING FROM THE USE OR HANDLING OF THIS PRODUCT, WHETHER OR NOT BASED IN CONTRACT, NEGLIGENCE, STRICT LIABILITY IN TORT OR OTHERWISE SHALL BE LIMITED, AT THE MANUFACTURER'S OPTION, TO REPLACEMENT OF, OR THE REPAYMENT OF THE PURCHASE PRICE FOR, THE QUANTITY OF PRODUCT WITH RESPECT TO WHICH DAMAGES ARE CLAIMED. IN NO EVENT SHALL MANUFACTURER OR SELLER BE LIABLE FOR SPECIAL, INDIRECT OR CONSEQUENTIAL DAMAGES RESULTING FROM THE USE OR HANDLING OF THIS PRODUCT.

NOTICE TO BUYER

Purchase of this material does not confer any rights under patents governing this product or the use thereof in countries outside of the United States.

Bayleton is a registered trademark of Mobay Chemical Corp.
Tilt is a registered trademark of Ciba-Geigy Corp.
Dithane is a registered trademark of Rohm and Haas Company
Manzate is a registered trademark of E.I. du Pont de Nemours & Company
CERONE is a registered trademark of Rhône-Poulenc

AG90036

THIS SPECIMEN LABEL IS INTENDED FOR USE ONLY AS A GUIDE IN PROVIDING GENERAL INFORMATION REGARDING THE DIRECTIONS, WARNINGS AND CAUTIONS ASSOCIATED WITH THE USE OF THIS PRODUCT. AS WITH ANY AGRICULTURAL CHEMICAL, ALWAYS FOLLOW THE LABEL INSTRUCTIONS ON THE PACKAGE BEFORE USING.

CHIPTOX®

MCPA SODIUM SALT HERBICIDE

CONTROLS WEEDS IN RICE, SMALL GRAINS, AND PEAS

ACTIVE INGREDIENT:
*Sodium salt of 2-methyl-4-chlorophenoxyacetic acid 24.0%
INERT INGREDIENT: 76.0%
*2-Methyl-4-chlorophenoxyacetic acid equivalent 21.7% or 2 pounds per gallon.

EPA Reg. No. 264-55AA   EPA Est. No. 264-MO-01

KEEP OUT OF REACH OF CHILDREN
DANGER PELIGRO

For PRODUCT USE Information Call 1-800-334-9745
For MEDICAL And TRANSPORTATION Emergencies ONLY Call 24 Hours A Day 1-800-334-7577

PRECAUTIONARY STATEMENTS

DANGER

HAZARDS TO HUMANS AND DOMESTIC ANIMALS

Corrosive. Causes eye damage. Harmful if swallowed. May be fatal if absorbed through the skin. Do not get in eyes, on skin or on clothing. Avoid breathing spray mist. Wear goggles or face shield when handling concentrate. In case of contact immediately flush eyes or skin with plenty of water. Get medical attention if irritation persists. Remove and wash contaminated clothing before reuse.

STATEMENT OF PRACTICAL TREATMENT

IF SWALLOWED: If patient is conscious and alert, give 2 to 3 glasses of water to drink. Do not induce vomiting. Get medical attention.

IF ON SKIN: Immediately wash skin with plenty of soap and water, if available, while removing contaminated clothing and shoes. Shoes and clothing contaminated by substantial spillage of concentrated product should be discarded in a manner which limits further exposure. Otherwise, wash clothing separately before reuse. Get medical attention.

IF INHALED: Remove victim to fresh air. If not breathing, give artificial respiration. Administer oxygen if necessary. Get medical attention.

IF IN EYES: Hold eyelids open and flush with a steady, gentle stream of water for at least 15 minutes. Get medical attention, preferably an ophthalmologist.

NOTE TO PHYSICIAN

No specific antidote is available.

All treatments should be based on observed signs and symptoms of distress in the patient. Overexposure to materials other than this product may have occurred.

This product is irritating to mucous membranes. If large amounts (greater than 1 ml/kg body weight) of the product have been ingested, the stomach should be evacuated by gastric intubation with the aid of a cuffed endotracheal tube to prevent exposure of the esophagus. After removal of stomach contents, wash stomach by instilling 30 to 50 g of activated charcoal in 3 to 4 ounces of water through the stomach tube and again remove stomach contents. Avoid oily laxatives.

ENVIRONMENTAL HAZARDS

Do not apply directly to water except as specified on this label. Do not contaminate water used for irrigation or domestic purposes by clearing equipment or disposal of wastes.

Do not use in a greenhouse. Avoid spray drift to susceptible plants such as cotton, beans, peas, and ornamentals as injury may result. Coarse sprays are less likely to drift.

Do not use same spray equipment for other purposes.

DIRECTIONS

IT IS A VIOLATION OF FEDERAL LAW TO USE THIS PRODUCT IN A MANNER INCONSISTENT WITH ITS LABELING.

Do not apply WEEDAR® Sodium MCPA through any type of irrigation system.

GENERAL INFORMATION

CHIPTOX® MCPA Sodium Salt Herbicide is effective on a wide range of broadleaf weeds and is useful for controlling these weeds in certain crops. Several crops, such as oats, rice, and small grains underseeded to legumes, are more tolerant of MCPA than of 2,4-D.

CHIPTOX® Herbicide is the sodium salt of MCPA and is readily dissolved in water, even in the hard waters found in some areas. Mix thoroughly while diluting this material and mix again if allowed to stand for an extended period of time before spraying.

COMPATIBILITY: If CHIPTOX® MCPA Sodium Salt Hebicide is to be tank mixed with other pesticides, compatibility should be tested prior to mixing. To test for compatibility, use a small container and mix a small amount (0.5 to 1 quart) of spray, combining all ingredients in the same ratio as the anticipated use. If any indications of physical incompatibility develop, do not use this mixture for spraying. Indications of incompatibility usually will appear within 5 to 15 minutes after mixing. Read and follow all directions and precautions on this label and on the labels of any product(s) for which a tank mixture is being considered.

To convert locally recommended application rates into amount of CHIPTOX® MCPA Sodium Salt Herbicide, use the following table:

Conversion of pound acid equivalent of MCPA to pints of CHIPTOX® MCPA Sodium Salt Herbicide

MCPA acid equivalent	1/4 lb.	3/8 lb.	1/2 lb.	3/4 lb.	1 lb.	1 1/4 lb.	1 1/2 lb
CHIPTOX® MCPA	1 pt.	1 1/2 pts.	2 pts.	3 pts.	4 pts.	5 pts.	6 pts.

Injury to crops may occur from this herbicide. If you are not prepared to accept some degree of crop injury, do not use this product.

Crop varieties vary in response to MCPA, and some may be easily injured. Apply CHIPTOX® MCPA Sodium Salt Herbicide only to varieties known to be tolerant to MCPA. If you are uncertain concerning tolerant varieties or local use situations that may affect crop tolerance to MCPA, contact your seed company or state Agricultural Extension Service for advice.

THE WEEDS CONTROLLED BY PROPERLY APPLIED TREATMENTS OF CHIPTOX® MCPA SODIUM SALT HERBICIDE INCLUDE:

common lambsquarters	pepperweeds
common cocklebur	purslanes
lambsquarters	ragweed (common and giant)
mustards (except blue mustard)	yellow rocket

RECOMMENDATIONS FOR APPLYING CHIPTOX® MCPA SODIUM SALT HERBICIDE

CEREAL CRAINS - POSTEMERGENCE - WHEAT, BARLEY, OATS, AND RYE (not underseeded with legumes)

Crop	Amount Per Acre	Directions
Annual weeds	1 to 2 pints*	Apply after grain is in the 4-leaf stage but not forming joints in the stem. *Do not spray grain in the boot stage to dough stage.*
Annual and biennial weeds	2 to 4 pints*	Apply after grain is fully tillered (usually about 4 to 8 inches high) but not forming joints in the stem. *Do not spray grain in the boot to dough stage.*
Perennial boradleaf weeds	3 to 4 pints*	
Emergency treatment in wheat Perennial broadleaf weeds	6 pints	Apply when weeds are approaching bud stage, *but do not spray grain during the boot to dough stage.* The 6 pint (1.5 pounts acid equivalent) per acre application can produce injury to wheat. Balance the severity of your weed problem against the possibility of crop damage. Where perennial weeds are scattered, spot treatment is suggested to minimize the extent of crop injury.

WHEAT, BARLEY, OATS, AND RYE (underseeded with legumes)

Crop	Amount Per Acre	Directions
CEREAL GRAINS underseeded with legumes (alfalfa, alsike, red and ladino clovers, trefoil)	1/2 to 1 pint	Treat after grain is tillered and is in the 4-leaf stage and when legume seedlings are 2 to 3 inches tall. Small grains should form a protective canopy over the legume seedlings. *Do not spray when grain is in the boot to dough stage.* Do not use more than 6 gallons of water per acre, higher volumes may result in injury to legumes.

* Use the lower rate if small annual and biennial weeds are the major problem. Use the higher rate if perennial weeds or annual and biennial weeds are present which are in the hard-to-kill categories as determined by local experience. The higher rates increase the risk of grain injury and should be used only where the weed control problem justifies the risk of grain damage.

RESTRICTIONS AND LIMITATIONS FOR USE ON SMALL GRAINS

- Use a minimum of 10 gallons of water per acre for ground application and 2 gallons of water per acre for aerial application, unless specific site directions listed above indicate otherwise.
- Do not allow livestock to forage or graze treated areas within 7 days of slaughter.

Crop	Amount Per Acre	Directions
Rice	3 to 5 pints	Treat 7 to 10 weeks after planting when rice is fully tillered and 6 to 8 inches above water. Do not spry after rice has reached the boot stage. Do not spray when temperatures ae over 90° F. Use the higher rate where sedges or bulrushes are the major problem. Use the lower rate for control of annual arrowhead, water plantain, and redstem. Use at least 3 to 5 gallons of water per acre for aerial application.

RESTRICTIONS AND LIMITATIONS FOR USE ON RICE

- Do not allow livestock animals to forage or graze treated areas within 7 days of slaughter.
- Do not grow crayfish or catfish in treated rice fields.

Crop	Amount Per Acre	Directions
Peas	1/2 to 1 1/2 pint	Treat when peas are 4 to 6 inches tall. Do not spray while peas are in blossom. Peas may show slight injury, but they usually recover a few days after treatment. Use the higher rate before buds form on Canada thistle to prevent from going to seed. Apply in 15 to 30 gallons of water per acre for ground application and 10 gallons of water per acre for aerial application.

RESTRICTION AND LIMITATIONS FOR USE ON PEAS

- Do not allow livestock to forage or graze treated fields. Do not feed treated vines to livestock.

STORAGE AND DISPOSAL

STORAGE

Do not store near fertilizers, seeds, insecticides, or fungicides. Do not contaminate water, food, or feed by storage or disposal.

PESTICIDE DISPOSAL

Pesticide wastes are acutely hazardous. Improper disposal of excess pesticide, spray mixture, or rinsate is a violation of Federal Law. If these wastes cannot be disposed of by use according to label instructions, contact your State Pesticide or Environmental Control Agency, or the Hazardous Waste representative at the nearest EPA Regional Office for guidance.

CONTAINER DISPOSAL

Triple rinse (or equivalent). Then offer for recycling or reconditioning, or puncture and dispose of in a sanitary landfill, or by other procedures approved by state and local authorities.

LIMITED WARRANTY AND DISCLAIMER

The manufacturer warrants that this product conforms to the chemical description on the label; that this product is reasonably fit for the purposes set forth in the directions for use when it is used in accordance with such directions; and that the directions, warning and other statements on this label are based upon responsible experts' evaluation of reasonable tests of effectiveness, of toxicity to laboratory animals and to plants, and of residues on food crops and upon reports of field experience. Tests have not been made on all varieties or in all states or under all conditions. THE MANUFACTURER NEITHER MAKES NOR INTENDS, NOR

DOES IT AUTHORIZE ANY AGENT OR REPRESENTATIVE TO MAKE, AND OTHER WARRANTIES, EXPRESS OR IMPLIED, AND IT EXPRESSLY EXCLUDES AND DISCLAIMS ALL IMPLIED WARRANTIES OF MERCHANTIBILITY OR FITNESS FOR A PARTICULAR PURPOSE.

THIS WARRANTY DOES NOT EXTEND TO, AND THE BUYER SHALL BE SOLELY RESPONSIBLE FOR, ANY AND ALL LOSS OR DAMAGE WHICH RESULTS FROM THE USE OF THIS PRODUCT IN ANY MANNER WHICH IS INCONSISTENT WITH THE LABEL DIRECTIONS, WARNINGS OR CAUTIONS.

BUYER'S EXCLUSIVE REMEDY AND MANUFACTURER'S OR SELLER'S EXCLUSIVE LIABILITY FOR ANY AND ALL CLAIMS, LOSSES, DAMAGES OR INJURIES RESULTING FROM THE USE OR HANDLING OF THIS PRODUCT, WHETHER OR NOT BASED IN CONTRACT, NEGLIGENCE, STRICT LIABILITY IN TORT OR OTHERWISE SHALL BE LIMITED, AT THE MANUFACTURER'S OPTION, TO REPLACEMENT OF, OR THE REPAYMENT OF THE PURCHASE PRICE FOR, THE QUANTITY OF PRODUCT WITH RESPECT TO WHICH DAMAGES ARE CLAIMED. IN NO EVENT SHALL MANUFACTURER OR SELLER BE LIABLE FOR SPECIAL INDIRECT OR CONSEQUENTIAL DAMAGES RESULTING FROM THE USE OR HANDLING OF THIS PRODUCT.

NOTICE TO BUYER

Purchase of this material does not confer any rights under parents governing this product or the use thereof in countries outside of the United States.

CHIPTOX is a registered trademark of Rhône-Poulenc

THIS SPECIMEN LABEL IS INTENDED FOR USE ONLY AS A GUIDE IN PROVIDING GENERAL INFORMATION REGARDING THE DIRECTIONS, WARNINGS AND CAUTIONS ASSOCIATED WITH THE USE OF THIS PRODUCT. AS WITH ANY AGRICULTURAL CHEMICAL, ALWAYS FOLLOW THE LABEL INSTRUCTIONS ON THE PACKAGE BEFORE USING.

AG90022

ENVERT 171®

WOODY PLANT HERBICIDE

A FORMULATION OF 2,4-D AND 2,4-DP FOR CREATING AN INVERT EMULSION OR CONVENTIONAL SPRAY. CONTROLS MIXED BRUSH ON UTILITY RIGHTS-OF-WAY, ROADSIDES, DRAINAGE DITCHBANKS, AND SIMILAR AREAS.

ACTIVE INGREDIENTS:	
*2,4-Dichlorophenoxyacetic acid, butoxyethanol ester	16.8%
**2-(2,4-Dichlorophenoxy) propionic acid, butoxyethanol ester	19.5%
INERT INGREDIENTS:	63.7%
TOTAL:	100.0%

*2,4-Dichlorophenoxyacetic acid equivalent 11.6% by weight or 0.95 pound per gallon.
Isomer specific by AOAC method No. 6. DO1-5

**2-(2,4-Dichlorophenoxy) propionic acid equivalent 13.7% by weight or 1.125 pounds per gallon.
Isomer specific by AOAC method No. 6 Dol-5

KEEP OUT OF REACH OF CHILDREN

CAUTION

EPA Reg. No. 264-223 EPA Est. No. 264-MO-01

For PRODUCT USE Information Call 1-800-334-9745
For MEDICAL and TRANSPORTATION Emergencies ONLY Call 24 Hours A Day 1-800-334-7577

PRECAUTIONARY STATEMENTS

CAUTION

HAZARDS TO HUMANS AND DOMESTIC ANIMALS

Harmful if swallowed or absorbed through skin. Avoid contact with skin, eyes or clothing.

Wash thoroughly with soap and water after handling.

STATEMENT OF PRACTICAL TREATMENT

IF SWALLOWED: Call a physician or Poison Control Center. If patient is conscious and alert, give 2-3 glasses of water. Do not induce vomiting. Get medical attention.

IF ON SKIN: Wash skin with plenty of soap and water, while removing contaminated clothing. Wash clothing separately before reuse.

IF IN EYES: Hold eyelids open and flush with a steady, gentle stream of water for at least 15 minutes. Get medical attention.

NOTE TO PHYSICIAN

No specific antidote is available. All treatments should be based on observed signs and symptoms of distress in the patient. Overexposure to materials other than this product may have occurred.

This product contains petroleum distillates. If large amounts (greater than 1 ml/kg body weight) of the product have been ingested, the stomach should be evacuated by gastric intubation with the aid of a cuffed endotracheal tube to prevent aspiration of petroleum distillates. After removal of stomach contents, wash stomach by instilling 30-50 grams of activated charcoal in 3-4 ounces of water through the stomach tube and again remove stomach contents. Avoid oily laxatives.

ENVIRONMENTAL HAZARDS

This product is toxic to fish and aquatic invertebrates.

Drift or runoff may adversely affect fish and non-target plants. Do not apply directly to water or wetlands (swamps, bogs, marshes and potholes). Do not contaminate water when disposing of equipment washwaters. Do not apply when weather conditions favor drift from treated areas. Do not use in or near a greenhouse. Do not use same spray equipment for other purposes.

USE PRECAUTIONS

Do not apply ENVERT® 171 Woody Plant Herbicide directly to, or allow spray mist to drift onto cotton, okra, grapes, tomatoes, fruit trees, vegetables, flowers or other desirable crop or ornamental plants which are susceptible to 2,4-D or 2,4-DP herbicides. Very small quantities of the 2,4-D or 2,4-DP can cause severe injury during the growing or dormant periods. At temperatures above 95°F vapors from ENVERT® 171 may injure susceptible plants.

Drift from aerial application may be reduced by (1) applying as near to the target as possible in order to obtain coverage; (2) by applying 12 or more gallons of spray per acre; (3) by using 20 pounds pressure or less at the nozzle tips; (4) by using nozzles which produce a coarse spray pattern; (5) by spraying only when the wind velocity is less than 5 miles per hour; and (6) by spraying when there is no possibility for a temperature inversion at time of spraying.

Use coarse sprays to minimize drift. Do not apply with hollow cone-type insecticide or other nozzles that produce fine spray droplets. Drift from ground application may be reduced by: (1) keeping spray as near to target as possible in order to obtain complete coverage; (2) by using no more than 20 pounds of pressure at the nozzle tips; and (3) by not spraying when the wind exceeds 5 miles per hour.

Do not apply near cotton, grapes, tomatoes or near desirable 2,4-D or 2,4-DP susceptible crop or ornamental vegetation.

Local conditions may affect the use of herbicides. Consult your State Agriculture Experiment Station, Farm Advisors, or Extension Weed Specialists for advice in selecting treatment from this label to best fit local conditions.

Be sure that use of the product conforms to all applicable laws, rules and regulations. Certain states have restrictions pertaining to application distances from susceptible crops. The applicator should become familiar with these laws, rules or regulations and follow them exactly.

DIRECTIONS FOR USE

IT IS A VIOLATION OF FEDERAL LAW TO USE THIS PRODUCT IN A MANNER INCONSISTENT WITH ITS LABELING.

GENERAL INFORMATION

ENVERT® 171 provides an effective woody plant herbicide in a "invert emulsion" form or as a conventional spray.

When mixed according to directions, ENVERT® 171 Woody Plant Herbicide makes an "invert emulsion," a thicker, more viscous spray mixture than standard water, oil or oil-water mixtures used in aerial or ground spraying.

To control mixed brush on utility rights-of-way, roadsides, drainage ditchbanks and similar areas, especially where conifers are a problem, spray in the spring after hardwoods have developed full sized leaves but before the spring growth of conifers has hardened off.

MIXING DIRECTIONS

Add the proper amount of ENVERT® 171 Woody Plant Herbicide to the mixing tank. While agitating, add the required amount of water. Always add water to ENVERT® 171, never the reverse. Mechanical paddle agitators that operate independently of the pump should be used. Re-circulate through the by-pass valves until the thick emulsion is obtained. To transfer the spray mixture from mixing tanks to aircraft tanks, use a positive displacement pump with 1-½ to 2 inch hose fittings on both the suction and discharge end.

DIRECTIONS For Air Application

APPLY through Rhône-Poulenc's SPRA-DISK® or other suitable equipment, mounted on a helicopter.

Where root suckering species such as sumac, sassafras, locust, and black gum predominate, mix 3 gallons of ENVERT® 171 Woody Plant Herbicide with 9 gallons of water for each acre to be sprayed.

Where harder to control species such as red maple, elm or oaks are present, mix 5 to 6 gallons of ENVERT® 171 Woody Plant Herbicide with 15 to 18 gallons of water for each acre to be sprayed.

Where brush is very dense and tall (25' to 30') or where conifers are present, better results will be obtained by increasing the rate and volume to 10 gallons of ENVERT® 171 Woody Plant Herbicide in 30 gallons of water per acre.

Always add water to ENVERT® 171 Woody Plant Herbicide; never the reverse.

Aerial spraying of ENVERT® 171 Woody Plant Herbicide is a specialized job. Only employ competent, reliable aerial applicators who have qualified technical guidance. Become familiar with state laws governing the application of herbicides. Do not apply ENVERT® 171 Woody Plant Herbicide in areas where possible drift may injure valuable crops or plants (See Precaution Statements).

Ground Application with Mechanical Invert Sprayers such as the WANNER Sprayer

Read sprayer operating manual thoroughly before operating sprayer. Be sure all strainers, screens, control orifices and spray tanks are clean, undamaged and correctly installed.

Conventional sprays rather than invert emulsions may be made when ratios of 11:1 and 15:1 (water to ENVERT® 171 Woody Plant Herbicide + kerosene or fuel oil) are mixed. Invert characteristics are provided when ratios of 7:1 (water to ENVERT® 171 + kerosene or fuel oil) are mixed.

Select the dilution rates, water orifice and chemical orifice from the following tables. Install the selected orifices and proceed with mixing.

Fill the concentrate tank with a mixture of ENVERT® 171 Woody Plant Herbicide and kerosene (or fuel oil) using 1 to 1 1/2 gallons kerosene or fuel oil for each gallon of ENVERT® 171.

Thoroughly pre-mix ENVERT® 171 Woody Plant Herbicide and kerosene or fuel oil before adding to the sprayer tank.

Fill the water tank with clean water, start up engine and pump as directed in the sprayer operating manual.

Set pressure regulator to apply spray mixture at 30-40 psi for boom spraying and at 100 psi for handgun application. Apply the spray to thoroughly cover plants to be controlled using total gallons spray per acre indicated in the following table. When treatment is directed toward scattered clumps of weeds or brush, total amount of spray applied per acre should be reduced proportional to the area actually treated.

MIXING RATIO AND ORIFICE SELECTION GUIDE FOR MECHANICAL INVERT SPRAYERS

APPLICATION	PUMP	CHEMICAL ORIFICE	WATER ORIFICE	CHEMICAL TANK RATIO ENVERT® 171: KEROSENE OR FUEL OIL	RATIO WATER: ENVERT® 171 + KEROSENE OR FUEL OIL MIX	TOTAL AMOUNT OF SPRAY TO APPLY PER TREATED ACRE
For light to medium density stands of susceptible Woody Plants and root suckering species such as Sumac, Locust, Sassafras and Black Gum.	A10FB (10 gallons) A20FB (20 gallons)	4916-80 4916-125 or 4916-156	5/16" 1/2" 9/16"	1:1 1:1 1:1	15:1 15:1 15:1	96 gallons
For general use for control of susceptible broadleaf weeds and light density stands of woody plants.	A10FB (10 gallons) A20FB (20 gallons)	4916-80 4916-125	5/16" 1/2"	2:3 2:3	15:1 15:1	50 to 100 gallons
For dense stands of brush or where species such as Red Maple, Elm or Oaks are present.	A10FB (10 gallons) A20FB (20 gallons)	4916-89 4916-125	5/16" 3/8"	2:3 2:3	11:1 11:1	150 to 180 gallons
For very dense and tall (25' to 30') brush or where conifers are to be controlled.	A10FB (10 gallons) A20FB (20 gallons)	4916-89 or 4916-120 4916-125	5/16" 5/16" 3/8"	1:1 2:3 1:1	11:1 7:1 11:1	240 gallons 200 gallons 240 gallons

CHEMICAL TANK MIXING TABLE FOR MECHANICAL INVERT SPRAYERS

TOTAL CAPACITY OF CHEMICAL TANK	AMOUNTS TO MIX RATIO FOR EACH TANK ENVERT® 171 : KEROSENE OR FUEL OIL 1:1	2:3
40 gallons	20 gallons ENVERT® 171 + 20 gallons kerosene or fuel oil	16 gallons ENVERT® 171 + 24 gallons kerosene or fuel oil
60 gallons	30 gallons ENVERT® 171 + 30 gallons kerosene or fuel oil	24 gallons ENVERT® 171 + 36 gallons kerosene or fuel oil
100 gallons	50 gallons ENVERT® 171 + 50 gallons kerosene or fuel oil	40 gallons ENVERT® 171 + 60 gallons kerosene or fuel oil

STORAGE AND DISPOSAL

STORAGE

Do not store near fertilizers, seeds, insecticides, or fungicides.

PESTICIDE DISPOSAL

Do not contaminate water, food or feed by storage or disposal. Open dumping is prohibited. Pesticide wastes are toxic. Improper disposal of excess pesticide, spray mixture, or rinsate is a violation of Federal Law. If these wastes cannot be disposed of by use according to label instructions, contact your State Pesticide or Environmental Control Agency, or the Hazardous Waste representative at the nearest EPA Regional Office for guidance.

CONTAINER DISPOSAL

Triple rinse (or equivalent). Then offer for recycling or reconditioning, or puncture and dispose of in a sanitary landfill, or by other procedures approved by state and local authorities.

LIMITED WARRANTY AND DISCLAIMER

The manufacturer warrants (a) that this product conforms to the chemical description on the label; (b) that this product is reasonably fit for the purposes set forth in the directions for use when it is used in accordance with such directions; and (c) that the directions, warnings and other statements on this label are based upon responsible experts' evaluation of reasonable tests of effectiveness, of toxicity to laboratory animals and to plants, and upon reports of field experience. Tests have not been made on all varieties or in all states or under all conditions. THE MANUFACTURER NEITHER MAKES NOR INTENDS, NOR DOES IT AUTHORIZE ANY AGENT OR REPRESENTATIVE TO MAKE ANY OTHER WARRANTIES, EXPRESS OR IMPLIED, AND IT EXPRESSLY EXCLUDES AND DISCLAIMS ALL IMPLIED WARRANTIES OF MERCHANTABILITY OR FITNESS FOR A PARTICULAR PURPOSE.

THIS WARRANTY DOES NOT EXTEND TO, AND THE BUYER SHALL BE SOLELY RESPONSIBLE FOR, ANY AND ALL LOSS OR DAMAGE WHICH RESULTS FROM THE USE OF THIS PRODUCT IN ANY MANNER WHICH IS INCONSISTENT WITH THE LABEL DIRECTIONS, WARNINGS, OR CAUTIONS.

BUYER'S EXCLUSIVE REMEDY AND MANUFACTURER'S OR SELLER'S EXCLUSIVE LIABILITY FOR ANY AND ALL CLAIMS, LOSSES, DAMAGES, OR INJURIES RESULTING FROM THE USE OR HANDLING OF THIS PRODUCT, WHETHER OR NOT BASED IN CONTRACT, NEGLIGENCE, STRICT LIABILITY IN TORT OR OTHERWISE SHALL BE LIMITED, AT THE MANUFACTURER'S OPTION, TO REPLACEMENT OF, OR THE REPAYMENT OF THE PUCHASE PRICE FOR, THE QUANTITY OF PRODUCT WITH RESPECT TO WHICH DAMAGES ARE CLAIMED. IN NO EVENT SHALL MANUFACTURER OR SELLER BE LIABLE FOR SPECIAL, INDIRECT OR CONSEQUENTIAL DAMAGES RESULTING FROM THE USE OR HANDLING OF THIS PRODUCT.

NOTICE TO BUYER

Purchase of this material does not confer any rights under patents governing this product or the use thereof in countries outside of the United States.

ENVERT is a registered trademark of Rhône-Poulenc

AG90013

EPA Approval: 4/11/86

THIS SPECIMEN LABEL IS INTENDED FOR USE ONLY AS A GUIDE IN PROVIDING GENERAL INFORMATION REGARDING THE DIRECTIONS, WARNINGS AND CAUTIONS ASSOCIATED WITH THE USE OF THIS PRODUCT. AS WITH ANY AGRICULTURAL CHEMICAL, ALWAYS FOLLOW THE LABEL INSTRUCTIONS ON THE PACKAGE BEFORE USING.

ESTERON* 99*

CONCENTRATE

For the Control of Many Broadleaf Weeds, Herbaceous Perennials and Woody Plants Susceptible to 2,4-D in Grass Pastures, Certain Crops and Non-Crop Areas.

ACTIVE INGREDIENT(S):

2,4-Dichlorophenoxyacetic Acid, Isooctyl Ester[1]	65.9%
INERT INGREDIENTS	34.1%
TOTAL	100.0%

2,4-Dichlorophenoxyacetic Acid Equivalent[1]: 43.7%–3.8 lb/gal

[1] Isomer Specific by AOAC Method No. 6.275-6.279 (13th Ed.)

EPA Reg. No. 464-349-264 EPA Est No.

KEEP OUT OF REACH OF CHILDREN

CAUTION

PRECAUCION: PRECAUCION AL USUARIO: Si usted no lee ingles, no use este producto hasta que la etiqueta le haya sido explicada ampliamente.

TRANSLATION: (TO THE USER: If you cannot read English, do not use this product until the label has been fully explained to you.)

For PRODUCT USE Information Call 1-800-334-9745
For MEDICAL and TRANSPORTATION Emergencies ONLY Call 24 Hours A Day 1-800-334-7577

PRECAUTIONARY STATEMENTS

CAUTION

HAZARDS TO HUMANS AND DOMESTIC ANIMALS

HARMFUL IF SWALLOWED OR ABSORBED THROUGH SKIN

MAY CAUSE ALLERGIC SKIN REACTION. Avoid breathing vapors or mists. Avoid contact with skin, eyes, or clothing. Wash thoroughly after handling and before eating or smoking. Remove contaminated clothing and wash before reuse. Do not apply this product in such a manner as to directly or through drift expose workers or other persons. The area being treated must be vacated by unprotected persons.

STATEMENT OF PRACTICAL TREATMENT

IN CASE OF CONTACT: immediately flush eyes or skin with plenty of water. Get medical attention if irritation persists.

IF SWALLOWED: DO NOT induce vomiting. Get medical attention. Contains petroleum distillate.

NOTE TO PHYSICIAN: May cause chemical pneumenitis if aspirated. If lavage is performed, suggest endotracheal and/or esophagoscopic control.

PHYSICAL OR CHEMICAL HAZARDS

Do Not Cut or Weld Container.

ENVIRONMENTAL HAZARDS

Do not apply directly to any body of water. Do not apply where runoff is likely to occur. Do not contaminate water when disposing of equipment washwaters. Do not contaminate irrigation ditches or water used for irrigation or domestic purposes.

NOTICE: Read the entire label.

Before buying or using this product, read LIMITED WARRANTY AND DISCLAIMER elsewhere on this label. If terms are not acceptable, return **unopened** package at once to seller for full refund of purchase price paid. Otherwise, use by the buyer or any other user constitutes acceptance of the terms under the LIMITED WARRANTY AND DISCLAIMER.

AGRICULTURAL CHEMICAL

Do Not Ship or Store with Food, Feeds, Drugs, or Clothing.

WEED LIST

ESTERON 99 CONCENTRATE is recommended for control of numerous broadleaf weeds and certain 2,4-D susceptible woody plants without injury to most established grasses. Species controlled include the following, plus many others:

beggarticks	milkvetch
bitterweed	morningglory, annual
blueweed, Texas	nettles
broomweed	onion, wild
buckbrush	pennycress (fanweed)
buckwheat, wild	pepperweed, field
burdock	pigweed(1)
burhead	plantains
carpetweed	poorjoe
catnip	rabbitbrush
chamise	radish, wild
chicory	ragweed
cocklebur	rape, wild
coffeeweed	redstem
cornflower	sage, coastal
coyotebrush	sagebrush, big
croton	sagebrush, sand
dandelion	salsify
docks	sand shinnery oak
dogfennel	shepherdspurse
elderberry galinsoga	sicklepod
garlic, wild	smartweed (annual)
goatsbeard	sneezeweed, bitter
hemp, wild	sowthistle, annual
jewelweed	spanishneedles
jimsonweed	sumac
ladysthumb	sunflower
lambsquarter	sweetclover
loco, bigbend	tansyragwort
mallow, Venice	thistle, bull
manzanita	thistle, musk
marshelder	wild mustard
thistle, Russian	willow
tumbleweed	witchweed
velvetleaf	wormwood
vervains	yellow rocket
vetch	yellow starthistle
water plantain	

(1) The control of "hybrid" pigweeds appears to be less satisfactory from 2,4-D products than formerly experienced on "non-hybrid" varieties. Since 2,4-D herbicides are not as effective on the "hybrid" pigweeds it is necessary to apply higher rates of 2,4-D for control, especially later in the growing season. Higher rates injure some crops, so less than satisfactory pigweed control may be experienced by the highest tolerated crop dosages.

Therefore, Rhône-Poulenc no longer includes pigweed among the species covered by the performance guarantee statements on the labels for ESTERON 99 CONCENTRATE. At this time, this disclaimer applies only to the High Plains of Texas and western Oklahoma, including the Panhandles. All other guarantees on these product labels are unchanged by this disclaimer.

DIRECTIONS FOR USE

IT IS A VIOLATION OF FEDERAL LAW TO USE THIS PRODUCT IN A MANNER INCONSISTENT WITH ITS LABELING.

DO NOT APPLY THIS PRODUCT THROUGH ANY TYPE OF IRRIGATION SYSTEM.

Apply ESTERON 99 CONCENTRATE as water or oil spray during warm weather when weeds or brush are actively growing. Application under drought conditions often will give poor results. Use low spray pressure to minimize drift. On cropland and along roadsides, do not exceed 20 psi pressure. Apply enough spray volume to provide uniform coverage of weeds and brush, usually 5 to 20 gallons per acre by ground equipment and 3 to 5 gallons by aircraft. Higher gallonage may be used if desired to improve spray coverage. Generally, the lower dosages recommended on this label will be satisfactory for young, succulent growth of sensitive weed species. For less sensitive species and under conditions where control is more difficult, the higher dosages will be needed. For crop uses, do not mix with oil or other adjuvants unless specifically recommended on this label. Deep-rooted perennial weeds such as Canada thistle and field bindweed and many woody plants usually require repeated applications for maximum control. **Do not apply ESTERON 99 CONCENTRATE where spray drift may contact nearby susceptible crops or other desirable plants, or may contaminate water for irrigation or domestic use. Read and follow all Use Precautions given on this label.**

NOTE: If there are uncertainties concerning special local use situations or specific crop variety tolerances to 2,4-D, consult your State Agricultural Experiment Station or local Extension Service weed specialists for advice.

TO PREPARE THE SPRAY

(1) Fill the spray tank about half full with water, then add the required amount of ESTERON 99 CONCENTRATE with agitation, and finally the rest of the water. NOTE: ESTERON 99 CONCENTRATE in water forms an emulsion which tends to separate unless the mixture is kept agitated. (2) If oil is added, first mix the ESTERON 99 CONCENTRATE and the oil and then add this mixture to the water. However, with adequate agitation, the oil can be added after the ESTERON 99 CONCENTRATE is mixed in the water. (3) If straight oil is used, a solution is formed and separation does not occur. Do not allow any water to get into the oil-herbicide mixture to avoid formation of an invert emulsion.

WEED CONTROL IN SMALL GRAINS NOT UNDERSEEDED WITH A LEGUME

NOTE: Do not permit dairy animals or meat animals being finished for slaughter to forage or graze treated grain fields within 2 weeks after treatment.

Spring and Winter Wheat, Barley, and Rye:

Apply 1/2 to 1 pint per acre. Spray when grain is in full tiller stage (usually 4 to 8 inches tall) but before the boot stage and boot to dough stage. For improved control of difficult weeds including wild garlic and wild onion or under dry or cool conditions, apply up to 2 pints per acre. Wild garlic and wild onion may not be killed but dockage should be reduced. Do not use higher rates unless possible crop injury will be acceptable. Consult State Agricultural Experiment Station or Extension Service Weed Specialists for recommendations or suggestions to fit local conditions.

Spring Seeded Oats:

Apply 1/2 pint per acre at the full tiller stage but before the early boot stage. Oats are less tolerant to 2,4-D than wheat or barley and are more likely to suffer some injury.

Fall Seeded Oats (Southern) Grown for Grain:

Apply 3/4 to 1 1/4 pints per acre after full tillering but before the early boot stage. Some difficult weeds may require higher rates for maximum control but crop injury may result. Do not spray during or immediately following cold weather.

Preharvest Treatment:

Apply 1 to 2 pints per acre when grains are in the hard dough stage to control large weeds that may interfere with harvest. Best results will be obtained when soil moisture is sufficient to cause succulent weed growth. NOTE: Do not feed treated straw to livestock.

WEED CONTROL IN CORN

Use one of the following programs. **Preemergence:** Apply 1 to 2 quarts per acre to soil anytime after planting but before corn emerges. Only emerged broadleaf weeds are likely to be controlled. Do not apply more than 1 quart per acre unless the increased risk of crop injury can be tolerated. Do not use on light sandy soil. **Emergence:** Apply 1 pint per acre just as corn plants are breaking ground. **Postemergence:** After emergence of corn use 1/2 pint per acre. Application of 3/4 to 1 pint per acre may be needed for maximum control of some weeds but such rates are more likely to injure the corn. If corn is over 8 inches tall, use drop nozzles to keep the spray off the corn foliage as much as possible. Do not apply from the tasseling to dough stage. Do not use with oil, atrazine or other adjuvants. Crop injury is more likely to occur if corn is growing rapidly under high temperature and high soil moisture conditions. To reduce breakage of stalks from temporary brittleness caused by 2,4-D, delay cultivation for 8 to 10 days after treatment. Do not forage or feed corn fodder for 7 days following application. NOTE: Hybrids vary in response to 2,4-D and some are easily injured. Spray only varieties known to be tolerant to 2,4-D. Contact seed company or your Agricultural Experiment Station or Extension Service weed specialists for this information.

Preharvest Corn Treatment:

After the hard dough or denting stage, apply 1 to 2 pints per acre by air or ground equipment to suppress perennial weeds, decrease weed seed production, and control tall weeds such as bindweed, cocklebur, dogbane, jimsonweed, ragweed, sunflower, velvetleaf and vines that interfere with harvesting. Do not forage or feed corn fodder for 7 days following application.

WEED CONTROL IN SORGHUM (MILO)

Apply 1/2 pint per acre when sorghum is 5 to 15 inches tall. A higher rate of 3/4 to 1 pint per acre may be needed to control some weeds but the chance for crop injury is likewise increased. Do not use with oil. Do not treat before the sorghum is 5 inches tall nor during the boot, flowering or early dough stages. If sorghum is taller than 8 inches, use drop nozzles to keep the spray off the foliage as much as possible. Temporary crop injury may occur under conditions of high soil moisture and high air temperatures. Varieties vary in tolerance to 2,4-D and some hybrids are quite sensitive. Spray only varieties known to be tolerant to 2,4-D. Contact seed company or your Agricultural Experiment Station or Extension Service weed specialists for this information.

WEED CONTROL IN GRASS SEED CROPS

Use 1 to 1 1/2 pints per acre in the amount of water required for uniform application by air or ground equipment. Apply to established stands in spring from the tiller to early boot stage. Do not spray in boot stage. New spring seedings may be treated with the lower rate after the grasses have at least five leaves. Perennial weed regrowth may be treated in the fall.

WEED AND BRUSH CONTROL IN RANGELAND, CONSERVATION RESERVE AND GRASS PASTURES

NOTE: Do not graze dairy animals on treated areas within 7 days after application. Do not use on bent grass, alfalfa, clover, or other legumes. Do not use on newly seeded areas until grass is well established. Do not use from early boot to milk stage where grass seed production is desired. For Conservation Reserve Land follow all applicable state and Federal regulations. Follow the most severe grazing restrictions imposed either by the pesticide label or by the USDA Acreage Conservation Reserve Program, whichever is longest.

Bitterweed, Broomweed, Croton, Docks, Kochia, Marshelder, Muskthistle and Other Broadleaf Weeds:

Use 2 quarts of ESTERON 99 CONCENTRATE per acre in the amount of water needed for uniform application. If the weeds are young and growing actively, 1 quart per acre will provide control of some species. Deeprooted perennial weeds may require repeated treatments in the same year or in subsequent years.

Wild Garlic and Wild Onion:

Apply 2 to 3 quarts per acre, making three applications (fall-spring-fall or spring-fall-spring) starting in late fall or early spring.

Weed Control in Newly Sprigged Coastal Bermudagrass:

Apply 1 to 2 quarts per acre preemergence and/or postemergence.

Sand Shinnery Oak and Sand Sagebrush:

On the oak, use 1 quart in 5 gallons of oil or in 4 gallons of water plus 1 gallon of oil per acre. Apply by aircraft between May 15 and June 15. On the sagebrush, use 1 quart in 3 gallons of oil per acre and apply by aircraft when foliage is fully expanded and the brush is actively growing.

Big Sagebrush and Rabbitbrush:

Use 2 to 3 quarts per acre in 2 to 3 gallons of oil or in 3 to 5 gallons of oil-water emulsion spray. For rabbitbrush, the 3 quart rate is usually required. Brush should be leafed out and growing actively when treated. Retreatment may be needed.

Chamise, Manzanita, Buckbrush, Coastal Sage, Coyotebrush and Certain Other Chaparral Species:

Use 2 to 3 quarts per acre in 5 to 10 gallons of water. One gallon of fuel oil may be included in the spray mixture for added effectiveness. Make applications by aircraft or ground equipment to obtain uniform spray coverage. For effective control, the brush must be fully leafed out and growing actively when sprayed. Retreatment may be needed.

WEED CONTROL IN FALLOW LAND AND CROP STUBBLE

Apply 1 to 2 pints per acre for control of annual and biennial broadleaf weeds. Use the higher rate on older drought stressed plants or hard to kill species. Apply 2 to 6 pints per acre for control of perennial broadleaf weeds or 4 to 6 pints for control of wild onion and wild garlic. Spray weeds in the bud to bloom stage or in good vegetative growth. Do not plant treated fallow land for three months or until chemical has disappeared from soil.

WEED CONTROL IN SUGARCANE

Preemergence: Apply 2 pints per acre before cane emerges to actively growing broadleaf weeds. **Postemergence:** Apply 2 to 4 pints per acre after cane emerges through layby. Use the higher rate for perennial weeds or hard to control species.

WOODY PLANT CONTROL IN NON-CROP AREAS

To control species susceptible to 2,4-D in rights-of-way, fencerows, roadsides, and along drainage ditch banks, spray brush up to 5 to 8 feet tall after spring foliage is well developed, using 3 to 4 quarts of ESTERON 99 CONCENTRATE in 100 gallons of water and wetting all parts of the brush including foliage, stems and bark. This may require up to 400 gallons of spray per acre for adequate coverage of solid stand of brush. Make application in such a way as to prevent drift of the spray off the area being treated. Spraying can be effective at any time up to 3 weeks before frost as long as the soil moisture is sufficient for active growth of the brush. Control will be less effective in midsummer during hot dry weather when soil moisture is deficient and plants are not actively growing. Oil or wetting agent may be added to the spray, if needed for increased effectiveness.

Forest Site Preparation: For control of susceptible broadleaf weeds and brush on sites to be planted in forests, use 1.5 to 8 quarts per acre of ESTERON 99 CONCENTRATE in sufficient spray volume for good plant coverage, usually 6 to 25 gallons. Applications can be made by air or ground (hand gun, boom, or powered knapsack sprayer). Two to eight quarts of diesel oil per acre or a suitable surfactant or penetrant may be added to improve brush control.

Forest Conifer Release: For applications in late winter or spring to control susceptible deciduous brush species, such as alder, willow, poplars, cascara, cherry, service cherry and vine maple during early growth and before conifer budbreak, use ESTERON 99 CONCENTRATE at rates up to 3 quarts per acre in diesel or stove oil by air or ground in sufficient spray volume for good plant coverage, usually 6 to 25 gallons. **Do not use in plantations where pine or larch are among the desired species.**

For treatment before conifer budbreak to control susceptible evergreen brush species, such as tanoak, mandrone, chinquapin, *ceanothus spp.* and manzanita or deciduous brush after leafout or broadleaf weeds, use ESTERON 99 CONCENTRATE at rates up to 3 quarts per acre alone or with 0.5 to 2.0 gallons per acre of diesel or similar oil or suggested rates of suitable surfactants or penetrants. After conifer budbreak, ESTERON 99 CONCENTRATE without oil, surfactant or penetrant can be used at rates up to 2 quarts per acre but may cause injury or suppression of the conifer growth. Use sufficient volume of spray for good coverage of brush, usually 6 to 25 gallons. Some species of pine may be seriously injured by treatment at these growth stages.

After conifer species such as white pine, ponderosa pine, jack pine, red pine, black spruce, white spruce, red spruce, and balsam fir cease growth and harden off and brush is still actively growing in late summer, 1.5 to 3.0 quarts of ESTERON 99 CONCENTRATE per acre in enough water to obtain good plant coverage may be applied by air or ground to control certain competing hardwood species such as alder, aspen, birch, hazel, and willow. Since this treatment may cause occasional conifer injury, do not use if such injury cannot be tolerated.

Directed Sprays In Conifer Plantations (including pine): Apply ESTERON 99 CONCENTRATE at any time brush or broadleaf weeds are susceptible by directing spray around the conifers to avoid contact of needles with injurious amounts of spray. Rates of ESTERON 99 CONCENTRATE are not to exceed 4 quarts per acre in oil, oil-water, or water carrier at 10 to 100 gallons per acre.

WEED CONTROL IN NON-CROP AREAS SUCH AS LAWNS, GOLF COURSES, CEMETERIES, PARKS, AIRFIELDS, ROADSIDES, VACANT LOTS, DRAINAGE DITCH BANKS

Apply 1 to 3 quarts of ESTERON 99 CONCENTRATE per acre in the amount of water needed for uniform application. Usually 2 quarts per acre provides good control under average conditions. Treat when weeds are young and growing well. Do not use on golf greens nor on dichondra or other broadleaf herbaceous ground covers. Do not use on creeping grasses such as bent and St. Augustine except for spot treating, and nor on newly seeded turf until grass is well established. Reseeding of treated areas should be delayed following treatment. With spring application, reseed in the fall; with fall application, reseed in the spring. Legumes are usually damaged or killed so do not treat areas where legumes are desired. Deeprooted perennial weeds may require repeated treatments in the same season or in subsequent years.

TULE (BULRUSH) AND OTHER RUSHES

Mix 2 quarts of ESTERON 99 CONCENTRATE and 1 gallon of diesel oil or kerosene, then add this mixture to 100 gallons of water. Spray to wet all foliage (400-800 gallons per acre). Addition of a wetting agent may be advisable. Apply in the spring during flower head emergence. Respray if needed when regrowth is 3 to 5 feet tall.

SPOT TREATMENT

To control broadleaf weeds in small non-cropland areas with a hand sprayer, use 1/4 pint of ESTERON 99 CONCENTRATE in 3 gallons of water and spray to thoroughly wet all weed foliage. Keep spray mixture agitated to prevent separation.

CONTROL OF WOODY WEEDS IN LOW-BUSH BLUEBERRY FIELDS IN MAINE

HOW TO USE:

Mount a drum 8 to 10 feet long or some other suitable length, and 1 1/2 to 2 feet in diameter on an axle such as an old hay rake frame. Cover the drum with water-absorbent yet tough cloth which will resist rapid wear and tear. Draw the cloth-covered drum across the blueberry field and at the same time spray evenly onto the full length of the top of the cloth-covered drum a spray mixture made by diluting 1 quart of ESTERON 99 CONCENTRATE in 50 gallons of water per acre. Have the drum mounted so that as it revolves on its axis it is high enough to miss most of the low bush blueberry stems, yet low enough to forcibly brush the spray-saturated cloth-covered drum against the higher woody weeds, principally sweet fern, wild cherry and poplar. Keep the cloth wet enough to provide top coverage of the weeds, yet not so wet as to allow runoff of the liquid which could cause injury to the blueberry plants.

WHEN TO USE:

Apply during June and July when weed tops have emerged sufficiently above the blueberry stems to allow treatment of the weeds and not the blueberry plants. Apply only during the year before the first burn. To use this method of weed control, two-year burns should be extended to three years. Caution: Do not allow the spray being applied to the cloth-covered drum to be directed onto the blueberries. Do not harvest-rake field during the herbicide treatment year or until a two-year interval thereafter.

REENTRY STATEMENT

Do not enter treated areas without protective clothing until sprays have dried. Because certain states may require more restrictive reentry intervals for various crops treated with this product, consult your State Department of Agriculture for further information. Written or oral warnings must be given to workers who are expected to be in a treated area or in an area about to be treated with this product. The front panel PRECAUTIONARY STATEMENTS should be read to workers as well as the instruction not to enter until sprays have dried. When oral warnings are given, warnings shall be given in a language customarily understood by workers. Oral warnings must be given if there is reason to believe that written warnings cannot be understood by workers. Written warnings must include the following information: "DANGER. Area treated with ESTERON 99 CONCENTRATE on (date of application). Do not enter without appropriate protective clothing until sprays have dried. (Insert here Statements of Practical Treatment as on front panel.)"

USE PRECAUTIONS

AVOID CONTACT WITH 2,4-D SUSCEPTIBLE CROPS AND OTHER DESIRABLE BROADLEAF PLANTS: Do not apply directly to or otherwise permit even minute amounts to contact cotton, grapes, tobacco, fruit trees, vegetables, flowers, ornamentals or other desirable plants susceptible to 2,4-D. Do not use in or near a greenhouse.

DO NOT APPLY IN THE VICINITY OF COTTON, GRAPES, TOBACCO, TOMATOES OR OTHER DESIRABLE 2,4-D SUSCEPTIBLE CROPS OR PLANTS. DO NOT SPRAY WHEN WIND IS BLOWING TOWARDS SUSCEPTIBLE CROPS OR ORNAMENTAL PLANTS.

AVOID SPRAY DRIFT: Applications should be made only when there is no hazard from spray drift since very small quantities of the spray, which may not be visible, may severely injure susceptible crops during both growing and dormant periods. Use coarse sprays to minimize drift since, under adverse weather conditions, fine spray droplets may drift a mile or more. A spray thickening agent such as NALCO-TROL, may be used with this product to aid in reducing spray drift. If used, follow all use recommendations and precautions on the product label.

GROUND EQUIPMENT: With ground equipment, spray drift can be lessened by keeping the spray boom as low as possible; by applying 20 gallons or more of spray per acre; by using no more than 20 pounds spraying pressure at large droplet producing nozzle tips; by spraying when wind velocity is low; and by stopping all spraying when wind exceeds 6 to 7 miles per hour. Do not apply with hollow cone-type insecticide or other nozzles that produce a fine-droplet spray.

DETERMINE AIR MOVEMENT AND DIRECTIONS BEFORE FOLIAR APPLICATION: Use a smoke generator or other means at or near the application site for the detection of air movement, air stability or temperature inversions. Such a condition exists when there is little or no wind and air temperature is lower near the ground than at higher levels. Use appropriate drift control measures or avoid application when smoke is moving toward nearby desirable susceptible plants or sensitive areas.

AERIAL APPLICATION: With aircraft, drift can be lessened by applying a coarse spray; by using no more than 20 pounds spray pressure at the nozzles; by using straight stream nozzles directed straight back; by using a spray boom no longer than 3/4 the wing or rotor span of the aircraft; and by spraying only when wind velocity is less than 6 mph.

Excessive amounts of this herbicide in the soil may temporarily inhibit seed germination or plant growth. Violent wind storms may move soil particles. If 2,4-D is on soil particles and they are blown onto the susceptible plants, visible symptoms may appear. Serious injury is unlikely. The hazard of movement of 2,4-D on dust is reduced if treated fields are irrigated or if rain occurs shortly after application.

AT HIGH TEMPERATURES, VAPORS FROM THIS PRODUCT MAY INJURE SUSCEPTIBLE PLANTS GROWING NEARBY.

To avoid injury to desirable plants, do not handle or apply other agricultural chemicals with the same equipment used for ESTERON 99 CONCENTRATE unless appropriately cleaned first. Local conditions may affect the use of herbicides. Consult your State Agricultural Experiment Station or Extension Service weed specialists for cleaning methods which are in compliance with local regulations and for advice in selecting treatments from this label to best fit local conditions. Be sure that use of this product conforms to all applicable regulations.

STORAGE AND DISPOSAL

Do not contaminate water, food or feed by storage or disposal.

STORAGE: Keep container tightly closed when not in use. This product can be stored in an unheated building.

PESTICIDE DISPOSAL: Pesticide wastes are toxic. Improper disposal of excess pesticide, spray mixture, or rinsate is a violation of Federal Law. If these wastes cannot be disposed of by use according to label instructions, contact your State Pesticide or Environmental Control Agency, or the Hazardous Waste representative at the nearest EPA Regional Office for guidance.

CONTAINER DISPOSAL: Triple rinse (or equivalent). Then offer for recycling or reconditioning, or puncture and dispose of in a sanitary landfill, or by other procedures approved by state and local authorities. **(Metal containers)** or

Triple rinse (or equivalent). Then dispose of in a sanitary landfill, or by incineration, or if allowed by local authorities, by burning. **(Plastic containers 5-gal or less)**

LIMITED WARRANTY AND DISCLAIMER

The manufacturer warrants (a) that this product conforms to the chemical description on the label; (b) that this product is reasonably fit for the purposes set forth in the directions for use when it is used in accordance with such directions; and (c) that the directions, warning and other statements on this label are based upon responsible experts' evaluation of reasonable tests of effectiveness, of toxicity to laboratory animals and to plants, and of residues on food crops and upon reports of field experience. Tests have not been made on all varieties or in all states or under all conditions. THE MANUFACTURER NEITHER MAKES NOR INTENDS, NOR DOES IT AUTHORIZE ANY AGENT OR REPRESENTATIVE TO MAKE, ANY OTHER WARRANTIES, EXPRESS OR IMPLIED, AND IT EXPRESSLY EXCLUDES AND DISCLAIMS ALL IMPLIED WARRANTIES OF MERCHANTABILITY OR FITNESS FOR A PARTICULAR PURPOSE.

THIS WARRANTY DOES NOT EXTEND TO, AND THE BUYER SHALL BE SOLELY RESPONSIBLE FOR, ANY AND ALL LOSS OR DAMAGE WHICH RESULTS FROM THE USE OF THIS PRODUCT IN ANY MANNER WHICH IS INCONSISTENT WITH THE LABEL DIRECTIONS, WARNINGS OR CAUTIONS.

BUYER'S EXCLUSIVE REMEDY AND MANUFACTURER'S OR SELLER'S EXCLUSIVE LIABILITY FOR ANY AND ALL CLAIMS, LOSSES, DAMAGES, OR INJURIES RESULTING FROM THE USE OR HANDLING OF THIS PRODUCT, WHETHER OR NOT BASED IN CONTRACT, NEGLIGENCE, STRICT LIABILITY IN TORT OR OTHERWISE SHALL BE LIMITED, AT THE MANUFACTURER'S OPTION, TO REPLACEMENT OF OR THE REPAYMENT OF THE PURCHASE PRICE FOR, THE QUANTITY OF PRODUCT WITH RESPECT TO WHICH DAMAGES ARE CLAIMED. IN NO EVENT SHALL MANUFACTURER OR SELLER BE LIABLE FOR SPECIAL, INDIRECT OR CONSEQUENTIAL DAMAGES RESULTING FROM THE USE OR HANDLING OF THIS PRODUCT.

NOTICE TO BUYER

Purchase of this material does not confer any rights under patents governing this product or the use thereof in countries outside of the United States.

Form No. AG90028

*ESTERON and 99 are trademarks of THE DOW CHEMICAL COMPANY

THIS SPECIMEN LABEL IS INTENDED FOR USE ONLY AS A GUIDE IN PROVIDING GENERAL INFORMATION REGARDING THE DIRECTIONS, WARNINGS AND CAUTIONS ASSOCIATED WITH THE USE OF THIS PRODUCT. AS WITH ANY AGRICULTURAL CHEMICAL, ALWAYS FOLLOW THE LABEL INSTRUCTIONS ON THE PACKAGE BEFORE USING.

ETHREL®

PLANT REGULATOR

For Flue Cured Tobacco
Aids Harvesting and Curing Efficiency
FOR COMMERCIAL OR AGRICULTURAL USE ONLY.
NOT FOR RESIDENTIAL USE.

ACTIVE INGREDIENT:	
Ethephon [(2-chloroethyl) phosphonic acid]*	21.7%
INERT INGREDIENTS:	78.3%
TOTAL:	100.0%

*Contains 2 pounds ethephon per gallon.

EPA Reg. No. 264-292 EPA Est. No. 264-PA-01

KEEP OUT OF REACH OF CHILDREN

WARNING AVISO

PRECAUCION AL USUARIO: Si usted no lee ingles, no use este producto hasta que le etiqueta haya sido explicado ampliamente.

For PRODUCT USE Information Call 1-800-334-9745
For MEDICAL and TRANSPORTATION Emergencies ONLY Call 24 Hours A Day 1-800-334-7577

PRECAUTIONARY STATEMENTS

WARNING

HAZARDS TO HUMANS AND DOMESTIC ANIMALS

HARMFUL IF SWALLOWED. May cause eye damage. Wear safety goggles when handling. Avoid contact with skin or clothing and do not inhale vapors as this product will irritate skin and mucous membranes.

STATEMENT OF PRACTICAL TREATMENT

IF SWALLOWED: Do not induce vomiting. Promptly drink a large quantity of milk, egg whites, gelatin solution or if these are not available, drink large quantities of water. Avoid alcohol. Call a physician immediately.

IF ON SKIN: Immediately wash with plenty of soap and water. Remove contaminated clothing and shoes. Wash clothing separately before reuse.

IF INHALED: Remove victim to fresh air. If not breathing, give artifical respiration. Administer oxygen if necessary. Get medical attention.

IF IN EYES: Hold eyelids open and flush with a steady, gentle stream of water for at least 15 minutes. Get medical attention, preferably an ophthalmologist.

NOTE TO PHYSICIAN

No specific antidote is available. All treatments should be based on observed signs and symptoms of distress in the patient. Overexposure to materials other than this product may have occurred.

Victims of severe overexposure, by inhalation, should be kept under medical observation for up to 72 hours for delayed onset of pulmonary edema. In a victim of overexposure by ingestion, careful gastric lavage is required due to the possibility of stomach or esophogeal perforation. This material is an acid, but the use of alkaline substances to neutralize it is contraindicated.

ENVIRONMENTAL HAZARDS

Do not contaminate water used for irrigation or domestic purposes. Avoid spray drift to nearby crops as this product may cause temporary modifications in plant growth.

Do not apply directly to water or wetlands (swamps, bogs, marshes, and potholes). Do not contaminate water when disposing of equipment washwaters.

Do not exceed the rate of ETHREL® Plant Regulator per acre per year recommended on this label.

GENERAL INFORMATION

A foliar spray of ETHREL® Plant Regulator hastens "yellowing" of mature tobacco and reduces curing time. ETHREL® increases the capacity of the curing barn by shortening the curing time and allows adjustments in harvest schedules. ETHREL® can be used as a directed spray to the bottom or middle portion of the tobacco plant or as an over-the-top spray. ETHREL® is not intended for use on immature tobacco.

USE PRECAUTIONS

Do not apply ETHREL® to immature leaves as this can result in unsatisfactory coloring, weight loss and reduced leaf quality.

Do not allow crop to over ripen in the field after using ETHREL®, since this may cause some reduction in yield and quality.

Treat only the acreage of tobacco which can be harvested and burned in one day. Do not plant another crop within 30 days after treatment.

Do not treat before anticipated major storm which could prevent harvest and result in crop loss.

Avoid spray drift to nearby crops as this product will cause modifications in plant growth. Plant injury or reduced yields will result.

Mix only the amount of spray you expect to use each day. Do not allow mixed solution to stand overnight.

Do not use surfactants or wetting agents with ETHREL® Plant Regulator.

Do not apply ETHREL® Plant Regulator if rain is expected within 4 hours.

DIRECTIONS FOR USE

IT IS A VIOLATION OF FEDERAL LAW TO USE THIS PRODUCT IN A MANNER INCONSISTENT WITH ITS LABELING.

Do not apply ETHREL® PLANT REGULATOR through any type of irrigation system.

GENERAL WORKER PROTECTION STATEMENTS

Do not apply this product in such a manner as to directly or through drift expose workers or other persons. The area being treated must be vacated by unprotected persons.

Mixers, loaders and applicators must wear a full face shield, long trousers, long sleeved shirt, gloves, and boots to avoid as much skin and eye contact as possible.

Do not enter treated areas within 24 hours after application unless protective clothing is worn. Because certain states may require more restrictive reentry intervals for various crops treated with this product, consult your State Department of Agriculture for further information.

Written or oral warnings must be given to workers who are expected to be in a treated area or in an area about to be treated with this product. Advise workers to stay out of fields during application and until sprays have dried. Regular long-sleeved work clothing should be worn when working in treated fields. See PRECAUTIONARY STATEMENTS; STATEMENT OF PRACTICAL TREATMENT and NOTE TO PHYSICIAN for information on accidental exposures. When oral warnings are given, warnings shall be given in a language customarily understood by workers. Oral warnings must be given if there is reason to believe that written warnings cannot be understood by workers. Written warnings must include the following information: appropriate signal word (WARNING), area treated with ETHREL® Plant Regulator, date of application, appropriate protective clothing, and reentry interval.

PROPER TIME FOR TREATMENT

Successful results with ETHREL® calls for treatment when leaves are **mature,** not overly rank green when sprayed. To easily determine the proper time for treatment and to determine the number of leaves per stalk that are ready for harvest, test spray several plants in more than one location in each field and observe their response. Mature leaves will begin yellowing in 24 to 72 hours. Test leaves that fail to yellow in 72 hours are not mature and are not ready for ETHREL® treatment. Wait a few days to permit further natural maturing, then make another test spray or "maturity" check.

When you have determined the number of leaves per plant that will color, you can determine the number of acres to treat in order to fill a barn. (See Supplemental Information.)

Prepare your test spray by mixing four teaspoons of ETHREL® in one quart of water. Spray each test plant with about 1 oz. of this mixture, covering all leaves with a fine mist. ETHREL® will not color immature sprayed leaves.

DIRECTED SPRAY APPLICATION

For a directed spray to the bottom or middle leaves of a tobacco plant, apply 2 quarts of ETHREL® in 50 to 60 gallons of water per acre. Drop nozzles similar to those used in corn can be used for directed application of ETHREL® to tobacco. Nozzles mounted on drop pipes plus slow tractor speeds will keep the boom from bouncing and insure proper placement of the directed spray. Use TG or OC spray tips designed to apply 50-60 gpa at 35 to 40 psi and at tractor speed of 2 to 3 mph. Direct ETHREL® in a fine spray to the mid or lower leaves to be treated using 2 nozzles per row; one on each side of the row dropped low enough to direct the spray to the leaves to be ripened and harvested. Thorough spray coverage is essential.

OVER-THE-TOP APPLICATION

Mix 2 to 4 quarts of ETHREL® with 40 to 60 gallons of water for each acre of tobacco to be treated.

Treat only when leaves remaining on the stalk are **mature.** To ensure remaining leaves are **mature,** test spray on several tobacco plants as described under the section on Proper Time of Treatment. Use the lower rate in a normally matured crop when experience indicates that minimum ripening inducement is required. Use the higher rate when the crop is heavy and has a tendency to be more rank or when temperatures are lower than normal. Always test spray to determine if the tobacco is mature enough to respond to treatment with ETHREL®.

Apply the over-the-top ETHREL® spray as a fine mist using three nozzles, (one nozzle tip over the center of the plant, and one on each side) so all leaves are covered thoroughly, similar to the application pattern of systemic sucker control agents. Use a spray pressure of 40 to 60 psi.

EQUIPMENT CLEANING

Because of the acidic nature of this product, prolonged exposure to spray deposits will damage acrylic plastics, certain paints, and metals.

Rinse thoroughly all exposed acrylic-plastic materials and painted surfaces with a detergent and water *within one hour* after exposure to spary deposits.

HARVESTING AND CURING

All mature, sprayed leaves will color 48 to 72 hours after ETHREL® application. The yellowing process is weather dependent; cool days will delay, while hot sunny weather can speed up the process. Harvest treated tobacco when leaves have reached the desired color intensity.

Directed Spray Application: With a directed spray be sure to harvest all leaves with 20% or more yellowing.

Over-The-Top Application: With an over-the-top application, harvesting can commence 24 hours after the ETHREL® application, depending on weather conditions or to avoid overripe tobacco. Tobacco harvested 24 hours after treatment will require more time in the coloring phase of the curing process.

Do not allow tobacco to become overripe before starting harvest. Leaf drop can occur when tobacco is overripe and left on the stalk. If a number of the bottom leaves have yellowed at the time of treatment, they should be harvested prior to spraying with ETHREL®.

Curing procedures are as much an art as a science and each cure must be judged on the basis of tobacco condition, interval between treatment and harvest, weather and type of curing facility before prescription temperature and ventilation schedules can be established. To obtain maximum quality, care must be taken to observe and control the curing process closely, especially during the late "coloring" and early "drying" stages of the leaf.

ETHREL® treated tobacco will have started the coloring process when harvested, reducing the time required in the coloring phase of curing. Treated tobacco should be dried faster. If tobacco leaves are green or contain some green when harvested, it may be necessary to color them for a few hours. If the leaves are completely yellow, the temperature and ventilation must be adjusted in a manner to dry the tobacco as fast as possible without scalding. Once the leaf is dried (3/4 dry), the grower should follow normal procedures for curing. Since ETHREL® treated leaves cure faster, treated and untreated leaves should not be cured together in the same barn.

SUPPLEMENTAL INFORMATION

NUMBER OF PLANTS PER ACRE AT DIFFERENT ROW WIDTHS AND DRILL SPACINGS

ROW WIDTH	SPACING IN DRILL						
	16 IN.	18 IN.	20 IN.	22 IN.	24 IN.	26 IN.	28 IN.
36 in.	10,890	9,680	8,712	7,920	7,260	6,701	6,222
38 in.	10,317	9,170	8,255	7,503	6,878	6,349	5,896
40 in.	9,801	8,712	7,841	7,128	6,535	6,032	5,601
42 in.	9,334	8,297	7,467	6,788	6,222	5,744	5,333
44 in.	8,912	7,920	7,128	6,480	5,940	5,483	5,091
46 in.	8,523	7,576	6,818	6,198	5,682	5,245	4,870
48 in.	8,167	7,260	6,534	5,940	5,445	5,026	4,667

NUMBER OF LEAVES PER ACRE AT DIFFERENT PLANT POPULATIONS

PLANTS PER ACRE	HEIGHT OF TOPPING (NUMBER OF LEAVES)					
	16	18	20	22	24	26
	LEAVES PER ACRE (TO NEAREST 100)					
9800	156,800	176,400	196,000	215,600	235,200	254,800

PLANTS PER ACRE	HEIGHT OF TOPPING (NUMBER OF LEAVES)					
	16	18	20	22	24	26
	LEAVES PER ACRE (TO NEAREST 100)					
9300	148,800	167,400	186,000	204,600	223,200	241,800
8900	142,400	160,200	178,000	195,800	213,600	231,400
8700	139,200	156,600	174,000	191,400	208,800	226,200
8500	136,000	153,000	170,000	187,000	204,000	221,000
8300	132,800	149,400	166,000	182,600	199,200	215,800
8200	131,200	147,600	164,000	180,400	196,800	213,200
7900	126,400	142,200	158,000	173,800	189,600	205,400
7800	124,800	140,400	156,000	171,600	187,200	202,800
7600	121,600	136,800	152,000	167,200	182,400	197,600
7500	120,000	135,000	150,000	165,000	180,000	195,000
7300	116,800	131,400	146,000	160,600	175,200	189,800
7100	113,600	127,800	142,000	156,200	170,400	184,600
6800	108,800	122,400	136,000	149,600	163,200	176,800
6500	104,000	117,000	130,000	143,000	156,000	169,000
6200	99,200	111,600	124,000	136,400	148,800	161,200
6000	96,000	108,000	120,000	132,000	144,000	156,000
5900	94,400	106,200	118,000	129,800	141,600	153,400
5700	91,200	102,600	114,000	125,400	136,800	148,200
5500	88,000	99,000	110,000	121,000	132,000	143,000
5400	86,400	97,200	103,000	118,800	129,600	140,400
5200	83,200	93,600	104,000	114,400	124,800	135,200
5000	80,000	90,000	100,000	110,000	120,000	130,000

For further help in determining the time of application, changes in curing procedure, and/or the required number of treated acres needed to fill one barn, contact your County Extension Agent, Area Tobacco Specialist, or your Rhône-Poulenc Ag Company Representative. (See supplemental information for guidance on the amount of tobacco needed to fill a barn).

STORAGE AND DISPOSAL

STORAGE

Do not store near fertilizers, seeds, insecticides or fungicides.

PESTICIDE DISPOSAL

Do not contaminate water, food, or feed by storage or disposal. Wastes resulting from the use of this product may be disposed of on site or at an approved waste disposal facility.

CONTAINER DISPOSAL

Do not reuse container. Triple rinse (or equivalent). Then offer for recycling or reconditioning, or puncture or dispose of in a sanitary landfill or incineration, or if allowed by state and local authorities, by burning. If burned, stay out of smoke.

LIMITED WARRANTY AND DISCLAIMER

The manufacturer warrants (a) that this product conforms to the chemical description on the label; (b) that this product is reasonably fit for the purposes set forth in the directions for use when it is used in accordance with such directions; and (c) that the directions, warnings and other statements on this label are based upon responsible experts' evaluation of reasonable tests of effectiveness, of toxicity to laboratory animals and to plants, and of residues on food crops and upon reports of field experience. Tests have not been made on all varieties or in all states or under all conditions. THE MANUFACTURER NEITHER MAKES NOR INTENDS, NOR DOES IT AUTHORIZE ANY AGENT OR REPRESENTATIVE TO MAKE, ANY OTHER WARRANTIES, EXPRESS OR IMPLIED, AND IT EXPRESSLY EXCLUDES AND DISCLAIMS ALL IMPLIED WARRANTIES OF MERCHANTABILITY OR FITNESS FOR A PARTICULAR PURPOSE.

THIS WARRANTY DOES NOT EXTEND TO, AND THE BUYER SHALL BE SOLELY RESPONSIBLE FOR, ANY AND ALL LOSS OR DAMAGE WHICH RESULTS FROM THE USE OF THIS PRODUCT IN ANY MANNER WHICH IS INCONSISTENT WITH THE LABEL DIRECTIONS, WARNINGS OR CAUTIONS.

BUYER'S EXCLUSIVE REMEDY AND MANUFACTURER'S OR SELLER'S EXCLUSIVE LIABILITY FOR ANY AND ALL CLAIMS, LOSSES, DAMAGES, OR INJURIES RESULTING FROM THE USE OR HANDLING OF THIS PRODUCT, WHETHER OR NOT BASED IN CONTRACT, NEGLIGENCE, STRICT LIABILITY IN TORT OR OTHERWISE SHALL BE LIMITED, AT THE MANUFACTURER'S OPTION, TO REPLACEMENT OF, OR THE REPAYMENT OF THE PURCHASE PRICE FOR, THE QUANTITY OF PRODUCT WITH RESPECT TO WHICH DAMAGES ARE CLAIMED. IN NO EVENT SHALL MANUFACTURER OR SELLER BE LIABLE FOR SPECIAL, INDIRECT OR CONSEQUENTIAL DAMAGES RESULTING FROM THE USE OR HANDLING OF THIS PRODUCT.

NOTICE TO BUYER

Purchase of this material does not confer any rights under patents governing this product or the use thereof in countries outside of the United States.

ETHREL is a registered trademark of RHONE-POULENC.

AG 90037

THIS SPECIMEN LABEL IS INTENDED FOR USE ONLY AS A GUIDE IN PROVIDING GENERAL INFORMATION REGARDING THE DIRECTIONS, WARNINGS AND CAUTIONS ASSOCIATED WITH THE USE OF THIS PRODUCT. AS WITH ANY AGRICULTURAL CHEMICAL, ALWAYS FOLLOW THE LABEL INSTRUCTIONS ON THE PACKAGE BEFORE USING.

ETHREL®

ETHEPHON PLANT REGULATOR

For Pineapple and Sugarcane
For Commercial or Agricultural Use Only
Not for Residential Use

ACTIVE INGREDIENT:	Percent by wt.
Ethephon (2-chloroethyl) phosphonic acid*	39.9%
INERT INGREDIENTS:	60.1%
TOTAL	100.0%

*This product contains 4 pounds ethephon per gallon.

EPA Reg. No. 264-257 EPA EST. NO. 264-PA-01

KEEP OUT OF REACH OF CHILDREN

DANGER PELIGRO

PRECAUCION AL USARIO: Si usted no lee ingles, no use este producto hasta que la informacion en la etiqueta le haya sido explicado ampliamente.

STATEMENT OF PRACTICAL TREATMENT

IF IN EYES: Hold eyelids open and flush with a steady, gentle stream of water for at least 15 minutes. Get medical attention, preferably an Ophthalmologist.

For **PRODUCT USE** Information Call 1-800-334-9745
For **MEDICAL** and **TRANSPORTATION** Emergencies ONLY Call 24 Hours A Day 1-800-334-7577

PRECAUTIONARY STATEMENTS

DANGER

HAZARDS TO HUMANS AND DOMESTIC ANIMALS

Corrosive. Causes eye damage and skin irritation. Do not get in eyes, on skin or on clothing. Wear goggles or face shield and rubber gloves when handling.

Harmful or fatal if swallowed. Avoid contamination of food.

STATEMENT OF PRACTICAL TREATMENT

If swallowed: Do not induce vomiting. Promptly drink a large quantity of milk, egg whites, gelatin solution, or if these are not available, drink large quantities of water. Avoid alcohol. Call a physician immediately.

IF ON SKIN: Immediately wash with plenty of soap and water. Remove contaminated clothing and shoes. Wash clothing separately before reuse.

IF INHALED: Remove victim to fresh air. If not breathing, give artificial respiration. Administer oxygen if necessary. Get medical attention.

NOTE TO PHYSICIAN

No specific antidote is available. All treatments should be based on observed signs and symptoms of distress in the patient. Overexposure to materials other than this product may have occurred.

Victims of severe overexposure, by inhalation, should be kept under medical observation for up to 72 hours for delayed onset of pulmonary edema. In a victim of overexposure by ingestion, careful gastric lavage is required due to the possibility of stomach or esophogeal perforation. This material is an acid, but the use of alkaline substances to neutralize it is contraindicated.

ENVIRONMENTAL HAZARDS

Avoid spray drift to nearby crops as this product may cause temporary modifications in plant growth. Do not contaminate water used for irrigation or domestic purposes.

Do not apply directly to water or wetlands (swamps, bogs, marshes, and potholes). Do not contaminate water when disposing of equipment washwaters.

Do not exceed the rate of ETHREL® EthephonPlant Regulator per acre year recommended on this label.

GENERAL INFORMATION

PINEAPPLES

ETHREL® Ethephon Plant Regulator for Pineapple and Sugarcane concentrates fruit maturity by stimulating uniform initiation of flower buds and by stimulating uniform shell coloration of pineapple fruit. When ETHREL® Ethephon Plant Regulator for Pineapple and Sugarcane is applied to pineapples, the chemical decomposes within the plant tissue to release ethylene, a naturally occurring plant hormone.

Sugarcane (For use in Hawaii Only)

ERTHREL® Ethephon Plant Regulator For Pineapple and Sugarcane applied just prior to floral induction will reduce or prevent tasseling (flowering). Flowering in sugarcane is associated with the formation of unusable pithy tissue, lower sucrose yields per unit of cane harvested and decreased processing efficiency.

DIRECTIONS FOR USE

IT IS A VIOLATION OF FEDERAL LAW TO USE THIS PRODUCT IN A MANNER INCONSISTENT WITH ITS LABELING.

Do not apply ETHREL® Ethephon Plant Regulator For Pineapple and Sugarcane through any type of irrigation system.

GENERAL WORKER PROTECTION STATEMENT

Do not apply this product in such a manner as to directly or through drift expose workers or other persons. The area being treated must be vacated by unprotected persons.

Mixers, loaders and applicators must wear a full face shield, long trousers, long sleeved shirt, gloves, and boots to avoid as much skin and eye contact as possible.

Do not enter treated fields within 24 hours after application unless protective clothing is worn. Because certain states may require more restrictive reentry intervals for various crops treated with this product, consult your State Department of Agriculture for further information. Written or oral warnings must be given to workers who are expected to be in a treated area or in an area about to be treated with this product. Advise workers to stay out of fields during application. Regular long-sleeved work clothing should be worn when working in treated fields. See Precautionary Statement, Statement of Practical Treatment and Note to Physician for information on accidental exposures. When oral warnings are given, warnings shall be given in a language customarily understood by workers. Oral warnings must be given if there is a reason to believe that written warnings cannot be understood by workers. Written warnings must include the following information: Appropriate signal word (DANGER), area treated with ETHREL® Ethephon Plant Regulator for Pineapple and Sugarcane, date of application, appropriate protective clothing, and reentry interval if applicable.

PINEAPPLE

Fruit Shell Coloration

Apply 1/2 to 1 quart per acre of ETHREL® Ethephon Plant Regulator for Pineapple and Sugarcane as a broadcast spray, thoroughly covering both foliage and fruit. Treat when first fruits begin to change shell color. Use the higher rate during periods of cool or cloudy weather when normal ripening has slowed. The spray volume should be 100 to 300 gallons of water per acre. Mature fruit should be harvested at optimum ripeness at least 48 hours after treatment.

Flower Bud Initiation

Apply 2 to 4 quarts of ETHREL® Ethephon Plant Regulator for Pineapple and Sugarcane per acre in 100 to 500 gallons of water. Treat vegetatively mature plants, generally 6 to 8 months prior to desired picking date. Proper rate will vary according to local growing conditions, varieties, plantation management practices, and time of year. Use the higher rate where earlier harvest is desired. Apply when pineapple foliage is dry.

Do not use treated plant foliage for forage or fodder within 7 days of treatment.

Do not harvest sugarcane treated with ETHREL® Ethephon Plant Regulator sooner than 2 days after the last application.

NOTE: Mix only the amount of spray to be used that day.

SUGARCANE (FOR USE IN HAWAII ONLY)

Apply ETHREL® Ethephon Plant Regulator for Pineapple and Sugarcane at 1 pint per acre in 7 to 14 gallons of water per acre, before flower induction takes place, usually between August 15 and September 15.

Apply by fixed wing aircraft or helicopter as a broadcast spray over the top of the cane, using equipment designed to give uniform coverage.

Harvest cane at desired maturity, generally 15 to 30 months after planting.

Do not harvest sugarcane treated with ETHREL® Plant Regulator sooner than 2 months after the last application.

STORAGE AND DISPOSAL

STORAGE

Do not contaminate water, food, or feed by storage or disposal.

PESTICIDE DISPOSAL

Pesticide wastes are acutely hazardous. Improper disposal of excess pesticide, spray mixture, or rinsate is a violation of Federal law. If these wastes cannot be disposed of by use according to label instructions, contact your State Pesticide or Environmental Control Agency or the Hazardous Waste representative at the nearest EPA Regional Office for guidance.

CONTAINER DISPOSAL

Triple rinse (or equivalent). Then offer for recycling or reconditioning, or puncture and dispose of in a sanitary landfill, or by incineration or if allowed by state and local authorities, by open burning. If burned, stay out of smoke.

LIMITED WARRANTY AND DISCLAIMER

The manufacturer warrants that this product conforms to the chemical description on the label; that this product is reasonably fit for the purposes set forth in the directions for use when it is used in accordance with such directions; and that the directions, warnings and other statements on this label are based upon responsible experts' evaluation of reasonable tests of effectiveness, of toxicity to laboratory animals and to plants, and of residues on food crops, and upon reports of field experience. Tests have not been made on all varieties or in all states or under all conditions. THE MANUFACTURER NEITHER MAKES NOR INTENDS, NOR DOES IT AUTHORIZE ANY AGENT OR REPRESENTATIVE TO MAKE, ANY OTHER WARRANTIES, EXPRESS OR IMPLIED, AND IT EXPRESSLY EXCLUDES AND DISCLAIMS ALL IMPLIED WARRANTIES OF MERCHANTABILITY OR FITNESS FOR A PARTICULAR PURPOSE.

THIS WARRANTY DOES NOT EXTEND TO, AND THE BUYER SHALL BE SOLELY RESPONSIBLE FOR, ANY AND ALL LOSS OR DAMAGE WHICH RESULTS FROM THE USE OF THIS PRODUCT IN ANY MANNER WHICH IS INCONSISTENT WITH THE LABEL DIRECTIONS, WARNINGS OR CAUTIONS.

BUYER'S EXCLUSIVE REMEDY AND MANUFACTURER'S OR SELLER'S EXCLUSIVE LIABILITY FOR ANY AND ALL CLAIMS, LOSSES, DAMAGES, OR INJURIES RESULTING FROM THE USE OR HANDLING OF THIS PRODUCT, WHETHER OR NOT BASED IN CONTRACT, NEGLIGENCE, STRICT LIABILITY IN TORT OR OTHERWISE SHALL BE LIMITED, AT THE MANUFACTURER'S OPTION, TO REPLACEMENT OF, OR THE REPAYMENT OF THE PURCHASE PRICE FOR, THE QUANTITY OF PRODUCT WITH RESPECT TO WHICH DAMAGES ARE CLAIMED. IN NO EVENT SHALL MANUFACTURER OR SELLER BE LIABLE FOR SPECIAL, INDIRECT OR CONSEQUENTIAL DAMAGES RESULTING FROM THE USE OR HANDLING OF THIS PRODUCT.

NOTICE TO BUYER

Purchase of this material does not confer any rights under patents governing this product or the use thereof in countries outside of the United States.

ETHREL is a registered trademark of Rhône-Poulenc

AG 90039

THIS SPECIMEN LABEL IS INTENDED FOR USE ONLY AS A GUIDE IN PROVIDING GENERAL INFORMATION REGARDING THE DIRECTIONS, WARNINGS AND CAUTIONS ASSOCIATED WITH THE USE OF THIS PRODUCT. AS WITH ANY AGRICULTURAL CHEMICAL, ALWAYS FOLLOW THE LABEL INSTRUCTIONS ON THE PACKAGE BEFORE USING.

ETHREL®

PLANT REGULATOR

For Commercial or Agricultural Use Only. Not for Residential Use.
For use on Tobacco, Tomatoes, Cherries, Grapes, Apples, Walnuts, Peppers, Blackberries and Cantaloupes.

ACTIVE INGREDIENT:	
Ethephon (2-chloroethyl)phosphonic acid*	21.7%
INERT INGREDIENTS:	78.3%
TOTAL:	100.0%

*Contains 2 pounds ethephon per gallon.

KEEP OUT OF REACH OF CHILDREN

WARNING AVISO

PRECAUCION AL USARIO: Si usted no lee ingles, no use este producto hasta que le etiqueta haya sido explicada ampliamente.

EPA Reg. No. 264-267 EPA Est. No. 264-PA-01

FOR PRODUCT USE INFORMATION CALL 1-800-334-9745
For MEDICAL And TRANSPORTATION EMERGENCIES ONLY CALL 1-800-334-7577

ETHREL® HELPS NATURE HELP YOU

ETHREL® Plant Regulator

* Promotes earlier, uniform coloring of mature FLUE-CURED TOBACCO.
* Promotes earlier coloration and maturity of TOMATO, APPLES, GRAPES, and PEPPERS
* Loosens CHERRIES, WALNUTS, and APPLES for earlier, more efficient harvest
* Promotes fruit abscission (slipping) in CANTALOUPES
* Increases dormant fruit bud hardiness and delayed spring bloom of SWEET CHERRIES in the Pacific Northwest
* Accelerates BLACKBERRY ripening and loosening

PRECAUTIONARY STATEMENTS

WARNING

HAZARDS TO HUMANS AND DOMESTIC ANIMALS

HARMFUL IF SWALLOWED. May cause eye damage. Do not get in eyes. Wear safety goggles or face shield when handling. Avoid contact with skin or clothing and do not inhale vapors as this product will irritate skin and mucous membranes. Avoid contamination of food. If taken internally call a physician and promptly drink large quantities of milk, egg white, gelatin solution or if these are not available, large quantities of water.

STATEMENT OF PRACTICAL TREATMENT

IF SWALLOWED: Do not induce vomiting. Promptly drink a large quantity of milk, egg whites, gelatin solution or if these are not available, drink large quantities of water. Avoid alcohol. Call a physician immediately.

IF ON SKIN: Immediately wash with plenty of soap and water. Remove contaminated clothing and shoes. Wash clothing separately before reuse.

IF INHALED: Remove victim to fresh air. If not breathing, give artificial respiration. Administer oxygen if necessary. Get medical attention.

IF IN EYES: Hold eyelids open and flush with a steady, gentle stream of water for at least 15 minutes. Get medical attention.

NOTE TO PHYSICIAN

No specific antidote is available. All treatments should be based on observed signs and symptoms of distress in the patient. Overexposure to materials other than this product may have occurred.

Victims of severe overexposure, by inhalation, should be kept under medical observation for up to 72 hours for delayed onset of pulmonary edema. In a victim of overexposure by ingestion, careful gastric lavage is required due to the possibility of stomach or esophogeal perforation. This material is an acid, but the use of alkaline substances to neutralize it is contraindicated.

ENVIRONMENTAL HAZARDS

Do not contaminate water used for irrigation or domestic purposes. Keep out of lakes, streams and ponds.

Do not apply directly to water or wetlands (swamps, bogs, marshes, and potholes). Do not contaminate water when disposing of equipment washwaters.

Avoid spray drift to nearby crops as this product will cause modifications in plant growth. Plant injury or reduced yields may result.

Do not plant another crop with 30 days after treatment.

IMPORTANT: Do not apply ETHREL® Plant Regulator through any type of irrigation system.

Do not use this product for purposes other than those listed on the label.

Do not exceed the rate of ETHREL® Plant Regulator per acre per year recommended on this label.

GENERAL INFORMATION

For local recommendations on rates, spray volumes, and spray equipment under varying temperature and rainfall conditions consult your Extension Pomologist, Horticultural Specialist, Rhone-Poulenc Ag Company Representative or Farm Advisor for his experience with this product in your area.

TOBACCO (FLUE-CURED) - A foliar spray of ETHREL® Plant Regulator promotes early, uniform "yellowing" of mature tobacco. ETHREL® reduces curing time, allowing more efficient use of curing barn space, and increased control over harvest schedules.

TOMATOES—FRESH MARKET IN CALIFORNIA—A foliar spray of ETHREL® Plant Regulator accelerates tomato ripening increasing the early yields of marketable ripe fruit.

TOMATOES—PROCESSED—ETHREL® Plant Regulator application starts the normal ripening process earlier and increases the rate of ripening of mature green fruit concentrating maturity. Concentrated fruit maturity enables high recoverable yields of ripe tomatoes in once-over harvest. Early concentrated maturity extends the normal harvest season and will help growers to schedule harvest and handling more efficiently.

CHERRIES—A foliar spray of ETHREL® Plant Regulator loosens fruit and promotes early uniform ripening, and reduces the shaker force needed to mechanically harvest fruit. These effects are important for increasing efficiency, recoverable yields, maintaining fruit quality and reducing tree injury. DO NOT USE ON CHERRIES IN CALIFORNIA. A fall application of ETHREL® to sweet cherries in the Pacific Northwest area increases dormant bud hardiness and delays bloom the following spring.

GRAPES - TABLE GRAPES IN CALIFORNIA AND ARIZONA—An application of ETHREL® Plant Regulator promotes early uniform color development on table grape varieties Cardinal, Emperor, Flame Seedless, Red Malaga, Queen and Tokay allowing more efficient harvests of quality fruit.

GRAPES - RAISIN PRODUCTION—A foliar spray of ETHREL® Plant Regulator will hasten maturity of Thompson Seedless grapes resulting in reduced acids, increased sugars and increased raisin quality.

APPLES—A foliar spray of ETHREL® Plant Regulator promotes fruit maturity and loosens apples, making harvest by hand or machine easier and more efficient. ETHREL® applied in combination with FRUITONE® N to control pre-harvest drop stimulates early development of red color and ripening without loosening fruit. Applied to young trees, ETHREL® will suppress vegetative growth and promote flower bud formation.

WALNUTS IN CALIFORNIA—A foliar spray of ETHREL® Plant Regulator will loosen walnuts for increased efficiency of mechanical harvest, promote earlier harvest, improve hull removal, and increase recoverable yields from a once-over harvest.

PEPPERS—A foliar spray of ETHREL® Plant Regulator promotes early, uniform ripening and coloring for more efficient packing and handling of fruit.

BLACKBERRIES IN OREGON AND WASHINGTON—A foliar spray of ETHREL® Plant Regulator will concentrate maturity and loosen fruit improving the harvesting efficiency while reducing cane injury from mechanical harvest.

CANTALOUPES IN CALIFORNIA, ARIZONA AND TEXAS—A foliar spray of ETHREL® Plant Regulator promotes abscission (slipping) of fruit allowing more efficient and economical harvesting.

USE PRECAUTIONS

Mix only the amount of spray you expect to use each day. Do not allow mixed solution to stand overnight.

Avoid spray drift to nearby crops as this product will cause modification in plant growth. Plant injury or reduced yields will result. Do not plant another crop within 30 days after treatment.

Do not use ETHREL® Plan Regulator with additives other than recommended on this label.

Avoid spills of concentrated product on spray equipment or airplane parts.

IMMEDIATELY RINSE ANY SPILLS WITH PLENTY OF WATER AS ETHREL® PLANT REGULATOR IS CORROSIVE.

EQUIPMENT CLEANING

Because of the acidic nature of this product, prolonged exposure to spray deposits will damage acrylic plastics, certain paints, and metals.

Rinse thoroughly all exposed acrylic-plastic materials and painted surfaces with a detergent and water within one hour after exposure to spray deposits.

DIRECTIONS FOR USE

IT IS A VIOLATION OF FEDERAL LAW TO USE THIS PRODUCT IN A MANNER INCONSISTENT WITH ITS LABELING.

GENERAL WORKER PROTECTION STATEMENTS

Do not apply this product in such a manner as to directly or through drift expose workers or other persons. The area being treated must be vacated by unprotected persons.

Mixers, loaders and applicators must wear a full face shield, long trousers, long sleeved shirt, gloves, and boots to avoid as much skin and eye contact as possible.

Do not enter treated fields within 24 hours after application unless protective clothing is worn. Because certain states may require more restrictive reentry intervals for various crops treated with this product, consult your State Department of Agriculture for further information. Written or oral warnings must be given to workers who are expected to be in a treated area or in an area about to be treated with this product. Advise workers to stay out of fields during application and until sprays have dried. See PRECAUTIONARY STATEMENTS, STATEMENT OF PRACTICAL TREATMENT, and NOTE TO PHYSICIAN for information on accidental exposure. When oral warnings are given, warnings shall be given in a language customarily understood by workers. Oral warnings must be given if there is reason to believe that written warnings cannot be understood by workers. Written warnings must include the following information: appropriate signal word (WARNING), area treated ETHREL® Plant Regulator, date of application, appropriate protective clothing, and reentry interval.

FLUE-CURED TOBACCO

A foliar spray of ETHREL® Plant Regulator promotes earlier, more uniform "yellowing" of mature tobacco leaf and reduces curing time. ETHREL® increases the capacity of the curing barn by shortening the curing time and allows adjustments in harvest schedules. ETHREL® can be used as a directed spray to the bottom or middle portion of the tobacco plant or as an over-the-top spray. ETHREL® is not intended for use on immature tobacco.

Important

Do not apply ETHREL® to immature leaves as this can result in unsatisfactory coloring, weight loss and reduced leaf quality.

Do not allow the crop to over ripen in the field after using ETHREL®, since this may cause some reduction in yield and quality.

Do not treat before anticipated major storm which could prevent harvest and result in crop loss.

Do not apply ETHREL® Plant Regulator if rain is expected within 4 hours.

PROPER TIME FOR TREATMENT

Successful results with ETHREL® calls for treatment when leaves are mature not overly rank green when sprayed. To easily determine the proper treatment timing and the number of leaves per stalk ready for harvest, test spray several plants in more than one location in each field and observe the response. Mature leaves will begin to yellow in 24 to 72 hours. Test leaves that fail to yellow in 72 hours are not mature and are not ready for ETHREL® treatment. Wait a few days to permit further natural maturing, then make another test spray or "maturity" check.

When you have confirmed the number of leaves per plant that will color, you can determine the number of acres to treat in order to fill a barn.

Prepare your test spray by mixing 4 teaspoons of ETHREL® in 1 quart of water. Spray each test plant with about 1 oz. of this mixture, covering all leaves with a fine mist. ETHREL® will not color immature sprayed leaves.

Directed Spray Application

For a directed spray to the bottom or middle leaves of a tobacco plant, apply 2 quarts of ETHREL® in 50 to 60 gallons of water per acre. Drop nozzles similar to those used in corn can be used for directed application of ETHREL® to tobacco. Nozzles mounted on drop pipes plus slow tractor speeds will keep the boom from bouncing and insure proper placement of the directed spray. Use TG or OC spray tips designed to apply 50 to 60 gpa at 35 to 40 psi and at tractor speed of 2 to 3 mph. Direct ETHREL® in a fine spray to the mid or lower leaves to be treated using 2 nozzles per row; one on each side of the row dropped low enough to direct the spray to the leaves to be ripened and harvested. Thorough spray coverage is essential.

Over-The-Top Application

Mix 2 to 4 quarts of ETHREL® with 40 to 60 gallons of water for each acre of tobacco to be treated.

Treat only when leaves remaining on the stalk are mature. To ensure remaining leaves are mature, test spray several tobacco plants as described under the section "Proper Time of Treatment". Use the lower rate in a normally mature crop when experience indicates that minimum ripening inducement is required. Use the higher rate when the crop is heavy and has a tendency to be more rank or when temperatures are lower than normal. Always test spray to determine if the tobacco is mature enough to respond to treatment with ETHREL®.

Apply over-the-top ETHREL® spray as a fine mist using three nozzles, (one nozzle tip over the center of the plant, and one on each side) so all leaves are covered thoroughly, similar to the application pattern of systemic sucker control agents. Use a spray pressure of 40 to 60 psi.

HARVESTING AND CURING

All mature, sprayed leaves will begin to color within 24 to 72 hours after ETHREL® application. The yellowing process is weather dependent; cool days will delay, while hot sunny weather can speed up the process. Harvest treated tobacco when leaves have reached the desired color intensity.

Directed Spray Application

With a directed spray be sure to havest all leaves with 20% of more yellowing.

Over-The-Top Application

With an over-the-top application, harvesting can commence 24 hours after ETHREL® application depending on weather conditions or to avoid over ripe tobacco. Tobacco harvested 24 hours after treatment will require more time in the coloring phase of the curing process.

Do not allow tobacco to become over ripe before starting harvest. Leaf drop can occur when tobacco is over ripe and left on the stalk. If a number of the bottom leaves have yellowed at the time of treatment, they should be harvested prior to spraying with ETHREL®.

Curing procedures are as much an art as a science and each cure must be judged on the basis of tobacco condition, interval between treatment and harvest, weather and type of curing facility before prescription temperature and ventilation schedules can be established. To obtain maximum quality, care must be taken to observe and control the curing process closely, especially during the late "coloring" and early "drying" stages of the leaf.

ETHREL® treated tobacco will have started the coloring process when harvested, reducing the time required in the coloring phase of curing. Treated tobacco should be dried faster. If tobacco leaves are green or contain some green when harvested, it may be necessary to color them for a few hours. If the leaves are completely yellow, temperature and ventilation must be adjusted in a manner to dry the tobacco as fast as possible without scalding. Once the leaf is dried (¾ dry), you should follow normal procedures for curing. Since ETHREL® treated leaves cure faster, treated and untreated leaves should not be cured together in the same barn.

TOMATOES

A foliar spray of ETHREL® Plant Regulator accelerates and concentrates tomato ripening, increasing recoverable fruit yield from a once-over harvest. Apply ETHREL® when desired tonnage of fruit has reached marketable size and maturity. Check fields closely for stage and degree of uniform maturity prior to application. Sample several plants throughout the field to determine proper spray date. Sort, weigh and calculate the percent fruit which is red and pink, including breakers. Fruit size alone is not an adequate indicator. Observe specific directions below to determine treatment stage for your situation. Walk treated fields frequently to evaluate crop condition. Maintain normal cultural practice between treatment and harvest. Timely harvest is essential for maximum return of high quality fruit. Consult processor representatives concerning delivery schedules and quotas prior to application of ETHREL®.

Consult your local Rhone-Poulenc AG Company Representative, Extension Horticultural Specialist or Farm Advisor for suggestions on the proper use of ETHREL® when fields have variable plant vigor due to differences in soil conditions or cultural practices and for rates of fruit ripening as affected by temperature, within the rate and timing limitations shown on the label.

Important

Do not apply before there is sufficient mature green fruit to produce the desired tonnage. ETHREL® will not ripen immature green fruit.

Some yellowing and general aging of foliage may be observed following treatment.

Do not treat plants with poor root systems or growing under stress due to poor soil conditions, drought, disease or insect damage. Treatment of weak plants will result in rapid loss of foliage cover increasing sunburn and sun scald potential, especially under high temperature conditions.

Do not treat when sustained temperatures avoe 105°F are anticipated.

Do not treat sensitive varieties during period when temperatures exceed 100°F. Foliar damage is more severe on certain varieties such a VF10, VF315, VF145, 21-4, and 13L. During periods of high temperatures use the lower rates on sensitive varieties.

Do not tank mix with sun protection products, sun protection whiteners, spray adjuvants or other additives.

Do not apply ETHREL® to more acreage than can be harvested in 2 to 3 days.

Do not apply ETHREL® to varieties known to soften rapidly or shatter when ripe.

Do not use on greenhouse tomatoes.

Do not harvest tomatoes for processing treated with ETHREL® Plant Regulator sooner than 14 days after last application.

Application

Apply recommended amount of ETHREL® in 20 to 100 gallons of water per acre for ground boom sprayers or in 10 to 12 gallons of water per acre for aerial application. Uniform ripening requires uniform spray coverage of foliage and fruit. Overdosing from overlapping swaths or spray nozzle patterns can cause severe foliage injury. Rates listed below are for broadcast coverage of the total area. If bands are sprayed over the row reduce the amount of ETHREL® used in proportion to the area actually treated.

FOR USE IN CALIFORNIA, TEXAS AND OTHER WESTERN AREAS HAVING SIMILAR VARIETIES AND CULTURES

Early and Midseason Tomato Crops

Apply 3¹/₄ pints of ETHREL® Plant Regulator per acre when 5% to 15% of the fruit in the field are red and pink (including breakers) and there is sufficient mature green fruit to produce the desired tonnage.

For early and midseason crops, particularly under high temperatures, rates as low as 1¹/₄ pints ETHREL® per acre have been effective for increasing recoverable yields and reducing sorting times with less vine damage. Expect little change in the harvest date, compared to untreated tomatoes. Harvest fruit at proper maturity generally 14 to 21 days after treatment.

Late Season or Coastal Crops

Apply 3¹/₄ to 6¹/₂ pints ETHREL® Plant Regulator per acre when 5% to 30% of fruit in the field are red and pink (including breakers) and there is sufficient mature green fruit to produce the desired tonnage. Maximum response is achieved when fruit to be treated is 5 to 15% pink and red. Use the higher rate of ETHREL® when nighttime temperatures are cool (below 65°F) or vegetative growth is dense.

Harvest maturity of late season crops can be expected 18 to 24 days after treatment. However, nighttime temperatures below 65°F retard natural color development and may extend the interval between treatment and harvest.

MIDWEST AND EASTERN TOMATO PRODUCING AREAS

Apply 3¹/₄ pints of ETHREL® Plant Regulator per acre when 5% to 30% of the fruit in the field are red and pink (including breakers) and there is sufficient mature green fruit to produce the desired tonnage. When average temperatures are cool (below 65°F) and vegetative growth is dense, apply 6¹/₂ pints of ETHREL® per acre when 5% to 30% of the fruit in the field are red and pink (including breakers).

For early and midseason crops, under high temperatures (above 85°F), rates as low as 1¹/₄ pints per acre of ETHREL® have been effective for increasing recoverable yields with less vine damage particularly on sensitive varieties - see above.

Harvest fruit at proper maturity, generally 14 to 21 days after treatment. Temperatures above 90°F may cause fruit to ripen sooner, whereas temperatures below 65°F retard natural color development and may extend the interval between treatment and harvest.

FRESH MARKET TOMATOES IN CALIFORNIA

A foliar spray of ETHREL® Plant Regulator accelerates and concentrates tomato ripening. Apply ETHREL® when desired tonnage of fruit has reached marketable size and maturity, generally 3 to 6 days before desired harvest date. Expect only one harvest from treated areas.

Apply a foliar spray of 3¹/₄ to 5 pints of ETHREL® per acre in 40 or more gallons of water for ground sprayers or 10 to 12 gallons of water per acre for aerial application. Use the higher rate and longer preharvest interval on late season crops, when temperatures are cool (generally below 85° F) and foliage is dense. Under high temperatures, rates as low as 1¹/₄ pints per acre have been effective while reducing foliage damage particularly on more sensitive varieties - see above.

Do not harvest fresh market tomatoes in California treated with ETHREL® Plant Regulator sooner than 3 days after last application.

CHERRIES

Fruit Loosening And Uniform Ripening

A foliar spray of ETHREL® Plant Regulator will promote fruit loosening and early uniform ripening. ETHREL® enhances separation of fruit from stems and reduces the shaking force required to harvest fruit. This can result in less risk of torn or bruised fruit and less tree damage. Check treated fruit frequently and harvest at optimum maturity. When programming harvest, spray a different orchard block each day and harvest blocks daily in the same sequence.

Important: Not for use in on cherries in California

Some gummosis of cherry trees is associated with treatment.

High temperatures during and after application may cause severe gummosis.

Do not treat trees that had severe gummosis the previous year as tree damage will result.

Do not apply to trees that are of low vigor or have experienced severe stress (such as winter injury, drought, or disease), since gummosis will be excessive, particularly when high temperatutres and/or drought follow treatment.

If possible, irrigate orchards to prevent drought stress resulting in gummosis.

Do not harvest cherries treated with ETHREL® Plant Regulator sooner than 7 days after last application.

Application

Do not treat too early as this can cause early fruit drop with stems attached. At the time of application all fruit including those inside the tree canopy should be in stage 3 as indicated by a rapid size increase and the presence of ground color (change from bright green to yellow background color). Treat when air temperatures are between 60 and 85°F. Do not treat when air temperatures exceed 85°F. Uniform coverage is important, concentrated sprays associated with erratic application can result in gummosis and tip dieback.

TART CHERRIES—Varieties such as Montmorency: Mix ¹/₃ pint of ETHREL® Plant Regulator in 100 gallons of water. Apply as a dilute spray with sufficient water for thorough, uniform coverage, generally 300 gallons per acre depending on the type and delivery of sprayer used and size of trees.

For concentrate sprayers, apply ²/₃ to 1 pint ETHREL® per acre in 50 to 100 gallons of water. Uniform coverage is essential for optimum response.

SWEET CHERRIES—(Varieties such as Windsor, Napoleon Royal Anne, Emperor Francis): Mix 1 to 1¹/₃ pints of ETHREL® Plant Regulator in 100 gallons of water. Apply as a dilute spray with sufficient water for thorough uniform coverage, generally 300 gallons per acre depending on the type and delivery of the sprayer and size of trees. Use low rates on the light or yellow colored varieties. Expect longer treatment to harvest interval with the lower rate.

For concentrate sprayers, apply 2 to 3 pints of ETHREL® per acre in 50 to 100 gallons of water per acre. Uniform coverage is essential for optimum response.

INCREASED DORMANT FRUIT BUD HARDINESS AND DELAYED SPRING BLOOM OF SWEET CHERRIES IN THE PACIFIC NORTHWEST

A fall application of ETHREL® Plant Regulator will increase fruit bud hardiness by decreasing the chance for winter injury and delaying bloom by 3 to 5 days which may help avoid frost injury. Treatment of early flowering varieties may delay bloom to better coincide with pollenation from other varieties.

IMPORTANT: Some early leaf yellowing and drop may be noted following application. Slight gummosis and individual fruit size reduction may be noted.

APPLICATION: Apply 3 pints of ETHREL® per acre in a dilute spray (300 to 400 gallons per acre). Treat in first two weeks of September.

GRAPES

TABLE GRAPES IN CALIFORNIA AND ARIZONA

An application of ETHREL® Plant Regulator promotes early uniform color development of colored table grape varieties including Cardinal, Flame Seedless, Red Malaga, Queen and Tokay allowing more efficient harvest of quality fruit.

Harvest fruit at desired maturity and quality, as indicated by sugar content, acidity and color. This is generally two weeks or more after treatment. Watch treated areas closely and harvest before berries become too dark.

For more information regarding local experiences with this product on grapes consult your Farm Advisor or Extension Viticulturist.

IMPORTANT

Some berry softening is associated with ETHREL® treatment of some varieties which may limit or influence storage considerations. Do not store Tokays.

Do not harvest grapes treated with ETHREL® Plant Regulator sooner than 14 days after last application.

APPLICATION: Apply 1 to 2 pints of ETHREL® per acre in sufficient water to wet vines and fruit clusters uniformly, generally 50 to 300 gallons using conventional ground sprayers. Rates higher than 1 pint/A may increase incidence of cracked fruit and should only be used in areas or during weather conditions where grapes have been especially hard-to-color in past seasons. Except Tokay treat all varieties when 5 to 30% of the berries are showing color. Treat Tokays when 5 to 15% of the berries are showing color. Aerial application of 1 to 2 pints of ETHREL® in 20 gallons of spray volume per acre of concentrate ground applications of 1 to 2 pints of ETHREL® in 25 to 50 gallons per acre may be used to treat Tokays.

THOMPSON SEEDLESS GRAPES FOR RAISIN PRODUCTION

A foliar spray of ETHREL® Plant Regulator will hasten maturity of Thompson Seedless grapes resulting in reduced acids, increased sugars and increased raisin quality.

For more information regarding local experience with this product on grapes consult your Farm Advisor or Extension Viticulturist.

APPLICATION Apply a foliar spray of 1 to 2 pints per acre ETHREL® Plant Regulator in 30 to 300 gpa of water at 5% berry softening. Do not apply to grapes which are under stress from insect damage or moisture stress. Observe treated vineyards closely and harvest at proper maturity as determined by sugar and acid levels.

APPLES

ETHREL® Plant Regulator provides several beneficial responses on apples. Depending on rate and time of application ETHREL® will: **1.)** promote fruit loosening, **2.)** promote uniform ripening and coloring without loosening, **3.)** promote thinning and return bloom, and **IV.)** increase flower bud development in young trees. For desired response follow specific recommendations below.

Important:

For best results, a thorough uniform spray coverage of leaves and fruit is needed. A wetting agent may improve spray coverage.

Do not graze or feed cover crops grown in treated apple orchards. Do not harvest apples treated with ETHREL® Plant Regulator sooner than 7 days after last application.

Fruit Loosening

Apply a foliar spray to apple trees 7 to 14 days prior to normal anticipated harvest. Pick apples when they reach the degree of loosening desired for mechanical or hand harvesting. Observe treated fruit closely for abscission response since temperatures above 75°F accelerate fruit loosening and promote preharvest drop.

Application

Use sufficient water for thorough uniform spray coverage, generally 50 to 500 gallons per acre depending on the type and delivery of sprayer used and size of trees.

Early and Midseason Maturing Varieties (such as Tydeman, Milton, Wealthy and McIntosh).

Apply 2½ pints of ETHREL® Plant Regulator per acre.

Late Maturing Varieties (such as Cortland, Red Delicious, Rome, Idared and Jonathan).

Apply 5 pints of ETHREL® Plant Regulator per acre.

Promotion of Uniform Ripening and Coloring of Red Varieties Without Loosening

Apply a foliar spray of ETHREL® Plant Regulator and a preharvest drop control chemical such as FRUITONE® brand N. Begin spraying 2 to 3 weeks before normal harvest period and about 1 to 2 weeks before desired harvest date.

Important: Fruit sized reduction may occur when used to obtain early maturity, especially if fruit is small at the time of treatment. Harvest fruit at proper maturity and placed in cold storage immediately. Observe fruit daily as the proper picking period is shorter with ETHREL® treated fruit than with untreated fruit. Do not allow fruit to become overripe on trees. Fruit intended for fresh market must be checked for quality and maturity. Color alone is not an adequate guide for fruit maturity. Use a pressure gauge or other suitable methods for determining internal maturity. Fruit over-mature when harvested and stored may soften sooner than untreated fruit. ETHREL® applied earlier than 3 weeks before normal anticipated harvest may result in reduced fruit quality and size. Do not apply ETHREL® to more acreage than can be harvested in 1 to 2 days. Treated fruit can be satisfactorily stored in a controlled atmosphere or regular cold air storage provided fruit is in good condition when put into storage.

APPLICATION

Treat when air temperatures are between 60°F and 90°F. However, application may be made at 50°F under rising temperature conditions. If daytime temperatures are high and nighttime temperatures are warm, color response from ETHREL® treatment will be reduced, but ripening and loosening effects will be accelerated. Cool weather may extend the interval between treatment and harvest. Apply as a normal dilute spray using sufficient water for thorough uniform coverage, generally 200 to 400 gallons per acre depending on size of trees. Since recommendations for chemical control of preharvest drop vary according to location and variety, follow directions on the FRUITONE® N label or consult your local extension Pomologist or Horticulturist for his recommendations.

Early or Midseason Maturing Varieties (varieties maturing with McIntosh or earlier).

Mix ½ to 1 pint of ETHREL® Plant Regulator in 100 gallons of water. Use the lower rate on apples intended for storage.

Late Maturing Varieties (varieties maturing later than McIntosh).

Mix 1 pint of ETHREL® Plant Regulator in 100 gallons of water.

THINNING AND RETURN BLOOM ON GOLDEN DELICIOUS

Golden Delicious is a difficult variety to adequately thin. Regulating cropping from one season to another is also a serious problem. Whenever a high percent of the spurs and lateral buds bloom in a single season severe alternate bearing can develop the following year.

To overcome this problem and maintain annual production, a series of sprays is usually required, starting with a bloom spray with Elgetol followed by AMID-THIN® W, ETHREL® or a combination of AMID-THIN® W plus ETHREL® applied 7 to 21 days after full bloom.

The program to use will depend on the amount of thinning required and the biennial bearing history of your orchard.

IMPORTANT: THE USE OF ETHREL® CAN RESULT IN OVER THINNING AND REDUCE FRUIT SIZE. AVOID DOUBLE COVERAGE AND USE ON LOW VIGOR TREES. IF ETHREL® HAS BEEN USED FOR THINNING DURING THE SPRING DO NOT APPLY A FALL TREATMENT FOR FRUIT LOOSENING.

There are many environmental factors which can affect thinning and return bloom. It is advisable to obtain experience under your conditions by initially treating only a small percent of your trees each year with any one program.

APPLICATION

Mix ½ to 1 pint of ETHREL® Plant Regulator per 100 gallons of water (2 to 4 pints per acre). Treat 10 to 20 days after full bloom. Apply as a normal dilute spray of 300 to 400 gallons per acre. For greater thinning apply AMID-THIN® W at 3 to 4 oz. plus ETHREL® at ½ to 1 pint per 100 gallons of spray solution. Treat 10 to 20 days after full bloom. Apply as normal dilute spray of 300 to 400 gallons per acre.

SPUR GOLDEN DELICIOUS

Thinning of Spur Golden Delicious with a heavy fruit set may require a higher rate of ETHREL® Plant Regulator of 1 to 1½ pints per acre per 100 gallons of water at the earlier timing for adequate thinning and return bloom.

INCREASED FLOWER BUD DEVELOPMENT ON YOUNG TREES

To increase flower bud development on young non-bearing trees, apply a foliar spray of ETHREL® Plant Regulator 3 to 4 weeks after bloom. On young trees just beginning to bear fruit, apply ETHREL® 5 to 6 weeks after full bloom after and June drop. Treatment can result in excessive fruit thinning and fruit size and yield reduction the year of usage. ETHREL® should reduce vegetative growth and increase flowering the following spring.

Important

User must consider the risk of excessive fruit thinning and fruit size reduction when using this treatment on bearing trees. Avoid double coverage and use on low vigor trees. Trees should be large enough to support a crop of apples before they are treated to initiate flower buds. Treat only vigorous trees since excessive growth reduction will result on weak trees. Treat when air temperatures are between 60°F and 90°F.

APPLICATION

Spray trees thoroughly and uniformly to the point of runoff. Use the higher rate on more vigorous trees.

Non-bearing trees: Treat 2 to 4 weeks after full bloom period. Mix 1 to 2 pints of ETHREL® Plant Regulator in 100 gallons of water (3 to 6 pints/acre).

Young trees just starting to bear: To help minimize fruit thinning, delay application 5 to 6 weeks after bloom and after June drop. Mix ½ to 1 pint of ETHREL® Plant Regulator in 100 gallons of water (1½ to 3 pints/A).

WALNUTS IN CALIFORNIA

ETHREL® Plant Regulator application promotes uniform hull split providing greater potential for increased recoverable yields of optimum quality walnuts with early harvest. Maximum benefit is obtained with application immediately after walnut packing tissue has turned brown.

For a programmed harvest, spray a different orchard block each day and harvest blocks daily in the same sequence starting 5 to 10 days after treatment.

Inspect treated orchards frequently and harvest as soon as adequate hull split occurs, generally 5 to 10 days after treatment. Cool temperatures increase the time requirement for optimum harvest response. As with early harvest, longer drying time may be required.

Important

Treatment prior to brown coloration of packing tissue may reduce nut quality. Some leaf drop is associated with treatment. Do not treat diseased, low vigor or moisture stressed trees as excessive leaf drop may occur.

Rates above that recommended are illegal and may result in tree injury (such as excessive defoliation, reduced catkin formation, twig dieback).

Do not harvest walnuts treated with ETHREL® Plant Regulator sooner than 5 days after last application.

APPLICATION

Apply 3 to 5 pints of ETHREL® per acre in ground spray volumes of 100 to 500 gallons per acre depending on spray equipment and size of trees. Treat when air temperatures are between 60°F and 90°F and after the packing tissue of all nuts inspected are completely brown. Use 5 pints per acre when minimum time interval between treatment and harvest is desired and with cool temperature harvest conditions.

Uniform spray coverage of walnut hulls is essential for maximum hull split and loosening response. Be sure equipment provides good penetration of spray to all parts of the tree. Check nozzle systems frequently to assure uniform distribution of spray droplets. Do not operate in excess of 3 mph. Use directional spray volume equipment when treating tall dense foliage trees.

PEPPERS

A foliar spray of ETHREL® Plant Regulator promotes early, uniform ripening and coloring for more efficient picking and handling of fruit. Apply ETHREL® to bell peppers when 10% of the fruit are red and chocolate and to pepper varieties, chili and pimento when 10 to 30% of the fruit are red and chocolate and there is sufficient mature green fruit to produce the desired tonnage.

ETHREL® will not ripen immature green fruit. Check several field locations to determine crop stage and degree of maturity. ETHREL® treatments may reduce total yields if applied too early or there is a lack of uniform, mature green fruit resulting from split fruit set or variable soil cultural practices. Harvest fruit when the desired color and maturity have developed, generally 14 or more days after treatment.

Important

Do not apply when prolonged average high temperatures of 95°F after treatment are expected. Do not treat when temperatures exceed 100°F. Applications made under high temperatures will accentuate fruit ripening, yellowing of foliage, defoliation and immature fruit abscission.

Do not treat when temperatures are below 60°F. Low temperatures after treatment may reduce or negate the effects of ETHREL®. Some yellowing and general aging of leaves will be noted after treatment.

Do not harvest peppers treated with ETHREL® Plant Regulator sooner than 5 days after last application.

NOTE: Under certain conditions tank mixtures of ETHREL® Plant Regulator with desiccants containing sodium chlorate could result in the formation of hypochlorous acids which on heating will emit toxic chloride fumes.

Application

Apply to 3 to 4 pints of ETHREL® in 40 to 100 gallons of water per acre with ground spray equipment. Use the higher rate when cool temperatures are anticipated (less than 65°F), when plants are growing vigorously or when foliage is dense to assure thorough wetting of all leaves and fruit.

BLACKBERRIES (CULTIVARS CHEHALEM, THORNLESS EVERGREEN AND BOYSENBERRY) IN OREGON AND WASHINGTON

Use ETHREL® Plant Regulator: 1) early in the harvest season to reduce number of pickings or 2) treat later in the harvest season for a once-over final picking. To loosen fruit and concentrate maturity, apply a foliar spray of ETHREL® Plant Regulator at least 3 days prior to the expected harvest when the air temperatures are between 60°F and 90°F and rain is not predicted within 24 hours.

Application

Do not spray more than can be harvested on the third day following treatment as considerable fruit may drop.

Use only at recommended times and rates on healthy, vigorous plants, as a reduction in berry size may occur. Do not use on damaged or diseased plants.

Do not harvest blackberries treated with ETHREL® Plant Regulator sooner than 3 days after last application.

Application

Mix 5 pints of ETHREL® per 100 gallons water and apply 100-175 gallons of spray mixture per acre.

CANTALOUPES IN CALIFORNIA, ARIZONA AND TEXAS

ETHREL® Plant Regulator promotes abscission (slipping) so fruit separates more easily from vines. This allows more efficient and economical harvesting. Abscission generally will occur two to five days after ETHREL® treatment. The effect will be faster at higher temperatures. Proper timing of ETHREL® treatment is critical, but may vary from season to season. Fruit quality, in terms of soluble solids and flesh color, does not improve following treatment. Therefore, ETHREL® should not be used until after fruit has developed marketable levels of soluble solids and flesh color. ETHREL® promotes abscission of immature fruit as well as marketable fruit. Thus, using ETHREL® too soon will result in poorly colored fruit with substandard soluble solids.

Observe treated fields frequently and pick fruit when market quality is reached. Fruit allowed to remain in the field too long will lose quality. Some yellowing or rapid aging of vines will be seen following treatment. Plants with low vigor will not respond properly. Treat each field only once with ETHREL®. Do not plant another crop within 30 days after treatment.

Do not harvest cantaloupes treated with ETHREL® Plant Regulator sooner than 2 days after last application.

For more specific directions under varying temperature and moisture conditions, consult your Extension Horticulturist or Farm Advisor for his experience with ETHREL® in your area.

CALIFORNIA AND ARIZONA

Treat only those fruit that have a fairly uniform fruit set, have vines in good condition, and have fruit with marketable soluble solids and internal flesh color. Do not treat fields where soluble solids are running less than 10%.

Apply 3 pints of ETHREL® Plant Regulator per acre in 40 to 80 gallons of water. Good spray coverage is important for producing uniform abscission. Be prepared to harvest abscised fruit two to five days after treatment. The exact pre-harvest interval will vary with temperature. Be sure to coordinate your harvesting schedule with your packer/shipper.

TEXAS

Treat only those fields that have a fairly uniform fruit set, have vines in good condition, and have fruit with marketable soluble solids and internal flesh color. Apply 3 pints of ETHREL® Plant Regulator per acre in 40 to 80 gallons of water with ground sprayers, or 10 to 12 gallons of water with aerial applications. ETHREL® should not be applied when night temperatures are below 60°F.

ETHREL® RATE GUIDE
Read Detailed Directions In This Label First

Crop	Situation	Amount of ETHREL®*	Volume (gallons/acre)
Tobacco	directed	4 pt/A	50 to 60
	over-top	4 to 8 pt/A	40 to 60
Tomatoes	early/midseason	1¼ to 3¼ pt/A	ground 20 to 100 air 10 to 12
	late/coastal	3¼ to 6½ pt/A	ground 20 to 100 air 10 to 12
	Midwest/east	1¼ to 6½ pt/A	ground 20 to 100 air 10 to 12
	fresh market	1¼ to 5 pt/A	ground 40 minimum air 10 to 12
Cherries	tart (dilute spray)	⅓ pt/100	300
	tart (conc spray)	⅔ to 1 pt/A	50 to100
	sweet (dilute spray)	1 to 1⅓ pt/100	300
	sweet (conc spray)	2 to 3 pt/A	50 to 100
	delay bloom	3 pt/A	300 to 400
Grapes	table	1 to 2 pt/A	25 to 300
	raisin	1 to 2 pt/A	30 to 300
Apples	**Fruit Loosening:**		
	early/mid season	2½ pt/A	50 to 100
	late season	5 pt/A	50 to 100
	Ripening:		
	early/mid season	½ to 1 pt/100	200 to 400
	late season	1 pt/100	200 to 400
	thinning/return	½ to 1½ pt/100	200 to 400
	bloom	1½ to 6 pt/A	100 to 300
	increase budding		
Walnuts		3 to 5 pt/A	100 to 500
Peppers		3 to 4 pt/A	40 to 100
Blackberries		5 pt/a00	100 to 175
Cantaloupes		3 pt/A	ground 40 to 80 air 10 to 12

***Amount of ETHREL® to use: pints per acre or pints per 100 gallons of water:**
1 Pint/Acre is Equivalent to ¼ lb AI/Acre.
2 Pints (1 Quart/Acre) is Equivalent to ½ Lb AI/Acre.
4 Pints (2 Quarts/Acre) is Equivalent to 1 Lb AI/Acre.

STORAGE AND DISPOSAL

STORAGE

Do not contaminate water, food, or feed by storage and disposal.

PESTICIDE DISPOSAL

Wastes resulting from the use of this product may be disposed of on site or at an approved waste disposal facility.

CONTAINER DISPOSAL

Triple rinse (or equivalent). Then offer for recycling or reconditioning or puncture and dispose of in a sanitary landfill or by incineration, or if allowed by state and local authorities, by burning. If burned, stay out of smoke.

LIMITED WARRANTY AND DISCLAIMER

The manufacturer warrants (a) that this product conforms to the chemical description on the label; (b) that this product is reasonably fit for the purposes set forth in the directions for use when it is used in accordance with such directions; and (c) that the directions, warnings and other statements on this label are based upon responsible experts' evaluation of reasonable tests of effectiveness, of toxicity to laboratory animals and to plants and of residues and upon reports of field experience. Tests have not been made on all varieties or in all states or under all conditions. THE MANUFACTURER NEITHER MAKES NOR INTENDS, NOR DOES IT AUTHORIZE ANY AGENT OR REPRESENTATIVE TO MAKE ANY OTHER WARRANTIES, EXPRESS OR IMPLIED, AND IT EXPRESSLY EXCLUDES AND DISCLAIMS ALL IMPLIED WARRANTIES OF MERCHANTABILITY OR FITNESS FOR A PARTICULAR PURPOSE.

THIS WARRANTY DOES NOT EXTEND TO, AND THE BUYER SHALL BE SOLELY RESPONSIBLE FOR, ANY AND ALL LOSS OR DAMAGE WHICH RESULTS FROM THE USE OF THIS PRODUCT IN ANY MANNER WHICH IS INCONSISTENT WITH THE LABEL DIRECTIONS, WARNINGS OR CAUTIONS.

BUYER'S EXCLUSIVE REMEDY AND MANUFACTURER'S OR SELLER'S EXCLUSIVE LIABILITY FOR ANY AND ALL CLAIMS, LOSSES, DAMAGES, OR INJURIES RESULTING FROM THE USE OR HANDLING OF THIS PRODUCT, WHETHER OR NOT BASED IN CONTRACT, NEGLIGENCE, STRICT LIABILITY IN TORT OR OTHERWISE SHALL BE LIMITED AT THE MANUFACTURER'S OPTION, TO REPLACEMENT OF, OR THE REPAYMENT OF THE PURCHASE PRICE FOR, THE QUANTITY OF PRODUCT WITH RESPECT TO WHICH DAMAGES ARE CLAIMED. IN NO EVENT SHALL MANUFACTURER OR SELLER BE LIABLE FOR SPE-

CIAL, INDIRECT OR CONSEQUENTIAL DAMAGES RESULTING FROM THE USE OR HANDLING OF THIS PRODUCT.

NOTICE TO BUYER

Purchase of this material does not confer any rights under patents governing this product or the use thereof in countries outside of the United States.

ETHREL is a registered trademark of Rhône-Poulenc

AG90038

THIS SPECIMEN LABEL IS INTENDED FOR USE ONLY AS A GUIDE IN PROVIDING GENERAL INFORMATION REGARDING THE DIRECTIONS, WARNINGS AND CAUTIONS ASSOCIATED WITH THE USE OF THIS PRODUCT. AS WITH ANY AGRICULTURAL CHEMICAL, ALWAYS FOLLOW THE LABEL INSTRUCTIONS ON THE PACKAGE BEFORE USING.

FOLEX® 6-EC

COTTON DEFOLIANT

Very Low Odor

FOR FAST AND EFFECTIVE COTTON DEFOLIATION

ACTIVE INGREDIENT:	
S,S,S-Tributyl phosphorotrithioate	70.5%
INERT INGREDIENTS:	29.5%
TOTAL	100.0%

Contains 6 pounds S,S,S-Tributyl phosphorotrithioate per U.S. gallon.

EPA Reg. No. 264-498 EPA Est. No.

KEEP OUT OF REACH OF CHILDREN

WARNING AVISO

PRECAUCION AL USUARIO: Si usted no puede leer o entender ingles, no use este producto hasta que la etiqueta le haya sido explicada ampliamente.

For **PRODUCT USE** Information Call 1-800-334-9745
For **MEDICAL** and **TRANSPORTATION Emergencies** ONLY Call 24 Hours A Day 1-800-334-7577

PRECAUTIONARY STATEMENTS

WARNING

HAZARDS TO HUMANS AND DOMESTIC ANIMALS

May be fatal if swallowed, inhaled, or absorbed through the skin. Do not breathe spray mist. Do not get in eyes, on skin, or on clothing. Wash with soap and warm water after handling and before eating, drinking or using tobacco.

Remove and wash contaminated clothing before reuse.

For protection during application, wear a hat, long sleeved shirt, and long legged trousers. In addition, mixer-loaders must wear rubber or neoprene gloves.

STATEMENT OF PRACTICAL TREATMENT

IF SWALLOWED: Call a physician or Poison Control Center. Vomiting should be induced. Drink 1 or 2 glasses of water and induce vomiting by giving one dose (1/2 oz. or 15 ml.) of Syrup of Ipecac. If vomiting does not occur within 10 to 20 minutes, administer a second dose. If Syrup of Ipecac is not available, induce vomiting by sticking finger down throat. Repeat until vomit is clear. Never give anything by mouth to an unconscious person.

IF ON SKIN: Wash with plenty of soap and water.

IF INHALED: Remove victim to fresh air. If not breathing, give artifical respiration preferably mouth to mouth. Get medical attention.

IF IN EYES: Flush eyes with plenty of water. Get medical attention if irritation persists.

NOTE TO PHYSICIAN

This product inhibits cholinesterase resulting in stimulation of the central nervous system, the parasympathetic nervous system and the somatic motor nerves. Poisoning with this product also results in cardiovascular and respiratory symptoms which must be treated as separate pathological entities apart from the cholinergic effects.

Use atropine sulfate to reverse cholinergic symptoms; and maintain a systematic, symptomatic treatment of the cardiovascular and respiratory effects, even though the cholinergic symptoms have ceased. Do not give central nervous system depressants. Watch for pulmonary edema, which may develop in serious cases of poisoning even after 12 hours. The use of oxygen is recommended in case of respiratory distress.

ENVIRONMENTAL HAZARDS

This product is toxic to fish.

Do not apply directly to water or wetlands. Drift and runoff from treated areas may be hazardous to aquatic organisms in neighboring areas. Do not contaminate water when disposing of equipment washwaters. Do not contaminate water by cleaning of equipment, or disposal of wastes. Thoroughly clean the spray equipment with a chemical cleaner before using for other products.

PHYSICAL OR CHEMICAL HAZARDS

Do not use or store near heat or open flame.

RESTRICTIONS

Do not apply FOLEX® 6-EC through any type of irrigation system.

Do not use on other crops used for food or forage. Avoid spray drift to susceptible plants other than cotton as this product may injure or defoliate other crops. (Coarse sprays are less likely to drift.)

Thoroughly clean the spray equipment with a chemical cleaner before using for other products. If chlorate type chemicals have been used in the same spray equipment, thoroughly wash spray tank and particularly dried deposits on aircraft fabric.

DIRECTIONS FOR USE

IT IS A VIOLATION OF FEDERAL LAW TO USE THIS PRODUCT IN A MANNER INCONSISTENT WITH ITS LABELING.

Apply this product only as specified on this label.

GENERAL INFORMATION

FOLEX® 6 EC is a cotton defoliant that is used to remove leaves from cotton plants prior to anticipated harvesting. Formulated as an emulsifiable concentrate, FOLEX® 6 EC contains 6 pounds active ingredient per gallon. FOLEX® 6 EC is most commonly mixed with water and applied to cotton using either ground or aerial equipment. However, white diesel oil may be substituted for water under adverse weather conditions. FOLEX® 6 EC is non-corrosive and will not damage metal parts normally used in applying spray chemicals.

FOLEX® 6 EC usually drops leaves in a green condition and will not appear effective until defoliation actually begins, approximately 3 days after application. By this time, the field will have an "off-cast" color. Under favorable conditions, satisfactory leaf drop will require 5 to 7 days. Adverse conditions, such as low temperatures and/or toughened plants (drought stressed, etc.) may extend this to 10 to 14 days.

FOLEX® 6 EC is effective when applied to cotton with a heavy dew. Once FOLEX® 6 EC has dried on the leaf, subsequent rainfall or dew does not adversely affect its activity. Application of FOLEX® 6 EC is not recommended when a heavy rainfall is expected within 1 hour after treatment. Conditions which delay absorption into the leaves are primarily those which cause the cotton leaves to be wilted, toughened, or leathery. When these conditions prevail, use white diesel oil instead of water, or make 2 applications 10 to 14 days apart.

FOLEX® 6 EC does not suppress second growth especially where rainfall follows defoliation. A second application of FOLEX® 6 EC will be effective in defoliation this second growth if excessive regrowth occurs between the first application and first and/or second harvest. FOLEX® 6 EC must be applied in sufficient water or diesel oil to give thorough coverage of the leaves. Larger droplets more effectively reach the lower leaves.

FOLEX® 6 EC may be used alone for bottom defoliation of the leaves on the lower portion of the plant as preconditioner prior to total defoilation, or as a total defoliation treatment. FOLEX® 6 EC may also be used in combination with Dropp (tank mix).

IMPORTANT: Observe all cautions and limitations on labels of all products used in mixtures specifically when FOLEX® 6 EC is tank mixed with products for use on cotton in areas where citrus is grown. Observe buffer zones and other restrictions on the FOLEX® 6 EC label and/or in the manufacturer's recommendations for FOLEX® 6 EC.

MIXING INSTRUCTIONS

FOLEX® 6 EC ALONE:

Fill the spray tank 1/2 to 3/4 full with clean water. Begin agitation and add the recommended amount of FOLEX® 6 EC. Add water to the spray tank to the desired level. Maintain sufficient agitation to ensure a uniform spray mixture during application. Follow same mixing instructions when diesel oil is substituted for water.

TANK MIXTURES: Fill the spray tank 1/2 to 3/4 full with clean water and begin agitation. When tank mixing with Dropp, add the recommended amount of Dropp first. After Dropp is thoroughly mixed with water, add the recommended amount of FOLEX® 6 EC. Add water to the spray tank to the desired level. This tank water and begin agitation. When tank mixing with Dropp, add the Maintain sufficient agitation while mixing and during application to ensure a uniform spray mixture. If spray mixture is allowed to remain without agitation for short periods of time, be sure to agitate until uniformly mixed before application. Do not use white diesel oil when FOLEX® 6 EC is used in a tank mixture with other products.

APPLICATION PROCEDURES

Do not apply FOLEX® 6 EC through any type of irrigation system.

FOLEX® 6 EC alone and in a tank mixture with Dropp can be applied by both ground and aerial equipment.

GROUND: Use a standard high clearance sprayer that provides a uniform and accurate application. Sprayer should be equipped with screens no finer than 50 mesh in the nozzle tips and in-line strainers. A spray volume of 10 to 25 gallons per acre is recommended when FOLEX® 6 EC is applied with water and 5 to 10 gallons per acre is recommended when FOLEX® 6 EC is applied with diesel oil.

AERIAL: Use aerial equipment equipped with a diaphragm-type nozzle that produces cone spray patterns with a maximum spray pressure of 40 psi at the nozzle tips. Nozzles should be located no farther than 3/4 the distance from the center of the aircraft to the end of the wing or rotor. The aircraft should discharge the spray a maximum of 10 feet above the crop. Do not apply when winds may cause drift to adjacent fields.

The Spray volume of 5 to 10 gallons per acre should be used when FOLEX® 6 EC alone may be applied in 3 to 5 gallons of spray volume per acre when using white diesel oil.

FOLEX® 6 EC RECOMMENDATIONS

PRODUCT	RATE	DIRECTIONS
FOLEX® 6 EC	1 to 1½ Pints/Acre	**Bottom Defoliation:** Apply when bottom bolls are mature or when youngest bolls cannot be dented by pressure between the thumb and the forefinger or cut through with a sharp knife. Direct spray to thoroughly cover the leaves on the desired portion of plants to be defoliated.
	1½ to 2 Pints/Acre	**Total Defoliation:** Apply when fiber quality of top bolls will not not be damaged by loss of top leaves or when top bolls are firm to thumb pressure. Application can be made up until first frost but not later than 7 days prior to cotton harvest.
	3 Pints/Acre	**Defoliation of Rank Cotton:** Apply when 50% or more of the bolls are open. Two applications can be made 2 to 6 weeks apart for complete defoliation. Applications can be made until first frost, but not later than 7 days prior to cotton harvest.

FOLEX® 6 EC + DROPP TANK MIX RECOMENDATIONS

Tank mixtures of FOLEX® 6 EC + DROPP are recommended for total defoliation when heavy regrowth control is required. Defoliation and inhibition of regrowth is obtained when the tank mixture is applied to mature cotton plants when 60% or more bolls are open. A second application of labeled rates of FOLEX® 6 EC + DROPP, or FOLEX® 6 EC alone may be made where necessary. FOLEX® 6 EC + Dropp tank mixtures are recommended for use in Alabama, Arkansas, Georgia, Louisiana, Mississippi, North Carolina, South Carolina, Tennessee and Texas.

PRODUCT	RATE	DIRECTIONS
FOLEX® 6 EC + Dropp 50WP	1 to 1½ Pints/Acre +	Mix FOLEX® 6 EC and Dropp with water and apply the tank mixture by ground or aerial equipment for total defoliation and regrowth control when minimum night temperature is above 65°F.

Use the high rate of Dropp when environmental conditions favor heavy regrowth. When minimum night temperature is below 60°F use FOLEX® 6 EC alone.

Do not add an adjuvant to FOLEX® 6 EC + Dropp tank mixtures or use higher recommended rates of either FOLEX® 6 EC or Dropp because desiccation may occur.

Refer to Dropp label for additional use directions and cautions when using tank mixtures of FOLEX® 6 EC with Dropp.

DEFOLIATION AND BOLL OPENING

The combination of FOLEX® 6 EC and PREP™ Ethephon Plant Regulator is used for total plant defoliation and to accelerate cotton boll opening. FOLEX® 6 EC and PREP™ Ethephon Plant Regulator may be applied as a tank mixture or used in 2 separate sequential applications. Tank mixture of FOLEX® 6 EC + PREP™ is recommended for use in all cotton growing states as is applications of FOLEX® 6 EC sequential to PREP™ (minimum of 4 days between applications). FOLEX® 6 EC + PREP™ Ethephon Plant Regulator tank mixtures or sequential applications should be applied to cotton when there is sufficient mature unopened bolls to produce the desired yield and when 30% to 60% of the cotton bolls are open. If used in rank cotton, a pretreatment with FOLEX® 6 EC (See FOLEX® 6 EC RECOMMENDATIONS for use rate) may be useful to ensure thorough coverage of unopened bolls by later sequential application of PREP™ Ethephon Plant Regulator. Refer to and observe all label use directions and precautions on the PREP™ label. Occassionally slight reduction on boll opening response has been observed when tank mixture of FOLEX® 6 EC + PREP™ Ethephon Plant Regulator have been used.

Always follow the mixing directions found on this label or on the appropriate spray adjuvant label. Contact your local Rhone-Pôulenc representative for additional information.

STORAGE AND DISPOSAL

STORAGE

Do not contaminate water, food, or feed by storage or disposal.

Store in cool, dry place, away from open flame and extreme heat, and in such a manner as to prevent cross contamination with other pesticides, fertilizers, food, and feed. Store in original container and out of the reach of children, preferably in a locked storage area.

PESTICIDE DISPOSAL

Pesticide wastes are toxic. Improper disposal of excess pesticide, spray mixture, or rinsate is a violation of Federal Law. If these wastes cannot be disposed of by use according to label instructions, contact your State Pesticide or Environmental Control Agency, or the Hazardous Waste representative at the nearest EPA Regional Office for guidance.

CONTAINER DISPOSAL

Triple rinse (or equivalent). Then offer for recycling or reconditioning, or puncture and dispose of in a sanitary landfill, or by other procedures approved by state and local authorities.

Handle an open container in a manner as to prevent spillage. If container is leaking, invert to prevent leakage. If the container is leaking or material spilled for any reason or cause, carefully dam-up spilled material to prevent runoff. Refer to PRECAUTIONARY STATEMENTS on label for hazards associated with the handling of this material. Do not walk through spilled material. Absorb spilled material with absorbing type compounds and dispose of as directed for pesticides above. In spill or leak incidents, keep unauthorized people away. You may contact Rhone-Poulenc for decontamination procedures or any other assistance that may be necessary. The RHONE-POULENC Emergency Telephone (24 Hours A Day in U.S.A.) 1-800-334-7577.

LIMITED WARRANTY AND DISCLAIMER

The Manufacturer warrants (a) that this product conforms to the chemical description on the label; (b) that this product is reasonably fit for the purposes set forth in the directions for use when it is used in accordance with such directions; (c) that the directions, warnings and other statements on this label are based upon responsible experts' evaluation of reasonable tests of effectiveness, of toxicity to laboratory animals and to plants; and of residues on food crops; and upon reports of field experience. Tests have not been made on all varieties or in all states or under all conditions. THE MANUFACTURER NEITHER MAKES NOR INTENDS, NOR DOES IT AUTHORIZE ANY AGENT OR REPRESENTATIVE TO MAKE, ANY OTHER WARRANTIES, EXPRESS OR IMPLIED, AND IT EXPRESSLY EXCLUDES AND DISCLAIMS ALL IMPLIED WARRANTIES OF MERCHANTABILITY OR FITNESS FOR A PARTICULAR PURPOSE.

THIS WARRANTY DOES NOT EXTEND TO, AND THE BUYER SHALL BE SOLELY RESPONSIBLE FOR, ANY AND ALL LOSS OR DAMAGE WHICH RESULTS FROM THE USE OF THIS PRODUCT IN ANY MANNER WHICH IS INCONSISTENT WITH THE LABEL DIRECTIONS, WARNINGS OR CAUTIONS.

BUYER'S EXCLUSIVE REMEDY AND MANUFACTURER'S OR SELLER'S EXCLUSIVE LIABILITY FOR ANY AND ALL CLAIMS, LOSSES, DAMAGES OR INJURIES RESULTING FROM THE USE OR HANDLING OF THIS PRODUCT, WHETHER OR NOT BASED IN CONTRACT, NEGLIGENCE, STRICT LIABILITY IN TORT OR OTHERWISE, SHALL BE LIMITED, AT THE MANUFACTURER'S OPTION, TO REPLACEMENT OF, OR THE REPAYMENT OF THE PURCHASE PRICE FOR, THE QUANITY OF PRODUCT WITH RESPECT TO WHICH DAMAGES ARE CLAIMED. IN NO EVENT SHALL MANUFACTURER OR SELLER BE LIABLE FOR SPECIAL, INDIRECT OR CONSEQUENTIAL DAMAGES RESULTING FROM THE USE OR HANDLING OF THIS PRODUCT.

NOTICE TO BUYER

Purchase of this material does not confer any rights under patents governing this product or the use thereof in countries outside of the United States.

Dropp is a trademark of Nor-Am Chemical Corporation
FOLEX is a registered trademark of Rhône-Poulenc
AG 90031

THIS SPECIMEN LABEL IS INTENDED FOR USE ONLY AS A GUIDE IN PROVIDING GENERAL INFORMATION REGARDING THE DIRECTIONS, WARNINGS AND CAUTIONS ASSOCIATED WITH THE USE OF THIS PRODUCT. AS WITH ANY AGRICULTURAL CHEMICAL, ALWAYS FOLLOW THE LABEL INSTRUCTIONS ON THE PACKAGE BEFORE USING.

Formula 40*

Herbicide

CONTAINS ALKANOLAMINE SALTS † OF 2,4-D

For the Selective Control of Many Broadleaf Weeds in Non-Crop Areas, Grass Pastures, Rangelands and in Certain Crops. Also for Control of Trees by Injection.

ACTIVE INGREDIENT(S):

Alkanolamine Salts (of the Ethanol and Isopropanol Series) of 2,4-Dichlorophenoxyacetic Acid**	56.5%
INERT INGREDIENTS	43.5%
TOTAL	100.0%

2,4-Dichlorophenoxyacetic Acid** Equivalent: 36.5% - 3.8 lb/gal

** Isomer Specific by AOAC Method No. 6.275-6.279 (13th Ed.)

† Salts are the least volatile forms of 2,4-D, however, the use precautions specified on this label must be followed to avoid possible injury to susceptible crops from spray drift.

E.P.A. Registration No. 464-1-264 E.P.A. Est. 464-MI-1

KEEP OUT OF REACH OF CHILDREN

DANGER PELIGRO

PRECAUCION AL USUARIO: Si usted no lee ingles no use este producto hasta que la informacion en la etiqueta le haya sido explicada ampliamente.

For PRODUCT USE Information Call 1-800-334-9745
For MEDICAL and TRANSPORTATION Emergencies ONLY Call 24 Hours A Day 1-800-334-7577

PRECAUTIONARY STATEMENTS

DANGER

HAZARDS TO HUMANS AND DOMESTIC ANIMALS

CAUSES IRREVERSIBLE EYE DAMAGE AND SKIN IRRITATION

Harmful if swallowed. Do not get in eyes, on skin, or on clothing. Wear goggles or face shield, or safety glasses when handling. Avoid contamination of food. Wash thoroughly after handling. Do not apply this product in such a manner as to directly through drift expose workers or other persons. The area being treated must be vacated by unprotected persons.

STATEMENT OF PRACTICAL TREATMENT

IN CASE OF CONTACT: immediately flush eyes with plenty of water for at least 15 minutes and get medical attention promptly. Flush skin with water and get medical attention if signs of irritation develop. Remove and wash contaminated clothing before reuse.

IF SWALLOWED: promptly drink a large quantity of milk, egg whites, gelatin solution or, if these are not available, drink large quantities of water. Induce vomiting by placing finger in back of throat. Never give anything by mouth to an unconscious person. Avoid alcohol. Call a physician.

PHYSICAL OR CHEMICAL HAZARDS

Do not cut or weld container

ENVIRONMENTAL HAZARDS

Do not apply directly to water. Do not contaminate water by cleaning of equipment or disposal of wastes. Do not contaminate irrigation ditches or water used for irrigation or domestic purposes.

AGRICULTURAL CHEMICAL: Do not ship or store with food, feeds, drugs, or clothing.

WEED LIST

FORMULA 40* Herbicide is recommended for control of numerous broadleaf weeds and certain 2,4-D susceptible woody perennials without injuring most grasses. Species controlled include the following plus many others:

bitterweed	plantains	mallow, Venice	velvetleaf
broomweed	poorjoe	marshelder	vetch
burdock	pusley, Florida	morningglory, annual	water plantain
carpetweed	ragweed, common	pennycress, (fanweed)	wild mustard
chicory	rape, wild	pennywort	witchweed
cocklebur	rocket, yellow	pepperweed, field	wormwood
coffeeweed	shepherdspurse	pigweed***	
croton	sicklepod		
dandelion	smartweed (annual)		
dock	sneezeweed, bitter		
galinsoga	sowthistle, annual		
hemp, wild	spanishneedles		
jewelweed	sunflower		
jimsonweed	sweetclover		
lambsquarters	thistle, bull		
loco, bigbend	thistle, musk		
lupines	thistle, Russian		

*** The control of "hybrid" pigweeds appears to be less satisfactory from 2,4-D products than formerly experienced on "non-hybrid" varieties. Since 2,4-D herbicides are not as effective on the "hybrid" pigweeds, it is necessary to apply higher rates of 2,4-D for control, especially later in the growing season. Higher rates injure some crops, so less than satisfactory pigweed control may be experienced by the highest tolerated crop dosages.

Therefore, Rhône-Poulenc no longer includes pigweed among the species covered by the performance guarantee statements on the labels for FORMULA 40* Herbicide. At this time, this disclaimer applies only to the High Plains of Texas and western Oklahoma, including the Panhandles. All other guarantees on these product labels are unchanged by this disclaimer.

DIRECTIONS FOR USE

IT IS A VIOLATION OF FEDERAL LAW TO USE THIS PRODUCT IN A MANNER INCONSISTENT WITH ITS LABELING.

Do not apply this product through any type of irrigation system.

Generally, the lower dosages given will be satisfactory for young, succulent growth of sensitive weed species. For less sensitive species and under conditions where control is more difficult, the higher dosages will be needed. Apply FORMULA 40* Herbicide during warm weather when weeds are young and growing actively. Use enough spray volume for uniform coverage by ground or air application. If only bands or rows are treated, leaving middles unsprayed, the dosage per crop acre is reduced proportionately. Do not apply where spray drift may be a problem due to proximity of susceptible crops or other desirable plants. Read and follow all Use Precautions given on this label.

To prepare the Spray, mix FORMULA 40* Herbicide **only** with water, unless otherwise directed on this label. Add about half the water to the mixing tank, then add the FORMULA 40* Herbicide with agitation, and finally the rest of the water with continuing agitation. Note: Adding oil, wetting agent or other surfactant to the spray may increase effectiveness on weeds, but also may reduce selectivity to crops resulting in crop damage.

Use with Liquid Fertilizer: FORMULA 40* Herbicide may be combined with liquid fertilizers suitable for foliar application to accomplish weeding and feeding of corn, small grains, sorghum and grass pastures in one operation. Use FORMULA 40* Herbicide in accordance with recommendations for these crops given in the following text and table. Use liquid fertilizer at rates recommended by supplier or local extension service specialist. To prepare the spray, FORMULA 40* Herbicide must first be premixed with water. For liquid nitrogen fertilizer, use a premix consisting of 1 part of FORMULA 40* Herbicide and 4 parts of water; for other liquid fertilizers, use 1 part FORMULA 40* Herbicide with 50 to 60 parts of water. Add the premix to the fertilizer while maintaining continuous agitation during both mixing and spraying operations. Apply the spray the same day it is prepared; **do not store. NOTE: Always premix FORMULA 40* Herbicide with water before adding to the liquid fertilizer.**

WEED CONTROL IN ASPARAGUS; See Table for recommended use rates. To control wild morningglory and certain other broadleaf weeds apply in about 60 gallons of water per acre for ground application and about 12 gallons per acre for aircraft application. Apply on actively growing weeds, usually in April or May. If spears are present, treat immediately after cutting. Make no more than 2 applications during the harvest season and these should be spaced at least one month apart. Spears contacted by the spray may be malformed and off-flavored. Malformed spears should be cut immediately and discarded. Post-harvest spraying should be only by ground rig using drop nozzles to avoid spraying the fern.

WEED CONTROL IN SMALL GRAINS NOT UNDERSEEDED WITH A LEGUME (Wheat, Barley, Rye, Oats, Millet): See Table for recommended use rates. Spray after grain begins tillering and before the boot stage (usually 4 to 8 inches tall) and weeds are small. **Do not apply before the tiller stage nor from early boot through the milk stage.** To control weeds that will interfere with harvest or to suppress perennial weeds, preharvest treatment can be applied when the grain is in the dough stage. Best results will be obtained when soil moisture is adequate for plant growth and weeds are growing well. Note: Do not permit dairy animals or meat animals being finished for slaughter to forage or graze treated grain fields within 2 weeks after treatment.

WEED CONTROL IN CORN: See Table for recommended use rates. **Preemergence** - Apply to soil anytime after planting but before corn emerges. Only emerged broadleafed weeds are likely to be controlled. Do not use on very light, sandy soil. **Emergence** - Apply just as corn plants are breaking ground. **Post-emergence** - Apply to emerged corn. When corn is over 8 inches tall, use drop nozzles to keep spray off corn foliage. Do not apply from tasseling to dough stage. Injury to corn is most likely to occur if FORMULA 40* Herbicide is applied when corn is growing rapidly under high temperature and high soil moisture conditions. In such situations, use the low rate of 1/2 pint per acre. After application, delay cultivation for 8 to 10 days to allow the corn to overcome any temporary brittleness. **NOTE:** Hybrids vary in tolerance to 2,4-D. Some are easily injured, especially sweet corn. Spray only varieties known to be tolerant to 2,4-D. Consult the seed company or your Agricultural Experiment Station or Extension Service Weed Specialist for this information before using. **Preharvest** - After the hard dough or denting stage, apply by air or ground equipment to suppress perennial weeds, decrease weed seed production, and control tall weeds such as bindweed, cocklebur, dogbane, jimsonweed, ragweed, sunflower, velvetleaf, and vines that interfere with harvesting. **NOTE: do not forage or feed corn fodder for 7 days following application.**

WEED CONTROL IN SORGHUM (MILO): See Table for recommended use rates. Treat only after the sorghum is 6 inches high and preferably before it is 15 inches high. Do not treat during the boot, tasseling or early dough stages. Reduce spray drift by keeping the boom and spray nozzles as low as possible. If crop is taller than 8 inches, use drop nozzles to keep the spray off the leaves. Temporary crop injury can be expected under conditions of high soil moisture and high air temperatures. If it is necessary to apply FORMULA 40* Herbicide under these conditions, use no more than 2/3 pint per acre. NOTE: Hybrids vary in tolerance to 2,4-D. Some are easily injured. Spray only varieties known to be tolerant to 2,4-D. Consult the seed company or your Agricultural Experiment Station or Extension Service Weed Specialist for this information.

WEED CONTROL IN RICE: See Table for recommended use rates. Apply in the late tillering stage of rice development, at the time of first joint development (first to second green ring), usually 6 to 9 weeks after emergence. Do not apply after panicle initiation, after rice internodes exceed 1/2 inch, at early seedling, early panicle, boot, flowering or early heading growth stages. Do not use in rice paddies where shellfish are of economic importance or where flood water is used for irrigation of other crops. **NOTE:** Some rice varieties under certain conditions can be injured by 2,4-D. Therefore before spraying consult local Extension Service or University specialists for appropriate rates and timing of 2,4-D sprays.

WEED CONTROL IN SUGAR CANE: See Table for recommended use rates. Use up to 4 applications per year in accordance with state recommendations. NOTE: Do not apply within 6 weeks before harvest.

WEED CONTROL IN STRAWBERRIES: See Table for recommended use rates. To control many broadleaf weeds in established or old strawberry plantings apply in 25 to 50 gallons of water per acre in early spring when strawberries are dormant or immediately after the last picking. Do not apply unless possible injury to the crop is acceptable. Follow recommendations of State Extension Weed or Horticultural Specialists to fit local conditions.

AMOUNT OF FORMULA 40 HERBICIDE TO USE IN CROPS
by air or ground application

NOTE: Do not apply when weather conditions favor drift from treated areas. Read complete directions and precautions before using.

	DOSAGE PER ACRE	
CROP	**Normal rates (usually safe to crop)**	**Higher rates for special situations[2] (more likely to injure crop)**
SMALL GRAINS		
Spring postemergence		
wheat, barley, rye,	2/3 to 1 1/3 pints	2 to 3 pints
millet oats	1/2 to 1 pint	1 1/2 to 2 pints
Preharvest(dough stage)		
wheat, barley, oats	1 to 2 pints	2 to 3 pints
CORN		
Preemergence	2 to 4 pints	
Emergence[1]	1 pint	1 1/2 pints
Postemergence[1]		
up to 8 inches tall	1/2 to 1 pint	1 1/2 to 2 1/2 pints
8 inches to tasseling (Use only directed spray)	1 pint	
Preharvest	1 to 2 pints	
SORGHUM(Milo)[1]		
Postemergence		
6 to 8 inches tall	2/3 pint	1 1/2 to 2 pints
8 to 15 inches tall (use only directed spray)	1 pint	
RICE	1 to 2 1/2 pints	2 to 3 pints
SUGARCANE		
Fall, after harvest or planting	2 to 4 pints	
Spring, once or twice before close-in	2 to 4 pints	
Summer, up to 6 weeks before harvest	2 1/2 pints	
ASPARAGUS	1 1/2 to 2 quarts	
STRAWBERRIES	1 to 1 1/2 quarts	

[1] Corn and sorghum varieties vary in tolerance to 2,4-D; some are easily injured. Before spraying, get information on 2,4-D tolerance of specific varieties and spray only those **known** to be resistant to 2,4-D injury. If plants are more than 8 inches tall, use directed spray and keep spray off corn and sorghum foliage.

[2] These higher rates may be needed to handle difficult weed problems in certain areas such as under dry conditions especially in western areas. **However, do not use unless possible crop injury will be acceptable.** Consult State Agricultural Experiment Station or Extension Service weed specialists for recommendations or suggestions to fit local conditions.

WEED CONTROL IN ESTABLISHED GRASS PASTURES, CONSERVATION RESERVE, AND RANGELANDS: Use at 2 to 4 pints per acre. Apply preferably when weeds are small and growing actively before the bud stage. Do not use from early boot to milk stage where grass seed production is desired. Do not use on newly seeded areas until grass is well established. Do not use on bentgrass, alfalfa, clover, or other legumes. When using on Conservation Reserve lands, follow all applicable state and Federal regulations. Follow the most severe grazing restrictions imposed either by the pesticide label or by the USDA Acreage Conservation Reserve Program, whichever is longest. NOTE: Do not graze dairy cattle in treated areas for 14 days after application. Remove meat animals from freshly treated areas for 7 days before slaughter. Withdrawal is not needed if 2 weeks or more have elapsed since application. Do not cut treated grass for hay within 30 days after application.

CONTROL OF SOUTHERN WILD ROSE, PERENNIAL PEPPERWEED, HOARY CRESS: On rangelands, forests, roadsides and fence rows use 1 gallon of FORMULA 40* Herbicide plus 4 to 8 fluid ounces of an agricultural surfactant per 100 gallons of water and spray thoroughly as soon as foliage is well-developed. Two or more treatments may be required. Apply at a maximum rate of 6 quarts per acre per application as a spot treatment so the overall rate is no more than 3 quarts per acre. See NOTE on Grazing Restrictions in grass pastures section above.

WEED CONTROL IN GRASS SEED CROPS: Use 1 to 4 pints per acre in spring or fall. Do not apply from early boot to the milk stage of the grass. Spray seedling grass only after the five-leaf stage, using 3/4 to 1 pint per acre to control small seedling weeds. After the grass is well established, higher rates, up to 4 pints, can be used to control hard-to-kill annual or perennial weeds. Do not use on bentgrass unless grass injury can be tolerated. See NOTE on Grazing Restrictions in grass pastures section above.

BROADLEAF WEED CONTROL IN NON-CROPLAND AREAS SUCH AS LAWNS, GOLF COURSES, CEMETERIES AND PARKS, AIRFIELDS, ROADSIDES, VACANT LOTS, DRAINAGE DITCH BANKS: Use 1 to 3 quarts of FORMULA 40* Herbicide per acre in the amount of water needed for uniform application. Treat when weeds are young and growing well. Usually 2 quarts per acre will provide adequate weed control. Do not use on dichondra or other herbaceous ground covers. Do not use on creeping grasses such as bent except for spot treating nor on freshly seeded turf until grass is well established. Reseeding of lawns should be delayed following treatment. With spring application, reseed in the fall; with fall application, reseed in the spring. Legumes are usually damaged or killed. Deep-rooted perennial weeds such as bindweed and Canada thistle may require repeated applications.

WEED CONTROL ON FALLOW LAND: Use 1 1/2 to 2 1/4 quarts per acre on annual broadleaf weeds and up to 3 quarts per acre on established perennial species, such as Canada thistle and field bindweed. Apply to actively growing weeds. Do not plant treated fallow land until three months after treatment, or until chemical has disappeared from soil.

WEED CONTROL IN STONE FRUIT AND NUT ORCHARDS: Apply 3 pints per acre for control of annual broadleaf weeds in the orchard floor. Use coarse sprays and low pressure in sufficient volume of water to obtain thorough wetting of weeds. Do not use in light sandy soil. Do not use in California. Do not apply to bare ground as injury may result. Do not apply immediately before irrigation and withhold irrigation for 2 days before and for 3 days after treatment. Do not allow spray to drift or contact foliage, fruit, stems, trunks of trees, or exposed roots as injury may result. Do not apply to newly established or young orchards. Trees must be at least 1 year old and in vigorous condition. Do not apply during bloom. Do not graze or feed cover crops from treated orchards. Do not make more than two applications per year. Do not harvest stone fruit within 40 days of application or harvest nuts within 60 days of application.

SPOT TREATMENT IN NON-CROP AREAS: To control broadleaf weeds in small areas with a hand sprayer, use 1/4 pint of FORMULA 40* Herbicide in 3 gallons of water and spray to thoroughly wet all foliage.

TREE INJECTION TREATMENT: To control unwanted hardwood trees such as elm, hickory, oaks and sweetgum in forest and other noncrop areas, apply FORMULA 40* Herbicide by injecting 1 ml of the undiluted product through the bark around the trunk at intervals of 1 to 3 inches between edges of the injector wounds. For harder to control species such as ash, maples and dogwood use 2 ml of undiluted FORMULA 40* Herbicide per injection site. Continuous cuts around the trunk often provide improved control. Also, cuts near the ground level may be more effective than at higher levels. Treatments can be made at any season; however, effectiveness may be less during winter months. Maples should not be treated during the spring sap flow.

USE PRECAUTIONS

AVOID CONTACT WITH 2,4-D SUSCEPTIBLE CROPS AND OTHER DESIRABLE BROADLEAF PLANTS: Do not apply directly to or otherwise permit even minute amounts to contact cotton, grapes, tobacco, fruit trees, vegetables, flowers, ornamentals or other desirable plants susceptible to 2,4-D. Do not use in or near a greenhouse.

DO NOT APPLY IN THE VICINITY OF COTTON, GRAPES, TOBACCO, TOMATOES OR OTHER DESIRABLE 2,4-D SUSCEPTIBLE CROPS OR PLANTS. DO NOT SPRAY WHEN WIND IS BLOWING TOWARDS SUSCEPTIBLE CROPS OR ORNAMENTAL PLANTS.

AVOID SPRAY DRIFT: Applications should be made only when there is no hazard from spray drift since very small quantities of the spray, which may not be visible, may severely injure susceptible crops during both growing and dormant periods. Use coarse sprays to minimize drift since, under adverse weather conditions, fine spray droplets may drift a mile or more. A spray thickening agent such as NALCO-TROL, may be used with this product to aid in reducing spray drift. If used, follow all use recommendations and precautions on the product label.

GROUND EQUIPMENT: With ground equipment, spray drift can be lessened by keeping the spray boom as low as possible; by applying 20 gallons or more of spray per acre; by using no more than 20 pounds spraying pressure at large droplet producing nozzle tips; by spraying when wind velocity is low; and by stopping all spraying when wind exceeds 6 to 7 miles per hour. Do not apply with hollow cone-type insecticide or other nozzles that produce a fine-droplet spray.

DETERMINE AIR MOVEMENT AND DIRECTIONS BEFORE FOLIAR APPLICATION: Use a smoke generator or other means at or near the application site for the detection of air movement, air stability or temperature inversions. Such a condition exists when there is little or no wind and air temperature is lower near the ground than at higher levels. Use appropriate drift control measures or avoid application when smoke is moving toward nearby desirable susceptible plants or sensitive areas.

AERIAL APPLICATION: With aircraft, drift can be lessened by applying a coarse spray; by using no more than 20 pounds spray pressure at the nozzles; by using straight stream nozzles directed straight back; by using a spray boom no longer than 3/4 the wing or rotor span of the aircraft; and by spraying only when wind velocity is less than 6 mph.

Excessive amounts of this herbicide in the soil may temporarily inhibit seed germination or plant growth. Violent wind storms may move soil particles. If 2,4-D is on soil particles and they are blown onto the susceptible plants, visible symptoms may appear. Serious injury is unlikely. The hazard of movement of 2,4-D on dust is reduced if treated fields are irrigated or if rain occurs shortly after application.

To avoid injury to desirable plants, do not handle or apply other agricultural chemicals with the same equipment used for FORMULA 40* Herbicide unless appropriately cleaned first. Local conditions may affect the use of herbicides. Consult your State Agricultural Experiment Station or Extension Service weed specialists for cleaning methods which are in compliance with local regulations and for advice in selecting treatments from this label to best fit local conditions. Be sure that use of this product conforms to all applicable regulations.

REENTRY STATEMENT

Do not enter treated areas without protective clothing until sprays have dried. Because certain states may require more restrictive reentry intervals for various crops treated with this product, consult your State Department of Agriculture for further information. Written or oral warnings must be given to workers who are expected to be in a treated area or in an area about to be treated with this product. The front panel PRECAUTIONARY STATEMENTS should be read to workers as well as the instruction not to enter until sprays have dried. When oral warnings are given, warnings shall be given in a language customarily understood by workers. Oral warnings must be given if there is reason to believe that written warnings cannot be understood by workers. Written warnings must include the following information: "DANGER. Area treated with FORMULA 40* Herbicide on (date of application). Do not enter without appropriate protective clothing until sprays have dried. (Insert here Statements of Practical Treatment as on front panel.)"

STORAGE AND DISPOSAL

Do not contaminate water, food or feed by storage or disposal.

STORAGE: Keep container tightly closed when not in use.

PESTICIDE DISPOSAL: Pesticide wastes are acutely hazardous. Improper disposal of excess pesticide, spray mixture, or rinsate is a violation of Federal Law. If these wastes cannot be disposed of by use according to label instructions, contact your State Pesticide or Environmental Control Agency, or the Hazardous Waste representative at the nearest EPA Regional Office for guidance.

CONTAINER DISPOSAL: Triple rinse (or equivalent). Then offer for recycling or reconditioning, or puncture and dispose of in a sanitary landfill, or by other procedures approved by state and local authorities. (Metal containers)

CONTAINER DISPOSAL: Triple rinse (or equivalent). Then dispose of in a sanitary landfill, or by incineration, or if allowed by local authorities, by burning. (Plastic containers 5-gal or less)

NOTE: FORMULA 40* Herbicide, exposed to subfreezing temperatures, should be warmed to at least 40°F and mixed thoroughly before using.

LIMITED WARRANTY AND DISCLAIMER

The manufacturer warrants (a) that this product conforms to the chemical description on the label; (b) that this product is reasonably fit for the purposes set forth in the directions for use when it is used in accordance with such directions; and (c) that the directions, warning and other statements on this label are based upon responsible experts' evaluation of reasonable tests of effectiveness, of toxicity to laboratory animals and to plants, and of residues on food crops and upon reports of field experience. Tests have not been made on all varieties or in all states or under all conditions. THE MANUFACTURER NEITHER MAKES NOR INTENDS, NOR DOES IT AUTHORIZE ANY AGENT OR REPRESENTATIVE TO MAKE, ANY OTHER WARRANTIES, EXPRESS OR IMPLIED, AND IT EXPRESSLY EXCLUDES AND DISCLAIMS ALL IMPLIED WARRANTIES OF MERCHANTABILITY OR FITNESS FOR A PARTICULAR PURPOSE.

THIS WARRANTY DOES NOT EXTEND TO, AND THE BUYER SHALL BE SOLELY RESPONSIBLE FOR, ANY AND ALL LOSS OR DAMAGE WHICH RESULTS FROM THE USE OF THIS PRODUCT IN ANY MANNER WHICH IS INCONSISTENT WITH THE LABEL DIRECTIONS, WARNINGS OR CAUTIONS.

BUYER'S EXCLUSIVE REMEDY AND MANUFACTURER'S OR SELLER'S EXCLUSIVE LIABILITY FOR ANY AND ALL CLAIMS, LOSSES, DAMAGES, OR INJURIES RESULTING FROM THE USE OR HANDLING OF THIS PRODUCT, WHETHER OR NOT BASED IN CONTRACT, NEGLIGENCE, STRICT LIABILITY IN TORT OR OTHERWISE SHALL BE LIMITED, AT THE MANUFACTURER'S OPTION, TO REPLACEMENT OF OR THE REPAYMENT OF THE PURCHASE PRICE FOR, THE QUANTITY OF PRODUCT WITH RESPECT TO WHICH DAMAGES ARE CLAIMED. IN NO EVENT SHALL MANUFACTURER OR SELLER BE LIABLE FOR SPECIAL, INDIRECT OR CONSEQUENTIAL DAMAGES RESULTING FROM THE USE OR HANDLING OF THIS PRODUCT.

NOTICE TO BUYER

Purchase of this material does not confer any rights under patents governing this product or the use thereof in countries outside of the United States.

*FORMULA 40 is a trademark of THE DOW CHEMICAL COMPANY

AG90019

THIS SPECIMEN LABEL IS INTENDED FOR USE ONLY AS A GUIDE IN PROVIDING GENERAL INFORMATION REGARDING THE DIRECTIONS, WARNINGS AND CAUTIONS ASSOCIATED WITH THE USE OF THIS PRODUCT. AS WITH ANY AGRICULTURAL CHEMICAL, ALWAYS FOLLOW THE LABEL INSTRUCTIONS ON THE PACKAGE BEFORE USING.

FRUITONE® N

TO CONTROL PRE-HARVEST DROP OF APPLES AND PEARS. ALSO USEFUL IN THINNING APPLES.

ACTIVE INGREDIENT:	
1-Naphthaleneacetic Acid, Sodium Salt*	3.5%
INERT INGREDIENTS:	96.5%
*1-Naphthaleneacetic acid equivalent 3.1%	
TOTAL:	100.0%

EPA Reg. No. 264-141 AA EPA Est. No. 57062-OH-001

KEEP OUT OF REACH OF CHILDREN

CAUTION

For PRODUCT USE Information Call 1-800-334-9745
For MEDICAL AND TRANSPORTATION Emergencies Information ONLY Call 24 Hours a Day 1-800-334-7577

PRECAUTIONARY STATEMENTS

CAUTION

HAZARDS TO HUMANS AND DOMESTIC ANIMALS

Harmful if swallowed. Avoid contact with skin, eyes, and clothing. Avoid breathing spray mist.

STATEMENT OF PRACTICAL TREATMENT

If Swallowed: If patient is conscious and alert, give 2-3 glasses of water or milk to drink. If available, give one tablespoon of Syrup of Ipecac to induce vomiting. Alternatively, induce vomiting by touching back of throat with finger. Do not repeat the Syrup of Ipecac. Do not make an unconscious person vomit. GET MEDICAL ATTENTION.

If On Skin: Immediately wash skin with plenty of soap and water, if available, while removing contaminated clothing and shoes. Wash clothing separately before reuse.

If Inhaled: Remove victim to fresh air. If not breathing, give artifical respiration. Administer oxygen if necessary. GET MEDICAL ATTENTION.

If In Eyes: Hold eyelids open and flush with a steady, gentle stream of water for at least 15 minutes. GET MEDICAL ATTENTION.

NOT TO PHYSICIAN

No specific antidote is available.

All treatments should be based on observed signs and symptoms of distress in the patient. Overexposure to materials other than this product may have occurred.

ENVIRONMENTAL HAZARDS

Do not apply directly to water or wetlands. Do not contaminate water when disposing of equipment washwaters. Do not contaminate irrigation ditches or water used for irrigation or domestic purposes. Do not apply when weather conditions favor drift from treated areas or use in a greenhouse. Do not apply FRUITONE® N through any type of irrigation system.

DIRECTIONS FOR USE

IT IS A VIOLATION OF FEDERAL LAW TO USE THIS PRODUCT IN A MANNER INCONSISTENT WITH ITS LABELING.

Do not apply FRUITONE® N through any type of irrigation system.

GENERAL WORKER PROTECTION STATEMENTS

Do not apply this product in such a manner as to directly or through drift expose workers or other persons. The area being treated must be vacated by unprotected persons.

Do not enter treated areas without protective clothing until sprays have dried. Because certain states may require more restrictive reentry intervals for various crops treated with this product, consult your State Department of Agriculture for further information. Written or oral warnings must be given to workers who are expected to be in a treated area or in an area about to be treated with this product. Advise workers to stay out of fields during application and until sprays have dried. Regular long-sleeved work clothing should be worn when working in treated fields. See PRECAUTIONARY STATEMENTS AND STATEMENTS OF PRACTICAL TREATMENT AND NOTE TO PHYSICIAN for information on accidential exposures. When oral warnings are given, warnings shall be given in a language customarily understood by workers. Oral warnings must be given if there is reason to believe that written warnings cannot be understood by workers. Written warnings must include the following information: appropriate signal word (CAUTION), area treated with, FRUITONE® N, date of application, appropriate protection clothing, and reentry interval if applicable.

GENERAL INFORMATION

Spraying just before harvest time (before pre-harvest drop commences) permits a longer harvest period, reduces wind losses, and decreases the need for spot picking. Be sure applications are completed in time to become effective before actual pre-harvest drop begins.

CONTROL PRE-HARVEST DROP

APPLES

With summer and early fall varieties during periods of warm weather, FRUITONE® N usually becomes effective in one to two days. For later varieties during cool weather, a slightly longer period is necessary. Healthy foliage, freedom from mites and frost damage is necessary for the most effective action from this product. For maximum effectiveness apply sprays when the temperature is 70° F or higher. Spray the entire tree canopy to "wet" assuring total complete coverage.

The following is offered only as a general guide. Variety and climatic conditions will affect the actual rate required. Generally in the Eastern U. S., a single spray of 5 to 10 ppm is recommended while in the Pacific Northwest rates of 15 to 20 ppm are usually required. Consult your state pomologist for specific recommendations.

For summer varieties such as Yellow Transparent and Wealthy, sprays should be applied when the first sound, uninjured fruit begin to drop. The effectiveness of this material is usually much longer with these varieties in hot weather. For summer varieties, one-half concentration (5 ppm) may be effective, especially with hot weather.

For late fall varieties such as McIntosh, Baldwin, and Red Delicious, application should be made as the first sound, uninjured fruit begin to drop. With a variety such as McIntosh, apply FRUITONE® N as the first sound, uninjured fruit begin to drop. If the area treated is not to be picked within 7 to 10 days after first treatment, we suggest a second application be applied to remaining unpicked fruit 5 to 6 days after the first application. With later varieties such as Baldwin and Red Delicious, one application of a 10 ppm spray at the time the first sound, uninjured fruit begin to drop is usually sufficient. Under cooler temperatures (below 70°F) higher rates may be required.

PEARS

For Bartlett pears apply a spray of 5 ppm when the first sound, uninjured fruit begin to drop. Consult your state pomologist for specific recommendations.

IMPORTANT: FRUITONE® N inhibits fruit drop, but does not effect normal ripening. The fruit must be picked at the proper stage of maturity, even though the hormone-like material is still effectively delaying excessive drop. Do not make more than 2 applications and do not apply closer than 5 days before harvest.

CHEMICAL THINNING

APPLES

Because of wide variations in varieties, climatic factors, tree vigor and fruit set rates for thinning can range from 2 to 20 ppm and application timing can range from petal fall to 21 days past full bloom.

The higher concentrations should be limited to vigorous trees with high fruit set potential. Use the lower concentrations for weaker, less vigorous trees with less food reserves and lower fruit set potential.

Midseason to late blooming varieties such as Red Delicious, Golden Delicious, and Romes should be treated 5 to 21 days after full bloom depending on locality and other factors.

For rates and application timing for your specific variety and location, consult your local extension pomologist for recommendations.

The preferred application temperature is between 70°F and 75°F. Application is not recommended below 60°F or above 80°F.

IMPORTANT: FRUITONE® N should not be used on early season varieties (those that ripen before McIntosh) such as Yellow Transparent, Early Williams Red, and Oldenburg, since severe injury or over-thinning may occur. AMID-THIN® W is recommended for these early season varieties.

SPRAY PREPARATION

One and one-quarter pounds of FRUITONE® N in 500 gallons of water gives a concentation of 10 ppm of naphthaleneacetic acid. See the spray preparation chart for other dilutions. If possible, sprays should be applied when the temperature is 70°F to 75°F. Spray the entire tree canopy to "wet" assuring uniform coverage.

SPRAY PREPARATION CHART

AMOUNT OF FRUITONE® N IN OUNCES AND POUNDS				
Gallons/ Acre	75	100	150	300
RATE ppm	OZS (LBS)	OZS (LBS)	OZS (LBS)	OZS (LBS)
2.0	0.6	0.8	1.2	2.4
5.0	1.4	2.1	3.0	6.1 (0.38)
7.5	2.2	3.0	4.5 (0.28)	9.0 (0.56)
10.0	3.0	4.0 (0.25)	6.1 (0.38)	12.0 (0.75)
20.0	6.1 (0.38)	8.0 (0.50)	12.0 (0.75)	24.0 (1.50)

STORAGE AND DISPOSAL

STORAGE

Do not store near fertilizers, seed, insecticides, or fungicides.

Do not contaminate water, food, or feed by storage or disposal.

PESTICIDE DISPOSAL

Open dumping is prohibited. Wastes resulting from the use of this product may be disposed of on site or at an approved waste disposal facility.

CONTAINER DISPOSAL

Do not reuse empty container. Triple rinse (or equivalent). Then offer for recycling or reconditioning, or puncture and dispose of in a sanitary landfill, or by other procedures approved by state and local authorities.

LIMITED WARRANTY AND DISCLAIMER

The manufacturer warrants that this product conforms to the chemical description on the label; that this product is reasonably fit for the purposes set forth in the directions for use when it is used in accordance with such directions; and that the directions, warnings and other statements on this label are based upon responsible experts' evaluation of reasonable tests of effectiveness, of toxicity to laboratory animals and to plants, and of residues on food crops and upon reports of field experience. Tests have not been made on all varieties or in all states or under all conditions. THE MANUFACTURER NEITHER MAKES NOR INTENDS, NOR DOES IT AUTHORIZE ANY AGENT OR REPRESENTATIVE TO MAKE, ANY OTHER WARRANTIES, EXPRESS OR IMPLIED, AND IT EXPRESSLY EXCLUDES AND DISCLAIMS ALL IMPLIED WARRANTIES OF MERCHANTABILITY OR FITNESS FOR A PARTICULAR PURPOSE.

THIS WARRANTY DOES NOT EXTEND TO, AND THE BUYER SHALL BE SOLELY RESPONSIBLE FOR, ANY AND ALL LOSS OR DAMAGE WHICH RESULTS FROM THE USE OF THIS PRODUCT IN ANY MANNER WHICH IS INCONSISTENT WITH THE LABEL DIRECTIONS, WARNINGS, OR CAUTIONS.

BUYER'S EXCLUSIVE REMEDY AND MANUFACTURER'S OR SELLER'S EXCLUSIVE LIABILITY FOR ANY AND ALL CLAIMS, LOSSES, DAMAGES OR INJURIES RESULTING FROM THE USE OR HANDLING OF THIS PRODUCT, WHETHER OR NOT BASED IN CONTRACT, NEGLIGENCE, STRICT LIABILITY IN TORT OR OTHERWISE SHALL BE LIMITED, AT THE MANUFACTURER'S OPTION, TO REPLACEMENT OF, OR THE REPAYMENT OF THE PURCHASE PRICE FOR, THE QUANTITY OF PRODUCT WITH RESPECT TO WHICH DAMAGES ARE CLAIMED. IN NO EVENT SHALL MANUFACTURER OR SELLER BE LIABLE FOR SPECIAL, INDIRECT OR CONSEQUENTIAL DAMAGES RESULTING FROM THE USE OR HANDLING OF THIS PRODUCT.

NOTICE TO BUYER

Purchase of this material does not confer any rights under patents governing this product or the use thereof in countries outside of the United States.

FRUITONE and AMID-THIN is a registered trademark of Rhône-Poulenc

AG 90041

THIS SPECIMEN LABEL IS INTENDED FOR USE ONLY AS A GUIDE IN PROVIDING GENERAL INFORMATION REGARDING THE DIRECTIONS, WARNINGS AND CAUTIONS ASSOCIATED WITH THE USE OF THIS PRODUCT. AS WITH ANY AGRICULTURAL CHEMICAL, ALWAYS FOLLOW THE LABEL INSTRUCTIONS ON THE PACKAGE BEFORE USING.

Fruitone® CPA

PINEAPPLE GROWTH REGULATOR

ACTIVE INGREDIENT:

Sodium 2-(3-chlorophenoxy) propionate*	8.0%
INERT INGREDIENTS:	92.0%
TOTAL	100.0%

Total 2-(3-chlorophenoxy) propionic acid equivalent 7.2% by weight or 75 grams per liter

E.P.A. Reg. No. 264-338 E.P.A. Est. No.

KEEP OUT OF REACH OF CHILDREN
CAUTION

For MEDICAL And TRANSPORTATION Emergencies ONLY Call 24 Hours A Day 1-800-334-7577

For **PRODUCT USE** Information Call 1-800-334-9745

PRECAUTIONARY STATEMENTS

CAUTION

HAZARDS TO HUMANS AND DOMESTIC ANIMALS

Keep out of reach of children. Harmful if swallowed. Avoid contact with skin, eyes or clothing.

ENVIRONMENTAL HAZARDS

Keep out of all bodies of water. Do not contaminate water by cleaning of equipment or disposal of wastes. Apply this product only as specified on this label. Do not apply FRUITONE® CPA through any type of irrigation system.

Avoid spray drift to susceptible plants as this product may injure beans, peas and ornamentals. Do not use the same spray equipment for other purposes unless thoroughly cleaned.

GENERAL INFORMATION

FRUITONE® CPA PINEAPPLE GROWTH REGULATOR may be applied to evaluate reduction of crown growth and the increase in size of pineapple fruit on both plant and ratoon crops. Reducing crown growth may be especially useful on fresh market fruit. Injury during the packing and shipping operation can be reduced due to smaller crown size. Shipping costs could also be lowered due to the reduction in weight of the nonedible crown. Increased fruit size may lead to larger yields of high grade slices and chunks for the processing market.

The application of FRUITONE® CPA PINEAPPLE GROWTH REGULATOR may result in delayed fruit maturity thus allowing for harvest scheduling.

DIRECTIONS FOR USE

IT IS A VIOLATION OF FEDERAL LAW TO USE THIS PRODUCT IN A MANNER INCONSISTENT WITH ITS LABELING.

Be sure that use of this product conforms to all applicable laws, rules and regulations. Certain states have restrictions pertaining to application distances from susceptible crops. The applicator should become familiar with these laws, rules or regulations and follow them exactly. Apply this only as specified on this label.

GENERAL WORKER PROTECTION STATEMENTS

Do not apply this product in such a manner as to directly or through drift expose workers or other persons. The area being treated must be vacated by unprotected persons.

Do not enter treated areas without protective clothing until sprays have dried. Because certain states may require more restrictive reentry intervals for various crops treated with this product, consult your State Department of Agriculture for further information.

Written or oral warnings must be given to workers who are expected to be in a treated area or in an area about to be treated with this product. Advise workers to stay out of fields during application and until sprays have dried. When oral warnings are given, warnings shall be given in a language customarily understood by workers. Oral warnings must be given if there is reason to believe that written warnings cannot be understood by workers. Written warnings must include the following information: appropriate signal word (CAUTION), area treated with FRUITONE® CPA Pineapple Growth Regulator, date of application, appropriate protection clothing and reentry interval if applicable.

APPLICATION INSTRUCTIONS

(A) Suggested Single Application Rates

RATE Grams a.i./acre	FRUITONE® CPA Pints/acre	WATER Gal./acre
50	1.4	250
100	2.75	250
150	4.0	250

The lower rate will be effective for yield increase while the higher rate will give maximum crown position and delayed fruit maturity.

Fill spray tank half full with water. With agitator running add the required amount of FRUITONE CPA according to the table. With continued agitation, add the remaining amount of water.

Apply as a broadcast spray over the top of the plants and fruit approximately 15 weeks prior to anticipated normal harvest. This should correspond to the stage when the last flowers on the fruit have shriveled and dried and when the pineapple crown is 3-5 cm. long.

(B) Sequential Applications

RATE Grams a.i./acre	FRUITONE CPA Pints/acre	WATER Gal./acre	TIME OF APPLICATION (Flower/fruit stage)
50	1.4	250	Mid-petal
+			
50	1.4	250	Dry petal + 10-14 days

As an **alternate** to the above referenced sequential treatment, 50 grams of FRUITONE CPA may be applied at dry petal with a second 50 grams applied 3 weeks later.

Follow mixing instructions given for Single Application. Apply as a broadcast spray over the top of plants and fruit.

Harvest only at optimum ripeness as evidenced by shell color development.

STORAGE AND DISPOSAL

STORAGE

Store above 40°F.

Do not store near fertilizers, seeds, insecticides or fungicides.
Do not contaminate water, food or feed by storage or disposal.

PESTICIDE DISPOSAL

Wastes resulting from the use of this product may be disposed of on site or at an approved waste disposal facility.

CONTAINER DISPOSAL

Triple rinse or equivalent. Then offer for recycling or reconditioning, or puncture and dispose of in a sanitary landfill or incineration, or if allowed by state and local authorities, by burning. If burned, stay out of smoke.

LIMITED WARRANTY AND DISCLAIMER

The manufacturer warrants that this product conforms to the chemical description on the label; that this product is reasonably fit for the purposes set forth in the directions for use when it is used in accordance with such directions; and that the directions, warnings and other statements on this label are based upon responsible experts' evaluation of reasonable tests of effectiveness of toxicity to laboratory animals and to plants, and of residues on food crops, and upon reports of field experience. Tests have not been made on all varieties or in all states or under all conditions. THE MANUFACTURER NEITHER MAKES NOR INTENDS, NOR DOES IT AUTHORIZE ANY AGENT OR REPRESENTATIVE TO MAKE ANY OTHER WARRANTIES, EXPRESS OR IMPLIED, AND IT EXPRESSLY EXCLUDES AND DISCLAIMS ALL IMPLIED WARRANTIES OF MERCHANTABILITY OR FITNESS FOR A PARTICULAR PURPOSE.

THIS WARRANTY DOES NOT EXTEND TO, AND THE BUYER SHALL BE SOLELY RESPONSIBLE FOR, ANY AND ALL LOSS OR DAMAGE WHICH RESULTS FROM THE USE OF THIS PRODUCT

IN ANY MANNER WHICH IS INCONSISTENT WITH THE LABEL DIRECTIONS, WARNINGS OR CAUTIONS.

BUYER'S EXCLUSIVE REMEDY AND MANUFACTURER'S OR SELLER'S LIABILITY FOR ANY AND ALL CLAIMS, LOSSES, DAMAGES, OR INJURIES RESULTING FROM THE USE OR HANDLING OF THIS PRODUCT, WHETHER OR NOT BASED IN CONTRACT, NEGLIGENCE, STRICT LIABILITY IN TORT OR OTHERWISE SHALL BE LIMITED, AT THE MANUFACTURER'S OPTION, TO REPLACEMENT OF, OR THE REPAYMENT OF THE PURCHASE PRICE FOR, THE QUANTITY OF PRODUCT WITH RESPECT TO WHICH DAMAGES OF THE PURCHASE PRICE FOR, THE QUANTITY OF PRODUCT WITH RESPECT TO WHICH DAMAGES ARE CLAIMED. IN NO EVENT SHALL MANUFACTURER OR SELLER BE LIABLE FOR SPECIAL INDIRECT OR CONSEQUENTIAL DAMAGES RESULTING FROM THE USE OR HANDLING OF THIS PRODUCT.

NOTICE TO BUYER

Purchase of this material does not confer any rights under patents governing this product or the use thereof in countries outside of the United States.

FRUITONE is a registered trademark of Rhône-Poulenc

AG 90040

THIS SPECIMEN LABEL IS INTENDED FOR USE ONLY AS A GUIDE IN PROVIDING GENERAL INFORMATION REGARDING THE DIRECTIONS, WARNINGS AND CAUTIONS ASSOCIATED WITH THE USE OF THIS PRODUCT, AS WITH ANY AGRICULTURAL CHEMICAL, ALWAYS FOLLOW THE LABEL INSTRUCTIONS ON THE PACKAGE BEFORE USING.

LARVIN® brand 3.2 Thiodicarb

INSECTICIDE/OVICIDE/Aqueous Flowable

ACTIVE INGREDIENT:

THIODICARB: Dimethyl N, N' [thiobis [(methylimino) carbonyloxy]] bis [ethanimidothioate]	32.5% by wt.
INERT INGREDIENTS:	67.5% by wt.

CONTAINS 3.2 pounds active thiodicarb per U.S gallon

EPA Reg. No.264-379 EPA Est. No. 264-MO-02

KEEP OUT OF REACH OF CHILDREN

WARNING AVISO

PRECAUCION AL USUARIO: Si usted no lee ingles, no use este producto hasta que la etiqueta le haya sido explicada ampliamente.

STATEMENT OF PRACTICAL TREATMENT

MAY BE FATAL IF SWALLOWED. MAY BE HARMFUL IF INHALED OR ABSORBED THROUGH THE SKIN. ANTIDOTE IS ATROPINE SULFATE. FOR ADDITIONAL PRECAUTIONARY STATEMENTS. READ THE ENTIRE LABEL BEFORE USING THIS THIS PRODUCT.

For PRODUCT USE Information Call 1-800-334-9745
For MEDICAL And TRANSPORTATION Emergencies ONLY Call 24 Hours A Day 1-800-334-7577

PRECAUTIONARY STATEMENTS

WARNING

HAZARDS TO HUMANS AND DOMESTIC ANIMALS

Illness may result rapidly following overexposure to this product. May be fatal if swallowed. May be harmful if inhaled or absorbed through the skin. May be irritating to eyes and may cause mild skin sensitization. Do not take internally. Avoid contact with skin, eyes or clothing. Avoid breathing spray mist. Keep away from domestic animals.

Wear long-sleeved clothing, full-length trousers, head covering and protective gloves while using this product. Remove contaminated clothing daily and wash before reuse. Wash exposed skin thoroughly with soap and water after use and before eating or using tobacco. Discontinue use if allergic reaction occurs. Do not eat or use tobacco while using this product.

SIGNS AND SYMPTOMS OF OVEREXPOSURE:

Abdominal cramps	Diarrhea
Excessive sweating	Muscle tremors
Pinpoint eye pupils	Vomiting
Blurred vision	Difficult breathing
Headache	Nausea
Salivation	Weakness
	Watery eyes

In severe cases, convulsions, unconsiousness and respiratory failure may occur.

ANTIDOTE STATEMENT

ANTROPINE SULFATE IS HIGHLY EFFECTIVE AS AN ANTIDOTE. See NOTE TO PHYSICIAN below.

STATEMENT OF PRACTICAL TREATMENT

GENERAL

Contact a physician immediately in all cases of suspected poisoning. If breathing stops, establish an airway, start artificial respiration, and provide oxygen. Make certain to remove all sources of continuing contamination. Remove clothing and wash skin and hair immediately with large amounts of water. Transport the patient to a physician or hospital immediately and SHOW A COPY OF THIS LABEL TO THE PHYSICIAN. If poisoning is suspected in animals, contact a veterinarian.

IF SWALLOWED: If patient is conscious and alert, give 2 to 3 glasses of water or milk to drink. If available, give one tablespoon of Syrup of Ipecac to induce vomiting. If vomiting does not occur within 20 minutes of the first dose, follow with a second and last dose of 1 teaspoon. Alternatively, induce vomiting by touching back of throat with finger. Do not make an unconscious person vomit. GET MEDICAL ATTENTION.

IF IN EYES: Hold eyelids open and flush with a steady, gentle stream of water for at least 15 minutes. GET MEDICAL ATTENTION.

IF ON SKIN: Immediately flush skin with plenty of water while removing contaminated clothing and shoes. Wash clothes separately before reuse.

IF INHALED: Remove to fresh air. If not breathing, give artificial respiration. Administer oxygen if necessary. GET MEDICAL ATTENTION.

NOTE TO PHYSICIAN

All treatments should be based on observed signs and symptoms of distress in the patient. Overexposure to materials other than this product may have occurred.

This product contains a methyl carbamate insecticide, which is cholinesterase inhibitor. Overexposure to this substance may cause toxic signs and symptoms due to stimulation of the cholinergic nervous system. These effects of overexposure are spontaneously and rapidly reversible.

Gastric lavage may be used if this product has been swallowed. LARVIN® poisoning may occur rapidly after ingestion and prompt removal of stomach contents is indicated.

Specific treatment consists of parenteral atropine sulfate. Caution should be maintained to prevent overatropinization. Mild cases may be given 1 to 2 mg intramuscularly every 10 minutes until full atropinization has been achieved and repeated thereafter whenever symptoms reappear. Severe cases should be given 2 to 4 mg intravenously every 10 minutes until fully atropinized, then intramuscularly every 30 to 60 minutes as needed to maintain the effect for at least 12 hours. Dosages for children should be appropriately reduced. Complete recovery from overexposure is to be expected within 24 hours.

Narcotics and other sedatives should not be used. Further, drugs like 2-PAM (pyridine-2-aldoxime methiodide) are NOT recommended unless organophosphate intoxication is also suggested.

To aid in confirmation of a diagnosis, urine samples should be obtained within 24 hours of exposure and immediately frozen. Analyses will be arranged by Rhone-Poulenc Ag Company.

Consultation on therapy can be obtained at all hours by calling the Rhone-Poulenc emergency number: 1-800-334-7577.

ENVIRONMENTAL HAZARDS

This product is toxic to fish, birds, invertebrates and other wildlife. Do not apply directly to lakes, streams and ponds. Do not contaminate water by cleaning of equipment or disposal of wastes. Do not apply when weather conditions favor drift from area treated. This product is toxic to bees exposed to direct treatment. Do not apply when bees are actively visiting the treatment area.

GENERAL INFORMATION

LARVIN® brand 3.2 Thiodicarb Insecticide/Ovicide is an aqueous flowable formulation that readily disperses in water for spraying by ground or air equipment. Directions on this label are based on tests and field experience relating to effectiveness, impact on environment and residues remaining in food and feed. READ THE ENTIRE LABEL AND OBSERVE ALL LABEL DIRECTIONS AND PRECAUTIONS BEFORE USE.

DIRECTIONS FOR USE

IT IS A VIOLATION OF FEDERAL LAW TO USE THIS PRODUCT IN A MANNER INCONSISTENT WITH ITS LABELING.

Do not use this product except as directed on this label or on other Rhone-Poulenc Ag Company supplemental labeling for this product. Applications at higher rates or at more frequent intervals than is stated on the label may result in illegal residues in crops, meat and milk.

NOTE: In case of minor spills or leaks, follow all precautions indicated above and clean up immediately. Soak up with sand, earth or other suitable material and dispose of wastes and broken or empty containers in a landfill approved for pesticide use.

GENERAL WORKER PROTECTION STATEMENTS

Do not apply this product in such a manner as to directly or through drift expose workers or other persons. The area being treated must be vacated by unprotected persons.

Do not enter treated areas without protective clothing until sprays have dried.

Because certain states may require more restrictive reentry intervals for various crops treated with this product, consult your State Department of Agriculture for further information.

Written or oral warnings must be given to workers who are expected to be in a treated area or in an area about to be treated with this product. Advise workers to stay out of fields during application and until sprays have dried. Regular long-sleeved work clothing should be worn when working in treated fields. See PRECAUTIONARY STATEMENT, STATEMENT OF PRACTICAL TREATMENT and NOTE TO PHYSICIAN for information on accidental exposures. When oral warnings are given, warnings shall be given in a language customarily understood by workers. Oral warnings must be given if there is reason to believe that written warnings cannot be understood by workers. Written warnings must include the following information: appropriate signal word (WARNING), area treated with LARVIN® brand 3.2 Thiodicarb Insecticide/Ovicide, date of application, appropriate protective clothing, appropriate reentry interval, if applicable.

GENERAL DIRECTIONS

SPRAY PREPARATION: TO ASSURE A UNIFORM PRODUCT, AGITATE OR SHAKE ALL CONTAINERS OF THIS PRODUCT PRIOR TO USE. Remove oil, rust, scale, pesticide residues and other foreign matter from sprayer and strainer. Flush with clean water. Use 50 mesh screens or equivalent slotted strainers in spray system. To prepare for spraying, fill tank to approximately 1/2 the needed volume of water. Add the required amount of this insecticide and mix thoroughly by mechanical or hydraulic agitation. Finish filling tank with water to desired volume and thoroughly mix. Do not store spray mixture for prolonged periods. If tank mixes are to be used, LARVIN® brand 3.2 Thiodicarb must be fully dispersed in water first, followed by the addition of the intended tank-mix material.

COMPATABILITY: Thiodicarb Insecticide when diluted with an equal volume of water, is physically compatible with a wide range of commonly used spray products, but the full range of compatabilities under local conditions is not known. Therefore, it is essential that before using LARVIN® brand 3.2 Thiodicarb Insecticide/Ovicide in any tank mixture the compatability of the mixture be established. Add a small amount of this product to an equal volume of water in a small container and then add the other pesticide or spray product and mix thoroughly. DO NOT USE MIXTURES THAT CURDLE, PRECIPITATE, OR GREASE. THIS PRODUCT IS NOT COMPATIBLE WITH NON-EMULSIFIED CROP OR VEGETABLE OIL DILUENTS. FOR BEST RESULTS, SPRAY MIXTURES SHOULD BE USED IMMEDIATELY AFTER MIXING WITH ADEQUATE AGITATION.

NOTE: Thiodicarb Insecticide is hydrolytically sensitive to degradation of active ingredient by strong acids, strong bases, and certain heavy metal oxides and salts of certain fungicides. Degradation can cause reduced effectiveness, plant phytotoxicity and increased handling hazards. DO NOT USE THIS PRODUCT IN SPRAY MIXTURES WITH THE ABOVE TYPE OF PRODUCTS INCLUDING MIXTURES WITH Maneb, Mancozeb (except Manzate 200 and Dithane M-45 wettable powders where mixtures are used immediately after mixing), Copper Count N-8, and Bordeaux mixture. Do not use a four way tank mixture of LARVIN® brand 3.2 + Pix® + chlordimeform + Guthion® or Parathion as this mixture is incompatible. DO NOT ADD LARVIN® brand 3.2 THIODICARB TO WATER WITH pH VALUES BELOW 3.0 OR ABOVE 8.5. If necessary, water should be buffered within this range before adding LARVIN® brand Thiodicarb.

DIRECTIONS FOR AERIAL OR GROUND SPRAY APPLICATION

APPLICATION TIMING: Begin application when insect populations reach recognized economic threshold levels. Consult the Cooperative Extension Service, professional consultants or other qualified authorities to determine appropriate threshold levels for treatment in your area.

APPLICATION INSTRUCTIONS: Apply a minimum finished spray volume of 2 gallons per acre by air or 5 gallons per acre by ground unless otherwise directed under crop specific directions. For best results, it is important to obtain thorough and uniform spray coverage of the plant. Repeat application as needed to maintain control. Use higher dosage rates for heavy infestations, large larvae, or dense foliage. The specific length of residual control depends on environmental factors, plant growth, dosage rate, and degree of insect infestation. Although thiodicarb has some contact activity, larvae not feeding on treated foliage will not be controlled. When banding, determine the amount of chemical to use per acre by dividing the band width by the row width and multiplying by the appropriate broadcast rate.

To clean the sprayer after use, drain and flush with water. Use rinsings on crop according to label instructions or dispose of in an approved manner (See STORAGE AND DISPOSAL).

DIRECTIONS FOR USE THROUGH SPRINKLER IRRIGATION SYSTEMS

Apply this product only through sprinkler irrigation systems including center pivot and solid set. Do not apply this product through any other type of irrigation system.

SPRAY PREPARATION: Remove scale, pesticide residues, and other foreign matter from the chemical tank and entire injector system. Flush with clean water.

APPLICATION INSTRUCTIONS: First prepare a suspension of LARVIN® brand 3.2 Thiodicarb Insecticide/Ovicide in a mix tank. Fill tank with 1/2 to 3/4 the desired amount of water. Start mechanical or hydraulic agitation. Add the required amount of LARVIN® 3.2, and then the remaining volume of water. (Suspension concentrations using the appropriate dosage per acre recommended on this label of LARVIN® per 1 to 4 gallons of water are recommended). Then set sprinkler to deliver 0.1 to 0.3 inch of water per acre. Start sprinkler and uniformly inject the suspension of LARVIN® into the irrigation water line so as to deliver the desired rate per acre. The suspension of LARVIN® should be injected with a positive displacement pump into the main line ahead of a right angle turn to insure adequate mixing. If you should have any other questions about calibration, you should contact State Extension Service specialists, equipment manufacturers or other experts.

NOTE: When treatment with LARVIN® brand 3.2 Thiodicarb Insecticide/Ovicide has been completed, further field irrigation over the treated area should be avoided for 24 to 48 hours to prevent washing the chemical off the crop.

GENERAL PRECAUTIONS FOR APPLICATIONS THROUGH SPRINKLER IRRIGATION SYSTEMS

Maintain continuous agitation in mix tank during mixing and application to assure a uniform suspension.

Greater accuracy in calibration and distribution will be achieved by injecting a larger volume of a more dilute solution per unit time.

The system must contain a functional check valve, vacuum relief valve, and low pressure drain appropriately located on the irrigation pipeline to prevent water source contamination from backflow. The pesticide injection pipeline must contain a functional, automatic, quick-closing check valve to prevent the flow of fluid back toward the injection pump. The pesticide injection pipeline must also contain a functional, normally closed solenoid-operated valve located on the intake side of the injection pump and connected to the system interlock to prevent fluid from being withdrawn from the supply tank when the irrigation system is either automatically or manually shutdown. The system must contain functional interlocking controls to automatically shut off the pesticide injection pump when the water pump motor stops. The irrigation line or water pump must include a functional pressure switch which will stop the water pump motor when the water pressure decreases to the point where pesticide distribution is adversely affected. Systems must use a metering pump, such as a positive displacement injection pump (e.g., diaphragm pump) effectively designed and constructed of materials that are compatible with pesticides and capable of being fitted with a system interlock. Do not apply when wind speed favors drift beyond the area intended for treatment.

Do not apply when wind speed favors drift, when system connection or fittings leak, when nozzles do not provide uniform distribution or when lines containing the product must be dismantled and drained.

Crop injury, lack of effectiveness, or illegal pesticide residues in the crop may result from non-uniform distribution of treated water.

Allow sufficient time for pesticide to be flushed through all lines and all nozzles before turning off irrigation water. A person knowledgeable of the chemigation system and responsible for its operation shall shut the system down and make necessary adjustments should the need arise.

Do not connect an irrigation system (including greenhouse systems) used for pesticide application to a public water system unless the label-prescribed safety devices for public water supplies are in place.

RECOMMENDED USES

COTTON

INTEGRATED PEST MANAGEMENT (IPM) USE OF THIS PRODUCT: LARVIN® Thiodicarb Insecticide/Ovicide has both contact and residual ovicidal activity which can be effectively utilized in IPM programs (See ovicide section). LARVIN® Thiodicarb Insecticide/Ovicide has been shown to leave substantial populations of beneficial insects and spiders after use. The lower rates allow for maximum beneficial survival and faster rebound of beneficial populations. Larvicidal and ovicidal control coupled with retention of beneficial insects and spiders can offer significant benefits to those producers utilizing integrated pest control programs.

SPRAY VOLUME FOR COTTON: LARVIN® brand 3.2 Thiodicarb Insecticide/Ovicide should be applied in a minimum finished spray volume of 2 gallons per acre by aircraft and 5 gallons per acre by ground equipment. Under extreme pest populations or heavy crop canopy; use a minimum spray volume of 5 gallons per acre by air and a minimum spray volume of 10 gallons per acre by ground.

SITE	PEST	DOSAGE PER ACRE POUNDS ACTIVE	DOSAGE PER ACRE FLUID OUNCES	SPECIFIC DIRECTIONS	USE RESTRICTIONS
COTTON (DO NOT USE IN CALIFORNIA)	Ovicide: Cotton bollworm Tobacco budworm	0.125 to 0.25	5.0 to 10.0	Apply on a 3 to 5 day schedule when moth flights begin or sufficient eggs are present. Thorough spray coverage is necessary as eggs must be contacted by LARVIN® for activity to occur. Eggs laid on untreated foliage or not contacted by LARVIN® sprays will not be controlled. If significant number of larvae survive, use higher recommended LARVIN® rates or 0.125 to 0.25 lbs ai/A of LARVIN® in tank mixture combinations with organophosphates or pyrethroids registered for use on cotton at their label recommended rates. When a larvicide is needed and maximum survival of beneficial insects and spiders is desired, use LARVIN® ovicide recommended rates in tank mix combinations with recommended rates of DIPEL® ES. Observe all restrictions and precautions which appear on the labels of all products to be tank mixed with LARVIN®.	Do not allow livestock to graze fields. Do not apply less than 28 days before harvest. In Arizona, apply before bolls begin to open and do not exceed 2 applications per season.
	Larvicide: Cotton bollworm Tobacco budworm	0.6 to 0.9	24.0 to 36.0	In this rate range, LARVIN® is a direct contact and residual contact ovicide, as well as an effective larvicide. Although LARVIN® has some contact activity, larvae must feed on treated foliage to be controlled. Use the higher rate when insect larvae populations are heavier or when infestations are predominantly large larvae (> ¼ inch in size).	
	Larvicide Armyworms (such as: Beet, Fall, etc.) Cabbage looper Cotton leaf perforator Cotton leafworm Soybean looper Fleahopper (suppression) Pink bollworm (suppression) Plant bugs (Suppression)	0.6 to 0.9	24.0 to 36.0	As pest pressure increases, reduce intervals between applications and/or increase dosage rate as indicated in specific directions below. A rate of 24 fluid ounces (0.36 lbs. active) should be used when: For low to moderate populations and when maximum survival of beneficial insects and spiders is desired. Rates at 30 to 60 fluid ounces (0.75 to 0.90 lbs. active) should be used when: Insect populations are heavy or infestation is predominantly large larvae. Although LARVIN® has some contact activity, larvae must feed on treated foliage to be controlled. Applications for suppression of Pink bollworm should be timed for the presence of adult moths.	Do not allow livestock to graze fields. Do not apply less than 28 days before harvest.
	Boll weevil (Suppression)	0.6 to 0.9	24.0 to 36.0	LARVIN® will reduce boll weevil populations comparable to most pyrethroids. LIGHT BOLL WEEVIL POPULATIONS When spraying LARVIN® on a 5 to 7 day worm control schedule, control of boll weevil will usually be adequate to maintain boll weevil populations below established economic thresholds. MODERATE TO HEAVY BOLL WEEVIL POPULATIONS: When spraying LARVIN® on a worm control schedule, intersprays of a registered boll weevil control product and/or tank mix of LARVIN® plus a registered boll weevil product should be used. A 3 to 5 day interval between sprays may be required to maintain boll weevil control under heavy infestations.	In Arizona, apply before bolls begin to open and do not exceed 2 applications per season.
	Cutworms	0.6	24.0	See "SPECIAL DIRECTIONS ON CUTWORMS"	
SOYBEAN (Standard Spray Volume Applications)	Podworms (Bollworm, Corn earworm Tobacco budworm) Armyworms (Such as: Beet, Fall, Southern, Yellow-striped, etc.) Green cloverworm Velvetbean caterpiller Woollybear caterpiller	0.25 to 0.4	10.0 to 16.0	Use the lower rates for low to moderate populations and maximum protection of beneficials. Apply in a minimum finished spray volume of 2 gallon per acre by air or 5 gallons per acre by ground. Under extreme pest populations, use a minimum spray volume of 5 gallons per acre by air and a minimum of 10 gallons per acre by ground.	Do not feed forage, hay, or straw to livestock. Do not apply less than 28 days before harvest.

SITE	PEST	DOSAGE PER ACRE POUNDS ACTIVE	FLUID OUNCES	SPECIFIC DIRECTIONS	USE RESTRICTIONS
	Soybean looper Cabbage looper Mexican bean beetle Bean leaf beetle Stinkbugs (Suppression) Three-Cornered Alfalfa hopper (Suppression)	0.45 to 0.75	18.0 to 30.0		
	Cutworms	See "SPECIAL DIRECTIONS ON "CUTWORMS"			
SOYBEAN (Low Spray Volume Applications)	Podworms (Bollworm, Corn earworm Tobacco budworm) Armyworms (Such as: Beet, Fall, Southern, etc.) Green cloverworm Velvetbean caterpiller Woollybear caterpiller	0.25 to 0.4	10.0 to 16.0	For best results, use the lower rates for low to moderate populations and maximum protection of beneficials. Apply in a minimum finished spray volume of 1 gallon per acre by aircraft. When applying at 1 gallon of spray volume per acre by air, use 1 quart of emulsified crop or vegetable oil (e.g., emulsified corn, soybean or cottonseed oil). Water alone may be used as a diluent for spray volumes of 2 gallons or higher per acre. TANK MIX DIRECTIONS FOR EMULSIFIED CROP OR VEGETABLE OILS: Fill spray tank with approximately 3/4 of the water required for the spray mixture. Add the required amount of LARVIN® 3.2 to the spray tank with the agitator running. Then add the desired amount of emulsified crop or vegetable oil. Continue agitation while adding the remainder of water.	FOR USE ONLY IN MARYLAND, VIRGINIA, NORTH CAROLINA, DELAWARE, SOUTH CAROLINA, GEORGIA, ALABAMA, MISSISSIPPI, LOUISIANA, ARKANSAS AND MISSOURI. Do not feed forage, hay, or straw to livestock. Do not apply less than 28 days before harvest.
SWEET CORN (Florida Fresh Market Only)	Corn earworm European corn borer Armyworms (Such as Fall, Beet, etc) Western bean cutworm	0.5 to 0.75	20 to 30	Whorl feeders: Spray as needed. As insect populations increase and/or when new plant growth is rapid, the spray interval should be shorter. For best results use sprays of 15 gallons or more per acre applied by ground equipment using solid or hollow cone nozzles directed into the plant whorl. Insects attacking silks and ears: Apply at 1-7 day intervals starting when silks first appear and continuing until harvest, if necessary. As insect populations increase and/or silk growth is rapid, the spray interval should be shorter. Use the higher dosage rate and apply at 1 to 2 day interval following periods of heavy moth flights. Follow local recommendations for proper timing. Apply in a minimum finished spray volume of 3 gallons per acre by air or 15 gallons per acre by ground. Foliage Feeders: Use at least 2 gallons finished spray volume per acre by air or 10 gallons per acre by ground.	After silk initiation, do not exceed 7.5 pounds of active ingredient (300 fluid ounces of this product) per acre per season. Do not allow livestock to graze treated field. Do not feed treated corn silage (green plant) or fodder to livestock. Ears may be harvested on day of last application.
	Cutworms			See "SPECIAL DIRECTIONS ON CUTWORMS."	

SPECIAL DIRECTIONS ON CUTWORMS

SITE	PEST	DOSAGE PER ACRE POUNDS ACTIVE	FLUID OUNCES	SPECIFIC DIRECTIONS	USE RESTRICTIONS
CROPS INDICATED ABOVE	Cutworms	0.5 to 0.75	20.0 to 30.0	Postemergence Rescue Treatments: For best results, apply as a broadcast spray when cutworms are actively feeding. If cutworms are feeding below the ground, control will be more difficult. Light cultivation with a rotary hoe before spraying may enhance control. If banding, use a minimum width of 10 inches over the row. When banding, determine the amount of chemical to use per acre, by dividing the band width by the row width and multiplying by the appropriate broadcast rate. Apply in a minimum finished spray volume of 3 gallons per acre by air or 15 gallons per acre by ground.	Refer to each specific crop for use restrictions.

FIELD USE CONVERSION TABLE

DOSAGE PER ACRE		ACRES TREATED PER VOLUME OF LARVIN® brand 3.2		
Pounds Active	Fluid Ounces	1 gallon	2.5 gallons	5 gallons
0.125	5.0	25.6	64.0	128.0
0.25	10.0	12.8	32.0	64.0
0.33	13.2	9.7	24.2	48.5
0.40	16.0	8.0	20.0	40.0
0.45	18.0	7.1	17.7	35.5
0.50	20.0	6.4	16.0	32.0
0.55	22.0	5.9	14.5	29.0
0.60	24.0	5.3	13.2	26.5
0.65	26.0	4.9	12.3	24.6
0.70	28.0	4.6	11.4	22.8

DOSAGE PER ACRE		ACRES TREATED PER VOLUME OF LARVIN® brand 3.2		
Pounds Active	Fluid Ounces	1 gallon	2.5 gallons	5 gallons
0.75	30.0	4.2	10.5	21.4
0.80	32.0	4.0	10.0	20.0
0.90	36.0	3.6	8.9	17.8
1.00	40.0	3.2	8.0	16.0

1 U.S. Gallon Equals 128 Fluid Ounces

STORAGE AND DISPOSAL

STORAGE: Do not store in or around the home. Store unused product in a cool, ventilated, dry, locked area. Do not allow prolonged storage in areas where temperatures frequently exceed 115°F (46°C). NEVER TRANSFER THIS PRODUCT TO ANOTHER CONTAINER FOR STORAGE.

If container is damaged, before clean-up, put on long-sleeved shirt, full-length trousers, head covering and protective gloves.

PESTICIDE DISPOSAL: Do not contaminate water, food or feed by storage or disposal. Contamination with this product will render water, food or feed unfit for human or animal consumption. Wastes resulting from the use of this product may be disposed of on site or at an approved waste disposal facility.

CONTAINER DISPOSAL: NEVER REUSE EMPTY CONTAINERS. Triple rinse or equivalent. Then offer for recycling or reconditioning or puncture and dispose of in a sanitary landfill or by incineration, or if allowed by state and local authorities, by burning. If container is burned, stay out of smoke.

ACCIDENTS: In case of a major spill of LARVIN® brand 3.2 Thiodicarb Insecticide/Ovicide: TELEPHONE (24 HOURS A DAY) IN THE U.S.A. 1-800-334-7577.

In case of minor spills or leaks follow all precautions indicated above and cleanup immediately. Soak up with sand, earth or other suitable material and dispose of wastes and broken or empty containers in a landfill approved for pesticide use.

LIMITED WARRANTY AND DISCLAIMER

The manufacturer warrants (a) that this product conforms to the chemical description on the label; (b) that this product is reasonably fit for the purposes set forth in the directions for use when it is used in accordance with such directions; and (c) that the directions, warnings and other statements on this label are based upon responsible experts' evaluation of reasonable tests of effectiveness, of toxicity to laboratory animals and to plants, and of residues on food crops, and upon reports of field experiences. Tests have not been made on all varieties or in all states or under all conditions. THE MANUFACTURER NEITHER MAKES NOR INTENDS, NOR DOES IT AUTHORIZE ANY AGENT OR REPRESENTATIVE TO MAKE, ANY OTHER WARRANTIES, EXPRESS OR IMPLIED, AND IT EXPRESSLY EXCLUDES AND DISCLAIMS ALL IMPLIED WARRANTIES OF MERCHANTABILITY OR FITNESS FOR A PARTICULAR PURPOSE.

THIS WARRANTY DOES NOT EXTEND TO, AND THE BUYER SHALL BE SOLELY RESPONSIBLE FOR, ANY AND ALL LOSS OR DAMAGE WHICH RESULTS FROM THE USE OF THIS PRODUCT IN ANY MANNER WHICH IS INCONSISTENT WITH THE LABEL DIRECTIONS, WARNINGS OR CAUTIONS.

BUYER'S EXCLUSIVE REMEDY AND MANUFACTURER'S OR SELLER'S EXCLUSIVE LIABILITY FOR ANY AND ALL CLAIMS, LOSSES, DAMAGES, OR INJURIES RESULTING FROM THE USE OR HANDLING OF THIS PRODUCT, WHETHER OR NOT BASED IN CONTRACT, NEGLIGENCE, STRICT LIABILITY IN TORT OR OTHERWISE, SHALL BE LIMITED, AT THE MANUFACTURER'S OPTION, TO REPLACEMENT OF, OR THE REPAYMENT OF THE PURCHASE PRICE FOR, THE QUANTITY OF PRODUCT WITH RESPECT TO WHICH DAMAGES ARE CLAIMED. IN NO EVENT SHALL MANUFACTURER OR SELLER BE LIABLE FOR SPECIAL, INDIRECT OR CONSEQUENTIAL DAMAGES RESULTING FROM THE USE OR HANDLING OF THIS PRODUCT.

NOTICE TO THE BUYER

Purchase of this material does not confer any rights under patents governing this product or the use thereof in countries outside of the United States.

LARVIN is a registered trademark of Rhône-Poulenc
US Patent No.: 4,382,957

AG 90047

EPA Approval 7/21/89

THIS SPECIMEN LABEL IS INTENDED FOR USE ONLY AS A GUIDE IN PROVIDING GENERAL INFORMATION REGARDING THE DIRECTIONS, WARNINGS AND CAUTIONS ASSOCIATED WITH THE USE OF THIS PRODUCT. AS WITH ANY AGRICULTURAL CHEMICAL, ALWAYS FOLLOW THE LABEL INSTRUCTIONS ON THE PACKAGE BEFORE USING.

Liquid AMIZINE® brand Herbicide

RESTRICTED USE PESTICIDE

FOR RETAIL SALE TO AND APPLICATION ONLY BY CERTIFIED APPLICATORS OR PERSONNEL UNDER THEIR DIRECT SUPERVISION

For Use On Substations, Storage Yards, Driveways, Parking Areas, Baseball Diamonds And Similar Areas.

ACTIVE INGREDIENTS:
Amitrole (3-amino-1,2,4-triazole) * 4.9%
Simazine (2-chloro-4,6-bis-ethylamino)-s-triazine) ** 9.9%
INERT INGREDIENTS 85.2%

* 0.44 pounds per gallon. ** 0.88 pounds per gallon

EPA Reg. No.: 264-196AA EPA Est. No.: 477-MD-01

KEEP OUT OF REACH OF CHILDREN
CAUTION

For PRODUCT USE Information Call 1-800-334-9745
For MEDICAL And TRANSPORTATION Emergencies ONLY Call 24 Hours A Day 1-800-334-7577

PRECAUTIONARY STATEMENTS

CAUTION

HAZARDS TO HUMANS AND DOMESTIC ANIMALS

Harmful if swallowed. Avoid contact with eyes, skin and clothing.

The use of this product may be hazardous to your health. This product contains amitrole which has been determined to cause cancer in laboratory animals. Wear a respirator during mixing and loading of this product.

Wear lightweight, waterproof clothing (jumpsuit or coverall and boots or shoes), gloves, and a wide-brimmed plastic or protective hat when applying this product to control dense, tall vegetation. Workers applying this product in other situations must wear lightweight waterproof gloves and boots (or shoes).

STATEMENT OF PRACTICAL TREATMENT

IF SWALLOWED: Induce vomiting immediately. Get medical attention.

IF ON SKIN: Wash thoroughly with soap and water. Remove contaminated clothing and wash before reuse.

IF IN EYES: Flush eyes with plenty of water.

NOTE TO PHYSICIAN

This product may affect thyroid function. Overexposure can be determined by monitoring thyroid function. If worker is under treatment for hypo- or hyperthyroidism, he should be removed from spray crew until the euthyroid state is achieved.

ENVIRONMENTAL HAZARDS

Avoid spray drift to susceptible plants such as cotton, beans, tomatoes and ornamentals. (Coarse sprays are less likely to drift.)

Do not use same spray equipment for other purposes. Do not contaminate water used for irrigation or domestic purposes. Do not use in a greenhouse.

Do not allow spray or spray drift to contaminate edible crops or water intended for irrigation, drinking or other domestic purposes.

Do not allow livestock to graze or feed in treated non-crop areas.

Do not apply directly to water or wetlands. Do not contaminate water by cleaning of equipment or disposal of wastes.

Simazine is a chemical which can travel (seep or leach) through soil and can contaminate groundwater which may be used as drinking water. Simazine has been found in groundwater as a result of agricultural use. Users are advised not to apply simazine where the water table (groundwater) is close to the surface and where the soils are very permeable, i.e., well-drained soils such as loamy sands. Your local agricultural agencies can provide further information on the type of soil in your area and the location of groundwater.

Do not use this product for purposes other than those recommended on this label.

DIRECTIONS FOR USE

IT IS A VIOLATION OF FEDERAL LAW TO USE THIS PRODUCT IN A MANNER INCONSISTENT WITH ITS LABELING.

Apply Liquid AMIZINE® Herbicide in spring or early summer before weeds are 3 to 4 inches tall. For best results on larger weeds, cut or mow them to ground line and remove. Treat 1 or 2 weeks later when regrowth appears. Use Liquid AMIZINE® early every spring to keep ground clear of broadleaf weeds and grasses.

For conventional high volume ground sprayers, mix 9 gallons of Liquid AMIZINE® Herbicide in a minimum of 50 gallons of water for each acre to be sprayed. Screens in nozzles as well as those in the suction hose should be no finer than 50 mesh. Use a Tee-Jet 8002 or larger fan type nozzle or its equivalent. For knap-sack sprayers, apply 1/5 gallon of Liquid AMIZINE® per 1,000 sq. ft. in sufficient water for good coverage: generally 2 to 3 gallons.

Liquid AMIZINE® Herbicide acts slowly but thoroughly. Treated weeds start to turn white or brown within a week and are usually dead in about three weeks. Small weeds growing actively are killed more rapidly than weeds in a dormant or dried condition.

In arid areas, Liquid AMIZINE® Herbicide should be watered into the ground if no rain falls within a week after treatment. Liquid AMIZINE® is non-flammable and does not stain metal, stone or concrete.

USING THE METER-MISER® SPRAY APPARATUS

The METER-MISER applicator is calibrated to accurately apply the spray mixture to the correct area, when pushed at such a speed that the spray pattern covers the entire width between the wheels. To avoid skips, walk in straight lines and over-lap wheel tracks about 3 inches.

Open valve after sprayer is in motion. Close valve before stopping sprayer.

Amounts to Use

Liquid AMIZINE®	Add Water to Make	Area Covered Square Feet
1 gal.	5 gal.	5000
4/5 gal.	4 gal.	4000
3/5 gal.	3 gal.	3000
2/5 gal.	2 gal.	2000
1/5 gal.	1 gal.	1000

MIXING SPRAYS (SHAKE CONTAINER BEFORE USING)

A. WHERE HOSE WATER UNDER PRESSURE IS AVAILABLE.
Half fill tank of METER-MISER applicator with water. Add 1 gallon of Liquid AMIZINE® Herbicide. Complete filling tank with water. Rock METER-MISER back and forth several times to assure thorough mixing and then spray.

B. WHERE HOSE WATER UNDER PRESSURE IS NOT AVAILABLE, OR LESS THAN A TANKFUL IS NEEDED.
Pour half the required water in a bucket. Mix in correct amount of Liquid AMIZINE® Herbicide. Add remaining water needed. Pour into METER-MISER tank. Rock sprayer back and forth several times to assure complete mixing and then spray.

C. DOUBLE SPRAY COVERAGE.
In areas where high volume sprays are desired, or to minimize the possibility of skipped streaks, use **double coverage.** Mix spray at half-strength. Cover the area twice, first from end to end, then from side to side.

IMPORTANT: Apply spray as soon as possible after mixing. Clean sprayer after use each day by flushing several times with clean water. Before using for weed control in turf, spray out a tankful of detergent solution to thoroughly clean tank and sprayer head. Also, clean wheels thoroughly.

Some of the Weeds Controlled Are

Blue grass	Pigweed
Canada thistle	Quackgrass
Chrysanthemum weed (mugwort)	Ragweed
Crabgrass	Smartweed
Dandelion	Snowthistle
Foxtail	Plantain
Kochia	Purslane
Lambsquarters	Velvetgrass
Nightshade	Wild oats
	Witchgrass

STORAGE AND DISPOSAL

STORAGE

Do not store near fertilizers, seeds, insecticides or fungicides.

KEEP FROM FREEZING.

PESTICIDE DISPOSAL

Do not contaminate water, food, or feed by storage or disposal. Pesticide wastes are toxic. Improper disposal of excess pesticide, spray mixture, or rinsate is a violation of Federal law. If these wastes cannot be disposed of by use according to label instructions, contact your State Pesticide or Environmental Control Agency or the Hazardous Waste representative at the nearest EPA Regional Office for guidance.

CONTAINER DISPOSAL

Triple rinse or equivalent. Then offer for recycling or reconditioning, or puncture and dispose of in a sanitary landfill, or by incineration, or if allowed by state and local authorities, by burning. If burned, stay out of smoke.

LIMITED WARRANTY AND DISCLAIMER

The manufacturer warrants (a) that this product conforms to the chemical description on the label; (b) that this product is reasonably fit for the purposes set forth in the directions for use when it is used in accordance with such directions; and (c) that the directions, warnings and other statements on this label are based upon responsible experts' evaluation of reasonable tests of effectiveness, of toxicity to laboratory animals and to plants, and upon reports of field experience. Tests have not been made on all varieties or in all states or under all conditions. THE MANUFACTURER NEITHER MAKES NOR INTENDS, NOR DOES IT AUTHORIZE ANY AGENT OR REPRESENTATIVE TO MAKE, ANY OTHER WARRANTIES, EXPRESS OR IMPLIED, AND IT EXPRESSLY EXCLUDES AND DISCLAIMS ALL IMPLIED WARRANTIES OF MERCHANTABILITY OR FITNESS FOR A PARTICULAR PURPOSE.

THIS WARRANTY DOES NOT EXTEND TO, AND THE BUYER SHALL BE SOLELY RESPONSIBLE FOR, ANY AND ALL LOSS OR DAMAGE WHICH RESULTS FROM THE USE OF THIS PRODUCT IN ANY MANNER WHICH IS INCONSISTENT WITH THE LABEL DIRECTIONS, WARNINGS, OR CAUTIONS.

BUYER'S EXCLUSIVE REMEDY AND MANUFACTURER'S OR SELLER'S EXCLUSIVE LIABILITY FOR ANY AND ALL CLAIMS, LOSSES, DAMAGES OR INJURIES RESULTING FROM THE USE OR HANDLING OF THIS PRODUCT, WHETHER OR NOT BASED IN CONTRACT, NEGLIGENCE, STRICT LIABILITY IN TORT OR OTHERWISE SHALL BE LIMITED, AT THE MANUFACTURER'S OPTION, TO REPLACEMENT OF, OR THE REPAYMENT OF THE PURCHASE PRICE FOR, THE QUANTITY OF PRODUCT WITH RESPECT TO WHICH DAMAGES ARE CLAIMED. IN NO EVENT SHALL MANUFACTURER OR SELLER BE LIABLE FOR SPECIAL, INDIRECT OR CONSEQUENTIAL DAMAGES RESULTING FROM THE USE OR HANDLING OF THIS PRODUCT.

NOTICE TO BUYER

Purchase of this material does not confer any rights under patents governing this product or the use thereof in countries outside of the United States.

AMIZINE & METER-MISER are registered trademarks of Rhône-Poulenc

Form No. AG 90004
Printed USA - 9/89

THIS SPECIMEN LABEL IS INTENDED FOR USE ONLY AS A GUIDE IN PROVIDING GENERAL INFORMATION REGARDING THE DIRECTIONS, WARNINGS AND CAUTIONS ASSOCIATED WITH THE USE OF THIS PRODUCT. AS WITH ANY AGRICULTURAL CHEMICAL, ALWAYS FOLLOW THE LABEL INSTRUCTIONS ON THE PACKAGE BEFORE USING.

MOCAP® 10% GRANULAR

NEMATICIDE INSECTICIDE

NOT FOR DOMESTIC USE

ACTIVE INGREDIENTS:
Ethoprop: (O-Ethyl S, S-Diproply Phosphorodithioate). 10%
INERT INGREDIENTS: 90%
TOTAL. 100%

EPA Reg. No. 264-465 EPA Est. No.

KEEP OUT OF REACH OF CHILDREN

WARNING AVISO

Precaucion al usuario: Si usted no lee ingles, no use este producto hasta que le etiqueta haya sido explicado ampliamente.

Trnaslation to the user: If you cannot read English, do not use this product until the label has been fully explained to you.

For PRODUCT USE Information Call 1-800-334-9745
For Medical and Transportation Emergencies ONLY CALL 24 Hours A Day 1-800-334-7577

PRECAUTIONARY STATEMENTS

WARNING

HAZARDS TO HUMANS AND DOMESTIC ANIMALS

May be fatal if swallowed, inhaled or absorbed through skin. Rapidly absorbed through skin. Wear protective clothing including rubber gloves. In case of contact, wash immediately with soap and water. Do not breathe fumes or dust. Wear a mask or pesticide respirator jointly approved by the Mining Enforcement & Safety Administration and the National Institute for Occupational Safety and Health. After use, wash hands, arms and face thoroughly with soap and warm water before eating or smoking. Wash all clothing with soap and hot water before reuse. Discard contaminated shoes. Do not contaminate feed or food.

Do not store in or around the home.

SYMPTOMS OF POISONING: Nausea, vomiting, abdominal cramps, diarrhea, excessive salivation, headache, dizziness, weakness, blurring or dimness of vision, excessive tearing, loss of muscular coordination, slurring of speech, twitching of muscles (especially of tongue and eyelids), mental confusion, disorientation, drowsiness, difficulty in breathing (chest tightness), runny nose.

STATEMENT OF PRACTICAL TREATMENT

IF SWALLOWED: Give 1 or 2 glasses of water and induce vomiting by touching back of throat with finger. Do not induce vomiting or give anything by mouth to an unconscious person. Call a physician, Poison Control Center, or hospital emergency room.

If not breathing, give mouth to mouth artificial respiration.

IN CASE OF SKIN CONTACT: Immediately remove all contaminated clothing, and wash skin with soap and water. Get medical attention. Wash clothing before reuse. Discard contaminated shoes.

FOR EYES: Flush with plenty of water for at least 15 minutes. Call a physician immediately in all cases of suspected poisoning.

IF INHALED: Remove victim to fresh air. If not breathing, give artificial respiration. Get medical attention. Persons providing first aid should prevent contamination of themselves.

NOTE TO PHYSICIAN

This product is a cholinesterase inhibitor. Severe symptoms and signs include diarrhea, pinpoint and non-reactive pupils, respiratory difficulty, pulmonary edema, cyanosis, loss of sphincter control, convulsions, and coma. Support respiration as needed. Measure should include removal of secretions, maintenance of a patent airway, and if necessary, artificial ventilation.

If cyanosis is absent, give ATROPINE 2-4 mg intravenously (0.05 mg/kg for children). Repeat atropine at 5-10 minute intervals until atropinization occurs (dry, flushed skin, tachycardia, pupillary dilatation), and maintain for 48 hours. If cyanotic, give initial atropine intramuscularly and start measures to improve ventilation.

Start 2-PAM (PROTOPAM, Ayerst) at the same time. Give 1-2 grams PROTOPAM (20-40 mg/kg for children) in 100cc saline over 15-30 minutes. If pulmonary edema is present, give intravenously slowly as a 5% solution in water over a period of at least 5 minutes. A second dose may given after one hour if muscle weakness presists. Additional doses may be given cautiously for persistent muscle weakness. May be given by intramuscular or subcutaneous routes if intravenous administration is not feasible.

In case of skin contact, wash patient with soap and water followed by wash with 95% ethyl alcohol. Keep patient under constant observation for 24-36 hours. Symptoms may persist for one month.

The use of theophylline, morphine, barbiturates, phenothiazines, reserpine, and succinyl choline is contraindicated.

ENVIRONMENTAL HAZARDS

This pesticide is toxic to aquatic organisms (fish and invertebrates), and wildlife. Birds feeding in treated areas may be killed. Do not apply directly to water or wetlands (swamps, bogs, marshes and potholes). Runoff from treated areas may be hazardous to aquatic organisms in neighboring areas. Cover or incorporate granules that are spilled during loading or are visible on soil surface in turn areas. Do not contaminate water when disposing of equipment wash water.

DIRECTIONS FOR USE

IT IS A VIOLATION OF FEDERAL LAW TO USE THIS PRODUCT IN A MANNER INCONSISTENT WITH ITS LABELING.

Unprotected persons may enter treated field after chemical has been mixed into the soil. Granules lying on the soil surface in turn areas at row ends should be incorporated to remove possible hazard to birds and other wildlife.

Do not apply in Long Island, New York.

INCORPORATION INFORMATION

MOCAP® 10% Granular Nematicide Insecticide can be incorporated with a rotary tiller, rotary hoe, springtooth harrow, by double-discing or with other equipment for mixing the chemical with the soil to the depth recommended. Follow any specific incorporating directions given under the crop heading.

If moisture is applied immediately after application to carry the treatment into the soil, only shallow incorporation (1/4 to 1/2 inch) with mechanical equipment such as drag chains is required.

NEMATODES CONTROLLED/SUPPRESSED

	Bananas & Plaintain	Cabbage	Field corn & Sweet corn	Cucumbers	Peanuts (planting)	Peanuts (pegging)	White (Irish) Potato	Sweet Potato	Snap beans & Lima beans	Soybeans	Sugarcane	Tobacco
BURROWING NEMATODES												
(*Radopholus similis*)	C											
DAGGER NEMATODES												
(*Xiphinema* spp.)			C							S		
CYST NEMATODES												
(*Heterodera glycines*)											S	
(*Heterodera cruciferae*)		C										
(*Heterodera schactii*)		C										
LESION NEMATODES												
(*Pratylenchus* spp.)	C	C	C	C	C	C	C		C	C	C	C
ROOT-KNOT NEMATODES												
(*Meloidogyne spp.*)	C									C	C	
(*Meloidogyne hapla*)		C	C	C	C	C	C	C	C			C
(*Meloidogyne incognita*)		C	C	C			C	C	C			C
(*Meloidogyne arenaria*)					C	C						
(*Meloidogyne javanica*)				C								
(*Meloidogyne chitwoodi*)							S					
LANCE NEMATODES												
(*Hoplolaimus spp.*)		S	C	S					S			
(*Hoplolaimus tylenchiformis*)				C								
(*Hoplolaimus columbus*)										S		
SPIRAL NEMATODES												
(*Helicotylenchus* spp.)	C		C	C			C	C	C	C		
(*Rotylenchus* spp.)			C									
STING NEMATODES												
(*Belonolaimus* spp.)			C		C	C						
(*Belonolaimus longicaudatus*)		C		C			C		C	C		
RING NEMATODES												
(*Criconemella* spp.)				C	C	C	C			C	C	
(*Criconemella ornata*)			C						C			
RENIFORM NEMATODES												

	Bananas & Plaintain	Cabbage	Field corn & Sweet corn	Cucumbers	Peanuts (planting)	Peanuts (pegging)	White (Irish) Potato	Sweet Potato	Snap beans & Lima beans	Soybeans	Sugarcane	Tobacco
(*Rotylenchulus reniformis*)								C				
STUNT NEMATODES												
(*Tylenchorhynchus spp.*)			C	C	C	C	C	C		C		C
(*Tylenchorhynchus claytoni*)									C			
(*Tylenchorhynchus martini*)											C	
STUBBY-ROOT NEMATODES												
(*Paratrichodorus spp.*)			C		C	C	C	C				
(*Trichodorus christiei*)		C		C						C	C	

C———Control
S———Suppression

INSECTS CONTROLLED/SUPPRESSED

	Bananas & Plaintain	Field corn & Sweet corn	Peanuts (pegging)	White (Irish) Potato	Sweet Potato	Sugarcane	Tobacco
CORN ROOTWORM							
(*Diabrotica virgifera virgifera*)		C					
(*Diabrotica longicornis barberi*)		C					
(*Diabrotica undecimpuctata howardi*)		C	C				
WIREWORMS							
(*Conoderus spp.*)							C
(*Conoderus vespertinus*)		C			C		
(*Conoderus falli*)		C					
(*Melanotus spp.*)		C			C		
(*Malanotus communis*)				C		C	
(*Limonius dubitan*)		C					
(*Limonius canus*)				C			
(*Limonius californicus*)				C			
(*Aeolus mellius*)		C					
(*Ctenicera spp.*)		C					
(*Ctenicera pruinina*)		C		C			
(*Eleodes spp.*)				C			
BLACK CUTWORM							
(*Agrotis ipsilon*)		C*					
SANDHILL CUTWORM							
(*Euxoa detersa*)		C*					
WHITE GRUBS							
(*Phyllophaga crinita*)		S					
(*Phyllophaga ephilida*)		S			C		
BANDED CUCUMBER BEETLE							
(*Diabrotica balteata*)					C		
FLEA BEETLE LARVAE							
(*Chaetocnema confinis*)					C		
(*Systena blanda*)					C		
LESSER CORNSTALK BORER							
(*Elasmopalpus lignosellus*)			S				
BANANA ROOT BORER							
(*Cosmopolites sordidus*)	C						

C———Controlled
S———Suppression

*** NOTE: Under heavy infestations of cutworms, enough cutworms may survive to cause some crop damage. If this occurs, it may be necessary to apply a rescue treatment with another registered pesticide to control the surviving worm.**

BANANA AND PLANTAIN

FOR NEMATODE AND BANANA ROOT BORER (*Cosmopolites sordidus*) CONTROL

Apply 60 grams of MOCAP® 10% Granular Nematicide–Insecticide in a radius of 3/4 meter around each producing stem or 2 ounces of MOCAP® 10% Granular in a radius of 30 inches. Before application, remove ground litter from area to be treated, and apply granules evenly. For best results, apply just before irrigation or a rainy period. If application is made during a dry period, mix granules into the top inch of soil with a rake. Repeat application in 6 months.

BEAN (SNAP AND LIMA)

FOR NEMATODE AND GARDEN SYMPHYLAN (*Scutigerella immaculata*) CONTROL

Row Treatment: Apply 20 to 30 pounds of MOCAP® 10% Granular Nematicide–Insecticide per acre (36 inch row spacing) or 1.4 to 2 pounds per 1,000 linear feet of row in band 12 to 15 inches wide on the row at-planting. Mix with the top 2 to 4 inches of soil. Use the higher rates in fields where nematode infestations have been especially severe. **Do not** use as a seed furrow treatment or allow granules to contact the seed.

FOR NEMATODE AND GARDEN SYMPHYLAN (*Scutigerella immaculata*) CONTROL

Broadcast Treatment: Apply 60 to 80 pounds per acre from 3 days before planting to at-planting time. Mix with the top 2 to 4 inches of soil right after application. Use the higher rates in fields where nematode infestations have been especially severe.

NOTE: Beans grown in treated soil may remain green longer than other beans and mature 1 to 2 weeks later.

CABBAGE

FOR NEMATODE AND GARDEN SYMPHYLAN (*Scutigerella immaculata*) CONTROL

Row Treatment: Apply 20 pounds of MOCAP® 10% Granular Nematicide Insecticide per acre (36 inch row spacing) or 1.4 pounds per 1,000 linear feet of row in a band 15 inches wide on the row at-planting. **Do not** use as a seed furrow treatment or allow granules to contact the seed.

FOR NEMATODE AND GARDEN SYMPHYLAN (*Scutigerella immaculata*) CONTROL

Broadcast Treatment: Apply 50 pounds of MOCAP® 10% Granular Nematicide–Insecticide per acre from 1 week before planting to at-planting time. Mix with the top 2 to 4 inches of soil right after application.

CORN (FIELD AND SWEET)

FOR CONTROL OF CORN ROOTWORMS, WIREWORMS, CUTWORMS, GARDEN SYMPHYLAN (*Scutigerella immaculata*), NEMATODES AND SUPPRESSION OF WHITE GRUBS:

Row Treatment: Use 12 ounces of MOCAP® 10% Granular Nematicide–Insecticide per 1,000 feet of row. Rate per acre depends on row spacing as shown below:

Row Spacing	40 Inches	38 Inches	36 Inches	30 Inches	20 Inches
Lbs. Per Acre	10	10.5	11.1	13.3	20

Apply the granules in a 6 to 7 inch band on the row over a closed seed furrow at-planting. Mix the granules with the top 1/2 inch of soil with drag chains, spring-tooth incorporators, or similar equipment.

FOR COMMONLY USED APPLICATORS, SEE CALIBRATIONS CHART BELOW

Suggested gauge settings for 12 ounces per 1,000 feet of row. (Do not change with row width). Be sure to calibrate your applicator under your operating conditions.

FOR APPLICATION RATE OF 12 OUNCES/1000 FEET OF ROW
SPEED IN MILES PER HOUR

BRAND OF APPLICATOR	2	3	4	5	6	7	8	9	10
Allis Chalmers	At all Speeds: GA9 = 11.35 Ounces . . . GA10 = 12.7 Ounces								
Gandy	19.1	23.5	26.7	25.9	32.1	34.5	36.6	38.7	40.8
International Harvester*	1/8.0	2/2.7	2/8.5	3/4.0	3/8.7	4/5.0	5/1.3	5/6.0	6/1.5
International Harvester (Cyclo)	1/5.7	1/9.3	2/4.7	3/0.5	3/5.1	4/0.6	4/5.8	5/1.5	5/7.0
John Deere**	1/29	2/6	2/13	2/19	2/27	3/1	3/6	3/11	3/15
John Deere (Max. Emerge.)	13	19	24	29	32	33	35	36	37
Noble	13	19	24	28	31	34	36	38	39
Noble (new)	9	11	14	17	20	22	25	27	29

* 1/2.0 means Gate 1. Dial 8.0

** 1/29 means Range 1. Notch 29

FOR CORN ROOTWORM CONTROL

Row Treatment: Use at the rates shown above for at-planting application. Apply at cultivation after plant emergence until layby.

On small corn, apply in a 6 to 7 inch band over the row. On taller corn, apply in bands 3 to 4 inches wide on each side of the corn, directing the granules toward the base of the plants. Cover the granules with soil by making application immediately ahead of cultivation equipment.

FOR CUTWORM CONTROL

Broadcast Treatment: Apply 30 pounds of MOCAP® 10% Granular Nematicide–Insecticide per acre. Apply 3 days before planting to at-planting time. Mix with the top 2 inches of soil right after application.

FOR NEMATODE CONTROL

Row Treatment: Apply 1 1/8 to 1 1/2 pound of MOCAP® 10% Granular Nematicide–Insecticide per 1000 feet of row in a band 12 to 15 inches wide on the row at-planting. Mix treatment with the top 2 to 4 inches of soil right after application. Rate per acre depends on row spacing as shown below:

Row Spacing	40 Inches	38 Inches	36 Inches	30 Inches	20 Inches
Lbs. Per Acre	15-20	15.5-21	16.3-22.2	19.6-26.6	30-40

FOR NEMATODE CONTROL

Broadcast Treatment: Apply 60 pounds of MOCAP® 10% Granular Nematicide–Insecticide per acre from 3 days before planting to at-planting time. Mix with the top 2 to 4 inches of soil right after application.

The treatments for nematode control will also control rootworms, wireworms, cutworms, garden symphylan (*Scutigerella immaculata*) and suppress white grubs.

CUCUMBER

FOR NEMATODE CONTROL

Row Treatment: Apply 20 pounds of MOCAP® 10% Granular Nematicide–Insecticide per acre (7-foot row spacing) or 3.2 pounds per 1000 linear feet of row in a band 12 to 15 inches wide on the row at or just before planting. Prepare bed and knock off top to provide a level area for treatment. Mix with the top 2 to 4 inches of soil right after application. **Do not** use as a seed furrow treatment or allow granules to contact the seed.

PEANUT

FOR NEMATODE CONTROL

Row Treatment: Use 20 to 40 pounds of MOCAP® 10% Granular Nematicide–Insecticide per acre (36-inch row spacing) or 1.4 to 2.8 pounds per 1000 linear feet of row. Apply in a band 12 inches wide on the row at planting and mix with the top 2 to 4 inches of soil. Use the higher rates of MOCAP® 10% Granular for control of peanut root-knot nematode (*M. arenaria*), and in fields where nematode infestations have been especially severe. **Do not** use as a seed furrow treatment or allow granules to contact the seed.

Broadcast Treatment: Apply 40 to 60 pounds of MOCAP® 10% Granular Nematicide Insecticide per acre. Make application from 1 week before planting to at-planting time. Mix with the top 2 to 4 inches of soil right after application. Use the higher rates of MOCAP® 10% Granular for control of peanut root-knot nematode (*M. arenaria*), and in fields where nematode infestations have been especially severe.

FOR NEMATODE AND CORN ROOTWORM CONTROL AND SUPPRESSION OF LESSER CORNSTALK BORER AND WHITE MOLD (*Sclerotium rolfsii*)

Row Treatment: Apply 30 pounds of MOCAP® 10% Granular Nematicide–Insecticide per acre (36 inch row spacing) or 2.1 pounds per 1000 linear feet of row in a band 15 to 18 inches wide on the row at pegging.

FOR CORN ROOTWORM CONTROL

Row Treatment: Apply 25 pounds of MOCAP® 10% Granular Nematicide–Insecticide per acre (36 inch row spacing) or 1.7 pounds per 1000 linear feet of row in a band 18 inches wide on the row at pegging.

NOTE: Pegging treatment should be incorporated lightly into the soil.

POTATOES

FOR CONTROL OF WIREWORMS AND GARDEN SYMPHYLAN (*Scutigerella immaculata*)

Row Treatment: Apply 30 pounds of MOCAP® 10% Granular Nematicide–Insecticide per acre (36 inch row spacing) or 2.1 pounds per 1000 linear feet of row in a band 12 inches wide on the row at-planting. Mix with the top 2 to 4 inches of soil.

The following broadcast treatment is recommended for fields where wireworm injury has been (or is suspected to be) especially severe (i.e., potatoes planted following sod, pasture or new ground).

FOR CONTROL OF WIREWORMS AND GARDEN SYMPHYLAN (*Scutigerella immaculata*)

Broadcast Treatment: Apply 40 to 60 pounds of MOCAP® 10% Granular Nematicide Insecticide anytime from 2 weeks before planting to at-planting time. Mix with the top 2 to 4 inches of soil right after application.

FOR CONTROL OF GARDEN SYMPHYLAN (*Scutigerella immaculata*) AND LESION, SPIRAL, STING, RING, STUNT, AND STUBBY-ROOT NEMATODES AND SUPPRESSION OF NORTHERN ROOT-KNOT AND SOUTHERN ROOT-KNOT NEMATODES

Row Treatment: Apply 30 pounds of MOCAP® 10% Granular Nematicide–Insecticide per acre (36 inch row spacing) or 2.1 pounds per 1000 linear feet of row in a band 12 inches wide on the row at-planting. Mix with the top 2 to 4 inches of soil.

FOR CONTROL OF GARDEN SYMPHYLAN (*Scutigerella immaculata*) AND NORTHERN ROOT-KNOT, SOUTHERN ROOT-KNOT, LESION, SPIRAL, STING, RING, STUNT, NEMATODES, AND SUPPRESSION OF COLUMBIA ROOT-KNOT NEMATODE

Broadcast Treatment: Apply 60 to 120 pounds of MOCAP® 10% Granular Nematicide–Insecticide per acre. Apply by ground or aerial application as close to planting as possible, but no earlier than 2 weeks before planting, or apply before potato emergence. Mix with the top 2 to 4 inches of soil right after application if applied before planting. If applied after planting, apply 1/4 to 1/2 inch of water to incorporate in soil. Uniformity of application is important for best results.

NOTE: For best suppression of Columbia Root-knot Nematodes, it is recommended that MOCAP® 10% Granular be applied following fall or spring application of a registered fumigant nematicide. **Do not** exceed a maximum of 12 lbs. active ingredient per acre per season on potatoes.

SOYBEANS

FOR NEMATODE CONTROL

Row Treatment: Apply 15 to 30 pounds of MOCAP® 10% Granular Nematicide–Insecticide per acre (36 inch row spacing) or 1.0 to 2.1 pounds per 1,000 linear feet in a band 12 to 15 inches wide on the row from 1 week before planting to at-planting time. Mix with the top 2 to 4 inches of soil right after application. User the higher rates in fields where nematode infestations have been especially severe. **Do not** use as a seed furrow treatment or allow spray to contact the seed.

NOTE: Soybeans treated with organophosphate chemicals such as MOCAP® may be more susceptible to injury from Metribuzin and Propanil herbicides.

SUGARCANE

FOR NEMATODE AND WIREWORM CONTROL

Furrow Treatment: Apply 20 to 40 pounds of MOCAP® 10% Granular Nematicide Insecticide per acre (6 foot row spacing) or 2¾ to 5½ pounds per 1,000 linear feet of row at-planting. Apply evenly in a band 12 to 15 inches wide, centered over the seed pieces in the opened furrow immediately before closing. Mix with the soil with a disc hiller or other suitable equipment as the cane is covered. Use the higher rates in fields where nematode and/or wireworm infestations have been especially severe.

SWEET POTATO

FOR NEMATODE, WIREWORM, WHITE GRUB, BANDED CUCUMBER BEETLE AND FLEA BEETLE LARVAE (Palestriped and Sweet Potato Flea Beetle) CONTROL

Row Treatment: Use 30 to 46 pounds of MOCAP® 10% Granular Nematicide–Insecticide per acre or 2.4 to 3.2 pounds per 1,000 linear feet of row (minimum 42 inch row spacing). Apply in a band 12 to 15 inches wide on the row 2 to 3 weeks before planting. Mix with the top 2 to 4 inches of soil right after application. Use the higher rates in fields where nematode and/or insect infestations have been especially severe.

Good results have been obtained by opening the row, applying MOCAP® 10% Granular in a 12 to 15 inch band on the row, bedding over and dragging.

FOR NEMATODE, WIREWORM, WHITE GRUB, BANDED CUCUMBER BEETLE AND FLEA BEETLE LARVAE (Palestriped and Sweet Potato Flea Beetle) CONTROL

Broadcast Treatment: Apply 60 to 80 pounds of MOCAP® 10% Granular Nematicide–Insecticide per acre 2 to 3 weeks before planting. Mix with the top 2 to 4 inches of soil right after application. Use double gang disc harrow or other equipment which will do a good job of mixing the MOCAP® 10% Granular with the soil. Use the higher rates in fields where nematode and/or insect infestations have been especially severe.

TOBACCO

FOR NEMATODE AND WIREWORM CONTROL

Row Treatment: Use 60 to 120 pounds of MOCAP® 10% Granular Nematicide–Insecticide per acre (42 inch row spacing) or 4.8 to 9.6 pounds per 1,000 linear feet of row. Apply after fertilizer has been listed or bedded in a band 18 to 24 inches wide on the row from 1 week before planting to at-planting time. Mix with the top 2 to 4 inches of soil right after application.

Use the higher rates in fields in where nematode injury to tobacco has been especially severe, or where soil is sandy.

FOR NEMATODE AND WIREWORM CONTROL

Broadcast Treatment: Apply 60 to 120 pounds of MOCAP® 10% Granular Nematicide–Insecticide per acre from 1 week before planting to at-planting time. Mix with the top 2 to 4 inches of soil right after application. Use a double gang disc harrow or other cultivating equipment that will do a good job of mixing MOCAP® 10% Granular with the soil. After mixing, apply fertilizer and, for best results, shape a wide bed with a tractor equipped with 4 disc hillers and wide sweeps for running out the middles. Be sure that only the treated soil (top 2 to 4 inches) is used to make beds.

Use the higher rates in fields where nematode injury to tobacco has been especially severe, or where the soil is sandy.

FOR WIREWORM CONTROL

Broadcast Treatment: Apply 20 pounds of MOCAP® 10% Granular Nematicide–Insecticide per acre from 2 weeks before planting to at-planting time. Mix with the top 2 to 4 inches of soil right after application. Use a disc harrow or other cultivating equipment that will do a good job of mixing MOCAP® 10% Granular with the soil.

Rhône-Poulenc does not recommend the use of MOCAP on tobacco in Florida.

STORAGE AND DISPOSAL

STORAGE

Do not store in or around the home. Do not contaminate water, food, or feed by storage or disposal.

PESTICIDE DISPOSAL

Pesticide wastes are acutely hazardous. Improper disposal of excess pesticide, granules mixture, or rinsate in a violation of Federal law. If these wastes cannot be disposed of by use according to label instructions, contact your State Pesticide or Environmental Control Agency, or the Hazardous Waste representative at the nearest EPA Regional Office for guidance.

CONTAINER DISPOSAL

Completely empty bag into application equipment. Then dispose of empty bag in a sanitary landfill or by incineration, or if allowed by State and local authorities, by burning. If burned, stay out of smoke.

LIMITED WARRANTY AND DISCLAIMER

The manufacturer warrants (a) that this product conforms to the chemical description on the label; (b) that this product is reasonably fit for the purposes set forth in the directions for use when it is used in accordance with such directions; and (c) that the directions, warnings, and other statements on this label are based upon responsible experts' evaluation of reasonable tests of effectiveness, of toxicity to laboratory animals and to plants, and residues on food crops, and upon reports of field experience. Tests have not been made on all varieties or in all states or under all conditions. THE MANUFACTURER NEITHER MAKES NOR INTENDS, NOR DOES IT AUTHORIZE ANY AGENT OR REPRESENTATIVE TO MAKE, ANY OTHER WARRANTIES, EXPRESSED OR IMPLIED, AND IT EXPRESSLY EXCLUDES AND DISCLAIMS ALL IMPLIED WARRANTIES OF MERCHANTABILITY OF FITNESS FOR A PARTICULAR PURPOSE.

THIS WARRANTY DOES NOT EXTEND TO, AND THE BUYER SHALL BE SOLELY RESPONSIBLE FOR, ANY AND ALL LOSS OR DAMAGE WHICH RESULTS FROM THE USE OF THIS PRODUCT IN ANY MANNER WHICH IS INCONSISTENT WITH THE LABEL DIRECTIONS, WARNINGS, OR CAUTIONS.

BUYER'S EXCLUSIVE REMEDY AND MANUFACTURER'S OR SELLER'S EXCLUSIVE LIABILITY FOR ANY AND ALL CLAIMS, LOSSES, DAMAGES, OR INJURIES RESULTING FROM THE USE OR HANDLING OF THIS PRODUCT, WHETHER OR NOT BASED IN CONTRACT, NEGLIGENCE, STRICT LIABILITY IN TORT OR OTHERWISE, SHALL BE LIMITED, AT THE MANUFACTURER'S OPTION, TO REPLACEMENT OF, OR THE REPAYMENT OF THE PURCHASE PRICE FOR, THE QUANTITY OF PRODUCT WITH RESPECT TO WHICH DAMAGES ARE CLAIMED. IN NO EVENT SHALL MANUFACTURER OR SELLER BE LIABLE FOR SPECIAL, INDIRECT OR CONSEQUENTIAL DAMAGES RESULTING FROM THE USE OR HANDLING OF THIS PRODUCT.

NOTICE TO BUYER

Purchase of this material does not confer any rights under patents governing this product or the use thereof in countries outside the United States.

MOCAP is a registered trademark of Rhône-Poulenc.

AG 90048 EPA Approval: 6/1/89

THIS SPECIMEN LABEL IS INTENDED FOR USE ONLY AS A GUIDE IN PROVIDING GENERAL INFORMATION REGARDING THE DIRECTIONS, WARNINGS AND CAUTIONS ASSOCIATED WITH THE USE OF THIS PRODUCT. AS WITH ANY AGRICULTURAL CHEMICAL, ALWAYS FOLLOW THE LABEL INSTRUCTIONS ON THE PACKAGE BEFORE USING.

MOCAP® 20% GRANULAR

NEMATICIDE-INSECTICIDE

RESTRICTED USE PESTICIDE

Acute Dermal Toxicity.

For retail sale to and use only by Certified Applicators or persons under the direct supervision of a Certified Applicator, and only for those uses covered by the Certified Applicator's certification.

ACTIVE INGREDIENTS:
Ethoprop: (O-Ethyl S, S-Dipropyl Phosphorodithioate) 20%
INERT INGREDIENTS: 80%
TOTAL: 100%
EPA Reg. No. 264-469 EPA Est. No.

KEEP OUT OF REACH OF CHILDREN

DANGER–POISON

PELIGRO

PRECAUCION AL USUARIO: Si usted no lee ingles, no use este producto hasta que la etiqueta haya sido explicada ampliamente.

TRANSLATION TO THE USER: If you cannot read English, do not use this product until the label has been fully explained to you.

STATEMENT OF PRACTICAL TREATMENT

IF SWALLOWED: Call a physician or Poison Control Center. Drink 1 to 2 glasses of water and induce vomiting by touching back of throat with finger. Do not induce vomiting or give anything to an unconscious person.

IF INHALED: Remove victim to fresh air and apply artificial respiration if indicated.

IF ON SKIN: Remove all contaminated clothing and wash skin with soap and water. Wash clothing before reuse. Discard contaminated shoes.

IF IN EYES: Flush with plenty of water for at least 15 minutes. Call a physician immediately.

For PRODUCT USE Information Call 1-800-334-9745
For MEDICAL And TRANSPORTATION Emergencies ONLY Call 24 Hours A Day 1-800-334-7577

PRECAUTIONARY STATEMENTS

DANGER

HAZARD TO HUMANS AND DOMESTIC ANIMALS

Poisonous if swallowed, inhaled or absorbed through skin. Rapidly absorbed through skin. Wear protective clothing. In case of contact, wash immediately with soap and water. Do not breathe fumes or dust. After use, wash hands, arms and face thoroughly with soap and warm water before eating or smoking. Wash all clothing with soap and hot water before reuse. Discard contaminated shoes. Do not contaminate feed or food.

Do not store in or around the home.

SYMPTOMS OF POISONING: Nausea, vomiting, abdominal cramps, diarrhea, excessive salivation, headache, dizziness, weakness, blurring or dimness of vision, excessive tearing, loss of muscular coordination, slurring of speech, twitching of muscles (especially of tongue and eyelids), mental confusion, disorientation, drowsiness, difficulty in breathing (chest tightness), runny nose.

STATEMENT OF PRACTICAL TREATMENT

IF SWALLOWED: Give 1 or 2 glasses of water to drink and induce vomiting by touching back of throat with finger. Do not induce vomiting or give anything by mouth to an unconscious person. Call a physician, Poison Control Center, or hospital emergency room. If not breathing, give mouth to mouth artificial respiration.

IN CASE OF SKIN CONTACT: Immediately remove all contaminated clothing, and wash skin with soap and water. Wash clothing before reuse. Discard contaminated shoes.

FOR EYES: Flush with plenty of water for at least 15 minutes. Call a physician immediately in all cases of suspected poisoning.

Persons providing first aid should prevent contamination of themselves.

NOTE TO PHYSICIAN: This product is a cholinesterase inhibitor. Severe symptoms and signs include diarrhea, pinpoint and non-reactive pupils, respiratory difficulty, pulmonary edema, cyanosis, loss of sphincter control, convulsions, and coma. Support respiration as needed. Measures should include removal of secretions, maintenance of a patent airway, and if necessary, artificial ventilation.

If cyanosis is absent, give ATROPINE 2-4 mg intravenously (0.05 mg/kg for children). Repeat atropine at 5-10 minute intervals until atropinization occurs (dry, flushed skin, tachycardia, pupillary dilation), and maintain for 48 hours. If cyanotic, give initial atropine intramuscularly and start measures to improve ventilation.

Start 2-PAM (PROTOPAM, Ayerst) at the same time. Give 1-2 grams PROTOPAM (20-40 mg/kg for children) in 100cc saline over 15-30 minutes. If pulmonary edema is present, give intravenously slowly as a 5% solution in water over a period of at least 5 minutes. A second dose may be given after one hour if muscle weakness persists. Additional doses may be given cautiously for persistent muscle weakness. May be given by intramuscular or subcutaneous routes if intravenous administration is not feasible.

In case of skin contact wash patient with soap and water followed by wash with 95% ethyl alcohol. Keep patient under constant observation for 24-36 hours. Symptoms may persist for one month.

The use of theophylline, morphine, barbiturates, phenothiazines, reserpine, and succinyl choline is contraindicated.

ENVIRONMENTAL HAZARDS

This pesticide is toxic to aquatic organisms (fish & invertebrates), and wildlife. Birds feeding in treated areas may be killed. Do not apply directly to water or wetlands (swamps, bogs, marshes and potholes). Runoff from treated areas may be hazardous to aquatic organisms in neighboring areas. Cover or incorporate granules that are spilled during loading or are visible on soil surface in turn areas. Do not contaminate water when disposing of equipment washwater.

DIRECTIONS FOR USE

It is a violation of Federal law to use this product in a manner inconsistent with its labeling.

Unprotected persons may enter treated field or area after chemical has been mixed into the soil. Granules lying on the soil surface in turn areas at row ends should be incorporated to remove possible hazard to birds and other wildlife.

Do not apply in Long Island, New York.

INCORPORATION INFORMATION

MOCAP® 20% Granular Nematicide-Insecticide can be incorporated with a rotary tiller, rotary hoe, springtooth harrow, by double-discing or with other equipment for mixing the chemical with the soil to the depth recommended. Follow any specific incorporating directions given under the crop heading.

If moisture is applied immediately after application to carry the treatment into the soil, only shallow incorporation (1/4 to 1/2 inch) with mechanical equipment such as drag chains is required.

NEMATODES CONTROLLED
FIELD CORN AND SWEET CORN

	6-7 INCH BAND 5 LBS./20G/A*	12-15 INCH BAND 7.5-10 LBS./20g/A*	BROADCAST 30 LBS./20G/A	POTATO	SUGARCANE
Dagger Nematodes					
(*Xiphenema spp.*)	C	C	C		
X. americanum	C	C	C		
Lance Nematodes					
Hoplolaimus spp.		S	S		
H. galeatus		S	S		
Lesion Nematodes					
Pratylenchus spp.	C	C	C	C	C
P. hexincisus	C				
P. zea		C	C		
P. penetrans	C	C	C		

	6-7 INCH BAND 5 LBS./20G/A*	12-15 INCH BAND 7.5-10 LBS./20g/A*	BROADCAST 30 LBS./20G/A	POTATO	SUGARCANE
Ring Nematodes					
Criconemoides spp.	S	S	S	C	C
C. ornatus	S	S	S		
Root-Knot Nematodes					
Meloidogyne spp.		C	C		C
M. hapla		C	C	C	
M. incognita		C	C	C	
M. Chitwoodi				S	
Spiral Nematodes					
Helicotylenchus spp.	C	C	C	C	
H. pseudorobustus		C	C		
H. dihystera		C	C		
Rotylenchus spp.		C	C		
Sting Nematodes					
Belonolaimus spp.		C	C		
B. longicaudatus		C	C	C	
Stubby-root Nematode					
Trichodorous spp.	C	C	C	C	
T. christiei	C	C	C		C
Stunt Nematodes					
Tylenchorhynchus spp.	C	C	C	C	
T. martini					C

C = Control S = Suppression

* Rates Based on 40-inch row spacing. Rates per acre for other row spacings are shown in the section below.

INSECTS CONTROLLED
FIELD CORN AND SWEET CORN

	6-7 INCH BAND 5 LB. 20G/A*	CULTIVATION APPLICATION 5 LB. 20G/A*	POTATO	SUGARCANE
Black Cutworm** (Agrotis ipsilon)	C			
Sandhill Cutworm** Euxoa detersa	C			
Western Corn Rootworm Diabrotica virgilera virgifera	C	C		
Northern Corn Rootworm *D. barberi*	C	C		
Southern Corn Rootworm D. undecimpunctata howardi	C	C		
Corn Wireworm Melanotus communis	C		C	C
Southern Potato Wireworm Conoderus falli	C			
Other Wireworms				
M. dietrichi	C			
M. depressus	C			
M. verberans	C			
M. lanei	C			
M. opacicollis	C			
M. similis	C			
Limonius dubitans	C			
Limonius californicus			C	
Limonius canus			C	
Ctenicera pruinina			C	
Elepdes spp.			C	
Aeolus mellillus	S			
White Grubs *Phyllophaga spp.*	S			

C = Control S = Suppression

* Rates based on 40-inch row spacing.

** Note: Under heavy infestations of cutworms, enough cutworms may survive to cause some crop damages. If this occurs, it may be necessary to apply a rescue treatment with another registered pesticide to control the surviving worms.

FIELD CORN AND SWEET CORN

NOTE TO USER: Make only one application of MOCAP® 20% Granular Nematicide-Insecticide per year.

FOR CONTROL OF CORN ROOTWORM, CORN WIREWORM, CUTWORMS, GARDEN SYMPHYLAN (Scutigerella immaculata), NEMATODES AND SUPPRESSION OF WHITE GRUBS: Row treatment: Use 6 ounces of MOCAP® 20% Granular Nematicide-Insecticide per 1,000 feet of row. Rate per acre depends on row spacing as shown below:

Row Spacing:	40″	38″	36″	30″	20″
Lbs. per acre:	5	5.3	5.6	6.7	10

Apply the granules in a 6 to 7 inch band over the row at planting. Mix the granules with the top 1/2 inch of soil with drag chains, spring-tooth incorporators, or similar equipment.

FOR COMMONLY USED APPLICATORS, SEE CALIBRATIONS CHART BELOW:

Suggested gauge settings for 6 ounces per 1,000 feet of row. (Do not change with row width.) Be sure to calibrate your applicator under your operating conditions.

FOR APPLICATION RATE OF 6 OUNCES/1,000 FEET OF ROW

BRAND OF APPLICATOR	SPEED IN MILES PER HOUR 2	3	4	5	6	7	8	9	10
Allis Chalmers	Gauge 5 for all speeds								
Old Gandy	11.4	14.6	16.8	18.6	20.2	21.8	23.4	25.2	26.8
New Gandy	11.8	14.9	17.0	18.9	20.5	21.9	23.3	24.7	26.0
International Harvester (old)	1/3.0	1/5.2	1/7.0	1/9.0	2/0.9	2/2.8	2/5.4	2/8.0	3/0.2
International Harvester (new)	1/3.2	1/5.3	1/7.1	1/9.0	2/0.8	2/2.9	2/4.9	2/6.8	2/8.6
John Deere (metal boxes)	1/16	1/18	1/20	1/32	2/1	2/2	2/2	2/3	2/3
John Deere (new) (1)	7	9	11	14	16	19	21	24	26
John Deere (newest) (2)	6	8	10	11	13	15	16	17	19
Noble (3)	5	7	9	12	14	16	18	20	22
Noble (new) (4)	4	5	7	9	11	13	15	16	17

(1)—Gauge Reads 10-20-30-40
(2)—Gauge Reads 5-15-25-35-45
(3)—Round hole in bottom of hopper
(4)—Rectangular hole in bottom of hopper

FOR CORN ROOTWORM CONTROL

Row treatment: Use the rates shown above for at-planting application. Apply at cultivation after plant emergence until layby.

On small corn, apply in a 6 to 7 inch band over the row. On taller corn, apply in bands 3 to 4 inches wide on each side of the corn, directing the granules toward the base of the plants. Cover the granules with soil by making application immediately ahead of cultivation equipment.

FOR CUTWORM CONTROL

Broadcast Treatment: Apply 15 pounds of MOCAP® 20% Granular Nematicide-Insecticide per acre from 3 days before planting to at-planting time. Mix treatment with the top 2 inches of soil right after application.

FOR NEMATODE CONTROL

Row Treatment: Apply 9 to 12 ounces of MOCAP® 20% Granular Nematicide-Insecticide per 1,000 feet of row in a band 12 to 15 inches wide on the row at-planting. Mix treatment with the top 2 to 4 inches of soil right after application. Rate per acre depends on row spacing as shown below:

Row Spacing:	40″	38″	36″	30″	20″
Lbs. Per Acre:	7.5-10	7.8-10.5	8.1-11.1	9.8-13.3	15-20

FOR NEMATODE CONTROL

Broadcast Treatment: Apply 30 pounds of MOCAP 20% Granular Nematicide-Insecticide per acre from 3 days before planting to at-planting time. Mix with the top 2 to 4 inches of soil right after application.

The treatments for nematode control will also control rootworms, wireworms, cutworms, garden symphylan (*Scutigerella immaculata*) and suppress white grubs.

POTATOES

WIREWORM AND GARDEN SYMPHYLAN (Scutigerella immaculata) CONTROL

Row Treatment: Apply 15 pounds of MOCAP® 20% Granular Nematicide-Insecticide per acre (36 inch row spacing) or 1.0 pound per 1,000 linear feet of row in a band 12 inches wide on the row at-planting. Mix with the top 2 to 4 inches of soil.

The following broadcast treatment is recommended for fields where wireworm injury has been (or is suspected to be) especially severe (i.e., potatoes planted following sod, pasture or new ground).

WIREWORMS AND GARDEN SYMPHYLAN (Scutigerella immaculata) CONTROL

Broadcast Treatment: Apply 20 to 30 pounds of MOCAP® 20% Granular Nematicide-Insecticide per acre. Mix with the top 2 to 4 inches of soil right after application.

FOR CONTROL OF GARDEN SYMPHYLAN (Scutigerella immaculata) AND LESION SPIRAL STING RING STUNT AND STUBBY-ROOT NEMATODES AND SUPPRESSION OF NORTHERN ROOT-KNOT AND SOUTHERN ROOT-KNOT NEMATODES

Row Treatment: Apply 15 pounds of MOCAP® 20% Granular Nematicide-Insecticide per acre (36 inch row spacing) or 1.0 pound per 1,000 linear feet of row in a band 12 inches wide on the row at-planting. Mix with the top 2 to 4 inches of soil.

FOR CONTROL OF GARDEN SUMPHYLAN (Scutigerella immaculata) AND NORTHERN ROOT-KNOT, SOUTHERN ROOT-KNOT, LESION, SPIRAL, STING, RING, STUNT NEMATODES AND SUPPRESSION OF COLUMBIA ROOT-KNOT NEMATODE

Broadcast Treatment: Use 30 to 60 pounds MOCAP® 20% Granular Nematicide-Insecticide per acre. Apply by ground or aerial application as close to planting as possible, but no earlier than 2 weeks before planting, or apply before potatoes emergence. Mix with the top 2 to 4 inches of soil after application if applied before planting. If applied after planting, apply 1/4 to 1/2 inch water to incorporate in soil. Uniformity of application is important for best results.

NOTE: For best suppression of Columbia Root-knot Nematodes, it is recommended that MOCAP® 20% Granular be applied following fall or spring application of a registered fumigant nematicide. **Do not** exceed a maximum of 12 lbs. active ingredient per acre per season on potatoes.

SUGARCANE

FOR NEMATODE AND WIREWORM CONTROL

Furrow treatment: Use 10 to 20 pounds of MOCAP® 20% Granular Nematicide-Insecticide per acre (6-foot row spacing) or 1.4 to 2.8 pounds per 1,000 linear feet of row at-planting. Apply evenly in a band 12 to 15 inches wide, centered over the seed pieces in the opened furrow, immediately before closing. Mix the soil with a disc hiller or other suitable equipment as the cane is covered. Use the higher rates in fields where nematode and/or wireworm infestations have been especially severe.

STORAGE AND DISPOSAL

STORAGE

Do not store in or around the home. Do not contaminate water, food, or feed by storage or disposal.

PESTICIDE DISPOSAL

Pesticide wastes are acutely hazardous. Improper disposal of excess pesticide is a violation of Federal law. If these wastes cannot be disposed of by use according to label instructions, contact your State Pesticide or Environmental Control Agency, or the Hazardous Waste representative at the nearest EPA Regional Office for guidance.

CONTAINER DISPOSAL

Completely empty bag into application equipment. Then dispose of empty bag in a sanitary landfill or by incineration, or if allowed by State and local authorities, by burning. If burned, stay out of smoke.

LIMITED WARRANTY AND DISCLAIMER

The manufacturer warrants (a) that this product conforms to the chemical description on the label; (b) that this product is reasonably fit for the purposes set forth in the directions for use when it is used in accordance with such directions; and (c) that the directions, warnings and other statements on this label are based upon responsible experts' evaluation of reasonable tests of effectiveness, of toxicity to laboratory animals and to plants, and of residues on food crops, and upon reports of field experience. Tests have not been made on all varieties or in all states or under all conditions. THE MANUFACTURER NEITHER MAKES NOR INTENDS NOR DOES IT AUTHORIZE ANY AGENT OR REPRESENTATIVE TO MAKE, ANY OTHER WARRANTIES, EXPRESS OR IMPLIED AND IT EXPRESSLY EXCLUDES AND DISCLAIMS ALL IMPLIED WARRANTIES OF MERCHANTIBILITY OR FITNESS FOR A PARTICULAR PURPOSE.

THIS WARRANTY DOES NOT EXTEND TO, AND THE BUYER SHALL BE SOLELY RESPONSIBLE FOR, ANY AND ALL LOSS OR DAMAGE WHICH RESULTS FROM THE USE OF THIS PRODUCT

IN ANY MANNER WHICH IS INCONSISTENT WITH THE LABEL DIRECTIONS, WARNINGS OR CAUTIONS.

BUYER'S EXCLUSIVE REMEDY AND MANUFACTURER'S OR SELLER'S EXCLUSIVE LIABILITY FOR ANY AND ALL CLAIMS, LOSSES, DAMAGES OR INJURIES RESULTING FROM THE USE OR HANDLING OF THIS PRODUCT WHETHER OR NOT BASED IN CONTRACT, NEGLIGENCE, STRICT LIABILITY IN TORT OR OTHERWISE, SHALL BE LIMITED, AT THE MANUFACTURER'S OPTION, TO REPLACEMENT OF, OR THE REPAYMENT OF THE PURCHASE PRICE FOR, THE QUANTITY OF PRODUCT WITH RESPECT TO WHICH DAMAGES ARE CLAIMED. IN NO EVENT SHALL MANUFACTURER OR SELLER BE LIABLE FOR SPECIAL, INDIRECT OR CONSEQUENTIAL DAMAGES RESULTING FROM THE USE OR HANDLING OF THIS PRODUCT.

NOTICE TO BUYER

Purchase of this material does not confer any rights under patents governing this product or the use thereof in countries outside of the United States.

MOCAP is a registered trademark of Rhône-Poulenc

AG 90049

EPA Approval: 6/1/89

THIS SPECIMEN LABEL IS INTENDED FOR USE ONLY AS A GUIDE IN PROVIDING GENERAL INFORMATION REGARDING THE DIRECTIONS, WARNINGS AND CAUTIONS ASSOCIATED WITH THE USE OF THIS PRODUCT. AS WITH ANY AGRICULTURAL CHEMICAL, ALWAYS FOLLOW THE LABEL INSTRUCTIONS ON THE PACKAGE BEFORE USING.

MOCAP® EC

NEMATICIDE-INSECTICIDE

RESTRICTED USE PESTICIDE

Acute Dermal Toxicity.

For retail sale to and use only by Certified Applicators or persons under the direct supervision of a Certified Applicator, and only for those uses covered by the Certified Applicator's certification.

ACTIVE INGREDIENTS:	
Ethoprop: (O-Ethyl S, S-Dipropyl Phosphorodithioate)	69.6%
INERT INGREDIENTS*:	30.4%
TOTAL:	100.0%

(Contains 6 pounds active ingredient per gallon)

EPA Reg. No. 264-458 EPA Est. No.

KEEP OUT OF REACH OF CHILDREN

DANGER POISON

DANGER–POISON

PELIGRO

PRECAUCION AL USUARIO: Si usted no lee ingles, no use este producto hasta que la etiqueta haya sido explicada ampliamente.

TRANSLATION TO THE USER: If you cannot read English, do not use this product until the label has been fully explained to you.

STATEMENT OF PRACTICAL TREATMENT

IF SWALLOWED: Call a physician or Poison Control Center. Drink 1 to 2 glasses of water and induce vomiting by touching back of throat with finger. Contains petroleum distillates. Avoid aspiration to lungs.

IF INHALED: Remove to fresh air and apply artificial respiration if needed.

IF ON SKIN: Remove all contaminated clothing and wash skin with soap and water. Wash clothing before reuse. Discard contaminated shoes. Get medical attention.

IF IN EYES: Flush with plenty of water for at least 15 minutes. Call a physician immediately.

FOR ADDITIONAL PRECAUTIONARY STATEMENTS INCLUDING NOTE TO PHYSICIAN:

See Inside Label

For PRODUCT USE Information Call 1-800-334-9745

For MEDICAL And TRANSPORTATION Emergencies ONLY Call 24 Hours A Day 1-800-334-7577

PRECAUTIONARY STATEMENTS

DANGER

HAZARD TO HUMANS AND DOMESTIC ANIMALS

Poisonous if swallowed, inhaled or absorbed through skin. Rapidly absorbed through skin. Wear waterproof protective clothing, rubber gloves and goggles. In case of contact, wash immediately with soap and water. Do not get into eyes, on skin or on clothing.

Do not breathe fumes or spray mist. Wear an AO R-6058 respirator with an R-58 cartridge, or equivalent, for protection during field handling and field exposures. After use, wash hands, arms and face thoroughly with soap and warm water before eating or smoking. Wash all clothing with soap and hot water before reuse. Discard contaminated shoes. Do not contaminate feed or food.

Do not store in or around the home.

SYMPTOMS OF POISONING: Nausea, vomiting, abdominal cramps, diarrhea, excessive salivation, headache, dizziness, weakness, blurring or dimness of vision, excessive tearing, loss of muscular coordination, slurring of speech, twitching of muscles (especially of tongue and eyelids), mental confusion, disorientation, drowsiness, difficulty in breathing (chest tightness), runny nose.

FIRST AID

IF SWALLOWED: Give 1 or 2 glasses of water and induce vomiting by touching back of throat with finger. Do not induce vomiting or give anything by mouth to an unconscious person. Call a physician, Poison Control Center, or hospital emergency room.

If not breathing, give mouth to mouth artificial respiration.

IF INHALED: Remove to fresh air and apply artificial respiration if needed.

IN CASE OF SKIN CONTACT: Immediately remove all contaminated clothing, and wash skin with soap and water. Wash clothing before reuse. Discard contaminated shoes.

FOR EYES—Flush with plenty of water for at least 15 minutes. Call a physician immediately in all cases of suspected poisoning.

Persons providing first aid should prevent contamination of themselves.

NOTE TO PHYSICIAN

This product is a cholinesterase inhibitor. Severe symptoms and signs include diarrhea, pinpoint and non-reactive pupils, respiratory difficulty, pulmonary edema, cyanosis, loss of sphincter control, convulsions, and coma.

Support respiration as needed. Measures should include removal of secretions, maintenance of a patent airway, and if necessary, artificial ventilation.

If cyanosis is absent, give ATROPINE 2-4 mg intravenously (0.05 mg/kg for children). Repeat atropine at 5-10 minute intervals until atropinization occurs (dry, flushed skin, tachycardia, pupillary dilatation), and maintain for 48 hours.

If cyanotic, give initial atropine intramuscularly and start measures to improve ventilation.

Start 2-PAM (PROTOPAM, Ayerst) at the same time. Give 1-2 grams PROTOPAM (20-40 mg/kg for children) in 100cc saline over 15-30 minutes. If pulmonary edema is present, give intravenously slowly as a 5% solution in water over a period of at least 5 minutes.

A second dose may be given after one hour if muscle weakness persists. Additional doses may be given cautiously for persistent muscle weakness. May be given by intramuscular or subcutaneous routes if intravenous administration is not feasible.

In case of skin contact wash patient with soap and water followed by wash with 95% ethyl alcohol. Keep patient under constant observation for 24-36 hours. Symptoms may persist for one month.

The use of theophylline, morphine, barbiturates, phenothiazines, reserpine, and succinyl choline is contraindicated.

PHYSICAL OR CHEMICAL HAZARDS

Do not use or store near heat or open flame.

ENVIRONMENTAL HAZARDS

This pesticide is toxic to aquatic organisms (fish and invertebrates), and extremely toxic to birds. *Cover or disc spill areas.* Birds in treated areas may be killed. Do not apply directly to water or wetlands (swamps, bogs, marshes and potholes). Drift and runoff from treated areas may be dangerous to aquatic organisms in neighboring areas. Do not contaminate water by cleaning of equipment washwaters.

Do not apply within 140 feet of inland freshwater habitats.

Along the Atlantic seaboard do not apply within 800 feet of brackish water habitats.

Do not apply when weather conditions favor drift from areas treated. Apply this product only as specified on this label.

DIRECTIONS FOR USE

It is a violation of Federal law to use this product in a manner inconsistent with its labeling.

Unprotected persons may enter treated field or area after chemical has been mixed into the soil.

Do not apply in Long Island, New York.

EQUIPMENT

MOCAP® EC spray mixture may affect some plastic materials after prolonged use. Spray tubing hoses and gaskets should be inspected frequently, particularly black-colored materials. Replace any material which is softened or swollen. Polyethylene, polypropylene, nylon and Teflon materials are recommended. Polyvinyl chloride (PVC) material is not recommended.

MIXING: Add the recommended amount of MOCAP® EC to the water in the spray tank and mix well. Use enough water in the sprayer to thoroughly and evenly cover the area to be treated. Agitate frequently during use.

INCORPORATION INFORMATION

MOCAP® EC Nematicide-Insecticide can be incorporated with a rotary tiller, rotary hoe, spring-tooth harrow, by double-discing or with other equipment for mixing the chemical with the soil to the depth recommended. Follow any specific incorporating directions given under the crop heading.

If moisture is applied immediately after application to carry the treatment into the soil, only shallow incorporation (¼ to ½ inch) with mechanical equipment such as drag chains is required.

SPECIFIC REQUIREMENTS FOR DRIP IRRIGATION SYSTEMS

Apply this product only through: sprinkler systems including: center pivot, lateral move, end tow, side (wheel) roll, traveler, big gun, solid set, or hand move; or drip (trickle) irrigation systems. Do not apply this product through any other type of irrigation system.

Crop injury, lack of effectiveness, or illegal pesticide residues in the crop can result from non-uniform distribution of treated water.

If you have questions about calibration, you should contact State Extension Service specialists, equipment manufacturers or other experts.

Do not connect an irrigation system (including greenhouse systems) used for pesticide application to a public water system unless the pesticide label prescribed safety devices for public water systems are in place.

A person knowledgeable of the chemigation system and responsible for its operation, or under the supervision of the responsible person, shall shut the system down and make necessary adjustments should the need arise.

Wear rubber gloves and boots when making adjustments or repairs on the chemigation system when MOCAP® EC is in the irrigation water.

POSTING OF AREAS TO BE CHEMIGATED IS REQUIRED WHEN:

1. Any part of a treated area is within 300 feet of sensitive areas such as residential areas, labor camps, businesses, day care centers, hospitals, in-patient clinics, nursing homes or any other public areas such as schools, parks, playgrounds, or other public facilities.
2. When the chemigated area is open to the public such as golf courses or retail greenhouses.

Posting must conform to the following requirements. Treated areas shall be posted with signs at all usual points of entry and along likely routes of approach from the listed sensitive areas. When there are no usual points of entry, signs must be posted in the corners of the treated areas and in any other location affording maximum visibility to sensitive areas. The printed side of the sign should face away from the treated area towards the sensitive area. The signs shall be printed in English. Signs must be posted prior to application and must remain posted until foliage has dried and soil surface water has disappeared. Signs may remain in place indefinitely as long as they are composed of materials to prevent deterioration and maintain legibility for the duration of the posting period.

All words shall consist of letters at least 2½ inches tall, and all letters and the symbol shall be a color which sharply contrasts with their immediate background. At the top of the sign shall be the words KEEP OUT, followed by an octagonal stop sign symbol at least 8 inches in diameter containing the word STOP. Below the symbol shall be the words PESTICIDES IN IRRIGATION WATER.

EQUIPMENT AND OPERATIONS REQUIREMENTS:

The system must contain a functional check valve, vacuum relief valve and low pressure drain appropriately located on the irrigation pipeline to prevent water source contamination from backflow.

The pesticide injection pipeline must contain a functional, automatic quick-closing check valve to prevent the flow of fluid back toward the injection pump.

The pesticide injection pipeline must also contain a functional, normally closed, solenoid-operated valve located on the intake side of the injection pump and connected to the system interlock to prevent fluid from being withdrawn from the supply tank when the irrigation system is either automatically or manually shut down.

The system must contain functional interlocking controls to automatically shut off the pesticide injection pump when the water pump motor stops.

The irrigation line or water pump must include a functional pressure switch which will stop the water pump motor when the water pressure decreases to the point where pesticide distribution is adversely affected.

Systems must use a metering pump such as a positive displacement injection pump (e.g. diaphragm pump) effectively designed and constructed of materials that are compatible with pesticides and capable of being fitted with a system interlock.

Do not apply when wind speed favors drift beyond the area intended for treatment.

The chemical injection tank should be cleaned before application to remove all fertilizer, pesticide and other foreign matter. The MOCAP® EC water mixture in the injection tank should be under constant agitation. Carefully calibrate the system before and during application. The drip irrigation system should provide uniform water flow and should have no leaks.

For sprinkler irrigation systems, MOCAP® EC or the MOCAP® EC water mixture should be injected continuously during water application in 7,000 to 28,000 gallons of water per acre.

INSECTS CONTROLLED/SUPPRESSED

	BANANAS & PLANTAIN	FIELD CORN & SWEET CORN	WHITE (IRISH) POTATO	SWEET POTATO	SUGARCANE	TOBACCO
Corn Rootworm						
(*Diabrotica virgifera virgifera*)		C				
(*Diabrotica longicornus barberi*)		C				
(*Diabrotica undecimpuctata howardi*)		C				
Wireworms						
(*Conoderus* spp.)						C
(*Conoderus vespertinus*)		C		C		
(*Conoderus falli*)		C				
(*Melanotus* spp.)		C		C		
(*Malanotus communis*)			C		C	
(*Limonius dubitan*)		C				
(*Limonius canus*)			C			
(*Limonius californicus*)			C			
(*Aeolus mellius*)		C				
(*Ctenicera* spp.)		C				
(*Ctenicera pruinina*)		C	C			
(*Eleodes* spp.)			C			
Black cutworm (*Agrotis ipsilon*)		C*				
Sandhill cutworm (*Euxoa detersa*)		C*				
White grubs (*Phyllophaga crinita*)		S				

	BANANAS & PLANTAIN	FIELD CORN & SWEET CORN	WHITE (IRISH) POTATO	SWEET POTATO	SUGARCANE	TOBACCO
(*Phyllophaga ephilida*)		S		C		
Banded cucumber beetle (*Diabrotica balteata*)				C		
Flea beetle larvae (*Chaetocnema confinis*)				C		
(*Systena blanda*)				C		
Banana root borer (*Cosmopolites sordidus*)	C					

C — Control
S — Suppression
*NOTE: Under heavy infestations of cutworms, enough cutworms may survive to cause some crop damage. If this occurs, it may be necessary to apply a rescue treatment with another registered pesticide to control the surviving worms.

NEMATODES CONTROLLED/SUPPRESSED

	BANANAS AND PLANTAIN	CABBAGE	FIELD CORN & SWEET CORN	CUCUMBERS	PEANUTS (PLANTING)	PINEAPPLE	WHITE (IRISH) POTATO	SWEET POTATO	SNAP BEANS & LIMA BEANS	SOYBEANS	SUGARCANE	TOBACCO
Burrowing Nematodes (*Radopholus similis*)	C											
Dagger Nematodes (*Xiphinema* spp.)			C							S		
Cyst Nematodes *Heterodera glycines*										S		
(*Heterodera cruciferae*)		C										
(*Heterodera schactii*)		C										
Lesion Nematodes (*Pratylenchus* spp.)	C	C	C	C	C		C		C	C	C	C
Root-Knot Nematodes (*Meloidogyne* spp.)	C									C	C	
(*Meloidogyne hapla*)		C	C	C	C		C	C	C			C
(*Meloidogyne incognita*)		C	C	C			C	C	C			C
(*Meloidogyne arenaria*)					C							
(*Meloidogyne javanica*)				C								
(*Meloidogyne chitwoodi*)							S					
Lance Nematodes (*Hoplolaimus* spp.)		S	C	S					S			
(*Hoplolaimus tylenchiformis*)				C								
(*Hoplolaimus columbus*)										S		
Spiral Nematodes (*Helicotylenchus* spp.)	C		C	C			C	C	C	C		
(*Rotylenchus* spp.)			C									
Sting Nematodes (*Belonolaimus* spp.)			C		C							
(*Belonolaimus longicaudatus*)		C		C			C		C	C		
Ring Nematodes (*Criconemoides* spp.)				C	C		C			C	C	
(*Criconemoides ornatus*)			C						C			
Reniform Nematodes (*Rotylenchulus reniformis*)						C		C				
Stunt Nematodes (*Tylenchorhynchus* spp.)			C	C	C		C	C		C		C
(*Tylenchorhynchus claytoni*)									C			
(*Tylenchorhynchus martini*)											C	
Stubby-root Nematodes (*Paratrichodorus* spp.)			C		C		C	C				
(*Trichodorus christiei*)		C		C						C	C	

C—Control S—Suppression

BANANA AND PLANTAIN

FOR NEMATODE AND BANANA ROOT BORER (*Cosmopolites sordidus*) CONTROL

Apply 8 ml of MOCAP® EC in a radius of ¾ meter around each producing stem. Before application, remove ground litter from area to be treated.

For best results, make application just before irrigation or a rainy period. If application is made during a dry period, mix MOCAP® EC into the top 2 cm of soil with a rake. Repeat application in 6 months.

BEAN (SNAP AND LIMA)

FOR NEMATODE AND GARDEN SYMPHYLAN ***(Scutigerella immaculata)*** **CONTROL**

Row treatment: Apply 1⅓ to 2 quarts of MOCAP® EC per acre (36 inch row spacing) or 2.0 to 4.4 fluid ounces per 1,000 linear feet of row in band 12 to 15 inches wide on the row at planting. Mix with the top 2 to 4 inches of soil. Use the higher rate in fields where nematode infestations have been especially severe. **Do not** use as a seed furrow treatment or allow spray to contact the seed.

FOR NEMATODE AND GARDEN SYMPHYLAN ***(Scutigerella immaculata)*** **CONTROL**

Broadcast treatment: Apply 1 to 1⅓ gallons of MOCAP® EC per acre from 3 days before planting to at-planting time. Mix with the top 2 to 4 inches of soil right after application. Use the higher rate in fields where nematode infestations have been especially severe.

NOTE: Beans grown in treated soil may remain green longer than other beans and mature 1 to 2 weeks later.

CABBAGE

FOR NEMATODE AND GARDEN SYMPHYLAN ***(Scutigerella immaculata)*** **CONTROL**

Row treatment: Apply 1⅓ quarts of MOCAP® EC per acre (36-inch row spacing) or 2.9 fluid ounces per 1,000 linear feet of row in a band 15 inches wide on the row at-planting. Mix with the top 2 to 4 inches of soil. **Do not** use as a seed furrow treatment or allow granules to contact the seed.

FOR NEMATODE AND GARDEN SYMPHYLAN ***(Scutigerella immaculata)*** **CONTROL**

Broadcast treatment: Apply 3⅓ quarts of MOCAP® EC per acre from 1 week before planting to at-planting time. Mix with the top 2 to 4 inches of soil right after application.

CITRUS SEEDLINGS

FOR NEMATODE CONTROL

Bare root dip treatment: Mix ½ pint (1 cupful) of MOCAP® EC with 50 gallons of water* (or 2 teaspoonfuls with 2 gallons of water). Dip the washed roots of citrus seedlings in the mixture so that only the roots are covered. Allow the roots to soak for 30 minutes then remove for planting.

(**MOCAP EC concentration is 900 parts per million in this mixture.*)

Treated plants should be tagged: "Warning. Wear rubber gloves when handling treated plants."

FOR NEMATODE CONTROL

Pot drench treatment: Mix ½ pint (1 cupful) of MOCAP® EC with 50 gallons of water (or 5 teaspoonfuls with 5 gallons of water). Pour enough of the mixture into each pot to completely drench the soil (about ½ pint for a 6-inch pot; 1 pint for an 8-inch pot).

CORN (FIELD AND SWEET)

FOR CONTROL OF CORN ROOTWORMS, WIREWORMS, CUTWORMS, GARDEN SYMPHYLAN ***(Scutigerella immaculata)*****, AND SUPPRESSION OF WHITE GRUBS:**

Row treatment: Apply 1⅔ ounces per 1,000 feet of row or ⅔ quarts MOCAP® EC per acre (40 inch row spacing) at planting. Apply the spray in a 6 to 7 inch band on the row over a closed seed furrow. Mix the spray with the top ½ inch with drag chains, spring-tooth incorporations, or similar equipment.

FOR CUTWORM CONTROL

Broadcast treatment: Apply 2 quarts of MOCAP® EC per acre from 3 days before planting to at-planting time. Mix with the top 2 inches of soil right after application.

FOR NEMATODE CONTROL

Row treatment: Apply 1 to 1⅓ quarts of MOCAP® EC per acre (40 inch row spacing) or 2.4 to 3.3 fluid ounces per 1,000 linear feet of row in a band 12 to 15 inches wide on the row at planting. Mix with the top 2 to 4 inches of soil right after application.

FOR NEMATODE CONTROL

Broadcast treatment: Apply 1 gallon of MOCAP® EC per acre from 3 days before planting to at-planting time. Mix with the top 2 to 4 inches of soil right after application.

The treatments for nematode control will also control corn rootworms, wireworms, cutworms, garden symphylan *(Scutigerella immaculata)* and suppress white grubs.

CUCUMBERS

FOR NEMATODE CONTROL

Row treatment: Apply 1⅓ quarts of MOCAP® EC per acre (7 foot row spacing) or 6.8 fluid ounces per 1,000 linear feet of row in a band 12 to 15 inches wide on the row at or just before planting. Prepare bed and knock off top to provide a level area for treatment. Mix with the top 2 to 4 inches of soil right after application. **Do not** use as a seed furrow treatment or allow spray to contact the seed.

ORNAMENTALS PLANTS

NOTE: (Aglaonema, Azalea, Boxwood, Bromeliads, Cacti, Caladium, Camellia, Cape Jasmine, Fern, Gardenia, Holly Philodendron, Pothos, Sansevieria and Yew).

FOR NEMATODE CONTROL

BARE ROOT AND TUBER DIP TREATMENT: Mix ½ pint (1 cupful) of MOCAP® EC with 50 gallons of water* (or 2 teaspoonfuls with 2 gallons of water). Dip the washed roots of plants in the mixture so that only the roots are covered. Allow the roots to soak for 30 minutes then remove for planting. For Leatherleaf Fern, mix ¼ pint (½ cupful) with 100 gallons of water (or 1 teaspoonful with 4 gallons of water), and dip washed roots for 15 minutes.

(MOCAP® EC concentration is 900 parts per million in this mixture.)*

TREATED PLANTS SHOULD BE TAGGED

"WARNING. Wear rubber gloves when handling treated plants".

FOR NEMATODE CONTROL

POT DRENCH TREATMENT: Mix ¼ pint (½ cupful) of MOCAP® EC with 50 gallons of water (or 1 teaspoonful with 2 gallons of water). Pour enough of the mixture into each pot to completely drench the soil (about ½ pint for 6 inch pot; 1 pint for an 8 inch pot). If the mixture contacts foliage wash it off immediately with fresh water to avoid foliage injury. For Leatherleaf Fern, mix ⅛ pint (¼ cupful) with 50 gallons of water (or 1 teaspoonful with 4 gallons of water). Apply after harvesting and following directions above.

FOR NEMATODE CONTROL

BED AND BENCH TREATMENT: Mix ½ to 1 pint (1 to 2 cupfuls) of MOCAP® EC with enough water to completely cover 1,000 sq. feet of bed or bench area (or 2 to 4 tablespoonfuls for 100 sq. feet). Wash the mixture off contacted foliage immediately to avoid foliage injury. After treatment apply enough water to soak the treatment 4 to 6 inches into the soil. For Leatherleaf Fern, use ⅓ pint (⅔ cupful) for 1,000 sq. feet (or 2 teaspoonfuls for 100 sq. feet). Apply after harvesting, and follow with enough water to soak the treatment 4 to 6 inches into the soil.

FOR NEMATODE CONTROL

FIELD NURSERY STOCK TREATMENT: Mix ½ pint (1 cupful) of MOCAP® EC with 100 gallons of water (or 2½ teaspoonfuls with 5 gallons of water). Apply 1 gallon of mixture to the soil in a square yard area around each plant. Wash the mixture off contacted foliage immediately to avoid injury. After treatment, apply enough additional water to wet the soil to a depth of 6 to 12 inches, depending upon depth of the roots.

NOTE: If plants other than those mentioned on this label require treatment, only a few should be treated until its effects can be determined. Some plants or varieties may show temporary growth retardation after treatment.

PEANUT

FOR NEMATODE CONTROL

Row treatment: Use 1⅓ to 2⅔ quarts of MOCAP® EC per acre (36 inch row spacing) or 2.9 to 5.9 fluid ounces per 1,000 linear feet of row. Apply in a band 12 inches wide on the row at planting and mix with the top 2 to 4 inches of soil. Use the higher rates of MOCAP® EC for control of peanut root-knot nematode *(M. arenaria)*, and in fields where nematode infestations have been especially severe. **Do not** use as a seed furrow treatment or allow granules to contact the seed.

FOR NEMATODE CONTROL

Broadcast treatment: Apply 3 to 4 quarts of MOCAP® EC per acre from 1 week before planting to at-planting time. Mix with the top 2 to 4 inches of soil right after application. Use the higher rates of MOCAP® EC for control of peanut root-knot nematode *(M. arenaria)*, and in fields where nematode infestations have been especially severe.

PINEAPPLE (Hawaii Only)

POSTPLANT APPLICATION THROUGH DRIP IRRIGATION SYSTEM

Plant Crop

Apply 1 to 2 gallons of MOCAP® EC Nematicide Insecticide per acre through drip irrigation immediately after planting. Dilute MOCAP® EC in 50 to 100 gallons of water before injection. Apply additional treatments every 2 months. Do not treat within 120 days before harvest.

Do not apply more than 16 gallons of MOCAP® EC per acre per plant crop.

Ratoon Crop

Apply 1 to 2 gallons of MOCAP® EC per acre through drip irrigation. Dilute MOCAP® EC in 50 to 100 gallons of water before injection. The first application may begin immediately after crop harvest. Apply additional treatments every 2 months. Do not treat within 120 days before harvest.

Do not apply more than 10 gallons of MOCAP® EC per acre to each ratoon crop.

SEE SPECIFIC REQUIREMENTS FOR DRIP IRRIGATION SYSTEMS IN EARLIER PORTION OF LABEL FOR GENERAL USE AND POSTING REQUIREMENTS.

Use the following sample calculations for MOCAP® EC Nematicide Insecticide applied through drip irrigation systems:

1. Acres watered by system: 25 acres
2. MOCAP® EC application rate per acre: 1 to 2 gallons MOCAP® EC per acre.
3. Diluent rate per acre: 50 gallons of water for each gallon of MOCAP® EC per acre.
4. Total gallons of material to be injected per acre:
   51 gallons for use rate of 1 gallon MOCAP® EC per acre.
   102 gallons for use rate of 2 gallons MOCAP® EC per acre.
5. Total gallons of material to be injected for whole field:
   1,275 gallons for use rate of 1 gallon MOCAP® EC per acre.
   2,550 gallons for use rate of 2 gallons MOCAP® EC per acre.
6. Hours of irrigation to apply 6,000 gallons of water per acre: 4 hours
7. Hours of irrigation to inject MOCAP® EC water mixture into the system: 3 hours
8. Injection rate per hour: 425 gallons to be injected per hour for use rate of 1 gallon MOCAP® EC per acre. 850 gallons to be injected per hour for use rate of 2 gallons of MOCAP® EC per acre.
9. Conversion of gallons of material to be injected per hour into fluid ounces per minute in order to calibrate and adjust the injector pump:
   906.5 fluid ounces to be injected per minute for use rate of 1 gallon of MOCAP® EC per acre.
   1,813.1 fluid ounces to be injected per minute for use rate of 2 gallons of MOCAP® EC per acre.

POTATOES

FOR WIREWORM AND GARDEN SYMPHYLAN (*Scutigerella immaculata*) CONTROL

Row treatment: Apply 2 quarts MOCAP® EC per acre or 4.4 fluid ounces per 1,000 linear feet of row in a band 12 inches wide on the row at planting. Mix with the top 2 to 4 inches of soil.

The following broadcast treatment is recommended for fields where wireworm injury has been (or is suspected to be) especially severe (i.e., potatoes planted following sod, pasture or new ground).

FOR WIREWORM AND GARDEN SYMPHYLAN (*Scutigerella immaculata*) CONTROL

Broadcast treatment: Apply 2²/₃ to 4 quarts of MOCAP® EC per acre. Mix treatment with the top 2 to 4 inches of soil right after application.

FOR CONTROL OF GARDEN SYMPHYLAN (SCUTIGERELLA IMMACULATA) AND LESION, SPIRAL, STING, RING, STUNT, AND STUBBY ROOT NEMATODES AND SUPPRESSION OF NORTHERN ROOT-KNOT AND SOUTHERN ROOT-KNOT NEMATODES:

Row treatment: Apply 2 quarts of MOCAP® EC per acre (36 inch row spacing) or 4.4 fluid ounces per 1,000 linear feet of row in a band 12 inches wide on the row at planting. Mix with the top 2 to 4 inches of soil.

FOR CONTROL OF GARDEN SYMPHYLAN (*Scutigerella immaculata*) AND NORTHERN ROOT-KNOT, SOUTHERN ROOT-KNOT, LESION, SPIRAL, STING, RING, STUNT, AND STUBBY-ROOT NEMATODES AND SUPPRESSION OF COLUMBIA ROOT-KNOT NEMATODES

Broadcast treatment: Apply 1 to 2 gallons MOCAP® EC per acre. Apply by ground application as close to planting as possible but no earlier than 2 weeks before planting or apply before potato emergence. Mix with the top 2 to 4 inches of soil right after application if applied before planting. If applied after planting, apply ¼ to ½ inch water to incorporate in soil. Uniformity of application is important for best results.

NOTE: For best suppression of Columbia root-knot nematodes, it is recommended that MOCAP® EC be applied following fall or spring application of a registered fumigant nematicide.

Do not exceed a maximum of 12 lbs. active ingredient per acre per season on potatoes.

SOYBEANS

FOR NEMATODE CONTROL

Row treatment: Apply 1 to 2 quarts of MOCAP® EC per acre (36 inch row spacing) or 2.2 to 4.4 fluid ounces per 1,000 linear feet in a band 12 to 15 inches wide on the row at planting. Mix with the top 2 to 4 inches of soil. Use the higher rates in fields where nematode infestations have been especially severe. **Do not** use as a seed furrow treatment or allow spray to contact the seed.

FOR NEMATODE CONTROL

Broadcast treatment: Apply 3 to 4 quarts of MOCAP® EC per acre from 1 week before planting to at-planting time. Mix with the top 2 to 4 inches of soil right after application. Use the higher rate in fields where nematode infestations have been especially severe.

TANKMIX WITH TREFLAN® HERBICIDE

MOCAP® EC may be tank-mixed with Treflan® (Trifluralin) Herbicide for control of the nematodes, grasses, and weeds named on the labels. Follow all directions and precautions on each label.

NOTE: Soybeans treated with organophosphate chemicals such as MOCAP® may be more susceptible to injury from metribuzin and propanil herbicides.

SUGARCANE

FOR NEMATODE AND WIREWORM CONTROL

Furrow Treatment: Apply 1⅓ to 2⅔ quarts of MOCAP® EC per acre at planting time (6 foot row spacing) or 5.0 to 11.8 fluid ounces per 1,000 linear feet of row at-planting. Apply evenly in a band 12 to 15 inches wide, centered over the seed pieces in the opened furrow immediately before closing. Mix with the soil with a disc hiller or other suitable equipment as the cane is covered. Use the higher rates in fields where nematode and/or wireworm infestations have been especially severe.

SWEET POTATO

FOR NEMATODE, WIREWORM, WHITE GRUB, BANDED CUCUMBER BEETLE AND FLEA BEETLE LARVAE (PALESTRIPED AND SWEET POTATO FLEA BEETLE) CONTROL

Row treatment: Use 2 to 2⅔ quarts of MOCAP® EC per acre or 5.1 to 6.9 fluid ounces per 1,000 linear feet of row (minimum 42 inch row spacing). Apply in a band 12 to 15 inches wide on the row 2 to 3 weeks before planting. Mix with the top 2 to 4 inches of soil right after application. Good results have been obtained by opening the row, applying MOCAP® EC in a 12 to 15 inch band on the row, bedding over and dragging. Use the higher rate in fields where nematode and/or insect infestations have been especially severe.

FOR NEMATODE, WIREWORM, WHITE GRUB, BANDED CUCUMBER BEETLE AND FLEA BEETLE LARVAE (PALESTRIPED AND SWEET POTATO FLEA BEETLE) CONTROL

Broadcast treatment: Apply 1 to 1⅓ gallons of MOCAP® EC per acre 2 to 3 weeks before planting. Mix with the top 2 to 4 inches of soil right after application. Use a double gang disc harrow or other equipment which will do a good job of mixing the MOCAP® EC with the soil. Use the higher rate in fields where nematode and/or insect infestations have been especially severe.

TOBACCO

FOR NEMATODE AND WIREWORM CONTROL

Row treatment: Apply 1 to 2 gallons of MOCAP® EC per acre (42 inch row spacing) or 10.3 to 20.6 fluid ounces per 1,000 linear feet of row. Apply after fertilizer has been listed or bedded. Make application in a band 18 to 24 inches wide on the row from one week before planting to at-planting time. Mix with the top 2 to 4 inches of soil right after application. Use the higher rates in fields where nematode injury to tobacco has been especially severe, or where the soil is sandy.

FOR NEMATODE AND WIREWORM CONTROL

Broadcast treatment: Apply 1 to 2 gallons of MOCAP® EC per acre from one week before planting to at-planting time. Mix the MOCAP® EC with the top 2 to 4 inches of soil right after application. Use a double gang disc harrow or other equipment which will do a good job of mixing the MOCAP® EC with the soil. After mixing, apply fertilizer and for best results, shape a wide bed with a tractor equipped with 4 disc hillers and wide sweeps for running out the middles. Be sure that only the treated soil (top 2 to 4 inches) is used to make beds. Use higher rates in fields where nematode injury to tobacco has been especially severe, or where the soil is sandy.

FOR WIREWORM CONTROL

Broadcast treatment: Apply 1⅓ quarts of MOCAP® EC per acre from 2 weeks before planting to at-planting time. Mix with the top 2 to 4 inches of soil right after application. Use a disc harrow or other cultivating equipment that will do a good job of mixing the MOCAP® EC with the soil.

FOR FLEA BEETLE CONTROL

Add ½ gallon (4 pints) per acre of Di-Syston* 8 Liquid Systemic Insecticide to the spray tank containing the MOCAP® EC mixture for nematode and wireworm control (see above). Observe all precautions on the label. Do not make more than one field application per crop season regardless of the formulation of Di-Syston used. (*Di-Syston is a registered trademark of Farbenfabriken Bayer A.G.)

Rhône-Poulenc does not recommend the use of MOCAP on tobacco in Florida.

STORAGE AND DISPOSAL

STORAGE

Do not contaminate water, food, or feed by storage or disposal.

PESTICIDE DISPOSAL

Pesticide wastes are acutely hazardous. Improper disposal of excess pesticide, spray mixture, or rinsate is a violation of Federal law. If these wastes cannot be disposed of by use according to label instructions, contact your State Pesticide or Environmental Control Agency, or the Hazardous Waste representative at the nearest EPA Regional Office for guidance.

CONTAINER DISPOSAL

Triple rinse (or equivalent). Then dispose of in a sanitary landfill, or by other procedures approved by state and local authorities.

LIMITED WARRANTY AND DISCLAIMER

The manufacturer warrants (a) that this product conforms to the chemical description on the label; (b) that this product is reasonably fit for the purposes set forth in the directions for use when it is used in accordance with such directions; and (c) that the directions, warnings and other statements on this label are based upon responsible experts' evaluation of reasonable tests of effectiveness, of toxicity to laboratory animals and to plants, and of residues on food crops, and upon reports of field experience. Tests have not been made on all varieties or in all states of under all conditions. THE MANUFACTURER NEITHER MAKES NOR INTENDS NOR DOES IT AUTHORIZE ANY AGENT OR REPRESENTATIVE TO MAKE, ANY OTHER WARRANTIES, EXPRESS OR IMPLIED AND IT EXPRESSLY EXCLUDES AND DISCLAIMS ALL IMPLIED WARRANTIES OF MERCHANTIBILITY OR FITNESS FOR A PARTICULAR PURPOSE.

THIS WARRANTY DOES NOT EXTEND TO, AND THE BUYER SHALL BE SOLELY REPONSIBLE FOR, ANY AND ALL LOSS OR DAMAGE WHICH RESULTS FROM THE USE OF THIS PRODUCT IN ANY MANNER WHICH IS INCONSISTENT WITH THE LABEL DIRECTIONS, WARNINGS OR CAUTIONS.

BUYER'S EXCLUSIVE REMEDY AND MANUFACTURER'S OR SELLER'S EXCLUSIVE LIABILITY FOR ANY AND ALL CLAIMS, LOSSES, DAMAGES OR INJURIES RESULTING FROM THE USE OR HANDLING OF THIS PRODUCT, WHETHER OR NOT BASED IN CONTRACT, NEGLIGENCE, STRICT LIABILITY IN TORT OR OTHERWISE, SHALL BE LIMITED, AT THE MANUFACTURER'S OPTION, TO REPLACEMENT OF, OR THE REPAYMENT OF THE PURCHASE PRICE FOR, THE QUANTITY OF PRODUCT WITH RESPECT TO WHICH DAMAGES ARE CLAIMED. IN NO EVENT SHALL MANUFACTURER OR SELLER BE LIABLE FOR SPECIAL, INDIRECT OR CONSEQUENTIAL DAMAGES RESULTING FROM THE USE OR HANDLING OF THIS PRODUCT.

NOTICE TO BUYER

Purchase of this material does not confer any rights under patents governing this product or the use thereof in countries outside of the United States.

MOCAP is a registered trademark of Rhône-Poulenc
AG90050

THIS SPECIMEN LABEL IS INTENDED FOR USE ONLY AS A GUIDE IN PROVIDING GENERAL INFORMATION REGARDING THE DIRECTIONS, WARNINGS AND CAUTIONS ASSOCIATED WITH THE USE OF THIS PRODUCT. AS WITH ANY AGRICULTURAL CHEMICAL, ALWAYS FOLLOW THE LABEL INSTRUCTIONS ON THE PACKAGE BEFORE USING.

MOCAP® PLUS

NEMATICIDE-INSECTICIDE FOR TOBACCO

RESTRICTED USE PESTICIDE

Acute human oral, dermal and inhalation toxicity.

For retail sale to and use only by Certified Applicators or persons under the direct supervision of a Certified Applicator, and only for those uses covered by the Certified Applicator's certification.

ACTIVE INGREDIENTS:	
Ethoprop: (O-Ethyl S, S-Dipropyl Phosphorodithioate)	10%
Disulfoton: (O, O-Diethyl S-[2-(ethylthio) ethyl] Phosphorodithioate)	5%
INERT INGREDIENTS:	85%
TOTAL:	100%

Contains 10% ETHOPROP® and 5% DISULFOTON

EPA Reg. No. 264-459 359-TN-1

KEEP OUT OF REACH OF CHILDREN

DANGER–POISON

PELIGRO

PRECAUCION AL USUARIO: Si usted no lee ingles, no use este producto hasta que la etiqueta haya sido explicada ampliamente.

TRANSLATION TO THE USER: If you cannot read English, do not use this product until the label has been fully explained to you.

STATEMENT OF PRACTICAL TREATMENT

IF SWALLOWED: Call a physician or Poison Control Center. Drink 1 to 2 glasses of water and induce vomiting by touching back of throat with finger. Do not induce vomiting or give anything to an unconscious person.

IF INHALED: Remove victim to fresh air and apply artificial respiration if indicated.

IF ON SKIN: Remove all contaminated clothing and wash skin with soap and water. Wash clothing before reuse. Discard contaminated shoes.

IF IN EYES: Flush with plenty of water for at least 15 minutes. Call a physician immediately.

For PRODUCT USE Information Call 1-800-334-9745
For MEDICAL And TRANSPORTATION Emergencies ONLY Call 24 Hours A Day 1-800-334-7577

PRECAUTIONARY STATEMENTS

DANGER

HAZARDS TO HUMANS AND DOMESTIC ANIMALS

Poisonous if swallowed, inhaled or absorbed through skin. Rapidly absorbed through skin. Wear waterproof protective clothing, natural rubber gloves and goggles. In case of contact, wash immediately with water.

Do not breathe fumes or dust. Wear an AO R-6058 respirator with an R-58 cartridge, or equivalent, for protection during field handling and field exposure. After use, wash hands, arms, and face thoroughly with soap and warm water before eating or smoking. Wash all contaminated clothing with soap and hot water before reuse.

Do not contaminate feed or food. Do not allow livestock to graze in treated area. Keep all unprotected persons and animals off treated area until chemical is mixed into the soil. Do not apply where runoff is likely to occur. Keep out of any body of water.

SYMPTOMS OF POISONING: A sense of "tightness" in chest, sweating, stomach pains, vomiting, diarrhea, contraction of pupils.

STATEMENT OF PRACTICAL TREATMENT

IF SWALLOWED: Call a physician or Poison Control Center. Drink 1 or 2 glasses of water and induce vomiting by touching back of throat with finger. Do not induce vomiting or give anything by mouth to an unconscious person.

IF ON SKIN: Immediately remove all contaminated clothing, and wash skin with soap and water. Wash clothing before reuse. Discard contaminated shoes.

IF IN EYES: Flush eyes with plenty of water. Call a physician immediately in all cases of suspected poisoning.

IF INHALED: Remove victim to fresh air.

NOTE TO PHYSICIAN

This product is a cholinesterase inhibitor. Warning symptoms include weakness, headaches, "tightness" of chest, blurred vision, non-reactive pinpoint pupils, salivation, sweating, nausea, vomiting, diarrhea, and abdominal cramps.

Administer atropine, 2 mg intramuscularly or intravenously. Repeat dose every 3 to 8 minutes until signs of atropinization occur. Give pralidoxime chloride (Protopam or 2-PAM), 2.5 gm in 100 cc of sterile water, intravenously, slow in 15 to 30 minutes or by deep intramuscular injection of 1 gram in 3 cc of distilled water. Repeat every half hour if respiration weakens or muscle fasciculation or convulsion recur. Clear chest by postural drainage. Artificial respiration may be necessary. Observe patient continuously for 48 hours.

ENVIRONMENTAL HAZARDS

This pesticide is toxic to aquatic organisms (fish & invertebrates), and wildlife. Birds feeding in treated areas may be killed. Do not apply directly to water or wetlands (swamps, bogs, marshes and potholes). Runoff from treated areas may be hazardous to aquatic organisms in neighboring areas. Cover or incorporate granules that are spilled during loading or are visible on soil surface in turn areas. Do not contaminate water when disposing of equipment wash water.

Do not apply within 140 feet of inland freshwater habitats.

Along the Atlantic seaboard, do not apply within 800 feet of brackish water habitats.

BEE CAUTION

This pesticide is toxic to bees exposed to direct application. Applications should be timed to coincide with periods of minimum bee activity, usually between late evening and early morning.

Contact your Cooperative Agricultural Extension Service or your local Rhône-Poulenc Ag Company representative for further information.

DIRECTIONS FOR USE

It is a violation of Federal law to use this product in a manner inconsistent with its labeling.

Do not apply in Long Island, New York.

REENTRY STATEMENT

Reentry into treated fields is prohibited within 24 hours of application. State or local use and reentry regulations may be more restrictive than those stated on this label.

Do not apply this product in such a manner as to directly or through drift expose workers or other persons. The area being treated must be vacated by unprotected persons.

Because certain states may require more restrictive reentry intervals for various crops treated with this product, consult your State Department of Agriculture for further information.

Do not allow reentry into treated fields within 24 hours of application unless appropriate protective clothing is worn. Written or oral warnings must be given to workers who are expected to be in a treated area or in an area about to be treated with this product. Oral warnings must be given if there is reason to believe that written warnings cannot be understood by workers and given in a language customarily understood by the workers.

Oral and written warnings must include the STATEMENT OF PRACTICAL TREATMENT and notification that unprotected persons may enter treated fields or area only after chemical has been mixed in the soil.

INCORPORATION INFORMATION

MOCAP® PLUS Nematicide-Insecticide can be incorporated with a rotary tiller, rotary hoe, spring-tooth harrow, by double-discing or with other equipment for mixing the chemical with the soil to the depth recommended. Follow any specific incorporating directions given under the heading.

If moisture is applied immediately after application to carry the treatment into the soil, only shallow incorporation (1/4 to 1/2 inch) with mechanical equipment such as drag chains is required.

Granules lying on the soil surface in turn areas at row ends should be incorporated to remove possible hazard to birds and other wildlife.

TOBACCO NEMATODES CONTROLLED	TOBACCO INSECTS CONTROLLED
Dagger nematode (*Xiphinema* spp.)	Wireworm (*Conoderus* spp.)
Lesion nematode (*Pratylenchus* spp.)	Tobacco flea beetle (*Epitrix hirtipennis*)
Root-knot nematode (*Meloidogyne hapla*) (*Meloidogyne incognita*)	Tobacco Aphid (*Myzus persicae*)
Stunt nematode (*Tylenchorhynchus* spp.)	

TOBACCO

FOR CONTROL OF NEMATODES, FLEA BEETLES, APHIDS AND WIREWORMS

Row treatment: Apply 60 pounds of MOCAP® PLUS Nematicide-Insecticide per acre (42-inch row spacing) or 4.8 pounds per 1000 linear feet of row. Apply after fertilizer has been listed or bedded. Make application in a band 18 inches wide on the row from 1 week before planting to at-planting time. Mix MOCAP® PLUS with the top 2 to 4 inches of soil right after application. Use a rotary tiller, rotary hoe or other cultivating equipment that will do a good job of mixing MOCAP® PLUS with the soil.

FOR CONTROL OF NEMATODES, FLEA BEETLES, APHIDS AND WIREWORMS

Broadcast treatment: Apply 60 to 80 pounds of MOCAP® PLUS Nematicide-Insecticide per acre from 1 week before planting to at-planting time. Mix with the top 2 to 4 inches of soil right after application. Use a double gang disc harrow or other equipment that will do a good job of mixing MOCAP® PLUS with the soil. After mixing, apply fertilizer and, for best results, shape a wide bed with a tractor equipped with 4 disc hillers and wide sweeps for running out the middles. Be sure that only the treated soil (top 2 to 4 inches) is used to make beds. Use the higher rate in fields where nematode injury to tobacco has been especially severe, or where the soil is sandy.

FOR WIREWORM CONTROL

Broadcast treatment: Apply 20 pounds of MOCAP® PLUS Nematicide-Insecticide per acre from 2 weeks before planting to at-planting time. Mix with the top 2 to 4 inches of soil right after application. Use a disc harrow or other cultivating equipment that will do a good job of mixing MOCAP® PLUS with the soil.

Rhône Poulenc Ag Company does not recommend the use of MOCAP on tobacco in Florida.

Do not plant any food crop in rotation after a field treatment with disulfoton unless it is a registered use for disulfoton.

STORAGE AND DISPOSAL

STORAGE

Do not contaminate water, food or feed by storage or disposal. Do not store in or around the home.

PESTICIDE DISPOSAL

Pesticide wastes are acutely hazardous. Improper disposal of excess pesticide, spray mixture, or rinsate is a violation of Federal law. If these wastes cannot be disposed of by use according to label instructions, contact your State Pesticide or Environmental Control Agency, or the Hazardous Waste representative at the nearest EPA Regional Office for guidance.

CONTAINER DISPOSAL

Completely empty bag into application equipment. Then dispose of empty bag in a sanitary landfill or by incineration, or, if allowed by State and local authorities, by burning. If burned, stay out of smoke.

LIMITED WARRANTY AND DISCLAIMER

The manufacturer warrants that this product conforms to the chemical description on the label; that this product is reasonably fit for the purposes set forth in the directions for use when it is used in accordance with such directions; that the directions, warnings and other statements on this label are based upon responsible experts' evaluation of reasonable tests of effectiveness, of toxicity to laboratory animals and to plants, and of residues on food crops, and upon reports of field experience. Tests have not been made on all varieties or in all states or under all conditions. THE MANUFACTURER NEITHER MAKES NOR INTENDS NOR DOES IT AUTHORIZE ANY AGENT OR REPRESENTATIVE TO MAKE, ANY OTHER WARRANTIES, EXPRESS OR IMPLIED AND IT EXPRESSLY EXCLUDES AND DISCLAIMS ALL IMPLIED WARRANTIES OF MERCHANTIBILITY OR FITNESS FOR A PARTICULAR PURPOSE.

THIS WARRANTY DOES NOT EXTEND TO, AND THE BUYER SHALL BE SOLELY RESPONSIBLE FOR, ANY AND ALL LOSS OR DAMAGE WHICH RESULTS FROM THE USE OF THIS PRODUCT IN ANY MANNER WHICH IS INCONSISTENT WITH THE LABEL DIRECTIONS, WARNINGS OR CAUTIONS.

BUYER'S EXCLUSIVE REMEDY AND MANUFACTURER OR SELLER'S EXCLUSIVE LIABILITY FOR ANY AND ALL CLAIMS, LOSSES, DAMAGES OR INJURIES RESULTING FROM THE USE OR HANDLING OF THIS PRODUCT, WHETHER OR NOT BASED IN CONTRACT, NEGLIGENCE, STRICT LIABILITY IN TORT OR OTHERWISE, SHALL BE LIMITED, AT THE MANUFACTURER'S OPTION, TO REPLACEMENT OF, OR, THE REPAYMENT OF THE PURCHASE PRICE FOR, THE QUANTITY OF PRODUCT WITH RESPECT TO WHICH DAMAGES ARE CLAIMED. IN NO EVENT SHALL MANUFACTURER OR SELLER BE LIABLE FOR SPECIAL, INDIRECT OR CONSEQUENTIAL DAMAGES RESULTING FROM THE USE OR HANDLING OF THIS PRODUCT.

NOTICE TO BUYER

Purchase of this material does not confer any rights under patents governing this product or the use thereof in countries outside of the United States.

MOCAP is a registered trademark of Rhône-Poulenc

AG90051 EPA Approval: 7/25/89

THIS SPECIMEN LABEL IS INTENDED FOR USE ONLY AS A GUIDE IN PROVIDING GENERAL INFORMATION REGARDING THE DIRECTIONS, WARNINGS AND CAUTIONS ASSOCIATED WITH THE USE OF THIS PRODUCT. AS WITH ANY AGRICULTURAL CHEMICAL, ALWAYS FOLLOW THE LABEL INSTRUCTIONS ON THE PACKAGE BEFORE USING.

MOCAP® PLUS 4-2 EC

NEMATICIDE-INSECTICIDE

RESTRICTED USE PESTICIDE

Acute human oral, dermal and inhalation toxicity.

For retail sale to and use only by Certified Applicators or persons under the direct supervision of a Certified Applicator, and only for those uses covered by the Certified Applicator's certification.

ACTIVE INGREDIENTS:	
Ethoprop: (O-Ethyl S, S-Dipropyl Phosphorodithioate)*	46.0%
Disulfoton: (O,O-Diethyl S-[2-(ethylthio) ethyl] Phosphorodithioate)**	23.0%
INERT INGREDIENTS:	31.0%
TOTAL	100.0%

*Equivalent to 4 pounds per gallon.
**Equivalent to 2 pounds per gallon.

EPA Reg. No. 264-464 EPA Est. No.

KEEP OUT OF REACH OF CHILDREN

DANGER POISON

PELIGRO

DANGER–POISON

PRECAUCION AL USUARIO: Si usted no lee ingles, no use este producto hasta que la etiqueta haya sido explicada ampliamente.

TRANSLATION TO THE USER: If you cannot read English, do not use this product until the label has been fully explained to you.

STATEMENT OF PRACTICAL TREATMENT

IF SWALLOWED: Drink 1 to 2 glasses of water and induce vomiting by touching back of throat with finger. Do not induce vomiting or give anything to an unconscious person. Call a physician, Poison Control Center or hospital emergency room.

If not breathing, give mouth to mouth artificial respiration.

IF ON SKIN: Immediately remove all contaminated clothing and wash skin with soap and water. Wash clothing before reuse. Discard contaminated shoes.

IF IN EYES: Flush with plenty of water for at least 15 minutes. Call a physician immediately in all cases of suspected poisoning.

Persons providing first aid should prevent contamination of themselves.

For PRODUCT USE Information Call 1-800-334-9745
For MEDICAL And TRANSPORTATION Emergencies ONLY Call 24 Hours A Day 1-800-334-7577

PRECAUTIONARY STATEMENTS

DANGER

HAZARDS TO HUMAN AND DOMESTIC ANIMALS

Fatal if swallowed, inhaled or absorbed through skin. Rapidly absorbed through skin. Wear water proof protective clothing, rubber gloves and goggles. In case of contact, wash immediately with water.

Do not breathe fumes or spray mist. Wear an AO R-6058 respirator with an R-58 cartridge, or equivalent, for protection during field handling and field exposure. After use, wash hands, arms, and face thoroughly with soap and warm water before eating or smoking. Wash all clothing with soap and hot water before reuse. Discard contaminated shoes. Do not contaminate feed or food.

Do not store in or around the home.

SYMPTOMS OF POISONING: Nausea, vomiting, abdominal cramps, diarrhea, excessive salivation, headache, dizziness, weakness, blurring or dimness of vision, excessive tearing, loss of muscular coordination, slurring of speech, twitching of muscles (especially of tongue and eyelids), mental confusion, disorientation, drowsiness, difficulty in breathing (chest tightness), runny nose.

STATEMENT OF PRACTICAL TREATMENT

IF SWALLOWED: Give 1 or 2 glasses of water and induce vomiting by touching back of throat with finger.

Do not induce vomiting or give anything by mouth to an unconscious person. Call a physician, Poison Control Center, or hospital emergency room.

If not breathing, give mouth to mouth artificial respiration.

IN CASE OF SKIN CONTACT: Immediately remove all contaminated clothing, and wash skin with soap and water. Wash clothing before reuse. Discard contaminated shoes.

IF IN EYES: Flush with plenty of water for at least 15 minutes. Call a physician immediately in all cases of suspected poisoning.

Persons providing first aid should prevent contamination to themselves.

NOTE TO PHYSICIAN

This product is a cholinesterase inhibitor. Severe symptoms and signs include diarrhea, pinpoint and non-reactive pupils, respiratory difficulty, pulmonary edema, cyanosis, loss of sphincter control, convulsions, and coma.

Solvent presents aspiration hazard. Gastric lavage may be indicated if material was taken internally. Support respiration as needed. Measures should include removal of secretions, maintenance of a patent airway, and if necessary, artificial ventilation.

If cyanosis is absent, give ATROPINE 2-4 mg intravenously (0.5 mg/kg for children). Repeat Atropine at 5 to 10 minute intervals until atropinization occurs (dry, flushed skin, tachycardia, pupillary dilation), and maintain for 48 hours. If cyanotic, give initial atropine intramuscularly and start measures to improve ventilation.

Start 2-PAM at the same time. Give 1 to 2 grams 2-PAM (20 to 40 mg/kg for children) in 100cc saline over 15 to 30 minutes. If pulmonary edema is present, give intravenously slowly as a 5% solution in water over 5 minutes.

A second dose may be given after 1 hour if muscle weakness persists. Additional doses may be given cautiously for persistent muscle weakness. May be given by intramuscular or subcutaneous routes if intravenous administration is not feasible.

In case of skin contact wash patient with soap and water followed by wash with 95% ethyl alcohol. Keep patient under constant observation for 24 to 36 hours. Symptoms may persist for 1 month.

ENVIRONMENTAL HAZARDS

This pesticide is toxic to aquatic organisms (fish & invertebrates), and extremely toxic to birds. *Cover or disc spill areas.* Birds in treated areas may be killed. Do not apply directly to water or wetlands (swamps, bogs, marshes and potholes). Drift and runoff from treated areas may be dangerous to aquatic organisms in neighboring areas. Do not contaminate water by cleaning of equipment wash waters.

Do not apply within 140 feet of inland freshwater habitats.

Along the Atlantic seaboard, do not apply with 800 feet of brackish water habitats.

PHYSICAL AND CHEMICAL HAZARDS

Do not use or store near heat or open flame.

BEE CAUTION

This pesticide is toxic to bees exposed to direct application. Applications should be timed to coincide with periods of minimum bee activity, usually between late evening and early morning. Contact your Cooperative Agricultural Extension Service or your local Rhône-Poulenc Ag Company representative for further information.

DIRECTIONS FOR USE

It is a violation of Federal law to use this product in a manner inconsistent with its labeling.

Do not apply in Long Island, New York.

Do not apply MOCAP® Plus 4-2 EC Nematicide-Insecticide through any type of irrigation system.

REENTRY STATEMENT

Reentry into treated fields is prohibited within 24 hours of application. State or local use and reentry regulations may be more restrictive than those stated on this label. Do not apply this product in such a manner as to directly or through drift expose workers or other persons. The area being treated must be vacated by unprotected persons. Because certain states may require more restrictive reentry intervals for various crops treated with this product, consult your State Department of Agriculture for further information.

Do not allow reentry into treated fields within 24 hours of application unless appropriate protective clothing is worn. Written or oral warnings must be given to workers who are expected to be in a treated area or in an area about to be treated with this product. Oral warnings must be given if there is reason to believe that written warnings cannot be understood by workers and given in a language customarily understood by the workers.

Oral and written warnings must include the STATEMENT OF PRACTICAL TREATMENT and notification that unprotected persons may enter treated fields or area only after chemical has been mixed in the soil.

EQUIPMENT: MOCAP® spray mixtures may affect some plastic materials after prolonged use. Spray tubing, hoses and gaskets should be inspected frequently, particularly black-colored materials. Replace any material which is softened or swollen. Polyethylene, polypropylene, nylon and teflon materials are recommended. Polyvinyl chloride (PVC) material is not recommended.

MIXING: Add the recommended amount to the water in the spray tank and mix well. Use enough water in the sprayer (usually 20 to 25 gallons) to thoroughly and evenly cover the area to be treated. Agitate frequently during use.

INCORPORATION INFORMATION

MOCAP® PLUS 4-2 EC Nematicide-Insecticide can be incorporated with a rotary tiller, rotary hoe, springtooth harrow, by double-discing or with other equipment for mixing the chemical with the soil to the depth recommended. Follow any specific incorporating directions given under the crop heading.

If moisture is applied immediately after application to carry the treatment into the soil, only shallow incorporation (1/4 to 1/2 inch) with mechanical equipment such as drag chains is required.

Tobacco Nematodes Controlled	Tobacco Insects Controlled
Dagger nematode (*Xiphinema* spp.)	Wireworm (*Conoderus* spp.)
Lesion nematode (*Pratylenchus* spp.)	Tobacco flea beetle (*Epitrix hirtipennis*)
Root-knot nematode (*Meloidogyne hapla*) (*Meloidogyne incognita*)	Tobacco Aphid (*Myzus persicae*)
Stunt nematode (*Tylenchorhynchus* spp.)	

TOBACCO

FOR CONTROL OF NEMATODES, FLEA BEETLES, APHIDS, AND WIREWORMS

Row Treatment: Use 1½ gallons of MOCAP® PLUS 4-2 EC Nematicide-Insecticide per acre (42 inch row spacing) or 15.4 fluid ounces per 1000 linear feet of row. Apply after fertilizer has been listed or bedded. Make application in a band 18 inches wide on the row from 1 week before planting to at-planting time. Mix MOCAP® PLUS 4-2 EC with the top 2 to 4 inches of soil right after application. Use a rotary tiller, rotary hoe or other cultivating equipment that will do a good job of mixing MOCAP® PLUS 4-2 EC with the soil.

FOR CONTROL OF NEMATODES, FLEA BEETLES, APHIDS, AND WIREWORMS

Broadcast Treatment: Apply 1½ to 2 gallons of MOCAP® PLUS 4-2 EC per acre from 1 week before planting to at-planting time. Mix with the top 2 to 4 inches of soil right after application. Use a double gang disc harrow or other equipment that will do a good job of mixing MOCAP® PLUS 4-2 EC with the soil. After mixing, apply fertilizer and, for best results, shape a wide bed with a tractor equipped with 4 disc hillers and wide sweeps for running out the middles. Be sure that only the treated soil (top 2 to 4 inches) is used to make beds. Use the higher rate in fields where nematode injury to tobacco has been especially severe, or where the soil is sandy.

FOR CONTROL OF NEMATODES, APHIDS, FLEA BEETLES, WIREWORMS, BLACK SHANK AND BLUE MOLD

Broadcast Treatment In Tank Mix With Ridomil® 2EC: Use 1½ to 2 gallons of MOCAP® PLUS 4-2 EC Nematicide-Insecticide mixed with 2 to 4 quarts of Ridomil® 2EC per acre. Mix the treatment with the top 2 to 4 inches of soil right after application, then form beds. Read both product labels for specific use and rate information. Follow all precautions, directions, and restrictions on the labels.

Rhône-Poulenc Ag Company does not recommend the use of MOCAP® PLUS 4-2 EC Nematicide-Insecticide on tobacco in Florida.

Do not plant any food crop in rotation after a field treatment with disulfoton unless it is a registered use for disulfoton.

STORAGE AND DISPOSAL

STORAGE

Do not contaminate water, food or feed by storage or disposal.

PESTICIDE DISPOSAL

Pesticide wastes are acutely hazardous. Improper disposal of excess pesticide, spray mixture, or rinsate is a violation of Federal law. If these wastes cannot be disposed of by use according to label instructions, contact your State Pesticide or Environmental Control Agency, or the Hazardous Waste representative at the nearest EPA Regional Office for guidance.

CONTAINER DISPOSAL

Triple rinse or equivalent. Then puncture and dispose of in a sanitary landfull, or by other procedures approved by state and local authorities.

LIMITED WARRANTY AND DISCLAIMER

The manufacturer warrants that this product conforms to the chemical description on the label; that this product is reasonably fit for the purposes set forth in the directions for use when it is used in accordance with such directions; and that the directions, warnings and other statements on this label are based upon responsible experts' evaluation of reasonable tests of effectiveness, of toxicity to laboratory animals and to plants, and of residues on food crops, and upon reports of field experience. Tests have not been made on all varieties or in all states or under all conditions. THE MANUFACTURER NEITHER MAKES NOR INTENDS NOR DOES IT AUTHORIZE ANY AGENT OR REPRESENTATIVE TO MAKE, ANY OTHER WARRANTIES, EXPRESS OR IMPLIED AND IT EXPRESSLY EXCLUDES AND DISCLAIMS ALL IMPLIED WARRANTIES OF MERCHANTIBILITY OR FITNESS FOR A PARTICULAR PURPOSE.

THIS WARRANTY DOES NOT EXTEND TO, AND THE BUYER SHALL BE SOLELY RESPONSIBLE FOR, ANY AND ALL LOSS OR DAMAGE WHICH RESULTS FROM THE USE OF THIS PRODUCT IN ANY MANNER WHICH IS INCONSISTENT WITH THE LABEL DIRECTIONS, WARNINGS OR CAUTIONS.

BUYER'S EXCLUSIVE REMEDY AND MANUFACTURER OR SELLER'S EXCLUSIVE LIABILITY FOR ANY AND ALL CLAIMS, LOSSES, DAMAGES OR INJURIES RESULTING FROM THE USE OR HANDLING OF THIS PRODUCT, WHETHER OR NOT BASED IN CONTRACT, NEGLIGENCE, STRICT LIABILITY IN TORT OR OTHERWISE, SHALL BE LIMITED, AT THE MANUFACTURER'S OPTION, TO REPLACEMENT OF OR THE REPAYMENT OF THE PURCHASE PRICE FOR, THE QUANTITY OF PRODUCT WITH RESPECT TO WHICH DAMAGES ARE CLAIMED. IN NO EVENT SHALL MANUFACTURER OR SELLER BE LIABLE FOR SPECIAL, INDIRECT OR CONSEQUENTIAL DAMAGES RESULTING FROM THE USE OR HANDLING OF THIS PRODUCT.

NOTICE TO BUYER

Purchase of this material does not confer any rights under patents governing this product or the use thereof in countries outside of the United States.

Ridomil® is a registered trademark of Ciba-Geigy for metalaxyl.
U.S. Patent No. 4,151,299
MOCAP is a registered trademark of Rhône-Poulenc
AG 90052

EPA Approval: 7/25/89

THIS SPECIMEN LABEL IS INTENDED FOR USE ONLY AS A GUIDE IN PROVIDING GENERAL INFORMATION REGARDING THE DIRECTIONS, WARNINGS AND CAUTIONS ASSOCIATED WITH THE USE OF THIS PRODUCT. AS WITH ANY AGRICULTURAL CHEMICAL, ALWAYS FOLLOW THE LABEL INSTRUCTIONS ON THE PACKAGE BEFORE USING.

NAA-800

FOR CONTROL OF PRE-HARVEST DROP OF APPLES AND PEARS

ACTIVE INGREDIENT:	
*Potassium salt of 1-Naphthaleneacetic acid	24.2%
INERT INGREDIENTS:	75.8%
TOTAL	100.0%

*1-Naphthaleneacetic acid equivalent 20.1% by weight or 800 grams per gallon.

EPA Reg. No. 264-142 EPA Est. No.

KEEP OUT OF REACH OF CHILDREN

WARNING AVISO

PRECAUCION AL USUARIO: Si usted no lee ingles, no use este producto hasta que la etiqueta haya sido explicada ampliamente.

For PRODUCT USE Information Call 1-800-334-9745
For MEDICAL AND TRANSPORTATION Emergencies ONLY Call 24 hours A Day 1-800-334-7577

PRECAUTIONARY STATEMENTS

WARNING

HAZARD TO HUMANS AND DOMESTIC ANIMALS

Harmful if swallowed. Causes substantial but temporary eye irritation. Do not get in eyes, on skin or on clothing. Wear safety goggles. Do not breathe spray mist. Wear full-length trousers, long-sleeve shirt and head covering. Do not eat, drink, or use tobacco while using. Change to clean clothing daily. Remove contaminated clothing and wash before reuse. Wash hands and face with soap and water after handling and before eating. Keep away from domestic animals.

STATEMENT OF PRACTICAL TREATMENT

GENERAL: Contact a physician in all cases of suspected poisoning. If poisoning is suspected in animals, contact a veterinarian.

IF SWALLOWED: If subject is conscious and alert, give 2 to 3 glasses of water or milk to drink. If available, give one tablespoon of Syrup of Ipecac to induce vomiting. Alternatively, induce vomiting by touching back of throat with finger. Do not repeat the Syrup of Ipecac. Do not make an unconscious person vomit. GET MEDICAL ATTENTION.

IN CASE OF EYE CONTACT: Hold eyelids open and flush with a steady, gentle stream of water for at least 15 minutes. GET MEDICAL ATTENTION.

IN CASE OF SKIN CONTACT: Wash with soap and plenty of water.

IF INHALED: Remove from contaminated atmosphere. Call a physician.

ENVIRONMENTAL HAZARDS

Do not apply directly to water. Do not contaminate water by cleaning of equipment or disposal of wastes. Do not contaminate irrigation ditches or water used for irrigation or domestic purposes. Do not apply when weather conditions favor drift from treated areas. Do not use in a greenhouse.

PHYSICAL/CHEMICAL HAZARDS

Flammable. Do not use or store near heat sources or open flame. Do not use in the vicinity of pilot lights.

DIRECTIONS FOR USE

IT IS A VIOLATION OF FEDERAL LAW TO USE THIS PRODUCT IN A MANNER INCONSISTENT WITH ITS LABELING.

Do not apply NAA-800 through any type of irrigation system.

GENERAL WORKER PROTECTION STATEMENTS

Do not apply this product in such a manner as to directly or through drift expose workers or other persons. The area treated must be vacated by unprotected persons.

Do not enter treated areas without protective clothing until sprays have dried.

Because certain states may require more restrictive reentry intervals for various crops treated with this product, consult your State Department of Agriculture for further information.

Written or oral warnings must be given to workers who are expected to be in a treated area or in an area about to be treated with this product. Advise workers to stay out of fields during application and until sprays have dried. Regular long-sleeved work clothing should be worn when working in treated fields. See PRECAUTIONARY STATEMENTS and STATEMENT OF PRACTICAL TREATMENT for information on accidental exposures. When oral warnings are given, warnings shall be given in a language customarily understood by workers. Oral warnings must be given if there is reason to believe that written warnings cannot be understood by workers. Written warnings must include the following information: Appropriate signal word (WARNING), area treated with NAA-800, date of application, appropriate clothing, and reentry interval (i.e., until sprays have dried).

FOR CONTROL OF PRE-HARVEST DROP OF APPLES & PEARS

DIRECTIONS - PRE-HARVEST DROP

NAA-800 contains 800 grams of naphthaleneacetic acid equivalent per gallon and has been formulated for aerial and ground applications to control pre-harvest drop of apples and pears.

It is important that aerial operators apply NAA-800 under conditions where drift can be minimized so that material is equally distributed over the area being sprayed.

Applications should be completed in time to become effective before actual pre-harvest drop begins. NAA-800 becomes effective in one or two days when applied to summer and early fall maturing varieties. For later varieties ripening during cooler weather, a slightly longer period is necessary. Healthy foliage, freed from mites and frost damage is necessary for the most effective action from this material. For maximum effectiveness, sprays should be applied when the temperature is 70° F or higher.

For aerial application use at least 5 gallons of water per acre and for ground applications use sufficient amounts of water to ensure good coverage.

In California, do not use this product for aerial application on pears.

AMOUNTS PER ACRE

APPLES

Proper timing of sprays to reduce pre-harvest drop is extremely important. Make applications as the first few sound, uninjured fruit begin to drop. With summer varieties, 1/4 pint NAA-800 per acre should be used, especially in hot weather.

For a variety such as McIntosh, 1/3 pint of NAA-800 per acre is suggested. A second application of the same amount can be applied 5 to 6 days after the first application. Do not harvest within 5 days of the last application.

With later varieties, such as Delicious, 1/3 pint of NAA-800 for one acre of apples is suggested.

PEARS

Make applications as the first few sound uninjured fruit begin to drop. Use 1/8 to 1/4 pint of NAA-800 per acre.

IMPORTANT: NAA-800 inhibits fruit drop, but does not effect normal ripening. The fruit must be picked at the proper stage of maturity even though NAA-800 is still effectively delaying excessive drop. Do not apply within 5 days of harvest. Do not make more than 2 applications per growing season.

STORAGE AND DISPOSAL

STORAGE

Do not store near fertilizers, seeds, insecticides or fungicides.

PESTICIDE DISPOSAL

Do not contaminate water, food or feed by storage or disposal. Wastes resulting from the use of this product may be disposed of on site or at an approved waste disposal facility.

CONTAINER DISPOSAL

Triple rinse (or equivalent). Then offer for recycling or reconditioning, or puncture and dispose of in a sanitary landfill, or incineration, or if allowed by state and local authorities, by burning. If container is burned, stay out of smoke.

LIMITED WARRANTY AND DISCLAIMER

The manufacturer warrants; that this product conforms to the chemical description on the label; that this product is reasonably fit for the purposes set forth in the directions for use when it is used in accordance with such directions; and that the directions, warnings and other statements on this label are based upon responsible experts' evaluation of reasonable tests of effectiveness of toxicity to laboratory animals and to plants, and of residues on food crops, and upon reports of field experience. Tests have not been made on all varieties or in all states or under all conditions. THE MANUFACTURER NEITHER MAKES NOR INTENDS, NOR DOES IT AUTHORIZE ANY AGENT OR REPRESENTATIVE TO MAKE ANY OTHER WARRANTIES, EXPRESS OR IMPLIED, AND IT EXPRESSLY EXCLUDES AND DISCLAIMS ALL IMPLIED WARRANTIES OF MERCHANTABILITY OR FITNESS FOR A PARTICULAR PURPOSE.

THIS WARRANTY DOES NOT EXTEND TO, AND THE BUYER SHALL BE SOLELY RESPONSIBLE FOR, ANY AND ALL LOSS OR DAMAGE WHICH RESULTS FROM THE USE OF THIS PRODUCT IN ANY MANNER WHICH IS INCONSISTENT WITH THE LABEL DIRECTIONS, WARNINGS OR CAUTIONS.

BUYER'S EXCLUSIVE REMEDY AND MANUFACTURER'S OR SELLER'S EXCLUSIVE LIABILITY FOR ANY AND ALL CLAIMS, LOSSES, DAMAGES OR INJURIES RESULTING FROM THE USE OR HANDLING OF THIS PRODUCT, WHETHER OR NOT BASED IN CONTRACT, NEGLIGENCE, STRICT LIABILITY IN TORT OR OTHERWISE, SHALL BE LIMITED, AT THE MANUFACTURER'S OPTION, TO REPLACEMENT OF, OR THE REPAYMENT OF THE PURCHASE PRICE FOR, THE QUANTITY OF PRODUCT WITH RESPECT TO WHICH DAMAGES ARE CLAIMED IN NO EVENT SHALL MANUFACTURER OR SELLER BE LIABLE FOR SPECIAL, INDIRECT OR CONSEQUENTIAL DAMAGES RESULTING FROM THE USE OR HANDLING OF THIS PRODUCT.

NOTICE TO BUYER

Purchase of this material does not confer any rights under patents governing this product or the use thereof in countries outside of the United States.

AG90042

THIS SPECIMEN LABEL IS INTENDED FOR USE ONLY AS A GUIDE IN PROVIDING GENERAL INFORMATION REGARDING THE DIRECTIONS, WARNINGS AND CAUTIONS ASSOCIATED WITH THE USE OF THIS PRODUCT. AS WITH ANY AGRICULTURAL CHEMICAL, ALWAYS FOLLOW THE LABEL INSTRUCTIONS ON THE PACKAGE BEFORE USING.

PREP™

PLANT REGULATOR FOR COTTON

FOR COMMERCIAL USE OR AGRICULTURAL USE ONLY.
NOT FOR RESIDENTIAL USE.

ACTIVE INGREDIENT:	
Ethephon (2-Chloroethyl) phosphonic acid*	55.4%
INERT INGREDIENTS:	44.6%
TOTAL	100.0%

*This product contains 6 pounds ethephon per gallon.

EPA Reg. No. 264-418 EPA Est. No. 264-PA-01

KEEP OUT OF REACH OF CHILDREN

DANGER PELIGRO

PRECAUCION AL USUARIO: Si usted no lee ingles, no use este producto hasta que la informacion en la etiqueta le haya sido explicado ampliamente.

STATEMENT OF PRACTICAL TREATMENT

IF IN EYES: Hold eyelids open and flush with a steady, gentle stream of water for at least 15 minutes. Get medical attention, preferably an Ophthalmologist.

For MEDICAL And TRANSPORTATION Emergencies ONLY Call 24 Hours A Day 1-800-334-7577

For PRODUCT USE Information Call 1-800-334-9745

The use of this product for a variety of plant growth regulation uses is protected by various United States and foreign patents including U.S. Patents 3,879,188 and 4,240,819. No license is granted to use this product in countries other than the United States or for any use not contemplated by this label. Liability for patent infringement may result from use or sale of this product outside the United States.

PRECAUTIONARY STATEMENTS

DANGER

HAZARDS TO HUMANS AND DOMESTIC ANIMALS

CORROSIVE. Causes eye and skin damage. May be fatal if absorbed through skin. Harmful if swallowed or inhaled. Avoid breathing spray mist. Do not get in eyes, on skin or on clothing. Wear full length trousers, long sleeved shirt, protective gloves, and goggles or face shield when handling. Change to clean work clothes daily. Wash thoroughly with soap and water after handling and before eating, or smoking. Remove and wash contaminated clothing before reuse. Keep away from domestic animals.

STATEMENT OF PRACTICAL TREATMENT

IF IN EYES: Hold eyelids open and flush with a gentle stream of water for at least 15 minutes. Get medical attention, preferably an ophthalmologist.

IF ON SKIN: Immediately wash skin with plenty of soap and water. Remove contaminated clothing and shoes. Wash clothing separately before reuse.

IF SWALLOWED: Do not induce vomiting. Promptly drink a large quantity of milk, egg whites, gelatin solution or if these are not available, drink large quantities of water. Avoid alcohol. Call a physician immediately.

IF INHALED: Remove victim to fresh air. If not breathing, give artificial respiration. Administer oxygen if necessary. Get medical attention.

NOTE TO PHYSICIAN

There is no specific antidote. All treatments should be based on observed signs and symptoms of distress in the patient. Overexposure to materials other than this product may have occurred.

Victims of severe overexposure, by inhalation, should be kept under medical observation for up to 72 hours for delayed onset of pulmonary edema. In a victim of overexposure by ingestion, careful gastric lavage is required due to the possibility of stomach or esophogeal perforation. This material is an acid, but use of alkaline substances to neutralize it is contraindicated.

ENVIRONMENTAL HAZARDS

Do not apply directly to water or wetlands (swamps, bogs, marshes, and potholes). Do not contaminate water used for irrigation or domestic purposes.

USE PRECAUTIONS

Do not apply PREP™ Plant Regulator for cotton through any type of irrigation system.

Avoid spray drift to nearby crops as this product will cause modifications in plant growth. Plant injury or reduced yields will result.

Do not plant another crop within 30 days after treatment.

Mix only the amount of spray you expect to use each day. Do not allow mixed solution to stand overnight.

DIRECTIONS FOR USE

IT IS A VIOLATION OF FEDERAL LAW TO USE THIS PRODUCT IN A MANNER INCONSISTENT WITH ITS LABELING.

Mixers, loaders and applicators must wear a full face shield, long trousers, long sleeved shirt, gloves and boots to avoid as much skin and eye contact as possible.

Do not enter treated fields within 24 hours after application unless protective clothing is worn.

GENERAL INFORMATION

A foliar spray of PREP™ Plant Regulator will accelerate opening of mature unopened cotton bolls and enhance defoliation which can result in earlier harvest with an increased recoverable yield. PREP™ Plant Regulator treatment allows increased efficiency from a once-over harvest. Some premature drop of small, immature bolls may be associated with treatment. Do not apply before sufficient, mature, unopened bolls have developed to produce the desired yield of cotton.

A boll is mature when it is too hard to be dented when squeezed between thumb and fingers, too hard to be sliced with a sharp knife, and when the seed coat becomes light brown in color.

For best results, thorough and uniform spray coverage is required. It is essential that the bolls are contacted by the spray to obtain optimum boll opening. Observe treated fields closely. Harvest at optimum boll opening and defoliation, generally 14 to 21 days after treatment. Sustained cool weather with average day and night temperatures of 65°F or lower after treatment with PREP™ Plant Regulator will delay the boll opening response.

DO NOT APPLY PREP™ PLANT REGULATOR IF RAIN IS EXPECTED WITHIN 6 TO 8 HOURS. Rainfall within 6 to 8 hours of application may reduce product performance.

Certified cotton seed growers should observe boll and seed maturity before treatment with PREP™ Plant Regulator. Bolls not completely mature might open with treatment of PREP™ Plant Regulator giving lower quality seed.

Spray deposits of PREP™ Plant Regulator, when left on aircraft acrylic plastic type windshields for longer than 1 hour, may react with this material, forming an opaque film or spots on the surface which cannot be removed by washing with water or any other cleaner. Rinse windshields with water immediately after each tankload of PREP™ Plant Regulator has been applied. Thoroughly rinse the aircraft with clean water after each day PREP™ Plant Regulator is applied.

PREP™ Plant Regulator may also affect certain paints, and if accidental deposits on painted objects such as cars, trucks, agricultural equipment, etc. occurs, these objects should be rinsed with water as soon as possible (within 1 hour) to prevent permanent staining of the painted surface.

SPRAY PREPARATION

Add 1/2 to 3/4 of the required amount of water to the spray tank. Start agitation. Add the required amount of PREP™ Plant Regulator, and the remaining amount of water. Prepare only as much spray solution as can be used on the day of mixing. Do not allow spray solution to stand overnight. Do not spill the concentrated product on spray equipment, or any airplane parts. ANY SPILLS SHOULD BE RINSED IMMEDIATELY WITH PLENTY OF WATER.

Use of a nurse tank is highly recommended for avoiding possible spills of concentrated formulation on spray equipment or any airplane parts.

EQUIPMENT CLEANING

Because of the acidic nature of this product, prolonged exposure to spray deposit will damage acrylic plastics, certain paints, and metals.

Rinse thoroughly with a detergent and water all exposed acrylic plastic-type materials (e.g., aircraft windshields), and painted surfaces **within an hour** after exposure to spray deposits.

At the end of each day, rinse thoroughly with a detergent and water all the metal parts of the aircraft and the associated spray equipment exposed to the spray deposits.

BOLL OPENING

RATES FOR BOLL OPENING

CONDITIONS EXPECTED 5-7 DAYS AFTER TREATMENT	PREP™ PINTS/ACRE	ONE GALLON OF PREP™ TREATS	SPRAY VOLUME GAL./ACRE GROUND	AIR
Hot and dry 80°F or higher	1 1/3 (1.0 lb a.i.)	6 Acres	15 to 50	2 to 5*
Dry with temperature 75°F to 80°F	2 (1.5 lb a.i.)	4 Acres	15 to 50	2 to 5*
Cool but higher than 65°F	2 2/3 (2.0 lb a.i.)	3 Acres	15 to 50	2 to 5*
Rank cotton	2 2/3 (2.0 lb a.i.)	3 Acres	25 to 50	5

*NOTE: For California and Arizona use a volume of no less than 5 gallons per acre for aerial applications

Time of Application (Boll Opening)

Apply when there are sufficient mature unopened bolls to produce the desired yield, but not later than when 60% of the bolls are opened (see General Information section on how to test for boll maturity). Do not tank mix PREP™ Plant Regulator with a desiccant if the cotton is to be spindle harvested.

PRECONDITIONING FOR DEFOLIATION

Rates for Preconditioning for Defoliation

CONDITIONS EXPECTED 5-7 DAYS AFTER TREATMENT	PREP™ PINTS/ACRE	ONE GALLON OF PREP™ TREATS	SPRAY VOLUME GAL./ACRE GROUND	AIR
Hot and dry 80°F or higher	⅔ (0.5 lb a.i)	12 Acres	15 to 50	2 to 5*
Cool but higher than 65° and also for rank cotton	1⅓ (1.0 lb a.i.)	6 Acres	15 to 50	2 to 5*

*NOTE: For California and Arizona use a volume of no less than 5 G.P.A. for aerial applications.

Time of Application (Preconditioning)

Apply PREP™ Plant Regulator 4 to 7 days prior to application of defoliant or tank mixed with the defoliant (see section on Compatibility). Do not tank mix PREP™ with desiccants unless plant desiccation is required.

Pre-Treatment With Defoliants Prior to PREP™ Plant Regulator

If the cotton is overly rank or laying down in the middles and good spray coverage of the bolls with PREP™ is difficult, a pretreatment with defoliants will be useful to improve boll coverage with PREP™. Use dosage rates of PREP™ Plant Regulator recommended for boll opening. Read and observe all appropriate label use directions and precautions for the defoliant used.

Do not use a defoliant before there are sufficient mature unopened bolls to produce the desired yield (see General Information section on How to Test for Boll Maturity).

COMPATIBILITY

PREP™ Plant Regulator is compatible with DEF, FOLEX, DROPP, HARVADE, METHYL PARATHION, GUTHION AND MALATHION, and may be applied in sequence or as a tank mixture (DO NOT TANK MIX WITH DESICCANTS IF COTTON IS TO BE SPINDLE HARVESTED). In some cases slight reduction in boll opening response has been observed when tank mixes with defoliants were used.

Good agitation in the spray tank is essential and a tank mixture should not be allowed to stand without agitation for more than 5 to 10 minutes. Read and observe all appropriate label use directions and precautions for the defoliants and insecticides used.

PRECAUTION

DO NOT TANK MIX PREP™ PLANT REGULATOR WITH DEFOLIANTS CONTAINING SODIUM CHLORATE BECAUSE THIS RESULTS IN THE FORMATION OF HYPOCHLOROUS ACIDS WHICH UPON HEATING EMIT TOXIC CHLORINE FUMES.

WHEN TO HARVEST

Do not harvest cotton sooner than 7 days after a treatment with PREP™ Plant Regulator. Observe the treated crop and harvest when optimum boll opening has been reached. Too early harvest might reduce the full advantage of the treatment and too late a harvest may result in reduced quality and loss of lint which will drop from the plant.

STORAGE AND DISPOSAL

STORAGE

Do not contaminate water, food, or feed by storage or disposal.

If container is broken or contents have spilled, follow all precautions indicated above and clean up immediately. Before cleaning up, put on full length trousers, long sleeved shirt, protective gloves, and goggles or face shield. Soak up spill with absorbent media such as sand, earth or other suitable material and dispose of waste at an approved waste disposal facility.

PESTICIDE DISPOSAL

Pesticide wastes are acutely hazardous. Improper disposal of excess pesticide, spray mixture, or rinsate is a violation of Federal law. If these wastes cannot be disposed of by use according to label instructions, contact your State Pesticide or Environmental Control Agency or the Hazardous Waste representative at the nearest EPA Regional Office for guidance.

CONTAINER DISPOSAL

Triple rinse (or equivalent) the empty containers. Then offer for recycling or reconditioning, or puncture and dispose of in a sanitary landfill, or by incineration or if allowed by state and local authorities, by burning. If container is burned, stay out of smoke.

LIMITED WARRANTY AND DISCLAIMER

The manufacturer warrants (a) that this product conforms to the chemical description on the label; (b) that this product is reasonably fit for the purposes set forth in the directions for use when it is used in accordance with such directions; and (c) that the directions, warning and other statements on this label are based upon responsible experts' evaluation of reasonable tests of effectiveness, of toxicity to laboratory animals and to plants, and of residues on food crops and upon reports of field experience. Tests have not been made on all varieties or in all states or under all conditions. THE MANUFACTURER NEITHER MAKES NOR INTENDS, NOR DOES IT AUTHORIZE ANY AGENT OR REPRESENTATIVE TO MAKE, ANY OTHER WARRANTIES, EXPRESS OR IMPLIED, AND IT EXPRESSLY EXCLUDES AND DISCLAIMS ALL IMPLIED WARRANTIES OF MERCHANTABILITY OR FITNESS FOR A PARTICULAR PURPOSE.

THIS WARRANTY DOES NOT EXTEND TO, AND THE BUYER SHALL BE SOLELY RESPONSIBLE FOR, ANY AND ALL LOSS OR DAMAGE WHICH RESULTS FROM THE USE OF THIS PRODUCT IN ANY MANNER WHICH IS INCONSISTENT WITH THE LABEL DIRECTIONS, WARNINGS OR CAUTIONS.

BUYER'S EXCLUSIVE REMEDY AND MANUFACTURER'S OR SELLER'S EXCLUSIVE LIABILITY FOR ANY AND ALL CLAIMS, LOSSES, DAMAGES, OR INJURIES RESULTING FROM THE USE OR HANDLING OF THIS PRODUCT, WHETHER OR NOT BASED IN CONTRACT, NEGLIGENCE, STRICT LIABILITY IN TORT OR OTHERWISE SHALL BE LIMITED, AT THE MANUFACTURER'S OPTION, TO REPLACEMENT OF, OR THE REPAYMENT OF THE PURCHASE PRICE FOR, THE QUANTITY OF PRODUCT WITH RESPECT TO WHICH DAMAGES ARE CLAIMED. IN NO EVENT SHALL MANUFACTURER OR SELLER BE LIABLE FOR SPECIAL, INDIRECT OR CONSEQUENTIAL DAMAGES RESULTING FROM THE USE OR HANDLING OF THIS PRODUCT.

NOTICE TO BUYER

Purchase of this material does not confer any rights under patents governing this product or the use thereof in countries outside of the United States.

PREP is a trademark of Rhône-Poulenc

AG90043

THIS SPECIMEN LABEL IS INTENDED FOR USE ONLY AS A GUIDE IN PROVIDING GENERAL INFORMATION REGARDING THE DIRECTIONS, WARNINGS AND CAUTIONS ASSOCIATED WITH THE USE OF THIS PRODUCT. AS WITH ANY AGRICULTURAL CHEMICAL, ALWAYS FOLLOW THE LABEL INSTRUCTIONS ON THE PACKAGE BEFORE USING.

RHOMENE®

MCPA Amine HERBICIDE

Controls weeds in flax, rice, peas, and small grains

ACTIVE INGREDIENT:

Dimethylamine salt of 2-methyl-4-chlorophenoxyacetic acid*	52.1%
INERT INGREDIENTS:	47.9%
TOTAL	100.0%

*2-Methyl-4-chlorophenoxyacetic acid equivalent 42.6% by weight or four pounds per gallon.

E.P.A. Reg. No. 264-47AA E.P.A. Est. No. 264-MO-01

KEEP OUT OF REACH OF CHILDREN

DANGER PELIGRO

For PRODUCT USE Information Call 1-800-334-9745
For MEDICAL and TRANSPORTATION Emergencies ONLY Call 24 Hours A Day 1-800-334-7577

PRECAUTIONARY STATEMENTS

DANGER

HAZARDS TO HUMANS AND DOMESTIC ANIMALS

Corrosive. Causes irreversible eye damage. Harmful if swallowed. May be fatal if absorbed through the skin. Avoid contact with eyes, skin or clothing. Avoid breathing vapors or spray mist. Remove and wash contaminated clothing before reuse. Wash thoroughly before eating or smoking.

STATEMENT OF PRACTICAL TREATMENT

IF SWALLOWED: If patient is conscious and alert, give 2 to 3 glasses of water to drink. Do not induce vomiting. Get medical attention.

IF ON SKIN: Wash skin with plenty of soap and water, if available, while removing contaminated clothing and shoes. Wash clothing separately before reuse.

IF IN EYES: Hold eyelids open and flush with a steady gentle stream of water for 15 minutes. Get medical attention, preferably an ophthalmologist.

IF INHALED: Remove victim to fresh air. If not breathing, give artificial respiration. Administer oxygen if necessary. GET MEDICAL ATTENTION.

NOTE TO PHYSICIAN

No specific antidote is available.

All treatments should be based on observed signs and symptoms of distress in the patient. Overexposure to materials other than this product may have occurred.

This product is irritating to mucous membranes. If large amounts (greater than 1 ml/kg body weight) of the product have been ingested, the stomach should be evacuated by gastric intubation with the aid of a cuffed endotracheal tube to prevent exposure of the esophagus. After removal of stomach contents, wash stomach by instilling 30-50 g. of activated charcoal in 3-4 ounces of water through the stomach tube and again remove stomach contents. Avoid oily laxatives.

ENVIRONMENTAL HAZARDS

Do not apply directly to water except as specified on this label. Do not contaminate water by cleaning of equipment or disposal of wastes. Avoid spray drift to susceptible plants such as cotton, beans, tomatoes, and ornamentals (Coarse sprays are less likely to drift). Do not apply when weather conditions favor drift from treated areas. Avoid use of small diameter spray nozzles. At high air or ground surface temperatures, vapors from this product may injure susceptible plants. Do not contaminate water used for irrigation or domestic purposes. Do not use the same spray equipment for other purposes unless thoroughly cleaned. Do not use in or near a greenhouse.

GENERAL WORKER PROTECTION STATEMENT

Do not apply this product in such a manner as to directly or through drift expose workers or other persons. The area treated must be vacated by unprotected persons.

Do not enter treated areas without protective clothing until sprays have dried.

Because certain states may require more restrictive reentry intervals for various crops treated with this product, consult your State Department of Agriculture for further information.

Written or oral warnings must be given to workers who are expected to be in treated area or in an area about to be treated with this product. Advise workers to stay out of fields during application and until sprays have dried. Regular long-sleeved work clothing should be worn when working in treated fields. See PRECAUTIONARY STATEMENTS, STATEMENT OF PRACTICAL TREATMENT and NOTE TO PHYSICIAN for information on accidental exposures. When oral warnings are given, warnings shall be given in a language customarily understood by workers. Oral warnings must be given if there is reason to believe that written warnings cannot be understood by workers. Written warnings must include the following information: Appropriate signal word (DANGER), area treated with RHOMENE® MCPA Amine Herbicide, date of application, appropriate protective clothing, and reentry interval (i.e., until sprays have dried).

DIRECTIONS FOR USE

IT IS A VIOLATION OF FEDERAL LAW TO USE THIS PRODUCT IN A MANNER INCONSISTENT WITH ITS LABELING.

Do not apply RHOMENE® MCPA Amine Herbicide through any type of irrigation system.

GENERAL INFORMATION

RHOMENE® Herbicide is effective on a large number of broadleaf weeds and is useful for controlling these weeds in certain crops. Several crops, such as flax, rice, oats and small grains underseeded to legumes, are more tolerant of MCPA than they are of 2,4-D.

RHOMENE® Herbicide is the amine form of MCPA that is easily dissolved in water, even in the hard waters found in some areas. Mix thoroughly while diluting this material and mix again if allowed to stand long before spraying.

COMPATIBILITY: If RHOMENE® Herbicide is to be tank mixed with other pesticides, compatibility should be tested prior to mixing. To test for compatibility, use a small container and mix a small amount (0.5 to 1 quart) of spray, combining all ingredients in the same ratio as the anticipated use. If any indications of physicial incompatibility develop, do not use this mixture for spraying. Indications of incompatibility usually will appear within 5 to 15 minutes after mixing. Read and follow all directions and precautions on this label and on the labels of any product(s) for which a tank mixture is being considered.

WEEDS CONTROLLED BY PROPERLY APPLIED TREATMENTS OF RHOMENE® HERBICIDE INCLUDE THE FOLLOWING:

- *Amaranthus* species (pigweeds)
- common burdock
- common cocklebur
- common lambsquarters
- dandelion
- field pennycress
- mustard (except blue mustard)
- pepperweeds (except perennial)
- plantains
- purslanes
- ragweed (common and giant)
- shepherdspurse
- thistles (bull and Russian)
- Yellow rocket

To convert locally recommended application rates into amount of RHOMENE® Herbicide, use the following table.

MCPA acid	1½ lb	1¼ lb	1 lb	¾ lb	½ lb	⅜ lb	¼ lb
RHOMENE® Herbicide	3 pts	2½ pts	2 pts	1½ pts	1 pt	¾ pt	½ pt

INJURY TO CROPS MAY OCCUR FROM THIS PESTICIDE. IF YOU ARE NOT PREPARED TO ACCEPT SOME DEGREE OF CROP INJURY, DO NOT USE THIS PRODUCT.

Crop varieties vary in response to MCPA, and some may be easily injured. Apply RHOMENE® Herbicide only to varieties known to be tolerant to MCPA. If you are uncertain concerning tolerant varieties or local use situations that may affect crop tolerance to MCPA, contact your seed company or state Agricultural Extension Service for advice.

RECOMMENDATIONS FOR APPLYING RHOMENE® MCPA AMINE HERBICIDE

CEREAL GRAINS: WHEAT, BARLEY, OATS, AND RYE (not underseeded with legumes)

CROP	AMOUNT RHOMENE® PER ACRE	DIRECTIONS
Annual Weeds	0.5 pint	Apply after grain is in the 4-leaf stage but not forming joints in the stem. *Do not spray in the boot to dough stage.*
Annual and Biennial weeds	1-2 pints*	Apply after grain is fully tillered (usually about 4 to 8 inches high), but not forming joints in the stem.
Perennial broadleaf weed	1.5 to 2 pints*	*Do not apply in the boot to dough stage.*
Emergency control in wheat Perennial broadleaf weed	3 pints	Apply when weeds are approaching bud stage, *but do not spray grain in the boot to dough stage.* The 3 pints (1.5 lb. acid equivalent) per acre application can produce injury to wheat. Balance the severity of your weed problem against the possibility of crop damage. Where perennial weeds are scattered, spot treatment is suggested to minimize the effect of crop injury.
WHEAT, BARLEY, OATS, and RYE: Underseeded with legumes: (alfalfa, alsike, red and ladino clovers, trefoil)	0.5 pint	Treat after grain is tillered and in the 4-leaf stage and when legumes are 2 to 3 inches tall. Small grains should form a protective canopy over the legume seedlings. Do not use more than 6 gallons of water per acre, higher volumes may result in injury to legumes. Do not apply to grain underseeded with vetch or sweet clover. Do not spray when grain is in the boot to dough stage.

*Use the lower rate if small annual and biennial weeds are the major problem. Use the higher rate if perennial weeds or annual and biennial weeds are present which are in the hard-to-kill categories as determined by local experience. The higher rates increase the risk of grain injury and should be used only where the weed control problem justifies the risk of grain damage.

RESTRICTIONS AND LIMITATIONS FOR USE ON SMALL GRAINS:

- Use a minimum of 10 gallons of water per acre for ground application and 2 gallons of water per acre for aerial application, unless specific site directions listed above indicate otherwise.
- Do not allow livestock to forage or graze treated areas within 7 days of slaughter.

CROP	AMOUNT RHOMENE® PER ACRE	DIRECTIONS
FLAX	⅓-½ pint	Apply when flax is 2 to 8 inches tall. Do not spray once flax has reached the bud stage. Use at least 8 to 10 gallons of water per acre for ground application and at least 3 to 5 gallons of water per acre for aerial application.
RICE	1½ - 2½ pints	Treat 7 to 10 weeks after planting when rice is fully tillered and 6 to 8 inches above water. **Do not spray**

CROP	AMOUNT RHOMENE® PER ACRE	DIRECTIONS
Rice (California only)	1 - 1½ pints	rice that has reached the boot stage. Use the higher rate where sedges, or bulrushes are the major problem. Use the lower rate for control of arrowhead, water plantain, and redstem. Do not spray when temperatures are over 90°F. Use at least 3 to 5 gallons of water per acre for aerial application. Treat when rice is in the 3 to 4-leaf stage, and has well established root systems, usually 18 to 24 days after planting. Applications made during hot weather or at the higher rate may cause temporary stunting of the rice. For best results, weeds should be exposed at time of application. If retreatment is necessary, usually 35 to 65 days after planting, the total amount per acre of RHOMENE® Herbicide applied should not exceed 2½ pints in one season. Use 8 to 10 gallons of water per acre for aerial application.

RESTRICTIONS AND LIMITATIONS FOR USE ON FLAX AND RICE:

Do not allow livestock animals to forage or graze treated areas within 7 days of slaugter.

Do not grow crayfish or catfish in treated rice fields.

CROP	AMOUNT RHOMENE® PER ACRE	DIRECTIONS
PEAS	½ to ¾ pint	Treat when peas are 4 to 6 inches tall. Do not spray while peas are in blossom. Peas may show slight injury, but they usually recover a few days after treatment. Use the higher rate before buds form on Canada thistle to prevent from going to seed. Do not spray peas that are stressed from lack of moisture or when temperatures are over 90°F. Apply in 15 to 30 gallons of water per acre for ground application and 10 gallons of water per acre for aerial application.

RESTRICTIONS AND LIMITATIONS FOR USE ON PEAS:

Do not allow livestock to forage or graze treated pea fields.

Do not feed treated vines to livestock.

STORAGE AND DISPOSAL

Do not store near food, feedstuffs, fertilizers, seeds, insecticides or fungicides. Do not contaminate water, food or feed by storage or disposal.

PESTICIDE DISPOSAL

Open dumping is prohibited. Pesticide wastes are acutely hazardous. Improper disposal of excess pesticide, spray mixture or rinsate is a violation of Federal law. If these wastes cannot be disposed of by use according to label instructions, contact your State Pesticide or Environmental Control Agency, or the Hazardous Waste representative at the nearest EPA Regional Office for guidance.

CONTAINER DISPOSAL

Triple rinse (or equivalent). Then offer for recycling or reconditioning, or puncture and dispose of in a sanitary landfill, or by incineration, or if allowed by state and local authorities, by burning. If burned, stay out of smoke.

LIMITED WARRANTY AND DISCLAIMER

The manufacturer warrants that this product conforms to the chemical description on the label, that this product is reasonably fit for the purposes set forth in the directions for use when it is used in accordance with such directions, and that the directions, warnings and other statements on this label are based upon responsible experts' evaluation of reasonable tests of effectiveness, of toxicity to laboratory animals and to plants, and of residues of food crops, and upon reports of field experience. Tests have not been made on all varieties or in all states or under all conditions. THE MANUFACTURER NEITHER MAKES NOR INTENDS, NOR DOES IT AUTHORIZE ANY AGENT OR REPRESENTATIVE TO MAKE, ANY OTHER WARRANTIES, EXPRESS OR IMPLIED, AND IT EXPRESSLY EXCLUDES AND DISCLAIMS ALL IMPLIED WARRANTIES OF MERCHANTABILITY OR FITNESS FOR A PARTICULAR PURPOSE.

THIS WARRANTY DOES NOT EXTEND TO, AND THE BUYER SHALL BE SOLELY RESPONSIBLE FOR, ANY AND ALL LOSS OR DAMAGE WHICH RESULTS FROM THE USE OF THIS PRODUCT IN ANY MANNER WHICH IS INCONSISTENT WITH THE LABEL DIRECTIONS, WARNINGS OR CAUTIONS.

BUYER'S EXCLUSIVE REMEDY AND MANUFACTURER'S OR SELLER'S EXCLUSIVE LIABILITY FOR ANY AND ALL CLAIMS, LOSSES, DAMAGES, OR INJURIES RESULTING FROM THE USE OR HANDLING OF THIS PRODUCT, WHETHER OR NOT BASED IN CONTRACT, NEGLIGENCE, STRICT LIABILITY IN TORT OR OTHERWISE, SHALL BE LIMITED, AT THE MANUFACTURER'S OPTION, TO REPLACEMENT OF, OR THE REPAYMENT OF THE PURCHASE PRICE FOR, THE QUANTITY OF PRODUCT WITH RESPECT TO WHICH DAMAGES ARE CLAIMED. IN NO EVENT SHALL MANUFACTURER OR SELLER BE LIABLE FOR SPECIAL, INDIRECT OR CONSEQUENTIAL DAMAGES RESULTING FROM THE USE OR HANDLING OF THIS PRODUCT.

NOTICE TO BUYER

Purchase of this material does not confer any rights under patents governing this product or the use thereof in countries outside of the United States.

RHOMENE is a registered trademark of Rhône-Poulenc

AG90021 EPA Approval: 7/25/87

THIS SPECIMEN LABEL IS INTENDED FOR USE ONLY AS A GUIDE IN PROVIDING GENERAL INFORMATION REGARDING THE DIRECTIONS, WARNINGS AND CAUTIONS ASSOCIATED WITH THE USE OF THIS PRODUCT. AS WITH ANY AGRICULTURAL CHEMICAL, ALWAYS FOLLOW THE LABEL INSTRUCTIONS ON THE PACKAGE BEFORE USING.

RHONOX®

MCPA Low Volatile Ester HERBICIDE

For the selective control of broadleaf weeds in small grains, rice, grasses, rangelands, pastures, and non-crop areas

ACTIVE INGREDIENT:

Isooctyl Ester of 2-Methyl-4-Chlorophenoxyacetic Acid*	68.7%
INERT INGREDIENTS:	31.3%
TOTAL	100.0%

*2-Methyl-4-Chlorophenoxyacetic Acid equivalent 44.1%, 3.7 lbs./gals. Isomer Specific by AOAC Method

EPA REG. NO. 228-156-264 EPA EST. NO. 228-IL-1

KEEP OUT OF REACH OF CHILDREN

WARNING

For PRODUCT USE Information Call 1-800-334-9745
For MEDICAL and TRANSPORTATION Emergencies ONLY Call 24 Hours A Day 1-800-334-7577

PRECAUTIONARY STATEMENTS

WARNING

HAZARDS TO HUMANS AND DOMESTIC ANIMALS

Do not get in eyes, on skin, or clothing. Causes eye and skin irritation. Avoid inhalation of spray mists. Harmful if swallowed. Wash thoroughly with soap and water after handling. Remove contaminated clothing and wash before reuse.

STATEMENT OF PRACTICAL TREATMENT

IF SWALLOWED: Drink large quantities of water. Do not induce vomiting. Do not give anything by mouth to an unconscious person. Get medical attention immediately.

IF IN EYES: Immediately flush eyes with plenty of water for 15 minutes and get medical attention at once.

IF ON SKIN: Wash promptly with soap and water and rinse thoroughly. Get medical attention.

ENVIRONMENTAL HAZARDS

Do not apply directly to water, wetlands or vegetables except as directed under crop uses. Do not contaminate water when disposing of equipment washwaters. Do not apply this product through any type of irrigation system. Do not contaminate water used for irrigation or domestic purposes. Spray equipment used in applying this product should be thoroughly cleaned before using for any other purpose. Use repeated flushing with soap and warm water or suitable chemical cleaner. It is best to use a separate sprayer for application of insecticides and fungicides. This product will kill or seriously injure many desirable forms of vegetation. Do not apply directly to flowers, fruits, grapes, ornamentals, cotton or other desirable plants. Vapors from this product may injure susceptible plants in the immediate vicinity. Do not apply when weather conditions favor drift from target area. Avoid use of small-diameter nozzles (Coarse sprays are less likely to drift). Excessive amounts of this product in the soil may temporarily inhibit seed germination and plant growth.

DIRECTIONS FOR USE

IT IS A VIOLATION OF FEDERAL LAW TO USE THIS PRODUCT IN A MANNER INCONSISTENT WITH ITS LABELING. READ ENTIRE LABEL BEFORE USING THIS PRODUCT.

EFFECTIVENESS ON WEEDS: Rhonox® Herbicide is effective on a large number of broadleaf weeds. On mustard, thistle, and white top or hoary cress; it has superiority over 2,4-D.

The following are some of the weeds killed.

Beggarstick	Honeysuckle
Burcucumber	Jimsonweed
Burdock	Knotweed
Buttercup	Kochia
Canada thistle	Lambsquarters (Goosefoot)
Carpetweed	Marshelder
Catsear	Mexican poppy
Chervil	Miner's lettuce
Cocklebur	Narrow leaf plantain
Croton	Nutgrass
Daisy	Pennycress
Dandelion	Pepperweed
Dock	Perennial morningglory
Dragon heat mint	Perennial ragweed
Dwarf nettle	Pigweed
Fanweed	Poison hemlock
Fat hen	Pickly lettuce
Fennel	Puncturevine
Field bindweed	Purslane
Galinsoga	Red root
Goatsbeard	Redstem
Goldenrod	Sheperdspurse
Halbertleaved salt bush	Sicklepod
Hempnettle	Small plantain
Hoary cress	Sneezeweed
Sowthistle	Wild jute
Spanish needle	Wild marigold
Stinging nettle	Wild petunia
Stinkweed	Wild radish
Sunflower	Wild sage
Thornapple	Witchweed
Tree-of-heaven	Yellow charlock mustard
Vetch	Yellow daisy
Whitebrush	Yellow rocket
Wild carrot	
Wild gooseberry	

SELECTIVE SPRAYING

NOTE: When using on grains and flax, do not forage or graze dairy and meat animals on treated areas within 7 days of slaughter. Also, except for small underseeded grains, use at least 8 to 10 gallons of water per acre for ground application and at least 3 to 5 gallons of water per acre for aerial application.

SMALL GRAINS: Wheat, Barley, Rye (For small grains underseeded with legumes, see the Small Grains paragraph below).

Spring planted—not seeded to legumes. Apply as a water mix spray by ground sprayer or airplane. Use 1/2 to 1 1/2 pints of this product per acre for the more susceptible weeds. Treat when fully tillered grains have reached 3 to 4 leaf stage up to boot stage (8 to 10 inches tall). Do not spray grain in the boot to dough stage. **Fall Planted—Fall application—not seeded to legumes.** The safest time to treat is when fall planted small grains are fully tillered and have reached 3 to 4 leaf to the boot stage (8 to 10 inches tall). Infestations of broadleaf weeds germinating in fall and winter may, however, justify treatment at an earlier stage. Fall planted wheat, barley, or rye can be sprayed with 1/2 to 1 pint of Rhonox® Herbicide per acre from the 3 to 4 leaf stage to the boot stage.

Spring application—Apply in early spring when annual broadleaf weeds are small and grain is fully tillered, but before grain is in the jointed stage.

SMALL GRAINS—Underseeded with Alfalfa, Birdsfoot Trefoil, Lespedeza, Red and White Clover. For emergency control of serious infestations of mustard, yellow rocket, and other susceptible broadleaf weeds, apply 1/4 to 1/2 pint in not more than 5 to 10 gallons of water per acre. Make application after cereal is well tillered (4 to 8 inches tall) and before reaching boot stage. The nurse crop and weeds should provide a protecting canopy which, together with the use of low gallonage applied at low pressure, will reduce the risk of damage to the legumes. Do not apply to small grains underseeded with vetch or sweet clover, which are very susceptible. There is also some risk to other legumes if only thinly protected by a canopy.

OATS: Spray after oats are fully tillered (4 to 6 inches tall) but before they are in the boot to dough stage. Apply the lower rate of 1/2 pint where annual weeds are the problem, the higher rate of 1 pint for perennial weeds.

RICE (California and other areas where recommended): For control of arrowhead, water plantain and red stem, apply 1 1/2 to 2 1/2 pints per acre. Use 2 1/2 pints per acre where sedge* nutgrass, curly indigo, and bullrush are also present. Treat only when rice is well established, 6 to 8 inches above water (no sooner than 55 and no later than 65 days after seeding), or when crop stems begin to elongate. Water should not be less than 2 to 3 inches deep at shallowest point. Do not spray after rice is in boot stage or when temperature is over 90°F. Do not grow crayfish or catfish in treated rice fields.

*In some cases, sedge is only partially controlled.

FLAX: Use 1/4 to 1/2 pint per acre. Use lower rate where susceptible weeds such as mustard are the main problem. Use higher rate when weeds such as pigweed, lambsquarters, stinkweed, ragweed, cocklebur are the problem. Apply only when weeds are up and when flax is 2 to 8 inches high and before it comes into bud. Treatment after early bud stage may result in severe damage. If Canada thistle is present, it may be necessary to go as high as 3/4 pint per acre to prevent seed head production. Some injury to the flax may result.

RANGE AND TIMBER LANDS: For control of Whitebrush—Use 1 1/4 quarts in mixture of 1 gallon of diesel oil and sufficient water to make 8 gallons of solution per acre. Apply in spring or fall under good moisture conditions, full leaf, before blossoms begin to fall.

GRASSES: Established Lawns, Golf Courses and similar Turf Grasses— Use 1 to 3 pints per acre in 10 to 100 gallons of water to give thorough coverage. Use higher rate for white top and Canada thistle. On small areas, use 1 fluid ounce (4 tablespoonsful) mixed in 3 to 5 gallons of water and apply uniformly over 1,000 square feet. Do not mow within 2 days before or after application. Spring and fall are best times to treat. Do not use on lawns of creeping grasses such as bent, except for spot spraying, nor on freshly seeded turf until the grass has become well established.

Established Grassland and Pastures—Use 1 to 3 pints per acre in sufficient water (10 to 100 gallons) in airplane or ground spray application and give thorough coverage. Use higher rate for white top, Canada thistle and other less easy-to-kill weeds; spray perennials in early bud to full bloom stage and regrowth in fall. Other weeds in spring or fall.

Grasses Grown for Seed—Use 1 to 2 pints per acre in 5 to 50 gallons of water by air or ground sprayer application. Use higher rate where weed stands are heavy. In established grasses, apply in spring before head comes into boot stage and on seedling grass after grass has tillered.

NOTE: For weed control in grasses, repeat treatment may be needed for less susceptible weeds. White clover and other legumes may be temporarily injured or killed. In some areas, bent, buffalo, carpet, centipede, dichondra and St. Augustine grasses may also be injured by the treatment.

NON-CROP SPRAYING: Canada thistle, White top and Meadow buttercup. (In Non-Crop Areas such as Roadsides, Fencerows, Rights-of-way and Similar Places): For spot treatment, use 1/4 pint to 3 to 4 gallons of water or 3/4 gallon per acre in 12 to 20 gallons of water to give coverage for most extensive areas. Spray to wet weeds thoroughly when in bud to early bloom and again on fall regrowth.

STORAGE AND DISPOSAL

STORAGE: Always store pesticides in a secured warehouse or storage building. Do not store near seeds, fertilizers, insecticides or fungicides. Containers should be opened in well ventilated areas. Keep container tightly sealed when not in use. Do not stack cardboard cases more than two pallets high. Do not contaminate water, food or feed by storage or disposal.

PESTICIDE DISPOSAL: Wastes resulting from the use of this product may be disposed of on site or an approved waste disposal facility. If container is damaged or if pesticide has leaked, contain all spillage. Absorb and clean up all spilled material with granules or sand.

CONTAINER DISPOSAL: Triple rinse (or equivalent). Then offer for recycling or reconditioning, or puncture and dispose of in a sanitary landfill, or by other procedures approved by State and Local authorities.

NOTICE: Local conditions may affect the use of this chemical. Consult State Agricultural Extension or Experiment Station weed specialists for specific recommendations for local weed problems and for information on possible lower dosages.

LIMITED WARRANTY AND DISCLAIMER

The manufacturer warrants (a) that this product conforms to the chemical description on the label; (b) that this product is reasonably fit for the purpose set forth in the directions for use when it is used in accordance with such directions; and (c) that the directions, warnings and other statements on this label are based upon responsible experts' evaluation of reasonable tests of effectiveness of toxicity to laboratory animals and to plants, and of residues on food crops and upon reports of field experience. Tests have not been made on all varieties or in all states or under all conditions. THE MANUFACTURER NEITHER MAKES NOR INTENDS, NOR DOES IT AUTHORIZE ANY AGENT OR REPRESENTATIVE TO MAKE ANY OTHER WARRANTIES EXPRESS OR IMPLIED, AND

IT EXPRESSLY EXCLUDES AND DISCLAIMS ALL IMPLIED WARRANTIES OF MERCHANTABILITY OR FITNESS FOR A PARTICULAR PURPOSE.

THIS WARRANTY DOES NOT EXTEND TO, AND THE BUYER SHALL BE SOLELY RESPONSIBLE FOR, ANY AND ALL LOSS OR DAMAGE WHICH RESULTS FROM THE USE OF THIS PRODUCT IN ANY MANNER WHICH IS INCONSISTENT WITH THE LABEL DIRECTIONS, WARNINGS, OR CAUTIONS.

BUYER'S EXCLUSIVE REMEDY AND MANUFACTURER'S OR SELLER'S EXCLUSIVE LIABILITY FOR ANY AND ALL CLAIMS, LOSSES, DAMAGES, OR INJURIES RESULTING FROM THE USE OR HANDLING OF THIS PRODUCT, WHETHER OR NOT BASED IN CONTRACT, NEGLIGENCE, STRICT LIABILITY IN TORT OR OTHERWISE SHALL BE LIMITED, AT THE MANUFACTURER'S OPTION, TO REPLACEMENT OF, OR THE REPAYMENT OF THE PURCHASE PRICE FOR, THE QUANTITY OF PRODUCT WITH RESPECT TO WHICH DAMAGES ARE CLAIMED. IN NO EVENT SHALL MANUFACTURER OR SELLER BE LIABLE FOR SPECIAL, INDIRECT OR CONSEQUENTIAL DAMAGES RESULTING FROM THE USE OR HANDLING OF THIS PRODUCT.

NOTICE TO BUYER

Purchase of this material does not confer any rights under patents governing this product or the use thereof in countries outside of the United States.

AG 90025

THIS SPECIMEN LABEL IS INTENDED FOR USE ONLY AS A GUIDE IN PROVIDING GENERAL INFORMATION REGARDING THE DIRECTIONS, WARNINGS AND CAUTIONS ASSOCIATED WITH THE USE OF THIS PRODUCT. AS WITH ANY AGRICULTURAL CHEMICAL, ALWAYS FOLLOW THE LABEL INSTRUCTIONS ON THE PACKAGE BEFORE USING.

ROVRAL®

FUNGICIDE

ACTIVE INGREDIENT:

Iprodione: 3-(3,5-dichlorophenyl)-N-(1-methylethyl)-2,4-dioxo-1-imidazolidinecarboxamide	50.0%
INERT INGREDIENTS:	50.0%
TOTAL:	100.0%

KEEP OUT OF REACH OF CHILDREN

CAUTION

EPA Reg. No. 264-453 EPA Est. No. 35982-TX-1

For PRODUCT USE Information Call 1-800-334-9745

For MEDICAL and TRANSPORTATION Emergencies ONLY Call 24 Hours A Day

1-800-334-7577

PRECAUTIONARY STATEMENTS

CAUTION

HAZARDS TO HUMANS AND DOMESTIC ANIMALS

Avoid contact with skin, eyes, or clothing. In case of contact immediately flush eyes or skin with plenty of water. Get medical attention if irritation persists.

STATEMENT OF PRACTICAL TREATMENT

IN CASE OF CONTACT: Wash skin with soap and water.

IF IN EYES: Flush with water for 15 minutes and get medical attention.

ENVIRONMENTAL HAZARDS

This pesticide is toxic to aquatic invertebrates. Do not apply directly to water or wetlands. Do not contaminate water when disposing of equipment washwater. Drift or run-off from treated areas are hazardous to aquatic invertebrates in neighboring areas.

GENERAL INFORMATION

ROVRAL® is a wettable powder formulation of iprodione for the control of certain diseases on Almonds and Stone Fruits (Apricots, Cherries, Nectarines, Peaches, Plums, and Prunes), Garlic, Lettuce, Grapes, Ginseng, Carrots, Broccoli, Beans, Peanuts, Potatoes, Caneberries, and Dry Bulb Onions.

NOTE TO USER: Do not graze animals in treated orchards. Do not feed cover crops grown in treated orchards to livestock.

DIRECTIONS FOR USE THROUGH SPRINKLER IRRIGATION SYSTEMS

Apply this product only through sprinkler irrigation systems including microjet, drip, solid set and center pivot. Do not apply this product through any other type of irrigation system.

SPRAY PREPARATION: Remove scale, pesticide residues, and other foreign matter from the chemical tank and entire injector system. Flush with clean water.

APPLICATION INSTRUCTIONS: First prepare a suspension of ROVRAL® Fungicide in a mix tank. Fill tank with 1/2 to 3/4 the desired amount of water. Start mechanical or hydraulic agitation. Add the required amount of ROVRAL® Fungicide, and then the remaining volume of water. (Suspension concentrations using the appropriate dosage per acre recommended on this label of ROVRAL® Fungicide per 1 to 4 gallons of water are recommended) Then set sprinkler to deliver 0.1 to 0.3 inch of water per acre. Start sprinkler and uniformly inject the suspension of ROVRAL® Fungicide into the irrigation water line so as to deliver the desired rate per acre. The suspension of ROVRAL® Fungicide should be injected with a positive displacement pump into the main line ahead of a right angle turn to insure adequate mixing. If you should have any other questions about calibration, you should contact State Extension Service specialists, equipment manufacturers or other experts.

NOTE: When treatment with ROVRAL® Fungicide has been completed, further field irrigation over the treated area should be avoided for 24 to 48 hours to prevent washing the chemical off the crop.

GENERAL PRECAUTIONS FOR APPLICATIONS THROUGH SPRINKLER IRRIGATION SYSTEMS

Maintain continuous agitation in mix tank during mixing and application to assure a uniform suspension. Greater accuracy in calibration and distribution will be achieved by injecting a larger volume of a more dilute solution per unit time.

Greater accuracy in calibration and distribution will be achieved by injecting a larger volume of a more dilute solution per unit time.

The system must contain a functional check valve, vacuum relief valve, and low pressure drain appropriately located on the irrigation pipeline to prevent water source contamination from backflow. The pesticide injection pipeline must contain a functional, automatic, quick-closing check valve to prevent the flow of fluid back toward the injection pump. The pesticide injection pipeline must also contain a functional, normally closed solenoid-operated valve located on the intake side of the injection pump and connected to the system interlock to prevent fluid from being withdrawn from the supply tank when the irrigation system is either automatically or manually shutdown. The system must contain functional interlocking controls to automatically shut off the pesticide injection pump when the water pump motor stops. The irrigation line or water pump must include a functional pressure switch which will stop the water pump motor when the water pressure decreases to the point where pesticide distribution is adversely affected. Systems must use a metering pump, such as a positive displacement injection pump (e.g., diaphragm pump) effectively designed and constructed of materials that are compatible with pesticides and capable of being fitted with a system interlock.

Do not apply when wind speed favors drift, when system connection or fittings leak, when nozzles do not provide uniform distribution or when lines containing the product must be dismantled and drained. Crop injury, lack of effectiveness, or illegal pesticide residues in the crop may result from nonuniform distribution of treated water.

Allow sufficient time for pesticide to be flushed through all lines and all nozzles before turning off irrigation water. A person knowledgeable of the chemigation system and responsible for its operation shall shut the system down and make necessary adjustments should the need arise.

Do not connect an irrigation system (including greenhouse systems) used for pesticide application to a public water system unless the label-prescribed safety devices for public water supplies are in place. **NOT FOR USE IN CALIFORNIA.**

DIRECTIONS FOR USE

IT IS A VIOLATION OF FEDERAL LAW TO USE THIS PRODUCT IN A MANNER INCONSISTENT WITH ITS LABELING.

REENTRY STATEMENT

Do not apply this product in such a manner as to directly or through drift expose workers or other persons. The area being treated must be vacated by unprotected persons. Do not enter treated areas without protective clothing until sprays have dried.

HOW TO USE ROVRAL® FUNGICIDE

Partially fill the spray tank with clean water. Measure the required amount of ROVRAL® Fungicide and pre-mix with a small volume of water, add this to the tank. Agitate to ensure thorough mixing while filling tank with remaining water. Maintain agitation during application and apply with properly calibrated application equipment. Do not allow spray mixture to stand overnight or for prolonged periods, as some chemical breakdown may occur, particularly in water with a high pH.

HOW MUCH ROVRAL® FUNGICIDE TO USE

The rate chosen will depend on the severity of disease pressure and weather conditions at the time of application (if conditions are severe, use the highest rate). Apply ROVRAL® Fungicide in accordance with the directions in the following tables.

LEAFY VEGETABLES

CROP	HOW TO USE	DISEASE	DOSAGE RATE LBS. PRODUCT/ACRE	WHEN TO APPLY
Lettuce*	Apply as a foliar spray in sufficient water to obtain thorough coverage (50 to 100 gallons per acre). Do not apply within 14 days of harvest. Do not make more than 3 applications of ROVRAL® Fungicide per season.	Lettuce Drop (*Sclerotinia sp.*) Bottom Rot (*Rhizoctonia solani*)	1.5 to 2.0	Application should be made with tractor mounted boom sprayers equipped with flat fan or hollow cone nozzles directed to ensure thorough coverage of the lower portion of the plants and the surrounding soil surface. Do not drench. The higher rate should be used under severe disease conditions. Apply at the 3 leaf stage of growth and again 10 days later. If conditions still favor disease development, a third application should be made 10 days after the second spray.

VEGETABLES

CROP	HOW TO USE	DISEASE	DOSAGE RATE LBS. PRODUCT/ ACRE	DOSAGE RATE GALS./ WATER PER ACRE	WHEN TO APPLY
Broccoli*	Application should be made with a tractor mounted boom sprayer with 2 flat fan nozzles per row (one on either side) directed at the base of the plant and the adjacent soil surface. Position nozzles to insure thorough coverage of the stem. Do not drench.	Black Leg (*Phoma lingiam*)	2.0 (1.0 Oz/1500 feet of row)	40-200	Apply immediately after thinning (2 to 4 leaf stage) as a directed spray to the base of the plant and the adjacent soil surface. If disease conditions persist or reoccur, a second application may be made up to the day of harvest.
Carrots*	Apply as a foliar spray, in sufficient water to obtain thorough coverage.	Alternaria Blight (*Alternaria dauci*)	1.0 to 2.0		Make the first application of ROVRAL® Fungicide as conditions become favorable for disease development in the Spring. Continue applications on a 7 to 14 day interval as long as conditions favor disease development. The higher rate and/or shorter spray interval should be used under severe disease conditions. ROVRAL® Fungicide should be applied in a minimum of 10 gallons per acre of solution, and may be applied up to the day of harvest. Do not make more than 8 applications of ROVRAL® per season.
Dry Bulb Onions*	A single, flat fan, or cone nozzle should be centered and adjusted to provide complete coverage of each row. Up to 10 applications can be made per season. Do not apply within 7 days of harvest. Apply ROVRAL® Fungicide using ground equipment in accordance with the directions in the table.	Botrytis Leaf Blight (*Botrytis squamosa*) Purple Blotch (*Alternaria pom*) Botrytis Neck Rot (*Botrytis allii*)	1.5	50 to 100	Apply as a foliar spray as soon as conditions become favorable for disease. Continue application on 7 to 14 day interval as long as conditions favor disease development. Under severe disease conditions, the shorter interval should be used.
Garlic*	Apply as an in furrow spray in sufficient water to obtain thorough coverage. **Using a 38 to 40 inch bed spacing.	White Rot (*Sclerotium cepivorum*)	4.0**	50 to 100	Apply ROVRAL® Fungicide in the furrow at planting. Position a single flat fan nozzle to treat the covering soil used to fill the furrow.

*NOTE TO USER: The following crops may be rotated after harvest: Garlic, Dry Bulb Onions, Broccoli, Lettuce, Peanuts, Carrots, Beans, and Potatoes.
The following crops may be rotated one month following the last iprodione application: Root Crops, Tomatoes, and Cotton.

BEANS (SNAP, DRY, AND LIMA)

CROP	HOW TO USE	DISEASE	DOSAGE RATE LBS. PRODUCT/ ACRE	DOSAGE RATE GALS./ WATER PER ACRE	WHEN TO APPLY
BEANS* (Snap, Dry, and Lima)	Apply ROVRAL® Fungicide using ground equipment with a spray pressure of 50 to 100 psi using three nozzles per row arranged with 1 directly over the row and 1 on each side of the row.	Gray Mold (*Botrytis spp.*) White Mold (*Sclerotinia spp.*)	1.5 to 2.0	20 to 40	Apply ROVRAL® as a foliar spray at early bloom (25 to 50% of plants with at least 1 open bloom) and, if conditions are still favorable for disease development, again at peak bloom (maximum number of open blooms). Under severe disease conditions, the higher rate should be used. Thorough coverage is essential for proper disease control. Do not allow foraging for 14 days after last application of ROVRAL® Fungicide. Do not feed snap bean hay to livestock. Do not feed dry bean hay to livestock until 45 days after last application. Do not feed succulent bean hay to livestock.

PEANUTS

CROP	HOW TO USE	DISEASE	DOSAGE RATE LBS. PRODUCT/ ACRE	DOSAGE RATE GALS./ WATER PER ACRE	WHEN TO APPLY
Peanuts*	Apply ROVRAL® Fungicide using a tractor mounted boom sprayer equipped with low pressure nozzles (e.g., 8008 LP, 8010 LP, and TK 7.5) that produce large droplets. A single nozzle should be centered and adjusted to provide complete coverage of the row.	Sclerotinia Blight (*Sclerotinia ap.*)	2.0	40	Make an initial application when conditions first become favorable for disease development. Up to 2 subsequent applications should be made at 4 week intervals. Do not apply within 10 days of harvest. NOT FOR USE IN CALIFORNIA.

POTATOES

CROP	HOW TO USE	DISEASE	DOSAGE RATE LBS. PRODUCT/ ACRE	DOSAGE RATE GALS./ WATER PER ACRE	WHEN TO APPLY
Potatoes*	A single, flat fan or cone nozzle should be centered and adjusted to provide complete coverage of each row. Up to 4 applications can be made per season. DO NOT APPLY WITHIN 14 DAYS OF HARVEST. Apply ROVRAL® Fungicide using ground application or chemigation, in sufficient water for thorough coverage (minimum of 10 GPA).	Early Blight (*Alternaria solani*) White Mold (*Sclerotinia sclerotiorum*)	2.0	Minimum 10	**For early blight,** begin applications when conditions first become favorable for disease development. Up to 3 subsequent applications can be applied at 7 to 10 day intervals or as required. **For white mold,** apply at first sign of disease or immediately prior to row closing and, if conditions are favorable for disease development, again 28 days later. NOT FOR USE IN CALIFORNIA.

RICE

CROP	HOW TO USE	DISEASE	DOSAGE RATE LBS. PRODUCT/ACRE	WHEN TO APPLY
RICE	Apply ROVRAL® Fungicide as a broadcast spray using aerial equipment. ROVRAL® Fungicide should be used with a minimum of 10 gallons of water per acre.	Sheath blight (*Rhizoctonia solani*)	1.0	ROVRAL® Fungicide may be applied twice per year. The first application should be made between joint movement and booting. The second application should be applied 2 weeks after the first application but not later than 75% heading. NOT FOR USE IN CALIFORNIA.

Do not apply in areas where catfish and crayfish are commercially cultivated.

*NOTE TO USER: The following crops may be rotated after harvest: Garlic, Dry Bulb Onions, Broccoli, Lettuce, Peanuts, Carrots, Beans, and Potatoes. The following crops may be rotated one month following the last iprodione application: Root Crops, Tomatoes, and Cotton.

CANEBERRIES

CROP	HOW TO USE	DISEASE	DOSAGE RATE: LBS. PRODUCT/ ACRE	DOSAGE RATE: GALS./ WATER PER ACRE	WHEN TO APPLY
Caneberries	Apply ROVRAL® Fungicide as a foliar spray with ground equipment in sufficient water to obtain thorough coverage at bloom and/or fruit (minimum of 100 gallons).	Botrytis Fruit Rot (*Botrytis cinerea*)	1.0 to 2.0	Minimum 100	Apply ROVRAL® Fungicide first at early bloom (5 to 10% bloom and make a repeat application again at full bloom. Up to 3 subsequent applications can be applied at 14 day intervals or as required. The final application can be made up to and including the day of harvest. NOT FOR USE IN CALIFORNIA.

STONE FRUITS AND NUTS

CROP	HOW TO USE	DISEASE	DOSAGE RATE: LBS./ 100 GAL	DOSAGE RATE: LBS./ ACRE	WHEN TO APPLY
Apricots Cherries Nectarines Peaches Plums Prunes	The rate of ROVRAL® Fungicide per 100 gallons are based on a standard of 400 gallons per acre dilute spray for mature trees. For less than mature trees, apply the rate per 100 gallons until run-off. If less than 400 gallons of spray solution per acre is applied to mature trees, refer to the rate per acre to ensure that the proper amount of material is applied. Apply ROVRAL® as an overall spray in sufficient water to obtain thorough coverage of bloom, foliage, and/or fruit (20 to 400 gallons per acre by ground equipment and in a minimum of 15 gals. per acre by air). Aerial applications are only recommended during the bloom period.	Brown Rot Blossom Blight (*Monilinia spp.*) Fruit Brown Rot (*Monilinia spp.*)	0.25 to 0.50	1.0 to 2.0	**Brown Rot Blossom Blight** Apply first at early bloom (approximately 5% bloom). If conditions are favorable for disease development, apply again at full bloom and at petal fall. **Fruit Brown Rot:** An application should be made whenever temperatures and moisture conditions favor disease infection in the 5 week period prior to harvest. If these conditions persist or reoccur, a second application should be made. This second application should be made no sooner than 7 days following the first pre-harvest application. Pre-harvest applications may be made up to and including the day of harvest. Do not apply more than 2.0 pints of ROVRAL® Fungicide per acre per application. Do not make more than 5 applications of ROVRAL® Fungicide per season.
Almonds	The rates of ROVRAL® Fungicide per 100 gallons are based on a standard of 400 gallons per acre dilute spray for mature trees. For less than mature trees, apply the rate per 100 gallons until run-off. If less than 400 gallons of spray solution per acre is applied, to mature trees, refer to the rate per acre to ensure that the proper amount of material is applied. Apply ROVRAL® Fungicide as an overall spray in sufficient water to obtain thorough coverage of bloom and foliage (in a minimum of 20 gals per acre by air). The use of aerial equipment is not recommended after petal fall.	Brown Rot Blossom Blight (*Monilinia laxa*) Shothole (*Stigmina carpophila*)	0.25	1.0	**Brown Rot Blossom Blight Shothole:** Apply first at pink bud and, if conditions favorable for disease development persist or reoccur up to 3 subsequent applications can be made at: 1) full bloom; 2) petal fall; and 3) up to 5 weeks after petal fall. Do not make more than 4 applications of ROVRAL® Fungicide per season.

NOTE TO USER: Do not graze animals in treated orchards. Do not feed cover crops grown in treated orchards to livestock.

GINSENG

CROP	HOW TO USE	DISEASE	DOSAGE RATE: LBS. PRODUCT/ ACRE	DOSAGE RATE: GALS./ WATER PER ACRE	WHEN TO APPLY
Ginseng	Apply as a foliar spray, in sufficient water to obtain thorough coverage, using ground equipment. ROVRAL® should be used as part of a complete program, including the use rate of alternate fungicides, for the control of Alternaria blight.	Alternaria Blight (*Alternaria panax*)	2.0	10 (Minimum)	Make the first application of ROVRAL® Fungicide as conditions become favorable for disease development in the Spring. Continue applications on a 7 day interval as long as conditions favor disease development. Do not make more than 12 applications of ROVRAL® Fungicide per season. Do not apply within 36 days of harvest. NOT FOR USE IN CALIFORNIA

GRAPES

CROP	HOW TO USE	DISEASE	DOSAGE RATE: LBS. PRODUCT/ ACRE	DOSAGE RATE: GALS./ WATER PER ACRE	WHEN TO APPLY
Grapes	ROVRAL® Fungicide is a flowable formulation for the control of Bunch Rot on Grapes.	Bunch Rot (*Botrytis sp.*)	1.5 to 2.0	50-200	Apply ROVRAL® Fungicide in accordance with directions provided in the table:

CROP	HOW TO USE	DISEASE	DOSAGE RATE LBS. PRODUCT/ ACRE	DOSAGE RATE GALS./ WATER PER ACRE	WHEN TO APPLY
	Apply as a foliar spray in sufficient water to obtain thorough coverage (50 to 200 gals per acre) The spray mixture should not be allowed to stand overnight as some breakdown of the product may occur from prolonged delays, particularly in water with a high pH. Maintain agitation during spray operations and apply with properly calibrated application equipment. Do not make more than 4 applications of ROVRAL® Fungicide per season.				SPRAY SCHEDULE 1) Early mid-bloom. 2) Prior to bunch closing. 3) Beginning of fruit ripening (veraison). 4) Final application prior to harvest as needed. The last application may be made up to and including the day of harvest. **Thorough Coverage Of The Bunches Is Essential.** Under severe disease conditions, the higher rate is recommended. ROVRAL® Fungicide must be used in conjuction with good cultural practices designed to minimize conditions conductive to Bunch Rot (*Botrytis cinerea*). It should be applied with equipment calibrated and adjusted to direct the spray at the bunches to ensure thorough bunch coverage.

NOTE TO USER: Do not graze animals in treated orchards. Do not feed cover crops grown in treated orchards to livestock.

STORAGE AND DISPOSAL

STORAGE

Do not contaminate water, food, or feed by storage or disposal.

PESTICIDE DISPOSAL

Wastes resulting from the use of this product may be disposed of on site or at an approved waste disposal facility.

CONTAINER DISPOSAL

Completely empty bag into application equipment. Then dispose of empty bag in sanitary landfill or by incineration, or if allowed by State and local authorities, by burning. If burned, stay out of smoke.

LIMITED WARRANTY AND DISCLAIMER

The manufacturer warrants (a) that this product conforms to the chemical description on the label; (b) that this product is reasonably fit for the purposes set forth in the directions for use when it is used in accordance with such directions; and (c) that the directions, warning and other statements on this label are based upon responsible experts' evaluation of reasonable tests of effectiveness, of toxicity to laboratory animals and to plants, and of residues on food crops and upon reports of field experience. Tests have not been made on all varieties or in all states or under all conditions. THE MANUFACTURER NEITHER MAKES NOR INTENDS NOR DOES IT AUTHORIZE ANY AGENT OR REPRESENTATIVE TO MAKE, ANY OTHER WARRANTIES, EXPRESS OR IMPLIED, AND IT EXPRESSLY EXCLUDES AND DISCLAIMS ALL IMPLIED WARRANTIES OF MERCHANTABILITY OR FITNESS FOR A PARTICULAR PURPOSE.

THIS WARRANTY DOES NOT EXTEND TO, AND THE BUYER SHALL BE SOLELY RESPONSIBLE FOR, ANY AND ALL LOSS OR DAMAGE WHICH RESULTS FROM THE USE OF THIS PRODUCT IN ANY MANNER WHICH IS INCONSISTENT WITH THE LABEL DIRECTIONS, WARNINGS OR CAUTIONS.

BUYER'S EXCLUSIVE REMEDY AND MANUFACTURER'S OR SELLER'S EXCLUSIVE LIABILITY FOR ANY AND ALL CLAIMS, LOSSES, DAMAGES, OR INJURIES RESULTING FROM THE USE OR HANDLING OF THIS PRODUCT, WHETHER OR NOT BASED IN CONTRACT, NEGLIGENCE, STRICT LIABILITY IN TORT OR OTHERWISE SHALL BE LIMITED, AT THE MANUFACTURER'S OPTION, TO REPLACEMENT OF OR THE REPAYMENT OF THE PURCHASE PRICE FOR, THE QUANTITY OF PRODUCT WITH RESPECT TO WHICH DAMAGES ARE CLAIMED, IN NO EVENT SHALL MANUFACTURER OR SELLER BE LIABLE FOR SPECIAL INDIRECT OR CONSEQUENTIAL DAMAGES RESULTING FROM THE USE OR HANDLING OF THIS PRODUCT.

NOTICE TO BUYER

Purchase of this material does not confer any rights under patents governing this product or the use thereof in countries outside the United States.

ROVRAL is a trademark of Rhône-Poulenc

AG90033

THIS SPECIMEN LABEL IS INTENDED FOR USE ONLY AS A GUIDE IN PROVIDING GENERAL INFORMATION REGARDING THE DIRECTIONS, WARNINGS AND CAUTIONS ASSOCIATED WITH THE USE OF THIS PRODUCT. AS WITH ANY AGRICULTURAL CHEMICAL, ALWAYS FOLLOW THE LABEL INSTRUCTIONS ON THE PACKAGE BEFORE USING.

ROVRAL® 4

FLOWABLE FUNGICIDE

NOT REGISTERED FOR USE IN CALIFORNIA

ACTIVE INGREDIENT:

Iprodione: 3-(3,5-dichlorophenyl)-N-(1-methylethyl)-2, 4-dioxo-1-imidazolidinecarboxamide*	41.6%
INERT INGREDIENTS:	58.4%
TOTAL:	100.0%

*Equivalent to 4 Lbs. Iprodione per gallon.

EPA Reg. No. 264-482 EPA Est. No.: 1812-GA-3

KEEP OUT OF REACH OF CHILDREN

CAUTION

For PRODUCT USE Information Call 1-800-334-9745
For MEDICAL And TRANSPORTATION Emergencies ONLY Call 24 Hours A Day 1-800-334-7577

PRECAUTIONARY STATEMENTS

CAUTION

HAZARDS TO HUMANS AND DOMESTIC ANIMALS

Avoid contact with skin, eyes, or clothing. In case of contact immediately flush eyes or skin with plenty of water. Get medical attention if irritation persists.

STATEMENT OF PRACTICAL TREATMENT

IN CASE OF CONTACT: Wash skin with soap and water.

IF IN EYES: Flush with water for 15 minutes and get medical attention.

ENVIRONMENTAL HAZARDS

This pesticide is toxic to aquatic invertebrates. Do not apply directly to water or wetlands. Do not contaminate water when disposing of equipment washwater. Drift or run-off from treated areas are hazardous to aquatic invertebrates in neighboring areas.

GENERAL INFORMATION: ROVRAL® is a flowable formulation of iprodione for the control of certain diseases on Stone Fruits, Nuts, Grapes, Leafy Vegetables, and Vegetables.

DIRECTIONS FOR USE

IT IS A VIOLATION OF FEDERAL LAW TO USE THIS PRODUCT IN ANY MANNER INCONSISTENT WITH ITS LABELING.

REENTRY STATEMENT

Do not apply this product in such a manner as to directly or through drift expose workers or other persons. The area being treated must be vacated by unprotected persons. Do not enter treated areas without protective clothing until sprays have dried.

Because certain states may require more restrictive reentry intervals for various crops treated with this product, consult your State Department of Agriculture for further information. Written or oral warnings must be given to workers who are expected to be in a treated area or in an area about to be treated with this product. Oral warnings must be given if there is reason to believe that written warnings cannot be understood by workers and given in a language customarily understood by the workers. Oral and written warning must include the STATEMENT OF PRACTICAL TREATMENT and REENTRY STATEMENTS as they appear on the label.

DIRECTIONS FOR USE THROUGH SPRINKLER IRRIGATION SYSTEMS

Apply this product only through sprinkler irrigation systems including microjet, drip, solid set and center pivot. Do not apply this product through any other type of irrigation system.

SPRAY PREPARATION: Remove scale, pesticide residues, and other foreign matter from the chemical tank and entire injector system. Flush with clean water.

APPLICATION INSTRUCTIONS: First prepare a suspension of ROVRAL® 4 Flowable Fungicide in a mix tank. Fill tank with 1/2 to 3/4 the desired amount of water. Start mechanical or hydraulic agitation. Add the required amount of ROVRAL® 4 Flowable Fungicide, and then the remaining volume of water. (Suspension concentrations using the appropriate dosage per acre recommended on this label of ROVRAL® 4 Flowable Fungicide per 1 to 4 gallons of water are recommended). Then set sprinkler to deliver 0.1 to 0.3 inch of water per acre. Start sprinkler and uniformly inject the suspension of ROVRAL® 4 Flowable Fungicide into the irrigation water line so as to deliver the desired rate per acre. The suspension of ROVRAL® 4 Flowable Fungicide should be injected with a positive displacement pump into the main line ahead of a right angle turn to insure adequate mixing. If you should have any other questions about calibration, you should contact State Extension Service specialists, equipment manufacturers or other experts.

NOTE: When treatment with ROVRAL® 4 Flowable Fungicide has been completed, further field irrigation over the treated area should be avoided for 24 to 48 hours to prevent washing the chemical off the crop.

GENERAL PRECAUTIONS FOR APPLICATIONS THROUGH SPRINKLER IRRIGATION SYSTEMS

Maintain continuous agitation in mix tank during mixing and application to assure a uniform suspension. Greater accuracy in calibration and distribution will be achieved by injecting a larger volume of a more dilute solution per unit time. The system must contain a functional check valve, vacuum relief valve, and low pressure drain appropriately located on the irrigation pipeline to prevent water source contamination from backflow. The pesticide injection pipeline must contain a functional, automatic, quick-closing check valve to prevent the flow of fluid back toward the injection pump. The pesticide injection pipeline must also contain a functional, normally closed solenoid-operated valve located on the intake side of the injection pump and connected to the system interlock to prevent fluid from being withdrawn from the supply tank when the irrigation system is either automatically or manually shutdown. The system must contain functional interlocking controls to automatically shut off the pesticide injection pump when the water pump motor stops. The irrigation line or water pump must include a functional pressure switch which will stop the water pump motor when the water pressure decreases to the point where pesticide distribution is adversely affected. Systems must use a metering pump, such as a positive displacement injection pump (e.g., diaphragm pump) effectively designed and constructed of materials that are compatible with pesticides and capable of being fitted with a system interlock. Do not apply when wind speed favors drift beyond the area intended for treatment.

Do not apply when wind favors drift, when system connection or fittings leak, when nozzles do not provide uniform distribution or when lines containing the product must be dismantled and drained. Crop injury, lack of effectiveness, or illegal pesticide residues in the crop may result from nonuniform distribution of treated water.

Allow sufficient time for pesticide to be flushed through all lines and all nozzles before turning off irrigation water. A person knowledgeable of the chemigation system and responsible for its operation shall shut the system down and make necessary adjustments should the need arise.

Do not connect an irrigation system (including greenhouse systems) used for pesticide application to a public water system unless the label-prescribed safety devices for public water supplies are in place.

HOW TO USE ROVRAL® 4 FLOWABLE FUNGICIDE

Partially fill the spray tank with clean water. Measure the required amount of ROVRAL® 4 Flowable Fungicide and pre-mix with a small volume of water, add this to the tank. Agitate to ensure thorough mixing while filling tank with remaining water. Maintain agitation during application and apply with properly calibrated application equipment. Do not allow spray mixture to stand overnight or for prolonged periods, as some chemical breakdown may occur, particularly in water with a high pH.

LEAFY VEGETABLES

CROP	HOW TO USE	DISEASE	DOSAGE RATE PINTS/ ACRE	WHEN TO APPLY
Head Lettuce*	Apply as a foliar spray in sufficient water to obtain thorough coverage (50 to 100 gallons per acre).	Lettuce Drop (Sclerotinia sp.) Bottom Rot (Rhizoctonia solani)	1.5 to 2.0	Application should be made with tractor mounted boom sprayers equipped with flat fan or hollow cone nozzles directed to ensure thorough coverage of the lower portion of the plants and the surrounding soil surface. Do not drench. The higher rate should be used under severe disease conditions. Apply at the 3 leaf stage of growth and again 10 days later. If conditions still favor disease development, a third application should be made 10 days after the second spray. Do not apply within 14 days of harvest. Do not make more than 3 applications of ROVRAL® 4 Flowable per season.

VEGETABLES

CROP	HOW TO USE	DISEASE	DOSAGE RATE PINTS/ ACRE	GALS/WATER PER ACRE	WHEN TO APPLY
Broccoli*	Application should be made with a tractor mounted boom sprayer with 2 flat fan nozzles per row (one on either side) directed at the base of the plant and the adjacent soil surface. Position nozzles to insure thorough coverage of the stem. Do not drench.	Black Leg (*Phoma lingiam*)	2.0		Apply immediately after thinning (2 to 4 leaf stage) as a directed spray to the base of the plant and the adjacent soil surface. If disease conditions persist or reoccur, a second application may be made up to the day of harvest. Apply in 40 to 200 gal. of spray solution per acre.

CROP	HOW TO USE	DISEASE	DOSAGE RATE PINTS/ ACRE	DOSAGE RATE GALS/WATER PER ACRE	WHEN TO APPLY
Dry Bulb Onions*	A single, flat fan, or cone nozzle should be centered and adjusted to provide complete coverage of each row. Up to 10 applications can be made per season. Do not apply within 7 days of harvest. Apply ROVRAL® 4 Flowable using ground equipment in accordance with the directions in the table.	Botrytis Leaf Blight (*Botrytis squamosa*) Purple Blotch (*Alternaria porri*) Botrytis Neck Rot (*Botrytis allii*)	1.5	50 to 100	Apply as a foliar spray as soon as conditions become favorable for disease. Continue application on 7 to 14 day interval as long as conditions favor disease development. Under disease conditions, the shorter interval should be used.

*** NOTE TO USER:** The following crops may be rotated after harvest: Garlic, Dry Bulb Onions, Broccoli, Lettuce, Peanuts, Carrots, Beans and Potatoes. The following crops may be rotated one month following the last iprodione application: Root Crops, Tomatoes, and Cotton.

GARLIC

CROP	HOW TO USE	DISEASE	DOSAGE RATE PINTS/ ACRE	DOSAGE RATE GALS./WATER PER ACRE	WHEN TO APPLY
Garlic*	Apply as an in furrow spray in sufficient water to obtain thorough coverage.	White Rot (*Sclerotium cepivorum*)	4.0**	50 to 100	Apply ROVRAL® 4 Flowable in the furrow at planting. Position a single flat fan nozzle to treat the covering soil used to fill the furrow. **Using a 38 to 40 inch bed spacing.

*** NOTE TO USER:** The following crops may be rotated after harvest: Garlic, Dry Bulb Onions, Broccoli, Lettuce, Peanuts, Carrots, Beans and Potatoes. The following crops may be rotated one month following the last iprodione application: Root Crops, Tomatoes, and Cotton.

PEANUTS

CROP	HOW TO USE	DISEASE	DOSAGE RATE PINTS/ ACRE	DOSAGE RATE GALS./WATER PER ACRE	WHEN TO APPLY
Peanuts*	Apply ROVRAL® 4 Flowable using a tractor mounted boom sprayer equipped with low pressure nozzles (e.g., 8008 LP, 8010 LP, and TK 7.5) that produce large droplets. A single nozzle should be centered and adjusted to provide complete coverage of the row.	Sclerotinia Blight (*Sclerotinia ap.*)	2.0	40	Make an initial application when conditions first become favorable for disease development. Up to 2 subsequent applications should be made at 4 week intervals. Do not apply within 10 days of harvest.

POTATOES

CROP	HOW TO USE	DISEASE	DOSAGE RATE PINTS/ ACRE	DOSAGE RATE GALS./WATER PER ACRE	WHEN TO APPLY
Potatoes*	A single, flat fan or cone nozzle should be centered and adjusted to provide complete coverage of each row. Up to 4 applications can be made per season. DO NOT APPLY WITHIN 14 DAYS OF HARVEST. Apply ROVRAL® Fungicide using ground application or chemigation, in sufficient water for thorough coverage (minimum of 10 gallons per acre).	Early Blight (*Alternaria solani*) White Mold (*Sclerotinia sclerotiorum*)	2.0	Minimum 10	**For early blight,** begin applications when conditions first become favorable for disease development. Up to 3 subsequent applications can be applied at 7 to 10 day intervals or as required. **For white mold,** apply at first sign of disease or immediately prior to row closing and, if conditions are favorable for disease development, again 28 days later. Do not apply within 14 days of harvest.

*** NOTE TO USER:** The following crops may be rotated after harvest: Garlic, Dry Bulb Onions, Broccoli, Lettuce, Peanuts, Carrots, Beans and Potatoes. The following crops may be rotated one month following the last iprodione application: Root Crops, Tomatoes, and Cotton.

STONE FRUITS AND NUTS

CROP	HOW TO USE	DISEASE	DOSAGE RATE PINTS/ 100 GAL.	DOSAGE RATE PINTS/ ACRE	WHEN TO APPLY
Apricots Cherries Nectarines Peaches Plums Prunes	Apply ROVRAL® as an overall spray in sufficient water to obtain thorough coverage of bloom, foliage, and/or fruit (20 to 400 gallons per acre by ground equipment and in a minimum of 15 gals. per acre by air). Aerial applications are only recommended during the bloom period.	Brown Rot Blossom Blight (*Monilinia* spp.) Fruit Brown Rot (*Monilinia* spp.)	0.25 to 0.50	1.0 to 2.0	The rate of ROVRAL® 4 Flowable per 100 gallons are based on a standard of 400 gallons per acre dilute spray for mature trees. For less than mature trees, apply the rate per 100 gallons until run-off. If less than 400 gallons of spray solution per acre is applied to mature trees, refer to the rate per acre to ensure that the proper amount of material is applied. **Brown Rot Blossom Blight:** Apply first at early bloom (approximately 5% bloom). If conditions are favorable for disease development, apply again at full bloom and at petal fall. **Fruit Brown Rot:** An application should be made whenever temperatures and moisture conditions favor disease infection in the 5 week period prior to harvest. If these conditions persist or reoccur, a second application should be made. This second application should be made no sooner than 7 days following the first pre-harvest application. Pre-harvest applications may be made up to and including the day of harvest. Do not apply more than 2.0 pints of ROVRAL® 4 Flowable per acre per application. Do not make more than 5 applications of ROVRAL® 4 Flowable per season.
Almonds	Apply ROVRAL® Flowable as an overall spray in sufficient water to obtain thorough coverage of bloom and foliage (100 to 400 gals. per acre by ground and in a minimum of 20 gals. per acre by air). The use of aerial equipment is not recommended after petal fall.	Brown Rot Blossom Blight (*Monilinia laxa*) Shothole (*Stigmina carpophila*)	0.25	1.0	The rates of ROVRAL® 4 Flowable per 100 gallons are based on a standard of 400 gallons per acre dilute spray for mature trees. For less than mature trees, apply the rate per 100 gallons until run-off. If less than 400 gallons of spray solution per acre is applied, to mature trees, refer to the rate per acre to ensure that the proper amount of material is applied. **Brown Rot Blossom Blight Shothole:** Apply first at pink bud and, if conditions favorable for disease development persist or reoccur up to 3 subsequent applications can be made at: 1) full bloom; 2) petal fall; and 3) up to 5 weeks after petal fall. Do not make more than 4 applications of ROVRAL® 4 Flowable per season.

NOTE TO USER: Do not graze animals in treated orchards. Do not feed cover crops grown in treated orchards to livestock.

GRAPES

CROP	HOW TO USE	DISEASE	DOSAGE RATE PINTS/ ACRE	WHEN TO APPLY
Grapes	ROVRAL® 4 Flowable is a flowable formulation for the control of Bunch Rot on Grapes. Apply as a foliar spray in sufficient water to obtain thorough coverage (50 to 200 gals. per acre). The spray mixture should not be allowed to stand overnight as some breakdown of the product may occur from prolonged delays, particularly in water with a high pH. Maintain agitation during spray operations and apply with properly calibrated application equipment.	Bunch Rot (*Botrytis* sp.)	1.5 to 2.0	Apply ROVRAL® 4 Flowable in accordance with directions provided in table: **SPRAY SCHEDULE** 1) Early mid-bloom. 2) Prior to bunch closing. 3) Beginning of fruit ripening (veraison). 4) Final application prior to harvest as needed. The last application may be made up to and including the day of harvest. **Thorough Coverage Of The Bunches Is Essential.** Under severe disease conditions, the higher rate is recommended. ROVRAL® 4 Flowable must be used in conjunction with good cultural practice designed to minimize conditions conductive to Bunch Rot (*Botrytis cinerea*). It should be applied with equipment calibrated and adjusted to direct the spray at the bunches to ensure thorough bunch coverage.

CROP	HOW TO USE	DISEASE	DOSAGE RATE PINTS/ ACRE	WHEN TO APPLY
				Do not make more than 4 applications of ROVRAL® 4 Flowable per season.

NOTE TO USER: Do not graze animals in treated orchards. Do not feed cover crops grown in treated orchards to livestock.

STORAGE AND DISPOSAL

STORAGE

Do not contaminate water, food, or feed by storage or disposal.

PESTICIDE DISPOSAL

Wastes resulting from the use of this product may be disposed of on site or at an approved waste disposal facility.

CONTAINER DISPOSAL

Triple rinse (or equivalent). Then dispose of empty bag in a sanitary landfill or by incineration, or by other procedures approved by State and local authorities.

LIMITED WARRANTY AND DISCLAIMER

The manufacturer warrants (a) that this product conforms to the chemical description on the label; (b) that this product is reasonably fit for the purposes set forth in the directions for use when it is used in accordance with such directions; and (c) that the directions, warning and other statements on this label are based upon responsible experts' evaluation of reasonable tests of effectiveness, of toxicity to laboratory animals and to plants, and of residues on food crops and upon reports of field experience. Tests have not been made on all varieties or in all states or under all conditions. THE MANUFACTURER NEITHER MAKES NOR INTENDS, NOR DOES IT AUTHORIZE ANY AGENT OR REPRESENTATIVE TO MAKE, ANY OTHER WARRANTIES, EXPRESS OR IMPLIED, AND IT EXPRESSLY EXCLUDES AND DISCLAIMS ALL IMPLIED WARRANTIES OF MERCHANTABILITY OR FITNESS FOR A PARTICULAR PURPOSE.

THIS WARRANTY DOES NOT EXTEND TO, AND THE BUYER SHALL BE SOLELY RESPONSIBLE FOR, ANY AND ALL LOSS OR DAMAGE WHICH RESULTS FROM THE USE OF THIS PRODUCT IN ANY MANNER WHICH IS INCONSISTENT WITH THE LABEL DIRECTIONS, WARNINGS OR CAUTIONS.

BUYER'S EXCLUSIVE REMEDY AND MANUFACTURER'S OR SELLER'S EXCLUSIVE LIABILITY FOR ANY AND ALL CLAIMS, LOSSES, DAMAGES, OR INJURIES RESULTING FROM THE USE OR HANDLING OF THIS PROUDCT, WHETHER OR NOT BASED IN CONTRACT, NEGLIGENCE, STRICT LIABILITY IN TORT OR OTHERWISE SHALL BE LIMITED, AT THE MANUFACTURER'S OPTION, TO REPLACEMENT OF OR THE REPAYMENT OF THE PURCHASE PRICE FOR, THE QUANTITY OF PRODUCT WITH RESPECT TO WHICH DAMAGES ARE CLAIMED. IN NO EVENT SHALL MANUFACTURER OR SELLER BE LIABLE FOR SPECIAL, INDIRECT OR CONSEQUENTIAL DAMAGES RESULTING FROM THE USE OR HANDLING OF THIS PRODUCT.

NOTICE TO BUYER

Purchase of this material does not confer any rights under patents governing this product or the use thereof in countries outside of the United States.

ROVRAL is a registered trademark of Rhône-Poulenc

AG 90034

EPA Approval: 5/31/89

THIS SPECIMEN LABEL IS INTENDED FOR USE ONLY AS A GUIDE IN PROVIDING GENERAL INFORMATION REGARDING THE DIRECTIONS, WARNINGS AND CAUTIONS ASSOCIATED WITH THE USE OF THIS PRODUCT. AS WITH ANY AGRICULTURAL CHEMICAL, ALWAYS FOLLOW THE LABEL INSTRUCTIONS ON THE PACKAGE BEFORE USING.

SEVIN® brand XLR PLUS

CARBARYL INSECTICIDE

For Agricultural or Commercial Use Only

ACTIVE INGREDIENT:	
Carbaryl (1-naphthyl N-methylcarbamate)	41.2% by wt.
INERT INGREDIENTS:	58.8% by wt.
TOTAL	100.0%

(Contains 4 Pounds Carbaryl Per Gallon)

E.P.A. Reg. No. 264-333 E.P.A. Est. No. 264-M0-02

KEEP OUT OF REACH OF CHILDREN

CAUTION

CUIDADO

PRECAUCION AL USUARIO: Si usted no lee ingles, no use este producto hasta que la etiqueta le haya sido explicada ampliamente.

For PRODUCT USE Information Call 1-800-334-9745
For MEDICAL and TRANSPORTATION Emergencies ONLY Call 24 Hours A Day 1-800-334-7577

PRECAUTIONARY STATEMENTS

CAUTION

HAZARDS TO HUMANS AND DOMESTIC ANIMALS

MAY BE HARMFUL IF SWALLOWED OR INHALED. OVEREXPOSURE MAY CAUSE: Salivation, watery eyes, pinpoint eye pupils, blurred vision, muscle tremors, difficult breathing, excessive sweating, abdominal cramps, nausea, vomiting, diarrhea, weakness, headache. IN SEVERE CASES CONVULSION, UNCONSCIOUSNESS AND RESPIRATORY FAILURE MAY OCCUR. SIGNS AND SYMPTOMS OF OVEREXPOSURE OCCUR RAPIDLY FOLLOWING EXPOSURE TO THIS PRODUCT.

Avoid breathing spray mist. Do not take internally. Avoid contact with eyes, skin or clothing. Wear regular long-sleeved work clothing and head covering. Change to clean clothing daily. Bathe and wash hair after each work day. Do not eat, drink or use tobacco while working with this product or spray solutions. Wash hands and face before eating, drinking or using tobacco. Keep out of reach of children and domestic animals.

ANTIDOTE STATEMENT

ATROPINE SULFATE IS HIGHLY EFFECTIVE AS AN ANTIDOTE. Narcotics and other sedatives should not be used. Further, drugs like 2-PAM (pyridine-2-aldoxime methiodide) are NOT recommended. See NOTE TO PHYSICIAN below.

STATEMENT OF PRACTICAL TREATMENT

GENERAL

Contact a physician immediately in all cases of suspected poisoning. Transport to a physician or hospital immediately and SHOW A COPY OF THIS LABEL TO THE PHYSICIAN. If poisoning is suspected in animals, contact a veterinarian.

IF SWALLOWED: Never give anything by mouth to an unconscious or convulsing person. If conscious and not convulsing, drink 1 to 2 glasses of water and induce vomiting by touching the back of the throat with finger.

IF IN EYES: Flush eyes with plenty of water. Get medical attention if irritation persists.

IF ON SKIN: Wash thoroughly with soap and water.

IF INHALED: Move from contaminated atmosphere and call a physician.

NOTE TO PHYSICIAN

Carbaryl is a carbamate insecticide, which is a cholinesterase inhibitor. Overexposure to this substance may cause toxic signs and symptoms due to stimulation of the cholinergic nervous system. These effects of overexposure are spontaneously and rapidly reversible. Gastric lavage may be used if this product has been swallowed. Carbaryl poisoning may occur rapidly after ingestion and prompt removal of stomach contents is indicated.

Specific treatment consists of parenteral atropine sulfate. Caution should be maintained to prevent overatropinization. Mild cases may be given 1 to 2 mg intramuscularly every 10 minutes until full atropinization has been achieved and repeated thereafter whenever symptoms reappear. Severe cases should be given 2 to 4 mg intravenously every 10 minutes until fully atropinized, then intramuscularly every 30 to 60 minutes as needed to maintain the effect for at least 12 hours. Dosages for children should be appropriately reduced. Complete recovery from overexposure is to be expected within 24 hours.

Narcotics and other sedatives should not be used. Further, drugs like 2-PAM (pyridine-2-aldoxime methiodide) are NOT recommended.

To aid in confirmation of a diagnosis, urine samples should be obtained within 24 hours of exposure and immediately frozen.

Consultation on diagnosis, therapy and analysis can be obtained at all hours by calling the Rhône-Poulenc emergency number 1-800-334-7577.

ENVIRONMENTAL HAZARDS

This product is extremely toxic to aquatic and estuarine invertebrates. Do not apply directly to water and wetlands, except under forest canopy and use on rice. Discharge from rice fields may kill aquatic and estuarine invertebrates. Do not apply when weather conditions favor drift from area treated. Do not contaminate water by cleaning equipment or disposal of wastes.

BEE CAUTION

This product is highly toxic to bees exposed to direct treatment on blooming crops or weeds. However, field studies have shown that SEVIN® brand XLR PLUS Carbaryl Insecticide is less hazardous to honey bees than other carbaryl products when direct application to bees is avoided and the spray residues have dried. For maximum honey bee hazard reduction, apply from late evening to early morning or when bees are not foraging. Do not apply this product or allow it to drift to blooming crops or weeds if bees are foraging in the treatment area. However, applications may be made during foraging periods if the beekeeper takes one of the following precautionary measures prior to bee flight activity on the day of treatment: (1) Confine the honey bees to the hive by covering the colony or screening the entrance or; (2) locate hives beyond bee flight range from the treated area. Precautionary measures may be discontinued after spray residues have dried. Contact your cooperative Agricultural Extension Service or your local Rhône-Poulenc Ag Company representative for further information.

DIRECTIONS FOR USE AS A WHEAT BRAN BAIT

FOR END USE ONLY. NOT FOR REPACKAGING.

FOR USE ONLY BY GOVERNMENT PERSONNEL OR PERSONS UNDER THEIR DIRECT SUPERVISION.

Mixing Instructions

Mix the appropriate amount of SEVIN® brand XLR PLUS Carbaryl Insecticide with wheat bran to make a carbaryl wheat bran bait containing 2% to 10% active carbaryl. For example, for a bait containing 5% carbaryl, mix 1 quart SEVIN® brand XLR PLUS Carbaryl Insecticide (contains 1 lb. active carbaryl) with each 19 pounds of wheat bran. Mix only the amount of bait necessary for each insect control program.

Storage Instructions

Store carbaryl bran baits in cool, dry area out of reach of children and animals. Do not contaminate water, food, or feed by storage or disposal.

NOTE: Carbaryl bran baits should only be stored temporarily while awaiting application.

Application Instructions

Applications may be made with ground equipment (hand cyclone spreader) or with aerial application equipment with a metered bait spreader attachment.

PASTURES, RANGELAND, WASTELAND, ROADSIDES

Use 0.50-1.50 lbs. active ingredient/acre for the control of grasshoppers and Mormon crickets. The lower rate is suggested for early instars on small plants or sparse vegetation. Use the higher rate for adults or dense vegetation. Use of low bait assay and higher rate is suggested for control of high grasshopper populations. Treatment may be repeated as necessary.

Preharvest interval is 0 days.

IMPORTANT: BEFORE USING SEVIN® BRAND XLR PLUS CARBARYL INSECTICIDE, READ AND FOLLOW CAREFULLY. OBSERVE THE PRECAUTIONARY STATEMENTS AND ALL OTHER INFORMATION APPEARING ON THE PRODUCT LABEL.

DIRECTIONS FOR USE

IT IS A VIOLATION OF FEDERAL LAW TO USE THIS PRODUCT IN A MANNER INCONSISTENT WITH ITS LABELING.

READ THIS LABEL BEFORE USE. STRICTLY OBSERVE LABEL DIRECTIONS AND CAUTIONS, AND APPLICABLE FEDERAL AND STATE REGULATIONS. DO NOT USE ON ANY CROP NOT LISTED ON THIS LABEL OR SUPPLEMENTAL LABELING AS ANY RESIDUES REMAINING MAY BE ILLEGAL OR HARMFUL.

GENERAL INFORMATION

SEVIN® brand XLR PLUS Carbaryl Insecticide is a suspension of microfine SEVIN® brand carbaryl insecticide in an aqueous medium. It readily disperses in water to form a spray which may be applied by air or ground.

GENERAL WORKER PROTECTION STATEMENT

Do not apply this product in such a manner as to directly or through drift expose workers or other persons. The area being treated must be vacated by unprotected persons.

Do not enter treated areas without protective clothing until sprays have dried.

Because certain states may require more restrictive reentry intervals for various crops treated with this product, consult your State Department of Agriculture for further information.

Written or oral warnings must be given to workers who are expected to be in a treated area or in an area about to be treated with this product. Advise workers to stay out of treated areas until spray deposits have dried. This includes regular long-sleeved and long-legged work clothing. See PRECAUTIONARY STATEMENTS, STATEMENT OF PRACTICAL TREATMENT and NOTE TO PHYSICIAN for additional information. When oral warnings are given, warnings shall be given in a language customarily understood by workers. Oral warnings must be given if there is reason to believe that written warnings cannot be understood by workers. Written warnings must include the following information: appropriate signal word (CAUTION), area treated with SEVIN® brand XLR PLUS Carbaryl Insecticide, date of application, appropriate clothing, and re-entry interval (i.e., until sprays have dried) and actions to be taken in case of accidental exposure.

PREHARVEST AND GRAZING USE INFORMATION AND LIMITATIONS

Tolerances established under the Federal Food, Drug and Cosmetic Act permit the sale of crops bearing probable carbaryl residues when this product is used in accordance with the label directions. If used as directed, treated forage may be grazed or used as feed for dairy and meat animals without causing illegal residues in meat or milk. This product may be applied up to and including the day of harvest or grazing of forage crops (except alfalfa). Application may be made without removing livestock from area being treated. Do not apply at greater rates or at more frequent intervals than is stated on the label. To do so may result in illegal residues in crops, meat, and milk.

Do not plant rotational food and feed crops not listed on this or other carbaryl labels in carbaryl treated soil.

Do not use reclaimed irrigation water from crops treated with carbaryl on crops for which carbaryl tolerances are not established.

PLANT RESPONSE PRECAUTIONS

To avoid possible injury to tender foliage, do not apply to wet foliage or during periods of high humidity.

Do not use on Boston Ivy, Virginia creeper and maidenhair fern as injury may result. Carbaryl may also injure Virginia and sand pines.

Carefully observe label instructions for apple thinning to avoid excessive thinning. Combinations with certain herbicides on rice and soybeans may be phytotoxic. Refer to specific directions for appropriate crop.

SPRAY PREPARATION

TO ASSURE A UNIFORM SUSPENSION, AGITATE, STIR OR RECIRCULATE ALL CONTAINERS OF THIS PRODUCT PRIOR TO USE. Remove oil, rust, scale, pesticide residues and other foreign matter from mix tanks and entire spray system. Flush with clean water. Fill spray or mix tank with 1/2 to 3/4 the desired amount of water. Start mechanical or hydraulic agitation. Slowly add the required amount of SEVIN® brand XLR PLUS Carbaryl Insecticide, and then the remaining volume of water. Include rinse water from container. Prepare only as much spray mixture as can be applied on the day of mixing. MAINTAIN CONTINUOUS AGITATION DURING MIXING AND APPLICATION TO ASSURE A UNIFORM SUSPENSION. DO NOT STORE SPRAY MIXTURE FOR PROLONGED PERIODS OR DEGRADATION OF CARBARYL MAY OCCUR. Local water conditions may also accelerate the degradation of spray mixtures containing carbaryl. See COMPATIBILITY STATEMENT below.

WASHOFF RESISTANCE AND COVERAGE

Dilution of 1 volume of SEVIN® brand XLR PLUS Carbaryl Insecticide with 1 volume of water provides maximum resistance to washoff by rainfall or overhead irrigation. Dilutions higher than 1 part SEVIN® brand XLR PLUS Carbaryl Insecticide to 39 parts water (1:39) are not recommended when washoff resistance is desired.

To achieve washoff resistance, SEVIN® brand XLR PLUS Carbaryl Insecticide must be diluted as stated above, and droplets must dry on the foliage. When atmospheric humidity is low, a drying time of at least two hours is generally adequate. Under high humidity a longer drying time is required. Washoff resistance cannot be expected if this product is applied to wet foliage and has not thoroughly dried prior to rainfall or overhead irrigation.

On all crops, use sufficient spray volume to obtain thorough coverage. Optimum pest control under certain crop, pest or climatic conditions may require spray gallonages higher than the 1:39 dilution. For example in hot, arid weather (low humidity), higher spray gallonage per acre may be required to compensate for loss from evaporation and insure thorough coverage. The total spray volume required for effective pest control can best be determined by previous experience, pest and crop conditions and local recommendations.

COMPATABILITY

SEVIN® brand XLR PLUS Carbaryl Insecticide, when diluted with at least an equal volume of water, is compatible with a wide range of pesticides. It is not compatible with diesel fuel, kerosene, fuel oil or aromatic solvents. If compatability with another product and the resulting crop response is unknown, the mixture should be tested on a small scale. Curdling, precipitation, greasing, layer formation or increased viscosity are symptoms of incompatibility. Incompatibility will reduce insect control and may cause application and handling difficulties or plant injury. Observe all cautions and limitations on labeling of all products used in mixtures. WHEN PREPARING COMBINATION SPRAYS, FIRST ADD "SEVIN®" BRAND XLR PLUS CARBARYL INSECTICIDE TO AT LEAST AN EQUAL VOLUME OF WATER, MIX THOROUGHLY, AND THEN ADD COMBINATION PRODUCTS TO THE MIXTURE. DO NOT APPLY TANK MIX COMBINATIONS UNLESS YOUR PREVIOUS EXPERIENCE INDICATES THE MIXTURE IS EFFECTIVE AND WILL NOT RESULT IN APPLICATION PROBLEMS OR PLANT INJURY.

Carbaryl is unstable under highly alkaline conditions and mixtures with strong bases, such as Bordeaux, lime-sulfur and casein-lime spreaders, will result in chemical degradation of the insecticide. Do not use this product in water with pH values above 8.0 unless a buffer is added. If necessary, water should be buffered to neutral (pH = 7.0) before adding this product to the spray tank. Overhead irrigation with alkaline or muddy water after application will also accelerate chemical degradation and may result in reduced insect control.

APPLICATION

On all crops use sufficient gallonage to obtain thorough and uniform coverage.

Observe crop label instructions for specific directions regarding spray volume where they occur. Calibrate spray equipment to deliver the required volume. The flow rate of this product diluted 1:1 with water is similar to water. Use 50 mesh slotted strainers in spray system and 25 mesh slotted strainers behind nozzles.

To clean spray system after use, drain and flush with a water and detergent mixture. Rinse thoroughly with clean water. Refer to the Storage and Disposal directions for disposal instructions.

INSECT CONTROL

Begin application when insect populations reach recognized economic threshold levels. Consult the Cooperative Extension Service, Professional Consultants or other qualified authorities to determine appropriate threshold levels for treatment in your area. Where a dosage range is indicated, use the lower rate on light to moderate infestations, young plants and early instars and the higher rate on heavy infestations, mature plants, advanced instars and adults. Thorough and uniform spray coverage is essential for effective control.

NOTE: All references to armyworm on the crops listed below refer to the species, *Pseudaletia unipuncta,* often called the "true armyworm". Except where indicated otherwise, this product is not registered for the control of other armyworm species. Regional differences have been noted in the susceptibility of certain strains of fall armyworm, Colorado potato beetle, spotted tentiform leafminer, and tobacco budworm (on cotton) to carbaryl. If local experience indicates inadequate control, use an alternative pesticide.

FORAGE, FIELD AND VEGETABLE CROPS

Apply in sufficient spray volume to obtain thorough coverage. Do not dilute greater than 1:39 (volume product: volume water) where washoff resistance is desired. Where maximum washoff resistance is needed, do not exceed a 1:1 dilution. To prepare small volumes of spray mixture use 1 tablespoonful (½ fluid ounce) of SEVIN® brand XLR PLUS Carbaryl Insecticide per gallon of water where rates of 1 quart per acre or 1 quart per 100 gallons are indicated in the tables below.

GRASSHOPPERS

CROP		QUARTS OF "SEVIN®" BRAND XLR PLUS/ACRE	PREHARVEST INTERVAL (DAYS)	SPECIFIC DIRECTIONS
All Forage, Field and Vegetable Crops		½ to 1½ Unless Noted Otherwise	See specific Forage, Field or Vegetable Crops	Use the lower rate for nymphs on small plants or sparse vegetation in wasteland, rangeland, ditchbanks and borders. Use the higher rate for mature grasshoppers or applications to dense vegetation or if extended residual control is desired. Be certain spray volumes are appropriate to assure adequate coverage.
Alfalfa Clovers Birdsfoot trefoil (DO NOT USE ON BIRDSFOOT TREFOIL IN CALIFORNIA)	Blister beetles Mexican bean beetle	½ to 1		OBSERVE BEE CAUTION Observe plant response precautions.
	Alfalfa caterpillar Bean leaf beetle Cucumber beetles Green cloverworm Japanese beetle Leafhoppers Potato Leafhopper Three cornered alfalfa hopper Thrips Velvetbean caterpillar	1	7 (Alfalfa)	Carbaryl may cause a temporary bleaching of tender alfalfa foliage. Apply only once per cutting for alfalfa at 1½ quarts.
	Armyworm Cloverhead weevil Corn earworm Egyptian alfalfa weevil larvae Essex skipper European alfalfa beetle Fall armyworm Lygus bugs Stink bugs Webworms Yellowstriped armyworm		0 (Clover & Birdsfoot trefoil)	On dense growth, use 25 to 40 gallons per acre with ground equipment to ensure adequate coverage. For alfalfa weevil larvae, if pretreatment damage is extensive, cut alfalfa and treat the stubble.
	Alfalfa weevil larvae West of the Rocky Mountains	1 to 1½		

CROP		QUARTS OF "SEVIN®" BRAND XLR PLUS/ACRE	PREHARVEST INTERVAL (DAYS)	SPECIFIC DIRECTIONS
	East of the Rocky Mountains	1½		
	Grasshoppers			Refer to the general GRASSHOPPER heading above.
Asparagus	Asparagus beetle	1 to 2	1	Treat ferns or brush growth. Do not treat more than once every 3 days.
	Apache cicada Asparagus beetle	2 to 4	Post harvest application only	
Beans (including blackeyed peas, cowpeas, southern peas, dry beans, green beans, lima beans, navy beans and snap beans) Lentils (DO NOT USE ON LENTILS IN CALIFORNIA)	Blister beetles Mexican bean beetle	½ to 1	0 (except cowpeas and lentils) 3 (cowpeas) 7 (lentils)	Observe plant response precautions.
	Alfalfa caterpillar Bean leaf beetles Cucumber beetles Flea beetles Green cloverworm Japanese beetle Leafhoppers Three cornered alfalfa hopper Thrips Velvetbean caterpillar Western bean cutworm	1		
	Grasshoppers	½ to 1½		Refer to general GRASSHOPPER heading above.
	Armyworm Corn earworm Cutworms European corn borer Fall armyworm Stink bugs Tarnished plant bug Webworms	1 to 1½		
	Alfalfa Looper	1½		
	Cowpea curculio	2		
	Corn earworm Limabean podborer Lygus bugs Stink bugs	2		CALIFORNIA ONLY
Cabbage Broccoli Brussels sprouts Cauliflower Kohlrabi	Flea beetles Harlequin bug	½ to 1	3	
	Armyworm Corn earworm Diamondback moth Fall armyworm Imported cabbage-worm	1 to 2		
Chinese cabbage Collards Hanover salad Horseradish Kale Mustard greens Radishes Rutabagas Turnips	Flea beetles Harlequin bug Leafhoppers	½ to 1	3 (horseradish, radishes, ruta-bagas and turnip roots) 14 (Chinese cab-bage, collards, Hanover salad, kale, mustard greens, and turnip tops)	
	Aster leafhopper	1 to 1½		
	Armyworm Corn earworm Fall armyworm Imported cabbageworm Stink bugs Tarnished plant bug	1 to 2		
Carrots Celery Parsley Parsnips (DO NOT USE ON CELERY IN CALIFORNIA)	Flea beetles Leafhoppers	½ to 1	0 (carrots) 3 (parsnips) 14 (celery & parsley)	
	Aster leafhopper Lygus bugs Spittlebugs	1 to 1½		
	Armyworm Corn earworm Fall armyworm Stink bugs Tarnished plant bug	1 to 2		
Corn (field, sweet, pop)	Armyworm Chinch bugs Corn earworm Corn rootworm adults Fall armyworm Flea beetles Japanese beetle Sap beetles Southwestern corn borer Leafhoppers	1 to 2	0	OBSERVE BEE CAUTION For insects attacking silks and ears apply at 1 to 6 day intervals starting when first silks appear and continuing until silks begin to dry. For larvae in whorl and foliage feeders, apply as necessary. Optimum timing and good coverage are essential for effective control. For optimum chinch bug control, apply at least 20 gallons of water per acre by ground and direct spray toward stalk to provide thorough coverage.

CROP			QUARTS OF "SEVIN®" BRAND XLR PLUS/ACRE	PREHARVEST INTERVAL (DAYS)	SPECIFIC DIRECTIONS
	Grasshoppers		½ to 1½		Refer to general GRASSHOPPER heading above.
	European Corn borer		1½ to 2		For optimum control, do not apply in less than 3 gallons of water per acre by air and 15 gallons of water per acre by ground
	Western bean cutworm		2		
	Cutworms		2 to 3		For optimum control, apply in a 12 inch band, over the row, using sufficient volume of water to obtain thorough coverage. For broadcast application use at least 20 gallons (ground) or 5 gallons (air) of water per acre.
Cotton	Cotton fleahopper Cotton leafworm Flea beetles	Striped blister beetle Thrips	½ to 1	7	
	Boll weevil Cotton bollworm Fall armyworm Leafrollers Leafhoppers	Tarnished plant bug Yellowstriped armyworm (cotton cutworm)	1 to 2		Treat on a 5 to 7 day schedule for as long as control is necessary. Mid and late season insect control. May be applied after bolls open.
	Lygus bugs		1 to 2		For light to moderate populations in Western irrigated cotton.
	Pink bollworm		1½ to 2½		Aphid populations will be suppressed by repeated applications of this insecticide.
	Cutworms Stink bugs	Saltmarsh caterpillar	2		
Cucumber Melons Pumpkin Squash	Pickleworm	Melonworm	½ to 1	0	Observe plant response precautions.
	Cucumber beetles Flea beetles	Leafhoppers Squash bugs	1		
Dandelion Endive (Escarole) Lettuce Salsify	Flea beetles Harlequin bug	Leafhoppers	½ to 1	3 (head lettuce & salsify roots)	Observe plant response precautions. Treat on a 5 to 7 day schedule after heads begin to form.
	Aster leafhopper Lygus bugs	Spittlebugs	1 to 1½	14 (dandelion, endive (escarole), leaf lettuce & salsify tops)	
	Armyworm Corn earworm Fall armyworm	Imported cabbageworm Stink bugs Tarnished plant bug	1 to 2		
Flax Proso Millet (DO NOT USE IN CALIFORNIA)	Armyworm		1 to 1½	42	Do not graze treated areas or harvest for dairy feed prior to crop maturity.
	Grasshopper		½ to 1½		Refer to the general GRASSHOPPER heading above.
Pasture	Armyworm Chinch bugs Essex skipper Fall armyworm Striped grass looper Thrips Range caterpillar Range crane fly		1 to 1½	0 (aerial application) 14 (ground applications)	To control thrips in grasses grown for seed, use high spray pressure to improve penetration into boot. Apply a minimum of 2 applications per year. Allow at least 14 days between applications. Do not allow foraging or cut for hay within 14 days of last application by ground. Aerially treated pastures may be grazed or cut for hay on day of treatment. Carefully mark swaths to avoid overapplication.
	Grasshoppers				Refer to the general GRASSHOPPER heading above.
Rangeland	Black grass bug Mormon cricket Range caterpillar Range crane fly		½ to 1	0	FOR AERIAL APPLICATION ONLY. Apply a maximum of 2 applications per year. Allow at least 14 days between applications.

CROP			QUARTS OF "SEVIN®" BRAND XLR PLUS/ACRE	PREHARVEST INTERVAL (DAYS)	SPECIFIC DIRECTIONS
					Carefully mark swaths to avoid overapplication.
	Grasshoppers				Refer to the general GRASSHOPPER heading above.
Wasteland, Rights-of Way, Hedgerows, Ditch banks, Roadsides	Black grass bug Mormon cricket Range caterpillar Range crane fly		½ to 1	0 (aerial application) 14 (ground application)	Apply a maximum of 2 applications per year. Allow at least 14 days between applications. Do not allow foraging or cut for hay within 14 days of last application by ground. Aerially treated areas may be grazed or cut for hay on day of treatment. Carefully mark swaths to avoid overapplication.
	Grasshoppers				Refer to the general GRASSHOPPER heading above.
Garden beet Spinach Swiss chard	Flea beetles Harlequin bug	Leafhoppers	½ to 1	3 (garden beet roots)	
	Aster leafhopper		1 to 1½	14 (garden beet tops, spinach, Swiss chard)	
	Armyworm Corn earworm Fall armyworm	Stink bugs Tarnished plant bug	1 to 2		
Okra	Corn earworm	Stink bugs	1 to 2	0	
Peanuts	Blister beetles	Mexican bean beetle	½ to 1	0	Observe plant response precautions.
	Alfalfa caterpillar Bean leaf beetle Cucumber beetles Green cloverworm Japanese beetle Leafhoppers	Rednecked peanutworm Three cornered alfalfa hopper Thrips Velvetbean caterpillar	1		
	Armyworm Corn earworm Fall armyworm	Stink bugs Webworms	1 to 1½		
	Whitefringed beetle adults Cutworms		2		
Peas	Colorado potato beetle Leafhoppers		1	3	
	Armyworm		1 to 1½		
	Alfalfa caterpillar Cutworms Pea leaf weevil	Pea weevil Yellowstriped armyworm	1½		
	Alfalfa looper		2½		WASHINGTON STATE ONLY
Potato Tomato Eggplant Pepper	Colorado potato beetle Flea beetles Leafhoppers		½ to 1	0	Thorough coverage is essential to effectively suppress stink bugs. When disease transmission is suspected, monitor fields following application and re-treat if reinfestation occurs.
	European corn borer Fall armyworm Lace bugs Stink bugs (suppression)	Tarnished plant bug Tomato fruitworm Tomato hornworm Tomato pinworm	1 to 2		
	Cutworms		2		
Prickly Pear cactus	Cochineal scale (crawlers)		2	1	Apply as needed at 7-10 day intervals. Do not make more than 7 applications per season.
Rice	Armyworm Chinch bugs	Fall armyworm Stink bugs	1 to 1½	14	MISSISSIPPI DELTA & TEXAS. CAUTION: May kill shrimp and crabs. Do not use in areas where these are important resources. Do not use on rice fields in which crayfish and catfish farming are included in the cultural practice. DO NOT APPLY PROPANIL HERBICIDES WITHIN 15 DAYS BEFORE OR AFTER APPLICATION OF THIS PRODUCT OR PLANT INJURY WILL RESULT
	Grasshoppers		½ to 1½		Refer to the general GRASSHOPPER heading above.

CROP	PEST		QUARTS OF "SEVIN®" BRAND XLR PLUS/ACRE	PREHARVEST INTERVAL (DAYS)	SPECIFIC DIRECTIONS
	Armyworm Leafhoppers	Tadpole shrimp	2		CALIFORNIA ONLY. For optimum tadpole shrimp control apply to water when pest first appears.
Sorghum (milo, grain sweet and hybrid)	Sorghum midge		¼ to 1	21 (grain) 0 (forage)	Direct spray into forming heads for optimum insect control. Treat for sorghum midge when 25 to 30 percent of heads have emerged from boot and are in bloom. Repeat application 3 to 5 days later if adults are still active. A third application may be necessary in late planted sorghum or if midge are abundant. For chinch bugs use high gallonage ground application directed at the base of plants.
	Armyworm Chinch bugs Corn earworm	Fall armyworm Stink bugs Webworms	1 to 2		
	Southwestern corn borer		1½		
	Cutworms		2		
	Grasshoppers		½ to 1½		Refer to the general GRASSHOPPER heading above.
Soybeans	Bean leaf beetle Blister beetle Cucumber beetles Grape colaspis Green cloverworm	Japanese beetle Mexican bean beetle Velvetbean caterpillar	½ to 1	0	DO NOT APPLY A COMBINATION OF THIS PRODUCT AND 2,4 DB HERBICIDES TO SOYBEANS AS CROP INJURY MAY RESULT. Use lower rates for light to moderate populations and smaller instars and to provide maximum survival of beneficial insects and spiders. Use the higher rates for heavy populations and larger instars.
	Corn earworm		½ to 1½		
	Grasshoppers		½ to 1½		Refer to general GRASSHOPPER heading above.
	Alfalfa caterpillar Leafhoppers	Three cornered alfalfa hopper Thrips	1		
	Armyworm Cutworms	Fall armyworm Webworms	1 to 1½		
	Painted lady (Thistle caterpillar) Saltmarsh caterpillar	Woollybear caterpillar Yellowstriped armyworm	1½ to 2		
Sugar beets	Armyworm Beet leaf beetle Fall armyworm	Flea beetles Leafhoppers Webworms	1 to 1½	14	
	Grasshopper		½ to 1½		Refer to general GRASSHOPPER heading above.
	Cutworms		1½		
Sunflower (DO NOT USE IN CALIFORNIA)	Cutworms		1½	60	
	Armyworm Fall armyworm Sunflower moth		1½ to 2		
	Stem weevil	Sunflower beetle	1 to 2		
	Grasshopper		½ to 1½		Refer to general GRASSHOPPER heading above.
Sweet Potato (DO NOT USE IN CALIFORNIA)	Corn earworm Cucumber beetles Flea beetles	Sweet potato hornworm Tortoise beetles	1 to 2	0	Apply as a foliar spray as needed.
	Yellowstriped armyworm		2		
	Sweet potato weevil		1 to 2		Full coverage of plants is essential. Use lower rate on young plants and higher rate on mature plants.
Tobacco Plant bed Treatment	Flea beetle		1	0	OBSERVE PLANT RESPONSE PRECAUTIONS. For flea beetle control, use 4 tsps. (0.7 fl. oz.) in 6 gallons of water and apply to 100 square yards.
	Green June beetle grubs		8		For green June beetle grub control mix 11 tablespoons (5.5 fl. oz.) in 50 to 100 gallons of water and apply to 100 square yards. Applications should be made to areas that larvae have uprooted by sprinkling mixture as a drench treatment.

CROP			QUARTS OF "SEVIN®" BRAND XLR PLUS/ACRE	PREHARVEST INTERVAL (DAYS)	SPECIFIC DIRECTIONS
Field Treatment	Budworms Fall armyworm Flea beetles Hornworms	Japanese beetle June beetle Suckfly	1 to 2		Use lower rate on young plants (up to knee height). Use at least 10 gallons of prepared spray per acre. Begin treatments when worms are small.
Wheat (including triticale) (DO NOT USE IN CALIFORNIA)	Flea beetles		1/2 to 1	0 (forage) 21 (grain)	
	Grasshoppers		1/2 to 1 1/2		Refer to the general GRASSHOPPER heading above.
	Cereal leaf beetle		1		Application is effective against eggs, larvae and adult of the cereal leaf beetle.
	Armyworm	Fall armyworm	1 to 1 1/2		

TREE FRUIT AND NUT CROPS

For dilute sprays apply the specified dosage per 100 gallons of water. For concentrate and aerial sprays, maintain the recommended rate per acre equivalent to that used in a dilute spray. The optimum spray gallonage will depend on tree size, density and stage of growth. Typical spray gallonages per acre range from, but not limited to, 200-600 gallons for dilute sprays, 20-100 gallons for concentrate sprays and 3-25 gallons of aerial sprays. Do not exceed maximum label rate per acre per application.

APPLE THINNING

CROP	QUARTS OF "SEVIN®" BRAND XLR PLUS/ 100 GAL.	PREHARVEST INTERVAL (DAYS)	SPECIFIC DIRECTIONS
Apples	1/4 to 1/2	1	OBSERVE BEE CAUTION. Apply 1 full coverage dilute spray between 10 and 25 days after full bloom. Factors such as tree age, variety, nutrition, previous crop, pruning, bloom and degree of set favor excessive fruit thinning with this product. Exercise caution to avoid possible yield reduction. Rates may vary depending on variety and local orchard conditions. Consult with your County Extension Service or other experts for advice on the proper use of this product.In Eastern apple growing areas, tank mix combinations of Carbaryl and Naphthaleneacetic Acid (NAA) or Naphthaleneacetamide (NAD) have successfully thinned several early-maturing, heavy-setting varieties, as well as hard-to-thin varieties such as Golden Delicious and Rhode Island Greening. The higher rate of Carbaryl and reduced rates of NAA or NAD are recommended for the combination. Also, a petal fall application of NAA or NAD followed 7 to 10 days later by an application of Carbaryl has improved thinning on these varieties. For easily thinned varieties including Cortland, Grimes, Johathan, McIntosh, Orleans, Rome Beauty, Puritan, Red Delicious, Winesap, Yellow Newton.
	1/2 to 1	1	For difficult to thin varieties including Baldwin, Ben Davis, Duchess, Early McIntosh, Golden Delicious, Lady Apple, Northern Spy, Rhode Island Greening, Steele Red, Turley, Wealthy, Yellow Transparent, and York Imperial.

CROP	INSECT		QUARTS OF SEVIN® BRAND XLR PLUS/ 100 GALS.	PREHARVEST INTERVAL (DAYS)	SPECIFIC DIRECTIONS
Almond	Peach twig borer San Jose scale Fruittree leafroller		1	28	OBSERVE BEE CAUTION Apply in "popcorn" or petal fall stages and again when the May brood of the peach twig borer begins to hatch or thereafter as needed.
	Navel orangeworm				Time early and mid season applications to correspond to moth flight peaks. Make a late season application at initiation of hull split or up to 10% hull split. Do not apply more than 5 quarts per acre.
Apples Pears	Apple aphid Apple maggot Apple rust mite Apple sucker Bagworms California pearslug (pear sawfly) Eastern tent caterpillar European apple sawfly Eyespotted bud moth Forbes scale Fruit tree leafroller Green fruitworm Gypsy moth	Lygus bugs Orange tortrix Oystershell scale Pear leaf blister mite Pear psylla Pear rust mite Periodical cicada Plum curculio Redbanded leaf-roller Rosy apple aphid San Jose scale Tarnished plant bug Tentiform leaf-miners Woolly apple	3/4 to 1	1	OBSERVE BEE CAUTION To avoid undesired apple thinning, delay use until at least 30 days after full bloom. For psylla control apply when eggs hatch or young nymphs are present. To control scale insects, apply when crawlers are present. Apply dilute sprays in 200-400 gallons per acre.

CROP	INSECT	QUARTS OF SEVIN® BRAND XLR PLUS/ 100 GALS.	PREHARVEST INTERVAL (DAYS)	SPECIFIC DIRECTIONS
	aphid Yellowheaded fireworm			
	Apple mealybug Apple aphid Codling moth White apple leafhopper	½		
Chestnuts (DO NOT USE IN CALIFORNIA)	Chestnut weevil	2 to 3	0	Make 4 applications at weekly intervals beginning in late July for adult chestnut weevil control. Last application should be made prior to shuck split.
Citrus Fruits (such as grapefruit, lemons, limes, oranges, tangelos, tangerines, citrus citron, kumquats and hybrids)	Avocado leafroller California orangedog Citrus cutworm Citrus root weevil Fruittree leafroller Orange tortrix Western tussock moth West Indian sugarcane borer (adults)	1	5	OBSERVE BEE CAUTION Do not apply more than 20 quarts of this product per acre per application.
	Black scale Brown soft scale California red scale Citricola scale Citrus snow scale Yellow scale	¾ to 1		May be mixed with petroleum oils commonly used on citrus.
Filbert	Filbert aphid Filbert leafroller Filbertworm	1	0	Apply when leafroller eggs are hatching. Repeat on first appearance of adult filbert moths and again 3 to 4 weeks later. Apply dilute sprays in 300-400 gallons per acre.
Olives	Olive scale	¾ to 1	0	For optimum scale control add 1½ gallons of summer oil and apply mixture when crawlers are present. Do not exceed 2 applications per year. Do not apply more than 15 quarts of this product per acre per application.
Peaches Apricots Nectarines	Apple pandemis Codling moth Cucumber beetles European earwig Fruittree leafroller Gypsy moth Japanese beetle June beetle Lecanium scales Lesser peachtree borer Olive scale Orange tortrix Oriental fruit moth Peach twig borer Periodical cicada Plum curculio Redbanded leaf-roller San Jose scale Tarnished plant bug Tussock moths Variegated leafroller	1	1 (peaches) 3 (apricots & nectarines)	Do not apply more than 6 quarts of this product per acre per application to apricots. For optimum scale control apply when crawlers are present. Spray limbs and trunk thoroughly at weekly intervals during moth flight. Apply dilute sprays in 200-400 gallons per acre.
Pecans	Black margined aphid Fall webworm Hickory shuck-worm Lesser webworm Pecan leaf phylloxera Pecan stem phylloxera Pecan nut casebearer Pecan spittlebug Pecan weevil Twig girdler Walnut caterpillar	1 to 2½	0	Do not apply more than 7.2 quarts of this product per acre per application. Apply dilute sprays in 200-400 gallons of water per acre.
Pistachios	Navel orangeworm	½ to 2	14	Apply dilute volumes of 150 to 300 gallons of mixed spray per acre for full coverage.
Plums Prunes Cherries	Codling moth Eastern tent caterpillar Orange tortrix Tussock moth	¾	1	Do not apply more than 6 quarts of this product per acre per application.
	Black cherry aphid Brown soft scale Cherry fruitworm Cherry maggot Eyespotted bud moth Forbes scale Fruittree leaf-roller Green fruitworm Gypsy moth Japanese beetle Lecanium scales Lesser peachtree borer Mealy plum aphid Oystershell scale Peach twig borer Plum curculio Prune leafhopper Redbanded leaf-roller Rose chafer San Jose scale Variegated leafroller	1		For optimum scale control apply when crawlers are present. For lesser peachtree borer control spray limbs and tree trunks thoroughly at weekly intervals during moth flight. Apply dilute sprays in 200-400 gallons of water per acre.

CROP	INSECT	QUARTS OF SEVIN® BRAND XLR PLUS/ 100 GALS.	PREHARVEST INTERVAL (DAYS)	SPECIFIC DIRECTIONS
Walnut	Calico scale European fruit lecanium Filbertworm Fruittree leafroller Frosted scale	½	0	Apply 1000 gallons of dilute spray per acre for mature trees.
	Codling moth	½	6	For codling moth apply first spray when average cross-sectional diameters of developing nuts are ½ to ¾ inch. Repeat during middle or late June as needed. Apply dilute sprays in 200-500 gallons of water per acre
	European earwig	2		Spray tree trunks to point of run-off.

SMALL FRUIT CROPS

Recommended dosages refer to quarts of SEVIN® brand XLR PLUS Carbaryl Insecticide per acre. The optimum spray gallonage will depend on plant size, density and stage of growth. Typical spray gallonage per acre range from 100-300 gallons for dilute sprays, 30-100 gallons for concentrate sprays and 5-25 gallons for aerial sprays. Do not exceed maximum label rate per acre per application.

CROP	INSECT	QUARTS OF "SEVIN®" BRAND XLR PLUS/ACRE	PREHARVEST INTERVAL (DAYS)	SPECIFIC DIRECTIONS
Blackberries Raspberries Dewberries (including boysenberries and loganberries)	European raspberry aphid Japanese beetle Leafhoppers Leafrollers Rose chafer Snowy tree cricket	1 to 2	7	
	Omnivorous leafroller Raspberry sawfly	2		CALIFORNIA ONLY
Blueberries	Blueberry maggot Cherry fruitworm Cranberry fruitworm European fruit lecanium Japanese beetle	1½ to 2	0	Apply 3 weeks before harvest and repeat as necessary.
Cranberries	Cutworms Cranberry fireworms Cranberry fruitworms Cranberry twig girdler Elm spanworm Japanese beetle Leafhoppers Rose chafer Spaganothus worm	1½ to 3	1	CAUTION: May kill shrimp and crabs. Do not use in areas where these are important resources. Apply in late bloom and as needed at 7 to 10 day intervals.
Strawberries	Flea beetles Meadow spittlebug (strawberry fruitworm) Omnivorous leaftier Strawberry clipper Strawberry bud weevil Strawberry leafroller Strawberry weevil	1 to 2	1	Carbaryl may injure Early Dawn and Sunrise varieties.

IMPORTED FIRE ANT CONTROL

SITE	AMOUNT OF "SEVIN®"BRAND XLR PLUS/ VOLUME OF WATER	SPECIFIC DIRECTIONS
Lawns, Cemeteries and Recreational Areas (including turf, golf courses, and parks) Pastures Rangeland Forested Lands and Wasteland	½ quarts/ 100 gallons or 1½ tablespoons/ gallon	Apply a total of 2 gallons of the diluted solution over the surface of each mound or at least 1 quart per 6 inches of mound diameter using a bucket, can or other appropriate equipment. Thoroughly wet mound and surrounding area to a 4 ft. diameter (12 sq. ft.). Do not disturb mounds prior to treatment. Pour solution from a height of about three feet to give sufficient force to break mound apex and flow into ant tunnels. For best results apply in cool weather, 65°-80°F or in early morning or late evening hours. Repeat application if mound activity resumes after 10 days. Treat new mounds as they appear. Pressurized sprays may disturb the ants and cause migration, reducing product effectiveness. DO NOT ALLOW PUBLIC USE OF TREATED AREAS DURING APPLICATIONS OR UNTIL SPRAYS HAVE DRIED.

SITE	AMOUNT OF "SEVIN®"BRAND XLR PLUS/ VOLUME OF WATER	SPECIFIC DIRECTIONS
Nursery Stock, Vegetable Transplants, Foliage Plants and Bedding Plants	1½ quarts/ 100 gallons	DO NOT USE ON ANY FOOD CROP NOT LISTED ON LABEL. Do not make more than one application, either as a root-dip or a drench treatment (applied to the point of saturation). Avoid contact with foliage and treat only the growing media when using on bedding plants.

ADULT MOSQUITO CONTROL

For dilute-spray ground applications to trees (including shade trees, shelter belts, forests, plantations, parks and recreational areas), ornamentals, woody plants and shrubs, apply the specified dosage per 100 gallons of water. For concentrate-spray ground applications, apply the specified dosage per acre in sufficient spray volume to provide thorough coverage. To prepare small volumes of spray, use 1 tablespoon (½ fluid ounce) of SEVIN® brand XLR PLUS Carbaryl Insecticide per gallon of water, where rates of 1 quart are indicated.

CROP	QUARTS OF "SEVIN®" BRAND XLR PLUS/100 GAL.	SPECIFIC DIRECTIONS
Pastures, Rangelands, Parks, Recreational Areas, Logging Camps, Military Posts and Adjacent Forested lands or Wastelands	¼ to 1 (½ to 1 tablespoonful per gallon)	DO NOT ALLOW PUBLIC USE OF TREATED AREAS DURING APPLICATION OR UNTIL SPRAYS HAVE DRIED. CAUTION: May kill shrimp and crabs. Do not use in areas where these are important resources. OBSERVE BEE CAUTION Treat shrubbery and areas where adult mosquitoes congregate. Treat when adult mosquitoes are active in early mornings or late evenings. Repeat at 7 to 10 day intervals. Use ¼ to ½ quart per 100 gallons in mistblowers, ½ to 1 quart per acre in aerial sprays and 1 quart per acre in low pressure ground sprayers.
	25	For residual control in subtropical regions apply 4 gallons of prepared spray per 2000 square feet of surface area. Repeat in 3 to 6 months or when necessary.

POULTRY ROOSTS AND BUILDINGS ONLY

POULTRY QUARTERS	QUARTS OF "SEVIN®" BRAND XLR PLUS/100 GAL.	SPECIFIC DIRECTIONS
Bed bugs Chicken mite Fleas Lice Northern fowl mite	4	DO NOT TREAT POULTRY OR GAME BIRDS. Apply 1 to 2 gallons of spray mixture per 1000 square feet of wall, litter or roost surface. Force spray into cracks. Repeat as needed. Avoid spraying nests, eggs and feeding and watering troughs.
Fowl tick	16	Ventilate while spraying. Do not treat premises within 7 days of slaughter.
Darkling Beetle (Lesser mealworms)	50	Apply 2 gallons of spray mixture per 1000 square feet of floor space or litter surface. Repeat as needed. Ventilate while spraying. Do not treat premises within 7 days of slaughter.

DIRECTIONS FOR USE THROUGH SPRINKLER IRRIGATION SYSTEMS

Apply this product only through sprinkler irrigation systems including center pivot and solid set. Do not apply this product through any other type of irrigation system.

SPRAY PREPARATION: First prepare a suspension of SEVIN® brand XLR PLUS Carbaryl Insecticide in a mix tank. Fill tank with ½ to ¾ the desired amount of water. Start mechanical or hydraulic agitation. Add the required amount of SEVIN® brand XLR PLUS, and then the remaining volume of water. (Suspension concentrations using the appropriate dosage per acre recommended on this label of SEVIN® BRAND XLR PLUS, per 1 to 4 gallons of water are recommended). Then set sprinkler to deliver 0.1 to 0.3 inch of water per acre. Start sprinkler and uniformly inject the suspension of SEVIN® brand XLR PLUS into the irrigation water line so as to deliver the desired rate per acre. The suspension of SEVIN® brand XLR PLUS should be injected with a positive displacement pump into the main line ahead of a right angle turn to insure adequate mixing. If you should have any other questions about calibration, you should contact State Extension Service specialists, equipment manufacturers or other experts.

NOTE: When treatment with SEVIN® brand XLR PLUS has been completed, further field irrigation over the treated area should be avoided for 24 to 48 hours to prevent washing the chemical off the crop.

GENERAL PRECAUTIONS FOR APPLICATIONS THROUGH SPRINKLER IRRIGATION SYSTEMS

Maintain continuous agitation in mix tank during mixing and application to assure a uniform suspension.

Greater accuracy in calibration and distribution will be achieved by injecting a larger volume of a more dilute solution per unit time.

The system must contain a functional check valve, vacuum relief valve, and low pressure drain appropriately located on the irrigation pipeline to prevent water source contamination from backflow. The pesticide injection pipeline must contain a functional, automatic, quick-closing check valve to prevent the flow of fluid back toward the injection pump. The pesticide injection pipeline must also contain a functional, normally closed solenoid-operated valve located on the intake side of the injection pump and connected to the system interlock to prevent fluid from beieng withdrawn from the supply tank when the irrigation system is either automatically or manually shutdown. The system must contain functional interlocking controls to automatically shut off the pesticide injection pump when the water pump motor stops. The irrigation line or water pump must include a functional pressure switch which will stop the water pump motor when the water pressure decreases to the point where pesticide distribution is adversely affected. Systems must use a metering pump, such as a positive displacement injection pump (e.g., diaphragm pump) effectively designed and constructed of materials that are compatible with pesticides and capable of being fitted with a system interlock. Do not apply when wind speed favors drift beyond the area intended for treatment.

Do not apply when wind speed favors drift, when system connection or fittings leak, when nozzles do not provide uniform distribution or when lines containing the product must be dismantled and drained.

Crop injury, lack of effectiveness, or illegal pesticide residues in the crop may result from nonuniform distribution of treated water.

Allow sufficient time for pesticide to be flushed through all lines and all nozzles before turning off irrigation water. A person knowledgeable of the chemigation system and responsible for its operation shall shut the system down and make necessary adjustments should the need arise.

Do not connect an irrigation system (including greenhouse systems) used for pesticide application to a public water system unless the label-prescribed safety devices for public water supplies are in place.

PEST CONTROL IN AND AROUND BUILDINGS

General Information

NOTE: Staining may occur on certain surfaces such as stucco, brick, cinder block, and wood. Therefore, applications of SEVIN® brand XLR PLUS Carbaryl Insecticide to surfaces where a noticeable residue or discoloration is objectionable should be avoided. Do not apply to carpets or draperies as staining may occur. Care should also be exercised to avoid spotting of wallpaper and fabrics.

Do not use this product in commercial food areas of food handling establishments, restaurants or other places where food is prepared or processed. Do not use in serving areas while food is exposed.

INDOORS

ANTS/CRICKETS/FIREBRATS/SILVERFISH: Mix 3 oz. this product per gallon of water and apply as fine, low pressure (20 psi) spot spray or as crack and crevice application to areas where these pests hide, such as baseboards, storage areas, closets, around water pipes, doors and windows, behind and under refrigerators, cabinets, sinks, stoves, dishwashers, hot water heaters, the underside of shelves, drawers and similar areas. For ants, apply to ant trails, around doors and windows and other places where ants enter premises.

BEES and WASPS: Mix 3 oz. this product per gallon of water and thoroughly spray nest and entrance and surrounding areas where insects alight. It is generally advisable to spray the nests in the evening when the insects are less active and have returned to the nest. For best results, check nest carefully one or two days after spraying to ensure complete kill, then remove and destroy nest to prevent emergence of newly hatched insects.

BROWN DOG TICKS and FLEAS: Mix 3 oz. this product per gallon of water and thoroughly spray infested areas such as nearby cracks and crevices, between and under cushions of upholstered furniture, along and behind baseboards, window and door frames and other areas where these pests may be present.

CARPENTER ANTS: Mix 3 oz. this product per gallon of water and apply to ant trails, around doors and windows and other places where ants enter premises. Where possible, apply this product directly to ant nest or infested wood.

CENTIPEDES/EARWIGS/MILLIPEDES/SCORPIONS: Mix 3 oz. SEVIN® brand XLR PLUS Carbaryl Insecticide per gallon of water and apply around water pipes, doors, and windows, and other places where these pests may enter premises. Spray baseboards, storage areas, garages, carports, basements and other areas where these pests are found.

COCKROACHES: Efficacy varies with species sensitivity. This product is generally not highly effective in controlling German cockroaches. However, the following may be controlled with the rates as stated. American roach, Australian roach, Brown roach, Smoky brown roach, and others. Mix 3 oz. this product per gallon of water and apply as a fine, low pressure (psi) spot spray or as a crack and crevice application to areas where these pests hide, such as baseboards, storage areas, closets, around water pipes, doors and windows, behind and under refrigerators, cabinets, sinks, stoves, dishwashers, hot water heaters, the underside of shelves, drawers and similar areas.

SPIDERS: Mix 3 oz. this product per gallon of water and apply to infested baseboards, window and door frames, corners, pipes, storage areas, attics and under eaves. Make spot applications to other areas where these pests are present.

OUTDOORS

PERIMETER TREATMENT: Residual spray for control of ants, bees and wasps, brown dog ticks, carpenter ants, centipedes, cockroaches, crickets, earwigs, firebrats and silverfish, fleas, millipedes, scorpions and spiders. Mix 16 oz. this product per 50 gallons of water and apply via power spray or other spray methods.

To help prevent infestations of buildings by the above pests, outside perimeter treatment should be in a band 6 to 10 feet wide and confined to shrub beds, foundation plantings and lawn or soil areas immediately adjacent to the structure. Direct application to structures should be minimal and restricted to cracks and crevices and other areas where insects tend to congregate.

STORAGE AND DISPOSAL

STORAGE

Store unused SEVIN® brand XLR PLUS Carbaryl Insecticide in original container only, in cool, dry area out of reach of children and animals. Do not store in areas where temperatures frequently exceed 100°F.

If container is damaged, before cleaning up, put on long-sleeved shirt, full length trousers, head covering, and protective gloves.

PESTICIDE DISPOSAL

Do not contaminate water, food or feed by storage or disposal. Open dumping is prohibited. Wastes resulting from the use of this product may be disposed of on site or at an approved waste disposal facility.

CONTAINER DISPOSAL

Triple rinse (or equivalent). Then offer for recycling or reconditioning, or dispose of in a sanitary landfill, or incineration, or if allowed by state and local authorities, by burning. If container is burned, stay out of smoke.

LIMITED WARRANTY AND DISCLAIMER

The manufacturer warrants (a) that this product conforms to the chemical description on the label; (b) that this product is reasonably fit for the purposes set forth in the directions for use when it is used in accordance with such directions; and (c) that the directions, warnings and other statements on this label are based upon responsible experts' evaluation of reasonable tests of effectiveness of toxicity to laboratory animals and to plants, and of residues on food crops, and upon reports of field experience. Tests have not been made on all varieties or in all states or under all conditions. THE MANUFACTURER NEITHER MAKES NOR INTENDS, NOR DOES IT AUTHORIZE ANY AGENT OR REPRESENTATIVE TO MAKE, ANY OTHER WARRANTIES, EXPRESS OR IMPLIED, AND IT EXPRESSLY EXCLUDES AND DISCLAIMS ALL IMPLIED WARRANTIES OF MERCHANTABILITY OR FITNESS FOR A PARTICULAR PURPOSE.

THIS WARRANTY DOES NOT EXTEND TO, AND THE BUYER SHALL BE SOLELY RESPONSIBLE FOR, ANY AND ALL LOSS OR DAMAGE WHICH RESULTS FROM THE USE OF THIS PRODUCT IN ANY MANNER WHICH IS INCONSISTENT WITH THE LABEL DIRECTIONS, WARNINGS OR CAUTIONS.

BUYER'S EXCLUSIVE REMEDY AND MANUFACTURER'S OR SELLER'S EXCLUSIVE LIABILITY FOR ANY AND ALL CLAIMS, LOSSES, DAMAGES, OR INJURIES RESULTING FROM THE USE OR HANDLING OF THIS PRODUCT, WHETHER OR NOT BASED IN CONTRACT, NEGLIGENCE, STRICT LIABILITY IN TORT OR OTHERWISE, SHALL BE LIMITED, AT THE MANUFACTURER'S OPTION, TO REPLACEMENT OF, OR THE REPAYMENT OF THE PURCHASE PRICE FOR, THE QUANTITY OF PRODUCT WITH RESPECT TO WHICH DAMAGES ARE CLAIMED. IN NO EVENT SHALL MANUFACTURER OR SELLER BE LIABLE FOR SPECIAL, INDIRECT OR CONSEQUENTIAL DAMAGES RESULTING FROM THE USE OR HANDLING OF THIS PRODUCT.

AG90053

EPA Approval 8/22/88

THIS SPECIMEN LABEL IS INTENDED FOR USE ONLY AS A GUIDE IN PROVIDING GENERAL INFORMATION REGARDING THE DIRECTIONS, WARNINGS AND CAUTIONS ASSOCIATED WITH THE USE OF THIS PRODUCT. AS WITH ANY AGRICULTURAL CHEMICAL, ALWAYS FOLLOW THE LABEL INSTRUCTIONS ON THE PACKAGE BEFORE USING.

SEVIN® BRAND 4 OIL

CARBARYL INSECTICIDE

For Agricultural or Commercial Use Only

ACTIVE INGREDIENTS:	
Carbaryl (1-naphthyl N-methylcarbamate)	46.1% by wt.
INERT INGREDIENTS:	53.9% by wt.
TOTAL	100.0%

This product contains petroleum distillates.

(Contains 4 pounds carbaryl per gallon)

E.P.A. Reg. No. 264-323 E.P.A. Est. No. 264-MO-02

KEEP OUT OF REACH OF CHILDREN

CAUTION

For MEDICAL And TRANSPORTATION Emergencies ONLY Call 24 Hours A Day 1-800-334-7577
For PRODUCT USE Information Call 1-800-334-9745

PRECAUTIONARY STATEMENTS

CAUTION

HAZARDS TO HUMANS AND DOMESTIC ANIMALS

MAY BE HARMFUL IF SWALLOWED. OVEREXPOSURE MAY CAUSE: Salivation, watery eyes, pinpoint eye pupils, blurred vision, muscle tremors, difficult breathing, excessive sweating, abdominal cramps, nausea, vomiting, diarrhea, weakness, headache. IN SEVERE CASES CONVULSION, UNCONSCIOUSNESS AND RESPIRATORY FAIL-

URE MAY OCCUR. SIGNS AND SYMPTOMS OCCUR RAPIDLY FOLLOWING OVEREXPOSURE TO THIS PRODUCT.

Avoid breathing of spray mist. Do not take internally. Avoid contact with eyes, skin or clothing. Wear regular long-sleeved work clothing and head covering when making overhead applications. Change to clean clothing daily. Bathe and wash hair after each work day. Do not eat, drink or use tobacco while working with this product or spray solutions. Wash hands and face before eating, drinking or using tobacco. Keep out of reach of children and domestic animals.

ANTIDOTE STATEMENT

ATROPINE SULFATE IS HIGHLY EFFECTIVE AS AN ANTIDOTE. Narcotics and other sedatives should not be used. Further, drugs like 2-PAM (pyridine-2-aldoxime methiodide) are NOT recommended. See NOTE TO PHYSICIAN below.

STATEMENT OF PRACTICAL TREATMENT

GENERAL

Contact a physician immediately in all cases of suspected poisoning. Transport patient to a physician or hospital immediately and SHOW A COPY OF THIS LABEL TO THE PHYSICIAN. If poisoning is suspected in animals, contact a veterinarian.

IF SWALLOWED: Call a physician or Poison Control Center immediately. Give 1 or 2 glasses of water or milk to drink. Do not induce vomiting, may cause aspiration hazard. GET MEDICAL ATTENTION.

IF IN EYES: Hold eyelids open and flush with a steady, gentle stream of water for at least 15 minutes. Get medical attention if irritation persists.

IF ON SKIN: Wash thoroughly with soap and water.

IF INHALED: Remove to fresh air. If not breathing, give artificial respiration. Administer oxygen if necessary. GET MEDICAL ATTENTION.

NOTE TO PHYSICIAN

Carbaryl is a carbamate insecticide, which is a cholinesterase inhibitor. Overexposure to this substance may cause toxic signs and symptoms due to stimulation of the cholinergic nervous system. These effects of overexposure are spontaneously and rapidly reversible. Gastric lavage may be used if this product has been swallowed. Carbaryl poisoning may occur rapidly after ingestion and prompt removal of stomach contents is indicated. PRECAUTION: To prevent aspiration of petroleum distillates, lay patient on side with head lower than waist.

Specific treatment consists of parenteral atropine sulfate. Caution should be maintained to prevent overatropinization. Mild cases may be given 1 to 2 mg intramuscularly every 10 minutes until full atropinization has been achieved and repeated thereafter whenever symptoms reappear. Severe cases should be given 2 to 4 mg intravenously every 10 minutes as needed until fully atropinized, then intramuscularly every 30 to 60 minutes to maintain the effect for at least 12 hours. Dosages for children should be appropriately reduced. Complete recovery from overexposure is to be expected within 24 hours.

Narcotics and other sedatives should not be used. Further, drugs like 2-PAM (pyridine-2-aldoxime methiodide) are NOT recommended.

To aid in confirmation of a diagnosis, urine samples should be obtained within 24 hours of exposure and immediately frozen. Analyses will be arranged by Rhône-Poulenc Ag Company.

Consultation on therapy can be obtained at all hours by calling the Union Carbide emergency number 1-800-334-7577.

ENVIRONMENTAL HAZARDS

This product is extremely toxic to aquatic and estuarine invertebrates. Do not apply directly to water and wetlands, except under the forest canopy. Do not apply when weather conditions favor drift from the area treated. Do not contaminate water by cleaning equipment or disposal of wastes. Do not contaminate water when disposing of equipment wash waters.

BEE CAUTION

MAY KILL HONEYBEES IN SUBSTANTIAL NUMBERS. This product is highly toxic to bees exposed to direct treatment on blooming crops or weeds. Do not apply this product or allow it to drift to blooming crops or weeds if bees are visiting the treatment area. Contact your Cooperative Agricultural Extension Service or your local Rhône-Poulenc Ag Company representative for further information.

DIRECTIONS FOR USE

IT IS A VIOLATION OF FEDERAL LAW TO USE THIS PRODUCT IN A MANNER INCONSISTENT WITH ITS LABELING.

GENERAL INFORMATION

SEVIN® brand 4-Oil Carbaryl Insecticide is a dispersion of finely ground technical carbaryl in a non-aromatic, low volatile oil. It is designed for air application as a low volume or ultra low volume spray. It may also be applied by ground mist blower or cold fogging, if diluted with diesel fuel, kerosene or #2 fuel oil. READ THIS LABEL BEFORE USE. USE IN STRICT ACCORDANCE WITH DIRECTIONS AND CAUTIONS.

GENERAL WORKER PROTECTION STATEMENT

Do not apply this product in such a manner as to directly or through drift expose workers or other persons. The area treated must be vacated by unprotected persons.

Do not enter treated areas without protective clothing until sprays have dried.

Because certain states may require more restrictive reentry intervals for various crops treated with this product, consult your State Department of Agriculture for further information.

Written or oral warnings must be given to workers who are expected to be in treated area or in an area about to be treated with this product. Advise workers to stay out of fields during application and until sprays have dried. Regular long-sleeved work clothing should be worn when working in treated fields. See PRECAUTIONARY STATEMENTS, STATEMENT OF PRACTICAL TREATMENT AND NOTE TO PHYSICIAN for information on accidental exposures. When oral warnings are given, warnings shall be given in a language customarily understood by workers. Oral warnings must be given if there is reason to believe that written warnings cannot be understood by workers. Written warnings must include the following information: Appropriate signal work (CAUTION), area treated with SEVIN® brand 4-Oil Carbaryl Insecticide, date of application, appropriate clothing, and re-entry interval (i.e., until sprays have dried).

PRE-HARVEST AND GRAZING USE INFORMATION AND LIMITATIONS

Tolerances established under the Federal Food, Drug and Cosmetic Act permit the sale of crops bearing probable carbaryl residues when this product is used in accordance with label directions. If used as directed, treated forage may be grazed or used as feed for dairy and meat animals without causing illegal residues in meat or milk. See specific crop directions for grazing and preharvest interval restrictions and limitations.

Do not plant rotational food and feed crops not listed on this or other carbaryl labels in carbaryl treated soil.

Do not use reclaimed irrigation water from crops treated with carbaryl on upland crops for which carbaryl tolerances are not established.

PLANT RESPONSE PRECAUTION

To avoid possible injury to tender foliage, do not apply to wet foliage or when rain or high humidity is expected during the next two days.

SPRAY PREPARATION

SEVIN® brand 4-Oil Carbaryl Insecticide thickens during storage. Roll or agitate drums or recirculate product in bulk tanks before use to reduce viscosity and restore homogeneity. SEVIN® brand 4-Oil Carbaryl Insecticide may be held prior to use and transferred in any materials of construction suitable for use with fuel oil, kerosene or diesel oil. Protect from water.

Before and after use, flush nurse tanks, pumps, lines, hoses and entire spray systems with diesel fuel, kerosene or #2 fuel oil until clean. Water, scale, rust and other residue must be removed from pumps, mixing and spray systems before use.

COMPATIBILITY

"SEVIN®" BRAND 4-OIL CARBARYL INSECTICIDE COMPATIBILITY INSECTICIDE IS NOT COMPATIBLE WITH AND SHOULD NOT BE MIXED WITH WATER, ALCOHOL OR AROMATIC SOLVENTS, but can be diluted with aliphatic oils (diesel fuel, kerosene or #2 fuel oil). Compatibility with other pesticides has not been thoroughly tested. DO NOT USE IN TANK MIX COMBINATIONS UNLESS YOUR EXPERIENCE INDICATES THAT THE MIXTURE IS EFFECTIVE AND WILL NOT RESULT IN PLANT INJURY OR MECHANICAL DIFFICULTY. For best mixing when preparing diluted SEVIN® brand 4-Oil Carbaryl Insecticide, add diluent oil to mix tank before adding proper volume of SEVIN® brand 4-Oil Carbaryl Insecticide. Provide complete recirculation of diluted suspensions once every 24 hours to maintain uniformity.

APPLICATION

On all crops use sufficient gallonage to obtain thorough and uniform coverage. Calibrate equipment to deliver the desired spray volume. Flat fan nozzles may be used to apply both undiluted and diluted SEVIN® brand 4-Oil Carbaryl Insecticide. Rotary atomizing and hollow cone nozzles are not recommended for applying undiluted SEVIN® brand 4-Oil Carbaryl Insecticide but can be used if 1 to 1 dilution is made. Use of a high-volume 50-mesh in-line strainer is suggested. Use of screens behind nozzles is not recommended. Apply by air undiluted, or diluted with kerosene, diesel fuel or #2 fuel oil. At temperatures below 50°F (10°C) dilution of 4 volumes of SEVIN® brand 4-Oil Carbaryl Insecticide with 1 volume of diluent oil is recommended to assure uniform flow and spray distribution. Dilutions greater than 1 to 1 by volume are not recommended; higher dilutions will reduce residual properties. Dilute 1 to 1 by volume when applying with ground equipment. Apply only when weather conditions are favorable. Wind and rising air currents may cause undesirable spray drift and reduce insect control.

DIRECTIONS FOR USE THROUGH SPRINKLER IRRIGATION SYSTEMS

Apply this product only through sprinkler irrigation systems including center pivot and solid set. Do not apply this product through any other type of irrigation system.

SPRAY PREPARATION: Remove scale, pesticide residues, and other foreign matter from the chemical tank and entire injector system. Flush with clean water.

APPLICATION INSTRUCTIONS: First prepare a suspension of SEVIN® brand 4 Oil Carbaryl Insecticide in a mix tank. Fill tank with 1/2 to 3/4 the desired amount of diesel. Start mechanical or hydraulic agitation. Add the required amount of SEVIN® brand 4 Oil, and then the remaining volume of diesel. (Suspension concentrations using the appropriate dosage per acre recommended on this label of SEVIN® brand 4 Oil per 1 to 4 gallons of diesel are recommended). Then set sprinkler to deliver 0.1 to 0.3 inch of water per acre. Start sprinkler and uniformly inject the suspension of SEVIN® brand 4 Oil into the irrigation water line so as to deliver the desired rate per acre. The suspension of SEVIN® brand 4 Oil should be injected with a positive displacement pump into the main line ahead of a right angle turn to insure adequate mixing. If you should have any other questions about calibration, you should contact State Extension Service specialists, equipment manufacturers or other experts.

NOTE: When treatment with SEVIN® brand 4 Oil has been completed, further field irrigation over the treated area should be avoided until foliage has dried to prevent washing the chemical off the crop.

GENERAL PRECAUTIONS FOR APPLICATIONS THROUGH SPRINKLER IRRIGATION SYSTEMS

Maintain continuous agitation in mix tank during mixing and application to assure a uniform suspension.

Greater accuracy in calibration and distribution will be achieved by injecting a larger volume of a more dilute solution per unit time.

The system must contain a functional check value, vacuum relief valve, and low pressure drain appropriately located on the irrigation pipeline to prevent water source contamination from backflow. The pesticide injection pipeline must contain a functional, automatic, quick-closing check value to prevent the flow of fluid back toward the injection pump. The pesticide injection pipeline must also contain a functional, normally closed solenoid-operated valve located on the intake side of the injection pump and connected to the system interlock to prevent fluid from being withdrawn from the supply tank when the irrigation system is either automatically or manually shutdown. The system must contain functional interlocking controls to automatically shut off the pesticide injection pump when the water pump motor stops. The irrigation line or water pump must include a functional pressure switch which will stop the water pump motor when the water pressure decreases to the point where pesticide distribution is adversely affected. Systems must use a metering pump, such as a positive displacement injection pump (e.g., diaphram pump) effectively designed and constructed of materials that are compatible with pesticides and capable of being fitted with a system interlock. Do not apply when wind speed favors drift beyond the area intended for treatment.

Do not apply when wind speed favors drift, when system connection or fittings leak, when nozzles do not provide uniform distribution or when lines containing the product must be dismantled and drained.

Crop injury, lack of effectiveness, or illegal pesticide residues in the crop may result from non-uniform distribution of treated water.

Allow sufficient time for pesticide to be flushed through all lines and all nozzles before turning off irrigation water. A person knowledgeable of the chemigation system and responsible for its operation shall shut the system down and make necessary adjustments should the need arise.

Do not connect an irrigation system (including greenhouse systems) used for pesticide application to a public water system unless the label-prescribed safety devices for public water supplies are in place.

INSECT CONTROL

Apply when insects or their damage appear. To maintain control repeat at 7 to 14 day intervals or as necessary unless a shorter interval is specified below. Where a dosage range is indicated, use lower rate on young plants and early instars and higher rate on mature plants, advanced instars and adults. Thorough and uniform spray coverage is essential for effective control.

SEVIN® brand 4-Oil Carbaryl Insecticide does not control spider mites. If spider mites are a problem, use a registered miticide.

Regional differences have been noted in the susceptibility of certain strains of fall armyworm to SEVIN® brand 4-Oil Carbaryl Insecticide. If local experience indicates inadequate control, use an alternative pesticide.

Follow directions for most effective use. DOSAGES LISTED REFER TO QUARTS OF SEVIN® BRAND 4-OIL CARBARYL INSECTICIDE PER ACRE, NOT TO VOLUME OF FINISHED SPRAY PER ACRE.

CAREFULLY CALIBRATE ALL APPLICATION EQUIPMENT.

ALFALFA INSECT CONTROL

For grasshoppers: Apply 1/2 to 1 1/2 quarts of this product per acre. To avoid possible injury to tender alfalfa foliage, do not apply to wet foliage or when rain or high humidity is expected during the next two days.

Apply only once per cutting.

Do not apply to alfalfa within 7 days of harvest.

CORN (field, pop) INSECT CONTROL

For corn rootworm beetles: Apply 1 quart of this product per acre as needed at first silking if populations are at economic levels and/or apply later when needed to control adult population buildup to reduce larval damage to next year's crop. Do not make more than two applications.

For western bean cutworm: Apply a single application of 2 quarts of this product plus 2 quarts diesel fuel, kerosene or #2 fuel oil per acre when an average of 1 in 7 plants have western bean cutworm egg masses or newly hatched larvae and 90 to 100% of the tassels have emerged. Application after 100% of the silks have emerged will reduce effectiveness.

For Japanese beetle: Apply 1 quart of this product per acre when adult beetles are present.

For grasshoppers: Apply 1/2 to 1 1/2 quarts of this product per acre. Lower rate is suggested for nymphs on small plants. Use 1 1/2 quarts when grasshoppers of foliage are mature and greater coverage is required.

FOREST INSECT CONTROL

For control of gypsy moth, fall and spring cankerworms, saddled prominent, forest tent caterpillar, elm spanworm, oak leafroller complex and Japanese beetle: Apply 1 quart of this product per acre when larvae are in early instars and leaves are at least 1/3 grown, or when adult Japanese beetles are present.

For spruce budworm and western spruce budworm: East of the Rocky Mountains apply 1/2 to 1 quart of this product per acre when a majority of the larvae are in the third through sixth instars. Lower rate is suggested for heavy spruce budworm infestations where 2 applications are necessary for adequate foliage protection. In the Rocky Mountains and west of the Rocky Mountains apply 1/2 to 1 quart of this product per acre when a majority of the larvae are in the third through fifth instars.

For Douglas-fir tussock moth: For maximum foliage protection apply 2 quarts of this product per acre when larvae are in first and second instars. Application to later instars will provide control but with reduced foliage protection.

For population control and foliage protection of high value trees using ground mist blower application: Apply 0.8 pint of spray mixture containing equal volumes of this product and diesel fuel, kerosene or #2 fuel oil per 20 to 30 foot tree (0.2 lb. ai/tree) when larvae are in early instars. Cover foliage thoroughly, but avoid runoff.

PASTURE INSECT CONTROL

For grasshoppers: Apply 3/8 quarts (12 fl. oz.) to 1 1/2 (48 fl. oz.) of this product per acre in a total spray volume of at least 20 fluid ounces per acre. For True armyworm, fall armyworm, chinch bugs, essex skipper, striped grass looper, thrips: Apply 1/2 to 1 quart of this product per acre. Use the lower rate for nymphs and immature insects on small plants or sparse vegetation. Apply the higher rate for adults or when vegetation is thick.

Apply a maximum of 2 applications per year. Allow at least 14 days between applications. Do not allow foraging or cut for hay within 14 days of last application by ground. Aerially treated pastures may be grazed or cut for hay on day of treatment.

Carefully mark swaths to avoid over-application.

RANGELAND INSECT CONTROL (FOR AERIAL APPLICATION ONLY)

For grasshoppers: Apply 3/8 quart (12 fl. oz.) to 1 quart (32 fl. oz.) of this product per acre in a total spray volume of at least 15 fluid ounces per acre. For black grass bugs, mormon cricket, range caterpillars, range crane fly: Apply 1/2 to 1 quart of this product per acre. Use the lower rate for nymphs and immature insects on small plants or sparse vegetation. Apply the higher rate for adults or when vegetation is thick.

Apply a maximum of 2 applications per year. Allow at least 14 days between applications.

Carefully mark swaths to avoid over-application.

WASTELANDS, RIGHTS-OF-WAY, HEDGEROWS, DITCHBANKS, ROADSIDES INSECT CONTROL

For grasshoppers: Apply 3/8 quart (12 fl. oz.) to 1 quart (32 fl. oz.) of this product per acre in a total volume spray of at least 20 fluid ounces per acre. For black grass bugs, mormon cricket, range caterpillar, range crane fly: Apply 1/2 to 1 quart of this product per acre. Use the lower rate for nymphs and immature insects on small plants or sparse vegetation. Apply the higher rate for adults or when vegetation is thick.

Apply a maximum of 2 applications per year. Allow at least 14 days between applications. Do not allow foraging or cut for hay within 14 days of last application by ground. Aerially treated areas may be grazed or cut for hay on day of treatment.

Carefully mark swaths to avoid over-application.

WHEAT INSECT CONTROL—(DO NOT USE IN CALIFORNIA).

Do not apply within 21 days of harvest of grain. No time limitation on green wheat used as pasture or forage.

For grasshoppers: Apply 1/2 to 1 1/2 quarts of this product per acre. Lower rate is suggested for nymphs on small plants. Use higher rate when grasshoppers are mature or vegetation is thick.

For cereal leaf beetle: Apply 1 quart of this product per acre. Application is effective against eggs, larvae and adults.

For armyworm and fall armyworm: Apply 1 to 1½ quarts of this product per acre.

DIRECTIONS FOR USE AS A WHEAT BRAN BAIT

FOR END USE ONLY. NOT FOR REPACKAGING.

FOR USE ONLY BY GOVERNMENT PERSONNEL OR PERSONS UNDER THEIR DIRECT SUPERVISION.

Mixing Instructions

Mix the appropriate amount of SEVIN® brand 4-Oil Carbaryl Insecticide with wheat bran to make carbaryl wheat bran bait containing 2% to 10% active carbaryl. For example, for a bait containing 5% carbaryl, mix 1 quart SEVIN® brand 4-Oil Carbaryl Insecticide (contains 1 lb. active carbaryl) with each 19 pounds of wheat bran. Mix only the amount of bait necessary for each insect control program.

Storage Instructions

Store carbaryl bran baits in cool, dry area out of reach children and animals. Do not contaminate water, food, or feed by storage or disposal.

NOTE: Carbaryl bran baits should only be stored temporarily while awaiting application.

Application Instructions

Applications may be made with ground equipment (hand cyclone spreader) or with aerial application equipment with a metered bait spreader attachment.

PASTURES, RANGELAND, WASTELAND, ROADSIDES

Use 0.50-1.50 lbs. active ingredient/acre for the control of grasshoppers and Mormon crickets. The lower rate is suggested for early instars on small plants or sparse vegetation. Use the higher rate for adults or dense vegetation. Use of low bait assay and higher rate is suggested for control of high grasshopper populations. Treatment may be repeated as necessary. Preharvest interval is 0 days.

STORAGE AND DISPOSAL

STORAGE

Store unused SEVIN® brand 4-Oil Carbaryl Insecticide in original container only, in cool, dry area out of reach of children and animals. Do not store in areas where temperatures frequently exceed 100°F.

PESTICIDE DISPOSAL

Do not contaminate water, food or feed by storage or disposal. Wastes resulting from the use of this product may be disposed of on site or at an approved waste disposal facility.

CONTAINER DISPOSAL

Decontaminate empty tanks. For drums: Triple rinse (or equivalent). Then offer for recycling or reconditioning, or puncture and dispose of in a sanitary landfill, or by other procedures approved by state and local authorities.

Consult Federal, State or local disposal authorities for approved alternative procedures.

LIMITED WARRANTY AND DISCLAIMER

The manufacturer warrants (a) that this product conforms to the chemical description on the label; (b) that this product is reasonably fit for the purposes set forth in the directions for use when it is used in accordance with such directions; and (c) that the directions, warnings and other statements on this label are based upon responsible experts' evaluation of reasonable tests of effectiveness, of toxicity to laboratory animals and to plants, and of residues on food crops, and upon reports of field experience. Tests have not been made on all varieties or in all states or under all conditions. THE MANUFACTURER NEITHER MAKES, NOR INTENDS, NOR DOES IT AUTHORIZE ANY AGENT OR REPRESENTATIVE TO MAKE, ANY OTHER WARRANTIES, EXPRESS OR IMPLIED, AND IT EXPRESSLY EXCLUDES AND DISCLAIMS ALL IMPLIED WARRANTIES OF MERCHANTABILITY OR FITNESS FOR A PARTICULAR PURPOSE.

THIS WARRANTY DOES NOT EXTEND TO, AND THE BUYER SHALL BE SOLELY RESPONSIBLE FOR, ANY AND ALL LOSS OR DAMAGE WHICH RESULTS FROM THE USE OF THIS PRODUCT IN ANY MANNER WHICH IS INCONSISTENT WITH THE LABEL DIRECTIONS, WARNINGS OR CAUTIONS.

BUYER'S EXCLUSIVE REMEDY AND MANUFACTURER'S OR SELLER'S EXCLUSIVE LIABILITY FOR ANY AND ALL CLAIMS, LOSSES, DAMAGES, OR INJURIES RESULTING FROM THE USE OR HANDLING OF THIS PRODUCT, WHETHER OR NOT BASED IN CONTRACT, NEGLIGENCE, STRICT LIABILITY IN TORT OR OTHERWISE, SHALL BE LIMITED, AT THE MANUFACTURER'S OPTION, TO REPLACEMENT OF, OR THE REPAYMENT OF THE PURCHASE PRICE FOR, THE QUANTITY OF PRODUCT WITH RESPECT TO WHICH DAMAGES ARE CLAIMED. IN NO EVENT SHALL MANUFACTURER OR SELLER BE LIABLE FOR SPECIAL, INDIRECT OR CONSEQUENTIAL DAMAGES RESULTING FROM THE USE OR HANDLING OF THIS PRODUCT.

AG 90054

SEVIN® is the registered trademark of RHÔNE-POULENC NEDERLAND B.V. for carbaryl insecticides.

EPA Approval 7/25/88

THIS SPECIMEN LABEL IS INTENDED FOR, USE ONLY AS A GUIDE IN PROVIDING GENERAL INFORMATION REGARDING THE DIRECTIONS, WARNINGS AND CAUTIONS ASSOCIATED WITH THE USE OF THIS PRODUCT. AS WITH ANY AGRICULTURAL CHEMICAL, ALWAYS FOLLOW THE LABEL INSTRUCTIONS ON THE PACKAGE BEFORE USING.

SEVIN® BRAND 80S

CARBARYL INSECTICIDE

For Agricultural or Commercial Use Only

NOT FOR RESIDENTIAL USE

ACTIVE INGREDIENT:	
Carbaryl (1-naphthyl N-methylcarbamate)	80% by wt.
INERT INGREDIENTS:	20% by wt.
TOTAL	100%

E.P.A. Reg. No. 264-316

E.P.A. Est. No.

KEEP OUT OF REACH OF CHILDREN

WARNING AVISO

PRECAUCION AL USARIO: Si usted no lee ingles, no use este producto hasta que la etiqueta le haya sido explicada ampliamente.

For MEDICAL and TRANSPORTATION Emergencies ONLY Call 24 Hours A Day 1-800-334-7577

For PRODUCT USE Information Call 1-800-334-9745

PRECAUTIONARY STATEMENTS

WARNING

HAZARDS TO HUMANS AND DOMESTIC ANIMALS

MAY BE FATAL IF SWALLOWED. Avoid breathing of dust or spray mist. Do not take internally. Avoid contact with eyes, skin or clothing. Wear regular long-sleeved work clothing. Change to clean clothing daily. Wash hands and face before eating. Wash thoroughly after handling.

ANTIDOTE STATEMENT

ATROPINE SULFATE IS HIGHLY EFFECTIVE AS AN ANTIDOTE. See NOTE TO PHYSICIAN below.

STATEMENT OF PRACTICAL TREATMENT

GENERAL

Contact a physician immediately in all cases of suspected poisoning. Transport to a physician or hospital immediately and SHOW A COPY OF THIS LABEL TO THE PHYSICIAN.

IF SWALLOWED: Never give anything by mouth to an unconscious or convulsing person. If conscious and not convulsing, drink 1 to 2 glasses of water and induce vomiting by touching the back of the throat with finger.

IF IN EYES: Flush eyes with plenty of water. Get medical attention if irritation persists.

IF ON SKIN: Wash thoroughly with soap and water.

IF INHALED: Move from contaminated atmosphere and call a physician.

NOTE TO PHYSICIAN

Carbaryl is a carbamate insecticide, which is a cholinesterase inhibitor. Overexposure to this substance may cause toxic signs and symptoms due to stimulation of the cholinergic nervous system. These effects of overexposure are spontaneously and rapidly reversible.

Specific treatment consists of parenteral atropine sulfate. Caution should be maintained to prevent over atropinization. Mild cases may be given 1 to 2 mg intramuscularly every 10 minutes until full atropinization has been achieved and repeated thereafter whenever symptoms reappear. Severe cases should be given 2 to 4 mg intravenously every 10 minutes until fully atropinized, then intramuscularly every 30 to 60 minutes to maintain the effect for at least 12 hours. Dosages for children should be appropriately reduced. Complete recovery from overexposure is to be expected within 24 hours.

Narcotics and other sedatives should not be used. Further, drugs like 2-PAM (pyridine-2-aldoxime methiodide) are NOT recommended.

To aid in confirmation of a diagnosis, urine samples should be obtained within 24 hours of exposure and immediately frozen. Analyses will be arranged by Rhône-Poulenc Ag Company.

Consultation on therapy can be obtained at all hours by calling the Rhône-Poulenc emergency number 1-800-334-7577

ENVIRONMENTAL HAZARDS

This product is extremely toxic to aquatic and estuarine invertebrates. Do not apply directly to water and wetlands, except under the forest canopy and use on rice. Discharge from rice fields may kill aquatic and estaurine invertebrates. Do not apply when weather conditions favor drift from the area treated. Do not contaminate water by cleaning equipment or disposal of wastes.

Do not contaminate water when disposing of equipment washwaters.

BEE CAUTION:

MAY KILL HONEYBEES IN SUBSTANTIAL NUMBERS.

This product is highly toxic to bees exposed to direct treatment or residues on blooming crops or weeds. Do not apply this product or allow it to drift to blooming crops or weeds if bees are visiting the treatment area. Contact your Cooperative Agricultural Extension Service or your local Rhône-Poulenc Ag Company representative for further information.

DIRECTIONS FOR USE

IT IS A VIOLATION OF FEDERAL LAW TO USE THIS PRODUCT IN A MANNER INCONSISTENT WITH ITS LABELING.

GENERAL INFORMATION

SEVIN® Brand 80S Carbaryl Insecticide is a dry powder for dispersion in water and application as an insecticidal spray in hydraulic sprayers, mist blowers, low gallonage ground equipment and aircraft. READ THIS LABEL BEFORE USE. STRICTLY OBSERVE LABEL DIRECTIONS AND CAUTIONS, AND APPLICABLE FEDERAL AND STATE REGULATIONS.

GENERAL WORKER PROTECTION STATEMENTS

Do not apply this product in such a manner as to directly or through drift expose workers or other persons. The area treated must be vacated by unprotected persons.

Do not enter treated areas without protective clothing until sprays have dried.

Because certain states may require more restrictive reentry intervals for various crops treated with this product, consult your State Department of Agriculture for further information.

Written or oral warnings must be given to workers who are expected to be in treated area or in an area about to be treated with this product. Advise workers to stay out of fields during application and until sprays have dried. Regular long-sleeved work clothing should be worn when working in treated fields. See PRECAUTIONARY STATEMENTS, STATEMENT OF PRACTICAL TREATMENT and NOTE TO PHYSICIAN for information on accidental exposures. When oral warnings are given, warnings shall be given in a language customarily understood by workers. Oral warnings must be given if there is reason to believe that written warnings cannot be understood by workers. Written warnings must include the following information: appropriate signal word (WARNING), area treated with SEVIN® brand 80S Carbaryl Insecticide, date of application, appropriate clothing, and re-entry interval (i.e., until sprays have dried).

PREHARVEST AND GRAZING USE INFORMATION AND LIMITATIONS

Tolerances established under the Federal Food, Drug and Cosmetic Act permit the sale of crops bearing probable carbaryl residues when this product is used in accordance with label directions. If used as directed, treated forage may be grazed or used as feed for dairy and meat animals without causing illegal residues in meat or milk. This product may be applied up to and including the day of harvest or grazing of forage crops. Application may be made without removing livestock from area being treated.

Do not plant rotational food and feed crops not listed on this or other carbaryl labels in carbaryl treated soil.

Do not use reclaimed irrigation water from crops treated with carbaryl on upland crops for which carbaryl tolerances are not established.

PLANT RESPONSE PRECAUTIONS

To avoid possible injury to tender foliage, do not apply to wet foliage or when rain or high humidity is expected during the next two days.

Do not use on Boston ivy, Virginia creeper and maidenhair fern as injury will result. During early season, it may also injure Virginia and sand pines.

Observe label instructions on apple thinning and on combinations with certain herbicides on rice and soybeans.

SPRAY PREPARATION

TO ASSURE A UNIFORM SUSPENSION, AGITATE, STIR OR RECIRCULATE ALL "SEVIN®" BRAND 80S CARBARYL INSECTICIDE SPRAY MIXTURES PRIOR TO AND DURING USE. Remove oil, rust, scale, pesticide residues and other foreign matter from mix tanks and entire spray system. Flush with clean water. Fill spray or mix tank with 1/2 to 3/4 the desired amount of water. Start mechanical or hydraulic agitation. Slowly add the required amount of SEVIN® brand 80S Carbaryl Insecticide, and then the remaining volume of water. Prepare only as much spray mixture as can be applied on the day of mixing. MAINTAIN CONTINUOUS AGITATION DURING MIXING AND APPLICATION TO ASSURE A UNIFORM SUSPENSION. DO NOT STORE SPRAY MIXTURES FOR PROLONGED PERIODS.

COMPATIBILITY

SEVIN® brand 80S Carbaryl Insecticide has been used without plant injury with most common insecticides, miticides, fungicides, nutrients, adjuvants and petroleum oil as used on citrus. If compatibility of SEVIN® brand 80S Carbaryl Insecticide with another product and the resulting crop response is unknown, it should be tested on a small scale. Curdling, precipitation, greasing, layer formation or increased viscosity are symptoms of incompatibility. Incompatibility will reduce insect control and may cause mechanical difficulties or plant injury. Observe all cautions and limitations on labeling of all products used in mixtures. WHEN PREPARING COMBINATION SPRAYS, FIRST ADD "SEVIN®" 80S CARBARYL INSECTICIDE TO AT LEAST 1/2 THE DESIRED AMOUNT OF WATER, MIX THOROUGHLY, AND THEN ADD COMBINATION PRODUCTS TO THE MIXTURE, AND THEN THE REMAINING VOLUME OF WATER. DO NOT APPLY TANK MIX COMBINATIONS UNLESS YOUR PREVIOUS EXPERIENCE INDICATES THE MIXTURE IS EFFECTIVE AND WILL NOT RESULT IN APPLICATION PROBLEMS, EXCESSIVE RESIDUES OR PLANT INJURY.

Unstable under highly alkline conditions. Not effective if used with alkaline materials such as Bordeaux, lime-sulfur and casein-lime spreaders.

APPLICATION

On all crops, use sufficient gallonage to obtain thorough and uniform coverage.

Calibrate spray equipment to deliver the required volume. Use 50 mesh slotted strainers in spray system and 25 mesh slotted strainers behind nozzles. Avoid applications just before rainfall as poor insect control may result. To clean spray system after use, drain and flush with water. Refer to the Storage and Disposal directions for proper disposal of rinse water.

INSECT CONTROL

Apply when insects or their damage appear. To maintain control repeat at 7 to 14 day intervals or as necessary unless a shorter interval is specified below. Where a dosage range is indicated, use lower rate on young plants and early instars and higher rate on mature plants, advanced instars and adults. Thorough and uniform spray coverage is essential for effective control.

SEVIN® brand 80S Carbaryl Insecticide does not control spider mites. If spider mites are a problem, use a registered miticide.

Regional differences have been noted in the susceptibility of certain strains of fall armyworm to SEVIN® brand 80S Carbaryl Insecticide. If your experience indicates inadequate control, use an alternative pesticide.

FORAGE, FIELD AND VEGETABLE CROPS

Use at least 1 gallon of finished spray per acre for aerial application and at least 3 gallons of finished spray per acre for concentrate ground application.

CROP	GRASSHOPPERS	POUNDS OF SEVIN® BRAND 80S/ACRE	PREHARVEST INTERVAL (DAYS)	SPECIFIC DIRECTIONS
All Forage, Field and Vegetable Crops		5/8 to 1 7/8 Unless Noted Otherwise	See Specific Forage, Field or Vegetable Crop	Use the Lower rate for nymphs on small plants or sparse vegetation in wasteland, rangeland, pastures, rights-of-way, ditchbanks, hedgerows and roadsides. Use the higher rate for adults or applications to dense vegetation.

CROP	INSECT	POUNDS OF SEVIN® BRAND 80S/ACRE	PREHARVEST INTERVAL (DAYS)	SPECIFIC DIRECTIONS
Alfalfa Clovers	Blister Beetles Mexican Bean beetle	2/3 to 1 1/4		Observe plant response precautions.
	Alfalfa caterpillar Bean leaf beetle Cucumber beetles Green cloverworm Japanese beetle Leafhoppers Three cornered alfalfa hopper Thrips Velvetbean caterpillar	1 1/4	7 (Alfalfa)	On dense growth use 25 to 40 gallons per acre with ground equipment to ensure adequate coverage.
	Armyworm Cloverhead weevil Corn earworm Cutworm Egyptian alfalfa weevil larvae Essex skipper European alfalfa beetle Fall armyworm Lygus bugs Webworms Yellowstriped armyworm	1 1/4 to 1 7/8	0 (Clovers)	For alfalfa weevil larvae, if pretreatment damage is extensive, cut alfalfa and treat the stubble. Apply only once per cutting.
	Alfalfa weevil larvae West of the Rocky Mountains	1/4 to 1 7/8		
	East of the Rocky Mountains	1 7/8		
Asparagus	Asparagus beetle	1 1/4 to 2 1/2	1	
	Apache cicada Asparagus beetle	2 1/2 to 5	Post harvest	Treat ferns or brush growth. Do not treat more than once every 3 days.
Beans (including blackeyed peas, cowpeas, crowder or southern peas, dry beans, green beans, lima beans, navy beans and snap beans)	Blister beetles Mexican bean beetle	2/3 to 1 1/4		Observe plant response precautions.
	Alfalfa caterpillar Bean leaf beetle Cucumber beetles Flea beetles Green cloverworm Japanese beetle Leafhoppers Three cornered alfalfa hopper Thrips Velvetbean caterpillar Western bean cutworm	1 1/4	0 (except cowpeas) 3 (cowpeas)	
	Armyworm Corn earworm Cutworms European corn borer Fall armyworm Stink bugs Tarnished plant bug Webworms	1 1/4 to 1 7/8		
	Cowpea curculio	2 1/2		
	Corn earworm Limabean pod borer Lygus bugs Stink bugs	2 1/2		CALIFORNIA ONLY
Cabbage Broccoli Brussels Sprouts Cauliflower Kohlrabi	Flea beetles Harlequin bug	2/3 to 1 1/4	3	
	Armyworm Corn earworm Fall armyworm Imported cabbageworm	1 1/4 to 2 1/2		
Chinese cabbage Collards Hanover salad Horseradish Kale Mustard greens Radishes Rutabagas Turnips	Fleabeetles Harlequin bug Leafhoppers	2/3 to 1 1/4	3 (Horseradish, radishes, rutabagas and turnip roots) 14 (Chinese cabbage, collards, Hanover salad, kale, mustard greens and turnip tops)	
	Aster Leafhopper	1 1/4 to 1 7/8		
	Armyworm Corn earworm Fall armyworm Imported cabbageworm Stink bugs Tarnished plant bug	1 1/4 to 2 1/2		
Carrots Celery Parsnips Parsley	Flea beetles Leafhoppers	2/3 to 1 1/4	0 (carrots) 3 (parsnips) 14 (celery & parsley)	DO NOT USE ON CELERY IN CALIFORNIA.
	Aster leafhopper Lygus bugs Spittlebugs	1 1/4 to 1 7/8		
	Armyworm Corn earworm Fall armyworm Stink bugs Tarnished plant bugs	1 1/4 to 2 1/2		

CROP	INSECT	POUNDS OF SEVIN® BRAND 80S/ACRE	PREHARVEST INTERVAL (DAYS)	SPECIFIC DIRECTIONS
Corn (field, sweet, pop)	Armyworm Chinch bugs Corn earworm Corn rootworm adults European corn borer Fall armyworm Flea beetles Japanese beetle Sap beetles Southwestern corn borer Leafhoppers	1¼ to 2½	0	OBSERVE BEE CAUTION: For insects attacking silks and ears apply at 1 to 6 day intervals starting when first silks appear and continuing until silks begin to dry. For larvae in whorl and foliage feeders, apply as necessary. For chinch bugs use high gallonage ground application directed at the base of plants. Optimum timing and good coverage are essential for effective control.
	Western bean cutworm	2½		Treat when infestation averages 15% and at 90% to 100% tassel emergence. Treatment after 100% silk emergence will reduce effectiveness.
	Cutworms	2½ to 8		Apply in a 12 inch band, using ¼ pound (4 ounces) per 1000 linear feet of row, in at least 15 gallons of water per acre. For broadcast application use up to 8 pounds in at least 20 gallons (ground) or 5 gallons (air) or water per acre.
Cotton	Cotton leafhopper Cotton leafworm Flea beetles Striped blister beetle Thrips	⅔ to 1¼	7	Early season insect control.
	Boll weevil Bollworms Cotton leafperforator Fall armyworm Leafrollers Leafhoppers Tarnished plant bug Yellowstriped armyworm (cotton cutworm)	1¼ to 2½	7	Treat on a 5 to 7 day schedule for as long as control is necessary. Mid and late season insect control. May be applied after bolls open.
	Lygus bugs	1¼ to 2½		For light to moderate populations in Western irrigated cotton.
	Pink bollworm	1⅞ to 3		Aphid populations will be suppressed by repeated applications of this insecticide.
	Stink bugs Saltmarsh caterpillar	2½		
Cucumber Melons Pumpkins Squash	Pickleworm Melonworm	⅔ to 1¼	0	Observe plant response precautions.
	Cucumber beetles Fleabeetles Leafhoppers Squash bugs	1¼		Avoid excessive applications.
Dandelion Endive (Escarole) Lettuce Salsify	Flea beetles Harlequin bug Leafhoppers	⅔ to 1¼	3 (head lettuce & salsify roots) 14 (dandelion, endive (escarole), leaf lettuce & salsify tops)	Observe plant response precautions. Treat on a 5 to 7 day schedule after heads begin to form.
	Aster leafhopper Lygus bugs Spittlebugs	1¼ to 1⅞		
	Armyworm Corn earworm Fall armyworm Imported cabbage worm Stink bugs Tarnished plant bug	1¼ to 2½		
Pasture	Armyworm Black grass bugs Chinch bugs Essex skipper Fall armyworm Range caterpillars Range crane fly Striped grass looper Thrips	1¼ to 1⅞	0 (aerial application) 14 (ground application)	To control thrips in grasses grown for seed use high spray pressure to improve penetration into boot. Apply a maximum of 2 applications per year. Allow at least 14 days between applications. Do not allow foraging or cut for hay within 14 days of last application by ground. Aerially treated pastures may be grazed or cut for hay on day of treatment. Carefully mark swaths to avoid over-application
	Grasshoppers			Refer to general GRASSHOPPER heading above.
Rangeland	Black grass bug Mormon cricket Range caterpillars Range crane fly	⅔ to 1¼	0	FOR AERIAL APPLICATION ONLY. Apply a maximum of 2 applications per year. Allow at least 14 days between applications.

CROP	INSECT		POUNDS OF SEVIN® BRAND 80S/ACRE	PREHARVEST INTERVAL (DAYS)	SPECIFIC DIRECTIONS
					Carefully mark swaths to avoid over-application.
	Grasshoppers				Refer to general GRASSHOPPER heading above.
Wasteland, Rights-of-Way, Hedgerows, Ditchbanks, Roadsides	Black grass bug Mormon cricket Range caterpillar Range crane fly		⅔ to 1¼	0 (aerial application) 14 (ground application)	Apply a maximum of 2 applications per year. Allow at least 14 days between applications. Do not allow foraging or cut for hay within 14 days of last application by ground. Aerially treated areas may be grazed or cut for hay on day of treatment. Carefully mark swaths to avoid over-applications.
	Grasshoppers				Refer to general GRASSHOPPER heading above.
Garden beet Spinach Swiss chard	Flea beetles Harlequin bug	Leafhoppers	⅔ to 1¼	3 (garden beet roots)	
	Aster leafhopper		1¼ to 1⅞		
	Armyworm Corn earworm Fall armyworm	Stink bugs Tarnished plant bug	1¼ to 2½	14 (garden beet tops, spinach, Swiss chard)	
Okra	Corn earworm	Stink bugs	1¼ to 2½	0	Treat on a 5 to 7 day schedule.
Peanuts	Blister beetles	Mexican bean beetle	⅔ to 1¼		Observe plant response precautions.
	Alfalfa caterpillar Bean leaf beetle Cucumber beetles Green cloverworm Japanese beetle Leafhoppers	Red-necked peanut worm Three cornered alfalfa hopper Thrips Velvetbean caterpillar	1¼	0	
	Armyworm Corn earworm Fall armyworm	Stinkbugs Webworms	1¼ to 1⅞		
	Whitefringed beetle (adults) Cutworms		2½		
Peas	Colorado potato beetle Leafhoppers		1¼	3	
	Armyworm		1¼ to 1⅞		
	Alfalfa caterpillar Cutworms Pea leaf weevil	Pea weevil Yellowstriped armyworm	1⅞		
Potato Tomato Eggplant Pepper	Colorado potato beetle Flea beetles Leafhoppers		⅔ to 1¼		
	European corn borer Fall armyworm Lace bugs Stink bugs	Tarnished plant bug Tomato fruit worm Tomato hornworm Tomato pinworm	1½ to 2½	0	
	Cutworms		2½		
Rice	Armyworm Chinch bugs	Fall armyworm Stink bugs	1¼ to 1⅞	14	CAUTION! May kill shrimp and crabs. Do not use in areas where these are important resources. Do not use on rice fields in which crayfish and/or catfish farming are included in the cultural practice. DO NOT APPLY PROPANIL HERBICIDES WITHIN 15 DAYS BEFORE OR AFTER APPLICATION OF THIS PRODUCT OR PLANT INJURY WILL RESULT. MISSISSIPPI DELTA & TEXAS.

CROP	INSECT		POUNDS OF SEVIN® BRAND 80S/ACRE	PREHARVEST INTERVAL (DAYS)	SPECIFIC DIRECTIONS
	Armyworm Leafhoppers	Tadpole shrimp	2 1/2		CALIFORNIA ONLY: For optimum tadpole shrimp control, apply to water when pest first appears.
Sorghums (milo, grain sorghum, sweet sorghum and hybrids)	Armyworm Chinch bugs Corn earworm	Fall armyworm Stink bugs Webworms	1 1/4 to 2 1/2	21 (grain) 0 (forage)	Direct spray into forming heads for optimum insect control. Treat for sorghum midge when 25 to 30 percent of heads have emerged from boot and are in bloom. Repeat application 3 to 5 days later if adults are still active. A third application may be necessary in late planted sorghum or if midge are abundant. For chinch bugs use high gallonage ground application directed at base of plants.
	Sorghum midge Southwestern corn borer		1 7/8		
	Cutworms		2 1/2		
Soybeans	Bean leaf beetle Cucumber beetles Green cloverworm	Mexican bean beetle Velvetbean caterpillar	2/3	0	DO NOT APPLY A COMBINATION OF THIS PRODUCT AND 2, 4-DB HERBICIDES TO SOYBEANS AS CROP INJURY MAY RESULT. Use lower rates for light to moderate populations and smaller instars.
	Corn earworm		2/3 to 1 1 1/4 to 1 7/8		
	Blister beetles Grape colaspis Mexican bean beetle		2/3 to 1 1/4		Use higher rates for heavy populations and larger instars.
	Alfalfa caterpillar Bean leaf beetle Cucumber beetles Green cloverworm Japanese beetle	Leafhoppers Three cornered alfalfa hopper Thrips Velvetbean caterpillar	1 1/4		
	Armyworm Cutworms	Fall armyworm Stink bugs Webworms	1 1/4 to 1 7/8		
	Painted lady (Thistle caterpillar)	Saltmarsh caterpillar Yellowstriped armyworm	2 1/2 to 3		
Sugar beets	Armyworm Beet leaf beetle Fall armyworm	Flea beetles Leafhoppers Webworms	1 1/4 to 1 7/8	14	
	Cutworms		1 7/8		
Sunflower	Cutworms		1 7/8	60	DO NOT USE IN CALIFORNIA. Do not allow animals to graze on treated crops.
	Armyworm Fall armyworm		1 7/8 to 2 1/2		
	Stem Weevil Sunflower beetle		1 1/4 to 2 1/2		
Sweet Potatoes	Sweet potato weevil		9 tablespoons/gallon.	0	DO NOT USE IN CALIFORNIA **Preplant Dip:** Just prior to planting, dip sweet potato cuttings in a suspension of this product (9 tablespoons/gallon of water).
			1 1/4 to 2 1/2		**Foliar Application:** Full coverage of plants is essential. Use lower rate on young plants and higher rate on mature plants.
	Corn earworm Cucumber beetles Flea beetles	Sweet potato hornworm Tortoise beetles	1 1/4 to 2 1/2		Apply as a foliar spray as needed.
Tobacco (In Plant Beds)	Tobacco flea beetle		2 1/2 Tbs/gal. or 1 1/4 lbs/ 50 gallons and apply 6 gal/100 sq. yards	0	To prevent plant injury, avoid excessive applications.
	Green June beetle grubs		2/3 lb./200 gal.		Apply only to areas that larvae have

CROP	INSECT	POUNDS OF SEVIN® BRAND 80S/ACRE	PREHARVEST INTERVAL (DAYS)	SPECIFIC DIRECTIONS
				uprooted by sprinkling as a drench with 50 to 100 gallons per 100 square yards.
(In Fields)	Budworms Fall armyworm Flea beetles Hornworms Japanese June beetles Suckfly	1 1/4 to 2 1/2		Use lower rate on young plants (up to knee height). Use at least 10 gallons of prepared spray per acre. Begin treatments when worms are small. If late season applications are necessary, allow three days before priming or cutting.
Wheat	Cereal leaf beetle	1 1/4	0 (Forage) 21 (grain)	DO NOT USE IN CALIFORNIA. Do not make more than two applications after grain heads emerge from boot. Application is effective against eggs larvae and adults.
	Armyworm Fall armyworm	1 1/4 to 1 7/8		

TREE FRUIT AND NUT CROPS

For dilute sprays apply the specified dosage per 100 gallons of water. For concentrate and aerial sprays increase the concentration of SEVIN® brand 80S Carbaryl Insecticide in the spray mixture to apply an amount per acre equivalent to that in a dilute spray. The optimum spray gallonage will depend on tree size, density and stage of growth. Typical spray gallonages per acre range from 200-300 gallons for dilute sprays, 30-100 gallons for concentrate sprays and 10-25 gallons for aerial sprays. Do not exceed maximum label rate per acre per application.

CROP	POUNDS OF SEVIN® BRAND 80S/100 GAL.	PREHARVEST INTERVAL (DAYS)	SPECIFIC DIRECTIONS
APPLE THINNING Apples only	1/3 to 2/3	1	Apply 1 full coverage dilute spray between 10 and 25 days after full bloom. If factors such as tree age, variety, nutrition, previous crop, pruning, bloom and degree of set favor excessive fruit thinning with this product, exercise caution to avoid possible yield reduction. Consult with your County Extension Service or other experts for advice on the proper use of this product. For easily thinned varieties including Cortland, Grimes, Jonathan, McIntosh, Orleans, Rome Beauty, Puritan, Red Delicious, Winesap, Yellow Newton.
	2/3 to 1 1/4		For difficult to thin varieties including Baldwin, Ben Davis, Duchess, Early McIntosh, Golden Delicious, Lady Apple, Northern Spy, Rhode Island Greening, Steele Red, Turley, Wealthy, Yellow Transparent and York Imperial.

CROP	INSECT	POUNDS OF SEVIN® BRAND 80S/100 GAL.	PREHARVEST INTERVAL (DAYS)	SPECIFIC DIRECTIONS
Almond	Peach twig borer San Jose scale Fruittree leafroller	1 1/4	28	Apply in "popcorn" or petal fall stages when the May brood of the peach twig borer begins to hatch or thereafter as needed.
	Navel orangeworm			Time early and mid season applications to correspond to moth flight peaks. Make a late season application at initiation of hull split or up to 10% hull split. Do not apply more than 6 1/4 pounds per acre.
Apples Pears	Apple aphid Apple rust mite Apple sucker Bagworms California pearslug (pear sawfly) Codling moth Eyespotted bud moth Green fruitworm Lecanium scales Lesser appleworm Lygus bugs Orange tortrix Oystershell scale Pear leaf blister mite Pear psylla Pear rust mite San Jose scale Tentiform leafminers Woolly apple aphid	1 to 1 1/4	1	WEST OF THE ROCKY MOUNTAINS. To avoid undesired apple thinning, delay use until at least 30 days after full bloom. For psylla control apply when eggs hatch or young nymphs are present.
	Apple mealybug Apple aphid Codling moth White apple leafhopper	2/3		EAST OF THE ROCKY MOUNTAINS.

CROP	INSECT	POUNDS OF SEVIN® BRAND 80S/100 GAL.	PREHARVEST INTERVAL (DAYS)	SPECIFIC DIRECTIONS
	Apple maggot Apple rust mite Bagworms Eastern tent caterpillar European apple sawfly Eyespotted bud moth Forbes scale Fruittree leafroller Green fruitworm Japanese beetle Lecanium scales Lesser appleworm Oystershell scale Pear psylla Pear leaf blister mite Pear rust mite Periodical cicada Plum curculio Redbanded leafroller Rosy apple aphid San Jose scale Tarnished plant bug Tentiform leafminers Woolly apple aphid Yellowheaded fireworm	1 1/4	1	To avoid undesired apple thinning, delay use until at least 30 days after full bloom. For optimum scale control apply when crawlers are present.
Chestnuts	Chestnut weevil	2 1/2 to 3 3/4	0	DO NOT USE IN CALIFORNIA. Make 4 applications at weekly intervals beginning in late July for adult chestnut weevil control. Last application should be made prior to shuck split.
Citrus Fruits (such as grapefruit, lemons, limes, oranges, tangelos, tangerines, citrus citron, kumquats and hybrids)	Avocado leafroller California orangedog Citrus cutworm Citrus root weevil Fruittree leafroller Orange tortrix Western tussock moth West Indian sugarcane borer (adults)	1 1/4	5	Do not apply more than 25 pounds of this product per acre per application. May be mixed with petroleum oils commonly used on citrus.
	Black scale Brown soft scale California red scale Citricola scale Citrus snow scale Yellow scale	1 to 1 1/4		
Filbert	Filbert aphid Filbert leafroller Filbertworm	1 1/4	0	Apply when leafroller eggs are hatching. Repeat on first appearance of adult filbert moths and again 3 to 4 weeks later.
Olives	Olive scale	1 to 1 1/4	0	For optimum scale control add 1 1/2 gallons of summer oil and apply mixture when crawlers are present. Do not exceed 2 applications per year. Do not apply more than 18 3/4 pounds of this product per acre per application.
Peaches Apricots Nectarines	Apple pandemis Codling moth Cucumber beetles European earwig Fruittree leafroller Orange tortrix Oriental fruit moth Peach twig borer Periodical cicada Plum curculio Redbanded leafroller	1 1/4	1 (peaches) 3 (apricots & nectarines)	Do not apply more than 7 1/2 pounds of this product per acre per application to apricots.
	Japanese beetle June beetles Lecanium scales Lesser peachtree borer Olive scale San Jose scale Tarnished plant bug Tussock moths Variegated leafroller	1 1/4		For optimum scale control apply when crawlers are present. Spray limbs and trunk thoroughly, weekly during moth flight.
Pecans	Black margined aphid Fall webworm Hickory shuckworm Lesser webworm Pecan leaf phylloxera Pecan weevil Twig girdler Walnut caterpillar Pecan nut casebearer Pecan spittlebug Pecan stem pyylopera	1 1/4 to 3	0	Do not apply more than 9 pounds of this product per acre per application.
Pistachios	Navel orangeworm	2/3 to 2 1/2	14	Apply dilute volumes of 150 to 300 gallons of mixed spray per acre for full coverage.

CROP	INSECT		POUNDS OF SEVIN® BRAND 80S/100 GAL.	PREHARVEST INTERVAL (DAYS)	SPECIFIC DIRECTIONS
Plums Prunes Cherries	Black cherry aphid Brown soft scale Cherry fruitworm Cherry maggot European earwig Eyespotted bud moth Forbes scale Fruittree leafroller Green fruitworm Japanese beetle Lecanium scales Lesser peach-tree borer	Mealy plum aphid Oystershell scale Peach twig borer Plum curculio Prune leafhopper Redbanded leafroller Rose chafer San Jose scale Variegated leafroller	1 1/4	1	Do not apply more than 7 1/2 pounds of this product per acre per application. For optimum scale control apply when crawlers are present. For lesser peachtree borer control, spray limbs and tree trunks thoroughly, weekly during moth flight.
	Codling moth Eastern tent caterpillar	Orange tortrix Tussock moth	1		
Walnut	Calico scale European fruit lecanium	Filbertworm Fruittree leafroller Frosted scale	2/3	0	Apply 1000 gallons of dilute spray per acre for mature trees.
	Codling moth				For codling moth apply first spray when average cross sectional diameters of developing nuts are 1/2 to 3/4 inch. Repeat during middle or late June as needed.
	European earwig		2 1/2		Spray tree trunks to point of run-off.

SMALL FRUIT CROPS

Recommended dosages refer to pounds of SEVIN® Brand 80S Carbaryl Insecticide per acre. The optimum spray gallonage will depend on plant size, density and stage of growth. Typical spray gallonages per acre range from 100-300 gallons for dilute sprays, 30-100 gallons for concentrate sprays and 10-25 gallons for aerial sprays. Do not exceed maximum label rate per acre per application.

CROP	INSECT		POUNDS OF SEVIN® BRAND 80S/ACRE	PRE-HARVEST INTERVAL (DAYS)	SPECIFIC DIRECTIONS
Blackberries Raspberries Dewberries (including boysenberries and loganberries)	European raspberry aphid Japanese beetle	Leafhoppers Leafrollers Rose chafer Snowy tree cricket	1 1/4 to 2 1/2	7	
	Omnivorous leafroller Raspberry sawfly		2 1/2		CALIFORNIA ONLY
Blueberries	Blueberry maggot Cherry fruitworm Cranberry fruitworm	European fruit lecanium Japanese beetle	1 7/8	0	Apply 3 weekds before harvest and repeat as necessary.
Cranberries	Cutworms Cranberry fireworms Cranberry fruitworms	Japanese beetle Leafhoppers Rose chafer	1 7/8 to 3 3/5	1	CAUTION! May kill shrimp and crabs. Do not use in areas where these are important resources. Apply in late bloom and as needed at 7 to 10 day intervals.
Grapes	European fruit lecanium Grape leaffolder Grape leafhopper	Western grapeleaf skeletonizer Western yellow-striped armyworm	1 1/4 to 2 1/2	7	Apply before first brood leaffolder larvae emerge from rolls.
	Cutworms Eight spotted forester Grape berry moth Japanese beetle June beetles	Orange tortrix Omnivorous leafroller Redbanded leafroller Saltmarsh caterpillar	2 1/2		

CROP	INSECT		POUNDS OF SEVIN® BRAND 80S/ACRE	PRE-HARVEST INTERVAL (DAYS)	SPECIFIC DIRECTIONS
Strawberries	Flea beetles Meadow spittlebug Omnivorous leaftier (strawberry fruitworm)	Strawberry leafroller Strawberry weevil	1 ½ to 2 ½	1	Carbaryl may injure Early Dawn and Sunrise varieties on the Delmarva Peninsula.

NONRESIDENTIAL TREES AND ORNAMENTALS

For dilute-spray ground applications to nonresidential trees (including shade trees, shelter belts, forests, plantations, parks, and recreational areas), nonresidential ornamentals, woody plants and shrubs, apply the specified dosage per 100 gallons of water. For concentrate-spray ground applications, apply the specified dosage per acre in sufficient spray volume to provide thorough coverage.

For aerial applications to forest trees (including shade trees, shelter belts, plantations, parks, and recreational areas) and commercially grown ornamentals, woody plants and shrubs, apply the specified dosage per acre in sufficient spray volume to provide thorough coverage. Avoid direct application to lakes, streams and ponds.

CROP	INSECT			POUNDS OF SEVIN® BRAND 80S/100 GAL.	SPECIFIC DIRECTIONS
GROUND APPLICATION: Trees (including shade trees, shelter belts, forests, plantations, parks and recreational areas). Ornamentals (including roses, flowers and other herbaceous plants), Woody Plants and Shrubs.	Ants Apple aphid Armyworm Azalea leafminer Bagworms Birch leafminer Blister beetle Boxelder bug Boxwood leafminer Brown tall moth Cankerworms Catalpa sphinx Chiggers Cooley spruce gall aphid Cutworms Cypress tip moth	Hackberry nipplegall maker Holly bud moth Holly leafminer Jackpine budworm Japanese beetle Jeffrey pine needleminer June beetles Lace bugs Leafhoppers Leafrollers Locust borer Maple leafcutter Mealybugs Mimosa webworm	Psyllids Puss caterpillar Redhumped oakworm Rose aphid Rose chafer Roseslug Saddled prominent Sawflies Scale insects Sowbugs Spiny elm caterpillar Springtails Spruce budworm	1 ¼	Observe plant response precautions. Apply dilute sprays to obtain thorough coverage of upper and lower leaf surfaces. To control scale insects, treat trunks, stems and twigs in addition to plant foliage. For optimum worm control, treat when in early instars. Addition of a sticker may improve residual control. Applications for control of Maple leafcutter on sugar maple should be made when larvae are in 2nd instar, after mining, and as cases are being formed.
AERIAL APPLICATION: Forest Trees (including shade trees, shelter belts, forests, planatations parks and recreational areas) and Commercially Grown Ornamentals (including roses, flowers and other herbaceous plants), Woody Plants and Shrubs	Douglas-fir tussock moth Eastern spruce gall aphid Elm leaf aphid Elm leaf beetle Elm spanworm Erlophyid mites European pine shoot moth Fall armyworm Flea beetles Fuller rose beetle Gall midges Gall wasps Green striped mapleworm Grasshoppers Gypsy moth	Nantucket pine tip moth Oak leafminers Oak leaf skeletonizer Oakworm complex Oleander caterpillar Olive ash borer Orange-striped oakworm Orange tortrix Periodical cicada Pine sawfly Pine spittlebug Pitch pine tip moth Plant bugs Poinsettia Hornworm	Spruce needleminer Subtropical pine tip moth Tent caterpillars Thorn bug Walnut caterpillar Webworms Western hemlock looper Western spruce budworm Willow leaf beetles Yellow poplar weevil		
	Ips engraver beetles Mountain pine beetle Roundheaded pine beetle Western pine beetle			25	Effective as a preventive treatment only. Repeat annually as required to prevent beetle attacks. Apply 1 gallon of spray per 50 square feet of bark in late May to early July, or prior to beetle attack. Treat tree trunks from ground level up, until trunk diameter is less than 5 inches.
	Elm bark beetle				Apply approximately 20 to 30 gallons of spray mixture for each 50 feet of elm tree for thorough coverage of all bark surfaces on trunk, limbs and twigs.

NONRESIDENTIAL TURF AND RECREATIONAL AREAS

CROP	INSECT		SPECIFIC DIRECTIONS
Turf grasses	Ants Armyworm Bluegrass billbug Centipedes Chiggers Chinch bugs Cutworms Earwigs Essex skipper European chafer Fall armyworm Fiery skipper Fleas Grasshoppers	June beetles Leafhoppers Lucerne moth Millipedes Mosquitoes Sod webworm (lawn moths) Sowbugs Springtails Ticks White grubs Yellowstriped armyworm	Use 4 ounces of SEVIN® brand 80S Carbaryl Insecticide per 1000 square feet (10 ¾ pounds per acre) of turf grass. Make applications in sufficient spray volume for thorough coverage and turf thatch penetration. Repeat treatment as necessary. For Armyworm, Cutworms, Fall Armyworm and Sod Webworm Control: Do not irrigate treated areas following insecticide application. For Chinch Bug Control: Treat entire turf grass area rather than just damaged areas. Irrigation of turf grass area before Insecticide application will aid in penetration into turf grass. For White Grub Control: Applications should be made when grubs are feeding near the soil surface, usually during late March through May, or July to early September, or as recommended by local Agricultural Extension Service agents. Water or irrigate turf grasses thoroughly soon after treatment.

IMPORTED FIRE ANT CONTROL

SITE	POUNDS OF SEVIN® BRAND 80S/ 100 GALLONS OF WATER		SPECIFIC DIRECTIONS
Cemeteries, and non-residential Recreational Area (including turf, golf courses, and parks) Pastures, Rangeland, Forested Lands, Wastelands	2		Apply a total of 2 gallons of the diluted solution over the surface of each mound or at least 1 quart per 6 inches of mound diameter using a bucket, can or other appropriate equipment. Thoroughly wet mound and surrounding area to a 4 ft. diameter (12 sq. ft.). Do not disturb mounds prior to treatment. Pour solution from a height of about three feet to give sufficient force to break mound apex and flow into ant tunnels. For best results, apply in cool weather 65-80°F, or in early morning or late evening hours. Repeat application if mound activity resumes after 10 days. Treat new mounds as they appear. Pressurized sprays may disturb the ants and cause migration, reduction in product effectiveness.

PEST CONTROL IN AND AROUND NON-RESIDENTIAL BUILDINGS LIMITED TO PEST CONTROL OPERATORS ONLY

INSECT	POUNDS OF SEVIN® BRAND 80S/100 GAL.	SPECIFIC DIRECTIONS
Ants Cockroaches	25 (25 tablespoonfuls/ gallon)	For use in and around non-residential buildings such as warehouses, barns, and municipal and recreational areas.
Brown dog tick Earwigs Fleas Millipedes Silverfish Spiders	12 ½ (12 ½ tablespoonfuls/ gallon)	Apply as a coarse wet spray or with a paint brush, and thoroughly treat cracks and crevices, baseboards, corners, walls, door sills, and window frames, ant trails, and other areas where insects tend to congregate. Do not use as a space spray. Repeat when necessary but not more often than twice per week. Do not use this product in commercial food areas of food handling establishments, restaurants, or other places where food is prepared or processed. Do not use in serving areas while food is exposed. Avoid application to surfaces where visible spray residues are objectionable.

POULTRY INSECT CONTROL

POULTRY	INSECT	SPECIFIC DIRECTIONS
Chickens Ducks, Geese, Game birds Pigeons, and Turkeys	Chicken mite Fleas Lice Northern fowl mite	For use as a direct spray on birds by: 1. Misting with Electric Fog Machine: Mix ⅜ pounds (6 ounces) of this product in 1 gallon of spray. Use 1 ½ gallons per 1000 hens in cages, on litter or on slatted floors. Repeat in 4 weeks if necessary. 2. Spraying with knapsack or cylinder type compressed air sprayers: Mix ¼ pound (4 ounces) of this product in 5 gallons of spray. Use 1 gallon per 100 hens in cages, on litter or on slatted floors. Repeat in 4 weeks if necessary. Direct mist spraying for chicken mite and fleas is a supplement to spraying roosts and buildings for control of these pests. Do not apply to poultry and game birds within 7 days of slaughter.

POULTRY QUARTERS	INSECT	SPECIFIC DIRECTIONS
Roosts and Buildings	Bedbugs Chicken mites Fleas Fowl tick Darkling Beetle (Lesser mealworms) Lice Northern fowl mite	Spray roosts and buildings with conventional power spray or knapsack equipment. For chicken mite, fleas and bedbugs, use 5 pounds of this product per 100 gallons of water. For fowl ticks, use 20 pounds of this product per 100 gallons of water. For lesser mealworms, use 62 ½ pounds per 100 gallons. Spray 1 to 2 gallons per 1000 square feet of wall, bedding, litter or roost surface. Force spray into cracks. Repeat as needed. Avoid spraying of nests, eggs and feeding and watering troughs. Ventilate while spraying. Do not treat premises within 7 days of slaughter.

DIRECTIONS FOR USE THROUGH SPRINKLER IRRIGATION SYSTEMS

Apply this product only through sprinkler irrigation systems including center pivot and solid set. Do not apply this product through any other type of irrigation system.

SPRAY PREPARATION: Remove scale, pesticide residues, and other foreign matter from the chemical tank and entire injector system. Flush with clean water.

APPLICATION INSTRUCTIONS

First prepare a suspension of SEVIN® brand 80S Carbaryl Insecticide in a mix tank. Fill tank with 1/2 to 3/4 the desired amount of water. Start mechanical or hydraulic agitation. Add the required amount of SEVIN® brand 80S, and then the remaining volume of water. (Suspension concentrations using the appropriate dosage per acre recommended on this label of SEVIN® brand 80S per 1 to 4 gallons of water are recommended). Then set sprinkler to deliver 0.1 to 0.3 inch of water per acre. Start sprinkler and uniformly inject the suspension of SEVIN® brand 80S into the irrigation water line so as to deliver the desired rate per acre. The suspension of SEVIN® brand 80S should be injected with a positive displacement pump into the main line ahead of a right angle turn to insure adequate mixing. If you should have any other questions about calibration, you should contact State Extension Service specialists, equipment manufacturers or other experts.

Note: When treatment with SEVIN® brand 80S has been completed, further field irrigation over the treated area should be avoided for 24 to 48 hours to prevent washing the chemical off the crop.

GENERAL PRECAUTIONS FOR APPLICATIONS THROUGH SPRINKLER IRRIGATION SYSTEMS

Maintain continuous agitation in mix tank during mixing and application to assure a uniform suspension.

Greater accuracy in calibration and distribution will be achieved by injecting a larger volume of a more dilute solution per unit time.

The system must contain a functional check value, vacuum relief valve, and low pressure drain appropriately located on the irrigation pipeline to prevent water source contamination from backflow. The pesticide injection pipeline must contain a functional, automatic, quick-closing check value to prevent the flow of fluid back toward the injection pump. The pesticide injection pipeline must contain a functional, automatic, quick-closing check valve to prevent the flow of fluid back toward the injection pump. The pesticide injection pipeline must also contain a functional, normally closed solenoid-operated valve located on the intake side of the injection pump and connected to the system interlock to prevent fluid from being withdrawn from the supply tank when the irrigation system is either automatically or manually shutdown. The system must contain functional interlocking controls to automatically shut off the pesticide injection pump when the water pump motor stops. The irrigation line or water pump must include a functional pressure switch which will stop the water pump motor when the water pressure decreases to the point where pesticide distribution is adversely affected. Systems must use a metering pump, such as a positive displacement injection pump (e.g., diaphram pump) effectively designed and constructed of materials that are compatible with pesticides and capable of being fitted with a system interlock. Do not apply when wind speed favors drift beyond the area intended for treatment.

Do not apply when wind speed favors drift, when system connection or fittings leak, when nozzles do not provide uniform distribution or when lines containing the product must be dismantled and drained.

Crop injury, lack of effectiveness, or illegal pesticide residues in the crop may result from non-uniform distribution of treated water.

Allow sufficient time for pesticide to be flushed through all lines and all nozzles before turning off irrigation water. A person knowledgeable of the chemigation system and responsible for its operation shall shut the system down and make necessary adjustments should the need arise.

Do not connect an irrigation system (including greenhouse systems) used for pesticide application to a public water system unless the label-prescribed safety devices for public water supplies are in place.

STORAGE AND DISPOSAL

STORAGE

Store unused SEVIN® brand 80S Carbaryl Insecticide in original container only, in cool, dry area out of reach of children and animals. Do not store in areas where temperatures frequently exceed 100°F.

PESTICIDE DISPOSAL

Do not contaminate water, food or feed by storage or disposal. Open dumping is prohibited. Wastes resulting from the use of this product may be disposed of on site or at an approved waste disposal facility.

CONTAINER DISPOSAL

Completely empty bag into application equipment. Then dispose of empty bag in a sanitary landfill or by incineration, or, if allowed by State and local authorities, by burning. If bag is burned, stay out of smoke.

LIMITED WARRANTY AND DISCLAIMER

The manufacturer warrants (a) that this product conforms to the chemical description on the label; (b) that this product is reasonably fit for the purposes set forth in the directions for use when it is used in accordance with such directions; and (c) that the directions, warnings and other statements on this label are based upon responsible experts' evaluation of reasonable tests of effectiveness, of toxicity to laboratory animals and to plants, and of residues on food crops, and upon reports of field experience. Tests have not been made on all varieties or in all states or under all conditions. THE MANUFACTURER NEITHER MAKES NOR INTENDS, NOR DOES IT AUTHORIZE ANY AGENT OR REPRESENTATIVE TO MAKE, ANY OTHER WARRANTIES, EXPRESS OR IMPLIED, AND IT EXPRESSLY EXCLUDES AND DISCLAIMS ALL IMPLIED WARRANTIES OF MERCHANTABILITY OR FITNESS FOR A PARTICULAR PURPOSE.

THIS WARRANTY DOES NOT EXTEND TO, AND THE BUYER SHALL BE SOLELY RESPONSIBLE FOR, ANY AND ALL LOSS OR DAMAGE WHICH RESULTS FROM THE USE OF THIS PRODUCT IN ANY MANNER WHICH IS INCONSISTENT WITH THE LABEL DIRECTIONS, WARNINGS OR CAUTIONS.

BUYER'S EXCLUSIVE REMEDY AND MANUFACTURER'S OR SELLER'S EXCLUSIVE LIABILITY FOR ANY AND ALL CLAIMS, LOSSES, DAMAGES, OR INJURIES RESULTING FROM THE USE OR HANDLING OF THIS PRODUCT, WHETHER OR NOT BASED IN CONTRACT, NEGLIGENCE, STRICT LIABILITY IN TORT OR OTHERWISE, SHALL BE LIMITED, AT THE MANUFACTURER'S OPTION, TO REPLACEMENT OF, OR THE REPAYMENT OF THE PURCHASE PRICE FOR, THE QUANTITY OF PRODUCT WITH RESPECT TO WHICH DAMAGES ARE CLAIMED. IN NO EVENT SHALL MANUFACTURER OR SELLER BE LIABLE FOR SPECIAL, INDIRECT OR CONSEQUENTIAL DAMAGES RESULTING FROM THE USE OR HANDLING OF THIS PRODUCT.

AG 90055 EPA Approval: 8/22/88

THE SPECIMEN LABEL IS INTENDED FOR USE ONLY AS A GUIDE IN PROVIDING GENERAL INFORMATION REGARDING THE DIRECTIONS, WARNINGS AND CAUTIONS ASSOCIATED WITH THE USE OF THIS PRODUCT. AS WITH ANY AGRICULTURAL CHEMICAL, ALWAYS FOLLOW THE LABEL INSTRUCTIONS PACKAGE BEFORE USING.

SEVIN® BRAND 50W

CARBARYL INSECTICIDE

FOR AGRICULTURAL OR COMMERCIAL USE ONLY
NOT FOR RESIDENTIAL USE

ACTIVE INGREDIENT:
Carbaryl (1-naphthyl N-methylcarbamate). . . . . 50% by wt.
INERT INGREDIENTS:. . . . . 50% by wt.

KEEP OUT OF REACH OF CHILDREN

WARNING ADVISO

PRECAUSION AL USARIO: Si usted no lee ingles, no use este producto hasta que la etiqueta le haya sido explicada ampliamente.

E.P.A. Reg. No. 264-314 E.P.A. EST. NO. 477-MD-01

For MEDICAL And TRANSPORTATION Emergencies ONLY Call 24 Hours A Day 1-800-334-7577
For Product Use Information Call 1-800-334-9745

PRECAUTIONARY STATEMENTS

WARNING

HAZARDS TO HUMANS AND DOMESTIC ANIMALS

MAY BE FATAL IF SWALLOWED.
MAY BE HARMFUL IF INHALED. OVEREXPOSURE MAY CAUSE: Salivation, watery eyes, pinpoint eye pupils, blurred vision muscle tremors, difficult breathing, excessive sweating, abdominal cramps' nausea, vomiting, diarrhea, weakness, headache. IN SEVERE CASES CONVULSION, UNCONSCIOUSNESS AND RESPIRATORY FAILURE MAY OCCUR. SIGNS AND SYMPTOMS OF OVEREXPOSURE OCCUR RAPIDLY FOLLOWING EXPOSURE TO THIS PRODUCT. Avoid breathing of dust and spray mist. Do not take internally. Avoid contact with eyes, skin or clothing. Wear regular long-sleeved work clothing and head covering. Change to clean clothing daily.
Bathe and wash hair after each work day. Do not eat, drink, or use tobacco while working with this product or spray solutions. Wash hands and face before eating, drinking or using tobacco. Keep out of reach of children and domestic animals.

ANTIDOTE STATEMENT

ATROPINE SULFATE IS HIGHLY EFFECTIVE AS AN ANTIDOTE.

Narcotics and other sedatives should not be used. Further, drugs like 2-PAM (pyridine-2-aldoxime methiodide) are not recommended. See NOTE TO PHYSICIAN below.

STATEMENT OF PRACTICAL TREATMENT

GENERAL

Contact a physician immediately in all cases of suspected poisoning. Transport to a physician or hospital immediately and SHOW A COPY OF THIS LABEL TO THE PHYSICIAN. If poisoning is expected in animals, contact a veterinarian.

IF SWALLOWED: Never give anything by mouth to an unconscious or convulsing person. If conscious and not convulsing, drink 1 to 2 glasses of water and induce vomiting by touching the back of the throat with finger.

IF IN EYES: Flush eyes with plenty of water. Get medical attention if irritation persists.

IF ON SKIN: Wash thoroughly with soap and water.

IF INHALED: Move from contaminated atmosphere and call a physician.

NOTE TO PHYSICIAN:

Carbaryl is a carbamate insecticide, which is a cholinesterase inhibitor. Overexposure to this substance may cause toxic signs and symptoms due to stimulation of the cholinergic nervous system. These effects of overexposure are spontaneously and rapidly reversible.

Gastric lavage may be used if this product has been swallowed. Carbaryl poisoning may occur rapidly after ingestion and prompt removal of stomach contents is indicated.

Specific treatment consists of parenteral atropine sulfate. Caution should be maintained to prevent overatropinization. Mild cases may be given 1 to 2 mg intramuscularly every 10 minutes until full atropinization has been achieved and repeated thereafter whenever symptoms reappear. Severe cases should be given 2 to 4 mg intravenously every 10 minutes until fully atropinized, then intramuscularly every 30 to 60 minutes, as needed, to maintain the effect for at least 12 hours. Dosages for children should be appropriately reduced. Complete recovery from overexposure is to be expected within 24 hours.

Narcotics and other sedatives should not be used. Further, drugs like 2-PAM (pyridine-2-aldoxime methiodide) are NOT recommended.

To aid in confirmation of a diagnosis, urine samples should be obtained within 24 hours of exposure and immediately frozen.

Consultation on diagnosis therapy and analysis can be obtained at all hours by calling the Rhône-Poulenc emergency phone number 1-800-334-7577

ENVIRONMENTAL HAZARDS

This product is extremely toxic to aquatic and estuarine invertebrates. Do not apply directly to water and wetlands, except under the forest canopy and use on rice. Discharge from rice fields may kill aquatic and estuarine invertebrates. Do not contaminate water by cleaning equipment or disposal of wastes. Do not apply when weather conditions favor drift from the area treated.

BEE CAUTION: MAY KILL HONEYBEES IN SUBSTANTIAL NUMBERS.

This product is highly toxic to bees exposed to direct treatment or residues on blooming crops or weeds. Do not apply this product or allow it to drift to blooming crops or weeds if bees are visiting the treatment area. Contact your Cooperative Extension Service or your local Rhône-Poulenc Ag Company representative for further information.

DIRECTIONS FOR USE

IT IS A VIOLATION OF FEDERAL LAW TO USE THIS PRODUCT IN A MANNER INCONSISTENT WITH ITS LABELING. READ THIS LABEL BEFORE USE. STRICTLY OBSERVE LABEL DIRECTIONS AND CAUTIONS, AND APPLICABLE FEDERAL AND STATE REGULATIONS.

DO NOT USE ON ANY CROP NOT LISTED ON THIS LABEL OR SUPPLEMENTAL LABELING AS ANY RESIDUES REMAINING MAY BE ILLEGAL AND/OR HARMFUL.

GENERAL INFORMATION

SEVIN® brand 50-W Carbaryl Insecticide is a dry powder for dispersion in water and application as an insecticide spray.

GENERAL WORKER PROTECTION STATEMENTS

Do not apply this product in such a manner as to directly or through drift expose workers or other persons. The area treated must be vacated by unprotected persons.

Do not enter treated areas without protective clothing until sprays have dried.

Because certain states may require more restrictive reentry intervals for various crops treated with this product, consult your State Department of Agriculture for further information.

Written or oral warnings must be given to workers who are expected to be in treated area or in area about to be treated with this product. Advise workers to stay out of fields during application and until sprays have dried. Regular long-sleeved work clothing should be worn when working in treated fields. See PRECAUTIONARY STATEMENTS. STATEMENT OF PRACTICAL TREATMENT and NOTE TO PHYSICIAN for additional information. When oral warnings are given, warnings shall be given in a language customarily understood by workers. Oral warnings must be given if there is reason to believe that written warnings cannot be understood by workers. Written warnings must include the following information: Appropriate signal word (WARNING), area treated with SEVIN® brand 50-W Carbaryl Insecticide, date of application, appropriate clothing, and reentry interval (i.e., until sprays have dried).

DIRECTIONS FOR USE THROUGH SPRINKLER IRRIGATION SYSTEMS

Apply this product only through sprinkler irrigation systems including center pivot and solid set. Do not apply this product through any other type of irrigation system.

SPRAY PREPARATION: First prepare a suspension of SEVIN® brand 50W Carbaryl Insecticide in a mix tank. Fill tank with 1/2 to 3/4 the desired amount of water. Start mechanical or hydraulic agitation. Add the required amount of SEVIN® brand 50W, and then the remaining volume of water. (Suspension concentrations using the appropriate dosage per acre recommended on this label of SEVIN® brand 50W per 1 to 4 gallons of water are recommended). Then sets sprinkler to deliver 0.1 to 0.3 inch of water per acre. Start sprinkler and uniformly inject the suspension of SEVIN® brand 50W into the irrigation water line so as to deliver the desired rate per acre. The suspension of SEVIN® brand 50W should be injected with a positive displacement pump into the main line ahead of a right angle turn to insure adequate mixing. If you should have any other questions about calibration, you should contact State Extension Service specialists, equipment manufacturers or other experts.

NOTE: When treatment with SEVIN® brand 50W has been completed, further field irrigation over the treated area should be avoided for 24 to 48 hours to prevent washing the chemical off the crop.

GENERAL PRECAUTIONS FOR APPLICATIONS THROUGH SPRINKLER IRRIGATION SYSTEMS

Maintain continuous agitation in mix tank during mixing and application to assure a uniform suspension.

Greater accuracy in calibration and distribution will be achieved by injecting a larger volume of a more dilute solution per unit time.

The system must contain a functional check valve, vacuum relief valve, and low pressure drain appropriately located on the irrigation pipeline to prevent water source contamination from backflow. The pesticide injection pipeline must contain a functional automatic, quick-closing check valve to prevent the flow of fluid back toward the injection pump. The pesticide injection pipeline must also contain a functional, normally closed solenoid-operated valve located on the intake side of the injection pump and connected to the system interlock to prevent fluid from being withdrawn from the supply tank when the irrigation system is either automatically or manually shutdown. The system must contain functional interlocking controls to automatically shut off the pesticide injection pump when the water pump motor stops. The irrigation line or water pump must include a functional pressure switch which will stop the water pump motor when the water pressure decreases to the point where pesticide distribution is adversely affected. Systems must use a metering pump, such as a positive displacement injection pump (e.g., diaphragm pump) effectively designed and constructed of materials that are compatible with pesticides and capable of being fitted with a system interlock. Do not apply when wind speed favors drift beyond the area intended for treatment.

Do not apply when wind speed favors drift, when system connection or fittings leak, when nozzles do not provide uniform distribution or when lines containing the product must be dismantled and drained.

Crop injury, lack of effectiveness, or illegal pesticide residues in the crop may result from non-uniform distribution of treated water.

Allow sufficient time for pesticide to be flushed through all lines and all nozzles before turning off irrigation water. A person knowledgeable of the chemigation system and responsible for its operation shall shut the system down and make necessary adjustments should the need arise.

Do not connect an irrigation system (including greenhouse systems) used for pesticide application to a public water system unless the label-prescribed safety devices for public water supplies are in place.

PRE-HARVEST AND GRAZING USE INFORMATION AND LIMITATIONS

Tolerances established under the Federal Food, Drug and Cosmetic Act permit the sale of crops bearing probable carbaryl residues when carbaryl is used in accordance with label directions.

If used as directed, treated forage may be grazed or used as feed for dairy and meat animals without resulting in illigal residues in meat and milk. This product may be applied up to and including the day of harvest or grazing of forage crops (except alfalfa). Application may be made without removing livestock from the area being treated. Do not apply at greater rates or at more frequent intervals than is stated on the label. To do so may result in illegal residues in crops, meat and milk.

Do not plant rotational food and feed crops not listed on this or other carbaryl labels in carbaryl treated soils.

Do not use reclaimed irrigation water from crops treated with carbaryl on crops for which carbaryl tolerances are not established.

PLANT RESPONSE PRECAUTIONS

To avoid possible injury on tender foliage, do not apply to wet foliage or during periods of high humidity.

Do not use on Boston Ivy, Virginia creeper and maidenhair fern as injury will result. Carbaryl may also injure Virginia and sand pines.

Carefully observe label instructions for apple thinning to avoid excessive thinning. Combinations with certain herbicides on rice and soybeans may be phytotoxic. Refer to specific directions for the appropriate crop.

SPRAY PREPARATION

TO ASSURE A UNIFORM SUSPENSION, AGITATE, STIR OR RECIRCULATE ALL "SEVIN®" BRAND 50-W CARBARYL INSECTICIDE SPRAY MIXTURES PRIOR TO AND DURING USE. Remove oil, rust, scale, pesticide residues and other foreign matter from mix tanks and entire spray system. Flush with clear water. Fill spray or mix tank with 1/2 to 3/4 the desired amount of water. Start mechanical or hydraulic agitation. Slowly add the required amount of SEVIN® brand 50-W Carbaryl Insecticide and then the remaining volume of water. Include rinse water from the container. Prepare only as much spray mixture as can be applied on the day of mixing. MAINTAIN CONTINUOUS AGITATION DURING MIXING AND APPLICATION TO ASSURE A UNIFORM SUSPENSION. DO NOT STORE SPRAY MIXTURE FOR PROLONGED PERIODS OR DEGRADATION OF CARBARYL MAY OCCUR. Local water conditions may accelerate the degradation of spray mixtures containing carbaryl. See COMPATABILITY STATEMENT below.

COMPATIBILITY

SEVIN® brand 50-W Carbaryl Insecticide has been effectively used without plant injury with most common insecticides, miticides, fungicides, nutrients, adjuvants, and with petroleum oil as used on citrus. If compatibiity with another product and the resulting crop response is unknown, it should be tested on a small scale. Curdling, precipitation, greasing, layer formation or increased viscosity are symptoms of incompatibility. Incompatibility will reduce insect control and may cause mechanical difficulties or plant injury. Observe all cautions and limitations on labeling of all products used in mixtures. WHEN PREPARING COMBINATION SPRAYS, FIRST ADD "SEVIN®" BRAND 50-W CARBARYL INSECTICIDE TO 1/2 THE DESIRED AMOUNT OF WATER, MIX THOROUGHLY, AND THEN ADD COMBINATION PRODUCTS TO THE MIXTURE. DO NOT APPLY TANK MIX COMBINATIONS UNLESS YOUR PREVIOUS EXPERIENCE INDICATES THE MIXTURE IS EFFECTIVE AND WILL NOT RESULT IN APPLICATION PROBLEMS, EXCESSIVE RESIDUES OR PLANT INJURY.

Carbaryl is unstable under highly alkaline conditions and mixtures with strong bases such as Bordeaux, lime-sulfur and casein-lime spreaders, will result in chemical degradation of the insecticide. Do not use this product in water wih pH values above 8.0 unless a buffer is added. If necessary, water should be buffered to neutral (pH=7.0) before adding this product to the spray tank. Overhead irrigation with alkaline or muddy water after application will also accelerate chemical degradation and may result in reduced insect control.

APPLICATION

On all crops use sufficient gallonage to obtain thorough and uniform coverage. Observe crop label instructions for specific directions regarding spray volume where they occur. Calibrate spray equipment to deliver the required volume. Use 50 mesh slotted strainers in spray system and 25 mesh slotted strainers behind nozzles. To clean spray system after use, drain and flush with water and detergent mixture. Refer to the Storage and Disposal directions for disposal instructions.

INSECT CONTROL

Begin application when insect populations reach recognized economic threshold levels. Consult the Cooperative Extension Service, Professional Consultants or other qualified authorities to determine appropriate threshold levels for treatment in your area. Where a dosage range is indicated, use the lower rate on light to moderate infestations, young plants and early instars and the higher rate on heavy infestations, mature plants, advanced instars and adults. Thorough and uniform spray coverage is essential for effective control.

NOTE: All references to aryworm on the crops listed below refer to the species. *Pseudaletia unipuncta*, often called the "true armyworm." Except where indicated otherwise, this product is not registered for the control of other armyworm species. Regional differences have been noted in the susceptibility of certain strains of fall armyworm. Colorado potato beetle, spotted tentiform leafminer, and tobacco budworm (on cotton) to carbaryl. If local experience indicates inadequate control, use an alternative pesticide.

FORAGE, FIELD AND VEGETABLE CROP INSECT CONTROL

Recommended dosages refer to pounds of SEVIN® brand 50-W Carbaryl Insecticide per acre. Where a dosage range is indicated, use lower rate on young plants, higher rate on mature plants. Use sufficient spray gallonage to obtain full coverage. This will usually vary from 3 to 40 gallons per acre depending on equipment, degree of insect infestation, and stage of crop growth. Keep spray mixture well agitated. Apply when insects or their damage appear and repeat at 7 to 10 day intervals or as necessary unless shorter interval is specified below. See specific directions below for grasshopper control.

GRASSHOPPERS

Apply 1 to 3 pounds as often as necessary. TO CONTROL GRASSHOPPERS ON THE FORAGE FIELD AND VEGETABLE CROPS, FOLLOW PREHARVEST LIMITATIONS GIVEN AFTER EACH CROP. Use 1 to 1½ pounds for nymphs on small plants or sparse vegetation in wasteland, rangeland, ditchbanks and borders. Use 2 to 3 pounds for mature grasshoppers, applications to dense vegetation or if extended residual control is desired. Be certain spray volumes are appropriate to assure adequate coverage.

ALFALFA, CLOVERS AND BIRDSFOOT TREFOIL (DO NOT USE ON BIRDSFOOT TREFOIL IN CALIFORNIA)

Use 1 to 2 pounds for blister beetles and Mexican bean beetle.

Use 2 pounds for alfalfa caterpillar, bean leaf beetle, cucumber beetles green cloverworm, Japanese beetle, leafhoppers, three-cornered alfalfa hopper, thrips, and velvetbean caterpillar.

Use 1 to 3 pounds for grasshopper control. Refer to the general GRASSHOPPER heading above.

Use 2 to 3 pounds for alfalfa blotch leafminer, armyworm, cloverhead weevil, corn earworm, cutworms, Egyptian alfalfa weevil larvae, essex skipper, European alfalfa beetle, fall armyworm, lygus bugs, stink bugs, webworms, and yellowstriped armyworm.

Use 2 to 3 pounds for alfalfa weevil larvae west of the Rocky Mountains and 3 pounds east of the Rocky Mountains. For alfalfa weevil larvae, if pretreatment damage is extensive, cut alfalfa and treat the stubble.

OBSERVE BEE CAUTION

Observe plant response precautions.

On dense growth, use 20 to 40 gallons per acre with ground equipment to ensure adequate coverage.

Carbaryl may cause temporary bleaching of tender alfalfa foliage.

Apply only once per cutting.

Do not apply within 7 days of harvest.

ASPARAGUS

Use 2 to 4 pounds for asparagus beetle on seedlings or spears.

Use 4 to 8 pounds for asparagus beetle and Apache cicada on ferns or brush growth in the post-harvest period.

Do not apply more often than once every 3 days.

Do not apply within 1 day of harvest.

BEANS (including blackeyed peas, cowpeas, crowder or southern peas, dry beans, green beans, lima beans, navy beans and snap beans)

LENTILS (DO NOT USE ON LENTILS IN CALIFORNIA)

Use 1 to 2 pounds for blister beetles and Mexican bean beetle.

Use 2 pounds for alfalfa caterpillar, bean leaf beetle, cucumber beetles, flea beetles, green cloverworm, Japanese beetle, leafhoppers, three-cornered alfalfa hopper, thrips, velvetbean caterpillar and western bean cutworm.

Use 1 to 3 pounds for grasshopper control. Refer to the general GRASSHOPPER heading above.

Use 2 to 3 pounds for armyworm, corn earworm, cutworms, European corn borer, fall armyworm, stink bugs, tarnished plant bug, and webworms.

Use 3 pounds for alfalfa looper.

Use 4 pounds for cowpea curculio.

In California, use 4 pounds for corn earworm, limabean pod borer, lygus bugs, and stink bugs.

Observe plant response precautions.

Do not appy to cowpeas within 3 days of harvest or to lentils within 0 days of harvest.

CABBAGE, BROCCOLI, BRUSSELS SPROUTS, CAULIFLOWER AND KOHLRABI

Use 1 to 2 pounds for flea beetles and harlequin bug.

Use 2 to 4 pounds for armyworm, fall armyworm, diamondback moth, imported cabbageworm and corn earworm.

Do not apply within 3 days of harvest.

CHINESE CABBAGE, COLLARDS, HANOVER SALAD, HORSERADISH, KALE, MUSTARD GREENS, RADISHES, RUTABAGAS AND TURNIPS.

Use 1 to 2 pounds for flea beetles, harlequin bug and leafhoppers.

Use 2 to 3 pounds for aster leafhopper.

Use 2 to 4 pounds for armyworm, fall armyworm, imported cabbageworm, corn earworm, tarnished plant bug and stink bugs.

Do not apply within 3 days of harvest of horseradish, radishes, rutabagas and turnips (roots).

Do not apply within 14 days of harvest of Chinese cabbage, collards, hanover salad, kale, mustard greens and turnips (tops).

CARROTS, CELERY, PARSNIPS AND PARSLEY (DO NOT USE ON CELERY IN CALIFORNIA)

Use 1 to 2 pounds for flea beetles and leafhoppers

Use 2 to 3 pounds in a 5 to 7 day schedule for aster leafhopper, lygus bugs, and spittlebugs.

Use 2 to 4 pounds for armyworm, fall armyworm, corn earworm, tarnished plant bug and stink bugs.

Treat on a 5 to 7 day schedule.

No time limitation on carrots.

Do not apply within 3 days of harvest of parsnips.

Do not apply within 14 days of harvest of parsley and celery.

CORN (FIELD, SWEET AND POP)

Use 2 to 4 pounds for armyworm, chinch bugs, corn earworm, corn rootworm adults, fall armyworm, flea beetles, Japanese beetle, sap beetles, Southwestern corn borer, and leafhoppers.
OBSERVE BEE CAUTION.

For insects attacking silks and ears, apply at 1 to 6 day intervals starting when first silks appear and continuing until silks begin to dry. For larvae in whorl and foliage feeders, apply as necessary. Optimum timing and good coverage are essential for effective control. For optimum chinch bug control, apply at least 20 gallons of water per acre by ground and direct spray toward stalk to provide thorough coverage.

Use 1 to 3 pounds for grasshopper control. Refer to the general GRASSHOPPER heading above.

Use 3 to 4 pounds for European corn borer control. For optimum control, do not apply in less than 3 gallons of water per acre by air and 15 gallons of water per acre by ground.

Use 4 to 6 pounds for cutworms. For optimum control, apply in a 12 inch band over the row, using sufficient volume of water to obtain thorough coverage. For broadcast application use at least 20 gallons (ground) or 5 gallons (air) of water per acre.

COTTON

Use 1 to 2 pounds for early season control of cotton fleahopper, cotton leafworm, flea beetles, striped blister beetle, and thrips.

Use 2 to 4 pounds for mid and late season control of boll weevil, cotton bollworms, fall armyworm, leafrollers, leafhoppers, tarnished plant bug, and yellowstriped armyworm (cotton cutworm). Treat on a 5 to 7 day schedule for as long as necessary. May be applied after bolls open.

Use 2 to 4 pounds for light to moderate populations of lygus bugs in Western irrigated cotton.

Use 3 to 5 pounds for pink bollworm.

Use 4 pounds for cutworms, stink bugs (except southern green stinkbug) and saltmarsh caterpillar.

Aphid populations will be suppressed by repeated applications of this insecticide.

Do not use this product within 7 days of harvest.

CUCUMBER, MELONS, PUMPKIN AND SQUASH

Use 1 to 2 pounds for pickleworm and melonworm, 2 pounds for cucumber beetles, flea beetles, leafhoppers and squash bug.

Observe plant response precautions.

DANDELION, ENDIVE (ESCAROLE), LETTUCE AND SALSIFY

Use 1 to 2 pounds for flea beetles, harlequin bug and leafhoppers.

Use 2 to 3 pounds for aster leafhoppers, lygus bugs, and spittlebugs.

Use 2 to 4 pounds for armyworm, fall armyworm, imported cabbageworm, corn earworm, tarnished plant bug and stink bugs.

Observe plant response precautions. Treat on a 5 to 7 day schedule after heads begin to form.

Do not apply within 3 days of harvest of head lettuce and salsify (roots).

Do not apply within 14 days of harvest of dandelion, endive (escarole), leaf lettuce and salsify (tops).

FLAX AND PROSO MILLET (DO NOT USE IN CALIFORNIA)

Use 2 to 3 pounds for armyworm control.

Use 1 to 3 pounds for grasshopper control. Refer to the general GRASSHOPPER heading above.

Do not graze treated areas or harvest for dairy feed prior to crop maturity. Do not apply within 42 days of harvest.

GARDEN BEET, SPINACH AND SWISS CHARD

Use 1 to 2 pounds for the flea beetles, harlequin bug and leafhoppers.

Use 2 to 3 pounds for aster leafhopper.

Use 2 to 4 pounds for armyworm, corn earworm, fall armyworm, tarnished plant bug and stink bugs.

Treat on a 5 to 7 day schedule.

Do not apply within 3 days of harvest of garden beets (roots).

Do not apply within 14 days of harvest of garden beets (tops), spinach and Swiss chard.

OKRA

Use 2 to 4 pounds for corn earworm and stink bugs. Treat on a 5 to 7 day schedule.

PASTURE

Use 2 to 3 pounds for armyworm, blackgrass bugs, chinch bugs, essex skipper, fall armyworm, range caterpillar, range crane fly, striped grass looper, and thrips. Use the lower rate for nymphs and immature insects on small plants or sparse vegetation. Apply the higher rate for adults or when vegetation is thick. For thrips control in grasses grown for seed, use high spray pressure to improve penetration into boot.
Use 1 to 3 pounds for grasshoppers. Refer to the general GRASSHOPPER heading above.

Apply a maximum of 2 applications per year. Allow at least 14 days between applications. Do not allow foraging or cut for hay within 14 days of last application by ground. Aerially treated pastures may be grazed or cut for hay on day of treatment.

Carefully mark swaths to avoid over-application.

PEANUTS

Use 1 to 2 pounds for blister beetles and Mexican bean beetle.
Use 2 pounds for alfalfa caterpillar, bean leaf beetle, cucumber beetles, green cloverworm, Japanese beetle, leafhoppers, red-necked peanutworm, three-cornered alfalfa hopper, thrips and velvetbean caterpillar.

Use 2 to 3 pounds for armyworm, fall armyworm, corn earworm, stink bugs and webworms.

Use 4 pounds for white fringed beetle adults and cutworms.

Observe plant response precautions.

PEAS

Use 2 pounds for leafhoppers and Colorado potato beetle.

Use 2 to 3 pounds for armyworm.

Use 3 pounds for alfalfa caterpillar, cutworms, pea leaf weevil, pea weevil, and yellow-striped armyworm.

Use 5 pounds for alfalfa looper in Washington State only.

Do not apply within 3 days of harvest.

POTATO, TOMATO, EGGPLANT AND PEPPER

Use 1 to 2 pounds for Colorado potato beetle, flea beetles and leafhoppers.

Use 2 to 4 pounds for European corn borer, fall armyworm, lace bugs, stink bugs, (suppression), tomato fruitworm, tomato hornworm, tomato pinworm and tarnished plant bug.

Use 4 pounds for cutworm control.

Thorough coverage is essential to effectively control stink bugs. When disease transmission is suspected, monitor fields following application and re-treat if reinfestation occurs.

RICE

CAUTION: May kill shrimp and crabs. Do not use in areas where these are important resources. Do not use on rice fields in which crayfish and/or catfish farming are included in the cultural practice.

In the Mississippi Delta and Texas, use 2 to 3 pounds for armyworm, fall armyworm, chinch bugs, and stink bugs.

DO NOT APPLY PROPANIL HERBICIDES WITHIN 15 DAYS BEFORE OR AFTER APPLICATION OF THIS PRODUCT OR PLANT INJURY WILL RESULT

Use 1 to 3 pounds for grasshoppers. Refer to the general GRASSHOPPER heading above.

In California use 4 pounds for armyworm, leafhoppers and tadpole shrimp. For optimum tadpole shrimp control apply to the water when the pest first appears.

Do not apply within 14 days of harvest.

RANGELAND (FOR AERIAL APPLICATION ONLY)

Use 1 to $1^{1}/_{2}$ pounds for black grass bug, mormon cricket, range caterpillar, and range crane fly. Use the lower rate for nymphs and immature insects on small plants or sparse vegetation. Apply the higher rate for adults or when vegetation is thin.

Use 1 to 3 pounds for grasshoppers. Refer to the general GRASSHOPPER heading above.

Apply a maximum of 2 applications per year. Allow at least 14 days between application.

Carefully mark swaths to avoid over-application.

SORGHUMS (milo, grain sorghum, sweet sorghum, and hybrids)

Use $1^{1}/_{2}$ to 2 pounds for sorghum midge control. Treat for sorghum midge when 25 to 30 percent of heads have emerged from boot and are in bloom. Repeat 3 to 5 days later if adults are still active. A third application may be necessary.
Use 2 to 4 pounds for armyworm, chinch bugs, corn earworm, fall armyworm, stink bugs (suppression of green stinkbugs), and webworms. For chinch bugs use high gallonage ground application directed to base of plants.

Use 3 pounds for Southwestern corn borer.

Use 4 pounds for cutworm control.

Use 1 to 3 pounds for grasshoppers. Refer to the general GRASSHOPPER heading above.

Direct spray into forming heads for optimum insect control.

Do not apply within 21 days of harvest of grain. No time limitations on forage.

SOYBEANS

DO NOT APPLY A COMBINATION OF THIS PRODUCT AND 2,4-DB HERBICIDES TO SOYBEANS AS CROP INJURY MAY RESULT

Use 1 to 2 pounds for bean leaf beetle, blister beetle, cucumber beetle, grape colaspis, green cloverworm, Japanese beetle, Mexican bean beetle and velvetbean caterpillar.

Use 1 to 3 pounds for corn earworm control.

Use 1 to 3 pounds for grasshoppers. Refer to the general GRASSHOPPER heading above.

Use 2 pounds for alfalfa caterpillar, leafhoppers, three cornered alfalfa hopper, and thrips.

Use 2 to 3 pounds for armyworm, cutworm, fall armyworms, and webworms.

Use 3 to 4 pounds for painted lady (thistle caterpillar), saltmarsh caterpillar, woolly bear caterpillar, and yellow striped armyworm.

Use lower rates for light to moderate populations and smaller instars and to provide maximum survival of beneficial insects and spiders. Use the higher rates for heavy infestations and larger instars.

SUGAR BEETS

Use 2 to 3 pounds for armyworms, beet leaf beetle, fall armyworm, flea beetles, leafhoppers and webworms.

Use 1 to 3 pounds for grasshoppers. Refer to the general GRASSHOPPER heading above.

Use 3 pounds for cutworms.

Do not apply within 14 days of harvest.

SUNFLOWER (DO NOT USE IN CALIFORNIA)

Use 3 pounds for cutworm control.

Use 3 to 4 pounds on armyworm, fall armyworm and sunflower moth.

Use 2 to 4 pounds on stem weevil and sunflower beetle.

Use 1 to 3 pounds for grasshoppers. Refer to the general GRASSHOPPER heading above.

Do not apply within 60 days of harvest.

SWEET POTATOES (DO NOT USE IN CALIFORNIA)

Apply as a foliar spray as needed.

Use 2 to 4 pounds for corn earworm, cucumber beetles, flea beatles, sweet potato hornworm, sweet potato weevil, and tortoise beetles.

For sweet potato weevil, full coverage of plants is essential. Use the lower rate on young plants and the higher rate on mature plants.

Use 4 pounds for yellowstriped armyworm control.

TOBACCO

PLANT BED TREATMENTS: For flea beetle, use 4 level tablespoonfuls per gallon or 2 pounds per 50 gallons, and apply 6 gallons of water per 100 square yards.

For green June beetle grubs, use 1 pound per 100 gallons of water when insects or their damage appear. Apply only to areas that larvae have uprooted by sprinkling as a drench with 50 to 100 gallons per 100 square yards. May be applied before or after seeding. Avoid excessive application as plant injury may result.

FIELD TREATMENT: For budworms, fall armyworm, flea beetles, Japanese beetle, June beetles, hornworms and suckfly, use 2 to 4 pounds. Use lower rate on young plants up to knee height. Use at least 10 gallons of prepared spray per acre. Begin treatments when worms are small. Avoid excessive application in bud of plant as injury may result.

WASTELANDS, RIGHTS-OF-WAY, HEDGEROWS, DITCHBANKS, ROADSIDES INSECT CONTROL Use 1 to 2 pounds for black grass bugs, mormon cricket, range caterpillars and range crane fly. Use the lower rate for nymphs and immature insects on small plants or sparse vegetation. Apply the higher rate for adults or when vegetation is thick.

Use 1 to 3 pounds for grasshoppers. Refer to the general GRASSHOPPER heading above.

Apply a maximum of 2 applications per year. Allow at least 14 days between applications. Do not allow foraging or cut for hay within 14 days of last application by ground. Aerially treated area may be grazed or cut for hay on day of treatment.

Carefully mark swaths to avoid over-application.

WHEAT (INCLUDING TRITICALE) (DO NOT USE IN CALIFORNIA)

Use 1 to 2 pounds for fleabeetle control.

Use 1 to 3 pounds for grasshoppers. Refer to the general GRASSHOPPER heading above.

Use 2 pounds on cereal leaf beetle. Application is effective against eggs, larvae and adult of the cereal leaf beetle.

Use 2 to 3 pounds on armyworm and fall armyworm.

Do not apply within 21 days of harvest of grain. No time limitations on forage.

TREE FRUIT AND NUT INSECT CONTROL

Recommended dosages refer to pounds of SEVIN® brand 50-W Carbaryl Insecticide per 100 gallons of dilute spray in conventional hydraulic-type sprayers. Apply dilute sprays for full coverage. Spray gallonage will vary according to tree size, density, planting distance and stage of growth. With concentrate spray, where lower volumes of spray are applied per acre, the concentration of SEVIN® brand 50-W Carbaryl Insecticide should be increased in the spray mixture so that the amount of SEVIN® brand 50-W Carbaryl Insecticide applied per acre is equivalent to a full-coverage dilute spray. Keep spray mixtures well agitated. Additional instructions appear with the crop categories below.

APPLE THINNING

OBSERVE BEE CAUTION.

Apply 1 full coverage dilute spray between 10 and 25 days after full bloom. Factors such as tree age, variety, nutrition, previous crop, pruning, bloom and degree of set favor execessive fruit thinning with this product. Exercise caution to avoid possible yield reduction. Rates may vary depending on variety and local orchard conditions.

Consult with your County Extention Service or other experts for advice on the proper use of this product.

In Eastern apple growing areas, tank mix combinations of carbaryl and naphthaleneacetic acid (NAA) or naphthaleneacetamide (NAD) have successfully thinned several early-maturing, heavy-setting varieties, as well as hard-to-thin varieties such as Golden Delicious and Rhode Island Greening. The higher rate of carbaryl and reduced rates of NAA or NAD are recommended for the combination. Also, a petal fall application of NAA or NAD followed 7 to 10 days later by an application of carbaryl has improved thinning on these varieties.

Concentration: Use 1/2 to 1 pound on easily thinned varieties. On hard-to-thin varieties, use 1 to 2 pounds.

Varietal Response: Easily-thinned varieties include Cortland, Grimes, Jonathan, McIntosh, Orleans, Rome Beauty, Puritan, Red Delicious, Winesap, and Yellow Newton. Difficult-to-thin varieties include Baldwin, Ben Davis, Dutchess, Early McIntosh, Golden Delicious, Lady Apple, Northern Spy, Rhode Island Greening, Steele Red, Turley, Wealthy, Yellow Transparent, and York Imperial.

ALMOND

OBSERVE BEE CAUTIONS.

Use 2 pounds for peach twig borer, San Jose scale and fruittree leafroller. Apply in "popcorn" or petal fall stages and again when the May brood of the peach twig borer begins to hatch or thereafter as needed.

Use 2 pounds for navel orangeworm. Time early and mid season applications to correspond to moth flight peaks. Make late season application at initiation of or up to 10% hull split.

Do not apply more than 10 pounds per acre.

Do not apply within 28 days of harvest.

AVOCADOS (FOR CALIFORNIA USE ONLY) FOR USE IN JAPANESE BEETLE AND GYPSY MOTH ERADICATION PROGRAMS IN CALIFORNIA ONLY.

Use 1 pound for Japanese beetle and gypsy moth control only. Use with ground equipment and spray to run off. Apply when insects or their damage appear. Do not make more than 1 application/season.

Do not apply within 1 day of harvest.

CHESTNUTS—DO NOT USE IN CALIFORNIA

Use 4 to 6 pounds for chestnut weevil.

Do not apply within 1 day of harvest.

Make 4 applications at weekly intervals beginning in late July for adult chestnut weevil control. Last application should be made prior to shuck split.

CITRUS FRUITS (such as grapefruit, lemons, limes, oranges, tangelos, tangerines, citrus citron, kumquats and hybrids).

Use 2 pounds for avocado leafroller, California orangedog, citrus cutworm, citrus root weevil, fruittree leafroller, orange tortrix, Western tussock moth and West Indian sugarcane borer (adults).

Use 1 1/2 to 2 pounds for black scale, brown soft scale, California red scale, citricola scale, citrus scale and yellow scale.

OBSERVE BEE CAUTION.

Do not apply more than 40 pounds of this product per acre per application.

To insure thorough coverage, do not apply less than 10 gallons of dilute spray mixture per mature tree.

May be mixed with petroleum oils commonly used on citrus.

Apply dilute sprays in 300 to 500 gallons of water.

Do not apply less than 10 gallons of dilute spray mixture per mature tree.

Do not apply within 5 days of harvest of these crops.

FILBERT

Use 2 pounds for filbert aphid, filbert leafroller and fibertworm. Apply when leafroller eggs are hatching. Repeat on first appearance of adult filbert moths and again 3 to 4 weeks later. Apply dilute sprays in 300 to 400 gallons of water per acre.

OLIVES

Use 1½ to 2 pounds for olive scale control. For optimum scale control add 1½ gallons of summer oil and apply mixture when crawlers are present. Do not exceed 2 applications per year.

Do not apply more than 30 pounds of this product per acre per application.

PEACHES, APRICOTS AND NECTARINES

Use 2 pounds for apple pandemis, codling moth, cucumber beetles, European earwig, Japanese beetle, June beetles, lecanium scale, lesser peachtree borer, orange tortrix, oriental fruit moth, peach twig borer, periodical cicada, plum curculio, redbanded leafroller, variegated leafroller, tarnished plant bug, tussock moths, San Jose scale and olive scale. For optimum scale control, apply when crawlers are present. Spray limbs and trunk thoroughly at weekly intervals during moth flight.

Do not apply more than 12 pounds of this product per acre per application on apricots.

Apply dilute sprays in 200 to 400 gallons of water per acre.

Do not apply within 1 day of harvest of peaches.

Do not apply within 3 days of harvest of apricots and nectarines.

PECANS

Use 2 to 5 pounds for black margined aphid, fall webworm, hickory shuckworm, lesser webworm, pecan leaf phylloxera, pecan nut casebearer, pecan spittlebug, pecan weevil, twig girdler, walnut caterpillar.

Do not apply more than 14⅖ pounds of this product per acre per application.

Apply dilute sprays in 200 to 500 gallons of water per acre.

PISTACHIOS

Use 1 to 4 pounds for navel orangeworm control. Apply dilute volumes of 150 to 300 gallons of mixed spray per acre for full coverage at onset of hull split. Do not use within 14 days of harvest.

PLUMS, PRUNES AND CHERRIES

Use 1½ pounds for eastern tent caterpillar, codling moth, orange tortrix, and tussock moths.

Use 2 pounds for black cherry aphid, brown soft scale, cherry maggots, cherry fruitworm, eyespotted bud moth, fruittree leafroller, redbanded leafroller, gypsy moth, Japanese beetle, lesser peachtree borer, peach twig borer, plum curculio, prune leafhopper, European earwig, green fruitworm, rose chafer, variegated leafroller, Forbes scale, lecanium scales, mealy plum aphid, oystershell scale, and San Jose scale. For optimum scale control, apply when crawlers are present. For lesser peachtree borer control spray limbs and trunk thoroughly, weekly during moth flight.

Do not apply more than 12 pounds of this product per acre per application.

Apply dilute sprays in 200 to 400 gallons of water per acre.

Do not apply within 1 day of harvest of these crops.

POME FRUITS: (including apples, pears, loquats, crabapples, oriental pears, and quince)

Application of this product within 30 days after full bloom may also provide apple thinning. To avoid this, delay use until at least 30 days after bloom.

Use 1½ to 2 pounds for apple aphid, apple maggot, apple rust mite, apple sucker, bagworms, California pearslug (Pear sawfly), Eastern tent caterpillar, European apple sawfly, eyespotted bud moth, Forbes scale, fruittree leafroller, green fruitworm, gypsy moth, lygus bugs, orange tortrix, oystershell scale, pear leaf blister mite, pear psylla, pear rust mite, periodical cicada, plum curculio, redbanded leafroller, rosy apple aphid, San Jose scale, tarnished plant bug, tentiform leafminers, wooly apple aphid, yellowheaded fireworm control. For psylla control apply when eggs hatch or young nymphs are present. To control scale insects, apply when crawlers are present.

Use 1 pound for apple mealybug, apple aphid, codling moth, white apple leafhopper.

Use 1¼ pounds for Japanese beetle. Apply as necessary with ground equipment at 7-14 day intervals.

Apply dilute sprays in 200 to 400 gallons of water per acre.

Do not apply within 1 day of harvest of apples and pears.

WALNUT

Use 1 pound for calico scale, codling moth, European fruit lecanium, filbertworm, fruittree leafroller, and frosted scale. Apply 1000 gallons of diluted spray per acre for mature trees. For codling moth, apply first spray when average cross-sectional diameters of developing nuts are ½ to ¾ inch. Repeat during middle or late June as needed.

Apply dilute sprays in 200 to 500 gallons of water per acre.

Use 4 pounds for European earwig. Spray tree trunks to point of run-off.

SMALL FRUIT INSECT CONTROL

Recommended dosage refers to pounds of SEVIN® brand 50-W Carbaryl Insecticide per acre. Keep spray mixtures well agitated. The optimum spray gallonage will depend on plant size, density and stage of growth. Typical spray gallonage per acre ranges from 100 to 300 gallons for dilute sprays, and 30 to 100 gallons for concentrate sprays and 5 to 25 gallons for aerial sprays. Do not exceed maximum label rate per acre per application.

BLACKBERRIES, RASPBERRIES AND DEWBERRIES (including Boysenberries and Loganberries).

Use 2 to 4 pounds for Japanese beetle, leafrollers, leafhoppers, European raspberry aphid, rose chafer, snowy tree cricket.

In California, use 4 pounds for omnivorous leafroller and raspberry sawfly.

Do not apply within 7 days of harvest of these crops.

BLUEBERRIES: Use 3 pounds for blueberry maggot, cherry and cranberry fruitworms, European fruit lecanium and Japanese beetle. Apply 3 weeks before harvest and repeat as necessary.

CRANBERRIES: Use 3 to 6 pounds for cutworms, cranberry fireworms, cranberry fruitworm, cranberry twig girdler, elm span, Japanese beetle, leafhoppers, rose chafer, and spaganothus worm. Apply as needed through prebloom at 7 to 10 day intervals. Resume applications as needed beginning at late bloom.

OBSERVE BEE CAUTION.

Do not apply within 1 day of harvest.

CAUTION: May kill shrimp and crabs. Do not use in areas where these are important resources.

GRAPES: Use 2 to 4 pounds for European fruit lecanium, grape leaffolder, grape leafhopper, western grapeleaf skeletonizer, and Western yellowstriped armyworm.

Use 4 pounds for grape berry moth, Japanese beetle, June beetles, orange tortrix, omnivorous leafroller, redbanded leafroller, cutworms, eight spotted forester, and saltmarsh caterpillar. Apply just before first brood leaffolder larvae emerge from rolls.

Do not apply within 7 days of harvest.

STRAWBERRIES: Use 2 to 4 pounds for flea beetles, meadow spittlebug, omnivorous leaftier (strawberry fruitworm), strawberry clipper, strawberry bud weavil, strawberry leafroller and strawberry weevil. Carbaryl may injure Early Dawn and Sunrise varieties. Do not apply within 1 day of harvest.

NON-RESIDENTIAL TREES AND ORNAMENTAL INSECT CONTROL

For dilute spray ground applications to nonresidential trees (including shade trees, sugar maple, shelter belts, forests, plantations, parks, and recreational areas), nonresidential ornamentals (including roses, flowers and other herbaceous plants), woody plants and shrubs, apply the specified dosage per 100 gallons of water. For concentrate-spray ground applications, apply the specified dosage per acre in sufficient spray volume to provide thorough coverage. Avoid direct application to lakes, streams and ponds.

Use 2 pounds for control of:

Ants	European pine	Mealybugs	Roseslug
Apple aphid	shoot moth	Mimosa webworm	Saddled prominent
Armyworm	Fall armyworm	Nantucket pine tip	Sawflies (exposed)
Azalea leafminer	Flea beetles	moth	Scale insects
Bagworms	Fuller rose beetle	Oak leafminers	Sowbugs
Birch leafminer	Gall midges	Oak leaf	Spiny elm
Blister beetle	Gall wasps	skeletonizer	caterpillar
Boxelder bug	Green striped	Oakworm complex	Springtails
Boxwood leafminer	mapleworm	Oleander caterpillar	
Brown tail moth	Grasshoppers	Olive ash borer	Spruce needleminer
Cankerworms	Gypsy moth	Orange-striped	Subtropical pine tip moth
Catalpa sphinx	Hackberry nipple-	oakworm	Tent caterpillars
Chiggers	gall maker	Orange tortrix	
	Holly bud moth	Periodical cicada	Thrips (exposed)
		Pine looper	
	Holly leafminer	Pine sawfly	Ticks
Cutworms	Jackpine	Pine spittlebug	Walnut caterpillar
	budworm		
Cypress tip moth	Japanese beetle	Pitch pine tip moth	Webworms
	Jeffrey pine	Plant bugs	Western hemlock
	needleminer	Poinsettia	looper
		hornworm	
Eastern spruce gall	June beetles	Psyllids	
adelgid	Lace bugs	Puss caterpillar	
Elm leaf aphid	Leafhoppers	Redhumped	Willow leaf
Elm leaf beetle	Leafrollers	oakworm	beetles
Elm spanworm	Locust borer	Rose aphid	Yellow popular
Eriophyid mites	Maple leafcutter	Rose chafer	weevil

USE 2 TO 4 POUNDS FOR CONTROL OF:

Balsam twig aphid	Douglas-fir	Western spruce

Cooley spruce gall adelgid	tussock moth Spruce budworm	budworm

Use the higher rate on large pest populations or advanced instars.

Apply dilute sprays to obtain thorough coverage of upper and lower leaf surfaces. To control scale insects, treat trunks, stems, and twigs in addition to plant foliage. For optimum worm control, treat when in early instars. Addition of a sticker may improve residual control.

Applications for control of Maple leafcutter on sugar maple should be made when larvae are in 2nd instar, after mining, and as cases are being formed.

Observe plant response precautions.

Use 32 pounds for control of elm bark beetle, Ips engraver beetles, mountain pine beetle, roundheaded pine beetle, spruce beetle and western pine beetle. Effective as a preventive treatment only. Repeat annually as required to prevent beetle attacks. Apply 1 gallon of spray per 50 square feet of bark prior to beetle flight or host-tree attack. Treat tree trunks from ground level up, until trunk diameter is less than 5 inches. For elm bark beetle, apply approximately 20 to 30 gallons of spray mixture for each 50 feet of elm tree for thorough coverage of all bark surfaces on trunks, limbs and twigs.

DO NOT ALLOW PUBLIC USE OF TREATED AREAS DURING APPLICATIONS OR UNTIL SPRAYS HAVE DRIED.

NON-RESIDENTIAL RECREATIONAL AREA INSECT CONTROL

For control of certain turfgrass pests, apply the recommended rates, using hose-end sprayers or pressure type equipment, in sufficient spray volume for thorough coverage and turf thatch penetration. Where a dosage range is indicated, use the higher rate on large pest populations, advanced instars and dense vegetation. Repeat treatment as necessary.

Use 3.0 ounces of SEVIN® brand 50W Carbaryl Insecticide per 1000 square feet (8 pounds per acre) of turfgrass for control of:

Ants	European chafer	June beetles	Sowbugs
Armyworm	Fall armyworm	Leafhoppers	Spittlebugs
Centipedes	Fiery skipper	Lucerne moth	Springtails
Chiggers	Grasshoppers	Millipedes	Ticks
Chinch bugs	Green June beetle grubs	Mosquitoes (adults)	Yellowstriped armyworm
Cutworms			
Earwigs			
Essex skipper			

For Armyworm, Cutworm and Fall Armyworm Control: Do not irrigate treated areas following insecticide application.

For Green June Beetle Grub Control: Make applications when grubs are feeding near the soil surface. Water or irrigate turf grasses soon after treatment.

Use 6.0 ounces of SEVIN® brand 50W Carbaryl Insecticide per 1000 square feet (16 pounds per acre) of turfgrass for control of:

Chinch bugs	White grubs (such as Japanese beetle
Blue grass billbug	Chafer beetle and
European crane fly	*Phyllophaga* spp. larvae)
Fleas	
Sod Webworm (lawn moths)	

For Chinch Bug Control: Treat entire grass area rather than just damaged areas. Irrigation of turfgrass area before treatment will aid in penetration of insecticide into turfgrass.

For European Crane Fly Control: Treatments should be applied in early spring, April 1 to April 15, or as recommended by local Agricultural Extension Service agents. Water or irrigate turf grasses thoroughly soon after treatment.

For Sod Webworm Control; Do not irrigate treated areas following insecticide application.

For White Grub Control: Applications should be made when grubs are feeding near the soil surface, usually during late March through May, or July to early September, or as recommended by local Agricultural Extension Service agents. Water or irrigate turf grasses thoroughly soon after treatment.

DO NOT ALLOW PUBLIC USE OF TREATED AREAS DURING APPLICATIONS OR UNTIL SPRAYS HAVE DRIED.

POULTRY INSECT CONTROL
CHICKENS, DUCKS, GAME BIRDS, GEESE, PIGEONS AND TURKEYS

Direct Mist Spray on Birds. Control northern fowl mite, chicken mite, lice and fleas by:

1. Misting with Electric Fog Machine. Mix 5/8 pound (10 ounces) of SEVIN® brand 50-W Carbaryl Insecticide in 1 gallon of spray. Use 1 1/2 gallons per 1000 hens in cages, on litter or slatted floor. Repeat in 4 weeks if necessary.

or

2. Spraying with knapsack or cylinder type compressed air sprayers. Mix 3/8 pound (6 ounces) of this product in 5 gallons of spray. Use 1 gallon per 100 hens in cages on litter or on slatted floor. Repeat in 4 weeks if necessary.

Direct mist spraying for chicken mite and fleas, as a supplement to spraying roosts and buildings for control of these pests.

For bedbugs, chicken mite, fleas, foul tick, lesser mealworms, lice and northern fowl mite; spray roosts and buildings with conventional power spray or knapsack equipment. For chicken mite, fleas and bedbugs, use 2 pounds per 25 gallons of water. For fowl ticks, use 8 pounds per 25 gallons of water. Spray 1 to 2 gallons per 1000 square feet of wall, bedding, litter or roost surfaces. Force spray into cracks and crevices. Repeat as needed. Ventilate while spraying. Avoid spraying of nests, eggs and feeding and watering troughs.

Do not apply to poultry and game birds or to their premises within 7 days of slaughter.

IMPORTED FIRE ANT CONTROL

Site	Pounds of "SEVIN®" BRAND 50-W/ volume of water	Specific Instructions
Cemetaries and nonresidential Recreational Areas (including turf, golf courses, and parks) Pastures, Rangeland, Forested Lands and Wastelands	3 pounds/ 100 gallons	Apply a total of 2 gallons of the diluted solution over the surface of each mound or at least 1 quart per 6 inches of mound diameter using a bucket, can or other appropriate equipment. Thoroughly wet mound and surrounding area to a 4 ft. diameter (12 sq. ft.). Do not disturb mounds prior to treatment. Pour solution from a height of about 3 ft. to give sufficient force to break mound apex and flow into ant tunnels. For best results apply in cool weather, 65-80°F, or in early morning or late evening hours. Repeat application if mound activity resumes after 10 days. Treat new mounds as they appear. Pressurized spray may disturb ants and causes migration, reducing product effectiveness. DO NOT ALLOW PUBLIC USE OF TREATED AREAS DURING APPLICATIONS OR UNTIL SPRAYS HAVE DRIED.

PEST CONTROL IN AND AROUND NONRESIDENTIAL BUILDINGS LIMITED TO PEST CONTROL OPERATORS ONLY

General Information

NOTE: Staining may occur on certain surfaces such as stucco, brick, cinder block and wood. Therefore, applications of this product to surfaces where a noticeable residue or discoloration is objectionable should be avoided. Do not apply to carpets or draperies as staining may occur. Care should also be exercised to avoid spotting of wallpaper and fabrics.

Do not use in this product in commercial food areas of food handling establishments, restaurants, or other places where food is prepared or processed. Do not use in serving areas while food is exposed.

INDOORS

ANTS/CRICKETS/FIREBATS/SILVERFISH

Mix 20 pounds in 100 gallons of water and apply as fine, low pressure (20 psi) spot spray or as crack and crevice application to areas where these pests hide, such as baseboards, storage areas, closets, around water pipes, doors and windows, behind and under refrigerators, cabinets, sinks, stoves, dishwashers, hot water heaters, the underside of shelves, drawers and similar areas. For ants, apply to ant trails, around doors and windows and other places where ants enter premises.

BEES AND WASPS

Mix 20 pounds in 100 gallons of water and thoroughly spray nest and entrance and surrounding areas where insects alight. It is generally advisable to spray the nests in the evening when the insects are less active and have returned to the nest. For best results, check nest carefully one or two days after spraying to ensure complete kill, then remove and destroy nest to prevent emergence of newly hatched insects.

BROWN DOG TICKS AND FLEAS

Mix 20 pounds in 100 gallons of water and thoroughly spray infested areas such as nearby cracks and crevices, between and under cushions of upholstered furniture, along and behind baseboards, window and door frames and other areas where these pests may be present.

CARPENTER ANTS

Mix 20 pounds in 100 gallons of water and apply to ant trails, around doors and windows and other places where ants enter premises. Where possible, apply this product directly to any nest or infested wood.

CENTIPEDES/EARWIGS/MILLIPEDES/SCORPIONS

Mix 20 pounds in 100 gallons of water and apply around water pipes, doors and windows, and other places where these pests may enter premises. Spray baseboards, storage areas, garages, carports, basements and other areas where these pests are found.

COCKROACHES

Efficacy varies with species sensitivity. This product is generally not highly effective in controlling German cockroaches. However, the following may be controlled with the rates as stated. American roach, Australian roach, Brown roach, Smoky brown roach, and others. Mix 20 pounds in 100 gallons of water and apply as a fine, low pressure (psi) spot spray or as crack and crevice application to areas where these pests hide, such as baseboards, storage areas, closets, around water pipes, doors and windows, behind and under refrigerators, cabinets, sinks, stoves, dishwashers, hot water heaters, the underside of shelves, drawers and similar areas.

SPIDERS

Mix 20 pounds in 100 gallons of water and apply to infested baseboards, window and door frames, corners, pipes, storage areas, attics and under eaves. Make spot applications to other areas where these pests are present.

OUTDOORS

PERIMETER TREATMENT

Residual spray for control of ants, bees and wasps, brown dog ticks, carpenter ants, centipedes, cockroaches, crickets, earwigs, firebrats and silverfish, fleas, millipedes, scorpions and spiders. Mix 10 pounds this product per 50 gallons of water and apply via power spray or other spray methods.

To help prevent infestations of buildings by the above pests, outside perimeter treatment should be in a band 6 to 10 feet wide and confined to shrub beds, foundation plantings and lawn or soil areas immediately adjacent to the structure. Direct application to structures should be minimal and restricted to cracks and crevices and other areas where insects tend to congregate.

STORAGE AND DISPOSAL

STORAGE

Store unused product in original container only, in cool, dry area out of reach of children and animals.

Do not store in areas where temperatures frequently exceed 100°F. If container is damaged, before cleaning up, put on long-sleeved shirt, full length trousers, head covering, and protective gloves.

PESTICIDE DISPOSAL

Do not contaminate water, food or feed by storage or disposal.

Open dumping is prohibited. Wastes resulting from the use of this product may be disposed of on site or at an approved waste disposal facility.

CONTAINER DISPOSAL

Completely empty bag into application equipment. Then dispose of bag in a sanitary landfill or by incineration, or, if allowed by State and local authorities, by burning. If bag is burned, stay out of smoke.

LIMITED WARRANTY AND DISCLAIMER

The manufacturer warrants (a) that this product conforms to the chemical description on the label; (b) that this product is reasonably fit for the purposes set forth in the directions for use when it is used in accordance with such directions; and (c) that the directions, warnings and other statements on this label are based upon responsible experts' evaluation of reasonable tests of effectiveness, of toxicity to laboratory animals and to plants, and of residues on food crops, and upon reports of field experience. Tests have not been made on all varieties or in all states or under all conditions. THE MANUFACTURER NEITHER MAKES NOR INTENDS, NO DOES IT AUTHORIZE ANY AGENT OR REPRESENTATIVE TO MAKE, ANY OTHER WARRANTIES, EXPRESS OR IMPLIED, AND IT EXPRESSLY EXCLUDES AND DISCLAIMS ALL IMPLIED WARRANTIES OF MERCHANTABILITY OR FITNESS FOR A PARTICULAR PURPOSE.

THIS WARRANTY DOES NOT EXTEND TO, AND THE BUYER SHALL BE SOLELY RESPONSIBLE FOR, ANY AND ALL LOSS OR DAMAGE WHICH RESULTS FROM THE USE OF THIS PRODUCT IN ANY MANNER WHICH IS INCONSISTENT WITH THE LABEL DIRECTIONS, WARNINGS OR CAUTIONS.

BUYER'S EXCLUSIVE REMEDY AND MANUFACTURER'S OR SELLER'S EXCLUSIVE LIABILITY FOR ANY AND ALL CLAIMS, LOSSES, DAMAGES, OR INJURIES RESULTING FROM THE USE OR HANDLING OF THIS PRODUCT, WHETHER OR NOT BASED IN CONTRACT, NEGLIGENCE, STRICT LIABILITY IN TORT OR OTHERWISE, SHALL BE LIMITED, AT THE MANUFACTURERS OPTION, TO REPLACEMENT OF, OR THE REPAYMENT OF THE PURCHASE PRICE FOR, THE QUANTITY OF PRODUCT WITH RESPECT TO WHICH DAMAGES ARE CLAIMED. IN NO EVENT SHALL MANUFACTURER OR SELLER BE LIABLE FOR SPECIAL, INDIRECT OR CONSEQUENTIAL DAMAGES RESULTING FROM THE USE OR HANDLING OF THIS PRODUCT.

SEVIN is the registered trademark of RHONE-POULENC NEDERLAND B.V. for carbaryl insecticides.

AG90056 EPA Approval: 12/8/88

THIS SPECIMEN LABEL IS INTENDED FOR USE ONLY AS A GUIDE IN PROVIDING GENERAL INFORMATION REGARDING THE DIRECTIONS, WARNINGS AND CAUTIONS ASSOCIATED WITH THE USE OF THIS PRODUCT. AS WITH ANY AGRICULTURAL CHEMICAL, ALWAYS FOLLOW THE LABEL INSTRUCTIONS ON THE PACKAGE BEFORE USING.

SEVIN® brand 4F

CARBARYL INSECTICIDE

For Agricultural Or Commercial Use Only

ACTIVE INGREDIENTS:	
Carbaryl (1-naphthyl N-methylcarbamate	43.0% by Wt.
INERT INGREDIENTS:	57.0% by Wt.
TOTAL	100%

(Contains 4 pounds Carbaryl per Gallon)

E.P.A. Reg No. 264-349 E.P.A. Est. No. 264-M0-02

KEEP OUT OF REACH OF CHILDREN

CAUTION CUIDADO

PRECAUCION AL USARIO: Si usted no lee ingles, no use este producto hasta que la etiqueta le haya sido explicada ampliamente.

For PRODUCT USE Information Call 1-800-334-9745
For MEDICAL and TRANSPORTATION Emergencies ONLY Call 24 Hours A Day 1-800-334-7577

PRECAUTIONARY STATEMENTS

CAUTION

HAZARDS TO HUMANS AND DOMESTIC ANIMALS

MAY BE HARMFUL IF SWALLOWED. OVEREXPOSURE MAY CAUSE: Salivation, watery eyes, pinpoint eye pupils, blurred vision, muscle tremors, difficult breathing, excessive sweating, abdominal cramps, nausea, vomiting, diarrhea, weakness, headache. IN SEVERE CASES CONVULSION, UNCONSCIOUSNESS AND RESPIRATORY FAILURE MAY OCCUR. SIGNS AND SYMPTOMS OCCUR RAPIDLY FOLLOWING OVEREXPOSURE TO THIS PRODUCT.

Avoid breathing spray mist. Do not take internally. Avoid contact with eyes, skin or clothing. Wear regular long-sleeved work clothing and head covering when making overhead applications. Change to clean clothing daily. Bathe and wash hair after each work day. Do not eat, drink or use tobacco while working with this product or spray solutions. Wash hands and face before eating, drinking or using tobacco. Keep out of reach of children and domestic animals.

ANTIDOTE STATEMENT

ATROPINE SULFATE IS HIGHLY EFFECTIVE AS AN ANTIDOTE. Narcotics and other sedatives should not be used. Further, drugs like 2-PAM (pyridine-2-aldoxime methiodide) are NOT recommended. See NOTE TO PHYSICIAN below.

STATEMENT OF PRACTICAL TREATMENT

GENERAL

Contact a physician immediately in all cases of suspected poisoning. Transport to a physician or hospital immediately and SHOW A COPY OF THIS LABEL TO THE PHYSICIAN. If poisoning is suspected in animals, contact a veterinarian.

IF SWALLOWED: Never give anything by mouth to an unconscious or convulsing person. If conscious and not convulsing, drink 1 to 2 glasses of water and induce vomiting by touching the back of the throat with finger.

IF IN EYES: Flush eyes with plenty of water. Get medical attention if irritation persists.

IF ON SKIN: Wash thoroughly with soap and water.

IF INHALED: Move from contaminated atmosphere and call a physician.

NOTE TO PHYSICIAN

Carbaryl is a carbamate insecticide, which is a cholinesterase inhibitor. Overexposure to this substance may cause toxic signs and symptoms due to stimulation of the cholinergic nervous system. These effects of overexposure are spontaneously and rapidly reversible. Gastric lavage may be used if this product has been swallowed. Carbaryl poisoning may occur rapidly after ingestion and prompt removal of stomach contents is indicated.

Specific treatment consists of parenteral atropine sulfate. Caution should be maintained to prevent overatropinization. Mild cases may be given 1 to 2 mg intramuscularly every 10 minutes until full atropinization has been achieved and repeated thereafter whenever symptoms reappear. Severe cases should be given 2 to 4 mg intravenously every 10 minutes until fully atropinized, then intramuscularly every 30 to 60 minutes as needed to maintain the effect for at least 12 hours. Dosages for children should be appropriately reduced. Complete recovery from overexposure is to be expected within 24 hours.

Narcotics and other sedatives should not be used. Further, drugs like 2-PAM (pyridine-2-aldoxime methoiodide) are NOT recommended.

To aid in confirmation of a diagnosis, urine samples should be obtained within 24 hours of exposure and immediately frozen.

Consultation on diagnosis, therapy and analysis can be obtained at all hours by calling the Rhône-Poulenc emergency number 1-800-334-7577.

ENVIRONMENTAL HAZARDS

This product is extremely toxic to aquatic and estuarine invertebrates. Do not apply directly to water and wetlands except under forest canopy and use on rice. Discharge from rice fields may kill aquatic and estuarine invertebrates. Do not apply when weather conditions favor drift from area treated. Do not contaminate water by cleaning equipment or disposal of wastes. Do not contaminate water when disposing of equipment washwaters.

BEE CAUTION: MAY KILL HONEYBEES IN SUBSTANTIAL NUMBERS.

This product is highly toxic to bees exposed to direct treatment on blooming crops or weeds. Do not apply this product or allow it to drift to blooming crops or weeds if bees are foraging in the treatment area. Contact your Cooperative Agricultural Extention Service or your local Rhône-Poulenc Ag Company representative for further information.

DIRECTIONS FOR USE

IT IS A VIOLATION OF FEDERAL LAW TO USE THIS PRODUCT IN A MANNER INCONSISTENT WITH ITS LABELING.

READ THIS LABEL BEFORE USE. STRICTLY OBSERVE LABEL DIRECTIONS AND CAUTIONS, AND APPLICABLE FEDERAL AND STATE REGULATIONS. DO NOT USE ON ANY CROP NOT LISTED ON THIS LABEL OR SUPPLEMENTAL LABELING AS ANY RESIDUES REMAINING MAY BE ILLEGAL OR HARMFUL.

GENERAL INFORMATION

SEVIN® brand 4F Carbaryl Insecticide is a suspension of microfine SEVIN® brand Carbaryl Insecticide in an aqueous medium. It readily disperses in water to form a spray which may be applied by ground or air and with hand or power-operated sprayers.

GENERAL WORKER PROTECTION STATEMENT

Do not apply this product in such a manner as to directly or through drift expose workers or other persons. The area being treated must be vacated by unprotected persons.

Do not enter treated areas without protective clothing until sprays have dried.

Because certain states may require more restrictive reentry intervals for various crops treated with this product, consult your State Department of Agriculture for further information.

Written or oral warnings must be given to workers who are expected to be in a treated area or in an area about to be treated with this product. Advise workers to stay out of treated areas until spray deposits have dried unless suitable clothing is worn. This includes regular long-sleeved (overhead applications) and long-legged work clothing. See PRECAUTIONARY STATEMENTS, STATEMENT OF PRACTICAL TREATMENT and NOTE TO PHYSICIAN for additional information. When oral warnings are given, warnings shall be given in a language customarily understood by workers. Oral warnings must be given if there is reason to believe that written warnings cannot be understood by workers. Written warnings must include the following information: appropriate signal word (CAUTION), area treated with SEVIN® brand 4F Carbaryl Insecticide, date of application, appropriate clothing, and reentry interval (i.e., until sprays have dried).

PREHARVEST AND GRAZING USE INFORMATION AND LIMITATIONS

Tolerances established under the Federal Food, Drug and Cosmetic Act permit the sale of crops bearing probable carbaryl residues when this product is used in accordance with the label directions. If used as directed, treated forage may be grazed or used as feed for dairy and meat animals without causing illegal residues in meat or milk. This product may be applied up to and including the day of harvest or grazing of forage crops (except alfalfa). Application may be made without removing livestock from area being treated. Do not apply at greater rates or at more frequent intervals than is stated on the label. To do so may result in illegal residues in crops, meat, and milk. Do not apply at greater rates or at more frequent intervals than are stated on the label.

Do not plant rotational food and feed crops not listed on this or other carbaryl labels in carbaryl treated soil.

Do not use reclaimed irrigation water from crops treated with carbaryl on crops for which carbaryl tolerances are not established.

PLANT RESPONSE PRECAUTIONS

To avoid possible injury to tender foliage, do not apply to wet foliage or during periods of high humidity.

Do not use on Boston ivy, Virginia creeper and maidenhair fern as injury may result. Carbaryl may also injure Virginia and sand pines.

Carefully observe label instructions for apple thinning to avoid excessive thinning. Combinations with certain herbicides on rice and soybeans may be phytotoxic. Refer to specific directions for appropriate crop.

SPRAY PREPARATION

TO ASSURE A UNIFORM SUSPENSION, AGITATE, STIR OR RECIRCULATE ALL CONTAINERS OF THIS PRODUCT PRIOR TO USE. Remove oil, rust, scale, pesticide residues and other foreign matter from mix tanks and entire spray system. Flush with clean water. Fill spray or mix tank with 1/2 to 3/4 the desired amount of water. Start mechanical or hydraulic agitation. Slowly add the required amount of SEVIN® brand 4F Carbaryl Insecticide, and then add the remaining volume of water. Include rinse water from container. Prepare only as much spray mixture as can be applied on the day of mixing. MAINTAIN CONTINUOUS AGITATION DURING MIXING AND APPLICATION TO ASSURE A UNIFORM SUSPENSION. DO NOT STORE SPRAY MIXTURE FOR PROLONGED PERIODS OR DEGRADATION OF CARBARYL MAY OCCUR. Local water conditions may also accelerate the degradation of spray mixtures containing carbaryl. See COMPATIBILITY STATEMENT below.

COMPATIBILITY

SEVIN® brand 4F Carbaryl Insecticide, when diluted with at least an equal volume of water, is compatible with a wide range of pesticides. It is not compatible with diesel fuel, kerosene, fuel oil or aromatic solvents. If compatibility with another product and the resulting crop response is unknown, the mixture should be tested on a small scale. Curdling, precipitation, greasing, layer formation or increased viscosity are syptoms of incompatibility. Incompatibility will reduce insect control and may cause application and handling difficulties or plant injury. Observe all cautions and limitations on labeling of all products used in mixtures. WHEN PREPARING COMBINATION SPRAYS, FIRST ADD SEVIN® BRAND 4F CARBARYL INSECTICIDE TO AT LEAST AN EQUAL VOLUME OF WATER, MIX THOROUGHLY, AND THEN ADD COMBINATION PRODUCTS TO THE MIXTURE. DO NOT APPLY TANK MIX COMBINATIONS UNLESS YOUR PREVIOUS EXPERIENCE INDICATES THE MIXTURE IS EFFECTIVE AND WILL NOT RESULT IN APPLICATION PROBLEMS OR PLANT INJURY.

Carbaryl is unstable under highly alkaline conditions and mixtures with strong bases, such as Bordeaux, lime-sulfur and casein-lime spreaders, will result in chemical degradation of the insecticide. Do not use this product in water with pH values above 8.0 unless a buffer is added. If necessary, water should be buffered to neutral (pH = 7.0) before adding this product to the spray tank.

Overhead irrigation with alkaline or muddy water after application will also accelerate chemical degradation and may result in reduced insect control.

APPLICATION

For all applications use sufficient gallonage to obtain thorough and uniform coverage. Observe label instructions for specific directions regarding spray volume where they occur. Calibrate spray equipment to deliver the required volume. Use 50 mesh slotted strainers in spray system and 25 mesh slotted strainers behind nozzles.

To clean spray system after use, drain and flush with a water and detergent mixture. Rinse thoroughly with clean water. Refer to the Storage and Disposal directions for disposal instructions.

INSECT CONTROL

Begin application when insect populations reach recognized economic threshold levels. Consult the Cooperative Extension Service, Professional Consultants or other qualified authorities to determine appropriate threshold levels for treatment in your area. To maintain control repeat at 7 to 14 day intervals or as necessary unless a shorter interval is specified below. Where a dosage range is indicated, use the lower rate on light to moderate infestations, young plants and early instars and the higher rate on heavy infestations, mature plants, advanced instars and adults. Thorough and uniform spray coverage is essential for effective control.

NOTE: All references to armyworm on the crops listed below refer to the species, *Pseudaletia unipuncta*, often called the "true armyworm". Except where indicated otherwise, this product is not registered for the control of other armyworm species. Regional differences have been noted in the susceptibility of certain strains of fall armyworm, Colorado potato beetle, spotted tentiform leafminer, and tobacco budworm (on cotton) to carbaryl. If local experience indicates inadequate control, use an alternative pesticide.

FORAGE, FIELD AND VEGETABLE CROPS

Apply in sufficient spray volume to obtain thorough coverage. Do not dilute greater than 1:39 (volume product: volume water) where washoff resistance is desired. Where maximum washoff resistance is needed, do not exceed a 1:1 dilution. To prepare small volumes of spray mixture use 1 tablespoonful (1/2 fluid ounce) of SEVIN® brand 4F Carbaryl Insecticide per gallon of water where rates of 1 quart per acre or 1 quart per 100 gallons are indicated in the tables below.

GRASSHOPPERS

CROP	QUARTS OF "SEVIN®" BRAND 4F/ ACRE	PREHARVEST INTERVAL (DAYS)	SPECIFIC DIRECTIONS
All Forage, Field and Vegetable Crops	½ to 1½ (Unless Noted Otherwise)	See specific Forage, Field or Vegetable Crops	Use the lower rate for nymphs on small plants or sparse vegetation in wasteland, rangeland, ditchbanks, rights-of-way, pastures, hedgerows, and roadsides. Use the higher rate for adult grasshoppers or applications to dense vegetation.

CROP	INSECT	QUARTS OF "SEVIN®" BRAND 4F/ ACRE	PREHARVEST INTERVAL (DAYS)	SPECIFIC DIRECTIONS
Alfalfa Clovers Birdsfoot trefoil (DO NOT USE ON BIRDSFOOT TREFOIL IN CALIFORNIA	Blister beetles Mexican bean beetle	½ to 1	7 (Alfalfa) 0 (Clover & Birdsfoot trefoil)	OBSERVE BEE CAUTION Observe plant response precautions. On dense growth use 25 to 40 gallons per acre with ground equipment to ensure adequate coverage. For alfalfa weevil larvae, if pretreatment damage is extensive, cut alfalfa and treat the stubble.
	Alfalfa caterpillar Bean leaf beetle Cucumber beetles Green cloverworm Japanese beetle Leafhoppers Three cornered alfalfa hopper Thrips Velvetbean caterpillar	1		
	Armyworm Cloverhead weevil Corn earworm Egyptian alfalfa weevil larvae Essex skipper European alfalfa beetle Fall armyworm Lygus bugs Stink bugs Webworms Yellowstriped armyworm	1 to 1½		Apply only once per cutting.
	Alfalfa weevil larvae West of the Rocky Mountains East of the Rocky Mountains	 1 to 1½ 1½		
Asparagus	Asparagus beetle	1 to 2	1	Treat ferns or brush growth. Do not treat more than once every 3 days.
	Apache cicada Asparagus beetle	2 to 4	Post harvest application only	
Beans (including blackeyed peas, cowpeas, southern peas, dry beans, green beans, lima beans, navy beans and snap beans) Lentils (DO NOT USE ON LENTILS IN CALIFORNIA)	Blister beetles Mexican bean beetle	½ to 1	0 (except cowpeas) 3 (cowpeas)	Observe plant response precautions.
	Alfalfa caterpillar Bean leaf beetles Cucumber beetles Flea beetles Green cloverworm Japanese beetle Leafhoppers Three cornered alfalfa hopper Thrips Velvetbean caterpillar Western bean cutworm	1		
	Grasshoppers	½ to 1½		Refer to general GRASSHOPPER heading above.
	Armyworm Corn earworm Cutworms European corn borer Fall armyworm Stink bugs Tarnished plant bug Webworms	1 to 1½		
	Alfalfa Looper	1½		
	Cowpea curculio	2		
	Corn earworm Limabean podborer Lygus bugs Stink bugs	2		CALIFORNIA ONLY
Cabbage Broccoli Brussels sprouts Cauliflower Kohlrabi	Flea beetles Harlequin bug	½ to 1	3	-
	Armyworm Corn earworm Diamondback moth Fall armyworm Imported cabbageworm	1 to 2		
Chinese cabbage Collards Hanover	Flea beetles Harlequin bug Leafhoppers	½ to 1	3 (horseradish, radishes, rutabagas and turnip roots)	

CROP	INSECT		QUARTS OF "SEVIN®" BRAND 4F/ ACRE	PREHARVEST INTERVAL (DAYS)	SPECIFIC DIRECTIONS
salad Horseradish Kale Mustard greens Radishes Rutabagas Turnips				14 (Chinese cabbage, collards, Hanover salad, kale, mustard greens, and turnip tops)	
	Aster leafhopper		1 to 1 ½		
	Armyworm Corn earworm Fall armyworm	Imported cabbageworm Stink bugs Tarnished plant bug	1 to 2		
Carrots Celery Parsley Parsnips (DO NOT USE ON CELERY IN CALIFORNIA)	Flea beetles	Leafhoppers	½ to 1	0 (carrots) 3 (parsnips) 14 (celery &parsley)	Treat on a 5 to 7 day schedule.
	Aster leafhopper Lygus bugs	Spittlebugs	1 to 1 ½		
	Armyworm Corn earworm Fall armyworm	Stink bugs Tarnished plant bug	1 to 2		
Corn (field, sweet, pop)	Armyworm Chinch bugs Corn earworm Corn rootworm adults Fall armyworm	Flea beetles Japanese beetle Sap beetles Southwestern corn borer Leafhoppers	1 to 2	0	OBSERVE BEE CAUTION For insects attacking silks and ears apply at 1 to 6 day intervals starting when first silks appear and continuing until silks begin to dry. For larvae in whorl and foliage feeders, apply as necessary. Optimum timing and good coverage are essential for effective control. For optimum chinch bug control, apply at least 20 gallons of water per acre by ground and direct spray toward stalk to provide thorough coverage.
	Grasshoppers		½ to 1 ½		Refer to general GRASSHOPPER heading above
	European Corn borer		1 ½ to 2		For optimum control, do not apply in less than 3 gallons of water per acre by air and 15 gallons of water per acre by ground.
	Western bean cutworm		2		
	Cutworms		2 to 3		For optimum control, apply in a 12 inch band, over the row, using sufficient volume of water to obtain thorough coverage. For broadcast application use at least 20 gallons (ground) or 5 gallons (air) of water per acre.
Cotton	Cotton fleahopper Cotton leafworm Flea beetles	Striped blister beetle Thrips	½ to 1	7	Treat on a 5 to 7 day schedule for as long as control is necessary. Mid and late season insect control. May be applied after bolls open.
	Boll weevil Cotton bollworm Fall armyworm Leafrollers Leafhoppers	Tarnished plant bug Yellowstriped armyworm (cotton cutworm)	1 to 2		
	Lygus bugs		1 to 2		For light to moderate populations in Western irrigated cotton.
	Pink bollworm		1½ to 2½		Aphid populations will be suppressed by repeated applications of this insecticide.
	Cutworms Stink bugs	Saltmarsh caterpillar	2		
Cucumber Melons Pumpkin Squash	Pickleworm	Melonworm	½ to 1	0	Observe plant response precautions.
	Cucumber beetles Flea beetles	Leafhoppers Squash bugs	1		
Dandelion Endive (Escarole) Lettuce Salsify	Flea beetles Harlequin bug	Leafhoppers	½ to 1	3 (head lettuce & salsify roots) 14 (dandelion, endive (escarole), leaf lettuce & salsify tops)	Observe plant response precautions. Treat on a 5 to 7 day schedule after heads begin to form.
	Aster leafhopper Lygus bugs	Spittlebugs	1 to 1½		
	Armyworm Corn earworm Fall armyworm	Imported cabbageworm Stink bugs	1 to 2		

CROP	INSECT	QUARTS OF "SEVIN®" BRAND 4F/ ACRE	PREHARVEST INTERVAL (DAYS)	SPECIFIC DIRECTIONS
	Tarnished plant bug			
Flax for Proso Millet (DO NOT USE IN CALIFORNIA	Armyworm	1 to 1½	42	Do not graze treated areas or harvest dairy feed prior to crop maturity.
	Grasshopper	½ to 1½		Refer to the general GRASSHOPPER heading above.
Garden beet Spinach Swiss chard	Flea beetles Harlequin bug Leafhoppers	½ to 1	3 (garden beet roots)	
	Aster leafhopper	1 to 1½	14 (garden beet tops, spinach, Swiss chard)	
	Armyworm Corn earworm Fall armyworm Stink bugs Tarnished plant bug	1 to 2		
Okra	Corn earworm Stink bugs	1 to 2	0	
Pasture	Armyworm Black grass bugs Chinch bugs Essex skipper Fall armyworm Range caterpiller Range crane fly Striped grass looper Thrips	1 to 1½	0 (aerial applications) 14 (ground applications)	To control thrips in grasses grown for seed, use high spray pressure to improve penetration into boot. Apply a minimum of 2 applications per year. Allow at least 14 days between applications. Do not allow foraging or cut for hay within 14 days of last application by ground. Aerially treated pastures may be grazed or cut for hay on day of treatment. Carefully mark swath to avoid overapplication.
	Grasshoppers	½ to 1½		Refer to the general GRASSHOPPER heading above.
Peanuts	Blister beetles Mexican bean beetle	½ to 1		Observe plant response precautions.
	Alfalfa caterpillar Bean leaf beetle Cucumber beetles Green cloverworm Japanese beetle Leafhoppers Rednecked peanutworm Three cornered alfalfa hopper Thrips Velvetbean caterpillar	1	0	
	Armyworm Corn earworm Fall armyworm Stink bugs Webworms	1 to 1½		
	Whitefringed beetle adults Cutworms	2		
Peas	Colorado potato beetle Leafhoppers	1	3	
	Armyworm	1 to 1½		
	Alfalfa caterpillar Cutworms Pea leaf weevil Pea weevil Yellowstriped armyworm	1½		
	Alfalfa looper	2½		WASHINGTON STATE ONLY
Potato Tomato Eggplant Pepper	Colorado potato beetle Flea beetles Leafhoppers	½ to 1		Thorough coverage is essential to effectively suppress stink bugs. When disease transmission is suspected, monitor fields following application and re-treat if reinfestation occurs.
	European corn borer Fall armyworm Lace bugs Stink bugs (suppression) Tarnished plant bug Tomato fruitworm Tomato hornworm Tomato pinworm	1 to 2	0	
	Cutworms	2		
Prickly Pear cactus	Cochineal scale (crawlers)	2	1	Apply as needed at 7-10 day intervals. Do not make more than 7 applications per season.
Rangeland	Black grass bug Mormon cricket Range caterpillars Range crane fly	½ to ¾	0	FOR AERIAL APPLICATIONS ONLY. Apply a maximum of 2 applications per year. Allow at least 14 days between applications. Carefully mark swaths to avoid overapplication.

CROP	INSECT		QUARTS OF "SEVIN®" BRAND 4F/ ACRE	PREHARVEST INTERVAL (DAYS)	SPECIFIC DIRECTIONS
	Grasshoppers		1 to 1½		Refer to the general GRASSHOPPER heading above.
Rice	Armyworm Chinch bugs	Fall armyworm Stink bugs	1 to 1½	14	MISSISSIPPI DELTA & TEXAS. CAUTION: May kill shrimp and crabs. Do not use in areas where these are important resources. Do not use on rice fields in which crayfish and catfish farming are included in the cultural practice. DO NOT APPLY PROPANIL HERBICIDES WITHIN 15 DAYS BEFORE OR AFTER APPLICATION OF THIS PRODUCT OR PLANT INJURY WILL RESULT
	Grasshoppers		½ to 1½		Refer to the general GRASSHOPPER heading above.
	Armyworm Leafhoppers	Tadpole shrimp	2		CALIFORNIA ONLY. For optimum tadpole shrimp control apply to water when pest first appears.
Sorghum (milo, grain sweet and hybrid)	Sorghum midge		¾ to 1	21 (grain) 0 (forage)	Direct spray into forming heads for optimum insect control. Treat for sorghum midge when 25 to 30 percent of heads have emerged from boot and are in bloom. Repeat application 3 to 5 days later if adults are still active. A third application may be necessary in late planted sorghum or if midge are abundant. For chinch bugs use high gallonage ground application directed at the base of plants. This product will not effectively control the Southern green stink bug.
	Armyworm Chinch bugs Corn earworm	Fall armyworm Stink bugs Webworms	1 to 2		
	Southwestern corn borer		1½		
	Cutworms		2		
	Grasshoppers		½ to 1½		Refer to the general GRASSHOPPER heading above.
Soybeans	Bean leaf beetle Blister beetle Cucumber beetles Grape colaspis Green cloverworm	Japanese beetle Mexican bean beetle Velvetbean caterpillar	½ to 1	0	DO NOT APPLY A COMBINATION OF THIS PRODUCT AND 2,4 DB HERBICIDES TO SOYBEANS AS CROP INJURY MAY RESULT. Use lower rates for light to moderate populations and smaller instars and to provide maximum survival of beneficial insects and spiders. Use the higher rates for heavy populations and larger instars.
	Corn earworm		½ - 1½		
	Grasshoppers		½ to 1½		Refer to general GRASSHOPPER heading above.
	Alfalfa caterpillar Leafhoppers	Three cornered alfalfa hopper Thrips	1		
	Armyworm Cutworms	Fall armyworm Webworms	1 to 1½		
	Painted lady (Thistle caterpillar) Saltmarsh caterpillar	Woollybear caterpillar Yellowstriped armyworm	1½ to 2		
Sugar beets	Armyworm Beet leaf beetle Fall armyworm	Flea beetles Leafhoppers Webworms	1 to 1½	14	
	Grasshopper		½ to 1½		Refer to general GRASSHOPPER heading above.
	Cutworms		1½		
Sunflower (DO NOT USE IN CALIFORNIA)	Cutworms		1½	60	
	Armyworm Fall armyworm Sunflower moth		1½ to 2		
	Stem weevil	Sunflower beetle	1 to 2		
	Grasshopper		½ to 1½		Refer to general GRASSHOPPER heading above.

CROP	INSECT	QUARTS OF "SEVIN®" BRAND 4F/ ACRE	PREHARVEST INTERVAL (DAYS)	SPECIFIC DIRECTIONS
Sweet Potato (DO NOT USE IN CALIFORNIA	Corn earworm, Cucumber beetles, Flea beetles, Sweet potato hornworm, Tortoise beetles	1 to 2	0	Apply as a foliar spray as needed.
	Yellowstriped armyworm	2		
	Sweet potato weevil	1 to 2		Full coverage of plants is essential. Use lower rate on young plants and higher rate on mature plants.
Tobacco Plant bed Treatment	Flea beetle	1	0	OBSERVE PLANT RESPONSE PRECAUTIONS. For flea beetle control, use 4 teaspoons (0.7 fl. oz.) in 6 gallons of water and apply to 100 square yards.
	Green June beetle grubs	8		For green June beetle grub control mix 11 tablespoons (5.5 fl. oz.) in 50 to 100 gallons of water and apply to 100 square yards. Applications should be made to areas that larvae have uprooted by sprinkling mixture as a drench treatment.
Field Treatment	Budworms, Fall armyworm, Flea beetle, Hornworms, Japanese beetle, June beetle, Suckfly	1 to 2		Use lower rate on young plants (up to knee height). Use at least 10 gallons of prepared spray per acre. Begin treatments when worms are small.
Wasteland, Rights-of Way, Hedgerows, Ditchbanks, Roadsides	Black grass bug, Mormon cricket, Range caterpillars, Range crane fly	½ to 1	0 (aerial applications) 14 (ground applications)	Apply a maximum of 2 applications per year. Allow at least 14 days between applications. Do not allow foraging or cut for hay within 14 days of last application by ground. Aerially treated areas may be grazed or cut for hay on day of treatment. Carefully mark swaths to avoid overapplication.
	Grasshoppers	½ to 1		Refer to the general GRASSHOPPER heading above.
Wheat (including triticale) (DO NOT USE IN CALIFORNIA)	Flea beetles	½ to 1	0 (forage) 21 (grain)	
	Grasshoppers	½ to 1½		Refer to the general GRASSHOPPER heading above.
	Cereal leaf beetle	1		Application is effective against eggs, larvae and adult of the cereal leaf beetle.
	Armyworm, Fall armyworm	1 to 1½		

TREE FRUIT AND NUT CROPS

Apply the recommended amount of SEVIN® brand 4F Carbaryl Insecticide in 100 gallons of water. Use hand or power-operated sprayer to wet leaves and branches to the dripping point.

CROP	QUARTS OF "SEVIN®" BRAND 4F/ 100 GAL.	PREHARVEST INTERVAL (DAYS)	SPECIFIC DIRECTIONS
Apples	¼ to ½	1	OBSERVE BEE CAUTION. Apply 1 full coverage dilute spray between 10 and 25 days after full bloom. Factors such as tree age, variety, nutrition, previous crop, pruning, bloom and degree of set favor excessive fruit thinning with the product. Exercise caution to avoid possible yield reduction. Rates may vary depending on variety and local orchard conditions. Consult with your County Extensions Service or other experts for advice on the proper use of this product. In Eastern apple growing areas, tank mix combinations of Carbaryl and Naphthaleneacetic Acid (NAA) or Naphthaleneacetamide (NAD) have successfully thinned several early-maturing, heavy-setting varieties, as well as hard-to-thin varieties such as Golden Delicious and Rhode Island Greening. The higher rate of Carbaryl and reduced rates of NAA or NAD followed 7 to 10 days later by an application of Carbaryl has improved thinning on these varieties. For easily thinned varieties including Cortland, Grimes, Johathan, McIntosh, Orleans, Rome Beauty, Puritan, Red Delicious, Winesap, Yellow Newton.
	½ to 1	1	For difficult to thin varieties including Baldwin, Ben Davis, Duchess, Early McIntosh, Golden Delicious, Lady Apple, Northern Spy, Rhode Island Greening, Steele Red, Turley, Wealthy, Yellow Transparent, and York Imperial

CROP	INSECT	QUARTS OF "SEVIN®" BRAND 4F/ 100 GALS.	PREHARVEST INTERVAL (DAYS)	SPECIFIC DIRECTIONS
Almond	Fruittree leafroller Peach twig borer San Jose scale	1	28	OBSERVE BEE CAUTION Apply in "popcorn" or petal fall stages and again when the May brood of the peach twig borer begins to hatch or thereafter as needed.
	Navel orangeworm			Time early and mid season applications to correspond to moth flight peaks. Make a late season application at initiation of hull split or up to 10% hull split.
Apples Pears	Apple aphid Apple maggot Apple rust mite Apple sucker Bagworms California pearslug (pear sawfly) Eastern tent caterpillar European apple sawfly Eyespotted bud moth Forbes scale Fruittree leafroller Green fruitworm Gypsy moth Japanese beetle Lecanium scales Lesser appleworm Lygus bugs Orange tortrix Oystershell scale Pearleaf blister mite Pear psylla Pear rust mite Periodical cicada Plum curculio Redbanded leaf-roller Rosy apple aphid San Jose scale Tarnished plant bug Tentiform leaf-miners Woolly apple aphid Yellowheaded fireworm	¾ to 1	1	OBSERVE BEE CAUTION To avoid undesired apple thinning, delay use until at least 30 days after full bloom. For psylla control apply when eggs hatch or young nymphs are present. To control scale insects, apply when crawlers are present.
	Apple mealybug Apple aphid Codling moth White apple leafhopper	½		
Chestnuts (Do not use in California)	Chestnut weevil	2 to 3	0	Make 4 applications at weekly intervals beginning in late July for adult chestnut weevil control. Last application should be made prior to shuck split.
Citrus Fruits (such as grapefruit, lemons, limes, oranges, tangelos, tangerines, citrus citron, kumquats and hybrids)	Avocado leafroller California orangedog Citrus cutworm Citrus root weevil Fruittree leafroller Orange tortrix Western tussock moth West Indian sugarcane borer (adults)	1	5	OBSERVE BEE CAUTION To insure thorough coverage, do not apply less than 10 gallons of dilute spray mixture per mature tree. May be mixed with petroleum oils commonly used on citrus.
	Black scale Brown soft scale California red scale Citricola scale Citrus snow scale Yellow scale	¾ to 1		
Filbert	Filbert aphid Filbert leafroller Filbertworm	1	0	Apply when leafroller eggs are hatching. Repeat on first appearance of adult filbert moths and again 3 to 4 weeks later. Apply dilute sprays in 300-400 gallons per acre.
Olives	Olive scale	¾ to 1	0	For optimum scale control add 1½ gallons of summer oil and apply mixture when crawlers are present. Do not exceed 2 applications per year.
Peaches Apricots Nectarines	Apple pandemis Codling moth Cucumber beetles European earwig Fruittree leafroller Gypsy moth Japanese beetle June beetle Lecanium scales Lesser peachtree borer Olive scale Orange tortrix Oriental fruit moth Peach twig borer Periodical cicada Plum curculio Redbanded leaf-roller San Jose scale Tarnished plant bug Tussock moths Variegated leafroller	1	1 (peaches) 3 (apricots & nectarines)	For optimum scale control apply when crawlers are present. Spray limbs and trunk thoroughly at weekly intervals during moth flight.

CROP	INSECT		QUARTS OF "SEVIN®" BRAND 4F/ 100 GALS.	PREHARVEST INTERVAL (DAYS)	SPECIFIC DIRECTIONS
Pecans	Black margined aphid Fall webworm Hickory shuckworm Lesser webworm Pecan leaf phylloxera	Pecan stem phylloxera Pecan nut casebearer Pecan spittlebug Pecan weevil Twig girdler Walnut caterpillar	1 to 2½	0	
Pistachios	Navel orangeworm		½ to 2	14	Apply dilute volumes of 150 to 300 gallons of mixed spray per acre for full coverage at onset of hull split.
Plums Prunes Cherries	Codling moth Eastern tent caterpillar	Orange tortrix Tussock moth	¾	1	For optimum scale control apply when crawlers are present. For lesser peachtree borer control spray limbs and tree trunks thoroughly at weekly intervals during moth flight.
	Black cherry aphid Brown soft scale Cherry fruitworm Cherry maggot European earwig Eyespotted bud moth Forbes scale Fruittree leafroller Green fruitworm Gypsy moth Japanese beetle	Lecanium scales Lesser peachtree borer Mealy plum aphid Oystershell scale Peach twig borer Plum curculio Prune leafhopper Redbanded leafroller Rose chafer San Jose scale Variegated leafroller	1		
Walnut	Calico scale European fruit lecanium	Filbertworm Fruittree leafroller Frosted scale	½	0	
	Codling moth		½		For codling moth apply first spray when average cross-sectional diameters of developing nuts are ½ to ¾ inch. Repeat during middle or late June as needed.
	Navel Orangeworm		1		For navel orangeworm, time application to coincide with peak egg laying activity at time of or following hull split.
	European earwig		2		Spray tree trunks to point of run-off.

SMALL FRUIT CROPS

Recommended dosages refer to quarts of SEVIN® brand 4F Carbaryl Insecticide per acre. The optimum spray gallonage will depend on plant size, density and stage of growth. Typical spray gallonage per acre range from 100-300 gallons for dilute sprays, 30-100 gallons for concentrate sprays and 5-25 gallons for aerial sprays. Do not exceed maximum label rate per acre per application.

CROP	INSECT		QUARTS OF "SEVIN®" BRAND 4F/ ACRE	PREHARVEST INTERVAL (DAYS)	SPECIFIC DIRECTIONS
Blackberries Raspberries Dewberries (including boysenberries and loganberries)	European raspberry aphid Japanese beetle Leafhoppers	Leafrollers Rose chafer Snowy tree cricket	1 to 2	7	
	Omnivorous leafroller Raspberry sawfly		2		CALIFORNIA ONLY
Blueberries	Blueberry maggot Cherry fruitworm Cranberry fruitworm	European fruit lecanium Japanese beetle	1½ to 2	0	Apply 3 weeks before harvest and repeat as necessary.
Cranberries	Cutworms Cranberry fireworms Cranberry fruitworms Cranberry twig girdler	Elm spanworm Japanese beetle Leafhoppers Rose chafer Spaganothus worm	1½ to 3	1	CAUTION: May kill shrimp and crabs. Do not use in areas where these are important resources. Apply in late bloom and as needed at 7 to 10 day intervals.
Grapes	European fruit lecanium Grape leaffolder Grape leafhopper	Western grapeleaf skeletonizer Western yellow-striped armyworm	1 to 2	7	Apply before first brood leaffolder larvae emerge from rolls.

CROP	INSECT		QUARTS OF "SEVIN®" BRAND 4F/ ACRE	PREHARVEST INTERVAL (DAYS)	SPECIFIC DIRECTIONS
	Cutworms Eight-spotted forester Grape berry moth Japanese beetles June beetles	Orange tortrix Omnivorous leafroller Redbanded leafroller Saltmarsh caterpillar	2		
Strawberries	Flea beetles Meadow spittlebug (strawberry fruitworm) Omnivorous leaftier	Strawberry bud weevil Strawberry leafroller Strawberry weevil Strawberry clipper	1 to 2	1	Carbaryl may injure Early Dawn and Sunrise varieties.

IMPORTED FIRE ANT CONTROL

SITE	AMOUNT OF SEVIN® BRAND 4F/ VOLUME OF WATER	SPECIFIC DIRECTIONS
Lawns, Cemeteries and Recreational Areas (including turf, golf courses, and parks) Pastures Rangeland Forested Lands and Wasteland	1½ quarts/diameter 100 gallons or 1½ tablespoons/ gallon	Apply a total of 2 gallons of the diluted solution over the surface of each mound or at least 1 quart per 6 inches of mound diameter using a bucket, can or other appropriate equipment. Thoroughly wet mound and surrounding area to a 4 ft. diameter (12 sq. ft.). Do not disturb mounds prior to treatment. Pour solution from a height of about three feet to give sufficient force to break mound apex and flow into ant tunnels. For best results apply in cool weather, 65-80 degrees F, or in early morning or late evening hours. Repeat application if mound activity resumes after 10 days. Treat new mounds as they appear. Pressurized sprays may disturb the ants and cause migration, reducing product effectiveness. DO NOT ALLOW PUBLIC USE OF TREATED AREAS DURING APPLICATIONS OR UNTIL SPRAYS HAVE DRIED.
Nursery Stock, Vegetable Transplants, Foliage Plants and Bedding Plants	1½ quarts/ 100 gallons	DO NOT USE ON ANY FOOD CROP NOT LISTED ON LABEL. Do not make more than one application, either as a root-dip or a drench treatment (applied to the point of saturation). Avoid contact with foliage and treat only the growing media when using on bedding plants.

ADULT MOSQUITO CONTROL

For dilute-spray ground applications to trees (including shade trees, shelter belts, forests, plantations, parks and recreational areas), ornamentals, woody plants and shrubs, apply the specified dosage per 100 gallons of water. For concentrate-spray ground applications, apply the specified dosage per acre in sufficient spray volume to provide thorough coverage. To prepare small volumes of spray, use 1 tablespoon (½ fluid ounce) of SEVIN® brand 4F Carbaryl Insecticide per gallon of water, where rates of 1 quart are indicated.

SITE	AMOUNT OF "SEVIN®" BRAND 4F/100 GAL.	SPECIFIC DIRECTIONS
Pastures, Rangelands, Parks, Recreational Areas, Logging Camps, Military Posts and Adjacent Forested lands or Wastelands	¼ to 1 (½ to 1 tablespoonful per gallon)	DO NOT ALLOW PUBLIC USE OF TREATED AREAS DURING APPLICATIONS OR UNTIL SPRAYS HAVE DRIED. CAUTION: May kill shrimp and crabs. Do not use in areas where these are important resources. OBSERVE BEE CAUTION Treat shrubbery and areas where adult mosquitoes congregate. Treat when adult mosquitoes are active in early mornings or late evenings. Repeat at 7 to 10 day intervals. Use ¼ to ½ quart per 100 gallons in mistblowers, ½ to 1 quart per acre in aerial sprays and 1 quart per acre in low pressure ground sprayers.
	25	For residual control in subtropical regions apply 4 gallons of prepared spray per 2000 square feet of surface area. Repeat in 3 to 6 months or when necessary.

POULTRY ROOSTS AND BUILDINGS ONLY

POULTRY QUARTERS	QUARTS OF "SEVIN®" BRAND 4F/100 GAL.	SPECIFIC DIRECTIONS
Bed bugs Chicken mite Fleas Lice Northern fowl mite	4	DO NOT TREAT POULTRY OR GAME BIRDS. Apply 1 to 2 gallons of spray mixture per 1000 square feet of wall, litter or roost surface. Force spray into cracks. Repeat as needed. Avoid spraying nests, eggs and feeding and watering troughs.
Fowl tick	16	Ventilate while spraying. Do not treat premises within 7 days of slaughter.

POULTRY QUARTERS	QUARTS OF "SEVIN®" BRAND 4F/100 GAL.	SPECIFIC DIRECTIONS
Darkling Beetle Lesser mealworms	50	Apply 2 gallons of spray mixture per 1000 square feet of floor space or litter surface. Repeat as needed. Ventilate while spraying. Do not treat premises within 7 days of slaughter.

DIRECTIONS FOR USE THROUGH SPRINKLER IRRIGATION SYSTEMS

Apply this product only through sprinkler irrigation systems including center pivot and solid set. Do not apply this product through any other type of irrigation system.

SPRAY PREPARATION: Remove scale, pesticide residues, and other foreign matter from the chemical tank and entire injector system. Flush with clean water.

APPLICATION INSTRUCTIONS: First prepare a suspension of SEVIN® brand 4F Carbaryl Insecticide in a mix tank. Fill tank with ½ to ¾ the desired amount of water. Start mechanical or hydraulic agitation. Add the required amount of SEVIN® brand 4F, and then the remaining volume of water. (Suspension concentrations using the appropriate dosage per acre recommended on this label of SEVIN® brand 4F per 1 to 4 gallons of water are recommended). Then, set sprinkler to deliver 0.1 to 0.3 inches of water per acre. Start sprinkler and then uniformly inject the suspension of SEVIN® brand 4F into the irrigation water line so as to deliver the desired rate per acre. The suspension of SEVIN® brand 4F should be injected with a positive displacement pump into the main line ahead of a right angle turn to insure adequate mixing. If you should have any other questions about calibration, you should contact State Extension Service specialists, equipment manufacturers or other experts.

NOTE: When treatment with SEVIN® brand 4F has been completed, further field irrigation over the treated area should be avoided for 24 to 48 hours to prevent washing the chemical off the crop.

GENERAL PRECAUTIONS FOR APPLICATIONS THROUGH SPRINKLER IRRIGATION SYSTEMS

Maintain Continuous Agitation in mix tank during mixing and application to assure a uniform suspension.

Greater accuracy in calibration and distribution will be achieved by injecting a larger volume of a more dilute solution per unit time.

The system must contain a functional check valve, vacuum relief valve, and low pressure drain appropriately located on the irrigation pipeline to prevent water source contamination from backflow. The pesticide injection pipeline must contain a functional, automatic, quick-closing check valve to prevent the flow of fluid back toward the injection pump. The pesticide injection pipeline must also contain a functional, normally closed solenoid-operated valve located on the intake side of the injection pump connected to the system interlock to prevent fluid from being withdrawn from the supply tank when the irrigation system is either automatically or manually shutdown. The system must contain functional interlocking controls to automatically shut off the pesticide injection pump when the water pump motor stops. The irrigation line or water pump must include a functional pressure switch which will stop the water pump motor when the water pressure decreases to the point where pesticide distribution is adversely affected. Systems must use a metering pump, such as a positive displacement injection pump (e.g., diaphragm pump) effectively designed and constructed of materials that are compatible with pesticides and capable of being fitted with a system interlock. Do not apply when wind speed favors drift beyond the area intended for treatment.

Do not apply when wind speed favors drift, when system connections or fittings leak, when nozzles do not provide uniform distribution or when lines containing the product must be dismantled and drained.

Crop injury, lack of effectiveness, or illegal pesticide residues in the crop may result from non-uniform distribution of treated water.

Allow sufficient time for pesticide to be flushed through all lines and all nozzles before turning off irrigation water. A person knowledgeable of the chemigation system and responsible for its operation shall shut the system down and make necessary adjustments should the need arise.

Do not connect an irrigation system (including greenhouse systems) used for pesticide application to a public water system unless the label-prescribed safety devices for public water supplies are in place.

STORAGE AND DISPOSAL

STORAGE

Store unused product in original container only, in cool, dry area out of reach of children and animals. Do not store in areas where temperatures frequently exceed 100°F.

If container is damaged, before cleaning up, put on long-sleeved shirt, full length trousers, head covering, and protective gloves.

PESTICIDE DISPOSAL

Do not contaminate water, food or feed by storage or disposal. Open dumping is prohibited. Wastes resulting from the use of this product may be disposed of on site or at an approved waste disposal facility.

CONTAINER DISPOSAL

Triple rinse (or equivalent). Then offer for recycling or reconditioning, or dispose of in a sanitary landfill, or incineration, or if allowed by state and local authorities, by burning. If container is burned, stay out of smoke.

LIMITED WARRANTY AND DISCLAIMER

The manufacturer warrants (a) that this product conforms to the chemical description on the label; (b) that this product is reasonably fit for the purposes set forth in the directions for use when it is used in accordance with such directions; and (c) that the directions, warnings and other statements on this label are based upon responsible experts' evaluation of reasonable tests of effectiveness of toxicity to laboratory animals and to plants, and of residues on food crops, and upon reports of field experience. Tests have not been made on all varieties or in all states or under all conditions. THE MANUFACTURER NEITHER MAKES NOR INTENDS, NOR DOES IT AUTHORIZE ANY AGENT OR REPRESENTATIVE TO MAKE, ANY OTHER WARRANTIES, EXPRESS OR IMPLIED, AND IT EXPRESSLY EXCLUDES AND DISCLAIMS ALL IMPLIED WARRANTIES OF MERCHANTABILITY OR FITNESS FOR A PARTICULAR PURPOSE.

THIS WARRANTY DOES NOT EXTEND TO, AND THE BUYER SHALL BE SOLELY RESPONSIBLE FOR, ANY AND ALL LOSS OR DAMAGE WHICH RESULTS FROM THE USE OF THIS PRODUCT IN ANY MANNER WHICH IS INCONSISTENT WITH THE LABEL DIRECTIONS, WARNINGS OR CAUTIONS.

BUYER'S EXCLUSIVE REMEDY AND MANUFACTURER'S OR SELLER'S EXCLUSIVE LIABILITY FOR ANY AND ALL CLAIMS, LOSSES, DAMAGES OR INJURIES RESULTING FROM THE USE OR HANDLING OF THIS PRODUCT, WHETHER OR NOT BASED IN CONTRACT, NEGLIGENCE, STRICT LIABILITY IN TORT OR OTHERWISE, SHALL BE LIMITED, AT THE MANUFACTURER'S OPTION, TO REPLACEMENT OF, OR THE REPAYMENT OF THE PURCHASE PRICE FOR, THE QUANTITY OF PRODUCT WITH RESPECT TO WHICH DAMAGES ARE CLAIMED. IN NO EVENT SHALL MANUFACTURER OR SELLER BE LIABLE FOR SPECIAL, INDIRECT OR CONSEQUENTIAL DAMAGES RESULTING FROM THE USE OR HANDLING OF THIS PRODUCT.

SEVIN is the registered trademark of RHONE-POULENC NEDERLAND B.V. for carbaryl insecticides.

AG 90057 EPA Approval: 7/19/88

THIS SPECIMEN LABEL IS INTENDED FOR USE ONLY AS A GUIDE IN PROVIDING GENERAL INFORMATION REGARDING THE DIRECTIONS, WARNINGS AND CAUTIONS ASSOCIATED WITH THE USE OF THIS PRODUCT. AS WITH ANY AGRICULTURAL CHEMICAL, ALWAYS FOLLOW THE LABEL INSTRUCTIONS ON THE PACKAGE BEFORE USING.

TACKLE® PLUS

Herbicide

FOR BROAD SPECTRUM WEED CONTROL IN SOYBEANS AND PEANUTS

A soluble liquid formulation containing:

ACTIVE INGREDIENT*:

Sodium salt of bentazon: 3-(1methyl[ethyl])-1-H-2,1,3-benzothiadiazin-4-(3H)-one, 2, 2-dioxide) 29.2%

Sodium salt of acifluorfen:.sodium 5-[2-chloro-4-(trifluoromethyl)-phenoxy]-2-nitrobenzoate 13.4%

INERT INGREDIENTS: 57.4%

Equivalent to 2.67 pounds per gallon bentazon and 1.33 pounds per gallon of sodium acifluorfen.

EPA Reg No. 7969-76-264 EPA Est. No.

KEEP OUT OF REACH OF CHILDREN

DANGER PELIGRO

PRECAUCION AL USUARIO: Si usted no lee ingles, no use este producto hasta que la etiqueta haya sido explicada ampliamente.

STATEMENT OF PRACTICAL TREATMENT

IF IN EYES: Flush eyes with large amounts of water for at least 15 minutes. Get medical attention.

IF SWALLOWED: Call a physician or Poison Control Center. Drink 1 or 2 glasses of water and induce vomiting by touching back of throat with finger. Do not induce vomiting or give anything by mouth to an unconscious person.

IF ON SKIN: Wash with plenty of soap and water. Consult a physician if irritation persists.

PRECAUTIONARY STATEMENTS

DANGER

HAZARDS TO HUMANS AND DOMESTIC ANIMALS

Causes irreversible eye damage. Harmful if swallowed, inhaled or absorbed through skin. Do not get in eyes. Wear goggles or face shield when handling. Avoid breathing vapors or spray mist and contact with skin or clothing. In case of contact, immediately remove contaminated clothing and shoes. Wash thoroughly with soap and water after handling. Wash contaminated clothing with soap and hot water before reuse. This product may cause an allergic skin reaction.

STATEMENT OF PRACTICAL TREATMENT

IF IN EYES: Flush eyes with large amounts of water for at least 15 minutes. Get medical attention.

IF SWALLOWED: Call a physician or Poison Control Center. Drink 1 or 2 glasses of water and induce vomiting by touching back of throat with finger. Do not induce vomiting or give anything by mouth to an unconscious person.

IF ON SKIN: Wash with plenty of soap and water. Get medical attention if irritation persists.

NOTE TO PHYSICIAN

Emesis is recommended.

ENVIRONMENTAL HAZARDS

Do not apply directly to water or wetlands. Do not contaminate water when disposing of equipment washwaters. Do not apply when weather conditions favor drift from target area.

DIRECTIONS FOR USE

IT IS A VIOLATION OF FEDERAL LAW TO USE THIS PRODUCT IN A MANNER INCONSISTENT WITH ITS LABELING.

RE-ENTRY AND WORKER PROTECTION STATEMENTS

Do not apply this product in such a manner as to directly or through drift, expose workers or other persons. The area being treated must be vacated by unprotected persons. Do not enter treated areas without protective clothing until sprays have dried. Because certain states may require more restrictive re-entry intervals for various crops treated with this product, consult your State Department of Agriculture for further information.

GENERAL INFORMATION

For weed control in soybeans and peanuts in the following states: AL, AR, DE, FL, GA, KY, LA, MD, MS, NC, NJ, NM, OK, PA, SC, TN, TX, VA, and in Southeast MO. (Jefferson County and south).

TACKLE® Plus Herbicide is intended fo selective post-emergence control of certain broadleaf weeds. TACKLE® Plus may provide partial control of some grasses. TACKLE® Plus is effective mainly through contact action; therefore, weeds must be thoroughly covered with spray. TACKLE® Plus may cause some soybean leaf-speckling and leaf-bronzing to occur under certain conditions. (See Restrictions and Limitations)

TIMING OF APPLICATON

Apply postemergence applications of TACKLE® Plus Herbicide early when weeds are small and actively growing and before weeds reach the minimum size listed in the APPLICATION RATE TABLE.

IN PEANUTS: Applications may be made from cracking through two expanded tetrafoliates leaves. Soybeans would generally be in the 2 to 3 trifoliate stage. Early application to weeds produces the most beneficial affect on weed control, and makes it easier to obtain thorough spray coverage. Delay in application which permits weeds to exceed the maximum size stated will result in inadequate control.

Cultivation before or during the application is not recommended. Cultivation may put weeds under stress, thus making control more difficult to obtain. Timely cultivation after application will usually assist in weed control.

WATER VOLUME AND SPRAY PRESSURE (Apply recommended rates of TACKLE® Plus Herbicide as follows):

Ground Equipment

Use a minimum of 20 gallons of water per broadcast acre and a minimum of 40 psi pressure (measured at the boom not at the pump or in the line). When crop and weed foliage is dense, use up to 50 gallons of water and up to 80 psi pressure. Use standard high pressure pesticide hollow cone or flat fan nozzles spaced 20 inches apart. Do not use flood, whirl chamber or controlled droplet applicator (CDA) nozzles.

Air Equipment

Use a minimum of 10 gallons of water per acre and a maximum of 40 psi pressure. Use only diaphragm-type nozzles producing cone or fan spray patterns.

Aerial application-special directions: To obtain uniform coverage and to avoid drift hazards, the following application equipment and practices should be used.

Nozzle Orientation

Nozzles must be oriented so as to discharge straight back with the air stream (opposite the direction of travel of the aircraft) and not more than 20 degrees downward. Nozzles must not be located further out than 3/4 the distance from the center of the aircraft to the end of the wing or rotor.

Water volume and spray pressure (See Air Equipment)

Do not apply TACKLE® Plus by aircraft when wind is blowing at a velocity above 10 mph. Coarse sprays (larger droplets) are less likely to drift.

Do not apply TACKLE® Plus by air if ornamentals or sensitive non-target crops, such as cotton, sugar beets, sunflowers, or okra are within 200 feet downwind.

Applicator must follow the most restrictive use precautions to avoid drift hazards including those found in this labeling as well as applicable state and local regulations and ordnances.

ADDITION OF OIL CONCENTRATE TO SPRAY TANK

A non-phytotoxic oil concentrate (commonly referred to as oil concentrate) should be added to the spray tank with TACKLE® Plus Herbicide. The oil concentrate must contain either a pertroleum or vegetable oil base and must meet the following criteria: 1) be non-phytotoxic: 2) contain only EPA exampt ingredients; 3) provide good mixing quality in the jar test (See below); and 4) be successful in local experience. The exact composition of suitable products will vary; however, vegetable and petroleum oil concentrates should contain emulsifiers which provide good mixing quality. For vegetable oil concentrates, it has been observed that highly refined vegetable oils are more satisfactory than unrefined vegetable oils.

For additional information, **see "Jar test for estimating suitability of oil concentrates"** at the end of this section.

With the addition of oil concentrate to TACKLE® Plus Herbicide on soybeans and peanuts, a leaf burn may occur; but all new growth is normal and crop vigor is not reduced. The potential for leaf burn is increased when relative humidity and temperatures are high. A few oil concentrates have exhibited excessive leaf burn.

Refer to your supplier for information concerning successful local experience prior to purchasing any oil concentrate.

Rate of Oil Concentrate

Ground application:	**1 pint/acre (maximum)**
Air application:	**1 pint/acre (maximum)**

MIXING/SPRAYING

Fill a thoroughly cleaned sprayer 1/2 to 2/3 full with clean water. Start agitation and add TACKLE® Plus Herbicide, allow to mix thoroughly; add oil concentrate and remaining volume of water. Maintain constant agitation during application.

Jar test for estimating suitability of oil concentrates

1. **Water supply.**
   Use only water from intended source and at the source temperature.
2. **Amount of water in jar.**
   Ground application: For 20 gallons/acre spray volume, use 3 1/3 cups or 800 ml of water.
   Air application: For 10 gallons/acre spray volume, use 1 2/3 cups or 400 ml of water. For other spray volumes, adjust proportionately to above.
3. **Amount of TACKLE® Plus Herbicide and oil concentrate to add.**
   Add TACKLE® Plus and oil concentrate at the rate of 1 teaspoon or 5 ml for each pint of recommended label rate.
4. **Add components in the following sequence, gently mixing between component additions.**
   a. TACKLE® Plus Herbicide
   b. Oil Concentrate
5. **Cap, jar, invert 10 cycles, let stand for 15 minutes, evaluate.**
6. **Evaluation.**
   An ideal tank mix combination will be uniform; thus, the suitability of the oil concentrate is questionable if any of the following are observed:
   a. Free oil (film or gloubles) at the surface.
   b. Flocculation of fine particles which may be suspended in the liquid or found as a precipitated layer at the bottom of the jar.
   c. Clabbering - thickening texture (coagulated) resembling yogurt or a curd-like texture as with cottage cheese.

An 80% active nonionic spray adjuvant may be used (in place of the oil concentrate) at the recommended rate of 1 pint per 100 gallons. For certain weeds, a higher spray adjuvant rate is recommended.

APPLICATION RATE TABLE - Soybeans And Peanuts

Apply TACKLE® Plus Herbicide when weeds are small and actively growing and before weeds reach the maximum size listed below. Such applications generally correspond to the soybean growth stages of unifoliate to 2 expanded trifoliate leaves; and in peanuts to 2 expanded tetrafoliate leaves. Soybeans and peanuts may experience yellowing, bronzing, speckling, or burning of leaves under certain conditions, but soybeans and peanut plants generally outgrow this condition within 10 days.

APPLICATION FOR WEED GROWTH STAGES
Soybeans And Peanuts

Weed Controlled	Leaf Stage	Max. Weight	TACKLE® Plus Rate/Acre	Oil Concentrate Rate/Acre
Anoda, Spurred*	Up to 4	2″		
Carpetweed	3″ diameter	2″		
Cocklebur	2 to 6	6″		
Crotalaria	Up to 6	6″		
Croton, Tropic	1 to 2	<2″		
Croton, Woolly	1 to 2	<2″		
Jimsonweed	Up to 6	6″		
Ladysthumb	Up to 6	6″		
Lambsquarters*	4 to 6	2″		
Mallow, Venice	Up to 6	2″		
Morningglories	Up to 4	4″	1½ Pint	1 Pint Maximum
Mustard, Wild	Up to 6	4″		
Nightshade, Black	Up to 6	2″		
Pigweed, Redroot	Up to 6	2″		
Pigweed, Smooth	Up to 6	3″		
Ragweed, Common	4 to 6	3″		
Ragweed, Giant	Up to 4	6″		
Redweed	2 to 4	3″		
Sesbania, Hemp	Up to 4	6″		
Sida, Prickly or Teaweed*	Up to 4	2″		
Smartweed, or Pennsylvania	Up to 6	6″		
Starbur, Bristly	4 to 6	3″		
Velvetleaf*	Up to 4	2″		
Waterhemp, Tall	Up to 6	3″		

* *Control may be inconsistent with this rate of TACKLE® Plus Herbicide. A later application of Basagran® Herbicide may be necessary (See Basagran® label).*

RESTRICTIONS AND LIMITATIONS

Do not apply more than a total of 1½ pints of TACKLE® Plus Herbicide per acre per season to soybeans or peanuts.

Do not apply TACKLE® Plus Herbicide to soybeans or peanuts that have been subject to stress conditions as hail damage, flooding, drought, injury from other herbicides or widely fluctuating temperatures, as crop injury may result.

Do not apply TACKLE® Plus Herbicide to soybeans or peanuts that show injury (leaf phytotoxicity and/or plant stunting) produced by any other prior herbicide applications, because this injury may be enhanced and/or prolonged.

Do not apply TACKLE® Plus Herbicide during prolonged periods of drought or during unseasonably cold weather, as unsatisfactory weed control may result.

Rainfall or overhead irrigation soon after application (within 8 hours) may nullify the effectiveness of TACKLE® Plus.

Do not mix or apply TACKLE® Plus Herbicide with any other pesticide or any fertilizer except as specifically recommended on this labeling.

Clean sprayer thoroughly prior to application of TACKLE® Plus Herbicide, particularly if a herbicide was used which has the potential to injure the crop to be sprayed.

In the case of crop failure, only soybeans or peanuts may be immediately replanted.

Crop rotation restriction: Root crops (such as carrots, turnips, sweet potatoes, etc.) must not be planted in fields treated with TACKLE® Plus for a period of 18 months following treatment.

Do not apply TACKLE® Plus Herbicide within 50 days of soybean harvest and 75 days of peanut harvest.

Do not use treated plants for feed or forage.

Do not make more than 1 application of TACKLE® Plus Herbicide per season.

Avoid drift to all other crops and non-target areas.

Do not apply TACKLE® Plus Herbicide through any type of irrigation system.

STORAGE AND DISPOSAL

Keep from freezing. Store above 32° F. Do not contaminate water, food, or feed by storage or disposal.

PESTICIDE DISPOSAL

Pesticide wastes are acutely hazardous. Improper disposal of excess pesticide spray mixture, or rinsate is a violation of Federal Law. If these wastes cannot be disposed of by use according to label instructions, contact your State Pesticide or Environmental Control Agency, or the Hazardous Waste representative at the nearest EPA Regional Office for guidance.

CONTAINER DISPOSAL

Triple rinse container (or equivalent). Do not reuse empty container. Puncture and dispose of in a sanitary landfill, or by incineration, or if allowed by state and local authorities, by burning. If burned, stay out of smoke.

IN CASE OF EMERGENCY

In case of large scale spillage regarding this product, call:

CHEMTREC - 1-800-424-9300
BASF Corporation - 1 (201) 263-3400
Rhône-Poulenc Ag Company - 1-800-334-7577

In case of medical emergency regarding this product, call:

1. Your local doctor for immediate treatment
2. Your local Poison Control Center (hospital)
3. BASF Corporation 1 (201) 263-3400

Steps to be taken in case material is released or spilled.

1. Dike and contain spill with inert material (sand, earth, etc.) and transfer liquid and solid diking material to separate containers for disposal.
2. Remove contaminated clothing and wash affected skin areas with water.
3. Wash clothing before reuse.
4. Keep spill out of all sewers and open bodied water.

LIMITED WARRANTY AND DISCLAIMER

The manufacturer warrants (a) that this product conforms to the chemical description on the label; (b) that this product is reasonably fit for the purposes set forth in the directions for use when it is used in accordance with such directions; and (c) that the directions, warning and other statements on this label are based upon responsible experts' evaluation of reasonable tests of effectiveness, of toxicity to laboratory animals and to plants, and of residues on food crops and upon reports of field experience. Tests have not been made on all varieties or in all states or under all conditions. THE MANUFACTURER NEITHER MAKES NOR INTENDS, NOR DOES IT AUTHORIZE ANY AGENT OR REPRESENTATIVE TO MAKE, ANY OTHER WARRANTIES, EXPRESS OR IMPLIED, AND IT EXPRESSLY EXCLUDES AND DISCLAIMS ALL IMPLIED WARRANTIES OF MERCHANTABILITY OR FITNESS FOR A PARTICULAR PURPOSE.

THIS WARRANTY DOES NOT EXTEND TO, AND THE BUYER SHALL BE SOLELY RESPONSIBLE FOR, ANY AND ALL LOSS OR DAMAGE WHICH RESULTS FROM THE USE OF THIS PRODUCT IN ANY MANNER WHICH IS INCONSISTENT WITH THE LABEL DIRECTIONS. WARNINGS OR CAUTIONS.

BUYER'S EXCLUSIVE REMEDY AND MANUFACTURER'S OR SELLER'S EXCLUSIVE LIABILITY FOR ANY AND ALL CLAIMS, LOSSES, DAMAGES, OR INJURIES RESULTING FROM THE USE OR HANDLING OF THIS PRODUCT, WHETHER OR NOT BASED IN CONTRACT, NEGLIGENCE, STRICT LIABILITY IN TORT OR OTHERWISE SHALL BE LIMITED, AT THE MANUFACTURER'S OPTION, TO REPLACEMENT OF OR THE REPAYMENT OF THE PURCHASE PRICE FOR, THE QUANTITY OF PRODUCT WITH RESPECT TO WHICH DAMAGES ARE CLAIMED. IN NO EVENT SHALL MANUFACTURER OR SELLER BE LIABLE FOR SPECIAL, INDIRECT OR CONSEQUENTIAL DAMAGES RESULTING FROM THE USE OR HANDLING OF THIS PRODUCT.

NOTICE TO BUYER

Purchase of this material does not confer any rights under patents governing this product or the use thereof in countries outside of the United States.

Rhône-Poulenc Ag Company

Form No. AG90015

THIS SPECIMEN LABEL IS INTENDED FOR USE ONLY AS A GUIDE IN PROVIDING GENERAL INFORMATION REGARDING THE DIRECTIONS, WARNINGS AND CAUTIONS ASSOCIATED WITH THE USE OF THIS PRODUCT. AS WITH ANY AGRICULTURAL CHEMICAL, ALWAYS FOLLOW THE LABEL INSTRUCTIONS ON THE PACKAGE BEFORE USING.

TEMIK® BRAND 15G

ALDICARB PESTICIDE

RESTRICTED USE PESTICIDE

FOR RETAIL SALE TO AND USE ONLY BY CERTIFIED APPLICATORS OR PERSONS UNDER THE DIRECT SUPERVISION OF A CERTIFIED APPLICATOR, AND ONLY FOR THOSE USES COVERED BY THE CERTIFIED APPLICATOR'S CERTIFICATION.

For Control of Certain Insects, Mites, and Nematodes

ACTIVE INGREDIENT: Aldicarb [2-methyl-2-(methylthio) propionaldehyde *0*-(methylcarbamoyl)oxime] 15%
INERT INGREDIENTS: 85%

TOTAL 100%

EPA Reg. No. 264-330 EPA Est. No. 10352-GA-1

KEEP OUT OF REACH OF CHILDREN

DANGER–POISON

PELIGRO

PRECAUCION AL USUARIO: Si usted no lee ingles, no use este producto hasta que le etiqueta haya sido explicada ampliamente.

TRANSLATION TO THE USER: If you cannot read English do not use this product until the label has been fully explained to you.

STATEMENT OF PRACTICAL TREATMENT

IF SWALLOWED: Drink 1 or 2 glasses of water and induce vomiting by touching back of throat with finger. Do not induce vomiting or give anything orally to an unconscious or convulsing person. Call a physician and follow General advice.

IF IN EYES: Flush eyes with plenty of water and get medical attention.

IF ON SKIN: Wash thoroughly with soap and water. Remove contaminated clothing and wash before reuse.

IF INHALED: Call a physician and follow General advice.

For PRODUCT USE Information Call 1-800-334-9745
For MEDICAL and TRANSPORTATION Emergencies ONLY Call 24 Hours A Day 1-800-334-7577

PRECAUTIONARY STATEMENTS

DANGER

HAZARDS TO HUMANS AND DOMESTIC ANIMALS

FATAL IF SWALLOWED. Causes Cholinesterase Inhibition. May Be Fatal Or Harmful By Skin Or Eye Contact Or By Breathing Dust. Rapidly Absorbed Through Skin Or Eyes. Do Not Get On Skin Or In Eyes. Do Not Breathe Dust. Keep Away From Domestic Animals.

Wear long-sleeved clothing, full-length trousers, head coverings, and protective gloves when handling. Always stand up-wind from hopper when loading. Wash hands and face before eating, drinking, or using tobacco. Bathe at the end of work day, washing entire body and hair with soap and water. Change clothing daily. Wash contaminated clothing in strong washing soda solution and rinse thoroughly before reusing.

SIGNS AND SYMPTOMS OF OVEREXPOSURE:

Salivation	Muscle tremor	Nausea
Watery eyes	Difficult breathing	Vomiting
Pinpoint eye pupils	Excessive sweating	Diarrhea
Blurred vision	Abdominal cramps	Weakness
Headache		

In severe cases, convulsions, unconsciousness, and respiratory failure may occur.

ANTIDOTE STATEMENT

ATROPINE SULFATE IS HIGHLY EFFECTIVE AS AN ANTIDOTE. See NOTE TO PHYSICIAN below.

STATEMENT OF PRACTICAL TREATMENT

GENERAL

Contact a physician immediately in all cases of suspected poisoning. Illness may be produced rapidly following overexposure to TEMIK® Aldicarb. If breathing stops, establish an airway and start artificial respiration, and provide oxygen. Make certain to remove all sources of continuing contamination. Remove clothing and wash skin and hair immediately with large amounts of water. Transport the patient to a physician or hospital immediately and SHOW A COPY OF THIS LABEL TO THE PHYSICIAN. If poisoning is suspected in animals, contact a veterinarian.

IF SWALLOWED: Drink 1 or 2 glasses of water and induce vomiting by touching back of throat with finger. Do not induce vomiting or give anything orally to an unconscious or convulsing person. Call a physician and follow General advice listed above.

IF IN EYES: Flush eyes with plenty of water and get medical attention.

IF ON SKIN: Wash thoroughly with soap and water. Remove contaminated clothing and wash before reuse.

IF INHALED: Call a physician and follow General advice listed above.

NOTE TO PHYSICIAN

TEMIK® brand aldicarb is a methyl carbamate insecticide which is a cholinesterase inhibitor. Overexposure to this substance may cause toxic signs and symptoms due to stimulation of the cholinergic nervous system. These effects of overexposure are spontaneously and rapidly reversible.

Gastric lavage may be used if this product has been swallowed. TEMIK® brand aldicarb poisoning may occur rapidly after ingestion and prompt removal of stomach contents is indicated.

Specific treatment consists of the administration of parenteral atropine sulfate. Caution should be exercised to prevent overatropinization. Mild cases may be given 1 to 2 mg intramuscularly every 10 minutes until full atropinization has been achieved and repeated thereafter whenever symptoms reappear. Severe cases should be given 2 to 4 mg intravenously every 10 minutes until the patient is fully atropinized, then intramuscularly every 30 to 60 minutes as needed to maintain the effect for at least 12 hours. Dosages for children should be appropriately reduced. Complete recovery from overexposure is to be expected within 24 hours.

Narcotics and other sedatives should not be used. Further, drugs such as (pyridine-2-aldoxime methiodide) are NOT recommended unless organophosphate intoxication is also suggested.

To aid in confirmation of a diagnosis, urine samples must be obtained within 24 hours of exposure and immediately frozen. Analyses will be arranged by Rhône-Poulenc Ag Company.

Consultation on therapy can be obtained at all hours by calling the Rhône Poulenc emergency number: 1-800-334-7577.

ENVIRONMENTAL HAZARDS

DECOMPOSITION AND MOVEMENT IN SOIL: This product is readily decomposed into harmless residues under most use conditions. However, a combination of permeable and acidic soil conditions, moderate to heavy irrigation and/or rainfall, use of 20 or more pounds per acre, and soil temperature below 50°F (10°C) at application time tend to reduce degradation and promote movement of residues to groundwater. If the above describes your local use conditions and groundwater in your area is used for drinking, do not use this product without first contacting Rhône-Poulenc at 1-800-334-7577.

TOXIC TO FISH, BIRDS, AND WILDLIFE: This product is toxic to fish, birds, aquatic invertebrates and other wildlife. Birds feeding on exposed granules in treated areas may be killed. Cover or incorporate granules in spill areas.

Run-off from treated areas may be hazardous to fish in neighboring areas. Do not apply directly to water or wetlands (swamps, bogs, marshes and potholes). Keep out of lakes, streams, and ponds. Do not contaminate water when disposing of equipment washwaters. Apply this product only as specified on this label.

NOTICE: Under the Endangered Species Act, it is a Federal Offense to use any pesticide in a manner that results in the death of a member of an endangered species.

This Act protects Attwater's Greater Prairie Chicken in the Texas counties of Aransas, Austin, Brazoria, Colorado, Galveston, Goliad, Harris, Refugio, and Victoria.

Prior to making applications in these counties, the user must determine that this species is not located in or immediately adjacent to the area to be treated. If the user is in doubt whether or not the above named endangered species may be affected, he should contact either the regional U.S. Fish & Wildlife Service office (Endangered Species Specialist) or personnel of the State Fish and Game office.

GENERAL DIRECTIONS FOR USE

GENERAL INFORMATION

READ ENTIRE LABEL BEFORE USING THIS PRODUCT.

TEMIK® brand 15G Aldicarb Pesticide controls certain insects, mites, and nematodes. When applied into moist soil at planting and/or postemergence the active ingredient aldicarb is rapidly absorbed by roots and translocated to all parts of the plant. Rainfall or irrigation soon after application will ensure prompt uptake of aldicarb. Control often lasts more than six weeks varying with growing conditions, rate of use, and pests. Crop yields are usually increased with treatments of TEMIK® brand 15G Aldicarb Pesticide.

GENERAL CAUTIONS

Use TEMIK® brand 15G Aldicarb Pesticide only in accordance with label directions, warnings, and cautions. DO NOT USE ON ANY CROP NOT LISTED ON THIS LABEL OR SUPPLEMENTAL LABELING

AS ANY RESIDUES REMAINING MAY BE ILLEGAL OR HARMFUL. DO NOT STORE OR USE IN OR AROUND THE HOME OR HOME GARDEN. Use high rate on heavy organic or clay soils. Do not exceed the maximum label rate. Applications at higher rates or at more frequent intervals than is stated on the label may result in illegal residues in crop, meat, and milk. Treatments in excess of 7 pounds per acre made directly in the seed furrow of cotton, sugar beets, soybeans, or drybeans may delay plant emergence and reduce plant stand.

Make side-dress applications close enough to plants to allow good uptake by the roots without injury to the plants from root pruning. In irrigated areas, follow application with irrigation within one week. If alternate furrows are irrigated after side-dress application, TEMIK® brand 15G Aldicarb Pesticide and water must be on the same side of the plant row.

Calibrate and adjust application equipment to insure proper rate and accurate placement. Clean application equipment thoroughly after use. For any left over material, see instructions for STORAGE AND DISPOSAL in this booklet.

Deep disc any spills at row ends immediately to prevent birds from feeding on exposed granules.

Do not wash, load, or empty application equipment near any well, as this practice is a potential source of ground water contamination.

Do not apply in Del Norte or Humboldt counties in California; Nassau or Suffolk Counties in New York; or in Curry County, Oregon.

COMPATIBILITY

Pesticidal activity of TEMIK® brand 15G Aldicarb Pesticide is not affected by normal applications of fertilizers or other pesticides. Its effectiveness may be reduced or lost if applied with alkaline materials such as lime. To minimize potential exposure hazards, do not mix TEMIK® aldicarb with other materials before application.

GENERAL WORKER PROTECTION STATEMENTS

Do not apply this product in such a manner as to directly expose workers or other persons. The area being treated must be vacated by unprotected persons.

Do not enter treated area without protective footwear. Persons coming in direct contact with wet treated soil after the initial irrigation or rainfall following treatment shall wear coveralls, rubber or neoprene boots, and gloves. Because certain states may require more restrictive re-entry intervals for various crops treated with this product, consult your State Department of Agriculture for further information. Written or oral warnings must be given to workers who are expected to be in a treated area or in an area about to be treated with this product. Advise workers to stay out of fields during application. Regular long-sleeved shirt, full-length trousers, and head covering should be worn when working in treated fields. See Precautionary Statement, Statement of Practical Treatment, and Note To Physician for information on accidental exposures. When oral warnings are given, warnings shall be given in a language customarily understood by workers. Oral warnings must be given if there is reason to believe that written warnings cannot be understood by workers. Written warnings must include the following information: appropriate signal word (DANGER-POISON), area treated with TEMIK® brand 15G Aldicarb Pesticide, date of application, appropriate protective clothing, and re-entry requirements.

PRE-HARVEST, GRAZING AND GROUNDWATER LIMITATIONS

To avoid excessive residues in or on:

COTTONSEED

- Do not make more than one 'at planting' application and one postemergence application per crop.
- Do not apply within 90 days of harvest.
- Do not feed cotton forage to livestock or allow livestock to graze in treated area.

POTATOES

- Do not make more than one application per crop.
- Do not harvest within 90 days of an at planting application or within 50 days of an 'after planting' application.

PEANUTS

- Do not make more than one application per crop.
- Do not harvest within 90 days of application.
- Do not hog-off treated fields.
- Do not feed hay or vines to livestock.

SUGAR BEETS

- Do not make more than one at planting application and two postemergence applications per crop.
- Do not exceed a total of 40 pounds per acre.
- Do not apply within 90 days of harvest.
- If tops are to be fed to livestock, do not apply within 120 days of harvest.
- Do not use tops as food for humans.

SOYBEANS

- Do not make more than one application per crop.
- Do not harvest within 90 days of application.
- Do not feed green forage, hay, or straw to livestock.

DRY BEANS

- Do not make more than one application per crop.
- Do not harvest within 90 days of application.
- Do not feed green forage, hay or straw to livestock.
- Do not use green pods as food for humans.

SUGARCANE AND SWEET POTATOES

- Do not make more than one application per crop.
- Do not harvest within 120 days of application.
- Do not feed sweet potato vines to livestock.

PECANS

- Do not make more than one application per year.
- Do not allow livestock to graze in treated areas.
- Do not harvest foliage or hay from treated areas.

SORGHUM

- Do not make more than one application per year.
- Do not harvest within 90 days of application.
- Do not feed green forage to livestock.

OTHER CROPS

- Do not plant any crop not listed on this label in soil treated with TEMIK® 15G Aldicarb Pesticide within six months after last application.

MEAT AND MILK

- Do not allow livestock to graze in treated areas before harvest.

GROUNDWATER

- Observe Environmental Hazard precautions regarding Decomposition and Movement in Soil, and carefully follow Directions For Use.
- Do not wash, load, or empty application requirement near any well, as this practice is a potential source of ground water contamination.
- Do not apply more than 33 lbs/A in the state of Florida.
- Do not apply in Del Norte or Humboldt counties in California; Nassau or Suffolk Counties in New York; or in Curry County, Oregon.
- Do not apply to any field more than once every two years in Maine and Wisconsin.

DIRECTIONS FOR USE AND RECOMMENDED APPLICATIONS

IT IS A VIOLATION OF FEDERAL LAW TO USE THIS PRODUCT IN A MANNER INCONSISTENT WITH ITS LABELING. To provide maximum performance and to minimize hazard to birds, work granules into or cover with soil to a depth of 2 inches or more. Deep disc any spills at row ends immediately to prevent birds from feeding on granules. When a range of rates is recommended, use the higher rate if pest infestations are expected to be severe.

CROP & TIME OF APPLICATION	PESTS CONTROLLED	POUNDS/ACRE	OUNCES/1000 FEET OF ROW	RECOMMENDED APPLICATION
COTTON		40" row spacing		
	Aphids, Thrips	3.5 to 5	4.5 to 6	
		(Except Texas Oklahoma and New Mexico)		
		2 to 5	2.5 to 6	
		(Texas, Oklahoma, and New Mexico Only)		
At planting	Fleahoppers, Leafminers, Mites, Overwintering boll weevil (adults feeding on foliage), Plant bugs, including Lygus	4 to 7	5 to 8.5	Drill granules just below seed line or place in seed furrow and cover with soil. If seeds and TEMIK are hill-dropped, TEMIK rates may be reduced by one-half.
	Nematodes	3.5 to 10 (Except Far West)	4.5 to 12	Drill granules just below seed line or place in seed furrow and cover with soil. If rate exceeds 7 pounds per acre apply granules in a 4 to 6-inch band and work into the soil or cover with soil; plant seed in or above treated zone.

CROP & TIME OF APPLICATION	PESTS CONTROLLED	POUNDS/ACRE	OUNCES/1000 FEET OF ROW	RECOMMENDED APPLICATION
	Nematodes	14 to 27 (Far West Only)	17 to 33	Apply granules in a 4 to 6-inch band 3 to 6 inches below seed line or cover with 3 to 6 inches of soil. Plant seed above treated zone.
At first squaring	Leafhoppers, Fleahoppers	7 to 14	8.5 to 17	Side-dress granules 8 to 16 inches to one side of the plant row and 2 to 6 inches deep (usually at or below bottom of water furrow).
	Mites, Boll weevil, Plant bugs, including Lygus	14 to 20	17 to 24.5	
From squaring through early bloom	Cotton leaf perforator	14	17	
	Whiteflies	14 to 20	17 to 24.5	
GRAIN SORGHUM		36" row spacing		
	Nematodes	7	7.5	
		3.5 to 7 (Texas and Oklahoma only)	4 to 7.5	
At Planting	Greenbug and Chinch Bug (early season control)	7	7.5	Apply granules in seed furrow and cover with soil.
POTATOES DO NOT APPLY IN SUFFOLK COUNTY LONG ISLAND, NEW YORK. In the following states apply at emergence only, before plants are 3 inches high: States of: Maine and Wisconsin Counties of: Hartford, CT Kent and New Castle, DE Franklin and Hampshire, MA Worchester, MD Atlantic, Burlington, Cucumberland, Monmouth and Salem, NJ Newport and Washington, RI Accomack and Northampton, VA		34" row spacing		
	Aphids	7 (Maine Only)	7	Apply granules in 4 to 6-inch band over row on top of hill. Cover granules with soil by cultivators or hilling discs.
	Colorado potato beetle Leafhoppers Aphids (except in Maine)	14	14.5	
IN ALL OTHER AREAS: At Planting Do not exceed a total of 20 lbs. per acre.	Flea beetles	20	21	Apply granules with seed pieces in planting furrow and cover with soil.
	Aphids, Colorado potato beetle, Leafhoppers	14 to 20	14.5 to 21	Apply as above, OR for furrow irrigation drill granules 2 to 4 inches on both sides of seed row and 1 to 2 inches below seed piece depth.
	Psyllids	14 (Arizona Only)	14.5	Apply as above.
	Nematodes (Except, Golden Nematode and Root-knot nematodes). Verticillium wilt has been suppressed where nematode populations have been controlled. Root-knot nematodes are suppressed. Despite suppression of Root-knot nematode, tuber quality may not be adequately protected. If Root-knot nematode is a severe economic pest, the use of other control measures should also be employed.	20	21	Apply granules with seed pieces in planting furrow and cover with soil, OR apply granules in an 6-inch band and work into the soil or cover with soil to a depth of 4 inches. Plant seed pieces in treated zone.
After Planting At emergence before plants are 3 inches high.	Aphids, Colorado potato beetle	14 to 20	14.5 to 21	Apply granules in a 4 to 6 inch band over on top of hill. Cover granules with soil by cultivator or hilling discs. Where furrow irrigation is used, side-dress granules 5 to 8 inches on both sides of plant row and 4 inches deep where pest populations are expected 2-3 weeks after plant emergence. Do not apply if at planting treatment was made.
PEANUTS		36" row spacing		
At planting	Thrips	7 to 14	7.5 to 15	Apply granules in seed furrow and cover with soil. In Southwest use high rate only.
Do not exceed a total of 20 lbs. per acre.	Nematodes	14 to 20	15 to 22	Apply granules in a 6 to 12-inch band and work into the soil or cover with soil to a depth of 2 to 4 inches. Plant seed in treated zone.
SUGAR BEETS		22" row spacing		

CROP & TIME OF APPLICATION	PESTS CONTROLLED	POUNDS/ACRE	OUNCES/1000 FEET OF ROW	RECOMMENDED APPLICATION
At planting or within one week before planting.	Nematodes	27 to 33	18 to 22	Apply granules in a 4 to 6-inch band and immediately work into the soil or cover with soil to a depth of 2 to 4 inches. Plant seed in or above treated zone, OR where furrow irrigation is employed for seed germination, drill granules 3 to 4 inches deep and 3 inches from seed row on water furrow side.
At planting	Aphids	7 to 14	4.5 to 9.5	Drill granules 1 to 3 inches below seed line. Granules can be placed in seed furrow if rate does not exceed 7 pounds per acre. Repeat applications will be required for continued protection against virus vectors (aphid and leafhoppers).
	Leafminers, Leafhoppers	14 to 20	9.5 to 13.5	
	Sugar beet root maggot	7 to 14	4.5 to 9.5	Apply granules in a 2 to 3-inch band over seed row and immediately work into the soil or cover with soil, OR where furrow irrigation is employed for seed germination, drill granules 2 inches deep and 2 inches from seed row on water furrow side. For the 7 lb. rate, apply granules in a 1 to 2 inch band in front of the press wheel as the furrow is closing.
At planting plus postemergence (split application)	Nematodes	14 to 20 at planting and 14 to 20 postemergence	9.5 to 13.5 at planting and 9.5 to 13.5 postemergence	AT PLANTING: Apply granules in a 4 to 6-inch band and immediately work into the soil or cover with soil to a depth of 2 to 4 inches. Plant seed in or above treated zone, OR where furrow irrigation is employed for seed germination, drill granules 3 to 4 inches deep and 3 inches from seed row on water furrow side. POSTEMERGENCE: Apply granules to both sides of plant row and immediately work into the soil or cover with soil, or for furrow irrigation side-dress granules 4 to 8 inches to water furrow side of plant row at furrow depth. Irrigate soon after application. Apply within 60 days after planting.
Postemergence	Sugar beet root maggot	7 to 14	4.5 to 9.5	Apply granules to both sides of plant row and immediately work into the soil or cover with soil, or for furrow irrigation side-dress granules 4 to 8 inches to water furrow side of plant row ant at Furrow depth. Irrigate soon after application.
Do not make any postemergence application if 27 to 33 lbs/acre were applied at planting or 1 week before planting.	Aphids	7 to 14	4.5 to 9.5	Apply as above. A repeat application may be required for continued protection against virus vectors (aphids, leafhoppers).
	Leafminers, Leafhoppers	14 to 20	9.5 to 13.5	
Do not make more than 1 at-planting application and 2 postemergence applications per crop. Do not exceed a total of 40 lbs per acre.	Nematodes	27	18	Apply as above. Apply within 60 days after planting.
SOYBEANS		30" to 48" row spacing		
At planting	Mexican bean beetle Thrips Suppression of three cornered alfalfa hopper	5 to 10	5.5 to 11	Drill granules 2 to 3 inches below seed line, OR 2 to 3 inches to the side of the seed row and 2 to 3 inches deep. Granules can be applied in seed furrow if rate does not exceed 5 pounds per acre.
Do not exceed a total of 20 lbs. per acre.	Nematodes	10 to 20 (Based on 36 inch row spacing)	11 to 22	Apply granules in a 6 to 8-inch band and immediately work into the soil or cover with soil. Plant seed in treated zone.
BEANS (Dried only) At planting		22" to 48" row spacing		
	Seedcorn maggot	3.5 to 5	4.0 to 5.5	Apply granules in seed furrow and cover with soil.
		(Michigan Only)		
	Aphids	5 to 7	5.5 to 7.5	Drill granules 2 to 3 inches below seed line OR, 2 to 3 inches to side of seed row and 2 to 3 inches deep, OR Granules can be placed in seed if rate does not exceed 5 pounds per acre.
	Leafhoppers, Mexican bean beetle, Mites	7 to 14		

CROP & TIME OF APPLICATION	PESTS CONTROLLED	POUNDS/ACRE	OUNCES/1000 FEET OF ROW	RECOMMENDED APPLICATION
Do not exceed a total of 14 lbs. per acre.	Nematodes	7 to 14	7.5 to 15	Apply granules in a 6-inch band and immediately work into the soil or cover with soil. Plant seed in treated zone. Where furrow irrigation is employed soon after planting, drill granules 3 to 4 inches deep and 3 inches from seed row on water furrow side.
SWEET POTATOES		48" row spacing		
At planting	Nematodes	10 - 20	15 to 30	Apply granules in a 12-inch band in opened row. Cover immediately with soil by hilling up 8 to 10 inches. Plant in center of treated zone.
SUGAR CANE		60" row spacing		
At planting	Nematodes	14 to 20	26 to 37 (Louisiana Only)	Apply granules in open row on top of newly planted cane and cover immediately with at least 6 inches of soil.
PECANS (Southeast Only) Producing trees During period from bud break to nut set, or about April 15 to May 15	Aphids, Mites	33 - 67 (except Florida)	Not Applicable	Apply as a 4 to 6-foot band along dripline on both sides of tree row by spreading granules uniformly and immediately working into soil, OR by shanking 2-3 inches into soil on 12 inch centers.
One to two weeks before bud break.	Suppression of pecan leaf phylloxera	33 (Florida only)		In Florida, apply between January 1 and April 30 only.
Newly transplanted trees (1-5 years old) During period of bud break, or about April 15 to May 15 (except Florida)	Aphids Mites	Dosage Ounces TEMIK 15G/Tree		Apply as a side-dress to individual trees by spreading the granules uniformly around the tree and immediately working into the soil to a depth of 2-3 inches.
One to two weeks before bud break.	Bud moth, Suppression of pecan leaf phylloxera	5 - 20	Not Applicable	In Florida, do not exceed 33 lbs/A and apply between January 1 and April 30 only.

SEE PREHARVEST AND GRAZING LIMITATIONS

PLANT GROWTH STIMULATION: In the absence of recognizable target pests and under certain growing conditions, stimulation of plant growth by TEMIK aldicarb on certain crops has been demonstrated under laboratory, greenhouse, and field conditions. Plants may be visibly taller, greener, and denser. Faster grow-off and an increase in fruiting rate and size often are results of the growth enhancement by TEMIK® brand aldicarb. However, such effects do not occur with all crops or under all conditions. Therefore, this product should not be used solely as a plant growth regulator, and its effect as one, should be considered a side benefit which may or may not occur as a consequence of its use.

TEMIK® BRAND 15G (GYPSUM) ALDICARB PESTICIDE CALIBRATION GUIDE
APPROXIMATE APPLICATOR SETTINGS* FOR TEMIK® BRAND 15G ALDICARB PESTICIDE (GYPSUM) AT 4, 5 & 6 MPH

NOTE: This calibration chart is applicable only to TEMIK® brand 15G Aldicarb Pesticide in this container which is formulated for use on a gypsum carrier. All rates are APPROXIMATIONS and must be confirmed by using a calibration tube (See Calibration Guide provided with tube). Calibration tubes are available from your TEMIK band 15G Aldicarb supplier.

POUNDS TEMIK® 15G PER ACRE FOR VARIOUS ROW SPACINGS					TYPE OF GRANULAR APPLICATOR									
ROW SPACING					GANDY GAUGE SETTING			JOHN DEERE (7100) SETTING			NOBLE E-Z FlOW GAUGE SETTING			CLAMPCO ⅜" FEEDER DIAL RPM
					MPH			MPH			MPH			MPH
22"	34"	36"	40"	48"	4	5	6	4	5	6	4	5	6	5
1.8	1.2	1.1	1	.8	7	8	9	3**	3**	4**	2	2	3	3
3.6	2.4	2.2	2	1.7	10	11	12	4**	5**	6**	3	4	5	6
5.5	3.5	3.3	3	2.5	12	13	14	6**	7**	8	5	6	7	8
7.3	4.7	4.4	4	3.3	13	15	16	7**	9	10	6	7	8	11
9.1	5.9	5.5	5	4.2	14	16	18	9	11	13	7	9	10	14
10.9	7.1	6.7	6	5.0	16	18	19	11	13	15	8	10	12	16
12.7	8.2	7.8	7	5.8	17	19	21	12	15	18	9	11	13	18
14.5	9.4	8.9	8	6.7	18	20	22	14	17	20	10	12	14	21
16.4	10.6	10.0	9	7.5	19	21	23	16	19	23	11	13	15	23
18.2	11.8	11.1	10	8.3	20	22	24	17	21	25	12	15	16	26
21.8	14.1	13.3	12	10.0	22	24	26	20	25	28	14	16	19	31
25.5	16.5	15.6	14	11.7	23	26	29	24	28	30	16	18	21	37
29.1	18.8	17.8	16	13.3	25	28	31	26	29	32	17	20	23	42
32.7	21.2	20.0	18	15.0	26	30	33	28	31	33	19	21	24	48
36.4	23.5	22.2	20	16.7	28	31	34	30	32	35	20	23	26	54
43.6	28.2	26.7	24	20.0	31	34	37	32	34	37	22	26	29	65
49.1	31.8	30.0	27	22.5	33	36	40	33	36	38	24	28	31	74

* All settings for one hopper box outlet per row
** For John Deere gauge setting below 8 there will be great variations in flow rate. A gauge setting of 6 is the lowest recommended setting.

NOTE: THIS CALIBRATION CHART WILL APPEAR ON THE 30 LBS POUR BOX.

TEMIK BRAND 15G (CORNCOB) ALDICARB PESTICIDE CALIBRATION GUIDE
APPROXIMATE APPLICATOR SETTINGS* FOR TEMIK® BRAND 15G ALDICARB PESTICIDE (CORNCOB) AT 4, 5, & 6 MPH

NOTE: This calibration chart is applicable only to TEMIK® brand 15G Aldicarb Pesticide in this container which is formulated for use on a corncob carrier. All rates are APPROXIMATIONS and must be confirmed by using a calibration tube (See Calibration Guide provided with tube). Calibration tubes are available from your TEMIK 15G Aldicarb supplier.

POUNDS TEMIK 15 G PER ACRE FOR VARIOUS ROW SPACINGS					TYPE OF GRANULAR APPLICATOR									
ROW SPACING					GANDY GAUGE SETTING			JOHN DEERE MAX EMERGE GAUGE SETTING**			NOBLE E-Z FLOW GAUGE SETTING			CLAMPCO ⅜" FEEDER DIAL RPM
					MPH			MPH			MPH			MPH
22"	34"	36"	40"	48"	4	5	6	4	5	6	4	5	6	5
1.8	1.2	1.1	1	.8	14	16	17	8	9	10	5	6	7	4
3.6	2.4	2.2	2	1.7	19	21	22	11	13	14	10	11	13	8
5.5	3.5	3.3	3	2.5	22	24	27	14	16	18	13	15	17	11
7.3	4.7	4.4	4	3.3	25	28	30	17	19	21	16	18	21	15
9.1	5.9	5.5	5	4.2	28	31	33	19	21	23	18	22	24	19
10.9	7.1	6.7	6	5.0	30	33	35	21	23	25	21	24	26	22
12.7	8.2	7.8	7	5.8	32	35	37	22	25	26	23	26	29	26
14.5	9.4	8.9	8	6.7	34	37	39	24	26	28	25	28	31	30
16.4	10.6	10.0	9	7.5	35	38	41	25	27	29	26	30	33	33
18.2	11.8	11.1	10	8.3	37	40	43	26	28	30	28	31	34	37
21.8	14.1	13.3	12	10.0	39	43	47	28	30	33	31	34	37	44
25.5	16.5	15.6	14	11.7	42	46	50	29	32	35	33	36	40	52
29.1	18.8	17.8	16	13.3	44	49	54	31	34	37	35	39	44	59
32.7	21.2	20.0	18	15.0	47	52	59	33	36	38	37	42		67
36.4	23.5	22.2	20	16.7	49	56	66	34	37	40	39	45		76
43.6	28.2	26.7	24	20.0	54	66		37	40	44	44			95
49.1	31.8	30.0	27	22.5	59			38	42					

* All settings for one hopper box outlet per row
** Odd Numbered Housing

STORAGE AND DISPOSAL

STORAGE

Store unused TEMIK® Brand Aldicarb pesticide in original container only, in well ventilated clean dry area out of reach of children and animals. Do not store in areas where temperature averages 115°F (46°C) or greater. Do not store in or around the home or home garden.

If container is broken, before cleaning up put on long-sleeved shirt, full-length trousers, head covering and rubber gloves. Do not get dust or granules on skin or in eyes. Do not breathe dust. Always stand upwind from spill when cleaning up. Sweep up and bury any small spills or excess TEMIK® Aldicarb at least 18 inches deep in soil, isolated from water supplies and food crops.

PESTICIDE DISPOSAL

Do not contaminate water, food, or feed by storage or dipsosal.

Open dumping is prohibited.

Pesticide wastes are acutely hazardous. Improper disposal of excess pesticide is a violation of federal law. If these wastes cannot be disposed of by use according to label instructions, contact your state pesticide or environmental control agency or the hazardous waste representative at the nearest EPA regional office for guidance.

CONTAINER DISPOSAL

Completely empty container into application equipment. Then dispose of empty container in a sanitary landfill or by incineration, or if allowed by state and local authorities, by open burning. If container is burned, stay away from and do not breathe or contact smoke.

In case of transportation or warehouse emergency involving a spill, fire, or exposure, call collect 1 (800) 334-7577 twenty-four hours a day in the USA.

LIMITED WARRANTY AND DISCLAIMER

The manufacturer warrants (a) that this product conforms to the chemical description on the label; (b) that this product is reasonably fit for the purposes set forth in the directions for use when it is used in accordance with such directions; and (c) that the directions, warnings and other statements on this label are based upon responsible experts' evaluation of reasonable tests of effectiveness, of toxicity to laboratory animals and to plants, and of residues on food crops, and upon reports of field experience. Tests have not been made on all varieties or in all states or under all conditions. THE MANUFACTURER NEITHER MAKES NOR INTENDS, NOR DOES IT AUTHORIZE ANY AGENT OR REPRESENTATIVE TO MAKE, ANY OTHER WARRANTIES, EXPRESS OR IMPLIED, AND IT EXPRESSLY EXCLUDES AND DISCLAIMS ALL IMPLIED WARRANTIES OF MERCHANTABILITY OR FITNESS FOR A PARTICULAR PURPOSE.

THIS WARRANTY DOES NOT EXTEND TO, AND THE BUYER SHALL BE SOLELY RESPONSIBLE FOR, ANY AND ALL LOSS OR DAMAGE WHICH RESULTS FROM THE USE OF THIS PRODUCT IN ANY MANNER WHICH IS INCONSISTENT WITH THE LABEL DIRECTIONS, WARNINGS OR CAUTIONS.

BUYER'S EXCLUSIVE REMEDY AND MANUFACTURER'S OR SELLER'S EXCLUSIVE LIABILITY FOR ANY CLAIMS, LOSSES, DAMAGES, OR INJURIES RESULTING FROM THE USE OR HANDLING OF THIS PRODUCT WHETHER OR NOT SUCH LIABILITY IS BASED IN CONTRACT, NEGLIGENCE, STRICT LIABILITY IN TORT OR OTHERWISE SHALL BE LIMITED, AT THE MANUFACTURER'S OPTION, TO REPLACEMENT OF, OR THE REPAYMENT OF THE PURCHASE PRICE FOR, THE QUANTITY OF PRODUCT WITH RESPECT TO WHICH DAMAGES ARE CLAIMED. IN NO EVENT SHALL MANUFACTURER OR SELLER BE LIABLE FOR SPECIAL, INDIRECT OR CONSEQUENTIAL DAMAGES RESULTING FROM THE USE OR HANDLING OF THIS PRODUCT.

TEMIK is a registered trademark of RHONE-POULENC for aldicarb pesticide.

AG 90057

EPA Approval: 2/7/89

THIS SPECIMEN LABEL IS INTENDED FOR USE ONLY AS A GUIDE IN PROVIDING GENERAL INFORMATION REGARDING THE DIRECTIONS, WARNINGS AND CAUTIONS ASSOCIATED WITH THE USE OF THIS PRODUCT. AS WITH ANY AGRICULTURAL CHEMICAL, ALWAYS FOLLOW THE LABEL INSTRUCTIONS ON THE PACKAGE BEFORE USING.

TEMIK® brand 15G

ALDICARB PESTICIDE

For Use on Citrus Only

RESTRICTED USE PESTICIDE

FOR RETAIL SALE TO AND USE ONLY BY CERTIFIED APPLICATORS OR PERSONS UNDER THE DIRECT SUPERVISION OF A CERTIFIED APPLICATOR, AND ONLY FOR THOSE USES COVERED BY THE CERTIFIED APPLICATOR'S CERTIFICATION.

For Control of Certain Insects, Mites, and Nematodes on Oranges, Grapefruit, Lemons, and Limes Only

ACTIVE INGREDIENT:
Aldicarb [2-methyl-2-(methylthio) propionaldehyde O-(methylcarbamoyl)oxime] 15%
INERT INGREDIENTS: 85%
TOTAL. 100%

EPA REG. NO. 264-417 EPA EST. NO. 264-GA-1

KEEP OUT OF REACH OF CHILDREN

 **DANGER–POISON** 

PELIGRO

PRECAUCION AL USUARIO: Si usted no lee ingles, no use este producto hasta que le etiqueta haya sido explicado ampliamente.

TRANSLATION TO THE USER: IF YOU CANNOT READ ENGLISH DO NOT USE THIS PRODUCT UNTIL THE LABEL HAS BEEN FULLY EXPLAINED TO YOU.

STATEMENT OF PRACTICAL TREATMENT

IF SWALLOWED: Drink 1 or 2 glasses of water and induce vomiting by touching back of throat with finger. Do not induce vomiting or give anything orally to an unconscious or convulsing person. Call a physician and follow General advice.

IF IN EYES: Flush eyes with plenty of water and get medical attention.

IF ON SKIN: Wash thoroughly with soap and water. Remove contaminated clothing and wash before reuse.

IF INHALED: Call a physician and follow General advice.

For PRODUCT USE Information Call 1-800-334-9745.

For MEDICAL and TRANSPORTATION EMERGENCIES ONLY Call 24 Hours A Day 1-800-334-7577

PRECAUTIONARY STATEMENTS

DANGER

HAZARDS TO HUMANS AND DOMESTIC ANIMALS

FATAL IF SWALLOWED. CAUSES CHOLINESTERASE INHIBITION. MAY BE FATAL OR HARMFUL BY SKIN OR EYE CONTACT OR BY BREATHING DUST. RAPIDLY ABSORBED THROUGH SKIN OR EYES. DO NOT GET ON SKIN OR IN EYES. DO NOT BREATHE THE DUST. KEEP AWAY FROM DOMESTIC ANIMALS.

WEAR LONG-SLEEVED CLOTHING, FULL-LENGTH TROUSERS, HEAD COVERING, AND PROTECTIVE GLOVES WHEN HANDLING. ALWAYS STAND UP-WIND FROM HOPPER WHEN LOADING. WASH HANDS AND FACE BEFORE EATING, DRINKING OR USING TOBACCO. BATHE AT THE END OF WORK DAY, WASHING ENTIRE BODY AND HAIR WITH SOAP AND WATER. CHANGE CLOTHING DAILY. WASH CONTAMINATED CLOTHING IN STRONG WASHING SODA SOLUTION AND RINSE THOROUGHLY BEFORE REUSING.

SIGNS AND SYMPTOMS OF OVEREXPOSURE:

Salivation	Muscle tremor	Nausea
Watery eyes	Difficult breathing	Vomiting
Pinpoint eye pupils	Excessive sweating	Diarrhea
Blurred vision	Abdominal cramps	Weakness
Headache		

In severe cases, convulsions, unconsciousness, and respiratory failure may occur.

ANTIDOTE STATEMENT

Atropine Sulfate is highly effective as an antidote. See NOTE TO PHYSICIAN below.

STATEMENT OF PRACTICAL TREATMENT

GENERAL

Contact a physician immediately in all cases of suspected poisoning. Illness may be produced rapidly following overexposure to TEMIK® Aldicarb. If breathing stops, establish an airway, start artificial respiration, and provide oxygen. Make certain to remove all sources of continuing contamination. Remove clothing and wash skin and hair immediately with large amounts of water. Transport the patient to a physician or hospital immediately and SHOW A COPY OF THIS LABEL TO THE PHYSICIAN. If poisoning is suspected in animals, contact a veterinarian.

IF SWALLOWED: Drink 1 to 2 glasses of water and induce vomiting by touching back of throat with finger. Do not induce vomiting or give anything orally to an unconscious or convulsing person. Call a physician and follow General advice listed above.

IF IN EYES: Flush eyes with plenty of water and get medical attention.

IF ON SKIN: Wash thoroughly with soap and water. Remove contaminated clothing and wash before reuse.

IF INHALED: Call physician and follow General advice listed above.

NOTE TO PHYSICIAN

TEMIK® brand aldicarb is a methyl carbamate insecticide which is a cholinesterase inhibitor. Overexposure to this substance may cause toxic signs and symptoms due to stimulation of the cholinergic nervous system. These effects of overexposure are spontaneously and rapidly reversible.

Gastric lavage may be used if this product has been swallowed. TEMIK® brand aldicarb poisoning may occur rapidly after ingestion and prompt removal of stomach contents is indicated.

Specific treatment consists of the administration of parenteral atropine sulfate. Caution should be exercised to prevent overatropinization. Mild cases may be given 1 to 2 mg intramuscularly every 10 minutes until full atropinization has been achieved and repeated thereafter whenever symptoms reappear. Severe cases should be given 2 to 4 mg intravenously every 10 minutes until the patient is fully atropinized, then intramuscularly every 30 to 60 minutes as needed to maintain the effect for at least 12 hours. Dosages for children should be appropriately reduced. Complete recovery from overexposure is to be expected within 24 hours.

Narcotics and other sedatives should not be used. Further, drugs such as (pyridine-2-aldoxime methiodide) are NOT recommended unless organophosphate intoxication is also suggested.

To aid in confirmation of a diagnosis, urine samples must be obtained within 24 hours of exposure and immediately frozen. Analyses will be arranged by Rhône-Poulenc Ag Company.

Consultation on therapy can be obtained at all hours by calling the Rhône-Poulenc emergency number: 1-(800)-334-7577.

ENVIRONMENTAL HAZARDS

DECOMPOSITION AND MOVEMENT IN SOIL: This product is readily decomposed into harmless residues under most use conditions. However, a combination of sandy and acidic soil conditions, moderate to heavy irrigation and/or rainfall, use of 20 or more pounds per acre, and soil temperature below 50°F (10°C) at application time tend to reduce degradation and promote movement of residues to groundwater. If the above describes your local use conditions and groundwater in your area is used for drinking, do not use this product without first contacting Rhône-Poulenc at 1-(800) 334-8577.

TOXIC TO FISH, BIRDS, AND WILDLIFE

This product is toxic to fish, birds, aquatic invertebrates, and other wildlife. Birds feeding on exposed granules in treated areas may be killed. Cover or incorporate granules in spill areas. Runoff from treated areas may be hazardous to fish in neighboring areas. Do not apply directly to water or wetlands (swamps, bogs, marshes, and potholes). Keep out of lakes, streams, and ponds. Do not contaminate water when disposing of equipment washwaters. Apply this product only as specified on this label.

NOTICE: Under the Endangered Species Act, it is a Federal Offense to use any pesticide in a manner that results in the death of a member of an endangered species.

This Act protects Attwater's Greater Prairie Chicken in the Texas counties of Aransas, Austin, Brazoria, Colorado, Galveston, Goliad, Harris, Refugio, and Victoria.

Prior to making applications in these counties, the user must determine that this species is not located in or immediately adjacent to the area to be treated. If the user is in doubt whether or not the above named endangered species may be affected, he should contact either the regional U.S. Fish & Wildlife Service office (Endangered Species Specialist) or personnel of the State Fish and Game office.

GENERAL DIRECTIONS FOR USE

GENERAL INFORMATION

READ ENTIRE LABEL BEFORE USING THIS PRODUCT.

This product controls certain insects, mites, and nematodes. When applied into moist soil the active ingredient aldicarb is rapidly absorbed by roots and translocated to all parts of the plant. Rainfall or irrigation soon after application will insure prompt uptake of aldicarb. Control often lasts more than six weeks varying with growing conditions, rate of use, and pests. Crop yields are usually increased with treatments of TEMIK® brand 15G Aldicarb Pesticide.

GENERAL CAUTIONS

Use TEMIK® brand 15G Aldicarb Pesticide only in accordance with label directions, warnings, and cautions. DO NOT USE ON ANY CROP NOT LISTED ON THIS LABEL OR SUPPLEMENTAL LABELING AS ANY RESIDUES REMAINING MAY BE ILLEGAL OR HARMFUL. DO NOT STORE OR USE IN OR AROUND THE HOME OR HOME GARDEN. Use high rate on heavy organic or clay soils. Do not exceed the maximum label rate. Applications at higher rates or at more frequent levels than is stated on the label may result in illegal residues in crop, meat, and milk.

Calibrate and adjust application equipment to insure proper rate and accurate placement. Clean application equipment thoroughly after use. For any left over material, see instructions for STORAGE AND DISPOSAL in this booklet.

Deep disc any spills at row ends immediately to prevent birds from feeding on exposed granules.

COMPATIBILITY

Pesticidal activity of TEMIK® brand 15G Aldicarb Pesticide is not affected by normal applications of fertilizers or other pesticides. Its effectiveness may be reduced or lost if applied with alkaline materials such as lime. To minimize potential exposure hazards, do not mix TEMIK® aldicarb with other materials before application.

GENERAL WORKER PROTECTION STATEMENTS

Do not apply this product in such a manner as to directly expose workers or other persons. The area being treated must be vacated by unprotected persons.

Do not enter treated areas without protective footwear. Persons coming in direct contact with wet treated soil after the initial irrigation or rainfall following treatment shall wear coveralls and rubber or neoprene boots, and gloves. Because certain states may require more restrictive re-entry intervals for various crops treated with this product, consult your State Department of Agriculture for further information. Written or oral warnings must be given to workers who are expected to be in a treated area or in an area about to be treated with this product. Advise workers to stay out of fields during application. Regular long-sleeved shirt, full-length trousers, and head covering should be worn when working in treated fields. See Precautionary Statements, Statement of Practical Treatment, and Note to Physician for information on accidental exposures. When oral warnings are given, warnings shall be given in a language customarily understood by workers. Oral warnings must be given if there is reason to believe that written warnings cannot be understood by workers. Written warnings must include the following information: Appropriate signal word (DANGER - POISON), area treated with TEMIK® brand 15G Aldicarb Pesticide, date of application, appropriate protective clothing, and re-entry requirements.

DIRECTIONS FOR USE ON CITRUS (ORANGES, GRAPEFRUIT, LEMONS, AND LIMES ONLY)

IT IS A VIOLATION OF FEDERAL LAW TO USE THIS PRODUCT IN A MANNER INCONSISTENT WITH ITS LABELING. To provide maximum performance and to minimize hazard to birds, work granules into or cover with soil to a depth of 2 inches or more. Deep disc any spills at row ends immediately to prevent birds from feeding on granules.

Pests Controlled	Pounds Per Acre (18 To 25 Ft Row Spacing)	Recommended Application
Citrus Nematode (Except Florida)	67 33 - 67 (Texas only)	Apply just prior to or during the spring flush of foliage growth in a band along the dripline on both sides of tree row. Spread granules uniformly and immediately work into the soil (preferred method); OR, shank granules 2 to 3 inches deep using 4 to 6 shanks on 12 inch centers. The width of each band should equal one fourth the tree row spacing.
Aphids Mites: Citrus Red Citrus Rust Texas Citrus Whiteflies	33 - 67 (Except Florida)	Apply as above, or apply in irrigation furrows using 2 shanks per furrow.
Aphids Mites: Citrus Rust Texas Citrus Whiteflies	33 (Florida only)	Apply as above between January 1 and April 30 only. See other ENVIRONMENTAL LIMITATIONS listed below.

PRE-HARVEST, GRAZING AND ENVIRONMENTAL LIMITATIONS

To avoid illegal residues:

- Do not make more than one application per year.
- Do not allow livestock to graze in treated areas.
- Do not harvest lemons within 30 days of application.
- Do not plant any crop not listed on this label in soil treated with TEMIK® 15G Aldicarb Pesticide within six months after last application.
- Do not apply this product in Suffolk County, Long Island, New York or Del Norte, California.
- In Florida, TEMIK® brand Aldicarb Pesticide cannot be applied within 300 feet of any drinking water well. In addition, TEMIK® aldicarb cannot be used in Florida citrus within 1000 feet of a drinking water well, when soils (such as Astatula soils) have a permeability rate greater than 20 inches per hour with an available water capacity less than 0.06 in all layers to a depth of 80 inches as identified by the U.S.D.A. Soil Conservation Service, unless it is known or reasonably believed based on authoritative sources that such wells are either cased to 100 feet below ground level or a minimum of 30 feet below the water table. The U.S.D.A. Soil Conservation Service which serves your county can tell you if the soils in your grove(s) fall within this category.

STORAGE AND DISPOSAL

STORAGE

Store unused TEMIK® brand Aldicarb Pesticide in original container only, in well ventilated clean dry area out of reach of children and animals. Do not store in areas where temperatures averages 115°F (46°C) or greater. Do not store in or around the home or home garden.

If container is broken, before cleaning up put on long-sleeved shirt, full-length trousers, head covering and rubber gloves. Do not get dust or granules on skin or in eyes. Do not breathe dust. Always stand upwind from spill when cleaning up. Sweep up and bury any small spills or excess TEMIK® Aldicarb at least 18 inches deep in soil, isolated from water supplies and food crops.

PESTICIDE DISPOSAL

Do not contaminate water, food, or feed by storage or disposal.

Open dumping is prohibited.

Pesticide wastes are acutely hazardous. Improper disposal of excess pesticide is a violation of Federal law. If these wastes cannot be disposed of by use according to label instructions, contact your State Pesticide or Environmental Control Agency, or The Hazardous Waste representative at the nearest EPA Regional Office for guidance.

CONTAINER DISPOSAL

Completely empty container into the application equipment. Then dispose of empty container in a sanitary landfill or by incineration, or if allowed by state and local authorities, by open burning. If container is burned, stay away from and do not breathe or contact smoke.

IN CASE OF A TRANSPORTATION OR WAREHOUSE EMERGENCY INVOLVING A SPILL, FIRE, OR EXPOSURE, CALL COLLECT 1-(800) 334-7577 TWENTY-FOUR HOURS A DAY IN THE U.S.A.

LIMITED WARRANTY AND DISCLAIMER

The manufacturer warrants (a) that this product conforms to the chemical description on the label; (b) that this product is reasonably fit for the purposes set forth in the directions for use when it is used in accordance with such directions; and (c) that the directions, warnings, and other statements on this label are based upon responsible experts' evaluation of reasonable tests of effectiveness, of toxicity to laboratory animals and to plants, and of residues on food crops, and upon reports of field experience. Tests have not been made on all varieties or in all states or under all conditions. THE MANUFACTURER NEITHER MAKES NOR INTENDS, NOR DOES IT AUTHORIZE ANY AGENT OR REPRESENTATIVE TO MAKE, ANY OTHER WARRANTIES, EXPRESS OR IMPLIED, AND IT EXPRESSLY EXCLUDES AND DISCLAIMS ALL IMPLIED WARRANTIES OF MERCHANTABILITY OR FITNESS FOR A PARTICULAR PURPOSE.

THIS WARRANTY DOES NOT EXTEND TO, AND THE BUYER SHALL BE SOLELY RESPONSIBLE FOR, ANY AND ALL LOSS OR DAMAGE WHICH RESULTS FROM THE USE OF THIS PRODUCT IN ANY MANNER WHICH IS INCONSISTENT WITH THE LABEL DIRECTIONS, WARNINGS, OR CAUTIONS.

BUYER'S EXCLUSIVE REMEDY AND MANUFACTURER'S OR SELLER'S EXCLUSIVE LIABILITY FOR ANY CLAIMS, LOSSES, DAMAGES, OR INJURIES RESULTING FROM THE USE OR HANDLING OF THIS PRODUCT WHETHER OR NOT SUCH LIABILITY IS BASED IN CONTRACT, NEGLIGENCE, STRICT LIABILITY IN TORT OR OTHERWISE SHALL BE LIMITED, AT THE MANUFACTURER'S OPTION TO REPLACEMENT OF, OR THE REPAYMENT OF THE PURCHASE PRICE FOR, THE QUANTITY OF PRODUCT WITH RESPECT TO WHICH DAMAGES ARE CLAIMED. IN NO EVENT SHALL MANUFACTURER OR SELLER BE LIABLE FOR SPECIAL, INDIRECT OR CONSEQUENTIAL DAMAGES RESULTING FROM THE USE OR HANDLING OF THIS PRODUCT.

NOTICE TO BUYER

Purchase of this material does not confer any rights under patents governing this product or the use thereof in countries outside of the United States.

TEMIK is the registered trademark of RHÔNE-POULENC for aldicarb pesticide.

AG 90059

EPA Approval: 2/5/87

THIS SPECIMEN LABEL IS INTENDED FOR USE ONLY AS A GUIDE IN PROVIDING GENERAL INFORMATION REGARDING THE DIRECTIONS, WARNINGS AND CAUTIONS ASSOCIATED WITH THE USE OF THIS PRODUCT. AS WITH ANY AGRICULTURAL CHEMICAL, ALWAYS FOLLOW THE LABEL INSTRUCTIONS ON THE PACKAGE BEFORE USING.

TEMIK® BRAND 15G

ALDICARB PESTICIDE

FOR SALE AND USE IN CALIFORNIA AND ARIZONA ONLY

RESTRICTED USE PESTICIDE

FOR RETAIL SALE TO AND USE ONLY BY CERTIFIED APPLICATORS OR PERSONS UNDER THE DIRECT SUPERVISION OF A CERTIFIED APPLICATOR, AND ONLY FOR THOSE USES COVERED BY THE CERTIFIED APPLICATOR'S CERTIFICATION.

For Control of Certain Insects, Mites, and Nematodes

ACTIVE INGREDIENT:
Aldicarb [2-methyl-2-(methylthio) propionaldehyde *O*-(methylcarbamoyl) oxime] 15%
INERT INGREDIENTS: 85%

TOTAL 100%

EPA Reg. No. 264-426 EPA Est. No. 10352-GA-01

For PRODUCT USE Information Call 1-800-334-9745
For MEDICAL and TRANSPORTATION Emergencies ONLY Call 24 Hours A Day 1-800-334-7577

KEEP OUT OF REACH OF CHILDREN

DANGER–POISON

PELIGRO

PRECAUCION AL USUARIO: Si usted no lee ingles no use este producto hasta que le etiqueto haya sido explicado ampliamente.

TRANSLATION TO THE USER: IF YOU CANNOT READ ENGLISH, DO NOT USE THIS PRODUCT UNTIL THE LABEL HAS BEEN FULLY EXPLAINED TO YOU.

STATEMENT OF PRACTICAL TREATMENT

IF SWALLOWED: Drink 1 or 2 glasses of water and induce vomiting by touching back of throat with finger. Do not induce vomiting or give anything orally to an unconscious or convulsing person. Call a physician and follow General advice.

IF IN EYES: Flush eyes with plenty of water and get medical attention.

IF ON SKIN: Wash thoroughly with soap and water. Remove contaminated clothing and wash before reuse.

IF INHALED: Call a physician and follow General advice.

PRECAUTIONARY STATEMENTS

DANGER

HAZARDS TO HUMANS AND DOMESTIC ANIMALS

FATAL IF SWALLOWED. Causes cholinesterase inhibition. May be fatal or harmful by skin or eye contact or by breathing dust. Rapidly absorbed through skin or eyes. Do not get on skin or in eyes. Do not breathe dust. Keep away from domestic animals.

Wear long-sleeved clothing, full-length trousers, head coverings, and protective gloves when handling. Always stand up-wind from hopper when loading. Wash hands and face before eating drinking or using tobacco. Bathe at the end of work day, washing entire body and hair with soap and water. Change clothing daily. Wash contaminated clothing in strong washing soda solution and rinse thoroughly before reusing.

SIGNS AND SYMPTOMS OF OVEREXPOSURE:

Salivation	Muscle tremor	Nausea
Watery eyes	Difficult breathing	Vomiting
Pinpoint eye pupils	Excessive sweating	Diarrhea
Blurred vision	Abdominal cramps	Weakness
Headache		

In severe cases, convulsions, unconsciousness, and respiratory failure may occur.

ANTIDOTE STATEMENT

ATROPINE SULFATE IS HIGHLY EFFECTIVE AS AN ANTIDOTE. See NOTE TO PHYSICIAN below.

STATEMENT OF PRACTICAL TREATMENT

GENERAL

Contact a physician immediately in all cases of suspected poisoning. Illness may be produced rapidly following overexposure to TEMIK® Aldicarb. If breathing stops, establish an airway and start artificial respiration, and provide oxygen. Make certain to remove all sources of continuing contamination. Remove clothing and wash skin and hair immediately with large amounts of water. Transport the patient to a physician or hospital immediately and SHOW A COPY OF THIS LABEL TO THE PHYSICIAN. If poisoning is suspected in animals, contact a veterinarian.

IF SWALLOWED: Drink 1 or 2 glasses of water and induce vomiting by touching back of throat with finger. Do not induce vomiting or give anything orally to an unconscious or convulsing person. Call a physician and follow General advice listed above.

IF IN EYES: Flush eyes with plenty of water and get medical attention.

IF ON SKIN: Wash thoroughly with soap and water. Remove contaminated clothing and wash before reuse.

IF INHALED: Call a physician and follow General advice listed above.

NOTE TO PHYSICIAN

TEMIK® brand aldicarb is a methyl carbamate insecticide which is a cholinesterase inhibitor. Overexposure to this substance may cause toxic signs and symptoms due to stimulation of the cholinergic nervous system. These effects of overexposure are spontaneously and rapidly reversible.

Gastric lavage may be used if this product has been swallowed. TEMIK® brand aldicarb poisoning may occur rapidly after ingestion and prompt removal of stomach contents is indicated.

Specific treatment consists of the administration of parenteral atropine sulfate. Caution should be exercised to prevent overatropinization. Mild cases may be given 1 to 2 mg intramuscularly every 10 minutes until full atropinization has been achieved and repeated thereafter whenever symptoms reappear. Severe cases should be given 2 to 4 mg intravenously every 10 minutes until the patient is fully atropinized, then intramuscularly every 30 to 60 minutes as needed to maintain the effect for at least 12 hours. Dosages for children should be appropriately reduced. Complete recovery from overexposure is to be expected within 24 hours.

Narcotics and other sedatives should not be used. Further, drugs such as (pyridine-2-aldoxime methiodide) are NOT recommended unless organophosphate intoxication is also suggested.

To aid in confirmation of a diagnosis, urine samples must be obtained within 24 hours of exposure and immediately frozen. Analyses will be arranged by Rhône-Poulenc Company.

Consultation on therapy can be obtained at all hours by calling the Rhône-Poulenc Company emergency number: 1-(800) 334-7577.

ENVIRONMENTAL HAZARDS

DECOMPOSITION AND MOVEMENT IN SOIL: This product is readily decomposed into harmless residues under most use conditions. However, a combination of permeable and acidic soil conditions, moderate to heavy irrigation and/or rainfall, use of 20 or more pounds per acre, and soil temperature below 50°F (10°C) at application time tend to reduce degradation and promote movement of residues to groundwater. If the above describes your local use conditions and groundwater in your area is used for drinking, do not use this product without first contacting Rhône-Poulenc at 1-(800) 334-8577.

TOXIC TO FISH, BIRDS, AND WILDLIFE: This product is toxic to fish, birds, aquatic invertebrates and other wildlife. Birds feeding on exposed granules in treated areas may be killed. Cover or incorporate granules in spill areas. Run-off from treated areas may be hazardous to fish in neighboring areas. Do not apply directly to water or wetlands (swamps, bogs, marshes and pot holes). Keep out of lakes, streams, and ponds. Do not contaminate water when disposing of equipment washwaters. Apply this product only as specified on this label.

GENERAL INFORMATION

READ ENTIRE LABEL BEFORE USING THIS PRODUCT.

TEMIK® brand 15G Aldicarb Pesticide controls certain insects, mites, and nematodes. When applied into moist soil at planting and/or postemergence the active ingredient aldicarb is rapidly absorbed by roots and translocated to all parts of the plant. Rainfall or irrigation soon after application will ensure prompt uptake of aldicarb. Control often lasts more than six weeks varying with growing conditions, rate of use, and pests. Crop yields are usually increased with treatments of TEMIK® brand 15G Aldicarb Pesticide.

GENERAL CAUTIONS

Use TEMIK® brand 15G Aldicarb Pesticide only in accordance with label directions, warnings, and cautions. DO NOT USE ON ANY CROP NOT LISTED ON THIS LABEL OR SUPPLEMENTAL LABELING AS ANY RESIDUES REMAINING MAY BE ILLEGAL OR HARMFUL. DO NOT STORE OR USE IN OR AROUND THE HOME OR HOME GARDEN. Use high rate on heavy organic or clay soils. Do not exceed the maximum label rate. Applications at higher rates or at more frequent intervals than is stated on the label may result in illegal residues in crop, meat, and milk. Treatments in excess of 7 pounds per acre made directly in the seed furrow of cotton, sugar beets, soybeans, or drybeans may delay plant emergence and reduce plant stand.

Make side-dress applications close enough to plants to allow good uptake by the roots without injury to the plants from root pruning. In irrigated areas, follow application with irrigation within one week. If alternate furrows are irrigated after side-dress application, TEMIK® brand 15G Aldicarb Pesticide and water must be on the same side of the plant row.

Calibrate and adjust application equipment to insure proper rate and accurate placement. Clean application equipment thoroughly after use. For any left over material, see instructions for STORAGE AND DISPOSAL in this booklet.

Deep disc any spills at row ends immediately to prevent birds from feeding on exposed granules.

Do not wash, load, or empty application equipment near any well, as this practice is a potential source of ground water contamination.

DO NOT APPLY IN DEL NORTE OR HUMBOLDT COUNTIES IN CALIFORNIA.

COMPATIBILITY

Pesticidal activity of TEMIK® brand 15G Aldicarb Pesticide is not affected by normal applications of fertilizers or other pesticides. Its effectiveness may be reduced or lost if applied with alkaline materials such as lime. To minimize potential exposure hazards, do not mix TEMIK® aldicarb with other materials before application.

GENERAL WORKER PROTECTION STATEMENTS

Do not apply this product in such a manner as to directly expose workers or other persons. The area being treated must be vacated by unprotected persons.

Do not enter treated area without protective footwear. Persons coming in direct contact with wet treated soil after the initial irrigation or rainfall following treatment shall wear coveralls, rubber or neoprene boots, and gloves. Because certain states may require more restrictive re-entry intervals for various crops treated with this product, consult your State Department of Agriculture for further information. Written or oral warnings must be given to workers who are expected to be in a treated area or in an area about to be treated with this product. Advise workers to stay out of fields during application. Regular long-sleeved shirt, full-length trousers, and head covering should be worn when working in treated fields. SEE PRECAUTIONARY STATEMENT, STATEMENT OF PRACTICAL TREATMENT, and NOTE TO PHYSICIAN for information on accidental exposures. When oral warnings are given, warnings shall be given in a language customarily understood by workers. Oral warnings must be given if there is reason to believe that written warnings cannot be understood by workers. Written warnings must include the following information: appropriate signal word (DANGER-POISON), area treated with TEMIK® brand 15G Aldicarb Pesticide, date of application, appropriate protective clothing, and re-entry requirements.

PRE-HARVEST, GRAZING AND GROUNDWATER LIMITATIONS

To avoid excessive residues in or on:

COTTONSEED

- Do not make more than one 'at planting' application and one post-emergence application per crop.
- Do not apply within 90 days of harvest.
- Do not feed cotton forage to livestock or allow livestock to graze in treated area.

POTATOES

- Do not make more than one application per crop.
- Do not harvest within 90 days of an 'at planting' application or within 50 days of an 'after planting' application.

PEANUTS

- Do not make more than one application per crop.
- Do not harvest within 90 days of application.
- Do not hog-off treated fields.
- Do not feed hay or vines to livestock.

SUGAR BEETS

- Do not make more than one 'at planting application' and two post-emergence applications per crop.
- Do not exceed a total of 40 pounds per acre.
- Do not apply within 90 days of harvest.
- If tops are to be fed to livestock, do not apply within 120 days of harvest.
- Do not use tops as food for humans.

SOYBEANS

- Do not make more than one application per crop.
- Do not harvest within 90 days of application.
- Do not feed green forage, hay, or straw to livestock.

DRY BEANS

- Do not make more than one application per crop.
- Do not harvest within 90 days of application.
- Do not feed green forage, hay or straw to livestock.
- Do not use green pods as food for humans.

SWEET POTATOES

- Do not make more than one application per crop.
- Do not harvest within 120 days of application.
- Do not feed sweet potato vines to livestock.

PECANS (Arizona only)

- Do not make more than one application per year.
- Do not allow livestock to graze in treated areas.
- Do not harvest foliage or hay from treated areas.

SORGHUM

- Do not make more than one application per year.
- Do not harvest within 90 days of application.
- Do not feed green forage to livestock.

OTHER CROPS

- Do not plant any crop not listed on this label in soil treated with TEMIK® 15G Aldicarb Pesticide within six months after last application.

MEAT AND MILK

- Do not allow livestock to graze in treated areas before harvest.

GROUNDWATER

- Observe Environmental Hazard precautions regarding Decomposition and Movement in Soil, and carefully follow Directions For Use.
- Do not wash, load, or empty application equipment near any well, as this pracice is a potential source of groundwater contamination.
- Do Not apply in Del Norte or Humboldt Counties, California.

DIRECTIONS FOR USE AND RECOMMENDED APPLICATIONS

IT IS A VIOLATION OF FEDERAL LAW TO USE THIS PRODUCT IN A MANNER INCONSISTENT WITH ITS LABELING. To provide maximum performance and to minimize hazard to birds, work granules into or cover with soil to a depth of 2 inches or more. Deep disc any spills at row ends immediately to prevent birds from feeding on granules. When a range of rates is recommended, use the higher rate if pest infestations are expected to be severe.

CROP & TIME OF APPLICATION	PESTS CONTROLLED	POUNDS/ACRE	OUNCES/1000 FEET OF ROW	RECOMMENDED APPLICATION
COTTON		40" row spacing		
At planting	Aphids, Thrips	3.5 to 5	4.5 to 6	Drill granules just below seed line or place in seed furrow and cover with soil. If seeds and TEMIK® are hill-dropped, TEMIK® rates may be reduced by one-half.
	Fleahoppers Leafminers, Mites, Overwintering boll weevil (adults feeding on foliage), Plant bugs, including Lygus	4 to 7	5 to 8.5	
	Nematodes	14 to 27	17 to 33	Apply granules in a 4 to 6-inch band 3 to 6 inches below seed line or cover with 3 to 6 inches of soil. Plant seed above treated zone.
At first squaring	Leafhoppers, Fleahoppers	7 to 14	8.5 to 17	Side-dress granules 8 to 16 inches to one side of the plant row and 2 to 6 inches deep (usually at or below bottom of water furrow).
	Mites, Boll weevil, Plant bugs, including Lygus	14 to 20	17 to 24.5	
From squaring through early	Cotton leaf perforator	14	17	

CROP & TIME OF APPLICATION	PESTS CONTROLLED	POUNDS/ACRE	OUNCES/1000 FEET OF ROW	RECOMMENDED APPLICATION
bloom	Whiteflies	14 to 20	17 to 24.5	
GRAIN SORGHUM At Planting		36" row spacing		Apply granules in seed furrow and cover with soil.
	Nematodes	7	7.5	
	Greenbug And Chinch Bug (early season control)	7	7.5	
POTATOES At-Planting		34" row spacing		
	Flea beetles	20	21	Apply granules with seed pieces in planting furrow and cover with soil.
	Aphids, Colorado potato beetle, Leafhoppers	14 to 20	14.5 to 21	Apply as above, OR for furrow irrigation drill granules 2 to 4 inches on both sides of seed row and 1 to 2 inches below seed piece depth.
	Psyllids	14 (Arizona Only)	14.5	Apply as above.
	Nematodes (Except Golden Nematode and Root-knot nematodes) Verticillium wilt has been suppressed where nematode populations have been controlled. Root-knot nematodes are suppressed. Despite suppression of Root-knot nematode, tuber quality may not be adequately protected. If Root-knot nematode is a severe economic pest, the use of other control measures should also be employed.	20	21	Apply granules with seed pieces in planting furrow and cover with soil, OR apply granules in an 8-inch band and work into the soil or cover with soil to a depth of 4 inches. Plant seed pieces in treated zone.
After Planting At emergence before plants are 3 inches high.	Aphids Colorado potato beetle	14 to 20	14.5 to 21	Apply granules in a 4 to 6 inch band over on top of hill. Cover granules with soil by cultivator or hilling discs. Where furrow irrigation is used, side-dress granules 5 to 8 inches on both sides of plant row and 4 inches deep where pest populations are expected 2 to 3 weeks after plant emergence. Do not apply if an at-planting treatment was made.
PEANUTS At planting		36" row spacing		
	Thrips	7 to 14	7.5 to 15	Apply granules in seed furrow and cover with soil. In Southwest use high rate only.
	Nematodes	14 to 20	15 to 22	Apply granules in a 6 to 12 inch band and work into the soil or cover with soil to a depth of 2 to 4 inches. Plant seed in treated zone.
SUGAR BEETS At planting or within one week **before planting.**		22" row spacing		
	Nematodes	27 to 33	18 to 22	Apply granules in a 4 to 6 inch band and immediately work into the soil or cover with soil to a depth of 2 to 4 inches. Plant seed in or above treated zone, OR Where furrow irrigation is employed for seed germination, drill granules 3 to 4 inches deep and 3 inches from seed row on water furrow side.
At planting	Aphids	7 to 14	4.5 to 9.5	Drill granules 1 to 3 inches below seed line. Granules can be placed in seed furrow if rate does not exceed 7 pounds per acre. Repeat applications will be required for continued protection against virus vectors (aphids and leafhoppers).
	Leafminers, Leafhoppers	14 to 20	9.5 to 13.5	
	Sugar beet root maggot	7 to 14	4.5 to 9.5	Apply granules in a 2 to 4-inch band over seed row and immediately work into the soil or cover with soil, OR Where furrow irrigation is employed for seed germination, drill granules 2 inches deep and 2 inches from seed row on water furrow side. For the 7 lb. rate, apply granules in a 1 to 2 inch band in front of the press wheel as the furrow is closing.
At planting plus postemergence (split applications)	Nematodes	14 to 20 at planting and postemergence	9.5 to 13.5 at planting and postemergence	AT PLANTING: Apply granules in a 4 to 6-inch band and immediately work into the soil or cover with soil to a depth of 2 to 4 inches. Plant seed in or above treated zone, OR Where furrow irrigation is employed for seed germination, drill granules 3 to 4 inches deep and 3 inches from seed row on water furrow side. POSTEMERGENCE: Apply granules to both sides of plant row and immediately work into the soil or cover with soil, OR For furrow irrigation, side-dress granules 4 to 8 inches to water furrow side of plant row at furrow depth. Irrigate soon after application. Apply within 60 days after planting.
Postemergence Do not make any postemergence application if 27 to 33 lbs/acre were used for nematode control at planting or 1 week before planting. Do not make more than 1 'at planting' application and 2 postemergence applications per crop.	Sugar beet root maggot	7 to 14	4.5 to 9.5	Apply granules to both sides of plant row and immediately work into the soil or cover with soil, OR For furrow irrigation, side-dress granules 4 to 8 inches to water furrow side or plant row and at furrow depth. Irrigate soon after application.
	Aphids	7 to 14	4.5 to 9.5	Apply as above. A repeat application may be required for continued protection against virus vectors (aphids, leafhoppers).
	Leafminers, Leafhoppers	14 to 20	9.5 to 13.5	
	Nematodes	27	18	Apply as above. Apply within 60 days after planting. Do not exceed a total of 40 lbs. per acre.
SOYBEANS		30"-48" row spacing		

CROP & TIME OF APPLICATION	PESTS CONTROLLED	POUNDS/ACRE	OUNCES/1000 FEET OF ROW	RECOMMENDED APPLICATION
At planting	Mexican bean beetle Thrips Suppression of three cornered alfalfa hopper	5 to 10	5.5 to 11	Drill granules 2 to 3 inches below seed line, OR two to 3 inches to the side of the seed row and 2 to 3 inches deep. Granules can be applied in seed furrow if rate does not exceed 5 pounds per acre.
	Nematodes	10 to 20	11 to 22	Apply granules in a 6 to 8-inch band and immediately work into the soil or cover with soil. Plant seed in treated zone.
BEANS (Dried only)		22 - 48" row spacing		
	Aphids	5 to 7	5.5 to 7.5	Drill granules 2 to 3 inches below seed line, OR 2 to 3 inches to side of seed row and 2 to 3 inches deep, OR Granules can be placed in seed furrow if rate does not exceed 5 lbs per acre.
	Leafhoppers Mexican bean beetle, Mites	7 to 14	7.5 to 15	
	Nematodes	7 to 14	7.5 to 15	Apply granules in a 6-inch band and immediately work into the soil or cover with soil. Plant seed in treated zone. OR Where furrow irrigation is employed soon after planting, drill granules 3 to 4 inches deep and 3 inches from seed row on water furrow side.
SWEET POTATOES At planting	Nematodes	48" row spacing 10 to 20	15 to 30	Apply granules in a 12-inch band in opened row. Cover immediately with soil by hilling up 8 to 10 inches. Plant in center of treated zone.
PECANS **Producing trees** During period from bud break to nut set	Aphids	33 to 47 (Arizona Only)	Not Applicable	Apply as a 4 to 6-foot band along dripline on both sides of tree row by spreading granules uniformly and immediately working into soil, or by shanking 2-3 inches into soil on 12 inch centers.
Newly transplanted trees (1-5 years old) During period of bud break	Aphids	Dosage Ounces TEMIK® 15G/Tree 5 to 20	Not Applicable	Apply as a side-dress to individual trees by spreading the granules uniformly around the tree and immediately working into the soil to a depth of 2-3 inches.

SEE PREHARVEST AND GRAZING LIMITATIONS ELSEWHERE IN THIS LABEL.

TEMIK® BRAND 15G (GYPSUM) ALDICARB PESTICIDE CALIBRATION GUIDE

APPROXIMATE APPLICATOR SETTINGS* FOR TEMIK® BRAND 15G ALDICARB PESTICIDE (GYPSUM) AT 4, 5 & 6 MPH

NOTE: This calibration chart is applicable only to TEMIK® brand 15G Aldicarb Pesticide in this container which is formulated for use on a gypsum carrier. All rates are APPROXIMATIONS and must be confirmed by using a calibration tube (See Calibration Guide provided with tube). Calibration tubes are available from your TEMIK® brand 15G Aldicarb supplier.

POUNDS TEMIK 15G PER ACRE FOR VARIOUS ROW SPACINGS					TYPE OF GRANULAR APPLICATOR									
ROW SPACING					GANDY GAUGE SETTING			JOHN DEERE (7100) SETTING			NOBLE E-Z FLOW GAUGE SETTING			CLAMPCO ⅜" FEEDER DIAL RPM
					MPH			MPH			MPH			MPH
22"	34"	36"	40"	48"	4	5	6	4	5	6	4	5	6	5
1.8	1.2	1.1	1	.8	7	8	9	3**	3**	4**	2	2	3	3
3.6	2.4	2.2	2	1.7	10	11	12	4**	5**	6**	3	4	5	6
5.5	3.5	3.3	3	2.5	12	13	14	6**	7**	8	5	6	7	8
7.3	4.7	4.4	4	3.3	13	15	16	7**	9	10	6	7	8	11
9.1	5.9	5.5	5	4.2	14	16	18	9	11	13	7	9	10	14
10.9	7.1	6.7	6	5.0	16	18	19	11	13	15	8	10	12	16
12.7	8.2	7.8	7	5.8	17	19	21	12	15	18	9	11	13	18
14.5	9.4	8.9	8	6.7	18	20	22	14	17	20	10	12	14	21
16.4	10.6	10.0	9	7.5	19	21	23	16	19	23	11	13	15	23
18.2	11.8	11.1	10	8.3	20	22	24	17	21	25	12	15	16	26
21.8	14.1	13.3	12	10.0	22	24	26	20	25	28	14	16	19	31
25.5	16.5	15.6	14	11.7	23	26	29	24	28	30	16	18	21	37
29.1	18.8	17.8	16	13.3	25	28	31	26	29	32	17	20	23	42
32.7	21.2	20.0	18	15.0	26	30	33	28	31	33	19	21	24	48
36.4	23.5	22.2	20	16.7	28	31	34	30	32	35	20	23	26	54
43.6	28.2	26.7	24	20.0	31	34	37	32	34	37	22	26	29	65
49.1	31.8	30.0	27	22.5	33	36	40	33	36	38	24	28	31	74

* All settings for one hopper box outlet per row

** For John Deere gauge setting below 8 there will be great variations in flow rate. A gauge setting of 6 is the lowest recommended setting.

TEMIK® BRAND 15G (CORNCOB) ALDICARB PESTICIDE CALIBRATION GUIDE

APPROXIMATE APPLICATOR SETTINGS* FOR TEMIK® BRAND 15G ALDICARB PESTICIDE (CORNCOB) AT 4, 5 & 6 MPH

NOTE: **This calibration chart is applicable only to TEMIK® brand 15G Aldicarb Pesticide in this container which is formulated for use on a corncob carrier. All rates are APPROXIMATIONS and must be confirmed by using a calibration tube (See Calibration Guide provided with tube). Calibration tubes are available from your TEMIK 15G Aldicarb supplier.**

POUNDS TEMIK® 15G PER ACRE FOR VARIOUS ROW SPACINGS					TYPE OF GRANULAR APPLICATOR GANDY			JOHN DEERE			NOBLE			CLAMPCO
ROW SPACING					GAUGE SETTING			MAX EMERGE GAUGE SETTING			E-Z FLOW GAUGE SETTING			½" FEEDER DIAL RPM
					MPH			MPH			MPH			MPH
22"	34"	36"	40"	48"	4	5	6	4	5	6	4	5	6	5
1.8	1.2	1.1	1	.8	14	16	17	8	9	10	5	6	7	4
3.6	2.4	2.2	2	1.7	19	21	22	11	13	14	10	11	13	8
5.5	3.5	3.3	3	2.5	22	24	27	14	16	18	13	15	17	11
7.3	4.7	4.4	4	3.3	25	28	30	17	19	21	16	18	21	15
9.1	5.9	5.5	5	4.2	28	31	33	19	21	23	18	22	24	19
10.9	7.1	6.7	6	5.0	30	33	35	21	23	25	21	24	26	22
12.7	8.2	7.8	7	5.8	32	35	37	22	25	26	23	26	29	26
14.5	9.4	8.9	8	6.7	34	37	39	24	26	28	25	28	31	30
16.4	10.6	10.0	9	7.5	35	38	41	25	27	29	26	30	33	33
18.2	11.8	11.1	10	8.3	37	40	43	26	28	30	28	31	34	37
21.8	14.1	13.3	12	10.0	39	43	47	28	30	33	31	34	37	44
25.5	16.5	15.6	14	11.7	42	46	50	29	32	35	33	36	40	52
29.1	18.8	17.8	16	13.3	44	49	54	31	34	37	35	39	44	59
32.7	21.2	20.0	18	15.0	47	52	59	33	36	38	37	42		67
36.4	23.5	22.2	20	16.7	49	56	66	34	37	40	39	45		76
43.6	28.2	26.7	24	20.0	54	66		37	40	44	44			95
49.1	31.8	30.0	27	22.5	59			38	42					

* All settings for one hopper box outlet per row

** Odd Number Housing

STORAGE AND DISPOSAL

STORAGE

STORE UNUSED TEMIK® brand ALDICARB PESTICIDE IN ORIGINAL CONTAINER ONLY, IN WELL VENTILATED CLEAN DRY AREA OUT OF REACH OF CHILDREN AND ANIMALS. DO NOT STORE IN AREAS WHERE TEMPERATURE AVERAGES 115°F (46°C) OR GREATER. DO NOT STORE IN OR AROUND THE HOME OR HOME GARDEN.

IF CONTAINER IS BROKEN, BEFORE CLEANING UP PUT ON LONG-SLEEVED SHIRT, FULL-LENGTH TROUSERS, HEAD COVERING AND RUBBER GLOVES. DO NOT GET DUST OR GRANULES ON SKIN OR IN EYES. DO NOT BREATHE DUST. ALWAYS STAND UPWIND FROM SPILL WHEN CLEANING UP. SWEEP UP AND BURY ANY SMALL SPILLS OR EXCESS TEMIK® ALDICARB AT LEAST 18 INCHES DEEP IN SOIL, ISOLATED FROM WATER SUPPLIES AND FOOD CROPS.

PESTICIDE DISPOSAL

DO NOT CONTAMINATE WATER, FOOD, OR FEED BY STORAGE OR DISPOSAL.

OPEN DUMPING IS PROHIBITED.

PESTICIDE WASTES ARE ACUTELY HAZARDOUS. IMPROPER DISPOSAL OF EXCESS PESTICIDE IS A VIOLATION OF FEDERAL LAW. IF THESE WASTES CANNOT BE DISPOSED OF BY USE ACCORDING TO LABEL INSTRUCTIONS, CONTACT YOUR STATE PESTICIDE OR ENVIRONMENTAL CONTROL AGENCY OR THE HAZARDOUS WASTE REPRESENTATIVE AT THE NEAREST EPA REGIONAL OFFICE FOR GUIDANCE.

CONTAINER DISPOSAL

COMPLETELY EMPTY CONTAINER INTO APPLICATION EQUIPMENT. THEN DISPOSE OF EMPTY CONTAINER IN A SANITARY LANDFILL OR BY INCINERATION, OR IF ALLOWED BY STATE AND LOCAL AUTHORITIES, BY OPEN BURNING. IF CONTAINER IS BURNED, STAY AWAY FROM AND DO NOT BREATHE OR CONTACT SMOKE.

IN CASE OF TRANSPORTATION OR WAREHOUSE EMERGENCY INVOLVING A SPILL, FIRE, OR EXPOSURE, CALL 1 (800) 334-7577 TWENTY-FOUR HOURS A DAY IN THE USA.

LIMITED WARRANTY AND DISCLAIMER

The manufacturer warrants (a) that this product conforms to the chemical description on the label; (b) that this product is reasonably fit for the purposes set forth in the directions for use when it is used in accordance with such directions; and (c) that the directions, warnings and other statements on this label are based upon responsible experts' evaluation of reasonable tests of effectiveness, of toxicity to laboratory animals and to plants, and of residues on food crops, and upon reports of field experience. Tests have not been made on all varieties or in all states or under all conditions. THE MANUFACTURER NEITHER MAKES NOR INTENDS, NOR DOES IT AUTHORIZE ANY AGENT OR REPRESENTATIVE TO MAKE, ANY OTHER WARRANTIES, EXPRESS OR IMPLIED, AND IT EXPRESSLY EXCLUDES AND DISCLAIMS ALL IMPLIED WARRANTIES OF MERCHANTABILITY OR FITNESS FOR A PARTICULAR PURPOSE.

THIS WARRANTY DOES NOT EXTEND TO, AND THE BUYER SHALL BE SOLELY RESPONSIBLE FOR, ANY AND ALL LOSS OR DAMAGE WHICH RESULTS FROM THE USE OF THIS PRODUCT IN ANY MANNER WHICH IS INCONSISTENT WITH THE LABEL DIRECTIONS, WARNINGS OR CAUTIONS.

BUYER'S EXCLUSIVE REMEDY AND MANUFACTURER'S OR SELLER'S EXCLUSIVE LIABILITY FOR ANY CLAIMS, LOSSES, DAMAGES, OR INJURIES RESULTING FROM THE USE OR HANDLING OF THIS PRODUCT WHETHER OR NOT SUCH LIABILITY IS BASED IN CONTRACT, NEGLIGENCE, STRICT LIABILITY IN TORT OR OTHERWISE SHALL BE LIMITED, AT THE MANUFACTURER'S OPTION, TO REPLACEMENT OF, OR THE REPAYMENT OF THE PURCHASE PRICE FOR, THE QUANTITY OF PRODUCT WITH RESPECT TO WHICH DAMAGES ARE CLAIMED. IN NO EVENT SHALL MANUFACTURER OR SELLER BE LIABLE FOR SPECIAL, INDIRECT OR CONSEQUENTIAL DAMAGES RESULTING FROM THE USE OR HANDLING OF THIS PRODUCT.

NOTICE TO BUYER

Purchase of this material does confer any rights under patents this product or the use thereof outside of the United States.

TEMIK is a registered trademark of RHÔNE-POULENC for aldicarb pesticide.

AG 90060

EPA Approval: 2/7/89

THIS SPECIMEN LABEL IS INTENDED FOR USE ONLY AS A GUIDE IN PROVIDING GENERAL INFORMATION REGARDING THE DIRECTIONS, WARNINGS AND CAUTIONS ASSOCIATED WITH THE USE OF THIS PRODUCT. AS WITH ANY AGRICULTURAL CHEMICAL, ALWAYS FOLLOW THE LABEL INSTRUCTIONS ON THE PACKAGE BEFORE USING.

TEMIK® BRAND TSX

GRANULAR PESTICIDE

RESTRICTED USE PESTICIDE

For retail sale to and use only by Certified Applicators or persons under the direct supervision of a Certified Applicator, and only for those uses covered by the Certified Applicator's certification.

For control of certain insects, mites, nematodes, and soil-borne seedling diseases of cotton.

ACTIVE INGREDIENTS:	
Aldicarb [2-methyl-2-(methylthio) propionaldehyde-O-methylcarbamoyl)oxime]	5.0%
Pentachloronitrobenzene (PCNB)	10.0%
5-Ethoxy-3-(trichloromethyl) 1,2,4-thiodiazole	2.5%
INERT INGREDIENTS:	82.5%
TOTAL	100.0%

EPA Reg. No. 264-319 EPA Est. No. 264-GA-1

KEEP OUT OF REACH OF CHILDREN

DANGER POISON

DANGER–POISON

PELIGRO

PRECAUCION AL USUARIO: Si usted no lee ingles, no use este producto hasta que le etiqueta le haya sido explicado ampliamente.

TRANSLATION TO THE USER: If you cannot read English, do not use this product until the label has been fully explained to you.

STATEMENT OF PRACTICAL TREATMENT

IF SWALLOWED: Drink 1 or 2 glasses of water and induce vomiting by touching back of throat with finger. Do not induce vomiting or give anything orally to an unconscious or convulsing person. Call a physician and follow General advice.

IF IN EYES: Flush eyes with plenty of water and get medical attention.

IF ON SKIN: Wash thoroughly with soap and water. Remove contaminated clothing and wash before reuse.

IF INHALED: Call a physician and follow General advice.

For MEDICAL and TRANSPORTATION Emergencies ONLY Call 24 Hours A Day 1-800-334-7577
For PRODUCT USE Information Call 1-800-334-9745

PRECAUTIONARY STATEMENTS

DANGER

HAZARDS TO HUMANS AND DOMESTIC ANIMALS

FATAL IF SWALLOWED. Causes cholinesterase inhibition. May be fatal or harmful by skin or eye contact or by breathing dust. Rapidly absorbed through skin or eyes. Do not get on skin or in eyes. Do not breathe dust. Keep away from domestic animals.

Wear long-sleeved clothing, full-length trousers, head coverings, and protective gloves when handling. Always stand up-wind from hopper when loading. Wash hands and face before eating, drinking or using tobacco. Bathe at the end of work day, washing entire body and hair with soap and water. Change clothing daily. Wash contaminated clothing in strong washing soda solution and rinse thoroughly before reusing.

SIGNS AND SYMPTOMS OF OVEREXPOSURE:

Salivation	Excessive sweating
Watery eyes	Abdominal cramps
Pinpoint eye pupils	Nausea
Blurred vision	Vomiting
Headache	Diarrhea
Muscle tremor	Weakness
Difficult breathing	

In severe cases, convulsions, unconsciousness, and respiratory failure may occur.

ANTIDOTE STATEMENT

ATROPINE SULFATE IS HIGHLY EFFECTIVE AS AN ANTIDOTE.

STATEMENT OF PRACTICAL TREATMENT

See NOTE TO PHYSICIAN below.

GENERAL

Contact a physician immediately in all cases of suspected poisoning. Illness may be produced rapidly following overexposure to TEMIK® Aldicarb. If breathing stops, establish an airway and start artificial respiration, and provide oxygen. Make certain to remove all sources of continuing contamination. Remove clothing and wash skin and hair immediately with large amounts of water. Transport the patient to a physician or hospital immediately and SHOW A COPY OF THIS LABEL TO THE PHYSICIAN. If poisoning is suspected in animals, contact a veterinarian.

NOTE TO PHYSICIAN

TEMIK® brand aldicarb is a methyl carbamate insecticide which is a cholinesterase inhibitor. Overexposure to this substance may cause toxic signs and symptoms due to stimulation of the cholinergic nervous system. These effects of overexposure are spontaneously and rapidly reversible.

Gastric lavage may be used if this product has been swallowed. TEMIK® brand aldicarb poisoning may occur rapidly after ingestion and prompt removal of stomach contents is indicated.

Specific treatment consists of the administration of parenteral atropine sulfate. Caution should be exercised to prevent overatropinization. Mild cases may be given 1 to 2 mg intramuscularly every 10 minutes until full atropinization has been achieved and repeated thereafter whenever symptoms reappear. Severe cases should be given 2 to 4 mg intravenously every 10 minutes until the patient is fully atropinized, then intramuscularly every 30 to 60 minutes as needed to maintain the effect for at least 12 hours. Dosages for children should be appropriately reduced. Complete recovery from overexposure is to be expected within 24 hours.

Narcotics and other sedatives should not be used. Further, drugs such as (pyridine-2-aldoxime methiodide) are NOT recommended unless organophosphate intoxication is also suggested.

To aid in confirmation of a diagnosis, urine samples must be obtained within 24 hours of exposure and immediately frozen. Analyses will be arranged by Rhône-Poulenc Ag Company.

Consultation on therapy can be obtained at all hours by calling the Rhône-Poulenc emergency number: 1-800-334-7577.

ENVIRONMENTAL HAZARDS

DECOMPOSITION AND MOVEMENT IN SOIL: This product is readily decomposed into harmless residues under most use conditions. However, a combination of permeable and acidic soil conditions, moderate to heavy irrigation and/or rainfall, use of 20 or more pounds per acre, and soil temperature below 50°F (10°C) at application time tend to reduce degradation and promote movement of residues to groundwater. If the above describes your local conditions and groundwater in your area is used for drinking, do not use this product without first contacting Rhône-Poulenc at 1-(800) 334-7577.

TOXIC TO FISH, BIRDS, AND WILDLIFE

This product is toxic to fish, birds, invertebrates and other wildlife. Birds feeding on exposed granules in treated areas may be killed. Cover or incorporate granules in spill areas.

Run-off from treated areas may be hazardous to fish in neighboring areas. Do not apply directly to water or wetlands (swamps, bogs, marshes, and potholes). Keep out of lakes, streams, and ponds. Do not contaminate water when disposing of equipment washwaters. Apply this product only as specified on this label.

NOTICE: Under the Endangered Species Act, it is a Federal Offense to use any pesticide in a manner that results in the death of a member of an endangered species.

This Act protects Attwater's Greater Prairie Chicken in the Texas counties of Aransas, Austin, Brazoria, Colorado, Galveston, Goliad, Harris, Refugio, and Victoria.

Prior to making applications in these counties, the user must determine that this species is not located in or immediately adjacent to the areas to be treated. If the user is in doubt whether or not the above named endangered species may be affected, he should contact either the regional U.S. Fish & Wildlife Service office (Endangered Species Specialist) or personnel of the State Fish and Game office.

GENERAL DIRECTIONS FOR USE

TEMIK® brand TSX Granular Pesticide controls certain nematodes, insects, mites, and soil-borne seedling diseases of cotton. When applied into moist soil at planting the active fungicide ingredients (Terraclor and Terrazole fungicides) protect the germinating seed and seedling from Rhizoctonia, Pythium, and Fusarium diseases. At the same time, the active nematicide ingredient (aldicarb) protects the developing root system from nematodes which damage the seedling directly and promote seedling disease. When the germinating plant emerges from the soil, the active insecticide ingredient (aldicarb) has moved upward from the roots to protect the developing stem and leaves from attack by early season mites and insects. Control of insects and mites often lasts more than six weeks, depending upon growing conditions, rate of use and pests. Cotton yields are usually increased with treatments of TEMIK® brand TSX Granular Pesticide.

GENERAL CAUTIONS

Use TEMIK® brand TSX Granular Pesticide only in accordance with label directions, warnings and cautions. Do not use on any crop other than cotton, as any residues remaining may be illegal or harmful. DO NOT STORE OR USE IN OR AROUND THE HOME OR HOME GARDEN.

Use high rate on heavy organic or clay soils. Do not over apply. Treatments in excess of 15 pounds per acre may delay emergence and reduce plant stand.

In irrigated areas, follow application with irrigation within one week.

Do not drill TEMIK® brand TSX Granular Pesticide if seed is hill-dropped.

Calibrate and adjust application equipment to insure proper rate and accurate placement. Clean application equipment thoroughly after use. Bury any excess material in soil (see **STORAGE AND DISPOSAL** instructions on FRONT PANEL of this bag).

Disc any spills at row ends immediately to prevent birds from feeding on exposed granules.

Do not wash, load, or empty application equipment near any well, as this practice is a potential source of groundwater contamination.

Do not apply this product in Del Norte, and Humboldt Counties in California; Nassau or Suffolk Counties in New York; or in Curry County, Oregon.

COMPATIBILITY

Pesticide activity of TEMIK® brand TSX Granular Pesticide is not affected by normal applications of fertilizers or other pesticides. It may be reduced or lost if applied with alkaline materials such as lime or fertilizers containing lime.

GENERAL WORKER PROTECTION STATEMENTS

Do not apply this product in such a manner as to directly expose workers or other persons. The area being treated must be vacated by unprotected persons.

Do not enter treated area without protective footwear. Persons coming in direct contract with wet treated soil after the initial irrigation or rainfall following treament shall wear coveralls, rubber or neoprene boots, and gloves. Because certain states may require more restrictive re-entry intervals for various crops treated with this product, consult your State Department of Agriculture for further information. Written or oral warnings must be given to workers who are expected to be in a treated area or in an area about to be treated with this product. Advise workers to stay out of fields during application. Regular long-sleeved shirt, full-length trousers, and head covering should be worn when working in treated fields. See Precautionary Statement, Statement of Practical Treament, and Note To Physician for information on accidental exposures. When oral warnings are given, warnings shall be given in a language customarily understood by workers. Oral warnings must be given if there is reason to believe that written warnings cannot be understood by workers. Written warnings must include the following information: appropriate signal word (DANGER-POISON), area treated with TEMIK® brand 15G Aldicarb Pesticide, date of application, appropriate protective clothing, and re-entry requirements.

PRE-HARVEST, GRAZING, AND GROUNDWATER LIMITATIONS

TO AVOID EXCESSIVE RESIDUES IN OR ON:

COTTONSEED

- Do not apply within 90 days of harvest.
- Do not feed cotton forage to livestock or allow livestock to graze in treated area.

OTHER CROPS

- Do not plant any crops in soil treated with TEMIK® brand TSX Granular Pesticide within 12 months of broadcast or banding applications unless PCNB is registered for use on those crops.

GROUNDWATER

- Observe ENVIRONMENTAL HAZARD precautions regarding DECOMPOSITION AND MOVEMENT IN SOIL, and carefully follow DIRECTIONS FOR USE.
- Do not wash, load, or empty application equipment near any well, as this practice is a potential source of groundwater contamination.
- Do not apply this product in Del Norte and Humboldt Counties in California; Nassau or Suffolk Counties in New York; or in Curry County, Oregon.

DIRECTIONS FOR USE AND RECOMMENDED APPLICATIONS

FOR CONTROL OF CERTAIN INSECTS, MITES, NEMATODES, AND SOIL-BORNE SEEDLING DISEASE OF COTTON.

IT IS A VIOLATION OF FEDERAL LAW TO USE THIS PRODUCT IN A MANNER INCONSISTENT WITH ITS LABELING.

To provide maximum performance and to minimize hazard to birds, work granules into or cover with soil to a depth of 2 inches or more. Deep disc any spills at row ends immediately to prevent birds from feeding on granules.

CROP AND TIME OF APPLICATION	PESTS CONTROLLED	POUNDS/ACRE	OZ./1000 FT/ROW	RECOMMENDED APPLICATION		
COTTON At planting	Aphids, mites, thrips, bollweevil (adults feeding on foliage)	(For example 40-inch row spacing)		Place granules in seed furrow and cover with soil. If seeds and TEMIK® TSX are hill-dropped, reduce rates of application of TEMIK TSX as indicated.		
	Nematodes (Except Far West	10-15	12.3 to 18.5	Hill Spacing	Hill Drop Rates	
					Pounds/Acre	Oz./1000 ft. of row
	Seedling Diseases complex (Rhizoctonia Phythium, Fusarium)			8-inch	10	12.5
				12-inch	8.5	10.5
				16-inch	7.5	9
				20-inch	6	7

STORAGE AND DISPOSAL

STORAGE

Store unused TEMIK® brand TSX Granular Pesticide in original container only in well ventilated clean dry area out of reach of children and animals. Do not store in areas where temperature averages 115°F or greater. Do not store in or around the home or home garden.

If container is broken, handle with rubber gloves. Do not get dust or granules on skin or in eyes. Do not breathe dust. Sweep up and bury any small spills or excess TEMIK® brand TSX at least 18 inches deep in soil isolated from water supplies and food crops.

PESTICIDE DISPOSAL

Do not contaminate water, food, or feed by storage or disposal. Open dumping is prohibited. Pesticide wastes are acutely hazardous. Improper disposal of excess pesticide is a violation of Federal law. If these wastes cannot be disposed of by use according to label instructions, contact your State Pesticide or Environmental Control Agency, or the Hazardous Waste representative at the nearest EPA Regional Office for guidance.

CONTAINER DISPOSAL

Completely empty container into application equipment. Then dispose of empty container in a sanitary landfill, or by incineration, or if allowed by state and local authorities, by open burning. If container is burned, stay away from and do not breathe or contact smoke.

IN CASE OF A TRANSPORTATION OR WAREHOUSE EMERGENCY INVOLVING A SPILL, FIRE, OR EXPOSURE CALL 1-800-344-7577 TWENTY-FOUR HOURS A DAY IN THE USA.

LIMITED WARRANTY AND DISCLAIMER

The manufacturer warrants (a) that this product conforms to the chemical description on the label; (b) that this product is reasonably fit for the purposes set forth in the directions for use when it is used in accordance with such directions and; (c) that the directions, warnings, and other statements on this label are based upon responsible experts' evaluation of reasonable tests of effectiveness, of toxicity to laboratory animals and to plants, and of residues on food crops, and upon reports of field experience. Tests have not been made on all varieties or in all states or under all conditions. THE MANUFACTURER NEITHER MAKES NOR INTENDS, NOR DOES IT AUTHORIZE ANY AGENT OR REPRESENTATIVE TO MAKE, ANY OTHER WARRANTIES, EXPRESS OR IMPLIED, AND IT EXPRESSLY EXCLUDES AND DISCLAIMS ALL IMPLIED WARRANTIES OF MERCHANTABILITY OR FITNESS FOR A PARTICULAR PURPOSE.

THIS WARRANTY DOES NOT EXTEND TO, AND THE BUYER SHALL BE SOLELY RESPONSIBLE FOR ANY AND ALL LOSS OR DAMAGE WHICH RESULTS FROM THE USE OF THIS PRODUCT IN ANY MANNER WHICH IS INCONSISTENT WITH THE LABEL DIRECTIONS, WARNINGS OR CAUTIONS.

BUYER'S EXCLUSIVE REMEDY AND MANUFACTURER'S OR SELLER'S EXCLUSIVE LIABILITY FOR ANY AND ALL CLAIMS, LOSSES, DAMAGES, OR INJURIES RESULTING FROM THE USE OR HANDLING OF THIS PRODUCT WHETHER OR NOT BASED IN CONTRACT, NEGLIGENCE, STRICT LIABILITY IN TORT OR OTHERWISE SHALL BE LIMITED, AT THE MANUFACTURER'S

OPTION, TO REPLACEMENT OF, OR THE REPAYMENT OF THE PURCHASE PRICE FOR, THE QUANTITY OF PRODUCT WITH RESPECT TO WHICH DAMAGES ARE CLAIMED. IN NO EVENT SHALL MANUFACTURER OR SELLER BE LIABLE FOR SPECIAL, INDIRECT OR CONSEQUENTIAL DAMAGES RESULTING FROM THE USE OR HANDLING OF THIS PRODUCT.

Contains TEMIK® brand ALDICARB, and TERRACLOR, and TERRAZOLE fungicides.

TEMIK is the registered trademark of RHÔNE-POULENC for aldicarb pesticide.

AG 90058

Printed USA-8/89

THIS SPECIMEN LABEL IS INTENDED FOR USE ONLY AS A GUIDE IN PROVIDING GENERAL INFORMATION REGARDING THE DIRECTIONS, WARNINGS AND CAUTIONS ASSOCIATED WITH THE USE OF THIS PRODUCT. AS WITH ANY AGRICULTURAL CHEMICAL, ALWAYS FOLLOW THE LABEL INSTRUCTIONS ON THE PACKAGE BEFORE USING.

THISTROL®

THISTLE HERBICIDE

Contains 4-(MCPB). Controls Canada Thistle, Other Broadleaf Weeds In Peas.

ACTIVE INGREDIENT
Sodium salt of 4-(2-methyl-4-chlorophenoxy) butyric acid* 23.5%
INERT INGREDIENTS: 76.5%

*4-(2-Methyl-4-chlorophenoxy) butyric acid equivalent 21.4% by weight or 2 pounds per gallon

E.P.A. Reg. No. 264-181-ZA EPA Est. No.

KEEP OUT OF REACH OF CHILDREN
CAUTION

For PRODUCT USE Information Call 1-800-334-9745
For MEDICAL And TRANSPORTATION Emergencies ONLY Call 24 Hours A Day 1-800-334-7577

PRECAUTIONARY STATEMENTS

CAUTION

HAZARD TO HUMANS AND DOMESTIC ANIMALS

Harmful if swallowed. Avoid contact with skin, eyes or clothing. In case of contact, wash with plenty of water.

ENVIRONMENTAL HAZARD

Avoid spray drift as this product may injure susceptible crops and plants such as cotton, beans, grape, tomatoes, and ornamentals. (Coarse sprays are less likely to drift.) Do not use same spray equipment for other purposes unless thoroughly cleaned prior to use.

Do not apply THISTROL® Thistle Herbicide through any type of irrigation system.

DIRECTIONS FOR USE

IT IS A VIOLATION OF FEDERAL LAW TO USE THIS PRODUCT IN A MANNER INCONSISTENT WITH ITS LABELING.

GENERAL WORKER PROTECTION STATEMENTS

Do not apply this product in such a manner as to directly or through drift expose workers or other persons. The area treated must be vacated by unprotected persons.

Do not enter treated areas without protective clothing until sprays have dried.

Because certain states may require more restrictive reentry intervals for various crops treated with this product, consult your State Department of Agriculture for further information.

Written or oral warnings must be given to workers who are expected to be in treated areas or in an area about to be treated with this product. Advise workers to stay out of fields during application and until sprays have dried. Regular long-sleeved work clothing should be worn when working in treated fields. See **Precautionary Statements** for information on accidental exposures. When oral warnings are given, warnings shall be given in a language customarily understood by workers. Oral warnings must be given if there is reason to believe that written warnings cannot be understood by workers. Written warnings must include the following information: Appropriate signal word (CAUTION), area treated with THISTROL® Thistle Herbicide, date of application, appropriate clothing, and reentry interval (i.e. until sprays have dried).

WHEN TO APPLY

Canada thistle growing in pea fields is a serious problem. It can reduce pea yields and the Canada thistle flower buds can contaminate the harvested peas, reducing market value and cash return to the grower. At rates used to control Canada thistle, THISTROL® Thistle Herbicide will control or suppress other broadleaf weeds such as lambsquarters, pigweed, smartweed, sow thistle, and annual morningglory.

Spray THISTROL® on thistles as soon as good shoot emergence has developed and when annual broadleaf weeds are less than 3 inches high. During this period, peas will have developed 6 to 12 nodes, depending on variety. Do not apply THISTROL® later than 3 nodes before pea flowering, as pea yields may be affected adversely. Do not spray peas under stress from lack of moisture, or when temperatures are over 90° F. Do not apply after pea flower buds appear.

AMOUNTS TO USE

WEEDS AND STAGE OF GROWTH	AMOUNT OF THISTROL PER ACRE
Young susceptible weeds less than 3 inches high (2 to 3 leaves)	2 to 3 pints
Canada thistle 4 to 10 inches high, or before thistle buds appear.	
Other susceptible weeds over 3 inches high	3 to 4 pints
Susceptible weeds requiring higher rates due to local conditions.	up to 6 pints

For ground equipment apply in 20 to 40 gallons of water per acre as a uniform, overall spray, using 40 psi maximum pressure.

For aerial application use 5 to 10 gallons of spray solution per acre.

Do not exceed the recommended dosage. Most pea varieties may be safely sprayed with THISTROL®; however, temporary twisting of some pea varieties may follow application. Check with state weed specialists or with local canning company representatives.

Do not feed treated peas or vines to livestock.

STORAGE AND DISPOSAL

STORAGE

Do not store near fertilizers, seeds, insecticides or fungicides. Protect from freezing. If allowed to freeze, hold in warm temperature for several days and remix before using.

PESTICIDE DISPOSAL

Do not contaminate water, food or feed by storage or disposal.
Wastes resulting from the use of this product may be disposed of on site or at an approved waste disposal facility.

CONTAINER DISPOSAL

Triple rinse (or equivalent). Then offer for recycling or reconditioning, or puncture and dispose of in a sanitary landfill, or incineration, or, if allowed by state and local authorities, by burning. If burned, stay out of smoke.

Do not use this product for purposes other than those recommended on the label.

LIMITED WARRANTY AND DISCLAIMER

The manufacturer warrants that this product conforms to the chemical description on the label; that this product is reasonably fit for the purposes set forth in the directions for use when it is used in accordance with such directions; and that the directions, warning and other statements on this label are based upon responsible experts' evaluation of reasonable tests of effectiveness, of toxicity to laboratory animals and to plants, and of residues on food crops and upon reports of field experience. Tests have not been made on all varieties or in all states or under all conditions. THE MANUFACTURER NEITHER MAKES NOR INTENDS, NOR DOES IT AUTHORIZE ANY AGENT OR REPRESENTATIVE TO MAKE, ANY OTHER WARRANTIES, EXPRESS OR IMPLIED, AND IT EXPRESSLY EXCLUDES AND DISCLAIMS ALL IMPLIED WARRANTIES OF MERCHANTABILITY OR FITNESS FOR A PARTICULAR PURPOSE.

THIS WARRANTY DOES NOT EXTEND TO, AND THE BUYER SHALL BE SOLELY RESPONSIBLE FOR, ANY AND ALL LOSS OR DAMAGE WHICH RESULTS FROM THE USE OF THIS PRODUCT IN ANY MANNER WHICH IS INCONSISTENT WITH THE LABEL DIRECTIONS, WARNINGS OR CAUTIONS.

BUYER'S EXCLUSIVE REMEDY AND MANUFACTURER'S OR SELLER'S EXCLUSIVE LIABILITY FOR ANY AND ALL CLAIMS, LOSSES, DAMAGES OR INJURIES RESULTING FROM THE USE OR HANDLING OF THIS PRODUCT, WHETHER OR NOT BASED IN CONTRACT, NEGLIGENCE, STRICT LIABILITY IN TORT OR OTHERWISE SHALL BE LIMITED, AT THE MANUFACTURER'S OPTION, TO REPLACEMENT OF, OR THE REPAYMENT OF THE PURCHASE PRICE FOR, THE QUANTITY OF PRODUCT WITH RESPECT TO WHICH DAMAGES ARE CLAIMED. IN NO EVENT SHALL MANUFACTURER OR SELLER BE LIABLE FOR SPECIAL, INDIRECT OR CONSEQUENTIAL DAMAGES RESULTING FROM THE USE OR HANDLING OF THIS PRODUCT.

NOTICE TO BUYER

Purchase of this material does not confer any rights under patents governing this product or the use thereof in countries outside of the United States.

THISTROL is a registered trademark of Rhône-Poulenc.

AG 90016

THIS SPECIMEN LABEL IS INTENDED FOR USE ONLY AS A GUIDE IN PROVIDING GENERAL INFORMATION REGARDING THE DIRECTIONS, WARNINGS AND CAUTIONS ASSOCIATED WITH THE USE OF THIS PRODUCT. AS WITH ANY AGRICULTURAL CHEMICAL, ALWAYS FOLLOW THE LABEL INSTRUCTIONS ON THE PACKAGE BEFORE USING.

TRE-HOLD®

BRAND TREE WOUND DRESSING

Healing Aid/Sprout Inhibitor

ACTIVE INGREDIENTS:	
1-Naphathaleneacetic acid, ethyl ester	1%
INERT INGREDIENTS:	99%
This product contains the toxic inert ingredient methylene chloride.	
TOTAL:	100.0%

EPA Reg. No. 264-431 EPA Est. No. 34631-PA-01

KEEP OUT OF REACH OF CHILDREN

CAUTION

For PRODUCT USE Information Call 1-800-334-9745
For MEDICAL And TRANSPORTATION Emergencies ONLY Call 24 Hours A Day 1-800-334-7577

PRECAUTIONARY STATEMENTS

CAUTION

HAZARDS TO HUMANS AND DOMESTIC ANIMALS

Harmful if swallowed. Avoid contact with eyes, skin and clothing. Avoid breathing spray mist.

STATEMENT OF PRACTICAL TREATMENT

IF SWALLOWED: If patient is conscious and alert, give 2 to 3 glasses of water to drink. Do not induce vomiting. GET MEDICAL ATTENTION. Swallowing is not an expected route of exposure, since this product is pressurized and applied as an aerosol spray.

IF ON SKIN: Cleanse skin with vegetable oil while removing exposed clothing. Wash skin with soap and water. Contaminated articles should be discarded.

IF INHALED: Remove victim to fresh air. If not breathing, give artifical respiration. Administer oxygen if necessary. GET MEDICAL ATTENTION.

IF IN EYES: Hold eyelids open and flush with a steady, gentle stream of water for at least 15 minutes. GET MEDICAL ATTENTION.

NOTE TO PHYSICIAN

No specific antidote is available.

All treatments should be based on observed signs and symptoms of distress in the patient. Overexposure to materials other than this product may have occurred.

GENERAL INFORMATION

Many trees produce vigorous unwanted sprouts around pruning cuts on branches, from the rootstock, and base of the trunk. These sprouts must be removed to promote and maintain desired plant form.

TRE-HOLD® Tree Wound Dressing treatment prevents or inhibits vegetative growth in woody plants after pruning. TRE-HOLD® is formulated in an aerosol container with an asphalt base which provides a protective barrier against disease and aids in the "healing" process. TRE-HOLD® is effective on bearing* and non-bearing fruit trees as well as many ornamental flowering and shade trees such as:

Apples*	American Elm	Silver Maple
Crabapples	Chinese Elm	Sugar Maple
Olives*	Sycamore	Water Oak
Pears*	Red Maple	

DIRECTIONS FOR USE

IT IS A VIOLATION OF FEDERAL LAW TO USE THIS PRODUCT IN A MANNER INCONSISTENT WITH ITS LABELING.

Do not use on citrus as the black asphalt base may enhance sun scald damage.

Shake the can well and spray the pruned surface from a distance of 10 to 20 inches.

BRANCH PRUNING AND WATER SPROUTS

Treat the pruning cuts and the surrounding bark after pruning branches or sprouts. Do not treat pruning cuts on olive branches.

For branch pruning: prune to a lateral branch as shown in the illustration. Stub cuts result in maximum resprouting and leave an unsightly stub which may die back.

For water sprouts: prune flush with the lateral branch.

TRUNK BASAL SPROUTS AND ROOT SUCKERS

Remove the existing sprouts and treat with TRE-HOLD® Tree Wound Dressing.

ON OLIVE TREES: First remove the existing sprouts and allow new sprouts to grow, and then treat the new sprouts with TRE-HOLD® before they become 10 inches long.

Do not treat fruit trees in blossom.

Spray pruned cuts and the surrounding bark only. Do not spray the leaves. Avoid spray drift to other plants as injury may result.

Do not treat more than once a year.

In thin bark species, sun scald may result if the number of leaves on trees or shrubs are greatly reduced from severe pruning.

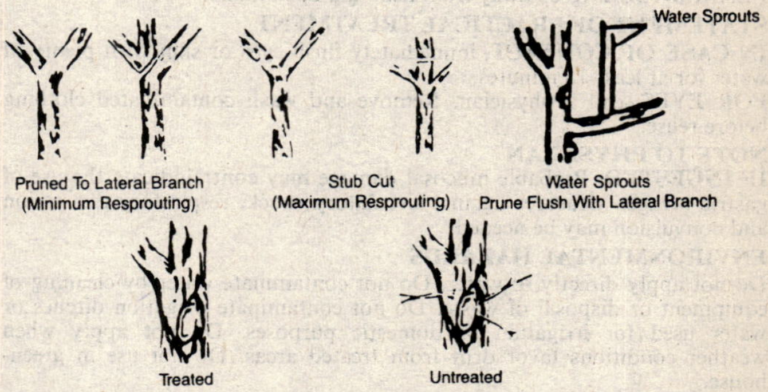

STORAGE AND DISPOSAL

Storage:

Do not store near fertilizers, seeds, insecticides, or fungicides. Avoid contamination of food or feed stuffs. Contents under pressure. Do not set on stove, radiator, or store where temperature exceeds 120°F as container may burst.

Pesticide and Container Disposal

Replace cap securely. Wrap original container in several layers of newspaper and discard in trash. Do not incinerate or puncture.

LIMITED WARRANTY AND DISCLAIMER

"NOTICE: BUYER SHALL BE SOLELY RESPONSIBLE FOR ANY AND ALL INJURY, LOSS OR DAMAGE WHICH RESULTS FROM THE USE OF THIS PRODUCT IN ANY MANNER WHICH IS INCONSISTENT WITH THE LABEL DIRECTIONS, WARNING OR CAUTIONS."

TREE-HOLD is a registered trademark of RHONE-POULENC.

AG 90044

THIS SPECIMEN LABEL IS INTENDED FOR USE ONLY AS A GUIDE IN PROVIDING GENERAL INFORMATION REGARDING THE DIRECTIONS, WARNINGS AND CAUTIONS ASSOCIATED WITH THE USE OF THIS PRODUCT. AS WITH ANY AGRICULTURAL CHEMICAL, ALWAYS FOLLOW THE LABEL INSTRUCTIONS ON THE PACKAGE BEFORE USING.

TRE-HOLD®

SPROUT INHIBITOR A112

CONTROLS SPROUT OR SUCKER GROWTH ON APPLE, PEAR AND OLIVE TREES.

FOR COMMERCIAL OR AGRICULTURAL USE. NOT FOR RESIDENTIAL USE.

ACTIVE INGREDIENT:
1-Naphthaleneacetic acid, ethyl ester* 15.1%
INERT INGREDIENTS: 84.9%

* 1-Naphthaleneacetic acid equivalent 1 lb./gal. or 13.2% by weight

E.P.A. Reg. No. 264-336 E.P.A. Est. No. 264-MO-01

KEEP OUT OF REACH OF CHILDREN

DANGER PELIGRO

PRECAUCION AL USUARIO: Si usted no lee ingles, no use este producto hasta que la etiqueta le haya sido explicada ampliamente.

For PRODUCT USE Information Call 1-800-334-9745
For MEDICAL and TRANSPORTATION EMERGENCIES ONLY Call 24 Hours A Day 1-800-334-7577

PRECAUTIONARY STATEMENTS

DANGER

HAZARD TO HUMANS AND DOMESTIC ANIMALS

Corrosive. Causes eye and skin irritation. Do not get in eyes, on skin or on clothing. Wear goggles or face shield and rubber gloves when handling. Harmful or fatal if swallowed. Avoid contamination of food. Flammable, keep away from heat and open flame.

PHYSICAL OR CHEMICAL HAZARD

FLAMMABLE: Keep away from heat and open flame.

STATEMENT OF PRACTICAL TREATMENT

IN CASE OF CONTACT, immediately flush eyes or skin with plenty of water for at least 15 minutes.
FOR EYES, call a physician. Remove and wash contaminated clothing before reuse.

NOTE TO PHYSICIAN

IF INGESTED: Probable mucosal damage may contraindicate the use of gastric lavage. Measures against circulatory shock, respiratory depression and convulsion may be needed.

ENVIRONMENTAL HAZARDS

Do not apply directly to water. Do not contaminate water by cleaning of equipment or disposal of waste. Do not contaminate irrigation ditches or water used for irrigation or domestic purposes. Do not apply when weather conditions favor drift from treated areas. Do not use in greenhouse.

Do not use the same spray equipment for other purpose unless thoroughly cleaned.

GENERAL INFORMATION

Many trees sprout vigorously from the rootstock, base of the trunks, scaffold limbs, and around pruning cuts on limbs. TRE-HOLD® Sprout Inhibitor A112 prevents or inhibits vegetative bud development in woody plants after pruning. TRE-HOLD® Sprout Inhibitor A 112 reduces resprouting on bearing and non-bearing apple, olive and pear trees and on ornamental olives, pears and crabapples.

DIRECTIONS FOR USE

IT IS A VIOLATION OF FEDERAL LAW TO USE THIS PRODUCT IN A MANNER INCONSISTENT WITH ITS LABELING. Do not apply TRE-HOLD® Sprout Inhibitor A 112 through any type of irrigation system.

TRE-HOLD® Sprout Inhibitor A112 is formulated for dilution in water and interior latex paint. After mixing, the solution should be applied to pruning cuts on the scaffold limbs, base of trunks and on rootstock suckers to control resprouting. Thorough coverage of treated area is essential for most effective sprout control.

GENERAL WORKER PROTECTION STATEMENTS

Do not apply this product in such a manner as to directly or through drift expose workers or other persons. The area being treated must be vacated by unprotected persons.

Do not enter treated areas without protective clothing until sprays have dried.

Because certain states may require more restrictive reentry intervals for various crops treated with this product, consult your State Department of Agriculture for further information. Written or oral warnings must be given to workers who are expected to be in a treated area or in an area about to be treated with this product. Advise workers to stay out of fields during application and until sprays have dried. Regular long-sleeved work clothing should be worn when working in treated fields. See PRECAUTIONARY STATEMENTS, STATEMENT OF PRACTICAL TREATMENT, and NOTE TO PHYSICIAN for information on accidental exposures. When oral warnings are given, warnings shall be given in a language customarily understood by workers. Oral warnings must be given if there is reason to believe that written warnings cannot be understood by workers. Written warnings must include the following information: appropriate signal word (DANGER), area treated with TRE-HOLD® Sprout Inhibitor A112, date of application, appropriate protective clothing, and reentry interval if applicable.

MIXING DIRECTIONS

TRE-HOLD® Sprout Inhibitor A112 must be diluted before use.

NEW PLANTINGS

For use on newly established plantings of apples and pears, dilute TRE-HOLD® Sprout Inhibitor A-112 to 0.5% solution.

Example for preparation of 0.5% dilute mixture

Amount of TRE-HOLD® SPROUT A 112	Amount of Water	Total Volume
1.25 fluid ounces	31.0 fluid ounces	1 quart
5.0 fluid ounces	123.0 fluid ounces	1 gallon
2.0 gallons	48.0 gallons	50 gallons

ESTABLISHED PLANTINGS

The following dilution volumes are examples for typical small and large scale needs. For most uses on established plantings, TRE-HOLD® Sprout Inhibitor A112 is diluted to a 1% concentration with water.

Examples for preparation of 1% dilute mixtures

Amount of TRE-HOLD® SPROUT A 112	Amount of Water	Total Volume
2.5 fluid ounces	29.5 fluid ounces	1 quart
10.0 fluid ounces	118 fluid ounces	1 gallon
4.0 gallons	46.0 gallons	50 gallons

For marking purposes, interior white latex paint may be substituted for an equal volume of water as indicated below. The white paint also provides protection from sun scald damage in thin barked trees or shrubs.

Example for preparations of 1% dilute mixtures with latex paint.

Amount of TRE-HOLD® SPROUT A 112	Amount of Latex Paint	Add Water to Bring Total Volume To
2.5 fluid ounces	4.0 to 15.0 fluid ounces	1 quart
10.0 fluid ounces	15 to 60 fluid ounces	1 gallon
4.0 gallons	5.0 to 20.0 gallons	50 gallons

Use the higher rate of interior white latex paint where bark damage from sun scald is likely to be most severe. Be sure the paint used is not harmful to trees.

APPLICATION INSTRUCTIONS—APPLES AND PEARS

NEW PLANTINGS—Apple and pear trees, healthy, vigorous and free from stress, may be treated after planting in the grove when the trees show established growth. Remove sprouts and follow the application procedures described below. Use only 0.5% dilution of TRE-HOLD® Sprout Inhibitor A112. Do not apply to non-established or stressed trees as injury and stand loss will result.

ESTABLISHED PLANTINGS

SCAFFOLD LIMBS

Prune existing sprouts and treat during the dormant season. Thoroughly cover area where existing sprouts were removed but restrict treatment to the cut surface and 2 to 3 inches of the surrounding area. Apply TRE-HOLD® Sprout Inhibitor A112 with a brush or small hand held pump-up type sprayer. Use low pressure (not to exceed 25 psi) to avoid fine spray particles from drifting or splashing on surrounding fruiting wood or buds. The use of interior white latex paint is required to help reflect sun light and cut down on high bark temperature.

Do not treat scaffold limbs after bud activity starts in the Spring. There has been some fruit size reduction on Golden Delicious and Anjou pears from excessive treatments to scaffold limbs on older, weaker trees. To minimize the chance of injury, limit treated areas to 10 percent of the total bark area of the tree. DO NOT treat weak trees, and DO NOT repeat applications on the same tree for at least one year.

TRUNK BASAL SPROUTS AND ROOT SUCKERS

Prune existing sprouts and apply TRE-HOLD® Sprout Inhibitor A-112 during the dormant season prior to green tip stage "or" during the summer pruning season when the new shoots are 6 to 12 inches in height. The later applications have been the most effective. On bearing trees do not treat suckers during bud swell, bloom or fruit set. This period is from the start of growth to 4 weeks after petal fall. Application during this time can result in excessive thinning.

Apply TRE-HOLD® Sprout Inhibitor A112 with a brush, small hand held sprayer or back pack sprayer. For large projects, a power operated herbicide type spray equipment with attached hand gun or trunk directed nozzle can be used. Maintain low spray pressure (not to exceed 25 psi) to avoid spray drift of fine spray particles.

PRUNING CUTS ON LARGE LATERAL OR UPRIGHT LIMBS

To control sprouts in the vicinity of pruning cuts, paint the 1% dilute mixture of TRE-HOLD® Sprout Inhibitor A112 on the bark in the immediate vicinity of the cut. Treat several inches of bark around the cut. DO NOT allow the mixture to splash or drip on to other parts of the tree. Treat only in the dormant season prior to bud activity.

Spray application is not recommended, due to the possibility of the spray contacting fruit buds.

OLIVES—TRUNK BASAL SPROUTS AND ROOT SUCKERS

(Do not treat scaffold branches or pruning cuts on olives.) Prune off existing sprouts or suckers and allow new sprouts to regrow. Treat with a dilute spray of TRE-HOLD® Sprout Inhibitor A112 when sprouts are no more than 10" tall. TRE-HOLD® Sprout Inhibitor A112 will control new sprout growth of olives, but will not control old woody sprouts. Do not treat during olive bloom or fruit set.

Limit treated areas to 10% of the plant's total bark surface.

STORAGE AND DISPOSAL

STORAGE

Do not use or store near heat or open flame.

PESTICIDE DISPOSAL

Do not contaminate water, food, or feed by storage or disposal. Pesticide wastes are acutely hazardous. Improper disposal of excess pesticide, spray mixture, or rinsate is a violation of Federal law. If these wastes cannot be disposed of by use according to label instructions, contact your State Pesticide or Environmental Control Agency or the Hazardous Waste representative at the nearest EPA Regional Office for guidance.

CONTAINER DISPOSAL

Triple rinse (or equivalent). Then offer for recycling or reconditioning, or puncture and dispose of in a sanitary landfill, or by other procedures approved by state and local authorities.

LIMITED WARRANTY AND DISCLAIMER

The manufacturer warrants that this product conforms to the chemical description on the label; that this product is reasonably fit for the purposes set forth in the directions for use when it is used in accordance with such directions; and that the directions, warnings and other statements on this label are based upon responsible experts' evaluation of reasonable tests of effectiveness, of toxicity to laboratory animals and to plants, and of residues on food crops and upon reports of field experience. Tests have not been made on all varieties or in all states or under all conditions. THE MANUFACTURER NEITHER MAKES NOR INTENDS, NOR DOES IT AUTHORIZE ANY AGENT OR REPRESENTATIVE TO MAKE ANY OTHER WARRANTIES, EXPRESS OR IMPLIED, AND IT EXPRESSLY EXCLUDES AND DISCLAIMS ALL IMPLIED WARRANTIES OF MERCHANTABILITY OR FITNESS FOR A PARTICULAR PURPOSE.

THIS WARRANTY DOES NOT EXTEND TO, AND THE BUYER SHALL BE SOLELY RESPONSIBLE FOR, ANY AND ALL LOSS OR DAMAGE WHICH RESULTS FROM THE USE OF THIS PRODUCT IN ANY MANNER WHICH IS INCONSISTENT WITH THE LABEL DIRECTIONS, WARNINGS OR CAUTIONS.

BUYER'S EXCLUSIVE REMEDY AND MANUFACTURER'S OR SELLER'S EXCLUSIVE LIABILITY FOR ANY AND ALL CLAIMS, LOSSES, DAMAGES OR INJURIES RESULTING FROM THE USE OR HANDLING OF THIS PRODUCT, WHETHER OR NOT BASED IN CONTRACT, NEGLIGENCE, STRICT LIABILITY IN TORT OR OTHERWISE SHALL BE LIMITED, AT THE MANUFACTURER'S OPTION, TO REPLACEMENT OF, OR THE REPAYMENT OF THE PURCHASE PRICE FOR, THE QUANTITY OF PRODUCT WITH RESPECT TO WHICH DAMAGES ARE CLAIMED. IN NO EVENT SHALL MANUFACTURER OR SELLER BE LIABLE FOR SPECIAL, INDIRECT OR CONSEQUENTIAL DAMAGES RESULTING FROM THE USE OR HANDLING OF THIS PRODUCT.

NOTICE TO BUYER

Purchase of this material does not confer any rights under patents governing this product or the use thereof in countries outside of the United States.

TRE-HOLD is a registered trademark of Rhône-Poulenc.

AG 90045

THIS SPECIMEN LABEL IS INTENDED FOR USE ONLY AS A GUIDE IN PROVIDING GENERAL INFORMATION REGARDING THE DIRECTIONS, WARNINGS AND CAUTIONS ASSOCIATED WITH THE USE OF THIS PRODUCT. AS WITH ANY AGRICULTURAL CHEMICAL, ALWAYS FOLLOW THE LABEL INSTRUCTIONS ON THE PACKAGE BEFORE USING.

TRE-HOLD®

SPROUT INHIBITOR FOR CITRUS

(Ready to Use Formulation)

Controls Sprouting on Non-Bearing Citrus and Non-Bearing Nectarines.

ACTIVE INGREDIENT:

Ethyl 1-naphthaleneacetate	1.15%
INERT INGREDIENTS:	98.85%
TOTAL	100.00%

This product contains the toxic inert ingredient formaldehyde.

EPA Reg. No. 264-397 EPA Est. No. 264-MO-01

KEEP OUT OF REACH OF CHILDREN

CAUTION

For **MEDICAL** and **TRANSPORTATION** Emergencies *ONLY* Call 24 Hours A Day 1-800-334-7577

For **PRODUCT** *USE* Information Call 1-800-334-9745

PRECAUTIONARY STATEMENTS

CAUTION

HAZARD TO HUMANS AND DOMESTIC ANIMALS

Harmful if swallowed or inhaled. Causes eye and skin irritation. Avoid contact with eyes, skin or clothing. Wear goggles or face shield and rubber gloves when handling. Wash thoroughly with soap and water after handling.

STATEMENT OF PRACTICAL TREATMENT

In case of eye contact: Flush with plenty of water for at least 15 minutes. Get medical attention if irritation persists.

If inhaled: Remove victim to fresh air. Get medical attention.

In case of skin contact: Wash with plenty of soap and water. Get medical attention if irritation persists. Remove and wash contaminated clothing before reuse.

ENVIRONMENTAL HAZARDS

Do not contaminate water when disposing of equipment washwaters. Do not apply directly to water or wetlands.

GENERAL INFORMATION

Proper pruning of young nonbearing trees promoted the growth and development of strong, well-spaced branches that can support more larger sized fruit, resist wind and ice damage, improve spray penetration for better insect and disease control and improve light penetration for better fruit set, size and color.

Citrus and nectarine trees sprout vigorously from the root stocks and pruning cuts made on hard wood branches.

TRE-HOLD® Sprout Inhibitor for citrus controls sprouting from pruning cuts, trunks, and root stocks. TRE-HOLD® Sprout Inhibitor for Citrus is a ready to use product formulated in a white latex paint to protect against sun scald damage. FOR NON-BEARING NECTARINES, the product is diluted—see dilution chart.

Thorough coverage immediately after pruning is necessary for maximum sprout inhibition. Do not allow TRE-HOLD® to contact buds or foliage as injury may result. Mix thoroughly before using.

DIRECTIONS FOR USE

IT IS A VIOLATION OF FEDERAL AND STATE LAW TO USE THIS PRODUCT IN A MANNER INCONSISTENT WITH ITS LABELING.

Do not apply TRE-HOLD® Sprout Inhibitor for Citrus through any type of irrigation system.

NON-BEARING CITRUS

Use the product as supplied (undiluted) and follow the procedures described below for treating sprouts on scaffold limbs, trunks and rootstocks.

NON-BEARING NECTARINES

For non-bearing nectarines first dilute the product to ¼ of the standard rate as shown in the following examples and follow the procedures described below for treating sprouts on scaffold limbs, trunks and rootstock.

Not for use on non-bearing nectarines in California.

PRODUCT DILUTION FOR NECTARINES

Amount of TRE-HOLD Sprout Inhibitor Paint	Amount of Water	Total Volume
1 quart	3 quarts	1 gallon
1 gallon	3 gallons	4 gallons

APPLICATION

Sprouting From Scaffold Limbs: Remove unwanted sprouts and thoroughly coat cut surfaces and surrounding bark.

Sprouting from Trunk: Remove trunk suckers and water sprouts and apply TRE-HOLD® to the entire trunk from ground line to scaffold limbs.

Rootstock: Citrus rootstock which is healthy, vigorous and free from stress may be treated. After planting apply TRE-HOLD® from the ground line to the first scaffold branches.

Do not allow the product to puddle at the base of the tree as injury and stand loss may occur.

Do not apply to rootstock if the bud union has not callused.

In temperate zones where freezing may injure scaffold branches, restrict treatment to the frost-free period to allow for a short period of growth between treatment and damaging freeze.

GENERAL WORKER PROTECTION STATEMENTS

Do not apply this product in such a manner as to directly or through drift expose workers or other persons. The area treated must be vacated by unprotected persons.

Do not enter treated areas without protective clothing until sprays have dried.

Because certain states may require more restrictive reentry intervals for various crops treated with this product, consult your State Department of Agriculture for further information.

Written or oral warnings must be given to workers who are expected to be in treated area or in an area about to be treated with this product. Advise workers to stay out of fields during application and until sprays have dried. Regular long-sleeved work clothing should be worn when working in treated fields. See Precautionary Statements and Statement Of Practical Treatment for information on accidental exposures. When oral warnings are given, warnings shall be given in a language customarily understood by workers. Oral warnings must be given if there is reason to believe that written warnings cannot be understood by workers. Written warnings must include the following information: Appropriate signal word (CAUTION), area treated with TRE-HOLD® Sprout Inhibitor for Citrus, date of application, appropriate clothing and reentry interval (i.e., until sprays have dried).

STORAGE AND DISPOSAL

STORAGE

Do not store near fertilizers, seeds, insecticides, or fungicides. Do not use or store near heat or open flame.

PESTICIDE DISPOSAL

Do not contaminate water, food, or feed by storage or disposal. Wastes resulting from the use of this product may be disposed of on site or at an approved waste disposal facility.

CONTAINER DISPOSAL

Do not reuse empty container. Triple rinse (or equivalent). Then offer for recycling or reconditioning, or puncture and dispose of in a sanitary landfill, or incineration, or, if allowed by state and local authorities, by burning. If container is burned, stay out of smoke.

LIMITED WARRANTY AND DISCLAIMER

The manufacturer warrants that this product conforms to the chemical description on the label; that this product is reasonably fit for the purposes set forth in the directions for use when it is used in accordance with such directions; and that the directions, warnings and other statements on this label are based upon responsible experts' evaluation of reasonable tests of effectiveness, of toxicity to laboratory animals and to plants, and of residues on food crops and upon reports of field experience. Test have not been made on all varieties or in all states or under all conditions. THE MANUFACTURER NEITHER MAKES NOR INTENDS, NOR DOES IT AUTHORIZE ANY AGENT OF REPRESENTATIVE TO MAKE, ANY OTHER WARRANTIES, EXPRESS OR IMPLIED, AND IT EXPRESSLY EXCLUDES AND DISCLAIMS ALL IMPLIED WARRANTIES OF MERCHANTABILITY OR FITNESS FOR A PARTICULAR PURPOSE.

THIS WARRANTY DOES NOT EXTEND TO, AND THE BUYER SHALL BE SOLELY RESPONSIBLE FOR, ANY AND ALL LOSS OR DAMAGE WHICH RESULTS FROM THE USE OF THIS PRODUCT IN ANY MANNER WHICH IS INCONSISTENT WITH THE LABEL DIRECTIONS, WARNINGS OR CAUTIONS.

BUYER'S EXCLUSIVE REMEDY AND MANUFACTURER'S OR SELLER'S EXCLUSIVE LIABILITY FOR ANY AND ALL CLAIMS, LOSSES, DAMAGES OR INJURIES RESULTING FROM THE USE OR HANDLING OF THIS PRODUCT, WHETHER OR NOT BASED IN CONTRACT, NEGLIGENCE, STRICT LIABILITY IN TORT OR OTHERWISE SHALL BE LIMITED, AT THE MANUFACTURER'S OPTION, TO REPLACEMENT OF, OR THE REPAYMENT OF THE PURCHASE PRICE FOR, THE QUANTITY OF PRODUCT WITH RESPECT TO WHICH DAMAGES ARE CLAIMED. IN NO EVENT SHALL MANUFACTURER OR SELLER BE LIABLE FOR SPECIAL, INDIRECT OR CONSEQUENTIAL DAMAGES RESULTING FROM THE USE OR HANDLING OF THIS PRODUCT.

NOTICE TO BUYER

Purchase of this material does not confer any rights under patents governing this product or the use thereof in countries outside the United States.

TRE-HOLD® is a registered trademark of RHÔNE-POULENC

AG 90046

THIS SPECIMEN LABEL IS INTENDED FOR USE ONLY AS A GUIDE IN PROVIDING GENERAL INFORMATION REGARDING THE DIRECTIONS, WARNINGS AND CAUTIONS ASSOCIATED WITH THE USE OF THIS PRODUCT. AS WITH ANY AGRICULTURAL CHEMICAL, ALWAYS FOLLOW THE LABEL INSTRUCTIONS ON THE PACKAGE BEFORE USING.

WEEDAR® EMULSAMINE® E-3

BROADLEAF HERBICIDE

The oil soluble amine form of 2,4-D

ACTIVE INGREDIENTS:

Dodecylamine salt of 2,4-dichlorophenoxyacetic acid*	47.3%
Tetradecylamine salt of 2,4-dichlorophenoxyacetic acid*	11.0%
INERT INGREDIENTS:	41.7%
TOTAL	100.0%

* Isomer specific by AOAC method No. 6.D01-5

* Total 2,4-dichlorophenoxyacetic acid equivalent 31.2% by weight or 2.7 pounds per gallon.

EPA Reg. No. 264-179AA EPA Est. 264-MO-1

KEEP OUT OF REACH OF CHILDREN

CAUTION

For PRODUCT USE Information Call 1-800-334-9745.
MEDICAL and TRANSPORTATION Emergencies ONLY Call 24 Hours A Day 1-800-334-7577

EPA Reg. No. 264-179 AA EPA Est. No. 264-MO-01

PRECAUTIONARY STATEMENTS

CAUTION

HAZARDS TO HUMANS AND DOMESTIC ANIMALS

Harmful if swallowed or absorbed by skin. Do not allow contact with skin or clothing. Causes irritation of eyes and skin. Do not get in eyes.

STATEMENT OF PRACTICAL TREATMENT

IF IN EYES: Flush eyes with plenty of water for at least 15 minutes and get medical attention.

IF ON SKIN: Wash exposed skin with plenty of soap and water.

ENVIRONMENTAL HAZARDS

Do not apply to water except as specified on this label.

This product is toxic to fish. Keep out of lakes, streams or ponds except for water hyacinth control.

Do not apply where runoff is likely to occur. Do not apply when weather conditions favor drift from treated areas.

Do not contaminate any body of water by washing of spray equipment or disposal of waste. Do not contaminate water used for irrigation or domestic purposes.

Avoid spray drift to susceptible plants such as cotton, beans, tomatoes, and ornamentals as this product may cause injury. Coarse sprays are less likely to drift.

Do not use the same spray equipment for other purposes. Do not use in or near a greenhouse.

GENERAL INFORMATION

Do not apply WEEDAR® Emulsamine® E-3 Broadleaf Herbicide through any type of irrigation system. Do not use this product for purposes other than those recommended on this label. Do not graze dairy animals on treated areas within 7 days after treatment.

WEEDAR® Emulsamine® E-3 Broadleaf Herbicide is a concentrated formulation of 2,4-D as the new oil soluble amine. WEEDAR® Emulsamine® E-3 Broadleaf Herbicide combines the effectiveness of low volatile ester forms of 2,4-D with maximum safety from volatility damage. Especially useful in areas of high summer temperatures.

CONTROLS THESE WEEDS

Annual and Biennial Weeds

- bull thistle
- burdock
- cocklebur
- coffeeweed
- French weed
- galinsoga
- jimsonweed
- kochia
- lambsquarters
- morningglory
- mustard (wild)
- peppergrass
- pigweed
- lettuce (wild)
- radish (wild)
- ragweed (common)
- Russian thistle
- sowthistle (common)
- velvetleaf

Perennial Weeds

- alkali mallow
- aster
- bindweed
- blue lettuce
- blueweed
- bur ragweed
- Canada thistle
- chicory
- dandelion
- docks
- goldenrod
- nettles
- orange hawkweed
- plantains
- povertyweed
- ragweed
- sowthistle
- stinging nettles
- toadflax
- wild garlic
- wild onion

MIXING SPRAYS

Fill the spray tank with half the required amount of water. Add the recommended amount of WEEDAR® Emulsamine® E-3 while agitating, then continue filling the spray tank with water.

WATER HYACINTH CONTROL IN STILL WATER (LAKES, PONDS, AND MARSHES)

WEEDAR® Emulsamine® E-3 Broadleaf Herbicide has proven highly effective for controlling water hyacinths, whether applied with air or boat equipment. In areas of dense stands of hyacinths, weed foliage intercepts and quickly absorbs WEEDAR® Emulsamine® E-3, producing a rapid browning which acts as a marker for later spraying.

Use in waters covered with dense stands of water hyacinths, which can intercept the spray material. Spray the weed mass only. When applied directly to open water, this product is toxic to fish. It should not be used for total water treatments of submersed weeds in ponds, lakes or streams.

Treatment of aquatic weeds can result in oxygen loss from decomposition of dead vegetation, fish injury or death may result. To reduce risk, spray only 1/3 to 1/2 the area in a single application allowing fish to move to untreated areas.

AERIAL APPLICATION

Use 1 1/3 gallons WEEDAR® Emulsamine® E-3 in 5 to 15 gallons of water to cover one surface acre. The use of the MICROFOIL™ boom will provide added drift control.

BOAT APPLICATION

Use 1 1/3 gallons WEEDAR® Emulsamine® E-3 in 50 to 100 gallons of water per acre. Uniform coverage is essential. Avoid submerging plants after treatment.

APPLICATION WITH LO-DRIFT® SPRAY ADDITIVE

For fixed wing aerial spraying, use 1 1/3 gallons WEEDAR® Emulsamine® E-3 plus 1 1/2 to 3 oz. LO-DRIFT in the 5 gallons of water per acre. If using DIRECTA-SPRAY® spray applicator use 1 1/3 gallons WEEDAR® Emulsamine® E-3 with LO-DRIFT® in 50-100 gallons of water. Use 1 pt. to 1 qt. LO-DRIFT® per 100 gallons of solution. Follow mixing directions on label for LO-DRIFT® Spray Additive.

DIRECTIONS FOR USE

IT IS A VIOLATION OF FEDERAL LAW TO USE THIS PRODUCT IN A MANNER INCONSISTENT WITH ITS LABELING.

Weeds In Non-Crop Areas	Amounts To Use	When To Spray
Lawns, Parks, Golf Courses, Cemeteries, and Similar Turf Areas	2 to 4 pints per acre	Apply in spring or fall, when annual weeds such as knot-weed are up but still small and in the seedling stage. To control dandelion, plantain and most other common weeds in turf, apply in spring or fall when these weeds are growing actively. Treat wild garlic or onion in three successive times (Example: spring, fall, spring). Do not apply to newly seeded areas until sod is formed and grass has been cut several times. Do not use WEEDAR® Emulsamine® E-3 Broadleaf Herbicide for susceptible Southern grasses such as St. Augustine. Bent grasses, dichondra or clover may be injured by this treatment. Do not graze dairy animals on treated areas within 7 days after treatment.
Fencerows, Roadsides*, Ditchbanks, and Vacant Lots	3 to 5 pints per acre	Apply when annual weeds are young and growing vigorously. For perennials, spray when actively growing and near the bud stage but before flowering. A second application is usually needed for best results on thistle, nettle, and other hard-to-control weeds.
Woody Plants* (vacant lots, fencerows, similar areas)	1 1/2 to 2 gallons per 100 gallons water	To control brush susceptible to 2,4-D such as willow, alder, buckbrush, sumac and elderberry; treat in spring or early summer. Wet thoroughly all parts of the brush, including foliage and stems, until spray starts to run off. For knapsack sprayers; use 2 fluid ounces of WEEDAR® Emulsamine® E-3 per gallon of water. Addition of 1 to 5 gallons of fuel oil may increase weed control.

* Use enough water for uniform coverage; 20-30 gallons per acre in low volume sprayers, 100-250 gallons per acre in high volume sprayers. Use the lower rate where annual weeds are the problem, the higher rate for perennials. Addition of 1 to 5 gallons of fuel oil may increase weed control.

Weeds In Cropped Areas	Amounts To Use	When To Spray
Wheat,* Barley and Rye * Treat fall-planted wheat in the spring.	1 to 2 pints per acre	Apply after grains are fully tillered (4 to 6 inches tall) but before they are in the boot stage. Apply when annual weeds are small and perennial weeds are in the bud stage if possible, but do not spray after grains are in the boot to dough stage. Do not use on grains undersown with legumes such as alfalfa or clovers, except where some legume injury can be tolerated. Do not forage or graze treated grain fields within 2 weeks after treatment nor feed treated straw to livestock.
Grain Sorghum	1 to 1 1/2 pints per acre	Treat when sorghum is 6 to 10 inches tall, using drop nozzles, when secondary roots are well established. **Do not apply from tassle to dough stage.**
Field Corn—post-emergence	2/3 pint per acre	Apply when weeds are small and corn is less than 10 inches tall. Avoid spraying just after corn leaves unfold as injury may occur. Use drop pipes to spray corn more than 10 inches tall. **Do not apply during or after the tassle to dough stage.**
Hayfields, Pastures and Rangelands	1/2 to 4 quarts per acre	Treat when weeds are growing actively. Do not apply when grass is in the boot to milk stage. Do not apply to recently seeded pastures until grass is well established. Most legumes will be injured or killed at rates recommended. Treat wild onion or garlic in early spring and fall and repeat for at least two seasons. Do not graze dairy animals for 7 days after spraying pastures with this product. For aerial application on grain, apply WEEDAR® Emulsamine® E-3 in 3-10 gallons of water per acre.

* Use low pressure sprayers (25 to 40 psi). Apply in minimum of 7 1/2 to 10 gallons water per acre. Use lower rate of WEEDAR® Emulsamine® E-3 where annual weeds are the problem, the higher rate for perennial weeds.

STORAGE AND DISPOSAL

STORAGE

Do not store near fertilizers, seeds, insecticides or fungicides. Do not contaminate water, food, or feed by storage or disposal.

PESTICIDE DISPOSAL

Pesticide wastes are toxic. Improper disposal of excess pesticide, spray mixture, or rinsate is a violation of Federal Law. If these wastes cannot be disposed of by use according to label instructions, contact your State Pesticide or Environmental Control Agency, or the Hazardous Waste representative at the nearest EPA Regional Office for guidance.

CONTAINER DISPOSAL

Triple rinse (or equivalent). Then offer for recycling or reconditioning, or puncture and dispose of in a sanitary landfill, or incinerate, or, if allowed by state and local authorities, by burning. If container is burned, stay out of smoke.

LIMITED WARRANTY AND DISCLAIMER

The manufacturer warrants that this product conforms to the chemical description on the label; that this product is reasonably fit for the purposes set forth in the directions for use when it is used in accordance with such directions, warnings and other statements on this label are based upon responsible experts' evaluation of reasonable tests of effectiveness of toxicity to laboratory animals and to plants, and of residues on food crops, and upon reports of field experience. Tests have not been made on all varieties or in all states or under all conditions. THE MANUFACTURER NEITHER MAKES NOR INTENDS, NOR DOES IT AUTHORIZE ANY AGENT OR REPRESENTATIVE TO MAKE ANY OTHER WARRANTIES, EXPRESS OR IMPLIED, AND IT EXPRESSLY EXCLUDES AND DISCLAIMS ALL IMPLIED WARRANTIES OF MERCHANTABILITY OR FITNESS FOR A PARTICULAR PURPOSE.

THIS WARRANTY DOES NOT EXTEND TO, AND THE BUYER SHALL BE SOLELY RESPONSIBLE FOR, ANY AND ALL LOSS OR DAMAGE WHICH RESULTS FROM THE USE OF THIS PRODUCT IN ANY MANNER WHICH IS INCONSISTENT WITH THE LABEL DIRECTIONS, WARNINGS OR CAUTIONS.

BUYER'S EXCLUSIVE REMEDY AND MANUFACTURER'S OR SELLER'S EXCLUSIVE LIABILITY FOR ANY AND ALL CLAIMS, LOSSES, DAMAGES, OR INJURIES RESULTING FROM THE USE OR HANDLING OF THIS PRODUCT, WHETHER OR NOT BASED IN CONTRACT, NEGLIGENCE, STRICT LIABILITY IN TORT OR OTHERWISE, SHALL BE LIMITED, AT THE MANUFACTURER'S OPTION, TO REPLACEMENT OF, OR THE REPAYMENT OF THE PURCHASE PRICE FOR, THE QUANTITY OF PRODUCT WITH RESPECT TO WHICH DAMAGES OF THE PURCHASE PRICE FOR, THE QUANTITY OF PRODUCT WITH RESPECT TO WHICH DAMAGES ARE CLAIMED. IN NO EVENT SHALL MANUFACTURER OR SELLER BE LIABLE FOR SPECIAL INDIRECT OR CONSEQUENTIAL DAMAGES RESULTING FROM THE USE OR HANDLING OF THIS PRODUCT

NOTICE TO BUYER

Purchase of this material does not confer any rights under patents governing this product or the use thereof in countries outside of the United States.

WEEDAR and EMULSAMINE are registered trademarks of Rhône-Poulenc

AG 90017

THIS SPECIMEN LABEL IS INTENDED FOR USE ONLY AS A GUIDE IN PROVIDING GENERAL INFORMATION REGARDING THE DIRECTIONS, WARNINGS AND CAUTIONS ASSOCIATED WITH THE USE OF THIS PRODUCT. AS WITH ANY AGRICULTURAL CHEMICAL, ALWAYS FOLLOW THE LABEL INSTRUCTIONS ON THE PACKAGE BEFORE USING.

WEEDAR 64®

BROADLEAF HERBICIDE

The 2,4-D Amine Weed Killer

ACTIVE INGREDIENT:	
Dimethylamine salt of 2,4-Dichlorophenoxyacetic acid*	46.8%
INERT INGREDIENTS:	53.2%
TOTAL	100.0%

* 2,4-Dichlorophenoxyacetic acid equivalent 38.9% by weight or 3.8 pounds per gallon.

* Isomer specific by AOAC method No. 6.D01-5

E.P.A. Reg. No. 264-2AA E.P.A. Est. No. 264-MO-01

KEEP OUT OF REACH OF CHILDREN

DANGER PELIGRO

PRACTICAL TREATMENT

IF IN EYES: Flush with water for at least 15 minutes and get medical attention

PRECAUCION AL USUARIO: Si usted no lee ingles, no use este producto hasta que la etiqueta le haya sido explicado ampliamente.

For PRODUCT USE Information Call 1-800-334-9745
For MEDICAL and TRANSPORTATION Emergencies Information ONLY Call 24 Hours A Day 1-800-334-7577

PRECAUTIONARY STATEMENTS

DANGER

HAZARDS TO HUMANS AND DOMESTIC ANIMALS

Harmful if swallowed. May be fatal if absorbed through the skin. Causes irreversible eye damage. Avoid breathing vapors or spray mist. Do not get in eyes, on skin, or on clothing. Wear goggles or safety glasses, protective clothing and rubber gloves. Wash thoroughly with soap and water after handling and before eating, drinking or using tobacco. Remove contaminated clothing and wash before reuse.

STATEMENT OF PRACTICAL TREATMENT

IF ON SKIN: Wash skin with plenty of soap and water. Remove contaminated clothing. Get medical attention.

IF SWALLOWED: If patient is conscious and alert, give 2-3 glasses of water or milk to drink. If available, give one tablespoon of Syrup of Ipecac to induce vomiting. Alternatively, induce vomiting by touching back of throat with finger. Do not repeat the Syrup of Ipecac. Do not make an unconscious person vomit. Get medical attention.

IF IN EYES: Flush with water for at least 15 minutes. Get medical attention, preferably an ophthalmologist.

IF INHALED: Move to an uncontaminated area. Get medical attention.

NOTE TO PHYSICIAN This product contains a phenoxy herbicidal chemical.There is no specific antidote. All treatments should be based on observed signs and symptoms of distress in the patient. Overexposure to materials other than this product may have occurred.

ENVIRONMENTAL HAZARDS

Do not contaminate water by cleaning equipment or disposal of wastes. Do not apply when weather conditions favor drift from treated areas. Do not use the same spray equipment for other purposes unless thoroughly cleaned.

Do not use in or near a greenhouse.

Do not contaminate water used for irrigation or domestic purposes (except as specifically recommended on this label) especially in areas where grapes, cotton, tomatoes or other susceptible plants are grown.

Do not treat irrigation ditches in areas where water will be used to overhead (sprinkler) irrigate susceptible crops especially grapes, tomatoes, tobacco and cotton.

Do not apply WEEDAR® 64 Broadleaf Herbicide directly to, or permit mist to drift onto cotton, okra, grapes, tomatoes, fruit trees, vegetables, flowers or other desirable crop or ornamental plants which are susceptible to 2,4-D herbicide. Do not apply near susceptible plants since very small quantities of the 2,4-D will cause severe injury during the growing or dormant periods. Crops contacted by WEEDAR® 64 Broadleaf Herbicide sprays or spray drift may be killed or suffer significant stand loss with extensive quality and yield reduction.

Do not apply when a temperature air inversion exists. Such a condition is characterized by little or no air movement and an increase in air temperature with an increase in height. In humid regions a fog or mist may form. An inversion may be detected by producing a smoke column and checking for a layering effect. If questions exist pertaining to the existence of an inversion, consult with local weather services before making an application.

Use coarse sprays to minimize drift. Do not apply with hollow cone type insecticide or other nozzles that produce fine spray droplets. Drift from aerial or ground application may be reduced by: (1) applying as near the target as possible in order to obtain coverage; (2) by increasing the volume of spray mix per acre; (3) by decreasing the pounds of pressure at the nozzle tips; (4) by using nozzles which produce a coarse spray pattern; (5) by not applying when wind is blowing toward susceptible crops or valuable plants.

DIRECTIONS FOR USE

IT IS A VIOLATION OF FEDERAL LAW TO USE THIS PRODUCT IN A MANNER INCONSISTENT WITH ITS LABELING.

Do not apply WEEDAR® 64 Broadleaf Herbicide through any type of irrigation system.

GENERAL WORKER PROTECTION STATEMENTS

Do not apply this product in such a manner as to directly or through drift expose workers or other persons. The area treated must be vacated by unprotected persons.

Do not enter treated areas without protective clothing until sprays have dried.

Because certain states may require more restrictive reentry intervals for various crops treated with this product, consult your State Department of Agriculture for further information.

Written or oral warnings must be given to workers who are expected to be in a treated area or in an area about to be treated with this product. Advise workers to stay out of fields during application and until sprays have dried. Regular long-sleeved work clothing should be worn when working in treated fields. See Precautionary Statements, Statement of Practical Treatment and Note to Physician for information on accidental exposures. When oral warnings are given, warnings shall be given in a language customarily understood by workers. Oral warnings must be given if there is reason to believe that written warnings cannot be understood by workers. Written warnings must include the following information: appropriate signal word (DANGER—PELIGRO), area treated with WEEDAR® 64 Broadleaf Herbicide, date of application, appropriate protective clothing, and reentry interval (i.e. until sprays have dried).

GENERAL INFORMATION

TO PREPARE A SPRAY:

Mix WEEDAR® 64 Broadleaf Herbicide only with water unless otherwise indicated on this label. Add about one-half the water to the mixing tank, then add WEEDAR® 64 with agitation, and finally the rest of the water with continuing agitation.

Note: Adding oil, wetting agent, or other surfactant to the spray may increase effectiveness on weeds but also may reduce selectivity to crops, resulting in crop damage.

COMPATIBILITY:

If WEEDAR® 64 is to be tank mixed with fertilizers or with other pesticides, compatibility should be tested prior to mixing. To test for compatibility, use a small container and mix a small amount (0.5 to 1 qt) of spray, combining all ingredients in the same ratio as the anticipated use. If any indications of physical incompatibility develop, do not use this mixture for spraying. Indications of incompatibility usually will appear within 5-15 minutes after mixing.

Read and follow all directions and precautions on this label and on the labels of any products for which a tank mixture is being considered.

Apply by air or ground equipment in sufficient gallonage to obtain adequate coverage, except as otherwise directed on this label.

Use 2 or more gallons of water per acre for aerial application and 10 or more gallons of water per acre for ground application.

Weeds controlled by properly applied treatments of WEEDAR® 64 include the following:

Annual and Biennial Weeds

*beggarticks	pepperweeds (except perennial)
bullthistle	**pigweeds (*Amaranthus* spp.)
coffeeweed	prickly lettuce
common cocklebur	ragweed (common or giant)
common burdock	rough fleabane
common evening primrose	*Russian thistle
common lambsquarters	Salsify (western or common)
hairy galinsoga	*smartweeds (annual species)
jimsonweed	sowthistles (annual or spiny)
*knotweed	sunflower
*mallow (venice or little)	*vervains
marshelder	vetches
morningglory (common, ivy, woolly)	*wild carrot
*musk thistle (***)	wild lettuce
mustards (except blue mustard)	wild parsnips

Perennial Weeds

*bindweed (hedge, field, European)	*hoary cress
blue lettuce	*ironweed
*Canada thistle	'Jerusalem-artichoke
catnip	*many flowered aster
chicory	*nettles (including stinging)
dandelion	*orange hawkweed
*docks	plantains
*dogbanes	sowthistle (perennial)
*goldenrod	*vervains
*ground ivy	*wild garlic
healall	*wild onion

* These species may require repeat applications and/or use of the higher rate recommended on this product label even under ideal conditions for application.

** Control of pigweeds in the High Plains area of Texas and Oklahoma may not be satisfactory with this product.

*** Not registered for control of musk thistle in California.

INJURY TO CROPS FROM THIS HERBICIDE MAY OCCUR. IF YOU ARE NOT PREPARED TO ACCEPT SOME DEGREE OF CROP INJURY DO NOT USE THIS PRODUCT.

Crop varieties vary in response to 2,4-D and some are easily injured. Apply WEEDAR® 64 Broadleaf Herbicide only to varieties known to be tolerant to 2,4-D. If you are uncertain concerning tolerant varieties or local use situations that may affect crop tolerance to 2,4-D consult your seed company, state Agricultural Extension Service or qualified crop consultant for advice.

Recommendations for Applying WEEDAR® 64 Broadleaf Herbicide

Crop	Amount of WEEDAR® 64 Per Acre	Directions
CEREAL GRAINS		
Postemergence WHEAT, BARLEY, RYE, OATS (not underseeded with legumes) Annual and Biennial weeds	½ to 2 pints*	Apply after grain is fully tillered (usually about 4-8" high) but not forming joints in the stem. **Do not spray grain in boot to dough stage.**
Perennial broadleaf weeds	1 to 2 pints	
Grain underseeded with legumes	¼-½ pints	Apply after grain is 8" tall. **Do not spray grain in boot to dough stage.** Do not spray alfalfa or sweet clover unless the infestation is severe and injury to these legumes can be tolerated.
Emergency weed control in wheat Perennial broadleaf weeds	3 pints*	Apply when weeds are approaching bud stage, after the grain dough stage. **Do not spray during the boot to dough stage.** The 3 pint per acre application can produce injury to wheat. Balance the severity of your weed problem against the possibility of crop damage. Where perennial weeds are scattered, spot treatment is suggested to minimize the extent of crop injury.

* Use the lower rate if small annual and biennial weeds are the major problem. Use the higher rate if perennial weeds or annual and biennial weeds are present which are in the hard to kill categories as determined by local experience. The higher rates increase the risk of grain injury and should be used only where the weed control problem justifies the grain damage risk. Do not apply WEEDAR® 64 to grain in the seedling stage.

RESTRICTIONS AND LIMITATIONS FOR USE ON CEREAL GRAINS

- **For aerial application** on grain, apply WEEDAR® 64 in 3 to 10 gallons of water per acre.
- For ground application use a minimum of 10 gallons of water per acre.
- Do not permit dairy animals or meat animals being finished for slaughter to forage treated grain fields within 2 weeks after treatment.
- Do not feed treated straw to livestock if an emergency treatment as described above is applied.

Crop	Amount of WEEDAR® 64 Per Acre	Directions
CORN (Field and Sweet)		
Preplant	1 to 2 pints	To control emerged broadleaf weed seedlings or existing cover crops prior to planting corn, apply 7 to 14 days before planting. Do not use on light, sandy soil, or where soil moisture is inadequate for normal weed growth. Use high rate for less susceptible weeds or cover crops such as alfalfa.
Preemergence	2 to 3 pints	Apply 3 to 5 days after planting but before corn emerges. Do not use on light, sandy soils or where soil moisture is low.
Postemergence Annual broadleaf weeds Perennial broadleaf weeds	½ to 1 pint 1 to 1 ½ pint	Apply when weeds are small and corn is less than 8 inches tall (to top of canopy). When corn is over 8 inches high, use drop nozzles and keep spray off foliage. Treat perennial weeds when they are in the bud to bloom stage. **Do not spray corn in the tassel to dough stage.** Corn treated with 2,4-D may become temporarily brittle. Winds or cultivation may cause stalk breakage during the period of time when the corn is brittle.
GRAIN SORGHUM (Milo)	1 pint	Apply when sorghum is 6 to 15 inches tall. If sorghum is taller than 8 inches to the top of the canopy, use drop nozzles and keep spray off the foliage. **Do not treat during the boot, flowering, or dough stages.**

RESTRICTIONS AND LIMITATIONS FOR USE ON CORN OR SORGHUM:

Crop	Amount of WEEDAR® 64 Per Acre	Directions
- Do not forage or feed fodder for 7 days following application		
RICE	1½ to 2½ pints	Apply when rice is in the late tillering stage of development, at the time of first joint development. Do not apply after panicle initiation, after rice internodes exceed one-half inch, at early seedling, early panicle, boot, or heading stages. Consult local University or Agricultural Extension Service Specialists for more specific information on rates and timings of application.
SUGARCANE Preemergence Postemergence	 4 pints 1½ to 2 pints	Apply before canes appear for control of emerged broadleaf weeds. Apply after cane emerges and through layby. **DO NOT USE IN CALIFORNIA.**
FALLOW LAND AND CROP STUBBLE: Annual broadleaf weeds	1 to 2 pints	Use the lower rate when weeds are small (2-3" tall) and actively growing. Use the higher rate on older and drought-stressed plants.
Biennial broadleaf weeds	2 to 4 pints	Spray while musk thistles or other biennial species are in the seedling to rosette stage, and before flower stalks become apparent. The lower rate can be used in spring during rosette stage. Use the highest rate, in fall or after flower stalks have developed.
Perennial broadleaf weeds	2 to 6 pints	Spray weeds in bud to bloom stage, or in good vegetative growth. Do not disturb treated areas for at least 2 weeks after treatment, or until weed tops are dead.
Wild garlic and onion in crop stubble	4 to 6 pints	Apply to new regrowth of wild garlic or onion which occurs in the fall following harvest of small grains, corn, or grain sorghum.
RESTRICTIONS AND LIMITATIONS FOR USE IN FALLOWLAND AND CROP STUBBLE: - Do not plant any crop for 3 months after treatment or until chemical has disappeared from the soil.		
ESTABLISHED GRASS PASTURES AND RANGELAND Annual broadleaf weeds Biennial and Perennial broadleaf weeds	 2 pints 2 to 4 pints	Apply when weeds are small and actively growing and prior to bud stage. Spray while musk thistles or other biennial species are in the seedling to rosette stage and before flower stalks become apparent. The lower rate can be used in the spring during rosette stage. Use the highest rate in the fall or after flower stalks have developed. Do not apply to newly seeded areas until grass is well established. Do not apply to grass in the early boot through milk stage if grass seed production is desired. Bentgrass and legumes may be injured by this treatment.
RESTRICTIONS AND LIMITATIONS FOR USE IN PASTURES AND RANGELANDS: - Do not graze (dairy) cattle in treated areas for 14 days after application.		
CONSERVATION RESERVE PROGRAM AREAS		
ANNUAL BROADLEAF WEEDS in young grasses in established grasses	 ½ to 1 pint ½ to 2 pints	Apply to actively growing annual broadleaf weeds. Use ½ to 1 pint when weeds are small; use higher rates on older weeds. Do not apply to young grasses with fewer than 6 leaves or prior to tillering, as excessive injury may result. Do not apply more than 1 pint until grasses are well established as excessive injury may result.
BIENNIAL AND PERENNIAL broadleaf weeds in established grasses	2 to 4 pints	Treat when biennial weeds are in the seedling to rosette stage and before flower stalks become apparent. Treat perennial weeds in the bud to bloom stage. Apply to actively growing weeds.
RESTRICTIONS AND LIMITATIONS FOR USE ON CONSERVATION RESERVE PROGRAM AREAS: - Use at least 2 gallons of water per acre by air and 5 gallons of water per acre by ground. - Do not harvest or graze treated Conservation Reserve Program areas. - Do not apply to grasses in the boot to dough stage if grass seed production is desired.		
GRASSES FOR SEED PRODUCTION	2 to 4 pints	Apply to established stands in spring from tiller to early boot stage. **Do not spray in boot stage.** New spring seedlings may be treated with the lower rate after grass seedlings have at least 5 leaves. Perennial weed regrowth may be treated in the fall. **DO NOT USE IN CALIFORNIA.** Restrictions and limitations for use on grasses for seed production. Do not graze dairy animals or cut forage for hay within 7 days of application.
NON-CROPLAND (Fencerows, roadsides, drainage ditches, golf courses, cemeteries, parks, turfgrass and other grass areas) Annual broadleaf weeds Biennial and Perennial broadleaf weeds	 2 to 4 pints 4 to 8 pints	Treat when weeds are young and actively growing. Perennial weeds should be near the bud stage but not flowering at application. Do not use on susceptible Southern grasses such as St. Augustine. Do not apply to newly seeded areas until grass is well established. Bentgrass, clover, legumes, and dichondria may be injured by this treatment.
SPOT TREATMENT IN NON-CROP AREAS		Mix 2 to 3 fluid ounces of WEEDAR® 64 Broadleaf Herbicide in 3 gallons of water. Wet all weeds and stems thoroughly. For best results, treat when weeds are actively growing.
RESTRICTIONS AND LIMITATIONS FOR USE ON NON-CROPLAND: - Do not graze dairy animals for 7 days following treatment - Use sufficient gallonage to obtain thorough and uniform coverage		
FORESTRY—TREE INJECTION For controlling species such as alder, aspen, birch, blackgum, cherry, oak, sweetgum, and tulip popular:		Make injections as near the root collar as possible, using one injection per inch of trunk dbh (4 ½ feet). For resistant species such as hickory, injections should overlap. For best results, injections should be made during the growing season, May 15 through October 15. **For Dilute Injection:** Mix 1 gallon of WEEDAR® 64 Broadleaf Herbicide in 19 gallons of water for dilute injections. **For Concentrate Injections:** Use 1 to 2 ml of concentrate WEEDAR® 64 Broadleaf Herbicide per injection. The injector bit must penetrate the inner bark.
STONE FRUIT AND NUT ORCHARDS Annual broadleaf weeds	 3 pints	For control of weeds on the orchard floor, apply using coarse sprays and low pressure in sufficient volume of water to obtain thorough wetting of weeds. Treat when weeds are

Crop	Amount of WEEDAR® 64 Per Acre	Directions
		small and actively growing. Do not use on light, sandy soil. **DO NOT USE IN CALIFORNIA**

RESTRICTIONS AND LIMITATIONS FOR USE IN STONE FRUIT AND NUT ORCHARDS:

- Do not apply to bare ground as injury may result
- Do not apply immediately before irrigation and withhold irrigation for 2 days before and for 3 days after treatment.
- Do not allow spray to drift to or contact foliage, fruit, stems, trunks of trees or exposed roots as injury may result.
- Do not apply to newly established or young orchards. Trees must be at least 1 year old and in vigorous condition.
- Do not apply during bloom
- Do not graze or feed cover crops from treated orchards.
- Do not make more than 2 applications per year.
- Do not harvest stone fruit within 40 days of application.
- Do not harvest nuts within 60 days of application.

Crop	Amount of WEEDAR® 64 Per Acre	Directions
WEEDS AND BRUSH IRRIGATION CANAL DITCHBANKS (Seventeen Western States: Arizona, California, Colorado, Idaho, Kansas, Montana, Nebraska, New Mexico, Nevada, North Dakota, Oklahoma, Oregon, South Dakota, Texas, Utah, Washington, and Wyoming).		For control of annual and perennial broadleaf weeds, apply 1 to 2 quarts of WEEDAR® 64 Broadleaf Herbicide per acre in approximately 20 to 100 gallons per acre. Treat when weeds are young and actively growing before the bud or early bloom stage. For harder-to-control weeds a repeat spray after 3 to 4 weeks using the same rates may be needed for maximum results. Apply no more than two treatments per season. For woody brush and patches of perennial broadleaf weeds, mix 1 gallon of WEEDAR® 64 Broadleaf Herbicide in 150 gallons of water. Wet foliage thoroughly using about 1 gallon of solution per square rod. **Spraying Instructions** Apply with low pressure (10 to 40 psi) power spray equipment mounted on a truck, tractor, or boat. Apply while traveling upstream to avoid accidental concentration of chemical into water. Spray when the air is fairly calm; 5 mph or less. Do not use on small canals (less than 10 cfs) where water will be used for drinking purposes. Boom spraying onto water surface must be held to a minimum and no cross-stream spraying to opposite banks should be permitted. When spraying shoreline weeds, allow no more than 2 foot overspray onto water with an average of less than 1-foot overspray to prevent introduction of greater than negligible amounts of chemical into the water. Do not allow dairy animals to graze on treated areas for at least 7 days after spraying. Water within treated banks should not be fished.

STORAGE AND DISPOSAL

STORAGE

Do not contaminate or store near food, feedstuff, fertilizers, seeds, insecticides, fungicides or other pesticides.

PESTICIDE DISPOSAL

Do not contaminate water, food or feed by storage or disposal. Open dumping is prohibited.

Pesticide wastes are acutely hazardous. Improper disposal of excess pesticide, spray mixture, or rinsate is a violation of Federal law. If these wastes cannot be disposed of by use according to label instructions, contact your State Pesticide or Environmental Control Agency, or the Hazardous Waste representative at the nearest EPA Regional Office for guidance.

CONTAINER DISPOSAL

Triple rinse or (equivalent). Then offer for recycling or reconditioning, or puncture and dispose of in a sanitary landfill, or incineration, or, if allowed by state and local authorities, by burning. If burned, stay out of smoke.

Local conditions may affect the use of herbicides. Consult your State Agriculture Experiment Station, Farm Advisors, or Extension Weed Specialists for advice in selecting treatment from this label to best fit local conditions.

Be sure that use of this product conforms to all applicable laws, rules and regulations. Certain states have restrictions pertaining to application distances from susceptible crops. The applicator should become familiar with these laws, rules or regulations and follow them exactly. Apply this product only as specified on this label.

LIMITED WARRANTY AND DISCLAIMER

The manufacturer warrants that this product conforms to the chemical description on the label; that this product is reasonably fit for the purposes set forth in the directions for use when it is used in accordance with such directions; and that the directions, warnings and other statements on this label are based upon responsible experts' evaluation of reasonable tests of effectiveness, of toxicity to laboratory animals and to plants, and of residues on food crops, and upon reports of field experience. Tests have not been made on all varieties or in all states or under all conditions. THE MANUFACTURER NEITHER MAKES NOR INTENDS, NOR DOES IT AUTHORIZE ANY AGENT OR REPRESENTATIVE TO MAKE, ANY OTHER WARRANTIES, EXPRESS OR IMPLIED, AND IT EXPRESSLY EXCLUDES AND DISCLAIMS ALL WARRANTIES OF MERCHANTABILITY OR FITNESS FOR A PARTICULAR PURPOSE.

THIS WARRANTY DOES NOT EXTEND TO, AND THE BUYER SHALL BE SOLELY RESPONSIBLE FOR, ANY AND ALL LOSS OR DAMAGE WHICH RESULTS FROM THE USE OF THIS PRODUCT IN ANY MANNER WHICH IS INCONSISTENT WITH THE LABEL DIRECTIONS, WARNINGS OR CAUTIONS.

BUYER'S EXCLUSIVE REMEDY AND MANUFACTURER'S OR SELLER'S EXCLUSIVE LIABILITY FOR ANY AND ALL CLAIMS, LOSSES, DAMAGES, OR INJURIES RESULTING FROM THE USE OR HANDLING OF THIS PRODUCT, WHETHER OR NOT BASED IN CONTRACT, NEGLIGENCE, STRICT LIABILITY IN TORT OR OTHERWISE SHALL BE LIMITED, AT THE MANUFACTURER'S OPTION, TO REPLACEMENT OF, OR THE REPAYMENT OF THE PURCHASE PRICE FOR, THE QUANTITY OF PRODUCT WITH RESPECT TO WHICH DAMAGES ARE CLAIMED. IN NO EVENT SHALL MANUFACTURER OR SELLER BE LIABLE FOR SPECIAL, INDIRECT OR CONSEQUENTIAL DAMAGES RESULTING FROM THE USE OR HANDLING OF THIS PRODUCT.

NOTICE TO BUYER

Purchase of this material does not confer any rights under patents governing this product or the use thereof in countries outside of the United States.

WEEDAR is the registered trademark of Rhône-Poulenc

AG90018

THIS SPECIMEN LABEL IS INTENDED FOR USE ONLY AS A GUIDE IN PROVIDING GENERAL INFORMATION REGARDING THE DIRECTIONS, WARNINGS AND CAUTIONS ASSOCIATED WITH THE USE OF THIS PRODUCT. AS WITH ANY AGRICULTURAL CHEMICAL, ALWAYS FOLLOW THE LABEL INSTRUCTIONS ON THE PACKAGE BEFORE USING.

WEEDAR® 64A

HERBICIDE

The 2,4-D Amine For Weed Control

ACTIVE INGREDIENT

Diethanolamine salt of 2, 4-Dichlorophenoxyacetic acid*	54.6%
INERT INGREDIENTS:	45.4%
TOTAL	100.0%

*2,4-Dichlorophenoxyacetic acid equivalent 37.0% by weight or 3.8 pounds per gallon.

*Isomer specific by AOAC method No. 6.D01-5

E.P.A. Reg. No. 264-143AA E.P.A. Est. No. 264-MO-01

KEEP OUT OF REACH OF CHILDREN
DANGER PELIGRO

PRACTICAL TREATMENT

IF IN EYES: Flush with water for at least 15 minutes and get medical attention.

PRECAUTION AL USUARIO: Si usted no lee ingles, no use este producto hasta que la etiqueta le haya sido explicado ampliamente.

For PRODUCT USE Information Call 1-800-334-9745
For MEDICAL and TRANSPORTATION Emergencies ONLY Call 24 Hours A Day 1-800-334-7577.

PRECAUTIONARY STATEMENTS

DANGER

HAZARDS TO HUMANS AND DOMESTIC ANIMALS

Harmful if swallowed. Harmful if absorbed through the skin. Causes irreversible eye damage. Avoid breathing vapors or spray mist. Do not get in eyes, on skin, or on clothing. Wear goggles or safety glasses. Wash thoroughly with soap and water after handling and before eating, drinking or using tobacco. Remove contaminated clothing and wash before reuse.

STATEMENT OF PRACTICAL TREATMENT

IF ON SKIN: Wash skin with plenty of soap and water. Remove contaminated clothing. GET MEDICAL ATTENTION.

IF SWALLOWED: If patient is conscious and alert, give 2-3 glasses of water or milk to drink. If available, give one tablespoon of Syrup of Ipecac to induce vomiting. Alternatively, induce vomiting by touching back of throat with finger. Do not repeat the Syrup of Ipecac. Do not make an unconscious person vomit. GET MEDICAL ATTENTION.

IF IN EYES: Flush with water for at least 15 minutes. Get medical attention PREFERABLY AN OPTHAMOLOGIST.

IF INHALED: Move to an uncontaminated area. Get medical attention.

NOTE TO PHYSICIAN

This product contains a phenoxy herbicidal chemical. There is no specific antidote. All treatments should be based on observed signs and symptoms of distress in the patient. Overexposure to materials other than this product may have occurred.

ENVIRONMENTAL HAZARDS

Do not contaminate water by cleaning equipment or disposal of wastes.

Do not apply when weather conditions favor drift from treated areas.

Do not use the same spray equipment for other purposes unless thoroughly cleaned.

Do not use in or near a greenhouse.

Do not apply directly to water except as specified in the rice directions for use.

Do not contaminate water used for irrigation or domestic purposes.

Do not apply WEEDAR® 64A Herbicide directly to, or otherwise permit it to come in contact with cotton, okra, grapes, tomatoes, fruit trees, vegetables, flowers or other desirable crop or ornamental plants which are susceptible to 2,4-D herbicide. Do not permit spray mist to drift onto susceptible plants since very small quantities of the 2,4-D can cause severe injury during the growing or dormant periods.

Use coarse sprays to minimize drift. Do not apply with hollow cone-type insecticide or other nozzles that produce fine spray droplets. Drift from ground application may be reduced by: (1) keeping the spray boom as near to the crop as possible in order to obtain complete coverage; (2) by applying 5 gallons or more of spray per acre; (3) by using no more than 20 pounds of pressure at the nozzle tips; and (4) by not spraying when the wind exceeds 5 miles per hour.

Drift from aerial application may be reduced by: (1) applying as near to the crop as possible in order to obtain coverage; (2) by applying 3 or more gallons of spray per acre; (3) by using 20 pounds pressure or less at the nozzle tips; (4) by using nozzles which produce a coarse spray pattern; (5) by spraying only when the wind velocity is less than 5 miles per hour; and (6) by spraying when there is no possibility for a temperature inversion at time of spraying.

Applications by aircraft, ground rig and hand sprayers should be carried out only when there is no hazard from spray drift. Do not apply near cotton, grapes, tomatoes, or near desirable crops or ornamental vegetation susceptible to 2,4-D herbicide.

At air temperatures above 95° F do not apply WEEDAR® 64A as vapors from this product may injure susceptible plants. Do not use the same spray equipment for other purposes.

DIRECTIONS FOR USE

IT IS A VIOLATION OF FEDERAL LAW TO USE THIS PRODUCT IN A MANNER INCONSISTENT WITH ITS LABELING.

Do not apply WEEDAR® 64A Herbicide through any type of irrigation system.

GENERAL WORKER PROTECTION STATEMENTS

Do not apply this product in such a manner as to directly or through drift expose workers or other persons. The area treated must be vacated by unprotected persons.

Do not enter treated areas without protective clothing until sprays have dried.

Because certain states may require more restrictive reentry intervals for various crops treated with this product, consult your State Department of Agriculture for further information.

Written or oral warnings must be given to workers who are expected to be in a treated area or in an area about to be treated with this product. Advise workers to stay out of fields during application and until sprays have dried. Regular long-sleeved work clothing should be worn when working in treated fields. See PRECAUTIONARY STATEMENTS, STATEMENT OF PRACTICAL TREATMENT and NOTE TO PHYSICIAN for information on accidental exposures. When oral warnings are given, warnings shall be given in a language customarily understood by workers. Oral warnings must be given if there is reason to believe that written warnings cannot be understood by workers. Written warnings must include the following information: appropriate signal word (DANGER - PELIGRO), area treated with WEEDAR® 64A Herbicide, date of application, appropriate protective clothing, and reentry interval (i.e. until sprays have dried).

GENERAL INFORMATION

For Ground Application

Apply WEEDAR® 64A Herbicide in 5 to 40 gallons of water per acre with the higher volume used to provide good coverage when weeds are dense.

For Aerial Application on Grain

Apply WEEDAR® 64A in 3 to 10 gallons of water per acre.

TO PREPARE A SPRAY: Mix WEEDAR® 64A only with water, unless otherwise directed on this label. Add about one-half the required amount of water to the spray tank, then add the WEEDAR® 64A with agitation. Add the remaining amount of water with continued agitation.

Note: Adding oil, wetting agent, or other surfactants may increase effectiveness on weeds but may reduce selectivity to crops, resulting in crop damage.

COMPATIBILITY: If WEEDAR® 64A is to be tank mixed with fertilizers or with other pesticides, compatibility should be tested prior to mixing. To test for compatibility, use a small container and mix a small amount (0.5 to 1 qt) of spray, combining all ingredients in the same ratio as the anticipated use. If any indications of physical incompatibility develop, do not use this mixture for spraying. Indications of incompatibility usually will appear within 5 to 15 minutes after mixing.

Read and follow all directions and precautions on this label and on the labels of any products for which a tank mixture is being considered.

USE WITH LIQUID FERTILIZER: WEEDAR® 64A may be combined with liquid fertilizers suitable for foliar applications to corn, small grains, sorghum, and grass pastures. Use WEEDAR® 64A in accordance with directions for these crops given in the following label. Use liquid fertilizer at rates recommended by local supplier or extension service specialist. To prepare the spray, WEEDAR® 64A must be premixed with water. For liquid nitrogen fertilizer, use a premix consisting of 1 part of WEEDAR® 64A and 4 parts of water; for other liquid fertilizers, use 1 part of WEEDAR® 64A with 50 to 60 parts water. Add the premix to the fertilizer while maintaining continuous agitation during both mixing and spraying operations. Apply the spray the same day it is prepared; **do not store.**

NOTE: Always premix WEEDAR® 64A Herbicide with water before adding to the liquid fertilizer.

WEEDS CONTROLLED by properly applied treatments of WEEDAR® 64A Herbicide include the following:

Annual And Biennial Weeds

*beggarticks	mustards (except blue mustard)
bullthistle	wild parsnip
common burdock	pepper weeds (except perennial)
common cocklebur	**pigweeds (*Amaranthus* spp.)
coffeeweed	prickly lettuce

fleabane (rough)
hairy galinsoga
jimsonweed
*knotweed
common lambsquarters
lettuce (wild)
*mallow (venice or little)
marshelder
morningglory (common, ivy, wooly)
*musk thistle (***)
common evening primrose
radish (wild)
ragweed (common or giant)
*Russian thistle
Salsify (western or common)
*smartweeds (annual species)
sowthistles (annual or spiny)
sunflower
*vervains
vetches
wild carrot

Perennial Weeds

artichoke (jerusalem)
*many-flowered aster
bindweed (hedge, field, European)
blue lettuce
Canada thistle
*ground ivy
healall
*hoary cress
*ironweed
nettles (including stinging)
*orange hawkneed
catnip
chicory
dandelion
*docks
*dogbanes
*goldenrod
plantains
sowthistle (perennial)
*Vervains
*wild garlic
*wild onion

* These species may require repeated applications and/or use of the higher rate recommended on this product label even under ideal conditions for application.
** Control of pigweeds in the High Plains area of Texas and Oklahoma may not be satisfactory with this product.
***** Not registered for musk thistle control in California.**

INJURY TO CROPS FROM THIS HERBICIDE MAY OCCUR. IF YOU ARE NOT PREPARED TO ACCEPT SOME DEGREE OF CROP INJURY DO NOT USE THIS PRODUCT.

Crop varieties vary in response to 2,4-D and some are easily injured. Apply WEEDAR® 64A only to varieties known to be tolerant to 2,4-D. If you are uncertain concerning tolerant varieties or local use situations that may affect crop tolerance to 2,4-D consult your seed company, state Agricultural Extension Service or qualified crop consultant for advice.

RECOMMENDATIONS FOR APPLYING WEEDAR® 64A BROADLEAF POSTEMERGENCE HERBICIDE
WEEDS IN CROPS - CEREAL GRAINS

CROP	AMOUNT OF WEEDAR® 64A PER ACRE	DIRECTIONS
Postemergence WHEAT, BARLEY, RYE, OATS (not underseeded with legumes)		
Annual and biennial weeds	1/2 to 2 pints*	Apply after grain is fully tillered (usually about 4 to 8 inches high) but not forming joints in the stem. **Do not spray grain in boot to dough stage.**
Perennial broadleaf weeds	1 to 2 pints*	
Grain underseeded with legumes.	1/4 to 1/2 pints	Apply after grain is 8" tall. **Do not spray grain in boot to dough stage.** Do not spray alfalfa or sweet clover unless the infestation is severe and injury to these legumes can be tolerated.
Emergency weed control in wheat Perennial broadleaf weeds	3 pints*	Apply when weeds are approaching bud stage, after the grain dough stage. **Do not spray during the boot to dough stage.** The 3 pint per acre application can produce injury to wheat. Balance the severity of your weed problem against the possibility of crop damage. Where perennial weeds are scattered, spot treatment is suggested to minimize the extent of crop injury.

* Use the lower rate if small annual and biennial weeds are the major problem. Use the higher rate if perennial weeds or annual and biennial weeds are present which are in the hard-to-kill categories as determined by local experience. The higher rates increase the risk of grain injury and should be used only where the weed control problem justifies the grain damage risk. Do not apply WEEDAR® 64A to grain in the seedling stage.

RESTRICTIONS AND LIMITATIONS FOR USE ON CEREAL GRAINS

- For aerial application on grain, apply WEEDAR® 64A in 3 to 10 gallons of water per acre. For ground application use a minimum of 10 gallons of water per acre.
- Do not permit dairy animals or meat animals being finished for slaughter to forage treated grain fields within 2 weeks after treatment.
- Do not feed treated straw to livestock if an emergency treatment as described above is applied.

CROP	AMOUNT OF WEEDAR® 64A PER ACRE	DIRECTIONS
ASPARAGUS	3-4 pints	Apply on actively growing weeds, usually in April or May. Use 60 gallons of water per acre for ground application and 12 gallons of water per acre for aerial application. If spears are present, treat immediately after cutting. Make no more than 2 applications during the harvest season; applications should be spaced at least one month apart. Spears contacted by the spray may be malformed and off-flavored. Malformed spears should be cut immediately and discarded. Post harvest applications should be made only by ground rig using drop nozzles to avoid spraying the fern. **DO NOT USE IN CALIFORNIA.**
CORN (FIELD AND SWEET) Preplant	1 to 2 pints	To control emerged broadleaf weed seedlings or existing cover crops prior to planting corn, apply 7 to 14 days before planting. Do not use on light, sandy soil, or where soil moisture is inadequate for normal weed growth. Use high rate for less susceptible weeds or cover crops such as alfalfa.
Preemergence	2 to 3 pints	Apply 3 to 5 days after planting but before corn emerges. Do not use on light, sandy soils or where soil moisture is inadequate.
Postemergence Annual broadleaf weeds	1/2 to 1 pint	Apply when weeds are small and corn is less than 8 inches tall (to top of canopy). When corn is over 8 inches tall, use drop nozzles and keep spray off foliage. Avoid spraying just after first corn leaves unfold, as injury may occur.
Perennial broadleaf weeds	1 to 1 1/2 pints	Apply when weeds are in bud to bloom stage. If corn is more than 8 inches tall, use drop nozzles and keep spray off foliage. **Do not spray when corn is in the tassel to dough stage.**
Preharvest	1 to 2 pints	After the hard dough (dent) stage, apply by air or ground equipment to suppress weeds that interfere with harvest or to suppress perennial weeds. **DO NOT USE IN CALIFORNIA**

RESTRICTIONS AND LIMITATIONS FOR USE ON CORN

- Do not forage or feed corn fodder for 7 days following application.
- Corn treated with 2,4-D may become temporarily brittle. Winds or cultivation may cause stalk breakage during the period of time when the corn is brittle.
- Do not spray corn in the tassel to dough stage.

CROP	AMOUNT OF WEEDAR® 64A PER ACRE	DIRECTIONS
SORGHUM (milo)	1 pint	Treat only when sorghum is 6 to 15 inches tall (to top of canopy). If sorghum is over 8 inches tall, use drop nozzles to keep spray off foliage. Do not treat during boot, flowering, or early dough stages.

RESTRICTIONS AND LIMITATIONS FOR USE ON SORGHUM
- Do not forage or feed fodder for seven days following application

CROP	AMOUNT OF WEEDAR® 64A PER ACRE	DIRECTIONS
RICE	1 ½ to 2 ½ pints	Apply when rice is in the late tillering stage of development, at the time of first joint development. Do not apply after panicle initiation, after rice internodes exceed one-half inch, at early seedling, early panicle, boot, or heading stages. Consult local university or Agricultural Extension Service Specialists for more specific information on rates and timing of application.

RESTRICTIONS AND LIMITATIONS FOR USE ON RICE
- Do not use in fields where shellfish are of economic importance or where flood water is used for irrigation of other crops.

CROP	AMOUNT OF WEEDAR® 64A PER ACRE	DIRECTIONS
STRAWBERRIES	2 to 3 pints	Apply only to established or old strawberry plantings. Use 25 to 50 gallons of water per acre in early spring when strawberries are dormant or immediately after the last picking. Do not apply unless some degree of crop injury is acceptable. Follow recommendations of State Extension Weed or Horticulture Specialists to fit local conditions. **DO NOT USE IN CALIFORNIA**

FALLOW LAND AND CROP STUBBLE

CROP	AMOUNT OF WEEDAR® 64A PER ACRE	DIRECTIONS
Annual broadleaf weeds	1 to 2 pints	Use the lower rate when weeds are small (2-3 inches) and actively growing. Use the higher rate on older or drought-stressed plants.
Biennial broadleaf weeds	2 to 4 pints	Spray while musk thistles or other biennial species are in the seedling to rosette stage and before flower stalks become apparent. The lower rate can be used in the spring during rosette stage. Use the highest rate in the fall or after flower stalks have developed.
Perennial broadleaf weeds	2 to 6 pints	Spray weeds in bud to bloom stage, or while in good vegetative growth. Do not disturb treated areas for at least 2 weeks after treatment or until weed tops are dead.
Wild garlic and onion in crop stubble	4-6 pints	Apply to new regrowth of wild garlic or onion which occurs in the fall following harvest of small grains, soybeans, corn or grain sorghum.

RESTRICTIONS AND LIMITATIONS FOR USE ON FALLOWLAND AND CROP STUBBLE
- Do not plant any crop for three months after treatment or until chemical has disappeared from the soil.

ESTABLISHED GRASS PASTURES, RANGELAND, AND CONSERVATION RESERVE PROGRAM AREAS:

CROP	AMOUNT OF WEEDAR® 64A PER ACRE	DIRECTIONS
Annual broadleaf weeds Biennial and Perennial broadleaf weeds	2 pints 2 to 4 pints	Apply when weeds are small and actively growing and prior to bud stage. Spray while musk thistles or other biennial species are in the seedling to rosette stage and before flower stalks become apparent. The lower rate can be used in the spring during rosette stage. Use the highest rate in the fall or after flower stalks have developed. Do not apply to newly seeded areas until grass is well established. Do not apply if grass is in the boot to milk stage if grass seed production is desired. Bentgrass and legumes may be injured by this treatment.

RESTRICTIONS AND LIMITATIONS FOR USE ON PASTURES AND RANGELANDS
- Do not graze (dairy) cattle in treated areas for 14 days after application.
- Remove meat animals from treated areas for 7 days prior to slaughter. Withdrawal is not needed if more than 2 weeks have elapsed since treatment.
- Do not cut treated grass for hay within 30 days after application

CONSERVATION RESERVE PROGRAM AREAS

CROP	AMOUNT OF WEEDAR® 64A PER ACRE	DIRECTIONS
Annual broadleaf weeds In young grass in established grasses	 ½ to 1 pint ½ to 2 pints	Apply to actively growing annual broadleaf weeds. Use ½ to 1 pint when weeds are small; use higher rates on older weeds. Do not apply to young grasses with fewer than 6 leaves or prior to tillering, as excessive injury may result. Do not apply more than 1 pint until grasses are well established as excessive injury may result.
Biennial and perennial broadleaf weeds in established grasses	2 to 4 pints	Treat when biennial weeds are in the seedling to rosette stage and before flower stalks become apparent. Treat perennial weeds in the bud to bloom stage. Apply to actively growing weeds.

RESTRICTIONS AND LIMITATIONS FOR USE ON CONSERVATION RESERVE PROGRAM AREAS
- Use at least 2 gallons of water per acre by air and 5 gallons of water per acre by ground.
- Do not harvest or graze treated Conservation Reserve Program areas.
- Do not apply to grasses in the boot to dough stage if grass seed production is desired.

CROP	AMOUNT OF WEEDAR® 64A PER ACRE	DIRECTIONS
GRASSES FOR SEED PRODUCTION	2-4 pints	Apply to established stands in spring from tiller to early boot stage. Do not spray in boot stage. New spring seedings may be treated with the lower rate after grass seedlings have at least 5 leaves. Perennial weed regrowth may be treated in the fall. **DO NOT USE IN CALIFORNIA.**

RESTRICTIONS AND LIMITATIONS FOR USE ON GRASSES FOR SEED PRODUCTION
- Do not graze dairy animals or cut forage for hay within 7 days of application.

NON-CROPLAND (FENCEROWS, ROADSIDES, DRAINAGE, DITCHES, GOLF COURSES, CEMETERIES, PARKS, TURFGRASS AND OTHER GRASS AREAS).

CROP	AMOUNT OF WEEDAR®64A PER ACRE	DIRECTIONS
Annual broadleaf weeds	2 to 4 pints	Apply lower rates when weeds are young and actively growing. Use the higher rates for perennial weeds and apply when weeds are near the bud stage. Do not apply to newly seeded areas until grass is well established. Where bentgrass predominates, make 2 applications of 1 pint per acre at 3-week intervals. Do not use on susceptible southern grasses such as St. Augustine. Bentgrass, dichondria, clovers, and legumes may be injured by this treatment. Repeated applications to deep rooted perennials may be required.
Biennial and Perennial broadleaf weeds	4 to 8 pints	
SPOT TREATMENT IN NON-CROP AREAS		Mix 2 to 3 fluid ounces of WEEDAR® 64A in 3 gallons of water. Thoroughly wet all stems and foliage. For best results, treat when weeds are actively growing.

RESTRICTIONS AND LIMITATIONS FOR USE ON NON-CROPLAND
- Do not graze dairy animals for seven days following application.
- Use sufficient gallonage for through and uniform coverage.

STONE FRUIT AND NUT ORCHARDS

CROP	AMOUNT OF WEEDAR® 64A PER ACRE	DIRECTIONS
Annual broadleaf weeds	3 pints	For control of weeds on the orchard floor, apply using coarse sprays and low pressure in sufficient volume of water to obtain thorough wetting of weeds. Treat when weeds are small and actively growing. Do not use on light, sandy soil. **DO NOT USE IN CALIFORNIA.**

RESTRICTIONS AND LIMITATIONS FOR USE IN STONE FRUIT AND NUT ORCHARDS
- Do not apply to bare ground as injury may result.
- Do not apply immediately before irrigation and withhold irrigation for 2 days before and for 3 days after treatment.
- Do not allow spray to drift or contact foliage, fruit, stems, trunks of trees or exposed roots as injury may result.
- Do not apply to newly established or young orchards. Trees must be at least 1 year old and in vigorous condition.
- Do not apply during bloom.
- Do not graze or feed cover crops from treated orchards.
- Do not make more than 2 applications per year.
- Do not harvest stone fruit within 40 days of application.
- Do not harvest nuts within 60 days of application.

STORAGE AND DISPOSAL

STORAGE

Do not contaminate or store near food, feedstuff, fertilizers, seeds, insecticides, fungicides or other pesticides.

PESTICIDE DISPOSAL

Do not contaminate water, food or feed by storage or disposal. Open dumping is prohibited.

Pesticide wastes are acutely hazardous. Improper disposal of excess pesticide, spray mixture or rinsate is a violation of Federal law. If these wastes cannot be disposed of by use according to label instructions, contact your State Pesticide or Environmental Control Agency, or the Hazardous Waste representative at the nearest EPA Regional Office for guidance.

CONTAINER DISPOSAL

Triple rinse (or equivalent). Then offer for recycling or reconditioning, or puncture and dispose of in a sanitary landfill, or incineration, or, if allowed by state and local authorities, by burning. If burned, stay out of smoke.

Local conditions may affect the use of herbicides. Consult your State Agriculture Experiment Station, Farm Advisors, or Extension Weed Specialists for advice in selecting treatment from this label to best fit local conditions.

Be sure that use of this product conforms to all applicable laws, rules and regulations. Certain states have restrictions pertaining to application distances from susceptible crops. The applicator should become familiar with these laws, rules or regulations and follow them exactly. Apply this product only as specified on this label.

LIMITED WARRANTY AND DISCLAIMER

The manufacturer warrants that this product conforms to the chemical description on the label; that this product is reasonably fit for the purposes set forth in the directions for use when it is used in accordance with such directions; and that the directions, warnings and other statements on this label are based upon responsible experts' evaluation of reasonable tests of effectiveness, of toxicity to laboratory animals and to plants, and of residues on food crops, and upon reports of field experience. Tests have not been made on all varieties or in all states or under all conditions. THE MANUFACTURER NEITHER MAKES NOR INTENDS, NOR DOES IT AUTHORIZE ANY AGENT OR REPRESENTATIVE TO MAKE, ANY OTHER WARRANTIES, EXPRESS OR IMPLIED, AND IT EXPRESSLY EXCLUDES AND DISCLAIMS ALL IMPLIED WARRANTIES OF MERCHANTABILITY OR FITNESS FOR A PARTICULAR PURPOSE.

THIS WARRANTY DOES NOT EXTEND TO, AND THE BUYER SHALL BE SOLELY RESPONSIBLE FOR, ANY AND ALL LOSS OR DAMAGE WHICH RESULTS FROM THE USE OF THIS PRODUCT IN ANY MANNER WHICH IS INCONSISTENT WITH THE LABEL DIRECTIONS, WARNINGS OR CAUTIONS.

BUYER'S EXCLUSIVE REMEDY AND MANUFACTURER'S OR SELLER'S EXCLUSIVE LIABILITY FOR ANY AND ALL CLAIMS, LOSSES, DAMAGES, OR INJURIES RESULTING FROM THE USE OR HANDLING OF THIS PRODUCT, WHETHER OR NOT BASED IN CONTRACT, NEGLIGENCE, STRICT LIABILITY IN TORT OR OTHERWISE SHALL BE LIMITED, AT THE MANUFACTURER'S OPTION, TO REPLACEMENT OF, OR THE REPAYMENT OF THE PURCHASE PRICE FOR, THE QUANTITY OF PRODUCT WITH RESPECT TO WHICH DAMAGES ARE CLAIMED. IN NO EVENT SHALL MANUFACTURER OR SELLER BE LIABLE FOR SPECIAL, INDIRECT OR CONSEQUENTIAL DAMAGES RESULTING FROM THE USE OR HANDLING OF THIS PRODUCT.

NOTICE TO BUYER

Purchase of this material does not confer any rights under patents governing this product or the use thereof in countries outside of the United States.

WEEDAR is the registered trademark of Rhône-Poulenc

AG90020

THIS SPECIMEN LABEL IS INTENDED FOR USE ONLY AS A GUIDE IN PROVIDING GENERAL INFORMATION REGARDING THE DIRECTIONS, WARNINGS AND CAUTIONS ASSOCIATED WITH THE USE OF THIS PRODUCT. AS WITH ANY AGRICULTURAL CHEMICAL, ALWAYS FOLLOW THE LABEL INSTRUCTIONS ON THE PACKAGE BEFORE USING.

WEEDONE® LV4

Broadleaf Postemergence Herbicide

The 2,4-D low volatile ester for agricultural weed control, conifer release, site preparation, turf and non-cropland weed control.

ACTIVE INGREDIENT:
*2,4-Dichlorophenoxyacetic acid, butoxyethyl ester 60.8%
INERT INGREDIENTS: 39.2%

*2,4-Dichlorophenoxyacetic acid equivalent 41.9% by weight or 3.8 pounds per gallon.

*Isomer specific by AOAC method No. 6.D01-5

KEEP OUT OF REACH OF CHILDREN
CAUTION

For PRODUCT USE Information Call 1-800-334-9745
For MEDICAL And TRANSPORTATION Emergencies ONLY Call 1-800-334-7577

E.P.A. Reg. No. 264-20ZA   E.P.A. Est. No. 264-MO-01

PRECAUTIONARY STATEMENTS

CAUTION

HAZARD TO HUMANS AND DOMESTIC ANIMALS

Harmful if swallowed or absorbed through skin. Do not get in eyes on skin or on clothing. Avoid breathing vapors or spray mist. Remove and wash contaminated clothing before reuse. Wash thoroughly before eating or smoking.

STATEMENT OF PRACTICAL TREATMENT

In case of contact wash skin with soap and water; for eyes flush with water for at least 15 minutes and get medical attention. If swallowed, do not induce vomiting. Get medical attention.

NOTE TO PHYSICIAN

Aspiration hazards may contraindicate the use of gastric lavage.

Never induce vomiting or give anything by mouth to an unconscious person. Get medical attention immediately.

ENVIRONMENTAL HAZARDS

This product is toxic to fish. Do not apply directly to water. Do not contaminate water by cleaning of equipment or disposal of wastes.

USE PRECAUTIONS

Do not apply WEEDONE® LV4 Broadleaf Postemergence Herbicide through any type of irrigation system.

Do not apply when weather conditions favor drift from treated areas.

Do not use the same spray equipment for other purposes unless thoroughly cleaned.

Do not use in or near a greenhouse.

Do not contaminate water used for irrigation or domestic purposes.

Do not apply WEEDONE® LV4 Broadleaf Postemergence Herbicide directly to, or permit spray mist to drift onto cotton, okra, grapes, tomatoes, fruit trees, vegetables, flowers or other desirable crop or ornamental plants which are susceptible to 2,4-D herbicide. Do not apply near susceptible plants since very small quantities of the 2,4-D will cause severe injury during the growing or dormant periods. Crops contacted by WEEDONE® LV4 Broadleaf Postemergence Herbicide sprays or spray drift may be killed or suffer significant stand loss with extensive quality and yield reduction.

Do not apply when a temperature air inversion exists. Such a condition is characterized by little or no air movement and an increase in air temperature with an increase in height. In humid regions a fog or mist may form. An inversion may be detected by producing a smoke column and checking for a layering effect. If questions exist pertaining to the existence of an inversion, consult with local weather services before making an application.

Use coarse sprays to minimize drift. Do not apply with hollow cone type insecticide or other nozzles that produce fine spray droplets.

Drift from **aerial or ground application** may be reduced by: (1) applying as near to the target as possible in order to obtain coverage; (2) by increasing the volume of spray mix per acre; (3) by decreasing the pounds of pressure at the nozzle tips; (4) by using nozzles which produce a coarse spray pattern; (5) by not applying when wind is blowing toward susceptible valuable plants.

At high air or ground surface temperatures, vapors from this product may injure susceptible plants.

DIRECTIONS FOR USE

IT IS A VIOLATION OF FEDERAL LAW TO USE THIS PRODUCT IN A MANNER INCONSISTENT WITH ITS LABELING.

GENERAL INFORMATION

WEEDONE® LV4 Broadleaf Postemergence Herbicide contains the butoxyethyl ester of 2,4-D, the original and one of the best low volatile esters. In cropland, WEEDONE® LV4 is more effective than amines for controlling hard-to-kill weeds such as bindweed, thistles, smartweeds, wild garlic, curled dock, tansy ragwort, and wild onions.

TO PREPARE A SPRAY

Add one-half the required amount of water to the spray tank, then add the WEEDONE® LV4 with agitation, and finally, the balance of the water with continued agitation. This material forms an emulsion in water, not a solution. This tends to separate on standing. Provide agitation to prevent such separation and insure uniform spray mixtures.

COMPATIBILITY

If WEEDONE® LV4 is to be tank mixed with fertilizers or with other pesticides, compatibility should be tested prior to mixing. To test for compatibility, use a small container and mix a small amount (0.5 to 1 qt) of spray, combining all ingredients in the same ratio as the anticipated use. If any indications of physical incompatibility develop, do not use this mixture for spraying. Indications of incompatibility usually will appear within 5-15 minutes after mixing. Read and follow the label of each tank-mix product used for precautionary statements, directions for use, geographic and other restrictions.

COMBINATION WITH LIQUID NITROGEN FERTILIZER: Use 1/2 to 3/4 pint WEEDONE® LV4 per acre for weeding and feeding corn, small grains, grasses for seed production or grass pastures as directed on this label. Use the fertilizer at rates recommended by supplier or Extension Service Specialist. Fill the spray tank about half full with the liquid fertilizer, then add the WEEDONE® LV4 with vigorous agitation, and complete filling the tank with fertilizer. Apply immediately and continue agitation in the spray tank during application. Application during very cold weather (near freezing) is not advisable. Do not allow mixture to stand overnight. Incompatibility may be encountered with some fertilizer brands or under some environmental conditions. If in doubt, test a small sample in the dilution ratio planned for application.

NOTE: Fertilizers can increase foliage contact burn of herbicides. Reducing the fertilizer rate and concentration will reduce the hazard of leaf burn.

WEEDS CONTROLLED BY PROPERLY APPLIED TREATMENTS OF WEEDONE LV-4 INCLUDE THE FOLLOWING:

ANNUAL AND BIENNIAL WEEDS

annual yellow sweetclover	pepperweeds
beggarticks*	(except perennial)
bull thistle	pigweeds**
coffeeweed	(*amaranthus* spp.)
common evening primrose	prickly* lettuce
common broomweed*	ragweed (common, giant)
common burdock	rough fleabane
common cocklebur	Russian thistle*
common lambsquarters	salsify (western or common)
croton (Texas or woolly)	smartweeds*
hairy galinsoga	(annual species)
jimsonweed	sowthistle (annual or spiny)
knotweed*	sunflower
mallow* (venice, dwarf, little)	velvetleaf
marshelder	vervains*
morning glory	vetches
(common, ivy, woolly)	wild carrot*
mustards	wild lettuce
(except blue mustard)	wild parsnips
	wild radish

PERENNIAL WEEDS

alfalfa*	healall
bindweed*	hoary cress*
(hedge, field and European)	Jerusalem-artichoke
blue lettuce	many-flowered aster*
Canada thistle*	nettles*
catnip	(including stinging)
chicory	plantains
dandelion	sowthistle (perennial)
docks*	tansy ragwort*
dogbanes*	vervains*
goldenrod*	western ironweed*
ground ivy*	wild garlic*
hawkweed*	wild onion*
(orange)	

* These species may require repeat applications and/or use of the higher rate recommended on this product label even under ideal conditions for application.

** Control of pigweeds in the High Plains area of Texas and Oklahoma may not be satisfactory with this product.

INJURY TO CROPS FROM THIS HERBICIDE MAY OCCUR. IF YOU ARE NOT PREPARED TO ACCEPT SOME DEGREE OF CROP INJURY DO NOT USE THIS PRODUCT.

Crop varieties vary in response to 2,4-D and some are easily injured. Apply WEEDONE® LV4 only to varieties known to be tolerant to 2,4-D. If you are uncertain concerning tolerant varieties or local use situations that may affect crop tolerance to 2,4-D consult your seed company, state Agricultural Extension Service or qualified crop consultant for advice.

RECOMMENDATIONS FOR APPLYING WEEDONE® LV4 BROADLEAF POSTEMERGENCE HERBICIDE

WEEDS IN CROPS	AMOUNT OF WEEDONE® LV4 PER ACRE	DIRECTIONS
CEREAL GRAINS Postemergence WHEAT, BARLEY, RYE (not underseeded with legumes) Annual and biennial weeds Perennial broadleaf weeds	 ½ to 2 pints* 1 to 2 pints*	Apply after grain is fully tillered (usually about 4 to 8" high) but not forming joints in the stem. **Do not spray grain in boot to dough stage.**
Wild onion or garlic	1½ to 2 pints*	Apply 1½ pints when grain is fully tillered and wild garlic or onion plants are small. Apply 2 pints after harvest in the crop stubble. For control of new fall growth of wild onion or garlic refer to fallowland use directions.
Emergency Weed Control in Wheat Perennial broadleaf weeds	3 pints*	Apply when weeds are approaching bud stage, but **do not spray grain during the boot to dough stage.** The 3 pint per acre application can produce injury to wheat. Balance the severity of your weed problem against the possibility of crop damage. Where perennial weeds are scattered, spot treatment is suggested to minimize the extent of crop injury.
Preharvest	1-2 pints*	Apply when grains are in the hard dough stage to suppress large weeds that may interfere with harvest. Best results will be obtained when soil moisture is sufficient to induce succulent weed growth.
TANK MIXTURES FOR CEREAL GRAINS WEEDONE LV4 and Buctril® ME4 (Tank Mixture on Wheat, Barley and Rye) Cereal areas except Washington Oregon Idaho Washington Oregon Idaho	½ to 1 pint Plus ½ to ¾ pint Buctril® ½ to 1 pint Plus ¾ to 1 pint Buctril®	Buctril® Broadleaf Herbicide will control some annual weeds that are resistant to WEEDONE® LV4 and may be tank mixed with WEEDONE® LV4 for broader spectrum weed control on small grains. TO PREPARE the spray: First mix WEEDONE® LV4 in water then add the BUCTRIL®. Use the higher rates for larger weeds or where weed growth is slow due to dry or cold weather. Apply before weeds are 6 inches high. Use 10-20 gallons total spray volume per acre with ground equipment or use 5-10 gallons total spray volume with air application. Use higher volume on larger weeds.

Read and follow the label of each tank-mix product used for precautionary statements, directions for use, geographic and other restrictions.

* Use the lower rate if small annual and biennial weeds are the major problem. Use the higher rate if perennial weeds or annual and biennial weeds are present which are in the hard-to-kill categories as determined by local experience. The higher rates increase the risk of grain injury and should be used only where the weed control problem justifies the risk of grain damage.

Do not apply to grain in the seedling stage.

RESTRICTIONS AND LIMITATIONS FOR USE ON SMALL GRAINS

- For aerial application on grain, apply WEEDONE® LV4 in 2 or more gallons of water per acre.
- For ground application, use a minimum of 10 gallons of water per acre.
- Do not mix with oil for crop uses.
- Do not permit dairy animals or meat animals being finished for slaughter to forage treated grain fields within 2 weeks after treatment.
- Do not feed treated straw to livestock if a preharvest treatment or emergency treatment as described above is applied.

FIELD CORN

WEEDS	AMOUNT OF WEEDONE® LV4 PER ACRE	DIRECTIONS
Preplant	1 to 2 pints	To control emerged broadleaf weed seedlings or existing cover crops prior to planting corn. Apply 7 to 14 days before planting. Do not use on light, sandy soil, or where soil moisture is inadequate for normal weed growth. Use high rate for control of less susceptible weeds or cover crops such as alfalfa.
Preemergence	2 to 3 pints	To control broadleaf weeds and suppress annual grasses, apply 3 to 5 days after planting, but before corn emerges. Use high rate on soil high in organic matter. Do not use on light, sandy soil, or where soil moisture is inadequate for normal weed growth.
Postemergence Annual broadleaf weeds Perennial broadleaf weeds	 ½ pint ¾ pint	Avoid spraying just after corn leaves unfold, as injury may occur. Apply when weeds are small and corn is less than 8" tall. If corn more than 8" tall must be sprayed, use drop pipes. Spray when weeds are in the bud to bloom stage. If corn is more than 8" tall, use drop pipes to keep spray off corn leaves. **Do not spray when corn is in tassel to dough stage.** 2,4-D may make corn brittle. Winds or cultivation may cause stalk breakage while brittle.
Preharvest	1 to 2 pints	After the hard dough (denting) stage, to suppress weeds that interfere with harvest such as bindweed, cocklebur, dogbane, jimsonweed, ragweed, sunflower, and velvetleaf, and to decrease production of weed seeds, spray with air or ground equipment. The high rate will be needed for tough weeds under stress.

RESTRICTIONS AND LIMITATIONS FOR USE ON FIELD CORN

- Do not forage or feed corn fodder for 7 days following application.

WEEDS	AMOUNT OF WEEDONE® LV4 PER ACRE	DIRECTIONS
GRAIN SORGHUM (milo) Annual broadleaf weeds Perennial broadleaf weeds	3/4 pint 1 1/4 pints	Treat when sorghum is 6 to 15" tall (to top of canopy). If sorghum is taller than 8" to top of canopy, use drop nozzles to keep spray off crop foliage. **Do not treat during the boot, flowering, or early dough stages.**

RESTRICTIONS AND LIMITATIONS FOR USE ON SORGHUM:
- Do not forage or feed corn fodder for 7 days following application.
- A minimum application volume of 5 gallons per acre by air or 10 gallons per acre by ground is recommended.

WEEDS	AMOUNT OF WEEDONE® LV4 PER ACRE	DIRECTIONS
GRASSES FOR SEED PRODUCTION	1 to 1 1/2 pints	Apply to established stands in spring from tiller to early boot stage. **Do not spray in boot stage.** New spring seedings may be treated with the lower rate after grass seedlings have at least 5 leaves. Perennial weed regrowth may be treated in the fall. DO NOT USE IN CALIFORNIA
SUGARCANE Preemergence Postemergence	2 pints 2 to 4 pints	Apply before canes appear for control of emerged broadleaf weeds. Apply after cane emerges and through layby. DO NOT USE IN CALIFORNIA

WEEDS AND BRUSH IN PASTURES, RANGELAND, AND FALLOW LAND

WEEDS	AMOUNT OF WEEDONE® LV4 PER ACRE	DIRECTIONS
Established Grass Pastures and Rangeland Annual broadleaf weeds Perennial and biennial broadleaf weeds	2 pints 3 to 4 pints	Apply when weeds are small and actively growing and prior to bud stage. Do not apply to newly seeded areas until grass is well established. Do not apply when grass is in the boot to milk stage if grass seed production is desired. Bentgrass and legumes may be injured by this treatment.
Sand sage, sagebrush, shinnery oak, and other 2,4-D susceptible woody species	1 to 4 quarts	Apply aerially to brush in 2 or more gallons of water per acre. One gallon fuel oil may be included in the mixture. Consult state or local brush control specialists for most effective rate, volume and timing of spray applications.
FALLOW LAND AND CROP STUBBLE		
Annual broadleaf weeds	1 to 2 pints	Use the lower rate when weeds are small (2 to 3" tall) and growing actively. Use the higher rate on older and drought to stressed plants.
Biennial broadleaf weeds	2 to 4 pints	Spray while musk thistles or other biennial species are in the seedling to rosette stage, and before flower stalks are initiated. The lower rate can be used in spring during rosette stage. In fall or after flower stalks have developed, use the highest rate.
Perennial broadleaf weeds	2-6 pints	Spray weeds in bud to bloom stage, or in good vegetative growth. Do not disturb treated area for at least 2 weeks after treatment, or until weed tops are dead.
Wild garlic and onion in crop stubble	4-6 pints	Apply to new regrowth of wild onion or garlic which occurs in the fall following harvest of small grains, soybeans, corn, or grain sorghum.

RESTRICTIONS AND LIMITATIONS FOR USE ON PASTURES, RANGELAND, AND FALLOW LAND:
- For ground application, use a minimum of 10 gallons of water per acre.
- For aerial application, use a minimum of 2 gallons of water per acre.
- Do not graze animals on treated areas within 7 days after treatment.
- Do not plant treated fallow land until three months after treatment or until chemical has disappeared from soil.

WEEDS IN ORNAMENTAL TURF AREAS

Golf Courses, Cemeteries, Parks, and Other Turf Areas		
Annual broadleaf weeds	2 pints	Apply any time weeds are growing vigorously. Cool-season weeds such as common chichweed, henbit and pepperweeds should be sprayed when they appear in fall or early spring. Warm-season weeds such as knotweed respond best when they are small. If there is additional germination, repeat treatment. Also controls cocklebur, galinsoga, lambsquarters, pigweed, ragweed, sunflower.
Perennial and biennial broadleaf weeds	3 to 4 pints	To control dandelion, plantain, and most other common broadleaf weeds in turf, apply in spring or fall when these weeds are growing actively. Treat wild garlic or onion in three successive seasons. (Example: spring, fall, spring). Do not apply to newly seeded areas until sod is formed and grass has been cut at least twice. Do not use WEEDONE® LV4 for susceptible southern grasses such as St. Augustine. Bentgrass and clover may be injured by this treatment. Also controls bindweed, chicory, docky, ground ivy, hawkweed, thistles, (Canada, musk, bull, and Scotch), smartweed, tansy ragwort, wild onion, wild garlic.

RESTRICTIONS AND LIMITATIONS FOR USE IN TURF AREAS
- For ground application in turf areas, use enough water for good coverage (minimum 10 gallons per acre).
- Do not graze animals on treated areas within 7 days after treatment.
- If needed, irrigate ornamental turf thoroughly before application. Spray when air is calm to avoid drift. Avoid treatment when temperatures exceed 85 degrees F or are expected to do so within 24 hours after application.

WEEDS AND BRUSH IN NON-CROP AREAS

Fencerows, Roadsides, Drainage Ditchbanks, Vacant Lots, Airfields, Railroad, Highway and Utility Rights-Of-Way, and Other Non-Crop Areas.

Annual broadleaf weeds	2 to 4 pints	Apply when most weeds are still young and growing vigorously.
Perennial and biennial broadleaf weeds	3 to 6 pints	Apply when weeds are actively growing and near the bud stage, but before flowering. For best results on tansy ragwort and musk thistle, treat in rosette stage, before bolting. A second application is usually needed for best results on thistle, nettle, and bindweed. Treat wild onion or garlic in early spring and in fall when they are young and growing actively.
Cattails and tules	$1^1/_2$ to $2^1/_2$ gallons	Mix 2 quarts WEEDONE® LV4 in 2 quarts kerosene or diesel oil then add this mixture to 100 gallons of water. Apply 300 to 500 gallons of spray per acre, depending on the stand. The addition of a wetting agent such as SURFEL Spray Adjuvant is suggested.
Woody plants	1 to 2 gallons	Apply in 30-100 gallons of spray per acre, preferably through low-volume equipment such as DIRECTA-SPRA, WOBBLER, MINI-WOBBLER, OR SPIROMETER. Treat when 2,4-D-susceptible species are in full leaf and growing actively. Repeated applications, oil or wetting agent, and/or high volume spraying (300 to 500 gallons of spray per acre) may be needed for acceptable control of some species or if brush growth is dense.

RESTRICTIONS AND LIMITATIONS FOR USE IN NON-CROP AREAS

- For aerial application to solid stands of susceptible brush, use 2 to 4 quarts in 3 to 12 gallons volume per acre. Two to 4 quarts of fuel oil may be included in this mixture.
- Applications will be less effective when conditions such as deficient soil moisture reduce brush growth.

TANK MIXTURES FOR NON-CROP AREAS
WEEDONE® LV4 and GARLON 4 or GARLON 3A TANK MIXTURES FOR NON-CROP AREAS.

WEEDS	AMOUNT OF WEEDONE® LV4 PER ACRE	DIRECTIONS
Read and observe all precautions and restrictions of labels of all products tank mixed with WEEDONE® LV4.		
Broadleaf Weed Control	2 to 4 pints WEEDONE® LV4 Plus 2 to 6 pints GARLON-4 or 3 to 8 pints GARLON-3A	For wider spectrum control of broadleaf weeds and woody plants. Apply as a broadcast spray in enough water to deliver 20 to 100 gallons total spray per acre. Apply when broadleaf weeds are actively growing.
Woody Plant Control Broadcast Foliar Spray	1 to 2 gallons WEEDONE® LV4 Plus $1^1/_2$ to 3 quarts GARLON-4 or 2-4 quarts GARLON-3A	Apply as a broadcast spray in enough water to deliver 20 to 100 gallons total spray per acre. Apply when woody plants are actively growing.
Woody Plant Control High Volume Leaf-Stem Treatment with Ground Equipment	1 to 8 quarts WEEDONE® LV4 Plus $1^1/_2$ to 12 pints GARLON-4 or 2 to 16 pints GARLON-3A	Mix 1 to 2 quarts WEEDONE LV-4 plus 1 $^1/_2$ to 3 pints GARLON 4 or 2 to 4 pints GARLON 3A in enough water to make 100 gallons of spray. Apply at a volume of 100 to 400 gallons of total spray per acre depending on size and density of woody plants. Thoroughly wet all leaves, stems and root collars of plants to be controlled.
Woody Plant Control Aerial Application (Helicopter only)	1 to 2 gallons WEEDONE® LV4 Plus 3 to 4 quarts GARLON-4 or 4 to 6 quarts GARLON-3A	Apply in a total spray volume of 10 to 30 gallons per acre using drift control equipment such as the MICROFOIL boom or an effective drift control agent such as LO-DRIFT spray additive. Use the higher rates and volumes when plants are dense or under drought conditions.

TANK MIXTURES FOR NON-CROP AREAS
WEEDONE® LV4 and BANVEL HERBICIDE TANK MIXTURES FOR NON-CROP AREAS.

WEEDS	AMOUNT OF WEEDONE® LV4 PER ACRE	DIRECTIONS
Annual Broadleaf Weeds	2 to 4 pints WEEDONE® LV4 plus $^1/_2$ to $1^1/_2$ pints BANVEL	For wider spectrum control of broadleaf weeds and woody plants. Apply as a broadcast spray in enough water to deliver 20 to 100 gallons total spray per acre. Apply when broadleaf weeds are actively growing. Use the higher rates when treating dense or tall vegetative growth.

WEEDS	AMOUNT OF WEEDONE® LV4 PER ACRE	DIRECTIONS
Perennial and biennial broadleaf weeds	3 to 6 pints WEEDONE® LV4 plus 1/2 to 6 pints BANVEL	Apply as a broadcast spray in enough water to deliver 20 to 100 gallons total spray per acre. Apply when broadleaf weeds are actively growing but prior to flowering. Use the lower rates for biennials less than 3 inches rosette diameter. Use the higer rates for perennial weeds or for biennial weeds past the 3 inch rosette stage.
Woody Plant Control Broadcast, high volume, stem foliage or aerial application	1 to 2 gallons WEEDONE® LV4 plus 2 to 8 quarts BANVEL	Apply as a broadcast spray in enough water to deliver 20 to 100 gallons total spray per acre or apply as a high volume stem foliage spray in enough volume to thoroughly wet leaves, stems and root collars (100 to 400 gallons per acre) or apply aerially in enough water to deliver total spray volume of 10 to 30 gallons per acre using drift control equipment such as the MICROFOIL® boom or an effective drift control agent such as LO-DRIFT spray additive. Use the higher rates and volumes when plants are dense or under drought conditions.

SMALL AREA APPLICATIONS FOR NON-CROP USE

For control of broadleaf weeds in small non-crop areas with hand held or back-pack sprayers mix 4 fluid ounces WEEDONE® LV4 per gallon of water. Thoroughly wet all weed foliage. Maintain agitation of mixture to prevent separation.

FOREST MANAGEMENT

Conifer Release

To control alder, apply 1 1/2 to 2 quarts WEEDONE® LV4 Broadleaf Postemergence Herbicide in 9 1/2 to 15 gallons water per acre as a foliage spray. Treat when 3/4 of the brush foliage has attained full size leaves and before new conifer growth reaches 2" in length. This is usually between early May and mid-June. Adjust treatment date depending on stage of growth of conifers and brush species. This may cause leader deformation on exposed firs, but they should overcome this during the second year after spraying.

To control tanoak, madrone, ceanothus, canyon live oak, and manzanita, and to release Douglas fir, hemlock, Sitka spruce or grand fir, apply 3 quarts WEEDONE® LV4 plus 3 quarts fuel oil in 8 1/2 to 15 gallons water per acre before new growth on Douglas fir is 2" long. To control manzanita and ceanothus in ponderosa pine, apply 3 quarts WEEDONE® LV4 before pine growth begins in spring.

To control hazel brush in the Lake states, apply 2 quarts WEEDONE® LV4 in 1 1/2 to 4 1/2 gallons water per acre when new shoot growth of hazel is complete (usually mid-July).

In northern areas, if possible conifer injury can be tolerated, 1 1/2 to 3 quarts WEEDONE® LV4 applied aerially in 8 to 25 gallons of water per acre after conifers such as jack pine, red pine, black spruce, and white spruce have hardened off (usually mid-July) will provide control of competing hardwoods including alder, aspen, birch, hazel, and willow.

SITE PREPARATION

Budbreak Spray: To control alder and other susceptible species before planting forest seedlings, apply 2 to 4 quarts WEEDONE® LV4 in 9 to 15 gallons fuel oil per acre after alder buds break, but before foliage is 1/4 full size.

Foliage spray: To control alder before planting forest tree seedlings, apply 2 quarts WEEDONE® LV4 plus 2 quarts fuel oil in 9 to 15 gallons water per acre after most alder leaves are full size. One-tenth gallon TRANS-VERT Herbicide Additive may be added for additional alder control. WEEDONE® LV4 in combination with TRANS-VERT is not registered for use in California.

Consult your regional or extension forester or state herbicide specialist for recommendations to fit local conditions.

WEEDONE, BUCTRIL ME4, SURFEL, DIRECTA-SPRA, WOBBLER, MINI WOBBLER, SPIROMETER, LO-DRIFT and TRANSVERT are registered trademarks of RHÔNE-POULENC AG COMPANY.

NALCO-TROL is a trademark of NALCO Chemical Company.

GARLON is a trademark of the DOW Chemical Co.

BANVEL is a trademark of SANDOZ Chemical Corporation.

STORAGE AND DISPOSAL

STORAGE

Do not store near food, feedstuff, fertilizers, seeds, insecticides, fungicides or other pesticides. Do not contaminate water, food or feed by storage or disposal.

PESTICIDE DISPOSAL

Pesticide wastes are toxic. Improper disposal of excess pesticide, spray mixture, or rinsate is a violation of Federal law. If these wastes cannot be disposed of by use according to label instructions, contact your State Pesticide or Environmental Control Agency or the Hazardous Waste representative at the nearest EPA Regional Office for guidance.

CONTAINER DISPOSAL

Triple rinse (or equivalent). Then offer for recycling or reconditioning, or puncture and dispose of in a sanitary landfill or by incineration, or if allowed by state and local authorities, by burning. If burned, stay out of smoke.

LIMITED WARRANTY AND DISCLAIMER

The manufacturer warrants that this product conforms to the chemical description on the label; that this product is reasonably fit for the purposes set forth in the directions for use when it is used in accordance with such directions; and that the directions, warnings and other statements on this label are based upon responsible experts' evaluation of reasonable tests of effectiveness, of toxicity to laboratory animals and to plants, and of residues on food crops and upon reports of field experience. Tests have not been made on all varieties or in all states or under all conditions. THE MANUFACTURER NEITHER MAKES NOR INTENDS, NOR DOES IT AUTHORIZE ANY AGENT OR REPRESENTATIVE TO MAKE, ANY OTHER WARRANTIES, EXPRESS OR IMPLIED, AND IT EXPRESSLY EXCLUDES AND DISCLAIMS ALL IMPLIED WARRANTIES OF MERCHANTABILITY OR FITNESS FOR A PARTICULAR PURPOSE.

THIS WARRANTY DOES NOT EXTEND TO, AND THE BUYER SHALL BE SOLELY RESPONSIBLE FOR, ANY AND ALL LOSS OR DAMAGE WHICH RESULTS FROM THE USE OF THIS PRODUCT IN ANY MANNER WHICH IS INCONSISTENT WITH THE LABEL DIRECTIONS, WARNINGS, OR CAUTIONS.

BUYER'S EXCLUSIVE REMEDY AND MANUFACTURER'S OR SELLER'S EXCLUSIVE LIABILITY FOR ANY AND ALL CLAIMS, LOSSES, DAMAGES, OR INJURIES RESULTING FROM THE USE OR HANDLING OF THIS PRODUCT, WHETHER OR NOT BASED IN CONTRACT, NEGLIGENCE, STRICT LIABILITY IN TORT OR OTHERWISE SHALL BE LIMITED, AT THE MANUFACTURER'S OPTION, TO REPLACEMENT OF, OR THE REPAYMENT OF THE PURCHASE PRICE FOR, THE QUANTITY OF PRODUCT WITH RESPECT TO WHICH DAMAGES ARE CLAIMED. IN NO EVENT SHALL MANUFACTURER OR SELLER BE LIABLE FOR SPECIAL INDIRECT OR CONSEQUENTIAL DAMAGES RESULTING FROM THE USE OR HANDLING OF THIS PRODUCT.

NOTICE TO BUYER

Purchase of this material does not confer any rights under patents governing this product or the use thereof in countries outside of the United States.

WEEDONE is a registered trademark of Rhône-Poulenc

AG90023

 EPA Approval: 12/2/86

THIS SPECIMEN LABEL IS INTENDED FOR USE ONLY AS A GUIDE IN PROVIDING GENERAL INFORMATION REGARDING THE DIRECTIONS, WARNINGS AND CAUTIONS ASSOCIATED WITH THE USE OF THIS PRODUCT. AS WITH ANY AGRICULTURAL CHEMICAL, ALWAYS FOLLOW THE LABEL INSTRUCTIONS ON THE PACKAGE BEFORE USING.

WEEDONE® LV6

EMULSIFIABLE BROADLEAF HERBICIDE

Weed And Brush Control In Small Grains, Field Corn, Rangeland, Pastures, Roadsides And Fencerows

ACTIVE INGREDIENT:	
2,4-Dichlorophenoxyacetic acid, butoxyethyl ester*	83.5%
INERT INGREDIENTS:	16.5%

*Isomer specific by AOAC method No. 6.DO1-5

2,4-Dichlorophenoxyacetic acid equivalent 5.7 lb./gal. or 57.5% wt.

KEEP OUT OF REACH OF CHILDREN

CAUTION

For PRODUCT USE Information Call 1-800-334-9745

For MEDICAL And TRANSPORTATION Emergencies ONLY Call 24 hours A day 1-800-334-7577

E.P.A. Reg. No. 264-271AA E.P.A. Est. No. 264-MO-01

PRECAUTIONARY STATEMENTS

CAUTION

HAZARD TO HUMANS AND DOMESTIC ANIMALS

Harmful if swallowed or absorbed through skin. Do not get in eyes, on skin or on clothing. Avoid breathing vapors or spray mist. Remove and wash contaminated clothing before reuse. Wash thoroughly before eating or smoking.

STATEMENT OF PRACTICAL TREATMENT

IF IN EYES: Hold eyelids open and flush with a steady gentle stream of water for at least 15 minutes. Get medical attention, preferable an ophthalmologist.

IF SWALLOWED: If patient is conscious and alert, give 2 to 3 glasses of water to drink. Do not induce vomiting. Get medical attention.

IF ON SKIN: Immediately wash skin with plenty of soap and water, if available, while removing contaminated clothing and shoes. Wash clothing separately before reuse.

IF INHALED: Remove victim to fresh air. If not breathing, give artificial respiration. Administer oxygen, if necessary. Get medical attention.

NOTE TO PHYSICIAN

No specific antidote is available.

All treatments should be based on observed signs and symptoms of distress in the patient. Overexposure to materials other than this product may have occurred.

This product contains petroleum distillates. If large amounts (greater than 1 ml/kg body weight) of the product have been ingested, the stomach should be evacuated by gastric intubation with the aid of a cuffed endotracheal tube to prevent aspiration of petroleum distillates. After removal of stomach contents, wash stomach by instilling 30 to 50 g. of activated charcoal in 3 to 4 ounces of water through the stomach tube and again remove stomach contents. Avoid oily laxatives.

ENVIRONMENTAL HAZARDS

This product is toxic to fish. Drift or run-off may adversely affect fish and non-target plants. Do not apply directly to water or wetlands (swamps, bogs, marshes and potholes). Do not contaminate water when disposing of equipment washwaters.

USE PRECAUTIONS

Do not apply WEEDONE® LV6 through any type of irrigation system.
Do not apply when weather conditions favor drift from treated areas.

Do not use the same spray equipment for other purposes unless thoroughly cleaned.

Do not use in or near a greenhouse.

Do not contaminate water used for irrigation or domestic purposes.

Do not apply WEEDONE® LV6 Emulsifiable Broadleaf Herbicide directly to, or permit spray mist to drift onto cotton, okra, grapes, tomatoes, fruit trees, vegetables, flowers or other desirable crop or ornamental plants which are susceptible to 2,4-D herbicide. Do not apply near susceptible plants since very small quantities of the 2,4-D will cause severe injury during the growing or dormant periods. Crops contacted by WEEDONE® LV6 Emulsifiable Broadleaf Herbicide sprays or spray drift may be killed or suffer significant stand loss with extensive quality and yield reduction.

Do not apply when a temperature air inversion exists. Such a condition is characterized by little or no air movement and an increase in air temperature with an increase in height. In humid regions a fog or mist may form. An inversion may be detected by producing a smoke column and checking for a layering effect. If questions exist pertaining to the existence of an inversion, consult with local weather services before making an application.

Use coarse sprays to minimize drift. Do not apply with hollow cone type insecticide or other nozzles that produce fine spray droplets.

Drift from aerial or ground application may be reduced by: (1) applying as near to the target as possible in order to obtain coverage; (2) by increasing the volume of spray mix per acre; (3) by decreasing the pounds of pressure at the nozzle tips; (4) by using nozzles which produce a coarse spray pattern; (5) by not applying when wind is blowing toward susceptible valuable plants.

At high air or ground surface temperatures, vapors from this product may injure susceptible plants.

DIRECTIONS FOR USE

IT IS A VIOLATION OF FEDERAL LAW TO USE THIS PRODUCT IN A MANNER INCONSISTENT WITH ITS LABELING.

GENERAL INFORMATION

WEEDONE® LV6 Emulsifiable Broadleaf Herbicide contains the butoxyethyl ester of 2,4-D, the original and one of the best low volatile esters. In cropland, WEEDONE® LV6 is more effective than amines for controlling hard-to-kill weeds such as bindweed, thistles, smartweeds, wild garlic, curled dock, tansy ragwort, and wild onions.

TO PREPARE A SPRAY: Add 1/2 the required amount of water to the spray tank, then add the WEEDONE® LV6 with agitation, and finally, the balance of the water with continued agitation. This material forms an emulsion in water, not a solution. This tends to separate on standing. Provide agitation to prevent such separation and insure uniform spray mixtures.

COMPATIBILITY: If Weedone® LV6 Emulsifiable Broadleaf Herbicide is to be tank mixed with fertilizers or with other pesticides, compatibility should be tested prior to mixing. To test for compatibility, use a small container and mix a small amount (0.5 to 1 qt) of spray, combining all ingredients in the same ratio as the anticipated use. If any indications of physical incompatibility develop, do not use this mixture for spraying. Indications of incompatibility usually will appear within 5 to 15 minutes after mixing.

Read and follow the label of each tank mix product used for precautionary statements, directions for use, geographic and other restrictions.

COMBINATION WITH LIQUID NITROGEN FERTILIZER

Use 1/3 to 1/2 pint WEEDONE® LV6 Emulsifiable Broadleaf Herbicide per acre for weeding and feeding corn, small grains, or grass pastures as directed on this label.

Use the fertilizer at rates recommended by supplier or Extension Service Specialist.

Fill the spray tank about half full with the liquid fertilizer, then add the WEEDONE® LV6 Emulsifiable Broadleaf Herbicide with vigorous agitation, and complete filling the tank with fertilizer. Apply immediately and continue agitation in the spray tank during application. Application during very cold weather (near freezing) is not advisable. Do not allow mixture to stand overnight. Incompatibility may be encountered with some fertilizer brands or under some environmental conditions. If in doubt, test a small sample in the dilution ratio planned for application.

NOTE: Fertilizers can increase foliage contact burn of herbicides. Reducing the fertilizer rate and concentration will reduce the hazard of leaf burn.

WEEDS CONTROLLED BY PROPERLY APPLIED TREATMENTS OF WEEDONE® LV6 INCLUDE THE FOLLOWING.

ANNUAL AND BIENNIAL WEEDS

annual yellow sweetclover	**pigweeds (Amaranthus spp.)
*beggarticks	*prickly lettuce
bull thistle	ragweed (common, giant)
coffeeweed	rough fleabane
*common broomweed	*Russian thistle
common burdock	salsify (western or common)
common cocklebur	*smartweeds (annual species)
common evening primrose	sowthistle (annual or spiny)
common lambsquarters	sunflower
croton (Texas or woolly)	velvetleaf
hairy galinsoga	*vervains
jimsonweed	vetches
*knotweed	*wild carrot
*mallow (venice, dwarf, little)	wild lettuce
marshelder	wild parsnips
morning glory (common, ivy, woolly)	wild radish
mustards (except blue mustard)	
pepperweeds (except perennial)	

Perennial weeds

*alfalfa	healall
*bindweed (hedge, field and European)	*hoary cress
blue lettuce	Jerusalem-artichole
*Canada thistle	*many-flowered aster
catnip	*nettles (including stinging)
chicory	plantains
dandelion	sowthistle (perennial)
*docks	*tansy ragwort
*dogbanes	*vervains
*goldenrod	*western ironweed
*ground ivy	*wild garlic
*hawkweed (orange)	*wild onion

* These species may require repeat applications and/or use of the higher rate recommended on this product label even under ideal conditions for applications.

** Control of pigweeds in the High Plains area of Texas and Oklahoma may not be satisfactory with this product.

INJURY TO CROPS FROM THIS HERBICIDE MAY OCCUR. IF YOU ARE NOT PREPARED TO ACCEPT SOME DEGREE OF CROP INJURY, DO NOT USE THIS PRODUCT.

Crop varieties vary in response to 2,4-D and some are easily injured. Apply WEEDONE® LV6 Emulsifiable Broadleaf Herbicide to varieties known to be tolerant to 2,4-D. If you are uncertain concerning tolerant varieties or local use situations that may affect crop tolerance to 2,4-D, consult your seed company, state Agricultural Extension Service or qualified crop consultant for advice.

RECOMMENDATIONS FOR APPLYING WEEDONE® LV6 EMULSIFIABLE BROADLEAF HERBICIDE CEREAL GRAINS

WHEAT, BARLEY, RYE (NOT UNDERSEEDED WITH LEGUMES)

WEEDS	AMOUNT OF WEEDONE® LV6 PER ACRE	DIRECTIONS
Postemergence Annual and Biennial Weeds Perennial broadleaf weeds	1/3 to 1 1/3 pints* 2/3 to 1 1/3 pints*	Apply after grain is fully tillered (usually about 4 to 8 inches high) but not forming joints in the stem. **Do not spray grain in boot to dough stage.**
Wild onion or garlic	1 to 1 1/3 pints*	Apply 1 pint when grain is fully tillered and wild garlic or onion plants are small. Apply 1 1/3 pints after harvest in the crop stubble. For control of new fall growth of wild onion or garlic refer to **FALLOWLAND USE DIRECTIONS.**
Emergency Weed Control In Wheat Perennial broadleaf weeds	2 pints*	Apply when weeds are approaching bud stage, **but do not spray grain during the boot to dough stage.** The 2 pints per acre application can produce injury to wheat. Balance the severity of your weed problem against the possibility of crop damage. Where perennial weeds are scattered, spot treatment is suggested to minimize the extent of crop injury.
Preharvest	2/3 to 1 1/3 pints	Apply when grains are in the hard dough stage to suppress large weeds that may interfere with harvest. Best results will be obtained when soil moisture is sufficient to induce succulent weed growth.

*Use the lower rate if small annual and biennial weeds are the major problem. Use the higher rate if perennial weeds or annual and biennial weeds are present which are in the hard-to-kill categories as determined by local experience. The higher rates increase the risk of grain injury and should be used only where the weed control problem justifies the risk of grain damage. Do not apply to grain in the seedling stage.

RESTRICTIONS AND LIMITATIONS FOR USE ON SMALL GRAINS

—For aerial application on grain: apply WEEDONE® LV6 Emulsifiable Broadleaf Herbicide in 2 or more gallons of water per acre.
—For ground application: use a minimum of 10 gallons of water per acre.
—Do not mix with oil for crop uses.
—Do not permit dairy animals or meat animals being finished for slaughter to forage treated grain fields within 2 weeks after treatment.
—Do not feed treated straw to livestock if a preharvest treatment or emergency treatment as described above is applied.

FIELD CORN

Preplant	2/3 to 1 1/3 pints	To control emerged broadleaf weed seedlings or existing cover crops prior to planting corn. Apply 7 to 14 days before planting. Do not use on light, sandy soil, or where soil moisture is inadequate for normal weed growth. Use high rate for control of less susceptible weeds or cover crops such as alfalfa.
Preemergence	1 1/3-2 pints	To control broadleaf weeds and suppress annual grasses, apply 3 to 5 days after planting, but before corn emerges. Use high rate on soil high in organic matter. Do not use on light, sandy soil, or where soil moisture is inadequate for normal weed growth.
Postemergence Annual broadleaf weeds	1/3 pint	Avoid spraying just after corn leaves unfold as injury may occur. Apply when weeds are small and corn is less than 8 inches tall. If corn more than 8" tall must be sprayed, use drop pipes.
Perennial broadleaf weeds	1/2 pint	**Spray when weeds are in the bud to bloom stage.** If corn is more than 8" tall, use drop pipes to keep spray off corn leaves. **Do not spray when corn is in tassel to dough stage.** 2,4-D may make corn brittle. Winds or cultivation may cause stalk breakage while brittle.
Preharvest	2/3 to 1 1/3 pints	After the hard dough (denting) stage, to suppress weeds that interfere with harvest such as bindweed, cocklebur, dogbane, jimsonweed, ragweed, sunflower, and velvetleaf, and to decrease production of weed seeds, spray with air or ground equipment. The high rate will be needed for tough weeds under stress.

RESTRICTIONS AND LIMITATIONS FOR USE ON FIELD CORN

—Do not forage or feed corn fodder for 7 days following application.

GRAIN SORGHUM (milo)

WEEDS	AMOUNT OF WEEDONE® LV6 PER ACRE	DIRECTIONS
Annual broadleaf weeds Perennial broadleaf weeds	1/2 pint 3/4 pint	Treat when sorghum is 6 to 15 inches tall (to top of canopy). If sorghum is taller than 8 inches to top of canopy, use drop nozzles to keep spray off crop foliage. **Do not treat during the boot, flowering, or early dough stages.**

RESTRICTIONS AND LIMITATIONS FOR USE ON SORGHUM

—Do not forage or feed fodder for 7 days following application.
—A minimum application volume of 5 gallons per acre by air or 10 gallons per acre by ground is recommended.

SUGARCANE

Preemergence Postemergence	1 1/3 pints 1 1/3 to 2 2/3 pints	Apply before canes appear for control of emerged broadleaf weeds. Apply after cane emerges and through layby. DO NOT USE IN CALIFORNIA

WEEDS AND BRUSH IN PASTURES, RANGELAND, AND FALLOW LAND
ESTABLISHED GRASS PASTURES AND RANGELAND

WEEDS	AMOUNT OF WEEDONE® LV6 PER ACRE	DIRECTIONS
Annual broadleaf weeds	1⅓ pints	Apply when weeds are small and actively growing and prior to bud stage. Do not apply to newly seeded areas until grass is well established. Do not apply when grass is in the boot to milk stage if grass seed production is desired. Bentgrass and legumes may be injured by this treatment.
Perennial and biennial broadleaf weeds	2 to 2⅔ pints	
Sand sage, sagebrush, shinnery oak, and other 2,4-D susceptible woody species	⅔ to 2⅔ quarts	Apply aerially to brush in 2 or more gallons of water per acre. One gallon fuel oil may be included in the mixture. Consult state or local brush control specialists for most effective rate, volume and timing of spray applications.

FALLOW LAND AND CROP STUBBLE

WEEDS	AMOUNT OF WEEDONE® LV6 PER ACRE	DIRECTIONS
Annual broadleaf weeds	⅔ to 1⅓ pints	Use the lower rate when weeds are small (2 to 3" tall) and growing actively. Use the higher rate on older and drought–stressed plants.
Biennial broadleaf weeds	⅔ to 1⅔ pints	Spray while musk thistles or other biennial species are in the seedling to rosette stage, and before flower stalks are initiated. The lower rate can be used in spring during rosette stage. In fall or after flower stalks have developed, use the highest rate.
Perennial broadleaf weeds	1⅓ to 4 pints	Spray weeds in bud to bloom stage, or in good vegetative growth. Do not disturb treated area for at least 2 weeks after treatment, or until weed tops are dead.
Wild garlic and onion	2⅔ to 4 pints	Apply to new regrowth of wild onion or garlic which occurs in the fall following harvest of small grains, soybeans, corn, or grain sorghum.

RESTRICTIONS AND LIMITATIONS FOR USE ON PASTURES, RANGELAND, AND FALLOW LAND

- For ground application, use a minimum of 10 gallons of water per acre.
- For aerial application, use a minimum of 2 gallons of water per acre.
- Do not graze animals on treated areas within 7 days after treatment.
- Do not plant treated fallow land until 3 months after treatment or until chemical has disappeared from soil.
- Do not cut grass for hay within 30 days after application.
- Do not permit dairy animals or meat animals being finished for slaughter to forage treated fields within 3 days of slaughter.

FENCEROWS, ROADSIDES, DRAINAGE DITCHBANKS, VACANT LOTS, AIRFIELDS, RAILROAD, HIGHWAY AND UTILITY RIGHTS-OF-WAY, AND OTHER NON-CROP AREAS.

WEEDS	AMOUNT OF WEEDONE® LV6 PER ACRE	DIRECTIONS
Annual broadleaf weeds	1⅓ to 2⅔ pints	Apply when most weeds are still young and growing vigorously.
Perennial and biennial broadleaf weeds	2 to 4 pints	Apply when weeds are actively growing and near the bud stage, but before flowering. For best results on tansy ragwort and musk thistle, treat in rosette stage, before bolting. A second application is usually needed for best results on thistle, nettle, and bindweed. Treat wild onion or garlic in early spring and in fall when they are young and growing actively.
Cattails and tules	1 to 1⅔ gallons	Mix 1 ⅓ quarts WEEDONE® LV6 in 2 quarts kerosene or diesel oil then add this mixture to 100 gallons of water. Apply 300 to 500 gallons of spray per acre, depending on the stand. The addition of a wetting agent such as SURFEL Spray Adjuvant is suggested.
Woody Plants	⅔ to 1⅓ gallons	Apply in 30 to 100 gallons of spray per acre, preferably through low-volume equipment such as DIRECTA-SPRA, WOBBLER, MINI-WOBBLER, OR SPIROMETER. Treat when 2,4-D-susceptible species are in full leaf and growing actively. Repeated applications, oil or wetting agent, and/or high volume spraying (300 to 500 gallons of spray per acre) may be needed for acceptable control of some species or if brush growth is dense.

RESTRICTIONS AND LIMITATIONS FOR USE IN NON-CROP AREAS

- For aerial application to solid stands of susceptible brush, use 1⅓ to 2⅔ quarts in 3-12 gallons volume per acre. Two to 4 quarts of fuel oil may be included in this mixture.
- Applications will be less effective when conditions such as deficient soil moisture reduce brush growth.

TANKS MIXTURES FOR NON-CROP AREAS
WEEDONE® LV6 and GARLON 4 OR GARLON
3A TANK MIXTURES FOR NON-CROP AREAS.

Read and follow the label of each tank-mix product used for precautionary statements, directions for use, geographic and other restrictions.

WEEDS	AMOUNT OF WEEDONE® LV6 PER ACRE	DIRECTIONS
Broadleaf Weed Control	1⅓ to 2⅔ pints WEEDONE® LV6 Plus 2 to 6 pints GARLON-4 or 3 to 8 pints GARLON-3A	For wider spectrum control of broadleaf weeds and woody plants. Apply as a broadcast spray in enough water to deliver 20 to 100 gallons total spray per acre. Apply when broadleaf weeds are actively growing.
Woody Plant Control Broadcast Foliar Spray	⅔ to 1⅓ gallons WEEDONE® LV6 Plus 1½ to 3 quarts GARLON-4 or 2 to 4 quarts GARLON-3A	Apply as a broadcast spray in enough water to deliver 20 to 100 gallons total spray per acre. Apply when wood plants are actively growing.
Woody Plant Control High Volume Leaf-Stem Treatment with Ground Equipment	⅔ to 5⅓ quarts WEEDONE® LV6 Plus 1½ to 12 pints GARLON-4 or 2 to 16 pints GARLON-3A	Mix ⅔ to 1⅓ to 2 quarts WEEDONE® LV6 plus 1½ to 3 pts. GARLON 4 or 2 to 4 pints GARLON 3A in enough water to make 100 gallons of spray. Apply at a volume of 100 to 400 gallons of total spray per acre depending on size and density of woody plants. Thoroughly wet all leaves, stems and root collars of plants to be controlled.
Woody Plant Control Aerial Application (Helicopter only)	⅔ to 1⅓ gallons WEEDONE® LV6 plus 3 to 4 quarts GARLON-4 or 4 to 6 quarts GARLON-3A	Apply in a total spray volume of 10 to 30 gallons per acre using drift control equipment such as the MICROFOIL boom or an effective drift control agent such as LO-DRIFT Spray Additive. Use the higher rates and volumes when plants are dense or under drought conditions.

WEEDONE LV6 and BANVEL Herbicide Tank Mixtures For Non-Crop Areas.

WEEDS	AMOUNT	DIRECTIONS
Annual broadleaf weeds	1⅓ to 2⅔ pints WEEDONE® LV6 plus ½ to 1½ pints BANVEL	For wider spectrum control of broadleaf weeds and woody plants. Apply as a broadcast spray in enough water to deliver 20 to 100 gallons total spray per acre. Apply when broadleaf weeds are actively growing. Use the higher rates when treating dense or tall vegetative growth.
Perennial and biennial broadleaf weeds	2 to 4 pints WEEDONE® LV6 plus ½ to 6 pints BANVEL	Apply as a broadcast spray in enough water to deliver 20 to 100 gallons total spray per acre. Apply when broadleaf weeds are actively growing but prior to flowering. Use the lower rates for biennials less than 3 inches rosette diameter. Use the higher rates for perennial weeds or for biennial weeds past the 3 inch rosette stage.
Woody Plant Control Broadcast, high volume, stem foliage or aerial application	⅔ to 1⅓ gallons WEEDONE® LV6 plus 2 to 8 quarters BANVEL	Apply as a broadcast spray in enough water to deliver 20 to 100 gallons total spray per acre or apply as a high volume stem foliage spray in enough volume to thoroughly wet leaves, stems and root collars (100 to 400 gallons per acre) or apply aerially in enough water to deliver total spray volume of 10 to 30 gallons per acre using drift control equipment such as the MICROFOIL Boom or an effective drift control agent such as LO-DRIFT spray additive. Use the higher rates and volumes when plants are dense or under drought conditions.

SMALL AREA APPLICATIONS FOR NON-CROP USE

For control of broadleaf weeds in small non-crop areas with hand held or back-pack sprayers mix 2 ⅔ fluid ounces WEEDONE® LV6 per gallon of water. Thoroughly wet all weed foliage. Maintain agitation of mixture to prevent separation.

WEEDONE®, SURFEL, DIRECTA-SPRA, WOBBLER, MINI WOBBLER, SPIROMETER, LO-DRIFT are registered trademarks of RHONE-POULENC AG COMPANY

GARLON is a trademark of the DOW Chemical Co.

BANVEL is a trademark of Sandoz Chemical Corporation

STORAGE AND DISPOSAL

STORAGE

Do not store near food, feedstuff, fertilizers, seeds, insecticides, fungicides or other pesticides. Do not contaminate water, food or feed by storage or disposal.

PESTICIDE DISPOSAL

Pesticide wastes are toxic. Improper disposal of excess pesticide, spray mixture, or rinsate is a violation of Federal law. If these wastes cannot be disposed of by use according to label instructions, contact your State Pesticide or Environmental Control Agency or the Hazardous Waste representative at the nearest EPA Regional Office for guidance.

CONTAINER DISPOSAL

Triple rinse (or equivalent). Then offer for recycling or reconditioning, or puncture and dispose of in a sanitary landfill, or incineration, or, if allowed by state and local authorities, by burning. If burned, stay out of smoke.

LIMITED WARRANTY AND DISCLAIMER

The manufacturer warrants that this product conforms to the chemical description on the label; that this product is reasonably fit for the purposes set forth in the directions for use when it is used in accordance with such directions; and that the directions, warnings and other statements on this label are based upon responsible experts' evaluation of reasonable tests of effectiveness, of toxicity to laboratory animals and to plants, and of residues on food crops, and upon reports of field experience. Tests have not been made on all varieties or in all states or under all conditions. THE MANUFACTURER NEITHER MAKES NOR INTENDS, NOR DOES IT AUTHORIZE ANY AGENT OR REPRESENTATIVE TO MAKE, ANY OTHER WARRANTIES, EXPRESS OR IMPLIED, AND IT EXPRESSLY EXCLUDES AND DISCLAIMS ALL IMPLIED WARRANTIES OF MERCHANTABILITY OR FITNESS FOR A PARTICULAR PURPOSE.

THIS WARRANTY DOES NOT EXTEND TO, AND THE BUYER SHALL BE SOLELY RESPONSIBLE FOR, ANY AND ALL LOSS OR

DAMAGE WHICH RESULTS FROM THE USE OF THIS PRODUCT IN ANY MANNER WHICH IS CONSISTENT WITH THE LABEL DIRECTIONS, WARNINGS, OR CAUTIONS.

BUYER'S EXCLUSIVE REMEDY AND MANUFACTURER'S OR SELLER'S EXCLUSIVE LIABILITY FOR ANY AND ALL CLAIMS, LOSSES, DAMAGES, OR INJURIES RESULTING FROM THE USE OR HANDLING OF THIS PRODUCT, WHETHER OR NOT BASED IN CONTRACT, NEGLIGENCE, STRICT LIABILITY IN TORT OR OTHERWISE SHALL BE LIMITED, AT THE MANUFACTURER'S OPTION, TO REPLACEMENT OF, OR THE REPAYMENT OF THE PURCHASE PRICE FOR, THE QUANTITY OF PRODUCT WITH RESPECT TO WHICH DAMAGES ARE CLAIMED. IN NO EVENT SHALL MANUFACTURER OR SELLER BE LIABLE FOR SPECIAL, INDIRECT OR CONSEQUENTIAL DAMAGES RESULTING FROM THE USE OR HANDLING OF THIS PRODUCT.

NOTICE TO BUYER

Purchase of this material does not confer any rights under patents governing this product or the use thereof in countries outside of the United States.

WEEDONE is a registered trademark of Rhône-Poulenc.

AG90024

THIS SPECIMEN LABEL IS INTENDED FOR USE ONLY AS A GUIDE IN PROVIDING GENERAL INFORMATION REGARDING THE DIRECTIONS, WARNINGS AND CAUTIONS ASSOCIATED WITH THE USE OF THIS PRODUCT. AS WITH ANY AGRICULTURAL CHEMICAL, ALWAYS FOLLOW THE LABEL INSTRUCTIONS ON THE PACKAGE BEFORE USING.

WEEDONE® 2,4-DP

WOODY PLANT HERBICIDE

Controls mixed brush on Highways, Railroads, and Utility Rights-of-way. Also controls solid stands of post, blackjack, sand shinnery oak, and sandsage.

ACTIVE INGREDIENT:

*2,4-Dichlorophenoxypropionic acid, butoxyethyl ester	59.1%
INERT INGREDIENTS	40.9%
TOTAL	100.0%

*2,4-Dichlorophenoxypropionic acid equivalent 3.7 lb./gal. or 41.5%/wt.

E.P.A. Reg. No. 264-231 ZA E.P.A. Est. No. 264-MO-01

KEEP OUT OF REACH OF CHILDREN

CAUTION

For PRODUCT USE Information Call 1-800-334-9745
For MEDICAL AND TRANSPORTATION Emergencies Only Call 24 Hours A Day 1-800-334-7577

PRECAUTIONARY STATEMENTS

CAUTION

HAZARDS TO HUMANS AND DOMESTIC ANIMALS

Harmful if swallowed. Avoid contact with eyes, skin, and clothing. Wash thoroughly with soap and water after handling.

STATEMENT OF PRACTICAL TREATMENT

IF SWALLOWED: Call a physician or Poison Control Center. If patient is conscious and alert, give 2-3 glasses of water to drink. Do not induce vomiting.

IF ON SKIN: Wash with plenty of soap and water. Remove and wash contaminated clothing before reuse.

IF IN EYES: Flush eyes with plenty of water. See a physician if irritation persists.

NOTE TO PHYSICIAN

There is no specific antidote. Treatment of exposure should be directed at the control of the symptoms and clinical condition.

Overexposure to materials other than this product may have occurred.

This product contains petroleum distillates. If large amounts (greater than 1 ml/kg body weight) of the product have been ingested, the stomach should be evacuated by gastric intubation with the aid of a cuffed endotracheal tube to prevent aspiration of petroleum distillates. After removal of stomach contents, wash stomach by instilling 30-50 grams of charcoal in 3-4 ounces of water through the stomach tube and again remove stomach contents. Avoid oily laxatives.

ENVIRONMENTAL HAZARD

This product is toxic to fish and aquatic invertebrates. Drift or runoff may adversely affect non-target plants. Do not apply directly to water or wetlands (swamps, bogs, marshes and potholes). Do not contaminate water when disposing of equipment washwaters.

Do not apply when runoff is likely to occur.

Do not apply WEEDONE® 2,4-DP Woody Plant Herbicide near, directly to, or permit spray mist to drift onto cotton, okra, grapes, tomatoes, fruit trees, vegetables, flowers, or other desirable crop or ornamental plants which are susceptible to 2,4-DP herbicide. Very small quantities of WEEDONE® 2,4-DP Woody Plant Herbicide will cause severe injury during the growing or dormant periods. Crops contacted by WEEDONE® 2,4-DP Woody Plant Herbicide sprays or spray drift may be killed or suffer significant stand loss with extensive quality and yield reduction. Do not apply when weather conditions favor drift from target areas.

Do not use the same spray equipment for other purposes unless thoroughly cleaned. Do not contaminate irrigation ditches or water used for domestic or irrigation purposes especially in areas where grapes, cotton, tomatoes, or other susceptible plants are grown. Do not use in a greenhouse.

Do not apply when a temperature air inversion exists. Such a condition is characterized by little or no air movement and an increase in air temperature with an increase in height. In humid regions a fog or mist may form. An inversion may be detected by producing a smoke column and checking for a layering effect. If questions exist pertaining to the existence of an inversion, consult with local weather services before making an application.

Use coarse sprays to minimize drift. Do not apply with hollow cone-type insecticide or other nozzles that produce fine spray droplets. Do not apply when any wind is blowing toward susceptible crops or valuable plants. Drift from *aerial application* may be reduced by: (1) applying as near to the target as possible in order to obtain coverage; (2) by using 20 pounds of pressure or less at the nozzle tips; (3) by using nozzles which produce a coarse spray pattern; (4) by spraying only when the wind velocity is less than 5 miles per hour; (5) by applying sprays at low pressure and high volume; (6) by applying 5 or more gallons of spray per acre.

Under very high temperatures, vapors from this product may injure susceptible plants in the immediate vicinity.

Drift from ground application may be reduced by: (1) keeping the spray as near to the target as possible in order to obtain complete coverage; (2) by applying 25 gallons or more of spray per acre; (3) by using no more than 20 pounds of pressure at the nozzle tips; (4) by using nozzles which produce a coarse spray pattern; and (5) by applying sprays at low pressure and high volume.

Under very high temperatures vapors from this product may injure susceptible plants in the immediate vicinity.

DIRECTIONS FOR USE

IT IS A VIOLATION OF FEDERAL LAW TO USE THIS PRODUCT IN A MANNER INCONSISTENT WITH ITS LABELING.

WEEDONE® 2,4-DP Woody Plant Herbicide will control mixed brush along utility rights-of-way, highways, railroads, drainage ditchbanks, and firebreaks; including post oak, blackjack oak, white oak, sand shinnery oak, red oak, pine, fir, spruce, black cherry, alder, willow, sandsage, elm, and similar species, as well as solid stands of post oak, blackjack oak, sand shinnery oak, and sandsage.

FOLIAGE STEM METHOD

This is the standard method for high volume sprays along fencerows, highways, and utility rights-of-way. Use it as a "first spray" on thick brush composed of mixed species. Apply spray to both foliage and stems of all plants from the time foliage is fully developed until plants begin to go dormant. For effective control, all leaves, stems, and suckers must be thoroughly wet to ground line. Some regrowth may be expected on resistant species, such as ash, maple, oak, and persimmon.

To control mixed brush, add 1 gallon WEEDONE® 2,4-DP Woody Plant Herbicide to 99 gallons of water. Mix thoroughly before spraying. Apply 100 to 300 gallons of spray mixture per acre.

Up to 5 gallons of oil per 100 gallons may be added to these spray mixtures.

PINE RELEASE

GROUND SPRAY

To control hickory, sweetgum, oaks, wax myrtle, chinquapin, boxelder, cherry, sumac, sassafras and similar species in pine plantations, apply a mixture of 1 gallon WEEDONE® 2,4-DP + 24 gallons of water to the foliage of competing hardwood brush species. Apply with a backpack or similar sprayer with a handheld wand and single nozzle. For best results, apply when hardwoods are in full leaf and actively growing and when soil moisture is high. Brush foliage should be thoroughly covered but short of runoff. Avoid spraying the planted pines, as some needle damage may result.

Application to a 1 to 3 year old pine plantation with 2-4' brush of average density should require 5 to 10 gallons of spray mixture per acre.

Brush species such as blackgum, water and willow oak or hophornbeam require thorough coverage for good results. Brush species such as red maple, ash or dogwood are not usually controlled by this treatment.

IMPORTANT: Before using WEEDONE® 2,4-DP Woody Plant Herbicide, read and carefully observe the Precautionary Statements and all other information appearing on the product label.

RAILROAD ON-TRACK APPLICATION

For use with the DIRECTA-SPRA® spray applicator mounted on Hi-Rail equipment, use 3 gallons WEEDONE® 2,4-DP Woody Plant Herbicide in 25-50 gallons of water per acre. For added drift control add 2-2.5 oz. Lo-Drift™ Spray Additive in each 25-50 gallons of water. If higher total spray volumes are needed, do not apply more than 3 gallons WEEDONE® 2,4-DP Woody Plant Herbicide per acre.

AERIAL APPLICATION—UTILITY RIGHTS-OF-WAY

For aerial application to mixed brush along utility rights-of-way, apply 2-3 gallons of WEEDONE® 2,4-DP Woody Plant Herbicide in 12 to 30 gallons of water per acre. One gallon of fuel oil may be added.

Apply the spray only through equipment designed to provide effective drift control. A helicopter mounted MICROFOIL® applicator or other equipment that provides equivalent drift control is recommended.

SOLID STANDS OF OAK OR SANDSAGE

Apply chemicals with fixed wing or helicopter aircraft using 5 gallons of spray mixture per acre. Spray when plants have just developed full sized leaves, when soil moisture is sufficient for good growing conditions, when relative humidity is high, and wind velocities are less than 5 miles per hour. Spray season normally runs from early May to mid-June in Texas, California, and New Mexico and from mid-May to early July in Oklahoma and northward.

AMOUNTS PER ACRE

Types of brush	WEEDONE 2,4-DP Qts.	Oil Gallons	Water Gallons	Remarks
Post & Blackjack Oak	2	1/2	4	Re-treat the 2nd year with 1 qt. per acre.
Sand shinnery Oak	1/2	1/2	4 3/8	Re-treat the 2nd year with 1/2 qt. per acre.
Sandsage	1/2	1/2	4 3/8	

Aerial spraying is a specialized job. Secure qualified technical guidance for each job and employ a competent, reliable applicator. Become familiar with state laws governing the application of herbicides. Do not use aerial sprays in areas where possible spray drift may injure valuable crops or plants.

TO PREPARE A SPRAY: Add this product to the proper amount of oil, if any, and mix thoroughly. Then pour this mixture into the required amount of water while agitating continuously. This material forms an emulsion in water—not a solution. Provide agitation to prevent separation and insure a uniform spray mixture.

AERIAL OR GROUND APPLICATION TO CONTROL MIXED BRUSH INCLUDING CONIFERS AND ROOT SUCKERING SPECIES ON UTILITY, RAILROAD AND HIGHWAY RIGHTS-OF-WAY AND FENCEROWS: Not for use in California.

Where red maple, conifers, and root-suckering species such as sassafras, sumac, black locust, or persimmon are the major problem, use 1 to 2 1/2 gallons WEEDONE® 2,4-DP Woody Plant Herbicide in combination with 1 to 2 1/2 gallons Tordon*K herbicide. Dilute with water and apply in a total spray volume of 15 to 50 gallons per acre. Use lower rates to control brush on sandy soils or coarse soils with susceptible species such as seedling aspen, cherry, willow, and sumac. Use higher rates to control brush on clay soils or fine soils or rocky terrain with more resistant species such as maple, oak, pine, and red cedar. When making the spray mixture, add WEEDONE® 2,4-DP to the required amount of water while agitating the mixture in the spray tank. Then add the TORDON* 101 or TORDON* K while continuing agitation. (Do not mix concentrates of WEEDONE® 2,4-DP with concentrates of TORDON* 101 or TORDON* K.) Aerial applications of the tank mixture should be made only with a helicopter mounted MICROFOIL® applicator or an equipment system providing equivalent drift control. Ground applications to control brush on railroad or utility rights-of-way should be made only with the DIRECTA-SPRA® spray applicator or an equipment system providing equivalent drift control. Do not use these tank mixtures on drainage ditchbanks or firebreaks.

NOTE: Observe all restrictions, precautions, and limitations on the labeling of each product used in tank mixtures.

STORAGE AND DISPOSAL

STORAGE

Do not store near other pesticides, seeds, fertilizers, food, or feedstuffs. Do not contaminate water, food, or feed by storage or disposal.

PESTICIDE DISPOSAL

Open dumping is prohibited. Wastes resulting from the use of this product may be disposed of on site or at an approved waste disposal facility.

CONTAINER DISPOSAL

Do not reuse container. Do not burn. Triple rinse or equivalent. Then offer for recycling or reconditioning, or puncture and dispose of in a sanitary landfill, or by other procedures approved by state and local authorities.

Local conditions may affect the use of herbicides. Consult your State Agricultural Experiment Station, Farm Advisors, or Extension Weed Specialist for advice in selecting treatment from this label to best fit local conditions.

Be sure that use of this product conforms to all applicable laws, rules and regulations. Certain states have restrictions pertaining to the application distances from susceptible crops. The applicator should become familiar with these laws, rules, or regulations and follow them exactly.

LIMITED WARRANTY

The manufacturer warrants that this product conforms to the chemical description on the label; that this product is reasonably fit for the purposes set forth in the directions for use when it is used in accordance with such directions; and that the directions, warnings and other statements on this label are based upon responsible experts' evaluation of reasonable tests of effectiveness of toxicity to laboratory animals and to plants, and of residues, and upon reports of field experience. Tests have not been made on all varieties or in all states or under all conditions. THE MANUFACTURER NEITHER MAKES NOR INTENDS, NOR DOES IT AUTHORIZE ANY AGENT OR REPRESENTATIVE TO MAKE, ANY OTHER WARRANTIES EXPRESS OR IMPLIED, AND IT EXPRESSLY EXCLUDES AND DISCLAIMS ALL IMPLIED WARRANTIES OF MERCHANTABILITY OR FITNESS FOR A PARTICULAR PURPOSE.

THIS WARRANTY DOES NOT EXTEND TO, AND THE BUYER SHALL BE SOLELY RESPONSIBLE FOR, ANY AND ALL LOSS OR DAMAGE WHICH RESULTS FROM THE USE OF THIS PRODUCT IN ANY MANNER WHICH IS INCONSISTENT WITH THE LABEL DIRECTIONS, WARNINGS OR CAUTIONS.

BUYER'S EXCLUSIVE REMEDY AND MANUFACTURER'S OR SELLER'S EXCLUSIVE LIABILITY FOR ANY AND ALL CLAIMS, LOSSES, DAMAGES, OR INJURIES RESULTING FROM THE USE OR HANDLING OF THIS PRODUCT, WHETHER OR NOT BASED IN CONTRACT, NEGLIGENCE, STRICT LIABILITY IN TORT OR OTHERWISE SHALL BE LIMITED. AT THE MANUFACTURER'S OPTION, TO REPLACEMENT OF, OR THE REPAYMENT OF THE PURCHASE PRICE FOR, THE QUANTITY OF PRODUCT WITH RESPECT TO WHICH DAMAGES ARE CLAIMED. IN NO EVENT SHALL MANUFACTURER OR SELLER BE LIABLE FOR SPECIAL, INDIRECT OR CONSEQUENTIAL DAMAGES RESULTING FROM THE USE OR HANDLING OF THIS PRODUCT.

WEEDONE® is a registered trademark of Rhône-Poulenc.

DIRECTA-SPRA® and MICRO-FOIL® are registered trademarks of Rhone-Poulenc.

*TORDON 101 and TORDON K are trademarks of the Dow Chemical Co.

AG90030

 EPA Approval 7/27/87

THIS SPECIMEN LABEL IS INTENDED FOR USE ONLY AS A GUIDE IN PROVIDING GENERAL INFORMATION REGARDING THE DIRECTIONS, WARNINGS AND CAUTIONS ASSOCIATED WITH THE USE OF THIS PRODUCT. AS WITH ANY AGRICULTURAL CHEMICAL, ALWAYS FOLLOW THE LABEL INSTRUCTIONS ON THE PACKAGE BEFORE USING.

WEEDONE® 170

WOODY PLANT HERBICIDE

Contains the butoxyethanol ester of 2,4-DP and 2,4-D. For controlling mixed brush on utility rights-of-way, along highways and solid stands of oak or elm.

ACTIVE INGREDIENTS:

*2-(2,4-Dichlorophenoxy)propionic acid, butoxyethanol ester . . .	29.3%
**2,4-Dichlorophenoxyacetic acid, butoxyethanol ester.	29.9%
INERT INGREDIENTS:	40.8%
TOTAL:	100.0%

*2-(2,4-Dichlorophenoxy)propionic acid equivalent 20.6% by weight or 1.85 pounds per gallon.

**2,4-Dichlorophenoxyacetic acid equivalent 20.6% by weight or 1.85 pounds per gallon.

Isomer Specific by AOAC Method No. 6. DO1-5.

EPA Reg. No. 264-222ZB EPA Est. No.

KEEP OUT OF REACH OF CHILDREN

CAUTION

For PRODUCT USE Information Call 1-800-334-9745

For MEDICAL and TRANSPORTATION Emergencies Information ONLY Call 1-800-334-7577

PRECAUTIONARY STATEMENTS

CAUTION
HAZARDS TO HUMANS AND DOMESTIC ANIMALS

Harmful if swallowed or absorbed through skin. Do not get in eyes, on skin or clothing. Avoid breathing vapors or spray mist. Remove and wash contaminated clothing before reuse. Wash thoroughly before eating or smoking.

STATEMENT OF PRACTICAL TREATMENT

If SWALLOWED: Call a physician or Poison Control Center. If patient is conscious and alert, give 2 to 3 glasses of water to drink. Do not induce vomiting.

IF ON SKIN: Wash skin with plenty of soap and water while removing contaminated clothing and shoes. Wash contaminated clothing separately before reuse.

IF IN EYES: Hold eyelids open and flush with a steady gentle stream of water for at least 15 minutes. Get medical attention.

NOTE TO PHYSICIAN: No specific antidote is available. All treatments should be based on observed signs and symptoms of distress in the patient. Overexposure to materials other than this product may have occurred.

This product contains petroleum distillates. If large amounts (greater than 1 ml/kg body weight) of the product have been ingested, the stomach should be evacuated by gastric intubation with the aid of a cuffed endotracheal tube to prevent aspiration of petroleum distillates. After removal of stomach contents, wash stomach by instilling 30-50 grams of activated charcoal in 3-4 ounces of water through the stomach tube and again remove stomach contents. Avoid oily laxatives.

ENVIRONMENTAL HAZARDS

This product is toxic to fish and aquatic invertebrates. Drift or runoff may adversely affect non-target plants. Do not apply directly to water or wetlands (swamp, bogs, marshes and potholes). Do not contaminate water when disposing of equipment washwaters. Do not apply when weather conditions favor drift from treated areas. Do not allow livestock to graze treated areas.

Do not use the same spray equipment for other purposes unless thoroughly cleaned. Do not contaminate water used for domestic or irrigation purposes. Do not use in or near a greenhouse.

Do not apply WEEDONE® 170 Woody Plant Herbicide directly to, or permit spray mist to drift onto cotton, okra, grapes, tomatoes, fruit trees, vegetables, flowers or other desirable crop or ornamental plants which are susceptible to 2,4-D or 2,4-DP herbicides. Do not apply near susceptible plants since very small quantities of the 2,4-D or 2,4-DP will cause severe injury during the growing or dormant periods. Crops contacted by WEEDONE® 170 sprays or spray drift may be killed or suffer significant stand loss with extensive quality and yield reduction.

Do not apply when a temperature air inversion exists. Such a condition is characterized by little or no air movement and an increase in air temperature with an increase in height. In humid regions a fog or mist may form. An inversion may be detected by producing a smoke column and checking for a layering effect. If questions exist pertaining to the existence of an inversion, consult with local weather services before making an application.

Use coarse sprays to minimize drift. Do not apply with hollow cone-type insecticide or other nozzles that produce fine spray droplets.

Drift from aerial or ground applications may be reduced by: (1) applying as near to the target as possible in order to obtain coverage; (2) by increasing the volume of spray mix per acre; (3) by decreasing the pounds of pressure at the nozzle tips; (4) by using nozzles which produce a coarse spray pattern; (5) by not applying when wind is blowing toward susceptible crop or valuable plants.

At high air or ground surface temperature, vapors from the product may injure susceptible plants.

GENERAL INFORMATION

WEEDONE® 170 Woody Plant Herbicide is a concentrated 50-50 combination of 2,4-D and 2,4-DP as the low-volatile butoxyethanol ester.

WEEDONE® 170 Woody Plant Herbicide is specifically designed for utilities and other industrial users to control woody plants.

WEEDONE® 170 Woody Plant Herbicide mixes with oil or kerosene and emulsifies in water.

WEEDONE® 170 Woody Plant Herbicide controls many noxious perennial weeds on uncropped land, such as along highways, drainage ditchbanks and fence rows.

WEEDONE® 170 Woody Plant Herbicide is effective for controlling:

Alder	Oaks
Aspen	Pine
Bigleaf Maple	Poison Ivy
Birch	Poison Oak
Blackberry	Poplar
Black Cherry	Red Elder
Blackjack Oak	Red Elm
Buckbrush	Red Maple
Black Locust	Salmonberry
Ceanothus	Sand Shinnery Oak
Chamise	Serviceberry
Coffeeberry	Snowberry
Currant	Spruce
Fir	Sycamore
Gooseberry	Tulip Poplar
Hemlock	Willow
Honeysuckle	Winged Elm
Manzanita	Yerba Santa
Multiflora Rose	(and many other species)

WEEDONE® 170 Woody Plant Herbicide in combination with TRANSVERT® is not registered for use in California.

DIRECTIONS FOR USE

IT IS A VIOLATION OF FEDERAL LAW TO USE THIS PRODUCT IN A MANNER INCONSISTENT WITH ITS LABELING.

TO PREPARE A SPRAY: Add one-half the required amount of oil or water to the spray tank, then add this product with agitation and finally the balance of the water or oil with continued agitation.

If this material is to be used in straight oil mixtures, do not let water get into it or the finished mixture.

This material forms an emulsion in water, not a solution. This tends to separate on standing. Provide agitation to prevent such separation and ensure a uniform mixture.

FOLIAGE STEM METHOD

This is the standard method for high volume sprays along fencerows, highways and rights-of-way. Use it as a **first spray** on thick brush composed of mixed species.

Apply spray to both foliage and stems of all plants from the time foliage is fully developed until plants begin to go dormant. For effective control all leaves, stems and suckers must be thoroughly wet to ground line. Some regrowth may be expected on resistant species, such as ash, maple, oak and persimmon.

Add 1 to 1½ gallons of WEEDONE® 170 to 100 gallons of water. Mix thoroughly before spraying. Use 200 to 600 gallons of spray mixture per acre depending on height and density of brush.

SPOT TREATMENT

For spot spraying with backpack sprayers mix 1 cup (8 oz.) WEEDONE® 170 Woody Plant Herbicide with 5 gallons of water. Wet brush stems and foliage thoroughly.

BASAL BARK METHOD

Thoroughly wet the base and root collar of all stems until spray collects around the root collar at ground line. This spray may be applied in any season. Apply WEEDONE® 170 Woody Plant Herbicide with the basal bark method on scattered brush or as a **second spray** on species resistant to a first foliage spray.

Mix 3 to 4 gallons of WEEDONE® 170 Woody Plant Herbicide to 100 gallons of oil. Apply with low volume knapsack sprayer or power equipment. Concentration will depend on species present, volume of spray used, and season applied. Use a coarse spray to avoid drift.

MODIFIED BASAL METHOD

Drench the base of plants, then wet the lower ⁴⁄₅ of remaining stems and leaves to run-off. Treat when brush is in full foliage. This method may be used where susceptible species have been controlled by prior sprays and resistant species such as maple and oak remain. Drenching the base of the plant and wetting all stems to run-off is essential for complete control.

Early season spraying: Add 1 to 1½ gallons of WEEDONE® 170 Woody Plant Herbicide to 10 gallons of diesel oil and mix thoroughly. Add this mixture to 89 gallons of water.

During dry weather or the latter part of spray season: Add 1½ gallons of WEEDONE® 170 Woody Plant Herbicide to 15 gallons of diesel oil and mix thoroughly. Add this mixture to 83½ gallons of water. Agitate thoroughly before spraying, particularly if material is allowed to stand for more than one hour after mixing.

CUT SURFACE

STUMP: This method can be used at any time of year, but it is more effective when applied as soon as possible after trees are cut. Spray the entire stump, particularly bark and exposed roots. A thorough drenching is essential for complete control.

Use this method after original or capital clearing. It is the first step toward a chemical brush control program on newly cleared highways and rights-of-way. Spray is most effective and economical on stumps 3 to 4 inches and larger.

Mix 3 to 4 gallons of WEEDONE® 170 Woody Plant Herbicide to 100 gallons of oil. Apply with low volume knapsack sprayer using a solid cone nozzle of medium orifice.

CONCENTRATE STUMP TREATMENT: For small (up to 3-inch diameter) stems, cut them as close to the ground as possible and apply undiluted WEEDONE® 170 directly from the can to the surface of the freshly cut stump.

FRILL: Make a frill by using an axe to cut overlapping notches in a continuous ring around the trunk near its base. Cut through the bark but do not remove chips. This method is recommended for cull trees 5 to 6 inches in diameter and larger. Treat freshly cut frills at any time of the year.

Mix 3 to 4 gallons of WEEDONE® 170 Woody Plant Herbicide in 100 gallons of oil. Pour in as much of this mixture as the frill will hold without wasting.

POWERED KNAPSACK BASAL SPRAY

Mix 8 to 14 gallons of WEEDONE® 170 Woody Plant Herbicide with fuel oil or kerosene to make 100 gallons of spray solution. Apply with a portable knapsack mistblower to lower brush stems. Apply spray to all sides of stems; good root collar coverage is essential. Run knapsack mistblower at 1/4 to 1/3 throttle for best spray delivery and coverage. For maximum drift control use a basal nozzle attachment. Do not raise spray nozzle above horizontal position.

FENCE ROW APPLICATION

To control mixed brush, perennial and annual broadleaf weeds use one of the application methods such as the foliage stem method described on this label. Some regrowth may be expected on resistant species such as ash, maple, oak and persimmon.

RAILROAD ON-TRACK APPLICATION

WEEDONE® 170 Woody Plant Herbicide in combination with TRANS-VERT® is not registered for use in California.

Use 1 gallon of WEEDONE® 170 and 1 quart of TRANS–VERT® herbicide additive in each 100 gallons of water and apply 100 to 300 gallons per acre to control mixed brush growing along railroad rights-of-way and under railroad communication lines. Apply from spray trains or Hi-Rail equipment only. Avoid contact with skin, eyes or clothing.

LOW VOLUME STEM FOLIAGE SPRAY (AERIAL OR GROUND)

WEEDONE® 170 Woody Plant Herbicide in combination with TRANS-VERT® is not registered for use in California.

Mixed Brush - Utility Rights-of-Way: Use 3 gallons of WEEDONE® 170 and 1/2 gallon TRANSVERT® mixed in water to make 12 to 15 gallons total volume per acre to control mixed brush growing on utility rights-of-way. Where red maple, conifers, ash, hickory or oak are the major problems, use 3 gallons of WEEDONE® 170 and 2/3 gallon TRANS–VERT® in 15 gallons total volume per acre.

Apply the spray only through equipment designed to provide effective drift control. A helicopter mounted MICROFOIL® applicator or other equipment that provides equivalent drift control is recommended for aerial sprays. A DIRECTA-SPRA® or mini-wobbler is recommended for ground sprays.

Solid Stands of Oak or Elm: Apply with fixed-wing or helicopter aircraft. Apply in the spring after hardwoods have just developed full sized leaves. The spray season normally runs from early May to mid-June in Texas and California, and from early May to early July in Oklahoma and northward. Spray when wind velocities are less than 5 miles per hour.

Brush	Add Water WEEDONE® 170	Oil Total	To Make
post, blackjack oak, winged elm	2/3 gallon	1 gallon	5 gallons
sand shinnery oak	1/2-1 quart	1 gallon	3 gallons

If necessary, repeat spray the second year with 1 quart WEEDONE® 170 per acre. For maximum control, use higher rate and repeat spray the second year using 1 quart WEEDONE® 170 per acre.

For aerial application to solid stands of brush in California use 2 to 4 quarts in 3 to 12 gallons volume per acre. Two to 4 quarts of fuel oil may be included in this mixture.

FORESTRY SITE PREPARATION - To reduce competition from mixed trees and brush before planting forest trees.

Apply one gallon WEEDONE® 170 Woody Plant Herbicide for easy to control species such as aspen, hazel, alder, sassafras, sumac, black locust, willow, and other similar species. Apply 2 gallons WEEDONE® 170 to control more difficult species such as oaks, sweet gum, black gum, hickory, and tulip poplar.

To control annual grasses and provide additional foliage brownout for prescribed foliage burning, add one quart of TRANS–VERT® per to the above spray mixtures.

Mix with water to make 10 gallons total solution when applying as an aerial spray or 20 gallons total solution when applying as a ground spray. Do not apply to established plantations as this spray mix will injure planted conifers.

PINE RELEASE - To control hardwood brush and release Northern conifers such as red pine Jack pine, white pine and white spruce.

Apply 2 - 4 qts. WEEDONE® 170 Woody Plant Herbicide as a broadcast spray in 10 gallons of water per acre. Make applications in midsummer after pine height growth is complete and the conifer buds are set. WEEDONE® 170 will control aspen, birch, cherry, alder, hazel, oaks and similar species.

WEEDONE® 170 will not, at these rates, provide satisfactory control of red maple, sugar maple, or ash.

AERIAL OR GROUND APPLICATION TO CONTROL MIXED BRUSH (Including conifers and root suckering species on utility, railroad and highway rights-of-way fencerows, and other non crop areas.)

WEEDONE® 170 And Tordon* Mixtures

Where red maple, conifers and root-suckering species such as sassafras, sumac, black locust or persimmon are the major problem; use 1 to 2 1/2 gallons WEEDONE® 170 in combination with 3 to 6 pints of Tordon* K Herbicide. Dilute with water and apply in a total spray volume of 15 to 50 gallons per acre.

Use lower rates to control brush on sandy soils or coarse soils with susceptible species such as seedling aspen, cherry, willow and sumac.

Use higher rates to control brush on clay soils or fine soils or rocky terrain with more resistant species such as maple, oak, pine and red cedar. When making the spray mixture, add WEEDONE® 170 to the required amount of water while agitating the mixture in the spray tank. Then add the Tordon 101 or Tordon® K while continuing agitation. (Do not mix concentrates of WEEDONE® 170 with concentrates of Tordon 101 or Tordon® K.)

Aerial applications of the tank mixture should be made only with a helicopter mounted MICROFOIL® applicator or an equipment system providing equivalent drift control. Do not use these tank mixtures on drainage ditchbanks or for site preparation in forest plantations.

Observe all restrictions, precautions and limitations on the labeling of each product used in tank mixtures.

*Tordon is a trademark of the Dow Chemical Company

WEEDONE® 170 WOODY PLANT HERBICIDE and Banvel* Herbicide Mixtures

HIGH VOLUME STEM FOLIAGE SPRAY

Mix 1 to 1 1/2 gallons WEEDONE® 170 with 1 to 2 quarts of Banvel 4WS per 100 gallons total spray mix. Use the low rate for easy to kill brush such as alder, aspen, cherry, sycamore, tulip poplar, and willows. Use the high rate to control species such as oak, ash, elm, pines, spruce, and fir.

LOW VOLUME STEM FOLIAGE SPRAY (Aerial or Ground)

Mix 2 to 3 gallons of WEEDONE® 170 with 1 1/2 gallons of Banvel herbicide in 15-30 gallons of water per acre to control mixed brush including conifers and ash. Observe all restrictions, precautions and limitations on the labelling of each product used in tank mixtures.

BANVEL is a trademark of the SANDOZ Chemical Company

STORAGE AND DISPOSAL

STORAGE

Do not contaminate or store near food, feedstuffs, fertilizers, seeds, insecticides or fungicides or other pesticides. Do not contaminate water, food, or feed by storage or disposal.

PESTICIDE DISPOSAL

Open dumping is prohibited. Pesticide wastes are toxic. Improper disposal of excess pesticide, spray mixture, or rinsate is a violation of Federal law. If these wastes cannot be disposed by use according to label instructions, contact your State Pesticide or Environmental Control Agency, or the Hazardous Waste representative at the nearest EPA Regional Office for guidance.

CONTAINER DISPOSAL

Triple rinse or equivalent. Then offer for recycling, or reconditioning, or puncture and dispose of in a sanitary landfill, or incineration, or, if allowed by state and local authorities, by burning. If container is burned, stay out of smoke.

LIMITED WARRANTY AND DISCLAIMER

The manufacturer warrants that this product conforms to the chemical description on the label; that this product is reasonably fit for the purposes set forth in the directions for use when it is used in accordance with such directions; and that the directions, warnings and other statements on this label are based upon responsible experts' evaluation of reasonable tests of effectiveness, of toxicity to laboratory animals and to plants, and of residues and upon reports of field experience. Tests have not been made on all varieties or in all states or under all conditions. THE MANUFACTURER NEITHER MAKES NOR INTENDS, NOR DOES IT AUTHORIZE ANY AGENT OR REPRESENTATIVE TO MAKE, ANY OTHER WARRANTIES, EXPRESS OR IMPLIED, AND IT EXPRESSLY EXCLUDES AND DISCLAIMS ALL IMPLIED WARRANTIES OF MERCHANTABILITY OR FITNESS FOR A PARTICULAR PURPOSE.

THIS WARRANTY DOES NOT EXTEND TO, AND THE BUYER SHALL BE SOLELY RESPONSIBLE FOR, ANY AND ALL LOSS OR

DAMAGE WHICH RESULTS FROM THE USE OF THIS PRODUCT IN ANY MANNER WHICH IS INCONSISTENT WITH THE LABEL DIRECTIONS, WARNINGS OR CAUTIONS.

BUYER'S EXCLUSIVE REMEDY AND MANUFACTURER'S OR SELLER'S EXCLUSIVE LIABILITY FOR ANY AND ALL CLAIMS, LOSSES, DAMAGES, OR INJURIES RESULTING FROM THE USE OR HANDLING OF THIS PRODUCT, WHETHER OR NOT BASED IN CONTRACT, NEGLIGENCE, STRICT LIABILITY IN TORT OR OTHERWISE SHALL BE LIMITED, AT THE MANUFACTURER'S OPTION, TO REPLACEMENT OF, OR THE REPAYMENT OF THE PURCHASE PRICE FOR, THE QUANTITY OF PRODUCT WITH RESPECT TO WHICH DAMAGES ARE CLAIMED. IN NO EVENT SHALL MANUFACTURE OR SELLER BE LIABLE FOR SPECIAL, INDIRECT OR CONSEQUENTIAL DAMAGES RESULTING FROM THE USE OR HANDLING OF THIS PRODUCT.

NOTICE TO BUYER

Purchase of this material does not confer any rights under patents governing this product or the use thereof in countries outside of the United States.

WEEDONE is a registered trademark of Rhône-Poulenc.

AG 90026

EPA approval: 2/27/87

THIS SPECIMEN LABEL IS INTENDED FOR USE ONLY AS A GUIDE IN PROVIDING GENERAL INFORMATION REGARDING THE DIRECTIONS, WARNINGS AND CAUTIONS ASSOCIATED WITH THE USE OF THIS PRODUCT. AS WITH ANY AGRICULTURAL CHEMICAL, ALWAYS FOLLOW THE LABEL INSTRUCTIONS ON THE PACKAGE BEFORE USING.

WEEDONE® 638

BROADLEAF HERBICIDE

A SPECIAL FORMULATION CONTAINING, 2,4-D ACID.

ACTIVE INGREDIENTS:	
*2,4-Dichlorophenoxyacetic acid	13.8%
**2,4-Dichlorophenoxyacetic acid, butoxyethyl ester	24.5%
INERT INGREDIENTS:	61.7%

*Isomer specific by AOAC method No. 6.DO1-5

**Total 2,4-dichlorophenoxyacetic acid equivalent 30.8% by weight or 2.8 pounds per gallon.

EPA Reg. No. 264-37AA EPA Est. No. 264-MO-01

KEEP OUT OF REACH OF CHILDREN

DANGER PELIGRO

PRECAUTION AL USUARIO: Si usted no lee ingles, no use este producto hasta que la etiqueta le haya sido explicado ampliamente.

For PRODUCT USE Information Call 1-800-334-9745
For MEDICAL And TRANSPORATION Emergencies ONLY Call 24 Hours a Day 1-800-334-7577

PRECAUTIONARY STATEMENTS

DANGER

HAZARDS TO HUMANS AND DOMESTIC ANIMALS

Causes irreversible eye damage. Do not get in eyes or on clothing. Wear goggles, face shield or safety glasses when handling concentrates. Wash thoroughly with soap and water after handling. Remove contaminated clothing and wash separately before reuse. Harmful if swallowed or absorbed through the skin. Avoid breathing spray mist.

STATEMENT OF PRACTICAL TREATMENT

IF IN EYES: Hold eyelid open and flush with a steady stream of water for at least 15 minutes. Get medical attention, preferably an OPHTHALMOLOGIST.

IF SWALLOWED: Give two or three glasses of water to drink. Do not induce vomiting. PRECAUTION: To prevent aspiration of swallowed product, lay patient on side with head lower than waist. Do not induce vomiting or give anything by mouth to an unconscious person. Get medical attention.

IF ON SKIN: Wash with plenty of soap and water, if available, while removing contaminated clothing.

NOTE TO PHYSICIAN

No specific antidote is available. All treatments should be based on observed signs and symptoms of distress in the patient. Overexposure to materials other than this product may have occurred.

This product contains petroleum distillates. If large amounts, greater than 1 mg/kg body weight have been ingested, the stomach should be evacuated by gastric intubation with the aid of a cuffed endotrachael tube to prevent aspiration of petroleum distillates. After removal of stomach contents, wash stomach by instilling 30-50 grams of activated charcoal in 3-4 ounces of water through the stomach tube and again remove stomach contents. Avoid oily laxatives.

ENVIRONMENTAL HAZARDS

This product is toxic to fish. Drift or runoff may adversely affect aquatic invertebrates and non-target plants. Do not apply directly to water or wetlands (swamps, bogs, marshes and potholes). Do not contaminate water when disposing of equipment washwaters.

Do not contaminate water used for irrigation or domestic purposes. Do not apply when weather conditions favor drift from treated areas. Avoid spray drift to susceptible plants such as cotton, beans, peas, ornamentals, and most vegetables, as injury may occur. Coarse sprays are less likely to drift.

Under very high temperature, vapors from this product may injure susceptible plants in the immediate vicinity. Do not use in or near a greenhouse.

DIRECTIONS FOR USE

IT IS A VIOLATION OF FEDERAL LAW TO USE THIS PRODUCT IN A MANNER INCONSISTENT WITH ITS LABELING.

Do not apply WEEDONE® 638 Broadleaf Herbicide through any type of irrigation system.

GENERAL INFORMATION

WEEDONE® 638 Broadleaf Herbicide is a special formulation containing 2,4-D acid. WEEDONE® 638 is more effective than 2,4-D amines for controlling hard-to-kill weeds such as field bindweed, Russian knapweed, Canada thistle, leafy spurge, cattails, tules and nutgrass.

WEEDONE® 638 Broadleaf Herbicide should be applied when daytime temperatures are lower than 80 to 85 degrees. Spraying when temperatures are higher may cause rapid foliar burn and decrease the effectiveness of WEEDONE® 638. Treatment at higher temperatures also increases the possibility of crop injury.

MIXING SPRAYS

Add the required amount of water to the spray tank. Start agitation. While agitating, slowly add WEEDONE® 638 Broadleaf Herbicide. Mix thoroughly. Whenever practical, maintain continuous agitation and use spray materials as soon as possible after mixing.

COMBINATION WITH NITROGEN SOLUTIONS

Use 14 fluid ounces (7/8 pint) WEEDONE® 638 Broadleaf Herbicide per acre, mixed with the amount of nitrogen solution required to apply the pounds of nitrogen you want per acre.

Pour liquid nitrogen into the spray tank, then add WEEDONE® 638 Broadleaf Herbicide. Mix thoroughly before spraying and maintain good by-pass agitation while spraying.

If WEEDONE® 638 is tank mixed with nitrogen compatibility should be tested prior to mixing. To test for compatibility, mix a small amount (1/2 - 1 qt.) of spray combining all ingredients in the same ratios you anticipate using. If any physical indications of incompatibility develop do not utilize the mixture for spraying.

GENERAL WORKER PROTECTION STATEMENTS

Do not apply this product in such a manner as to directly or through drift expose workers or other persons. The area being treated must be vacated by unprotected persons.

Do not enter treated areas without protective clothing until sprays have dried. Because certain states may require more restrictive reentry intervals for various crops treated with this product, consult your State Department of Agriculture for further information. Written or oral warnings must be given to workers who are expected to be in a treated area or in an area about to be treated with this product. Advise workers to stay out of fields during application and until sprays have dried. Regulation long sleeved work clothing should be worn when working in treated fields. See PRECAUTIONARY STATEMENT, STATEMENT OF PRACTICAL TREATMENT and NOTE TO PHYSICIAN. When oral warnings are given, warnings shall be given in a language customarily understood by workers. Oral warnings must be given if there is reason to believe that written warnings cannot be understood by workers. Written warnings must include the following information: appropriate signal word (DANGER), area treated with WEEDONE® 638 Broadleaf Herbicide, date of application, appropriate protection clothing, and reentry interval if applicable.

WEEDS CONTROLLED BY PROPERLY APPLIED TREATMENTS OF WEEDONE® 638 INCLUDE THE FOLLOWING PLUS MANY OTHERS:

ANNUAL AND BIENNIAL WEEDS

annual yellow sweetclover
*beggarticks
bull thistle
coffeeweed
*common broomweed
common burdock
common cocklebur
common eveningprimrose
common lambsquarters
croton (Texas or woolly)
galinsoga
jimsonweed
*knotweed
*mallow (venice or little)
marshelder
morningglory (common, ivy, woolly)
*musk thistle***
mustards (except blue mustard)
pepperweeds (except perennial)
**pigweeds (*Amaranthus* spp.)
*prickly lettuce
ragweed (common or giant)
rough fleabane
*Russian thistle
*salsify (western or common)
*smartweeds (annual species)
sowthistle (annual or spiny)
sunflower
velvetleaf
*vervains
veches
*wild carrot
wild lettuce
wild parsnip
wild radish

PERENNIAL WEEDS

*alfalfa
*Austrian fieldcress
*bindweed (hedge, field and European)
blue lettuce
*Canada thistle
catnip
chicory
dandelion
*docks
*dogbanes
*goldenrod
*ground ivy
healall
*hoary cress
Jerusalem-artichoke
*many-flowered aster
*nettles (including stinging)
*orange hawkweed
plantains
sowthistle (perennial)
*tansy ragwort
Texas blueweed
*vervains
*western ironweed
*wild garlic
*wild onion

*These species may require repeat applications and/or use of higher rate recommended on this product label even under ideal conditions for application.
**Control of pigweeds in the High Plains area of Texas and Oklahoma may not be satisfactory with this product.
***Not registered for control of musk thistle in California.

INJURY TO CROPS FROM THIS HERBICIDE MAY OCCUR. IF YOU ARE NOT PREPARED TO ACCEPT SOME DEGREE OF CROP INJURY DO NOT USE THIS PRODUCT.

Crop varieties vary in response to 2,4-D and some are easily injured. Apply WEEDONE® 638 only to varieties known to be tolerant to 2,4-D. If you are uncertain concerning tolerant varieties or local use situations that may affect crop tolerance to 2,4-D consult your seed company or state Agricultural Extension Service for advice.

RECOMMENDATIONS FOR APPLYING WEEDONE® 638 BROADLEAF HERBICIDE

CEREAL GRAINS, WHEAT, BARLEY, (NOT UNDERSEEDED WITH LEGUMES)

WEEDS IN CROPS	AMOUNT OF WEEDONE® 638 PER ACRE	DIRECTIONS
Postemergence Annual and biennial weeds Perennial broadleaf weeds	1 pint 2 pints*	Apply after grain is fully tillered (usually about 4-8" high) but not forming joints in the stem. Apply when weeds are near bud stage. **Do not spray grain in boot to dough stage.**
Wild onion or garlic	1½-2 pints*	Apply 1½ pints when grain is fully tillered and wild garlic or onion plants are small. Apply 2 pints after harvest in the crop stubble. For control of new fall growth of wild onion or garlic refer to FALLOWLAND use directions.
Emergency Weed Control in Wheat Perennial broadleaf weeds	3 pints*	Apply when weeds are approaching bud stage, but **do not spray grain during the boot or dough stage.** The 3 pint per acre application can produce injury to wheat. Balance the severity of your weed problem against the possibility of crop damage. Where perennial weeds are scattered, spot treatment is suggested to minimize the extent of crop injury.
Preharvest	1-2 pints*	Apply when grains are in the hard dough stage to suppress large weeds that may interfere with harvest. Best results will be obtained when soil moisture is sufficient to induce succulent weed growth. DO NOT USE IN CALIFORNIA.

* Use the lower rate if small annual and biennial weeds are the major problem. Use the higher rate if perennial weeds or annual and biennial weeds are present which are in the hard-to-kill categories as determined by local experience. The higher rates increase the risk of grain injury and should be used only where weed control problems justify the risk of grain damage. Do not apply to grain in the seedling stage.

RESTRICTIONS AND LIMITATIONS FOR USE ON CEREAL GRAINS

- For aerial application on grain, apply WEEDONE® 638 in 2 or more gallons of water per acre. For ground application, use a minimum of 10 gallons of water per acre.
- Do not mix with oil for crop uses.
- Do not permit dairy animals or meat animals being finished for slaughter to forage treated grain fields within 2 weeks after treatment.
- Do not feed treated straw to livestock if a preharvest treatment or emergency treatment as described above is applied.

FIELD CORN

WEEDS IN CROPS	AMOUNT OF WEEDONE® 638 PER ACRE	DIRECTIONS
Preplant	1½ to 2½ pts	To control emerged broadleaf weed seedlings or existing cover crops prior to planting corn, apply 7 to 14 days befor planting. Do not use on light, sandy soil, or where soil moisture is inadequate for normal weed growth. Use high rate for control of less susceptible weeds or cover crops such as alfalfa.
Postemergence Annual broadleaf weeds	⅔ pint	Avoid spraying just after the first corn leaves unfold, as injury may occur. Apply when weeds are small and corn is less than 8" tall (to top of canopy). If corn more than 8" tall must be sprayed, use drop pipes.
Perennial broadleaf weeds	1 pint	Spray when weeds are in the bud to bloom stage. If corn is more than 8" tall, use drop pipes to keep spray off corn leaves. **Do not spray in tassel to dent stage.** 2,4-D may make corn brittle. Winds or cultivation may cause stalk breakage while brittle.

FIELD CORN		
WEEDS IN CROPS	AMOUNT OF WEEDONE® 638 PER ACRE	DIRECTIONS
Preharvest	1½ to 2½ pints	After the dent stage to suppress weeds that interfere with harvest, such as bindweed, cocklebur, dogbane, jimsonweed, ragweed, sunflower, and velvetleaf, and to decrease production of weed seeds, spray with air or ground equipment. The high rate will be needed for weeds under stress. DO NOT USE IN CALIFORNIA.

RESTRICTIONS AND LIMITATIONS FOR USE ON FIELD CORN

- Do not forage or feed corn fodder for 7 days following application.
- A minimum application volume of 5 gallons per acre by air or 10 gallons per acre by ground is recommended.

GRAIN SORGHUM (milo)		
WEEDS IN CROPS	AMOUNT OF WEEDONE® 638 PER ACRE	DIRECTIONS
Annual broadleaf weeds Perennial broadleaf weeds	1 pint 1½ pints	Treat when sorghum is 6 to 15" tall. If sorghum is taller than 8" (to top of canopy) use drop nozzles to keep spray off crop foliage. Do not treat during the boot, flowering or early dough stages.

RESTRICTIONS AND LIMITATIONS FOR USE ON SORGHUM

- Do not forage or feed sorghum fodder for 7 days following application.
- A minimum application volume of 5 gallons per acre by air or 10 gallons per acre by ground is recommended.

WEEDS AND BRUSH IN PASTURES, RANGELAND, TURF AREAS, FALLOW LAND AND CONSERVATION RESERVE PROGRAM AREAS RANGELAND, PASTURES, GOLF COURSES, CEMETERIES, PARKS, AND OTHER TURF AREAS		
WEEDS IN CROPS	AMOUNT OF WEEDONE® 638 PER ACRE	DIRECTIONS
Annual broadleaf weeds	2½ pints	Apply in spring or fall when annual weeds such as knotweed or common chickweed are up but still small and in or near the seedling stage, and growing vigorously. Also controls cocklebur, galinsoga, lambsquarters, mustards, pigweed, ragweed, sunflower.
Perennial and biennial broadleaf weeds	3 to 5 pints	To control dandelion, plantain, and most other common broadleaf weeds in turf, apply in spring or fall when these weeds are growing actively. Treat wild garlic or onion in three successive seasons. (Example: spring, fall, spring).

RESTRICTIONS AND LIMITATIONS FOR USE ON RANGELAND, PASTURES, TURF AREAS

- A minimum application volume of 25 gallons per acre is recommended for use on turf.
- A minimum application volume of 2 gallons by air or 5 gallons by ground is recommended for pastures.
- Do not apply to newly seeded areas until sod is formed and grass has been cut a few times. Do not use WEEDONE® 638 for susceptible southern grasses such as St. Augustine. Bentgrass and clover may be injured by this treatment.
- Do not graze animals on treated areas within 7 days of treatment.
- Do not cut forage for hay within 30 days of treatment.
- Do not permit dairy animals or meat animals being finished for slaughter to forage treated fields within 3 days of slaughter.

CONSERVATION RESERVE PROGRAM AREAS		
WEEDS IN CROPS	AMOUNT OF WEEDONE® 638 PER ACRE	DIRECTIONS
Annual broadleaf weeds in young grasses	⅔ to 1⅓ pints ⅔ to 2⅔ pints	Apply to actively growing annual broadleaf weeds. Use ⅔ to 1⅓ pint when weeds are small; use higher rates on older weeds. Do not apply to young grasses with fewer than 6 leaves or prior to tillering, as excessive injury may result. Do not apply more than 1⅓ pint until grasses are well established as excessive injury may result.
Biennial and perennial broadleaf weeds in established grasses	2⅔ to 5 5⅓ pints	Treat when biennial weeds are in the seedling to rosette stage and before flower stalks become apparent. Treat perennial weeds in the bud to bloom stage. Apply to actively growing weeds.

RESTRICTIONS AND LIMITATIONS FOR USE ON CONSERVATION RESERVE PROGRAM AREAS

- Use at least 2 gallons of water per acre by air and 5 gallons of water per acre by ground.
- Do not harvest or graze treated Conservation Reserve Program areas.
- Do not apply to grasses in the boot to dough stage if grass seed production is desired.

FALLOW LAND AND CROP STUBBLE		
WEEDS IN CROPS	AMOUNT OF WEEDONE® 638 PER ACRE	DIRECTIONS
Annual broadleaf weeds	1⅓ to 2⅔ pints	Use the lower rate when weeds are small (2-3" tall) and growing actively. Use the higher rate on older and drought-stressed plants, except as directed in the special weed problems section of this label.
Biennial broadleaf weeds	2⅔ to 5⅓ pints	Spray while musk thistles or other biennial species are in the seedling to rosette stage, and before flower stalks become apparent. The lower rate can be used in spring during rosette stage. In fall or after flower stalks have developed, use the highest rate, except as directed in the special weed problems section of this label.

FALLOW LAND AND CROP STUBBLE

WEEDS IN CROPS	AMOUNT OF WEEDONE® 638 PER ACRE	DIRECTIONS
Perennial broadleaf weeds	1½ to 4 quarts	Spray weeds in bud to bloom stage, or in good vegetative growth. Do not disturb treated area for at least 2 weeks after treatment, or until weed tops are dead, except as directed in the special weed problems section of this label.
Wild garlic and onion in crop stubble	2⅔ to 4 quarts	Apply to new regrowth of wild onion or garlic which occurs in the fall following harvest of small grains, soybeans, corn, or grain sorghum.

RESTRICTIONS AND LIMITATIONS FOR USE ON FALLOW LAND AND CROP STUBBLE

- A minimum spray volume of 5 gallons per acre by air or 10 gallons per acre by ground is recommended.
- Do not plant treated fallow land until 3 months after treatment or until chemical has disappeared from soil.
- Do not graze dairy animals on treated areas within 7 days of treatment.

WEEDS AND BRUSH IN NON-CROP AREAS

FENCEROWS, ROADSIDES, DRAINAGE DITCHBANKS, VACANT LOTS, AIRFIELDS, RAILROAD, HIGHWAY AND UTILITY RIGHTS-OF-WAY, AND OTHER NON-CROP AREAS.

WEEDS IN CROPS	AMOUNT OF WEEDONE® 638 PER ACRE	DIRECTIONS
Annual broadleaf weeds	1⅓ to 2⅔ quarts	Apply when most weeds are still young and growing vigorously, except as directed in the special weed problems section of this label.
Perennial and biennial broadleaf weeds	2 to 4 quarts	Apply when weeds are actively growing and near the bud stage, but before flowering. For best results on tansy ragwort and musk thistle, treat in rosette stage, before bolting. A second application is usually needed for best results on thistle, nettle, and bindweed. Treat wild onion or garlic in early spring and in fall when they are young and growing actively, except as directed in the special weed problems section of this label.

SMALL AREA APPLICATIONS FOR NON-CROP USE

For control of broadleaf weeds in small non-crop areas with hand-held or back-pack sprayers mix 5 fluid ounces WEEDONE® 638 per gallon of water. Thoroughly wet all weed foliage. Maintain agitation of mixture.

SPECIAL WEED PROBLEMS

DIRECTIONS FOR SPECIAL WEED PROBLEMS IN PASTURES, FALLOW LAND, NON CROP AND CONSERVATION RESERVE PROGRAM AND CONSERVATION RESERVE ACREAGE.

To control the following special weed problems in fallow land, non-crop areas or pastures apply the rates indicated at the recommended application time in sufficient spray volume by air or ground to obtain thorough and uniform coverage. Thorough and uniform coverage may require high volume ground spraying at 50 gallons per acre or more or aerial spraying at 10 gallons per acre or more when weed growth is dense. Adequate soil moisture for weed growth is essential for good control.

WEEDS	AMOUNT WEEDONE® 638	WHEN TO SPRAY
musk thistle	1½ to 2 quarts	rosette stage spring or fall, use higher rate in fall.
field bindweed	1½ to 2 quarts	bud to full bloom
hoary cress	2 quarts	bud to full bloom
leafy spurge, russian knapweed	2 quarts	full bloom
Canada thistle, camelthorn, lotus	2 quarts	bud to early bloom in spring or summer
Canada thistle	1½ to 2½ quarts	6 to 8 inches in fall, prior to a killing frost with good soil moisture. Do not till treated area within 7 days of treatment.
cattails, tule (hard-stem bulrush)	3 quarts	late spring to early summer when 4 to 6 feet tall; repeat in fall and following spring.
tansy ragwort	3 quarts	rosette stage up to early bloom
nutgrass	3 quarts	6 to 8 inches high, before bloom, repeat when regrowth is 6 to 8 inches high.

Refer to the restrictions and limitations for use on these sites.

STORAGE AND DISPOSAL STATEMENT

STORAGE

Do not store near fertilizers, seeds, insecticides, or fungicides. Do not contaminate water, food, or feed by storage or disposal.

PESTICIDE DISPOSAL

Open dumping is prohibited. Pesticide wastes are acutely hazardous. Improper disposal of excess pesticide, spray mixture or rinsate is a violation of Federal law. If these wastes cannot be disposed of by use according to label instructions, contact your State Pesticide or Environmental Control Agency or the Hazardous Waste representative at the nearest EPA Regional office for guidance.

CONTAINER DISPOSAL

Triple rinse (or equivalent). Then offer for recycling or reconditioning, or puncture and dispose of in a sanitary landfill, or incineration, or, if allowed by state and local authorities, by burning. If burned, stay out of smoke.

LIMITED WARRANTY AND DISCLAIMER

The manufacturer warrants (a) that this product conforms to the chemical description on the label; (b) that this product is reasonably fit for the purposes set forth in the directions for use when it is used in accordance with such directions; and (c) that the directions, warnings and other statements on this label are based upon responsible experts' evaluation of reasonable tests of effectiveness, of toxicity to laboratory animals and to plans, and of residues on food crops, and upon reports of field experience. Tests have not been made on all varieties or in all states or under all conditions. THE MANUFACTURER NEITHER MAKES NOR INTENDS, NOR DOES IT AUTHORIZE ANY AGENT OR REPRESENTATIVE TO MAKE, ANY OTHER WARRANTIES, EXPRESS OR IMPLIED, AND IT EXPRESSLY EXCLUDES AND DISCLAIMS ALL IMPLIED WARRANTIES OF MERCHANTABILITY OR FITNESS FOR A PARTICULAR PURPOSE.

THIS WARRANTY DOES NOT EXTEND TO, AND THE BUYER SHALL BE SOLELY RESPONSIBLE FOR, ANY AND ALL LOSS OR DAMAGE WHICH RESULTS FROM THE USE OF THIS PRODUCT IN ANY MANNER WHICH IS INCONSISTENT WITH THE LABEL DIRECTIONS, WARNINGS OR CAUTIONS.

BUYER'S EXCLUSIVE REMEDY AND MANUFACTURER'S OR SELLER'S EXCLUSIVE LIABILITY FOR ANY AND ALL CLAIMS, LOSSES, DAMAGES, OR INJURIES RESULTING FROM THE USE OR HANDLING OF THIS PRODUCT, WHETHER OR NOT BASED IN CONTRACT, NEGLIGENCE, STRICT LIABILITY IN TORT OR OTHERWISE, SHALL BE LIMITED, AT THE MANUFACTURER'S OPTION, TO REPLACEMENT OF, OR THE REPAYMENT OF THE PURCHASE PRICE FOR, THE QUANTITY OF PRODUCT WITH RESPECT TO WHICH DAMAGES ARE CLAIMED. IN NO EVENT SHALL MANUFACTURED OR SELLER BE LIABLE FOR SPE-

CIAL, INDIRECT OR CONSEQUENTIAL DAMAGES RESULTING FROM THE USE OR HANDLING OF THIS PRODUCT.

NOTICE TO BUYER

Purchase of this material does not confer any rights under patents governing this product or the use thereof in countries outside of the United States.

WEEDONE is a registered trademark of Rhône-Poulenc.

EPA Approval 2/27/87

AG90027

THIS SPECIMEN LABEL IS INTENDED FOR USE ONLY AS A GUIDE IN PROVIDING GENERAL INFORMATION REGARDING THE DIRECTIONS, WARNINGS AND CAUTIONS ASSOCIATED WITH THE USE OF THIS PRODUCT. AS WITH ANY AGRICULTURAL CHEMICAL, ALWAYS FOLLOW THE LABEL INSTRUCTIONS ON THE PACKAGE BEFORE USING.

WEEDONE® CB brand

READY-TO-USE WOODY PLANT HERBICIDE

For Controlling Woody Plants In Forest Management Programs And On Fencerows, Roadsides, Drainage Ditchbanks, Railroads, Utility Right-Of-Ways And Other Non-Crop Areas.

ACTIVE INGREDIENTS:

*2-(2,4-Dichlorophenoxy)propionic acid, butoxyethyl ester	11.7%
**2,4-Dichlorophenoxyacetic acid, butoxyethyl ester	11.9%
INERT INGREDIENTS:	76.4%

*2-(2,4-Dichlorophenoxy)propionic acid equivalent 8.2% by weight or 2/3 pounds per gallon.

**2,4-Dichlorophenoxyacetic acid equivalent 8.2% by weight or 2/3 pounds per gallon.

Isomer specific by AOAC method No. 6.DO1-5.

Contains Petroleum Distillates

EPA Reg. No. 264-393 EPA Est. No.

KEEP OUT OF REACH OF CHILDREN

WARNING AVISO

PRECAUCION AL USUARIO: Si usted ne lee ingles, no use este producto hasta que la etiqueta le haya side explicada ampliamente.

For **PRODUCT USE** Information Call 1-800-334-9745
For **MEDICAL and TRANSPORTATION Emergencies** Information **ONLY** Call 24 Hours A Day 1-800-334-7577

PRECAUTIONARY STATEMENTS

WARNING

HAZARDS TO HUMANS AND DOMESTIC ANIMALS

Harmful if swallowed, or inhaled, or absorbed through the skin. Causes skin and eye irritation. Do not get in eyes, on skin or clothing. Avoid breathing spray mist. Wear protective clothing, including long-sleeve shirt, full length trousers, and rubber gloves when handling. Wash thoroughly with soap and water after handling and before eating or smoking. Remove contaminated clothing and wash before reuse.

STATEMENT OF PRACTICAL TREATMENT

IN CASE OF SKIN CONTACT: Wash with plenty of soap and water, if available, while removing contaminated clothing. Get medical attention if irritation develops.

IN CASE OF EYE CONTACT: Hold eyelids open and flush with a steady, gentle stream of water for at least 15 minutes. Get medical attention.

IF SWALLOWED: Drink promptly 2-3 glasses of milk, large quantities of egg whites, gelatin solution, or if these are not available, drink 2-3 glasses of water. Avoid alcohol. Get medical attention. Do not induce vomiting. Do not give anything by mouth to an unconscious person.

NOTE TO PHYSICIAN

No specific antidote is available. Treat symptomatically. Overexposure to materials other than this product may have occurred.

This product contains petroleum distillates. If large amounts (greater than 1 ml/kg body weight) of the product have been ingested, the stomach should be evacuated by gastric intubation with the aid of a cuffed endotracheal tube to prevent aspiration of petroleum distillates. After removal of stomach contents, wash stomach by instilling 30 to 50 grams of activated charcoal in 3 to 4 ounces of water through the stomach tube and again remove stomach contents. Avoid oily laxatives.

ENVIRONMENTAL HAZARDS

This product is toxic to fish and aquatic invertebrates. Drift or runoff may adversely affect fish and non-target plants. Do not apply directly to water or wetlands (swamps, bogs, marshes, and potholes).

Do not contaminate water when disposing of equipment washwaters. Do not graze meat or dairy animals on treated areas.

Do not apply when weather conditions favor drift. Do not use the same spray equipment for other purposes unless thoroughly cleaned. Do not contaminate water used for domestic or irrigation purposes. Do not use in or near a greenhouse.

Do not apply WEEDONE® CB brand Ready-To-Use Woody Plant Herbicide directly to or permit spray mist to drift onto cotton, okra, grapes, tomatoes, fruit trees, vegetables, flowers, or other desirable crop or ornamental plants which are susceptible to 2,4-D or 2,4-DP herbicides. Do not allow spray to contact susceptible plants since very small quantities of the 2,4-D or 2,4-DP will cause severe injury during the growing or dormant periods. Crops contacted by WEEDONE® CB brand Ready-To-Use Woody Plant Herbicide sprays or spray drift may be killed or suffer significant stand loss with extensive quality and yield reduction.

Vapors from this product may injure highly sensitive plants already growing in the immediate vicinity or planted there two to three weeks after spraying. Do not use in or near a greenhouse.

Drift may be reduced by: (1) applying as near to the target as possible; (2) by decreasing the pounds of pressure at the nozzle tips; (3) by using nozzles which produce a coarse spray pattern; (4) by not applying when wind is blowing toward susceptible crops or valuable plants.

GENERAL INFORMATION

WEEDONE® CB brand READY-TO-USE WOODY PLANT HERBICIDE is a ready-to-apply formulation that can be used in basal, stump and stem applications.

WEEDONE® CB is a ready-to-use product, applied undiluted, using the special application equipment described on the label. It controls over 100 major woody plant species, including oak, hickory, multiflora rose, sweetgum and blackgum when applied to the lower 18 to 24 inches of the plant stem. During the growing season, vegetation near treated stems may be discolored and even killed. WEEDONE® CB brand READY-TO-USE WOODY PLANT HERBICIDE normally allows ground cover plants to become re-established in the treated area.

WEEDONE® CB brand READY-TO-USE WOODY PLANT HERBICIDE can be applied in both dormant and growing seasons. This year-round capability lengthens treatment programs and improves crew utilization.

WEEDONE® CB brand READY-TO-USE WOODY PLANT HERBICIDE is ready-to-apply and requires no mixing with oil, thus reducing application time.

WEEDONE® CB brand READY-TO-USE WOODY PLANT HERBICIDE will not freeze under normal storage or use conditions.

WEEDONE® CB brand READY-TO-USE WOODY PLANT HERBICIDE gives effective control of numerous unwanted woody plant species.

Species controlled by WEEDONE® CB brand READY-TO-USE WOODY PLANT HERBICIDE include:

Ailanthus (tree-of-heaven)	*Ailanthus altissima*
Alder	*Alnus* sp.
Apple seedlings	*Malus* sp.
Aralia (devils-walkingstick)	*Aralia spinosa*
Ash	*Fraxinus* sp.
Ash, green	*Fraxinus pennsylvanica*
*Ash, white	*Fraxinus americana*
Aspen	*Populus* sp.
Aspen, big-tooth	*Populus grandidentata*
Aspen, quaking	*Populus tremuloides*
Balm-of-Gilead	*Populus gileadensis*
Bayberry, northern	*Myrica pennsylvanica*
Beautybush	*Kolkwitzia amabilis*
Beech	*Fagus* sp.
Birch	*Betula* sp.
*Birch, gray	*Betula populifolia*
Birch, paper	*Betula papyrifera*
*Birch, river	*Betula nigra*
Blackberry	*Rubus* sp.
Boxelder	*Acer negundo*
Buckeye	*Aesculus* sp.
Buckeye, Ohio	*Aesculus glabra*
Butternut	*Juglans cinerea*
Briars	*Rosa* sp.
Camphor tree	*Cinnamonum camphor*
Catalpa, western	*Catalpa speciosa*
Cherry	*Prunus virginiana*

*Cherry, black	*Prunus serotina*
Cherry, pin	*Prunus pensylvanica*
Chokecherry, common	*Prunus virginiana*
Cottonwood	*Populus* sp.
Dogwood	*Cornus* sp.
Elder	*Sambucus* sp.
Elder, Pacific red	*Sambucus callicarpa*
Elm, American	*Ulmus americana*
Elm, Chinese	*Ulmus parvifolia*
Elm, red	*Ulmus fulva*
Elm, winged	*Ulmus alata*
Fir	*Abies* sp.
Fir, Balsam	*Abies balsamea*
Fir, Douglas	*Pseudotsuga menziesii*
Grape	*Vitis* sp.
Grape, Fox	*Vitis labrusca*
Gum, black	*Nyssa sylvatica*
Gum, sweet	*Liquidambar styraciflua*
Hackberry	*Celtis occidentalis*
Hardhack	*Spiraea tomentosa*
Hawthorn	*Crataegus* sp.
Hawthorn, English	*Crataegus oxyacantha*
Hazel, beaked	*Corylus cornuta*
Hickory	*Carya* sp.
Hickory, pignut	*Carya glabra*
Hickory, shagbark	*Carya ovata*
Honeysuckle	*Lonicera japonica*
Hornbeam	*Carpinus* sp.
Huckleberry	*Vaccinium* sp.
Ironwood	*Ostrya virginiana*
Ivy, English	*Hedera helix*
Juniper, common	*Juniperus communis*
Laurel, sheep	*Kalmia angustifolia*
Locust	*Robinia* sp.
Locust, black	*Robinia pseudoacacia*
Magnolia (Sweet bay)	*Magnolia virginiana*
*Maple	*Acer* sp.
*Maple, bigleaf	*Acer macrophyllum*
*Maple, red	*Acer rubum*
*Maple, sugar	*Acer saccharum*
Mulberry	*Morus* sp.
Oak	*Quercus* sp.
Oak, bur	*Quercus macrocarpa*
Oak, chestnut	*Quercus prinus*
Oak, northern red	*Quercus rubra*
Oak, pin	*Quercus palustris*
Oak, southern red	*Quercus falcata*
Oak, water	*Quercus nigra*
Oak, white	*Quercus alba*
Orange, sour	*Citrus aurantium*
Osageorange	*Maclura pomifera*
Persimmon, common	*Diospyros virginiana*
*Pine, shortleaf **	*Pinus echinata*
*Pine, Virginia **	*Pinus virginiana*
Pine, jack	*Pinus banksiana*
Pine, red	*Pinus resinosa*
Pine, white	*Pinus strobus*
Poison-ivy	*Rhus radicans*
Poison-oak	*Rhus toxicodendron*
Poplar	*Populus* sp.
Poplar, balsam	*Populus balsamifera*
Prickly-ash	*Zanthoxylum americanum*
Prickly pear	*Opuntia* sp.
Raspberry,	*Rubus* sp.
Redcedar, eastern	*Juniperus virginiana*
Rhododendron, Canadian	*Rhododendron canadense*
Rose	*Rosa* sp.
Rose, multiflora	*Rosa multiflora*
Sassafras	*Sassafras albidum*
Serviceberry (shadbush)	*Amelanchier* sp.
Sourwood	*Oxydendrum arboreum*
Spruce, white	*Picea glauca*
Sumac	*Rhus* sp.
Sumac, staghorn	*Rhus typhina*
Sumac, poison	*Rhus vernix*
Swamp cyrilla	*Cyrilla racemiflora*
Sweetfern	*Comptonia peregrina*
Sycamore, American	*Platanus occidentalis*
Tallowtree, Chinese	*Sapium sebiferum*
Tulip-popular (yellow poplar)	*Liriodendron tulipifera*
Walnut	*Juglans* sp.
Walnut, black	*Juglans nigra*
Wax myrtle, (southern bayberry)	*Myrica cerifera*
Willow	*Salix* sp.
Willow, Bebb	*Salix bebbiana*
Witchhazel	*Hamamelis* sp.
Withe-rod	*Virburnum cassinoides*

*Treat at least the lower 36 inches of stem as specified in Directions for Use. **Not effective on southern pines larger than 2½" in stem diameter (DBH)

FLORIDA DRAINAGE DITCHBANK SPECIES CONTROLLED BY WEEDONE® CB BRAND READY-TO-USE WOODY PLANT HERBICIDE INCLUDE THE FOLLOWING:

Lantana, largeleaf	*Lantana camara*
Sea Myrtle	*Baccharis* sp.
Brazilian peppertree	*Schinus erebinthifolius*
Australian-pine	*Casuarina equisetifolia*
Cabbage palmetto	*Sabal palmetto*

DIRECTIONS FOR USE

IT IS A VIOLATION OF FEDERAL LAW TO USE THIS PRODUCT IN A MANNER INCONSISTENT WITH ITS LABELING.

Apply this product only as specified on this label.

Be sure that use of this product conforms to all applicable laws, rules and regulations. Certain states have restrictions pertaining to application distances from susceptible crops. The applicator should become familiar with these laws, rules or regulations and follow them exactly. Do not dilute WEEDONE® CB brand READY-TO-USE WOODY PLANT HERBICIDE with water. Do not apply WEEDONE® CB brand READY-TO-USE WOODY PLANT HERBICIDE when stems are wet. Rainfall which follows application of WEEDONE® CB brand READY-TO-USE WOODY PLANT HERBICIDE to dry stems does not normally reduce brush control.

Do not graze meat or dairy animals on treated area.

FOR CONTROLLING WOODY PLANTS IN FOREST MANAGEMENT PROGRAMS AND ON FENCEROWS, ROADSIDES, DRAINAGE DITCHBANKS, RAILROADS, UTILITY RIGHT-OF-WAYS AND OTHER NON-CROP AREAS.

Average application volume of WEEDONE® CB brand READY-TO-USE WOODY PLANT HERBICIDE is 3 gallons per acre, although actual amount used will vary from 1 to 5 gallons depending on the amount and size of brush in the treated area. Total spray volume should be approximately 10% of that required for a conventional basal treatment. Do not exceed a total application volume of 20 gallons WEEDONE® CB brand READY-TO-USE WOODY PLANT HERBICIDE per acre.

BASAL BARK TREATMENT

To control general mixed brush up to 4 inches DBH (trunk diameter at breast height) including oak, hickory, sweetgum, blackgum, and many others, apply WEEDONE® CB brand READY-TO-USE WOODY PLANT HERBICIDE undiluted to the lower 18 to 24 inches of the plant stem. Treat all around the stem including the root collar area and any exposed roots. It is not necessary for WEEDONE® CB brand READY-TO-USE WOODY PLANT HERBICIDE to run down or puddle in the root collar area.

To control Southern pines less than 2½ inches DBH, increase treatment height to cover the lower 36 inches of stem bark.

To control ash, cherry, or maple increase treatment height to cover the lower 36 inches of stem bark.

WEEDONE® CB brand READY-TO-USE WOODY PLANT HERBICIDE basal bark treatment will control top growth of root-suckering species such as aspen, sumac, Sassafras, persimmon and black locust, but regrowth from lateral roots may occur the following growing season. Fall and winter treated stems may briefly leaf out in the following spring before final kill. WEEDONE® CB brand READY-TO-USE WOODY PLANT HERBICIDE will not usually control woody stems larger than 4" DBH. For large stems, cut the tree and treat the stump as directed in the section on stump treatment.

STUMP TREATMENT

Control stems larger than 4 inches DBH, by cutting the tree and thoroughly treating the bark and root collar area of the remaining stump. The stump can be cut at any height - low or high as long as the remaining wood, bark and root collar area is thoroughly treated.

Cutting the stump as close to the ground as possible will reduce the amount of WEEDONE® CB brand READY-TO-USE WOODY PLANT HERBICIDE needed for control.

Stumps do not require immediate treatment permitting separate cut and spray operations for greater efficiency.

FENCEROW TREATMENT

To control brush in fencerows apply WEEDONE® CB brand READY-TO-USE WOODY PLANT HERBICIDE to brush or stumps as directed above. Do not allow WEEDONE® CB brand READY-TO-USE WOODY PLANT HERBICIDE spray or drift to contact crops, forage plants or ornamentals.

DRAINAGE DITCHBANK TREATMENT

WEEDONE® CB brand READY-TO-USE WOODY PLANT HERBICIDE effectively removes unwanted brush along ditchbanks that interferes with the growth of bank-holding perennial grasses. WEEDONE® CB brand READY-TO-USE WOODY PLANT HERBICIDE is an excellent management tool, allowing you to selectively remove unwanted brush while leaving desired species. Do not allow spray or spray drift to contact water.

To control brush on drainage ditchbanks apply WEEDONE® CB brand READY-TO-USE WOODY PLANT HERBICIDE to brush or stumps as directed above. Do not allow WEEDONE® CB brand READY-TO-USE WOODY PLANT HERBICIDE spray or spray drift to contact crops, forage plants or ornamentals.

TANK MIXTURES WITH OTHER HERBICIDES

When mixing WEEDONE® CB brand READY-TO-USE WOODY PLANT HERBICIDE with other herbicides, follow all directions, precautions and limitations on both product labels.

WEEDONE CB/Garlon 4

To improve control on (resistant) difficult-to-control species such as red maple, white ash, and black cherry, apply a mixture of 7 parts WEEDONE® CB brand WOODY PLANT HERBICIDE and 1 part Garlon 4 Herbicide (7 pints + 1 pint). Cover the lower 18" to 24" of brush stem thoroughly as with other WEEDONE® CB brand READY-TO-USE WOODY PLANT HERBICIDE applications.

Add the required amount of Garlon 4 to the WEEDONE® CB brand READY-TO-USE WOODY PLANT HERBICIDE in the sprayer tank. Shake or stir the sprayer to thoroughly mix the two ingredients. Once mixed, the Garlon 4 and WEEDONE® CB brand READY-TO-USE WOODY PLANT HERBICIDE tank mixture should not need additional agitation.

WEEDONE CB/Tordon®

To improve control of pine and root-suckering species such as sassafras, black locust or sumac apply a mixture of 7 parts WEEDONE® CB brand READY-TO-USE WOODY PLANT HERBICIDE and 1 part Tordon® Herbicide. Cover the lower 18" to 24" of brush stem thoroughly as with other WEEDONE® CB brand READY-TO-USE WOODY PLANT HERBICIDE applications.

Add the required amount of Tordon® Herbicide to the WEEDONE® CB brand READY-TO-USE WOODY PLANT HERBICIDE in the sprayer tank. Shake or stir the sprayer to thoroughly mix the two ingredients and continue to agitate the sprayer every few minutes while spraying.

WEEDONE® CB brand READY-TO-USE WOODY PLANT HERBICIDE tank mixtures with Tordon® Herbicide will separate on standing. Use entire spray mixture, do not allow this tank mixture to remain in the sprayer overnight. Do not apply this tank mixture to drainage ditch banks.

WEEDONE® CB/Banvel OS

To improve control of pines, ash, persimmon, osageorange and certain other difficult-to-control species, apply a mixture of 7 parts WEEDONE® CB brand READY-TO-USE WOODY PLANT HERBICIDE and 1 part Banvel OS Herbicide. Cover the lower 18" to 24" of brush stem thoroughly as with other WEEDONE® CB brand READY-TO-USE WOODY PLANT HERBICIDE applications.

Add the required amount of Banvel OS to the WEEDONE® CB brand READY-TO-USE WOODY PLANT HERBICIDE in the sprayer tank. Shake or stir the sprayer to thoroughly mix the two ingredients. Once mixed, the Banvel OS and WEEDONE® CB brand READY-TO-USE WOODY PLANT HERBICIDE tank mixture should not need additional agitation.

Banvel Herbicide can be substituted for an equal amount of Banvel OS, but a suitable emulsifier or compatibility agent must be added to enable the Banvel Herbicide to disperse in the WEEDONE® CB brand READY-TO-USE WOODY PLANT HERBICIDE. Test suitable mixtures before using. Banvel Herbicide/WEEDONE® CB brand READY-TO-USE WOODY PLANT HERBICIDE tank mixtures will separate on standing, do not allow this tank mixture to remain in the sprayer overnight or for any prolonged period without agitation. Do not apply this tank mixture to drainage ditchbanks.

APPLICATION EQUIPMENT

WEEDONE® CB brand READY-TO-USE WOODY PLANT HERBICIDE can be applied from a hand pump back pack sprayer or a hand-carried pump-up pressurized sprayer such as Chapin Model 152, Sears 1443, or Birchmeier Iris 10, Cooper Pegler Model CP 15 or Solo models 425/475 (equipped with Aflas gaskets).

For most efficient use of the spray mixture, apply with a gun such as the B & G Extenda-wand equipped with a Spraying System 5500 Y series adjustable ConeJet nozzle. Equip all nozzles with a 200 mesh screen.

An alternate system which will reduce but not eliminate dripping between trigger releases is a 200 mesh TeeJet strainer and check valve number 4193A-SS having a spring opening that operates at 10 psi with the Spraying Systems 5500 Y series adjustable ConeJet nozzle mounted on a conventional basal spray wand.

Since WEEDONE® CB brand READY-TO-USE WOODY PLANT HERBICIDE is applied as a concentrate, the most efficient nozzle size will vary with the diameter of the brush stems to be treated:

> For average brush of 1-3 inches DBH, or for heavy stands of brush with most stems less than 1 inch DBH use the Spraying Systems 5500 adjustable ConeJet nozzle with a Y2 orifice.
>
> For scattered stands or large trees or brush or to treat large stumps, use the Spraying Systems 5500 adjustable ConeJet nozzle with a Y3 orifice.
>
> For small stems less than 1 inch diameter or to limit amount of WEEDONE® CB brand READY-TO-USE WOODY PLANT HERBICIDE applied, use the Spraying Systems 5500 adjustable ConeJet nozzle with a Y1 orifice.

Spray pressure of 15 psi is optimum for most use. Do not use spray pressure higher than 20 psi.

EQUIPMENT MAINTENANCE

WEEDONE® CB brand READY-TO-USE WOODY PLANT HERBICIDE may affect O rings and other seals in certain sprayer units. The most effective method of minimizing this problem is to drain the sprayer completely every evening. WEEDONE® CB brand READY-TO-USE WOODY PLANT HERBICIDE is a concentrate and can be returned to the original container.

If daily applicator draining cannot be performed, prolonged exposure of specific wetted rubber parts, starting after two weeks, will result in reduced applicator reliability. Viton, Nitrile Rubber, Buna N., Neoprene and EPDM are particularly susceptible to swelling after prolonged exposure to WEEDONE® CB brand READY-TO-USE WOODY PLANT HERBICIDE. Tests have shown Aflas-Poly (tetrafluorethylene-co-propylene) to have excellent long-term resistance to WEEDONE® CB brand READY-TO-USE WOODY PLANT HERBICIDE. Solo Inc., Newport News Virginia, telephone (804) 245-4228, currently provides Aflas replacement gaskets (part # 4900337) for the Solo Model 425 Applicator. Other applicator manufacturers are currently evaluating other alternative rubbers having long-term resistance to WEEDONE® CB brand READY-TO-USE WOODY PLANT HERBICIDE.

To prevent time loss in the field, have available replacement parts whenever you start out to spray.

Sprayers can be cleaned of WEEDONE® CB brand READY-TO-USE WOODY PLANT HERBICIDE at the end of the season, or prior to shifting to other herbicides, by first rinsing with kerosene, then following with a triple rinse of hot water and detergent. At the end of the spray season replace all O rings and/or seals showing signs of deterioration. Such a maintenance program will prolong the life of your equipment.

STORAGE AND DISPOSAL

STORAGE

Do not store near fertilizers, seeds, insecticides, or fungicides.

PESTICIDE DISPOSAL

Do not contaminate water, food, or feed by storage or disposal. Pesticide wastes are toxic. Improper disposal of excess pesticide, spray mixture, or rinsate is a violation of Federal law. If these wastes cannot be disposed of by use according to label instructions, contact your State Pesticide or Environmental Control Agency or the Hazardous Wastes representative at the nearest EPA Regional Office for guidance.

CONTAINER DISPOSAL

Triple rinse (or equivalent). Then offer for recycling or reconditioning, or puncture and dispose of in a sanitary landfill or by other procedures approved by state and local procedures.

LIMITED WARRANTY AND DISCLAIMER

The manufacturer warrants (a) that this product conforms to the chemical description on the label; (b) that this product is reasonably fit for the purposes set forth in the directions for use when it is used in accordance with such directions; and (c) that the directions, warning and other statements on this label are based upon responsible experts' evaluation of reasonable tests of effectiveness, of toxicity to laboratory animals and to plants, and of residues on food crops and upon reports of field experience. Tests have not been made on all varieties or in all states or under all conditions. THE MANUFACTURER NEITHER MAKES NOR INTENDS, NOR DOES IT AUTHORIZE ANY AGENT OR REPRESENTATIVE TO MAKE, ANY OTHER WARRANTIES, EXPRESS OR IMPLIED, AND IT EXPRESSLY EXCLUDES AND DISCLAIMS ALL IMPLIED WARRANTIES OF MERCHANTABILITY OR FITNESS FOR A PARTICULAR PURPOSE.

THIS WARRANTY DOES NOT EXTEND TO, AND THE BUYER SHALL BE SOLELY RESPONSIBLE FOR, ANY AND ALL LOSS OR

DAMAGE WHICH RESULTS FROM THE USE OF THIS PRODUCT IN ANY MANNER WHICH IS INCONSISTENT WITH THE LABEL DIRECTIONS, WARNINGS OR CAUTIONS.

BUYER'S EXCLUSIVE REMEDY AND MANUFACTURER'S OR SELLER'S EXCLUSIVE LIABILITY FOR ANY AND ALL CLAIMS, LOSSES, DAMAGES, OR INJURIES RESULTING FROM THE USE OR HANDLING OF THIS PRODUCT, WHETHER OR NOT BASED IN CONTRACT, NEGLIGENCE, STRICT LIABILITY IN TORT OR OTHERWISE SHALL BE LIMITED, AT THE MANUFACTURER'S OPTION, TO REPLACEMENT OF OR THE REPAYMENT OF THE PURCHASE PRICE FOR, THE QUANTITY OF PRODUCT WITH RESPECT TO WHICH DAMAGES ARE CLAIMED. IN NO EVENT SHALL MANUFACTURER OR SELLER BE LIABLE FOR SPECIAL, INDIRECT OR CONSEQUENTIAL DAMAGES RESULTING FROM THE USE OR HANDLING OF THIS PRODUCT.

NOTICE TO BUYER

Purchase of this material does not confer any rights under patents governing this product or the use thereof in countries outside of the United States.

WEEDONE® is a registered trademark of Rhône-Poulenc

Garlon and Tordon are registered trademarks of the Dow Chemical Company.

Banvel is a registered trademark of Sandoz.

Chapin is a registered trademark of R.E. Chapin Mfg. Works, Inc.

Birchmeier Iris is a registered trademark of Birchmeier & Company.

CP-15 is a registered trademark of Cooper Pegler and Company, Ltd.

Solo 425 is a registered trademark of Solo Incorporated.

Viton and Neoprene are registered trademarks of E.I. duPont de Nemours and Company and Copolymer Rubber and Chemical.

ConeJet and TeeJet are registered trademarks of Spraying Systems Company.

AG90029

EPA Approval 10/27/86

THIS SPECIMEN LABEL IS INTENDED FOR USE ONLY AS A GUIDE IN PROVIDING GENERAL INFORMATION REGARDING THE DIRECTIONS, WARNINGS AND CAUTIONS ASSOCIATED WITH THE USE OF THIS PRODUCT. AS WITH ANY AGRICULTURAL CHEMICAL, ALWAYS FOLLOW THE LABEL INSTRUCTIONS ON THE PACKAGE BEFORE USING.

Rohm and Haas Company

Independence Mall West
Philadelphia, PA 19105
(215) 592-3000

DIKAR®

FUNGICIDE-MITICIDE

NOT FOR SALE, DISTRIBUTION OR USE IN ARIZONA AND CALIFORNIA.

For Agricultural Use Only Or Only For Sale To, Use And Storage By Service Persons

ACTIVE INGREDIENTS:

A coordination product of zinc ion and manganese ethylene bisdithiocarbamate	72.0%
In which the ingredients are	
Manganese++	14.4%
Zinc++	1.8%
Ethylene bisdithiocarbamate	55.8%
2.4 Dinitro-6-octyl* phenyl crotonate	4.6%
2.6 Dinitro-4-octyl* phenyl crotonate Nitrooctyl* phenols (principally dinitro)	0.1%
INERT INGREDIENTS:	23.3%
TOTAL	100.0%

*A mixture of 1-methylheptyl l-ethylhexyl and 1-propylpentyl isomers

EPA Reg. No. 707-93

KEEP OUT OF REACH OF CHILDREN

CAUTION

NOTICE: Before using this product, read the entire Precautionary Statements, Conditions of Sale and Warranty, Directions for Use, Use Restrictions, and Storage and Disposal Instructions. If the Conditions of Sale and Warranty are not acceptable, return the product unopened within thirty days of purchase to the place of purchase.

USE OF THIS PRODUCT MAY BE HAZARDOUS TO YOUR HEALTH. THIS PRODUCT HAS BEEN DETERMINED TO CAUSE BIRTH DEFECTS IN LABORATORY ANIMALS.

PRECAUTIONARY STATEMENTS

HAZARDS TO HUMANS AND DOMESTIC ANIMALS

Harmful if swallowed, absorbed through the skin or inhaled. Causes moderate eye, skin, nose and throat irritation. Avoid contact with skin, eyes and clothing. Avoid breathing dust or spray mist.

Wash thoroughly with soap and water after handling, before eating or smoking and before using the toilet.

PROTECTIVE CLOTHING AND EQUIPMENT consisting of a long-sleeved shirt and long pants or a coverall that covers all parts of the body except the head, hands and feet; chemical-resistant gloves; shoes, socks and goggles or a face shield must be worn during handling.

See Re-entry and Worker Protection Statements on this label for additional precautions.

STATEMENTS OF PRACTICAL TREATMENT

IF SWALLOWED: Dilute by giving 2 glasses of water to drink and call a physician. Never give anything by mouth to an unconscious person.

IF INHALED: Move subject to fresh air.

IF ON SKIN: Wash exposed areas with plenty of soap and water. If irritation persists, call a physician.

IF IN EYES: Flush eyes with plenty of water for at least 15 minutes. If irritation persists, call a physician.

ENVIRONMENTAL HAZARDS

This pesticide is toxic to fish. Drift and runoff from treated areas may be hazardous to aquatic organisms in neighboring areas. Do not apply directly to water or wetlands (swamps, bogs, marshes and potholes). Do not contaminate water when disposing of equipment washwaters or disposing of wastes.

DIRECTIONS FOR USE

It is a violation of Federal law to use this product in a manner inconsistent with its labeling.

RE-ENTRY AND WORKER PROTECTION STATEMENTS

Do not enter treated areas without protective clothing until sprays have dried. After (sprays have dried/dusts have settled/vapors have dispersed, as applicable) **do not enter or allow entry into treated areas until the 24-hour re-entry interval has expired** unless wearing the personal protective equipment listed on the label. Because certain states may require more restrictive re-entry intervals for various crops treated with this product, consult your State Department of Agriculture for further information. Do not apply this product in such a manner as to directly or through drift expose workers or other persons. The area being treated must be vacated by unprotected persons.

Written or oral warnings must be given to workers who are expected to be in a treated area or in an area about to be treated with this product. Oral warnings must be given if there is reason to believe that written warnings cannot be understood by workers. When oral warnings are given, warnings shall be given in a language customarily understood by workers. Written warnings must include the following information:

CAUTION

(Insert area or field description) treated with DIKAR on (insert date of application). Do not enter treated areas without protective clothing until sprays have dried. After (sprays have dried/dusts have settled/vapors have dispersed, as applicable) **do not enter or allow entry into treated areas until the 24-hour re-entry interval has expired** unless wearing the personal protective equipment listed on the label. In case of accidental exposure: Call a doctor (physician), clinic or hospital immediately. Explain that the victim has been exposed to DIKAR and describe his condition. For further information see the "Statements of Practical Treatment."

Oral warnings must include the same information that is in the written warnings.

WORKER SAFETY RULES—Keep all unprotected persons, children, livestock and pets away from treated area or where there is danger of drift. Do not rub eyes or mouth with hands (see "Statements of Practical Treatment").

HANDLE THIS PRODUCT ONLY WHEN WEARING THE FOLLOWING PROTECTIVE CLOTHING AND EQUIPMENT: a long-sleeved shirt and long pants or a coverall that covers all parts of the body except the head, hands, and feet, chemical-resistant gloves, shoes, socks, and goggles or a face shield.

PERSONAL PROTECTIVE EQUIPMENT—For Mixers, Loaders, Applicators and Early Re-entry Workers.

During the mixing and loading of DIKAR fungicide, wear a protective suit of one or two pieces that covers all parts of the body except the head, hands, and feet. Wear chemical-resistant gloves, chemical-resistant apron, and chemical-resistant shoes, shoe coverings, or boots. Wear goggles or a face shield. A hood or hat may also be worn.

When used on field cucurbits, grapes and pears, application MUST be made from a vehicle with a completely enclosed cab. All vents and windows of the cab should remain closed during application. During application from a vehicle with a completely enclosed cab, wear a long-sleeved shirt and long pants. A clean set of protective clothing and equipment as defined here must be available in case you must exit the cab in treated areas for repairs or other reasons. Always wear protective gloves upon exiting the cab. Remove protective clothing and equipment before re-entering the cab to prevent cab contamination.

When used on apples, it is recommended that application be made from a vehicle with a completely enclosed cab. When used on apples, if application from a vehicle with a completely enclosed cab is not possible or during repair and cleanup of equipment or re-entry into treated areas prior to the 24-hour re-entry period, wear a protective suit of one or two pieces that covers all parts of the body except the hands, feet, and head. Wear chemical-resistant gloves and chemical-resistant shoes, shoe coverings or boots. A hood or hat may also be worn. During airblast or mist blower application a protective hood or wide brim hat must be worn.

WARNING: Wearing the protective suit for the prolonged periods required for application may produce heat stress at temperatures above 85°F on a cloudy day or 80°F on a sunny day. Application of this product must be conducted from enclosed cab vehicles when the temperature exceeds 85°F (80°F on sunny days) or delayed until the temperature drops.

IMPORTANT! Before removing gloves, wash with soap and water. Always wash hands, face, and arms with soap and water before eating, smoking or drinking. Always wash hands and arms before using the toilet. After application of the pesticide is completed, remove all clothes and shoes. Shower using soap and water. Redress only in clean clothes. Do not use contaminated clothing. Wash non-disposable protective clothing and protective equipment with soap or detergent and water after each use. Personal and non-disposable protective clothing worn during use must be laundered separately from household articles. Clothing and protective equipment heavily contaminated or drenched with DIKAR must be destroyed according to state and local regulations.

HEAVILY CONTAMINATED OR DRENCHED CLOTHING CANNOT BE ADEQUATELY DECONTAMINATED.

During aerial application, human flaggers are prohibited unless in totally enclosed vehicles.

CHEMIGATION

DO NOT APPLY THIS PRODUCT THROUGH ANY TYPE OF IRRIGATION SYSTEM.

GENERAL USE INFORMATION

DIKAR is a broad-spectrum protectant fungicide and mite suppressant recommended for use on specific fruit and vegetable crops. Optimum disease control and mite suppression is achieved when a regularly scheduled preventative spray program is followed. The addition of an agricultural surfactant will improve fungicide performance by providing a more uniform spray deposit, increased foliar redistribution, and fungicide retention during periods of wet weather.

NOTE: Under conditions of high temperature and humidity, DIKAR fungicide-miticide can injure the blossoms and foliage of some fruit and vegetable crops. **Do not apply to sensitive plant species when air temperature exceeds or is expected to exceed 90°F.**

USE RATE DETERMINATION

Carefully read, understand, and follow label use rates and restrictions.

Under low disease conditions, minimum label rates can be used while maximum label rates and shortened spray schedules are recommended for severe or threatening disease conditions.

For proper application, determine the number of acres to be treated, the recommended label use rate, and the gallonage to be applied per acre. Prepare only the amount of spray solution required to treat the measured acreage. Careful calibration of spray equipment is recommended prior to use.

When applied by hand sprayers, 1 pound DIKAR fungicide-miticide per 100 gallons per acre is equivalent to 1 level tablespoon per gallon spray solution.

MIXING

Slowly place into spray tank as it is being filled or thoroughly premix in a nurse tank for concentrate or aircraft sprayers. Add other co-applied fungicides, insecticides, growth regulators, micronutrients, and spray adjuvants after DIKAR fungicide-miticide.

When preparing spray solutions for use in a hand sprayer, premix as a slurry in a small container, and then add to sprayer containing 1/3 to 1/2 the desired final water volume. Use spray solutions immediately and do not leave standing overnight.

COMPATIBILITY

DIKAR fungicide-miticide is compatible with most commonly used agricultural fungicides, insecticides, and growth regulators except spray oils, calcium arsenate, dinitro acaracides and mildewcides, and phosphorodithioate acaracides. When preparing tank mixes, user should consult spray compatibility charts or State Extension Service Specialists prior to actual use.

SPRAY ADJUVANTS

The addition of agricultural surfactants to DIKAR fungicide-miticide sprays, will improve initial spray deposits, fungicide redistribution, and weatherability. The following spray adjuvants have been especially formulated to optimize the performance of foliar-applied agricultural chemicals:

TRITON AG-98®	A low foam, nonionic, general purpose spray adjuvant. May be used in dilute or concentrate sprays applied by aircraft or ground equipment.
TRITON® B-1956®	A water-dispersible, resin-based nonionic surfactant which resists re-wetting and removal by rain. Effective with dilute sprays applied by ground equipment.
TRITON CS-7®	A spreader-binder designed specifically for use in concentrate and low volume sprays applied by aircraft or ground equipment.

Place DIKAR fungicide-miticide into suspension prior to adding an adjuvant to the spray mixture. Read and carefully observe the precautionary statements and all other information appearing on both product labels prior to spray preparation.

DISEASE MONITORING

DIKAR is a broad-spectrum, protectant fungicide and mite suppressant. If not applied on a routine protectant spray schedule, crops should be scouted on a weekly basis. Application should be made at the recommended label use rate and spray schedule, at the first sign of or report of disease or mites in the area, or during environmental conditions favorable for disease or mite development.

If the spray schedule is extended beyond the schedules recommended in the use directions, and powdery mildew and mites are especially severe, 1 or 2 sprays of KARATHANE® LC or KARATHANE WD agricultural fungicide-miticide should be used between or with the regular DIKAR fungicide-miticide application for added powdery mildew and mite control.

RESTRICTIONS

Users should carefully read, understand, and follow all use restrictions prior to using DIKAR fungicide-miticide.

FRUITS

CROP	DISEASES CONTROLLED	RATE OF DIKAR FUNGICIDE-MITICIDE PER APPLICATION LB./A	LB./100 GALS.	REMARKS (ALSO REFER TO DIRECTIONS FOR USE)	RESTRICTIONS
Grapes	Black rot Bunch rot Deadarm Downy mildew Mite suppression Powdery mildew	1½ to 4		Apply in sufficient water to provide thorough coverage starting when new shoots are ½ to 1½ inches long. Repeat when shoots are 3 to 5 inches long, when shoots are 8 to 10 inches long, and then at 10 to 14 day intervals until fruit is set. **NOTE: Do not use DIKAR fungicide-miticide on grape varieties known to be sensitive to KARATHANE agricultural fungicide-miticide.** For further information, consult State Agricultural Experiment Station or State Extension Service specialists.	**Application MUST be made from an enclosed cab. Refer to Re-entry and Worker Protection Statements.** Do not apply within 66 days of harvest.

STORAGE AND DISPOSAL

STORAGE: Keep away from fire and sparks. Store in a cool, dry, well ventilated area. Do not allow to become wet or overheated in storage: decomposition, impaired activity, or fire may result. Keep container closed when not in use.

Pallets of containers should not be stacked more than three high. Provide access aisle for each two rows. Decomposition produces a foul odor; if observed, check for hot containers and remove immediately to open areas for disposal.

PESTICIDE DISPOSAL: Wastes resulting from the use of this product may be disposed of on-site or at an approved waste disposal facility.

CONTAINER DISPOSAL: Completely empty bag into application equipment. Then dispose of it in a sanitary landfill or by incineration, or, if allowed by state and local authorities, by burning. If burned, stay out of smoke.

STEPS TO BE TAKEN IN CASE MATERIAL IS RELEASED OR SPILLED: Sweep or shovel into containers for disposal or reworking. Keep dusting to a minimum. Flush contaminated area with a large amount of water to a chemical or sanitary sewer containing a settling pit.

CONDITIONS OF SALE AND WARRANTY

Rohm and Haas warrants that the product conforms to its chemical description and is reasonably fit for the purpose stated on the label only when used in accordance with label directions under normal conditions of use. **ROHM AND HAAS MAKES NO OTHER EXPRESS OR IMPLIED WARRANTIES EITHER OF MERCHANTABILITY OR FITNESS FOR A PARTICULAR USE.** Handling, storage and use of the product by Buyer or User are beyond the control of Rohm and Haas and Seller. Risks such as crop injury, ineffectiveness or other unintended consequences resulting from, but not limited to, weather or soil conditions, presence of other materials, disease, pests, drift to other crops or property or failure to follow label directions will be assumed by the Buyer or User. **IN NO CASE WILL ROHM AND HAAS OR SELLER BE HELD LIABLE FOR CONSEQUENTIAL, SPECIAL OR INDIRECT DAMAGES RESULTING FROM THE HANDLING, STORAGE OR USE OF THIS PRODUCT.**

DIKAR, KARATHANE, TRITON, AG-98, B-1956 and CS-7 are registered trademarks of the Rohm and Haas Company.

8118-B2

DITHANE® DF

AGRICULTURAL FUNGICIDE

ACTIVE INGREDIENTS
MANCOZEB
A coordination product of zinc ion and manganese ethylene bisdithiocarbamate 75%
In which the ingredients are:
Manganese++ 15.00%
Zinc++ 1.87%
Ethylene bisdithiocarbamate ion ($C_4H_6N_2S_4$) 58.13%
INERT INGREDIENTS 25%

TOTAL 100%

EPA Reg. No. 707-180

KEEP OUT OF REACH OF CHILDREN

CAUTION

NOTICE: Before using this product, read the entire Precautionary Statements, Conditions of Sale and Warranty, Directions for Use, Use Restrictions and Storage and Disposal Instructions. If the Conditions of Sale and Warranty are not acceptable, return the product unopened within thirty days of purchase to the place of purchase.

PRECAUTIONARY STATEMENTS

HAZARDS TO HUMANS AND DOMESTIC ANIMALS

May cause irritation of nose, throat, eyes and skin. Do not breathe dust or spray mist. **PROTECTIVE CLOTHING AND EQUIPMENT** consisting of a long-sleeved shirt and long pants or a coverall that covers all parts of the body except the head, hands and feet; chemical-resistant gloves; shoes, socks and goggles or a face shield must be worn during mixing and loading. During mixing and loading, a chemical-resistant apron must also be worn.

STATEMENTS OF PRACTICAL TREATMENT

IF SWALLOWED: Dilute by giving 2 glasses of water to drink and call a physician. Never give anything by mouth to an unconscious person.

IF INHALED: Move subject to fresh air.

IF IN EYES: Flush eyes with large amounts of water for at least 15 minutes. Consult a physician if irritation persists.

IF ON SKIN: Wash affected skin areas with soap and water.

ENVIRONMENTAL HAZARDS

This pesticide is toxic to fish. Drift and runoff from treated areas may be hazardous to aquatic organisms in neighboring areas. Do not apply directly to water or wetlands (swamps, bogs, marshes and potholes). Do not contaminate water when disposing of equipment washwaters or disposing of wastes. For Seed Treatment Products — cover or incorporate spilled treated seed.

DIRECTIONS FOR USE

It is a violation of Federal law to use this product in a manner inconsistent with its labeling.

RE-ENTRY AND WORKER PROTECTION STATEMENTS

WORKER SAFETY RULES—Keep all unprotected persons, children, livestock and pets away from treated area or where there is danger of drift.

Do not rub eyes or mouth with hands. See Statements of Practical Treatment.

PERSONAL PROTECTIVE EQUIPMENT — For Mixers, Loaders, Applicators and Early Re-entry Workers.

HANDLE THIS PRODUCT ONLY WHEN WEARING THE FOLLOWING PROTECTIVE CLOTHING AND EQUIPMENT: a long-sleeved shirt and long pants or a coverall that covers all parts of the body except the head, hands, and feet; chemical-resistant gloves; shoes, socks, and goggles or a face shield. During mixing and loading, a chemical-resistant apron must also be worn.

During application from a tractor with a completely enclosed cab with positive pressure filtration, or aerially with an enclosed cockpit, a long-sleeved shirt and long pants may be worn in place of the above protective clothing. Chemical-resistant gloves must be available in the cab or cockpit and worn while exiting.

IMPORTANT! Before removing gloves, wash them with soap and water. Always wash hands, face, and arms with soap and water before eating, smoking or drinking. Always wash hands and arms with soap and water before using the toilet.

After work take off all clothes and shoes. Shower using soap and water. Wear only clean clothes. Do not use contaminated clothing. Wash protective clothing and protective equipment with soap and water after each use. Personal clothing worn during use must be laundered separately from household articles. Clothing and protective equipment heavily contaminated or drenched with DITHANE must be destroyed according to state and local regulations.

HEAVILY CONTAMINATED OR DRENCHED CLOTHING CANNOT BE ADEQUATELY DECONTAMINATED.

During aerial application, human flaggers are prohibited unless in totally enclosed vehicles.

RE-ENTRY AND WORKER PROTECTION STATEMENTS

RE-ENTRY STATEMENTS

Written or oral warnings must be given to workers who are expected to be in a treated area or in an area about to be treated with this product. When oral warnings are given, warnings shall be given in a language customarily understood by workers. Oral warnings must be given if there is reason to believe that written warnings cannot be understood by workers. Warnings must include the following information:

CAUTION. Area treated with DITHANE on (date of application). Do not enter without appropriate protective clothing until sprays have dried for turf uses or within **24 hours** of application for agricultural uses. In case of contact, flush skin or eyes with plenty of water; for eyes, consult a physician if irritation persists.

For Agricultural Uses

After sprays have dried, **do not enter or allow entry into treated areas until the 24-hour re-entry interval has expired** unless wearing the personal protective equipment listed on this label.

For Turf Uses

Do not enter treated areas without protective clothing until sprays have dried.

GENERAL USE INFORMATION

DITHANE DF is a broad-spectrum protectant fungicide recommended for outdoor or greenhouse grown crops. Optimum disease control is achieved when the fungicide is applied in a regularly scheduled preventive spray program. The addition of an agricultural surfactant will improve fungicide performance by providing a more uniform spray deposit, increased foliar redistribution, and improved fungicide retention during periods of wet weather.

USE RATE DETERMINATION

Carefully read, understand, and follow label use rates and restrictions.

Under low disease conditions, minimum label rates per application can be used while maximum label rates and shortened spray schedules are recommended for severe or threatening disease conditions.

For proper application, determine the number of acres to be treated, the recommended label use rate, and the gallonage to be applied per acre. Prepare only the amount of spray solution required to treat the measured acreage. Careful calibration of spray equipment is recommended prior to use.

When applied by hand sprayers, 1 pound DITHANE DF fungicide per 100 gallons per acre is equivalent to 1 level tablespoon per gallon spray solution.

MIXING

Slowly place into spray tank as it is being filled or thoroughly premix in a nurse tank for concentrate or aircraft sprayers. Add other co-applied fungicides, insecticides, growth regulators, micronutrients, and spray adjuvants after DITHANE DF fungicide has been placed into suspension.

When preparing spray solutions for use in a hand sprayer, premix as a slurry in a small container, and then add to sprayer containing 1/3 to 1/2 the desired final water volume.

COMPATIBILITY

DITHANE DF fungicide is compatible with most commonly used agricultural fungicides, insecticides, and growth regulators. When preparing tank mixes, user should consult spray compatibility charts or State Cooperative Extension Service Specialists prior to actual use.

SPRAY ADJUVANTS

The addition of agricultural surfactants to DITHANE DF fungicide sprays will improve initial spray deposits, fungicide redistribution, and weatherability. The following spray adjuvants have been especially formulated to optimize the performance of foliar-applied agricultural chemicals:

TRITON® B-1956®	A water-dispersible, resin-based nonionic surfactant which resists re-wetting and removal by rain. Effective with dilute sprays applied by ground equipment.
TRITON CS-7®	A spreader-binder designed specifically for use in concentrate and low volume sprays applied by aircraft or ground equipment.

Place DITHANE DF fungicide into suspension prior to adding an adjuvant to the spray mixture. Read and carefully observe the precautionary statements and all other information appearing on both product labels prior to spray preparation.

APPLICATION

Ground

Thorough coverage foliar sprays generally result in optimum disease control. To achieve good coverage use proper spray pressure, gallonage per acre, nozzles (generally hollow cone), discs (generally D-5 to D-7), nozzle spacing, and tractor speed. Consult spray nozzle and accessory catalogues for specific information on proper equipment calibration.

Hand Sprayers

Thoroughly spray plant foliage until runoff.

Aerial

A uniform initial spray deposit over the crop canopy generally results in optimum disease control. Each aircraft should be prechecked for droplet size, uniformity of spray pattern, swath width, and spray volume.

Nozzle selection — Hollow cone brass nozzles with a D-series orifice disc and core (whirlplate) are recommended. Nozzles should point straight down or slightly backward.

Swath width — For most field and vegetable crops, swaths just beyond the wingspan of 36 to 40 feet for light aircraft and up to 45 feet for heavier aircraft are suggested. Optimum swath for helicopters is usually 5 to 10 feet beyond normal boom length.

Spray volume — On vegetable and field crops, 2 to 3 gallons of spray per acre are generally optimum; orchards and vineyards can be handled with spray volumes of 5 gallons per acre. Some tall or dense foliage crops, requiring greater penetration to the lower leaf surface, will require higher spray volumes. **Do not use less than 5 gallons per acre in California.**

Altitude — For most crops, the spray boom should be positioned 5 to 10 feet above the crop canopy.

Flagging — Swaths should be marked at the end of the field with permanent flags or by a flagman. Swaths should be measured accurately with a chain or other device except when rows can be accurately counted.

CHEMIGATION USE DIRECTIONS

Sprinkler Irrigation:

DITHANE DF fungicide must be applied on a regular protectant fungicide schedule, **not an irrigation schedule.** If irrigation cycles are less frequent than recommended DITHANE DF fungicide application intervals, ground or aerial applications must supplement chemigation applications to achieve adequate disease control.

Apply DITHANE DF fungicide only through sprinkler irrigation systems including center-pivot, lateral move, end tow, side (wheel) roll, traveler, solid set, or hand move irrigation systems. Do not apply product through any other type of irrigation system.

Lack of fungicidal effectiveness or illegal pesticide residues in the crop can result from nonuniform distribution of treated water.

If you have questions about calibration, you should contact State Extension Service specialists, equipment manufacturers or other experts.

Do not connect an irrigation system (including greenhouse systems) used for pesticide application to a public water system unless the pesticide label-prescribed safety devices for public water systems are in place.

A person knowledgeable of the chemigation system and responsible for its operation, or under the supervision of the responsible person, shall shut the system down and make necessary adjustments should the need arise.

Before applying DITHANE DF fungicide through sprinkler irrigation equipment, the chemigation system must meet the following specifications:

- Public water system means a system for the provision to the public of piped water for human consumption if such system has at least 15 service connections or regularly serves an average of at least 25 individuals daily at least 60 days out of the year.
- Chemigation systems connected to public water systems must contain a functional reduced-pressure zone (RPZ), backflow preventer or the functional equivalent in the water supply line upstream from the point of pesticide introduction. As an option to the RPZ, the water from the public water system should be discharged into a reservoir tank prior to pesticide introduction. There shall be a complete physical break (air gap) between the outlet end of the fill pipe and the top or overflow rim of the reservoir tank of at least twice the inside diameter of the fill pipe.
- Systems not connected to a public water supply must contain a functional check valve, vacuum relief valve, and low-pressure drain appropriately located in the irrigation pipeline to prevent water source contamination from back flow.
- The pesticide injection pipeline must contain a functional, automatic, quick-closing check valve to prevent the flow of fluid back toward the injection pump.
- The pesticide injection pipeline must also contain a functional, normally closed, solenoid-operated valve located on the intake side of the injection pump and connected to the system interlock to prevent fluid from being withdrawn from the supply tank when the irrigation system is either automatically or manually shut down.
- The system must contain functional interlocking controls to automatically shut off the pesticide injection pump when the water pump motor stops, or in cases where there is no water pump, when the water pressure decreases to the point where pesticide distribution is adversely affected.
- The irrigation line or water pump must include a functional pressure switch which will stop the water pump motor when the water pressure decreases to the point where pesticide distribution is adversely affected.
- Systems must use a metering pump, such as a positive displacement injection pump (e.g., diaphragm pump) effectively designed and constructed of materials that are compatible with pesticides and capable of being fitted with a system interlock.
- Do not apply when wind speed favors drift beyond the area intended for treatment.

Center-pivot, Lateral Move, End Tow, and Traveler Irrigation Equipment: (Use only with electric or oil hydraulic drive systems which provide a uniform water distribution.)

- Determine size of area to be treated.
- Determine the time required to apply no more than 1/4 inch water (6,750 gallons water per acre) over the area to be treated when the system and injection equipment are operated at normal pressures recommended by the equipment manufacturer. Run system at 80 to 95% of manufacturer's rated capacity.
- Using only water, determine the injection pump output when operated at normal line pressure.
- Determine the amount of DITHANE DF fungicide required to treat area.
- Add the required amount of DITHANE DF fungicide and sufficient water to meet the injection time requirements of the solution tank.
- Maintain constant solution tank agitation during the injection period.
- Stop injection equipment after treatment is completed. Continue to operate the system until DITHANE DF fungicide solution has cleared the sprinkler head.

Solid-set, Side (wheel) Roll, and Hand Move Irrigation Equipment:

- Determine acreage covered by sprinkler.
- Fill injector solution tank with water and adjust flow rate to use contents over a 10 to 30 minute interval.
- Determine the amount of DITHANE DF fungicide required to treat area.
- Add the required amount of DITHANE DF fungicide into the same quantity of water used to calibrate the injection equipment.
- Maintain constant solution tank agitation during the injection period.
- Operate system at normal pressures recommended by the manufacturer of the injection equipment and used for the time interval established during calibration.
- Inject DITHANE DF fungicide at the end of the irrigation cycle or as a separate application to maximize foliar fungicide retention.
- Stop injection equipment after treatment is completed. Continue to operate the system until DITHANE DF fungicide solution has cleared the last sprinkler head.

DISEASE MONITORING

DITHANE DF fungicide is a broad-spectrum, protectant fungicide. If not applied on a routine protectant spray schedule, crops should be scouted on a weekly basis. Fungicide application should be made, at the recommended label use rate and spray schedule, at the first sign of disease, report of disease in the area, or during environmental conditions favorable for disease development.

RESTRICTIONS

Users should carefully read, understand, and follow all use restrictions prior to using DITHANE DF fungicide.

VEGETABLES

CROP	DISEASES CONTROLLED	RATE OF DITHANE DF PER APPLICATION LB./A	REMARKS (Also Refer To Directions For Use)	RESTRICTIONS
Asparagus	Rust Cercospora leaf spot	2	Start applications when rust first appears and repeat at 10 day intervals. Four applications are usually sufficient.	Apply only on asparagus ferns after spears have been harvested.

CROP	DISEASES CONTROLLED	RATE OF DITHANE DF PER APPLICATION LB./A	REMARKS (Also Refer To Directions For Use)	RESTRICTIONS
Corn (sweet, popcorn)	Common rust Helminthosporium leaf blights	1½	Use sufficient water for thorough coverage. Start applications when disease first appears and repeat at 4 to 7 day intervals. The addition of TRITON surfactants to spray solutions will improve performance.	Do not apply within 7 days of harvest. Do not feed treated forage to livestock.
Onions (dry bulb)	Botrytis leaf blight Downy mildew Neck rot Purple blotch	3	Follow a protective spray schedule starting when diseases are first reported in the area and repeat at 7 day intervals throughout the season. The addition of a TRITON surfactant to spray solutions will improve performance. Do not allow spray or drift to contact bulbs after lifting from soil.	Do not apply within 7 days of harvest.
(furrow drench)	Smut		Apply 3 lbs. per acre as a furrow drench at time of planting onion seeds. Use 75 to 125 gallons of water per acre.	Do not use more than 2.4 lbs. actual per acre (29,000 linear feet of furrow) with an 18 inch row spacing. Do not use in California.
Potatoes	Early blight Late blight	1 to 2	Begin application when plants are 4 to 6 inches high by applying 1 lb. per acre. As the vines increase in size, apply 1½ to 2 lbs. per acre on a 7 to 10 day spray schedule throughout the season. Reduce the spray interval under heavy disease conditions or weather conditions favorable for rapid disease development. The addition of a TRITON surfactant to spray solutions will improve performance.	
Seedpiece (treatment)	Fusarium decay Seedborne common scab		Dip whole or cut potato tubers in 1⅓ lbs. DITHANE DF fungicide per 50 gallons of water. Place treated tubers in a clean container following treatment and plant as soon as possible. Spread treated seedpieces in a cool place if held before planting.	Do not use treated seed potatoes for food or feed purposes.
Tomatoes	Anthracnose Early blight Gray leaf spot Late blight Leaf mold Septoria leaf spot	1½ to 3¼	Start applications when seedlings emerge or transplants are set and repeat at 7 day intervals throughout the season. The addition of a TRITON surfactant to spray solutions will improve performance.	Do not apply within 5 days of harvest.

FRUITS

CROP	DISEASES CONTROLLED	RATE OF DITHANE DF PER APPLICATION LB./A	LB./100 GALS.	REMARKS (Also Refer To Directions For Use)	RESTRICTIONS
Bananas	Sigatoka	2 to 4	—	Apply when leaves first appear and repeat every 2 to 3 weeks or as required. Use sufficient water to provide adequate coverage. The addition of a TRITON surfactant to spray solutions will improve performance.	
Cranberries	Fruit rot	3 to 6	—	Start applications at mid-bloom and repeat at 7 to 10 day intervals as required.	Do not apply within 30 days of harvest.
Grapes (California only)	Black rot Bunch rot Deadarm Downy mildew	1½ to 4	—	Apply in sufficient water to provide thorough coverage starting when new shoots are ½ to 1½ inches long. Repeat when shoots are 3 to 5 inches long, when shoots are 8 to 10 inches long, and then at 10 to 14 day intervals until fruit is set. For late season control of black rot, deadarm and downy mildew the use of other approved and recommended fungicides is suggested.	In Calif only, Do not apply after fruit is set. In other areas Do not apply within 66 days of harvest.

FIELD CROPS

CROP	DISEASES CONTROLLED	RATE OF DITHANE DF PER APPLICATION LB./A	REMARKS (Also Refer To Directions For Use)	RESTRICTIONS
Oats	Refer to wheat			
Peanuts	Cercospora leaf spot Rust	1 to 2	Start application when disease first appears or is reported in area. Repeat sprays at 10 to 14 day intervals. Reduce sprays to a 7 day interval during humid weather.	Do not apply within 14 days of harvest. Do not feed treated vines to livestock.
Sugar Beets	Cercospora leaf spot	1½ to 2	Start application when disease first threatens and repeat every 7 to 10 days as needed. The addition of a TRITON surfactant to spray solutions will improve performance.	Do not apply within 14 days of harvest. Do not feed treated tops to livestock.

CROP	DISEASES CONTROLLED	RATE OF DITHANE DF PER APPLICATION LB./A	REMARKS (Also Refer To Directions For Use)	RESTRICTIONS
Wheat	Helminthosporium blights and leaf spot Leaf rust Septoria leaf spot Septoria glume blotch Tan spot	2	Start application at onset of disease or when plants are in the tillering to jointing stage and repeat at 7 to 10 day intervals. The addition of TRITON CS-7 to spray solutions will improve performance.	Do not make more than three applications during the season. Do not apply within 26 days after application. Do not graze livestock in treated areas within 26 days after application.

SEED TREATMENT

CROP	DISEASES CONTROLLED	SEED TREATMENT RATE FOR DITHANE DF PER APPLICATION OZ./BU.	OZ./100 LB.	REMARKS (Also Refer To Directions For Use)	RESTRICTIONS
GENERAL USE INSTRUCTIONS				**Seeds to be treated should be cleaned and well cured prior to treatment. DITHANE DF fungicide may be applied to dry seed with conventional slurry seed treating equipment. For best results, the seed must be completely and uniformly covered with fungicide. For seed treatment, a dye must be added to DITHANE DF fungicide which will impart an unnatural color to the seed.**	**Treated seed should be labeled, "Must not be used for food, feed, or oil purposes."**
Corn (field)	Damping-off Seed rots Seedling blights	1.6 to 3.2	2.9 to 5.8		
Cotton (acid delinted)	Damping-off Seedling blights	—	3.2		
(reginned)	Damping-off Seedling blights	—	6.4		
Flax	Damping-off Seed rots Seedling blights	2.1 to 4.3	3.8 to 7.6		
Peanuts (shelled)	Damping-off Seed rots Seedling blights	2.1 to 4.3	8.5 to 17.0		
Rice	Damping-off Seed rots Seedling blights	—	2.1 to 4.3	Apply before, during or after soaking in water.	
Safflower	Seedborne rust (*Puccinia carthami*)		2.1		
Sorghum	Covered kernel smut Damping-off Seed rots Seedling blights	1.6 to 2.7	2.9 to 4.8		
Tomatoes	Damping-off Seed rots Seedling blights		8.5		

MISCELLANEOUS

CROP	DISEASES CONTROLLED	RATE OF DITHANE DF PER APPLICATION	REMARKS (Also Refer To Directions For Use)
Asparagus crowns	Crown rot	1 lb. per 100 gals.	Place loosely packed crowns into a burlap bag and soak, with gentle agitation, in the fungicide solution for 5 minutes. Remove bag, drain well, and plant crowns as soon as possible. A tank large enough to hold a single burlap bag will treat 2 bags of crowns. Clean dipping suspension should then be prepared in a clean tank. Dirty crowns should be pre-washed to remove excess soil.
Caprifig	Assorted molds Endosepsis (Fusarium)	1 lb. per 25 gals.	Prepare mamme figs by making a shallow cut through the eye and then hand dividing to avoid wasp injury. Submerge mamme figs in the fungicide suspension for a minimum of 15 minutes. The fungicide suspension should be stirred frequently to prevent settling out. Fresh dipping solutions should be used after treating 4 or 5 batches of figs. After treatment, figs should be drained prior to placement in trees.

TURF

CROP	DISEASES CONTROLLED	RATE OF DITHANE DF PER APPLICATION OZ./1000 SQ. FT.	REMARKS (Also Refer To Directions For Use)	RESTRICTIONS
GENERAL USE INSTRUCTIONS			**Start application when grass greens-up in spring or when disease first appears and repeat at 7 to 14 day intervals or until disease threat is past.** **When conditions are especially favorable for disease development, apply maximum fungicide use rate on a 7 day spray schedule.** **Apply in sufficient water to provide adequate coverage.**	**Do not graze treated areas. Do not feed clippings to livestock.**
Assorted Grasses	Helminthosporium melting-out Rust (leaf, stem, stripe)	4		
	Copper spot Fusarium blight Powdery mildew Red thread Slime mold	4 to 8		
	Algae	6		
	Dollar spot	6 to 8		
	Rhizoctonia brown patch	4	Apply on a 7 day spray schedule.	
	Pythium blight	8	Begin applications at first sign of disease and repeat applications at 5 day intervals, or more frequently, if conditions are especially favorable for disease development.	
	Fusarium snow mold	6 to 8	Apply at 2 to 6 week intervals during winter.	

ORNAMENTALS

GENERAL USE INSTRUCTIONS Neither the manufacturer nor the seller has determined the effects of using DITHANE DF fungicide on ornamentals not specified on this label.

Prior to any large-scale application on such plants, the user should determine the effects of DITHANE DF fungicide by testing a small section of the type of plants to be treated. User assumes all risks arising out of application to unlabeled plants. The Conditions of Sale and Warranty apply to all uses.

For outdoor or greenhouse use, apply the equivalent of 1½ lbs. DITHANE DF fungicide per 100 gals dilute spray. The addition of TRITON B-1956 to spray solutions will improve performance.

Begin spraying when plants are well leafed out or at first sign of disease, in a full coverage spray at 7 to 10 day intervals throughout season or follow State Extension Service recommendations for disease control on the following ornamental plants:

CROP	DISEASES CONTROLLED	REMARKS (Also Refer To Directions For Use)	RESTRICTIONS
African violet	Botrytis blight		
Anthurium	Anthracnose, spadix rot		
Arborritae	Cercospora blight		
Ash, mountain	Entomosporium leaf spot Guignardia leaf blotch		
Ash, white	Anthracnose Cylindrosporium leaf spot		
Aster, perennial	Puccinia rusts		
Aucuba, japonica	Alternaria leaf spot Anthracnose		
Azalea	Cylindrocladium rot Petal blight Phytophthora twig and bud blight	Apply in a full coverage spray, 2 to 3 times a week, while flowers are opening. Direct spray into flowers and thoroughly spray ground under bushes.	
Begonia	Botrytis blight		
Buffaloberry	Cylindrosporium leaf spot		Do not use fruit for food or feed purposes.
Camellias	Petal blight	Refer to azalea.	
Carnation	Rust Septoria leaf spot		
Cedar, red (Juniper)	Cercospora blight Phomopsis blight		
Chrysanthemum	Ascochyta blight Botrytis petal spot Rust	Apply twice weekly during blooming period.	

CROP	DISEASES CONTROLLED	REMARKS (Also Refer To Directions For Use)	RESTRICTIONS
Conifers	Lophodermium needle cast Pine gall rust Scirrhia brown spot	Begin application in spring or early summer before infection occurs. Repeat after heavy rains and at two-week intervals as long as needed.	
Cordyline	Cercospora leaf spot		
Crabapple (Ornamental)	Cedar-apple rust Scab Sphaeropsis leaf spot		
Cypress, Arizona (Cupressus sp.)	Cercospora blight Monochaetia canker		
Dahlia	Botrytis blight		
Dieffenbachia	Leptosphaeria brown spot		
Dogwood, flowering	Anthracnose Elsinoe leaf spot Septoria leaf spot	Apply when buds begin to open, when bracts have fallen, 4 weeks later and again in late summer after flower buds for next season have formed.	
Dracaena	Fusarium leaf spot		
Elm	Black leaf spot		
Euonymus	Anthracnose		
Fatsia	Anthracnose		
Fern	Rhizoctonia blight		
Ficus	Cercospora leaf spot		
Firethorn (Pyracantha)	Fusicladium scab		
Fir, Douglas	Swiss needle cast		
Fuchsia	Botrytis blight Rust		
Geranium	Rust		
Gladiolus	Botrytis blossom blight Curvularia leaf spot	Make regular weekly applications starting before diseases appear and increase to 2 or 3 applications per week during periods of heavy disease and during rainy weather. On flower spikes, reduce spray concentration to 3/4 lb. per 100 gals.	
Hawthorn	Cedar-apple rust Fabraea leaf spot Frogeye leaf spot Hawthorn rust Scab		
Holly	Purple spot		
Hollyhock	Anthracnose Cercospora leaf spot Puccinia rusts		
Honeysuckle	Herpobasidium blight		
Horsechestnut, Buckeye	Alternaria leaf spot Guignardia leaf blotch		
Hydrangea	Botrytis blight Cercospora leaf spot		
Iris	Didymellina leaf spot Mystrosporium ink spot		
Juniper	Phomopsis blight		
Laurel, mountain	Cercospora leaf spot Petal blight	Refer to azalea.	
Ligustrum	Cercospora leaf spot		
Lily	Botrytis blight		
Magnolia	Gloeosporium leaf spot		
Maple	Alternaria leaf spot Phyllosticta leaf spot		Do not use on Sugar maples intended for the production of maple syrup.
Marigold	Botrytis blossom blight	Do not use on French dwarf double or Signet type marigolds .	
Narcissus	Botrytis blight (fire) Smoulder		
Oak	Actinopelte leaf spot Taphrina leaf blister		
Orchid (Dendrobium)	Botrytis blossom blight		
Pachysandra	Volutella blight	Use a drenching spray of 2 lbs. per 50 gallons of water per 5,000 sq. ft. of bed. Start application at first sign of disease and apply at least 5 applications at 10 to 14 day intervals.	
Pansy	Anthracnose		
Peony	Botrytis blossom blight Phytophthora blight	Apply in early spring and early fall, drenching soil around plants as well as the foliage. Promptly destroy all infected plant parts.	
Peperomia	Cercospora leaf spot		
Philodendron	Dactylaria leaf spot Phytophthora leaf spot		

CROP	DISEASES CONTROLLED	REMARKS (Also Refer To Directions For Use)	RESTRICTIONS
Photinia	Entomosporium leaf spot		
Pleomele	Fusarium leaf spot		
Poinsettia	Sphaceloma scab		
Rhododendron	Cercospora leaf spot Discosia leaf spot Petal blight	Refer to azalea.	
Rose	Black spot Cercospora leaf spot Rust		
Schefflera	Alternaria blight		
Skunkbush, sumac	Cylindrosporium leaf spot		
Snapdragon	Rust		
Statice	Cercospora frogeye		
Syngonium	Cephalosporium leaf spot		
Tulip	Botrytis blight (fire)		
Venus, flytrap	Anthracnose		
Viburnum	Downy mildew Ramularia leaf spot		
Walnut	Anthracnose		Do not use treated walnuts for food or feed purposes.
Zinnia	Alternaria leaf blight		

FUNGICIDE TANK MIXES

The following tank mix recommendations are made for the improved control of specific diseases. Before mixing DITHANE DF fungicide with any other fungicide, read and carefully observe the precautionary and all other information appearing on both product labels. User must abide by the most restrictive limitations appearing on the product labels.

CROP	CO-APPLIED FUNGICIDE	FOR IMPROVED CONTROL OF:	USE RATE PER APPLICATION	DITHANE DF USE RATE PER APPLICATION	REMARKS (ALSO REFER TO DIRECTIONS FOR USE)
Grapes	BAYLETON 50 WP	Powdery mildew	2 to 6 ozs./A	1½ to 4 lbs./A	
	KARATHANE WD		1 to 2½ lbs./A		Do not use in California.
	KARATHANE LC		4 to 8 fl. ozs./100 gals.		
	BENLATE 50 WP*		¾ to 1 lb./A		
Peanuts	BENLATE 50 WP*	Ascochyta web blotch Early leaf spot Late leaf spot	4 ozs./A	1 to 2 lbs./A	
	TOPSIN-M70 WP*				
Potatoes	Triphenyl Tin Hydroxide (TPTH)	Early blight	½ to full label use rate	1½ to 2 lbs./A	
Roses, Flowers, Ornamentals, and shade trees	BENLATE 50 WP*	Leaf spots Powdery mildew	½ to 1 lb./100 gals.	1½ lb./100 gals.	
	TOPSIN-M70 WP*				
	KARATHANE WD	Mite suppression Powdery mildew	6 to 8 ozs./100 gals.		
Sugar beets	BENLATE 50 WP*	Cercospora leaf spot	4 ozs./A	1½ to 2 lbs./A	
	TOPSIN-M70 WP*				
	MERTECT 340-F*		6 fl. ozs./A		
	Triphenyl Tin Hydroxide (TPTH)		½ to full label rate		
Tomatoes	Fixed Coppers	Bacterial leaf spot	Manufacturer's label use rate	1½ to 3¼ lbs./A	
	DYRENE 50 WP	Gray leaf spot	½ to 1 lb./A		
	BENLATE 50 WP*	Botrytis gray mold Sclerotinia white mold			
Wheat	BAYLETON 50 WP	Powdery mildew Rust (leaf, stem, stripe)	2 to 4 ozs./A	2 lbs./A	

* Do not use if fungal resistance to this fungicide is present in field to be treated or has been reported in the area.

STORAGE AND DISPOSAL

STORAGE: Keep away from fire and sparks. Store in a cool, dry, well-ventilated area. Do not allow to become wet or overheated in storage: decomposition, impaired activity, or fire may result. Keep container closed when not in use.

Pallets of containers should not be stacked more than three high. Provide access aisle for each two rows. Decomposition produces a foul odor; if observed, check for hot containers and immediately remove to open areas for disposal.

PESTICIDE DISPOSAL: Wastes resulting from the use of this product may be disposed of on site or at an approved waste disposal facility.

CONTAINER DISPOSAL: Completely empty bag into application equipment. Then dispose of it in a sanitary landfill or by incineration, or, if allowed by state and local authorities, by burning. If burned, stay out of smoke.

STEPS TO BE TAKEN IN CASE MATERIAL IS RELEASED OR SPILLED: Sweep or shovel into containers for disposal or reworking. Keep dusting to a minimum. Flush contaminated area with a large amount of water to a chemical or sanitary sewer containing a settling pit. Refer to Precautionary Statements.

CONDITIONS OF SALE AND WARRANTY

Rohm and Haas warrants that the product conforms to its chemical description and is reasonably fit for the purpose stated on the label only when used in accordance with label directions under normal conditions of use. **ROHM AND HAAS MAKES NO OTHER EXPRESS OR IMPLIED WARRANTIES EITHER OF MERCHANTABILITY OR FITNESS FOR A PARTICULAR USE.** Handling, storage and use of the product by Buyer or User are beyond the control of Rohm and Haas and Seller. Risks such as crop injury, ineffectiveness or other unintended consequences resulting from, but not limited to, weather or soil conditions, presence of other materials, disease, pests, drift to other crops or property or failure to follow label directions will be assumed by the Buyer or User. **IN NO CASE WILL ROHM AND HAAS OR SELLER BE HELD LIABLE FOR CONSEQUENTIAL, SPECIAL OR INDIRECT DAMAGES RESULTING FROM THE HANDLING, STORAGE OR USE OF THIS PRODUCT.**

DITHANE, KARATHANE, TRITON, AG-98, B-1956 and CS-7 are registered trademarks of the Rohm and Haas Company.
BAYLETON and DYRENE are registered trademarks of the parent company of Farbenfabriken Bayer GmbH, Leverkusen.
BENLATE is a registered trademark of E.I. duPont de Nemours and Co., Inc.
MERTECT is a registered trademark of Merck and Co., Inc.
TOPSIN-M is a registered trademark of Nippon Soda Co., Ltd.
8170-B2

DITHANE F-45®

FLOWABLE MANCOZEB AGRICULTURAL FUNGICIDE

ACTIVE INGREDIENT:		
MANCOZEB		
A coordination product of zinc ion and manganese ethylene bisdithiocarbamate		37%*
In which the ingredients are:		
Manganese++	7.4%	
Zinc++	0.9%	
Ethylene bisdithiocarbamate ion ($C_4H_6N_2S_4$)	28.7%	
INERT INGREDIENTS:		63%
TOTAL		100%

*Equivalent to 4 lbs. active ingredient per gallon.

EPA Reg. No. 707-156

NOTICE: Before using this product, read the entire Precautionary Statements, Conditions of Sale and Warranty, Directions for Use, Use Restrictions, and Storage and Disposal Instructions. If the Conditions of Sale and Warranty are not acceptable, return the product unopened within thirty days of purchase to the place of purchase.

KEEP OUT OF REACH OF CHILDREN
CAUTION

PRECAUTIONARY STATEMENTS

HAZARDS TO HUMANS AND DOMESTIC ANIMALS

May cause irritation of nose, throat, eyes and skin. Avoid contact with skin, eyes or clothing. Avoid breathing spray mist. In case of contact immediately flush eyes or skin with plenty of water. Get medical attention if irritation persists.

PROTECTIVE CLOTHING AND EQUIPMENT consisting of a long-sleeved shirt and long pants or a coverall that covers all parts of the body except the head, hands and feet; chemical-resistant gloves; shoes, socks and goggles or a face shield must be worn during mixing and loading. During mixing and loading, a chemical-resistant apron must also be worn.

STATEMENTS OF PRACTICAL TREATMENT

IF SWALLOWED: Dilute by giving 2 glasses of water to drink and call a physician. Never give anything by mouth to an unconscious person.

IF INHALED: Move subject to fresh air.

IF IN EYES: Flush eyes with large amounts of water for at least 15 minutes. Consult a physician if irritation persists.

IF ON SKIN: Wash affected skin areas with soap and water.

ENVIRONMENTAL HAZARDS

This pesticide is toxic to fish. Drift and runoff from treated areas may be hazardous to aquatic organisms in neighboring areas. Do not apply directly to water or wetlands (swamps, bogs, marshes and potholes). Do not contaminate water when disposing of equipment washwaters or disposing of wastes. For Seed Treatment Products—cover or incorporate spilled treated seed.

PHYSICAL OR CHEMICAL HAZARDS

Keep away from fire and sparks. Keep from freezing.

DIRECTIONS FOR USE

It is a violation of Federal law to use this product in a manner inconsistent with its labeling.

RE-ENTRY AND WORKER PROTECTION STATEMENTS

WORKER SAFETY RULES—Keep all unprotected persons, children, livestock and pets away from treated area or where there is danger of drift. Do not rub eyes or mouth with hands. See Statements of Practical Treatment.

PERSONAL PROTECTIVE EQUIPMENT—For Mixers, Loaders, Applicators and Early Re-entry Workers.

HANDLE THIS PRODUCT ONLY WHEN WEARING THE FOLLOWING PROTECTIVE CLOTHING AND EQUIPMENT: a long-sleeved shirt and long pants or a coverall that covers all parts of the body except the head, hands, and feet; chemical-resistant gloves; shoes, socks, and goggles or a face shield. During mixing and loading, a chemical-resistant apron must also be worn.

During application from a tractor with a completely enclosed cab with positive pressure filtration, or aerially with an enclosed cockpit, a long-sleeved shirt and long pants may be worn in place of the above protective clothing. Chemical-resistant gloves must be available in the cab or cockpit and worn while exiting.

IMPORTANT! Before removing gloves, wash them with soap and water. Always wash hands, face, and arms with soap and water before eating, smoking or drinking. Always wash hands and arms with soap and water before using the toilet.

After work take off all clothes and shoes. Shower using soap and water. Wear only clean clothes. Do not use contaminated clothing. Wash protective clothing and protective equipment with soap and water after each use. Personal clothing worn during use must be laundered separately from household articles. Clothing and protective equipment heavily contaminated or drenched with DITHANE must be destroyed according to state and local regulations.

HEAVILY CONTAMINATED OR DRENCHED CLOTHING CANNOT BE ADEQUATELY DECONTAMINATED.

During aerial application, human flaggers are prohibited unless in totally enclosed vehicles.

RE-ENTRY STATEMENTS

Written or oral warnings must be given to workers who are expected to be in a treated area or in an area about to be treated with this product. When oral warnings are given, warnings shall be given in a language customarily understood by workers. Oral warnings must be given if there is reason to believe that written warnings cannot be understood by workers. Warnings must include the following information:

CAUTION

Area treated with DITHANE on (date of application). Do not enter without appropriate protective clothing until sprays have dried for turf uses or within **24 hours** of application for agricultural uses. In case of contact, flush skin or eyes with plenty of water; for eyes, consult a physician if irritation persists.

For Agricultural Uses

After sprays have dried, **do not enter or allow entry into treated areas until the 24-hour re-entry interval has expired** unless wearing the personal protective equipment listed on this label.

For Turf Uses

Do not enter treated areas without protective clothing until sprays have dried.

GENERAL USE INFORMATION

DITHANE F-45 is a broad-spectrum protectant fungicide recommended for outdoor or greenhouse grown crops. Optimum disease control is achieved when the fungicide is applied in a regularly scheduled preventive spray program. The addition of an agricultural surfactant will improve fungicide performance by providing a more uniform spray deposit, increased foliar redistribution, and improved fungicide retention during periods of wet weather.

USE RATE DETERMINATION

Carefully read, understand, and follow label use rates and restrictions.

Under low disease conditions, minimum label rates per application can be used while maximum label rates and shortened spray schedules are recommended for severe or threatening disease conditions.

For proper application, determine the number of acres to be treated, the recommended label use rate, and the gallonage to be applied per acre. Prepare only the amount of spray solution required to treat the measured acreage. Careful calibration of spray equipment is recommended prior to use.

If only a portion of the container's contents are to be used, thoroughly shake the container prior to measuring. When small quantities of spray solution are being prepared for use in hand or power sprayers, the following conversion table should be followed (rates are based on dilute thorough coverage sprays):

Recommended Label Use Rate Per* Acre or 100 Gals.*	Fluid Ounces DITHANE F-45 Flowable Agricultural Fungicide Required For:			
	10 Gals.	5 Gals.	2 Gals.	1 Gal.
0.5 qt.	1.6	0.8	0.3	0.2
1.0 qt.	3.2	1.6	0.7	0.3
1.5 qts.	4.8	2.4	1.1	0.4
2.0 qts.	6.4	3.2	1.3	0.6
2.5 qts.	8.0	4.0	1.6	0.8
3.0 qts.	9.6	4.8	2.0	0.9
3.5 qts.	11.2	5.6	2.3	1.1
4.0 qts.	12.8	6.4	2.6	1.3

1 cup = 8 fluid ounces or 237 milliliters
1 fluid ounce = 2 tablespoons or 30 milliliters
1 tablespoon = 3 teaspoons or 15 milliliters
* Dilute thorough coverage sprays.

MIXING

Slowly place into spray tank as it is being filled or thoroughly premix in a nurse tank for concentrate or aircraft sprayers. Add other co-applied fungicides, insecticides, growth regulators, micronutrients, and spray adjuvants after DITHANE F-45 flowable fungicide has been placed into suspension.

When preparing spray solutions for use in a hand sprayer, premix as a slurry in a small container, and then add to sprayer containing 1/3 to 1/2 the desired final water volume.

COMPATIBILITY

DITHANE F-45 flowable fungicide is compatible with most commonly used agricultural fungicides, insecticides, and growth regulators. When preparing tank mixes, user should consult spray compatibility charts or State Cooperative Extension Service Specialists prior to actual use.

SPRAY ADJUVANTS

The addition of agricultural surfactants to DITHANE F-45 flowable fungicide sprays will improve initial spray deposits, fungicide redistribution, and weatherability. The following spray adjuvants have been especially formulated to optimize the performance of foliar-applied agricultural chemicals:

TRITON® B-1956® A water-dispersible, resin-based nonionic surfactant which resists re-wetting and removal by rain. Effective with dilute sprays applied by ground equipment.

TRITON CS-7® A spreader-binder designed specifically for use in concentrate and low volume sprays applied by aircraft or ground equipment.

Place DITHANE F-45 flowable fungicide into suspension prior to adding an adjuvant to the spray mixture. Read and carefully observe the precautionary statements and all other information appearing on both product labels prior to spray preparation.

APPLICATION

Ground

Thorough coverage foliar sprays generally result in optimum disease control. To achieve good coverage use proper spray pressure, gallonage per acre, nozzles (generally hollow cone), discs (generally D-5 to D-7), nozzle spacing, and tractor speed. Consult spray nozzle and accessory catalogues for specific information on proper equipment calibration.

Hand sprayers

Thoroughly spray plant foliage until run-off.

Aerial

A uniform initial spray deposit over the crop canopy generally results in optimum disease control. Each aircraft should be prechecked for droplet size, uniformity of spray pattern, swath width, and spray volume.

Nozzle selection—Hollow cone brass nozzles with a D series orifice disc and core (whirlplate) are recommended. Nozzles should point straight down or slightly backward.

Swath width—For most field and vegetable crops, swaths just beyond the wingspan of 36 to 40 feet for light aircraft and up to 45 feet for heavier aircraft are suggested. Optimum swath for helicopters is usually 5 to 10 feet beyond normal boom length.

Spray volume—On vegetable and field crops, 2 to 3 gallons of spray per acre are generally optimum; orchards and vineyards can be handled with spray volumes of 5 gallons per acre. Some tall or dense foliage crops, requiring greater penetration to the lower leaf surface, will require higher spray volumes. **Do not use less than 5 gallons per acre in California.**

Altitude—For most crops, the spray boom should be positioned 5 to 10 feet above the crop canopy.

Flagging—Swaths should be marked at the end of the field with permanent flags or by a flagman. Swaths should be measured accurately with a chain or other device except when rows can be accurately counted.

CHEMIGATION USE DIRECTIONS

Sprinkler Irrigation

DITHANE F-45 fungicide must be applied on a regular protectant fungicide schedule, **not an irrigation schedule.** If irrigation cycles are less frequent than recommended DITHANE F-45 fungicide application intervals, ground or aerial applications must supplement chemigation applications to achieve adequate disease control.

Apply DITHANE F-45 fungicide only through sprinkler irrigation systems including center-pivot, lateral move, end tow, side (wheel) roll, traveler, solid set, or hand move irrigation systems. Do not apply product through any other type of irrigation system.

Lack of fungicidal effectiveness or illegal pesticide residues in the crop can result from nonuniform distribution of treated water.

If you have questions about calibration, you should contact State Extension Service specialists, equipment manufacturers or other experts.

Do not connect an irrigation system (including greenhouse systems) used for pesticide application to a public water system unless the pesticide label-prescribed safety devices for public water systems are in place.

A person knowledgeable of the chemigation system and responsible for its operation, or under the supervision of the responsible person, shall shut the system down and make necessary adjustments should the need arise.

Before applying DITHANE F-45 fungicide through sprinkler irrigation equipment, the chemigation system must meet the following specifications:

- Public water system means a system for the provision to the public of piped water for human consumption if such system has at least 15 service connections or regularly serves an average of at least 25 individuals daily at least 60 days out of the year.
- Chemigation systems connected to public water systems must contain a functional reduced-pressure zone (RPZ), backflow preventer or the functional equivalent in the water supply line upstream from the point of pesticide introduction. As an option to the RPZ, the water from the public water system should be discharged into a reservoir tank prior to pesticide introduction. There shall be a complete physical break (air gap) between the outlet end of the fill pipe and the top or overflow rim of the reservoir tank of at least twice the inside diameter of the fill pipe.
- Systems not connected to a public water supply must contain a functional check valve, vacuum relief valve, and low-pressure drain appropriately located in the irrigation pipeline to prevent water source contamination from back flow.
- The pesticide injection pipeline must contain a functional, automatic, quick-closing check valve to prevent the flow of fluid back toward the injection pump.
- The pesticide injection pipeline must also contain a functional, normally closed, solenoid-operated valve located on the intake side of the injection pump and connected to the system interlock to prevent fluid from being withdrawn from the supply tank when the irrigation system is either automatically or manually shut down.
- The system must contain functional interlocking controls to automatically shut off the pesticide injection pump when the water pump motor stops, or in cases where there is no water pump, when the water pressure decreases to the point where pesticide distribution is adversely affected.
- The irrigation line or water pump must include a functional pressure switch which will stop the water pump motor when the water pressure decreases to the point where pesticide distribution is adversely affected.
- Systems must use a metering pump, such as a positive displacement injection pump (e.g., diaphragm pump) effectively designed and constructed of materials that are compatible with pesticides and capable of being fitted with a system interlock.
- Do not apply when wind speed favors drift beyond the area intended for treatment.

Center-pivot, Lateral Move, End Tow, and Traveler Irrigation Equipment: (Use only with electric or oil hydraulic drive systems which provide a uniform water distribution.)

- Determine size of area to be treated.
- Determine the time required to apply no more than 1/4 inch water (6,750 gallons water per acre) over the area to be treated when the system and injection equipment are operated at normal pressures recommended by the equipment manufacturer. Run system at 80 to 95% of manufacturer's rated capacity.
- Using only water, determine the injection pump output when operated at normal line pressure.

- Determine the amount of DITHANE F-45 fungicide required to treat area.
- Add the required amount of DITHANE F-45 fungicide and sufficient water to meet the injection time requirements of the solution tank.
- Maintain constant solution tank agitation during the injection period.
- Stop injection equipment after treatment is completed. Continue to operate the system until DITHANE F-45 fungicide solution has cleared the sprinkler head.

Solid-set, Side (wheel) Roll, and Hand Move Irrigation Equipment:

- Determine acreage covered by sprinkler.
- Fill injector solution tank with water and adjust flow rate to use contents over a 10 to 30 minute interval.
- Determine the amount of DITHANE F-45 fungicide required to treat area.
- Add the required amount of DITHANE F-45 fungicide into the same quantity of water used to calibrate the injection equipment.
- Maintain constant solution tank agitation during the injection period.
- Operate system at normal pressures recommended by the manufacturer of the injection equipment and used for the time interval established during calibration.
- Inject DITHANE F-45 fungicide at the end of the irrigation cycle or as a separate application to maximize foliar fungicide retention.
- Stop injection equipment after treatment is completed. Continue to operate the system until DITHANE F-45 fungicide solution has cleared the last sprinkler head.

DISEASE MONITORING

DITHANE F-45 flowable fungicide is a broad-spectrum, protectant fungicide. If it is not applied on a routine protectant spray schedule, crops should be scouted on a weekly basis. Fungicide application should be made, at the recommended label use rate and spray schedule, at the first sign of disease, report of disease in the area, or during environmental conditions favorable for disease development.

RESTRICTIONS

Users should carefully read, understand, and follow all use restrictions prior to using DITHANE F-45 flowable fungicide.

VEGETABLES

CROP	DISEASES CONTROLLED	RATE OF DITHANE F-45 FLOWABLE PER APPLICATION QTS./A	REMARKS (ALSO REFER TO DIRECTIONS FOR USE)	RESTRICTIONS
Asparagus	Cercospora leaf spot Rust	1.6	Start applications when rust first appears and repeat at 10 day intervals. Four applications are usually sufficient.	Apply only on asparagus ferns after spears have been harvested.
Corn (sweet, popcorn)	Common rust Helminthosporium leaf blight	1.2	Use sufficient water for thorough coverage. Start applications when disease first appears and repeat at 4 to 7 day intervals. The addition of a TRITON surfactant to spray solutions will improve performance.	Do not apply within 7 days of harvest. Do not feed treated forage to livestock.
Onions (dry bulb)	Botrytis leaf blight Downy mildew Neck rot Purple blotch	2.4	Follow a protective spray schedule starting when diseases are first reported in the area and repeat at 7 day intervals throughout the season. The addition of a TRITON surfactant to spray solutions will improve performance. Do not allow spray or drift to contact bulbs after lifting from soil.	Do no apply within 7 days of harvest.
(furrow drench)	Smut		Apply 2.4 qts. per acre as a furrow drench at time of planting onion seeds. Use 75 to 125 gallons water per acre.	Do not use more than 2.4 qts. actual per acre (29,000 linear feet of furrow). Do not use in California.
Potatoes	Early blight Late blight	0.8 to 1.6	Begin application when plants are 4 to 6 inches high by applying 0.8 qt. per acre. As the vines increase in size, apply 1.2 to 1.6 qts. per acre. The addition of a TRITON surfactant to spray solutions will improve performance.	
Seedpiece (treatment)	Fusarium decay Seedborne common scab		Dip whole or cut potato tubers in 1 qt. DITHANE F-45 flowable fungicide per 50 gallons of water. Place treated tubers in a clean container following treatment and plant as soon as possible. Spread treated seedpieces in a cool place if held before planting.	Do not use treated seed potatoes for food or feed purposes.
Tomatoes	Anthracnose Early blight Gray leaf spot Late blight Leaf mold Septoria leaf spot	1.2 to 2.4	Start applications when seedlings emerge or transplants are set and repeat at 7 day intervals throughout the season. The addition of a TRITON surfactant to spray solutions will improve performance.	Do not apply within 5 days of harvest.

FRUITS

CROP	DISEASES CONTROLLED	RATE OF DITHANE F-45 FLOWABLE PER APPLICATION QTS./A	QTS./100 GALS.	REMARKS (ALSO REFER TO DIRECTIONS FOR USE)	RESTRICTIONS
Bananas	Sigatoka	1.6 to 3.2		Apply when leaves first appear and repeat every 2 to 3 weeks or as required. Use sufficient water to provide adequate coverage. The addition of a TRITON surfactant to spray solutions will improve performance.	
Cranberries	Fruit rot	2.4 to 4.8		Start applications at mid-bloom and repeat at 7 to 10 day intervals as required.	Do not apply within 30 days of harvest.

CROP	DISEASES CONTROLLED	RATE OF DITHANE F-45 FLOWABLE PER APPLICATION QTS./A	QTS./100 GALS.	REMARKS (ALSO REFER TO DIRECTIONS FOR USE)	RESTRICTIONS
Grapes	Black rot Bunch rot Deadarm Downy mildew	1.2 to 3.2		Apply in sufficient water to provide thorough coverage starting when new shoots are 1/2 to 1 1/2 inches long. Repeat when shoots are 3 to 5 inches long, when shoots are 8 to 10 inches long, and then at 10 to 14 day intervals until fruit is set or 66 days before harvest. For late season control of black rot, deadarm and downy mildew the use of other approved and recommended fungicides is suggested.	In California do not apply after fruit is set. In other area do not apply within 66 days of harvest.

FIELD CROPS

CROP	DISEASES CONTROLLED	RATE OF DITHANE F-45 FLOWABLE PER APPLICATION QTS./A	REMARKS (ALSO REFER TO DIRECTIONS FOR USE)	RESTRICTIONS
Peanuts	Cercospora leaf spot Rust	0.8 to 1.6	Start applications when disease first appears or is reported in area. Repeat sprays at 10 to 14 day intervals. Reduce sprays to a 7 day interval during humid weather.	Do not apply within 14 days of harvest. Do not feed treated vines to livestock.
Sugar Beets	Cercospora leaf spot	1.2 to 1.6	Start application when disease first threatens and repeat every 7 to 10 days as needed. The addition of a TRITON surfactant to spray solutions will improve performance.	Do not apply within 14 days of harvest. Do not feed treated tops to livestock.
Wheat	Helminthosporium leaf spot Leaf rust Septoria glume blotch Septoria leaf spot Tan spot	1.6	Start application at onset of disease or when plants are in the tillering to jointing stage and repeat at 7 to 10 day intervals. The addition of TRITON CS-7 to spray solutions will improve performance.	Do not make more than three applications during the season. Do not apply within 26 days of harvest. Do not graze livestock in treated areas within 26 days after application.

SEED TREATMENT

CROP	DISEASES CONTROLLED	RATE OF DITHANE F-45 FLOWABLE PER APPLICATION FL. OZ./BU.	FL. OZ./100 LB.	REMARKS (ALSO REFER TO DIRECTIONS FOR USE)	RESTRICTIONS
GENERAL USE INSTRUCTIONS				**Seeds to be treated should be cleaned and well cured prior to treatment.** **DITHANE F-45 flowable fungicide may be applied to dry seed with conventional slurry or mist seed treating equipment or as a planter-box application. For best results, the seed must be completely and uniformly covered with fungicide.** **For seed treatment, a dye must be added to DITHANE F-45 flowable fungicide that will impart an unnatural color to the seed.**	**Treated seed should be labeled, "Must not be used for food, feed, or oil purposes."**
Corn (field)	Damping-off Seed rots Seedling blights	2.4 to 4.8	4.3 to 8.6		
Cotton (acid delinted)	Damping-off Seedling blights		4.8		
(reginned)	Damping-off Seedling blights		9.6		
Flax	Damping-off Seed rots Seedling blights	3.2 to 6.4	5.7 to 11.3		
Peanuts (shelled)	Damping-off Seed rots Seedling blights	3.2 to 6.4	12.8 to 25.6		
Rice	Damping-off Seed rots Seedling blights		3.2 to 6.4	Apply before, during or after soaking in water.	
Safflower	Seedborne rust (Puccinia carthami)		3.2		

CROP	DISEASES CONTROLLED	RATE OF DITHANE F-45 FLOWABLE PER APPLICATION FL. OZ./BU.	FL. OZ./100 LB.	REMARKS (ALSO REFER TO DIRECTIONS FOR USE)	RESTRICTIONS
Sorghum	Covered kernel smut Damping-off Seed rots Seedling blights	2.4 to 4.0	4.3 to 7.2		
Tomatoes	Damping-off Seed rots Seedling blights		12.8		

MISCELLANEOUS

CROP	DISEASES CONTROLLED	RATE OF DITHANE F-45 FLOWABLE PER APPLICATION	REMARKS (ALSO REFER TO DIRECTIONS FOR USE)
Asparagus crowns	Crown rot	0.8 qts. per 100 gals.	Place loosely packed crowns into a burlap bag and soak, with gentle agitation, in the fungicide solution for 5 minutes. Remove bag, drain well, and plant crowns as soon as possible. A tank large enough to hold a single burlap bag will treat 2 bags of crowns. Clean dipping suspension should then be prepared in a clean tank. Dirty crowns should be pre-washed to remove excess soil.
Caprifig	Endosepsis (Fusarium) Assorted molds	0.8 qts. per 25 gals.	Prepare mamme figs by making a shallow cut through the eye and then hand dividing to avoid wasp injury. Submerge mamme figs in the fungicide suspension for a minimum of 15 minutes. The fungicide suspension should be stirred frequently to prevent settling out. Fresh dipping solutions should be used after treating 4 or 5 batches of figs. After treatment, figs should be drained prior to placement in trees.

TURF

CROP	DISEASES CONTROLLED	RATE OF DITHANE F-45 FLOWABLE PER APPLICATION OZ/1000 SQ. FT.	REMARKS (ALSO REFER TO DIRECTIONS FOR USE)	RESTRICTIONS
GENERAL USE INSTRUCTIONS			**Start application when grass greens-up in spring or when disease first appears and repeat at 7 to 14 day intervals or until disease threat is past.** **When conditions are especially favorable for disease development, apply maximum fungicide use rate on a 7 day spray schedule.** **Apply in sufficient water to provide adequate coverage.**	**Do not graze treated areas.** **Do not feed clippings to livestock.**
Assorted Grasses	Helminthosporium melting-out Rust (leaf, stem, stripe)	6.4		
	Copper spot Fusarium blight Red thread Slime mold	6.4 to 12.8		
	Algae	9.6		
	Dollar spot	9.6 to 12.8		
	Rhizoctonia brown patch	6.4	Apply on a 7 day spray schedule.	
	Pythium blight	12.8	Apply at 5 day intervals, or more frequently, if conditions are especially favorable for disease development.	
	Fusarium snow mold	9.6 to 12.8	Apply at 2 to 6 week intervals during winter.	

ORNAMENTALS

CROP	DISEASES CONTROLLED	REMARKS (ALSO REFER TO DIRECTIONS FOR USE)	RESTRICTIONS
GENERAL USE INSTRUCTIONS Neither the manufacturer nor the seller has determined the effects of using DITHANE F-45 flowable fungicide on ornamentals not specified on this label. **Prior to any large-scale application on such plants, the user should determine the effects of DITHANE F-45 flowable fungicide by testing a small section of the type of plants to be treated. User assumes all risks arising out of application to unlabeled plants. The Conditions of Sale and Warranty apply to all uses.** **For outdoor or greenhouse use apply the equivalent of 1.2 qts. DITHANE F-45 flowable fungicide per 100 gals. dilute spray. The addition of TRITON B-1956 to spray solutions will improve performance.** **Begin spraying when plants are well leafed out or at first sign of disease, in a full coverage spray at 7 to 10 day intervals throughout season or follow State Extension Service recommendations for disease control on the following ornamental plants:**			

CROP	DISEASES CONTROLLED	REMARKS (ALSO REFER TO DIRECTIONS FOR USE)	RESTRICTIONS
African violet	Botrytis blight		
Anthurium	Anthracnose, spadix rot		
Arborvitae	Cercospora blight		
Ash, mountain	Entomosporium leaf spot Guignardia leaf blotch		
Ash, white	Anthracnose Cylindrosporium leaf spot		
Aster, perennial	Puccinia rusts		
Aucuba, japonica	Alternaria leaf spot Anthracnose		
Azalea	Cylindrocladium rot Petal blight Phytophthora twig and bud blight	Apply in a full coverage spray, 2 to 3 times a week, while flowers are opening. Direct spray into flowers and thoroughly spray ground under bushes.	
Begonia	Botrytis blight		
Buffaloberry	Cylindrosporium leaf spot		Do not use fruit for food or feed purposes.
Camellias	Petal blight	Refer to azalea.	
Carnation	Rust Septoria leaf spot		
Cedar, red (Juniper)	Cercospora blight Phomopsis blight		
Chrysanthemum	Ascochyta blight Botrytis petal spot Rust	Apply twice weekly during blooming period.	
Conifers	Lophodermium needle cast Pine gall rust Scirrhia brown spot	Begin application in spring or early summer before infection occurs. Repeat after heavy rains and at two-week intervals as long as needed.	
Cordyline	Cercospora leaf spot		
Crabapple (Ornamental)	Cedar-apple rust Scab Sphaeropsis leaf spot		
Cypress Arizona (Cupressus sp.)	Cercospora blight Monochaetia canker		
Dahlia	Botrytis blight		
Dieffenbachia	Leptosphaeria brown spot		
Dogwood, flowering	Anthracnose Elsinoe leaf spot Septoria leaf spot	Apply when buds begin to open, when bracts have fallen, 4 weeks later and again in late summer after flower buds for next season have formed.	
Dracaena	Fusarium leaf spot		
Elm	Black leaf spot		
Euonymus	Anthracnose		
Fatsia	Anthracnose		
Fern	Rhizoctonia blight		
Ficus	Cercospora leaf spot		
Firethorn (Pyracantha)	Fusicladium scab		
Fir, Douglas	Swiss needle cast		
Fuchsia	Botrytis blight Rust		
Geranium	Rust		
Gladiolus	Botrytis blossom blight Curvularia leaf spot	Make regular weekly applications starting before diseases appear and increase to 2 or 3 applications per week during periods of heavy disease and during rainy weather. On flower spikes, reduce spray concentration to 0.6 qts. per 100 gals.	
Hawthorn	Cedar-apple rust Fabraea leaf spot Frogeye leaf spot Hawthorn rust Scab		
Holly	Purple spot		
Hollyhock	Anthracnose Cercospora leaf spot Puccinia rusts		
Honeysuckle	Herpobasidium blight		
Horsechestnut, buckeye	Alternaria leaf spot Guignardia leaf blotch		
Hydrangea	Botrytis blight Cercospora leaf spot		
Iris	Didymellina leaf spot Mystrosporium ink spot		
Juniper	Phomopsis blight		

CROP	DISEASES CONTROLLED	REMARKS (ALSO REFER TO DIRECTIONS FOR USE)	RESTRICTIONS
Laurel, mountain	Cercospora leaf spot Petal blight	Refer to azalea.	
Ligustrum	Cercospora leaf spot		
Lily	Botrytis blight		
Magnolia	Gloeosporium leaf spot		
Maple	Alternaria leaf spot Phyllosticta leaf spot		Do not use on Sugar maples intended for the production of maple syrup.
Marigold	Botrytis blossom blight	Do not use on French dwarf or Signet type marigold seedlings.	
Narcissus	Botrytis blight (fire) Smoulder		
Oak	Actinopelte leaf spot Taphrina leaf blister		
Orchid (Dendrobium)	Botrytis blossom blight		
Pachysandra	Volutella blight	Use a drenching spray of 1.6 qts. per 50 gallons of water per 5,000 sq. ft. of bed. Start application at first sign of disease and apply at least 5 applications at 10 to 14 day intervals.	
Pansy	Anthracnose		
Peony	Botrytis blossom blight Phytophthora blight	Apply in early spring and early fall, drenching soil around plants as well as the foliage. Promptly destroy all infected plant parts.	
Peperomia	Cercospora leaf spot		
Philodendron	Dactylaria leaf spot Phytophthora leaf spot		
Photinia	Entomosporium leaf spot		
Pleomele	Fusarium leaf spot		
Poinsettia	Sphaceloma scab		
Rhododendron	Cercospora leaf spot Discosia leaf spot Petal blight	Refer to azalea.	
Rose	Black spot Cercospora leaf spot Rust		
Schefflera	Alternaria blight		
Skunkbush, sumac	Cylindrosporium leaf spot		
Snapdragon	Rust		
Statice	Cercospora frogeye		
Syngonium	Cephalosporium leaf spot		
Tulip	Botrytis blight (fire)		
Venus, flytrap	Anthracnose		
Viburnum	Downy mildew Ramularia leaf spot		
Walnut	Anthracnose		Do not use treated walnuts for food or feed purposes.
Zinnia	Alternaria leaf blight		

FUNGICIDE TANK MIXES:
The following tank mix recommendations are made for the improved control of specific diseases. Before mixing DITHANE F-45 flowable with any other fungicide, read and carefully observe the precautionary and all other information appearing on both product labels. User must abide by the most restrictive limitations appearing on the product labels.

CROP	CO-APPLIED FUNGICIDE	FOR IMPROVED CONTROL OF:	USE RATE PER APPLICATION	DITHANE F-45 FLOWABLE USE RATE PER APPLICATION	REMARKS (Also Refer To Directions For Use)
Grapes	BAYLETON 50 WP	Powdery mildew	2 to 6 ozs./A	1.2 to 3.2 qts./A	
	KARATHANE WD		1 to 2½ lbs./A		Do not use in California.
	KARATHANE LC		4 to 8 fl. ozs./100 gals.		
	BENLATE 50 WP*		¾ to 1 lb./A		
Peanuts	BENLATE 50 WP*	Ascochyta web blotch Early leaf spot Late leaf spot	4 ozs./A	0.8 to 1.6 qts./A	
	TOPSIN-M70 WP*	Early leaf spot			
Potatoes	Triphenyl Tin Hydroxide (TPTH)	Early blight	½ to full label use rate	1.2 to 1.6 qts./A	
Sugar Beets	BENLATE 50 WP*	Cercospora leaf spot	4 ozs./A	1.2 to 1.6 qts./A	
	TOPSIN-M70 WP*				
	MERTECT 340-F*		6 fl. ozs./A		

CROP	CO-APPLIED FUNGICIDE	FOR IMPROVED CONTROL OF:	USE RATE PER APPLICATION	DITHANE F-45 FLOWABLE USE RATE PER APPLICATION	REMARKS (Also Refer To Directions For Use)
	Triphenyl Tin Hydroxide (TPTH)		½ to full label rate		
Tomatoes	Fixed Coppers	Bacterial leaf spot	Manufacturer's label use rate	1.2 to 2.4 qts./A	
	DYRENE 50 WP	Gray leaf spot	½ to 1 lb./A		
	BENLATE 50 WP*	Botrytis gray mold Sclerotinia white mold			
Wheat	BAYLETON 50 WP	Powdery mildew Rust (leaf, stem, stripe)	2 to 4 ozs./A	1.6 qts./A	

* Do not use if fungal resistance to this fungicide is present in field to be treated or has been reported in the area.

STORAGE AND DISPOSAL

STORAGE: Keep away from fire and sparks. Store in a cool, well ventilated area, but not below 32°F. Do not allow to become overheated in storage. This may bring on chemical changes which will impair the fungicidal effectiveness of DITHANE F-45 flowable fungicide. Keep container closed when not in use.

PESTICIDE DISPOSAL: Wastes resulting from the use of this product may be disposed of on-site or at an approved waste disposal facility.

CONTAINER DISPOSAL

Plastic Containers—Triple rinse (or equivalent). Then puncture and dispose of in a sanitary landfill, by incineration, or, if allowed by State and local authorities, by burning. If burned, stay out of smoke. **Fiber Drum (with liner)**—Completely empty liner by shaking and tapping sides and bottom to loosen clinging particles. Dispose of liner in a sanitary landfill or by incineration if allowed by State and local authorities. If drum cannot be reused, dispose of in the same manner.

Metal Containers—Triple rinse (or equivalent) then offer for recycling or reconditioning or puncture and dispose of in a sanitary landfill or by other procedures approved by State and local authorities.

Bulk Containers—Drain thoroughly and return to specified destination for cleaning and re-use.

STEPS TO BE TAKEN IN CASE MATERIAL IS RELEASED OR SPILLED: Dike and contain the spill. Transfer liquid and solid diking material to separate containers for recovery or disposal. Flush contaminated area with a large amount of water to a chemical or sanitary sewer containing a settling pit. Remove contaminated clothing and wash affected skin areas with soap and water. Wash clothing before reuse. Keep the solids out of the muncipal sewers and open bodies of water. Refer to Precautionary Statements.

CONDITIONS OF SALE AND WARRANTY

Rohm and Haas warrants that the product conforms to its chemical description and is reasonably fit for the purpose stated on the label only when used in accordance with label directions under normal conditions of use. **ROHM AND HAAS MAKES NO OTHER EXPRESS OR IMPLIED WARRANTIES EITHER OF MERCHANTABILITY OR FITNESS FOR A PARTICULAR USE.** Handling, storage and use of the product by Buyer or User are beyond the control of Rohm and Haas and Seller. Risks such as crop injury, ineffectiveness or other unintended consequences resulting from, but not limited to, weather or soil conditions, presence of other materials, disease, pests, drift to other crops or property or failure to follow label directions will be assumed by the Buyer or User. **IN NO CASE WILL ROHM AND HAAS OR SELLER BE HELD LIABLE FOR CONSEQUENTIAL, SPECIAL OR INDIRECT DAMAGES RESULTING FROM THE HANDLING, STORAGE OR USE OF THIS PRODUCT.**

DITHANE, F-45, KARATHANE, TRITON, AG-98, B-1956 and CS-7 are registered trademarks of the Rohm and Haas Company.

BAYLETON and DYRENE are registered trademarks of the parent company of Farbenfabriken Bayer GmbH, Leverkusen.

BENLATE is a registered trademark of E.I. duPont de Nemours and Co., Inc.

MERTECT is a registered trademark of Merck and Co., Inc.

TOPSIN-M is a registered trademark of Nippon Soda Co., Ltd.

9646-B7

DITHANE M-45™

AGRICULTURAL FUNGICIDE

ACTIVE INGREDIENTS:
MANCOZEB

A coordination product of zinc ion and manganese ethylene bisdithiocarbamate	80%
In which the ingredients are:	
Manganese++	16%
Zinc++	2%
Ethylene bisdithiocarbamate ion ($C_4H_6N_2S_4$)	62%
INERT INGREDIENTS:	20%
TOTAL	100%

EPA Reg. No. 707-78

NOTICE: Before using this product, read the entire Precautionary Statements, Conditions of Sale and Warranty, Directions for Use, Use Restrictions and Storage and Disposal Instructions. If the Conditions of Sale and Warranty are not acceptable, return the product unopened within thirty days of purchase to the place of purchase.

KEEP OUT OF REACH OF CHILDREN

CAUTION

PRECAUTIONARY STATEMENTS

HAZARDS TO HUMANS AND DOMESTIC ANIMALS

May cause irritation of nose, throat, eyes and skin. Do not breathe dust or spray mist. **PROTECTIVE CLOTHING AND EQUIPMENT** consisting of a long-sleeved shirt and long pants or a coverall that covers all parts of the body except the head, hands and feet; chemical-resistant gloves; shoes, socks and goggles or a face shield must be worn during mixing and loading. During mixing and loading, a chemical-resistant apron must also be worn.

STATEMENTS OF PRACTICAL TREATMENT

IF SWALLOWED: Dilute by giving 2 glasses of water to drink and call a physician. Never give anything by mouth to an unconscious person.

IF INHALED: Move subject to fresh air.

IF IN EYES: Flush eyes with large amounts of water for at least 15 minutes. Consult a physician if irritation persists.

IF ON SKIN: Wash affected skin areas with soap and water.

ENVIRONMENTAL HAZARDS

This pesticide is toxic to fish. Drift and runoff from treated areas may be hazardous to aquatic organisms in neighboring areas. Do not apply directly to water or wetlands (swamps, bogs, marshes and potholes). Do not contaminate water when disposing of equipment washwaters or disposing of wastes. For Seed Treatment Products—cover or incorporate spilled treated seed.

DIRECTIONS FOR USE

It is a violation of Federal law to use this product in a manner inconsistent with its labeling.

RE-ENTRY AND WORKER PROTECTION STATEMENTS

WORKER SAFETY RULES—Keep all unprotected persons, children, livestock and pets away from treated area or where there is danger of drift.

Do not rub eyes or mouth with hands. See Statements of Practical Treatment.

PERSONAL PROTECTIVE EQUIPMENT—For Mixers, Loaders, Applicators and Early Re-entry Workers.

HANDLE THIS PRODUCT ONLY WHEN WEARING THE FOLLOWING PROTECTIVE CLOTHING AND EQUIPMENT: a long-sleeved shirt and long pants or a coverall that covers all parts of the body except the head, hands, and feet; chemical-resistant gloves; shoes, socks, and goggles or a face shield. During mixing and loading, a chemical-resistant apron must also be worn.

During application from a tractor with a completely enclosed cab with positive pressure filtration, or aerially with an enclosed cockpit, a long-sleeved shirt and long pants may be worn in place of the above protective clothing. Chemical-resistant gloves must be available in the cab or cockpit and worn while exiting.

IMPORTANT! Before removing gloves, wash them with soap and water. Always wash hands, face, and arms with soap and water before eating, smoking or drinking. Always wash hands and arms with soap and water before using the toilet.

After work take off all clothes and shoes. Shower using soap and water. Wear only clean clothes. Do not use contaminated clothing. Wash protective clothing and protective equipment with soap and water after each use. Personal clothing worn during use must be laundered separately from household articles. Clothing and protective equipment heavily contaminated or drenched with DITHANE must be destroyed according to state and local regulations.

HEAVILY CONTAMINATED OR DRENCHED CLOTHING CANNOT BE ADEQUATELY DECONTAMINATED.

During aerial application, human flaggers are prohibited unless in totally enclosed vehicles.

RE-ENTRY STATEMENTS

Written or oral warnings must be given to workers who are expected to be in a treated area or in an area about to be treated with this product. When oral warnings are given, warnings shall be given in a language customarily understood by workers. Oral warnings must be given if there is reason to believe that written warnings cannot be understood by workers. Warnings must include the following information: "**CAUTION.** Area treated with DITHANE on (date of application). Do not enter without appropriate protective clothing until sprays have dried for turf uses or within **24 hours** of application for agricultural uses. In case of contact, flush skin or eyes with plenty of water; for eyes, consult a physician if irritation persists."

For Agricultural Uses

After sprays have dried, **do not enter or allow entry into treated areas until the 24-hour re-entry interval has expired** unless wearing the personal protective equipment listed on this label.

For Turf Uses

Do not enter treated areas without protective clothing until sprays have dried.

GENERAL USE INFORMATION

DITHANE M-45 is a broad-spectrum protectant fungicide recommended for the control of fruit, field, vegetable, and ornamental crop diseases. Optimum disease control is achieved when the fungicide is applied in a regularly scheduled preventive spray program. The addition of an agricultural surfactant will improve fungicide performance by providing a more uniform spray deposit, increased foliar redistribution, and improved fungicide retention during periods of wet weather.

USE RATE DETERMINATION

Carefully read, understand, and follow label use rates and restrictions.

Under low disease conditions, minimum label rates per application can be used while maximum label rates and shortened spray schedules are recommended for severe or threatening disease conditions.

For proper application, determine the number of acres to be treated, the recommended label use rate, and the gallonage to be applied per acre. Prepare only the amount of spray solution required to treat the measured acreage. Careful calibration of spray equipment is recommended prior to use.

When applied by hand sprayers, 1 pound DITHANE M-45 fungicide per 100 gallons per acre is equivalent to 1 level tablespoon per gallon spray solution.

MIXING

Slowly place into spray tank as it is being filled or thoroughly premix in a nurse tank for concentrate or aircraft sprayers. Add other co-applied fungicides, insecticides, growth regulators, micronutrients, and spray adjuvants after DITHANE M-45 fungicide has been placed into suspension.

When preparing spray solutions for use in a hand sprayer, premix as a slurry in a small container, and then add to sprayer containing 1/3 to 1/2 the desired final water volume.

COMPATIBILITY

DITHANE M-45 fungicide is compatible with most commonly used agricultural fungicides, insecticides, and growth regulators. When preparing tank mixes, user should consult spray compatibility charts or State Cooperative Extension Service Specialists prior to actual use.

SPRAY ADJUVANTS

The addition of agricultural surfactants to DITHANE M-45 fungicide sprays will improve initial spray deposits, fungicide redistribution, and weatherability. The following spray adjuvants have been especially formulated to optimize the performance of foliar-applied agricultural chemicals:

TRITON® B-1956®	A water-dispersible, resin-based nonionic surfactant which resists re-wetting and removal by rain. Effective with dilute sprays applied by ground equipment.
TRITON CS-7®	A spreader-binder designed specifically for use in concentrate and low volume sprays applied by aircraft or ground equipment.

Place DITHANE M-45 fungicide into suspension prior to adding an adjuvant to the spray mixture. Read and carefully observe the precautionary statements and all other information appearing on both product labels prior to spray preparation.

APPLICATION

Ground

Thorough coverage foliar sprays generally result in optimum disease control. To achieve good coverage use proper spray pressure, gallonage per acre, nozzles (generally hollow cone), discs (generally D-5 to D-7), nozzle spacing, and tractor speed. Consult spray nozzle and accessory catalogues for specific information on proper equipment calibration.

Hand Sprayers

Thoroughly spray plant foliage until runoff.

Aerial

A uniform initial spray deposit over the crop canopy generally results in optimum disease control. Each aircraft should be prechecked for droplet size, uniformity of spray pattern, swath width, and spray volume.

Nozzle selection—Hollow cone brass nozzles with a D series orifice disc and core (whirlplate) are recommended. Nozzles should point straight down or slightly backward.

Swath width—For most field and vegetable crops, swaths just beyond the wingspan of 36 to 40 feet for light aircraft and up to 45 feet for heavier aircraft are suggested. Optimum swath for helicopters is usually 5 to 10 feet beyond normal boom length.

Spray volume—On vegetable and field crops, 2 to 3 gallons of spray per acre are generally optimum; orchards and vineyards can be handled with spray volumes of 5 gallons per acre. Some tall or dense foliage crops, requiring greater penetration to the lower leaf surface, will require higher spray volumes. **Do not use less than 5 gallons per acre in California.**

Altitude—For most crops, the spray boom should be positioned 5 to 10 feet above the crop canopy.

Flagging—Swaths should be marked at the end of the field with permanent flags or by a flagman. Swaths should be measured accurately with a chain or other device except when rows can be accurately counted.

CHEMIGATION USE DIRECTIONS

Sprinkler Irrigation

DITHANE M-45 fungicide must be applied on a regular protectant fungicide schedule, **not an irrigation schedule.** If irrigation cycles are less frequent than recommended DITHANE M-45 fungicide application intervals, ground or aerial applications must supplement chemigation applications to achieve adequate disease control.

Apply DITHANE M-45 fungicide only through sprinkler irrigation systems including center-pivot, lateral move, end tow, side (wheel) roll, traveler, solid set, or hand move irrigation systems. Do not apply product through any other type of irrigation system.

Lack of fungicidal effectiveness or illegal pesticide residues in the crop can result from nonuniform distribution of treated water.

If you have questions about calibration, you should contact State Extension Service specialists, equipment manufacturers or other experts.

Do not connect an irrigation system (including greenhouse systems) used for pesticide application to a public water system unless the pesticide label-prescribed safety devices for public water systems are in place.

A person knowledgeable of the chemigation system and responsible for its operation, or under the supervision of the responsible person, shall shut the system down and make necessary adjustments should the need arise.

Before applying DITHANE M-45 fungicide through sprinkler irrigation equipment, the chemigation system must meet the following specifications:

- Public water system means a system for the provision to the public of piped water for human consumption if such system has at least 15 service connections or regularly serves an average of at least 25 individuals daily at least 60 days out of the year.
- Chemigation systems connected to public water systems must contain a functional reduced-pressure zone (RPZ), backflow preventer or the functional equivalent in the water supply line upstream from the point of pesticide introduction. As an option to the RPZ, the water from the public water system should be discharged into a reservoir tank prior to pesticide introduction. There shall be a complete physical break (air gap) between the outlet end of the fill pipe and the top or overflow rim of the reservoir tank of at least twice the inside diameter of the fill pipe.
- Systems not connected to a public water supply must contain a functional check valve, vacuum relief valve, and low-pressure drain appropriately located in the irrigation pipeline to prevent water source contamination from back flow.

- The pesticide injection pipeline must contain a functional, automatic, quick-closing check valve to prevent the flow of fluid back toward the injection pump.
- The pesticide injection pipeline must also contain a functional, normally closed, solenoid-operated valve located on the intake side of the injection pump and connected to the system interlock to prevent fluid from being withdrawn from the supply tank when the irrigation system is either automatically or manually shut down.
- The system must contain functional interlocking controls to automatically shut off the pesticide injection pump when the water pump motor stops, or in cases where there is no water pump, when the water pressure decreases to the point where pesticide distribution is adversely affected.
- The irrigation line or water pump must include a functional pressure switch which will stop the water pump motor when the water pressure decreases to the point where pesticide distribution is adversely affected.
- Systems must use a metering pump, such as a positive displacement injection pump (e.g., diaphragm pump) effectively designed and constructed of materials that are compatible with pesticides and capable of being fitted with a system interlock.
- Do not apply when wind speed favors drift beyond the area intended for treatment.

Center-pivot, Lateral Move, End Tow, and Traveler Irrigation Equipment: (Use only with electric or oil hydraulic drive systems which provide a uniform water distribution.)

- Determine size of area to be treated.
- Determine the time required to apply no more than 1/4 inch water (6,750 gallons water per acre) over the area to be treated when the system and injection equipment are operated at normal pressures recommended by the equipment manufacturer. Run system at 80 to 95% of manufacturer's rated capacity.
- Using only water, determine the injection pump output when operated at normal line pressure.
- Determine the amount of DITHANE M-45 fungicide required to treat area.
- Add the required amount of DITHANE M-45 fungicide and sufficient water to meet the injection time requirements of the solution tank.
- Maintain constant solution tank agitation during the injection period.
- Stop injection equipment after treatment is completed. Continue to operate the system until DITHANE M-45 fungicide solution has cleared the sprinkler head.

Solid-set, Side (wheel) Roll, and Hand Move Irrigation Equipment:

- Determine acreage covered by sprinkler.
- Fill injector solution tank with water and adjust flow rate to use contents over a 10 to 30 minute interval.
- Determine the amount of DITHANE M-45 fungicide required to treat area.
- Add the required amount of DITHANE M-45 fungicide into the same quantity of water used to calibrate the injection equipment.
- Maintain constant solution tank agitation during the injection period.
- Operate system at normal pressures recommended by the manufacturer of the injection equipment and used for the time interval established during calibration.
- Inject DITHANE M-45 fungicide at the end of the irrigation cycle or as a separate application to maximize foliar fungicide retention.
- Stop injection equipment after treatment is completed. Continue to operate the system until DITHANE M-45 fungicide solution has cleared the last sprinkler head.

DISEASE MONITORING

DITHANE M-45 fungicide is a broad-spectrum, protectant fungicide. If not applied on a routine protectant spray schedule, crops should be scouted on a weekly basis. Fungicide application should be made, at the recommended label use rate and spray schedule, at the first sign of disease, report of disease in the area, or during environmental conditions favorable for disease development.

RESTRICTIONS

Users should carefully read, understand, and follow all use restrictions prior to using DITHANE M-45 fungicide.

VEGETABLES

CROP	DISEASES CONTROLLED	RATE OF DITHANE M-45 PER APPLICATION LB./A	REMARKS (ALSO REFER TO DIRECTIONS FOR USE)	RESTRICTIONS
Asparagus	Cercospora leaf spot Rust	2	Start applications when rust first appears and repeat at 10 day intervals. Four applications are usually sufficient.	Apply only on asparagus ferns after spears have been harvested.
Corn (sweet, popcorn)	Common rust Helminthosporium leaf blight	1½	Use sufficient water for thorough coverage. Start applications when disease first appears and repeat at 4 to 7 day intervals. The addition of a TRITON surfactant to spray solutions will improve performance.	Do not apply within 7 days of harvest. Do not feed treated forage to livestock.
Onions (dry bulb)	Botrytis leaf blight Downy mildew Purple blotch Neck rot	3	Follow a protective spray schedule starting when diseases are first reported in the area and repeat at 7 day intervals throughout the season. The addition of a TRITON surfactant to spray solutions will improve performance. Do not allow spray or drift to contact bulbs after lifting from soil.	Do not apply within 7 days of harvest.
(furrow drench)	Smut		Apply 3 lbs. per acre as a furrow drench at time of planting onion seeds. Use 75 to 125 gallons water per acre.	Do not use more than 2.4 lbs. actual per acre (29,000 linear feet of furrow). Do not use in California.
Potatoes	Early blight Late blight	1 to 2	Begin application when plants are 4 to 6 inches high by applying 1 lb. per acre. As the vines increase in size, apply 1½ to 2 lbs. per acre. The addition of a TRITON surfactant to spray solutions will improve performance.	
Seedpiece (treatment)	Fusarium decay Seedborne common scab		Dip whole or cut potato tubers in 1¼ lbs. DITHANE M-45 fungicide per 50 gallons of water. Place treated tubers in a clean container following treatment and plant as soon as possible. Spread treated seedpieces in a cool place if held before planting.	Do not use treated seed potatoes for food or feed purposes.
Tomatoes	Anthracnose Early blight Gray leaf spot Late blight Leaf mold Septoria leaf spot	1½ to 3	Start applications when seedlings emerge or transplants are set and repeat at 7 day intervals throughout the season. The addition of a TRITON surfactant to spray solutions will improve performance.	Do not apply within 5 days of harvest.

FRUITS AND NUTS

CROP	DISEASES CONTROLLED	RATE OF DITHANE M-45 PER APPLICATION LB./A	LB./100 GALS.	REMARKS (ALSO REFER TO DIRECTIONS FOR USE)	RESTRICTIONS
Bananas	Sigatoka	2 to 4		Apply when leaves first appear and repeat every 2 to 3 weeks or as required. Use sufficient water to provide adequate coverage. The addition of a TRITON surfactant to spray solutions will improve performance.	
Cranberries	Fruit rot	3 to 6		Start applications at mid-bloom and repeat at 7 to 10 day intervals as required.	Do not apply within 30 days of harvest.
Grapes	Black rot Bunch rot Deadarm Downy mildew	1½ to 4		Apply in sufficient water to provide thorough coverage starting when new shoots are ½ to 1½ inches long. Repeat when shoots are 3 to 5 inches long, when shoots are 8 to 10 inches long, and then at 10 to 14 day intervals until fruit is set or 66 days before harvest. For late season control of black rot, deadarm and downy mildew the use of other approved and recommended fungicides is suggested.	In Calif. do not apply after fruit is set. In other areas do not apply within 66 days of harvest.

FIELD CROPS

CROP	DISEASES CONTROLLED	RATE OF DITHANE M-45 PER APPLICATION LB./A	REMARKS (ALSO REFER TO DIRECTIONS FOR USE)	RESTRICTIONS
Peanuts	Cercospora leaf spot Rust	1 to 2	Start applications when disease first appears or is reported in area. Repeat sprays at 10 to 14 day intervals. Reduce sprays to a 7 day interval during humid weather.	Do not apply within 14 days of harvest. Do not feed treated vines to livestock.
Sugar Beets	Cercospora leaf spot	1½ to 2	Start application when disease first threatens and repeat every 7 to 10 days as needed. The addition of a TRITON surfactant to spray solutions will improve performance.	Do not apply within 14 days of harvest. Do not feed treated tops to livestock.
Wheat	Helminthosporium leaf spot Leaf rust Septoria leaf spot Septoria glume blotch Tan spot	2	Start application at onset of disease or when plants are in the tillering to jointing stage and repeat at 7 to 10 day intervals. The addition of TRITON CS-7 to spray solutions will improve performance.	Do not make more than three applications during the season. Do not apply within 26 days of harvest. Do not graze livestock in treated areas within 26 days after application.

SEED TREATMENT

CROP	DISEASES CONTROLLED	RATE OF DITHANE M-45 PER APPLICATION OZ./BU.	OZ./100 LB.	REMARKS (ALSO REFER TO DIRECTIONS FOR USE)	RESTRICTIONS
GENERAL USE INSTRUCTIONS				**Seeds to be treated should be cleaned and well cured prior to treatment. DITHANE M-45 fungicide may be applied to dry seed with conventional slurry or mist seed treating equipment or as a planter-box application. For best results, the seed must be completely and uniformly covered with fungicide. For seed treatment, a dye must be added to DITHANE M-45 fungicide that will impart an unnatural color to the seed.**	**Treated seed should be labeled, "Must not be used for food, feed, or oil purposes."**
Barley	Covered smut Damping-off False loose smut Seed rots Seedling blights	1.3 to 2.0	2.7 to 4.2	For planter box treatment only.	
Corn (field)	Damping-off Seed rots Seedling blights	1.5 to 3.0	2.7 to 5.4		
Cotton (acid delinted)	Damping-off Seedling blights		3.0		
(reginned)	Damping-off Seedling blights		6.0		
Flax	Damping-off Seed rots Seedling blights	2.0 to 4.0	3.6 to 7.1		

CROP	DISEASES CONTROLLED	RATE OF DITHANE M-45 PER APPLICATION OZ./BU.	OZ./100 LB.	REMARKS (ALSO REFER TO DIRECTIONS FOR USE)	RESTRICTIONS
Oats	Damping-off Seed rots Seedling blights Smuts	1.3 to 2.0	4.0 to 6.3	For planter box treatment only.	
Peanuts (shelled)	Damping-off Seed rots Seedling blights	2.0 to 4.0	8.0 to 16.0		
Rice	Damping-off Seed rots Seedling blights		2.0 to 4.0	Apply before, during or after soaking in water.	
Rye	Bunt Damping-off Seed rots Seedling blights	1.3 to 2.0	2.3 to 3.6	For planter box treatment only.	
Safflower	Seedborne rust (Puccinia carthami)		2.0		
Sorghum	Covered kernel smut Damping-off Seed rots Seedling blights	1.5 to 2.5	2.7 to 4.5		
Tomatoes	Damping-off Seed rots Seedling blights		8.0		
Wheat	Bunt Damping-off Seed rots Seedling blights	1.3 to 2.0	2.2 to 3.3	For planter box treatment only.	

MISCELLANEOUS

CROP	DISEASES CONTROLLED	RATE OF DITHANE M-45 PER APPLICATION	REMARKS (ALSO REFER TO DIRECTIONS FOR USE)
Asparagus crowns	Crown rot	1 lb. per 100 gals.	Place loosely packed crowns into a burlap bag and soak, with gentle agitation, in the fungicide solution for 5 minutes. Remove bag, drain well, and plant crowns as soon as possible. A tank large enough to hold a single burlap bag will treat 2 bags of crowns. Clean dipping suspension should then be prepared in a clean tank. Dirty crowns should be pre-washed to remove excess soil.
Caprifig	Assorted molds Endosepsis (Fusarium)	1 lb. per 25 gals.	Prepare mamme figs by making a shallow cut through the eye and then hand dividing to avoid wasp injury. Submerge mamme figs in the fungicide suspension for a minimum of 15 minutes. The fungicide suspension should be stirred frequently to prevent settling out. Fresh dipping solutions should be used after treating 4 or 5 batches of figs. After treatment, figs should be drained prior to placement in trees.

TURF

CROP	DISEASES CONTROLLED	RATE OF DITHANE M-45 PER APPLICATION OZ./1000 SQ. FT.	REMARKS (ALSO REFER TO DIRECTIONS FOR USE)	RESTRICTIONS
GENERAL USE INSTRUCTIONS			**Start application when grass greens-up in spring or when disease first appears and repeat at 7 to 14 day intervals or until disease threat is past.** **When conditions are especially favorable for disease development, apply maximum fungicide use rate on a 7 day spray schedule.** **Apply in sufficient water to provide adequate coverage.**	**Do not graze treated areas.** **Do not feed clippings to livestock.**
Assorted Grasses	Helminthosporium melting-out Rust (leaf, stem, stripe)	4		
	Copper spot Fusarium Blight Red thread Slime mold	4 to 8		
	Algae	6		
	Dollar spot	6 to 8		

CROP	DISEASES CONTROLLED	RATE OF DITHANE M-45 PER APPLICATION OZ./1000 SQ. FT.	REMARKS (ALSO REFER TO DIRECTIONS FOR USE)	RESTRICTIONS
	Rhizoctonia brown patch	4	Apply on a 7 day spray schedule.	
	Pythium blight	8	Apply at 5 day intervals, or more frequently, if conditions are especially favorable for disease development.	
	Fusarium snow mold	6 to 8	Apply at 2 to 6 week intervals during winter.	

ORNAMENTALS

GENERAL USE INSTRUCTIONS Neither the manufacturer nor the seller has determined the effects of using DITHANE M-45 fungicide on ornamentals not specified on this label.

Prior to any large-scale application on such plants, the user should determine the effects of DITHANE M-45 fungicide by testing a small section of the type of plants to be treated. User assumes all risks arising out of application to unlabeled plants. The Conditions of Sale and Warranty apply to all uses.

For outdoor or greenhouse use apply the equivalent of 1½ lbs. DITHANE M-45 fungicide per 100 gals. dilute spray. The addition of TRITON B-1956 to spray solutions will improve performance.

Begin spraying when plants are well leafed out or at first sign of disease, in full coverage spray at 7 to 10 day intervals throughout season or follow State Extension Service recommendations for disease control on the following ornamental plants:

CROP	DISEASES CONTROLLED	REMARKS (ALSO REFER TO DIRECTIONS FOR USE)	RESTRICTIONS
African violet	Botrytis blight		
Anthurium	Anthracnose, spadix rot		
Arborvitae	Cercospora blight		
Ash, mountain	Entomosporium leaf spot Guignardia leaf blotch		
Ash, white	Anthracnose Cylindrosporium leaf spot		
Aster, perennial	Puccinia rusts		
Aucuba, japonica	Alternaria leaf spot Anthracnose		
Azalea	Cylindrocladium rot Petal blight Phytophthora twig and bud blight	Apply in a full coverage spray, 2 to 3 times a week, while flowers are opening. Direct spray into flowers and thoroughly spray ground under bushes.	
Begonia	Botrytis blight		
Buffaloberry	Cylindrosporium leaf spot		Do not use fruit for food or feed purposes.
Camellias	Petal blight	Refer to azalea.	
Carnation	Rust Septoria leaf spot		
Cedar, red (Juniper)	Cercospora blight Phomopsis blight		
Chrysanthemum	Ascochyta blight Botrytis petal spot Rust	Apply twice weekly during blooming period.	
Conifers	Lophodermium needle cast Pine gall rust Scirrhia brown spot	Begin application in spring or early summer before infection occurs. Repeat after heavy rains and at two week intervals as long as needed.	
Cordyline	Cercospora leaf spot		
Crabapple (Ornamental)	Cedar-apple rust Scab Sphaeropsis leaf spot		
Cypress Arizona (Cupressus sp.)	Cercospora blight Monochaetia canker		
Dahlia	Botrytis blight		
Dieffenbachia	Leptosphaeria brown spot		
Dogwood, flowering	Anthracnose Elsinoe leaf spot Septoria leaf spot	Apply when buds begin to open, when bracts have fallen, 4 weeks later and again in late summer after flower buds for next season have formed.	
Dracaena	Fusarium leaf spot		
Elm	Black leaf spot		

CROP	DISEASES CONTROLLED	REMARKS (ALSO REFER TO DIRECTIONS FOR USE)	RESTRICTIONS
Euonymus	Anthracnose		
Fatsia	Anthracnose		
Fern	Rhizoctonia blight		
Ficus	Cercospora leaf spot		
Firethorn (Pyracantha)	Fusicladium scab		
Fir, Douglas	Swiss needle cast		
Fuchsia	Botrytis blight Rust		
Geranium	Rust		
Gladiolus	Botrytis blossom blight Curvularia leaf spot	Make regular weekly applications starting before diseases appear and increase to 2 to 3 applications per week during periods of heavy disease and during rainy weather. On flower spikes, reduce spray concentration to ¾ lb. per 100 gals.	
Hawthorn	Cedar-apple rust Fabraea leaf spot Frogeye leaf spot Hawthorn rust Scab		
Holly	Purple spot		
Hollyhock	Anthracnose Cercospora leaf spot Puccinia rusts		
Honeysuckle	Herpobasidium blight		
Horsechestnut, Buckeye	Alternaria leaf spot Guignardia leaf blotch		
Hydrangea	Botrytis blight Cercospora leaf spot		
Iris	Didymellina leaf spot Mystrosporium ink spot		
Juniper	Phomopsis blight		
Laurel, mountain	Cercospora leaf spot Petal blight	Refer to azalea.	
Ligustrum	Cercospora leaf spot		
Lily	Botrytis blight		
Magnolia	Gloeosporium leaf spot		
Maple	Alternaria leaf spot Phyllosticta leaf spot		Do not use on Sugar maples intended for the production of maple syrup.
Marigold	Botrytis blossom blight	Do not use on French dwarf double or Signet type marigold seedlings.	
Narcissus	Botrytis blight (fire) Smoulder		
Oak	Actinopelte leaf spot Taphrina leaf blister		
Orchid (Dendrobium)	Botrytis blossom blight		
Pachysandra	Volutella blight	Use a drenching spray of 2 lbs. per 50 gallons of water per 5,000 sq. ft. of bed. Start application at first sign of disease and apply at least 5 applications at 10 to 14 day intervals.	
Pansy	Anthracnose		
Peony	Botrytis blossom blight Phytophthora blight	Apply in early spring and early fall, drenching soil around plants as well as the foliage. Promptly destroy all infected plant parts.	
Peperomia	Cercospora leaf spot		
Philodendron	Dactylaria leaf spot Phytophthora leaf spot		

CROP	DISEASES CONTROLLED	REMARKS (ALSO REFER TO DIRECTIONS FOR USE)	RESTRICTIONS
Photinia	Entomosporium leaf spot		
Pleomele	Fusarium leaf spot		
Poinsettia	Sphaceloma scab		
Rhododendron	Cercospora leaf spot Discosia leaf spot Petal blight	Refer to azalea.	
Rose	Black spot Cercospora leaf spot Rust		
Schefflera	Alternaria blight		
Skunkbush, sumac	Cylindrosporium leaf spot		
Snapdragon	Rust		
Statice	Cercospora frogeye		
Syngonium	Cephalosporium leaf spot		
Tulip	Botrytis blight (fire)		
Venus, flytrap	Anthracnose		
Viburnum	Downy mildew Ramularia leaf spot		
Walnut	Anthracnose		Do not use treated walnuts for food or feed purposes.
Zinnia	Alternaria leaf blight		

FUNGICIDE TANK MIXES:
The following tank mix recommendations are made for the improved control of specific diseases. Before mixing DITHANE M-45 with any other fungicide, read and carefully observe the precautionary and all other information appearing on both product labels. User must abide by the most restrictive limitations appearing on the product labels.

CROP	CO-APPLIED FUNGICIDE	FOR IMPROVED CONTROL OF:	USE RATE PER APPLICATION	DITHANE M-45 USE RATE PER APPLICATION	REMARKS (ALSO REFER TO DIRECTIONS FOR USE)
Grapes	BAYLETON 50 WP	Powdery mildew	2 to 6 ozs./A	1½ to 4 lbs./A	
	KARATHANE WD		1 to 2½ lbs./A		Do not use in California.
	KARATHANE LC		4 to 8 fl. ozs./100 gals.		
	BENLATE 50 WP*		¾ to 1 lb./A		
Peanuts	BENLATE 50 WP*	Ascochyta web blotch Early leaf spot Late leaf spot	4 ozs./A	1 to 2 lbs./A	
	TOPSIN-M70 WP*	Early leaf spot			
Potatoes	Triphenyl Tin Hydroxide (TPTH)	Early blight	½ to full label use rate	1½ to 2 lbs./A	
Sugar Beets	BENLATE 50 WP*	Cercospora leaf spot	4 ozs./A	1½ to 2 lbs./A	
	TOPSIN-M70 WP*				
	MERTECT 340-F*		6 fl. ozs./A		
	Triphenyl Tin Hydroxide (TPTH)		½ to full label use rate		
Tomatoes	FIXED COPPERS	Bacterial leaf spot	Manufacturer's label use rate	1½ to 3 lbs./A	
	DYRENE 50 WP	Gray leaf spot	½ to 1 lb./A		
	BENLATE 50 WP*	Botrytis gray mold Sclerotinia white mold			
Wheat	BAYLETON 50 WP	Powdery mildew Rust (leaf, stem stripe)	2 to 4 ozs./A	2 lbs./A	

* Do not use if fungal resistance to this fungicide is present in field to be treated or has been reported in the area.

STORAGE AND DISPOSAL

STORAGE: Keep away from fire and sparks. Store in a cool, dry, well ventilated area. Do not allow to become wet or overheated in storage: decomposition, impaired activity, or fire may result. Keep container closed when not in use. Pallets of containers should not be stacked more than three high. Provide access aisle for each two rows. Decomposition produces a foul odor; if observed, check for hot containers and immediately remove to open areas for disposal.

PESTICIDE DISPOSAL: Wastes resulting from the use of this product may be disposed of on-site, or at an approved waste disposal facility.

CONTAINER DISPOSAL: Completely empty bag into application equipment. Then dispose of it in a sanitary landfill or by incineration, or, if allowed by state and local authorities, by burning. If burned, stay out of smoke.

STEPS TO BE TAKEN IN CASE MATERIAL IS RELEASED OR SPILLED: Sweep or shovel into containers for disposal or reworking. Keep dusting to a minimum. Flush contaminated area with a large amount of water to a chemical or sanitary sewer containing a settling pit. Refer to Precautionary Statements.

CONDITIONS OF SALE AND WARRANTY

Rohm and Haas warrants that the product conforms to its chemical description and is reasonably fit for the purpose stated on the label only when used in accordance with label directions under normal conditions of use. **ROHM AND HAAS MAKES NO OTHER EXPRESS OR IMPLIED WARRANTIES EITHER OF MERCHANTABILITY OR FITNESS FOR A PARTICULAR USE.** Handling, storage and use of the product by Buyer or User are beyond the control of Rohm and Haas and Seller. Risks such as crop injury, ineffectiveness or other unintended consequences resulting from, but not limited to, weather or soil conditions, presence of other materials, disease, pests, drift to other crops or property or failure to follow label directions will be assumed by the Buyer or User. **IN NO CASE WILL ROHM AND HAAS OR SELLER BE HELD LIABLE FOR CONSEQUENTIAL, SPECIAL OR INDIRECT DAMAGES RESULTING FROM THE HANDLING, STORAGE OR USE OF THIS PRODUCT.**

DITHANE, KARATHANE, TRITON, AG-98, B-1956 AND CS-7 are registered trademarks of the Rohm and Haas Company.
BAYLETON and DYRENE are registered trademarks of the parent company of Farbenfabriken Bayer GmbH, Leverkusen.
BENLATE is a registered trademark of E.I. duPont de Nemours and Co., Inc.
MERTECT is a registered trademark of Merck and Co., Inc.
TOPSIN-M is a registered trademark of Nippon Soda Co., Ltd.

8120-B8

FORE®

TURF AND ORNAMENTAL FUNGICIDE

ACTIVE INGREDIENTS:	
MANCOZEB	
A coordination product of zinc ion and manganese ethylene bisdithiocarbamate	80%
In which the ingredients are:	
Manganese++	16%
Zinc++	2%
Ethylene bisdithiocarbamate ion ($C_4H_6N_2S_4$)	62%
INERT INGREDIENTS:	20%
TOTAL	100%

EPA Reg. No. 707-87

NOTICE: Before using this product, read the entire Precautionary Statements, Conditions of Sale and Warranty, Directions for Use, Use Restrictions and Storage and Disposal Instructions. If the Conditions of Sale and Warranty are not acceptable, return the product unopened within thirty days of purchase to the place of purchase.

KEEP OUT OF REACH OF CHILDREN
CAUTION

PRECAUTIONARY STATEMENTS

HAZARDS TO HUMANS AND DOMESTIC ANIMALS

May cause irritation of nose, throat, eyes and skin. Do not breathe dust or spray mist. **PROTECTIVE CLOTHING AND EQUIPMENT** consisting of a long-sleeved shirt and long pants or a coverall that covers all parts of the body except the head, hands and feet; chemical resistant gloves; shoes, socks and goggles or a face shield must be worn during mixing and loading. During mixing and loading, a chemical-resistant apron must also be worn.

STATEMENTS OF PRACTICAL TREATMENT

IF SWALLOWED: Dilute by giving 2 glasses of water to drink and call a physician. Never give anything by mouth to an unconscious person.

IF INHALED: Move subject to fresh air.

IF IN EYES: Flush eyes with large amounts of water for at least 15 minutes. Consult a physician if irritation persists.

IF ON SKIN: Wash affected skin areas with soap and water.

ENVIRONMENTAL HAZARDS

This product is toxic to fish. Drift and runoff from treated areas may be hazardous to aquatic organisms in neighboring areas. Do not apply directly to water or wetlands (swamps, bogs, marshes and potholes). Do not contaminate water when disposing of equipment washwaters or disposing of wastes.

CONDITIONS OF SALE AND WARRANTY

Rohm and Haas warrants that the product conforms to its chemical description and is reasonably fit for the purpose stated on the label only when used in accordance with label directions under normal conditions of use. **ROHM AND HAAS MAKES NO OTHER EXPRESS OR IMPLIED WARRANTIES EITHER OF MERCHANTABILITY OR FITNESS FOR A PARTICULAR USE.** Handling, storage and use of the product by Buyer or User are beyond the control of Rohm and Haas and Seller. Risks such as crop injury, ineffectiveness or other unintended consequences resulting from, but not limited to, weather or soil conditions, presence of other materials, disease, pests, drift to other crops or property or failure to follow label directions will be assumed by the Buyer or User. **IN NO CASE WILL ROHM AND HAAS OR SELLER BE HELD LIABLE FOR CONSEQUENTIAL, SPECIAL OR INDIRECT DAMAGES RESULTING FROM THE HANDLING, STORAGE OR USE OF THIS PRODUCT.**

DIRECTIONS FOR USE

It is a violation of Federal law to use this product in a manner inconsistent with its labelling.

RE-ENTRY AND WORKER PROTECTION STATEMENTS (APPLIES ONLY TO USES ON SOD FARMS AND COMMERCIAL NURSERIES)

Do not enter treated areas without protective clothing until sprays have dried. After (sprays have dried/dusts have settled/vapors have dispersed, as applicable) **do not enter or allow entry into treated areas until the 24-hour re-entry interval has expired** unless wearing the personal protective equipment listed on the label. Because certain states may require more restrictive re-entry intervals for various crops treated with this product, consult your State Department of Agriculture for further information. Do not apply this product in such a manner as to directly or through drift expose workers or other persons. The area being treated must be vacated by unprotected persons.

Written or oral warnings must be given to workers who are expected to be in a treated area or in an area about to be treated with this product. Oral warnings must be given if there is reason to believe that written warnings cannot be understood by workers. When oral warnings are given, warnings shall be given in a language customarily understood by workers. Written warnings must include the following information.

CAUTION

(Insert area or field description) treated with FORE turf and ornamental fungicide on (insert date of application). Do not enter treated areas without protective clothing until sprays have dried. After (sprays have dried/dusts have settled/vapors have dispersed, as applicable) **do not enter or allow entry into treated areas until the 24-hour re-entry interval has expired** unless wearing the personal protective equipment listed on the label. In case of accidental exposure: Call a doctor (physician), clinic or hospital immediately. Explain that the victim has been exposed to FORE and describe his condition. For further information see the "Statements of Practical Treatment."

Oral warnings must include the same information that is in the written warnings.

WORKER SAFETY RULES—Keep all unprotected persons, children, livestock and pets away from treated area or where there is danger of drift. Do not rub eyes or mouth with hands (See "Statements of Practical Treatment").

PERSONAL PROTECTIVE EQUIPMENT—For Mixers, Loaders, Applicators and Early Re-entry Workers.

HANDLE THIS PRODUCT ONLY WHEN WEARING THE FOLLOWING PROTECTIVE CLOTHING AND EQUIPMENT: a long-sleeved shirt and long pants or a coverall that covers all parts of the body except the head, hands, and feet; chemical-resistant gloves; shoes, socks, and goggles or a face shield. During mixing and loading, a chemical-resistant apron must also be worn.

During application from a tractor with a completely enclosed cab with positive pressure filtration, or aerially with an enclosed cockpit, a long-sleeved shirt and long pants may be worn in place of the above protective clothing. Chemical-resistant gloves must be available in the cab or cockpit and worn while exiting.

IMPORTANT! Before removing gloves, wash them with soap and water. Always wash hands, face, and arms with soap and water before eating, smoking or drinking. Always wash hands and arms with soap and water before using the toilet.

After work take off all clothes and shoes. Shower using soap and water. Wear only clean clothes. Do not use contaminated clothing. Wash protective clothing and protective equipment with soap and water after each use. Personal clothing worn during use must be laundered separately from household articles. Clothing and protective equipment heavily contaminated or drenched with mancozeb must be destroyed according to state and local regulations.

HEAVILY CONTAMINATED OR DRENCHED CLOTHING CANNOT BE ADEQUATELY DECONTAMINATED.

During aerial application, human flaggers are prohibited unless in totally enclosed vehicles.

CHEMIGATION

Do not apply this product through any type of irrigation system.

GENERAL USE INFORMATION

FORE turf and ornamental fungicide is a unique and different broad-spectrum protectant fungicide which, when used according to label directions, provides high levels of disease control. This fungicide is safe for use on turf grasses and most ornamentals.

Optimum disease control is achieved when the fungicide is applied in a regularly scheduled preventative spray program. The addition of a spreader-sticker will improve fungicide performance by providing a uniform spray deposit, increased foliar redistribution, and fungicide retention during periods of wet weather.

USE RATE DETERMINATION

Carefully read, understand, and follow label rates and use restrictions.

Under low disease conditions, minimum label rates per application can be used while maximum label rates and shortened spray schedules are recommended for severe or threatening disease conditions.

For proper application to turf, determine the square footage to be treated, divide footage by 1000 and multiply by the recommended fungicide use rate per 1000 sq. ft., and then determine the amount of water required to provide adequate coverage. When treating ornamentals, determine the recommended fungicide use rate and the spray gallonage required to provide a thorough coverage. Prepare only the amount of spray solution required to treat the desired area.

When FORE fungicide is applied by a hand sprayer, 1 pound per 100 gallons or acre is equivalent to 1 level tablespoon per gallon spray solution.

MIXING

Slowly place into spray tank as it is being filled or thoroughly premix in a nurse tank for concentrate sprayers. Add other co-applied fungicides, insecticides, growth regulators, micronutrients, and spray adjuvants after FORE fungicide has been placed into suspension.

When preparing spray solutions for use in a hand sprayer, premix as a slurry in a small container, and then add to sprayer containing $^1/_3$ to $^1/_2$ the desired final water volume.

COMPATIBILITY

FORE fungicide is compatible with most commonly used agricultural fungicides, insecticides, and growth regulators. When preparing tank mixes, user should consult spray compatibility charts or State Extension Service Specialists prior to actual use.

The physical spray compatibility of specific tank mixes can be checked by preparing a sample suspension in a glass container and observing the suspension for any adverse effects which might interfere with proper spray application.

SPRAY ADJUVANTS

The addition of agricultural surfactants to FORE fungicide sprays will improve initial spray deposits, fungicide redistribution, and weatherability. The following spray adjuvants have been especially formulated to optimize the performance of foliar-applied agricultural chemicals:

TRITON AG-98® — A low foam nonionic general purpose spray adjuvant. May be used in dilute or concentrate sprays applied by aircraft or ground equipment.

TRITON® B-1956® — A water-dispersible, resin-based nonionic surfactant which resists re-wetting and removal by rain. Effective with dilute sprays applied by ground equipment.

TRITON CS-7® — A spreader-binder designed specifically for use in concentrate and low volume sprays applied by aircraft or ground equipment.

Place FORE fungicide into suspension prior to adding an adjuvant to the spray mixture. Read and carefully observe the precautionary statements and all other information appearing on both product labels prior to spray preparation.

APPLICATION

FORE is a protectant fungicide. Optimum disease control is achieved with thorough coverage sprays applied on a regular basis.

DISEASE SCOUTING

FORE fungicide is a broad spectrum protectant fungicide. If not applied on a routine protectant spray schedule, turf and ornamental plants should be frequently observed for disease signs or symptoms. Fungicide application should be made, at the recommended label use rate and spray schedule, when disease is first observed, reported in local area, or during environmental conditions favorable for disease development.

RESTRICTIONS

Users should carefully read, understand and follow all use restrictions prior to using FORE fungicide.

TURF

CROP	DISEASES CONTROLLED	RATE OF FORE FUNGICIDE PER APPLICATION OZ./1000 SQ. FT.	REMARKS (ALSO REFER TO DIRECTIONS FOR USE)	RESTRICTIONS
GENERAL USE INSTRUCTIONS			**Start application when grass greens-up in spring or when disease first appears, and repeat at 7 to 14 day intervals or until disease threat is past.** **When conditions are especially favorable for disease development, apply maximum fungicide use rate on a 7 day spray schedule.** **Apply in sufficient water to provide adequate coverage.**	**Do not graze treated areas. Do not feed clippings to livestock.**
Assorted Grasses	Helminthosporium melting-out	4		
	Rust (leaf, stem, stripe)	4		
	Copper spot Fusarium blight Red thread Slime mold	4 to 8		
	Algae	6		
	Dollar spot	6 to 8		
	Rhizoctonia brown patch	4	Apply on a 7 day spray schedule.	
	Pythium blight	8	Apply at 5 day intervals, or more frequently, if conditions are especially favorable for disease development.	
	Fusarium snow mold	6 to 8	Apply at 2 to 6 week intervals during winter.	

CROP	DISEASES CONTROLLED	RATE OF FORE FUNGICIDE PER APPLICATION OZ./1000 SQ. FT.	REMARKS (ALSO REFER TO DIRECTIONS FOR USE)	RESTRICTIONS

ORNAMENTALS

CROP	DISEASES CONTROLLED	REMARKS (ALSO REFER TO DIRECTIONS FOR USE)	RESTRICTIONS
GENERAL USE INSTRUCTIONS Neither the manufacturer nor the seller has determined the effects of using FORE fungicide on ornamentals not specified on this label. Prior to any large-scale application on such plants, the user should determine the effects of FORE fungicide by testing a small section of the type of plants to be treated. User assumes all risks arising out of application to unlabeled plants. The Conditions of Sale and Warranty apply to all uses. Apply 1½ pounds FORE fungicide per 100 gallons spray. The addition of TRITON B-1956 to spray solutions will improve performance. Begin spraying when plants are well leafed out or at first sign of disease, in a full coverage spray at 7 to 10 day intervals throughout season or follow State Extension Service recommendations for disease control on the following ornamental plants:			
African violet	Botrytis blight		
Anthurium	Anthracnose, spadix rot		
Arborvitae	Cercospora blight		
Ash, mountain	Entomosporium leaf spot Guignardia leaf blotch		
Ash, white	Anthracnose Cylindrosporium leaf spot		
Aster, perennial	Puccinia rusts		
Aucuba, japonica	Anthracnose Alternaria leaf spot		
Azalea	Cylindrocladium rot Petal blight Phytophthora twig and bud blight	Apply in a full coverage spray, 2 to 3 times a week, while flowers are opening. Direct spray into flowers and thoroughly spray ground under bushes.	
Begonia	Botrytis blight		
Buffaloberry	Cylindrosporium leaf spot		Do not use fruit for food or feed purposes.
Camellias	Petal blight	Refer to azalea.	
Carnation	Septoria leaf spot Rust	Refer to azalea.	
Cedar, red (Juniper)	Cercospora blight Phomopsis blight		
Chrysanthemum	Ascochyta blight Botrytis petal spot Rust	Apply twice weekly during blooming period.	
Conifers	Lophodermium needle cast Pine gall rust Scirrhia brown spot	Begin application in spring or early summer before infection occurs. Repeat after heavy rains and at two week intervals as long as needed.	
Cordyline	Cercospora leaf spot		
Crabapple (Ornamental)	Cedar-apple rust Scab Sphaeropsis leaf spot		
Cypress, Arizona (Cupressus sp.)	Cercospora blight Monochaetia canker		
Dahlia	Botrytis blight		
Dieffenbachia	Leptosphaeria brown spot		
Dogwood, flowering	Anthracnose Elsinoe leaf spot Septoria leaf spot	Apply when buds begin to open, when bracts have fallen, 4 weeks later, and again in late summer after flower buds for next season have formed.	
Dracaena	Fusarium leaf spot		
Elm	Black leaf spot		
Euonymus	Anthracnose		
Fatsia	Anthracnose		
Fern	Rhizoctonia blight		
Ficus	Cercospora leaf spot		
Firethorn (Pyracantha)	Fusicladium scab		
Fir, Douglas	Swiss needle cast		

CROP	DISEASES CONTROLLED	REMARKS (ALSO REFER TO DIRECTIONS FOR USE)	RESTRICTIONS
Fuchsia	Botrytis blight Rust		
Geranium	Rust		
Gladiolus	Botrytis blossom blight Curvularia leaf spot	Make regular weekly applications starting before diseases appear and increase to 2 or 3 applications per week during periods of heavy disease and during rainy weather. On flower spikes, reduce spray concentration to $^3/_4$ lb. per 100 gals.	
Hawthorn	Cedar-apple rust Fabraea leaf spot Frogeye leaf spot Hawthorn rust Scab		
Holly	Purple spot		
Hollyhock	Anthracnose Cercospora leaf spot Puccinia rusts		
Honeysuckle	Herpobasidium blight		
Horsechestnut, Buckeye	Alternaria leaf spot Guignardia leaf blotch		
Hydrangea	Botrytis blight Cercospora leaf spot		
Iris	Didymellina leaf spot Mystrosporium ink spot		
Juniper	Phomopsis blight		
Laurel, mountain	Cercospora leaf spot Petal blight	Refer to azalea.	
Ligustrum	Cercospora leaf spot		
Lily	Botrytis blight		
Magnolia	Gloeosporium leaf spot		
Maple	Alternaria leaf spot Phyllosticta leaf spot		Do not use on Sugar maples intended for the production of maple syrup.
Marigold	Botrytis blossom blight		
Narcissus	Botrytis blight (fire) Smoulder		
Oak	Actinopelte leaf spot Taphrina leaf blister		
Orchid (Dendrobium)	Botrytis blossom blight		
Pachysandra	Volutella blight	Use a drenching spray of 2 lbs. per 50 gallons of water per 5,000 sq. ft. of bed. Start application at first sign of disease and apply at least 5 applications at 10 to 14 day intervals.	
Pansy	Anthracnose		
Peony	Botrytis blossom blight Phytophthora blight	Apply in early spring and early fall, drenching soil around plants as well as the foliage. Promptly destroy all infected plant parts.	
Peperomia	Cercospora leaf spot		
Philodendron	Dactylaria leaf spot Phytophthora leaf spot		
Photinia	Entomosporium leaf spot		
Pleomele	Fusarium leaf spot		
Poinsettia	Sphaceloma scab		
Rhododendron	Cercospora leaf spot Discosia leaf spot Petal blight	Refer to azalea.	
Rose	Black spot Cercospora leaf spot Rust		
Schefflera	Alternaria blight		
Skunkbush, sumac	Cylindrosporium leaf spot		
Snapdragon	Rust		
Statice	Cercospora frogeye		
Syngonium	Cephalosporium leaf spot		
Tulip	Botrytis blight (fire)		
Venus, flytrap	Anthracnose		
Viburnum	Downy mildew Ramularia leaf spot		
Walnut	Anthracnose		Do not use treated walnuts for food or feed purposes.
Zinnia	Alternaria leaf blight		

STORAGE AND DISPOSAL

STORAGE: Keep away from fire and sparks. Store in a cool, dry, well ventilated area. Do not allow to become wet or overheated in storage: decomposition, impaired activity, or fire may result. Keep container closed when not in use.

Pallets of containers should not be stacked more than three high. Provide access aisle for each two rows. Decomposition produces a foul odor; if observed, check for hot containers and immediately remove to open areas for disposal.

PESTICIDE DISPOSAL: Wastes resulting from the use of this product may be disposed of on-site or at an approved waste disposal facility.

CONTAINER DISPOSAL

Paper Bags—Completely empty bag into application equipment. Then dispose of it in a sanitary landfill or by incineration, or, if allowed by State and local authorities, by burning. If burned, stay out of smoke.

Fiber Drums with liners—Do not re-use empty containers. Completely empty liner by shaking and tapping sides and bottom to loosen clinging particles. Empty residue into application equipment. Then dispose of drum and liner in a sanitary landfill or by incineration, if allowed by State and local authorities.

STEPS TO BE TAKEN IN CASE MATERIAL IS RELEASED OR SPILLED: Sweep or shovel into containers for disposal or reworking. Keep dusting to a minimum. Flush contaminated area with a large amount of water to a chemical or sanitary sewer containing a settling pit. REFER TO PRECAUTIONARY STATEMENTS.

FORE®, TRITON®, AG-98®, B-1956® and CS-7® are registered trademarks of Rohm and Haas Company.

8107-B3

GOAL® 1.6E

HERBICIDE

USE DIRECTIONS FOR:
ARTICHOKES (GLOBE)
BROCCOLI/CABBAGE/CAULIFLOWER
CITRUS (NONBEARING)
COFFEE
CONIFER
COTTON
FALLOW BED
GUAVA
JOJOBA
MINT
ONIONS
TREEFRUIT/NUT/VINE

ACTIVE INGREDIENT:	
Oxyfluorfen	
2-chloro-1-(3-ethoxy-4-nitrophenoxy) 4-(trifluoromethyl) benzene	19.4%*
INERT INGREDIENTS:	80.6%
TOTAL	100%

*Equivalent to 1.6 lbs. active ingredient per gallon.

EPA REG. NO. 707-174

Before using this product, read the entire Precautionary Statements, Conditions of Sale and Warranty, Directions for Use, Use Restrictions and Storage and Disposal Instructions. If the Conditions of Sale and Warranty are not acceptable, return the product unopened within thirty days of purchase to the place of purchase.

KEEP OUT OF REACH OF CHILDREN
WARNING AVISO

PRECAUTIONARY STATEMENTS

HAZARDS TO HUMANS AND DOMESTIC ANIMALS

PRECAUCION AL USUARIO: Si usted no lee inglés, no use este producto hasta que la etiqueta haya sido explicada ampliamente.

Causes substantial but temporary eye injury and skin irritation. Do not get in eyes, on skin, or on clothing. Wear goggles or face shield. This material may be harmful if swallowed. Wash thoroughly with soap and water after handling. Remove contaminated clothing and wash before reuse.

STATEMENTS OF PRACTICAL TREATMENT

IF IN EYES: Flush eyes with a large amount of water for at least 15 minutes. Get prompt medical attention.

IF ON SKIN: Wash thoroughly with plenty of soap and water. If irritation develops, get medical attention.

IF SWALLOWED: Do not induce vomiting. Dilute by giving 2 glasses of water to drink. Call a physician. Do not induce vomiting or give anything by mouth to an unconscious person.

IF INHALED: Move victim to fresh air.

NOTE TO PHYSICIAN: Contains petroleum distillates. Aspiration hazard may contraindicate the use of gastric lavage.

ENVIRONMENTAL HAZARDS

Do not apply directly to water or wetlands. Do not contaminate water by cleaning of equipment or disposal of wastes.

This product is highly toxic to aquatic invertebrates, aquatic plants, wildlife and fish. Use with care when applying in areas frequented by wildlife or adjacent to any body of water or wetland area. Do not apply when weather conditions favor drift or erosion from target areas.

PHYSICAL OR CHEMICAL HAZARDS

Combustible.

Do not use or store near heat or open flame.

CONDITIONS OF SALE AND WARRANTY

Rohm and Haas warrants that the product conforms to its chemical description and is reasonably fit for the purpose stated on the label only when used in accordance with label directions under normal conditions of use. **ROHM AND HAAS MAKES NO OTHER EXPRESS OR IMPLIED WARRANTIES EITHER OF MERCHANTABILITY OR FITNESS FOR A PARTICULAR USE.** Handling, storage and use of the product by Buyer or User are beyond the control of Rohm and Haas and Seller. Risks such as crop injury, ineffectiveness or other unintended consequences resulting from, but not limited to, weather or soil conditions, presence of other materials, disease, pests, drift to other crops or property or failure to follow label directions will be assumed by the Buyer or User. **IN NO CASE WILL ROHM AND HAAS OR SELLER BE HELD LIABLE FOR CONSEQUENTIAL, SPECIAL OR INDIRECT DAMAGES RESULTING FROM THE HANDLING, STORAGE OR USE OF THIS PRODUCT.**

RE-ENTRY AND WORKER PROTECTION STATEMENTS

Do not enter treated areas without protective clothing until sprays have dried. Because certain states may require more restrictive re-entry intervals for various crops treated with this product, consult your State Department of Agriculture for further information. Do not apply this product in such a manner as to directly or through drift, expose workers or other persons. The area being treated must be vacated by unprotected persons.

Written or oral warnings must be given to workers who are expected to be in a treated area or in an area about to be treated with this product. Oral warnings must be given if there is reason to believe that written warnings cannot be understood by workers. When oral warnings are given, warnings shall be given in a language customarily understood by workers. Written warnings must include the following information:

WARNING

(Insert area or field description) treated with GOAL 1.6E herbicide on (insert date of application). Do not enter treated areas without protective clothing until sprays have dried. In case of accidental exposure: Call a doctor (physician), clinic or hospital immediately. Explain that the victim has been exposed to GOAL 1.6E herbicide and describe his condition. For further information see the "Statement of Practical Treatment."

Oral warnings must include the same information that is in the written warnings.

STORAGE AND DISPOSAL

Do not contaminate water, food or feed by storage or disposal.

STORAGE: Keep from freezing. Store above 32°F.

PESTICIDE DISPOSAL: Pesticide wastes are toxic. Improper disposal of excess pesticide, spray mixture or rinsate is a violation of Federal Law. If these wastes cannot be disposed of by use according to label instructions, contact your State Pesticide or Environmental Control Agency, or the Hazardous Waste representative at the nearest EPA Regional Office for guidance.

CONTAINER DISPOSAL

Plastic Containers: Triple rinse (or equivalent). Puncture and dispose of in a sanitary landfill, or incinerate, or, if allowed by State and local authorities, by burning. If burned, stay out of smoke.

Metal Containers: Triple rinse (or equivalent). Then offer for recycling or reconditioning, or puncture and dispose of in a sanitary landfill, or by other procedures approved by State and local authorities.

STEPS TO BE TAKEN IN CASE MATERIAL IS RELEASED OR SPILLED: Ventilate area. Avoid breathing vapors. Use MSHA/NIOSH self-contained breathing apparatus or airline respirator for large spills in confined areas. Dike the spill with inert material (sand, earth, etc.) and transfer the liquid or solid diking material to separate containers for recovery or disposal. Remove contaminated clothing promptly and wash exposed skin areas with soap and water. Wash clothing before re-use. Keep spill out of all sewers and bodies of water.

CHEMIGATION

Do not apply this product through any type of irrigation system.

GENERAL CLASSIFICATION

GOAL 1.6E herbicide is an emulsifiable concentrate containing 1.6 lbs. of active ingredient per gallon. It is useful in preemergence, postemergence or post-directed applications.

GENERAL USE INFORMATION

CULTURAL CONSIDERATIONS

In order for GOAL 1.6E herbicide to provide maximum preemergence activity:

Prior to application, the bed or soil surface should be smooth and free of crop and weed trash (decaying leaves, clippings, dead weeds, etc.). Leaves and trash may be removed by blowing the area to be treated or by thoroughly mixing the trash into the soil through cultivation prior to herbicide application.

After application, at least one-quarter inch (1/4 inch) of irrigation or rainfall should occur within 3 or 4 weeks after application. The best results from GOAL 1.6E herbicide are from applications to established beds or soil surfaces that are left undisturbed during the time period for which weed control is desired. Cultural practices that result in redistribution or disturbance of the soil surface after treatment will decrease the herbicidal effectiveness of GOAL 1.6E. Cutting water furrows or cultivations that mix untreated soil into treated areas will also reduce the effectiveness of the treatment.

MIXING DIRECTIONS

Fill the spray tank at least one-third full of clean water. With the pump and agitator running, add the recommended amount of herbicides to the spray tank. The order of addition to the spray tank should be wettable powders first, flowables second and liquids last. Complete filling of the spray tank with water. For all applications of GOAL 1.6E herbicide (except onions) where postemergence weed control is desired, add 2 to 4 pints of TRITON AG-98® (or comparable 80% active nonionic surfactant cleared for application to growing crops) per each 100 gallons of spray. The addition of 4 pints of TRITON AG-98 per 100 gallons of spray is recommended to enhance postemergence activity when hard water (greater than 600 ppm) is used as carrier. Maintain agitation until spraying is complete.

Spray equipment should be calibrated carefully before each use.

Dosages listed on this label are for broadcast application. For banded application, the amount of GOAL 1.6E herbicide used per acre should be reduced according to the following formula:

$$\frac{\text{Band Width (in inches)}}{\text{Row Width (in inches)}} \times \text{Rate per Broadcast Acre} = \text{Amount Needed per Acre for Banded Application}$$

ARTICHOKES (GLOBE)

POST-DIRECTED SPRAY

GENERAL INFORMATION

GOAL 1.6E is an effective herbicide for postemergence and preemergence control of certain broadleaf weeds in artichokes. GOAL 1.6E herbicide should be directed towards the winter ditch, levees or flat rows between the artichoke rows. Artichoke fronds receiving accidental spray or drift will be injured. Over-the-top applications may exhibit severe injury to the foliage and flower bud and are not recommended.

DOSAGE

GOAL 1.6E herbicide is recommended as a post-directed application at 5 to 10 pints (1.0 to 2.0 lb. active) per acre. Optimum control is achieved when two applications of GOAL 1.6E herbicide are applied at 5 pints (1.0 lb. active) per acre. The initial application should be made to susceptible weed seedlings (up to 8 leaf stage). It is recommended that a second application be made 8 to 10 weeks later. Good results may be achieved when a single application of 10 pints (2.0 lb. active) of GOAL 1.6E herbicide is applied to susceptible weed seedlings (up to 8 leaf stage). Do not apply more than 10 pints (2.0 lb. active) of GOAL 1.6E herbicide per treated acre per season as a result of a single application or multiple applications. Do not apply within 5 days of harvest.

WEEDS CONTROLLED POSTEMERGENCE

CHEESEWEED (MALVA)	OXALIS (BERMUDA BUTTERCUP)
GROUNDSEL, COMMON	SHEPHERDSPURSE
MUSTARD, COMMON YELLOW	SOWTHISTLE, ANNUAL
NETTLE, BURNING	

WEEDS CONTROLLED PREEMERGENCE

CHEESEWEED (MALVA)	*OXALIS (BERMUDA BUTTERCUP)
GROUNDSEL, COMMON	SHEPHERDSPURSE
LAMBSQUARTERS, COMMON	SOWTHISTLE, ANNUAL
MUSTARD, COMMON YELLOW	

* Suppression.

TIMING AND METHOD OF APPLICATION

Treatments should be made October 1 through February 15 (after completion of the ditching operation).

GOAL 1.6E herbicide should be applied in a minimum of 40 gallons of water per acre depending upon density of emerged weeds. Spray volume should be increased as weed height and density increase. Use a low-pressure sprayer equipped with flat fan nozzles. Spray equipment should be calibrated carefully before each use. Spray should be directed towards the winter ditch, levees or flat rows between the artichoke rows. ARTICHOKE FRONDS RECEIVING ACCIDENTAL SPRAY OR DRIFT WILL BE INJURED.

ARTICHOKES (GLOBE) SPECIFIC USE RESTRICTIONS

In addition to the following, also observe GENERAL USE RESTRICTIONS listed at the end of this label.

- Do not apply more than 10 pints (2.0 lb. active) of GOAL 1.6E herbicide per treated acre per season as a result of a single application or multiple applications.
- Do not apply GOAL 1.6E herbicide within 5 days of harvest.
- DO NOT APPLY GOAL 1.6E HERBICIDE TO GLOBE ARTICHOKE PLANTINGS BEFORE OCTOBER 1 OR AFTER FEBRUARY 15.
- Avoid direct spray or drift contact of GOAL 1.6E herbicide with artichoke flowers or buds as severe injury may result.
- Do not apply GOAL 1.6E herbicide to artichoke plantings within 60 days after cutting back or transplanting.

BROCCOLI/CABBAGE/CAULIFLOWER

PRETRANSPLANT (PREPLANT) APPLICATION FOR PREEMERGENCE BROADLEAF WEED CONTROL

NOT FOR USE IN CALIFORNIA

GENERAL INFORMATION

GOAL 1.6E is a selective herbicide for preemergence control of certain annual broadleaf weeds. Applications must be made after completion of soil preparation but *prior* to transplanting of broccoli, cabbage or cauliflower plants. Transplanting should be completed with minimal soil disturbance. Treated soil surfaces should be left undisturbed after transplanting to obtain greatest benefit of GOAL 1.6E herbicide on susceptible annual broadleaf weeds during the time period for which weed control is desired. However, timely cultivations after weed emergence will assist in weed control.

Pretransplant applications of GOAL 1.6E herbicide in broccoli, cabbage and cauliflower may result in a temporary initial crop response (leaf cupping or crinkling). Crop response may be enhanced if crop leaves come in direct contact with treated soil. Crops rapidly outgrow this condition and develop normally.

Severe crop response may result from the use of transplants that are under stress due to temperature, disease, fertilizer, nematodes, insects, pesticides or storage conditions. The use of young (less than 5 weeks old), extremely succulent transplants grown in containers, less than 1 inch square, may increase the severity of crop injury. Hardening off, increasing the age of transplants or increasing the size of the rooting container will lessen the possibility and/or severity of crop injury.

DOSAGE

GOAL 1.6E herbicide is recommended for use at 1.25 to 2.5 pints (0.25 to 0.5 lb. active) per broadcast acre. The lower rate (1.25 pint per acre) is recommended for preemergence weed control on coarse texture soils with less than 1% organic matter. The 2.5 pint per acre rate is recommended for preemergence weed control on medium to fine texture soils or soils containing greater than 1% organic matter.

GOAL 1.6E herbicide will assist in early season annual grass control. However, GOAL 1.6E herbicide must not be a basic portion of the grass herbicide program. A planned herbicide program for preemergence or postemergence grass control is recommended. Research has shown that severe crop injury can occur if GOAL 1.6E herbicide is applied to a field that has had an acetanilide herbicide (Dual®, Lasso®, Ramrod®) application during the current growing season, therefore, it is not recommended.

WEEDS CONTROLLED*

CARPETWEED	PURSLANE, COMMON
PIGWEED, REDROOT	SMARTWEED, PENNSYLVANIA

GOAL 1.6E herbicide at the rate of 1.25 to 2.5 pints per acre may provide partial control or suppression of galingsoga, common lambsquarters and wild mustard.

*Applications of GOAL 1.6E herbicide to muck soils may result in partial control or suppression of the weeds listed.

METHOD OF APPLICATION

GOAL 1.6E herbicide should be thoroughly mixed with clean water at recommended concentrations, and applied in a minimum of 20 gallons of water per acre. Use conventional ground spray equipment with flat fan nozzles at 20 to 40 psi. Do not exceed 40 psi. Accurately calibrate spray equipment prior to each use.

Thoroughly flush the spray equipment (tank, hose, pump, boom) with water before and after each use. Residual GOAL 1.6E herbicide remaining in spray equipment may damage other crops.

AVOID DRIFT TO ALL OTHER CROPS AND NON-TARGET AREAS. DO NOT APPLY WHEN WEATHER CONDITIONS FAVOR DRIFT. GOAL 1.6E HERBICIDE IS PHYTOTOXIC TO PLANT FOLIAGE.

BROCCOLI/CABBAGE/CAULIFLOWER
SPECIFIC USE RESTRICTIONS

In addition to the following, also observe GENERAL USE RESTRICTIONS listed at the end of this label.

- Do not apply more than 2.5 pints (0.5 lb. active) of GOAL 1.6E herbicide per treated acre per season.
- Do not apply GOAL 1.6E herbicide preemergence to direct seeded broccoli, cabbage or cauliflower.
- Do not apply GOAL 1.6E herbicide post-transplant or postemergence (over the top) to broccoli, cabbage or cauliflower.
- For field use only. Do not apply GOAL 1.6E herbicide in an enclosed greenhouse structure as injury to plant foliage may result.

CITRUS (NONBEARING)

CALAMONDIN, CHIRONJA, CITRUS CITRON, GRAPEFRUIT, KUMQUAT, LEMON, LIME, MANDARIN, PUMMELO, SATSUMA MANDARIN, SOUR ORANGE, SWEET ORANGE, TANGELO, TANGERINE, TANGOR

FOR USE ONLY IN PERMANENTLY ESTABLISHED GROVES IN ARIZONA, CALIFORNIA, FLORIDA, LOUISIANA AND TEXAS

GENERAL INFORMATION

GOAL 1.6E is effective as a preemergence and/or postemergence herbicide when used alone or in recommended tank mix combinations, for the control of certain annual broadleaf weeds in nonbearing citrus plantings. GOAL 1.6E herbicide may be applied to newly planted trees or to young trees that will not bear fruit within one year.

The most effective postemergence weed control is achieved when GOAL 1.6E herbicide is applied to seedling weeds at the recommended growth stage. For postemergence control of certain grassy and broadleaf weeds, a tank mix of GOAL 1.6E herbicide with Gramoxone Super® or Roundup® can be used.

For residual grass control in citrus, a tank mixture of GOAL 1.6E herbicide with Devrinol®, simazine, Solicam® or Surflan® can be used. Contact herbicides such as Gramoxone Super or Roundup may also be added to the tank mixture.

Check individual product labels to determine suitability and use rates for various crops.

GOAL 1.6E HERBICIDE USED ALONE

GEOGRAPHIC USE DIRECTIONS
ARIZONA AND CALIFORNIA

DOSAGE

GOAL 1.6E herbicide is recommended for postemergence control at 2.5 to 10 pints (0.5 to 2.0 lb. active) per broadcast acre. For preemergence control of susceptible weeds, use 6 to 10 pints (1.2 to 2.0 lb. active) per broadcast acre.

WEEDS CONTROLLED POSTEMERGENCE (weeds up to 4 inches high) — Apply 2.5 to 10 pints (0.5 to 2.0 lb. active) of GOAL 1.6E herbicide per broadcast acre. Applications to weeds beyond this 4-inch stage may result in partial control.

CHEESEWEED (MALVA)	MINERSLETTUCE
FIDDLENECK, COAST	NETTLE, BURNING
*FILAREE, BROADLEAF	PIGWEED, REDROOT
*FILAREE, REDSTEM	REDMAIDS
*FILAREE, WHITESTEM	SHEPHERDSPURSE
GROUNDSEL, COMMON	SOWTHISTLE, ANNUAL
HENBIT	

*GOAL 1.6E herbicide at the 10 pint rate (2.0 lb. active) will provide control of filaree not exceeding the 4-inch stage. Applications to filaree beyond the 4-inch stage may result in partial control.

WEEDS CONTROLLED PREEMERGENCE—Apply 6 to 10 pints (1.2 to 2.0 lb. active) of GOAL 1.6E herbicide per broadcast acre.

BURCLOVER	LETTUCE, PRICKLY
CHEESEWEED (MALVA)	PIGWEED, REDROOT
FIDDLENECK, COAST	PURSLANE, COMMON
FILAREE, BROADLEAF	REDMAIDS
FILAREE, REDSTEM	ROCKET, LONDON
FILAREE, WHITESTEM	SHEPHERDSPURSE
GROUNDSEL, COMMON	SOWTHISTLE, ANNUAL
HENBIT	SPURGE, PROSTRATE
KNOTWEED, PROSTRATE	SPURGE, SPOTTED
LAMBSQUARTERS, COMMON	

FLORIDA, LOUISIANA AND TEXAS

DOSAGE

GOAL 1.6E herbicide is recommended for postemergence control at 2.5 to 10 pints (0.5 to 2.0 lb. active) per broadcast acre. For preemergence control of susceptible weeds, GOAL 1.6E herbicide is recommended at 10 pints (2.0 lb. active) per broadcast acre.

WEEDS CONTROLLED POSTEMERGENCE—Apply 2.5 to 10 pints (0.5 to 2.0 lb. active) of GOAL 1.6E herbicide per broadcast acre. The lower rate is recommended for the control of susceptible seedling weeds in the early postemergence stage, up to the 4 leaf stage. The higher rate (2.0 lb. active) should be used for weeds up to the 6 leaf stage. Applications to weeds beyond the 6 leaf stage may result in partial control.

BALSAMAPPLE	PEPPERWEED, VIRGINIA
*CUDWEED, NARROWLEAF	PIGWEED, REDROOT
**EVENINGPRIMROSE, CUTLEAF	POINSETTIA, WILD
GROUNDCHERRY, CUTLEAF	PURSLANE, COMMON
GROUNDCHERRY, WRIGHT	PUSLEY, FLORIDA
LAMBSQUARTERS, COMMON	SIDA, PRICKLY (TEAWEED)
MORNINGGLORY, ANNUAL	SMARTWEED, PENNSYLVANIA
NIGHTSHADE, AMERICAN BLACK	SOWTHISTLE, ANNUAL
NIGHTSHADE, BLACK	

WEEDS CONTROLLED PREEMERGENCE—Apply 10 pints (2.0 lb. active) of GOAL 1.6E herbicide per broadcast acre.

CUDWEED, NARROWLEAF	POINSETTIA, WILD
**EVENINGPRIMROSE, CUTLEAF	PUSLEY, FLORIDA
GROUNDCHERRY, CUTLEAF	SIDA, PRICKLY (TEAWEED)
LAMBSQUARTERS, COMMON	SMARTWEED, PENNSYLVANIA
NIGHTSHADE, AMERICAN BLACK	SOWTHISTLE, ANNUAL
NIGHTSHADE, BLACK	SPURGE, PROSTRATE
PEPPERWEED, VIRGINIA	SPURGE, SPOTTED
PIGWEED, REDROOT	

*Maximum 0.5 inch diameter.
**Highest rate and/or multiple applications may be required for acceptable control. Do not apply more than 20 pints (4.0 lb. active) per broadcast acre during any 12-month period as a result of multiple applications.

ALL STATES—ARIZONA, CALIFORNIA, FLORIDA, LOUISIANA AND TEXAS

TIMING AND METHOD OF APPLICATION

GOAL 1.6E herbicide should be directed to the soil and the base of trees. Avoid direct spray contact on the citrus foliage. Use a low-pressure sprayer equipped with a breakaway boom and flat fan nozzles. An off-center (OC) nozzle positioned at the end of the boom may be desired.

SPRAY VOLUME:

Weed Stage	Gallons of Water per Acre
Preemergence	40 or more
Postemergence	
up to 4 inch or 4 leaf stage	40 or more
exceeding 4 inch or 4 leaf stage	100 or more

TANK MIXES WITH GOAL 1.6E HERBICIDE

IMPORTANT: Read and observe all label directions before using. When tank mixing, always read all individual manufacturers' labels. In interpreting all labels for the tank mix, the most restrictive situations must apply.

DOSAGE

For preemergence control of susceptible grassy and broadleaf weeds in citrus plantings, a tank mixture of GOAL 1.6E herbicide with Devrinol, simazine, Solicam or Surflan can be applied. Apply at the recommended rates and growth stages to susceptible weed species in a manner described on the respective labels.

For postemergence control of susceptible grassy and broadleaf weeds, a tank mixture of Gramoxone Super or Roundup with GOAL 1.6E herbicide or combinations of GOAL 1.6E herbicide plus Devrinol, simazine, Solicam or Surflan can be used. Apply at the recommended rates and growth stages to susceptible weed species in a manner described on the respective labels.

WEEDS CONTROLLED

In addition to the weeds controlled by GOAL 1.6E herbicide used alone, control of susceptible weeds listed on the respective labels for the following products is also obtained:

Devrinol 50-WP	simazine*
Gramoxone Super	Solicam
Roundup	Surflan

*In addition, provides preemergence control of horseweed (marestail).

CITRUS (NONBEARING)
SPECIFIC USE RESTRICTIONS

In addition to the following, also observe the GENERAL USE RESTRICTIONS listed at the end of this label.

- Apply GOAL 1.6E herbicide only to nonbearing citrus trees.
- Do not apply more than 10 pints of GOAL 1.6E herbicide (2.0 lb. active) per broadcast acre in a single application or more than 20 pints (4.0

lb. active) per broadcast acre during any 12-month period as a result of multiple applications.

- GOAL 1.6E herbicide or any of the combinations recommended on this label should only be applied to healthy growing trees.
- Do not apply during periods of new foliage growth. Applications should be made after foliage has fully expanded and hardened off.
- Direct spray toward the base of trees. Avoid direct spray contact on the citrus foliage.

COFFEE

BEARING AND NONBEARING COFFEE IN HAWAII

GENERAL INFORMATION

GOAL 1.6E is effective as a preemergence herbicide when used alone for the control of certain annual broadleaf weeds in bearing and nonbearing coffee plantings. Preemergence control is most effective when spray is applied to clean, weed-free established soil surfaces. Treated berms or soil surfaces should not be disked or disturbed in any manner as the herbicidal effectiveness of GOAL 1.6E may be decreased. Seedling weeds are controlled as they come in contact with soil applied herbicide during emergence.

For postemergence control of certain grassy broadleaf weeds, a tank mixture of either Gramoxone Super or Roundup with GOAL 1.6E herbicide can be used. The most effective postemergence weed control is achieved when tank mix combinations of GOAL 1.6E herbicide with Gramoxone Super or Roundup are applied to seedling weeds (less than four inches in height).

DOSAGE

For preemergence control of susceptible weeds, GOAL 1.6E herbicide is recommended at 2.5 to 10 pints (0.5 to 2.0 lb. active) per broadcast acre as a preemergence application directed to the orchard floor beneath coffee plants, or at a dosage of up to 5 pints per broadcast acre as a pre-transplant application.

For directed spray applications, coffee transplants must be healthy, and of suitable size for field transplanting. Avoid spray contact with coffee foliage as injury may result.

For postemergence control of annual broadleaf and grassy weed species, a tank mix of GOAL 1.6E herbicide with Roundup or Gramoxone Super may be applied as a directed spray. GOAL 1.6E herbicide may be tank mixed at 2.5 to 10 pints per broadcast acre according to weeds present and growth stage. The lower rate (0.5 lb. active) will assist in the control of seedling weeds less than four inches in height. The higher rate (2.0 lb. active) is recommended for control of seedling weeds less than four inches in height and will provide additional preemergence control of susceptible weeds. Refer to the Gramoxone Super or Roundup label for susceptible weed species and recommended use rates for control.

WEEDS CONTROLLED PREEMERGENCE

AGERATUM	PURSLANE, COMMON
BUTTONWEED	SPURGE, GARDEN
CROTALARIA	

TIMING AND METHOD OF APPLICATION

DO NOT APPLY PREPLANT OR PREEMERGENCE TO DIRECT SEEDED COFFEE

Treatments should only be applied to healthy coffee stock (as determined by standard commercial growing practices). Care must be taken to prevent direct spray contact with foliage. Coffee foliage receiving accidental spray or drift may be injured.

As a preemergence or postemergence treatment to weeds, apply in a minimum of 30 gallons of water per acre. Use higher volumes to assure adequate coverage in high densities of emerged weeds or heavy trash. GOAL 1.6E herbicide should be directed to the soil and the base of the tree. Use of a low pressure sprayer equipped with a breakaway boom and flat fan or off-center (OC) nozzles is recommended. Spray equipment should be calibrated carefully before each use.

COFFEE

SPECIFIC USE RESTRICTIONS

In addition to the following, also observe GENERAL USE RESTRICTIONS listed at the end of this label.

- Do not apply preplant or preemergence to direct seeded coffee.
- Direct spray toward the base of the trees. Avoid direct plant contact.
- GOAL 1.6E herbicide or any of the combinations recommended on this label should be applied to only healthy growing trees/transplants under standard commercial growing practices.
- Do not apply more than 10 pints (2.0 lb. active) per broadcast acre of GOAL 1.6E herbicide in a single application or 30 pints (6.0 lb. active) per broadcast acre per year.
- Do not apply GOAL 1.6E herbicide within one (1) day of harvesting.
- Applications of GOAL 1.6E herbicide during periods of rapid new foliage growth may cause injury.

CONIFER SEEDBEDS, TRANSPLANTS AND CONTAINER STOCK

GENERAL INFORMATION

GOAL 1.6E is effective as a preemergence and/or postemergence herbicide for the control of certain annual grassy and broadleaf weeds in conifer seedbeds, transplant and conifer stock. The most effective postemergence weed control is achieved when GOAL 1.6E herbicide is applied to seedling weeds less than 4 inches in height. Preemergence control is most effective when spray is applied to clean weed-free soil surfaces. Treated soil surfaces should not be disturbed as the herbicidal effectiveness of GOAL 1.6E may be decreased. Seedling weeds are controlled during emergence as they come in contact with the soil-applied herbicide.

IMPORTANT: Some varieties or cultivars of conifers and ornamental species listed may be susceptible to GOAL 1.6E herbicide. Care should be taken to insure that the particular variety to be sprayed with GOAL 1.6E herbicide is tolerant. It is suggested that unfamiliar species be tested in limited areas prior to application for preemergence and postemergence weed control.

WEEDS CONTROLLED

When GOAL 1.6E herbicide is applied preemergence or postemergence at recommended dosages and weed stages, the following grasses and broadleaf weeds are controlled:

*BARNYARDGRASS	*MORNINGGLORY, TALL
BEDSTRAW, CATCHWEED	MUSTARD, BLUE
BITTERCRESS, LESSER	MUSTARD, TUMBLE
*BLUEGRASS, ANNUAL	MUSTARD, WILD
BUCKWHEAT, WILD	NETTLE, BURNING
BURCLOVER	NIGHTSHADE, BLACK
CARPETWEED	NIGHTSHADE, HAIRY
*CLOVER, RED	OATS, WILD
*CLOVER, WHITE	ORACH, RED
COCKLEBUR, COMMON	PEPPERWEED, YELLOWFLOWER
*CRABGRASS, LARGE	PIGWEED, PROSTRATE
*FIDDLENECK, COAST	PIGWEED, REDROOT
FILAREE, BROADLEAF	PIMPERNEL, SCARLET
FILAREE, REDSTEM	PURSLANE, COMMON
FIREWEED (FROM SEED)	REDMAIDS
FLIXWEED	ROCKET, LONDON
*FOXTAIL, GIANT	SANDSPURRY, RED
*GOOSEGRASS	*SHEPHERDSPURSE
GROUNDCHERRY CUTLEAF	SIDA, PRICKLY
GROUNDCHERRY, WRIGHT	SMARTWEED PENNSYLVANIA
GROUNDSEL, COMMON	SORREL, RED (FROM SEED)
HENBIT	SOWTHISTLE, ANNUAL
JIMSONWEED	SPEEDWELL, BIRDSEYE
KNOTWEED, PROSTRATE	**SPURGE, PROSTRATE
LADYSTHUMB	**SPURGE, SPOTTED
LAMBSQUARTERS COMMON	SPURRY, CORN
LETTUCE, PRICKLY	TANSYMUSTARD
MALLOW, LITTLE	**THISTLE, BULL
MAYWEED	THISTLE, RUSSIAN
MINERSLETTUCE	VELVETLEAF
*MORNINGGLORY, IVYLEAF	**WITCHGRASS
	WOODSORREL, YELLOW

* Highest rate and/or multiple applications may be required for acceptable control.
** Preemergence control only.

GOAL 1.6E herbicide is most effective when applied preemergence to annual grasses. Postemergence applications should be made to seedling grasses not exceeding the 2-leaf stage. The addition of 0.25% (2 pints/100 gallons of spray solution) of TRITON AG-98 or comparable 80% active nonionic spray surfactant, cleared for application on growing crops, enhances the GOAL 1.6E herbicide activity on emerged weeds.

CONIFER SEEDBEDS

To assist in the establishment of conifer seedbeds, a preemergence application should be made after seeding but prior to conifer emergence. Postemergence applications should not be made until a minimum of 5 weeks after emergence of the conifer seedlings. Additional care should be taken if cool, cloudy weather occurs during emergence to make certain that seedlings have hardened off prior to spraying.

Conifers are tolerant to preemergence and postemergence applications of GOAL 1.6E herbicide. Applied postemergence, GOAL 1.6E herbicide will provide both postemergence and residual preemergence control of many broadleaf weeds and annual grass species.

CONIFER SPECIES

GOAL 1.6E herbicide may be applied to conifer seedbeds of numerous species including the following:

DOUGLAS-FIR	*Pseudotsuga menziesii*
FIR	
FRASER	*Abies fraseri*

GRAND	*Abies grandis*
NOBLE	*Abies procera*
HEMLOCK	
EASTERN HEMLOCK	*Tsuga canadensis*
PINE	
AUSTRIAN	*Pinus nigra*
EASTERN WHITE	*Pinus strobus*
HIMALAYAN	*Pinus wallichiana*
JACK	*Pinus banksiana*
LOBLOLLY	*Pinus taeda*
LODGEPOLE	*Pinus contorta*
LONGLEAF	*Pinus palustris*
MONTEREY	*Pinus radiata*
MUGHO	*Pinus mugo*
PONDEROSA	*Pinus ponderosa*
SCOTCH	*Pinus sylvestris*
SHORTLEAF	*Pinus echinata*
SLASH	*Pinus elliottii*
VIRGINIA	*Pinus virginiana*
SPRUCE	
BLUE	*Picea pungens*
DWARF ALBERTA	*Picea glauca Conica*
NORWAY	*Picea abies*
SITKA	*Picea sitchensis*

PREEMERGENCE DOSAGE

Apply 1.25 to 5 pints (0.25 to 1.0 lb. active) of GOAL 1.6E herbicide per broadcast acre as a preemergence application. Where grassy weeds are present, a minimum rate of 2.5 pints (0.5 lb. active) of GOAL 1.6E herbicide per broadcast acre is suggested. In known areas of high weed competition, 5 pints (1.0 lb. active) of GOAL 1.6E herbicide per broadcast acre are recommended.

TIMING AND METHOD OF APPLICATION

Apply recommended dosage after seeding but prior to conifer emergence. GOAL 1.6E herbicide should be thoroughly mixed with clean water at recommended concentration and applied at 20 to 40 psi in a minimum of 20 gallons of water per treated acre. Broadcast to beds and irrigate prior to weed emergence with 1/2 to 3/4 inch of sprinkler irrigation.

POSTEMERGENCE DOSAGE

Apply 1.25 to 2.5 pints (0.25 to 0.5 lb. active) of GOAL 1.6E herbicide per broadcast acre with each postemergence application. Two or three postemergence applications may be necessary for season-long weed control.

TIMING AND METHOD OF APPLICATION

Apply recommended dosage to seedbeds no sooner than 5 weeks after emergence of conifer seedlings. Additional care should be taken if cool, cloudy weather occurs during emergence to make certain that seedlings have hardened off prior to spraying. Application should be made to seedling weeds (less than 4 inches in height). GOAL 1.6E herbicide should be thoroughly mixed with clean water at recommended concentration and applied as a broadast application at 20 to 40 psi in a minimum of 20 gallons of water per treated acre.

CONIFER TRANSPLANTS AND CONTAINER STOCK (INCLUDES 2-0 SEEDLING AND CHRISTMAS TREE PLANTINGS)

Many container-grown conifers and conifer transplants are tolerant to preemergence and postemergence applications of GOAL 1.6E herbicide. Applied postemergence, GOAL 1.6E herbicide will provide both postemergence and preemergence control of many broadleaf weeds and grasses. Postemergence applications should be applied before bud break or after foliage has had an opportunity to harden off. Conifers may be transplanted from seedbeds and sprayed directly providing bud break has not occurred.

The following conifer species in addition to species listed under the CONIFER SEEDBED section have been shown to be tolerant to GOAL 1.6E herbicide.

ARBORVITAE
  Thuja occidentalis
  Thuja orientalis
JUNIPER
  Juniperus chinensis
  Juniperus horizontalis
  Juniperus procumbens
  Juniperus sabina
  Juniperus scopulorum
RED CEDAR
  Juniperus virginiana
WESTERN HEMLOCK
  Tsuga heterophylla
YEW
  Taxus species

DOSAGE

For preemergence or postemergence weed control apply 5 to 10 pints (1.0 to 2.0 lb. active) of GOAL 1.6E herbicide per broadcast acre.

TIMING AND METHOD OF APPLICATION

For optimum weed control, preemergence applications should be made immediately after transplanting seedlings or to weed-free container stock. Postemergence applications should be made to weeds less than 4 inches in height. Two applications may be necessary, in fall-transplanted conifer fields, for season-long weed control. The addition of 0.25% (2 pints/100 gals. of spray solution) of TRITON AG-98 or comparable 80% active nonionic surfactant, cleared for application to growing crops, enhances GOAL 1.6E herbicide activity on emerged weeds. GOAL 1.6E herbicide must be applied only to conifer transplants prior to bud break or after foliage has had an opportunity to harden off. Thoroughly mix with clean water at recommended concentration and apply at 20 to 40 psi in a minimum of 20 gallons of water per treated acre. Spray over the top of transplants. Heavy rainfall immediately following application to emerged weeds may reduce effectiveness.

POAST® HERBICIDE TANK MIX IN CONIFER PLANTATIONS

FOR USE ONLY IN THE STATES OF ALABAMA, GEORGIA, KENTUCKY, NORTH CAROLINA, SOUTH CAROLINA, TENNESSEE, VIRGINIA AND WEST VIRGINIA

GENERAL INFORMATION

GOAL 1.6E is effective as a preemergence and/or postemergence herbicide for the control of certain annual grassy and broadleaf weeds in conifer plantations.

The tank mix of GOAL 1.6E herbicide with Poast will provide additional control of certain annual and perennial grass weeds (see Poast label for weeds).

The following plants are tolerant to a tank mix of GOAL 1.6E herbicide plus Poast:

COMMON NAME	SCIENTIFIC NAME
FIR, FRASER	*Abies Fraseri*
*HEMLOCK, CANADA	*Tsuga canadensis*
PINE, VIRGINIA	*Pinus virginiana*
PINE, WHITE	*Pinus strobus*
SPRUCE, NORWAY	*Picea abies*

*Canada Hemlock has a prolonged period of bud break and new growth, thus, directed applications are recommended during this period.

Broadleaf weeds controlled: See GOAL 1.6E herbicide label.

Grasses controlled: See Poast herbicide label.

GOAL 1.6E herbicide and Poast rates: A maximum of 10 pints per acre per season of GOAL 1.6E herbicide may be tank mixed with Poast. A maximum of 2.5 pints of Poast may be tank mixed with GOAL 1.6E herbicide.

See GOAL 1.6E herbicide label for minimum recommended rates of GOAL 1.6E herbicide and see Poast label for minimum recommended rates of Poast. Two or three applications may be needed for season-long control. In some cases, reduced grass control with Poast may be experienced when tank mixed with GOAL 1.6E herbicide.

TIMING

Applications should be made when weeds are actively growing but before conifer bud break or after conifer foliage has had an opportunity to harden off. Broadleaf weeds must be within the height limitations indicated on the GOAL 1.6E herbicide label.

METHOD OF APPLICATION

Fill the spray tank at least one-third full of clean water and add the recommended amounts of GOAL 1.6E herbicide and Poast while the pump and agitator are running. Complete filling of the spray tank with water. Apply at 20 to 40 psi in a minimum of 20 gallons of water per treated acre. Heavy rainfall immediately following application to emerged weeds may reduce effectiveness.

IMPORTANT: When tank mixing, always read all individual manufacturers' labels. In interpreting all labels for the tank mixture, the most restrictive situations must apply.

CONIFER—SPECIFIC USE RESTRICTIONS

In addition to the following, also observe GENERAL USE RESTRICTIONS listed at the end of this label.

- NOT FOR CONIFER RELEASE IN FOREST MANAGEMENT PROGRAMS OR FOR FOREST REGENERATION APPLICATIONS.
- Do not apply GOAL 1.6E herbicide in an enclosed greenhouse structure as injury to plant foliage may result.
- Do not store or transport treated container stock in an enclosed structure until completion of 4 irrigations (minimum 21 days) as injury to nonlabeled plants may occur.
- Always apply GOAL 1.6E herbicide only to healthy conifer stock. Do not apply GOAL 1.6E herbicide to conifers that are under stress from excessive fertilizer or soil salts, disease, nematodes, frost, drought, flooding, previously applied pesticides, soil insects, or winter injury, as severe injury may result.

COTTON

POST-DIRECTED SPRAY
GENERAL INFORMATION

GOAL 1.6E is a selective herbicide for use as a post-directed application for broadleaf weed control in cotton. Cotton leaves that are accidentally sprayed will exhibit necrotic spotting and may drop from the plant, therefore, care must be exercised to avoid spray contact with the cotton leaves. Crop response may be enhanced if applications are made when excessive soil moisture is present or if rainfall occurs following application. Cotton will outgrow this condition and continue to develop normally.

DOSAGE

GOAL 1.6E herbicide is recommended as a post-directed application at 1.25 to 2.5 pints (0.25 to 0.5 lb. active) per acre. Optimum control is achieved when 2.5 pints of GOAL 1.6E herbicide (0.5 lb. active) per acre are applied to weed seedlings not exceeding 4 true leaves. Effective control of succulent weed seedlings in the 2 to 3 leaf stage can usually be obtained when 1.25 pints of GOAL 1.6E herbicide (0.25 lb. active) per acre are applied. See "Mixing Directions" for surfactant recommendation. Weeds should be in the seedling stage, young and actively growing. Do not count cotyledon leaves.

WEEDS CONTROLLED POSTEMERGENCE

When GOAL 1.6E herbicide is applied as a post-directed application at the recommended weed stage and dosage in cotton, the following weeds are controlled.

COCKLEBUR, COMMON	NIGHTSHADE, HAIRY
CROTON, TROPIC	PIGWEED, REDROOT
GROUNDCHERRY, CUTLEAF	*POINSETTIA, WILD
GROUNDCHERRY, WRIGHT	PURSLANE, COMMON
JIMSONWEED	SESBANIA, HEMP
LAMBSQUARTERS, COMMON	**SICKLEPOD
MORNINGGLORY, ANNUAL (UP TO 6 LEAF)	*SIDA, PRICKLY (TEAWEED)
NIGHTSHADE, AMERICAN BLACK	SMARTWEED, PENNSYLVANIA
NIGHTSHADE, BLACK	VELVETLEAF

* Multiple applications may be required for acceptable control.
** Post-direct applications of GOAL 1.6E herbicide will kill or suppress seedlings not exceeding the one true leaf stage.

TIMING
SOUTHERN COTTON

ALABAMA, ARKANSAS, GEORGIA, LOUISIANA, MISSISSIPPI, MISSOURI, NEW MEXICO, NORTH CAROLINA, OKLAHOMA, SOUTH CAROLINA, TENNESSEE AND TEXAS

Cotton plant height must be a minimum 6 inches or greater. Application to cotton plants less than 6 inches tall may result in severe crop injury and is not recommended. In cotton 6 to 8 inches tall, GOAL 1.6E herbicide must be applied using rigid precision ground sprayer equipment. The use of spray shields is recommended to avoid spray contact with cotton foliage. Use branch lifters or shields if excessive spray contact on larger cotton plants (8 inches or greater) cannot be avoided by the directed spray.

WESTERN COTTON
ARIZONA AND CALIFORNIA

Cotton plant height must be a minimum 6 inches or greater. Application to cotton plants less than 6 inches tall may result in severe crop injury and is not recommended. In cotton 6 to 8 inches tall, GOAL 1.6E herbicide must be applied using **rigid** precision ground sprayer equipment. The use of spray shields is recommended to avoid spray contact with cotton foliage. Use branch lifters or shields if excessive spray contact on larger cotton plants (8 inches or greater) cannot be avoided by the directed spray. To obtain the maximum benefit of postemergence activity, encourage weed emergence by irrigating prior to spraying. Irrigate immediately following herbicide application to obtain greatest benefit of preemergence activity from GOAL 1.6E herbicide on nightshade and groundcherry species.

METHOD OF APPLICATION
SOUTHERN AND WESTERN COTTON

Accurate, uniform placement of GOAL 1.6E herbicide spray is essential for effective weed control and to minimize cotton injury. As a directed postemergence application, GOAL 1.6E herbicide should be applied at 20 to 25 psi using 20 to 40 gallons of spray on a broadcast acre basis. Do **not** exceed 25 psi. Spray should be directed towards the base of the cotton plant. Cotton foliage receiving accidental spray or drift may be injured. Weeds should be in the seedling stage, young and actively growing.

GOAL 1.6E herbicide can be applied using a post-direct spray rig with only 2 flat fan nozzles per row, 1 nozzle on each side of the row. Additional care should be taken when adjusting sprayer prior to application. For best coverage, it is suggested to use 4 flat fan nozzles per row, 2 nozzles on each side of the row. The 2 forward nozzles should point forward and downward while the rear nozzles should point to the rear and downward. With either sprayer system, nozzles should be adjusted to cover the weed foliage with minimum contact to the cotton plant. **Do not use cone nozzles.**

TANK MIXES
WITH GOAL 1.6E HERBICIDE

IMPORTANT: Read and observe all label directions before using. When tank mixing, always read all individual manufacturers' labels. In interpreting all labels for tank mixtures, the most restrictive situations must apply.

DOSAGE

For postemergence control of susceptible grassy and broadleaf weeds in cotton, a tank mixture of GOAL 1.6E herbicide with either Bladex®, Karmex®, or MSMA or Probe can be applied as a post-directed application. Apply at the recommended rates and growth stages to susceptible weed species in a manner described on the respective labels.

Rate of Product Per Broadcast Acre

Product	Rate Per Acre
Bladex 4L	1.2 to 2 pts.
Karmex 80W	0.25 to 0.5 lb.
MSMA 6.6	2.4 pts.
Probe	0.66 to 2.0 lbs.

COTTON — SOUTHERN AND WESTERN
SPECIFIC USE RESTRICTIONS

In addition to the following, also observe GENERAL USE RESTRICTIONS listed at the end of this label.

- SOUTHERN COTTON: Do not apply more than 2.5 pints (0.5 lb. active) per broadcast acre of GOAL 1.6E herbicide per season as a result of a single application or multiple applications.
- WESTERN COTTON: Do not apply more than 2.5 pints (0.5 lb. active) of GOAL 1.6E herbicide per broadcast acre in a single application, or more than a total of 5.0 pints (1.0 lb. active) of Goal 1.6E herbicide per broadcast acre per season as a result of multiple applications.
- Do not apply within 75 days of harvest.

COTTON FALLOW BED AIR APPLICATION

FOR USE ONLY IN THE FOLLOWING CALIFORNIA COUNTIES: FRESNO, KERN, KINGS, MADERA, MERCED, TULARE. TREATMENTS MUST BE MADE OCTOBER 1 THROUGH MARCH 31.

GENERAL INFORMATION

GOAL 1.6E herbicide is effective as a preemergence herbicide when used alone or as a postemergence herbicide when applied in a tank mix combination with Roundup or Gramoxone Super for the control of winter annual broadleaf weeds in fallow beds. Do not apply within 14 days of planting. The fallow bed should be worked with a Lilliston, or similar incorporation tool, to a depth of at least 2 inches prior to planting. It is important to thoroughly break the soil surface prior to planting; weed control should not be expected following breaking of the soil surface.

EXERCISE EXTREME CARE TO AVOID HERBICIDE CONTACT WITH ANY DESIRABLE DORMANT OR NON-DORMANT CROP, PLANT, TREE OR VEGETATION, AS SEVERE INJURY MAY RESULT.

Cotton can be planted 14 days or more following a GOAL 1.6E herbicide fallow bed application.

GOAL 1.6E HERBICIDE USED ALONE
DOSAGE

Goal 1.6E herbicide may be applied at 1.25 to 2.5 pints (0.25 to 0.5 lb. active) per broadcast acre. The lower rate (1.25 pints per acre) should provide up to 4 weeks of preemergence control of susceptible weeds and provide postemergence activity on susceptible weeds (up to 4 leaf stage). The higher rate (2.5 pints per acre) should provide preemergence control of susceptible weeds for up to 8 weeks and postemergence activity on susceptible weeds up to 6 leaf stage. Best preemergence control is achieved when irrigation or rainfall occurs within 3 or 4 weeks following application.

WEEDS CONTROLLED

GOAL 1.6E herbicide will provide preemergence and postemergence* control of the following weeds when used at recommended dosages and weed stage.

CHEESEWEED (MALVA)	MUSTARD, BLACK
FIDDLENECK, COAST	NETTLE, BURNING
FILAREE, BROADLEAF	REDMAIDS
FILAREE, REDSTEM	ROCKET, LONDON
GROUNDSEL, COMMON	SHEPHERDSPURSE
HENBIT	SOWTHISTLE, ANNUAL
MINERSLETTUCE	

* Thorough spray coverage is essential to maximize the postemergence activity of GOAL 1.6E herbicide. For postemergence control, a tank mixture of GOAL 1.6E herbicide with Roundup or Gramoxone Super is recommended.

AERIAL APPLICATION

GOAL 1.6E herbicide should be applied using swirl jet or hollow cone nozzles and a spray pressure less than 40 psi to deliver a minimum spray volume of 10 gallons of water per acre (minimum 5 gallons for GOAL 1.6E/Roundup tank mix). Application should be made at a height of 6 to 10 feet above the soil surface. It is suggested that the nozzles on the spray boom should not be placed any closer to the wing or rotor tips than ¼ of the span; this will minimize the formation of spray or wing tip vortice roll. Nozzles should be spaced and positioned to produce a uniform spray pattern and to minimize or eliminate the formation of droplets 100 microns or less in diameter.

AVOID DRIFT

WHEN APPLYING TO FALLOW BEDS, EXTREME CARE MUST BE EXERCISED TO PREVENT SPRAY DRIFT WHICH COULD RESULT IN DAMAGE TO OTHER CROPS OR DESIRABLE VEGETATION. USE THE FOLLOWING GUIDELINES WHEN AERIAL APPLICATIONS ARE TO BE MADE.

1. Do not apply when the wind direction is not stable, when inversion conditions exist, or when wind velocity exceeds 10 mph.

2. When wind speeds are 5 mph or less, maintain a minimum downwind buffer zone of at least 1/2 mile from all crops and desirable vegetation, except the following:

   Maintain a minimum downwind buffer zone of:

   - 150 feet from dormant tree fruit, dormant vines and overwintering sugarbeets.
   - 650 feet from garlic, jojoba, legumes, onions, pastures, small grains, seedling sugarbeets, and vegetable fallow beds.

3. When wind speeds are between 5 to 10 mph, downwind buffer zones in excess of those listed above are suggested.

4. For upwind and side borders, maintain a minimum buffer zone of 150 feet from any vegetable fallow bed, crop, or desirable vegetation.

The use of a drift control agent may be required by local regulations. However, the drift control agent may decrease the weed control activity.

IMPORTANT

Aerial applicators must be familiar with the EPA-registered label and follow the use precautions. Spraying GOAL 1.6E herbicide in a manner other than as recommended is done at the user's risk. Users are responsible for all loss or damage that results from such spraying. In addition, aerial applicators should follow all applicable state and local regulations and ordinances. In interpreting the label and local regulations, the most restrictive situations should apply to avoid drift hazards.

GOAL 1.6E HEBICIDE—TANK MIXES

DOSAGE

GOAL 1.6E herbicide can be tank mixed with Roundup or Gramoxone Super to obtain postemergence control of annual grassy weeds, volunteer grains and broadleaf weeds.

GOAL 1.6E/ROUNDUP TANK MIX

Tank mix 1.25 to 2.5 pints (0.25 to 0.5 lb. active) of GOAL 1.6E herbicide with 0.75 to 1.0 pint (0.38 to 0.5 lb. active) of Roundup for each acre treated. Apply in a minimum of 5 gallons of water per acre at the recommended rates and growth stages to susceptible weed species in a manner consistent with the respective labels.

GOAL 1.6E/GRAMOXONE SUPER TANK MIX

Tank mix 1.25 to 2.5 pints (0.25 to 0.5 lb. active) of GOAL 1.6E herbicide with 1.5 to 5 pints (0.28 to .93 lb. active) of Gramoxone Super for each acre treated. Apply in a minimum of 10 gallons of water per acre at the recommended rates and growth stages to susceptible weed species in a manner consistent with the respective labels.

COTTON FALLOW BED AIR APPLICATION

SPECIFIC USE RESTRICTIONS

In addition to the following, also observe GENERAL USE RESTRICTIONS listed at the end of this label.

- Read and observe all label directions before using. When tank mixing, always read all individual manufacturers' labels. In interpreting all labels for the tank mixture, the most restrictive situations must apply.
- DO NOT AERIALLY APPLY GOAL 1.6E HERBICIDE (EITHER ALONE OR IN ANY COMBINATION RECOMMENDED ON THIS LABEL) TO FALLOW BEDS BEFORE OCTOBER 1 OR AFTER MARCH 31.
- Do not apply more than 2.5 pints (0.5 lb. active) of GOAL 1.6E herbicide per acre, per fallow season.
- Do not apply GOAL 1.6E herbicide within 14 days of planting.

COTTON FALLOW BED GROUND APPLICATION

GENERAL INFORMATION

GOAL 1.6E is effective as a preemergence and/or postemergence herbicide when used alone or in a tank mix combination with Roundup or Gramoxone Super for the control of winter annual broadleaf weeds in fallow beds. Do not apply within 14 days of planting. The fallow bed should be worked with a Lilliston, or similar incorporation tool, to a depth of a least 2 inches prior to planting. It is important to thoroughly break the soil surface prior to planting; weed control should not be expected following breaking of the soil surface.

Cotton can be planted 14 days or more following a GOAL 1.6E herbicide fallow bed application.

GOAL 1.6E HERBICIDE USED ALONE

DOSAGE

GOAL 1.6E herbicide should be used at 1.25 to 2.5 pints (0.25 to 0.5 lb. active) per broadcast acre. The lower rate (1.25 pints per acre) should provide postemergence control of susceptible weeds (up to 4 leaf stage) and provide up to 4 weeks of preemergence control of susceptible weeds. The higher rate (2.5 pints per acre) should provide postemergence control of susceptible weeds (up to 6 leaf stage) and preemergence control of susceptible weeds for up to 8 weeks. Best preemergence control is achieved when pre-irrigation or rainfall occurs within 3 or 4 weeks following application.

WEEDS CONTROLLED

GOAL 1.6E herbicide will provide preemergence and postemergence control of the following weeds when used at recommended dosages and weed stage.

CHEESEWEED (MALVA)	MINERSLETTUCE
FIDDLENECK, COAST	MUSTARD, BLACK
FILAREE, BROADLEAF	NETTLE, BURNING
FILAREE, REDSTEM	REDMAIDS
GROUNDSEL, COMMON	ROCKET, LONDON
HENBIT	SOWTHISTLE, ANNUAL

GOAL 1.6E is a contact herbicide, therefore, coverage is essential for acceptable postemergence control. If dense weed populations, oversized weed seedlings, volunteer grains, annual grasses or unfavorable environmental conditions exist, a tank mixture of GOAL 1.6E herbicide with Gramoxone Super or Roundup for postemergence control is recommended.

METHOD OF APPLICATION

GOAL 1.6E herbicide should be applied in a minimum of 20 gallons of water per acre. The volume of water used should be increased as the weeds become taller and more dense. Use a low pressure sprayer equipped with flat fan nozzles. Spray equipment should be calibrated carefully before each use.

TANK MIXES WITH GOAL 1.6E HERBICIDE

IMPORTANT: Read and observe all label directions before using. When tank mixing, always read all individual manufacturers' labels. In interpreting all labels for the tank mix, the most restrictive situations must apply.

DOSAGE

GOAL 1.6E herbicide can be tank mixed with Gramoxone Super or Roundup to obtain postemergence control of annual grassy weeds. Tank mix 1.25 to 2.5 pints (0.25 to 0.5 lb. active) of GOAL 1.6E herbicide with 1.5 to 5 pints (0.28 to .93 lb. active) of Gramoxone Super or 0.75 to 1.0 pint (0.38 to 0.5 lb. active) of Roundup for each acre treated. Refer to the "Preplant or Preemergence Uses" section on the Gramoxone Super label or the FALLOW AND REDUCED TILLAGE SYSTEM section on the Roundup label for specific use directions and restrictions. Apply at the recommended rates and growth stages to susceptible weed species in a manner described on the respective labels.

COTTON FALLOW BED GROUND APPLICATION

SPECIFIC USE RESTRICTIONS

In addition to the following, also observe GENERAL USE RESTRICTIONS listed at the end of this label.

- Do not apply more than 2.5 pints (0.5 lb. active) of GOAL 1.6E herbicide per acre, per fallow season.
- Do not apply GOAL 1.6E herbicide within 14 days of planting.

GUAVA

FOR USE ONLY IN HAWAII

GENERAL INFORMATION

GOAL 1.6E is effective as a preemergence herbicide when used alone for the control of certain annual broadleaf weeds in bearing and nonbearing guava plantings.

For postemergence control of certain grassy and broadleaf weeds, a tank mixture of either Gramoxone Super or Roundup with GOAL 1.6E herbicide can be applied to seedling weeds.

Check individual product labels to determine suitability and use rates for crop.

GOAL 1.6E HERBICIDE USED ALONE

DOSAGE

GOAL 1.6E herbicide is recommended for postemergence control of susceptible weeds at 2.5 to 10 pints (0.5 to 2.0 lb. active) per broadcast acre.

For preemergence control of susceptible weeds, use 6 to 10 pints (1.2 to 2.0 lb. active) of GOAL 1.6E herbicide per broadcast acre.

WEEDS CONTROLLED POSTEMERGENCE

Apply 2.5 to 10 pints (0.5 to 2.0 lb. active) of GOAL 1.6E herbicide per broadcast acre. Applications to weeds beyond the 4 leaf stage may result in partial control.

PURSLANE, COMMON	SPURGE, GARDEN

WEEDS CONTROLLED PREEMERGENCE

Apply 6 to 10 pints (1.2 to 2.0 lb. active) of GOAL 1.6E herbicide per broadcast acre.

AGERATUM	PURSLANE, COMMON
BUTTONWEED	SPURGE, GARDEN
CROTALARIA	

TIMING AND METHOD OF APPLICATION

Treatments should be applied only to healthy guava trees. Care must be taken to prevent direct spray or drift from contacting green stems, fruit or foliage, as injury may result. Applications should be made only after new foliage has hardened off, or injury may result.

As a preemergence or postemergence treatment to weeds, apply in a minimum of 15 gallons of water per acre. Use higher volumes to assure adequate coverage in high densities of emerged weeds or heavy trash. GOAL 1.6E herbicide should be directed to the soil and the base of the tree. Use of a low pressure sprayer equipped with a breakaway boom and flat fan or off-center (OC) nozzles is recommended. An off-center nozzle positioned at the end of the boom may be desired. Spray shields are suggested for use in young trees.

TANK MIXES WITH GOAL 1.6E HERBICIDE

IMPORTANT: Read and observe all label directions before using. When tank mixing, always read all individual manufacturers' labels. In interpreting all labels for the tank mixture, the most restrictive situations must apply.

DOSAGE

For postemergence control of susceptible grassy and broadleaf weeds in guava plantings, a tank mixture of GOAL 1.6E herbicide with either Gramoxone Super or Roundup can be used. Apply at recommended rates and growth stages to susceptible weed species in a manner described on the respective labels.

WEEDS CONTROLLED POSTEMERGENCE

In addition to the weeds controlled by GOAL 1.6E herbicide used alone, control of susceptible weeds listed on the respective labels for the following products is also obtained: Gramoxone Super Roundup

GUAVA—SPECIFIC USE RESTRICTIONS

In addition to the following, also observe GENERAL USE RESTRICTIONS listed at the end of this label.

- Do not apply more than 10 pints (2.0 lb. active) per broadcast acre of GOAL 1.6E herbicide in a single application or more than 20 pints (4.0 lb. active) per season.
- Do not apply GOAL 1.6E herbicide within 1 day of harvest.
- Direct spray toward the base of the trees. Avoid direct plant contact.
- GOAL 1.6E herbicide or any of the combinations recommended on this label should be applied to only healthy growing trees.
- GOAL 1.6E applications should be made only after new foliage has hardened off.

JOJOBA

GENERAL INFORMATION

GOAL 1.6E is a selective herbicide for postemergence and preemergence control of certain broadleaf weeds in jojoba. GOAL 1.6E herbicide should be post-directed to the base of the jojoba plant to avoid possible phytotoxicity to the jojoba foliage. Over-the-top applications may exhibit burning, crinkling or bronzing of jojoba foliage, particularly to the youngest leaves, flowers or buds present at the time of application.

DOSAGE

GOAL 1.6E herbicide is recommended for postemergence and preemergence control of susceptible seedling weeds (up to 12 inches in height) at 10 pints (2.0 lb. active) per broadcast acre. For optimum residual control, apply during the fall or winter.

For early postemergence control of susceptible seedling weeds (less than 8 inches in height) apply GOAL 1.6E herbicide at a rate of 5 pints (1.0 lb. active) per broadcast acre.

WEEDS CONTROLLED POSTEMERGENCE

FIDDLENECK, COAST	MINERSLETTUCE
**FILAREE, BROADLEAF	NETTLE, BURNING
**FILAREE, REDSTEM	*PIGWEED, REDROOT
**FILAREE, WHITESTEM	REDMAIDS
*GROUNDSEL, COMMON	SHEPHERDSPURSE
HENBIT	SOWTHISTLE, ANNUAL
MALLOW, LITTLE (MALVA, CHEESEWEED)	

WEEDS CONTROLLED PREEMERGENCE

BURCLOVER	LETTUCE, PRICKLY
FIDDLENECK, COAST	MALLOW, LITTLE (MALVA, CHEESEWEED)
FILAREE, BROADLEAF	PIGWEED, REDROOT
FILAREE, REDSTEM	PURSLANE, COMMON
FILAREE, WHITESTEM	REDMAIDS
GROUNDSEL, COMMON	ROCKET, LONDON
HENBIT	SHEPHERDSPURSE
KNOTWEED, PROSTRATE	SOWTHISTLE, ANNUAL
LAMBSQUARTERS, COMMON	

*Highest rate may be required for acceptable postemergence control.

**GOAL 1.6E herbicide at the 10 pint rate (2.0 lb. active) will provide control of filaree not exceeding the 4 inch stage. Applications to filaree beyond the 4 inch stage may result in partial control.

TIMING AND METHOD OF APLICATION

Apply the first application of GOAL 1.6E herbicide after jojoba plants have grown to a minimum 6 inch height or greater. Additional applications should be applied as needed for post and preemergence weed control. Weed height should not exceed 12 inches or unsatisfactory weed control may result.

GOAL 1.6E herbicide should be applied in a minimum spray volume of 40 gallons of water per acre depending upon density of emerged weeds. Spray volume should be increased as weed height and density increase. Use a low-pressure sprayer equipped with flat fan nozzles. Spray equipment should be calibrated carefully before each use.

JOJOBA—SPECIFIC USE RESTRICTIONS

In addition to the following, also observe GENERAL USE RESTRICTIONS listed at the end of this label.

- Avoid direct spray or drift contact of GOAL 1.6E herbicide with jojoba flowers or buds as severe injury may result.
- Do not apply more than 10 pints (2.0 lb. active) per broadcast acre in a single application.

MINT
SPEARMINT/PEPPERMINT

FOR USE ONLY IN IDAHO, OREGON AND WASHINGTON

GENERAL INFORMATION

GOAL 1.6E is a selective herbicide for the control of certain annual grasses and broadleaf weeds in spearmint and peppermint grown in Idaho, Oregon and Washington. Applications should only be made to spearmint and peppermint that are dormant.

METHOD OF APPLICATION

Application must be made prior to new spring growth or severe crop injury may result. GOAL 1.6E herbicide should be thoroughly mixed with clean water at recommended concentration and applied at 20 to 40 psi in 20 to 40 gallons of water per acre.

WEEDS CONTROLLED

When GOAL 1.6E herbicide is applied as a dormant application at recommended dosages in spearmint and peppermint, the following annual weeds are controlled:

BEDSTRAW, CATCHWEED	NIGHTSHADE, HAIRY
*BLUEGRASS ANNUAL	*OATS, WILD
FLIXWEED	ORACH, RED
GROUNDSEL, COMMON	PEPPERWEED, YELLOWFLOWER
LAMBSQUARTERS, COMMON	PIGWEED, REDROOT
LETTUCE, PRICKLY (CHINA LETTUCE)	*RYEGRASS, ITALIAN
MUSTARD, BLUE (PURPLE MUSTARD)	SHEPERDSPURSE
	SOWTHISTLE, ANNUAL
MUSTARD, TUMBLE (JIM HILL MUSTARD)	TANSYMUSTARD
	THISTLE, RUSSIAN

* Control of annual grasses is best obtained when GOAL 1.6E herbicide is applied prior to emergence. Postemergence control of winter annual grasses is generally unsatisfactory if applications are made after the 1 to 2 leaf stage.

WESTERN OREGON

PEPPERMINT (WILLAMETTE VALLEY)

Apply 2.5 to 3.75 pints (0.5 to 0.75 lb. active) of GOAL 1.6E herbicide from November through February to dormant peppermint only. Treatments in January or February generally provide better residual preemergence control of annual broadleaf weeds. Full season weed control should not be expected from this treatment.

- DO NOT APPLY GOAL 1.6E HERBICIDE IN THE WILLAMETTE VALLEY TO MINT THAT HAS BEEN PLOWED.

OREGON AND WASHINGTON (EAST OF CASCADES) AND WESTERN IDAHO

SPEARMINT AND PEPPERMINT

Apply 5 to 10 pints (1 to 2 lb. active) of GOAL 1.6E herbicide from December through March to dormant mint only. Later winter applications will provide maximum activity on summer weeds. Summer grass control may be inconsistent. For best results, fall plowed fields should be harrowed to provide a smooth surface prior to application. Plowed fields should not be harrowed after GOAL 1.6E herbicide has been applied, as soil disturbance will decrease the herbicidal effectiveness. In furrow-irrigated fields, corrugating must be done prior to application. Corrugating after application can cover treated rows with untreated soil, resulting in poor weed control.

MINT (SPEARMINT AND PEPPERMINT)

SPECIFIC USE RESTRICTIONS

In addition to the following, also observe GENERAL USE RESTRICTIONS listed at the end of this label.

- Apply GOAL 1.6E herbicide only to healthy spearmint and peppermint. Do not apply to spearmint or peppermint that has been weakened by disease, nematodes, soil insects, or winter injury, as severe injury may result.

ONIONS

GENERAL INFORMATION

GOAL 1.6E is a selective herbicide for postemergence application to direct-seeded and transplanted onions for early postemergence control of certain annual broadleaf and grass weeds. Initial spray application should be made only when the onions have reached the development stage specified in the DOSAGE section and the SPECIFIC USE RESTRICTIONS section of this label. On onion transplants spray as soon after transplanting as practical. GOAL 1.6E herbicide can cause necrotic lesions, twisting, pigtailing or stunting of the onion plants. Injury will be more severe if applications are made immediately following or during cool, wet weather and/or if applications are made prior to the development stage of the onion plants as specified in the DOSAGE section and the SPECIFIC USE RESTRICTIONS section of this label.

DOSAGE

SEEDED ONIONS

NORTHEASTERN STATES

CONNECTICUT, MAINE, MASSACHUSETTS, NEW HAMPSHIRE, NEW JERSEY, NEW YORK, RHODE ISLAND AND VERMONT

GOAL 1.6E herbicide is recommended for postemergence control at 2.4 to 4.8 fluid ounces (0.03 to 0.06 lb. active) per acre when applied postemergence to seeded onions that have *at least three (3) true leaves.* Multiple treatments at the aforementioned rate may be applied. Do not apply more than 2.5 pints (0.5 lb. active) per broadcast acre of GOAL 1.6E herbicide as a result of multiple applications in one season.

WESTERN STATES

ARIZONA, CALIFORNIA, COLORADO, IDAHO, NEVADA, NEW MEXICO, OREGON, TEXAS, UTAH AND WASHINGTON

GOAL 1.6E herbicide is recommended for postemergence control at 0.6 to 1.25 pints (0.12 to 0.25 lb. active) per acre when applied postemergence to onions that have *at least two (2) true leaves.* Multiple treatments at the aforementioned rates may be applied. Do not apply more than 2.5 pints (0.5 lb. active) per broadcast acre of GOAL 1.6E herbicide in one season.

ALL OTHER STATES

GOAL 1.6E herbicide is recommended for postemergence control at 0.6 pints (0.12 lb. active) per acre when applied postemergence to onions that have *at least two (2) true leaves.* Multiple treatments at the aforementioned rates may be applied. Do not apply more than 2.5 pints (0.5 lb. active) per broadcast acre of GOAL 1.6E herbicide in one season.

TRANSPLANTED ONIONS (ALL STATES)

Transplanted onions are most tolerant of a GOAL 1.6E herbicide application immediately after transplanting. For all states except the Northeastern states listed under DOSAGE—SEEDED ONIONS section, an application of up to 2.5 pints (0.5 lb. active) per acre within two days after transplanting may be made. If less than 2.5 pints per acre is applied, a second application can be made two weeks or more after transplanting. Do not exceed the maximum use rate of 2.5 pints (0.5 lb. active) of GOAL 1.6E herbicide per acre per season. For transplanted onions in the Northeastern states, see DOSAGE—SEEDED ONIONS section.

For transplanted onions apply the same rates within two days after transplanting.

WEEDS CONTROLLED

GOAL 1.6E herbicide will provide postemergence control of the following weeds when applied at the recommended dosage and leaf stage (2 to 4 leaves):

CANARYGRASS (ANNUAL)	*PIGWEED, REDROOT
EVENINGPRIMROSE, CUTLEAF	PUNCTUREVINE
	*PURSLANE, COMMON
GROUNDSEL, COMMON	ROCKET, LONDON
MALLOW, LITTLE (MALVA)	SAGE, LANCELEAF
NIGHTSHADE BLACK	*SHEPHERDSPURSE
*PIGWEED, PROSTRATE	SOWTHISTLE, ANNUAL

* Specific weeds controlled at rates recommended for use in Northeastern States (see DOSAGE section).

TIMING AND METHOD OF APPLICATION

For best postemergence control of susceptible weeds, apply when the weeds are in the 2 to 4 leaf stage. Application of GOAL 1.6E herbicide after the weeds exceed the maximum leaf stage may result in reduced weed control. More than one postemergence application may be necessary to control subsequent weed flushes.

GOAL 1.6E herbicide should be thoroughly mixed with clean water at recommended concentrations, and applied in a minimum of 40 gallons of water per acre. Use conventional ground spray equipment with flat fan spray nozzles, at 20 to 40 psi. Accurately calibrate spray equipment prior to each use. Avoid drift to all other crops and non-target areas. Thoroughly flush the spray equipment (tank, hose, pump, boom) with water before and after each use. Residual GOAL 1.6E herbicide remaining in spray equipment may damage other crops.

ONIONS—SPECIFIC USE RESTRICTIONS

In addition to the following, also observe GENERAL USE RESTRICTIONS listed at the end of this label.

- In all states except Northeastern states, do not start spraying until the onions *have two (2) fully developed true leaves.* In the Northeastern states (Connecticut, Maine, Massachusetts, New Hampshire, New Jersey, New York, Rhode Island, Vermont), do not start spraying until the onions *have three (3) fully developed true leaves.* Applications made prior to the recommended onion development stage may result in serious injury and is not recommended.
- Do not apply more than a total of 2.5 pints (0.5 lb. active) per acre of GOAL 1.6E herbicide during one use season.
- Do not apply within 60 days of harvest.
- Use only on dry bulb onions.
- Do not apply to onions grown for seed, except as specified on other approved supplemental labeling.
- Do not mix GOAL 1.6E herbicide with oils, surfactants, liquid fertilizers or pesticides except as specified on other approved Rohm and Haas supplemental labeling.
- Do not apply to onion plants that are under stress due to drought, flooding, excessive fertilizer or soil salts, wind injury, hail, frost damage, injury from previously applied pesticides, or injury due to insects or diseases.

TREEFRUIT/NUT/VINE

DORMANT APPLICATION

ALMOND, APPLE, APRICOT, AVOCADO, BEECHNUT, BRAZIL NUT, BUTTERNUT, CASHEW, CHERRY, CHESTNUT, CHINQUAPIN, CRAB APPLE, DATE, FIG, FILBERT, GRAPES, HICKORY NUT, KIWI, LOQUAT, MACADAMIA NUT, NECTARINE, OLIVES, PEACH, PEAR, PECAN, PISTACHIO, PLUM, POMEGRANATES, PRUNE, QUINCE, WALNUT

GENERAL INFORMATION

GOAL 1.6E herbicide is effective as a preemergence and/or postemergence herbicide when used alone or in recommended combinations, for the control of certain annual broadleaf weeds, in certain bearing and non-bearing treefruit, nut or vine plantings.

The most effective postemergence weed control is achieved when GOAL 1.6E herbicide is applied to seedling weeds. For postemergence control of certain grassy and broadleaf weeds, a tank mixture of GOAL 1.6E herbicide with either Gramoxone Super or Roundup can be used.

For residual grass control in dormant treefruit, nut or vine plantings, a tank mixture of GOAL 1.6E herbicide with Devrinol, Karmex, KERB®, simazine, Solicam or Surflan can be applied. Contact herbicides such as Gramoxone Super or Roundup may also be added to the tank mixture.

Check individual product labels to determine suitability and use rates for various crops.

GOAL 1.6E HERBICIDE USED ALONE

GEOGRAPHIC USE DIRECTIONS

ARIZONA AND CALIFORNIA

DOSAGE

GOAL 1.6E herbicide is recommended for postemergence control at 2.5 to 10 pints (0.5 to 2.0 lb active) per broadcast acre.

WEEDS CONTROLLED POSTEMERGENCE—(weeds up to 4 inches high) Apply 2.5 to 10 pints (0.5 to 2.0 lb. active) of GOAL 1.6E herbicide per broadcast acre. Applications to weeds beyond this 4 inch stage may result in partial control.

CHEESEWEED (MALVA)	MINERSLETTUCE
FIDDLENECK, COAST	NETTLE, BURNING
*FILAREE, BROADLEAF	PIGWEED, REDROOT
*FILAREE, REDSTEM	REDMAIDS
*FILAREE, WHITESTEM	SHEPHERDSPURSE
GROUNDSEL, COMMON	SOWTHISTLE, ANNUAL
HENBIT	

* GOAL 1.6E herbicide at the 10 pint rate (2.0 lb. active) will provide control of filaree not exceeding the 4 inch stage. Applications to filaree beyond the 4 inch stage may result in partial control.

WEEDS CONTROLLED PREEMERGENCE—Apply 6 to 10 pints (1.2 to 2.0 lb. active) of GOAL 1.6E herbicide per broadcast acre.

BURCLOVER	LAMBSQUARTERS, COMMON
CHEESEWEED (MALVA)	LETTUCE, PRICKLY
FIDDLENECK, COAST	PIGWEED, REDROOT
FILAREE, BROADLEAF	PURSLANE, COMMON
FILAREE, REDSTEM	REDMAIDS
FILAREE, WHITESTEM	ROCKET, LONDON
GROUNDSEL, COMMON	SHEPHERDSPURSE
HENBIT	SOWTHISTLE, ANNUAL
KNOTWEED, PROSTRATE	

ALL OTHER STATES (EXCEPT CALIFORNIA AND ARIZONA)

DOSAGE

GOAL 1.6E herbicide is recommended for postemergence control at 2.5 to 10 pints (0.5 to 2.0 lb. active) per broadcast acre. For preemergence control of susceptible weeds, use 6 to 10 pints (1.2 to 2.0 lb. active) per broadcast acre.

WEEDS CONTROLLED POSTEMERGENCE—Apply 2.5 to 10 pints (0.5 to 2.0 lb. active) of GOAL 1.6E herbicide per broadcast acre. The lower rate is recommended for the control of susceptible seedling weeds in the early postemergence stage up to the 4 leaf stage. The higher rate (2.0 lb. active) should be used for weeds up to the 6 leaf stage. Applications to weeds beyond the 6 leaf stage may result in partial control.

BALSAMAPPLE	NIGHTSHADE, BLACK
COCKLEBUR, COMMON	PEPPERWEED, VIRGINIA
* CUDWEED, NARROWLEAF	PIGWEED, REDROOT
** EVENINGPRIMROSE, CUTLEAF	POINSETTIA, WILD
GROUNDCHERRY, CUTLEAF	PURSLANE, COMMON
GROUNDCHERRY, WRIGHT	SESBANIA, HEMP
JIMSONWEED	SHEPHERDSPURSE
LAMBSQUARTERS, COMMON	SIDA, PRICKLY (TEAWEED)
MORNINGGLORY, ANNUAL	SMARTWEED, PENNSYLVANIA
NIGHTSHADE, AMERICAN BLACK	SOWWHISTLE, ANNUAL
	VELVETLEAF

* Maximum 0.5 inch diameter.

** Highest rate and/or multiple applications may be required for acceptable control. Do not apply more than 10 pints (2.0 lb. active) per broadcast acre of GOAL 1.6E herbicide in one season.

WEEDS CONTROLLED PREEMERGENCE—Apply 6 to 10 pints (1.2 to 2.0 lb. active) of GOAL 1.6E herbicide per broadcast acre.

CAMPHORWEED	PEPPERWEED, VIRGINIA
CUDWEED, NARROWLEAF	PIGWEED, REDROOT
*EVENINGPRIMROSE, CUTLEAF	POINSETTIA, WILD
	SIDA, PRICKLY
GROUNDCHERRY CUTLEAF	SMARTWEED, PENNSYLVANIA
JIMSONWEED	
LAMBSQUARTERS, COMMON	SOWWHISTLE, ANNUAL
NIGHTSHADE, AMERICAN BLACK	SPURGE, PROSTRATE
	SPURGE, SPOTTED
NIGHTSHADE, BLACK	VELVETLEAF

* Highest rate and/or multiple applications may be required for acceptable control. Do not apply more than 10 pints (2.0 lb. active) per broadcast acre of GOAL 1.6E herbicide in one season.

ALL STATES

TIMING AND METHOD OF APPLICATION

In Arizona and California, applications of GOAL 1.6E herbicide should be made October 1 through February 15. In Coachella Valley (California), do not apply to grapes after February 1.

In all states, dormancy determines the suitability and safety of using GOAL 1.6E herbicide in treefruit, nut and vine plantings. Do not apply GOAL 1.6E herbicide after buds start to swell or when fruit/nuts are present.

As a preemergence treatment, apply a minimum of 40 gallons of water per acre. Use higher volumes to assure adequate coverage in high densities of emerged weeds or heavy trash. Best preemergence results are achieved when spray is applied to a relatively weed-free established berm or soil surface. GOAL 1.6E herbicide should be directed to the soil and the base of dormant trees or vines. Use a low-pressure sprayer equipped with a breakaway boom and flat fan nozzles. An off center (OC) nozzle positioned at the end of the boom may be desired. See SPECIFIC USE RESTRICTIONS for GOAL 1.6E herbicide application on dormant tree or vine plantings.

SPRAY VOLUME:

Weed Stage	Gallons of Water per Acre
Preemergence	40 or more
Postemergence (up to 4 inch or 4 leaf stage)	40 or more
Exceeding 4 inch or 4 leaf stage	100 or more

TANK MIXES WITH GOAL 1.6E HERBICIDE

IMPORTANT: Read and observe all label directions before using. When tank mixing, always read all individual manufacturers' labels. In interpreting all labels for the tank mixture, the most restrictive situations must apply.

DOSAGE

For preemergence control of susceptible grassy and broadleaf weeds in certain dormant treefruit, nut crops and vines, a tank mixture of GOAL 1.6E herbicide with Devrinol, Karmex, KERB, simazine, Solicam or Surflan can be applied. Apply at the recommended rates and growth stages to susceptible weed species in a manner described on the respective labels.

For postemergence control of susceptible grassy and broadleaf weeds in certain dormant treefruit, nut crops and vines, a tank mixture of Gramoxone Super or Roundup with GOAL 1.6E herbicide or combinations of GOAL 1.6E herbicide plus Devrinol, Karmex, KERB, simazine, Solicam or Surflan with either Gramoxone Super or Roundup can be used. Apply at the recommended rates and growth stages to susceptible weed species in a manner described on the respective labels.

WEEDS CONTROLLED

In addition to the weeds controlled by GOAL 1.6E herbicide used alone, control of susceptible weeds listed on the respective labels for the following products is also obtained:

Devrinol	simazine*
Gramoxone Super	Solicam
Karmex	Surflan
KERB	

* In addition, simazine provides preemergence control of horseweed (marestail).

TREEFRUIT/NUT/VINE

SPECIFIC USE RESTRICTIONS

In addition to the following, also observe GENERAL USE RESTRICTIONS listed at the end of this label.

- Apply only to dormant fruit or nut trees or vines. Dormancy determines the suitability and safety of using GOAL 1.6E herbicide on treefruit, nuts and vines. Do not apply after buds start to swell or when foliage or fruit/nuts are present.
- **IN ARIZONA AND CALIFORNIA, DO NOT APPLY GOAL 1.6E HERBICIDE (EITHER ALONE OR IN ANY COMBINATION RECOMMENDED ON THIS LABEL) TO TREE OR VINE PLANTINGS BEFORE OCTOBER 1 OR AFTER FEBRUARY 15. DO NOT APPLY GOAL 1.6E HERBICIDE (EITHER ALONE OR IN ANY COMBINATION RECOMMENDED ON THIS LABEL) TO GRAPES IN THE COACHELLA VALLEY (CALIFORNIA) BEFORE OCTOBER 1 OR AFTER FEBRUARY 1.**
- Do not apply more than 10 pints (2.0 lb. active) per broadcast acre of GOAL 1.6E herbicide in one season.
- Do not apply to grapes or kiwi established less than 3 years unless vines are on a trellis wire a minimum of 3 feet above the soil surface.
- Do not apply to grapes or kiwi which are not staked or trellised unless grapes are free-standing.
- GOAL 1.6E herbicide or any of the combinations recommended on this label should be applied to only healthy growing trees or vines.
- Direct spray toward the base of tree or vines. Avoid direct plant contact.

SPECIFIC USE RESTRICTIONS FOR INDIVIDUAL CROPS ARE FOUND UNDER DIRECTIONS FOR USE IN EACH CROP GROUP SECTION.

USE RESTRICTIONS THAT APPLY TO ALL REGISTERED APPLICATIONS ARE LISTED BELOW:

GENERAL USE RESTRICTIONS

- Read and observe all label directions before using. When tank mixing, always read all individual manufacturers' labels. In interpreting all labels for the tank mixture, the most restrictive situations must apply.
- Do not contaminate irrigation water or water used for domestic purposes.
- Do not use any plants treated with GOAL 1.6E herbicide for feed or forage.
- Do not feed or allow animals to graze on any areas treated with GOAL 1.6E herbicide.
- GOAL 1.6E herbicide should be applied only by ground application equipment except as specifically directed on this label or on other approved Rohm and Haas Company supplemental labeling.
- Do not apply when weather conditions favor drift. Avoid drift to all non-target areas. GOAL 1.6E herbicide is phytotoxic to plant foliage.
- Thoroughly flush spray equipment (tank, pump, hoses and boom) with clean water before and after each use. Residual GOAL 1.6E herbicide remaining in spray equipment may damage other crops. To assist removal of GOAL 1.6E herbicide residues in spray equipment, TRITON AG-98 OR TRITON CS-7® may be added at the rate of 1 quart per 100 gallons of water during flushing.
- Use GOAL 1.6E herbicide only for recommended purposes and at recommended rates.
- Do not treat ditch banks or waterways with GOAL 1.6E herbicide.
- CHEMIGATION: Do not apply this product through any type of irrigation system.
- **ROTATION CROP RESTRICTIONS**

Do not rotate to small-grain crops (includes barley, buckwheat, corn, proso millet, pearl millet, oats, popcorn, rice, rye, sorghum, triticale, wheat, wild rice) within 10 months following a GOAL treatment.

Do not direct seed any crops, other than GOAL-labeled crops, within 60 days following a GOAL treatment.

Do not transplant seedling crops, other than GOAL-labeled crops, within 30 days following a GOAL treatment. **IMPORTANT:** TREATED SOIL MUST BE THOROUGHLY INCORPORATED TO A DEPTH OF 4 INCHES AFTER HARVEST (OR ABANDONING) OF TREATED CROP BUT PRIOR TO PLANTING OF THE ROTATIONAL CROP.

FAILURE TO ACHIEVE THIS THOROUGH AND COMPLETE INCORPORATION OR TO FOLLOW THE REQUIRED MINIMUM PLANT-BACK INTERVAL MAY RESULT IN CROP INJURY, STAND REDUCTION AND/OR VIGOR REDUCTION OF THE PLANT-BACK CROP. See specific fallow bed labeling regarding crop planting information for applications of GOAL 1.6E made to a fallow bed or fallow field.

WEED CONTROL LIST

AGERATUM	*Ageratum conyzoides*
BALSAMAPPLE	*Momordica charantia*
BARNYARDGRASS (WATERGRASS)	*Echinochloa crus-galli*
BEDSTRAW, CATCHWEED	*Galium aparine*
BITTERCRESS, LESSER	*Cardamine oligosperma*
BLUEGRASS, ANNUAL	*Poa annua*
BLUEGRASS, KENTUCKY	*Poa pratensis*
BROME, RIPGUT	*Bromus rigidus*
BUCKWHEAT, WILD	*Polygonum convolvulus*
BURCLOVER	*Medicago hipida*
BUTTONWEED	*Borreria laevis*
CAMPHORWEED	*Heterotheca subaxillaris*
CANARYGRASS (ANNUAL)	*Phalaris canariensis*
CARPETWEED	*Mollugo verticillata*
CHEESEWEED (MALVA)	*Malva parviflora*
CHICKWEED, COMMON	*Stellaria media*
CHICKWEED, STICKY	*Cerastium viscosum*
CLOVER, RED	*Trifolium pratense*
CLOVER, WHITE	*Trifolium repens*
COCKLEBUR, COMMON	*Xanthium pensylvanicum*
CRABGRASS, LARGE (HAIRY)	*Digitaria sanguinalis*
CRABGRASS, SMOOTH	*Digitaria ischaemum*
CROTALARIA	*Crotalaria species*
CROTON, TROPIC	*Croton glandulosus*
CUDWEED, NARROWLEAF	*Gnaphalium falcatum*
CUPGRASS, SOUTHWESTERN	*Eriochloa gracilus*
EVENINGPRIMROSE, CUTLEAF	*Oenothera laciniata*
FIDDLENECK COAST	*Amsinckia intermedia*
FILAREE, BROADLEAF	*Erodium botrys*
FILAREE, REDSTEM	*Erodium cicutarium*
FILAREE, WHITESTEM	*Erodium moschatum*
FIREWEED (FROM SEED)	*Epilobium angustifolium*
FLIXWEED	*Descurainia sophia*
FOXTAIL, GIANT	*Setaria faberi*
GOOSEGRASS	*Eleusine indica*
GROUNDCHERRY, CUTLEAF	*Physalis angulata*
GROUNDCHERRY, WRIGHT	*Physalis wrightii*
GROUNDSEL, COMMON	*Senecio vulgaris*
HENBIT	*Lamium amplexicaule*
HORSEWEED (MARESTAIL)	*Conyza canadensis*
JIMSONWEED	*Datura stramonium*
KNOTWEED, PROSTRATE	*Polygonum aviculare*
LADYSTHUMB (SMARTWEED)	*Polygonum persicaria*
LAMBSQUARTERS, COMMON	*Chenopodium album*
LETTUCE, PRICKLY (CHINA LETTUCE)	*Lactuca serriola*
MALLOW, LITTLE (MALVA)	*Malva parviflora*
MAYWEED (DOG FENNEL)	*Anthemis cotula*
MINERSLETTUCE	*Montia perfoliata*
MORNINGGLORY SPECIES, ANNUAL	*Ipomoea species*
MORNINGGLORY, TALL	*Ipomea purpurea*
MUSTARD, BLACK	*Brassica nigra*
MUSTARD, BLUE (PURPLE MUSTARD)	*Chorispora tenella*
MUSTARD, COMMON YELLOW	*Brassica campestris*
MUSTARD, TUMBLE (JIM HILL MUSTARD)	*Sisymbrium altissimum*
MUSTARD, WILD	*Brassica kaber*
NETTLE, BURNING	*Urtica urens*
NIGHTSHADE, AMERICAN BLACK	*Solanum nodiflorum*
NIGHTSHADE, BLACK	*Solanum nigrum*
NIGHTSHADE, HAIRY	*Solanum sarachoides*
OATS, WILD	*Avena fatua*
ORACH RED	*Atriplex rosea*
ORCHARDGRASS	*Dactylis glomerata*
OXALIS (BERMUDA BUTTERCUP)	*Oxalis pes-caprae*
PEPPERWEED, VIRGINIA	*Lepidium virginicum*
PEPPERWEED, YELLOWFLOWER	*Lepidium perfoliatum*
PIGWEED, PROSTRATE	*Amaranthus blitoides*
PIGWEED, REDROOT	*Amaranthus retroflexus*
PIMPERNEL, SCARLET	*Anagallis arvensis*
PINEAPPLEWEED	*Matricaria matricarioides*
POINSETTIA, WILD	*Euphorbia heterophylla*
PUNCTUREVINE	*Tribulus terrestris*
PURSLANE, COMMON	*Portulaca oleracea*
PUSLEY, FLORIDA	*Richardia scabra*
QUACKGRASS	*Agropyron repens*
REDMAIDS	*Calandrinia caulescens*
ROCKET, LONDON	*Sisymbrium iria*
RYEGRASS, ITALIAN	*Lolium multiflorum*
RYEGRASS, PERENNIAL	*Lolium perene*
SAGE, LANCELEAF	*Salvia lanceplata*
SANDSPURRY	*Spergularia rubra*
SESBANIA, HEMP	*Sesbania exaltata*
SHEPHERDSPURSE	*Capsella bursa-pastoris*
SICKLEPOD	*Cassia obtusifolia*
SIDA, PRICKLY (TEAWEED)	*Sida spinosa*
SMARTWEED, PENNSYLVANIA	*Polygonum pensylvanicum*
SORREL, RED (FROM SEED)	*Rumex acetosella*
SOWTHISTLE, COMMON	*Sonchus oleraceus*
SPEEDWELL, BIRDSEYE	*Veronica persica*
SPURGE, GARDEN	*Euphorbia hirta*
SPURGE, PROSTRATE	*Euphorbia supina*
SPURGE, SPOTTED	*Euphorbia maculat*
SPURRY, CORN	*Spergula arvensis*
TANSYMUSTARD	*Descurainia pinnata*
THISTLE, BULL	*Cirsium vulgare*
THISTLE, RUSSIAN	*Salsola kali*
VELVETLEAF	*Albutilon theophrasti*
WITCHWEED	*Striga asiatica*
WITCHGRASS	*Panicum capillare*
WOODSORREL, COMMON YELLOW	*Oxalis stricta*

SPECIAL LOCAL NEED REGISTRATIONS FOR GOAL 1.6E

State	Use	SLN No.
California	*Noncrop Uses (Fence Rows, etc.)*	*CA-830065*
Hawaii	*Macadamia Nuts*	*HI-840006*
Idaho	*Cottonwood*	*ID-860015*
Indiana	*Mint*	*IN-840003*
Michigan	*Mint*	*MI-840011*
North Carolina	*Clary Sage*	*NC-830023*
North Carolina	*Witchweed on Corn*	*NC-880004*
North Carolina	*Witchweed on Corn*	*NC-850004*
Oregon	*Cottonwood*	*OR-850021*
Washington	*Cottonwood*	*WA-850023*
Wisconsin	*Mint*	*WI-840002*

NOTE: Supplemental Labeling is available for the following uses: Corn (USDA Witchweed eradication program in North and South Carolina), Onions Grown for Seed. Please contact the GOAL Marketing Manager, Rohm and Haas Company, Independence Mall West, Philadelphia, PA 19105 (215) 592-3292.

GOAL, KERB, TRITON, AG-98 and CS-7 are registered trademarks of Rohm and Haas Company.
LASSO, RAMROD and ROUNDUP are registered trademarks of Monsanto Company.
DUAL is a registered trademark of Ciba-Geigy Corporation.
DEVRINOL and GRAMOXONE are registered trademarks of ICI Americas Inc.
BLADEX, KARMEX are registered trademarks of E.I. duPont de Nemours and Co.
POAST is a registered trademark of BASF-AG Corp.
PROBE is a registered trademark of Sandoz Ltd.
SOLICAM is a registered trademark of Sandoz, Inc.
SURFLAN is a registered trademark of Elanco Products Company.

9541-B5 Made in USA

KARATHANE Liquid Concentrate

AGRICULTURAL FUNGICIDE AND MITICIDE

NOT FOR SALE, DISTRIBUTION OR USE IN ARIZONA AND CALIFORNIA.

FOR AGRICULTURAL USE ONLY OR ONLY FOR SALE TO, USE, AND STORAGE BY SERVICEPERSONS.

ACTIVE INGREDIENTS:	
2,4-Dinitro-6-octyl* phenyl crotonate	37.2%**
2,6-Dinitro-4-octyl* phenyl crotonate	
Nitrooctyl* phenols (principally dinitro)	0.2%
INERT INGREDIENTS:	62.6%
TOTAL	100.0%

*A mixture of 1-methylheptyl, 1-ethylhexyl, and 1-propylpentyl isomers.
**Equivalent to 48% or 4 lbs. Technical Karathane.

EPA Reg. No. 707-71

KEEP OUT OF REACH OF CHILDREN
WARNING AVISO

PRECAUCION AL USUARIO: Si usted no lee inglés, no use este producto hasta que la etiqueta haya sido explicada ampliamente.

NOTICE: Before using this product, read the entire Precautionary Statements, Conditions of Sale and Warranty, Directions for Use, Use Restrictions and Storage and Disposal Instructions. If the Conditions of Sale and Warranty are not acceptable, return the product unopened within thirty days of purchase to the place of purchase.

USE OF THIS PRODUCT MAY BE HAZARDOUS TO YOUR HEALTH. THIS PRODUCT HAS BEEN DETERMINED TO CAUSE BIRTH DEFECTS IN LABORATORY ANIMALS.

PRECAUTIONARY STATEMENTS

HAZARDS TO HUMANS AND DOMESTIC ANIMALS

Harmful if swallowed, absorbed through the skin or inhaled. Causes severe eye irritation. May cause allergic skin reaction in susceptible individuals. Avoid breathing vapors or spray mist. Do not get in eyes, or on skin, or on clothing. Wash thoroughly with soap and water after handling and before eating or smoking.

When mixing/loading and spraying wear midforearm-to-elbow-length chemical-resistant gloves, boots or overshoes, one-piece overalls that have long sleeves and long pants, and face shield or goggles. Mixer/loaders should, in addition, wear a chemical-resistant apron.

Instead of clothing and equipment specified above, the applicator can use an enclosed tractor cab with filtered air supply. See Re-entry and Worker Protection Statements for additional precautions.

STATEMENT OF PRACTICAL TREATMENT

IF SWALLOWED — Dilute by giving 2 glasses of water to drink. Call a physician. Never give anything by mouth to an unconscious person.

NOTE TO PHYSICIAN — Because of the presence of xylene (petroleum distillates) it may not be advisable to induce emesis. Careful gastric lavage may be indicated depending on the extent of involvement.

IF INHALED — Remove the victim to fresh air. Treat the person symptomatically. Call a physician.

IF ON SKIN — Wash exposed areas with plenty of soap and water. If irritation persists, call a physician.

IF IN EYES — Flush eyes with plenty of water for at least 15 minutes. If irritation persists, call a physician.

ENVIRONMENTAL HAZARDS

This product is toxic to fish. Do not apply directly to water or wetlands. Do not apply when weather conditions favor drift from areas treated. Do not apply where runoff is likely to occur. Do not contaminate water by cleaning of equipment or disposal of wastes. Apply this product only as specified on this label.

PHYSICAL OR CHEMICAL HAZARDS

FLAMMABLE

Do not use, pour, spill or store near heat or open flame. Do not use in the vicinity of pilot lights.

DIRECTIONS FOR USE

It is a violation of Federal law to use this product in a manner inconsistent with its labeling.

RE-ENTRY AND WORKER PROTECTION STATEMENTS

Do not enter treated areas without protective clothing until sprays have dried. Because certain states may require more restrictive re-entry intervals for various crops treated with this product, consult your State Department of Agriculture for further information. Do not apply this product in such a manner as to directly or through drift, expose workers or other persons. The area being treated must be vacated by unprotected persons.

Written or oral warnings must be given to workers who are expected to be in a treated area or in an area about to be treated with this product. Oral warnings must be given if there is reason to believe that written warnings cannot be understood by workers. When oral warnings are given, warnings shall be given in a language customarily understood by workers. Written warnings must include the following information:

WARNING

(Insert area or field description) treated with KARATHANE LIQUID CONCENTRATE fungicide-miticide on (insert date of application). Do not enter treated areas without protective clothing until sprays have dried. In case of accidental exposure: Call a doctor (physician), clinic or hospital immediately. Explain that the victim has been exposed to KARATHANE LIQUID CONCENTRATE fungicide-miticide and describe his condition. For further information see the "Statement of Practical Treatment."

Oral warnings must include the same information that is in the written warnings.

PERSONAL PROTECTIVE EQUIPMENT—For Mixers, Loaders, when used on Apples, Field Cucurbits, Grapes and Pears.

During the mixing and loading of KARATHANE LIQUID CONCENTRATE, wear a protective suit of one or two pieces that covers all parts of the body except the head, hands, and feet. Wear chemical-resistant gloves, chemical-resistant apron, and chemical-resistant shoes, shoe coverings, or boots. Wear goggles or a face shield. A hood or hat may also be worn.

When used on field cucurbits, grapes and pears, application MUST be made from a vehicle with a completely enclosed cab. All vents and windows of the cab should remain closed during application. During application from a vehicle with a completely enclosed cab, wear a long-sleeved shirt and long pants. A clean set of protective clothing and equipment as defined here must be available in case you must exit the cab in treated areas for repairs or other reasons. Always wear protective gloves upon exiting the cab. Remove protective clothing and equipment before re-entering the cab to prevent cab contamination.

When used on apples, it is recommended that application be made from a vehicle with a completely enclosed cab. When used on apples, if application from a vehicle with a completely enclosed cab is not possible or during repair and cleanup of equipment or re-entry into treated areas prior to drying of the spray, wear a protective suit of one or two pieces that covers all parts of the body except the hands, feet, and head. Wear chemical-resistant gloves and chemical-resistant shoes, shoe coverings or boots. A hood or hat may also be worn. During airblast or mist blower application a protective hood or wide brim hat must be worn.

WARNING: Wearing the protective suit for the prolonged periods required for application may produce heat stress at temperatures above 85°F on a cloudy day or 80°F on a sunny day. Application of this product must be conducted from enclosed cab vehicles when the temperature exceeds 85°F (80°F on sunny days) or delayed until the temperature drops.

IMPORTANT: After application of the pesticide is completed, remove all clothes and shoes. Shower using soap and water. Redress only in clean clothes. Do not use contaminated clothing. Wash non-disposal protective clothing and protective equipment with soap or detergent and water after each use. Personal and non-disposable protective clothing worn during use must be laundered separately from household articles. Clothing or protective equipment heavily contaminated or drenched with KARATHANE LIQUID CONCENTRATE must be destroyed according to state or local regulations.

HEAVILY CONTAMINATED OR DRENCHED CLOTHING CANNOT BE ADEQUATELY DECONTAMINATED.

CHEMIGATION

Do not apply this product through any type of irrigation system.

GENERAL USE INFORMATION

KARATHANE LIQUID CONCENTRATE fungicide-miticide will protect against and eradicate powdery mildew when used according to the label directions.

KARATHANE LIQUID CONCENTRATE fungicide-miticide also provides effective control or suppression of the following mites, including those that may be resistant to other miticides:

Citrus Red	European Red	Plum Nursery
Clover	Peach Silver	Red Spider (two-spotted)

KARATHANE LIQUID CONCENTRATE fungicide-miticide possesses short residual activity and is effective only against motile mite forms. Although of short residual activity, effective reduction of motile mite populations may provide economic mite control for many weeks. When used alone, application should be repeated every 7 to 10 days to kill mites emerging from eggs unaffected by previous sprays. Alternating sprays or tank-mix combinations with other miticides will delay the development of mite resistance to other miticides and improve initial mite kill.

USE RATE DETERMINATION

Carefully read, understand, and follow label rates and use limitations.

Label use rates are for single strength (dilute) sprays. KARATHANE LIQUID CONCENTRATE fungicide-miticide may be used in concentrate or airblast sprayers but correct use rates must be maintained.

When spray volumes of less than 100 gallons are required to provide a thorough full coverage spray, the following table may be used to determine the proper use rate:

Recommended KARATHANE LIQUID CONCENTRATE Label Use Rate Per 100 Gals. Spray	Milliliters KARATHANE LIQUID CONCENTRATE Required in Varied Single-Strength Spray Gallonages*				
(Oz.)	1 gal.	2 gals.	3 gals.	5 gals.	10 gals.
3	0.9	1.8	2.6	4.4	8.9
4	1.2	2.4	3.5	5.9	11.8
6	1.8	3.6	5.4	9.0	18.0
8	2.4	4.8	7.0	11.8	23.6

* 15 milliliters = 1 teaspoon

COMPATIBILITY

KARATHANE LIQUID CONCENTRATE fungicide-miticide is compatible with DIKAR®, DITHANE®, captan, ferbam, glyodin, and thiram fungicides and with most insecticides and miticides including azinphosmethyl, demeton, and parathion. Exceptions are noted under specific use directions. **Do not use in combination with phosphate insecticides or nutritional sprays in California.** When preparing tank mixes, user should consult current spray compatibility charts or State Cooperative Extension Service Specialists prior to actual use.

SPRAY ADJUVANTS

The addition of agricultural surfactants to KARATHANE LIQUID CONCENTRATE fungicide-miticide sprays will improve initial spray deposits, fungicide redistribution, and weatherability. The following spray adjuvants have been especially formulated to optimize the performance of foliar-applied agricultural chemicals:

TRITON AG-98® A low foam, nonionic, general purpose spray adjuvant. May be used in dilute or concentrate sprays applied by aircraft or ground equipment.

TRITON® B-1956® A water-dispersible, resin-based nonionic surfactant which resists re-wetting and removal by rain. Effective with dilute sprays applied by ground equipment.

TRITON CS-7® A spreader-binder designed specifically for use in concentrate and low volume sprays applied by aircraft or ground equipment.

Place KARATHANE LIQUID CONCENTRATE fungicide-miticide into suspension prior to adding an adjuvant. Read and carefully observe the precautionary statements and all other information appearing on both product labels prior to spray preparation.

APPLICATION

Thorough full coverage foliar sprays generally result in optimum powdery mildew and mite control. To achieve good foliar coverage use proper spray pressure, gallonage per acre, spray nozzles, discs, nozzle spacing, and tractor speed. Consult spray nozzle and accessory catalogues for specific information on proper equipment calibration. When applied with hand pressurized sprayers, thoroughly spray plant foliage until runoff.

PEST MONITORING

KARATHANE LIQUID CONCENTRATE fungicide-miticide should be applied when the first signs of powdery mildew are noted or when adult mites are observed on plant foliage.

RESTRICTIONS

Users should carefully read, understand, and follow all use restrictions prior to using KARATHANE LIQUID CONCENTRATE.

FRUITS

CROP	CONTROL OF:	RATE OF KARATHANE LIQUID CONCENTRATE PER APPLICATION: OZ./100 GAL.	OZ./A	TIMING AND REMARKS	RESTRICTIONS
Apples (orchard)	Powdery mildew	4 to 6	20 to 48	For powdery mildew begin application at delayed dormant to pink stage and continue sprays at 7 to 14 day intervals until terminal growth is completed or mildew activity has ceased. Applications should not be more than 14 days apart. Control measures are needed from the time buds break or at first signs of mildew activity, and should be continued through terminal growth and new bud formation to reduce overwintering infection in the buds. From 4 to 8 applications are usually required per season depending on variety, location and disease severity. Thorough coverage is necessary. When mildew growth is heavy add a spreader-sticker for thorough wetting of mildew spores and mycelium. Certain spreader-stickers will reduce the deposit on maturing fruit. TRITON B-1956 is recommended at 1 to 2 ozs. per 100 gals. of dilute spray.	Refer to Re-entry and Worker Protection Statements. Do not apply when day temperatures are expected to exceed 90°F. Do not use with oil or oil-based pesticides or apply to foliage within 30 days after an oil spray application. Do not use within 21 days of harvest. Do not graze treated areas to livestock or feed cover crops grown in treated orchards to animals.
	Clover mites European red mites Red spider (two-spotted) mites			For mites, apply at first signs of an infestation and continue on a 7 to 14 day schedule as long as necessary. Do not use more than ½ lb. of 25% wettable parathion with KARATHANE LIQUID CONCENTRATE from pink through second cover or apply in separate applications. Other phosphate insecticides such as azinphos methyl, demeton, and malathion, appear to be compatible with KARATHANE LIQUID CONCENTRATE.	
Apples (nursery)	Powdery mildew Red spider mites Clover mites	4	—	On non-bearing nursery stock, apply at first sign of powdery mildew and/or mites, at 7 to 10 day intervals or as necessary.	Refer to Re-entry and Worker Protection Statements. Do not graze treated areas to livestock or feed cover crops grown in treated areas to animals.
Apricots	Powdery mildew Clover mites European red mites Peach silver mites Plum rust mites Two-spotted spider mites	4	20 to 32	Begin application at pink bud and petal fall and in subsequent spring sprays if disease persists, and post harvest.	**Do not apply within 45 days of harvest.** **Do not graze treated areas to livestock or feed cover crops grown in treated orchards to animals.** **Avoid drift to bearing trees.**
Cherries (nursery)	Refer to Apples (nursery)				Do not apply to trees that will bear fruit within one year of treatment.
Citrus (nursery)	Citrus red mites	4 to 8	—	Apply two sequential sprays: the first when mites appear and the second 10 to 14 days later.	Apply only to nursery trees or to young trees that will not bear fruit within one year. Avoid drift to bearing trees.
Grapes	Powdery mildew	4 to 8	12 to 24	Begin applications when mildew first appears (usually during early stages of foliage growth) and repeat at 10 to 14 day intervals. From 2 to 6 applications may be required.	**Application MUST be made from an enclosed cab. Refer to Re-entry and Worker Protection Statements.**

CROP	CONTROL OF:	RATE OF KARATHANE LIQUID CONCENTRATE PER APPLICATION OZ./100 GAL.	OZ./A	TIMING AND REMARKS	RESTRICTIONS
					Do not use within 21 days of harvest.
Peaches (orchard)	Refer to Apricots				
Peaches (nursery)	Refer to Apples (nursery)				
Pears	Refer to Apples (orchard)	4 to 6	20		**Application MUST be made from an enclosed cab. Refer to Re-entry and Worker Protection Statements.**

VEGETABLES

CROP	CONTROL OF:	RATE OF KARATHANE LIQUID CONCENTRATE PER APPLICATION OZ./100 GAL.	OZ./A	TIMING AND REMARKS	RESTRICTIONS
Cantaloupes	Powdery mildew	3 to 4	3 to 6	Apply when powdery mildew signs are first observed and repeat at 7 to 10 day intervals. From 3 to 10 applications may be required.	**Application MUST be made from an enclosed cab. Refer to Re-entry and Worker Protection Statements.** Do not use within 7 days of harvest.
Cucumbers	Powdery mildew	3 to 4	3 to 6	Apply in a light, thorough spray. Begin when mildew first appears and repeat at 5 to 7 day intervals or as necessary. Spray a maximum of 3 to 4 times to an individual fruit. Do not apply when temperatures exceed 90°F.	
Melons	Refer to Cantaloupes				
Pumpkins	Refer to Cantaloupes				
Squash	Refer to Cantaloupes			On summer squash, spray a maximum of 3 to 4 times to an individual fruit.	

STORAGE AND DISPOSAL

STORAGE — Do not store at temperatures above 90°F for periods longer than 48 hours.

PESTICIDE DISPOSAL — Do not contaminate water, food or feed by storage or disposal. Pesticide wastes are toxic. Improper disposal of excess pesticide, spray mixture or rinsate is a violation of Federal law. If these wastes cannot be disposed of by use according to label instructions, contact your State Pesticide or Environmental Control Agency, or the Hazardous Waste representative at the nearest EPA Regional Office for guidance.

CONTAINER DISPOSAL — Triple rinse (or equivalent). Then puncture and dispose of in a sanitary landfill, or by incineration, or if allowed by state and local authorities, by burning. If burned, stay out of smoke.

STEPS TO BE TAKEN IN CASE MATERIAL IS RELEASED OR SPILLED — Eliminate ignition sources. Ventilate area. Avoid breathing vapors. Use approved self-contained breathing apparatus or air respirators for large spills in confined areas. Dike or contain liquid. Scoop or shovel into a suitable container for recovery or disposal. Remove contaminated clothing promptly and wash exposed skin areas with soap and water. Wash clothing before reuse. Keep spill out of all sewers and open bodies of water. REFER TO PRECAUTIONARY STATEMENTS.

CONDITIONS OF SALE AND WARRANTY

Rohm and Haas warrants that the product conforms to its chemical description and is reasonably fit for the purpose stated on the label only when used in accordance with label directions under normal conditions of use. **ROHM AND HAAS MAKES NO OTHER EXPRESS OR IMPLIED WARRANTIES EITHER OF MERCHANTABILITY OR FITNESS FOR A PARTICULAR USE.** Handling, storage and use of the product by Buyer or User are beyond the control of Rohm and Haas and Seller. Risks such as crop injury, ineffectiveness or other unintended consequences resulting from, but not limited to, weather or soil conditions, presence of other materials, disease, pests, drift to other crops or property or failure to follow label directions will be assumed by the Buyer or User. **IN NO CASE WILL ROHM AND HAAS OR SELLER BE HELD LIABLE FOR CONSEQUENTIAL, SPECIAL OR INDIRECT DAMAGES RESULTING FROM THE HANDLING, STORAGE OR USE OF THIS PRODUCT.**

DIKAR, DITHANE, KARATHANE, TRITON, AG-98, B-1956 AND CS-7 are registered trademarks of the Rohm and Haas Company.

9180-B5

KARATHANE® WD

AGRICULTURAL FUNGICIDE AND MITICIDE

NOT FOR SALE, DISTRIBUTION OR USE IN ARIZONA AND CALIFORNIA

FOR AGRICULTURAL USE ONLY OR ONLY FOR SALE TO, USE, AND STORAGE BY SERVICEPERSONS

ACTIVE INGREDIENTS:	
2,4-Dinitro-6-octyl* phenyl crotonate	
2,6-Dinitro-4-octyl* phenyl crotonate	19.4%
Nitrooctyl* phenol (principally dinitro)	0.1%
INERT INGREDIENTS:	80.50%
TOTAL	100.00%

*A mixture of 1-methylheptyl, 1-ethylhexyl, and 1-propylpentyl isomers

EPA Reg. No. 707-53

KEEP OUT OF REACH OF CHILDREN

CAUTION

NOTICE: Before using this product, read the entire Precautionary Statements, Conditions of Sale and Warranty, Directions for Use, Use Restrictions and Storage and Disposal Instructions. If the Conditions of Sale and Warranty are not acceptable, return the product unopened within thirty days of purchase to the place of purchase.

USE OF THIS PRODUCT MAY BE HAZARDOUS TO YOUR HEALTH. THIS PRODUCT HAS BEEN DETERMINED TO CAUSE BIRTH DEFECTS IN LABORATORY ANIMALS.

PRECAUTIONARY STATEMENTS

HAZARDS TO HUMANS AND DOMESTIC ANIMALS

Harmful if swallowed, absorbed through the skin or inhaled. Causes moderate eye, skin, nose and throat irritation. May cause allergic skin reaction in susceptible individuals. Avoid contact with skin, eyes, and clothing. Avoid breathing dust or spray mist. Wash thoroughly with soap and water after handling and before eating or smoking.

When mixing/loading and spraying wear midforearm-to-elbow-length chemical-resistant gloves, boots or overshoes, one-piece overalls that have long sleeves and long pants, and face shield or goggles. Mixer/loaders should, in addition, wear a chemical-resistant apron.

Instead of clothing and equipment specified above, the applicator can use an enclosed tractor cab with filtered air supply. See Re-entry and Worker Protection Statements for additional precautions.

STATEMENT OF PRACTICAL TREATMENT

IF SWALLOWED: Drink 1 to 2 glasses of water. Call a physician. Never give anything by mouth to an unconscious person.

NOTE TO PHYSICIAN: Administer sedatives if necessary to control apprehension and excitement. Be prepared to counteract respiratory depression and hypotension which may occur following the administration of anti-convulsants.

IF INHALED: Remove victim to fresh air. Treat the victim symptomatically. If irritation persists, call a physician.

IF ON SKIN: Wash exposed areas with plenty of soap and water. If irritation persists, call a physician.

IF IN EYES: Flush eyes with plenty of water for at least 15 minutes. If irritation persists, call a physician.

ENVIRONMENTAL HAZARDS

This product is toxic to fish. Do not apply directly to water or wetlands. Do not apply where runoff is likely to occur. Do not apply when weather conditions favor drift from areas treated. Do not contaminate water by cleaning of equipment or disposal of wastes. Apply this product only as specified on this label.

DIRECTIONS FOR USE

It is a violation of Federal law to use this product in a manner inconsistent with its labeling.

RE-ENTRY AND WORKER PROTECTION STATEMENTS

Do not enter treated areas without protective clothing until sprays have dried. Because certain states may require more restrictive re-entry intervals for various crops treated with this product, consult your State Department of Agriculture for further information. Do not apply this product in such a manner as to directly or through drift, expose workers or other persons. The area being treated must be vacated by unprotected persons.

Written or oral warnings must be given to workers who are expected to be in a treated area or in an area about to be treated with this product. Oral warnings must be given if there is reason to believe that written warnings cannot be understood by workers. When oral warnings are given, warnings shall be given in a language customarily understood by workers. Written warnings must include the following information:

CAUTION

(Insert area or field description) treated with KARATHANE WD on (insert date of application). Do not enter treated areas without protective clothing until sprays have dried. In case of accidental exposure: Call a doctor (physician), clinic or hospital immediately. Explain that the victim has been exposed to KARATHANE WD and describe his condition. For further information see the "Statement of Practical Treatment."

Oral warnings must include the same information that is in the written warnings.

PERSONAL PROTECTIVE EQUIPMENT—For Mixers and Loaders, when used on Apples, Field Cucurbits, Grapes and Pears.

During the mixing and loading of KARATHANE WD, wear a protective suit of one or two pieces that covers all parts of the body except the head, hands, and feet. Wear chemical-resistant gloves, chemical-resistant apron, and chemical-resistant shoes, shoe coverings, or boots. Wear goggles or a face shield. A hood or hat may also be worn.

When used on field cucurbits, grapes and pears, application MUST be made from a vehicle with a completely enclosed cab. All vents and windows of the cab should remain closed during application. During application from a vehicle with a completely enclosed cab, wear a long-sleeved shirt and long pants. A clean set of protective clothing and equipment as defined here must be available in case you must exit the cab in treated areas for repairs or other reasons. Always wear protective gloves upon exiting the cab. Remove protective clothing and equipment before re-entering the cab to prevent cab contamination.

When used on apples, it is recommended that application be made from a vehicle with a completely enclosed cab. When used on apples, if application from a vehicle with a completely enclosed cab is not possible or during repair and cleanup of equipment or re-entry into treated areas prior to drying of the spray, wear a protective suit of one or two pieces that covers all parts of the body except the hands, feet, and head. Wear chemical-resistant gloves and chemical-resistant shoes, shoe coverings or boots. A hood or hat may also be worn. During airblast or mist blower application a protective hood or wide brim hat must be worn.

WARNING: Wearing the protective suit for the prolonged periods required for application may produce heat stress at temperatures above 85°F on a cloudy day or 80°F on a sunny day. Application of this product must be conducted from enclosed cab vehicles when the temperature exceeds 85°F (80°F on sunny days) or delayed until the temperature drops.

IMPORTANT: After application of the pesticide is completed, remove all clothes and shoes. Shower using soap and water. Redress only in clean clothes. Do not use contaminated clothing. Wash non-disposable protective clothing and protective equipment with soap or detergent and water after each use. Personal and non-disposable protective clothing worn during use must be laundered separately from household articles. Clothing or protective equipment heavily contaminated or drenched with KARATHANE WD must be destroyed according to state or local regulations.

HEAVILY CONTAMINATED OR DRENCHED CLOTHING CANNOT BE ADEQUATELY DECONTAMINATED.

CHEMIGATION

Do not apply this product through any type of irrigation system.

GENERAL USE INFORMATION

KARATHANE WD fungicide-miticide will protect against and eradicate powdery mildew when used according to the label directions.

KARATHANE WD fungicide-miticide also provides effective control or suppression of the following mites, including those that may be resistant to other miticides:

Citrus Red	European Red	Plum Nursery
Clover	Peach Silver	Red Spider (two-spotted)

KARATHANE WD fungicide-miticide possesses short residual activity and is effective only against motile mite forms. Although of short residual activity, effective reduction of motile mite populations may provide economic mite control for many weeks. When used alone, application should be repeated every 7 to 10 days to kill mites emerging from eggs and unaffected by previous sprays. Alternating sprays or tank mix combinations with other miticides will delay the development of mite resistance to other miticides and improve initial mite kill.

USE RATE DETERMINATION

Carefully read, understand, and follow label rates and use limitations.

Label use rates are for single strength (dilute) sprays. KARATHANE WD fungicide-miticide may be used in concentrate or airblast sprayers but correct use rates must be maintained.

When spray volumes of less than 100 gallons are required to provide a thorough full coverage spray, the following table may be used to determine the proper use rate:

Recommended KARATHANE WD Fungicide-Miticide Label Use Rate Per 100 Gals. Spray (Oz.)	Teaspoons KARATHANE WD Fungicide-Miticide Required in Varied Single Strength Spray Gallonages*				
	1 gal.	2 gals.	3 gals.	5 gals.	10 gals.
4	1/2	1	1 1/2	2 1/2	5
6	3/4	1 1/2	2 1/4	3 3/4	7 1/2
8	1	2	3	5	10
12	1 1/2	3	4 1/2	7 1/2	15
16	2	4	6	10	20

* 1 tablespoon = 3 teaspoons

MIXING

Slowly place KARATHANE WD fungicide-miticide into spray tank as it is being filled or premix as a slurry prior to adding to spray tank. Add other co-applied fungicides, miticides, insecticides, growth regulators, micronutrients, and spray adjuvants after KARATHANE WD fungicide-miticide has been placed into suspension.

COMPATIBILITY

KARATHANE WD fungicide-miticide is compatible with DIKAR®, DITHANE®, captan, ferbam, glyodin, and thiram fungicides and with most insecticides and miticides including azinphos-methyl, demeton, and parathion. Exceptions are noted under specific use directions. When preparing tank mixes, user should consult current spray compatibility charts or State Extension Service Specialists prior to actual use.

SPRAY ADJUVANTS:

The addition of agricultural surfactants to KARATHANE WD fungicide-miticide sprays will improve initial spray deposits, redistribution, and weatherability. The following spray adjuvants have been especially formulated to optimize the performance of foliar-applied agricultural chemicals:

TRITON AG-98® — A low foam, nonionic, general purpose spray adjuvant. May be used in dilute or concentrate sprays applied by aircraft or ground equipment.

TRITON® B-1956® — A water-dispersible, resin-based nonionic surfactant which resists rewetting and removal by rain. Effective with dilute sprays applied by ground equipment.

TRITON CS-7® A spreader-binder designed specifically for use in concentrate and low volume sprays applied by aircraft or ground equipment.

Place KARATHANE WD fungicide-miticide into suspension prior to adding an adjuvant to the spray mixture. Read and carefully observe the precautionary statements and all other information appearing on both product labels prior to spray preparation.

APPLICATION

Thorough, full-coverage foliar sprays generally result in optimum powdery mildew and mite control. To achieve good foliar coverage use proper spray pressure, gallonage per acre, spray nozzles, discs, nozzle spacing, and tractor speed. Consult spray nozzle and accessory catalogues for specific information on proper equipment calibration.

When applied with hand-pressurized sprayers, thoroughly spray plant foliage until runoff.

PEST MONITORING

KARATHANE WD fungicide-miticide should be applied when the first signs of powdery mildew are noted or when mites are observed on plant foliage.

RESTRICTIONS

Users should carefully read, understand, and follow all use restrictions prior to using KARATHANE WD fungicide-miticide.

FRUITS

		RATE OF KARATHANE WD FUNGICIDE-MITICIDE PER APPLICATION			
CROP	**CONTROL OF**	**OZ./100 GAL.**	**LB./A**	**TIMING AND REMARKS**	**RESTRICTIONS**
Apples (orchard)	Powdery mildew	8 to 12	2½ to 3½	For powdery mildew begin application at delayed dormant to pink stage and continue sprays at 7 to 14 day intervals until terminal growth is completed or mildew activity has ceased. Applications should not be more than 14 days apart. Control measures are needed from the time buds break or at first signs of mildew activity, and should be continued through terminal growth and new bud formation to reduce overwintering infection in the buds. From 4 to 8 applications are usually required per season depending on variety, location and disease severity. Thorough coverage is necessary. In the East, ¼ lb. per 100 gals. water may be used on varieties (such as Delicious, McIntosh and Stayman) which are slightly to moderately susceptible to powdery mildew, and on susceptible varieties under light infection conditions. Apply every 5 to 7 days as needed or until terminal growth is completed. When mildew growth is heavy add a spreader-sticker for thorough wetting of mildew spores and mycelium. Certain spreader-stickers will reduce the deposit on maturing fruit. TRITON B-1956 is recommended at 1 to 2 ozs. per 100 gals. of dilute spray.	**Refer to Re-entry and Worker Protection Statements.** Do not use within 21 days of harvest. Do not graze livestock in treated areas or feed cover crops grown in treated orchards to animals.
	Clover mites European red mites Red spider (two-spotted) mites			For mites, apply at first signs of an infestation and continue on a 7 to 14 day schedule as long as necessary. Do not use more than ½ lb. of 25% wettable parathion with KARATHANE WD fungicide-miticide from pink through second cover or apply in separate applications. Other phosphate insecticides such as azinphosmethyl, demeton, and malathion, are compatible with KARATHANE WD fungicide-miticide.	
Apples (nursery)	Clover mites Powdery mildew Red spider mites	8	—	On non-bearing nursery stock, apply at first sign of powdery mildew and/or mites, at 7 to 10 day intervals or as necessary.	**Refer to Re-entry and Worker Protection Statements.** Do not graze livestock in treated areas or feed cover crops grown in treated areas to animals.
Apricots	Clover mites European red mites Peach silver mites Plum rust mites Powdery mildew Two-spotted spider mites	8 to 12	2½ to 3¼	Begin application at pink bud and petal fall and in subsequent spring sprays if disease persists, and post harvest.	Do not apply within 45 days of harvest. Do not graze livestock in treated areas or feed cover crops grown in treated orchards to animals.
Cherries (nursery)	Refer to Apples (nursery)				Do not apply to trees that will bear fruit within one year of treatment.
Citrus (nursery)	Citrus red mites	8	—	First apply when mites are noted and repeat 10 to 14 days later.	Apply only to nursery trees or to young trees that will not bear fruit within one year.
Grapes	Mites Powdery mildew	8	1 to 2½	Begin applications when mildew first appears (usually during early stages of foliage growth) and repeat at 10 to 14 day intervals. From 2 to 6 applications may be required.	**Applications MUST be made from an enclosed cab. Refer to Re-entry and Worker Protection Statements.** Do not use within 21 days of harvest.
Peaches (orchard)	Refer to Apricots				
Peaches (nursery)	Refer to Apples (nursery)				
Pears	Refer to Apples (orchard)				**Applications MUST be made from an enclosed cab. Refer to Re-entry and Worker Protection Statements.**

CROP	CONTROL OF	RATE OF KARATHANE WD FUNGICIDE-MITICIDE PER APPLICATION OZ./100 GAL.	LB./A	TIMING AND REMARKS	RESTRICTIONS
VEGETABLES					
Cantaloupes	Powdery mildew	6 to 8	⅓ to ¾	Apply when powdery mildew signs are first observed and repeat at 7 to 10 day intervals. From 3 to 10 applications may be required. On cucumbers and summer squash, spray a maximum of 3 to 4 times to an individual set of fruit.	**Applications MUST be made from an enclosed cab. Refer to Re-entry and Worker Protection Statements.** **Do not use within 7 days of harvest.**
Cucumbers (field)	Refer to Cantaloupes				
Cucumbers (green-house)	Powdery mildew	4 to 8	—	Apply in a light thorough spray. Begin when mildew first appears and repeat at 5 to 7 day intervals or as necessary. Do not apply when temperatures exceed 90°F.	Do not use within 7 days of harvest.
Melons	Refer to Cantaloupes				
Pumpkins	Refer to Cantaloupes				
Squash	Refer to Cantaloupes				

ORNAMENTALS

CROP	CONTROL OF	TIMING AND REMARKS
African violets Asters Begonias Calendulas Chrysanthemums Cinerarias Dahlias Delphiniums Euonymus Gardenia Hydrangeas Larkspur Lilacs	Clover mites Powdery mildew Red spider mites	GENERAL USE DIRECTIONS: **Greenhouse**—Use 4 to 6 ozs./100 gals. water or 2 level tablespoons per 12½ gals. water. Begin applications when first signs of powdery mildew or mites are observed and apply thorough coverage sprays at 7 to 10 day intervals. **Outdoors**—Use 6 to 8 ozs./100 gals. water or 2 level teaspoons per 3 gals. water. Begin applications when first signs of mildew are observed and apply thorough coverage sprays at 5 to 7 day intervals or as necessary. Do not apply when greenhouse or outdoor temperatures are expected to reach or exceed 90°F.
Roses (greenhouse)		Apply in a light misting spray when signs of powdery mildew or mites are first observed and repeat at 5 to 7 day intervals or as long as necessary. The use of TRITON B-1956 will improve uniformity of spray deposits. Do not apply if temperatures are expected to reach or exceed 90°F.
Roses (outdoors)		Use includes multiflora roses. Apply 6 to 8 ozs. per 100 gals. when first signs of powdery mildew or mites are observed and repeat at 7 to 10 day intervals. Use thorough full coverage sprays. The addition of TRITON B-1956 will improve uniformity of spray deposits.

STORAGE AND DISPOSAL

STORAGE: Do not store at temperatures above 90°F for periods longer than 48 hours.

PESTICIDE DISPOSAL: Do not contaminate water, food or feed by storage or disposal. Pesticide wastes are toxic. Improper disposal of excess pesticide, spray mixture or rinsate is a violation of Federal law. If these wastes cannot be disposed of by use according to label instructions, contact your State Pesticide or Environmental Control Agency, or the Hazardous Waste Representative at the nearest EPA Regional Office for guidance.

CONTAINER DISPOSAL

PAPER BAGS: Completely empty bag into application equipment. Then dispose of in a sanitary landfill or by incineration, or, if allowed by State and local authorities, by burning. If burned, stay out of smoke.

FIBER DRUMS: Complete empty liner by shaking and tapping sides and bottom to loosen clinging particles. Empty residue into application equipment. Then dispose of drum and liner in a sanitary landfill, or by incineration if allowed by State and local authorities.

STEPS TO BE TAKEN IN CASE MATERIAL IS RELEASED OR SPILLED: Scoop or shovel solid material into a suitable container for recovery or disposal. Remove contaminated clothing promptly and wash exposed skin areas with soap and water. Wash clothing before reuse. Keep spill out of all sewers and open bodies of water. Refer to Precautionary Statements.

CONDITIONS OF SALE AND WARRANTY

Rohm and Haas warrants that the product conforms to its chemical description and is reasonably fit for the purpose stated on the label only when used in accordance with label directions under normal conditions of use. **ROHM AND HAAS MAKES NO OTHER EXPRESS OR IMPLIED WARRANTIES EITHER OF MERCHANTABILITY OR FITNESS FOR A PARTICULAR USE.** Handling, storage and use of the product by Buyer or User are beyond the control of Rohm and Haas and Seller. Risks such as crop injury, ineffectiveness or other unintended consequences resulting from, but not limited to, weather or soil conditions, presence of other materials, disease, pests, drift to other crops or property or failure to follow label directions will be assumed by the Buyer or User. **IN NO CASE WILL ROHM AND HAAS OR SELLER BE HELD LIABLE FOR CONSEQUENTIAL, SPECIAL OR INDIRECT DAMAGES RESULTING FROM THE HANDLING, STORAGE OR USE OF THIS PRODUCT.**

DIKAR, DITHANE, KARATHANE, TRITON, AG-98, B-1956 and CS-7 are registered trademarks of the Rohm and Haas Company.

8157-B3

KELTHANE® MF

AGRICULTURAL MITICIDE EMULSIFIABLE CONCENTRATE

ACTIVE INGREDIENT:

Dicofol	
1,1-Bis(chlorophenyl) 2,2,2-trichloroethanol	42%*
INERT INGREDIENTS:	58%
TOTAL	100%

* Equivalent to 4 lbs. active ingredient per gallon

EPA Reg. No. 707-202

NOTICE: Before using this product, read the entire Precautionary Statements, Conditions of Sale and Warranty, Directions for Use, Use Precautions and Storage and Disposal Instructions. If the Conditions of Sale and Warranty are not acceptable, return the product unopened within thirty days of purchase to the place of purchase.

KEEP OUT OF REACH OF CHILDREN CAUTION

PRECAUTIONARY STATEMENTS

HAZARDS TO HUMANS AND DOMESTIC ANIMALS

Harmful if swallowed, absorbed through the skin or inhaled. Causes moderate eye, skin, nose and throat irritation. Avoid contact with skin, eyes and clothing. **Skin contact with this pesticide may be hazardous; wear chemical-resistant gloves when mixing, loading, or applying this product.** This product is a skin sensitizer. Repeated exposure may cause allergic reaction in some individuals. If allergic reaction should occur, discontinue use and see a physician. Avoid breathing vapors or spray mist.

Wash thoroughly with soap and water after handling and before eating or smoking. Remove contaminated clothing and wash before reuse.

STATEMENT OF PRACTICAL TREATMENT

IF SWALLOWED: Drink 1 or 2 glasses of water and induce vomiting by touching back of throat with finger, or, call a physician and have the stomach emptied immediately. Do not induce vomiting or give anything by mouth to an unconscious person.

NOTE TO PHYSICIAN: Because this product contains a chlorinated hydrocarbon insecticide, vomiting or careful gastric lavage is recommended. Epinephrin or other adrenergic amines can cause myocardial irritability.

IF INHALED: Remove victim to fresh air. Treat the victim symptomatically. Call a physician if problem continues.

IF ON SKIN: Wash exposed areas with plenty of soap and water. If irritation persists, call a physician.

IF IN EYES: Flush eyes with plenty of water for at least 15 minutes. Call a physician if irritation persists.

ENVIRONMENTAL HAZARDS

This product is toxic to fish. Do not apply directly to lakes, streams or ponds. Do not apply where runoff is likely to occur. Do not apply when weather conditions favor drift from areas treated. Do not contaminate water by cleaning of equipment or disposal of wastes. Apply this product only as specified on this label.

PHYSICAL OR CHEMICAL HAZARDS

FLAMMABLE

Do not use, pour, spill or store near heat or open flame.

CONDITIONS OF SALE AND WARRANTY

Rohm and Haas warrants that the product conforms to its chemical description and is reasonably fit for the purpose stated on the label only when used in accordance with label directions under normal conditions of use. **ROHM AND HAAS MAKES NO OTHER EXPRESS OR IMPLIED WARRANTIES EITHER OF MERCHANTABILITY OR FITNESS FOR A PARTICULAR USE.** Handling, storage and use of the product by Buyer or User are beyond the control of Rohm and Haas and Seller. Risks such as crop injury, ineffectiveness or other unintended consequences resulting from, but not limited to, weather or soil conditions, presence of other materials, disease, pests, drift to other crops or property or failure to follow label directions will be assumed by the Buyer or User. **IN NO CASE WILL ROHM AND HAAS OR SELLER BE HELD LIABLE FOR CONSEQUENTIAL, SPECIAL OR INDIRECT DAMAGES RESULTING FROM THE HANDLING, STORAGE OR USE OF THIS PRODUCT.**

DIRECTIONS FOR USE

It is a violation of Federal law to use this product in a manner inconsistent with its labeling.

RE-ENTRY AND WORKER PROTECTION STATEMENTS

Do not enter treated areas without protective clothing until sprays have dried. Because certain states may require more restrictive re-entry intervals for various crops treated with this product, consult your State Department of Agriculture for further information. Do not apply this product in such a manner as to directly or through drift expose workers or other persons. The area being treated must be vacated by unprotected persons.

Written or oral warnings must be given to workers who are expected to be in a treated area or in an area about to be treated with this product. Oral warnings must be given if there is reason to believe that written warnings cannot be understood by workers. When oral warnings are given, warnings shall be given in a language customarily understood by workers. Written warnings must include the following information:

CAUTION

(Insert area or field description) treated with KELTHANE MF on (insert date of application). Do not enter treated areas without protective clothing until sprays have dried. In case of accidental exposure: Call a doctor (physician), clinic or hospital immediately. Explain that the victim has been exposed to KELTHANE MF and describe his condition. For further information see the "Statement of Practical Treatment."

Oral warnings must include the same information that is in the written warnings.

CHEMIGATION

Do not apply this product through any type of irrigation system.

GENERAL USE INFORMATION

KELTHANE MF Miticide is an emulsifiable formulation that gives high initial kill against most species of agricultural mites. KELTHANE MF is a specific miticide and will not kill bees and beneficial insect predators, when used as recommended.

KELTHANE MF Miticide is compatible with most other commonly used insecticides and fungicides. It should not be combined with lime. KELTHANE MF Miticide may be applied in dilute or in concentrate ground sprays. For aerial sprays use 3 to 10 gallons of water per acre to provide thorough coverage of all plant parts. Coverage on the underside of leaves is especially important. For further details see specific recommendations on this label, then consult current State Agricultural Experiment Station spray schedules or State Extension Service Specialists for additional details on compatibility, dosage, timing and uses.

VEGETABLES AND FIELD CROPS

BEANS (dry, green, lima)—Pacific, tropical and two-spotted mites. Use 3/4 to 1 pint KELTHANE MF Miticide per acre. Apply at first signs of mite buildup and repeat as necessary. Do not apply within 7 days of harvest. Do not feed treated vines (or crop residues) to meat or dairy animals.

BEANS (dry, green, lima) California only—Pacific, tomato russet, tropical and two-spotted mites. Use 3 pints KELTHANE MF Miticide per acre. Apply at first signs of mite buildup and repeat as necessary. Do not apply within 21 days of harvest. Do not feed treated vines (or crop residues) to meat or dairy animals.

COTTON—Desert, Pacific and two-spotted mites. Use 1 to 1 1/2 quarts KELTHANE MF Miticide per acre, depending on size of plant and density of foliage. Apply in sufficient water to provide thorough coverage. Begin application when mites are first noticed and repeat if necessary but do not apply within 14 days of harvest. Do not feed cotton stalks or trash to meat or dairy animals.

HOPS—Pacific, tropical and two-spotted mites. Use 2 to 2 1/3 pints KELTHANE MF Miticide per acre. Apply at first signs of mite buildup and repeat as necessary. Do not apply within 7 days of harvest. Do not feed treated vines (or crop residues) to meat or dairy animals.

MINT—Peppermint and spearmint for commercial oil production. For control of two-spotted mites use KELTHANE MF Miticide at 1 3/4 to 2 1/2 pints per acre in sufficient water for thorough coverage. Make application when mites become active. Do not apply within 30 days of harvest and do not make more than one application per crop per season. Do not feed treated fresh or spent hay to livestock.

FRUITS AND NUTS

CITRUS (Grapefruit, Kumquats, Lemons, Limes, Oranges, Tangelos, Tangerines)—Citrus flat, citrus red (purple), citrus rust, six-spotted, Texas citrus and Yuma mites. Apply KELTHANE MF Miticide in full coverage sprays using 1 pint per 100 gallons of dilute spray (2 gallons maximum/acre). If concentrate sprays are used adjust dosage accordingly. Apply when mites appear in threatening numbers and repeat as required for satisfactory control. Do not apply within 7 days of harvest.

GRAPES—For clover (brown almond), European red, McDaniel, Pacific, two-spotted, Willamette and yellow mites. Use 1 1/3 to 2 1/3 pints KELTHANE MF Miticide per acre. Apply just after bloom or whenever mites appear and repeat if necessary. Do not apply within 7 days of harvest.

STRAWBERRIES—Six-spotted, strawberry (Atlantic) and two-spotted mites. Use 1 quart KELTHANE MF Miticide per acre. Apply when mites appear and repeat as necessary but do not apply within 2 days of harvest, or if temperature exceeds 90°F.

WALNUTS—For clover (brown almond), European red, McDaniel, Pacific, two-spotted, Willamette and yellow mites. Use a full coverage spray of 3/4 to 1 pint KELTHANE MF Miticide per 100 gallons dilute spray (1 gallon maximum/acre). Apply just after bloom or whenever mites appear and repeat if necessary. Do not apply within 14 days of harvest.

NOTE: KELTHANE MF Miticide may be used without any spreader-sticker. When necessary to improve the spray coverage on foliage difficult to wet and to smooth out spray deposits on maturing fruit, TRITON B-1956® is recommended. Other suitable pesticide spreader-stickers may also be used except in situations where TRITON B-1956 is recommended as the only acceptable spreader-sticker.

STORAGE AND DISPOSAL

STORAGE—Keep from freezing. Store at 32°F-110°F.

PESTICIDE DISPOSAL—Do not contaminate water, food, or feed by storage or disposal. **Pesticide wastes are toxic. Improper disposal of excess pesticide, spray mixture, or rinsate is a violation of Federal law. If these wastes cannot be disposed of by use according to label instructions, contact your State Pesticide or Environmental Control Agency, or the Hazardous Waste representative at the nearest EPA Regional Office for guidance.**

CONTAINER DISPOSAL—Triple rinse (or equivalent). Then puncture and dispose of in a sanitary landfill, or by incineration, or if allowed by state and local authorities, by burning. If burned, stay out of smoke.

STEPS TO BE TAKEN IN CASE MATERIAL IS RELEASED OR SPILLED: Eliminate ignition sources. Ventilate area. Avoid breathing vapors. Wear respirator suitable for vapor concentration encountered (MSHA/NIOSH approved or equivalent). Dike the spill with inert material and transfer the liquid or solid diking materials to separate containers. Wash clothing before reuse. REFER TO PRECAUTIONARY STATEMENTS.

10490-B0

KELTHANE, Triton and B-1956 are registered trademarks of Rohm and Haas Company.

KELTHANE® 35

AGRICULTURAL MITICIDE WETTABLE POWDER

FOR AGRICULTURAL USE ONLY OR ONLY FOR SALE TO, USE, AND STORAGE BY SERVICEPERSONS

ACTIVE INGREDIENT:	
Dicofol 1,1-Bis(chlorophenyl)-2,2,2-trichloroethanol	35%
INERT INGREDIENTS:	65%
TOTAL	100%

EPA Reg. No. 707-205

KEEP OUT OF REACH OF CHILDREN

WARNING AVISO

Before using this product, read the entire Precautionary Statements, Conditions of Sale and Warranty, Directions for Use, Use Restrictions and Storage and Disposal Instructions on side and back panel. If the Conditions of Sale and Warranty are not acceptable, return the product unopened within thirty days of purchase to the place of purchase.

Precaucion Al Usuario: Si usted no lee inglés, no use este producto hasta que la etiqueta haya sido explicada ampliamente.

PRECAUTIONARY STATEMENTS

HAZARDS TO HUMANS AND DOMESTIC ANIMALS

May cause eye injury. Harmful if swallowed, inhaled, or absorbed through the skin. Do not get in eyes or on clothing. Avoid breathing dust or spray mist. Frequent or repeated exposure to this product may cause allergic reactions in some people. **Skin contact with this pesticide may be hazardous: wear chemical-resistant gloves when mixing, loading, or applying this product.**

Wash thoroughly with soap and water after handling and before eating or smoking. Remove contaminated clothing and wash before reuse.

STATEMENT OF PRACTICAL TREATMENT

IF SWALLOWED: Call a physician or Poison Control Center immediately. Drink 1 or 2 glasses of water. Do not use milk, cream or other liquids containing vegetable or animal fats. Induce vomiting by touching back of throat with finger. Do not induce vomiting or give anything by mouth to an unconscious or convulsing person.

NOTE TO PHYSICIAN: Because this product is a chlorinated miticide vomiting is recommended. The administration of milk or other fat-based demulcents which might enhance absorption is to be avoided. Epinephrin or other adrenergic amines can cause myocardial irritability in persons poisoned with chlorinated hydrocarbons. **IF INHALED:** Remove victim to fresh air. Treat the victim symptomatically. If irritation persists call a physician. **IF ON SKIN:** Wash exposed areas with plenty of soap and water. If irritation persists, call a physician. **IF IN EYES:** Flush eyes with plenty of water for at least 15 minutes. If irritation persists call a physician.

ENVIRONMENTAL HAZARDS

This product is toxic to fish. Do not apply directly to lakes, streams or ponds. Do not contaminate water by cleaning of equipment or disposal of wastes. Do not apply where runoff is likely to occur. Do not apply when weather conditions favor drift from treated areas. Apply this product only as specified on this label.

CONDITIONS OF SALE AND WARRANTY

Rohm and Haas warrants that the product conforms to its chemical description and is reasonably fit for the purpose stated on the label only when used in accordance with label directions under normal conditions of use. **ROHM AND HAAS MAKES NO OTHER EXPRESS OR IMPLIED WARRANTIES EITHER OF MERCHANTABILITY OR FITNESS FOR A PARTICULAR USE.** Handling, storage and use of the product by Buyer or User are beyond the control of Rohm and Haas and Seller. Risks such as crop injury, ineffectiveness or other unintended consequences resulting from, but not limited to, weather or soil conditions, presence of other materials, disease, pests, drift to other crops or property or failure to follow label directions will be assumed by the Buyer or User. IN NO CASE WILL ROHM AND HAAS OR SELLER BE HELD LIABLE FOR CONSEQUENTIAL, SPECIAL OR INDIRECT DAMAGES RESULTING FROM THE HANDLING, STORAGE OR USE OF THIS PRODUCT.

RE-ENTRY AND WORKER PROTECTION STATEMENTS

Do not enter treated areas without protective clothing until sprays have dried. Because certain states may require more restrictive re-entry intervals for various crops treated with this product consult your State Department of Agriculture for further information. Do not apply this product in such a manner as to directly or through drift, expose workers or other persons. The area being treated must be vacated by unprotected persons.

Written or oral warnings must be given to workers who are expected to be in a treated area or in an area about to be treated with this product. Oral warnings must be given if there is reason to believe that written warnings cannot be understood by workers. When oral warnings are given, warnings shall be given in a language customarily understood by workers. Written warnings must include the following information:

WARNING

(Insert area or field description) treated with KELTHANE 35 miticide on (insert date of application). Do not enter treated areas without protective clothing until sprays have dried. In case of accidental exposure: Call a doctor (physician), clinic or hospital immediately. Explain that the victim has been exposed to KELTHANE 35 miticide and describe his condition. For further information see the "Statement of Practical Treatment."

Oral warnings must include the same information that is in the written warnings.

STORAGE AND DISPOSAL

PESTICIDE DISPOSAL: Do not contaminate water, food, or feed by storage or disposal. Open dumping is prohibited. Pesticide wastes are toxic. Improper disposal of excess pesticide, spray mixture, or rinsate is a violation of Federal law. If these wastes cannot be disposed of by use according to label instructions, contact your State Pesticide or Environmental Control Agency or the Hazardous Waste representative at the nearest EPA Regional Office for guidance.

CONTAINER DISPOSAL: Completely empty bag into application equipment. Then dispose of empty bag in a sanitary landfill or by incineration or, if allowed by State and local authorities, by burning. If burned, stay out of smoke.

STEPS TO BE TAKEN IN CASE MATERIAL IS RELEASED OR SPILLED: Sweep or shovel solid material into a suitable container for recovery or disposal. Remove contaminated clothing promptly and wash exposed skin areas with soap and water. Wash clothing before reuse. Keep spill out of all sewers and open bodies of water. REFER TO PRECAUTIONARY STATEMENTS.

CHEMIGATION

Do not apply this product through any type of irrigation system.

DIRECTIONS FOR USE

It is a violation of Federal law to use this product in a manner inconsistent with its labeling.

KELTHANE 35 miticide gives high initial kill against most species of agricultural mites including European red, two-spotted (red spider), six-spotted, Pacific, privet, Schoene, yellow (carpini), McDaniel, Willamette, spruce, desert and tropical mites. KELTHANE 35 is a specific miticide and is harmless to predacious and other beneficial insects.

KELTHANE 35 miticide can be applied alone or it can be applied with various other pesticides in recommended spray schedules. For further details, see recommendations below and consult current State Agricultural

Experiment Station spray schedules, or State Extension Service Specialists for additional details on compatibility, dosage, timing and uses. KELTHANE 35 miticide is believed to be compatible with other insecticides and fungicides in common use.

For proper application, mix KELTHANE 35 miticide with water and apply as a wet spray with suitable spray equipment. The following chart is offered as a guide in preparing dilute, high-gallonage sprays. For concentrate ground sprayers, the amount of water used should be reduced in accordance with instructions from the sprayer manufacturer, but the amount of KELTHANE 35 miticide applied per acre should be the same whether dilute or concentrate sprays are used.

To avoid excessive residues at harvest follow the use recommendations. Do not exceed the maximum dosage per acre per application, or the time restriction between last application and harvest for each crop listed in the table below:

Summary of KELTHANE 35 Miticide Dosage per Acre and Limitations Between Last Application and Harvest

Crop	KELTHANE 35/lbs. per Acre Each Application Small Plants	Large Plants	Mixing Rate lbs. per 100 gal. Dilute Spray*	Days Before Harvest
Apples, crabapples, pears, quince	6	10-12	1-1⅓	7
Cantaloupe, cucumbers, melons, pumpkins, squash, watermelons	1	1⅔	1	2
Grapes	1½	3½	1-1⅓	7
Strawberries (see use instructions)	1	6¾	1-1⅓	2
Ornamentals, flowers, nursery stock, shade trees	—	—	1-1⅓	—
Buildings (see use directions) Clover mites on buildings	—	—	1-1⅓	—

* Adjust water gallonage if concentrate ground sprays are used. Amount of spray applied per acre must be adjusted to the size and number of trees or plants per acre and the foliage density.

FRUITS AND NUTS

APPLES, PEARS, CRABAPPLES, QUINCES—To control two-spotted, Schoene, McDaniel, yellow (carpini), Willamette, Pacific and apple rust mite and to clean up late spring European red mite and clover mite, apply at petal fall to second cover or whenever mite population appears. Make thorough full-coverage application. Repeat as necessary during the season for adequate control. To avoid possible injury do not use glyodin fungicide with KELTHANE 35 miticide in late cover sprays.

GRAPES—For brown almond (clover), two-spotted, European red, Pacific, McDaniel, Willamette and yellow mites, apply just after bloom or whenever mites appear and repeat if necessary.

STRAWBERRIES—Two-spotted, strawberry (Atlantic), and six-spotted mite, use 1 to 3 lbs. KELTHANE 35 miticide per acre. Apply when mites appear and repeat as necessary.

For Cyclamen mite use 4 to 6¾ lbs. KELTHANE 35 miticide per acre. Apply 2 or 3 thorough high-pressure, drenching sprays (400 gallons per acre) at 10 to 20 day intervals when new growth starts in spring, or use as a spot treatment in infested areas during production season.

VEGETABLES AND FIELD CROPS

CUCUMBERS, CANTALOUPES, MELONS, WATERMELONS, PUMPKINS, WINTER AND SUMMER SQUASH.—For two-spotted, Pacific, tropical and tomato russet mite, apply at first signs of mite buildup and repeat as necessary for adequate control. Do not feed treated vines, forage or crop residues to meat or dairy animals. In California mite control on cantaloupe, cucumbers, melons, pumpkins, squash, and watermelons is inadequate when vine size exceeds 14 inch runner length. Apply with ground rig only in 40 to 100 gallons of water spray mix. Use 2 qts. TRITON B-1956® per 100 gals. spray mix.

FOR USE ON

CLOVER MITES ON BUILDINGS—Use 1 to 1⅓ pounds KELTHANE 35 miticide per 100 gallons of water. Thoroughly spray outside walls, foundations and windowsills, and plants and lawn at base of infested buildings.

LAWN AND TURF GRASSES—Clover, brown and Bermuda grass mites. Use 1 to 1⅓ pounds KELTHANE 35 miticide per 100 gallons of water (1½ tablespoons per gallon). Apply at 2½ gallons per 1000 sq. ft. Make application when mites appear, covering the entire lawn area with a thorough spray and repeat in 10 to 14 days if necessary.

ORNAMENTALS, FLOWERS, NURSERY STOCK AND SHADE TREES—For control of mites on greenhouse or outdoor plants use 1 to 1⅓ pounds KELTHANE 35 miticide per 100 gallons (1½ tablespoons per gallon of water). Apply when mites are present and repeat as necessary.

NOTE: KELTHANE 35 miticide may be used without any spreader-sticker. When necessary to improve the spray coverage on foliage difficult to wet and to smooth out spray deposits on maturing fruit, TRITON B-1956 is recommended. Other suitable pesticide spreader-stickers may also be used, except in situations where TRITON B-1956 is recommended as the only acceptable spreader-sticker.

8114-D18-US

KERB® 50-W

SELECTIVE HERBICIDE IN WATER SOLUBLE POUCHES

RESTRICTED USE PESTICIDE

Because pronamide has produced tumors in laboratory animals, this product is for retail sale to and use only by Certified Applicators or persons under their direct supervision, and only for those uses covered by the Certified Applicator's certification.

USE DIRECTIONS FOR:

Artichokes (Globe) California Only • Blueberry • Fallow Land (Idaho, Oregon and Washington Only) • Legumes for Forage and Seed • Lettuce, Endive and Escarole • Ornamental Bermudagrass Turf • Raspberry and Blackberry (Oregon and Washington Only) • Rhubarb (Oregon and Washington Only) • Tree Fruit (Apple, Cherry, Nectarine, Peach, Pear, Plum, Prune) and Grapes • Woody Ornamentals, Nursery Stock and Christmas Trees

ACTIVE INGREDIENT:

Pronamide 3,5-dichloro-N-(1, 1-dimethyl-2-propynyl)-benzamide	50.0%
INERT INGREDIENTS:	50.0%
TOTAL	100.0%

EPA Reg. No. 707-159

KEEP OUT OF REACH OF CHILDREN
CAUTION

NOTICE: Before using this product, read the entire Precautionary Statements, Conditions of Sale and Warranty, Directions for Use, Use Restrictions and Storage and Disposal Instructions. If the Conditions of Sale and Warranty are not acceptable, return the product unopened within thirty days of purchase to the place of purchase.

PRECAUTIONARY STATEMENTS

HAZARDS TO HUMANS AND DOMESTIC ANIMALS

Harmful if swallowed. Take special care to avoid contact with eyes, skin or clothing. Do not breathe the dust or spray mist. Wash clothing and gloves after use.

STATEMENTS OF PRACTICAL TREATMENT

IF SWALLOWED: Dilute by giving 2 glasses of water to drink and call a physician. Never give anything by mouth to an unconscious person.

NOTE TO PHYSICIAN: Emesis is recommended.

IF INHALED: Move subject to fresh air.

IF ON SKIN: Wash affected skin areas with soap and water. Consult a physician if irritation persists.

IF IN EYES: Flush eyes with large amounts of water for at least 15 minutes. Consult a physician if irritation persists.

PROTECTIVE CLOTHING: When applying this product, wear midforearm waterproof gloves, long-sleeved shirts and long pants, preferably one-piece (coveralls). Hand-spray or hand-spreaders also require the use of waterproof boots or shoe coverings. Wash nondisposable gloves, boots and shoe coverings thoroughly with soap and water before removing.

Protective clothing/equipment is not needed during application if an enclosed tractor cab with filtered air supply or enclosed cockpit is used.

Any article of clothing worn while handling product must be cleaned before reusing. Clothing should be laundered separately from household articles. Clothing which has been drenched or heavily contaminated should be disposed of in accordance with state or local regulations.

ENVIRONMENTAL HAZARDS

Do not apply directly to water. Do not contaminate water when disposing of equipment washwaters.

CONDITIONS OF SALE AND WARRANTY

Rohm and Haas warrants that the product conforms to its chemical description and is reasonably fit for the purpose stated on the label only when used in accordance with label directions under normal conditions of use. **ROHM AND HAAS MAKES NO OTHER EXPRESS OR IMPLIED WARRANTIES EITHER OF MERCHANTABILITY OR FITNESS FOR A PARTICULAR USE.** Handling, storage and use of the product by Buyer or User are beyond the control of Rohm and Haas and Seller. Risks such as crop injury, ineffectiveness or other unintended consequences resulting from, but not limited to, weather or soil conditions, presence of other materials, disease, pests, drift to other crops or property or failure to follow label directions will be assumed by the Buyer or User. **IN NO CASE WILL ROHM AND HAAS OR SELLER BE HELD LIABLE FOR CONSEQUENTIAL, SPECIAL OR INDIRECT DAMAGES RESULTING FROM THE HANDLING, STORAGE OR USE OF THIS PRODUCT.**

DIRECTIONS FOR USE

It is a violation of Federal law to use this product in a manner inconsistent with its labeling.

USE RESTRICTIONS: Hand-spray applications of pronamide may be made only to ornamentals and nursery stock.

CHEMIGATION

DO NOT APPLY THIS PRODUCT THROUGH ANY TYPE OF IRRIGATION SYSTEM.

RE-ENTRY AND WORKER PROTECTION STATEMENTS

Do not enter treated areas without protective clothing until sprays have dried. Because certain states may require more restrictive re-entry intervals for various crops treated with this product, consult your State Department of Agriculture for further information. Do not apply this product in such a manner as to directly or through drift, expose workers or other persons. The area being treated must be vacated by unprotected persons.

Written or oral warnings must be given to workers who are expected to be in a treated area or in an area about to be treated with this product. Oral warnings must be given if there is reason to believe that written warnings cannot be understood by workers. When oral warnings are given, warnings shall be given in a language customarily understood by workers. Written warnings must include the following information:

CAUTION

(Insert area or field description) treated with KERB 50-W on (insert date of application). Do not enter treated areas without protective clothing until sprays have dried. In case of accidental exposure: Call a doctor (physician), clinic or hospital immediately. Explain that the victim has been exposed to KERB 50-W and describe his condition. For further information see the "Statements of Practical Treatment."

Oral warnings must include the same information that is in the written warnings.

GENERAL INFORMATION

KERB 50-W selective herbicide is formulated as a wettable powder containing 50% active ingredient. It is effective for the control of a wide range of grasses and certain broadleaf weeds. The product is a soil active herbicide with uptake by sensitive weeds occurring through the roots. Before using this herbicide for a specific crop use, study the following general use information which provides important instructions for the safe and effective application of the product.

WEED SPECTRUM

KERB 50-W herbicide is recommended for both preemergence and early postemergence control of winter annual and perennial grasses and chickweed and for preemergence control only of certain other broadleaf weeds and grasses listed.

WEEDS CONTROLLED BOTH PREEMERGENCE AND EARLY POSTEMERGENCE

BARLEY, FOXTAIL	*Hordeum jubatum*
BARLEY, VOLUNTEER	*Hordeum vulgare*
BENTGRASS	*Agrostis species*
BLUEGRASS, ANNUAL	*Poa annua*
BLUEGRASS, BULBOUS	*Poa bulbosa*
BLUEGRASS, KENTUCKY	*Poa pratensis*
BROME, DOWNY (CHEATGRASS)	*Bromus tectorum*
CHICKWEED, COMMON	*Stellaria media*
CHICKWEED, MOUSE-EAR	*Cerastium vulgatum*
FESCUE, TALL	*Festuca arundinaceae*
GOATGRASS, JOINTED	*Aegilops cylindrica*
OAT, VOLUNTEER	*Avena sativa*
OAT, WILD	*Avena fatua*
ORCHARDGRASS	*Dactylis glomerata*
QUACKGRASS	*Agropyron repens*
RYE, VOLUNTEER	*Secale cereale*
RYEGRASS, ITALIAN	*Lolium multiflorum*
RYEGRASS, PERENNIAL	*Lolium perenne*
VELVETGRASS	*Holcus lanatus*
WHEAT, VOLUNTEER	*Triticum aestivum*

WEEDS CONTROLLED ONLY PREEMERGENCE

BARNYARDGRASS	*Echinochloa crus-galli*
CANARYGRASS	*Phalaris canariensis*
CARPETWEED	*Mollugo verticillata*
CRABGRASS, LARGE	*Digitaria sanguinalis*
DODDER, FIELD	*Cuscuta campestris*
FOXTAIL, YELLOW	*Setaria lutescens*
GOOSEGRASS	*Eleusine indica*
GOOSEFOOT, NETTLELEAF	*Chenopodium murale*
HENBIT	*Lamium amplexicaule*
KNOTWEED, PROSTRATE	*Polygonum aviculare*
LAMBSQUARTERS, COMMON	*Chenopodium album*
LOVEGRASS	*Eragrostis diffusa*
ROCKET, LONDON	*Sisymbrium irio*
MALLOW, LITTLE (CHEESEWEED)	*Malva parviflora*
MORNINGGLORY, ANNUAL	*Ipomoea purpurea*
MUSTARD, WILD	*Brassica kaber*
NETTLE, BURNING	*Urtica urens*
NIGHTSHADE, BLACK	*Solanum nigrum*
NIGHTSHADE, HAIRY	*Solanum sarachoides*
PANICUM, FALL	*Panicum dichotomiflorum*
PURSLANE, COMMON	*Portulaca oleracea*
SHEPHERDSPURSE	*Capsella bursa-pastoris*
SMARTWEED, PALE	*Polygonum lapathifolium*
SORREL, RED (from seed)	*Rumex acetosella*
TOMATO, VOLUNTEER	*Solanum esculentum*
RADISH, WILD	*Raphanus sativus*

NOTE: The weed species controlled by KERB 50-W herbicide are dependent on the rate used, specific crop culture involved, and the associated conditions of temperature, soil type and moisture availability. Refer to specific crop use directions for weed species controlled.

DOSAGE

The rate of KERB 50-W herbicide required will vary depending on the crop culture involved and weed species to be controlled. See specific crop use directions for all dosage recommendations. All dosage recommendations listed in this label are in terms of pounds KERB 50-W herbicide per broadcast acre. For banded application, the amount of KERB 50-W herbicide used per acre should be reduced to the following formula:

$$\frac{\text{Band width (in.)}}{\text{Row width (in.)}} \times \text{Rate per Acre Broadcast} = \text{Amount needed per Acre for Band Application}$$

TIMING AND APPLICATION

Unless specific directions are given under the crop to be treated, KERB 50-W herbicide should be applied in the fall or early winter, when temperatures do not exceed 55°F, BUT PRIOR TO SOIL FREEZE-UP. Best weed control results when KERB 50-W herbicide is applied preemergence to the weeds and when application is followed by rainfall or irrigation to move the product into the root zone of the germinating weeds.

KERB 50-W herbicide should be thoroughly mixed in clean water at the recommended concentration and applied uniformly as a spray. For ground application, use a conventional low pressure herbicide sprayer equipped with flat fan nozzles spaced and calibrated to uniformly deliver 20 to 50 gallons of spray per acre. For aerial applications apply in a coarse droplet spray at 5 to 10 gallons per acre. Accurately calibrate spray equipment prior to each use.

EFFECT OF SOIL TYPE, MOISTURE AND TEMPERATURE

KERB 50-W herbicide is most active in coarse to medium textured soils of low organic matter and is relatively inactive in peat or muck soils or mineral soils high in organic matter content at rates recommended in this label. Herbicidal activity is best in soils containing less than 4 percent organic matter. Use in soils with higher organic matter may result in inconsistent or incomplete weed control.

The herbicidal activity of KERB 50-W is mainly through root absorption in sensitive weed species. Rain, melting snow or irrigation is ESSENTIAL following treatment to move KERB 50-W herbicide into the root zone of germinating weeds.

Under field conditions, KERB 50-W herbicide will remain relatively stable with little loss of herbicidal activity when soil temperatures are less than 55°F. As soil temperatures increase, degradation of the active ingredient takes place. KERB 50-W herbicide may degrade rather quickly if left exposed on the soil surface in warm weather. If KERB 50-W herbicide is applied when air temperatures exceed 85°F, it should be soil incorporated to a shallow depth (top two to three inches) or watered into the soil as soon as possible.

CULTURAL CONSIDERATIONS

For best results apply KERB 50-W herbicide to a trash free soil surface. Clean cultivation before application is preferable, but not necessary. To obtain optimum weed control in areas not clean cultivated, the area to be treated should be free of surface litter (dead or decaying crop and weed debris, mowing clippings, etc.). Trash free areas create ideal conditions for rapid movement of KERB 50-W into the weed root zone following rain or irrigation.

ROTATION CROP PLANTING INFORMATION

Follow the directions given below when rotation crops will be planted to areas previously treated with KERB 50-W herbicide:

1. Whether KERB 50-W herbicide is bed-topped, banded or broadcast, the beds should be knocked down and the field cross-disced before rotation crops other than artichokes, lettuce, endive or escarole are planted.
2. Where the KERB 50-W herbicide treatment is to be followed by a rotation crop within six months of application, bed-topped or banded applications are suggested.
3. Where small grains or grasses will be planted within one year to areas where KERB 50-W herbicide is to be applied at a dosage rate greater than 1.0 pound of product per acre, KERB 50-W herbicide should not be broadcast. Treatments should be limited to band or bed-top applications only.
4. Where rotation crops are to follow within one year of the KERB 50-W herbicide treatment, the following rotation plan applies:

Amount of KERB 50-W Applied/Planted Acre[2]	Waiting Period in Months Before Planting the Crops Indicated[1]			
	Group A	Group B	Group C	Group D
1.0 lb.	0	3	3	6
2.0 lb.	0	3	4	9
3.0 lb.	0	4	6	9
4.0 lb.	0	5	7	12

[1] Crop Group A — Artichokes, lettuce, endive, escarole

Crop Group B — Beans, corn, cotton, sorghum, carrots, celery

Crop Group C — Broccoli, cabbage, cauliflower, cucurbits, spinach, sugarbeets, onions, tomatoes

Crop Group D — Wheat, barley, oats, grasses

NOTE: "Crops not listed in Crop Groups A, B, C or D (above) should not be planted back sooner than 12 months after the last KERB 50-W herbicide application regardless of the rate of KERB 50-W used."

[2] Planted Acre refers to Acreage planted. KERB at 4.0 pounds of product per treated acre on two 5 inch treated bands (40 inch centers) is the equivalent of one pound KERB 50-W per planted acre. KERB at 4.0 pounds of product as a broadcast treatment (total acreage sprayed) is the equivalent of 4 pounds of KERB 50-W per planted acre.

ARTICHOKES (GLOBE)

(CALIFORNIA ONLY)

GENERAL INFORMATION

KERB 50-W is a selective herbicide for the control of susceptible weeds in either established (ratoon) or transplanted globe artichokes.

WEEDS CONTROLLED

KERB 50-W herbicide is effective at 4.0 to 8.0 pounds of product (2.0 to 4.0 pounds active ingredient) per treated acre for the preemergence control of the following weeds:

BARLEY, VOLUNTEER	HENBIT	NIGHTSHADE, HAIRY
BLUEGRASS, ANNUAL	KNOTWEED, PROSTRATE	OAT, VOLUNTEER
CHICKWEED, COMMON	MALLOW, LITTLE (CHEESEWEED)	OAT, WILD
CHICKWEED, MOUSE-EAR		RYEGRASS, ITALIAN
FOXTAIL, YELLOW	MUSTARD, WILD	WHEAT, VOLUNTEER
GOOSEFOOT, NETTLELEAF	NETTLE, BURNING	

Crop	Weeds	KERB 50-W Rate (Per Broadcast Acre)* Dependable Rainfall or Overhead Irrigation	Less Dependable Rainfall or Furrow Irrigation	Comments
Globe Artichokes (established ratoon)	Susceptible annual grasses, volunteer grains and broad-leaf weeds	4.0 lbs.	Not recommended	Sandy soils, sandy loams and silt loams
		8.0 lbs.	Not recommended	Silt, silty clay loams, clay loams and clay soils
Globe Artichokes (newly transplanted crowns)	Susceptible annual grasses, volunteer grains and broad-leaf weeds	4.0 lbs.	Not recommended	All soil types except peat and muck soils

*Dosage recommendations listed on this label are in terms of pounds KERB 50-W per acre broadcast application. For banded treatments down artichoke rows or between rows, the amount of KERB 50-W used per acre should be reduced according to the following formula:

$$\frac{\text{Band width (in.)}}{\text{Row width (in.)}} \times \text{Rate per Acre Broadcast} = \text{Amount needed per Acre for Band Application}$$

DOSAGE AND TIMING RECOMMENDATIONS

Established Ratoon Artichokes

Apply KERB 50-W in a single postemergence application to the crop after tillage operations are completed and shoot regrowth of the artichokes has occurred. Application timing of KERB 50-W should be preemergence to the weeds and before new artichoke leaves are greater than 14 to 16 inches long. Apply KERB 50-W in a banded treatment over the crop row at the rate of 4.0 to 8.0 pounds of product per broadcast acre (see dosage rate for soil type in chart). A second application of KERB 50-W at the same rate may be applied in a banded treatment directed to the untreated soil surface between the artichoke rows after the ditching operation is completed later in the season.

Transplanted Artichoke Crowns

Apply KERB 50-W in a single application after transplanting the crowns but before new shoots have developed 3 to 4 new leaves. KERB 50-W should be applied preemergence to the weeds and banded over the crop row at the rate of 4.0 pounds of product per broadcast acre. Do not use higher rates of KERB 50-W than 4.0 pounds per acre in one season. A second application of KERB 50-W at the same rate may be applied in a banded treatment directed to the untreated soil surface between the artichoke rows after the ditching operation is completed later in the season.

APPLICATION

KERB 50-W herbicide may be applied by aircraft or ground sprayer for preemergence control of susceptible grasses and broadleaf weeds in established ratoon artichokes or transplanted artichoke crowns.

Aerial — Mix the recommended amount of KERB 50-W in a minimum of 10 gallons of water per acre for aerial application. Avoid drift to all other crops and non-target areas.

Ground — Mix the recommended amount of KERB 50-W herbicide in clean water and apply uniformly with a ground sprayer in 20 to 50 gallons of water per acre. Reduce dosage and volume accordingly for banded treatments. Use a standard low pressure herbicide sprayer equipped with flat fan nozzles that give uniform spray distribution.

MOISTURE AND IRRIGATION REQUIREMENTS

Moisture is necessary to activate KERB 50-W in the soil and move it into the root zone of germinating weeds. In artichoke culture natural rainfall or supplementary overhead irrigation within 1 to 3 days after the application of KERB 50-W herbicide is essential for effective weed control. For best results use overhead sprinkler irrigation equipment to irrigate the field with 1 to 2 inches of water after application of KERB 50-W herbicide.

EFFECT OF SOIL TYPE

KERB 50-W is not recommended for use on highly organic or muck soils because herbicidal activity is lowered significantly in these soils. Follow dosage rates suggested in the dosage recommendations chart according to the soil type for established and transplanted artichokes.

ROTATION CROPS

Artichokes are generally long-term perennial crops. In the event that artichokes are discontinued and a rotational crop will be planted within one year where KERB 50-W was applied at the rate of 4.0 pounds of product per acre, follow the crop rotational plan outlined on this label for lettuce. If artichokes are discontinued where KERB 50-W was applied at a dosage rate greater than 4.0 pounds of product per acre, do not plant back to crops other than artichokes for a period of 18 months.

ARTICHOKES — SPECIFIC USE RESTRICTIONS

- Do not apply more than 8.0 pounds of KERB 50-W herbicide to established artichokes or more than 4.0 pounds to newly transplanted artichokes per acre broadcast application.
- Do not make more than one application to the artichoke row in one season.

- Do not harvest artichokes within 60 days of the first application (over the plant rows) or within 1 day of the second application (to the untreated area between rows after ditching).

BLUEBERRIES

GENERAL INFORMATION

KERB 50-W is a selective herbicide recommended for fall and winter applications to established blueberries for both preemergence and postemergence control of winter annual and perennial grasses and chickweed and preemergence control of certain broadleaf weeds.

DOSAGE RECOMMENDATIONS

KERB 50-W herbicide is recommended at the rate of 2.0 to 4.0 pounds of product (1.0 to 2.0 pounds active ingredient) per acre broadcast application. The rate will depend on the weed species present. Follow the weed control recommendations listed in the chart below:

Weeds Controlled		Pounds KERB 50-W Per Acre[1] Dependable Rainfall or Overhead Irrigation[2]
BLUEGRASS ANNUAL BROME, DOWNY (CHEATGRASS) CHICKWEED	OAT, WILD SORREL, RED (from seed)	2.0
BENTGRASS BLUEGRASS, KENTUCKY FESCUE, TALL ORCHARDGRASS	QUACKGRASS RYEGRASS, PERENNIAL VELVETGRASS	4.0

[1] Dosage rates recommended are in pounds of KERB 50-W per acre broadcast application. Reduce rates accordingly for banded applications.

[2] For effective weed control, rainfall or overhead irrigation is essential following the application of KERB 50-W.

CROP TOLERANCE

Established blueberry plants are tolerant to recommended rates of KERB 50-W herbicide. Do not apply KERB 50-W to newly transplanted blueberries until roots are well established.

TIMING AND APPLICATION

Apply KERB 50-W herbicide in a single application during the fall or early winter months, but prior to soil freeze-up and snow cover. Optimum herbicidal activity occurs when applications are made under cool temperature conditions (55°F or less) and are followed by rainfall or overhead irrigation.

Mix the recommended amount of KERB 50-W in clean water and apply uniformly with a low pressure ground sprayer in 20 to 50 gallons of water per acre.

BLUEBERRIES — SPECIFIC USE RESTRICTIONS

- Do not make more than one application of KERB 50-W herbicide per year and do not apply KERB 50-W at rates higher than 4.0 pounds product (2.0 pounds active ingredient) per acre.

FALLOW LAND
(IDAHO, OREGON AND WASHINGTON ONLY)

GENERAL INFORMATION

KERB 50-W herbicide is a selective herbicide for use on fallow land to control certain winter annual grasses and volunteer grains.

Use of the product is restricted to summer fallow land which will be planted back the following year to winter wheat, barley or oats.

Use rates and plant-back restrictions for KERB 50-W herbicide should be strictly followed or severe injury to the following small grain crop may occur.

KERB 50-W is a soil active herbicide that is absorbed through weed roots. In order to achieve good weed control, sufficient soil moisture from rain, melting snow or irrigation is required following treatment to move KERB 50-W herbicide into the soil to the weed root zone.

Do not use any fall tillage prior to or after the application of KERB 50-W herbicide to fallow land, since weed control will be reduced. To ensure best weed control results with KERB 50-W, use a straw spreader when combining to eliminate chaff rows.

WEEDS CONTROLLED

KERB 50-W herbicide will provide preemergence and postemergence control of the following weeds when used according to label directions:

BARLEY, VOLUNTEER	GOATGRASS, JOINTED
BLUEGRASS, BULBOUS	RYE, VOLUNTEER
BROME, DOWNY (CHEATGRASS)	WHEAT, VOLUNTEER

DOSAGE AND TIMING

Apply KERB 50-W herbicide in a single application at a rate of 0.5 to 1.0 pound (0.25 to 0.50 pound active ingredient) per acre. For optimum weed control, apply KERB 50-W herbicide between mid-October and mid-December, prior to soil freeze-up. Applications made outside these dates could result in poor weed control.

APPLICATION

Mix the recommended amount of KERB 50-W herbicide in clean water and apply uniformly in a broadcast application. For ground application KERB 50-W herbicide should be applied in 10 to 20 gallons of water per acre with a conventional low pressure herbicide sprayer equipped with flat fan nozzles. For aerial applications apply in a coarse droplet spray at 5 to 10 gallons per acre. Spray equipment should be carefully calibrated before each use.

ROTATION CROPS

Winter wheat, barley and oats can be planted into fallow land treated with KERB 50-W herbicide 9 months (270 days) or more after application.

FALLOW LAND — SPECIFIC USE RESTRICTIONS

- Do not plant winter wheat, barley or oats within 270 days of the application of KERB 50-W herbicide.
- Do not use any fall tillage prior to or after the application of KERB 50-W herbicide.

TANK MIXTURE WITH GLEAN® HERBICIDE FOR FALLOW LAND (IDAHO, OREGON AND WASHINGTON ONLY)

For control or suppression of certain broadleaf weeds not controlled by KERB 50-W herbicide in fallow land areas, Glean herbicide may be tank mixed with KERB 50-W. See Glean label for list of weeds claimed.

DOSAGE

Use KERB 50-W herbicide at the rate of 0.5 to 0.75 pound (0.25 to 0.38 lb. active ingredient) per acre tank mixed with Glean herbicide at the dosage rate recommended on the Glean label for reduced tillage fallow weed control.

Important: Before applying a tank mixture of KERB 50-W and Glean herbicides on fallow land areas, read and observe all label directions for each product. Use rates and plant back restrictions should be strictly followed or severe injury to the following small grain crops may occur. In interpreting product labels for tank mixtures, the most restrictive situations must apply.

For crop rotation flexibility, do not use Glean herbicide on all your fallow acreage.

LEGUMES GROWN FOR FORAGE AND SEED
ALFALFA, CLOVER, BIRDSFOOT TREFOIL, CROWN VETCH AND SAINFOIN

GENERAL INFORMATION

KERB 50-W is a selective herbicide recommended for fall or winter applications to alfalfa, clover, birdsfoot trefoil, crown vetch and sainfoin for both preemergence and postemergence control of susceptible winter annual and perennial grasses and for preemergence control of certain broadleaf weeds.

DOSAGE

KERB 50-W herbicide is recommended at the rate of 1.0 to 4.0 pounds of product (0.5 to 2.0 pounds active ingredient) per broadcast acre application. The required rate will depend on the weed species present as well as the type of irrigation used or the dependability of rainfall following application. The effective rate will be higher in low rainfall areas or where furrow irrigation is used than in areas of dependable rainfall or where overhead irrigation is practiced. Follow the weed control recommendations given in the chart below for fall or winter applications of KERB 50-W herbicide:

Pounds KERB 50-W Per Acre Broadcast Application

Weeds Controlled		Dependable Rainfall or Overhead Irrigation	Low Rainfall or Furrow Irrigation
Apply Preemergence or Postemergence to These Weeds:			
BARLEY, FOXTAIL BLUEGRASS, ANNUAL BROME, DOWNY (CHEATGRASS) CHICKWEED	GRAIN, VOLUNTEER OAT, WILD RYEGRASS, ITALIAN	1.0-1.5	1.5-2.0
BLUEGRASS, KENTUCKY ORCHARDGRASS RYEGRASS, PERENNIAL		1.5-2.0	2.0-3.0
QUACKGRASS		2.0-3.0	3.0-4.0
Apply Preemergence Only to These Weeds: SORREL, RED (from Seed)		1.5-2.0	2.0-3.0
MUSTARD, WILD	ROCKET, LONDON	3.0	4.0

RADISH, WILD	SHEPHERDSPURSE		

NOTE: For control of spring germinating dodder refer to specific instructions under Spring Use Directions for Established Alfalfa.

TIMING AND APPLICATION

KERB 50-W herbicide should be applied during the fall or winter months. Optimum herbicidal activity occurs when applications are made under cool temperature conditions (55°F or less) and are followed by rainfall or overhead irrigation. Applications should always be made BEFORE SOIL FREEZE-UP.

Applications are recommended postemergence to established, actively growing or dormant forage legumes or to new plantings after the legume has reached the trifoliate leaf stage. In established forage legume stands, applications should be made after the last cutting when the weather and soil temperatures are cool. In fall-seeded forage legumes, applications should be made after legumes have reached the trifoliate leaf stage. In spring-seeded forage legumes, applications of KERB 50-W herbicide should be made the following fall or early winter to control winter annual and perennial grasses. Do not use KERB 50-W herbicide as a preplant or preemergence treatment or before the trifoliate leaf stage of the legume has developed in new plantings as injury to the legume stand may result. Remove or disperse trash, crop residues and ashes before treatment.

Mix the recommended amount of KERB 50-W herbicide in clean water and apply uniformly with a ground sprayer at 20 to 50 gallons per acre. Use a conventional herbicide sprayer equipped with flat fan nozzles at 40 to 60 psi.

ROTATION CROPS

Where rotation crops are to follow within one year of the KERB 50-W herbicide treatment to alfalfa, clover, birdsfoot trefoil, crown vetch or sainfoin, follow the directions given in the General Information section of this label under ROTATION CROP PLANTING INFORMATION.

SPECIFIC USE RESTRICTIONS — ALFALFA, CLOVER, BIRDSFOOT TREFOIL, CROWN VETCH AND SAINFOIN

- Do not use more than 4.0 pounds KERB 50-W herbicide per acre per season.
- Do not graze or harvest for forage or dehydration within the following intervals after application.

Alfalfa — below 3.0 pounds KERB 50-W/A (West of Mississippi River)	— 25 days
Alfalfa — 3.0 to 4.0 pounds KERB 50-W/A (West of Mississippi River)	— 45 days
Clover, birdsfoot trefoil, crown vetch, sainfoin (entire U.S.) and alfalfa (East of Mississippi River) up to 4.0 pounds KERB 50-W/A	— 120 days

SPRING USE DIRECTIONS FOR ESTABLISHED ALFALFA —

DODDER CONTROL IN ALFALFA SEED CROPS ONLY IN CALIFORNIA, IDAHO, NEVADA, OREGON, UTAH AND WASHINGTON

GENERAL INFORMATION

For effective control KERB 50-W herbicide must be moved into the soil either by rainfall or irrigation before the germination of dodder. Preferably, irrigation should be made within 1 to 3 days following the KERB 50-W application, but can be delayed up to 2 weeks if necessary provided that irrigation precedes dodder germination. If irrigation of the field treated with KERB 50-W herbicide must be delayed, a light mechanical incorporation (maximum 1 inch depth) should follow the KERB 50-W application and the field irrigated within two weeks.

When using flood type or overhead sprinkler irrigation systems the amount of irrigation following the KERB 50-W application should not exceed one inch of water. Excess irrigation following the KERB 50-W application and prior to germination of dodder may decrease the effectiveness of KERB 50-W herbicide.

DOSAGE AND TIMING

For effective control, KERB 50-W herbicide must be applied before dodder germinates. Follow directions given below depending on method of irrigation used:

Furrow Irrigation — Apply KERB 50-W at the rate of 3.0 to 4.0 pounds of product (1.5 to 2.0 pounds active ingredient) per acre. Incorporate lightly at time of application and irrigate within seven days.

Flood Irrigation — Apply KERB 50-W at the rate of 3.0 pounds of product (1.5 pounds active ingredient) per acre. Flood field with 0.5 to 1.0 inch of water within 1 to 3 days after application.

Overhead Sprinkler Irrigation — Use same directions as given above for flood irrigation.

Excessive amounts of irrigation water following KERB 50-W herbicide application may adversely affect the herbicidal activity.

NOTE: If the legume crop treated with KERB 50-W herbicide should be winter killed or otherwise destroyed, do not plant the field in the spring to small grain or grass crops since these crops may be injured from residual KERB 50-W in the soil. For additional information see Rotation Crop Planting Chart in the General Information section of this label.

LETTUCE, ENDIVE AND ESCAROLE

GENERAL INFORMATION

KERB 50-W herbicide is a selective herbicide for the control of certain annual grasses and broadleaf weeds in direct seeded or transplanted lettuce, endive and escarole.

WEEDS CONTROLLED

KERB 50-W herbicide is effective at 2.0 to 4.0 pounds of product (1.0 to 2.0 pounds active ingredient) per treated acre for the preemergence control of the following weeds:

GRASSES		
BARLEY, FOXTAIL	CANARYGRASS	OATS, VOLUNTEER
BARLEY, VOLUNTEER	CRABGRASS	PANICUM, FALL
BARNYARDGRASS	FOXTAIL, YELLOW	RYEGRASS, ITALIAN
BLUEGRASS, ANNUAL	GOOSEGRASS	RYE, VOLUNTEER
BROME, DOWNY (CHEATGRASS)	LOVEGRASS	WHEAT, VOLUNTEER

BROADLEAF WEEDS		
CARPETWEED	LAMBSQUARTERS, COMMON	PURSLANE, COMMON
CHICKWEED, COMMON	MORNINGGLORY, ANNUAL	ROCKET, LONDON
GOOSEFOOT, NETTLELEAF	MUSTARD, WILD	SHEPHERDSPURSE
	NETTLE, BURNING	SMARTWEED, PALE
HENBIT	NIGHTSHADE, BLACK	TOMATO, VOLUNTEER
KNOTWEED	NIGHTSHADE, HAIRY	

DOSAGE

KERB 50-W herbicide is recommended at the rate of 2.0 to 4.0 pounds of product (1.0 to 2.0 pounds active ingredient) per acre broadcast application. The dosage rate required is dependent on soil texture and method of irrigation. At rates recommended on this label, KERB 50-W is not recommended for weed control on highly organic (peat and muck) soils. For lettuce, endive and escarole follow the dosage recommendations listed in chart below:

Pounds KERB 50-W Per Acre Broadcast Application[1]

Weeds	Dependable Rainfall or Overhead Irrigation	Less Dependable Rainfall or Furrow Irrigation	Soil Texture Group[2]
Susceptible annual grasses and	2.0 - 3.0 (Surface application)	3.0 - 4.0 (Soil incorporation)	Coarse and medium textured soils
broadleaf weeds	3.0 - 4.0 (Surface application)	4.0 (Soil incorporation)	Fine textured soils

[1] Reduce dosage rate accordingly for banded applications.
[2] Soil Texture Group
Coarse: sand, loamy sand, sandy loam
Medium: loam, silt loam, silt, sandy clay loam
Fine: silty clay loam, clay loam, sandy clay, silty clay, clay

CROP TOLERANCE

Most varieties of lettuce are highly tolerant of the recommended rates of KERB 50-W herbicide. Do not use more than 3 pounds KERB 50-W on Val Temp, Grande Verde and Prima Verde varieties of crisp lettuce, or on endive and escarole.

TIMING AND APPLICATION

KERB 50-W herbicide can be applied either pre-plant, post-plant or postemergence to lettuce in banded, bed-topped or broadcast applications. Most applications will be made preemergence to the lettuce just before or after planting and preemergence to the weeds. Applications can be made before or after lettuce thinning but should be made prior to weed emergence. Do not apply KERB 50-W herbicide to lettuce within 55 days of harvest and do not make more than one application to each crop of lettuce.

Mix the recommended amount of KERB 50-W herbicide in clean water and apply uniformly with a ground sprayer in 20 to 50 gallons of water per treated acre. Reduce dosage and volume accordingly for banded treatments. Use a standard low pressure sprayer equipped with flat fan nozzles that provide uniform spray distribution.

APPLICATION-MOISTURE REQUIREMENTS

KERB 50-W herbicide acts mainly through root absorption, therefore it is necessary to move KERB 50-W herbicide into the root zone of germinating weeds to provide effective control. This can be accomplished by overhead sprinkler irrigation, by rainfall or by shallow mechanical incorporation.

Sprinkler Irrigation

KERB 50-W herbicide can be applied to the soil surface without mechanical incorporation after planting or transplanting if overhead irrigation is used. An initial irrigation of 1 to 2 inches should promptly follow the application of KERB 50-W herbicide, especially in hot weather.

Applications Dependent on Natural Rainfall
In areas of dependable natural rainfall, KERB 50-W herbicide can be applied as a surface treatment preemergence to the weeds. Applications to direct seeded or transplanted lettuce are most successful when followed by 1/2 to 1 inch of rainfall within two to three days after application.

Furrow Irrigation — Mechanical Incorporation
Where rainfall is not dependable or supplementary overhead irrigation is not used, shallow pre-plant incorporation is recommended. PTO driven incorporators or rolling cultivators that thoroughly mix KERB 50-W herbicide into the top 2 inches of soil are suggested.

Incorporation should be simultaneous or immediately after application of KERB 50-W herbicide, especially in hot weather. Irrigation should be started as soon as possible.

Where furrow irrigation is used, spray application and mechanical incorporation should be made after beds have been formed. KERB 50-W herbicide will not be as effective if disced in prior to bed shaping.

Hoeing, thinning or shallow cultivation of soil treated with KERB 50-W herbicide will not destroy its herbicidal activity.

TEMPERATURE
KERB 50-W herbicide is not highly volatile, but it may degrade rather quickly if left exposed on the soil surface in warm weather. If applied when air temperatures exceed 85°F it should be shallow incorporated or watered into the soil as soon as possible, preferably within 1 or 2 days.

ROTATION CROPS
Where rotation crops are to follow within one year of the KERB 50-W herbicide treatment to lettuce, endive or escarole, follow the directions given in the General Information section of this label under ROTATION CROP PLANTING INFORMATION.

LETTUCE — SPECIFIC USE RESTRICTIONS • Do not apply KERB 50-W herbicide to lettuce varieties which will be harvested less than 55 days after treatment. • Do not apply more than one application of KERB 50-W herbicide to each crop of lettuce.

ORNAMENTAL BERMUDAGRASS

GROWN FOR TURF OR SEED
GENERAL INFORMATION
KERB 50-W is a selective herbicide recommended for the preemergence and postemergence control of annual bluegrass *(Poa annua)* in ornamental bermudagrass *(Cynodon dactylon)* turf or bermudagrass grown for seed.

WEED CONTROL
KERB 50-W herbicide will control annual bluegrass for pregermination and seedling stages through tillering, heading and seed formation. KERB 50-W herbicide acts slowly on seedling to mature annual bluegrass. Following application of KERB 50-W annual bluegrass may first become dark green and then gradually turn yellow and die over a 3 to 5 week period.

For effective control of annual bluegrass in bermudagrass turf or seed crops, moisture is necessary to move KERB 50-W herbicide in the weed root zone. Refer to the Moisture Requirement section of this label for details.

DOSAGE AND TIMING
KERB 50-W herbicide is recommended at the rate of 1.0 to 3.0 pounds of product (0.5 to 1.5 pounds active ingredient) per acre broadcast application. The dosage rate required is dependent on the growth stage of annual bluegrass at time of application. Follow the dosage rate and timing recommendations given below:

Annual Bluegrass Growth Stage	Pounds KERB 50-W Per Acre Broadcast Application
Preemergence or early postemergence	1.0-2.0*
Postemergence-early tillering to heading	1.5-2.0
Postemergence-seed forming stages	2.0-3.0

*Use the higher rate when longer preemergence residual control is desired.

APPLICATION
Mix the recommended amount of KERB 50-W herbicide in clean water and apply uniformly with a low pressure ground sprayer in 20 to 50 gallons of water per acre. The sprayer should be equipped with flat fan nozzles, spaced to provide uniform distribution without skips or excessive overlapping of spray patterns.

NOTE: Avoid spraying on fairways or approaches that may drain onto bentgrass greens or to areas overseeded with sensitive cool season grasses.

MOISTURE REQUIREMENTS
KERB 50-W herbicide acts mainly through root absorption in sensitive weed species. If no rainfall occurs within a day or so of the application, a light overhead irrigation should be made to move the chemical into the weed root zone. Avoid heavy irrigations of more than 1 inch to reduce the possibility of excess washing or leaching of the chemical from the area of application.

ORNAMENTAL BERMUDAGRASS TURF OR SEED CROP — SPECIFIC USE RESTRICTIONS
- Do not apply KERB 50-W herbicide to areas that are to be overseeded with susceptible cool season grasses within 90 days of treatment.
- Do not apply KERB 50-W to dicondra, perennial bluegrass, annual and perennial ryegrasses, fescues and bentgrasses.
- KERB 50-W herbicide is not recommended for use on golf greens.
- Do not graze treated areas and do not feed clippings to livestock.

RASPBERRIES AND BLACKBERRIES
(OREGON AND WASHINGTON ONLY)
GENERAL INFORMATION
KERB 50-W is a selective herbicide recommended for fall and winter applications to established raspberries and blackberries for both preemergence and postemergence control of certain winter annual and perennial grasses.

DOSAGE
KERB 50-W herbicide is recommended at the rate of 2.0 to 6.0 pounds of product (1.0 to 3.0 pounds active ingredient) per acre broadcast application. The rate will depend on the weed species present and the soil texture of the site being treated. Follow the weed control recommendations listed in the chart below:

Pounds of KERB 50-W Per Acre[1]

Weeds Controlled	Dependable Rainfall or Overhead Irrigation[2]	Comments
BLUEGRASS, ANNUAL	2.0 - 4.0	Use low rates on light to medium soils and high rates on heavy soils
RYEGRASS, PERENNIAL	4.0 - 6.0	
QUACKGRASS	4.0 - 6.0	

[1] Dosage rates recommended are in pounds of KERB 50-W per acre broadcast application. Reduce rates accordingly for banded applications.
[2] For effective weed control, rainfall or overhead irrigation is essential following the application of KERB 50-W.

CROP TOLERANCE
Established canefruit are tolerant to recommended rates of KERB 50-W herbicide. Newly transplanted raspberries or blackberries should be well rooted and transplanted for at least 3 months prior to the application of KERB 50-W herbicide.

TIMING AND APPLICATION
Apply KERB 50-W herbicide only during the fall or winter months. For optimum results, apply KERB 50-W during November or December. Do not make applications when the ground is frozen. Mix the recommended amount of KERB 50-W in clean water and apply uniformly with a low pressure ground sprayer in 20 to 50 gallons of water per acre.

RASPBERRIES AND BLACKBERRIES — SPECIFIC USE RESTRICTIONS
- Do not make more than one application of KERB 50-W herbicide per year and do not apply KERB 50-W at rates higher than 6.0 pounds product (3.0 pounds active ingredient) per acre.

RHUBARB
(OREGON AND WASHINGTON ONLY)
GENERAL INFORMATION
KERB 50-W is a selective herbicide recommended for fall and winter applications to established rhubarb for both preemergence and postemergence control of winter annual and perennial grasses and chickweed and preemergence control of certain broadleaf weeds.

DOSAGE
KERB 50-W herbicide is recommended at the rate of 2.0 to 4.0 pounds of product (1.0 to 2.0 pounds active ingredient) per acre broadcast application. The rate will depend on the weed species present. Follow the weed control recommendations listed in the chart below:

Weeds Controlled		Pounds KERB 50-W Per Acre[1] Dependable Rainfall or Overhead Irrigation[2]
BLUEGRASS, ANNUAL BROME, DOWNY (CHEATGRASS) CHICKWEED	OAT, WILD SORREL, RED (from seed)	2.0
BENTGRASS[3] BLUEGRASS, KENTUCKY FESCUE, TALL[3] ORCHARDGRASS[3]	QUACKGRASS RYEGRASS, PERENNIAL VELVETGRASS[3]	4.0

[1] Dosage rates recommended are in pounds of KERB 50-W per acre broadcast application. Reduce rates accordingly for banded applications.
[2] For effective weed control, rainfall or overhead irrigation is essential following the application of KERB 50-W.
[3] KERB 50-W at the rate of 4.0 pounds product per acre may only provide partial control to these weeds.

CROP TOLERANCE

Established rhubarb plants, in a dormant growth condition, are tolerant to recommended rates of KERB 50-W herbicide. Do not apply KERB 50-W to newly transplanted rhubarb or to rhubarb during the active growing stage.

TIMING AND APPLICATION

Apply KERB 50-W herbicide in a single application during the fall or winter months as a broadcast surface application to dormant rhubarb. Optimum herbicidal activity occurs when applications are made after soil temperatures drop to 55°F or less and are followed by rainfall or overhead irrigation. Applications must be made prior to soil freeze-up and snow cover.

Mix the recommended amount of KERB 50-W in clean water and apply uniformly with a low-pressure ground sprayer in 20 to 50 gallons of water per acre.

RHUBARB — SPECIFIC USE RESTRICTIONS

- Do not apply KERB 50-W herbicide to rhubarb within 38 days of harvest.
- Use of KERB 50-W in rhubarb is restricted to Oregon and Washington only.
- Do not make more than one application of KERB 50-W herbicide per year to rhubarb and do not apply KERB 50-W at rates higher than 4.0 pounds product (2.0 pounds active ingredient) per acre.

TREE FRUIT AND GRAPES (APPLE, CHERRY, NECTARINE, PEACH, PEAR, PLUM, PRUNE)

GENERAL INFORMATION

KERB 50-W is a selective herbicide for use in directed spray applications for the control of winter annual and perennial grasses and certain broadleaf weeds in non-bearing and bearing apples, cherries, nectarines, peaches, pears, plums, prunes and grape plantings.

WEED CONTROL

KERB 50-W herbicide is effective at 2.0 to 8.0 pounds of product (1.0 to 4.0 pounds active ingredient) per treated acre for the preemergence and postemergence control of susceptible winter annual and perennial grasses and chickweed and for the preemergence control only of other broadleaf weeds listed on this label. Refer to chart in dosage rate section below for specific weeds controlled.

DOSAGE AND TIMING

KERB 50-W herbicide is recommended in a single, directed application to labeled fruit trees and grape plantings at dosage rates of 2.0 to 8.0 pounds of product (1.0 to 4.0 pounds active ingredient) per treated acre. Application of KERB 50-W should be in the fall, after the fruit is harvested, but prior to leaf drop and soil freeze-up.

The dosage rate required for effective weed control will depend on the weed species present and the soil texture of the area being treated. Follow the specific rate recommendations given in the chart below for the use of KERB 50-W herbicide in labeled fruit trees and grapes:

Pounds KERB 50-W Per Acre —		Dependable Rainfall or Overhead Irrigation		
Weeds Controlled		Soil Texture Group*		
		Coarse	Medium	Fine
BLUEGRASS, ANNUAL BROME, DOWNY (CHEATGRASS) CHICKWEED GRAIN VOLUNTEER	OAT, WILD RYEGRASS, ITALIAN SORREL, RED (from seed)	2.0	3.0	4.0
BLUEGRASS, KENTUCKY FESCUE, TALL ORCHARDGRASS	QUACKGRASS RYEGRASS, PERENNIAL	3.0-4.0	4..0-6.0	6.0-8.0

*Soil Texture Group — Coarse: sand, loamy sand, sandy loam — Medium: loam, silt loam, silt, sandy clay loam — Fine: silty clay loam, clay loam, sandy clay, silty clay, clay

APPLICATION

Mix the recommended amount of KERB 50-W herbicide in clean water and apply uniformly in 40 to 50 gallons of water per acre. Use of a low pressure ground sprayer equipped with a breakaway boom and flat fan or off-center (OC) nozzles is recommended. KERB 50-W should be directed to the soil and the base of trees and vines.

NOTE: Dosage recommendations listed on this label are for surface broadcast application. For banded treatments, the amount of KERB 50-W used per acre should be reduced according to the following formula:

$$\frac{\text{Band width (in.)}}{\text{Row width (in.)}} \times \text{Rate per Acre Broadcast} = \text{Amount Needed per Acre for Band Application}$$

KERB 50-W herbicide should not be soil incorporated.

CROP TOLERANCE

When used as recommended, established non-bearing or bearing fruit trees and grapes listed on this label are very tolerant to KERB 50-W herbicide. KERB 50-W herbicide should not be applied to seedling trees or vines less than 1 year old or to fall-transplanted stock transplanted less than 1 year or to spring-transplanted stock transplanted less than 6 months.

CULTURAL CONSIDERATIONS

KERB 50-W herbicide acts mainly through root absorption in sensitive weed species. Dependable rainfall or overhead irrigation is essential following the application for effective weed control. Trash-free areas create ideal conditions for rapid movement of KERB 50-W herbicide into the weed root zone following rain or irrigation. Clean cultivation before application is preferable but not necessary.

To obtain optimum weed control in areas not clean cultivated, the area to be treated should be free of surface litter (dead or decaying weeds, leaves, mowing clippings, etc). If area to be treated is under a mixed grass or weed sod, it should be mowed and the clippings removed.

TREE FRUIT (APPLE, CHERRY, NECTARINE, PEACH, PEAR, PLUM, PRUNE) AND GRAPE PLANTINGS-SPECIFIC USE RESTRICTIONS

- Do not feed or allow livestock to graze areas treated with KERB 50-W herbicide.
- Do not apply KERB 50-W herbicide to labeled fruit trees or grapes at rates higher than 8.0 pounds product (4.0 pounds active ingredient) per acre and make only one application per season.

WOODY ORNAMENTALS, NURSERY STOCK AND CHRISTMAS TREES

GENERAL INFORMATION

KERB 50-W is a selective herbicide recommended for fall applications to established woody ornamentals, nursery stock and Christmas trees for the control of winter annual and perennial grasses and certain broadleaf weeds.

CROP TOLERANCE

At recommended rates of KERB 50-W herbicide the following trees and shrubs are tolerant to topical applications made in the fall:

ASH	FIRETHORN	MOUNTAIN ASH
ARBORVITAE	FLOWERING CHERRY	MOUNTAIN LAUREL
AZALEA	FLOWERING CRABAPPLE	OHIO BUCKEYE
BARBERRY	FLOWERING QUINCE	OAK
BASSWOOD	FORSYTHIA	PINE
BEECH	GINKO	POPLAR
BIRCH	HAWTHORN	PRIVET
BOXWOOD	HEMLOCK	RHODODENDRON
BRADFORD PEAR	HOLLY	SPIREA
CEDAR	HONEY LOCUST	SPRUCE
COTONEASTER	JUNIPER	SWEETGUM
DOGWOOD	LILAC	SYCAMORE
DOUGLAS FIR	LINDEN	TULIPTREE
EASTERN REDBUD	LONDON PLANE	VIBURNUM
ELM	MAGNOLIA	WALNUT
EUONYMUS	MAPLE	WILLOW
FIR	MOCK ORANGE	YEW

KERB 50-W herbicide should only be used on established trees and woody ornamentals. KERB 50-W herbicide is not recommended for use on seedling trees or shrubs less than one year old or to fall transplanted stock transplanted less than one year or to spring transplanted stock transplanted less than six months.

WEED CONTROL

KERB 50-W herbicide is effective in fall applications at the rate of 2.0 to 4.0 pounds of product (1.0 to 2.0 pounds active ingredient) per treated acre for the preemergence and postemergence control of susceptible winter annual and perennial grasses and chickweed and for preemergence control only of other broadleaf weeds listed on this label. Refer to chart in dosage rate section below for specific weeds controlled.

DOSAGE AND TIMING

KERB 50-W herbicide is recommended in a single, fall application, either directed or topically applied, to woody ornamentals, nursery stock or Christmas trees at the rate of 2.0 to 4.0 pounds of product (1.0 to 2.0 pounds active ingredient) per treated acre. Application of KERB 50-W herbicide should be made in the fall prior to leaf drop and soil freeze-up. For control of winter annual or perennial grasses or chickweed, applications can be made either preemergence or postemergence to the weeds. For control of other labeled broadleaf weeds, preemergence applications must be used to achieve control.

The dosage rate required will depend on the weed species present in the area to be treated. Follow the weed control recommendations given in the chart below:

Weeds Controlled		Pounds KERB 50-W Per Acre Broadcast Application
BARLEY, FOXTAIL BLUEGRASS, ANNUAL	GRAIN VOLUNTEER RYEGRASS, ITALIAN	2.0

Weeds Controlled		Pounds KERB 50-W Per Acre Broadcast Application
BROME, DOWNY (CHEATGRASS) CHICKWEED	SORREL, RED (from seed)	
MUSTARD, WILD ROCKET, LONDON	SHEPHERDSPURSE	3.0
BLUEGRASS, KENTUCKY ORCHARDGRASS QUACKGRASS	RYEGRASS, PERENNIAL	4.0

APPLICATION

Mix the recommended amount of KERB 50-W herbicide in clean water and apply uniformly in 20 to 50 gallons per acre. Use a low pressure ground sprayer equipped with flat fan nozzles spaced to provide uniform distribution.

Dosage recommendations listed on this label are for surface broadcast application. For banded treatments down the row, the amount of KERB 50-W used per acre should be reduced according to the following formula:

$$\frac{\text{Band width (inches)}}{\text{Row width (inches)}} \times \text{Rate per Acre Broadcast} = \text{Amount Needed per Acre for Band Application}$$

KERB 50-W herbicide should not be soil incorporated.

NOTE: Most ornamental turf grass species and ground covers are sensitive to KERB 50-W. Care should be exercised to avoid contact of KERB 50-W herbicide with these plants from either direct application, spray drift or from applications to areas that may drain onto established ornamental turf and ground cover.

SOIL AND MOISTURE REQUIREMENTS

KERB 50-W herbicide is most active in coarse to medium textured soils of low organic matter and is relatively inactive in peat or muck soils or mineral soils high in organic matter content at rates recommended in this label. Herbicidal activity is best in soils containing less than 4 percent organic matter. Use in soils of higher organic matter content may result in inconsistent or incomplete weed control.

KERB 50-W herbicide acts mainly through root absorption in sensitive weed species. Dependable rainfall or overhead irrigation is essential following application for effective weed control.

WOODY ORNAMENTALS, NURSERY STOCK AND CHRISTMAS TREES

SPECIFIC USE RESTRICTIONS • Apply KERB 50-W herbicide in the fall prior to soil freeze-up. • Do not soil incorporate KERB 50-W herbicide.

STORAGE AND DISPOSAL

PROHIBITIONS: Do not contaminate water, food or feed by storage or disposal. Open dumping is prohibited.

STORAGE: Store in a cool, dry place but not below 32°F (0°C). Do not remove package from container except for immediate use.

PESTICIDE DISPOSAL: Wastes resulting from the use of this product may be disposed of on-site or at an approved waste disposal facility.

CONTAINER DISPOSAL: Completely empty bag into application equipment. Then dispose of in a sanitary landfill or by incineration, or, if allowed by State and local authorities, by burning. If burned, stay out of smoke.

STEPS TO BE TAKEN IN CASE MATERIAL IS RELEASED OR SPILLED: Sweep up and place in containers for recovery or disposal. Keep dusting to a minimum. Wear MSHA/NIOSH respirator for dusts encountered. Flush area with water. Keep spill out of all sewers and open bodies of water. Protective Clothing — See Precautionary Statements.

KERB is a registered trademark of the Rohm and Haas Company.
GLEAN is a registered trademark of E. I. duPont de Nemours & Company, Inc.

8116-B9

NOVA® 40W

AGRICULTURAL FUNGICIDE IN WATER-SOLUBLE POUCHES

NOT FOR USE IN ALASKA, ARIZONA, CALIFORNIA, HAWAII, IDAHO, MONTANA, NEVADA, OREGON, UTAH, WASHINGTON AND WYOMING.

ACTIVE INGREDIENT
MYCLOBUTANIL
a-butyl-*a*-(4-chlorophenyl)-1H-1,2-4-triazole-1-propanenitrile. . . . 40%
INERT INGREDIENTS 60%
TOTAL 100%

 EPA Reg. No. 707-221

Notice: Before using this product, read the entire Precautionary Statements, Conditions of Sale and Warranty, Directions for Use, Use Restrictions and Storage and Disposal Instructions. If the Conditions of Sale and Warranty are not acceptable, return the product unopened within thirty days of purchase to the place of purchase.

KEEP OUT OF REACH OF CHILDREN WARNING AVISO

Precaucion al Usuario: Si usted no lee inglés, no use este producto hasta que la etiqueta haya sido explicada ampliamente.

PRECAUTIONARY STATEMENTS

HAZARDS TO HUMANS AND DOMESTIC ANIMALS

Causes eye irritation. Harmful if swallowed, inhaled, or absorbed through the skin. Do not swallow, get in eyes or on skin, or breathe spray mist. Wear goggles, mid-forearm to elbow-length chemical-resistant gloves, hat, long-sleeved shirt and long pants when mixing, loading and applying this product. Contaminated clothing must be laundered separately in hot water before using. Wash thoroughly after handling and before eating or smoking.

STATEMENT OF PRACTICAL TREATMENT

IF IN EYES: Flush eyes with a large amount of water for at least 15 minutes. Call a physician if irritation persists.

IF INHALED: Move victim to fresh air.

IF ON SKIN: Wash affected skin thoroughly with soap and water.

IF SWALLOWED: Dilute by giving two glasses of water to drink and call a physician. Never give anything by mouth to an unconscious person.

ENVIRONMENTAL HAZARDS

Do not apply directly to water or wetlands. Do not contaminate water when disposing of equipment washwaters. Do not apply when weather conditions favor drift or runoff from areas treated.

CONDITIONS OF SALE AND WARRANTY

Rohm and Haas warrants that the product conforms to its chemical description and is reasonably fit for the purpose stated on the label only when used in accordance with label directions under normal conditions of use. **ROHM AND HAAS MAKES NO OTHER EXPRESS OR IMPLIED WARRANTIES EITHER OF MERCHANTABILITY OR FITNESS FOR A PARTICULAR USE.** Handling, storage and use of the product by Buyer or User are beyond the control of Rohm and Haas and Seller. Risks such as crop injury, ineffectiveness or other unintended consequences resulting from, but not limited to, weather or soil conditions, presence of other materials, disease, pests, drift to other crops or property or failure to follow label directions will be assumed by the Buyer or User. **IN NO CASE WILL ROHM AND HAAS OR SELLER BE HELD LIABLE FOR CONSEQUENTIAL, SPECIAL OR INDIRECT DAMAGES RESULTING FROM THE HANDLING, STORAGE OR USE OF THIS PRODUCT.**

DIRECTIONS FOR USE

It is a violation of Federal law to use this product in a manner inconsistent with its labeling.

CHEMIGATION—Do not apply this product through any type of irrigation system.

RE-ENTRY AND WORKER PROTECTION STATEMENTS

Do not enter treated areas without protective clothing until sprays have dried. Because certain states may require more restrictive re-entry intervals for various crops treated with this product, consult your State Department of Agriculture for further information. Do not apply this product in such a manner as to directly or through drift expose workers or other persons. The area being treated must be vacated by unprotected persons.

Written or oral warnings must be given to workers who are expected to be in a treated area or in an area about to be treated with this product. Oral warnings must be given if there is reason to believe that written warnings cannot be understood by workers. When oral warnings are given, warnings shall be given in a language customarily understood by workers. Written warnings must include the following information:

WARNING

(Insert area or field description) treated with NOVA 40W fungicide on (insert date of application). Do not enter treated area without protective clothing until sprays have dried. In case of accidental exposure, call a doctor (physician), clinic or hospital immediately. Explain that the victim has been exposed to NOVA 40W fungicide and describe his condition. For further information see the "Statement of Practical Treatment."

Oral warnings must include the same information that is in the written warnings.

GENERAL INFORMATION

NOVA 40W fungicide is a systemic, protectant and curative fungicide recommended for the control of specific diseases mentioned on this label. Optimum disease control is achieved when the fungicide is applied in a regularly scheduled preventive spray program.

USE RATE DETERMINATION —Carefully read, understand and follow label use rates and restrictions.

Under low disease conditions, minimum label use rates per application can be used while maximum label rates and shortened spray schedules are recommended for severe or threatening disease conditions.

For proper application, determine the number of acres to be treated, the recommended label use rate and the gallonage to be applied per acre. Prepare only the amount of spray solution required to treat the measured acreage. Careful calibration of spray equipment is recommended prior to use.

HANDLING—The enclosed pouches of NOVA 40W fungicide are water-soluble. **Do not** allow pouches to become wet prior to adding to the spray tank. **Do not** handle the pouches with wet hands or wet gloves. Always reseal overwrap bag to protect remaining unused pouches. Do not remove water-soluble pouches from overwrap except to add directly to the spray tank.

MIXING—Always place NOVA 40W fungicide into solution prior to adding co-applied agricultural chemicals registered for use on apples and grapes. Add the required number of unopened pouches as determined by the dosage recommendations, into the spray tank with agitation. Depending on the water temperature and the degree of agitation, the pouches should dissolve completely within approximately ten minutes from the time they are added to the water.

COMPATIBILITY—NOVA 40W fungicide is compatible with most commonly used agricultural fungicides, insecticides, growth regulators, micronutrients and spray adjuvants. When preparing tank mixes, user should consult spray compatibility charts or State Cooperative Extension Service Specialists prior to actual use.

CAUTION: Do not use PVA packets in a tank mix with products that contain Boron or release free chlorine. The resultant reaction of PVA and Boron or free chlorine is a plastic which is not soluble in water or solvents such as diesel oils, kerosene, gasoline or alcohol.

Do not attempt to use the PVA packets directly in diesel oils or summer spray type oils as in ULV or LV uses. PVA packets are water soluble not oil soluble.

STORAGE AND DISPOSAL

Store in a cool dry place, but not below 32°F. Do not remove package from container except for immediate use.

PESTICIDE DISPOSAL: Do not contaminate water, food or feed by storage or disposal. Pesticide wastes are toxic. Improper disposal of excess pesticide, spray mixture, or rinsate is a violation of Federal law. If these wastes cannot be disposed of by use according to label instructions, contact your State Pesticide or Environmental Control Agency, or the Hazardous Waste representative at the nearest EPA regional office for guidance.

CONTAINER DISPOSAL: Completely empty container into application equipment. Dispose of in a sanitary landfill or by incineration or, if allowed by state and local authorities, by burning. If burned, stay out of smoke.

STEPS TO BE TAKEN IN CASE MATERIAL IS RELEASED OR SPILLED: Wear eye protection. Wear protective clothing. Spray water on powder and dust. Scoop or shovel solid material into a suitable container for recovery or disposal. Keep dust to a minimum. Flush contaminated area with a large amount of water to a chemical or sanitary sewer containing a settling pit. Remove contaminated clothing promptly and wash affected skin areas with soap and water. Thoroughly launder clothing before reuse. Refer to Precautionary Statements.

USE DIRECTIONS FOR USE ON APPLES — Best control of labeled diseases is achieved when the fungicide is applied on a 7- to 10-day application schedule.

Label use recommendations (ounces product per 100 gallons spray) are based on dilute sprays with a 400-gallon per acre base. Orchard tree heights, tree spacing and pruning practices vary. The following summary table provides additional guidance in calculating proper use rates for well-pruned orchards:

Tree Height (feet)	Approximate Gallons per Acre for Dilute Sprays on Well-Pruned Trees	Ounces NOVA 40W Fungicide per 10,000 cu. ft. of Canopy Volume	Ounces NOVA 40W Fungicide per Acre*
10	200	0.1 to 0.2	2.5 to 5.0
15	300	0.1 to 0.2	3.75 to 7.5
20	400	0.1 to 0.2	5.0 to 10.0
25	500	0.1 to 0.2	6.25 to 12.5
30	600	0.1 to 0.2	7.5 to 15.0

*Note recommended use rates for the control of specific diseases. For orchards that receive minimal pruning, increase the per acre use rate by 1.4 times.

Disease	Rate of NOVA 40W oz./100 gals.[a]	Use Recommendations	Restrictions
Scab Prebloom	1.25 to 2.0 (0.5 to 0.8 active)	Begin application at green tip. During periods favorable for primary scab development, use high rate and a 7-day application schedule.	Do not apply within 14 days of harvest. Do not apply more than 5 pounds (2 lb. active) NOVA 40W fungicide per acre per season.
Postbloom	1.25 to 2.0 (0.5 to 0.8 active)	Use NOVA 40W in a tank mixture with the recommended use rate of DITHANE M-45™ fungicide or another protectant fungicide registered for use on apples for improved fruit scab and summer disease control.	Do not graze livestock in treated areas or feed cover crops grown in treated areas to livestock. Application by ground equipment only.
Post-infection	2.0 (0.8 active)	NOVA 40W fungicide provides 96-hour post-infection control or curative activity and suppresses sporulation of established lesions when used in two or more applications. Apply as soon as possible after an infection period. Follow with standard protectant spray schedule.	
Powdery Mildew	1.25 to 2.5 (0.5 to 1.0 active)	Begin application at tight cluster and continue through the second cover spray. Additional sprays beyond second cover may be needed on susceptible varieties or under heavy disease pressure. Use high label rate if powdery mildew was present in previous years.	
Rusts	1.25 to 2.0 (0.5 to 0.8 active)	Begin application at pink stage and continue through the second cover spray.	

[a]Based on dilute sprays with a 400-gallon per acre base.

USE DIRECTIONS FOR USE ON GRAPES — Thorough spray coverage is essential for good disease control. NOVA 40W fungicide should be uniformly applied with a ground sprayer using 50 or more gallons of water per acre.

Disease	Rate of NOVA 40W oz. per Acre	Use Recommendations	Restrictions
Black Rot	3 to 5 (1.2 to 2.0 active)	Begin application when new vines are 1 to 3 inches in length and continue application on a 14- to 21-day schedule. Use higher rate under heavy disease pressure.	Do not apply more than 1.5 lbs. (0.6 lb. A.I.) per acre per year. Do not apply within 14 days of harvest.
Powdery Mildew	3 to 5 (1.2 to 2.0 active)	Begin application at prebloom (12- to 18-inch canes) and continue application on a 14- to 21-day interval. Use higher rate on susceptible varieties or under heavy disease pressure.	

DITHANE and NOVA are registered trademarks of Rohm and Haas Company.

10491-B2 6-3046 Made in Italy by Rohm and Haas Italia S.p.A.
March 1990

PROSTAR™ 4E

HERBICIDE

ACTIVE INGREDIENT

propanil
3',4'-dichloropropionanilide 43.5%*
INERT INGREDIENTS 56.5%
TOTAL 100.0%

* Equivalent to 4 lbs. active ingredient per gallon. This product contains the toxic inert ingredient Isophorone

EPA Reg. No. 707-109

PROSTAR 4E is a trademark of Rohm and Haas Company.

NOTICE: Before using this product read the entire Precautionary Statements, Conditions of Sale and Warranty, Directions for Use, Use Restrictions and Storage and Disposal Instructions. If the Conditions of Sale and Warranty are not acceptable, return the product unopened within thirty days of purchase to the place of purchase.

KEEP OUT OF REACH OF CHILDREN

WARNING

PRECAUTIONARY STATEMENTS

HAZARDS TO HUMANS AND DOMESTIC ANIMALS

IRRITATING TO EYES AND SKIN
HARMFUL IF SWALLOWED OR ABSORBED THROUGH SKIN
MAY BE HARMFUL IF INHALED

Do not get in eyes or on skin. Wear eye and skin protection when handling. Do not take internally. Avoid breathing spray mist.

STATEMENT OF PRACTICAL TREATMENT

IF IN EYES: Flush eyes with large amounts of water for at least 15 minutes. Consult a physician if irritation persists.

IF ON SKIN: Wash affected area with soap and water.

IF SWALLOWED: Dilute by giving 2 glasses of water to drink and call a physician. Never give anything by mouth to an unconscious person.

NOTE TO PHYSICIAN: Emesis is recommended.

IF INHALED: Remove victim to fresh air. Treat the person symptomatically. If irritation persists, call a physician.

ENVIRONMENTAL HAZARDS

This pesticide is toxic to fish. Drift and runoff from treated areas may be hazardous to aquatic organisms in neighboring areas. Do not apply directly to water except as specified on this label. Do not contaminate water when disposing of equipment washwaters.

Water drained from treated rice fields must not be used to irrigate other crops or released within ½ mile upstream of a potable water intake in flowing water (e.g., river, stream, etc.) or within ½ mile of a potable water intake in a standing body of water, such as a lake, pond or reservoir.

CONDITIONS OF SALE AND WARRANTY

Rohm and Haas warrants that the product conforms to its chemical description and is reasonably fit for the purpose stated on the label only when used in accordance with label directions under normal conditions of use. **ROHM AND HAAS MAKES NO OTHER EXPRESS OR IMPLIED WARRANTIES EITHER OF MERCHANTABILITY OR FITNESS FOR A PARTICULAR USE.** Handling, storage and use of the product by Buyer or User are beyond the control of Rohm and Haas and Seller. Risks, such as crop injury, ineffectiveness or other unintended consequences resulting from, but not limited to, weather or soil conditions, presence of other materials, disease, pests, drift to other crops or property or failure to follow label directions, will be assumed by the Buyer or User. **IN NO CASE WILL ROHM AND HAAS OR SELLER BE HELD LIABLE FOR CONSEQUENTIAL, SPECIAL OR INDIRECT DAMAGES RESULTING FROM THE HANDLING, STORAGE OR USE OF THIS PRODUCT.**

DIRECTIONS FOR USE

(FOR RICE GROWN IN SOUTHERN UNITED STATES ONLY)

It is a violation of Federal law to use this product in a manner inconsistent with its labeling.

RE-ENTRY AND WORKER PROTECTION STATEMENTS

Do not enter treated areas without protective clothing until sprays have dried. Because certain states may require more restrictive re-entry intervals for various crops treated with this product, consult your State Department of Agriculture for further information. Do not apply this product in such a manner as to directly or through drift expose workers or other persons. The area being treated must be vacated by unprotected persons.

CHEMIGATION

DO NOT APPLY THIS PRODUCT THROUGH ANY TYPE OF IRRIGATION SYSTEM.

GENERAL INFORMATION

PROSTAR 4E herbicide is a selective postemergence herbicide for use in rice only for control of the following weeds:

BARNYARDGRASS (WATERGRASS) *Echinochloa crus-galli* *Echinochloa colonum*	FOXTAIL *Setaria* spp.	PANICUM, TEXAS *Panicum Texanum*
	GOOSEGRASS *Eleusine indica*	PARAGRASSS *Panicum purpurascens*
BEAKRUSH (SPEARHEAD) *Rhynchospora corniculata*	GULF COCKSPUR *Echinochloa crus-pavonis*	PIGWEED, REDROOT *Amaranthus retroflexus*
CRABGRASS SPECIES *Digitaria* spp.	HEMP SESBANIA (COFFEE BEAN) *Sesbania exaltata*	REDWEED *Melochia corchorifolia*
CROTON, WOOLLY *Croton capitatus*	HOORAHGRASS *Fimbristylis miliaceae*	SIGNALGRASS, BROADLEAF *Bracharia platyphylla*
DOCK, CURLY *Rumex crispus*	MEXICANWEED *Caperonia castanaefolia*	SPIKERUSH (WIREGRASS) *Eleocharis* spp.

PROSTAR 4E herbicide is an emulsifiable concentrate containing 4 pounds active ingredient per U.S. gallon. The product is not a hormone-type herbicide, but kills susceptible weeds by direct contact action. For this reason, thorough coverage of emerged weeds is essential for best results. Only weeds that have emerged and are exposed at time of application will be controlled. The product has no preemergence or residual herbicidal activity. Apply PROSTAR 4E herbicide only to fields that have been drained of flood water. The product is most effective if applied when susceptible grasses and broadleaf weeds are small and growing actively under favorable soil moisture and weather conditions. Early weed control removes competition, saves moisture and generally contributes to increased yields.

USE RESTRICTIONS

Do not apply to any crop other than rice. PROSTAR 4E herbicide injures most crops except cereal grains and perennial grasses. Avoid drift or accidental application from turning aircraft on cotton, soybeans, corn, safflower, seedling legumes, vegetables, orchards, vineyards, gardens, shrubs, and ornamentals. Once applied, it does not release fumes hazardous to nearby crops.

TIMING AND DOSAGE RECOMMENDATIONS

Treat grassy and weedy fields when a satisfactory stand of rice that will tolerate flooding is established. The amount of PROSTAR 4E herbicide to apply depends primarily upon the stage and growth condition of the grasses. The growth stage of the rice is also a factor in dosage and timing limitations so as to avoid the possibility of excessive residues.

For best results apply PROSTAR 4E herbicide at the rate of 3 to 4 quarts (3 to 4 lbs. active) per acre when the grasses are actively growing in the 1 to early 4 leaf stage. This rate will also control many seedling broadleaf and aquatic weeds. Generally this will be 15 to 25 days after planting of the rice. In order to insure satisfactory weed control, do not apply less than 3 quarts of PROSTAR 4E herbicide per acre in a single spray application.

Apply PROSTAR 4E herbicide at the rate of 4 to 6 quarts (4 to 6 lbs. active) per acre to actively growing grasses in the 4 to 6 leaf and early tillering stage or when they are in the 2 to 4 leaf stage but stressed under dry soil conditions. Generally this will be 20 to 30 days after planting of the rice.

EMERGENCY TREATMENT: Apply PROSTAR 4E herbicide at the rate of 5 to 6 quarts (5 to 6 lbs. active) in 15 gallons of spray per acre for emergency control of older tillering grass. Generally this will be 30 to 40 days after planting. If the field is already flooded, the water should be lowered or drained before spraying to expose more of the grass and weeds. Emergency treatment should be considered as a salvage operation only and cannot be relied upon for total control of grass and weeds.

TO AVOID EXCESSIVE RESIDUES AT HARVEST, DO NOT APPLY AFTER THE END OF TILLERING FOR THE RICE VARIETY BEING TREATED. DO NOT APPLY MORE THAN A MAXIMUM OF SIX POUNDS ACTIVE INGREDIENT PER ACRE IN A SINGLE APPLICATION OR EXCEED EIGHT POUNDS ACTIVE INGREDIENT PER ACRE TOTAL DOSAGE PER SEASON.

APPLICATION EQUIPMENT

Aircraft — Fixed wing aircraft or helicopters should have well-designed spray systems that produce a uniform pattern of medium-fine spray droplets. Apply PROSTAR 4E herbicide on small grass in no less than 10 gallons of total spray per acre with boom-nozzle sprayers. Increase volume to 12 to 15 gallons per acre for larger or denser stands of grass or during periods of low humidity.

The optimum effective spray swath width depends on operating conditions and type of aircraft being used. For uniform spray coverage with fixed wing aircraft, do not exceed a spray swath width 10 percent greater than the wingspan or the length of the boom in helicopters. Measure the swaths accurately for flagging.

Ground Sprayers — Use standard low-pressure herbicide sprayers equipped with boom and flat fan nozzles. Use nozzle sizes that deliver a medium-fine droplet in 15 to 20 gallons total spray per acre at 40 to 50 p.s.i. and at ground speeds not in excess of 3 to 4 m.p.h. Avoid raising boom too high. Spray patterns should meet uniformly.

Flush all equipment with clear water after each day's use. Clean all equipment, including nurse tanks used for PROSTAR 4E herbicide, with detergent wash followed by a water rinse, BEFORE AND AFTER spraying other pesticides or other crops. Applicators and flagmen should avoid contact with spray mist from PROSTAR 4E herbicide or any pesticides and should wear protective clothing and goggles. Wash thoroughly after exposure.

CROP TOLERANCE AND GROWING CONDITIONS

All leading commercial varieties of rice are exceptionally tolerant to PROSTAR 4E herbicide. A temporary yellowing or tip burn of rice may be noted after treatment, but new growth is normal. Severe leaf burn and partial killing of rice may occur if the product is applied when rice is under stress and in a weakened growth condition due to disease or insect infestations, excessive soil salts, overwatering, or prolonged drought and extremely hot weather. Growers are cautioned not to spray under such conditions and/or when maximum daily temperatures have been or are expected to go above 100°F.

EFFECT OF CLIMATIC CONDITIONS AND CULTURAL PRACTICES ON WEED CONTROL

Field and Seedbed Preparation

Fields should be accurately leveled and contoured and have well-prepared seedbeds free of clods. This encourages uniform and rapid emergence of rice, grass and broadleaf weeds, and permits better timing and coverage of PROSTAR 4E herbicide sprays resulting in optimum weed control.

Water Management

Before application of PROSTAR 4E herbicide, drained or dry planted fields should be flushed as often as needed to prevent drying and crusting. Flushing encourages uniform emergence and vigorous growth of grass, broadleaf weeds and rice which is essential for best results. Flush fields in sufficient time so that weeds and rice are actively growing at time of treatment. Make sure the field is drained prior to treatment so that grasses and broadleaf weeds are fully exposed. Weeds that are partially submerged in standing water at time of application will not be satisfactorily controlled.

After treatment, treated fields should always be flooded before a second infestation of grass has a chance to develop. To prevent more grass from germinating after treatment, fields should be flooded within 24 hours after spraying, or as soon as possible after 24 hours.

Temperature

The temperature a few days before and after applying PROSTAR 4E herbicide has an important bearing on the weed-killing activity. The activity increases as daily maximum temperatures increase above 75°F and decreases as the daily maximum temperatures decline below 75°F. Do not apply PROSTAR 4E herbicide when maximum temperatures have been or are expected to stay below 65°F or to go above 100°F. Low temperature at time of application is not so important as long as it warms up later during the day.

Relative Humidity and Rain

Grasses and weeds are more responsive to PROSTAR 4E herbicide during periods of high humidity when the foliage is moist or covered by dew. When the humidity is very low some of the spray may crystalize in the air. When this condition exists, increase spray volume to 12-15 gallons per acre for best results. Do not spray when rain threatens within six hours to avoid loss of the spray deposit before adsorption by the grass.

Wind

Avoid applications when the wind speed exceeds 10 m.p.h. because of the drift hazard to sensitive crops and the possibility of uneven (streaked) application.

COMPATIBILITY WITH OTHER CHEMICALS

Tank-mix applications of PROSTAR 4E herbicide with other herbicides, insecticides, spray adjuvants, or liquid fertilizers may reduce crop tolerance and/or weed control or impair mixing properties. Use of these products in tank-mix applications with PROSTAR 4E herbicide is done at the user's risk.

Insecticides

Severe injury or kill of rice plants may result from tank-mix combinations or separate sprays of PROSTAR 4E herbicide and certain insecticides. Do not combine PROSTAR 4E herbicide with carbamate insecticides such as carbaryl (Sevin, etc.), methomyl (Lannate, Nudrin, etc.) or organophosphorus insecticides such as parathion, methyl parathion, Guthion, malathion, Systox, EPN, Phosphamidon, etc. Do not apply any of the above insecticides to rice fields within 14 days before or after PROSTAR 4E herbicide. Do not use carbamate or systemic organophosphorus insecticides on rice fields to be treated with PROSTAR 4E herbicide.

Do not apply PROSTAR 4E herbicide to rice fields that were planted with rice seed treated with bird repellents containing methiocarb such as Mesurol, Borderland Red, etc. Consult local Extension specialist for current recommendations of approved insecticides on rice.

STORAGE AND DISPOSAL

STORAGE: Do not use, pour, spill or store near heat or open flame. Ground all metal containers when transferring product. Protect from freezing. If stored below 32°F and crystals form, warm to 72°F for 24 hours, periodically shaking container to reconstitute.

PESTICIDE DISPOSAL: Do not contaminate water, food, or feed by storage or disposal. Pesticide wastes are toxic. Improper disposal of excess pesticide, spray mixture, or rinsate is a violation of Federal Law. If these wastes cannot be disposed of by use according to label instructions, contact your State Pesticide or Environmental Control Agency, or the Hazardous Waste representative at the nearest EPA regional office for guidance.

CONTAINER DISPOSAL

Metal Containers — Triple rinse (or equivalent) then offer for recycling or reconditioning or puncture and dispose of in a sanitary landfill or by other procedures approved by State and local authorities. **Bulk Containers** — Drain thoroughly and return to specified destination for cleaning and reuse.

STEPS TO BE TAKEN IN CASE MATERIAL IS RELEASED OR SPILLED: Eliminate ignition sources. Ventilate area. Avoid breathing vapors. Use MSHA/NIOSH self-contained breathing apparatus or air mask for large spills in confined areas. Dike the spill with inert material (sand, earth, fuller's earth, etc.) and if appropriate transfer the liquid and solid diking material to separate containers for recovery or disposal. Remove contaminated clothing promptly and wash affected skin areas with soap and water. Wash clothing before reuse. Keep out of sewers and open bodies of water. REFER TO PRECAUTIONARY STATEMENTS.

10499-R1

RALLY™ 40W

AGRICULTURAL FUNGICIDE IN WATER-SOLUBLE POUCHES

FOR USE ONLY IN ALASKA, IDAHO, MONTANA, NEVADA, OREGON, UTAH, WASHINGTON AND WYOMING.

ACTIVE INGREDIENT	
MYCLOBUTANIL	
a-butyl-*a*-(4-chlorophenyl)-1H-1,2-4-triazole-1-propanenitrile	40%
INERT INGREDIENTS	60%
TOTAL	100%

 EPA Reg. No. 707-221

Notice: Before using this product, read the entire Precautionary Statements, Conditions of Sale and Warranty, Directions for Use, Use Restrictions and Storage and Disposal Instructions. If the Conditions of Sale and Warranty are not acceptable, return the product unopened within thirty days of purchase to the place of purchase.

KEEP OUT OF REACH OF CHILDREN
WARNING AVISO

Precaucion al Usuario: Si usted no lee inglés, no use este producto hasta que la etiqueta haya sido explicada ampliamente.

PRECAUTIONARY STATEMENTS

HAZARDS TO HUMANS AND DOMESTIC ANIMALS

Causes eye irritation. Harmful if swallowed, inhaled, or absorbed through the skin. Do not swallow, get in eyes or on skin, or breathe spray mist. Wear goggles, mid-forearm to elbow-length chemical-resistant gloves, hat, long-sleeved shirt and long pants when mixing, loading and applying this product. Contaminated clothing must be laundered separately in hot water before using. Wash thoroughly after handling and before eating or smoking.

STATEMENTS OF PRACTICAL TREATMENT

IF IN EYES: Flush eyes with a large amount of water for at least 15 minutes. Call a physician if irritation persists.

IF INHALED: Move victim to fresh air.

IF ON SKIN: Wash affected skin thoroughly with soap and water.

IF SWALLOWED: Dilute by giving two glasses of water to drink and call a physician. Never give anything by mouth to an unconscious person.

ENVIRONMENTAL HAZARDS

Do not apply directly to water or wetlands. Do not contaminate water when disposing of equipment washwaters. Do not apply when weather conditions favor drift or runoff from areas treated.

CONDITIONS OF SALE AND WARRANTY

Rohm and Haas warrants that the product conforms to its chemical description and is reasonably fit for the purpose stated on the label only when used in accordance with label directions under normal conditions of use. **ROHM AND HAAS MAKES NO OTHER EXPRESS OR IMPLIED WARRANTIES EITHER OF MERCHANTABILITY OR FITNESS FOR A PARTICULAR USE.** Handling, storage and use of the product by Buyer or User are beyond the control of Rohm and Haas and Seller. Risks, such as crop injury, ineffectiveness or other unintended consequences resulting from, but not limited to, weather or soil conditions, presence of other materials, disease, pests, drift to other crops or property or failure to follow label directions, will be assumed by the Buyer or User. **IN NO CASE WILL ROHM AND HAAS OR SELLER BE HELD LIABLE FOR CONSEQUENTIAL, SPECIAL OR INDIRECT DAMAGES RESULTING FROM THE HANDLING, STORAGE OR USE OF THIS PRODUCT.**

DIRECTIONS FOR USE

It is a violation of Federal law to use this product in a manner inconsistent with its labeling.

CHEMIGATION—Do not apply this product through any type of irrigation system.

RE-ENTRY AND WORKER PROTECTION STATEMENTS

Do not enter treated areas without protective clothing until sprays have dried. Because certain states may require more restrictive re-entry intervals for various crops treated with this product, consult your State Department of Agriculture for further information. Do not apply this product in such a manner as to directly or through drift expose workers or other persons. The area being treated must be vacated by unprotected persons.

Written or oral warnings must be given to workers who are expected to be in a treated area or in an area about to be treated with this product. Oral warnings must be given if there is reason to believe that written warnings cannot be understood by workers. When oral warnings are given, warnings shall be given in a language customarily understood by workers. Written warnings must include the following information:

WARNING

(Insert area or field description) treated with RALLY 40W fungicide on (insert date of application). Do not enter treated area without protective clothing until sprays have dried. In case of accidental exposure, call a doctor (physician), clinic or hospital immediately. Explain that the victim has been exposed to RALLY 40W fungicide and describe his condition. For further information see the "Statements of Practical Treatment."

Oral warnings must include the same information that is in the written warnings.

GENERAL INFORMATION

RALLY 40W fungicide is a systemic, protectant and curative fungicide recommended for the control of specific diseases mentioned on this label. Optimum disease control is achieved when the fungicide is applied in a regularly scheduled preventive spray program.

USE RATE DETERMINATION—Carefully read, understand and follow label use rates and restrictions.

Under low disease conditions, minimum label use rates per application can be used while maximum label rates and shortened spray schedules are recommended for severe or threatening disease conditions.

For proper application, determine the number of acres to be treated, the recommended label use rate and the gallonage to be applied per acre. Prepare only the amount of spray solution required to treat the measured acreage. Careful calibration of spray equipment is recommended prior to use.

HANDLING—The enclosed pouches of RALLY 40W fungicide are water-soluble. **Do not** allow pouches to become wet prior to adding to the spray tank. **Do not** handle the pouches with wet hands or wet gloves. Always reseal overwrap bag to protect remaining unused pouches. Do not remove water-soluble pouches from overwrap except to add directly to the spray tank.

MIXING—Always place RALLY 40W fungicide into solution prior to adding co-applied agricultural chemicals registered for use on apples and grapes. Add the required number of unopened pouches, as determined by the dosage recommendations, into the spray tank with agitation. Depending on the water temperature and the degree of agitation, the pouches should dissolve completely within approximately ten minutes from the time they are added to the water.

COMPATIBILITY—RALLY 40W fungicide is compatible with most commonly used agricultural fungicides, insecticides, growth regulators, micronutrients and spray adjuvants. When preparing tank mixes, user should consult spray compatibility charts or State Cooperative Extension Service Specialists prior to actual use.

CAUTION: Do not use PVA packets in a tank mix with products that contain Boron or release free chlorine. The resultant reaction of PVA and Boron or free chlorine is a plastic which is not soluble in water or solvents such as diesel oils, kerosene, gasoline or alcohol.

Do not attempt to use the PVA packets directly in diesel oils or summer-spray-type oils as in ULV or LV uses. PVA packets are water-soluble not oil-soluble.

STORAGE AND DISPOSAL

Store in a cool dry place, but not below 32°F. Do not remove package from container except for immediate use.

PESTICIDE DISPOSAL: Do not contaminate water, food or feed by storage or disposal. Pesticide wastes are toxic. Improper disposal of excess pesticide, spray mixture, or rinsate is a violation of Federal law. If these wastes cannot be disposed of by use according to label instructions, contact your State Pesticide or Environmental Control Agency, or the Hazardous Waste representative at the nearest EPA regional office for guidance.

CONTAINER DISPOSAL: Completely empty container into application equipment. Dispose of in a sanitary landfill or by incineration or, if allowed by state and local authorities, by burning. If burned, stay out of smoke.

STEPS TO BE TAKEN IN CASE MATERIAL IS RELEASED OR SPILLED: Wear eye protection. Wear protective clothing. Spray water on powder and dust. Scoop or shovel solid material into a suitable container for recovery or disposal. Keep dust to a minimum. Flush contaminated area with a large amount of water to a chemical or sanitary sewer containing a settling pit. Remove contaminated clothing promptly and wash affected skin areas with soap and water. Thoroughly launder clothing before reuse. Refer to Precautionary Statements.

USE DIRECTIONS FOR USE ON APPLES — Best control of labeled diseases is achieved when the fungicide is applied on a 7- to 10-day application schedule.

Label use recommendations (ounces product per 100 gallons spray) are based on dilute sprays with a 400-gallon per acre base. Orchard tree heights, tree spacing and pruning practices vary. The following summary table provides additional guidance in calculating proper use rates for well-pruned orchards:

Tree Height (feet)	Approximate Dilute Sprays on Well-Pruned Trees (Gals. per Acre)	Rally 40W Fungicide (Oz./10,000 cu. ft. of canopy volume	Rally 40W Fungicide (Oz./A.)*
10	200	0.1 to 0.2	2.5 to 5.0
15	300	0.1 to 0.2	3.75 to 7.5
20	400	0.1 to 0.2	5.0 to 10.0
25	500	0.1 to 0.2	6.25 to 12.5
30	600	0.1 to 0.2	7.5 to 15.0

*Note recommended use rates for the control of specific diseases. For orchards that receive minimal pruning, increase the per acre use rate by 1.4 times.

Disease	Rate of RALLY 40W (oz./100 gals.[a])	Use Recommendations	Restrictions
Scab Prebloom	1.25 to 2.0 (0.5 to 0.8 active)	Begin application at green tip. During periods favorable for primary scab development, use high rate and a 7-day application schedule.	Do not apply within 14 days of harvest. Do not apply more than 5 pounds (2 lb. active)

Disease	Rate of RALLY 40W (oz./100 gals.[a])	Use Recommendations	Restrictions
			RALLY 40W fungicide per acre per season.
Postbloom	1.25 to 2.0 (0.5 to 0.8 active)	Use RALLY 40W in a tank mixture with the recommended use rate of DITHANE M-45™ fungicide or another protectant fungicide registered for use on apples for improved fruit scab and summer disease control.	Do not graze livestock in treated areas or feed cover crops grown in treated areas to livestock. Application by ground equipment only.
Post-infection	2.0 (0.8 active)	RALLY 40W fungicide provides 96-hour post-infection control or curative activity and suppresses sporulation of established lesions when used in two or more applications. Apply as soon as possible after an infection period. Follow with standard protectant spray schedule.	
Powdery Mildew	1.25 to 2.5 (0.5 to 1.0 active)	Begin application at tight cluster and continue through the second cover spray. Additional sprays beyond second cover may be needed on susceptible varieties or under heavy disease pressure. Use high label rate if powdery mildew was present in previous years.	
Rusts	1.25 to 2.0 (0.5 to 0.8 active)	Begin application at pink stage and continue through the second cover spray.	

[a]Based on dilute sprays with a 400-gallon per acre base.

USE DIRECTIONS FOR USE ON GRAPES — Thorough spray coverage is essential for good disease control. RALLY 40W fungicide should be uniformly applied with a ground sprayer using 50 or more gallons of water per acre.

Disease	Rate of RALLY 40W (oz./A.)	Use Recommendations	Restrictions
Black Rot	3 to 5 (1.2 to 2.0 active)	Begin application when new vines are 1 to 3 inches in length and continue application on a 14- to 21-day schedule. Use higher rate under heavy disease pressure.	Do not apply more than 1.5 lbs. (0.6 lb. A.I.) per acre per year. Do not apply within 14 days of harvest.
Powdery Mildew	3 to 5 (1.2 to 2.0 active)	Begin application at prebloom (12- to 18-inch canes) and continue application on a 14- to 21-day interval. Use higher rate on susceptible varieties or under heavy disease pressure.	

DITHANE and RALLY are trademarks of Rohm and Haas Company.

10339-B2 6-2865 Made in Italy by Rohm and Haas Italia S.p.A.
March 1990

RALLY™ 40W

AGRICULTURAL FUNGICIDE IN WATER-SOLUBLE POUCHES

FOR USE ONLY IN ARIZONA, CALIFORNIA AND HAWAII.

ACTIVE INGREDIENT
MYCLOBUTANIL
a-butyl-*a*-(4-chlorophenyl)-1H-1,
2-4- triazole-1-propanenitrile 40%
INERT INGREDIENTS 60%
TOTAL 100%

 EPA Reg. No. 707-215

Notice: Before using this product, read the entire Precautionary Statements, Conditions of Sale and Warranty, Directions for Use, Use Restrictions and Storage and Disposal Instructions. If the Conditions of Sale and Warranty are not acceptable, return the product unopened within thirty days of purchase to the place of purchase.

KEEP OUT OF REACH OF CHILDREN WARNING AVISO

Precaucion al Usuario: Si usted no lee inglés, no use este producto hasta que la etiqueta haya sido explicada ampliamente.

PRECAUTIONARY STATEMENTS

HAZARDS TO HUMANS AND DOMESTIC ANIMALS

Causes eye irritation. Harmful if swallowed, inhaled, or absorbed through the skin. Do not swallow, get in eyes or on skin, or breathe spray mist. Wear goggles, mid-forearm to elbow-length chemical-resistant gloves, hat, long-sleeved shirt and long pants when mixing, loading and applying this product. Contaminated clothing must be laundered separately in hot water before using. Wash thoroughly after handling and before eating or smoking.

STATEMENTS OF PRACTICAL TREATMENT

IF IN EYES: Flush eyes with a large amount of water for at least 15 minutes. Call a physician if irritation persists.

IF INHALED: Move victim to fresh air.

IF ON SKIN: Wash affected skin thoroughly with soap and water.

IF SWALLOWED: Dilute by giving two glasses of water to drink and call a physician. Never give anything by mouth to an unconscious person.

ENVIRONMENTAL HAZARDS

Do not apply directly to water or wetlands. Do not contaminate water when disposing of equipment washwaters. Do not apply when weather conditions favor drift or runoff from areas treated.

CONDITIONS OF SALE AND WARRANTY

Rohm and Haas warrants that the product conforms to its chemical description and is reasonably fit for the purpose stated on the label only when used in accordance with label directions under normal conditions of use. **ROHM AND HAAS MAKES NO OTHER EXPRESS OR IMPLIED WARRANTIES EITHER OF MERCHANTABILITY OR FITNESS FOR A PARTICULAR USE.** Handling, storage and use of the product by Buyer or User are beyond the control of Rohm and Haas and Seller. Risks, such as crop injury, ineffectiveness or other unintended consequences resulting from, but not limited to, weather or soil conditions, presence of other materials, disease, pests, drift to other crops or property or failure to follow label directions, will be assumed by the Buyer or User. **IN NO CASE WILL ROHM AND HAAS OR SELLER BE HELD LIABLE FOR CONSEQUENTIAL, SPECIAL OR INDIRECT DAMAGES RESULTING FROM THE HANDLING, STORAGE OR USE OF THIS PRODUCT.**

DIRECTIONS FOR USE

It is a violation of Federal law to use this product in a manner inconsistent with its labeling.

CHEMIGATION—Do not apply this product through any type of irrigation system.

RE-ENTRY AND WORKER PROTECTION STATEMENTS

Do not enter treated areas without protective clothing until sprays have dried. Because certain states may require more restrictive re-entry intervals for various crops treated with this product, consult your State Department of Agriculture for further information. Do not apply this product in such a manner as to directly or through drift expose workers or other persons. The area being treated must be vacated by unprotected persons.

Written or oral warnings must be given to workers who are expected to be in a treated area or in an area about to be treated with this product. Oral warnings must be given if there is reason to believe that written warnings cannot be understood by workers. When oral warnings are given, warnings shall be given in a language customarily understood by workers. Written warnings must include the following information:

WARNING

(Insert area or field description) treated with RALLY 40W fungicide on (insert date of application). Do not enter treated area without protective clothing until sprays have dried. In case of accidental exposure, call a doctor (physician), clinic or hospital immediately. Explain that the victim has been exposed to RALLY 40W fungicide and describe his condition. For further information see the "Statements of Practical Treatment."

Oral warnings must include the same information that is in the written warnings.

GENERAL INFORMATION

RALLY 40W fungicide is a systemic, protectant and curative fungicide recommended for the control of specific diseases mentioned on this label. Optimum disease control is achieved when the fungicide is applied in a regularly scheduled preventive spray program.

USE RATE DETERMINATION—Carefully read, understand and follow label use rates and restrictions.

Under low disease conditions, minimum label use rates per application can be used while maximum label rates and shortened spray schedules are recommended for severe or threatening disease conditions.

For proper application, determine the number of acres to be treated, the recommended label use rate and the gallonage to be applied per acre. Prepare only the amount of spray solution required to treat the measured acreage. Careful calibration of spray equipment is recommended prior to use.

HANDLING—The enclosed pouches of RALLY 40W fungicide are water-soluble. **Do not** allow pouches to become wet prior to adding to the spray tank. **Do not** handle the pouches with wet hands or wet gloves. Always reseal overwrap bag to protect remaining unused pouches. Do not remove water-soluble pouches from overwrap except to add directly to the spray tank.

MIXING—Always place RALLY 40W fungicide into solution prior to adding co-applied agricultural chemicals registered for use on apples and grapes. Add the required number of unopened pouches, as determined by the dosage recommendations, into the spray tank with agitation. Depending on the water temperature and the degree of agitation, the pouches should dissolve completely within approximately ten minutes from the time they are added to the water.

COMPATIBILITY—RALLY 40W fungicide is compatible with most commonly used agricultural fungicides, insecticides, growth regulators, micronutrients and spray adjuvants. When preparing tank mixes, user should consult spray compatibility charts or State Cooperative Extension Service Specialists prior to actual use.

CAUTION: Do not use PVA packets in a tank mix with products that contain Boron or release free chlorine. The resultant reaction of PVA and Boron or free chlorine is a plastic which is not soluble in water or solvents such as diesel oils, kerosene, gasoline or alcohol.

Do not attempt to use the PVA packets directly in diesel oils or summer-spray-type oils as in ULV or LV uses. PVA packets are water-soluble not oil-soluble.

STORAGE AND DISPOSAL

Store in a cool dry place, but not below 32°F. Do not remove package from container except for immediate use.

PESTICIDE DISPOSAL: Do not contaminate water, food or feed by storage or disposal. Pesticide wastes are toxic. Improper disposal of excess pesticide, spray mixture, or rinsate is a violation of Federal law. If these wastes cannot be disposed of by use according to label instructions, contact your State Pesticide or Environmental Control Agency, or the Hazardous Waste representative at the nearest EPA regional office for guidance.

CONTAINER DISPOSAL: Completely empty container into application equipment. Dispose of in a sanitary landfill or by incineration or, if allowed by state and local authorities, by burning. If burned, stay out of smoke.

STEPS TO BE TAKEN IN CASE MATERIAL IS RELEASED OR SPILLED: Wear eye protection. Wear protective clothing. Spray water on powder and dust. Scoop or shovel solid material into a suitable container for recovery or disposal. Keep dust to a minimum. Flush contaminated area with a large amount of water to a chemical or sanitary sewer containing a settling pit. Remove contaminated clothing promptly and wash affected skin areas with soap and water. Thoroughly launder clothing before reuse. Refer to Precautionary Statements.

USE DIRECTIONS FOR USE ON APPLES — Best control of labeled diseases is achieved when the fungicide is applied on a 7- to 10-day application schedule.

Label use recommendations (ounces product per 100 gallons spray) are based on dilute sprays with a 400-gallon per acre base. Orchard tree heights, tree spacing and pruning practices vary. The following summary table provides additional guidance in calculating proper use rates for well-pruned orchards:

Tree Height (feet)	Approximate Gallons per Acre for Dilute Sprays on Well-Pruned Trees	Ounces RALLY 40W Fungicide per 10,000 cu. ft. of Canopy Volume	Ounces RALLY 40W Fungicide per Acre*
10	200	0.1 to 0.2	2.5 to 5.0
15	300	0.1 to 0.2	3.75 to 7.5
20	400	0.1 to 0.2	5.0 to 10.0
25	500	0.1 to 0.2	6.25 to 12.5
30	600	0.1 to 0.2	7.5 to 15.0

*Note recommended use rates for the control of specific diseases. For orchards that receive minimal pruning, increase the per acre use rate by 1.4 times.

Disease	Rate of RALLY 40W oz./100 gals.[a]	Use Recommendations	Restrictions
Scab Prebloom	1.25 to 2.0 (0.5 to 0.8 active)	Begin application at green tip. During periods favorable for primary scab development, use high rate and a 7-day application schedule.	Do not apply within 14 days of harvest. Do not apply more than 5 pounds (2 lb. active) RALLY 40W fungicide per acre per season.
Postbloom	1.25 to 2.0 (0.5 to 0.8 active)	Use RALLY 40W in a tank mixture with the recommended use rate of DITHANE M-45™ fungicide or another protectant fungicide registered for use on apples for improved fruit scab and summer disease control.	Do not graze livestock in treated areas or feed cover crops grown in treated areas to livestock. Application by ground equipment only.
Post-infection	2.0 (0.8 active)	RALLY 40W fungicide provides 96-hour post-infection control or curative activity and suppresses sporulation of established lesions when used in two or more applications. Apply as soon as possible after an infection period. Follow with standard protectant spray schedule.	
Powdery Mildew	1.25 to 2.5 (0.5 to 1.0 active)	Begin application at tight cluster and continue through the second cover spray. Additional sprays beyond second cover may be needed on susceptible varieties or under heavy disease pressure. Use high label rate if powdery mildew was present in previous years.	
Rusts	1.25 to 2.0 (0.5 to 0.8 active)	Begin application at pink stage and continue through the second cover spray.	

[a]Based on dilute sprays with a 400-gallon per acre base.

USE DIRECTIONS FOR USE ON GRAPES — Thorough spray coverage is essential for good disease control. RALLY 40W fungicide should be uniformly applied with a ground sprayer using 50 or more gallons of water per acre.

Disease	Rate of RALLY 40W oz. per acre	Use Recommendations	Restrictions
Black Rot	3 to 5 (1.2 to 2.0 active)	Begin application when new vines are 1 to 3 inches in length and continue application on a 14- to 21-day schedule. Use higher rate under heavy disease pressure.	Do not apply more than 1.5 lbs. (0.6 lb. A.I.) per acre per year. Do not apply within 14 days of harvest.
Powdery Mildew	3 to 5 (1.2 to 2.0 active)	Begin application at prebloom (12- to 18-inch canes) and continue application on a 14- to 21-day interval. Use higher rate on susceptible varieties or under heavy disease pressure.	

DITHANE and RALLY are trademarks of Rohm and Haas Company.

8190-B2 6-2865 Made in Italy by Rohm and Haas Italia S.p.A.
March 1990

STAM® M-4

HERBICIDE

ACTIVE INGREDIENT
propanil
3',4'-dichloropropionanilide 43.5%*
INERT INGREDIENTS 56.5%

TOTAL 100.0%

*Equivalent to 4 lbs. active ingredient per gallon

This product contains the toxic inert ingredient ISOPHORONE

EPA Reg. No. 707-109

KEEP OUT OF REACH OF CHILDREN

WARNING

Notice: Before using this product, read the entire Precautionary Statements, Conditions of Sale and Warranty, Directions for Use, Use Restrictions and Storage and Disposal Instructions on back panel. If the Conditions of Sale and Warranty are not acceptable, return the product unopened within thirty days of purchase to the place of purchase.

PRECAUTIONARY STATEMENTS

HAZARDS TO HUMANS AND DOMESTIC ANIMALS

WARNING

IRRITATING TO EYES AND SKIN

HARMFUL IF SWALLOWED OR ABSORBED THROUGH SKIN

MAY BE HARMFUL IF INHALED

Do not get in eyes or on skin. Wear eye and skin protection when handling. Do not take internally. Avoid breathing spray mist.

STATEMENT OF PRACTICAL TREATMENT

IF IN EYES: Flush eyes with large amounts of water for at least 15 minutes. Consult a physician if irritation persists.

IF ON SKIN: Wash affected area with soap and water.

IF SWALLOWED: Dilute by giving 2 glasses of water to drink and call a physician. Never give anything by mouth to an unconscious person.

NOTE TO PHYSICIAN: Emesis is recommended.

IF INHALED: Remove victim to fresh air. Treat the person symptomatically. If irritation persists, call a physician.

ENVIRONMENTAL HAZARDS

This pesticide is toxic to fish. Drift and runoff from treated areas may be hazardous to aquatic organisms in neighboring areas. Do not apply directly to water except as specified on this label. Do not contaminate water when disposing of equipment washwaters. Water drained from treated rice fields must not be used to irrigate other crops or released within 1/2 mile upstream of a potable water intake in flowing water (e.g., river, stream, etc.) or within 1/2 mile of a potable water intake in a standing body of water, such as a lake, pond or reservoir.

CONDITIONS OF SALE AND WARRANTY

Rohm and Haas warrants that the product conforms to its chemical description and is reasonably fit for the purpose stated on the label only when used in accordance with label directions under normal conditions of use. **ROHM AND HAAS MAKES NO OTHER EXPRESS OR IMPLIED WARRANTIES EITHER OF MERCHANTABILITY OR FITNESS FOR A PARTICULAR USE.** Handling, storage and use of the product by Buyer or User are beyond the control of Rohm and Haas and Seller. Risks such as crop injury, ineffectiveness or other unintended consequences resulting from, but not limited to, weather or soil conditions, presence of other materials, disease, pests, drift to other crops or property or failure to follow label directions will be assumed by the Buyer or User. **IN NO CASE WILL ROHM AND HAAS OR SELLER BE HELD LIABLE FOR CONSEQUENTIAL, SPECIAL OR INDIRECT DAMAGES RESULTING FROM THE HANDLING, STORAGE OR USE OF THIS PRODUCT.**

DIRECTIONS FOR USE

(FOR RICE GROWN IN SOUTHERN UNITED STATES ONLY.)

It is a violation of Federal law to use this product in a manner inconsistent with its labeling.

RE-ENTRY AND WORKER PROTECTION STATEMENTS

Do not enter treated areas without protective clothing until sprays have dried. Because certain states may require more restrictive re-entry intervals for various crops treated with this product, consult your State Department of Agriculture for further information. Do not apply this product in such a manner as to directly or through drift expose workers or other persons. The area being treated must be vacated by unprotected persons.

CHEMIGATION

DO NOT APPLY THIS PRODUCT THROUGH ANY TYPE OF IRRIGATION SYSTEM.

GENERAL INFORMATION

STAM M-4 is a selective postemergence herbicide for use in rice only for the control of the following weeds:

BARNYARDGRASS (WATERGRASS)
Echinochloa crus-galli
Echinochloa colonum
BEAKRUSH (SPEARHEAD)
Rhynchospora corniculata
CRABGRASS SPECIES
Digitaria spp.
CROTON, WOOLLY
Croton capitatus
DOCK, CURLY
Rumex crispus
FOXTAIL
Setaria spp.
GOOSEGRASS
Eleusine indica
GULF COCKSPUR
Echinochloa crus-pavonis
HEMP SESBANIA (COFFEE BEAN)
Sesbania exaltata
HOORAHGRASS
Fimbristylis miliaceae
MEXICANWEED
Caperonia castanaefolia
PANICUM, TEXAS
Panicum Texanum
PARAGRASS
Panicum purpurascens
PIGWEED, REDROOT
Amaranthus retroflexus
REDWEED
Melochia corchorifolia
SIGNALGRASS, BROADLEAF
Bracharia platyphylla
SPIKERUSH (WIREGRASS)
Eleocharis spp.

STAM M-4 herbicide is an emulsifiable concentrate containing 4 pounds active ingredient per U.S. gallon. The product is not a hormone-type herbicide, but kills susceptible weeds by direct contact action. For this reason, thorough coverage of emerged weeds is essential for best results. Only weeds which have emerged and are exposed at time of application will be controlled. The product has no preemergence or residual herbicidal activity. Apply STAM M-4 herbicide only to fields which have been drained of flood water. The product is most effective if applied when susceptible grasses and broadleaf weeds are small and growing actively under favorable soil moisture and weather conditions. Early weed control removes competition, saves moisture and generally contributes to increased yields.

USE RESTRICTIONS

Do not apply to any crop other than rice. STAM M-4 herbicide injures most crops except cereal grains and perennial grasses. Avoid drift or accidental application from turning aircraft on cotton, soybeans, corn, safflower, seedling legumes, vegetables, orchards, vineyards, gardens, shrubs, and ornamentals. Once applied, it does not release fumes hazardous to nearby crops.

TIMING AND DOSAGE RECOMMENDATIONS

Treat grassy and weedy fields when a satisfactory stand of rice that will tolerate flooding is established. The amount of STAM M-4 herbicide to apply depends primarily upon the stage and growth condition of the grasses. The growth stage of the rice is also a factor in dosage and timing limitations so as to avoid the possibility of excessive residues.

For best results apply STAM M-4 herbicide at the rate of 3 to 4 quarts (3 to 4 lbs. active) per acre when the grasses are actively growing in the 1 to early 4 leaf stage. This rate will also control many seedling broadleaf and aquatic weeds. Generally this will be 15 to 25 days after planting of the rice. In order to insure satisfactory weed control, do not apply less than 3 quarts of STAM M-4 herbicide per acre in a single spray application.

Apply STAM M-4 herbicide at the rate of 4 to 6 quarts (4 to 6 lbs. active) per acre to actively growing grasses in the 4 to 6 leaf and early tillering stage or when they are in the 2 to 4 leaf stage but stressed under dry soil conditions. Generally this will be 20 to 30 days after planting of the rice.

EMERGENCY TREATMENT: Apply STAM M-4 herbicide at the rate of 5 to 6 quarts (5 to 6 lbs. active) in 15 gallons of spray per acre for emergency control of older tillering grass. Generally this will be 30 to 40 days after planting. If the field is already flooded, the water should be lowered or drained before spraying to expose more of the grass and weeds. Emergency treatment should be considered as a salvage operation only and cannot be relied upon for total control of grass and weeds.

TO AVOID EXCESSIVE RESIDUES AT HARVEST, DO NOT APPLY AFTER THE END OF TILLERING FOR THE RICE VARIETY BEING TREATED. DO NOT APPLY MORE THAN A MAXIMUM OF SIX POUNDS ACTIVE INGREDIENT PER ACRE IN A SINGLE APPLICATION OR EXCEED EIGHT POUNDS ACTIVE INGREDIENT PER ACRE TOTAL DOSAGE PER SEASON.

APPLICATION EQUIPMENT

AIRCRAFT — Fixed wing aircraft or helicopters should have well designed spray systems that produce a uniform pattern of medium fine spray droplets. Apply STAM M-4 herbicide on small grass in no less than 10 gallons of total spray per acre with boom-nozzle sprayers. Increase volume to 12 to 15 gallons per acre for larger or denser stands of grass or during periods of low humidity.

The optimum effective spray swath width depends on operating conditions and type of aircraft being used. For uniform spray coverage with fixed wing aircraft, do not exceed a spray swath width 10 percent greater than the wingspan or the length of the boom in helicopters. Measure the swaths accurately for flagging.

Ground Sprayers — Use standard low pressure herbicide sprayers equipped with boom and flat fan nozzles. Use nozzle sizes that deliver a medium fine droplet in 15 to 20 gallons total spray per acre at 40 to 50 p.s.i. and at ground speeds not in excess of 3 to 4 m.p.h. Avoid raising boom too high. Spray patterns should meet uniformly.

Flush all equipment with clear water after each day's use. Clean all equipment including nurse tanks used for STAM M-4 herbicide with a detergent wash followed by a water rinse, BEFORE AND AFTER spraying other pesticides or other crops. Applicators and flagmen should avoid contact with spray mist from STAM M-4 herbicide or any pesticides and should wear protective clothing and goggles. Wash thoroughly after exposure.

CROP TOLERANCE AND GROWING CONDITIONS

All leading commercial varieties of rice are exceptionally tolerant to STAM M-4 herbicide. A temporary yellowing or tip burn of rice may be noted after treatment, but new growth is normal. Severe leaf burn and partial killing of rice may occur if the product is applied when rice is under stress and in a weakened growth condition due to disease or insect infestations, excessive soil salts, overwatering, or prolonged drought and extremely hot weather. Growers are cautioned not to spray under such conditions and/or when maximum daily temperatures have been or are expected to go above 100°F.

EFFECT OF CLIMATIC CONDITIONS AND CULTURAL PRACTICES ON WEED CONTROL

Field and Seedbed Preparation

Fields should be accurately leveled and contoured and have well prepared seedbeds free of clods. This encourages uniform and rapid emergence of rice, grass and broadleaf weeds and permits better timing and coverage of STAM M-4 herbicide sprays resulting in optimum weed control.

Water Management

Before application of STAM M-4 herbicide, drained or dry planted fields should be flushed as often as needed to prevent drying and crusting. Flushing encourages uniform emergence and vigorous growth of grass, broadleaf weeds and rice which is essential for best results. Flush fields in sufficient time so that weeds and rice are actively growing at time of treatment. Make sure the field is drained prior to treatment so that grasses and broadleaf weeds are fully exposed. Weeds that are partially submerged in standing water at time of application will not be satisfactorily controlled.

After treatment, treated fields should always be flooded before a second infestation of grass has a chance to develop. To prevent more grass from germinating after treatment, fields should be flooded within 24 hours after spraying, or as soon as possible after 24 hours.

Temperature

The temperature a few days before and after applying STAM M-4 herbicide has an important bearing on the weed killing activity. The activity increases as daily maximum temperatures increase above 75°F and decreases as the daily maximum temperatures decline below 75°F. Do not apply STAM M-4 herbicide when maximum temperatures have been or are expected to stay below 65°F or to go above 100°F. Low temperature at time of application is not so important as long as it warms up later during the day.

Relative Humidity and Rain

Grasses and weeds are more responsive to STAM M-4 herbicide during periods of high humidity when the foliage is moist or covered by dew. When the humidity is very low some of the spray may crystalize in the air. When this condition exists, increase spray volume to 12-15 gallons per acre for best results. Do not spray when rain threatens within six hours to avoid loss of the spray deposit before adsorption by the grass.

Wind

Avoid applications when the wind speed exceeds 10 m.p.h. because of drift hazard to sensitive crops and the possibility of uneven (streaked) application.

COMPATIBILITY WITH OTHER CHEMICALS

Tank-mix applications of STAM M-4 herbicide with other herbicides, insecticides, spray adjuvants, or liquid fertilizers may reduce crop tolerance and/or weed control or impair mixing properties. Use of these products in tank-mix applications with STAM M-4 herbicide is done at the user's risk.

Insecticides

Severe injury or kill of rice plants may result from tank-mix combinations or separate sprays of STAM M-4 herbicide and certain insecticides. Do not combine STAM M-4 herbicide with carbamate insecticides such as carbaryl (Sevin, etc.), methomyl (Lannate, Nudrin, etc.) or organophosphorus insecticides such as parathion, methyl parathion, Guthion, malathion, Systox, EPN, Phosphamidon, etc. Do not apply any of the above insecticides to rice fields within 14 days before or after STAM M-4 herbicide. Do not use carbamate or systemic organophosphorus insecticides on rice fields to be treated with STAM M-4 herbicide.

Do not apply STAM M-4 herbicide to rice fields which were planted with rice seed treated with bird repellents containing methiocarb such as Mesurol, Borderland Red, etc. Consult local Extension specialist for current recommendations of approved insecticides on rice.

STORAGE AND DISPOSAL

STORAGE: Do not use, pour, spill or store near heat or open flame. Ground all metal containers when transferring product. Protect from freezing. If stored below 32°F and crystals form, warm to 72°F for 24 hours, periodically rolling drum to reconstitute.

PESTICIDE DISPOSAL: Do not contaminate water, food, or feed by storage or disposal. Pesticide wastes are toxic. Improper disposal of excess pesticide, spray mixture, or rinsate is a violation of Federal Law. If these wastes cannot be disposed of by use according to label instructions, contact your State pesticide or Environmental Control Agency, or the Hazardous Waste representative at the nearest EPA regional office for guidance.

CONTAINER DISPOSAL: **Metal Containers**—Triple rinse (or equivalent). Then offer for recycling or reconditioning, or puncture and dispose of in a sanitary landfill, or by other procedures approved by state and local authorities. **Bulk Containers**—Drain thoroughly and return to specified destination for cleaning and reuse.

STEPS TO BE TAKEN IN CASE MATERIAL IS RELEASED OR SPILLED: Eliminate ignition sources. Ventilate area. Avoid breathing vapors. Use MSHA/NIOSH self-contained breathing apparatus or air mask for large spills in confined areas. Dike the spill with inert material (sand, earth, fuller's earth, etc.) and if appropriate transfer the liquid and solid diking material to separate containers for recovery or disposal. Remove contaminated clothing promptly and wash affected skin areas with soap and water. Wash clothing before reuse. Keep out of all sewers and open bodies of water. REFER TO PRECAUTIONARY STATEMENTS.

9234-B4

STAM® is a registered trademark of Rohm and Haas Company.

STAMPEDE™ CM

HERBICIDE

ACTIVE INGREDIENTS:		48%
propanil (3′,4′-Dichloropropionanilide)	33%*	
2-Methyl-4-chlorophenoxyacetic acid, isooctyl ester	15%**	
INERT INGREDIENTS:		52%
TOTAL		100%

This product contains the toxic inert ingredient isophorone.

*Equivalent to 3 lbs. active ingredient per gallon.

**Equivalent to 1.4 lbs. per gallon of 2-Methyl-4-chlorophenoxyacetic acid, isooctyl ester (0.85 lbs. 2-Methyl-4-chlorophenoxyacetic acid equivalent).

EPA Reg. No. 707-182

NOTICE: Before using this product, read the entire Precautionary Statements, Conditions of Sale and Warranty, Directions for Use, Use Restrictions and Storage and Disposal Instructions. If the Conditions of Sale and Warranty are not acceptable, return the product unopened within thirty days of purchase to the place of purchase.

KEEP OUT OF REACH OF CHILDREN
CAUTION

PRECAUTIONARY STATEMENTS

HAZARDS TO HUMANS AND DOMESTIC ANIMALS

Harmful if swallowed, inhaled, or absorbed through the skin. Avoid breathing vapors or spray mist. Avoid contact with skin, eyes or clothing. Remove and wash clothing before reuse.

STATEMENT OF PRACTICAL TREATMENT

IF IN EYES — Flush eyes with plenty of water for at least 15 minutes. Call a physician if irritation persists.

IF INHALED — Move subject to fresh air.

IF ON SKIN — Wash affected area with soap and water.

IF SWALLOWED — Dilute by giving 2 glasses of water to drink and call a physician. Never give anything by mouth to an unconscious person.

NOTE TO PHYSICIAN — Emesis is recommended.

ENVIRONMENTAL HAZARDS

This pesticide is toxic to fish. Drift and runoff from treated areas may be hazardous to aquatic organisms in neighboring areas. Do not apply directly to water or wetlands (swamps, bogs, marshes, and potholes.) Do not contaminate water when disposing of equipment washwaters.

PHYSICAL OR CHEMICAL HAZARDS

COMBUSTIBLE

Do not use, pour, or store near heat or open flame.

CONDITIONS OF SALE AND WARRANTY

Rohm and Haas warrants that the product conforms to its chemical description and is reasonably fit for the purpose stated on the label only when used in accordance with label directions under normal conditions of use. **ROHM AND HAAS MAKES NO OTHER EXPRESS OR IMPLIED WARRANTIES EITHER OF MERCHANTABILITY OR FITNESS FOR A PARTICULAR USE.** Handling, storage and use of the product by Buyer or User are beyond the control of Rohm and Haas and Seller. Risks such as crop injury, ineffectiveness or other unintended consequences resulting from, but not limited to, weather or soil conditions, presence of other materials, disease, pests, drift to other crops or property or failure to follow label directions will be assumed by the Buyer or User. **IN NO CASE WILL ROHM AND HAAS OR SELLER BE HELD LIABLE FOR CONSEQUENTIAL, SPECIAL OR INDIRECT DAMAGES RESULTING FROM THE HANDLING, STORAGE OR USE OF THIS PRODUCT.**

DIRECTIONS FOR USE

It is a violation of Federal law to use this product in a manner inconsistent with its labeling.

RE-ENTRY AND WORKER PROTECTION STATEMENTS

Do not enter treated areas without protective clothing until sprays have dried. Because certain states may require more restrictive re-entry intervals for various crops treated with this product, consult your State Department of Agriculture for further information. Do not apply this product in such a manner as to directly or through drift, expose workers or other persons. The area being treated must be vacated by unprotected persons.

CHEMIGATION

DO NOT APPLY THIS PRODUCT THROUGH ANY TYPE OF IRRIGATION SYSTEM.

GENERAL INFORMATION

STAMPEDE CM is a selective postemergence herbicide for the control of green and yellow foxtail (wild millet, pigeongrass) and specific broadleaf weeds in hard red spring wheat, spring barley and durum wheat in North Dakota, South Dakota, Minnesota and Montana.

STAMPEDE CM is a contact herbicide, therefore, thorough coverage of emerged susceptible weeds is essential for acceptable postemergence control. STAMPEDE CM has no preemergence or residual herbicidal activity. The most effective control is achieved when foxtail (pigeongrass) and susceptible broadleaf weeds are small and growing actively under favorable soil moisture and weather conditions.

WEEDS CONTROLLED POSTEMERGENCE

Weeds Controlled		Seedling Stage*
Pigeongrass		
Foxtail, Green	*Setaria viridis*	1 to 3 leaf
Foxtail, Yellow	*Setaria lutescens*	1 to 3 leaf
Buckwheat, Wild	*Polygonum convolvulus*	1 to 4 leaf
Kochia**	*Kochia scoparia*	1 to 4 leaf
Lambsquarters, Common	*Chenopodium album*	1 to 4 leaf
Mustard, Wild	*Brassica kaber*	1 to 4 leaf
Pigweed, Prostrate	*Amaranthus blitoides*	1 to 4 leaf
Pigweed, Redroot	*Amaranthus retroflexus*	1 to 4 leaf

*Refers to weed seedling stages for control. Pigeongrass greater than the 3 leaf stage or listed broadleaf weeds beyond the 4 leaf stage will not be controlled.

**Suppression — may cause spotting, stunting or death of kochia not exceeding the 4 leaf stage.

DOSAGE RATE RECOMMENDATIONS — SPRING BARLEY, DURUM AND HARD RED SPRING WHEAT

STAMPEDE CM herbicide is recommended for control of susceptible weed seedlings when applied as a single postemergence application at the rate of 2.5 to 3 pints per acre. The lower rate (2.5 pints/A) should be applied when weed seedlings are growing under favorable soil moisture conditions. The higher rate (3 pints/A) is recommended when heavy weed pressure is present or less than favorable soil moisture conditions exist. Do not apply STAMPEDE CM at rates higher than 3 pints per acre.

Do not apply STAMPEDE CM when soil moisture levels are deeper than 1.5 inches from the soil surface and active plant growth stops.

APPLICATION TIMING

WEED CONTROL — STAMPEDE CM herbicide is most effective on controlling foxtail if spray applications are made when the majority of the emerged foxtail are in the 2 to 3 LEAF STAGE. The effectiveness of STAMPEDE CM to provide control declines rapidly as the 4th leaf of the foxtail begins to emerge. Since foxtail at these early stages is often less than one inch tall, close inspection of the field is essential to determine correct application timing. Fields should be checked frequently for emerging pigeongrass starting 2 to 2½ weeks after planting. The best time to apply STAMEDE CM often will be 10 to 17 days after crop emergence when the cereal grain is in the 3 to 4 LEAF STAGE.

CROP TOLERANCE

STAMPEDE CM herbicide causes temporary yellowing or tip browning to the leaves of the grain crop 2 to 5 days after application. The effects are temporary and usually will disappear 10 to 14 days after application. New leaves will have normal green color. Application of STAMPEDE CM under adverse growing conditions may result in greater crop injury and slower recovery. Care should be taken not to spray STAMPEDE CM on grain crops that are under growth stress caused by drought, flooding, excess soil salts, hail damage, disease or extreme temperatures. Do not apply STAMPEDE CM if frost is expected within 24 hours or when temperatures are above 85°F, especially with drying winds.

Applications of STAMPEDE CM herbicide should be made only to actively growing (healthy) grain crops in the seedling stages recommended below:

Crop	Seedling Stage*
Hard Red Spring Wheat	2 to 5 leaf
Durum Wheat	2 to 4 leaf
Spring Barley	2 to 4 leaf

*Refers to crop seedling stages for best tolerance. To avoid crop injury do not apply STAMPEDE CM herbicide to labeled grain crops beyond the recommended seedling stage.

MIXING AND EQUIPMENT

STAMPEDE CM herbicide is an emulsifiable concentrate that mixes readily with water. With the pump and agitator running, add the recommended amount of STAMPEDE CM into a partially filled spray tank. Complete filling of the spray tank to the desired level. Spray water should be clean and a minimum temperature of 50°F or warmer when mixed. STAMPEDE CM may be applied with conventional low-pressure herbicide sprayers.

GROUND APPLICATION

Field Sprayers: Spra-Coupe, tractor-drawn or truck-mounted ground sprayers, floaters (Big A, Terragator, etc.).

STAMPEDE CM herbicide should be thoroughly mixed with clean water, at recommended concentrations and applied in a minimum of 10 gallons of water per acre. Use conventional ground spray equipment with flat fan nozzles at a minimum spray pressure of 40 psi. Recommended ground speeds should not exceed 8 mph (1st or 2nd gears only) for Spra-Coupe, 5 mph for tractor-drawn sprayers and 10 mph for truck-mounted sprayers. Hollow cone or flood jet nozzles are not recommended on these sprayers.

AERIAL APPLICATION

STAMPEDE CM herbicide can be applied either by fixed wing aircraft or helicopter. Make applications at the recommended dosage and timing in a minimum of 5 gallons of water per acre. Aircraft should be equipped with a well designed spray distribution system that is adjusted and operated to provide a uniform pattern of medium-sized droplets (400 micron size) distributed over an optimum width swath. Only hollow cone nozzles with cores are recommended for STAMPEDE CM aerial application. Nozzles should be angled 0° to 20° downward and to the rear, depending on air speed. Nozzles should not be placed within 2 to 3 feet of the wing tips in fixed wing aircraft and extras should be placed on the right side of the fuselage. Use a spray pump of at least 1½ to 2 inches in diameter and operate at 20 to 50 psi. Application should be made with boom height 6 to 10 feet above crop. The optimum or effective spray swath width depends on operating conditions and type of aircraft being used. For uniform spray coverage with fixed wing aircraft do not exceed a spray swath width 10%

greater than the wingspan or the length of the boom in helicopters. Measure the swaths accurately for flagging.

DO NOT APPLY STAMPEDE CM HERBICIDE IN AERIAL APPLICATIONS WITHIN 600 FEET OF ANY PERMANENT BODY OF WATER OR WHEN WIND SPEED EXCEEDS 10 MPH.

CALIBRATION

Sprayers should be checked carefully for pressure, ground speed, and uniformity of spray pattern, and calibrated accurately for spray volume.

CLEANING SPRAYERS

Thoroughly flush spray equipment (tank, pump, hoses and boom) with clean water before and after each use. Repeat this procedure at the end of each day's spraying and before changing to or from another pesticide. This cleaning operation of tank and lines is especially important if insecticides, such as SEVIN®, or any of the organophosphorus insecticides are used in the sprayer prior to or after STAMPEDE CM. Interaction of these insecticides and STAMPEDE CM can injure the crop severely. See COMPATIBILITY WITH OTHER CHEMICALS section.

EFFECT OF CLIMATIC CONDITIONS AND CULTURAL PRACTICES ON WEED CONTROL

Field and Seedbed Preparation

Fields under conventional tillage should have well prepared seedbeds, free of large clods, to encourage more uniform and rapid emergence of the crop, foxtail, and broadleaf weeds. This permits better timing of sprays. In no-till grain fields, pigeongrass and other weeds may emerge rapidly and very thickly because of the firm seedbed. Monitoring of no-till fields should begin early to assure proper timing of sprays.

Soil Moisture and Rainfall

Weed control with STAMPEDE CM herbicide is strongly influenced by soil moisture. Weed control and crop tolerance is best when adequate moisture is available for active weed and crop growth. The performance of STAMPEDE CM is reduced substantially during periods of prolonged dry weather. STAMPEDE CM IS NOT RECOMMENDED FOR USE WHEN SOIL MOISTURE LEVELS ARE DEEPER THAN 1½ INCHES FROM THE SOIL SURFACE AND ACTIVE PLANT GROWTH STOPS.

STAMPEDE CM herbicide enters the leaf rapidly. Spraying of weeds at the correct stage should not be postponed even if showery weather is expected; however, a heavy rain of 1 inch or more, within 4 hours after application, may reduce control. Fields may be sprayed when the plants are dry or wet with dew or rain.

TEMPERATURE

Providing soil moisture is not a limiting factor; the herbicidal activity of STAMPEDE CM herbicide improves as daily maximum temperatures increase. Best weed control occurs when the daily maximum temperatures are 65°F and above. Do not apply STAMPEDE CM when daily maximum temperatures will remain below 50°F or when they are expected to exceed 85°F.

COMPATIBILITY WITH OTHER CHEMICALS

Foliar-Applied Chemicals

If another herbicide is required, a 3 day interval should be allowed between the application of STAMPEDE CM herbicide and the other herbicide.

Severe injury or kill of spring barley, durum or hard red spring wheat plants may result from tank mix or separate applications of STAMPEDE CM herbicide and certain insecticides. Grain crops that have been or will be treated with STAMPEDE CM should not be treated with carbamate insecticides such as carbaryl (Sevin, etc.), methomyl (Lannate®, Nudrin®, etc.) or organophosphorus insecticides such as methyl parathion, Guthion®, etc.

If a foliar insecticide is necessary, malathion can be safely applied 14 days before or after the STAMPEDE CM herbicide treatment.

Soil-Applied Chemicals

Do not spray STAMPEDE CM herbicide on spring barley, durum or hard red spring wheat if the field was treated the previous year with soil-applied systemic organophosphorus insecticides (Counter®, Disyston®, Thimet®, and others).

Do not spray STAMPEDE CM herbicide on spring barley, durum or hard red spring wheat that has been treated at planting time with soil-applied systemic insecticides e.g., Disyston, Furadan®, Thimet, and others.

Grain crops protected with maneb/lindane seed dressings may be treated with STAMPEDE CM herbicide.

USE RESTRICTIONS

- Do not apply STAMPEDE CM herbicide to any crops other than spring barley, durum or hard red spring wheat.
- Do not mix or apply STAMPEDE CM herbicide with any other pesticide, spray adjuvant or with fertilizer except as specifically recommended on this label.
- Do not apply STAMPEDE CM herbicide under windy conditions that will allow spray drift to adjacent susceptible crops such as sunflowers, soybeans, sugarbeets, potatoes, forage legumes, gardens, orchards, and shelter belts.
- DO NOT GRAZE TREATED CROP OR CUT FOR GREEN CHOP FEED.

STORAGE AND DISPOSAL

STORAGE — Do not store at temperatures below 15°F. If stored below 15°F and crystals form, warm to 60°F for 24 hours, periodically shaking container to reconstitute.

PESTICIDE DISPOSAL — Do not contaminate water, food or feed by storage or disposal. Pesticide wastes are toxic. Improper disposal of excess pesticide, spray mixture or rinsate is a violation of Federal law. If these wastes cannot be disposed of by use according to label instructions, contact your State Pesticide or Environmental Control Agency, or the Hazardous Waste representative at the nearest EPA Regional Office for guidance.

CONTAINER DISPOSAL — Triple rinse (or equivalent). Then puncture and dispose of in a sanitary landfill, or by incineration, or if allowed by state and local authorities, by burning. If burned, stay out of smoke.

STEPS TO BE TAKEN IN CASE MATERIAL IS RELEASED OR SPILLED — Eliminate ignition sources. Ventilate area. Avoid breathing vapors. Use MSHA/NIOSH self-contained breathing apparatus or air masks for large spills in confined areas. Dike the spill with inert material (sand, earth, fuller's earth, etc.) and if appropriate, transfer the liquid and solid diking material to separate containers for recovery or disposal. Keep spill out of all sewers and open bodies of water. Refer to Precautionary Statements.

STAMPEDE™ is a trademark of Rohm and Haas Company.
COUNTER® and THIMET® are registered trademarks of American Cyanamid Co.
DISYSTON® is a registered trademark of Bayer AG.
FURADAN® is a registered trademark of FMC Corp.
GUTHION® is a registered trademark of Mobay Chemical Corp.
LANNATE® and NUDRIN® are registered trademarks of E.I. duPont de Nemours & Co.
SEVIN® is a registered trademark of Rhone Poulenc Inc.

10012-B4

STREL™ 4E
HERBICIDE

ACTIVE INGREDIENT

propanil 3',4'-dichloropropionanilide	43.5%*
INERT INGREDIENTS	56.5%
TOTAL	100.0%

* Equivalent to 4 lbs. active ingredient per gallon. This product contains the toxic inert ingredient Isophorone

EPA Reg. No. 707-109

NOTICE: Before using this product read the entire Precautionary Statements, Conditions of Sale and Warranty, Directions for Use, Use Restrictions and Storage and Disposal Instructions. If the Conditions of Sale and Warranty are not acceptable, return the product unopened within thirty days of purchase to the place of purchase.

KEEP OUT OF REACH OF CHILDREN
WARNING

PRECAUTIONARY STATEMENTS

HAZARDS TO HUMANS AND DOMESTIC ANIMALS

IRRITATING TO EYES AND SKIN
HARMFUL IF SWALLOWED OR ABSORBED THROUGH SKIN
MAY BE HARMFUL IF INHALED

Do not get in eyes or on skin. Wear eye and skin protection when handling. Do not take internally. Avoid breathing spray mist.

STATEMENT OF PRACTICAL TREATMENT

IF IN EYES: Flush eyes with large amounts of water for at least 15 minutes. Consult a physician if irritation persists.

IF ON SKIN: Wash affected area with soap and water.

IF SWALLOWED: Dilute by giving 2 glasses of water to drink and call a physician. Never give anything by mouth to an unconscious person.

NOTE TO PHYSICIAN: Emesis is recommended.

IF INHALED: Remove victim to fresh air. Treat the person symptomatically. If irritation persists, call a physician.

ENVIRONMENTAL HAZARDS

This pesticide is toxic to fish. Drift and runoff from treated areas may be hazardous to aquatic organisms in neighboring areas. Do not apply directly to water except as specified on this label. Do not contaminate water when disposing of equipment washwaters.

Water drained from treated rice fields must not be used to irrigate other crops or released within ½ mile upstream of a potable water intake in flowing water (e.g., river, stream, etc.) or within ½ mile of a potable water intake in a standing body of water, such as a lake, pond or reservoir.

CONDITIONS OF SALE AND WARRANTY

Rohm and Haas warrants that the product conforms to its chemical description and is reasonably fit for the purpose stated on the label only when used in accordance with label directions under normal conditions of use. **ROHM AND HAAS MAKES NO OTHER EXPRESS OR IMPLIED WARRANTIES EITHER OF MERCHANTABILITY OR FITNESS FOR A PARTICULAR USE.** Handling, storage and use of the product by Buyer or User are beyond the control of Rohm and Haas and Seller. Risks, such as crop injury, ineffectiveness or other unintended consequences resulting from, but not limited to, weather or soil conditions, presence of other materials, disease, pests, drift to other crops or property or failure to follow label directions, will be assumed by the Buyer or User. **IN NO CASE WILL ROHM AND HAAS OR SELLER BE HELD LIABLE FOR CONSEQUENTIAL, SPECIAL OR INDIRECT DAMAGES RESULTING FROM THE HANDLING, STORAGE OR USE OF THIS PRODUCT.**

DIRECTIONS FOR USE

(FOR RICE GROWN IN SOUTHERN UNITED STATES ONLY)

It is a violation of Federal law to use this product in a manner inconsistent with its labeling.

RE-ENTRY AND WORKER PROTECTION STATEMENTS

Do not enter treated areas without protective clothing until sprays have dried. Because certain states may require more restrictive re-entry intervals for various crops treated with this product, consult your State Department of Agriculture for further information. Do not apply this product in such a manner as to directly or through drift expose workers or other persons. The area being treated must be vacated by unprotected persons.

CHEMIGATION

DO NOT APPLY THIS PRODUCT THROUGH ANY TYPE OF IRRIGATION SYSTEM.

GENERAL INFORMATION

STREL 4E herbicide is a selective postemergence herbicide for use in rice only for control of the following weeds:

BARNYARDGRASS (WATERGRASS) Echinochloa crus-galli Echinochloa colonum	FOXTAIL Setaria spp. GOOSEGRASS Eleusine indica	PANICUM, TEXAS Panicum Texanum PARAGRASSS Panicum purpurascens
BEAKRUSH (SPEARHEAD) Rhynchospora corniculata	GULF COCKSPUR Echinochloa crus-pavonis	PIGWEED, REDROOT Amaranthus retroflexus
CRABGRASS SPECIES Digitaria spp.	HEMP SESBANIA (COFFEE BEAN) Sesbania exaltata	REDWEED Melochia corchorifolia
CROTON, WOOLLY Croton capitatus	HOORAHGRASS Fimbristylis miliaceae	SIGNALGRASS, BROADLEAF Bracharia platyphylla
DOCK, CURLY Rumex crispus	MEXICANWEED Caperonia castanaefolia	SPIKERUSH (WIREGRASS) Eleocharis spp.

STREL 4E herbicide is an emulsifiable concentrate containing 4 pounds active ingredient per U.S. gallon. The product is not a hormone-type herbicide, but kills susceptible weeds by direct contact action. For this reason, thorough coverage of emerged weeds is essential for best results. Only weeds that have emerged and are exposed at time of application will be controlled. The product has no preemergence or residual herbicidal activity. Apply STREL 4E herbicide only to fields that have been drained of flood water. The product is most effective if applied when susceptible grasses and broadleaf weeds are small and growing actively under favorable soil moisture and weather conditions. Early weed control removes competition, saves moisture and generally contributes to increased yields.

USE RESTRICTIONS

Do not apply to any crop other than rice. STREL 4E herbicide injures most crops, except cereal grains and perennial grasses. Avoid drift or accidental application from turning aircraft on cotton, soybeans, corn, safflower, seedling legumes, vegetables, orchards, vineyards, gardens, shrubs, and ornamentals. Once applied, it does not release fumes hazardous to nearby crops.

TIMING AND DOSAGE RECOMMENDATIONS

Treat grassy and weedy fields when a satisfactory stand of rice that will tolerate flooding is established. The amount of STREL 4E herbicide to apply depends primarily upon the stage and growth conditions of the grasses. The growth stage of the rice is also a factor in dosage and timing limitations so as to avoid the possibility of excessive residues.

For best results apply STREL 4E herbicide at the rate of 3 to 4 quarts (3 to 4 lbs. active) per acre when the grasses are actively growing in the 1 to early 4 leaf stage. This rate will also control many seedling broadleaf and aquatic weeds. Generally this will be 15 to 25 days after planting of the rice. In order to insure satisfactory weed control, do not apply less than 3 quarts of STREL 4E herbicide per acre in a single spray application.

Apply STREL 4E herbicide at the rate of 4 to 6 quarts (4 to 6 lbs. active) per acre to actively growing grasses in the 4 to 6 leaf and early tillering stage or when they are in the 2 to 4 leaf stage but stressed under dry soil conditions. Generally this will be 20 to 30 days after planting of the rice.

EMERGENCY TREATMENT: Apply STREL 4E herbicide at the rate of 5 to 6 quarts (5 to 6 lbs. active) in 15 gallons of spray per acre for emergency control of older tillering grass. Generally this will be 30 to 40 days after planting. If the field is already flooded, the water should be lowered or drained before spraying to expose more of the grass and weeds. Emergency treatment should be considered as a salvage operation only and cannot be relied upon for total control of grass and weeds.

TO AVOID EXCESSIVE RESIDUES AT HARVEST, DO NOT APPLY AFTER THE END OF TILLERING FOR THE RICE VARIETY BEING TREATED. DO NOT APPLY MORE THAN A MAXIMUM OF SIX POUNDS ACTIVE INGREDIENT PER ACRE IN A SINGLE APPLICATION OR EXCEED EIGHT POUNDS ACTIVE INGREDIENT PER ACRE TOTAL DOSAGE PER SEASON.

APPLICATION EQUIPMENT

Aircraft — Fixed wing aircraft or helicopters should have well-designed spray systems that produce a uniform pattern of medium-fine spray droplets. Apply STREL 4E herbicide on small grass in no less than 10 gallons of total spray per acre with boom-nozzle sprayers. Increase volume to 12 to 15 gallons per acre for larger or denser stands of grass or during periods of low humidity.

The optimum effective spray swath width depends on operating conditions and type of aircraft being used. For uniform spray coverage with fixed wing aircraft, do not exceed a spray swath width 10 percent greater than the wingspan or the length of the boom in helicopters. Measure the swaths accurately for flagging.

Ground Sprayers — Use standard low-pressure herbicide sprayers equipped with boom and flat fan nozzles. Use nozzle sizes that deliver a medium-fine droplet in 15 to 20 gallons total spray per acre at 40 to 50 p.s.i. and at ground speeds not in excess of 3 to 4 m.p.h. Avoid raising boom too high. Spray patterns should meet uniformly.

Flush all equipment with clear water after each day's use. Clean all equipment, including nurse tanks used for STREL 4E herbicide, with detergent wash followed by a water rinse, BEFORE AND AFTER spraying other pesticides or other crops. Applicators and flagmen should avoid contact with spray mist from STREL 4E herbicide or any pesticides and should wear protective clothing and goggles. Wash thoroughly after exposure.

CROP TOLERANCE AND GROWING CONDITIONS

All leading commercial varieties of rice are exceptionally tolerant to STREL 4E herbicide. A temporary yellowing or tip burn of rice may be noted after treatment, but new growth is normal. Severe leaf burn and partial killing of rice may occur if the product is applied when rice is under stress and in a weakened growth condition due to disease or insect infestations, excessive soil salts, overwatering, or prolonged drought and extremely hot weather. Growers are cautioned not to spray under such conditions and/or when maximum daily temperatures have been or are expected to go above 100°F.

EFFECT OF CLIMATIC CONDITIONS AND CULTURAL PRACTICES ON WEED CONTROL

Field and Seedbed Preparation

Fields should be accurately leveled and contoured and have well-prepared seedbeds free of clods. This encourages uniform and rapid emergence of rice, grass and broadleaf weeds and permits better timing and coverage of STREL 4E herbicide sprays resulting in optimum weed control.

Water Management

Before application of STREL 4E herbicide, drained or dry planted fields should be flushed as often as needed to prevent drying and crusting. Flushing encourages uniform emergence and vigorous growth of grass, broadleaf weeds and rice which is essential for best results. Flush fields in sufficient time so that weeds and rice are actively growing at time of treatment. Make sure the field is drained prior to treatment so that grasses and broadleaf weeds are fully exposed. Weeds that are partially submerged in standing water at time of application will not be satisfactorily controlled.

After treatment, treated fields should always be flooded before a second infestation of grass has a chance to develop. To prevent more grass from germinating after treatment, fields should be flooded within 24 hours after spraying, or as soon as possible after 24 hours.

Temperature

The temperature a few days before and after applying STREL 4E herbicide has an important bearing on the weed-killing activity. The activity increases as daily maximum temperatures increase above 75°F and decreases as the daily maximum temperatures decline below 75°F. Do not apply

STREL 4E herbicide when maximum temperatures have been or are expected to stay below 65°F or to go above 100°F. Low temperature at time of application is not so important as long as it warms up later during the day.

Relative Humidity and Rain

Grasses and weeds are more responsive to STREL 4E herbicide during periods of high humidity when the foliage is moist or covered by dew. When the humidity is very low some of the spray may crystalize in the air. When this condition exists, increase spray volume to 12-15 gallons per acre for best results. Do not spray when rain threatens within six hours to avoid loss of the spray deposit before adsorption by the grass.

Wind

Avoid applications when the wind speed exceeds 10 m.p.h. because of the drift hazard to sensitive crops and the possibility of uneven (streaked) application.

COMPATIBILITY WITH OTHER CHEMICALS

Tank-mix applications of STREL 4E herbicide with other herbicides, insecticides, spray adjuvants, or liquid fertilizers may reduce crop tolerance and/or weed control or impair mixing properties. Use of these products in tank-mix applications with STREL 4E herbicide is done at the user's risk.

Insecticides

Severe injury or kill of rice plants may result from tank-mix combinations or separate sprays of STREL 4E herbicide and certain insecticides. Do not combine STREL 4E herbicide with carbamate insecticides such as carbaryl (Sevin, etc.), methomyl (Lannate, Nudrin, etc.) or organophosphorus insecticides such as parathion, methyl parathion, Guthion, malathion, Systox, EPN, Phosphamidon, etc. Do not apply any of the above insecticides to rice fields within 14 days before or after STREL 4E herbicide. Do not use carbamate or systemic organophosphorus insecticides on rice fields to be treated with STREL 4E herbicide.

Do not apply STREL 4E herbicide to rice fields that were planted with rice seed treated with bird repellents containing methiocarb such as Mesurol, Borderland Red, etc. Consult local Extension specialist for current recommendations of approved insecticides on rice.

STORAGE AND DISPOSAL

STORAGE: Do not use, pour, spill or store near heat or open flame. Ground all metal containers when transferring product. Protect from freezing. If stored below 32°F and crystals form, warm to 72°F for 24 hours, periodically shaking container to reconstitute.

PESTICIDE DISPOSAL: Do not contaminate water, food, or feed by storage or disposal. Pesticide wastes are toxic. Improper disposal of excess pesticide, spray mixture, or rinsate is a violation of Federal Law. If these wastes cannot be disposed of by use according to label instructions, contact your State Pesticide or Environmental Control Agency, or the Hazardous Waste representative at the nearest EPA regional office for guidance.

CONTAINER DISPOSAL

Metal Containers — Triple rinse (or equivalent) then offer for recycling or reconditioning or puncture and dispose of in a sanitary landfill or by other procedures approved by State and local authorities. **Bulk Containers** — Drain thoroughly and return to specified destination for cleaning and reuse.

STEPS TO BE TAKEN IN CASE MATERIAL IS RELEASED OR SPILLED: Eliminate ignition sources. Ventilate area. Avoid breathing vapors. Use MSHA/NIOSH self-contained breathing apparatus or air mask for large spills in confined areas. Dike the spill with inert material (sand, earth, fuller's earth, etc.) and if appropriate transfer the liquid and solid diking material to separate containers for recovery or disposal. Remove contaminated clothing promptly and wash affected skin areas with soap and water. Wash clothing before reuse. Keep out of sewers and open bodies of water. REFER TO PRECAUTIONARY STATEMENTS.

1380-B1 MADE IN U.S.A.

TRITON AG-44M®

COMPATIBILITY & BUFFERING (ACIDIFYING) AGENT

Principal Functioning Agents:	
Alkylaryl polyoxyethylene glycol phosphate ester surfactants, solubilizers	82.5%
Constituents ineffective as spray adjuvants	17.5%
TOTAL	100.0%

Calif. Reg. No. 707-50040-AA

KEEP OUT OF REACH OF CHILDREN

DANGER

NOTICE: Before using this product, read the entire Precautionary Statements, Conditions of Sale and Warranty, Directions for Use, and Storage and Disposal Instructions on side panels. If the Conditions of Sale and Warranty are not acceptable, return the product unopened within thirty days of purchase to the place of purchase.

PRECAUTIONARY STATEMENTS

PRECAUCION AL USUARIO:
Si usted no lee inglés, no use este producto hasta que la etiqueta haya sido explicada ampliamente.

HAZARDS TO HUMANS AND DOMESTIC ANIMALS

DANGER

CORROSIVE — CAUSES EYE AND SKIN DAMAGE

HARMFUL IF SWALLOWED, INHALED, OR ABSORBED

DO NOT TAKE INTERNALLY

CAUSES ALLERGIC SKIN REACTION

COMBUSTIBLE

Do not get in eyes, on skin or clothing.
Wear chemical splash goggles and face shield, and impervious rubber gloves when handling.
Avoid breathing vapor or mist.
Use with adequate ventilation.
Keep container closed when not in use.
Keep away from heat and flame.
Wash thoroughly after handling.
Remove contaminated clothing promptly.
Thoroughly launder clothing before reuse.

STATEMENT OF PRACTICAL TREATMENT

IF IN EYES: Flush eyes with a large amount of water for at least 15 minutes. Get prompt medical attention.

IF ON SKIN: Flood exposed skin areas with water. Wash affected skin areas with soap and water. If irritation develops, get prompt medical attention.

IF SWALLOWED: Dilute by giving two glasses of water to drink. Get prompt medical attention. Do not induce vomiting or give anything by mouth to an unconscious person.

IF INHALED: Move subject to fresh air. If not breathing, give artificial respiration and get prompt medical attention. If breathing is difficult, give oxygen.

CONDITIONS OF SALE AND WARRANTY

Rohm and Haas warrants that the product conforms to its chemical description and is reasonably fit for the purpose stated on the label only when used in accordance with label directions under normal conditions of use. **ROHM AND HAAS MAKES NO OTHER EXPRESS OR IMPLIED WARRANTIES EITHER OF MERCHANTABILITY OR FITNESS FOR A PARTICULAR USE.** Handling, storage and use of the product by Buyer or User are beyond the control of Rohm and Haas and Seller. Risks such as crop injury, ineffectiveness or other unintended consequences resulting from, but not limited to, weather or soil conditions, presence of other materials, disease, pests, drift to other crops or property or failure to follow label directions will be assumed by the Buyer or User. **IN NO CASE WILL ROHM AND HAAS OR SELLER BE HELD LIABLE FOR CONSEQUENTIAL, SPECIAL OR INDIRECT DAMAGES RESULTING FROM THE HANDLING, STORAGE OR USE OF THIS PRODUCT.**

DIRECTIONS FOR USE

TRITON AG-44M is a spray tank additive for use as a compatibility and/or buffering (acidifying) agent. As a compatibility agent, TRITON AG-44M will improve the stability and compatibility of pesticides (alone or in tank mixtures) when using liquid fertilizers as a spray carrier. As a buffering agent, TRITON AG-44M will reduce the pH of alkaline waters to maximize the effectiveness of pesticides that are affected by high pH waters. In addition, the use of TRITON AG-44M will assist in improving the wetting characteristics of the spray solution.

BUFFERING

TRITON AG-44M will reduce the pH of alkaline waters to maximize the effectiveness of pesticides that are affected by high pH waters. A low-pH solution will enhance the performance of many agricultural chemicals, but can be detrimental to others. Refer to the pesticide label.

USE RATE

Determine the initial pH of the spray solution. If a lower pH is recommended for the pesticide to be applied, add TRITON AG-44M at the rate of three fluid ounces per 100 gallons of spray solution. Recheck the pH. If further reduction of the pH is required, add one ounce of TRITON AG-44M per 100 gallons of spray solution until the desired pH range is achieved.

Many factors can influence the pH of the spray solution, therefore, it is recommended that the pH be monitored during the addition of TRITON AG-44M to avoid overadjustment of the spray solution pH.

COMPATIBILITY

When used as a compatibility agent, TRITON AG-44M will improve the stability and compatibility of pesticides (alone or in tank mixtures) when using liquid fertilizers as a spray carrier. TRITON AG-44M should, under most conditions, be added at the suggested use rates to the spray solution (liquid fertilizer) in the spray tank and thoroughly mixed prior to the addition of the pesticide. The required amount of pesticide should be added with agitation and allowed to mix for several minutes before application. If the pesticide to be used is a wettable powder or flowable formulation, the pesticide should be slurried with TRITON AG-44M before addition to the spray tank. The use of a premix with TRITON AG-44M is recommended when a pesticide formulation is added to high-phosphate liquid fertilizers.

Always check the compatibility of liquid fertilizer/pesticide mixtures in small containers before application because variations in fertilizers may affect the use rate of TRITON AG-44M compatibility agent.

- **Wettable Powder or Flowable Formulations**—Place the pesticide formulation into suspension and then add TRITON AG-44M. Add liquid fertilizer to the suspension contained in the spray or nurse tank, or add as a premix to a spray or nurse tank fertilizer solution.
- **Liquid Formulations**—Add TRITON AG-44M to the spray solution (liquid fertilizer) in the spray or nurse tank, at the suggested label use rates. Thoroughly mix prior to the addition of the liquid pesticide formulation.
- **Phosphate Fertilizer**—TRITON AG-44M should be premixed with the pesticide formulation prior to addition to high-phosphate liquid fertilizers.
- **Unknown Compatibility**—If the compatibility of the intended pesticide combination or liquid fertilizer/pesticide mixture is unknown, prepare sample mixtures in small containers for observation and adjustment of the TRITON AG-44M use rate. (See "Premix Test Procedure" section.)

USE RATES

One to four pints of TRITON AG-44M per 100 gallons of spray solution are normally required to improve the stability of spray mixtures:

High-Nitrogen Liquid Fertilizers and Pesticides	1-3 pints per 100 gallons of fertilizer
Mixed Liquid Fertilizers and Pesticides	2-3 pints per 100 gallons of fertilizer
Liquid Fertilizers and Multiple Pesticide Combinations	2-4 pints per 100 gallons of fertilizer

The addition of TRITON AG-44M as a compatibility agent will also reduce the pH of the spray solution. A low pH will enhance the performance of many agricultural chemicals, but can be detrimental to others. Refer to the pesticide label.

Agitation—Continuous spray or nurse tank agitation is recommended. If spray solution has been allowed to stand, agitate the spray solution prior to application.

PREMIX TEST PROCEDURE

To determine the exact amount of TRITON AG-44M needed and the proper order of addition for each tank mix component, proceed as follows:

1. Pour one pint (473 ml) of the spray solution (liquid fertilizer) into a suitable container.
2. Add one-eighth (1/8) to one-half (1/2) teaspoon (0.6 to 2.4 ml) of TRITON AG-44M to the spray solution and thoroughly mix.
3. Add the recommended amount of pesticide to the spray solution mixture, agitate and observe stability. Stability for 60 minutes or more indicates a proper use level.
4. When wettable powder/flowable formulations are added to high-phosphate solutions, TRITON AG-44M should be added to the pesticide first, and the resultant mixture added to the liquid fertilizer before observing stability.

CONVERSION FACTORS:

- One teaspoon (4.9 ml) of liquid pesticide per one pint of spray solution (liquid fertilizer) equals one gallon pesticide per 100 gallons spray solution.
- One-eighth (1/8) teaspoon (0.6 ml) TRITON AG-44M compatibility agent per one pint of spray solution (liquid fertilizer) equals one pint TRITON AG-44M per 100 gallons spray solution.
- One teaspoon (0.56 grams) of dry pesticide per one pint of spray solution (liquid fertilizer) equals one pound pesticide per 100 gallons spray solution.

NOTE: These directions for use do not constitute a recommendation to use a pesticide in any manner inconsistent with the product's EPA-registered label.

SPILLS AND LEAKS

Eliminate ignition sources. Ventilate area. Avoid breathing vapor. Wear respirator suitable for vapor concentration encountered (MSHA/NIOSH-approved or equivalent). Wear full protective equipment. (Acid-resistant clothing, gloves and boots, chemical splash goggles and face shield.) Dike and contain spill with inert material (e.g., sand, earth). Floor may be slippery; use care to avoid falling. Remove contaminated clothing promptly and wash affected skin areas with soap and water. Thoroughly launder clothing before reuse. Do not take clothing home to be laundered. Transfer liquid to containers for recovery or disposal and solid diking material to separate containers for disposal. Keep spills and cleaning runoffs out of municipal sewers and open bodies of water. When discarded, this material is a hazardous waste. RCRA#D-001 (ignitable); reportable quantities 100 lbs. (40 CFR 302) "Superfund." Incinerate liquid and contaminated diking material, in accordance with local, state and federal regulations.

IN CASE OF FIRE

Use water, dry chemical, CO_2 or alcohol foam extinguishers.

Wear self-contained breathing apparatus (pressure-demand, MSHA/NIOSH-approved or equivalent) and full protective gear.

Container Disposal—Empty containers may contain product residue (vapor/liquid) and can be hazardous. Observe all label precautions. Containers should be triple rinsed (or equivalent), punctured and then disposed of in a sanitary landfill or by incineration, if permitted by state and local authorities.

TRITON AG-44M is a registered trademark of Rohm and Haas Company

9809-R3

6-1902

TRITON AG-98®

NO FOAM
SPREADER ACTIVATOR

PRINCIPAL FUNCTIONING AGENTS:

Alkylaryl polyoxyethylene glycols	80%
Constituents ineffective as spray adjuvants	20%
TOTAL	100%

Calif. Reg. No. 707-50036-AA

KEEP OUT OF REACH OF CHILDREN
WARNING

Before using this product, read the entire Precautionary Statements, Conditions of Sale and Warranty, Directions for Use, and Storage and Disposal Instructions on back panel. If the Conditions of Sale and Warranty are not acceptable, return the product unopened within thirty days of purchase to the place of purchase.

Triton AG-98 is a registered trademark of Rohm and Haas Company.

PRECAUTIONARY STATEMENTS

HAZARDS TO HUMANS AND DOMESTIC ANIMALS

PRECAUCION AL USUARIO: Si usted no lee inglés, no use este producto hasta que la etiqueta haya sido explicada ampliamente.

Avoid contact with eyes, skin and clothing as severe eye irritation and possible eye damage, or skin irritation may result. Harmful if swallowed. Vapor harmful if inhaled. Avoid breathing vapor or mist. Use with adequate ventilation. Do not take internally. Wear chemical splash goggles and impervious rubber gloves when handling. Wash thoroughly after handling. Remove and launder clothing before reuse. Keep container closed when not in use. Combustible — Keep away from heat, sparks and flame.

STATEMENT OF PRACTICAL TREATMENT

IF IN EYES. Flush eyes with a large amount of water for at least 15 minutes. Get prompt medical attention. **IF ON SKIN:** Wash skin thoroughly with soap and water. Remove and wash clothing before reuse. **IF SWALLOWED:** Dilute by giving two glasses of water to drink. Get prompt medical attention. Do not induce vomiting or give anything by mouth to an unconscious person. **IF INHALED:** Move subject to fresh air. Give artificial respiration if breathing has stopped; give oxygen if breathing is difficult, and get prompt medical attention.

SPILLS AND LEAKS: Eliminate ignition sources. Ventilate area. Avoid breathing vapor. Floor may be slippery; use care to avoid falling. Dike and contain spill with inert material (e.g., sand, earth). Transfer liquid to containers for recovery or disposal and solid diking material to separate containers for disposal. Remove contaminated clothing promptly and wash affected skin areas with soap and water. Keep spills and cleaning runoffs out of municipal sewers and open bodies of water.

IN CASE OF FIRE: Use water, dry chemical, CO_2 or alcohol foam extinguishers.

Wear respirator (pressure-demand, self-contained breathing apparatus, MSHA/NIOSH-approved or equivalent) and full protective gear.

CONTAINER DISPOSAL: Empty containers may contain product residue (vapor and/or liquid) and can be hazardous. Observe all label precautions. Containers should be triple rinsed (or equivalent), punctured and then disposed of in a sanitary landfill or by incineration, if permitted by state and local authorities.

CONDITIONS OF SALE AND WARRANTY

Rohm and Haas warrants that the product conforms to its chemical description and is reasonably fit for the purpose stated on the label only when used in accordance with label directions under normal conditions of use. **ROHM AND HAAS MAKES NO OTHER EXPRESS OR IMPLIED WARRANTIES EITHER OF MERCHANTABILITY OR FITNESS FOR A PARTICULAR USE.** Handling, storage and use of the product by Buyer or User are beyond the control of Rohm and Haas and Seller. Risks such as crop injury, ineffectiveness or other unintended consequences resulting from, but not limited to, weather or soil conditions, presence of other materials, disease, pests, drift to other crops or property or failure to follow label directions will be assumed by the Buyer or User. **IN NO CASE WILL ROHM AND HAAS OR SELLER BE HELD LIABLE FOR CONSEQUENTIAL, SPECIAL OR INDIRECT DAMAGES RESULTING FROM THE HANDLING, STORAGE OR USE OF THIS PRODUCT.**

DIRECTIONS FOR USE

SHAKE WELL BEFORE USING

TRITON AG-98 is a general purpose low foam nonionic spray adjuvant for use on plants to improve the effectiveness of insecticides, herbicides (contact and aquatic), fungicides, and defoliants. TRITON AG-98 increases contact activity and improves spray penetration by providing more uniform spray coverage without excessive runoff of the spray.

HERBICIDES

TRITON AG-98 is recommended for use at a rate of 0.5 to 2 pints per 100 gallons of spray with most postemergence herbicides that recommend the use of a nonionic surfactant (unless otherwise directed by the herbicide manufacturer).

DEFOLIANTS

(Use with magnesium and sodium chlorate and with chlorate/borate defoliants and herbicides.)

TRITON AG-98 should be applied at a rate of 2 pints per each 100 gallons of dilute spray when spray application is at the rate of 10 or more gallons per acre. Use 4 pints of TRITON AG-98 per 100 gallons of spray when the application rate is less than 10 gallons per acre.

INSECTICIDES, FUNGICIDES, ACARICIDES

TRITON AG-98 may be used at a rate of 2 to 8 ounces per 100 gallons of spray to enhance the wetting characteristics of an insecticide, fungicide or acaricide application.

SOIL PENETRANT

TRITON AG-98 should be applied at a rate of 8 ounces per 1,000 square feet of lawn. Applications should be made in a minimum spray volume of 12 gallons of water per acre.

NOTE: Due to variations in materials, application equipment, waters and other conditions, the optimum use rate may vary. Sufficient wetting agent should be added so that the spray uniformly wets the surface being treated and allows an even deposit of the active chemical.

8550-R7

TRITON® B-1956®

SPREADER STICKER

PRINCIPAL FUNCTIONING AGENTS:

Modified phthalic glycerol alkyd resin	77%
Constituents ineffective as spray adjuvants	23%
TOTAL	100%

KEEP OUT OF REACH OF CHILDREN

CAUTION

Before using this product, read the entire Precautionary Statements, Conditions of Sale and Warranty and Directions for Use, and Disposal Instructions on side and back panels. If the Conditions of Sale and Warranty are not acceptable, return the product unopened within thirty days of purchase to the place of purchase.

Calif. Reg. No. 707-50032-AA

PRECAUTIONARY STATEMENTS

HAZARDS TO HUMANS AND DOMESTIC ANIMALS

PRECAUCION AL USUARIO: Si usted no lee inglés, no use este producto hasta que la etiqueta haya sido explicada ampliamente.

CONTAINS CHLORINATED SOLVENT

Avoid contact with eyes, skin and clothing, as severe eye irritation or skin irritation may result.
Harmful if swallowed.
Vapor harmful if inhaled.
Avoid breathing vapor or mist.
Use with adequate ventilation.
Do not take internally.
Wear chemical splash goggles and impervious rubber gloves when handling.
Wash thoroughly after handling.
Remove and launder clothing before reuse.
Keep container closed when not in use.

STATEMENT OF PRACTICAL TREATMENT

IF IN EYES: Flush eyes with a large amount of water for at least 15 minutes. Get prompt medical attention.

IF ON SKIN: Wash skin thoroughly with soap and water. Remove and wash clothing before reuse.

IF SWALLOWED: Dilute by giving two glasses of water to drink. Get prompt medical attention. Do not induce vomiting or give anything by mouth to an unconscious person. See "Note to Physician."

IF INHALED: Move subject to fresh air. Give artificial respiration if breathing has stopped; give oxygen if breathing is difficult, and get prompt medical attention. See "Note to Physician."

Note to Physician: Overexposure to many chlorinated solvents, especially if accompanied by anoxia, can temporarily increase cardiac irritability. Maintain adequate oxygenation until recovery. Avoid sympathomimetic amines, such as epinephrine, which may precipitate arrhythmias.

SPILLS AND LEAKS: Wear respirator suitable for vapor concentration encountered (MSHA/NIOSH-approved or equivalent). Wear protective clothing. Dike and contain spill with inert material (e.g., sand, earth). Transfer liquid to containers for recovery or disposal and solid diking material to separate containers for disposal. Remove contaminated clothing promptly and wash affected skin areas with soap and water. Keep spills and cleaning runoffs out of municipal sewers and open bodies of water.

IN CASE OF FIRE: Use water, dry chemical, or CO_2 extinguishers. Wear respirator, self-contained breathing apparatus (pressure-demand, MSHA/NIOSH-approved or equivalent) and full protective gear. Use water spray to cool containers.

CONTAINER DISPOSAL: Empty containers may contain product residue (vapor/or liquid) and can be hazardous. Observe all label precautions. Containers should be triple rinsed (or equivalent), punctured and then disposed of in a sanitary landfill or by incineration, if permitted by state and local authorities.

CONDITIONS OF SALE AND WARRANTY

Rohm and Haas warrants that the product conforms to its chemical description and is reasonably fit for the purpose stated on the label only when used in accordance with label directions under normal conditions of use. **ROHM AND HAAS MAKES NO OTHER EXPRESS OR IMPLIED WARRANTIES EITHER OF MERCHANTABILITY OR FITNESS FOR A PARTICULAR USE.** Handling, storage and use of the product by Buyer or User are beyond the control of Rohm and Haas and Seller. Risks such as crop injury, ineffectiveness or other unintended consequences resulting from, but not limited to, weather or soil conditions, presence of other materials, disease, pests, drift to other crops or property or failure to follow label directions will be assumed by the Buyer or User. **IN NO CASE WILL ROHM AND HAAS OR SELLER BE HELD LIABLE FOR CONSEQUENTIAL, SPECIAL OR INDIRECT DAMAGES RESULTING FROM THE HANDLING, STORAGE OR USE OF THIS PRODUCT.**

DIRECTIONS FOR USE

TRITON B-1956 is a water-dispersible, resin-based nonionic surfactant (not a water-soluble detergent-type wetting agent) that resists re-wetting and removal by rain. TRITON B-1956 is effective as a spreading, depositing and sticking agent to improve the performance of INSECTICIDES, MITICIDES, FUNGICIDES, DEFOLIANTS, and HERBICIDES.

TRITON B-1956 is beneficial when used on smooth and waxy or fuzzy (hairy) foliage and fruit. TRITON B-1956 will also improve spray coverage on fruit or foliage that is covered by dust or honeydew, webbed by spider mites or infected with powdery mildew and on weeds and brush. TRITON B-1956 will assist in the spray deposit/penetration on hard-to-wet insects such as mealy bugs and scales.

Growers are urged to observe closely the spray deposit pattern during spraying and after it has dried. If the pattern is blotchy and uneven, more TRITON B-1956 is needed. If the spray is running to lower leaf margins, "eyebrowing," or collecting in drop spots on fruit, the amount of TRITON B-1956 should be reduced.

DOSAGE

Recommendations for use of TRITON B-1956 depend on the crop, the pest, the kind of spray equipment, volume, pressure and droplet size of the spray and the pesticide or mixture being used.

As a general guide, 1 to 2 ounces are used on moderately easy-to-wet plants, 2 to 4 ounces on hard-to-wet plants, and 4 to 8 ounces on very hard-to-wet plants per 100 gallons of dilute full-volume spray. As sprays are concentrated to 2X, 10X or 20X, the amount of TRITON B-1956 should be increased in about the same proportion as the pesticide. The chart below will serve as a general guide for some of the more common types of spray application.

GALLONS PER ACRE

	Field and row crops	Vineyards, hop yards	Tree fruits and nuts	TRITON B-1956 per 100 gallon spray
Dilute	100-160	150-300	600-1000	1-2 oz.
2X	50-80	75-150	300-500	2-4 oz.
3X	35-55	50-100	200-330	3-6 oz.
8X	12-20	20-40	75-125	8-16 oz.
10X	10-16	15-30	60-100	10-20 oz.
20X	5-8	8-16	30-50	1 pt.-1 qt.
30X	3-5	5-10	20-30	1½ pt.-1½ qt.

MISCELLANEOUS USES

To improve the wetting properties of insecticidal sprays on livestock, especially in hard water areas, use 1 to 2 ounces TRITON B-1956 per 100 gallons of spray.

MIXING

TRITON-B 1956 should be thoroughly dispersed in water before being added to the spray tank.

COMPATIBILITY

TRITON B-1956 can be used with wettable powder, emulsifiable concentrate or solution formulations of most pesticides now on the market.

TRITON® and B-1956® are registered trademarks of Rohm and Haas Company.

9606-R3

TRITON CS-7®

SPREADER BINDER

PRINCIPAL FUNCTIONING AGENTS:

Blend of alkylaryl polyethoxylate and sodium salt of alkylsulfonatedalkylate 60%

Constituents ineffective as spray adjuvants 40%

TOTAL 100%

Calif. Reg. No. 707-50033-AA

KEEP OUT OF REACH OF CHILDREN

WARNING

Before using this product, read the entire Precautionary Statements, Conditions of Sale and Warranty, Directions for Use, and Storage and Disposal Instructions on back panel. If the Conditions of Sale and Warranty are not acceptable, return the product unopened within thirty days of purchase to the place of purchase.

Triton CS-7 is a registered trademark of Rohm and Haas Company.

PRECAUTIONARY STATEMENTS

HAZARDS TO HUMANS AND DOMESTIC ANIMALS

PRECAUCION AL USUARIO: Si usted no lee inglés, no use este producto hasta que la etiqueta haya sido explicada ampliamente.

Avoid contact with eyes, skin and clothing as severe eye irritation and possible eye damage, or skin irritation may result. Harmful if swallowed. Vapor harmful if inhaled. Avoid breathing vapor or mist. Use with adequate ventilation. Do not take internally. Wear chemical splash goggles and impervious rubber gloves when handling. Wash thoroughly after handling. Remove and launder clothing before reuse. Keep container closed when not in use. Combustible — Keep away from heat, sparks and flame.

STATEMENT OF PRACTICAL TREATMENT

IF IN EYES: Flush eyes with a large amount of water for at least 15 minutes. Get prompt medical attention. **IF ON SKIN:** Wash skin thoroughly with soap and water. Remove and wash clothing before reuse. **IF SWALLOWED:** Dilute by giving two glasses of water to drink. Get prompt medical attention. Do not induce vomiting or give anything by mouth to an unconscious person. **IF INHALED:** Move subject to fresh air. Give artificial respiration if breathing has stopped; give oxygen if breathing is difficult, and get prompt medical attention.

SPILLS AND LEAKS: Eliminate ignition sources. Ventilate area. Avoid breathing vapor. Wear respirator suitable for vapor concentration encountered (MSHA/NIOSH-approved or equivalent). Floor may be slippery, use care to avoid falling. Dike and contain spill with inert material (e.g., sand, earth). Scoop or shovel solid material into a suitable container for recovery or disposal. Transfer liquid to containers for recovery or disposal and solid diking material to separate containers for disposal. Remove contaminated clothing promptly and wash affected skin areas with soap and water. Keep spills and cleaning runoffs out of municipal sewers and open bodies of water.

IN CASE OF FIRE: Use water, dry chemical, CO_2 or alcohol foam extinguishers.

Wear respirator (pressure-demand, self-contained breathing apparatus, MSHA/NIOSH-approved or equivalent) and full protective gear.

CONTAINER DISPOSAL: Empty containers may contain product residue (vapor and/or liquid) and can be hazardous. Observe all label precautions. Containers should be triple rinsed (or equivalent), punctured and then disposed of in a sanitary landfill or by incineration, if permitted by state and local authorities.

CONDITIONS OF SALE AND WARRANTY

Rohm and Haas warrants that the product conforms to its chemical description and is reasonably fit for the purpose stated on the label only when used in accordance with label directions under normal conditions of use. **ROHM AND HAAS MAKES NO OTHER EXPRESS OR IMPLIED WARRANTIES EITHER OF MERCHANTABILITY OR FITNESS FOR A PARTICULAR USE.** Handling, storage and use of the product by Buyer or User are beyond the control of Rohm and Haas and Seller. Risks such as crop injury, ineffectiveness or other unintended consequences resulting from, but not limited to, weather or soil conditions, presence of other materials, disease, pests, drift to other crops or property or failure to follow label directions will be assumed by the Buyer or User. **IN NO CASE WILL ROHM AND HAAS OR SELLER BE HELD LIABLE FOR CONSEQUENTIAL, SPECIAL OR INDIRECT DAMAGES RESULTING FROM THE HANDLING, STORAGE OR USE OF THIS PRODUCT.**

DIRECTIONS FOR USE

TRITON CS-7 is a spray adjuvant for use in concentrate sprays of insecticides or fungicides applied by aircraft or low volume spray equipment. TRITON CS-7 is also useful when applied to ornamental turf grass areas as a soil penetrant and dew suppressant.

SPRAY ADJUVANT

TRITON CS-7 is a spreader-binder designed specifically for use as a spray adjuvant in concentrate and low volume sprays. TRITON CS-7 is suggested for use on all crops receiving ground or aerial applications of an insecticide or fungicide.

When applying low volume or concentrate spray applications, complete coverage of the foliage is largely dependent on good initial spray coverage and redeposition of the spray deposit over the leaf surface. TRITON CS-7 provides uniform spray deposits and significantly aids in redistribution of the spray deposit and so extends the effective area of protection with minimum loss during weathering.

Add TRITON CS-7 to the nurse tank or just prior to complete filling of the spray tank and after the fungicide or insecticide has been added and is well dispersed.

USE RATE

Add TRITON CS-7 at a rate of 1 to 2 pints per 100 gallons of spray mix. Highly concentrated spray mixes require the 2 pint use rate.

This suggested rate is considered adequate for most uses. Some pesticide formulations or some crops may require more or less TRITON CS-7 than the above. Check the deposit after the initial spray load and adjust the concentration if necessary.

SOIL PENETRANT AND DEW SUPPRESSANT

When applied as a dew suppressant on ornamental turf grass. TRITON CS-7 is suggested for use at an application rate of 4 ounces per 1,000 square feet. For spot applications or use as a soil penetrant, the suggested use rate is 8 ounces per 1,000 square feet. Applications should be made in a minimum spray volume of 12 gallons of water per acre.

Sprays should be applied uniformly over the turf area to insure uniform distribution of TRITON CS-7 to the grass blades. Experience will indicate when repeat applications are needed and the best appropriate dosage of TRITON CS-7 to use in individual cases.

5820-R11

Sandoz Crop Protection Corporation

1300 East Touhy Avenue
Des Plaines, IL 60018
(312) 699-1616

BANVEL®

HERBICIDE

ACTIVE INGREDIENTS:

Dimethylamine salt of dicamba (3,6-dichloro-o-anisic acid)*	48.2%
Dimethylamine salts of related acids	12.0%
INERT INGREDIENTS:	39.8%
TOTAL	100.0%

*This product contains 40.0% 3,6-dichloro-o-anisic acid (dicamba) or 4 pounds per gallon.

EPA Reg. No. 55947-1 EPA Est. No. 55947-TX-1

KEEP OUT OF REACH OF CHILDREN

WARNING

PRECAUTIONARY STATEMENTS

HAZARDS TO HUMANS AND DOMESTIC ANIMALS

WARNING

Causes eye irritation. Do not get in eyes, on skin, or on clothing. Harmful if swallowed. Avoid breathing spray mist. Wash thoroughly after handling. In case of contact, wash skin with soap and water; for eyes, flush with water for 15 minutes and get medical attention.

ENVIRONMENTAL HAZARDS

Keep out of lakes, streams or ponds. Do not contaminate water by cleaning of equipment or disposal of wastes. Apply this product only as directed on label.

DIRECTIONS FOR USE

It is a violation of Federal Law to use this product in a manner inconsistent with its labeling.

For additional copies of the Specimen Label, write to the manufacturer.

**BEFORE USING BANVEL HERBICIDE
READ AND FOLLOW THE PRECAUTIONS**

IMPORTANT

The following directions apply to all uses of BANVEL® Herbicide. Additional precautions and restrictions will be found in each specific use section.

Do not contaminate irrigation ditches or water used for domestic purposes.

SENSITIVE CROP PRECAUTIONS: BANVEL Herbicide may cause injury to desirable trees and plants, particularly beans, cotton, flowers, fruit trees, grapes, ornamentals, peas, potatoes, soybeans, sunflowers, tobacco, tomatoes and other broadleaf plants when contacting their roots, stems or foliage. These plants are most sensitive to BANVEL Herbicide during their development or growing stage. FOLLOW THE PRECAUTIONS LISTED BELOW WHEN USING BANVEL HERBICIDE.

- Do not treat areas where either possible downward movement into the soil or surface washing may cause contact of BANVEL Herbicide with the roots of desirable plants such as trees and shrubs.
- Avoid making applications when spray particles may be carried by air currents to areas where sensitive crops and plants are growing. Do not spray near sensitive plants if wind is gusty or in excess of 5 mph and moving in the direction of nearby sensitive crops. However, always make applications when there is some air movement to determine the direction and distance of possible spray drift. Leave an adequate buffer zone between area to be treated and sensitive plants. Coarse sprays are less likely to drift out of the target area than fine sprays. Agriculturally approved drift-reducing additives may be used.
- Do not apply BANVEL Herbicide in the vicinity of sensitive crops when the temperature on the day of application is expected to exceed 85°F as drift is more likely to occur.
- Do not use aerial equipment to apply BANVEL Herbicide when sensitive crops and plants are growing in the vicinity of area to be treated.
- To avoid injury to desirable plants, equipment used to apply BANVEL Herbicide should be thoroughly cleaned (see PROCEDURE FOR CLEANING SPRAY EQUIPMENT) before reusing to apply any other chemicals.

All crop uses of BANVEL Herbicide are intended for a normal growing interval between planting and harvest. No crop rotation restrictions exist if normal harvest of treated crop has occurred. If this interval is shortened, such as in cover crops that will be plowed under, do not follow up with the planting of a sensitive crop.

Crops growing under stress conditions such as drought, poor fertility, or foliar damage due to hail, wind or insects, can exhibit various injury symptoms that may be more pronounced if herbicides are applied.

Consult your local or state authorities for possible application restrictions and advice concerning these and other special local use situations. Tank mix recommendations are for use ony in states where the tank mix product and application site are registered.

Do not apply this product through any type of irrigation system.

PROCEDURE FOR CLEANING SPRAY EQUIPMENT

The steps listed below are suggested for thorough cleaning of spray equipment following applications of BANVEL Herbicide or tank mixes of BANVEL Herbicide plus 2,4-D amine.

1) Hose down thoroughly the inside as well as outside surfaces of equipment while filling the spray tank half full of water. Flush by operating sprayer until the system is purged of the rinse water.
2) Fill tank with water while adding 1 quart of household ammonia for every 25 gallons of water. Operate the pump to circulate the ammonia solution through the sprayer system for 15 to 20 minutes and discharge a small amount of the ammonia solution through the boom and nozzles. Let the solution stand for several hours, preferably overnight.
3) Flush the solution out of the spray tank through the boom.
4) Remove the nozzles and screens and flush the system with two full tanks of water.

The steps listed below are suggested for thorough cleaning of spray equipment used to apply BANVEL Herbicide as a tank mix with wettable powders (WP), emulsifiable concentrates (EC), or other types of water-dispersible formulations. BANVEL Herbicide tank mixes with water-dispersible formulations require the use of a water/detergent rinse.

5) Complete step 1.
6) Fill tank with water while adding 2 lbs. of detergent for every 40 gallons of water. Operate the pump to circulate the detergent solution through the sprayer system for 5 to 10 minutes and discharge a small amount of the solution through the boom and nozzles. Let the solution stand for several hours, preferably overnight.
7) Flush the detergent solution out of the spray tank through the boom.
8) Repeat step 1, and follow with steps 2, 3, and 4.

MIXING AND APPLICATION

UNLESS OTHERWISE SPECIFIED UNDER THE INDIVIDUAL USE HEADINGS OF THIS BOOKLET, THE FOLLOWING DIRECTIONS APPLY TO ALL CROP AND NON-CROP USES OF BANVEL HERBICIDE. REFER TO INDIVIDUAL USE SECTIONS FOR ADDITIONAL PRECAUTIONS, RESTRICTIONS, APPLICATION RATES AND TIMINGS.

BANVEL Herbicide is a water-soluble formulation that can be applied using water or sprayable fluid fertilizer as the carrier. If a fluid fertilizer is to be used, a compatibility test (see COMPATIBILITY TEST) should be made prior to tank mixing.

Ground or aerial application equipment which will give good spray coverage of weed foliage should be used. HOWEVER, DO NOT USE AERIAL APPLICATION EQUIPMENT IF SENSITIVE CROPS ARE GROWING IN THE VICINITY OF THE AREA TO BE TREATED.

Apply 5 to 50 gallons of diluted spray per treated acre when using ground application equipment, or 3 to 10 gallons of diluted spray per treated acre when using aerial application equipment. Use the higher level of the listed spray volumes when treating dense or tall vegetation. Use coarse sprays.

Select nozzles designed to produce minimal amounts of fine spray particles. Spray with nozzles as close to the weeds as is practical for good weed coverage.

BANVEL Herbicide should not be applied during periods of gusty wind or when wind is in excess of 15 mph as uneven spray coverage may occur.

Avoid disturbing (e.g. cultivating or mowing) treated areas for at least 7 days following application.

BAND TREATMENTS

BANVEL Herbicide may be applied as a band treatment. Use the formulas below to determine the appropriate rate and volume per treated acre.

$$\frac{\text{Band width in inches}}{\text{Row width in inches}} \times \text{Broadcast RATE per treated acre} = \text{Band RATE per treated acre}$$

$$\frac{\text{Band width in inches}}{\text{Row width in inches}} \times \begin{array}{c}\text{Broadcast}\\ \text{VOLUME}\\ \text{per treated acre}\end{array} = \begin{array}{c}\text{Band VOLUME}\\ \text{per treated acre}\end{array}$$

COMPATIBILITY TEST

Before mixing in the spray tank, it is advisable to test compatibility by mixing all components in a small container in proportionate quantities (see following table).

Amount of Herbicide to Add to One Pint of Spray Carrier (Assuming Volume is 25 Gallons per Acre)

Herbicide Formulations	Rate Per Acre	Level Teaspoons
Dry	1 lb.	1½
Liquid	1 pt.	½

If herbicide(s) do not ball-up or form flakes, sludge, gels, oily films or layers, or other precipitates, then the tested spray mix is compatible. Usually, incompatibility in any of the above described forms will occur within 5 minutes after mixing.

If components are incompatible, the use of a compatibility agent is recommended. Rerun the above COMPATIBILITY TEST with a suitable compatibility agent (¼ teaspoon is equivalent to 2 pints per 100 gallons of fluid fertilizer).

GENERAL WEED LIST

This is a general list of weeds which may be treated with BANVEL Herbicide in accordance with this label as recommended under the rates and timing sections of the Individual Use Headings. Proper usage of this product will give control or growth suppression of many ANNUAL, BIENNIAL, and PERENNIAL broadleaf weeds, and many WOODY brush and vine species including:

ANNUALS

Amaranth, Spiny
  (Spiny Pigweed)
Aster, Slender,
Bedstraw
Beggarweed, Florida
Broomweed, Common
Buckwheat, Wild
Buffalobur
Burclover, California
Burcucumber
Buttercup, Roughseed
Carpetweed
Catchfly, Nigthflowering
Chamomile, Corn
Chickweed, Common
Clovers (Annual)
Cockle, Corn
Cockle, Cow
Cocklebur, Common
Croton, Tropic
Croton, Woolly
Cucumber, Wild
Daisy, English
Eveningprimrose,
  Cutleaf
Fleabane, Annual
Goosefoot, Nettleleaf
Henbit
Jimsonweed
Knotweed
Kochia
Ladysthumb
Lambsquarter, Common
Lambsquarters
  (triazine resistant)
Lettuce, Prickly
Mallow, Common
Mallow, Venice
Mayweed
Morningglory, Ivyleaf
Morningglory, Tall
Mustard, Tansy
Mustard, Wild
Mustard (Yellowtops)
Nightshade, Black
Pennycress, Field
  (Fanweed,
  Frenchweed,
  Stinkweed)
Pepperweed, Virginia
  (Peppergrass)
Pigweed, Prostrate
Pigweed, Redroot
  (Carelessweed)
Pigweed, Rough
Pigweed, Smooth
Pigweed
  (triazine resistant)
Pigweed, Tumble
Poorjoe
Puncturevine
Purslane, Common
Pusley, Florida
Radish, Wild
Ragweed, Common
Ragweed, Giant
  (Buffaloweed)
Rubberweed, Bitter
  (Bitterweed)
Sesbania, Hemp
Shepherdpurse
Sicklepod
Sida, Prickly
  (Teaweed)
Smartweed, Green
Smartweed, Pennsylvania
Sneezeweed, Bitter
Sowthistle, Annual
Sowthistle, Spiny
Spikeweed, Common
Spurge, Prostrate
Spurry, Corn
Starbur, Bristly
Sumpweed, Rough
Sunflower, Common
  (Wild)
Sunflower, Volunteer
Thistle, Russian
Velvetleaf
Waterhemp
Waterprimrose, Winged
Wormwood, Annual

BIENNIALS

Burdock, Common
Carrot, Wild
  (Queen Anne's Lace)
Cockle, White
Eveningprimrose,
  Common
Geranium, Carolina
Gromwell
Knapweed, Diffuse
Knapweed, Spotted
Mallow, Dwarf
Plantain, Bracted
Ragwort, Tansy
Starthistle, Yellow
Sweetclover
Teasel
Thistle, Bull
Thistle, Milk
Thistle, Musk
Thistle, Plumeless

PERENNIALS

* Alfalfa
Artichoke, Jerusalem
Aster, Spiny
Aster, Whiteheath
Bedstraw, Smooth
Bindweed, Field
Bindweed, Hedge
Blueweed, Texas
* Bursage
  (Bur Ragweed)
  (Lakeweed)
  (Povertyweed)
Bursage, Woolyleaf
  (Lakeweed)
* Buttercup, Tall
Campion Bladder
Chickweed, Field
Chickweed, Mouseear
Chicory
* Clover, Hop
* Dandelion, Common
* Dock, Broadleaf
  (Bitterdock)
Dock, Curly
Dogbane, Hemp
* Dogfennel
  (Cypressweed)
Fern, Bracken
Garlic, Wild
Goldenrod, Canada
Goldenrod, Missouri
Goldenweed, Common
Hawkweed
Horsenettle, Carolina
Ironweed
Knapweed, Black
Knapweed, Russian
Mare's Tail
  (Horseweed)
Milkweed, Climbing
Milkweed, Common
Milkweed, Honeyvine
Milkweed,
  Western Whorled
Nettle, Stinging
Nightshade, Silverleaf
  (White Horsenettle)
Onion, Wild
* Plantain, Broadleaf
* Plantain, Buckhorn
Pokeweed
Ragweed, Western
Redvine
Smartweed, Swamp
Snakeweed, Broom
*Sorrel, Red
  (Sheep Sorrel)
Sowthistle
Sowthistle, Perennial
Spurge, Leafy
Sundrop, Halfshrub
  (Eveningprimrose)
Thistle, Canada
Toadflax, Dalmatian
Trumpetcreeper
  (Buckvine)
Vetch
Waterhemlock
Waterprimrose,
  Creeping
* Woodsorrel,
  Common Yellow
Wormwood, Common
Wormwood, Louisiana
* Yankeeweed
Yarrow, Common

* Noted perennials may be controlled using BANVEL Herbicide at rate lower than those recommended for other listed perennial weeds. (See application rates and timing on pages 4, 5, 7, 8, 10, 11, 13, 14 and 15.)

WOODY

Alder
Ash
Aspen
Basswood
Beech
Birch
Blackberry
Blackgum
Cedar
Cherry
Chinquapin
Cottonwood
Creosotebush
Cucumbertree
Dewberry
Dogwood
Elm
Grape
Hawthorn (Thornapple)
Hemlock
Hickory
Locust, Black
Maple
Mesquite
Oak
Oak, Poison
Olive, Russian
Persimmon, Eastern
Pine
Plum, Sand (Wild Plum)
Poplar
Rabbitbrush
Redcedar, Eastern
Rose, McCartney
Rose, Multiflora
Sagebrush, Fringed
Sassafras
Serviceberry
Spicebush
Spruce
Sumac
Sweetgum

Honeylocust
Honeysuckle
Hornbeam
Huckleberry
Huisache
Ivy, Poison
Kudzu
Sycamore
Tarbush
Willow
Witchhazel
Yaupon
Yucca

FIELD, SEED* AND SILAGE CORN

IMPORTANT

Observe all precautions, mixing and application instructions as well as the following:

* Do not apply BANVEL Herbicide to seed corn without first verifying with your local seed corn company (supplier) the Banvel selectivity on your inbred line. This precaution will help avoid potential injury of sensitive varieties.

BANVEL is not registered for use on sweet corn and popcorn.

Direct chemical contact with corn seed must be avoided. If corn seeds are less than 1½ inches below the surface, delay application until corn has emerged.

Up to 2 applications of BANVEL Herbicide may be made during a growing season. Do not exceed a total of 1½ pints of BANVEL Herbicide per treated acre per crop year. Allow two weeks between applications of BANVEL Herbicide. See appropriate section for rate information. For combination options or sequential treatments, refer to appropriate section.

Applications of BANVEL Herbicide to corn during periods of rapid growth may result in temporary leaning. Corn will usually become erect within 3 to 7 days. Cultivation should be delayed until after corn is growing normally to avoid breakage.

Do not use adjuvants containing penetrants such as petroleum and crop oils after crop emergence.

Prior to the ensilage (milk) stage of the crop, do not harvest or graze corn for dairy or beef feed.

Use coarse sprays to avoid potential herbicide drift. Select nozzles which are designed to produce minimal amounts of fine spray particles such as raindrops, LP flat fans or large capacity flood nozzles such as D10, TK10 or greater capacity tips. Keep the spray pressure at or below 20 psi and the spray volume at or above 20 gpa, unless otherwise required by the manufacturer of drift-reducing nozzles. An agriculturally approved drift control agent may be added to further reduce the potential for drift.

Several synthetic pyrethroid insecticides are labeled for tankmix applications of BANVEL. Refer to their label for specific recommendations.

WEEDS CONTROLLED

BANVEL Herbicide will control many ANNUAL broadleaf weeds or give growth suppression of many PERENNIAL broadleaf weeds commonly found in corn. (Refer to the GENERAL WEED LIST on pages 3-4.)

For best performance, make application when weeds have emerged and are actively growing.

Preemergence control of cocklebur, velvetleaf, and jimsonweed may be reduced if conditions such as low temperature or lack of soil moisture cause delayed or deep germination of weeds.

Adding an agriculturally approved surfactant to BANVEL Herbicide will improve wild mustard, common sunflower, or velvetleaf control when weeds are over 5 inches tall or growing under drought stress at time of application. Tank mixing BANVEL Herbicide plus 2,4-D will improve control from a late postemergence application, but 2,4-D may cause brittleness to corn. Refer to tank mix treatments.

PREPLANT/PREEMERGENCE IN NO TILLAGE CORN

Applications of BANVEL Herbicide may be made before, during, or after planting to emerged and actively growing broadleaf weeds. Apply BANVEL Herbicide at 1 pint per treated acre on medium or fine textured soils containing 2% or greater organic matter. This rate also applies to soils containing greater than 8% organic matter. Use ½ pint per treated acre on coarse textured soils (sand, sandy loam, and loamy sand).

When planting into a legume sod (e.g., alfalfa or clover), apply BANVEL Herbicide after 4-6 inches of regrowth has occurred.

PREEMERGENCE IN CONVENTIONAL OR REDUCED TILLAGE CORN

BANVEL Herbicide may be applied after planting and prior to corn emergence. Application at 1 pint per treated acre may be made to medium or fine textured soils which contain 2% or greater organic matter. DO NOT apply to coarse textured soils.

Preemergence application of BANVEL Herbicide does not require mechanical incorporation to become active. A shallow mechanical incorporation is recommended if application is not followed by adequate rainfall or sprinkler irrigation. Avoid tillage equipment (e.g., drags, harrows) which concentrate treated soil over seed furrow.

EARLY POSTEMERGENCE (ALL TILLAGE SYSTEMS)

(Spike to 5th true leaf stage)

BANVEL Herbicide at 1 pint per treated acre may be applied during the period from corn emergence through the 5th true leaf stage or 8 inches tall, whichever comes first. Reduce the rate to ½ pint per treated acre if the crop is beyond the 8 inch stage or growing on coarse textured soils (sand, sandy loam, loamy sand).

LATE POSTEMERGENCE (ALL TILLAGE SYSTEMS)

(8″ to 36″ tall corn)

Application of BANVEL Herbicide at ½ pint per treated acre may be made from 8 to 36 inch tall corn or 15 days before tassel emergence, whichever comes first.

Make directed spray application when: (1) corn leaves prevent proper spray coverage; (2) sensitive crops are growing nearby; (3) tank mixing with 2,4-D; or (4) surfactants are added to the spray mixture.

DO NOT apply BANVEL Herbicide when soybeans are growing nearby if any of these conditions exist:

—corn is more than 24 inches tall

—soybeans are more than 10 inches tall

—soybeans have begun to bloom

OVERLAY (SEQUENTIAL) TREATMENTS

BANVEL Herbicide may be applied to ground previously treated with one or more of the following herbicides.

atrazine
Bexton®/Ramrod (propachlor)
Bicep®
Bladex® (cyanazine)
Bronco®
Dual® (metolachlor)
Eradicane® (EPTC)
Extrazine®
Gramoxone® (paraquat)
Lasso® (alachor)
Marksman®
Princep® (simazine)
Prowl® (pendimethalin)
Roundup® (glyphosate)
Sutan + ®/Genate® (butylate)

READ AND FOLLOW LABEL DIRECTIONS FOR EACH OF THE ABOVE PRODUCTS.

TANK MIX TREATMENTS FOR CORN

BANVEL Herbicide may be tank mixed with one or more of the following herbicides for control of grasses or additional broadleaf weeds. Read and follow the label of each tank mix product used for precautionary statement, directions for use, rates and timings, and other restrictions.

RATES AND TIMINGS

BANVEL PLUS	PREPLANT PREEMERGENT (NO TILLAGE CORN)	PREEMERGENT (CONVENTIONAL OR REDUCED TILLAGE CORN)	EARLY POST-EMERGENT (ALL TILLAGE SYSTEMS)	LATE POSTEMERGENT (ALL TILLAGE SYSTEMS)	ADDITIONAL DIRECTIONS
Atrazine	1¼-4 lbs. ai/A	1¼-4 lbs. ai/A	1¼-4 lbs. ai/A	1¼-4 lbs. ai/A	Application may be made before grasses are 1½ inches tall.
BLADEX®	1¼-4 lbs. ai/A	1¼-4 lbs. ai/A	1¼-2 lbs. ai/A (use the 80W, 90DF formulations only.)	-	Application may be made before grasses are 1½ inches tall, and before corn is beyond the 4 true leaf stage.

RATES AND TIMINGS					
BANVEL PLUS	PREPLANT PREEMERGENT (NO TILLAGE CORN)	PREEMERGENT (CONVENTIONAL OR REDUCED TILLAGE CORN)	EARLY POST-EMERGENT (ALL TILLAGE SYSTEMS)	LATE POSTEMERGENT (ALL TILLAGE SYSTEMS)	ADDITIONAL DIRECTIONS
DUAL®	-	1½-3 lbs. ai/A (use only fine or medium textured soils with 2½% or greater organic matter)	1½-3 lbs. ai/A	-	Application may be made after planting and until grasses reach the 2 leaf stage and before corn is greater than 3 inches tall.
LASSO®	1½-4 lbs. ai/A (use only on fine textured soils greater than 2½% organic matter)	1½-4 lbs. ai/A (use only on fine textured soils with greater than 2½% organic matter)	1½-4 lbs. ai/A	-	Application may be made before grasses reach the 2 leaf stage and before corn is greater than 3 inches tall.
PARAQUAT	¼-1lb. ai/A	1¼-1 lb. ai/A	-	-	Application may be made to emerged weeds but prior to corn emergence.
PRINCEP®	2.0-3.0 lbs. ai/A	2.0-3.0 lbs. ai/A	-	-	Application may be made prior to weed and corn emergence.
PROWL®	-	1.0-2.0 lbs. ai/A	-	-	Application may be made immediately after planting but prior to weed and corn emergence.
ROUNDUP®	1.0-3.0 lbs. ai/A	1.0-3.0 lbs. ai/A	-	-	Application may be made to emerged weeds but prior to corn emergence.
2,4-D	¼-½ lbs. ai/A	¼-½ lbs. ai/A	Not recommended	⅛ lb. ai/A	Drop pipes are to be used when corn reaches the 8" growth stage. Keeping the spray off the corn leaves and out of the whorl will reduce the likelihood of crop injury and improve spray coverage of weed foliage.

GRAIN SORGHUM (Milo)

IMPORTANT

Observe all precautions, including the reference to crops growing under stress.

Read and follow mixing and application instructions.

Applications of BANVEL Herbicide to sorghum during periods of rapid growth may result in temporary leaning of plants or rolling of leaves. These effects are usually outgrown within 10 to 14 days.

Do not graze or feed treated sorghum forage or silage prior to mature grain stage.

Do not apply BANVEL Herbicide to sorghum grown for seed production.

Make no more than one application per growing season.

Delay harvest until 30 days after treatment.

WEEDS CONTROLLED

BANVEL Herbicide, when applied at the recommended rate for grain sorghum, will control many actively growing ANNUAL broadleaf weeds and will reduce competition from established PERENNIAL broadleaf weeds as well as control their seedlings. (Refer to GENERAL WEED LIST).

RATES AND TIMINGS

BANVEL Herbicide may be applied to emerged and actively growing weeds at least 15 days prior to planting. Postemergence application of BANVEL Herbicide must be made after sorghum is in the 3 leaf stage but before sorghum is 15 inches tall. For best performance, make applications when sorghum is in the 3-5 leaf stage and weeds are small (less than 3 inches tall). Use drop pipes (drop nozzles) if crop is taller than 8 inches. Keeping the spray off the sorghum leaves and out of the whorl will reduce the likelihood of crop injury and improve spray coverage of weed foliage.

BROADCAST RATE PER TREATED ACRE:

½ pint (¼ lb. a.i.)

TANK MIX TREATMENT

BANVEL plus Atrazine

For improved control of emerged, actively growing broadleaf weeds including triazine resistant species and added supression of perennial broadleaf weeds, tank mix ½ pint BANVEL Herbicide with 0.5 to 1.25 lbs. a.i. atrazine per treated acre. For control of grasses (less than 1.5 inches tall), tank mix ½ pint BANVEL Herbicide with 2-3 lbs. a.i. atrazine per treated acre. For best performance and minimal crop injury, make application when sorghum is 3-8" tall and when broadleaf weeds are small (less than 6 inches tall). The atrazine rate will depend upon soil texture and length of residual weed control desired.

READ AND FOLLOW THE LABEL OF EACH TANK MIX PRODUCT USED FOR PRECAUTIONARY STATEMENTS, DIRECTIONS FOR USE, APPLICATION RATES AND TIMINGS AND OTHER RESTRICTIONS.

OVERLAY (SEQUENTIAL) TREATMENTS

BANVEL Herbicide may be applied to ground previously treated with one or more of the following herbicides:

Herbicide	Maximum rate per treated acre (lbs. a.i.)
alachlor (Lasso®) (Screen®-treated seed)	4
atrazine	3
metolachlor (Dual®) (Concep®- treated seed)	2.5
propachlor (Ramrod®)	5
propazine (Milogard®)	3.2

PREHARVEST USES

For Use Only in the States of Texas and Oklahoma

Application of BANVEL Herbicide may be made any time after the sorghum has reached the soft dough stage of development for suppression of weeds. An agriculturally approved surfactant may be used to improve performance.

BROADCAST RATE PER TREATED ACRE:

½ pint (¼ lb. a.i.)

SMALL GRAINS (NOT UNDERSEEDED TO LEGUMES)

IMPORTANT

Observe all precautions. Read and follow mixing and application instructions.

If small grains are grown for pasture only, refer to the PASTURE, RANGELAND and NON-CROPLAND section.

Do not graze or harvest for livestock feed prior to crop maturity.

WEEDS CONTROLLED

BANVEL Herbicide, when applied at recommended rates, will control ANNUAL broadleaf weeds commonly found in small grains, such as:

Buckwheat, Wild	Pigweed, Redroot
Chamomile, Corn	(Carelessweed)

Cockle, Corn
Cockle, Cow
Cocklebur, Common
Henbit
Knotweed
Kochia
Ladysthumb
Lambsquarters, Common
Mallow, Common
Nightshade, Black
Pennycress, Field
Pigweed, Rough
Pigweed, Tumble
Ragweed, Common
Ragweed, Giant (Buffaloweed)
Smartweed, Green
Smartweed, Pennsylvania
Sowthistle, Annual
Sunflower, Common (Wild)
Sunflower, Volunteer
Thistle, Russian
Velvetleaf

BANVEL Herbicide and BANVEL Herbicide tank mixes will reduce competition from established PERENNIAL broadleaf weeds and control their seedlings. (Refer to GENERAL WEED LIST).

THE SPECIAL USE TANK MIX FOR FALL SEEDED WHEAT ONLY allows a higher rate of 2,4-D to be used in combination with BANVEL Herbicide. This tank mix treatment may be used for improved performance of difficult-to-control weeds including:

*Fiddleneck (Tarweed)	Gromwell
Garlic, Wild	*Onion, Wild

* Spring applications may not control weeds that develop in the fall. For fall applications, refer to the BETWEEN CROPPING APPLICATIONS section.

RATES AND TIMINGS

Application of BANVEL Herbicide may be made before, during or after planting to emerged and actively growing weeds. See specific crop for timing restrictions. For best performance, make application when weeds are in the 2-3 leaf stage and rosettes are less than 2 inches across. Use the higher level of listed rate ranges when treating more mature weeds or dense vegetative growth. Surfactant should not be used when applying BANVEL Herbicide alone or in a tank mix on small grains except when tank mixing with chlorsulfuron (Glean), metsulfuron (Ally) or Finesse.

FALL SEEDED WHEAT

BANVEL HERBICIDE MUST BE APPLIED TO FALL SEEDED WHEAT PRIOR TO THE JOINTING STAGE.

BROADCAST RATE PER TREATED ACRE:

1/4 pint (1/8 lb. a.i.)

TANK MIX TREATMENTS

For control of grasses or additional broadleaf weeds, BANVEL Herbicide may be tank mixed with the following herbicides. Read and follow the label of each tank mix product used for precautionary statements, directions for use, weeds controlled and geographic and other restrictions.

BROADCAST RATE PER TREATED ACRE:

Apply 1/8-1/4 pint (1/16-1/8 lb. a.i.) BANVEL Herbicide with:

	Amount Product*	lb. a.i.
2,4-D	1/2-3/4 pint	1/4-3/8
MCPA	1/2-3/4 pint	1/4-3/8
bromoxynil		
(Brominal ME4,	1/2-3/4 pint	1/4-3/8
Buctril)	1-1 1/2 pint	1/3-3/8
bromoxynil + MCPA		
(Bronate,	1 pint	1/4 + 1/4
Brominal Plus)	1 pint	1/4 + 1/4
metsulfuron (Ally®)****	1/10 wt.oz.	0.0038
metribuzin		
(Sencor® 75DF**, Lexone® DF™)	1/6-1/3 lb.	1/8-1/4
(Sencor® 4**, Lexone® 4L)	1/4-1/2 pint	1/8-1/4
(Lexone® 50WP)	1/4-1/2 lb.	1/8-1/4
chlorsulfuron***		
(Glean® 75DF)	1/6-1/2 wt. oz.	0.008-0.024
diuron (Karmex® 80WP)	1/2-2 lbs.	2/5-1 3/5
terbutryn (Igran® 80W)	1 1/2-2 3/4 lbs.	1 1/5-2 1/5
Finesse®****	0.3--0.5 wt.oz.	

* Based on 4 pounds per gallon formulations of MCPA and 2,4-D.
** Application may be made after fall seeded wheat has started to grow and has a well established secondary root system or is beginning to tiller, but prior to the jointing stage.
*** When making tank mix applications with Glean, add a surfactant of at least 80% active ingredient at the rate of 1-2 quarts/100 gallons of spray or not more than 1/4-1/2% by volume. Use the higher rate of surfactant when using the lower rate ranges of the tank mix and/or when treating more mature weeds or dense vegetative growth.
**** Use 0.25% surfactant per treated acre for improved control of broadleaf weeds.

SPECIAL USE TANK MIX FOR FALL SEEDED WHEAT ONLY

BANVEL	1/4 pint	1/8 lb. a.i.
plus	plus	plus
2,4-D amine	1 to 2 pints	1/2 to 1 lb. a.i.
or	or	or
2,4-D ester	1 to 1 1/2 pints	1/2 to 3/4 lb. a.i.

Note: Do not use unless possible crop injury will be acceptable.

SPRING SEEDED WHEAT

BANVEL HERBICIDE MUST BE APPLIED BEFORE SPRING SEEDED WHEAT EXCEEDS THE 5 LEAF STAGE.

BROADCAST RATE PER TREATED ACRE:

1/4 pint (1/8 lb.a.i.)

TANK MIX TREATMENTS

For control of additional broadleaf weeds, BANVEL Herbicide may be tank mixed with the following herbicides. Read and follow the label of each tank mix product used for precautionary statements, directions for use, weeds controlled and geographic and other restrictions.

BROADCAST RATE PER TREATED ACRE:

APPLY 1/8-1/4 pint (1/16-1/8 lb. a.i.) BANVEL Herbicide with:

Herbicide	Amount product*	lb. a.i.
2,4-D	1/2-3/4 pint	1/4-3/8
MCPA	1/2-3/4 pint	1/4-3/8
bromoxynil		
(Brominal,	1/2-3/4 pint	1/4-3/8
Buctril)	1-1 1/2 pint	1/4-3/8
chlorsulfuron**		
(Glean® 75DF)	1/6-1/2 wt. oz.	0.008-0.024
Finesse®***	0.3-0.5 wt. oz.	-
metsulfuron (Ally®)***	1/10 wt. oz.	0.0038

* Based on 4 pounds per gallon formulations of MCPA and 2,4-D.
** When making tank mix applications with Glean, add a surfactant of at least 80% active ingredient at the rate of 1-2 quarts per 100 gallons of spray or not more than 1/4-1/2% by volume. Use the higher rate of surfactant when using the lower rate ranges of the tank mix and/or when treating more mature weeds or dense vegetative growth.
*** Use 0.25% surfactant per treated acre for improved control of broadleaf weeds.

FALL SEEDED BARLEY

BANVEL HERBICIDE MUST BE APPLIED TO FALL SEEDED BARLEY PRIOR TO THE JOINTING STAGE.

Note: For fall barley varieties that are seeded during the winter months or later, follow the rates and timings given for Spring Seeded Barley.

BROADCAST RATE PER TREATED ACRE:

1/4 pint (1/8 lb. a.i.)

TANK MIX TREATMENTS

For control of additional broadleaf weeds, BANVEL Herbicide may be tank mixed with the following herbicides. Read and follow the label of each tank mix product used for precautionary statements, directions for use, weeds controlled and geographic and other restrictions.

BROADCAST RATE PER TREATED ACRE:

Apply 1/8-1/4 pint (1/16-1/8 lb.a.i.) BANVEL Herbicide with:

Herbicide	Amount product*	lb. a.i.
2,4-D	1/2 pint	1/4
MCPA	1/2-3/4 pint	1/4-3/8
chlorsulfuron**		
Glean® 75DF)	1/6-1/2 wt. oz.	0.008-0.024
metsulfuron (Ally®)***	1/10 wt. oz.	0.0038
metribuzin		
Sencor 4	1/2-1 pint	1/4-1/2
Sencor 75 DF	1/3-2/3 pound	1/4-1/2

* Based on 4 pounds per gallon formulations of MCPA and 2,4-D.
** When making tank mix applications with Glean, add a surfactant of at least 80% active ingredient at the rate of 1-2 quarts/100 gallons of spray or not more than 1/4-1/2% by volume. Use the highest rate of surfactant when using the lower rate ranges of the tank mix and/or when treating more mature weeds or dense vegetative growth.
*** Use 0.25% surfactant per treated acre for improved control of broadleaf weeds.

SPRING SEEDED BARLEY

BANVEL HERBICIDE MUST BE APPLIED BEFORE SPRING SEEDED BARLEY EXCEEDS THE 3 LEAF STAGE.

BROADCAST RATE PER TREATED ACRE:

3/16 pint (3/32 lb. a.i.)

TANK MIX TREATMENTS

For control of additional broadleaf weeds, BANVEL Herbicide may be tank mixed with the following herbicides. Read and follow the label of each tank mix product used for precautionary statements, directions for use, weeds controlled and geographic and other restrictions.

BROADCAST RATE PER TREATED ACRE:

Apply 1/8-3/16 pint (1/16-3/32 lb. a.i.) BANVEL Herbicide with:

Herbicide	Product amount*	lb. a.i.
MCPA	1/2 pint	1/4
metribuzin		
Sencor 4	1/2 pint	1/4

Herbicide	Product amount*	lb. a.i.
Sencor 75DF	1/3 pound	1/4
chlorsulfuron** (Glean® 75DF)	1/6-1/2 wt. oz.	0.008-0.024
metsulfuron (Ally®)***	1/10 wt. oz.	0.0038

* Based on 4 pounds per gallon formulations of MCPA.

** When making tank mix applications with Glean, add a surfactant of at least 80% active ingredient at the rate of 1-2 quarts/100 gallons of spray or not more than 1/4-1/2% by volume. Use the highest rate of surfactant when using the lower rate ranges of the tank mix and/or when treating more mature weeds or dense vegetative growth.

*** Use 0.25% surfactant for improved control of broadleaf weeds.

FALL AND SPRING SEEDED OATS

BANVEL HERBICIDE MUST BE APPLIED BEFORE SPRING SEEDED OATS EXCEED THE 5 LEAF STAGE. APPLICATIONS TO FALL SEEDED OATS MUST BE MADE PRIOR TO THE JOINTING STAGE.

BROADCAST RATE PER TREATED ACRE:

1/4 pint (1/8 lb. a.i.)

TANK MIX TREATMENTS

For control of additional broadleaf weeds, BANVEL Herbicide may be tank mixed with the following herbicides. Read and follow the label of each tank mix product used for precautionary statements, directions for use, weeds controlled and geographic and other restrictions.

BROADCAST RATE PER TREATED ACRE:

Apply 1/8-1/4 pint (1/16-1/8 lb. a.i.) BANVEL Herbicide with:

Herbicide	Amount product*	lb.a.i.
MCPA	1/2-3/4 pint	1/4-3/8

* Based on 4 pounds per gallon formulations of MCPA.

SUGARCANE

IMPORTANT

Observe all precautions. Read and follow mixing and application instructions.

Consult your local or state authorities for possible application restrictions, especially concerning aerial applications and advice concerning special local use situations.

WEEDS CONTROLLED

BANVEL Herbicide, when applied at recommended rates, will control many ANNUAL, BIENNIAL and PERENNIAL broadleaf weeds commonly found in sugarcane. (Refer to GENERAL WEED LIST).

RATES AND TIMINGS

Application of BANVEL Herbicide may be made any time after weeds have emerged and are actively growing but before the close-in stage of sugarcane. Application rates and timings of BANVEL Herbicide are given below. Use the higher level of listed rate ranges when treating dense vegetative growth.

Weed Stage & Type	Broadcast Rate Per Treated Acre	
	Product amount	lbs.a.i
Annual		
Small, actively growing	1/2-1 pt.	1/4-1/2
Established weed growth	1-1 1/2 pts.	1/2-3/4
Biennial	1-2 pints	1/2-1
Perennial		
Noted (*pg 3 &4) Perennials	2-4 pts.	1-2
Other Perennials	4-6 pts.	2-3†

†Application made over the top of actively growing sugarcane may result in crop injury.

When possible, direct the spray beneath the sugarcane canopy in order to minimize the likehood of crop injury. The use of directed sprays will also aid in maximizing spray coverage of weed foliage.

Retreatments may be made as needed, however, do not exceed a total of 6 pints (3 lbs. a.i.) of BANVEL Herbicide per treated acre during a growing season.

TANK MIX TREATMENTS

BANVEL Herbicide may be tank mixed with one or more of the following herbicides for control of grasses or additional broadleaf weeds. Read and follow the label of each tank mix product used for precautionary statements, directions for use, rates and timings, weeds controlled, geographic and other restrictions.

Herbicide	Rates per treated acre (lbs. a.i.)
ametryn (Evik®)	2/5 to 8
asulam (Asulox®)	2 to 3 1/3
atrazine	2/5 to 4
dalapon (Dalapon®)	3 1/2 to 8 1/2
2,4-D	1/2 to 3*

* Application of BANVEL Herbicide plus 2,4-D tank mix at the higher listed rate ranges may result in crop injury.

PASTURE, RANGELAND AND NON-CROPLAND AREAS

BANVEL Herbicide is recommended for use on pasture rangeland, general farmstead weed and brush control and for use on non-cropland areas such as fence rows, roadways, rights-of-way (utility, railroad, highway, pipeline), non-selective forest brush control (including site preparation), wasteland and other non-cropland areas.

IMPORTANT

Observe all precautions. Read and follow mixing and application instructions.

BANVEL Herbicide uses described in this section also pertain to small grains such as barley, oats, rye or wheat grown for pasture use only.

NEWLY SEEDED AREAS, including small grains such as barley, oats, rye or wheat grown for pasture, may be severely injured if rates of BANVEL Herbicide are applied in excess of those listed for control of ANNUAL weeds.

ESTABLISHED GRASS CROPS growing under stress can exhibit various injury symptoms that may be more pronounced if herbicides are applied. Furthermore, rates of BANVEL Herbicide in excess of 2 quarts (2 lbs. a.i.) per treated acre may cause temporary injury to many grass species.

Bentgrass, carpetgrass, buffalograss and St. Augustine grass may be injured at rates exceeding 1 pint BANVEL Herbicide (1/2 lb. a.i.) per treated acre. Usually colonial bentgrasses are more tolerant than creeping types. Velvetgrasses are most easily injured. Treatments will kill or injure alfalfa, clovers, lespedeza, wild winter peas, vetch and other legumes.

REMOVE MEAT ANIMALS FROM TREATED AREAS 30 DAYS PRIOR TO SLAUGHTER.

THERE IS NO WAITING PERIOD BETWEEN TREATMENT AND GRAZING FOR NON-LACTATING ANIMALS.

TIMING RESTRICTIONS FOR LACTATING DAIRY ANIMALS FOLLOWING TREATMENT

BANVEL Herbicide Rate Per Treated Acre	Days Before Grazing	Days Before Hay Harvest
Up to 1 pint (1/2 lb. a.i.)	7 days	37 days
Up to 1 quart (1 lb. a.i.)	21 days	51 days
Up to 2 quarts (2 lbs. a.i.)	40 days	70 days
Up to 8 quarts (8 lbs. a.i.)	60 days	90 days

Note: Observe all precautions and restrictions on labels of products used in tank mixtures.

MIXING AND APPLICATION

BANVEL Herbicide can be applied using water, oil in water emulsions (including invert systems), or sprayable fluid fertilizer as a carrier. A COMPATIBILITY TEST should be made prior to tank mixing.

To prepare oil in water emulsions, half-fill spray tank with water plus appropriate amount of herbicide. With continuous agitation, slowly add a premix of oil (such as diesel oil or fuel oil) plus a suitable emulsifier to spray tank. Complete filling of spray tank with water. Maintain vigorous agitation during spray operation to prevent oil and water from forming separate layers.

BANVEL Herbicide may be applied broadcast using either ground or aerial application equipment. When using ground equipment, apply 5 to 600 gallons of diluted spray per treated acre. Volume of spray applied will depend on the height, density, and type of weeds or brush being treated and on the type of equipment being used. When using aerial equipment apply 3 to 40 gallons of diluted spray per treated acre.

BANVEL Herbicide may be applied to individual clumps or small areas (SPOT TREATMENT) of undesirable vegetation using handgun or similar types of application equipment. Apply diluted sprays to allow complete wetting (up to run-off) of foliage and stems.

ACCUTROL® Herbicide adjuvant or other spray additives (emulsifiers, surfactants, wetting agents, drift control agents, or penetrants) may be used for wetting, penetration, or drift control. Spray additives must be agriculturally approved when used in pasture applications. If spray additives are used, read and follow all use recommendations and precautions on product label.

WEEDS CONTROLLED

BANVEL Herbicide, when applied at recommended rates, will give control of many ANNUAL, BIENNIAL, and PERENNIAL broadleaf weeds, and many WOODY brush and vine species commonly found in pasture, rangeland and non-cropland areas. (Refer to GENERAL WEED LIST). Noted (*) PERENNIAL weeds may be controlled with lower rates of either BANVEL Herbicide or BANVEL Herbicide plus 2,4-D. See RATES AND TIMINGS below.

RATES AND TIMINGS

Application rates and timing of BANVEL Herbicide are given below. Use the higher level of listed rate ranges when treating dense or tall vegetative growth.

Weed Storage & Type	Broadcast rate per treated acre	
	Product amount	lbs. a.i.
Annual		
Small, actively growing	½-1 pt.	¼-½
Established weed growth	1-1½ pts.	½-¾
†Biennial		
Rosette diameter		
Less than 3 inches	½-1 pt.	¼-½
3 inches or more	1-2 pts.	½-1
Bolting	2-3 pts.	1-1½
Perennial		
Suppression or top growth control	½-1 qt.	½-1
Noted(*) Perennials	1-2 qts.	1-2
Other perennials	2-4 qts.	2-4
Dense stands	4-6 qts.	4-6
Woody Brush & Vines		
Foliage Suppression	½-1 qt.	½-1
Stems	1-2 qts.	1-2
Stems and Stem Sprouts	½-1 gal.	2-4
Stems and Root Sprouts	1-2 gals.	4-8

† For best performance, make application when BIENNIAL WEEDS are in the rosette stage. Retreatments may be made as needed; however, do not exceed a total of 2 gallons (8 lbs. a.i.) of BANVEL Herbicide per treated acre during a growing season.

TANK MIX TREATMENTS

READ AND FOLLOW THE LABEL OF EACH TANK MIX PRODUCT USED FOR PRECAUTIONARY STATEMENTS, DIRECTIONS FOR USE, APPLICATION RATES AND OTHER RESTRICTIONS. BANVEL Herbicide may be tank mixed with one or more of the following herbicides for control of grasses, additional broadleaf weeds, and woody brush and vines.

Herbicide	Rates per treated acre (lbs. a.i.)
Pasture, rangeland, and non-cropland use:	
atrazine	½ to 2
diuron (Karmex®)	4 to 48
glyphosate (Roundup®)	¾ to 3¾
simazine (Princep®)	5 to 40
paraquat	½ to 1
triclopyr (Garlon®)	¾ to 9
2,4-D	¼ to 6
Rangeland and non-cropland use only:	
picloram (Tordon®)	¼ to 3
2,4,5-T	¼ to 6
Non-cropland use only:	
amitrole	2 to 8
atratol (Atratol®)	4⅘ to 40
bromacil (Hyvar®)	1½ to 24
dalapon (Dalapon®)	4¼ to 12¾
diquat	½
fosamine ammonium (Krenite®)	6 to 12
hexazinone (Velpar®)	2 to 12
MSMA	2
prometon (Pramitol®)	10 to 60
sulfometuron methyl (Oust™)	0.14 to 0.56
tebuthiuron (Spike®)	1 to 16
2,4-DP (Weedone®)	½ to 11

Due to the variations that may occur in formulated products and specific use ingredients (e.g. water supplies), a COMPATIBILITY TEST is recommended prior to actual tank mixing.

CUT SURFACE TREE TREATMENTS

BANVEL Herbicide may be applied as a cut surface treatment for control of unwanted trees and prevention of sprouts of cut trees. A mix of 1 part BANVEL Herbicide with 1 to 3 parts water should be used in application. Use the lower dilution when treating difficult-to-control species.

- FRILL OR GIRDLE TREATMENTS*: Make a continuous cut or a series of overlapping cuts using an axe to girdle tree trunk. Spray or paint cut surface with the BANVEL Herbicide/water mix.
- STUMP TREATMENTS; Spray or paint freshly cut surface with the BANVEL Herbicide/water mix. The area adjacent to the bark should be thoroughly wet.

* Note: For more rapid foliar effects, 2,4-D may be added to the BANVEL Herbicide/water mix.

DORMANT APPLICATIONS FOR CONTROL OF MULTIFLORA ROSE

BANVEL Herbicide can be applied when plants are dormant as an undiluted SPOT-CONCENTRATE directly to the soil or as a LO-OIL BASAL BARK treatment using an oil-water emulsion solution.

MIXING AND APPLICATION

SPOT-CONCENTRATE applications of BANVEL Herbicide should be applied directly to the soil as close as possible to the root crown but within 6-8 inches of the crown. On sloping terrain, application should be made to the uphill side of the crown. Do not make application when snow or water prevents applying BANVEL Herbicide directly to the soil.

LO-OIL BASAL BARK applications of BANVEL Herbicide should be applied to the basal stem region from the ground line up to a height of 12-18 inches. Spray until runoff, with special emphasis on covering the root crown. For best results, make application when plants are dormant. Do not make application after bud break or when plants are showing signs of active growth. Do not make application when snow or water prevents applying BANVEL Herbicide to the ground line.

Note: To prepare oil-in-water emulsions, half fill spray tank with water plus appropriate amount of herbicide. With continuous agitation, slowly add a premix of oil plus a suitable emulsifier to the spray tank. Complete filling of spray tank with water. Maintain vigorous agitation during spray operation to prevent oil and water from forming separate layers.

RATES AND TIMINGS

Application rates of BANVEL Herbicide are given below:

Spot Concentrate Treatment

Canopy Diameter (Feet)	BANVEL Herbicide (Ounces)
5	¼
10	1
15	2¼

Do not exceed a total of 2 gallons (8 lbs. a.i.) of BANVEL Herbicide per acre per year.

LO-OIL BASAL BARK TREATMENT

Mix the appropriate amount of BANVEL Herbicide with the appropriate amount of water, emulsifier and No. 2 diesel fuel to obtain the volume of spray desired. See table below. Refer to MIXING and APPLICATION section before mixing. Do not exceed 30 gallons of spray solution per acre per year.

Volume Of Spray Solution Desired (Gal.)	Ounces			
	Water	Emulsifier	BANVEL Herbicide	No. 2 Diesel Fuel
1	100* (3 qts.)	½	8	20
2	200 (6 qts.)	1	16	40
5	500 (3.75 gals.)	2½	40	100
10	1000 (7.5 gals.)	5	80	200

*Conversion: 100 ounces = 3 quarts

CONSERVATION RESERVE PROGRAM (CRP) ACRES

Important

Before using this product, read and carefully observe all applicable directions, restrictions, and precautionary statements on the EPA-registered container label and the BANVEL Herbicide Directions for Use booklet (especially pages 4-5 and 14-21) attached to the container.

Newly Seeded Areas—seedling grasses or small grains such as barley, oats, rye, wheat, or other grass species grown as a cover crop may be severely injured if rates of BANVEL Herbicide are applied in excess of those listed for control of ANNUAL weeds. **Do not apply BANVEL Herbicide to seedling grasses until grasses exceed the three-leaf stage of growth.**

Established grass stands growing under stress can exhibit various injury symptoms that may be more pronounced if herbicides are applied. Bentgrass, carpetgrass, smooth brome, buffalograss and St. Augustine grass may be injured at rates exceeding 1 pint of BANVEL Herbicide (0.5 lb a.i.) per treated acre. Depending on the rate used, treatment with BANVEL Herbicide will injure or kill alfalfa, clovers, lespedeza, wild winter peas, vetch or other legumes.

Mixing And Application

BANVEL Herbicide can be applied using water, oil in water emulsions (including invert systems), or sprayable fluid fertilizer as a carrier. A COMPATIBILITY TEST (page 9 of the BANVEL Herbicide Booklet) should be made prior to tank mixing.

To prepare oil in water emulsions, half-fill the spray tank with water plus the appropriate amount of herbicide. With continuous agitation, slowly add a premix of oil (such as diesel oil or fuel oil) plus a suitable emulsifier to the spray tank. Complete filling of spray tank with water. Maintain vigorous agitation during spray operation to prevent oil and water from forming separate layers.

BANVEL Herbicide may be applied broadcast using either ground or aerial equipment. When using ground equipment, apply 5-50 gallons of diluted spray per treated acre. The volume of spray applied will depend on the height, density, and types of weeds being treated and the type of equipment being used. When using aerial equipment, apply 3-10 gallons of diluted spray per treated acre.

BANVEL Herbicide may be applied to individual clumps or small areas (SPOT TREATMENT) of weeds using a handgun or similar application equipment. Apply diluted sprays to allow complete wetting (up to runoff) of foliage and stems.

Agriculturally approved drift-reducing additives or surfactants may be used. If spray additives are used, read and follow all use recommendations and precautions on product label. Do not use adjuvants containing penetrants such as petroleum and crop oils after cover crop emergence.

DIRECTIONS FOR USE

BANVEL Herbicide, when applied at recommended rates, will control many annual and biennial weeds and provide top growth control of many perennial weeds. Many of the key broadleaf weed species controlled or suppressed (perennials) are listed below.

Fo perennial broadleaf suppression in established grass or in the fallow (stubble) or cover crop period prior to seeding a grass crop, apply 1 quart (1 lb a.i.) of BANVEL Herbicide per treated acre. Established grass is defined as: Grass that is treated the season after planting.

Annual Weeds Controlled

Buckwheat, Wild	Pigweed, Rough
Chamomile, Corn	Pigweed, Tumble
Cockle, Corn	Ragweed, Common
Cockle, Cow	Ragweed, Giant (Buffaloweed)
Cocklebur, Common	Smartweed, Green
Henbit	Smartweed, Pennsylvania
Knotweed	Sowthistle, Annual
Kochia	Sunflower, Common (Wild)
Ladysthumb	Sumflower, Volunteer
Lambsquarters, Common	Thistle, Russian
Mallow, Common	Velvetleaf
Nightshade, Black	
Pennycress, Field	
Pigweed, Redroot	

Biennial Weeds Controlled

Knapweed, Diffuse	Thistle, Bull
Knapweed, Spotted	Thistle, Musk
Starthistle, Yellow	Thistle, Plumeless
Sweetclover	

Perennials (Suppression or Top Growth Control)

Alfalfa (Volunteer)	Garlic, Wild
Artichoke, Jerusalem	Horsenettle, Carolina
Bindweed, Field	Knapweed, Russian
Bindweed, Hedge	Nightshade, Silverleaf
Blueweed, Texas	Redvine
Bursage (Bur Ragweed) (Povertyweed) (Lakeweed)	Smartweed, Swamp
Dandelion, Common	Spurge, Leafy
Dock, Curly	Sowthistle, Perennial
Dogbane, Hemp	Thistle, Canada
	Trumpetcreeper (Buckvine)

Rates and Timing

Application rates and timing of BANVEL Herbicide treatment are given below. Use the higher rate of the rate range when vegetation is either dense or tall, or when weeds are growing under stressed conditions such as drought or cool temperature.

The addition of an agriculturally approved surfactant will improve wetting and coverage of weed foliage and improve control of drought stressed weeds.

Weed Type[1] & Stage	Broadcast Rate Per Treated Acre: Amount of Formulated BANVEL Herbicide	Equivalent lbs a.i.
	pints	
Annuals		
- Small actively growing	1/4 to 1	1/8 to 1/2
-Established weed growth	1	1/2
Biennials[2]		
-Rosette diameter		
a) less than 3 inches	1/2 to 1	1/4 to 1/2
b) 3 inches or greater	1 to 2	1/2 to 1
c) bolting biennial	2 to 3	1 to 1 1/2
Perennials[2]		
Suppression/Control	2 to 4	1 to 2

[1] For best results, treat Biennial weeds with BANVEL Herbicide when they are in the rosette stage of growth. Retreatments may be made as needed; however, DO NOT EXCEED A TOTAL OF 2 QUARTS (2 lbs a.i.) of BANVEL Herbicide per treated acre during a growing season.

[2] Biennial and Perennial weeds will require follow-up (sequential) treatments for seedling control and escapes.

Tank Mix Treatments

Read and follow the label of each tank mix product used for precautionary statements, directions for use, application rates, and other restrictions.

BANVEL Herbicide may be tank mixed with one or more of the following herbicides for control of grasses or additional broadleaf weeds.

Tank Mix Herbicide	Rates Per Treated Acre (lbs a.i.)
Atrazine*	1/2 to 2
chlorsulfuron (Glean 75DF)**	0.008 tp 0.024
Glyphosate (Roundup)*	1/2 to 2
2,4-D	1/4 to 6
metsulfuron (Ally)***	0.0038
Paraquat*	1/2 to 1

* Preplant application

** When making tank mix applications with Glean, add a surfactant of at least 80 percent active ingredient at the rate of 1-2 quarts/100 gallons of spray or not more than 1/4-1/2 percent by volume. Use the highest rate of surfactant when using the lower rate ranges of the tank mix and/or when treating more mature weeds or dense vegetative growth

*** Use 0.25 percent surfactant per treated acre for improved control of broadleaf weeds.

Caution

When used as a preplant application for control of existing weeds, allow 45 days per pint of BANVEL Herbicide applied per acre before seeding grass west of the Mississippi River, or 20 days per pint east of the Mississippi River. Exclude days when the ground is frozen. **Do not apply prior to planting of grass-legume mixtures.**

ASPARAGUS

For Use Only in the States of California, Oregon and Washington

IMPORTANT

Observe all precautions. Read and follow mixing and application instructions.

If spray contacts emerged spears, crooking (twisting) of some spears may result. If such crooking occurs, discard affected spears.

Do not harvest prior to 24 hours after treatment.

Make only one application per season.

RATES AND TIMINGS

Apply BANVEL Herbicide to emerged and actively growing weeds in 40 to 60 gallons of diluted spray per treated acre immediately after cutting the field, but at least 24 hours before the next cutting.

Weeds	Rate Per Treated Acre
Mustard, Black Pigweed, Redroot (carelessweed) Sowthistle, Annual *Thistle, Canada Thistle, Russian	1/2 to 1 pt. (1/4-1/2 lb. a.i.)
*Bindweed, Field Chickweed, Common Goosefoot, Nettleleaf Radish, Wild Thistle, Milk	1 pt. (1/2 lb. a.i.)

BANVEL Herbicide plus 2,4-D herbicide tank mixture may be used for improved control of noted (*) weeds. READ AND FOLLOW 2,4-D PRODUCT LABELING FOR PRECAUTIONARY STATEMENTS, DIRECTIONS FOR USE, APPLICATION RATES AND TIMINGS, AND OTHER RESTRICTIONS.

TURF AND LAWNS

Including Golf Course Fairways, Aprons, Tees and Rough.

IMPORTANT

Observe all precautions. Read and follow mixing and application instructions.

To avoid injury to newly seeded grasses, application of BANVEL Herbicide should be delayed until after the second mowing. Furthermore, application rates in excess of 1 pint (1/2 lb. a.i.) per treated acre may cause noticeable stunting or discoloration of sensitive grass species such as bentgrass, carpetgrass, buffalograss, and St. Augustine grass.

In areas where roots of sensitive plants extend, do not apply in excess of 1/4 pint (1/8 lb. a.i.) of BANVEL Herbicide per treated acre on coarse textured (sandy-type) soils, or in excess of 1/2 pint (1/4 lb. a.i.) per treated acre on fine textured (clayey-type) soils. Do not make repeat applications in these areas for 30 days and until previous applications of BANVEL Herbicide have been activated in the soil by rain or irrigation.

WEEDS CONTROLLED

BANVEL Herbicide, when applied at recommended rates, will give control of many ANNUAL, BIENNIAL, and noted (*) PERENNIAL broadleaf weeds commonly found in turf. BANVEL Herbicide will also give growth suppression of many other listed PERENNIAL broadleaf weeds and WOODY brush and vine species. (Refer to GENERAL WEED LIST).

MIXING AND APPLICATION

Apply 30 to 200 gallons of diluted spray per treated acre (3 qts. to 4 1/4 gals. per 1,000 sq. ft.), depending on density or height of weeds treated and on the type of equipment used.

RATES AND TIMINGS

Use the higher level of listed rate ranges when treating dense vegetative growth.

Weed Stage & Type	BANVEL Herbicide		
	Pints per Treated Acre	Lbs. a.i. Per Treated Acre	Teaspoons per 1000 sq. ft.
Annual			
Small, actively growing	1/2-1	1/4-1/2	1-2 1/4
Established weed growth	1-1 1/2	1/2-3/4	2 1/4-3 1/4
Biennial			
Rosette diameter			
Less than 3 inches	1/2-1	1/4-1/2	1-2 1/4
3 inches or more	1-2	1/2-1	2 1/4-4 1/2
Perennials and Woody			
Brush and Vines	1-2	1/2-1	2 1/4-4 1/2

For best performance, apply when weeds are emerged and actively growing.

Retreatments may be made as needed; however, do not exceed a total of 2 pints (1 lb. a.i.) BANVEL Herbicide per treated acre during a growing season.

TANK MIX TREATMENTS

READ AND FOLLOW THE LABEL OF EACH TANK MIX PRODUCT USED FOR PRECAUTIONARY STATEMENTS, DIRECTIONS FOR USE, APPLICATION RATES AND TIMINGS AND OTHER RESTRICTIONS.

Tank mix treatments of BANVEL Herbicide may be made with 2,4-D, MCPA, MCPP, or bromoxynil for control of additional weeds listed on the tank mix product label.

Apply 1/5 to 1/2 pint (1/10-1/4 lb. a.i.) of BANVEL Herbicide per treated acre with 1/2 to 1 1/2 lbs. acid equivalent of 2,4-D, MCPA, or MCPP, or with 3/8 to 1/2 lb. a.i. of bromoxynil. Use the higher level of the listed rate ranges when treating established weeds. Repeat treatments may be made as needed; however, do not exceed 2 pints (1 lb. a.i.) of BANVEL Herbicide per treated acre during the growing season.

GRASS SEED CROPS

Perennial Grasses such as Bermuda grass, Bluegrass, Lawntype Fescue and Ryegrass

IMPORTANT

Observe all precautions. Read and follow mixing and application instructions.

Refer to the PASTURE, RANGELAND, AND NONCROPLAND IMPORTANT section for possible grazing and feeding restrictions.

Do not use on bentgrass unless possible crop injury can be tolerated.

RATES AND TIMINGS

Apply 1/2 to 2 pints (1/4-1 lb. a.i.) of BANVEL Herbicide in 5 to 40 gallons of diluted spray per treated acre after weeds have emerged and are actively growing for control of broadleaf weeds such as:

Alfalfa	Cockle, White
*Bindweed, Field	Dock, Curly
Catchfly, Nightflowering	*Knapweed, Russian
Chamomile, Corn	Knotweed
Chickweed, Common	Sorrel, Red (Sheep Sorrel)
Chickweed, Mouseear	Starwort, Little
Clover	*Thistle, Canada

*Top growth control only.

Use 1/2 to 1 pint (1/4-1/2 lb. a.i.) of BANVEL Herbicide per treated acre on SEEDLING GRASS after the crop reaches the 3-5 leaf stage. Up to 2 pints (1 lb. a.i.) of BANVEL Herbicide per treated acre may be used on well-established PERENNIAL grass. DO NOT APPLY AFTER THE GRASS SEED CROP BEGINS TO JOINT.

For control of ANNUAL GRASS WEEDS such as:

- Brome, Downy (Cheatgrass)
- Brome, Ripgut
- Fescue, Rattail

Apply 2 to quarts (2-4 lbs. a.i.) of BANVEL Herbicide per treated acre in the fall or late summer after harvest and burning of established grass seed crops. Applications should be made within 3-14 days following first irrigation and before weeds have more than 2 leaves.

BETWEEN CROPPING APPLICATIONS (BCA) FOR BROADLEAF WEED CONTROL

IMPORTANT

Observe all precautions. Read and follow mixing and application instructions.

WEEDS CONTROLLED

BANVEL Herbicide, when applied at the recommended rates, will control many ANNUAL and BIENNIAL broadleaf weeds. (Refer to GENERAL WEED LIST). In addition, BANVEL Herbicide will control the following PERENNIAL broadleaf weeds:

*Alfalfa	*Dogbane, Hemp
Artichoke, Jerusalem	Garic, Wild**
Bindweed, Field	Horsenettle, Carolina
Bindweed, Hedge	Nightshade, Silverleaf
Blueweed, Texas	Redvine
*Bursage (Bur Ragweed) (Povertyweed) (Lakeweed)	Smartweed, Swamp
*Dandelion, Common	*Sowthistle, Perennial
*Dock, Curly	Thistle, Canada**
	Trumpetcreeper (Buckvine)

Noted (*) perennials may be controlled using BANVEL Herbicide at rates lower than those recommended for other listed perennial weeds. (See RATES AND TIMINGS, page 14.)

** SPECIAL TANK MIX TREATMENTS, page 14, for specific control program.

RATES AND TIMINGS

Apply BANVEL Herbicide as a broadcast or spot treatment to emerged and actively growing weeds after crop harvest and before a killing frost. Agriculturally approved spray additives, such as surfactants or oils, may be used to enhance spray coverage and the herbicide's penetration of weed foliage. See ROTATIONAL CROPS for recommended interval between application and planting to prevent crop injury.

For best performance, make application when ANNUAL weeds are less than 6 inches tall, when BIENNIAL weeds are in the rosette stage, and to PERENNIAL weed regrowth in late summer or fall following a mowing or tillage treatment. Most effective control of upright perennial broadleaf weeds, such as Canada thistle and Jerusalem artichoke, occurs if application is made when the majority of weeds is 8 inches or taller. Viney perennial broadleaf weeds, such as field bindweed and hedge bindweed, are best controlled when weeds are in or beyond the full bloom stage.

Avoid disturbing treated areas for at least 7 days following application. Treatments may not kill weeds which develop from seed or underground plant parts, such as rhizomes or bulblets, after the effective period for BANVEL Herbicide. For seedling control, a follow-up program or other cultural practices could be instituted (refer to corn, sorghum and wheat in-crop uses of BANVEL Herbicide).

Weed Stage & Type	BANVEL Herbicide per treated acre	
	Product amount	lbs. a.i.
Annual	1/2-1 pt.	1/4-1/2
Biennial	1-2 pts.	1/2-1
Perennial		
Perennial suppression	1-2 pts.	1/2-1
Noted (*) perennials	2-4 pts.	1-2
Other perennials	4 pts.	2

Retreatments may be made as needed; however, do not exceed a total of 4 pints (2 lbs. a.i.) of BANVEL Herbicide per treated acre during any given fallow period.

TANK MIX TREATMENTS

BANVEL Herbicide may be tank mixed with one or more of the following herbicides for control of grasses or additional broadleaf weeds. Read and follow the label of each tank mix product used for precautionary statements, directions for use, rates and timings, weeds controlled and geographic or other restrictions.

Herbicide	Rate per treated acre (lbs. a.i.)
ANNUAL WEED CONTROL	
atrazine	1/2 to 3
chlorsulfuron* (Glean®)	0.016 to 0.024 (1/3-1/2 wt. oz. product)
cyanazine (Bladex®)	1 3/5 to 3 1/5
glyphosate (Roundup®)	1/4 to 1/2
metribuzin (Sencor®, Lexone®)	1/3 to 3/4
paraquat	1/2 to 1
2,4-D	1/4 to 1/2
PERENNIAL WEED CONTROL	
glyphosate	1 to 2

Herbicide	Rate per treated acre (lbs. a.i.)
2,4-D	1 to 2

*When making tank mix applications with Glean, add a surfactant of at least 80% active ingredient at the rate of 1-2 quarts/100 gallons of spray or not more than 1/4-1/2% by volume. Use the highest rate of surfactant when using the lower rate ranges of the tank mix and/or when treating more mature weeds or dense vegetative growth.

SPECIAL TANK MIX TREATMENTS

For suppression of perennial weeds, apply 0.25-0.50 lb. a.i. BANVEL Herbicide with 0.25-0.50 lb. a.i. Roundup herbicide per treated acre.

For wild garlic control, apply 1 pint (1/2 lb. a.i.) BANVEL Herbicide with 1 1/2 lbs. acid equivalent 2,4-D low volatile ester per treated acre. Apply when wild garlic is 4 to 8 inches tall.

For Canada thistle control, use BANVEL Herbicide or BANVEL Plus Roundup Herbicide tank mix treatments.

Application may be made during fallow periods for control of volunteer barley, bulbous bluegrass, downy brome, jointed goatgrass, common rye and volunteer wheat when they are actively growing. Use 1 pint BANVEL Herbicide with 1/2-3/4 lb. Kerb 50W (0.25-0.38 lb. a.i.). Fall seeded wheat may be planted 9 months or more after application. For best performance, make application between mid-October and mid-December, prior to soil freeze up.

During fallow periods, apply BANVEL Herbicide plus Landmaster Herbicide to give improved control of kochia, wild buckwheat, prickly lettuce, field bindweed and Canada thistle. Use 4-8 fluid ounces of BANVEL Herbicide plus 40-54 fluid ounces of Landmaster Herbicide for annual weed control or 8-16 fluid ounces of BANVEL Herbicide plus 40-54 fluid ounces of Landmaster Herbicide for perennial weed suppression.

ROTATIONAL CROPS

The following recommendations are based on BANVEL Herbicide use rates up to 4 pints (2 lbs. a.i.) per treated acre.

CORN and SORGHUM may be planted in the spring following applications made during the previous year.

SOYBEANS may be planted in the spring following applications made during the previous year. If less than 1 inch of rainfall occurs between application and first killing frost, treated areas should be cultivated to allow herbicide to come in contact with moist soil. Cultivation may take place before or immediately after ground thaw.

Soybean injury may occur if the interval between application and planting is less than specified. In areas with greater than 30 inches of rainfall, delay planting for 30 days per pint of BANVEL Herbicide per treated acre. In areas with less than 30 inches of rainfall, delay planting for 45 days per pint of BANVEL Herbicide per treated acre. Exclude days when ground is frozen.

WHEAT may be planted in the fall or spring following applications. Also, spot applications may be made any time prior to crop emergence if crop injury can be tolerated in treated areas. Wheat injury may occur if the interval between application and planting is less than specified.

East of the Mississippi River, the interval is 20 days per pint of BANVEL Herbicide per treated acre. Exclude days when ground is frozen.

West of the Mississippi River, the interval is 45 days per pint of BANVEL Herbicide per treated acre. Exclude days when ground is frozen.

Following a normal harvest of corn, sorghum, soybeans, or wheat, any rotational crop may be planted. If the interval before harvest is shortened, such as when cover crops will be plowed under, do not follow up with the planting of a sensitive crop.

CONTROL OF PERENNIAL BROADLEAF WEEDS IN CROPLAND (SPOT APPLICATION ONLY) For Use Only in the States of Idaho, Montana, Nevada, Oregon, Utah, and Washington.

IMPORTANT

Observe all precautions. Read and follow mixing and application instructions.

Do not treat subirrigated cropland or areas where the soil remains saturated with water throughout the year.

Make only one application on BANVEL Herbicide per year.

WEEDS CONTROLLED

BANVEL Herbicide, when applied at recommended rates, will control many broadleaf weeds including:

Bindweed, Field	Knapweed, Russian
Dock, Broadleaf (Bitterdock)	Ragwort, Tansy
Dock, Curly	Spurge, Leafy
Knapweed, Black	Thistle, Canada

RATES AND TIMINGS

BANVEL Herbicide may be applied at any time following a crop harvest to stubble fallow or other cropland. Application should be made when weeds are actively growing and prior to a killing frost.

Apply 4 to 6 quarts (4-6 lbs. a.i.) of BANVEL Herbicide per treated acre. Application may be made up to one month prior to the planting of wheat.

Note: Do not use unless injury to wheat or rotated barley will be acceptable.

Barley, oats, corn, sorghum (milo), annual or perennial grass crops may be planted into treated areas one year after application. Crops grown for seed (other than perennial grass seed) should not be planted into treated areas until three years after application. Do not plant broadleaf crops such as alfalfa, beans, peas, potatoes, or sugarbeets into treated areas until two years after application.

In most cases, treatments will not kill perennial weed seedlings which germinate from seed one or two years after treatment. Once the effect of the chemical has been lost, a follow-up program for seedling control or other cultural practices should be instituted.

WIPER APPLICATION USES

IMPORTANT

Observe all precautions.

BANVEL Herbicide may be applied through wiper application equipment to control or suppress actively growing broadleaf weeds, brush and vines. Use a solution containing 1 part BANVEL Herbicide to 1 part water. Do not contact desirable vegetation with herbicide solution. Wiper application should only be made to crops (including pastures) and non-cropland areas described in this label with the exception of Grain Sorghum (Milo).

24(c) SPECIAL LOCAL NEED LABELING

24(c) SLN #50
BANVEL® Herbicide
Annual Weed Control in Millet (Nebraska) . . . .

24(c) SLN #61
BANVEL® Herbicide
Preharvest Application in Fall Seeded Wheat
(Oklahoma) . . . .

24(c) SLN #66
BANVEL® Herbicide
Preharvest Application in Fall Seeded Wheat
(Kansas) . . . .

24(c) SLN #86
BANVEL® Herbicide
For control of Annual and Perennial Broadleaf Weeds Such as Field Bindweed in Between Cropping Applications to Wheat
(Oklahoma) . . . .

24(c) SLN #89
BANVEL® Herbicide
For Control of Annual and Perennial Broadleaf Weeds such as Field Bindweed in Between Cropping Applications to Wheat
(Texas) . . . .

24(c) SLN #90
BANVEL® Herbicide
Wild Buckwheat Control in Fall Seeded Wheat (Oklahoma) . . . .

24(c) SLN #91
BANVEL® Herbicide
Small Grain Aerial Application (Colorado, Montana, No. Dakota, So. Dakota and Wyoming) . . . .

2(ee) RECOMMENDATIONS

2(ee) Recommendation #73
BANVEL® Herbicide
Tank Mix With Roundup® Herbicide for Preplant Applications to Spring Barley and Spring Wheat (Specific Counties in South Idaho) . . . .

2(ee) Recommendation #88
BANVEL® Herbicide
Between Cropping Applications (BCA)
BANVEL® Plus Landmaster® Tank Mix For Use in Fallow Rotated to Wheat (Montana and North Dakota) . . . .

2(ee) Recommendation #93
BANVEL® Herbicide
Between Cropping Applications (BCA) BANVEL® Plus Landmaster® Tank Mix for Use in Fallow Rotated to Wheat (Montana, No. Dakota, Wyoming, Washington, Idaho, Utah and Nebraska) . . . .

STORAGE AND DISPOSAL

PROHIBITIONS

Do not contaminate water, food or feed by storage or disposal.

STORAGE

Store in original container in well-ventilated area separately from fertilizer, feed and foodstuffs. Avoid cross-contamination with other pesticides. Spillage or leakage should be contained and absorbed with clay granules, sawdust, or equivalent material for disposal.

PESTICIDE DISPOSAL

Triple rinse pesticide from containers and use rinsates in the pesticide application. Wastes which cannot be used according to label instructions may be disposed of on site or at an approved waste disposal facility.

CONTAINER DISPOSAL

Plastic or Metal: After triple rinsing (or equivalent), offer for recycling or reconditioning, or puncture and dispose of in a sanitary landfill, or by other procedures approved by state and local authorities, such as burning of plastic containers. If burned, stay out of smoke.

LIMITED WARRANTY AND LIABILITY

Notice: Read this Limited Warranty and Liability before buying or using this product. If the terms are not acceptable, return it at once, unopened.

It is critical that this product be used and mixed only as specified on the label. The laws of a State may make some or all of this paragraph inapplicable or may give you rights in addition to your rights hereunder. Except to the extent prohibited by applicable law, the exclusive remedy of the user or buyer and the limit of liability of this Company or any other Seller for any and all losses, personal injuries or damages resulting from the use of this product, shall be the purchase price paid by the user or buyer for the quantity of product involved. Except to the extent prohibited by State Law, there is no warranty, and this Company and other Sellers disclaim all liability for losses, personal injury or damages: (i) arising from any use of this product in a manner or for a purpose not recommended in its label directions or from mixing this product before use with any substance except as recommended by the product's label; (ii) arising from handling or storage in violation of label instructions; (iii) for all indirect, special or consequential damages; (iv) when not reported to this Company within one year of discovery; and (v) arising from product not used within the label-designated shelf life or four years from the date of purchase, whichever first occurs. THERE ARE NO IMPLIED WARRANTIES AND NO WARRANTIES OF MERCHANTABILITY OR FITNESS.

Prowl® is a registered trademark of American Cyanamid Company.
Atratrol®, Bicep®, Concep®, Dual®, Evik®, Igran®, Milogard®, Pramitol®, and Princep® are registered trademarks of Ciba-Geigy Corporation, Agricultural Division.
Bexton®, Dowpon®, Garlon®, and Tordon® are registered trademarks of Dow Chemical Company.
Dalapon® is a registered trademark of SDS Biotech Corp.
Ally®, Bladex®, Extrazine®, Finesse®, Glean®, Hyvar®, Karmex®, Krenite®, Lexone®, and Velpar® are registered trademarks of E.I. duPont de Nemours & Co., Inc.
Oust™ is a trademark of E.I. duPont de Nemours & Co., Inc.
Spike® is a registered trademark of Elanco Products Company.
Weedone® is a registered trademark of Union Carbide Argicultural Products Company, Inc.
Sencor® is a registered trademark of Mobay Chemical Corporation.
Bronco®, Landmaster®, Lasso®, Ramrod®, Roundup®, and Screen® are registered trademarks of Monsanto Agricultural Products Company.
Eradicane® and Sutan +® are registered trademarks of Stauffer Chemical Company.
Asulox®, Brominal®, and Buctril® are registered trademarks of Rhone-Poulenc Chemical Company.
Gramoxone® is a registered trademark of ICI Agrichemicals, Inc.
Banvel® Herbicide, Marksman® Herbicide and Accutrol® are registered trademarks of Sandoz Ltd.

January 1989

BANVEL® SGF™

HERBICIDE

ACTIVE INGREDIENT

Sodium salt of dicamba (3,6-dichloro-o-anisic acid)*	23.15%
Sodium salts of related acids	5.79%
INERT INGREDIENTS:	71.06%
TOTAL	100.00%

*This product contains 21.06% 3,6-dichloro-o-anisic acid (dicamba) or 2 pounds per gallon.

EPA Reg. No. 55947-28 EPA Est. No. 55947-TX-1

KEEP OUT OF REACH OF CHILDREN

WARNING

PRECAUTIONARY STATEMENTS

HAZARDS TO HUMANS AND DOMESTIC ANIMALS

WARNING

Causes eye irritation. Do not get in eyes, on skin, or on clothing. Harmful if swallowed. Avoid breathing spray mist. Wash thoroughly after handling. In case of contact, wash skin with soap and water; for eyes, flush with water for 15 minutes and get medical attention.

ENVIRONMENTAL HAZARDS

Keep out of lakes, streams or ponds. Do not contaminate water by cleaning of equipment or disposal of wastes. Apply this product only as directed on label.

DIRECTIONS FOR USE

It is a violation of Federal Law to use this product in a manner inconsistent with its labeling.

For additional copies of the Specimen Label, write to the manufacturer.

BEFORE USING BANVEL SGF READ AND FOLLOW THE PRECAUTIONS

IMPORTANT

The following directions apply to all uses of BANVEL SGF. Additional precautions and restrictions will be found in each specific use section.

Do not contaminate irrigation ditches or water used for domestic purposes.

SENSITIVE CROP PRECAUTIONS

BANVEL SGF may cause injury to desirable trees and plants, particularly beans, cotton, flowers, fruit trees, grapes, ornamentals, peas, potatoes, soybeans, sunflowers, tobacco, tomatoes and other broadleaf plants when contacting their roots, stems or foliage. These plants are most sensitive to BANVEL SGF during their development or growing stage. FOLLOW THE PRECAUTIONS LISTED BELOW WHEN USING BANVEL SGF.

- Do not treat areas where either possible downward movement into the soil or surface washing may cause contact of BANVEL SGF with the roots of desirable plants such as trees and shrubs.
- Avoid making applications when spray particles may be carried by air currents to areas where sensitive crops and plants are growing. Do not spray near sensitive plants if wind is gusty or in excess of 5 mph and moving in the direction of nearby sensitive crops. However, always make applications when there is some air movement to determine the direction and distance of possible spray drift. Leave an adequate buffer zone between area to be treated and sensitive plants. Coarse sprays are less likely to drift out of the target area than fine sprays. Agriculturally approved drift-reducing additives may be used.
- Do not apply BANVEL SGF in the vicinity of sensitive crops when the temperature on the day of application is expected to exceed 85°F as drift is more likely to occur.
- Do not use aerial equipment to apply BANVEL SGF when sensitive crops and plants are growing in the vicinity of the area to be treated.
- To avoid injury to desirable plants, equipment used to apply BANVEL SGF should be thoroughly cleaned (see PROCEDURE FOR CLEANING SPRAY EQUIPMENT) before reusing to apply any other chemicals.

All crop uses of BANVEL SGF are intended for a normal growing interval between planting and harvest. No crop rotation restrictions exist if normal harvest of treated crop has occurred. If this interval is shortened, such as in cover crops that will be plowed under, do not follow up with the planting of a sensitive crop.

Crops growing under stress conditions such as drought, poor fertility, or foliar damage due to hail, wind or insects, can exhibit various injury symptoms that may be more pronounced if herbicides are applied.

Consult your local or state authorities for possible application restrictions and advice concerning these and other special local use situations. **Tank mix recommendations are for use only in states where the tank mix product and application site are registered.**

Do not apply this product through any type of irrigation system.

PROCEDURE FOR CLEANING SPRAY EQUIPMENT

The steps listed below are suggested for thorough cleaning of spray equipment following applications of BANVEL SGF or tank mixes of BANVEL SGF or tank mixes of BANVEL SGF plus 2,4-D amine.

1) Hose down thoroughly the inside as well as outside surfaces of equipment while filling the spray tank half full of water. Flush by operating sprayer until the system is purged of the rinse water.
2) Fill tank with water while adding 1 quart of household ammonia for every 25 gallons of water. Operate the pump to circulate the ammonia solution through the sprayer system for 15 to 20 minutes and discharge a small amount of the ammonia solution through the boom and nozzles. Let the solution stand for several hours, preferably overnight.
3) Flush the solution out of the spray tank through the boom.
4) Remove the nozzles and screens and flush the system with two full tanks of water.

The steps listed below are suggested for thorough cleaning of spray equipment used to apply BANVEL SGF as a tank mix with wettable powders (WP) emulsifiable concentrates (EC), or other types of water-dispersible formulations. BANVEL SGF tank mixes with water-dispersible formulations require the use of a water/detergent rinse.

5) Complete step 1.
6) Fill tank with water while adding 2 lbs. of detergent for every 40 gallons of water. Operate the pump to circulate the detergent solution through the sprayer system for 5 to 10 minutes and discharge a small amount of the solution through the boom and nozzles. Let the solution stand for several hours, preferably overnight.
7) Flush the detergent solution out of the spray tank through the boom.

8) Repeat step 1, and follow with steps 2, 3, and 4.

MIXING AND APPLICATION

UNLESS OTHERWISE SPECIFIED UNDER THE INDIVIDUAL USE HEADINGS OF THIS BOOKLET, THE FOLLOWING DIRECTIONS APPLY TO ALL CROP AND NON-CROP USES OF BANVEL SGF. REFER TO INDIVIDUAL USE SECTIONS FOR ADDITIONAL PRECAUTIONS, RESTRICTIONS, APPLICATION RATES AND TIMINGS.

BANVEL SGF is a water-soluble formulation that can be applied using water or sprayable fluid fertilizer as the carrier. If a fluid fertilizer is to be used, a compatibility test (see COMPATIBILITY TEST) should be made prior to tank mixing.

Ground or aerial application equipment which will give good spray coverage of weed foliage should be used. HOWEVER, DO NOT USE AERIAL APPLICATION EQUIPMENT IF SENSITIVE CROPS ARE GROWING IN THE VICINITY OF THE AREA TO BE TREATED.

Apply 5 to 50 gallons of diluted spray per treated acre when using ground application equipment, or 3 to 10 gallons of diluted spray per treated acre when using aerial application equipment. Use the higher level of the listed spray volumes when treating dense or tall vegetation. Use coarse sprays.

Select nozzles designed to produce minimal amounts of fine spray particles. Spray with nozzles as close to the weeds as is practical for good weed coverage.

Avoid disturbing (e.g. cultivating or mowing) treated areas for at least 7 days following application.

COMPATIBILITY TEST

Before mixing in the spray tank, it is advisable to test compatibility by mixing all components in a small container in proportionate quantities (see following table).

Amount of Herbicide to Add to One Pint of Spray Carrier (Assuming Volume is 25 Gallons per Acre)

Herbicide Formulations	Rate Per Acre	Level Teaspoons
Dry	1 lb.	1½
Liquid	1 pt.	½

If herbicide(s) do not ball-up or form flakes, sludge, gels, oily films or layers, or other precipitates, then the tested spray mix is compatible. Usually, incompatibility in any of the above described forms will occur within 5 minutes after mixing.

If components are incompatible, the use of a compatibility agent is recommended. Rerun the above COMPATIBILITY TEST with a suitable compatibility agent (¼ teaspoon is equivalent to 2 pints per 100 gallons of fluid fertilizer).

SMALL GRAINS (NOT UNDERSEEDED TO LEGUMES)

IMPORTANT

Observe all precautions. Read and follow cleaning, mixing and application instructions.

If small grains are grown for pasture only, refer to the BANVEL Herbicide Label under the PASTURE, RANGELAND and NON-CROP-LAND section.

Do not graze or harvest for livestock feed prior to crop maturity.

WEEDS CONTROLLED

BANVEL SGF, when applied at recommended rates, will control ANNUAL broadleaf weeds commonly found in small grains, such as:

Buckwheat, Wild	Pigweed, Redroot (Carelessweed)
Chamomile, Corn	Pigweed, Rough
Cockle, Corn	Pigweed, Tumble
Cockle, Cow	Ragweed, Common
Cocklebur, Common	Ragweed, Giant (Buffaloweed)
Henbit	Smartweed, Green
Knotweed	Smartweed, Pennsylvania
Kochia	Sowthistle, Annual
Ladysthumb	Sunflower, Common (Wild)
Lambsquarters, Common	Sunflower, Volunteer
Mallow, Common	Thistle, Russian
Nightshade, Black	Velvetleaf
Pennycress, Field	

BANVEL SGF and BANVEL SGF tank mixes will reduce competition from established PERENNIAL broadleaf weeds and control their seedlings.

THE SPECIAL USE OF TANK MIX FOR FALL SEEDED WHEAT ONLY allows a higher rate of 2,4-D to be used in combination with BANVEL SGF. This tank mix treatment may be used for improved performance of difficult-to-control weeds including:

*Fiddleneck (Tarweed)	Gromwell
Garlic, Wild	*Onion, Wild

*Spring applications may not control weeds that develop in the fall. For fall applications, refer to the BETWEEN CROPPING APPLICATIONS section.

RATES AND TIMINGS

Application of BANVEL SGF may be made before, during or after planting to emerged and actively growing weeds. See specific crop for timing restrictions. For best performance, make application when weeds are in the 2-3 leaf stage and rosettes are less than 2 inches across. Use the higher level of listed rate ranges when treating more mature weeds or dense vegetative growth. Surfactant should not be used when applying BANVEL SGF alone or in a tank mix on small grains except when tank mixing with registered sulfonylurea small grain herbicides such as chlorsulfuron (Glean), metsulfuron (Ally) or Finesse.

FALL SEEDED WHEAT

BANVEL SGF MUST BE APPLIED TO FALL SEEDED WHEAT PRIOR TO THE JOINTING STAGE.

BROADCAST RATE PER TREATED ACRE:

½ pint (⅛ lb. a.i.)

TANK MIX TREATMENTS

For control of grasses or additional broadleaf weeds, BANVEL SGF may be tank mixed with the following herbicides. Read and follow the label of each tank mix product used for precautionary statements, directions for use, weeds controlled and geographic and other restrictions.

BROADCAST RATE PER TREATED ACRE:

Apply ¼-½ pint (1/16-⅛ lb. a.i.) BANVEL SGF with:

Herbicide	Product Amount	lb. a.i.
2,4-D*	½-¾ pint	¼-⅜
MCPA	½-¾ pint	¼-⅜
bromoxynil (Buctril)	1-1½ pint	⅓-⅜
bromoxynil + MCPA (Bronate),	1 pint	¼ + ¼
metsulfuron (Ally)***	1/10 wt.oz.	0.0038
metribuzin (Sencor 75DF**, Lexone DF™)	⅙-⅓ lb.	⅛-¼
(Sencor 4**, Lexone 4L)	¼-½ pint	⅛-¼
(Lexone 50WP)	¼-½ lb.	⅛-¼
chlorsulfuron*** (Glean 75DF)	⅙-1.2 wt. oz.	0.008-0.024
diuron (Karmex 80WP)	½-2 lbs.	⅖-1⅗
Finesse***	0.3-0.5 wt.oz.	-

* Based on 4 pounds per gallon formulations of MCPA and 2,4-D.

** Application may be made after fall seeded wheat has started to grow and has a well established secondary root system or is beginning to tiller, but prior to the jointing stage.

*** When making tank mix applications with Glean, Ally, or Finesse add a surfactant of at least 80% active ingredient at the rate of 1-2 quarts/100 gallons of spray or not more than ¼-½% by volume. Use the higher rate of surfactant when using the lower rate ranges of the tank mix and/or when treating more mature weeds or dense vegetative growth.

SPECIAL USE TANK MIX FOR FALL SEEDED WHEAT ONLY

BANVEL SGF	½ pint	⅛ lb. a.i.
plus	plus	plus
2,4-D amine	1 to 2 pints	½ to 1 lb. a.i.
or	or	or
2,4-D ester	1 to 1½ pints	½ to ¾ lb. a.i.

Note: Do not use unless possible crop injury will be acceptable.

SPRING SEEDED WHEAT

BANVEL SGF MUST BE APPLIED BEFORE SPRING SEEDED WHEAT EXCEEDS THE 5 LEAF STAGE.

BROADCAST RATE PER TREATED ACRE:

½ pint (⅛ lb. a.i.)

TANK MIX TREATMENTS

For control of additional broadleaf weeds, BANVEL SGF may be tank mixed with the following herbicides. Read and follow the label of each tank mix product used for precautionary statements, directions for use, weeds controlled and geographic and other restrictions.

BROADCAST RATE PER TREATED ACRE:

Apply ¼-½ pint (1/16- ⅛ lb. a.i.) BANVEL SGF with:

Herbicide	Product Amount	lb. a.i.
2,4-D*	½-¾ pint	¼-⅜
MCPA	½-¾ pint	¼-⅜
bromoxynil (Brominal,	½-¾ pint	¼-⅜
Buctril)	1-1½ pint	¼-⅜
chlorsulfuron** (Glean 75DF)	⅙-½ wt. oz.	0.008-0.024

Herbicide	Product Amount	lb. a.i.
Finesse**	0.3-0.5 wt. oz.	-
metsulfuron (Ally)**	1/10 wt. oz.	0.0038

*Based on 4 pounds per gallon formulations of MCPA and 2,4-D.

**When making tank mix applications with Glean, Ally, or Finesse, add a surfactant of at least 80% active ingredient at the rate of 1-2 quarts per 100 gallons of spray or not more than 1/4-1/2% by volume. Use the higher rate of surfactant when using the lower rate ranges of the tank mix and/or when treating more mature weeds or dense vegetative growth.

FALL SEEDED BARLEY

BANVEL SGF MUST BE APPLIED TO FALL SEEDED BARLEY PRIOR TO THE JOINTING STAGE.

Note: For fall barley varieties that are seeded during the winter months or later, follow the rates and timings given for Spring Seeded Barley.

BROADCAST RATE PER TREATED ACRE:

1/2 pint (1/8 lb. a.i.)

TANK MIX TREATMENTS

For control of additional broadleaf weeds, BANVEL SGF may be tank mixed with the following herbicides. Read and follow the label of each tank mix product used for precautionary statements, directions for use, weeds controlled and geographic and other restrictions.

BROADCAST RATE PER TREATED ACRE:

Apply 1/4-1/2 pint (1/16- 1/8 lb. a.i.) BANVEL SGF with:

Herbicide	Product Amount	lb. a.i.
2,4-D*	1/2 pint	1/4
MCPA	1/2-3/4 pint	1/4-3/8
chlorsulfuron**		
(Glean 75DF)	1/6-1/2 wt. oz.	0.008-0.024
metsulfuron (Ally)**	1/10 wt. oz.	0.0038
metribuzin		
Sencor 4	1/2-1 pint	1/4-1/2
Sencor 75DF	1/3-2/3 pound	1/4-1/2

*Based on 4 pounds per gallon formulations of MCPA and 2,4-D.

**When making tank mix applications with Glean or Ally, add a surfactant of at least 80% active ingredient at the rate of 1-2 quarts/100 gallons of spray or not more than 1/4-1/2% by volume. Use the higher rate of surfactant when using the lower rate ranges of the tank mix and/or when treating more mature weeds or dense vegetative growth.

SPRING SEEDED BARLEY

BANVEL SGF MUST BE APPLIED BEFORE SPRING SEEDED BARLEY EXCEEDS THE 3 LEAF STAGE.

BROADCAST RATE PER TREATED ACRE:

3/8 pint (3/32 lb. a.i.)

TANK MIX TREATMENTS

For control of additional broadleaf weeds, BANVEL SGF may be tank mixed with the following herbicides. Read and follow the label of each tank mix product used for precautionary statements, directions for use, weeds controlled and geographic and other restrictions.

BROADCAST RATE PER TREATED ACRE:

Apply 1/4-3/8 pint (1/16-3/32 lb. a.i.) BANVEL SGF with:

Herbicide	Product Amount	lb. a.i.
MCPA*	1/2 pint	1/4
metribuzin		
Sencor 4	1/2 pint	1/4
Sencor 75DF	1/3 pound	1/4
chlorsulfuron**		
(Glean 75DF)	1/6-1/2 wt. oz.	0.008-0.024
metsulfuron (Ally)***	1/10 wt. oz.	0.0038

*Based on 4 pounds per gallon formulations of MCPA.

**When making tank mix applications with Glean or Ally, add a surfactant of at least 80% active ingredient at the rate of 1-2 quarts/100 gallons of spray or not more than 1/4-1/2% by volume. Use the higher rate of surfactant when using the lower rate ranges of the tank mix and/or when treating more mature weeds or dense vegetative growth.

FALL AND SPRING SEEDED OATS

BANVEL SGF MUST BE APPLIED BEFORE SPRING SEEDED OATS EXCEED THE 5 LEAF STAGE. APPLICATIONS TO FALL SEEDED OATS MUST BE MADE PRIOR TO THE JOINTING STAGE.

BROADCAST RATE PER TREATED ACRE:

1/2 pint (1/8 lb. a.i.)

TANK MIX TREATMENTS

For control of additional broadleaf weeds, BANVEL SGF may be tank mixed with the following herbicides. Read and follow the label of each tank mix product used for precautionary statements, directions for use, weeds controlled and geographic and other restrictions.

BROADCAST RATE PER TREATED ACRE:

Apply 1/4-1/2 pint (1/16-1/8 lb. a.i.) BANVEL SGF with:

Herbicide	Product Amount	lb. a.i.
MCPA*	1/2-3/4 pint	1/4-3/8

*Based on 4 pounds per gallon formulations of MCPA.

BETWEEN CROPPING APPLICATIONS (BCA) FOR BROADLEAF WEED CONTROL

IMPORTANT

Observe all precautions. Read and follow mixing and application instructions.

WEEDS CONTROLLED

BANVEL SGF, when applied at the recommended rates, will control many ANNUAL and BIENNIAL broadleaf weeds. In addition, BANVEL SGF will control the following PERENNIAL broadleaf weeds:

*Alfalfa	*Dogbane, Hemp
Artichoke, Jerusalem	Garic, Wild**
Bindweed, Field	Horsenettle, Carolina
Bindweed, Hedge	Nightshade, Silverleaf
Blueweed, Texas	Redvine
*Bursage	Smartweed, Swamp
(Bur Ragweed)	*Sowthistle, Perennial
(Povertyweed)	Thistle, Canada**
(Lakeweed)	Trumpetcreeper
*Dandelion, Common	(Buckvine)
*Dock, Curly	

Noted (*) perennials may be controlled using BANVEL SGF at rates lower than those recommended for other listed perennial weeds. (See RATES AND TIMINGS.)

**SPECIAL TANK MIX TREATMENTS for specific control program.

RATES AND TIMINGS

Apply BANVEL SGF as a broadcast or spot treatment to emerged and actively growing weeds after crop harvest and before a killing frost. Agriculturally approved spray additives, such as surfactants or oils, may be used to enhance spray coverage and the herbicide's penetration of weed foliage. See ROTATIONAL CROPS for recommended interval between application and planting to prevent crop injury.

For best performance, make application when ANNUAL weeds are less than 6 inches tall, when BIENNIAL weeds are in the rosette stage, and to PERENNIAL weed regrowth in late summer or fall following a mowing or tillage treatment. Most effective control of upright perennial broadleaf weeds, such as Canada thistle and Jerusalem artichoke, occurs if application is made when the majority of weeds are 8 inches or taller. Viney perennial broadleaf weeds, such as field bindweed and hedge bindweed, are best controlled when weeds are in or beyond the full bloom stage.

Avoid disturbing treated areas for at least 7 days following application. Treatments may not kill weeds which develop from seed or underground plant parts, such as rhizomes or bulblets, after the effective period for BANVEL SGF. For seedling control, a follow-up program or other cultural practices could be instituted (refer to pages 3-4) for small grain in-crop uses of BANVEL SGF.

Weed Stage & Type	BANVEL SGF per treated acre	
	Product amount	lbs. a.i.
Annual	1-2 pts.	1/4-1/2
Biennial	2-4 pts.	1/2-1
Perennial		
Perennial suppression	2-4 pts.	1/2-1
Noted (*) perennials	4-8 pts.	1-2
Other perennials	8 pts.	2

Retreatments may be made as needed; however, do not exceed a total of 8 pints (2 lbs. a.i.) of BANVEL SGF per treated acre during any given fallow period.

TANK MIX TREATMENTS

BANVEL SGF may be tank mixed with one or more of the following herbicides for control of grasses or additional broadleaf weeds. Read and follow the label of each tank mix product used for precautionary statements, directions for use, rates and timings, weeds controlled and geographic or other restrictions.

Herbicide	Rate per treated acre (lbs. a.i.)
ANNUAL WEED CONTROL	
atrazine	1/2 to 3
chlorsulfuron* (Glean)	0.016 to 0.024
	(1/3-1/2 wt. oz. product)
cyanazine (Bladex)	1 2/5 to 3 1/5
glyphosate (Roundup)	1/4 to 1/2
metribuzin (Sencor, Lexone)	1/3 to 3/4
paraquat	1/2 to 1
2,4-D	1/4 to 1/2

Herbicide	Rate per treated acre (lbs. a.i.)
PERENNIAL WEED CONTROL	
glyphosate	1 to 2
2,4-D	1 to 2

*When making tank mix applications with Glean, add a surfactant of at least 80% active ingredient at the rate of 1-2 quarts/100 gallons of spray or not more than 1/4-1/2% by volume. Use the highest rate of surfactant when using the lower rate ranges of the tank mix and/or when treating more mature weeds or dense vegetative growth.

SPECIAL TANK MIX TREATMENTS

For suppression of perennial weeds, apply 0.25-0.50 lb. a.i. BANVEL SGF with 0.25-0.50 lb. a.i. Roundup herbicide per treated acre.

For wild garlic control, apply 2 pints (1/2 lb. a.i.) BANVEL SGF with 1 1/2 lbs. acid equivalent 2,4-D low volatile ester per treated acre. Apply when wild garlic is 4 to 8 inches tall.

For Canada thistle control, use BANVEL SGF or BANVEL SGF plus Roundup Herbicide tank mix treatments.

Application may be made during fallow periods for control of volunteer barley, bulbous bluegrass, downy brome, jointed goatgrass, common rye and volunteer wheat when they are actively growing. Use 2 pints BANVEL SGF with 1/2- 3/4 lb. Kerb 50W (0.25-0.38 lb. a.i.). Fall seeded wheat may be planted 9 months or more after application. For best performance, make application between mid-October and mid-December, prior to soil freeze up.

During fallow periods, apply BANVEL SGF plus Landmaster II Herbicide to give improved control of kochia, wild buckwheat, prickly lettuce, field bindweed and Canada thistle. Use 4-8 fluid ounces of BANVEL SGF and 40-54 fluid ounces of Landmaster II Herbicide for annual weed control or 8-16 fluid ounces of BANVEL SGF plus 40-54 fluid ounces of Landmaster II Herbicide for perennial weed suppression.

ROTATIONAL CROPS

The following recommendations are based on BANVEL SGF use rates up to 8 pints (2 lbs. a.i.) per treated acre.

CORN and SORGHUM may be planted in the spring following applications made during the previous year.

SOYBEANS may be planted in the spring following applications made during the previous year. If less than 1 inch of rainfall occurs between application and first killing frost, treated areas should be cultivated to allow herbicide to come in contact with moist soil. Cultivation may take place before or immediately after ground thaw.

Soybean injury may occur if the interval between application and planting is less than specified. In areas with greater than 30 inches of rainfall, delay planting for 30 days per quart of BANVEL SGF per treated acre. In areas with less than 30 inches of rainfall, delay planting for 45 days per quart of BANVEL SGF per treated acre. Exclude days when ground is frozen.

WHEAT may be planted in the fall or spring following applications. Also, spot applications may be made any time prior to crop emergence if crop injury can be tolerated in treated areas. Wheat injury may occur if the interval between application and planting is less than specified.

East of the Mississippi River, the interval is 20 days per quart of BANVEL SGF per treated acre. Exclude days when ground is frozen.

West of the Mississippi River, the interval is 45 days per quart of BANVEL SGF per treated acre. Exclude days when ground is frozen.

Following a normal harvest of corn, sorghum, soybeans, or wheat, any rotational crop may be planted. If the interval before harvest is shortened, such as when cover crops will be plowed under, do not follow up with the planting of a sensitive crop.

CONTROL OF PERENNIAL BROADLEAF WEEDS IN CROPLAND (SPOT APPLICATION ONLY)

For Use Only in the States of Idaho, Montana, Nevada, Oregon, Utah, and Washington.

IMPORTANT

Observe all precautions. Read and follow mixing and application instructions.

Do not treat subirrigated cropland or areas where the soil remains saturated with water throughout the year.

Make only one application of BANVEL SGF per year.

WEEDS CONTROLLED

BANVEL SGF, when applied at recommended rates, will control many broadleaf weeds including:

Bindweed, Field	Knapweed, Russian
Dock, Broadleaf (Bitterdock)	Ragwort, Tansy
Dock, Curly	Spurge, Leafy
Knapweed, Black	Thistle, Canada

RATES AND TIMINGS

BANVEL SGF may be applied at any time following a crop harvest to stubble fallow or other cropland. Application should be made when weeds are actively growing and prior to a killing frost.

Apply 8 to 12 quarts (4-6 lbs. a.i.) of BANVEL SGF per treated acre. Application may be made up to one month prior to the planting of wheat.

Note: Do not use unless injury to wheat or rotated barley will be acceptable.

Barley, oats, corn, sorghum (milo), annual or perennial grass crops may be planted into treated areas one year after application. Crops grown for seed (other than perennial grass seed) should not be planted into treated areas until three years after application. Do not plant broadleaf crops such as alfalfa, beans, peas, potatoes, or sugarbeets into treated areas until two years after application.

In most cases, treatments will not kill perennial weed seedlings which germinate from seed one or two years after treatment. Once the effect of the chemical has been lost, a follow-up program for seedling control or other cultural practices should be instituted.

WIPER APPLICATION USES

IMPORTANT

Observe all precautions.

BANVEL SGF may be applied through wiper application equipment to control or suppress actively growing broadleaf weeds, brush and vines. Use a solution containing 2 parts BANVEL SGF to 1 part water. Do not contact desirable vegetation with herbicide solution. Wiper application should only be made to crops and non-cropland areas described in this label.

STORAGE AND DISPOSAL

PROHIBITIONS

Do not contaminate water, food or feed by storage or disposal.

STORAGE

Store in original container in a well-ventilated area separately from fertilizer, feed and foodstuffs. Avoid cross-contamination with other pesticides. Spillage or leakage should be contained and absorbed with clay granules, sawdust, or equivalent material for disposal.

PESTICIDE DISPOSAL

Triple rinse pesticide from containers and use rinsates in the pesticide application. Wastes which cannot be used according to label instructions may be disposed of on site or at an approved waste disposal facility.

CONTAINER DISPOSAL

Plastic or Metal: After triple rinsing (or equivalent), offer for recycling or reconditioning, or puncture and dispose of in a sanitary landfill, or by other procedures approved by state and local authorities, such as burning of plastic containers. If burned, stay out of smoke.

LIMITED WARRANTY AND LIABILITY

Notice: Read this Limited Warranty and Liability before buying or using this product. If terms are not acceptable, return it at once, unopened.

It is critical that this product be used and mixed only as specified on the label. The laws of a State may make some or all of this paragraph inapplicable or may give you rights in addition to your rights hereunder. Except to the extent prohibited by applicable law, the exclusive remedy of the user or buyer and the limit of liability of this Company or any other Seller for any and all losses, personal injuries or damages resulting from the use of this product, shall be the purchase price paid by the user or buyer for the quantity of product involved. Except to the extent prohibited by State Law, there is no warranty, and this Company and other Sellers disclaim all liability for losses, personal injury or damages: (i) arising from any use of this product in a manner or for a purpose not recommended in its label directions or from mixing this product before use with any substance except as recommended by the product's label; (ii) arising from handling or storage in violation of label instructions; (iii) for all indirect, special or consequential damages; (iv) when not reported to this Company within one year of discovery; and (v) arising from product not used within the label-designated shelf life or four years from the date of purchase, whichever occurs first. THERE ARE NO IMPLIED WARRANTIES AND NO WARRANTIES OF MERCHANTABILITY OR FITNESS.

BANVEL® SGF™ is a trademark of Sandoz Ltd.

March 1989

EVITAL®

A GRANULAR HERBICIDE FOR CRANBERRIES

For the control of certain annual and perennial weeds in cranberries grown in New Jersey, Massachusetts, Oregon, Washington, and Wisconsin.

ACTIVE INGREDIENTS:
norflurazone [4-chloro-5-(methylamino)-2-(alpha, alpha, alpha-trifluoro-m- tolyl)-3(2H)- pyridazinone]. 5%
INERT INGREDIENTS:. 95%

TOTAL. 100%

E.P.A. Reg. No. 55947-57 E.P.A. Est. No. 39578-TX-1
U.S. Pat. No. 3,644,355 and 3,834,889

KEEP OUT OF REACH OF CHILDREN
CAUTION

In case of skin contact, wash thoroughly with soap and water. Wear clean clothes and wash before reuse. Wash hands thoroughly with soap and water before eating, drinking or smoking.

STORAGE AND DISPOSAL

Do not contaminate water, food or feed by storage or disposal.

Pesticide Disposal: Wastes resulting from the use of this product may be disposed of on site or at an approved waste disposal facility.

Container Disposal: Completely empty bag into applicator equipment. Then dispose of empty bag in a sanitary landfill or by incineration, or, if allowed by state and local authorities, by burning. If burned, stay out of smoke.

Do not contaminate streams, ponds, or watershed areas with unused materials, or with hopper washings when cleaning equipment.

DIRECTIONS FOR USE

It is a violation of Federal law to use this product in a manner inconsistent with its labeling.

Apply EVITAL as a single application in the early spring after removal of winter flood and before weed growth resumes or in the fall after harvest at least two weeks before winter flood. Only one application of EVITAL per year (12 months) can be made.

Apply EVITAL granules with any suitable calibrated ground or aerial equipment to ensure uniform coverage. Flooding after application may reduce weed control. Injury to the cranberry plant may occur where applications overlap or where water stands in low wet areas of a treated bog.

WARRANTY AND CONDITIONS OF SALE

Sandoz Crop Protection warrants that the composition of this product conforms to the description on the label and is reasonable fit for the purpose stated on the label when used in accordance with directions under normal conditions of use. Sandoz Crop Protection makes no other warranty, express or implied, concerning the use of this product other than as indicated on the label. Buyer assumes all risks of use, storage or handling of this material in strict accordance with directions given on the label. This statement supersedes all other statements and should be incorporated into new and existing label revisions. (See attached memo.)

Application Rates Per Acre

80 - 120 LBS.	120 - 160 LBS.
Barnyardgrass	Broomsedge
Needlegrass	Bog rush
Rice cutgrass (sicklegrass)	Fall panicum
Smokegrass	Little bluestem (Poverty grass)
	Nutsedge *(Cyperus dentatus)*
160 LBS.	Red root
Star grass	Redtop grass
Switch grass	Spike rush
Wool grass	Summer grass

The rate of application will depend on many factors. Several are listed below.

Use Lower Rates Where One or More of the Following Exist:	Use Higher Rates Where One or More of the Following Exist:
Open, sanded, well maintained bogs	A mat of vine and weed stems overlay the soil
Sprinkler irrigation	No sprinkler irrigation
Sandy soils, low in organic matter	Peat soils, high in organic matter
Perennial weed seedlings and young plants	Established perennial weed clumps
Stressed cranberry plants and sensitive varieties (Stevens and McFarlin)	Vigorous cranberry plants

NOTE: The following rates per acre should not be exceeded in any one year:

In a newly planted bog -80 lbs.
In an established bog - 160 lbs.

2-87

JAVELIN® WG

BIOLOGICAL INSECTICIDE FOR CONTROL OF INSECT PESTS OF VEGETABLES, FRUIT AND FIELD CROPS

ACTIVE INGREDIENT:
Bacillus thuringiensis Variety *Kurstaki*, potency of 52,863 Spodoptera Units (at least 60 million viable spores) per milligram*. 6.4%
INERT INGREDENTS:. 93.6%

TOTAL. 100.0%

*Equivalent to 24 billion Spodoptera Units per pound (14.4 billion International Units per pound).

EPA Reg. No. 55947-136 EPA Est. No. 55947-CA-2

KEEP OUT OF REACH OF CHILDREN
CAUTION

STORE IN A COOL PLACE

DIRECTIONS FOR USE

It is a violation of Federal Law to use this product in a manner inconsistent with its labeling.

JAVELIN® WG is a biological insecticide specific for the control of lepidopteran larvae (see Recommended Application Rates section).

Fill spray or mixing tank ³/₄ full. Turn on agitation and pour JAVELIN WG into water while maintaining continuous agitation. Add other spray material (if any) and add balance of water. Agitate as necessary to maintain suspension. Do not allow diluted sprays to remain in the tank for more than 48 hours. JAVELIN WG is formulated to provide desirable coverage and stickability on leaf surfaces. Additional adjuvants, spreaders or stickers may be added to improve product performance, especially under heavy dew or rainy conditions. Combinations with commonly used insecticides, fungicides or other spray tank adjuvants are generally not deleterious to JAVELIN WG if the mix is used promptly.

Do not apply this product through any type of irrigation system.

GROUND APPLICATION

Use recommended amount of JAVELIN WG in a minimum of 20 gallons of water per acre depending on type of crop and requirements of state regulations.

AERIAL APPLICATION

Use recommended amount of JAVELIN WG in at least 5 gallons of water per acre. Apply early morning or evening when air is calm.

RECOMMENDED APPLICATION RATES

CROP

Asparagus, Bok Choy, Broccoli (except Raab), Brussel Sprouts, Cabbage, Cauliflower, Celery, Collards, Dry Beans and Peas, Dry Bulb Onions, Field Corn, Grapes, Kale, Lettuce (Endive, Romaine, Head Lettuce, Escarole, Butter Crunch, etc.), Melons (Cantaloupe, Crenshaw, Honeydew, Muskmelon, Watermelon), Napa, Peas, Peppers, Potatoes, Radish, Strawberries, Sugar Beets, Sweet Corn, Swiss Chard, Tomatoes

Pest	Rate Lbs/Acre	Pre-Harvest Interval
Armyworm	1/2-1 1/4	0 Days
Diamondback Moth	1/2	
Imported Cabbageworm	1/4-1/2	
Loopers	1/2-1	
Crop		
Cotton		
Pest	**Rate Lbs/Acre**	**Pre-Harvest Interval**
Looper	1/4-1/2	0 Days
Tobacco Budworm*	1/2-1 1/4	
Cotton Bollworm*	1/2-1 1/4	
Armyworm	1/2-1 1/4	
Crop		

Pest	Rate Lbs/Acre	Pre-Harvest Interval
Alfalfa (Hay and Seed), Hay Crops & other Forage Crops.		
Pest	**Rate Lbs/Acre**	**Pre-Harvest Interval**
Alfalfa Caterpillar	1/4-1/2	
Armyworm**	1/2-1 1/4	0 Days
Looper	1/4-1/2	
Crop		
Citrus		
Pest	**Rate Lbs/Acre**	**Pre-Harvest Interval**
Orangedog	1/4-1/2	
Citrus Cutworm	1/2-1 1/4	0 Days

SPECIAL INSTRUCTIONS

For most consistent control apply at first sign of newly hatched worms (1st and 2nd instar larvae).

Reapply as necessary under a pest management program that includes close scouting.

If rapid knockdown of heavy worm or nonlepidopteran populations is necessary, include an effective contact insecticide in combination with JAVELIN WG.

For heavy worm infestations use the higher JAVELIN WG rate. During situations of dense foliage and rapid growth, increasing water carrier volumes will provide more complete crop coverage and improve JAVELIN WG performance.

* For the suppression of light to moderate infestations, apply at first sign of egg laying or newly hatched worms (1st instar larvae).

** Under intense methods of western alfalfa production, conditions of mixed or overlapping generations of armyworm may exist. In this situation, it is important to monitor insect numbers and make JAVELIN WG applications when populations reach 10 worms per 180° sweep. Most consistent control will be obtained with applications at the 1 1/4 lb. rate when worms are small—first and second instar.

STORAGE AND DISPOSAL

STORAGE

Store in a cool place. Activity may be impaired by storage at prolonged temperatures above 90°F. Protect from freezing.

PESTICIDE DISPOSAL

Pesticide, spray mixture or rinse water that cannot be used according to label instruction must be disposed of according to Federal, state or local procedures.

CONTAINER DISPOSAL

Do not contaminate water, food or feed by storage or disposal. Completely empty container by shaking and tapping sides and bottom to loosen clinging particles. Empty residue into application equipment. Then dispose of container in a sanitary landfill or by incineration if allowed by state and local authorities.

LIMITED WARRANTY AND LIABILITY

NOTICE: Read this Limited Warranty and Liability before buying or using this product. If terms are not acceptable, return it at once, unopened.

It is critical that this product be used and mixed only as specified on the label. The laws of a State may make some or all of this paragraph inapplicable or may give you rights in addition to your rights hereunder. Except to the extent prohibited by applicable law, the exclusive remedy of the user or buyer and the limit of liability of this Company or any other Seller for any and all losses, personal injuries or damages resulting from the use of this product, shall be the purchase price paid by the user or buyer for the quantity of product involved. Except to the extent prohibited by State Law, there is no warranty, and this Company and other Sellers disclaim all liability for losses, personal injury or damages: (i) arising from any use of this product in a manner or for a purpose not recommended in its label directions, or from mixing this product before use with any substance except as recommended by the product's label; (ii) arising from handling or storage in violation of label instructions; (iii) for all indirect, special or consequential damages; (iv) when not reported to this Company within one year of discovery; and (v) arising from product not used within the label-designated shelf life or four years from the date of purchase, whichever occurs first. There are no implied warranties and no warranties of merchantability or fitness.

JAVELIN® WG is a Registered Trademark of Sandoz Ltd.

SP-376

August 1988

MARKSMAN®

HERBICIDE

ACTIVE INGREDIENTS:

Potassium salt of dicamba (3,6-dichloro-o-anisic acid)*	13.42%
Potassium salts of related acids	3.35%
Atrazine**	22.23%
INERT INGREDIENTS:	61.00%
TOTAL	100.00%

*Equivalent to 11.45% 3,6-dichloro-o-anisic acid (dicamba) or 1.1 pounds/gallons.

**Equivalent to 22.23% 2-chloro-4-ethylamino-6-isopropyl/amino-s-triazine (atrazine) or 2.1 pounds/gallons.

EPA Reg. No. 55947-39 EPA Est. No. 11773-1A-1

KEEP OUT OF REACH OF CHILDREN

CAUTION

SHAKE WELL BEFORE USING

PRECAUTIONARY STATEMENTS

HAZARDS TO HUMANS AND DOMESTIC ANIMALS

CAUTION

Harmful if swallowed. Avoid breathing spray mist. Avoid contact with skin, eyes, or clothing. Wash thoroughly after handling. In case of contact, wash skin with soap and water; for eyes flush with water for 15 minutes and get medical attention.

ENVIRONMENTAL HAZARDS

Keep out of lakes, streams or ponds. This pesticide is toxic to aquatic invertebrates. Do not apply directly to water or wetlands. Run off and drift from treated areas may be hazardous to aquatic organisms in neighboring areas. Do not contaminate water by cleaning of equipment or disposal of wastes.

DIRECTIONS FOR USE

It is a violation of Federal Law to use this product in a manner inconsistent with its labeling.

Do not apply this product through any type of irrigation system.

APPLICATION FOR WEED CONTROL IN FIELD, SEED*, AND SILAGE CORN

IMPORTANT

* Do not apply MARKSMAN Herbicide to seed corn without first verifying with your local seed corn company (supplier) the selectivity of MARKSMAN on your inbred line. This precaution will help avoid potential injury of sensitive varieties. Also follow this precaution before application to any specialty type corn.

MARKSMAN is not registered for use on sweet corn and popcorn.

Do not contaminate irrigation ditches or water used for domestic purposes.

Atrazine leaches readily and accepted label rates have been found to result in contamination of water supplies by way of ground-water. Therefore, users are advised to avoid use of atrazine in well drained soils, particularly areas having high ground-water tables.

Surfactants (nonionic) added to a spray mixture may improve control of drought stressed weeds but do not apply any adjuvant which contains penetrants such as petroleum and crop oils-after corn emergence, crop injury could result.

A maximum of two applications of MARKSMAN Herbicide may be made up to the 5th true leaf stage. Do not exceed a total of 5 1/4 pints MARKSMAN Herbicide per treated acre per crop year. Consult rates and timing section and do not exceed rates of application for any given time of application.

BANVEL® Herbicide can be applied before or after a MARKSMAN treatment. The combined total of the BANVEL active ingredient (dicamba) - regardless if applied as BANVEL or MARKSMAN - must not exceed a total equivalent of 0.75 lbs. per treated acre per crop year. A gallon of MARKSMAN contains 1.1 lbs. of dicamba (.14 lbs./pt.), and a gallon of BANVEL contains 4.0 lbs. of dicamba (0.5 lbs./pt.). A gallon of MARKSMAN also contains 2.1 lbs. of atrazine as an active ingredient (.26 lbs./pt.) - consider this when reading the next section on rotations.

ROTATIONAL CROPS

MARKSMAN contains some atrazine (.26 lb./pint) so follow these directions: (1) If treated crop is destroyed due to weather conditions such as hail, flooding, etc., only corn or sorghum can be planted as a replacement until the folowing year. If corn is replanted, do not apply MARKSMAN or BANVEL until after emergence. Consult label of each product for application directions and do not exceed the maximum yearly use rate for MARKSMAN or BANVEL. If sorghum is the replanted crop only BANVEL is registered for postemergence application - also follow label directions and don't exceed maximum yearly use rate. (2) If applied after June 10, do not rotate with crops other than corn or sorghum the next year, or injury to the rotated crop may occur. (3) In the High Plains and intermountain areas of the West, where rainfall is sparse and erratic or where irrigation is required, use only when corn or sorghum is to follow corn, or when a crop of untreated corn or sorghum is to precede other rotational crops. (4) Plant only corn, sorghum, soybeans or peanuts the year following treatment. For soils containing a calcareous surface layer such as those found in eastern parts of the Dakotas, KS., western MN. or NE., injury may occur to soybeans planted the year following application. (5) Small grains may be planted 10 months following treatement and all other crops not previous designated to follow a treated corn crop may be planted 18 months after application.

SENSITIVE CROP PRECAUTIONS

MARKSMAN Herbicide may cause injury to desirable broadleaf plants and trees, when contacting their roots, stems or foliage. To avoid potential off-target herbicide movement:

- Do not apply MARKSMAN in general vicinity of tobacco and tomatoes.
- Use nozzles designed to produce large spray droplets such as Delavan Raindrops Spraying Systems XR flat fans, or large capacity flood nozzles, such as Delavan D10, Spraying Systems TK10-or greater capacity.
- Use a spray pressure of 20 psi or less, 10 gpa spray volume or more.
- MARKSMAN Herbicide is intended for early season application. Although off-target movement is usually not significant at that time, Sandoz Crop Protection Corporation does recommend certain precautions be taken. Do not make application when high winds are moving towards sensitive crops, inversions are present or high temperatures are expected on the day of application.

Corn growing under stress-including but not limited to moisture, low temperatures, drought, hail and poor fertility can exhibit various injury symptoms that may be more pronounced if herbicides are applied.

Consult your local or state authorities for possible application restrictions and advice concerning these and other special local use situations.

MIXING AND APPLICATION

MARKSMAN Herbicide is a water-dispersible formulation that can be applied using water or sprayable fluid fertilizer as the carrier. Use of sprayable fluid fertilizer carrier after the corn emerges may result in crop injury. If a fluid fertilizer is to be used, a compatibility test (see COMPATIBILITY TEST) should be made prior to tank mixing.

Ground or aerial application equipment, which will give good spray coverage of weed foliage, should be used. However, do not use aerial application equipment if broadleaf crops are growing in the vicinity of the area to be treated.

Before making application, be sure all equipment is clean and not contaminated with any other materials, or crop injury or sprayer clogging may result.

Apply 10 to 50 gallons of diluted spray per treated acre when using ground application equipment, or 3 to 10 gallons of diluted spray per treated acre when using aerial application equipment. Use the higher level of the listed spray volumes when treating dense or tall vegetation. Use coarse sprays. Thoroughly clean equipment following applications of MARKSMAN Herbicide. Any tank mix with a water-dispersible formulation will require the use of a water/detergent rinse.

Avoid cultivating treated acres for at least seven days following postemergence application.

BAND TREATMENTS

MARKSMAN Herbicide may be applied as a band treatment. Use the formulas below to determine the appropriate rate and volume per treated acre.

$$\frac{\text{Band width in inches}}{\text{Row width in inches}} \times \text{Broadcast RATE per treated acre} = \text{Band RATE per treated acre}$$

$$\frac{\text{Band width in inches}}{\text{Row width in inches}} \times \text{Broadcast VOLUME per treated acre} = \text{Band VOLUME per treated acre}$$

COMPATIBILITY TEST

Before mixing in the spray tank, it is advisable to test compatibility by mixing all components in a small container in proportionate quantities (see following table).

Amount of Herbicide to Add to One Pint of Spray Carrier (Assuming Volume is 25 Gallons Per Acre)

Herbicide Formulations	Rate Per Acre	Level Teaspoons
Dry	1 lb.	$1^1/_2$
Liquid	1 pt.	$^1/_2$

If herbicide(s) do not ball-up or form flakes, sludge, gels, oily films or layers, or other precipitates, then the tested spray mix is compatible. Usually, incompatibility in any of the above described forms will occur within 5 minutes after mixing.

If components are incompatible, the use of a compatibility agent is recommended. Rerun the above COMPATIBILITY TEST with a suitable compatibility agent ($^1/_4$ teaspoon is equivalent to 2 pints per 100 gallons of fluid fertilizer).

WEEDS CONTROLLED

MARKSMAN Herbicide, when applied at recommended rates, will control many ANNUAL broadleaf weeds and give growth suppression of many PERENNIAL broadleaf weeds commonly found in corn.

ANNUALS

Buckwheat, Wild	Pigweed, Rough
Burcucumber	Pigweed, Smooth
Chickweed, Common	Pigweed (Triazine Resistant)
Clover, (Annual)	Pigweed, Tumble
Cocklebur, Common	Puncturevine
Cucumber, Wild	Purslane, Common
Jimsonweed	Ragweed, Giant (Buffaloweed)
Kochia	Sicklepod
Kochia (Triazine Resistant)	Sida, Prickly (Teaweed)
Ladysthumb	Smartweed, Green
Lambsquarter Common	Smartweed, Pennsylvania
Lambsquarter, (Triazine Resistant)	Spanishneedles
Mallow, Common	Spurge, Prostrate
Mallow, Venice	Sunflower, Common (Wild)
Morningglory, Ivyleaf	Sunflower, Volunteer
Morningglory, Tall	Tansymustard
Mustard, Wild	Thistle, Russian
Mustard (Yellowtops)	Velvetleaf
Nightshade, Black	
Pigweed, Prostrate	
Pigweed, Redtop (Carelessweed)	

PERENNIALS

Alfalfa	Dock, Curly
Artichoke, Jerusalem	Dogbane, Hemp
Bindweed, Field	Horsenettle, Carolina
Bindweed, Hedge	Lespedeza
Clovers (Perennials)	Milkweed, Common
Dandelion, Common	Smartweed, Swamp
Dock, Broadleaf (Bitterdock)	Thistle, Canada
	Vetch

RATES AND TIMINGS

PREPLANT/PREEMERGE IN NO TILLAGE CORN

Applications of MARKSMAN may be made before, during, or after planting for control of emerged and actively growing broadleaf weeds.

See Application Table below for use directions.

Application Table

Soil Texture	Organic Matter	MARKSMAN RATE
Medium or Fine (a)	2% or above (including soil above 8%)	$3^1/_2$ pints
Medium or Fine	Below 2%	(c)
Coarse (b)	2% or above	2 pints
Coarse	Below 2%	(c)

(a) Fine texture - silty clay, silty clay loam, clay.
(a) Medium texture - loam, silt loam, silt, sandy clay and sandy clay loam.
(b) Coarse texture - sandy, sandy loam and loamy sand.
(c) Delay application until after corn emergence - see EARLY POSTEMERGENCE SECTION for instructions.

The addition of an agriculturally approved surfactant, crop oil, or petroleum oil to the spray mixture may improve weed control - oil or other such penetrants are not to be applied after corn emergence.

When planting into a legume sod (e.g., alfalfa or clover), apply MARKSMAN Herbicide after 4-6 inches of regrowth has occurred.

PREEMERGENCE IN CONVENTIONAL OR REDUCED TILLAGE CORN

MARKSMAN Herbicide may be applied after planting and prior to corn emergence. Application of $3^1/_2$ pints per treated acre should be made to medium or fine textured soils which contain 2% or greater organic matter. DO NOT apply to coarse soils or any soil with less than 2% organic matter.

MARKSMAN is registered for control of a wide range of broadleaf weeds. For grass control MARKSMAN must be tank mixed with an appropriate herbicide registered to control the spectrum of grasses present. (See TANKMIX TREATMENTS).

Make applications of MARKSMAN to corn covered with at least 1½ inches of soil - that is free of soil clods - and under conditions that favor normal seed emergence. During unfavorable growing conditions (cold, wet soil), direct herbicide contact with seed or other improper crop management practices could impair seed development and result in reduced corn population. Avoidance of these stresses will allow for normal plant development.

Do not incorporate preemergence applications of MARKSMAN. In dry weather a rotary hoe can be used to break soil crust but do not disturb the soil more than one-half inch deep. Avoid tillage equipment (e.g., drags, harrows) which concentrate treated soil over seed furrow, as seed damage could result.

SEE TANK MIX SECTION FOR GRASS AND ADDITIONAL BROADLEAF WEED CONTROL.

EARLY POST EMERGENCE
(ALL TILLAGE SYSTEMS)

(Emergence to 5th true leaf stage)

MARKSMAN Herbicide at 3½ pints per treated acre may be applied during the period from corn emergence through the 5th leaf stage. For corn growing on coarse textured soils (sand, sandy loam, loamy sand), reduce the rate to 2 pints per treated acre. For drought stressed weeds the addition of an agriculturally approved nonionic surfactant to the spray mixture may improve weed control. However, do not use adjuvants containing penetrants such as petroleum and crop oils after crop emergence - injury may result.

OVERLAY
(SEQUENTIAL TREATMENTS)

MARKSMAN Herbicide may be applied to ground previously treated with one or more of the following herbicides:

Herbicide	Maximum rate per treated acre (lbs. a.i.)
Alachlor (Lasso®)	4
Alachlor + glyphosphate premix (Bronco®)	5
Atrazine (see Important Section-Rotational Crops)	3
Butylate (Sutan + ®)	6
Cyanazine (Bladex®)	4
EPTC (Eradicane®)	6
Glyphosate (Roundup®)	5
Metolachlor (Dual®)	3
Paraquat	1
Pendimethalin (Prowl®)	2
Propachlor (Bexton®, Ramrod®)	6

Read and follow the label of each of the above products for precautionary statements, directions for use and other restrictions.

TANK MIX TREATMENTS

MARKSMAN Herbicide will control emerged and actively growing broadleaf weeds when applied prior to, during, or after planting, but before the corn exceeds the 5 leaf stage. MARKSMAN Herbicide may be tank mixed with one or more of the following herbicides for control of grasses or additional broadleaf weeds. When tank mixing MARKSMAN with other products, read the label of each tank mix partner for precautionary statements, directions for use and other restrictions. Also, read the MARKSMAN label.

MARKSMAN Plus Bladex Application may be made before grasses are 1½ inches tall and the corn is not beyond the four true leaf stage. Use 1.25-4.0 lbs. a.i. Bladex per treated acre for preemergence and 1.25-2.0 lbs. a.i. for postemergence treatments. AFTER CORN EMERGENCE, USE ONLY THE BLADEX 80W FORMULATIONS.

MARKSMAN Plus Dual Application may be made until grasses reach the two-leaf stage and before corn is greater than 3 inches tall. Applications prior to crop emergence should only be applied on medium to fine textured soils containing 2.5% or more organic matter. Use 1.5-2.5 lbs. a.i. Dual per treated acre.

MARKSMAN Plus Lasso Application may be made after planting until grasses reach the two-leaf stage and before corn is greater than 3 inches tall. Applications prior to crop emergence should only be made to fine textured soils containing more than 2.5% organic matter. Apply 2.5 lbs. a.i. Lasso per treated acre.

MARKSMAN Plus Paraquat Application may be made to emerged weeds, but before corn emerges. Use 0.25-1.0 lb. a.i. paraquat per treated acre.

MARKSMAN Plus Prowl Application may be made immediately after planting but before weeds or crop emerge. Do not apply following preemerge Prowl applications. Apply only to medium soils with more than 3% organic matter and fine soils with 2.0% or more organic matter. Prowl rate is 1.5 lb. a.i. per treated acre.

MARKSMAN Plus Roundup Application may be made to emerged weeds, but before corn emerges. Use 1.0-3.0 lbs. a.i. Roundup per treated acre.

MARKSMAN Plus Atrazine For additional broadleaf weed control, apply after planting, but before the corn exceeds the 5 leaf stage. See RATES AND TIMINGS section for soil type and organic matter restrictions, plus the IMPORTANT section for rotational restrictions. Refer to table below for rates.

Soil Texture	MARKSMAN Rate	Maximum Atrazine Rate Per Treated Acre
Medium	3½ pts.	1.5 lbs. a.i.
Fine	3½ pts.	2.1 lbs. a.i.

Peat, muck, and high organic clay soils - do not apply preemerge.

Post-Harvest Weed Control on Fallow Ground in Wheat/Fallow/Wheat Rotation, Or A Wheat/Corn or Sorghum/Fallow Rotation

MARKSMAN 3.2FL Applied To Fallow Ground

CAUTION

This product contains atrazine (0.26 lbs a.i./pint) and dicamba (0.138 lbs a.i./pint) - therefore some rotational crop restrictions and sensitive crop precautions may apply - refer to the MARKSMAN label for restrictions or precautions that may pertain to use on fallow ground in either cropping system covered in this supplemental labeling.

Surfactants (nonionic), crop oil concentrate, or fluid fertilizers added to the spray mixture may improve weed control, particularly on weeds growing under stressed conditions such as drought.

IMPORTANT

Do not contaminate irrigation ditches or water used for domestic purposes.

Atrazine leaches readily and accepted label rates may result in contamination of groundwater supplies by way of leaching. Users are advised, therefore, to avoid application of atrazine on coarse, well-drained soils, particularly in areas with a high groundwater table.

Risk of atrazine carry-over exists with use on fallow ground if not used according to local University Experiment Station herbicide use recommendations and the atrazine label. Consult with your local Experiment Station personnel for the correct recommended atrazine use rate for your area and soil.

Mixing And Application

MARKSMAN Heribicide is a water-dispersible formulation that can be applied in 10 to 50 gallons of water as the spray carrier per treated acre using ground equipment. Apply 3 to 10 gallons of diluted spray per treated acre when using aerial application equipment. Read and follow the "SENSITIVE CROP PRECAUTIONS" section of this label, particularly if MARKSMAN is aerially applied. Provide sufficient agitation during mixing to maintain a uniform spray mixture.

Tank Mixes

To ensure a uniform mixture when MARKSMAN Herbicide is tankmixed with one or more other products, follow this procedure: Fill the spray tank approximately one-third full with water and with the agitator operating, add the recommended amount of ingredients using the following order: dry formulations (e.g. wettable powders, dry flowables) first, and liquid suspensions (e.g. flowables) next. Mix thoroughly and fill the tank to one-half full with continuous agitation. Add Emulsifiable Concentrate formulations last while maintaining agitation and then complete filling the spray tank with water. If a surfactant is to be used, add it last. If a drift control agent is to be used, follow the directions for mixing on the specific product label.

Sensitive Crop Precautions

MARKSMAN Herbicide may cause injury to desirable broadleaf plants and trees when contacting their roots, stems, and/or foliage. To avoid potential off-target herbicide movement:

- Do not apply MARKSMAN in the general vicinity of tobacco and tomatoes, or other highly sensitive plants.
- Use nozzles designed to produce large spray droplets.
- Use a spray pressure of 30 psi or less, and 10 gpa or more spray volume.
- Do not make applications when high winds are moving towards sensitive crops, plants, or trees.

Consult your local or state authorities for possible other application restrictions and/or advice.

Annuals Preemergence

Cheatgrass	Downy Brome
Kochia	Jointed Goatgrass
Lambsquarters, Common	Rye
Mustard	Volunteer Barley
Thistle, Russian	Volunteer Wheat
Wild Lettuce	Wild Oats

Annuals Postemergence

Buckwheat, Wild	Pigweed, Redroot
Burcucumber	Rough, Smooth
Chickweed, Common	Puncturevine
Clovers (Annual)	Purslane, Common
Cucumber, Wild	Ragweed, Common
Jimsonweed	Sicklepod
Kochia	Sida, Prickly
Lambsquarters, Common	Smartweed, Green & Pennsylvania
Mallow, Common & Venice	Spanish Needles
Morningglory, Tall Ivyleaf	Spurge, Prostrate
Mustard, Wild Yellowtops & Tansy	Thistle, Russian
Nightshade, Black	Velvetleaf

Perennial & Biennial Broadleaf Weeds
Postemergence

MARKSMAN will provide suppression or burndown of perennial or biennial broadleaf species.

Triazine resistant species are also controlled.

Wheat/Fallow/Wheat

MARKSMAN USE IN: CO, KS, NE, OK, SD, TX, & WY

For preemergence or postemergence control or suppression of the weed species listed in this label, apply MARKSMAN 3.2FL as a broadcast treatment alone or tankmixed with appropriate herbicides on all soils except those listed under the "Precautions" section of this label. A split application of MARKSMAN can be made after wheat harvest, but only in the summer to fall timeframe - the maximum labeled rate of 3.5 pints/A should not be exceeded, a split treatment should not be made in the spring, and all appropriate crop rotation restrictions must be followed.

Application Timings

Preemergence

Apply MARKSMAN 3.2FL as a broadcast preemergence treatment at the rates and timing shown in the table below. For best results, MARKSMAN should be applied within 10 days following wheat harvest.

Application Timing	MARKSMAN Application Rate
	(pints/Acre)
SUMMER-Fall (June - November)	2.0-3.5

To obtain effective control of listed annual broafleaf weeds, preemergence applications must be made prior to weed emergence.

If actively growing weeds are present at the time of application, refer to the "Postemergence" section of this label for use directions.

Active Ingredients Per Pint Of MARKSMAN 3.2FL

MARKSMAN	Dicamba	+	Atrazine
		lbs a.i.	
2.0 pints	0.28	+	0.53
2.5 pints	0.34	+	0.66
3.0 pints	0.41	+	0.79
3.5 pints	0.48	+	0.92

Postemergence

MARKSMAN can be applied in either the Summer or Fall following wheat harvest to emerged actively growing weeds. For best performance, an agriculturally approved spray additive, such as a surfactant, crop oil concentract, or fluid fertilizer is recommended to enhance spray coverage and herbicide penetration of the weed foliage. For best performance, apply MARKSMAN to ANNUAL weeds that are less than four inches tall, and at least 6 hours prior to anticipated precipitation.

For control of PERENNIAL of BIENNIAL weed species on fallow ground, MARKSMAN Herbicide can be tankmixed with BANVEL Herbicide. Refer to the Banvel Herbicide label under "Between Cropping Applications" (BCA) use directions.

Rotational Crops

The application rates and timings in this label pertain only to a cropping system of WHEAT/FALLOW/WHEAT or WHEAT/CORN or SORGHUM/FALLOW. If any other crop is to be substituted for wheat, corn, sorghum or the fallow period, the crop rotation restrictions on the MARKSMAN-CORN label must be adhered to.

Precautions

To avoid crop injury:

1) Do not use on sand soils
2) Do not treat erodable hillsides, calliche, and rocky outcroppings, or exposed calcareous subsoil.
3) Do not treat soils of the Rosebud and Canyon series in Western NE and adjoining counties in CO and WY.
4) Do not treat soils with calcareous surface layers.
5) Avoid overlap.

Wheat/Corn or Sorghum/Fallow

MARKSMAN USE IN: CO, KS, NE, OK & TX

Application Timings

Preemergence of Postemergence

For control of annual broadleaf or grass weeds following wheat harvest and in the following corn or sorghum crop when grown under minimum tillage, apply 2 to 11 pints/Acre of MARKSMAN as a broadcast treatment to wheat stubble immediately following wheat harvest. For best results, apply MARKSMAN within 10 days following wheat harvest and at least 6 hours prior to anticipated precipitation.

Plant corn or sorghum the following spring with minimum soil disturbance. Use a surface planter or a planter leaving a shallow furrow. If weeds are present at planting, remove them with a sweep plow or other suitable implement before planting.

An additional 1 pint of BANVEL Herbicide can be tankmixed with MARKSMAN for suppression of broadleaf perennial species if present at the time of application. Refer to the "Tankmix" section of this label for other herbicides that may be used in combination with MARKSMAN for control of grasses or additional broadleaf weeds. When tankmixing MARKSMAN with other products, read the label and follow the directions of each tankmix partner.

PRECAUTIONS

1) Use only on silt loam soil or finer-textured soils, or crop injury may result; 2) Wheat/corn or sorghum/fallow cropping sequence must be followed; 3) Do not apply following sorghum harvest.

NOTE

To avoid illegal residues, do not graze or feed forage from treated area to livestock. To avoid illegal residues and crop injury, do not plant any crop other than those listed on this supplemental or the MARKSMAN label within 18 months following treatment.

A split application of MARKSMAN may be made after wheat harvest, but only during the summer to fall timeframe - the maximum labeled rate of 11 pints/ A should not be exceeded, a split treatment should not be made in the spring, and all appropriate crop rotation restrictions must be followed.

NOTE

In areas where a higher ratio of atrazine to dicamba is needed (i.e. a 4 parts atrazine to 1 part dicamba ratio in OK & TX), atrazine can be tankmixed with MARKSMAN. Do not exceed the 11 pints/A rate of MARKSMAN or the equivalent 6 pints/A rate of atrazine 4L, and read the precautions concerning soil textures, split applications, and cropping sequence.

Example: Six (6) pints/A of MARKSMAN applied delivers 0.825 lbs. a.i./A of Dicamba plus 1.5 lbs a.i./A of atrazine.
Tankmixing an additional 3 pints/A of atrazine 4L (equal to 1.5 lbs a.i./A) to the 6 pints/A of MARKSMAN will provide approximately a 4.1 ratio of atrazine to dicamba (3.1 lbs a.i./A of atrazine plus 0.825 lbs a.i./A of dicamba).

MARKSMAN 3.2FL Tankmixes - Post Harvest - Fallow

Wheat/Fallow/Wheat or Wheat/Corn-Sorghum/ Fallow

2, 4-D

Apply 0.125 to 1.0 lb a.i./A as a tank mixture to MARKSMAN for improved postemergence burndown of annual or perennial broadleaf weeds. Burndown activity will particularly be enhanced on weeds growing under drought conditions or weeds that have been "topped" during the harvest operation.

Roundup or Roundup RT

Apply as a tank mixture to MARKSMAN for postemergence control of additional grass weeds or broadleaf weeds at a maximum rate of 16 ozs/ Acre. Refer to the Roundup label for use directions, precautions, and limitations.

Landmaster BW

Apply as a tank mixture for postemergence control of certain broadleaf and grass weeds at a rate of 27 to 54 ozs/Acre. Refer to the Landmaster label for use directions, precautions, weeds controlled, and limitations.

Landmaster II

Apply as a tank mixture for postemergence control of certain broadleaf and grass weeds at a rate of 27 to 54 ozs/Acre. Refer to the Landmaster II label for use directions, precautions, weeds controlled, and limitations.

Fallow Master

Apply as a tank mixture for postemergence control of certain broadleaf and grass weeds at 32 to 44 ozs/acre. Refer to the Fallow Master label for use directions, precautions, and limitations.

Gramoxone, Cyclone, Paraquat

Use as a tank mixture for additional postemergence control of certain broadleaf and grass weeds at 0.38 to 0.5 lb a.i./A. Refer to the Gramoxone label for use directions, precautions, and limitations.

Glean

Apply as a tankmix for additional preemergence or postemergence control of certain broadleaf weeds at 1/6 to 1/2 oz of product/A. Refer to the Glean label for use directions, precautions, and limitations.

Ally

Apply as a tankmix for additional preemergence or postemergence control of certain broadleaf weeds at 1/10 oz product/A. Refer to the Ally label for use directions, precautions, and limitations.

Command

Apply as a tankmix for additional preemergence control of certain broadleaf and grass weeds at 0.5 to 1.0 lba a.i./A. Refer to the Command label for use directions, precautions, and limitations.

STORAGE AND DISPOSAL

PROHIBITIONS

Do not contaminate water, food, or feed by storage or disposal. Open dumping is prohibited.

PESTICIDE DISPOSAL

Pesticide, spray mixture or rinsate that cannot be used according to label instructions must be disposed of according to Federal, State or local procedures under Subtitle C of the Resource Conservation and Recovery Act.

CONTAINER DISPOSAL

For Plastic Containers: Triple rinse (or equivalent). Then offer for recycling or reconditioning or puncture and dispose of in a sanitary landfill, or incineration, or if allowed by state and local authorities, by burning. If burned, stay out of smoke.

GENERAL

Consult Federal, State, or local disposal authorities for approved alternative procedures such as limited burning.

LIMITED WARRANTY AND LIABILITY

NOTICE: Read this Limited Warranty and Liability before buying or using this product. If terms are not acceptable, return it at once, unopened.

It is critical that this product be used and mixed only as specified on the label. The laws of a State may make some or all of this paragraph inapplicable or may give you rights in addition to your rights hereunder. Except to the extent prohibited by applicable law, the exclusive remedy of the user or buyer and the limit of liability of this Company or any other Seller for any and all losses, personal injuries or damages resulting from the use of this product, shall be the purchase price paid by the user or buyer for the quantity of product involved. Except to the extent prohibited by State Law, there is no warranty, and this Company and other Sellers disclaim all liability for losses, personal injury or damages: (i) arising from any use of this product in a manner or for a purpose not recommended in its label directions or from mixing this product before use with any substance except as recommended by the product's label; (ii) arising from handling or storage in violation of label instructions; (iii) for all indirect, special or consequential damages; (iv) when not reported to this Company within one year of discovery; and (v) arising from product not used within the label-designated shelf life or four years from the date of purchase, whichever occurs first. THERE ARE NO IMPLIED WARRANTIES AND NO WARRANTIES OF MERCHANTABILITY OR FITNESS.

Registered Trademarks

MARKSMAN® and BANVEL® are registered trademarks of Sandoz Ltd.

Ally®, Bladex® and Glean® are registered trademarks of E.I. DuPont de Nemours & Company (Inc.)
Bexton® is a registered trademark of Dow Chemical Company.
Bronco®, Landmaster BW®, Landmaster II®, Lasso®, Ramrod®, Roundup® and Roundup RT® are registered trademarks of Monsanto Agricultural Products Company.
Command® is a registered trademark of FMC Corporation.
Dual® is a registered trademark of Ciba-Geigy Corporation, Agricultural Division.
Eradicane® and Sutan +® are registered trademarks of I.C.I. Chemical Company.
Fallow Master® is a registered trademark of Sandoz, Ltd. and used under license by Monsanto Ag. Products Co.
Gramoxone® is a registered trademark of ICI Agrichemicals, Inc.
Prowl® is a registered trademark of American Cyanamid Company.

March 1989

MARKSMAN®
Herbicide

SUPPLEMENTAL LABELING

Grain Sorghum

EPA Reg. No. 55947-39

Directions For Use

It is a violation of Federal law to use this product in a manner inconsistent with its labeling.

Important

Before using this product, read and carefully observe all applicable directions, restrictions, and precautionary statements on the EPA registered container label and the MARKSMAN® Herbicide Directions For Use booklet attached to the container. Also, READ AND FOLLOW THE LABEL OF EACH TANK MIXED PRODUCT USED FOR PRECAUTIONARY STATEMENTS, DIRECTIONS FOR USE, WEEDS CONTROLLED, GEOGRAPHIC AND OTHER RESTRICTIONS.

Applications of MARKSMAN Herbicide to sorghum during periods of rapid growth may result in temporary leaning of plants or rolling of leaves. These effects are usually outgrown within 10-14 days.

On coarse soils, injury to sorghum may occur if heavy rain immediately follows application.

Rainfall or irrigation occurring within 6 hours after application may reduce the effectiveness of MARKSMAN Herbicide.

Do not apply to furrow planted sorghum until level (plowed in).

Make no more than one application of MARKSMAN Herbicide per growing season.

Delay harvest until 30 days after treatment.

Do not apply MARKSMAN Herbicide to sorghum grown for seed production.

Do not graze or feed treated sorghum forage or silage prior to mature grain stage.

Weeds Controlled

MARKSMAN Herbicide, when applied at the recommended rates and timings for grain sorghum, will control many ANNUAL broafleaf weeds (e.g. pigweed) and will reduce the competition from established PERENNIAL broadleaf weeds as well as control their seedlings.

For best performance, make application when weeds are small (less than 3 inches tall) and actively growing.

Refer to the WEEDS CONTROLLED LIST in the MARKSMAN Herbicide Directions For Use booklet for specific weeds.

Rates And Timings

MARKSMAN Herbicide may be applied postemergence in grain sorghum (milo). Applications may be made to sorghum once it has 3 leaves up until it has 5 leaves.

BROADCAST RATE PER ACRE:

1½ to 2 pints

Use 1½ pints MARKSMAN Herbicide for control of pigweed spp. that are less than 3 inches tall and are actively growing. Use 2 pints of MARKSMAN Herbicide for control of all other listed broadleaf weeds.

Do not add surfactant or crop oil.

Overlay (Sequential) Treatments

MARKSMAN Herbicide may be applied to ground previously treated with one or more of the following herbicides:

Herbicide	Maximum rate per treated acre (lbs. a.i.)
alachlor (Lasso®) (Screen®-treated seed)	4
atrazine	2.25
metolachlor (Dual®) (Concep®-treated seed)	2.50
propachlor (Ramrod®)	5

Tank Mix Treatments

MARKSMAN Herbicide may be tank mixed with the following herbicide for added residual or for grass control in sorghum. Read and follow the label of each product used for precautionary statements, directions for use, rates and timings, and other restrictions.

MARKSMAN PLUS ATRAZINE: For use only on fine or medium soils. Add 0.50 lb. a.i. atrazine per treated acre for added residual broadleaf weed control. Add 1.50-2.25 lbs. a.i. atrazine per treated acre for control of emerged grasses less than 1½ inches tall. No crop oil or surfactant is to be added to this combination - crop damage could result. Atrazine carryover may injure small grains and broadleaf crops if the total rate of atrazine exceeds the rate recommended for that specific geographic area or crop rotation.

This labeling must be in the possession of the user at the time of pesticide application.

Registered Trademarks

MARKSMAN® Herbicide is a registered trademark of Sandoz Ltd.
Lasso®, Ramrod® and Screen® are registered trademarks of Monsanto Agricultural Company.
Concep® and Dual® are registered trademarks of Ciba-Geigy Corporation, Agricultural Division

MARKSMAN® is a Registered Trademark of Sandoz Ltd.

May 1989

PROBE®
HERBICIDE

A 75% WDG FORMULATION

ACTIVE INGREDIENTS:

Methazole*	75.0%
Related Compounds	3.9%
INERT INGREDIENTS:	21.1%
TOTAL	100.0%

*2-(3,4-dichlorophenyl) -4-methyl-1,2,4-oxadiazolidine-3,5-dione.

EPA Reg. No. 55947-23 EPA Est. No. 33560-TN-1

KEEP OUT OF REACH OF CHILDREN
WARNING

PRECAUTIONARY STATEMENTS

HAZARDS TO HUMANS AND DOMESTIC ANIMALS

WARNING

Harmful if swallowed, inhaled or absorbed through the skin or eyes. Do not get in eyes, on skin or on clothing. Avoid breathing dust or spray mist. Keep out of the reach of children. Avoid contamination of feed and foodstuff. In case of skin contact, wash immediately with soap and water. Skin contact may result in acne-form reaction. For eyes, flush with plenty of water for at least 15 minutes and get medical attention. Wear impervious gloves and protective clothing when handling. Wash hands, arms and face thoroughly with soap and water before eating or smoking. Wash all contaminated clothing with detergent and hot water before reuse.

Do not contaminate domestic or irrigation waters. Observe all precautions on labels of all products used in tank mixtures.

ENVIRONMENTAL HAZARDS

Do not apply directly to water or wetlands. Do not contaminate water by cleaning of equipment or disposal of wastes. Do not apply when weather conditions favor drift from areas treated. Do not apply where runoff is likely to occur.

IMPORTANT

The following directions apply to all uses of PROBE Herbicide. Additional precautions and restrictions will be found in each specific use section.

Do not graze or feed forage from treated areas to livestock.

Crops growing under stress conditions, such as drought, disease, poor fertility or foliar damage, due to hail, wind, sand blasting, or insects, can exhibit various injury symptoms that may be more pronounced if herbicides are applied.

Do not rotate with any other crops within one year of treatment, except cover crops, which must be plowed under and not grazed.

Consult your local or state authorities for possible application restrictions and advice concerning these and other special local use situations.

Do not apply this product through any type of irrigation system.

DIRECTIONS FOR USE

PROBE Herbicide is a selective herbicide for use in cotton. Directed applications can be made early postemergence (to cotton 3-6 inches tall) or late postemergence (to cotton over 6 inches tall). Post directed applications may be made at lay-by.

MIXING AND APPLICATION

PROBE Herbicide should be applied in water solutions. Premix the proper amount of PROBE Herbicide with a small amount of water. Add this mixture to the spray tank already filled with 3/4 of the required volume of water. Maintain agitation while filling and spraying. If spray mixture is allowed to settle at any time, thorough agitation is required to resuspend the mixture before spraying is resumed. Keep by-pass line on or near bottom of tank to minimize foaming. Ground equipment with 50 mesh or coarser screens in the nozzles and in-line strainers should be used.

PROBE Herbicide may be applied as a directed broadcast or band treatment. Apply the equivalent of 20-40 gallons of water carrier per acre on a broadcast basis.

Thoroughly clean the sprayer immediately after use by flushing the system with water containing a detergent.

BAND TREATMENTS

PROBE Herbicide may be applied as a band treatment. Use the formulas below to determine the appropriate rate and volume per treated acre.

$$\frac{\text{Band Width in inches}}{\text{Row Width in Inches}} \times \text{Broadcast Rate per treated acre} = \text{Band Rate per treated acre}$$

$$\frac{\text{Band Width in inches}}{\text{Row Width in Inches}} \times \text{Broadcast Volume per treated acre} = \text{Band Volume per treated acre}$$

COMPATIBILITY TEST

Before mixing in the spray tank, it is advisable to test compatibility by mixing all components in a small container in proportionate quantities (see following table).

Amount of Herbicide to Add to One Pint of Spray Carrier (Assuming Volume is 25 Gallons per Acre)

Herbicide Formulations	Rate Per Acre	Level Teaspoons
Dry	1 lb.	1½
Liquid	1 pt.	½

If herbicide(s) do not ball-up or form flakes, sludge, gels, oily films or layers, or other precipitates, then the tested spray mix is compatible. Usually incompatibility in any of the above described forms will occur within 5 minutes after mixing.

If components are incompatible, the use of an agriculturally approved compatibility agent, such as Compex®, is recommended. Rerun the above Compatibility Test with a suitable compatibility agent (¼ teaspoon is equivalent to 2 pints per 100 gallons of spray carrier).

GENERAL WEED LIST

When applied according to label directions, PROBE Herbicide will control the following broadleaf weeds.

Anoda, Spurred	Purslane
Cocklebur, Common	Sesbania, Hemp
Lambsquarters, Common	Sicklepod
Morningglory (Annual)	Sida, Prickly (Teaweed)
Pigweed	Spurge, Prostrate
	Spurge, Spotted

RATES AND TIMINGS

For cotton grown in Texas, Oklahoma and areas west thereof, only 2 applications per season may be made. In other states, 4 applications per season may be made.

PREEMERGENCE WEED CONTROL IN COTTON

For use only in the states of Arizona, New Mexico, Oklahoma and Texas.

APPLICATION DIRECTIONS

PROBE Herbicide may be applied immediately after planting but before cotton emergence. If rainfall (or irrigation) is not anticipated before weed emergence, shallow incorporation may be used for activation. The following table gives the recommended rate based on soil texture and organic matter content. Use the appropriate level of the listed rate range based on specific soil organic matter content of area to be treated.

BROADCAST RATE PER TREATED ACRE
(pounds of Probe Herbicide)

SOIL TEXTURE	2% or less Organic Matter	greater than 2% but less than 4% Organic Matter
COARSE SOILS sand, loamy sand, and sand loam	1.5-2.0	2.0-2.5
MEDIUM SOILS loam, silt loam, silt, sandy clay loam, and sandy clay	2.0-2.5	2.5-3.3
FINE SOILS silty clay loam, clay loam, silty clay, and clay	2.5-3.3	NOT RECOMMENDED

PROBE HERBICIDE IS NOT RECOMMENDED FOR PREEMERGENCE USE ON SOILS CONTAINING GREATER THAN 4% ORGANIC MATTER.

POSTEMERGENCE USES

PROBE Herbicide may be applied as a post-directed spray when cotton is 3 inches or greater in height. In addition, PROBE Herbicide may be applied post-directed up to layby. Do not exceed 1.0 lb. of PROBE Herbicide on cotton less than 6 inches in height. Where initial application is made to dense weed populations, or where stress conditions exist, a follow-up application may be necessary for optimum control. Apply only as a directed spray to the base of the cotton plants whereby contact with the cotton foliage is minimized and weed coverage is maximized. For best performance, make application when broadleaf weeds are small (2 inches in height or less). Repeat postemergence applications may be made as needed, however, do not exceed a total of 4.0 lbs. of product (3.0 lbs. active) per cropping season.

Weed Height (inches)	Broadcast Rate Per Acre (pounds of Probe Herbicide)
cotyledon to 3	0.66-2.0

For lay-by treatment, refer to the following table:

Weed Height (inches)	Broadcast Rate Per Acre (pounds of Probe Herbicide)
cotyledon to 3	2.0

Use the higher level of the listed rate ranges when weeds approach 3 inches or when weeds are drought stressed. Use an agriculturally approved non-ionic surfactant with all PROBE Herbicide postemergence applications.

POSTEMERGENCE TANKMIX TREATMENTS

PROBE Herbicide may be tank mixed with one or more of the following herbicides for control of grasses and improved control of broadleaf weeds. Read and follow the label of each tank mix product used for precautionary statements, directions for use, rates and timings, and other restrictions. Use an agriculturally approved non-ionic surfactant with all PROBE Herbicide postemergence tank mix applications.

PROBE Herbicide plus DSMA

Apply after cotton is 3 inches high and before the first bloom. Do not make more than two applications. Use 3.6 lbs. a.i. DSMA per treated acre.

PROBE Herbicide plus MSMA

Apply after cotton is three inches high and before the first bloom. Do not make more than two applications. Use 2.0 lbs. a.i. MSMA per treated acre.

PROBE Herbicide plus cyanazine (Bladex®)

Apply after the cotton is six inches high. Use 0.6 to 1.0 lb. a.i. cyanazine per treated acre.

OVERLAY (SEQUENTIAL) TREATMENT

PROBE Herbicide may be applied to growth previously treated with one or more of the following herbicides.

Herbicide	Maximum Rate Per Treated Acre (lbs. a.i.)
Diuron (Karmex®, other name brands)	1.6
Fluometuron (Cotoran®)	2.0
Norflurazon (Zorial®)	2.0
Pendimethalin (Prowl®)	1.5
Prometryn (Caparol®)	2.8
Trifluralin (Treflan®, other name brands)	1.0

Read and follow the label of each of the above products for precautionary statements, directions for use and other restrictions.

STORAGE AND DISPOSAL

PROHIBITIONS

Do not contaminate water, food or feed by storage or disposal.

STORAGE

Store in original container in a well ventilated area separately from fertilizer, feed and foodstuffs. Avoid cross-contamination with other pesticides. Spillage or leakage should be contained, moistened, and carefully swept up avoiding inhalation of dust. Wash area of spill with detergent and water.

PESTICIDE DISPOSAL

Wastes resulting from the use of this product may be disposed of on site or at an approved disposal facility.

CONTAINER DISPOSAL

Plastic containers: Triple rinse (or equivalent). Then offer for recycling or reconditioning, or puncture and dispose of in a sanitary landfill, or incineration, or, if allowed by State and local authorities, by burning. If burned, stay out of smoke. Plastic bags: Completely empty bag into application equipment. Then dispose of empty bag in a sanitary landfill or by incineration, or, if allowed by State and local authorities, by burning. If burned, stay out of smoke.

LIMITED WARRANTY AND LIABILITY

Notice: Read this Limited Warranty and Liability before buying or using this product. If terms are not acceptable, return it at once, unopened.

It is critical that this product be used and mixed only as specified on the label. The laws of a State may make some or all of this paragraph inapplicable or may give you rights in addition to your rights hereunder. Except to the extent prohibited by applicable law, the exclusive remedy of the user or buyer and the limit of liability of this Company or any other Seller for any and all losses, personal injuries or damages resulting from the use of this product, shall be the purchase price paid by the user or buyer for the quantity of product involved. Except to the extent prohibited by State Law, there is no warranty, and this Company and other Sellers disclaim all liability for losses, personal injury or damages: (i) arising from any use of this product in a manner or for a purpose not recommended in its label directions, or from mixing this product before use with any substance except as recommended by the product's label; (ii) arising from handling or storage in violation of label instructions; (iii) for all indirect, special or consequential damages; (iv) when not reported to this Company within one year of discovery; and (v) arising from product not used within the label-designated shelf life or four years from the date of purchase, whichever occurs first. THERE ARE NO IMPLIED WARRANTIES AND NO WARRANTIES OF MERCHANTABILITY OR FITNESS.

PROBE® is a Registered Trademark of Sandoz Ltd.

Prowl® is registered trademark of American Cyanamid Company
Caparol® and Cotoron® are registered trademarks of Ciba-Geigy Corporation, Agricultural Division
Bladex® and Karmex® are registered trademarks of E.I. duPont de Nemours and Co., Inc.
Treflan® is a registered trademark of Elanco Products Company, (a division of Eli Lilly and Co).
Zorial® is a registered trademark of Sandoz Ltd.
Compex® is a registered trademark of Kalo, Inc.

March 1989

SOLICAM® DF

Herbicide for control of grass and broadleaf weeds in tree fruit and nut crops, and citrus.

ACTIVE INGREDIENT: norflurazon [4-chloro-5-(methylamino)-2-(alpha, alpha, alpha-trifluoro-*m*-tolyl)-3(2*H*)-pyridazinone]	80.0%
INERT INGREDIENTS	20.0%
TOTAL	100.0%

EPA Reg. No. 55947-78 EPA Est. No. 55618-SC-001

SOLICAM® is a Registered Trademark of Sandoz Ltd.

KEEP OUT OF REACH OF CHILDREN
CAUTION

PRECAUTIONARY STATEMENTS

HAZARDS TO HUMANS

CAUTION

Harmful if swallowed or absorbed through the skin. Avoid contact with skin, eyes or clothing. In case of skin or eye contact, flush with plenty of water.

STATEMENT OF PRACTICAL TREATMENT

If swallowed

Call a physician or poison control center. Drink 1 or 2 glasses of water and induce vomiting by touching back of throat with finger. Do not induce vomiting or give anything by mouth to an unconscious person.

If on skin

Wash with plenty of soap and water. Get medical attention if irritation persists.

If in eyes

Flush eyes with plenty of water. Get medical attention if irritation persists.

ENVIRONMENTAL HAZARDS

Do not apply directly to any body of water. Do not apply when weather conditions favor run-off or drift from treated areas. Do not contaminate water when disposing of equipment washwaters.

WEEDS CONTROLLED

SOLICAM DF at recommended rates controls the following weeds:

Grass Weeds

Annual bluegrass	*Guineagrass (seedling)
Annual ryegrass	Italian ryegrass
Annual sedge	Johnsongrass (seedling)
Bahiagrass (seedling)	*Natalgrass
Barnyardgrass	*Pangolagrass
Broadleaf signalgrass	*Sandbur
Bromus spp.	***Sandspur
Cheat	Southwestern cupgrass
Crabgrass	Six weeks grama grass
*Crowfootgrass	Tall Fescue
***Downy brome	Texas panicum
Fall panicum	*Vaseygrass
Feather fingergrass	Wild barley
Foxtails	Wild onion
*Goosegrass	Witchgrass

Broadleaf Weeds

Annual bursage	*Goldenrod
***Black mustard	**Little mallow (cheeseweed)
*Camphorweed	London rocket
Carolina geranium	Pineapple weed
Chickweed	Prostate spurge
*Common ragweed	Puncturevine
Cudweed	Redmaids
*Dayflower	Russian thistle
Dog fennel	Shepherdspurse
False dandelion (smooth	

cat's ear)	Stinging nettle
Fiddleneck	***Tumble mustard
**Filaree (redstem and whitestem)	***Virginia pepperweed
	Wild buckwheat
***Flixweed	Wild geranium

SOLICAM DF at recommended rates suppresses the following weeds:

Bermudagrass	Henbit
Common purslane	Horseweed (marestail)
Nutsedge	Pigweeds (redroot, tumble and green amaranth)
Orchardgrass	
Quackgrass	Plaintains (bracted and buckhorn)
Rhizome Johnsongrass	
*Torpedograss	Poorjoe
Western muhly (wirestem muhly)	Silverleaf nightshade (white horsenettle)
Common lambsquarters	Sowthistle
*Florida pusley	Flax-leaved fleabane
Groundsel	

* When applied at the higher rates recommended for weed control in Florida citrus.

** Treat prior to germination and incorporate with water on coarse and medium soils for adequate control.

*** Registration pending in California.

DIRECTIONS FOR USE

It is a violation of Federal law to use this product in a manner inconsistent with its labeling.

GENERAL INFORMATION

SOLICAM DF is a herbicide for control of certain grasses and broadleaf weeds in tree and vine crops, including citrus.

SOLICAM DF must be moved into the weed seed germination zone to be effective. If no rainfall occurs within four weeks after application, the product must be activated by flood irrigation or sprinkler irrigation. SOLICAM DF alone has no post-emergence activity and will not control established weeds.

Do not graze livestock on treated areas. Do not feed or graze cover crops grown in treated orchards.

Do not use SOLICAM DF in deciduous crops (except apples) in Arizona, or the Coachella, Imperial, Palo Verde and Tehachapi Valleys of California. Do not use in stone fruits on the western slope of Colorado.

APPLICATION DIRECTIONS

Apply SOLICAM DF before weeds emerge. The soil surface must be free of weeds and relatively free of plant residues. The soil should be settled and firm at the time of application.

Apply as a directed spray. Avoid contact with fruit or foliage. Make only one application per year unless otherwise indicated. Do not apply when nuts or fruit are on the ground at harvest.

Equipment

SOLICAM DF should be applied using an accurately calibrated fixed boom sprayer. Filters with screen sizes of 50 mesh or larger should be used. Supplemental applications may be made in citrus using ring drench techniques or chemigation through low volume sprinkler or drip irrigation systems. Chemigation can only be used in citrus crops.

Mixing Instructions

Clean and calibrate the sprayer. Add SOLICAM DF to the spray tank already filled with the required volume of water. This will eliminate or minimize foaming. Maintain agitation while filling and spraying. If a bypass line is used, discharge at the bottom of the tank to further minimize foaming.

Do not allow SOLICAM DF spray mixture to remain in the spray tank overnight, because settling could occur and resuspension would be difficult.

Tank Mixes

SOLICAM DF is approved as a tank mix with Roundup®, Goal®, Surflan®, Princep®, Caliber® and Karmex®. Always predetermine the compatibility of labeled tank mixes by mixing small proportional quantities in advance. Products should be added to the spray tank in the following order:

1. Wettable powders and water dispersible granules. Wettable powders should be premixed in a small amount of water. Water dispersible granulars should be added during filling. Allow both to disperse before other products are added. The order of addition is not important.
2. Flowable liquids.
3. Emulsifiable concentrates.
4. Surfactants cleared for application to growing crops. Begin adding wettable powders, flowable liquids, emulsifiable concentrates and surfactants after the spray tank is 3/4 full. Continue agitation during the addition of all the materials and while filling and spraying.

Broadcast Rates

Unless otherwise directed, use 2.5-5 lbs. of SOLICAM DF per acre as a directed spray in 20-100 gallons of water.

Use the lower rates on coarse soils. Use the higher rates on medium to fine soils or soils having organic matter content of greater than 3%.

Soil Texture	Soil Type	SOLICAM DF Rate of Application
Coarse	sand, loamy sand, sandy loam	2.5 lbs.
Medium	loam, silt loam, silt sandy clay loam	3.75 lbs.
Fine	sandy clay, clay loam, silty clay loam, silty clay, clay	5.0 lbs.

Strip or Band Treatment

When applying a strip treatment of SOLICAM DF, the following formula may be used to calculate the amount per acre:

$$\frac{\text{Width of strip in feet}}{\text{Distance between rows in feet}} \times \text{Pounds per acre for broadcast treatment} = \text{Pounds per acre for strip treatment}$$

CROP RECOMMENDATIONS

DECIDUOUS TREE AND VINE CROPS

Apply SOLICAM DF as directed on sandy loam or finer soil types from fall to early spring, before weeds emerge. For almonds, it is not recommended that SOLICAM DF be applied after bud break on sandy loam soils. In California and the Pacific Northwest, trees must be established at least 18 months in the field. Do not use on coarse-textured soil such as sand, loamy sand, and gravelly sand.

Almonds - Preharvest Application

Apply SOLICAM DF broadcast or band as directed on sandy loam or finer soil types prior to harvest for clean orchard floors. Activate SOLICAM DF with a minimum of one-two inches of irrigation water. Trees must be established at least 18 months in the field. Do not use on coarse-textured soil such as sand, loamy sand, and gravelly sand.

Grapes

Apply SOLICAM DF as directed on sandy loam or finer soil types from fall to early spring. It is not recommended that SOLICAM DF be applied after bud break on sandy loam soils. Vines must be established at least two years in the field. Do not use on nursery stock. In California, do not use on any counties south of Monterey, Kings and Tulare counties. Do not use on coarse-textured soil such as sand, loamy sand, and gravelly sand in all areas.

Apples and Peaches

Apply SOLICAM DF as directed. The product may be used on newly planted apples. Peaches must be established at least 6 months in the field. However, in California and the Pacific Northwest, peach trees must be established at least 18 months in the field. The soil around the trees should be firm and free of depressions in which rain or irrigation water could accumulate. Higher rates of 3.75-5 lbs. per acre may be used in Coastal Plains soils of the Southeast.

Apricots, Filberts, Nectarines, Pears, Plums, Prunes and Walnuts

Apply SOLICAM DF as directed. Trees must be established at least 18 months in the field in California and the Pacific Northwest. Elsewhere, trees must be established at least 12 months in the field.

Pecans

Apply SOLICAM DF as directed to bearing and non-bearing pecans. The product may also be used on newly set pecan trees. Higher rates of 2.5-5 lbs. per acre may be used in Coastal Plains soils of the Southeast.

Blueberries

Apply SOLICAM DF as a directed spray from fall to early spring before weeds emerge. The soil surface must be free of weeds and relatively free of plant residues at the time of application. Make only one application per year. SOLICAM DF may be used on newly set or established blueberry plantings.

Raspberries, Blackberries

Apply SOLICAM DF as a directed spray according to label directions from fall to early spring before weeds emerge. Apply when the crop is dormant. The soil surface must be free of weeds and relatively free of plant residues at the time of application. Make only one application per year. Application of SOLICAM DF may result in temporary bleaching or chlorosis of the leaves from which the plant will recover.

CITRUS CROPS

Apply SOLICAM DF as directed any time during the year according to local practice. SOLICAM DF may be used on bearing and non-bearing citrus. It may also be used on field grown nursery stock. Do not use on germinating citrus seed beds. Do not use where citrus is interplanted with palm trees. Do not harvest citrus crops treated at greater than 4 lbs. per acre prior to 12 weeks after SOLICAM DF treatment.

California and Arizona Citrus Crops

Apply 2.5-5 lbs. of SOLICAM DF per acre as directed. Higher rates of 3.75-5 lbs. per acre can also be used in coarse-textured soils and incorporated to a depth of not more than 3 to 4 inches for increased activity against less susceptible weeds.

Florida and Texas Citrus Crops

For oranges, grapefruit, lemons, limes, and tangerines, apply 5-10 lbs. of SOLICAM DF per acre unless otherwise directed. Use the same rate for all coarse-textured soils. SOLICAM DF may be applied up to two times per year to newly planted or established trees. Apply no more than 10 lbs. total SOLICAM DF per acre per year.

Ring Drench Application (Florida Citrus only)

Apply SOLICAM DF to newly planted citrus as a ring drench treatment at the rate of 10 lbs. product broadcast per acre. Make only one application per year. Consult the following table for the amount of SOLICAM DF to add to a 500-gallon water tank for various diameter rings.

Ounces of SOLICAM per Tank

	3 ft. ring	4 ft. ring	5 ft. ring
3 gals./tree (167 trees/tank)	4.3	7.6	12.0
5 gals./tree (100 trees/tank)	2.6	4.6	7.2
7 gals./tree (71 trees/tank)	1.8	3.3	5.2
10 gals./tree (50 trees/tank)	1.3	2.3	3.6

Injection into Crop Irrigation Systems

(Citrus Crops Only)

Low volume sprinkler - 4 to 50 gallons per hour (gph) per emitter, drip 0.5-3 gph per emitter. Point of application should be above ground.

Irrigation system should run a sufficient amount of time prior to SOLICAM DF injection to have all emitters functioning properly. After system is operating properly, length of injection should be such that at one period of time during the injection, the first and last emitters in the system contain SOLICAM DF treated water. Add SOLICAM DF to the supply tank already filled with the volume of water required for the injection period (this should be at least four (4) gallons for each pound of SOLICAM DF used). Maintain proper agitation in SOLICAM DF injection tank. SOLICAM DF should be mixed in clean water and injected down-line from filters. Following SOLICAM DF injection, system should be flushed for a period of time sufficient to clear the line of SOLICAM DF. (If SOLICAM DF application is made during a normal irrigation cycle, injection should be made during the late stage.)

Apply this product only through low volume sprinkler (micro sprinkler), and drip (trickle) irrigation systems. Do not apply this product through any other type of irrigation system.

Crop injury, lack of effectiveness, or illegal pesticide residues in the crop can result from non-uniform distribution of treated water. If you have questions about calibration, you should contact State Extension Service specialists, equipment manufacturers or other experts. Do not connect an irrigation system used for pesticide application to a public water system unless the pesticide label-prescribed safety devices for public water systems are in place. A person knowledgeable of the chemigation system and responsible for its operation, or under the supervision of the responsible person, shall shut the system down and make necessary adjustments should the need arise.

Calibration

Calculation of use rate is based on *wetted area* around emitters - *NOT* on tree acres. To determine correct amount of SOLICAM DF, use the following formula:

1. Treated area per each emitter = A
   A = 3.14 × (radius × radius)
   Example: If the average distance from emitter to perimeter of wetted area, measured one inch below soil surface - 13 inches, then
   A = 3.14 × (13″ × 13″)
   A = 3.14 × (169″)
   A = 530.7 square inches.
2. The area in square feet wet in each acre = B
   $B = \frac{A \times \text{emitters/acre}}{144}$
   Example: If there are 300 emitters per acre, then
   $B = \frac{530.7 \times 300}{144}$ = B = 1105.6 square feet wetted per acre.
3. The total area (in square feet) wet by your system = C
   C = B × acres covered by system
   Example: If the system covers 20 acres, then
   C = 1105.6 square feet per acre × 20 acres
   = 22,112 square feet wetted by system.
4. Amount of SOLICAM DF to inject = S
   Rate per treated acre of Solicam DF = R

$S = \frac{C}{43,560} \times R$ = pounds of SOLICAM DF

Example: If the desired application rate per treated acre is 1.5 lbs. of Solicam DF, then

$S = \frac{22,112}{43,560} \times 1.5 = S$ = 0.76 pounds of SOLICAM DF should be injected into the system.

(Note: This will change the amount of product from 1.0 to 1.5 lbs. of SOLICAM DF per acre.)

Application of SOLICAM DF through irrigation systems should be used as a supplemental weed control practice. The addition of SOLICAM DF through irrigation systems should help prevent breakthrough weeds at irrigation point providing the application is made before weed seeds germinate.

PRECAUTIONS FOR ALL SPRINKLER OR DRIP CHEMIGATION APPLICATIONS

1. The system must contain a functional check valve, vacuum relief valve, and low pressure drain appropriately located on the irrigation pipeline to prevent water source contamination from backflow.
2. The pesticide injection pipeline must contain a functional, automatic, quick-closing check valve to prevent the flow of fluid back toward the injection pump.
3. The pesticide injection pipeline must also contain a functional, normally closed solenoid-operated valve located on the intake side of the injection pump and connected to the system interlock to prevent fluid from being withdrawn from the supply tank when the irrigation system is either automatically or manually shut down.
4. The system must contain functional interlocking controls to automatically shut off the pesticide injection pump when the water pump motor stops, or in cases where there is no water pump, when the water pressure decreases to the point where pesticide distribution is adversely affected.
5. The irrigation line or water pump must include a functional pressure switch which will stop the water pump motor when the water pressure decreases to the point where pesticide distribution is adversely affected.
6. Systems must use a metering pump, such as a positive displacement injection pump (e.g., diaphragm pump) effectively designed and constructed of materials that are compatible with pesticides and capable of being fitted with a system interlock.
7. Do not apply when wind speed favors drift beyond the area intended for treatment.
8. Application when drift may occur, such as from windy conditions, or when systems joints and connections are leaking, or when nozzles are not providing uniform distribution, may cause crop injury.
9. Application should be directed in such a way that SOLICAM DF not come into contact with foliage.

ADDITIONAL PRECAUTIONS FOR CHEMIGATION SYSTEMS CONNECTED TO PUBLIC WATER SYSTEMS

1. Public water system means a system for the provision to the public of piped water for human consumption if such system has at least 15 service connections or regularly serves an average of at least 25 individuals daily at least 60 days out of the year.
2. Chemigation systems connected to public water systems must contain a functional, reduced-pressure zone, backflow preventer (RPZ) or the functional equivalent in the water supply line upstream from the point of pesticide introduction. As an option to the RPZ, the water from the public water system should be discharged into a reservoir tank prior to pesticide introduction. There shall be a complete physical break (air gap) between the outlet end of the fill pipe and the top or overflow rim of the reservoir tank of at least twice the inside diameter of the fill pipe.

SOLICAM DF TANK MIX COMBINATIONS

The following tank mix combinations are recommended for improved control of weeds in tree and vine crops, including citrus. No label dosage rates may be exceeded and the more restrictive label limitations and precautions should be observed. Before applying these combinations, read thoroughly all precautionary statements and other information in both labels. Refer to the Mixing Instructions section of this label for specific tank mixing directions. SOLICAM DF may not be tank mixed with any product containing a prohibition against such mixing.

SOLICAM DF plus Gramoxone (paraquat)

This tank mix combination is recommended for kill of existing vegetation and for residual preemergence control of susceptible broadleaf and grassy weeds in all tree crops listed on this label.

Apply 2.5-5 lbs. of SOLICAM DF per acre and 0.5-1 lb a.i. of Gramoxone (paraquat) per acre in 50-200 gals. of water per acre. Apply to the orchard floor avoiding contact with fruit, foliage, or stems. Add an agriculturally approved nonionic surfactant such as Ortho X-77 at the rate of 1/2 pint per

100 gallons of spray. Apply when the weeds and grasses are succulent and new growth is 1-6 inches tall. Apply this combination only once per year.

Florida citrus only - Apply 5-10 lbs. of SOLICAM DF per acre and 0.5-1 lb a.i./A of Gramoxone per acre in 50-200 gals. of water per acre. This combination may be applied up to two times per year. Apply no more than 10 lbs. total SOLICAM DF per acre per year.

Do not spray under windy conditions, and use shield for young trees. Refer to the SOLICAM DF and paraquat labels for further directions, cautions, warnings, precautions, and notes.

SOLICAM DF plus Roundup

This tank mix combination is recommended for control of existing annual and perennial weeds and for residual preemergence control of susceptible broadleaf and grassy weeds in all tree crops listed on this label.

Apply 2.5-5 lbs. of SOLICAM DF per acre as recommended on its label and 1-5 qts. of Roundup per acre as recommended on its label. Apply to the orchard floor in 20-40 gals. of water per acre, avoiding contact with fruit, foliage or other tree parts. Apply when the weeds are actively growing.

Florida citrus only - Apply 5-10 lbs. of SOLICAM DF per acre and 1-5 qts. of Roundup per acre in 20-40 gals. of water per acre. This combination may be applied up to two times per year. Apply no more than 10 lbs. total SOLICAM DF per acre per year.

Do not apply under windy conditions and use a shield for young trees. Use extreme care when applying this combination. Refer to the SOLICAM DF and Roundup labels for further directions, cautions, warnings, precautions, and notes.

SOLICAM DF plus Princep Caliber 90, Princep 80W or Registered Simazine 80W or 90 Products

This tank mix combination is recommended for improved control of broadleaf weeds, and more specifically the following: Clover, Cutleaf Eveningprimrose, Dandelion, Henbit, Horseweed (Marestail), Lambsquarters, Pigweeds, Puncturevine.

This combination may be used in the following crops: Almonds, Apples, Cherries, Filberts, Grapefruit, Lemons, Oranges, Peaches, Pears, Pecans, Plums, Walnuts.

Apply the appropriate rates in the following table as a tank mix in 20-100 gals. water per acre. Apply as a directed spray to the soil surface before weeds emerge, avoiding contact with fruit, foliage, or stems.

Soil Type	SOLICAM DF	Pounds Product per Acre Princep Caliber 90	Simazine 80W
Medium	2.5	1.125	1.25
Fine	3.75	2.25	2.5

Do not use this combination on coarse soils. Use only in those crops and locations specified in both the SOLICAM DF and Simazine 80W labels.

Florida only (grapefruit and oranges)—Apply 4-5 lbs. of SOLICAM DF per acre and 3.6-4.5 lbs. Princep Caliber 90 or 4-5 lbs. Simazine 80W per acre in 20-100 gals. of water per acre. Use the same rates in all coarse-textured soils. Apply this combination only once per year.

SOLICAM DF plus Simazine plus Gramoxone (paraquat)

0.5-1 lb a.i./A of Gramoxone may be added to the SOLICAM DF plus Simazine tank mix combinations to kill existing vegetation. Add an agriculturally approved nonionic surfactant such as Ortho X-77 at the rate of 1/2 pint per 100 gals. of spray. Refer to directions for the SOLICAM DF plus Simazine combinations for recommended crops, application rates and restrictions.

Apply to the orchard floor in at least 50 gals. of water per acre, avoiding contact with fruit, foliage or stems. Apply when the weeds are succulent and new growth is 1-6 inches tall.

Do not spray under windy conditions and use a shield for young trees. Refer to the product labels for further directions, warnings, precautions, and notes.

SOLICAM DF plus Simazine plus Roundup

1-5 qts. of Roundup may be added to the SOLICAM DF plus Simazine tank mix combinations to control existing annual and perennial weeds. Refer to directions for the SOLICAM DF plus Simazine combinations for recommended crops, application rates and restrictions.

Apply in 20-40 gals. of water per acre, avoiding contact with fruit, foliage or stems. Apply when the weeds are actively growing. Use extreme care when applying this combination. Refer to the individual product labels for further directions, warnings, precautions, and notes.

SOLICAM DF plus Goal

This tank mix combination is recommended for control of certain existing weeds, including Malva and Filaree, and improved residual control of annual broadleaf weeds.

This combination may be used in the following crops: Almonds, Apricots, Cherries, Grapes (California only), Nectarines, Peaches, Pears, Plums Prunes, Walnuts.

Apply 2.5-5 lbs. of SOLICAM DF per acre and 2-4 qts. of Goal 2E or 2.5-5 qts. of Goal 1.6E per acre in at least 50 gals. of water per acre. Add an agriculturally approved nonionic surfactant such as Triton AG-98 or Ortho X-77 at the rate of 1 qt. per 100 gals. of spray. Apply to the orchard floor avoiding contact with fruit, foliage or other tree parts. Refer to the Goal label for directions regarding application timing. Do not apply under windy conditions.

Apply this combination only once per year. Use only in those locations specified in both SOLICAM DF and Goal labels.

SOLICAM DF plus Surflan plus Gramoxone (paraquat)

This tank mix combination is recommended for kill of existing vegetation and improved residual control of certain broadleaf weeds, including Purslane, Lambsquarters, and Pigweeds.

This combination may be used in the following crops: Apples, Almonds, Cherries, Filberts, Grapefruit, Lemons, Oranges, Peaches, Pears, Plums, English Walnuts.

Apply 2.5-5 lbs. of SOLICAM DF per acre, 0.5-1 lb a.i. of Gramoxone per acre, and 2-4 qts. of Surflan A.S. or 2 2/3-5 1/3 lbs. of Surflan 75W per acre in at least 50 gals. of water. Add an agriculturally approved nonionic surfactant such as Ortho X-77 at the rate of 1/2 pint per 100 gals. of spray. Apply to the orchard floor avoiding contact with fruit, foliage or stems. Apply when the weeds are succulent and new growth is 1-6 inches tall.

Do not spray under windy conditions and use a shield for young trees. Refer to the SOLICAM DF, paraquat and Surflan labels for further directions, warnings, precautions, and notes.

SOLICAM DF plus Karmex or Other Diuron 80WP Products

There may be instances where the combination with Diuron will improve overall weed control in sandy loam, medium or fine soils. SOLICAM DF at labeled rates is safe in citrus crops in all soil types when used alone.

This combination may be used in the following crops: Apples, Citrus, Peaches, Pears, Pecans, Walnuts.

Apply in the appropriate rates in the following table as a tank mix in 20-100 gals. water per acre. Apply as directed to the soil surface before weeds emerge, avoiding contact with fruit, foliage and stems.

Soil Type	Pounds Product Per Acre SOLICAM DF	Diuron 80WP
Sandy loam	1.75	1
Medium	2.5	1.5
Fine	3.75	2

Do not use this combination on sand, loamy sand, gravelly soils or exposed subsoils, or in soils with less than 1% organic matter (0.5% for pecans). Use only in those crops and locations specified in both the SOLICAM DF and Diuron labels. In California and Arizona use this combination in citrus only.

STORAGE AND DISPOSAL

Storage: Do not contaminate water, food or feed by storage or disposal.

Pesticide Disposal: Wastes resulting from the use of this product may be disposed of on site or at any approved waste disposal facility.

Container Disposal: Completely empty container into application equipment. Then dispose of empty container in a sanitary landfill or by incineration, or, if allowed by state and local authorities, by burning. If burned, stay out of smoke.

LIMITED WARRANTY AND LIABILITY

NOTICE: Read this Limited Warranty and Liability before buying or using this product. If terms are not acceptable, return it at once, unopened.

It is critical that this product be used and mixed only as specified on the label. The laws of a State may make some or all of this paragraph inapplicable or may give you rights in addition to your rights hereunder. Except to the extent prohibited by applicable law, the exclusive remedy of the user or buyer and the limit of liability of this Company or any other Seller for any and all losses, personal injuries or damages resulting from the use of this product, shall be the purchase price paid by the user or buyer for the quantity of product involved. Except to the extent prohibited by State Law, there is no warranty, and this Company and other Sellers disclaim all liability for losses, personal injury or damages: (i) arising from any use of this product in a manner or for a purpose not recommended in its label directions or from mixing this product before use with any substance except as recommended by the product's label; (ii) arising from handling or storage in violation of label instructions; (iii) for all indirect, special or consequential damages; (iv) when not reported to this Company within one year of discovery; and (v) arising from product not used within the label-designated shelf life or four years from the date of purchase, whichever occurs first. THERE ARE NO IMPLIED WARRANTIES AND NO WARRANTIES OF MERCHANTABILITY OR FITNESS.

SOLICAM® is a Registered Trademark of Sandoz Ltd.
Gramoxone is a Registered trademark of ICI Americas Co.
Roundup is a Registered trademark of Monsanto Co.
AG-98 and Goal are Registered trademarks of Rohm and Haas Co.
Ortho X-77 is a Registered trademark of Chevron Chemical Co.
Surflan is a Registered trademark of Elanco Products Co.
Princep, Caliber are Registered trademarks of CIBA-GEIGY Corp.
Karmex is a Registered trademark of Dupont.
Sandox Crop Protection Corp. 1988

September 1988

THIOLUX®

DRY FLOWABLE SULFUR

ACTIVE INGREDIENT:	
Sulfur	80%
INERT INGREDIENTS	20%
TOTAL	100%

EPA Reg. No. 55947-48

IMPORTANT

Before using this product, read and carefully observe all applicable directions, restrictions, and precautionary statements on the EPA-registered container label attached to the container.

RE-ENTRY STATEMENT

Do not apply this product in such a manner as to directly or through drift expose workers or other persons. The area being treated must be vacated by unprotected persons. Because certain states may require more restrictive reentry intervals for various crops treated with this product, consult your State Department of Agriculture for further information. Written or oral warnings must be given to workers who are expected to be in a treated area or in an area about to be treated with this product. Do not enter treated areas for 24 hours unless appropriate protective clothing is worn.

Written or oral warnings must be given to workers who are expected to be in a treated area or in an area about to be treated with this product. When oral warnings are given, warnings shall be given in a language customarily understood by workers. Oral warnings must be given if there is reason to believe that written warnings cannot be understood by workers. Written warnings must include the following information. "Area treated with THIOLUX® on (date). Do not enter without appropriate protective clothing for 24 hours."

DIRECTIONS FOR USE

It is a violation of Federal Law to use this product in a manner inconsistent with its labeling.

Pour needed amount into spray tank under agitation. If necessary, add a few drops of oil to avoid foaming.

USE PRECAUTIONS

Some crops are susceptible to injury from sulfur under certain climatic conditions. The user is advised not to use on any crop unless local use has proven sulfur to be safe in that locality.

Sulfur may burn foliage or cause fruit injury when temperatures are excessively high. Do not apply to or allow spray drift on sulfur sensitive plants such as Anjou pears, apricots, cranberries, or spinach. Do not apply with or following an oil application for 21 to 60 days. Do not apply oil for 14 days following treatment, except for dormant sprays on pears. Follow all other label directions or precautions as indicated.

When growing crops for processing consult processor before applying sulfur.

DIRECTIONS FOR DILUTION

Dilute Application

Ground Application—Apply specified rate in 20 to 60 gallons of water per acre. Orchard Application—Apply specified rate in 100 to 800 gallons of water per acre.

Concentrate Application

Ground Application—Apply specified rate in not less than 5 gallons of water per acre. Orchard Application—Apply specified rate in 20 to 100 gallons of water per acre. Special concentrate equipment is necessary for these uses.

Air Application

Apply specified rate in 3 to 20 gallons of water per acre. Orchard rates should not be applied in less than 10 gallons of water per acre.

Use lower recommended rates when using concentrate equipment. Use higher rates when using dilute applications.

Crops

Alfalfa, Cotton: Atlantic mite, Red spider mite - Apply 3 to 5 lbs. per acre as needed.

Almonds, Cherries, Nectarines, Peaches, Plums, Prunes: Brown rot blossom blight, Cherry leafspot, Powdery mildew, Scab, Leafspot, Silver mite, Flat mite, Red spider mite - Apply 10 to 30 lbs. per acre. Apply at bloom or early petal fall. Repeat as necessary, usually 10 to 14 days after a period of wet weather. For brown rot of fruit, apply when fruit starts to ripen.

Apples, Pears: Powdery mildew, Scab, Blister mite, Red spider mite - Apply 10 to 30 lbs. per acre. Do not apply to Anjou pears. Apply pre-bloom and petal fall. Apply with cover sprays throughout the season.

Asparagus: Rust - Apply 10 to 30 lbs. per acre. Use after cutting stops. Irrigate and cultivate before applying the sulfur. Repeat at 7 to 10 day intervals throughout the season.

Avocados: Brown mite - Apply 10 to 30 lbs. per acre as needed.

Barley, Corn, Oats, Rye, Sorghum, Soybeans, Wheat: For suppression of the Red spider mite, Two-spotted mite, Grass banks mite, and Pacific mite; for control of the Atlantic mite (small grain only) and for control of Powdery mildew (soybeans only) - Apply 6 lbs. per acre. For soybeans, apply 7.5 to 12.5 lbs. per acre. Apply when disease first appears and repeat as necessary or with the regularly scheduled insect program.

Beans, Peas, Potatoes, Rutabagas, Turnips: Powdery mildew, Rust, Red spider mite - Apply 3 to 10 lbs. per acre. Start in early leaf stage. Repeat application every 10 to 14 days with or without pesticides immediately after a rain.

Broccoli, Brussels Sprouts, Cabbage, Cauliflower, Collards, Kale, Peppers: Powdery mildew, Rust, Red spider mite - Apply 3 to 5 lbs. per acre. Start in early leaf stage. Repeat application every 10 to 14 days with or without pesticides immediately after a rain.

Blackberries, Boysenberries, Gooseberries, Loganberries: Powdery mildew - Apply 6 to 15 lbs. per acre. Begin when new shoots are 6 inches long and before blossoms open or when disease first appears. Repeat at 7 to 10 day intervals or as necessary.

Blueberries, Currants, Dewberries, Huckleberries: Powdery mildew - Apply 5 to 10 lbs. per acre. Begin when new shoots are six inches long and before blossoms open or when disease first appears. Repeat at 7 to 10 day intervals or as necessary.

Carrots, Garlic, Onions: Petrobia mite and Powdery mildew - Apply 3 to 10 lbs. per acre when the disease first appears and repeat as needed.

Citrus: Flat mite, Silver mite, Rust mite, Bud mite, Red spider mite, Six spotted mite, Thrips, Yuma Spider mite - Apply 8 to 20 lbs. per acre. Apply in any spray—NOT CONTAINING OIL—or following oil by 14 days.

Cucurbits: For the control of Powdery mildew apply 4 to 6 lbs. per acre when the disease first appears and repeat as necessary. Application intervals throughout the growing season 10 to 14 days.

Grapes: Powdery mildew, Bud mite, Blister mite, Red spider mite - Apply 6 lbs. per acre. Apply in late dormant or with cutworm spray, early shoot period and through season on 14 day schedule or with pesticides.

Walnuts: Pacific mite, Almond mite, Two-spotted mite, Red spider mite, - Apply 10 to 20 lbs. per acre and repeat as necessary. Apply with mite sprays during season to increase the effectiveness. CAUTION: Certain varieties of walnuts are sensitive to sulfur sprays under certain conditions. Do not apply unless the varieties are known to be tolerant of sulfur.

Mangoes: Powdery mildew - Apply 10 to 30 lbs. per acre. Begin before bloom. Repeat in bloom, after fruit set and 3 weeks later or as necessary.

Peanuts: Rust, Red spider mite, Leaf spot - Apply 3 to 5 lbs. per acre. Start in early leaf stage. Repeat applications as necessary, usually 10 to 24 days.

Peppermint, Spearmint: Powdery mildew - Apply 4 to 6 pounds per acre when mint is 5-6" tall or when disease appears. Repeat twice, at 30 day intervals. Do not apply within 30 days of harvest.

Raspberries: Powdery mildew - Apply 6 to 15 lbs. per acre. Spray each week starting at first bloom color. Continue until all fruit is set, then repeat as weather requires.

Strawberries: Powdery mildew, Red spider mite - Apply 5 to 10 lbs. per acre. Start in early leaf stage, repeat application with or without pesticides. For Powdery mildew, begin at first signs of disease and repeat as necessary. Do not use on sulfur sensitive varieties.

Sugar Beets: Powdery mildew, Red spider mite - Apply 5 to 10 lbs. per acre at 10 to 30 day intervals with or without other pesticides as soon as the first symptoms appear.

Tomatoes: Russet mite, Powdery mildew - Apply 3 to 10 lbs. per acre. Begin as soon as first symptoms appear. Thorough coverage is required.

Ornamentals

Apply 3 to 10 lbs. per acre by ground or 3 to 20 lbs. per acre by air in aqueous spray ensuring complete plant coverage. Begin when disease first appears and repeat at 5 to 10 day intervals.

Asters, Carnations: For control of Powdery mildew.

Calendula, Cherry laurels, Chrysanthemums, Hollyhock, Smilax: For control of Leaf spot and Powdery mildew.

Clematis, Columbine, Ivy (English), Laurel, Petunia, Sage: For control of Leaf spot.

Cosmos, Crepe Myrtle, Dahlia, Delphinium, Gladiolus, Ligustrum, Sunflowers, Sweet Peas, Violets, Zinnias: For control of Powdery mildew.

Roses: For control of Powdery mildew and Black spot.

Nutrient Use

THIOLUX applications will provide sulfur for the plants' nutrient requirements and should be considered in the total plant nutrient applications.

THIOLUX can be used as a micronutrient supplement by either foliar or soil application.

THIOLUX is recommended on the following crops: Alfalfa, almonds, avocados, apples, barley, all beans, broccoli, cabbage, cauliflower, carrots, citrus, corn, grapes, lawns, melons, nectarines, pears, peaches, peppers, plums, prunes, potatotes, peanuts, sorghum, soybeans, sugar beets, strawberries, tomatoes, turnips, walnuts, watermelons, wheat and most other crops. Consult your State Agricultural Experimental Station or Extension Specialist for advice in selecting treatments to best fit local conditions.

LIMITED WARRANTY AND LIABILITY

Notice: Read this Limited Warranty and Liability before buying or using this product. If terms are not acceptable, return it at once, unopened.

It is critical that this product be used and mixed only as specified on the label. The laws of a State may make some or all of this paragraph inapplicable or may give you rights in addition to your rights hereunder. Except to the extent prohibited by applicable law, the exclusive remedy of the user or buyer and the limit of liability of this Company or any other Seller for any and all losses, personal injuries or damages resulting from the use of this product, shall be the purchase price paid by the user or buyer for the quantity of product involved. Except to the extent prohibited by State Law, there is no warranty and this Company and other Sellers disclaim all liability for losses, personal injury or damages; (i) arising from any use of this product in a manner or for a purpose not recommended in its label directions, or from mixing this product before use with any substance except as recommended by the product's label; (ii) arising from handling or storage in violation of label instructions; (iii) for all indirect, special or consequential damages; (iv) when not reported to this Company within one year of discovery. THERE ARE NO IMPLIED WARRANTIES AND NO WARRANTIES OF MERCHANTABILITY OR FITNESS.

TRIDENT™

INSECTICIDE

BIOLOGICAL INSECTICIDE FOR CONTROL OF COLORADO POTATO BEETLE

ACTIVE INGREDIENT:

Bacillus thuringiensis variety **tenebrionis** Potency of 3200 tenebrionis units per milligram*	0.32%
INERT INGREDIENTS:	99.68%
TOTAL	100.00%

*Equivalent to 12.6 billion tenebrionis units per gallon.

EPA Reg. No. 55947-131 EPA Est. No. 55947-CA-2

NOT REGISTERED IN CALIFORNIA

KEEP OUT OF REACH OF CHILDREN

CAUTION

PRECAUTIONARY STATEMENTS

HAZARDS TO HUMANS

Avoid inhalation and contact with eyes, skin and especially open cuts or wounds. When handling this product, wear gloves, dust mask or equivalent pulmonary tract protection, and goggles.

In case of contact with skin, eyes, open cuts or wounds, flush the contaminated area with water. If irritation occurs and persists, get medical attention.

ENVIRONMENTAL HAZARDS

Do not apply directly to water or wetlands. Do not contaminate water when disposing of equipment washwaters.

DIRECTIONS FOR USE

It is a violation of Federal law to use this product in a manner inconsistent with its labeling.

Do not apply this product through any type of irrigation system.

GENERAL

TRIDENT™ is a biological insecticide specific for the control of Colorado Potato Beetle (Leptinotarsa decemlineata) on potatoes. TRIDENT, while equally effective controlling chemical resistant and nonresistant varieties of Colorado Potato Beetle, is not harmful to beneficial insects.

TRIDENT should be applied against young larvae, hatchling to 1/4 inch in length. TRIDENT attacks the larval gut and must be ingested by the insect to be effective. Thorough plant coverage is essential for best results. After eating TRIDENT, larvae stop feeding within a few hours and die within 2-5 days.

Since the active ingredient, Bacillus thuringiensis is exempt from tolerance requirements, TRIDENT may be applied up to the day of harvest.

MIXING

Always shake or stir TRIDENT thoroughly before use. Add recommended amount of TRIDENT to 3/4 volume of required spray before filling tank. Maintain agitation. Do not allow diluted sprays to remain in the tank for more than 72 hours. TRIDENT is formulated to provide desirable coverage and adherence to leaf surfaces. Additional adjuvants, spreaders, or stickers may be added to improve product performance, especially under heavy dew or rainy conditions. Combinations with commonly used insecticides, fungicides or other spray tank adjuvants are generally not deleterious to TRIDENT if the mix is used promptly. Before mixing in the spray tank, it is advisable to test physical compatibility by mixing all components in a small container in proportionate quantities.

APPLICATION

VOLUME

For ground equipment use the recommended amount of TRIDENT in a minimum of 20 gallons of water per acre. Thorough coverage of all foliage is essential for good Colorado Potato Beetle control and dictates the minimum volume necessary.

RATE

Apply 4 to 6 quarts of TRIDENT for control of young larval populations of Colorado Potato Beetles, hatchling to 1/4 inch in length. Use the higher rate against heavy Colorado Potato Beetle infestations or where it is difficult to get good coverage. Reapply as necessary under a pest management program that includes close scouting.

A 3-quart rate of TRIDENT may be used following a previous, effective application of a contact insecticide and when light populations of young Colorado Potato Beetle larvae are present.

SPECIAL INSTRUCTIONS

Addition of a spreader sticker, such as pinolene to TRIDENT has demonstrated improved control of young beetle larvae.

TRIDENT is most effective against young larvae. Early application is essential for maximum effects. When heavy infestations of older larval (greater than 1/4 inch) or adult Colorado Potato Beetle occur, use an effective contact insecticide with rapid knockdown activity.

TRIDENT will not provide control of potato insect pests other than Colorado Potato Beetle. For other pests use an appropriately labeled insecticide to achieve control.

STORAGE AND DISPOSAL

Do not contaminate water, feed or food by storage or disposal.

STORAGE

Store in a cool place. Activity may be impaired by storage at temperatures above 90°F. Do not contaminate water, food, or feed by storage or disposal.

PESTICIDE DISPOSAL

Pesticide, spray mixture or rinse water that cannot be used according to label instructions must be disposed of according to federal, state or local procedures.

CONTAINER DISPOSAL

Triple rinse or equivalent. Then offer for recycling or reconditioning, or puncture and dispose of in a sanitary landfill, or if allowed by state and local authorities, by burning. If burned, stay out of smoke.

LIMITED WARRANTY AND LIABILITY

NOTICE: Read this Limited Warranty and Liability before buying or using this product. If terms are not acceptable, return it at once, unopened.

It is critical that this product be used and mixed only as specified on the label. The laws of the State may make some or all of this paragraph inapplicable or may give you rights in addition to your rights hereunder. Except to the extent prohibited by applicable law, the exclusive remedy of the user or buyer and the limit of liability of this Company or any other Seller for any and all losses, personal injuries or damages resulting from the use of this product, shall be the purchase price paid by the user or buyer for the quantity of product involved. Except to the extent prohibited by State Law, there is no warranty, and this Company and other Sellers disclaim all liability for losses, personal injury or damages: (i) arising from any use of this product in a manner or for a purpose not recommended in its label directions, or from mixing this product before use with any substance except as recommended by the product's label; (ii) arising from handling or storage in violation of label instructions; (iii) for all indirect, special or consequential damages; (iv) when not reported to this Company within one year of discovery; and (v) arising from product not used within the label-designated shelf life or two years from the date of purchase, whichever occurs first. There are no implied warranties and no warranties of merchantability or fitness.

TRIDENT™ is a Trademark of Sandoz Ltd.

January 1989

TRIDENT™ BIOLOGICAL INSECTICIDE

SUPPLEMENTAL LABELING

FOR CONTROL OF COLORADO POTATO BEETLE ON POTATOES, TOMATOES, AND EGGPLANT

Not For Sale or Distribution After August 30, 1990
Not For Use After August 30, 1990

ACTIVE INGREDIENT:

Bacillus thuringiensis variety tenebrionis Potency of 3200 tenebrionis units/mg*	0.32%
INERT INGREDIENTS:	99.68%
TOTAL	100.00%

*Equivalent to 12.6 billion tenebrionis units per gallon.

EPA Reg. No. 55947-131

EPA Est. No. 55947-CA-2

KEEP OUT OF REACH OF CHILDREN
CAUTION

Precautionary Statements
Hazards To Humans

Avoid inhalation, contact with eyes, skin, and especially open cuts or wounds.

When handling this product, wear gloves, goggles, and dust mask or equivalent nose and mouth protective covering device.

In case of contact with skin, eyes or open wounds, flush contaminated area with water. If irritation occurs and persists, get medical attention.

Environmental Hazards

Keep out of lakes, ponds or streams. Do not contaminate water when disposing of equipment wash waters.

Directions For Use

It is a violation of Federal law to use this product in a manner inconsistent with its labeling.

Do not apply this product through any type of irrigation system.

General: TRIDENT is a biological insecticide specific for the control of Colorado potato beetle (Leptinotarsa decemlineata) larvae on potatoes, tomatoes, and eggplants. TRIDENT, while equally effective in controlling chemical-resistant populations of Colorado potato beetles, is not harmful to beneficial insects.

Initiate TRIDENT application as soon as eggs begin to hatch. TRIDENT is most effective on young Colorado potato beetle larvae, in the first and second instars or up to 1/4 inch in length. TRIDENT attacks the larval gut and must be ingested by the insect to be effective. Thorough plant coverage is essential for best results. After eating TRIDENT, larvae stop feeding within a few hours and die within 2 - 4 days.

Since the active ingredient, Bacillus thuringiensis, is exempt from tolerance requirements, TRIDENT may be applied up to the day of harvest.

Mixing: Always shake or stir TRIDENT thoroughly before use. Add recommended amount of TRIDENT to 3/4 volume of required spray before filling tank. Maintain agitation. Do not allow diluted sprays to remain in the tank for more than 48 hours. TRIDENT is formulated to provide desirable coverage and adherence to leaf surfaces. Additional adjuvants, spreaders, or stickers may be added to improve product performance, especially under heavy dew or rainy conditions. Combinations with commonly used insecticides, fungicides or other spray tank adjuvants are generally not deleterious to TRIDENT if the mix is used promptly. Before mixing in the spray tank, it is advisable to test physical compatibility by mixing all components in a small container in proportionate quantities.

Application

Volume: For ground equipment use the recommended amount of TRIDENT in a minimum of 20 gallons of water per acre. For application by air, use the recommended rate of TRIDENT in a minimum of 3 gallons of water per acre. Thorough coverage of all foliage is essential for good Colorado potato beetle control and dictates the minimum volume necessary.

Rate: With ground equipment, apply 3 to 6 quarts per acre of TRIDENT for control of young larval populations of Colorado potato beetles, hatchling to 1/4 inch in length. Use the 3 quart per acre rate only when light, uniform populations of larvae are present. For aerial application, use TRIDENT at 4 to 6 quarts per acre. Use the higher rate against heavy Colorado potato beetle infestations or where it is difficult to get good coverage. Reapply as necessary under a pest management program that includes close scouting. In periods of heavy egg hatching, the application interval should be shortened to 4 to 5 days.

Early morning application of TRIDENT may enhance Colorado potato beetle control.

Special Instructions: Addition of a spreader sticker, such as pinolene to TRIDENT improves control of young beetle larvae.

TRIDENT is most effective against young larvae. Early application is essential for maximum effect. When infestations of older larval Colorado potato beetle occur, use an effective contact insecticide with rapid knockdown activity.

TRIDENT will not provide control of potato insect pests other than Colorado potato beetle. For other pests use an appropriately labeled insecticide to achieve control.

March 1989

WEEDMASTER®

HERBICIDE

ACTIVE INGREDIENTS:

Dimethylamine salt of dicamba (3,6-dichloro-o-anisic acid)*	12.4%
Dimethylamine salts of related acids	3.1%
Dimethylamine salt of 2,4 dichloro-phenoxyacetic acid**	35.7%
INERT INGREDIENTS:	48.8%
TOTAL	100.0%

*Equivalent to 10.3% 3,6-dichloro-o-anisic acid
**Equivalent to 29.6% 2,4-dichlorophenoxyacetic acid
**Isomer specific by AOAC method 6.DO1-5

Contains 1 lb. Dicamba and 2.87** lbs. 2,4-D Per Gallon

EPA Reg. No. 55947-24 EPA Est. No. 55947-TX-1

KEEP OUT OF REACH OF CHILDREN
CAUTION

SHAKE WELL BEFORE USING

PRECAUTIONARY STATEMENTS

HAZARDS TO HUMANS AND DOMESTIC ANIMALS

CAUTION

Harmful if swallowed. Avoid breathing spray mist. Avoid contact with skin, eyes or clothing. In case of contact, wash skin with soap and water; for eyes, flush with water for 15 minutes and get medical attention.

ENVIRONMENTAL HAZARDS

Keep out of lakes, streams or ponds. Do not contaminate water by cleaning of equipment or disposal of wastes.

Refer to use restrictions to protect ENDANGERED SPECIES.

DIRECTIONS FOR USE

It is a violation of Federal law to use this product in a manner inconsistent with its labeling.

Do not apply this product through any type of irrigation system.

For additional copies of the Specimen Label, write to the manufacturer.

BEFORE USING WEEDMASTER HERBICIDE READ AND FOLLOW THE PRECAUTIONS

IMPORTANT

The following directions apply to all uses of WEEDMASTER Herbicide.

Do not contaminate irrigation ditches or water used for domestic purposes.

SENSITIVE CROP PRECAUTIONS

WEEDMASTER Herbicide may cause injury to desirable trees and plants, particularly beans, cotton, flowers, fruit trees, grapes, ornamentals, peas, potatoes, soybeans, sunflowers, tobacco, tomatoes and other broadleaf plants when contacting their roots, stems or foliage. These plants are most sensitive to WEEDMASTER Herbicide during their development or growing stage. FOLLOW THE PRECAUTIONS LISTED BELOW WHEN USING WEEDMASTER HERBICIDE.

- Do not treat areas where either possible downward movement into the soil or surface washing may cause contact of WEEDMASTER Herbicide with the roots of desirable plants such as trees and shrubs.
- Avoid making applications when spray particles may be carried by air current to areas where sensitive crops and plants are growing. Do not spray near sensitive plants if wind is gusty or in excess of 5 mph and moving in the direction of nearby sensitive crops. However, always make applications when there is some air movement to determine the direction and distance of possible spray drift. Leave an adequate buffer zone between area to be treated and sensitive plants. Coarse sprays are less likely to drift out of the target area than fine sprays. Agriculturally approved drift-reducing additives may be used.
- Do not use aerial equipment to apply WEEDMASTER Herbicide when sensitive crops and plants are growing in the vicinity of area to be treated.
- To avoid injury to desirable plants, equipment used to apply WEEDMASTER Herbicide should be thoroughly cleaned (see PROCEDURE FOR CLEANING SPRAY EQUIPMENT) before reusing to apply any other chemicals.

Crops growing under stress conditions such as drought, poor fertility, or foliar damage due to hail, wind or insects, can exhibit various injury symptoms that may be more pronounced if herbicides are applied.

Consult your local state authorities for possible application restrictions and advice concerning these and other special local use situations.

MIXING AND APPLICATION

WEEDMASTER Herbicide is a water-soluble formulation that may be applied using water.

WEEDMASTER Herbicide may be applied broadcast using either ground or aerial equipment. When using ground equipment, apply 5 to 40 gallons of diluted spray per treated acre. Volume of spray applied will depend on the height, density, and type of weeds or brush being treated and on the equipment being used. When using aerial equipment, apply 3 to 10 gallons of diluted spray per treated acre.

WEEDMASTER Herbicide may be applied to individual clumps or small areas (SPOT TREATMENT) of undesirable vegetation using handgun or similar types of application equipment. Apply diluted sprays to allow complete wetting (up to runoff) of foliage and stems. For knapsack or other small capacity sprayers utilize the table below to calculate material needed. (The below table is based on the assumption that the spot treatment rate equates to 60 gallons per acre on a broadcast basis).

Knapsack Sprayer Dilution Instructions

Sprayer capacity (Gallons of Water)	Ounces* of Weedmaster to add per filled tank
1 gallon	1 oz.
2½ gallons	2.5 oz.
3 gallons	3 oz.
5 gallons	5 oz.

The addition of a surfactant can help improve control. Add ½% (.005) by volume. For example, 5 gallons (40 pts./640 oz.) of herbicide solution would require .2 pt. (3.2 oz.) of surfactant.

* 1 fluid ounce = 2 tablespoons and 1 cup (liquid) = 16 tablespoons

Select nozzles designed to produce minimal amounts of fine spray particles. Spray with nozzles as close to the weeds as is practical for good weed coverage.

WEEDMASTER Herbicide should not be applied during periods of gusty wind or when wind is in excess of 15 mph as uneven spray coverage may occur.

Avoid disturbing (e.g. cultivating or mowing) treated areas for at least 7 days following application.

PROCEDURE FOR CLEANING SPRAY EQUIPMENT

The steps listed below are suggested for thorough cleaning of spray equipment following applications of WEEDMASTER Herbicide.

1) Hose down thoroughly the inside as well as outside surfaces of equipment while filling the spray tank half full of water. Flush by operating sprayer until the system is purged of the rinse water.
2) Fill tank with water while adding 1 quart of household ammonia for every 25 gallons of water. Operate the pump to circulate the ammonia solution through the sprayer system for 15 to 20 minutes and discharge a small amount of the ammonia solution through the boom and nozzles. Let the solution stand for several hours, preferably overnight.
3) Flush the solution out of the spray tank through the boom.
4) Remove the nozzles and screens and flush the system with two full tanks of water.

PASTURE RANGELAND AND NONCROPLAND

IMPORTANT

Do not use on bent, susceptible grass pastures, lespedeza, wild winter peas, vetch, clover and alfalfa pastures as injury will occur. For pasture renovation, wait 3 weeks per quart per acre of WEEDMASTER Herbicide used before interseeding or injury may occur.

If grasses are grown for seed or for seed-down purposes, do not apply when grass is in the boot to milk stage. Do not apply to newly seeded areas or after heading begins.

Do not graze meat animals in treated areas within 30 days of slaughter.

Do not graze lactating dairy animals within 7 days of treatment.

Treated-grasses may be harvested for dry hay, however, do not harvest within 37 days of treatment.

Do not use seed from treated grasses for feed or food purposes.

Adjuvants or other spray additives (emulsifiers, surfactants, wetting agents, drift control agents, or penetrants) may be used for wetting, penetration or drift control. Spray additives must be agriculturally approved when used in pasture applications. If spray additives are used, read and follow all use recommendations and precautions on product label.

WEEDS CONTROLLED, RATES AND TIMINGS

Application rates and timings of WEEDMASTER Herbicide are given below. When perennial weeds are reaching maturity, mowing and allowing some regrowth will enhance control.

WEED STAGE/APPLICATION STAGE

WEED	WEEDMASTER RATE (PINTS/ACRE)					
	0.5	1.0	1.5	2.0	3.0	4.0
Beebalm, Spotted				pre-bloom	post bloom	
Bindweed, Field						actively growing
Broomweed	1-3"	3"-branching		branching		after branching
Buckeye spp.*					full leaf	
Buffalobur				first flower	late flower	mature
Burdock Spp.		pre-flower				
Buttercup Spp.		pre-flower		post-flower		
Chickweed, Common		seedling	1-3"	over 3"		
Chicory				rosette	early bolting	
Clover, Bur			prior to flowering	mature		
Cocklebur, Common		1-6"	6-12"	12"-30"		
Coreopsis, Plains		1-6"	6-12"	flowering	prior to	
Croton, Woolly (Doveweed, Goatweed)	1-4"	4-12"	12-30"			
Dandelion, Common		rosette		bolting		
Dewberry, Southern*						spring or fall
Dock, Curly			prior to bolting		after bolting	
Dogfennel (Cypressweed, Rosinweed)				10-"15"		
Elderberry**						actively growing
Eupatorium, Late (Thoroughwort)				10-15"		
Fleabane, Annual		1-4"	4-8"	8"		
Goldenrod, Missouri				3" to flower		
Goldenweed, Common						actively growing
Groundsel, Texas (Yellowtop)		rosette	post bolting			
Henbit			pre-flower		flower	
Honeysuckle, Hairy					spring or fall	
Horsenettle, Carolina*						full flower
Ivy, Poison				spring or fall		
Knapweed, Black**						actively growing
Knapweed, Russian**						actively growing

WEED	WEEDMASTER RATE (PINTS/ACRE)					
	0.5	1.0	1.5	2.0	3.0	4.0
Knapweed, Spotted						actively growing
Knotweed Spp.		3" runners		3" runners		
Kochia		1-6"	6-10"	10-20"		
Lambsquarters, Common		1-6"	6-10"	10-20"		
Morningglory, Tall		pre-flower		post-flower		
Mustards, Annual			rosette			
Nightshade, Silverleaf*						full flower early berry
Pennycress, Field				rosette		
Pepperweed, Virginia		1-3"	3-6"	after branching		full leaf
Persimmon, Eastern*						actively growing
Pigweed, Redroot (Carelessweed)		3"	3-10"	10"		
Poorjoe		prior to flowering		mature		
Pivet**						actively growing
Rabbitbrush**						actively growing
Ragweed, Common	1-3"	3-6"	6-10"	10"		
Ragwort, Tansy				rosette		
Redvine**						actively growing
Sage, Fringed**						actively growing
Sneezeweed, Bitter (Bitterweed)		1-4"	prior to flowering	flower		
Smartweed, Pennsylvania		4"				
Smartweed, Perennial						actively growing
Sorrel, Red (Sour Dock, Sheep Sorrell)		rosette	bolting	flower		
Sowthistle, Perennial**						actively growing
Spurge, Leafy**						actively growing
Sunflower		1-3"	3-6"	6-24"		
Thistle, Bull			rosette	bolting		
Thistle, Canada**						actively growing
Thistle, Musk			rosette,	bolting		
Thistle, Plumeless			rosette	bolting		
Velvetleaf		6"	6-20"	20"		
Vetch, Hairy		1-4"	4-8"	8" full flower		
Yankeeweed (Rosinweed)				10-18"		

* may require repeat applications.
** recommended rate will provide Top Growth Suppression only.

BETWEEN CROPPING APPLICATION (BCA) FOR BROADLEAF WEED CONTROL IN CROPLAND ROTATED TO WHEAT

IMPORTANT

Before using this product, read and carefully observe all applicable directions, restrictions, and precautionary statements on the EPA-registered container label and the WEEDMASTER Herbicide Directions For Use booklet.

WEEDS CONTROLLED

WEEDMASTER Herbicide, when applied at the recommended rates, will control the ANNUAL and BIENNIAL weeds and suppress the PERENNIAL weeds listed below.

Annuals
Buckwheat, Wild
Cockle, Cow
Cocklebur, Common
Knotweed
Kochia
Lambsquarters, Common
Mallow, Common
Mustards
Nightshade, Black
Pigweed, Redroot (Carelessweed)
Pigweed, Rough
Purslane, Common

Annuals (cont.)
Ragweed, Common
Smartweed, Pennsylvania
Sowthistle, Annual
Sunflower
Tansymustard
Thistle, Russian
Velvetleaf

Biennials
Carrot, Wild
Ragwort, Tansy
Thistle, Bull
Thistle, Musk
Thistle, Plumeless
Star Thistle, Yellow

Perennials
Bindweed, Field
Dock, Curly
Thistle, Canada

RATES AND TIMINGS

Application may be made to fallow land, wheat stubble or land to be rotated to wheat. Application should be made to emerged and actively growing weeds. Use higher rate when treating dense vegetative growth. Avoid disturbing treated areas for seven days following application.

Wheat injury may occur if the interval between application and planting is less than 10 days for each pint per acre of WEEDMASTER Herbicide used. Exclude days when ground is frozen.

Weed Type & Stage	Broadcast Rate Per Treated Acre
Annual	**Amount**
small, actively growing (less than 4 inches)	1-1½ pints
established weed growth (greater than 4 inches)	1½-3 pints
Biennial	
rosette diameter	
3 inches or less	1½-2 pints
3 inches or more	2-4 pints
greater than 4 inches, tillering, bolted or flowering	4 pints
Perennial	
suppression or top growth control	2-4 pints
seasonal control	4-8 pints

Add 0.5% v/v of an agriculturally approved surfactant to WEEDMASTER Herbicide when used alone or in a tank mix. The addition of surfactant will enhance spray coverage and the herbicide's penetration and weed foliage.

Retreatments may be made as needed; however, do not exceed a total of 8 pints of WEEDMASTER Herbicide per treated acre.

TANK MIX TREATMENTS

WEEDMASTER Herbicide may be tank mixed with one or more of the following herbicides for control of grasses or additional broadleaf weeds. Read and follow the label of each tank mix product used for precautionary statements, directions for use, rates and timings, weeds controlled, geographic or other restrictions. Add 0.5% v/v of an agriculturally approved surfactant to all tank mixes.

Herbicide	Rate Per Treated Acre (lbs. a.i)
atrazine	1/2 to 3
chlorsulfuron (Glean®)	0.016 to 0.024
cyanazine	1 3/5 to 3 1/5
glyphosphate (Roundup®)	1/4 to 2
metribuzin	1/3 to 3/4
paraquat	1/2 to 1

SUGARCANE

IMPORTANT

Consult your local or state authorities for possible application restrictions, especially concerning aerial applications and advice concerning special local use situations.

WEEDS CONTROLLED

WEEDMASTER Herbicide, when applied at recommended rates, will control the ANNUAL and PERENNIAL broadleaf weeds commonly found in sugarcane, such as:

Annuals

Burclover, California	Fleabane, Annual
Buttercup (Ranunculus Spp.)	Geranium, Carolina
Chickweed, Common	Henbit
Clovers, Annual	Lettuce, Prickly
Cocklebur, Common	Pepperweed, Virginia
Eveningprimrose, Cutleaf	Pigweed, Prostrate
Pigweed, Redroot (Carelessweed)	Shepherdspurse
Pigweed, Smooth	Sida, Prickly (Teaweed)
Purslane, Common	Sowthistle, Annual
Ragweed, Common	Spurge, Prostrate

Perennials

Aster Spp.	Garlic, Wild
Bindweed, Field	Goldenrod Spp.
Chickweed, Mouseear	Groundsel, Cressleaf
Dandelion, Common	Onion, Wild
Dock, Curly	Sowthistle, Perennial
Dogfennel	

APPLICATION DIRECTIONS

DO NOT USE AERIAL APPLICATION EQUIPMENT IF SENSITIVE CROPS ARE GROWING IN THE VICINITY OF THE AREA TO BE TREATED. When band treatments are desired apply an equivalent rate and volume per acre by using the formula below:

$$\frac{\text{Band width in inches}}{\text{Row width in inches}} \times \text{Broadcast RATE per treated acre} = \text{Band RATE per treated acre}$$

$$\frac{\text{Band width in inches}}{\text{Row width in inches}} \times \text{Broadcast VOLUME per treated acre} = \text{Band VOLUME per treated acre}$$

Application of WEEDMASTER Herbicide can be made any time after the weeds have emerged and are actively growing but prior to the close-in stage of sugarcane. When possible, direct the spray beneath the sugarcane canopy in order to minimize the likelihood of crop injury. The use of directed sprays will also aid in maximizing spray coverage of weed foliage. Application rates and timings are given on the following page. Use the higher level of listed rate ranges when treating dense vegetative growth.

- For control of listed ANNUAL broadleaf weeds, apply 1 quart of WEEDMASTER Herbicide per treated acre.
- For suppression of listed PERENNIALS, apply 1 to 2 quarts of WEEDMASTER Herbicide per treated acre.
- For control of listed PERENNIAL broadleaf weeds, apply 2 to 3 quarts of WEEDMASTER Herbicide per treated acre.

Retreatments may be made as needed, however, do not exceed 8 quarters of WEEDMASTER Herbicide per treated acre during a growing season.

TANK MIX TREATMENTS

WEEDMASTER Herbicide may be tank mixed with one or more of the following herbicides for control of grasses or additional broadleaf weeds. Read and follow the label of each tank mix product used for precautionary statements, directions for use, rates and timings, weeds controlled, geographic and other restrictions.

Herbicide	Rates Per Treated Acre (lbs. a.i.)
ametryn (Evik®)	2/5 to 8
asulam (Asulox®)	2 to 3 1/3
atrazine	2/5 to 4
dalapon (Dalapon®)	3 1/2 to 8 1/2
metribuzine (Sencor®, Lexone®)	0.75 to 3.0
terbacil (Sinbar®)	1.6 to 3.2

* based on 4 pounds per gallon formulations of MCPA and 2 pounds per gallon formulation bromoxynil.

STORAGE AND DISPOSAL

PROHIBITIONS

Do not contaminate water, food or feed by storage or disposal.

STORAGE

Store in original container in a well ventilated area separately from fertilizer, feed and food stuffs. Avoid cross-contamination with other pesticides. Spillage or leakage should be contained and absorbed with clay granules, saw dust, or equivalent material for disposal.

PESTICIDE DISPOSAL

Pesticide wastes may be toxic. Triple rinse pesticide from containers and use rinsate in the pesticide application. Improper disposal of excess pesticide, mixture, or rinsate is a violation of Federal Law. If these wastes cannot be disposed of by use according to label instructions, contact your State Pesticide or Environmental Control Agency, or the Hazardous Waste representative at the nearest EPA Regional Office for guidance.

CONTAINER DISPOSAL

Plastic or Metal: After triple rinsing (or equivalent), offer for recycling or reconditioning, or puncture and dispose of in a sanitary landfill, or by other procedures approved by state and local authorities, such as burning of plastic containers. If burned stay out of smoke.

LIMITED WARRANTY AND LIABILITY

NOTICE: Read this Limited Warranty before buying or using this product. If terms are not acceptable, return it at once, unopened.

It is critical that this product be used and mixed only as specified on the label. The laws of a State may make some or all of this paragraph inapplicable or may give you rights in addition to your rights hereunder. Except to the extent prohibited by applicable law, the exclusive remedy of the user or buyer and the limit of liability of this Company or any other Seller for any and all losses, personal injuries or damages resulting from the use of this product, shall be the purchase price paid by the user or buyer for the quantity of product involved. Except to the extent prohibited by State Law, there is no warranty, and this Company and other Sellers disclaim all liability for losses, personal injury or damages: (i) arising from any use of this product in a manner or for a purpose not recommended in its label directions, or from mixing this product before use with any substance except as recommended by the products label; (ii) arising from handling or storage in violation of label instructions; (iii) for all indirect, special or consequential damages; (iv) when not reported to this Company within one year of discovery; and (v) arising from product not used within the label-designated shelf life or four years from the date of purchase, whichever occurs first. THERE ARE NO IMPLIED WARRANTIES AND NO WARRANTIES OF MERCHANTABILITY OR FITNESS.

WEEDMASTER® is a Registered Trademark of Sandoz Ltd.

Glean®, Lexone®, Karemex®, and Sinbar® are registered trademarks of E.I. du-Pont de Nemours & Co., Inc.

Roundup® is a registered trademark of Monsanto Agricultural Products Company.

Sencor® is a registered trademark of Mobay Chemical Corporation.

Evik® and Igran® are registered trademarks of Ciba Geigy Corporation, Agricultural Division.

Asulox® is a registered trademark of Rhone-Poulenc Inc.

Dalapan® is a registered trademark of SDS Biotech Corporation.

May 1989

ZORIAL® RAPID 80™

HERBICIDE

A DRY FLOWABLE HERBICIDE FOR CONTROL OF GRASS AND BROADLEAF WEEDS IN COTTON AND SOYBEANS.

ACTIVE INGREDIENT:
Norflurazon [4-chloro-5-(methylamino)-2-(alpha, alpha, alpha-trifluoro-*m*-tolyl)-3(2*H*)-pyridazinone] 80.0%
INERT INGREDIENTS: 20.0%
TOTAL 100.0%

EPA Reg No. 55947-77 EPA Est. No. 55618-SC-001

KEEP OUT OF REACH OF CHILDREN
CAUTION

PRECAUTIONARY STATEMENTS

ENVIRONMENTAL HAZARDS

Do not apply directly to any body of water. Do not apply when weather conditions favor runoff or drift from treated areas. Do not contaminate water when disposing of equipment washwaters.

WEEDS CONTROLLED

ZORIAL® RAPID 80™ Herbicide at recommended rates is effective for:

A. Controlling the following weeds in cotton or soybeans:

Barnyardgrass	Pigweeds
Crabgrass	Prickly Sida
Foxtail	Panicum Spp.
Goosegrass[1]	Spotted Spurge[2]
Signalgrass	Spurred Anoda (Cottonweed)
Johnsongrass (seedling)	
Carpetweed	Velvetleaf
Florida Pusley	Tropic Croton[3]
Pennsylvania Smartweed	Purslane

[1] Includes DNA resistant goosegrass.
[2] Use the ZORIAL® SPLIT™ or preemergence for improved control.
[3] Use higher rates on Coastal Plains soils only, for improved control.

B. Suppressing the following weeds in cotton or soybeans:

Bermudagrass	Hemp Sesbania
Johnsongrass (rhizome)	Kochia
Nutsedge (yellow and purple)	Morningglory
	Russian Thistle
Lanceleaf Sage (Mintweed)	Sicklepod
	Other Smartweeds
Cocklbur	Venice Mallow
Common Ragweed	

LOCATION OF USE

For use in cotton and soybeans in the states of Alabama, Arkansas, North Florida, Georgia, Kentucky, Louisiana, Mississippi, Missouri, North Carolina, Oklahoma, South Carolina, Tennessee, Texas and Virginia.

DIRECTIONS FOR USE

It is a violation of Federal Law to use this product in a manner inconsistent with its labeling.

Do not apply this product through any type of irrigation system.

GENERAL INFORMATION

ZORIAL RAPID 80 is a herbicide for control of certain grasses and broadleaf weeds in cotton and soybeans. Apply ZORIAL before weeds emerge. ZORIAL can be applied preplant incorporated, preemergence surface, or as a split application in areas with 40 inches or more annual rainfall. In areas with less than 40 inches rainfall, apply as a preplant incorporated or split application only.

The soil surface should be free of clods and existing weeds prior to application.

If replanting of cotton or soybeans is necessary, a light reworking of the topsoil will retain the herbicidal activity of ZORIAL.

Peanuts can be planted as a replacement crop 30 days after ZORIAL application. Prior to planting, a light reworking of the topsoil is recommended to remove existing plant material and to retain herbicidal activity.

Do not graze treated cotton fields with livestock or feed treated cotton forage to livestock. Cover crops planted after harvest should be plowed under and not grazed or harvested. To minimize injury to small grain cover crops, they should be broadcast planted rather than drilled.

Quantity of ZORIAL applied in any one year must not exceed the broadcast rate in pounds per acre for a given soil type.

For one year following ZORIAL application, treated acreage may only be rotated to cotton, soybeans and peanuts.

Crop Safety

If the cotton or soybean plant is stressed during early development, application of ZORIAL at the label rate may result in temporary bleaching or chlorosis of the leaves from which the plant will recover.

MIXING INSTRUCTIONS

Clean and calibrate the sprayer. ADD ZORIAL RAPID 80 to the spray tank already filled with 1/4 of the required volume of the water. Maintain agitation while filling and spraying. If a by-pass line is used, discharge at the bottom of the tank to minimize foaming.

Do not allow ZORIAL RAPID 80 spray mixture to remain in the spray tank overnight because settling could occur and resuspension would be difficult.

Tank Mixes ZORIAL RAPID 80 is approved as a tank mix with Prowl®, Treflan®, Sencor®, Lexone®, Cotoran®, Dual®, Lasso®, Lorox®, and Bladex®. Always predetermine the compatibility of labeled tank mixes by mixing small proportional quantities in advance.

Products should be added to the spray tank in the following order:

1. Wettable powders and water dispersible granules. Wettable powders should be premixed in a small amount of water. Water dispersible granulars should be added during filling. Allow both to disperse before other products are added. The order of addition is not important.
2. Flowable liquids.
3. Emulsifiable concentrates.
4. Surfactants.

Begin adding wettable powders, flowable liquids, emulsifiable concentrates, and surfactants after the spray tank is 3/4 full. Continue agitation during the addition of all the materials and while filling and spraying.

APPLICATION DIRECTIONS

Ground Application Apply the recommended rate of ZORIAL in 10-20 gallons of water per acre. Apply accurately with a fixed boom sprayer. Equip sprayer with 8002 to 8006 fan-jet nozzle tips or eqivalent for broadcast application. Use 8002E to 8006E nozzle tips or equivalent for band applications. Use screen size 50 mesh or larger.

Aerial Application Apply the recommended rate of ZORIAL in 5 or more gallons of water per acre. To minimize drift, use ZORIAL with a drift control agent such as NalcoTrol® and Ortho-Trol®. Do not apply during periods of gusty winds or winds in excess of 5 mph. Aerial spray equipment, flying speed and flying height should be carefully adjusted to provide a uniform application to the soil surface. Use markers to avoid overlapping and possible crop injury.

Preplant incorporated application (PPI) Apply ZORIAL at the recommended rate from 30 days to just prior to planting. In West Texas and Southwest Oklahoma, ZORIAL may be applied up to 60 days prior to planting. Incorporate not deeper than 2-3 inches with commonly used equipment. Incorporation should occur within one week after application.

Preemergence surface application (PRE) Apply ZORIAL at the recommended rate at the time of planting and preemergence to weeds.

Split application-THE ZORIAL® SPLIT™ Apply ZORIAL preplant incorporated, as indicated above, at one half the recommended rate. Apply the other half preemergence surface as recommended above.

Band Treatment When applying a band treatment of ZORIAL RAPID 80 the following formula may be used to calculate the amount per acre:

$$\frac{\text{Band width in inches}}{\text{Row width in inches}} \times \text{Rate in lbs per acre broadcast treatment} = \text{Pounds per acre for band treatment}$$

SOYBEAN INFORMATION

Do not use soybeans in West Texas or Southwest Oklahoma. Do not use when the soil type is sand, loamy sand, or sandy loam except the Coastal Plains soils of the Southeast. Do not use the PPI treatment in soybeans unless as part of the split application treatment. Do not use PPI treatments in Coastal Plains soils of the Southeast. Soybeans should be planted to a depth of 1-2 inches and soil firmly packed above and around the seed bed.

Broadcast rates for soybeans in pounds of ZORIAL RAPID 80 per acre:

Soil Type	PRE	Split PPI	Split PRE
Coastal Plains soils (SE)	1.25		
loam, silty loam, silt, sandy clay loam, sandy clay	1.9	.95	.95
clay loam, silty clay loam, silty clay, clay	2.5	1.25	1.25

ZORIAL RAPID 80 COMBINATIONS IN SOYBEANS

ZORIAL RAPID 80 plus Sencor or Lexone Tank Mix

This combination is recommended for improved control of cocklebur, morning glory, hemp sesbania, common ragweed and Venice mallow.

Apply a tank mix of ZORIAL plus Sencor or Lexone preemergence surface at the rates per acre (broadcast) recommended in the following table. These rates represent 1/2 the normal ZORIAL rate and 1/2 to 1/3 the normal metribuzin rate. Apply using 10-20 gallons of water per acre.

Broadcast Rates	ZORIAL RAPID 80 Pounds Per Acre	Sencor DF/ Lexone DF: Pounds Per Acre		
		Organic Matter		
Soil Type		Less than 2%	2-4%	Over 4%
Coastal Plains soils (SE only)	1.25	do not use	0.25-0.50	0.33-0.66
Medium	1.25-1.50	0.25-0.66	0.33-0.83	0.425-1
Fine (heavy)	1.25-1.50*-1.90	0.33-0.83	0.425-1	0.50-1.17
Mississippi Delta only	1.25-1.50*-1.90	0.50-1	0.58-1.17	0.665-1.3

* Optimum rate

When using Sencor 4 or Lexone 4L convert to equivalent rates. The amount of Sencor DF or Lexone DF equals two-thirds the amount of Sencor 4 or Lexone 4L. For example, one pint of Sencor 4 or Lexone 4L equals two-thirds pound Sencor DF or Lexone DF.

COTTON INFORMATION

Broadcast rates for cotton in pounds of ZORIAL RAPID 80 per acre:

Soil Type	PPI	PRE	Split PPI	Split PRE
sand, loamy sand, sandy loam	1.25	1.25	.625	.625
Coastal Plains soils (SE)*	**	1.9	.95	.95
loam, silty loam, silt, sandy clay loam, sandy clay	1.9	1.9	.95	.95
clay loam, silty clay, loam, silty clay, clay	2.5	2.5	1.25	1.25

* For improved control of hard to control weeds like pigweed, cocklebur and tropic croton.

** Do not use 1.9 lbs. PPI on Coastal Plains soils.

West Texas and Southwest Oklahoma Only Do not use on sandy soils (soils with greater than 70% sand content). Prior to application of ZORIAL, completely destroy the seed beds of the previous year by "cross-discing" Apply ZORIAL PPI only. For best results, incorporate ZORIAL on flat disced soil at labeled rates for control spurred anoda, annual grasses and suppression of yellow and purple nutsedge. To prevent removal of the herbicide application by wind, immediate incorporation is recommended. Soybeans can be planted as a replacement crop 45 days after the first planting of cotton. Do not plant peanut as replacement crop. Do not use ZORIAL on furrow planted cotton or on glandless cotton varieties.

STORAGE AND DISPOSAL

STORAGE

Do not contaminate water, food or feed by storage or disposal.

PESTICIDE DISPOSAL

Wastes resulting from the use of this product may be disposed of on site or at an approved waste disposal facility.

CONTAINER DISPOSAL

Triple rinse (or equivalent). Then puncture and dispose of in a sanitary landfill, or by incineration, or, if allowed by state and local authorities, by burning. If burned, stay out of smoke.

LIMITED WARRANTY AND LIABILITY

Notice: Read this Limited Warranty and Liability before buying or using this product. If terms are not acceptable, return at once, unopened.

It is critical that this product be used and mixed only as specified on the label. The laws of a State may make some or all of this paragraph inapplicable or may give you rights in addition to your rights hereunder. Except to the extent prohibited by applicable law, the exclusive remedy of the user or buyer and the limit of liability of this Company or any other Seller for any and all losses, personal injuries or damages resulting from the use of this product, shall be the purchase price paid by the user or buyer for the quantity of product involved. Except to the extent prohibited by State Law, there is no warranty, and this Company and other Sellers disclaim all liabilty for losses, personal injury or damages: (i) arising from any use of this product in manner or for a purpose not recommended in its label directions or from mixing this product before use with any substance except as recommended by the product's label; (ii) arising from handling or storage in violation of label instruction; (iii) for all indirect, special or consequential damages; (iv) when not reported to this Company within one year of discovery; and (v) arising from product not used within the label-designated shelf life or four years from the date of purchase, whichever occurs first. THERE ARE NO IMPLIED WARRANTIES AND NO WARRANTIES OF MERCHANTABILITY OR FITNESS.

ZORIAL®, ZORIAL® RAPID 80™ and ZORIAL® SPLIT™ are Trademarks of Sandoz Ltd.
Nalco/Trol® is a Registered Trademark of Nalco Chemical Co.
Ortho-Trol® is a Registered Trademark of Chevron Chemical Co.
Sencor® is Registered Trademark of Farben Fabriken Bayer GmbH. Leverkusen
Lorox® and Lexone® are Registered Trademarks of E.I. DuPont de Nemours and Co.
Prowl® is a Registered Trademark of American Cyanimid
Cotoran® and Dual® are Registered Trademarks of Ciba-Geigy
Lasso® is a Registered Trademark of Monsanto
Bladex® is a Registered Trademark of Shell Chemical Co.
Treflan® is a Registered Trademark of Elanco.

March 1989

Uniroyal Chemical Company, Inc.

Crop Protection
World Headquarters
Middlebury, CT 06749
(203) 573-2000

ALANAP® -L

KILLS BROADLEAF WEEDS

COMPOSITION

ACTIVE INGREDIENTS:	(% by weight)
Sodium salt of naptalam (sodium 2-[(1-naphthalenylamino) carbonyl] benzoate)	23.7%
INERT INGREDIENTS:	76.3%
TOTAL:	100.0%

Each gallon contains two pounds of active ingredient.

EPA REG. NO. 400–49 12–15–87

KEEP OUT OF REACH OF CHILDREN

WARNING

STATEMENT OF PRACTICAL TREATMENT

IF IN EYES: Flush with plenty of water for fifteen minutes. Call a physician.

IF SWALLOWED: Call a physician or Poison Control Center. Drink two large glasses of water and induce vomiting.

IF INHALED: See a physician if an abnormal reaction occurs.

IF ON SKIN: Wash thoroughly with soap and water.

PRECAUTIONARY STATEMENTS

HAZARDS TO HUMANS

WARNING

Causes eye irritation. Do not get in eyes, on skin or on clothing. Harmful if swallowed.

ENVIRONMENTAL HAZARDS

Do not apply directly to water. Do not contaminate water by cleaning of equipment or disposal of wastes.

STORAGE AND DISPOSAL

Do not contaminate water, food or feed by storage or disposal.

STORAGE: Store in a dry location. ALANAP-L is normally a solution but it may crystallize in the cold (32°F.). Gentle warming (60°F.) will redissolve any crystals. Mild agitation, such as rolling, shaking, or stirring before use is recommended.

PESTICIDE DISPOSAL: Wastes resulting from the use of this product may be disposed of on site or at an approved disposal facility.

CONTAINER DISPOSAL: Triple rinse (or equivalent). Then offer for recycling or reconditioning, or puncture and dispose of in a sanitary landfill, or by other procedures approved by state and local authorities.

EQUIPMENT CLEANING: Thoroughly rinse spray tank, hose and nozzles with water after application. Use ALANAP-L in metal or fiberglass tanks. ALANAP-L can be removed from metal or fiberglass tanks by washing thoroughly with ammonia water and rinsing several times with water containing a good detergent.

DIRECTIONS FOR USE

It is a violation of Federal law to use this product in a manner inconsistent with its labeling.

GENERAL INSTRUCTIONS

DO NOT APPLY THIS PRODUCT THROUGH ANY TYPE OF IRRIGATION SYSTEM.

DO NOT HARVEST SOYBEANS EARLIER THAN 60 DAYS POST-TREATMENT.

DO NOT GRAZE OR FEED FORAGE OR HAY FROM TREATED SOYBEANS OR PEANUTS TO LIVESTOCK.

OBSERVE ALL CAUTIONS AND LIMITATIONS ON LABELING OF ALL PRODUCTS USED IN MIXTURES.

PREEMERGENCE

1. PREPARATION OF SOIL

Area to be treated must be free of all weeds at time of application. ALANAP-L usually will not control established weeds. To insure maximum weed control, it is important to have a relatively smooth soil surface. If the area to be treated has been plowed, the soil clods should be broken up. If weather factors make replanting necessary after ALANAP-L has been applied, thoroughly disk the area prior to replanting.

2. METHOD OF APPLICATION

a. Uniform coverage of soil surface is essential.
b. Apply ALANAP-L within 48 hours of planting.
c. Rotary hoe may be used to destroy small weeds without imparing herbicide activity if dry weather follows ALANAP-L treatment.
d. If irrigation is used best results are obtained when water is applied after spraying ALANAP-L.
e. Avoid spray drift on areas planted to such sensitive crops as beets, tomatoes and spinach.
f. Do not use ALANAP-L on silt loam soils of extremely fine texture.

POSTEMERGENCE

Uniform coverage of foliage is essential.
Do not apply if rain is expected within 6 hours of application.

WEEDS CONTROLLED:

BROADLEAF WEEDS: *Common chickweed, common lambsquarters, common ragweed, common purslane, velvetleaf, common cocklebur, field and hedge bindweed, cutleaf groundcherry, white mustard, shepherds purse, redroot pigweed, hairy galinsoga and carpetweed (black and hairy nightshade are controlled in California).*

NOTES:

1. **If heavy rains immediately follow ALANAP-L application, weed control normally will not be impaired. However, some stunting of growth or even some stand reduction may be noted, especially in the lighter soils temperature may accentuate this condition. With the return of good growing conditions the stunted plants will usually recover and maintain a normal growth pattern.**
2. **Do not mix ALANAP-L with liquid fertilizer solutions since solid precipitates may form.**

CUCURBITS

Cantaloupe, Cucumber, Muskmelon and Watermelon

Apply ALANAP-L as a preemergence spray immediately after seeding. A second application postemergence to the crop may be made about a month after planting just before the crop is starting to vine, but before weeds have emerged. ALANAP-L can also be used as a postemergence spray immediately after transplanting crops. ALANAP-L may be used as a tank mix with bensulfide [PREFAR®]. See specific directions.

1. PREEMERGENCE

ALANAP-L: Use 6–8 quarts of ALANAP-L in 20–40 gallons of water per acre (broadcast). On light sandy soils reduce the rate to 4–6 quarts of ALANAP-L.

2. IRRIGATED MELONS (Preplant or Preemergence):

Prepare and shape seed beds. Band treat on the bed with the recommended rate (see above use directions for ALANAP-L alone or use directions for Prefar plus ALANAP-L) incorporate to 1–2½ inches. Plant, then irrigate as needed. On sandy loam or lighter soils ALANAP-L should not be used with sprinkler irrigation.

3. POSTEMERGENCE:

ALANAP-L may be used postemergence at up to 8 quarts per acre after plants are ready to vine.

NOTES:

1. **Do not use ALANAP-L on vine crops of any kind when growing conditions are adverse namely in early spring when weather is cold and wet. Under such conditions ALANAP-L may cause stunting.**
2. **Do not apply postemergence to vine crops when temperature is 100° F. or above.**

ALANAP PLUS PREFAR®:

DIRECTIONS FOR USE PRE-PLANT OR PREEMERGENCE: Cantaloupe, cucumber, muskmelon and watermelon—Apply 1–1½ gallons per acre of PREFAR 4E (4 to 6 lbs. a.i.) plus 1 to 2 gallons per acre of ALANAP-L (2 to 4 lbs. a.i.) in sufficient water (10 to 40 gal./acre) to provide uniform distribution. Incorporate lightly (½ to 1 inch) into the soil using suitable equipment just prior to planting. Use lower rates on light (sandy) soils and higher rates on heavier (clay) soils. Reduce the rate proportionately for band applications. **For irrigated melons see treating sequence given above.**

BROADLEAF WEEDS CONTROLLED:

PREFAR plus ALANAP-L in tank mix combination will control: *common chickweed, common lambsquarters, common ragweed, common purslane, velvetleaf, common cocklebur, field and hedge bindweed, cutleaf groundcherry, white mustard, shepherdspurse, redroot pigweed, hairy galinsoga, carpetweed.*

ANNUAL GRASSES CONTROLLED:

Smooth and large crabgrass, foxtails, barnyard grass, goosegrass, fall panicum and *red sprangletop.*

IMPORTANT - Follow all general directions for use and precautions listed above on the ALANAP-L label, as well as those listed on the PREFAR label.
PREFAR is a registered trademark of ICI AMERICAS INC.

NURSERY STOCK

A. For established Woody Plants (e.g. elm, arborvitae, taxus, pine, spruce, maple, cedar, cottonwood, plum, thuja, picea, lilac, privet, philadelphus, azalea, holly, rhododendron, juniper.)

1. Apply ALANAP-L on a broadcast basis or as a band spray centering on the row. Additional applications may be made as required.
2. Direct spray so as to strike the nursery stock about 2–3 inches or less above the ground.

B. For small Woody Transplants (2 methods suggested).

1. Apply ALANAP-L prior to transplanting. Care should be taken not to disturb much of the treated area.
2. Apply ALANAP-L as directed spray after transplanting.

Use 12 quarts of ALANAP-L in 20–40 gallons of water per acre (broadcast). On light sandy soils the rate should be reduced to 8 quarts. If applying on silt or clay soils increase the rate to 16 quarts.

NOTE: Avoid application to foliage or nursery plantings. Do not use on herbaceous [non-woody] plants.

PEANUTS (except California)

ALANAP-L is an effective preemergent herbicide for controlling annual broadleaf weeds. It is particularly effective for weeds that are deep seeded in peanuts. ALANAP-L may be used following a preplant incorporated grass treatment. For maximum weed control, use ALANAP-L in a tank mix combination with LASSO® or DUAL® 8E. If rainfall does not occur within 5-7 days, a soil surface incorporation is recomended.

WEEDS CONTROLLED

ALANAP-L plus LASSO:

Common chickweed, common lambsquarters, common ragweed, common purslane, velvetleaf, common cocklebur, field and hedge bindweed, cutleaf groundcherry, white mustard, shepherdspurse, redroot pigweed, hairy galinsoga and carpetweed, barnyardgrass, crabgrass, giant foxtail, green foxtail (robust purple, robust white), yellow foxtail, goosegrass, fall panicum, red rice, broadleaf signalgrass, witchgrass, black nightshade, hairy nightshade, Florida pusley, yellow nutsedge.

Partial Control: Florida beggarweed, annual morningglory.

ALANAP-L plus DUAL 8E:

Common chickweed, common lambsquarters, common ragweed, common purslane, velvetleaf, common cocklebur, field and hedge bindweed, cutleaf groundcherry, white mustard, shepherdspurse, redroot pigweed, hairy galinsoga and carpetweed, barnyardgrass, crabgrass, crowfootgrass, fall panicum, foxtail millet, giant foxtail, goosegrass, green foxtail, prairie cupgrass, red rice, signalgrass (Brachiaria), southwestern cupgrass, witchgrass, yellow foxtail, yellow nutsedge, black nightshade, Florida pusley.

Partial Control: Florida beggarweed, annual morningglory.

ALANAP-L alone:

Common chickweed, common lambsquarters, common ragweed, common purslane, velvetleaf, common cocklebur, field and hedge bindweed, cutleaf groundcherry, white mustard, shepherdspurse, redroot pigweed, hairy galinsoga and carpetweed.

APPLICATION TIMING AND RATE

ALANAP-L should be sprayed uniformly on the soil prior to crop emergence (immediately after planting through ground cracking).

ALANAP-L plus LASSO (liquid):

Use ALANAP-L at 4 quarts plus LASSO at 2 quarts in 20-40 gallons of water per acre broadcast.

ALANAP-L plus DUAL 8E:

Use ALANAP-L at 4 quarts plus DUAL 8E at 2 pints in 20-40 gallons of water per acre.

ALANAP-L alone:

Use ALANAP-L alone at 4-8 quarts in 20-40 gallons of water per acre broadcast. On light, sandy soils the rate should be reduced to 4 quarts.
LASSO is a Registered Trademark of Monsanto Agricultural Products Co.
DUAL is a Registered Trademark of Ciba-Geigy Corp.

SOYBEANS

Use ALANAP-L as a preemergence spray immediately after seeding, or as an overtop application when soybeans are about 14" tall to 60 days before harvest.

PREEMERGENCE THROUGH CRACKING:

Use 8 quarts of ALANAP-L in 20–40 gallons of water per acre (broadcast). On light sandy soils the rate should be reduced to 4 quarts. A band treatment may also be used.

NOTES: Do not apply ALANAP-L to soil area where crop seeds are exposed or shallow planted. The usual depth of planting to insure a stand is sufficient providing that the crop seeds are well covered with soil.

POSTEMERGENCE:
ALANAP-L Butyrac® 200 or Butoxone®

Application Timing: Apply on soybeans that are 14" tall but before 60 days of harvest.

Ground Application: Apply 2–3 quarts of ALANAP-L plus 3–4 ounces of Butyrac*200 or Butoxone** in 10–20 gallons of water per acre (broadcast rate) on soybeans. The spray boom should be positioned 18"–24" above the tops of the soybean plants or weeds and be equipped with hollow cone nozzles. Maintain a high spray pressure (40–50 psi) during application. This insures a fine spray mist for better overall coverage.

Aerial Application: Apply 2–3 quarts of ALANAP-L plus 3–4 ounces of Butyrac 200 or Butoxone in a minimum of 5 gallons of water per acre on soybeans. Nozzle size and orientation should provide maximum droplet breakup without causing excessive drift. Do not apply by air within 50 feet of susceptible crops such as cotton, tomatoes and tobacco. For best results, use 2 quarts of ALANAP-L plus 3 ounces of Butyrac 200 or Butoxone on weeds that are 10–12 inches tall such as *common cocklebur, giant ragweed, wild mustard or sunflower.* However, if these weeds are taller than 12 inches, use 3 quarts of ALANAP-L plus 4 ounces of Butyrac 200 or Butoxone. ALANAP-L plus Butyrac 200 or Butoxone if applied according to directions will control or suppress growth of above weeds reducing their competition to the soybean crop.

NOTES:

1. **Use surfactant [nonionic 80–20 etc.] or COC at manufacturers recommended dosage.**
2. **Avoid application of this tank mix to drought stressed soybeans.**
3. **Do not apply the tank mix if rain is expected within 6 hours after application.**
4. **Applications of this tank mix should not be made when weather conditions favor drift.**
5. **Observe all cautions and limitations on the Butyrac 200 or Butoxone labels.**

* Butyrac 200 contains 2 lbs/gallons of 4–(2,4-dicholorophenoxy) butyric acid and is a registered trademark of Union Carbide.
** Butoxone contains 1.75 lbs/gallon of 4–(2,4-dicholorophenoxy) butyric acid and is a registered trademark of Rhone-Poulenc, Inc.

IMPORTANT NOTICE

Seller warrants that this product conforms to its chemical description and is reasonably fit for the purposes stated on the label when used in accordance with the directions and instructions specified on the label under normal conditions of use, **but neither this warranty nor any other warranty of merchantability or fitness for a particular purpose, express or implied, extends to the use of this product, contrary to label instructions, or under abnormal conditions, or under conditions not reasonably foreseeable to seller, and buyer assumes the risk of any such use.**

APRON®-TERRACLOR®

DUST SEED TREATMENT FUNGICIDE

Gustafson

ACTIVE INGREDIENTS:	
Pentachloronitrobenzene (PCNB)	25.00%
METALAXYL:N–(2,6–dimethylphenyl)– N-methoxyacetyl) analine methyl ester	6.25%
INERT INGREDIENTS:	78.75%

EPA Reg. No. 7501-126

KEEP OUT OF REACH OF CHILDREN CAUTION

See additional precautionary statements

PRECAUTIONARY STATEMENTS

HAZARDS TO HUMANS AND DOMESTIC ANIMALS
CAUTION

Harmful if swallowed, absorbed through skin or inhaled. Avoid contact with skin, eyes or clothing. Avoid breathing dust. Wash thoroughly with soap and water after handling. Remove contaminated clothing and wash before reuse. Protective clothing including dust mask, gloves, long-sleeved shirt, long pants, socks and shoes should be worn during mixing/loading procedures.

STATEMENT OF PRACTICAL TREATMENT

If swallowed: Call a physician or Poison Control Center. Drink 1 or 2 glasses of water and induce vomiting by touching back of throat with finger. Do not induce vomiting or give anything by mouth to an unconscious person.

If in eyes: Flush eyes with plenty of water. Call a physician if irritation persists.

If on skin: Wash with plenty of soap and water. Call a physician if irritation occurs.

If inhaled: Move to fresh air. Administer CPR if not breathing.

ENVIRONMENTAL HAZARDS

This product is toxic to fish and aquatic organisms. Cover or incorporate treated seeds. Do not contaminate water when disposing of equipment washwaters.

DIRECTIONS FOR USE

It is a violation of Federal Law to use this product in a manner inconsistent with its labeling.

STORAGE AND DISPOSAL

STORAGE: Store in a cool, dry place. Do not store in direct sunlight. Protect from extreme temperatures.

PROHIBITIONS: Do not contaminate water, food or feed by storage or disposal. Open dumping is prohibited. Do not reuse container.

PESTICIDE DISPOSAL: Pesticide wastes are toxic. Improper disposal of excess pesticide, spray mixture, or rinsate is a violation of Federal Law. If these wastes cannot be disposed of by use according to label instructions, contact your State Pesticide or Environmental Control Agency, or the Hazardous Waste representative at the nearest EPA Regional Office for guidance.

CONTAINER DISPOSAL: Completely empty container into application equipment. Then offer for recycling or reconditioning, or puncture and dispose of in a sanitary landfill, or incineration, or, if allowed by state and local authorities, by burning. If burned, stay out of smoke.

NOTE: Metalaxyl is a systemic fungicide having a specific mode of action. Metalaxyl could be subject to development of resistant strains of fungi. Development of resistance cannot be predicted. Therefore, Gustafson, Inc. cannot assume liability for crop damage resulting from resistant strains of fungi. Consult with your State Agricultural Experiment Station or Extention Service Specialist for guidance or ways to control any possible metalaxyl resistant strains of fungi which may occur.

RECOMMENDED APPLICATIONS

APRON-TERRACLOR will control damping off, seed and seeding rot diseases caused by Pythium spp. and Rhizoctonia solani. APRON-TERRACLOR will also provide early season control of Phytophthor in soybeans.

PLANTER BOX TREATMENT

Apply at the specified rate and premix with the seed directly in the planter box at planting. A non-oil lubricant has also been added to help seeds flow more easily and evenly from the planter box. Use the higher rate of application in fields with a history of severe disease pressure.

SEED TREATER APPLICATION

APRON-TERRACLOR may be applied at planting time, utilizing the Gustafson CF–80° treater to maximize seed coverage. ("Calibrate the Gustafson CF–80 as per the manufacturer's instructions.) Do not use the CF–80 for applications to peanuts.

CROPS	RATE	DISEASE
Cotton	8 oz/cwt.	*Pythium* and *Rhizoctonia* Seedling Disease Complex, Common Smut or Bunt (Wheat), Covered Smut (Barley)
Peanuts	4 to 8 oz/cwt.	
Barley, Rice	2 to 4 oz/bu.	
Beans, Corn, Wheat	2 oz./bu.	
Soybeans, Peas	4 oz./bu.	*Pythium* and *Rhizoctonia* Seedling Disease Complex, Early Season *Phytophthora*

USE RESTRICTIONS

Do not use treated seed for food, feed or oil purposes.
All seed treated with this product (except by the planter box method) must be colored with an EPA approved dye (e.g., those listed in 40 CFR 180.1001) which imparts an unnatural color to the seed.

NOTICE—READ CAREFULLY
LIMITED WARRANTY STATEMENT

Gustafson warrants that this product conforms to the chemical description on the label and is reasonably fit for purposes stated on the label only when used in accordance with directions and instructions specified on the label. Gustafson neither makes, nor authorizes any agent or representative to make any other warranty of FITNESS or MERCHANTABILITY, guarantee or representation, express or implied, concerning this material. In the event of any breach of this limited warranty, Gustafson shall not be liable for consequential damages.

Gustafson (logo) is a Reg. TM of Gustafson, Inc.

Terraclor is a Reg. TM of Uniroyal Chemical Company, Inc.

Apron® is a Trademark of Ciba Geigy for Metalaxyl.

L157–1/88

This is a specimen label. Uniroyal is not responsible for the accuracy of the information contained herein. As labels are subject to revisions, always carefully read and follow the label on the product container.

B-NINE® SP

CONTROLS HEIGHT, PROMOTES FLOWERING OF PLANTS

COMPOSITION

ACTIVE INGREDIENT:	(% by weight)
Daminozide [butanedioic acid mono (2,2- dimethylhydrazide)]. . .	85.0%
INERT INGREDIENTS:	15.0%
TOTAL.	100.0%

EPA REG. NO. 400-110 003 6-20-89

B-NINE is a Registered Trademark of UNIROYAL CHEMICAL COMPANY, INC.

KEEP OUT OF REACH OF CHILDREN

CAUTION

STATEMENT OF PRACTICAL TREATMENT

IF SWALLOWED: Drink promptly a large quantity of milk, egg white or gelatin solution; if not available drink large amounts of water. Avoid alcohol. Call a physician immediately.

IF ON EYES OR SKIN: Wash with plenty of water for at least 15 minutes. For eyes, call a physician. Remove and wash contaminated clothing before use.

IF INHALED: and abnormal symptoms develop see a physician.

PRECAUTIONARY STATEMENTS

HAZARDS TO HUMANS
CAUTION

Harmful if absorbed through skin. Avoid contact with skin, eyes or clothing. Do not allow spray solution to touch skin. Wear protective clothing to protect hands, arms and legs, both when diluting and applying this product. Avoid breathing dusts. Wear a protective mask during mixing/loading operations. Wash hands and face thoroughly with soap and water after use or before eating or smoking.

ENVIRONMENTAL HAZARDS

Do not apply directly to water or wetlands. Do not discharge into lakes, streams, ponds or public waters unless in accordance with NPDES permit. For guidance, contact your Regional Office of the EPA. Do not contaminate water when disposing of equipment washwaters.

PHYSICAL OR CHEMICAL HAZARDS

Do not use or store near heat or open flame.

DIRECTIONS FOR USE

Each formulator is responsible for obtaining EPA registrations for its formulated product(s).

It is a violation of Federal Law to use this product in a manner inconsistent with its labeling.

REENTRY STATEMENT

Do not reenter treated areas until 24 hours after application, unless wearing protective clothing.

STORAGE AND DISPOSAL

Do not contaminate water, food or feed by storage or disposal.

STORAGE

Store in a dry location. Use spray and stock solutions within 24 hours. Do not store solutions. Immediate use is appropriate, if another component is added to spray mixture.

PESTICIDE DISPOSAL

Wastes resulting from the use of this product may be disposed of on site or at an approved waste disposal facility.

CONTAINER DISPOSAL

Plastic: Triple rinse (or equivalent). Then offer for recycling or reconditioning, or puncture and dispose of in a sanitary landfill, or incineration, or, if allowed by state and local authorities, by burning. If burned, stay out of smoke.

Paper: Completely empty bag into application equipment. Then dispose of empty bag in a sanitary landfill or by incineration, or if allowed by State and local authorities, by burning. If burned, stay out of smoke.

NOT FOR REFORMULATION

APPLICATION NOTES

1. Do not add additional wetting agent or mix with insecticides or fungicides as burning of leaves may result.
2. Best results are obtained if plants are watered before spraying. Foliage should be dry when spray is applied. Wilted plants do not readily absorb B-NINE SP and poor results might be obtained if wilted plants are

sprayed. Do not syringe for 18 to 24 hours after spraying to allow the chemical to enter the plant.
3. Do not handle treated plants until sprays have dried.
4. To measure B-NINE SP, use the **large end** of the scoop provided.

B-NINE® SP Dilution Table

% B-NINE SP Desired	Level Scoops* Per Gallons** Of Solution		Weight Measure For One (1) Gallon Of Solution	
	Scoops	Gallons	Ounces (avoirdupois)	Grams
0.15	1	3/4	0.24	6.7
0.25	2	1	0.40	11.1
0.37	3	1	0.60	16.7
0.50	4	1	0.80	22.3
0.75	6	1	1.20	33.4
1.00	8	1	1.60	44.5

* ONE LARGE LEVEL SCOOP = TWO LEVEL TEASPOONS

** Approximate dosage only. Use weight measurements wherever possible.

NOTE: Stock solutions prepared from B-NINE SP should not be held for more than 24 hours.

Place the amount of B-NINE SP shown into an empty one-gallon container. Then add enough warm water with stirring to make one gallon of solution. As B-NINE SP is soluble in water, no additional mixing is required.

GENERAL INFORMATION

B-NINE SP is dissolved in warm water then sprayed on the plant foliage using standard spray equipment. B-NINE SP passes into the leaf, then moves within the plant to reduce stem elongation. Treated plants develop a shorter, more compact habit of growth when compared to untreated plants. (See dilution table for converting percent concentration to dosage of B-NINE SP per gallon.)

DIRECTIONS FOR USE

Do not apply this product through any type of irrigation system.

Pot Chrysanthemums

To produce compact potted, branched (pinched) plants, spray to runoff 2 weeks after pinching with a 0.25% solution of B-NINE SP. If additional retardation is desired, a second application of 0.25% B-NINE SP may be applied 3 weeks after the first application. It is suggested that a "MEDIUM" schedule calling for shading and pinching on the same day be used for "TALL" varieties. This will eliminate the open, relatively leafless appearance often associated with the use of delayed pinch schedules.

Cut Chrysanthemums

To retard neck stretching just below the flower on potted and cut flowers, spray the top 1/3 of the foliage to runoff with 0.25% B-NINE SP two days after disbudding.

Chrysanthemum Cuttings

(Do Not use is California)

At certain times of the year or with certain tall or long internoded varieties, the conventional spray applications described above may not be adequate. Additional applications of B-NINE SP may be useful as pre-plant dips and post-plant sprays as follows:

PRE-PLANT DIP - Use 0.10% (1,000 ppm)

Rooted cutting: Hold a handful of cuttings by the root system; dip in the B-NINE SP solution just long enough to thoroughly wet all leaves and stems, remove and pot. Allow the foliage to dry before watering-in. Normally, by the time you are at the end of a bed, you can start watering-in those plants first potted.

Unrooted cutting: Remove unrooted cuttings from the box. Dip them in the B-NINE SP solution, remove and place in a flat. Cover with cheesecloth to avoid dehydration and place in the cooler overnight. Stick the next day and go on to a normal misting program.

POST PLANT SPRAY - Spray rooted cuttings as soon as the cuttings establish themselves (approximately 3-5 days after potting). Concentration will vary from 0.125% (1,250 ppm) to 0.25% (2,500 ppm), depending on variety and time of year.

Azaleas - Greenhouse Plants

To promote development of additional buds on plants grown for forcing, spray when new growth following final shaping is 1-2 inches with either:
1. Two applications of 0.15% B-NINE SP at one-week intervals.
2. One application of 0.25% solution of B-NINE SP.

Azaleas - Nursery Plants

To produce compact plants for landscape use, spray the foliage to runoff with a 0.37% solution of B-NINE SP during the fourth week in July. The final shaping of the plants should be done by early July to permit adequate regrowth prior to spraying.

Hydrangeas - Summer Growth

For those greenhouse varieties that produce excessive growth during the summer following pinching, spray to runoff with a 0.75% solution of B-NINE SP when the regrowths are 1-2 inches in length, usually in late July, but not after the first week in August.

Hydrangeas - Spring Growth

To retard spring forcing growth, spray with a 0.5% solution of B-NINE SP when the new growth has just started to unfold and four to five pairs of leaves are visible. Apply 2-4 weeks after the start of forcing but not later than this time.

Bedding Plants

Excessive stem elongation of most bedding plants is prevented by a B-NINE SP spray made 2-3 weeks after transplanting, or when the plants begin to elongate (stretch). Use a 0.5% concentration sprayed to the drip point is particularly effective on the following species: Ageratum*, Asters, Begonia*, Cosmos, Dahlia*, Dusty Miller*, Marigolds, Petunias, Phlox*, Salvia, Verbena* and Zinnias. In the sunbelt, B-NINE SP sprayed at 0.25% is effective for controlling elongation of Celosia*, Crossandra*, Exacum*, and Vinca*. Two or more applications may be necessary on any of the above species if new growth begins to stretch.

* DO NOT USE ON THESE SPECIES IN CALIFORNIA.

TO REDUCE LEAF INJURY OF PETUNIAS CAUSED BY AIR POLLUTANT, OZONE AND SULFUR DIOXIDE make a second application of 0.5% B-NINE SP one week after the first. B-NINE SP will not protect against leaf damage caused by the air pollutant PAN. Growth retardation will be more pronounced and there will be a delay in flowering from the second B-NINE SP application. Plants sprayed for pollution protection will be no shorter than those sprayed for compact growth.

Foliage Plants

(Do not use in California)

In the sunbelt, B-NINE SP sprayed at 0.25% concentration to the drip point is effective for controlling elongation (stretching) and promoting improved appearance and coloration of certain foliage species such as **Radermachera** (China Doll). Two or more applications may be necessary if new growth begins to stretch or for enhanced coloration.

Poinsettias

B-NINE SP reduces the elongation (stretching of poinsettias). For optimum growth control, a B-NINE SP spray should be applied to shoot breaks that are 1½-2" long. The dosage range is 0.2 - 0.3% (2,000 - 3,000 ppm). Use the 0.2% dilution in areas north of the sunbelt and 0.3% in the sunbelt. A repeat application may be necessary for optimum plant size.

B-NINE SP Plus CYCOCEL® - Sunbelt Region: A tank mix combination of B-NINE SP plus CYCOCEL has been shown to provide optimum retardation without marginal chlorosis. The tank mixture recommended is B-NINE SP at 0.2% - 0.25% plus CYCOCEL at 0.15%. Spray this dilution on shoot breaks that are 1½-2" long. Under prevailing high temperature conditions, that often exist in the sunbelt, the 0.25% dilution should be used with 0.15% CYCOCEL. A repeat application may be desirable if excessive elongation occurs. A slight reduction of bract size may occur on Hegg cultivars. No reduction has been observed on V-14 Glory or V-10 Amy cultivars.

Northern Belt Region: The tank mixture recommended is B-NINE SP at 0.25% (2,500 ppm) plus CYCOCEL at 0.15% (1,500 ppm). One gallon of tank mix should be applied to 200 square feet of area. Application should be made when axillary shoots (shoot breaks) are 1½ to 2 inches long. Only one application should be made due to the extreme effectiveness of the tank mix. The tank mix should not be applied after September 30.

Gardenias

To produce compact plants for spring bloom, spray with 0.5% B-NINE SP when plants are about 2/3 of final marketing size. This is from mid-August to mid-September for the Florida area and during the first two weeks of December for other areas.

IMPORTANT NOTICE

Seller warrants that this product conforms to its chemical description and is reasonably fit for the purposes stated on the label when used in accordance with the directions and instructions specified on the label under normal conditions of use. **But neither this warranty nor any other warranty of merchantability or fitness for a particular purpose, express or implied, extends to the use of this product, contrary to label instructions, or under abnormal conditions, or under conditions not reasonably foreseeable to seller, and buyer assumes the risk of any such use.**

CYCOCEL is a registered trademark of AMERICAN CYANAMID CO.

CASORON® 2G

CONTROLS WEEDS AND GRASSES IN ROSES, EVERGREENS AND WOODY ORNAMENTALS

COMPOSITION

ACTIVE INGREDIENT:	(% by weight)
Dichlobenil (2,6-dichlorobenzonitrile)	2.0%
INERT INGREDIENTS:	98.0%
TOTAL:	100.0%

EPA REG. NO. 400-170

2-25-88

CASORON is a Registered Trademark of Duphar B.V., Weesp, the Netherlands

KEEP OUT OF REACH OF CHILDREN

CAUTION

STATEMENT OF PRACTICAL TREATMENT

IF SWALLOWED: Call a physician or poison control center immediately. Drink promptly large quantities of water.
IF INHALED: Remove to fresh air; if abnormal symptoms develop, contact a physician.
IF ON SKIN: Wash affected area with plenty of soap and water.
IF IN EYES: Flush eyes with plenty of water for at least 15 minutes. Contact a physician.

PRECAUTIONARY STATEMENTS

HAZARDS TO HUMANS AND DOMESTIC ANIMALS

CAUTION

Harmful if swallowed. Avoid breathing of dust. Avoid contact with eyes or skin.

ENVIRONMENTAL HAZARDS

Do not apply directly to water or wetlands (swamps, bogs, marshes, and potholes). Do not contaminate water when disposing of equipment washwaters. Cover, collect, or incorporate granules spilled on the soil surface.

DIRECTIONS FOR USE

It is a violation of Federal law to use this product in a manner inconsistent with its labeling.

STORAGE AND DISPOSAL

Do not contaminate water, food or feed by storage or disposal.

STORAGE—Store in a dry location and keep container tightly closed. Do not store with propagative structures such as seed, bulbs, tubers, nursery stock, etc., or with food or feed products.

DISPOSAL—Do not reuse empty container. Securely wrap original container in several layers of newspaper and discard in trash.

WORKERS SAFETY RULES

IMPORTANT! Always wash hands, face, and arms with soap and water before smoking, eating, drinking, or toileting. Keep all unprotected persons, children, livestock, and pets away from treated area or where there is a danger of drift. Do not rub eyes or mouth with hands. If you feel sick in any way. STOP work and get help right away. See Statement of Practical Treatment.

BEFORE APPLICATION

Remove existing weeds.
Thoroughly work and smooth the area to be treated.
Carefully measure area to be treated and determine the number of square feet.
If area to be treated is large, it should be divided into smaller sections, and each treated individually.

APPLICATION AND RATE

Shake CASORON 2G uniformly on the soil at the rate of 3/4 ounce (2 level tablespoons) for each 10 sq. ft. to be treated (i.e. 20 oz. treats 270 sq. ft.) Rates higher than those indicated may cause injury. More accurate and uniform application can be obtained by applying one-half of the required material over the entire treatment area and repeating the application with the remaining material.

CASORON 2G should not be used on sloping areas as run-off may result in injury to lawn or other plants in adjacent areas.

As a guide for the proper application rate and uniformity of distribution of the granules, consult the following diagram:

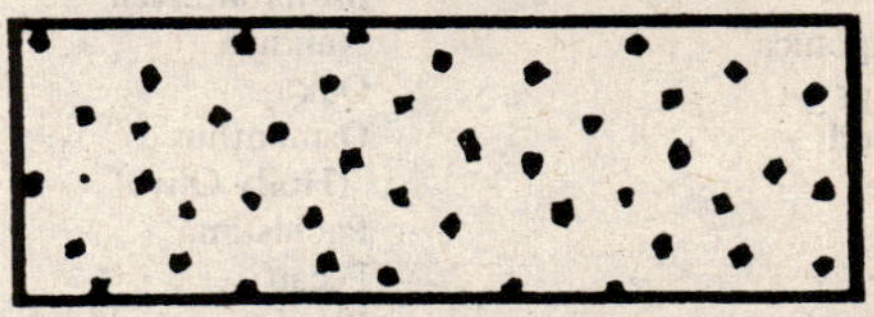

AFTER APPLICATION

Unless the soil is wet, sprinkle lightly to moisten the soil 1/4 to 1/2 inch deep. Do not flood; avoid puddling and washing.

Avoid disturbing soil in the treated area as long as weed control is being maintained.

USE PRECAUTIONS

Use only on the ornamental plants listed on the label. Do not apply to transplant or cutting beds or to newly set transplants, and do not seed or transplant into treated areas. **Do not use in greenhouses.** Do not use on extremely light sandy soils. Do not use on plants, any part of which is used for human or animal consumption.

WEEDS CONTROLLED

CASORON 2G when applied to the soil as directed will give selective control of annual grassy and broadleaved weeds and certain perennial weeds such as:

Annual Bluegrass	Knotweed
Bluegrass	Lambsquarters
Bull Thistle	Milkweed Vine (Morrenia odorata)
Camphorweed	Natalgrass
Carpetweed	Peppergrass
Chickweed	Pineapple Weed
Citron Melon	Wild Mustard
Coffeeweed	Plantain
Crabgrass	Purslane
Cudweed	Ragweed
Dandelion	Redroot Pigweed
Evening primrose	Rosarypea
Fiddleneck	Russian Thistle
Florida Purslane	Shepherdspurse
Pursley	Smartweed
Foxtail	Spanishneedles
Gisekia	Spurge
Groundsel	Teaweed
Henbit	Texas Panicum (Hurrahgrass)
Horsetail	Yellow Woodsorrel
Jerusalemoak	
Goosefoot	

WOODY ORNAMENTALS

CASORON 2G IS TO BE APPLIED ONLY TO ESTABLISHED WOODY PLANTS—THOSE WHICH HAVE BEEN PLANTED FOR AT LEAST FOUR MONTHS. CASORON 2G MAY BE USED FOR WEED CONTROL ON THE FOLLOWING PLANTS:

Amur Corktree	Juniper
Arbor Vitae	Kinnikinnick
Ash	Laurel
Azalea[1]	Leucothoe
Barberry	Lilac
Beauty Bush	Linden
Birch	Locust
Boxelder	Magnolia
Boxwood	Maple
Camellia	Mock Orange
Caragana	Mountain Ash
Cleyera japonica	Nandina
Cotoneaster	Oak
Cottonwood	Osmanthus
Crabapple	(Holly Olive)
Deutzia	Pachistima
Dogwood	Pecan
Eastern Red Cedar	Photinia
Elm	Pittosporum
Elwood Cypress	Poplar
English Walnut	Privet
Euonymus	Pyracantha
Filbert	Rhododendron
Flowering Almond	Rockrose
Flowering Quince	Rose
Forsythia	Russian Olive
Gardenia	Spirea
Goldenraintree	Squawcarpet
Hackberry	Weigela
Heather	Willow
Holly[2]	Yew (Taxus)
Honeysuckle	
Ivy (Hedera helix)	

[1]Kurume, Mollis, Hardy hybrid types and hardly native species only.
[2]Except I. crenata, I. rotunda, I. vomitoria.

IMPORTANT NOTICE

Seller warrants that this product conforms to its chemical description and is reasonably fit for the purposes stated on the label when used in accordance with the directions and instructions specified on the label under normal conditions of use, **but neither this warranty nor any other warranty of merchantability or fitness for a particular purpose, express or implied, extends to the use of this product, contrary to label instructions, or under abnormal conditions, or under conditions not reasonably foreseeable to seller, and buyer assumes the risk of any such use.**

CASORON® 4G

DICHLOBENIL WEED AND GRASS KILLER

COMPOSITION

ACTIVE INGREDIENT:	(% by weight)
Dichlobenil (2,6-dichlorobenzonitrile)	4.0%
INERT INGREDIENTS:	96.0%
TOTAL	100.0%
EPA REG. NO. 400-168	001 4-3-89

KEEP OUT OF REACH OF CHILDREN

CAUTION

STATEMENT OF PRACTICAL TREATMENT

IF SWALLOWED: Call a physician or poison control center immediately. Drink promptly a large quantity of water.

IF INHALED: Remove to fresh air; if abnormal symptoms develop, contact a physician.

IF ON SKIN: Wash affected area with plenty of soap and water.

IF IN EYES: Flush eyes with plenty of water for at least 15 minutes. Contact a physician.

CASORON is a Registered Trademark of Duphar B.V., Weesp, the Netherlands

PRECAUTIONARY STATEMENTS

HAZARDS TO HUMANS AND DOMESTIC ANIMALS

CAUTION

Harmful if swallowed. Avoid breathing of dust. Avoid contact with eyes or skin.

ENVIRONMENTAL HAZARDS

Do not contaminate water when disposing of equipment washwaters. Cover, collect, or incorporate granules spilled on the soil surface.

DIRECTIONS FOR USE

It is a violation of Federal Law to use this product in a manner inconsistent with its labeling.

STORAGE AND DISPOSAL

Do not contaminate water, food or feed by storage or disposal.

STORAGE—Store in a dry location. Do not store with propagative structures such as seed, bulbs, tubers, nursery stock, etc., or with food or feed products.

PESTICIDE DISPOSAL—Wastes resulting from the use of this product may be disposed of on site or at an approved wastes disposal facility.

CONTAINER DISPOSAL—Completely empty bag into application equipment. Then dispose of empty bag in a sanitary landfill or by incineration, or if allowed by State and local authorities, by burning. If burned, stay out of smoke.

WORKERS SAFETY RULES

IMPORTANT! Always wash hands, face, and arms with soap and water before smoking, eating, drinking, or toileting. Keep all unprotected persons, children, livestock, and pets away from treated area or where there is a danger of drift. Do not rub eyes or mouth with hands. If you feel sick in any way, STOP work and get help right away. See Statement of Practical Treatment.

PERENNIAL WEED CONTROL

CASORON 4G is recommended for control of certain perennial weeds in established woody ornamentals, nuts and fruit and for control in non-crop areas such as roadways, around buildings and fences, railroad and utility rights-of-way and installations.

WEEDS CONTROLLED:

Artemisia	Russian Knap weed
Canada Thistle	Smooth Brome
Curly Dock	Timothy
Falsedandelion (Catsear)	Wild Artichoke
Fescue	Wild Aster
Leafy Spurge	Wild Carrot
Orchardgrass	Yellow Rocket
Quackgrass	

SOIL SURFACE TREATMENT: Apply CASORON 4G as a soil surface treatment in late fall from November 15 to February 15. Old weed growth should not be removed before application. Do not make surface application to areas which have been cultivated during the fall or summer prior to application.

INCORPORATED TREATMENT: Apply CASORON 4G and incorporate immediately after application in late fall or very early spring before May 1.

ANNUAL WEED CONTROL

CASORON 4G is recommended for control of annual grassy and broadleaved weeds and certain perennial weeds such as:

Annual Bluegrass (*Poa annua*)
Bluegrass
Bull Thistle
Camphorweed
Carpetweed
Chickweed
Citron Melon
Coffeeweed
Crabgrass
Cudweed
Dandelion
Dogfennel
Evening Primrose
Fiddleneck
Florida Purslane (Pusley)
Foxtail
Gisekia
Groundsel
Henbit
Horsetail
Jerusalemoak
Goosefoot
Knotweed
Lambsquarter
Maypops
Milkweed Vine (*Morrenia odorata*)
Miners Lettuce
Natalgrass
Old Witchgrass
Peppergrass
Pineapple Weed
Plantain
Purslane
Ragweed
Red Deadnettle
Redroot Pigweed
Rosarypea
Russian Thistle
Shepherdspurse
Smartweed
Spanishneedles
Spurge
Teaweed
Texas Panicum (Hurrah grass)
Wild Barley
Wild Mustard
Wild Radish
Yellow Woodsorrel

Apply CASORON 4G either in early spring before seeds of annual weeds germinate or after cultivation has removed the growing weeds. Distribute granules uniformly over the soil surface. If treatment is followed by application of 1/2 to 1 inch of water by overhead irrigation, the lower rates are recommended. Shallow incorporation immediately after application is recommended in southern areas. In other areas, shallow incorporation or sprinkler irrigation is suggested when applications are made during periods of high temperature.

NOTE: Use of CASORON 4G on plants not listed on the label may result in damage.

FRUIT AND NUT ORCHARDS AND NURSERIES

CASORON 4G may be used on the following crops.

A. **Apple, Blueberry, Cherry, Grape, Nectarine, Peach, Pear, Prune, Plum, Filbert:** May be used on bearing, non-bearing and nursery stock. Do not apply until 4 weeks after transplanting. Do not make application within one month of harvest of nuts or stone fruits. Apply 100 to 150 pounds per acre for annual weed control based on area actually treated. For control of perennial weeds apply 150 pounds per acre. Follow general directions above.

B. **Blackberry and Raspberry:** For control of annual weeds follow general directions above except use only on established plantings and do not apply more than 100 pounds per acre. Do not apply during new shoot emergence.

Do not apply CASORON 4G during 6 months following possible freeze damage on any citrus. Do not allow livestock to graze treated orchards.

Do not plant rotational crops, on which dichlobenil is not registered, in treated soil within one year after application.

WOODY ORNAMENTALS AND SHELTERBELTS (ESTABLISHED AND IN NURSERIES)

For control of annual weeds CASORON 4G may be used on the following plants. Apply 100 to 150 pounds of CASORON 4G per acre, based on area actually treated. Do not apply until 4 weeks after transplanting. Follow general directions above.

For control of perennial weeds listed, apply 150 to 200 pounds per acre. Do not apply more than 150 pounds per acre on Azalea, Boxwood, Euonymus, Forsythia, Heather, Holly, Ivy, Lilac, Leucothoe. Do not apply more than 150 pounds per acre on plantings less than 1 year old. Follow general directions listed above.

Amur Corktree
Arbor Vitae
Ash
Azalea[1]
Barberry
Beauty Bush
Birch
Boxelder
Boxwood
Camellia
Caragana
Cleyera japonica
Cotoneaster
Cottonwood
Crabapple
Deutzia
Dogwood
Eastern Red Cedar
Elm
Elwood Cypress
English Walnut
Euonymus
Filbert
Flowering Almond
Flowering Quince
Forsythia
Gardenia
Goldenraintree
Hackberry
Heather
Holly (Ilex spp)[2]
Honeysuckle
Ivy (Hedera helix)
Juniper
Kinnikinnick
Laurel
Leucothoe
Lilac
Linden
Locust
Magnolia
Maple
Mock Orange
Mountain Ash
Nandina
Oak
Osmanthus (Holly-olive)
Pachistima
Pecan
Photinia
Pines[3]
Pittosporum
Poplar
Privet
Pyracantha
Rhododendron
Rockrose
Rose
Russian Olive
Spirea
Squawcarpet
Weigela
Willow
Yew (Taxus)

[1] Kurume, Mollis, Hardy hybrid types and hardy native Azalea species only.
[2] Except Ilex crenata, I. rotunda, I. vomitoria.
[3] Pine species are sensitive to injury from CASORON 4G if applied within two years after transplanting. Do not use on pines in California.

USE PRECAUTIONS

AVOID APPLICATIONS WHICH ALLOW CASORON 4G GRANULES TO LODGE IN THE FOLIAGE, PARTICULARLY IN CONIFER TYPE SPECIES, SUCH AS PINES, YEWS, ETC.

Do not allow CASORON 4G granules to accumulate and be in contact with trunk at or below soil line.

Do not use in seed beds, transplant or cutting beds or greenhouses.

Do not apply until 6 months after rooting of cuttings in the field, or after planting rooted cuttings.

Do not apply CASORON 4G to fruit stock within 3 months prior to or following grafting or budding of root stocks or planting of new grafts.

Do not apply more than the maximum recommended amount of CASORON 4G per growing season.

Do not use on "St. Lucie fine sand," "Arzell fine sand" or other light sand soil.

Do not graze livestock in treated areas.

WEED CONTROL IN CRANBERRIES

Control of perennial and annual, broadleaved and grassy weeds (including rushes and sedges) in cranberries can be obtained by used of CASORON 4G. The following weeds are controlled:

Sensitive Fern	Cotton grass
Bracken Fern	Needle grass
Royal Fern	Oniongrass
Hair Cap Moss	Arrowleaved Tear Thumb
Common Horsetail	Beggarsticks
Water Horsetail (pipes)	Knotweed
	Loosestrife
Rush (*Juncus* spp.)	Marsh St. Johnswort
Dodder	Tideland clover
Bluejoint grass	Ragweed
Rattlesnake grass (Manna grass)	Sorrell
	Wild Strawberry
Summer grass	Aster
Velvetgrass	Buckbean
Bent Grasses	Hawkweed
Little Hairgrass	Western Lilaeopsis
Crabgrass	Marsh Pea
Rice Cutgrass	Plantain
Bunch grass	Smartweed
Muskrat grass	(Marsh pepper
Nutsedge (Nutgrass)	Pennsylvania
Short Wiregrass	Spotted
Wideleaf grass	Swamp
Stargrass	Water)
Woolgrass	

Application should be made in early spring while perennial weeds are still dormant and annual weeds have not started to germinate or in late fall after crop has been harvested. Apply pre-bloom or post-harvest.

Spring Treatment

AREAS OTHER THAN PACIFIC NORTHWEST: Apply 100 pounds CASORON 4G per acre. Do not make spring application if treatment was made the previous fall. See specific directions for spring treatment in the Pacific Northwest.

PACIFIC NORTHWEST: Two equal applications may be made in early spring with an interval of 3 to 6 weeks between treatments. Total application should be 100 pounds per acre.

Fall Treatment

In all growing areas apply 100 to 150 pounds of CASORON 4G per acre in the fall after crop has been harvested.
Do not make fall application if treatment was made the previous spring. Use high rate only on very weedy bogs, or on bogs abandoned because of weeds.

USE PRECAUTIONS

Distribute CASORON 4G uniformly. Avoid over-application which may result from over-lapping during treatment period. Temporary reddening of cranberry plants may occur especially with late spring application or when applied on sandy bogs. Do not apply after cranberry plants start to bloom as yields may be reduced. Do not use on young beds or newly sanded beds or on bogs prior to or immediately after mowing for vines.

OTHER USES

NUTSEDGE CONTROL: To control nutsedge (nutgrass) on mineral soils in areas on which containers of nursery stock are to be grown or held—and also for control in non-crop areas—apply 250 to 500 pounds of CASORON 4G per acre and incorporate immediately to a depth of 4 to 6 inches. Seasonal control of annual weeds will also be obtained. Containers of plants listed above may be placed on treated soil immediately. Do not transplant into treated soil.

UNDER ASPHALT: For general weed control under asphalt (in areas such as roadways, parking lots, recreational areas), after final grade is achieved, apply 250 to 300 lbs. per acre, ($5^3/_4$ to 7 lbs. per 1,000 sq. ft.). Treated area should be covered with asphalt as soon as possible. For resurfacing work, existing weeds should be sprayed with a contact herbicide prior to CASORON 4G application.

IMPORTANT NOTICE

Seller warrants that this product conforms to its chemical description and is reasonably fit for the purposes stated on the label when used in accordance with the directions and instructions specified on the label under normal conditions of use, **but neither this warranty nor any other warranty of merchantability or fitness for a particular purpose, express or implied, extends to the use of this product, contrary to label instructions, or under abnormal conditions, or under conditions not reasonably foreseeable to seller, and buyer assumes the risk of any such use.**

CASORON® 10G

AQUATIC AND INDUSTRIAL HERBICIDE

COMPOSITION

ACTIVE INGREDIENT:	(% by weight)
Dichlobenil (2,6-dichlorobenzonitrile)	10.0%
INERT INGREDIENTS:	90.0%
TOTAL	100.0%

EPA REG. NO. 400-178 2-22-88

CASORON is a Registered Trademark of Duphar B.V., Weesp, the Netherlands

KEEP OUT OF REACH OF CHILDREN

CAUTION

STATEMENT OF PRACTICAL TREATMENT

IF SWALLOWED: Call a physician or poison control center immediately. Drink promptly a large quantity of water.
IF INHALED: Remove to fresh air; if abnormal symptoms develop, contact a physician.
IF ON SKIN: Wash affected area with plenty of soap and water.
IF IN EYES: Flush eyes with plenty of water for at least 15 minutes. Contact a physician.

PRECAUTIONARY STATEMENTS

HAZARDS TO HUMANS AND DOMESTIC ANIMALS
CAUTION

Harmful if swallowed. Avoid breathing of dust. Avoid contact with eyes or skin.

ENVIRONMENTAL HAZARDS

Do not contaminate untreated water when disposing of equipment wash-waters. Treatment of weed areas can result in oxygen loss from decomposition of dead weeds. This loss can cause fish suffocation. Therefore treat only $^1/_3$ of the weed areas at a time and wait 14 days between treatments. Consult your State Environmental Regulatory Agency concerning the need for a permit before applying this product to public waters. Do not apply to water which will be used for crop irrigation, for livestock water or for human consumption. Do not use fish from treated water for food or feed for a period of 90 days after application. Do not use in commercial fish or shellfish water. Cover, collect, or incorporate granules spilled on the soil surface.

DIRECTIONS FOR USE

It is a violation of Federal Law to use this product in a manner inconsistent with its labeling.

STORAGE AND DISPOSAL

Do not contaminate water, food or feed by storage or disposal.

STORAGE—Store in a dry location. Do not store with propagative structures such as seed, bulbs, tubers, nursery stock, etc., or with food or feed products.

PESTICIDE DISPOSAL—Wastes resulting from the use of this product may be disposed of on site or at an approved waste disposal facility.

CONTAINER DISPOSAL—Completely empty bag into application equipment. Then dispose of empty bag in a sanitary landfill or by incineration, or if allowed by State and local authorities, by burning. If burned, stay out of smoke.

WORKERS SAFETY RULES

IMPORTANT! Always wash hands, face, and arms with soap and water before smoking, eating, drinking, or toileting. Keep all unprotected persons, children, livestock, and pets away from treated area or where there is a danger of drift. Do not rub eyes or mouth with hands. If you feel sick in any way, STOP work and get help right away. See Statement of Practical Treatment.

AQUATIC WEED CONTROL

CASORON 10G is recommended for control of rooted aquatic weeds in non-flowing water, such as ponds, reservoirs and lakes.

Apply CASORON 10G in water bodies operated under the supervision of a government agency, corporate or land development ponds or lakes, providing the user and/or applicator has control of the uses of water.

CASORON 10G may be applied either through the water column, to exposed pond bottoms after drawdown, or to shorelines at low water. CASORON 10G does not harm fish when used as directed.

CASORON 10G applied as directed below will give season long control of several species of weeds including the following:

*Alligatorweed	*Bladderwort (*Utricularia* spp.)
*Cattails (*Typha* spp.)	Chara
Coontail	*Eelgrass (*Vallisneria*)
Elodea (*Egeria* sp.)	*Eurasian watermilfoil
*Hydrilla	Naiads
Northern watermilfoil	*Pennywort
Pondweeds (*potamogeton* spp.)	*Purple Loosestrife
*Spatterdock (*Nuphar* sp. yellow waterlily, cowlily)	*Waterprimrose
*White waterlily	*Widgeongrass

*DO NOT USE IN CALIFORNIA FOR THESE WEEDS

SUBMERGED WEED CONTROL—Elodea, Naiads, Chara, watermilfoils, Pondweeds, Hydrilla, (*Potamogeton spp.*) and Widgeongrass: Apply CASORON 10G during early spring either before growth begins, or during the period of early rapid growth. As CASORON 10G controls weeds only where applied, uniform distribution of the granules over the area to be treated is essential. It is particularly suited for spot treatments, treatment around boat docks, and swimming areas. Submerged weeds must be rooted to the hydrosoil for adequate residual control.

Treatment of Exposed Bottoms and Shorelines at low water

Apply CASORON 10G at a rate of 70 to 100 pounds per acre. Complete and uniform coverage of the area is necessary for control. Use the lower rate if the soil is wet at the time of application, or if the water is less than 1 foot deep. In the case of treatment after drawdown, the pond should be refilled promptly.

Treatment Through Water

Distribute CASORON 10G uniformly over the surface of the water using a rate of 100 to 150 pounds per acre of surface area. Apply 100 to 125 pounds per acre when the water is less than 4 feet deep, or when light weed populations are expected. Apply 125 to 150 pounds per acre when applying through water deeper than 4 feet or if heavy weed populations are expected. Application should be made during periods of calm when wave action is at a minimum.

LILY CONTROL: White waterlily and spatterdock (yellow waterlily, cowlily, pondlily) should be treated in spring after the new leaves reach the surface of the water. **In southern areas** treatment may be made in Spring or after plants are more mature—June or July. Use application rates listed above.

ALLIGATORWEED CONTROL: Timing is very important in control of this weed. Apply 40 to 80 pounds of CASORON 10G per acre in early spring after growth begins and not later than full-size leaf stage. The minimum rates of 40 to 60 pounds per acre should be used when new growth is one-fourth to one-half inch in length. Treatments made earlier or later than this optimum growth stage, should be at rates of 60 to 80 pounds per acre. Treat only alligatorweed which is firmly rooted.

PURPLE LOOSESTRIFE CONTROL: Apply CASORON 10G at the rate of 150 lb./A during early spring before growth begins.

APPLICATION EQUIPMENT

CASORON 10G should be distributed by use of an applicator such as a hand cyclone type spreader. If areas to be treated are small, satisfactory application can be made by hand, spreading the granules with a small scoop. If large areas are to be treated such as along shorelines, aerial equipment can be employed. Boat mounted granular applicators are effective for evenly distributing CASORON 10G along shorelines and boat dock areas. CASORON 10G is designed to sink rapidly in water.

INDUSTRIAL WEED CONTROL

CASORON 10G will give nonselective weed control for an extended (12 month) period of time when applied as directed. The weeds listed below can be controlled in industrial and noncropland areas such as railroads, utility, highway, and pipeline rights-of-way, non-surfaced roadways, utility plant sites and substations, petroleum tank farms, pumping installations, fencerows, storage areas, lumber yard and non-irrigation ditch banks and similar areas.

WEEDS AND GRASSES		RATES	DIRECTIONS
Perennial Weeds: Artemisia (Mugwort) Aster Bindweeds Bluegrass Canada Thistle Curly Dock False Dandelion Fescue Florida Betony	Leafy Spurge Orchardgrass Quackgrass Red Sorrel Russian Knapweed Timothy Wild Carrot Yellow Rocket	120-200 lbs/Acre ($2^{3}/_{4}$ lbs - $4^{1}/_{2}$ lbs. per 1,000 sq. ft.)	For non-selective weed control CASORON 10G should be applied at the rate of 120 to 200 pounds per acre ($2^{2}/_{3}$ to $4^{1}/_{2}$ pounds per thousand square feet, $4^{1}/_{2}$ to $7^{1}/_{2}$ ounces per 100 square feet or $^{3}/_{4}$ to $1^{1}/_{4}$ pounds per square rod). The lower rates may be used if the weed infestation is primarily annuals. **For perennial weeds** and in dry soil areas the higher rates should be used. Dead weeds from previous seasons growth do not need to be removed and soil should not be cultivated prior to application.
Annual Grass: Annual Bluegrass Crabgrass Foxtail Foxtail Barley	Natalgrass Texas Panicum Virginia Pepperweed Witchgrass		Make application during coldest weather when plants are fully dormant or in the winter rosette stage.
Annual Broadleaf Weeds: Bull Thistle Camphorweed Carpetweed Chickweed Citronmelon Coffeeweed Cudweed Dandelion Dog Fennel Eveningprimrose Fiddleneck Florida Purslane Gisekia Groundsel Henbit Horsetail Jerusalemoak Goosefood Knotweed Lambsquarters	Maypops Miners Lettuce Pineappleweed Plantain Prickly sida Purple Deadnettle Purslane Ragweed Redroot Pigweed Rosary Pea Russian Thistle Shepherdspurse Smartweed Spanish needles Spurge Stranglervine Wild Mustard Wild Radish Yellow Woodsorrel		**For annual weed control** CASORON 10G should be applied prior to germination or when new plants are very small. For best results CASORON 10G should be applied during the season of lowest temperatures. Apply November 1 to March 1 in most areas or until April 1 in the northern tier of states. In areas of limited rainfall apply just prior to or during rainy season. For prompt and effective results rainfall or other precipitation should occur within a few days of application. Do not apply if air temperature is expected to rise above 70°F within the following week. If application is made when temperature is above 70°F, material should be watered in promptly or lightly incorporated into soil, or covered with mulch. Do not seed or transplant into treated area for 24 months.

OTHER USES

UNDER ASPHALT: For general weed control under asphalt (in areas such as roadways, parking lots, recreational areas), after final grade is achieved, apply 100 to 120 lbs. per acre. (2.3 to 2.8 lbs. per 1,000 sq. ft.). Treated area should be covered with asphalt as soon as possible. For resurfacing work, existing weeds should be sprayed with a contact herbicide prior to CASORON 10G application.

USE PRECAUTIONS

This product is not selective in action and may destroy all types of vegetation. The treated area may be totally or partially nonproductive for one or more years. Apply only to areas where complete plant control is desired. Do not apply to areas on slope above lawns or susceptible crops as run-off water from treated area could cause injury. Do not graze livestock in treated areas.

IMPORTANT NOTICE

Seller warrants that this product conforms to its chemical description and is reasonably fit for the purposes stated on the label when used in accordance with the directions and instructions specified on the label under normal conditions of use, **but neither this warranty nor any other warranty of merchantability or fitness for a particular purpose, express or implied, extends to the use of this product, contrary to label instructions, or under abnormal conditions, or under conditions not reasonably foreseeable to seller, and buyer assumes the risk of any such use.**

CASORON® 50W

DICHLOBENIL WEED AND GRASS KILLER

COMPOSITION

ACTIVE INGREDIENT:	(% by weight)
Dichlobenil(2,6-dichlorobenzonitrile)	50.0%
INERT INGREDIENTS:	50.0%
TOTAL	100.0%
EPA REG. NO. 400-169	001 1-27-89

KEEP OUT OF REACH OF CHILDREN

CAUTION

STATEMENT OF PRACTICAL TREATMENT

IF SWALLOWED: Call a physician or poison control center immediately. Drink an eight ounce glass of water containing 2 tablespoons of activated charcoal.

IF INHALED: Remove to fresh air; if abnormal symptoms develop, contact a physician.

IF ON SKIN: Wash affected area with plenty of soap and water.

IF IN EYES: Flush eyes with plenty of water for at least 15 minutes. Contact a physician.

CASORON is a Registered Trademark of Duphar B.V., Weesp, the Netherlands

PRECAUTIONARY STATEMENTS

HAZARDS TO HUMANS AND DOMESTIC ANIMALS

CAUTION

Harmful if swallowed. Avoid breathing of dust. Avoid contact with eyes or skin.

ENVIRONMENTAL HAZARDS

Do not apply directly to water or wetlands (swamps, bogs, marshes, and potholes). Do not contaminate water when disposing of equipment washwaters. Cover, collect, or incorporate granules spilled on the soil surface.

DIRECTIONS FOR USE

It is a violation of Federal Law to use this product in a manner inconsistent with its labeling.

Do not apply this product through any type of irrigation system.

STORAGE AND DISPOSAL

Do not contaminate water, food or feed by storage or disposal.

STORAGE—Store in a dry location. Do not store with propagative structures such as seed, bulbs, tubers, nursery stock, etc., or with food or feed products.

PESTICIDE DISPOSAL—Wastes resulting from the use of this product may be disposed of on site or at an approved wastes disposal facility.

CONTAINER DISPOSAL—Completely empty bag into application equipment. Then dispose of empty bag in a sanitary landfill or by incineration, or if allowed by State and local authorities, by burning. If burned, stay out of smoke.

WORKERS SAFETY RULES

IMPORTANT! Always wash hands, face, and arms with soap and water before smoking, eating, drinking, or toileting. Keep all unprotected persons, children, livestock, and pets away from treated area or where there is a danger of drift. Do not rub eyes or mouth with hands. If you feel sick in any way, STOP work and get help right away. See Statement of Practical Treatment.

DIRECTIONS FOR USE

It is a violation of Federal law to use this product in a manner inconsistent with its labeling. Do not apply this product through any type of irrigation system.

ANNUAL WEED CONTROL

CASORON 50W is recommended for control of annual grassy and broadleaved weeds and certain perennial weeds such as:

Annual Bluegrass (*Poa annua*)	Milkweed Vine (*Morrenia odorata*)
Bluegrass	Miners lettuce
Bull Thistle	Natalgrass
Camphorweed	Old Witchgrass
Carpetweed	Peppergrass
Chickweed	Pineapple Weed
Citron Melon	Plantain
Coffeeweed	Purslane
Crabgrass	Ragweed
Cudweed	Red Deadnettle
Dandelion	Redroot Pigweed
Dogfennel	Rosarypea
Evening Primrose	Russian Thistle
Fiddleneck	Shepherdspurse
Florida Purslane (Pusley)	Smartweed
Foxtail	Spanishneedles
Gisekia	Spurge
Groundsel	Teaweed
Henbit	Texas Panicum (Hurrahgrass)
Horsetail	
Jerusalemoak	Wild Barley
Goosefoot	Wild Mustard
Knotweed	Wild Radish
Lambsquarters	Yellow Woodsorrel
Maypops	

Apply CASORON 50W over-all or as a directed spray either in early spring before seeds of annual weeds germinate or after cultivation has removed the growing weeds. Apply in sufficient water for uniform coverage (at least 50 gallons per acre). If treatment is followed by application of 1/2 to 1 inch of water by overhead irrigation, the lower rates are recommended. Shallow incorporation immediately after application is recommended in southern areas. In other areas, shallow incorporation or sprinkler irrigation is suggested when applications are made during periods of high temperature. CASORON 50W may be used for weed control on the plants listed below.

USE PRECAUTIONS

Do not use in seed beds, transplant or cutting beds or greenhouses. Do not apply this product through any type of irrigation system. Do not apply until 6 months after rooting of cuttings in the field, or after planting rooted cuttings. Do not apply more than the maximum recommended amount of CASORON 50W per growing season. Do not use on "St. Lucie fine sand," "Arzell fine sand" or other light sand soil. Do not graze livestock in treated areas.

WOODY ORNAMENTALS AND SHELTERBELTS (ESTABLISHED AND IN NURSERIES)

CASORON 50W may be used for weed control on the plants listed below. Apply 8 to 12 pounds per acre, based on area actually treated. Do not apply until 4 weeks after transplanting. Following general directions above.

Amur Corktree	Ivy (*Hedera helix*)
Arbor Vitae	Juniper
Ash	Kinnikinnick
Azalea[1]	Laurel
Barberry	Leucothoe
Beauty Bush	Lilac
Birch	Linden
Boxelder	Locust
Boxwood	Magnolia
Camellia	Maple
Caragana	Mock Orange
Cleyera japonica	Mountain Ash
Cotoneaster	Nandina
Cottonwood	Oak
Crabapple	Osmanthus (Holly-olive)
Deutzia	Pachistima
Dogwood	Pecan
Eastern Red Ceder	Photina
Elm	Pittosporum
Elwood Cypress	Poplar
English Walnut	Privet
Euonymus	Pyracantha
Filbert	Rhododendron
Flowering Almond	Rockrose
Flowering Quince	Rose
Forsythia	Russian Olive
Gardenia	Spirea
Goldenraintree	Squawcarpet
Hackberry	Weigela
Heather	Willow
Holly (Ilex spp.)[2]	Yew (Taxus)
Honeysuckle	

[1] Kurume, Mollis, Hardy hybrid types and hardy native Azalea species only.
[2] Except Ilex crenata, I. rotunda, I. vomitoria.

OTHER USES

UNDER ASPHALT: For general weed control under asphalt (in areas such as roadways, parking lots, recreational areas): after final grade is achieved, apply 20-24 lbs. in 75 to 100 gals. of water per acre. (7½ to 9 oz. per 1,000 sq. ft. in 1½ to 2½ gals. water). Treated area should be covered with asphalt as soon as possible. For re-surfacing work, existing weeds should be sprayed with a contact herbicide prior to CASORON 50W application.

IMPORTANT NOTICE

Seller warrants that this product conforms to its chemical description and is reasonably fit for the purposes stated on the label when used in accordance with the directions and instructions specified on the label under normal conditions of use, **but neither this warranty nor any other warranty of merchantability or fitness for a particular purpose, express or implied, extends to the use of this product, contrary to label instructions, or under abnormal conditions, or under conditions not reasonably foreseeable to seller, and buyer assumes the risk of any such use.**

COMITE®

AGRICULTURAL MITICIDE RECOMMENDED FOR AGRICULTURAL USE ONLY

COMPOSITION

ACTIVE INGREDIENTS:	(% by weight)
Propargite [2–(p-tert-butylphenoxy) cyclohexyl 2-propynyl sulfite]**	73.0%
INERT INGREDIENTS:	27.0%
TOTAL:	100.0%

**Contains 6.55 pounds OMITE per gallon

EPA REG. NO. 400-104 004 2-21-89

COMITE is a Registered Trademark of UNIROYAL CHEMICAL COMPANY, INC.

KEEP OUT OF REACH OF CHILDREN
DANGER
PELIGRO

PRECAUCION AL USUARIO: Si usted no lee ingles, no use este producto hasta que le etiqueta haya sido explicado ampliamente.

STATEMENT OF PRACTICAL TREATMENT

IF ON SKIN: Wash thoroughly with soap and water.

IF IN EYES: Immediately flush eyes with plenty of water. See a physician.

IF SWALLOWED: Call a physician or Poison Control Center immediately. Do not induce vomiting. Drink an eight ounce glass of water containing 2 tablespoons of activated charcoal. If this is not available, drink a large quantity of whole or condensed milk. Avoid alcohol.

IF INHALED: Remove person to fresh air. Apply artificial respiration if symptoms indicate. Call a physician.

PRECAUTIONARY STATEMENTS

HAZARDS TO HUMANS

DANGER

Corrosive, causes eye damage. May be fatal if inhaled. Harmful if swallowed or absorbed through skin. Do not breathe vapors or spray mist. Do not get in eyes, on skin or on clothing. Wash hands and face thoroughly with soap and water after use and before eating or smoking. (See use restrictions for protective clothing and clean up requirements).

ENVIRONMENTAL HAZARDS

This product is toxic to fish.

Do not apply directly to water or wetlands (swamps, bogs, marshes, and potholes). Drift or runoff from treated areas may be hazardous to aquatic organisims in neighboring areas. Do not contaminate water when disposing of equipment wash waters.

PHYSICAL OR CHEMICAL HAZARDS

Flammable, keep away from heat and open flame.

DIRECTIONS FOR USE

It is a violation of Federal law to use this product in a manner inconsistent with its labeling.

Do not apply this product through any type of irrigation system.

STORAGE AND DISPOSAL

Do not contaminate water, food or feed by storage or disposal

PESTICIDE DISPOSAL: Pesticide wastes are acutely hazardous. Improper disposal of excess pesticide, spray mixture or rinsate is a violation of Federal Law. If these wastes cannot be disposed of by use according to label instructions, contact your State Pesticide or Environmental Control Agency, or the Hazardous Waste representative at the nearest EPA Regional Office for guidance.

CONTAINER DISPOSAL: Triple rinse (or equivalent). Then offer for recylcing or reconditioning, or puncture and dispose of in a sanitary landfill, or by other procedures approved by state and local authorities.

REENTRY STATEMENT—EXCEPT FOR HARVESTING

Do not apply this product in such a manner as to directly or through drift expose workers or other persons. The area being treated must be vacated by unprotected persons.

Do not enter treated areas without protective clothing until sprays have dried.

Protective clothing requirements are outlined below for the indicated reentry.

CROP	EARLY[1, 2] ENTRY WORKER	TIME FOR REENTRY: LOW EXPOSURE ACTIVITIES IRRIGATOR, TRACTOR DRIVER, SCOUT[3]	TIME FOR REENTRY: ALL OTHER[3, 4, 5] ACTIVITES
All labeled crops	when sprays have dried	48 hours	7 days

1. Early entry is defined as any worker who is directed to enter a renetry restricted area prior to the expiration of any reentry interval that applies to that area. This will include those activities necessitated by unexpected events such as weather, pest conditions, equipment failure, etc. and does not include routine cultural practices.
2. These workers will wear protective clothing to include **at least:**
   — a hat or other suitable head covering
   — long-sleeved shirt and long-legged trousers or a coverall type garment (closely woven fabric covering the body **including** arms and legs).
   — chemical resistant gloves.
   — shoes and socks.
   NOTE: Occupants of enclosed cab vehicles must wear normal work attire.
3. These workers will wear normal work attire to include **at least:**
   — a hat or other suitable head covering.
   — long-sleeved shirt and long-legged trousers
   — shoes and socks
4. All other activities include those which have high foliar exposure such as pruners, cane turners and tippers (table grapes), etc.
5. See use directions for time of harvest after last application.

Because certain states may require more restrictive reentry intervals for various crops treated with this product, consult your State Department of Agriculture for further information.

Written or oral warnings must be given to the workers who are expected to be in a treated area or in an area to be treated with this product. When oral warnings are given, warnings shall be given in a language customarily understoood by workers. Oral warnings must be given if there is reason to believe that written warnings cannot be understood by workers. Written or oral warnings must include the following information:

"DANGER, Area to be treated with Propargite on (date of application). Do not enter without appropriate protective clothing until sprays have dried. Tractor drivers and irrigators and scouts wearing normal work attire may enter after 48 hours. Do not enter treated areas without protective clothing to perform activities that involve substantial or prolonged exposure with treated foliage for (enter number of days)."

If accidentally exposed, consult the Statement of Practical Treatment portion of this label for first aid recommendations.

USE RESTRICTIONS

Mixer/loaders must wear goggles or a face shield, chemical-resistant apron, long-sleeved shirt, long pants, and mid-forearm to elbow length chemical-resistant gloves. Applicators, if applying more than 2 pints of COMITE per acre in air blast equipment to citrus must be in an enclosed cab. At rates of 2 pints or less and if not in an enclosed cab, applicators must wear goggles, wide brimmed hat, a long-sleeved shirt and long pants, and chemical-resistant gloves while applying this pesticide.

Any article of clothing worn while applying product must be cleaned before re-use. Clothing should be laundered separately from household articles. Clothing that has been drenched or has otherwise absorbed concentrated pesticide must be disposed of in a sanitary landfill, incinerated, or burned if allowed by State and local authorities.

Do not apply this product through any type of irrigation system.

Do not mix COMITE with alkaline materials (such as lime, Bordeaux mixture or lime sulfur) or with materials containing a large amount of petroleum type solvents.

Do not use in spray solution above pH 10.

Do not plant any food or feed crop in rotation within 6 months after last application of propargite unless the crop is a registered use for propargite.

Cotton - Apply only before bolls open. Do not feed treated foliage or cotton trash to livestock.

Beans - Do not feed or forage treated bean vines or bean trash after harvesting.

Peanuts - Do not graze or feed livestock on treated areas or cut treated forage for hay.

GENERAL INSTRUCTIONS

COMITE is a liquid emulsifiable concentrate for preparing sprays to control the banks grass, Pacific spider, strawberry spider, and two-spotted spider mites.

COMITE is not systemic in action; therefore complete coverage of both upper and lower leaf surfaces and of fruit is necessary for effective control.

After three-quarters full of water, add recommended dosage [from table] of COMITE to the spray tank. Fill tank, agitate and spray thoroughly to cover foliage and fruit for best results.

DOSAGE INSTRUCTIONS

CROP	MITES CONTROLLED	TIMING OF APPLICATION	AMOUNT OF COMITE PER ACRE	GALLONS OF SPRAY SOLUTION PER ACRE: Ground Application	GALLONS OF SPRAY SOLUTION PER ACRE: Aerial Application	TOTAL NUMBER OF SPRAYS PER YEAR	EARLIEST HARVEST DAYS AFTER APPLICATION
COTTON (California & Arizona only)	two-spotted spider Pacific spider strawberry spider	Early[1]	1 pint	15[2] to 30 gallons	Not recommended	3	Before bolls open
		Mid Season to Layby	1½ to 2 pints	25[2] to 40 gallons	—		
			2 pints	—	5 to 15 gallons		
		Layby	2 pints	25[2] to 50 gallons	5 to 15 gallons		
		Layby to Boll Opening	2 pints	—	5 to 15 gallons		
COTTON (East of the Rocky Mts.)	two-spotted spider Pacific spider Strawberry spider	Early[1]	1 pint	15[2] to 30 gallons	Not recommended	3	Before bolls open
		Mid Season to Layby	1 to 2 pints	25[2] to 40 gallons	5 to 15 gallons		
		Layby to Boll Opening	2 pints	25[2] to 50 gallons	5 to 15 gallons		
FIELD CORN[3]	two-spotted spider Banks grass		2 to 3 pints	minimum of 20 gallons	minimum of 2 gallons[4]	1	30
SWEET CORN[3] (California only)	two-spotted spider Banks grass		2 to 3 pints	minimum of 20 gallons	minimum of 2 gallons	1	30

CROP	MITES CONTROLLED	TIMING OF APPLICATION	AMOUNT OF COMITE PER ACRE	GALLONS OF SPRAY SOLUTION PER ACRE Ground Application	Aerial Application	TOTAL NUMBER OF SPRAYS PER YEAR	EARLIEST HARVEST DAYS AFTER APPLICATION
GRAIN SORGHUM[5] (East of Rocky Mountains)	two-spotted spider Banks grass		1½ to 2 pints	minimum of 20 gallons	minimum of 5 gallons	1	30[6]
PEANUTS	two-spotted spider		2 pints	minimum of 20 gallons	minimum of 5 gallons	1	14
POTATOES (Pacific North west Only)	two-spotted spider		1½ to 2½ pints[7]	20 to 50 gallons	minimum of 10 gallons	2	14
BEANS[8] (green, lima & dry) (West of Rocky Mts.)	two-spotted spider		2 to 3 pints	minimum of 20 gallons	minimum of 5 gallons	2	succulent - 7 dry - 28

APPLICATION NOTES

COTTON

1. COMITE may be phytotoxic to young cotton under certain conditions. Age and climatic conditions result in a physiological hardening when plants are generally between 6 and 12 inches tall. Applications can begin after plants are hardened.
2. COMITE may be used in concentrate sprayers designed for concentrate row crop applications. Maintain COMITE dosage rate per acre.

CORN

3. Treat early. Best results are achieved when mite populations are low, before the mites have damaged the plants. Treatment is recommended when mite populations are just beginning to build, and the corn is still small enough (usually 3 to 4 feet) to allow good coverage. In California, ensure applications are made only when corn leaves are dry. This is especially true for white corn varieties which are the most susceptible to plant injury.
3. In Kansas and Colorado, use a minimum of 2 gallons spray solution per acre. In California, best results are obtained from applying 10–20 gallons of spray solution per acre. In Texas and New Mexico and other states, use a minimum of 5 gallons spray solution per acre.

GRAIN SORGHUM

5. Due to the large number of varieties of sorghum, it is recommended that a small test plot be treated to check phytotoxicity before using COMITE on a large area.
6. SILAGE STAGE: Grain should not be harvested within 60 days of spraying COMITE.

POTATOES

7. The 2½ pints per acre rate should be used when the crop is under extreme mite pressure.

BEANS

8. Do not use on varieties grown for fresh market or on varieties whose value is linked to pod appearance as pods may be spotted.

IMPORTANT NOTICE

Seller warrants that this product conforms to its chemical description and is reasonably fit for the purposes stated on the label when used in accordance with the directions and instructions specified on the label under normal conditions of use, **but neither this warranty nor any other warranty of merchantability or fitness for a particular purpose, express or implied, extends to the use of this product, contrary to label instructions, or under abnormal conditions, or under conditions not reasonably foreseeable to seller, and buyer assumes the risk of any such use.**

COMITE®

AGRICULTURAL MITICIDE FOR CITRUS

COMPOSITION

ACTIVE INGREDIENTS:	(% by weight)
Propargite [2-(p-tert-butylphenoxy) cyclohexyl 2-propynyl sulfite]*.	73.0%
INERT INGREDIENTS:	27.0%
TOTAL:	100.0%

* Contains 6.55 pounds OMITE per gallon

EPA REG. NO. 400-104 003 2-21-89

KEEP OUT OF REACH OF CHILDREN

DANGER

PELIGRO

PRECAUCION AL USUARIO: Si usted no lee ingles, no use este producto hasta que le etiqueta haya sido explicado ampliamente.

STATEMENT OF PRACTICAL TREATMENT

IF ON SKIN: Wash thoroughly with soap and water.

IF IN EYES: Immediately flush eyes with plenty of water. See a physician.

IF SWALLOWED: Call a physician or Poison Control Center immediately. Do not induce vomiting. Drink an eight ounce glass of water containing 2 tablespoons of activated charcoal. If this is not available, drink a large quantity of whole or condensed milk. Avoid alcohol.

IF INHALED: Remove person to fresh air. Apply artificial respiration if symptoms indicate. Call a physician.

COMITE is a Registered Trademark of UNIROYAL CHEMICAL COMPANY, INC.

PRECAUTIONARY STATEMENTS

HAZARDS TO HUMANS

DANGER

Corrosive, causes eye damage. May be fatal if inhaled. Harmful if swallowed or absorbed through skin. Do not breathe vapors or spray mist. Do not get in eyes, on skin or on clothing. Wear goggles or face shield when handling. Wash hands and face thoroughly with soap and water after use and before eating and smoking. (See use restrictions for protective clothing and clean up requirements).

ENVIRONMENTAL HAZARDS

This product is toxic to fish. Do not apply directly to water or wetlands (swamps, bogs, marshes, and potholes). Drift or runoff from treated areas may be hazardous to aquatic organisms in neighboring areas. Do not contaminate water when disposing of equipment washwaters.

PHYSICAL OR CHEMICAL HAZARDS

Flammable, keep away from heat and open flame.

DIRECTIONS FOR USE

It is a violation of Federal law to use this product in a manner inconsistent with its labeling.

Do not apply this product through any type of irrigation system.

STORAGE AND DISPOSAL

Do not contaminate water, food or feed by storage or disposal

PESTICIDE DISPOSAL: Pesticide wastes are acutely hazardous. Improper disposal of excess pesticide, spray mixture or rinsate is a violation of Federal Law. If these wastes cannot be disposed of by use according to label instructions, contact your State Pesticide or Environmental Control Agency, or the Hazardous Waste representative at the nearest EPA Regional Office for guidance.

CONTAINER DISPOSAL: Triple rinse (or equivalent). Then offer for recycling or reconditioning, or puncture and dispose of in a sanitary landfill, or by other procedures approved by state and local authorities.

REENTRY STATEMENT—EXCEPT FOR HARVESTING

Do not apply this product in such a manner as to directly or through drift expose workers or other persons. The area being treated must be vacated by unprotected persons.

Do not enter treated areas without protective clothing until sprays have dried.

Protective clothing requirements are outlined below for the indicated reentry.

Crop	Time For Reentry Early[1,2] Entry Worker	Low Exposure Activities Irrigator, Tractor Driver, Scout[3]	All Other[3, 4, 5] Activities
Oranges & Grapefruit	when sprays have dried	48 hours	21 days

1. Early entry is defined as any worker who is directed to enter a reentry restricted area prior to the expiration of any reentry interval that applies to that area. This

will include those activities necessitated by unexpected events such as weather, pest conditions, equipment failure, etc. and does not include routine cultural practices.

2. These workers will wear protective clothing to include **at least:**
   — a hat or other suitable head covering
   — long-sleeved shirt and long-legged trousers or a coverall type garment (closely woven fabric covering the body **including** arms and legs).
   — chemical resistant gloves.
   — shoes and socks.
   NOTE: Occupants of enclosed cab vehicles must wear normal work attire.
3. These workers will wear normal work attire to include **at least:**
   — a hat or other suitable head covering.
   — long-sleeved shirt and long-legged trousers
   — shoes and socks
4. All other activities include those which have high foliar exposure such as pruners, can turners and tippers (table grapes), etc.
5. See directions for time of harvest after last application.

Because certain states may require more restrictive reentry intervals for various crops treated with this product, consult your State Department of Agriculture for further information.

Written or oral warnings must be given to the workers who are expected to be in a treated area or in an area to be treated with this product. When oral warnings are given, warnings shall be given in a language customarily understood by workers. Oral warnings must be given if there is reason to believe that written warnings cannot be understood by workers. Written or oral warnings must include the following information:

"DANGER, Area to be treated with Propargite on (date of application). Do not enter without appropriate protective clothing until sprays have dried. Tractor drivers and irrigators and scouts wearing normal work attire may enter after 48 hours. Do not enter treated areas without protective clothing to perform activities that involve substantial or prolonged exposure with treated foliage for [enter number of days]." If accidentally exposed, consult the Statement of Practical Treatment portion of this label for first aid recommendations.

USE RESTRICTIONS

Mixer/loaders must wear goggles or a face shield, chemical-resistant apron, long-sleeved shirt, long pants, and mid-forearm to elbow length chemical-resistant gloves. Applicators, if applying more than 2 pints of COMITE per acre in air blast equipment to citrus must be in an enclosed cab. At rates of 2 pints or less and if not in an enclosed cab, applicators must wear goggles, wide brimmed hat, a long-sleeved shirt and long pants, and chemical-resistant gloves while applying this pesticide.

Any article of clothing worn while applying product must be cleaned before re-use. Clothing should be laundered separately from household articles. Clothing that has been drenched or has otherwise absorbed concentrated pesticide must be disposed of in a sanitary landfill, incinerated, or burned if allowed by State and local authorities.

Do not apply this product through any type of irrigation system.

Do not mix COMITE with alkaline materials (such as lime, Bordeaux mixture or lime sulfur) or with materials containing a large amount of petroleum type solvents.

Do not use in spray solution above pH 10.

Do not plant any food or feed crop in rotation within 6 months after last application of propargite unless the crop is a registered use for propargite.

Do not graze or feed livestock on cover crops grown among the trees listed on this label.

GENERAL INSTRUCTIONS

COMITE is a liquid emulsifiable concentrate for preparing sprays to control the citrus red, citrus rust, false spider, six-spotted spider, Texas citrus and broad spider mites.

COMITE is not systemic in action; therefore complete coverage of both upper and lower leaf surfaces and of fruit is necessary for effective control.

After three-quarters full of water, add recommended dosage (from table) of COMITE to the spray tank. Fill tank, agitate and spray thoroughly to cover foliage and fruit for best results.

DOSAGE INSTRUCTIONS

CROP	MITES CONTROLLED	AMOUNT OF COMITE PER ACRE	GALLONS OF SPRAY SOLUTION PER ACRE GROUND APPLICATION	AERIAL APPLICATION	TOTAL NUMBER OF SPRAYS PER YEAR	DAYS BEFORE HARVEST (LAST SPRAY)
ORANGES & GRAPEFRUIT Florida & Texas only	citrus red Texas citrus six-spotted citrus rust false spider broad mite	2-3 pints	minimum of 25 gallons	minimum of 10 gallons	2	21

DOSAGE INSTRUCTIONS

CROP	MITES CONTROLLED	AMOUNT OF COMITE PER ACRE	GALLONS OF SPRAY SOLUTION PER ACRE GROUND APPLICATION	AERIAL APPLICATION	TOTAL NUMBER OF SPRAYS PER YEAR	EARLIEST HARVEST DAYS AFTER APPLICATION
ORANGES & GRAPEFRUIT Florida & Texas only	citrus red Texas citrus six-spotted citrus rust false spider broad mite	2-3 pints	minimum of 25 gallons	minimum of 10 gallons	2	21

APPLICATION NOTES

1. Apply in sufficient water for thorough coverage.
2. Leaf distortion and/or fruit spotting may occur when used in the spring or it may occur if COMITE is tank mixed with oil or applied within two weeks prior to or following an oil application.

IMPORTANT NOTICE

Seller warrants that this product conforms to its chemical description and is reasonably fit for the purposes stated on the label when used in accordance with the directions and instructions specified on the label under normal conditions of use, **but neither this warranty nor any other warranty of merchantability or fitness for a particular purpose, express or implied, extends to the use of this product, contrary to label instructions, or under abnormal conditions, or under conditions not reasonably foreseeable to seller, and buyer assumes the risk of any such use.**

DIMILIN® 25W

INSECT GROWTH REGULATOR

RESTRICTED USE PESTICIDE

Due to toxicity to aquatic invertebrate animals.
For retail sale to and use only by Certified Applicators, or persons under their direct supervision, and only for those uses covered by the Certified applicator's certification.

COMPOSITION

ACTIVE INGREDIENT:	(% by weight)
N-[[(4-Chlorophenyl)amino]carbonyl]-2,6- difluorobenzamide*	25.0%
INERT INGREDIENTS:	75.0%
TOTAL	100.0%

*Diflubenzuron: U.S. Patent Numbers 3,748,356; 3,933,908; 3,989,842; 4,013,717; 4,110,469; 4,399,152 and other patents pending.

EPA Reg. No. 37100–8–400 5-24-88

KEEP OUT OF REACH OF CHILDREN
CAUTION

DIMILIN® is a Registered Trademark of Duphar B. V. Weesp, Holland

PRECAUTIONARY STATEMENTS

HAZARDS TO HUMANS
CAUTION

Avoid contact with skin. Avoid contamination of feed and foodstuffs.

ENVIRONMENTAL HAZARDS

This pesticide is extremely toxic to crab, shrimp and other aquatic invertebrates. Do not apply directly to water or wetlands, (swamps, bogs, marshes, and potholes), except under the forest canopy when aerially applied to control forest pests. Drift or runoff from treated areas may be hazardous to aquatic organisms in neighboring areas. Do not contaminate water when disposing of equipment washwaters.

DIRECTIONS FOR USE

It is a violation of Federal Law to use this product in a manner inconsistent with its labeling.

Do not apply this product through any type of irrigation system.

STORAGE AND DISPOSAL

Do not contaminate water, food or feed by storage or disposal.

STORAGE—Store in a dry location.

PESTICIDE DISPOSAL—Wastes resulting from the use of this product may be disposed of on site or at an approved waste disposal facility.

CONTAINER DISPOSAL—Completely empty bag into application equipment. Then dispose of empty bag in a sanitary landfill or by incineration, or if allowed by State and local authorities, by burning. If burned, stay out of smoke.

DIRECTIONS FOR USE

FORESTS, TREES AND SHRUBS

DIMILIN 25W will control larvae of gypsy moth, forest tent caterpillar, Nantucket pine tipmoth, and tussock moth in forests, trees and shrubs.

Action of DIMILIN 25W is slow because it disrupts the normal molting process of insect larvae. In most instances, several days may be needed before full effect can be seen. DIMILIN 25W applied at label rates does not affect bees or other beneficial insects.

DIMILIN 25W may be used to protect trees and shrubs such as in:

- Forests
- Residential, municipal and shade tree areas
- Recreational, areas such as campgrounds, golf courses, parks, parkways
- Ornamental, shade tree, and forest nurseries
- Forest plantings
- Shelterbelts
- Rights of way and other easements

INSECT	RATE DIMILIN 25W PER ACRE	TIMING	SPRAY VOLUME PER ACRE-GALLONS[1,2] AERIAL	GROUND AIR BLAST	GROUND HYDRAULIC
Gypsy Moth[3]	1-4 oz.	Early instar (1st, 2nd or 3rd) preferred, but prior to full leaf expansion.	0.5-2.5	5-20	100-400
Forest Tent Caterpillar	2-8 oz.	Early instar (1st, 2nd or 3rd) preferred, but prior to full leaf expansion.	1.0-2.0	5-20	100-400
Nantucket Pine Tip Moth	4 oz.	Early instar or when 75% of pupal cases are empty.	2.0-5.0	5-30[4]	100-400
Tussock Moth	4-8 oz.	Early instar	1.0-2.0	5-20	100-400

Do not make more than one application per year.

1. Uniform coverage of the foliage is essential. The higher volumes are recommended for mature or very large or dense tree stands or high population pressures. The higher water volumes in aerial application should be used when conditions of application are less than ideal and/or after larvae have reached 3rd instar.
2. Continuous agitation during mixing and application is required to maintain suspension of DIMILIN 25W. Do not use equipment without adequate agitation.
3. Gypsy Moth–To provide maximum protection from defoliation and reduce egg mass deposition, treatment can begin at 5 to 20% leaf expansion provided egg hatch is underway. Because of its slowness of action, applications later than 3rd instar may result in reduced foliage protection. Higher rates of DIMILiN 25W should be used if larvae have reached 3rd or 4th instar.
4. Use 5 to 15 gallons per acre if trees are less than 3½ feet tall. Use 15 to 30 gallons per acre for taller trees. Application should be made at peak emergence which can be determined by twig sampling, phermone traps, degree days, etc.

QUARANTINE PROGRAMS (Gypsy Moth)

For use in Quarantine programs conducted by State Cooperators as well as USDA personnel of both Plant Protection and Quarantine, APHIS and the U.S. Forest Service. For use in eradication of isolated infestations, make two applications of 1 to 2 ounces of DIMILIN 25W per acre 7-14 days apart. For use in quarantine programs involving the movement of nursery stock from infested to non-infested areas, make two applications of 1 to 2 ounces of DIMILIN 25W per acre 7-14 days apart on nursery stock.

MUSHROOMS

DIMILIN 25W will control larvae of sciarid flies in mushroom growing facilities. DIMILIN 25W in the mushroom growing media will prevent the development of the larval stages of the sciarids. This effectively stops reproduction in the growing medium and prevents damage to the mushrooms. Because of its unique type of activity, do not expect immediate reductions in adult fly populations. DIMILIN 25W does not directly affect adults, but kills the larvae in the growing medium.

Compost treatment: Apply 2.4 to 4 pounds of DIMILIN 25W per 1000 square feet to the compost at spawning time by thorough incorporation such as with a spawning machine. This is equivalent to 30 to 50 ppm active ingredient assuming a compost wet weight of 40 pounds per cubic foot.

Casing treatment: Apply 13.5 ounces of DIMILIN 25W per 1000 square feet at the time of casing by thorough incorporation into the casing or in sufficient water to obtain a thorough drench. This is equivalent to a rate of 30 ppm active ingredient assuming a casing weight of 6700 pounds per 1000 square feet.

SOYBEANS (except California)

DIMILIN 25W will control larvae of velvetbean caterpillar, Mexican bean beetle and green cloverworm. Apply DIMILIN 25W at the rate of 2 to 4 ounces (0.125 to 0.25 pounds) per acre. Make application when larvae are small (less than 0.5 inches) to give greater control and minimum insect damage to leaves. Repeat application if damaging numbers reappear.

DIMILIN 25W may be applied at the lower rate (2 ounces) to prevent velvetbean caterpillar build-up when the vegetative growth of soybeans is completed and as pod formation begins. Consult local Extension Service regarding infestation levels requiring treatment.

Aerial Application: Apply recommended amount of DIMILIN 25W in sufficient water (1 to 3 gallons per acre) to achieve uniform coverage of foliage.

Ground Application: Apply recommended amount of DIMILIN 25W in 9 to 35 gallons of water per acre to give uniform coverage.

Do not make more than two applications per season. Do not apply within 21 days of harvest. Do not for cut hay nor allow milk or meat animals to graze treated fields.

DIMILIN 25W inhibits the molting process of larvae, therefore, it does not provide immediate kill. From 3 to 5 days may be required before populations are reduced. Do not rotate crops other than soybeans or cotton until six (6) months following last application.

Restrictions on rotational crops: Do not plant food or feed crops in DIMILIN 25W treated soils within 6 months following last application, unless DIMILIN 25W is authorized for use on these crops.

COTTON

BEET ARMYWORM AND FALL ARMYWORM: Apply 4 to 8 ounces of DIMILIN 25W per acre (in California, do not exceed 4 ounces per acre per application). Use sufficient water to assure adequate coverage. 1 to 2 quarts of emulsifiable vegetable or paraffinic crop oil may be used to enhance canopy penetration. Application should be made during early stages of armyworm larval development. Based on DIMILIN's unique mode of action, full mortality effects may be delayed for 3 to 5 days following application. Do not apply more than 1.5 pounds of DIMILIN 25W per acre during one growing season. Do not apply after bolls begin to open.

Aerial Application: Apply in 2 to 3 gallons total volume per acre.

Ground Application: Apply in 5 to 30 gallons of total volume per acre to give uniform coverage.

BOLL WEEVIL: DIMILIN 25W will control boll weevil through reproduction suppression. Apply 4 to 8 ounces (0.25 to 0.5 pounds) of DIMILIN 25W per acre per treatment in combination with 2 to 4 quarts of emulsifiable vegetable or paraffinic crop oil. Consult your supplier or Uniroyal representative for oil specifications. In California do not apply in oil without water, do not exceed 4 ounces per acre per application.

Aerial Application: Apply in 2 to 3 gallons total volume per acre.

Ground Application: Apply in 5 to 30 gallons of total volume per acre to give uniform coverage.

MIXING DIRECTIONS IF USED WITH WATER:

1. Fill tank with half of the required amount of water.

2. Begin agitation and add required amount of DIMILIN 25W.
3. Continue agitation while adding remainder of water.
4. Add proper quantity of oil slowly. To avoid formation of an invert emulsion use at least 2 parts water for each part oil.

MIXING DIRECTIONS IF USED WITHOUT WATER:

1. **Prepare slurry.** In a container outside the spray tank, prepare a slurry in the proportion of 5 pounds DIMILIN 25W to 1 gallon of oil. Mix until smooth.
2. Slowly add slurry to spray tank containing balance of oil, using vigorous agitation.
3. Do not dilute with water.

Aerial or ground applications: Spray should be applied with aerial or ground equipment designed or modified to insure full uniform coverage of the entire plant. Equipment adjusted to provide droplets with a diameter 150 to 220 microns is recommended. Provide agitation prior to, during, and after blending and while applying.

For maximum control of boll weevil reproduction, make first application at pinhead square stage of cotton growth when overwintering boll weevils are entering the fields. Repeat treatments should allow at least 5 to 7 day intervals between applications. Do not make more than 6 applications and do not apply more than 1.5 pounds of DIMILIN 25W per acre during one growing season. Do not apply after bolls begin to open. Do not rotate crops other than cotton or soybeans until six (6) months following last application of DIMILIN 25W.

DIMILIN 25W does not kill the adult boll weevils, however, eggs deposited by treated female weevils will not hatch, thus limiting reproduction. The control of egg hatch and larval development within the square prevents its shedding and will then allow normal boll development. After initial treatment of the female weevil, 7 to 10 days are required before non-hatching eggs are laid.

When used for boll weevil control, DIMILIN 25W allows normal build-up of beneficial insects which may aid in control of bollworm and budworm. Emulsifiable concentrate insecticide formulations used in tank mixes, in the presence of oil, may result in phytoxicity. Care should be taken where such mixture is used.

Restrictions on rotational crops: Do not plant food or feed crops in DIMILIN 25W treated soils within 6 months following last application, unless DIMILIN 25W is authorized for use on these crops.

IMPORTANT NOTICE—Seller warrants that this product conforms to its chemical description and is reasonably fit for the purposes stated on the label when used in accordance with the directions and instructions specified on the label under normal conditions of use, **but neither this warranty nor any other warranty of merchantability or fitness for a particular purpose, express or implied, extends to the use of this product, contrary to label instructions, or under abnormal conditions, or under conditions not reasonably foreseeable to seller, and buyer assumes the risk of any such use.**

DIMILIN® 25W

INSECT GROWTH REGULATOR

RESTRICTED USE PESTICIDE

Due to toxicity to aquatic invertebrate animals. For retail sale to and use only by Certified Applicators, or persons under their direct supervision, and only for those uses covered by the Certified applicator's certification.

COMPOSITION

ACTIVE INGREDIENT:	(% BY WEIGHT)
N-[[(4-Chlorophenyl)amino]carbonyl]-2,6-difluorobenzamide*	25.0%
INERT INGREDIENTS:	75.0%
TOTAL	100.0%

* Diflubenzuron: U.S. Patent Numbers 3,748,356; 3,933,908; 3,989,842; 4,013,717; 4,110,469; 4,399,152 and other patents pending.

EPA Reg. No. 37100-8-400 EPA Est. No.

DIMILIN® is a Registered Trademark of Duphar B.V. Weesp, Holland

Lot No. 4-1-87B

KEEP OUT OF REACH OF CHILDREN

CAUTION

See left side panel for additional precautionary statement.

PRECAUTIONARY STATEMENTS

HAZARDS TO HUMANS

CAUTION

Avoid contact with skin. Avoid contamination of feed and foodstuffs.

ENVIRONMENTAL HAZARDS

This pesticide is extremely toxic to crab, shrimp and other aquatic invertebrates. Do not apply directly to water or wetlands, except under the forest canopy when used to control forest pests. Drift or runoff from treated areas may be hazardous to aquatic organisms in neighboring areas. Do not contaminate water by cleaning of equipment or disposal of wastes.

DIRECTIONS FOR USE

It is a violation of Federal Law to use this product in a manner inconsistent with its labeling. Do not apply this product through any type of irrigation system.

STORAGE AND DISPOSAL

Do not contaminate water, food or feed by storage or disposal.

STORAGE—Store in a dry location.

PESTICIDE DISPOSAL—Wastes resulting from the use of this product may be disposed of on site or at an approved waste disposal facility.

CONTAINER DISPOSAL—Triple rinse (or equivalent). Then offer for recycling or reconditioning, or puncture and dispose of in a sanitary landfill, by incineration or, if allowed by State and local authorities, by burning. If burned, stay out of smoke.

DIRECTIONS FOR USE

TREES AND SHRUBS

NANTUCKET PINE TIPMOTH: Apply 4 ounces DIMILIN 25W per acre in the early larval stages of development, preferably at the beginning of the second tipmoth generation at egg hatch. Do not make more than one application per year.

Aerial Application: Apply the recommended amount in 0.5 to 2 gallons of water per acre for uniform coverage.

Ground Application: Apply recommended amount in sufficient water for uniform, full coverage of foliage (10 to 200 gallons per acre).

For use in Christmas tree plantations and forested areas.

ORNAMENTALS

For control of beet armyworm on field or greenhouse grown chrysanthemums apply 0.5 to 1 pound of DIMILIN 25W per acre in a dilute spray not to exceed 200 gallons of water per acre. Begin applications when larvae appear and repeat at 2 week intervals as required. The insect dies during molting following contact and full effect will not be seen for 3 to 5 days following application.

IMPORTANT NOTICE—Seller warrants that this product conforms to its chemical description and is reasonably fit for the purposes stated on the label when used in accordance with the directions and instructions specified on the label under normal conditions of use, **but neither this warranty nor any other warranty of merchantability or fitness for a particular purpose, express or implied, extends to the use of this product, contrary to label instructions, or under abnormal conditions, or under conditions not reasonably foreseeable to seller, and buyer assumes the risk of any such use.**

HARVADE® -5F

HARVEST GROWTH REGULANT

FOR COTTON

COMPOSITION

ACTIVE INGREDIENT:	(% by weight)
Dimethipin (2,3-Dihydro-5,6-dimethyl-1, 4-dithiin 1,1,4,4-tetraoxide)* **	48.0%
INERT INGREDIENTS:	52.0%
TOTAL	100.0%

*U.S. Patent No. 3,920,438

**Contains 4.9 pounds of active ingredient per gallon.

EPA REG. NO. 400–155 003 8-4-89

KEEP OUT OF REACH OF CHILDREN

CAUTION

STATEMENT OF PRACTICAL TREATMENT

IF SWALLOWED: Drink two large glasses of water and induce vomiting. Contact a physician.

IF ON SKIN: Wash affected area with plenty of soap and water.

IF IN EYES: Flush eyes with plenty of water for at least 15 minutes. Contact a physician.

IF INHALED: Remove to fresh air; if abnormal symptoms develop, contact a physician.

HARVADE is a Registered Trademark of UNIROYAL CHEMICAL COMPANY, INC.

SHAKE WELL BEFORE USING

PRECAUTIONARY STATEMENTS

HAZARDS TO HUMANS

CAUTION

Harmful if swallowed or inhaled. Do not breathe vapors or spray mist. Avoid contact with skin, eyes or clothing. Wear a long sleeve shirt and impermeable gloves during mixing, loading, flagging, or application of HARVADE-5F. In case of contact, immediately flush eyes or skin with plenty of water. Get medical attention if irritation persists. Keep away from food or food products.

Wash hands and face thoroughly with soap and water after use and before eating or smoking.

ENVIRONMENTAL HAZARDS

Do not apply directly to water. Do not contaminate water when disposing of equipment washwaters. Do not apply when weather conditions favor drift from treated area.

DIRECTIONS FOR USE

It is violation of Federal law to use this product in a manner inconsistent with its labeling.

Do not apply this product through any type of irrigation system.

STORAGE AND DISPOSAL

Do not contaminate water, food or feed by storage or disposal.

PESTICIDE DISPOSAL: Wastes resulting from the use of this product may be disposed of on site or at an approved waste disposal facility.

CONTAINER DISPOSAL: Plastic containers. Triple rinse (or equivalent). Then offer for recycling or reconditioning, or puncture and dispose of in a sanitary landfill, or incineration, or if allowed by State and local authorities, by burning. If burned, stay out of smoke.

MIXING INSTRUCTIONS

SHAKE CONTAINER VIGOROUSLY BEFORE USE: Fill the spray tank with 2/3 of the required amount of water. Add the desired amount of HARVADE-5F, the selected crop oil concentrate and remaining required amount of water. The HARVADE-5F and crop oil concentrate will suspend easily with minimum agitation. To assure that container content is emptied—TRIPLE RINSE INTO SPRAY TANK.

DIRECTIONS FOR USE

GENERAL INFORMATION

HARVADE-5F is a plant growth regulant that effects certain plant processes which lead to defoliation in cotton. Since HARVADE-5F must enter the plant through the leaves for optimum activity, it must be applied in a manner that will allow maximum penetration through the leaf surface. Complete coverage of the entire plant is essential for optimum effectiveness. **HARVADE-5F MUST BE USED WITH ONE PINT/A OF THE DESIGNATED CROP OIL CONCENTRATES.** HARVADE-5F will improve the harvest operation by also desiccating most mature morningglory species. Desiccation/defoliation of other weed species such as sicklepod, croton, and teaweed has been observed. Good coverage of these above weeds is necessary for optimum results.

Application Equipment:

Nearly all spray application equipment is suitable for applying HARVADE-5F if fine spray droplets can be deposited. If aerial equipment is used, it should be fitted with a fixed boom set at 45-90° relative to the trailing edge of wing and flown 10–12 ft. above the crop canopy.

Compatibility:

HARVADE-5F is compatible with most commonly used cotton desiccants, defoliants, and insecticides. HARVADE-5F has been shown to be incompatible with EPN when used as a tank mix. Consult your supplier or Uniroyal for details.

Application Timing

Optimum defoliation will be obtained when HARVADE-5F is applied to mature cotton plants; i.e. 70% or more of the bolls open, no active growth, lower leaves have purple tinge. Apply 7–14 days prior to anticipated harvest.

Application Rate

Air Application:

Mix 8 fl. oz. of HARVADE-5F plus one pint of the crop oil concentrate designated below in a minimum of 5 gallons of spray solution per acre. If cotton is rank or conditions favor regrowth, see application notes.

Ground Application:

Mix 8 fl. oz. of HARVADE-5F plus one pint of the crop oil concentrates designated below in 10–20 gallons of spray solution per acre. If cotton is rank or conditions favor regrowth, see application notes.

One gallon of HARVADE-5F will treat 16 acres of cotton, when applied at the 8 fl. oz. rate.

Application Notes:

Partial defoliation from the first application will most likely result from one or more of the following conditions:

- Rank cotton foliage is dense enough to inhibit spray droplets from reaching bottom leaves.
- Plants are actively growing, with new growth and fruiting evident at the top.
- Bolls are not 70% open.
- Cotton is drought stressed. Under this condition, applications should be made before mid morning or at late afternoon.
- Night temperatures are below 60°F for 3–4 nights before or after application.
- Rainfall occurs within 6 hours after application.

Under these conditions a second application of 6 oz/A may be required 5–7 days after initial application. If only one application is desired on rank or actively growing cotton, a mixture of HARVADE-5F with a phosphate defoliant has been shown to be advantageous.

TANK MIX COMBINATIONS

HARVADE-5F plus PREP®*:

HARVADE-5F may be applied with 1 1/3–2 2/3 pints of PREP® to enhance boll opening and control regrowth. Apply this tank mixture when the bolls are 60–70% open.

HARVADE-5F plus GRAMOXONE® SUPER**:

On rain-belt cotton if a high percentage of cotton leaves are still actively growing, then 4-6 fl. oz./A of GRAMOXONE® SUPER should be added to the labeled rate of HARVADE-5F for more complete defoliation.

HARVADE-5F plus DROPP® 50WP***:

Under conditions that promote excessive regrowth, a tank mixture of HARVADE-5F at 6.5 oz/A with one pint of crop oil plus DROPP 50WP at 0.1-0.2 lb/A provides excellent regrowth control. This combination also provides defoliation when night time temperatures are below 60°-70°F for 2-3 nights preceeding or following application.

*PREP® is the registered trademark of Union Carbide Corporation

**GRAMOXONE® SUPER is the registered trademark of ICI Americas, Inc.

***DROPP® is the registered trademark of NOR-AM Chemcal Co.

CROP OIL CONCENTRATES REQUIRED WITH HARVADE-5F

Any petroleum based crop oil concentrate (83-17) or similar to those listed below **must be added at 1-2 pints per acre:** Agridex[1], Prime Oil[2], Super Oil[3], Surf Oil[4], Peptoil[5], Clean Crop Oil Concentrate[6] and Super Savol[7]. These crop oil concentrates are trademarks of [1] Helena Chemical Company,[2] Terra International Inc.,[3] Red Panther Cooperative, [4] Cleveland Chemical Company, [5] Drexel Chemical Co., [6] United Agri-Products, Inc. and [7] Leffingwell.

USE PRECAUTIONS

1. Do not plant rotational crops within 6 months after application.
2. Keep HARVADE drift from non-target areas.

a. Do not apply HARVADE-5F directly to, or permit mist to drift onto citrus, broccoli, almonds, grapes, peppers, tomatoes, lettuce, alfalfa, sugarbeets or wheat. These crops may sustain significant injury if contacted with HARVADE-5F.

b. Do not apply HARVADE-5F to, or permit mist to drift onto soil that will subsequently be seeded to peppers, tomatoes, lettuce, broccoli, alfalfa or sugarbeets. Significant crop injury may occur from such an application.

c. Do not apply HARVADE-5F to, or permit mist to drift onto irrigation water that will subsequently be applied to peppers, tomatoes or lettuce.

IMPORTANT NOTICE

Seller warrants that this product conforms to its chemical description and is reasonably fit for the purposes stated on the label when used in acordance with the directions and instructions specified on the label under normal conditions of use, **but neither this warranty nor any other warranty of merchantability or fitness for a particular purpose, express or implied, extends to the use of this product, contrary to label instructions, or under abnormal conditions, or under conditions not reasonably foreseeable to seller, and buyer assumes the risk of any such use.**

OMITE® -30W

AGRICULTURAL MITICIDE

RECOMMENDED FOR AGRICULTURAL USE ONLY

COMPOSITION

ACTIVE INGREDIENTS:	(% by weight)
Propargite [2-(p-tert-butylphenoxy) cyclohexyl 2-propynyl sulfite].	30.0%
INERT INGREDIENTS:	70.0%
TOTAL:	100.0%

EPA REG. NO. 400-82 003 3-21-89

OMITE is a Registered Trademark of UNIROYAL CHEMICAL COMPANY, INC.

KEEP OUT OF REACH OF CHILDREN

DANGER

PELIGRO

PRECAUCION AL USUARIO: Si usted no lee ingles, no use este producto hasta que le etiqueta haya sido explicado ampliamente.

STATEMENT OF PRACTICAL TREATMENT

IF ON SKIN: Wash thoroughly with soap and water.

IF IN EYES: Immediately flush eyes with plenty of water. See a physician.

IF SWALLOWED: Call a physician or Poison Control Center immediately. Drink promptly a large quantity of milk, egg whites or gelatin solution. If these are not available, drink large quantities of water. Avoid alcohol.

IF INHALED: Remove person to fresh air. Apply artificial respiration if symptoms indicate. Call a physician.

PRECAUTIONARY STATEMENTS

HAZARDS TO HUMANS

DANGER

Corrosive, causes eye damage. Harmful if swallowed, inhaled or absorbed through skin. Do not breathe vapors or spray mist. Do not get in eyes, on skin or on clothing. Wash hands and face thoroughly with soap and water after use and before eating or smoking. (See use restrictions for protective clothing and clean up requirements)

ENVIRONMENTAL HAZARDS

This pesticide is toxic to fish. Do not apply directly to water or wetlands (swamps, bogs, marshes, and potholes). Drift or runoff from treated areas may be hazardous to aquatic organisms in neighboring areas. Do not contaminate water when disposing of equipment wash waters.

DIRECTIONS FOR USE

It is a violation of Federal law to use this product in a manner inconsistent with its labeling.

Do not apply this product through any type of irrigation system.

STORAGE AND DISPOSAL

Do not contaminate water, food or feed by storage or disposal.

STORAGE: Store in a dry location.

PESTICIDE DISPOSAL: Pesticide wastes are acutely hazardous. Improper disposal of excess pesticide, spray mixture or rinsate is a violation of Federal Law. If these wastes cannot be disposed of by use according to label instructions, contact your State Pesticide or Environmental Control Agency, or the Hazardous Waste representative at the nearest EPA Regional Office guidance.

CONTAINER DISPOSAL: Completely empty bag into application equipment. Then dispose of empty bag in a sanitary landfill or by incineration, or if allowed by State and local authorities, by burning. If burned, stay out of smoke.

REENTRY STATEMENT—EXCEPT FOR HARVESTING

Do not apply this product in such a manner as to directly or through drift expose workers or other persons. The area being treated must be vacated by unprotected persons.

Do not enter treated area until sprays have dried.

Protective clothing requirements are outlined below for the indicated reentry.

Crop	TIME FOR REENTRY Early[1, 2] Entry Worker	Low Exposure Activities Irrigator, Tractor Driver, Scout[3]	All Other[3, 4, 5] Activities
Strawberries	when sprays have dried	48 hours	3 days
Grapes-wine & raisin	when sprays have dried	48 hours	14 day
Grapes-table	when sprays have dried	48 hours	21 days
Citrus - 10.5 lbs/per acre or less	when sprays have dried	48 hours	21 days
Citrus - at rates more than 10.5 lbs.	when sprays have dried	48 hours	28 days
All other crops	when sprays have dried	48 hours	7 days

1. Early entry is defined as any worker who is directed to enter a reentry restricted area prior to the expiration of any reentry interval that applies to that area. This will include those activities necessitated by unexpected events such as weather, pest conditions, equipment failure, etc. and does not include routine cultural practices.
2. These workers will wear protective clothing to include **at least:**
   —a hat or other suitable head covering
   —long-sleeved shirt and long-legged trousers or a coverall type garment (closely woven fabric covering the body **including** arms and legs).
   —chemical resistant gloves.
   —shoes and socks.
   NOTE: Occupants of enclosed cab vehicles must wear normal work attire.
3. These workers will wear normal work attire to include **at least:**
   —a hat or other suitable head covering.
   —long-sleeved shirt and long-legged trousers
   —shoes and socks
4. All other activities include those which have high foliar exposure such as pruners, cane turners and tippers (table grapes), etc.
5. See use directions for time of harvest after last application.

Because certain states may require more restrictive reentry intervals for various crops treated with this product, consult your State Department of Agriculture for further information.

Written or oral warnings must be given to the workers who are expected to be in a treated area or in an area to be treated with this product. When oral warnings are given, warnings shall be given in a language customarily understood by workers. Oral warnings must be given if there is reason to believe that written warnings cannot be understood by workers. Written or oral warnings must include the following information:

"DANGER, Area to be treated with Propargite on (date of application). Do not enter without appropriate protective clothing until sprays have dried." Tractor drivers and irrigators and scouts wearing normal work attire may enter after 48 hours. Do not enter treated areas to perform activities that involve substantial or prolonged contact with treated foliage for 7 days."

If accidentally exposed, consult the Statement of Practical Treatment portion of this label for first aid recommendations.

USE RESTRICTIONS

Mixer/loaders must wear goggles or a face shield, chemical-resistant apron, long-sleeved shirt, long pants, and mid-forearm to elbow length chemical-resistant gloves.

Applicators, if applying more than 10 lbs. of OMITE-30W per acre in air blast equipment must be in an enclosed cab. At rates of 10 lbs. or less and if not in an enclosed cab, applicators must wear goggles, wide brimmed hat, a long-sleeved shirt and long pants, and chemical-resistant gloves while applying this pesticide.

Any article of clothing worn while applying product must be cleaned before re-use. Clothing should be laundered separately from household articles. Clothing that has been drenched or has otherwise absorbed concentrated pesticide must be disposed of in a sanitary landfill, incinerated, or burned if allowed by State and local authorities.

Do not apply this product through any type of irrigation system.

Do not plant any food or feed crop in rotation within 6 months after last application of propargite unless the crop is a registered use for propargite.

Do not feed livestock on cover crops grown among the trees listed on this label.

DIRECTIONS FOR USE

OMITE-30W is a dry wettable powder for preparing sprays to control brown almond, citrus red, European red, McDaniel, Pacific spider, peach silver, strawberry spider, two-spotted spider mites and Willamette mites.

OMITE-30W is not systemic in action; therefore complete coverage of both upper and lower leaf surfaces and of fruit is necessary for effective control. Always use at least the minimum recommended amount of OMITE-30W per acre. OMITE-30W can be used in concentrate sprays, however, do not reduce the recommended rate per acre.

OMITE-30W performance is best when day temperatures average above 70°F.

DOSAGE INSTRUCTIONS

CROP	MITES CONTROLLED	AMOUNT OMITE-30W PER ACRE	MINIMUM GALLONS SPRAY SOLUTION PER ACRE GROUND	AERIAL	TOTAL NUMBER OF SPRAYS PER YEAR	EARLIEST HARVEST DAYS AFTER APPLICATION
ALMONDS (CA & AZ)	clover European red two-spotted spider strawberry spider Pacific spider	5-15 lbs.	50	—	2	28

CROP	MITES CONTROLLED	AMOUNT OMITE-30W PER ACRE	MINIMUM GALLONS SPRAY SOLUTION PER ACRE GROUND	AERIAL	TOTAL NUMBER OF SPRAYS PER YEAR	EARLIEST HARVEST DAYS AFTER APPLICATION
APPLES[1]	European red two-spotted spider Pacific spider McDaniel	5-12 lbs.	500-800	—	3	7
APRICOTS (West of Rocky Mts.)	Pacific spider two-spotted spider	5-9 lbs.	50	—	2	14
FIGS	two-spotted spider Pacific spider strawberry spider	3-6 lbs.	50	—	2	14
GRAPES 7-Day reentry Interval (West of Rocky Mts.)	Pacific spider strawberry spider Willamette two-spotted spider	5-9 lbs.	40	—	2	21
HOPS[1]	two-spotted spider	5 lbs.	100	—	2	14
LEMONS[2] (CA & AZ)	citrus red	10 lbs.	100	—	2	21
		11-15 lbs.	100	—	2	28
NECTARINES (West of Rocky Mts.)	Pacific spider brown almond	5-9 lbs.*	50	20 gal.	2	14
PEACHES	European red Pacific spider two-spotted spider strawberry spider McDaniel peach silver	5-15 lbs.**	50	20 gal.	2	14
PEANUTS	two-spotted spider	3-5 lbs.	20	—	2	14
PLUMS PRUNES	European red two-spotted spider Pacific spider	5-15 lbs.**	50	20 gal.	2	14
ROSES[3] (field grown)	two-spotted spider Pacific spider	2-5 lbs.	50	—	—	—
STRAWBERRIES[4] (West of Rocky Mts.)	Pacific spider strawberry spider two-spotted spider	5-10 lbs.	100	—	3 per 40 day harvest season	3
STRAWBERRIES[4] (LA & FL)	Pacific spider strawberry spider two-spotted spider	2½ lbs.	100	—	3 per 21 day harvest season	3
WALNUTS (CA only)	European red two-spotted spider	7-22 lbs.	100	10	2	14

*** Use 7½-9 pounds of OMITE-30W per acre when applied by air.**
****Use 7½-10 pounds of OMITE-30W per acre when applied by air.**

FOR USE AFTER HARVEST AND ON NON-BEARING TREES ONLY

Non-bearing trees are those that will not bear fruit within one year of Omite Application

DOSAGE INSTRUCTIONS

Crop	Mites Controlled	Amount OMITE-30W Per Acre	Minimum Gallons Spray Solution Per Acre	Total Number of Sprays Per Year
NAVEL ORANGES[2] (West of Rocky Mts.)	citrus red	10-15 lbs.	100	1
GRAPEFRUIT[2] (West of Rocky Mts.)	citrus red	10-15 lbs.	100	1
CHERRIES (West of Rocky Mts.)	European red two-spotted spider Pacific spider McDaniel	6 lbs.	50	—
PEARS	European red two-spotted spider	5-7½ lbs.	50	For use after harvest only.

APPLICATION NOTES:

1. Apples and Hops: Leaf injury and/or fruit spotting may occur if the following conditions are not observed:
   Do not apply OMITE-30W with Diazinon WP if air temperatures are expected to exceed 95°F the day of spraying apples and hops.
   Do not apply in combination with Guthion emulsifiable concentrate.
   Do not apply OMITE-30W with foliar spray oils because of incompatibility. See Compatibilities or contact your supplier or Uniroyal for further information.
2. Citrus: OMITE-30W may cause some leaf burn on new growth (feather growth) of lemons.
   Foliar oil applications must not follow OMITE-30W treatments within 30 days.
   OMITE-30W when used alone may follow foliar oils at an interval of greater than 40 days after an oil application.
   Do not use on navel oranges or grapefruit after new fruit is exposed in the spring as fruit spotting may occur.
3. Roses: 1 lb. of OMITE-30W per 100 gallons of water equals 1 ounce (28.4 grams) per 6.25 gallons. Because cultivars and cultural practices differ widely, the user should initially treat only a small portion of his crop to confirm plant safety under his conditions. Repeat applications of OMITE-30W are normally necessary for extended control.
4. Strawberries: Do not combine with other chemicals or spray adjuvants, when using on strawberries, as leaf and fruit injury may occur.
   California: Do not apply at day temperatures above 80°F as fruit burn may occur.

COMPATIBILITIES—Except as noted OMITE-30W is compatible with many pesticides (See application notes). Check with your Supplier or Uniroyal for details. Observe all cautions and limitations of labeling of all products used in mixtures.

- Do not use additional spray adjuvants.
- OMITE-30W is not compatible with alkaline materials such as lime, lime sulfur or Bordeaux. The effectiveness of OMITE-30W will be reduced.
- Do not apply in combination with foliar spray oils as phytotoxicity may result.

USE RESTRICTIONS—Do not feed livestock on cover crops grown among the trees listed on this label.

IMPORTANT NOTICE

Seller warrants that this product conforms to its chemical description and is reasonably fit for the purposes stated on the label when used in accordance with the directions and instructions specified on the label under normal conditions of use, **but neither this warranty nor any other warranty of merchantability or fitness for a particular purpose, express or implied, extends to the use of this product, contrary to label instructions, or under abnormal conditions, or under conditions not reasonably foreseeable to seller, and buyer assumes the risk of any such use.**

OMITE®-30W

THIS BAG CONTAINS 2 x 2½ LB. WATER SOLUBLE BAGS SEE FOLDER FOR USE INSTRUCTIONS—ENCLOSED IN BOTTOM FOLD

AGRICULTURAL MITICIDE WEST OF THE ROCKY MOUNTAINS

RECOMMENDED FOR AGRICULTURAL USE ONLY

COMPOSITION

ACTIVE INGREDIENTS:	(% by weight)
Propargite [2-(p-tert-butylphenoxy) cyclohexyl 2-propynyl sulfite]	30.0%
INERT INGREDIENTS:	70.0%
Total:	100.0%

EPA REG. NO. 400–82 003 2–21–89

OMITE is a Registered Trademark of UNIROYAL CHEMICAL COMPANY, INC.

KEEP OUT OF REACH OF CHILDREN

DANGER

PELIGRO

PRECAUCION AL USUARIO: Si usted no lee ingles, no use este producto hasta que le etiqueta haya sido explicado ampliamente.

STATEMENT OF PRACTICAL TREATMENT

IF ON SKIN: Wash thoroughly with soap and water.

IF IN EYES: Immediately flush eyes with plenty of water. See a physician.

IF SWALLOWED: Drink promptly a large quantity of milk, egg whites or gelatin solution. If these are not available, drink large quantities of water. Avoid alcohol. Call a physician or Poison Control Center immediately.

IF INHALED: Remove person to fresh air. Apply artificial respiration if symptoms indicate. Call a physician.

AVOID EXPOSURE TO DUST

PRECAUTIONARY STATEMENTS

HAZARDS TO HUMANS

DANGER

Corrosive, causes eye damage. Harmful if swallowed, inhaled or absorbed through skin. Do not breathe vapors or spray mist. Do not get in eyes, on skin or on clothing. Wear goggles or face shield when handling. Wash hands and face thoroughly with soap and water after use and before eating or smoking. [See use restrictions for protective clothing and clean up requirements]

ENVIRONMENTAL HAZARDS

This pesticide is toxic to fish. Do not apply directly to water or wetlands [swamps, bogs, marshes, and potholes]. Drift or runoff from treated areas may be hazardous to aquatic organisms in neighboring areas. Do not contaminate water when disposing of equipment wash waters.

PHYSICAL OR CHEMICAL HAZARDS

Flammable, keep away from heat and open flame.

DIRECTIONS FOR USE

It is a violation of Federal law to use this product in a manner inconsistent with its labeling.

Do not apply this product through any type of irrigation system.

STORAGE AND DISPOSAL

Do not contaminate water, food or feed by storage or disposal.

STORAGE: Store in a dry location.

PESTICIDE DISPOSAL: Pesticide wastes are acutely hazardous. Improper disposal of excess pesticide, spray mixture or rinsate is a violation of Federal Law. If these wastes cannot be disposed of by use according to label instructions, contact your State Pesticide or Environmental Control Agency, or the Hazardous Waste representative at the nearest EPA Regional Office for guidance.

CONTAINER DISPOSAL: Completely empty bag into application equipment. Then dispose of empty bag in a sanitary landfill or by incineration, or if allowed by State and local authorities, by burning. If burned, stay out of smoke.

REENTRY STATEMENT - EXCEPT FOR HARVESTING

Do not apply this product in such a manner as to directly or through drift expose workers or other persons. The area being treated must be vacated by unprotected persons.

Do not enter treated areas until sprays have dried.

Protective clothing requirements are outlined below for the indicated reentry.

	Time for Reentry		
Crop	Early[1, 2] Entry Worker	Low Exposure Activities Irrigator, Tractor Driver, Scout[3]	All Other[3, 4, 5] Activities
Strawberries	when sprays have dried	48 hours	3 days
Grapes-wine & raisin	when sprays have dried	48 hours	14 days
Grapes-table	when sprays have dried	48 hours	21 days
Citrus - 10.5 lbs/per acre or less	when sprays have dried	48 hours	21 days
Citrus - at rates more than 10.5 lbs.	when sprays have dried	48 horus	28 days
All other crops	when sprays have dried	48 hours	7 days

1. Early entry is defined as any worker who is directed to enter a reentry restricted area prior to the expiration of any reentry interval that applies to that area. This will include those activities necessitated by unexpected events such as weather, pest conditions, equipment failure, etc. and does not include routine cultural practices.
2. These workers will wear protective clothing to include **at least:**
   — a hat or other suitable head covering
   — long-sleeved shirt and long-legged trousers or a coverall type garment (closely woven fabric covering the body **including** arms and legs).
   — chemical resistant gloves.
   — shoes and socks.
   NOTE: Occupants of enclosed cab vehicles must wear normal work attire.
3. These workers will wear normal work attire to include **at least:**
   — a hat or other suitable head covering.
   — long-sleeved and long-legged trousers
   — shoes and socks
4. All other activities include those which have high foliar exposure such as pruners, cane turners and tippers (table grapes), etc.
5. See use directions for time of harvest after last application.

Because certain states may require more restrictive reentry intervals for various crops treated with this product, consult your State Department of Agriculture for further information.

Written or oral warnings must be given to the workers who are expected to be in a treated area or in an area to be treated with this product. When oral warnings are given, warnings shall be given in a language customarily understood by workers. Oral warnings must be given if there is reason to believe that written warnings cannot be understood by workers. Written or oral warnings must include the following information:

"DANGER, Area to be treated with Propargite on [date of application]. Do not enter without appropriate protective clothing until sprays have dried. Tractor drivers and irrigators and scouts wearing normal work attire may enter after 48 hours. Do not enter treated areas without protective clothing to perform activities that involve substantial or prolonged exposure with treated foliage for [enter number of days]."

If accidentally exposed, consult the Statement of Practical Treatment portion of this label for first aid recommendations.

USE DIRECTIONS

Mixer/loaders must wear goggles or a face shield, chemical-resistent apron, long-sleeved shirt, long pants, and mid-forearm to elbow length chemical-resistant gloves.

Applicators, if applying more than 10 lbs. of OMITE-30W per acre in air blast equipment must be in an enclosed cab. At rates of 10 lbs. or less and if not in an enclosed cab, applicators must wear goggles, wide brimmed hat, a long-sleeved shirt and long pants, and chemical-resistant gloves while applying this pesticide.

Any article of clothing worn while applying product must be cleaned before re-use. Clothing should be laundered separately from household articles. Clothing that has been drenched or has otherwise absorbed concentrated pesticide must be disposed of in a sanitary landfill, incinerated, or burned if allowed by State and local authorities.

Do not apply this product through any type of irrigation system.

Do not plant any food or feed crop in rotation within 6 months after last application of propargite unless crop is a registered use for propargite.

Do not feed livestock on cover crops grown among the trees listed on this label.

GENERAL INSTRUCTIONS

A dry wettable powder for preparing sprays to control brown almond, citrus red, clover, European red, McDaniel, Pacific spider, peach silver, strawberry spider, two-spotted spider and Willamette mites.

OMITE-30W is not systemic in action; therefore complete coverage of both upper and lower leaf surfaces and of fruit is necessary for effective control. Always use at least the minimum recommended amount of OMITE-30W per acre. OMITE-30W can be used in concentrate sprays.

OMITE-30W performance is best when day temperatures average above 70°F.

PRECAUCION AL USUARIO:

Si usted no puede lee ingles no use este producto hasta que la etiqueta le haya sido explicada ampliamente.

MIXING INSTRUCTIONS

PACKAGE CONTAINS OMITE-30W IN A WATER SOLUBLE BAG-ENTIRE CONTENTS DISSOLVE IN WATER. After opening package, immediately dump entire contents into spray tank. Do not touch bag: handling may cause breakage. THIS BAG CONTAINS 2x2½ POUND WATER SOLUBLE BAGS.

DOSAGE INSTRUCTIONS [West of the Rocky Mountains]

CROP	MITES CONTROLLED	NO. OF 2½LB. WATER SOLUBLE BAGS PER ACRE, PER TREATMENT	GALLONS SPRAY SOLUTION PER ACRE Ground	Aerial	TOTAL NUMBER OF SPRAYS PER YEAR	EARLIEST HARVEST DAYS AFTER APPLICATION
ALMONDS (CA & AZ)	clover European red two-spotted spider strawberry spider Pacific spider	2–6	500–1500	—	2	28
APPLES	European red two-spotted spider Pacific spider McDaniel	2–5	500–800	—	3	7
APRICOTS	Pacific spider two-spotted spider	2–4	300–600	—	2	14
FIGS	two-spotted spider Pacific spider strawberry spider	1–2	200–400	—	2	14
GRAPES 7-Day reentry Interval (see note 10)	Pacific spider strawberry spider Willamette two-spotted spider	2–4	350–600	—	2	21
HOPS	two-spotted spider	2	200	—	2	14
LEMONS (CA & AZ)	citrus red	4–6	600–1500	—	2	21
		5–6	600–1500	—	2	28
NECTARINES	Pacific spider brown almond	2–4*	300–600	Minimum 20 gal.	2	14
PEACHES	European red Pacific spider two-spotted spider strawberry spider McDaniel peach silver	2–6**	300–1000	Minimum 20 gal.	2	14
PLUMS PRUNES	European red two-spotted spider Pacific spider	2–6**	300–1000	Minimum 20 gal.	2	14
ROSES (field grown)	two-spotted spider Pacific spider	1–2	50–300	—	—	—
STRAWBERRIES (See application notes below)	Pacific spider strawberry spider two-spotted spider	2–4	100–200	—	3†† (per crop season)	3
WALNUTS (CA only)	European red two-spotted spider	3–9	500–1500	—	2	14

††Do not apply more than 3 applications of OMITE in any 40 day period during the harvesting season. *Use 7½–9 pounds of OMITE-30W per acre when applied by air. **Use 7½–10 pounds of OMITE-30W per acre when applied by air.

APPLICATION NOTES:

1. Complete and uniform coverage is necessary for optimum results.
2. Do not reduce the recommended rate per acre when applying OMITE-30W in concentrate sprays.
3. Do not combine with other chemicals or spray adjuvants, when using on strawberries, as leaf and fruit injury may occur.
4. California: Do not apply at day temperatures above 80°F as fruit burn may occur. (Strawberries only).
5. OMITE-30W may cause some leaf burn on new growth (feather growth) of lemons.
6. Omite performance is best when daily temperatures average above 70°F.
7. Leaf injury and/or fruit spotting may occur, if the following conditions are not observed:
   — Do not apply OMITE-30W with DIAZINON-WP if air temperatures are expected to exceed 95°F the day of spraying apples or hops.
   — OMITE-30W is not compatible with foliar spray oils. (Further information available from supplier or Uniroyal).
8. OMITE-30W is not compatible with alkaline materials such as lime, lime sulfur or Bordeaux. The effectiveness of OMITE-30W will be reduced.

Post Harvest Only

Crop	Mites Controlled	No. of 2½ lb. Water Soluble Bags per Acre per Treatment	Gallons Spray Solution per acre	Total Number of Sprays per Year	Last Spray Days Before Harvest
NAVEL ORANGES	citrus red	4–6	600–1500	1	SEE CAUTION BELOW
GRAPE-FRUIT	citrus red	4–6	600–1500	1	SEE CAUTION BELOW

CAUTION: DO NOT USE ON NAVEL ORANGES OR GRAPEFRUIT AFTER NEW FRUIT IS EXPOSED IN THE SPRING AS FRUIT SPOTTING MAY OCCUR.

FOR POST HARVEST AND NON-BEARING TREES ONLY

Non-bearing trees are those that will not bear fruit within one year of Omite Application.

DOSAGE INSTRUCTIONS [WEST OF THE ROCKY MOUNTAINS]

Crop	Mites Controlled	No. of 2½ lb. Water Soluble Bags per Acre per Treatment	Gallons Spray Solution per acre	Total Number of Sprays per Year	Last Spray Days Before Harvest
CHERRIES	European red two-spotted spider Pacific spider McDaniel	2	400	—	—
PEARS	European red two-spotted spider	2–3	FOR POST HARVEST TREES ONLY 300–500		

COMPATIBILITIES—Except as noted above OMITE-30W is compatible with many pesticides. (See application note 7). Check with your supplier or Uniroyal for details. Observe all cautions and limitations of labeling of all products used in mixtures.

IMPORTANT NOTICE

Seller warrants that this product conforms to its chemical description and is reasonably fit for the purposes stated on the label when used in accordance with the directions and instructions specified on the label under normal conditions of use, **but neither this warranty nor any other warranty of merchantability or fitness for a particular purpose, express or implied, extends to the use of this product, contrary to label instructions, or under abnormal conditions, or under conditions not reasonably foreseeable to seller, and buyer assumes the risk of any such use.**

OMITE® -CR

AGRICULTURAL MITICIDE IN WATER SOLUBLE BAGS

RECOMMENDED FOR AGRICULTURAL USE ONLY

NOT FOR USE IN CALIFORNIA

COMPOSITION

ACTIVE INGREDIENTS:	(% by weight)
Propargite [2-(p-tert-butylphenoxy) cyclohexyl 2-propynyl sulfite]	30.0%
INERT INGREDIENTS:	70.0%
TOTAL:	100.0%

EPA REG. NO. 400-82 002 2-21-89

KEEP OUT OF REACH OF CHILDREN

DANGER
PELIGRO

PRECAUCION AL USUARIO: Si usted no lee ingles, no use este producto hasta que le etiqueta haya sido explicado ampliamente.

OMITE is a Registered Trademark of UNIROYAL CHEMICAL COMPANY, INC.

STATEMENT OF PRACTICAL TREATMENT

IF ON SKIN: Wash thoroughly with soap and water.

IF IN EYES: Immediately flush eyes with plenty of water. See a physician.

IF SWALLOWED: Drink promptly a large quantity of milk, egg whites or gelatin solution. If these are not available, drink large quantities of water. Avoid alcohol. Call a physician or Poison Control Center immediately.

IF INHALED: Remove person to fresh air. Apply artificial respiration if symptoms indicate. Call a physician.

AVOID EXPOSURE TO DUST

PRECAUTIONARY STATEMENTS

HAZARDS TO HUMANS

DANGER

Corrosive, causes eye damage. May be fatal if inhaled. Harmful if swallowed, or absorbed through skin. Do not breathe vapors or spray mist. Do not get in eyes, on skin or on clothing. Wear goggles or face shield when handling. Wash hands and face thoroughly with soap and water after use and before eating or smoking. (See use restrictions for protective clothing and clean-up requirements)

ENVIRONMENTAL HAZARDS

This pesticide is toxic to fish. Do not apply directly to water or wetlands (swamps, bogs, marshes, and potholes). Drift or runoff from treated areas may be hazardous to aquatic organisms in neighboring areas. Do not contaminate water when disposing of equipment wash waters.

DIRECTIONS FOR USE

It is a violation of Federal law to use this product in a manner inconsistent with its labeling.

Do not apply this product through any type of irrigation system.

STORAGE AND DISPOSAL

Do not contaminate water, food or feed by storage or disposal.

STORAGE: Store in a dry location.

PESTICIDE DISPOSAL: Pesticide wastes are acutely hazardous. Improper disposal of excess pesticide, spray mixture or rinsate is a violation of Federal Law. If these wastes cannot be disposed of by use according to label instructions, contact your State Pesticide or Environmental Control Agency, or the Hazardous Waste representative at the nearest EPA Regional Office for guidance.

CONTAINER DISPOSAL: Completely empty bag into application equipment. Then dispose of empty bag in a sanitary landfill or by incineration, or if allowed by State and local authorities, by burning. If burned, stay out of smoke.

REENTRY STATEMENT—EXCEPT FOR HARVESTING

Do not apply this product in such a manner as to directly or through drift expose workers or other persons. The area being treated must be vacated by unprotected persons.

Do not enter treated areas without protective clothing until sprays have dried.

Protective clothing requirements are outlined below for the indicated reentry.

Crop	Time For Reentry: Early (1, 2) Entry Worker	Low Exposure Activities Irrigator, Tractor Driver, Scout (3)	All Other (3, 4, 5)
Strawberries	when sprays have dried	48 hours	3 days
Citrus	when sprays have dried	48 hours	28 days
All other crops	when sprays have dried	48 hours	7 days

1. Early entry is defined as any worker who is directed to enter a reentry restricted area prior to the expiration of any reentry interval that applies to that area. This will include those activities necessitated by unexpected events such as weather, pest conditions, equipment failure, etc. and does not include routine cultural practices.
2. These workers will wear protective clothing to include **at least:**
   — a hat or other suitable head covering
   — long-sleeved shirt and long-legged trousers or a coverall type garment (closely woven fabric covering the body **including** arms and legs).
   — chemical resistant gloves.
   — shoes and socks.
   NOTE: Occupants of enclosed cab vehicles must wear normal work attire.
3. These workers will wear normal work attire to include **at least:**
   — a hat or other suitable head covering.
   — long-sleeved shirt and long-legged trousers
   — shoes and socks
4. All other activities include those which have high foliar exposure such as pruners, can turners and tippers (table grapes), etc.
5. See use directions for time of harvest after last application.

Because certain states may require more restrictive reentry intervals for various crops treated with this product, consult your State Department of Agriculture for further information.

Written or oral warnings must be given to the workers who are expected to be in a treated area or in an area to be treated with this product. When oral warnings are given, warnings shall be given in a language customarily understood by workers. Oral warnings must be given if there is reason to believe that written warnings cannot be understood by workers. Written or oral warnings must include the following information:

"DANGER, Area to be treated with Propargite on (date of application). Do not enter without appropriate protective clothing until sprays have dried. Tractor drivers and irrigators and scouts wearing normal work attire may enter after 48 hours. Do not enter treated areas without protective clothing to perform activities that involve substantial or prolonged exposure with treated foliage for (enter number of days)." If accidentally exposed, consult the Statement of Practical Treatment portion of this label for first aid recommendations.

USE RESTRICTIONS

Mixer/loaders must wear goggles or a face shield, chemical-resistant apron, long-sleeved shirt, long pants, and mid-forearm to elbow length chemical-resistant gloves.

Applicators, if applying more than 10 lbs. of OMITE CR per acre in airblast equipment must be in an enclosed cab. At rates of 10 lbs. or less and if not in an enclosed cab, applicators must wear goggles, wide brimmed hat, a long-sleeved shirt and long pants, and chemical-resistant gloves while applying this pesticide.

Any article of clothing worn while applying product must be cleaned before re-use. Clothing should be laundered separately from household articles. Clothing that has been drenched or has otherwise absorbed concentrated pesticide must be disposed of in a sanitary landfill, incinerated, or burned if allowed by State and local authorities.

Do not apply this product through any type of irrigation system.

Do not plant any food or feed crop in rotation within 6 months after last application of propargite unless the crop is a registered use for propargite.

Do not feed livestock on cover crops grown among the trees listed on this label.

GENERAL INSTRUCTIONS

OMITE-CR is a dry wettable powder for preparing sprays to control citrus red, clover, European red, McDaniel, Pacific spider, strawberry spider, and two-spotted spider mites.

OMITE-CR is not systemic in action; therefore complete coverage of both upper and lower leaf surfaces and of fruit is necessary for effective control. Always use at least the minimum recommended amount of OMITE-CR per acre. OMITE-CR can be used in concentrate sprays.

AL USUARIO: Si usted no puede leer o entender ingles no use este producto hasta que la etiqueta le haya sido explicada ampliamente.

TO THE USER: If you cannot read or understand English, do not use this product until the label has been fully explained to you.

MIXING INSTRUCTIONS

This package contains 2 × 1½ pound water soluble bags of OMITE-CR. After opening package:

1. Do not open or perforate the inner water soluble bags of OMITE-CR. The water soluble bags and contents will dissolve in water.
2. Immediately dump entire contents into spray tank.
3. Do not touch water soluble bags as handling may cause breakage.
4. Do not begin spraying until the bag is completely dissolved and the powder is thoroughly dispersed in the spray tank.
5. If only one water soluble bag is used, store the remaining bag in the original package in a dry location away from heat or cold.

DOSAGE INSTRUCTIONS

CROP	MITES CONTROLLED	NUMBER OF 1½ POUND *WATER SOLUBLE BAGS* PER ACRE	GALLONS SPRAY SOLUTION PER ACRE GROUND	GALLONS SPRAY SOLUTION PER ACRE AERIAL	TOTAL NUMBER OF SPRAYS PER YEAR	EARLIEST HARVEST DAYS AFTER APPLICATION
ALMONDS (CA only)	clover European red two-spotted spider strawberry spider Pacific spider	3-10	500-1500	—	2	28
APPLES	European red two-spotted spider Pacific spider McDaniel	3-8	500-800	—	3	7
HOPS	two-spotted spider	3	200	—	2	14
LEMONS (AZ only) (See application note 3, 5 & 6)	citrus red	7-10	600-1500	—	2	28
STRAWBERRIES (See application notes 1 & 2)†	Pacific spider strawberry spider two-spotted spider	3-7	100-200	—	3†	3
ROSES (field grown)	two-spotted spider Pacific spider	1-3	50-300	—	—	—

† Do not apply more than 3 applications of OMITE-CR in any 40 day period during the harvesting season.

APPLICATION NOTES:

1. Do not combine with other chemicals or spray adjuvants, when using on strawberries, as leaf and fruit injury may occur.
2. California: Do not apply at day temperatures above 80°F as fruit burn may occur. (Strawberries only).
3. OMITE-CR may cause some leaf burn on new growth (feather growth) of lemons.
4. Omite performance is best when daily temperatures average above 70°F.
5. OMITE-CR when used alone may follow foliar oils at an interval of greater than 40 days after an oil application.
6. Foliar oil applications must not follow OMITE-CR treatments within 30 days.

NOTE: This is a Federal label. This label is not registered in the state of California.

FOR POST HARVEST AND NON-BEARING TREES ONLY

Non-bearing trees are those that will not bear fruit within one year of Omite Application.

Dosage Instructions

Crop	Mites Controlled	Number of 1½ Pound Water Soluble Bags Per Acre	Gallons Spray Solution Per Acre	Total Number of Sprays Per Year	Last Spray Days Before Harvest
CHERRIES	European red two-spotted spider Pacific spider McDaniel	4	400	—	—

USE RESTRICTIONS—Do not feed livestock on cover crops grown among the trees listed on this label.

CAUTION:

A. Leaf injury and/or fruit spotting may occur if the following conditions are not observed.
   1. Do not apply OMITE-CR with Diazinon-WP if air temperatures are expected to exceed 95°F the day of spraying apples and hops.
   2. OMITE-CR is not compatible with foliar spray oils. (Further information from supplier or Uniroyal).

B. OMITE-CR is not compatible with alkaline materials such as lime, lime sulfur or Bordeaux. The effectiveness of OMITE-CR will be reduced.

COMPATIBILITIES—Except as noted above OMITE-CR is compatible with many pesticides. (See caution statements). Check with your supplier or Uniroyal for details. Observe all cautions and limitations of labeling of all products used in mixtures.

IMPORTANT NOTICE

Seller warrants that this product conforms to its chemical description and is reasonably fit for the purposes stated on the label when used in accordance with the directions and instructions specified on the label under normal conditions of use, **but neither this warranty nor any other warranty of merchantability or fitness for a particular purpose, express or implied, extends to the use of this product, contrary to label instructions, or under abnormal conditions, or under conditions not reasonably foreseeable to seller, and buyer assumes the risk of any such use.**

OMITE®-CR

COLLATERAL LABEL
FOR CALIFORNIA ONLY
AGRICULTURAL MITICIDE
IN WATER SOLUBLE BAGS

RECOMMENDED FOR AGRICULTURAL USE ONLY

COMPOSITION

ACTIVE INGREDIENTS:	(% by weight)
Propargite [2-(p-tert-butylphenoxy) cyclohexyl 2-propynyl sulfite]*	30.0%
INERT INGREDIENTS:	70.0%
TOTAL:	100.0%

EPA REG. NO. 400-82 003 3-21-89

OMITE is a Registered Trademark of UNIROYAL CHEMICAL COMPANY, INC.

KEEP OUT OF REACH OF CHILDREN

DANGER
PELIGRO

PRECAUCION AL USUARIO: Si usted no lee ingles, no use este producto hasta que le etiqueta haya sido explicado ampliamente.

STATEMENT OF PRACTICAL TREATMENT

IF ON SKIN: Wash thoroughly with soap and water.

IF IN EYES: Immediately flush eyes with plenty of water. See a physician.

IF SWALLOWED: Drink promptly a large quantity of milk, egg whites or gelatin solution. If these are not available, drink large quantities of water. Avoid alcohol. Call a physician or Poison Control Center immediately.

IF INHALED: Remove person to fresh air. Apply artificial respiration if symptoms indicate. Call a physician.

AVOID EXPOSURE TO DUST

PRECAUTIONARY STATEMENTS

HAZARDS TO HUMANS

DANGER

Corrosive, causes eye damage. Harmful if swallowed, inhaled or absorbed through skin. Do not breathe vapors or spray mist. Do not get in eyes, on skin or on clothing. Wash hands and face thoroughly with soap and water after use and before eating or smoking. [See use restrictions for protective clothing and clean up requirements]

ENVIRONMENTAL HAZARDS

This pesticide is toxic to fish. Do not apply directly to water or wetlands [swamps, bogs, marshes, and potholes]. Drift or runoff from treated areas may be hazardous to aquatic organisms in neighboring areas. Do not contaminate water when disposing of equipment wash waters.

DIRECTIONS FOR USE

It is a violation of Federal law to use this product in a manner inconsistent with its labeling.

Do not apply this product through any type of irrigation system.

STORAGE AND DISPOSAL

Do not contaminate water, food or feed by storage or disposal.

STORAGE: Store in a dry location.

PESTICIDE DISPOSAL: Pesticide wastes are acutely hazardous. Improper disposal of excess pesticide, spray mixture or rinsate is a violation of Federal Law. If these wastes cannot be disposed of by use according to label instructions, contact your State Pesticide or Environmental Control Agency, or the Hazardous Waste representative at the nearest EPA Regional Office for guidance.

CONTAINER DISPOSAL: Completely empty bag into application equipment. Then dispose of empty bag in a sanitary landfill or by incineration, or if allowed by State and local authorities, by burning. If burned, stay out of smoke.

REENTRY STATEMENT - EXCEPT FOR HARVESTING

Do not apply this product in such a manner as to directly or through drift expose workers or other persons. The area being treated must be vacated by unprotected persons.

Do not enter treated areas without protective clothing until sprays have dried.

Protective clothing requirements are outlined below for the indicated reentry.

TIME FOR REENTRY			
CROP	EARLY[1, 2] ENTRY WORKER	LOW EXPOSURE ACTIVITIES IRRIGATOR, TRACTOR DRIVER, SCOUT[3]	ALL OTHER[3, 4, 5] ACTIVITIES
Citrus	when sprays have dried	48 hours	28 days

1. Early entry is defined as any worker who is directed to enter a reentry restricted area prior to the expiration of any reentry interval that applies to that area. This will include those activities necessitated by unexpected events such as weather, pest conditions, equipment failure, etc. and does not include routine cultural practices.
2. These workers will wear protective clothing to include **at least:**
   — a hat or other suitable head covering
   — long-sleeved shirt and long legged trousers or a coverall type garment (closely woven fabric covering the body **including** arms and legs).
   — chemical resistant gloves.
   — shoes and socks.
   NOTE: Occupants of enclosed cab vehicles must wear normal work attire.
3. These workers will wear normal work attire to include **at least:**
   — a hat or other suitable head covering.
   — long-sleeved shirt and long-legged trousers
   — shoes and socks
4. All other activities include those which have high foliar exposure such as pruners, cane turners and tippers (table grapes), etc.
5. See use directions for time of harvest after last application.

Because certain states may require more restrictive reentry intervals for various crops treated with this product, consult your State Department of Agriculture for further information.

Written or oral warnings must be given to the workers who are expected to be in a treated area or in an area to be treated with this product. When oral warnings are given, warnings shall be given in a language customarily understood by workers. Oral warnings must be given if there is reason to believe that written warnings cannot be understood by workers. Written or oral warnings must include the following information:

"DANGER, Area to be treated with Propargite on [date of application]. Do not enter without appropriate protective clothing until sprays have dried. Tractor drivers and irrigators and scouts wearing normal work attire may enter after 48 hours. Do not enter treated areas without protective clothing to perform activities that involve substantial or prolonged exposure with treated foliage for [enter number of days]."

If accidentally exposed, consult the Statement of Practical Treatment portion of this label for first aid recommendations.

USE RESTRICTIONS

Mixer/loaders must wear goggles or a face shield, chemical-resistant apron, long-sleeved shirt, long pants, and mid-forearm to elbow length chemical-resistant gloves.

Applicators, if applying more than 10 lbs. of OMITE CR per acre in airblast equipment must be in an enclosed cab. At rates of 10 lbs. or less and if not in an enclosed cab, applicators must wear goggles, wide brimmed hat, a long-sleeved shirt and long pants, and chemical-resistant gloves while applying this pesticide.

Any article of clothing worn while applying product must be cleaned before re-use. Clothing should be laundered separately from household articles. Clothing that has been drenched or has otherwise absorbed concentrated pesticide must be disposed of in a sanitary landfill, incinerated, or burned if allowed by State and local authorities.

Do not apply this product through any type of irrigation system.

Do not plant any food or feed crop in rotation within 6 months after last application of propargite unless the crop is a registered use for propargite.

Do not feed livestock on cover crops grown among the trees listed on this label.

Foliar oil applications must not follow OMITE-CR treatments within 30 days.

OMITE-CR when used alone may follow foliar oils at an interval of greater than 40 days after an oil application.

GENERAL INSTRUCTIONS

OMITE-CR is a dry wettable powder for preparing sprays to control citrus red and two-spotted spider mites.

OMITE-CR is not systemic in action; therefore complete coverage of both upper and lower leaf surfaces and of fruit is necessary for effective control. Always use at least the minimum recommended amount of OMITE-CR pere acre. OMITE-CR can be used in concentrae sprays, but no less than 100 gallons per acre.

AL USUARIO:

Si usted no puede leer o entender ingles no use este producto hasta que la etiqueta le haya sido explicada ampliamente.

TO THE USER:

If you cannot read or understand English, do not use this product until the label has been fully explained to you.

MIXING INSTRUCTIONS

This package contains 2x1½ pound water soluble bags of OMITE-CR. After opening package:

1. Do not open or perforate the inner water soluble bags of OMITE-CR. The water soluble bags and contents will dissolve in water.
2. Immediately dump entire contents into spray tank.
3. Do not touch water soluble bags as handling may cause breakage.
4. Do not begin spraying until the bag is completely dissolved and the powder is thoroughly dispersed in the spray tank.
5. If only one water soluble bag is used, store the remaining bag in the original package in a dry location away from heat or cold.

Dosage Instructions

Crop	Mites Controlled	Pounds OMITE-CR Per Acre [No. of 1½ Lb. *Water Soluble* Bags Per Acre]	Gallons Spray Solution Per Acre Ground	Total Number of Sprays Per Year	Earliest Harvest Days After Application
ORANGES[1,2] GRAPEFRUIT LEMONS (CA only)	citrus red two-spotted spider	10.5 lbs. (7 bags)	1000 gals.	2[4]	42 days

APPLICATION NOTES:

1. OMITE-CR may cause some leaf burn on new growth (feather growth) of lemons.
2. Omite performance is best when daily temperatures average above 70°F.
3. Even though there is a 7 day PHI, harvesting cannot take place until the reentry interval expires.
4. Second treatment must be made no closer than 42 days after the first.

CAUTION:

OMITE-CR is not compatible with foliar spray oils. (Further information from supplier or Uniroyal).

OMITE-CR is not compatible with alkaline materials such as lime, lime sulfur or Bordeaux. The effectiveness of OMITE-CR will be reduced.

COMPATIBILITIES —Except as noted above OMITE-CR is compatible with many pesticides. (See caution statements). Check with your supplier or Uniroyal for details. Observe all cautions and limitations of labeling of all products used in mixtures.

IMPORTANT NOTICE

Seller warrants that this product conforms to its chemical description and is reasonably fit for the purposes stated on the label when used in accordance with the directions and instructions specified on the label under normal conditions of use, **but neither this warranty nor any other warranty of merchantability or fitness for a particular purpose, express or implied, extends to the use of this product, contrary to label instructions, or under abnormal conditions, or under conditions not reasonably foreseeable to seller, and buyer assumes the risk of any such use.**

OMITE®-6E

AGRICULTURAL MITICIDE

RECOMMENDED FOR AGRICULTURAL USE ONLY

COMPOSITION

ACTIVE INGREDIENTS:	(% by weight)
Propargite 2-(p-tert-butylphenoxy) cyclohexyl 2-propynyl sulfite*.	68.1%
INERT INGREDIENTS:	31.9%
TOTAL:	100.0%

*6 lbs. Active per Gallon

EPA REG. NO. 400-89 003 3-31-89

KEEP OUT OF REACH OF CHILDREN

WARNING

AVISO

PRECAUCION AL USUARIO: Si usted no lee ingles, no use este producto hasta que le etiqueta haya sido explicado ampliamente.

STATEMENT OF PRACTICAL TREATMENT

IF ON SKIN: Wash thoroughly with soap and water.

IF IN EYES: Immediately flush eyes with plenty of water. See a physician.

IF SWALLOWED: Call a physician or poison control center immediately. Drink a large quantity of milk, egg whites or gelatin solution. If these are not available, drink large quantities of water. Avoid alcohol.

IF INHALED: Remove person to fresh air. Apply artificial respiration if symptoms indicate. Call a physician.

OMITE is a Registered Trademark of UNIROYAL CHEMICAL COMPANY, INC.

PRECAUTIONARY STATEMENTS

HAZARDS TO HUMANS

WARNING

Causes substantial but temporary eye injury. May be fatal if absorbed through skin. Harmful if swallowed or inhaled. Do not breathe vapors or spray mist. Do not get in eyes, on skin or on clothing. Wash hands and face thoroughly with soap and water after use and before eating, drinking, or smoking. (See use restrictions for protective clothing and clean up requirements).

ENVIRONMENTAL HAZARDS

This pesticide is toxic to fish. Do not apply directly to water or wetlands (swamps, bogs, marshes, and potholes). Drift or runoff from treated areas may be hazardous to aquatic organisms in neighboring areas. Do not contaminate water when disposing of equipment washwaters.

PHYSICAL OR CHEMICAL HAZARDS

Flammable, keep away from heat and open flame.

DIRECTIONS FOR USE

It is a violation of Federal law to use this product in a manner inconsistent with its labeling.

STORAGE AND DISPOSAL

Do not contaminate water, food or feed by storage or disposal.

PESTICIDE DISPOSAL: Wastes resulting from the use of this product may be disposed of on site or at an approved waste disposal facility.

CONTAINER DISPOSAL: Triple rinse (or equivalent). Then offer for recycling or reconditioning, or puncture and dispose of in a sanitary landfill, or if allowed by state and local authorities, by burning. If burned, stay out of smoke.

REENTRY STATEMENT

Do not apply this product in such a manner as to directly or through drift expose workers or other persons. The area being treated must be vacated by unprotected persons.

Do not enter treated areas without protective clothing until sprays have dried.

Protective clothing requirements are outlined below for the indicated reentry.

Crop	Time For Reentry		
	Early[1,2] Entry Worker	Low Exposure Activities Irrigator, Tractor Driver, Scout[3]	All Other[3,4,5] Activities
All labelled crops	when sprays have dried	48 hours	7 days

1. Early entry is defined as any worker who is directed to enter a reentry restricted area prior to the expiration of any reentry interval that applies to that area. This will include those activities necessitated by unexpected events such as weather, pest conditions, equipment failure, etc. and does not include routine cultural practices.
2. These workers will wear protective clothing to include **at least:**
   — a hat or other suitable head covering
   — long-sleeved shirt and long-legged trousers or a coverall type garment (closely woven fabric covering body **including** arms and legs).
   — chemical resistant gloves.
   — shoes and socks.
   NOTE: Occupants of enclosed cab vehicles must wear normal work attire.
3. These workers will wear normal work attire to include **at least:**
   — a hat or other suitable head covering.
   — long-sleeved shirt and long-legged trousers
   — shoes and socks
4. All other activities include those which have high foliar exposure such as pruners, cane turners and tippers (table grapes), etc.
5. See use directions for time of harvest after last application.

Because certain states may require more restrictive reentry intervals for various crops treated with this product, consult your State Department of Agriculture for further information.

Written or oral warnings must be given to workers who are expected to be in a treated area or in an area about to be treated with this product. When oral warnings are given, warnings shall be given in a language customarily understood by workers. Oral warnings must be given if there is reason to believe that written warnings cannot be understood by workers. Written or oral warnings must include the following information:

"WARNING Area to be treated with Propargite on (date of application). Do not enter without appropriate protective clothing until sprays have dried. Tractor drivers and irrigators and scouts wearing normal work attire may enter after 48 hours. Do not enter treated areas without protective clothing to perform activities that involve substantial or prolonged exposure with treated foliage for (enter number of days)." If accidently exposed, consult the Statement of Practical Treatment portion of the pesticide label for first aid recommendations.

USE RESTRICTIONS

Mixer/loaders must wear goggles or a face shield, chemical-resistant apron, long-sleeved shirt, long pants, and mid-forearm to elbow length chemical-resistant gloves.

Applicators, if applying more than 4 pints of OMITE-6E (64 oz.) per acre in air blast equipment must be in an enclosed cab. At rates of 4 pints or less and if not in an enclosed cab, applicators must wear goggles, wide brimmed hat, a long-sleeved shirt and long pants, and chemical-resistant gloves while applying this pesticide.

Any article of clothing worn while applying product must be cleaned before re-use. Clothing should be laundered separately from household articles. Clothing that has been drenched or has otherwise absorbed concentrated pesticide must be disposed of in a sanitary landfill, incinerated, or burned if allowed by State and local authorities.

Do not use water leaving propargite treated fields to irrigate crops used for food or feed that are not registered for use with propargite.

Do not graze or feed livestock on cover crops grown among the trees and vines listed on this label. Do not feed hop refuse or cones to livestock. However, spent hops from the extract operation may be fed. Do not feed mint spent hay to livestock.

Do not plant any food or feed crop in rotation within 6 months after last application of propargite unless the crop is a registered use for propargite.

DIRECTIONS FOR USE

OMITE®-6E is a liquid emulsifiable concentrate for preparing sprays to control clover, European red, Pacific spider, southern red, strawberry spider, and two-spotted spider mites.

OMITE®-6E is not systemic in action; therefore complete coverage of both upper and lower leaf surfaces and of fruit is necessary for effective control.

Add recommended dosage (from table) of OMITE®-6E to the spray tank, agitate and spray thoroughly to cover foliage and fruit for best results.

COMPATIBILITIES: OMITE®-6E is compatible with most pesticides (see caution statements). Check with your supplier for details. Observe all cautions and limitations on labeling of all products used in mixtures.

CAUTION:

A. Leaf injury may occur if the following conditions are not observed.
   1. Do not apply OMITE®-6E with Diazinon WP if air temperatures are expected to exceed 95°F. the day of spraying hops.
   2. OMITE®-6E is not compatible with foliar spray oils. (Further information available from supplier).
   3. Do not apply OMITE®-6E to cranberries if air temperatures are above 85°F.

B. OMITE®-6E is not compatible with alkaline materials such as lime, lime sulfur or Bordeaux. The effectiveness of OMITE will be reduced.

DOSAGE INSTRUCTIONS

See dosage instructions chart and application notes.

AL USUARIO:

Si usted no puede leer o entender ingles no use este producto hasta que la etiqueta le haya sido explicada ampliamente.

TO THE USER:

If you cannot read or understand English, do not use this product until the label has been fully explained to you.

DOSAGE INSTRUCTIONS

CROP	MITES CONTROLLED	AMOUNT PER ACRE	GALLONS OF SPRAY SOLUTION PER ACRE: GROUND APPLICATION	GALLONS OF SPRAY SOLUTION PER ACRE: AERIAL APPLICATION	NUMBER OF SPRAYS PER YEAR	DAYS BEFORE HARVEST (LAST SPRAY)
Almonds (California & Arizona only)	Clover European red strawberry spider two-spotted spider Pacific spider	3-4 pints	minimum of 50 gallons	15[1]	2	28
Apricots (California only)	two-spotted spider Pacific spider	2-3 pints		minimum of 10 gallons	2	post-harvest only
Cranberries[3] (Massachusetts only)	southern red mite	30-60[4,5] fluid ounces	300		2	14
Field Corn (California only)	two-spotted spider	2 pints	20-50	minimum of 10 gallons	1	56
Hops	two-spotted spider	2 pints	200		2	14
Mint	two spotted spider	2-3 pints	20-50	minimum of 10 gallons	2	14
Potatoes (Pacific Northwest only)	two-spotted spider	2-3 pints	20-50	minimum of 10 gallons	2	14
Walnuts	two-spotted spider European red	2-9 pints	500-1500	minimum of 20 gallons[2]	2	14

APPLICATION NOTES—ALMONDS

1. One application of OMITE-6E may be applied by air to almonds over one application of OMITE-30W.

APPLICATION NOTES—WALNUTS

2. Use 3 to 4 pints of OMITE-6E in a minimum of 20 gallons of water per acre when applied by air.

APPLICATION NOTES—CRANBERRIES

3. OMITE-6E may be applied through a sprinkler irrigation system to cranberries only.
4. 30-60 ounces OMITE-6E may be applied to cranberries once prior to bloom followed by a second application of 30 ounces anytime prior to 14 days of harvest.
5. Regardless of the size of the irrigation system, if OMITE-6E is used at the recommended rate per acre, the ratio of diluent to product in the slurry tank should be approximately 4:1.

IRRIGATION USE PRECAUTIONS FOR CRANBERRIES

A. Apply this product only through sprinkler systems, including center pivot, lateral move, end tow, side (wheel) roll, traveler, big gun, solid set, or hand move irrigation sytems. Do not apply this product through any other type of irrigation system.

B. Crop injury, lack of effectiveness, or illegal pesticide residues in the crop can result from nonuniform distribution of treated water.

C. If you have any questions about calibration, you should contact State Extension Service specialists, equipment manufacturers or other experts.

D. Do not connect an irrigation system (including greenhouse systems) used for pesticide application to a public water system unless the pesticide label prescribed safety devices for public water systems are in place.

E. A person knowledgeable of the chemigation system and responsible for its operation, or under the supervision of the responsible person, shall shut the system down and make the necessary adjustments should the need arise.

F. The system must contain a functional check valve, vacuum relief valve, and low pressure drain appropriately located on the irrigation pipeline to prevent water source contamination from backflow.
G. The pesticide injection pipeline must contain a functional, automatic, quick-closing check valve to prevent the flow of fluid back toward the injection pump.
H. The pesticide injection pipeline must also contain a functional, normally closed, solenoid-operated valve located on the intake side of the injection pump and connected to the system interlock to prevent fluid from being drawn from the supply tank when the irrigation system is either automatically or manually shut down.
I. The system must contain functional interlocking controls to automatically shut off the pesticide injection pump when the water pump motor stops.
J. The irrigation line or water pump must include a functional pressure switch which will stop the water pump motor when the water pressure decreases to the point where pesticide distribution is adversely affected.
K. Systems must use a metering pump, such as a positive displacement injection pump (e.g., diaphragm pump) effectively designed and constructed of materials that are compatible with pesticides and capable of being fitted with a system interlock.
L. Do not apply when wind speed favors drift beyond the area intended for treatment.
M. Constant agitation must be maintained in the chemical supply tank during the entire period of insecticide application.
N. Inject the product with a positive replacement pump into the main line ahead of a right angle turn, to insure adequate mixing.
O. Application of more than label recommended quantities of irrigation water per acre may result in decreased product performance by removing the chemical from the zone of effectiveness.
P. Do not apply when system connections or fittings leak, when nozzles do not provide uniform distribution or when lines containing the product must be dismantled and drained.
Q. Greater accuracy in calibration and distribution will be achieved by injecting a larger volume of more dilute mixture per hour. Pesticide should be applied continuously for the duration of the water addition.
R. Where sprinkler irrigation patterns do not overlap sufficiently unacceptable mite control may result. Where sprinkler distribution patterns overlap excessively crop injury may result.

IMPORTANT NOTICE

Seller warrants that this product conforms to its chemical description and is reasonably fit for the purposes stated on the label when used in accordance with the directions and instructions specified on the label under normal conditions of use, **but neither this warranty nor any other warranty of merchantability or fitness for a particular purpose, express or implied, extends to the use of this product, contrary to label instructions, or under abnormal conditions, or under conditions not reasonably foreseeable to seller, and buyer assumes the risk of any such use.**

ORNAMITE™

MITICIDE FOR ORNAMENTALS

FOR COMMERCIAL USE ONLY

COMPOSITION

ACTIVE INGREDIENTS:	(% by weight)
Propargite [2-(p-tert-butylphenoxy) cyclohexyl 2-propynyl sulfite]*	30.0%
INERT INGREDIENTS:	70.0%
TOTAL:	100.0%

EPA REG. NO. 400-82 12-17-87

ORNAMITE is a Registered Trademark of UNIROYAL CHEMICAL COMPANY, INC.

KEEP OUT OF REACH OF CHILDREN

DANGER PELIGRO

PRECAUCION AL USUARIO: Si usted no lee ingles, no use este producto hasta que le etiqueta haya sido explicado ampliamente.

STATEMENT OF PRACTICAL TREATMENT

IF SWALLOWED: Drink promptly a large quantity of milk, egg whites or gelatin solution. If these are not available, drink large quantities of water. Avoid alcohol. Call a physician or Poison Control Center immediately.

IF INHALED: See a physician if an abnormal reaction occurs.

IF ON SKIN: Wash thoroughly with soap and water.

IF IN EYES: Immediately flush eyes with plenty of water. See a physician.

AVOID EXPOSURE TO DUST

PRECAUTIONARY STATEMENTS

HAZARDS TO HUMANS

DANGER

Corrosive, causes eye damage. May be fatal if inhaled. Harmful if swallowed, or absorbed through skin. Do not breathe vapors or spray mist. Do not get in eyes, on skin or on clothing. Wear goggles or face shield when handling. Wash hands and face thoroughly with soap and water after use and before eating or smoking.

ENVIRONMENTAL HAZARDS

This pesticide is toxic to fish and shrimp. Keep out of lakes, streams or ponds. Do not apply where runoff is likely to occur. Do not apply when weather conditions favor drift from areas treated. Do not contaminate water by cleaning of equipment or disposal of wastes.

DIRECTIONS FOR USE

It is a violation of Federal law to use this product in a manner inconsistent with its labeling.

STORAGE AND DISPOSAL

Do not contaminate water, food or feed by storage or disposal.

STORAGE: Store in a dry location.

PESTICIDE DISPOSAL: Pesticide wastes are acutely hazardous. Improper disposal of excess pesticide, spray mixture or rinsate is a violation of Federal Law. If these wastes cannot be disposed of by use according to label instructions, contact your State Pesticide or Environmental Control Agency, or the Hazardous Waste representative at the nearest EPA Regional Office for guidance.

CONTAINER DISPOSAL: Completely empty bag into application equipment. Then dispose of empty bag in a sanitary landfill or by incineration, or if allowed by State and local authorities, by burning. If burned, stay out of smoke.

REENTRY STATEMENT

Do not apply this product in such a manner as to directly or through drift expose workers or other persons. The area being treated must be vacated by unprotected persons.

Do not enter treated areas without protective clothing until sprays have dried.

Because certain states may require more restrictive reentry intervals, consult your State Department of Agriculture for further information.

Written or oral warnings must be given to the workers who are expected to be in a treated area or in an area about to be treated with this product. When oral warnings are given, warnings shall be given in a language customarily understood by workers. Oral warnings must be given if there is reason to believe that written warnings cannot be understood by workers. Written or oral warnings must include the following information:

"DANGER, Area to be treated with propargite on [date of application]. "Do not enter without appropriate protective clothing until sprays have dried." If accidentally exposed, consult the Statement of Practical Treatment portion of the pesticide label for first aid recommendations."

GENERAL INSTRUCTIONS

ORNAMITE is a dry wettable powder for preparing sprays to control Pacific spider, strawberry spider, and two-spotted spider mites.

ORNAMITE is not systemic in action; therefore complete coverage of both upper and lower leaf surfaces and of fruit is necessary for effective control. Always use at least the minimum recommended amount of ORNAMITE per acre. ORNAMITE can be used in concentrate sprays.

AL USARIO:

Si usted no puede leer o entender ingles no use este producto hasta que la etiqueta le haya sido explicada ampliamente.

TO THE USER:

If you cannot read or understand English, do not use this product until the label has been fully explained to you.

MIXING INSTRUCTIONS

This package contains $4 \times ^1/_2$ pound water soluble bags. After opening package:

1. Do not open or perforate the inner water soluble bags. The water soluble bags and contents will dissolve in water.
2. Immediately dump entire contents into spray tank.
3. Do not touch water soluble bags as handling may cause breakage.
4. Do not begin spraying until the bag is completely dissolved and the powder is thoroughly dispersed in the spray tank.
5. If only one water soluble bag is used, store the remaining bag in the original package in a dry location away from heat or cold.

DOSAGE INSTRUCTIONS

CROP	MITES CONTROLLED	NUMBER OF ½ POUND *WATER SOLUBLE BAGS* PER ACRE	GALLONS SPRAY SOLUTION PER ACRE GROUND	AERIAL	TOTAL NUMBER OF SPRAYS PER YEAR	LAST SPRAY DAYS BEFORE HARVEST
ORNAMENTALS (field, nursery, greenhouse)	two-spotted spider	3 See application note 2	Spray to wet	—	—	—
ROSES (greenhouse grown)	two-spotted spider Pacific spider strawberry spider	3 See application note 2	150	—	—	—
ROSES (field grown)	two-spotted spider Pacific spider	3-9 See application note 2	50-300	—	—	—

APPLICATION NOTES:

1. ORNAMITE performance is best when daily temperatures average above 70°F.
2. Ornamentals - 3 — ½ lb. bags of ORNAMITE per 150 gallons of water equals 1 ounce (28.4 grams) per 6.25 gallons. The user should initially treat only a small portion of his crop to confirm plant safety under his conditions.

USE RESTRICTIONS

Do not use this product through any type of irrigation system.

CAUTION:

ORNAMITE is not compatible with foliar spray oils. Leaf injury may occur. (Further information from supplier or Uniroyal)

ORNAMITE is not compatible with alkaline materials such as lime, lime sulfur or Bordeaux. The effectiveness of ORNAMITE will be reduced.

COMPATIBILITIES — ORNAMITE is compatible with many pesticides. Check with your supplier or Uniroyal for details. Observe all cautions and limitations of labeling of all products used in mixtures.

IMPORTANT NOTICE — Seller warrants that this product conforms to its chemical description and is reasonably fit for the purposes stated on the label when used in accordance with the directions and instructions specified on the label under normal conditions of use, **but neither this warranty nor any other warranty of merchantability or fitness for a particular purpose, express or implied, extends to the use of this product, contrary to label instructions, or under abnormal conditions, or under conditions not reasonably foreseeable to seller, and buyer assumes the risk of any such use.**

PLANTVAX®-75W

SYSTEMIC FUNGICIDE FOR RUST CONTROL OF GERANIUMS IN GREENHOUSES

COMPOSITION

ACTIVE INGREDIENT:	(% by weight)
oxycarboxin (5,6-dihydro 2-methyl-1, 4-oxathiin-3-carboxanilide-4, 4-dioxide)	75.0%
INERT INGREDIENTS:	25.0%
TOTAL	100.0%

EPA REG. NO. 400-144 7-8-87

PLANTVAX is a Registered Trademark of Uniroyal Chemical Company, Inc.

KEEP OUT OF REACH OF CHILDREN
CAUTION

STATEMENT OF PRACTICAL TREATMENT

IF SWALLOWED: Induce vomiting and see a physician.
IF INHALED: See a physician if an abnormal reaction occurs.
IF ON SKIN: Wash thoroughly with soap and water.
IF IN EYES: Immediately flush eyes with plenty of water. See a physician.
SEE LEFT SIDE PANEL FOR ADDITIONAL PRECAUTIONARY STATEMENT.

PRECAUTIONARY STATEMENTS

HAZARDS TO HUMANS
CAUTION

Harmful if swallowed, inhaled or absorbed through the skin. Avoid breathing vapors or spray mist. Avoid contact with skin, eyes or clothing. Wash hands and face thoroughly with soap and water after use or before eating or smoking.

PHYSICAL OR CHEMICAL HAZARDS

Do not use or store near heat or open flame.

STORAGE AND DISPOSAL

Do not contaminate water, food or feed by storage or disposal.

STORAGE—Store in a dry location.

PESTICIDE DISPOSAL—Wastes resulting from the use of this product may be disposed of on site or at an approved waste disposal facility.

CONTAINER DISPOSAL—Completely empty bag into application equipment. Then dispose of empty bag in a sanitary landfill, or by incineration, or if allowed by State and local authorities, by burning. If burned, stay out of smoke.

DIRECTIONS FOR USE

It is a violation of Federal law to use this product in a manner inconsistent with its labeling.

Do not apply this product through any type of irrigation system.

PLANTVAX-75W is a wettable powder suitable for use as a spray when thoroughly mixed with water. Make a thin paste using a portion of the required water and the recommended amount of PLANTVAX-75W. Stir this paste by adding additional water. Provide enough agitation to maintain a uniform chemical suspension.

For the control of rust on geraniums. When first signs of rust disease appear, apply PLANTVAX-75W at a concentration of 1.6 to 2.4 ounces per 10 gallons of water at approximately 2 week intervals, using 2 to 4 sprays per season. In spite of the systemic effect of the material, the spray application should be made evenly. The necessary spray volume depends on the growth stage and should be made so that the plants are thoroughly covered.

Do not exceed recommended rates or number of applications, as plant injury may occur. Keep away from food and food products.

COMPATIBILITY

Do not mix PLANTVAX-75W with insecticides and acaricides as plant injury may occur.

IMPORTANT NOTICE

Seller warrants that this product conforms to its chemical description and is reasonably fit for the purposes stated on the label when used in accordance with the directions and instructions specified on the label under normal conditions of use, **but neither this warranty nor any other warranty of merchantability or fitness for a particular purpose, express or implied, extends to the use of this product, contrary to label instructions, or under abnormal conditions, or under conditions not reasonably foreseeable to seller, and buyer assumes the risk of any such use.**

RELDAN® 3%

DUST INSECTICIDE

Gustafson

ACTIVE INGREDIENTS:	
Chlorpyrifos-methyl [0.0-dimethyl 0–(3,5,6–trichloro–2–pyridyl) phosphorothloate	3%
INERT INGREDIENTS:	97%

EPA Reg. No. 7501–99 U.S. Patent No. 3,244,586

KEEP OUT OF REACH OF CHILDREN
CAUTION

STATEMENT OF PRACTICAL TREATMENT

If Swallowed: Do not induce vomiting. Call a physician immediately.

If Inhaled: Remove victim to fresh air. Apply respiration, if indicated. Call a physician.

If On Skin: Remove contaminated clothing and immediately wash with soap and water.

If In Eyes: Flush eyes with plenty of water and call a physician.

NOTE TO PHYSICIAN:

Chlorpyrifos-methyl is a cholinesterase inhibitor. Treat symptomatically. Atropine by injection is the preferable antidote. Oximes such as 2–PAM may or may not be therapeutic, but it is recommended they not be used in place of atropine.

SEE ADDITIONAL PRECAUTIONARY STATEMENTS

PRECAUTIONARY STATEMENTS

HAZARDS TO HUMANS AND DOMESTIC ANIMALS

CAUTION

CAUSES EYE IRRITATION—MAY CAUSE ALLERGIC SKIN REACTION

HARMFUL IF SWALLOWED

Do not get in eyes, on skin, or on clothing. Remove contaminated clothing. Wash thoroughly after handling. Wear goggles or face shield and rubber gloves whenever eye or skin exposure is likely to occur and wash before reuse. Avoid breathing dust. Keep container closed. Handle dust in a ventilated area.

ENVIRONMENTAL HAZARDS

RELDAN Insecticide is toxic to fish, birds and other wildlife. Do not apply directly to water. Do not discharge directly or indirectly into surface waters. Do not contaminate water by cleaning of equipment or disposal of wastes.

STORAGE AND DISPOSAL

Do not contaminate water, food or feed by storage or disposal.

STORAGE: Store in a cool, dry location. Do not store in direct, hot sunlight.

PESTICIDE DISPOSAL: Wastes resulting from the use of this product may be disposed of on-site or at an approved waste disposal facility.

CONTAINER DISPOSAL: Completely empty bag into application equipment. Then dispose of empty bag in a sanitary landfill or by incineration, or, if allowed by State and Local authorities, by burning. If burned, stay out of smoke.

DIRECTIONS FOR USE

It is a violation of Federal law to use this product in a manner inconsistent with its labeling.

STORED GRAIN: For protection of barley, oats, rice, sorghum and wheat against injury from stored grain weevils, moths, beetles and mealworms including granary weevil, rice weevil, red flour beetle, confused flour beetle, saw-toothed grain beetle, Indian meal moth, and Angoumois grain moth. Apply to the grain at the rate of 10 lbs. of product per 1000 bu. as the grain is being loaded or turned into final storage. To protect stored grains from attack by Indian meal moth, apply to the surface of clean or infested grain at the rate of 7 lbs. per 1000 sq. ft. of grain surface area. Apply evenly. Repeat if necessary.

HOW TO APPLY TO SMALL GRAINS

Apply Gustafson RELDAN 3% Powder Insecticide when grain is in the truck or wagon prior to binning. To determine the amount in the truck or wagon, multiply the length by width by depth (all in feet) and multiply this total by 0.8 to get the number of bushels. Measure out the correct amount of Gustafson RELDAN 3% Powder Insecticide (10 lbs. per 1000 bu.) and spread evenly over the surface of the grain in the truck or wagon. Cut into the grain with a shovel. Coating will become mixed with grain as it is dumped from the truck or wagon into the hopper. Further mixing will occur when the grain is augered or blown into the bin. Mixing is completed as the coating and grain roll down the cone shaped pile in the bin.

NOTICE—READ CAREFULLY
LIMITED WARRANTY STATEMENT

Gustafson warrants that this product conforms to the chemical description on the label and is reasonably fit for purposes stated on the label only when used in accordance with directions and instructions specified on the label. Gustafson neither makes, nor authorizes any agent or representative to make any other warranty of FITNESS or MERCHANTABILITY, guarantee or representation, expressed or implied, concerning this material. In the event of any breach by Gustafson of this limited warranty, Gustafson shall not be liable for consequential damages.

Gustafson (logo) is a Reg. TM of Gustafson, Inc.
RELDAN is a Reg. TM of The Dow Chemical Company

L131–2/87

This is a specimen label. Uniroyal is not responsible for the accuracy of the information contained herein. As labels are subject to revisions, always carefully read and follow the label on the product container.

RESCUE™

POSTEMERGENT HERBICIDE

FOR SOYBEANS

COMPOSITION

ACTIVE INGREDIENT:	(percent by weight)
Sodium salt of naptalam (N-1-Naphthylphthalamic acid)*	23.7%
Sodium salt of 4-(2,4-dichlorophenoxy) butyric acid**	.6%
INERT INGREDIENTS:	75.7%
TOTAL	100.0%

*Equivalent to 2 lbs. of the sodium salt of naptalam per gallon.
**Equivalent to 0.06 lbs 4-(2,4-DB) acid per gallon.

EPA REG. NO. 400-166 3-10-87A

RESCUE™ is a trademark of UNIROYAL CHEMICAL COMPANY, INC.

KEEP OUT OF REACH OF CHILDREN

WARNING

STATEMENT OF PRACTICAL TREATMENT

IF SWALLOWED: Drink two large glasses of water and induce vomiting. Contact a physician.

IF ON SKIN: Wash affected area with plenty of soap and water.

IF IN EYES: Flush eyes with plenty of water for at least 15 minutes. Contact a physician.

IF INHALED: Remove to fresh air; if abnormal symptoms develop, contact a physician.

PRECAUTIONARY STATEMENTS

HAZARDS TO HUMANS

WARNING

Causes eye irritation. Do not get in eyes. Harmful if inhaled or absorbed through the skin. Avoid breathing spray mist. Avoid contact with skin or clothing.

ENVIRONMENTAL HAZARDS

Do not apply directly to water. Do not contaminate water by cleaning of equipment or disposal of wastes. Do not apply when weather favors drift from treated area.

DIRECTIONS FOR USE

It is a violation of Federal Law to use this product in a manner inconsistent with its labeling.

Do not apply this product through any type of irrigation system.

STORAGE AND DISPOSAL

Do not contaminate water, food or feed by storage and disposal.

STORAGE: RESCUE is normally a solution but it may crystallize in the cold (32°F.). Gentle warming (60°F.) will redissolve crystals. Mild agitation, such as rolling, shaking, or stirring before use is recommended.

PESTICIDE DISPOSAL: Wastes resulting from the use of this product may be disposed of on site or at an approved waste disposal facility.

CONTAINER DISPOSAL: Triple rinse (or equivalent). Then offer for recycling or reconditioning, or puncture and dispose of in a sanitary landfill, or by other procedures approved by state and local authorities.

EQUIPMENT CLEANING: Thoroughly rinse spray tank, hose and nozzles with water after application. Use RESCUE in metal or fiberglass tanks. RESCUE can be removed from metal or fiberglass tanks by washing thoroughly with ammonia water and rinsing several times with water containing a good detergent.

USE DIRECTIONS

Do not apply this product through any type of irrigation system.

GENERAL

RESCUE is a selective herbicide recommended for mid-to-late season postemergence weed control in soybeans. RESCUE is a systemic herbicide and therefore requires 20-30 days to obtain maximum effect. The reduction of weed competition to the soybeans begins immediately after RESCUE is applied even though the weeds are not killed immediately. RESCUE may be tank mixed with BLAZER® 2L or TACKLE® to stop growth faster and to reduce competition from additional labelled weeds that may also be present. Weed size will determine the most effective rate for the tank mix of RESCUE plus BLAZER 2L or TACKLE. Control with this mixture will decrease with increasing weed size and density of the soybean canopy. Thorough coverage of the weed is essential for maximum control. DO NOT APPLY TO STRESSED SOYBEAN PLANTS.

WEEDS CONTROLLED: When applied according to label instructions, RESCUE or the tank mix with BLAZER 2L or TACKLE will control or suppress the following broadleaf weeds, reducing their competition to the soybean crop.

NOTE: Read carefully the result which can be expected from the use of RESCUE or the tank mix with BLAZER 2L or TACKLE on individual weed species. The various responses are coded as follows:

A Stem twisting
B Growth termination
C Reduction of viable seed
D Eventual desiccation if soybeans canopy over weeds
E Reduce harvest losses
F Faster growth termination and better weed desiccation
G Plants 24"-36" will be suppressed and may not die unless soybeans canopy over them
H Terminate growth and flowering
I Reduce competition to crop

WEED	GROWTH STAGE FOR OPTIMUM RESULTS	RESCUE APPLICATION RATE/A		BLAZER 2L or TACKLE APPLICATION RATE/A	IMPORTANT COMMENTS	EXPECTED RESULTS
Cocklebur	Up to 12" (but before flowering begins)	2-3 qt		—	Use 2 qt rate in midsouth and when open canopy exists	A, B, C, D, E
		2 qt	+	1.0 pt	Use the 1 pt rate of BLAZER 2L or TACKLE where heavy infestations occur	A, B, C, E, F
	Up to 24" (but before flowering begins)	3 qt		—		A, B, C, D, G
		3 qt	+	1 pt		A, B, C, G, F
Giant Ragweed	Up to 18" (but before flowering begins)	3 qt		—		A, B, D, E
	Up to 36" (but before flowering begins)	2-3 qt	+	1 pt	Use 3 qts of RESCUE where heavy infestations occur or when soybeans cover middles	A, B, E, F
Sunflower	Up to 36" but before early flower bud	3 qts		—	A non-ionic surfactant is recommended for optimum control at 1 pt/25 gal of spray	A, B, D, E
Marshelder	12" thru early flower bud	3 qts		—		A, D, E
Morningglory (annual tall, ivy leaf, entire leaf)	Up to 8 leaf	2 qt	+	1.5 pt		A, F, H
	8 leaf but before vining	3 qt	+	1.5 pt	Spray solutions should be applied before canopy closure of crop to facilitate good coverage	A, F, H
Common Ragweed	Up to 24"	3 qt	+	1 pt		E, F, I
	24" or taller but before flowering	3 qt.	+	1.5 pt		B, E, G, I
Jimsonweed and/or Pigweed	Up to 18"	2 qt	+	1 pt		D, E, F, I
	18" or taller but before flowering	3 qt	+	1 pt		A, E, F, I

APPLICATION TIMING

RESCUE may be applied to determinate or indeterminate soybeans which are at least 14" tall or if blooming has begun.

Upper Midwest and High Plains (Minnesota, North Dakota, South Dakota): RESCUE applications should be delayed until the Group O, I or II soybeans have begun to bloom.

GROUND APPLICATION

Apply the suggested label rates indicated in the above table in 10-25 gallons of water per acre. NON-PHYTOTOXIC SPRAY ADJUVANTS SUCH AS A NON-IONIC SPRAY SURFACTANT SUCH AS TRITON AG98 OR CROP OIL CONCENTRATE SHOULD BE ADDED TO THE SPRAY MIX AT THE MANUFACTURERS RECOMMENDED RATE. Soybean leaf response will be increased by the addition of a crop oil concentrate with the BLAZER 2L or TACKLE tank mix. Fertilizer solutions used as spray adjuvants such as an adjuvant grade of 10-34-0 may be used at 1 qt per acre. A SPRAY SOLUTION CONTAINING 10-34-0 MUST BE APPLIED WITHIN 24 HOURS OF MIXING. The spray boom should be positioned 18-24" above the tops of soybean plants or weeds and be equipped with standard high pressure hollow cone or flat fan nozzles (no greater than 20" spacing) that provide thorough, uniform coverage of the weed foliage. Maintain a high pressure (40-50 psi) during application to insure better overall coverage.

SPOT SPRAYERS: RESCUE may be applied through spot spraying equipment to control or suppress the above labeled weeds. The use of a spray solution containing 7.5% RESCUE (1½ gal. RESCUE in 20 gallons of water plus surfactant/adjuvant) has obtained the best results. Applications should be made when the weeds are actively growing with additional recommendations stated in the broadcast section. A flat fan or hollow cone nozzle tip should be used with the handgun sprayer to ensure complete coverage of the target weed.

AERIAL APPLICATION

Apply the suggested label rates of RESCUE or tank mixtures indicated in the weed table, in a minimum of 5 gallons of water per acre. A NON-IONIC SURFACTANT SUCH AS TRITON AG98 OR CROP OIL CONCENTRATE SHOULD BE USED AT THE MANUFACTURERS RECOMMENDED RATE. Soybean leaf response will be increased by the addition of a crop oil concentrate with the BLAZER 2L or TACKLE tank mix. Fertilizer solutions used as adjuvants such as an adjuvant grade of 10-34-0 may be used at 1 qt per acre. THE SPRAY SOLUTION CONTAINING 10-34-0 MUST BE APPLIED WITHIN 24 HOURS OF MIXING. Nozzle size and orientation should provide maximum droplet breakup without causing excessive drift.

ATTENTION: IMPORTANT APPLICATION NOTES

1. AVOID APPLICATIONS TO DROUGHT, HEAT OR DISEASE STRESSED SOYBEANS.
2. Less activity should be expected if applied on stressed weeds.
3. If rain occurs within 6 hours after application effectiveness may be reduced.
4. Drift-applications should not be made when weather conditions favor drift. Do not apply by air within 500 feet of susceptible crops such as cotton, tomatoes, tobacco or sunflower.
5. Soybean tolerance—Most soybeans are tolerant to RESCUE when used according to label instructions. Some upper soybean plant drooping and leaf wrinkling or twisting may occur under certain conditions, however, soybeans will usually outgrow this condition and continue to develop normally. Soybean height reduction may occur due to the shortening of the soybean stem internodes and has been shown to have no direct relationship to yield. Mitchell, Corsoy and Forrest varieties are more sensitive to RESCUE than other varieties.
6. DO NOT GRAZE OR FEED FORAGE FROM TREATED SOYBEANS TO LIVESTOCK.
7. Do not harvest soybeans earlier than 60 days post treatment.
8. Do not mix BLAZER 2L or TACKLE Herbicides with liquid fertilizers or pesticides except as specifically directed on this label or on other approved Rohm & Haas Company and Rhone-Poulenc Inc. supplemental labeling.
9. Before applying a tank mix of RESCUE and BLAZER 2L or TACKLE, read both labels and follow precautionary statements on each label.

®BLAZER is a Registered Trademark of Rohm & Haas Company.
®TACKLE is a Registered Trademark of Rhone-Poulenc Inc.

IMPORTANT NOTICE

Seller warrants that this product conforms to its chemical description and is reasonably fit for the purposes stated on the label when used in accordance with the directions and instructions specified on the label under normal conditions of use, **but neither this warranty nor any other warranty of merchantability or fitness for a particular purpose, express or implied, extends to the use of this product, contrary to label instructions, or under abnormal conditions, or under conditions not reasonably foreseeable to seller, and buyer assumes the risk of any such use.**

ROYAL MH-30®

GROWTH RETARDANT WITH SORBATRAN*

CONTROLS TOBACCO SUCKER GROWTH

COMPOSITION

ACTIVE INGREDIENTS:	(% by weight)
Potassium salt of 1,2-dihydro-3,6-pyridazinedione	21.7%
INERT INGREDIENTS:	78.3%
TOTAL:	100.0%

(One gallon contains 1.5 pounds as maleic hydrazide)

EPA REG. NO. 400-84 001 3-29-89

KEEP OUT OF REACH OF CHILDREN

CAUTION

STATEMENT OF PRACTICAL TREATMENT

IF IN EYES: Flush eyes with plenty of water. Call a physician if irritation persists.

IF ON SKIN: Wash with plenty of soap and water.

MH-30 is a Registered Trademark of UNIROYAL CHEMICAL COMPANY, INC.

PRECAUTIONARY STATEMENTS

HAZARDS TO HUMANS

CAUTION

Wear a long-sleeved shirt, long pants, and chemical resistant gloves at all times when mixing, loading, handling, or applying this chemical. Causes eye irritation. Avoid contact with skin, eyes or clothing. Wash thoroughly with soap and water after handling.

ENVIRONMENTAL HAZARDS

Do not apply directly to water or wetlands [swamps, bogs, marshes, and potholes]. Do not contaminate water when disposing of equipment washwaters.

DIRECTIONS FOR USE

It is a violation of Federal law to use this product in a manner inconsistent with its labeling.

Do not apply this product through any type of irrigation system.

STORAGE AND DISPOSAL

Do not contaminate water, food or feed by storage and disposal.

STORAGE: The active ingredient in ROYAL MH-30 will separate from the surfactant at temperatures above 134°F. If containers are exposed to excessive heating, the contents can be remixed by lowering the temperature of the entire formulation below 134°F. and applying agitation. Always shake or stir well before use.

PESTICIDE DISPOSAL: Pesticide wastes are toxic. Improper disposal of excess pesticide, spray mixture or rinsate is a violation of Federal Law. If these wastes cannot be disposed of by use according to label instructions, contact your State Pesticide or Environmental Control Agency, or the Hazardous Waste representative at the nearest EPA Regional Office for guidance.

CONTAINER DISPOSAL: Triple rinse (or equivalent). Then offer for recycling or reconditioning, or puncture and dispose of in a sanitary landfill, or incineration, or if allowed by State and local authorities, by burning. If burned stay out of smoke.

EQUIPMENT CLEANING: ROYAL MH-30 is not corrosive to spray equipment. Following applications, all spray equipment should be thoroughly rinsed with water.

REENTRY STATEMENT

Do not apply this product in such a manner as to directly or through drift expose workers or other persons. The area being treated must be vacated by unprotected persons.

Do not enter treated areas without protective clothing until sprays have dried.

Because certain states may require more restrictive reentry intervals for various crops treated with this product, consult your State Department of Agriculture for further information.

Written or oral warnings must be given to workers who are expected to be in a treated area or in an area about to be treated with this product. When oral warnings are given, warnings shall be given in a language customarily understood by workers. Oral warnings must be given if there is reason to believe that written warnings cannot be understood by workers. Written or oral warnings must include the following information:

"CAUTION" "Area to be treated with ROYAL MH-30 on (date of application). Do not enter without appropriate protective clothing until sprays have dried. If accidentally exposed, consult the Statement of Practical Treatment portion of the pesticide label for first aid recommendations."

NOTES

1. Even though ROYAL MH-30 is non-volatile no drift should be permitted onto nearby crops. Particular care should be taken to prevent drift when aerial applications are being made.
2. To control tobacco suckers apply ROYAL MH-30 at the proper time. Too early spraying may prevent normal development of top leaves. Too late spraying will allow some sucker growth. FULL FLOWER is the stage when 90% of your plants have begun to shed their first flowers.
3. Rains occuring within 6 hours after spraying may reduce effectiveness. Should wash off occur within 6 hours, a single repeat application may be made.
4. Follow dosage recommendations carefully and spray field uniformly.
5. Compatibility studies with ROYAL MH-30 and tobacco insecticides have not been completed. Separate application is recommended.
6. Do not use on tobacco plants grown for seed production.
7. Do not spray tobacco in the middle of hot, dry days when leaves are severely wilted. For best results spray during the early morning, all day on cloudy days, avoiding the late afternoon or night applications. (Use a nozzle delivering a fine mist spray.)
8. Closely hand-suckered tobacco or tobacco treated with ROYAL MH-30 will not ripen as rapidly as poorly suckered tobacco. Also, under some conditions ROYAL MH-30 treated tips have a tendency to turn yellow prematurely. This is not harmful and does not mean that your tobacco is ripe. Therefore, do not harvest tobacco treated with ROYAL MH-30 by using time and color of the leaves as your only guide to ripeness. Use other characteristics indicative of ripeness such as:
   a. Fading of color from midrib.
   b. Change in apparent leaf texture.
9. If tobacco plants have been grown under drought conditions, absorption of ROYAL MH-30 will be slow and results may not be as good as plants treated under normal conditions.
10. Too little ROYAL MH-30 will result in less than the desired control, too much reduces quality of tips.
11. Use only one application of ROYAL MH-30 per season.
12. Allow at least 7 days after application before harvest.
13. **Do not apply this product through any type of irrigation system.**

TOBACCO

DIRECTIONS FOR USE

To prevent sucker development on flue cured, burley, dark fire, Maryland and cigar (Pennsylvania, and Wisconsin) tobacco.

ROYAL MH-30 is a potassium salt formulation of maleic hydrazide used to prevent sucker development in tobacco. Since ROYAL MH-30 is absorbed and moves inside the tobacco plant, spraying the whole plant is not necessary. Best results are obtained by spraying the upper 1/3 to 1/2 of the plant. It is very important to apply equal amounts of ROYAL MH-30 to each plant at the correct time to obtain satisfactory sucker control.

Under favorable conditions, ROYAL MH-30 kills tobacco suckers in a manner similar to contact agents through systemic action.

TIMING

For flue cured, burley, dark fire and Maryland tobacco

1. If all plants flower at about the same time, proceed as follows: As soon as the plants are in FULL FLOWER, top, pull all suckers from early maturing plants and spray ROYAL MH-30 uniformly onto the entire field within 24 hours after topping. (See notes 1-12).
2. If there is a wide range in the time of flowering, the following procedure should be used:
   a. Go through the field and top the few early flowering plants.
   b. Wait until the remaining plants are in FULL FLOWER, top them, remove all suckers from the entire field, and apply ROYAL MH-30 within 24 hours after topping. (See notes 1-12).

Tip leaves to be harvested should be at least 6 to 8 inches long at time of spraying.

For Pennsylvania cigar tobacco

1. If your field flowers uniformly:
   a. Top when about 1/4 to 1/2 of your plants show one or more open flowers on each head.
   b. Wait three days after topping.
   c. Spray ROYAL MH-30. For best results ROYAL MH-30 should be sprayed on dry or nearly dry leaves. Avoid spraying on leaves covered with dew.
2. If your field doesn't flower uniformly:
   a. When flowering in your field is particularly uneven you should top when about 50 to 60 percent of the plants show one or more open flowers per head.
   b. Wait three days after topping and remove by hand any suckers more than two inches long.
   c. Spray ROYAL MH-30 over your entire field.

For Wisconsin cigar tobacco

1. Top the field when most of the plants in your field are in full flower.
2. Remove, by hand, any suckers larger than one or two inches.
3. Spray within 24 hours after topping and removal of suckers.

HOW TO SPRAY

ROYAL MH-30 may be applied with either a conventional sprayer, or with a compressed air shoulder sprayer.

1. ROYAL MH-30 can be applied by spraying one or more rows at a time, wetting upper 1/2 to 1/3 of plant.
2. Compressed air shoulder sprayer with nozzle spraying a wide cone. Spray one side of each tobacco row while walking at a constant rate. Use up and down motion of spray nozzle to moisten upper 6 to 8 leaves of each plant. Apply equal amounts of spray to each plant.

RATE OF APPLICATION

Apply the indicated amounts of ROYAL MH-30 in 40 to 50 gallons of water per acre. Apply at 40 to 50 psi.

For flue cured, burley, cigar (Pennsylvania, Wisconsin) and Maryland Tobacco

Apply 1.5 to 2 pints of ROYAL MH-30 per 1000 plants. Example: use 1.5 to 2 gallons of ROYAL MH-30 for 8000 plants per acre.

For dark tobacco

Apply 3.0 pints of ROYAL MH-30 per 1000 plants. Example: use 3.0 gallons ROYAL MH-30 for 8000 plants per acre.

NOTE: ALWAYS SHAKE OR STIR WELL BEFORE USE.

IMPORTANT NOTICE

Seller warrants that this product conforms to its chemical description and is reasonably fit for the purposes stated on the label when used in accordance with the directions and instructions specified on the label under normal conditions of use, **but neither this warranty nor any other warranty of merchantability or fitness for a particular purpose, express or implied, extends to the use of this product, contrary to label instructions, or under abnormal conditions, or under conditions not reasonably foreseeable to seller, and buyer assumes the risk of any such use.**

ROYAL MH-30®

CONTROLS SPROUTING OF POTATOES AND ONIONS
CONTROLS VOLUNTEER POTATOES
CONTROLS SIZE AND SHAPE OF POTATOES

COMPOSITION

ACTIVE INGREDIENTS:	(% by weight)
Potassium salt of 1,2-dihydro-3,6-pyridazinedione	21.7%
INERT INGREDIENTS:	78.3%
TOTAL	100.0%

(One gallon contains 1.5 pounds as maleic hydrazide)

EPA REG. NO. 400-84 001 3-29-89

KEEP OUT OF REACH OF CHILDREN

CAUTION

STATEMENT OF PRACTICAL TREATMENT

IF IN EYES: Flush eyes with plenty of water. Call a physician if irritation persists.

IF ON SKIN: Wash with plenty of soap and water.

MH-30 is a Registered Trademark of UNIROYAL CHEMICAL COMPANY, INC.

PRECAUTIONARY STATEMENTS

HAZARDS TO HUMANS

CAUTION

Wear a long-sleeved shirt, long pants, and chemical resistant gloves at all times when mixing, loading, handling, or applying this chemical. Causes eye irritation. Avoid contact with skin, eyes or clothing. Wash thoroughly with soap and water after handling.

ENVIRONMENTAL HAZARDS

Do not apply directly to water or wetlands (swamps, bogs, marshes, and potholes). Do not contaminate water when disposing of equipment washwaters.

DIRECTIONS FOR USE

It is a violation of Federal law to use this product in a manner inconsistent with its labeling.

Do not apply this product through any type of irrigation system.

STORAGE AND DISPOSAL

Do not contaminate water, food or feed by storage and disposal.

STORAGE: The active ingredient in ROYAL MH-30 will separate from the surfactant at temperatures above 134°F. If containers are exposed to excessive heating, the contents can be remixed by lowering the temperature of the entire formulation below 134°F. and applying agitation. Always shake or stir well before use.

PESTICIDE DISPOSAL: Pesticide wastes are toxic. Improper disposal of excess pesticide, spray mixture or rinsate is a violation of Federal Law. If these wastes cannot be disposed of by use according to label instructions, contact your State Pesticide or Environmental Control Agency, or the Hazardous Waste representative at the nearest EPA Regional Office for guidance.

CONTAINER DISPOSAL: Triple rinse (or equivalent). Then offer for recycling or reconditioning, or puncture and dispose of in a sanitary landfill, or incineration, or if allowed by State and local authorities, by burning. If burned stay out of smoke.

EQUIPMENT CLEANING: ROYAL MH-30 is not corrosive to spray equipment. Following applications, all spray equipment should be thoroughly rinsed with water.

REENTRY STATEMENT

Do not apply this product in such a manner as to directly or through drift expose workers or other persons. The area being treated must be vacated by unprotected persons.
Do not enter treated areas without protective clothing until sprays have dried.

Because certain states may require more restrictive reentry intervals for various crops treated with this product, consult your State Department of Agriculture for further information.

Written or oral warnings must be given to workers who are expected to be in a treated area or in an area about to be treated with this product. When oral warnings are given, warnings shall be given in a language customarily understood by workers. Oral warnings must be given if there is reason to believe that written warnings cannot be understood by workers. Written or oral warnings must include the following information:

"CAUTION" "Area to be treated with ROYAL MH-30 on (date of application). Do not enter without appropriate protective clothing until sprays have dried. If accidentally exposed, consult the Statement of Practical Treatment portion of the pesticide label for first aid recommendations."

NOTES

1. Follow dosage recommendations carefully and spray field uniformly.
2. Do not use on potato plants grown for seed production.
3. Even though Royal MH-30 is non-volatile no drift should be permitted onto nearby crops. Particular care should be taken to prevent drift when aerial applications are being made.
4. Spray potatoes or onions at least 24 hours before rain or applying overhead irrigation. Rain or overhead irrigation occurring within 24 hours after application will reduce effectiveness.
5. At least a 2-week interval should be allowed between a ROYAL MH-30 application to potatoes and the use of a vine killer.
6. Uneven application may result in budend cracking and/or elephant hide on potato tubers.
7. To obtain desired results plants should not be stressed from either too little or excessive water before or after application.
8. **IMPORTANT: Geographical response of potato plants.**
   EAST OF THE ROCKY MOUNTAINS
   Occasional injury to the terminal growth of potato foliage has been observed following the application of ROYAL MH-30. However, trials and usage indicate that this does not adversely affect the yield or the quality of the potato tubers at harvest if the chemical is applied according to label recommendations.
   WEST OF THE ROCKY MOUNTAINS
   Reduced sizing, yield, tuber cracking and/or "elephant hide" disorders have been observed in some instances following application of ROYAL MH-30 on potatoes grown under irrigation in the Northwest. At this time, the association, if any, between these disorders and ROYAL

MH-30 use is not clear or confirmed. Until such time as the association or the lack of association can be determined, ROYAL MH-30 is not recommended for use on irrigated potatoes, and use on any potatoes, irrigated or not, is at the user's risk. Uniroyal will not be responsible for any loss incurred by users should any of these disorders occur after the application of ROYAL MH-30.

9. Do not use this product in any type of irrigation system.

DIRECTIONS FOR USE

ONIONS
TO CONTROL SPROUTING OF ONIONS IN STORAGE

Timing
ROYAL MH-30 should be applied when the bulbs are fully mature, have at least 5-8 green leaves, and necks soft enough for the tops to fall if they have not already done so. The spray is often applied when 50 percent of the tops have fallen but all tops are still green. This condition usually occurs about two weeks to 10 days before harvest time. Green tops are essential for the absorption of ROYAL MH-30.

RATE OF APPLICATION
NOTE: Uniform and complete coverage is essential for good sprout control.

Ground Application
Use $1^1/_3$ gallons of ROYAL MH-30 in a minimum of 30 gallons of water per acre.

Aerial Application
Use $1^1/_3$ gallons of ROYAL MH-30 in a minimum of 10 gallons of water per acre. Do not fly at heights of less than 8-12 feet. Maintaining this height will give more uniform coverage and aid in preventing an excessive build-up of chemical on the left side of the treated swath. These heights should be reduced as cross-wind velocities exceed 2-4 MPH.

Spray Adjuvants
West of Rocky Mountains: Use Regulaid or X-77 at the rate of 1 quart per 100 gallons of spray solution in arid sections of the U.S. Either product will help provide better coverage and aid in translocation of the ROYAL MH-30. East of Rocky Mountains: use of surfactants is normally not recommended.

Temperature
ROYAL MH-30 should be applied at temperatures below 85°F or at temperatures below 80°F if the temperature is expected to exceed 85°F. ROYAL MH-30 is water soluble and under high temperature and low humidity conditions will crystalize on the leaf surface and not be absorbed into the plant.

CAUTION
Avoid spraying edible onions too early. Spongy bulbs may result if spraying is done earlier than two weeks before maturity.

Varieties
Use of ROYAL MH-30 should be combined with selection of a good-keeping variety. Do not apply to onion varieties with poor-keeping quality. ROYAL MH-30 will not make poor-keeping varieties equal to good-keeping varieties.

POTATOES
BENEFITS

Sprout Control
ROYAL MH-30 will control sprouting of potatoes in storage. However, ROYAL MH-30 will not make a poor-keeping potato such as Norgold or Norchip into a good-keeping potato.

Reduction of Volunteer Potatoes
ROYAL MH-30 will reduce the number of volunteer potatoes the following spring because of its ability to control sprouting. If applied at the proper time it will also reduce the number of undersize potatoes left in the field, which also reduces the potential for volunteers.

May Improve Grade
ROYAL MH-30 may improve the grade of potatoes, especially Russet Burbank variety, if the application is made prior to any secondary tuber formation.

Undersize Potatoes
ROYAL MH-30 will help reduce any additional "sets" of potatoes, thus decreasing the number of undersize potatoes in the field at harvest.

Reduces Shrinkage
ROYAL MH-30 applications can reduce shrinkage in storage up to 15 percent.

DIRECTIONS FOR USE

Timing
Timing is important. To allow translocation of chemical down to the tubers ROYAL MH-30 should be applied after tubers are formed and while vines are still green and are free from insects, diseases or other damage. Time of application may occur as early as 2-3 weeks past full bloom when the primary and secondary inflorescences have lost most of their flowers and a few third order inflorescences and buds are present. This occurs when a few lower leaves show a yellow color. Application prior to this time may result in reduced size of potatoes and phytotoxicity. The following tuber sizes may also be used as a guide, provided the above conditions have been met.

Russet Burbank
The smallest tubers that are required to reach a minimum size of 4-6 oz. must be $1^1/_2$-2" in diameter (and elongating) at time of spraying.

Norgold or Norchip
The smallest tubers that are required to reach maximum size must be 2" or more in diameter at time of spraying.

Other Varieties
The smallest tubers that are required to reach maximum size, should be $1^1/_2$" in diameter at time of spraying.

Rate of Application
NOTE: Uniform and complete coverage is essential for satisfactory results.

Ground
Use 2 gallons of ROYAL MH-30 in a minimum of 30 gallons of water per acre.

Aerial
Use 2 gallons of ROYAL MH-30 per acre.
West of the Rocky Mountains: Use in a minimum of 10 gallons of water per acre.
East of the Rocky Mountains: Use in a sufficient quantity of water to gain complete uniform coverage.
Do not fly at heights of less than 8-12 feet. Maintaining this height will give more uniform coverage and aid in preventing an excessive build up of chemical on the left side of the treated swath. These heights should be reduced as cross-wind velocities exceed 2-4 MPH.

Spray Adjuvants
West of Rocky Mountains: Use Regulaid or X-77 at the rate of 1 quart per 100 gallons of spray solution in arid sections of the U.S. Either product will help provide better coverage and aid in translocation of the ROYAL MH-30. East of the Rocky Mountains: Use of surfactants is normally not recommended.

Temperature
ROYAL MH-30 should be applied at temperatures below 85°F or at temperatures below 80°F if the temperature is expected to exceed 85°F. ROYAL MH-30 is water soluble and under high temperature and low humidity conditions will crystalize on the leaf surface and not be absorbed into the plant.

Note: Always Shake or Stir Well Before Use.

IMPORTANT NOTICE
Seller warrants that this product conforms to its chemical description and is reasonably fit for the purposes stated on the label when used in accordance with the directions and instructions specified on the label under normal conditions of use, **but neither this warranty nor any other warranty of merchantability or fitness for a particular purpose, express or implied, extends to the use of this product, contrary to label instructions, or under abnormal conditions, or under conditions not reasonably foreseeable to seller, and buyer assumes the risk of any such use.**

ROYAL MH-30® SG

WATER SOLUBLE GRANULE GROWTH RETARDANT

FOR POTATOES AND ONIONS

COMPOSITION

ACTIVE INGREDIENTS:	(% by weight)
Potassium salt of 1,2-dihydro-3,6-pyridazinedione	80.0%**
INERT INGREDIENTS:	20.0%
TOTAL:	100.0%

* Contains 60% by weight of maleic hydrazide

EPA REG. NO. 400-165 002 8-29-89

KEEP OUT OF REACH OF CHILDREN

CAUTION

STATEMENT OF PRACTICAL TREATMENT

IF IN EYES: Flush with plenty of water. Call a physician if irritation persists.

IF ON SKIN: Wash with plenty of soap and water. Get medical attention.

IF INHALED: Remove to fresh air.

MH-30 is a Registered Trademark of UNIROYAL CHEMICAL COMPANY, INC.

PRECAUTIONARY STATEMENTS

HAZARDS TO HUMANS

CAUTION

Wear a long-sleeved shirt, long pants, and chemical resistant gloves at all times when mixing, loading, handling, or applying this chemical. Causes moderate eye injury. Avoid contact with skin, eyes or clothing. Wash thoroughly with soap and water after handling.

ENVIRONMENTAL HAZARDS

Do not apply directly to water or wetlands (swamps, bogs, marshes, and potholes). Do not contaminate water when disposing of equipment washwaters.

DIRECTIONS FOR USE

It is a violation of Federal law to use this product in a manner inconsistent with its labeling.
Do not apply this product through any type of irrigation system.

STORAGE AND DISPOSAL

Do not contaminate water, food or feed by storage and disposal.

STORAGE: Store in a dry location.

PESTICIDE DISPOSAL: Pesticide wastes are toxic. Improper disposal of excess pesticide, spray mixture, or rinsate is a violation of Federal Law. If these wastes cannot be disposed of by use according to label instructions, contact your State Pesticide or Environmental Control Agency, or the Hazardous Waste representative at the nearest EPA Regional Office for guidance.

EQUIPMENT CLEANING: ROYAL MH-30SG is not corrosive to spray equipment. Following applications, all spray equipment should be thoroughly rinsed with water.

CONTAINER DISPOSAL:

Fiber carton with liners or bags: Completely empty liner or bag by shaking and tapping sides and bottom to loosen clinging particles. Empty residue into application equipment. Then dispose of liner or bag in a sanitary landfill, or by incineration if permitted by State and local authorities. Dispose of carton in the same manner.

REENTRY STATEMENT

Do not apply this product in such a manner as to directly or through drift expose workers or other persons. The area being treated must be vacated by unprotected persons.
Do not enter treated areas without protective clothing until sprays have dried.

Because certain states may require more restrictive reentry intervals for various crops treated with this product, consult your State Department of Agriculture for further information.

Written or oral warnings must be given to workers who are expected to be in a treated area or in an area about to be treated with this product. When oral warnings are given, warnings shall be given in a language customarily understood by workers. Oral warnings must be given if there is reason to believe that written warnings cannot be understood by workers. Written or oral warnings must include the following information:

"CAUTION" "Area to be treated with ROYAL MH-30SG on [date of application]. Do not enter without appropriate protective clothing until sprays have dried. If accidentally exposed, consult the Statement of Practical Treatment portion of the pesticide label for first aid recommendations."

NOTES

1. Do not apply this product through any type of irrigation system.
2. Follow dosage recommendations carefully and spray field uniformly.
3. Do not use on potato plants grown for seed production.
4. Even though ROYAL MH-30SG is non-volatile no drift should be permitted onto nearby crops. Particular care should be taken to prevent drift when aerial applications are being made.
5. Spray potatoes or onions at least 24 hours before rain or applying overhead irrigation. Rain or overhead irrigation occurring within 24 hours after application will reduce effectiveness.
6. At least a 2-week interval should be allowed between a ROYAL MH-30SG application to potatoes and the use of a vine killer.
7. Uneven application may result in budend cracking and/or elephant hide on potato tubers.
8. To obtain desired results plants should not be stressed from either too little or excessive water before or after application.
9. **IMPORTANT: Geographical response of potato plants.**
   EAST OF THE ROCKY MOUNTAINS
   Occasional injury to the terminal growth of potato foliage has been observed following the application of ROYAL MH-30SG. However, trials and usage indicate that this does not adversely affect the yield or the quality of the potato tubers at harvest if the chemical is applied according to label recommendations.
   WEST OF THE ROCKY MOUNTAINS
   Reduced sizing, yield, tuber cracking and/or "elephant hide" disorders have been observed in some instances following application of ROYAL MH-30SG on potatoes grown under irrigation in the Northwest. At this time, the association, if any, between these disorders and ROYAL MH-30SG use is not clear or confirmed. Until such time as the association or lack of association can be determined, ROYAL MH-30SG is not recommended for use on irrigated potatoes, and use on any potatoes, irrigated or not, is at the user's own risk. Uniroyal will not be responsible for any loss incurred by users should any of these disorders occur after the application of ROYAL MH-30SG.
10. Do not add any extra wetting agents or commercial spray adjuvants to ROYAL MH-30SG spray solutions except as specified under potatoes and onions.
11. Do not handle treated plants until sprays have dried.

MIXING INSTRUCTIONS

NOTE: ROYAL MH-30SG must be dissolved first if it is to be tank mixed with an insecticide or fungicide.

1. Be sure sprayer is clean and not contaminated with any material.
2. Fill tank ¾ full with clean water, and start agitation.
3. Pour ROYAL MH-30SG slowly from bag into tank. Avoid dumping all contents at once.
4. Allow 2-3 minutes agitation for dissolving.
5. Add compatible tank mix insecticide or fungicide if needed.
6. Complete filling tank.

DIRECTIONS FOR USE

ONIONS
TO CONTROL SPROUTING OF ONIONS IN STORAGE

Timing

ROYAL MH-30SG should be applied when the bulbs are fully mature, have at least 5-8 green leaves, and necks soft enough for the tops to fall if they have not already done so. The spray is often applied when 50 percent of the tops have fallen but all tops are still green. This condition usually occurs about two weeks to 10 days before harvest time. Green tops are essential for the absorption of ROYAL MH-30SG.

RATE OF APPLICATION

NOTE: Uniform and complete coverage is essential for good sprout control.

Ground Application

Use 3⅓ pounds of ROYAL MH-30SG in a minimum of 30 gallons of water per acre.

Aerial Application

Use 3⅓ pounds of ROYAL MH-30SG in a minimum of 10 gallons of water per acre. Do not fly at heights of less than 8-12 feet. Maintaining this height will give more uniform coverage and aid in preventing an excessive build-up of chemical on the left side of the treated swath. These heights should be reduced as cross-wind velocities exceed 2-4 MPH.

Spray Adjuvants

West of Rocky Mountains: Use Regulaid, X-77 or R11 at the rate of 1 quart per 100 gallons of spray solution in arid sections of the U.S. These products will help provide better coverage and aid in translocation of the ROYAL MH-30SG. East of Rocky Mountains: use of surfactants is normally not recommended.

Temperature

ROYAL MH-30SG should be applied at temperatures below 85°F or at temperatures below 80°F if the temperature is expected to exceed 85°F. ROYAL MH-30SG is water soluble and under high temperature and low humidity conditions will crystalize on the leaf surface and not be absorbed into the plant.

CAUTION

Avoid spraying edible onions too early. Spongy bulbs may result if spraying is done earlier than two weeks before maturity.

Varieties

Use of ROYAL MH-30SG should be combined with selection of a good-keeping variety. Do not apply to onion varieties with poor-keeping quality. ROYAL MH-30SG will not make poor-keeping varieties equal to good-keeping varieties.

POTATOES
TO IMPROVE QUALITY AND TO CONTROL SPROUTING IN STORAGE

BENEFITS

Sprout Control

ROYAL MH-30SG will control sprouting of potatoes in storage. However, ROYAL MH-30SG will not make a poor-keeping potato such as Norgold or Norchip into a good-keeping potato.

Reduction of Volunteer Potatoes

Royal MH-30SG will reduce the number of volunteer potatoes the following spring because of its ability to control sprouting. If applied at the proper time it will also reduce the number of undersize potatoes left in the field, which also reduces the potential for volunteers.

May Improve Grade

ROYAL MH-30SG may improve the grade of potatoes, especially Russet Burbank variety, if the application is made prior to any secondary tuber formation.

Undersize Potatoes

ROYAL MH-30SG will help reduce any additional "sets" of potatoes, thus decreasing the number of undersize potatoes in the field at harvest.

Reduces Shrinkage

ROYAL MH-30SG applications can reduce shrinkage in storage up to 15 percent.

DIRECTIONS FOR USE

Timing

Timing is important. To allow translocation of chemical down to the tubers ROYAL MH-30SG should be applied after tubers are formed and while vines are still green and are free from insects, diseases or other damage. Time of application may occur as early as 2-3 weeks past full bloom when the primary and secondary inflorescences have lost most of their flowers and a few third order inflorescences and buds are present.
This occurs when a few lower leaves show a yellow color. Application prior to this time may result in reduced size of potatoes and phytotoxicity. The following tuber sizes may also be used as a guide, provided the above conditions have been met.

Russet Burbank

The smallest tubers that are required to reach a minimum size of 4-6 oz. must be 1½-2" in diameter (and elongating) at time of spraying.

Norgold or Norchip

The smallest tubers that are required to reach maximum size must be 2" or more in diameter at time of spraying.

Other Varieties

The smallest tubers that are required to reach maximum size, should be 1½" in diameter at time of spraying.

Rate of Application

NOTE: Uniform and complete coverage is essential for satisfactory results.

Ground

Use 5 pounds of Royal MH-30SG in a minimum of 30 gallons of water per acre.

Aerial

Use 5 pounds of ROYAL MH-30SG per acre.
West of the Rocky Mountains: Use in a minimum of 10 gallons of water per acre.
East of the Rocky Mountains: Use in a sufficient quantity of water to gain complete uniform coverage.
Do not fly at heights of less than 8-12 feet. Maintaining this height will give more uniform coverage and aid in preventing an excessive build up of chemical on the left side of the treated swath. These heights should be reduced as cross-wind velocities exceed 2-4 MPH.

Spray Adjuvants

West of Rocky Mountains: Use Regulaid, X-77 or R11 at the rate of 1 quart per 100 gallons of spray solution in arid sections of the U.S. These products will help provide better coverage and aid in translocation of the ROYAL MH-30SG.
East of the Rocky Mountains: Use of surfactants is normally not recommended.

Temperature

ROYAL MH-30SG should be applied at temperatures below 85°F or at temperatures below 80°F if the temperature is expected to exceed 85°F later that day. ROYAL MH-30SG is water soluble and under high temperature and low humidity conditions will crystalize on the leaf surface and not be absorbed into the plant.

IMPORTANT NOTICE

Seller warrants that this product conforms to its chemical description and is reasonably fit for the purposes stated on the label when used in accordance with the directions and instructions specified on the label under normal conditions of use, **but neither this warranty nor any other warranty of merchantability or fitness for a particular purpose, express or implied, extends to the use of this product, contrary to label instructions, or under abnormal conditions, or under conditions not reasonably foreseeable to seller, and buyer assumes the risk of any such use.**

ROYAL SLO-GRO®

GROWTH RETARDANT —WITH SORBATRAN

Controls Growth of Grass, Trees, Shrubs and Ivy

COMPOSITION

ACTIVE INGREDIENTS:	(% by weight)
Potassium salt of 1,2-dihydro-3,6-pyridazinedione	21.7%
INERT INGREDIENTS:	78.3%
TOTAL	100.0%

(One gallon contains 1.5 pounds as maleic hydrazide)

EPA REG. NO. 400-94 001 3-29-89
ROYAL SLO-GRO is a Registered Trademark of UNIROYAL CHEMICAL COMPANY, INC.

KEEP OUT OF REACH OF CHILDREN

CAUTION

STATEMENT OF PRACTICAL TREATMENT

IF IN EYES: Flush eyes with plenty of water. Call a physician if irritation persists.

IF ON SKIN: Wash with soap and water. Get medical attention.

PRECAUTIONARY STATEMENTS

HAZARDS TO HUMANS
CAUTION

Wear a long-sleeved shirt, long pants, and chemical resistant gloves at all times when mixing, loading, handling, or applying this chemical. Causes eye irritation. Avoid contact with skin, eyes or clothing. Wash thoroughly with soap and water after handling.

ENVIRONMENTAL HAZARDS

Do not apply directly to water or wetlands (swamps, bogs, marshes, and potholes). Do not contaminate water when disposing of equiment washwaters.

DIRECTIONS FOR USE

It is a violation of Federal law to use this product in a manner inconsistent with its labeling.
Do not apply this product through any type of irrigation system.

STORAGE AND DISPOSAL

Do not contaminate water, food or feed by storage and disposal.

STORAGE

The active ingredient in ROYAL SLO-GRO will separate from the surfactant at temperatures above 134°F. If containers are exposed to excessive heating, the contents can be remixed by lowering the temperature of the entire formulation below 134°F, and applying agitation. Always shake or stir well before use.

PESTICIDE DISPOSAL

Pesticide wastes are toxic. Improper disposal of excess pesticide, spray mixture or rinsate is a violation of Federal Law. If these wastes cannot be disposed of by use according to label instructions contact your State Pesticide or Environmental Control Agency, or the Hazardous Waste representative at the nearest EPA Regional Office for guidance.

EQUIPMENT CLEANING

ROYAL SLO-GRO is not corrosive to spray equipment. Following applications, all spray equipment should be thoroughly rinsed with water.

CONTAINER DISPOSAL

Triple rinse (or equivalent). Then offer for recycling or reconditioning, or puncture and dispose of in a sanitary landfill, or incineration, or, if allowed by state and local authorities, by burning. If burned, stay out of smoke.

DIRECTIONS FOR USE

ROYAL SLO-GRO goes into solution readily. After initial mixing, the spray solution requires no additional agitation.

MIXING INSTRUCTIONS

1. Be sure sprayer is clean and not contaminated with any material.
2. Fill tank ¾ full with clean water, and start agitation.
3. Pour ROYAL SLO-GRO into tank. Avoid dumping all contents at once.
4. Mild agitation is recommended.
5. Add compatible tank mix insecticide or fungicide if needed. (ROYAL SLO-GRO should be added before an insecticide or fungicide).
6. Complete filling tank.

APPLICATION NOTES

1. Follow dosage recommendations carefully and spray field uniformly.
2. Apply only to green vigorous plants.
3. Even though ROYAL SLO-GRO is non-volatile no drift should be permitted onto nearby crops.

4. All turf areas treated should contain well established perennial grasses at least three years old.
5. Do not add any extra wetting agents or commercial spray adjuvants to ROYAL SLO-GRO spray solutions.
6. Do not handle treated plants until sprays have dried.
7. Time treatment to allow a minimum rainfree period of at least 12 hours after application to insure complete absorption of the chemical.
8. **Do not apply this product through any type of irrigation system.**

DO NOT PASTURE LIVESTOCK IN, OR CUT HAY FROM, TREATED AREAS.

TURF
TO CONTROL THE GROWTH OF PERENNIAL GRASSES ON COMMERCIAL TURF LOCATIONS

BENEFITS

ROYAL SLO-GRO is used mostly as a substitute for mechanical mowing on various commercial turf locations such as hard-to-mow areas along highways, airport property, military installations, Federal, state and municipal properties, industrial areas, golf course roughs and similar areas. It can be used for grass inhibition on all turf areas except on fine lawns (where esthetic appearance is more important than reducing a maintenance chore) and on those areas under heavy foot traffic (such as golf course fairways and greens, except for *Poa Annua* control).

Perennial grasses effectively retarded by ROYAL SLO-GRO are: bluegrasses, fescues, bromegrass, orchard grass, quackgrass, perennial rye, ryegrass and bahiagrass (not for use on bahiagrass in California). Bentgrass can be inhibited, but often shows discoloration effects. ROYAL SLO-GRO will injure St. Augustine grass and should not be used on this species.

The growth regulant action of ROYAL SLO-GRO reduces or prevents seed stalk and seed head formation and slows leaf growth. If the application is timed properly, no significant growth will occur for several weeks after the treatment. As the effect gradually "wears off", the turf may grow to 6-12 inches in height by the end of the growing season.

DIRECTIONS FOR USE

APPLICATION TIMING
SPRING

Apply spring treatments when grass is 2-3 inches tall for the best growth control. This timing usually occurs just before and during early dandelion or forsythia bloom. If this timing period is missed and the grass reaches a height of 6 inches or more, spray and then mow 5 to 7 days later. UNDER NO CONDITIONS SHOULD GRASS BE CUT SHORTER THAN 3 to 4 INCHES. The grass growth rate is greatly reduced by a ROYAL SLO-GRO treatment. A "short cut" may produce a scalped or light brown effect, especially if dry weather follows. **To extend the growth control period of bahiagrass, a repeat application may be necessary in the late spring or approximately one month after initial application.**

AUTUMN

ROYAL SLO-GRO may also be applied late in the growing season to reduce grass growth the following spring. Treatment should be made while grass is still green but before it becomes dormant (usually during October). An additional benefit of Fall treatment is control of wild onion, garlic and biennial type weeds such as dandelion and plantain. Since grass growth is inhibited the following spring, the area will "green up" about two weeks later than untreated turf.

One application per year either Spring or Fall should be used. At either time, the area to be treated must be green and actively growing. Turf to be treated must be free of leaf cover or other debris which would prevent direct contact of the spray with the grass.

RATE OF APPLICATION

Apply 3 to 4 gallons of ROYAL SLO-GRO in 30-50 gallons of water per acre. Use the higher rate for fall applications and when a longer period of retardation is desired.

On grass where broadleaf growth is a problem, ROYAL SLO-GRO may be used in a tank-mix combination with low volatile ester or amine formulations of 2,4-D. One pound of 2,4-D per acre should be used with ROYAL SLO-GRO in most spring applications. Follow dosage and precautionary information of the 2,4-D label. If weeds have not emerged at time of ROYAL SLO-GRO application a separate, later spray of 2,4-D is recommended to control these weeds.

APPLICATION EQUIPMENT:

Applications may be made with standard booms using flat fan type nozzle tips in the 80° series, i.e. 8003, 8004. These tips should be placed on 10 inch centers and turned 45° out from the boom line. This spacing allows 50% overlap when the boom is 18-20 inches above the grass and a uniform spray pattern. Applications may also be made with off center nozzle systems. Accurate calibration is required using low speed speedometers. Spraying should not be done on excessively windy days.

SPECIAL GRASS AREAS

GOLF COURSE FAIRWAYS: ROYAL SLO-GRO may be used to reduce *Poa annua* (annual blue grass) in golf course fairways. Recommended procedure is to first mow area twice in normal sequence (usually 5 to 6 days apart). When third mowing is needed and **before** first *Poa annua* seed heads appear - spray 2 qts. of ROYAL SLO-GRO in 30-40 gallons of water per acre. Do not use over 40 gallons per acre as effectiveness may be reduced. The effect of this treatment should be evident in 8-10 days showing up as a reduction in *Poa annua* reseeding with little retardation of desirable grass growth.

OVERSEEDING RYEGRASS IN BERMUDAGRASS: Excessive growth of Bermuda grass on golf greens and fairways is often a problem in the fall after verticutting, fertilization, overseeding and top dressing. Apply ROYAL SLO-GRO in late September or early October to inhibit the growth of Bermuda grass and enable ryegrass or other grasses to produce a good stand. Use 1½ gallons of ROYAL SLO-GRO in 50 gallons of water per acre. ROYAL SLO-GRO should be applied with a calibrated boom. The Bermuda grass must be in good condition. Overseed no sooner than 48 hours after spraying.

HOME LAWNS: ROYAL SLO-GRO is not recommended as an overall treatment for prime lawns or other fine turf areas. ROYAL SLO-GRO can be used as a band or edge treatment of lawns where it is difficult to trim mechanically. Examples are: along walls, around trees, rocks, etc.

SPECIAL NOTE: Because of the difficulties in applying ROYAL SLO-GRO at a uniform rate and dosage to small areas, some color modification may occur. Any slightly abnormal color of the treated area is a temporary effect. At times, the treated grass may be a greener color than untreated turf. Do not apply ROYAL SLO-GRO during the summer or other than times permanent grasses are dormant under drought conditions.

SHRUBS AND DECIDUOUS TREES
TO RETARD THE GROWTH OF CERTAIN ORNAMENTALS

BENEFITS

ROYAL SLO-GRO is used to suppress excessive vegetative growth and reduce the need to mechanically prune or shear.

DIRECTIONS FOR USE

APPLICATION TIMING:

The best procedure is to apply to plants that have been previously pruned into the desired shape. Some regrowth should be allowed before treatment to hide fresh cut ends of limbs or stems to prevent a barren appearance of the treated plant. ROYAL SLO-GRO should only be used on vigorous, healthy plants.

The following **trees** can be treated with ROYAL SLO-GRO:

Acacia, Black	Linden
Alder	Liquid Amber (Sweet Gum)
Ash	Madrone
Bay (California Laurel)	Manzanita
Birch	Maple
Box-Elder	Mulberry
Brevillea Silk Oak	Oak
Buckeye, California	Pine, Monterey
Catalpa	Plane (Sycamore)
Cypress, Monterey	Poplar
Dogwood	Redbud, Western
Elderberry	Redwood
Elm	Walnut
Eucalyptus	Wax-Myrtle, California
Fir, Douglas	Willow, Black

ROYAL SLO-GRO will effectively retard excessive vegetative growth on the following **shrubs:**

Cissus	Pittosporum
Eugenia	Privet
Forsythia	Pyracantha
Honeysuckle	Viburnum
Myrtus	Xylosoma

ROYAL SLO-GRO may be used to inhibit the growth of **Hahn** and **Algerian Ivy.** A special use for ROYAL SLO-GRO is for weed control in **Ice Plant.** It may be used for both emerged broadleaf weed and annual grass control without injury to the herbaceous growth.

All areas to be inhibited should be sprayed to drip-point covering all surfaces of leaves and stems. If only one section of a tree is to be controlled, spray just that section and normal growth will continue in the unsprayed sections. Do not spray where wind velocity exceeds 15 mph. Drift should not go on nearby crops.

Best results are obtained when spray is applied to green, vigorous, new growth. On trees, new leaves should be fully expanded when treatment is made. ROYAL SLO-GRO should be used once a year - in spring, after new growth has started, or in early summer - usually after pruning when new regrowth is about 2 to 4 inches in length.

RATE OF APPLICATION:
ROYAL SLO-GRO is recommended for all uses at a rate of 1½- 2 gallons per 100 gallons of water. Use the higher rate on shrubs and trees that are vigorous.

FALLOW LAND
TO CONTROL SILVERLEAF NIGHTSHADE
(DO NOT USE IN ARIZONA AND CALIFORNIA)

BENEFITS
Silverleaf nightshade is a troublesome perennial weed that over winters and may compete with subsequent crops and cultivations. ROYAL SLO-GRO may be applied to Silverleaf nightshade in the fall after the seasons's crop is harvested. The effect on the weed obtained is no subsequent growth and eventual kill the following spring.

DIRECTIONS FOR USE

RATE OF APPLICATION:
Apply 6 gallons of ROYAL SLO-GRO in 10-20 gallons of water per acre.

APPLICATION TIMING:
Apply to Silverleaf nightshade during periods of active growth but after crop harvest. It may be applied to Silverleaf nightshade anytime during bloom.

NOTE: Do not graze treated areas or feed treated forage to livestock. Do not plant food or feed crops within 6 months of after application.

IMPORTANT NOTICE
Seller warrants that this product conforms to its chemical description and is reasonably fit for the purposes stated on the label when used in accordance with the directions and instructions specified on the label under normal conditions of use, **but neither this warranty nor any other warranty of merchantability or fitness for a particular purpose, express or implied, extends to the use of this product, contrary to label instructions, or under abnormal conditions, or under conditions not reasonably foreseeable to seller, and buyer assumes the risk of any such use.**

SULV™

HERBICIDE 2,4-D 3.8 LBS./GALLON

READY-TO-USE-DO NOT DILUTE FOR AERIAL APPLICATION
FOR DISTRIBUTION AND USE ONLY WITHIN STATES LISTED IN ATTACHED BOOKLET

COMPOSITION

ACTIVE INGREDIENTS:	(% by weight)
Dimethylamine salt of 2,4-dichlorophenoxyacetic acid	46.6%* **
INERT INGREDIENTS:	53.4%
Total:	100.0%

*This product contains 3.8 lbs. 2,4-dichlorophenoxyacetic acid per gallon or 38.7%
**Isomer specific by AOAC Method No. 6 D01–5.

24[c] Registrant 9–22–88

KEEP OUT OF REACH OF CHILDREN
DANGER

STATEMENT OF PRACTICAL TREATMENT

IF IN EYES: Flush with plenty of water and get medical attention.

IF SWALLOWED: Drink one or two glasses of water and induce vomiting by placing finger in back of throat and get medical attention. Never give anything by mouth to an unconscious person.

IF ON SKIN: Wash with plenty of soap and water and get medical attention if irritation persists.

PRECAUTIONARY STATEMENTS

HAZARDS TO HUMANS
DANGER
Corrosive, causes irreversible eye damage. Harmful if swallowed or absorbed through skin. Avoid contact with skin, eyes or clothing. Wear goggles or face shield when handling. Wash thoroughly after handling.

ENVIRONMENTAL HAZARDS
Do not apply where runoff is likely to occur. Do not apply directly to any body of water. Do not apply when weather conditions favor drift from treated areas. Do not contaminate water by cleaning of equipment or disposal of wastes.

DIRECTIONS FOR USE
It is a violation of Federal Law to use this product in a manner inconsistent with its labeling. Use this product only as specified on this label. Many states have regulations on Phenoxy Herbicide applications. Check with your State Department of Agriculture for current information.

Under no circumstances should this herbicide product or any 2,4-D weed killer be used in the vicinity of cotton, tomatoes, garden crops, grapes, ornamentals, or other susceptible crops or severe damage may result. Do not apply on windy days.

Do not use equipment used in applying this product or any 2,4-D weed killer to apply insecticides, fungicides or other material to susceptible crops.

Do not apply this product through any type of irrigation system.

STORAGE AND DISPOSAL

PROHIBITIONS: Do not contaminate water, food or feed by storage or disposal.

PESTICIDE DISPOSAL: Pesticide wastes are acutely hazardous. Improper disposal of excess pesticide spray mixture, or rinsate is a violation of Federal Law. If these wastes cannot be disposed of according to label instructions, contact your State Pesticide or Environmental Control Agency, or the Hazardous Waste representative at the nearest EPA Regional Office for guidance.

CONTAINER DISPOSAL: Triple rinse (or equivalent). Then offer for recycling or reconditioning or puncture and dispose of in a sanitary landfill, or by other procedures approved by state and local authorities.

AUTOMOBILE FINISH PRECAUTION
Undiluted spray droplets may damage automobile finishes. Cars should not be sprayed. If accidental exposure does occur, the car should be washed before product dries.

SEE DIRECTIONS FOR YOUR STATE ONLY

24[c] State Labels
FOR USE AND DISTRIBUTION ONLY WITHIN
WASHINGTON WA–790065 OREGON OR–790026

AERIAL APPLICATION - READY-TO-USE, DO NOT DILUTE
GROUND APPLICATION —Mix with water, 2–5 gallons total solution per acre.

USE SITES: PASTURE, RANGELAND, NON-CROP [Including fence rows, utility rights-of-way, ditch banks, field borders, etc.] AND FALLOW LAND [used in rotation to grow wheat, barley, rye or oats].

PESTS CONTROLLED

Field Bindweed (perennial morningglory): Apply up to 2 to 3 quarts of DED-WEED SULV per acre. Summer fallow applications should be made in early bloom of field bindweed. For treatment during cropping year, make application after harvest to field bindweed in standing stubble. Do not make application within 30 days of planting.

Canada Thistle: Apply 1½–2 quarts per acre. For best results make application to Canada Thistle at approximately 6" to 8" of growth or in the bud stage prior to flowering in non-tilled areas. If recently tilled, delay application as long as possible to allow rhyzome fragments to emerge or early bloom.

Musk Thistle: Apply 1 to 2 quarts DED-WEED SULV per acre. Apply in spring when plants are growing vigorously and/or in the fall when plants are in the rosette stage.

APPLICATION**

Aerial - (fixed wing or rotary wing)-Aircraft specifications: **Boom width should not exceed ¾ the length of the aircraft wingspan. Do not exceed 25 psi nozzle pressure.** Number of nozzles required to obtain desired volume per acre is dependent on swath width and speed of aircraft. Nozzles should be positioned between 135° and 175° from direction of flight for fixed wing. DO NOT APPLY THROUGH BECO-MIST nozzle systems. Maintain aircraft altitude of 6–10 feet during application. See manufacturers technical bulletin regarding nozzling and method of application specifications.

Ground - Apply in water 2 to 5 gallons total solution per acre with conventional equipment. Use nozzle systems capable of spraying correct gallonage; 25 psi is recommended.

Do not apply when grass is in boot to milk stage. Do not apply to seedling grasses, or after heading begins.

**Timing of applications may vary - your State Cooperative Extension service may have information on the correct application time for your area.

RESTRICTIONS

1. **Pastures/Rangeland:** Do not graze dairy cattle in treated areas for 14 days after application. Remove meat animals from freshly treated areas for 7 days before slaughter. Withdrawal is not needed if 2 weeks or more have elapsed since application. Do not cut treated grass for hay within 30 days after application.
2. **Fallow land:** Do not plant any crop until 3 months after treatment or until chemical has disappeared from soil.

SEE DIRECTIONS FOR YOUR STATE ONLY

24[c] State Labels
FOR USE AND DISTRIBUTION ONLY WITHIN
Montana MT–800006
Idaho ID–800047
North Dakota ND–800010
Texas TX–820011
Missouri MO–800023
Utah UT–830010
New Mexico NM–820001

Nevada NV–830007
Oklahoma OK–820008

AERIAL APPLICATION - READY-TO-USE, DO NOT DILUTE

GROUND APPLICATION — Mix with water, 1–5 gallons total solution per acre.

USE SITES: PASTURE, RANGELAND, NON-CROP [including fence rows, utility rights-of-way, ditch banks, field borders, etc.] AND FALLOW LAND [used in rotation to grow wheat, barley, rye or oats].

PESTS CONTROLLED

Field Bindweed (perennial morningglory): Apply up to 2 to 3 quarts of DED-WEED SULV per acre. Summer fallow applications should be made in early bloom of field bindweed. For treatment during cropping year, make application after harvest to field bindweed in standing stubble. Do not make application within 30 days of planting.

Canada Thistle: Apply 1½-2 quarts per acre. For best results make application to Canada Thistle at approximately 6" to 8" of growth or in the bud stage prior to flowering in non-tilled areas. If recently tilled, delay application as long as possible to allow rhyzome fragments to emerge or early bloom.

Musk Thistle: Apply 1 to 2 quarts DED-WEED SULV per acre. Apply in spring when plants are growing vigorously and/or in the fall when plants are in the rosette stage.

APPLICATION**

Aerial (fixed wing or rotary wing)-Aircraft specifications: **Boom width should not exceed ¾ the length of the aircraft wingspan. Do not exceed 25 psi nozzle pressure.** Number of nozzles required to obtain desired volume per acre is dependent on swath width and speed of aircraft. Nozzles should be positioned between 135° and 175° from direction of flight for fixed wing. DO NOT APPLY THROUGH BECO-MIST nozzle systems. Maintain aircraft altitude of 6–10 feet during application. See manufacturers technical bulletin regarding nozzling and method of application specifications.

Ground - Apply in water 1 to 5 gallons total solution per acre with conventional equipment. Use nozzle systems capable of spraying correct gallonage; 25 psi is recommended.

Do not apply when grass is in boot to milk stage. Do not apply to seedling grasses, or after heading begins.

**Timing of applications may vary - your State Cooperative Extension service may have information on the correct application time for your area.

RESTRICTIONS

1. **Pastures/Rangeland:** Do not graze dairy cattle in treated areas for 14 days after application. Remove meat animals from freshly treated areas for 7 days before slaughter. Withdrawal is not needed if 2 weeks or more have elapsed since application. Do not cut treated grass for hay within 30 days after application.
2. **Fallow land:** Do not plant any crop until 3 months after treatment or until chemical has disappeared from soil.

CROP STATE LABELS

24[c] State Labels

FOR USE AND DISTRIBUTION ONLY WITHIN

Oregon OR–790026
Montana MT–800006
Idaho ID–800047
North Dakota ND–800010
Texas TX–820011

See Directions For Your State Only

WEED CONTROL IN CROPS

AERIAL APPLICATION ON GRAIN CROPS - Mix with water for a minimum total solution per acre of 2 quarts. No additives are necessary.

GROUND APPLICATION ON GRAIN CROPS - Mix with water, 1 to 5 gallons total solution per acre. To control broadleaf weeds such as: Carpetweed, Cocklebur, Common radish, Cow cockle, Florida pusley, Kochia, Lambsquarter, Morningglory (seedling), Mustards, Pigweed, Russian thistle, Canada thistle (seedling), Wild turnip, Musk thistle (seedling). Apply as directed below:

Crop	Amount SULV	Timing	Remarks
WHEAT (Fall or Spring Sown) BARLEY RYE	¾ to 1½ pints per acre*	Apply full tiller to boot stage and after milk stage to harvest	Do not treat in fall. Do not apply from boot to milk stage or in seedling stages.

*Up to 2 pints may be used at dough stage.
Do not permit dairy animals or meat animals being finished for slaughter to forage or graze treated grain fields within two weeks of treatment.

24[c] State Labels

FOR USE AND DISTRIBUTION ONLY WITHIN

Washington WA–810002

WEED CONTROL IN CROPS

AERIAL APPLICATION ON GRAIN CROPS - Mix with water for a minimum total solution per acre of 2 quarts. No additives are necessary.

GROUND APPLICATION ON GRAIN CROPS - Mix with water, 1 to 5 gallons total solution per acre. To control broadleaf weeds such as: Carpetweed, Cocklebur, Common radish, Cow cockle, Florida pusley, Kochia, Lambsquarter, Morningglory (seedling), Mustards, Pigweed, Russian thistle, Canada thistle (seedling), Wild turnip, Musk thistle (seedling). Apply as directed below:

Crop	Amount SULV	Timing	Remarks
WHEAT (Fall or Spring Sown) BARLEY RYE	¾ to 1½ pints per acre*	Apply full tiller to boot stage (Do not apply during boot stage) and after milk stage to harvest	Do not treat in fall. Do not apply from boot to milk stage or in seedling stages.

*Up to 2 pints may be used at dough stage.
Do not permit dairy animals or meat animals being finished for slaughter to forage or graze treated grain fields within two weeks of treatment.

24[c] State Labels

FOR USE AND DISTRIBUTION ONLY WITHIN

Oklahoma OK–820008

WEED CONTROL IN CROPS

AERIAL APPLICATION ON GRAIN CROPS - Mix with water for a minimum total solution per acre of 2 quarts. No additives are necessary.

GROUND APPLICATION ON GRAIN CROPS - Mix with water, 1 to 5 gallons total solution per acre. To control broadleaf weeds such as: Carpetweed, Cocklebur, Common radish, Cow cockle, Florida pusley, Kochia, Lambsquarter, Morningglory (seedling), Mustards, Pigweed, Russian thistle, Canada thistle (seedling), Wild turnip, Musk thistle (seedling). Apply as directed below:

Crop	Amount SULV	Timing	Remarks
WHEAT (Fall or Spring Sown) BARLEY RYE	¾ to 1½ pints per acre*	Apply in the spring when wheat is fully tillered, but before jointing	Do not treat in fall. Do not apply from boot to milk stage or in seedling stages.

*Up to 2 pints may be used at dough stage.
Do not permit dairy animals or meat animals being finished for slaughter to forage or graze treated grain fields within two weeks of treatment.

IMPORTANT NOTICE

Seller warrants that this product conforms to its chemical description and is reasonably fit for the purposes stated on the label when used in accordance with the directions and instructions specified on the label under normal conditions of use, **but neither this warranty nor any other warranty of merchantability or fitness for a particular purpose, express or implied, extends to the use of the product, contrary to label instructions, or under abnormal conditions, or under conditions not reasonably foreseeable to seller, and buyer assumes the risk of any such use.**

TERRACLOR®

2 LB. EMULSIFIABLE SOIL FUNGICIDE

COMPOSITION

ACTIVE INGREDIENT:	(% by weight)
Pentachloronitrobenzene (PCNB)	23.8%
INERT INGREDIENTS:	76.2%
TOTAL	100.0%

Contains 2 lbs. of Pentachloronitrobenzene per U.S. Gallon at 68°F.

EPA REG. NO. 400-400 8-30-87

Terraclor is a Registered Trademark of UNIROYAL CHEMICAL COMPANY, INC.

KEEP OUT OF REACH OF CHILDREN

CAUTION

STATEMENT OF PRACTICAL TREATMENT

IF IN EYES: Flush with plenty of water. Get medical attention.

IF ON SKIN: Wash with plenty of soap and water. Get medical attention.

IF SWALLOWED: Drink promptly a large quantity of milk, egg whites, gelatin solution or, if these are not available, drink large quantities of water. Avoid alcohol. Get medical attention.

PRECAUTIONARY STATEMENTS

HAZARDS TO HUMANS AND DOMESTIC ANIMALS

CAUTION

This material may cause skin irritation. Avoid contact with eyes or clothing. Wash thoroughly with soap and water after handling. Use with adequate ventilation. Do not contaminate feed or foodstuffs. The following protective clothing are required during mixing/loading operations: respirator, gloves, long-sleeve shirt, long pants, socks and shoes.

ENVIRONMENTAL HAZARDS

This pesticide is toxic to fish and aquatic organisms. Drift and runoff from treated areas may be hazardous to fish and aquatic organisms in adjacent aquatic sites. Do not apply directly to water or wetlands (swamps, bogs, marshes and potholes). Do not contaminate water or wetlands by cleaning of equipment or disposal of wastes.

PHYSICAL OR CHEMICAL HAZARDS

Flammable. Do not use or store near heat sources or open flame. Do not use in the vicinity of pilot lights.

STORAGE AND DISPOSAL

Do not contaminate water, food or feed by storage or disposal.

STORAGE: Protect from temperatures below 20°F.

PESTICIDE DISPOSAL: Pesticide wastes are toxic. Improper disposal of excess pesticide, spray mixture, or rinsate is a violation of Federal Law. If these wastes cannot be disposed of by use according to label instructions, contact your State Pesticide or Environmental Control Agency, or the Hazardous Waste representative at the nearest EPA Regional Office for guidance.

CONTAINER DISPOSAL: Triple rinse (or equivalent). Then offer for recycling or reconditioning, or puncture and dispose of in a sanitary landfill, or by other procedures approved by State and local authorities.

DIRECTIONS FOR USE

It is a violation of Federal law to use this product in a manner inconsistent with its labeling.

Do not apply this product through any type of irrigation system.

	RATES OF TERRACLOR 2 LB. EMULSIFIABLE			
CROPS (DISEASE)	GALS./ACRE	FL. OZ./1000 FEET OF ROW	APPLICATION	REMARKS
DRY, SUCCULENT & SNAP BEANS (Root and Stem Rot-Rhizoctonia solani)	$1/2$ to $3/4$ gallons Use in 8 to 10 gallons water (36 inch row spacing - bush type beans; 62 inch row spacing - pole beans) In California and Arizona: 3 to 4 pts. in 8 to 10 gals. (spacings as above)	Bush: 4.4 to 6.6 Pole: 7.6 to 11.3 Bush: 2.2 to 4.4 Pole: 3.8 to 5.6	Spray planting furrow and covering soil at planting time.	Avoid spraying directly to seed; otherwise delayed emergence may occur. Use lower rate on lighter soils.
(White Mold - Sclerotinia sclerotiorum)	1 gallon Use in 8 to 10 gals. water (36 inch row spacing - bush type beans; 62 inch row spacing - pole beans)	Bush: 8.8 Pole: 15	Spray an 8 inch wide band centered on the row immediately after or at time of seeding. If disease is severe, application may be repeated at 2 to 3 week intervals using 1 to 2 nozzles per row and directing spray at base of plants.	Remarks as above. Do not apply after first bloom. Soil should remain undisturbed after treatment. Do not feed treated vines to livestock.
Preharvest Intervals: Dry beans - Do not apply within 45 days of harvest; Succulent beans - Do not apply within 35 days of harvest; Snap beans - Do not apply within 14 days of harvest.				
COTTON (Damping off - Rhizoctonia solani)	$1/2$ to 1 gallon Use in 15 gals. water (40 inch row spacing)	4.8 to 9.7	Spray on seed and surrounding soil at planting time.	Do not allow the feeding or grazing of cotton foliage by livestock.
GARLIC (White Rot - Sclerotium cepivorum)	10 gallons Use in 100 gals. water (24 inch row spacing)	58.7	Apply at planting. Attach a suitable rig to the machine planter and spray 8-12 inch bands using 2 nozzles per planting furrow. Direct the front nozzle to spray the bottom of the open furrow and the "seed" as it is dropped. Direct the rear nozzle to spray the covering soil.	
POTATOES (Rhizoctonia solani)	9 to $12\frac{1}{2}$ gallons Use in sufficient water to total 25 gals final spray volume	—	Apply as a broadcast to the soil surface prior to planting and thoroughly mix (disc and cross disc) to a depth of 4 inches. A drag harrow is usually attached behind the discs on the last discing.	It is very important to thoroughly mix the TERRACLOR into the tuber-forming zone of the soil. It is desirable to make treatment when the soil is slightly moist.
	5 gallons Use in sufficient water to total 20 gallons final spray volume (42 inch row spacing)	51.6	Direct the spray solution to several points in the row to insure optimum treatment of the stem, stolen and tuber-forming zones, using three nozzles per row. Place one fan nozzle ahead of the furrow opening discs and two cone spray nozzles above and slightly ahead of the closing discs. Select nozzles that will deliver approximately $1/3$ of the recommended spray volume per acre through each nozzle at the desired planting speed and pressure.	
Rotation Restriction: Do not plant root crops in PCNB treated fields within 12 months of broadcast and banding applications unless PCNB is registered for use on those crops.				

IMPORTANT NOTICE

Seller warrants that this product conforms to its chemical description and is reasonably fit for the purposes stated on the label when used in accordance with the directions and instructions specified on the label under normal conditions of use, **but neither this warranty nor any other warranty of merchantability or fitness for a particular purpose, express or implied, extends to the use of this product, contrary to label instructions, or under abnormal conditions, or under conditions not reasonably foreseeable to seller, and buyer assumes the risk of any such use.**

TERRACLOR® 6.5%
PLUS

Di-Syston® 6.5%
GRANULAR
SOIL FUNGICIDE — SYSTEMIC INSECTICIDE

RESTRICTED USE PESTICIDE

Acute Human Oral, Dermal, and Inhalation Toxicity.

For retail sale to and application only by certified applicators or personnel under their direct supervision and only for those uses covered by the certified applicator's certification.

COMPOSITION

ACTIVE INGREDIENT:	(% BY WEIGHT)
Pentachloronitrobenzene (PCNB)	6.5%
Disulfoton: O,O-Diethyl S-[2-(ethylthio) ethyl] phosphorodithioate**	6.5%
INERT INGREDIENTS:	87.0%
TOTAL:	100.0%
EPA REG. NO. 400-411	2-2-88

**U.S. Patent No. 2,759,010.

KEEP OUT OF REACH OF CHILDREN
PELIGRO

DANGER–POISON

POISONOUS IF SWALLOWED, INHALED OR ABSORBED THROUGH THE SKIN. RAPIDLY ABSORBED THROUGH SKIN. REPEATED INHALATION OR SKIN CONTACT MAY, WITHOUT SYMPTOMS PROGRESSIVELY INCREASE SUSCEPTIBILITY TO DISULFOTON POISONING. ATROPINE AND 2-PAM ARE ANTIDOTES FOR DI-SYSTON. CALL A PHYSICIAN IMMEDIATELY IN ALL CASES OF SUSPECTED POISONING

PRECAUTION USUARIO: Si usted no lee ingles, no use este producto hasta que le etiqueta haya sido explicado ampliamente.

STATEMENT OF PRACTICAL TREATMENT

IF SWALLOWED: Call a physician or Poison Control Center immediately. Drink one or two glasses of water and induce vomiting by touching back of throat with finger. Repeat until vomit fluid is clear. Do not induce vomiting or give anything by mouth to an unconscious or convulsing person.

IF INHALED: Remove victim to fresh air. Apply artificial respiration if indicated.

IF ON SKIN: Remove contaminated clothing and wash affected areas with soap and water.

IF IN EYES: Flush with plenty of water. Call a physician immediately.

NOTE TO PHYSICIAN: Warning symptoms include weakness, headache, tightness in chest, blurred vision, non-reactive pinpoint pupils, salivation, sweating, nausea, vomiting, diarrhea and abdominal cramps.

ANTIDOTE: Administer atropine sulfate in large therapeutic doses. Repeat as necessary to the point of tolerance.

2-PAM is also antidotal and may be administered in conjunction with atropine.

Compound inhibits cholinesterase resulting in stimulation of the central nervous system, the parasympathetic nervous system, and the somatic motor nerves. Do not give morphine. Watch for pulmonary edema, which may develop in serious cases of poisoning even after 12 hours. At first sign of pulmonary edema, the patient should be placed in an oxygen tent and treated symptomatically. Repeated exposure to cholinestrase inhibitors may, without warning, cause increasing susceptibility to very small doses of any cholinesterase inhibitor. Allow no further exposure until cholinesterase regeneration has taken place as determined by blood tests.

TERRACLOR is a Registered Trademark of UNIROYAL CHEMICAL COMPANY, INC.

DI-SYSTON is a Registered Trademark of the Parent Company of Farbenfabriken Bayer GmbH, Leverkusen.

PRECAUTIONARY STATEMENTS

HAZARDS TO HUMANS

DANGER

Poisonous if swallowed, inhaled, or absorbed through the skin. Rapidly absorbed through the skin. Do Not Breathe Dust. Do Not Get In Eyes, On Skin, Or Clothing. In case of contact, immediately remove contaminated clothing and wash skin thoroughly with soap and water. Wear freshly laundered long-sleeve work clothes daily. Wash thoroughly with soap and water before eating or smoking. Bathe at the end of the work day and change outer clothing. While transferring material from package to equipment and during mixing/loading/application operations, wear a hat, goggles, face shield or equivalent, gloves (natural rubber), long-sleeved shirt, long pants, socks, shoes, and a pesticide respirator jointly approved by the Mine Safety Health Administration and the National Institute for Occupational Safety and Health under the provisions of 30 CFR Part 11. Rubber gloves should be washed with soap and water after each use. Do not wear the same gloves for other work. Destroy and replace gloves frequently. Keep all unprotected persons out of the operating area.

ENVIRONMENTAL HAZARDS

This pesticide is toxic to fish and wildlife. Do not apply directly to water or wetlands. Drift and run-off from treated areas may be hazardous to fish in neighboring areas. Do not contaminate water by cleaning of equipment or disposal of wastes. Cover or incorporate spills.

This pesticide is toxic to bees exposed to direct application. Applications should be timed to coincide with periods of minimum bee activity, usually between late evening and early morning.

DIRECTIONS FOR USE

It is a violation of Federal law to use this product in a manner inconsistent with its labeling.

STORAGE AND DISPOSAL

Do not contaminate water, food or feed by storage or disposal.

STORAGE: Store in a dry location.

PESTICIDE DISPOSAL: Pesticide wastes are acutely hazardous. Improper disposal of excess pesticide, spray mixture, or rinsate is a violation of Federal Law. If these wastes cannot be disposed of by use according to label instructions, contact your State Pesticide or Environmental Control Agency, or the Hazardous Waste representative at the nearest EPA Regional Office for guidance.

CONTAINER DISPOSAL: Completely empty bag into application equipment. Then dispose of empty bag in a sanitary landfill or by incineration, or if allowed by State and local authorities, by burning. If burned, stay out of smoke.

REENTRY STATEMENT

Do not apply this product in such a manner as to directly or through drift expose workers or other persons. The area being treated must be vacated by unprotected persons.

Do not allow reentry into treated fields within 24 hours of application unless appropriate protective clothing is worn.

Because certain states may require more restrictive reentry intervals for various crops treated with this product consult your State Department of Agriculture for further information.

Written or oral warnings must be given to workers who are expected to be in a treated area or in an area about to be treated with this product. When oral warnings are given if there is reason to believe that written warnings cannot be understood by workers. Written or oral warnings must include the following information:

"DANGER" Area to be treated with TERRACLOR with Di-Syston on [date of application]. Do not enter without appropriate protective clothing until 24 hours after application. If accidently exposed, consult the Statement of Practical Treatment portion of the pesticide label for first aid recommendations."

DIRECTIONS FOR USE

It is a violation of Federal law to use this product in a manner inconsistent with its labeling.

COTTON

For control of damping-off (Rhizoctonia solani) and early season control of thrips, aphids and mites, apply at planting time as a soil in-furrow application, near seeds and over covering soil while using a suitable applicator properly calibrated to assure accurate placement and proper dosage levels.

Application Method	Lbs./drill row per acre	Lbs./1300 linear feet of row based on 40 inch spacing
DRILL PLANTING	15	1.5
HILL-DROP PLANTING	5	0.5

Soil temperature, moisture and tilth should be favorable for good germination and emergence. Systemic insecticides, however, can possibly damage seed germination or stunting of seedlings under adverse conditions such as extremely cool or wet or extremely dry weather. Such damage may be more pronounced in light, sandy soils or when used in conjunction with certain pre-emergence herbicides.

USE PRECAUTIONS

While emptying bags into equipment, pour downwind and allow as little free fall as possible. Do not pour at face level and do not allow dust to reach the breathing zone.

Do not plant root crops in PCNB treated fields within 12 months of broadcast and banding applications unless PCNB is registered for use on those crops.

IMPORTANT NOTICE

Seller warrants that this product conforms to its chemical description and is reasonably fit for the purposes stated on the label when used in accordance with the directions and instructions specified on the label under normal conditions of use, **but neither this warranty nor any other warranty of merchantability or fitness for a particular purpose, express or implied, extends to the use of this product, contrary to label instructions, or under abnormal conditions, or under conditions not reasonably foreseeable to seller, and buyer assumes the risk of any such use.**

TERRACLOR®

10% GRANULAR SOIL FUNGICIDE

COMPOSITION

ACTIVE INGREDIENT:	(% by weight)
Pentachloronitrobenzene	10%
INERT INGREDIENTS:	90%
TOTAL	100%

EPA REG. NO. 400-402 10-1-88

TERRACLOR is a Registered Trademark of UNIROYAL CHEMICAL COMPANY, INC.

KEEP OUT OF REACH OF CHILDREN

CAUTION

STATEMENT OF PRACTICAL TREATMENT

IF IN EYES: Flush eyes with plenty of water. Get medical attention.

IF ON SKIN: Wash with plenty of water. Get medical attention.

IF SWALLOWED: Drink promptly a large quantity of milk, egg whites, gelatin solution or, if these are not available, drink large quantities of water. Avoid alchohol. Get medical attention.

PRECAUTIONARY STATEMENTS

HAZARDS TO HUMANS

CAUTION

Harmful if swallowed. Avoid inhaling dust. This material may cause skin irritation. Wash hands and face thoroughly after using. The following protective clothing are required during mixing/loading operations: gloves, a respirator, long-sleeved shirt, long pants, socks and shoes.

ENVIRONMENTAL HAZARDS

This product is toxic to fish and aquatic organisms. Cover, collect, or incorporate granules. Do not contaminate water by cleaning of equipment or disposal of wastes.

DIRECTIONS FOR USE

It is a violation of Federal law to use this product in a manner inconsistent with its labeling.

STORAGE AND DISPOSAL

Do not contaminate water, food or feed by storage or disposal.

STORAGE: Store in a dry location.

PESTICIDE DISPOSAL: Pesticide wastes are toxic. Improper disposal of excess pesticide, spray mixture, or rinsate is a violation of Federal Law. If these wastes cannot be disposed of by use according to label instructions, contact your State Pesticide or Environmental Control Agency, or the Hazardous Waste representative at the nearest EPA Regional Office for guidance.

CONTAINER DISPOSAL: Completely empty bag into application equipment. Then dispose of empty bag in a sanitary landfill or by incineration, or if allowed by State and local authorities, by burning. If burned, stay out of smoke.

RATE OF TERRACLOR 10% GRANULAR

CROP	DISEASE	POUNDS/ACRE	POUNDS/1000 FEET OF ROW	RECOMMENDED APPLICATION
COTTON	Damping-off, Seed and Seedling Rots Rhizoctonia solani	10-20 (40 inch row spacings)	¾ to 1½	In-furrow Application: Apply over seed and covering soil at planting time, using suitable application equipment. Do not allow feeding or grazing of cotton foliage by livestock.
BEANS	Root and Stem Rot- Rhizoctonia solani	10 to 15 (36 inch row spacing - bush type; 62 inch row spacing - pole beans)	Bush: ¾ to 1 Pole: 1¼ to 1¾	In-furrow Application: Direct the material into the planting furrow and covering soil at planting time. NOTE: Do not feed treated vines to livestock.
CABBAGE BRUSSELS SPROUTS, BROCCOLI	Clubroot- Plasmodiphora brassicae	200 (40 inch row spacing)	15	Row Application: Apply in a 12 to 15 inch band, by means of a suitable applicator, and rototil to a depth of 4 to 6 inches, immediately prior to planting.
		300	—	Broadcast Application: Apply to the soil surface prior to planting. Disc and cross disc to a depth of 4 to 6 inches.
	For control of clubroot, it is very important to thoroughly mix TERRACLOR with the soil.			
CAULIFLOWER	Clubroot- Plasomidophora brassicae	200 (48 inch row (spacing)	18.3	Row Application: Apply in a 12 to 15 inch band, by means of a suitable applicator, and rototil to a depth of 4 to 6 inches, immediately prior to planting. It is very important to thoroughly mix TERRACLOR with the soil.
		300	—	Broadcast Application: Apply to the soil surface prior to planting. Disc and cross disc to a depth of 4 to 6 inches.

CROP	DISEASE	POUNDS/ACRE	POUNDS/1000 FEET OF ROW	RECOMMENDED APPLICATION
	Wire Stem or Black Root Corticium solani	110 to 150 (48 inch row spacing)	10 to 13.7	Row Application: Apply in an 8 inch band centered on the row, immediately prior to seeding. For smaller areas apply 12 lbs. per 1,000 sq. ft. Rotoil to a depth of 2 inches.
PEANUTS	Southern blight Sclerotium rolfsii	36 inch row spacing 100	6.9	Surface Band Treatment: Apply in an 8 to 12 inch band, centered on the row, at pegging time.
		100	6.9	Pre-plant Soil Mix: Apply in an 8 to 12 inch band, centered on the row, and mix to a depth of 2 to 4 inches. Application may be made simultaneously with planting, using suitable equipment.
		33 per application	2.3 pounds per application	Soil Mix Cultivation: Apply in split applications during three cultivations. Attach two delivery tubes per row ahead of inside cultivator sweeps (one on each side) and adjust so as to treat a total band of 12 inches wide, centered on the row. Do not pile untreated soil on the crown of the plant.
		50	3.4 pounds per application	Split Pegging Applications: Apply as a 12 to 18 inch band centered over the row follow by shallow incorporation if possible. The initial application should be timed to coincide with early pegging. A second application at the same rate may be made if disease symptoms reappear. Depending on weather, this usually occurs 4 to 5 weeks after the first application.
	NOTE: Do not exceed 10 lbs. actual PCNB per acre in any one season, regardless of the application methods used. Do not feed treated vines to livestock. Do not allow hogging down of peanuts.			
POTATOES	Rhizoctonia solani	180 to 250	—	Broadcast Application: Spread evenly over the soil surface prior to planting. Disc and cross disc to depth of 4 to 6 inches.
		100 (36 inch row spacing)	6.8	Row Application: Apply in-furrow and thoroughly work into top 4 to 6 inches of the soil. It is essential that the fungicide is worked into the tuber forming zone of the soil. It is desirable to make treatment when the soil is slightly moist.
	Rotation Restriction: Do not plant root crops in PCNB treated fields within 12 months of broadcast and banding applications unless PCNB is registered for use on those crops.			

IMPORTANT NOTICE

Seller warrants that this product conforms to its chemical description and is reasonably fit for the purposes stated on the label when used in accordance with the directions and instructions specified on the label under normal conditions of use, **but neither this warranty nor any other warranty of merchantability or fitness for a particular purpose, express or implied, extends to the use of this product, contrary to label instructions, or under abnormal conditions, or under conditions not reasonably foreseeable to seller, and buyer assumes the risk of any such use.**

TERRACLOR®

75% WETTABLE POWDER TURF, ORNAMENTAL & VEGETABLE SOIL FUNGICIDE

COMPOSITION

ACTIVE INGREDIENT:	(% by weight)
Pentachloronitrobenzene	75%
INERT INGREDIENTS	25%
TOTAL	100%

EPA REG. NO. 400-399 12-4-87G

Terraclor is a Registered Trademark of Uniroyal Chemical Company, Inc.

KEEP OUT OF REACH OF CHILDREN

CAUTION

STATEMENT OF PRACTICAL TREATMENT

IF SWALLOWED: Call a physician or Poison Control Center immediately. Drink 1 or 2 glasses of water and induce vomiting by touching back of throat with finger, or blunt object. Do not induce vomiting or give anything by mouth to an unconscious person.

IF IN EYES: Flush with plenty of water. Get medical attention if irritation persists.

IF ON SKIN: Wash with plenty of soap and water. Get medical attention if irritation persists.

IF INHALED: Remove to fresh air. See a physician if an abnormal reaction occurs.

PRECAUTIONARY STATEMENTS

HAZARDS TO HUMANS AND DOMESTIC ANIMALS CAUTION

Harmful if swallowed. Avoid inhaling dust. May cause skin irritation. Wash hands and face thoroughly after using. The following protective clothing are required during mixing/loading operations: gloves, long-sleeved shirt, long pants, socks and shoes. Wash thoroughly all contaminated clothing before reuse.

ENVIRONMENTAL HAZARDS

This pesticide is toxic to fish and aquatic organisms. Drift and runoff from treated areas may be hazardous to fish and aquatic organisms in adjacent aquatic sites. Do not apply directly to water or wetlands (swamps, bogs, marshes, and potholes). Do not contaminate water when disposing of equipment washwaters.

STORAGE AND DISPOSAL

Do not contaminate water, food or feed by storage or disposal.

STORAGE: Store in a dry location.

PESTICIDE DISPOSAL: Pesticide wastes are toxic. Improper disposal of excess pesticide, spray mixture, or rinsate is a violation of Federal Law. If these wastes cannot be disposed of by use according to label instructions, contact your State Pesticide or Environmental Control Agency, or the Hazardous Waste representative at the nearest EPA Regional Office for guidance.

CONTAINER DISPOSAL:

Paper Bags: Completely empty bag into application equipment. Then dispose of bag in a sanitary landfill or by incineration, of if allowed by State and local authorities, by burning. If burned, stay out of smoke.

Cardboard Canisters: [Commercial] Completely empty canisters into application equipment. Then dispose of empty container in a sanitary landfill or by incineration, or if allowed by State and local authorities, by burning. If burned, stay out of smoke.

Household: Do not reuse empty container. Securely wrap original container in several layers of newspaper and discard in trash.

DIRECTIONS FOR USE

IT IS A VIOLATION OF FEDERAL LAW TO USE THIS PRODUCT IN A MANNER INCONSISTENT WITH ITS LABELING. DO NOT APPLY THIS PRODUCT THROUGH ANY TYPE OF IRRIGATION SYSTEM. DO NOT PLANT ROOT CROPS IN PCNB TREATED FIELDS WITHIN 12 MONTHS OF BROADCAST AND BANDING

APPLICATIONS UNLESS PCNB IS REGISTERED FOR USE ON THOSE CROPS.

GENERAL INFORMATION

TERRACLOR 75%WP is a protective fungicide recommended for control of certain soil borne diseases of vegetable and ornamental crops. Best results are achieved when used in a preventative program following recommended rates and application directions. Carefully read, understand and follow label directions.

VEGETABLES

TERRACLOR 75%WP should be applied as a preventative fungicide to vegetable crops. It is important when using TERRACLOR 75%WP as a protective fungicide that material be placed into the soil. This can be accomplished as a soil drench, mechanical incorporation or as an in-furrow application. Under certain crop use (snap & dry beans), repeated applications may be necessary. See use and rate recommendations for specific crop use recommendations and diseases controlled by TERRACLOR 75%WP.

Crop	Diseases Controlled
BEANS (snap & dry)	Root & Stem Rot (Rhizoctonia solani) White Mold (Sclerotium sclerotiorum)
PEPPERS, TOMATOES	Southern Blight (Sclerotium rolfsii)
POTATOES	Stem & Root Rot (Rhizoctonia solani)
GARLIC	White Rot (Sclerotium capivorum)
BROCCOLI BRUSSELS SPROUTS CABBAGE CAULIFLOWER	Wire Stem or Black Root (Corticium solani) Club Root (Plasmodiophora brassicae)

RATE OF TERRACLOR 75%WP

CROP	DISEASE	POUNDS/ACRE	OUNCE/1000 FEET OF ROW	RECOMMENDED APPLICATION
COTTON	Damping off (Rhizoctonia solani)	1⅓ to 2¾ (40 inch row spacings) Apply in at least 15 gals. of water	1.6 to 3.4	Apply specified dosage as a directed in-furrow spray to the seed and surrounding soil at planting time. Do not allow the feeding or grazing of cotton foliage by livestock.
PEANUTS	Southern stem and root rot (Sclerotium rolfsii)	36 inch row spacing 13⅓ Apply in 8 to 10 gallons of water	14.7	PRE-PLANT SOIL MIX: Apply specified dosage as a spray in an 8 to 12 inch band centered on the row and incorporate to a depth of 2 to 4 inches. Application may be made simultaneously with planting using suitable application equipment.
				AT-PEGGING BAND: Apply specified dosage as a spray in a 12 inch band centered on the row so as to cover the crown and pegging zones of the plant. If the row middles are cultivated at the same time, attach two spray nozzles (one to each side) - ahead of inside cultivator sweeps. Do not move any untreated soil into the area (if necessary remove inside sweep).
		4⅓	4.8	AT CULTIVATION: Apply specified dosage at cultivation. Do not apply more than 3 times per crop season. Attach two spray nozzles per row of inside cultivator sweeps (one on each side) and adjust so as to treat a total band of 12 inches centered on the row. Do not pile untreated soil on crown of plant.
	NOTE: Do not exceed 10 lbs. actual PCNB per acre in any one season, regardless of application methods used. Do not apply within 45 days of harvest. Do not feed treated vines to livestock. Do not allow hogging down of peanuts.			
BEANS	Root and Stem Rot (Rhizoctonia solani)	1⅓ to 2 Apply in 8 to 10 gallons of water. (36 inch row spacing—bush type beans: 62 inch row spacing—pole beans)	Bush 1.4 to 2.2 Pole: 2.5 to 3.8	Apply as a directed spray in the seed furrow and to the covering soil at planting time.
BEANS (Snap and Dry)	White Mold (Sclerotinia sclerotiorum)	**spacing as above** 2¾ Apply in 8 to 10 gallons of water.	Bush: 3 Pole: 5.2	Apply as an 8 inch band centered on the row at or immediately after planting. If the disease problem is severe, applications may be repeated at 2 to 3 week intervals using directed spray application with one or two nozzles per row aimed at the soil at the base of the plants. Soil should remain undisturbed after application.
	Do not apply after first bloom. Do not exceed 10 lbs. actual PCNB per acre in any one season, regardless of method of application. Do not feed treated vines to livestock. Preharvest Intervals: Dry beans—Do not apply within 45 days of harvest: Succulent beans—Do not apply within 35 days of harvest. Snap beans—Do not apply within 14 days of harvest.			
BROCCOLI BRUSSELS SPROUTS CABBAGE CAULIFLOWER	Clubroot (plasmodiophora brassicae)	2 to 6 pounds Apply in 100 gallons of water. Use ½ to ¾ pint of this solution per plant.		Transplant Solution: The applied dosage of TERRACLOR should be based on the severity of the disease infection. Maintain agitation to hold the powder in suspension.
		30 Apply in 25 gallons of water (40 inch row spacings—broccoli, brussels sprouts and cabbage; 48 inch row spacings—cauliflower).	Broccoli, Brussels Sprouts, Cabbage: 36.6 Cauliflower: 44	Band Application: Spray as a 12 to 15 inch band centered on the row and incorporate to a depth of 4 to 6 inches immediately prior to planting.
		40 Apply in 30 gallons of water	—	Broadcast Application: Apply as a preplant broadcast spray prior to planting. Thoroughly incorporate to a depth of 4 to 6 inches using a disc or other suitable equipment.
	NOTE: For effective control of clubroot in broccoli, brussels sprouts, cabbage and cauliflower, it is extremely important to thoroughly mix TERRACLOR with the soil.			
	Wire Stem or Black Root (Corticium solani)	15 to 20 Apply in 50 gallons of water	—	Broadcast Drench Application: Apply as a soil drench at the time of or immediately after seeding.
		For smaller areas use 1 level tablespoon per gallon of water as a soil drench per 50 sq. ft. of seed bed. Apply with water can or similar equipment.		

CROP	DISEASE	POUNDS/ACRE	OUNCE/1000 FEET OF ROW	RECOMMENDED APPLICATION
		10 to 15 Apply in 35 gallons water. (40 inch row spacings—broccoli, brussels sprouts and cabbage; 48 inch row spacings—cauliflower)	Broccoli, Brussels Sprouts, Cabbage: 12.2 to 18.3 Cauliflower: 14.6 to 22	Row Drench Treatment: Spray as an 8 inch band centered on the row at the time of or immediately after seeding.
GARLIC	White rot (Sclerotium capivorum)	27½ Apply in 100 gallons water. (24 inch row spacings)	20	In-Furrow Application: Apply as an in-furrow spray at planting time. Attach a spray rig to the machine planter and spray 8-12 inch bands using 2 nozzles per furrow. Direct front nozzle to spray the "seed" as it is dropped: direct the rear nozzle to spray the covering soil. Use good mechanical agitation in the spray tank to maintain a uniform suspension.
		13½ pounds per 1000 lbs. of "seed" cloves.		Clove Treatment: Mix in sufficient water to apply dosage as a uniform mist-spray for complete coverage of treated cloves. For optimum results add a commercial sticker to the spray solution.
PEPPERS TOMATOES	Southern Blight (Sclerotium rolfsii)	3 to 5 pounds in 100 gallons of water. Use ½ pint of this solution per plant.		Apply at the time of transplanting. The solution should be agitated often to maintain a uniform suspension and assure proper suspension and assure proper dosage.
		10 Apply in 100 gallons of water. (36 inch row spacing—peppers; 72 inch row spacings—flat tomatoes: 48 inch row spacings—staked tomatoes)	Peppers: 11 Flat Tomatoes: 22 Staked Tomatoes: 14.6	Apply as an in-furrow spray to the open "V" trench, just prior to planting. When cultivating, set plows as flat as possible to avoid getting untreated soil against stems of plants.
POTATOES	Rhizoctonia solani	24 to 33 Apply in 25 gallons water	—	Broadcast Application: Apply as a preplant broadcast spray prior to planting. Thoroughly incorporate to a depth of 4 to 6 inches using a disc or other suitable equipment.
		13½ Apply in 20 gallons water. (42 inch row spacings)	17.4	In-furrow Application: Spray several points in the planting furrow so as to evenly treat all the soil that surrounds and covers the seed piece and potato forming zone. For best results, attach 3 nozzles to the planter; one fan nozzle ahead of the opening discs (Approximately 5 gallons of spray per acre); one cone nozzle placed above and slightly ahead of the closing discs (approximately 10 gallons of spray per acre); one fan nozzle to rear of closing discs (approximately 5 gallons of spray per acre). Spray pattern should treat a 12 to 14 inch band centered on the row.

ROTATION RESTRICTION: Do not plant crops in PCNB treated fields within 12 months of broadcast and banding applications unless PCNB is registered for use on those crops.

GENERAL INFORMATION

TERRACLOR 75%WP is a preventative fungicide recommended for control of certain soil borne diseases of turf and certain ornamental crops. Best results are achieved when used in preventative program following recommended rates and application directions. Carefully read, understand and follow label directions.

TURF

TERRACLOR 75%WP is a fungicide for the prevention and control of certain soil borne diseases on golf courses, sod farms, home lawns and institutional areas where turf is grown.

DISEASES CONTROLLED:

	Fungi
Brown Patch	Rhizoctonia solani
Helminthosporium Leaf Spot*	Helminthosporium spp
Dollar spot*	Sclerotinia homoecarpa
Gray Snow Mold	Typhula spp
Pink Snow Mold	Fusarium nivale

Best results are obtained when TERRACLOR 75%WP is used as a preventative application before diseases appear, however, if you do not apply as a preventative treatment, TERRACLOR 75%WP should be applied at first sign of the disease.

Areas treated with TERRACLOR 75%WP should be watered following application to move material down to the soil level. If treated area is subjected to unusually heavy rainfall, or flooded or if disease is severe or reappears, the area should be treated again. See use and rate recommendations for intervals for retreating.

USE & RATE RECOMMENDATION FOR TURF:

DISEASE/CROP	RATE/1,000 SQ. FT.	REMARKS
Brown Patch (Rhizoctonia solani) St. Augustine, Bermuda, and similar warm climate grasses	16 oz. in 10-15 gallons water	1. Apply one application in the spring or fall or at the 1st indication of disease. 2. If disease is severe or reappears, retreat 3-4 weeks later.
Dichondra	16 oz. in 40 gallons water	1. Treat as above
Bluegrass, Rye Grass, Fescue, Bent and similar cool climate grasses	3-4 oz. in 3-6 gallons water	1. Apply at first sign of disease. 2. Repeat every 7-10 days during warm damp weather.
Helminthosporium Leaf Spot* (Helminthosporium spp) **Dollar Spot*** (Sclerotinia homoecarpa)	7-10 oz. in 5-10 gallons water	1. Preventative applications should be in spring or fall of each year. 2. For control of existing infections, apply at higher rate as soon as symptoms appear. 3. If disease is severe or reappears, area should be retreated 3-4 weeks later.
Gray Snow Mold (Typhula spp) **Pink Snow Mold** (Fusarium nivale)	8 oz. in 5-10 gallons water	1. First application should be made prior to first snowfall.

* DO NOT USE IN CALIFORNIA

Treated area should be watered lightly as soon as possible following treatment to wash chemical down to the soil level.

Do not graze treated areas or feed grass clippings to livestock.

Under certain growing conditions, a temporary discoloration of the grass occasionally may occur. This causes no harm and will disappear in a short time.

ORNAMENTALS

TERRACLOR 75%WP can be used as a soil treatment for protection against a variety of soil borne diseases in nursery, greenhouse and field grown ornamentals.
It is important that TERRACLOR 75%WP be placed or mixed in the soil for best results.
See use and rate recommendations for specific plant and disease recommendations.

DISEASES CONTROLLED

	Fungi
Root and stem rots	Rhizoctonia spp.
Botrytis storage rot	Botrytis spp.
Neck dry rot	Stomatinia gladioli
Bulb dry rot	Sclerotium gladioli
Needle blight	Dothistromia pini
Petal blight	Ovulinia azalea
Bulb and stem rot	Sclerotium rolfsii
Flower blight	Sclerotium camelliae
Crown rot	Pellicularia rolfsii
Black rot	Sclerotinia sclerotiorum

PLANT TOLERANCE

With specific plants not listed below we recommend a preliminary trial to determine plant tolerance.

USE AND RATE RECOMMENDATION:

TERRACLOR 75%WP AS SOIL DRENCH AT TIME OF SEEDING AND TRANSPLANTING IN NURSERY AND GREENHOUSES FOR RHIZOCTONIA CONTROL.

Plants	Use and Rates
Tropical Foliage Plants Ajuga, Aphelandra, Dieffenbachia, Dracena, Ficus, Fittonia, Gynura, Hoya, Peperomia, Pothos, Schefflera, Syngonium, Tradescantia, Similar foliage plants **One repeat application can be made 4 to 6 weeks later, if necessary.** **Do not apply to Philodendron or Pilea or injury may occur.**	Use 4 to 8 oz. in 100 gallons of water and apply to 400-800 sq. ft. of bed or bench area. Coverage: For soil depths of 4 inches or less, apply 100 gallons to 800 sq. ft. For soil depths of more than 4 inches apply 100 gallons to 400 sq. ft.
Bedding and Flowering Plants Carnation, Chrysanthemum, Dahlia, Geranium, Marigold, Petunia, Poinsettia, Portulaca, Salvia, Vinca, Zinnia, Similar plants	Use 4 oz. of TERRACLOR 75%WP with 100 gallons of water and apply to 800 sq. ft. of flat or bench area. Apply at seeding or transplanting.
Herbaceous and Woody Ornamentals **Herbaceous**: Kalanchoe, Lily **Woody**: Azalea, Boxwood, English Ivy, Holly, Juniper, Pine, Rhododendron Similar herbaceous and woody plants	Use 4 to 8 oz. of TERRACLOR 75%WP with 100 gallons of water and apply to 400-800 sq. ft. of bed or bench area.

Solution of TERRACLOR 75%WP and water should be continuously agitated to assure uniform application of the fungicide.

USE AND RATE RECOMMENDATIONS:

TERRACLOR 75%WP AS A BROADCAST INCORPORATION PRIOR TO PLANTING FIELD GROWN CROPS

CROP	DISEASE	RATE OF TERRACLOR 75%WP (LBS. PER 1,000 SQ. FT. AREA)	APPLICATION
Easter Lillies	Rhizoctonia Root Rot	$3^1/_4$ to $6^1/_2$	Apply specified rate in sufficient water to insure uniform ground coverage prior to planting and thoroughly incorporate to a depth of 6–7 inches and see note below.
Iris, Hyascinth, Narcissus, Tulips	Crown Rot, Black Rot		
		Use lower rates for light soils and higher rate for heavy soils.	
Dutch Iris	Bulb Rot, Stem Rot (Sclerotium rolfsii)	$3^1/_4$ to $4^1/_2$	Apply specified rate in sufficient water to insure uniform ground coverage prior to planting and thoroughly incorporate to a depth of 2-3 inches.
Gladiolus	Neck or Bulb Dry Rot (Stomatinia or Sclerotinia gladioli)	$3^2/_3$	Apply specified rate in sufficient water to insure uniform ground coverage prior to planting and thoroughly incorporate to a depth of 4-6 inches. Bulbs may be planted immediately after incorporation.
Calendula, Larkspur, Snapdragon, Sweet Pea	Stem Rot (Rhizoctonia)	$3^1/_2$	Apply specified rate in sufficient water to insure uniform ground coverage prior to planting and thoroughly incorporate to a depth of 4 inches.
Southern Pine Seedlings	Rhizoctonia Needle Blight	50 (lbs. per acre)	Apply specified rate per acre in 100 gallons of water to insure uniform ground coverage. An additional $1^1/_2$ inch of

TERRACLOR 75%WP AS A BROADCAST INCORPORATION PRIOR TO PLANTING FIELD GROWN CROPS

CROP	DISEASE	RATE OF TERRACLOR 75%WP (LBS. PER 1,000 SQ. FT. AREA)	APPLICATION
			water per acre should be applied by irrigation prior to planting. Seed may be planted immediately after irrigation or within one week of this date.
Solution of TERRACLOR 75%WP and water should be continuously agitated to assure uniform application of the fungicide.			

TERRACLOR 75%WP AS GROUND SPRAY ON ESTABLISHED WOODY ORNAMENTALS

CROP	DISEASE	RATE OF TERRACLOR 75%WP (LBS. PER 150 SQ. FT. AREA)	APPLICATION
Azaleas Camellia	Azalea petal blight (Ovulinia Azalea) Camellia flower blight	1.0 1.0	Apply specified rate per 150 sq. ft. in sufficient water to assure uniform ground coverage. Apply to the ground beneath the bushes and immediate surrounding area. Begin application prior to opening of buds and repeat every 3-4 weeks during bloom season.
Solution of TERRACLOR 75%WP and water should be continuously agitated to assure uniform application of fungicide.			

TERRACLOR 75%WP AS A DIP APPLICATION

CROP	DISEASE	RATE OF TERRACLOR 75%WP (LBS.)	APPLICATION
Iris Hyascinth Narcissus Tulips	Crown Rot Black Rot	2	Dip bulbs for 5 minutes in 75% concentration which is obtained by adding 2 lbs. of TERRACLOR 75%WP to 3.2 gallons of water. Add 1% sticker to solution and maintain good agitation.
Easter Lillies	Root Rot (Rhizoctonia)	4-6	Mix TERRACLOR 75%WP in 100 gallons of water and dip bulbs for 5-15 minutes. A sticking agent may be added. Maintain good agitation in dip tanks.
Roses	Botrytis Storage Rot	2	Mix TERRACLOR 75%WP in 100 gallons of water and dip dormant roses prior to storage. A spray application of the same solution may be substituted for dipping. Maintain good agitation.

TERRACLOR 75%WP AS BAND APPLICATION PRIOR TO PLANTING FIELD GROWN CROPS

CROP	DISEASE	RATE OF TERRACLOR 75%WP (LBS. PER 1,000 LINEAR FT. OF ROW)	APPLICATION
Gladiolus	Neck or Bulb Dry Rot (Stomatinia or Sclerotinia gladioli)	5½	Apply as a 12 inch band centered on the row at planting time. Incorporate to depth of 4 inches. Bulb may be planted immediately after incorporation.
Dutch Iris	Bulb Rot Stem Rot (Sclerotium rolfsii)	½	Apply as a 12 inch band centered on the row at planting time. Follow with 3 additional applications at the same rate at 15 day intervals.

TERRACLOR 75%WP AS IN FURROW APPLICATION AT PLANTING

CROP	DISEASE	RATE OF TERRACLOR 75%WP (LBS. PER 1,000 LINEAR FT. OF ROW)	APPLICATION
Easter Lillies Iris Hyascinth Narcissus Tulips	Rhizoctonia Root Rot Crown Rot Black Rot	4½ to 6	Apply in 5-10 gallons of water. Direct spray into open furrow to thoroughly cover entire furrow and bulbs. Close furrow in normal manner. See notes below.
Use lower rates for lighter soils and higher rates for heavier soils. For Black rot control, use high rates. Solution of TERRACLOR 75%WP and water should be agitated to assure uniform application of the fungicide.			

IMPORTANT NOTICE

Seller warrants that this product conforms to its chemical description and is reasonably fit for the purposes stated on the label when used in accordance with the directions and instructions specified on the label under normal conditions of use, **but neither this warranty nor any other warranty of merchantability or fitness for a particular purpose, express or implied, extends to the use of this product, contrary to label instructions, or under abnormal conditions, or under conditions not reasonably foreseeable to seller, and buyer assumes the risk of any such use.**

TERRACLOR®

with MOCAP® 10 Plus 3

GRANULAR

Pegging time treatment for nematodes, corn rootworm and whitemold control on peanuts.

COMPOSITION

ACTIVE INGREDIENT:	(% by weight)
Pentachloronitrobenzene (PCNB)	10%
Ethoprop: (O-Ethyl-S, S-Dipropyl Phosphorodithioate)	3%
INERT INGREDIENTS:	87%
TOTAL	100%

EPA REG. NO. 264-475-400

TERRACLOR is a Registered Trademark of Uniroyal Chemical Company, Inc.

Mocap is a Registered Trademark of Rhone-Poulenc Ag company.

Contains MOCAP® brand ethoprop nematicide/insecticide manufactured and supplied by Rhone-Poulenc Ag Company.

KEEP OUT OF REACH OF CHILDREN

WARNING

AVISO

PRECAUCION AL USUARIO: Si usted no lee ingles, no use este producto hasta que le etiqueta haya sido explicado ampliamente.

TRANSLATION TO THE USER: If you cannot read English, do not use this product until the label has been fully explained to you.

STATEMENT OF PRACTICAL TREATMENT

IF SWALLOWED: Call a physician or Poison Control Center. Drink 1 or 2 glasses of water and induce vomiting by touching back of throat with finger. Do not induce vomiting or give any thing by mouth to an unconscious person.

IF IN EYES: Poisonous if in eyes. Flush eyes with plenty of water for at least 15 minutes. Call a physician immediately.

IF ON SKIN: Immediately remove all contaminated clothing. Wash skin with soap and water. Discard contaminated shoes.

IF INHALED: Remove victim to fresh air and apply artificial respiration if indicated.

NOTE TO PHYSICIAN: This product is a cholinesterase inhibitor. Warning symptoms include: weakness, headaches, tightness of chest, blurred vision, non-reactive pupils, salivation, sweating, nausea, vomiting, diarrhea, and abdominal cramps.

Administer Atropine, 2 mg. intramuscularly or intravenously. Repeat dose every 3 to 8 minutes until signs of atropinization occur. Give pralidoxime chloride (PROTOPAM or 2-PAM), 2.5 gm in 100 cc of sterile water, intravenously, slowly in 15 to 30 minutes or by deep intramuscular injection of 1 gram in 3cc of distilled water. Repeat every half-hour if respiration weakens or muscle fasciculation or convulsion recur. Clear chest by postural drainage. Artificial respiration may be necessary. Observe patient contінously for 48 hours. Allow no further exposure until cholinesterase regeneration has taken place as determined by blood tests. Repeated exposure to cholinesterase inhibitors may, without warning, cause increasing susceptibility to very small doses of any cholinesterase inhibitor.

PRECAUTIONARY STATEMENTS

HAZARDS TO HUMANS AND DOMESTIC ANIMALS

WARNING

May be fatal if absorbed through skin. Rapidly absorbed through skin. May cause skin irritation. This product is a possible skin sensitizer. If an allergic reaction occurs after repeated exposure, discontinue use and consult a physician. Harmful if swallowed or inhaled. Do not get in eyes, on skin or on clothing. Wear rubber gloves, a long-sleeved shirt, long pants, socks and shoes. IN case of contact, wash thoroughly with soap and water. Do not breathe fumes or dust. After use wash hands, arms and face thoroughly with soap and warm water before eating or smoking. Wash all contaminated clothing with soap and hot water before reuse. Do not contaminate feed or food. Do not allow livestock to feed in treated areas.

SYMPTOMS OF POISONING: A sense of [tightness] in chest, sweating, stomach pains, vomiting, diarrhea, contraction of pupils.

ENVIRONMENTAL HAZARDS

This pesticide is toxic to aquatic organisms [fish and invertebrates], and wildlife. Birds feeding in treated areas may be killed. Do not apply directly to water or wetlands [swamps, bogs, marshes and potholes). Runoff from treated areas may be hazardous to aquatic organisms in neighboring areas. Cover or incorporate granules that are spilled during loading or are visible on soil surface in turn areas. Do not contaminate water when disposing of equipment washwaters.

DIRECTIONS FOR USE

It is a violation of Federal law to use this product in a manner inconsistent with its labeling.

Unprotected persons may enter treated areas after chemical has been mixed into the soil.

STORAGE AND DISPOSAL

STORAGE: Do not contaminate water, food or feed by storage or disposal. Do not store in or around the home.

PESTICIDE DISPOSAL: Pesticide wastes are acutely hazardous. Improper disposal of excess pesticide, spray mixture, or rinsate is a violation of Federal Law. If these wastes cannot be disposed of by use according to label instructions, contact your State Pesticide or Environmental Control Agency, or the Hazardous Waste representative at the nearest EPA Regional Office for guidance.

CONTAINER DISPOSAL: Completely empty bag into application equipment. Then dispose of empty bag in a sanitary landfill or by incineration, or if allowed by State and local authorities, by burning. If burned, stay out of smoke.

DIRECTIONS FOR USE

PEANUTS

Row Treatment At Early Pegging For Control Of Nematodes, Corn rootworm, and White mold (Southern Stem Rot)

Use 100 pounds of TERRACLOR® with MOCAP® per acre at early pegging. Apply in a band 12 to 15 inches wide, centered on the row (36-inch row spacing), or 6.9 pounds per 1000 linear feet of row. Do not apply when plants are wet. Incorporate lightly into the soil.

NOTE: Do not apply more than 10 pounds of pentachloronitrobenzene (PCNB) in any form in any one season. Do not hog down peanuts. Do not plant any root crop in rotation on PCNB treated soil except potatoes.

LIMITED WARRANTY AND DISCLAIMER

The manufacturer warrants (a) that this product conforms to the chemical description on the label; (b) that this product is reasonably fit for the purposes set forth in the directions for use when it is used in accordance with such directions; and (c) that the directions, warnings and other statements of this label are based upon responsible expert's evaluation of reasonable tests of effectiveness of toxicity to laboratory animals and to plants, and of residues on food crops, and upon reports of field experience. Tests have not been made on all varieties or in all states under all conditions. THE MANUFACTURER NEITHER MAKES NOR INTENDS, NOR DOES IT AUTHORIZE ANY AGENT OR REPRESENTATIVE TO MAKE ANY OTHER WARRANTIES, EXPRESS OR IMPLIED, AND IT EXPRESSLY EXCLUDES AND DISCLAIMS ALL IMPLIED WARRANTIES OF MERCHANTABILITY OR FITNESS FOR A PARTICULAR PURPOSE.

THIS WARRANTY DOES NOT EXTEND TO, AND THE BUYER SHALL BE SOLELY RESPONSIBLE FOR ANY AND ALL LOSS OR DAMAGE WHICH RESULTS FROM THE USE OF THIS PRODUCT IN ANY MANNER WHICH IS INCONSISTENT WITH THE LABEL DIRECTIONS, WARNINGS OR CAUTIONS.

BUYER'S EXCUSIVE REMEDY AND MANUFACTURER'S OR SELLER'S EXCLUSIVE LIABILITY FOR ANY AND ALL CLAIMS, LOSSES, DAMAGES OR INJURIES RESULTING FROM THE USE OR HANDLING OF THIS PRODUCT, WHETHER OR NOT BASED IN CONTRACT, NEGLIGENCE, STRICT LIABILITY IN TORT OR OTHERWISE, SHALL BE LIMITED, AT THE MANUFACTURER'S OPTION TO REPLACEMENT OF, OR THE REPAYMENT OF THE PURCHASE PRICE FOR THE QUALITY OF PRODUCT WITH RESPECT TO WHICH DAMAGES ARE CLAIMED. IN NO EVENT SHALL MANUFACTURER OR SELLER BE LIABLE FOR SPECIAL, INDIRECT OR CONSEQUENTIAL DAMAGES RESULTING FROM THE USE OR HANDLING OF THIS PRODUCT.

TERRACLOR Super X®

EMULSIFIABLE SOIL FUNGICIDE

Controls Rhizoctonia, Pythium, and Fusarium Soil-Borne Seedling Diseases of Cotton

COMPOSITION

ACTIVE INGREDIENT:	(% by weight)
Pentachloronitrobenzene (PCNB)	23.2%
5-Ethoxy-3-(trichloromethyl)-1,2,4-thiadiazole	5.8%
INERT INGREDIENTS:	71.0%
TOTAL	100.0%

Contains 2 lbs. of Pentachloronitrobenzene and 0.5 lb. of 5-Ethoxy-3-trichloromethyl-1,2,4-thiadiazole per U.S. gallon at 68°F.

EPA REG. NO. 400-405 12-9-87A

TERRACLOR Super X is a Registered Trademark of UNIROYAL CHEMICAL COMPANY, INC.

KEEP OUT OF REACH OF CHILDREN

WARNING

STATEMENT OF PRACTICAL TREATMENT

IF IN EYES: Flush with plenty of water. Get medical attention.

IF ON SKIN: Wash with plenty of soap and water. Get medical attention.

IF SWALLOWED: Call a physician or Poison Control Center. Drink an eight ounce glass of water containing 2 tablespoons of activated charcoal.

PRECAUTIONARY STATEMENTS

HAZARDS TO HUMANS
WARNING

Harmful if swallowed. Do not inhale spray mist. Do not get on skin or in eyes. May cause skin irritation. Skin contact may be harmful. The following protective clothing are required during mixing/loading operations: respirator, gloves, long-sleeved shirt, long pants, socks and shoes. Wash thoroughly all exposed skin. In case of eye contact, flush with plenty of water. Wash thoroughly all contaminated clothing before reuse.

ENVIRONMENTAL HAZARDS

This pesticide is toxic to fish and aquatic organisms. Drift and runoff from treated areas may be hazardous to fish and aquatic organisms in adjacent aquatic sites. Do not apply directly to water or wetlands (swamps, bogs, marshes, and potholes). Do not contaminate wtaer when disposing of equipment washwaters.

PHYSICAL OR CHEMICAL HAZARDS

Flammable. Do not use or store near heat sources or open flame. Do not use in the vicinity of pilot lights.

DIRECTIONS FOR USE

It is a violation of Federal law to use this product in a manner inconsistent with its labeling.

STORAGE AND DISPOSAL

Do not contaminate water, food or feed by storage or disposal.

STORAGE: Protect from temperatures below 20°F.

PESTICIDE DISPOSAL: Pesticide wastes are toxic. Improper disposal of excess pesticide, spray mixture, or rinsate is a violation of Federal Law. If these wastes cannot be diposed of by use according to label instructions, contact your State Pesticide or Environmental Control Agency, or the Hazardous Waste representative at the nearest EPA Regional Office for guidance.

CONTAINER DISPOSAL: Triple rinse (or equivalent). Then offer for recycling or reconditioning, or puncture and dispose of in a sanitary landfill, or by other procedures approved by State and local authorities.

DIRECTIONS FOR USE

Do not apply this product through any type of irrigation system.

COTTON
(Seedling Disease Complex - Rhizoctonia, Pythium and Fusarium)

SOIL IN-FURROW APPLICATION: At planting time, use 2 to 3 quarts. per acre (13,000 linear ft. based on 40 in. rows) mixed with 5 to 15 gallons of water. Use the higher treatment rate under conditions of expected heavy disease pressure. Spray all the soil that surrounds and covers the seed. Two nozzles work best; center the front nozzle over the open furrow to spray soil around the seed and the rear nozzle to spray the covering soil as it fills the furrow. Do not allow the feeding or grazing of cotton foliage by livestock. Do not feed gin waste to livestock.

ROTATION RESTRICTION: Do not plant root crops in PCNB treated fields within 12 months of application unless PCNB is registered for use on those crops.

IMPORTANT NOTICE

Seller warrants that this product conforms to its chemical description and is reasonably fit for the purposes stated on the label when in accordance with the directions and instructions specified on the label under normal conditions of use, **but neither this warranty nor any other warranty of merchantability or fitness for a particular purpose, express or implied, extends to the use of this product, contrary to label instructions, or under abnormal conditions, or under conditions not reasonably foreseeable to seller, and buyer assumes the risk of any such use.**

TERRACLOR Super X®

GRANULAR SOIL FUNGICIDE

FOR COTTON

COMPOSITION

ACTIVE INGREDIENT:	(% by weight)
Pentachloronitrobenzene (PCNB)	10.0%
5-Ethoxy-3-(trichloromethyl)-1,2,4-thiadiazole	2.5%
INERT INGREDIENTS:	87.5%
TOTAL:	100%

EPA REG. NO. 400-406 12-8-87A

TERRACLOR Super X is a Registered Trademark of UNIROYAL CHEMICAL COMPANY, INC.

KEEP OUT OF REACH OF CHILDREN

WARNING

STATEMENT OF PRACTICAL TREATMENT

IF IN EYES: Flush eyes with plenty of water. Get medical attention.

IF ON SKIN: Wash with plenty of water. Get medical attention.

IF SWALLOWED: Call a physician or Poison Control Center. Drink an eight ounce glass of water containing 2 tablespoons of activated charcoal.

PRECAUTIONARY STATEMENTS

HAZARDS TO HUMANS
WARNING

Harmful if swallowed. Do not inhale dust. Do not get on skin or in eyes. May cause skin irritation. Contact may be harmful. The following protective clothing are required during mixing/loading operations: respirator, gloves, long-sleeved shirt, long pants, socks and shoes. Wash thoroughly all exposed skin. In case of eye contact, flush with plenty of water. Wash thoroughly all contaminated clothing before reuse.

ENVIRONMENTAL HAZARDS

This pesticide is toxic to fish and aquatic organisms. Cover, collect or incorporate treated granules. Do not contaminate water when disposing of equipment washwaters.

DIRECTIONS FOR USE

It is a violation of Federal law to use this product in a manner inconsistent with its labeling.

STORAGE AND DISPOSAL

Do not contaminate water, food or feed by storage or disposal.

STORAGE: Store in a dry location.

PESTICIDE DISPOSAL: Pesticide wastes are toxic. Improper disposal of excess pesticide, spray mixture, or rinsate is a violation of Federal Law. If these wastes cannot be disposed of by use according to label instructions, contact your State Pesticide or Environmental Control Agency, or the Hazardous Waste representative at the nearest EPA Regional Office for guidance.

CONTAINER DISPOSAL: Completely empty bag into application equipment. Then dispose of empty bag in a sanitary landfill or by incineration, or if allowed by State and local authorities, by burning. If burned, stay out of smoke.

DIRECTIONS FOR USE

COTTON
(Seedling Disease Complex -Rhizoctonia, Pythium and Fusarium)

SOIL IN-FURROW APPLICATION: At planting time, use 10 to 15 lbs. per acre (per 13,000 linear ft. based on 40 in. rows). Apply over the seed and to the covering soil, using suitable equipment. Do not allow the feeding or grazing of cotton foliage by livestock.

ROTATION RESTRICTION: Do not plant root crops in PCNB treated fields within 12 months of application unless PCNB is registered for use on those crops.

IMPORTANT NOTICE

Seller warrants that this product conforms to its chemical description and is reasonably fit for the purposes stated on the label when used in accordance with the directions and instructions specified on the label under normal conditions of use, **but neither this warranty nor any other warranty of merchantability or fitness for a particular purpose, express or implied, extends to the use of this product, contrary to label instructions, or under abnormal conditions, or under conditions not reasonably foreseeable to seller, and buyer assumes the risk of any such use.**

TERRACLOR Super X®

GRANULAR SOIL FUNGICIDE WITH DI-SYSTON® SYSTEMIC INSECTICIDE

RESTRICTED USE PESTICIDE

Acute Human Oral, Dermal, and Inhalation Toxicity.
For retail sale to and application only be certified applicators or personnel under their direct supervision and only for those uses covered by the certified applicator's certification.

COMPOSITION

ACTIVE INGREDIENT:	(% by weight)
Pentachloronitrobenzene (PCNB)	6.50%
5-Ethoxy-3-(trichloromethyl)-1,2,4-thiadiazole	1.63%
Disulfoton: 0,0-Diethyl S-[2-(ethylthio) ethyl] phosphorodithioate*	6.50%
INERT INGREDIENTS:	85.37%
TOTAL:	100.0%

*U.S. Patent No. 2,759,010.

EPA REG. NO. 400-408 6-2-87A

TERRACLOR Super X is a Registered Trademark of UNIROYAL CHEMICAL COMPANY, INC.
DI-SYSTON is a Registered Trademark of the Parent Company of Farbenfabriken Bayer GmbH, Leverkusen.

KEEP OUT OF REACH OF CHILDREN
PELIGRO
DANGER—POISON

 **DANGER–POISON** 

POISONOUS IF SWALLOWED, INHALED OR ABSORBED THROUGH THE SKIN. RAPIDLY ABSORBED THROUGH SKIN. REPEATED INHALATION OR SKIN CONTACT MAY, WITHOUT SYMPTOMS PROGRESSIVELY INCREASE SUSCEPTIBILITY TO DISULFOTON POISONING. ATROPINE AND 2-PAM ARE ANTIDOTES FOR DI-SYSTON. CALL A PHYSICIAN IMMEDIATELY IN ALL CASES OF SUSPECTED POISONING

PRECAUCION AL USUARIO: Si usted no lee ingles, no use este producto hasta que le etiqueta haya sido explicado ampliamente.

STATEMENT OF PRACTICAL TREATMENT

IF SWALLOWED: Call a physician or Poison Control Center immediately. Drink one or two glasses of water and induce vomiting by touching back of throat with finger. Repeat until vomit fluid is clear. Do not induce vomiting or give anything by mouth to an unconscious or convulsing person.

IF INHALED: Remove victim to fresh air. Apply artificial respiration if indicated.

IF ON SKIN: Remove contaminated clothing and wash affected areas with soap and water.

IF IN EYES: Flush with plenty of water. Call a physician immediately.

NOTE TO PHYSICIAN: Warning symptoms include weakness, headache, tightness in chest, blurred vision, non-reactive pinpoint pupils, salivation, sweating, nausea, vomiting, diarrhea and abdominal cramps.

ANTIDOTE: Administer atropine sulfate in large therapeutic doses. Repeat as necessary to the point of tolerance. 2-PAM is also antidotal and may be administered in conjunction with atropine. Compound inhibits cholinesterase resulting in stimulation of the central nervous system, the parasympathetic nervous system, and the somatic motor nerves. Do not give morphine. Watch for pulmonary edema, which may develop in serious cases of poisoning even after 12 hours. At first sign of pulmonary edema, the patient should be placed in an oxygen tent and treated symptomatically. Repeated exposure to cholinesterase inhibitors may, without warning, cause increasing susceptability to very small doses of any cholinesterase inhibitor. Allow no further exposure until cholinesterase regeneration has taken place as determined by blood tests.

PRECAUTIONARY STATEMENTS

HAZARDS TO HUMANS
DANGER

Poisonous if swallowed, inhaled, or absorbed through the skin. Rapidly absorbed through the skin. Do Not Breathe Dust. Do Not Get In Eyes, On Skin, Or Clothing. In case of contact, immediately remove contaminated clothing and wash skin thoroughly with soap and water. Wear freshly laundered long-sleeve work clothes daily. Wash thoroughly with soap and water before eating or smoking. Bathe at the end of the work day and change outer clothing.

While transferring material from package to equipment, and during mixing/loading/application operations, wear a hat, goggles, face shield or equivalent, gloves [natural rubber], long-sleeved shirt, long pants, socks, shoes, and a pesticide respirator jointly approved by the Mine Safety Health Administration and by the National Institute for Occupational Safety and Health under the provisions of 30 CFR Part 11. Rubber gloves should be washed with soap and water after each use. Do not wear the same gloves for other work. Destroy and replace gloves frequently. Keep all unprotected persons out of the operating area.

ENVIRONMENTAL HAZARDS

This pesticide is toxic to fish and wildlife. Do not apply directly to water or wetlands. Drift and run-off from treated areas may be hazardous to fish in neighboring areas. Do not contaminate water by cleaning of equipment or disposal of wastes. Cover or incorporate spills.

This pesticide is toxic to bees exposed to direct application. Applications should be timed to coincide with periods of minimum bee activity, usually between late evening and early morning.

STORAGE AND DISPOSAL

Do not contaminate water, food or feed by storage or disposal.

STORAGE: Store in a dry location.

PESTICIDE DISPOSAL: Pesticide wastes are acutely hazardous. Improper disposal of excess pesticide, spray mixture, or rinsate is a violation of Federal Law. If these wastes cannot be disposed of by use according to label instructions, contact your State Pesticide or Environmental Control Agency, or the Hazardous Waste representative at the nearest EPA Regional Office for guidance.

CONTAINER DISPOSAL: Completely empty bag into application equipment. Then dispose of empty bag in a sanitary landfill or by incineration, or if allowed by State and local authorities, by burning. If burned, stay out of smoke.

REENTRY STATEMENT

Do not apply this product in such a manner as to directly or through drift expose workers or other persons. The area being treated must be vacated by unprotected persons.

Do not allow reentry into treated fields within 24 hours of application unless appropriate protective clothing is worn.

Because certain states may require more restrictive reentry intervals for various crops treated with this product consult your State Department of Agriculture for further information.

Written or oral warnings must be given to workers who are expected to be in a treated area or in an area about to be treated with this product. When oral warnings are given, warnings shall be in a language customarily understood by workers. Oral warnings must be given if there is reason to believe that written warnings cannot be understood by workers. Written or oral warnings must include the following information:

"DANGER" "Area to be treated with TERRACLOR Super X with Di-Syston on (date of application). Do not enter without appropriate protective clothing until 24 hours after application. If accidently exposed, consult the Statement of Practical Treatment portion of the pesticide label for first aid recommendations."

DIRECTIONS FOR USE

It is a violation of Federal law to use this product in a manner inconsistent with its labeling.

COTTON

For control of Seedling Disease Complex-Rhizoctonia, Pythium and Fusarium and for early season control of thrips, aphids and mites.

Apply as a soil in-furrow application at planting time with a granular pesticide applicator properly calibrated to assure accurate placement and proper dosage.

Application Method	Lbs./drill row per acre	Lbs./1300 linear feet of row based on 40 inch spacing
DRILL PLANTING	15	1.5
HILL-DROP PLANTING	5	0.5

Soil temperature, moisture and tilth should be favorable for good germination and emergence. Systemic insecticides can possibly damage seed germination when placed on or very near the seed, however, the addition of a fungicide such as Terraclor Super X, considerably lessens this hazard. Do not allow the feeding or grazing of cotton foliage by livestock. Do not plant root crops in treated fields within 12 months of application.

IMPORTANT NOTICE

Seller warrants that this product conforms to its chemical description and is reasonably fit for the purposes stated on the label when used in accordance with the directions and instructions specified on the label under normal conditions of use, **but neither this warranty nor any other warranty of merchantability or fitness for a particular purpose, express or implied, extends to the use of this product, contrary to label instructions, or under abnormal conditions, or under conditions not reasonably foreseeable to seller, and buyer assumes the risk of any such use.**

TERRA-COAT® LT-2N

SEED TREATMENT FUNGICIDE WITH DYE

Gustafson

ACTIVE INGREDIENTS: (% by weight)
Pentachloronitrobenzene (PCNB) 23.7%
INERT INGREDIENTS: 76.3%
TOTAL 100.0%

This product contains the toxic inert ingredient Rhodamine B.

Contains 2 lbs. of Pentachloronitrobenzene per U.S. Gallon at 68°F (20°C)

EPA Reg. No. 7501-55

KEEP OUT OF REACH OF CHILDREN CAUTION

PRECAUTIONARY STATEMENTS

HAZARDS TO HUMANS AND DOMESTIC ANIMALS

CAUTION

Harmful if swallowed, inhaled or absorbed through skin. Avoid contact with skin, eyes or clothing. Wash thoroughly with soap and water after handling. During mixing/loading procedures, wear respirator, gloves, long-sleeved shirt, long pants, socks and shoes. Remove contaminated clothing and wash before use. Do not inhale spray mist. Do not contaminate feed or foodstuffs.

STATEMENT OF PRACTICAL TREATMENT

If in eyes: Flush with plenty of water. Get medical attention.

If on skin: Wash with plenty of soap and water. Get medical attention.

If swallowed: Drink promptly a large quantity of milk, egg whites, gelatin solution or, if these are not available, drink large quantities of water. Avoid alcohol. Get medical attention. Do not induce vomiting.

ENVIRONMENTAL HAZARDS

This product is toxic to fish and aquatic organisms. Cover or incorporate treated seeds. Do not contaminate water or wetlands by cleaning of equipment or disposal of wastes.

PHYSICAL OR CHEMICAL HAZARDS

Do not use, pour, spill or store near heat or open flame.

STORAGE AND DISPOSAL

STORAGE: Store in a cool place. Do not store in direct sunlight. Protect from temperatures below 20°F.

PROHIBITIONS: Do not contaminate water, food or feed by storage or disposal. Open dumping is prohibited. Do not re-use empty container.

PESTICIDE DISPOSAL: Pesticide wastes are toxic. Improper disposal of excess pesticide, spray mixture, or rinsate is a violation of Federal Law. If these wastes cannot be disposed of by use according to label instructions, contact your State Pesticide or Environmental Control Agency, or the Hazardous Waste representative at the nearest EPA Regional Office for guidance.

CONTAINER DISPOSAL: Triple rinse (or equivalent) and offer for recycling or reconditioning, or puncture and dispose of in a sanitary landfill or by other procedures approved by State and local authorities.

DIRECTIONS FOR USE

It is a violation of Federal Law to use this product in a manner inconsistent with its labeling.

SEED TREATMENTS—Seed should be sound and well cured prior to treatment otherwise damage may occur. Apply GUSTAFSON TERRA-COAT LT-2N directly to seed at dosage rates indicated below, using suitable treating equipment. Note: It may be necessary to add up to equal parts of water to assure optimum seed coat coverage.

Planting Seeds	Application Rates in Fluid Ounces*	Seedling Diseases Complex
BARLEY	2-4 oz./bushel	Covered Smut
OATS		Oat Smut
WHEAT	2 oz./bushel	Common smut or Bunt and Damping-off caused by Rhizoctonia
COTTON	Acid Delinted 12 oz./100 lbs. Mechanically delinted 16 oz./100 lbs.	Seed rot and early stages of Preemergence damping-off caused by Rhizoctonia.

If the history of disease has been severe, then this treatment on cottonseeds may give only temporary protection and a subsequent soil application may be required to control later stages of post-emergence damping-off.

* 29.6 cc is equivalent to one fluid ounce.

IMPORTANT: Federal Law requires that bags containing treated seed shall be labeled with the following information.

This seed has been treated with Pentachloronitrobenzene

CAUTION

DO NOT USE FOR FEED, FOOD OR OIL PURPOSES

LIMITED WARRANTY STATEMENT

Treatment of highly mechanically damaged seed, or seed of known low vigor and poor quality, except for the purposes of curative control of existing disease or insect pests, may result in reduced germination and/or reduction of seed and seedling vigor. Treat and conduct germination tests on a small portion of seed before committing the total seed lot to a selected chemical treatment. Due to seed quality conditions beyond the control of Gustafson, Inc., no claims are made to guarantee germination of carryover seed.

Gustafson warrants that this product conforms to the chemical description on the label and is reasonably fit for the purposes stated on the label only when used in accordance with directions and instructions specified on the label. Gustafson neither makes, nor authorizes any agent or representative to make any other warranty of FITNESS or MERCHANTABILITY, guarantee or representation, expressed or implied, concerning this material. In the event of any breach of this limited warranty, Gustafson shall not be liable for consequential damages.

Gustafson (logo) is a Reg. TM of Gustafson, Inc.

Terra-Coat is a Reg. TM of Uniroyal Chemical Company, Inc.

L45-9/87

TERRAZOLE®

25% EMULSIFIABLE ORNAMENTAL FUNGICIDE

COMPOSITION

ACTIVE INGREDIENT: (% by weight)
5-Ethoxy-3-(trichloromethyl)-1,2,4-thiadiazole* 25.0%
INERT INGREDIENTS:** 75.0%
TOTAL 100.0%

* Contains 2 lbs. 5-Ethoxy-3-(trichloromethyl)-1,2,4-thiadiazole per U.S. Gallon at 68°F.

** Contains Petroleum Distillates

EPA REG. NO. 400-417 2-22-88A

Terrazole® is a Registered Trademark of Uniroyal Chemical Company, Inc.

KEEP OUT OF REACH OF CHILDREN DANGER!

STATEMENT OF PRACTICAL TREATMENT

IF IN EYES: Hold eyelids open and flush with a steady stream of water for 15 minutes. Get medical attention.

IF SWALLOWED: Do not induce vomiting. Drink promptly a large quantity of milk, egg whites, gelatin solution or if these are not available, drink large quantities of water. Avoid alcohol. Obtain prompt medical attention.

IF ON SKIN: Wash with plenty of soap and water. Get medical attention if irritation persists.

IF INHALED: Remove victim to fresh air. If not breathing, give artificial respiration, preferably mouth-to-mouth. Get medical attention.

NOTE TO ATTENDING PHYSICIAN: Probable mucosal damage and the presence of petroleum distillates in the formulation may contraindicate the use of gastric lavage.

PRECAUTIONARY STATEMENTS

HAZARDS TO HUMANS AND DOMESTIC ANIMALS

DANGER

Corrosive, causes irreversible eye damage. Harmful if swallowed. Do not take internally. Do not inhale vapors or spray mist. Do not get in eyes or on clothing. Wear goggles or face shield. May cause skin irritation. Wash thoroughly with soap and water after handling. Remove contaminated clothing and wash before reuse.

ENVIRONMENTAL HAZARDS

Do not apply directly to water. Do not contaminate water when disposing of equipment washwaters.

PHYSICAL OR CHEMICAL HAZARDS

Flammable. Do not use or store near heat source or open flame. Do not use in the vicinity of pilot lights.

DIRECTIONS FOR USE

It is a violation of Federal law to use this product in a manner inconsistent with its labeling.

Do not apply this product through any type of irrigation system.

STORAGE AND DISPOSAL

Do not contaminate water, food or feed by storage or disposal.

PESTICIDE DISPOSAL

Pesticide wastes are acutely hazardous. Improper disposal of excess pesticide, spray mixture, or rinsate is a violation of Federal Law. If these wastes cannot be disposed of by use according to label instructions, contact your State Pesticide or Environmental Control Agency, or the Hazardous Waste representative at the nearest EPA Regional Office for guidance.

CONTAINER DISPOSAL

Triple rinse (or equivalent). Then offer for recycling or reconditioning, or puncture and dispose of in a sanitary landfill, or by other procedures approved by State and local authorities.

DIRECTIONS FOR USE

Do not apply this product through any type of irrigation system.

Terrazole is useful as a soil fungicide for control of damping-off, root and stem diseases caused by Pythium and Phytophthora.

Use only at recommended rates as overdosage may be harmful to sensitive plants. Do not mix with other chemicals, other than Terraclor® unless prior use has proven compatibility.

RECOMMENDATIONS

For bedding and bench plantings use with water, as a soil drench at time of seeding and transplanting in commercial nurseries and greenhouses. The drench solution should be continually agitated to assure uniform application of the fungicidal material. To improve penetration of the fungicide, the drenching should be followed with an equal quantity of water applied either by sprinkling or by irrigation.

BEDDING PLANTS Ajuga, Alyssum, Aster, Calendula, Celosia, Cineraria, Chrysanthemum, Coleus, Dahlia, Dianthus, Dusty Miller, Larkspur, Marigold, Pansy, Petunia, Phlox, Salvia, Snapdragon, Verbena, Vinca, Zinnia	**BEDDING PLANTS:** (Soil 2" to 3" Deep) - Add 3 to 4 ozs. to 100 gallons of water and apply to 400 sq. ft. of bed area. Repeat application may be made at one to two month intervals, if necessary.
FOLIAGE PLANTS Aglaonema, Anthurium, Areca Palm, Brassala, Cactus, Caladium, Calathea, Chamaedorea, Chinese Evergreen, Crassula, Dieffenbachia, Gynura, Hedera, Hoya, Maranta, Nephrolepis, Nephthytis, Peperomia, Philodendron, Pilea, Pothos, Sansevieria, Stromanthe, Zygocactus **HERBACEOUS PLANTS** African Violet, Gerbera, Chrysanthemum, Kalanchoe, Easter Lily, Poinsettia, Geranium, Sinningia **WOODY PLANTS** Azalea, Boxwood, Euonymus, Ficus, Juniper, Maple, Pieris, Podocarpus, Red Pine, Rhododendron	**TROPICAL FOLIAGE, HERBACEOUS and WOODY PLANTS:** (Peat and high organic growth media 6" deep) **Bed and Bench Plantings:** Add 4 to 8 ozs. in 100 gallons of water and apply to 400 sq. ft of bed or bench area. Repeat application at two to three month intervals to maintain disease control. **Pot Plantings:** Use 1/2 pint solution per 6" pot or in sufficient amount to saturate the soil.

IMPORTANT NOTICE

Seller warrants that this product conforms to its chemical description and is reasonably fit for the purposes stated on the label when used in accordance with the directions and instructions specified on the label under normal conditions of use, **but neither this warranty nor any other warranty of merchantability or fitness for a particular purpose, express or implied, extends to the use of this product, contrary to label instructions, or under abnormal conditions, or under conditions not reasonably foreseeable to seller, and buyer assumes the risk of any such use.**

TERRAZOLE®

35% WETTABLE POWDER
TURF & ORNAMENTAL FUNGICIDE

COMPOSITION

ACTIVE INGREDIENT:	(% by weight)
5-Ethoxy-3-(trichloromethyl)-1,2,4-thiadiazole*	35%
INERT INGREDIENTS:	65%
TOTAL:	100%

EPA REG. NO. 400-416 2-24-88A

TERRAZOLE is a registered trademark of Uniroyal Chemical Company, Inc.

KEEP OUT OF REACH OF CHILDREN
WARNING

STATEMENT OF PRACTICAL TREATMENT

IF SWALLOWED: Call a physician or Poison Control Center. Drink 1 or 2 glasses of water and induce vomiting by touching back of throat with finger, or if available, by administering Syrup of Ipecac. Do not induce vomiting or give anything by mouth to an unconscious person.

IF IN EYES: Flush with plenty of water. Get medical attention if irritation persists.

IF ON SKIN: Wash with plenty of soap and water. Get medical attention if irritation persists.

IF INHALED: Remove to fresh air.

PRECAUTIONARY STATEMENTS

HAZARDS TO HUMANS AND DOMESTIC ANIMALS

WARNING

Harmful if swallowed. Do not take internally. Do not inhale spray mist. Causes eye irritation. Do not get on skin or in eyes. May cause skin irritation. Wash thoroughly all exposed skin. Repeated contact may be harmful. Wash thoroughly all contaminated clothing before reuse.

ENVIRONMENTAL HAZARDS

Do not apply directly to water or wetlands. Do not contaminate water by cleaning of equipment washwaters.

DIRECTIONS FOR USE

It is a violation of Federal law to use this product in a manner inconsistent with its labeling.

Do not apply this product through any type of irrigation system.

STORAGE AND DISPOSAL

Do not contaminate water, food or feed by storage or disposal.

PESTICIDE DISPOSAL:

Wastes resulting from the use of this product may be disposed of on site or at an approved wastes disposal facility.

CONTAINER DISPOSAL:

Paper Bags: Completely empty bag into application equipment. Then dispose of empty bag in a sanitary landfill, or by incineration, or if allowed by State and local authorities, by burning. If burned, stay out of smoke.

Plastic Containers: Triple rinse (or equivalent). Then offer for recycling or reconditioning, or puncture and dispose of in a sanitary landfill, or incineration, or, if allowed by State and local authorities, by burning. If burned, stay out of smoke.

DIRECTIONS FOR USE

It is a violation of Federal law to use this product in a manner inconsistent with its labeling.

Do not apply this product through any type of irrigation system.

Terrazole 35% WP is useful as a soil fungicide for control of damping-off, root and stem diseases caused by Pythium and Phytophthora. Use only at recommended rates as overdosage may be harmful to sensitive plants. Do not mix with chemicals other than Terraclor, unless prior use has proven compatibility.

LAWNS AND TURF

Terrazole 35% WP controls **Pythium** blight and **Pythium** damping-off in turf. Apply as a spray to established turf to control **Pythium** blight, or apply as a spray at time of seeding to control **Pythium** damping-off.

Within the rate range given for turf, use the lower rate for the shortest interval listed and the higher rate for the longest interval. Under severe disease conditions, use the highest rate and shortest interval specified.

ESTABLISHED TURF:

Pythium Blight: Mix 4 to 8 ozs. in 5 gallons of water and apply to 1,000 sq. ft. of established ornamental turf. Begin applications at first sign of disease and retreat in 5 to 10 days depending upon climatic conditions to maintain disease control.

NEWLY SEEDED AREAS:

Pythium Damping-off and **Pythium** Blight: On newly seeded areas mix 6 to 8 ozs. in 5 to 10 gallons of water and apply to 1,000 sq. ft. of area immediately after seeding. Retreat at 5 to 7 day intervals to maintain a preventive disease control program.

Do not allow animals to graze on treated areas. Do not feed clippings from treated areas to animals.

ORNAMENTALS

For Bedding and Bench Plantings: Use Terrazole 35% WP mixed with water, as a soil drench at time of seeding and transplanting in nurseries and greenhouses. The drench solution should be continually agitated to assure uniform application of the fungicide material. To improve penetration of the fungicide, the drenching should be followed with an equal quantity of water, applied either by sprinkling or irrigation.

BEDDING PLANTS Ageratum, Alyssum, Aster, Balsam, Begonia, Calendula, Carnation, Celosia, Coleus, Dahlia, Dianthus, Foxglove, Impatiens, Larkspur, Marigold, Pansy, Petunia, Portulaca, Salvia, Snapdragon, Stokesia, Verbena, Vinca, Zinnia	**BEDDING PLANTS:** (Soil 2" to 3" deep) - Mix 4-6 ozs. TERRAZOLE 35% WP with 100 gals. of water and apply to 800 sq. ft. of bed area. (1 gal. per 8 sq. ft.) and use at time of seeding and transplanting. Irrigate immediately with additional water equal to at least half the volume of the fungicidal drench for improved soil penetration of the fungicide. Re-treat at 4-8 week intervals if additional applications are necessary.
FOLIAGE PLANTS Aglaonema, Anthurium, Brassaia, Caladium, Chinese Evergreen, Dieffenbachia, Echeveria, Nephthytis, Peperomia, Pilea, Pothos, Philodendron, Zygocactus	**BENCH PLANTINGS OF TROPICAL FOLIAGE PLANTS (peat and other high organic growth media 6" deep)** Mix 8-12 ozs.of TERRAZOLE 35% WP with 100 gals of water and apply to 400 sq. ft of bench area. Irrigate immediately with additional water equal to at least half the volume of the fungicidal drench for improved soil penetration of the fungicide. Retreat at 2-3 month intervals to maintain disease control. **BED GROWN PLANTS** Mix 3 to 10 ozs. of TERRAZOLE 35% WP with 100 gals. of water per 400 sq. ft. or apply in sufficient amount to saturate the soil mixture. Re-treat at 4 to 12 week intervals if necessary.
CONTAINER AND BED GROWN HERBACEOUS African Violet, Begonia, Calceolaria, Carnation, Chrysanthemum, Daphne, Easter Lily, Geranium, Gerbera, Gypsophilla, Heather, Orchids, Sinningia, Statice	**CONTAINER AND BEDGROWN PLANTS** Mix 3 to 10 ozs. of TERRAZOLE 35% WP with 100 gals of water per 400 sq. ft. or apply in sufficient amount to saturate the soil mixture, such as 1/2 pint per 6" pot. Irrigate immediately with additional water equal to at least half the volume of the fungicide drench for improved soil penetration of the fungicide. Retreat at 4-12 week intervals if necessary.
WOODY Arborvitae, Arctostaphylus, Camellia, Carissa, Ceonothus, Fir, Hemlock, Juniper, Laurel, Leucothoe, Ligustrum, Magnolia, Norfolk Island Pine, Poinsettia, Rhamnus, Rhododendron, Southern Pine, Western Pine, Yew	

DRY SOIL MIX for Plants Listed above. Mix 1½ to 3 ozs. TERRAZOLE 35% WP with each cu. yd. of soil mixture. Ingredients for the soil mix should be free of large lumps and dry enough to prevent agglomeration during mixing. (Improved uniformity of blending will be assured if the TERRAZOLE 35% WP is pre-mixed with five times its volume of dry soil.) Re-treat if needed as indicated under appropriate soil drench recommendations.

IMPORTANT NOTICE

Seller warrants that this product conforms to its chemical description and is reasonably fit for the purposes stated on the label when used in accordance with the directions and instructions specified on the label under normal conditions of use, **but neither this warranty nor any other warranty of merchantability or fitness for a particular purpose, express or implied, extends to the use of this product, contrary to label instructions, or under abnormal conditions, or under conditions not reasonably foreseeable to seller, and buyer assumes the risk of any such use.**

TURFCIDE®

10% GRANULAR
TURF & ORNAMENTAL FUNGICIDE

COMPOSITION

ACTIVE INGREDIENT:	(% by weight)
Pentachloronitrobenzene (PCNB)	10.0%
INERT INGREDIENTS:	90.0%
TOTAL:	100.0%

EPA REG. NO. 400-407 002 2-29-88

TURFCIDE is a Registered Trademark of Uniroyal Chemical Company, Inc.

KEEP OUT OF REACH OF CHILDREN
CAUTION

STATEMENT OF PRACTICAL TREATMENT

IF IN EYES: Flush eyes with plenty of water. Get medical attention.

IF ON SKIN: Wash with plenty of water. Get medical attention.

IF SWALLOWED: Drink promptly a large quantity of milk, egg whites, gelatin solution or, if these are not available, drink large quantities of water. Avoid alcohol. Get medical attention.

PRECAUTIONARY STATEMENTS

HAZARDS TO HUMANS AND DOMESTIC ANIMALS

CAUTION

Harmful if swallowed, inhaled or absorbed through the skin. Avoid breathing dust. This material may cause skin irritation. Avoid contact with eyes, skin or clothing.

Homeowners: Avoid contact with skin by wearing the following protective clothing: gloves, long-sleeve shirt, long pants, socks and shoes. Commercial Applicators: The following protective clothing are required during mixing/loading operations: gloves, long-sleeve shirt, long pants, socks and shoes. Wash hands thoroughly after using.

Wash contaminated clothing before reuse. Do not handle or apply product with bare hands.

ENVIRONMENTAL HAZARDS

This pesticide is toxic to fish and aquatic organisms. Cover, collect, or incorporate spilled granules. Do not contaminate water when disposing of equipment washwaters. Do not apply directly adjacent to potable water supplies.

STORAGE AND DISPOSAL

Do not contaminate water, food or feed by storage or disposal.

STORAGE

Store in a dry location.

PESTICIDE DISPOSAL

Pesticide wastes are toxic. Improper disposal of excess pesticide, spray mixture or rinsate is a violation of Federal Law. If these wastes cannot be disposed of by use according to label instructions, contact your State Pesticide or Environmental Control Agency, or the Hazardous Waste representative at the nearest EPA Regional Office for guidance.

CONTAINER DISPOSAL

Completely empty bag into application equipment. Then dispose of empty bag in a sanitary landfill or by incineration, or if allowed by State and local authorities, by burning. If burned, stay out of smoke.
HOUSEHOLD: Do not reuse empty container. Securely wrap original container in several layers of newspaper and discard in trash.

DIRECTIONS FOR USE

It is a violation of Federal Law to use this product in a manner inconsistent with its labeling.

LAWNS AND TURF

Brown patch-Rhizoctonia solani: For Southern or warm season lawns such as St. Augustine, Bermuda and similar grasses: Make one application in the fall or the spring at first indication of infection. Apply 7 1/2 lbs. per 1,000 sq. ft. Following application, the treated areas should be lightly watered to insure movement through the grass to soil level. If the treated area is subjected to unusually heavy rainfall, or flooded, or if the disease is severe or reappears, the area should be retreated 3 to 4 weeks later. For Dichondra: Treat as above, using 7 1/2 lbs. per 1,000 sq. ft. For Northern or cool season lawn such as Rye Grass, Fescue, Bluegrass, Bent or similar grasses: Apply 2 to 2 1/2 lbs. per 1,000 sq. ft. at first sign of disease. Repeat every 7 to 10 days during warm, damp weather. Water treated area lightly to wash chemical down to soil. Under certain growing conditions, a temporary discoloration of the grass occasionally may occur. This causes no harm and will disappear in a short time.

Helminthosporium leaf spot and dollar spot: Apply at rate of 5 to 7 1/2 lbs. per 1,000 sq. ft. Preventive applications should be made in spring or fall of each year. For control of existing infections, apply at the higher rate as soon as symptoms are first noted. Following applications, the treated area should be lightly watered to insure movement through the grass to soil level. If the treated area is subjected to unusually heavy rainfall, or flooded, or if the disease is severe or reappears, the area should be retreated 3 to 4 weeks later.

Snow mold (Typhula incarnata, Fusarium nivale): Apply at rate of 5 to 7 1/2 lbs. per 1,000 sq. ft. using a granular lawn spreader or comparable equipment. Application should be made prior to first snowfall.

Do not graze treated areas or feed clippings to livestock.

Spreader Setting Number Per 1000 Sq. Ft.

Spreader Make	2 Lbs.	2 1/2 Lbs.	5 Lbs.	7 1/2 Lbs.
CYCLONE	3 1/2	3 3/4	6	7 1/2
SCOTT	5 1/8	5 5/8	6 1/2	7 1/2
SEARS	6 1/4	6 1/2	7 1/2	9

These settings apply to 20/40 mesh granular carriers. The spreader settings listed above are approximate. Accurately weigh a few pounds into your spreader and apply to measured area. Then adjust spreader setting, if necessary, in order to assure correct rate of application.

ORNAMENTALS

AFRICAN VIOLETS, BEGONIAS, CARNATIONS, CHRYSANTHEMUMS, POINSETTIAS, SNAPDRAGONS, and similar bedding plants (Stem Rot-Rhizoctonia): Apply 15 lbs. per 1,000 sq. ft. by spreading evenly on the soil surface and thoroughly mixing into the top 2 inches of soil. Apply to a well prepared seed bed.

BULBOUS IRIS, HYACINTH, NARCISSUS, AND TULIP (Crown Rot and Black Rot): Apply broadcast 1,000 to 2,000 lbs. per acre (2 to 4 lbs. per 100 sq. ft.) and mix thoroughly in upper 6 to 7 inches of soil by double discing or rotovating.
NOTE: Use higher rate for heavier soils and for Black Rot control.

CALENDULA, LARKSPUR, SNAPDRAGON, AND SWEET PEA (Stem Rot-Sclerotinia): Apply 1,000 lbs. per acre broadcast one week prior to planting. For smaller areas apply 2 lbs. per 100 sq. ft. Spread on the soil surface and uniformly mix in the soil to a depth of 4 inches. Terraclor was developed for this use in Charleston, South Carolina area.

DUTCH IRIS (Bulb and stem rot caused by Sclerotium rolfsii): Broadcast: Apply 1,000 to 1,500 lbs. per acre (2 to 4 lbs. per 100 sq. ft.) prior to planting and mix with the top 2 or 3 inches of soil by discing or roto-tilling.
ROW: Apply 60 lbs. per 14,500 linear feet of row, in a 12 inch band, centered on the row, at planting time. Follow with three additional applications at this rate at 15 day intervals.

EASTER LILIES (Rhizoctonia root rot): Broadcast: Apply 1,000 to 2,000 lbs. per acre (2 to 4 lbs. per 100 sq. ft.) to soil surface prior to planting and mix thoroughly in upper 6 to 7 inches by double discing or rotovating. Note: Use the lower rate of application for light soils and higher rate of application for heavier soils.

GLADIOLUS (Neck or Bulb Dry Rot-Stromatinia or Sclerotinia gladioli): Broadcast: At planting time apply 1,200 lbs. per acre (1 1/2 lbs. per 100 sq. ft.) to soil surface and mix thoroughly in upper 4 to 6 inches by double discing or rotovating. (Bulbs may be planted immediately.)

Do not plant vegetable root crops in PCNB treated fields within 12 months of broadcast and banding applications unless PCNB is registered for use on those crops.

IMPORTANT NOTICE

Seller warrants that this product conforms to its chemical description and is reasonably fit for the purposes stated on the label when used in accordance with the directions and instructions specified on the label under normal conditions of use, **but neither this warranty nor any other warranty of merchantability or fitness for a particular purpose, express or implied, extends to the use of this product, contrary to label instructions, or under abnormal conditions, or under conditions not reasonably foreseeable to seller, and buyer assumes the risk of any such use.**

TURFCIDE®

2 LB. EMULSIFIABLE
TURF & ORNAMENTAL FUNGICIDE

COMPOSITION

ACTIVE INGREDIENT:	(% by weight)
Pentachloronitrobenzene (PCNB)	24%
INERT INGREDIENTS:	76%
TOTAL	100.0%

Contains 2 lbs. of Pentachloronitrobenzene per U.S. Gallon at 68°F.

EPA REG. NO. 400-404 12-3-87

Turfcide is a Registered Trademark of UNIROYAL CHEMICAL COMPANY, INC.

KEEP OUT OF REACH OF CHILDREN
CAUTION

STATEMENT OF PRACTICAL TREATMENT

IF IN EYES: Flush eyes with plenty of water. Get medical attention.
IF ON SKIN: Wash with plenty of water. Get medical attention.
IF SWALLOWED: Drink promptly a large quantity of milk, egg whites, gelatin solution or, if these are not available, drink large quantities of water. Avoid alcohol. Get medical attention.

PRECAUTIONARY STATEMENTS

HAZARDS TO HUMANS AND DOMESTIC ANIMALS

CAUTION

This material may cause skin irritation. Avoid contact with eyes or clothing. Wash thoroughly with soap and water after handling. Use with adequate ventilation. Do not contaminate feed or foodstuffs.
Homeowners: Avoid contact with skin by wearing the following protective clothing: gloves, long-sleeve shirt, long pants, socks and shoes. Commercial Applicators: The following protective clothing are required during mixing/loading operations: respirator, gloves, long-sleeve shirt, long pants, socks and shoes. Wash hands thoroughly after using.

ENVIRONMENTAL HAZARDS

This pesticide is toxic to fish and aquatic organisms. Drift and runoff from treated areas may be hazardous to fish and aquatic organisms in adjacent aquatic sites. Do not apply directly to water or wetlands (swamps, bogs, marshes and potholes). Do not contaminate water or wetlands by cleaning of equipment or disposal of wastes.

PHYSICAL OR CHEMICAL HAZARDS

Flammable. Do not use or store near heat sources or open flame. Do not use in the vicinity of pilot lights.

STORAGE AND DISPOSAL

Do not contaminate water, food or feed by storage or disposal.

STORAGE: Protect from temperatures below 20°F.

PESTICIDE DISPOSAL:

Pesticide wastes are toxic. Improper disposal of excess pesticide, spray mixture, or rinsate is a violation of Federal Law. If these wastes cannot be disposed of by use according to label instructions, contact your State Pesticide or Environmental Control Agency, or the Hazardous Waste representative at the nearest EPA Regional Office for guidance.

CONTAINER DISPOSAL:

Triple rinse (or equivalent). Then offer for recycling or reconditioning, or puncture and dispose of in a sanitary landfill, or by other procedures approved by State and local authorities.

HOUSEHOLD:

Do not reuse empty container. Securely wrap original container in several layers of newspaper and discard in trash.

DIRECTIONS FOR USE

It is a violation of Federal law to use this product in a manner inconsistent with its labeling.

Do not apply this product through any type of irrigation system.

LAWNS AND TURF

LAWNS (Brown Patch-Rhizoctonia solani): For Southern or warm season lawns such as St. Augustine, Bermuda and similar grasses: Make one application in the fall or spring at first indication of infection. Mix 1 qt. in sufficient water (10-15 gals.) to cover 1,000 sq. ft. Apply by means of a sprinkler can, sprayer or any other convenient method of obtaining even coverage. Following application, the treated area should be lightly watered to insure movement through the grass to the soil level. If the treated area is subjected to unusually heavy rainfall, or flooded, or if the disease is severe or reappears, the area should be retreated 3-4 weeks later. Under certain growing conditions, a temporary discoloration of the grass occasionally may occur. This causes no harm and will disappear in a short time.

Helminthosporium leaf spot and dollar spot*: Apply at rate of 1 to 1 1/2 qts. in sufficient water (10-15 gals.) per 1,000 sq. ft. Apply by means of a sprinkler can, sprayer or any other convenient method of obtaining even coverage. Preventative applications should be made in spring or fall of each year. For control of existing infections, apply at the higher rate as soon as symptoms are first noted. Following applications, the treated area should be lightly watered to insure movement through the grass to soil level. If the treated area is subjected to unusually heavy rainfall, or flooded, or if the disease is severe or reappears, the area should be retreated 3 to 4 weeks later.

Snow mold* (Typhula incarnata, Fusarium nivale): Apply at rate of 1 to 1 1/2 qts. in sufficient water (10-15 gals.) per 1,000 sq. ft. Apply by means of a sprinkler can, sprayer or any other convenient method of obtaining even coverage. Application should be made prior to first snowfall.
Do not graze treated areas or feed clippings to livestock.

***DO NOT USE IN CALIFORNIA.**

ORNAMENTALS

AZALEA AND CAMELLIA (Azalea Petal blight and Camellia flower blight): Apply as a drench under plants at a rate of 1 1/2 qts. in 30 gallons of water per 150 sq. ft. Begin application prior to opening of buds and repeat every 3 to 4 weeks through bloom period.

Do not plant root crops in PCNB treated fields within 12 months of broadcast and banding applications unless PCNB is registered for use on these crops.

IMPORTANT NOTICE

Seller warrants that this product conforms to its chemical description and is reasonably fit for the purposes stated on the label when used in accordance with the directions and instructions specified on the label under normal conditions of use **but neither this warranty nor any other warranty of merchantability or fitness for a particular purpose, express or implied, extends to the use of this product, contrary to label instructions, or under abnormal conditions, or under conditions not reasonably foreseeable to seller, and buyer assumes the risk of any such use.**

VITAVAX® -200

FLOWABLE FUNGICIDE (VITAVAX WITH THIRAM)

COMPOSITION

ACTIVE INGREDIENTS:	(% by weight)
Carboxin (5,6-dihydro-2-methyl-N-phenyl-1,4-oxathiin-3-carboxamide)*	17.0%
Thiram (tetramethylthiuram disulfide)	17.0%
INERT INGREDIENTS:	66.0%
TOTAL:	100.0%

This product contains the toxic inert ingredient Rhodamine B dye.

Contains 1.67 pounds carboxin and 1.67 pounds thiram per U.S. gallon.

EPA REG. NO. 400-112 10-10-88

VITAVAX is a Registered Trademark of UNIROYAL CHEMICAL COMPANY, INC.

KEEP OUT OF REACH OF CHILDREN
CAUTION

STATEMENT OF PRACTICAL TREATMENT

IF IN EYES: Immediately flush eyes with plenty of water. See a physician.

IF SWALLOWED: Call a physician or Poison Control Center immediately. Drink 1 or 2 glasses of water and induce vomiting by touching back of throat with finger. Do not induce vomiting or give anything by mouth to an unconscious person.

IF INHALED: See a physician if an abnormal reaction occurs.

IN ON SKIN: Wash thoroughly with soap and water.

SYMPTOMS OF POISONING: No specific symptoms. Acute poisoning is accompanied by general depression and illness. Consumption of alcoholic beverages with exposure to Thiram may cause flushing, breathing difficulty, nausea, vomiting and low blood pressure. In case of poisoning, call a physician immediately. Have patient lie down and keep quiet.

NOTE TO PHYSICIAN: No specific antidote is available. Treat symptomatically. In the presence of consumed alcohol, thiram responds similarly to Disulfiram.

PRECAUTIONARY STATEMENTS

HAZARDS TO HUMANS
CAUTION

Causes eye and skin irritation. Harmful if swallowed, inhaled or absorbed through skin. Do not get in eyes, on skin, or on clothing. Wear goggles or face shield and rubber gloves when handling. Avoid inhaling of spray mist. Do not use or store near food or feed.

During commercial exposure in seed treatment and loading operations, wear goggles, clean rubber gloves, protective clothing, and a pesticide respirator jointly approved by the Mine Safety and Health Administration (formerly the U.S. Bureau of Mines) and by the National Institute of Occupational Safety and Health under provisions of 30 CFR, Part 11. Wash hands, arms and face thoroughly with soap and water after handling and before eating or smoking. Routinely shower or bathe after work; wash all clothing with soap and water before reusing. Consumption of alcoholic beverages increases the toxic effects of thiram.

ENVIRONMENTAL HAZARDS

This pesticide product is toxic to fish. Do not contaminate water when disposing of equipment washwaters. Drift or runoff from treated areas may be hazardous to aquatic organisms in neighboring areas. Do not apply directly to water.

PHYSICAL OR CHEMICAL HAZARDS

Do not use or store near heat or open flame.

STORAGE AND DISPOSAL

Do not contaminate water, food or feed by storage or disposal.

STORAGE: Store in a cool, dry location. Do not store in direct sunlight. Protect from extreme temperatures.

PESTICIDE DISPOSAL: Pesticide wastes are toxic. Improper disposal of excess pesticide, spray mixture or rinsate is a violation of Federal Law. If these wastes cannot be disposed of by use according to label instructions, contact your State Pesticide or Environmental Control Agency, or the Hazardous Waste representative at the nearest EPA Regional Office for guidance.

CONTAINER DISPOSAL: Triple rinse (or equivalent). Then offer for recycling or reconditioning, or puncture and dispose of in a sanitary landfill, or by other procedures approved by State and local authorities.

DIRECTIONS FOR USE

It is a violation of Federal law to use this product in a manner inconsistent with its labeling. Do not apply this product through any type of irrigation system.

SEED TREATMENT: VITAVAX-200 Flowable Fungicide is a ready-to-seed treatment. Application should be made using standard mechanical, slurry or mist-type seed treatment equipment. Uniform application to seed is necessary to insure seed safety and best disease control. Seed should be sound and well cured prior to treatment. With high seed temperatures and when treating at low relative humidity, dilution with clean water will assist in achieving adequate seed coverage. Do not exceed 4 parts water per 1 part product. Dilution of product requires mechanical agitation.

COMPATIBILITY

VITAVAX-200 Flowable Fungicide should not be tank mixed with any pesticide unless adequate physical compatibility and phytotoxicity tests have been completed and proved to be satisfactory.

GENERAL: VITAVAX-200 Flowable Fungicide combines the systemic action of carboxin with the surface action of thiram to reduce seed rot/seedling blight diseases. It is particularly effective against **Rhizoctonia solani** and shows good activity various **Helminthosporium, Fusarium,** and **Pythium** species. This often results in increased and more uniform stands of seedlings with higher yield potential. Use the higher rate under conditions of expected heavy disease pressure.

BARLEY, WHEAT AND OATS

In addition to the above, loose smut **(Ustilago nuda),** and covered smut **(Ustilago hordei),** are controlled in barley and loose smut **(Ustilago tritici),** and common bunt **(Tilletia caries)** are controlled on wheat by VITAVAX-200 Flowable Fungicide.

Use 3-4 fluid ounces of VITAVAX-200 Flowable Fungicide per 100 pounds of seed. VITAVAX-200 Flowable Fungicide should be applied as a slurry treatment by machine.

RICE

VITAVAX-200 Flowable Fungicide is particularly effective against **Rhizoctonia solani** and **helminthosporium oryzae.** Use 3-4 fluid ounces of VITAVAX-200 Flowable Fungicide per 100 pounds of seed applied as a slurry treatment by machine. The higher rate is recommended for control of **Helminthosporium oryzae.**

SOYBEANS

Use 2 fluid ounces of VITAVAX-200 Flowable Fungicide per bushel of seed or 4 fluid ounces per 100 pounds of seed. VITAVAX-200 Flowable Fungicide should be applied as a slurry treatment by machine.

COTTON

VITAVAX-200 Flowable Fungicide provides **systemic** control of seed and seedling diseases of cotton. Seed rot, damping off and sore-shin due to **Rhizoctonia solani** are reduced. VITAVAX-200 Flowable Fungicide can be used on both machine and acid delinted cottonseed at the rate of 6.4 to 8 fl. oz. per 100 pounds of seed. VITAVAX-200 Flowable Fungicide should be applied as a slurry treatment by machine.

USE RESTRICTIONS

Do not use treated seed for food, feed or oil purposes. Do not graze or feed livestock on treated areas for six weeks after planting. Care must be exercised in the handling of treated seed. Do not graze or feed livestock on vines grown from treated seed.

USE PRECAUTION

When using formulations that do not contain dye, to comply with 21 CFR Chapter 1, Section 2.25, all grain seed treated with an economic poison must be colored with an approved dye (40 CFR 180.1001) to distinguish and prevent subsequent and inadvertent use as a food for man or feed for animals.

SEED LABELING

1. "This seed is treated with carboxin and thiram [VITAVAX-200 Flowable Fungicide] at the manufacturer's recommended rate. Do not use treated seed for food, feed, or oil purposes." Add appropriate information for any other chemical treatments. Consult the manufacturer for additional tagging requirements.
2. "Planting Note: Seed treated with VITAVAX-200 Flowable Fungicide does not flow as freely as untreated seed. The drill should therefore be adjusted to compensate for this restricted seed flow to prevent lower than desired seed rates."
3. "Augers used for handling treated seed should not be used to move seed for feed, food or oil processing."
4. "Do not re-use bags from treated seed to handle food or feed products.

NOTICE - READ CAREFULLY
LIMITED WARRANTY STATEMENT

Treatment of highly mechanically damaged seed or seed of low vigor and poor quality, except for the purposes of curative control of existing disease or insect pests, may result in reduced germination and or reduction of seed and seedling vigor. Treat and conduct germination tests on a small portion of seed before commiting the total seed lot to a selected chemical treatment. Due to seed quality conditions beyond the control of Uniroyal, no claims are made to guaranteed germination of carry-over seed.

Uniroyal warrants that this product conforms to the chemical description on the label and is reasonably fit for the purposes stated on the label only when used in accordance with directions and instructions specified on the label. Uniroyal neither makes, nor authorizes any agent or representative to make any other warranty of FITNESS or MERCHANTABILITY, guarantee or representation, expressed or implied, concerning this material. In the event of any breach of this limited warranty, Uniroyal shall not be liable for consequential damages.

VITAVAX®

POUR ON
FLOWABLE FUNGICIDE
HOPPER BOX SEED TREATMENT

Gustafson

ACTIVE INGREDIENTS: (% by weight)	
Carboxin (5,6–dihydro–2–methyl–N–phenyl–1, 4–oxathiin–3–carboxamide)	5.7%
Thiram (tetramethylthiuram disulfide)	5.7%
INERT INGREDIENTS:	88.6%
TOTAL:	100.0%

EPA Reg. No. 400–156–7501

This product contains the toxic inert ingredient Rhodamine B dye.

KEEP OUT OF REACH OF CHILDREN
CAUTION

STATEMENT OF PRACTICAL TREATMENT

IF IN EYES: Immediately flush eyes with plenty of water. See a physician.

IF SWALLOWED: Call a physician or Poison Control Center immediately. Drink 1 or 2 glasses of water and induce vomiting by touching back of throat with finger or blunt object. Do not induce vomiting or give anything by mouth to an unconscious person.

IF INHALED: See a physician if an abnormal reaction occurs.

IF ON SKIN: Wash thoroughly with soap and water.

SEE ADDITIONAL PRECAUTIONARY STATEMENT.

PRECAUTIONARY STATEMENTS

HAZARDS TO HUMANS

CAUTION

Causes eye and skin irritation. Harmful if swallowed, inhaled or absorbed through skin. Do not get in eyes, or skin, or on clothing. Wear goggles or face shield and rubber gloves when handling. Avoid inhaling of spray mist. Do not use or store near food or feed.

During commercial exposure in seed treatment and loading operations, wear goggles, clean rubber gloves, protective clothing, and a pesticide respirator jointly approved by the Mine Safety and Health Administration (formerly the U.S. Bureau of Mines) and by the National Institute for Occupational Safety and Health under provisions of 30 CFR, Part 11. Wash hands, arms and face thoroughly with soap and water after handling and before eating or smoking. Routinely shower or bathe after work; wash all clothing with soap and water before reusing. Consumption of alcoholic beverages increases the toxic effects of thiram.

ENVIRONMENTAL HAZARDS

This product is toxic to fish. Do not contaminate water by cleaning of equipment or disposal of wastes. Drift or runoff from treated areas may be hazardous to aquatic organisms in neighboring areas. Do not apply directly to water.

PHYSICAL OR CHEMICAL HAZARDS

Do not use or store near heat or open flame.

DIRECTIONS

It is a violation of Federal law to use this product in a manner inconsistent with its labeling.

STORAGE AND DISPOSAL

Do not contaminate water, food or feed by storage or disposal.

PESTICIDE DISPOSAL: Pesticide wastes are toxic. Improper disposal of excess pesticide, spray mixture or rinsate is a violation of Federal Law. If these wastes cannot be disposed of by use according to label instructions, contact your State Pesticide or Environmental Control Agency, or the Hazardous Waste representative at the nearest EPA Regional Office for guidance.

CONTAINER DISPOSAL: Triple rinse (or equivalent). Then offer for recycling or reconditioning, or puncture and dispose.

DIRECTIONS FOR USE

It is a violation of Federal law to use this product in a manner inconsistent with its labeling.

Do not apply this product through any type of irrigation system. VITAVAX POUR-ON Flowable Fungicide is a ready-to-use seed treatment. VITAVAX POUR-ON Flowable Fungicide combines the systemic action of carboxin with the surface action of thiram to control various seed and seedling diseases.

Mix well before use: Apply VITAVAX POUR-ON Flowable Fungicide directly to seed in hopper box. Partially fill hopper box with a premeasured amount of seed. Apply the proper amount of VITAVAX POUR-ON Flowable Fungicide evenly over the surface of the seed. Mix with a stick or paddle until a uniform color is imparted to all seed. Repeat procedure until hopper box is filled.

For mechanical seed treaters apply VITAVAX POUR-ON Flowable Fungicide at a dosage of 12 fl. oz. per 100 pounds of seed directly to a premeasured amount of seed and mix thoroughly until all seed have a uniform color.

SOYBEANS AND DRY BEANS

For control of **Rhizoctonia solani,** use 6 fl. oz. of VITAVAX POUR-ON Flowable Fungicide per bushel of seed or 12 fl. oz. per 100 pounds of seed. On soybeans, do not graze or feed livestock on forage and hay on treated areas. On dry beans, do not allow livestock to graze or feed on bean forage until 60 days after treated seed is planted.

RICE

VITAVAX POUR-ON Flowable Fungicide is particularly effective against **Rhizoctonia solani** and **Helminthosporium oryzae.** Use 9–12 fluid ounces of VITAVAX POUR-ON Flowable Fungicide per 100 pounds of seed applied as a pour-on treatment or by machine. The higher rate is recommended for control of **Helminthosporium oryzae.**

BARLEY, OATS, WHEAT AND TRITICALE

VITAVAX POUR-ON Flowable Fungicide provides control of the following diseases:

Barley - loose smut **(Ustilago nuda)** covered smut **(U. hordei)**

Oats - loose smut **(U. avenae)** covered smut **(U. kolleri)**

Wheat - loose smut **(U. tritici)** common bunt **(Tilletia caries)**

Triticale - seed decay, damping off, and seedling blight.

Use at the rate of 9–12 fluid ounces per 100 pounds of seed as a pour on treatment. Uniform coverage is essential for product performance. Do not graze or feed livestock on treated areas for six weeks after planting.

COMPATIBILITY

VITAVAX POUR-ON Flowable Fungicide should not be tank mixed with any pesticide unless adequate physical compatibility and phytotoxicity tests have been completed and proved to be satisfactory.

USE RESTRICTIONS

This product contains a dye which imparts an unnatural color to the seed to help prevent the inadvertent use of treated seed for oil purposes or as food for man or feed for animals. Treated seed must not be used for or mixed with food or animal feed, or processed for oil. Treat only those seeds needed for immediate use, minimizing the interval between treatment and planting. Do not store excess treated seeds beyond planting time.

SEED LABELING

Treated seed must not be used or mixed with food or animal feed, or processed for oil. Seed commercially treated with VITAVAX POUR-ON Flowable Fungicide must be labeled in accord with all applicable requirements of the Federal Seed Act. On oats, barley, wheat and triticale, do not graze or feed livestock on treated areas for six weeks after planting. On soybeans, do not graze or feed livestock on forage and hay in treated areas. On dry beans, do not allow livestock to graze or feed on bean forage until 60 days after treated seed is planted. Add appropriate information for any other chemical treatments.

WARRANTY STATEMENT

Gustafson warrants that this product conforms to the chemical description on the label and is reasonably fit for purposes stated on the label only when used in accordance with directions and instructions specified on the label. Gustafson neither makes, nor authorizes any agent or representative to make any other warranty of FITNESS or MERCHANTABILITY, guarantee or representation, express or implied, concerning this material.

Gustafson (logo) is a Reg. TM of Gustafson, Inc.

VITAVAX is a Reg. Trademark of Uniroyal Chemical Company, Inc.

L136–10/87

This is a specimen label. Uniroyal is not responsible for the accuracy of the information contained herein. As labels are subject to revisions, always carefully read and follow the label on the product container.

Valent USA Corporation

1333 North California Boulevard
P.O. Box 8025
Walnut Creek, CA 94596
(415) 256-2700

BOLERO® 8EC

ACTIVE INGREDIENT:	BY WT.
*Thiobencarb	84%
INERT INGREDIENTS:	16%

*S-[(4-chlorophenyl)methyl] diethylcarbamothioate

Contains 8 pounds S-[(4-chlorophenyl)methyl] diethylcarbamothioate per gallon.

BOLERO®—Reg. TM of Kumiai Chemical Industry Co., Ltd. for thiobencarb herbicide.

Product 5281 Made in U.S.A. Form 201-C

EPA Reg. No. 62499-20-59639

NET CONTENTS 5 GALLONS

KEEP OUT OF REACH OF CHILDREN

CAUTION

PRECAUTIONARY STATEMENTS

HAZARDS TO HUMANS & DOMESTIC ANIMALS

CAUTION

Causes eye irritation. Harmful if swallowed. Avoid contact with eyes, skin and clothing. Avoid breathing spray mist.

STATEMENT OF PRACTICAL TREATMENT

In case of eye contact, wash eyes immediately with plenty of water. See a doctor if irritation persists. If swallowed, give a large amount of water to drink, make person vomit, and call a doctor.

Note to Physicians: Emergency Information—call (415) 233-3737.

ENVIRONMENTAL HAZARDS

This product is toxic to shrimp. Do not apply directly to lakes, streams, ponds, tidal marshes and estuaries. Do not contaminate water by cleaning of equipment or disposal of wastes.

The use of BOLERO 8 EC on rice is restricted to protect the endangered fat pocketbook pearly mussel *(Potamilus capax)* and its habitat. Use is prohibited in the following areas of Arkansas.

Mississippi County—Within the basin that drains directly into the Right Hand Chute of Little River, south of Big Lake National Wildlife Refuge.

Poinsett County—Between Crowley's Ridge and the levee east of the Right Hand Chute of Little River and the St. Francis Floodway. Use is also prohibited west of Rt. 140 and north of Rt. 63 at the siphon near Marked Tree. Except that the prohibited area does not include the area bounded by Arkansas highway 373 on the west, highway 63 on the east and highway 14 on the south.

Cross, St. Francis and Lee Counties—Between Crowley's Ridge and the levee east of the Right Hand Chute of Little River and the St. Francis Floodway as far south as the confluence of L'Anguille River (Lee County).

PHYSICAL OR CHEMICAL HAZARDS

Do not use or store near heat or open flame.

DIRECTIONS FOR USE

It is a violation of Federal Law to use this product in a manner inconsistent with its labeling.

READ ENTIRE LABEL. USE STRICTLY IN ACCORDANCE WITH PRECAUTIONARY STATEMENTS AND DIRECTIONS, AND WITH APPLICABLE STATE AND FEDERAL REGULATIONS.

STORAGE AND DISPOSAL

PROHIBITIONS

Do not contaminate water, food or feed by storage, disposal or cleaning of equipment. Open dumping is prohibited.

STORAGE

Keep pesticide in original container.
Do not put concentrate or dilute into food or drink containers.
Store in cool, dry place.
Protect from excessive heat.
For help with any spill, leak, fire or exposure involving this material, call day or night (415) 233-3737.

PESTICIDE DISPOSAL

Wastes resulting from the use of this product may be disposed of on site or at an approved waste disposal facility.

CONTAINER DISPOSAL

Triple rinse (or equivalent). Do not reuse container. Offer for recycling or reconditioning, or puncture and dispose of in a sanitary landfill, or dispose of by other procedures allowed by State and local authorities.

DIRECTIONS

Do not apply this product through any type of irrigation system.

WEED CONTROL IN DRY SEEDED RICE IN TEXAS, LOUISIANA, MISSISSIPPI AND ARKANSAS.

Application Directions—For Dry Seeded Rice in Texas, Louisiana, Mississippi and Arkansas.

Late Preemergence Application (Prior to rice and grass emergence)—For control of Watergrass (Echinochloa spp.) and Sprangletop (Leptochloa spp.). Apply BOLERO 8 EC 1 to 5 days prior to rice emergence; generally this would be 5 to 9 days after planting of the rice. Apply 4 pts. of BOLERO 8 EC per acre after soils have been sealed by rainfall or by flushing. Apply to well prepared seedbeds free of large clods. Soil should be wet at time of application. Do not apply BOLERO 8 EC as a preemergence treatment to cracked soil.

Early Postemergence Application—For control of emerged Watergrass. Apply 4 pts. of BOLERO 8 EC per acre when Watergrass has developed no further than the two-leaf stage. Rice may be fully emerged at time of application. Soil should be wet at time of application.

Do not apply BOLERO 8 EC to stressed rice. Do not apply BOLERO 8 EC to fields with exposed seed as exposed seed will be killed. Do not apply BOLERO 8 EC to second crop (stubble crop) rice.

Water Management—After application, flush the fields as often as necessary to prevent crusting and drying of the soil. Fields should be flooded as soon as the rice plants will tolerate permanent flooding.

APPLICATION EQUIPMENT

Aircraft: Apply BOLERO 8 EC in no less than 10 gals. spray mix per acre.

Ground Sprayers: Apply in 10 to 20 gals. of total spray mix per acre.

Precautions: Do not apply more than 4 lbs. active ingredient per acre per year. Do not plant subsequent crops in treated fields within 6 months of last application. Do not use in fields where fall farming of crayfish will be practiced. Do not use adjacent to catfish ponds. In Texas and other areas where double cropping is the agricultural practice, do not apply to a second stubble rice crop. Water drained directly from treated fields must not be used to irrigate other crops.

Application to stressed rice can result in stand reduction, chlorosis, growth inhibition, delayed maturity and/or leaf desiccation. Stress factors include but are not limited to the following: Daily temperatures below 65° F., or above 95° F., problem soils, (i.e., Zn deficiency, high salt content, high pH.), excessive moisture, (i.e., above field capacity while rice seed is germinating), drought conditions, poor field drainage or deep water after application.

Do not apply under conditions involving possible drift to food, forage or other plantings that might be damaged or the crops thereof rendered unfit for sale, use or consumption.

IMPORTANT INFORMATION. READ BEFORE USING PRODUCT.

LIMITED WARRANTY: 1. Valent U.S.A. Corporation warrants that this product conforms to the chemical description on the label and is reasonably fit for use under average conditions when used strictly in accordance with the directions on the labeling. Valent does not make nor authorizes any agent or representative to make, any other warranty, guarantee or representation, express or implied, concerning this product. Specifically, NO IMPLIED WARRANTY OF MERCHANTABILITY OR FITNESS FOR ANY PARTICULAR PURPOSE IS MADE.

2. Critical and unforeseeable factors beyond Valent's control prevent it from eliminating all risks in connection with the use of this product. Such risks include, but are not limited to, damage to plants and crops to which product is applied, lack of complete control over the handling and application of this product, and damage caused by drift to other plants or crops. Such risks occur even though the product is reasonably fit under average conditions for the uses stated on the labeling and even though label directions are followed. Buyer and user acknowledge and assume all risks and liability (except those assumed by Valent under 1 above) resulting from handling, storage, and use of the product.

3. Precautions stated on the labeling should be followed to avoid hazardous exposure to the product.

BOLERO® 10 G

HERBICIDE

ACTIVE INGREDIENT:	BY WT.
*Thiobencarb	10%
INERT INGREDIENTS:	90%

*S-[(4-chlorophenyl)methyl] diethylcarbamothioate

BOLERO®—Reg. TM of Kumiai Chemical Industry Co., Ltd. for thiobencarb herbicide.

Product 5298 Made in U.S.A. Form 100-C

EPA Reg. No. 62499-19-AA-59639

NET WEIGHT 50 POUNDS

KEEP OUT OF REACH OF CHILDREN

CAUTION

PRECAUTIONARY STATEMENTS

HAZARDS TO HUMANS & DOMESTIC ANIMALS

CAUTION

Causes eye irritation. Avoid contact with eyes or clothing. Avoid breathing dust. Use with proper ventilation or wear a Mine Safety & Health Act and National Institute of Occupational Safety and Health approved pesticide respirator. Wash thoroughly with soap and water after handling.

STATEMENT OF PRACTICAL TREATMENT

If in eyes, flush eyes immediately with fresh water for 15 minutes. If irritation continues, see a doctor. If inhaled, remove person from exposure area.

Note to Physicans: Emergency Information—call (415) 233-3737.

ENVIRONMENTAL HAZARDS

This product is toxic to fish and shrimp. Do not apply directly to lakes, streams, ponds, tidal marshes and estuaries. Do not contaminate water when disposing of equipment washwaters.

DIRECTIONS FOR USE

It is a violation of Federal Law to use this product in a manner inconsistent with its labeling.

READ ENTIRE LABEL. USE STRICTLY IN ACCORDANCE WITH PRECAUTIONARY STATEMENTS AND DIRECTIONS, AND WITH APPLICABLE STATE AND FEDERAL REGULATIONS.

STORAGE AND DISPOSAL

PROHIBITIONS

Do not contaminate water, food or feed by storage, disposal or cleaning of equipment. Open dumping is prohibited.

STORAGE

Keep pesticide in original container.
Do not put concentrate or dilute into food or drink containers.
Store in cool, dry place.
Protect from excessive heat.
For help with any spill, leak, fire or exposure involving this material, call day or night (415) 233-3737.

PESTICIDE DISPOSAL

Wastes resulting from the use of this product may be disposed of on site or at an approved waste disposal facility.

CONTAINER DISPOSAL

Completely empty bag into application equipment. Do not reuse container. Dispose of empty bag in a sanitary landfill or by incineration, or, if allowed by State and local authorities, by burning. If burned, stay out of smoke.

DIRECTIONS

CROP	WEEDS	AMOUNT BOLERO 10 G PER ACRE	FURTHER USE INSTRUCTIONS
Rice (California Only)	Barnyardgrass (Watergrass) Sprangletop Smallflower Umbrellaplant	40 lbs.	Apply after flooding when rice is in the expanded two-leaf stage (first open leaf). Do not apply if rice exhibit symptoms of stress. The variety Cal Pearl may react adversely to some stress conditions more than other varieties. Barnyardgrass beyond the two-leaf stage and Smallflower Umbrellaplant beyond the three-leaf stage may not be controlled. Smallflower Umbrellaplant must be submerged for optimum control. Cold weather during the planting season may result in decreased weed control due to prolonged period of weed seed germination. Do not drain field for a minimum of four days after a BOLERO application. Water level in the checks should be maintained at 3 to 4 inches with no exposed soil. Water flow should be maintained at a minimum during and up to four days after application for good weed control.
			California may require special water management practices to minimize off-site water movement of BOLERO residues; consult your State Department of Food and Agriculture or County Agricultural Commissioner for further information.
			Precautions: Do not make more than one application per year. Do not plant subsequent crops in treated fields within 6 months of last application. Do not use in fields where fall farming of crayfish will be practiced. Do not use adjacent to catfish ponds. Water drained directly from treated fields must not be used to irrigate other crops.

IMPORTANT INFORMATION. READ BEFORE USING PRODUCT.

LIMITED WARRANTY: 1. Valent U.S.A. Corporation warrants that this product conforms to the chemical description on the label and is reasonably fit for use under average conditions when used strictly in accordance with the directions on the labeling. Valent does not make nor authorizes any agent or representative to make, any other warranty, guarantee or representation, express or implied, concerning this product. Specifically, NO IMPLIED WARRANTY OF MERCHANTABILITY OR FITNESS FOR ANY PARTICULAR PURPOSE IS MADE.

2. Critical and unforeseeable factors beyond Valent's control prevent it from eliminating all risks in connection with the use of this product. Such risks include, but are not limited to, damage to plants and crops to which product is applied, lack of complete control over the handling and application of this product, and damage caused by drift to other plants or crops. Such risks occur even though the product is reasonably fit under average conditions for the uses stated on the labeling and even though label directions are followed. Buyer and user acknowledge and assume all risks and liability (except those assumed by Valent under 1 above) resulting from handling, storage, and use of the product.

3. Precautions stated on the labeling should be followed to avoid hazardous exposure to the product.

COBRA™
HERBICIDE

ACTIVE INGREDIENT:	BY WT.
Lactofen	23.2%
* Contains 2 lbs. active ingredient per gallon	
INERT INGREDIENTS:	76.8%

* 1,(carboethoxy) ethyl 5-[2-chloro-4-(trifluoromethyl) phenoxy]-2-nitrobenzoate

EPA Reg. No. 59639–34

Valent U.S.A. Corporation © 1989

Product 5623 Form 131-B Made in U.S.A.

DANGER

KEEP OUT OF REACH OF CHILDREN

SEE NEXT PAGE FOR ADDITIONAL PRECAUTIONARY STATEMENTS

Read all Precautionary Statements, Directions for Use, Conditions of Sale, Storage and Disposal Statements and Environment Hazards Statements before using this product. It is a violation of Federal Law to use this product in a manner inconsistent with the label.

PRECAUTIONARY STATEMENTS

HAZARDS TO HUMANS & DOMESTIC ANIMALS

DANGER

Corrosive. Causes skin burns and irreversible eye damage. Harmful if swallowed, inhaled, or absorbed through skin. Do not get in eyes or on skin or clothing. Wear goggles or face shield, coveralls or tightly woven long-sleeved shirt, and rubber gloves when handling. Avoid breathing vapor or spray mist. Wash thoroughly with soap and water after handling. Remove contaminated clothing and wash before reuse.

This product contains lactofen, which has been determined to cause tumors in laboratory animals (mouse, rat). Risks can be reduced by closely following use directions and precautions, and by wearing the protective clothing specified elsewhere on this label.

NOTE TO PHYSICIANS: Emergency Information—call (415) 233-3737.

STATEMENT OF PRACTICAL TREATMENT

IF IN EYES, flush with plenty of water for at least 15 minutes. Get medical attention immediately.

IF SWALLOWED, DO NOT INDUCE VOMITING. Drink promptly a large quantity of milk, egg whites, gelatin solution, or, if these are not available, large quantities of water. Avoid alcohol. Get medical attention immediately.

IF ON SKIN, wash with plenty of soap and water. Get medical attention.

ENVIRONMENTAL HAZARDS

This pesticide is toxic to fish. Do not apply directly to water or wetlands (swamps, bogs, marshes, and potholes). Drift and runoff from treated areas may be hazardous to aquatic organisms in neighboring areas. Do not contaminate water by cleaning of equipment or disposal of waste. Do not apply when weather conditions favor drift from target area.

PHYSICAL OR CHEMICAL HAZARDS

Do not use or store near heat or open flame.

DIRECTIONS FOR USE

It is a violation of Federal Law to use this product in a manner inconsistent with its labeling.

READ ENTIRE LABEL AND PAMPHLET. USE STRICTLY IN ACCORDANCE WITH PRECAUTIONARY STATEMENTS AND DIRECTIONS, AND WITH APPLICABLE STATE AND FEDERAL REGULATIONS.

Do not apply this product through any type of irrigation system.

WORKER PROTECTION AND REENTRY

Do not treat areas while unprotected humans or domestic animals are present in the treatment area. Do not allow entry into treated areas without protective clothing until the spray has dried.

TABLE OF CONTENTS

DIRECTIONS FOR USE IN SOYBEANS

GENERAL INFORMATION

COBRA™ is a selective, broad spectrum herbicide for postemergence control of susceptible broadleaf weeds. COBRA is formulated as an emulsifiable concentrate containing 2 lbs. of active ingredient per gallon. COBRA works primarily through contact action. Good coverage of young, actively growing weeds is essential for maximum weed control. The use of an EPA approved spray adjuvant is required. Refer to the label section on ADJUVANTS AND ADDITIVES for specific recommendations.

When COBRA is applied post emergent, a portion of the spray solution may contact the soil surface. If soil moisture conditions are favorable for **preemergence activity** following the application, suppressed germination of small-seeded broadleaf weeds, such as nightshade species, pigweed species, and prickly sida may be expected for a 2 to 3 week period. The presence of excessive crop or weed foliage at the time of application will reduce the amount of herbicide spray contacting the soil surface and will reduce the level of soil activity.

A temporary crop response should be expected following a post-emergence application of COBRA. Soybean leaves which are open at the time of application will show some burn, bronzing and speckling. Trifoliate soybean leaves which have emerged but are unopened at the time of application may appear cupped at the tip and/or crinkled along the edges of the leaf. Soybeans quickly outgrow all initial herbicide effects. When COBRA is used as directed, under commercial condition **soybean yields will not be affected. Under conditions of normal weed growth COBRA is rainfast 30 minutes after application.**

PRECAUTIONS

Apply COBRA once per growing season, not later than 90 days before harvest. Do not graze animals on green forage or stubble. Do not utilize hay or straw for animal feed or bedding.

APPLICATION

TIMING OF APPLICATION

For best results, COBRA should be applied to small actively growing weeds that are not larger than indicated in Table 1 of this label. Normally this occurs 14 to 21 days after planting or last field preparation. When soybeans are at the first to second trifoliate leaf stage.

Soybeans at or larger than the third trifoliate stage may interfere with the spray pattern and reduce coverage of the weed leaves. **Do not apply COBRA when** the soybeans or the weeds are under stress.

Applying COBRA under conditions that do not promote active weed growth will reduce herbicide effectiveness. These conditions include drought, excessive water, extremes in temperature, and low humidity. Weeds under stress tend to "harden off" and become less susceptible to herbicidal action.

Do not cultivate prior to or during application. Do not generate excessive dust while spraying. Excessively dusty conditions may interfere with the coverage of the weed leaf surface by the spray solution. A timely cultivation approximately one week after application will assist in weed control. Under drilled/solid-seeded soybeans cropping systems, a dense crop canopy develops more rapidly than conventional row spacings. The crop canopy may restrict penetration of the herbicide spray pattern and reduce coverage of the weed foliage. Applications should be made when soybeans are at the first trifoliate leaf stage usually 7 to 10 days after planting. Delaying application beyond the first trifoliate leaf stage may result in unsatisfactory weed control. For improved weed coverage and canopy penetration in drilled or solid seeded soybean, the higher range of spray volume and pressure is required. See the BROADCAST GROUND APPLICATIONS section of this label for application information.

BROADCAST GROUND APPLICATION

COBRA and COBRA tank mixes can be applied by ground equipment using standard commercial sprayers equipped with flat fan or hollow cone nozzles. COBRA is a contact herbicide. Special attention should be given to preparing and operating the sprayer to assure proper coverage of the weed leaf surface.

Use COBRA on a broadcast basis in **20 to 30 gals. of water per acre** and a spray pressure of **40 to 60 PSI measured at the boom.** Apply COBRA using a flat fan or hollow cone nozzle designed to deliver the desired spray pressure and spray volume. Avoid use of flat fan nozzles larger than 8006 (or equivalent) because they do not break up spray patterns into small enough droplets to provide adequate weed coverage for foliar herbicides. Spray nozzles should be centered at 20 and spacing to provide adequate coverage. Ground speed should not exceed 10 mph to provide proper spray coverage. Boom height, ground speed, and pressure recommendations should not exceed those recommended by the spray nozzle manufacturer for the type and size of nozzle being used. Improper use of the selected spray nozzle will adversely affect the spray pattern, prevent proper coverage of weed leaf surface, and reduce weed control. Refer to the manufacturer's spray chart for nozzle selection and operating information.

REFER TO THE AERIAL APPLICATION SECTION FOR SPECIFIC INSTRUCTIONS FOR AERIAL APPLICATION.

DO NOT USE THE FOLLOWING DELIVERY SYSTEMS TO APPLY COBRA HERBICIDE:

1 — Flood Nozzles
2 — Control Droplet Applicators (CDA)
3 — Flat Fan Nozzles Larger than 8006
4 — Spray rigs which utilize wheel driven pumps

BAND APPLICATION

Row banding equipment should be adjusted to provide maximum coverage of weeds in the row. Base the band use of COBRA from a broadcast use rate of 20 to 30 gallons of water per acre by reducing the spray gallons in proportion to the area actually treated. The spray pressure should be 40 to 60 PSI measured at the boom. A minimum of two nozzles per row is required to provide optimum coverage of the weed foliage.

DO NOT make band applications while cultivating or create excessive dust while spraying. Excessively dusty conditions will interfere with proper coverage of the weed leaf surface, thereby reducing contact activity.

LOW VOLUME GROUND APPLICATION
SPRA-COUPE™, HAGIE™

Application of COBRA and COBRA tank mixes using a low volume application require a minimum of 10 gals. of spray solution per acre.

A minimum of 20 gals. of spray solution per acre is required in North Dakota, South Dakota and Minnesota.

Applications at less than 10 gals. per acre will provide inconsistent weed control. The spray pressure at the boom should be between 40 to 60 PSI. Flat fan nozzles are recommended at 20 inch spacing for proper spray coverage. To provide adequate coverage flat fan nozzles larger than 8006 (or equivalent) should not be used. Height of the spray boom should be adjusted so as not to exceed the manufacturer's recommendation for proper coverage by the spray nozzle being used. Maximum speed of operation should not exceed 10 mph as spray coverage of weed foliage may be adversely affected.

Crop oil concentrate is required at the rate of 1 qt. per acre to achieve optimum control with low volume applications.

AERIAL APPLICATION

To obtain satisfactory weed control with aerial applications of COBRA, uniform coverage must be obtained. Do not spray when drift is possible or when wind velocity is more than 5 mph. Avoid spraying COBRA within 200 ft. of dwellings, or adjacent sensitive crops such as ornamentals, cotton, tobacco, or sorghum. To obtain satisfactory application and minimize drift, the following directions must be observed:

Volume and Pressure: Use COBRA in **5 to 10 gals. per acre water** and a maximum spray pressure of **40 PSI.** Applications at less than 5 gals. per acre will provide inadequate control. The higher gallons applications generally afford more consistent weed control.

Nozzle and Nozzle Orientation: Use nozzles which produce flat or hollow cone spray patterns. Use non-drip type nozzles such as diaphragm-type nozzles to avoid unwanted discharge of spray solution. The nozzles must be directed toward the rear of the aircraft, at an angle between 0 and 15 downward, do not place nozzles on the outer 25% of wings or rotors.

CROP OIL CONCENTRATE: With aerial applications the use of crop oil concentrate at the rate of **1 qt. per acre** in the spray solution is required. **Do not use surfactants.**

IMPORTANT: Use of less than 1 qt. of crop oil concentrate per acre or use of an adjuvant other than crop oil concentrate will result in unsatisfactory weed control.

ADJUVANTS AND ADDITIVES

Control of weeds by COBRA Herbicide at or near the maximum growth stage indicated in Table 1 of this label is enhanced by the addition of an adjuvant to the spray mixture. Use an EPA approved petroleum-based crop oil concentrate, or 28% N liquid fertilizer, or non-ionic surfactant, such as VALENT X-77® Spreader. Mixing and compatibility qualities should be verified by a jar test. The addition of a spray adjuvant may intensify the initial crop response.

Crop Oil Concentrate: Crop oil concentrate is the preferred adjuvant for COBRA for weed control over a wide spectrum of application conditions. Higher levels of crop response are also generally observed with the use of a crop oil concentrate; however soybeans quickly outgrow all initial herbicide effects. The rate of crop oil concentrate will depend on the environmental conditions preceding the application and the weed species and weed size at the time of application. If environmental conditions are good and weeds are growing vigorously, use of the low rate of crop oil concentrate is recommended. The higher rate is required when the weeds are under environmental stress such as low temperature, low humidity, or low soil moisture. Refer to the adjuvant table for the recommended rate of crop oil concentrate by rate and environmental conditions.

IMPORTANT: For aerial applications and low volume ground application, the addition of 1 qt. per acre of crop oil concentrate is required.

28% N Liquid Fertilizer: A 28% N liquid fertilizer may be substituted for crop oil concentrate in the spray solution when weeds are actively growing. Refer to the adjuvant and additive table for the recommended rate of 28% N liquid fertilizer and additives by state and environmental conditions.

Surfactant: When soil moisture is good and weeds are actively growing, the addition of an EPA approved non-ionic surfactant such as VALENT X-77 Spreader at the rate of 2 pts. per 100 gals. of spray solution will improve the wetting and contact activity of COBRA Herbicide. Refer to the adjuvant table for the recommended rate of non-ionic surfactant by state and environmental conditions.

Table 1.
RELATIVE HUMIDITY ADJUVANT RECOMMENDATIONS

	PERCENT RELATIVE HUMIDITY		
	100%, 80% (High)	80%, 60% (Medium)	60%, Less (Low)
NON-IONIC SURFACTANT SOLUTION (ALL STATES)	2 pts./100 gals.	Not Recommended	Not Recommended
CROP OIL CONCENTRATE/ ACRE* (AL, DE, FL, GA, MD, NC, PA, SC, VA)	1/4 to 1/2 pt.	1/2 pt.	1/2 pt.
(ALL OTHER STATES)	1/2 pt.	1 pt.	1 pt.
28% NITROGEN/ ACRE**	1 gal.	1 gal. + X-77 Spreader at 2 pts. 100 gals.	1 gal. + crop oil concentrate at 1 pt./acre

* 1 pt./acre required to control velvetleaf, wild sunflower, and morningglory species.
** 1 gallon 28% N + X-77 at 2 pts./100 gals. required to control velvetleaf.

MIXING

Add about 1/2 of the required amount of water to the spray tank and begin agitation. Add the required amount of COBRA Herbicide and mix thoroughly. Then add the correct amount of adjuvant and the remaining water. Maintain agitation during filling and spraying to insure a uniform spray mixture.

WEEDS CONTROLLED BY COBRA HERBICIDE

Identify your weed species as early as possible and determine the stage of growth by counting the true leaves. Ignore the cotyledon (seed leaves) when you count.

Use Table 1 of this label to determine the maximum weed leaf stage at which you must apply COBRA for effective weed control of the species desired.

Table 2.
WEEDS CONTROLLED BY COBRA HERBICIDE IN SOYBEANS
Weed Growth Stages
12.5 fl. oz. per acre

Common Name	Maximum Leaf Stages
Balloonvine	4
Beggarsticks Devils	6
Bristly Starbur	4
Buffalobur	4
Burcucumber	4
Carpetweed	8 in. diameter
Cocklebur, Common	6
Copperleafs	
Hophornbeam	6
Virginia	4
Devils Claw	4

WEEDS CONTROLLED BY COBRA HERBICIDE IN SOYBEANS Weed Growth Stages 12.5 fl. oz. per acre	
Common Name	Maximum Leaf Stages
Eclipta	6
Florida Beggarweed	2
Florida Pusley	6
Groundcherry	
Cutleaf	6
Lanceleaf	6
Hairy Galinsoga	4
Hemp Sesbania	6
Jimsonweed	4
Mexican weed	4
Morningglories	
Cypressvine	4
*Entireleaf	2
*Ivyleaf	2
*Palmleaf	4
*Pitted	4
*Purple Moonflower	4
*Smallflower	4
*Tall	4
Mustard Wild	6
Nightshades	
Black	6
Eastern Black	6
Hairy	4
Pigweeds	
Palmer Amaranth	6
Prostrate	6
Redroot	6
Smooth	6
Spiny Amaranth	6
Poorjoe	6
Prickly Sida (Teaweed)	4
Puncturevine	1.5 in. diameter
Purslane Common	8 in. diameter
Ragweeds	
Common	6
Giant	4
Showy Crotalaria	4
Smell Melon	6
Spurges	
Prostrate	1 in. diameter
Spotted	4
Tall Waterhemp	4
Tropic Croton	4
**Velvetleaf	4
Venice Mallow	4
Wild Poinsettia	4
Wild Sunflower	2
Witchweed	6-8 in. prior to bloom
Woolly Croton	4

* Crop oil concentrate is required at the rate of 1 pt./acre for contol.

** Crop oil concentrate at 1 pt./acre or 28% N liquid fertilizer at 1 gal./acre plus non-ionic surfactant X-77 at 2 pts./100 gals. of solution is required for control.

Table 3.

WEEDS SUPPRESSED BY COBRA Herbicide Weeds Growth Stages 12.5 fl. oz. per acre	
Common Name	Maximum Leaf States
*Coffee Senna	2
Pennsylvania Smartweed	4
*Sicklepod	2
Spurred Anoda	2

* Suppression may be improved when COBRA is applied following a preemergence application of Lasso® or Dual® at the recommended rates. Only suppression of growth and not acceptable commercial control may be expected when these weeds are treated with COBRA.

The use of a crop oil concentrate at 1 pt. per acre is required for suppression of these weeds.

Cultivation 1 week to 10 days after treatment will usually aid in obtaining satisfactory suppression of these weeds.

Table 4.

PERENNIAL WEEDS SUPPRESSED BY COBRA Herbicide Weed Growth Stages 12.5 fl. oz. per acre	
Common Name	Maximum Leaf Stages
Bigroot Morningglory (Wild Sweet Potato)	6 Leaf
Canada Thistle	6 Leaf
Milkweeds	
Climbing	6 Leaf
Common	6 Leaf
Redvine	6 Leaf
Swamp Smartweed	6 Leaf
Trumpetcreeper	6 Leaf

The addition of crop oil concentrate at 1 pt./acre is required to burn back existing above-ground vegetation and may retard the growth of new foliage.

TANK MIX WITH BASAGRAN® HERBICIDE FOR CONTROL OF BROADLEAF WEEDS IN SOYBEANS GENERAL INFORMATION

For best results the COBRA plus BASAGRAN tank mix should be applied to actively growing weeds.

Do not apply COBRA tank mixtures during periods ehn the soybeans and the weeds are under stress or when conditions do not favor active weed growth. Refer to the COBRA Soybean Label for application timing.

Table 5.
COBRA Herbicide & BASAGRAN Herbicide Tank Mix - Soybeans - Rate and Time of Application Table.

Product	Product Rate	Weeds Controlled/ Weed Size	Adjuvant
COBRA	12.5 fl. oz./A	Refer to Table 2	1 pt. non-ionic surfactant per 100 gals. water. Surfactant must have at least 80% active ingredient. Do Not Use crop oil concentrate or vegetable oil.
plus	plus		
		Maximum leaf stage	
BASAGRAN	1 pt.	Pigweed	
		Smooth	6
		Redroot	6
		Tall Waterhemp	6
		Cocklebur	6
		Smartweed	4
		Velvetleaf	6
		Wild Sunflower	4
	2 pt.	Cocklebur	10
		Common Lambsquarter	4-8
		Smartweed, Pennsylvania	10
		Wild Sunflower	6

MIXING INSTRUCTIONS

Maintain agitation throughout the spray application. Failure to agitate the spray volume may result in improper mixing of the herbicides and unsatisfactory weed control. Mixing and compatibility qualities should be verified by a jar test.

COBRA Herbicide Plus BASAGRAN Herbicide Tank Mix: Add 1/2 of the required water to the spray tank and begin agitation. Add the required amount of BASAGRAN and mix thoroughly. Then add the required amount of COBRA and continue mixing. Finally, add the required amount of crop oil concentrate or the 28% N liquid fertilizer and the remaining water.

Use Restrictions for COBRA Herbicide Plus BASAGRAN Herbicide Tank Mix:

1. Always read and follow all label directions when using any pesticide along or in tank mix combinations. **The most restrictive labeling applies when using a tank mix.**
2. Do not apply the COBRA plus BASAGRAN tank mix less than 90 days before harvesting soybeans.
3. Do not graze treated fields or harvest for forage hay.
4. Do not apply the COBRA plus BASAGRAN tank mix if rain is expected within 8 hours of application or unsatisfactory weed control may result.
5. Do not apply COBRA plus BASAGRAN tank mix by air.

TANK MIX WITH CLASSIC® HERBICIDE FOR CONTROL OF BROADLEAF WEEDS IN SOYBEANS GENERAL INFORMATION

For best results, the COBRA plus CLASSIC tank mix should be applied to actively growing weeds. Do not apply the COBRA and CLASSIC tank mix during periods when the soybeans and the weeds are under stress or when conditions do not favor active weed growth. Refer to the COBRA soybean label for application timing.

Table 6.
COBRA Herbicide & CLASSIC Herbicide Tank Mix - Soybeans - Rate and Time of Application Table.

Product	Product Rate	Weeds Controlled/ Weed Size		Adjuvant
COBRA	10 to 12.5 oz./A according to weed species and size on Table 1	Refer to Table 2		2 pts non-ionic surfactant per 100 gals. water. Surfactant must have at least 80% active ingredient. Do Not Use crop oil concentrate or vegetable oil.
plus	plus			
		Maximum leaf stage		
CLASSIC	0.25 to 38 oz.	Pigweed		
		Smooth	6	
		Redroot	6	
		Tall Water Hemp	6	
		Cocklebur	8	
		Florida Beggar-weed	6	
		Mustard	6	
		Smartweed, Pennsylvania	4	
		Sunflower	6	
plus	plus			
	0.38	Yellow Nutsedge	6	
		Velvetleaf*	5	
plus	plus			
	.5	Sicklepod**	3	

* Add 1 gal. 28% liquid fertilizer per acre and 2 pts. non-ionic surfactant per 100 gals. water.
** Required 12.5 oz. of COBRA for control.

MIXING INSTRUCTIONS

Add 1/2 of the required water to the spray tank and begin agitation. Add the required amount of CLASSIC and mix thoroughly. Then add the required amount of COBRA continue mixing. Finally, add the required amount of non-ionic surfactant and 28% N liquid fertilizer and the remaining water.

Maintain agitation throughout the spray application. Failure to agitate the spray mixture may result in improper mixing of the herbicides and unsatisfactory weed control.

Use Restrictions for COBRA Herbicide plus CLASSIC Herbicide Tank Mix:

1. Always read and follow all label directions when using any pesticide alone or in tank mix combinations. **The most restrictive labeling applies when using a tank mix.**
2. Do not apply the COBRA plus CLASSIC tank mix less than 90 days before harvesting soybeans.
3. Do not graze animals on green forage or stubble. Do not utilize hay or straw for animal feed or bedding.
4. Do not apply the COBRA plus CLASSIC tank mix if rain is expected within 1 hour of application or unsatisfactory weed control may result.
5. Refer to the CLASSIC label for restrictions on soil pH, minimum recropping interval and rotational guidelines.
6. Do not apply the COBRA plus CLASSIC tank mix by air.

TANK MIX WITH 2,4-DB PRODUCTS FOR CONTROL OF BROADLEAF WEEDS IN SOYBEANS
GENERAL INFORMATION

Use only the amine formulations of 2,4-DB registered for use on soybeans. For best results, the COBRA plus 2,4-DB tank mix should be applied to actively growing weeds. Do not apply COBRA and 2,4-DB product tank mixtures during periods when the soybeans and the weeds are under stress or when conditions do not favor active weed growth. Refer to the COBRA soybean label for application timing.

The use of a 2,4-DB tank mix may cause crop injury.

Table 7.
COBRA Herbicide & 2,4-DB Tank Mix - Soybeans - Rate and Time of Application Table.

Product	Product Rate	Weeds Controlled/ Weed Size		Adjuvant
COBRA	12.5 fl. oz./A	Refer to Table 2		1 pt. non-ionic surfactant per 100 gals. water. Surfactant must have at least 80% active ingredient. Do Not Use crop oil concentrate or vegetable oil.
plus	plus			
		Maximum leaf stage		
2,4-DB	2 oz./A	Morningglories	6	
		Sicklepod	2	

MIXING INSTRUCTIONS

Add 1/2 of the required water to the spray tank and begin agitation. Add the required amount of 2,4-DB and mix thoroughly. Then add the required amount of COBRA and continue mixing. Finally, add the required amount of non-ionic surfactant such as VALENT X-77 Spreader and the remaining water.

Maintain agitation throughout the spray application. Failure to agitate the spray mixture may result in improper mixing of the herbicides and unsatisfactory weed control.

Use Restrictions for COBRA Herbicide plus 2,4-DB Tank Mix:

1. Always read and follow all label directions when using any pesticide alone or in tank mix combinations. **The most restrictive labeling applies when using a tank mix.**
2. Do not apply the COBRA plus 2,4-DB tank mix by air.
3. Do not apply the COBRA plus 2,4-DB tank mix less than 90 days before harvesting soybeans.
4. Do not graze animals on green forage or stubble. Do not utilize hay or straw for animal feed or bedding.
5. Do not apply the COBRA plus 2,4-DB tank mix if rain is expected within 2 hours of application or unsatisfactory weed control may result.
6. Do not allow 2,4-DB to come into contact with cotton, okra, grapes, tomatoes, fruit trees, vegetables, flowers, or other desirable crop or ornamental plants. Do not permit spray mist to drift onto susceptible plants since very small quantities of 2,4-DB can cause severe injury during growing or dormant periods.

TANK MIX WITH SCEPTER™ HERBICIDE FOR CONTROL OF BROADLEAF WEEDS IN SOYBEANS
GENERAL INFORMATION

For best results, the COBRA plus SCEPTER tank mix should be applied to actively growing weeds.

Do not apply COBRA and SCEPTER tank mix during periods when the soybeans and the weeds are under stress or when conditions do not favor active weed growth.

Refer to the COBRA soybean label for application timing.

Table 8.
COBRA Herbicide & SCEPTER Herbicide Tank Mix - Soybeans - Rate and Time of Application Table.

Product	Product Rate	Weeds Controlled/ Weed Size		Adjuvant
COBRA	12.5 fl. oz./A	Refer to Table 2		1 pt. non-ionic surfactant per 100 gals. water. Surfactant must have at least 80% active ingredient. Do Not Use crop oil concentrate or vegetable oil.
plus	plus			
		Maximum leaf stage		
SCEPTER	4 oz.	Cocklebur	12	
		Pigweed	12	

MIXING INSTRUCTIONS

Add 1/2 of the required water to the spray tank and begin agitation. Add the required amount of SCEPTER and mix thoroughly. Then add the required amount of COBRA and continue mixing. Finally, add the required amount of non-ionic surfactant and the remaining water.

Maintain agitation throughout the spray application. Failure to agitate the spray mixture may result in improper mixing of the herbicides and unsatisfactory weed control.

Use Restrictions for COBRA Herbicide plus SCEPTER Herbicide Tank Mix:

1. Always read and follow all label directions when using any pesticide alone or in tank mix combinations. **The most restrictive labeling applies when using a tank mix.**
2. Do not apply the COBRA plus SCEPTER tank mix by air.
3. Do not graze animals forage or stubble. Do not utilize hay or straw for animal feed or bedding.

4. Do not apply the COBRA plus SCEPTER tank mix if rain is expected within 2 hours of application or unsatisfactory weed control may result.
5. Refer to the SCEPTER label for regional use and rotational crop restrictions.

DIRECTIONS FOR USE IN SOUTHERN PINE SEEDBEDS
GENERAL INFORMATION

COBRA is a selective herbicide for use in southern pine seedbeds, when used according to this label.

Pine seedlings will tolerate postemergence treatments when applications are made following complete emergence of the stand.

COBRA may be used on the pine species listed in Table A "Tolerant Species" in the following states:

Alabama	Georgia	Mississippi	South Carolina
Arkansas	Kentucky	Missouri	Tennessee
Delaware	Louisiana	North Carolina	Texas
Florida	Maryland	Oklahoma	Virginia

Occasionally **slight needle burn will be observed on the youngest growth** following application. **New growth will be normal** and the seedlings will continue vigorous growth under favorable environmental condition.

Crop tolerance is greatest if applications are made following complete stand emergence. Some forking and stunting of seedling may result if COBRA is applied to newly emerged seedlings. For maximum crop safety wait until the seedlings have initiated primary shoot growth before applying COBRA.

APPLICATION

Thoroughly mix COBRA with clean water and apply at 30 to 50 PSI in 20 to 40 gals. per acre. Flat fan or hollow cone nozzles are recommended.

Applications made at less than 20 gals. per acre or less than 30 PSI will not provide complete coverage of the weeds and will result in incomplete control. Do not make aerial applications of COBRA to pine seedbeds.

Table A.

COBRA Herbicide - Southern Pine Seedling - Rate and Time of Application Table.

Product	Rate	Weeds Controlled	Tolerant Species	Adjuvant
COBRA	6.5% oz./A every 1 week (maximum 4 applications)* or 12.5 oz./A every 2 weeks (maximum 2 applications)*	Black Nightshade Carpetweed Common Dayflower Common Ragweed Dog Fennel Eclipta Florida Beggarweed Florida Pusley Hairy Galinsoga Mayweed Morningglory species Mustard species Prickly Sida Pigweed species Prostrate Spurge Poorjoe Showy Crotalaria Sicklepod Spotted Spurge Tropic Croton Witchweed Yellow Woodsorrel	Eastern white pine (pinus strobus) Loblolly Pine (pinus taeda) Sand Pine (pinus clausa) Shortleaf Pine (pinus echinata) Slash Pine (pinus piliottii) Virginia Pine (pinus virginiana)	½ pt./A crop oil concentrate** or 1 qt./A non-ionic surfactant per 100 gals. (if increased post-emergent activity is desired)

* Do not apply more than 12.5 fl. oz. per acre within any 2 week period. The total amount of COBRA used per season must not exceed 26 fl. oz. per acre (0.416 active ingredients).

** The addition of crop oil concentrate to COBRA may cause some needle burn. **DO NOT APPLY WITH SPRAY ADJUVANTS UNTIL PRIMARY SHOOT GROWTH HAS BEGUN.**

APPENDIX

The following are scientific names for the weeds listed on this label:

Common Name	Scientific Name
Balloonvine	Cardiospermum halicacabum
Beggarticks, Devils	Bidens frondosa
Bigroot Morningglory	Ipomoea pandurata
Bristly Starbur	Acanthospermum hispidum
Buffalobur	Solanum rostratum
Burcucumber	Sicyos angulatus
Canada Thistle	Cirsium arvense
Carpetweed	Mollugo verticillata
Cocklebur, Common	Xanthium strumarium
Coffee Senna	Cassia occidentalis
Copperleafs	
Hophornbeam	Acalypha ostryaefolia
Virginia	Acalypha virginica
Devils Claw	Proboscidea louisianica
Eclipta	Eclipta prostrata
Florida Beggarweed	Desmodium tortuosum
Florida Pusley	Richardia scabra
Groundcherry	
Cutleaf	Physalis angulata
Lanceleaf	Physalis lanceifolia
Hairy Galinsoga	Galinsoga ciliata
Hemp Sesbania	Sesbania exaltata
Jimsonweed	Datura stramonium
Mexicanweed	Caperonia castaniifolia
Milkweeds	
Climbing	Sarcostemma cynanchoides
Common	Asclepias syriaca
Morningglories	
Cypressvine	Ipomoea quamoclit
Entireleaf	Ipomoea hederacea var. integriuscula
Ivyleaf	Ipomoea hederacea
Palmleaf	Ipomoea wrightii
Pitted	Ipomoea lacunosa
Purple Moonflower	Ipomoea turbinata
Smallflower	Jacquemontia tamnifolia
Tall	Ipomoea purpurea
Mustard, Wild	Sinapis arvensis
Nightshades	
Black	Solanum nigrum
Eastern Black	Solanum pytcanthum
Hairy	Solanum sarrachoides
Pennsylvania Smartweed	Polygonum pensylvanicum
Pigweeds	
Palmer Amaranth	Amaranthus palmeri
Prostrate	Amaranthus blitoides
Redroot	Amaranthus retroflexus
Smooth	Amaranthus hybridus
Spiny Amaranth	Amaranthus spinosus
Poorjoe	Diodia teres
Prickly Sida (Teaweed)	Sida spinosa
Puncturevine	Tribulus terrestris
Purslane, Common	Portulaca oleracea
Ragweeds	
Common	Ambrosia artemisiifolia
Giant	Ambrosia trifida
Redvine	Brunnichia ovata
Sicklepod	Cassia obtusifolia
Showy Crotalaria	Crotalaria spectabilis
Smell melon	Cucumis melo
Spurges	
Prostrate	Euphorbia humistrata
Spotted	Euphorbia maculata
Spurred Anoda	Anoda cristata
Swamp Smartweed	Polygonum coccineum
Tall Waterhemp	Amaranthus tuberculatos
Tropic Croton	Croton glandulosus
Trumpetcreeper	Campsis radicans
Velvetleaf	Abutilon theophrasti
Venice mallow	Hibiscus trionum
Wild Poinsettia	Euphorbia heterophylla
Wild Sunflower	Helianthus annuus
Witchweed	Striga asiatica
Woolly Croton	Croton capitatus

STORAGE AND DISPOSAL

PROHIBITIONS

Do not contaminate water, food, or feed by storage, disposal, or cleaning of equipment.

STORAGE

Store in cool, dry place.
Keep pesticide in original container.
Do not put concentrate or dilute into food or drink containers.
Not for use or storage in or around the home.
For help with any spill, leak, fire or exposure involving this material, call day or night (415) 233-3737.

PESTICIDE DISPOSAL
This product is acutely hazardous. Improper disposal of excess pesticide, spray mixture, or rinsate is a violation of Federal law. If these wastes cannot be disposed of by use according to label instructions, contact your State Pesticide or Environmental Control Agency, or the Hazardous Waste representative at the nearest EPA Regional Office for guidance.

CONTAINER DISPOSAL
Triple rinse (or equivalent). Do not reuse container. Offer for recycling or reconditioning, or puncture and dispose of in a sanitary landfill, or incineration, or, if allowed by state and local authorities, by burning. If burned, stay out of smoke.

IMPORTANT INFORMATION.
READ BEFORE USING PRODUCT.

LIMITED WARRANTY: 1. Valent U.S.A. Corporation warrants that this product conforms to the chemical description on the label and is reasonably fit for use under average conditions when used strictly in accordance with the directions on the labeling. Valent does not make nor authorizes any agent or representative to make, any other warranty, guarantee or representation, express or implied, concerning this product. Specifically, NO IMPLIED WARRANTY OF MERCHANTABILITY OR FITNESS FOR ANY PARTICULAR PURPOSE IS MADE.

2. Critical and unforeseeable factors beyond Valent's control prevent it from eliminating all risks in connection with the use of this product. Such risks include, but are not limited to, damage to plants and crops to which product is applied, lack of complete control over the handling and application of this product, and damage caused by drift to other plants or crops. Such risks occur even though the product is reasonably fit under average conditions for the uses stated on the labeling and even though label directions are followed. Buyer and user acknowledge and assume all risks and liability (except those assumed by Valent under 1 above) resulting from handling, storage, and use of the product.

3. Precautions stated on the labeling should be followed to avoid hazardous exposure to the product.

COBRA™ — TM of Chevron Chemical Company for lactofen herbicide.
BASAGRAN® — Reg. TM of BASF Corp. for bentazon herbicide.
CLASSIC® — Reg. TM of E.I. du Pont de Nemours & Co., Inc. for chlorimuron ethyl herbicide.
SCEPTER® — TM of American Cyanamid Co. for imaziquin herbicide
SPRA-COUPE™ — TM of Melroe Company-Fargo, North Dakota
X-77® — Reg. TM of Chevron Chemical Company.
HAGIE™ — TM of Hagie Manufacturing Company.
DUAL® — Reg. TM of Ciba-Giegy Corp. for metolachlor herbicide.
LASSO® — Reg. TM of Monsanto Agri. Co for alachlor herbicide.

DIBROM® 8 Emulsive

NALED INSECTICIDE

ACTIVE INGREDIENTS:	BY WT.
*Naled	58%
Light Aromatic Petroleum Distillate	20%
INERT INGREDIENTS:	22%

*1,2-dibromo-2,2-dichloroethyl dimethyl phosphate

Contains 8 pounds technical naled per gallon equivalent to 7.2 pounds naled.

DIBROM® Reg.-TM of Chevron Chemical Company for naled insecticide

Product 2738 Made in U.S.A. Form 333-C

EPA Reg. No. 59639-15-AA

Valent U.S.A. Corporation © 1989

NET CONTENTS 1 GALLON

KEEP OUT OF REACH OF CHILDREN

DANGER —PELIGRO

DO NOT TAKE INTERNALLY. DO NOT GET IN EYES. DO NOT GET ON SKIN.

PRECAUCION AL USUARIO: Si usted no lee ingles, no use este producto hasta que la etiqueta le haya sido explicada ampliamente.

PRECAUTIONARY STATEMENTS

HAZARDS TO HUMANS & DOMESTIC ANIMALS

DANGER

CORROSIVE. Causes irreversible eye and skin damage. May be fatal if swallowed, inhaled or absorbed through the skin. May cause allergic skin reaction. Do not get in eyes, on skin or on clothing. When handling this material wear impervious protective clothing including gloves, apron, overshoes, chemical goggles and face shield. Do not breathe vapor or spray mist. Wear a MSHA/NIOSH approved pesticide respirator. Wash thoroughly with soap and water after handling and before eating, drinking, or using tobacco. Remove contaminated clothing and wash before reuse.

Keep children and animals off treated areas. (Does not apply to uses for adult mosquito control or fly control or rangeland pests).

STATEMENT OF PRACTICAL TREATMENT

If in eyes, hold eyelids open and flush with a steady, gentle stream of water for 15 minutes. Get medical attention immediately.

If inhaled, remove victim to fresh air. If not breathing, give artificial respiration and get medical attention immediately.

If swallowed, drink 1 or 2 glasses of water (or milk) and induce vomiting by touching the back of the throat with finger. If possible contact a physician, Poison Control Center or emergency center before inducing vomiting. Do not induce vomiting or give anything by mouth to an unconscious person. Take person and product container to the nearest medical emergency treatment center.

If on skin, wash with plenty of soap and water. Get medical attention immediately.

Note to Physicians: Emergency Information - call (415) 233-3737. Naled is a cholinesterase inhibitor. Measurement of blood cholinesterase activity may be useful in monitoring exposure. If signs of cholinesterase inhibitation appear, atropine sulfate is antidotal. 2-PAM (Protopam) is also antidotal and may be used in conjunction with atropine, but should not be used alone.

ENVIRONMENTAL HAZARDS

This pesticide is toxic to fish, aquatic invertebrates, and wildlife. Do not apply directly to water or wetlands (swamps, bogs, marshes and potholes). Runoff from treated areas may be hazardous to aquatic organisms in neighboring areas. Do not contaminate water by cleaning of equipment or disposal of wastes.

This product is highly toxic to bees exposed to direct treatment on blooming crops or weeds. Do not apply this product or allow to drift to blooming crops or weeds while bees are actively visiting the treatment area.

PHYSICAL OR CHEMICAL HAZARDS

COMBUSTIBLE, KEEP OUT OF REACH OF CHILDREN.

DO NOT USE OR STORE near heat, open flame, sparks or hot surfaces. **USE ONLY IN WELL VENTILATED AREA.** Keep container closed. Clean up spills immediately. Liquid evaporates and forms vapor (fumes) which can catch fire and burn with explosive violence. Invisible vapor spreads easily and can be set on fire by many sources such as pilot lights, welding equipment and electrical motors and switches. Fire hazard is greater as liquid temperature rises above 85°F.

DO NOT heat this container. Replace cap or bung. Emptied container still contains hazardous or explosive vapor or liquid.

DIRECTIONS FOR USE

It is a violation of Federal Law to use this product in a manner inconsistent with its labeling.

READ ENTIRE LABEL. USE STRICTLY IN ACCORDANCE WITH PRECAUTIONARY STATEMENTS AND DIRECTIONS, AND WITH APPLICABLE STATE AND FEDERAL REGULATIONS.

REENTRY STATEMENT

(Reentry Statement does not apply to uses for adult mosquito or fly control or rangeland pests.)

Keep all unprotected persons out of operating areas or vicinity where there may be danger of drift.

Do not enter treated areas prior to 24 hours after application without protective clothing.

Certain states may require more restrictive reentry intervals; consult your State Department of Agriculture for further information.

Written or oral warnings regarding use of protective clothing and accidental exposure must be given to workers who are expected to be in treated areas or in areas about to be treated.

USE DESIGNATED AMOUNT OF PRODUCT IN THE FOLLOWING GALLONS WATER PER ACRE UNLESS OTHERWISE NOTED:

GROUND APPLICATION—Vegetable and Field Crops—30 to 250 gals. **Cotton**—3 to 20 gals. **Deciduous Fruit Crops**—60 to 800 gals. **Citrus**—400 to 3,000 gals. **(On Young Fruit Plantings**—Use a minimum of 15 gals. water per acre.)

AIR APPLICATION—Field Crops—3 to 10 gals. A short residual contact insecticide for kill of insect and mite adults and larvae present at the time of application.

Begin application at first sign of insects. Repeat as necessary. Do not apply within 4 days of harvest unless otherwise noted. Do not use with highly alkaline materials such as lime or bordeaux.

Do not spray on Nectarines. Ornamental Cherries, Liquidambar or Chrysanthemums as injury may occur.

MIXING DIRECTIONS: Add to spray tank when 1/4 to 1/2 full with agitator running and maintain agitation until spraying is completed. Make new dilution for each use.

Wash metal equipment thoroughly after use to avoid corrosion.

DIRECTIONS

Do not apply this product through any type of irrigation system.

ALMONDS: Peach Twig Borer—Apply only once during the dormant or delayed-dormant period. Apply 1 pt. in combination with 1½ to 2 gals. supreme type oil or 2 to 3 gals. dormant oil spray per 100 gals. water in high gallonage spray equipment. Thorough coverage is required. Observe oil spray precautions.

BEANS, LIMA BEANS AND PEAS (DRY AND SUCCULENT FORM): Alfalfa Looper, Aphids, Spider Mites, Lygus, Leafminer, Loopers, Nitidulids, Leafhoppers—1 pt. per acre in water to cover. 100 gals. per acre is preferable. For Large Loopers, Stink Bugs—1½ pts. per acre in water to cover. Aerial Application (California Only)—1 pt. per acre in minimum of 10 gals of water per acre by air. Begin application at first sign of insects. Make only one application per 7 day period. Do not apply within 1 day of harvest. Do not feed treated vines.

BROCCOLI, CABBAGE, CAULIFLOWER, BRUSSELS SPROUTS, KALE, COLLARDS: Imported Cabbageworm, Diamond-Back Caterpillars, Aphids—1 pt. per acre **Looper**—2 pts. per acre. Do not apply within 1 day of harvest.

CELERY: Aphids, Loopers, Armyworms, Salt Marsh Caterpillars, Leafminers—Use ground application only with fine mist spray—1 pt. in 50 to 75 gals. water per acre on plants less than ¾ grown. On larger plants use 1½ pts. in 100 to 150 gals. water per acre. Do not apply within 1 day of harvest.

COTTON: Lygus—½ to 1 pt. per acre by air or ground. **Leaf Perforator, Fleahoppers, Armyworms, Bollworms, Loopers and For Suppression of Mites**—1 pt. per acre by air or ground. Apply at 7 to 10 day intervals while pests persist. **FOR HEAVY POPULATION OF LISTED INSECTS**—Use ½ pt. per acre by air or ground in control programs with other insecticides registered for use on cotton. Do not apply after first bolls open. Do not graze livestock in treated fields.

GRAPES: (Ground Use Only) Drosophila, Leafhoppers—½ to ⅔ pt. per 100 gals. water (100 to 130 gals. diluted spray per acre depending on size of vines.) Use only ½ pt. per 100 gals. water on Muscat varieties. Do not apply within 1 day of harvest. **NOTE:** Do not use after first bloom forms on grapes for fresh fruit and cannery markets as fruit russeting will occur. Do not enter vineyards for 24 hours following treatment.

ORANGES, LEMONS, GRAPEFRUIT, TANGERINES: Aphids, Soft Brown Scale Crawlers, Mites, Citrus Cutworms, Leafroller (Tortrix)—1 pt. to 100 gals. water. **Thrips (Bloom)**—½ pt. per 100 gals. water. Use up to 7 days of harvest. Do not enter citrus groves for 24 hours following treatment.

SAFFLOWER (CALIFORNIA AND ARIZONA ONLY): Aphids, Thrips, Lygus—¾ pt. per acre. **By Air**—In 3 to 10 gals. spray per acre. Do not apply within 30 days of harvest.

STRAWBERRIES: Leafroller, Spider Mites, Aphids, Omnivorous Leaftier, Spittlebugs, Thrips, Lygus—1 pt. per acre. May be used 1 day before harvest.

SUGAR BEETS: Spider Mites, Leafhoppers—1 pt. per acre by air or ground. **By Air**—Use in 1 to 5 gals. water per acre. Do not apply within 2 days of harvest.

TOBACCO: Hornworm—1 pt. to 100 gals. water, using up to 200 gals. diluted spray per acre. **Flea Beetles**—1 pt. per acre in water to cover. Apply at first sign of insects. Repeat as necessary.

WALNUTS: Aphids—½ to 1 pt. to 100 gals. water. **Aerial Application (California only)—Aphids, Red Humped Caterpillar**—Apply 1½ to 2 pts. per acre in 10 to 20 gals. of water by air. Do not apply within 10 days of harvest. Do not graze livestock in treated orchards.

GREENHOUSE, VAPOR TREATMENT OF ROSES, OTHER ORNAMENTAL PLANTS: Spider Mites, Adult Whiteflies, Aphids, Leafrollers, Mealybugs—1 fl. oz. per 10,000 cu. ft. when plants are dry. Apply undiluted on heat pipes (or in open pans on hot plates). Apply continuously or to intermittent sections of pipe on each side and on 1 or more of the pipes through center of each range depending upon width. Application to cold pipes should be followed by immediately heating pipes to 160°F. In steam heated houses, close vents for at least 1 hour after pipes are hot. With hot water systems, close house for at least 3 hours. Houses may remain closed overnight following treatment. When application is made to hot pipes, wear goggles and approved respirator. Ventilate house before workers re-enter. Wear respirator when entering treated houses to ventilate. Protect hives of bees in houses by covering with plastic or rubber sheet during treatment and until house is ventilated.

Spider Mites (Resistant and Non-resistant)—Make 3 to 4 treatments at 3 or 4 day intervals. Repeat as necessary. **Adult Whiteflies**—Make 2 applications at 7 day intervals. Repeat as necessary. **Aphids**—Apply when aphids appear. Repeat as necessary. **NOTE:** Avoid overtreatment and direct application to plants as injury may result. White Butterfly and Golden Rapture Roses, Pink Champagne Chrysanthemums, Green Wandering Jew, Poinsettias and Dutchman's Pipe may be injured by DIBROM vapor.

FOREST AND SHADE TREES: CONIFERS (Arborvitae, Douglas Fir, Fir, Hemlock, Juniper, Pine, Spruce): Suppression of Two-Spot and Red Spider Mites, Aphids, Spittlebugs, Zimmerman Moths, Juniper Webworms, Tussock Moths, Spruce Budworms, Adult Sierra Fir and Western Hemlock Bark Borers. BROADLEAF TREES (Ash, Birch, Black Walnut, Box Elder, Dogwood, Elm, Evergreen Pear, Flowering Plum and Quince, Locust, Magnolia, Maple, Oak, Sycamore, Willow): Suppression of Two-Spot and Red Spider Mites, Aphids, Leafminers, Thrips, Elm Leaf Beetles, California Oakworms (California only), Spring and Fall Cankerworms, Fall Webworm, Oak Webworm, and Tussock Moths—1 pt. in 100 gals. water. Use when infestation is evident, repeat as necessary.

FRUIT FLIES (Drosophila sp.) in and around Food Processing Plants, Loading Docks, Cull Piles and Refuse Areas, Cider Mills—Use 2 oz. in 2½ gals. water (5 pints to 100 gals.) as a coarse spray to walls, floors, doorways, windows, refuse and cull piles where insects congregate. Apply every 5 to 7 days as necessary. Do not apply to cull fruit or refuse piles to be fed to livestock. Avoid contamination of feeds, foodstuffs and food processing machinery. Do not apply when plants are in operation or when foods are present or exposed. Do not spray surfaces which will come into contact with foods. Cover food containers during spraying.

GNAT, STABLE FLY (Dog Flies), AND HOUSE FLY CONTROL—RESIDENTIAL AREAS, CROPS LISTED ON LABEL, SWAMPS, WOODLAND: Consult your State Fish and Game Agency before applying this product. It is not necessary to avoid farm buildings. Make applications during peak of infestation and repeat as necessary. See crop recommendations for use limitations near harvest. This rate of application will kill shrimp. Do not apply to tidal or marsh waters which are important shrimp producing areas. **AIRCRAFT: East**—Apply 0.8 to 4 fl. oz. per acre diluted with water. Apply 2 to 8 qts. of diluted spray per acre. **West**—Apply 1.6 to 4 fl. oz. per acre diluted with water. Apply 2 to 8 qts. of diluted spray per acre. **MIST OR COLD FOG:** Use 3 to 5 qts. per 100 gals. water. Calibrate equipment (rate of travel and output) to apply 0.1 to 0.25 lb. technical naled per acre.

ADULT MOSQUITO CONTROL—RESIDENTIAL AREAS, CROPS LISTED ON LABEL, SWAMPS, WOODLAND AND LIVESTOCK PASTURES INCLUDING DAIRY CATTLE: Consult your State Fish and Game Agency before applying this product. Make applications during peak of infestation and repeat as necessary. See crop recommendations for use limitations near harvest. Treat shrubbery and vegetation where mosquitoes may rest. Shrubbery and vegetation around stagnant pools, marshy areas, ponds and shore lines may be treated. Direct application to water is prohibited. This rate of application will kill shrimp. Do not apply to tidal or marsh waters which are important shrimp producing areas. **AIRCRAFT: East**—Apply 0.8 to 1.6 fl. oz. per acre diluted with water. Apply 2 to 8 qts. of diluted spray per acre. **West**—Apply 1.6 fl. oz. per acre diluted with water. Apply 2 to 8 qts. of diluted spray per acre. **MIST OR COLD FOG:** Use 3 to 5 qts. per 100 gals. water. Calibrate equipment (rate of travel and output) to apply 0.1 lb. technical naled per acre. Do not make more than one application every 10 days to pastures or other areas where lactating dairy animals are being held.

REDUCTION OF LIVESTOCK PESTS—Gnat, Stable Fly (Dog Fly), Housefly, Mosquito; Horn Fly (Central and North Central U.S.) IN CORRALS, HOLDING PENS, FEEDLOTS AND RANGELAND CONTAINING DAIRY AND BEEF CATTLE, HOGS, SHEEP AND HORSES: AERIAL APPLICATION: Apply 4 fl. oz. diluted with 8 fl. oz. of water (total of 12 fl. oz.) per acre (0.25 lb. technical naled per acre). **GROUND APPLICATION:** Use 5 pts. per 100 gals. water. Apply with a mist blower or similar equipment as a space treatment. Calibrate equipment (rate of travel and output) to apply 0.25 lb. technical naled per acre (approximately 5 gals. of diluted spray).

CLOVER MITES, ROACHES, EARWIGS AND LEAFHOPPERS—Outside Dwellings and in Lawns: Use 1 fl. oz. in 3 gals. water. Apply to turf and to soil surfaces around flowers, shrubs, and trees for general pest cleanup.

GRASSHOPPERS, RANGE CATERPILLARS: Rangeland—½ to ¾ pt. per acre by air or ground. **Young Grasshoppers**—½ pt. per acre by air or ground. **Armyworms**—1 pt. per acre by air or ground. **By Air**—Use in 1 to 5 gals. water per acre. Apply when hoppers hatch is completed. Repeat as necessary. Animals may be present during treatment. Do not graze lactating dairy animals on treated areas.

Do not apply under conditions involving possible drift to food, forage or other plantings that might be damaged or the crops thereof rendered unfit for sale, use or consumption.

STORAGE AND DISPOSAL

PROHIBITIONS

Do not contaminate water, food or feed by storage, disposal or cleaning of equipment.
Open dumping is prohibited.

STORAGE

Store in cool, dry place.
Keep pesticide in original container.
Do not put concentrate or dilute into food or drink containers.
Not for use or storage in or around the home.
Do not store diluted spray.
For help with any spill, leak, fire or exposure involving this material, call day or night (415) 233-3737.

PESTICIDE DISPOSAL
This product is acutely hazardous. Improper disposal of excess pesticide, spray mixture, or rinsate is a violation of Federal Law. If these wastes cannot be disposed of by use according to label instructions, contact your State Pesticide or Environmental Control Agency, or the Hazardous Waste representative at the nearest EPA Regional Office for guidance.

CONTAINER DISPOSAL
Triple rinse (or equivalent). Do not reuse container. Offer for recycling or reconditioning, or puncture and dispose of in a sanitary landfill, or incineration, or, if allowed by State and local authorities, by burning. If burned, stay out of smoke.

IMPORTANT INFORMATION.
READ BEFORE USING PRODUCT.

LIMITED WARRANTY: 1. Valent U.S.A. Corporation warrants that this product conforms to the chemical description on the label and is reasonably fit for use under average conditions when used strictly in accordance with the directions on the labeling. Valent does not make nor authorize any agent or representative to make, any other warranty, guarantee or representation, express or implied, concerning this product. Specifically, NO IMPLIED WARRANTY OF MERCHANTABILITY OR FITNESS FOR ANY PARTICULAR PURPOSE IS MADE.

2. Critical and unforeseeable factors beyond Valent's control prevent it from eliminating all risks in connection with the use of this product. Such risks include, but are not limited to, damage to plants and crops to which the product is applied, lack of complete control over the handling and application of this product, and damage caused by drift to other plants or crops. Such risks occur even though the product is reasonably fit under average conditions for the uses stated on the labeling and even though label directions are followed. Buyer and user acknowledge and assume all risks and liability (except those assumed by Valent under 1 above) resulting from handling, storage, and use of the product.

3. Precautions stated on the labeling should be followed to avoid hazardous exposure to the product.

DIBROM®
14 Concentrate

(INSECTICIDE)

ACTIVE INGREDIENT:	BY WT.
*Naled	85%
INERT INGREDIENTS:	15%

*1,2-dibromo-2,2-dichloroethyl dimethyl phosphate

Contains 14 pounds technical naled per gallon equivalent to 12.6 pounds naled.

DIBROM®-Reg. TM of Chevron Chemical Company for naled insecticide.

Product 3290 Made in U.S.A. Form 105-D

EPA Reg. No. 59639-19-AA

NET CONTENTS 1 GALLON (3.785 LITRES)

KEEP OUT OF REACH OF CHILDREN

DANGER

DO NOT TAKE INTERNALLY. DO NOT GET IN EYES. DO NOT GET ON SKIN.

PRECAUTIONARY STATEMENTS

HAZARDS TO HUMANS & DOMESTIC ANIMALS

DANGER

CORROSIVE. Causes irreversible eye and skin damage. May be fatal if swallowed, inhaled or absorbed through the skin. May cause allergic skin reaction. Do not get in eyes, on skin, or on clothing. When handling this material wear impervious protective clothing including gloves, apron, overshoes, chemical goggles and face shield. Do not breathe vapor or spray mist. Wear a MSHA/NIOSH approved pesticide respirator. Wash thoroughly with soap and water after handling and before eating, drinking, or using tobacco. Remove contaminated clothing and wash before reuse.

STATEMENT OF PRACTICAL TREATMENT

If in eyes, hold eyelids open and flush with a steady, gentle stream of water for 15 minutes. Get medical attention immediately.

If inhaled, remove victim to fresh air. If not breathing, give artificial respiration and get medical attention immediately.

If swallowed, drink 1 or 2 glasses of water (or milk) and induce vomiting by touching the back of the throat with finger. If possible contact a physician, Poison Control Center or emergency center before inducing vomiting. Do not induce vomiting or give anything by mouth to an unconscious person. Take person and product container to the nearest medical emergency treatment center.

If on skin, wash with plenty of soap and water. Get medical attention immediately.

Note to Physicians: Emergency Information - call (415) 233-3737. Naled is a cholinesterase inhibitor. Measurement of blood cholinesterase activity may be useful in monitoring exposure. If signs of cholinesterase inhibition appear, atropine sulfate is antidotal. 2-PAM (Protopam) is also antidotal and may be used in conjunction with atropine, but should not be used alone.

ENVIRONMENTAL HAZARDS

This pesticide is toxic to fish, aquatic invertebrates, and wildlife. Do not apply directly to water except when used over water as labeled for adult mosquito, blackfly or housefly control. Runoff from treated areas may be hazardous to aquatic organisms in neighboring areas. Do not contaminate water when disposing of equipment washwaters.

This product is highly toxic to bees exposed to direct treatment on blooming crops or weeds. Do not apply this product or allow it to drift to blooming crops or weeds while bees are actively visiting the treatment area.

PHYSICAL OR CHEMICAL HAZARDS

COMBUSTIBLE. KEEP OUT OF REACH OF CHILDREN.

DO NOT USE OR STORE near heat, open flame, sparks or hot surfaces. **USE ONLY IN WELL VENTILATED AREA.** Keep container closed. Clean up spills immediately. Liquid evaporates and forms vapor (fumes) which can catch fire and burn with explosive violence. Invisible vapor spreads easily and can be set on fire by many sources such as pilot lights, welding equipment and electrical motors and switches. Fire hazard is greater as liquid temperature rises above 85°F.

DO NOT heat this container. Replace cap or bung. Emptied container still contains hazardous or explosive vapor or liquid.

DIRECTIONS FOR USE

It is a violation of Federal Law to use this product in a manner inconsistent with its labeling.

READ ENTIRE LABEL. USE STRICTLY IN ACCORDANCE WITH PRECAUTIONARY STATEMENTS, AND DIRECTIONS, AND WITH APPLICABLE STATE AND FEDERAL REGULATIONS.

DIBROM 14 Concentrate is a special formulation for use undiluted or diluted in diesel oil to be applied for the control of mosquitoes, houseflies and certain other nuisance insects. **This product cannot be diluted with water.** If crystallization occurs, warm at 70° until crystals disappear.

SPECIAL MIXING DIRECTIONS: The use of DIBROM 14 Concentrate and some grades of diesel fuel may result in precipitates which cause nozzle clogging and operational difficulties. To prevent this, DIBROM 14 Concentrate should be used with Valent Additive. See Valent Additive label for recommended rates.

A technical bulletin, Operating Manual For Ultra Low Volume Application of DIBROM 14 Concentrate, is available for detailed information on equipment for ULV applications.

DIRECTIONS

ADULT MOSQUITO CONTROL in Residential Areas, Municipalities, Tidal Marshes, Swamps, Woodlands, Livestock Pastures, Feed Lots and Pastures including Dairy Cattle. It is not necessary to avoid farm buildings, dairy barns, feed or forage areas. Treat shrubbery and vegetation where mosquitoes may rest. Shrubbery and vegetation around stagnant pools, marshy areas, swamps, residential areas, municipalities, woodlands, pastures, farm buildings and feed lots may be treated.

ULTRA LOW VOLUME AIRCRAFT APPLICATION: Apply 1/2 to 1 fl. oz. of undiluted product per acre. Use the 1 oz. rate where heavy vegetation exists, i.e., woodlands, etc. Aircraft should be equipped with a corrosion resistant spray system. Large droplets may damage certain automobile finishes. Flat fan nozzles should be positioned 45° forward and only on aircraft flying 100 MPH or faster. Aircraft flying at slower speeds should be equipped with atomizing nozzles. No more than 5% of the droplets should be larger than 80 microns. Spray during periods when wind is 10 mph or less and when thermal activity is low.

AIRPLANE APPLICATION: Dilute 50 to 100 oz (1.6 to 3.1 qts.) DIBROM 14 Concentrate in 100 gals. No. 2 Fuel Oil or diesel oil. Apply at the rate of 1 gal. diluted material per acre. This is equivalent to 0.05 to 0.1 lb. actual DIBROM per acre or dilute 1.0 fl. oz. DIBROM 14 Concentrate with HAN (Heavy Aromatic Naphtha) to make 16 fl. oz. Apply at the rate of 16 fl. oz. of dilute mixture per acre. This is equivalent to 0.10 lb. actual DIBROM per acre.

THERMAL FOG APPLICATION: Dilute 3.1 qts. (100 fl. oz.) to 99 gals. No. 2 Fuel Oil or diesel oil or 10 oz. per 10 gals. oil. Apply either at the rate of 40 gals. per hour at 5 mph vehicle speed, 80 gals./hour (10 mph), or 120 gals./hour (15 mph), applying swath 300 to 400 feet wide.

ULTRA LOW VOLUME NON-THERMAL GROUND APPLICATION: Dilute 2 qts. (64 fl. oz.) of DIBROM 14 Concentrate in soybean oil or HAN to make 5 gals. of dilute solution. Apply at the rate of 3 to 6 fl. oz. per minute at a vehicle speed of 5 mph, applying a swath 300 ft. wide; at 10 mph. deliver 6 to 12 fl. oz. per minute and at 15 mph deliver 9 to 18 fl. oz. per minute. This is equivalent to 0.01 to 0.02 lb. actual DIBROM per acre. Use the high rate and slower speed to penetrate dense vegetation. Maximum effect is produced by particles 11–20 microns in size. Mass median diameter of the droplets produced should not exceed 15 microns and no droplet should be larger than 50 microns. Excessive pressure and over atomization result in poor mosquito control. To determine droplet size, treat glass microscope slides with a thin film of General Electric SC-87 Dri Film (a silicone aerosol product). This is accomplished by waving slides rapidly, 3 to 6 feet in front of aerosol discharge. Place prepared slides in a spray deposit area far enough away from discharge to avoid saturation. Using an ocular micrometer, at 450x magnification, measure a reasonable sample (about 100) of droplet diameters. Multiply droplet diameter by a 0.42 spread factor to determine original droplet size. Proceed to obtain droplet size range and median diameter.

GROUND ULV CONCENTRATE APPLICATION: Apply DIBROM 14 Concentrate undiluted at the rate of 0.6 fl. oz. per minute at 5 mph; 1.2 fl. oz. per minute at 10 mph; and 1.8 fl. oz. per minute at 15 mph, applying a 300 ft. swath. These flow rates are equivalent to 0.02 lb. actual DIBROM per acre. CAUTION—SPECIAL INSTRUCTIONS: This application can be made only under the following conditions: 1. Application can be made only by appropriately licensed and certified personnel with public agencies or private contractors operating in conjunction with or under contract with such government agencies. 2. Application in calm air conditions is to be avoided. 3. Application is not to be made in the immediate vicinity of pedestrians. 4. Vehicles used to apply must be air conditioned, and equipped with automatic speed control flow device.

BLACKFLIES, HOUSEFLIES, GNATS, CERTAIN OTHER NUISANCE INSECTS—SMALL FLYING MOTHS, CRANE FLIES, MIDGES, ADULT STABLE FLIES (Dog Flies) in Residential Areas, Municipalities, Woodlands, Livestock Pastures, Feed Lots and Pastures including dairy cattle. It is not necessary to avoid farm buildings, dairy barns, feed or forage areas.

THERMAL FOG APPLICATION: Dilute 1 gal. to 99 gals. of No. 2 Fuel Oil or diesel oil or 13 oz. per 10 gals. oil. Apply at a rate of 40 gals. per hour output at an average vehicle speed of 5 miles per hour, applying swath 300 to 400 feet wide.

AIRPLANE APPLICATION: Dilute 100 to 230 fl. oz. (3/4 to 1 3/4 gals.) DIBROM 14 Concentrate in 100 gals. No. 2 Fuel Oil or diesel oil. **Smaller insects—Gnats and Midges:** 3/4 to 1 gal. **Houseflies, Flying Moths:** 1 to 1 3/4 gals. Apply 1 gal. diluted material per acre. This is equivalent to 0.1 to 0.2 lb. actual DIBROM per acre. **Suppression of Blackflies:** Dilute 1.5 fl. oz. DIBROM 14 Concentrate with 14.5 fl. oz. HAN. Apply at the rate of 16 fl. oz. diluted mix per acre. This is equivalent to 0.16 lb. actual DIBROM per acre.

HORN FLIES (Range Cattle): AIRPLANE APPLICATION: Apply 2 to 4 fl. oz. DIBROM 14 Concentrate undiluted per acre. Use higher rate for heavier fly populations. This is equivalent to 0.2–0.4 lb. actual DIBROM per acre.

SUPPRESSION OF DEER FLIES: AIRPLANE APPLICATION: Apply 1 to 2 fl. oz. DIBROM 14 Concentrate undiluted per acre. Use higher rate in denser vegetation. This is equivalent to 0.1–0.2 lb. actual DIBROM per acre.

Note: Concentrate may corrode metal spray or mixing equipment. Flush all equipment thoroughly with aromatic solvent after use.

Do not apply under conditions involving possible drift to food, forage or other plantings that might be damaged or the crops thereof rendered unfit for sale, use or consumption.

STORAGE AND DISPOSAL

PROHIBITIONS

Do not contaminate water, food or feed by storage, disposal or cleaning of equipment.

Open dumping is prohibited.

STORAGE

Keep pesticide in original container.

Do not put concentrate or dilute into food or drink containers.

For help with any spill, leak, fire or exposure involving this material, call day or night (415) 233-3737.

PESTICIDE DISPOSAL

This product is acutely hazardous. Improper disposal of excess pesticide, spray mixture, or rinsate is a violation of Federal Law. If these wastes cannot be disposed of by use according to label instructions, contact your State Pesticide or Environmental Control Agency, or the Hazardous Waste representative at the nearest EPA Regional Office for guidance.

CONTAINER DISPOSAL

Triple rinse (or equivalent). Do not reuse container. Dispose of in a sanitary landfill or by other procedures allowed by State and local authorities.

IMPORTANT INFORMATION.
READ BEFORE USING PRODUCT.

LIMITED WARRANTY: 1. Valent U.S.A. Corporation warrants that this product conforms to the chemical description on the label and is reasonably fit for use under average conditions when used strictly in accordance with the directions on the labeling. Valent does not make nor authorizes any agent or representative to make, any other warranty, guarantee or representation, express or implied, concerning this product. Specifically, NO IMPLIED WARRANTY OF MERCHANTABILITY OR FITNESS FOR ANY PARTICULAR PURPOSE IS MADE.

2. Critical and unforeseeable factors beyond Valent's control prevent it from eliminating all risks in connection with the use of this product. Such risks include, but are not limited to, damage to plants and crops to which product is applied, lack of complete control over the handling and application of this product, and damage caused by drift to other plants or crops. Such risks occur even though the product is reasonably fit under average conditions for the uses stated on the labeling and even though label directions are followed. Buyer and user acknowledge and assume all risks and liability (except those assumed by Valent under 1 above) resulting from handling, storage, and use of the product.

3. Precautions stated on the labeling should be followed to avoid hazardous exposure to the product.

Diquat

Herbicide–H/A

TO PREVENT ACCIDENTAL POISONING, NEVER PUT INTO FOOD, DRINK OR OTHER CONTAINERS AND USE STRICTLY IN ACCORDANCE WITH ENTIRE LABEL AND PAMPHLET.

ACTIVE INGREDIENT:	BY WT.
Diquat dibromide [6, 7-dihydrodipyrido (1,2-a:2′,1′-c) pyrazinediium dibromide]	35.3%
INERT INGREDIENTS:	64.7%

Contains 2 lbs. diquat cation per gal. as 3.73 lbs. salt per gal.

X-77®-Reg. TM of Chevron Chemical Company

Product 1000-2 Made in U.S.A. Form 166-D

EPA Reg. No. 62499-4-ZA-59639

NET CONTENTS 1 GALLON

KEEP OUT OF REACH OF CHILDREN

WARNING—AVISO

SEE BELOW FOR ADDITIONAL PRECAUTIONARY STATEMENTS

PRECAUCION AL USUARIO: Si usted no lee ingles, no use este producto hasta que la etiqueta le haya sido explicada ampliamente.

PRECAUTIONARY STATEMENTS

HAZARDS TO HUMANS & DOMESTIC ANIMALS

WARNING

May be fatal if swallowed, inhaled or absorbed through skin. Causes substantial but temporary eye injury. Causes skin irritation. Contact with irritated skin, or a cut, or repeated contact with intact skin may result in poisoning. Do not get in eyes or on skin or clothing. Do not breathe spray mist. Wash thoroughly with soap and water after handling and before eating, drinking, or using tobacco. Remove contaminated clothing and wash before reuse. Wear face shield or goggles; protective clothing; rubber gloves, rubber apron, and shoe coverings impermeable to diquat when handling concentrate. When spraying, wear waterproof footwear and clothing, except when using aquatic subsurface application methods. Do not feed forage from treated crops to livestock. Keep livestock and pets out of treated fields and crop areas.

STATEMENT OF PRACTICAL TREATMENT

IF SWALLOWED: IMMEDIATELY give water or milk to drink and induce vomiting by inserting finger in throat. Take person and product container to the nearest hospital or physician fast. PROMPT TREATMENT IS ESSENTIAL TO COUNTERACT POISONING and should be initiated before signs and symptoms of injury appear. **Note to Physicians:** call the Poison Information Center (415) 233-3737 at any hour to obtain toxicology information and a diquat analysis. To be effective, treatment for diquat poisoning must begin **IMMEDIATELY.** Treatment consists of binding diquat in the gut with suspensions of activated charcoal or bentonite clay, administration of cathartics to enhance elimination, and removal of diquat from the blood by charcoal hemoperfusion or continuous hemodialysis.

SKIN CONTACT: IMMEDIATELY wash with soap and water. Remove and wash all contaminated clothing with soap and hot water before reuse. See a doctor if diquat contacts a skin cut, abrasion or area of irritation.

EYE CONTACT: IMMEDIATELY wash eyes with water for at least 15 minutes and get medical attention.

INHALATION: IMMEDIATELY get away from spray mist. Stop and check spray procedure. See a doctor if irritation persists.

WORKER SAFETY RULES

Keep all unprotected persons and children away from treated area or where there may be drift. Do not rub eyes or mouth with hands. If you feel sick in any way, STOP work and get help right away (See Practical Treatment Section).

ENVIRONMENTAL HAZARDS (AQUATIC AND NON-AQUATIC USES)

This pesticide is toxic to wildlife. Do not apply directly to water except as specified on this label. **For Non-Aquatic Use** do not apply directly to water or wetlands (swamps, bogs, marshes and potholes). Do not contaminate water when disposing of equipment washwaters. Do not use treated water for animal consumption, spraying or irrigation within 14 days after treatment. Do not use treated water for drinking purposes until 14 days after treatment. Treatment of dense weed areas can result in oxygen loss from decomposition of dead weeds. This loss can cause fish suffocation. Therefore, treat only 1/3 to 1/2 of the dense weed areas at a time and wait 14 days between treatments. DO NOT APPLY TO MUDDY WATER. Avoid creating muddy water during application. Do not apply under conditions of high wind and wave action. Do not apply where plants are covered with mud deposits. Consult your State Fish and Game Agencies before applying this product to public waters.

DIRECTIONS FOR USE

It is a violation of Federal Law to use this product in a manner inconsistent with its labeling.

READ ENTIRE LABEL AND PAMPHLET. USE STRICTLY IN ACCORDANCE WITH PRECAUTIONARY STATEMENTS AND DIRECTIONS, AND WITH APPLICABLE STATE AND FEDERAL REGULATIONS.

REENTRY STATEMENT

Keep all unprotected persons out of operating areas or vicinity where there may be drift.

For terrestrial uses, do not enter or allow entry of agricultural or maintenance personnel into treated areas within 24 hours after treatment unless protective clothing is worn. When contacting sprayed vegetation wet with spray, dew or rain, wear waterproof footwear and clothing.

For aquatic uses, do not enter treated areas while treatments are in progress.

Do not allow swimming in treated areas within 24 hours after treatment.

Certain states may require more restrictive reentry intervals; consult your State Department of Agriculture for further information.

Written or oral warnings regarding use of protective clothing and accidental exposure must be given to workers who are expected to be in treated areas or in areas about to be treated. Oral warnings must be given if there is reason to believe that written warnings cannot be understood by workers.

When oral warnings are given, warnings shall be given in a language customarily understood by workers and must indicate the specific period of time that treated areas may not be entered without protective clothing. Oral warnings must also include appropriate first aid instructions in case of accidental exposure. These instructions are given in the STATEMENT OF PRACTICAL TREATMENT at the beginning of this pamphlet.

Written warnings must include the following information, "WARNING. Area treated with Valent Diquat Herbicide-H/A on (date of application). Do not enter treated areas without appropriate protective clothing within 24 hours after treatment (or other reentry interval if your State has a more restrictive interval for this product). If accidental exposure occurs, follow the instructions below." (Written warnings must include the STATEMENT OF PRACTICAL TREATMENT given at the beginning of this pamphlet.)

DIRECTIONS

Valent Diquat Herbicide-H/A is a non-volatile herbicidal chemical for use in the desiccation of certain plants in preharvest application in order to facilitate both mechanical and hand harvesting. It is also recommended for use as a general weed killer and as an aquatic weed killer. Absorption and herbicidal action is usually quite rapid with effects visible in a few days. Do not use on lawns. Rinse all spray equipment thoroughly with water after use. AVOID SPRAY DRIFT to crops which may be damaged during application. Use Valent X-77 Spreader (non-ionic). Valent X-77 Spreader is recommended, but a similar and suitable labeled non-ionic surfactant may be substituted. Do not use dirty or muddy water for Diquat dilution as Diquat will be inactivated.

NON-CROP OR NON-PLANTED AREAS: Rights-Of-Way, Including Railroads, Highways, Roads, Dividers and Medians, Pipelines, Public Utility Lines, Including Pumping Stations, Transformer Stations and Substations. Around Electric Utilities, Commercial Buildings, Manufacturing Plants, Storage Yards, Rail Yards, Fence Lines and Parkways, Edges and Non-Flooded Portions of Ponds, Lakes and Ditches. Also Around Ornamental Gardens, Walkways, Patios, Beneath Greenhouse Benches, Along Driveways and Around Golf Courses. To Kill Undesirable Above Ground Grass and Broadleaf Weed Growth—1 to 2 qts. plus 8 to 16 oz. Valent X-77 Spreader (non-ionic) per 100 gals. water (4 teaspoonfuls Valent Diquat Herbicide-H/A plus 1/2 teaspoonful Valent X-77 Spreader (non-ionic) to 1 gal. water). Apply for full coverage and thorough weed contact. Apply to young weeds since control decreases as weeds mature. Retreatment may be necessary to control grasses and established weeds. Avoid spray contact with foliage of food crops or ornamental plants.

ALFALFA, CLOVER, GRAIN SORGHUM, SOYBEANS (Seed Crops Only): 1 1/2 to 2 pts. per acre. **On Thin Stands of Seed Alfalfa**—1 pt. per acre. **Preharvest Desiccation of Foliage of Listed Crops**—Use specified dosage in 15 to 30 gals. water by ground spray application and 5 to 10 gals. water by air. On Alfalfa and Clover, desiccation is complete in 3 to 10 days. On Grain Sorghum, apply within 1 to 2 weeks of harvest and when seeds have not more than 30% moisture. On Soybeans, apply 1 week before harvest. Do not graze or feed treated forage to livestock. Do not use seed from treated plants for food, feed or oil purposes.

POTATOES (Eastern): For Desiccation of Potato Plants to Facilitate Harvest—Apply 1 pt. per acre in 20 to 100 gals. of water with ground equipment or in 5 to 10 gals. with aerial equipment. Use higher volumes (40 to 100 gals.) for thorough coverage of heavy vine growth or when hollow cone type nozzles are used. Make a second application if necessary to obtain additional desiccation where vine growth is dense. Allow a minimum of 5 days between applications. Do not apply to drought stressed potatoes. Applications should be made at least 7 days before harvest. Add Valent X-77 Spreader (non-ionic) at 8 to 16 fl. oz. per 100 gals. of spray mix.

POTATOES (Western): For Desiccation of Potato Plants to Facilitate Harvest:—Apply 1 pt. per acre in 20 to 100 gals. of water with ground equipment or in 5 to 10 gals. with aerial equipment. Use sufficient water for thorough coverage. Make a second application if necessary to obtain additional desiccation where vine growth is dense. Allow a minimum of 5 days between applications. Do not apply to drought stressed potatoes. Application should be made at least 7 days before harvest. Add Valent X-77 Spreader (non-ionic) at 8 to 16 fl. oz. per 100 gals of spray mix.

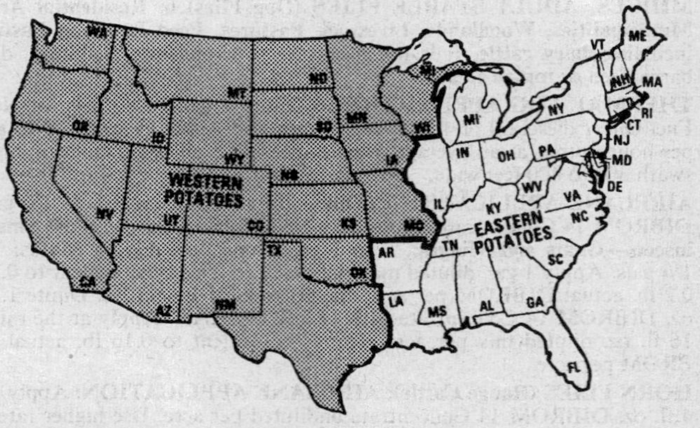

AQUATIC USE DIRECTIONS

For application only to ponds, lakes, and drainage ditches where there is little or no outflow of water and which are totally under the control of the product's user. Do not use the treated water for animal consumption, spraying, irrigation, or domestic purposes for 14 days after treatment.

WEED SPECIES	GALS./SURFACE ACRE
SUBMERSED WEEDS (Infesting Still Ponds, Lakes, Ditches or Portions Thereof):	
Bladderwort (Utricularia spp.)	1–2
Coontail (Ceratophyllum demersum)	2
Elodea (Elodea spp.)	2
Naiad (Najas spp.)	1
Pondweeds (Potamogeton spp.)†	2
Watermilfoil (Myriophyllum spp.)	1–2

† Except P. robbinsii; also Richardson Pondweed (P. richardsonii) in Minnesota.

APPLICATION: In early season where submersed weed growth has not reached the surface apply Valent Diquat Herbicide-H/A by pouring directly from the container into the water while moving slowly over the water surface in a boat. Distribute evenly over infested areas in strips 40 feet apart. In mixed weed population, use the high rate of application as indicated by weeds present.

In late season or where submersed weed growth has reached the water surface throughout the area to be treated apply Valent Diquat Herbicide-H/A by pouring as above or injecting a diluted (10–20 to 1) solution below the water surface in strips no more than 20 feet apart. Use the high rate of application as indicated by the weeds present. Where algae are present along with the submersed weeds, pre-treatment with Copper Sulphate at recommended rates is advised for best results with Valent Diquat Herbicide-H/A.

FLOATING WEEDS

Pennywort (Hydrocotyle umbellata)	1/2–3/4
Salvinia (Salvinia rotundifolia)	1/2–3/4
Waterhyacinth (Eichhornia crassipes)	1/2–3/4
Waterlettuce (Pistia Stratiotes)	1/2–3/4

APPLICATION: 150 to 200 gals. of water plus 1 pt. Valent X-77 Spreader (non-ionic) should be applied as an overall spray for control of the above weeds. For aerial application to Waterlettuce, use 7.5 gals. of water per acre plus 1 pt. Valent X-77 Spreader (non-ionic) per 100 gals. spray solution. The high rates of Valent Diquat Herbicide-H/A should be used for late season application.

Duckweed (Lemna spp.) 1

APPLICATION: For control, apply Valent Diquat Herbicide-H/A as an overall spray in 50 to 150 gals. of water plus 1 pt. Valent X-77 Spreader (non-ionic) per acre. Care should be taken to thoroughly cover all plants on water and on damp marginal areas. Reinfestation of Duckweed occurs readily from untreated areas. Retreatment with Valent Diquat Herbicide-H/A may be necessary to obtain season long control.

WEED SPECIES	GALS./100 GALS. WATER
MARGINAL WEEDS (Infesting the Edges and Other Portions of Ponds, Lakes and Ditches):	
Cattails (Typha spp.)	1.0

APPLICATION: For top kill, apply Valent Diquat Herbicide-H/A in 100 gals. water plus 1 pt. Valent X-77 Spreader (non-ionic) for full coverage and thorough weed contact. Repeat treatment as necessary to control regrowth. For best results apply before flowering.

ALGAE: Valent Diquat Herbicide-H/A will control certain filamentous algae. Spirogyra spp. and Pithophora spp.: Apply 0.5 to 1.5 p.p.m. Valent Diquat Herbicide-H/A for control. Application methods as for submersed weeds.

Formula For Parts Per Million (PPM) Recommendations

Gals. to use = p.p.m. recommended × acreage of water × average depth of water × 1.4. For example—½ p.p.m. recommendations; two acre pond; 4 feet average depth; $\frac{1}{2} \times 2 \times 4 \times 1.4 = 5.6$ gals. required.

Do not apply under conditions involving possible drift to food, forage or other plantings that might be damaged or the crops thereof rendered unfit for sale, use or consumption.

STORAGE AND DISPOSAL

PROHIBITIONS

Do not contaminate water, food or feed by storage, disposal or cleaning of equipment.

STORAGE

Keep pesticide in original container.
Do not put concentrate or dilute into food or drink containers.
Do not contaminate feed, foodstuffs or drinking water.
Do not store or transport near feed or food.
Store at temperature above 32°F.
For help with any spill, leak, fire or exposure involving this material, call day or night (415) 233-3737.

PESTICIDE DISPOSAL

This product is acutely hazardous. Improper disposal of excess pesticide, spray mixture, or rinsate is a violation of Federal Law. If these wastes cannot be disposed of by use according to label instructions, contact your State Pesticide or Environmental Control Agency, or the Hazardous Waste representative at the nearest EPA Regional Office for guidance.

CONTAINER DISPOSAL

Triple rinse (or equivalent). Do not reuse container. Incinerate, burn, or puncture and dispose of in a sanitary landfill, or dispose of by other procedures allowed by State and local authorites. If burned, stay out of smoke.

IMPORTANT INFORMATION. READ BEFORE USING PRODUCT.

LIMITED WARRANTY: 1. Valent U.S.A. Corporation warrants that this product conforms to the chemical description on the label and is reasonably fit for use under average conditions when used strictly in accordance with the directions on the labeling. Valent does not make nor authorizes any agent or representative to make, any other warranty, guarantee or representation, express or implied, concerning this product. Specifically, NO IMPLIED WARRANTY OF MERCHANTABILITY OR FITNESS FOR ANY PARTICULAR PURPOSE IS MADE.

2. Critical and unforeseeable factors beyond Valent's control prevent it from eliminating all risks in connection with the use of this product. Such risks include, but are not limited to, damage to plants and crops to which product is applied, lack of complete control over the handling and application of this product, and damage caused by drift to other plants or crops. Such risks occur even though the product is reasonably fit under average conditions for the uses stated on the labeling and even though label directions are followed. Buyer and user acknowledge and assume all risks and liability (except those assumed by Valent under 1 above) resulting from handling, storage, and use of this material.

3. Precautions stated on the labeling should be followed to avoid hazardous exposure to the product.

MONITOR®
4 Spray
(INSECTICIDE)

RESTRICTED USE PESTICIDE

For retail sale to and use only by Certified Applicators or persons under their direct supervision and only for those uses covered by the Certified Applicator's certification.

ACTIVE INGREDIENT	BY WT.
*Methamidophos	40.0%
INERT INGREDIENTS	60.0%

*O,S-dimethyl phosphoramidothioate

Contains 4 pounds active per gallon at 68°F.

MONITOR®-Reg. TM of Chevron Chemical Company for methamidophos insecticide.

Product 5197 Form 110-D Made in U.S.A.

EPA Reg. No. 62499-17-AA-59639

NET CONTENTS 30 GALLONS

KEEP OUT OF REACH OF CHILDREN
DANGER —PELIGRO

PRECAUCION AL USUARIO: Si usted no lee ingles, no use este producto hasta que la etiqueta le haya sido explicada ampliamente.

PRECAUTIONARY STATEMENTS

HAZARDS TO HUMANS & DOMESTIC ANIMALS

DANGER

Fatal if swallowed, inhaled or absorbed through skin! Do not breathe vapor or spray mist. Wear a MSHA/NIOSH approved pesticide respirator. Rapidly absorbed through skin! Do not get in eyes, on skin, or on clothing. Wear protective clothing, rubber gloves and goggles. Wash thoroughly with soap and water after handling and before eating, drinking, or using tobacco. Remove contaminated clothing and wash before reuse.

STATEMENT OF PRACTICAL TREATMENT

ANTIDOTE:

If Swallowed.—Drink 1 or 2 glasses of water or milk and induce vomiting by touching back of throat with finger. If possible contact a physician, Poison Control Center, or emergency center before inducing vomiting. Do not induce vomiting or give anything by mouth to an unconscious person. Take person and product container to the nearest emergency treatment center.

If Inhaled—Remove victim to fresh air. If not breathing, give artificial respiration and get medical attention immediately.

If On Skin—Remove contaminated clothing immediately and wash skin thoroughly with soap and water. Get medical attention immediately.

If In Eyes—Hold eyelids open and flush with a steady, gentle stream of water for 15 minutes. Get medical attention immediately.

Note to Physicians: Emergency Information - call (415) 233-3737. Methamidophos is a cholinesterase inhibitor. Measurement of blood cholinesterase activity may be useful in monitoring exposure. If signs of cholinesterase inhibition appear, atropine sulfate is antidotal. 2-PAM (Protopam) is also antidotal and may be used in conjunction with atropine, but should not be used alone.

ENVIRONMENTAL HAZARDS

This product is extremely toxic to birds and other wildlife. Birds and other wildlife in treated areas may be killed. Do not apply directly to water bodies or wetlands (e.g., lakes, streams, ponds, canals).

This product is highly toxic to bees exposed to direct treatment or residues on blooming crops or weeds. Do not apply this product or allow it to drift to blooming crops or weeds if bees are visiting the treatment area.

PHYSICAL OR CHEMICAL HAZARDS

COMBUSTIBLE. DO NOT USE OR STORE near flame, sparks or hot surfaces. **USE ONLY IN WELL VENTILATED AREA.** Keep container closed. Clean up spills immediately. Liquid evaporates and forms vapor (fumes) which can catch fire and burn with explosive violence. Invisible vapor spreads easily and can be set on fire by many sources such as pilot lights, welding equipment and electrical motors and switches. Fire hazard is greater as liquid temperature rises above 85°F.

DO NOT weld, heat or drill this container. Replace cap or bung. Emptied container still contains hazardous or explosive vapor or liquid.

DIRECTIONS FOR USE

It is a violation of Federal Law to use this product in a manner inconsistent with its labeling.

READ ENTIRE LABEL. USE STRICTLY IN ACCORDANCE WITH PRECAUTIONARY STATEMENTS AND DIRECTIONS AND WITH APPLICABLE STATE AND FEDERAL REGULATIONS.

REENTRY STATEMENT

Keep all unprotected persons out of operating areas or vicinity where there may be danger of drift.

Do not enter treated areas without protective clothing until sprays have dried.

Certain states may require more restrictive reentry intervals; consult your State Department of Agriculture for further information.

Written or oral warnings regarding use of protective clothing and accidental exposure must be given to workers who are expected to be in treated areas or in areas about to be treated. Oral warnings must be given if there is reason to believe that written warnings cannot be understood by workers.

When oral warnings are given, warnings shall be given in a language customarily understood by workers and must indicate the specific period of time that treated areas may not be entered without protective clothing. Oral warnings must also include appropriate first aid instructions in case of accidental exposure. These instructions are given in the STATEMENT OF PRACTICAL TREATMENT at the beginning of this label.

Written warnings must include the following information, "DANGER. Area treated with MONITOR 4 Spray on (date of application). Do not enter treated areas without appropriate protective clothing until sprays have dried (or other reentry interval if your State has a more restrictive interval for this product). If accidental exposure occurs, follow the instructions below." (Written warnings must include the STATEMENT OF PRACTICAL TREATMENT given at the beginning of this label).

DIRECTIONS

Apply to potatoes only by sprinkler irrigation systems. Do not apply by chemigation to any other crop, or this crop using any other type of irrigation system.

USE DESIGNATED AMOUNT OF PRODUCT IN THE FOLLOWING GALLONS OF WATER PER ACRE UNLESS OTHERWISE NOTED:

GROUND APPLICATION—25 to 125 gals. water.

AIR APPLICATION—3 to 10 gals. water.

CABBAGE (Includes Tight Heading Varieties of Chinese Cabbage), BROCCOLI, BRUSSELS SPROUTS, CAULIFLOWER: Cabbage Looper, Imported Cabbageworm, Diamondback Moth Larvae, Aphids: 1 to 2 pts. (0.5 to 1.0 lb. active) per acre. **In A Preventive Program**—Beginning when the plants are small, use the 1 pt. rate (0.5 lb. active) and apply at 7 day intervals. **For Clean-Up of Existing Populations**—Use 1½ to 2 pts. (0.75 to 1.0 lb. active) per acre. Do not apply to cabbage later than 35 days before the first harvest, to Brussels sprouts later than 14 days before the first harvest, or to cauliflower later than 28 days before the first harvest. For broccoli, do not apply the 1 pt. (0.5 lb. active) rate later than 14 days before the first harvest; do not apply the higher rates later than 21 days before the first harvest.

CELERY (Florida Only): Dipterous Leafminers—1 to 2 pts. (0.5 to 1.0 lb. active) per acre. Up to 10 applications may be made at 7 to 10 day intervals using ground equipment only. Do not apply within 21 days of harvest. Plants should be field trimmed (tops removed) before shipping or use. Celery trimmings cannot be fed to livestock and treated celery cannot be dried for celery flakes.

COTTON: Cabbage Looper, Beet Armyworm, Lygus Bugs (California and Arizona Only), Aphids, Thrips, Mites: 1 to 2 pts. (0.5 to 1.0 lb. active) per acre. **Whiteflies:** ½ to 1 pt. (0.25 to 0.5 lb. active) per acre. Apply as needed up to 50 days before harvest and before the bolls open. For whiteflies, a second application may be needed for heavy infestations. Do not feed gin trash to livestock or allow animals to graze on treated areas.

POTATOES: Aphids, Potato Tuberworm, Cabbage Looper, Colorado Potato Beetle (Except Long Island, New York), Flea Beetles, Potato Leafhopper—1½ to 2 pts. (0.75 to 1.0 lb. active) per acre. Apply in a 7 to 10 day preventive program or as necessary. Resistance to Colorado potato beetle has occurred in other areas. Consult your local extension service for details. Do not apply later than 14 days before harvest.

CHEMIGATION-POTATOES

SPRINKLER IRRIGATION APPLICATION TO POTATOES:

This product may only be applied through sprinkler irrigation systems including center pivot, lateral move, end tow, side (wheel) roll, travelers, big gun, solid set, or hand move. Do not apply this product through any other type of irrigation system. Crop injury, lack of effectiveness, or illegal pesticide residues in the crop may result from non-uniform distribution of treated water.

If you have questions about calibration, you should contact State Extension Service specialists, equipment manufacturers or other experts.

Do not connect an irrigation system (including greenhouse systems) used for pesticide application to a public water system unless the label-prescribed safety devices for public water supplies are in place.

A person knowledgeable of the chemigation system and responsible for its operation shall shut the system down and make necessary adjustments should the need arise.

The system must contain a functional check valve, vacuum relief valve, and low pressure drain appropriately located on the irrigation pipeline to prevent water source contamination from backflow.

The pesticide injection pipeline must contain a functional, automatic, quick-closing check valve to prevent the flow of fluid back toward the injection pump.

The pesticide injection pipeline must also contain a functional, normally closed solenoid-operated valve located on the intake side of the injection pump and connected to the system interlock to prevent fluid from being withdrawn from the supply tank when the irrigation system is either automatically or manually shut down.

The system must contain functional interlocking controls to automatically shut off the pesticide injection pump when the water pump motor stops.

The irrigation line or water pump must include a functional pressure switch which will stop the water pump motor when the water pressure decreases to the point where pesticide distribution is adversely affected.

Systems must use a metering pump, such as a positive displacement injection pump (e.g., diaphragm pump) effectively designed and constructed of materials that are compatible with pesticides and capable of being fitted with a system interlock.

Do not apply when wind speed favors drift beyond the area intended for treatment.

Solid Set Systems: Apply specified dosage for the entire length of the irrigation period or for a 30 to 60 minute period at the end of a regular irrigation set or as a 30 to 60 minute injection as a separate application not associated with a regular irrigation. Allow time for all lines to flush the pesticide through all nozzles before turning off irrigation water. To ensure the lines are flushed and free of remaining pesticide, a dye indicator may be injected into the lines to mark the end of the application period. See **NOTE.**

Center Pivot Systems: Inject the specified dosage per acre continuously for one complete revolution of the system. See **NOTE.**

NOTE: Constant agitation must be maintained in the chemical supply tank during the entire period of insecticide application. Inject the product with a positive displacement pump into the main line ahead of a right angle turn to insure adequate mixing.

Application of more than label recommended quantities of irrigation water per acre may result in decreased product performance by removing the chemical from the zone of effectiveness.

POSTING

Posting of areas to be chemigated is required when 1) any part of a treated area is within 300 feet of sensitive areas such as residential areas, labor camps, businesses, day care centers, hospitals, in-patient clinics, nursing homes or any public areas such as schools, parks, playgrounds, or other public facilities not including public roads, or 2) when the chemigated area is open to the public such as golf courses or retail greenhouses.

Posting must conform to the following requirements. Treated areas shall be posted with signs at all usual points of entry and along likely routes of approach from the listed sensitive areas. When there are no usual points of entry, signs must be posted in the corners of the treated areas and in any other location affording maximum visibility to sensitive areas. The printed side of the sign should face away from the treated area towards the sensitive area. The signs shall be printed in English. Signs must be posted prior to application and must remain posted until foliage has dried and soil surface water has disappeared. Signs may remain in place indefinitely as long as they are composed of materials to prevent deterioration and maintain legibility for the duration of the posting period.

All words shall consist of letters 2½ inches tall, and all letters and the symbol shall be a color which sharply contrasts with their immediate background. At the top of the sign shall be the words KEEP OUT, followed by an octagonal stop sign symbol at least 8 inches in diameter containing the word STOP. Below the symbol shall be the words PESTICIDES IN IRRIGATION WATER.

SUGAR BEETS (Arizona and California Only): Green Peach Aphid: 1 to 2 pts. (0.5 to 1.0 lb. active) per acre. Repeat at 14 day intervals as needed. Do not apply more than 6 applications of MONITOR 4 Spray per crop season. Do not apply within 30 days of harvest.

SUGAR BEETS (Seed Crop Only—California and Arizona Only): Aphids, Mites—Apply 1 to 2 pts. per acre. Use a minimum of 15 gals. water per acre with ground equipment or a minimum of 3 gals. water per acre with aircraft equipment. Repeat as necessary. Do not apply within 30 days of harvest of roots or tops for food/feed use. Do not use seed from treated plants for food or feed.

Do not apply under conditions involving drift to food, forage or other plantings that might be damaged or the crops thereof rendered unfit for sale, use or consumption.

STORAGE AND DISPOSAL

PROHIBITIONS

Do not contaminate water, food or feed by storage, disposal or cleaning of equipment.
Open dumping is prohibited.

STORAGE

Keep pesticide in original container. Do not put concentrate or dilute into food or drink containers. Not for use or storage in or around the home. Do not store next to feed or food, or transport in or on vehicles containing foodstuffs or feeds. Store in a cool, dry place. Protect from heat. For help with any spill, leak, fire or exposure involving this material, call day or night (415) 233-3737.

PESTICIDE DISPOSAL

This product is acutely hazardous. Improper disposal of excess pesticide, spray mixture, or rinsate is a violation of Federal Law. If these wastes cannot be disposed of by use according to label instructions, contact your State Pesticide or Environmental Control Agency, or the Hazardous Waste representative at the nearest EPA Regional Office for guidance.

CONTAINER DISPOSAL

Triple rinse (or equivalent). Do not reuse container. Offer for recycling or reconditioning, or puncture and dispose of in a sanitary landfill, or dispose of by other procedures allowed by State and local authorities.

IMPORTANT INFORMATION.
READ BEFORE USING PRODUCT.

LIMITED WARRANTY: 1. Valent U.S.A. Corporation warrants that this product conforms to the chemical description on the label and is reasonably fit for use under average conditions when used strictly in accordance with the directions on the labeling. Valent does not make nor authorizes any agent or representative to make, any other warranty, guarantee or representation, express or implied, concerning this product. Specifically, NO IMPLIED WARRANTY OF MERCHANTABILITY OR FITNESS FOR ANY PARTICULAR PURPOSE IS MADE.
2. Critical and unforeseeable factors beyond Valent's control prevent it from eliminating all risks in connection with the use of this product. Such risks include, but are not limited to, damage to plants and crops to which product is applied, lack of complete control over the handling and application of this product, and damage caused by drift to other plants or crops. Such risks occur even though the product is reasonably fit under average conditions for the uses stated on the labeling and even though label directions are followed. Buyer and user acknowledge and assume all risks and liability (except those assumed by Valent under 1 above) resulting from handling, storage, and use of the product.
3. Precautions stated on the labeling should be followed to avoid hazardous exposure to the product.

ORTHENE® 75 S
Soluble Powder

(INSECTICIDE)

ACTIVE INGREDIENT	BY WT.
*Acephate (O,S-Dimethyl acetylphosphoramidothioate)	75%
INERT INGREDIENTS	25%

*ORTHENE®, Acephate U.S. Pat. No. 3,716,600

Product 5332 Form 114-E Made In U.S.A.

EPA Reg. No. 59639-26-AA

KEEP OUT OF REACH OF CHILDREN
CAUTION

PRECAUTIONARY STATEMENTS

HAZARDS TO HUMANS & DOMESTIC ANIMALS

CAUTION

Harmful if swallowed. Causes eye irritation. Avoid contact with eyes, skin and clothing. Avoid breathing dust or spray mist. Do not allow children or pets to come into contact with treated areas until sprays have dried.

STATEMENT OF PRACTICAL TREATMENT

If swallowed drink 1 to 2 glasses of water (or milk) and induce vomiting by touching the back of the throat with finger. If possible contact a physician, Poison Control Center, or emergency center before inducing vomiting. Do not induce vomiting or give anything by mouth to an unconscious person. Take person and product container to the nearest emergency treatment center. In case of eye contact, wash eyes with fresh water for 15 minutes. If irritation continues, see a doctor. For skin contact, wash skin with plenty of soap and water. If inhaled, remove person from exposure area.

Note to Physicians: Emergency Information—call (415) 233-3737. Acephate is a cholinesterase inhibitor. If signs of cholinesterase inhibition appear, atropine is antidotal. 2-PAM may also be used in conjunction with atropine but should not be used alone.

WORKER SAFETY

USE ONLY WHEN WEARING THE FOLLOWING PROTECTIVE CLOTHING AND EQUIPMENT DURING MIXING/LOADING, REPAIR, AND CLEANING OF MIXING, LOADING, AND APPLICATION EQUIPMENT, AND DISPOSAL OF THE PESTICIDE: AND [areas for which a reentry interval has been established] EARLY REENTRY INTO TREATED AREAS: chemical-resistant gloves, long sleeved shirt, long legged pants, shoes, and socks. If overhead exposure is likely, such as during flagging, wear a hood or wide brimmed hat. **WORK SAFETY:** Always wash hands, face, and arms with soap and water before smoking, eating, drinking or toileting. **AFTER WORK:** Before removing gloves, wash them with soap and water., Take off all work clothes and shoes. Shower using soap and water. Wear only clean clothes when leaving job. Do not wear contaminated clothing. Personal and protective clothing worn during work must be laundered separately from household articles. Store protective clothing separately from personal clothing. Clean or launder protective clothing after each use.

ENVIRONMENTAL HAZARDS

This pesticide is toxic to birds. Applications may adversely affect birds in rangeland treatment areas. Do not apply directly to water or wetlands. Cover or soil-incorporate spills.

This product is highly toxic to bees exposed to direct treatment or residues on blooming crops or weeds. Do not apply this product or allow it to drift to blooming crops or weeds if bees are visiting the treatment area.

DIRECTIONS FOR USE

It is a violation of Federal Law to use this product in a manner inconsistent with its labeling.

READ ENTIRE LABEL AND PAMPHLET. USE STRICTLY IN ACCORDANCE WITH PRECAUTIONARY STATEMENTS AND DIRECTIONS AND WITH APPLICABLE STATE AND FEDERAL REGULATIONS.

REENTRY STATEMENT

Keep all unprotected persons out of operating areas or vicinity where there may be drift.

Do not enter treated areas without protective clothing until sprays have dried.

Certain states may require more restrictive reentry intervals; consult your State Department of Agriculture for further information.

Written or oral warnings regarding use of protective clothing and accidental exposure must be given to workers who are expected to be in treated areas or in areas about to be treated.

DIRECTIONS

Apply to cranberries only by sprinkler irrigation system.

Do not apply by chemigation to any other crop, or this corp using any other type of irrigation system.

Table of Equivalents

Amount of ORTHENE 75 S Soluble Powder	1⅓ lbs.	1 lb.	⅔ lb.	4 oz.	2 oz.
Amount Active Contained	1 lb.	¾ lb.	½ lb.	⅕ lb.	⅒ lb.

COTTON

CROPS	INSECTS	AMOUNT ORTHENE 75 S PER ACRE	FURTHER USE INSTRUCTIONS	DAYS TO HARVEST
Cotton	Thrips	2 oz. to 4 oz.	Apply in water at 3 to 10 gals. spray per acre by air (minimum 5 gals./acre in CA, AZ) or 10 to 25 gals. spray per acre by ground.	21
	Plantbugs (Lygus)	1/3 to 1 1/3 lbs. 2/3 to 1 1/3lbs. (CA, AZ)		
	Fleahopper	1/3 lb.	Apply when egges or insects first appear. Repeat as necessary to maintain control.	
	Cotton Aphid Whitefly	2/3 to 1 1/3 lbs.	**Lygus**—Use higher rate for lygus adults that have migrated into cotton.	
	Armyworms Cabbage looper	1 1/3 lbs.	**Bollworm and Tobacco budworm**—Early season light infestation use 2/3 lb. per acre. Mid and late season moderate to severe infestations use 1 to 1 1/3 lbs. per acre. Apply when eggs appear and repeat at a 3 to 7 day spray interval.	
	Bollworm Tobacco Budworm	2/3 to 1 1/3 lbs. (East of Rockies) 1 1/3 lbs. (CA, AZ)		
	Stinkbugs (CA, AZ)	1 lb.	**Pink bollworm**—Apply when insects appear and repeat at a 5 to 7 day spray interval.	
	Pink Bollworm (CA, AZ)	1 1/3 lbs.	Do not apply within 21 days of harvest. Do not feed gin trash to livestock or allow animals to graze on treated areas.	

COTTON TANK MIXES

CROPS	INSECTS	AMOUNT ORTHENE 75 S + LORSBAN® 4E PER ACRE	FURTHER USE INSTRUCTIONS	DAYS TO HARVEST
Cotton (CA, AZ)	Armyworms Bollworm Cabbage Looper Cotton Aphid Cutworms Fleahopper Grasshoppers Pink Bollworm Plant Bugs (Lygus, Mirids) Salt Marsh Caterpillar Thrips Tobacco Budworm Whitefly	1/3 to 1 1/3 lbs. + 1 to 2 pts.	Do not feed gin trash or treated forage to livestock. Do not allow livestock to graze on treated areas.	21

CROPS	INSECTS	AMOUNT ORTHENE 75 S + PYDRIN® 2.4 EC PER ACRE	FURTHER USE INSTRUCTIONS	DAYS TO HARVEST
Cotton (Except CA, AZ)	Tobacco Budworm	2/3 lb. + 2 2/3 fl. oz.	Begin applications when eggs or insects appear and apply as needed to maintain control Apply in 3 to 5 gals. of water with aerial equipment or in 10 to 25 gals. of water with ground equipment. Do not feed gin trash to livestock or allow animals to graze treated areas. Do not apply more than two (2) lbs. active ingredient of PYDRIN to a crop in a growing season. All applicable directions, restrictions and precautions on the EPA registered labels are to be followed.	21

COTTON SEED HOPPER BOX TREATMENT (EXCEPT AZ, CA AND THE BLACK LANDS OF TEXAS)

CROP	INSECTS	AMOUNT ORTHENE 75 S SOLUBLE POWDER PER ACRE	FURTHER USE INSTRUCTIONS
Cotton	Thirps Cotton Aphids Black (Greasy) Cutworm	3 to 4 oz.	Apply 3 to 4 oz. ORTHENE 75 S Soluble Powder per acre in the planter box as a dry powder. Even coverage is required for acceptable insect control. To obtain even coverage, fill planter box 1/2 full of seed, add the proper amount of ORTHENE 75 S Soluble Powder and mix thoroughly with a stick. Add the remainder of the seed to the planter box and the additional required amount of ORTHENE 75 S Soluble Powder followed by thorough mixing. For improved mixing, premix seed with ORTHENE 75 S Soluble Powder in a suitable container, then pour treated seed into

CROP	INSECTS	AMOUNT ORTHENE 75 S SOLUBLE POWDER PER ACRE	FURTHER USE INSTRUCTIONS
			planter box. Loss of ORTHENE 75 S Soluble Powder will occur if air planters are used. Plants grown from improperly covered seed will not exhibit insect control. Do not use treated seed for food or feed purposes or process for oil.

FIELD CROPS

CROPS	INSECTS	AMOUNT ORTHENE 75 S PER ACRE	FURTHER USE INSTRUCTIONS	DAYS TO HARVEST
Peppermint Spearmint	Alfalfa Looper Cutworms Aphids	1⅓ lbs.	Apply in water at 5 to 10 gals. spray per acre by air or 20 to 100 gals. spray per acre by ground. Apply when eggs or insects first appear. Make one repeat application, if necessary, to maintain control. Do not apply more than 2⅔ lbs. per acre per season.	14
Soybeans	Grasshoppers	⅓ to ⅔ lb.	Apply in water at 2 to 10 gals. spray per acre by air or 10 to 50 gals. spray per acre by ground. Repeat as necessary to maintain control. Do not graze or cut vines for hay or forage.	14
	Soybean Looper Cabbage Looper Green Cloverworm Velvetbean caterpillar Bean leaf beetle Mexican bean beetle	⅔ to 1⅓ lbs.		
	Silver Spotted Skipper (except CA)	1 lb.		
	Stink bugs Armyworms Threecornered alfalfa hopper Corn Earworm	1 to 1⅓ lbs.		
Peanuts	Grasshoppers	⅓ to ⅔ lb.	Apply in water at 10 to 50 gals. spray per acre by ground or in 5 to 10 gals. spray per acre by air. Begin applications when insects first appear and repeat as needed to maintain control. Do not feed treated forage or hay to livestock or allow animals to graze treated areas.	14 (of digging)
	Thrips	⅔ to 1 lb.		
	Corn earworm Fall armyworm Leafhoppers Loopers Velvetbean caterpillar	1 to 1⅓ lbs.		

FRUIT CROPS

CROPS	INSECTS	AMOUNT ORTHENE 75 S PER ACRE	FURTHER USE INSTRUCTIONS	DAYS TO HARVEST
Cranberries	Fireworms Spanworms Sparganothis	1⅓ lbs.	Apply in water by air, ground or with sprinklers. Use a minimum of 2 gals. spray per acre by air. Use sufficient water to give thorough coverage with ground or sprinkler equipment. **Note:** Limit to one application per growing season. Do not apply from start of bloom until all berries set.	90

SPRINKLER IRRIGATION APPLICATION TO CRANBERRIES: This product may only be applied through sprinkler irrigation systems including center pivot, lateral move, end tow, side (wheel) roll, travelers, big gun, solid set, or hand move. Do not apply this product through any other type of irrigation system.

Crop injury, lack of effectiveness, or illegal pesticide residues in the crop may result from non-uniform distribution of treated water. If you have questions about calibration, you should contact State Extension Service specialists, equipment manufacturers or other experts.

Do not connect an irrigation system (including greenhouse systsms) used for pesticide application to a public water system unless the label-prescribed safety devices for public water supplies are in place. A person knowledgeable of the chemigation system and responsible for its operation shall shut the system down and make necessary adjustments should the need arise.

The system must contain a functional check valve, vacuum relief valve, and low pressure drain appropriately located on the irrigation pipeline to prevent water source contamination from backflow. The pesticide injection pipeline must contain a functional, automatic, quick-closing check valve to prevent the flow of fluid back toward the injection pump.

The pesticide injection pipeline must also contain a functional, normally closed solenoid-operated valve located on the intake side of the injection pump and connected to the system interlock to prevent fluid from being withdrawn from the supply tank when the irrigation system is either automatically or manually shut down. The system must contain functional interlocking controls to automatically shut off the pesticide injection pump when the water pump motor stops.

The irrigation line or water pump must include a functional pressure switch which will stop the water pump motor when the water pressure decreases to the point where pesticide distribution is adversely affected. Systems must use a metering pump, such as a positive displacement injection pump (e.g., diaphragm pump) effectively designed and constructed of materials that are compatible with pesticides and capable of being fitted with a system interlock.

Do not apply when wind speed favors drift beyond the area intended for treatment.

Solid Set Systems: Apply specified dosage for the entire length of the irrigation period or for a 30 to 60 minute period at the end of a regular irrigation set or as a 30 to 60 minute injection as a separate application not associated with a regular irrigation. Allow time for all lines to flush the pesticide through all nozzles before turning off irrigation water. To ensure the lines are flushed and free of remaining pesticide, a dye indicator may be injected into the lines to mark the end of the application period. See NOTE.

Center Pivot Systems: Inject the specified dosage per acre continuously for one complete revolution of the system. See NOTE.

NOTE: Constant agitation must be maintained in the chemical supply tank during the entire period of insecticide application. Inject the product with a positive displacement pump into the main line ahead of a right angle turn to insure adequate mixing.

Application of more than label recommended quantities of irrigation water per acre may result in decreased product performance by removing the chemical from the zone of effectiveness.

VEGETABLE CROPS				
CROPS	**INSECTS**	**AMOUNT ORTHENE 75 S PER ACRE**	**FURTHER USE INSTRUCTIONS**	**DAYS TO HARVEST**
Beans and Lima Beans (Dry and Succulent Forms)	Fleahoppers Grasshoppers	⅓ to ⅔ lb.	Apply in water at a minimum of 2 gals. spray per acre by air or 20 to 100 gals. spray per acre by ground. Apply when eggs or insects first appear. Repeat at 7 to 10 day intervals as necessary. Severe insect infestations, use higher rates. Do not feed treated vines to livestock.	14 (snapbeans or dry beans) 0 (lima beans—succulent form)
	Aphids Bean leaf beetle Bean leafroller Cabbage looper Cutworms Green Cloverworm Leafhoppers Mexican bean beetle Plantbugs (Lygus) Soybean looper Thrips Whitefly	⅔ to 1⅓ lbs.		
	Armyworms Corn earworm European corn borer	1 to 1⅓ lbs.		
Brussels Sprouts Cauliflower	Green peach aphid	⅔ to 1⅓ lbs.	Apply in water at 25 to 150 gals. spray per acre by ground or a minimum of 5 gals. spray per acre by air. Use the high rate when heavy infestations of aphids are present. Begin applications when insects or eggs first appear. Repeat application as needed. Do not apply more than 8 lbs. per acre per season to Brussels Sprouts. Do not feed trimmings to livestock or allow animals to graze in treated areas.	14
	Cabbage looper Diamondback moth Imported Cabage worm	1⅓ lbs.		
Celery	Green peach aphid	⅔ to 1⅓ lbs.	Apply in water at 50 to 100 gals. spray per acre by ground or in a minimum of 5 gals. per acre by air. Apply when eggs or insects appear and repeat at 3 to 10 day intervals as necessary to maintain control. Trim plants (tops removed) before shipment or use. Do not use trimming (tops) for food or feed.	21
	Beet and Fall armyworms Cabbage looper	1⅓ lbs.		
Head Lettuce (Crisphead Type Only)	Green peach aphid	⅔ to 1⅓ lbs.	Apply in water at 5 to 10 gals. spray per acre by air or 10 to 60 gals. spray per acre (broadcast) by ground. Apply as necessary to maintain desired control. Do not apply more than 6⅔ lbs. per acre per season. Do not apply after first head begins to form to crops which germinate from mid-September through November in desert areas of AZ and CA. Do not feed trimmings to livestock or allow animals to graze in treated areas.	21 Spring. summer and early fall crops in all areas and winter crops in Florida and Texas, late fall crops in Arizona, winter crops in Arizona and California.
	Cabbage looper Armyworms	1⅓ lbs.		
Peppers (non-bell type)	Aphids	⅔ lbs.	For use in Midwestern, Eastern States and Puerto Rico only. Apply in water at 40 to 150 gals. spray per acre with ground equipment. Repeat at 7 to 10 day intervals as necessary. Do not exceed 7 applications per season.	7
Sweet peppers (Bell type)	Grasshoppers	⅓ to ⅔ lb.	Apply in water at a minimum of 3 gals. spray per acre by air or 25 to 150 gals. spray per acre by ground. Apply when eggs or insects appear. Repeat as necessary until insects have been reduced below economic numbers.	7
	Cabbage looper Green peach aphid Tobacco hornworm	⅔ to 1⅓ lbs.		
	European corn borer	1 to 1⅓ lbs.		

PASTURE, RANGELAND, WASTELAND				
CROPS	INSECTS	AMOUNT ORTHENE 75 S PER ACRE	FURTHER USE INSTRUCTIONS	DAYS TO HARVEST
Pasture Rangelend Wasteland	Grasshoppers Mormon Crickets Black Grass Bugs	⅛ to ⅙ lb. (0.094 to 0.125 lb. active ingredient)	Apply in water at a minimum of ½ gal. spray per acre by air or 10 to 20 gals. spray per acre by ground. Use the higher rate in thicker vegetation or on larger insects. ORTHOTROL or NALCO-TROL® may be added to aid in controlling drift and reducing evaporation of aerial sprays. Do not make more than one application per season. Do not apply when lactating dairy cattle are present. Do not pasture or feed treated hay to lactating dairy cattle within 21 days after application. Remove meat animals from treated areas at least one day before slaughter if they were present at application or grazed treated areas within 21 days after application.	21 for hay
	Imported Fire Ants	**Mound Treatment—Dry Method:** Evenly distribute 1 to 2 teaspoons over the mound. **Drench Method:** Mix 1 oz. in 5 gals. of water. Apply 1 gal. of mix to each mound area by sprinkling the mound until it is wet and treat a four (4) foot diamter circle around the mound.	For best results apply the material in the early morning or late afternoon when the ants are most active. Applications made under prolonged hot and dry conditions may be ineffective due to the location of the ants deep within the nest. Grass in treated area may be injured. Treat maximum of 13 mounds per acre. Do not treat mound more than once per season. Do not apply when lactating dairy cattle are present. Do not pasture or feed hay to lactating dairy cattle within 21 days after application. Remove meat animals from treated areas at least 1 day before slaughter if they were present at application or grazed treated areas within 21 days after application.	21
Non-Crop Areas (field borders, fencerows, roadsides, ditch-banks and borrow pits)	Grasshoppers	⅓ lb. (.25 lb. active Ingredient)	For early to mid-season application to grasshoppers in field borders, fencerows roadsides, ditchbanks, and borrow pits. Apply in 1 to 5 gals. of water by air, or in 10 to 20 gals. of water with ground equipment. Use the higher volumes when spraying heavier foliage. Do not graze or feed vegetation cut from treated areas.	

Do not apply under conditions involving possible drift to food, forage or other plantings that might be damaged or the crops thereof rendered unfit for sale, use or consumption.

NOTE: This product is sold by weight and package is full when packed but due to fluffy nature this material is inclined to settle.

STORAGE AND DISPOSAL

PROHIBITIONS

Do not contaminate water, food or feed by storage, disposal or cleaning of equipment.
Open dumping is prohibited.

STORAGE

Keep pesticide in original container.
Do not put concentrate or dilute into food or drink containers.
Store in cool, dry place. Protect from excessive heat.
Do not contaminate food or foodstuffs.
Do not store or transport near feed or food.
For help with any spill, leak, fire or exposure involving this material, call day or night (415) 233-3737.

PESTICIDE DISPOSAL

Wastes resulting from the use of this product may be disposed of on site or at an approved waste disposal facility.

CONTAINER DISPOSAL

Completely empty canister into application equipment. Do not reuse container. Dispose of empty canister in a sanitary landfill or by incineration, or if allowed by State and local authorities, by burning. If burned, stay out of smoke.

IMPORTANT INFORMATION READ BEFORE USING PRODUCT

LIMITED WARRANTY: 1. Valent U.S.A. Corporation warrants that this product conforms to the chemical description on the label and is reasonably fit for use under average conditions when used strictly in accordance with the directions on the labeling. Valent does not make nor authorizes any agent or representative to make, any other warranty, guarantee or representation, express or implied, concerning this product. Specifically, NO IMPLIED WARRANTY OF MERCHANTABILITY OR FITNESS FOR ANY PARTICULAR PURPOSE IS MADE.
2. Critical and unforeseeable factors beyond Valent's control prevent it from eliminating all risks in connection with the use of this product. Such risks include, but are not limited to, damage to plants and crops to which product is applied, lack of complete control over the handling and application of this product, and damage caused by drift to other plants or crops. Such risks occur even though the product is reasonably fit under average conditions for the uses stated on the labeling and even though label directions are followed. Buyer and user acknowledge and assume all risks and liability (except those assumed by Valent under 1 above) resulting from handling, storage, and use of the product.
3. Precautions stated on the labeling should be followed to avoid hazardous exposure to the product.

LORSBAN®—Reg. TM of Dow Chemical Company for chlorpyrifos insecticide
NALCO-TROL®—Reg. TM of Nalco Chemical Co.
PYDRIN®—Reg. TM of E.I. du Pont de Nemours & Co., Inc. for fenvalerate insecticide
ORTHENE®—Reg. TM of Chevron Chemical Company for acephate insecticide.
ORTHOTROL®—Reg. TM of Chevron Chemical Company.

ORTHENE®

80 SEED PROTECTANT (INSECTICIDE)

A SYSTEMIC INSECTICIDE FOR TREATMENT OF COTTON SEEDS TO HELP REDUCE INJURY TO COTTON SEEDLINGS FROM THRIPS, COTTON APHIDS, CUTWORMS AND SEED CORN MAGGOT

ACTIVE INGREDIENT	BY WT.
*Acephate (O, S-Dimethyl acetylphosphor-amidothioate)	80%
INERT INGREDIENTS	20%

*ORTHENE®, Acephate
U.S. Pat. No. 3,716,600

EPA Reg. No. 59639-29-AA

Product 5315 Form 115-D Made in U.S.A.
ORTHENE® Reg. TM of Chevron Chemical Company for acephate insecticide
ORTHOCIDE®. Reg. TM of Chevron Chemical Company for captan fungicide

KEEP OUT OF REACH OF CHILDREN

CAUTION

SEE SIDE PANEL FOR ADDITIONAL PRECAUTIONARY STATEMENTS.

DOES NOT CONTAIN DYE

PRECAUTIONARY STATEMENTS

HAZARDS TO HUMANS & DOMESTIC ANIMALS

CAUTION

Harmful if swallowed. Causes eye irritation. Avoid contact with eyes, skin and clothing. Avoid breathing dust or spray mist. Wash hands thoroughly after handling.

STATEMENT OF PRACTICAL TREATMENT

If swallowed drink 1 or 2 glasses of water (or milk) and induce vomiting by touching the back of the throat with finger. If possible contact a physician, Poison Control Center, or emergency center before inducing vomiting. Do not induce vomiting or give anything by mouth to an unconscious person. Take person and product container to the nearest emergency treatment center. In case of eye contact, wash eyes with fresh water for 15 minutes. If irritation continues, see a doctor. For skin contact, wash skin with plenty of soap and water. If inhaled, remove person from exposure area.

Note to Physicians: Emergency Information - call (415) 233-3737. Acephate is a cholinesterase inhibitor. If signs of cholinesterase inhibition appear, atropine is antidotal. 2-PAM may also be used in conjunction with atropine but should not be used alone.

WORKER SAFETY

USE ONLY WHEN WEARING THE FOLLOWING PROTECTIVE CLOTHING AND EQUIPMENT DURING MIXING/LOADING, REPAIR, AND CLEANING OF MIXING, LOADING, AND APPLICATION EQUIPMENT, AND DISPOSAL OF THE PESTICIDE: chemical-resistant gloves, long sleeved shirt, long legged pants, shoes, and socks. **WORK SAFETY:** Always wash hands, face, and arms with soap and water before smoking, eating, drinking or toileting. **AFTER WORK:** Before removing gloves, wash them with soap and water. Take off all work clothes and shoes. Shower using soap and water. Wear only clean clothes when leaving job. Do not wear contaminated clothing. Personal and protective clothing worn during work must be laundered separately from household articles. Store protective clothing separately from personal clothing. Clean or launder protective clothing after each use.

ENVIRONMENTAL HAZARDS

This pesticide is toxic to birds. Do not contaminate water by cleaning of equipment or disposal of wastes. Dispose of excess treated seeds or wastes by burial away from water supplies. Do not allow treated seed to be exposed on soil surfaces. Cover or soil-incorporate spills.

DIRECTIONS FOR USE

It is a violation of Federal Law to use this product in a manner inconsistent with its labeling.

READ ENTIRE LABEL. USE STRICTLY IN ACCORDANCE WITH PRECAUTIONARY STATEMENTS AND DIRECTIONS, AND WITH APPLICABLE STATE AND FEDERAL REGULATIONS.

USE PRECAUTION

When using formulations that do not contain dye, to comply with 21 CFR Chapter 1, Section 2.5, all seed treated with a pesticide must be colored to distinguish and prevent subsequent inadvertent use as a food for man or feed for animals.

ORTHENE 80 Seed Protectant is a water soluble systemic insecticide which helps to reduce injury to cotton seedlings by several insects.

Thrips and Cotton Aphids

ORTHENE 80 Seed Protectant gives effective reduction of these pests up to three weeks after planting.

Black (Greasy) Cutworm

ORTHENE 80 Seed Protectant gives effective reduction of Black (Greasy) Cutworm damage from planting through the 3 to 4 leaf stage of development. When planting seed into fields where large cutworms are present (5th instar and larger) economic damage may occur.

Seedcorn Maggot

ORTHENE 80 Seed Protectant effectively reduces feeding by Seedcorn Maggot on germinating and emerging cotton seedlings.

DIRECTIONS

COTTONSEED: Thrips, Cotton Aphids, Black (Greasy) Cutworm, Seedcorn Maggot — Use 8 oz. per 100 lbs. seed. Can be mixed in the slurry tank with most of the fungicide seed treaters in common use. ORTHENE 80 Seed Protectant can also be applied as a separate treatment using enough water to give adequate coverage of the seed. ORTHENE 80 Seed Protectant dissolves quickly in water with a minimum of agitation.

Observe all cautions and limitations on labeling of all products used in mixtures.

Treated seed must not be used for or mixed with food or animal feed, or processed for oil. ORTHENE 80 Seed Protectant treated seeds sold or shipped as feed or foodstuffs may be considered adulterated under state and federal law. **Seed commercially treated with ORTHENE 80 Seed Protectant should be labeled as follows: "TREATED SEED. DO NOT USE FOR FOOD, FEED OR OIL."**

STORAGE AND DISPOSAL

PROHIBITIONS

Do not contaminate water, food or feed by storage, disposal or cleaning of equipment.
Open dumping is prohibited.

STORAGE

Keep pesticide in original container.
Do not put concentrate or dilute into food or drink containers.
Store in cool, dry place.
Protect from excessive heat.
Do not store or transport near feed or food.
For help with any spill, leak, fire or exposure involving this material, call day or night (415) 233-3737.

PESTICIDE DISPOSAL

Wastes resulting from the use of this product may be disposed of on site or at an approved waste disposal facility.

CONTAINER DISPOSAL

Completely empty liner by shaking and tapping sides and bottom to loosen clinging particles. Empty residue into application equipment. Do not reuse container. Dispose of liner in a sanitary landfill or by incineration if allowed by State and local authorities. If drum is contaminated and cannot be reused, dispose of in the same manner.

IMPORTANT INFORMATION
READ BEFORE USING PRODUCT

LIMITED WARRANTY: 1. Valent U.S.A. Corporation warrants that this product conforms to the chemical description on the label and is reasonably fit for use under average conditions when used strictly in accordance with the directions on the labeling. Valent does not make nor authorizes any agent or representative to make, any other warranty, guarantee or representation, express or implied, concerning this product. Specifically, NO IMPLIED WARRANTY OF MERCHANTABILITY OR FITNESS FOR ANY PARTICULAR PURPOSE IS MADE.

2. Critical and unforeseeable factors beyond Valent's control prevent it from eliminating all risks in connection with the use of this product. Such risks include, but are not limited to, damage to plants and crops to which product is applied, lack of complete control over the handling and application of this product, and damage caused by drift to other plants or crops. Such risks occur even though the product is reasonably fit under average conditions for the uses stated on the labeling and even though label directions are followed. Buyer and user acknowledge and assume all risks and liability (except those assumed by Valent under 1 above) resulting from handling, storage, and use of the product.

3. Precautions stated on the labeling should be followed to avoid hazardous exposure to the product.

ORTHENE® 90 S

(INSECTICIDE)

ACTIVE INGREDIENT	BY WT.
*Acephate (O,S-Dimethyl acetylphosphoramidothioate)	90%
INERT INGREDIENTS	10%

*ORTHENE®, Acephate U.S. Pat. No. 3,716,600

Product 5593 Form 325-B Made in U.S.A.

EPA Reg. 59639-33-AA

KEEP OUT OF REACH OF CHILDREN

CAUTION

SEE BACK PANEL FOR ADDITIONAL PRECAUTIONARY STATEMENTS

NET WEIGHT 20 POUNDS

PRECAUTIONARY STATEMENTS

HAZARDS TO HUMANS AND DOMESTIC ANIMALS

CAUTION

Harmful if swallowed. Causes eye irritation. Avoid contact with eyes, skin and clothing. Avoid breathing dust or spray mist. Wash hands thoroughly after handling.

STATEMENT OF PRACTICAL TREATMENT

If swallowed drink 1 or 2 glasses of water (or milk) and induce vomiting by touching the back of the throat with finger. If possible contact a physician, Poison Control Center, or emergency center before inducing vomiting. Do not induce vomiting or give anything by mouth to an unconscious person. Take person and product container to the nearest emergency treatment center. In case of eye contact, wash eyes with fresh water for 15 minutes. If irritation continues, see a doctor. For skin contact, wash skin with plenty of soap and water. If inhaled, remove person from exposure area.

Note to Physicians: Emergency Information — call (415) 233-3737. Acephate is a cholinesterase inhibitor. If signs of cholinesterase inhibition appear, atropine is antidotal. 2-PAM may also be used in conjunction with atropine but should not be used alone.

WORKER SAFETY

USE ONLY WHEN WEARING THE FOLLOWING PROTECTIVE CLOTHING AND EQUIPMENT DURING MIXING/LOADING, APPLICATION, REPAIR, AND CLEANING OF MIXING, LOADING, AND APPLICATION EQUIPMENT, DISPOSAL OF THE PESTICIDE, AND [areas for which a reentry interval has been established] EARLY REENTRY INTO TREATED AREAS: chemical-resistant gloves, long sleeved shirt, long legged pants, shoes, and socks. If overhead exposure is likely, such as during flagging, wear a hood or wide brimmed hat. **WORK SAFETY:** Always wash hands, face, and arms with soap and water before smoking, eating, drinking or toileting. **AFTER WORK:** Before removing gloves, wash them with soap and water. Take off all work clothes and shoes. Shower using soap and water. Wear only clean clothes when leaving job. Do not wear contaminated clothing. Personal and protective clothing worn during work must be laundered separately from household articles. Store protective clothing separately from personal clothing. Clean or launder protective clothing after each use.

ENVIRONMENTAL HAZARDS

This pesticide is toxic to birds. Do not apply directly to water or wetlands (swamps, bogs, marshes and potholes). Cover or soil-incorporate spills.

This product is highly toxic to bees exposed to direct treatment or residues on blooming crops or weeds. Do not apply this product or allow it to drift to blooming crops or weeds if bees are visiting the treatment area.

DIRECTIONS FOR USE

It is a violation of Federal law to use this product in a manner inconsistent with its labeling.

READ ENTIRE LABEL. USE STRICTLY IN ACCORDANCE WITH PRECAUTIONARY STATEMENTS AND DIRECTIONS, AND WITH APPLICABLE STATE AND FEDERAL REGULATIONS.

REENTRY STATEMENT

Keep all unprotected persons out of operating areas or vicinity where there may be drift.

Reentry into treated areas is prohibited for 24 hours after the end of application unless the protective clothing specified elsewhere on this label is worn.

Certain states may require more restrictive reentry intervals; consult your State Department of Agriculture for further information.

Written or oral warnings regarding use of protective clothing and accidental exposure must be given to workers who are expected to be treated in areas or in areas about to be treated.

STORAGE AND DISPOSAL

PROHIBITIONS

Do not contaminate water, food or feed by storage, disposal or cleaning of equipment.

Open dumping is prohibited.

STORAGE

Keep pesticide in original container.

Do not put concentrate or dilute into food or drink containers.

Store in cool, dry place. Protect from excessive heat.

Do not contaminate food or foodstuffs.

Do not store or transport near feed or food.

For help with any spill, leak, fire or exposure involving this material, call day or night (415) 233-3737.

PESTICIDE DISPOSAL

Wastes resulting from the use of this product may be disposed of on site or at an approved waste disposal facility.

CONTAINER DISPOSAL

Completely empty bag into application equipment. Do not reuse container.

Dispose of empty bag in a sanitary landfill or by incineration, or, if allowed by State and local authorities, by burning.

If burned, stay out of smoke.

DIRECTIONS

Do not apply this product through any type of irrigation system.

COTTON

CROPS	INSECTS	LABEL RATE* (LBS. TO ACRES	FURTHER USE INSTRUCTIONS	DAYS TO HARVEST
Cotton	Thrips	1 lb. to 5 Acres (3.2 oz. per acre)	Apply in water at 3 to 10 gals. spray per acre by air (minimum 5 gals./acre in CA, AZ) or 10 to 25 gals. spray per acre by ground.	21
	Plantbugs (Lygus)	1 lb. to 4 Acres up to 1 lb. to 1 Acre (0.25 to 1 lb. per acre)	Apply when eggs or insects first appear. Repeat as necessary to maintain control. **Lygus**—Use higher rate for lygus adults that have migrated into cotton.	
		1 lb. to 2 Acres up to 1 lb. to 1 Acre (CA, AZ) (0.5 lb. to 1 lb. per acre)		
	Fleahopper	1 lb. to 4 Acres (0.25 lb. per acre)	**Bollworm and Tobacco Budworm**—Early season light infestation use 1 lb. to 2 acres. Mid and late season moderate to severe infestations use 1 lb. to 1 acre. Apply when eggs appear and repeat at a 3 to 7 day interval.	
	Cotton Aphid Whitefly	1 lb. to 2 Acres up to 1 lb. to 1 Acre (0.5 to 1 lb. per acre)		

CROPS	INSECTS	LABEL RATE* (LBS. TO ACRES	FURTHER USE INSTRUCTIONS	DAYS TO HARVEST
	Armyworms Cabbage Looper	1 lb. to 1 Acre (1 lb. per acre)	**Pink Bollworm**—Apply when insects appear and repeat at a 5 to 7 day spray interval.	
	Bollworm Tobacco Budworm	1 lb. to 2 Acres up to 1 lb. to 1 Acre (East of Rockies) (0.5 lb. to 1 lb. per acre)	Do not feed gin trash to livestock or allow animals to graze on treated areas.	
		1 lb. to 1 Acre (CA, AZ) (1 lb. per acre)		
	Stinkbugs (CA, AZ)	1 lb. to 1.25 Acre (0.8 lb. per acre)		
	Pink Bollworm (CA, AZ)	1 lb. to 1 Acre (1 lb. per acre)		

* Where a rate is given, use the higher rate when large (late instar) larvae are present or where heavy insect pressure exists.

COTTON TANK MIXES

CROPS	INSECTS	AMOUNT ORTHENE 90 S + LORSBAN® 4E PER ACRE	FURTHER USE INSTRUCTIONS	DAYS TO HARVEST
Cotton (CA, AZ)	Armyworms Bollworm Cabbage Looper Cotton Aphid Cutworms Fleahopper Grasshoppers Pink Bollworm Plant Bugs (Lygus, Mirids) Salt Marsh Caterpillar Thrips Tobacco Budworm Whitefly	4.44 oz. to 17.77 oz. + 1 to 2 pts.	Do not feed gin trash or treated forage to livestock. Do not allow livestock to graze on treated areas.	21

CROPS	INSECTS	AMOUNT ORTHENE 90 S + PYDRIN® 2,4EC PER ACRE	FURTHER USE INSTRUCTIONS	DAYS TO HARVEST
Cotton (Except CA, AZ)	Tobacco Budworm	8.93 oz. + 2²/₃ fl. oz.	Begin applications when eggs or insects appear and apply as needed to maintain control. Apply in 3 to 5 gals. of water with aerial equipment or in 10 to 25 gals. of water with ground equipment. Do not feed gin trash to livestock or allow animals to graze on treated areas. Do not apply more than two (2) lbs. active ingredient of PYDRIN to a crop in a growing season. All applicable directions, restrictions and precautions on the EPA registered labels are to be followed.	21

COTTON SEED HOPPER BOX TREATMENT (EXCEPT AZ, CA AND THE BLACK LANDS OF TEXAS)

CROPS	INSECTS	AMOUNT ORTHENE 90 S PER ACRE	FURTHER USE INSTRUCTIONS
Cotton	Thrips Cotton Aphids Black (Greasy) Cutworm	2.5 to 3.25 oz.	Apply 2.5 to 3.25 oz. ORTHENE 90S per acre in the planter box as a dry powder. Even coverage is required for acceptable insect control. To obtain even coverage, fill planter box ½ full of seed, add the proper amount of ORTHENE 90 S and mix thoroughly with a stick. Add the remainder of the seed to the planter box and the additional required amount of ORTHENE 90 S followed by thorough mixing. For improved mixing, premix seed with ORTHENE 90 S in a suitable container, then pour treated seed into planter box. Loss of ORTHENE 90 S will occur if air planters are used. Plants grown from improperly covered seed will not exhibit insect control. Do not use treated seed for food or feed purposes or process for oil.

Do not apply under conditions involving possible drift to food, forage or other plantings that might be damaged or the crops thereof rendered unfit for sale, use or consumption.

NOTE: This product is sold by weight and package is full when packed, but due to fluffy nature this material is inclined to settle.

IMPORTANT INFORMATION. READ BEFORE USING PRODUCT.

LIMITED WARRANTY: 1. Valent U.S.A. Corporation warrants that this product conforms to the chemical description on the label and is reasonably fit for use under average conditions when used strictly in accordance with the directions on the labeling. Valent does not make nor authorizes any agent or representative to make, any other warranty, guarantee or representation, express or implied, concerning this product. Specifically, NO IMPLIED WARRANTY OF MERCHANTABILITY OR FITNESS FOR ANY PARTICULAR PURPOSE IS MADE.

2. Critical and unforeseeable factors beyond Valent's control prevent it from eliminating all risks in connection with the use of this product. Such risks include, but are not limited to, damage to plants and crops to which product is applied, lack of complete control over the handling and application of this product, and damage caused by drift to other plants or crops. Such risks occur even though the product is reasonably fit under average conditions for the uses stated on the labeling and even though label directions are followed. Buyer and user acknowledge and assume all risks and liability (except those assumed by Valent under 1 above) resulting from handling, storage, and use of the product.

3. Precautions stated on the labeling should be followed to avoid hazardous exposure to the product.

LORSBAN® — Reg. TM of Dow Chemical Company for chlorpyrifos insecticide.
PYDRIN® — Reg. TM of E. I. du Pont de Nemours & Co., Inc. for fenvalerate insecticide.
ORTHENE ® — Reg. TM of Chevron Chemical Company for acephate insecticide.
VALENT U.S.A. Corporation © 1989

ORTHENE® Specialty Concentrate

(INSECTICIDE)

ACTIVE INGREDIENT	BY WT.
Acephate (O,S-dimethyl acetylphosphoramidothioate)	97%
INERT INGREDIENTS	3%

ORTHENE®, Acephate U.S. Pat. No. 3,716,600

ORTHENE®- Reg. TM of Chevron Chemical Company for acephate insecticide

Product 5478 Form 112-C
Made in U.S.A.

EPA Reg. No. 59639-30-AA

CAUTION
KEEP OUT OF REACH OF CHILDREN

PRECAUTIONARY STATEMENTS

HAZARDS TO HUMANS & DOMESTIC ANIMALS

CAUTION

Harmful if swallowed. Avoid contact with eyes, skin and clothing. Avoid breathing dust, sprays, or vapors. Wash thoroughly with soap and water after handling. Remove and launder contaminated clothing before reuse.

STATEMENT OF PRACTICAL TREATMENT

If swallowed contact physician, Poison Control Center, or emergency clinic before inducing vomiting. Give water or milk to drink and induce vomiting by touching back of throat with finger, or by administering syrup of ipecac. Do not induce vomiting or give anything by mouth to an unconscious person. If medical advice cannot be obtained, take person and product container to the nearest medical emergency treatment facility. In case of eye contact, wash eyes immediately with fresh water for at least 15 minutes. See a doctor if irritation persists. If inhaled, remove person from exposure area. For skin contact, wash skin with plenty of soap and water.

Note to Physicians: Emergency Information—call (415) 233-3737.

Acephate is a cholinesterase inhibitor. If signs of cholinesterase inhibition appear, atropine is antidotal. 2-PAM may also be used in conjunction with atropine but should not be used alone.

WORKER SAFETY

USE ONLY WHEN WEARING THE FOLLOWING PROTECTIVE CLOTHING AND EQUIPMENT DURING MIXING/LOADING, REPAIR AND CLEANING OF MIXING, LOADING AND APPLICATION EQUIPMENT, AND DISPOSAL OF THE PESTICIDE: AND [areas for which a reentry interval has been established] EARLY REENTRY INTO TREATED AREAS: chemical-resistant gloves, long sleeved shirt, long legged pants, shoes and socks. If overhead exposure is likely such as during flagging, wear a hood or wide brimmed hat. **WORK SAFETY:** Always wash hands, face and arms with soap and water before smoking, eating, drinking or toileting. **AFTER WORK:** Before removing gloves, wash them with soap and water. Take off all work clothes and shoes. Shower using soap and water. Wear only clean clothes when leaving job. Do not wear contaminated clothing. Personal and protective clothing worn during work must be laundered separately from household articles. Store protective clothing separately from personal clothing. Clean or launder protective clothing after each use.

ENVIRONMENTAL HAZARDS

This pesticide is toxic to birds. Applications may adversely affect birds in treatment areas. Do not apply directly to water or wetlands except under forest canopy. Cover or soil-incorporate spills.

DIRECTIONS FOR USE

It is a violation of Federal Law to use this product in a manner inconsistent with its labeling.

READ ENTIRE LABEL. USE STRICTLY IN ACCORDANCE WITH PRECAUTIONARY STATEMENTS AND DIRECTIONS, AND WITH APPLICABLE STATE AND FEDERAL REGULATIONS.

STORAGE AND DISPOSAL

PROHIBITIONS

Do not contaminate water, food or feed by storage, disposal or cleaning of equipment.
Open dumping is prohibited.

STORAGE

Keep pesticide in original container.
Do not put concentrate or dilute into food or drink containers.
Store in cool, dry place.
Protect from excessive heat.
Do not contaminate food or foodstuffs.
Do not store or transport near feed or food.
For help with any spill, leak, fire or exposure involving this material, call day or night (415) 233-3737.

PESTICIDE DISPOSAL

Wastes resulting from the use of this product may be disposed of on site or at an approved waste disposal facility.

CONTAINER DISPOSAL

Completely empty liner by shaking and tapping sides and bottom to loosen clinging particles. Empty residue into application equipment. Do not reuse container. Dispose of liner in a sanitary landfill or by incineration if allowed by State and local authorities. If drum is contaminated and cannot be reused, dispose of in the same manner.

DIRECTIONS

Do not apply this product through any type of irrigation system.

PASTURES, RANGELAND AND WASTELAND: Grasshoppers, Mormon Crickets and Black Grass Bugs—Apply 0.094 to 0.125 lb. active ingredient per acre in 1 qt. water by air. ORTHOTROL or NALCO-TROL® may be added to aid in controlling drift and reducing evaporation. Use the higher rate of ORTHENE in thicker vegetation or on larger insects. Do not apply when lactating dairy cattle are present. Do not pasture or feed treated hay to lactating dairy cattle within 21 days after application. Do not remove hay from treated area within 21 days after application. Remove meat animals from treated areas at least 1 day before slaughter if they were present at application or grazed treated areas within 21 days after application. Do not make more than 1 application per season.

NALCO-TROL®-Reg. TM of Nalco Chemical Co.

FOREST USES

To be used by or under supervision of personnel involved in state or federal control programs.

For Some Foliage Protection Against Gypsy Moth Larvae in Infested Forests—Apply with aircraft equipment using .52 lb. to .77 lb. (0.5 to 0.75 lb. active) in 1/2 gal. of water per acre. Apply after foliage expansion and insect hatch have begun. Use the low rate for suppression of light to moderate populations. Use the high rate for heavy populations and where extended residual control is desired. Prepare solution in mixing tank with good agitation prior to filling aircraft.

Aerial Applications for Suppression of Larval Populations of Douglas Fir Tussock Moth—Apply .77 lb. (0.75 lb. active) in 1 gal. of water per acre.

Aerial (Helicopter) Application for Suppression of Larval Populations of Western Spruce Budworm—Apply .52 lb. (0.5 lb. active) in 1 gal. water per acre.

Aerial Application for Suppression of Larval Populations of Spruce Budworm—Apply .52 lb. (0.5 lb. active) in 1/4 to 1/2 gal. water per acre.

Do not apply under conditions involving possible drift to food, forage or other plantings that might be damaged or the crops thereof rendered unfit for sale, use or consumption.

IMPORTANT INFORMATION: READ BEFORE USING PRODUCT.

LIMITED WARRANTY: 1. Valent U.S.A. Corporation warrants that this product conforms to the chemical description on the label and is reasonably fit for use under average conditions when used strictly in accordance with the directions on the labeling. Valent does not make nor authorizes any agent or representative to make, any other warranty, guarantee or representation, express or implied, concerning this product. Specifically, NO IMPLIED WARRANTY OF MERCHANTABILITY OR FITNESS FOR ANY PARTICULAR PURPOSE IS MADE.

2. Critical and unforeseeable factors beyond Valent's control prevent it from eliminating all risks in connection with the use of this product. Such risks include, but are not limited to, damage to plants and crops to which product is applied, lack of complete control over the handling and application of this product, and damage caused by drift to other plants or crops. Such risks occur even though the product is reasonably fit under average conditions for the uses stated on the labeling and even though label directions are followed. Buyer and user acknowledge and assume all risks and

liability (except those assumed by Valent under 1 above) resulting from handling, storage, and use of the product.

3. Precautions stated on the labeling should be followed to avoid hazardous exposure to the product.

ORTHENE®
Tobacco Insect Spray

(INSECTICIDE)

ACTIVE INGREDIENT	BY WT.
*Acephate (O,S-dimethyl acetylphosphoramidothioate)	75%
INERT INGREDIENTS.	25%

*ORTHENE®, Acephate U.S. Pat No. 3,716,600

ORTHENE®- Reg. TM of Chevron Chemical Company for acephate insecticide

Product 5248 Form 134-D Made in U.S.A.

EPA Reg. No. 59639-27

CAUTION
KEEP OUT OF REACH OF CHILDREN

PRECAUTIONARY STATEMENTS

HAZARDS TO HUMANS & DOMESTIC ANIMALS

CAUTION

Harmful is swallowed. Causes eye irritation. Avoid contact with eyes, skin and clothing. Avoid breathing dust or spray mist. Wash hands thoroughly after handling.

STATEMENT OF PRACTICAL TREATMENT

If swallowed drink 1 or 2 glasses of water (or milk) and induce vomiting by touching the back of the throat with finger. If possible contact a physician, Poison Control Center, or emergency center before inducing vomiting. Do not induce vomiting or give anything by mouth to an unconcious person. Take person and product container to the nearest emergency treatment center. In case of eye contact, wash eyes with fresh water for 15 minutes. If irritation continues, see a doctor. For skin contact, wash skin with plenty of soap and water. If inhaled, remove person from exposure area.

Note to Physicians: Emergency Information - call (415) 233-3737. Acephate is a cholinesterase inhibitor. If signs of cholinesterase inhibition appear, atropine is antidotal. 2-PAM may also be used in conjunction with atropine but should not be used alone.

WORKER SAFETY

USE ONLY WHEN WEARING THE FOLLOWING PROTECTIVE CLOTHING AND EQUIPMENT DURING MIXING/LOADING, REPAIR AND CLEANING OF MIXING, LOADING AND APPLICATION EQUIPMENT, AND DISPOSAL OF THE PESTICIDE: AND [areas for which a reentry interval has been established] EARLY REENTRY INTO TREATED AREAS: chemical-resistant gloves, long sleeved shirt, long legged pants, shoes and socks. If overhead exposure is likely such as during flagging, wear a hood or wide brimmed hat. **WORK SAFETY:** Always wash hands, face and arms with soap and water before smoking, eating, drinking or toileting. **AFTER WORK:** Before removing gloves, wash them with soap and water. Take off all work clothes and shoes. Shower using soap and water. Wear only clean clothes when leaving job. Do not wear contaminated clothing. Personal and protective clothing worn during work must be laundered separately from household articles. Store protective clothing after each use.

ENVIRONMENTAL HAZARDS

This pesticide is toxic to birds. Do not apply directly to water or wetlands. Cover or soil-incorporate spills.

This product is highly toxic to bees esposed to direct treatment or residues on blooming crops or weeds. Do not apply this product or allow it to drfit to blooming crops or weeds if bees are visiting the treatment area.

DIRECTIONS FOR USE

It is a violation of Federal Law to use this product in a manner inconsistent with its labeling.

READ ENTIRE LABEL. USE STRICTLY IN ACCORDANCE WITH PRECAUTIONARY STATEMENTS AND DIRECTIONS, AND WITH APPLICABLE STATE AND FEDERAL REGULATIONS.

REENTRY STATEMENT

Keep all unprotected persons out of operating areas or vicinity where there may be drift.

Do not enter treated areas without protective clothing until sprays have dried.

Certain states may require more restrictive reentry intervals; consult your State Department of Agriculture for further information.

Written or oral warnings regarding use of protective clothing and accidental exposure must be given to workers who are expected to be in treated areas or in areas about to be treated.

STORAGE AND DISPOSAL

PROHIBITIONS

Do not contaminate water, food, or feed by storage, disposal or cleaning of equipment. Open dumping is prohibited.

STORAGE

Keep pesticide in original container.
Do not put concentrate or dilute into food or drink containers.
Store in a cool, dry place.
Protect from excessive heat.
Do not store or transport near feed or food.
For help with any spill, leak, fire or exposure involving this material, call day or night (415) 233-3737.

PESTICIDE DISPOSAL

Wastes resulting from the use of this product may be disposed of on site or at an approved waste disposal facility.

CONTAINER DISPOSAL

Completely empty canister into application equipment. Do not reuse container. Dispose of empty canister in a sanitary landfill or by incineration, or if allowed by State and local authorities, by burning. If burned, stay out of smoke.

DIRECTIONS

Do not apply this product through any type of irrigation system.

TRANSPLANT WATER APPLICATION

CROP	PEST	LBS. ORTHENE Tobacco INSECT SPRAY PER ACRE
Tobacco	Flea Beetle, Green Peach Aphid, Cutworm	1

Provides control of early season flea beetles, green peach aphids and cutworms for approximately 3 to 4 weeks after transplanting. For later season control of these insects, apply a foliar spray of ORTHENE Tobacco Insect Spray.

Apply in a minimum of 100 gallons of transplant water per acre.

DO NOT APPLY MORE THAN 1 LB. ORTHENE TOBACCO INSECT SPRAY PER ACRE AS A TRANSPLANT WATER APPLICATION AS SOME PHYTOTOXICITY MAY OCCUR.

FOLIAR APPLICATION

CROP	PEST	LBS. ORTHENE TOBACCO INSECT SPRAY PER ACRE
Tobacco (Flue-Cured, Air-Cured, Dark Fire-Cured)	Grasshoppers, Green Peach Aphid, Flea Beetle	1/3 to 2/3
	Hornworm	2/3
	Tobacco Asphid	2/3 to 1
	Budworm, Cabbage Looper, Cutworm	1

Apply in 10 to 50 gals. water per acre with ground equipment or a minimum of 3 gals. per acre by air. Apply on a 7 day schedule or as needed. Do not apply within 3 days of harvest. Do not re-enter sprayed areas until the spray deposit has dried.

PLANT BED APPLICATION

CROP	PEST	LBS. ORTHENE Tobacco INSECT SPRAY PER ACRE
Tobacco Bed Treatment	Flea Beetle, Cutworm, Green Peach Aphid	1

Apply to foliage at the equivalent of 1 tablespoon in 1 gal. of water per every 1,000 sq. ft. of bed. Apply evenly to insure thorough coverage.

SOIL APPLICATION

CROP	PEST	AMOUNT ORTHENE Tobacco INSECT SPRAY
Tobacco	Imported Fire Ants	**Mound Treatment—Dry Method:** Evenly distribute 1 to 2 teaspoons over the mound. **Drench Method:** Mix 1 oz. in 5 gals. of water. Apply 1 gal. of mix to each mound area by sprinkling the mound until it is wet and treat a four (4) foot diameter circle around the mound.

For best results apply the material in the early morning or late afternoon when the ants are most active. Applications made under prolonged hot and dry conditions may be ineffective due to the location of the ants deep within the nest. Treat maximum of 13 mounds per acre. Do not treat more than once per season. Do not apply within 3 days of harvest.

Do not apply under conditions involving possible drift to food, forage or other plantings that might be damaged or the crops thereof rendered unfit for sale, use or consumption.

NOTE: This product is sold by weight and package is full when packed, but due to fluffy nature this material is inclined to settle.

IMPORTANT INFORMATION.
READ BEFORE USING PRODUCT.

LIMITED WARRANTY: 1. Valent U.S.A. Corporation warrants that this product conforms to the chemical description on the label and is reasonably fit for use under average conditions when used strictly in accordance with the directions on the labeling. Valent does not make nor authorizes any agent or representative to make, any other warranty, guarantee or representation, express or implied, concerning this product. Specifically, NO IMPLIED WARRANTY OF MERCHANTABILITY OR FITNESS FOR ANY PARTICULAR PURPOSE IS MADE.

2. Critical and unforeseeable factors beyond Valent's control prevent it from eliminating all risks in connection with the use of this product. Such risks include, but are not limited to, damage to plants and crops to which the product is applied, lack of complete control over the handling and application of this product, and damage caused by drift to other plants or crops. Such risks occur even though the product is reasonably fit under average conditions for the uses stated on the labeling and even though label directions are followed. Buyer and user acknowledge and assume all risks and liability (except those assumed by Valent under 1 above) resulting from handling, storage, and use of the product.

3. Precautions stated on the labeling should be followed to avoid hazardous exposure to the product.

ORTHENE® Turf, Tree & Ornamental Spray

(INSECTICIDE)

ACTIVE INGREDIENT	BY WT.
*Acephate (O,S-Di-methyl acetylphos-phoramidothioate)	75%
INERT INGREDIENTS	25%

*ORTHENE®, Acephate

EPA Reg. No. 59639-26-ZA
U.S. Pat. No. 3,716,600

Product 5332 Form 113-D Made in U.S.A.

ORTHENE® - Reg. TM of Chevron Chemical Company for acephate insecticide.

KEEP OUT OF REACH OF CHILDREN

CAUTION

SEE BELOW FOR ADDITIONAL PRECAUTIONARY STATEMENTS.

PRECAUTIONARY STATEMENTS

HAZARDS TO HUMANS & DOMESTIC ANIMALS

CAUTION

Harmful if swallowed. Causes eye irritation. Avoid contact with eyes, skin and clothing. Avoid breathing dust or spray mist. Wash hands thoroughly after handling.

Do not allow children or pets to come into contact with treated areas until sprays have dried.

STATEMENT OF PRACTICAL TREATMENT

If swallowed drink 1 or 2 glasses of water (or milk) and induce vomiting by touching the back of the throat with finger. If possible contact a physician, Poison Control Center, or emergency center before inducing vomiting. Do not induce vomiting or give anything by mouth to an unconscious person. Take person and product container to the nearest emergency treatment center. In case of eye contact, wash eyes with fresh water for 15 minutes. If irritation continues, see a doctor. For skin contact, wash skin with plenty of soap and water. If inhaled, remove person from exposure area.

Note to Physicians: Emergency Information — call (415) 233-3737. Acephate is a cholinesterase inhibitor. If signs of cholinesterase inhibition appear, atropine is antidotal. 2-PAM may also be used in conjunction with atropine but should not be used alone.

WORKER SAFETY

USE ONLY WHEN WEARING THE FOLLOWING PROTECTIVE CLOTHING AND EQUIPMENT DURING MIXING/LOADING, REPAIR, AND CLEANING OF MIXING, LOADING, AND APPLICATION EQUIPMENT, AND DISPOSAL OF THE PESTICIDE: AND [areas for which a reentry interval has been established] EARLY REENTRY INTO TREATED AREAS: chemical-resistant gloves, long sleeved shirt, long legged pants, shoes, and socks. If overhead exposure is likely, such as during flagging, wear a hood or wide brimmed hat. **WORK SAFETY:** Always wash hands, face, and arms with soap and water before smoking, eating, drinking or toileting. **AFTER WORK:** Before removing gloves, wash them with soap and water. Take off all work clothes and shoes. Shower using soap and water. Wear only clean clothes when leaving job. Do not wear contaminated clothing. Personal and protective clothing worn during work must be laundered separately from household articles. Store protective clothing separately from personal clothing. Clean or launder protective clothing after each use.

ENVIRONMENTAL HAZARDS

This pesticide is toxic to birds. Applications may adversely affect birds in rangeland treatment areas. Do not apply directly to water or wetlands. Cover or soil-incorporate spills.

This product is highly toxic to bees exposed to direct treatment or residues on blooming crops or weeds. Do not apply this product or allow it to drift to blooming crops or weeds if bees are visiting the treatment area.

DIRECTIONS FOR USE

It is a violation of Federal law to use this product in a manner inconsistent with its labeling. READ ENTIRE LABEL AND PAMPHLET. USE STRICTLY IN ACCORDANCE WITH PRECAUTIONARY STATEMENTS AND DIRECTIONS AND WITH APPLICABLE STATE AND FEDERAL REGULATIONS.

REENTRY STATEMENT

Keep all unprotected persons out of operating areas or vicinity where there may be drift.

Do not enter treated areas without protective clothing until sprays have dried.

Certain states may require more restrictive reentry intervals; consult your State Department of Agriculture for further information.

Written or oral warnings regarding use of protective clothing and accidental exposure must be given to workers who are expected to be in treated areas or in areas about to be treated.

DIRECTIONS

Do not apply this product through any type of irrigation system.

Table of Equivalents

Amount of ORTHENE Turf, Tree & Ornamental Spray	1⅓ lbs.	1 lb.	⅔ lb.	4 oz.	2oz.
Amount Active Contained	1 lb.	¾ lb.	½ lb.	⅕ lb.	⅒ lb.

SPECIALTY USES

CROPS	INSECTS	AMOUNT ORTHENE TURF, TREE & ORNAMENTAL SPRAY PER ACRE	FURTHER USE INSTRUCTIONS	DAYS TO HARVEST
Non-Crop Areas	Imported Fire Ants	**Mound Treatment— Dry Method:** Evenly distribute 1 to 2 teaspoons over the mound. **Drench Method:** Mix 1 oz. in 5 gals. of water. Apply 1 gal. of mix to each mound area by sprinkling the mound until it is wet and treat a four (4) foot diameter circle around the mound.	For best results apply the material in the early morning or late afternoon when the ants are most active. Applications made under prolonged hot and dry conditions may be ineffective due to the location of the ants deep within the nest. Grass in treated area may be injured. Do not treat mound more than once per season.	

PLANTS,	INSECTS	AMOUNT ORTHENE TURF, TREE & ORNA-MENTAL SPRAY PER 100 GALS.	TIME OF APPLICATION	RECOMMENDED APPLICATION
Container Grown Nursery Stock (Arborvitae Azelea Camelia Rhododendron Roses Viburnum Yew)	Black Vine Weevil Strawberry Root Weevil	1 lb. (16 oz.)	Application should be made by mid-September for greenhouse stock and by mid-October for outdoor stock. Consult your local county extension agent for information on the identification and control of root weevils on ornamentals.	Apply the specified amount of ORTHENE Turf, Tree & Ornamental Spray per 100 gals. of solution so as to thoroughly drench the root system.

SMALL SPRAY EQUIPMENT

PREPARATION OF ORTHENE SPRAY MIX USING A 16 OZ. MEASURING CUP.

LABEL RECOMMENDATION ORTHENE IN 100 GALS. WATER	EQUIVALENT ORTHENE IN 20 GALS. WATER*	EQUIVALENT ORTHENE IN 1 GAL. WATER (TSP./GAL.)
1/3 lb.	2 1/2 oz.	1
2/3 lb.	5 oz.	2
1 lb.	7 1/2 oz.	3
1 1/3 lbs.	10 oz.	4

* Use in 16 oz. measuring cup.

TREES & SHRUBS

PLANTS	INSECTS	AMOUNT ORTHENE TURF, TREE & ORNA-MENTAL SPRAY PER 100 GALS.	TIME OF APPLICATION	RECOMMENDED APPLICATION
Trees and Shrubs (except Flowering Crabapple see below)	Aphids Bagworms Birch Leafminer Tent Caterpillar* Lace Bugs Leafrollers	1/3 lb. (5.3 oz)	As the insects begin to appear.	Apply the specified amount of ORTHENE Turf, Tree & Ornamental Spray in 100 gals. water with a hydraulic sprayer. The addition of a suitable sticker improves control of Gypsy Moth larvae.
	Douglas Fir Tussock Moth Larvae Gypsy Moth Larvae* Webworms	2/3 lb. (10.5 oz.)	As the insects begin to appear.	*Mist blower application. Adjust rates to 1 1/3 lbs. per 100 gals. water for Gypsy Moth control and 1.0 lb. per 100 gals. water for Tent Caterpillar control.
	Scales (Crawlers)	2/3 lb. (10.5 oz.)	As crawlers begin to appear. Repeat applications, at a 2 week or more interval, may be necessary where there is continuous crawler production.	Do not apply to Huckleberry, Balm of Gilead, Cottonwood, Lombardy Poplar and Viburnum suspensum. Nursery crops. Before treating large plantings, spray only a few plants and observe two weeks for phytotoxicity.
Trees and Shrubs	Ponderosa Pine Needle Miner	2/3 lb. (10.5 oz.)	Time of application is important. Consult your Farm advisor or County Extension Agent.	Apply the specified amount of ORTHENE Turf, Tree & Ornamental Spray in 100 gals. water with a hydraulic sprayer as a full coverage spray.
	Grasshoppers	2/3 lb. (10.5 oz.)	As the grasshoppers begin to appear.	
	California Oakworm Cankerworms (Spring and Fall)	1/3 to 2/3 lb. (5.3 to 10.5 oz.)	As the insects begin to appear. Use the higher amount when the larger larvae are present.	Apply the specified amount of ORTHENE Turf, Tree & Ornamental Spray in 100 gals. water with a hydraulic sprayer.
	Nantucket Pine Tip Moth Larvae	1 lb. (16 oz.)	Time of appliction is important. Consult your Farm Advisor or County Extension Agent. Repeat applications will be required for subsequent generations.	
	Root Weevil Adults	1 lb. (16 oz.)	Apply when first feeding damage occurs. Repeat applications, at four week intervals until the first heavy frost, may be necessary for complete foliage protection.	Apply the specified amount of ORTHENE Turf, Tree & Ornamental Spray in 100 gals. water with hydraulic sprayer as a full coverage spray.

PLANTS	INSECTS	AMOUNT ORTHENE TURF, TREE & ORNAMENTAL SPRAY PER 100 GALS.	TIME OF APPLICATION	RECOMMENDED APPLICATION
	Box Elder Bugs Sawflies Budworms Leafhoppers	1 lb. (16 oz.)	As the insects begin to appear.	Apply the specified amount of ORTHENE Turf, Tree & Ornamental Spray in 100 gals. water with a hydraulic sprayer as a full coverage spray.
	Japanese Beetle	1⅓ lbs. (21.3 oz.)	As the Janpense Beetles begin to appear. Repeat applications, at 2 week intervals, may be necessary.	
	Elm Leaf Beetle (larvae)	1⅓ lb. (21.3 oz.)	As the larvae begin to appear. ORTHENE will not prevent Elm Leaf Beetle eggs from hatching.	Apply the specified amount of ORTHENE Turf, Tree & Ornamental Spray in 100 gals. water with a hydraulic sprayer as a full coverage spray.
Douglas Fir Christmas Trees	Douglas Fir Needle Midge	⅔ lb. (10.5 oz.)	Application should be made no more than 2 weeks prior to bud burst. For additional pest management information, consult your county extension service.	Apply the specified amount of ORTHENE Turf, Tree & Ornamental Spray in not less than 2 gals. of spray per acre by air or in 100 gals. of spray per acre by ground.
Flowering Crabapples	Aphids Tent Caterpillars Leafrollers	⅓ lb. (5.3 oz.)	As the insects begin to appear.	Apply the sepcified amount of ORTHENE Turf, Tree & Ornamental Spray in 100 gals. water with a hydraulic sprayer. Do not apply more often than 3 times in a growing season at a 4 week interval. **Caution:** Phytotoxicity has occurred on the following Crabapple varieties: Hopa, Ichonoski, Malusfloribunda, Pink Perfection, Red Wine and Snow Cloud.

COMMERCIAL TURFGRASS, LAWNS, AND OTHER RECREATIONAL TURFGRASS AREAS
DO NOT ALLOW LIVESTOCK TO GRAZE TREATED AREAS. DO NOT FEED TREATED GRASS TO LIVESTOCK.

PLANTS	INSECTS	AMOUNT ORTHENE TURF, TREE & ORNAMENTAL SPRAY PER 1,000 SQ. FT.	TIME OF APPLICATION	RECOMMENDED APPLICATION
Turfgrass (Residential & Commercial)	Fall Armyworm Yellow Striped Armyworm Southern Armyworm	½ to 1.2 oz. (1⅓ to 3¼ lbs. per acre)	As the insects appear. A repeat application at 2 week intervals may be necessary.	Apply the specified amount of ORTHENE Turf, Tree & Ornamental Spray per 1,000 sq. ft. Use a minimum of 5 gals. water per 1,000 sq. ft. to obtain good coverage.
	Cutworm	1.2 to 2.4 oz. (3¼ to 6⅔ lbs. per acre)		
	Chinch bugs	1.2 to 2.4 oz (3-¼ to 6-⅔ lbs per acre)	Apply as needed for adult population knockdown (10 to 14) days).	Apply the specified amount of ORTHENE Turf, Tree & Ornamental Spray per 1,000 sq. ft. Use 1 to 15 gals water per 1,000 sq. ft. to obtain good coverage.
	Fleas	1.2 to 2.4 oz	Apply as needed for knockdown (7 to 10 days) only of existing adult populations.	
	Sod Webworm (Crambus spp.)	½ to 1 oz. (1⅓ to 2¾ lbs. per acre)	As sod webworms begin to appear. Use the higher amount when quick knockdown is needed or with heavy infestations. Repeat application may be necessary. Do not repeat at more than 1 week intervals.	Apply the specified amount of ORTHENE Turf, Tree & Ornamental Spray per 1,000 sq. ft. Use 1 to 15 gals. water per 1,000 sq. ft. to obtain good coverage.
	Leafhopper	1 oz. (2¾ lbs. per acre)	As the leafhoppers begin to appear. A repeat application at 1 week intervals may be necessary.	Apply the specified amount of ORTHENE Turf, Tree & Ornamental Spray per 1,000 sq. ft. Use 1 to 15 gals. water per 1,000 sq. ft. to obtain good coverage.
	Mole Crickets (Except CA)	1.0 to 1.9 oz. (2.66-5.2 lbs. per acre)	As mole crickets begin to appear. For knockdown of existing populations, more than one application may be required throughout the growing season. For heavy infestations, use the higher dosage rate.	Apply the specified amount of ORTHENE Turf, Tree & Ornamental Spray per 1,000 sq. ft. Use 1 to 15 gals. water per 1,000 sq. ft. to obtain good coverage. Apply during late afternoon or early evening hours and after an irrigation. Do not irrigate after application.

PLANTS	INSECTS	AMOUNT ORTHENE TURF, TREE & ORNAMENTAL SPRAY PER 1,000 SQ. FT.	TIME OF APPLICATION	RECOMMENDED APPLICATION
	Greenbug (Schizaphis graminun) Grasshoppers	½ oz. (1⅓ lbs. per acre)	Apply when insects or their damage first appears. Repeat as necessary. Application is not to be repeated at more than 1 week intervals.	Apply the specified amount of ORTHENE Turf, Tree & Ornamental Spray. Use 4 gals. of water per 1,000 sq. ft. to obtain good coverage. Do not mow turfgrass for at least 24 hours after application.
Dichondra	Cutworm Flea Beetle Southern Armyworm Yellow Striped Armyworm	1.0 to 1.9 oz. (2.7 to 5.2 lbs. per acre)	As the insects appear. Repeat at 2 week intervals as necessary.	Apply the specified amount of ORTHENE Turf, Tree & Ornamental Spray per 1,000 sq. ft. Use a minimum of 15 gals. of water per 1,000 sq. ft. to obtain good coverage.

COMMERCIAL TURFGRASS, LAWNS, AND OTHER RECREATIONAL TURFGRASS AREAS DO NOT ALLOW LIVESTOCK TO GRAZE TREATED AREAS. DO NOT FEED TREATED GRASS TO LIVESTOCK

PLANTS	INSECTS	AMOUNT ORTHENE TURF, TREE & ORNAMENTAL SPRAY	TIME OF APPLICATION	RECOMMENDED APPLICATION
Turfgrass	Imported Fire Ants	**Mound Treatment—Dry Method:** Evenly distribute 1 to 2 teaspoons over the mound. **Drench Method:** Mix 1 oz. in 5 gals. of water. Apply 1 gal. of mix to each mound area by sprinkling the mound until it is wet and treat a four (4) foot diameter circle around the mound.	For best results apply the material in the early morning or late afternoon when the ants are most active. Applications made under prolonged hot and dry conditions may be ineffective due to the location of the ants deep within the nest.	Apply the specified amount of ORTHENE Turf, Tree & Ornamental Spray as directed. Grass in treated areas may be injured. Do not treat mound more than once per season.

OUTDOOR WASP AND PERIMETER SPRAY

LOCATION	INSECTS	AMOUNT ORTHENE TURF, TREE & ORNAMENTAL SPRAY PER GALLON	TIME OF APPLICATION	RECOMMENDED APPLICATION
Outdoor wasp and perimeter area.	Wasps	1.6 oz. (0.10 lbs.)	Treat early or late in the day, as wasps are generally less active during these times.	Apply the specified amount of ORTHENE Turf, Tree & Ornamental Spray per each gallon of water used. Apply as a spot treatment to the nest, nest entrance, and surrounding areas where the wasps alight.
	Cockroaches, Ants, Pillbugs, Earwigs, (Perimeter Treatment)	1.6 oz (0.10 lbs.)	As the insects appear.	Apply specified amount of ORTHENE Turf, Tree & Ornamental Spray per each gallon of water used. Apply to a band of soil 6 to 10 feet adjacent to the structure and to a height of 2 to 3 feet on the foundation where pests may be active or may find entrance. Also apply as a residental spray or with a paint brush to surfaces of buildings, window frames, shutters, entry-ways, screens, eaves patios, garages, carports, around garbage areas and other areas where these pests congregate.

OUTDOOR FLORAL CROPS

PLANTS	INSECTS	AMOUNT ORTHENE TURF, TREE & ORNAMENTAL SPRAY PER 100 GALS.	TIME OF APPLICATION	RECOMMENDED APPLICATION
Daisies Dahlias Easter Lily Gladioli Gyposophila Pachysandra Peony Roses Sedum Statice Strawflower Yarrow Zinnia	Aphids Thrips Lygus	⅔ lb. (10.5 oz.)	As insects begin to appear. Repeat applications may be necessary.	Apply the specified amount of ORTHENE Turf, Tree & Ornamental Spray in 100 gals. water with a hydraulic sprayer as a full coverage spray. Multiple applications may cause a slight tip burn or marginal leaf necrosis on some varieties. Test on a few plants to determine varietal susceptibility.
Roses Boston Ivy	Japanese Beetle	1⅓ lbs. (21.3 oz.)	As the Japanese Beetles begin to appear. Repeat applications at 2 week intervals may be necessary.	

GREENHOUSE FLORAL AND FOLIAGE PLANT CROPS

PLANTS	INSECTS	AMOUNT ORTHENE TURF, TREE & ORNAMENTAL SPRAY PER 100 GALS.	TIME OF APPLICATION	RECOMMENDED APPLICATION
Roses	Leafrollers	2/3 to 1 lb. (10.5 to 16 oz.)	As leafrollers begin to appear. Use the higher amount when the large larvae are present.	Apply the specified amount of ORTHENE Turf, Tree & Ornamental Spray in 100 gals. water with a hydraulic sprayer as a full coverage spray.
Foliage Plants Orchids Anthuriums Cacti Poinsettia	Aphids	1/3 lb. (5.3 oz.)	As aphids begin to appear.	Apply the specified amount of ORTHENE Turf, Tree & Ornamental Spray in 100 gals. water with a hydraulic sprayer as a full coverage spray. The addition of a wetting agent may be required on difficult to wet foliage.
	Mealybugs Thrips Whiteflies	2/3 lb. (10.5 oz.)	As the insects begin to appear. A repeat application, at a 2 week interval, may be necessary for control of mealybugs and white flies.	
	Scales (Crawlers)	2/3 lb. (10.5 oz.)	As crawlers begin to appear. Repeat applications, at a 2 week or more interval, may be necessary where there is continuous crawler production.	**Caution:** Phytotoxicity has occurred on the following foliage plants: Bletchum gibbum, Cissus antarctica, Ficus triangularis, Fittonia verschaffeltii, Maranta leuconeura kerchoveana, Pachystachya lutea, Plectranthus australis, Polypodium aureus, Polystichum, Pteris ensiformis, Tolmica menziesii. Before treating large plantings spray only a few plants and observe 2 weeks for varietal phytotoxicity.
Roses Carnations Chrysanthemums	Aphids Thrips	2/3 lb. (10.5 oz.)	As aphids begin to appear. As thrips begin to appear or at the tight flower bud stage. Repeat applications may be necessary.	Apply the specified amount of ORTHENE Turf, Tree & Ornamental Spray in 100 gals. water with a hydraulic sprayer as a full coverage spray. Do not apply more often than once every 28 days to Carnations and Chrysantheumum. **Caution:** Phytotoxicity has occurred on the following Chrysanthemum varieties: Albatross, Bonnie Jean, Dixie, Garland, Gent, Iceberg, Pride, Showoff, Statesman, Tally Ho, Westward Ho, and Wild Honey. Before treating large Chrysanthemum plantings, spray only a few plants and observe two weeks for varietal phytotoxicity.

Do not apply under conditions involving possible drift to food, forage or other plantings that might be damaged or the crops thereof rendered unfit for sale, use or consumption.

STORAGE AND DISPOSAL

PROHIBITIONS: Do not contaminate water, food or feed by storage, disposal or cleaning of equipment. Open dumping is prohibited.

STORAGE: Keep pesticide in original container. Do not put concentrate or dilute into food or drink containers. Store in cool, dry place. Protect from excessive heat. Do not contaminate food or foodstuffs. Do not store or transport near feed or food. For help with any spill, leak, fire or exposure involving this material, call day or night (415) 233-3737.

PESTICIDE DISPOSAL: Wastes resulting from the use of this product may be disposed of on site or at an approved waste disposal facility.

CONTAINER DISPOSAL: Completely empty canister into application equipment. Do not reuse container. Dispose of empty canister in a sanitary landfill or by incineration, or if allowed by State and local authorities, by burning. If burned, stay out of smoke.

Note: This product is sold by weight and package is full when packed, but due to fluffy nature this material is inclined to settle.

IMPORTANT INFORMATION. READ BEFORE USING PRODUCT.

LIMITED WARRANTY: 1. Valent U.S.A. Corporation warrants that this product conforms to the chemical description on the label and is reasonably fit for use under average conditions when used strictly in accordance with the directions on the labeling. Valent does not make nor authorizes any agent or representative to make, any other warranty, guarantee or representation, express or implied, concerning this product. Specifically, NO IMPLIED WARRANTY OF MERCHANTABILITY OR FITNESS FOR ANY PARTICULAR PURPOSE IS MADE.

2. Critical and unforeseeable factors beyond Valent's control prevent it from eliminating all risks in connection with the use of this product. Such risks include, but are not limited to, damage to plants and crops to which product is applied, lack of complete control over the handling and application of this product, and damage caused by drift to other plants or crops. Such risks occur even though the product is reasonably fit under average conditions for the uses stated on the labeling and even though label directions are followed. Buyer and user acknowledge and assume all risks and liability (except those assumed by Valent under 1 above) resulting from handling, storage, and use of the product.

3. Precautions stated on the labeling should be followed to avoid hazardous exposure to the product.

VOLCK® Supreme Spray

(INSECTICIDE-MITICIDE)

A Superior Oil Spray
For Western Agriculture

ACTIVE INGREDIENT	BY WT.
Petroleum Oil	98%
INERT INGREDIENTS	2%

Petroleum Oil Min. Unsulfonated Residue—99%

Petroleum Oil Classification—Unclassified.

Product 2675 Made in U.S.A. Form 178-F

VOLCK® Reg. TM of Chevron Chemical Company

EPA Reg. No. 59639-20-AA

NET CONTENTS __ GALLONS

KEEP OUT OF REACH OF CHILDREN

CAUTION

PRECAUTIONARY STATEMENTS

HAZARDS TO HUMANS & DOMESTIC ANIMALS

CAUTION

Keep out of reach of children.

Note to Physicians: Emergency Information—call (415) 233-3737.

ENVIRONMENTAL HAZARDS

When this product is combined with parathion emulsive (4 lbs. per gal.) or Malathion, it is toxic to fish. Do not apply directly to water.

PHYSICAL OR CHEMICAL HAZARDS

Do not weld, heat or drill this container. Residues may ignite with explosive violence if heated sufficiently.

DIRECTIONS FOR USE

It is a violation of Federal Law to use this product in a manner inconsistent with its labeling.

READ ENTIRE LABEL. USE STRICTLY IN ACCORDANCE WITH PRECAUTIONARY STATEMENTS AND DIRECTIONS, AND WITH APPLICABLE STATE AND FEDERAL REGULATIONS.

REENTRY STATEMENT

Keep all unprotected persons out of operating areas or vicinity where there may be drift.

Do not enter treated areas without protective clothing until sprays have dried. Certain states may require more restrictive reentry intervals; consult your State Department of Agriculture for further information.

Written or oral warnings regarding use of protective clothing and accidental exposure must be given to workers who are expected to be in treated areas or in areas about to be treated.

STORAGE AND DISPOSAL

PROHIBITIONS

Do not contaminate water, food, or feed by storage, disposal or cleaning of equipment.

Open dumping is prohibited.

STORAGE

Keep pesticide in original container. Do not put concentrate or dilute into food or drink containers. Protect from water contamination during storage and transfer. Contamination with water may make this product unfit for its intended use. Always check emulsification (see MIXING DIRECTIONS) before use.

PESTICIDE DISPOSAL

Wastes resulting from the use of this product may be disposed of on site or at an approved waste disposal facility.

CONTAINER DISPOSAL

Triple rinse (or equivalent). Offer for recycling or reconditioning, or puncture and dispose of in a sanitary landfill, or by other procedures approved by state and local authorities.

CONTAINER REUSE BY MANUFACTURER

When empty, replace bungs and drum head and return to manufacturer for refilling with VOLCK® Supreme Spray.

DIRECTIONS

VOLCK Supreme Spray is an especially refined Phytonomic Oil Contact Insecticide. It is for use on plants in foliage or the dormant stage. When using this product in combination with other pesticides as directed on this label, observe appropriate cautions, limitations for the products involved.

For foliage applications, the spray tank must be free of any sulfur residue before using VOLCK Supreme Spray. Do not combine sulfur sprays with VOLCK Supreme Spray on foliage and do not follow a previous application of sulfur or apply sulfur following VOLCK Supreme Spray until sufficient time has elapsed for safety (at least 30 days). If in doubt, spray a portion of one tree previously treated with sulfur, using 1 qt. to 25 gals. water. If sufficient time has not elapsed, burn will show in several days. Do not use VOLCK Supreme Spray when temperatures are above 90° F. When temperatures drop below 32° F. (usually nighttime lows), do not spray tree and vine crops until daytime temperatures have warmed to a point above 35° F. and are free from moisture due to melting frost or heavy dews. Do not use on foliage closer than 60 days after or 90 days before using ORTHOCIDE captan or folpet products. Do not use in combination with SEVIN® (Carbaryl) on crops other than citrus. Application to trees weakened by disease, drought, drying winds, or high nitrogen applications may result in oil injury.

AMOUNTS GIVEN ARE FOR 100 GALLONS OF DILUTED SPRAY.

ALFALFA (Seed Crop Only) Spider Mites: Use 1 qt. per acre in 20 to 30 gals. of water by ground application or 10 to 15 gals. by air with recommended rates of KELTHANE® EC.

ALMONDS: Summer Use: Clover (Brown Almond Mite, European Red Mite and Two-Spotted Spider Mite—Concentrated Sprayers—Use 4 to 6 gals. VOLCK Supreme Spray in 60 to 150 gals. water per acre. Do not apply to trees lacking adequate mositure.

ALMONDS, APRICOTS, CHERRIES, NECTARINES, PEACHES, PLUMS, PRUNES—Dormant or Delayed Dormant: Brown Apricot Scale, Black Scale, San Jose Scale, Olive Scale (Parlatoria Scale), Oleander Scale, Frosted Scale, Italian Pear Scale, European Fruit Lecanium, Putnam Scale, Parlatoria, Soft Brown Scale, Certain Other Scale Insects, Eggs of Aphids, Red Spider Mites—1½ to 2 gals. Summer Use—Red Spider Mites—3 to 4 qts.; Summer Stages of Brown Apricot Scale, San Jose Scale—1 gal.

APPLES: Rust Mite—Delayed Dormant—1½ to 2 gals. Summer Use—1 to 1½ gals.

APPLES, PEARS: Dormant or Delayed Dormant (Except California); San Jose Scale, Italian Pear Scale, Eggs of Aphids or European Red Mites—Concentrated Sprayers only—For Application of 100 Gals., or More, Dilute Spray Per Acre—Use 5 gals. VOLCK Supreme Spray per acre. **For Applications of 50 to 100 Gals., Dilute Spray Per Acre**—Use 4 gals. VOLCK Supreme Spray per acre. **For Applications of 25 to 50 Gals., Dilute Spray Per Acre**—Use 3 gals. of VOLCK Supreme Spray per acre.

CAUTION: Multiple concentrate applications may cause wood injury. Double applications can occur on the ends of the row during turning—turn off sprayer to avoid multiple applications.

Equipment must be capable of giving thorough coverage for maximum insecticidal effectiveness. If San Jose Scale is a moderate to severe problem, use dilute application. Concentrate applications may delay bud development. The following conditions may cause oil injury; cool, damp, or extremely dry weather; broken emulsions and/or faulty application. Follow State recommendations and cautions.

APPLES, PEARS: Dormant or Delayed Dormant: San Jose Scale, Italian Pear Scale, Eggs of Aphids, Red Spider Mites—1½ to 2 gals. Summer Use: Red Spider Mites—3 to 4 qts. (Concentrated Sprayers—Pears (Calif. only)—4 to 6 gals. in 90 to 120 gals. of water per acre). Summer Forms of San Jose and Italian Pear Scale—1 to 1½ gals.

FIGS: Fig Scale—Dormant or Delayed Dormant—Use 3 gals. VOLCK Supreme Spray. Severe Infestation—Use 2 gals. VOLCK Supreme Spray.

GRAPES: Mealybug—Dormant-Use 4 to 6 gals. in 200 to 300 gals. of water per acre.

GRAPEFRUIT, LEMONS, VALENCIA ORANGES: Citrus Red Mite—Concentrate Spray Applications (mist sprayer)—Use 5 to 15 gals. in 50 to 100 gals. water per acre. Apply to Valencia oranges only during the period from mid-July through August. Do not apply to lemons within 60 days of harvest. Concentrate spray applications require that the spray droplets be discharged into an air stream pattern of high velocity (above 100 mph) and with sufficient volume to carry the spray through the tree. Ground speed should not exceed 1½ mph. Observe all label cautions, relative to the use of spray oils on citrus. Use only equipment capable of applying concentrate sprays.

LEMONS, GRAPEFRUIT: Red Scale, Yellow Scale, Purple Scale, Black Scale, Red Spider Mites—1 to 1¼ gals. VOLCK Supreme Spray. For best results, spray during winter and spring months when Red Scale are more easily controlled and trees generally less reactive to oil sprays. Do not use within 8 weeks of harvest on Lemons.

OLIVES: Parlatoria Scale, Spider Mites, Oleander Scale, Black Scale—Use 1½ gals. VOLCK Supreme Spray. Apply in the prebloom or in postbloom up to buckshot.

BLACK SCALE (California only)—Use ½ gal. per 100 gals. of water. Apply 1,200 to 1,800 gals. of dilute spray per acre. Make applications between July 1 and 31. Higher concentrations of VOLCK Supreme Spray than noted above may cause phytotoxicity.

NAVEL ORANGES: Red Scale, Black Scale, Yellow Scale, Purple Scale, Red Spider Mites—3 qts. VOLCK Supreme Spray. Do not use on Navel Oranges after Sept. 1.

VALENCIA ORANGES: Red Scale, Black Scale, Yellow Scale, Purple Scale, Red Spider Mites—1 gal. applying not later than Sept. 1.

PEARS: Pear Psylla—Delayed Dormant: 2 gals. Summer Use: 1 to 1½ gals. Blister Mite—Delayed Dormant and Fall Spray—2 gals.

ORNAMENTALS and SHADE TREES (Conifers (see Note), Holly, Pyracantha, Sycamores, Hawthorne, Flowering Peach, Flowering Cherry, Apple, Magnolia, Maple, Dogwood, Oak, Privet, Birch): Soft Brown Scale, Oyster Shell Scale, Brown Apricot Scale, Black Scale, San Jose Scale, Red Spider Mites and Eggs—1 gal. to 100 gals. water. Dormant Spray on Deciduous Varieties—2 gals. per 100 gals. water. Dormant Spray on Other Varieties—1 gal. per 100 gals. water. NOTE: May cause discoloration of blue spruce. Do not use on conifers or ferns not tolerant to oil sprays.

WALNUTS: Late Spring to Mid-Summer Only: Summer Stages of Frosted Scale, European Fruit Scale, San Jose and other Soft-Shelled Scale, Red Spider Mites—2 to 4 qts. Do not apply after husk split.

CONCENTRATE APPLICATION

Crop	Growth Stage	Gals. Form Per Acre	Gals. Spray Per Acre	Insects
Pears	Dormant & Delayed Dormant	4-6	40-125	San Jose Scale, Pear Psylla, European Red Mite Eggs

CONCENTRATE APPLICATION

Crop	Growth Stage	Gals. Form Per Acre	Gals. Spray Per Acre	Insects
Almonds, Apricots, Cherries, Peaches & Nectarines	Dormant & Delayed Dormant	4-6	40-125	San Jose Scale, Brown Apricot Scale, Walnut Scale, Parlatoria Scale, Peach Twig Borer, Mite Eggs
Prunes & Plums	Dormant & Delayed Dormant	3-5	40-125	Parlatoria (Olive) Scale, San Jose Scale, European Red Mite Eggs, Brown Mite Eggs, Apricot Scale, Black Scale, Mealy Plum Aphid Eggs, Plum Aphid Eggs

AIR APPLICATION

Apply quantity of product shown for each listed crop for control of listed insects in sufficient water to make a minimum of 20 gals. dilute spray per acre by air application. Follow timing shown for each crop as shown on container label. **Use only as emergency application when soil conditions do not permit regular ground application.**

CAUTION: Do not apply VOLCK Supreme Spray Concentrate to Queen Ann variety or other oil sensitive varieties as phytotoxicity to buds and new wood will result.

GENERAL CAUTIONS: Observe on all above crops and uses—oil effectiveness in dormant and delayed dormant applications is determined by thoroughness of coverage. Concentrate applications may reduce coverage and effectiveness. Oil applications made to trees that are weak or suffering from moisture stress or when slow drying of spray occurs may cause bud, twig or leaf injury. Phytotoxicity studies have not been made on all fruit varieties, or in all growing areas in California for concentrate applications. Concentrate spraying is a more exacting technique of spray application than dilute. Concentrate spray equipment must be properly calibrated and operated to obtain desired results.

MIXING DIRECTIONS:

Check this product for proper emulsification by mixing a small quantity of it with water in a small container before filling the entire spray tank. Correct emulsification is indicated when a uniform mixure, with a blue color, results on shaking. If this mixture is not a milky emulsion after shaking, do not use the remaining product.

With agitator running, start filling spray tank with water. When combining with wettable powder insecticide or fungicide products, add these first when tank is 1/3 filled. Add VOLCK Supreme Spray when tank is about 1/2 full and while there is sufficient agitation of the spray to ensure complete emulsification.

Do not apply under conditions involving possible drift to food, forage or other plantings that might be damaged or the crops thereof rendered unfit for sale, use or consumption.

IMPORTANT INFORMATION.
READ BEFORE USING PRODUCT.

LIMITED WARRANTY: 1. Valent U.S.A. Corporation warrants that this product conforms to the chemical description on the label and is reasonably fit for use under average conditions when used strictly in accordance with the directions on the labeling. Valent does not make nor authorizes any agent or representative to make, any other warranty, guarantee or representation, express or implied, concerning this product. Specifically, NO IMPLIED WARRANTY OF MERCHANTABILITY OR FITNESS FOR ANY PARTICULAR PURPOSE IS MADE.

2. Critical and unforeseeable factors beyond Valent's control prevent it from eliminating all risks in connection with the use of this product. Such risks include, but are not limited to, damage to plants and crops to which product is applied, lack of complete control over the handling and application of this product, and damage caused by drift to other plants or crops. Such risks occur even though the product is reasonably fit under average conditions for the uses stated on the labeling and even though label directions are followed. Buyer and user acknowledge and assume all risks and liability (except those assumed by Valent under 1 above) resulting from handling, storage, and use of the product.

3. Precautions stated on the labeling should be followed to avoid hazardous exposure to the product.

KELTHANE®—Reg. TM of Rohm & Haas Co.
SEVIN® is the registered trademark of Rhône-Poulenc for Carbaryl insecticide.

CPCR 6th Edition Supplement

AMERICAN CYANAMID COMPANY

New Registered label

TRI-SCEPT®

HERBICIDE

FOR USE IN SOYBEANS

ACTIVE INGREDIENTS:

Ammonium salt of imazaquin (2-[4,5-dihydro-4-methyl-4-(1-methylethyl)-5-oxo-1*H*-imidazol-2-yl]-3-quinolinecarboxylic acid)*	5.09%
trifluralin: α,α,α-trifluoro-2,6-dinitro-N,N- dipropyl-p-toluidine	28.99%
INERT INGREDIENTS:	65.92%
TOTAL	100.00%

Contains petroleum distillates

*Equivalent to 4.83%, 2-[4,5-dihydro-4-methyl-4-(1-methylethyl)5-oxo-1*H*-imidazol-2-yl]-3-quinolinecarboxylic acid.

TRI-SCEPT contains 3.0 pounds of active ingredients per gallon (0.43 pounds active ingredient of imazaquin and 2.57 pounds active ingredient of trifluralin per gallon).

EPA Reg. No. 241-307

® Registered Trademark of American Cyanamid Company

D41 11/89

KEEP OUT OF REACH OF CHILDREN

WARNING!/¡AVISO!

PRECAUCION AL USUARIO: Si usted no lee inglés, no use este producto hasta que la etiqueta le haya sido explicada ampliamente.

STATEMENT OF PRACTICAL TREATMENT

If in eyes: Hold eyelids open and flush with a steady, gentle stream of water for 15 minutes. Get medical attention if irritation persists.

If on skin: Wash with plenty of soap and water. Get medical attention if irritation persists.

If inhaled: Remove victim to fresh air.

In case of emergency endangering life or property involving this product, call collect, day or night, area code 201-835-3100.

PRECAUTIONARY STATEMENTS

HAZARDS TO HUMANS

WARNING!

Causes substantial but temporary eye injury. Harmful if absorbed through the skin or inhaled.

DO NOT get in eyes or on clothing. Wear goggles or face shield. Avoid breathing spray mist. Avoid contact with skin. Wash skin thoroughly with soap and water after handling. Remove contaminated clothing and wash before reuse.

ENVIRONMENTAL HAZARDS

This product is toxic to fish. DO NOT apply directly to water or wetlands (swamps, bogs, or marshes). Drift or runoff from treated areas may be hazardous to aquatic organisms in neighboring areas. DO NOT contaminate water when disposing of equipment washwaters.

DIRECTIONS FOR USE

It is a violation of Federal law to use this product in a manner inconsistent with its labeling.

This labeling must be in the possession of the user at the time of pesticide application.

Observe all cautions and limitations in this leaflet and on the labels of products used in combination with TRI-SCEPT. Do not use TRI-SCEPT other than in accordance with the instructions set forth on this label. The use of TRI-SCEPT not consistent with this label can result in injury to crops, animals, or persons. Keep container closed to avoid spills and contamination.

DO NOT apply this product through any type of irrigation system.

STORAGE AND DISPOSAL

STORAGE: KEEP FROM FREEZING, DO NOT STORE BELOW 32°F. SHAKE WELL BEFORE USING. DO NOT contaminate water, food, or feed by storage or disposal.

PESTICIDE DISPOSAL: Wastes resulting from the use of this product may be disposed of on site or at an approved waste disposal facility.

CONTAINER DISPOSAL: Triple rinse (or equivalent). Then offer for recycling or reconditioning, or puncture and dispose of in a sanitary landfill, or by incineration, or, if allowed by state and local authorities, by burning. If burned, stay out of smoke.

DISCLAIMER

The label instructions for the use of this product reflect the opinion of experts based on field use and tests. The directions are believed to be reliable and should be followed carefully. However, it is impossible to eliminate all risks inherently associated with use of this product. Crop injury, ineffectiveness or other unintended consequences may result because of such factors as weather conditions, presence of other materials, or the use or application of the product contrary to label instructions, all of which are beyond the control of American Cyanamid Company. All such risks shall be assumed by the user.

American Cyanamid Company shall not be responsible for losses or damages resulting from use of this product in any manner not set forth on this label. User assumes all risks associated with the use of this product in any manner not specifically set forth on this label.

American Cyanamid Company warrants only that the material contained herein conforms to the chemical description on the label and is reasonably fit for the use therein described when used in accordance with the directions for use, subject to the risks referred to above.

Any damages arising from a breach of this warranty shall be limited to direct damages, and shall not include consequential commercial damages such as loss of profits or values, etc. Cyanamid's liability will be limited to a refund of the purchase price of the TRI-SCEPT. In no case shall Cyanamid or the seller be liable for consequential, special or indirect damages resulting from the use or handling of this product.

American Cyanamid Company makes no other express or implied warranty, including any other express or implied warranty of FITNESS or MERCHANTABILITY. USER assumes the risk of any use contrary to label instructions, or under abnormal conditions, or under conditions not reasonably foreseeable by American Cyanamid Company.

GENERAL INFORMATION

For broad spectrum control of grass and broadleaf weeds in soybeans, apply TRI-SCEPT as a preplant incorporated treatment.

After TRI-SCEPT is applied, some susceptible weeds emerge, growth stops, and then the weeds either die or are not competitive with the crop.

A timely cultivation may aid in the control of certain weeds or improve general weed control when adequate moisture is not received after application. Cultivation should be shallow.

TRI-SCEPT reaches the growing points of susceptible weeds either by direct contact in the soil, or by root uptake and rapid translocation to the growing points. Therefore, adequate soil moisture is important for optimum TRI-SCEPT activity. The amount of rainfall or irrigation required following application depends on existing soil moisture, soil texture and organic matter content. Sufficient water to moisten the soil to a depth of 2 inches is normally adequate. If adequate moisture is not received within 7 days after treatment, then a cultivation is recommended to control escaped weeds. When adequate moisture is received after dry conditions, TRI-SCEPT will provide residual control of susceptible germinating weeds; activity on established weeds will depend on the weed species and the location of its root system in the soil.

TRI-SCEPT soil treatment followed by a SCEPTER® or SCEPTER® O.T.® postemergence treatment will control certain problem weeds (see SEQUENTIAL PROGRAM section). Use sequential treatments in recommended states only.

Occasionally, intemode shortening of soybean plants may be observed with TRI-SCEPT applications. This has no effect on soybean yields.

There should be an interval of 90 days between the TRI-SCEPT application and soybean harvest.

Do not graze or feed treated soybean forage, hay, or straw to livestock.

Use of TRI-SCEPT herbicide in accordance with label directions is expected to result in normal growth of rotational crops in most situations; however, various environmental and agronomic factors make it impossible to eliminate all risks associated with the use of this product and, therefore, rotational crop injury is always possible.

NOTE: DO NOT apply TRI-SCEPT postemergence to soybeans as crop injury may occur.

DO NOT use TRI-SCEPT other than in accordance with the instructions set forth on this label.

DO NOT use TRI-SCEPT in any tank mixes other than those specifically provided in this label.

Crops other than soybeans, such as cotton, corn, grain sorghum, rice and vegtables, may be injured by spray drift or other indirect contact with TRI-SCEPT.

To avoid injury to sensitive crops from spray drift, follow all use directions and precautions in Spraying Instructions on page 2.

To avoid injury to sensitive crops, spray equipment used for TRI-SCEPT applications must be drained and thoroughly cleaned with water before being used to apply other products to these crops.

Apply TRI-SCEPT prior to July 1 in the Northern Use Area, as defined in the Use Area section of this label.

See page 5 for Rotational Crop Restrictions.

Replanting: If replanting is necessary in a field previously treated with TRI-SCEPT, the field may be replanted to soybeans. Rework the soil no deeper than the treated zone. DO NOT apply a second treatment of TRI-SCEPT.

MIXING INSTRUCTIONS

Fill the spray tank one-half to three-quarters full with clean water or liquid fertilizer. While agitating add the required amount of product and then fill the remainder of the tank with water or liquid fertilizer. Maintain agitation while spraying to ensure a uniform spray mixture.

When tank mixing TRI-SCEPT with recommended herbicides, add the TRI-SCEPT to the spray tank first and make sure it is thoroughly mixed before adding the other herbicide.

SPRAYING INSTRUCTIONS

NOTE: DO NOT make applications when spray may be carried by wind to sensitive crops.

Avoid overlaps when spraying.

GROUND APPLICATIONS:

Uniformly apply with properly calibrated ground equipment in 10 to 40 gallons of water, or 20 or more gallons of liquid fertilizer per acre. A spray pressure of 20 to 40 psi is recommended.

DO NOT apply with ground equipment when wind velocity is greater than 10 mph.

AERIAL APPLICATIONS:

Uniformly apply with properly calibrated aerial equipment in 5 or more gallons of water per acre.

To avoid injury to sensitive crops from drift, aerial applicators must adhere to the following SPECIAL AERIAL USE DIRECTIONS AND PRECAUTIONS:

- — Use nozzles which produce a coarse spray.
- — Nozzle height above ground must be a maximum of 10 feet.
- — Nozzles must be pointed toward the rear of the aircraft. The downward angle of the nozzle should not be greater than 20 degrees.
- — To minimize wing-tip vortex roll nozzles or spray boom must not be located any closer to end of wing or rotor than three-fourths the distance from the center of the aircraft.
- — Use a maximum spray pressure of 40 psi.
- — A buffer zone must be established between the area to be sprayed and sensitive crops.
- — DO NOT spray when wind velocity is greater than 5 mph.

Applicator is responsible for any loss or damage which results from spraying TRI-SCEPT in a manner other than recommended in this label. In addition, applicator must follow all applicable state and local regulations and ordinances in regard to spraying.

APPLICATIONS WITH LIQUID FERTILIZERS

TRI-SCEPT can be applied with liquid fertilizers, alone or in combination with PROWL, trifluralin, or Command 4EC. Follow all TRI-SCEPT label recommendations regarding incorporation, timing of application, special instructions and precautions. Apply treatments in 20 or more gallons of liquid fertilizer per acre with ground equipment. A spray pressure of 20 to 40 psi is recommended.

All individual state regulations relating to fluid fertilizer mixing, registration, labelling and application are the responsibility of the individual and/or company selling the TRI-SCEPT/liquid fertilizer mixture.

LIQUID FERTILIZER COMPATIBILITY DETERMINATIONS

If a liquid fertilizer and herbicide(s) mixture separates in the spray tank, clogged equipment and uneven application can result, which can cause poor weed control and crop injury. Always predetermine the compatibility of TRI-SCEPT alone or with PROWL, trifluralin, or Command 4EC in the specific liquid fertilizer to be used according to the following directions:

1. Add 1 pint of fertilizer to each of 2 one-quart jars.
2. Add ½ teaspoon of adjuvant* to one jar.
3. (a) When using TRI-SCEPT alone, add to each jar the correct amount of TRI-SCEPT as specified in the table below.
   (b) When using TRI-SCEPT plus PROWL, trifluralin or Command, first add the specified quantity of TRI-SCEPT, then add the correct amount of PROWL, trifluralin or Command.
4. Close both jars and shake thoroughly for 10 seconds. Let them stand for 30 minutes and then observe the results. Look for signs of separation, an oily layer or globules, sludge, flakes or other precipitates.
5. Determine compatibility
   (a) If the mixture without adjuvant does not separate, use this mixture in your spray tank.
   (b) If the mixture with adjuvant does not separate, but the one without adjuvant separates, use the adjuvant mixture in your spray tank. Add the adjuvant to the liquid fertilizer as directed on the manufacturer's label.
   (c) If either mixture separates, but mixes readily with shaking, the mixture can be used providing good agitation is maintained in the spray tank.
   (d) If separation of the mixture occurs, and agitation and/or adjuvant does not correct the problem, DO NOT use the herbicide(s) in that specific liquid fertilizer.

* Compex-Kalo Laboratories, Inc. Kansas City, MO; Sponto 168 D-Witco Chemical Co., Houston TX; Unite-Hopkins Agricultural Chemical Co., Madison, WI; or other comparable adjuvants.

Gallons of Liquid Fertilizer to be Applied per Acre	Teaspoons of Specified Herbicide to be Added To 1 Pint of Liquid Fertilizer Solution**			
	TRI-SCEPT	PROWL	Trifluralin	Command 4EC
20	1½	⅔	⅔	⅓
30	1	½	⅓	¼
40	¾	⅓	⅓	⅙

** Based on highest per acre rate used in the TRI-SCEPT label for each of those products.

PREPLANT INCORPORATED APPLICATIONS WITH DRY BULK FERTILIZERS

TRI-SCEPT may be impregnated on dry bulk fertilizers. When applied as directed, TRI-SCEPT/dry bulk fertilizer mixtures provide weed control equal to that provided by the same rates of TRI-SCEPT applied in water or liquid fertilizer.

Follow all TRI-SCEPT label recommendations regarding incorporation, special instructions and precautions. Apply TRI-SCEPT/dry bulk fertilizer mixtures only with ground equipment.

All individual state regulations relating to dry bulk fertilizer blending, registration, labeling, and application are the responsibility of the individual and/or company selling the TRI-SCEPT/dry bulk fertilizer mixture.

A minimum of 200 pounds and a maximum of 450 pounds of dry bulk fertilizer impregnated with the recommended amount of TRI-SCEPT must be applied per acre.

DO NOT impregnate TRI-SCEPT onto coated ammonium nitrate or limestone because these materials will not absorb the herbicide. Dry fertilizer blends containing mixtures of ammonium nitrate or limestone may be impregnated with TRI-SCEPT. A minimum of 200 pounds of impregnated dry bulk fertilizer, excluding the weight of ammonium nitrate or limestone, must be applied per acre.

Use the following table to determine the amount of TRI-SCEPT to be impregnated on a ton of dry bulk fertilizer based on the rate of fertilizer which will be applied per acre.

RATE CHART FOR IMPREGNATION OF DRY BULK FERTILIZER WITH TRI-SCEPT

(Pints of TRI-SCEPT per Ton of Fertilizer)

TRI-SCEPT Rate Per Acre	Fertilizer Rate lbs/acre	Pints per Ton
2.33 Pints	200	23⅓
	250	18⅔
	300	15½
	350	13⅓
	400	11⅔
	450	10⅓

For those rates not listed in this table, calculate the pints of TRI-SCEPT to be impregnated on a ton of dry bulk fertilizer using the following formula:

$$\frac{2000}{\text{Pounds of dry fertilizer per acre}} \times \text{2.33 pints of TRI-SCEPT per acre (recommended rate)} = \text{Pints of TRI-SCEPT per ton of fertilizer}$$

To impregnate TRI-SCEPT on bulk fertilizer, use a closed rotary-drum mixer or other commonly used dry bulk fertilizer blender equipped with suitable spray equipment. Spray nozzles must be placed to provide uniform coverage of TRI-SCEPT onto the fertilizer during mixing.

If trifluralin or PROWL is to be combined with TRI-SCEPT prior to impregnation, premix the trifluralin or PROWL with an equal volume of water before adding it to the TRI-SCEPT. DO NOT mix undiluted trifluralin or PROWL with TRI-SCEPT.

Apply the TRI-SCEPT/dry bulk fertilizer mixture with an accurately calibrated dry fertilizer spreader. The TRI-SCEPT/dry bulk fertilizer mixture must be spread uniformly on the soil surface. Uneven spreading can cause poor weed control and crop injury.

Refer to Preplant Incorporated Application section of this label for incorporation directions.

USE RATE

Apply TRI-SCEPT at a rate of 2.33 pints per acre.

APPLICATION INSTRUCTIONS

PREPLANT INCORPORATED APPLICATIONS:

Apply TRI-SCEPT before planting and thoroughly incorporate. Incorporation implements should be set to thoroughly incorporate TRI-SCEPT into the top 2 inches of soil. A second pass must be made at an angle to the first pass to ensure thorough incorporation. PTO driven equipment should be set to cut 2 inches deep and operated one time at 4 mph or less. Incorporate into the soil within 24 hours after application. TRI-SCEPT may be applied during land preparation up to 30 days prior to planting or immediately prior to planting in Alabama, Arkansas, Florida, Georgia, Louisiana, Mississippi, North Carolina, Oklahoma, South Carolina, Tennessee and Texas only. For the remaining Southern and Northern Use Areas, TRI-SCEPT may be applied during land preparation up to 45 days prior to planting or immediately prior to planting. Adequate moisture is required for activation of TRI-SCEPT.

If soybeans are planted on beds, apply and incorporate after bed formation using PTO-driven equipment or rolling cultivator.

WEEDS CONTROLLED

The following grass and broadleaf weeds are controlled by recommended treatments of TRI-SCEPT:

GRASSES

Barnyardgrass (*Echinochloa crus-galli*)
Crabgrass (*Digitaria* spp.)
Crowfootgrass (*Dactyloctenium aegyptium*)
Cupgrass, Woolly (*Eriochloa villosa*)
Foxtail
- Giant (*Setaria faberi*)
- Green (*Setaria viridis*)
- Robust (*Setaria* sp.)
- Yellow (*Setaria glauca*)

Goosegrass (*Eleusine indica*)
Johnsongrass, seedling (*Sorghum halepense*)
Panicum
- Fall* (*Panicum dichotomiflorum*)
- Texas (*Panicum texanum*)

Sandbur, Field (*Cenchrus incertus*)
Shattercane* (*Sorghum bicolor*)
Signalgrass, Broadleaf (*Brachiaria platyphylla*)

BROADLEAF WEEDS

Bristly Starbur (*Acanthospermum hispidum*)
Burcucumber (*Sicyos angulatus*)
Carpetweed (*Mollugo verticillata*)
Cocklebur, Common (*Xanthium strumarium*)
Jimsonweed (*Datura stramonium*)
Kochia (*Kochia scoparia*)
Lambsquarters, Common (*Chenopodium album*)
Mallow, Venice (*Hibiscus trionum*)
Morningglory
- Palm Leaf (*Ipomoea wrightii*)
- Pitted (*Ipomoea lacunosa*)
- Smallflower (*Jacquemontia tamnifolia*)

Mustard Species (*Brassica* spp.)
Nightshade, Eastern Black (*Solanum ptycanthum*)
Pigweed
- Palmer (*Amaranthus palmeri*)
- *Redroot (Amaranthus retroflexus*)
- Smooth (*Amaranthus hybridus*)
- Spiny (*Amaranthus spinosus*)
- Waterhemp, Tall (*Amaranthus tuberculatus*)

Poinsettia, Wild (*Euphorbia heterophylla*)
Puncturevine (*Tribulus terrestris*)
Purslane (*Portulaca oleracea*)
Pusley, Florida (*Richardia scabra*)
Ragweed
- Common (*Ambrosia artemisiifolia*)
- Giant (*Ambrosia trifida*)

Redweed (*Melochia corchorifolia*)
Sida, Prickly (Teaweed)(*Sida spinosa*)
Smartweed
- Ladysthumb (*Polygonum persicaria*)
- Pennsylvania (*Polygonum pensylvanicum*)

Sunflower, Common (*Helianthus annuus*)
Velvetleaf (*Abutilon theophrasti*)

TRI-SCEPT will aid in the control and reduce competition from:

Morningglory
- Entireleaf (*Ipomoea hederacea var. intergriuscula*)
- Ivyleaf (*Ipomoea hederacea*)
- Tall (*Ipomoea purpurea*)

Nutsedge, Yellow (*Cyperus esculentus*)
Spurge, Spotted (*Euphorbia maculata*)
Texasweed (*Caperonia palustris*)

*See TRI-SCEPT plus trifluralin or PROWL® herbicide Tank Mixture section below for use directions.

® Registered Trademark of Amercian Cyanamid Company

TRI-SCEPT ROTATIONAL CROP RESTRICTIONS

The following rotational crops may be planted after applying TRI-SCEPT in soybeans:

1. Wheat may be planted in the fall, four months after TRI-SCEPT application in soybeans. Rice may be planted in the spring of the year following TRI-SCEPT application. (See Northern Use Area Restrictions*).
2. Eleven months after TRI-SCEPT application:
   Barley (See Northern Use Area Restrictions*)
   Field corn (See Northern Use Area Restrictions* or Southern Use Area Restrictions**)
   Edible Beans
   Grain sorghum
   Oats (See Northern Use Area Restrictions*) (***)
   Peanuts
   Tobacco
3. Sugarbeets may be planted 28 months after TRI-SCEPT application in soybeans

*** NORTHERN USE AREA RESTRICTIONS:**

For the Northern Use Area as defined in the USE AREA section of the label, DO NOT plant field corn, wheat, barley, or oats as rotational crops in the fall or in the spring of the year following TRI-SCEPT application.

**** SOUTHERN USE AREA RESTRICTIONS:**

For the Southern Use Area, as defined in the USE AREA section of this label; field corn may be planted as a rotational crop in the spring of the year following TRI-SCEPT application unless extreme drought conditions develop or less than 15 inches of rainfall or irrigation is received from the date of application through October of the year of application. For eastern Oklahoma, Arkansas, the bootheel of Missouri, Tennessee, Northern Carolina and states south: field corn may be planted in the spring of the year following TRI-SCEPT application unless extreme drought conditions develop and less than 15 inches of rainfall or irrigation is received within 6 months following the date of application. If postemergence applications of SCEPTER or SCEPTER O.T. follow a TRI-SCEPT treatment (refer to Sequential Program section of this label for use directions), the 6 month interval begins after the last postemergence application.

Use of TRI-SCEPT herbicide in accordance with label directions is expected to result in normal growth of rotational crops in most situations; however, various environmental and agronomic factors make it impossible to eliminate all risks associated with the use of this product and, therefore, rotational crop injury is always possible.

ONLY rotational crops harvested at maturity may be used for feed or food.

DO NOT plant other rotational crops prior to 18 months after last TRI-SCEPT application.

DO NOT apply products containing chlorimuron ethyl (e.g. Classic, Canopy, Gemini, Lorox Plus, Preview[1], etc.), imazaquin (e.g. SCEPTER®, ALA-SCEPT® SQUADRON®, etc. [refer to the SEQUENTIAL PROGRAM section of this label for exceptions]), or imazethapyr (e.g. PURSUIT®, PURSUIT® PLUS, etc.) the same year as TRI-SCEPT or injury to follow crops may occur.

DO NOT graze or feed treated soybean forage, hay, or straw to livestock.

[1] Classic, Canopy, Gemini, Lorox Plus, and Preview are trademarks of E.I. duPont de Nemours and Company.

HERBICIDE COMBINATIONS

TRI-SCEPT should be used only in accordance with the directions on this label. Do not use TRI-SCEPT in any tank mixes other than those specifically provided in this label.

TRI-SCEPT Plus trifluralin or PROWL® Tank Mixture

Trifluralin or PROWL must be tank mixed with TRI-SCEPT and applied preplant incorporated if infestations of shattercane or fall panicum are anticipated.

For shattercane control, add trifluralin 4 EC or PROWL at a rate of 0.5 pint per acre on medium textured soils or 1.0 pint per acre on fine textured soils. For control of fall panicum, add 0.5 pint per acre of trifluralin or PROWL for all soil types.

Observe all precautions and limitations on the trifluralin label.

Observe all precautions and limitations on the PROWL label.

TRI-SCEPT PLUS Command[2] 4 EC Preplant Incorporated Tank Mixture

Observe all precautionary statements on the Command label before using. Use sprayers equipped with nozzles that provide accurate and uniform application. TRI-SCEPT plus Command 4EC preplant incorporated tank mixture may be used when heavy infestations of velvetleaf are anticipated.

Apply 2.33 pints per acre of TRI-SCEPT plus 0.25 to 0.5 pint (4 to 8 oz.) per acre of Command 4EC to the soil surface and uniformly incorporate. Refer to the Command 4EC label for incorporation instructions. DO NOT apply aerially or through irrigation systems. DO NOT apply this tank mixture to overly moist or wet soils.

NOTE: The use of an agriculturally approved drift reducing agent is required at spray volumes of 10 to 15 gallons per acre.

Command 4EC is a volatile compound. Off-site movement of spray drift or vapors of Command can cause foliar whitening or yellowing of some plants. Prior to using Command, read and strictly follow all precautions and application instructions on the Command 4EC label.

When applied as directed, tank mixtures of TRI-SCEPT plus Command control heavy infestations of velvetleaf as well as the weeds listed in the WEEDS CONTROLLED section of this label. Use the high rate of Command when severe infestations of velvetleaf are anticipated.

In the event of a crop loss due to weather conditions, soybeans can be replanted. DO NOT work the soil deeper than 2 inches.

[2] Command is a trademark of FMC Corporation

SEQUENTIAL PROGRAM

TRI-SCEPT Followed by SCEPTER® herbicide Postemergence

A preplant incorporated application of TRI-SCEPT may be followed by a single postemergence application of SCEPTER to control certain problem weeds.

Apply SCEPTER at 1/3 to 2/3 pints per acre as a sequential postemergence treatment following a preplant incorporated treatment of TRI-SCEPT. The sequential postemergence treatment should be applied before weed species exceed the 1 to 2 true leaf stage.

Follow use directions for postemergence treatments on the SCEPTER label.

This sequential application can be used in Alabama, Arkansas, Florida, Georgia, Louisiana, Mississippi, North Carolina, Oklahoma, South Carolina, Tennessee, and Texas only.

When applied as directed, TRI-SCEPT preplant incorporated followed by SCEPTER postemergence at a rate of 2/3 pint per acre will control the following broadleaf weeds:

- Beggarweed, Florida *(Desmodium tortuosum)*
- Mexicanweed *(Caperonia castanifolia)*
- Sicklepod *(Cassia obtusifolia)*
- Texasweed *(Caperonia palustris)*

TRI-SCEPT Followed by SCEPTER® O.T.® Postemergence

A preplant incorporated application of TRI-SCEPT may be followed by a postemergence application of SCEPTER O.T. to control morningglories and hemp sesbania *(Sesbania exaltata)*.

Apply SCEPTER O.T. at 1 pint per acre as a sequential postemergence treatment following a preplant incorporated application of TRI-SCEPT. The total amount of SCEPTER O.T. applied in one season should not exceed 2 pints per acre when following TRI-SCEPT.

Follow use directions in the SCEPTER O.T. label. See the SCEPTER O.T. label for weed height limitations.

This sequential application can be used in Alabama, Arkansas, Florida, Georgia, Louisiana, Mississippi, North Carolina, Oklahoma, South Carolina, Tennessee, and Texas only.

For postemergence control of other weeds following a soil applications of TRI-SCEPT, refer to the SCEPTER or SCEPTER O.T. labels in the Postemergence Applications sections for use rates, weeds controlled, and growth stages.

USE AREA

TRI-SCEPT can be applied only in the states or parts of states shaded in the following map:

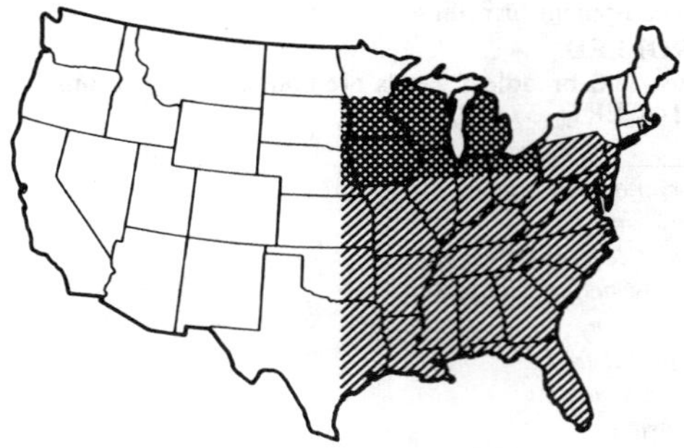

= Northern Use Area (includes: Nebraska, east of U.S. 81, north of U.S. 34; South Dakota, east of U.S. 81; Illinois, north of S. R. 116 for area west of Peoria, north of U.S. 24 for area east of Peoria; Indiana, north of U.S. 24 for area west of Huntington, north of U.S. 224 for area east of Huntington; Ohio, north of U.S. 224; and all other states or portions of states shaded in the Northern use area map).

= Southern Use Area (includes: Kansas, east of U.S. 81; and Cloud, Ellsworth, Harvey, Jewell, Lincoln, McPherson, Mitchell, Ottawa, Republic, Saline, Sedgewick and Sumner counties; southern portions of Illinois, Indiana, Ohio and eastern Nebraska not covered by Northern Use Area description; and all other states or portions of states shaded in the Southern Use Area map).

NOTE: See ROTATIONAL CROP RESTRICTIONS section for recommendations applying to the Northern vs. Southern Use Areas.

NOTE: See SEQUENTIAL PROGRAM section of this label for geographical restrictions for this method of application.

CIBA-GEIGY CORPORATION

The following revised product labels appear:
Ridomil 2E, Supracide® 2E, Tilt

RIDOMIL® 2E

FUNGICIDE

For the control of certain diseases in various crops caused by the Oomycete class of fungi

ACTIVE INGREDIENT:
Metalaxyl:*N*-(2,6-dimethyl-phenyl)-*N*-(methloxyacetyl) alanine methyl ester 25.1%
INERT INGREDIENTS: 74.9%

TOTAL 100.0%

EPA Reg. No. 100-607 EPA Est. 100-AL-1

RIDOMIL 2E contains 2 lbs. active ingredient per gallon.

KEEP OUT OF REACH OF CHILDREN

WARNING/AVISO

PRECAUCIÓN AL USUARIO: Si usted no lee inglés, no use este producto hasta que la etiqueta haya sido explicado ampliamente.

See additional precautionary statements inside booklet.

See directions for use inside booklet.

RIDOMIL® trademark of CIBA-GEIGY for metalaxyl
U.S. Patent No. 4,151,299

DIRECTIONS FOR USE AND CONDITIONS OF SALE AND WARRANTY

IMPORTANT: Read the entire **Directions for Use** and the **Conditions of Sale and Warranty** before using this product. If terms are not acceptable, return the unopened product container at once.

CONDITIONS OF SALE AND WARRANTY

The **Directions for Use** of this product reflect the opinion of experts based on field use and tests. The directions are believed to be reliable and should be followed carefully. However, it is impossible to eliminate all risks inherently associated with the use of this product. Crop injury, ineffectiveness, or other unintended consequences may result because of such factors as weather conditions, presence of other materials, or the manner of use or application all of which are beyond the control of CIBA-GEIGY or the Seller. All such risks shall be assumed by the Buyer.

CIBA-GEIGY warrants that this product conforms to the chemical description on the label and is reasonably fit for the purposes referred to in the **Directions for Use** subject to the inherernt risks referred to above. **CIBA-GEIGY makes no other express or implied warranty of Fitness or Merchantability or any other express or implied warranty. In no case shall CIBA-GEIGY or the Seller be liable for consequential, special, or indirect damages resulting from the use or handling of this product.** CIBA-GEIGY and the Seller offer the product, and the Buyer and user accept it, subject to the foregoing **Conditions of Sale and Warranty,** which may be varied only by agreement in writing signed by a duly authorized representative of CIBA-GEIGY.

DIRECTIONS FOR USE

It is a violation of federal law to use this product in a manner inconsistent with its labeling.

FAILURE TO FOLLOW DIRECTIONS FOR USE AND PRECAUTIONS ON THIS LABEL MAY RESULT IN CROP INJURY OR POOR DISEASE CONTROL.

Do not apply this product in such manner as to directly, or through drift, expose workers or other persons, except those knowingly involved in the application. The area being treated must be vacated by unprotected persons.

REENTRY STATEMENTS

Do not enter treated areas without protective clothing until sprays have dried.

Because certain states may require more restrictive reentry intervals for various crops treated with this product, consult your State Department of Agriculture for further information.

Written or oral warnings must be given to workers who are expected to be in the treated area or in an area about to be treated with this product. Oral warnings must be given which inform workers of areas or fields that may not be entered without specific protective clothing until sprays have dried and appropriate actions to take in case of accidental exposure, as described under **Precautionary Statements** at the back of this label. When oral warnings are given, warnings shall be given in a language customarily understood by workers. Oral warnings must be given if there is reason to believe that written warnings cannot be understood by workers. Written warnings must include the following information: "WARNING. Area treated with RIDOMIL 2E on *(date of application)*. Do not enter without appropriate protective clothing until sprays have dried. Flush eyes or skin with plenty of water. Call a physician if irritation persists. Remove contaminated clothing and wash before reuse."

GENERAL INFORMATION

RIDOMIL is a systemic fungicide for use on selected crops to control certain diseases caused by members of the Oomycete class of fungi. Other fungicides must be used to control diseases incited by other classes of fungi.

THIS PRODUCT IS NOT TO BE USED IN FOLIAR APPLICATIONS UNLESS SPECIFIED IN THIS LABEL OR IN SOLUTIONS USED TO DIP PLANTS. DO NOT USE IN NURSERIES, TURF, OR LANDSCAPE PLANTINGS UNLESS SPECIFIED IN THIS LABEL.

Note: RIDOMIL is a systemic fungicide having a specific mode of action and could be subject to development of resistant strains of fungi. Development of resistance cannot be predicted. Therefore, CIBA-GEIGY cannot assume liability for crop damage resulting from resistant strains of fungi. If treatment is not effective following the use of RIDOMIL as recommended, a resistant strain of fungi may be present. If the treatment is ineffective due to the presence of a RIDOMIL resistant strain of fungi, neither RIDOMIL nor any other fungicide with similar action will effectively control that disease. Consideration should then be given to the prompt use of other types of suitable fungicides. Consult with your State Agriculture Experiment Station or Extension Service Specialist for guidance on your particular crop and disease control situation.

Do not make foliar applications to field grown tobacco, or other crops since this practice may encourage more rapid development of resistance.

To avoid spray drift, do not apply under windy conditions. Avoid spray overlap, as crop injury may result.

Mixing Instructions

Add one-fourth to one-half of the required amount of water to the spray tank, add the proper amount of RIDOMIL 2E, then add the rest of the water. When tank mixing other products with RIDOMIL 2E, follow the proper sequence of adding products to the spray tank. Wettable powders or water dispersible granules should be added to the water in the tank first, followed by flowable products, with emulsifiable concentrates, such as RIDOMIL 2E, added last. Provide sufficient agitation during mixing and application to maintain a uniform emulsion.

RIDOMIL 2E is usually compatible with Balan®, Bravo®, Dasanit® + Di-Syston®, Dithane® M-22, Dithane M-45, Dasanit + Nemacur®, Dasanit, D·z·n®, Difolatan®, Enide®, Furadan®, Manzate®, Manzate 200, Mocap®, Mocap Plus 4-2EC, Paarlan®, and Tillam®.

To assure the compatibility of RIDOMIL 2E with these and other products, pour the products into a small container of water in the correct proportions. After thorough mixing, let stand for five minutes. If the combination remains mixed, or can be remixed readily, the mixture is compatible.

Calculate the amount of RIDOMIL 3E needed for band treatment by the formula:

$$\frac{\text{band width in inches}}{\text{row spacing in inches}} \times \text{broadcast rate per acre} = \text{amount needed per acre of field}$$

For banded applications, the area actually is the area covered by the band, not total cropland area planted. Some row-crop recommendations are based on the treating in-the-row and these rates generally are specified as amounts (ounces) of product per certain row length (often 1,000 feet). Other express rates as amount per treated acre which means the total area treated with the pesticide. If rates are expressed as amount per treated acre and banded applications are used, the amount of pesticide used per acre will be proportionately less. The formula above can be used to calculate the amount of RIDOMIL 2E needed per acre of crop when banded applications are made.

APPLICATION INSTRUCTIONS

Apply RIDOMIL 2E by ground or air in sufficient water or liquid fertilizer to provide uniform coverage of the soil surface. Apply in a minimum of 20 gallons/A for ground applications and 5 gallons/A by air. Refer to the specific crop directions for use for application recommendations.

BEFORE TANK MIXING RIDOMIL 2E WITH OTHER REGISTERED PRODUCTS FOR ANY USE ON THIS LABEL, READ THE LABEL OF THE TANK MIX PARTNER TO BE CERTAIN IT IS LABELED FOR USE ON THE PARTICULAR CROP AND THAT USE PATTERNS ARE COMPATIBLE WITH THOSE OF RIDOMIL 2E.

Application Through Irrigation Systems

RIDOMIL 2E, alone or in combination with other fungicides which are registered for application through irrigation systems, may be applied through irrigation systems. Apply this product only through center pivot, solid set, moving wheel, micro-sprinkler or drip irrigation systems. Do not apply this product through any other type of irrigation system. Crop injury, lack of effectiveness, or illegal pesticide residues in the crop can result from nonuniform distribution of treated water. If you have questions about calibrations, you should contact State Extension Service specialists, equipment manufacturers or other experts. Do not connect an irrigation system (including greenhouse systems) used for pesticide application to a public water system unless the pesticide label-prescribed safety devices for public water systems are in place. A person knowledgeable of the chemigation system and responsible for its operation, or under the supervision of the responsible person, shall shut the system down and make necessary adjustments should the need arise.

Operating Instructions

1. The system must contain a functional check valve, vacuum relief valve, and low pressure drain appropriately located on the irrigation pipeline to prevent water-source contamination from backflow.
2. The pesticide injection pipeline must contain a functional, automatic, quick-closing check valve to prevent the flow of fluid back toward the injection pump.
3. The pesticide injection pipeline must also contain a functional, normally closed, solenoid-operated valve located on the intake side of the injection pump and connected to the system interlock to prevent fluid from being withdrawn from the supply tank when the irrigation system is either automatically or manually shut down.
4. The system must contain functional interlocking controls to automatically shut off the pesticide injection pump when the water pump motor stops.
5. The irrigation line or water pump must include a functional pressure switch which will stop the water pump motor when the water pressure decreases to the point where pesticide distribution is adversely affected.
6. Systems must use a metering pump, such as a positive injection pump (e.g., diaphragm pump) effectively designed and constructed of materials that are compatible with pesticides and capable of being fitted with a system interlock.
7. Do not apply when wind speed favors drift beyond the area intended.

Application Instructions

RIDOMIL 2E must be applied on the schedule specified in the specific crop use recommendations, not according to the irrigation schedule.

With the exception of avocados and citrus, RIDOMIL 2E has not been sufficiently tested when applied through irrigation systems to assure consistent product performance for all labeled uses. The following calibration and application techniques are provided for user reference, but do not constitute a warranty of fitness for application through sprinkler or drip irrigation equipment. Users should check with state and local regulatory agencies for potential use restrictions before applying any agricultural chemical or drip irrigation equipment.

Note: Do not inject RIDOMIL 2E at full strength or deterioration of valves and seals may occur. Use a dilution ratio of at least 10 parts water to 1 part RIDOMIL 2E. RIDOMIL 2E is corrosive to many seal materials. Leather seals are best. EPDM or silicone rubber seals can be used, but should be replaced once a year. Do not use Viton, Buna-N, Neoprene or PVC seals.

Center Pivot Irrigation Equipment (Use only with drive systems which provide uniform water distribution)

- Determine the size of the area to be treated.
- Determine the time required to apply ½-1 inch of water over the area to be treated when the system and injection equipment are operated at normal pressures recommended by the equipment manufacturer. Run the system at 80-95% of the manufacturer's rated capacity.
- Using water, determine the injection pump output when operated at normal line pressure.
- Determine the amount of RIDOMIL 2E required to treat the area covered by the irrigation system.
- Add the required amount of RIDOMIL 2E and sufficient water to meet the injection time requirements to the solution tank.
- Maintain constant solution tank agitation during the injection period.
- Continue to operate the system until the RIDOMIL 2E solution has cleared the sprinkler head.

Solid Set and Moving Wheel Irrigation Equipment

- Determine the acreage covered by the sprinklers.
- Fill injector solution tank with water and adjust flow rate to use the contents over a 20- to 30-minute interval.
- Determine the amount of RIDOMIL 2E required to treat the area covered by the irrigation system.
- Add the required amount of RIDOMIL 2E into the same quantity of water used to calibrate the injection period.
- Operate the system at the same pressure and time interval established during the calibration.
- Inject RIDOMIL 2E at the end of the irrigation cycle in ½-1 inch of water or as a separate application to maximize the effectiveness of the fungicide.
- Stop injection equipment after treatment is completed. Continue to operate the system until the RIDOMIL 2E solution has cleared the last sprinkler head.

Micro Sprinkler or Drip Irrigation Systems

General Instructions

- Each run of the irrigation system must be calibrated separately to determine the time it takes water to move through the system and to make sure all emitters in the system are putting out the same amount of water.
- Only pressure injection or venturi equipment is recommended.
- Figure the area to be treated in each irrigation run.
- Measure the output of each of the emitters or drip tubes closest to and farthest from the injector site.
- For calibration, substituite a concentrated detergent (such as Wisk) or a soluble fertilizer for the RIDOMIL 2E in the injector tank. The detergent will bubble as it leaves the emitters. The time period over which bubbles occur should be checked for both the closest and farthest emitters. If these times are not within two minutes of each other, adjust the dilution ratio and/or the injection rate.
- If a soluble fertilizer is used, measure the time intervals with a salt bridge. If a drip system is being calibrated, substitute soluble fertilizer for the RIDOMIL 2E in the injector and measure the time intervals with a salt bridge.

Step-by-Step Instructions

- Before starting to calibrate, operate the system until all the emitters are putting out at equal flow rates or until the system is operating at full pressure.
- Make up an indicator solution of detergent or fertilizer, using the same ratio to be used when mixing RIDOMIL 2E.
- Set the injector to apply the indicator solution at the injection rate to be used in the actual RIDOMIL 2E application.
- Attach a 5-inch length of flexible tubing over the emitter closest to the injection point, another length over the emitter farthest away. Both emitters should be monitored to determine the time intervals that the indicator solutions are observed.
- Begin injecting the indicator solution. Direct the flow from the tubes at the emitters into a small container. Begin timing when the indicator solution is first detected, stop timing when the indicator solutions are no longer detected.
- If the period of detection of the indicator solution between the two emitters are within two minutes of each other, comparable coverage will be obtained. If they are not, make adjustments by increasing the dilution ratio, using more water per part of RIDOMIL 2E, or adjust the injector to a slower flow rate.
- Once the system is calibrated, dilute the needed amount of RIDOMIL 2E with water usiing a minimum of 10 parts water to 1 part of RIDOMIL 2E.
- Do not begin to inject RIDOMIL 2E into the system until all emitters are producing equal flow rates, or until the system is at full pressure.
- Inject the RIDOMIL 2E into the system at the end of the irrigation in ½-1 inch of irrigation water.

APPLES (BEARING AND NONBEARING TREES)

Use of RIDOMIL 2E will aid in the control of crown, collar, and root rot caused by *Phytophthora* spp. when used in conjunction with good cultural practices and rootstocks that are most tolerant to the disease. RIDOMIL 2E applications should be made before symptoms appear, especially in areas of the orchard favorable for disease development. RIDOMIL 2E will not revitalize trees showing moderate to severe disease symptoms.

Broadcast Spray or Banded Applications: Apply 2-4 gals./treated acre as a broadcast spray (6-12 fl. oz./1,000 sq. ft.) in sufficient water to obtain thorough coverage. The treated area is based on the area under the tree canopy or the area of the sprayed row (similar to herbicide strips). Soil surface sprays of RIDOMIL 2E will not be effective until the fungicide is moved into the root zone by rainfall or irrigation. Applications should be made in early spring before growth starts and in the fall after harvest but before the ground freezes up.

Drench: Mix 1 qt. of RIDOMIL 2E with 100 gals. of water. Apply the amount of diluted mixture indicated in the table below around the trunk of each tree. Applications should be made in early spring before growth starts and in the fall after harvest but before the ground freezes up. On new plantings, delay the first application until two weeks after planting.

Note: To determine trunk diameter, measure the trunk 12 inches above the soil line.

Trunk Diameter	Qts. of Diluted Mixture/Tree
<1 inch	1 qt.
1-3 inches	2 qts.
3-5 inches	3 qts.
>5 inches	4 qts.

Notes: (1) Do not dip roots of trees in or spray bare roots with solutions containing RIDOMIL 2E. (2) Do not graze or feed cover crops in treated orchards, or illegal residues may occur.

ASPARAGUS

RIDOMIL 2E will control crown rot and spear rot caused by *Phytophthora* spp.

Apply 2 qts./A in a minimum of 10 gals. of water over the beds.

Cutting Beds: Apply 30 to 60 days before the first cutting. For additional control, make another application just before harvest.

New Planting: Apply after planting seedlings or after covering one-year old crowns.

AVOCADOS

Root Rot—*Phytophthora cinnamomi*

Begin applications at the start of the growing season or at transplanting. Two additional applications should be made at three-month intervals. Applications are not needed during the winter months of November through February. RIDOMIL 2E may be applied as a sleeve drench at the time of transplanting, as a soil surface spray under sprinkler irrigation system, as a directed spray under drip emitters or injected into the irrigation water.

Sleeve Drench: Mix 1 fl. oz. of RIDOMIL 2E with 18 gals. of water. At time of transplanting, drench the roots inside the sleeve with 1 qt. of RIDOMIL 2E solution per tree.

Note: The sleeve drench will not replace the soil surface for long term control of root rot.

Sprinkler Irrigation: Apply as a soil surface spray under the canopy of the tree in sufficient water to obtain uniform coverage. See the following table for the amount of RIDOMIL 2E to use based on the diameter of the tree canopy. Start applications at the beginning of the growing season or at transplanting and continue at three-month intervals.

Drip Irrigation: Apply the recommended amount of RIDOMIL 2E (see table) to the soil directly under the emitter at each tree. If there is more than one emitter per tree, distribute the total amount of RIDOMIL 2E needed among the emitters.

Injection into Irrigation Water (Sprinkle or Drip Irrigation Only): Inject RIDOMIL 2E into the irrigation water at a rate of 2-4 fl. oz./1,000 gals. (3¾-7½ ppm active ingredient) at each irrigation. If RIDOMIL 2E is not applied at each irrigation, use the table below to determine how much RIDOMIL 2E should be injected into the irrigation water. If RIDOMIL is to be used more frequently than every three months, adjust the rates so that no more than the specified amount is applied during each three-month period. See the **General Information** section of this label for further instructions and precautions when making applications through irrigation systems.

Diameter of Tree Canopy (Ft.)	Amount of RIDOMIL 2E Per Ten Trees/3 Months
2	½-1 fl. oz.
5	3-6 fl. oz.
10	13-26 fl. oz.
15	29-58 fl. oz.
20	51-103 fl. oz.

Notes: (1) For best results, use RIDOMIL 2E as soon as soil tests indicate the presence of Phytophthora. (2) For new planting, the use of Phytophthora resistant rootstocks with RIDOMIL 2E is recommended. Mature trees in moderate to advanced stages of decline cannot be cured with RIDOMIL 2E. (3) Do not apply more than 24 gals./A of RIDOMIL 2E/year or make an application within 28 days of harvest, or illegal residues may result.

CITRUS

Includes grapefruits, lemons, limes, oranges, tangelos, tangerines, citrus citron, kumquats and hybrids of these.

Use RIDOMIL 2E on citrus for control of citrus foot rot, root rot, and trunk cankers caused by *Phytophthora* spp. RIDOMIL 2E can be applied as a topical canker application and as a soil application, as a spray through sprinkler or drip irrigation systems. If trees are on a drip irrigation system, distribute the amount of RIDOMIL 2E needed per tree (see tables) to the soil directly under the drip emitters at each tree. If there is more than one emitter per tree, distribute the total amount of RIDOMIL 2E needed among the emitters.

Note: Where nematodes are a problem, best results can be achieved if effective EPA-registered nematicides are used. Nematicides can be used in combination or in sequence with RIDOMIL applications.

Citrus in Nurseries (Arizona, California, and Florida Only)

Make the first application of RIDOMIL 2E at the time of planting. Make repeat applications at three-month intervals during the period when trees are actively growing. For banded applications, use a band wide enough to cover the root systems of the plants. Do not apply RIDOMIL 2E solutions to bare roots.

Soil Drench: Apply 4-6 fl. oz./100 gals. of water as a drench over the row at the rate of 100-250 gals./1,000 feet of row. The width of the drench treatment should be wide enough to cover thr root systems of the plants. Follow with a ½-1 inch irrigation over the treated area.

Soil Surface Spray: Apply 2 gals./treated acre in a broadcast or banded surface spray to seedbeds, liners, or bedded stock in sufficient water to obtain thorough coverage. If applications are banded, the treatment area should be wide enough to cover the root systems of the plants. Follow the applications with a ½-1 inch irrigation over the treated area.

Citrus Resets or New Plantings (Arizona, California, and Florida Only)

Make the first application of RIDOMIL 2E to citrus resets or new plantings at the time of transplanting. Make up to three repeat applications per year at three-month intervals or when rot growth flushes occur.

Water Ring Drench: Mix 4-6 fl. oz./100 gals. of water. Apply 5 gals. of mix around the base of each tree within the watering ring.

Soil Surface Spray: Apply 1-2 gals./treated acre (3-6 fl. oz./1,000 sq. ft.) in sufficient water to obtain uniform coverage of the soil surface. Apply spray to the soil surface beneath the tree canopy or apply through irrigation water. If natural rainfall is not expected within three days of a soil surface application, irrigate with ½-1 inch water over the treated area. See instructions below for application through irrigation water.

Established Plantings

Soil Applications (Florida Only): Apply 1-2 gals./treated acre (3-6 fl. oz./1,000 sq. ft.) as a soil surface spray under the canopy of the trees. Apply in sufficient water to provide uniform coverage or apply through irrigation water. See instructions below for application through irrigation water. Two additional applications may be made per year at three-month intervals. Use the table as a guide for treating small numbers of trees.

Diameter of Tree Canopy (Ft.)	Fl. Oz. RIDOMIL 2E per Ten Trees
5	½-1
10	2½-5
15	5-10
20	10-20

Soil Surface Spray (Arizona and California Only): Make one application in March-April. One or two additional applications per year can be made at three-month intervals or to coincide with flushes of root growth. Use the following table to determine the proper rate based on tree size and the number of applications per year. For applications based on broadcast rates, use RIDOMIL 2E at 4 gal./A (12 fl. oz./1,000 sq. ft.) when three applications are planned and at 5.5 gals./A (16 fl. oz./1,000 sq. ft.) when two applications are planned. Apply in sufficient water to provide uniform coverage or apply through irrigation water. See instructions below for application through irrigation water.

Diameter of Tree Canopy (Ft.)	Fl. Oz. RIDOMIL 2E per Ten Trees	
	2 Applications/yr.	3 Applications/yr.
5	3½	2¼
10	13	9
15	29	21
20	51	37

Trunk Spray for Control of Gummosis Caused by Phytophthora spp. (Arizona, California and Texas only): Add 1 gallon of RIDOMIL 2E to 15 gallons of water and spray the surface of the trunks using enough spray to thoroughly wet the cankers. Up to three applications per year may be made.

Notes: (1) In Arizona and California, do not make trunk spray applications if 3 soil applications are planned or illegal residues may result. (2) Do not apply more than 24 lbs. ai/A of metalaxyl per year (12 gals. of RIDOMIL 2E), or illegal residues may result.

Application Through Irrigation Water (Sprinkler or Drip Irrigation Only): See comments and precautions in the **General Information** section of this label. Inject RIDOMIL 2E into the irrigation water at rates specified in the directions.

COLE CROPS (BROCCOLI, CABBAGE, AND CAULIFLOWER)

Includes broccoli, Chinese broccoli (gai lon, white flowering broccoli), cabbage, Chinese cabbage (tight-heading varieties only), and cauliflower only.

RIDOMIL 2E applied as a soil application at planting will control damping-off caused by *Pythium* spp. and basal stem rot caused by *Phytophthora* spp. Applications may be made preplant incorporated or as a soil surface spray after planting.

Preplant Incorporated Application: Apply 4-8 pts./treated acre as a broadcast application in sufficient water or liquid fertilizer to provide uniform coverage and incorporate in the top two inches of soil. For control of Pythium damping-off use only 1-2 pts./A. For banded applications, a 7-inch band is recommended. Use the formula in the **General Information** section of this label to calculate the amount of RIDOMIL 2E needed per acre.

Surface Application: Apply 4-8 pts./treated acre at planting in sufficient water or liquid fertilizer to provide uniform coverage. For banded applications, a 7-inch band is recommended. Use the formula in the **General Information** section of this label to calculate the amount of RIDOMIL 2E needed per acre. If natural rainfall is not expected before the seeds start germinating, RIDOMIL 2E should be incorporated mechanically before planting or be moved into the seed zone after planting with ½-1 inch sprinkler irrigation.

Notes: (1) Do not use RIDOMIL 2E for disease control in greenhouse crops. (2) Do not dip plants in solutions containing RIDOMIL 2E, or crop injury may occur.

COTTON

See Rots and Seedling Diseases of Cotton Caused by Pythium spp.

Apply ¼-½ pt.13,000 linear feet of row (0.3-0.6 fl. oz./1,000 linear feet) as an in-furrow spray in 5-15 gals. of water or liquid fertilizer at planting. Mount the spray nozzle so the spray is directed into the furrow over the seed just before the seeds are covered.

For control of Pythium and Rhizoctonia apply 2-4 qts. Terraclor® 2E or 1⅓-2¾ lbs. Terraclor 75W 13,000 linear feet of row in tank mixture with RIDOMIL 2E or use labeled rates of RIDOMIL 2E PC® 11G.

Note: When RIDOMIL 2E is applied with Terraclor 2E or Terraclor 75W, observe all precautions and restrictions that appear on the Terraclor 2E or Terraclor 75W labels.

CUCURBIT VEGETABLES

Includes balsam pear (bitter melon), Chinese waxgourd (Chinese preserving melon), citron melon, cucumber, gherkin, edible gourds, cantaloupe, casaba, crenshaw, honeydew melon, honey balls, mango melon, muskmelon, Persian melon, pumpkin, summer squash, winter squash, watermelon, and cucurbit hybrids only.

RIDOMIL 2E applied at planting will provide control of damping-off and cottony leak caused by *Pythium* spp. Applications may be made preplant incorporated or as a soil surface spray after planting.

Preplant Incorporated Application: Apply 4-8 pts./treated acre as a broadcast soil application in sufficient water or liquid fertilizer to provide uniform coverage and incorporate in the top two inches of soil. For banded applications, a 7-inch band is recommended. Use the formula in the **General Information** section of this label to calculate the amount of RIDOMIL 2E needed per acre.

Surface Application: Apply 4-8 pts./treated acre at planting in sufficient water or liquid fertilizer to provide uniform coverage. For banded applications, a 7-inch band is recommended. Use the formula in the **General Information** section of this label to calculate the amount of RIDOMIL 2E needed per acre. If natural rainfall is not expected before the seeds start germinating, RIDOMIL 2E should be incorporated mechanically before planting or be moved into the seed zone after planting with ½-1 inch sprinkler irrigation.

Notes: (1) Do not use RIDOMIL 2E for disease control in greenhouse crops. (2) Do not dip plants in solutions containing RIDOMIL 2E, or crop injury may occur.

DECIDUOUS FRUITS AND NUTS IN ORCHARDS (NONBEARING)

Phytophthora Diseases

Use of RIDOMIL 2E will aid in the control of crown, collar and root rot of deciduous fruit and nut trees caused by *Phytophthora* spp. when used in conjunction with good cultural practices and rootstocks that are most tolerant to the disease. Applications should be made before symptoms appear, especially in areas favorable for disease development. RIDOMIL 2E will not revitalize trees showing moderate to severe disease symptoms.

Make the first application at the time of planting (see Notes below.). Additional applications should be made at three-month intervals during the time when conditions are favorable for disease development. For established plantings, make the first application in the spring before the growth starts.

Apply 2-4 gals./treated acre (6-12 fl. oz./1,000 sq. ft.) in sufficient water to obtain thorough coverage of the soil under the canopy of the trees. For banded applications, use the formula in the **General Information** section of this label to calculate the amount of RIDOMIL 2E needed per acre. Use the high rate in areas known to be infested with Phytophthora. Soil surface sprays of RIDOMIL 2E will not be effective until the fungicide is moved into the root zone by rainfall or irrigation.

Notes: (1) Do not apply to plantings that will bear harvestable fruit within 12 months of the last application, or illegal residues may result. To avoid injury, (2) Do not dip the roots of trees in RIDOMIL 2E solutions, spray the roots, or concentrate it around the tree trunks; (3) Do not apply to trees under stress; (4) In California, do not apply RIDOMIL 2E to newly planted trees within 90 days of planting; and (5) When applying RIDOMIL 2E to apples, see the **Apples** section of this label for specific recommendations.

HOPS

Downy Mildew *(Pseudoperonospora humuli):* Apply 1 qt./A of hops in a minimum of 20 gals. of water or liquid fertilizer to the soil surface over the crowns after pruning, but before training. Early application before shoots are six inches long is preferable.

Notes: (1) Do not apply after training or make more than one application per year, and (2) Do not feed hop refuse to livestock, or illegal residues may result.

LEAFY VEGETABLES (HEAD LETTUCE AND SPINACH ONLY)

RIDOMIL 2E applied as a soil application will control damping-off caused by *Pythium* spp. in head lettuce and spinach and white rust (*Albugo occidentalis*) in spinach. Applications may be made banded over the row, preplant incorporated, or injected with liquid fertilizer.

Preplant Incorporated Application: Apply 4-8 pts./treated acre as a broadcast soil application in sufficient water or liquid fertilizer to provide uniform coverage and incorporate in the top two inches of soil. For banded applications, a 7-inch band is recommended. Use the formula in the **General Information** section of this label to calculate the amount of RIDOMIL 2E needed per acre.

Surface Application: Apply 4-8 pts./treated acre at planting in sufficient water or liquid fertilizer to provide uniform coverage. For banded applications, a 7-inch band is recommended. Use the formula in the **General Information** section of this label to calculate the amount of RIDOMIL 2E needed per acre. If natural rainfall is not expected before the seeds start germinating, RIDOMIL 2E should be incorporated mechanically before planting or be moved into the seed zone after planting with ½-1 inch sprinkler irrigation.

White Rust Control (Spinach Only): Apply 1 pt./A or crop as a shanked in application in water or liquied fertilizer. Make the first additional application 40-50 days after planting or after the first cutting. One other application may be made after the next cutting. Use sufficient mechanical or bypass agitation to keep the RIDOMIL 2E mixed with the water or fertilizer.

Notes: (1) The additional applications of RIDOMIL 2E noted above and made after each cutting by shanking the fungicide into the beds along with liquid fertilizer provide continuing control of white rust. However, white rust can only be controlled in a preventative disease control program that begins with an application RIDOMIL 2E to the soil at planting. **If RIDOMIL 2E is not used at planting, do not use RIDOMIL 2E at any other time throughout the season.** Do not apply RIDOMIL 2E in foliar applications or in situations where white rust infections are already established. The use of RIDOMIL 2E in curative applications greatly increases the risk of the fungus developing resistance to metalaxyl. The development of resistance will destroy the effectiveness of RIDOMIL 2E in controlling white rust. (2) Do not harvest spinach within 21 days of a RIDOMIL 2E application. (3) Do not use RIDOMIL 2E for disease control in greenhouse crops.

LEGUME VEGETABLES (SUCCULENT OR DRIED)

Includes field beans, French beans, kidney beans, lima beans, mung beans, navy beans, pinto beans, runner beans, snap beans, wax beans, broad beans (fava beans), chickpeas (garbanzo beans), lentils, lupines (sweet, white sweet, white and grain), garden peas, field peas, sugar peas, southern peas (blackeyed peas, crowder peas, cowpeas, catjang), and edible soybeans, etc.

Pythium Damping-Off and Root Rot: RIDOMIL 2E applied at planting will control damping-off and root rot caused by *Pythium* spp. Applications may be made preplant incorporated, or as a soil surface spray after planting.

Preplant Incorporated Application: Apply 2-4 pts./treated acre as a broadcast soil application in sufficient water or liquid fertilizer to provide uniform coverage and incorporate in the top two inches of soil. For banded applications, a 7-inch band is recommended. Use the formula in the **General Information** section of this label to calculate the amount of RIDOMIL 2E needed per acre.

Surface Application: Apply 2-4 pts./treated acre at planting in sufficient water or liquid fertilizer to provide uniform coverage. For banded applications, a 7-inch band is recommended. Use the formula in the **General Information** section of this label to calculate the amount of RIDOMIL 2E needed per acre. If natural rainfall is not expected before the seeds start germinating, RIDOMIL 2E should be incorporated mechanically before planting or be moved into the seed zone after planting with ½-1 inch sprinkler irrigation.

Note: Do not use RIDOMIL 2E for disease control in greenhouse crops.

ONIONS—DRY BULB, GREEN, AND ONIONS GROWN FOR SEED

RIDOMIL 2E applied at planting will control damping-off caused by *Pythium* spp. Applications may be made preplant incorporated or as a soil surface spray after planting.

Preplant Incorporated Application: Apply 2-4 pts./treated acre as a broadcast soil application in sufficient water or liquid fertilizer to provide uniform coverage and incorporate in the top two inches of soil. For banded applications, a 7-inch band is recommended. Use the formula in the **General Information** section of this label to calculate the amount of RIDOMIL 2E needed per acre.

Surface Application: Apply 2-4 pts./treated acre at planting in sufficient water or liquid fertilizer to provide uniform coverage. For banded applications, a 7-inch band is recommended. Use the formula in the **General Information** section of this label to calculate the amount of RIDOMIL 2E needed per acre. If natural rainfall is not expected before the seeds start germinating, RIDOMIL 2E should be incorporated mechanically before planting or be moved into the seed zone after planting with ½-1 inch sprinkler irrigation.

Note: Do not use RIDOMIL 2E for disease control in greenhouse crops.

PEANUTS

RIDOMIL 2E is a soil-applied systemic fungicide for use in control of seedling and pod diseases of peanuts incited by *Pythium* spp.

Seedling Diseases: Apply 1 pt./A of crop as an application to the seed in-furrow or in a 7-inch band at the time of planting. For in-furrow applications, position the spray so the fungicide is mixed with the soil covering the seed. Avoid spraying the seed directly with the spray solution or crop injury may occur.

Pod Rot: Apply 1-2 qts./A at early pod set or pegging through the overhead irrigation systems. See the **General Information** section of this label for further instructions and precautions when making applications through irrigation systems.

Notes: (1) Where pathogens other than *Pythium* spp. are present, use fungicides that control those diseases in combination with RIDOMIL 2E. (2) Where the predominant pod rot pathogens are *Pythium* spp. and *Rhizoctonia* spp.: use RIDOMIL PC 11G at labeled rates.

PEPPERS AND EGGPLANT

Soil applications of RIDOMIL 2E will provide control of Pythium damping-off and crown rot caused by *Phytophthora caspsici.* RIDOMIL 2E must be applied to the soil before the plants are infected with Phytophthora to obtain satisfactory disease control.

Apply 4-8 pts./treated acre (broadcast rate) at the time of planting in sufficient water or liquid fertilizer to provide uniform coverage. If rainfall is not expected before the plants begin growth, RIDOMIL 2E should be incorporated mechanically before planting or be moved into the root zone after planting with ½-1 inch or sprinkler irrigation water. For banded applications, a 12-16 inch band is recommended. After the initial application, two supplemental post-directed applications at 4 pts./treated acre (broadcast rate) should be made at 30-day intervals. The spray should be directed at the base of the plants and cover 6-10 inches of soil on either side of the plants. Use the formula in the **General Information** section of this label to calculate the amount of RIDOMIL 2E needed per acre.

Precautions: (1) RIDOMIL 2E may cause some yellowing of the pepper leaves. (2) Plants already infected with Phytophthora cannot be cured with RIDOMIL 2E applications. (3) The foliar blight phase of Phytophthora cannot be controlled with foliar applications of RIDOMIL 2E .

Notes: (1) Do not apply within 7 days of harvest, and (2) Do not apply more than 12 pts. of RIDOMIL 2E per acre of crop per season, or illegal residues may result.

PINEAPPLE

RIDOMIL 2E , applied as a "seed piece" dip, provides effective control of heart rot disease of pineapple caused by *Phytophthora* spp.

Apply RIDOMIL 2E as a crown dip before planting at the rate of 1-2 qts./100 gals. of water. Use 75 to 100 gals. of dip solution per planted acre, depending on crown size, plant density and dipping techniques.

Note: If there is a crop failure within one year of planting treated crowns, do not harvest plant material for animal feed.

RASPBERRIES

RIDOMIL 2E is a soil-applied systemic fungicide for use in the control of Phytophthora root rot.

Apply 1 pt./1,000 linear ft. or row to the soil surface in a three-foot band over the row (14.5 pts./treated acre on broadcast basis). Make one application in the spring and another in the fall after harvest.

Note: Do not apply RIDOMIL 2E within 45 days before harvest or illegal residues may result.

SOYBEANS

RIDOMIL 2E is a soil-applied systemic fungicide for use in the control of Phytophthora root and stem rot and Pythium damping-off. RIDOMIL 2E may be applied broadcast, banded or in the seed furrow before the seeds are covered. The seed furrow applications will provide more consistent results if rain is not expected before the seeds germinate.

For best results against Phytophthora root and stem rot, use RIDOMIL 2E with soybean varieties that have some tolerance to the races of Phytophthora present in the field. The higher rate of RIDOMIL 2E should be used in areas with a history of heavy Phytophthora damage. Under heavy late season Phytophthora pressure, RIDOMIL 2E may not provide complete control.

Surface Applications: For full season control, apply 5 pts./treated acre in sufficient water or liquid fertilizer to provide uniform coverage at the time of planting. For banded applications, a 7-inch band is recommended. Use the formula in the **General Information** section of this label to calculate the amount of RIDOMIL 2E needed per acre. For early to mid season control, apply 1½-3 pts./treated acre.

In-Furrow Applications: For full season control, apply 1¹⁄₁₀ fl. oz./1,000 linear feet of row as an in-furrow spray in 5-10 gals. of water or liquid fertilizer at planting. Position the spray so the fungicide is mixed with the soil covering the seed. Avoid spraying the seed directly with the spray solution or crop injury may occur. Use sufficient water or liquid fertilizer to provide uniform coverage. Use the table below to determine the amount of RIDOMIL 2E needed per acre based on row spacing. For early to mid season control, apply ³⁄₁₀-⁶⁄₁₀ fl. oz./1,000 linear feet of row.

Suggested Rates Per Acre According to Row Spacing

Row Spacing	Linear Feet of Soybean Row Per Acre	Fluid Ounces of RIDOMIL 2E Needed Per Acre — Rate Desired: ³⁄₁₀ fl. oz.	⁶⁄₁₀ fl. oz.	1¹⁄₁₀ fl. oz.
38 in.	13,756	4	8	15
36 in.	14,520	4½	9	16
30 in.	17,424	5½	11	19
24 in.	21,780	6½	13	24
20 in.	26,136	8	16	28

Note: RIDOMIL 2E is specific for Pythium and Phytophthora and will not control other diseases that may attack soybeans.

TOBACCO

RIDOMIL 2E is a soil-applied systemic fungicide for use in the field before transplanting for control of black shank (*Phytophthora parasitica,* var. *Nicotianae*) and blue mold (*Peronospora tabacina*) on all types of tobacco. RIDOMIL 2E may also be applied in tobacco plant beds for control of blue mold and damping-off caused by *Pythium* spp. For control of anthracnose and other tobacco diseases, use fungicides that control those diseases.

Tobacco Plant Bed

For control of blue mold and damping-off caused by *Pythium* spp., apply RIDOMIL 2E only as a preplant broadcast soil application at the rate of 1 qt./treated acre in 50 gals. of water or liquid fertilizer (1 fl. oz. or 2 tablespoons/150 sq. yds. of bed in 2 gals. of water or liquid fertilizer). Apply before or at the time of seeding to the surface of the plant bed and lightly incorporate or follow with ½ inch sprinkler irrigation water. In Pennsylvania, use 2 qts./treated acre (2 fl. oz. or 4 tablespoons/150 sq. yds. of bed in 2 gals. of water or liquid fertilizer).

Notes: (1) Do not use RIDOMIL 2E for disease control in greenhouse crops. (2) Do not dip plants in solutions containing RIDOMIL 2E , or crop injury may occur.

Field Planted Tobacco

Blue Mold: Apply RIDOMIL 2E as a broadcast soil application and incorporate in the top 2-4 inches of soil before forming beds. For flue-cured tobacco, use 1-2 qts./treated acre, depending on disease pressure and length of control desired. Under low disease pressure or for early season control, use 1 qt./treated acre. For burley and other tobacco types, use 2 qts./treated acre.

Note: For best control of blue mold, use RIDOMIL 2E in the plant bed and follow with an application in the field prior to transplanting. Do not use RIDOMIL 2E in transplant water or in foliar applications on field tobacco because of the potential for the development of resistance in the blue mold fungus, crop injury or potential excessive residues in the crop.

Black Shank: Use RIDOMIL 2E as a broadcast soil application and incorporate in the top 2-4 inches of soil before forming beds. Apply RIDOMIL 2E using conventional ground application equipment in sufficient water or liquid/dry fertilizer to provdie uniform coverage. Use the table below to determine the amount of RIDOMIL 2E needed per acre depending on the black shank severity.

Type of Tobacco	Disease Level in Field	Rate of RIDOMIL 2E Per Acre
Flue-Cured	Low to Moderate (Less than 6% Disease)	2 qts.
	High (More than 6% Disease)	4 qts.*
Burley and Other**	Low to Moderate (Less than 6% Disease)	4 qts.
	High (More than 6% Disease)	6 qts.

* Florida and Georgia—Use 6 qts./treated acre of RIDOMIL 2E in fields with heavy black shank levels (greater than 6%).

** Pennsylvania—Do not use RIDOMIL 2E for black shank control.

Lay-by Application: For prolonged control of blue mold or black shank in field planted tobacco, make a supplemental application of 1 qt./acre of crop as a soil application at the last cultivation. Position the nozzles so that the spray is deposited under the plants and is covered with soil by the cultivator. Do not make this lay-by application if more than 2 qts./A of RIDOMIL 2E were applied at transplanting or if no RIDOMIL 2E was applied at transplanting.

Notes: (1) For best results against black shank, use RIDOMIL 2E with tobacco varieties that have some resistance to black shank and use crop rotation. In fields where there is a history of severe black shank, use the highest rate and plant a variety that is resistant to the race of Phythphthora present in the field. (Burley L8 hybrids are only resistant to Phytophthora Race O.) (2) RIDOMIL 2E is not recommended for use in high black shank areas on highly susceptible flue-cured varieties such Hicks, Virginia Gold or White Gold. (3) Failure to adequately control nematodes in fields treated with RIDOMIL 2E may result in poor control of black shank.

TOMATOES

Soil applications of RIDOMIL 2E at planting will provide control of damping-off caused by *Pythium* spp. and soil applications applied 4-12 weeks before harvest under the vines will reduce fruit and root rot caused by *Pythium* spp. and *Phytophthora* spp.

Damping-Off (*Pythium* spp.): Apply 4-8 pts./treated acre in sufficient water or liquid fertilizer to provide uniform coverage at the time of planting. If natural rainfall is not expected before the seeds start germinating. RIDOMIL 2E should be incorporated mechanically before planting or be moved into the seed zone after planting with ½-1 inch sprinkler irrigation. For banded applications, a 7-inch band is recommended. Use the formula in the **General Information** section of this label to calculate the amount of RIDOMIL 2E needed per acre.

Fruit and Root Rot (*Pythium* spp. and *Phytophthora* spp.): Additional applications can be made as soil surface applications under the vines to provide control of root and fruit rot. If only one application is planned, apply 4 pts./treated acre under the vines 4 weeks after the planting application to 4 weeks before harvest, depending on the expected occurrence of the diseases. If root and fruit rot are problems throughout the growing season or the occurrence of diseases cannot be predicted, make two additional applications at 2 pts./treated acre under the vines, one 4-6 weeks after the planting application and another 4 weeks before harvest. Use sufficient water to obtain thorough coverage. If rainfall is not expected within three days of the application, follow as soon as possible with ½-1 inch overhead irrigation. Use the formula shown in the **General Information** section to determine the amount of RIDOMIL 2E needed per acre.

RIDOMIL can be injected into the beds with liquid fertilizer. For calculating the amount of RIDOMIL 2E needed per acre, apply the same amount as would be applied on a 7-inch band at 4 pts./treated acre. Keep RIDOMIL 2E suspended in the fertilizer solution with mechanical or bypass agitation.

Notes: To avoid illegal residues. (1) Do not apply more than 12 pts./treated acre season and (2) Do not apply within 7 days of harvest. (3) Do not use RIDOMIL 2E for disease control in greenhouse crops.

VEGETABLE BEDDING PLANTS

Includes broccoli, Chinese broccoli (gai lon, white flowering broccoli), cabbage, Chinese cabbage (bok choy and napa), cauliflower, cucurbit vegetables, head lettuce, spinach, legume vegetables, onion and tomatoes.

Pythium Damping-Off: Apply 2-4 pts./treated acre as a preplant broadcast spray in 50 gals. of water or liquid fertilizer (1-2 fl. oz. or 2-4 tablespoons 150 sq. yds. of bed in 2 gals. water or liquid fertilizer) before or at the time of seeding to the surface of the beds. If natural rainfall is not expected before the seeds start germinating. RIDOMIL 2E should be incorporated mechanically before planting or be moved in the seed zone after planting with ½-1 inch sprinkler irrigation.

Note: In Georgia and Florida, do not use RIDOMIL 2E for disease control in vegetable plants grown in an enclosed environment, i.e., greenhouse, glasshouse, etc.

REPLANTING

If replanting is necessary, additional applications of RIDOMIL 2E or RIDOMIL 5G can be made provided that the total amount of metalaxyl applied does not exceed the maximum allowed for the specific crop. The maximum amount of metalaxyl applied to the soil per crop cycle (combined plants, failed and replanted) is as follows.

To determine the rate per application, follow the directions for use on a particular crop.

Failed and Replanted Crops	Maximum Lbs. Metalaxyl	Maximum Amount of RIDOMIL 2E or 5G
Asparagus	2	4 qts. 2E
Cole Crops (Broccoli, Cabbage, Cauliflower)		
Cotton	2	4 qts. 2E or 40 lbs. 5G
Cucurbit Vegetables**		
Leafy Vegetables (Head Lettuce and Spinach)	2	4 qts. 2E
Legume Vegetables***	1	2 qts. 2E
Onions		
Peanuts	1	2 qts. 2E or 20 lbs. 5G
Peppers/Eggplant	3	6 qts. 2E
Pineapples	2	4 qts. 2E
Soybeans	2	4 qts. 2E or 40 lbs. 5G
Tobacco	3	6 qts. 2E
Tomatoes		

Rotation (Plantback) Restrictions

Rotation Crop	Planting Time From Last RIDOMIL 2E Application
Apples, Asparagus Avocados	
Broccoli	
Cabbage, Cauliflower, Chinese Broccoli, (gai lon, white flowering broccoli), Chinese Cabbage (tight-heading varieties only), Citrus, Cucurbit Vegetables**	
Deciduous Fruits and Nuts*	
Eggplant	0 days
Head Lettuce, Hops	
Legume Vegetables***	
Onions	
Peanuts, Peppers, Pineapples, Potatoes	
Raspberries	
Soybeans, Spinach	
Tobacco, Tomatoes	
Wheat	14 days
Corn	12 months
Root Crops	
Crops **Not Intended** for Food or Feed	0 days
All Other Crops **Intended** for Food or Feed	12 months

* These crops and other perennial crops may be planted immediately following last application of RIDOMIL 2E provided they will not bear harvestable fruit within 12 months.

** **Cucurbit Vegetables:** Includes balsam pear (bitter melon), Chinese waxgourd (Chinese preserving melon), citron melon, cucumber, gherkin, edible gourds, cantaloupe, casaba, crenshaw, honeydew melon, honey balls, mango melon, muskmelon, Persian melon, pumpkin, summer squash, winter squash, watermelon, and cucurbit hybrids.

*** **Legume Vegetables (Succulent or Dried):** Includes beans, French beans, kidney beans, lima beans, mung beans, navy beans, pinto beans, runner beans, snap beans, wax beans, broad beans (fava beans), chickpeas (garbanzo beans), lentils, lupines, (sweet, white sweet, white and grain), garden peas, field peas, sugar peas, southern peas (blackeyed peas, crowder peas, cowpeas, catjang) and edible soybeans.

STORAGE AND DISPOSAL

Store at temperatures above 40°F. Crystals may form at lower storage temperatures. If this occurs, place the product in a warm room (68°F or above) and roll or shake the container at frequent intervals until all crystals are dissolved.

Pesticide Disposal

Do not contaminate water, food or feed by storage or disposal. Open dumping is prohibited. Wastes resulting from the use of this product are acutely toxic. Improper disposal of unused pesticide, spray mixture, or rinsate is a violation of federal law. Pesticide, spray mixture, or rinsate that cannot be used according to label instructions must be disposed of according to federal, state or local procedures. For guidance in proper disposal methods, contact your State Pesticide or Environmental Control Agency, or the Hazardous Waste representative at the nearest EPA Regional Office.

Container Disposal

Do not reuse empty container. Triple rinse (or equivalent) and offer for recycling or reconditioning, or puncture and dispose of in a sanitary landfill, or by incineration, or by open burning, if allowed by state and local authorities. If burned, keep out of smoke.

For minor spills, leaks, etc., follow all precautions indicated on this label and clean up immediately. Take special care to avoid contamination of equipment and facilities during cleanup and disposal of wastes. In the event of a major spill, fire, or other emergency, call 1-800-888-8372 day or night.

PRECAUTIONARY STATEMENTS

HAZARDS TO HUMANS AND DOMESTIC ANIMALS

WARNING

Causes eye injury. Do not get in eyes. Avoid contact with skin or clothing. Harmful if swallowed, inhaled or absorbed through the skin. Avoid breathing vapors or spray mist.

Statement of Practical Treatment

If in eyes: Immediately flush with plenty of water. Get medical attention.

If on skin: Wash thoroughly wiith soap and water. Get medical attention if irritation occurs.

If swallowed: Immediately contact a physician, hospital, or local Poison Control Center. Do not induce vomiting.

If inhaled: Move victim to fresh air. If not breathing, give artificial respiration (preferably mouth-to-mouth) and get medical attention.

Notes to Physician: If RIDOMIL 2E is ingested, lavage stomach. A slurry of activated charcoal in water can be left in the stomach. Give a saline laxative and supportive therapy.

ENVIRONMENTAL HAZARDS

Do not apply directly to water or wetlands (swamps, bogs, marshes, and potholes). Apply only as specified on this label. Do not apply when weather conditions favor drift from treated areas. Do not contaminate water when disposing of equipment washwaters.

PHYSICAL OR CHEMICAL HAZARDS

Do not use or store near heat or open flame.

Balan® trademark of Elanco Products for benefin
Bravo® trademark of Fermenta Plant Protection Company for chlorothalonil
Dasanit® trademark of the Parent Company of Farbenfabriken Bayer GmbH, Leverkusen for fensulfothion
Difolatan® trademark of Chevron Chemical Company for captafol
Di-Syston® trademark of the Parent Company of Farbenfabriken Bayer GmbH, Leverkusen for disulfoton
Dithane® trademark of Rohm and Haas Company for maneb/mancozeb
D·z·n® trademark of CIBA-GEIGY for diazinon
Enide® trademark of Tuco Division of the Upjohn Company for diphenamid
Furadan® trademark of FMC Corporation for carbofuran
Manzate® trademark of E.I. duPont de Nemours and Company for maneb/mancozeb
Mocap® trademark of Rhone Poulenc, Inc. for ethoprop
Nemacur® trademark for the Parent Company of Farbenfabriken Bayer GmbH, Leverkusen, for fenamiphos
Paarlan® trademark of Elanco Products for isopropalin
RIDOMIL® trademark of CIBA-GEIGY for metalaxyl
U.S. Patent No. 4,151,299
Terraclor® trademark of Uniroyal, Inc., for PCNB
Tillam® trademark of Stauffer Chemical Company for pebulate

CGA 65L14M 099

This is a specimen label which was not prepared by CIBA-GEIGY. CIBA-GEIGY is not responsible for the accuracy of the information contained herein. As labels are subject to revision, always carefully read and follow the label on the product container.

SUPRACIDE® 2E

INSECTICIDE-MITICIDE

RESTRICTED USE PESTICIDE

FOR RETAIL SALE TO AND USE ONLY BY CERTIFIED APPLICATORS OR PERSONS UNDER THEIR DIRECT SUPERVISION, AND ONLY FOR THOSE USES COVERED BY THE CERTIFIED APPLICATOR'S CERTIFICATION.

For control of certain insects of alfalfa; artichokes; certain citrus, fruits and nuts; cotton; nursery stock; olives; safflowers; sunflowers; sorghum; and tobacco

ACTIVE INGREDIENT:
Methidathion: O,O-dimethyl phosphorodithioate, S-ester with 4-(mercaptomethyl)-2-methoxy-Δ^2-1,3,4-thiadiazolin-5-one 24.4%
INERT INGREDIENTS: 75.6%
TOTAL 100.0%

EPA Reg. No. 100-501 EPA Est. 100-AL-1

SUPRACIDE 2E contains 2 lbs. active ingredient per gallon.

KEEP OUT OF REACH OF CHILDREN

DANGER/ PELIGRO

PRECAUCIÓN AL USUARIO: Si usted no lee inglés, no use este producto hasta que la etiqueta haya sido explicado ampliamente.

See additional precautionary statements at end of label booklet.

See directions for use inside booklet.

SUPRACIDE® trademark of CIBA-GEIGY for methidathion

CGA 30L23F 059

DIRECTIONS FOR USE AND CONDITIONS OF SALE AND WARRANTY

IMPORTANT: Read the entire **Directions for Use** and the **Conditions of Sale and Warranty** before using this product.

CONDITIONS OF SALE AND WARRANTY

The **Directions for Use** of this product reflect the opinion of experts based on field use and tests. The directions are believed to be reliable and should be followed carefully. However, it is impossible to eliminate all risks inherently associated with use of this product. Crop injury, ineffectiveness, or other unintended consequences may result because of such factors as weather conditions, presence of other materials, or the manner of use or application all of which are beyond the control of CIBA-GEIGY or the Seller. All such risks shall be assumed by the Buyer.

CIBA-GEIGY warrants that this product conforms to the chemical description on the label and is reasonably fit for the purposes referred to in the **Directions for Use** subject to the inherent risks referred to above. **CIBA-GEIGY makes no other express or implied warranty of Fitness or Merchantability or any other express or implied warranty. In no case shall CIBA-GEIGY or the Seller be liable for consequential, special, or indirect damages resulting from the use or handling of this product.** CIBA-GEIGY and the Seller offer this product, and the Buyer and user accept it, subject to the foregoing **Conditions of Sale and Warranty,** which may be varied only by agreement in writing signed by a duly authorized representative of CIBA-GEIGY.

DIRECTIONS FOR USE

It is a violation of federal law to use this product in a manner inconsistent with its labeling.

FAILURE TO FOLLOW THE DIRECTIONS FOR USE AND PRECAUTIONS ON THIS LABEL MAY RESULT IN POOR INSECT AND MITE CONTROL, CROP INJURY, AND/OR ILLEGAL RESIDUES.

GENERAL INFORMATION

SUPRACIDE 2E is an emulsifiable concentrate which when applied according to the Directions for Use will control certain insect pests and mites on agricultural crops.

Wear protective clothing as described in the USDA/EPA Guide for Commercial Applicators during the preparation of the spray mixture and during application.

Do not apply this product in such a manner as to directly or through drift expose workers or other persons except those knowingly involved in the application. Do not treat while unprotected humans or domestic animals are present in the treatment area.

REENTRY STATEMENT

Two days must elapse between the time of application and the time a worker reenters the treated area when SUPRACIDE 2E is applied at rates up to 2 lbs. ai/A.

Fourteen days must elapse between the time of application and the time a worker reenters the treated area when SUPRACIDE 2E is applied at rates greater than 2 lbs. ai/A and up to the maximum label rate of 10 lbs. ai/A. Do not apply Supracide at a rate exceeding 10 lbs. ai/A.

If a worker chooses to reenter a treated field prior to the allowable time of reentry, protective clothing consisting of a protective suit worn over normal work attire, chemical resistant gloves, chemical resistant shoes, shoe covers or boots, a NIOSH or MSHA approved respiratory protection device, goggles or face shield must be worn. Protective clothing must be removed daily and laundry washed, keeping the clothing separate from regularly worn clothing. Fresh protective clothing must be worn each day that the worker enters a field prior to the allowable time.

Because certain states may require more restrictive reentry intervals for various crops treated with this product, consult your State Department of Agriculture for further information.

Written or oral warnings must be given to workers who are expected to be in a treated area or in an area about to be treated with this product. Oral warnings must inform workers of areas or fields that may not be entered without specific protective clothing until the intervals stated above have elapsed, and appropriate actions to take in case of accidental exposure, as described under **Precautionary Statements** on this label. When oral warnings are given, warnings shall be given in a language customarily understood by workers. Oral warnings must be given if there is reason to believe that written warnings cannot be understood by workers. Written warnings must include the following information: "DANGER. Area treated with SUPRACIDE 2E on (date of application). Do not enter without appropriate protective clothing until (date based on intervals stated above). In case of accidental exposure, flush eyes or skin with plenty of water. Call physician if irritation persists. Remove and wash contaminated clothing before reuse."

Mixing Procedures

(1) Be sure sprayer is clean; if it is contaminated with any material, crop injury or sprayer clogging may result. (2) Fill tank 1/4-1/2 full with clean water, and start agitation. (3) Be certain that the agitation system is working properly and creates a rippling or rolling action on the liquid surface. (4) Add SUPRACIDE to the tank. Then add tank mix ingredients. (5) Complete filling tank, maintaining sufficient agitation at all times to ensure surface action. This applies to both spray and nurse tanks. (6) Do not leave tank mixtures in the tank without adequate agitation. Always maintain agitation to avoid separation and buildup of residues. If a residue starts to build up in tank, drain and clean with strong detergent and ammonia solution and rinse thoroughly with clean water. (7) Immediately after use, clean sprayer thoroughly by flushing system with water containing a detergent and rinse with clean water.

Chemigation: Do not apply this product through any type of irrigation system.

Alfalfa (Pure Stands or Stands Containing Clover or Grass) and Pure Stands of Timothy Grass

Apply at the rates listed below in a minimum of 10 gals. of water per acre with ground equipment or 2 gals. of water per acre by air. For alfalfa weevil and Egyptian alfalfa weevil control, make foliar applications to standing alfalfa when 20-30% of the growing tips show insect feeding damage. Make stubble applications immediately following a cutting. For other insects listed on alfalfa, apply when insects first appear. Make no more than one stubble application and no more than one foliage application per cutting. Use the lower rate for light infestations and the higher rate for moderate to heavy infestations.

Table 1: Alfalfa—Rates Per Acre

Insects Controlled	Northeastern States	Southern States	California	Pacific Intermountain States Excluding CA
Alfalfa weevil	2–4 pts.	3–4 pts.	2 pts.	2–4 pts.
Aphids	2–4 pts.	3–4 pts.	2 pts.*	2–4 pts.
Egyptian alfalfa weevil	—	—	2 pts.*	2–4pts.
Grass scale *Eriococcus insignis*	—	—	—	4 pts. (Washington only)
Lygus	—	—	2 pts.	2–4 pts.
Leafhoppers	2–4 pts.	2–4 pts.	2 pts.	2–4 pts.
Spittlebugs	2–4 pts.	—	—	—

* Egyptian alfalfa weevil and pea aphid: (California only) Use 1–2 pts. where integrated control is practiced. Apply as above. When Egyptian alfalfa weevil larval counts exceed 20 per sweep, use the 2 pt. rate only.

Note: To avoid injury to pollinating insects, do not apply during bloom. To avoid illegal residues, do not apply within 14 days of cutting, grazing, or feeding to livestock.

Almonds

Dilute and Concentrate Sprays

Peach twig borer, San Jose scale, brown almond mite: Apply during the dormant or delayed dormant period, or as a cover spray. Mix a dilute spray at the rate of 1–2 pts. per 100 gals. of water and apply 300–600 gals. per acre. If applying as a concentrate spray*, apply 3–12 pts. per acre in a minimum of 50 gals. of water per acre. For rates less than 2 pts. of Supracide per 100 gals., tank mix with EPA-registered oil.

When concentrate applications are made, proper equipment must be used to obtain uniform coverage.

Aerial Sprays

Peach twig borer, San Jose scale: Apply during the dormant or delayed dormant period and use 6–12 pts. per acre as an aerial spray* in a minimum of 20 gals. of spray per acre. Spray swath should be no wider than the distance between the tree rows or the length of the spray boom.

EPA-registered oils may be added to the SUPRACIDE spray mixture in the dormant/delayed dormant period. Do not use more than 1.5 gals of oil per 100 gals. spray as a dilute spray or 6 gals. of oil per acre as a concentrate spray or 5 gals of oil per acre as an aerial spray.

* The rate of SUPRACIDE for concentrate and aerial sprays is based on a dilute spray with conventional ground equipment. A dilute spray is considered to be 300–600 gals. per acre. This amounts to a minimum of 6 pts. and a maximum of 12 pts. of SUPRACIDE per acre.

Note: To avoid illegal residues, (1) Do not make more than one application during the dormant or delayed dormant period nor more than one cover spray each season. (2) Do not apply within 80 days of harvest. (3) Do not graze treated areas or feed cover crops grown in treated areas to livestock.

Precaution: Apply the dormant/delayed dormant application before any blossoms open, or injury may occur. Do not combine with oils in the cover spray, or injury or excessive residues may occur. The cover spray of SUPRACIDE may be slightly phytotoxic on some almond varieties.

Artichokes

Apply in a minimum of 50 gals. of water with ground equipment or a minimum of 10 gals. of water by air. Begin applications when insects appear in newly planted fields or when shoots emerge following complete cutback of the plants. Continue applications at 14 day intervals until buds begin to appear. Do not plant crops except those listed on this label within one year after last application.

Artichoke plume moth, chrysanthemum leafminer: Apply 4 pts. per acre.

Note: To avoid illegal residues, (1) Do not apply after buds begin to form. (2) Avoid application under conditions which favor drift to adjacent artichoke fields where buds are present. (3) Do not exceed a total of 8 applications per crop cycle.

Cotton

Apply in a minimum of 10 gals. of water with ground equipment or in a minimum of 3 gals. of water by air unless otherwise specified.

Spider mites, boll weevil, bollworm: Apply 4 pts. per acre. Repeat applications at 5–7 day intervals if necessary.

Lygus bug: (Arizona, California only) Apply 2–4 pts. per acre. Use the low rate for light to moderate infestations and the high rate for moderate to heavy infestations. Repeat applications if necessary.

Pink bollworm: (Arizona, California only) Apply 2 pts. per acre. Repeat applications at 5–7 day intervals if necessary.

Bandedwing whitefly: (mid-South and southeast United States only) Apply 1–2 pts. per acre. Use the low rate for light to moderate infestations and the high rate for moderate to heavy infestations. Repeat applications if necessary.

Note: To avoid illegal residues, (1) Do not apply more than 16 pts. per acre during any one growing season. (2) Do not apply within 14 days of harvest. (3) Do not apply more than the 2 pt. per acre rate to **open** bolls. (4) Do not graze treated cotton plants or feed gin trash to livestock.

Deciduous Fruit

Dilute and Concentrate Sprays

See Table 2 for specific crops and insects. Apply during the dormant or delayed dormant period. Use a rate of 1–2 pts. per 100 gals. of water as a dilute spray (300–600 gals. per acre) or a rate of 3–12 pts. per acre as a concentrate spray* (a minimum of 50 gals. of water per acre). The lower rates may be used for light infestations and the higher rates for moderate to heavy infestations. For rates less than 2 pts. of SUPRACIDE per 100 gals., tank mix with an EPA-registered oil.

When concentrate applications are made, proper equipment must be used to obtain uniform coverage.

Table 2: Deciduous Fruits

Crop	Insects
Apples	San Jose scale, rosy apple aphid
Cherries, pears	San Jose scale
Nectarines, peaches, plums, prunes	Peach twig borer, San Jose scale, Green peach aphid[1]
Apricots	Peach twig borer
Pears†	Pear psylla[2]

† California only

[1] For green peach aphid, use 8 pts. SUPRACIDE + 4 gals. oil per acre or 12 pts. SUPRACIDE per acre alone.

[2] For pear psylla, use 8–12 pts. SUPRACIDE + 4–6 gals. oil in a minimum of 100 gals. total spray per acre.

Aerial Sprays

See Table 3 for specific crops and insects. Apply during the dormant or delayed dormant period and use 3–12 pts. per acre as an aerial spray* in a minimum of 20 gals. total spray per acre. Spray swath should be no wider than the distance between the tree rows or the length of the spray boom. The lower rates may be used for light infestations and the higher rates for moderate to heavy infestations. For rates less than 2 pts. of SUPRACIDE per 100 gals., tank mix with an EPA-registered oil.

Table 3: Deciduous Fruits (Continued)

Crop	Insects
Apples	San Jose scale, rosy apple aphid
Peaches, plums, prunes, nectarines, apricots	Peach twig borer, San Jose scale
Cherries, pears	San Jose scale

EPA-registered oils may be added to the SUPRACIDE spray mixture in the dormant/delayed dormant period. Do not use more than 1.5 gals. of oil per 100 gals. spray as a dilute spray or 6 gals. of oil per acre as a concentrate spray or 5 gals. of oil per acre as an aerial spray.

* The rate of SUPRACIDE for concentrate and aerial sprays is based on a dilute spray with conventional ground equipment. A dilute spray is considered to be 300-600 gals. per acre. This amounts to a minimum of 3 pts. and a maximum of 12 pts. of SUPRACIDE per acre.

Note: To avoid illegal residues, (1) Apply before any blossoms open or injury may occur. (2) Do not graze treated areas or feed cover crops grown in treated areas to livestock.

Precaution: Do not use SUPRACIDE on Red Beaut plums.

Grapefruits, Lemons, Oranges and Other Citrus Fruits

SUPRACIDE Alone

Apply the recommended rates anytime except during the bloom period, or within 14 days of harvest. Preferred timing is during the immediate post-bloom period, **before fruit becomes infested.** Use the low rate for light infestations and the high rate for moderate to heavy infestations. Thorough coverage is essential for good control.

Make no more than 2 applications per growing season and allow at least 45 days between applications. Severe infestations may require 2 treatments.

California red scale: (Arizona and California only) Apply 1-2 pts. per 100 gals. of water.

Black scale: (California only) Apply 1-2 pts. per 100 gals. of water.

Chaff, Glovers, purple, snow, yellow scale, and mealybug: (Florida only) Apply 1/2 -2 pts. per 100 gals. of water.

Black scale, whiteflies, and citrus blackfly: (Florida only) Apply 1-2 pts. per 100 gals. of water.

Brown soft, California red, chaff scale, and mealybug: (Texas only) Apply 1/2-2 pts. per 100 gals. of water.

The minimum amount of SUPRACIDE applied per acre is dependent upon the number of trees per acre and upon tree height. See example below.

Example: Mature trees of normal foliage require a minimum of 5 gals. of spray plus one additional gallon for each foot of tree height for adequate coverage. Eighty trees per acre 20 feet tall require $80 \times 25 = 2{,}000$ gals. of dilute spray per acre.

Note: To avoid illegal residues, (1) Do not apply more than 40 pts. per acre. (2) Do not graze treated areas. (3) Make no more than 2 applications per growing season nor apply within 14 days of harvest.

SUPRACIDE Plus Oil

In Florida and Texas, EPA-registered citrus spray oils at rates of 2-4 qts. per 100 gals. of water may be tank mixed with SUPRACIDE following the rates and timing listed above for SUPRACIDE alone.

In California, EPA-registered citrus spray oils at rates listed below may be tank mixed with SUPRACIDE. See the above directions for rates and timing. A rate of 1/2 pt. of SUPRACIDE per 100 gals. of water may be used if tank mixed with oil.

Type of oil	Amount per 100 gals. of water
Light-medium grade	1.6 - 1.8 gals.
Medium grade	1.6 - 1.8 gals.
Narrow-range 415 grade	1.2 - 1.4 gals.

For further details, consult the current *University of California Citrus Treatment Guide.*

Note: To avoid illegal residues, (1) Do not apply more than 40 pts. of SUPRACIDE per acre in any one season. (2) Do not graze treated areas. (3) Make no more than 2 applications with oil per growing season nor apply within 60 days of harvest. (4) Lemons only, do not exceed 10 pts. per acre or apply more than once per season when tank mixed with oil.

Mangoes

Apply during the postharvest to bloom stage with ground equipment in 100 gals. of spray mixture per acre. Spray foliage to thoroughly wet upper and lower leaf surfaces. Repeat as needed at 3 week intervals.

False oleander scale, *Pseudaulacaspis cockerelli* (cooley): Apply 1/2-1 pt. and 0.5% oil per acre.

Note: To avoid illegal residues, (1) Do not exceed 5 applications per season. (2) Do not graze or feed cover crops from treated groves.

Nursery Stock (woody ornamental or herbaceous plants commercially grown for transplanting)

Apply at the rate listed below in a thorough cover spray to runoff. Thorough coverage and timing of application are essential for effective control.

To control scale insects on nursery stock, including both potted and field grown plants, apply two pints per 100 gals. of water once per season after sale crawlers have hatched. Monitor populations closely to determine when crawlers are present.

Because of changing cultivars in the nursery industry, test each new cultivar for phytotoxicity by treating one plant with the recommended rate at least 2 days prior to treatment of the remaining plants.

Olives

Apply once during the postharvest or prebloom season at the rate listed below as a full cover spray in a minimum of 400 gals. of water per acre. Supreme or superior type oil may be added to the spray mixture at the rate of 1-1.5 gals. per 100 gals. of water.

Black scale: Apply 2 pts./100 gals.

Note: To avoid illegal residues, (1) Do not apply more than 12 pts. per acre per year. (2) Do not apply after buds break. (3) Do not graze treated areas or feed cover crops grown in treated areas to livestock.

Pecans (Southeastern U.S. Only)

Apply the appropriate rate listed below as a foliar or full cover spray in sufficient water for thorough coverage.

Yellow pecan aphid: Apply 2 pts. per 100 gals. and repeat as necessary at 14-21 day intervals.

Hickory shuckworm and pecan serpentine leafminer: Apply 2-4 pts. per 100 gals. when insects first appear or leafminer damage first appears and repeat as necessary at 14-21 day intervals. Use the 4 pt. per 100 gals. rate when infestations are heavy.

Precaution: Do not tank mix SUPRACIDE with Cyprex® as phytotoxicity may result.

Note: To avoid illegal residues, (1) Do not apply within 60 days of harvest. (2) Do not graze treated areas.

Safflowers

Apply in a minimum of 20 gals. of water by ground equipment or in a minimum of 5 gals. of water by air.

Lygus bugs and thrips: Apply 2 pts. per acre. Begin applications either at bud stage, 12-14 days before onset of bloom, or as infestations develop. Additional applications can be made on a 7-14 day schedule or as further infestations develop.

Note: To avoid illegal residues, (1) Do not apply within 28 days of harvest. (2) Do not make more than three applications per year. (3) Do not graze treated areas.

Sunflowers

Apply 2 pts. per acre. Make ground application in a minimum of 10 gals. of water per acre; aerial application in a minimum of 5 gals. of water per acre. Thorough coverage of the heads is necessary. One to three applications may be made on a 7-day schedule. The number of applications will be determined by the severity of the infestations.

Stem weevil*: Begin applications in early season at vegetative stage 2.3 (third alternate leaf developed) to vegetative stage 3.1 (inflorescence head visible but tightly surrounded by young leaves).

Sunflower moth, seed weevil**, banded sunflower moth, and sunflower maggot complex: Begin applications at the onset of bloom as the infestations develop.

Sunflower seed midge: Begin applications during early bud development.

* *Cylindrocopturus adspersus* and *Apion occidentalis.*

** *Smicronyx fulvus and Smicronyx sordidus.*

Note: To avoid illegal residues, (1) Do not apply within 50 days of harvest. (2) Do not graze treated areas or feed treated forage to livestock.

Grain Sorghum Grown for Forage and Grain (Colorado, Kansas, Oklahoma, and Texas Only)

Apply in a minimum of 10 gals. of water per acre with ground equipment or 2 gals. of water per acre by air. Up to three applications may be made on a 10-14 day schedule. The number of applications will be determined by the severity of the infestation.

Banks grass mite, sorghum greenbug: Apply 2 pts. per acre.

Precaution: Application may result in phytotoxicity to certain sorghum varieties. This phytotoxicity is a temporary spotting, reddening or chlorosis of the leaves.

Note: To avoid illegal residues, do not apply within 30 days of harvesting or feeding to livestock.

Tobacco

Apply at the rates listed below in a minimum of 25 gals. of water per acre when insects appear. Apply 1-3 repeat applications on a 7-14 day interval as needed.

Flea beetles, hornworms: Apply 2-3 pts. per acre.

Budworms: Apply 4 pts. per acre.

Precaution: When applying SUPRACIDE with hand sprayers, avoid concentrating the spray in the bud area since phytotoxicity may occur. Do not apply SUPRACIDE to tobacco plants which are under stress as injury may occur.

Note: Do not apply in mixture with other pesticides or within 3 days of tobacco harvest.

Walnuts

Dilute and Concentrate Sprays

Walnut scale, frosted scale, Italian pear scale: Apply during the dormant or delayed dormant period or as a cover spray when crawlers appear.

Codling moth, walnut aphid: Apply as a cover spray according to moth activity or when aphids appear.

Use a rate of 2 pts. per 100 gals. of water as a dilute spray (400 gals. per acre), or a rate of 8 pts. per acre as a concentrate spray* (a minimum of 100 gals. of water per acre).

When concentrate applications are made, proper equipment must be used to obtain uniform coverage. If scale infestations are severe, dilute applications should be used.

Aerial Sprays

Walnut scale, frosted scale: Apply during the dormant or delayed dormant period and use 8 pts. per acre as an aerial spray* in a minimum of 20 gals. of spray per acre. Spray swath should be no wider than the distance between the tree rows or the length of the spray boom.

EPA-registered oils may be added to the SUPRACIDE spray mixture in the dormant/delayed dormant period. Do not use more than 1 gal. of oil per acre in concentrate and aerial sprays or 4 gals. of oil per acre in dilute sprays.

*The rate of SUPRACIDE for concentrate and aerial sprays is based on a dilute spray with conventional ground equipment. A dilute spray is considered to be 400 gals. per acre. This amounts to 8 pts. of SUPRACIDE per acre.

Note: To avoid illegal residues, (1) Make no more than one application during the dormant or delayed dormant period nor more than two cover sprays each season. (2) Do not apply within 7 days of harvest. (3) Do not graze treated areas or feed cover crops grown in treated areas to livestock.

Precaution: Do not add oil to SUPRACIDE sprays to walnuts in southern or central California or to trees that may be subject to stress due to drought, drying winds, disease, or severe insect infestations.

STORAGE AND DISPOSAL

Do not contaminate water, food, or feed by storage or disposal.

Storage: Store at temperatures above 40°F. Crystals may form at lower storage temperatures. If this occurs, place the product in a warm room (68°F or above) and roll or shake the container at frequent intervals until all crystals are dissolved.

Pouring the product through a 50-mesh or finer screen will indicate adequacy of or need to repeat above procedure. Do not store near food, feed, or fertilizer. Do not use or store in or around the home.

Pesticide and Container Disposal: Pesticide wastes are acutely hazardous. Improper disposal of unused pesticide, spray mixture, or rinse water is a violation of federal law. If these wastes cannot be used according to label instructions, contact your State Pesticide or Environmental Control Agency, or the Hazardous Waste representative at the nearest EPA Regional Office for guidance in proper disposal methods.

Triple rinse container (or equivalent) and offer for recycling or reconditioning, or puncture and dispose of in a sanitary landfill, by incineration, or if allowed by state and local authorities, by burning. Stay out of smoke from burning containers.

For minor spills, leaks, etc., follow all precautions indicated on this label and clean up immediately. Take special care to avoid contamination of equipment and facilities during cleanup procedures and disposal of wastes. In the event of a major spill, fire or other emergency, call 1-800-888-8372 day or night.

PRECAUTIONARY STATEMENTS

HAZARDS TO HUMANS AND DOMESTIC ANIMALS

DANGER

Corrosive. Causes irreversible eye damage. May cause skin irritation. Do not get in eyes, on skin, or on clothing. Wear goggles or face shield and rubber gloves when mixing. May be harmful or fatal if swallowed or absorbed through skin. Do not breathe vapor or spray mist. Prolonged or repeated inhalation may be harmful. Wear a mask or pesticide respirator jointly approved by the Mining Enforcement Safety Administration (MESA) or the National Institute for Occupational Safety and Health (NIOSH) during periods of prolonged or repeated use. Wash thoroughly with soap and water after handling and before eating, drinking, or smoking. Remove contaminated clothing and wash before reuse or if not washable, discard. Keep out of reach of children and domestic animals.

STATEMENT OF PRACTICAL TREATMENT

If in eyes: Flush with plenty of water. Get medical attention immediately.

If swallowed: Call a physician or poison control center immediately. DO NOT INDUCE VOMITING UNLESS NO OTHER TREATMENT IS AVAILABLE. Vomiting may cause aspiration pneumonia. If it is necessary to induce vomiting, do so under medical supervision. Give victim one or two glasses of water and insert finger in back of throat. Repeat until vomit fluid is clear. Do not induce vomiting or give anything by mouth to an unconscious person.

If on skin: Remove contaminated clothing and wash skin with plenty of soap and water. Get medical attention.

If inhaled: Remove victim to fresh air. If not breathing, give artificial respiration, preferably mouth-to-mouth unless exposure was due to swallowing; then use cardiopulmonary resuscitation (CPR). Get medical attention.

Note to Physician: Lavage stomach with care to prevent aspiration into lung. After lavage, instill and leave 30 gm. activated charcoal in 3-4 oz. water. Atropine sulfate and 2-PAM are specific antidotes. Morphine, theophylline, aminophylline, and large amounts of intravenous fluids are contraindicated.

ENVIRONMENTAL HAZARDS

This pesticide is toxic to fish and wildlife. Do not apply directly to water or wetlands (swamps, bogs, marshes, and potholes). Drift and runoff from treated areas may be hazardous to aquatic organisms in neighboring areas. Do not contaminate water by cleaning of equipment or disposal of wastes.

This pesticide is highly toxic to bees exposed to direct treatment or to residues remaining on the treated area. Do not apply when bees are actively visiting the crop, cover crop, or weeds blooming in the treatment area. Applications should be timed to provide the maximum possible interval between treatment and the next period of bee activity.

PHYSICAL OR CHEMICAL HAZARDS

Do not use, pour, spill, or store near heat or open flame.

SUPRACIDE® trademark of CIBA-GEIGY for methidathion
Cyprex® trademark of American Cyanamid for dodine

CGA 30L23F 059

This is a specimen label which was not prepared by CIBA-GEIGY. CIBA-GEIGY is not responsible for the accuracy of the information contained herein. As labels are subject to revision, always carefully read and follow the label on the product container.

TILT®

FUNGICIDE

For control of certain diseases in cereals, grasses grown for seed, and sugarcane

ACTIVE INGREDIENT:
1-[[2-(2,4-dichlorophenyl)-4-propyl-1,3-dioxolan-2-yl]methyl]-1*H*-1,2,4-triazole 41.8%
INERT INGREDIENTS: 58.2%
TOTAL 100.0%

EPA Reg. No. 100-617 EPA Est. 100-AL-1

KEEP OUT OF REACH OF CHILDREN.

WARNING

See additional precautionary statements and directions for use inside booklet.

DIRECTIONS FOR USE AND CONDITIONS OF SALE AND WARRANTY

IMPORTANT: Read the entire **Directions for Use** and the **Conditions of Sale and Warranty** before using this product.

CONDITIONS OF SALE AND WARRANTY

The **Directions for Use** of this product reflect the opinion of experts based on field use and tests. The directions are believed to be reliable and should be followed carefully. However, it is impossible to eliminate all risks inherently associated with use of this product. Crop injury, ineffectiveness, or other unintended consequences may result because of such factors as weather conditions, presence of other materials, or the manner of use or application all of which are beyond the control of CIBA-GEIGY or the Seller. All such risks shall be assumed by the Buyer.

CIBA-GEIGY warrants that this product conforms to the chemical description on the label and is reasonably fit for the purposes referred to in the **Directions for Use** subject to the inherent risks referred to above. **CIBA-GEIGY makes no other express or implied warranty of Fitness or Merchantability or any other express or implied warranty. In no case shall CIBA-GEIGY or the Seller by liable for consequential, special, or indirect damages resulting from the use or handling of this product.** CIBA-GEIGY and the Seller offer this product, and the Buyer and user accept it, subject to the foregoing **Conditions of Sale and Warranty,** which may be varied only by agreement in writing signed by a duly authorized representative of CIBA-GEIGY.

DIRECTIONS FOR USE

It is a violation of federal law to use this product in a manner inconsistent with its labeling.

GENERAL INFORMATION

TILT is a broad spectrum fungicide for the control of certain diseases in wheat, barley, rye, rice, sugarcane, pecans, and in grasses grown for seed.

Note: Do not use in nurseries, on turf, or landscape plantings.

FAILURE TO FOLLOW DIRECTIONS AND PRECAUTIONS ON THIS LABEL MAY RESULT IN CROP INJURY, POOR DISEASE CONTROL, OR ILLEGAL RESIDUES.

Spray Equipment

Thorough coverage is necessary to provide good disease control.

To avoid spray drift, do not apply when conditions favor drift beyond the target area. Avoid spray overlap, as crop injury may occur.

Equip sprayers with nozzles that provide accurate and uniform application. Be certain that nozzles are the same size and uniformly spaced across the boom. Calibrate sprayer before use.

Use a pump with capacity to (1) maintain 35-40 psi at nozzles, and (2) provide sufficient agitation in tank to keep mixture in suspension—this requires recirculation of 10 percent of tank volume per minute. Use a jet agitator or liquid sparge tube for agitation. Do not air sparge.

Although TILT is an emulsifiable concentrate, it is suggested screens be used to protect the pump and to prevent nozzles from clogging. Screens placed on suction side of pump should be *16-mesh or coarser.* Do not place a screen in the recirculation line. Use 50-mesh or coarser screens between the pump and boom, and where required, at the nozzles. Check nozzle manufacturer's recommendations.

For more information on spray equipment and calibration, consult sprayer manufacturers and state recommendations. For specific local directions on spray schedules, consult the current state agricultural experiment station recommendations.

To determine the physical compatibility of TILT with other products, use a jar test. Using a quart jar, add the proportionate amounts of the products to 1 qt. of water. Add wettable powders and water dispersible granular products first, then liquid flowables, and emulsifiable concentrates last. After thoroughly mixing, let stand for at least 5 minutes. If the combination remains mixed or can be remixed readily, it is physically compatible. Once compatibility has been proven, use the same procedure for adding required ingredients to the spray tank.

Aerial Application: For those crops where aerial applications are indicated, apply in a minimum of 5 gals. of water/A unless specified otherwise. Avoid applications under conditions when uniform coverage cannot be obtained or when excessive spray drift may occur. Do not apply directly to humans or animals. Flagmen and loaders should wash thoroughly before eating, drinking, or smoking, and at the end of each day's operation.

Ground Application: Apply TILT by ground equipment in a minimum of 15 gals. of water/A

Chemigation: Do not apply this product through any type of irrigation system.

Mixing Procedures

(1) Wear rubber gloves and long-sleeved shirt when mixing and loading this product. (2) Be sure sprayer is clean; if it is contaminated with any material, crop injury or sprayer clogging may result. (3) Fill tank 1/4-1/2 full with clean water, and start agitation. (4) Be certain that the agitation system is working properly and creates a rippling or rolling action on the liquid surface. (5) Add product to the tank. For wheat, barley, rye, and rice, custom applicators must use a closed system for mixing and loading this product. (6) Complete filling tank, maintaining sufficient agitation at all times to ensure surface action. This applies to both the spray and nurse tanks. (7) Do not leave tank mixtures in the tank without adequate agitation. Always maintain agitation to avoid separation and buildup of residues. If a residue starts to build up in tank, drain and clean with strong detergent and ammonia solution, and rinse thoroughly with clean water. (8) Immediately after use, clean sprayer thoroughly by flushing system with water containing a detergent and rinse with clean water.

Rotational Crops

To avoid possible illegal residues, (1) Do not double-crop treated acreage when TILT is applied to the first crop unless the second crop appears on this label. (2) Do not graze or feed forage, fodder, or straw from rotational crops planted in the fall or the spring following treatment.

CEREALS

Wheat, Barley, Rye

In wheat, barley, and rye, TILT controls rusts (*Puccinia* spp.), powdery mildew (*Erysiphe* spp.), leaf blight and glume blotch (*Septoria* spp.), tan spot (*Pyrenophora tritici-repentis*), Helminthosporium leaf blight, barley scald (*Rhynchosporium secalis*) and net blotch (*Pyrenophora teres*). Apply 4 fl. oz. of TILT/A by ground or aerial application. Highest yields are normally obtained when TILT is applied to the emerging flag leaf (Feekes growth stage 8). Do not apply after this growth stage to avoid possible illegal residues. TILT may be applied earlier if disease symptoms appear (especially applicable to barley). Make only one application per season.

For control of foot rot (*Pseudocercosporella* spp.), apply 4 fl. oz. of TILT/A plus half rates of other EPA-registered fungicides such as Benlate®, Mertect®, and Topsin® M. Apply at tillering but before elongation has occurred.

Do not graze or feed livestock treated forage or cut the green crop for hay or silage. After harvest the straw may be used for bedding.

Rice

In rice, TILT controls sheath blight (*Rhizoctonia solani*), brown leaf spot (*Helminthosporium oryzae*), narrow brown leaf spot and brown blotch (*Cereospora oryzae*), leaf smut (*Entyloma oryzae*) and sheath spot (*Rhizoctonia oryzae*). TILT also suppresses stem rot (*Sclerotium oryzae*). Apply TILT on either of the following schedules as an aerial spray in 5-10 gals. of water/A.

A. 6 fl. oz./A at first internode elongation (up to 2-inch panicle) and repeat at swollen boot. Make the second application 10-14 days after the first application, but before the boot splits and head emerges. TILT provides best control of sheath blight when the first application is applied at disease appearance in the field. The first application is recommended when 5 percent or fewer of the tillers are infected.

B. 10 fl. oz./A at first internode elongation (up to 2-inch panicle). The 10 oz. rate is recommended if greater than 10 percent of the tillers are infected with sheath blight. If disease reappears, use another registered fungicide for the second application.

Note: To avoid possible illegal residues (1) Do not apply to stubble or ratoon crop rice; (2) Do not use in rice fields where commercial farming of crayfish will be practiced; (3) Do not drain water from treated rice fields into ponds used for commercial catfish farming; (4) Do not use water drained from treated fields to irrigate other crops; (5) Do not use in California.

Endangered Species Restrictions: The use of TILT (propiconazole) on rice is restricted to protect the endangered fat pocketbook pearly mussel (*Potamilus capax*) and its habitat. Use is prohibited in the following areas of Arkansas:

Mississippi Country—Within the basin that drains directly into the Right Hand Chute of Little River, south of Big Lake National Wildlife Refuge.

Poinsett County—Between Crowley's Ridge and the levee east of the Right Hand Chute of Little River and the St. Francis Floodway. Use is also prohibited west of Rt. 140 and north of Rt. 63 at the siphon near Marked Tree. Except that the prohibited area does not include the area bounded by Arkansas highway 373 on the west, highway 63 on the east and highway 14 on the south.

Cross, St. Francis, and Lee Counties: Between Crowley's Ridge and the levee east of the Right Hand Chute of Little River and the St. Francis Floodway as far south as the confluence of L'Anguille River (Lee County).

Grasses Grown for Seed (Nebraska, Oregon, Washington, Idaho, and Minnesota only)

TILT controls rust (*Puccinia* spp.), powdery mildew, and Selenophoma stem eyespot in perennial ryegrass, fescues, bluegrass, orchardgrass, and wheatgrasses.

Mix 4-8 fl. oz. of TILT (maximum 4 fl. oz. on bluegrass) in a minimum of 20 gals. of water/A for ground application, or in a minimum of 10 gals. of water/A for aerial application. Apply when powdery mildew and Selenophoma infections or rust pustules are noticeable and increasing in number, in late spring or early summer. Repeat at 14-21 day intervals. To maximize control under severe rust pressure, use the higher rate of 8 fl. oz./A (except on bluegrass) and make applications at 14-day intervals until the seed is mature. Do not apply more than 32 fl. oz. of TILT per acre per growing cycle. Make the last application at least 20 days before seed matures. For bluegrass, it is important to begin applications early in the growing season.

Note: To avoid possible illegal residues do not:

(1) feed hay cut within 20 days of the last application or (2) graze treated areas within 140 daysof the last application.

Sugarcane (Hawaii only)

For control of pineapple disease of sugarcane (*Ceratocystis paradoxa*) apply TILT to cut seed pieces using 0.75 fl. oz. (22 ml) per 100 gals. water (1:17,000). Treatments can be applied in either a cold or hot water dip. Do not use treated seed pieces for food or feed purposes. Dispose of spent dip solution according to state and federal regulations.

Cold Water Dip—Immerse seed pieces to give thorough wetting, remove and allow to drain.

Hot Water Dip—Maintain water temperature at 52°C. Soak the seed pieces for 20-30 minutes, remove and allow to drain.

STORAGE AND DISPOSAL

Store at temperatures above 0° F. Do not contaminate water, food or feed by storage or disposal. Pesticide wastes may be toxic. Improper disposal of unused pesticide, spray mixture, or rinse water is a violation of federal law. If these wastes cannot be used according to label instructions, contact your State Pesticide or Environmental Control Agency, or the Hazardous Waste representative at the nearest EPA Regional Office for guidance in proper disposal methods.

Triple rinse container (or equivalent). Then offer for recycling or reconditioning, or puncture and dispose of in a sanitary landfill, by incineration, or, if allowed by state and local authorities, by burning. Stay out of smoke from burning containers.

For minor spills, leaks, etc., follow all precautions indicated on this label and clean up immediately. Take special care to avoid contamination of equipment and facilities during cleanup and disposal of wastes. In the event of a major spill, fire, or other emergency, call 1-800-888-8372 day or night.

PRECAUTIONARY STATEMENTS

HAZARDS TO HUMANS AND DOMESTIC ANIMALS

WARNING

This product may cause skin sensitization reactions in certain individuals. Wear rubber gloves and long-sleeved shirt when mixing or handling this product. Remove contaminated clothing and wash before reuse.

Causes eye irritation. Do not get in eyes, on skin, or on clothing. Harmful if swallowed, inhaled, or absorbed through skin. Wash thoroughly with soap and water after handling. Avoid breathing spray mist.

First Aid: If in eyes, flush with plenty of water and get medical attention if irritation persists. If on skin, wash with plenty of soap and water. Get medical attention if irritation persists. If swallowed, drink promptly a large quantity of milk, egg whites, or gelatin solution; if these are not available, drink large quantities of water. Never give anything by mouth to an unconscious person. Do not induce vomiting. Call a physician or Poison Control Center immediately. In case of inhalation exposure, move to fresh air.

Note to Physician: There is no specific antidote for this product. Induce emesis or lavage stomach, taking care to avoid aspiration of stomach contents into lungs. Give a saline laxative and supportive therapy.

ENVIRONMENTAL HAZARDS

This pesticide is toxic to fish. Do not apply directly to water. Do not contaminate water by cleaning of equipment or disposal of wastes. Do not contaminate water when disposing of equipment wash waters. Refer to product labeling for use restrictions to protect endangered species.

PHYSICAL OR CHEMICAL HAZARDS

Do not use, pour, spill or store near heat or open flame.

Store at temperatures above 0°F.

Do not apply this product through any type of irrigation system.

Benlate® trademark of E. I. duPont de Nemours & Co. (Inc.) for benomyl
Mertect® trademark of Merck and Co., Inc. for thiabendazole
TILT® trademark of CIBA-GEIGY

U.S. Patent No. 4,079,062

Topsin® trademark of Pennwalt Corporation for thiophanate-methyl

CGA 72L1K 069

This is a specimen label which was not prepared by CIBA-GEIGY. CIBA-GEIGY is not responsible for the accuracy of the information contained herein. As labels are subject to revision, always carefully read and follow the label on the product container.

THE DOW CHEMICAL COMPANY

The following revised product labels appear: Lorsban 4E, Lorsban 50W, Pathway, Tordon RTU.

LORSBAN* 4E

INSECTICIDE

For Control of Various Insects Infesting Certain Field, Fruit, Nut, and Vegetable Crops

ACTIVE INGREDIENT(S):

Chlorpyrifos [O,O-diethyl O-(3,5,6-trichloro-2-pyridinyl) phosphorothioate]	40.7%
INERT INGREDIENTS:	59.3%
TOTAL	100.0%

Contains 4 pounds of chlorpyrifos per gallon

E.P.A Registration No. 464-448 E.P.A. Est. 464-MI-1 *(MM), 464-CA-1 WP

*Superscript used corresponds to letter lot in no.

KEEP OUT OF REACH OF CHILDREN

WARNING

AVISO:

PRECAUCIÓN AL USUARIO: Si usted no lee inglés, no use este producto hasta que la etiqueta le haya sido explicada ampliamente.

PRECAUTIONARY STATEMENTS

HAZARDS TO HUMANS AND DOMESTIC ANIMALS

MAY BE FATAL IF SWALLOWED • HARMFUL IF ABSORBED THROUGH SKIN • CAUSES SUBSTANTIAL BUT TEMPORARY EYE INJURY • CAUSES SKIN IRRITATION

Avoid Breathing Vapor Or Spray Mist • Do Not Get In Eyes, On Skin, Or On Clothing • Wear Protective Clothing, Rubber Gloves, And Goggles • Wash Thoroughly With Soap And Water After Handling And Before Eating Or Smoking • Immediately Remove Contaminated Clothing And Wash Before Reuse • Destroy Contaminated Shoes • Do Not Apply This Product In Such A Manner As To Directly Or Through Drift Expose Workers Or Other Persons • The Area Being Treated Must Be Vacated By Unprotected Persons • Keep Away From Food, Feedstuffs And Water Supplies

STATEMENTS OF PRACTICAL TREATMENT:

If Swallowed: Do not induce vomiting. Contains aromatic petroleum solvent. Call a physician or poison control center immediately.

If In Eyes: Flush with plenty of water for at least 15 minutes. Get medical attention.

If On Skin: Wash with plenty of soap and water. Get medical attention if irritation persists.

If Inhaled: Remove to fresh air if effects occur. Get medical attention.

NOTE TO PHYSICIANS: Chlorpyrifos is a cholinesterase inhibitor. Treat symptomatically. If exposed, plasma and red blood cell cholinesterase tests may indicate significance of exposure (baseline data are useful). Atropine, only by injection, is the preferable antidote. Oximes, such as 2-PAM/protopam, may be therapeutic if used early; however, use only in conjunction with atropine. In case of severe acute poisoning, use antidote immediately after establishing an open airway and respiration.

PHYSICAL OR CHEMICAL HAZARDS

COMBUSTIBLE • Do Not Use or Store Near Heat or Open Flame • Do Not Cut or Weld Container.

ENVIRONMENTAL HAZARDS

This pesticide is toxic to birds and wildlife, and extremely toxic to fish and aquatic organisms. Do not apply directly to water. Drift and runoff from treated areas may be hazardous to aquatic organisms in adjacent aquatic sites. Cover or incorporate spills. Do not contaminate water by cleaning of equipment or disposal of waste. This product is highly toxic to bees exposed to direct treatment or residues on blooming crops or weeds. Avoid use when bees are actively foraging. Protective information may be obtained from your Cooperative Agricultural Extension Service.

NOTICE

Read and understand the entire label before using. Use only according to label directions.

Before buying or using this product, read "WARRANTY LIMITATIONS AND DISCLAIMER" elsewhere on this label. If terms are not acceptable, return **unopened** package at once to seller for full refund of purchase price paid. Otherwise, use by the buyer or any other user constitutes acceptance of the terms under WARRANTY LIMITATIONS AND DISCLAIMER.

IN CASE OF AN EMERGENCY endangering life or property involving this product, call collect 517-636-4400

AGRICULTURAL CHEMICAL

Do Not Ship or Store with Food, Feeds, Drugs, or Clothing

REENTRY STATEMENT

Do not allow reentry into treated areas within 24 hours unless protective clothing is worn. Protective clothing consists of at least a hat or other suitable head covering, a long-sleeved shirt and long-legged trousers or a coverall type garment (all of closely woven fabric covering the body, including the arms and legs), shoes and socks. Because certain states may require more restrictive reentry intervals for various crops treated with this product, consult your State Department of Agriculture for further information.

Written or oral warnings must be given to workers who are expected to be in a treated area or in an area about to be treated with this product. These oral warnings shall inform workers of areas or fields that may not be entered without protective clothing and actions to take in case of accidental exposure as shown by the Statement of Practical Treatment section on this label. When oral warnings are given, warnings shall be given in a language customarily understood by workers. Written warnings must include the following information: WARNING. Area treated with LORSBAN 4E insecticide on (insert date of application). Do not enter without protective clothing for 24 hours. In case of accidental exposure (insert actions to take as shown by the Statement of Practical Treatment section on this label).

Do not use on turf or for structural pest control. Do not formulate this product into other end use products.

DIRECTIONS FOR USE

It is a violation of Federal law to use this product in a manner inconsistent with its labeling.

GENERAL INFORMATION

LORSBAN 4E insecticide forms an emulsion when diluted with water and is suitable for use in all conventional spray equipment.

MIXING INSTRUCTIONS

To prepare the spray, add a portion of the required amount of water to the spray tank and with agitation add the LORSBAN 4E. Complete filling the tank with the balance of water needed. Maintain sufficient agitation during both mixing and application to ensure uniformity of the spray mixture.

LORSBAN 4E insecticide may also be used in tank mixtures with certain herbicides and/or with non-pressure fertilizer solutions as recommended under specific crop use directions. Note that LORSBAN 4E may be inactivated by sulfur containing fertilizer solutions. Prepare tank mixtures in the same manner as recommended above for use of LORSBAN 4E alone. When tank mixtures of LORSBAN 4E and herbicides are involved, add wettable powders first, flowables second, and emulsifiable concentrates last. Where a fertilizer solution is involved, it is strongly recommended that a fertilizer pesticide compatibility agent such as Unite or Compex be used. Maintain constant agitation during both mixing and application to ensure uniformity of the spray mixture. Do not allow spray mixtures to stand overnight.

NOTE: Test compatibility of the intended tank mixture before adding LORSBAN 4E to the spray or mix tank. Add proportionate amounts of each ingredient to a pint or quart jar, cap, shake, and let set 15 minutes. Formation of precipitates that do not readily redisperse indicates an incompatible mixture that should not be used.

SPRINKLER IRRIGATION

LORSBAN 4E may be applied by sprinkler irrigation for the following crop uses:

Alfalfa, Citrus Orchard Floors, Field Corn, Popcorn, Sweet Corn, Cotton, Cranberries, Sorghum, and Soybeans

See the use sections for the individual crops for further application information. Do not apply this product to the above listed crops through any other type of irrigation system. Do not apply this product by chemigation to any other crop.

The following use directions are to be followed when LORSBAN 4E is applied through sprinkler irrigation systrems. Thoroughly clean the injection system and tank of any fertilizer or chemical residues, and dispose of the residues according to state and federal laws. Flush the injector with soap and water. Determine the amount of insecticide needed to cover the desired acreage. Pump the required LORSBAN 4E insecticide into the steel tank, start mechanical or hydraulic agitation, and add in order the non-emulsifiable oil and/or water. Continually agitate the mixture containing LORSBAN 4E. Set the sprinkler system to deliver the desired inches of water per acre. Start the water pump and sprinkler, and let the system achieve the desired pressure and speed before starting the injector. Start the injector and calibrate the injector system according to section 14 (below). The mixture containing LORSBAN 4E must be injected continuously and uniformly into the irrigation water line as the sprinkler is moving. This procedure is necessary to deliver the desired rate per acre in a uniform manner. When the application is finished, allow the entire irrigation and injector system to be thoroughly flushed clean before stopping the system.

The following use precautions will result in a safe and successful application of mixtures containing LORSBAN 4E.

1. Apply this product only through sprinkler irrigation systems including center pivot, lateral move, end tow, side (wheel) roll, traveler, big gun, solid set, micro sprinkler, or hand move. Do not apply this product through any other type of irrigation system.
2. Crop injury, lack of effectiveness, or illegal pesticide residues in the crop can result from nonuniform distribution of treated water.

3. If you have questions about calibration, you should contact State Extension Service specialists, equipment manufacturers or other experts.
4. Do not connect an irrigation system (including greenhouse systems) used for pesticide application to a public water system.
5. A person knowledgeable of the chemigation system and responsible for its operation, or under the supervision of the responsible person, shall shut the system down and make necessary adjustments should the need arise.
6. The system must contain a functional check valve, vacuum relief valve, and low pressure drain appropriately located on the irrigation pipeline to prevent water source contamination from backflow. Refer to the American Society of Agricultural Engineer's Engineering Practice 409 for more information.
7. The pesticide injection pipeline must contain a functional, automatic, quick-closing check valve to prevent the flow of fluid back toward the injection pump.
8. The pesticide injection pipeline must also contain a functional, normally closed, solenoid-operated valve located on the intake side of the injection pump and connected to the system interlock to prevent fluid from being withdrawn from the supply tank when the irrigation system is either automatically or manually shut down.
9. The system must contain functional interlocking controls to automatically shut off the pesticide injection pump when the water pump motor stops.
10. The irrigation line or water pump must include a functional pressure switch which will stop the water pump motor when the water pressure decreases to the point where pesticide distribution is adversely affected.
11. Systems must use a metering pump, such as a positive displacement injection pump (e.g., diaphragm pump) effectively designed and constructed of materials that are compatible with pesticides and capable of being fitted with a system interlock. The metering pump must provide a greater pressure than that of the irrigation system at the point of injection. The pump must meet Section 675 for "Electrically Driven or Controlled Irrigation Machines" NEC 70 and must contain Viton or Teflon seals.
12. To insure uniform mixing of the insecticide into the water line, inject the mixture through a nozzle placed in the fertilizer injection port or just ahead of an elbow or tee in the irrigation line so that the turbulence created at those points will assist in mixing. It is suggested that the injection point be higher than the insecticide tank to prevent siphoning.
13. The steel tank holding the insecticide mixture should be large enough to allow the system to complete a revolution with one filling. It should be free of rust, fertilizer, sediment, and foreign material, and equipped with an in-line strainer situated between the tank and the injector pump.
14. In order to calibrate the irrigation system and injector to apply the mixture containing LORSBAN 4E, determine the following: 1) Calculate the number of acres irrigated by the system, 2) Set the irrigation rate and determine the number of minutes for the system to cover the intended treatment area, 3) Calculate the total gallons of insecticide mixture needed to cover the desired acreage. Divide the total gallons of insecticide mixture needed by the number of minutes to cover the treatment area. This value equals the gallons per minute output that the injector must deliver. Convert the gallons per minute to milliliters or ounces per minute. Calibrate the injector pump with the system in operation at the desired irrigation rate. It is suggested that the injector pump be calibrated at least twice before operation, and the system should be monitored during operation.
15. Do not apply when wind speed favors drift beyond the area intended for treatment. End guns must be turned off during the application, if they irrigate nontarget areas.
16. Do not allow irrigation water to collect or runoff and pose a hazard to livestock, wells, or adjoining crops.
17. Allow foliage to dry before reentering the field.
18. Do not apply through sprinkler systems which deliver a low coefficient of uniformity such as certain water drive units.

USE AND DOSAGE RECOMMENDATIONS

ALFALFA: Use LORSBAN 4E to control the following pests at the dosages indicated by application as a broadcast, foliar spray:

Aphids (suppression)	1/2 pint per acre;
Corn rootworm adults (spotted cucumber beetle), grasshoppers	1/2 to 1 pint per acre;
Alfalfa blotch leafminer, alfalfa looper, alfalfa weevil larvae and adults, armyworms, cutworms, Egyptian alfalfa weevil larvae and adults, plant bugs, leafhoppers, spittlebugs	1 to 2 pints per acre.

Note: Use higher rates to control spotted alfalfa aphid in California and Nevada. Stubble spray may be applied to control leafhopper in the Northeast.

Mix the required dosage with enough water to ensure thorough coverage of crop foliage and apply using aerial (fixed-wing or helicopter) or power-operated ground spray equipment. For aerial application use 2 to 5 gallons of water per acre. For best coverage when using ground application, a minimum of 20 gallons of water per acre with hollow cone nozzles is recommended. Control may be reduced at low spray volumes under high temperature and wind conditions. Treat when field counts or crop injury indicates that damaging pest populations are developing or present; however, do not apply more than once per crop cutting. Some reduction in insect control may be evident under excessively cool conditions. For Egyptian alfalfa weevil control in California, apply the specified dosage in a minimum of 5 gallons of water per acre when larvae are actively feeding and populations reach 15 to 20 larvae per 180 degree sweep with a 15-inch diameter net.

LORSBAN 4E may also be applied through **sprinkler irrigation systems** as a postemergence broadcast application to control the above listed foliar pests. For best results, use the recommended rate of LORSBAN 4E per acre. Maintain vigorous tank agitation to assure uniformity of the application throughout the injection period. See the Sprinkler Irrigation section of this label for further information.

LORSBAN 4E insecticide should not be tank-mixed with pesticides, surfactants, or fertilizer formulations unless prior use has shown the combination noninjurious under your current conditions of use. Some phytotoxic symptoms may be observed on young, tender rapidly growing alfalfa when treated with LORSBAN 4E. Alfalfa will outgrow the symptoms and no yield loss should be expected.

This product is highly toxic to bees exposed to direct treatment on alfalfa. Do not apply if nearby bees are clustered outside of hives and bees are foraging. Protective information may be obtained from your Agricultural Extension Service.

Restrictions: *Do not cut or graze treated alfalfa within 7 days after application of 1/2 pint of LORSBAN 4E per acre, within 14 days after application of 1 pint per acre, or within 21 days after application of rates above 1 pint per acre. Do not make more than 4 applications per year or apply more than once per crop cutting.*

ASPARAGUS: Use LORSBAN 4E to control cutworms, asparagus aphids, and asparagus beetles by application at the rate of 2 pints per acre. Mix the specified dosage in sufficient water to ensure thorough coverage of treated plants and apply as a broadcast, foliar spray. For cutworms, it is preferable to apply LORSBAN 4E when the soil is moist and worms are active on or near the soil surface. Applications may be made during the fern stage for control of asparagus beetles and asparagus aphids when field counts or crop injury indicates that damaging pest populations are developing or present.

Restrictions: *Do not make more than one preharvest application per season or apply within one day of harvest. Do not make more than two postharvest applications during the fern stage. Based on available residue data, the use of LORSBAN 4E on asparagus is limited to the Midwest and Pacific Northwest.*

CHERRIES: Use LORSBAN 4E for the control of lesser peachtree borer, greater peachtree borer, and American plum borer by application as a trunk spray. Mix 1 1/2 to 3 quarts of LORSBAN 4E with 100 gallons of water and apply as a coarse, low-pressure spray to give uniform coverage of tree trunks and lower limbs. Make a second application two weeks after the first one and a third application after harvest. Avoid contact with foliage in sweet cherries as premature leaf drop may result. Consult your State Agricultural Experiment Station or Extension Service Specialist for proper time to treat your area.

In addition, one of the three allowable applications per year may be applied as a dormant spray for control of San Jose scale, peach twig borer, and climbing cutworms. For control of these pests tank mix 1/2 to 1 pint of LORSBAN 4E with 1 to 2 gallons of a petroleum oil recommended for dormant use in 100 gallons of water and spray the entire tree by application to runoff using ground spray equipment. For low volume (concentrate) sprays (40 to 100 gallons of spray mixture per acre) use the same amounts of LORSBAN 4E and spray oil per acre required for application as a dilute spray and apply in a manner that will ensure thorough coverage of the trees. Use the higher dosage of LORSBAN 4E for severe infestations. Use oil as recommended by your State Agricultural Experiment Station or Extension Service Specialist.

Restrictions: *Make only three applications per year. Do not apply within 6 days before harvest. Do not allow meat or dairy animals to graze in treated orchards.*

CITRUS FRUITS: Use LORSBAN 4E insecticide at the rates indicated to control the following pests. Use the lower rates for light infestations and increase the dosage for heavier infestations.

CROP	PEST	DOSAGE LORSBAN 4E FOR SPRAY COVERAGE: DILUTE (PINTS/100 GAL. WATER)	DOSAGE LORSBAN 4E FOR SPRAY COVERAGE: DILUTE OR CONCENTRATE (PINTS/ACRE)	GALLONS OF SPRAY PER ACRE FOR DILUTE APPLICATIONS[3]	REMARKS
Grapefruit Lemons Oranges and Other Citrus Fruits	Aphids, Katydids Lepidopterous larvae Cutworms Fruittree leafroller Orange tortrix Western tussock moth	Outside Coverage Spray (OC) ½ to 1	2 to 7	200 to 700	Use directions apply in all states including California
	Citrus thrips[1], Mealy bugs[1] Scale insects[2] Black scale Brown soft scale California red scale Snow scale, Chaff scale	½ to 1	2 to 7	200 to 700	Use directions apply in all states except California (See footnotes 1 & 2)
	Citrus rust mite		4 to 7	100 to 700	Use directions apply in Florida, only.

[1] In California apply as Outside Coverage (OC) using 6 to 12 pints of LORSBAN 4E per acre in 100 to 300 gallons of finished spray per acre.

[2] In California apply as Thorough Coverage (TC) using 8 to 12 pints of LORSBAN 4E per acre. For dilute sprays apply 1200 to 2400 gallons of spray per acre but do not apply more than 12 pints of LORSBAN 4E per acre or use less than ½ pint of LORSBAN 4E per 100 gallons of finished spray per acre.

[3] Gallonage per acre based on average size of mature trees. Adjust to tree size per acre.

For dilute applications, use the specified dosage per 100 gallons of water for the spray type indicated for pest to be controlled. For **concentrate sprays,** use the same amount of LORSBAN 4E per acre that would be needed for application as a dilute spray, using appropriate application/coverage procedures. A petroleum spray oil recommended for use on citrus may be added to **dilute spray mixtures only** at a rate of up to 1.8 gallons per 100 gallons of water to improve control of aphids, mealybugs, scale insects, and thrips. Treat when insects become a problem or in accordance with the local spray schedule recommended by your State Extension Service Specialist. Do not apply when temperature exceeds 95°F. LORSBAN 4E is highly toxic to bees exposed to direct treatment and should not be applied when bees are actively visiting the area. Apply late in evening or early morning to groves where honey bees are expected to visit.

LORSBAN 4E may be applied in tank mixtures with ethion, dicofol, or Vendex. See the Mixing Instructions section of this label for further instructions. Read and carefully follow all applicable directions, restrictions, and precautions on labeling for the other products used in combination with LORSBAN 4E.

Precautions: Observe local use directions for tank mix combinations especially in regard to applications of LORSBAN 4E plus spray oil. Consult with a County Farm Advisor, County Agent, Extension Service personnel, or Agricultural Commissioner for such information regarding a given locality. In California LORSBAN 4E should not be used in combination with spray oil when temperatures are expected to exceed 95°F the day of application or for several consecutive days thereafter. Do not apply when trees are stressed by drought or high temperature. LORSBAN 4E should not be tank mixed with Difolatan 80 Sprills as crop injury may occur.

Restrictions: Do not apply more than 2 applications per fruit year or more than 15 pints of LORSBAN 4E insecticide per acre per fruit year. Do not make second application within 30 days of the first application. Do not treat within 21 days before harvest for applications up to 7 pints of LORSBAN 4E per acre nor within 35 days for application of rates above 7 pints per acre. Do not pick fruit or do other work involving contact with trees within 2 days after treatment. Do not allow livestock to graze in treated areas.

CITRUS ORCHARD FLOORS: Use LORSBAN 4E to control red imported fire ants and other ant species by applying the specified dose in 25 or more gallons of water with ground application equipment that will uniformly apply the spray to the orchard floor. To control foraging ants and suppress mounds, apply LORSBAN 4E to the orchard floor at the rate of ¾ to 1 quart per acre. Retreat as needed. For best insect control, uniform coverage of the orchard floor is necessary. Do not apply where weed growth or other obstructions would impede uniform coverage of the orchard floor. Do not apply in tank mixtures with Evik herbicide. Foliar applications of LORSBAN 4E may be made in addition to the orchard floor treatments.

LORSBAN 4E may also be applied to citrus orchard floors through sprinkler irrigation systems only if the system uniformly covers the soil surface at the base of the tree. For best results, use the recommended amount of LORSBAN 4E per acre. See the Sprinkler Irrigation section of this label for further information.

Restrictions: Do not apply more than 10 quarts of LORSBAN 4E per acre per season. Do not apply last treatment within 28 days before harvest. Do not allow livestock to graze in treated areas. In Florida, do not make more than 3 applications per season.

CRANBERRIES: Use LORSBAN 4E by application as a broadcast, foliar spray to control brown spanworm, cranberry fruitworm, cranberry weevil, cutworms, fireworms, and Sparganothis fruitworms at the rate of 3 pints per acre. Mix the specified dosage in enough water to ensure thorough coverage and apply no less than 5 gallons of spray per acre when using aerial equipment or no less than 100 gallons of spray per acre when using ground equipment. For weevil control, apply once at flower bud development (late May, early June) and, if weevils are present, once after 100% bloom (early to mid July). For other insects, treat when field counts indicate damaging insect populations are developing or present. Apply only after the winter flood has been removed. To avoid pesticide contamination of flood waters, make no applications while bogs are flooded.

LORSBAN 4E may also be applied through sprinkler irrigation systems to control the above listed pests. For best results, use the recommended rate of LORSBAN 4E per acre. Maintain vigorous tank agitation to assure uniformity of the application throughout the injection period. See the Sprinkler Irrigation section of this label for further information.

Restrictions: Do not make more than two applications per year or apply within 60 days before harvest.

FIELD CORN, POPCORN, SWEET CORN: For use to control cutworms, armyworms, corn earworm, corn rootworm adults, chinch bugs, grasshoppers, wireworms, flea beetle larvae and adults, aphids, billbugs, grubs, western bean cutworm, corn borers, symphylans, common stalk borer, and lesser cornstalk borer.

Preplant Incorporation Treatment: Use LORSBAN 4E at the following rates by application in sufficient water to the soil surface and incorporate into the soil:

Cutworms, symphylans	2 to 4 pints per acre;
Wireworms, billbugs, flea beetle larvae, grubs, seed corn maggots, seed corn beetle	4 pints per acre.
Lesser cornstalk borer, corn rootworm larvae	6 pints per acre.

NOTE: For suppression of nematodes in North Carolina, South Carolina, and Virginia, apply 4 to 6 pints of LORSBAN 4E per acre.

Use recommended rate in not less than 10 gallons of water per acre and apply as a broadcast spray to the soil surface using suitable power-operated ground spray equipment. On the same day of treatment, incorporate the insecticide into the top 2 to 4 inches of soil using a disc, field cultivator or equivalent equipment.

LORSBAN 4E insecticide may also be applied in tank mixtures with non-pressure fertilizer solutions and/or with Bladex, Eradicane, Sutan, Lasso, Dual, and atrazine herbicides. See the Mixing Instructions section of this label for further information. Read and carefully follow all applicable directions, restrictions, and precautions on labeling for the other products used in combination with LORSBAN 4E.

Preplant, At-Plant or Preemergence Treatment in Conservation Tillage: Use LORSBAN 4E at the following rates by application in sufficient water to surface trash and exposed soil:

Cutworms, armyworms	1 to 2 pints per acre.

Use recommended rate in not less than 20 gallons of water per acre and apply as a broadcast spray using suitable power-operated ground spray equipment. Use higher rates for residual control.

LORSBAN 4E may also be applied in tank mixtures with non-pressure fertilizer solutions and/or with paraquat and Roundup. See the Mixing Instructions section of this label for further information. Read and carefully follow all applicable directions, restrictions, and precautions on labeling for the other products used in combination with LORSBAN 4E.

Cultivation Time Treatment: Use LORSBAN 4E at the rate of 2 pints per acre to control corn rootworm larvae. Apply LORSBAN 4E as a water emulsion on both sides of the row at the base of the plants just ahead of the cultivator shovels. Cover the insecticide with soil around the brace roots. The best time to apply a basal treatment of a soil insecticide with cultivation is near the beginning of egg hatch. A cultivation application of LORSBAN 4E may be made in addition to an at planting application of LORSBAN 15G granular insecticide.

Postemergence Treatment: Use LORSBAN 4E at the following rate by application in sufficient water to ensure thorough coverage of treated plants:

Grasshoppers	½ to 1 pint per acre;

Armyworms, chinch bugs, aphids, corn rootworm adults, webworms, western bean cutworm, European corn borer (see note) . . . . 1 to 2 pints per acre;
Southwestern corn borer, corn earworm . . . . . . 1½ to 2 pints per acre;
Cutworms, billbugs, lesser cornstalk borer, flea beetle adults, common stalk borer . . . . 2 to 3 pints per acre.

Note: The recommended dosage will control silk clipping by corn rootworm adults. For European corn borer control, use 1½ to 2 pints per acre when application is made with power-operated ground and aerial equipment and 1 to 2 pints per acre when application is made through a sprinkler irrigation system.

Treat when field counts indicate that pests are or may become a problem. For best billbug, chinch bug, and flea beetle control, apply with sufficient water to ensure a minimum spray volume of 20 to 40 gallons per acre and 40 psi using **ground spray equipment.** On corn less than 6 inches tall, apply the insecticide spray in a 9 to 12 inch wide band over the row. On corn greater than 6 inches tall, apply the insecticide spray using drop nozzles directed to the base of the plant. Do not reduce the dosage for banded or directed applications. Concentrate the full labeled dosage rate in the treated zone. When chinch bugs continue to immigrate to corn over a prolonged period or under extreme pressure, a second application of LORSBAN 4E may be needed.

For cutworm, webworm, western bean cutworm, armyworm, aphid, European and southwestern corn borer, grasshopper, lesser cornstalk borer, corn rootworm adult, corn earworm, and common stalk borer control, apply as a broadcast spray using either aerial(fixed-wing or helicopter) or power-operated ground spray equipment. For aerial application use 2 to 5 gallons of spray per acre. Control may be reduced at low spray volumes under high temperature and wind conditions. For cutworms, it is preferable to apply LORSBAN 4E when soil is moist and worms are active on or near the soil surface. If ground is dry, cloddy or crusty at time of treatment, worms may be protected from the spray and effectiveness will be reduced. If such conditions exist, shallow incorporation using a rotary hoe or other suitable equipment immediately before or soon after treatment may improve control. Consult your Agricultural Experiment Station or Extension Service specialist for additional information concerning control practices in your area. For webworm control, shallow incorporation using a rotary hoe or other suitable equipment immediately before or soon after treatment is necessary. For first-generation European corn borer control, treat when 25 and 50 percent of the corn plants show pinhole feeding or leaf-feeding scars. Ground applications should be directed into the corn whorls. For second-generation European corn borer control, treat when field counts of egg masses indicate pests are or may become a problem. For southwestern corn borer control, treat when field counts of egg masses indicate pests are or may become a problem. A second application may be applied 10 to 14 days later, if needed due to reinfestation. For common stalk borer control, treat approximately 11 days after application of Roundup herbicide or after complete burndown with paraquat herbicide (3 to 5 days). Do not use LORSBAN 4E in combination with the burndown herbicide for control of common stalk borer.

LORSBAN 4E insecticide may also be applied through **sprinkler irrigation systems** as a postemergence broadcast application to control the above listed foliar insects. For best results, use the recommended rate of LORSBAN 4E in a tank mix with 2 pints per acre of non-emulsifiable oil. Maintain vigorous tank agitation to assure uniformity of the LORSBAN 4E plus oil mixture throughout the injection period. LORSBAN 4E may also be applied through sprinkler irrigation systems at the rate of 2 to 3 pints per acre to control corn rootworm larvae. Time application to coincide with the appearance of the second instar larvae. Maintain vigorous tank agitation to assure uniformity of the application throughout the injection period. Apply with enough water to wet the root zone to the depth control is needed. Under saturated soil conditions, allow enough soil drying to occur so that an application using a minimum water rate will not produce runoff. Consult University Extension personnel or other experienced consultants to determine the need to treat and to aid in application timing. See the Sprinkler Irrigation section of this label for further information.

Restrictions: Do not apply within 35 days before harvest of grain. Do not apply more than a total of 15 pints of LORSBAN 4E per acre per season. Do not allow livestock to graze in treated areas nor harvest treated corn silage as feed for meat or dairy animals within 14 days after last treatment. Do not feed treated corn fodder to meat or dairy animals within 35 days after last treatment.

SWEET CORN Grown Only in Florida and Georgia: Use LORSBAN 4E to control infestations of beet armyworm, fall armyworm and corn earworm by application as a broadcast, foliar spray at the rate of 1 to 2 pints per acre. Mix the specified dosage in enough water to ensure thorough coverage and apply using suitable aerial or ground spray equipment. For aerial application, use at least 2 gallons of spray per acre. Treat when field counts indicate damaging pest populations are developing or present. Re-treat as necessary to maintain control but do not apply more than twenty-two 1-pint or eleven 2-pint treatments per season.

LORSBAN 4E insecticide may also be applied through **sprinkler irrigation systems** as a postemergence broadcast application to control the above listed foliar insects. For best results, use the recommended rate of LORSBAN 4E in a tank mix with 2 pints per acre of non-emulsifiable oil. Maintain vigorous tank agitation to assure uniformity of the LORSBAN 4E plus oil mixture throughout the injection period. See the Overhead Sprinkler Irrigation section of this label for further information.

Restrictions: Do not apply more than 22 pints of LORSBAN 4E per acre per season. Do not harvest corn ears, allow livestock to graze in treated areas, nor feed treated silage, fodder, or grain to meat or dairy animals within 21 days after treatment. Do not use in conjunction with postplant broadcast, foliar applications of LORSBAN 15G granular insecticide.

COTTON: Use LORSBAN 4E for control of the following pests in all states except Arizona and California at the dosages indicated.

Cotton fleahopper, plant bugs (*Lygus, Mirids*) . . 3/8 to 1 pint per acre;
Cotton aphid, fall armyworm, grasshoppers, thrips, yellow-striped armyworm . . . . 1/2 to 1 pint per acre;
Spider mites . . . . 1 pint per acre;
Beet armyworm, bollworms (*Heliothis spp.*), boll weevil, cutworms, pink bollworm, salt marsh caterpillar . . . . 1½ to 2 pints per acre.

NOTE: The recommended dosage rate of 3/8 pint per acre will not achieve the high degree of control of the higher label rate, but will minimize the damage done by plant bugs and cotton fleahopper and allow the beneficial insects to survive, build up, and be available to aid in the control of bollworms infesting cotton.

Use LORSBAN 4E for control of the following pests in Arizona and California at the dosages indicated:

Armyworms, cotton aphid, cotton fleahopper, *Lygus*, salt marsh caterpillar, thrips . . . . 1 to 2 pints per acre;
Bollworms, (*Heliothis spp.*), boll weevil, cutworms, pink bollworm . . . . 2 pints per acre.

NOTE: The 2 pint rate will aid in the suppression of cotton leafperforator and spider mites.

Mix the required dosage with sufficient water to ensure thorough coverage of plants and apply using aerial or power-operated ground spray equipment. For aerial application, use at least 1 gallon of spray per acre. Treat when field counts indicate damaging insect populations are developing or present. Re-treat as necessary to maintain control.

LORSBAN 4E insecticide may also be applied through **sprinkler irrigation systems** as a postemergence broadcast application to control the above listed foliar pests. For best results, use the recommended rate of LORSBAN 4E per acre. Maintain vigorous tank agitation to assure uniformity of the application throughout the injection period. See the Sprinkler Irrigation section of this label for further information.

For effective control of spider mites when large numbers of eggs are present, apply a second spray 3 to 5 days after initial treatment to control newly hatched nymphs.

For best results on bollworms, it is suggested that fields be scouted twice per week and treatments made when worms are 1/4 inch or less in length. The following table illustrates the size of worms in relation to age and stage of development (instar) as a guide to timing of treatments for best control.

From the table it can be seen that a scouting schedule of only once per week will not be satisfactory since the worms may be too big to control effectively by the seventh or eighth day.

Timing for the best worm control

	Age (days)	Size	Instar
Get the worms at this stage	Hatch	1/16"	Hatch
	3	3/32"	I
	5	9/32"	II
	6	7/16"	III
	8	11/16"	IV

Proper application techniques help to ensure thorough spray coverage and correct dosage and are thus important in obtaining good control of pests. Consider these suggestions when applying LORSBAN 4E insecticide on cotton.

Aerial Application: Shorten boom length to avoid spray entering the vortices at the wing tips. Swath width should be reduced when wind direction is the same as direction of spraying.

The proper nozzle arrangement and swath width to avoid skips and vortices effect can be checked out by flying over a paper tape (adding machine paper) using water with or without soluble dye. (The dye gives a permanent record.)

Flying at a height of 5-15 feet above the target results in the best coverage.

Nozzle orientation of the boom is important. More break-up occurs when nozzles are pointed straight down versus the straight back position. Desired droplet size (100-200 microns) can be obtained by angling the nozzles somewhere in this range.

Marking of swath by flagging or permanent markers is essential.

Ground Application: Orient the boom and nozzles so that uniform coverage is obtained. The swath width should not be wider than the boom; drift spray is wasted spray so do not depend on it. Use flat fan or disc-core hollow cone nozzles with maximum spacing of 20 inches and spray pressure of 40-60 psi with a droplet size of 100-200 microns.

Restrictions: Do not apply within 14 days before harvest. Do not allow livestock to graze in treated areas. Do not feed gin trash or treated forage to livestock.

LORSBAN 4E insecticide may also be used in tank mixtures in all states to control armyworms, bollworms ***(Heliothis spp.),*** **boll weevil, cotton fleahopper, plant bugs** ***(Lygus, mirids),*** **and spider mites** as indicated in the following table.

Tank Mixture	Rate/Acre
LORSBAN 4E plus Ambush 2E	1 to 2 pints plus 1/5 to 2/5 pint
LORSBAN 4E plus Pounce 3.2E	1 to 2 pints plus 1/8 to 1/4 pint
LORSBAN 4E plus Pydrin 2.4E	1 to 2 pints plus 1/6 to 1/3 pint
LORSBAN 4E plus Bolstar 6E	1/2 to 1 pint plus 2/3 to 1 1/3 pints
LORSBAN 4E plus methyl parathion[1] Plus EPN[1]	1 pint plus 0.6 lb a.i plus 0.6 lb a.i.

[1] Use only emulsifiable formulations containing these active ingredients (a.i.).

Mix the tank mixture components in sufficient water to ensure thorough coverage of plants (at least 1 gallon of spray per acre) and apply using aerial spray equipment. Treat when field counts indicate damaging insect populations are developing or present. Re-treat as necessary to maintain control.

In California and Arizona, use LORSBAN 4E at 1 to 2 pints per acre plus Monitor 4E at 1 to 2 pints per acre or LORSBAN 4E at 1 to 2 pints per acre plus Orthene 75S at 0.33 to 1.3 pounds per acre for control of armyworms, bollworms, cabbage looper, cotton aphids, cotton fleahopper, cutworms, grasshopper, omnivorous leafroller, pink bollworm, plant bugs (*Lygus Mirids*), saltmarsh caterpillar, thrips, and whitefly (suppression). Apply in 10 to 25 gallons of water per acre with ground spray equipment or in 5 to 10 gallons of water per acre with aerial spray equipment. Treat when field counts indicate that damaging insect populations are developing or present. For tank mixtures which include Monitor 4E, apply before 65 percent of the bolls have opened. Re-treat as necessary to maintain control.

Restrictions: Make no applications of these tank mixtures closer to harvest than the longest pre-harvest interval shown for any of the products in the tank mixture. Do not allow livestock to graze in treated areas. Do not feed gin trash or treated forage to livestock. Workers entering the field within 48 hours after application should wear protective clothing.

FIGS: Use LORSBAN 4E at the rate of 2 quarts per acre for control of driedfruit beetle by application in sufficient water to the soil surface followed by incorporation into the top 3 inches of soil. Apply to fig orchard soil as a dormant application in late winter prior to beetle emergence and prior to leaf formation.

Restrictions: Make only one application per year. Do not apply within 7 months of harvest. Based on available residue data, use LORSBAN 4E on figs is restricted to California.

GRAPES: Use LORSBAN 4E insecticide for control of grape root borer by application just before the pest emerges from the soil. Mix 4 1/2 pints of LORSBAN 4E with 100 gallons of water and apply 2 quarts of the diluted spray mixture to the soil surface on a 15-square foot area around the base of each vine. Do not allow spray to contact fruit or foliage.

Restrictions: Do not make more than one application per season or apply within 35 days before harvest. Based upon available residue data, the use of LORSBAN 4E in grapes is restricted to states East of the Rocky Mountains.

MINT: Use LORSBAN 4E by application as a broadcast, foliar spray to control cutworms at the rate of 2 to 4 pints per acre and mint root borer at the rate of 4 pints per acre. Mix the specified dosage in water to give no less than 10 gallons of spray per acre and apply using ground spray equipment. For cutworm control, treat during May and June when field counts indicate damaging insect populations are developing or present. When larvae are less than 3/4 inch in length, use the 2 pint rate. When larvae are 3/4 inch or more in length, use the higher rate. **Make only one application during the growing season. Do not apply within 90 days before harvest.** For **mint root borer** control, apply post-harvest when field counts indicate damaging insect populations are developing or present. Follow treatment with approximately 1 acre inch of sprinkler irrigation immediately after application to incorporate the insecticide into the soil. **Make only one post-harvest application per season.**

NECTARINES, PEACHES: Use LORSBAN 4E insecticide for the control of peach tree borers by application as a trunk spray before newly hatched borers enter the trees. Mix 3 quarts of LORSBAN 4E with 100 gallons of water and apply as a coarse, low-pressure spray to give uniform coverage of tree trunks. Thoroughly wet all bark areas from ground level to scaffold limbs. Do not allow spray to contact fruit. Consult your State Agricultural Experiment Station's or Extension Service Specialist's written recommendations for proper time to treat in your area.

LORSBAN 4E may also be used as a preplant dip application for non-bearing peach trees at the equivalent application rate of 3 quarts per 100 gallons of water for control of peach tree borer. Dip trees several inches above the grafting bud scar and plant immediately or allow to dry before returning to storage. Do not allow peach trees to remain in contact with the dip solution.

Restrictions: Make only one application per season. Do not apply within 14 days before harvest. Do not allow meat or dairy animals to graze in treated orchards.

ONIONS (dry bulb): Use LORSBAN 4E insecticide to control onion maggot by application as an in-furrow drench. Apply LORSBAN 4E at the rate of 1.1 fluid ounce per 1000 linear feet of row at an 18-inch row spacing. Use a minimum of 40 gallons of total drench per acre. Incorporate to a depth of 1 to 2 inches.

Restrictions: Do not make more than one application per year.

PEANUTS: For suppression of wireworms, apply LORSBAN 4E at a rate of 4 pints per acre as a preplant broadcast spray to the soil surface followed by immediate soil incorporation to a depth of 3 to 4 inches. Use a minimum of 100 gal. of total spray per acre.

Restrictions: Do not make more than one application per season. Do not harvest within 21 days after treatment. Do not feed treated peanut forage or hay to meat or dairy animals.

SORGHUM: Use LORSBAN 4E Insecticide for control of the following pests at the dosages indicated:

Sorghum midge . . . . (Apply when 30 to 50 percent of the seed heads are in bloom, repeat at 3-day intervals if necessary.);	1/2 pint per acre
Greenbugs, grasshoppers, yellow sugar cane aphid, and other aphids	1/2 to 1 pint per acre;
Chinch bugs, lesser cornstalk borer . . . . (Apply as a directed spray toward the base of the plant using power-operated ground spray equipment with sufficient water to ensure coverage of an 8 to 12 inch band centered on the row. On plants less than 6 inches high, apply an 8 to 12 inch band over the row. Do not reduce the dosage for banded or directed applications. Concentrate the full labeled dosage rate in the treated zone.);	1 to 2 pints per acre
Webworms	1 pint per acre;
Armyworms, corn earworm, cutworms	1 to 2 pints per acre;
European and southwestern corn borer	1 1/2 to 2 pints per acre.

Mix the specified dosage in enough water to ensure thorough coverage and apply using suitable aerial or ground spray equipment. To minimize chemical injury, do not apply LORSBAN 4E to drought stressed grain sorghum within 3 days following irrigation or rain except where the product is applied in irrigation water.

LORSBAN 4E insecticide may also be applied through **sprinkler irrigation systems** as a postemergence broadcast application to control the above listed foliar pests. For best results, use the recommended rate of LORSBAN 4E per acre. Maintain vigorous tank agitation to assure uniformity of the application throughout the injection period. See the Sprinkler Irrigation section of this label for further information.

Restrictions: The treated crop is not to be used for grain forage, fodder, hay, or silage within 30 days after application of 1 pint of LORSBAN 4E per acre or within 60 days after application of rates above 1 pint per acre. Do not treat sweet varieties of sorghum. Do not apply more than 3 pints of LORSBAN 4E per acre per season.

SOYBEANS: For use to control armyworms, bean leaf beetle, corn earworm, cutworms, European corn borer, grasshoppers, green cloverworm, lesser cornstalk borer, Mexican been beetle, saltmarsh caterpillar and other woollybears, southern green stink bug, spider mites, and velvetbean caterpillar.

Soil Treatment: Use LORSBAN 4E at the rate of 1 to 2 pints per acre to control cutworms and lesser cornstalk borer. Mix the specified dosage in a minimum of 10 gallons of spray per acre and apply to the soil surface using suitable ground spray equipment. Equivalent rates of insecticide spray required per 100 feet of row for various row spacing are given in the accompanying table. For at-plant treatments apply the insecticide over the row in a 4 to 6-inch band in front of the planter shoe or press wheel or after the press wheel followed by a drag chain for light incorporation. Do not apply as an in-furrow treatment. For postemergence rescue treatments, apply as a directed spray in a 9 to 12-inch band at the base of the plant. To plants under 6 inches high apply over the top in a 6- to 12-inch band. Treat when field counts or conditions indicate that pests are or may become a problem.

	Fluid Ounces of Spray Required Per 100 Feet of Row for Various Row Spacing			
Volume of Spray Per Acre	36″	32″	28″	24″
10 Gallons	8.8	7.9	6.9	5.9
15 Gallons	13.2	11.8	10.3	8.8
20 Gallons	17.6	15.7	13.7	11.8

Foliar Treatment: Use LORSBAN 4E at the following rate by application in sufficient water to ensure thorough coverage of treated plants:

European corn borer, southern green stink bug . . 2 pints per acre
Bean leaf beetle, cutworms, corn earworm, salt-marsh caterpillar and other woollybears . . . . . . 1 to 2 pints per acre;
Mexican bean beetle, armyworms . . . . 1 to 1½ pints per acre;
Velvetbean caterpillar, grasshoppers, green clover-worm, spider mites . . . . ½ to 1 pint per acre.

Apply as a broadcast spray using either aerial or ground equipment when field counts indicate damaging insect populations are developing or present; retreat as necessary to maintain control. For effective control of spider mites when large numbers of eggs are present, apply a second spray 3 to 5 days after initial treatment to control newly-hatched nymphs. **On determinate soybeans do not apply more than one application after pod set.**

LORSBAN 4E insecticide may also be applied through **sprinkler irrigation systems** as a postemergence broadcast application to control the above listed foliar pests. For best results, use the recommended rate of LORSBAN 4E per acre. Maintain vigorous tank agitation to assure uniformity of the application throughout the injection period. See the Sprinkler Irrigation section of this label for further information.

Restrictions: *Do not apply more than 6 pints of LORSBAN 4E per acre or 3 pounds of chlorpyrifos (active ingredient) per acre per season. Do not apply last treatment within 28 days before harvest nor apply last two treatments closer than 14 days apart. Do not allow livestock to graze in treated areas or otherwise feed treated soybean forage, hay, and straw to meat or dairy animals.*

STRAWBERRIES: Use LORSBAN 4E insecticide by application as a broadcast foliar spray to control strawberry bud weevil at the rate of 1 quart per acre. Apply in a minimum of 40 gallons of spray per acre when buds first appear and 10 to 14 days later. Do not apply after berries start to form or when berries are present. LORSBAN 4E should not be tank mixed with pesticides, surfactants, or fertilizer formulations unless prior use has shown the combination noninjurious under your current conditions of use. Phytotoxicity may occur when LORSBAN 4E is applied to strawberries experiencing high temperature and drought stress.

Restrictions: *For pre-bloom use only. Do not make more than two applications per season or apply within 21 days before harvest.*

SUNFLOWERS: For use to control cutworms, sunflower beetle larvae and adults, stem weevil, sunflower moth, banded sunflower moth, woollybears, seed weevil, and grasshoppers.

Preplant Incorporation Treatment: Use LORSBAN 4E insecticide at the following rates by application in sufficient water to the soil surface and incorporate into the soil:

Cutworms . . . . 2 to 4 pints per acre.

Use recommended rate in not less than 10 gallons of water per acre and apply as a broadcast spray to the soil surface using suitable power-operated ground spray equipment. On the same day of treatment, incorporate the insecticide into the top 2 to 4 inches of soil using a disc, field cultivator or equivalent equipment.

Postemergence Treatment: Use LORSBAN 4E for control of the following pests at the dosage indicated by application in sufficient water to ensure thorough coverage of treated plants:

Cutworms . . . . 2 to 3 pints per acre;
Sunflower beetle larvae and adults, stem weevil, sunflower moth, banded sunflower moth, woollybears, and seed weevil . . . . 1 to 1½ pints per acre;
Grasshoppers . . . . 1 pint per acre.

Apply as a broadcast spray using either aerial (fixed-wing or helicopter) or power-operated ground spray equipment when field counts indicate that pests are or may become a problem. For cutworm control, a second treatment may be made 7 to 10 days later, if needed. For stem weevil control, optimal treatment time is within 5 to 7 days after adult weevils begin to appear. For sunflower moth control, make first application during early 1 to 5 percent bloom stage. A second treatment may be made 7 days later, if needed. For seed weevil control, treat when field counts indicate there are 10 to 12 adults per plant for oil crops and 1 to 3 adults per plant on confectionary crops. Additional treatments should be made at successive 7 to 10 day intervals if field counts indicate need to re-treat. For sunflower beetle larvae or adult control, treat when field counts indicate there are 10 larvae or 1 to 2 adults per seedling. Additional treatments may be made at successive 7 to 10 day intervals if field counts indicate need to retreat.

Restrictions: *Do not apply more than 9 pints of LORSBAN 4E per acre per season. Do not apply within 42 days before harvest. Do not allow livestock to graze in treated areas.*

SUGAR BEETS: Use LORSBAN 4E by application as a broadcast, foliar spray to control beet armyworms at a rate of 1½ to 2 pints per acre and cutworms at a rate of 2 pints per acre. Treat when field counts indicate that damaging insect populations are or may be a problem. Mix the specified dosage with 2 to 5 gallons of water per acre and apply using suitable aerial spray equipment or with 10 to 30 gallons of water per acre when using power-operated ground spray equipment. Re-treat as necessary to maintain control.

Restrictions: *Do not apply within 30 days before harvest of beet roots and tops. Do not apply more than a total of 8 pints of LORSBAN 4E per acre per season. Do not allow livestock to graze in treated areas nor harvest treated beet tops as feed for meat or dairy animals within 30 days after last treatment.*

SWEET POTATOES: Use LORSBAN 4E to reduce the feeding damage caused by populations of *Conderus* wireworm, *Systena* flea beetle and the sweet potato flea beetle. Apply at the rate of 4 pints per acre as a broadcast (overall) spray to the soil surface followed by incorporation. Mix the specified dosage with enough water to obtain uniform coverage and apply as a coarse spray using suitable ground spray equipment. Incorporate the insecticide to a depth of 4 to 6 inches as soon as possible after application by using a rotary hoe, disc cultivator, or other suitable incorporation equipment. Plant the crop in the usual manner no later than 14 days after treatment (any delay in planting will reduce the length of time that LORSBAN 4E will protect against feeding damage). LORSBAN 4E will not control false wireworms or white fringe beetle or other grubs that attack sweet potatoes.

Restrictions: *Do not make more than one application per season. Do not harvest within 125 days of treatment.*

TOBACCO: Use LORSBAN 4E for preplant treatment to control larvae of cutworms, flea beetles, mole crickets, root maggots, and wireworms. Apply 2 to 3 quarts of LORSBAN 4E per acre in not less than 10 gallons of water as a broadcast (overall) spray to the soil surface one week before transplanting. Immediately following application, incorporate the insecticide into the soil to a depth of 2 to 4 inches using suitable equipment. The application of LORSBAN 4E will also suppress the movement of imported fire ants into treated fields.

To control the above insects and low to moderate populations of rootknot nematodes in North Carolina, South Carolina, and Virginia, use LORSBAN 4E at the rate of 5 quarts per acre. To control the above insects and moderate populations of rootknot nematodes in all tobacco growing regions, use LORSBAN 4E in a tank mix with Nemacur 3 at the rate of 2 quarts of LORSBAN 4E plus 4 quarts of Nemacur 3 per acre. Read and carefully follow all applicable directions, restrictions, and precautions on labeling for Nemacur 3 used in combination with LORSBAN 4E. Apply the specified dosage in not less than 10 gallons of water as a broadcast (overall) spray to the soil surface 24 to 48 hours before bedding and transplanting. Immediately following application, incorporate into the soil to a depth of at least 4 inches using suitable equipment. Where the nematode species *Meloidogyne arenaria* or *M. javanica* are present or high populations of *M. incognita,* apply TELONE* II soil fumigant at the recommended label rate.

Restrictions: *Do not make more than one application per season.*

TREE FRUITS: Use LORSBAN 4E as a dormant or delayed dormant spray at the rates indicated to control the following insects on the crops listed. While LORSBAN 4E may be used without oil, oil is recommended for additional pest control.

Crop	Insect	Dosage LORSBAN 4E
Apples	Rosy apple aphid San Jose scale Lygus Pandemis Leafroller Climbing cutworms	½ to 1 pint
Pears	San Jose scale Climbing cutworms Pear psylla adults	
Plums, Prunes	San Jose scale Mealy plum aphid Climbing cutworms Peach twig borer	
Almonds Peaches, Nectarines	San Jose scale Peach twig borer Climbing cutworms	

Tank mix the specified dosage with 1 to 2 gallons of a petroleum spray oil recommended for dormant use in 100 gallons of water and spray the entire tree by application to runoff using suitable ground spray equipment. (SEE PRECAUTIONS FOR USE IN CALIFORNIA.) For low volume (concentrate) sprays (40 to 100 gallons of spray mixture per acre), use the same amounts of LORSBAN 4E and spray oil per acre required for application as a dilute spray and apply in a manner that will ensure thorough coverage of the trees. Use the higher dosage of LORSBAN 4E for severe infestations. Use oil as recommended by your State Agricultural Experiment Station or Extension Service Specialist.

Precautions: Because cold or dry conditions may cause LORSBAN 4E plus oil sprays to infuse trees resulting in bud damage or drop, do not apply until winter rains or irrigation has replenished soil moisture such that bark and twigs are not desiccated.

Additional Precautions Specific to California: Use a minimum of 250 gallons of total spray volume per acre. Do not use more than 4 pints of LORSBAN 4E per acre. Do not use more than 4 gallons of spray oil per acre on almonds, peaches, or nectarines. Do not use any adjuvants or surfactants in addition to or as a substitute for a petroleum spray oil in a tank mix with LORSBAN 4E. Do not apply on almonds in the following counties in California: Butte, Colusa, Glenn, Solano, Sutter, Tehama, Yolo, and Yuba.

Restrictions: *Make only one application during the dormant season. Do not allow meat or dairy animals to graze in treated orchards.*

TREE NUTS: Use LORSBAN 4E at the dosages indicated by application as a foliar spray to control pests listed in the following table. Mix the required dosage in sufficient water to ensure thorough and complete coverage of the foliage and crop and apply as a concentrate or dilute spray using conventional, power-operated spray equipment. For dilute sprays applied to tree nut crops, mix the required dosage in sufficient water to allow for spray to runoff. For concentrate sprays, apply an equivalent amount of LORSBAN 4E per acre. Treat when pests appear or in accordance with local conditions. To suppress pecan leaf scorch mite on pecans, use in a preventative program. For aerial application on pecans, apply 2 to 4 pints of LORSBAN 4E in 5 to 15 gallons of spray volume per acre. Insect control by aerial application may be less than control by ground application because of less coverage. For ant control on pecans, do not apply where weed growth or other obstructions would impede uniform coverage of the orchard floor. For best pecan *Phylloxera* control, use 2 applications 7 to 10 days apart starting at bud swell and ending by early leaf expansion. Consult your State Agricultural Experiment Station, Certified Pest Control Advisor, or Extension Service specialist for specific use information in your area.

Crop	Insects Controlled	Dosage LORSBAN 4E	Restrictions
Almonds	Navel orangeworm Peach twig borer San Jose scale	4 pints per acre	Make no more than 3 applications per season on almonds and filberts and no more than 2 applications per season on walnuts. Do not apply within 14 days of harvest. Do not allow livestock to graze in treated orchards.
Filberts	Eyespotted bud moth Filbert aphid Filbert leafroller Filbertworm Obliquebanded leafroller Omniverous leaftier Winter moth	3 to 4 pints per acre	
Walnuts	Coding moth Walnut scale	4 pints per acre	
Pecans	Black pecan aphid Hickory shuckworm Pecan leaf scorch mite (suppression)	2 pints per 100 gallon	Make no more than 5 applications per season. Do not apply within 28 days of harvest. Do not allow livestock to graze in treated orchards. For dilute spray apply 200 to 600 gallons per acre. Make no applications of tank mixtures closer to harvest than the longest pre-harvest interval shown for any of the products in the tank mixture.
	Pecan nut casebearer Phylloxera spp. Fall webworm Spittlebug	1 pint per 100 gallon	
	Fire ants and other ant species (orchard floor)	2 pints per acre as a ground spray	
	Yellow pecan aphid complex	1 pint of LORSBAN 4E plus $2\frac{2}{3}$ ounce of Pydrin 2.4E or 2.6 ounce of Cymbush 3E or 3 ounce of Ammo 2.5E per 100 gallon	

VEGETABLES: Use LORSBAN 4E to control root maggots on the following crops at the dosages indicated:

Cauliflower . . . . 1.6 to 2.4 fl oz per 1000 linear feet of row;

Broccoli, Brussels Sprouts, Cabbage, Chinese Cabbage, Collards, Kale, Kohlrabi, Turnips . . . . . . 1.6 to 2.75 fl oz per 1000 linear feet of row.

For **direct seeded crops** apply the specified dosage in a water-based spray as a 4-inch wide band over the row at planting time, behind the planter shoe and in front of the press wheel to achieve shallow incorporation. Apply in a minimum of 40 gallons of total spray per acre.

For **transplanted crops,** apply LORSBAN 4E as a water-based spray directed to the base of the plants immediately after setting. Use a minimum of 40 gallons of total spray per acre. Do not add any additional adjuvants, surfactants, or spreader stickers. **Do not apply as a foliage application.**

Restrictions: Do not apply more than 2 pints of LORSBAN 4E to cauliflower planted in 40-inch rows. Use proportional amounts for other row spacings not to exceed 4 pints of LORSBAN 4E per acre. Do not apply more than 2.6 pints of LORSBAN 4E per acre to broccoli, Brussels sprouts, cabbage, Chinese cabbage, collards, kale, kohlrabi, and turnips planted in 40-inch rows or more than $4\frac{1}{2}$ pints of LORSBAN 4E per acre to these crops in 20-inch rows (or two rows per bed). Use proportional amounts for other row spacings not to exceed $4\frac{1}{2}$ pints of LORSBAN 4E per acre. Do not make more than one application per season or apply within 30 days before harvest.

Radishes . . . . 1.0 fl oz per 1000 linear feet of row.

Apply the specified dosage as a water-based drench in the seed furrows with the seed at planting time. Use a minimum of 40 gallons of total drench per acre.

Restrictions: Do not apply more than $5\frac{1}{2}$ pints of LORSBAN 4E per acre or make more than one application per season.

Rutabagas . . . . 1.6 to 3.3 fl oz per 1000 linear feet of row.

Apply the specified dosage in a water-based spray as a 4-inch wide band over the row at planting time, behind the planter shoe and in front of the press wheel to achieve shallow incorporation. Use a minimum of 40 gallons of total spray per acre.

Restrictions: Do not apply more than $4\frac{1}{2}$ pints of LORSBAN 4E per acre or make more than one application per season. Do not use rutabaga tops for food or feed purposes.

To avoid phytotoxicity in vegetable crops, do not mix with other pesticide products or treat plants that are under extreme heat and drought stress.

STORAGE AND DISPOSAL

Do not contaminate water, food or feed by storage or disposal. Open dumping is prohibited.

Storage: Store in original container in secured dry storage area. Prevent cross-contamination with other pesticides and fertilizers. Do not store above 100°F for extended periods of time. Storage below 20°F may result in formation of crystals. If product crystallizes, store at 50 to 70°F and agitate to redissolve crystals. If container is damaged or spill occurs, use product immediately or dispose of product and damaged container as indicated below.

Pesticide Disposal: Pesticide wastes are toxic. Improper disposal of excess pesticide, spray mixture, or rinsate is a violation of Federal law. If these wastes cannot be disposed of by use according to label instructions, contact your State Pesticide or Environmental Control Agency, or the Hazardous Waste representative at the nearest EPA Regional Office for guidance.

Container Disposal for Refillable Containers: Replace the dry disconnect cap, if applicable, and seal all openings which have been opened during use. Return the empty container to a collection site designated by the Dow Chemical Company. If the container has been damaged and cannot be returned according to the recommended procedures, contact the Dow Chemical Customer Service Center at 1-800-258-1470 to obtain proper handling instructions.

OR

Container Disposal for Non-Refillable Containers: Triple rinse (or equivalent). Then offer for recycling or reconditioning, or puncture and dispose of in a sanitary landfill, or by other procedures approved by state and local authorities. For reusable portable containers, do not rinse or put any other material into container. When returning reusable portable container, seal all openings and return to owner, **or** triple rinse (or equivalent). Then puncture and dispose of in a sanitary landfill, or by incineration, or, if allowed by state and local authorities, by burning. If burned, stay out of smoke.

ATTENTION
Do Not Cut or Weld Container

WARRANTY LIMITATIONS AND DISCLAIMER

The Dow Chemical Company warrants that this product conforms to the chemical description on the label and is reasonably fit for the purposes stated on the label when used in strict accordance with the directions therein under normal conditions of use. THIS IS THE ONLY WARRANTY MADE ON THIS PRODUCT. NO OTHER EXPRESS AND NO IMPLIED WARRANTY OF MERCHANTABILITY OR FITNESS FOR A PARTICULAR PURPOSE IS MADE OUTSIDE OF THIS LABEL. Therefore, neither this warranty nor any other warranty of merchantability or fitness for a particular purpose, express or implied, extends to the use of this product contrary to label instructions (including conditions noted on the label, such as unfavorable temperatures, soil conditions, etc.), under abnormal conditions (such as excessive rainfall, drought, tornadoes, hurricanes, etc.) or under conditions not reasonably foreseeable to or beyond the control of seller.

When buyer or user suffers losses or damages resulting from the use or handling of this product (including claims based on contract, negligence, strict liability, or other legal theories), buyer or user must promptly notify in writing The Dow Chemical Company of any claims to be eligible to receive either remedy given below. The EXCLUSIVE REMEDY OF THE BUYER OR USER and the LIMIT OF LIABILITY of The Dow Chemical Company or any other seller will be one of the following, at the election of The Dow Chemical Company: (1) **Refund of purchase price** paid by buyer or user for product bought, **or** (2) **Replacement** of amount of product used. The seller **will not be liable for consequential or incidental damages or losses.**

The terms of this Warranty Limitations And Disclaimer cannot be varied by any written or verbal statements or agreements. Any employee or sales agent of the seller is not authorized to vary or exceed the terms of this Warranty Limitations And Disclaimer in any manner.

SPECIAL LOCAL NEED REGISTRATIONS

General Information and Mixing Instructions. Use LORSBAN 4E by mixing with water to form an emulsion spray. To prepare the spray, start to fill the spray tank with the required amount of water, add the recommended amount of LORSBAN 4E, and complete filling the tank with the balance of the water needed. Maintain sufficient agitation during mixing and application to ensure uniformity of the spray mixture. Apply LORSBAN 4E with the amounts of water recommended in each SLN. LORSBAN 4E should not be tank mixed with pesticides, surfactants, or fertilizer formulations unless prior use has shown the combination noninjurous under local conditions of use. Consult your Cooperative Agricultural Extension Service for additional information concerning control practices in your area.

CARROTS GROWN FOR SEED: For use only within **Washington** (EPA SLN No. WA-850025) to control cutworms and lygus, apply LORSBAN 4E with the amounts of water recommended below. Consult your Cooperative Agricultural Extension Service for additional information concerning cutworms and lygus control practices in your area. Use LORSBAN 4E by application as a foliar, broadcast spray at the rate of 1 quart per acre. Mix the specified dosage in sufficient water to ensure thorough coverage of treated areas. For **aerial application** use at lest 5 gallons of water per acre. Make application after carrots have started to bolt and when field counts or crop injury indicate that damaging pest populations are developing or present.

Restrictions: Do not allow livestock to graze in treated areas. Do not feed treated carrot cuttings or seed screenings to livestock.

This product is highly toxic to bees exposed to direct treatment or residue on treated crops or weeds. Do not apply when bees are in the field. Application must be made after bees have ceased flying for the day. Consult your Cooperative Agricultural Extension Service for additional information concerning protection of pollinating bees.

CLOVER GROWN FOR SEED For use only within **Oregon** (EPA SLN No. OR-850032) to control garden symphylans, armyworms, cutworms, aphids, weevils, lygus and spittlebugs, apply LORSBAN 4E with the amounts of water recommended below. Consult your Cooperative Agricultural Extension Service for additional information concerning garden symphylan, armyworm, cutworm, aphid, weevil, lygus, and spittlebug control practices in your area. For **garden symphylan control** use LORSBAN 4E as a preplant, broadcast application to soil surfaces at the rate of 2 quarts per acre. Mix the specified dosage in at least 10 gallons of water per acre to ensure thorough coverage of the treated soil. On the same day of treatment, incorporate the insecticide into the top 2 to 4 inches of soil using a disc, field cultivator, or equivalent equipment. For **established planting,** apply LORSBAN 4E at the rate of 2 quarts per acre as a broadcast, foliar spray before spring growth. Apply in at least 40 gallons of water to ensure thorough coverage and penetration. Incorporate to a depth of 2 to 4 inches using a disc or other suitable equipment. For **armyworm, cutworm, weevil, lygus and spittlebug control,** use LORSBAN 4E as a broadcast, foliar application at the rate of 2 pints per acre. For aphid control, use LORSBAN 4E as a broadcast, foliar application at the rate of 1 to 2 pints per acre. Mix the specified dosage in sufficient water to ensure thorough coverage of treated areas. Treat when field counts or crop injury indicates that damaging pest populations are developing or present. For **best cutworm control** soil should be moist and worms active on or near the soil surface at the time of treatment.

Restrictions: Do not allow livestock to graze in treated areas. Do not feed treated clover cuttings or seed screenings to livestock or use hay for livestock bedding.

This product is highly toxic to bees exposed to direct treatment or residues on treated crops or weeds. Do not apply to clover in bloom. Evening or night time applications when forages are not present reduce exposure hazards. Consult your Cooperative Agricultural Extension Service for additional information concerning protection of pollinating bees.

FIELD AND SWEET CORN: For use only within **Florida** (sweet corn only) EPA SLN No. FL-780046) and Georgia (EPA SLN No. GA-790019) to control beet armyworms, fall armyworm, and corn earworm larvae and **Colorado** (sweet corn only) (EPA SLN No. CO-810017), **Maryland** (sweet corn only) (EPA SLN No. MD-840004), **New Jersey** (sweet corn only) (EPA SLN No. NJ-840008), **Pennsylvania** (sweet corn only) (EPA SLN No. PA-840008), and **Delaware** (sweet corn only) (EPA SLN No. DE-840003) to control armyworms and corn earworm larvae, use 1 pint of LORSBAN 4E in sufficient water to ensure thorough coverage of plants and apply using either aerial (fixed-wing or helicopter) or power-operated ground spray equipment. For aerial application use at least 3 gallons of water per acre. Treat when field counts indicate that worms are or may become a problem. Repeat the application as needed, but do not exceed 13 applications per crop season.

Restrictions: Do not treat within 7 days before harvest. Do not allow livestock to graze in treated areas. Do not feed treated corn silage, forage, or fodder to meat or dairy animals.

For use only within **Florida** (field and sweet corn grown only for planting seed) (EPA SLN No. FL-770025) to control beet armyworm, fall armyworm, and corn earworm, use 1 to 2 pints of LORSBAN 4E per acre in sufficient water to ensure thorough coverage of plants and apply using either aerial (fixed-wing or helicopter) or power-operated ground spray equipment. For aerial application use at least 3 gallons of water per acre. Treat when field counts indicate that worms are or may become a problem. Repeat the application as needed.

Restrictions: Apply only to corn to be harvested for seed. Do not allow livestock to graze in treated areas. Do not feed treated corn grain, silage, or forage to meat or dairy animals.

COTTON: For use only within **Alabama** (EPA SLN No. AL-850003) used alone or in tank mixtures with pyrethroid insecticides is recommended for control and suppression of aphids, fall armyworm, and spider mites infesting cotton. Use LORSBAN 4E at a rate of 1/2 to 1 pint per acre and pyrethroid insecticides at the dosage rate shown on their product labels. Mix the required dosage of LORSBAN 4E or LORSBAN 4E and pyrethroid insecticide in sufficient cottonseed oil or soybean oil to give a finished spray volume 1 of quart per acre. Apply a minimum of 1 quart of spray per acre with aircraft calibrated to give adequate coverage of plants. If the aircraft is calibrated to deliver higher volumes, use **emulsifiable** vegetable oil at a rate of 1 pint per acre with the above dosage of insecticide and add enough water for 2 gallons of finished spray per acre. Treat when field count indicate damaging insect populations are developing or present. However, for fall armyworm control, use only as part of a regularly scheduled program (5 to 7 days apart).

Restrictions: Make no applications of these tank mixtures closer to harvest than the longest pre-harvest interval shown for any of the products in the tank mixture. Do not allow livestock to graze in treated areas. Do not feed gin trash or treated forage to livestock.

For use only within **Arkansas** (EPA SLN No. AR-870001) to suppress infestations of plantbugs and fleahoppers and preserve populations of beneficial arthropods in cotton. Use LORSBAN 4E at the rate of 1/4 pint per acre. Mix with water to form an emulsion spray. Apply the specified dosage in sufficient water to ensure thorough coverage using aerial or ground spary equipment. Treat when field counts indicate damaging density levels of plantbugs or fleahoppers are present.

Restrictions: Do not apply within 14 days before harvest. Do not allow livestock to graze in treated areas or feed gin trash or treated forage to livestock.

For use only within **Mississippi** (EPA SLN No. MS-840004) to control larvae of wireworms and cutworms, apply 2 pints of LORSBAN 4E per acre for wireworm control and 1 to 2 pints of LORSBAN 4E per acre for cutworm control in not less than 10 gallons of water in a 16 to 18 inch band over the row prior to or at planting. Immediately following application incorporate the insecticide into the soil to a depth of 3 to 4 inches using suitable equipment.

Restrictions: Do not make more than one application per season.

For use only within **Mississippi** (EPA SLN No. MS-830025) to control and suppress spider mites, use 1/2 to 1 pint of LORSBAN 4E per acre in tank mixtures with pyrethroid insecticides at the dosage rate shown on their product label and sufficient cottenseed oil or soybean oil to give a finished spray volume of 1 quart per acre. Apply a minimum of 1 quart of spray per acre with aircraft calibrated to give adequate coverage of plants. Treat when field counts indicate damaging insect populations are developing or present.

Restrictions: Make no applications of these tank mixtures closer to harvest than the longest preharvest interval shown for any of the products in the tank mixture. Do not allow livestock to graze in treated areas. Do not feed gin trash or treated forage to livestock.

GRAPES: For use only within **Michigan** (EPA SLN No. MI-840002) and **Missouri** (EPA SLN No. MO-870004) to control climbing cutworms, use 1 quart of LORSBAN 4E per acre in a minimum of 25 gallons of water and apply as a prebloom, dilute, or concentrate spray. LORSBAN 4E should not be tank mixed with pesticides, surfactants, or fertilizer formulations unless prior use has shown the combination noninjurious under your conditions of use. Consult your State Agricultural Station or Extension Service Specialist for proper time to treat in your area.

Restrictions: Do not make more than one application per season. Do not use in conjunction with a soil surface spray application of LORSBAN 4E for grape root borer control.

For use only with **Tennessee** (EPA SLN No. TN-880002) and **Georgia** (EPA SLN No. GA-860001) to control grape root borer by split soil applications, mix 2 1/4 pints of LORSBAN 4E with 100 gallons of water and apply 2 quarts of the diluted spray mixture to the soil surface on a 15 square foot area around the base of each vine. Do not allow spray to contact fruit or foilage. Make the first application before the pest emerges (at least 35 days prior to harvest) followed by a second application after harvest. Consult with local extension entomologists for optimum applicatin timing based on pest emergence.

Restrictions: Do not make more than 2 applications per season or apply within 35 days before harvest.

NON-BEARING GRAPES: For use only within **Idaho** (EPA SLN No. ID-820018), **Oregon** (EPA SLN No. OR-820001), and **Washington** (EPA SLN No. WA-810036) to control cutworms, apply 1 quart of LORSBAN 4E per acre as a broadcast spray using either aerial or power-operated ground spray equipment. For aerial application use at least three gallons of spray per acre. If sprinkler irrigation is available, apply 1/4 to 1/2 inch of water after application to help incorporate the material. Treat when cutworms first become active and when field counts indicate damaging insect populations are developing or present. Retreat as necessary to maintain control. Consult your State Agricultural Experiment Station or Extension Service specialist for additional information concerning cutworm control practices in your area.

Restrictions: Do not harvest grapes in the same growing season as application.

MINT: For use only within **Oregon** (EPA SLN No. OR-8500034) by application as a preplant, broadcast spray to soil surfaces at the rate of 2 quarts per acre to control garden symphylans infesting mint. Mix the specified dosage in at least 10 gallons of water per acre to ensure thorough coverage of treated soil. On the same day of treatment, incorporate the insecticide into the top 2 to 4 inches of soil using a disc, field cultivator, or equivalent equipment.

Restrictions: Do not make more than one preplant, incorporation application in the spring. Do not use in conjunction with a broadcast, foliar spray application of LORSBAN 4E for cutworm control.

ONIONS (Dry Bulb): For use only within **Michigan** (EPA SLN No. MI-850006) for early season application on onions (dry bulb), to control Onion maggot. Use LORSBAN 4E at the rate of 1 quart per acre (equivalent to one pound of chlorpyrifos per acre) and apply as a directed spray to the base of onion seedlings or transplants during peak egg laying by the onion maggot adult. Proper timing of the insecticide application is essential for satisfactory control. Use a minimum of 100 gallons of total spray per acre for thorough wetting of the soil at the base of the plants.

Restrictions: Do not make more than one directed spray application if LORSBAN 4E or LORSBAN 15G were used at planting. Do not use within 60 days of harvest or apply more than 2 pounds of chlorpyrifos per acre per growing season.

PEANUTS: For use only within **GEORIGA** (EPA SLN No. GA-870006) use LORSBAN 4E at the rate of 4 pints per acre to suppress wireworms. Apply the specified dosage in not less than 10 gallons of water as a preplant broadcast spray to the soil surface using suitable power-operated ground application equipment. Immediately following application, incorporate the insecticide into the soil to a depth of 3 to 4 iches using suitable equipment.

Restrictions: Do not make more than one application per season. Do not apply a postplant directed spray following a preplant broadcast spray.

PERENNIAL GRASS SEED CROPS: For use only within **Idaho** (EPA SLN No. ID-830005) and **Oregon** (EPA SLN No. OR-830007) to control billbugs, cutworms, and aphids and **Washington** (EPA SLN No. WA-830009) to control cutworms and aphids, use 1 quart of LORSBAN 4E per acre in sufficient water to ensure good coverage and apply as a broadcast spray using power-operated ground application equipment. For improved results, the thatch should be moist at the time of application and the insecticide washed into the thatch immediately after application. If sprinkler irrigation is available, apply 1/4 to 1/2 inch of water after application to help incorporate the material. Otherwise, apply to early morning dew-covered field or during or immediately following a light rain. For billbug control, apply in early April when adult billbugs are active. For cutworm control, apply in either spring or fall when larvae or larval damage first appear. Fall applications should follow sufficient irrigation and/or rainfall. Heavy carbon residue from field burning will prevent penetration of LORSBAN 4E into the plant crown where the insects feed. Retreat as necesary to maintain control. To control aphids, apply 1 to 2 pts of LORSBAN 4E per acre as a broadcast spray using power-operated ground application equipment. Treat as colonies begin forming before heading. Consult your Cooperative Agricultural Extension Service for additional information concerning billbug, cutworm, and aphid control practices in your area. This product is highly toxic to bees exposed to direct treatment or residues on treated crops or weeds. Protective information may be obtained from your Cooperative Agricultural Extension Service.

Restrictions: Apply only to perennial grass seed crops. Do not allow livestock to graze in treated areas. Do not feed treated grass cuttings or seed screenings to livestock nor use hay for livestock bedding.

For use only within **South Dakota** (EPA SLN No. SD-870006) to control billbugs and cutworms infesting perennial grass and seed crops. Apply 1 quarts of LORSBAN 4E per acre as a broadcast spray using power-operated ground application equipment. Apply LORSBAN 4E with enough water to ensure good coverage. For best results, use 20 gpa of spray mixture. For improved results, the thatch should be moist at the time of application and the insecticide washed into the thatch immediately after application. If sprinkler irrigation is available, apply 1/4 to 1/2 inches of water after application to help incorporate the material. Otherwise, apply to early morning dew-covered fields or during or immediately following a light rain. For **billbug control,** apply in early April when adult billbugs are active. For **cutworm control,** apply in either spring or fall when larvae or larval damage first appear. Fall applications should follow sufficient irrigation and/or rainfall. Heavy carbon residue from field burning will prevent penetration of LORSBAN 4E into the plant crown where the insects feed. Retreat as necessary to maintain control. To control aphids, apply 1 to 2 pts of LORSBAN 4E per acre as a broadcast spray using power-operated ground application equipment. Treat as colonies begin forming before heading. Consult your Cooperative Agricultural Extension Service for additional information concerning billbug, cutworm, and aphid control practices in your area.

For aerial application, follow above dosage recommendations and use two to five gallons of water per acre. Control may be reduced at low spray volumes under high temperature and wind conditions.

Restrictions: Apply only to perennial grass seed crops. Do not allow livestock to graze in treated areas. Do not feed treated grass cuttings or seed screenings to livestock nor use hay for livestock bedding.

RADISHES GROWN FOR SEED: For use only within **Oregon** (EPA SLN No. OR-850031) use LORSBAN 4E as a preplant, broadcast application to soil surface to control garden symphylan at the rate of 2 quarts per acre. Mix the specified dosage in at least 10 gallons of water per acre to ensure thorough coverage of the treated soil. On the same day of treatment, incorporate the herbicide into the top 2 to 4 inches of soil using a disc, field cultivator, or equivalent equipment.

For **cutworm control,** use LORSBAN 4E as a broadcast application to soil surface at the rate of 1 quart per acre. Mix the specified dosage in at least 10 gallons of water per acre to ensure thorough coverage of treated soil. Treat in the spring when cutworm larvae begin to appear. It is preferable to apply LORSBAN 4E when the soil is moist and worms are active on or near the soil surface.

Restrictions: Do not allow livestock to graze in treated areas. Do not feed treated radish cuttings or seed screenings to livestock.

This product is highly toxic to bees exposed to direct treatment or residue on treated crops or weeds. Do not apply to radishes in bloom. Evening or night time applications when foragers are not present in the fields reduce exposure hazards. Consult your Cooperative Agricultural Extension Service for additional information concerning protection of pollinating bees.

STRAWBERRIES: For use only within **Idaho** (EPA SLN No. ID-830028), **Oregon** (EPA SLN No. OR-830017), and **Washington** (EPA SLN No. WA-830033) to control garden symphylans, use 2 quarts of LORSBAN 4E per acre with enough water to ensure good coverage of the soil surface and apply pre-plant as a broadcast spray to the soil surface using suitable power-operated ground application equipment. On the same day of treatment, incorporate the insecticide into the top 2 to 4 inches of soil using a disc, field cultivator or equivalent equipment. LORSBAN 4E should not be tank mixed with pesticides, surfactants, or fertilizer formulations unless prior use has shown the combination noninjurious under local conditions of use. Consult your Cooperative Agricultural Extension Service for additional information concerning garden symphylan control practices in your area.

Restrictions: Apply only as a soil pre-plant incorporation application in the spring for protection of strawberries the following year.

For use only within **Oregon** (EPA SLN No. OR-840021) to control strawberry crown moth, apply 1 quart of LORSBAN 4E per acre as a directed spray to crowns of strawberry plants immediately after harvest and after plants are topped. Apply in a minimum of 100 gallons of spray per acre. **Do not** sprinkle irrigate for one weed following application. A second application may be required two to two and one half weeks later. Phytotoxicity may occur when LORSBAN 4E is applied to plants experiencing high temperature and water stress. LORSBAN 4E should not be tank mixed with pesticides, surfactants, or fertilizer formulations unless prior use has shown the combination noninjurious under local conditions. Consult your Cooperative Agricultural Extension Service for additional information concerning strawberry crown moth control practices in your area.

Restrictions: Use postharvest only. Do not make more than two applications per year.

SUGAR BEETS GROWN FOR SEED: For use only within **Idaho** (EPA SLN No. ID-830021) and **Oregon** (EPA SLN No. OR-830018) to control garden symphylans, use 2 quarts of LORSBAN 4E per acre with enough water to ensure good coverage of the soil surface and apply pre-plant as a broadcast spray to the soil surface using surface using suitable power-operated ground application equipment. LORSBAN 4E should not be tank mixed with pesticides, surfactants, or fertilizer formulations unless prior use has shown the combination noninjurious under local conditions of use. Consult your Cooperative Agricultural Extension Service for additional information concerning garden symphylan control practices in your area.

Restrictions: Apply only as a soil pre-plant incorporation application in the fall for protection of sugar beets the following year.

01545-L1 N1189

*Trademark of THE DOW CHEMICAL COMPANY

LORSBAN* 50W

WETTABLE POWDER INSECTICIDE

For Control of Various Insects on Certain Fruit and Nut Crops

ACTIVE INGREGIENT(S):
Chlorpyrifos [O,O-diethyl O-(3,5,6-trichloro-2-pyridinyl) phosphorothioate] 50%
INERT INGREDIENTS: 50%
TOTAL 100%

E.P.A. Registration No. 464–552 E.P.A. Est. 447–TX–1

KEEP OUT OF REACH OF CHILDREN

WARNING

AVISO:

PRECAUCIÓN AL USUARIO: Si usted no lee inglés, no use este producto hasta que la etiqueta la haya sido explicada ampliamente.

PRECAUTIONARY STATEMENTS

HAZARDS TO HUMANS AND DOMESTIC ANIMALS

MAY BE FATAL IF SWALLOWED • HARMFUL IF ABSORBED THROUGH SKIN OR INHALED • CAUSES EYE IRRITATION

Avoid Contact With Eyes, Skin, Or Clothing • Avoid Breathing Dust Or Spray Mist • Wash Thoroughly With Soap And Water After Handling And Before Eating Or Smoking • When Loading The Spray Tank Or Handling The Powder, Wear A Pesticide Respirator Jointly Approved By The Mining Enforcement And Safety Administration (Formerly The U.S. Bureau Of Mines) And By The National Institute For Occupational Safety And Health Under The Provisions Of 30 CFR Part II • Wear Protective Clothing Such As Long Pants, Long Sleeve Shirt, Gloves, Hat, And Boots When Loading The Spray Tank Or Handling The Powder And During Application Of The Spray Mixture •Work To Windward To Stay Out Of Spray Drift Or Mist • Remove Contaminated Clothing And Wash Before Reuse • Do Not Apply This Product In Such A Manner As To Directly Or Through Drift Expose Workers Or Other Persons • The Area Being Treated Must Be Vacated By Unprotected Persons • Keep Away From Food, Feedstuffs And Water Supplies

STATEMENT OF PRACTICAL TREATMENT

If Swallowed: Call a physician or Poison Control Center. Drink 1 or 2 glasses of water and induce vomiting by touching back of throat with finger. Do not induce vomiting or give anything by mouth to an unconscious person.

If On Skin: Wash with plenty of soap and water. Get medical attention if irritation persists.

If in Eyes: Flush with plenty of water for 5 minutes. Get medical attention if irritation persists.

If Inhaled: Remove to fresh air if effects occur. Get medical attention.

NOTE TO PHYSICIANS: Chlorpyrifos is a cholinesterase inhibitor. Treat symptomatically. If exposed, plasma and red blood cell cholinesterase tests may indicate significance of exposure (baseline data are useful). Atropine, only by injection, is the preferable antidote. Oximes, such as 2–PAM/protopam, may be therapeutic if used early; however, use only in conjunction with atropine. In case of severe acute poisoning, use antidote immediately after establishing an open airway and respiration.

ENVIRONMENTAL HAZARDS

This pesticide is toxic to birds and wildlife, and extremely toxic to fish and aquatic organisms. Do not apply directly to water. Drift and runoff from treated areas may be hazardous to aquatic organisms in adjacent aquatic sites. Cover or incorporate spills. Do not contaminate water when disposing of equipment washwaters. This product is highly toxic to bees exposed to direct treatment or residues on blooming crops or weeds. Avoid use when bees are actively foraging of if bees are likely to come in contact with wet or re-wetted treated surfaces. Protective information may be obtained from your Coorporative Extension Service. This product may pose a hazard to certain Federally endangered fish species. Refer to the Endangered Species section of this label for any geographic restrictions.

NOTICE

Read and understand the entire label before using. Use only according to label directions.

Before buying or using this product, read "WARRANTY LIMITATIONS AND DISCLAIMER" elsewhere on this label. If terms are not acceptable, return **unopened** package at once to seller for full refund of purchase price paid. Otherwise, use by the buyer or any other user constitutes acceptance of the terms under WARRANTY LIMITATIONS AND DISCLAIMER.

IN CASE OF AN EMERGENCY endangering life or property involving this product, call collect 517–636–4400

AGRICULTURAL CHEMICAL

Do Not Ship or Store with Food, Feeds, Drugs, or Clothing

ENDANGERED SPECIES RESTRICTIONS

The use of LORSBAN 50W may pose a hazard to certain Federally endangered fish species. This product **cannot** be used on sour cherry trees in the following countries:

COLORADO: Delta, Larimer and Mesa

OHIO: Washington

UTAH: Emery, Grand, Utah and Washington

WISCONSIN: Iowa, Pierce, Richland and Vernon

DIRECTIONS FOR USE

It is a violation of federal law to use this product in a manner inconsistent with its labeling.

GENERAL INFORMATION

LORSBAN 50W insecticide is a wettable powder formulation designed for use as a cover spray to control various pests. The pests controlled are listed in the accompanying table. LORSBAN 50W is compatible with insecticides, miticides, and fungicides commonly recommended in orchard spray schedules except for alkaline materials such as Bordeaux mixture and lime.

Test compatibility of the intended tank mixture before adding LORSBAN 50W to the spray or mix tank. Add proportionate amounts of each ingredient to a pint or quart jar, cap, shake, and let set 15 minutes. Formation of precipitates that do not readily redisperse indicates an incompatible mixture that should not be used.

MIXING INSTRUCTIONS

LORSBAN 50W mixes readily with water to form a suspension. Mix the required amount of LORSBAN 50W with a small quantity of water and add this premix while filling the spray tank. Alternatively, fill the tank with the required amount of water and then add the premix. Maintain sufficient agitation during both mixing and application to ensure uniformity of the spray mixture. LORSBAN 50W contains a wetting agent. Do not use additional wetting agents, spreaders or stickers.

USE AND DOSAGE RECOMMENDATIONS

Use LORSBAN 50W at the dosages indicated by application as a foliar spray to control pests listed in the following table. Where a rate range is listed in the use directions, use the higher rate for more severe infestations. Mix the required dosage in sufficient water to ensure adequate coverage and apply as a concentrate or dilute spray using conventional, power-operated ground spray equipment of by aircraft (fixed-wing or helicopter). For dilute sprays applied to tree fruits and nuts, mix the required dosage in sufficient water to allow for spray to run off. For concentrate sprays, apply an equivalent amount of LORSBAN 50W per acre. Treat when pests appear or in accordance with local conditions. Mow or beat down orchard blooming weed cover crops before applying spray to tree fruit crops to reduce hazard to bees. This step is especially important in relation to early Spring cover sprays when morning conditions may be wet from dew. LORSBAN 50W will perform best in an almond or walnut pest control program that includes winter sanitation (removal of mummy nuts) and regular monitoring of target pests. To control apple aphid infesting apples, spray every 10 to 14 days on a preventative schedule. To suppress pecan leaf scorch mite on pecans, use in a preventative program. For best pecan *Phylloxera* control, use two applications 7 to 10 days apart starting at bud swell and ending by early leaf expansion. For ant control on pecans, do not apply where weed growth or other obstructions would impede uniform coverage of the orchard floor. Consult your State Agricultural Experiment Station, Certified Pest Control Advisor, or Extension Service specialist for specific use information in your area.

LORSBAN 50W may also be used on almonds and walnuts as a dormant/delayed dormant treatment applied at the dosages indicated in a dilute or concentrate spray. Proper equipment must be used to obtain complete coverage of the entire bark area. If scale infestations are severe, dilute sprays should be used to ensure complete coverage. Use oil as recommended by your state Agricultural Experiment Station, State Extension Service Specialist, or Certified Pest Control Advisor. Do not use oil with LORSBAN 50W on walnuts. Because dry or cold conditions may cause LORSBAN 50W plus oil sprays to infuse trees resulting in bud damage or drop, do not apply until winter rains or irrigation has replenished soil moisture such that bark and twigs are not dessicated.

DOSAGE AND USE RECOMMENDATIONS
FOR LORSBAN 50W WETTABLE POWDER INSECTICIDE

APPLES: Use LORSBAN 50W to control the pests listed in the following table at the rates indicated.

GEOGRAPHIC LOCATION	PESTS CONTROLLED	DOSAGE RECOMMENDATIONS	REMARKS
Restrictions (All use Areas) Make no more than 8 applications per season. Do not apply last treatment within 28 days before harvest or apply last two treatments closer than 21 days apart. Do not allow livestock to graze in treated orchards. **Additional Restrictions** (by Geographic Location) In California: Do not do work involving contact with trees within 2 days after treatment.			
All Areas	Apple aphid[1], Apple maggot, Climbing cutworm, Codling moth, Dogwood borer (trunk spray), European apply sawfly, Eyespotted bud moth, European corn borer, Fruittree leafroller, Green fruitworm, Lesser appleworm, Obliquebanded leafroller, Orange tortrix, Pandemis leafroller, Oriental fruit moth, Periodical cicada, Plum curculio, Redbanded leafroller, Rosy apple aphid, San Jose scale, Tarnished plant bug, Tufted apple bud moth, Variegated leafroller, Woolly apple aphid.	3 lb/acre	[1] For apple aphid control LORSBAN 50W should not be used alone as a rescue treatment.
Connecticut, Delaware, Maine, Maryland, Massachusetts, Michigan, New Hampshire, New Jersey, New York, N. Carolina, Ohio, Pennsylvania, Rhode Island, South Carolina, Vermont, Virginia, West Virginia.	Apple aphid[2], Apple maggot[3], Codling moth, Dogwood borer (trunk spray)[4], European apply sawfly[5], European corn borer[6], European red mite[7], Eyespotted bud moth, Fruittree leafroller, Green fruitworm, Lesser appleworm, Obliquebanded leafroller[6], Oriental fruit moth, Periodical cicada[4], Plum curclio[8], Potato leafhopper, Redbanded leafroller, Rosy apple aphid, San Jose scale, Spotted tentiform leafminer[9], Tarnished plant bug, Tufted apple bug moth[10], Variegated leafroller, White apple leafroller[9], Wooly apple aphid	2-3 lb/acre (8-12 oz/100 gal dilute)[11] OR A tank mix of 1½ lbs/acre (6 oz/100 gal dilute) with one of the following: 1½ pints/acre LANNATE 1.8L (6 fl oz/100 gal dilute)[11] 6 oz/acre LANNATE 90SP (1½ oz/100 gal dilute)[11] 2 lbs/acre IMIDAN 50W (8 oz/100 gal dilute)[11] 1½ lbs./acre GUTHION 35W (6 oz/100 gal dilute)[11] 2 pints/acre PENNCAP M (8 fl oz/100 gal dilute)[11]	[2] For apple aphid control, use 2-3 lbs/acre LORSBAN 50W or a tank mix with LANNATE 1.8L or LANNATE 90SP as recommended. LORSBAN 50W should not be used alone or in a tank mix as a rescue treatment. Treatments should be applied at regular intervals (1-2 weeks apart). [3] For apple maggot control in Michigan or in other areas after July, use 3 lbs/acre LORSBAN 50W or a tank mix with IMIDAN 50W, GUTHION 35W or PENNCAP M as recommended. [4] No tank mix recommended. [5] For European apple sawfly control, use 2-3 lbs/acre LORSBAN 50W or a tank mix with IMIDAN 50W, GUTHION or PENNCAP M as recommended. [6] For control of European corn borer or obliquebanded leafroller, use 2-3 lbs/acre LORSBAN 50W or a tank mix with LANNATE 1.8L or LANNATE 90SP as recommended. [7] European red mite suppression with LORSBAN 50W is achieved when a regularly scheduled preventative spray program is followed or when used with other mite management practices such as biological control. [8] For plum curculio control in Michigan, use 3 lbs/acre LORSBAN 50W or a tank mix with IMIDAN 50W, GUTHION 35W or PENNCAP M as recommended. For other areas, use 2-3 lbs/acre LORSBAN 50W or one of the above-mentioned tank mixes. [9] For control of spotted tentiform (leafminer or white apple) leafhopper, use a tank mix with LANNATE 1.8L or LANNATE 90SP as recommended. LORSBAN 50W alone is not recommended for control of these pests. [10] For tufted apple bud moth control, use 3 lbs/acre LORSBAN 50W or a tank mix with LANNATE 1.8L OR LANNATE 90SP as recommended. [11] Based on 400 gal dilute/acre.

CROP	INSECTS CONTROLLED	DOSAGE LORSBAN 50W	RESTRICTIONS
Almonds	Navel orangeworm, Peach twig borer, San Jose scale	4 lb per acre	Make only one dormant/delayed dormant spray application and no more than three foliar spray applications on almonds per season, one dormant/delayed dormant spray application and no more than two foliar spray applications on walnuts per season, and no more than three foliar spray applications on filberts per season. Do not apply within 14 days of harvest. Do not allow livestock to graze in treated orchards. Do not apply as a dormant spray application on almonds in the following counties in California: Butte, Colusa, Glenn, Solano, Sutter, Tehama, Yolo and Yuba.
Walnuts	Codling moth, Walnut scale, Walnut husk fly	4 lb per acre	
Filberts	Eyespotted bud moth, Filbert aphid, Filbert leafroller, Filbertworm Obliquebanded leafroller, Omniverous leaftier, Winter moth	3 to 4 lb per acre	
Pecans	Black pecan aphid, Hickory shuckworm, Pecan leaf scorch mite (suppression)	2 lb per 100 gallons	Make no more than 5 applications per season. Do not apply within 28 days of harvest. Do not allow livestock to graze in treated orchards.
	Pecan nut casebearer, Phylloxera spp., Fall webworm, Spittlebug	1 lb per 100 gallons	
	Fire ants and other ant species (orchard floor)	2 lb per acre	

CROP	INSECTS CONTROLLED	DOSAGE LORSBAN 50W	RESTRICTIONS
Brassica (cole) Broccoli Brussels sprouts Cabbage Cauliflower Collards Kale Kohlrabi	Cabbage aphid Cutworms Imported cabbageworm Striped flea beetle (adult)	2 lb per acre	Make no more than 6 applications per season. Do not apply within 21 days before harvest. To avoid phytotoxicity in vegetable crops, do not mix with other pesticide products or treat plants that are under extreme heat and drought stress.
Sour Cherries	Borers American plum Lesser peachtree Pacific flatheaded Peach twig Peachtree Shothole Climbing cutworm Green fruitworm Leafrollers Fruittree Obliquebanded Pandemis Redbanded Lesser appleworm Tarnished plant bug Moth Eyespotted bud Mineola Scale European fruit lecanium San Jose Stink bug	2 to 3 lb per acre	Make no more than 8 applications per season. Do not apply within 14 days of harvest. Do not allow livestock to graze in treated orchard. In California do not do work involving contact with trees within 2 days after treatment. Do not use on sweet cherries. Refer to the Endangered Species Section of the label for geographic restrictions on the use of LORSBAN 50W on sour cherries.
	Black cherry aphid Cherry fruitfly Plum curculio Rose chafer	3 lb per acre	

REENTRY STATEMENT

Do not allow reentry into treated areas within 24 hours unless protective clothing is worn. Protective clothing consists of at least a hat or other suitable head covering, a long-sleeved shirt and long-legged trousers or a coverall type garment (all of closely woven fabric covering the body, including the arms and legs), shoes and socks. Because certain states may require more restrictive reentry intervals for various crops treated with this product, consult your state Department of Agriculture for further information.

Written or oral warnings must be given to workers who are expected to be in a treated area or in an area about to be treated with this product. These oral warnings shall inform workers of areas or fields that may not be entered without protective clothing and actions to take in case of accidental exposure as shown by the Statement of Practical Treatment section on this label. When oral warnings are given, warnings shall be given in a language customarily understood by workers. Written warnings must include the following information: WARNING. Area treated with LORSBAN 50W wettable powder insecticide on (insert date of application). Do not enter without protective clothing for 24 hours. In case of accidental exposure (insert actions to take as shown by the Statement of Practical Treatment section on this label).

STORAGE AND DISPOSAL

Do not contaminate water, food or feed by storage or disposal. Open dumping is prohibited.

Pesticide Disposal: Wastes resulting from the use of this product may be disposed of on site or at an approved waste disposal facility.

Container Disposal: Completely empty bag into application equipment. Then dispose of empty bag in a sanitary landfill or by incineration, or, if allowed by state and local authorities, by burning. If burned, stay out of smoke.

WARRANTY LIMITATIONS AND DISCLAIMER

The Dow Chemical Company warrants that this product conforms to the chemical description on the label and is reasonably fit for the purposes stated on the label when used in strict accordance with the directions therein under normal conditions of use. THIS IS THE ONLY WARRANTY MADE ON THIS PRODUCT. NO OTHER EXPRESS AND NO IMPLIED WARRANTY OF MERCHANTABILITY OR FITNESS FOR A PARTICULAR PURPOSE IS MADE OUTSIDE OF THIS LABEL. Therefore, neither this warranty nor any other warranty of merchantability, or fitness for a particular purpose, express or implied, extends to the use of this product contrary to label instructions (including conditions noted on the label, such as unfavorable temperatures, soil conditions, etc.) under abnormal conditions (such as excessive rainfall, drought, tornadoes, hurricanes, etc.) or under conditions not reasonably foreseeable to or beyond the control of seller.

When buyer or user suffers losses or damages resulting from the use or handling of this product (including claims based on contract, negligence, strict liability, or other legal theories), buyer or user must promptly notify in writing The Dow Chemical Company of any claims to be eligible to receive either remedy given below. The EXCLUSIVE REMEDY OF THE BUYER OR USER and the LIMIT OF LIABILITY of The Dow Chemical Company or any other seller will be one of the following, at the election of The Dow Chemical Company:

(1) **Refund of purchase price** paid by buyer or user for product bought, **or**
(2) **Replacement** of amount of product used.

The seller **will not be liable for consequential or incidental damages or losses**

The terms of this Warranty Limitations and Disclaimer cannot be varied by any written or verbal statements or agreements. Any employee or sales agent of the seller is not authorized to vary or exceed the terms of this Warranty Limitations And Disclaimer in any manner.

45386–L1 J989

* Trademark of THE DOW CHEMICAL COMPANY

PATHWAY*

HERBICIDE

For Controlling Unwanted Trees Via Cut Surface Treatments

Active Ingredient(s):
Picloram (4-amino-3,5,6-trichloropicolinic acid),
as the triisopropanolamine salt. 5.4%
2,4-Dichlorophenoxyacetic Acid, as the triisopropanolamine salt . 20.9%
Inert Ingredient(s): 73.7%

Acid Equivalents: Picloram - 3.0% 2,4-Dichlorophenoxyacetic acid - 11.2%

E.P.A. Registration No. 464-510
E.P.A. Est. 464-MI-1

KEEP OUT OF REACH OF CHILDREN

WARNING

AVISO:

PRECAUCION AL USUARIO: Si usted no lee inglés, no use este producto hasta que la etiqueta le haya sido explicada ampliamente.

PRECAUTIONARY STATEMENTS

HAZARDS TO HUMANS AND DOMESTIC ANIMALS

CAUSES SUBSTANTIAL BUT TEMPORARY EYE INJURY • HARMFUL IF SWALLOWED, INHALED OR ABSORBED THROUGH SKIN

Do Not Get In Eyes Or On Clothing • Wear Goggles, Face Shield Or Safety Glasses When Handling • Avoid Contact With Skin • Wash Thoroughly With Soap And Water After Handling • Remove Contaminated Clothing And Wash Before Reuse • Do Not Drink Solution • Avoid Breathing Spray Mist

STATEMENT OF PRACTICAL TREATMENT: If in eyes, flush with plenty of water for at least 15 minutes. Get medical attention. **If on skin,** wash with plenty of soap and water. Get medical attention if irritation persists. **Contains ethylene glycol. If swallowed,** induce vomiting immediately by giving two glasses of water and sticking finger down throat. Do not induce vomiting or give anything by mouth to an unconscious person.

PHYSICAL OR CHEMICAL HAZARDS

COMBUSTIBLE • Do Not Use or Store Near Heat or Open Flame. • Do Not Cut or Weld Container.

ENVIRONMENTAL HAZARDS

Do not apply directly to water or wetlands (swamps, bogs, marshes and potholes). Do not contaminate water used for irrigation or domestic purposes by cleaning of equipment or disposal of wastes. Do not allow run-off or spray to contaminate wells, irrigation ditches or any body of water used for irrigation or domestic purposes. Do not make application when circumstances favor movement from treatment site. See additional precautionary statements elsewhere on this label.

NOTICE

Read the entire label. Use only according to label directions.

Before buying or using this product, read "WARRANTY LIMITATIONS AND DISCLAIMER" elsewhere on this label. If terms are not acceptable, return **unopened** package at once to seller for full refund of purchase price paid. Otherwise, use by the buyer or any other user constitutes acceptance of the terms under "WARRANTY LIMITATIONS AND DISCLAIMER."

IN CASE OF AN EMERGENCY endangering life or property involving this product, call collect 517-636-4400

AGRICULTURAL CHEMICAL

Do Not Ship or Store with Food, Feeds, Drugs, or Clothing

GENERAL INFORMATION

PATHWAY Herbicide is a **ready-to-use product** that is effective in cut surface applications for killing unwanted trees and preventing undesirable sprouting of cut trees in forest and other non-crop areas such as fence rows and right-of-ways. Among the many species that can be controlled are: **ailanthus, alder, aspen, birch, cedar, cherry, dogwood, elm, firs, green ash, gum, hawthorn, hickory, hornbeam, maples, oaks, pecan, persimmon, serviceberry, sourwood, sweetbay.**

USE DIRECTIONS

It is a violation of Federal law to use this product in a manner inconsistent with its labeling.

Use undiluted as indicated below.

TREE INJECTION TREATMENT: Use 1 milliliter of undiluted PATHWAY Herbicide solution through the bark completely around the tree trunk at intervals of 2 to 3 inches between edges of the injector wounds. Make injections near ground level when using the tree injector or 2 to 4 feet above the ground when using a HYPOHATCHET Injector or similar device. Treatments can be made in any season. Maples should not be treated during the spring sap flow. With some difficult to control species such as dogwood, hickory, sugar maple, bigleaf maple, tanoak, and some firs, application to a continuous cut rather than to spaced cuts may be more uniformly effective.

FRILL OR GIRDLE TREATMENT: Make a single hack girdle or "frill" of overlapping axe cuts through the bark completely around the tree as close to the ground as feasible. Spray or paint the injured surface with undiluted PATHWAY Herbicide, using enough volume to wet treated areas.

STUMP TREATMENT: Spray or paint the cut surfaces of freshly cut stumps and stubs with undiluted PATHWAY Herbicide. The cambium area next to the bark is the most vital area to wet.

USE PRECAUTIONS

Use this product only as specified on this label. PATHWAY Herbicide is highly active as a herbicide, is water soluble, can move with surface runoff water, and can remain phytotoxic for a year or more if it gets into the soil. Untreated trees occasionally are affected by root uptake of herbicide leaches into the soil or excreted by the roots of nearby treated trees. If such injury cannot be tolerated, do not use PATHWAY close to desirable trees.

Do not contaminate cropland, water, or irrigation ditches.

Do Not Contaminate Water Intended for Irrigation or Domestic Purposes. To avoid injury to crops or other desirable plants, do not treat or allow spray drift or run-off to fall onto banks or bottoms of irrigation ditches, either dry or containing water, or other channels that carry water that may be used for irrigation or domestic purposes. Do not apply to snow or frozen ground.

Water contaminated with residues of picloram from drainage ditch bank uses shall not be used to irrigate crops which are not registered for use with this chemical.

Do not rotate food or feed crops on treated land if they are not registered for use with picloram until an adequately sensitive bioassay or chemical test shows that no detectable picloram is present in the soil.

Do not apply on or in the vicinity of sensitive crops of desirable broadleaf plants. Small amounts can cause injury to plants.

Do not allow careless application or spray drift. Do not permit any spray or spray drift to contact desirable plants. Use only gravity or very low spray pressure.

Do not remove product from original container except to apply as indicated under USE DIRECTIONS.

Do not store near food, feedstuff, fertilizer, seeds, insecticides, fungicides, or other pesticides.

PRECAUTIONARY STATEMENTS

ENVIRONMENTAL HAZARDS

Picloram is a chemical which can travel (seep or leach) through soil and under certain conditions contaminate groundwater which may be used for irrigation and drinking purposes. Users are advised not to apply picloram where soils have a rapid to very rapid permeability throughout the profile (such as loamy sand to sand) and the water table of an underlying aquifer is shallow **or** to soils containing sinkholes over limestone bedrock, severely fractured surfaces, and substrates which would allow direct introduction into an aquifer. Your local agricultural agencies can provide further information on the type of soil in your area and the location of groundwater.

An aquifer is defined as "an underground, saturated, permeable, geologic formation capable of producing significant quantities of water to a well or spring. It is the ability of the saturated zone, or portion of that zone, to yield water which makes it an aquifer" (American Chemical Society, 1983).

STORAGE AND DISPOSAL

Do not contaminate water, food or feed by storage or disposal.

Pesticide Disposal: Pesticide wastes are toxic. Improper use of excess pesticide, spray mixture, or rinsate is a violation of Federal law. If these wastes cannot be disposed of by use according to label instructions, contact your State Pesticide or Environmental Control Agency, or the Hazardous Waste representative at the nearest EPA Regional Office for guidance.

Rinse application equipment after use, at least three times with water, and dispose of rinse water in a non-cropland area away from water supplies.

Container Disposal: Do not reuse container. Triple rinse (or equivalent). Puncture and dispose of in a sanitary landfill, or by incineration, or, if allowed by state and local authorities, by burning. If burned, stay out of smoke.

General: Consult federal, state or local disposal authorities for approved alternative procedures.

Be sure that use of this product conforms to all application regulations.

WARRANTY LIMITATIONS AND DISCLAIMER

The Dow Chemical Company warrants that this product conforms to the chemical description on the label and is reasonably fit for the purposes stated on the label when used in strict accordance with the directions therein under normal conditions of use. THIS IS THE ONLY WARRANTY MADE ON THIS PRODUCT. NO OTHER EXPRESS AND NO IMPLIED WARRANTY OF MERCHANTABILITY OR FITNESS FOR A PARTICULAR PURPOSE IS MADE OUTSIDE OF THIS LABEL. Therefore, neither this warranty nor any other warranty of merchantability or fitness for a particular purpose, express or implied, extends to the use of this product contrary to label instructions (including conditions noted on the label, such as unfavorable temperatures, soil conditions, etc.), under abnormal conditions (such as excessive rainfall, drought, tornadoes, hurricanes, etc.) or under conditions not reasonably foreseeable to or beyond the control of seller.

When buyer or user suffers losses or damages resulting from the use or handling of this product (including claims based on contract, negligence, strict liability, or other legal theories), buyer or user must promptly notify in writing The Dow Chemical Company of any claims to be eligible to receive either remedy given below. THE EXCLUSIVE REMEDY OF THE BUYER OR USER and the LIMIT OF LIABILITY of The Dow Chemical Company or any other seller will be one of the following, at the election of The Dow Chemical Company:

(1) **Refund of purchase price** paid by buyer or user for product bought, **or**

(2) **Replacement** of amount of product used.

The seller **will not be liable for consequential or incidental damages or losses.**

The terms of this Warranty Limitations And Disclaimer cannot be varied by any written or verbal statements or agreements. Any employee or sales agent of the seller is not authorized to vary or exceed the terms of this Warranty Limitations And Disclaimer in any manner.

11387-L1 A1089

*Trademark of THE DOW CHEMICAL COMPANY

TORDON* RTU

HERBICIDE

For Controlling Unwanted Trees Via Cut Surface Treatments

ACTIVE INGREDIENT(S):

Picloram (4-amino-3,5,6-trichloropicolinic acid), as the triisopropanol-amine salt	5.4%
2,4-Dichlorophenoxyacetic Acid, as the triisopropanol-amine salt	20.9%
INERT INGREDIENTS:	73.7%
TOTAL	100.0%

Acid Equivalents:
Picloram - 3.0%
2,4-Dichlorophenoxyacetic acid - 11.2%

E.P.A Registration No. 464-510 E.P.A. Est. 464-MI-1

KEEP OUT OF REACH OF CHILDREN

WARNING

AVISO:

PRECAUCIÓN AL USUARIO: Si usted no lee inglés, no use este producto hasta que la etiqueta le haya sido explicada ampliamente.

PRECAUTIONARY STATEMENTS

HAZARDS TO HUMANS AND DOMESTIC ANIMALS

CAUSES SUBSTANTIAL BUT TEMPORARY EYE INJURY • HARMFUL IF SWALLOWED, INHALED OR ABSORBED THROUGH SKIN

Do Not Get In Eyes Or On Clothing • Wear Goggles, Face Shield Or Safety Glasses When Handling • Avoid Contact With Skin • Wash Thoroughly With Soap And Water After Handling • Remove Contaminated Clothing And Wash Before Reuse • Do Not Drink Solution • Avoid Breathing Spray Mist

STATEMENT OF PRACTICAL TREATMENT

If in eyes, flush with plenty of water for at least 15 minutes. Get medical attention. **If on skin,** wash with plenty of soap and water. Get medical attention if irritation persists. **Contains ethylene glycol. If swallowed,** induce vomiting immediately by giving two glasses of water and sticking finger down throat. Do not induce vomiting or give anything by mouth to an unconscious person.

PHYSICAL OR CHEMICAL HAZARDS

COMBUSTIBLE • Do Not Use or Store Near Heat or Open Flame • Do Not Cut or Weld Container.

ENVIRONMENTAL HAZARDS

Do not apply directly to water or wetlands (swamps, bogs, marshes, and potholes) Do not contaminate water used for irrigation or domestic purposes by cleaning of equipment or disposal of wastes. Do not allow run-off or spray to contaminate wells, irrigation ditches or any body of water used for irrigation or domestic purposes. Do not make application when circumstances favor movement from treatment site. See additional precautionary statements elsewhere on this label.

NOTICE

Read the entire label. Use only according to label directions.

Before buying or using this product, read "WARRANTY LIMITATIONS AND DISCLAIMER" elsewhere on this label. If terms are not acceptable, return **unopened** package at once to seller for full refund of purchase price paid. Otherwise, use by the buyer or any other user constitutes acceptance of the terms under "WARRANTY LIMITATIONS AND DISCLAIMER."

IN CASE OF AN EMERGENCY endangering life or property involving this product, call collect 517–636–4400

AGRICULTURAL CHEMICAL

Do Not Ship or Store with Food, Feeds, Drugs, or Clothing

GENERAL INFORMATION

TORDON RTU Herbicide is a **ready-to-use product** that is effective in cut surface applications for killing unwanted trees and preventing undesirable sprouting of cut trees in forest and other non-crop areas such as fence rows and right-of-ways. Among the many species that can be controlled are: **ailanthus, alder, aspen, birch, cedar, cherry, dogwood, elm, firs, green ash, gum, hawthorn, hickory, hornbeam, maples, oaks, pecan, persimmon, serviceberry, sourwood, sweetbay.**

USE DIRECTIONS

It is a violation of Federal law to use this product in a manner inconsistent with its labeling.

Use undiluted as indicated below.

TREE INJECTION TREATMENT: Use 1 milliliter of undiluted TORDON RTU Herbicide solution through the bark completely around the tree trunk at intervals of 2 to 3 inches between edges of the injector wounds. Make injections near ground level when using the tree injector or 2 to 4 feet above the ground when using a HYPOHATCHET Injector or similar device. Treatments can be made in any season. Maples should not be treated during the spring sap flow. With some difficult to control species such as dogwood, hickory, sugar maple, bigleaf maple, tanoak, and some firs, application to a continuous cut rather than to spaced cuts may be more uniformly effective.

FRILL OR GIRDLE TREATMENT: Make a single hack girdle or "frill" of overlapping axe cuts through the bark completely around the tree as close to the ground as feasible. Spray or paint the injured surface with undiluted TORDON RTU Herbicide, using enough volume to wet treated areas.

STUMP TREATMENT: Spray or paint the cut surfaces of freshly cut stumps and stubs with undiluted TORDON RTU Herbicide. The cambium area next to the bark is the most vital area to wet.

USE PRECAUTIONS

Use this product only as specified on this label. TORDON RTU Herbicide is highly active as a herbicide, is water soluble, can move with surface run-off water, and can remain phytotoxic for a year or more if it gets into the soil. Untreated trees occasionally are affected by root uptake of herbicide leaches into the soil or excreted by the roots of nearby treated trees. If such injury cannot be tolerated, do not use TORDON RTU close to desirable trees.

Do not contaminate cropland, water, or irrigation ditches.

Do Not Contaminate Water Intended for Irrigation or Domestic Purposes. To avoid injury to crops or other desirable plants, do not treat or allow spray drift or run-off to fall onto banks or bottoms of irrigation ditches, either dry or containing water, or other channels that carry water that may be used for irrigation or domestic purposes. Do not apply to snow or frozen ground.

Water contaminated with residues of picloram from drainage ditch bank uses shall not be used to irrigate crops which are not registered for use with this chemical.

Do not rotate food or feed crops on treated land if they are not registered for use with picloram until an adequately sensitive bioassay or chemical test shows that no detectable picloram is present in the soil.

Do not apply on or in the vicinity of sensitive crops of desirable broadleaf plants. Small amounts can cause injury to plants.

Do not allow careless application or spray drift. Do not permit any spray or spray drift to contact desirable plants. Use only gravity flow or very low spray pressure.

Do not remove product from original container except to apply as indicated under USE DIRECTIONS.

Do not store near food, feedstuff, fertilizer, seeds, insecticides, fungicides, or other pesticides.

PRECAUTIONARY STATEMENTS

ENVIRONMENTAL HAZARDS

Picloram is a chemical which can travel (seep or leach) through soil and under certain conditions has the potential to contaminate groundwater which may be used for irrigation and drinking purposes. Users are advised not to apply picloram where soils have a rapid to very rapid permeability throughout the profile (such as loamy sand to sand) and the water table of an underlying aquifer is shallow **or** to soils containing sinkholes over limestone bedrock, severely fractured surfaces, and substrates which would allow direct introduction into an aquifer. Your local agricultural agencies can provide further information on the type of soil in your area and the location of groundwater.

An aquifer is defined as "an underground, saturated, permeable, geologic formation capable of producing significant quantities of water to a well or spring. It is the ability of the saturated zone, or portion of that zone, to yield water which makes it an aquifer" (American Chemical Society, 1983).

STORAGE AND DISPOSAL

Do not contaminate water, food or feed by storage or disposal.

Pesticide Disposal: Pesticide wastes are toxic. Improper use of excess pesticide, spray mixture, or rinsate is a violation of Federal law. If these wastes cannot be disposed of by use according to label instructions, contact your State Pesticide or Environmental Control Agency, or the Hazardous Waste representative at the nearest EPA Regional Office for guidance.

Rinse application equipment after use, at least three times with water, and dispose of rinse water in a non-cropland area away from water supplies.

Container Disposal: Do not reuse container. Triple rinse (or equivalent). Puncture and dispose of in a sanitary landfill, or by incineration, or, if allowed by state and local authorities, by burning. If burned, stay out of smoke.

General: Consult federal, state or local disposal authorities for approved alternative procedures.

Be sure that use of this product conforms to all application regulations.

WARRANTY LIMITATIONS AND DISCLAIMER

The Dow Chemical Company warrants that this product conforms to the chemical description on the label and is reasonably fit for the purposes stated on the label when used in strict accordance with the directions therein under normal conditions of use. THIS IS THE ONLY WARRANTY MADE ON THIS PRODUCT. NO OTHER EXPRESS AND NO IMPLIED WARRANTY OF MERCHANTABILITY OR FITNESS FOR A PARTICULAR PURPOSE IS MADE OUTSIDE OF THIS LABEL. Therefore, neither this warranty nor any other warranty of merchantability or fitness for a particular purpose, express or implied, extends to the use of this product contrary to label instructions (including conditions noted on the label, such as unfavorable temperatures, soil conditions, etc.), under abnormal conditions (such as excessive rainfall, drought, tornadoes, hurricanes, etc.) or under conditions not reasonably foreseeable to or beyond the control of seller.

When buyer or user suffers losses or damages resulting from the use or handling of this product (including claims based on contract, negligence, strict liability, or other legal theories), buyer or user must promplty notify in writing The Dow Chemical Company of any claims to be eligible to receive either remedy given below. The EXCLUSIVE REMEDY OF THE BUYER OR USER and the LIMIT OF LIABILITY of The Dow Chemical Company or any other seller will one of the following, at the election of The Dow Chemical Company: (1) **Refund of purchase price** paid by buyer or user for product bought, **or** (2) **Replacement** of amount of product used. The seller **will not be liable for consequential or incidental damages or losses.**

The terms of this Warranty Limitations And Disclaimer cannot be varied by any written or verbal statements or agreements. Any employee or sales agent of the seller is not authorized to vary or exceed the terms of this Warranty Limitations And Disclaimer in any manner.

87244-L2 C1089

*Trademark of THE DOW CHEMICAL COMPANY

E.I. DU PONT DE NEMOURS & CO. (INC.)

The following revised product labels appear: Canopy, Escort, Express, Gemini, Harmony Extra, Lorox Plus, Manzate, Preview, Telar

CANOPY®

HERBICIDE DISPERSIBLE GRANULES

	BY WEIGHT
ACTIVE INGREDIENTS:	75.0%
Metribuzin: [4-Amino-6-1(1,1-dimethylethyl)-3-(methylthio)-1,2,4-triazin-5(4H)-one]	64.3%
Chlorimuron Ethyl: Ethyl 2-[[[[(4-chloro-6-methoxypyrimidin-2-yl) amino]carbonyl]amino]sulfonyl]benzoate	10.7%
INERT INGREDIENTS:	25.0%
TOTAL	100.0%

EPA Reg. No. 352-444

U.S. Pats. 4,346,220, 4,394,506, 4,547,215 & 4,645,530

KEEP OUT OF REACH OF CHILDREN

PRECAUTIONARY STATEMENTS

HAZARDS TO HUMANS AND DOMESTIC ANIMALS

CAUTION! HARMFUL IF SWALLOWED. CAUSES MODERATE EYE INJURY.

Avoid breathing dust or spray mist. Avoid contact with skin, eyes and clothing.

In case of contact with eyes, immediately flush with plenty of water. Get medical attention if irritation persists. Wash thoroughly with soap and water after handling. Remove and wash contaminated clothing before re-use.

STATEMENT OF PRACTICAL TREATMENT

IF IN EYES: Flush eyes with plenty of water. Call a physician if irritation persists.

IF SWALLOWED: Call a physician or Poison Control Center. Drink 1 or 2 glasses of water and induce vomiting by touching back of throat with finger. Do not induce vomiting or give anything by mouth to an unconscious person.

For medical emergencies involving this product, call toll free 1-800-441-3637.

ENVIRONMENTAL HAZARDS

Do not apply directly to water or wetlands. Do not contaminate water by cleaning of equipment or disposal of wastes.

IMPORTANT

Injury to or loss of desirable trees or vegetation may result from failure to observe the following: Do not apply or drain or flush equipment on or near desirable trees or other plants, or on areas where their roots may extend, or in locations where the chemical may be washed or moved into contact with their roots. Do not use on lawns, walks, driveways, tennis courts or similar areas. Prevent drift of spray to desirable plants. Do not contaminate any body of water. Keep from contact with fertilizers, insecticides, fungicides and seeds during storage.

PRIOR TO USING "CANOPY" HERBICIDE, CONSIDERATION SHOULD BE GIVEN TO CROP ROTATION PLANS. Crops other than soybeans may be extremely sensitive to low concentrations of "CANOPY" remaining in the soil the next planting season. Choice of rotation crop is restricted following application of "CANOPY". (See "ROTATIONAL CROP GUIDELINES".) "CANOPY" should not be applied on soils above pH 7.0 except in those states listed in Group III of the "ROTATIONAL CROP GUIDELINES" where application can be made to soils with a pH as high as 7.5.

Thoroughly clean all traces of "CANOPY" from application equipment immediately after use and prior to spraying crops other than soybeans. Failure to remove even small amounts of "CANOPY" from application equipment may result in injury to subsequently sprayed crops.

Metribuzin is a chemical which can travel (seep or leach) through soil and can contaminate groundwater which may be used as drinking water. Metribuzin has been found in groundwater as a result of agricultural use. Users are advised not to apply metribuzin where the water table (groundwater) is close to the surface and where the soils are very permeable, i.e., well drained soils such as loamy sands. Your local agricultural agencies can provide further information on the type of soil in your area and the location of groundwater.

GENERAL INFORMATION

Du Pont "CANOPY" Herbicide is a dispersible granule formulation to be mixed with water and sprayed for selective preemergence and preplant incorporated weed control in soybeans. When applied according to the instructions on this label, it will control many broadleaf weeds and provide partial control of nutsedge.

Preemergence and preplant incorporated applications of "CANOPY" require rainfall or sprinkler irrigation to activate the herbicide. Degree of control and duration of effect depend on: rate used, weed spectrum, growing conditions at and following time of treatment, soil pH, texture, organic matter, moisture and precipitation.

Because most crops are highly sensitive to "CANOPY", all direct or indirect contact (such as spray drift) to crops or to land scheduled to be planted to crops other than soybeans should be avoided.

BIOLOGICAL ACTIVITY

"CANOPY" rapidly inhibits the growth of susceptible weeds. Following application of preplant incorporation or preemergence treatment, susceptible weeds may germinate and emerge, but growth then ceases and leaves become yellow 3-5 days after emergence. Death of leaf tissue and growing point will follow in some species while others will remain green but stunted and noncompetitive. "CANOPY" provides partial control of some annual grasses when used preplant or preemergence but other products may be needed to ensure adequate grass control.

DIRECTIONS FOR USE

It is a violation of federal law to use this product in a manner inconsistent with its labeling. Do not apply this product through any type of irrigation system.

EQUIPMENT—SPRAY VOLUMES

Apply uniformly by ground equipment with a properly calibrated low pressure (20 to 40 psi) fixed-boom sprayer equipped with fan-type nozzles and screens no finer than 50 mesh. Use 10 to 40 gals of water per acre. Continuous agitation in the spray tank is required to keep the material in suspension. Avoid overlap and shut off spray booms while starting, turning, slowing or stopping, as injury to the crop may result.

EARLY PREPLANT OR PREPLANT INCORPORATED (See Rate Chart)

"CANOPY" can be applied at planting or up to 30 days prior to planting either preemergence or preplant incorporated. If incorporated, uniformly incorporate the treatment no deeper than the top 1-2" of soil prior to planting soybeans. Use equipment suited to proper incorporation into the top 1-2 inches of soil, eg: Do-all, field cultivator, roterra. "CANOPY" can be tank mixed with "Command"[10] and applied preplant incorporated. Refer to the "Command" label for specific use instructions. If tank-mixed with a grass herbicide such as a DNA like "Treflan"[1], "Sonalan[1]", or acetanilide such as "Dual"[2], or "Lasso"[3], follow label instructions for proper incorporation of the grass herbicide into the top 1 to 2" of soil. Improper soil incorporation can result in erratic weed control or crop injury. Rainfall* (~1") is required to activate preplant incorporated herbicides such as "CANOPY" before weed emergence.

"CANOPY" + "Treflan," "Prowl,"[4] "Dual" or "Sonalan" — Apply within 2 weeks of planting.
"CANOPY" + "Treflan" - (Early PPI) — Apply within 30 days of planting.
"CANOPY" + "Lasso" — Apply within 7 days of planting.
"CANOPY" + "Command" — Apply within 30 days of planting.

PREEMERGENCE (See Rate Chart)

Preemergence application: Apply before soybeans emerge. Do not spray overtop of emerged soybeans, as severe injury or death of crop will occur.

Preemergence activity of "CANOPY" requires rainfall* to activate. Best results are obtained if treatment is made to moist soil and followed by rainfall or irrigation (~1") before weeds germinate. On dry soil, more moisture is required for activation (1 to 2") before weed emergence. If moisture is insufficient to activate the herbicide, a rotary hoeing or shallow cultivation should be made after emergence of crop while weeds are small enough to be controlled by mechanical means. "CANOPY" can be applied preemergence in combination with "Lasso," "Dual," "Prowl" "Surflan" or following the use of a preplant grass material such as "Treflan" or "Sonalan". Consult labels for rates and use instructions.

*Rainfall required to activate ppi or pretreatments depends on soil moisture when activating moisture is received. Several small rainfalls of less than 1/4" are not as beneficial as one large rainfall of 1/2 inch or more for activation.

MINIMUM TILLAGE OR NO-TILL (See Rate Chart)

"CANOPY" applied preemergence to soybeans but postemergence to broadleaf weeds and small annual grasses will provide some burndown of existing weeds as well as normal preemergence broadleaf weed control. When used for burndown, add 0.25% v/v (1 qt/100 gal) surfactant such as Surfactant WK or crop oil concentrate at 1% v/v (1 qt./100 gal.). For added postemergence control, especially on grasses, "CANOPY" can be applied with "Roundup"[3], paraquat, or "Bronco"[3]. For added preemergence grass control, "CANOPY" can be mixed with "Dual", "Surflan" or "Lasso". For added postemergence broadleaf weed control, consult labels for rates and use instructions. Maintain constant agitation of spray mixture.

RATE CHART

Apply as follows:

Soil Texture		Broadcast Rate (Ounces Per Acre) Organic Matter Percent in Soil ½-3%	Percent in Soil 3-5%
Coarse:	Loamy sand Sandy loam	6	8
Medium:[a]	Loam Silt loam* Silt Sandy clay loam	8	10
Fine:	Silty clay loam Clay loam Clay	10	12

[a] Medium soils in southeast KS, southwest MO and northeast OK with ½-3% O.M., use 5-6 oz/A.

* Silt loam soils in Tennessee and Kentucky use 6-8 oz/A.

Do not apply to soil with less than ½% O.M.

On soils with higher organic matter or heavy pressure from large, deep-germinating weed seeds, an additional 15% more material should be added to the recommended rate for each soil type.

For band applications, use proportionately less than those listed above.

Plant seed 1 ½" to 2" deep on a flat or raised seedbed only or crop injury may occur.

Soybean stunting may occur if excessive rainfall occurs after application but before soybeans germinate. Injury is more prevalent under poor drainage or compacted conditions or when soil is saturated for long periods of time. Soybeans rapidly outgrow stunting once favorable growing conditions return.

Do not apply "CANOPY" to soils with pH greater than 7.0 except in those states as directed in the "ROTATIONAL CROP GUIDELINES" section. Soybean varieties such as Altona, Coker 6955 and 156, Govan, Semmes, Tracy, Vansoy, Terra Vig 505 and 606, AP 71, NKS 1884, Agropro 55, Asgrow 6520, Maple Amber, Portage and Vinton 81 are sensitive to "CANOPY." Injury may occur if "CANOPY" is used on these varieties. Before use on any soybean variety, tolerance to "CANOPY" must first be determined. Injury to soybeans may occur if "CANOPY" is used on soils having a calcareous surface layer or pH greater than 7.5, or if used in conjunction with soil-applied organic phosphate pesticides such as "Dasanit"[5], "Disyston"[5], "Mocap"[6], "Nemacur"[5], "Thimet"[4], parathion, or "Lorsban"[7].

NOTE: Injury may occur if atrazine was applied on the soil the year before use of "CANOPY". Seedling disease, cold weather, deep planting (more than 2"), excessive moisture, high salt concentration, or drought may weaken soybean seedlings and increase possibility of crop injury.

Do not apply to Black Belt Soils of Alabama and Mississippi with a soil pH greater than 7.0 or history of nutrient deficiency such as iron chlorosis, as injury may occur.

WEEDS CONTROLLED

The following weeds will be controlled at the above rates both preplant incorporated and preemergence:

*Cocklebur	Pigweed
Florida beggarweed	Palmer
Hemp sesbania	Redroot
Hophornbeam copperleaf	Smooth
Jimsonweed	Spiny amaranth
Lambsquarters	Tall waterhemp
Mustard, wild	Poinsettia (wild)
*Morningglory, annual	Prickly sida (teaweed)
Ivyleaf	Purslane, common
Entireleaf	Ragweed, common
Pitted	*Ragweed, giant
Smallflower	*Sicklepod
Tall	Smartweed (annual)
	Spotted spurge
	Sunflower
	Velvetleaf

PARTIAL CONTROL:

Barnyardgrass	Johnsongrass, seedling
Yellow nutsedge	Mexicanweed
Purple nutsedge	Burcucumber
Crabgrass	Panicum, Texas and fall
Foxtail, species	Signalgrass, broadleaf

* Large-seeded weeds, germinating deep in the soil such as morningglory, sicklepod, cocklebur and giant ragweed or other weeds which may emerge at various times during the growing season may require a cultivation or an application of Du Pont "Classic" Herbicide for season-long control.

NOTE: Consult "Lasso", "Dual", "Prowl", "Surflan", "Treflan", "Sonalan", or "Command" labels for additional weeds controlled when "CANOPY" is used in conjunction with these herbicides.

SPRAY TANK PREPARATION

Fill tank ¼ full with water. Start agitation system, add "CANOPY" and continue adding water. Add separately each additional component of any tank mix while adding water. Continue agitation throughout. If poor mixing should occur with any component, premix the component with two parts water before adding to the spray tank.

A fertilizer solution may be used in the spray mixture. Small quantities should be tested for compatibility by the following procedures before full-scale mixing.

1. Put 1 pint of fertilizer solution in a quart jar.
2. Mix 2 teaspoons "CANOPY" with 2 tablespoons of water; mix thoroughly and add to fertilizer solution.
3. Close jar and shake well.
4. If other herbicides are to be used in the mixture, premix 2 teaspoons of wettable powder or 1 teaspoon of liquid with 2 tablespoons of water; add to "CANOPY"/fertilizer solution mixture.
5. Close jar and shake well.
6. Watch mixture for several seconds; check again in 30 minutes.
7. If mixture does not separate, foam, gel, or become lumpy, it may be used.
8. Mixing ability may be improved by adding compatibility agents such as Kalo Laboratories' "Compex"[8] or Witco Chemicals' "Sponto"[9] 168D. Follow directions on container.

Provided the above procedure shows the mixture to be compatible, prepare the tank mixture as follows: Add the fertilizer solution to the spray tank first, with the agitator running, add the required amount of "CANOPY" and thoroughly mix. For tank mixtures with other herbicides, follow directions above. For tank mixtures with other herbicides, all applicable directions, restrictions and precautions for the additional herbicides are also to be followed.

Use "CANOPY" spray preparations the same day as mixed or product degradation may occur. Thoroughly reagitate and remix before using, if allowed to settle. When tank mixing with other herbicides, all applicable directions, restrictions and precautions for the additional herbicides are also to be followed.

ROTATIONAL CROP GUIDELINES

The minimum length of time in months before "CANOPY" treated soil can be planted with major crops is given below. Choice of rotational crop is determined by geographical location (state), rate, soil pH and time interval elapsed since last application (either band or broadcast). Guidelines are divided into three groups based on these criteria.

Group I

- States of AL, AR, FL, GA, LA, MS or TX - Soil pH 7.0 or less
- In southeast KS, southwest MO and northeast OK - Soil pH 7.0 or less and use rates of 5.0-6.0 ounces per acre

Group II

- States of DE, KY, MD, MO Bootheel, NJ, NC, PA, SC, TN VA or WV - Soil pH 6.8 or less

  AND

  —Use rate 8-10 oz./A

Group III

- All other states not listed in Groups I or II—Soil pH 7.0 or less OR
- States of AL, AR, FL, GA, LA, MS or TX—Soil pH 7.1-7.5 OR
- States of DE, KY, MD, MO Bootheel, NJ, NC, PA, SC, TN, VA or WV

  —Soil pH 6.9-7.5 OR
  —Fields not meeting rate requirements in Group II.

Minimum length of time in months before planting recrop:

Crops	Group I	Group II	Group III
Soybeans	Anytime	Anytime	Anytime
Barley	4	4	4
Ryegrass	4	4	4
Wheat	4	4	4
Alfalfa	10	10	18
Clover	12	12	18
Field Corn	9	10	18
Cotton	10	10	18
Peanuts	12	12	18
Rice	10	10	18
Sorghum	10	10	18
Tobacco (Transplant)	10	10	18
Tomato (Transplant)	10	10	18

All other crops require a minimum 18 month rotational interval and a successful field bioassay.**

The suggested interval before initiating a field bioassay is 10 months for Group I and 15 months for Groups II and III.

**The season before planting any crop not listed above, a successful field bioassay must be completed. The field bioassay will detect small quantities of "CANOPY" which can remain in the soil and injure rotational crops. A successful field bioassay means growing to maturity a test strip of the crop(s) intended for production the following year. The test strip should cross the whole field including knolls and low areas.

Combination Treatment: When recommended tank mix combinations are used, consult the product labels for replanting and rotational crop instructions and observe the directions that are the most restrictive.

REDUCED RATE FOR BROADLEAF WEED CONTROL IN SOYBEANS IN THE STATES OF IA, IL, IN, KS, MI, MO, MN, NE, OK, OH AND WI

DuPont "CANOPY" Herbicide is recommended preemergence at reduced rates in the states of IA, IL, IN, KS, MI, MO, MN, NE, OK, OH AND WI for broadleaf weed control in soybeans. This treatment will control those weeds listed on the "CANOPY" Herbicide product label.

DIRECTIONS FOR USE

Apply "CANOPY" early preplant, preplant incorporated or preemergence at reduced rates (see rate chart below) as directed on the product label. "CANOPY" can be applied early preplant, preplant incorporated or preemergence in combination with other herbicides such as Lasso, Dual, Prowl, Surflan, Treflan, Sonalan, Commence and Command. Consult the product label for rates, recropping and use instructions of these products. Follow the product label with the most restrictive recropping instructions when using "CANOPY" in combination with other herbicides.

RATE CHART

Reduced rate of "CANOPY" in IA, IL, IN, KS, MI, MO, MN, NE, OK, OH, AND WI	
	Broadcast Rate (Ounces per acre)
Soil Texture	½-4% Organic Matter
Coarse: Loamy sand, Sandy loam	5
Medium: Loam, Silt loam, Silt, Sandy clay loam	6
Fine: Silty clay loam, Clay loam, Clay	6-7

- Do not apply to soils with greater than 6.8 pH.
- Do not apply to soils with less than ½% organic matter or greater than 4% organic matter.

Rotational Crop Guidelines

The following "CANOPY" rates alone and followed by a postemergence "Classic" Herbicide application may be recropped as indicated in the chart below. For "CANPOY" rates other than those listed or for more than one "Classic" application, refer to the "CANOPY" and "Classic" Herbicide labels for rotational crop information.

- "CANOPY" rate of 5 ounces per acre and a single application of "Classic" Herbicide postemergence of up to ¾ ounce per acre.
- "CANOPY" rate of 6 ounces per acre and a single application of "Classic" Herbicide postemergence of no more than ⅔ ounce per acre.
- "CANOPY" rate of 7 ounces per acre and no "Classic" Herbicide postemergence.

Crop	Recropping Interval in Months
Soybeans	Anytime
Barley	4
Ryegrass	4
Wheat	4
Alfalfa	10
Cotton	10
Clover	12
Field Corn	10
Dry Beans	12
Sorghum	12
Rice	10
Tobacco (transplant)	10
Tomato (transplant)	10

REDUCED RATE FOLLOWED BY "CLASSIC" HERBICIDE FOR BROADLEAF WEED CONTROL IN SOYBEANS

DuPont "CANOPY" Herbicide is recommended preplant incorporated or preemergence at reduced rates followed by "Classic" Herbicide postemergence for broadleaf weed control in soybeans. This treatment will control those weeds listed on the "CANOPY" and "Classic" Herbicide product labels.

DIRECTIONS FOR USE

Apply "CANOPY" preplant incorporated or preemergence at reduced rates (see rate chart below) as directed on the product label. Follow the "CANOPY" preplant incorporated or preemergence reduced rate application with a postemergence application of "Classic" Herbicide at its labeled rate when weeds are small and actively growing (usually 14–28 days after preplant incorporated or premergence application of "CANOPY" reduced rate). "CANOPY" can be applied in tank mix with Lasso, Dual, Prowl, Surflan, Treflan, Sonalan, and Command. Read and follow the produce label for use of these products.

"CANOPY" Reduced Rate Followed by "Classic"	
	Broadcast Rate (Ounces Per Acre)
Soil Texture	½–4% Organic Matter
COARSE; Loamy Sand, Sandy Loam	4–6
MEDIUM; Loam, Siltloam, Silt, Sandy Clay loam	4–6
FINE; Silty Clay Loam, Clay Loam, Clay	6–8

- Do not exceed 7 ounces "CANOPY" per acre or apply to soils with greater than 6.8 pH in the states of IA, IL, IN, KS, MI, MO, MN, NE, OK, OH and WI.
- Do not apply to soils with less than ½% organic matter or greater than 4% organic matter.

Rotational Crop Guidelines

(A) States of IA, IL, IN, KS, MI, MO, MN, NE, OK, OH and WI

The following "CANOPY" rates followed by a postemergence "Classic" Herbicide application may be recropped as indicated in the chart below. For "CANOPY" rates other than those listed or for more than one "Classic" Herbicide labels for rotational crop information.

- "CANOPY" rate of 4–5 ounces per acre and a single application of "Classic" Herbicide postemergence of up to ¾ ounce per acre.
- "CANOPY" rate of 6 ounces per acre and a single application of "Classic" Herbicide postemergence of no more than ⅔ ounce per acre.
- "CANOPY" rate of 7 ounces per acre and No "Classic" Herbicide postemergence.

Crop	Recropping Interval in Months
Soybeans	Anytime
Barley	4
Ryegrass	4
Wheat	4
Alfalfa	10
Cotton	10
Clover	12
Field Corn	10
Dry Beans	12
Sorghum	12
Rice	10
Tobacco (transplant)	10
Tomato (transplant)	10

(B) ALL STATES NOT LISTED ABOVE FOLLOW ROTATIONAL AND SOIL pH RESTRICTIONS ON THE "CANOPY" AND "CLASSIC" HERBICIDE LABELS.

SPRAYER CLEANUP

To avoid subsequent injury to crops other than soybeans, immediately after spraying and prior to spraying other crops, thoroughly remove all traces of "CANOPY" from mixing and spray equipment. For more information, consult supplemental bulletin on sprayer cleanout following use of Du Pont "Classic", "CANOPY" or "Gemini" Herbicides.

PRECAUTIONS

Do not apply "CANOPY" Herbicide to soils with a pH greater than 7.0 except in those states in Group III of the "ROTATIONAL CROP GUIDELINES" where applications can be made to soils with a pH as high as 7.5.

CAUTION: Soil pH varies greatly even within the same field; pH variations as much as 2 pH units are common. Determine soil pH by laboratory analysis using a 1:1 soil:water suspension.

Do not graze treated fields or harvest for forage or hay.

Use of "CANOPY" may affect rotational crops. Follow "ROTATIONAL CROP GUIDELINES".

Do not use on sand, muck, or peat soils, nor on soils with less than ½% organic matter, as crop injury may occur.

Do not apply "CANOPY" after the soybean crop has emerged.

Do not apply to Black Belt Soils of Alabama and Mississippi with soil pH greater than 7.0 or on soils with a history of nutrient deficiency such as iron chlorosis or injury may occur.

If "Scepter"[4], "Pursuit"[4] or any product containing the active ingredients, imazaquin or imazethapyr (such as "Squadron"[4], "Partner"[3]) is applied the same year as "CANOPY", do not plant crops other than soybeans for at least 15 months from the last application. Crops listed in the "'CANOPY' Rotational Crop Guidelines" may be planted after 15 months; all other crops require a successful field bioassay.

Do not apply in land that has been or will be treated with "Ally", "Glean" or "Finesse" herbicide in the states of Nebraska, Kansas, North Dakota and South Dakota.

STORAGE AND DISPOSAL

Storage: Store product in original container only, away from other pesticides, fertilizer, food or feed.

Pesticide Disposal: Do not contaminate water, food or feed by storage or disposal. Wastes resulting from the use of this product may be disposed of on site or at an approved waste disposal facility.

Container Disposal: Triple rinse (or equivalent), then offer for recycling or reconditioning, or puncture and dispose of in a sanitary landfill, or by incineration, or, if allowed by state and local authorities, by burning. If burned, stay out of smoke.

NOTICE TO BUYER: Purchase of this material does not confer any rights under patents of countries outside of the United States.

[1] Trademark of Elanco Products Co., Indianapolis, Indiana

[2] Trademark of Ciba-Geigy Corp., Greensboro, North Carolina

[3] Trademark of Monsanto Agricultural Products Co., St. Louis, Missouri
[4] Trademark of American Cyanamid Corp., Princeton, New Jersey
[5] Trademark of Mobay Chemical Corp., Kansas City, Missouri
[6] Trademark of Rhone-Poulenc, Inc., Monmouth Junction, New Jersey
[7] Trademark of Dow Chemical Co., Midland, Michigan
[8] Trademark of Kalo Agricultural Chemicals, Inc., Overland Park, Kansas
[9] Trademark of Witco Chemical Corp., Houston, Texas
[10] Trademark of FMC Corp., Philadelphia, Pennsylvania

NOTICE OF WARRANTY

Du Pont warrants that this product conforms to the chemical description on the label thereof and is reasonably fit for the purpose stated on such label only when used in accordance with the directions under normal use conditions. It is impossible to eliminate all risks inherently associated with the use of this product. Crop injury, ineffectiveness, or other unintended consequences may result because of such factors as weather conditions, presence of other materials, or the manner of use or application, all of which are beyond the control of Du Pont. In no case shall Du Pont be liable for consequential, special or indirect damages resulting from the use or handling of this product. All such risks are assumed by the buyer. DU PONT MAKES NO WARRANTIES OF MERCHANTABILITY OR FITNESS FOR A PARTICULAR PURPOSE NOR ANY OTHER EXPRESS OR IMPLIED WARRANTY EXCEPT AS STATED ABOVE.

AG-3225 8098/8089 4/21/88

This is a specimen label. DuPont is not responsible for the accuracy of the information contained herein. As labels are subject to revision, always carefully read and follow the label on the product container.

ESCORT®

HERBICIDE
DRY FLOWABLE

ACTIVE INGREDIENT	BY WEIGHT
Metsulfuron Methyl [Methyl 2-[[[[(4-methoxy-6-methyl-1,3,5-triazin-2-yl)-amino]carbonyl]amino]sulfonyl]benzoate]	60%
INERT INGREDIENTS:	40%
TOTAL	100%

PACKAGE SIZE 8 - 8 OUNCE BOTTLES/CARTON

EPA Reg. No. 352-439

U.S. Patent 4,383,113

KEEP OUT OF REACH OF CHILDREN
CAUTION

PRECAUTIONARY STATEMENTS

HAZARDS TO HUMANS AND DOMESTIC ANIMALS

CAUTION!

Harmful if absorbed through skin. Causes eye irritation. Avoid contact with skin, eyes or clothing. Avoid breathing dust or spray mist. Wash thoroughly with soap and water after handling. Remove and wash contaminated clothing before reuse.

STATEMENT OF PRACTICAL TREATMENT

In case of contact with eyes, immediately flush with plenty of water. If on skin, wash with plenty of soap and water. Get medical attention if irritation persists.

For medical emergencies involving this product, call toll free 1-800-441-3637.

ENVIRONMENTAL HAZARDS

Keep out of any body of water. Do not contaminate water by cleaning of equipment or disposal of waste.

IMPORTANT

DO NOT USE ON FOOD OR FEED CROPS. Injury to or loss of desirable trees or other plants may result from failure to observe the following: Do not apply "ESCORT" Herbicide (except as recommended), or drain or flush equipment on or near desirable trees or other plants, or on areas where their roots may extend or in locations where the chemical may be washed or moved into contact with their roots.

Do not use on lawns, walks, driveways, tennis courts or similar areas. Keep from contact with fertilizers, insecticides, fungicides and seeds.

Do not apply to any body of water, including streams, irrigation water or wells. Do not apply where runoff water may flow onto agricultural land, as injury to crops may result.

Do not apply "ESCORT" during periods of intense rainfall or to water-saturated soils, or to frozen ground as off-target movement may occur.

Do not apply to impervious substrates such as paved or highly compacted surfaces nor to frozen ground as off-target movement will occur.

Do not apply this product through any type of irrigation system.

Do not allow spray to drift onto adjacent crops or other desirable plants or trees as injury may occur.

In areas where sensitive crops are grown, make applications to soils whose surface has been settled by rain. Do not treat powdery dry soil or light sandy soils, when there is little likelihood of rainfall soon after treatment, as wind may cause off-target movement.

Following an "ESCORT" application, do not use sprayer for application to crops. This is extremely important, as low rates of "ESCORT" can kill or severely injure most crops.

Do not use this product in California nor in the following counties of Colorado: Saguache, Rio Grande, Alamosa, Costilla and Conejos.

GENERAL INFORMATION

Du Pont "ESCORT" Herbicide is a dispersible granule to be mixed in water and applied as a foliar spray for control of unwanted woody plants and annual and perennial broadleaf weeds, basal soil treatment of multiflora rose, for preemergence and postemergence industrial weed control, and for the growth suppression of certain desirable grasses in noncrop areas such as airports, highways, roadsides, utility and pipeline rights-of-ways, petroleum tank farms, storage areas, pumping installations, plant sites, industrial turf, roadside turf, railroad rights-of-way and fencerows. "ESCORT" rapidly stops growth of susceptible plant species; however, typical symptoms (discoloration) may not be noticeable for several weeks after application, depending on growing conditions and susceptibility. Weeds absorb "ESCORT" through both the foliage and roots resulting in visual effects progressing from growth inhibition to reddish-purplish coloration, chlorosis, necrosis, vein discoloration and death of terminal. Initial effects are usually seen 2 or 3 weeks following application; however, the final effects on weeds are evident at about 4 to 6 weeks after application. On brush, ultimate effects occur during the season after treatment.

Warm, moist conditions following treatment promote the activity of "ESCORT", while cold, dry conditions may reduce or delay activity. Weeds and brush hardened off by cold weather or drought stress may not be controlled. Degree of control and duration of effect are dependent on the rate used, sensitivity and size of target species, as well as soil moisture and soil temperature. "ESCORT" is noncorrosive, nonflammable, nonvolatile and does not freeze.

Following use of "ESCORT" on some naturally-occurring biotypes,*, certain weeds listed on this label may not be effectively controlled by this product.

To delay the occurrence of resistant biotypes*, use "ESCORT" in tank mixes and/or sequential treatments with other registered herbicides effective on the same broadleaf weed species. Do not let weed escapes go to seed; time postemergence treatments before seed formation. Respray problem areas in a timely and effective manner using a herbicide with a different mode of action**.

*Biotypes are naturally-occurring individuals of the species which have a slightly different genetic makeup. Resistant biotypes may look exactly the same as susceptible biotypes. Herbicide resistant biotypes are able to survive a use rate several times higher than needed to control susceptible biotypes.

**Mode of action is the chemical interaction that interrupts a biological process necessary for plant growth and development.

Consult your chemical supplier, applicator, consultant, appropriate state agricultural extension service representative or your local Du Pont representatives for specific recommendations.

DIRECTIONS FOR USE

It is a violation of federal law to use this product in a manner inconsistent with its labeling.

"ESCORT" Herbicide should be used only in accordance with recommendations on this label or in separate published Du Pont recommendations available through local suppliers.

Du Pont will not be responsible for losses or damages resulting from the use of this product in any manner not specifically recommended by Du Pont. User assumes all risks associated with such nonrecommended use.

Du Pont "ESCORT" Herbicide is recommended as a postemergence application for control of brush and certain perennial broadleaf weeds and preemergence and postemergence for weed control on noncropland. "ESCORT" is also recommended for use on established, unimproved, industrial turf for weed control. On fescue/bluegrass turf, "ESCORT" is recommended for growth suppression and seedhead inhibition.

For tank mixes, use the most restrictive limitations from labeling of both products.

NON-CROP (INDUSTRIAL) WEED CONTROL

"ESCORT" is recommended for use for general weed control on non-crop, industrial sites such as airports, military installations, fence rows, roadsides and associated rights-of-ways, lumberyards, petroleum tank farms, pipeline and utility rights-of-ways, pumping installations, railroads, storage areas, plant sites and other similar areas including governmental and private lands.

When applied at lower dosage rates (as listed below), "ESCORT" provides short term control of weeds listed; usage at higher rates increases length and degree of weed control. Combinations with other herbicides such as 2,4-D, dicamba, "Karmex" DF, "Hyvar" X and "Krovar" I DF increases the spectrum of weeds controlled. Bare ground objectives can be achieved with higher rates of "ESCORT" plus residual-type companion herbicides.

In states of Montana, North Dakota and South Dakota, "ESCORT" is for use only on railraod rights-of-ways.

"ESCORT" may be applied preemergence or postemergence to the weeds, but for best results, apply postemergence to young, actively growing weeds (less than 2" tall or 2" across) at any time of the year, except when ground is frozen.

"ESCORT" should be used during seasons when rainfall normally occurs as moisture is required to move "ESCORT" into the root zone of weeds before they develop an established root system. Preemergence treatments control or suppress weeds through root uptake, while postemergence works through root uptake and foliar uptake. Under limited rainfall conditions, "ESCORT" may not provide satisfactory weed control.

Treated soils should be left undisturbed to reduce the potential for "ESCORT" movement by soil erosion due to wind or water. Injury to crops may occur when treated soil is blown or moved onto land used to produce corps.

Rate selection is based on weed species, weed size and soil texture. Use the higher rates on established plants and on fine textured soils, and lower rates on smaller weeds and coarse textured soils.

Since the degree of control and duration of effect will vary with the amount of chemical applied, soil texture, soil pH, soil organic matter, weed size, rainfall and other conditions, it is suggested that users limit their first use to small areas.

Apply "ESCORT" in a minimum of 10 gallons of water per acre. Spray volumes of 20 to 40 gallons per acre and pressures of 25 to 35 psi are preferred. Application may also be made with a hand gun using 100 to 300 gallons water per acre (as needed for coverage of weed growth).

Use the following rates of "ESCORT" (including surfactant at 0.25% to 0.5% by volume) to control weeds listed:

1/3 to 1/2 ounce "ESCORT"/acre*

Annual broomweed	Corn Cockle	Plantain
Annual sowthistle	Cow Cockle	Prickly lettuce*
Aster	Crown vetch	Redroot pigweed
Bahiagrass	Dandelion	Redstem filaree
Beebalm	Dogfennel	Rough Fleabane
Bittercress	False chamomile	Shepheardspurse
Bitter sneezeweed	Fiddleneck tarweed	Smallseed falseflax
Blackeyed susan	Field pennycress	Smooth pigweed
Blue Mustard	Flixweed	Sweet clover
Bur buttercup	Goldenrod	Tansymustard
Chicory	Gumweed	Treacle mustard
Clover	Henbit	Tumble Mustard
Cocklebur	Lambsquarters	Wild carrot
Goldenrod	Marestail	Wild lettuce
Common chickweed	Maximillion sunflower	Wild mustard
Common groundsel	Miners lettuce	Wooly croton
Common mullein	Pennsylvania smartweed	Wood sorrell
Common purslane	Plains coreopsis	Yankeeweed
Common yarrow		
Conical catchfly		

* Except resistant biotypes

1/2 to 1 ounce "ESCORT"/acre

Blackberry	Dyer's wood	Hoary cress (whitetop)
Broom snakeweed	Kochia*	Mutliflora rose
Buckhorn plantain	Common Sunflower	Rosering gaillardia
Canada thistle**	Prostrate knotweed	
Curly dock		

1 to 2 ounces* "ESCORT"/acre

Bull thistle	Russian thistle*	Salsify
Field bindweed**	Scouringrush	Musk thistle

3 to 4 ounces

Kudzu

* Except resistant biotypes
** Suppression

SPECIAL WEED PROBLEMS

Kochia, Russian Thistle and Prickly Lettuce:
Since these weeds have biotypes resistant to "ESCORT", tank mixture combinations should be used involving herbicides with different modes of action such as "Karmex" DF, "Hyvar" X, and "Krovar" I DF. Where resistance is known to occur, these weeds should not be allowed to form mature seed but should be treated postemergence with herbicides registered for control of these weeds such as 2, 4-D and dicamba.

TURF (INDUSTRIAL, UNIMPROVED ONLY)

Selective Control of Weeds in Turf
"ESCORT" is recommended for use on unimproved industrial turf where certain grasses are well-established and desired as ground cover. An application of "ESCORT" may cause temporary discoloration of the grasses.

Applications can be made anytime, except when ground is frozen. Apply uniformly with properly calibrated ground equipment, using 10 to 100 gallons of spray solution per acre. Sprayer pressures of 25 to 35 psi are preferred.

Avoid overlapping and shut off spray booms while starting, turning, slowing or stopping when spraying must be taken to prevent drift; add a drift control agent to reduce drift use at least 10 gals water per acre.

NOTE: Do not apply "ESCORT" to turf that is under stress from drought, insects, disease, cold temperatures or poor fertility, as injury may result.

Do not apply to turf less than 1 year old.
Do not use on bahiagrass, as severe injury will result.
Do not exceed 1/2 oz "ESCORT" per acre within 9 months. Use the lower rates for minimum effect (chlorosis) of turf, particularly fescue.

FESCUE/BLUEGRASS: Apply 1/4 to 1/2 ounce "ESCORT" per acre, but do not exceed 1/2 ounce per acre within a nine-month period.

CRESTED WHEATGRASS AND SMOOTH BROME: Apply 1/4 to 1 ounce "ESCORT" per acre for weed control in these grasses. Some chlorosis or stunting may occur following the application.

BERMUDAGRASS: Apply 1/4 to 2 ounces "ESCORT" per acre for weed control.

Some chlorosis or stunting may occur following the application.
These treatments will control the following weeds:

1/4 to 1/3 ounce "ESCORT"/acre

Bittercress	Chickweed	Redstem filaree
Bur buttercup	Chicory	Wild carrot
Blue mustard	Field pennycress	

1/3 to 2 ounces "ESCORT"/acre

(See listing under "NON-CROP WEED CONTROL")

BLUE GAMA, BLUESTEMS (BIG, LITTLE, PLAINS, SAND, WW SPAR), BROMEGRASSES (MEADOW), BUFFALOGRASS, GREEN SPRANGLETOP, INDIANGRASS, KLEINGRASS, LOVEGRASSES, (ATHERSTONE, SAND, WEEPING, WILMAN), ORCHARDGRASS, SIDEOATS GRAMA, SWITCHGRASS (BLACKWELL), WHEATGRASS (BLUEBUNCH, INTERMEDIATE, PUBESCENT SIBERIAN, SLENDER, STREAMBANK, TALL, THICKSPIKE, WESTERN), AND WILDRYE GRASS (RUSSIAN) In states of CO, KS, NE, NM, OK, UT & TX.

Apply 1/10 oz "ESCORT" in a minimum of 10 gals water per acre; add 0.25% surfactant by volume where weeds have emerged.

New Seedings

Apply preplant or preemergence where soil (seed bed) has been cultivated. Do not treat orchardgrass or Russian wildrye grass.

Established grasses

Treat when weeds are in the seedling stage.

WEEDS CONTROLLED

Bur buttercup (testiculate)	Kochia*-**	Prickly lettuce*
Common purslane	Lambsquarters**- common and slimleaf	Russian thistle* **
Common sunflower**	Marestail**	Snow speedwell
Cutleaf eveningprimrose**	Pigweed-redroot and tumble	Tansymustard**
Flixweed**		Trumble mustard (Jim Hill)

* Except resistant biotypes
** Weed suppression is a visual reduction in weed competition (reduced population or vigor) as compared to untreated area. Degree of suppression will vary with size of weed and environmental conditions following treatment.

GROWTH SUPPRESSION AND SEEDHEAD INHIBITION ("Chemical Mowing")

Fescue/Bluegrass:
"ESCORT" is recommended for use on unimproved turf in noncropland areas for growth suppression and seedhead inhibition of well-established turf. Apply 1/4 to 1/2 ounce "ESCORT" anytime from after green-up until seed stalk (boot) appearance. For improved performance, apply 1/4 to 1/3 ounce "ESCORT" plus 1/8 to 1/4 pint "Embark" 2-S[2] per acre. Addition of a nonionic surfactant at 0.25% by volume (1 qt per 100 gals) provides maximum performance, but may temporarily increase chlorosis of turf.

BRUSH CONTROL

Foliar applications should be made after brush is fully leafed until the beginning of fall leaf coloration. Complete coverage of all foliage and stems is required for brush control. Effectiveness may be reduced if rainfall occurs within 24 hours after application.

High Volume Foliar Ground Applications:
Apply as a full coverage spray to foliage, stems and limbs using up to 400 gallons of total spray per acre depending on plant species, height and density of growth and on the type of spray equipment used. On tall, dense brush, it is often necessary to sray from both sides to obtain adequate coverage. Use the following amounts of "ESCORT" per 100 gallons of water (include surfactant at 1 quart/100 gallons) and spray to runoff for control of:

1/2 to 1 Ounce*

Ash	Cherry	Salmonberry
Aspen	Elm	Thimbleberry
Blackberry	Multiflora rose	Willow

1 to 2 Ounces*

Elder	Ocean Spray (Holodiscus)	Red Maple
Oaks		Cottonwood

* Where mixed stands occur, use the higher rate for the most difficult-to-control species.

For broader spectrum of control, "ESCORT" may be tank mixed with Du Pont "Krenite" S, "Arsenal"[1], triglopyr or glyphosate using labeled use rates for each product.

"ESCORT" at 1/2 to 1 oz plus "Krenite" S at 1 1/2 to 3 gals (plus surfactant) per 100 gals water provides improved control of black cherry and red maple; do not exceed 6 gals "Krenite" S per acre.

Low Volume Foliar Ground Applications:
Apply 1 to 3 ounces "ESCORT" per acre in 25 to 50 gals of water plus 0.25% surfactant by volume for control of ash, cherry, elm, winged elm, black locust, red oak, tuliptree and mulberry.

For improved control of black cherry and red maple, apply 1/2 to 2 ozs "ESCORT" plus 1-1/2 to 3 gals "Krenite" S (plus surfactant) per acre.

Spotgun Basal Soil Treatment:
For control of multiflora rose, prepare a spray suspension of "ESCORT" using 1 ounce product per gallon of water. Mix vigorously until "ESCORT" is dispersed and agitate periodically while applying the spray suspension.

Apply the spray preparation with an exact-delivery hand gun applicator. This equipment delivers a thin stream of a predetermined volume when triggered. Apply at the rate of 4 milliters for each 2 feet of rose canopy diameter. Direct the treatment to the soil within 2 feet of the stem union of the rose to be controlled. When treating large plants and more than one delivery of "ESCORT" dispersion is required, make applications on opposite sides of the plant.

Applications should be made from early spring to summer, when the ground is not frozen.

LOBLOLLY PINE PLANTATIONS
(AL, AR, FL, GA, KY, LA, MS, NC, SC, TN, TX AND WV ONLY)

"ESCORT" is recommended for control of certain weeds in established (at least 2 years old) plantations of loblolly pine (only).

For control of blackberry, apply 1/2 to 1 oz "ESCORT" per acre in a minimum of 10 gals water per acre; add surfactant at 0.25% by volume (1 qt/100 gals water). Treat plants when they have reached full leaf but before leaf tissue has hardened in the fall. Treatment will also control many annual broadleaf weeds (see Weed Control).

For broader spectrum brush control "ESCORT" should be tank-mixed with "Velpar" L (or "Velpar") as directed on package labeling.

NOTE: Do not apply where conifers are suffering from loss of vigor caused by insects, disease, drought, winter damage, or other stresses, as injury may result. Excessive injury may result to conifers less than 4 years of age from transplanting on coarse-textured soils, and less than 3 years of age from transplanting on medium and fine-textured soils. Use on gravely or rocky soils and exposed subsoils may cause conifer injury or mortality. Poor weed and brush control may occur from application made when the soil is saturated with water and rain is imminent within 24 hours. Do not use on poorly drained or marshy sites.

SPRAY PREPARATION

Fill the spray tank half full of water and, with agitator running, add the proper amount of "ESCORT". Finish adding the required amount of water. Continuous agitation is required to keep the product in suspension. For postemergence applications, a nonionic surfactant of at least 80% active ingredient may be added to improve wetting and/or contact activity. To minimize drift, a drift control agent may be added at the recommended label rate. "ESCORT" spray preparations may degrade in acidic solutions if not used in 24 hours; they are stable in alkaline solutions. Thoroughly reagitate before using. Tank mixes with other registered herbicides should be tested for compatibility before full scale mixing.

Sprayer Agitation: Use mechanical or bypass agitation to thoroughly mix the spray solution. After initial mixing, do not use excessive agitation when using a surfactant as foaming problems may result. To avoid foaming problems, use antifoam agents.

Sprayer Cleanup: Immediately after spraying thoroughly remove all traces of "ESCORT" from mixing and spray equipment as follows:

1. Drain tank, then flush tank, boom and hoses with clean water for a minimum of 10 minutes.
2. Fill the tank with clean water, then add 1/2 gallon chlorine bleach (containing 5 1/4% sodium hypochlorite) per 100 gallons of water. Turn on sprayer long enough to flush through boom and hoses. Stop spraying, but keep agitator working in the tank for 15 minutes, then drain.
3. Repeat Step 2.
4. Nozzles and screens should be removed and cleaned separately. To remove traces of chlorine bleach, rinse the tank thoroughly with clean water and flush through hoses and boom.

CAUTION: Do not use chlorine bleach with ammonia. All traces of liquid fertilizer containing ammonia, ammonium nitrate or ammonium sulphate must be rinsed with water from the mixing and application equipment before adding chlorine bleach solution. Failure to do so will release a gas with a musty chlorine odor which can cause eye, nose, throat and lung irritation. Do not clean equipment in an enclosed area.

NOTICE TO BUYER: Purchase of this material does not confer any rights under patents of countries outside of the United States.

STORAGE AND DISPOSAL

Storage: Store product in original container, away from other pesticides, fertilizer, food or feed.

Pesticide Disposal: Do not contaminate water, food or feed by storage, disposal or cleaning of equipment. Wastes resulting from the use of this product may be disposed of on site or at an approved waste disposal facility.

Container disposal: Triple rinse (or equivalent) the container and then offer for recycling or reconditioning, or puncture and dispose of in a sanitary landfill, or by incineration, or, if allowed by state and local authorities, by burning. If burned, stay out of smoke.

NOTICE OF WARRANTY

Du Pont warrants that this product conforms to the chemical description on the label thereof and is reasonably fit for purposes stated on such label only when used in accordance with directions under normal use conditions. It is impossible to eliminate all risks inherently associated with the use of this product. Injury to adjacent crops, ineffectiveness or other unintended consequences may result because of such factors as weather conditions, presence of other material, or the manner of use or application, all of which are beyond the control of Du Pont. In no case shall Du Pont be liable for consequential, special or indirect damages resulting from the use or handling of this product. All such risks shall be assumed by the buyer. DU PONT MAKES NO WARRANTIES OF MERCHANTABILITY OR FITNESS FOR A PARTICULAR PURPOSE NOR ANY OTHER EXPRESS OR IMPLIED WARRANTY EXCEPT AS STATED ABOVE.

[1] Registered trademark of American Cyanamid Company.
[2] Registered trademark of 3M Agricultural Products.

AG-3287 8109 10/3/89.

This is a specimen label. DuPont is not responsible for the accuracy of the information contained herein. As labels are subject to revision, always carefully read and follow the label on the product container.

EXPRESS®

HERBICIDE
DRY FLOWABLE

ACTIVE INGREDIENT:	BY WEIGHT
Methyl 2-[[[[N-(4-methoxy-6-methyl-1,3,5-triazin-2-yl)-methylamino]carbonyl]amino]sulfonyl]benzoate	75%
INERT INGREDIENTS	25%
TOTAL	100%

EPA Reg. No. 352-509 U.S. Patents 4,383,113 & 4,740,234

KEEP OUT OF REACH OF CHILDREN
CAUTION

PRECAUTIONARY STATEMENTS

HAZARDS TO HUMANS AND DOMESTIC ANIMALS

CAUTION!

Avoid contact with skin, eyes and clothing. In case of contact with eyes, immediately flush with plenty of water. Get medical attention if irritation persists. Wash thoroughly after handling. Remove contaminated clothing and wash before reuse.

For medical emergencies involving this product, call toll free 1-800-441-3637.

ENVIRONMENTAL HAZARDS

Do not apply directly to water or wetlands (swamps, bogs, marshes and potholes). Do not contaminate water when disposing of equipment wash-waters.

DIRECTIONS FOR USE

It is a violation of federal law to use this product in a manner inconsistent with its labeling.

Do not apply this product through any type of irrigation system.

"EXPRESS" Herbicide should be used only in accordance with recommendations on this label or in separately published Du Pont recommendations.

Du Pont will not be responsible for losses or damages resulting from the use of this product in any manner not specifically recommended by Du Pont.

IMPORTANT INFORMATION

Injury to or loss of desirable trees or vegetation may result from failure to observe the following: Do not apply or drain or flush equipment on or near desirable trees or other plants, or on areas where their roots may extend, or in locations where the chemical may be washed or moved into contact with their roots. Do not use on lawns, walks, driveways, tennis courts or similar areas. Prevent drift of spray to desirable plants. Do not contaminate any body of water. Thoroughly clean all traces of Du Pont "EXPRESS" Herbicide from application equipment immediately after use. (Refer to "SPRAYER CLEANUP" section of this label.) Failure to follow these procedures may result in injury to subsequently sprayed crops.

GENERAL INFORMATION

Du Pont "EXPRESS" Herbicide is recommended for selective postemergence control of certain broadleaf weeds in wheat (including durum) and barley. Do not use "EXPRESS" on any other crop. "EXPRESS" is a dry flowable granule to be mixed in water or other recommended carrier and applied as a uniform broadcast spray. It is noncorrosive, nonflammable, nonvolatile and does not freeze.

Best results are obtained when "EXPRESS" is applied to young, actively-growing weeds. The use rate will depend on weed spectrum and size of weed at time of application. The degree of control and duration of effect are dependent on rate used, sensitivity and size of target weed and environmental conditions at the time of and following application.

"EXPRESS" stops growth of susceptible weeds rapidly. However, typical symptoms of dying weeds (discoloration) may not be noticeable for 1-3 weeks after application (2-5 weeks for wild garlic) depending on the environmental conditions and weed susceptibility. Warm, moist conditions following treatment promote the activity of "EXPRESS", while cold, dry conditions delay the activity. Weeds hardened-off by cold weather or drought stress will be less susceptible.

A vigorous growing crop will aid weed control by shading and providing competition for weeds. However, a dense crop canopy at time of application can intercept spray and result in reduced weed control. Weeds may not be adequately controlled in areas of thin crop stand or seeding skips.

Naturally occurring weed biotypes* that are resistant to "Glean", "Finesse" or "Ally" will also be resistant to "EXPRESS". In areas where these weed biotypes are known to exist, only spray "EXPRESS" in tank mixtures with other broadleaf herbicides having a different mode of action** such as: 2,4-D, "Banvel"[1]***, "Buctril"[2], "Bronate"[2], "Curtail"[3], MCPA, or "Lexone". Refer to TANK MIXTURES.

* Biotypes are naturally-occurring individuals of the species which have a slightly different genetic makeup. Resistant biotypes may look exactly the same as susceptible biotypes. Herbicide-resistant biotypes are able to survive a use rate several times higher than needed to control susceptible biotypes.

** Mode of action is the chemical interaction that interrupts a biological process necessary for plant growth and development.

*** Tank mixes with "Banvel" may result in reduced control of some broadleaf weeds.

Read and follow all appropriate sections of label, including precautions, before using this product.

WEED CONTROL IN WHEAT AND BARLEY

CROP STAGE AT APPLICATION

Make applications after the crop is in the 2-leaf stage, but before the flag leaf is visible. Do not harvest sooner than 45 days after the last application of "EXPRESS".

APPLICATION RATES

Apply 1/6 to 1/3 ounce "EXPRESS" per acre to wheat (including durum) or barley for control or partial control of the weeds listed below.

Use 1/3 ounce "EXPRESS" per acre when weed infestation is heavy and predominantly consists of those weeds listed under partial control, or when application timing and environmental conditions are marginal (refer to APPLICATION AND ENVIRONMENTAL CONDITIONS FOR OPTIMUM PERFORMANCE).

Use 1/6 to 1/4 ounce "EXPRESS" per acre when weed infestation is light and predominantly consists of those weeds listed under weeds controlled, and when optimum application conditions occur.

Sequential treatments of "EXPRESS" may be made provided the total amount of "EXPRESS" applied to the crop does not exceed 1/3 ounce per acre; for example, 1/6 ounce followed by 1/6 ounce. The final application must be made before the flag leaf is visible. Do not harvest sooner than 45 days after the last application of "EXPRESS".

WEEDS CONTROLLED

Black mustard	Kochia
Blue/purple mustard	Mayweed chamomile
Bushy wallflower/Treacle mustard	Miners lettuce
Coast fiddleneck	Pineappleweed
Common chickweed	Prickly lettuce
Common groundsel	Russian thistle
Common lambsquarters	Slimleaf lambsquarters
Corn spurry	Tarweed fiddleneck
False chamomile	Wild chamomile
Field pennycress	Wild mustard
Flixweed	

PARTIAL CONTROL*

Annual sowthistle	Redroot pigweed
Canada thistle**	Shepherdspurse
Common sunflower	Tansymustard
Hairy nightshade	Tumble/Jim Hill mustard
Henbit	Wild buckwheat
Pennsylvania smartweed	Wild garlic
Prostrate knotweed	Wild radish
Redmaids	

*Partial Control: A visual reduction of weed population as well as a significant loss of vigor for individual weed plants. For best results, use 1/3 ounce "EXPRESS" per acre.
**See SPECIFIC WEED PROBLEMS for more information.

SPECIFIC WEED PROBLEMS

Canada thistle: For best results apply 1/3 oz. per acre plus surfactant in the spring when all thistles have emerged above the soil, are actively growing, and are 4 to 8 inches tall with 2 to 6 inches of new growth.

APPLICATION AND ENVIRONMENTAL CONDITIONS FOR OPTIMUM PERFORMANCE

Crop Stage: Make applications after the crop is in the 2-leaf stage, but before the flag leaf is visible. Do not harvest sooner than 45 days after the last application of "EXPRESS".

Since thorough coverage is required, avoid crop canopy obstruction of the spray contacting the weed foliage.

Crop Competition: A vigorous growing crop will aid weed control by shading and providing competition for weeds. However, a dense crop canopy at time of application can intercept spray and result in reduced weed control. Weeds may not be adequately controlled in areas of thin crop stand or seeding skips.

Pest Stage: Since "EXPRESS" has very little or no soil activity, only those weeds that have germinated above the soil surface will be controlled. Consequently, application of "EXPRESS" should be made when all or the majority of weeds have germinated. Annual broadleaf weeds should be past the cotyledon stage, actively growing, and less than 4 inches tall or across. Wild garlic plants should be 6 to 12 inches tall with 4 to 6 inches of new growth. See SPECIFIC WEED PROBLEMS for more information on Canada thistle.

Application: Foliar absorption is the primary means of "EXPRESS" uptake by plants; therefore, thorough spray coverage of weeds is essential.

Weather: Conditions which are conducive to healthy, actively growing plants optimize "EXPRESS" weed control performance. Ideal conditions include warm temperatures (70 Deg.F. or more) and adequate soil moisture before, during and immediately after application.

Avoid making applications of "EXPRESS" to weeds when rainfall is threatening. Rainfall immediately after treatment can wash "EXPRESS" off weed foliage and result in reduced weed control effectiveness. Several hours of dry weather are needed to allow "EXPRESS" to be absorbed by weed foliage.

Surfactant: Use only EPA approved surfactants authorized for use on food crops.

Water Spray Carrier - Unless specified otherwise, add a nonionic surfactant of at least 80% active ingredient strength at 0.25% vol/vol (1 quart per 100 gallons of spray solution).

Liquid N Carrier - Always use surfactant when applying "EXPRESS" in liquid nitrogen fertilizer. When "EXPRESS" is applied using liquid nitrogen fertilizer solution as a spray carrier, early, temporary, crop yellowing and stunting may occur. Use 0.125 - 0.25% vol/vol surfactant (1 pint - 1 quart per 100 gallons of spray solution) when applying "EXPRESS" in liquid nitrogen fertilizer solution. Refer to TANK MIXTURES for surfactant information when tank mixing with other products.

EQUIPMENT-SPRAY VOLUMES

Prior to spraying "EXPRESS", thoroughly clean sprayer (tanks, booms, hoses, nozzles, mixing cones, tanks, etc.).

Apply using properly calibrated air or ground equipment. Select a spray volume and delivery system that will insure thorough coverage and a uniform spray pattern. Avoid overlapping, and shut off spray booms while starting, turning, slowing or stopping, or injury to the crop may result.

Do not use equipment and/or spray volumes that will cause spray to drift onto nontarget sites. Do not make applications during weather conditions which cause spray to drift onto nontarget sites. Refer to CAUTION-AVOID SPRAY DRIFT.

Do not apply "EXPRESS" through any type of irrigation system.

Spray Equipment: Refer to specific manufacturer's recommendations for additional information on spray volume (GPA), pressure, speed, nozzle types and arrangements, nozzle heights above the target area, etc., for respective application equipment.

Ground Application: For optimum spray distribution and thorough coverage, use flat fan or low volume flood nozzles. For flat fan nozzles, do not use less than 5 gallon spray volume per acre (GPA).

For flood nozzles on 30-inch spacing, use not less than 10 GPA and no larger than TK10 or equivalent and not less than 30 psi. On 40-inch nozzle spacings, use not less than 13 GPA or not less than 20 GPA when nozzles are on a 60-inch spacing. 100% overlapping of nozzle spray pattern is essential for 30, 40 and 60-inch spacings.

When using liquid nitrogen solution as a spray carrier, for flood nozzles use not less than 30 GPA and no larger than TK20. See statements on liquid N solutions below for additional information.

"Raindrop" nozzles[4] are not recommended for "EXPRESS" applications as weed control performance may be reduced.

Use 50-mesh screens or larger.

Aerial Application: Use orifice discs, cores and nozzle types and arrangements that will provide for optimum spray distribution and maximum coverage at 3 to 5 GPA. Do not apply during inversion conditions, when winds are gusty, or when other conditions will favor poor coverage and/or off target spray movement.

For aerial application in the state of Washington, refer to and follow the directions on the Washington special local need label, *"Express" Herbicide Aerial Application To Wheat and Barley in the State of Washington.*

Liquid Nitrogen Fertilizer Solutions: "EXPRESS" can be added directly to liquid N solutions (for example 28-0-0, 32-0-0). Use 0.125 - 0.25% vol/vol surfactant (1 pint - 1 quart per 100 gallons of spray solution) when applying "EXPRESS" in liquid nitrogen fertilizer solution. The addition of liquid nitrogen fertilizer to the spray solution, or the use of liquid nitrogen fertilizer as the total carrier of the spray solution will increase the weight of the spray solution as compared to water (see table below). Consequently, appropriate conversions of spray nozzles and/or pressure must be made in order to maintain proper spray volumes. Refer to the conversion chart below for guidance (taken from Spray Systems Company Catalog 39, page 2).

Influence of adding liquid 28% N fertilizer to spray solution weight per gallon:

% of Spray Solution		
Water	**28% N**	**Weight of Spray Solution (Lbs/Gallon)**
100	0	8.34
50	50	9.5
0	100	10.6

SPRAYING SYSTEMS CONVERSION CHART

SPRAY SOLUTIONS OTHER THAN WATER--NEW METHOD

Since all the tabulations are based on spraying water, which weighs 8.34 lbs per USA gallon, conversion factors must be used when spraying solutions which are heavier or lighter than water. To determine the proper size nozzle for the solution to be sprayed, first multiply the desired GPM or GPA of solution by the water rate conversion factor. Then use the new converted GPM or GPA rate to select the proper size nozzle.

Example: 10 GPA (28%N) × 1.13 = 11.3 GPA (water)

Weight of Solution	Specific Gravity	Conversion Factors
7.0 lbs per gallon	0.84	0.92
8.0 lbs per gallon	0.96	0.98
8.34 lbs per gallon	1.00	1.00
9.0 lbs per gallon	1.08	1.04
10.0 lbs per gallon	1.20	1.10
10.65 lbs per gallon-28% nitrogen	1.28	1.13
11.0 lbs per gallon	1.32	1.15
12.0 lbs per gallon	1.44	1.20
14.0 lbs per gallon	1.68	1.30

Published in Spraying Systems Company Catalog 39, page 2.

Agitation: Continuous agitation is required to keep "EXPRESS" in suspension.

CAUTION - AVOID SPRAY DRIFT

Follow these practices to minimize drift.

Do not allow spray from either ground or aerial equipment to drift onto adjacent crops or land, as even small amounts will injure other plants. When spraying near adjacent, sensitive crops or plants, do everything possible to reduce spray drift. This includes:

- Stop spraying wind speed becomes excessive. DO NOT SPRAY IF WIND SPEED IS 10 MPH OR GREATER. Spray drift can occur at wind speeds less than 10 MPH. If sensitive crops or plants are downwind, extreme caution must be used even in relatively low wind conditions! DO NOT SPRAY IF WINDS ARE GUSTY.
- High temperatures, drought, and low relative humidity increase the possibility of harmful spray drift. EXTREME CAUTION MUST BE USED WHEN THESE CONDITIONS ARE PRESENT AND SENSITIVE CROPS OR PLANTS ARE NEARBY, REGARDLESS OF WINDSPEED.
- Do not apply when temperature inversion exists. An inversion is characterized by low air movement and an increase in air temperature with an increase in altitude. In humid regions, a fog or mist may form. An inversion may be detected by producing a smoke column and checking for a layering effect. Smoke-producing devices on aircraft are recommended. If not sure whether inversion conditions are present, consult with local weather services before making an application.
- Postemergence grass herbicides (such as "Avenge"[5] and "Assert"[5]) are often applied using high pressure. When "EXPRESS" is tank mixed with these products, do not exceed 40 psi.
- Drift from aerial or ground equipment may be further reduced by:

  1. Using large droplet size sprays to minimize drift. DO NOT APPLY WITH HOLLOW-CONE INSECTICIDE NOZZLES ON GROUND EQUIPMENT. Do not use nozzles that produce small droplets, such as Sprayfoil[6] or airblast-type nozzles. Nozzles should be oriented at an angle between straight down and straight back for ground applications. For aerial applications orient nozzles straight back along the windstream. If using flood-type nozzles on aircraft, orient them so spray is produced in direction of the airstream.
  2. Increasing volume of spray mix per acre (for example, minimum GPA by air, 10 GPA by ground) by using higher flow rate nozzles.
  3. Reducing pressure (PSI). DO NOT EXCEED 40 PSI when applying "EXPRESS". [Vehicle speed must also be reduced to maintain spray mix volume per acre.] Consult manufacturer's catalogs for details on correct calibration.
  4. Apply as close to target plants as possible while still maintaining a good spray pattern.

TANK MIXTURES

2,4-D (amine or ester) or MCPA (amine or ester) - Use "EXPRESS" plus 1/8 to 3/8 lb active ingredient 2,4-D or MCPA (ester formulations have provided best results). Surfactant may be added at 0.125 - 0.25% vol/vol (1 to 2 pints per 100 gallons of spray volume); however, the addition of surfactant may increase the chance of crop injury. Use the 1 to 2 pint rate of surfactant with 1/8 lb active ingredient of 2,4-D or MCPA. Use the 1 pint rate of surfactant with 1/4 to 3/8 lb active ingredint of 2,4-D or MCPA. Higher rates of 2,4-D or MCPA may be used, but do not exceed highest rate allowed on the label.

Always mix "EXPRESS" in water prior to adding 2,4-D or MCPA and surfactant. Read and follow all label instructions on timing, precautions and warnings for these herbicides prior to using these tank mixtures.

"EXPRESS" may be tank mixed with other suitable registered herbicides (for example "Lexone") to control weeds other than those listed. Read and follow all manufacturer's label recommendations for the companion herbicide. If these recommendations conflict with this "EXPRESS" label , do not use as a tank mix with "EXPRESS".

"EXPRESS" will not control wild oats or other grasses. For wild oat control, "EXPRESS" can be tank mixed with "Avenge" or "Assert". When tank mixing "EXPRESS" and "Assert", ALWAYS include another broadleaf weed herbicide with a different mode of action, for example: MCPA ester, "Bronate". DO NOT tank mix with "Hoelon"[7] 3EC as grass weed control may be reduced.

Always mix "EXPRESS" in water PRIOR to adding other products, including surfactants. When mixing "EXPRESS" in liquid nitrogen fertilizer solutions (for example 28-0-0, 32-0-0), simply add the "EXPRESS" directly to the liquid fertilizer with the agitator running.

Refer to SPRAY PREPARATION for further information.

Tank mixes of "EXPRESS" plus "Banvel" may result in reduced control of some broadleaf weeds.

Tank mixes of "EXPRESS" plus metribuzin may result in reduced control of wild garlic.

"EXPRESS" may be tank mixed or used sequentially with insecticides registered for use on cereal grains. However, under certain conditions, (drought stress, crop in 2-4 leaf stage) tank mixes or sequential applications of "EXPRESS" plus organophosphate insecticides (such as parathion) may produce temporary crop yellowing or, in severe cases, crop injury. Limit first use to a small area before treating large areas.

DO NOT USE "EXPRESS" PLUS MALATHION OR "LORSBAN" 3 as crop injury will result

SPRAY PREPARATION

Mix the proper amount of "EXPRESS" into the necessary volume of water in the spray tank with the agitator running, then add the companion herbicide to the tank after all the "EXPRESS" is in suspension. ALWAYS MIX "EXPRESS" IN WATER FIRST, PRIOR TO ADDING OTHER PRODUCTS IN THE SAME SPRAY TANK.

Use the spray preparation of "EXPRESS" within 24 hours as product degradation may occur. If spray preparation is left standing, thoroughly agitate before reusing.

Surfactant: Unless specified otherwise, add a nonionic surfactant of at least 80% active ingredient strength at 0.25% vol/vol (1 quart per 100 gallons of spray solution). Use only EPA approved surfactants authorized for use on food crops.

Liquid Nitrogen Fertilizer: Liquid nitrogen fertilizer solutions such as 28-0-0 or 32-0-0, may be used as a carrier in place of water. To apply "EXPRESS" with liquid nitrogen fertilizer, simply add the "EXPRESS" directly to the liquid fertilizer with the agitator running (for at least 3 to 5 minutes). For fertilizer solutions that contain more than just nitrogen, for example 10-34-0, slurry the desired amount of "EXPRESS" in a clean bucket using water until a flowable mixture is produced. Add this slurry to the agitating spray tank of liquid nitrogen fertilizer solution. Thoroughly rinse all of the "EXPRESS" slurry into the spray tank. Do not use "EXPRESS" in liquid fertilizer solutions of less than pH 3. Run a tank mix compatibility test before mixing "EXPRESS" in fertilizer solution.

Use 0.125 - .025% vol/vol surfactant (1 pint - 1 quart per 100 gallons of spray solution) when applying "EXPRESS" in liquid nitrogen fertilizer solution.

Use of "EXPRESS" and a surfactant in liquid nitrogen fertilizer solutions may cause early, temporary crop injury (discoloration and stunting).

When the spray solution contains liquid nitrogen fertilizer, the weight per gallon of solution varied significantly from the weight of water (8.34 lbs per USA gallon). Consequently, liquid nitrogen fertilizer spray solutions must use the appropriate conversion in order to insure proper spray volume. See EQUIPMENT-SPRAY VOLUMES for further information.

CROP ROTATION

Any crop may be planted 60 days after the application of "EXPRESS".

SPRAYER CLEANUP

To avoid subsequent injury to crops other than wheat or barley, immediately after spraying thoroughly remove all traces of "EXPRESS" from mixing and spray equipment as follows:

1. Drain tank; rinse interior surfaces of tank; then flush tank, boom and hoses with clean water for a minimum of 5 minutes.
2. Fill the tank full with clean water, then add cleaning solution*. Flush solution through boom, hoses and nozzles, then allow to sit for 15 minutes with agitation running; then drain.
3. Repeat Step 2.
4. Repeat Step 1.
5. Nozzles and screens should then be removed and cleaned separately. To remove traces of cleaning solution, rinse the tank thoroughly with clean water and flush through hoses and boom.
6. Flush boom and hoses with clean water for 5 minutes just prior to using the sprayer for the first time after the "EXPRESS" application.

*Use any one of the following cleaning solutions:

1. One gallon ammonia (containing 3% active) per 100 gallons of water.
2. Nutra-Sol (carefully read and follow Nutra-Sol label directions).
3. Loveland Spray Tank Cleaner 9 (carefully read and follow Loveland Spray Tank Cleaner label directions).
4. Tank-Aid 10 (carefully read and follow Tank-Aid label directions).

PRECAUTIONS

The total rate of "EXPRESS" cannot exceed 1/3 ounce product per acre applied to any one crop during one growing season.

Do not harvest sooner than 45 days after the last application of "EXPRESS".

Do not graze or feed forage or hay from treated areas to livestock (harvested straw may be used for bedding and/or feed).

Do not plant to any crop other than wheat or barley for 60 days after application of "EXPRESS".

Do not apply to wheat or barley that is stressed by severe winter, drought, water-saturated soil, disease or insect damage as crop injury may result. Under certain conditions such as prolonged cool weather (daily high temperature less than 50 Deg.F.) or wide fluctuations in day/night temperatures just prior to or soon after treatment, temporary yellowing and/or crop stunting may occur. Risk of injury is greatest when crop is in the 2 to 5-leaf stage.

Do not apply to wheat or barley crops underseeded with another crop.

"EXPRESS" is only registered on wheat and barley. Do not use on any other crop.

For ground applications applied to weeds when dry, dusty field conditions exist, control of weeds in wheel track areas may be reduced.

The "EXPRESS" volumetric measuring cylinder is to be used only as a guide, since the degree of accuracy varies by plus or minus 7.5%. For more precise measurement, use scales calibrated in ounces.

STORAGE AND DISPOSAL

STORAGE: Store product in original container only, away from other pesticides, fertilizer, food or feed.

PRODUCT DISPOSAL: Do not contaminate water, food or feed by storage or disposal. Waste resulting from the use of this product may be disposed of on site or at an approved waste disposal facility.

CONTAINER DISPOSAL: Triple-rinse (or equivalent). Then offer for recycling or reconditioning, or puncture and dispose of in a sanitary landfill, or by incineration, or, if allowed by state and local authorities, by burning. If burned, stay out of smoke.

1 Registered trademark of Sandoz Crop Protection Corporation.
2 Registered trademark of Rhone - Poulenc Ag Company.
3 Registered trademark of the Dow Chemical Company.
4 Registered trademark of Delavan Corporation.
5 Registered trademark of American Cyanamid Company.
6 Registered trademark of D. AND W. INDUSTRIES.
7 Registered trademark of Hoechst-Roussel Agri-Vet Company.
8 Sold by Thomas G. Kilfoil, San Bruno, CA.
9 Registered trademark of Loveland Industries, Inc.
10 Manufactured for Cornbelt Chemical Company

NOTICE TO BUYER: Purchase of this material does not confer any rights under patents of countries outside of the United States.

NOTICE OF WARRANTY

Du Pont warrants that this product conforms to the chemcial description on the label thereof and is reasonably fit for purposes stated on such label only when used in accordance with directions under normal use conditions. It is impossible to eliminate all risks inherently associated with the use of this product. Crop injury, ineffectiveness, or other unintended consquences may result because of such factors as weather conditions, presence of other materials, or the manner of use or application, all of which are beyond the control of Du Pont. In no case shall Du Pont be liable for consequential, special or indirect damages resulting from the use of handling of this product. All such risks shall be assumed by the buyer. DU PONT MAKES NO WARRANTIES OF MERCHANTABILITY OR FITNESS FOR A PARTICULAR PURPOSE NOR ANY OTHER EXPRESSED OR IMPLIED WARRANTY EXCEPT AS STATED ABOVE.

AG - 3270 8079/8109 6/30/89

This is a specimen label. DuPont is not responsible for the accuracy of the information contained herein. As labels are subject to revision, always carefully read and follow the label on the product container.

GEMINI®

HERBICIDE
DISPERSIBLE GRANULES

ACTIVE INGREDIENTS: 60%
Linuron: [3-(3,4-dichlorophenyl)-1-methoxy-1-methylurea] . . 55.4%
Chlorimuron Ethyl: Ethyl 2-[[[[(4-chloro-6-methoxy-pyrimidin-2-yl)amino]- carbonyl]amino]sulfonyl]benzoate . . . 4.6%
INERT INGREDIENTS: 40%

TOTAL. 100%

EPA Reg. No. 352-544

U.S. Pat. 4,394,506 & 4,547,215

KEEP OUT OF REACH OF CHILDREN
CAUTION

PRECAUTIONARY STATEMENTS

HAZARDS TO HUMANS AND DOMESTIC ANIMALS

CAUTION!

HARMFUL IF SWALLOWED OR ABSORBED THROUGH SKIN. CAUSES MODERATE INJURY. AVOID CONTACT WITH SKIN, EYES AND CLOTHING. WASH THOROUGHLY WITH SOAP AND WATER AFTER HANDLING.

In case of eye contact, flush with plenty of water and get medical attention if irritation persists.

In case of skin contact, wash with soap and water and get medical attention if irritation persists. Wash thoroughly with soap and water after handling. Remove and wash contaminated clothing before reuse.

If swallowed, call a physician or Poison Control Center. Drink 1 or 2 glasses of water and induce vomiting by touching back of throat with finger. Do not induce vomiting or give anything by mouth to an unconscious person.

For medical emergencies involving this product, call toll free 1-800-441-3637.

ENVIRONMENTAL HAZARDS

Do not apply directly to water or wetlands (swamps, bogs, marshes and potholes). Do not contaminate water by cleaning of equipment or disposal of wastes.

IMPORTANT

When mixing/loading and spraying, wear a wide-brimmed hat, mid-forearm to elbow length chemical resistant gloves, and one-piece overalls which have long sleeves and long pants. Overalls should be constructed of finely woven fabric as specified in the USDA/EPA Guide for Commercial Applicators.

Any article of clothing worn while applying product must be cleaned before reusing. Clothing which has been drenched or has otherwise absorbed concentrated pesticide must be disposed of in a sanitary landfill, or incinerated, or, if allowed by state and local authorities, by burning.

Instead of clothing and equipment specified above, the applicator can use an enclosed tractor cab which provides a filtered air supply (as specified by Taschenberg and Bourke, 1975).

Injury to or loss of desirable trees or vegetation may result from failure to observe the following: Do not apply or drain or flush equipment on or near desirable trees or other plants, or on areas where their roots may extend, or in locations where the chemical may be washed or moved into contact with their roots. Do not use on lawns, walks, driveways, tennis courts, or similar areas. Prevent drift of spray to desirable plants. Do not contaminate any body of water. Keep from contact with fertilizers, insecticides, fungicides, and seeds during storage.

PRIOR TO USING "GEMINI", CONSIDERATION SHOULD BE GIVEN TO CROP ROTATION PLANS. Crops other than soybeans can be extremely sensitive to low concentrations of "Gemini" remaining in the soil the next planting season. Choice of rotational crop is restricted following application of "Gemini". (See "ROTATIONAL CROP GUIDELINES".)

Thoroughly clean all traces of "Gemini" from application equipment immediately after use and prior to spraying crops other than soybeans. Failure to remove even small amounts of "Gemini" from application equipment may result in injury to subsequently sprayed crops.

GENERAL INFORMATION

Du Pont "Gemini" Herbicide is a dispersible granule formulation to be mixed with water and sprayed for selective preemergence weed control in soybeans. When applied according to the instructions on this label, it will control many broadleaf weeds and provide partial control of nutsedge and annual grasses.

Preemergence and preplant Incorporated applications of "Gemini" require rainfall or sprinkler irrigation to activate the herbicide. Degree of control and duration of effect depend on: rate used, weed spectrum, growing conditions at and following time of treatment, soil pH, texture, organic matter, moisture, and precipitation.

Because most crops are highly sensitive to "Gemini", all direct or indirect contact (such as spray drift) to crops or to land scheduled to be planted to crops other than soybeans should be avoided.

BIOLOGICAL ACTIVITY

"Gemini" rapidly inhibits the growth of susceptible weeds. Following activation of preplant incorporated or preemergence treatment, susceptible weeds may germinate and emerge, but growth then ceases and leaves become yellow 3-5 days after emergence. Death of leaf tissue and growing point will follow in some species while others remain green but stunted and noncompetitive. "Gemini" provides partial control of some annual grasses when used preplant or preemergence but other products may be needed to ensure adequate grass control.

DIRECTIONS FOR USE

It is a violation of federal law to use this product in a manner inconsistent with its labeling.

Do not apply this product through any type of irrigation system.

EQUIPMENT - SPRAY VOLUMES

Apply uniformly by ground equipment with a properly calibrated low pressure (20 to 40 psi) fixed-boom sprayer equipped with fan-type nozzles and screens no finer than 50 mesh. Use 10 to 40 gals. of water per acre.

Continuous agitation in the spray tank is required to keep the material in suspension. Avoid overlap, and shut off spray booms while starting, turning, slowing or stopping, or injury to the crop may result.

EARLY PREPLANT OR PREPLANT INCORPORATED (See rate chart)

"Gemini" may be applied at planting or up to 30 days prior to planting either preemergence or preplant incorporated. If incorporated, "Gemini" should be uniformly incorporated no deeper than the top 1-2" of soil prior to planting soybeans. Use equipment suited for proper incorporation into the top 1-2 inches of soil e.g.: Do-all, field cultivator, Roterra. If tank-mixed with a grass herbicide, such as a DNA like "Treflan"[4], "Sonalan"[4], or acetanilide such as "Dual"[2], or "Lasso"[1], follow label instructions for proper incorporation of the grass herbicide into the top 1 to 2" of soil. Improper soil incorporation can result in erratic weed control or crop injury. Rainfall* (~1") is required to activate preplant incorporated herbicides such as "Gemini" before weed emergence.

Use of a disk for incorporation is not recommended.

"Gemini" + "Treflan", "Prowl", "Dual" or "Sonalan" - Apply within 2 weeks of planting.

"Gemini" + "Lasso" - Apply within 7 days of planting.

PREEMERGENCE (See rate chart)

Preemergence application: Apply before soybeans emerge. Do not spray over top of emerged soybeans as severe injury or death of crop will occur.

Preemergence activity of "Gemini" requires rainfall* to activate. Best results are obtained if treatment is made to moist soil and followed by rainfall or irrigation (~1") before weeds germinate. On dry soil more moisture is required for activation (1 to 2") before weed emergence. If moisture is insufficient to activate the herbicide, a rotary hoeing or shallow cultivation should be made after emergence of crop while weeds are small enough to be controlled by mechanical means. "Gemini" can be applied preemergence in combination with "Lasso", "Dual", "Prowl"[3], "Surflan"[4], or following the use of a preplant grass material such as "Treflan" or "Sonalan." Consult labels for rates and use instructions.

* Rainfall required to activate ppi or pre treatments depends on soil moisture when activating moisture is received. Several small rainfalls of less than 1/4" are not as beneficial as one large rain of 1/2 inch or more for activation.

MINIMUM TILLAGE OR NO-TILL (See rate chart)

"Gemini" applied preemergence to soybeans but postemergence to broadleaf weeds and small annual grasses will provide some burndown of existing weeds as well as normal preemergence broadleaf control. When used for burndown add 0.25% (1 qt./100 gal) surfactant such as "Surfactant WK" or crop oil concentrate at 1% v/v (1 gal/100 gal). For added postemergence control especially on grasses, "Gemini" can be applied with "Roundup"[1], paraquat, or "Bronco"[1]. For added preemergence grass control "Gemini" can be mixed with "Dual", "Surflan" or "Lasso." Consult labels for rates and use instructions. Maintain constant agitation of spray mixture.

RATE CHART

Apply as follows:

Soil Texture (1/2 - 3% Organic Matter)		Rate (Ounces Per Acre)
Coarse:	Loamy sand	12-16
Medium:	Loam Silty loam Silt	16-20
Fine:	Silty Clay loam Clay loam Clay	20-24

Do not use on sand or loamy sand nor on any soil with less than 1/2 percent organic matter.

REDUCED RATE FOLLOWED BY "CLASSIC" HERBICIDE FOR BROADLEAF WEED CONTROL IN SOYBEANS

"Gemini" Herbicide is recommended preplant incorporated or preemergence at reduced rates followed by "Classic" Herbicide postemergence for broadleaf weed control in soybeans. This treatment will control those broadleaf weeds listed on the "Gemini" and "Classic" Herbicide product labels.

Apply "Gemini" preplant incorporated or preemergence at 8-10 ounces product per acre to any soil texture with 1/2 to 3% organic matter as directed on the product label. Follow the "Gemini" preplant incorporated or preemergence reduced rate with a postemergence application of "Classic" Herbicide at its labeled rate when weeds are small and actively growing. (Usually 14-28 days after preplant incorporated or preemergence application of "Gemini" reduced rate.)

For grass weed control "Gemini" can be tank mixed with Dual, Lasso, Sonalon, Treflan or Surflan. Consult labels for rates and use instructions of these products.

Do not apply "Gemini" on soils with pH greater than 7.0 except in those states as directed in the "ROTATIONAL CROP GUIDELINES" where application can be made to soils with a pH up to 7.5.

For band applications use proportionately less than listed in rate chart.

Plant seed 1 1/2 to 2" deep on a flat or raised seedbed only, or crop injury may occur.

Soybean stunting may occur if excessive rainfall occurs after application. Injury is more prevalent under poor drainage or compacted conditions or when soil is saturated for long periods of time. Soybeans rapidly outgrow stunting once favorable growing conditions return.

Injury to soybeans may occur if "Gemini" is used on soils having a calcareous surface layer or a pH of 7.0 or higher, or if used in conjunction with soil-applied insecticides. Seedling disease, cold or hot weather, deep planting (more than 2"), excessive moisture, high salt concentration, or drought may weaken soybean seedlings and increase possibility of crop injury. Do not apply to Black Belt Soils of Alabama and Mississippi with a soil pH greater than 7.0 or a history of nutrient deficiency such as iron chlorosis, as injury may occur.

Replanting Soybeans

If initial seeding fails to produce a stand, treated fields may be replanted to soybeans. Do not rework soil; do not retreat field with a second application as injury to the crop may result. Do not replant treated areas to any crop other than soybeans within four months after treatment as injury to subsequent crops may result.

WEEDS CONTROLLED

The following weeds will be controlled at the above rates:

Carpetweed	Mustards	Prickly sida (teaweed)
Cocklebur	Pigweed species	Purslane
Florida beggarweed	Palmer	Ragweed, common
Jimsonweed	Redroot	*Ragweed, giant
Lambsquarters	Smooth	Smartweed
*Morningglory, annual	Spiny amaranth	Sunflower
Ivyleaf	Tall waterhemp	Velvetleaf
Entireleaf		
Pitted		
Smallflower		
Tall		

Partial Control:

Annual grasses: Burcucumber, Hemp sesbania, Purple nutsedge, Sicklepod*, Wild poinsettia, Yellow nutsedge

* Large seeded weeds germinating deep in the soil such as morningglory and giant ragweed may require a cultivation or an application of Du Pont "Classic" Herbicide postemergence for season long control. Consult "Classic" label for application directions, recropping intervals and use restrictions.

NOTE: Consult "Lasso", "Dual", "Prowl", "Sonalan", "Surflan", or "Treflan" labels for additional weeds controlled when "Gemini" is used in conjunction with these herbicides.

SPRAY TANK PREPARATION

Fill tank 1/4 full with water. Start agitation system, add "Gemini" and continue adding water. Add separately each additional component of any tank mix while adding water. Continue agitation throughout. If poor mixing should occur with any component, premix the component with two parts water before adding to the spray tank.

A fertilizer solution may be used in the spray mixture. Small quantities should be tested for compatibility by the following procedures before full scale mixing.

1. Put 1 pint of fertilizer solution in a quart jar.
2. Mix 2 teaspoons "Gemini" with 2 tablespoons of water; mix thoroughly and add to fertilizer solution.
3. Close jar and shake well.
4. If other herbicides are to be used in the mixture, premix 2 teaspoons of wettable powder or 1 teaspoon of liquid with 2 tablespoons of water; add to "Gemini"/fertilizer solution mixture.
5. Close jar and shake well.
6. Watch mixture for several seconds; check again in 30 minutes.
7. If mixture does not separate, foam, gel, or become lumpy, it may be used.
8. Mixing ability may be improved by adding compatibility agents such as Kalo Laboratories' "Compex"[5] or Witco Chemicals' "Sponto"[6] 168D. Follow directions on container.

Provided the above procedure shows the mixture to be compatible, prepare the tank mixture as follows: Add the fertilizer solution to the spray tank first, with the agitator running, add the required amount of "Gemini" and thoroughly mix. For tank mixtures with other herbicides, follow directions above.

When tank mixing with other herbicides, all applicable directions, restrictions and precautions for the additional herbicides are also to be followed. Use "Gemini" spray preparations the same day as mixed or product degradation may occur. Thoroughly reagitate before using if allowed to settle.

RECROP INTERVAL FOR "GEMINI" HERBICIDE

Crops that can be planted after using "Gemini" are listed below. Choice of rotational crop is determined by geographical location (state), soil pH, rate and time interval elapsed since last application.

ROTATIONAL CROP GUIDELINES

GROUP I

States of: AL, AR, FL, GA, LA, MS, TX

■ Soil pH 7.0 or less

GROUP II

States of: DE, KY, MD, MO Bootheel, NJ, NC, PA, SC, TN, VA, WV

■ Soil pH 6.8 or less

AND

■ Use rate 16-22 oz/A

GROUP III

■ All other states not listed in Group I or II

■ Soil pH 7.0 or less OR

States of: AL, AR, FL, GA, LA, MS, TX

■ Soil pH 7.1-7.5 OR

States of: DE, KY, MD, MO Bootheel, NJ, NC, PA, SC, TN, VA, WV

■ Soil pH 6.9-7.5 OR Fields not meeting rate requirement in Group II

TIME INTERVAL IN MONTHS BEFORE PLANTING ROTATIONAL CROP

Crops	Group I	Group II	Group III
Soybeans	Anytime	Anytime	Anytime
Cereal grains	4	4	4
Field Corn	9	10	18
Clover	10	12	18
Cotton	10	10	18
Sorghum	10	10	18
Tobacco	10	10	18
Tomato (transplant)	10	10	18

All other crops require a minimum 18 month rotational interval and a successful field bioassay[a].

Suggested interval (months) before initiating bioassay is 10 months for Group I, 12 months for Group II and 15 months for Group III bioassay interval.

[a] Field Bioassay - The field bioassay will detect small quantities of "Gemini" which can remain in the soil and injure rotational crops. A successful field bioassay must be completed the season before planting any crop not listed above. A successful field bioassay means growing to maturity a test strip of the crop(s) intended for production the following year. The test strip should cross the whole field, including knolls and low areas.

SPRAYER CLEANUP

To avoid subsequent injury to crops other than soybeans, immediately after spraying and prior to spraying other crops, thoroughly remove all traces of "Gemini" Herbicide from mixing and spray equipment. For more information, consult supplemental bulletin on sprayer cleanout following use of Du Pont "Classic", "Canopy" or "Gemini" Herbicides.

PRECAUTIONS

Do not apply "Gemini" to soils with a pH greater than 7.0, except for those states listed in Group III of the "ROTATIONAL CROP GUIDELINES" where "Gemini" can be applied to soils up to 7.5 pH.

CAUTION: Soil pH varies greatly even within the same field. pH variations as much as 2 pH units are common. Determine soil pH by laboratory analysis using a 1:1 soil:water suspension.

Do not graze treated fields or harvest for forage or hay.

Do not use on sand, muck, or peat soils, nor on soils with less than 1/2% organic matter as crop injury may result.

Do not apply "Gemini" after the soybean crop has emerged.

Use of "Gemini" may affect rotational crops. Follow "ROTATIONAL CROP GUIDELINES".

Do not apply to Black Belt Soils of Alabama and Mississippi with soil pH greater than 7.0, or on soils with a history of nutrient deficiencies such as iron chlorosis or injury may occur.

When recommended tank mix combinations are used, consult the product label for replanting and recropping instructions and observe the directions that are the most restrictive.

If "Scepter"[3], "Pursuit"[3] or any product containing the active ingredients, imazaquin or imazethapyr (such as "Squadron"[1], "Partner"[5]) is applied the same year as "Gemini", do not plant crops other than soybeans for at least 15 months from the last application. Crops listed in the "Gemini" Rotational Crop Guidelines" may be planted after 15 months; all other crops require a successful field bioassay.

Do not apply to land that has been or will be treated with "Ally", "Finesse" or "Glean" herbicides in the states of Nebraska, Kansas, North Dakota and South Dakota.

[1] Trademark of Monsanto Agricultural Products Co., St. Louis, MO
[2] Trademark of Ciba-Geigy Corp., Greensboro, NC
[3] Trademark of American Cyanamid Corp., Princeton, NJ
[4] Trademark of Elanco Products Co., Indianapolis, IN
[5] Trademark of Kalo Agricultural Chemicals, Inc., Overland Park, KS
[6] Trademark of Witco Chemical Corp., Houston, TX

NOTICE TO BUYER: Purchase of this material does not confer any rights under patents of countries outside of the United States.

STORAGE AND DISPOSAL

Storage: Store product in original container only, away from other pesticides, fertilizer, food, or feed.

Disposal: Do not contaminate water, food, or feed by storage or disposal. Wastes resulting from the use of this product may be disposed of on site or at an approved waste disposal facility.

Container Disposal: Triple rinse (or equivalent). Then offer for recycling or reconditioning, or puncture and dispose of in a sanitary landfill, or incineration, or, if allowed by state and local authorities, by burning. If burned, stay out of smoke.

NOTICE OF WARRANTY

Du Pont warrants that this product conforms to the chemical description on the label thereof and is reasonably fit for the purposes stated on such label only when used in accordance with the directions under normal use conditions. It is impossible to eliminate all risks inherently associated with the use of this product. Crop injury, ineffectiveness, or other unintended consequences may result because of such factors such as weather conditions, presence of other materials, or the manner of use or application, all of which are beyond the control of Du Pont. In no case shall Du Pont be liable for consequential, special or indirect damages resulting from the use or handling of this product. All such risks are assumed by the buyer. DU PONT MAKES NO WARRANTIES OF MERCHANTABILITY OR FITNESS FOR A PARTICULAR PURPOSE NOR ANY OTHER EXPRESS OR IMPLIED WARRANTY EXCEPT AS STATED ABOVE.

AG-3279 8099/8099 9/6/89

This is a specimen label. DuPont is not responsible for the accuracy of the information contained herein. As labels are subject to revision, always carefully read and follow the label on the product container.

HARMONY® EXTRA

HERBICIDE

DRY FLOWABLE

ACTIVE INGREDIENTS	BY WEIGHT
Methyl-3-[[[[(4-methoxy–6–methyl–1,3,5–triazin–2–yl) amino] carbonyl]amino]sulfonyl]–2–thiophenecarboxylate	50%
Methyl 2–[[[[N–(4–methoxy–6–methyl–1,3,5–triazin–2–yl) methylamino] carbonyl]amino]sulfonyl]benzoate	25%
INERT INGREDIENTS	25%
TOTAL	100%

EPA Reg. No. 352–538 U.S. Patents 4,481,029 & 4,383,113 & 4,740,234

KEEP OUT OF REACH OF CHILDREN

CAUTION

PRECAUTIONARY STATEMENT

HAZARDS TO HUMANS AND DOMESTIC ANIMALS

CAUTION! Avoid contact with skin, eyes and clothing. In case of contact with eyes, immediately flush with plenty of water. Get medical attention if irritation persists. Wash thoroughly after handling. Remove contaminated clothing and wash before reuse.

For medical emergencies involving this product, call toll free 1–800–441–3637.

ENVIRONMENTAL HAZARDS

Do not apply directly to water or wetlands (swamps, bogs, marshes, and potholes). Do not contaminate water wnen disposing of equipment wash-waters.

DIRECTIONS FOR USE

It is a violation of federal law to use this product in a manner inconsistent with its labeling.

Do not apply this product through any type of irrigation system.

"HARMONY" EXTRA HERBICIDE should be used only in accordance with recommendations on this label or in separately published Du Pont recommendations.

Du Pont will not be responsible for losses or damages resulting from the use of this product in any manner not specifically recommended by Du Pont.

IMPORTANT INFORMATION

Injury to or loss of desirable trees or vegetation may result from failure to observe the following: Do not apply or drain or flush equipment on or near desirable trees or other plants, or on areas where their roots may extend, or in locations where the chemical may be washed or moved into contact with their roots. Do not use on lawns, walks, driveways, tennis courts or similar areas. Prevent drift of spray to desirable plants. Do not contaminate any body of water. Thoroughly clean all traces of Du Pont "Harmony" Extra Herbicide from application equipment immediately after use. (Refer to "SPRAYER CLEANUP" section of this label.) Failure to follow these procedures may result in injury to subsequently sprayed crops.

GENERAL INFORMATION

Du Pont "HARMONY EXTRA herbicide is recommended for selective postemergence control of certain broadleaf weeds in wheat (including durum) and barley. Do not use "HARMONY" EXTRA on any other crop. "HARMONY EXTRA" is a dry flowable granule to be mixed in water or other recommended carrier and applied as a uniform broadcast spray. It is noncorrosive, nonflammable, nonvolatile and does not freeze.

Best results are obtained when "HARMONY" EXTRA is applied to young, actively growing weeds. The use rate will depend on weed spectrum and size of weed at time of application. The degree of control and duration of effect are dependent on rate used, sensitivity and size of target weed and environmental conditions at the time of and following application.

"HARMONY" EXTRA stops growth of susceptible weeds rapidly. However, typical symptoms of dying weeds (discoloration) may not be noticeable for 1–3 weeks after application (2–5 weeks for wild garlic) depending on the environmental conditions and weed susceptibility. Warm, moist conditions following treatment promote the activity of "HARMONY" EXTRA, while cold, dry conditions delay the activity. Weeds hardened–off by cold weather or drought stress will be less susceptible.

A vigorous growing crop will aid weed control by shading and providing competition for weeds. However, a dense crop canopy at time of application can intercept spray and result in reduced weed control. Weeds may not be adequately controlled in areas of thin crop stand or seeding skips.

Naturally-occurring weed biotypes* that are resistant to Du Pont "Glean" Herbicide, Du Pont "Finesse" Herbicide or Du Pont "Ally" Herbicide will also be resistant to "Harmony" Extra. In areas where these weed biotypes are known to exist, only spray "HARMONY" EXTRA in tank mixtures with other broadleaf herbicides having a different mode of action** such as: 2,4–D, Banvel[1]***, Buctril[2], Bronate[2], Curtail[3], MCPA or Du Pont "Lexone" DF Herbicide. Refer to TANK MIXTURES.

* Biotypes are naturally-occurring individuals of the species which have a slightly different genetic makeup. Resistant biotypes may look exactly the same as susceptible biotypes. Herbicide resistant biotypes are able to survive a use rate several times higher than needed to control susceptible biotypes.

** Mode of action is the chemical interaction that interrupts a biological process necessary for plant growth and development.

*** Tank mixes with "Banvel" may result in reduced control of some broadleaf weeds.

Read and follow all appropriate sections of label, including precautions, before using this product.

WEED CONTROL IN WHEAT AND BARLEY

CROP STAGE AT APPLICATION

For winter wheat, make applications after the crop is in the 2-leaf stage, but before the 3rd node is detectable.

For spring wheat and barley, make applications after the crop is in the 2-leaf stage, but before the 1st node is detectable.

APPLICATION RATES

Apply 0.5 ounce "HARMONY" EXTRA per acre to wheat (including durum) or barley for control or partial control of the weeds listed below.

Use 0.6 ounce "HARMONY" EXTRA per acre when weed infestation is heavy and predominately consists of those weeds listed under partial control, or when application timing and environmental conditions are marginal (refer to APPLICATION AND ENVIRONMENTAL CONDITIONS FOR OPTIMUM PERFORMANCE).

Use 0.3 to 0.4 ounce "HARMONY" EXTRA per acre when weed infestation is light and predominately consists of those weeds listed under weeds controlled, and when optimum application conditions occur.

Do not use less than 0.3 ounce "HARMONY" EXTRA per acre.

Sequential treatments of "HARMONY" EXTRA may be made provided the total amount of "HARMONY" EXTRA applied to the crop does not exceed 1.0 ounce per acre; for example, 0.5 ounce followed by 0.5 ounce. Do not make the first application until after winter wheat is in the 2-leaf stage and make the last application before the 3rd node is detectable. Do not make the first application until after spring wheat or barley is in the 2-leaf stage and make the last application before the 1st node is detectable.

WEEDS CONTROLLED

Annual knawel	Mouseear chickweed
Annual sowthistle	Mouseearcress
Black mustard	Narrowleaf lambsquarters
Blue/Purple mustard	Nightflowering catchfly
Broadleaf dock	Pennsylvania smartweed
Bur buttercup	Pineappleweed
Bushy wallflower/Treacle mustard	Prickly lettuce
Clasping pepperweed	Prostrate knotweed
Coast fiddleneck	Prostrate pigweed
Common buckwheat	Redmaids
Common chickweed	Redroot pigweed
Common groundsel	Russian thistle
Common lambsquarters	Scentless chamomile/mayweed
Common radish	Shepherdspurse
Common sunflower	Slimleaf lambsquarters
Corn chamomile	Smallflower buttercup
Corn gromwell*	Smallseed falseflax
Corn spurry	Stinking chickweed
Cowcockle	Stinking mayweed/Dogfennel
Curly dock	Swinecress
False chamomile	Tansymustard
Field chickweed	Tarweed fiddleneck
Field pennycress	Tumble/Jim Hill mustard
Flixweed	Volunteer lentils
Green smartweed	Volunteer peas
Henbit	Volunteer sunflower
Kochia	Wild buckwheat*
Ladysthumb	Wild chamomile
London rocket	Wild garlic*
Marshelder	Wild mustard
Mayweed chamomile	Wild radish*
Miners lettuce	

PARTIAL CONTROL**

Canada thistle*	Common vetch
Carolina geranium	Cutleaf eveningprimrose
Catchweed bedstraw	Hairy nightshade
Common cocklebur	Little mallow
Common mallow	

*See SPECIFIC WEED PROBLEMS for more information.

**Partial control: A visual reduction of weed population as well as a significant loss of vigor for individual weed plants. For best results, use 0.5 or 0.6 ounce "HARMONY" EXTRA per acre.

SPECIFIC WEED PROBLEMS

Canada thistle: For best results apply 0.6 ounce "HARMONY" EXTRA per acre plus surfactant in the spring when all thistles have emerged above the soil, are actively growing, and are 4 to 8 inches tall with 2 to 6 inches of new growth. Canada thistle control will be improved by using "HARMONY" EXTRA in combination with 2,4–D (refer to TANK MIXTURES).

Corn gromwell, Wild buckwheat: For best results, use 0.5 to 0.6 ounce "HARMONY" EXTRA per acre plus surfactant.

Wild garlic: For best results, apply 0.5 to 0.6 ounce "HARMONY" EXTRA per acre plus surfactant when wild garlic plants are less than 12 inches tall with 2 to 4 inches of new growth. For severe infestations, use the 0.6 ounce per acre rate of "HARMONY" EXTRA. Control may be reduced when plants are hardened-off by cold weather and/or drought stress. Control is enhanced when applications are made during warm temperatures to actively growing wild garlic plants. Typical symptoms of dying wild garlic plants (discoloration and collapse) may not be noticeable for 2–5 weeks.

Thorough spray coverage of all garlic plants is essential.

Wild radish: For best results, apply 0.5 to 0.6 ounce "HARMONY" EXTRA per acre plus surfactant either in the fall or spring to wild radish rosettes less than 6" in diameter. Applications made later than 30 days after weed emergence will result in partial control. Fall applications should be made prior to hardening-off of plants.

APPLICATION AND ENVIRONMENTAL CONDITIONS FOR OPTIMUM PEFORMANCE

Crop Stage: For winter wheat, make applications after the crop is in the 2-leaf stage, but before the 3rd node is detectable.

For spring wheat and barley, make applications after the crop is in the 2-leaf stage, but before the 1st node is detectable.

Since thorough coverage is required, avoid crop canopy obstruction of the spray contacting the weed foliage.

Crop Competition: A virgorous growing crop will aid weed control by shading and providing competition to the weeds. However, a dense crop canopy at time of application can intercept spray and result in reduced weed control. Weeds may not be adequately controlled in areas of thin crop stand or seeding skips.

Pest Stage: Since "HARMONY" EXTRA has very little or no soil activity, only those weeds that have germinated above the soil surface will be controlled. Consequently, application of "HARMONY" EXTRA should be made when all or the majority of weeds have germinated. Annual broadleaf weeds should be past the cotyledon stage, actively growing, and less than 4 inches tall or across. Wild garlic plants should be less than 12 inches tall with 2 to 4 inches of new growth. See SPECIFIC WEED PROBLEMS for more information on Canada thistle, corn gromwell, wild buckwheat, wild garlic, and wild radish.

Application: Foliar absorption is the primary means of "HARMONY" EXTRA uptake by plants; therefore, thorough spray coverage of all target weeds is essential.

Weather: Conditions which are conducive to healthy, actively growing plants optimize "HARMONY" EXTRA weed control performance. Ideal conditions include warm temperatures and adequate soil moisture before, during and immediately after application.

Avoid making applications of "HARMONY" EXTRA to weeds when rainfall is threatening. Rainfall immediately after treatment can wash "HARMONY" EXTRA off weed foliage and result in reduced weed control effectiveness. Several hours of dry weather are needed to allow "HARMONY" EXTRA to be absorbed by weed foliage.

Surfactant: Use only EPA approved surfactants authorized for use on food crops.

Water Spray Carrier - Unless specified otherwise, always add a nonionic surfactant of at least 80% active ingredient strength at 0.25% vol/vol (1 quart per 100 gallons of spray solution).

Liquid N Carrier - Surfactant use is recommended when applying "HARMONY" EXTRA in liquid nitrogen fertilizer. When "HARMONY" EXTRA is applied using liquid nitrogen fertilizer solution as a spray carrier, early, temporary, crop yellowing and stunting may occur. These symptoms will be more noticeable when surfactant is used. Use 0.06/0.25% vol/vol surfactant (1/2 pint - 1 quart per 100 gallons of spray solution) when applying "HARMONY" EXTRA in liquid nitrogen fertilizer solution. Refer to TANK MIXTURES for surfactant information when tank mixing with other products. If surfactant is not used when applying "HARMONY" EXTRA in liquid nitrogen fertilizer, weed control may be significantly reduced.

EQUIPMENT-SPRAY VOLUMES

Prior to spraying "HARMONY" EXTRA, thoroughly clean sprayer (tanks, booms, hoses, nozzles, mixing cones, all mixing equipment, etc.)

Apply using properly calibrated air or ground equipment. Select a spray volume and delivery system that will insure thorough coverage and a uniform spray pattern. Avoid overlapping, and shut off spray booms while starting, turning, slowing or stopping, or injury to the crop may result.

Do not use equipment and/or spray volumes that will cause spray to drift onto non-target sites. Do not make applications during weather conditions which cause spray to drift onto non-target sites. Refer to CAUTION-AVOID SPRAY DRIFT.

Do not apply "HARMONY" EXTRA through any type of irrigation system.

Spray Equipment: Refer to specific manufacturer's recommendations for additional information on spray volume (GPA), pressure, speed, nozzle types and arrangements, nozzle heights above the target area, etc., for respective application equipment.

Ground Application: For optimum spray distribution and thorough coverage, use flat fan or low volume flood nozzles. For flat fan nozzles, do not use less than 5 gallon spray volume per acre (GPA).

For flood nozzles on 30-inch spacing, use not less than 10 GPA and no larger than TK10 or equivalent and not less than 30 psi. On 40-inch nozzle spacings, use not less than 13 GPA or not less than 20 GPA when nozzles are on a 60-inch spacing. 100% overlapping of nozzle spray pattern is essential for 30, 40 and 60-inch spacings.

When using liquid nitrogen solution as a spray carrier, for flood nozzles use not less than 30 GPA and no larger than TK20. See statements on liquid N solutions below for additional information.

"Raindrop"[4] nozzles are not recocmmended for "HARMONY" EXTRA applications as weed control performance may be reduced.

Use 50-mesh screens or larger.

Aerial Application: Use nozzle types and arrangements that will provide for optimum spray distribution and maximum coverage at 3 to 5 GPA. Do not apply during inversion conditions, when winds are gusty, or when other conditions will favor poor coverage and/or off target spray movement.

For aerial application in the state of Washington, refer to and follow the directions on the Washington special local need label, "Harmony" Extra Herbicide Aerial Application To Wheat And Barley In the State Of Washington.

Liquid Nitrogen Fertilizer Solutions: Always slurry "HARMONY" EXTRA in clean water PRIOR to mixing in liquid N solutions. Surfactant use is recommended when applying "HARMONY" EXTRA in liquid nitrogen fertilizer. If surfactant is not used, weed control may be significantly reduced. When "HARMONY" EXTRA is applied using liquid nitrogen fertilizer solution as a carrier, early, temporary crop yellowing and stunting may occur. The symptoms will be more noticeable when surfactant is used. Use 0.06/0.25% vol/vol surfactant (1/2 pint - 1 quart per 100 gallons of spray solution) when applying "HARMONY" EXTRA in liquid nitrogen fertilizer solution.

The addition of liquid nitrogen fertilizer to the spray solution, or the use of liquid nitrogen fertilizer as the total carrier of the spray solution will increase the weight of the spray solution as compared to water (see table below). Consequently, appropriate conversions of spray nozzles and/or pressure must be made in order to maintain proper spray volumes. Refer to the conversion chart below for guidance (taken from Spray Systems Company Catalog 39, page 2).

Influence of adding liquid 28% N fertilizer to spray solution weight per gallon:

% of Spray Solution		
Water	**28% N**	**Weight of Spray Solution (Lbs/Gallon)**
100	0	8.34
50	50	9.5
0	100	10.6

SPRAYING SYSTEMS CONVERSION CHART

SPRAYING SOLUTIONS OTHER THAN WATER—NEW METHOD

Since all the tabulations in the Spray Systems Catalog are based on spraying water, which weighs 8.34 lbs. per USA gallon, conversion factors must be used when spraying solutions which are heavier or lighter than water. To determine the proper size nozzle for the solution to be sprayed, first multiply the desired GPM or GPA of solution by the water rate conversion factor. Then use the new converted GPM or GPA rate to select the proper size nozzle.

Example: 10 GPA (28%N) X 1.13 = 11.3 GPA (water)

Weight of Solution		Specific Gravity	Conversion Factors
7.0	lbs per gallon	0.84	0.92
8.0	lbs per gallon	0.96	0.98
8.34	lbs per gallon-Water	1.00	1.00
9.0	lbs per gallon	1.08	1.04
10.0	lbs per gallon	1.20	1.10
10.65	lbs per gallon-28% nitrogen	1.28	1.13
11.0	lbs per gallon	1.32	1.15
12.0	lbs per gallon	1.44	1.20
14.0	lbs per gallon	1.68	1.30

Published in Spraying Systems Company Catalog 39, page 2.

Agitation: Continuous agitation is required to keep "HARMONY" EXTRA in suspension.

CAUTION - AVOID SPRAY DRIFT

Follow these practices to minimize drift.

Do not allow spray from either ground or aerial equipment to drift onto adjacent crops or land, as even small amounts will injur other plants. When spraying, do everything possible to reduce spray drift. This includes:

- Stop spraying if wind speed becomes excessive. DO NOT SPRAY IF WIND SPEED IS 10 MPH OR GREATER. Spray drift can occur at wind speeds less than 10 MPH. If sensitive crops or plants are downwind, extreme caution must be used even in relatively low wind conditions! DO NOT SPRAY IF WINDS ARE GUSTY.
- High temperatures, drought, and low relative humidity increase the possibility of spray drift. EXTREME CAUTION MUST BE USED WHEN THESE CONDITIONS ARE PRESENT AND SENSITIVE CROPS OR PLANTS ARE NEARBY, REGARDLESS OF WIND SPEED.
- Do not apply when a temperature inversion exists. An inversion is characterized by low air movement and an increase in air temperature with an increase in altitude. In humid regions, a fog or mist may form. An inversion may be detected by producing a smoke column and checking for a layering effect. Smoke-producing devices on aircraft are

recommended. If not sure whether inversion conditions are present, consult with local weather serevices before making an application.

- Postemergence grass herbicides (such as Avenge[5] and Assert[5]) are often applied using high pressure. When "HARMONY" EXTRA is tank mixed with these products, do not exceed 40 psi.
- Drift from aerial or ground equipment may be further reduced by:
  1. Using large droplet size sprays to minimize drift. DO NOT APPLY WITH HOLLOW-CONE INSECTICIDE NOZZLES ON GROUND EQUIPMENT. Do not use nozzles that produce fine droplets, such as Sprayfoil[6] or airblast-type nozzles. Nozzles should be oriented at an angle between straight down and straight back for ground applications. For aerial applications orient nozzles straight back along the windstream. If using flood-type nozzles on aircraft, orient them so spray is produced in direction of the airstream.
  2. Increasing volume of spray mix per acre (for example, minimum 5 GPA by air, 10 GPA by ground) by using higher flow rate nozzles.
  3. Reducing pressure (PSI). DO NOT EXCEED 40 PSI when applying "HARMONY" EXTRA. [Vehicle speed must also be reduced to maintain spray mix volume per acre.] Consult manufacturer's catalogs for details on correct calibration.
  4. Apply as close to target plants as possible while still maintaining a good spray pattern.

TANK MIXTURES

2,4–D (amine or ester) or MCPA (amine or ester) - Use "HARMONY" EXTRA plus 1/8 to 3/8 lb active ingredient 2,4–D or MCPA (ester formulations have provided best results). Surfactant may be added at 0.125–0.25% vol/vol (1 pint to 1 quart per 100 gallons of spray volume); however, the addition of surfactant may increase the chance of crop injury. Use the 1 pint to 1 quart rate of surfactant with 1/8 lb active ingredient rate of 2,4–D or MCPA. Use the 1 pint rate of surfactant with 1/4 to 3/8 lb active ingredient of 2,4–D or MCPA. Higher rates of 2,4–D or MCPA may be used, but do not exceed highest rate allowed on the label.

Always mix "HARMONY" EXTRA in water prior to adding 2,4–D or MCPA. Always add surfactant last. Read and follow all label instructions on timing, precautions and warnings for these herbicides prior to using these tank mixtures.

"HARMONY" EXTRA may be tank mixed with other suitable registered herbicides (for example "Lexone") to control weeds other than those listed. Read and follow all manufacturer's label recommendations for the companion herbicide. If these recommendations conflict with this "HARMONY" EXTRA label or supplemental "HARMONY" EXTRA labels, do not use as a tank mix with "HARMONY" EXTRA.

"HARMONY" EXTRA will not control wild oats or other grasses. For wild oat control, "HARMONY" EXTRA can be tank mixed with "Avenge" or "Assert". When tank mixing "HARMONY" EXTRA and "Assert", ALWAYS include another broadleaf weed herbicide with a different mode of action, for example: 2,4–D ester, MCPA ester, "Buctril" or "Bronate". DO NOT tank mix with Hoelon[7] 3EC as grass weed control may be reduced.

Always mix "HARMONY" EXTRA in water PRIOR to adding other products, including surfactants. When mixing "HARMONY" EXTRA in liquid nitrogen fertilizer solutions, first mix the "HARMONY" EXTRA in clean water and then add the "HARMONY" EXTRA and water slurry to the liquid nitrogen solution. Refer to SPRAY PREPARATION for further information.

Tank mixes of "HARMONY" EXTRA plus "Banvel" may result in reduced control of some broadleaf weeds.

Tank mixes of "HARMONY" EXTRA plus metribuzin may result in reduced control of wild garlic.

HARMONY" EXTRA may be tank mixed or used sequentially with fungicides or insecticides registered for use on cereal grains. However, under certain conditions, (drought stress, crop in 2–4 leaf stage) tank mixes or sequential applications of "HARMONY" EXTRA plus organophosphate insecticides (such as parathion) may produce temporary crop yellowing or, in severe cases, crop injury. Limit first use to a small area before treating large areas.

DO NOT USE "HARMONY" EXTRA PLUS MALATHION OR "LORSBAN"[3] as crop injury will result.

SPRAY PREPARATION, ADDITIVES, PRODUCT MEASUREMENT, SURFACTANT AND LIQUID NITROGEN FERTILIZER

Spray Preparation: Mix the proper amount of "HARMONY" EXTRA into the necessary volume of water in the spray tank with the agitator running, then add the companion herbicide to the tank after all the "HARMONY" EXTRA is in suspension. ALWAYS MIX "HARMONY" EXTRA IN WATER FIRST, PRIOR TO ADDING OTHER PRODUCTS IN THE SAME SPRAY TANK.

Use the spray preparation of "HARMONY" EXTRA within 24 hours as product degradation may occur. If spray preparation is left standing, thoroughly agitate before reusing.

Additives: Do not use with spray additives that alter the pH of the spray solution below pH 5.0 or above pH 9.0, as rapid product degradation can occur. Spray solutions of pH 6.0 - 8.0 allow for optimum stability of "HARMONY" EXTRA.

Product Measurement: The "HARMONY" EXTRA volumetric measuring cylinder is to be used as a guide, since the degree of accuracy varies by plus or minus 7.5%. For more precise measurement, use scales calibrated in ounces.

Surfactant: Unless specified otherwise, add a nonionic surfactant of at least 80% active ingredient strength at 0.25% vol/vol (1 quart per 100 gallons of spray solution). Use only EPA approved surfactants authorized for use on food crops. Refer to TANK MIXTURES and EQUIPMENT - SPRAY VOLUMES for variations on surfactant rate.

Liquid Nitrogen Fertilizer: Liquid nitrogen fertilizer solutions may be used as a carrier in place of water. Slurry the desired amount of "HARMONY" EXTRA in a clean bucket using water until a flowable mixture is produced. Add this slurry to the agitating spray tank of liquid nitrogen fertilizer solution. Thoroughly rinse all of the "HARMONY" EXTRA slurry into the spray tank. Do not use "HARMONY" EXTRA in liquid fertilizer solutions of less than pH 5.0. Run a tank mix compatibility test before mixing "HARMONY" EXTRA in fertilizer solution.

Use 0.06/0.25% vol/vol surfactant (1/2 pint - 1 quart per 100 gallons of spray solution) when applying "HARMONY" EXTRA in liquid nitrogen fertilizer solution.

Use of "HARMONY" EXTRA and a surfactant in liquid nitrogen fertilizer solutions may cause early, temporary crop injury (discoloration and stunting).

When the spray solution contains liquid nitrogen fertilizer, the weight per gallon of solution varies significantly from the weight of water (8.34 lbs per USA gallon). Consequently, liquid nitrogen fertilizer spray solutions must use the appropriate conversion in order to insure proper spray volume. See EQUIPMENT-SPRAY VOLUMES for further information.

CROP ROTATION

Any crop may be planted 60 days after the application of "HARMONY" EXTRA.

SPRAYER CLEANUP

To avoid subsequent injury to crops other than wheat or barley, immediately after spraying thoroughly remove all traces of "HARMONY" EXTRA from mixing and spray equipment as follows:

1. Drain tank; rinse interior surfaces of tank; then flush tank, boom and hoses with clean water for a minimum of 5 minutes.
2. Fill the tank full with clean water, then add cleaning solution*. Flush solution through boom, hoses and nozzles, then allow to sit for 15 minutes with agitation running; then drain.
3. Repeat Step 2.
4. Repeat Step 1.
5. Nozzles and screens should then be removed and cleaned separately. To remove traces of cleaning solution, rinse the tank thoroughly with clean water and flush through hoses and boom.
6. Flush boom and hoses with clean water for 5 minutes just prior to using the sprayer for the first time after the "HARMONY" EXTRA application.

*Use any one of the following cleaning solutions:

1. One gallon ammonia (containing 3% active) per 100 gallons of water.
2. Nutra-Sol[8] (carefully read and follow Nutra-Sol label directions).
3. Loveland Spray Tank Cleaner[9] (carefully read and follow Loveland Spray Tank Cleaner label directions).
4. Tank-Aid[10] (carefully read and follow Tank-Aid label directions).

PRECAUTIONS

The total rate of "HARMONY" EXTRA cannot exceed 1.0 ounce product per acre applied to any one crop during one growing season.

Do not graze or feed forage or hay from treated areas to livestock (harvested straw may be used for bedding and/or feed).

Do not plant to any crop other than wheat or barley for 60 days after application of "HARMONY" EXTRA.

Do not apply to wheat or barley that is stressed by severe winter, drought, water-saturated soil, disease or insect damage as crop injury may result. Under certain conditions such as prolonged cool weather (daily high temperature less than 50 Deg.F.) or wide fluctuations in day/night temperatures just prior to, during or soon after treatment, temporary yellowing and/or crop stunting may occur. Risk of injury is greatest when crop is in the 2 to 5-leaf stage.

Do not apply to wheat or barley crops underseeded with another crop.

"HARMONY" EXTRA is only registered on wheat and barley. Do not use on any other crop.

For ground applications applied to weeds when dry, dusty field conditions exist, control of weeds in wheel track areas may be reduced.

STORAGE AND DISPOSAL

STORAGE: Store product in original container only, away from other pesticides, fertilizer, food or feed.

PRODUCT DISPOSAL: Do not contaminate water, food or feed by storage or disposal. Waste resulting from the use of this product may be disposed of on site or at an approved waste disposal facility.

CONTAINER DISPOSAL: Triple-rinse (or equivalent) the container. Then offer for recycling or reconditioning, or puncture and dispose of in a sanitary landfill, or by incineration, or, if allowed by state and local authorities, by burning. If burned, stay out of smoke.

NOTICE TO BUYER: Purchase of this material does not confer any rights under patents of countries outside of the United States.

NOTICE OF WARRANTY

Du Pont warrants that this product conforms to the chemical description on the label thereof and is reasonably fit for purposes stated on such label only when used in accordance with directions under normal use conditions. It is impossible to eliminate all risks inherently associated with the use of this product. Crop injury, ineffectiveness, or other unintended consequences may result because of such factors as weather conditions, presence of other materials, or the manner of use or application, all of which are beyond the control of Du Pont. In no case shall Du Pont be liable for consequential, special or indirect damages resulting from the use or handling of this product. All such risks shall be assumed by the buyer. DU PONT MAKES NO WARRANTIES OF MERCHANTABILITY OR FITNESS FOR A PARTICULAR PURPOSE NOR ANY OTHER EXPRESSED OR IMPLIED WARRANTY EXCEPT AS STATED ABOVE.

[1] Registered trademark of Sandoz Crop Protection Corporation.
[2] Registered trademark of Rhone-Poulenc Ag Company.
[3] Registered trademark of The Dow Chemical Company.
[4] Registered trademark of Delavan Corporation.
[5] Registered trademark of American Cyanamid Company.
[6] Registered trademark of D. AND W. INDUSTRIES.
[7] Registered trademark of Hoechst-Roussel Agri-Vet Company.
[8] Sold by Thomas G. Kilfoil, San Bruno, CA.
[9] Registered trademark of Loveland Industries, Inc.
[10] Manufactured for Cornbelt Chemical Company.

AG–3277 8089/8109 8/3/89

This is a specimen label. DuPont is not responsible for the accuracy of the information contained herein. As labels are subject to revision, always carefully read and follow the label on the product container.

LOROX PLUS™

HERBICIDE
DISPERSIBLE GRANULES

FOR USE ON SOYBEANS ONLY

	BY WEIGHT
ACTIVE INGREDIENTS:	60%
Linuron: [3-(3,4-dichlorophenyl)-1-methoxy-1-methylurea] . 56.9%	
Chlorimuron Ethyl: Ethyl 2-[[[[(4-chloro-6-methoxypyrimidin-2-yl)amino]carbonyl]-amino]sulfonyl]benzoate 3.1%	
INERT INGREDIENTS:	40%
TOTAL	100%

EPA Reg. No. 352-543
U.S. Pats. 4,394,506 & 4,547,215

KEEP OUT OF REACH OF CHILDREN

PRECAUTIONARY STATEMENTS

HAZARDS TO HUMANS AND DOMESTIC ANIMALS

CAUTION!

HARMFUL IF SWALLOWED OR ABSORBED THROUGH SKIN. CAUSES MODERATE EYE INJURY.

AVOID CONTACT WITH SKIN, EYES AND CLOTHING. AVOID BREATHING DUST OR SPRAY MIST. WASH THOROUGHLY WITH SOAP AND WATER AFTER HANDLING. REMOVE AND WASH CONTAMINATED CLOTHING BEFORE REUSE.

STATEMENT OF PRACTICAL TREATMENT

If in eyes: Flush with plenty of water and get medical attention if irritation persists.

If on skin: Wash with soap and water and get medical attention if irritation persists.

If swallowed: Call a physician or Poison Control Center. Drink 1 or 2 glasses of water and induce vomiting by touching back of throat with finger. Do not induce vomiting or give anything by mouth to an unconscious person.

For medical emergencies involving this product, call toll free 1-800-441-3637.

ENVIRONMENTAL HAZARDS

Do not apply directly to water or wetlands (swamps, bogs, marshes and potholes). Do not contaminate water by cleaning of equipment or disposal of wastes.

IMPORTANT

When mixing/loading and spraying, wear a wide-brimmed hat, mid-forearm to elbow length chemical resistant gloves, and one-piece overalls which have long sleeves and long pants. Overalls should be constructed of finely woven fabric as specified in the USDA/EPA Guide for Commercial Applicators.

Any article of clothing worn while applying product must be cleaned before reusing. Clothing which has been drenched or has otherwise absorbed concentrated pesticide must be disposed of in a sanitary landfill, or incinerated, or, if allowed by state and local authorities, by burning.

Instead of clothing and equipment specified above, the applicator can use an enclosed tractor cab which provides a filtered air supply (as specified by Taschenberg and Bourke, 1975).

Injury to or loss of desirable trees or vegetation may result from failure to observe the following: Do not apply or drain or flush equipment on or near desirable trees or other plants, or on areas where their roots may extend, or in locations where the chemical may be washed or moved into contact with their roots. Do not use on lawns, walks, driveways, tennis courts, or similar areas. Prevent drift of spray to desirable plants. Do not contaminate any body of water. Keep from contact with fertilizers, insecticides, fungicides, and seeds during storage.

Prior to using "Lorox Plus", consideration should be given to crop rotation plans. Crops other than soybeans can be extremely sensitive to low concentrations of "Lorox Plus" remaining in the soil the next planting season. Choice of rotational crop is restricted following application of "Lorox Plus". (See "ROTATIONAL CROP GUIDELINES").

Do not apply if soil pH is greater than 6.8.

"Lorox Plus" is registered for use on soybeans only. Use on crops other than soybeans will result in severe crop injury or crop death.

GENERAL INFORMATION

Du Pont "Lorox Plus" Herbicide is a dispersible granule formulation to be mixed with water and sprayed for selective preemergence weed control in soybeans. When applied according to the instructions on this label, it will control many broadleaf weeds and provide partial control of nutsedge.

Preemergence applications of "Lorox Plus" require rainfall or sprinkler irrigation to activate the herbicide. Degree of control and duration of effect depend on: rate used, weed spectrum, growing conditions at and following time of treatment, soil pH, texture, organic matter, moisture and precipitation.

Because most crops are highly sensitive to "Lorox Plus", all direct or indirect contact (such as spray drift) to crops or to land scheduled to be planted to crops other than soybeans should be avoided.

BIOLOGICAL ACTIVITY

"Lorox Plus" rapidly inhibits the growth of susceptible weeds. Following preemergence treatment, susceptible weeds may germinate and emerge, but growth then ceases and leaves become yellow 3-5 days after emergence. Death of leaf tissue and growing point will follow in some species while others will remain green but stunted and noncompetitive. "Lorox Plus" provides partial control of some annual grasses, but other products may be needed to ensure adequate grass control.

DIRECTIONS FOR USE

It is a violation of federal law to use this product in a manner inconsistent with the terms of the label.

Do not apply this product through any type of irrigation system.

EQUIPMENT-SPRAY VOLUMES

Apply uniformly by ground equipment with a properly calibrated low pressure (20 to 40 psi) fixed-boom sprayer equipped with fan-type nozzles or other carefully calibrated nozzles and screens no finer than 50 mesh. Use 10 to 40 gals of water per acre.

Continuous agitation in the spray tank is required to keep the material in suspension. Avoid overlap and shut off spray booms while starting, turning, slowing or stopping, as injury to the crop may result.

EARLY PREPLANT OR PREPLANT INCORPORATED (See Rate Chart)

"Lorox Plus" can be applied at planting or up to 30 days prior to planting either preemergence or preplant incorporated. If incorporated, "Lorox Plus" should be uniformly incorporated no deeper than the top 1-2" of soil prior to planting soybeans. Use equipment suited for proper incorporation into the top 1-2 inches of soil e.g.: Do-all, field cultivator or Roterra. If tank-mixed with a grass herbicide, such as a DNA like "Treflan"[4], "Sonalan"[4], or acetanilide such as "Dual"[2], or "Lasso", follow label instructions for proper incorporation of the grass herbicide into the top 1 to 2" of soil. Improper soil incorporation can result in erratic weed control or crop injury. Rainfall* (~1") is required to activate preplant incorporated herbicides such as "Lorox Plus" before weed emergence.

Use of a disk for incorporation is not recommended.

PREEMERGENCE USE ON SOYBEANS (See Rate Chart)

Apply before soybeans emerge. Do not spray over top of emerged soybeans or severe crop injury or crop death will occur.

Preemergence activity of "Lorox Plus" requires rainfall or irrigation to activate. Best results are obtained if treatment is made to moist soil and followed by rainfall or irrigation (1/2" to 1") before weeds germinate. On dry soil, chemical activation requires more moisture (1-2"). If moisture is insufficient to activate the herbicide, a rotary hoeing or shallow cultivation should be made after emergence of crop while weeds are small enough to be controlled by mechanical means. "Lorox Plus" can be applied preemergence in combination with "Lasso", "Dual", "Prowl"[3], "Surflan"[4] or following the use of a preplant grass material such as "Treflan" "Prowl" or "Sonalan". Consult labels for rates and use instructions.

SOYBEANS — MINIMUM TILLAGE OR NO-TILL (See Rate Chart)

"Lorox Plus" applied preemergence to soybeans but postemergence to broadleaf weeds and small annual grasses will provide limited burndown of existing weeds. When used for burndown add 0.25% V/V surfactant (1 qt/100 gal) such as Surfactant WK or crop oil concentrate at 1% V/V (1 qt/100 gal). For added postemergence control, especially on grasses, "Lorox Plus" can be applied with "Roundup"[1], paraquat or "Bronco"[1]. For added preemergence control, especially on grasses, "Lorox Plus" can be mixed with "Dual", "Lasso", or "Surflan". Consult labels for rates and use instructions. Maintain constant agitation of spray mixture.

RATE CHART

Apply within 2 weeks of planting as follows:

Soil Texture (1/2-3% Organic Matter*)	Rate (Ounces Per Acre)
Coarse: Loamy sand, Sandy loam	12-14
Medium: Loam, Silty loam, Silt	14-16
Fine: Silty clay loam, Clay loam, Clay	16-18

* Do not apply to soil with less than 1/2% O.M.

Do not apply "Lorox Plus" to soils with pH greater than 6.8.

Dosage rates listed are broadcast rates. For band applications use proportionally less.

Plant seed 1 1/2" to 2" deep on a flat or raised seedbed only, or crop injury may occur.

Soybean stunting may occur if excessive rainfall occurs after application but before soybeans germinate. Injury is more prevalent under poor drainage or compacted conditions or when soil is saturated for long periods of time. Soybeans rapidly outgrow stunting once favorable growing conditions return.

NOTE: Seedling disease, cold weather, deep planting (more than 2"), excessive moisture, high salt concentration or drought may weaken soybean seedlings and increase the possibility of crop injury.

WEEDS CONTROLLED

The following weeds will be controlled at the above rates preemergence:

Carpetweed	Pigweed species	Purslane
Cocklebur*	Palmer amaranth	Ragweed, common
Eastern black nightshade**	Redroot	Smartweeds, annual
	Smooth	Spotted spurge
Jimsonweed	Spiny amaranth	Velvetleaf
Lambsquarters	Tall waterhemp	
Mustards	Prickly sida (teaweed)	
	Partial Control:	
Burcucumber	Morningglory, annual*	Nutsedge species
Crabgrass species	Entireleaf	Panicum species
Eastern black nightshade	Ivyleaf	Ragweed, giant*
		Sunflower*
Foxtail species	Pitted	
Johnsongrass, seedling	Tall	

* Large seeded weeds, germinating deep in the soil, such as morningglory, sunflower, cocklebur and giant ragweed or weeds with subsequent flushes may require a cultivation or an application of Du Pont "Classic" Herbicide postemergent for season-long control. Consult "Classic" label for application directions, recropping intervals and use restrictions. NOTE: Consult "Lasso", "Dual", "Prowl", "Treflan", "Sonalan" or "Surflan" labels for additional weeds controlled when "Lorox Plus" is used in conjunction with these herbicides.

** Requires the addition of "Lasso".

SPRAY TANK PREPARATION

Fill tank 1/4 full with water. Start agitation system, add "Lorox Plus" and continue adding water. Add separately each additional component of any tank mix while adding water. Continue agitation throughout. If poor mixing should occur with any component, premix the component with two parts water before adding to the spray tank. A fertilizer solution may be used in the spray mixture. Small quantities should be tested for compatibility by the following procedures before full scale mixing.

1. Put 1 pint of fertilizer solution in quart jar.
2. Mix 2 teaspoons "Lorox Plus" with 2 tablespoons of water; mix thoroughly and add to fertilizer solution.
3. Close jar and shake well.
4. If other herbicides are to be used in the mixture, premix 2 teaspoons of wettable powder or 1 teaspoon of liquid with 2 tablespoons of water; add to "Lorox Plus"/fertilizer solution mixture.
5. Close jar and shake well.
6. Watch mixture for several seconds; check again in 30 minutes.
7. If mixture does not separate, foam, gel or become lumpy, it may be used.
8. Mixing ability may be improved by adding compatibility agents such as Kalo Laboratories "Compex"[5] or Witco Chemicals "Sponto"[6] 168D. Follow directions on container.

Provided the above procedure shows the mixture to be compatible, prepare the tank mixture as follows: Add the fertilizer solution to the spray tank first; with the agitator running, add the required amount of "Lorox Plus" and thoroughly mix. For tank mixtures with other herbicides, follow directions above.

When tank mixing with other herbicides, all applicable directions, restrictions and precautions for the additional herbicides are also to be followed. Use "Lorox Plus" spray preparations the same day as mixed or product degradation may occur. Thoroughly reagitate before using if allowed to settle.

ROTATIONAL CROP GUIDELINES

Crops that can be planted after using "Lorox Plus" are listed below. Choice of rotational crop is determined by time interval elapsed since last application. Prior to 18 months following application, do not rotate crops used for food or feed, which are not registered for use with "Lorox Plus" or Du Pont "Lorox" Herbicide onto areas previously treated with these chemicals.

Crops	Time Interval in Months Before Replanting
Soybeans	anytime
Wheat, Barley, Oats, Rye	4
Field Corn	10**
Clover	12
Sorghum	10**
Tobacco (transplant)	10**
Tomato (transplant)	10**

All Other Crops require a successful Field Bioassay***

** For applications of "Lorox Plus" after June 15, extend recrop interval one month.

***Field Bioassay - The season before planting any crop not listed above, a successful field bioassay must be completed. The field bioassay will detect small quantities of "Lorox Plus" which can remain in the soil and injure rotational crops. A successful field bioassay means growing to maturity a test strip of the crop(s) intended for production the following year. The test strip should cross the whole field including knolls and low areas.

Combination Treatment: When recommended tank mix combinations are used, consult the product label for replanting and recropping instructions and observe the directions that are the most restrictive.

REDUCED RATE FOLLOWED BY "CLASSIC" HERBICIDE FOR BROADLEAF WEED CONTROL IN SOYBEANS

DuPont "Lorox" Plus Herbicide is recommended preplant incorporated or preemergence at reduced rates followed by "Classic" Herbicide postemergence for broadleaf weed control in soybeans. This treatment will control those broadleaf weeds listed on the "Lorox" Plus and "Classic" Herbicide product labels.

Apply "Lorox" Plus preplant incorporated or preemergence at 12 ounces product per acre to any soil texture with 1/2 to 3% organic matter as directed on the product label. Follow the "Lorox" Plus preplant incorporated or preemergence reduced rate with a postemergence application of "Classic" Herbicide at its labeled rate when weeds are small and actively growing. (Usually 14-28 days after preplant incorporated or preemergence application of "Lorox" Plus reduced rate.)

For grass weed control "LOROX" Plus can be tank mixed with Dual, Lasso, Sonalan, Treflan, and Sonalan. Consult labels for rates and use instructions of these products.

"LOROX PLUS" HERBICIDE PLUS "PREVIEW" HERBICIDE TANK MIX AT REDUCED RATES FOR WEED CONTROL IN SOYBEANS IN THE STATES OF OH, IN, IL, IA AND MI.

"Lorox Plus" Herbicide and "Preview" Herbicide at Reduced Rates, Tank Mixed for Weed Control in Soybeans in the States of OH, IN, IL, IA AND MI.

DuPont "Lorox Plus" Herbicide and "Preview" Herbicide are recommended at reduced rates as a tank mixture for control of many annual broadleaf weeds and grasses in soybeans. This tank mixture will control those weeds listed on the "Lorox Plus" label.

DIRECTIONS FOR USE

Apply "Lorox Plus" and "Preview" tank mixture as a preplant incorporated (1-2 inches) or preemergence treatment. Follow the "Lorox Plus" package label for preplant incorporation guidelines and recommended equipment. Make a single application using 10-40 gallons of water per acre in ground equipment.

Choose the proper rates for your soil type from the rate chart below.

Soil Texture	Product Rate Per Acre (Ounces Per Acre)	
	"Lorox" Plus	"Preview"
Coarse: Loamy sand, Sandy loan	6	3
Medium: Loam, Silty loam, Silt	7	3.5
Fine: Silty clay loan, Clay loam, Clay	8	4

Do not use on soils with less than 1/2% or more than 3.5% organic matter.
Do not use on muck or peat soils

Important: Read and follow all use precautions, restrictions and recropping guidelines on the "Lorox" Plus and "Preview" product labels.

SPRAYER CLEANUP

To avoid subsequent injury to crops other than soybeans, immediately after spraying and prior to spraying other crops, thoroughly remove all traces of "Lorox Plus" from mixing and spray equipment. Be careful to rinse both the spray tank, boom and clean nozzles and screens. For more information, consult supplemental bulletin on sprayer cleanout following use of Du Pont "Classic", "Canopy", "Gemini", "Preview" or "Lorox Plus" Herbicides.

PRECAUTIONS

Do not apply "Lorox Plus" Herbicide to soils with a pH greater than 6.8.

CAUTION: Soil pH varies greatly even within the same field. pH variations as much as 2 pH units are common when the soil has a calcareous sub soil. Determine soil pH by laboratory analysis using a 1:1 soil:water suspension.

Do not graze treated fields or harvest for forage or hay.

Use of "Lorox Plus" may affect rotational crops. Follow "ROTATIONAL CROP GUIDELINES."

Do not use on sand, muck, or peat soils, nor on soils with less than 1/2% organic matter, as crop injury may occur.

Do not apply "Lorox Plus" after the soybean crop has emerged.

If "Scepter"[3], "Pursuit"[3] or any product containing the active ingredients, imazaquin or imazethapyr (such as "Squadron"[3], "Partner"[3]) is applied the same year as "Lorox Plus", do not plant crops other than soybeans for at least 15 months from the last application. Crops listed in the " 'Lorox Plus' Rotational Crop Guidelines" may be planted after 15 months; all other crops require a successful field bioassay.

Do not apply to land that has been or will be treated with "Glean", "Ally" or "Finesse" herbicides in the States of Nebraska, Kansas, North Dakota and South Dakota.

STORAGE AND DISPOSAL

Storage: Store product in original container only, away from other pesticides, fertilizer, food or feed.

Pesticide Disposal: Do not contaminate water, food or feed by storage or disposal. Wastes resulting from the use of this product may be disposed of on site or at an approved waste disposal facility.

Container Disposal: Triple rinse (or equivalent). Then offer for recycling or reconditioning, or puncture and dispose of in a sanitary landfill, or incineration, or, if allowed by state and local authorities, by burning. If burned, stay out of smoke.

NOTICE OF WARRANTY

Du Pont warrants that this product conforms to the chemical description on the label thereof and is reasonably fit for the purposes stated on such label only when used in accordance with the directions under normal conditions. It is impossible to eliminate all risks inherently associated with the use of this product. Crop injury, ineffectiveness or other unintended consequences may result because of such factors as weather conditions, presence of other materials, or the manner of use or application, all of which are beyond the control of Du Pont. In no case shall Du Pont be liable for consequential, special or indirect damages resulting from the use or handling of this product. All such risks are assumed by the buyer. DU PONT MAKES NO WARRANTIES OF MERCHANTABILITY OR FITNESS FOR A PARTICULAR PURPOSE NOR ANY OTHER EXPRESS OR IMPLIED WARRANTY EXCEPT AS STATED ABOVE.

[1]Registered Trademark of Monsanto Agricultural Products Co., St. Louis, MO
[2]Registered Trademark of Ciba-Geigy Corp., Greensboro, NC
[3]Registered Trademark of American Cyanamid Corp., Princeton, NJ
[4]Registered Trademark of Elanco Products Co. Indianapolis, IN
[5]Registered Trademark of Kalo Agricultural Chemicals Inc., Overland Park, KS
[6]Registered Trademark of Witco Chemical Corp., Houston, TX

NOTICE TO BUYER: Purchase of this material does not confer any rights under patents of countries outside of the United States.

AG-3280 8099/8099 9/6/89

This is a specimen label. DuPont is not responsible for the accuracy of the information contained herein. As labels are subject to revision, always carefully read and follow the label on the product container.

MANZATE 200 DF

Because the new Manzate 200 DF label had not received EPA approval at press time, it is not included in this supplement. Refer to page 869 in this edition for the current label approved as of November 15, 1989.

PREVIEW®

HERBICIDE
DISPERSIBLE GRANULES

	BY WEIGHT
ACTIVE INGREDIENT:	75%
Metribuzin [4-Amino-6-(1,1-dimethylethyl)-3-(methylthio)-1,2,4-triazin-5(4H)-one]	68.5%
Ethyl 2-[[[[(4-chloro-6-methoxypyrimidin-2-yl)-amino]carbonyl]amino]sulfonyl]benzoate	6.5%
INERT INGREDIENTS:	25%
TOTAL	100%

EPA Reg. No. 352-448
U.S. Patent 4,346,220; 4,547,215; 4,394,506 & 4,645,530

KEEP OUT OF REACH OF CHILDREN

PRECAUTIONARY STATEMENTS

HAZARDS TO HUMANS AND DOMESTIC ANIMALS

CAUTION!

HARMFUL IF SWALLOWED. CAUSES MODERATE EYE INJURY.

Avoid breathing dust or spray mist. Avoid contact with eyes. Wash thoroughly with soap and water after handling. Remove and wash contaminated clothing before reuse.

STATEMENT OF PRACTICAL TREATMENT

If in eyes: Flush eyes with plenty of water. Call a physician if irritation persists.

If swallowed: Call a physician or Poison Control Center. Drink 1 or 2 glasses of water and induce vomiting by touching back of throat with finger. Do not induce vomiting or give anything by mouth to an unconscious person.

For medical emergencies involving this product, call toll free 1-800-441-3637.

ENVIRONMENTAL HAZARDS

Do not apply directly to water or wetlands. Do not contaminate water by cleaning of equipment or disposal of wastes.

IMPORTANT

Injury to or loss of desirable trees or vegetation may result from failure to observe the following: Do not apply or drain or flush equipment on or near desirable trees or other plants, or on areas where their roots may extend, or in locations where the chemical may be washed or moved into contact with their roots. Do not use on lawns, walks, driveways, tennis courts or similar areas. Prevent drift of spray to desirable plants. Do not contaminate any body of water. Keep from contact with fertilizers, insecticides, fungicides and seeds during storage.

Thoroughly clean all traces of "PREVIEW" from application equipment immediately after use and prior to spraying crops other than soybeans. Failure to remove even small amounts of "PREVIEW" from application equipment may result in injury to subsequently sprayed crops.

Prior to using "PREVIEW", consideration should be given to crop rotation plans. Crops other than soybeans can be extremely sensitive to low concentrations of "PREVIEW" remaining in the soil the next planting season. Choice of rotational crop is restricted following application of "PREVIEW". (See "ROTATIONAL CROP GUIDELINES".)

Do not apply if soil pH is greater than 6.8.

Metribuzin is a chemical which can travel (seep or leach) through soil and can contaminate groundwater which may be used as drinking water. Metribuzin has been found in groundwater as a result of agricultural use. Users are advised not to apply metribuzin where the water table (groundwater) is close to the surface and where the soils are very permeable; i.e., well drained soils such as loamy sands. Your local agricultural agencies can provide further information on the type of soil in your area and the location of groundwater.

GENERAL INFORMATION

Du Pont "PREVIEW" Herbicide is a dispersible granule formulation to be mixed with water and sprayed for selective preemergence and preplant incorporated weed control in soybeans. When applied according to the instructions on this label, it will control many broadleaf weeds and provide partial control of nutsedge.

Preplant incorporated and preemergence applications of "PREVIEW" require rainfall or sprinkler irrigation to activate the herbicide. Degree of control and duration of effect depend on: Rate used, weed spectrum, growing conditions at and following time of treatment, soil pH, texture, organic matter, moisture and precipitation.

Because most crops are highly sensitive to "PREVIEW", all direct or indirect contact (such as spray drift) to crops or to land scheduled to be planted to crops other than soybeans should be avoided.

BIOLOGICAL ACTIVITY

"PREVIEW" rapidly inhibits the growth of susceptible weeds. Following activation of preplant incorporation or preemergence treatment, susceptible weeds may germinate and emerge, but growth then ceases and leaves become yellow 3-5 days after emergence. Death of leaf tissue and growing point will follow in some species while others will remain green but stunted and noncompetitive. "PREVIEW" provides partial control of some annual grasses when used preplant or preemergence but other products may be needed to ensure adequate grass control.

DIRECTIONS FOR USE

It is a violation of federal law to use this product in a manner inconsistent with its labeling.

Do not apply this product through any type of irrigation system.

EQUIPMENT-SPRAY VOLUMES

Apply uniformly by ground equipment with a properly calibrated low pressure (20 to 40 psi) fixed-boom sprayer equipped with fan-type nozzles or other carefully calibrated nozzles and screens no finer than 50 mesh. Use 10 to 40 gals of water per acre.

Continuous agitation in the spray tank is required to keep the material in suspension. Avoid overlap and shut off spray booms while starting, turning, slowing or stopping, as injury to the crop may result.

EARLY PREPLANT OR PREPLANT INCORPORATED
(See rate chart)

"PREVIEW" can be applied at planting or up to 30 days prior to planting either preemergence or preplant incorporated. If incorporated, uniformly incorporate the treatment no deeper than the top 1-2" of soil prior to planting soybeans. Use equipment suited to proper incorporation into the top 1-2 inches of soil, e.g.: Do-all, field cultivator, roterra. "PREVIEW" can be tank mixed with "Command"[10] and applied preplant incorporated. Refer to the "Command" label for specific use instructions. If tank-mixed with a grass herbicide such as a DNA like "Treflan"[1], "Sonalan"[1], or acetanilide such as "Dual"[2], or "Lasso"[3], follow label instructions for proper incorporation of the grass herbicide into the top 1 to 2 inches of soil. Improper soil incorporation can result in erratic weed control or crop injury. Rainfall* (~1") is required to activate preplant incorporated herbicides such as "PREVIEW" before weed emergence.

"PREVIEW" + "Treflan", "Prowl"[4], "Dual" or "Sonalan" — Apply within 2 weeks of planting.
"PREVIEW" + "Treflan" (Early PPI) — Apply within 30 days of planting.
"PREVIEW" + "Lasso" — Apply within 7 days of planting.
"PREVIEW" + "Command" — Apply within 30 days of planting.

PREEMERGENCE (See rate chart)

Preemergence application: Apply before soybeans emerge. Do not spray overtop of emerged soybeans, as severe injury or death of crop will occur.

Preemergence activity of "PREVIEW" requires rainfall* to activate. Best results are obtained if treatment is made to moist soil and followed by rainfall or irrigation (~1") before weeds germinate. On dry soil, more moisture is required for activation (1 to 2") before weed emergence. If moisture is insufficient to activate the herbicide, a rotary hoeing or shallow cultivation should be made after emergence of crop while weeds are small enough to be controlled by mechanical means. "PREVIEW" can be applied preemergence in combination with "Lasso", "Dual", "Prowl", "Surflan" or following the use of a preplant grass material such as "Treflan" or "Sonalan". Consult labels for rates and use instructions.

*Rainfall required to activate ppi or pre treatments depends on soil moisture when activating moisture is received. Several small rainfalls of less than 1/4" are not as beneficial as one large rain of 1/2 inch or more for activation.

MINIMUM TILLAGE OR NO-TILL (See rate chart)

"PREVIEW" applied preemergence to soybeans but postemergence to broadleaf weeds and small annual grasses will provide some burndown of existing weeds as well as normal preemergence broadleaf weed control. When used for burndown, add 0.25% v/v (1 qt./100 gal.) surfactant such as Surfactant WK or crop oil concentrate at 1% v/v (1 qt./100 gal.). For added postemergence control, especially on grasses, "PREVIEW" can be applied with "Roundup"[3], paraquat, or "Bronco"[3]. For added preemergence grass control, "PREVIEW" can be mixed with "Dual", "Surflan" or "Lasso". Consult labels for rates and use instructions. Maintain constant agitation of spray mixture.

RATE CHART

Apply as follows:

Soil Texture	Ounces of "Preview" Per Acre	
	Percent Organic Matter in Soil*	
	½-3%	3-5%
Coarse: Loamy sand, Sandy loam	6	7
Medium:**Loam, Silty loam, Silt	7	8
Fine: Silty clay loam, Clay loam, Clay	8	9-10

* Do not apply to soil with less than ½% O.M.
Do not apply "PREVIEW" to soils with pH greater than 6.8.
** For medium soils 1½ to 3% O.M. in MO and KS, use 7-8 oz./A.

REDUCED RATE FOLLOWED BY "CLASSIC" HERBICIDE FOR BROADLEAF WEED CONTROL IN SOYBEANS

DuPont "PREVIEW" Herbicide is recommended preplant incorporated or preemergence at reduced rates followed by "Classic" Herbicide postemergence for broadleaf weed control in soybeans. This treatment will control these broadleaf weeds listed on the PREVIEW and "Classic" Herbicide product labels.

Apply "PREVIEW" at reduced rates (see rate chart below) preplant incorporated or preemergence as directed on the product label. Follow the "PREVIEW" preplant incorporated or preemergence reduced rate with a postemergence application of "Classic" Herbicide at its labeled rate when weeds are small and actively growing. (Usually 14-28 days after preplant incorporated or preemergence application of "Preview" reduced rate).

For grass weed control "PREVIEW" can be applied in tank mix with "Lasso", "Dual", "Prowl", "Treflan" or "Sonalan". Read and follow the product label for use of these products.

"PREVIEW" Reduced Rate Chart

Soil Texture	Ounces of "Preview" Per Acre	
	Percent Organic Matter in Soil*	
	½-3%	3-5%
Coarse: Loamy sand, Sandy loam:sl	5	6
Medium: Loam, Silty loam, Silt	6	7
Fine: Silty clay loam, Clay loam, Clay	7	8

* Do not apply to soil with less than ½% O.M.
Do not apply "PREVIEW" to soils with pH greater than 6.8.

Dosage rates listed are broadcast rates. For band applications use proportionally less.

Plant seed 1 1/2" to 2" deep on a flat or raised seedbed only, or crop injury may occur.

Soybean stunting may occur if rainfall is excessive during soybean germination and emergence. Injury is more prevalent under poor drainage or compacted conditions or when soil is saturated for long periods of time. Soybeans outgrow stunting once favorable growing conditions return.

Soybean varieties such as Altona, Coker 6995 and 156, Govan, Semmes, Tracy, Vansoy, Terra Vig 505 and 606, AP 71, NKS 1884, Agropro 55, Asgrow 6520, Maple Amber, Portage, and Vinton 81 are more sensitive to "PREVIEW". Injury may occur if "PREVIEW" is used on these varieties. Before use on any soybean variety, tolerance to "PREVIEW" must first be determined. Injury to soybeans may occur if "PREVIEW" is used on soils having a calcareous surface layer, or if used in conjunction with soil-applied organic phosphate pesticides such as "Dasanit"[5], "Disyston"[5], "Mocap"[6], Nemacur[5], Thimet[4], parathion or "Lorsban"[7].

NOTE: Injury may occur if atrazine was applied on the soil the year before use of "PREVIEW". Seedling disease, cold weather, deep planting (more than 2"), excessive moisture, high salt concentration or drought may weaken soybean seedlings and increase the possibility of crop injury.

WEEDS CONTROLLED

The following weeds will be controlled at the above rates both preplant incorporated and preemergence:

Carpetweed	Pigweed species	Purslane
Cocklebur*	Palmer amaranth	Ragweed, common
Hophornbeam copperleaf	Redroot	Smartweeds, annual
	Smooth	Spotted spurge
Jimsonweed	Spiny amaranth	Sunflower*
Lambsquarters	Tall waterhemp	Velvetleaf
Mustards	Prickly sida (teaweed)	

PARTIAL CONTROL

Burcucumber	Morningglory, annual*	Nutsedge species
Crabgrass species	Tall	Panicum species
Eastern black nightshade	Ivyleaf	Ragweed, giant*
Foxtail species	Pitted	
Johnsongrass, seedling	Entireleaf	

*Large seeded weeds, germinating deep in the soil, such as morningglory, sunflower, cocklebur and giant ragweed or weeds with subsequent flushes may require a cultivation or postemergence application of Du Pont "Classic" Herbicide for season-long control. Consult "Classic" label for application directions and use restrictions.

NOTE: Consult "Lasso", "Dual", "Prowl", "Treflan", "Sonalan", "Surflan", or "Command" labels for additional weeds controlled when "PREVIEW" is used in conjunction with these herbicides.

SPRAY TANK PREPARATION

Fill tank 1/4 full with water. Start agitation system, add "PREVIEW" and continue adding water. Add separately each additional component of any tank mix while adding water. Continue agitation throughout. If poor mixing should occur with any component, premix the component with two parts water before adding to the spray tank. A fertilizer solution may be used in the spray mixture. Small quantities should be tested for compatibility by the following procedures before full scale mixing.

1. Put 1 pint of fertilizer solution in quart jar.
2. Mix 2 teaspoons "PREVIEW" with 2 tablespoons of water; mix thoroughly and add to fertilizer solution.
3. Close jar and shake well.
4. If other herbicides are to be used in the mixture, premix 2 teaspoons of wettable powders or 1 teaspoon of liquids with 2 tablespoons of water; add to "PREVIEW"/fertilizer solution mixture.
5. Close jar and shake well.
6. Watch mixture for several seconds; check again in 30 minutes.
7. If mixture does not separate, foam, gel or become lumpy, it may be used.
8. Mixing ability may be improved by adding compatibility agents such as Kalo Laboratories "Compex"[8] or Witco Chemicals "Sponto"[9] 168D. Follow directions on container.

Provided the above procedure shows the mixture to be compatible, prepare the tank mixture as follows: Add the fertilizer solution to the spray tank first; with the agitator running, add the required amount of "PREVIEW" and thoroughly mix. For tank mixtures with other herbicides, follow directions above.

Use "PREVIEW" spray preparations the same day as mixed or product degradation may occur. Thoroughly reagitate before using if allowed to settle.

When tank mixing with other herbicides, all applicable directions, restrictions and precautions for the additional herbicides are also to be followed.

ROTATIONAL CROP GUIDELINES

Crops that can be planted after using "PREVIEW" are listed below. Choice of rotational crop is determined by time interval elapsed since last application.

Crops	Time Interval in Months Before Replanting
Soybeans	anytime
Wheat, Barley, Ryegrass	4
Field Corn	10**
Alfalfa	10
Dry beans	12
Sorghum	12
Clover	12
Tobacco (transplant)	10
Tomato (transplant)	10

All other crops require a successful field bioassay***
The suggested interval before initiating a field bioassay is 10 months.

**For Applications of "PREVIEW" made after June 15, extend recrop interval 1 month.

***Field Bioassay - The season before planting any crop not listed above, a successful field bioassay must be completed. The field bioassay will detect small quantities of "PREVIEW" which can remain in the soil and injure rotational crops. A successful field bioassay means growing to maturity a test strip of the crop(s) intended for production the following year. The test strip should cross the whole field including knolls and low areas.

Combination Treatment: When recommended tank mix combinations are used, consult the product label for replanting and recropping instructions and observe the directions that are the most restrictive.

SPRAYER CLEANUP

To avoid subsequent injury to crops other than soybeans, immediately after spraying and prior to spraying other crops, thoroughly remove all traces of "PREVIEW" from mixing and spray equipment. Be careful to rinse both the spray tank, boom and clean nozzles and screens. For more information, consult supplemental bulletin on sprayer cleanout following use of Du Pont "Classic", "Canopy", "Gemini", "PREVIEW" or "Lorox Plus" Herbicides.

PRECAUTIONS

Do not apply "PREVIEW" Herbicide to soils with a pH greater than 6.8.

CAUTION: Soil pH varies greatly even within the same field. pH variations as much as 2 pH units are common when the soil has a calcareous sub soil. Determine soil pH by laboratory analysis using a 1:1 soil:water suspension.

Do not graze treated fields or harvest for forage or hay.

Use of "PREVIEW" may affect rotational crops. Follow "ROTATIONAL CROP GUIDELINES".

Do not use on sand, muck, or peat soils, nor on soils with less than 1/2% organic matter, as crop injury may occur.

Do not apply "PREVIEW" after the soybean crop has emerged.

If "Scepter"[4], "Pursuit"[4] or any product containing the active ingredients imazaquin or imazethapyr (such as "Squadron"[4], "Partner"[3]) is applied the same year as "PREVIEW", do not plant crops other than soybeans for at least 15 months from the last application. Crops listed in the " 'PREVIEW' Rotational Crop Guidelines" may be planted after 15 months; all other crops require a successful field bioassay.

Do not apply to land that has been or will be treated with "Ally", "Glean" or "Finesse" herbicides in the states of Nebraska, Kansas, North Dakota and South Dakota.

STORAGE AND DISPOSAL

Storage: Store product in original container only, away from other pesticides, fertilizer, food or feed.

Pesticide Disposal: Do not contaminate water, food or feed by storage or disposal. Wastes resulting from the use of this product may be disposed of on site or at an approved waste disposal facility.

Container Disposal: Triple rinse (or equivalent). Then offer for recycling or reconditioning, or puncture and dispose of in a sanitary landfill, or incineration, or, if allowed by state and local authorities, by burning. If burned, stay out of smoke.

NOTICE TO BUYER: Purchase of this material does not confer any rights under patents of countries outside of the United States.

NOTICE OF WARRANTY

Du Pont warrants that this product conforms to the chemical description on the label thereof and is reasonably fit for the purposes stated on such label only when used in accordance with the directions under normal use conditions. It is impossible to eliminate all risks inherently associated with the use of this product. Crop injury, ineffectiveness or other unintended consequences may result because of such factors as weather conditions, presence of other materials, or the manner of use or application, all of which are beyond the control of Du Pont. In no case shall Du Pont be liable for consequential, special or indirect damages resulting from the use or handling of this product. All such risks are assumed by the buyer. DU PONT MAKES NO WARRANTIES OF MERCHANTABILITY OR FITNESS FOR A PARTICULAR PURPOSE NOR ANY OTHER EXPRESS OR IMPLIED WARRANTY EXCEPT AS STATED ABOVE.

1Trademark of Elanco Products Co., Indianapolis, IN
2Trademark of Ciba-Geigy Corp., Greensboro, NC
3Trademark of Monsanto Agricultural Products Co., St. Louis, MO
4Trademark of American Cyanamid Corp., Princeton, NJ
5Trademark of Mobay Chemical Corp., Kansas City, MO
6Trademark of Rhone-Poulenc, Inc., Monmouth Junction, NJ
7Trademark of Dow Chemical Co., Midland, MI
8Trademark of Kalo Agricultural Chemicals, Inc., Overland Park, KS
9Trademark of Witco Chemical Corp., Houston, TX
10Trademark of FMC Corp., Philadelphia, PA

AG-3229 8098/8089 4/11/88

This is a specimen label. DuPont is not responsible for the accuracy of the information contained herein. As labels are subject to revision, always carefully read and follow the label on the product container.

TELAR®

HERBICIDE
DISPERSIBLE GRANULE

ACTIVE INGREDIENT:	BY WEIGHT
Chlorsulfuron	
2-Chloro-N-[(4-methoxy-6-methyl-1,3,5-triazin-2-yl)- aminocarbonyl]benzenesulfonamide	75%
INERT INGREDIENTS:	25%
TOTAL	100%

EPA Reg. No. 352-404

U.S. Pat. 4,127,405

KEEP OUT OF REACH OF CHILDREN
CAUTION

PRECAUTIONARY STATEMENTS

HAZARDS TO HUMANS AND DOMESTIC ANIMALS

CAUTION!

MAY IRRITATE EYES, NOSE, THROAT AND SKIN.

Avoid breathing dust or spray mist. Avoid contact with skin, eyes and clothing.

In case of contact with eyes, immediately flush with plenty of water. Get medical attention if irritation persists. Wash thoroughly after handling. Remove and wash contaminated clothing before reuse.

For medical emergencies involving this product, call toll free 1-800-441-3637.

ENVIRONMENTAL HAZARDS

Do not apply directly to any body of water. Do not contaminate water by cleaning of equipment or disposal of wastes.

IMPORTANT

DO NOT USE ON FOOD OR FEED CROPS. Injury to or loss of desirable trees or other plants may result from failure to observe the following: Do not apply "TELAR" (except as recommended), or drain or flush equipment on or near desirable trees or other plants, or on areas where their roots may extend or in locations where the chemical may be washed or moved into contact with their roots.

Do not use on lawns, walks, driveways, tennis courts or similar areas. Keep from contact with fertilizers, insecticides, fungicides and seeds.

Do not treat powdery, dry soil and light, sandy soils when there is little likelihood of rainfall soon after treatment; in areas where sensitive crops are grown, do not treat before soil has been settled by rain as off-target movement may occur. Do not apply to impervious substrates such as paved or highly compacted surfaces nor to frozen ground as off-target movement will occur.

Do not apply in or on irrigation ditches or canals including their outer banks.

Do not apply to any body of water. Do not apply where runoff water may flow onto agricultural land, as injury to crops may result. Do not apply "TELAR" during periods of intense rainfall or to soils saturated with water as off-target movement may occur.

Do not apply this product through any type of irrigation system.

Do not allow spray to drift onto adjacent crops or other desirable plants or trees as injury may occur.

Following a "TELAR" application, do not use sprayer for application to crops. This is extremely important, as low rates of "TELAR" can kill or severely injure most crops.

Do not use this products in the following counties of Colorado: Saguache, Rio Grande, Alamosa, Costilla and Conejos.

GENERAL INFORMATION

Du Pont "TELAR" Herbicide is a dispersible granule to be mixed in water and applied as a spray for control of many broadleaf weeds on non-cropland. It is noncorrosive, nonflammable, nonvolatile and does not freeze.

"TELAR" may be applied preemergence or postemergence to weeds; best results are obtained if application is made before or during early stages of weed growth. "TELAR" should be used during seasons when rainfall occurs as moisture is required to move "TELAR" into the root zone of weeds before they develop an established root system. Preemergence treatments control or suppress weeds through root uptake while postemergence works through root uptake and foliar uptake. Under limited rainfall conditions, "TELAR" may not provide satisfactory control of hard-to-kill perennials and annuals.

"TELAR" is absorbed by both roots and foliage of weeds, resulting in visual effects progressing from growth inhibition to reddish-purplish coloration, chlorosis, necrosis, vein discoloration and death of terminals. Initial effects are usually seen 2 or 3 weeks following application; however, the final effects are evident at about 4 to 6 weeks after application. Warm, moist conditions following treatment promote the activity of "TELAR" while cold, dry conditions delay the activity of "TELAR." Weeds hardened off by cold weather or drought stress may not be fully controlled.

Degree of control and duration of effect depend on rate used, weed spectrum, weed size, growing conditions at and following time of treatment, soil pH, precipitation, and soil organic matter.

Following use of "TELAR", some naturally-occurring biotypes* of certain weeds listed on this label may not be effectively controlled by this product.

To delay the occurrence of resistant biotypes, use "TELAR" in tank mixes and/or sequential treatments with other registered hjerbicides effective on the same broadleaf weed species. Do not let weed escapes go seed; time postemergence treatments before seed formation. Respray problem areas in a timely and effective manner using a broadleaf herbicide having a different mode of action**.

* Biotypes are naturally-occurring individuals of the species which have a slightly different genetic makeup. Resistant biotypes may look exactly the same as susceptible biotypes. Herbicide resistant biotypes are able to survive a use rate several times higher than needed to control susceptible biotypes.

** Mode of action is the chemical interaction that interrupts a biological process necessary for plant growth and development.

Consult your supplier, applicator, consultant, appropriate state agricultural extension service representative or your local Du Pont representatives for specific recommendations.

DIRECTIONS FOR USE

It is a violation of federal law to use this product in a manner inconsistent with its labeling.

"TELAR" Herbicide should be used only in accordance with recommendations on this label or in separate published Du Pont recommendations available through local suppliers.

Du Pont will not be responsible for losses or damages resulting from the use of this product in any manner not specifically recommended by Du Pont. User assumes all risks associated with such nonrecommended use.

Du Pont "TELAR" Herbicide is recommended for preemergence and postemergence control of many broadleaf weeds on noncropland areas. For best results, apply postemergence to young, acitvely growing weeds (less than 2" tall or 2" across) at any time of the year, except when ground is frozen.

Treated soils should be left undisturbed to reduce the potential for "TELAR" movement by soil erosion due to wind or water. Injury to crops may occur when treated soil is blown or moved onto land used to produce crops.

Rate selection is based on weed species, weed size and soil texture. Use the higher rates on established plants and on fine textured soils and lower rates on smaller weeds and coarse textured soils.

For tank mixes use the most restrictive limitations from labeling of both products.

NON-CROP (INDUSTRIAL) WEED CONTROL

"TELAR" is recommended for use for weed control on noncrop, industrial sites such as airports, military installations, fence rows, roadsides and associated rights-of-ways, lumberyards, petroleum tank farms, pipeline and utility rights-of-ways, pumping installations, railroads, storage areas, plant sites and other similar areas including governmental and private lands.

When applied at lower dosage rates (as listed below), "TELAR" provides short term control of weeds listed; usage at higher rates increases length and degree of weed control. Combinations with other herbicides increases the spectrum of weeds controlled and bare ground objectives can be achieved with higher rates of "TELAR" plus residual type companion herbicides.

In states of Montana, North Dakota and South Dakota, "TELAR" is for use only on railroad rights-of-ways.

Apply "TELAR" in 10 to 40 gals water at 1/4 to 3 oz per acre per year to control many annual, biennial and perennial broadleaf weeds in non-cropland areas. The addition of a surfactant at 1 qt./100 gals water for postemergence applications is recommended to enhance activity. For handgun application, mix 1 oz "TELAR" per 100 gals water and apply up to 300 gals spray mix per acre including surfactant as described above.

Since the degree of control and duration of effect will vary with the amount of chemical applied, soil texture, soil pH, soil organic matter, weed size, rainfall and other conditions, it is suggested that users limit their first use to small areas.

WEEDS CONTROLLED AND USE RATES

1/4 to 1/2 ozs per acre

Annual sowthistle	London Rocket
Blue mustard	Mayweed
Common chickweed	Miners lettuce
Common speedwell	Pineappleweed
Conical catchfly	Prostrate pigweed
Fiddleneck (tarweed)	Redroot Pigweed
Field Pennycress	Shepherdspurse
Fixweed*	Smooth pigweed
Hempnettle	Treacle mustard
Henbit	Tumble mustard (Jim Hill)
Lixweed	Wild mustard

1/2 to 1 oz per Acre

Bouncingbet	Goldenrod
Bur beakchervil	Groundsel
Buttercup	Musk Thistle
Canada Thistle*†	Sweet Clover*
Common lanbsquarters	Tumble mustard
Common sunflower	Turkey mullein*
Common speedwell*	Whitetop (hoary cress)†
Dandelion*	Wild parsnip

* Partial control only.
† Prebloom to bloom and fall rosette are recommended timings. Surfactant is required at 1 to 2 qts/100 gallons water.

1 to 3 ozs per Acre

Annual ryegrass (Lolium spp)*	Foxtail (Setaria spp)*
Aster	Horsetail (Equisetum spp)
Bedstraw	Kochia**
Black mustard	Prickly lettuce**
Burclover	Prostrate knotweed*
Common cinquefoil	Puncturevine
Common mallow	Red clover
Common mullein	Russian thistle**
Common ragweed*	Russian Knapweed†
Common yarrow	Scouringrush (Equisetum spp)
Corn spurry	Tansymustard
Cow cockle	White clover
Curly dock	Wild carrot
Dyer's woad	Wild garlic/wild onion
Flixweed	Yellow Starthistle*

* Partial control only.
** Except resistant biotypes.
† Prebloom to bloom and fall rosette are recommended timings. Surfactant is required at 1 to 2 qts/100 gallons water.

TANK MIXTURES WITH OTHER HERBICIDES

For broader spectrum and improved preemergence to early postemergence week control, tank mix "TELAR" with "Karmex" DF, "Krovar" I DF, dicamba, 2,4-D or glyphosate.

Do not tank mix "Telar" with Du Pont "Hyvar" XL Herbicide.

SPECIAL WEED PROBLEMS

Kochia, Russian Thistle and Prickly Lettuce: Since these weeds have biotypes resistance to "TELAR", tank mixture combinations should be used involving herbicides with different modes of action such as "Karmex" DF, "Hyvar" X, "Krovar" I DF. Where resistance is known, weeds should not be allowed to form mature seed but should be treated postemergence with herbicides registered for control of these weeds such as 2,4-D and dicamba.

Turf (Industrial, Unimproved Only): "TELAR" is recommended for use on unimproved industrial turf, on roadsides, and on other noncropland areas where bahiagrass, bermudagrass, bluegrass, fescue, wheatgrass and smooth brome are well established, and desired as ground covers. Apply "TELAR" at the rate of 1/4 to 1 oz per acre per year to bahiagrass, bermudagrass, bluegrass and wheatgrass and at the rate of 1/4 to 1/2 oz per acre per year to fescue and smooth brome. The higher rates and/or the addition of surfactant at 1 qt/100 gals water may result in temporary chlorosis of desirable grasses; application of "TELAR" may also suppress grass growth and inhibit grass seedhead formation. Desirable grasses should be well established at time of application or top kill and stand reduction may occur. For best results, treat turf at green-up and prior to formation of seed heads (boots). "TELAR" controls weeds listed at appropriate rates under NON-CROP (INDUSTRIAL) WEED CONTROL.

For growth suppression and seedhead inhibition ("chemical mowing") in mixed established stands of fescue/bluegrass, apply 1/4 oz "TELAR" plus 1/4 to 1/2 pt "Embark" 2-S[1] per acre. Time treatment to green-up until seedhead appearance.

In the Pacific Northwest for growth suppression of fescue, annual bluegrass, perennial ryegrass,smooth bromegrass, orchardgrass and reed canarygrass, apply 1/2 oz "TELAR" plus 1/2 to 1 pt. "Embark" 2-S per acre during the active vegetative growth stages of turf. Addition of a non-ionic surfactant at 1/4% by volume (1 quart/100 gallons) will improve performance, but may temporarily increase chlorosis of desirable grasses.

Note: Do not apply "TELAR" or "TELAR" plus "Embark" to turf that is under stress from drought, insects, disease, cold temperatures or poor fertility, as injury may result. Do not apply to turf less than 1 year old. Do not use "TELAR" plus "Embark" on bahiagrass turf, as severe injury will result. Do not exceed 1/2 ounce "TELAR" per acre within a 12-month period.

[1]Registered trademark 3M Corporation

For weed control established in bentgrasses, blue grama, bluestems (big, little plains, sand, ww spar), bromegrasses (meadow, smooth), buffalograss, galleta, green needlegrass, green sprangetop, indiangrass, indian ricegrass, kleingrass, lovegrasses (sand, weeping), orchardgrass, prairie sandreed, sand dropseed, sheep fescue, sideoats grams, switchgrass, wheatgrasses (crested, intermediate pubescent, slender, streambank, tall thickspike, western) and wildrye grasses (beardless, russian), apply 1/2 oz "Telar" in a minimum of 10 gals water per acre; add 0.25% surfactant by volume where emerged seedling weeds are present. Where new seeding is desired, planting in cultivated areas is permissible 6 months after treatment. Treatment controls or suppresses weeds as shown under Weeds Control.

SPRAY PREPARATION: Fill the spray tank half full of water and with agitator running, add the proper amount of "TELAR." Finish adding the required amount of water. Continuous agitation is required to keep the product in suspension. For postemergence applications, a nonionic surfactant of at least 80% active ingredient may be added to improve wetting and/or contact activity. To minimize drift, a drift control agent may be added at the recommended label rate. "TELAR" spray preparations may degrade in acidic solutions if not used in 24 hours; they are stable in alkaline solutions. Thoroughly reagitate before using.

APPLICATION TECHNIQUE: Before spraying, calibrate the equipment to determine the quantity of water necessary to uniformly cover the vegetation and soil in a measured area to be treated. Use a sprayer properly calibrated to a constant speed of travel and rate of delivery. Avoid overlapping and shut off spray booms while starting, turning, slowing or stopping when spraying industrial turf as injury to the desired species may result. Extreme care must be taken to prevent drift.

"TELAR" may also be applied by handgun.

SPRAY VOLUME: Apply "TELAR" in ground equipment. Select a spray volume that will assure thorough coverage and uniform spray pattern. Do not use less than 10 gallons of water per acre. Spray volumes of 20 to 40 gallons per acre are preferred. Spray pressure of 25 to 35 psi are adequate.

For handgun application, use 100 to 300 gals spray per acre.

AGITATION: Use mechanical or bypass agitation to thoroughly mix the spray solution. After initial mixing, do not use excessive agitation when using a surfactant as foaming problems may result. To avoid foaming problems, use antifoam agents.

SPRAYER CLEANUP: Immediately after spraying thoroughly remove all traces of "TELAR" from mixing and spray equipment as follows:

1. Drain tank, then flush tank, boom and hoses with clean water for a minimum of 10 minutes.
2. Fill the tank with clean water, then add 1/2 gallon chlorine bleach (containing 5 1/4% sodium hypochlorite) per 100 gallons of water. Turn on sprayer long enough to flush through boom and hoses. Stop spraying but keep agitator working in the tank for 15 minutes, then drain.
3. Repeat Step 2.
4. Nozzles and screeens should be removed and cleaned separately. To remove traces of chlorine bleach, rinse the tank thoroughly with clean water and flush through hoses and boom.

NOTE: This sprayer cleanup procedure is only effective for "TELAR" and for general uses specified under "Directions for Use"; sprayer should not be used on plants such as fine turf and ornamentals and other desirable plants.

STORAGE AND DISPOSAL

STORAGE: Store product in original container only, away from other pesticides, fertilizer, food or feed.

PRODUCT DISPOSAL: Do not contaminate water, food or feed by storage or disposal. Wastes resulting from the use of this product may be disposed of on site or at an approved waste disposal facility or on non-crop sites as previously recommended.

CONTAINER DISPOSAL: Triple rinse (or equivalent) the container. Then offer for recycling or reconditioning, or puncture and dispose of in a sanitary landfill, or incineration, or, if allowed by state and local authorities, by burning. If burned, stay out of smoke.

NOTICE TO BUYER: Purchase of this material does not confer any rights under patents of countries outside of the United States.

NOTICE OF WARRANTY: Du Pont warrants that this product conforms to the chemical description on the label thereof and is reasonably fit for purposes stated on such label only when used in accordance with directions under normal use conditions. It is impossible to eliminate all risks inherently associated with the use of this product. Injury to adjacent crops, ineffectiveness or other unintended consequences may result because of such factors as weather conditions, presence of other materials, or the manner of use or application, all of which are beyond the control of Du Pont. In no case shall Du Pont be liable for consequential, special or indirect damages resulting from the use or handling of this product. All such risks shall be assumed by the buyer. DU PONT MAKES NO WARRANTIES OF MERCHANTABILITY OR FITNESS FOR A PARTICULAR PURPOSE NOR ANY OTHER EXPRESS OR IMPLIED WARRANTY EXCEPT AS STATED ABOVE.

AG-3288 8109 10/3/89

HOECHST-ROUSSEL AGRI-VET COMPANY

For Hoelon 3EC Herbicide as found on page 1106, the following supplement should be included as part of the label:

Supplemental Labeling

HOELON 3EC HERBICIDE

For Postemergence Control of Goosegrass (Silver Crabgrass) in Bermudagrass Turf

EPA REG. NO. 8340-20-54382 **EPA SLN NO. FL 890030**

FOR DISTRIBUTION AND USE ONLY WITHIN THE STATE OF FLORIDA UNDER A SPECIAL LOCAL NEED AUTHORIZATION UNDER FIFRA SECTION 24(c)

FOR USE ON GOLF COURSES AND SOD FARMS ONLY

RESTRICTED USE PESTICIDE

For retail sale to and use only by certified applicators or persons under their direct supervision and only for those uses covered by the certified applicator's certification.

PRECAUTIONARY STATEMENTS

DANGER

Important

Before using this product, read and carefully observe all applicable directions, restrictions and precautionary statements on the EPA approved Hoelon® 3EC Herbicide container and use directions labels.

HAZARDS TO HUMANS AND DOMESTIC ANIMALS

Causes eye and skin damage. Do not get in eyes or on skin or clothing. Wear goggles and respirator when handling. Avoid contact or inhalation of spray mist. Wash thoroughly after using and change clothing. Do not take internally.

The following protective apparel must be worn during all loading and mixing operations and when applied by ground equipment: impermeable rubber gloves (such as neoprene or PVC), waterproof boots, impermeable pants and shirts, goggles and a cartridge type respirator. Aerial pilots are not to be involved in the mixing and loading unless the aforementioned equipment is worn. Pilots must wear a cartridge type respirator during application unless the aircraft is equipped with an air filter system.

The Environmental Protection Agency has concluded that diclofop-methyl, the active ingredient in this product, produced tumors in laboratory mice. This effect was not observed in other species tested. The user must read and follow all precautionary statements on this label. Give particular attention to protective clothing requirement.

ENVIRONMENTAL HAZARDS

This pesticide is toxic to fish. Use with care when applying in areas adjacent to any body of water. Keep out of lakes, streams, ponds, drainage basins, tidal marshes and estuaries. Do not apply when weather conditions favor runoff or drift. Avoid direct application or drift of spray material to water surfaces. Do not apply within 100 feet of aquatic habitat. Do not contaminate arable land and/or water when disposing of equipment washwaters.

STATEMENT OF PRACTICAL TREATMENT

In case of skin or eye contact, immediately flush with plenty of water for at least 15 minutes.

If swallowed, DO NOT INDUCE VOMITING, but seek medical attention immediately.

Note to Physician: Aspiration hazard may contraindicate the use of gastric lavage. For emergency assistance call (201) 231-2000.

STORAGE AND DISPOSAL

Do not contaminate water, food, or feed by storage or disposal.

Do not use or store near heat or open flame.

Pesticide Disposal: Pesticide wastes are toxic. Improper disposal of excess pesticide, spray mixture, or rinsate is a violation of Federal Law. If these wastes cannot be disposed of by use according to label instructions, contact your State Pesticide or Environmental Control Agency, or the hazardous waste representative at the nearest EPA Regional Office for guidance.

Container Disposal: Triple rinse (or equivalent). Then offer for recycling or reconditioning, or puncture and dispose of in a sanitary landfill, or by other procedures approved by state and local authorities.

Do not store below 20°F.

Important Notice: Disclaimer

Read "IMPORTANT NOTICE: DISCLAIMER" before buying or using. If terms are not acceptable, return at once unopened. HOECHST-ROUSSEL AGRI-VET COMPANY warrants only that the product conforms to the chemical description on the label and is reasonably fit for the purpose stated on the label when used in accordance with the directions under normal conditions of use. This warranty does not extend to the use of this product contrary to label instructions or under abnormal conditions, or under conditions not reasonably foreseeable to HOECHST-ROUSSEL AGRI-VET COMPANY and user assumes the risk of any such use. HOECHST-ROUSSEL AGRI-VET COMPANY MAKES NO OTHER WARRANTY, EXPRESS OR IMPLIED, INCLUDING ANY IMPLIED WARRANTY OF FITNESS FOR A PARTICULAR PURPOSE OR OF MERCHANTABILITY. In no case shall HOECHST-ROUSSEL AGRI-VET COMPANY be liable for consequential, special, indirect or incidental damages resulting from the use or handling of this product. The aforegoing conditions of sale and warranty can be varied only by an agreement in writing signed by a duly authorized representative of HOECHST-ROUSSEL AGRI-VET COMPANY.

GENERAL INFORMATION

Hoelon® 3EC Herbicide is an emulsifiable concentrate (EC) formulation containing three pounds of active ingredient per gallon. Hoelon 3EC Herbicide will control goosegrass/crowfootgrass/silver crabgrass *(Eleusine indica)* in established bermudagrass *(Cynodon dactylon)* turf. Hoelon 3EC Herbicide will not control broadcast weeds or sedges.

Hoelon 3EC Herbicide is absorbed primarily through the foliage. Thorough coverage is important. Visual injury on goosegrass is evident approximately two to three weeks after application (dependent upon environmental conditions).

DIRECTIONS FOR USE

It is a violation of Federal Law to use this product in a manner inconsistent with its labeling. This label must be in the possession of the user at the time of application.

The following use recommendation is provided as a special local need authorized for use in Florida under FIFRA Section 24(c).

Hoelon 3EC Herbicide can be applied as a broadcast application to all cultivars of bermudagrass turf only in the state of Florida. Avoid applications to other turf species as severe turf injury may occur.

Application information

Hoelon 3EC Herbicide only controls goosegrass that has emerged at the time of application. Young actively growing goosegrass plants are more easily controlled than larger plants. Refer to the application rate and time chart on this Supplemental Label for proper application timing and rates. Avoid applications to turfgrass under stress due to lack of moisture, chemical injury, insects, disease, or temperature extremes.

The addition of a non-ionic surfactant at 0.25% by volume may improve the control of goosegrass under less than optimum growing conditions.

Pressurized Hydraulic Sprayer Application

Apply Hoelon 3 EC Herbicide by means of a pressurized hydraulic sprayer using 30-60 PSI and 0.7 to 1.4 gallons of finished spray per 1,000 ft. (20-40 gallons per acre). Thorough spray coverage is essential for optimum results. Flat fan nozzles are recommended.

Applications should be made using a uniform spray pattern with a minimum overlap. Overdosing may result in turf injury.

Mowing Information

For optimum results, spray Hoelon 3 EC Herbicide when maximum leaf area is present. Avoid mowing treated areas for at least 36 hours to allow time for penetration into the grass weed. Hoelon 3EC Herbicide is most effective on closely mowed goosegrass but not less than 5/16" cutting height. The control of goosegrass in higher mowed turf such as in "rough" areas is less effective. The effectiveness of Hoelon 3EC Herbicide is reduced when grass clippings interfere with thorough spray coverage of the target weed.

Time of Application and Rate Chart Hoelon 3EC Herbicide

Grassy Weed	Stage of Growth	Fl. Oz. Per 1,000 Ft.^{2}	Fl. Oz. Per Acre
Goosegrass	1-3 Leaf	0.75	32.6
	3-4 Leaf	1.00	43.5
	4 Leaf - 1 Tiller	1.25	54.4
	1-2 Tiller	1.50	65.3

Special Notes

1. Avoid application to turfgrass under drought stress or stress from insects, diseases or nematodes.
2. Rainfall or irrigation within one hour of application may cause a reduction in grass control. Allow spray deposit to dry before irrigation.
3. DO NOT tank mix Hoelon 3EC Herbicide with any other pesticide or liquid fertilizer as reduced goosegrass control may occur.
4. DO NOT overseed treated area for at least six (6) months from time of application.
5. Aply Hoelon 3EC Herbicide only to established (stolons 4 inches in length) bermudagrass turf.
6. DO NOT apply more than a total of 1.5 fluid ounces per 1000 ft.^{2} (65.3 fluid ounces/A) per year.
7. DO NOT apply Hoelon 3EC Herbicide through any type of irrigation system.

ALL APPLICABLE DIRECTIONS, RESTRICTIONS AND PRECAUTIONS ON THE EPA REGISTERED LABEL ARE TO BE FOLLOWED.

This labeling must be in the possession of the user at the time of pesticide application.

TH8901042 (9/89)

"This is a specimen label. Hoechst-Roussel Agri-Vet Company is not responsible for the accuracy of the information contained herein, as labels are subject to revision. Always read and follow the use instructions on the container label."

NOR-AM CHEMICAL COMPANY

The following revised product labels appear:
Apollo®, Carzol SP

APOLLO®

OVICIDE MITICIDE

A Suspension Concentrate (SC) Formulation for Mite Control on Pears

ACTIVE INGREDIENT:	Percent by Weight
3,6-bis (2-chlorophenyl)-1,2,4,5-tetrazine.	42%
INERT INGREDIENTS:	58%
TOTAL.	100%

Contains 1 pound of active ingredient per quart.

EPA Registration No. 45639-135

PRECAUCION AL USUARIO: Si usted no lee ingles, no use este, producto hasta que la etiqueta haya sido explicada ampliamente.

KEEP OUT OF REACH OF CHILDREN

CAUTION

STATEMENT OF PRACTICAL TREATMENT

IF IN EYES: Flush eyes with plenty of water. Call a physician if irritation persists.

IF ON SKIN: Wash with plenty of soap and water. Get medical attention if irritation persists.

PRECAUTIONARY STATEMENTS

HAZARDS TO HUMANS AND DOMESTIC ANIMALS

CAUTION

Avoid contact with skin, eyes or clothing. In case of skin contact, carefully remove contaminated clothing and wash affected area with soap and water. In case of eye contact, flush eyes with plenty of water. Get medical attention if irritation persists.

Wash hands and exposed skin before eating, drinking or smoking and after spraying or handling. Wash all contaminated clothing thoroughly before reuse.

ENVIRONMENTAL HAZARDS

Do not apply directly to water. Do not apply when weather conditions favor drift from treated area. Do not contaminate water when disposing of equipment wash waters.

IN CASE OF FIRE, LEAKY OR DAMAGED CONTAINERS, OR OTHER EMERGENCY, REPORT AT ONCE BY TOLL-FREE TELEPHONE TO: 800-424-9300.

IMPORTANT—Read the entire Directions for Use and Disclaimer of Warranties before using this product.

DIRECTIONS FOR USE

It is a violation of federal law to use this product in a manner inconsistent with its labeling.

Do not apply this product in such a manner as to directly or through drift expose workers or other persons. The area being treated must be vacated by unprotected persons.

REENTRY STATEMENT

Do not enter treated areas without protective clothing until sprays have dried. Because certain states may require more restrict reentry intervals for various crops treated with this product, consult your State Department of Agriculture for further information.

Written or oral warnings must be given to workers who are expected to be in a treated area or in an area about to be treated with this product. When oral warnings are given, warnings shall be given in a language customarily understood by workers. Oral warnings must be given if there is reason to believe that written warnings cannot be understood by workers. Written warnings must include the following information:

CAUTION: Area treated with APOLLO® SC on (date of application). Do not enter without appropriate protective clothing until leaves are completely dry. In case of accidental exposure, see Statement of Practical Treatment on the label.

STORAGE AND DISPOSAL

Prohibitions: Do not contaminate water, food or feed by storage or disposal.

Storage: Store in original container, tightly closed, in a safe place.

Pesticide Disposal: Wastes resulting from the use of this product may be disposed of on site or at an approved waste disposal facility.

Container Disposal: Triple rinse (or equivalent). Then offer for recycling or reconditioning, or puncture and dispose of in a sanitary landfill, or incineration, or if allowed by state and local authorities, by burning. If burned, stay out of smoke.

GENERAL INFORMATION

APOLLO® SC is a specific miticide which acts as an ovicide and on the young motile stages. APOLLO SC is relatively non-toxic to beneficial insects and mites. It has good residual activity on European red mites, two-spotted spider mites, McDaniel mites and yellow spider mites when applied according to label directions.

USE ON PEARS

TARGET SPECIES

European red mite	*Panonychus ulmi*
Two-spotted spider mite	*Tetranychus urticae*
McDaniel mite	*Tetranychus mcdanieli*
Yellow spider mite	*Eotetranychus carpini boreallis*

APOLLO SC may be applied by ground using either dilute or concentrate spray equipment. Do not apply by air.

Concentrate Application (Less than 400 gallons per acre): Apply 4–8 ounces of APOLLO SC per acre.

Dilute Application (400 gallons per acre or more): Use 1–2 ounces of APOLLO SC per 100 gallons of spray. Do not apply less than 4 ounces of APOLLO SC per acre.

Recommendations for Use

Because APOLLO® SC acts primarily against the egg and larval stages, apply APOLLO SC at the first sign of mite activity. If the mite population is approaching damaging levels or if pear rust mite is present, combine APOLLO SC with an appropriate adulticide such as CARZOL® SP. When tank mixing APOLLO SC with an adulticide, use label rates for both products.

IMPORTANT—Do not apply within 21 days of harvest. Do not graze treated areas. Do not apply this product through any type of irrigation system.

DO NOT APPLY MORE THAN ONE APPLICATION OF APOLLO SC PER SEASON.

MANAGING MITE RESISTANCE

Repeated use of miticides with a similar mode of action may lead to buildup of resistant strains of mites. For good resistance management practices, NOR-AM Chemical Company recommends that the ovicide/laracide APOLLO SC be applied only once per year. If APOLLO SC has been applied, do not apply another ovicide such as Savey[1] in the same year. If Savey has been applied, do not apply APOLLO SC in the same year. If additional miticide applications are needed, use a miticide with a different mode of action such as the adulticide CARZOL® SP.

[1] Savey is a registered trademark of E.I. DuPont de Nemours and Company, Inc.

IMPORTANT: READ BEFORE USE

By using this product, user accepts the following conditions, warranty, disclaimer of warranties and limitations of liability.

CONDITIONS: The directions for use of this product are believed to be adequate and should be followed carefully. However, because of extreme weather and soil conditions, manner of use and other factors beyond NOR-AM Chemical Company's control, it is impossible for NOR-AM to eliminate all risks associated with the use of this product. As a result, crop injury or ineffectiveness is always possible.

DISCLAIMER OF WARRANTIES: THERE ARE NO WARRANTIES, EXPRESS OR IMPLIED, OF MECHANTABILITY OR OF FITNESS FOR A PARTICULAR PURPOSE OR OTHERWISE, WHICH EXTEND BEYOND THE STATEMENTS MADE ON THIS LABEL. No agent of NOR-AM Chemical Company is authorized to make any warranties beyond those contained herein or to modify the warranties contained herein. UNDER NO CIRCUMSTANCES SHALL NOR-AM CHEMICAL COMPANY BE LIABLE FOR ANY INCIDENTAL OR CONSEQUENTIAL DAMAGES RESULTING FROM FAILURE TO USE THIS PRODUCT IN ACCORDANCE WITH LABEL DIRECTIONS.

LIMITATIONS OF LIABILITY: In no event shall NOR-AM Chemical Company's liability for damages arising out of the use of this product exceed the purchase price of the product used.

530-A-1B-5M Rev. 11/89

This is a specimen label. NOR-AM Chemical Company is not responsible for the accuracy of the information contained herein. As labels are subject to revision, always carefully read and follow the label on the product container.

CARZOL® SP

MITICIDE/INSECTICIDE

FOR AGRICULTURAL USE ONLY

ACTIVE INGREDIENT:	Percent by Weight
Formetanate hydrochloride*	92%
*N,N-dimethyl-N'[3-[[(methylamino)carbonyl]oxy]phenyl] methanimidamide monohydrochloride	
INERT INGREDIENTS	8%
TOTAL	100%

EPA Reg. No. 45639-74   U.S. Pat. Nos. 3,336,186; Re 27,347; 3,542,853

PRECAUCION AL USUARIO: Si usted no lee ingles, no use este producto hasta que la etiqueta haya sido explicada ampliamente.

KEEP OUT OF REACH OF CHILDREN

DANGER–POISON

PELIGRO—VENENO

STATEMENT OF PRACTICAL TREATMENT

IF SWALLOWED—DRINK 1 OR 2 GLASSES OF WATER AND INDUCE VOMITING BY TOUCHING BACK OF THROAT WITH FINGER. DO NOT INDUCE VOMITING OR GIVE ANYTHING BY MOUTH TO AN UNCONSCIOUS PERSON. GET MEDICAL ATTENTION.

IF ON SKIN—WASH IMMEDIATELY WITH SOAP AND WATER.

IF IN EYES—FLUSH WITH WATER FOR AT LEAST 15 MINUTES AND GET MEDICAL ATTENTION.

NOTICE TO PHYSICIAN—CARZOL® SP IS A CARBAMATE WHICH ACTS AS A CHOLINESTERASE INHIBITOR. SPECIFIC ANTIDOTE—ATROPINE. DO NOT USE 2-PAM.

DO NOT REUSE THIS BOX

PRECAUTIONARY STATEMENTS

DANGER

HAZARDS TO HUMANS AND DOMESTIC ANIMALS

Fatal if swallowed. Do not breathe spray mist. Can cause irreversible eye damage. May cause skin sensitization. Do not get in eyes, on skin or on clothing. Wear goggles, face shield or safety glasses. Wear a pesticide respirator approved jointly by the Mining Enforcement and Safety Administration (formerly U.S. Bureau of Mines) and by the National Institute for Occupational Safety and Health under the provisions of 30 CFR Part II. Wash thoroughly with soap and water after handling and before eating or smoking. Remove contaminated clothing and wash before reuse.

Do not apply this product in such a manner as to directly or through drift expose workers or other persons. The area being treated must be vacated by unprotected persons.

ENVIRONMENTAL HAZARDS

This pesticide is toxic to wildlife. Do not apply directly to water or wetlands. Do not contaminate water by cleaning of equipment or disposal of wastes. Cover or incorporate spills. This product is toxic to bees exposed to direct treatment. Do not apply this product while bees are actively visiting the treatment area. Do not allow animals to graze in orchard areas treated with CARZOL SP.

STORAGE AND DISPOSAL

PESTICIDE DISPOSAL—

Do not contaminate water, food, or feed by storage or disposal. Open dumping is prohibited.

Pesticide wastes are acutely hazardous. Improper disposal of excess pesticide, spray mixture, or rinsate is a violation of Federal Law. If these wastes cannot be disposed of by use according to label instructions, contact your State Pesticide or Environmental Control Agency, or the Hazardous Waste representative at the nearest EPA Regional Office for guidance.

CONTAINER DISPOSAL—

Completely empty box into application equipment. Then dispose of empty box in a sanitary landfill or by incineration, or, if allowed by State and local authorities, by burning. If burned, stay out of smoke.

IN CASE OF FIRE, LEAKY OR DAMAGED CONTAINERS, OR OTHER EMERGENCY, REPORT AT ONCE BY TOLL-FREE TELEPHONE TO: 800-424-9300.

DIRECTIONS FOR USE

It is a violation of Federal law to use this product in a manner inconsistent with its labeling.

REENTRY STATEMENT

Do not enter treated areas for 24 hours unless protective clothing is worn.

Because certain states may require more restrictive reentry intervals for various crops treated with this product, consult your State Department of Agriculture for further information.

Written or oral warnings must be given to workers who are expected to be in a treated area or in an area about to be treated with this product. (Indicate specific oral warnings which inform workers of areas or fields that may not be entered without specific protective clothing, period of time field must be vacated and appropriate actions to take in case of accidental exposure.) When oral warnings are given, warnings shall be given in a language customarily understood by workers. Oral warnings must be given if there is reason to believe that written warnings cannot be understood by workers. Written warnings must include the following information:

DANGER: Area treated with CARZOL SP on (date of application). Do not enter without appropriate protective clothing for 24 hours or until leaves are completely dry. In case of accidental exposure see Statement of Practical Treatment on the label.

CARZOL SP is completely soluble in water. Preparation of a premix of slurry before adding to the spray tank is not necessary. Add the desired amount of CARZOL SP directly to the spray tank $\frac{1}{2}$ full of water with the agitator running. Then add the remainder of the water.

NOTE: CARZOL SP is not stable in alkaline water. Do not prepare the spray mixture until just before application. Do not prepare more spray mixture than will be used within 4 hours.

USE PRECAUTIONS

To avoid excessive residues, observe the following limitations:

- CITRUS: Do not apply more than 5 lbs. CARZOL SP per acre in a 12-month period. Allow 30 days between applications.
  NOTE: If brown soft scale is present in the grove, a suitable scallicide should be applied after the CARZOL SP application.
- APPLES, PEARS: Do not apply more than 4 lbs. CARZOL SP per acre in one crop season.
- PEACHES: Do not apply more than $2\frac{1}{2}$ lbs. CARZOL SP per acre in one crop season.
- NECTARINES: Do not apply more than 4 lbs. CARZOL SP per acre in one crop season.
- PLUMS, PRUNES: Do not apply more than 3 lbs. CARZOL SP per acre in one crop season.
- ALFALFA (Seed Crop Only): Use this product only on pure stands of alfalfa to be harvested for seed. Do not allow spray to drift onto adjacent crops. Do not spray while bees are visiting the treatment area. Do not allow livestock to graze treated fields. Do not use treated crops for livestock feed. Crop residues are not to be fed to livestock or baled for such use. Fields treated with CARZOL SP must be cleared after harvest by burning or by disking into the soil. CARZOL SP does not control aphids. If aphids appear, add an aphicide to the spray.
- Do not apply this product through any type of irrigation system.

To avoid excessive residues observe the pre-harvest intervals shown in the table.
For other use limitations see Use Precautions.

CROP	PEST	APPLICATION
ORANGES, LEMONS, TANGERINES, TANGELOS AND GRAPEFRUIT (Arizona only)	THRIPS	Apply $\frac{1}{2}$ to $1\frac{1}{2}$ lbs. of CARZOL SP per acre as outside coverage spray when thrips appear. For aerial application, use a minimum of 10 gallons of water per acre. Apply as needed up to a limit of 5 lbs. CARZOL SP per acre per 12 month period. Allow 7 days before harvest.
ORANGES, LEMONS TANGERINES, TANGELOS AND GRAPEFRUIT (California only)	THRIPS	Apply $\frac{1}{2}$ to $1\frac{1}{2}$ lbs. of CARZOL SP per acre as outside coverage spray when thrips appear. For aerial application, use a minimum of 10 gallons of water per acre. Apply no more than twice per season. Allow 7 days before harvest.
ORANGES, LEMONS, TANGERINES AND GRAPEFRUIT (Florida)	APOPKA WEEVIL CITRUS ROOT WEEVIL (BLUE-GREEN WEEVIL)	Apply 1 lb. CARZOL SP as a concentrate spray of 100 gallons per acre, or less. Repeat application as needed. Allow 7 days before harvest.

CROP	PEST	APPLICATION
	FULLER'S ROSE BEETLE LITTLE LEAF NOTCHER	
ORANGES, LEMONS, TANGERINES AND GRAPEFRUIT (Florida)	RUST MITE	Apply 1 lb. CARZOL SP per acre. Repeat application as needed. Allow 7 days before harvest.
ORANGES, LEMONS, TANGERINES AND GRAPEFRUIT (Texas)	RUST MITE	Use 2–4 oz. CARZOL SP per 100 gallons of water when mites appear. For full cover dilute spray, apply 500 gallons of water per acre. Repeat application as needed. Allow 7 days before harvest.
APPLES	TENTIFORM LEAF MINER	Use 4–8 oz. CARZOL SP per 100 gallons of water at pink and petal fall. For full cover dilute spray, apply 400 to 800 gallons of water per acre depending on the size of the trees. For concentrate spray, apply 1–4 lbs. CARZOL SP per acre in at least 50 gallons of water. Allow 7 days before harvest.
APPLES	WHITE APPLE LEAFHOPPER	Use 2–4 oz. CARZOL SP per 100 gallons when leafhoppers appear. For full cover dilute spray, apply 400 gallons of water per acre. For concentrate spray, apply 1/2–1 lb. CARZOL SP per acre in at least 50 gallons of water. Repeat application as needed. Allow 7 days before harvest.
APPLES AND PEARS	EUROPEAN RED MITE, TWO-SPOTTED MITE	Use 4–8 oz. CARZOL SP per 100 gallons of water when mites appear. For full cover dilute spray, apply 400 to 800 gallons of water per acre depending on the size of the trees and severity of the mite infestation. For concentrate spray, apply 1–4 lbs. CARZOL SP per acre in at least 50 gallons of water. Repeat application as needed. Allow 7 days before harvest.
APPLES (Also see below for Arizona only)	THRIPS	Use 4–5 oz. CARZOL SP per 100 gallons of water at bloom and repeat as necessary through petal fall. For full cover, dilute spray, apply 400 gallons of water per acre. For concentrate spray or dilute application of less than 400 gallons, apply 1 to 1 1/4 lbs. CARZOL SP per acre in a minimum of 50 gallons of water per acre. Apply at night or while bees are not working in blooms. Allow 7 days before harvest.
PEARS	PEAR RUST MITE	Use 4 oz. CARZOL SP per 100 gallons of water. Apply post bloom to harvest at first indication of rust mite infestation. For full cover dilute spray, apply 400–800 gallons of water per acre depending on the size of the trees. For concentrate spray, apply 1–2 lbs. CARZOL SP per acre in at least 50 gallons of water. Repeat application as needed. Allow 7 days before harvest.
PEACHES AND NECTARINES	EUROPEAN RED MITE, McDANIEL MITE, TWO-SPOTTED MITE	Use 4 oz. CARZOL SP per 100 gallons of water when mites appear. For full cover dilute spray, apply 400 to 500 gallons of water per acre depending on the size of the trees. For concentrate spray, apply 1–1 1/4 lbs. CARZOL SP per acre in at least 50 gallons of water. Repeat application as needed. Allow 21 days before harvest for peaches, 14 days for nectarines.
PEACHES	LYGUS BUGS, STINK BUGS	Use 4 oz. CARZOL SP per 100 gallons of water at first sign of infestation. For full cover dilute spray, apply 400–500 gallons of water per acre. For concentrate spray, apply 1–1 1/4 lbs. CARZOL SP per acre. Repeat application as needed. Allow 21 days before harvest.
NECTARINES	THRIPS	Use 4–6 oz. CARZOL SP per 100 gallons of water when thrips infest blossoms through petal fall. For full cover dilute spray, apply 400 to 500 gallons of water per acre depending on the size of the trees. For concentrate or aerial spray, apply 1–2 lbs. CARZOL SP per acre. Apply at night or while bees are not working in blooms. Repeat application as needed. Allow 14 days before harvest.
PLUMS AND PRUNES	EUROPEAN RED MITE, McDANIEL MITE, TWO-SPOTTED MITE	Use 4 oz. CARZOL SP per 100 gallons of water when mites appear. For full cover dilute spray, apply 400 to 600 gallons of water per acre depending on the size of the trees. For concentrate spray, apply 1–1 1/2 lbs. CARZOL SP per acre in at least 50 gallons of water Repeat application as needed. Allow 7 days before harvest.
ALFALFA (seed crop only in California, Oregon, Washington, Idaho and Nevada. Use pre-bloom only on alfalfa in all areas except California.)	LYGUS BUGS	Use 1/2–1 lb. CARZOL SP per acre. Apply when lygus bugs reach damaging numbers. Repeat application as needed. Consult your State Agricultural Experiment Station or State Extension Service Specialist for proper timing.
	CONSPERSE STINK BUGS (California only)	Use 1 lb. CARZOL SP per acre for knockdown of a heavy nymphal population. A lygus control program will control stink bugs. When applied by air use a minimum of 5 gallons per acre. When applied by ground use a minimum of 20 gallons per acre.
	TWO-SPOTTED MITE	Use 1/2–1 lb. CARZOL SP per acre. Apply when mites appear. When applied by air use a minimum of 10 gallons of spray per acre. When applied by ground use a minimum of 20 gallons per acre. Repeat application as needed.

SPECIAL LOCAL NEED REGISTRATION—EPA SLN NO. AZ-880009

APPLES (Arizona only)	THRIPS	CARZOL SP is used for the control of thrips on apples by aerial application at rates of 1 to 1 1/4 lbs. CARZOL SP per acre with a dilution rate of 10 gallons of water per acre minimum. Apply 2-4 times at 4-6 day intervals between bloom and petal fall. Do not apply more than 4 lbs. CARZOL SP per acre in one crop season. Apply at night or while bees are not working in blooms. Do not re-enter treated areas until leaves are completely dry or until 24 hours after application. Allow 7 days before harvest.

SPECIAL LOCAL NEED REGISTRATION—EPA SLN NO. AZ-890009

ORANGES, LEMONS TANGERINES, TANGELOS AND GRAPEFRUIT (Arizona only)	THRIPS	Apply 1/2 to 1 1/2 lbs. of CARZOL SP per acre as outside coverage spray when thrips appear. For aerial applications, use a minimum of 10 gallons of water per acre. Apply as needed up to a limit of 5 lbs. CARZOL SP per acre per 12 month period. Allow 7 days before harvest. Do not reenter treated areas for 48 hours unless protective clothing is worn. Allow 14 days between applications.

IMPORTANT: READ BEFORE USE

By using this product, user accepts the following conditions, warranty, disclaimer of warranties and limitations of liability.

CONDITIONS: The directions for use of this product are believed to be adequate and should be followed carefully. However, because of extreme weather and soil conditions, manner of use and other factors beyond NOR-AM Chemical Company's control, it is impossible for NOR-AM to eliminate all risks associated with the use of this product. As a result, crop injury or ineffectiveness is always possible.

DISCLAIMER OF WARRANTIES: THERE ARE NO WARRANTIES, EXPRESS OR IMPLIED, OF MERCHANTABILITY OR OF FITNESS FOR A PARTICULAR PURPOSE OR OTHERWISE, WHICH EXTEND BEYOND THE STATEMENTS MADE ON THIS LABEL. No agent of NOR-AM Chemical Company is authorized to make any warranties beyond those contained herein or to modify the warranties contained herein. UNDER NO CIRCUMSTANCES SHALL NOR-AM CHEMICAL COMPANY BE LIABLE FOR ANY INCIDENTAL OR CONSEQUENTIAL DAMAGES RESULTING FROM FAILURE TO USE THIS PRODUCT IN ACCORDANCE WITH LABEL DIRECTIONS.

LIMITATIONS OF LIABILITY: In no event shall NOR-AM Chemical Company's liability for damages arising out of the use of this product exceed the purchase price of the product used.

525-1B-C-10M-Rev. 10/89

This is a specimen label. NOR-AM Chemical Company is not responsible for the accuracy of the information contained herein. As labels are subject to revision, always carefully read and follow the label on the product container.